*Marks'*
# Standard Handbook
# for Mechanical Engineers

## OTHER McGRAW-HILL HANDBOOKS OF INTEREST

*American Institute of Physics* · American Institute of Physics Handbook
*American Society of Mechanical Engineers* · ASME Handbooks:

    Engineering Tables           Metals Engineering—Processes
    Metals Engineering—Design    Metals Properties

*Beeman* · Industrial Power Systems Handbook
*Brady and Clauser* · Materials Handbook
*Brater and King* · Handbook of Hydraulics
*Burington* · Handbook of Mathematical Tables and Formulas
*Burington and May* · Handbook of Probability and Statistics with Tables
*Cagle* · Handbook of Adhesive Bonding
*Callender* · Time-Saver Standards for Architectural Design Data
*Carrier Air Conditioning Company* · Handbook of Air Conditioning System Design
*Carroll* · Industrial Instrument Servicing Handbook
*Considine* · Chemical and Process Technology Encyclopedia
*Considine* · Encyclopedia of Instrumentation and Control
*Considine* · Energy Technology Handbook
*Considine* · Process Instruments and Controls Handbook
*Considine and Ross* · Handbook of Applied Instrumentation
*Crocker and King* · Piping Handbook
*Davis* · Handbook of Applied Hydraulics
*DeChiara and Callender* · Time-Saver Standards for Building Types
*Dudley* · Gear Handbook
*Emerick* · Heating Handbook
*Fink* · Electronics Engineer's Handbook
*Fink and Carroll* · Standard Handbook for Electrical Engineers
*Harris* · Handbook of Noise Control
*Harris and Crede* · Shock and Vibration Handbook
*Hicks* · Standard Handbook of Engineering Calculations
*Juran* · Quality Control Handbook
*Kallen* · Handbook of Instrumentation and Controls
*Korn and Korn* · Mathematical Handbook for Scientists and Engineers
*LeGrand* · The New American Machinists' Handbook
*Machol* · System Engineering Handbook
*Manas* · National Plumbing Code Handbook
*Mantell* · Engineering Materials Handbook
*Maynard* · Industrial Engineering Handbook
*Merritt* · Standard Handbook for Civil Engineers
*Morrow* · Maintenance Engineering Handbook
*Parmley* · Standard Handbook of Fastening and Joining
*Peckner* · Handbook of Stainless Steels
*Perry* · Chemical Engineers' Handbook
*Perry* · Engineering Manual
*Rohsenow and Hartnett* · Handbook of Heat Transfer
*Rothbart* · Mechanical Design and Systems Handbook
*Shand* · Glass Engineering Handbook
*Smeaton* · Motor Application and Maintenance Handbook
*Smeaton* · Switchgear and Control Handbook
*Society of Manufacturing Engineers:*

    Die Design Handbook       Handbook of Fixture Design
    Manufacturing Planning and   Tool and Manufacturing Engineers
      Estimating Handbook         Handbook

*Staniar* · Plant Engineering Handbook
*Streeter* · Handbook of Fluid Dynamics
*Truxal* · Control Engineers' Handbook

# Marks'
# Standard Handbook for Mechanical Engineers

*Revised by a staff of specialists*

**THEODORE BAUMEISTER**   *Editor-in-Chief*

Stevens Professor Emeritus of Mechanical Engineering,
Columbia University in the City of New York

**EUGENE A. AVALLONE**   *Associate Editor*

Consulting Engineer; Professor of Mechanical Engineering,
The City College of the City University of New York

**THEODORE BAUMEISTER III**   *Associate Editor*

Consultant, Information Systems Department,
E. I. du Pont de Nemours & Co.

*Eighth Edition*

**McGRAW-HILL BOOK COMPANY**

New York  St. Louis  San Francisco  Auckland  Bogotá
Düsseldorf  Johannesburg  London  Madrid
Mexico  Montreal  New Delhi  Panama
Paris  São Paulo  Singapore
Sydney  Tokyo  Toronto

Library of Congress Cataloged The First Issue
of this title as follows:

Standard handbook for mechanical engineers. 1st– ed.;
1916–
New York, McGraw-Hill.
v. Illus. 18–24 cm.
Title varies: 1916–58: Mechanical engineers' handbook.
Editors: 1916–51, L. S. Marks.—1958– T. Baumeister.
Includes bibliographies.

1. Mechanical engineering—Handbooks, manuals, etc. I. Marks,
Lionel Simeon, 1871– ed. II. Baumeister, Theodore, 1897–
ed. III. Title: Mechanical engineers' handbook.

TJ151.S82 502'.4'621 16–12915
Library of Congress [r68z²60–2]
ISBN 0-07-004123-7

## MARKS' STANDARD HANDBOOK FOR MECHANICAL ENGINEERS

891011 KPKP 876543

| First Edition | Third Edition | Fifth Edition | Seventh Edition |
| --- | --- | --- | --- |
| *Eleven Printings* | *Seven Printings* | *Seven Printings* | *Fifteen Printings* |
| Second Edition | Fourth Edition | Sixth Edition | |
| *Seven Printings* | *Thirteen Printings* | *Eight Printings* | |

*The editors for this book were Harold B. Crawford and Lester Strong
and the production supervisor was Teresa F. Leaden.
It was set in Janson by University Graphics, Inc.*

*Printed and bound by Kingsport Press, Inc.*

*The editors and the publishers will be grateful to readers who notify them
of any inaccuracy or important omission in this book*

# Contents

*For the detailed contents of any section consult the title page of that section.*
*The alphabetical index follows Section 19.*

*List of Contributors    ix*
*Preface to the Eighth Edition    xiii*
*Preface to the First Edition    xv*
*Symbols and Abbreviations    xvii*


**1. Mathematical Tables and Measuring Units** ............... **1-1**
  **Mathematical Tables** ..................................... **1-1**
  **Measuring Units** ........................................ **1-31**

**2. Mathematics** ........................................... **2-1**
  **Arithmetic** ............................................. **2-2**
  **Geometry and Mensuration** ............................... **2-7**
  **Algebra** ................................................ **2-15**
  **Trigonometry** ........................................... **2-25**
  **Analytical Geometry** .................................... **2-30**
  **Differential and Integral Calculus** .................... **2-42**
  **Graphical Representation of Functions** .................. **2-54**
  **Vector Analysis** ........................................ **2-63**
  **Computers** .............................................. **2-64**

**3. Mechanics of Solids and Fluids** ........................ **3-1**
  **Mechanics of Solids** .................................... **3-2**
  **Friction** ............................................... **3-24**
  **Mechanics of Fluids** .................................... **3-33**

**4. Heat** .................................................. **4-1**
  **Thermal Properties of Substances and Thermodynamics** ..... **4-2**
  **Transmission of Heat by Conduction and Convection** ....... **4-59**
  **Radiant-Heat Transfer** .................................. **4-70**

**5. Strength of Materials** ................................. **5-1**
  **Mechanical Properties of Materials** ..................... **5-2**
  **Mechanics of Materials** ................................. **5-16**
  **Pipeline Flexure Stresses Caused by Expansion or Movement of**
  **Supports** ............................................... **5-61**
  **Vibration** .............................................. **5-67**
  **Nondestructive Testing** ................................. **5-76**

**6. Materials of Engineering** .............................. **6-1**
  **General Properties of Materials** ........................ **6-3**
  **Iron and Steel** ......................................... **6-12**

Iron and Steel Castings ............................................. 6-47
Nonferrous Metals ................................................. 6-59
Corrosion ......................................................... 6-106
Paints and Protective Coatings .................................... 6-118
Wood .............................................................. 6-122
Nonmetallic Materials ............................................. 6-138
Cement, Mortar, and Concrete ..................................... 6-177
Water ............................................................. 6-189
Lubricants and Lubrication ........................................ 6-196

7. Fuels and Furnaces ................................................. 7-1
Fuels ............................................................. 7-2
Carbonization of Coal and Gas Making ............................. 7-37
Combustion Furnaces .............................................. 7-47
Incineration ...................................................... 7-53
Electric Furnaces and Ovens ...................................... 7-60

8. Machine Elements .................................................. 8-1
Mechanism ........................................................ 8-3
Machine Elements ................................................. 8-9
Gearing ........................................................... 8-98
Fluid-Film Bearings ............................................... 8-120
Bearings with Rolling Contact .................................... 8-136
Packings and Seals ............................................... 8-143
Pipe and Pipe Fittings ............................................ 8-147
Preferred Numbers ................................................ 8-201

9. Power Generation .................................................. 9-1
Sources of Energy ................................................ 9-3
Steam Boilers ..................................................... 9-7
Steam Engines .................................................... 9-35
Steam Turbines ................................................... 9-38
Power-Plant Heat Exchangers ..................................... 9-59
Internal-Combustion Engines ...................................... 9-78
Gas Turbines ..................................................... 9-114
Nuclear (Atomic) Power ........................................... 9-119
Hydraulic Turbines ............................................... 9-136
Direct Energy Conversion ......................................... 9-149
Power Miscellany ................................................. 9-153

10. Materials Handling ................................................ 10-1
Methods of Moving Materials ...................................... 10-2
Lifting ............................................................ 10-4
Dragging, Pulling, and Pushing ................................... 10-21
Carrying and Lifting .............................................. 10-25
Conveying and Continuous Flow ................................... 10-40

11. Transportation .................................................... 11-1
Automobiles ...................................................... 11-3
Railway Engineering .............................................. 11-19
Marine Engineering ............................................... 11-35
Air Resistance of Trains, Automobiles, and Ships ................. 11-57
Aeronautics ...................................................... 11-58
Jet Propulsion and Aircraft Propellers ............................ 11-84

Astronautics ......................................................... 11-104
Pipeline Transmission ............................................. 11-130

## 12. Building Construction and Equipment ..................... 12-1
Industrial Plants ..................................................... 12-2
Structural Design of Buildings ................................... 12-11
Reinforced-Concrete Design and Construction ...................... 12-46
Heating, Ventilation, and Air Conditioning ......................... 12-62
Illumination ........................................................ 12-116
Sound, Noise, and Ultrasonics ..................................... 12-136

## 13. Shop Processes ......................................... 13-1
Foundry Practice and Equipment .................................... 13-2
Working Metals and Plastics ........................................ 13-9
Welding ............................................................. 13-27
Metal-Removal Processes and Equipment ............................ 13-48
Surface-Texture Designation, Production, and Control ............. 13-73
Woodcutting Tools and Machines .................................... 13-80

## 14. Pumps and Compressors .................................. 14-1
Pumps ............................................................... 14-2
Centrifugal and Axial Pumps ....................................... 14-15
Compressors ........................................................ 14-30
High-Vacuum Pumps .................................................. 14-44
Fans ................................................................ 14-49

## 15. Electrical and Electronics Engineering .................. 15-1
Electrical Engineering ............................................. 15-2
Electronics ........................................................ 15-84

## 16. Instruments and Controls ............................... 16-1
Instruments ......................................................... 16-2
Automatic Controls ................................................. 16-25
Surveying .......................................................... 16-42

## 17. Industrial Engineering ................................. 17-1
Industrial Economics and Management ............................... 17-2
Cost Accounting .................................................... 17-10
Engineering Statistics and Quality Control ........................ 17-19
Time and Motion Study .............................................. 17-26
Cost of Electric Power ............................................. 17-33

## 18. Environmental Control .................................. 18-1

## 19. Refrigeration, Cryogenics, Optics, Patents, Trademarks, and Copyrights, and Miscellaneous ..................... 19-1
Mechanical Refrigeration ........................................... 19-3
Cryogenics ......................................................... 19-23
Optics ............................................................. 19-40
Patents, Trademarks, and Copyrights ............................... 19-41
Miscellaneous ...................................................... 19-45

*Index follows Section 19*

# Contributors

**Abraham Abramowitz** *Consulting Engineer; Professor Emeritus of Electrical Engineering, The City College, The City University of New York.* (ILLUMINATION)

**Ashok Ahuja** *Senior Power Generation Engineer. General Electric Co. Cost of Electric Power.*

**Anders T. Anderson** *Commander, U.S. Navy (CDR, USN), SSBN Project Officer, Submarine Logistics Division, Naval Sea Systems Command.* (MARINE ENGINEERING)

**William Antis** *Technical Director, Maynard Research Council Incorporated, Pittsburgh, P.A.* (TIME AND MOTION STUDY)

**Stephen D. Antolovitch** *Professor of Materials Science, College of Engineering, University of Cincinnati.* (ASTRONAUTICS)

**Neil A. Armstrong** *University Professor of Aerospace Engineering, University of Cincinnati.* (ASTRONAUTICS)

**Eugene A. Avallone** *Professor of Mechanical Engineering, The City College, The City University of New York.* (MACHINE ELEMENTS)

**Hilary E. Bacon** *Consulting Chemical Engineer and Partner, Sheppard T. Powell Associates.* (CORROSION)

**Frederick G. Baily** *Manager, Turbine Application Engineering, Large Steam Turbine-Generator Department, General Electric Co.* (STEAM TURBINES)

**Antonio F. Baldo** *Professor of Mechanical Engineering, The City College, The City University of New York.* (NONMETALLIC MATERIALS)

**Heard K. Baumeister** *Senior Engineer, International Business Machines Corporation.* (MECHANISM. OPTICS)

**Theodore Baumeister, III** *Consultant, Informations Systems Department, E. I. du Pont de Nemours & Co.* (MATHEMATICAL TABLES)

**Howard S. Bean** *late Physicist, National Bureau of Standards.* (MEASURING UNITS. GENERAL PROPERTIES OF MATERIALS)

**E. R. Behnke** *Product Manager, CM Chain Division, Columbus McKinnon Corp.* (MATERIALS HANDLING)

**R. R. Bennett** *Chief Consulting Mechanical Engineer, Ebasco Services, Inc.* (SOURCES OF ENERGY)

**C. H. Berry** *late Gordon McKay Professor of Mechanical Engineering, Emeritus, Harvard University; and late Professor of Mechanical Engineering, Northeastern University.* (PREFERRED NUMBERS)

**\*William Bollay** *President, Aerophysics Development Corp.* (AIR RESISTANCE OF TRAINS, AUTOMOBILES, AND SHIPS)

**Jay A. Bolt** *Professor of Mechanical Engineering, University of Michigan.* (INTERNAL-COMBUSTION ENGINES)

**F. G. Brickwedde** *Evan Pugh Research Professor Emeritus of Physics, The Pennsylvania State University.* (CRYOGENICS)

**Charles Willers Briggs** *late Consulting Engineer.* (IRON AND STEEL CASTINGS)

**James A. Broadston** *Chief Engineer, Surface Checking Gage Co.* (SURFACE-TEXTURE DESIGNATION, PRODUCTION, AND CONTROL)

**Frederick W. Buse** *Chief Engineer, Standard Pump-Aldrich Division, Ingersoll-Rand Co.* (PUMPS)

**Benson Carlin** *President, O. E. M. Medical, Inc.* (SOUND, NOISE, AND ULTRASONICS)

**C. L. Carlson** *Fellow Engineer, Research Labs, Westinghouse Electric Corporation.* (NONFERROUS METALS)

**David E. Cole** *Associate Professor of Mechanical Engineering, University of Michigan.* (INTERNAL-COMBUSTION ENGINES)

**J. J. Cornish, III** *Director of Engineering, Lockheed-Georgia Company* (AERONAUTICS)

**Robert T. Corry** *Associate Professor of Mechanical Engineering, Polytechnic Institute of New York.* (INSTRUMENTS)

**E. V. Crane** *late Chief Engineer and Research Director, E. W. Bliss Co.* (WORKING METALS AND PLASTICS)

**Marcelo Crespo da Silva** *Assistant Professor of Aerospace Engineering, University of Cincinnati.* (ASTRONAUTICS)

**Glenn H. Damon** *Staff Coordinator, U.S. Department of the Interior, Bureau of Mines, Mining Research Health and Safety.* (ROCKET FUELS)

**B. B. Dayton** *Consulting Physicist, East Flat Rock, NC.* (HIGH-VACUUM PUMPS)

**Carl H. de Zeeuw** *Professor, Department of Wood Products Engineering, College of Environmental Science and Forestry, State University of New York, Syracuse.* (WOOD)

**Donald D. Dodge** *Controls Sect. Supervisor, Alternate Engines Research Dept., Ford Motor Co.* (NONDESTRUCTIVE TESTING)

**Leander Economides** *Consulting Engineer, New York, NY.* (HEATING, VENTILATION, AND AIR CONDITIONING)

**F. J. Edeskuty** *Associate Group Leader—Cryogenics, Los Alamos Scientific Laboratory.* (CRYOGENICS)

**Odin Elnan** *Professor of Aerospace Engineering, University of Cincinnati.* (ASTRONAUTICS)

**Vivian F. Estcourt** *Consulting Engineer, Bechtel Power Corporation.* (GEOTHERMAL POWER)

**George H. Ewing** *Senior Vice President, Texas Eastern Gas Pipeline Co* (PIPELINE TRANSMISSION)

**Erich A. Farber** *Professor and Research Professor of Mechanical Engineering; Director, Solar Energy and Energy Conversion Laboratory, University of Florida.* (HOT-AIR ENGINES)

**J. F. Farnsworth** *Manager Gasification, Engineering and Construction Division, Koppers Company, Inc.* (CARBONIZATION OF COAL AND GAS MAKING)

**Edward A. Fenton** *Consulting Engineer, Miami, FL.* (WELDING)

**Arthur J. Fiehn** *Vice President, Power Technology, Burns and Roe, Inc.* (COST OF ELECTRIC POWER)

**Edward W. Fisher** *Product Specialist, Retired, Garlock, Inc.* (PACKINGS AND SEALS)

**\*Philip Franklin** *late Professor of Mathematics, Massachusetts Institute of Technology.* (MATHEMATICAL TABLES. MATHEMATICS)

**Dudley D. Fuller** *Stevens Professor of Mechanical Engineering, Columbia University.* (FRICTION, FLUID FILM BEARINGS)

**W. Lucas Gaillard** *Practicing Surveyor, State of South Carolina.* (SURVEYING)

**W. Lucas Gaillard, Jr.** *Practicing Surveyor, State of South Carolina.* (SURVEYING)

**William L. Gamble** *Professor of Civil Engineering, University of Illinois at Urbana-Champaign.* (CEMENT, MORTAR, AND CONCRETE. REINFORCED-CONCRETE DESIGN AND CONSTRUCTION)

**John E. Gray** *President, International Energy Associates, Ltd.* [NUCLEAR (ATOMIC) POWER]

**George A. Hawkins** *Vice President Emeritus for Academic Affairs, Purdue University.* (THERMAL PROPERTIES OF SUBSTANCES AND THERMODYNAMICS)

**Harold V. Hawkins** *Manager, Product Standards and Services, Columbus McKinnon Corp.* (MATERIALS HANDLING. PIPELINE FLEXURE STRESSES)

*\*Contributions by authors whose names are marked with an asterisk were made for the previous edition and have been revised or rewritten by others for this edition. The stated professional position in these cases is that held by the author at the time of his or her contribution.*

ix

**Hoyt C. Hottel** *Professor Emeritus, Massachusetts Institute of Technology.* (RADIANT-HEAT TRANSFER)

**William L. Hughes** *Professor of Electrical Engineering, Oklahoma State University.* (WIND POWER)

**John R. Immer** *President, Work Saving International, Washington, D.C.* (INDUSTRIAL PLANTS)

**Byron M. Jones** *Manager, Analysis and Test Department, Bucyrus-Erie Company.* (ELECTRONICS)

**Robert Jorgensen** *Chief Engineer, Air Handling Division, Buffalo Forge Company.* (FANS)

**Serope Kalpakjian** *Professor of Mechanical Engineering, Illinois Institute of Technology.* (MATERIAL-REMOVAL PROCESSES AND EQUIPMENT)

**I. J. Karassik** *Vice President and Chief Consulting Engineer, Worthington Pump Corp.* (CENTRIFUGAL AND AXIAL PUMPS)

**Robert W. Kennard** *Manager, Systems Engineering, Engineering Department, E. I. du Pont de Nemours & Co.* (ENGINEERING STATISTICS AND QUALITY CONTROL)

**George W. Kessler** *retired Vice President, The Babcock & Wilcox Company.* (STEAM BOILERS)

**Reno C. King** *Associate Dean, Queensborough Community College, Bayside, NY.* (PIPE AND PIPE FITTINGS)

**Ezra S. Krendel** *Professor of Operations Research and Statistics, University of Pennsylvania.* (MAN- AND ANIMAL-GENERATED POWER)

**Robert J. Kroll** *Professor of Aerospace Engineering, University of Cincinnati.* (ASTRONAUTICS)

**W. C. Krutzsch** *Manager, Product Development, Engineered Pump Division, Worthington Pump Corp. (U.S.A).* (CENTRIFUGAL AND AXIAL PUMPS)

**P. G. Kuchuris, Jr.** *Market Planning Manager, International Harvester Co.* (MATERIALS HANDLING)

**L. D. Kunsman** *Fellow Engineer, Research Labs, Westinghouse Electric Corp.* (NONFERROUS METALS)

**Walter A. La Pierre** *Consulting Engineer in Private Practice, Westfield, NJ.* (ELECTRICAL ENGINEERING)

**W. E. Lewis** *Retired Manager-Field Sales, Lectromelt Furnace Division, McGraw-Edison Co.* (ELECTRIC FURNACES AND OVENS)

**Anna Kazanjian Longobardo** *Manager of Plans and Programs, Systems Management Division, Sperry Rand Corp.* (AUTOMATIC CONTROLS)

**Guy S. Longobardo** *Corporate Staff, I. B. M. Corp.* (AUTOMATIC CONTROLS)

**Carl R. Loper, Jr.** *Professor, Minerals and Metals Engineering, University of Wisconsin.* (FOUNDRY PRACTICE AND EQUIPMENT)

**J. T. McCartney** *Research Physicist, Pittsburgh Energy Research Center, Energy Research and Development Administration, Pittsburgh, PA.* (COAL)

**Harold E. McGannon** *Technical Writer and Editor.* (IRON AND STEEL)

**Charles W. MacGregor** *Formerly Engineering Consultant and Manager of Advanced Technology, Systems Development Division, International Business Machines Corporation, Endicott, NY. Presently Consulting Engineer in Private Practice, Cohasset, MA.* (MECHANICAL PROPERTIES OF MATERIALS)

**Dennis K. McLaughlin** *Associate Professor of Mechanical Engineering, Oklahoma State University.* (WIND POWER)

**C. J. Manney** *Consultant, Columbus McKinnon Corp.* (MATERIALS HANDLING)

**Wilbur D. Marsh** *Senior Application Engineer, Electric Utility Systems Engineering Department, General Electric Co.* (DIRECT ENERGY CONVERSION)

**Adolph Matz** *Professor Emeritus of Accounting, The Wharton School, University of Pennsylvania.* (COST ACCOUNTING)

**M. H. Mawhinney** *Consulting Engineer, Salem, OH.* (COMBUSTION FURNACES)

**Sherwood B. Menkes** *Professor of Mechanical Engineering, The City College, The City University of New York.* (FLYWHEEL ENERGY STORAGE)

**George W. Michalec** *Assistant to Vice President, Operations Technology, American Can Co. Formerly Professor and Associate Dean of Engineering & Science, Stevens Institute of Technology.* (GEARING)

**Cort L. Miller** *Engineering Associate, Eastman Kodak Company, Kodak Park, Engineering Division.* (BEARINGS WITH ROLLING CONTACT)

**William Mirsky** *Professor of Mechanical Engineering, University of Michigan.* (INTERNAL-COMBUSTION ENGINES)

**Reeves Morrisson** *Corporate Technical Staff, United Technologies Corporation.* (GAS TURBINES)

**George J. Moshos** *Chairman, Department of Computer and Information Science, New Jersey Institute of Technology.* (COMPUTERS)

**James W. Murdock** *Consulting Engineer; Associate Professor, Mechanical Engineering and Mechanics Department, Drexel University.* (MECHANICS OF FLUIDS)

**John Nagy** *Physical Scientist, U.S. Department of the Interior, Mining Enforcement and Safety Administration, Pittsburgh Technical Support Center.* (DUST EXPLOSIONS IN BUILDINGS)

**Harold M. Nelson** *Managing Editor, Automotive Industries.* (AUTOMOBILES)

**Carl L. Newman** *Vice President, Engineering, Consolidated Edison Co. of New York, Inc.* [NUCLEAR (ATOMIC) POWER]

**Edward Taylor Newton** *Counsellor at Law; Senior Partner, Newton, Hopkins and Ormsby, Atlanta, GA.* (PATENTS, TRADEMARKS, AND COPYRIGHTS)

**Joseph P. Nicoletti** *Vice President, John A. Blume & Associates.* (STRUCTURAL DESIGN OF BUILDINGS)

**B. W. Niebel** *Professor of Industrial Engineering, The Pennsylvania State University.* (INDUSTRIAL ECONOMICS AND MANAGEMENT)

**N. J. Palladino** *Dean of Engineering, Pennsylvania State University.* [NUCLEAR (ATOMIC) POWER]

**John F. Partridge** *Manager, Industrial Engineering, Consolidated Rail Corp.* (RAILWAY ENGINEERING)

**Donald J. Patterson** *Associate Professor of Mechanical Engineering, University of Michigan.* (INTERNAL-COMBUSTION ENGINES)

**Richard W. Perkins** *Professor of Mechanical and Aerospace Engineering, Syracuse University.* (WOODCUTTING TOOLS AND MACHINES)

**Pascal M. Rapier** *Consulting Engineer, Richmond, CA.* (ENVIRONMENTAL CONTROL)

**Bruce A. Reese** *Professor and Head, School of Aeronautics and Astronautics, Purdue University.* (JET PROPULSION AND AIRCRAFT PROPELLERS)

**Frank A. Ritchings** *Vice President, Ebasco Services, Inc., New York.* (SOURCES OF ENERGY)

**Louis H. Roddis, Jr.** *Consulting Engineer.* [NUCLEAR (ATOMIC) POWER]

**Kenneth A. Roe** *Chairman and President, Burns and Roe, Inc.* (WATER. ENVIRONMENTAL CONTROL)

**Fred P. Roslyn** *Consulting Engineer.* (MECHANICAL REFRIGERATION)

**Ivan L. Ross** *International Manager, Chain Conveyor Division, ACCO.* (MATERIALS HANDLING)

**Adel F. Sarofim** *Professor of Chemical Engineering, Massachusetts Institute of Technology.* (RADIANT-HEAT TRANSFER)

**M. D. Schlesinger** *Deputy Director, Energy Research and Development Administration, Pittsburgh, PA.* (FUELS)

**Kenneth A. Smith** *Professor of Chemical Engineering, Massachusetts Institute of Technology.* (TRANSMISSION OF HEAT BY CONDUCTION AND CONVECTION)

**Lawrence Sobel** *Principal Engineer, Westinghouse Electric Corporation (formerly with the University of Cincinnati).* (ASTRONAUTICS)

**Joseph R. Spencer** *Senior Design Engineer, Heat Transfer Engineering, Westinghouse Electric Corporation.* (POWER-PLANT HEAT EXCHANGER)

**Robert F. Steidel, Jr.** *Professor of Mechanical Engineering, University of California, Berkeley* (MECHANICS OF SOLIDS)

**Richard M. Stephani** *Senior Design Engineer, Heat Transfer Engineering, Westinghouse Electric Corporation.* (POWER-PLANT HEAT EXCHANGERS)

**George G. Sward** *Retired Consultant, National Paint and Coatings Association.* (PAINTS AND PROTECTIVE COATINGS)

**John Symonds** *Fellow Engineer, Oceanic Division, Westinghouse Electric Corporation.* (MECHANICAL PROPERTIES OF MATERIALS)

**Widen Tabakoff** *Professor of Aerospace Engineering, University of Cincinnati.* (ASTRONAUTICS)

**William T. Thomson** *Emeritus Professor of Engineering, University of California, Santa Barbara.* (VIBRATION)

**Robert W. Van Dolah** *Research Director, U.S. Dept. of the Interior, Pittsburgh Mining & Research Center.* (EXPLOSIVES)

**Charles O. Velzy** *Secretary and Treasurer, Charles R Velzy Associates, Inc.* (INCINERATION)

**Charles R Velzy**  *President, Charles R Velzy Associates, Inc.* (INCINERATION)

**Harry Verakis**  *Chemist, U.S. Department of the Interior, Mining Enforcement and Safety Administration, Pittsburgh Technical Support Center.* (DUST EXPLOSIONS IN BUILDINGS)

**Joseph P. Vidosic**  *Regents' Professor Emeritus of Mechanical Engineering, Georgia Institute of Technology.* (MECHANICS OF MATERIALS)

**Fred J. Villforth, Jr.**  *Technologist, Beacon Research Laboratories, Research and Development Department, Texaco, Inc.* (LUBRICANTS AND LUBRICATION)

**C. C. Ward**  *Retired Chemical Engineer, Bartlesville Energy Research Center, U.S. Bureau of Mines.* (PETROLEUM AND OTHER LIQUID FUELS. GASEOUS FUELS)

**Michael W. Washo**  *Engineer, Eastman Kodak Company, Kodak Park, Engineering Division.* (BEARINGS WITH ROLLING CONTACT)

**Warren G. Whippen**  *Manager, Product Development, Allis Chalmers, York Division.* (HYDRAULIC TURBINES)

**Glenn C. Williams**  *Professor of Chemical Engineering, Massachusetts Institute of Technology.* (TRANSMISSION OF HEAT BY CONDUCTION AND CONVECTION)

**K. D. Williamson, Jr.**  *Associate Group Leader—Controlled Thermonuclear Research (Cryogenics), Los Alamos Scientific Laboratory.* (CRYOGENICS)

**David E. Wolfson**  *Retired Chemical Engineer, Pittsburgh Energy Research Center, U.S. Bureau of Mines.* (COKE)

**Arthur R. Worster**  *Manager, Engineering, Air Power Division, Ingersoll-Rand Co.* (COMPRESSORS)

**John I. Yellott**  *College of Architecture, Arizona State University.* (SOLAR ENERGY)

**Maurice J. Zucrow**  *Professor Emeritus, Purdue University.* (JET PROPULSION AND AIRCRAFT PROPELLERS)

# Preface to the Eighth Edition

In the preparation of the eighth edition of "Marks," the editors had several major objectives. First, to modernize and update the contents as required; second, to introduce the SI system of units where applicable; and third, to recast the Handbook into a two-column format to enhance the visual presentation of textual matter and thereby increase its readability and usefulness.

It is recognized that, at this time, the U. S. Customary System (USCS) of units cannot be completely replaced by the International System (SI). A transition from USCS to SI will proceed at a rational pace to accommodate the needs of the profession, of industry, of the polity, and of the citizenry. The transition period most likely will be long and complex, and duality of units probably will be demanded for several decades. Incorporation of dual units to the maximum extent possible will make the eighth edition useful for many years to professional engineers and to students.

Established practices and new concepts are the warp and woof which not only constitute the fabric of the profession, but also serve to keep it alive and progressive. For example, increased recognition of the social implications of engineering has resulted in a new section on environmental control.

The editors were cognizant of the competing requirements to offer the user the broad spectrum of information that has made the Handbook so useful for over sixty years, and yet to keep the size of the one volume within reason. This was achieved through the enthusiastic and timely cooperation of contributors, reviewers, and publisher.

Last, the Handbook is ultimately the responsibility of the editors. Meticulous care has been exercised to avoid errors, but, if any are inadvertently included, the editors will appreciate being informed so that they may be eliminated from subsequent printings of this edition.

*Willtown Bluff, SC*  
*Chauncey, NY*  
*Newark, DE*

THEODORE BAUMEISTER  
EUGENE A. AVALLONE  
THEODORE BAUMEISTER III

# Preface to the First Edition*

This Handbook is intended to supply both the practicing engineer and the student with a reference work which is authoritative in character and which covers the field of mechanical engineering in a comprehensive manner. It is no longer possible for a single individual or a small group of individuals to have so intimate an acquaintance with any major division of engineering as is necessary if critical judgment is to be exercised in the statement of current practice and the selection of engineering data. Only by the co-operation of a considerable number of specialists is it possible to obtain the desirable degree of reliability. This Handbook represents the work of fifty specialists.

Each contributor is to be regarded as responsible for the accuracy of his section. The number of contributors required to ensure sufficiently specialized knowledge for all the topics treated is necessarily large. It was found desirable to enlist the services of thirteen specialists for an adequate handling of the "Properties of Engineering Materials." Such topics as "Automobiles," "Aeronautics," "Illumination," "Patent Law," "Cost Accounting," "Industrial Buildings," "Corrosion," "Air Conditioning," "Fire Protection," "Prevention of Accidents," etc., though occupying relatively small spaces in the book, demanded each a separate writer.

A number of the contributions which deal with engineering practice, after examination by the Editor-in-Chief, were submitted by him to one or more specialists for criticism and suggestions. Their co-operation has proved of great value in securing greater accuracy and in ensuring that the subject matter does not embody solely the practice of one individual but is truly representative.

An accuracy of four significant figures has been assumed as the desirable limit; figures in excess of this number have been deleted, except in special cases. In the mathematical tables only four significant figures have been kept.

The Editor-in-Chief desires to express here his appreciation of the spirit of co-operation shown by the Contributors and of their patience in submitting to modifications of their sections. He wishes also to thank the Publishers for giving him complete freedom and hearty assistance in all matters relating to the book from the choice of contributors to the details of typography.

*Cambridge, Mass.*
April 23, 1916

LIONEL S. MARKS

*Excerpt.

# Symbols and Abbreviations

For symbols of chemical elements, see Sec. 6; for abbreviations applying to metric weights and measures and SI units, Sec. 1; SI unit prefixes are listed on p. 1–34.

Pairs of parentheses, brackets, etc., are frequently used in this work to indicate corresponding values. For example, the statement that "the cost per kw of a 30,000-kw plant is $86; of a 15,000-kw plant, $98; and of an 8,000-kw plant, $112," is condensed as follows: The cost per kw of a 30,000 (15,000) [8,000]-kw plant is $86 (98) [112].

In the citation of references readers should always attempt to consult the latest edition of referenced publications.

| | | | |
|---|---|---|---|
| A or Å | Angstrom unit $= 10^{-10}$m; $3.937 \times 10^{-11}$ in | AMCA | Air Moving & Conditioning Assoc., Inc. |
| A | mass number $= N + Z$; ampere | amu | atomic mass unit |
| AA | arithmetical average | AN | ammonium nitrate (explosive); Army-Navy Specification |
| AAA | Am. Automobile Assoc. | | |
| AAR | Assoc. of Am. Railroads | AN-FO | ammonium nitrate-fuel oil (explosive) |
| AAS | Am. Astronautical Soc. | ANC | Army-Navy Civil Aeronautics Committee |
| ABAI | Am. Boiler & Affiliated Industries | ANS | Am. Nuclear Soc. |
| abs | absolute | ANSI | American National Standards Institute |
| a.c. | aerodynamic center | antilog | antilogarithm of |
| a-c, ac | alternating current | API | Am. Petroleum Inst. |
| ACI | Am. Concrete Inst. | approx | approximately |
| ACM | Assoc. for Computing Machinery | APWA | Am. Public Works Assoc. |
| ACRMA | Air Conditioning and Refrigerating Manufacturers Assoc. | AREA | Am. Railroad Eng. Assoc. |
| | | ARI | Air Conditioning and Refrigeration Inst. |
| ACS | Am. Chemical Soc. | ARS | Am. Rocket Soc. |
| ACSR | aluminum cable steel-reinforced | ASCE | Am. Soc. of Civil Engineers |
| ACV | air cushion vehicle | ASHRAE | Am. Soc. of Heating, Refrigerating, and Air Conditioning Engineers |
| A.D. | anno Domini (in the year of our Lord) | | |
| AEC | Atomic Energy Commission (U.S.) | ASLE | Am. Soc. of Lubricating Engineers |
| a-f, af | audio frequency | ASM | Am. Soc. for Metals |
| AFBMA | Anti-friction Bearings Manufacturers' Assoc. | ASME | Am. Soc. of Mechanical Engineers |
| AFS | Am. Foundrymen's Soc. | ASST | Am. Soc. for Steel Treating |
| AGA | Am. Gas Assoc. | ASTM | Am. Soc. for Testing and Materials |
| AGMA | Am. Gear Manufacturers' Assoc. | ASTME | Am. Soc. of Tool & Manufacturing Engineers |
| ahp | air horsepower | atm | atmosphere |
| AIChE | Am. Inst. of Chemical Engineers | *Auto. Ind.* | Automotive Industries (New York) |
| AIEE | Am. Inst. of Electrical Engineers (see IEEE) | avdp | avoirdupois |
| AIME | Am. Inst. of Mining Engineers | avg, ave | average |
| AIP | Am. Inst. of Physics | AWG | Am. Wire Gage |
| AISC | American Institute of Steel Construction, Inc. | AWPA | Am. Wood Preservation Assoc. |
| AISE | Am. Iron & Steel Engineers | AWS | American Welding Soc. |
| AISI | Am. Iron and Steel Inst. | AWWA | American Water Works Assoc. |
| a.m. | ante meridiem (before noon) | b | barns |
| a-m, am | amplitude modulation | bar | barometer |
| *Am. Mach.* | Am. Machinist (New York) | B&S | Brown & Sharp (gage); Beams and Stringers |
| AMA | Automobile Manufacturers' Assoc.; Acoustical Materials Assoc. | bbl | barrels |
| | | B.C. | before Christ |

| | | | | |
|---|---|---|---|---|
| B.C.C. | body centered cubic | | coth | hyperbolic cotangent of |
| Bé | Baumé (degrees) | | $coth^{-1}$ | inverse hyperbolic cotangent of |
| B.G. | Birmingham gage (hoop and sheet) | | covers | coversed sine of |
| bgd | billions of gallons per day | | c.p. | circular pitch; center of pressure |
| BHN | Brinnell Hardness Number | | cp | candle power |
| bhp | brake horsepower | | *cp* | coef of performance |
| BLC | boundary layer control | | CP | chemically pure |
| B.M. | board measure; bench mark | | CPH | close packed hexagonal |
| bmep | brake mean effective pressure | | cpm, cycles/ | |
| B of M, | | | min | cycles per minute |
| BuMines | Bureau of Mines | | cps, cycles/s | cycles per second |
| BOD | biochemical oxygen demand | | CSA | Canadian Standards Assoc. |
| bp | boiling point | | csc | cosecant of |
| Bq | bequerel | | $csc^{-1}$ | arc whose cosecant is (see $cos^{-1}$) |
| bsfc | brake specific fuel consumption | | csch | hyperbolic cosecant of |
| BSI | British Standards Inst. | | $csch^{-1}$ | inverse hyperbolic cosecant of |
| Btu | British thermal units | | cu | cubic |
| Btuh, Btu/h | Btu per hr | | cyl | cylinder |
| bu | bushels | | db, dB | decibel |
| *Bull.* | Bulletin | | d-c, dc | direct current |
| Buweaps | Bureau of Weapons, U.S. Navy | | def | definition |
| BWG | Birmingham wire gage | | deg | degrees |
| c | velocity of light | | diam. (dia) | diameter |
| °C | degrees Celsius (centigrade) | | DO | dissolved oxygen |
| C | coulomb | | $D_2O$ | deuterium (heavy water) |
| CAB | Civil Aeronautics Board | | d.p. | double pole |
| CAGI | Compressed Air & Gas Inst. | | DP | Diametral pitch |
| cal | calories | | DPH | diamond pyramid hardness |
| C-B-R | chemical, biological & radiological (filters) | | DST | daylight saving time |
| CBS | Columbia Broadcasting System | | $d^2tons$ | breaking strength, $d$ = chain wire diam, in. |
| cc, cm³ | cubic centimeters | | DX | direct expansion |
| CCR | critical compression ratio | | *e* | base of Napierian logarithmic system |
| c to c | center to center | | | $(=2.7182+)$ |
| cd | candela | | EAP | equivalent air pressure |
| c.f. | centrifugal force | | EDR | equivalent direct radiation |
| *cf.* | confer (compare) | | EEI | Edison Electric Inst. |
| cfh, ft³/h | cubic feet per hour | | eff | efficiency |
| cfm, ft³/min | cubic feet per minute | | e.g. | exempli gratia (for example) |
| C.F.R. | Cooperative Fuel Research | | ehp | effective horsepower |
| cfs, ft³/s | cubic feet per second | | EHV | extra high voltage |
| cg | center of gravity | | *El. Wld.* | Electrical World (New York) |
| cgs | centimeter-gram-second | | elec | electric |
| *Chm. Eng.* | Chemical Eng'g (New York) | | elong | elongation |
| chu | centigrade heat unit | | emf | electromotive force |
| C.I. | cast iron | | *Engg.* | Engineering (London) |
| cir | circular | | *Engr.* | The Engineer (London) |
| cir mil | circular mils | | ENT | emergency negative thrust |
| cm | centimeters | | EP | extreme pressure (lubricant) |
| *CME* | Chartered Mech. Engr (IMechE) | | ERDA | Energy Research & Development Administration (successor to AEC; see also NRC) |
| C.N. | cetane number | | | |
| coef | coefficient | | Eq. | equation |
| COESA | U.S. Committee on Extension to the Standard Atmosphere | | est | estimated |
| | | | etc. | et cetera (and so forth) |
| col | column | | et seq. | et sequens (and the following) |
| colog | cologarithm of | | eV | electron volts |
| const | constant | | evap | evaporation |
| cos | cosine of | | exp | exponential function of |
| $cos^{-1}$ | arc whose cosine is, inverse cosine of | | exsec | exterior secant of |
| cosh | hyperbolic cosine of | | ext | external |
| $cosh^{-1}$ | inverse hyperbolic cosine of | | °F | degrees Fahrenheit |
| cot | cotangent of | | F | farad |
| $cot^{-1}$ | arc whose cotangent is (see $cos^{-1}$) | | FAA | Federal Aviation Agency |

| | | | |
|---|---|---|---|
| F.C. | fixed carbon, % | ICE | Inst. of Civil Engineers |
| FCC | Federal Communications Commission; Federal Construction Council | ICI | International Commission on Illumination |
| F.C.C. | face-centered-cubic (alloys) | I.C.T. | International Critical Tables |
| ff. | following (pages) | I.D., ID | inside diameter |
| fhp | friction horsepower | i.e. | id est (that is) |
| Fig. | figure | IEC | International Electrotechnical Commission |
| F.I.T. | Federal income tax | IEEE | Inst. of Electrical & Electronics Engineers (successor to AIEE, *q.v.*) |
| f-m, fm | frequency modulation | | |
| F.O.B. | free on board (cars) | IES | Illuminating Engineering Soc. |
| FP | fore perpendicular | i-f, if | intermediate frequency |
| FPC | Federal Power Commission | IGT | Inst. of Gas Technology |
| fpm, ft/min | feet per minute | ihp | indicated horsepower |
| fps | foot-pound-second system | IMechE | Inst. of Mechanical Engineers |
| ft/s | feet per second | imep | indicated mean effective pressure |
| F.S. | Federal Specifications | Imp | Imperial |
| FSB | Federal Specifications Board | in., in | inches |
| fsp | fiber saturation point | in.·lb, in·lb | inch-pounds |
| ft | feet | INA | Inst. of Naval Architects |
| fc | foot candles | *Ind. & Eng. Chem.* | Industrial & Eng'g Chemistry (Easton, Pa) |
| fL | foot lamberts | | |
| ft-lb | foot-pounds | int | internal |
| *g* | acceleration due to gravity | i-p, ip | intermediate pressure |
| g | grams | ipm, in/min | inches per minute |
| gal | gallons | ipr | inches per revolution |
| gc | gigacycles per sec | IPS | iron pipe size |
| GCA | ground-controlled approach | IRE | Inst. of Radio Engineers (see IEEE) |
| g·cal | gram-calories | IRS | Internal Revenue Service |
| gd | Gudermannian of | ISO | International Organization for Standardization |
| G.E. | General Electric Co. | isoth | isothermal |
| GEM | ground effect machine | ISTM | International Soc. for Testing Materials |
| GFI | gullet feed index | IUPAC | International Union of Pure & Applied Chemistry |
| G.M. | General Motors Co. | | |
| GMT | Greenwich Mean Time | J | joule |
| GNP | gross national product | J&P | joists & planks |
| gpcd | gallons per capita day | *Jour.* | Journal |
| gpd | gallons per day; grams per denier | JP | jet propulsion fuel |
| gpm, gal/min | gallons per minute | *k* | isentropic exponent; conductivity |
| gps, gal/s | gallons per second | K | degrees Kelvin (Celsius abs) |
| gpt | grams per tex | K | Knudsen number |
| H | henry | kB | kilo Btu (1000 Btu) |
| h | Planck's constant = $6.624 \times 10^{-27}$erg-sec | kc | kilocycles |
| ℏ | Planck's constant, $ℏ = h/2\pi$ | kcps | kilocycles per sec |
| HEPA | high efficiency particulate matter | kg | kilograms |
| h-f, hf | high frequency | kg·cal | kilogram-calories |
| hhv | high heat value | kg·m | kilogram-meters |
| horiz | horizontal | kip | 1000 lb or 1 kilo-pound |
| hp | horsepower | kips | thousands of pounds |
| h-p | high-pressure | km | kilometers |
| *HPAC* | Heating, Piping, & Air Conditioning (Chicago) | kmc | kilomegacycles per sec |
| | | kmcps | kilomegacycles per sec |
| hp·hr | horsepower-hour | kpsi | thousands of pounds per sq in |
| hr, h | hours | ksi | one kip per sq in, 1000 psi (lb/in²) |
| HSS | high speed steel | kts | knots |
| H.T. | heat-treated | kVA | kilovolt-amperes |
| HTHW | high temperature hot water | kW | kilowatts |
| Hz | hertz = 1 cycle/s (cps) | kWh | kilowatt-hours |
| IACs | International Annealed Copper Standard | l < L | litres < lamberts |
| IAeS | Institute of Aerospace Sciences | £ | Laplace operational symbol |
| ibid. | ibidem (in the same place) | lb | pounds |
| ICAO | International Civil Aviation Organization | L.B.P. | length between perpendiculars |
| ICC | Interstate Commerce Commission | lhv | low heat value |
| | | lim | limit |

| | |
|---|---|
| lin | linear |
| ln | Napierian logarithm of |
| loc. cit. | loco citato (place already cited) |
| log | common logarithm of |
| LOX | liquid oxygen explosive |
| l-p, lp | low pressure |
| LPG | liquified petroleum gas |
| lpw, lm/W | lumens per watt |
| lx | lux |
| L.W.L | load water line |
| lm | lumen |
| m | metres |
| M | thousand; Mach number; moisture, % |
| mA | milliamperes |
| *Machy.* | Machinery (New York) |
| max | maximum |
| MBh | thousands of Btu per hr |
| mc | megacycles per sec |
| m.c. | moisture content |
| Mcf | thousand cubic feet |
| mcps | megacycles per sec |
| *Mech. Eng.* | Mechanical Eng'g (ASME) |
| mep | mean effective pressure |
| METO | maximum, except during take-off |
| meV | million electron volts |
| MF | maintenance factor |
| mhc | mean horizontal candles |
| mi | mile |
| MIL-STD | U.S. Military Standard |
| min | minutes; minimum |
| mip | mean indicated pressure |
| MKS | meter-kilogram-second system |
| MKSA | meter-kilogram-second-ampere system |
| mL | millilamberts |
| ml | millilitre = 1.000027 cm³ |
| mlhc | mean lower hemispherical candles |
| mm | millimetres |
| mm-free | mineral matter free |
| mmf | magnetomotive force |
| mol | mole |
| mp | melting point |
| MPC | maximum permissible concentration |
| mph, mi/h | miles per hour |
| MRT | mean radiant temperature |
| ms | manuscript; milliseconds |
| msc | mean spherical candles |
| MSS | Manufacturers Standardization Soc. of the Valve & Fittings Industry |
| Mu | micron, micro |
| MW | megawatts |
| MW day | megawatt day |
| MWT | mean water temperature |
| *n* | polytropic exponent |
| *N* | number (in mathematical tables) |
| N | number of neutrons; newton |
| N$_s$ | specific speed |
| NA | not available |
| NAA | National Assoc. of Accountants |
| NACA | National Advisory Committee on Aeronautics (see NASA) |
| NACM | National Assoc. of Chain Manufacturers |
| NASA | National Aeronautics and Space Administration |
| nat. | natural |
| NBC | National Broadcasting Company |
| NBFU | National Board of Fire Underwriters |
| NBS | National Bureau of Standards |
| NCN | nitrocarbonitrate (explosive) |
| NDHA | National District Heating Assoc. |
| NEC | National Electrical Code |
| NEMA | National Electrical Manufacturers Assoc. |
| NFPA | National Fire Protecion Assoc. |
| NLGI | National Lubricating Grease Institute |
| nm | nautical miles |
| No. (Nos.) | number(s) |
| NPSH | net positive suction head |
| NRC | Nuclear Regulator Commission (successor to AEC; see also ERDA) |
| NTP | normal temperature and pressure |
| O.D., OD | outside diameter (pipes) |
| O.H | open-hearth (steel) |
| O.N. | octane number |
| op. cit. | opere citato (work already cited) |
| OSHA | Occupational Safety & Health Administration |
| OSW | Office of Saline Water |
| OTS | Office of Technical Services, U.S. Dept. of Commerce |
| oz | ounces |
| p. (pp.) | page (pages) |
| Pa | pascal |
| P.C. | propulsive coefficient |
| PE | polyethylene |
| PEG | polyethylene glycol |
| P.E.L. | proportional elastic limit |
| PETN | an explosive |
| pf | power factor |
| PFI | Pipe Fabrication Inst. |
| PIV | peak inverse voltage |
| p.m. | post meridiem (after noon) |
| PM | preventive maintenance |
| P.N. | performance number |
| ppb | parts per billion |
| PPI | plan position indicator |
| ppm | parts per million |
| press | pressure |
| *Proc.* | Proceedings |
| PSD | power spectral density, g²/cps |
| psi, lb/in² | lb per sq in |
| psia | lb per sq in. abs |
| psig | lb per sq in. gage |
| pt | point; pint |
| PVC | polyvinyl chloride |
| Q | 10¹⁸ Btu |
| qt | quarts |
| q.v. | quod vide (which see) |
| r | roentgens |
| *R* | gas constant |
| R | deg Rankine (Fahrenheit abs); Reynolds number |
| rad | radius; radiation absorbed dose; radian |
| RBE | see rem |
| R-C | resistor-capacitor |

| | |
|---|---|
| RCA | Radio Corporation of America |
| R&D | research & development |
| RDX | cyclonite, a military explosive |
| rem | Roentgen equivalent man (formerly RBE) |
| rev | revolutions |
| r-f, rf | radio frequency |
| RMA | Rubber Manufacturers Assoc. |
| rms | square root of mean square |
| rpm, r/min | revolutions per minute |
| rps, r/s | revolutions per second |
| RSHF | room sensible heat factor |
| ry. | railway |
| $s$ | entropy |
| s | seconds |
| S | sulphur, %; siemens |
| SAE | Soc. of Automotive Engineers |
| sat | saturated |
| SBI | Steel Boiler Inst. |
| scfm | standard cu ft per min |
| SCR | silicon controlled rectifier |
| sec | secant of |
| $\sec^{-1}$ | arc whose secant is (see $\cos^{-1}$) |
| Sec. | Section |
| sech | hyperbolic secant of |
| $\text{sech}^{-1}$ | inverse hyperbolic secant of |
| segm | segment |
| SE No. | steam emulsion number |
| sfc | specific fuel consumption, lb per hphr |
| sfm | surface feet per minute |
| sfpm | surface ft per minute |
| shp | shaft horsepower |
| SI | International System of Units (Le Système International d'Unites) |
| sin | sine of |
| $\sin^{-1}$ | arc whose sine is (see $\cos^{-1}$) |
| sinh | hyperbolic sine of |
| $\sinh^{-1}$ | inverse hyperbolic sine of |
| SME | Society of Manufacturing Engineers (successor to ASTME) |
| SNAME | Soc. of Naval Architects and Marine Engineers |
| SP | static pressure |
| sp | specific |
| specif | specification |
| sp gr | specific gravity |
| sp ht | specific heat |
| spp | species unspecified (botanical) |
| SPS | standard pipe size |
| sq | square |
| sr | steradian |
| SSF | sec Saybolt Furol |
| SSU | seconds Saybolt Universal (same as SUS) |
| std | standard |
| SUS | Saybolt Universal seconds (same as SSU) |
| SWG | Standard (British) wire gage |
| T | tesla |
| TAC | Technical Advisory Committee on Weather Design Conditions (ASHRAE) |
| tan | tangent of |
| $\tan^{-1}$ | arc whose tangent is (see $\cos^{-1}$) |
| tanh | hyperbolic tangent of |

| | |
|---|---|
| $\tanh^{-1}$ | inverse hyperbolic tangent of |
| TDH | total dynamic head |
| TEL | tetraethyl lead |
| temp | temperature |
| THI | temperature-humidity (discomfort) index |
| thp | thrust horsepower |
| TNT | trinitrotoluol (explosive) |
| torr | = 1 mm Hg = 1.332 millibars (1/760) atm = (1.013250/760) dynes per $cm^2$ |
| TP | total pressure |
| tph | tons per hour |
| tpi | turns per in |
| TR | transmitter-receiver |
| Trans. | Transactions |
| T.S. | tensile strength; tensile stress |
| tsi | tons per sq in |
| ttd | terminal temperature difference |
| UHF | ultra high frequency |
| UKAEA | United Kingdom Atomic Energy Authority |
| UL | Underwriters' Laboratory |
| ult | ultimate |
| UMS | universal maintenance standards |
| USAF | U.S. Air Force |
| USCG | U.S. Coast Guard |
| USCS | U.S. Commercial Standard; U.S. Customary System |
| USDA | U.S. Dept. of Agriculture |
| USFPL | U.S. Forest Products Laboratory |
| USGS | U.S. Geologic Survey |
| USHEW | U.S. Dept. of Health, Education & Welfare |
| USN | U.S. Navy |
| USP | U.S. Pharmacopoeia |
| USPHS | U.S. Public Health Service |
| USS | United States Standard |
| USSG | U.S. Standard Gage |
| UTC | Coordinated Universal Time |
| V | volt |
| VCF | visual confort factor |
| VCI | visual comfort index |
| VDI | Verein Deutscher Ingenieure |
| vel | velocity |
| vers | versed sine of |
| vert | vertical |
| VHF | very high frequency |
| VI | viscosity index |
| viz. | videlicet (namely) |
| V.M. | volatile matter, % |
| vol | volume |
| VP | velocity pressure |
| vs | versus |
| W | watt |
| Wb | weber |
| W&M | Washburn & Moen wire gage |
| w.g. | water gage |
| WHO | World Health Organization |
| W.I. | wrought iron |
| W.P.A. | Western Pine Assoc. |
| wt | weight |
| yd | yards |
| Y.P. | yield point |
| yr | year(s) |

| Y.S. | yield strength; yield stress | $\sigma$, s | Boltzmann constant |
|---|---|---|---|
| z | atomic number; figure of merit | $\mu$ | micro ($=10^{-6}$), as in $\mu$s |
| *Zeit.* | Zeitschrift | $\mu$m | micrometer (micron) $= 10^{-6}$ m ($10^{-3}$ mm) |
| $\Delta$ | mass defect | $\Omega$ | ohm |
| $\mu$c | microcurie | | |

## MATHEMATICAL SIGNS AND SYMBOLS

+ plus (sign of addition)

+ positive

− minus (sign of subtraction)

− negative

$\pm$ ($\mp$) plus or minus (minus or plus)

$\times$ times, by (multiplication sign)

· multiplied by

$\div$ sign of division

/ divided by

: ratio sign, divided by, is to

:: equals, as (proportion)

< less than

> greater than

<< much less than

>> much greater than

= equals

$\equiv$ identical with

~ similar to

$\approx$ approximately equals

$\cong$ approximately equals, congruent

$\leqq$ equal to or less than

$\geqq$ equal to or greater than

$\pm$ $\neq$ not equal to

$\rightarrow$ $\doteq$ approaches

$\propto$ varies as

$\infty$ infinity

$\sqrt{\phantom{x}}$ square root of

$\sqrt[3]{\phantom{x}}$ cube root of

$\therefore$ therefore

$\parallel$ parallel to

() [] {} parentheses, brackets and braces; quantities enclosed by them to be taken together in multiplying, dividing, etc.

$\overline{AB}$ length of line from $A$ to $B$

$\pi$ pi, $= 3.14159 +$

° degrees

′ minutes

″ seconds

$\angle$ angle

$dx$ differential of $x$

$\Delta$ (delta) difference

$\Delta x$ increment of $x$

$\partial u/\partial x$ partial derivative of $u$ with respect to $x$

$\int$ integral of

$\int_b^a$ integral of, between limits $a$ and $b$

$\oint$ line integral around a closed path

$\Sigma$ (sigma) summation of

$f(x)$, $F(x)$ functions of $x$

$\exp x = e^x$ [e $= 2.71828^x$ (base of natural, or Napierian, logarithms)]

$\nabla$ del or nabla, vector differential operator

$\nabla^2$ Laplacian operator

$\pounds$ Laplace operational symbol

4! factorial $4 = 1 \times 2 \times 3 \times 4$

$|x|$ absolute value of $x$

$\dot{x}$ first derivative of $x$ with respect to time

$\ddot{x}$ second derivative of $x$ with respect to time

$\mathbf{A} \times \mathbf{B}$ vector product; magnitude of $\mathbf{A}$ times magnitude of $\mathbf{B}$ times sine of the angle from $\mathbf{A}$ to $\mathbf{B}$; $AB$ sin $\overline{AB}$

$\mathbf{A} \cdot \mathbf{B}$ scalar product; magnitude of $\mathbf{A}$ times magnitude of $\mathbf{B}$ times cosine of the angle from $\mathbf{A}$ to $\mathbf{B}$; $AB$ cos $\overline{AB}$

*Marks'*
# Standard Handbook
# for Mechanical Engineers

# Mathematical Tables and Measuring Units

BY

**PHILIP FRANKLIN,** *Late Professor of Mathematics, Massachusetts Institute of Technology.*
**HOWARD S. BEAN,** *Late Physicist, National Bureau of Standards.*

**MATHEMATICAL TABLES**
**by Philip Franklin (revised by Theodore Baumeister, III)**

| | |
|---|---|
| Circles (Areas, Segments, etc.) | 1-2 |
| Regular Polygons | 1-8 |
| Binomial Coefficients | 1-8 |
| Common Logarithms | 1-9 |
| Degrees and Radians | 1-11 |
| Trigonometric Functions | 1-13 |
| Exponentials | 1-19 |
| Natural (Napierian) Logarithms | 1-20 |
| Hyperbolic Functions | 1-22 |
| Multiples of 0.4343 and 2.3026 | 1-24 |
| Residuals and Probable Errors | 1-25 |
| Compound Interest and Annuities | 1-26 |
| Decimal Equivalents | 1-30 |

**MEASURING UNITS**
**By Howard S. Bean**

| | |
|---|---|
| U.S. Customary System (USCS) | 1-31 |
| Metric System | 1-32 |
| The International System of Units (SI) | 1-32 |
| SI Base Units and Symbols | 1-33 |
| Conversion Table to SI | 1-34 |
| Systems of Units | 1-34 |
| Temperature | 1-40 |
| Terrestrial Gravity | 1-40 |
| Mohs Scale of Hardness | 1-40 |
| Acceleration of Gravity | 1-40 |
| Time | 1-41 |
| Density and Specific Gravity | 1-42 |
| Specific Gravity, API and Baumé | 1-42 |
| Conversion and Equivalency Tables | 1-44 |

## MATHEMATICAL TABLES

### By Philip Franklin (Revised by Theodore Baumeister, III)

REFERENCES FOR MATHEMATICAL TABLES: Allen, "Six-Place Tables," McGraw-Hill. Comrie, "Chambers Shorter Six-Figure Mathematical Tables," Chemical Publishing. Dwight, "Mathematical Tables of Elementary and Some Higher Mathematical Functions," McGraw-Hill. Dwight, "Tables of Integrals and Other Mathematical Data," Macmillan. Jahnke and Emde, "Tables of Functions," B. G. Teubner, Leipzig, or Dover. Peirce-Foster, "A Short Table of Integrals," Ginn. "Mathematical Tables from Handbook of Chemistry and Physics," Chemical Rubber Co. "Handbook of Mathematical Functions," NBS.

## Circumferences of Circles by Hundredths

| D | 0 | 1 | 2 | 3 | 4 | 5 | 6 | 7 | 8 | 9 | Avg diff |
|---|---|---|---|---|---|---|---|---|---|---|---|
| 1.0 | 3.142 | 3.173 | 3.204 | 3.236 | 3.267 | 3.299 | 3.330 | 3.362 | 3.393 | 3.424 | 31 |
| .1 | 3.456 | 3.487 | 3.519 | 3.550 | 3.581 | 3.613 | 3.644 | 3.676 | 3.707 | 3.738 | |
| .2 | 3.770 | 3.801 | 3.833 | 3.864 | 3.896 | 3.927 | 3.958 | 3.990 | 4.021 | 4.053 | |
| .3 | 4.084 | 4.115 | 4.147 | 4.178 | 4.210 | 4.241 | 4.273 | 4.304 | 4.335 | 4.367 | |
| .4 | 4.398 | 4.430 | 4.461 | 4.492 | 4.524 | 4.555 | 4.587 | 4.618 | 4.650 | 4.681 | |
| 1.5 | 4.712 | 4.744 | 4.775 | 4.807 | 4.838 | 4.869 | 4.901 | 4.932 | 4.964 | 4.995 | |
| .6 | 5.027 | 5.058 | 5.089 | 5.121 | 5.152 | 5.184 | 5.215 | 5.246 | 5.278 | 5.309 | |
| .7 | 5.341 | 5.372 | 5.404 | 5.435 | 5.466 | 5.498 | 5.529 | 5.561 | 5.592 | 5.623 | |
| .8 | 5.655 | 5.686 | 5.718 | 5.749 | 5.781 | 5.812 | 5.843 | 5.875 | 5.906 | 5.938 | |
| .9 | 5.969 | 6.000 | 6.032 | 6.063 | 6.095 | 6.126 | 6.158 | 6.189 | 6.220 | 6.252 | |
| 2.0 | 6.283 | 6.315 | 6.346 | 6.377 | 6.409 | 6.440 | 6.472 | 6.503 | 6.535 | 6.566 | |
| .1 | 6.597 | 6.629 | 6.660 | 6.692 | 6.723 | 6.754 | 6.786 | 6.817 | 6.849 | 6.880 | |
| .2 | 6.912 | 6.943 | 6.974 | 7.006 | 7.037 | 7.069 | 7.100 | 7.131 | 7.163 | 7.194 | |
| .3 | 7.226 | 7.257 | 7.288 | 7.320 | 7.351 | 7.383 | 7.414 | 7.446 | 7.477 | 7.508 | |
| .4 | 7.540 | 7.571 | 7.603 | 7.634 | 7.665 | 7.697 | 7.728 | 7.760 | 7.791 | 7.823 | |
| 2.5 | 7.854 | 7.885 | 7.917 | 7.948 | 7.980 | 8.011 | 8.042 | 8.074 | 8.105 | 8.137 | |
| .6 | 8.168 | 8.200 | 8.231 | 8.262 | 8.294 | 8.325 | 8.357 | 8.388 | 8.419 | 8.451 | |
| .7 | 8.482 | 8.514 | 8.545 | 8.577 | 8.608 | 8.639 | 8.671 | 8.702 | 8.734 | 8.765 | |
| .8 | 8.796 | 8.828 | 8.859 | 8.891 | 8.922 | 8.954 | 8.985 | 9.016 | 9.048 | 9.079 | |
| .9 | 9.111 | 9.142 | 9.173 | 9.205 | 9.236 | 9.268 | 9.299 | 9.331 | 9.362 | 9.393 | |
| 3.0 | 9.425 | 9.456 | 9.488 | 9.519 | 9.550 | 9.582 | 9.613 | 9.645 | 9.676 | 9.708 | 31 |
| .1 | 9.739 | 9.770 | 9.802 | 9.833 | 9.865 | 9.896 | 9.927 | 9.959 | 9.990 | 10.022 | 3 |
| .1 | | | | | | | | | | 10.02 | |
| .2 | 10.05 | 10.08 | 10.12 | 10.15 | 10.18 | 10.21 | 10.24 | 10.27 | 10.30 | 10.34 | |
| .3 | 10.37 | 10.40 | 10.43 | 10.46 | 10.49 | 10.52 | 10.56 | 10.59 | 10.62 | 10.65 | |
| .4 | 10.68 | 10.71 | 10.74 | 10.78 | 10.81 | 10.84 | 10.87 | 10.90 | 10.93 | 10.96 | |
| 3.5 | 11.00 | 11.03 | 11.06 | 11.09 | 11.12 | 11.15 | 11.18 | 11.22 | 11.25 | 11.28 | |
| .6 | 11.31 | 11.34 | 11.37 | 11.40 | 11.44 | 11.47 | 11.50 | 11.53 | 11.56 | 11.59 | |
| .7 | 11.62 | 11.66 | 11.69 | 11.72 | 11.75 | 11.78 | 11.81 | 11.84 | 11.88 | 11.91 | |
| .8 | 11.94 | 11.97 | 12.00 | 12.03 | 12.06 | 12.10 | 12.13 | 12.16 | 12.19 | 12.22 | |
| .9 | 12.25 | 12.28 | 12.32 | 12.35 | 12.38 | 12.41 | 12.44 | 12.47 | 12.50 | 12.53 | |
| 4.0 | 12.57 | 12.60 | 12.63 | 12.66 | 12.69 | 12.72 | 12.75 | 12.79 | 12.82 | 12.85 | |
| .1 | 12.88 | 12.91 | 12.94 | 12.97 | 13.01 | 13.04 | 13.07 | 13.10 | 13.13 | 13.16 | |
| .2 | 13.19 | 13.23 | 13.26 | 13.29 | 13.32 | 13.35 | 13.38 | 13.41 | 13.45 | 13.48 | |
| .3 | 13.51 | 13.54 | 13.57 | 13.60 | 13.63 | 13.67 | 13.70 | 13.73 | 13.76 | 13.79 | |
| .4 | 13.82 | 13.85 | 13.89 | 13.92 | 13.95 | 13.98 | 14.01 | 14.04 | 14.07 | 14.11 | |
| 4.5 | 14.14 | 14.17 | 14.20 | 14.23 | 14.26 | 14.29 | 14.33 | 14.36 | 14.39 | 14.42 | |
| .6 | 14.45 | 14.48 | 14.51 | 14.55 | 14.58 | 14.61 | 14.64 | 14.67 | 14.70 | 14.73 | |
| .7 | 14.77 | 14.80 | 14.83 | 14.86 | 14.89 | 14.92 | 14.95 | 14.99 | 15.02 | 15.05 | |
| .8 | 15.08 | 15.11 | 15.14 | 15.17 | 15.21 | 15.24 | 15.27 | 15.30 | 15.33 | 15.36 | |
| .9 | 15.39 | 15.43 | 15.46 | 15.49 | 15.52 | 15.55 | 15.58 | 15.61 | 15.65 | 15.68 | |
| 5.0 | 15.71 | 15.74 | 15.77 | 15.80 | 15.83 | 15.87 | 15.90 | 15.93 | 15.96 | 15.99 | 3 |
| .1 | 16.02 | 16.05 | 16.08 | 16.12 | 16.15 | 16.18 | 16.21 | 16.24 | 16.27 | 16.30 | |
| .2 | 16.34 | 16.37 | 16.40 | 16.43 | 16.46 | 16.49 | 16.52 | 16.56 | 16.59 | 16.62 | |
| .3 | 16.65 | 16.68 | 16.71 | 16.74 | 16.78 | 16.81 | 16.84 | 16.87 | 16.90 | 16.93 | |
| .4 | 16.96 | 17.00 | 17.03 | 17.06 | 17.09 | 17.12 | 17.15 | 17.18 | 17.22 | 17.25 | |
| 5.5 | 17.28 | 17.31 | 17.34 | 17.37 | 17.40 | 17.44 | 17.47 | 17.50 | 17.53 | 17.56 | |
| .6 | 17.59 | 17.62 | 17.66 | 17.69 | 17.72 | 17.75 | 17.78 | 17.81 | 17.84 | 17.88 | |
| .7 | 17.91 | 17.94 | 17.97 | 18.00 | 18.03 | 18.06 | 18.10 | 18.13 | 18.16 | 18.19 | |
| .8 | 18.22 | 18.25 | 18.28 | 18.32 | 18.35 | 18.38 | 18.41 | 18.44 | 18.47 | 18.50 | |
| .9 | 18.54 | 18.57 | 18.60 | 18.63 | 18.66 | 18.69 | 18.72 | 18.76 | 18.79 | 18.82 | |
| 6.0 | 18.85 | 18.88 | 18.91 | 18.94 | 18.98 | 19.01 | 19.04 | 19.07 | 19.10 | 19.13 | |
| .1 | 19.16 | 19.20 | 19.23 | 19.26 | 19.29 | 19.32 | 19.35 | 19.38 | 19.42 | 19.45 | |
| .2 | 19.48 | 19.51 | 19.54 | 19.57 | 19.60 | 19.63 | 19.67 | 19.70 | 19.73 | 19.76 | |
| .3 | 19.79 | 19.82 | 19.85 | 19.89 | 19.92 | 19.95 | 19.98 | 20.01 | 20.04 | 20.07 | |
| .4 | 20.11 | 20.14 | 20.17 | 20.20 | 20.23 | 20.26 | 20.29 | 20.33 | 20.36 | 20.39 | |
| 6.5 | 20.42 | 20.45 | 20.48 | 20.51 | 20.55 | 20.58 | 20.61 | 20.64 | 20.67 | 20.70 | |
| .6 | 20.73 | 20.77 | 20.80 | 20.83 | 20.86 | 20.89 | 20.92 | 20.95 | 20.99 | 21.02 | |
| .7 | 21.05 | 21.08 | 21.11 | 21.14 | 21.17 | 21.21 | 21.24 | 21.27 | 21.30 | 21.33 | |
| .8 | 21.36 | 21.39 | 21.43 | 21.46 | 21.49 | 21.52 | 21.55 | 21.58 | 21.61 | 21.65 | |
| .9 | 21.68 | 21.71 | 21.74 | 21.77 | 21.80 | 21.83 | 21.87 | 21.90 | 21.93 | 21.96 | |
| 7.0 | 21.99 | 22.02 | 22.05 | 22.09 | 22.12 | 22.15 | 22.18 | 22.21 | 22.24 | 22.27 | |
| .1 | 22.31 | 22.34 | 22.37 | 22.40 | 22.43 | 22.46 | 22.49 | 22.53 | 22.56 | 22.59 | |

**Circumferences by Hundredths (*continued*)**

| D | 0 | 1 | 2 | 3 | 4 | 5 | 6 | 7 | 8 | 9 | Diff Av |
|---|---|---|---|---|---|---|---|---|---|---|---|
| .2 | 22.62 | 22.65 | 22.68 | 22.71 | 22.75 | 22.78 | 22.81 | 22.84 | 22.87 | 22.90 | |
| .3 | 22.93 | 22.97 | 23.00 | 23.03 | 23.06 | 23.09 | 23.12 | 23.15 | 23.18 | 23.22 | |
| .4 | 23.25 | 23.28 | 23.31 | 23.34 | 23.37 | 23.40 | 23.44 | 23.47 | 23.50 | 23.53 | |
| 7.5 | 23.56 | 23.59 | 23.62 | 23.66 | 23.69 | 23.72 | 23.75 | 23.78 | 23.81 | 23.84 | |
| .6 | 23.88 | 23.91 | 23.94 | 23.97 | 24.00 | 24.03 | 24.06 | 24.10 | 24.13 | 24.16 | |
| .7 | 24.19 | 24.22 | 24.25 | 24.28 | 24.32 | 24.35 | 24.38 | 24.41 | 24.44 | 24.47 | |
| .8 | 24.50 | 24.54 | 24.57 | 24.60 | 24.63 | 24.66 | 24.69 | 24.72 | 24.76 | 24.79 | |
| .9 | 24.82 | 24.85 | 24.88 | 24.91 | 24.94 | 24.98 | 25.01 | 25.04 | 25.07 | 25.10 | |
| 8.0 | 25.13 | 25.16 | 25.20 | 25.23 | 25.26 | 25.29 | 25.32 | 25.35 | 25.38 | 25.42 | |
| .1 | 25.45 | 25.48 | 25.51 | 25.54 | 25.57 | 25.60 | 25.64 | 25.67 | 25.70 | 25.73 | |
| .2 | 25.76 | 25.79 | 25.82 | 25.86 | 25.89 | 25.92 | 25.95 | 25.98 | 26.01 | 26.04 | |
| .3 | 26.08 | 26.11 | 26.14 | 26.17 | 26.20 | 26.23 | 26.26 | 26.30 | 26.33 | 26.36 | |
| .4 | 26.39 | 26.42 | 26.45 | 26.48 | 26.52 | 26.55 | 26.58 | 26.61 | 26.64 | 26.67 | |
| 8.5 | 26.70 | 26.73 | 26.77 | 26.80 | 26.83 | 26.86 | 26.89 | 26.92 | 26.95 | 26.99 | |
| .6 | 27.02 | 27.05 | 27.08 | 27.11 | 27.14 | 27.17 | 27.21 | 27.24 | 27.27 | 27.30 | |
| .7 | 27.33 | 27.36 | 27.39 | 27.43 | 27.46 | 27.49 | 27.52 | 27.55 | 27.58 | 27.61 | |
| .8 | 27.65 | 27.68 | 27.71 | 27.74 | 27.77 | 27.80 | 27.83 | 27.87 | 27.90 | 27.93 | |
| .9 | 27.96 | 27.99 | 28.02 | 28.05 | 28.09 | 28.12 | 28.15 | 28.18 | 28.21 | 28.24 | |
| 9.0 | 28.27 | 28.31 | 28.34 | 28.37 | 28.40 | 28.43 | 28.46 | 28.49 | 28.53 | 28.56 | |
| .1 | 28.59 | 28.62 | 28.65 | 28.68 | 28.71 | 28.75 | 28.78 | 28.81 | 28.84 | 28.87 | |
| .2 | 28.90 | 28.93 | 28.97 | 29.00 | 29.03 | 29.06 | 29.09 | 29.12 | 29.15 | 29.19 | |
| .3 | 29.22 | 29.25 | 29.28 | 29.31 | 29.34 | 29.37 | 29.41 | 29.44 | 29.47 | 29.50 | |
| .4 | 29.53 | 29.56 | 29.59 | 29.63 | 29.66 | 29.69 | 29.72 | 29.75 | 29.78 | 29.81 | |
| 9.5 | 29.85 | 29.88 | 29.91 | 29.94 | 29.97 | 30.00 | 30.03 | 30.07 | 30.10 | 30.13 | |
| .6 | 30.16 | 30.19 | 30.22 | 30.25 | 30.28 | 30.32 | 30.35 | 30.38 | 30.41 | 30.44 | |
| .7 | 30.47 | 30.50 | 30.54 | 30.57 | 30.60 | 30.63 | 30.66 | 30.69 | 30.72 | 30.76 | |
| .8 | 30.79 | 30.82 | 30.85 | 30.88 | 30.91 | 30.94 | 30.98 | 31.01 | 31.04 | 31.07 | |
| .9 | 31.10 | 31.13 | 31.16 | 31.20 | 31.23 | 31.26 | 31.29 | 31.32 | 31.35 | 31.38 | |
| 10.0 | 31.42 | | | | | | | | | | |

## Explanation of Table of Circumferences

This table gives the product of $\pi$ times any number $D$ from 1 to 10; that is, it is a table of multiples of $\pi$. ($D$ = diameter.)

Moving the decimal point *one* place in column $D$ is equivalent to moving it *one* place in the body of the table.

$$\text{Circumference} = \pi \times \text{diam} = 3.141593 \times \text{diam}$$

Conversely,

$$\text{Diameter} = \frac{1}{\pi} \times \text{circum} = 0.31831 \times \text{cirum}$$

**Areas of Circles by Hundredths**

| D | 0 | 1 | 2 | 3 | 4 | 5 | 6 | 7 | 8 | 9 | Avg diff |
|---|---|---|---|---|---|---|---|---|---|---|---|
| 1.0 | 0.785 | 0.801 | 0.817 | 0.833 | 0.849 | 0.866 | 0.882 | 0.899 | 0.916 | 0.933 | 16 |
| .1 | 0.950 | 0.968 | 0.985 | 1.003 | 1.021 | 1.039 | 1.057 | 1.075 | 1.094 | 1.112 | 18 |
| .2 | 1.131 | 1.150 | 1.169 | 1.188 | 1.208 | 1.227 | 1.247 | 1.267 | 1.287 | 1.307 | 20 |
| .3 | 1.327 | 1.348 | 1.368 | 1.389 | 1.410 | 1.431 | 1.453 | 1.474 | 1.496 | 1.517 | 21 |
| .4 | 1.539 | 1.561 | 1.584 | 1.606 | 1.629 | 1.651 | 1.674 | 1.697 | 1.720 | 1.744 | 23 |
| 1.5 | 1.767 | 1.791 | 1.815 | 1.839 | 1.863 | 1.887 | 1.911 | 1.936 | 1.961 | 1.986 | 24 |
| .6 | 2.011 | 2.036 | 2.061 | 2.087 | 2.112 | 2.138 | 2.164 | 2.190 | 2.217 | 2.243 | 26 |
| .7 | 2.270 | 2.297 | 2.324 | 2.351 | 2.378 | 2.405 | 2.433 | 2.461 | 2.488 | 2.516 | 27 |
| .8 | 2.545 | 2.573 | 2.602 | 2.630 | 2.659 | 2.688 | 2.717 | 2.746 | 2.776 | 2.806 | 29 |
| .9 | 2.835 | 2.865 | 2.895 | 2.926 | 2.956 | 2.986 | 3.017 | 3.048 | 3.079 | 3.110 | 31 |
| 2.0 | 3.142 | 3.173 | 3.205 | 3.237 | 3.269 | 3.301 | 3.333 | 3.365 | 3.398 | 3.431 | 32 |
| .1 | 3.464 | 3.497 | 3.530 | 3.563 | 3.597 | 3.631 | 3.664 | 3.698 | 3.733 | 3.767 | 34 |
| .2 | 3.801 | 3.836 | 3.871 | 3.906 | 3.941 | 3.976 | 4.011 | 4.047 | 4.083 | 4.119 | 35 |
| .3 | 4.155 | 4.191 | 4.227 | 4.264 | 4.301 | 4.337 | 4.374 | 4.412 | 4.449 | 4.486 | 37 |
| .4 | 4.524 | 4.562 | 4.600 | 4.638 | 4.676 | 4.714 | 4.753 | 4.792 | 4.831 | 4.870 | 38 |
| 2.5 | 4.909 | 4.948 | 4.988 | 5.027 | 5.067 | 5.107 | 5.147 | 5.187 | 5.228 | 5.269 | 40 |
| .6 | 5.309 | 5.350 | 5.391 | 5.433 | 5.474 | 5.515 | 5.557 | 5.599 | 5.641 | 5.683 | 42 |
| .7 | 5.726 | 5.768 | 5.811 | 5.853 | 5.896 | 5.940 | 5.983 | 6.026 | 6.070 | 6.114 | 43 |
| .8 | 6.158 | 6.202 | 6.246 | 6.290 | 6.335 | 6.379 | 6.424 | 6.469 | 6.514 | 6.560 | 45 |
| .9 | 6.605 | 6.651 | 6.697 | 6.743 | 6.789 | 6.835 | 6.881 | 6.928 | 6.975 | 7.022 | 46 |
| 3.0 | 7.069 | 7.116 | 7.163 | 7.211 | 7.258 | 7.306 | 7.354 | 7.402 | 7.451 | 7.499 | 48 |
| .1 | 7.548 | 7.596 | 7.645 | 7.694 | 7.744 | 7.793 | 7.843 | 7.892 | 7.942 | 7.992 | 49 |
| .2 | 8.042 | 8.093 | 8.143 | 8.194 | 8.245 | 8.296 | 8.347 | 8.398 | 8.450 | 8.501 | 51 |
| .3 | 8.553 | 8.605 | 8.657 | 8.709 | 8.762 | 8.814 | 8.867 | 8.920 | 8.973 | 9.026 | 53 |
| .4 | 9.079 | 9.133 | 9.186 | 9.240 | 9.294 | 9.348 | 9.402 | 9.457 | 9.511 | 9.566 | 54 |
| 3.5 | 9.621 | 9.676 | 9.731 | 9.787 | 9.842 | 9.898 | 9.954 | 10.010 | | | 56 |
| | | | | | | | | 10.01 | 10.07 | 10.12 | 6 |
| .6 | 10.18 | 10.24 | 10.29 | 10.35 | 10.41 | 10.46 | 10.52 | 10.58 | 10.64 | 10.69 | 6 |
| .7 | 10.75 | 10.81 | 10.87 | 10.93 | 10.99 | 11.04 | 11.10 | 11.16 | 11.22 | 11.28 | |
| .8 | 11.34 | 11.40 | 11.46 | 11.52 | 11.58 | 11.64 | 11.70 | 11.76 | 11.82 | 11.88 | |
| .9 | 11.95 | 12.01 | 12.07 | 12.13 | 12.19 | 12.25 | 12.32 | 12.38 | 12.44 | 12.50 | |
| 4.0 | 12.57 | 12.63 | 12.69 | 12.76 | 12.82 | 12.88 | 12.95 | 13.01 | 13.07 | 13.14 | 7 |
| .1 | 13.20 | 13.27 | 13.33 | 13.40 | 13.46 | 13.53 | 13.59 | 13.66 | 13.72 | 13.79 | |
| .2 | 13.85 | 13.92 | 13.99 | 14.05 | 14.12 | 14.19 | 14.25 | 14.32 | 14.39 | 14.45 | |
| .3 | 14.52 | 14.59 | 14.66 | 14.73 | 14.79 | 14.86 | 14.93 | 15.00 | 15.07 | 15.14 | |
| .4 | 15.21 | 15.27 | 15.34 | 15.41 | 15.48 | 15.55 | 15.62 | 15.69 | 15.76 | 15.83 | |
| 4.5 | 15.90 | 15.98 | 16.05 | 16.12 | 16.19 | 16.26 | 16.33 | 16.40 | 16.47 | 16.55 | |
| .6 | 16.62 | 16.69 | 16.76 | 16.84 | 16.91 | 16.98 | 17.06 | 17.13 | 17.20 | 17.28 | |
| .7 | 17.35 | 17.42 | 17.50 | 17.57 | 17.65 | 17.72 | 17.80 | 17.87 | 17.95 | 18.02 | |
| .8 | 18.10 | 18.17 | 18.25 | 18.32 | 18.40 | 18.47 | 18.55 | 18.63 | 18.70 | 18.78 | 8 |
| .9 | 18.86 | 18.93 | 19.01 | 19.09 | 19.17 | 19.24 | 19.32 | 19.40 | 19.48 | 19.56 | |
| 5.0 | 19.63 | 19.71 | 19.79 | 19.87 | 19.95 | 20.03 | 20.11 | 20.19 | 20.27 | 20.35 | 8 |
| .1 | 20.43 | 20.51 | 20.59 | 20.67 | 20.75 | 20.83 | 20.91 | 20.99 | 21.07 | 21.16 | |
| .2 | 21.24 | 21.32 | 21.40 | 21.48 | 21.57 | 21.65 | 21.73 | 21.81 | 21.90 | 21.98 | |
| .3 | 22.06 | 22.15 | 22.23 | 22.31 | 22.40 | 22.48 | 22.56 | 22.65 | 22.73 | 22.82 | |
| .4 | 22.90 | 22.99 | 23.07 | 23.16 | 23.24 | 23.33 | 23.41 | 23.50 | 23.59 | 23.67 | 9 |
| 5.5 | 23.76 | 23.84 | 23.93 | 24.02 | 24.11 | 24.19 | 24.28 | 24.37 | 24.45 | 24.54 | |
| .6 | 24.63 | 24.72 | 24.81 | 24.89 | 24.98 | 25.07 | 25.16 | 25.25 | 25.34 | 25.43 | |
| .7 | 25.52 | 25.61 | 25.70 | 25.79 | 25.88 | 25.97 | 26.06 | 26.15 | 26.24 | 26.33 | |
| .8 | 26.42 | 26.51 | 26.60 | 26.69 | 26.79 | 26.88 | 26.97 | 27.06 | 27.15 | 27.25 | |
| .9 | 27.34 | 27.43 | 27.53 | 27.62 | 27.71 | 27.81 | 27.90 | 27.99 | 28.09 | 28.18 | |
| 6.0 | 28.27 | 28.37 | 28.46 | 28.56 | 28.65 | 28.75 | 28.84 | 28.94 | 29.03 | 29.13 | 10 |
| .1 | 29.22 | 29.32 | 29.42 | 29.51 | 29.61 | 29.71 | 29.80 | 29.90 | 30.00 | 30.09 | |
| .2 | 30.19 | 30.29 | 30.39 | 30.48 | 30.58 | 30.68 | 30.78 | 30.88 | 30.97 | 31.07 | |
| .3 | 31.17 | 31.27 | 31.37 | 31.47 | 31.57 | 31.67 | 31.77 | 31.87 | 31.97 | 32.07 | |
| .4 | 32.17 | 32.27 | 32.37 | 32.47 | 32.57 | 32.67 | 32.78 | 32.88 | 32.98 | 33.08 | |
| 6.5 | 33.18 | 33.29 | 33.39 | 33.49 | 33.59 | 33.70 | 33.80 | 33.90 | 34.00 | 34.11 | |
| .6 | 34.21 | 34.32 | 34.42 | 34.52 | 34.63 | 34.73 | 34.84 | 34.94 | 35.05 | 35.15 | |
| .7 | 35.26 | 35.36 | 35.47 | 35.57 | 35.68 | 35.78 | 35.89 | 36.00 | 36.10 | 36.21 | 11 |
| .8 | 36.32 | 36.42 | 36.53 | 36.64 | 36.75 | 36.85 | 36.96 | 37.07 | 37.18 | 37.28 | |
| .9 | 37.39 | 37.50 | 37.61 | 37.72 | 37.83 | 37.94 | 38.05 | 38.16 | 38.26 | 38.37 | |
| 7.0 | 38.48 | 38.59 | 38.70 | 38.82 | 38.93 | 39.04 | 39.15 | 39.26 | 39.37 | 39.48 | |
| .1 | 39.59 | 39.70 | 39.82 | 39.93 | 40.04 | 40.15 | 40.26 | 40.38 | 40.49 | 40.60 | |

## Areas of Circles by Hundredths (*continued*)

| D | 0 | 1 | 2 | 3 | 4 | 5 | 6 | 7 | 8 | 9 | Avg diff |
|---|---|---|---|---|---|---|---|---|---|---|---|
| .2 | 40.72 | 40.83 | 40.94 | 41.06 | 41.17 | 41.28 | 41.40 | 41.51 | 41.62 | 41.74 | |
| .3 | 41.85 | 41.97 | 42.08 | 42.20 | 42.31 | 42.43 | 42.54 | 42.66 | 42.78 | 42.89 | 12 |
| .4 | 43.01 | 43.12 | 43.24 | 43.36 | 43.47 | 43.59 | 43.71 | 43.83 | 43.94 | 44.06 | |
| 7.5 | 44.18 | 44.30 | 44.41 | 44.53 | 44.65 | 44.77 | 44.89 | 45.01 | 45.13 | 45.25 | |
| .6 | 45.36 | 45.48 | 45.60 | 45.72 | 45.84 | 45.96 | 46.08 | 46.20 | 46.32 | 46.45 | |
| .7 | 46.57 | 46.69 | 46.81 | 46.93 | 47.05 | 47.17 | 47.29 | 47.42 | 47.54 | 47.66 | |
| .8 | 47.78 | 47.91 | 48.03 | 48.15 | 48.27 | 48.40 | 48.52 | 48.65 | 48.77 | 48.89 | |
| .9 | 49.02 | 49.14 | 49.27 | 49.39 | 49.51 | 49.64 | 49.76 | 49.89 | 50.01 | 50.14 | |
| 8.0 | 50.27 | 50.39 | 50.52 | 50.64 | 50.77 | 50.90 | 51.02 | 51.15 | 51.28 | 51.40 | 13 |
| .1 | 51.53 | 51.66 | 51.78 | 51.91 | 52.04 | 52.17 | 52.30 | 52.42 | 52.55 | 52.68 | |
| .2 | 52.81 | 52.94 | 53.07 | 53.20 | 53.33 | 53.46 | 53.59 | 53.72 | 53.85 | 53.98 | |
| .3 | 54.11 | 54.24 | 54.37 | 54.50 | 54.63 | 54.76 | 54.89 | 55.02 | 55.15 | 55.29 | |
| .4 | 55.42 | 55.55 | 55.68 | 55.81 | 55.95 | 56.08 | 56.21 | 56.35 | 56.48 | 56.61 | |
| 8.5 | 56.75 | 56.88 | 57.01 | 57.15 | 57.28 | 57.41 | 57.55 | 57.68 | 57.82 | 57.95 | |
| .6 | 58.09 | 58.22 | 58.36 | 58.49 | 58.63 | 58.77 | 58.90 | 59.04 | 59.17 | 59.31 | 14 |
| .7 | 59.45 | 59.58 | 59.72 | 59.86 | 59.99 | 60.13 | 60.27 | 60.41 | 60.55 | 60.68 | |
| .8 | 60.82 | 60.96 | 61.10 | 61.24 | 61.38 | 61.51 | 61.65 | 61.79 | 61.93 | 62.07 | |
| .9 | 62.21 | 62.35 | 62.49 | 62.63 | 62.77 | 62.91 | 63.05 | 63.19 | 63.33 | 63.48 | |
| 9.0 | 63.62 | 63.76 | 63.90 | 64.04 | 64.18 | 64.33 | 64.47 | 64.61 | 64.75 | 64.90 | |
| .1 | 65.04 | 65.18 | 65.33 | 65.47 | 65.61 | 65.76 | 65.90 | 66.04 | 66.19 | 66.33 | 15 |
| .2 | 66.48 | 66.62 | 66.77 | 66.91 | 67.06 | 67.20 | 67.35 | 67.49 | 67.64 | 67.78 | |
| .3 | 67.93 | 68.08 | 68.22 | 68.37 | 68.51 | 68.66 | 68.81 | 68.96 | 69.10 | 69.25 | |
| .4 | 69.40 | 69.55 | 69.69 | 69.84 | 69.99 | 70.14 | 70.29 | 70.44 | 70.58 | 70.73 | |
| 9.5 | 70.88 | 71.03 | 71.18 | 71.33 | 71.48 | 71.63 | 71.78 | 71.93 | 72.08 | 72.23 | |
| .6 | 72.38 | 72.53 | 72.68 | 72.84 | 72.99 | 73.14 | 73.29 | 73.44 | 73.59 | 73.75 | |
| .7 | 73.90 | 74.05 | 74.20 | 74.36 | 74.51 | 74.66 | 74.82 | 74.97 | 75.12 | 75.28 | |
| .8 | 75.43 | 75.58 | 75.74 | 75.89 | 76.05 | 76.20 | 76.36 | 76.51 | 76.67 | 76.82 | |
| .9 | 76.98 | 77.13 | 77.29 | 77.44 | 77.60 | 77.76 | 77.91 | 78.07 | 78.23 | 78.38 | 16 |

### Explanation of Table of Areas of Circles

Moving the decimal point *one* place in column $D$ is equivalent to moving it *two* places in the body of the table. ($D$ = diameter.)

$$\text{Area of circle} = \frac{\pi}{4} \times (\text{diam}^2) = 0.785398 \times (\text{diam}^2)$$

Conversely,

$$\text{Diam} = \sqrt{\frac{4}{\pi}} \times \sqrt{\text{area}} = 1.128379 \times \sqrt{\text{area}}$$

### Segments of Circles, Given h/c

Given: $b$ = height; $c$ = chord. To find the diameter of the circle, the length of arc, or the area of the segment, form the ratio $b/c$, and find from the table the value of (diam/$c$), (arc/$c$), or (area/$bc$); then, by a simple multiplication,

$$\text{diam} = c \times (\text{diam}/c)$$
$$\text{arc} = c \times (\text{arc}/c)$$
$$\text{area} = b \times c \times (\text{area}/bc)$$

The table gives also the angle subtended at the center, and the ratio of $b$ to $D$.

| $\dfrac{h}{c}$ | $\dfrac{\text{Diam}}{c}$ | Diff | $\dfrac{\text{Arc}}{c}$ | Diff | $\dfrac{\text{Area}}{h \times c}$ | Diff | Central angle, $v$ | Diff | $\dfrac{h}{\text{Diam}}$ | Diff |
|---|---|---|---|---|---|---|---|---|---|---|
| .00 | | | 1.000 | 0 | .6667 | 0 | 0.00° | 458 | .0000 | 4 |
| 1 | 25.010 | 12490 | 1.000 | 1 | .6667 | 2 | 4.58 | 458 | .0004 | 12 |
| 2 | 12.520 | *4157 | 1.001 | 1 | .6669 | 2 | 9.16 | 457 | .0016 | 20 |
| 3 | 8.363 | *2073 | 1.002 | 2 | .6671 | 4 | 13.73 | 457 | .0036 | 28 |
| 4 | 6.290 | *1240 | 1.004 | 3 | .6675 | 5 | 18.30 | 454 | .0064 | 35 |
| .05 | 5.050 | *823 | 1.007 | 3 | .6680 | 6 | 22.84° | 453 | .0099 | 43 |
| 6 | 4.227 | *586 | 1.010 | 3 | .6686 | 7 | 27.37 | 451 | .0142 | 50 |
| 7 | 3.641 | *436 | 1.013 | 4 | .6693 | 8 | 31.88 | 448 | .0192 | 58 |
| 8 | 3.205 | *337 | 1.017 | 4 | .6701 | 9 | 36.36 | 446 | .0250 | 64 |
| 9 | 2.868 | *268 | 1.021 | 5 | .6710 | 10 | 40.82 | 442 | .0314 | 71 |
| .10 | 2.600 | *217 | 1.026 | 6 | .6720 | 11 | 45.24° | 439 | .0385 | 77 |
| 1 | 2.383 | *180 | 1.032 | 6 | .6731 | 12 | 49.63 | 435 | .0462 | 83 |
| 2 | 2.203 | *150 | 1.038 | 6 | .6743 | 13 | 53.98 | 432 | .0545 | 88 |
| 3 | 2.053 | *127 | 1.044 | 7 | .6756 | 14 | 58.30 | 427 | .0633 | 94 |
| 4 | 1.926 | *109 | 1.051 | 8 | .6770 | 15 | 62.57 | 423 | .0727 | 99 |
| .15 | 1.817 | *94 | 1.059 | 8 | .6785 | 16 | 66.80° | 418 | .0826 | 103 |
| 6 | 1.723 | *82 | 1.067 | 8 | .6801 | 17 | 70.98 | 413 | .0929 | 107 |
| 7 | 1.641 | *72 | 1.075 | 9 | .6818 | 18 | 75.11 | 409 | .1036 | 111 |
| 8 | 1.569 | *63 | 1.084 | 10 | .6836 | 19 | 79.20 | 403 | .1147 | 116 |
| 9 | 1.506 | 56 | 1.094 | 9 | .6855 | 20 | 83.23 | 399 | .1263 | 116 |
| .20 | 1.450 | 50 | 1.103 | 11 | .6875 | 21 | 87.21° | 392 | .1379 | 120 |
| 1 | 1.400 | 44 | 1.114 | 10 | .6896 | 22 | 91.13 | 387 | .1499 | 123 |
| 2 | 1.356 | 39 | 1.124 | 12 | .6918 | 23 | 95.00 | 381 | .1622 | 124 |
| 3 | 1.317 | 35 | 1.136 | 11 | .6941 | 24 | 98.81 | 375 | .1746 | 127 |
| 4 | 1.282 | 32 | 1.147 | 12 | .6965 | 24 | 102.56 | 370 | .1873 | 127 |
| .25 | 1.250 | 28 | 1.159 | 12 | .6989 | 25 | 106.26° | 364 | .2000 | 128 |
| 6 | 1.222 | 26 | 1.171 | 13 | .7014 | 27 | 109.90 | 358 | .2128 | 130 |
| 7 | 1.196 | 23 | 1.184 | 13 | .7041 | 27 | 113.48 | 352 | .2258 | 129 |
| 8 | 1.173 | 21 | 1.197 | 14 | .7068 | 28 | 117.00 | 345 | .2387 | 130 |
| 9 | 1.152 | 19 | 1.211 | 14 | .7096 | 29 | 120.45 | 341 | .2517 | 130 |
| .30 | 1.133 | 17 | 1.225 | 14 | .7125 | 29 | 123.86° | 334 | .2647 | 130 |
| 1 | 1.116 | 15 | 1.239 | 15 | .7154 | 31 | 127.20 | 328 | .2777 | 129 |
| 2 | 1.101 | 13 | 1.254 | 15 | .7185 | 31 | 130.48 | 322 | .2906 | 128 |
| 3 | 1.088 | 13 | 1.269 | 15 | .7216 | 32 | 133.70 | 316 | .3034 | 128 |
| 4 | 1.075 | 11 | 1.284 | 16 | .7248 | 32 | 136.86 | 311 | .3162 | 127 |
| .35 | 1.064 | 10 | 1.300 | 16 | .7280 | 34 | 139.97° | 305 | .3289 | 125 |
| 6 | 1.054 | 8 | 1.316 | 16 | .7314 | 34 | 143.02 | 299 | .3414 | 124 |
| 7 | 1.046 | 8 | 1.332 | 17 | .7348 | 35 | 146.01 | 293 | .3538 | 123 |
| 8 | 1.038 | 7 | 1.349 | 17 | .7383 | 36 | 148.94 | 288 | .3661 | 122 |
| 9 | 1.031 | 6 | 1.366 | 17 | .7419 | 36 | 151.82 | 282 | .3783 | 119 |
| .40 | 1.025 | 5 | 1.383 | 18 | .7455 | 37 | 154.64° | 277 | .3902 | 119 |
| 1 | 1.020 | 5 | 1.401 | 18 | .7492 | 38 | 157.41 | 271 | .4021 | 116 |
| 2 | 1.015 | 4 | 1.419 | 18 | .7530 | 38 | 160.12 | 266 | .4137 | 115 |
| 3 | 1.011 | 3 | 1.437 | 18 | .7568 | 39 | 162.78 | 261 | .4252 | 112 |
| 4 | 1.008 | 2 | 1.455 | 19 | .7607 | 40 | 165.39 | 256 | .4364 | 111 |
| .45 | 1.006 | 3 | 1.474 | 19 | .7647 | 40 | 167.95° | 251 | .4475 | 109 |
| 6 | 1.003 | 1 | 1.493 | 19 | .7687 | 41 | 170.46 | 245 | .4584 | 107 |
| 7 | 1.002 | 1 | 1.512 | 19 | .7728 | 41 | 172.91 | 241 | .4691 | 105 |
| 8 | 1.001 | 1 | 1.531 | 20 | .7769 | 42 | 175.32 | 237 | .4796 | 103 |
| 9 | 1.000 | 0 | 1.551 | 20 | .7811 | 43 | 177.69 | 231 | .4899 | 101 |
| .50 | 1.000 | | 1.571 | | .7854 | | 180.00° | | .5000 | |

*Interpolation may be inaccurate at these points.

## Segments of Circles, Given h/D

Given: $b$ = height; $D$ = diameter of circle. To find the chord, the length of arc, or the area of the segment, form the ratio $b/D$, and find from the table the value of (chord/$D$), (arc/$D$), or (area/$D^2$); then by a simple multiplication,

$$\text{chord} = D \times (\text{chord}/D)$$
$$\text{arc} = D \times (\text{arc}/D)$$
$$\text{area} = D^2 \times (\text{area}/D^2)$$

The table gives also the angle subtended at the center, the ratio of the arc of the segment to the whole circumference, and the ratio of the area of the segment to the area of the whole circle.

| $\dfrac{h}{D}$ | $\dfrac{\text{Arc}}{D}$ | Diff | $\dfrac{\text{Area}}{D^2}$ | Diff | Central angle, $v$ | Diff | $\dfrac{\text{Chord}}{D}$ | Diff | $\dfrac{\text{Arc}}{\text{Circum}}$ | Diff | $\dfrac{\text{Area}}{\text{Circle}}$ | Diff |
|---|---|---|---|---|---|---|---|---|---|---|---|---|
| .00 | 0.000 | 2003 | .0000 | 13 | 0.00° | 2296 | .0000 | *1990 | .0000 | *638 | .0000 | 17 |
| 1 | .2003 | *835 | .0013 | 24 | 22.96 | *956 | .1990 | *810 | .0638 | *265 | .0017 | 31 |
| 2 | .2838 | *644 | .0037 | 32 | 32.52 | *738 | .2800 | *612 | .0903 | *205 | .0048 | 39 |
| 3 | .3482 | *545 | .0069 | 36 | 39.90 | *625 | .3412 | *507 | .1108 | *174 | .0087 | 47 |
| 4 | .4027 | *483 | .0105 | 42 | 46.15 | *553 | .3919 | *440 | .1282 | *154 | .0134 | 53 |
| .05 | .4510 | *439 | .0147 | 45 | 51.68° | *504 | .4359 | *391 | .1436 | *139 | .0187 | 58 |
| 6 | .4949 | *406 | .0192 | 50 | 56.72 | *465 | .4750 | *353 | .1575 | *130 | .0245 | 63 |
| 7 | .5355 | *380 | .0242 | 52 | 61.37 | *435 | .5103 | *323 | .1705 | 121 | .0308 | 67 |
| 8 | .5735 | *359 | .0294 | 56 | 65.72 | *411 | .5426 | *298 | .1826 | 114 | .0375 | 71 |
| 9 | .6094 | *341 | .0350 | 59 | 69.83 | *391 | .5724 | *276 | .1940 | 108 | .0446 | 74 |
| .10 | .6435 | *326 | .0409 | 61 | 73.74° | *374 | .6000 | *258 | .2048 | 104 | .0520 | 78 |
| 1 | .6761 | *314 | .0470 | 64 | 77.48 | *359 | .6258 | *241 | .2152 | 100 | .0598 | 82 |
| 2 | .7075 | *302 | .0534 | 66 | 81.07 | *347 | .6499 | *227 | .2252 | 96 | .0680 | 84 |
| 3 | .7377 | *293 | .0600 | 68 | 84.54 | *335 | .6726 | *214 | .2348 | 93 | .0764 | 87 |
| 4 | .7670 | *284 | .0668 | 71 | 87.89 | *326 | .6940 | *201 | .2441 | 91 | .0851 | 90 |
| .15 | .7954 | 276 | .0739 | 72 | 91.15° | 316 | .7141 | *191 | .2532 | 88 | .0941 | 92 |
| 6 | .8230 | 270 | .0811 | 74 | 94.31 | 309 | .7332 | *181 | .2620 | 86 | .1033 | 94 |
| 7 | .8500 | 263 | .0885 | 76 | 97.40 | 302 | .7513 | *171 | .2706 | 83 | .1127 | 97 |
| 8 | .8763 | 258 | .0961 | 78 | 100.42 | 295 | .7684 | 162 | .2789 | 82 | .1224 | 99 |
| 9 | .9021 | 252 | .1039 | 79 | 103.37 | 289 | .7846 | 154 | .2871 | 81 | .1323 | 101 |
| .20 | 0.9273 | 248 | .1118 | 81 | 106.26° | 284 | .8000 | 146 | .2952 | 79 | .1424 | 103 |
| 1 | 0.9521 | 243 | .1199 | 82 | 109.01 | 279 | .8146 | 139 | .3031 | 77 | .1527 | 104 |
| 2 | 0.9764 | 240 | .1281 | 84 | 111.89 | 274 | .8285 | 132 | .3108 | 76 | .1631 | 107 |
| 3 | 1.0004 | 235 | .1365 | 84 | 114.63 | 271 | .8417 | 125 | .3184 | 75 | .1738 | 108 |
| 4 | 1.0239 | 233 | .1449 | 86 | 117.34 | 266 | .8542 | 118 | .3259 | 74 | .1846 | 109 |
| .25 | 1.0472 | 229 | .1535 | 88 | 120.00° | 263 | .8660 | 113 | .3333 | 73 | .1955 | 111 |
| 6 | 1.0701 | 227 | .1623 | 88 | 122.63 | 260 | .8773 | 106 | .3406 | 72 | .2066 | 112 |
| 7 | 1.0928 | 224 | .1711 | 89 | 125.23 | 256 | .8879 | 101 | .3478 | 72 | .2178 | 114 |
| 8 | 1.1152 | 222 | .1800 | 90 | 127.79 | 254 | .8980 | 95 | .3550 | 70 | .2292 | 115 |
| 9 | 1.1374 | 219 | .1890 | 92 | 130.33 | 251 | .9075 | 90 | .3620 | 70 | .2407 | 116 |
| .30 | 1.1593 | 217 | .1982 | 92 | 132.84° | 249 | .9165 | 85 | .3690 | 69 | .2523 | 117 |
| 1 | 1.1810 | 215 | .2074 | 93 | 135.33 | 247 | .9250 | 80 | .3759 | 69 | .2640 | 119 |
| 2 | 1.2025 | 214 | .2167 | 93 | 137.80 | 245 | .9330 | 74 | .3828 | 68 | .2759 | 119 |
| 3 | 1.2239 | 212 | .2260 | 95 | 140.25 | 242 | .9404 | 70 | .3896 | 67 | .2878 | 120 |
| 4 | 1.2451 | 210 | .2355 | 95 | 142.67 | 241 | .9474 | 65 | .3963 | 67 | .2998 | 121 |
| .35 | 1.2661 | 209 | .2450 | 96 | 145.08° | 240 | .9539 | 61 | .4030 | 67 | .3119 | 122 |
| 6 | 1.2870 | 208 | .2546 | 96 | 147.48 | 238 | .9600 | 56 | .4097 | 66 | .3241 | 123 |
| 7 | 1.3078 | 206 | .2642 | 97 | 149.86 | 237 | .9656 | 52 | .4163 | 66 | .3364 | 123 |
| 8 | 1.3284 | 206 | .2739 | 97 | 152.23 | 235 | .9708 | 47 | .4229 | 65 | .3487 | 124 |
| 9 | 1.3490 | 204 | .2836 | 98 | 154.58 | 235 | .9755 | 43 | .4294 | 65 | .3611 | 124 |
| .40 | 1.3694 | 204 | .2934 | 98 | 156.93° | 233 | .9798 | 39 | .4359 | 65 | .3735 | 125 |
| 1 | 1.3898 | 203 | .3032 | 98 | 159.26 | 233 | .9837 | 34 | .4424 | 65 | .3860 | 126 |
| 2 | 1.4101 | 202 | .3130 | 99 | 161.59 | 231 | .9871 | 31 | .4489 | 64 | .3986 | 126 |
| 3 | 1.4303 | 202 | .3229 | 99 | 163.90 | 232 | .9902 | 26 | .4553 | 64 | .4112 | 126 |
| 4 | 1.4505 | 201 | .3328 | 100 | 166.22 | 230 | .9928 | 22 | .4617 | 64 | .4238 | 126 |
| .45 | 1.4706 | 201 | .3428 | 99 | 168.52° | 230 | .9950 | 18 | .4681 | 64 | .4364 | 127 |
| 6 | 1.4907 | 201 | .3527 | 100 | 170.82 | 230 | .9968 | 14 | .4745 | 64 | .4491 | 127 |
| 7 | 1.5108 | 200 | .3627 | 100 | 173.12 | 229 | .9982 | 10 | .4809 | 64 | .4618 | 127 |
| 8 | 1.5308 | 200 | .3727 | 100 | 175.41 | 230 | .9992 | 6 | .4873 | 63 | .4745 | 128 |
| 9 | 1.5508 | 200 | .3827 | 100 | 177.71 | 229 | .9998 | 2 | .4936 | 64 | .4873 | 127 |
| .50 | 1.5708 | | .3927 | | 180.00° | | 1.0000 | | .5000 | | .5000 | |

*Interpolation may be inaccurate at these points.

**Regular Polygons**

$n$ = number of sides

$v$ = 360°/$n$ = angle subtended at the center by one side

$a$ = length of one side = $R\left(2\sin\dfrac{v}{2}\right) = r\left(2\tan\dfrac{v}{2}\right)$

$R$ = radius of circumscribed circle = $a\left(\tfrac{1}{2}\csc\dfrac{v}{2}\right) = r\left(\sec\dfrac{v}{2}\right)$

$r$ = radius of inscribed circle = $R\left(\cos\dfrac{v}{2}\right) = a\left(\tfrac{1}{2}\cot\dfrac{v}{2}\right)$

Area = $a^2\left(\tfrac{1}{4}\,n\cot\dfrac{v}{2}\right) = R^2(\tfrac{1}{2}\,n\sin v) = r^2\left(n\tan\dfrac{v}{2}\right)$

| $n$ | $v$ | $\dfrac{\text{Area}}{a^2}$ | $\dfrac{\text{Area}}{R^2}$ | $\dfrac{\text{Area}}{r^2}$ | $\dfrac{R}{a}$ | $\dfrac{R}{r}$ | $\dfrac{a}{R}$ | $\dfrac{a}{r}$ | $\dfrac{r}{R}$ | $\dfrac{r}{a}$ |
|---|---|---|---|---|---|---|---|---|---|---|
| 3 | 120° | 0.4330 | 1.299 | 5.196 | 0.5774 | 2.000 | 1.732 | 3.464 | 0.5000 | 0.2887 |
| 4 | 90° | 1.000 | 2.000 | 4.000 | 0.7071 | 1.414 | 1.414 | 2.000 | 0.7071 | 0.5000 |
| 5 | 72° | 1.721 | 2.378 | 3.633 | 0.8507 | 1.236 | 1.176 | 1.453 | 0.8090 | 0.6882 |
| 6 | 60° | 2.598 | 2.598 | 3.464 | 1.0000 | 1.155 | 1.000 | 1.155 | 0.8660 | 0.8660 |
| 7 | 51°.43 | 3.634 | 2.736 | 3.371 | 1.152 | 1.110 | 0.8678 | 0.9631 | 0.9010 | 1.038 |
| 8 | 45° | 4.828 | 2.828 | 3.314 | 1.307 | 1.082 | 0.7654 | 0.8284 | 0.9239 | 1.207 |
| 9 | 40° | 6.182 | 2.893 | 3.276 | 1.462 | 1.064 | 0.6840 | 0.7279 | 0.9397 | 1.374 |
| 10 | 36° | 7.694 | 2.939 | 3.249 | 1.618 | 1.052 | 0.6180 | 0.6498 | 0.9511 | 1.539 |
| 12 | 30° | 11.20 | 3.000 | 3.215 | 1.932 | 1.035 | 0.5176 | 0.5359 | 0.9659 | 1.866 |
| 15 | 24° | 17.64 | 3.051 | 3.188 | 2.405 | 1.022 | 0.4158 | 0.4251 | 0.9781 | 2.352 |
| 16 | 22°.50 | 20.11 | 3.062 | 3.183 | 2.563 | 1.020 | 0.3902 | 0.3978 | 0.9808 | 2.514 |
| 20 | 18° | 31.57 | 3.090 | 3.168 | 3.196 | 1.013 | 0.3129 | 0.3168 | 0.9877 | 3.157 |
| 24 | 15° | 45.58 | 3.106 | 3.160 | 3.831 | 1.009 | 0.2611 | 0.2633 | 0.9914 | 3.798 |
| 32 | 11°.25 | 81.23 | 3.121 | 3.152 | 5.101 | 1.005 | 0.1960 | 0.1970 | 0.9952 | 5.077 |
| 48 | 7°.50 | 183.1 | 3.133 | 3.146 | 7.645 | 1.002 | 0.1308 | 0.1311 | 0.9979 | 7.629 |
| 64 | 5°.625 | 325.7 | 3.137 | 3.144 | 10.19 | 1.001 | 0.0981 | 0.0983 | 0.9968 | 10.18 |

**Binomial Coefficients**

(For table giving binomial coefficients for fractional values of $n$, see Sec. 2, Algebra)

$(n)_0 = 1 \qquad (n)_1 = n \qquad (n)_2 = \dfrac{n(n-1)}{1\times 2} \qquad (n)_3 = \dfrac{n(n)-1)(n-2)}{1\times 2\times 3} \qquad$ etc. $\qquad$ in general

$(n)_r = \dfrac{n(n-1)(n-2)\ldots(n-[r-1])}{1\times 2\times 3\ldots\times r} \qquad$ Other notations: $nC_r = \begin{pmatrix} n \\ r \end{pmatrix} = (n)_r$

| $n$ | $(n)_0$ | $(n)_1$ | $(n)_2$ | $(n)_3$ | $(n)_4$ | $(n)_5$ | $(n)_6$ | $(n)_7$ | $(n)_8$ | $(n)_9$ | $(n)_{10}$ | $(n)_{11}$ | $(n)_{12}$ | $(n)_{13}$ |
|---|---|---|---|---|---|---|---|---|---|---|---|---|---|---|
| 1 | 1 | 1 | | | | | | | | | | | | |
| 2 | 1 | 2 | 1 | | | | | | | | | | | |
| 3 | 1 | 3 | 3 | 1 | | | | | | | | | | |
| 4 | 1 | 4 | 6 | 4 | 1 | | | | | | | | | |
| 5 | 1 | 5 | 10 | 10 | 5 | 1 | | | | | | | | |
| 6 | 1 | 6 | 15 | 20 | 15 | 6 | 1 | | | | | | | |
| 7 | 1 | 7 | 21 | 35 | 35 | 21 | 7 | 1 | | | | | | |
| 8 | 1 | 8 | 28 | 56 | 70 | 56 | 28 | 8 | 1 | | | | | |
| 9 | 1 | 9 | 36 | 84 | 126 | 126 | 84 | 36 | 9 | 1 | | | | |
| 10 | 1 | 10 | 45 | 120 | 210 | 252 | 210 | 120 | 45 | 10 | 1 | | | |
| 11 | 1 | 11 | 55 | 165 | 330 | 462 | 462 | 330 | 165 | 55 | 11 | 1 | | |
| 12 | 1 | 12 | 66 | 220 | 495 | 792 | 924 | 792 | 495 | 220 | 66 | 12 | 1 | |
| 13 | 1 | 13 | 78 | 286 | 715 | 1287 | 1716 | 1716 | 1287 | 715 | 286 | 78 | 13 | 1 |
| 14 | 1 | 14 | 91 | 364 | 1001 | 2002 | 3003 | 3432 | 3003 | 2002 | 1001 | 364 | 91 | 14 |
| 15 | 1 | 15 | 105 | 455 | 1365 | 3003 | 5005 | 6435 | 6435 | 5005 | 3003 | 1365 | 455 | 105 |

For $n = 14$, $(n)_{14} = 1$; for $n = 15$, $(n)_{14} = 15$, and $(n)_{15} = 1$.

**Common Logarithms**

| Number | 0 | 1 | 2 | 3 | 4 | 5 | 6 | 7 | 8 | 9 | Avg diff |
|---|---|---|---|---|---|---|---|---|---|---|---|
| 1.0 | 0.0000 | 0043 | 0086 | 0128 | 0170 | 0212 | 0253 | 0294 | 0334 | 0374 | |
| 1.1 | 0414 | 0453 | 0492 | 0531 | 0569 | 0607 | 0645 | 0682 | 0719 | 0755 | |
| 1.2 | 0792 | 0828 | 0864 | 0899 | 0934 | 0969 | 1004 | 1038 | 1072 | 1106 | |
| 1.3 | 1139 | 1173 | 1206 | 1239 | 1271 | 1303 | 1335 | 1367 | 1399 | 1430 | |
| 1.4 | 1461 | 1492 | 1523 | 1553 | 1584 | 1614 | 1644 | 1673 | 1703 | 1732 | |
| 1.5 | 1761 | 1790 | 1818 | 1847 | 1875 | 1903 | 1931 | 1959 | 1987 | 2014 | |
| 1.6 | 2041 | 2068 | 2095 | 2122 | 2148 | 2175 | 2201 | 2227 | 2253 | 2279 | |
| 1.7 | 2304 | 2330 | 2355 | 2380 | 2405 | 2430 | 2455 | 2480 | 2504 | 2529 | |
| 1.8 | 2553 | 2577 | 2601 | 2625 | 2648 | 2672 | 2695 | 2718 | 2742 | 2765 | |
| 1.9 | 2788 | 2810 | 2833 | 2856 | 2878 | 2900 | 2923 | 2945 | 2967 | 2989 | |
| 2.0 | 0.3010 | 3032 | 3054 | 3075 | 3096 | 3118 | 3139 | 3160 | 3181 | 3201 | 21 |
| 2.1 | 3222 | 3243 | 3263 | 3284 | 3304 | 3324 | 3345 | 3365 | 3385 | 3404 | 20 |
| 2.2 | 3424 | 3444 | 3464 | 3483 | 3502 | 3522 | 3541 | 3560 | 3579 | 3598 | 19 |
| 2.3 | 3617 | 3636 | 3655 | 3674 | 3692 | 3711 | 3729 | 3747 | 3766 | 3784 | 18 |
| 2.4 | 3802 | 3820 | 3838 | 3856 | 3874 | 3892 | 3909 | 3927 | 3945 | 3962 | 17 |
| 2.5 | 3979 | 3997 | 4014 | 4031 | 4048 | 4065 | 4082 | 4099 | 4116 | 4133 | 17 |
| 2.6 | 4150 | 4166 | 4183 | 4200 | 4216 | 4232 | 4249 | 4265 | 4281 | 4298 | 16 |
| 2.7 | 4314 | 4330 | 4346 | 4362 | 4378 | 4393 | 4409 | 4425 | 4440 | 4456 | 16 |
| 2.8 | 4472 | 4487 | 4502 | 4518 | 4533 | 4548 | 4564 | 4579 | 4594 | 4609 | 15 |
| 2.9 | 4624 | 4639 | 4654 | 4669 | 4683 | 4698 | 4713 | 4728 | 4742 | 4757 | 15 |
| 3.0 | 0.4771 | 4786 | 4800 | 4814 | 4829 | 4843 | 4857 | 4871 | 4886 | 4900 | 14 |
| 3.1 | 4914 | 4928 | 4942 | 4955 | 4969 | 4983 | 4997 | 5011 | 5024 | 5038 | 14 |
| 3.2 | 5051 | 5065 | 5079 | 5092 | 5105 | 5119 | 5132 | 5145 | 5159 | 5172 | 13 |
| 3.3 | 5185 | 5198 | 5211 | 5224 | 5237 | 5250 | 5263 | 5276 | 5289 | 5302 | 13 |
| 3.4 | 5315 | 5328 | 5340 | 5353 | 5366 | 5378 | 5391 | 5403 | 5416 | 5428 | 13 |
| 3.5 | 5441 | 5453 | 5465 | 5478 | 5490 | 5502 | 5514 | 5527 | 5539 | 5551 | 12 |
| 3.6 | 5563 | 5575 | 5587 | 5599 | 5611 | 5623 | 5635 | 5647 | 5658 | 5670 | 12 |
| 3.7 | 5682 | 5694 | 5705 | 5717 | 5729 | 5740 | 5752 | 5763 | 5775 | 5786 | 12 |
| 3.8 | 5798 | 5809 | 5821 | 5832 | 5843 | 5855 | 5866 | 5877 | 5888 | 5899 | 11 |
| 3.9 | 5911 | 5922 | 5933 | 5944 | 5955 | 5966 | 5977 | 5988 | 5999 | 6010 | 11 |
| 4.0 | 0.6021 | 6031 | 6042 | 6053 | 6064 | 6075 | 6085 | 6096 | 6107 | 6117 | 11 |
| 4.1 | 6128 | 6138 | 6149 | 6160 | 6170 | 6180 | 6191 | 6201 | 6212 | 6222 | 10 |
| 4.2 | 6232 | 6243 | 6253 | 6263 | 6274 | 6284 | 6294 | 6304 | 6314 | 6325 | 10 |
| 4.3 | 6335 | 6345 | 6355 | 6365 | 6375 | 6385 | 6395 | 6405 | 6415 | 6425 | 10 |
| 4.4 | 6435 | 6444 | 6454 | 6464 | 6474 | 6484 | 6493 | 6503 | 6513 | 6522 | 10 |
| 4.5 | 6532 | 6542 | 6551 | 6561 | 6571 | 6580 | 6590 | 6599 | 6609 | 6618 | 10 |
| 4.6 | 6628 | 6637 | 6646 | 6656 | 6665 | 6675 | 6684 | 6693 | 6702 | 6712 | 10 |
| 4.7 | 6721 | 6730 | 6739 | 6749 | 6758 | 6767 | 6776 | 6785 | 6794 | 6803 | 9 |
| 4.8 | 6812 | 6821 | 6830 | 6839 | 6848 | 6857 | 6866 | 6875 | 6884 | 6893 | 9 |
| 4.9 | 6902 | 6911 | 6920 | 6928 | 6937 | 6946 | 6955 | 6964 | 6972 | 6981 | 9 |

*(Avg diff column for rows 1.0–1.9:* See pages 1-40 to 1-42*)*

$$\log \pi = 0.4971 \qquad \log \pi/2 = 0.1961 \qquad \log \pi^2 = 0.9943 \qquad \log \sqrt{\pi} = 0.2486$$
$$\log e = 0.4343 \qquad \log (0.4343) = 0.6378 - 1$$

These two pages give the common logarithms of numbers between 1 and 10, correct to four places. Moving the decimal point $n$ places to the right [or left] in the number is equivalent to adding $n$ [or $-n$] to the logarithm. Thus, $\log 0.017453 = 0.2419 - 2$, which may always be written $\overline{2}.2419$ or $8.2419 - 10$. For procedures involved in finding the number corresponding to a given logarithm and for graphical representations of functions, see Sec. 2.

$$\log (ab) = \log a + \log b \qquad\qquad \log (a^N) = N \log a$$

$$\log \left(\frac{a}{b}\right) = \log a - \log b \qquad\qquad \log \left(\sqrt[v]{a}\right) = \frac{1}{N} \log a$$

**Common Logarithms** (continued)

| Num-ber | 0 | 1 | 2 | 3 | 4 | 5 | 6 | 7 | 8 | 9 | Avg diff |
|---|---|---|---|---|---|---|---|---|---|---|---|
| 5.0 | 0.6990 | 6998 | 7007 | 7016 | 7024 | 7033 | 7042 | 7050 | 7059 | 7067 | 9 |
| 5.1 | 7076 | 7084 | 7093 | 7101 | 7110 | 7118 | 7126 | 7135 | 7143 | 7152 | 8 |
| 5.2 | 7160 | 7168 | 7177 | 7185 | 7193 | 7202 | 7210 | 7218 | 7226 | 7235 | 8 |
| 5.3 | 7243 | 7251 | 7259 | 7267 | 7275 | 7284 | 7292 | 7300 | 7308 | 7316 | 8 |
| 5.4 | 7324 | 7332 | 7340 | 7348 | 7356 | 7364 | 7372 | 7380 | 7388 | 7396 | 8 |
| 5.5 | 7404 | 7412 | 7419 | 7427 | 7435 | 7443 | 7451 | 7459 | 7466 | 7474 | 8 |
| 5.6 | 7482 | 7490 | 7497 | 7505 | 7513 | 7520 | 7528 | 7536 | 7543 | 7551 | 8 |
| 5.7 | 7559 | 7566 | 7574 | 7582 | 7589 | 7597 | 7604 | 7612 | 7619 | 7627 | 8 |
| 5.8 | 7634 | 7642 | 7649 | 7657 | 7664 | 7672 | 7679 | 7686 | 7694 | 7701 | 7 |
| 5.9 | 7709 | 7716 | 7723 | 7731 | 7738 | 7745 | 7752 | 7760 | 7767 | 7774 | 7 |
| 6.0 | 0.7782 | 7789 | 7796 | 7803 | 7810 | 7818 | 7825 | 7832 | 7839 | 7846 | 7 |
| 6.1 | 7853 | 7860 | 7868 | 7875 | 7882 | 7889 | 7896 | 7903 | 7910 | 7917 | 7 |
| 6.2 | 7924 | 7931 | 7938 | 7945 | 7952 | 7959 | 7966 | 7973 | 7980 | 7987 | 7 |
| 6.3 | 7993 | 8000 | 8007 | 8014 | 8021 | 8028 | 8035 | 8041 | 8048 | 8055 | 7 |
| 6.4 | 8062 | 8069 | 8075 | 8082 | 8089 | 8096 | 8102 | 8109 | 8116 | 8122 | 7 |
| 6.5 | 8129 | 8136 | 8142 | 8149 | 8156 | 8162 | 8169 | 8176 | 8182 | 8189 | 7 |
| 6.6 | 8195 | 8202 | 8209 | 8215 | 8222 | 8228 | 8235 | 8241 | 8248 | 8254 | 7 |
| 6.7 | 8261 | 8267 | 8274 | 8280 | 8287 | 8293 | 8299 | 8306 | 8312 | 8319 | 6 |
| 6.8 | 8325 | 8331 | 8338 | 8344 | 8351 | 8357 | 8363 | 8370 | 8376 | 8382 | 6 |
| 6.9 | 8388 | 8395 | 8401 | 8407 | 8414 | 8420 | 8426 | 8432 | 8439 | 8445 | 6 |
| 7.0 | 0.8451 | 8457 | 8463 | 8470 | 8476 | 8482 | 8488 | 8494 | 8500 | 8506 | 6 |
| 7.1 | 8513 | 8519 | 8525 | 8531 | 8537 | 8543 | 8549 | 8555 | 8561 | 8567 | 6 |
| 7.2 | 8573 | 8579 | 8585 | 8591 | 8597 | 8603 | 8609 | 8615 | 8621 | 8627 | 6 |
| 7.3 | 8633 | 8639 | 8645 | 8651 | 8657 | 8663 | 8669 | 8675 | 8681 | 8686 | 6 |
| 7.4 | 8692 | 8698 | 8704 | 8710 | 8716 | 8722 | 8727 | 8733 | 8739 | 8745 | 6 |
| 7.5 | 8751 | 8756 | 8762 | 8768 | 8774 | 8779 | 8785 | 8791 | 8797 | 8802 | 6 |
| 7.6 | 8808 | 8814 | 8820 | 8825 | 8831 | 8837 | 8842 | 8848 | 8854 | 8859 | 6 |
| 7.7 | 8865 | 8871 | 8876 | 8882 | 8887 | 8893 | 8899 | 8904 | 8910 | 8915 | 6 |
| 7.8 | 8921 | 8927 | 8932 | 8938 | 8943 | 8949 | 8954 | 8960 | 8965 | 8971 | 6 |
| 7.9 | 8976 | 8982 | 8987 | 8993 | 8998 | 9004 | 9009 | 9015 | 9020 | 9025 | 5 |
| 8.0 | 0.9031 | 9036 | 9042 | 9047 | 9053 | 9058 | 9063 | 9069 | 9074 | 9079 | 5 |
| 8.1 | 9085 | 9090 | 9096 | 9101 | 9106 | 9112 | 9117 | 9122 | 9128 | 9133 | 5 |
| 8.2 | 9138 | 9143 | 9149 | 9154 | 9159 | 9165 | 9170 | 9175 | 9180 | 9186 | 5 |
| 8.3 | 9191 | 9196 | 9201 | 9206 | 9212 | 9217 | 9222 | 9227 | 9232 | 9238 | 5 |
| 8.4 | 9243 | 9248 | 9253 | 9258 | 9263 | 9269 | 9274 | 9279 | 9284 | 9289 | 5 |
| 8.5 | 9294 | 9299 | 9304 | 9309 | 9315 | 9320 | 9325 | 9330 | 9335 | 9340 | 5 |
| 8.6 | 9345 | 9350 | 9355 | 9360 | 9365 | 9370 | 9375 | 9380 | 9385 | 9390 | 5 |
| 8.7 | 9395 | 9400 | 9405 | 9410 | 9415 | 9420 | 9425 | 9430 | 9435 | 9440 | 5 |
| 8.8 | 9445 | 9450 | 9455 | 9460 | 9465 | 9469 | 9474 | 9479 | 9484 | 9489 | 5 |
| 8.9 | 9494 | 9499 | 9504 | 9509 | 9513 | 9518 | 9523 | 9528 | 9533 | 9538 | 5 |
| 9.0 | 0.9542 | 9547 | 9552 | 9557 | 9562 | 9566 | 9571 | 9576 | 9581 | 9586 | 5 |
| 9.1 | 9590 | 9595 | 9600 | 9605 | 9609 | 9614 | 9619 | 9624 | 9628 | 9633 | 5 |
| 9.2 | 9638 | 9643 | 9647 | 9652 | 9657 | 9661 | 9666 | 9671 | 9675 | 9680 | 5 |
| 9.3 | 9685 | 9689 | 9694 | 9699 | 9703 | 9708 | 9713 | 9717 | 9722 | 9727 | 5 |
| 9.4 | 9731 | 9736 | 9741 | 9745 | 9750 | 9754 | 9759 | 9763 | 9768 | 9773 | 5 |
| 9.5 | 9777 | 9782 | 9786 | 9791 | 9795 | 9800 | 9805 | 9809 | 9814 | 9818 | 5 |
| 9.6 | 9823 | 9827 | 9832 | 9836 | 9841 | 9845 | 9850 | 9854 | 9859 | 9863 | 4 |
| 9.7 | 9868 | 9872 | 9877 | 9881 | 9886 | 9890 | 9894 | 9899 | 9903 | 9908 | 4 |
| 9.8 | 9912 | 9917 | 9921 | 9926 | 9930 | 9934 | 9939 | 9943 | 9948 | 9952 | 4 |
| 9.9 | 9956 | 9961 | 9965 | 9969 | 9974 | 9978 | 9983 | 9987 | 9991 | 9996 | 4 |

**Degrees and Minutes Expressed in Radians**

| Degrees | | | | | | Hundredths | | | | Minutes | |
|---|---|---|---|---|---|---|---|---|---|---|---|
| 1° | .0175 | 61° | 1.0647 | 121° | 2.1118 | 0°.01 | .0002 | 0°.51 | .0089 | 1' | .0003 |
| 2 | .0349 | 2 | 1.0821 | 2 | 2.1293 | 2 | .0003 | 2 | .0091 | 2' | .0006 |
| 3 | .0524 | 3 | 1.0996 | 3 | 2.1468 | 3 | .0005 | 3 | .0093 | 3' | .0009 |
| 4 | .0698 | 4 | 1.1170 | 4 | 2.1642 | 4 | .0007 | 4 | .0094 | 4' | .0012 |
| 5° | .0873 | 65° | 1.1345 | 125° | 2.1817 | .05 | .0009 | .55 | .0096 | 5' | .0015 |
| 6 | .1047 | 6 | 1.1519 | 6 | 2.1991 | 6 | .0010 | 6 | .0098 | 6' | .0017 |
| 7 | .1222 | 7 | 1.1694 | 7 | 2.2166 | 7 | .0012 | 7 | .0099 | 7' | .0020 |
| 8 | .1396 | 8 | 1.1868 | 8 | 2.2340 | 8 | .0014 | 8 | .0101 | 8' | .0023 |
| 9 | .1571 | 9 | 1.2043 | 9 | 2.2515 | 9 | .0016 | 9 | .0103 | 9' | .0026 |
| 10° | .1745 | 70° | 1.2217 | 130° | 2.2689 | 0°.10 | .0017 | 0°.60 | .0105 | 10' | .0029 |
| 1 | .1920 | 1 | 1.2392 | 1 | 2.2864 | 1 | .0019 | 1 | .0106 | 11' | .0032 |
| 2 | .2094 | 2 | 1.2566 | 2 | 2.3038 | 2 | .0021 | 2 | .0108 | 12' | .0035 |
| 3 | .2269 | 3 | 1.2741 | 3 | 2.3213 | 3 | .0023 | 3 | .0110 | 13' | .0038 |
| 4 | .2443 | 4 | 1.2915 | 4 | 2.3387 | 4 | .0024 | 4 | .0112 | 14' | .0041 |
| 15° | .2618 | 75° | 1.3090 | 135° | 2.3562 | .15 | .0026 | .65 | .0113 | 15' | .0044 |
| 6 | .2793 | 6 | 1.3265 | 6 | 2.3736 | 6 | .0028 | 6 | .0115 | 16' | .0047 |
| 7 | .2967 | 7 | 1.3439 | 7 | 2.3911 | 7 | .0030 | 7 | .0117 | 17' | .0049 |
| 8 | .3142 | 8 | 1.3614 | 8 | 2.4086 | 8 | .0031 | 8 | .0119 | 18' | .0052 |
| 9 | .3316 | 9 | 1.3788 | 9 | 2.4260 | 9 | .0033 | 9 | .0120 | 19' | .0055 |
| 20° | .3491 | 80° | 1.3963 | 140° | 2.4435 | 0°.20 | .0035 | 0°.70 | .0122 | 20' | .0058 |
| 1 | .3665 | 1 | 1.4137 | 1 | 2.4609 | 1 | .0037 | 1 | .0124 | 21' | .0061 |
| 2 | .3840 | 2 | 1.4312 | 2 | 2.4784 | 2 | .0038 | 2 | .0126 | 22' | .0064 |
| 3 | .4014 | 3 | 1.4486 | 3 | 2.4958 | 3 | .0040 | 3 | .0127 | 23' | .0067 |
| 4 | .4189 | 4 | 1.4661 | 4 | 2.5133 | 4 | .0042 | 4 | .0129 | 24' | .0070 |
| 25° | .4363 | 85° | 1.4835 | 145° | 2.5307 | .25 | .0044 | .75 | .0131 | 25' | .0073 |
| 6 | .4538 | 6 | 1.5010 | 6 | 2.5482 | 6 | .0045 | 6 | .0133 | 26' | .0076 |
| 7 | .4712 | 7 | 1.5184 | 7 | 2.5656 | 7 | .0047 | 7 | .0134 | 27' | .0079 |
| 8 | .4887 | 8 | 1.5359 | 8 | 2.5831 | 8 | .0049 | 8 | .0136 | 28' | .0081 |
| 9 | .5061 | 9 | 1.5533 | 9 | 2.6005 | 9 | .0051 | 9 | .0138 | 29' | .0084 |
| 30° | .5236 | 90° | 1.5708 | 150° | 2.6180 | 0°.30 | .0052 | 0°.80 | .0140 | 30' | .0087 |
| 1 | .5411 | 1 | 1.5882 | 1 | 2.6354 | 1 | .0054 | 1 | .0141 | 31' | .0090 |
| 2 | .5585 | 2 | 1.6057 | 2 | 2.6529 | 2 | .0056 | 2 | .0143 | 32' | .0093 |
| 3 | .5760 | 3 | 1.6232 | 3 | 2.6704 | 3 | .0058 | 3 | .0145 | 33' | .0096 |
| 4 | .5934 | 4 | 1.6406 | 4 | 2.6878 | 4 | .0059 | 4 | .0147 | 34' | .0099 |
| 35° | .6109 | 95° | 1.6581 | 155° | 2.7053 | .35 | .0061 | .85 | .0148 | 35' | .0102 |
| 6 | .6283 | 6 | 1.6755 | 6 | 2.7227 | 6 | .0063 | 6 | .0150 | 36' | .0105 |
| 7 | .6458 | 7 | 1.6930 | 7 | 2.7402 | 7 | .0065 | 7 | .0152 | 37' | .0108 |
| 8 | .6632 | 8 | 1.7104 | 8 | 2.7576 | 8 | .0066 | 8 | .0154 | 38' | .0111 |
| 9 | .6807 | 9 | 1.7279 | 9 | 2.7751 | 9 | .0068 | 9 | .0155 | 39' | .0113 |
| 40° | .6981 | 100° | 1.7453 | 160° | 2.7925 | 0°.40 | .0070 | 0°.90 | .0157 | 40' | .0116 |
| 1 | .7156 | 1 | 1.7628 | 1 | 2.8100 | 1 | .0072 | 1 | .0159 | 41' | .0119 |
| 2 | .7330 | 2 | 1.7802 | 2 | 2.8274 | 2 | .0073 | 2 | .0161 | 42' | .0122 |
| 3 | .7505 | 3 | 1.7977 | 3 | 2.8449 | 3 | .0075 | 3 | .0162 | 43' | .0125 |
| 4 | .7679 | 4 | 1.8151 | 4 | 2.8623 | 4 | .0077 | 4 | .0164 | 44' | .0128 |
| 45° | .7854 | 105° | 1.8326 | 165° | 2.8798 | .45 | .0079 | .95 | .0166 | 45' | .0131 |
| 6 | .8029 | 6 | 1.8500 | 6 | 2.8972 | 6 | .0080 | 6 | .0168 | 46' | .0134 |
| 7 | .8203 | 7 | 1.8675 | 7 | 2.9147 | 7 | .0082 | 7 | .0169 | 47' | .0137 |
| 8 | .8378 | 8 | 1.8850 | 8 | 2.9322 | 8 | .0084 | 8 | .0171 | 48' | .0140 |
| 9 | .8552 | 9 | 1.9024 | 9 | 2.9496 | 9 | .0086 | 9 | .0173 | 49' | .0143 |
| 50° | .8727 | 110° | 1.9199 | 170° | 2.9671 | 0°.50 | .0087 | 1°.00 | .0175 | 50' | .0145 |
| 1 | .8901 | 1 | 1.9373 | 1 | 2.9845 | | | | | 51' | .0148 |
| 2 | .9076 | 2 | 1.9548 | 2 | 3.0020 | | | | | 52' | .0151 |
| 3 | .9250 | 3 | 1.9722 | 3 | 3.0194 | | | | | 53' | .0154 |
| 4 | .9425 | 4 | 1.9897 | 4 | 3.0369 | | | | | 54' | .0157 |
| 55° | .9599 | 115° | 2.0071 | 175° | 3.0543 | | | | | 55' | .0160 |
| 6 | .9774 | 6 | 2.0246 | 6 | 3.0718 | | | | | 56' | .0163 |
| 7 | .9948 | 7 | 2.0420 | 7 | 3.0892 | | | | | 57' | .0166 |
| 8 | 1.0123 | 8 | 2.0595 | 8 | 3.1067 | | | | | 58' | .0169 |
| 9 | 1.0297 | 9 | 2.0769 | 9 | 3.1241 | | | | | 59' | .0172 |
| 60° | 1.0472 | 120° | 2.0944 | 180° | 3.1416 | | | | | 60' | .0175 |

Arc 1° = 0.0174533; arc 1' = 0.000290888; arc 1" = 0.00000484814; 1 radian = 57°.295780 = 57° 17'.7468 = 57° 17' 44",806.

## Radians Expressed in Degrees

| Rad | Deg | Rad | Deg | Rad | Deg | Rad | Deg | Rad | Deg |
|---|---|---|---|---|---|---|---|---|---|
| 0.01 | 0°.57 | .64 | 36°.67 | 1.27 | 72°.77 | 1.90 | 108°.86 | 2.53 | 144°.96 |
| 2 | 1°.15 | .65 | 37°.24 | 8 | 73°.34 | 1 | 109°.43 | 4 | 145°.53 |
| 3 | 1°.72 | 6 | 37°.82 | 9 | 73°.91 | 2 | 110°.01 | 2.55 | 146°.10 |
| 4 | 2°.29 | 7 | 38°.39 | 1.30 | 74°.48 | 3 | 110°.58 | 6 | 146°.68 |
| .05 | 2°.86 | 8 | 38°.96 | 1 | 75°.06 | 4 | 111°.15 | 7 | 147°.25 |
| 6 | 3°.44 | 9 | 39°.53 | 2 | 75°.63 | 1.95 | 111°.73 | 8 | 147°.82 |
| 7 | 4°.01 | .70 | 40°.11 | 3 | 76°.20 | 6 | 112°.30 | 9 | 148°.40 |
| 8 | 4°.58 | 1 | 40°.68 | 4 | 76°.78 | 7 | 112°.87 | 2.60 | 148°.97 |
| 9 | 5°.16 | 2 | 41°.25 | 1.35 | 77°.35 | 8 | 113°.45 | 1 | 149°.54 |
| .10 | 5°.73 | 3 | 41°.83 | 6 | 77°.92 | 9 | 114°.02 | 2 | 150°.11 |
| 1 | 6°.30 | 4 | 42°.40 | 7 | 78°.50 | 2.00 | 114°.59 | 3 | 150°.69 |
| 2 | 6°.88 | .75 | 42°.97 | 8 | 79°.07 | 1 | 115°.16 | 4 | 151°.26 |
| 3 | 7°.45 | 6 | 43°.54 | 9 | 79°.64 | 2 | 115°.74 | 2.65 | 151°.83 |
| 4 | 8°.02 | 7 | 44°.12 | 1.40 | 80°.21 | 3 | 116°.31 | 6 | 152°.41 |
| .15 | 8°.59 | 8 | 44°.69 | 1 | 80°.79 | 4 | 116°.88 | 7 | 152°.98 |
| 6 | 9°.17 | 9 | 45°.26 | 2 | 81°.36 | 2.05 | 117°.46 | 8 | 153°.55 |
| 7 | 9°.74 | .80 | 45°.84 | 3 | 81°.93 | 6 | 118°.03 | 9 | 154°.13 |
| 8 | 10°.31 | 1 | 46°.41 | 4 | 82°.51 | 7 | 118°.60 | 2.70 | 154°.70 |
| 9 | 10°.89 | 2 | 46°.98 | 1.45 | 83°.08 | 8 | 119°.18 | 1 | 155°.27 |
| .20 | 11°.46 | 3 | 47°.56 | 6 | 83°.65 | 9 | 119°.75 | 2 | 155°.84 |
| 1 | 12°.03 | 4 | 48°.13 | 7 | 84°.22 | 2.10 | 120°.32 | 3 | 156°.42 |
| 2 | 12°.61 | .85 | 48°.70 | 8 | 84°.80 | 1 | 120°.89 | 4 | 156°.99 |
| 3 | 13°.18 | 6 | 49°.27 | 9 | 85°.37 | 2 | 121°.47 | 2.75 | 157°.56 |
| 4 | 13°.75 | 7 | 49°.85 | 1.50 | 85°.94 | 3 | 122°.04 | 6 | 158°.14 |
| .25 | 14°.32 | 8 | 50°.42 | 1 | 86°.52 | 4 | 122°.61 | 7 | 158°.71 |
| 6 | 14°.90 | 9 | 50°.99 | 2 | 87°.09 | 2.15 | 123°.19 | 8 | 159°.28 |
| 7 | 15°.47 | .90 | 51°.57 | 3 | 87°.66 | 6 | 123°.76 | 9 | 159°.86 |
| 8 | 16°.04 | 1 | 52°.14 | 4 | 88°.24 | 7 | 124°.33 | 2.80 | 160°.43 |
| 9 | 16°.62 | 2 | 52°.71 | 1.55 | 88°.81 | 8 | 124°.90 | 1 | 161°.00 |
| .30 | 17°.19 | 3 | 53°.29 | 6 | 89°.38 | 9 | 125°.48 | 2 | 161°.57 |
| 1 | 17°.76 | 4 | 53°.86 | 7 | 89°.95 | 2.20 | 126°.05 | 3 | 162°.15 |
| 2 | 18°.33 | .95 | 54°.43 | 8 | 90°.53 | 1 | 126°.62 | 4 | 162°.72 |
| 3 | 18°.91 | 6 | 55°.00 | 9 | 91°.10 | 2 | 127°.20 | 2.85 | 163°.29 |
| 4 | 19°.48 | 7 | 55°.58 | 1.60 | 91°.67 | 3 | 127°.77 | 6 | 163°.87 |
| .35 | 20°.05 | 8 | 56°.15 | 1 | 92°.25 | 4 | 128°.34 | 7 | 164°.44 |
| 6 | 20°.63 | 9 | 56°.72 | 2 | 92°.82 | 2.25 | 128°.92 | 8 | 165°.01 |
| 7 | 21°.20 | 1.00 | 57°.30 | 3 | 93°.39 | 6 | 129°.49 | 9 | 165°.58 |
| 8 | 21°.77 | 1 | 57°.87 | 4 | 93°.97 | 7 | 130°.06 | 2.90 | 166°.16 |
| 9 | 22°.35 | 2 | 58°.44 | 1.65 | 94°.54 | 8 | 130°.63 | 1 | 166°.73 |
| .40 | 22°.92 | 3 | 59°.01 | 6 | 95°.11 | 9 | 131°.21 | 2 | 167°.30 |
| 1 | 23°.49 | 4 | 59°.59 | 7 | 95°.68 | 2.30 | 131°.78 | 3 | 167°.88 |
| 2 | 24°.06 | 1.05 | 60°.16 | 8 | 96°.26 | 1 | 132°.35 | 4 | 168°.45 |
| 3 | 24°.64 | 6 | 60°.73 | 9 | 96°.83 | 2 | 132°.93 | 2.95 | 169°.02 |
| 4 | 25°.21 | 7 | 61°.31 | 1.70 | 97°.40 | 3 | 133°.50 | 6 | 169°.60 |
| .45 | 25°.78 | 8 | 61°.88 | 1 | 97°.98 | 4 | 134°.07 | 7 | 170°.17 |
| 6 | 26°.36 | 9 | 62°.45 | 2 | 98°.55 | 2.35 | 134°.65 | 8 | 170°.74 |
| 7 | 26°.93 | 1.10 | 63°.03 | 3 | 99°.12 | 6 | 135°.22 | 9 | 171°.31 |
| 8 | 27°.50 | 1 | 63°.60 | 4 | 99°.69 | 7 | 135°.79 | 3.00 | 171°.89 |
| 9 | 28°.07 | 2 | 64°.17 | 1.75 | 100°.27 | 8 | 136°.36 | 1 | 172°.46 |
| .50 | 28°.65 | 3 | 64°.74 | 6 | 100°.84 | 9 | 136°.94 | 2 | 173°.03 |
| 1 | 29°.22 | 4 | 65°.32 | 7 | 101°.41 | 2.40 | 137°.51 | 3 | 173°.61 |
| 2 | 29°.79 | 1.15 | 65°.89 | 8 | 101°.99 | 1 | 138°.08 | 4 | 174°.18 |
| 3 | 30°.37 | 6 | 66°.46 | 9 | 102°.56 | 2 | 138°.66 | 3.05 | 174°.75 |
| 4 | 30°.94 | 7 | 67°.04 | 1.80 | 103°.13 | 3 | 139°.23 | 6 | 175°.33 |
| .55 | 31°.51 | 8 | 67°.61 | 1 | 103°.71 | 4 | 139°.80 | 7 | 175°.90 |
| 6 | 32°.09 | 9 | 68°.18 | 2 | 104°.28 | 2.45 | 140°.37 | 8 | 176°.47 |
| 7 | 32°.66 | 1.20 | 68°.75 | 3 | 104°.85 | 6 | 140°.95 | 9 | 177°.04 |
| 8 | 33°.23 | 1 | 69°.33 | 4 | 105°.42 | 7 | 141°.52 | 3.10 | 177°.62 |
| 9 | 33°.80 | 2 | 69°.90 | 1.85 | 106°.00 | 8 | 142°.09 | 1 | 178°.19 |
| .60 | 34°.38 | 3 | 70°.47 | 6 | 106°.57 | 9 | 142°.67 | 2 | 178°.76 |
| 1 | 34°.95 | 4 | 71°.05 | 7 | 107°.14 | 2.50 | 143°.24 | 3 | 179°.34 |
| 2 | 35°.52 | 1.25 | 71°.62 | 8 | 107°.72 | 1 | 143°.81 | 4 | 179°.91 |
| 3 | 36°.10 | 6 | 72°.19 | 9 | 108°.29 | 2 | 144°.39 | 3.15 | 180°.48 |

### Interpolation

| | |
|---|---|
| .0002 | 0°.01 |
| 04 | .02 |
| 06 | .03 |
| 08 | .05 |
| .0010 | 0°.06 |
| 12 | .07 |
| 14 | .08 |
| 16 | .09 |
| 18 | .10 |
| .0020 | 0°.11 |
| 22 | .13 |
| 24 | .14 |
| 26 | .15 |
| 28 | .16 |
| .0030 | 0°.17 |
| 32 | .18 |
| 34 | .19 |
| 36 | .21 |
| 38 | .22 |
| .0040 | 0°.23 |
| 42 | .24 |
| 44 | .25 |
| 46 | .26 |
| 48 | .28 |
| .0050 | 0°.29 |
| 52 | .30 |
| 54 | .31 |
| 56 | .32 |
| 58 | .33 |
| .0060 | 0°.34 |
| 62 | .36 |
| 64 | .37 |
| 66 | .38 |
| 68 | .39 |
| .0070 | 0°.40 |
| 72 | .41 |
| 74 | .42 |
| 76 | .44 |
| 78 | .45 |
| .0080 | 0°.46 |
| 82 | .47 |
| 84 | .48 |
| 86 | .49 |
| 88 | .50 |
| .0090 | 0°.52 |
| 92 | .53 |
| 94 | .54 |
| 96 | .55 |
| 98 | .56 |

### Multiples of $\pi$

| | | |
|---|---|---|
| 1 | 3.1416 | 180° |
| 2 | 6.2832 | 360° |
| 3 | 9.4248 | 540° |
| 4 | 12.5664 | 720° |
| 5 | 15.7080 | 900° |
| 6 | 18.8496 | 1080° |
| 7 | 21.9911 | 1260° |
| 8 | 25.1327 | 1440° |
| 9 | 28.2743 | 1620° |
| 10 | 31.4159 | 1800° |

## Natural Sines and Cosines

### Natural Sines at intervals of 0°.1, or 6′

| Deg | °.0 = (0′) | °.1 (6′) | °.2 (12′) | °.3 (18′) | °.4 (24′) | °.5 (30′) | °.6 (36′) | °.7 (42′) | °.8 (48′) | °.9 (54′) | | | Avg diff |
|---|---|---|---|---|---|---|---|---|---|---|---|---|---|
| | | | | | | | | | | | 0.0000 | 90° | |
| 0° | 0.0000 | 0017 | 0035 | 0052 | 0070 | 0087 | 0105 | 0122 | 0140 | 0157 | 0175 | 89 | 17 |
| 1 | 0175 | 0192 | 0209 | 0227 | 0244 | 0262 | 0279 | 0297 | 0314 | 0332 | 0349 | 88 | 17 |
| 2 | 0349 | 0366 | 0384 | 0401 | 0419 | 0436 | 0454 | 0471 | 0488 | 0506 | 0523 | 87 | 17 |
| 3 | 0523 | 0541 | 0558 | 0576 | 0593 | 0610 | 0628 | 0645 | 0663 | 0680 | 0698 | 86 | 17 |
| 4 | 0698 | 0715 | 0732 | 0750 | 0767 | 0785 | 0802 | 0819 | 0837 | 0854 | 0.0872 | 85 | 17 |
| 5 | 0.0872 | 0889 | 0906 | 0924 | 0941 | 0958 | 0976 | 0993 | 1011 | 1028 | 1045 | 84 | 17 |
| 6 | 1045 | 1063 | 1080 | 1097 | 1115 | 1132 | 1149 | 1167 | 1184 | 1201 | 1219 | 83 | 17 |
| 7 | 1219 | 1236 | 1253 | 1271 | 1288 | 1305 | 1323 | 1340 | 1357 | 1374 | 1392 | 82 | 17 |
| 8 | 1392 | 1409 | 1426 | 1444 | 1461 | 1478 | 1495 | 1513 | 1530 | 1547 | 1564 | 81 | 17 |
| 9 | 1564 | 1582 | 1599 | 1616 | 1633 | 1650 | 1668 | 1685 | 1702 | 1719 | 0.1736 | 80° | 17 |
| 10° | 0.1736 | 1754 | 1771 | 1788 | 1805 | 1822 | 1840 | 1857 | 1874 | 1891 | 1908 | 79 | 17 |
| 11 | 1908 | 1925 | 1942 | 1959 | 1977 | 1994 | 2011 | 2028 | 2045 | 2062 | 2079 | 78 | 17 |
| 12 | 2079 | 2096 | 2113 | 2130 | 2147 | 2164 | 2181 | 2198 | 2215 | 2233 | 2250 | 77 | 17 |
| 13 | 2250 | 2267 | 2284 | 2300 | 2317 | 2334 | 2351 | 2368 | 2385 | 2402 | 2419 | 76 | 17 |
| 14 | 2419 | 2436 | 2453 | 2470 | 2487 | 2504 | 2521 | 2538 | 2554 | 2571 | 0.2588 | 75 | 17 |
| 15 | 0.2588 | 2605 | 2622 | 2639 | 2656 | 2672 | 2689 | 2706 | 2723 | 2740 | 2756 | 74 | 17 |
| 16 | 2756 | 2773 | 2790 | 2807 | 2823 | 2840 | 2857 | 2874 | 2890 | 2907 | 2924 | 73 | 17 |
| 17 | 2924 | 2940 | 2957 | 2974 | 2990 | 3007 | 3024 | 3040 | 3057 | 3074 | 3090 | 72 | 17 |
| 18 | 3090 | 3107 | 3123 | 3140 | 3156 | 3173 | 3190 | 3206 | 3223 | 3239 | 3256 | 71 | 17 |
| 19 | 3256 | 3272 | 3289 | 3305 | 3322 | 3338 | 3355 | 3371 | 3387 | 3404 | 0.3420 | 70° | 16 |
| 20° | 0.3420 | 3437 | 3453 | 3469 | 3486 | 3502 | 3518 | 3535 | 3551 | 3567 | 3584 | 69 | 16 |
| 21 | 3584 | 3600 | 3616 | 3633 | 3649 | 3665 | 3681 | 3697 | 3714 | 3730 | 3746 | 68 | 16 |
| 22 | 3746 | 3762 | 3778 | 3795 | 3811 | 3827 | 3843 | 3859 | 3875 | 3891 | 3907 | 67 | 16 |
| 23 | 3907 | 3923 | 3939 | 3955 | 3971 | 3987 | 4003 | 4019 | 4035 | 4051 | 4067 | 66 | 16 |
| 24 | 4067 | 4083 | 4099 | 4115 | 4131 | 4147 | 4163 | 4179 | 4195 | 4210 | 0.4226 | 65 | 16 |
| 25 | 0.4226 | 4242 | 4258 | 4274 | 4289 | 4305 | 4321 | 4337 | 4352 | 4368 | 4384 | 64 | 16 |
| 26 | 4384 | 4399 | 4415 | 4431 | 4446 | 4462 | 4478 | 4493 | 4509 | 4524 | 4540 | 63 | 16 |
| 27 | 4540 | 4555 | 4571 | 4586 | 4602 | 4617 | 4633 | 4648 | 4664 | 4679 | 4695 | 62 | 16 |
| 28 | 4695 | 4710 | 4726 | 4741 | 4756 | 4772 | 4787 | 4802 | 4818 | 4833 | 4848 | 61 | 15 |
| 29 | 4848 | 4863 | 4879 | 4894 | 4909 | 4924 | 4939 | 4955 | 4970 | 4985 | 0.5000 | 60° | 15 |
| 30° | 0.5000 | 5015 | 5030 | 5045 | 5060 | 5075 | 5090 | 5105 | 5120 | 5135 | 5150 | 59 | 15 |
| 31 | 5150 | 5165 | 5180 | 5195 | 5210 | 5225 | 5240 | 5255 | 5270 | 5284 | 5299 | 58 | 15 |
| 32 | 5299 | 5314 | 5329 | 5344 | 5358 | 5373 | 5388 | 5402 | 5417 | 5432 | 5446 | 57 | 15 |
| 33 | 5446 | 5461 | 5476 | 5490 | 5505 | 5519 | 5534 | 5548 | 5563 | 5577 | 5592 | 56 | 15 |
| 34 | 5592 | 5606 | 5621 | 5635 | 5650 | 5664 | 5678 | 5693 | 5707 | 5721 | 0.5736 | 55 | 14 |
| 35 | 0.5736 | 5750 | 5764 | 5779 | 5793 | 5807 | 5821 | 5835 | 5850 | 5864 | 5878 | 54 | 14 |
| 36 | 5878 | 5892 | 5906 | 5920 | 5934 | 5948 | 5962 | 5976 | 5990 | 6004 | 6018 | 53 | 14 |
| 37 | 6018 | 6032 | 6046 | 6060 | 6074 | 6088 | 6101 | 6115 | 6129 | 6143 | 6157 | 52 | 14 |
| 38 | 6157 | 6170 | 6184 | 6198 | 6211 | 6225 | 6239 | 6252 | 6266 | 6280 | 6293 | 51 | 14 |
| 39 | 6293 | 6307 | 6320 | 6334 | 6347 | 6361 | 6374 | 6388 | 6401 | 6414 | 0.6428 | 50° | 13 |
| 40° | 0.6428 | 6441 | 6455 | 6468 | 6481 | 6494 | 6508 | 6521 | 6534 | 6547 | 6561 | 49 | 13 |
| 41 | 6561 | 6574 | 6587 | 6600 | 6613 | 6626 | 6639 | 6652 | 6665 | 6678 | 6691 | 48 | 13 |
| 42 | 6691 | 6704 | 6717 | 6730 | 6743 | 6756 | 6769 | 6782 | 6794 | 6807 | 6820 | 47 | 13 |
| 43 | 6820 | 6833 | 6845 | 6858 | 6871 | 6884 | 6896 | 6909 | 6921 | 6934 | 6947 | 46 | 13 |
| 44 | 6947 | 6959 | 6972 | 6984 | 6997 | 7009 | 7022 | 7034 | 7046 | 7059 | 0.7071 | 45° | 12 |
| 45° | 0.7071 | | | | | | | | | | | | |

| | °.9 = (54′) | °.8 (48′) | °.7 (42′) | °.6 (36′) | °.5 (30′) | °.4 (24′) | °.3 (18′) | °.2 (12′) | °.1 (6′) | °.0 (0′) | | Deg |
|---|---|---|---|---|---|---|---|---|---|---|---|---|

For graphs, see Sec. 2.                    **Natural Cosines**

**Natural Sines & Cosines** *(continued)*

**Natural Sines at intervals of 0°.1, or 6′**

| Deg | °.0 =(0′) | °.1 (6′) | °.2 (12′) | °.3 (18′) | °.4 (24′) | °.5 (30′) | °.6 (36′) | °.7 (42′) | °.8 (48′) | °.9 (54′) | | Avg diff |
|---|---|---|---|---|---|---|---|---|---|---|---|---|
| | | | | | | | | | | | 0.7071 | 45° | |
| 45° | 0.7071 | 7083 | 7096 | 7108 | 7120 | 7133 | 7145 | 7157 | 7169 | 7181 | 7193 | 44 | 12 |
| 46 | 7193 | 7206 | 7218 | 7230 | 7242 | 7254 | 7266 | 7278 | 7290 | 7302 | 7314 | 43 | 12 |
| 47 | 7314 | 7325 | 7337 | 7349 | 7361 | 7373 | 7385 | 7396 | 7408 | 7420 | 7431 | 42 | 12 |
| 48 | 7431 | 7443 | 7455 | 7466 | 7478 | 7490 | 7501 | 7513 | 7524 | 7536 | 7547 | 41 | 12 |
| 49 | 7547 | 7559 | 7570 | 7581 | 7593 | 7604 | 7615 | 7627 | 7638 | 7649 | 0.7660 | 40° | 11 |
| 50° | 0.7660 | 7672 | 7683 | 7694 | 7705 | 7716 | 7727 | 7738 | 7749 | 7760 | 7771 | 39 | 11 |
| 51 | 7771 | 7782 | 7793 | 7804 | 7815 | 7826 | 7837 | 7848 | 7859 | 7869 | 7880 | 38 | 11 |
| 52 | 7880 | 7891 | 7902 | 7912 | 7923 | 7934 | 7944 | 7955 | 7965 | 7976 | 7986 | 37 | 11 |
| 53 | 7986 | 7997 | 8007 | 8018 | 8028 | 8039 | 8049 | 8059 | 8070 | 8080 | 8090 | 36 | 10 |
| 54 | 8090 | 8100 | 8111 | 8121 | 8131 | 8141 | 8151 | 8161 | 8171 | 8181 | 0.8192 | 35 | 10 |
| 55 | 0.8192 | 8202 | 8211 | 8221 | 8231 | 8241 | 8251 | 8261 | 8271 | 8281 | 8290 | 34 | 10 |
| 56 | 8290 | 8300 | 8310 | 8320 | 8329 | 8339 | 8348 | 8358 | 8368 | 8377 | 8387 | 33 | 10 |
| 57 | 8387 | 8396 | 8406 | 8415 | 8425 | 8434 | 8443 | 8453 | 8462 | 8471 | 8480 | 32 | 9 |
| 58 | 8480 | 8490 | 8499 | 8508 | 8517 | 8526 | 8536 | 8545 | 8554 | 8563 | 8572 | 31 | 9 |
| 59 | 8572 | 8581 | 8590 | 8599 | 8607 | 8616 | 8625 | 8634 | 8643 | 8652 | 0.8660 | 30° | 9 |
| 60° | 0.8660 | 8669 | 8678 | 8686 | 8695 | 8704 | 8712 | 8721 | 8729 | 8738 | 8746 | 29 | 9 |
| 61 | 8746 | 8755 | 8763 | 8771 | 8780 | 8788 | 8796 | 8805 | 8813 | 8821 | 8829 | 28 | 8 |
| 62 | 8829 | 8838 | 8846 | 8854 | 8862 | 8870 | 8878 | 8886 | 8894 | 8902 | 8910 | 27 | 8 |
| 63 | 8910 | 8918 | 8926 | 8934 | 8942 | 8949 | 8957 | 8965 | 8973 | 8980 | 8988 | 26 | 8 |
| 64 | 8988 | 8996 | 9003 | 9011 | 9018 | 9026 | 9033 | 9041 | 9048 | 9056 | 0.9063 | 25 | 7 |
| 65 | 0.9063 | 9070 | 9078 | 9085 | 9092 | 9100 | 9107 | 9114 | 9121 | 9128 | 9135 | 24 | 7 |
| 66 | 9135 | 9143 | 9150 | 9157 | 9164 | 9171 | 9178 | 9184 | 9191 | 9198 | 9205 | 23 | 7 |
| 67 | 9205 | 9212 | 9219 | 9225 | 9232 | 9239 | 9245 | 9252 | 9259 | 9265 | 9272 | 22 | 7 |
| 68 | 9272 | 9278 | 9285 | 9291 | 9298 | 9304 | 9311 | 9317 | 9323 | 9330 | 9336 | 21 | 6 |
| 69 | 9336 | 9342 | 9348 | 9354 | 9361 | 9367 | 9373 | 9379 | 9385 | 9391 | 0.9397 | 20° | 6 |
| 70° | 0.9397 | 9403 | 9409 | 9415 | 9421 | 9426 | 9432 | 9438 | 9444 | 9449 | 9455 | 19 | 6 |
| 71 | 9455 | 9461 | 9466 | 9472 | 9478 | 9483 | 9489 | 9494 | 9500 | 9505 | 9511 | 18 | 6 |
| 72 | 9511 | 9516 | 9521 | 9527 | 9532 | 9537 | 9542 | 9548 | 9553 | 9558 | 9563 | 17 | 5 |
| 73 | 9563 | 9568 | 9573 | 9578 | 9583 | 9588 | 9593 | 9598 | 9603 | 9608 | 9613 | 16 | 5 |
| 74 | 9613 | 9617 | 9622 | 9627 | 9632 | 9636 | 9641 | 9646 | 9650 | 9655 | 0.9659 | 15 | 5 |
| 75 | 0.9659 | 9664 | 9668 | 9673 | 9677 | 9681 | 9686 | 9690 | 9694 | 9699 | 9703 | 14 | 4 |
| 76 | 9703 | 9707 | 9711 | 9715 | 9720 | 9724 | 9728 | 9732 | 9736 | 9740 | 9744 | 13 | 4 |
| 77 | 9744 | 9748 | 9751 | 9755 | 9759 | 9763 | 9767 | 9770 | 9774 | 9778 | 9781 | 12 | 4 |
| 78 | 9781 | 9785 | 9789 | 9792 | 9796 | 9799 | 9803 | 9806 | 9810 | 9813 | 9816 | 11 | 3 |
| 79 | 9816 | 9820 | 9823 | 9826 | 9829 | 9833 | 9836 | 9839 | 9842 | 9845 | 0.9848 | 10° | 3 |
| 80° | 0.9848 | 9851 | 9854 | 9857 | 9860 | 9863 | 9866 | 9869 | 9871 | 9874 | 9877 | 9 | 3 |
| 81 | 9877 | 9880 | 9882 | 9885 | 9888 | 9890 | 9893 | 9895 | 9898 | 9900 | 9903 | 8 | 3 |
| 82 | 9903 | 9905 | 9907 | 9910 | 9912 | 9914 | 9917 | 9919 | 9921 | 9923 | 9925 | 7 | 2 |
| 83 | 9925 | 9928 | 9930 | 9932 | 9934 | 9936 | 9938 | 9940 | 9942 | 9943 | 9945 | 6 | 2 |
| 84 | 9945 | 9947 | 9949 | 9951 | 9952 | 9954 | 9956 | 9957 | 9959 | 9960 | 0.9962 | 5 | 2 |
| 85 | 0.9962 | 9963 | 9965 | 9966 | 9968 | 9969 | 9971 | 9972 | 9973 | 9974 | 9976 | 4 | 1 |
| 86 | 9976 | 9977 | 9978 | 9979 | 9980 | 9981 | 9982 | 9983 | 9984 | 9985 | 9986 | 3 | 1 |
| 87 | 9986 | 9987 | 9988 | 9989 | 9990 | 9990 | 9991 | 9992 | 9993 | 9993 | 9994 | 2 | 1 |
| 88 | 9994 | 9995 | 9995 | 9996 | 9996 | 9997 | 9997 | 9997 | 9998 | 9998 | 0.9998 | 1 | 0 |
| 89 | 0.9998 | 9999 | 9999 | 9999 | 9999 | 0000 | 0000 | 0000 | 0000 | 0000 | 1.0000 | 0° | 0 |
| 90° | 1.0000 | | | | | | | | | | | | |

| | °.9 =(54′) | °.8 (48′) | °.7 (42′) | °.6 (36′) | °.5 (30′) | °.4 (24′) | °.3 (18′) | °.2 (12′) | °.1 (6′) | °.0 (0′) | Deg |
|---|---|---|---|---|---|---|---|---|---|---|---|

**Natural Cosines**

## Natural Tangents and Cotangents

### Natural Tangents at intervals of 0°.1, or 6′

| Deg | °.0 =(0′) | °.1 (6′) | °.2 (12′) | °.3 (18′) | °.4 (24′) | °.5 (30′) | °.6 (36′) | °.7 (42′) | °.8 (48′) | °.9 (54′) | | | Avg diff |
|---|---|---|---|---|---|---|---|---|---|---|---|---|---|
| | | | | | | | | | | | 0.0000 | 90° | |
| 0° | 0.0000 | 0017 | 0035 | 0052 | 0070 | 0087 | 0105 | 0122 | 0140 | 0157 | 0175 | 89 | 17 |
| 1 | 0175 | 0192 | 0209 | 0227 | 0244 | 0262 | 0279 | 0297 | 0314 | 0332 | 0349 | 88 | 17 |
| 2 | 0349 | 0367 | 0384 | 0402 | 0419 | 0437 | 0454 | 0472 | 0489 | 0507 | 0524 | 87 | 17 |
| 3 | 0524 | 0542 | 0559 | 0577 | 0594 | 0612 | 0629 | 0647 | 0664 | 0682 | 0699 | 86 | 18 |
| 4 | 0699 | 0717 | 0734 | 0752 | 0769 | 0787 | 0805 | 0822 | 0840 | 0857 | 0.0875 | 85 | 18 |
| 5 | 0.0875 | 0892 | 0910 | 0928 | 0945 | 0963 | 0981 | 0998 | 1016 | 1033 | 1051 | 84 | 18 |
| 6 | 1051 | 1069 | 1086 | 1104 | 1122 | 1139 | 1157 | 1175 | 1192 | 1210 | 1228 | 83 | 18 |
| 7 | 1228 | 1246 | 1263 | 1281 | 1299 | 1317 | 1334 | 1352 | 1370 | 1388 | 1405 | 82 | 18 |
| 8 | 1405 | 1423 | 1441 | 1459 | 1477 | 1495 | 1512 | 1530 | 1548 | 1566 | 1584 | 81 | 18 |
| 9 | 1584 | 1602 | 1620 | 1638 | 1655 | 1673 | 1691 | 1709 | 1727 | 1745 | 0.1763 | 80° | 18 |
| 10° | 0.1763 | 1781 | 1799 | 1817 | 1835 | 1853 | 1871 | 1890 | 1908 | 1926 | 1944 | 79 | 18 |
| 11 | 1944 | 1962 | 1980 | 1998 | 2016 | 2035 | 2053 | 2071 | 2089 | 2107 | 2126 | 78 | 18 |
| 12 | 2126 | 2144 | 2162 | 2180 | 2199 | 2217 | 2235 | 2254 | 2272 | 2290 | 2309 | 77 | 18 |
| 13 | 2309 | 2327 | 2345 | 2364 | 2382 | 2401 | 2419 | 2438 | 2456 | 2475 | 2493 | 76 | 18 |
| 14 | 2493 | 2512 | 2530 | 2549 | 2568 | 2586 | 2605 | 2623 | 2642 | 2661 | 0.2679 | 75 | 19 |
| 15 | 0.2679 | 2698 | 2717 | 2736 | 2754 | 2773 | 2792 | 2811 | 2830 | 2849 | 2867 | 74 | 19 |
| 16 | 2867 | 2886 | 2905 | 2924 | 2943 | 2962 | 2981 | 3000 | 3019 | 3038 | 3057 | 73 | 19 |
| 17 | 3057 | 3076 | 3096 | 3115 | 3134 | 3153 | 3172 | 3191 | 3211 | 3230 | 3249 | 72 | 19 |
| 18 | 3249 | 3269 | 3288 | 3307 | 3327 | 3346 | 3365 | 3385 | 3404 | 3424 | 3443 | 71 | 19 |
| 19 | 3443 | 3463 | 3482 | 3502 | 3522 | 3541 | 3561 | 3581 | 3600 | 3620 | 0.3640 | 70° | 20 |
| 20° | 0.3640 | 3659 | 3679 | 3699 | 3719 | 3739 | 3759 | 3779 | 3799 | 3819 | 3839 | 69 | 20 |
| 21 | 3839 | 3859 | 3879 | 3899 | 3919 | 3939 | 3959 | 3979 | 4000 | 4020 | 4040 | 68 | 20 |
| 22 | 4040 | 4061 | 4081 | 4101 | 4122 | 4142 | 4163 | 4183 | 4204 | 4224 | 4245 | 67 | 21 |
| 23 | 4245 | 4265 | 4286 | 4307 | 4327 | 4348 | 4369 | 4390 | 4411 | 4431 | 4452 | 66 | 21 |
| 24 | 4452 | 4473 | 4494 | 4515 | 4536 | 4557 | 4578 | 4599 | 4621 | 4642 | 0.4663 | 65 | 21 |
| 25 | 0.4663 | 4684 | 4706 | 4727 | 4748 | 4770 | 4791 | 4813 | 4834 | 4856 | 4877 | 64 | 21 |
| 26 | 4877 | 4899 | 4921 | 4942 | 4964 | 4986 | 5008 | 5029 | 5051 | 5073 | 5095 | 63 | 22 |
| 27 | 5095 | 5117 | 5139 | 5161 | 5184 | 5206 | 5228 | 5250 | 5272 | 5295 | 5317 | 62 | 22 |
| 28 | 5317 | 5340 | 5362 | 5384 | 5407 | 5430 | 5452 | 5475 | 5498 | 5520 | 5543 | 61 | 23 |
| 29 | 5543 | 5566 | 5589 | 5612 | 5635 | 5658 | 5681 | 5704 | 5727 | 5750 | 0.5774 | 60° | 23 |
| 30° | 0.5774 | 5797 | 5820 | 5844 | 5867 | 5890 | 5914 | 5938 | 5961 | 5985 | 6009 | 59 | 24 |
| 31 | 6009 | 6032 | 6056 | 6080 | 6104 | 6128 | 6152 | 6176 | 6200 | 6224 | 6249 | 58 | 24 |
| 32 | 6249 | 6273 | 6297 | 6322 | 6346 | 6371 | 6395 | 6420 | 6445 | 6469 | 6494 | 57 | 25 |
| 33 | 6494 | 6519 | 6544 | 6569 | 6594 | 6619 | 6644 | 6669 | 6694 | 6720 | 6745 | 56 | 25 |
| 34 | 6745 | 6771 | 6796 | 6822 | 6847 | 6873 | 6899 | 6924 | 6950 | 6976 | 0.7002 | 55 | 26 |
| 35 | 0.7002 | 7028 | 7054 | 7080 | 7107 | 7133 | 7159 | 7186 | 7212 | 7239 | 7265 | 54 | 26 |
| 36 | 7265 | 7292 | 7319 | 7346 | 7373 | 7400 | 7427 | 7454 | 7481 | 7508 | 7536 | 53 | 27 |
| 37 | 7536 | 7563 | 7590 | 7618 | 7646 | 7673 | 7701 | 7729 | 7757 | 7785 | 7813 | 52 | 28 |
| 38 | 7813 | 7841 | 7869 | 7898 | 7926 | 7954 | 7983 | 8012 | 8040 | 8069 | 8098 | 51 | 28 |
| 39 | 8098 | 8127 | 8156 | 8185 | 8214 | 8243 | 8273 | 8302 | 8332 | 8361 | 0.8391 | 50° | 29 |
| 40° | 0.8391 | 8421 | 8451 | 8481 | 8511 | 8541 | 8571 | 8601 | 8632 | 8662 | 8693 | 49 | 30 |
| 41 | 8693 | 8724 | 8754 | 8785 | 8816 | 8847 | 8878 | 8910 | 8941 | 8972 | 9004 | 48 | 31 |
| 42 | 9004 | 9036 | 9067 | 9099 | 9131 | 9163 | 9195 | 9228 | 9260 | 9293 | 9325 | 47 | 32 |
| 43 | 9325 | 9358 | 9391 | 9424 | 9457 | 9490 | 9523 | 9556 | 9590 | 9623 | 0.9657 | 46 | 33 |
| 44 | 0.9657 | 9691 | 9725 | 9759 | 9793 | 9827 | 9861 | 9896 | 9930 | 9965 | 1.0000 | 45° | 34 |
| 45° | 1.0000 | | | | | | | | | | | | |

| | °.9 =(54′) | °.8 (48′) | °.7 (42′) | °.6 (36′) | °.5 (30′) | °.4 (24′) | °.3 (18′) | °.2 (12′) | °.1 (6′) | °.0 (0′) | Deg |

For graphs, see Sec. 2.        **Natural Cotangents**

**Natural Tangents and Cotangents** *(continued)*

Natural Tangents at intervals of 0°.1, or 6'

| Deg | °.0 =(0') | °.1 (6') | °.2 (12') | °.3 (18') | °.4 (24') | °.5 (30') | °.6 (36') | °.7 (42') | °.8 (48') | °.9 (54') | | Avg diff |
|---|---|---|---|---|---|---|---|---|---|---|---|---|
| | | | | | | | | | | | 1.0000 | 45° |
| 45° | 1.0000 | 0035 | 0070 | 0105 | 0141 | 0176 | 0212 | 0247 | 0283 | 0319 | 0355 | 44 | 35 |
| 46 | 0355 | 0392 | 0428 | 0464 | 0501 | 0538 | 0575 | 0612 | 0649 | 0686 | 0724 | 43 | 37 |
| 47 | 0724 | 0761 | 0799 | 0837 | 0875 | 0913 | 0951 | 0990 | 1028 | 1067 | 1106 | 42 | 38 |
| 48 | 1106 | 1145 | 1184 | 1224 | 1263 | 1303 | 1343 | 1383 | 1423 | 1463 | 1504 | 41 | 40 |
| 49 | 1504 | 1544 | 1585 | 1626 | 1667 | 1708 | 1750 | 1792 | 1833 | 1875 | 1.1918 | 40° | 41 |
| 50° | 1.1918 | 1960 | 2002 | 2045 | 2088 | 2131 | 2174 | 2218 | 2261 | 2305 | 2349 | 39 | 43 |
| 51 | 2349 | 2393 | 2437 | 2482 | 2527 | 2572 | 2617 | 2662 | 2708 | 2753 | 2799 | 38 | 45 |
| 52 | 2799 | 2846 | 2892 | 2938 | 2985 | 3032 | 3079 | 3127 | 3175 | 3222 | 3270 | 37 | 47 |
| 53 | 3270 | 3319 | 3367 | 3416 | 3465 | 3514 | 3564 | 3613 | 3663 | 3713 | 3764 | 36 | 49 |
| 54 | 3764 | 3814 | 3865 | 3916 | 3968 | 4019 | 4071 | 4124 | 4176 | 4229 | 1.4281 | 35 | 52 |
| 55 | 1.4281 | 4335 | 4388 | 4442 | 4496 | 4550 | 4605 | 4659 | 4715 | 4770 | 4826 | 34 | 55 |
| 56 | 4826 | 4882 | 4938 | 4994 | 5051 | 5108 | 5166 | 5224 | 5282 | 5340 | 5399 | 33 | 57 |
| 57 | 5399 | 5458 | 5517 | 5577 | 5637 | 5697 | 5757 | 5818 | 5880 | 5941 | 6003 | 32 | 60 |
| 58 | 6003 | 6066 | 6128 | 6191 | 6255 | 6319 | 6383 | 6447 | 6512 | 6577 | 6643 | 31 | 64 |
| 59 | 1.6643 | 6709 | 6775 | 6842 | 6909 | 6977 | 7045 | 7113 | 7182 | 7251 | 1.7321 | 30° | 67 |
| 60° | 1.732 | 1.739 | 1.746 | 1.753 | 1.760 | 1.767 | 1.775 | 1.782 | 1.789 | 1.797 | 1.804 | 29 | 7 |
| 61 | 1.804 | 1.811 | 1.819 | 1.827 | 1.834 | 1.842 | 1.849 | 1.857 | 1.865 | 1.873 | 1.881 | 28 | 8 |
| 62 | 1.881 | 1.889 | 1.897 | 1.905 | 1.913 | 1.921 | 1.929 | 1.937 | 1.946 | 1.954 | 1.963 | 27 | 8 |
| 63 | 1.963 | 1.971 | 1.980 | 1.988 | 1.997 | 2.006 | 2.014 | 2.023 | 2.032 | 2.041 | 2.050 | 26 | 9 |
| 64 | 2.050 | 2.059 | 2.069 | 2.078 | 2.087 | 2.097 | 2.106 | 2.116 | 2.125 | 2.135 | 2.145 | 25 | 9 |
| 65 | 2.145 | 2.154 | 2.164 | 2.174 | 2.184 | 2.194 | 2.204 | 2.215 | 2.225 | 2.236 | 2.246 | 24 | 10 |
| 66 | 2.246 | 2.257 | 2.267 | 2.278 | 2.289 | 2.300 | 2.311 | 2.322 | 2.333 | 2.344 | 2.356 | 23 | 11 |
| 67 | 2.356 | 2.367 | 2.379 | 2.391 | 2.402 | 2.414 | 2.426 | 2.438 | 2.450 | 2.463 | 2.475 | 22 | 12 |
| 68 | 2.475 | 2.488 | 2.500 | 2.513 | 2.526 | 2.539 | 2.552 | 2.565 | 2.578 | 2.592 | 2.605 | 21 | 13 |
| 69 | 2.605 | 2.619 | 2.633 | 2.646 | 2.660 | 2.675 | 2.689 | 2.703 | 2.718 | 2.733 | 2.747 | 20° | 14 |
| 70° | 2.747 | 2.762 | 2.778 | 2.793 | 2.808 | 2.824 | 2.840 | 2.856 | 2.872 | 2.888 | 2.904 | 19 | 16 |
| 71 | 2.904 | 2.921 | 2.937 | 2.954 | 2.971 | 2.989 | 3.006 | 3.024 | 3.042 | 3.060 | 3.078 | 18 | 17 |
| 72 | 3.078 | 3.096 | 3.115 | 3.133 | 3.152 | 3.172 | 3.191 | 3.211 | 3.230 | 3.251 | 3.271 | 17 | 19 |
| 73 | 3.271 | 3.291 | 3.312 | 3.333 | 3.354 | 3.376 | 3.398 | 3.420 | 3.442 | 3.465 | 3.487 | 16 | 22 |
| 74 | 3.487 | 3.511 | 3.534 | 3.558 | 3.582 | 3.606 | 3.630 | 3.655 | 3.681 | 3.706 | 3.732 | 15 | 24 |
| 75 | 3.732 | 3.758 | 3.785 | 3.812 | 3.839 | 3.867 | 3.895 | 3.923 | 3.952 | 3.981 | 4.011 | 14 | 28 |
| 76 | 4.011 | 4.041 | 4.071 | 4.102 | 4.134 | 4.165 | 4.198 | 4.230 | 4.264 | 4.297 | 4.331 | 13 | 32 |
| 77 | 4.331 | 4.366 | 4.402 | 4.437 | 4.474 | 4.511 | 4.548 | 4.586 | 4.625 | 4.665 | 4.705 | 12 | 37 |
| 78 | 4.705 | 4.745 | 4.787 | 4.829 | 4.872 | 4.915 | 4.959 | 5.005 | 5.050 | 5.097 | 5.145 | 11 | 44 |
| 79 | 5.145 | 5.193 | 5.242 | 5.292 | 5.343 | 5.396 | 5.449 | 5.503 | 5.558 | 5.614 | 5.671 | 10° | 53 |
| 80° | 5.671 | 5.730 | 5.789 | 5.850 | 5.912 | 5.976 | 6.041 | 6.107 | 6.174 | 6.243 | 6.314 | 9 | |
| 81 | 6.314 | 6.386 | 6.460 | 6.535 | 6.612 | 6.691 | 6.772 | 6.855 | 6.940 | 7.026 | 7.115 | 8 | |
| 82 | 7.115 | 7.207 | 7.300 | 7.396 | 7.495 | 7.596 | 7.700 | 7.806 | 7.916 | 8.028 | 8.144 | 7 | |
| 83 | 8.144 | 8.264 | 8.386 | 8.513 | 8.643 | 8.777 | 8.915 | 9.058 | 9.205 | 9.357 | 9.514 | 6 | |
| 84 | 9.514 | 9.677 | 9.845 | 10.02 | 10.20 | 10.39 | 10.58 | 10.78 | 10.99 | 11.20 | 11.43 | 5 | |
| 85 | 11.43 | 11.66 | 11.91 | 12.16 | 12.43 | 12.71 | 13.00 | 13.30 | 13.62 | 13.95 | 14.30 | 4 | |
| 86 | 14.30 | 14.67 | 15.06 | 15.46 | 15.90 | 16.35 | 16.83 | 17.34 | 17.89 | 18.46 | 19.08 | 3 | |
| 87 | 19.08 | 19.74 | 20.45 | 21.20 | 22.02 | 22.90 | 23.86 | 24.90 | 26.03 | 27.27 | 28.64 | 2 | |
| 88 | 28.64 | 30.14 | 31.82 | 33.69 | 35.80 | 38.19 | 40.92 | 44.07 | 47.74 | 52.08 | 57.29 | 1 | |
| 89 | 57.29 | 63.66 | 71.62 | 81.85 | 95.49 | 114.6 | 143.2 | 191.0 | 286.5 | 573.0 | ∞ | 0° | |
| 90° | ∞ | | | | | | | | | | | | |

| | °.9 =(54') | °.8 (48') | °.7 (42') | °.6 (36') | °.5 (30') | °.4 (24') | °.3 (18') | °.2 (12') | °.1 (6') | °.0 (0') | Deg |
|---|---|---|---|---|---|---|---|---|---|---|---|

**Natural Cotangents**

**Natural Secants and Cosecants**

**Natural Secants at intervals of 0°.1, or 6′**

| Deg | °.0 =(0′) | °.1 (6′) | °.2 (12′) | °.3 (18′) | °.4 (24′) | °.5 (30′) | °.6 (36′) | °.7 (42′) | °.8 (48′) | °.9 (54′) | | | Avg diff |
|---|---|---|---|---|---|---|---|---|---|---|---|---|---|
| | | | | | | | | | | | 1.0000 | 90° | |
| 0° | 1.0000 | 0000 | 0000 | 0000 | 0000 | 0000 | 0001 | 0001 | 0001 | 0001 | 0002 | 89 | 0 |
| 1 | 0002 | 0002 | 0002 | 0003 | 0003 | 0003 | 0004 | 0004 | 0005 | 0006 | 0006 | 88 | 0 |
| 2 | 0006 | 0007 | 0007 | 0008 | 0009 | 0010 | 0010 | 0011 | 0012 | 0013 | 0014 | 87 | 1 |
| 3 | 0014 | 0015 | 0016 | 0017 | 0018 | 0019 | 0020 | 0021 | 0022 | 0023 | 0024 | 86 | 1 |
| 4 | 0024 | 0026 | 0027 | 0028 | 0030 | 0031 | 0032 | 0034 | 0035 | 0037 | 1.0038 | 85 | 1 |
| 5 | 1.0038 | 0040 | 0041 | 0043 | 0045 | 0046 | 0048 | 0050 | 0051 | 0053 | 0055 | 84 | 2 |
| 6 | 0055 | 0057 | 0059 | 0061 | 0063 | 0065 | 0067 | 0069 | 0071 | 0073 | 0075 | 83 | 2 |
| 7 | 0075 | 0077 | 0079 | 0082 | 0084 | 0086 | 0089 | 0091 | 0093 | 0096 | 0098 | 82 | 2 |
| 8 | 0098 | 0101 | 0103 | 0106 | 0108 | 0111 | 0114 | 0116 | 0119 | 0122 | 0125 | 81 | 3 |
| 9 | 0125 | 0127 | 0130 | 0133 | 0136 | 0139 | 0142 | 0145 | 0148 | 0151 | 1.0154 | 80° | 3 |
| 10° | 1.0154 | 0157 | 0161 | 0164 | 0167 | 0170 | 0174 | 0177 | 0180 | 0184 | 0187 | 79 | 3 |
| 11 | 0187 | 0191 | 0194 | 0198 | 0201 | 0205 | 0209 | 0212 | 0216 | 0220 | 0223 | 78 | 4 |
| 12 | 0223 | 0227 | 0231 | 0235 | 0239 | 0243 | 0247 | 0251 | 0255 | 0259 | 0263 | 77 | 4 |
| 13 | 0263 | 0267 | 0271 | 0276 | 0280 | 0284 | 0288 | 0293 | 0297 | 0302 | 0306 | 76 | 4 |
| 14 | 0306 | 0311 | 0315 | 0320 | 0324 | 0329 | 0334 | 0338 | 0343 | 0348 | 1.0353 | 75 | 5 |
| 15 | 1.0353 | 0358 | 0363 | 0367 | 0372 | 0377 | 0382 | 0388 | 0393 | 0398 | 0403 | 74 | 5 |
| 16 | 0403 | 0408 | 0413 | 0419 | 0424 | 0429 | 0435 | 0440 | 0446 | 0451 | 0457 | 73 | 5 |
| 17 | 0457 | 0463 | 0468 | 0474 | 0480 | 0485 | 0491 | 0497 | 0503 | 0509 | 0515 | 72 | 6 |
| 18 | 0515 | 0521 | 0527 | 0533 | 0539 | 0545 | 0551 | 0557 | 0564 | 0570 | 0576 | 71 | 6 |
| 19 | 0576 | 0583 | 0589 | 0595 | 0602 | 0608 | 0615 | 0622 | 0628 | 0635 | 1.0642 | 70° | 7 |
| 20° | 1.0642 | 0649 | 0655 | 0662 | 0669 | 0676 | 0683 | 0690 | 0697 | 0704 | 0711 | 69 | 7 |
| 21 | 0711 | 0719 | 0726 | 0733 | 0740 | 0748 | 0755 | 0763 | 0770 | 0778 | 0785 | 68 | 7 |
| 22 | 0785 | 0793 | 0801 | 0808 | 0816 | 0824 | 0832 | 0840 | 0848 | 0856 | 0864 | 67 | 8 |
| 23 | 0864 | 0872 | 0880 | 0888 | 0896 | 0904 | 0913 | 0921 | 0929 | 0938 | 0946 | 66 | 8 |
| 24 | 0946 | 0955 | 0963 | 0972 | 0981 | 0989 | 0998 | 1007 | 1016 | 1025 | 1.1034 | 65 | 9 |
| 25 | 1.1034 | 1043 | 1052 | 1061 | 1070 | 1079 | 1089 | 1098 | 1107 | 1117 | 1126 | 64 | 9 |
| 26 | 1126 | 1136 | 1145 | 1155 | 1164 | 1174 | 1184 | 1194 | 1203 | 1213 | 1223 | 63 | 10 |
| 27 | 1223 | 1233 | 1243 | 1253 | 1264 | 1274 | 1284 | 1294 | 1305 | 1315 | 1326 | 62 | 10 |
| 28 | 1326 | 1336 | 1347 | 1357 | 1368 | 1379 | 1390 | 1401 | 1412 | 1423 | 1434 | 61 | 11 |
| 29 | 1434 | 1445 | 1456 | 1467 | 1478 | 1490 | 1501 | 1512 | 1524 | 1535 | 1.1547 | 60° | 11 |
| 30° | 1.1547 | 1559 | 1570 | 1582 | 1594 | 1606 | 1618 | 1630 | 1642 | 1654 | 1666 | 59 | 12 |
| 31 | 1666 | 1679 | 1691 | 1703 | 1716 | 1728 | 1741 | 1753 | 1766 | 1779 | 1792 | 58 | 13 |
| 32 | 1792 | 1805 | 1818 | 1831 | 1844 | 1857 | 1870 | 1883 | 1897 | 1910 | 1924 | 57 | 13 |
| 33 | 1924 | 1937 | 1951 | 1964 | 1978 | 1992 | 2006 | 2020 | 2034 | 2048 | 2062 | 56 | 14 |
| 34 | 2062 | 2076 | 2091 | 2105 | 2120 | 2134 | 2149 | 2163 | 2178 | 2193 | 1.2208 | 55 | 15 |
| 35 | 1.2208 | 2223 | 2238 | 2253 | 2268 | 2283 | 2299 | 2314 | 2329 | 2345 | 2361 | 54 | 15 |
| 36 | 2361 | 2376 | 2392 | 2408 | 2424 | 2440 | 2456 | 2472 | 2489 | 2505 | 2521 | 53 | 16 |
| 37 | 2521 | 2538 | 2554 | 2571 | 2588 | 2605 | 2622 | 2639 | 2656 | 2673 | 2690 | 52 | 17 |
| 38 | 2690 | 2708 | 2725 | 2742 | 2760 | 2778 | 2796 | 2813 | 2831 | 2849 | 2868 | 51 | 18 |
| 39 | 2868 | 2886 | 2904 | 2923 | 2941 | 2960 | 2978 | 2997 | 3016 | 3035 | 1.3054 | 50° | 19 |
| 40° | 1.3054 | 3073 | 3093 | 3112 | 3131 | 3151 | 3171 | 3190 | 3210 | 3230 | 3250 | 49 | 20 |
| 41 | 3250 | 3270 | 3291 | 3311 | 3331 | 3352 | 3373 | 3393 | 3414 | 3435 | 3456 | 48 | 21 |
| 42 | 3456 | 3478 | 3499 | 3520 | 3542 | 3563 | 3585 | 3607 | 3629 | 3651 | 3673 | 47 | 22 |
| 43 | 3673 | 3696 | 3718 | 3741 | 3763 | 3786 | 3809 | 3832 | 3855 | 3878 | 3902 | 46 | 23 |
| 44 | 3902 | 3925 | 3949 | 3972 | 3996 | 4020 | 4044 | 4069 | 4093 | 4118 | 1.4142 | 45° | 24 |
| 45° | 1.4142 | | | | | | | | | | | | |

| | °.9 =(54′) | °.8 (48′) | °.7 (42′) | °.6 (36′) | °.5 (30′) | °.4 (24′) | °.3 (18′) | °.2 (12′) | °.1 (6′) | °.0 (0′) | Deg |
|---|---|---|---|---|---|---|---|---|---|---|---|

For graphs, see Sec. 2.                                          **Natural Cosecants**

**Natural Secants and Cosecants (continued)**

**Natural Secants at intervals of 0°.1, or 6'**

| Deg | °.0 =(0') | °.1 (6') | °.2 (12') | °.3 (18') | °.4 (24') | °.5 (30') | °.6 (36') | °.7 (42') | °.8 (48') | °.9 (54') | | | Avg diff |
|---|---|---|---|---|---|---|---|---|---|---|---|---|---|
| | | | | | | | | | | | 1.4142 | 45° | |
| 45° | 1.4142 | 4167 | 4192 | 4217 | 4242 | 4267 | 4293 | 4318 | 4344 | 4370 | 4396 | 44 | 25 |
| 46 | 4396 | 4422 | 4448 | 4474 | 4501 | 4527 | 4554 | 4581 | 4608 | 4635 | 4663 | 43 | 27 |
| 47 | 4663 | 4690 | 4718 | 4746 | 4774 | 4802 | 4830 | 4859 | 4887 | 4916 | 4945 | 42 | 28 |
| 48 | 4945 | 4974 | 5003 | 5032 | 5062 | 5092 | 5121 | 5151 | 5182 | 5212 | 5243 | 41 | 30 |
| 49 | 5243 | 5273 | 5304 | 5335 | 5366 | 5398 | 5429 | 5461 | 5493 | 5525 | 1.5557 | 40° | 31 |
| 50° | 1.5557 | 5590 | 5622 | 5655 | 5688 | 5721 | 5755 | 5788 | 5822 | 5856 | 5890 | 39 | 33 |
| 51 | 5890 | 5925 | 5959 | 5994 | 6029 | 6064 | 6099 | 6135 | 6171 | 6207 | 6243 | 38 | 35 |
| 52 | 6243 | 6279 | 6316 | 6353 | 6390 | 6427 | 6464 | 6502 | 6540 | 6578 | 6616 | 37 | 37 |
| 53 | 6616 | 6655 | 6694 | 6733 | 6772 | 6812 | 6852 | 6892 | 6932 | 6972 | 7013 | 36 | 40 |
| 54 | 7013 | 7054 | 7095 | 7137 | 7179 | 7221 | 7263 | 7305 | 7348 | 7391 | 1.7434 | 35 | 42 |
| 55 | 1.7434 | 7478 | 7522 | 7566 | 7610 | 7655 | 7700 | 7745 | 7791 | 7837 | 7883 | 34 | 45 |
| 56 | 7883 | 7929 | 7976 | 8023 | 8070 | 8118 | 8166 | 8214 | 8263 | 8312 | 8361 | 33 | 48 |
| 57 | 8361 | 8410 | 8460 | 8510 | 8561 | 8612 | 8663 | 8714 | 8766 | 8818 | 8871 | 32 | 51 |
| 58 | 8871 | 8924 | 8977 | 9031 | 9084 | 9139 | 9194 | 9249 | 9304 | 9360 | 1.9416 | 31 | 54 |
| 59 | 1.9416 | 9473 | 9530 | 9587 | 9645 | 9703 | 9762 | 9821 | 9880 | 9940 | 2.0000 | 30° | 58 |
| 60° | 2.000 | 2.006 | 2.012 | 2.018 | 2.025 | 2.031 | 2.037 | 2.043 | 2.050 | 2.056 | 2.063 | 29 | 6 |
| 61 | 2.063 | 2.069 | 2.076 | 2.082 | 2.089 | 2.096 | 2.103 | 2.109 | 2.116 | 2.123 | 2.130 | 28 | 7 |
| 62 | 2.130 | 2.137 | 2.144 | 2.151 | 2.158 | 2.166 | 2.173 | 2.180 | 2.188 | 2.195 | 2.203 | 27 | 7 |
| 63 | 2.203 | 2.210 | 2.218 | 2.226 | 2.233 | 2.241 | 2.249 | 2.257 | 2.265 | 2.273 | 2.281 | 26 | 8 |
| 64 | 2.281 | 2.289 | 2.298 | 2.306 | 2.314 | 2.323 | 2.331 | 2.340 | 2.349 | 2.357 | 2.366 | 25 | 8 |
| 65 | 2.366 | 2.375 | 2.384 | 2.393 | 2.402 | 2.411 | 2.421 | 2.430 | 2.439 | 2.449 | 2.459 | 24 | 9 |
| 66 | 2.459 | 2.468 | 2.478 | 2.488 | 2.498 | 2.508 | 2.518 | 2.528 | 2.538 | 2.549 | 2.559 | 23 | 10 |
| 67 | 2.559 | 2.570 | 2.581 | 2.591 | 2.602 | 2.613 | 2.624 | 2.635 | 2.647 | 2.658 | 2.669 | 22 | 11 |
| 68 | 2.669 | 2.681 | 2.693 | 2.705 | 2.716 | 2.729 | 2.741 | 2.753 | 2.765 | 2.778 | 2.790 | 21 | 12 |
| 69 | 2.790 | 2.803 | 2.816 | 2.829 | 2.842 | 2.855 | 2.869 | 2.882 | 2.896 | 2.910 | 2.924 | 20° | 13 |
| 70° | 2.924 | 2.938 | 2.952 | 2.967 | 2.981 | 2.996 | 3.011 | 3.026 | 3.041 | 3.056 | 3.072 | 19 | 15 |
| 71 | 3.072 | 3.087 | 3.103 | 3.119 | 3.135 | 3.152 | 3.168 | 3.185 | 3.202 | 3.219 | 3.236 | 18 | 16 |
| 72 | 3.236 | 3.254 | 3.271 | 3.289 | 3.307 | 3.326 | 3.344 | 3.363 | 3.382 | 3.401 | 3.420 | 17 | 18 |
| 73 | 3.420 | 3.440 | 3.460 | 3.480 | 3.500 | 3.521 | 3.542 | 3.563 | 3.584 | 3.606 | 3.628 | 16 | 21 |
| 74 | 3.628 | 3.650 | 3.673 | 3.695 | 3.719 | 3.742 | 3.766 | 3.790 | 3.814 | 3.839 | 3.864 | 15 | 24 |
| 75 | 3.864 | 3.889 | 3.915 | 3.941 | 3.967 | 3.994 | 4.021 | 4.049 | 4.077 | 4.105 | 4.134 | 14 | 27 |
| 76 | 4.134 | 4.163 | 4.192 | 4.222 | 4.253 | 4.284 | 4.315 | 4.347 | 4.379 | 4.412 | 4.445 | 13 | 31 |
| 77 | 4.445 | 4.479 | 4.514 | 4.549 | 4.584 | 4.620 | 4.657 | 4.694 | 4.732 | 4.771 | 4.810 | 12 | 36 |
| 78 | 4.810 | 4.850 | 4.890 | 4.931 | 4.973 | 5.016 | 5.059 | 5.103 | 5.148 | 5.194 | 5.241 | 11 | 43 |
| 79 | 5.241 | 5.288 | 5.337 | 5.386 | 5.436 | 5.487 | 5.540 | 5.593 | 5.647 | 5.702 | 5.759 | 10° | 52 |
| 80° | 5.759 | 5.816 | 5.875 | 5.935 | 5.996 | 6.059 | 6.123 | 6.188 | 6.255 | 6.323 | 6.392 | 9 | |
| 81 | 6.392 | 6.464 | 6.537 | 6.611 | 6.687 | 6.765 | 6.845 | 6.927 | 7.011 | 7.097 | 7.185 | 8 | |
| 82 | 7.185 | 7.276 | 7.368 | 7.463 | 7.561 | 7.661 | 7.764 | 7.870 | 7.979 | 8.091 | 8.206 | 7 | |
| 83 | 8.206 | 8.324 | 8.446 | 8.571 | 8.700 | 8.834 | 8.971 | 9.113 | 9.259 | 9.411 | 9.567 | 6 | |
| 84 | 9.567 | 9.728 | 9.895 | 10.07 | 10.25 | 10.43 | 10.63 | 10.83 | 11.03 | 11.25 | 11.47 | 5 | |
| 85 | 11.47 | 11.71 | 11.95 | 12.20 | 12.47 | 12.75 | 13.03 | 13.34 | 13.65 | 13.99 | 14.34 | 4 | |
| 86 | 14.34 | 14.70 | 15.09 | 15.50 | 15.93 | 16.38 | 16.86 | 17.37 | 17.91 | 18.49 | 19.11 | 3 | |
| 87 | 19.11 | 19.77 | 20.47 | 21.23 | 22.04 | 22.93 | 23.88 | 24.92 | 26.05 | 27.29 | 28.65 | 2 | |
| 88 | 28.65 | 30.16 | 31.84 | 33.71 | 35.81 | 38.20 | 40.93 | 44.08 | 47.75 | 52.09 | 57.30 | 1 | |
| 89 | 57.30 | 63.66 | 71.62 | 81.85 | 95.49 | 114.6 | 143.2 | 191.0 | 286.5 | 573.0 | ∞ | 0° | |
| 90° | ∞ | | | | | | | | | | | | |
| | °.9 =(54') | °.8 (48') | °.7 (42') | °.6 (36') | °.5 (30') | °.4 (24') | °.3 (18') | °.2 (12') | °.1 (6') | °.0 (0') | | Deg | |

**Natural Cosecants**

**Exponentials** [$e^n$ and $e^{-n}$]

| n | $e^n$ | Diff | n | $e^n$ | Diff | n | $e^n$ | n | $e^{-n}$ | Diff | n | $e^{-n}$ | n | $e^{-n}$ |
|---|---|---|---|---|---|---|---|---|---|---|---|---|---|---|
| 0.00 | 1.000 | 10 | 0.50 | 1.649 | 16 | 1.0 | 2.718* | 0.00 | 1.000 | −10 | 0.50 | .607 | 1.0 | .368* |
| .01 | 1.010 | 10 | .51 | 1.665 | 16 | .1 | 3.004 | .01 | .990 | −10 | .51 | .600 | .1 | .333 |
| .02 | 1.020 | 10 | .52 | 1.682 | 17 | .2 | 3.320 | .02 | .980 | −10 | .52 | .595 | .2 | .301 |
| .03 | 1.030 | 11 | .53 | 1.699 | 17 | .3 | 3.669 | .03 | .970 | −10 | .53 | .589 | .3 | .273 |
| .04 | 1.041 | 10 | .54 | 1.716 | 17 | .4 | 4.055 | .04 | .961 | −10 | .54 | .583 | .4 | .247 |
| 0.05 | 1.051 | 11 | 0.55 | 1.733 | 18 | 1.5 | 4.482 | 0.05 | .951 | −9 | 0.55 | .577 | 1.5 | .223 |
| .06 | 1.062 | 11 | .56 | 1.751 | 17 | .6 | 4.953 | .06 | .942 | −10 | .56 | .571 | .6 | .202 |
| .07 | 1.073 | 11 | .57 | 1.768 | 18 | .7 | 5.474 | .07 | .932 | −9 | .57 | .566 | .7 | .183 |
| .08 | 1.083 | 10 | .58 | 1.786 | 18 | .8 | 6.050 | .08 | .923 | −9 | .58 | .560 | .8 | .165 |
| .09 | 1.094 | 11 | .59 | 1.804 | 18 | .9 | 6.686 | .09 | .914 | −9 | .59 | .554 | .9 | .150 |
| 0.10 | 1.105 | 11 | 0.60 | 1.822 | 18 | 2.0 | 7.389 | 0.10 | .905 | −9 | 0.60 | .549 | 2.0 | .135 |
| .11 | 1.116 | 11 | .61 | 1.840 | 19 | .1 | 8.166 | .11 | .896 | −9 | .61 | .543 | .1 | .122 |
| .12 | 1.127 | 12 | .62 | 1.859 | 19 | .2 | 9.025 | .12 | .887 | −9 | .62 | .538 | .2 | .111 |
| .13 | 1.139 | 11 | .63 | 1.878 | 18 | .3 | 9.974 | .13 | .878 | −9 | .63 | .533 | .3 | .100 |
| .14 | 1.150 | 12 | .64 | 1.896 | 20 | .4 | 11.02 | .14 | .869 | −8 | .64 | .527 | .4 | .0907 |
| 0.15 | 1.162 | 12 | 0.65 | 1.916 | 19 | 2.5 | 12.18 | 0.15 | .861 | −9 | 0.65 | .522 | 2.5 | .0821 |
| .16 | 1.174 | 11 | .66 | 1.935 | 19 | .6 | 13.46 | .16 | .852 | −8 | .66 | .517 | .6 | .0743 |
| .17 | 1.185 | 12 | .67 | 1.954 | 20 | .7 | 14.88 | .17 | .844 | −9 | .67 | .512 | .7 | .0672 |
| .18 | 1.197 | 12 | .68 | 1.974 | 20 | .8 | 16.44 | .18 | .835 | −8 | .68 | .507 | .8 | .0608 |
| .19 | 1.209 | 12 | .69 | 1.994 | 20 | .9 | 18.17 | .19 | .827 | −8 | .69 | .502 | .9 | .0550 |
| 0.20 | 1.221 | 13 | 0.70 | 2.014 | 20 | 3.0 | 20.09 | 0.20 | .819 | −8 | 0.70 | .497 | 3.0 | .0498 |
| .21 | 1.234 | 12 | .71 | 2.034 | 20 | .1 | 22.20 | .21 | .811 | −8 | .71 | .492 | .1 | .0450 |
| .22 | 1.246 | 13 | .72 | 2.054 | 21 | .2 | 24.53 | .22 | .803 | −8 | .72 | .487 | .2 | .0408 |
| .23 | 1.259 | 12 | .73 | 2.075 | 21 | .3 | 27.11 | .23 | .795 | −8 | .73 | .482 | .3 | .0369 |
| .24 | 1.271 | 13 | .74 | 2.096 | 21 | .4 | 29.96 | .24 | .787 | −8 | .74 | .477 | .4 | .0334 |
| 0.25 | 1.284 | 13 | 0.75 | 2.117 | 21 | 3.5 | 33.12 | 0.25 | .779 | −8 | 0.75 | .472 | 3.5 | .0302 |
| .26 | 1.297 | 13 | .76 | 2.138 | 22 | .6 | 36.60 | .26 | .771 | −8 | .76 | .468 | .6 | .0273 |
| .27 | 1.310 | 13 | .77 | 2.160 | 21 | .7 | 40.45 | .27 | .763 | −7 | .77 | .463 | .7 | .0247 |
| .28 | 1.323 | 13 | .78 | 2.181 | 22 | .8 | 44.70 | .28 | .756 | −8 | .78 | .458 | .8 | .0224 |
| .29 | 1.336 | 14 | .79 | 2.203 | 23 | .9 | 49.40 | .29 | .748 | −7 | .79 | .454 | .9 | .0202 |
| 0.30 | 1.350 | 13 | 0.80 | 2.226 | 22 | 4.0 | 54.60 | 0.30 | .741 | −8 | 0.80 | .449 | 4.0 | .0183 |
| .31 | 1.363 | 14 | .81 | 2.248 | 22 | .1 | 60.34 | .31 | .733 | −7 | .81 | .445 | .1 | .0166 |
| .32 | 1.377 | 14 | .82 | 2.270 | 23 | .2 | 66.69 | .32 | .726 | −7 | .82 | .440 | .2 | .0150 |
| .33 | 1.391 | 14 | .83 | 2.293 | 23 | .3 | 73.70 | .33 | .719 | −7 | .83 | .436 | .3 | .0136 |
| .34 | 1.405 | 14 | .84 | 2.316 | 24 | .4 | 81.45 | .34 | .712 | −7 | .84 | .432 | .4 | .0123 |
| 0.35 | 1.419 | 14 | 0.85 | 2.340 | 23 | 4.5 | 90.02 | 0.35 | .705 | −7 | 0.85 | .427 | 4.5 | .0111 |
| .36 | 1.433 | 15 | .86 | 2.363 | 24 |  |  | .36 | .698 | −7 | .86 | .423 |  |  |
| .37 | 1.448 | 14 | .87 | 2.387 | 24 | 5.0 | 148.4 | .37 | .691 | −7 | .87 | .419 | 5.0 | .00674 |
| .38 | 1.462 | 15 | .88 | 2.411 | 24 | 6.0 | 403.4 | .38 | .684 | −7 | .88 | .415 | 6.0 | .00248 |
| .39 | 1.477 | 15 | .89 | 2.435 | 25 | 7.0 | 1097. | .39 | .677 | −7 | .89 | .411 | 7.0 | .000912 |
| 0.40 | 1.492 | 15 | 0.90 | 2.460 | 24 | 8.0 | 2981. | 0.40 | .670 | −6 | 0.90 | .407 | 8.0 | .000335 |
| .41 | 1.507 | 15 | .91 | 2.484 | 25 | 9.0 | 8103. | .41 | .664 | −7 | .91 | .403 | 9.0 | .000123 |
| .42 | 1.522 | 15 | .92 | 2.509 | 26 | 10.0 | 22026. | .42 | .657 | −6 | .92 | .399 | 10.0 | .000045 |
| .43 | 1.537 | 16 | .93 | 2.535 | 25 | $\pi/2$ | 4.810 | .43 | .651 | −7 | .93 | .395 | $\pi/2$ | .208 |
| .44 | 1.553 | 15 | .94 | 2.560 | 26 | $2\pi/2$ | 23.14 | .44 | .644 | −6 | .94 | .391 | $2\pi/2$ | .0432 |
| 0.45 | 1.568 | 16 | 0.95 | 2.586 | 26 | $3\pi/2$ | 111.3 | 0.45 | .638 | −7 | 0.95 | .387 | $3\pi/2$ | .00898 |
| .46 | 1.584 | 16 | .96 | 2.612 | 26 | $4\pi/2$ | 535.5 | .46 | .631 | −6 | .96 | .383 | $4\pi/2$ | .00187 |
| .47 | 1.600 | 16 | .97 | 2.638 | 26 | $5\pi/2$ | 2576. | .47 | .625 | −6 | .97 | .379 | $5\pi/2$ | .000388 |
| .48 | 1.616 | 16 | .98 | 2.664 | 27 | $6\pi/2$ | 12392. | .48 | .619 | −6 | .98 | .375 | $6\pi/2$ | .000081 |
| .49 | 1.632 | 17 | .99 | 2.691 | 27 | $7\pi/2$ | 59610. | .49 | .613 | −6 | .99 | .372 | $7\pi/2$ | .000017 |
| 0.50 | 1.649 |  | 1.00 | 2.718 |  | $8\pi/2$ | 286751. | 0.50 | 0.607 |  | 1.00 | .368 | $8\pi/2$ | .000003 |

*Note: Do not interpolate in this column.

$e = 2.71828$    $1/e = 0.367879$    $\log_{10} e = 0.4343$    $1/(0.4343) = 2.3026$

$\log_{10}(0.4343) = \bar{1}.6378$    $\log_{10}(e^n) = n(0.4343)$

For graphical representations see Sec. 2.

### Natural Logarithms

| | $n$ | $n\,(2.3026)$ | $n\,(0.6974-3)$ |
|---|---|---|---|
| | 1 | 2.3026 | 0.6974–3 |
| | 2 | 4.6052 | 0.3948–5 |
| | 3 | 6.9078 | 0.0922–7 |
| These two pages give the natural or Napierian loga- | 4 | 9.2103 | 0.7897–10 |
| rithms (ln) of numbers between 1 and 10, correct to four | 5 | 11.5129 | 0.4871–12 |
| places. Moving the decimal point $n$ places to the right [or | 6 | 13.8155 | 0.1845–14 |
| left] in the number is equivalent to adding $n$ times 2.3026 | 7 | 16.1181 | 0.8819–17 |
| [or $n$ times $\bar{3}.6974$] to the logarithm.  Base $e = 2.71828+$ | 8 | 18.4207 | 0.5793–19 |
| | 9 | 20.7233 | 0.2767–21 |

| Number | 0 | 1 | 2 | 3 | 4 | 5 | 6 | 7 | 8 | 9 | Avg diff |
|---|---|---|---|---|---|---|---|---|---|---|---|
| 1.0 | 0.0000 | 0100 | 0198 | 0296 | 0392 | 0488 | 0583 | 0677 | 0770 | 0862 | 95 |
| 1.1 | 0953 | 1044 | 1133 | 1222 | 1310 | 1398 | 1484 | 1570 | 1655 | 1740 | 87 |
| 1.2 | 1823 | 1906 | 1989 | 2070 | 2151 | 2231 | 2311 | 2390 | 2469 | 2546 | 80 |
| 1.3 | 2624 | 2700 | 2776 | 2852 | 2927 | 3001 | 3075 | 3148 | 3221 | 3293 | 74 |
| 1.4 | 3365 | 3436 | 3507 | 3577 | 3646 | 3716 | 3784 | 3853 | 3920 | 3988 | 69 |
| 1.5 | 0.4055 | 4121 | 4187 | 4253 | 4318 | 4383 | 4447 | 4511 | 4574 | 4637 | 65 |
| 1.6 | 4700 | 4762 | 4824 | 4886 | 4947 | 5008 | 5068 | 5128 | 5188 | 5247 | 61 |
| 1.7 | 5306 | 5365 | 5423 | 5481 | 5539 | 5596 | 5653 | 5710 | 5766 | 5822 | 57 |
| 1.8 | 5878 | 5933 | 5988 | 6043 | 6098 | 6152 | 6206 | 6259 | 6313 | 6366 | 54 |
| 1.9 | 6419 | 6471 | 6523 | 6575 | 6627 | 6678 | 6729 | 6780 | 6831 | 6881 | 51 |
| 2.0 | 0.6931 | 6981 | 7031 | 7080 | 7129 | 7178 | 7227 | 7275 | 7324 | 7372 | 49 |
| 2.1 | 7419 | 7467 | 7514 | 7561 | 7608 | 7655 | 7701 | 7747 | 7793 | 7839 | 47 |
| 2.2 | 7885 | 7930 | 7975 | 8020 | 8065 | 8109 | 8154 | 8198 | 8242 | 8286 | 44 |
| 2.3 | 8329 | 8372 | 8416 | 8459 | 8502 | 8544 | 8587 | 8629 | 8671 | 8713 | 43 |
| 2.4 | 8755 | 8796 | 8838 | 8879 | 8920 | 8961 | 9002 | 9042 | 9083 | 9123 | 41 |
| 2.5 | 0.9163 | 9203 | 9243 | 9282 | 9322 | 9361 | 9400 | 9439 | 9478 | 9517 | 39 |
| 2.6 | 9555 | 9594 | 9632 | 9670 | 9708 | 9746 | 9783 | 9821 | 9858 | 9895 | 38 |
| 2.7 | 0.9933 | 9969 | *0006 | *0043 | *0080 | *0116 | *0152 | *0188 | *0225 | *0260 | 36 |
| 2.8 | 1.0296 | 0332 | 0367 | 0403 | 0438 | 0473 | 0508 | 0543 | 0578 | 0613 | 35 |
| 2.9 | 0647 | 0682 | 0716 | 0750 | 0784 | 0818 | 0852 | 0886 | 0919 | 0953 | 34 |
| 3.0 | 1.0986 | 1019 | 1053 | 1086 | 1119 | 1151 | 1184 | 1217 | 1249 | 1282 | 33 |
| 3.1 | 1314 | 1346 | 1378 | 1410 | 1442 | 1474 | 1506 | 1537 | 1569 | 1600 | 32 |
| 3.2 | 1632 | 1663 | 1694 | 1725 | 1756 | 1787 | 1817 | 1848 | 1878 | 1909 | 31 |
| 3.3 | 1939 | 1969 | 2000 | 2030 | 2060 | 2090 | 2119 | 2149 | 2179 | 2208 | 30 |
| 3.4 | 2238 | 2267 | 2296 | 2326 | 2355 | 2384 | 2413 | 2442 | 2470 | 2499 | 29 |
| 3.5 | 1.2528 | 2556 | 2585 | 2613 | 2641 | 2669 | 2698 | 2726 | 2754 | 2782 | 28 |
| 3.6 | 2809 | 2837 | 2865 | 2892 | 2920 | 2947 | 2975 | 3002 | 3029 | 3056 | 27 |
| 3.7 | 3083 | 3110 | 3137 | 3164 | 3191 | 3218 | 3244 | 3271 | 3297 | 3324 | 27 |
| 3.8 | 3350 | 3376 | 3403 | 3429 | 3455 | 3481 | 3507 | 3533 | 3558 | 3584 | 26 |
| 3.9 | 3610 | 3635 | 3661 | 3686 | 3712 | 3737 | 3762 | 3788 | 3813 | 3838 | 25 |
| 4.0 | 1.3863 | 3888 | 3913 | 3938 | 3962 | 3987 | 4012 | 4036 | 4061 | 4085 | 25 |
| 4.1 | 4110 | 4134 | 4159 | 4183 | 4207 | 4231 | 4255 | 4279 | 4303 | 4327 | 24 |
| 4.2 | 4351 | 4375 | 4398 | 4422 | 4446 | 4469 | 4493 | 4516 | 4540 | 4563 | 23 |
| 4.3 | 4586 | 4609 | 4633 | 4656 | 4679 | 4702 | 4725 | 4748 | 4770 | 4793 | 23 |
| 4.4 | 4816 | 4839 | 4861 | 4884 | 4907 | 4929 | 4951 | 4974 | 4996 | 5019 | 22 |
| 4.5 | 1.5041 | 5063 | 5085 | 5107 | 5129 | 5151 | 5173 | 5195 | 5217 | 5239 | 22 |
| 4.6 | 5261 | 5282 | 5304 | 5326 | 5347 | 5369 | 5390 | 5412 | 5433 | 5454 | 21 |
| 4.7 | 5476 | 5497 | 5518 | 5539 | 5560 | 5581 | 5602 | 5623 | 5644 | 5665 | 21 |
| 4.8 | 5686 | 5707 | 5728 | 5748 | 5769 | 5790 | 5810 | 5831 | 5851 | 5872 | 20 |
| 4.9 | 5892 | 5913 | 5933 | 5953 | 5974 | 5994 | 6014 | 6034 | 6054 | 6074 | 20 |

$\ln x = (2.3026)\log_{10}x \qquad \log_{10} x = (0.4343)\ln x$

where $2.3026 = \ln 10$ and $0.4343 = \log_{10} e$. For graphs, see Sec. 2.

**Natural Logarithms (*continued*)**

| Num-ber | 0 | 1 | 2 | 3 | 4 | 5 | 6 | 7 | 8 | 9 | Avg diff |
|---|---|---|---|---|---|---|---|---|---|---|---|
| 5.0 | 1.6094 | 6114 | 6134 | 6154 | 6174 | 6194 | 6214 | 6233 | 6253 | 6273 | 20 |
| 5.1 | 6292 | 6312 | 6332 | 6351 | 6371 | 6390 | 6409 | 6429 | 6448 | 6467 | 19 |
| 5.2 | 6487 | 6506 | 6525 | 6544 | 6563 | 6582 | 6601 | 6620 | 6639 | 6658 | 19 |
| 5.3 | 6677 | 6696 | 6715 | 6734 | 6752 | 6771 | 6790 | 6808 | 6827 | 6845 | 18 |
| 5.4 | 6864 | 6882 | 6901 | 6919 | 6938 | 6956 | 6974 | 6993 | 7011 | 7029 | 18 |
| 5.5 | 1.7047 | 7066 | 7084 | 7102 | 7120 | 7138 | 7156 | 7174 | 7192 | 7210 | 18 |
| 5.6 | 7228 | 7246 | 7263 | 7281 | 7299 | 7317 | 7334 | 7352 | 7370 | 7387 | 18 |
| 5.7 | 7405 | 7422 | 7440 | 7457 | 7475 | 7492 | 7509 | 7527 | 7544 | 7561 | 17 |
| 5.8 | 7579 | 7596 | 7613 | 7630 | 7647 | 7664 | 7681 | 7699 | 7716 | 7733 | 17 |
| 5.9 | 7750 | 7766 | 7783 | 7800 | 7817 | 7834 | 7851 | 7867 | 7884 | 7901 | 17 |
| 6.0 | 1.7918 | 7934 | 7951 | 7967 | 7984 | 8001 | 8017 | 8034 | 8050 | 8066 | 16 |
| 6.1 | 8083 | 8099 | 8116 | 8132 | 8148 | 8165 | 8181 | 8197 | 8213 | 8229 | 16 |
| 6.2 | 8245 | 8262 | 8278 | 8294 | 8310 | 8326 | 8342 | 8358 | 8374 | 8390 | 16 |
| 6.3 | 8405 | 8421 | 8437 | 8453 | 8469 | 8485 | 8500 | 8516 | 8532 | 8547 | 16 |
| 6.4 | 8563 | 8579 | 8594 | 8610 | 8625 | 8641 | 8656 | 8672 | 8687 | 8703 | 15 |
| 6.5 | 1.8718 | 8733 | 8749 | 8764 | 8779 | 8795 | 8810 | 8825 | 8840 | 8856 | 15 |
| 6.6 | 8871 | 8886 | 8901 | 8916 | 8931 | 8946 | 8961 | 8976 | 8991 | 9006 | 15 |
| 6.7 | 9021 | 9036 | 9051 | 9066 | 9081 | 9095 | 9110 | 9125 | 9140 | 9155 | 15 |
| 6.8 | 9169 | 9184 | 9199 | 9213 | 9228 | 9242 | 9257 | 9272 | 9286 | 9301 | 15 |
| 6.9 | 9315 | 9330 | 9344 | 9359 | 9373 | 9387 | 9402 | 9416 | 9430 | 9445 | 14 |
| 7.0 | 1.9459 | 9473 | 9488 | 9502 | 9516 | 9530 | 9544 | 9559 | 9573 | 9587 | 14 |
| 7.1 | 9601 | 9615 | 9629 | 9643 | 9657 | 9671 | 9685 | 9699 | 9713 | 9727 | 14 |
| 7.2 | 9741 | 9755 | 9769 | 9782 | 9796 | 9810 | 9824 | 9838 | 9851 | 9865 | 14 |
| 7.3 | 1.9879 | 9892 | 9906 | 9920 | 9933 | 9947 | 9961 | 9974 | 9988 | *0001 | 13 |
| 7.4 | 2.0015 | 0028 | 0042 | 0055 | 0069 | 0082 | 0096 | 0109 | 0122 | 0136 | 13 |
| 7.5 | 2.0149 | 0162 | 0176 | 0189 | 0202 | 0215 | 0229 | 0242 | 0255 | 0268 | 13 |
| 7.6 | 0281 | 0295 | 0308 | 0321 | 0334 | 0347 | 0360 | 0373 | 0386 | 0399 | 13 |
| 7.7 | 0412 | 0425 | 0438 | 0451 | 0464 | 0477 | 0490 | 0503 | 0516 | 0528 | 13 |
| 7.8 | 0541 | 0554 | 0567 | 0580 | 0592 | 0605 | 0618 | 0631 | 0643 | 0656 | 13 |
| 7.9 | 0669 | 0681 | 0694 | 0707 | 0719 | 0732 | 0744 | 0757 | 0769 | 0782 | 12 |
| 8.0 | 2.0794 | 0807 | 0819 | 0832 | 0844 | 0857 | 0869 | 0882 | 0894 | 0906 | 12 |
| 8.1 | 0919 | 0931 | 0943 | 0956 | 0968 | 0980 | 0992 | 1005 | 1017 | 1029 | 12 |
| 8.2 | 1041 | 1054 | 1066 | 1078 | 1090 | 1102 | 1114 | 1126 | 1138 | 1150 | 12 |
| 8.3 | 1163 | 1175 | 1187 | 1199 | 1211 | 1223 | 1235 | 1247 | 1258 | 1270 | 12 |
| 8.4 | 1282 | 1294 | 1306 | 1318 | 1330 | 1342 | 1353 | 1365 | 1377 | 1389 | 12 |
| 8.5 | 2.1401 | 1412 | 1424 | 1436 | 1448 | 1459 | 1471 | 1483 | 1494 | 1506 | 12 |
| 8.6 | 1518 | 1529 | 1541 | 1552 | 1564 | 1576 | 1587 | 1599 | 1610 | 1622 | 12 |
| 8.7 | 1633 | 1645 | 1656 | 1668 | 1679 | 1691 | 1702 | 1713 | 1725 | 1736 | 11 |
| 8.8 | 1748 | 1759 | 1770 | 1782 | 1793 | 1804 | 1815 | 1827 | 1838 | 1849 | 11 |
| 8.9 | 1861 | 1872 | 1883 | 1894 | 1905 | 1917 | 1928 | 1939 | 1950 | 1961 | 11 |
| 9.0 | 2.1972 | 1983 | 1994 | 2006 | 2017 | 2028 | 2039 | 2050 | 2061 | 2072 | 11 |
| 9.1 | 2083 | 2094 | 2105 | 2116 | 2127 | 2138 | 2148 | 2159 | 2170 | 2181 | 11 |
| 9.2 | 2192 | 2203 | 2214 | 2225 | 2235 | 2246 | 2257 | 2268 | 2279 | 2289 | 11 |
| 9.3 | 2300 | 2311 | 2322 | 2332 | 2343 | 2354 | 2364 | 2375 | 2386 | 2396 | 11 |
| 9.4 | 2407 | 2418 | 2428 | 2439 | 2450 | 2460 | 2471 | 2481 | 2492 | 2502 | 11 |
| 9.5 | 2.2513 | 2523 | 2534 | 2544 | 2555 | 2565 | 2576 | 2586 | 2597 | 2607 | 10 |
| 9.6 | 2618 | 2628 | 2638 | 2649 | 2659 | 2670 | 2680 | 2690 | 2701 | 2711 | 10 |
| 9.7 | 2721 | 2732 | 2742 | 2752 | 2762 | 2773 | 2783 | 2793 | 2803 | 2814 | 10 |
| 9.8 | 2824 | 2834 | 2844 | 2854 | 2865 | 2875 | 2885 | 2895 | 2905 | 2915 | 10 |
| 9.9 | 2925 | 2935 | 2946 | 2956 | 2966 | 2976 | 2986 | 2996 | 3006 | 3016 | 10 |
| 10.0 | 2.3026 | | | | | | | | | | |

Moving the decimal point $n$ places to the right [or left] in the number requires adding $n$ times 2.3026 [or $n$ times (0.6974−3)] in the body of the table. See auxiliary table of multiples on top of the preceding page.

**Hyperbolic Sines** [sinh $x = \frac{1}{2}(e^x - e^{-x})$]

| $x$ | 0 | 1 | 2 | 3 | 4 | 5 | 6 | 7 | 8 | 9 | Avg diff |
|-----|---|---|---|---|---|---|---|---|---|---|----------|
| 0.0 | .0000 | .0100 | .0200 | .0300 | .0400 | .0500 | .0600 | .0701 | .0801 | .0901 | 100 |
| 1 | .1002 | .1102 | .1203 | .1304 | .1405 | .1506 | .1607 | .1708 | .1810 | .1911 | 101 |
| 2 | .2013 | .2115 | .2218 | .2320 | .2423 | .2526 | .2629 | .2733 | .2837 | .2941 | 103 |
| 3 | .3045 | .3150 | .3255 | .3360 | .3466 | .3572 | .3678 | .3785 | .3892 | .4000 | 106 |
| 4 | .4108 | .4216 | .4325 | .4434 | .4543 | .4653 | .4764 | .4875 | .4986 | .5098 | 110 |
| 0.5 | .5211 | .5324 | .5438 | .5552 | .5666 | .5782 | .5897 | .6014 | .6131 | .6248 | 116 |
| 6 | .6367 | .6485 | .6605 | .6725 | .6846 | .6967 | .7090 | .7213 | .7336 | .7461 | 122 |
| 7 | .7586 | .7712 | .7838 | .7966 | .8094 | .8223 | .8353 | .8484 | .8615 | .8748 | 130 |
| 8 | .8881 | .9015 | .9150 | .9286 | .9423 | .9561 | .9700 | .9840 | .9981 | 1.012 | 138 |
| 9 | 1.027 | 1.041 | 1.055 | 1.070 | 1.085 | 1.099 | 1.114 | 1.129 | 1.145 | 1.160 | 15 |
| 1.0 | 1.175 | 1.191 | 1.206 | 1.222 | 1.238 | 1.254 | 1.270 | 1.286 | 1.303 | 1.319 | 16 |
| 1 | 1.336 | 1.352 | 1.369 | 1.386 | 1.403 | 1.421 | 1.438 | 1.456 | 1.474 | 1.491 | 17 |
| 2 | 1.509 | 1.528 | 1.546 | 1.564 | 1.583 | 1.602 | 1.621 | 1.640 | 1.659 | 1.679 | 19 |
| 3 | 1.698 | 1.718 | 1.738 | 1.758 | 1.779 | 1.799 | 1.820 | 1.841 | 1.862 | 1.883 | 21 |
| 4 | 1.904 | 1.926 | 1.948 | 1.970 | 1.992 | 2.014 | 2.037 | 2.060 | 2.083 | 2.106 | 22 |
| 1.5 | 2.129 | 2.153 | 2.177 | 2.201 | 2.225 | 2.250 | 2.274 | 2.299 | 2.324 | 2.350 | 25 |
| 6 | 2.376 | 2.401 | 2.428 | 2.454 | 2.481 | 2.507 | 2.535 | 2.562 | 2.590 | 2.617 | 27 |
| 7 | 2.646 | 2.674 | 2.703 | 2.732 | 2.761 | 2.790 | 2.820 | 2.850 | 2.881 | 2.911 | 30 |
| 8 | 2.942 | 2.973 | 3.005 | 3.037 | 3.069 | 3.101 | 3.134 | 3.167 | 3.200 | 3.234 | 33 |
| 9 | 3.268 | 3.303 | 3.337 | 3.372 | 3.408 | 3.443 | 3.479 | 3.516 | 3.552 | 3.589 | 36 |
| 2.0 | 3.627 | 3.665 | 3.703 | 3.741 | 3.780 | 3.820 | 3.859 | 3.899 | 3.940 | 3.981 | 39 |
| 1 | 4.022 | 4.064 | 4.106 | 4.148 | 4.191 | 4.234 | 4.278 | 4.322 | 4.367 | 4.412 | 44 |
| 2 | 4.457 | 4.503 | 4.549 | 4.596 | 4.643 | 4.691 | 4.739 | 4.788 | 4.837 | 4.887 | 48 |
| 3 | 4.937 | 4.988 | 5.039 | 5.090 | 5.142 | 5.195 | 5.248 | 5.302 | 5.356 | 5.411 | 53 |
| 4 | 5.466 | 5.522 | 5.578 | 5.635 | 5.693 | 5.751 | 5.810 | 5.869 | 5.929 | 5.989 | 58 |
| 2.5 | 6.050 | 6.112 | 6.174 | 6.237 | 6.300 | 6.365 | 6.429 | 6.495 | 6.561 | 6.627 | 64 |
| 6 | 6.695 | 6.763 | 6.831 | 6.901 | 6.971 | 7.042 | 7.113 | 7.185 | 7.258 | 7.332 | 71 |
| 7 | 7.406 | 7.481 | 7.557 | 7.634 | 7.711 | 7.789 | 7.868 | 7.948 | 8.028 | 8.110 | 79 |
| 8 | 8.192 | 8.275 | 8.359 | 8.443 | 8.529 | 8.615 | 8.702 | 8.790 | 8.879 | 8.969 | 87 |
| 9 | 9.060 | 9.151 | 9.244 | 9.337 | 9.431 | 9.527 | 9.623 | 9.720 | 9.819 | 9.918 | 96 |
| 3.0 | 10.02 | 10.12 | 10.22 | 10.32 | 10.43 | 10.53 | 10.64 | 10.75 | 10.86 | 10.97 | 11 |
| 1 | 11.08 | 11.19 | 11.30 | 11.42 | 11.53 | 11.65 | 11.76 | 11.88 | 12.00 | 12.12 | 12 |
| 2 | 12.25 | 12.37 | 12.49 | 12.62 | 12.75 | 12.88 | 13.01 | 13.14 | 13.27 | 13.40 | 13 |
| 3 | 13.54 | 13.67 | 13.81 | 13.95 | 14.09 | 14.23 | 14.38 | 14.52 | 14.67 | 14.82 | 14 |
| 4 | 14.97 | 15.12 | 15.27 | 15.42 | 15.58 | 15.73 | 15.89 | 16.05 | 16.21 | 16.38 | 16 |
| 3.5 | 16.54 | 16.71 | 16.88 | 17.05 | 17.22 | 17.39 | 17.57 | 17.74 | 17.92 | 18.10 | 17 |
| 6 | 18.29 | 18.47 | 18.66 | 18.84 | 19.03 | 19.22 | 19.42 | 19.61 | 19.81 | 20.01 | 19 |
| 7 | 20.21 | 20.41 | 20.62 | 20.83 | 21.04 | 21.25 | 21.46 | 21.68 | 21.90 | 22.12 | 21 |
| 8 | 22.34 | 22.56 | 22.79 | 23.02 | 23.25 | 23.49 | 23.72 | 23.96 | 24.20 | 24.45 | 24 |
| 9 | 24.69 | 24.94 | 25.19 | 25.44 | 25.70 | 25.96 | 26.22 | 26.48 | 26.75 | 27.02 | 26 |
| 4.0 | 27.29 | 27.56 | 27.84 | 28.12 | 28.40 | 28.69 | 28.98 | 29.27 | 29.56 | 29.86 | 29 |
| 1 | 30.16 | 30.47 | 30.77 | 31.08 | 31.39 | 31.71 | 32.03 | 32.35 | 32.68 | 33.00 | 32 |
| 2 | 33.34 | 33.67 | 34.01 | 34.35 | 34.70 | 35.05 | 35.40 | 35.75 | 36.11 | 36.48 | 35 |
| 3 | 36.84 | 37.21 | 37.59 | 37.97 | 38.35 | 38.73 | 39.12 | 39.52 | 39.91 | 40.31 | 39 |
| 4 | 40.72 | 41.13 | 41.54 | 41.96 | 42.38 | 42.81 | 43.24 | 43.67 | 44.11 | 44.56 | 43 |
| 4.5 | 45.00 | 45.46 | 45.91 | 46.37 | 46.84 | 47.31 | 47.79 | 48.27 | 48.75 | 49.24 | 47 |
| 6 | 49.74 | 50.24 | 50.74 | 51.25 | 51.77 | 52.29 | 52.81 | 53.34 | 53.88 | 54.42 | 52 |
| 7 | 54.97 | 55.52 | 56.08 | 56.64 | 57.21 | 57.79 | 58.37 | 58.96 | 59.55 | 60.15 | 58 |
| 8 | 60.75 | 61.36 | 61.98 | 62.60 | 63.23 | 63.87 | 64.51 | 65.16 | 65.81 | 67.47 | 64 |
| 9 | 67.14 | 67.82 | 68.50 | 69.19 | 69.88 | 70.58 | 71.29 | 72.01 | 72.73 | 73.46 | 71 |
| 5.0 | 74.20 | | | | | | | | | | |

If $x > $ sinh $x = \frac{1}{2}(e^x)$ and $\log_{10}$ sinh $x = (0.4343)x + 0.6990) - 1$, correct to four significant figures. For graphs, see Sec. 2.

**Hyperbolic Cosines** $[\cosh x = \tfrac{1}{2}(e^x + e^{-x})]$

| $x$ | 0 | 1 | 2 | 3 | 4 | 5 | 6 | 7 | 8 | 9 | Avg diff |
|---|---|---|---|---|---|---|---|---|---|---|---|
| 0.0 | 1.000 | 1.000 | 1.000 | 1.000 | 1.001 | 1.001 | 1.002 | 1.002 | 1.003 | 1.004 | 1 |
| 1 | 1.005 | 1.006 | 1.007 | 1.008 | 1.010 | 1.011 | 1.013 | 1.014 | 1.016 | 1.018 | 2 |
| 2 | 1.020 | 1.022 | 1.024 | 1.027 | 1.029 | 1.031 | 1.034 | 1.037 | 1.039 | 1.042 | 3 |
| 3 | 1.045 | 1.048 | 1.052 | 1.055 | 1.058 | 1.062 | 1.066 | 1.069 | 1.073 | 1.077 | 4 |
| 4 | 1.081 | 1.085 | 1.090 | 1.094 | 1.098 | 1.103 | 1.108 | 1.112 | 1.117 | 1.122 | 5 |
| 0.5 | 1.128 | 1.133 | 1.138 | 1.144 | 1.149 | 1.155 | 1.161 | 1.167 | 1.173 | 1.179 | 6 |
| 6 | 1.185 | 1.192 | 1.198 | 1.205 | 1.212 | 1.219 | 1.226 | 1.233 | 1.240 | 1.248 | 7 |
| 7 | 1.255 | 1.263 | 1.271 | 1.278 | 1.287 | 1.295 | 1.303 | 1.311 | 1.320 | 1.329 | 8 |
| 8 | 1.337 | 1.346 | 1.355 | 1.365 | 1.374 | 1.384 | 1.393 | 1.403 | 1.413 | 1.423 | 10 |
| 9 | 1.433 | 1.443 | 1.454 | 1.465 | 1.475 | 1.486 | 1.497 | 1.509 | 1.520 | 1.531 | 11 |
| 1.0 | 1.543 | 1.555 | 1.567 | 1.579 | 1.591 | 1.604 | 1.616 | 1.629 | 1.642 | 1.655 | 13 |
| 1 | 1.669 | 1.682 | 1.696 | 1.709 | 1.723 | 1.737 | 1.752 | 1.766 | 1.781 | 1.796 | 14 |
| 2 | 1.811 | 1.826 | 1.841 | 1.857 | 1.872 | 1.888 | 1.905 | 1.921 | 1.937 | 1.954 | 16 |
| 3 | 1.971 | 1.988 | 2.005 | 2.023 | 2.040 | 2.058 | 2.076 | 2.095 | 2.113 | 2.132 | 18 |
| 4 | 2.151 | 2.170 | 2.189 | 2.209 | 2.229 | 2.249 | 2.269 | 2.290 | 2.310 | 2.331 | 20 |
| 1.5 | 2.352 | 2.374 | 2.395 | 2.417 | 2.439 | 2.462 | 2.484 | 2.507 | 2.530 | 2.554 | 23 |
| 6 | 2.577 | 2.601 | 2.625 | 2.650 | 2.675 | 2.700 | 2.725 | 2.750 | 2.776 | 2.802 | 25 |
| 7 | 2.828 | 2.855 | 2.882 | 2.909 | 2.936 | 2.964 | 2.992 | 3.021 | 3.049 | 3.078 | 28 |
| 8 | 3.107 | 3.137 | 3.167 | 3.197 | 3.228 | 3.259 | 3.290 | 3.321 | 3.353 | 3.385 | 31 |
| 9 | 3.418 | 3.451 | 3.484 | 3.517 | 3.551 | 3.585 | 3.620 | 3.655 | 3.690 | 3.726 | 34 |
| 2.0 | 3.762 | 3.799 | 3.835 | 3.873 | 3.910 | 3.948 | 3.987 | 4.026 | 4.065 | 4.104 | 38 |
| 1 | 4.144 | 4.185 | 4.226 | 4.267 | 4.309 | 4.351 | 4.393 | 4.436 | 4.480 | 4.524 | 42 |
| 2 | 4.568 | 4.613 | 4.658 | 4.704 | 4.750 | 4.797 | 4.844 | 4.891 | 4.939 | 4.988 | 47 |
| 3 | 5.037 | 5.087 | 5.137 | 5.188 | 5.239 | 5.290 | 5.343 | 5.395 | 5.449 | 5.503 | 52 |
| 4 | 5.557 | 5.612 | 5.667 | 5.723 | 5.780 | 5.837 | 5.895 | 5.954 | 6.013 | 6.072 | 58 |
| 2.5 | 6.132 | 6.193 | 6.255 | 6.317 | 6.379 | 6.443 | 6.507 | 6.571 | 6.636 | 6.702 | 64 |
| 6 | 6.769 | 6.836 | 6.904 | 6.973 | 7.042 | 7.112 | 7.183 | 7.255 | 7.327 | 7.400 | 70 |
| 7 | 7.473 | 7.548 | 7.623 | 7.699 | 7.776 | 7.853 | 7.932 | 8.011 | 8.091 | 8.171 | 78 |
| 8 | 8.253 | 8.335 | 8.418 | 8.502 | 8.587 | 8.673 | 8.759 | 8.847 | 8.935 | 9.024 | 86 |
| 9 | 9.115 | 9.206 | 9.298 | 9.391 | 9.484 | 9.579 | 9.675 | 9.772 | 9.869 | 9.968 | 95 |
| 3.0 | 10.07 | 10.17 | 10.27 | 10.37 | 10.48 | 10.58 | 10.69 | 10.79 | 10.90 | 11.01 | 11 |
| 1 | 11.12 | 11.23 | 11.35 | 11.46 | 11.57 | 11.69 | 11.81 | 11.92 | 12.04 | 12.16 | 12 |
| 2 | 12.29 | 12.41 | 12.53 | 12.66 | 12.79 | 12.91 | 13.04 | 13.17 | 13.31 | 13.44 | 13 |
| 3 | 13.57 | 13.71 | 13.85 | 13.99 | 14.13 | 14.27 | 14.41 | 14.56 | 14.70 | 14.85 | 14 |
| 4 | 15.00 | 15.15 | 15.30 | 15.45 | 15.61 | 15.77 | 15.92 | 16.08 | 16.25 | 16.41 | 16 |
| 3.5 | 16.57 | 16.74 | 16.91 | 17.08 | 17.25 | 17.42 | 17.60 | 17.77 | 17.95 | 18.13 | 17 |
| 6 | 18.31 | 18.50 | 18.68 | 18.87 | 19.06 | 19.25 | 19.44 | 19.64 | 19.84 | 20.03 | 19 |
| 7 | 20.24 | 20.44 | 20.64 | 20.85 | 21.06 | 21.27 | 21.49 | 21.70 | 21.92 | 22.14 | 21 |
| 8 | 22.36 | 22.59 | 22.81 | 23.04 | 23.27 | 23.51 | 23.74 | 23.98 | 24.22 | 24.47 | 23 |
| 9 | 24.71 | 24.96 | 25.21 | 25.46 | 25.72 | 25.98 | 26.24 | 26.50 | 26.77 | 27.04 | 26 |
| 4.0 | 27.31 | 27.58 | 27.86 | 28.14 | 28.42 | 28.71 | 29.00 | 29.29 | 29.58 | 29.88 | 29 |
| 1 | 30.18 | 30.48 | 30.79 | 31.10 | 31.41 | 31.72 | 32.04 | 32.37 | 32.69 | 33.02 | 32 |
| 2 | 33.35 | 33.69 | 34.02 | 34.37 | 34.71 | 35.06 | 35.41 | 35.77 | 36.13 | 36.49 | 35 |
| 3 | 36.86 | 37.23 | 37.60 | 37.98 | 38.36 | 38.75 | 39.13 | 39.53 | 39.93 | 40.33 | 39 |
| 4 | 40.73 | 41.14 | 41.55 | 41.97 | 42.39 | 42.82 | 43.25 | 43.68 | 44.12 | 44.57 | 43 |
| 4.5 | 45.01 | 45.47 | 45.92 | 46.38 | 46.85 | 47.32 | 47.80 | 48.28 | 48.76 | 49.25 | 47 |
| 6 | 49.75 | 50.25 | 50.75 | 51.26 | 51.78 | 52.30 | 52.82 | 53.35 | 53.89 | 54.43 | 52 |
| 7 | 54.98 | 55.53 | 56.09 | 56.65 | 57.22 | 57.80 | 58.38 | 58.96 | 59.56 | 60.15 | 58 |
| 8 | 60.76 | 61.37 | 61.99 | 62.61 | 63.24 | 63.87 | 64.52 | 65.16 | 65.82 | 66.48 | 64 |
| 9 | 67.15 | 67.82 | 68.50 | 69.19 | 69.89 | 70.59 | 71.30 | 72.02 | 72.74 | 73.47 | 71 |
| 5.0 | 74.21 | | | | | | | | | | |

If $x > 5$, $\cosh x = \tfrac{1}{2}(e^x)$ and $\log_{10} \cosh x = (0.4343)x + 0.6990 - 1$, correct to four significant figures. For graphs, see Sec. 2.

**Hyperbolic Tangents** $[\tanh x = (e^x - e^{-x})/(e^x + e^{-x}) = \sinh x/\cosh x]$

| $x$ | 0 | 1 | 2 | 3 | 4 | 5 | 6 | 7 | 8 | 9 | Avg diff |
|-----|---|---|---|---|---|---|---|---|---|---|----------|
| 0.0 | .0000 | .0100 | .0200 | .0300 | .0400 | .0500 | .0599 | .0699 | .0798 | .0898 | 100 |
| 1 | .0997 | .1096 | .1194 | .1293 | .1391 | .1489 | .1587 | .1684 | .1781 | .1878 | 98 |
| 2 | .1974 | .2070 | .2165 | .2260 | .2355 | .2449 | .2543 | .2636 | .2729 | .2821 | 94 |
| 3 | .2913 | .3004 | .3095 | .3185 | .3275 | .3364 | .3452 | .3540 | .3627 | .3714 | 89 |
| 4 | .3800 | .3885 | .3969 | .4053 | .4137 | .4219 | .4301 | .4382 | .4462 | .4542 | 82 |
| 0.5 | .4621 | .4700 | .4777 | .4854 | .4930 | .5005 | .5080 | .5154 | .5227 | .5299 | 75 |
| 6 | .5370 | .5441 | .5511 | .5581 | .5649 | .5717 | .5784 | .5850 | .5915 | .5980 | 67 |
| 7 | .6044 | .6107 | .6169 | .6231 | .6291 | .6352 | .6411 | .6469 | .6527 | .6584 | 60 |
| 8 | .6640 | .6696 | .6751 | .6805 | .6858 | .6911 | .6963 | .7014 | .7064 | .7114 | 52 |
| 9 | .7163 | .7211 | .7259 | .7306 | .7352 | .7398 | .7443 | .7487 | .7531 | .7574 | 45 |
| 1.0 | .7616 | .7658 | .7699 | .7739 | .7779 | .7818 | .7857 | .7895 | .7932 | .7969 | 39 |
| 1 | .8005 | .8041 | .8076 | .8110 | .8144 | .8178 | .8210 | .8243 | .8275 | .8306 | 33 |
| 2 | .8337 | .8367 | .8397 | .8426 | .8455 | .8483 | .8511 | .8538 | .8565 | .8591 | 28 |
| 3 | .8617 | .8643 | .8668 | .8693 | .8717 | .8741 | .8764 | .8787 | .8810 | .8832 | 24 |
| 4 | .8854 | .8875 | .8896 | .8917 | .8937 | .8957 | .8977 | .8996 | .9015 | .9033 | 20 |
| 1.5 | .9052 | .9069 | .9087 | .9104 | .9121 | .9138 | .9154 | .9170 | .9186 | .9202 | 17 |
| 6 | .9217 | .9232 | .9246 | .9261 | .9275 | .9289 | .9302 | .9316 | .9329 | .9342 | 14 |
| 7 | .9354 | .9367 | .9379 | .9391 | .9402 | .9414 | .9425 | .9436 | .9447 | .9458 | 11 |
| 8 | .9468 | .9478 | .9488 | .9498 | .9508 | .9518 | .9527 | .9536 | .9545 | .9554 | 9 |
| 9 | .9562 | .9571 | .9579 | .9587 | .9595 | .9603 | .9611 | .9619 | .9626 | .9633 | 8 |
| 2.0 | .9640 | .9647 | .9654 | .9661 | .9668 | .9674 | .9680 | .9687 | .9693 | .9699 | 6 |
| 1 | .9705 | .9710 | .9716 | .9722 | .9727 | .9732 | .9738 | .9743 | .9748 | .9753 | 5 |
| 2 | .9757 | .9762 | .9767 | .9771 | .9776 | .9780 | .9785 | .9789 | .9793 | .9797 | 4 |
| 3 | .9801 | .9805 | .9809 | .9812 | .9816 | .9820 | .9823 | .9827 | .9830 | .9834 | 4 |
| 4 | .9837 | .9840 | .9843 | .9846 | .9849 | .9852 | .9855 | .9858 | .9861 | .9863 | 3 |
| 2.5 | .9866 | .9869 | .9871 | .9874 | .9876 | .9879 | .9881 | .9884 | .9886 | .9888 | 2 |
| 6 | .9890 | .9892 | .9895 | .9897 | .9899 | .9901 | .9903 | .9905 | .9906 | .9908 | 2 |
| 7 | .9910 | .9912 | .9914 | .9915 | .9917 | .9919 | .9920 | .9922 | .9923 | .9925 | 2 |
| 8 | .9926 | .9928 | .9929 | .9931 | .9932 | .9933 | .9935 | .9936 | .9937 | .9938 | 1 |
| 2.9 | .9940 | .9941 | .9942 | .9943 | .9944 | .9945 | .9946 | .9947 | .9949 | .9950 | 1 |
| 3. | .9951 | .9959 | .9967 | .9973 | .9978 | .9982 | .9985 | .9988 | .9990 | .9992 | 4 |
| 4. | .9993 | .9995 | .9996 | .9996 | .9997 | .9998 | .9998 | .9998 | .9999 | .9999 | 1 |
| 5. | .9999 | If $x > 5$, $\tanh x = 1.0000$ to four decimal places. | | | | | Graphs, Sec. 2. | | | | |

**Multiples of 0.4343** $(0.43429448 = \log_{10} e)$

| $x$ | 0 | 1 | 2 | 3 | 4 | 5 | 6 | 7 | 8 | 9 |
|-----|---|---|---|---|---|---|---|---|---|---|
| 0. | 0.0000 | 0.0434 | 0.0869 | 0.1303 | 0.1737 | 0.2171 | 0.2606 | 0.3040 | 0.3474 | 0.3909 |
| 1. | 0.4343 | 0.4777 | 0.5212 | 0.5646 | 0.6080 | 0.6514 | 0.6949 | 0.7383 | 0.7817 | 0.8252 |
| 2. | 0.8686 | 0.9120 | 0.9554 | 0.9989 | 1.0423 | 1.0857 | 1.1292 | 1.1726 | 1.2160 | 1.2595 |
| 3. | 1.3029 | 1.3463 | 1.3897 | 1.4332 | 1.4766 | 1.5200 | 1.5635 | 1.6069 | 1.6503 | 1.6937 |
| 4. | 1.7372 | 1.7806 | 1.8240 | 1.8675 | 1.9109 | 1.9543 | 1.9978 | 2.0412 | 2.0846 | 2.1280 |
| 5. | 2.1715 | 2.2149 | 2.2583 | 2.3018 | 2.3452 | 2.3886 | 2.4320 | 2.4755 | 2.5189 | 2.5623 |
| 6. | 2.6058 | 2.6492 | 2.6926 | 2.7361 | 2.7795 | 2.8229 | 2.8663 | 2.9098 | 2.9532 | 2.9966 |
| 7. | 3.0401 | 3.0835 | 3.1269 | 3.1703 | 3.2138 | 3.2572 | 3.3006 | 3.3441 | 3.3875 | 3.4309 |
| 8. | 3.4744 | 3.5178 | 3.5612 | 3.6046 | 3.6481 | 3.6915 | 3.7349 | 3.7784 | 3.8218 | 3.8652 |
| 9. | 3.9087 | 3.9521 | 3.9955 | 4.0389 | 4.0824 | 4.1258 | 4.1692 | 4.2127 | 4.2561 | 4.2995 |

**Multiples of 2.3026** $(2.3025851 = \ln 10 = 1/0.4343)$

| $x$ | 0 | 1 | 2 | 3 | 4 | 5 | 6 | 7 | 8 | 9 |
|-----|---|---|---|---|---|---|---|---|---|---|
| 0. | 0.0000 | 0.2303 | 0.4605 | 0.6908 | 0.9210 | 1.1513 | 1.3816 | 1.6118 | 1.8421 | 2.0723 |
| 1. | 2.3026 | 2.5328 | 2.7631 | 2.9934 | 3.2236 | 3.4539 | 3.6841 | 3.9144 | 4.1447 | 4.3749 |
| 2. | 4.6052 | 4.8354 | 5.0657 | 5.2959 | 5.5262 | 5.7565 | 5.9867 | 6.2170 | 6.4472 | 6.6775 |
| 3. | 6.9078 | 7.1380 | 7.3683 | 7.5985 | 7.8288 | 8.0590 | 8.2893 | 8.5196 | 8.7498 | 8.9801 |
| 4. | 9.2103 | 9.4406 | 9.6709 | 9.9011 | 10.131 | 10.362 | 10.592 | 10.822 | 11.052 | 11.283 |
| 5. | 11.513 | 11.743 | 11.973 | 12.204 | 12.434 | 12.664 | 12.894 | 13.125 | 13.355 | 13.585 |
| 6. | 13.816 | 14.046 | 14.276 | 14.506 | 14.737 | 14.967 | 15.197 | 15.427 | 15.658 | 15.888 |
| 7. | 16.118 | 16.348 | 16.579 | 16.809 | 17.039 | 17.269 | 17.500 | 17.730 | 17.960 | 18.190 |
| 8. | 18.421 | 18.651 | 18.881 | 19.111 | 19.342 | 19.572 | 19.802 | 20.032 | 20.263 | 20.493 |
| 9. | 20.723 | 20.954 | 21.184 | 21.414 | 21.644 | 21.875 | 22.105 | 22.335 | 22.565 | 22.796 |

## Standard Distribution of Residuals (See Sec. 2)

$a$ = any positive quantity
$y$ = the number of residuals which are numerically $< a$
$r$ = the probable error of a single observation
$n$ = number of observations

| $\dfrac{a}{r}$ | $\dfrac{y}{n}$ | Diff |
|---|---|---|
| 0.0 | .000 | |
| 1 | .054 | 54 |
| 2 | .107 | 53 |
| 3 | .160 | 53 |
| 4 | .213 | 53 |
| | | 51 |
| 0.5 | .264 | |
| 6 | .314 | 50 |
| 7 | .363 | 49 |
| 8 | .411 | 48 |
| 9 | .456 | 45 |
| | | 44 |
| 1.0 | .500 | |
| 1 | .542 | 42 |
| 2 | .582 | 40 |
| 3 | .619 | 37 |
| 4 | .655 | 36 |
| | | 33 |
| 1.5 | .688 | |
| 6 | .719 | 31 |
| 7 | .748 | 29 |
| 8 | .775 | 27 |
| 9 | .800 | 25 |
| | | 23 |
| 2.0 | .823 | |
| 1 | .843 | 20 |
| 2 | .862 | 19 |
| 3 | .879 | 17 |
| 4 | .895 | 16 |
| | | 13 |
| 2.5 | .908 | |
| 6 | .921 | 13 |
| 7 | .931 | 10 |
| 8 | .941 | 10 |
| 9 | .950 | 9 |
| | | 7 |
| 3.0 | .957 | |
| 1 | .963 | 6 |
| 2 | .969 | 6 |
| 3 | .974 | 5 |
| 4 | .978 | 4 |
| | | 4 |
| 3.5 | .982 | |
| 6 | .985 | 3 |
| 7 | .987 | 2 |
| 8 | .990 | 3 |
| 9 | .991 | 1 |
| | | 2 |
| 4.0 | .993 | |
| | | 6 |
| 5.0 | .999 | |

## Factors for Computing Probable Error (See Sec. 2)

| $n$ | Bessel $\dfrac{0.6745}{\sqrt{(n-1)}}$ | Bessel $\dfrac{0.6745}{\sqrt{n(n-1)}}$ | Peters $\dfrac{0.8453}{\sqrt{n(n-1)}}$ | Peters $\dfrac{0.8453}{n\sqrt{n-1}}$ |
|---|---|---|---|---|
| 2 | .6745 | .4769 | .5978 | .4227 |
| 3 | .4769 | .2754 | .3451 | .1993 |
| 4 | .3894 | .1947 | .2440 | .1220 |
| 5 | .3372 | .1508 | .1890 | .0845 |
| 6 | .3016 | .1231 | .1543 | .0630 |
| 7 | .2754 | .1041 | .1304 | .0493 |
| 8 | .2549 | .0901 | .1130 | .0399 |
| 9 | .2385 | .0795 | .0996 | .0332 |
| 10 | .2248 | .0711 | .0891 | .0282 |
| 11 | .2133 | .0643 | .0806 | .0243 |
| 12 | .2034 | .0587 | .0736 | .0212 |
| 13 | .1947 | .0540 | .0677 | .0188 |
| 14 | .1871 | .0500 | .0627 | .0167 |
| 15 | .1803 | .0465 | .0583 | .0151 |
| 16 | .1742 | .0435 | .0546 | .0136 |
| 17 | .1686 | .0409 | .0513 | .0124 |
| 18 | .1636 | .0386 | .0483 | .0114 |
| 19 | .1590 | .0365 | .0457 | .0105 |
| 20 | .1547 | .0346 | .0434 | .0097 |
| 21 | .1508 | .0329 | .0412 | .0090 |
| 22 | .1472 | .0314 | .0393 | .0084 |
| 23 | .1438 | .0300 | .0376 | .0078 |
| 24 | .1406 | .0287 | .0360 | .0073 |
| 25 | .1377 | .0275 | .0345 | .0069 |
| 26 | .1349 | .0265 | .0332 | .0065 |
| 27 | .1323 | .0255 | .0319 | .0061 |
| 28 | .1298 | .0245 | .0307 | .0058 |
| 29 | .1275 | .0237 | .0297 | .0055 |
| 30 | .1252 | .0229 | .0287 | .0052 |
| 31 | .1231 | .0221 | .0277 | .0050 |
| 32 | .1211 | .0214 | .0268 | .0047 |
| 33 | .1192 | .0208 | .0260 | .0045 |
| 34 | .1174 | .0201 | .0252 | .0043 |
| 35 | .1157 | .0196 | .0245 | .0041 |
| 36 | .1140 | .0190 | .0238 | .0040 |
| 37 | .1124 | .0185 | .0232 | .0038 |
| 38 | .1109 | .0180 | .0225 | .0037 |
| 39 | .1094 | .0175 | .0220 | .0035 |
| 40 | .1080 | .0171 | .0214 | .0034 |
| 45 | .1017 | .0152 | .0190 | .0028 |
| 50 | .0964 | .0136 | .0171 | .0024 |
| 55 | .0918 | .0124 | .0155 | .0021 |
| 60 | .0878 | .0113 | .0142 | .0018 |
| 65 | .0843 | .0105 | .0131 | .0016 |
| 70 | .0812 | .0097 | .0122 | .0015 |
| 75 | .0784 | .0091 | .0113 | .0013 |
| 80 | .0759 | .0085 | .0106 | .0012 |
| 85 | .0736 | .0080 | .0100 | .0011 |
| 90 | .0715 | .0075 | .0094 | .0010 |
| 95 | .0696 | .0071 | .0089 | .0009 |
| 100 | .0678 | .0068 | .0085 | .0008 |

## Compound Interest. Amount of a Given Principal

The amount $A$ at the end of $n$ years of a given principal $P$ placed at compound interest today is $A = P \times x$ or $A = P \times y$, according as the interest (at the rate of $r$ percent per annum) is compounded annually, or continuously; the factor $x$ or $y$ being taken from the following tables.

Values of $x$ (interest compounded annually; $A = P \times x$)

| Years | $r = 2$ | 3 | 4 | 5 | 6 | 7 | 8 | 10 | 12 |
|---|---|---|---|---|---|---|---|---|---|
| 1 | 1.0200 | 1.0300 | 1.0400 | 1.0500 | 1.0600 | 1.0700 | 1.0800 | 1.1000 | 1.1200 |
| 2 | 1.0404 | 1.0609 | 1.0816 | 1.1025 | 1.1236 | 1.1449 | 1.1664 | 1.2100 | 1.2544 |
| 3 | 1.0612 | 1.0927 | 1.1249 | 1.1576 | 1.1910 | 1.2250 | 1.2597 | 1.3310 | 1.4049 |
| 4 | 1.0824 | 1.1255 | 1.1699 | 1.2155 | 1.2625 | 1.3108 | 1.3605 | 1.4641 | 1.5735 |
| 5 | 1.1041 | 1.1593 | 1.2167 | 1.2763 | 1.3382 | 1.4026 | 1.4693 | 1.6105 | 1.7623 |
| 6 | 1.1262 | 1.1941 | 1.2653 | 1.3401 | 1.4185 | 1.5007 | 1.5869 | 1.7716 | 1.9738 |
| 7 | 1.1487 | 1.2299 | 1.3159 | 1.4071 | 1.5036 | 1.6058 | 1.7138 | 1.9487 | 2.2107 |
| 8 | 1.1717 | 1.2668 | 1.3686 | 1.4775 | 1.5938 | 1.7182 | 1.8509 | 2.1436 | 2.4760 |
| 9 | 1.1951 | 1.3048 | 1.4233 | 1.5513 | 1.6895 | 1.8385 | 1.9990 | 2.3579 | 2.7731 |
| 10 | 1.2190 | 1.3439 | 1.4802 | 1.6289 | 1.7908 | 1.9672 | 2.1589 | 2.5937 | 3.1058 |
| 11 | 1.2434 | 1.3842 | 1.5395 | 1.7103 | 1.8983 | 2.1049 | 2.3316 | 2.8531 | 3.4785 |
| 12 | 1.2682 | 1.4258 | 1.6010 | 1.7959 | 2.0122 | 2.2522 | 2.5182 | 3.1384 | 3.8960 |
| 13 | 1.2936 | 1.4685 | 1.6651 | 1.8856 | 2.1329 | 2.4098 | 2.7196 | 3.4523 | 4.3635 |
| 14 | 1.3195 | 1.5126 | 1.7317 | 1.9799 | 2.2609 | 2.5785 | 2.9372 | 3.7975 | 4.8871 |
| 15 | 1.3459 | 1.5580 | 1.8009 | 2.0789 | 2.3966 | 2.7590 | 3.1722 | 4.1772 | 5.4736 |
| 16 | 1.3728 | 1.6047 | 1.8730 | 2.1829 | 2.5404 | 2.9522 | 3.4259 | 4.5950 | 6.1304 |
| 17 | 1.4002 | 1.6528 | 1.9479 | 2.2920 | 2.6928 | 3.1588 | 3.7000 | 5.0545 | 6.8660 |
| 18 | 1.4282 | 1.7024 | 2.0258 | 2.4066 | 2.8543 | 3.3799 | 3.9960 | 5.5599 | 7.6900 |
| 19 | 1.4568 | 1.7535 | 2.1068 | 2.5270 | 3.0256 | 3.6165 | 4.3157 | 6.1159 | 8.6128 |
| 20 | 1.4859 | 1.8061 | 2.1911 | 2.6533 | 3.2071 | 3.8697 | 4.6610 | 6.7275 | 9.6463 |
| 25 | 1.6406 | 2.0938 | 2.6658 | 3.3864 | 4.2919 | 5.4274 | 6.8485 | 10.835 | 17.000 |
| 30 | 1.8114 | 2.4273 | 3.2434 | 4.3219 | 5.7435 | 7.6123 | 10.063 | 17.449 | 29.960 |
| 40 | 2.2080 | 3.2620 | 4.8010 | 7.0400 | 10.286 | 14.974 | 21.725 | 45.259 | 93.051 |
| 50 | 2.6916 | 4.3839 | 7.1067 | 11.467 | 18.420 | 29.457 | 46.902 | 117.39 | 289.00 |
| 60 | 3.2810 | 5.8916 | 10.520 | 18.679 | 32.988 | 57.946 | 101.26 | 304.48 | 897.60 |

This table is computed from the formula $x = [1 + (r/100)]^n$.

Values of $y$ (interest compounded continuously: $A = P \times y$)

| Years | $r = 2$ | 3 | 4 | 5 | 6 | 7 | 8 | 10 | 12 |
|---|---|---|---|---|---|---|---|---|---|
| 1 | 1.0202 | 1.0305 | 1.0408 | 1.0513 | 1.0618 | 1.0725 | 1.0833 | 1.1052 | 1.1275 |
| 2 | 1.0408 | 1.0618 | 1.0833 | 1.1052 | 1.1275 | 1.1503 | 1.1735 | 1.2214 | 1.2712 |
| 3 | 1.0618 | 1.0942 | 1.1275 | 1.1618 | 1.1972 | 1.2337 | 1.2712 | 1.3499 | 1.4333 |
| 4 | 1.0833 | 1.1275 | 1.1735 | 1.2214 | 1.2712 | 1.3231 | 1.3771 | 1.4918 | 1.6161 |
| 5 | 1.1052 | 1.1618 | 1.2214 | 1.2840 | 1.3499 | 1.4191 | 1.4918 | 1.6487 | 1.8221 |
| 6 | 1.1275 | 1.1972 | 1.2712 | 1.3499 | 1.4333 | 1.5220 | 1.6161 | 1.8221 | 2.0544 |
| 7 | 1.1503 | 1.2337 | 1.3231 | 1.4191 | 1.5220 | 1.6323 | 1.7507 | 2.0138 | 2.3164 |
| 8 | 1.1735 | 1.2712 | 1.3771 | 1.4918 | 1.6161 | 1.7507 | 1.8965 | 2.2255 | 2.6117 |
| 9 | 1.1972 | 1.3100 | 1.4333 | 1.5683 | 1.7160 | 1.8776 | 2.0544 | 2.4596 | 2.9447 |
| 10 | 1.2214 | 1.3499 | 1.4918 | 1.6487 | 1.8221 | 2.0138 | 2.2255 | 2.7183 | 3.3201 |
| 11 | 1.2461 | 1.3910 | 1.5527 | 1.7333 | 1.9348 | 2.1598 | 2.4109 | 3.0042 | 3.7434 |
| 12 | 1.2712 | 1.4333 | 1.6161 | 1.8221 | 2.0544 | 2.3164 | 2.6117 | 3.3201 | 4.2207 |
| 13 | 1.2969 | 1.4770 | 1.6820 | 1.9155 | 2.1815 | 2.4843 | 2.8292 | 3.6693 | 4.7588 |
| 14 | 1.3231 | 1.5220 | 1.7507 | 2.0138 | 2.3164 | 2.6645 | 3.0649 | 4.0552 | 5.3656 |
| 15 | 1.3499 | 1.5683 | 1.8221 | 2.1170 | 2.4596 | 2.8577 | 3.3201 | 4.4817 | 6.0496 |
| 16 | 1.3771 | 1.6161 | 1.8965 | 2.2255 | 2.6117 | 3.0649 | 3.5966 | 4.9530 | 6.8210 |
| 17 | 1.4049 | 1.6653 | 1.9739 | 2.3396 | 2.7732 | 3.2871 | 3.8962 | 5.4739 | 7.6906 |
| 18 | 1.4333 | 1.7160 | 2.0544 | 2.4596 | 2.9447 | 3.5254 | 4.2207 | 6.0496 | 8.6711 |
| 19 | 1.4623 | 1.7683 | 2.1383 | 2.5857 | 3.1268 | 3.7810 | 4.5722 | 6.6859 | 9.7767 |
| 20 | 1.4918 | 1.8221 | 2.2255 | 2.7183 | 3.3201 | 4.0552 | 4.9530 | 7.3891 | 11.023 |
| 25 | 1.6487 | 2.1170 | 2.7183 | 3.4903 | 4.4817 | 5.7546 | 7.3891 | 12.182 | 20.086 |
| 30 | 1.8221 | 2.4596 | 3.3201 | 4.4817 | 6.0496 | 8.1662 | 11.023 | 20.086 | 36.598 |
| 40 | 2.2255 | 3.3201 | 4.9530 | 7.3891 | 11.023 | 16.445 | 24.533 | 54.598 | 121.51 |
| 50 | 2.7183 | 4.4817 | 7.3891 | 12.182 | 20.086 | 33.115 | 54.598 | 148.41 | 403.43 |
| 60 | 3.3201 | 6.0496 | 11.023 | 20.086 | 36.598 | 66.686 | 121.51 | 403.43 | 1339.4 |

Formula: $y = e^{(r/100) \times n}$.

## Principal Which Will Amount to a Given Sum

The principal $P$, which, if placed at compound interest today, will amount to a given sum $A$ at the end of $n$ years is $P = A \times x'$ or $P = A \times y'$, according as the interest (at the rate of $r$ percent per annum) is compounded annually, or continuously; the factor $x'$ or $y'$ being taken from the following tables.

### Values of $x'$ (interest compounded annually: $P = A \times x'$)

| Years | $r = 2$ | 3 | 4 | 5 | 6 | 7 | 8 | 10 | 12 |
|---|---|---|---|---|---|---|---|---|---|
| 1 | .98039 | .97087 | .96154 | .95238 | .94340 | .93458 | .92593 | .90909 | .89286 |
| 2 | .96117 | .94260 | .92456 | .90703 | .89000 | .87344 | .85734 | .82645 | .79719 |
| 3 | .94232 | .91514 | .88900 | .86384 | .83962 | .81630 | .79383 | .75131 | .71178 |
| 4 | .92385 | .88849 | .85480 | .82270 | .79209 | .76290 | .73503 | .68301 | .63552 |
| 5 | .90573 | .86261 | .82193 | .78353 | .74726 | .71299 | .68058 | .62092 | .56743 |
| 6 | .88797 | .83748 | .79031 | .74622 | .70496 | .66634 | .63017 | .56447 | .50663 |
| 7 | .87056 | .81309 | .75992 | .71068 | .66506 | .62275 | .58349 | .51316 | .45235 |
| 8 | .85349 | .78941 | .73069 | .67684 | .62741 | .58201 | .54027 | .46651 | .40388 |
| 9 | .83676 | .76642 | .70259 | .64461 | .59190 | .54393 | .50025 | .42410 | .36061 |
| 10 | .82035 | .74409 | .67556 | .61391 | .55839 | .50835 | .46319 | .38554 | .32197 |
| 11 | .80426 | .72242 | .64958 | .58468 | .52679 | .47509 | .42888 | .35049 | .28748 |
| 12 | .78849 | .70138 | .62460 | .55684 | .49697 | .44401 | .39711 | .31863 | .25668 |
| 13 | .77303 | .68095 | .60057 | .53032 | .46884 | .41496 | .36770 | .28966 | .22917 |
| 14 | .75788 | .66112 | .57748 | .50507 | .44230 | .38782 | .34046 | .26333 | .20462 |
| 15 | .74301 | .64186 | .55526 | .48102 | .41727 | .36245 | .31524 | .23939 | .18270 |
| 16 | .72845 | .62317 | .53391 | .45811 | .39365 | .33873 | .29189 | .21763 | .16312 |
| 17 | .71416 | .60502 | .51337 | .43630 | .37136 | .31657 | .27027 | .19784 | .14564 |
| 18 | .70016 | .58739 | .49363 | .41552 | .35034 | .29586 | .25025 | .17986 | .13004 |
| 19 | .68643 | .57029 | .47464 | .39573 | .33051 | .27651 | .23171 | .16351 | .11611 |
| 20 | .67297 | .55368 | .45639 | .37689 | .31180 | .25842 | .21455 | .14864 | .10367 |
| 25 | .60953 | .47761 | .37512 | .29530 | .23300 | .18425 | .14602 | .09230 | .05882 |
| 30 | .55207 | .41199 | .30832 | .23138 | .17411 | .13137 | .09938 | .05731 | .03338 |
| 40 | .45289 | .30656 | .20829 | .14205 | .09722 | .06678 | .04603 | .02209 | .01075 |
| 50 | .37153 | .22811 | .14071 | .08720 | .05429 | .03395 | .02132 | .00852 | .00346 |
| 60 | .30478 | .16973 | .09506 | .05354 | .03031 | .01726 | .00988 | .00328 | .00111 |

Formula: $x' = [1 + (r/100)]^{-n} = 1/x$.

### Values of $y'$ (interest compounded continuously: $P = A \times y'$)

| Years | $r = 2$ | 3 | 4 | 5 | 6 | 7 | 8 | 10 | 12 |
|---|---|---|---|---|---|---|---|---|---|
| 1 | .98020 | .97045 | .96079 | .95123 | .94176 | .93239 | .92312 | .90484 | .88692 |
| 2 | .96079 | .94176 | .92312 | .90484 | .88692 | .86936 | .85214 | .81873 | .78663 |
| 3 | .94176 | .91393 | .88692 | .86071 | .83527 | .81058 | .78663 | .74082 | .69768 |
| 4 | .92312 | .88692 | .85214 | .81873 | .78663 | .75578 | .72615 | .67032 | .61878 |
| 5 | .90484 | .86071 | .81873 | .77880 | .74082 | .70469 | .67032 | .60653 | .54881 |
| 6 | .88692 | .83527 | .78663 | .74082 | .69768 | .65705 | .61878 | .54881 | .48675 |
| 7 | .86936 | .81058 | .75578 | .70469 | .65705 | .61263 | .57121 | .49659 | .43171 |
| 8 | .85214 | .78663 | .72615 | .67032 | .61878 | .57121 | .52729 | .44933 | .38289 |
| 9 | .83527 | .76338 | .69768 | .63763 | .58275 | .53259 | .48675 | .40657 | .33960 |
| 10 | .81873 | .74082 | .67032 | .60653 | .54881 | .49659 | .44933 | .36788 | .30119 |
| 11 | .80252 | .71892 | .64404 | .57695 | .51685 | .46301 | .41478 | .33287 | .26714 |
| 12 | .78663 | .69768 | .61878 | .54881 | .48675 | .43171 | .38289 | .30119 | .23693 |
| 13 | .77105 | .67706 | .59452 | .52205 | .45841 | .40252 | .35345 | .27253 | .21014 |
| 14 | .75578 | .65705 | .57121 | .49659 | .43171 | .37531 | .32628 | .24660 | .18637 |
| 15 | .74082 | .63763 | .54881 | .47237 | .40657 | .34994 | .30119 | .22313 | .16530 |
| 16 | .72615 | .61878 | .52729 | .44933 | .38289 | .32628 | .27804 | .20190 | .14661 |
| 17 | .71177 | .60050 | .50662 | .42741 | .36059 | .30422 | .25666 | .18268 | .13003 |
| 18 | .69768 | .58275 | .48675 | .40657 | .33960 | .28365 | .23693 | .16530 | .11533 |
| 19 | .68386 | .56553 | .46767 | .38674 | .31982 | .26448 | .21871 | .14957 | .10228 |
| 20 | .67032 | .54881 | .44933 | .36788 | .30119 | .24660 | .20190 | .13534 | .09072 |
| 25 | .60653 | .47237 | .36788 | .28650 | .22313 | .17377 | .13534 | .08208 | .04979 |
| 30 | .54881 | .40657 | .30119 | .22313 | .16530 | .12246 | .09072 | .04979 | .02732 |
| 40 | .44933 | .30119 | .20190 | .13534 | .09072 | .06081 | .04076 | .01832 | .00823 |
| 50 | .36788 | .22313 | .13534 | .08208 | .04979 | .03020 | .01832 | .00674 | .00248 |
| 60 | .30119 | .16530 | .09072 | .04979 | .02732 | .01500 | .00823 | .00248 | .00075 |

Formula: $y' = e^{-(r/100)\times n} = 1/y$.

## Amount of an Annuity

The amount $S$ accumulated at the end of $n$ years by a given annual payment $Y$ set aside at the end of each year is $S = Y \times v$, where the factor $v$ is to be taken from the following table (interest at $r$ percent per annum, compounded annually)

### Values of $v$

| Years | $r = 2$ | 3 | 4 | 5 | 6 | 7 | 8 | 10 | 12 |
|---|---|---|---|---|---|---|---|---|---|
| 1 | 1.0000 | 1.0000 | 1.0000 | 1.0000 | 1.0000 | 1.0000 | 1.0000 | 1.0000 | 1.0000 |
| 2 | 2.0200 | 2.0300 | 2.0400 | 2.0500 | 2.0600 | 2.0700 | 2.0800 | 2.1000 | 2.1200 |
| 3 | 3.0604 | 3.0909 | 3.1216 | 3.1525 | 3.1836 | 3.2149 | 3.2464 | 3.3100 | 3.3744 |
| 4 | 4.1216 | 4.1836 | 4.2465 | 4.3101 | 4.3746 | 4.4399 | 4.5061 | 4.6410 | 4.7793 |
| 5 | 5.2040 | 5.3091 | 5.4163 | 5.5256 | 5.6371 | 5.7507 | 5.8666 | 6.1051 | 6.3528 |
| 6 | 6.3081 | 6.4684 | 6.6330 | 6.8019 | 6.9753 | 7.1533 | 7.3359 | 7.7156 | 8.1152 |
| 7 | 7.4343 | 7.6625 | 7.8983 | 8.1420 | 8.3938 | 8.6540 | 8.9228 | 9.4872 | 10.089 |
| 8 | 8.5830 | 8.8923 | 9.2142 | 9.5491 | 9.8975 | 10.260 | 10.637 | 11.436 | 12.300 |
| 9 | 9.7546 | 10.159 | 10.583 | 11.027 | 11.491 | 11.978 | 12.488 | 13.579 | 14.776 |
| 10 | 10.950 | 11.464 | 12.006 | 12.578 | 13.181 | 13.816 | 14.487 | 15.937 | 17.549 |
| 11 | 12.169 | 12.808 | 13.486 | 14.207 | 14.972 | 15.784 | 16.645 | 18.531 | 20.655 |
| 12 | 13.412 | 14.192 | 15.026 | 15.917 | 16.870 | 17.888 | 18.977 | 21.384 | 24.133 |
| 13 | 14.680 | 15.618 | 16.627 | 17.713 | 18.882 | 20.141 | 21.495 | 24.523 | 28.029 |
| 14 | 15.974 | 17.086 | 18.292 | 19.599 | 21.015 | 22.550 | 24.215 | 27.975 | 32.393 |
| 15 | 17.293 | 18.599 | 20.024 | 21.579 | 23.276 | 25.129 | 27.152 | 31.772 | 37.280 |
| 16 | 18.639 | 20.157 | 21.825 | 23.657 | 25.673 | 27.888 | 30.324 | 35.950 | 42.753 |
| 17 | 20.012 | 21.762 | 23.698 | 25.840 | 28.213 | 30.840 | 33.750 | 40.545 | 48.884 |
| 18 | 21.412 | 23.414 | 25.645 | 28.132 | 30.906 | 33.999 | 37.450 | 45.599 | 55.750 |
| 19 | 22.841 | 25.117 | 27.671 | 30.539 | 33.760 | 37.379 | 41.446 | 51.159 | 63.440 |
| 20 | 24.297 | 26.870 | 29.778 | 33.066 | 36.786 | 40.995 | 45.762 | 57.275 | 72.052 |
| 25 | 32.030 | 36.459 | 41.646 | 47.727 | 54.865 | 63.249 | 73.106 | 98.347 | 133.33 |
| 30 | 40.568 | 47.575 | 56.085 | 66.439 | 79.058 | 94.461 | 113.28 | 164.49 | 241.33 |
| 40 | 60.402 | 75.401 | 95.026 | 120.80 | 154.76 | 199.64 | 259.06 | 442.59 | 767.09 |
| 50 | 84.579 | 112.80 | 152.67 | 209.35 | 290.34 | 406.53 | 573.77 | 1163.9 | 2400.0 |
| 60 | 114.05 | 163.05 | 237.99 | 353.58 | 533.13 | 813.52 | 1253.2 | 3034.8 | 7471.6 |

Formula: $v = \{[1 + (r/100)]^n - 1\} \div (r/100) = (x - 1) \div (r/100)$.

## Annuity Which Will Amount to a Given Sum (Sinking Fund)

The annual payment $Y$ which, if set aside at the end of each year, will amount with accumulated interest to a given sum $S$ at the end of $n$ years is $Y = S \times v'$, where the factor $v'$ is given below (interest at $r$ percent per annum, compounded annually)

### Values of $v'$

| Years | $r = 2$ | 3 | 4 | 5 | 6 | 7 | 8 | 10 | 12 |
|---|---|---|---|---|---|---|---|---|---|
| 1 | 1.0000 | 1.0000 | 1.0000 | 1.0000 | 1.0000 | 1.0000 | 1.0000 | 1.0000 | 1.0000 |
| 2 | .49505 | .49261 | .49020 | .48780 | .48544 | .48309 | .48077 | .47619 | .47170 |
| 3 | .32675 | .32353 | .32035 | .31721 | .31411 | .31105 | .30803 | .30211 | .29635 |
| 4 | .24262 | .23903 | .23549 | .23201 | .22859 | .22523 | .22192 | .21547 | .20923 |
| 5 | .19216 | .18835 | .18463 | .18097 | .17740 | .17389 | .17046 | .16380 | .15741 |
| 6 | .15853 | .15460 | .15076 | .14702 | .14336 | .13980 | .13632 | .12961 | .12323 |
| 7 | .13451 | .13051 | .12661 | .12282 | .11914 | .11555 | .11207 | .10541 | .09912 |
| 8 | .11651 | .11246 | .10853 | .10472 | .10104 | .09747 | .09401 | .08744 | .08130 |
| 9 | .10252 | .09843 | .09449 | .09069 | .08702 | .08349 | .08008 | .07364 | .06768 |
| 10 | .09133 | .08723 | .08329 | .07950 | .07587 | .07238 | .06903 | .06275 | .05698 |
| 11 | .08218 | .07808 | .07415 | .07039 | .06679 | .06336 | .06008 | .05396 | .04842 |
| 12 | .07456 | .07046 | .06655 | .06283 | .05928 | .05590 | .05270 | .04676 | .04144 |
| 13 | .06812 | .06403 | .06014 | .05646 | .05296 | .04965 | .04652 | .04078 | .03568 |
| 14 | .06260 | .05853 | .05467 | .05102 | .04758 | .04434 | .04130 | .03575 | .03087 |
| 15 | .05783 | .05377 | .04994 | .04634 | .04296 | .03979 | .03683 | .03147 | .02682 |
| 16 | .05365 | .04961 | .04582 | .04227 | .03895 | .03586 | .03298 | .02782 | .02339 |
| 17 | .04997 | .04595 | .04220 | .03870 | .03544 | .03243 | .02963 | .02466 | .02046 |
| 18 | .04670 | .04271 | .03899 | .03555 | .03236 | .02941 | .02670 | .02193 | .01794 |
| 19 | .04378 | .03981 | .03614 | .03275 | .02962 | .02675 | .02413 | .01955 | .01576 |
| 20 | .04116 | .03722 | .03358 | .03024 | .02718 | .02439 | .02185 | .01746 | .01388 |
| 25 | .03122 | .02743 | .02401 | .02095 | .01823 | .01581 | .01368 | .01017 | .00750 |
| 30 | .02465 | .02102 | .01783 | .01505 | .01265 | .01059 | .00883 | .00608 | .00414 |
| 40 | .01656 | .01326 | .01052 | .00828 | .00646 | .00501 | .00386 | .00226 | .00130 |
| 50 | .01182 | .00887 | .00655 | .00478 | .00344 | .00246 | .00174 | .00086 | .00042 |
| 60 | .00877 | .00613 | .00420 | .00283 | .00188 | .00123 | .00080 | .00033 | .00013 |

Formula: $v' = (r/100) \div \{[1 + (r/100)]^n - 1\} = 1/v$.

## Present Worth of an Annuity

The capital $C$ which, if placed at interest today, will provide for a given annual payment $Y$ for a term of $n$ years before it is exhausted is $C = Y \times w$, where the factor $w$ is given below (interest at $r$ percent per annum, compounded annually)

Values of $w$

| Years | $r = 2$ | 3 | 4 | 5 | 6 | 7 | 8 | 10 | 12 |
|---|---|---|---|---|---|---|---|---|---|
| 1 | .98039 | .97087 | .96154 | .95238 | .94340 | .93458 | .92593 | .90909 | .89286 |
| 2 | 1.9416 | 1.9135 | 1.8861 | 1.8594 | 1.8334 | 1.8080 | 1.7833 | 1.7355 | 1.6901 |
| 3 | 2.8839 | 2.8286 | 2.7751 | 2.7232 | 2.6730 | 2.6243 | 2.5771 | 2.4869 | 2.4018 |
| 4 | 3.8077 | 3.7171 | 3.6299 | 3.5460 | 3.4651 | 3.3872 | 3.3121 | 3.1699 | 3.0373 |
| 5 | 4.7135 | 4.5797 | 4.4518 | 4.3295 | 4.2124 | 4.1002 | 3.9927 | 3.7908 | 3.6048 |
| 6 | 5.6014 | 5.4172 | 5.2421 | 5.0757 | 4.9173 | 4.7665 | 4.6229 | 4.3553 | 4.1114 |
| 7 | 6.4720 | 6.2303 | 6.0021 | 5.7864 | 5.5824 | 5.3893 | 5.2064 | 4.8684 | 4.5638 |
| 8 | 7.3255 | 7.0197 | 6.7327 | 6.4632 | 6.2098 | 5.9713 | 5.7466 | 5.3349 | 4.9676 |
| 9 | 8.1622 | 7.7861 | 7.4353 | 7.1078 | 6.8017 | 6.5152 | 6.2469 | 5.7590 | 5.3282 |
| 10 | 8.9826 | 8.5302 | 8.1109 | 7.7217 | 7.3601 | 7.0236 | 6.7101 | 6.1446 | 5.6502 |
| 11 | 9.7868 | 9.2526 | 8.7605 | 8.3064 | 7.8869 | 7.4987 | 7.1390 | 6.4951 | 5.9377 |
| 12 | 10.575 | 9.9540 | 9.3851 | 8.8633 | 8.3838 | 7.9427 | 7.5361 | 6.8137 | 6.1944 |
| 13 | 11.348 | 10.635 | 9.9856 | 9.3936 | 8.8527 | 8.3577 | 7.9038 | 7.1034 | 6.4235 |
| 14 | 12.106 | 11.296 | 10.563 | 9.8986 | 9.2950 | 8.7455 | 8.2442 | 7.3667 | 6.6282 |
| 15 | 12.849 | 11.938 | 11.118 | 10.380 | 9.7122 | 9.1079 | 8.5595 | 7.6061 | 6.8109 |
| 16 | 13.578 | 12.561 | 11.652 | 10.838 | 10.106 | 9.4466 | 8.8514 | 7.8237 | 6.9740 |
| 17 | 14.292 | 13.166 | 12.166 | 11.274 | 10.477 | 9.7632 | 9.1216 | 8.0216 | 7.1196 |
| 18 | 14.992 | 13.754 | 12.659 | 11.690 | 10.828 | 10.059 | 9.3719 | 8.2014 | 7.2497 |
| 19 | 15.678 | 14.324 | 13.134 | 12.085 | 11.158 | 10.336 | 9.6036 | 8.3649 | 7.3658 |
| 20 | 16.351 | 14.877 | 13.590 | 12.462 | 11.470 | 10.594 | 9.8181 | 8.5136 | 7.4694 |
| 25 | 19.523 | 17.413 | 15.622 | 14.094 | 12.783 | 11.654 | 10.675 | 9.0770 | 7.8431 |
| 30 | 22.396 | 19.600 | 17.292 | 15.372 | 13.765 | 12.409 | 11.258 | 9.4269 | 8.0552 |
| 40 | 27.355 | 23.115 | 19.793 | 17.159 | 15.046 | 13.332 | 11.925 | 9.7791 | 8.2438 |
| 50 | 31.424 | 25.730 | 21.482 | 18.256 | 15.762 | 13.801 | 12.233 | 9.9148 | 8.3045 |
| 60 | 34.761 | 27.676 | 22.623 | 18.929 | 16.161 | 14.039 | 12.377 | 9.9672 | 8.3240 |

Formula: $w = \{1 - [1 + (r/100)]^{-n}\} \div [r/100] = v/x$.

## Annuity Provided for by a Given Capital

The annual payment $Y$ provided for for a term of $n$ years by a given capital $C$ placed at interest today is $Y = C \times w'$ (interest at $r$ percent per annum, compounded annually; the fund supposed to be exhausted at the end of the term)

Values of $w'$

| Years | $r = 2$ | 3 | 4 | 5 | 6 | 7 | 8 | 10 | 12 |
|---|---|---|---|---|---|---|---|---|---|
| 1 | 1.0200 | 1.0300 | 1.0400 | 1.0500 | 1.0600 | 1.0700 | 1.0800 | 1.1000 | 1.1200 |
| 2 | .51505 | .52261 | .53020 | .53780 | .54544 | .55309 | .56077 | .57619 | .59170 |
| 3 | .34675 | .35353 | .36035 | .36721 | .37411 | .38105 | .38803 | .40211 | .41635 |
| 4 | .26262 | .26903 | .27549 | .28201 | .28859 | .29523 | .30192 | .31547 | .32923 |
| 5 | .21216 | .21835 | .22463 | .23097 | .23740 | .24389 | .25046 | .26380 | .27741 |
| 6 | .17853 | .18460 | .19076 | .19702 | .20336 | .20980 | .21632 | .22961 | .24323 |
| 7 | .15451 | .16051 | .16661 | .17282 | .17914 | .18555 | .19207 | .20541 | .21912 |
| 8 | .13651 | .14246 | .14853 | .15472 | .16104 | .16747 | .17401 | .18744 | .20130 |
| 9 | .12252 | .12843 | .13449 | .14069 | .14702 | .15349 | .16008 | .17364 | .18768 |
| 10 | .11133 | .11723 | .12329 | .12950 | .13587 | .14238 | .14903 | .16275 | .17698 |
| 11 | .10218 | .10808 | .11415 | .12039 | .12679 | .13336 | .14008 | .15396 | .16842 |
| 12 | .09456 | .10046 | .10655 | .11283 | .11928 | .12590 | .13270 | .14676 | .16144 |
| 13 | .08812 | .09403 | .10014 | .10646 | .11296 | .11965 | .12652 | .14078 | .15568 |
| 14 | .08260 | .08853 | .09467 | .10102 | .10758 | .11434 | .12130 | .13575 | .15087 |
| 15 | .07783 | .08377 | .08994 | .09634 | .10296 | .10979 | .11683 | .13147 | .14682 |
| 16 | .07365 | .07961 | .08582 | .09227 | .09895 | .10586 | .11298 | .12782 | .14339 |
| 17 | .06997 | .07595 | .08220 | .08870 | .09544 | .10243 | .10963 | .12466 | .14046 |
| 18 | .06670 | .07271 | .07899 | .08555 | .09236 | .09941 | .10670 | .12193 | .13794 |
| 19 | .06378 | .06981 | .07614 | .08275 | .08962 | .09675 | .10413 | .11955 | .13576 |
| 20 | .06116 | .06722 | .07358 | .08024 | .08718 | .09439 | .10185 | .11746 | .13388 |
| 25 | .05122 | .05743 | .06401 | .07095 | .07823 | .08581 | .09368 | .11017 | .12750 |
| 30 | .04465 | .05102 | .05783 | .06505 | .07265 | .08059 | .08883 | .10608 | .12414 |
| 40 | .03656 | .04326 | .05052 | .05828 | .06646 | .07501 | .08386 | .10226 | .12130 |
| 50 | .03182 | .03887 | .04655 | .05478 | .06344 | .07246 | .08174 | .10086 | .12042 |
| 60 | .02877 | .03613 | .04420 | .05283 | .06188 | .07123 | .08080 | .10033 | .12013 |

Formula: $w' = [r/100] \div \{1 - [1 + (r/100)]^{-n}\} = 1/w = v' + (r/100)$.

### Decimal Equivalents

**From minutes and seconds into decimal parts of a degree**

| ′ | 0°. | ″ | 0°. |
|---|---|---|---|
| 0′ | 0°.0000 | 0″ | 0°.0000 |
| 1 | .0167 | 1 | .0003 |
| 2 | .0333 | 2 | .0006 |
| 3 | .05 | 3 | .0008 |
| 4 | .0667 | 4 | .0011 |
| 5′ | .0833 | 5″ | .0014 |
| 6 | .10 | 6 | .0017 |
| 7 | .1167 | 7 | .0019 |
| 8 | .1333 | 8 | .0022 |
| 9 | .15 | 9 | .0025 |
| 10′ | 0°.1667 | 10″ | 0°.0028 |
| 1 | .1833 | 1 | .0031 |
| 2 | .20 | 2 | .0033 |
| 3 | .2167 | 3 | .0036 |
| 4 | .2333 | 4 | .0039 |
| 15′ | .25 | 15″ | .0042 |
| 6 | .2667 | 6 | .0044 |
| 7 | .2833 | 7 | .0047 |
| 8 | .30 | 8 | .005 |
| 9 | .3167 | 9 | .0053 |
| 20′ | 0°.3333 | 20″ | 0°.0056 |
| 1 | .35 | 1 | .0058 |
| 2 | .3667 | 2 | .0061 |
| 3 | .3833 | 3 | .0064 |
| 4 | .40 | 4 | .0067 |
| 25′ | .4167 | 25″ | .0069 |
| 6 | .4333 | 6 | .0072 |
| 7 | .45 | 7 | .0075 |
| 8 | .4667 | 8 | .0078 |
| 9 | .4833 | 9 | .0081 |
| 30′ | 0°.50 | 30″ | 0°.0083 |
| 1 | .5167 | 1 | .0086 |
| 2 | .5333 | 2 | .0089 |
| 3 | .55 | 3 | .0092 |
| 4 | .5667 | 4 | .0094 |
| 35′ | .5833 | 35″ | .0097 |
| 6 | .60 | 6 | .01 |
| 7 | .6167 | 7 | .0103 |
| 8 | .6333 | 8 | .0106 |
| 9 | .65 | 9 | .0108 |
| 40′ | 0°.6667 | 40″ | 0°.0111 |
| 1 | .6833 | 1 | .0114 |
| 2 | .70 | 2 | .0117 |
| 3 | .7167 | 3 | .0119 |
| 4 | .7333 | 4 | .0122 |
| 45′ | .75 | 45″ | .0125 |
| 6 | .7667 | 6 | .0128 |
| 7 | .7833 | 7 | .0131 |
| 8 | .80 | 8 | .0133 |
| 9 | .8167 | 9 | .0136 |
| 50′ | 0°.8333 | 50″ | 0°.0139 |
| 1 | .85 | 1 | .0142 |
| 2 | .8667 | 2 | .0144 |
| 3 | .8833 | 3 | .0147 |
| 4 | .90 | 4 | .015 |
| 55′ | .9167 | 55″ | .0153 |
| 6 | .9333 | 6 | .0156 |
| 7 | .95 | 7 | .0158 |
| 8 | .9667 | 8 | .0161 |
| 9 | .9833 | 9 | .0164 |
| 60′ | 1.00 | 60″ | 0°.0167 |

**From decimal parts of a degree into minutes and seconds (exact values)**

| 0°. | ′ ″ | 0°. | ′ ″ |
|---|---|---|---|
| 0°.00 | 0′ | 0°.50 | 30′ |
| 1 | 0′ 36″ | 1 | 30′ 36″ |
| 2 | 1′ 12″ | 2 | 31′ 12″ |
| 3 | 1′ 48″ | 3 | 31′ 48″ |
| 4 | 2′ 24″ | 4 | 32′ 24″ |
| 0°.05 | 3′ | 0°.55 | 33′ |
| 6 | 3′ 36″ | 6 | 33′ 36″ |
| 7 | 4′ 12″ | 7 | 34′ 12″ |
| 8 | 4′ 48″ | 8 | 34′ 48″ |
| 9 | 5′ 24″ | 9 | 35′ 24″ |
| 0°.10 | 6′ | 0°.60 | 36′ |
| 1 | 6′ 36″ | 1 | 36′ 36″ |
| 2 | 7′ 12″ | 2 | 37′ 12″ |
| 3 | 7′ 48″ | 3 | 37′ 48″ |
| 4 | 8′ 24″ | 4 | 38′ 24″ |
| 0°.15 | 9′ | 0°.65 | 39′ |
| 6 | 9′ 36″ | 6 | 39′ 36″ |
| 7 | 10′ 12″ | 7 | 40′ 12″ |
| 8 | 10′ 48″ | 8 | 40′ 48″ |
| 9 | 11′ 24″ | 9 | 41′ 24″ |
| 0°.20 | 12′ | 0°.70 | 42′ |
| 1 | 12′ 36″ | 1 | 42′ 36″ |
| 2 | 13′ 12″ | 2 | 43′ 12″ |
| 3 | 13′ 48″ | 3 | 43′ 48″ |
| 4 | 14′ 24″ | 4 | 44′ 24″ |
| 0°.25 | 15′ | 0°.75 | 45′ |
| 6 | 15′ 36″ | 6 | 45′ 36″ |
| 7 | 16′ 12″ | 7 | 46′ 12″ |
| 8 | 16′ 48″ | 8 | 46′ 48″ |
| 9 | 17′ 24″ | 9 | 47′ 24″ |
| 0°.30 | 18′ | 0°.80 | 48′ |
| 1 | 18′ 36″ | 1 | 48′ 36″ |
| 2 | 19′ 12″ | 2 | 49′ 12″ |
| 3 | 19′ 48″ | 3 | 49′ 48″ |
| 4 | 20′ 24″ | 4 | 50′ 24″ |
| 0°.35 | 21′ | 0°.85 | 51′ |
| 6 | 21′ 36″ | 6 | 51′ 36″ |
| 7 | 22′ 12″ | 7 | 52′ 12″ |
| 8 | 22′ 48″ | 8 | 52′ 48″ |
| 9 | 23′ 24″ | 9 | 53′ 24″ |
| 0°.40 | 24′ | 0°.90 | 54′ |
| 1 | 24′ 36″ | 1 | 54′ 36″ |
| 2 | 25′ 12″ | 2 | 55′ 12″ |
| 3 | 25′ 48″ | 3 | 55′ 48″ |
| 4 | 26′ 24″ | 4 | 56′ 24″ |
| 0°.45 | 27′ | 0°.95 | 57′ |
| 6 | 27′ 36″ | 6 | 57′ 36″ |
| 7 | 28′ 12″ | 7 | 58′ 12″ |
| 8 | 28′ 48″ | 8 | 58′ 48″ |
| 9 | 29′ 24″ | 9 | 59′ 24″ |
| 0°.50 | 30′ | 1°.00 | 60′ |

| 0°. | ″ |
|---|---|
| 0°.000 | 0″.0 |
| 1 | 3″.6 |
| 2 | 7″.2 |
| 3 | 10″.8 |
| 4 | 14″.4 |
| 0°.005 | 18″ |
| 6 | 21″.6 |
| 7 | 25″.2 |
| 8 | 28″.8 |
| 9 | 32″.4 |
| 0°.010 | 36″ |

**Common fractions**

| 8 ths | 16 ths | 32 nds | 64 ths | Exact decimal values |
|---|---|---|---|---|
| | | | 1 | .01 5625 |
| | 1 | | 2 | .03 125 |
| | | | 3 | .04 6875 |
| | | 1 | 4 | .06 25 |
| | | | 5 | .07 8125 |
| | 3 | | 6 | .09 375 |
| | | | 7 | .10 9375 |
| 1 | 2 | 4 | 8 | .12 5 |
| | | | 9 | .14 0625 |
| | | 5 | 10 | .15 625 |
| | | | 11 | .17 1875 |
| | 3 | 6 | 12 | .18 75 |
| | | | 13 | .20 3125 |
| | | 7 | 14 | .21 875 |
| | | | 15 | .23 4375 |
| 2 | 4 | 8 | 16 | .25 |
| | | | 17 | .26 5625 |
| | | 9 | 18 | .28 125 |
| | | | 19 | .29 6875 |
| | 5 | 10 | 20 | .31 25 |
| | | | 21 | .32 8125 |
| | | 11 | 22 | .34 375 |
| | | | 23 | .35 9375 |
| 3 | 6 | 12 | 24 | .37 5 |
| | | | 25 | .39 0625 |
| | | 13 | 26 | .40 625 |
| | | | 27 | .42 1875 |
| | 7 | 14 | 28 | .43 75 |
| | | | 29 | .45 3125 |
| | | 15 | 30 | .46 875 |
| | | | 31 | .48 4375 |
| 4 | 8 | 16 | 32 | .50 |
| | | | 33 | .51 5625 |
| | | 17 | 34 | .53 125 |
| | | | 35 | .54 6875 |
| | 9 | 18 | 36 | .56 25 |
| | | | 37 | .57 8125 |
| | | 19 | 38 | .59 375 |
| | | | 39 | .60 9375 |
| 5 | 10 | 20 | 40 | .62 5 |
| | | | 41 | .64 0625 |
| | | 21 | 42 | .65 625 |
| | | | 43 | .67 1875 |
| | 11 | 22 | 44 | .68 75 |
| | | | 45 | .70 3125 |
| | | 23 | 46 | .71 875 |
| | | | 47 | .73 4375 |
| 6 | 12 | 24 | 48 | .75 |
| | | | 49 | .76 5625 |
| | | 25 | 50 | .78 125 |
| | | | 51 | .79 6875 |
| | 13 | 26 | 52 | .81 25 |
| | | | 53 | .82 8125 |
| | | 27 | 54 | .84 375 |
| | | | 55 | .85 9375 |
| 7 | 14 | 28 | 56 | .87 5 |
| | | | 57 | .89 0625 |
| | | 29 | 58 | .90 625 |
| | | | 59 | .92 1875 |
| | 15 | 30 | 60 | .93 75 |
| | | | 61 | .95 3125 |
| | | 31 | 62 | .96 875 |
| | | | 63 | .98 4375 |

# MEASURING UNITS

## By Howard S. Bean

REFERENCES: "International Critical Tables," McGraw-Hill. "Smithsonian Physical Tables," Smithsonian Institution. "Landolt-Börnstein: Zahlenwerte und Funktionen aus Physik, Chemie, Astronomie, Geophysik und Technik," Springer. "Handbook of Chemistry and Physics," Chemical Rubber Co. "Units of Weights and Measures; Definitions and Tables of Equivalents," Misc. Pub. 286, NBS. "Units and Systems of Weights and Measures; Their Origin, Development, and Present Status," Circ. 570 NBS. "Weights and Measures Standards of the United States, a Brief History," Misc. Pub. 247, NBS. "Standard Frequencies and Time Signals," Misc. Pub. 236, NBS. "Standard Time," Code of Federal Regulations, Title 49. "Fluid Meters, Their Theory and Application," 6th ed. chaps. 1–2, ASME, 1971. H. E. Huntley, "Dimensional Analysis," Richard & Co., New York, 1951. "U.S. Standard Atmosphere, 1962," Government Printing Office. "Policy for NBS Usage of SI Units," and "SI Units," reprints *Tech. News Bull.*, Jan. 1971, Mar. 1971, NBS. NBS Frequency and Time Broadcast Services, Special Publication 236, 1972 ed. Public Law 89-387, "Uniform Time Act of 1966." "Metric Conversion Act of 1975." ASTM E380-74; "Metric Practice Guide." E. A. Mechtly, "The International System of Units—Physical Constants and Conversion Factors," National Aeronautics and Space Administration, Publication SP-7012.

### U.S. CUSTOMARY SYSTEM (USCS)

The USCS is the system of units most commonly used for measures of weight and length. They are identical for practical purposes with the corresponding English units, but the capacity measures differ from those now in use in the British Empire, the U.S. gallon being defined as 231 cu in and the bushel as 2,150.42 cu in, whereas the corresponding British Imperial units are, respectively, 277.42 cu in, and 2,219.36 cu in (1 Imp gal = 1.2 U.S. gal, approx; 1 Imp bu = 1.03 U.S. bu, approx).

### U.S. Customary Units

#### Units of Length

| | |
|---|---|
| 12 inches | = 1 foot |
| 3 feet | = 1 yard |
| 5½ yards = 16½ feet | = 1 rod, pole, or perch |
| 40 poles = 220 yards | = 1 furlong |
| 8 furlongs = 1,760 yards = 5,280 feet | = 1 mile |
| 3 miles | = 1 league |
| 4 inches | = 1 hand |
| 9 inches | = 1 span |

#### Nautical Units

| | |
|---|---|
| 6,076.11549 feet | = 1 international nautical mile |
| 6 feet | = 1 fathom |
| 120 fathoms | = 1 cable length |
| 1 nautical mile per hr | = 1 knot |

#### Surveyor's or Gunter's Units

| | |
|---|---|
| 7.92 inches | = 1 link |
| 100 links = 66 ft = 4 rods | = 1 chain |
| 80 chains | = 1 mile |
| 33⅓ inches | = 1 vara (Texas) |

#### Units of Area

| | |
|---|---|
| 144 square inches | = 1 square foot |
| 9 square feet | = 1 square yard |
| 30¼ square yards | = 1 square rod, pole, or perch |
| 160 square rods = 10 square chains = 43,560 square feet = 5,645 sq varas (Texas) | = 1 acre |
| 640 acres = 1 square mile = | 1 "section" of U.S. government surveyed land |
| 1 circular inch = area of circle 1 inch in diameter | = 0.7854 sq in |
| 1 square inch | = 1.2732 circular inches |
| 1 circular mil | = area of circle 0.001 in in diam |
| 1,000,000 cir mils | = 1 circular inch |

#### Units of Volume

| | |
|---|---|
| 1,728 cubic inches | = 1 cubic foot |
| 231 cubic inches | = 1 gallon |
| 27 cubic feet | = 1 cubic yard |
| 1 cord of wood | = 128 cubic feet |
| 1 perch of masonry | = 16½ to 25 cu ft |

#### Liquid or Fluid Measurements

| | |
|---|---|
| 4 gills | = 1 pint |
| 2 pints | = 1 quart |
| 4 quarts | = 1 gallon |
| 7.4805 gallons | = 1 cubic foot |

(There is no standard liquid barrel; by trade custom, 1 bbl of petroleum oil, unrefined = 42 gal. The capacity of the common steel barrel used for refined petroleum products and other liquids is 55 gal.)

#### Apothecaries' Liquid Measurements

| | |
|---|---|
| 60 minims | = 1 liquid dram or drachm |
| 8 drams | = 1 liquid ounce |
| 16 ounces | = 1 pint |

#### Water Measurements

The **miner's inch** is the quantity of water that will pass through an orifice 1 sq in in cross section under a head of 4 to 6½ in, as fixed by statutes, and varies from 1/40 to 1/50 cu ft per sec. The units now most in use are 1 cu ft per sec and 1 gal per sec, the U.S. Reclamation Service employing the former. See Sec. 3.

#### Dry Measures

| | |
|---|---|
| 2 pints | = 1 quart |
| 8 quarts | = 1 peck |
| 4 pecks | = 1 bushel |
| 1 std bbl for fruits and vegetables | = 7,056 cu in or 105 dry qt, struck measure |

#### Shipping Measures

| | |
|---|---|
| 1 Register ton | = 100 cu ft |
| 1 U.S. shipping ton | = 40 cu ft |
| | = 32.14 U.S. bu or 31.14 Imp bu |

| | |
|---|---|
| 1 British shipping ton | = 42 cu ft |
| | = 32.70 Imp bu or 33.75 U.S. bu |

### Board Measurements

$$1 \text{ board foot} = \begin{cases} 144 \text{ cu in} = \text{volume of board} \\ 1 \text{ ft sq and 1 in thick} \end{cases}$$

The international log rule, based upon $\frac{1}{4}$ in kerf, is expressed by the formula

$$X = 0.904762(0.22D^2 - 0.71D)$$

where $X$ is the number of board feet in a 4-ft section of a log and $D$ is the top diam in in. In computing the number of board feet in a log, the taper is taken at $\frac{1}{2}$ in per 4 ft linear, and separate computation is made for each 4 ft section.

### Weights
(The grain is the same in all systems)

#### Avoirdupois Weights

| | |
|---|---|
| 16 drams = 437.5 grains | = 1 ounce |
| 16 ounces = 7,000 grains | = 1 pound |
| 100 pounds | = 1 cental |
| 2,000 pounds | = 1 short ton |
| 2,240 pounds | = 1 long ton |
| 1 std lime bbl, small | = 180 lb net |
| 1 std lime bbl, large | = 280 lb net |
| Also (in Great Britain): | |
| 14 pounds | = 1 stone |
| 2 stone = 28 pounds | = 1 quarter |
| 4 quarters = 112 pounds | = 1 hundredweight (cwt) |
| 20 hundredweight | = 1 long ton |

#### Troy Weights

| | |
|---|---|
| 24 grains | = 1 pennyweight (dwt) |
| 20 pennyweights = 480 grains | = 1 ounce |
| 12 ounces = 5,760 grains | = 1 pound |

**1 assay ton** = 29,167 milligrams, or as many milligrams as there are troy ounces in a ton of 2,000 lb avoirdupois. Consequently, the number of milligrams of precious metal yielded by an assay ton of ore gives directly the number of troy ounces that would be obtained from a ton of 2,000 lb avoirdupois.

#### Apothecaries' Weights

| | |
|---|---|
| 20 grains | = 1 scruple Э |
| 3 scruples = 60 grains | = 1 dram Ʒ |
| 8 drams | = 1 ounce Ʒ |
| 12 ounces = 5,760 grains | = 1 pound |

#### Weight for Precious Stones
1 carat = 200 milligrams
(Used by almost all important nations)

#### Circular Measures

| | |
|---|---|
| 60 seconds | = 1 minute |
| 60 minutes | = 1 degree |
| 90 degrees | = 1 quadrant |
| 360 degrees | = circumference |
| 52.2957795 degrees (=57°17′44.806″) | = 1 radian (or angle having arc of length equal to radius) |

## METRIC SYSTEM

In the United States the name **"metric system"** of length and mass units is commonly taken to refer to a system that was developed in France about 1800. The unit of length was equal to 1/10,000,000 of a quarter meridian (north pole to equator) and named the **metre**. A cube 1/10th metre on a side was the **litre,** the unit of volume. The mass of water filling this cube was the **kilogram,** or standard of mass; i.e., 1 litre of water = 1 kilogram of mass. Metal bars and weights were constructed conforming to these prescriptions for the metre and kilogram. One bar and one weight were selected to be the primary representations. The kilogram and the metre are now defined independently, and the litre, although for many years defined as the volume of a kilogram of water at the temperature of its maximum density, 4°C, and under a pressure of 76 cm of mercury, is now equal to 1 cubic decimeter.

In 1866, the U.S. Congress formally recognized metric units as a legal system, thereby making their use permissible in the United States. In 1893, the Office of Weights and Measures (now the National Bureau of Standards), by executive order, fixed the values of the U.S. yard and pound in terms of the meter and kilogram, respectively, as 1 yard = 3,600/3,937 m; and 1 lb = 0.453 592 4277 kg. By agreement in 1959 among the **national standards** laboratories of the **English-speaking nations,** the relations in use now are: 1 yd = 0.9144 m, whence 1 in = 25.4 mm exactly; and 1 lb = 0.453 592 37 kg, or 1 lb = 453.59 g (nearly).

## THE INTERNATIONAL SYSTEM OF UNITS (SI)

In October 1960, the Eleventh General (International) Conference on Weights and Measures redefined some of the original metric units and expanded the system to include other physical and engineering units. This expanded system is called, in French, **Le Système International d'Unités** (abbreviated **SI**), and in English, **The International System of Units.**

The **Metric Conversion Act of 1975** codifies the voluntary conversion of the U.S. to the SI system. It is expected that in time all units in the United States will be in SI form. For this reason, additional tables of units, prefixes, equivalents, and conversion factors are included below.

SI consists of **seven base** units, **two supplementary** units, a series of **derived units** consistent with the base and supplementary units, and a series of approved prefixes for the formation of multiples and submultiples of the various units (see tables below). Multiple and submultiple prefixes in steps of 1,000 are recommended. (See ASTM E380-74 for further details.)

**Base** and **supplementary units** are **defined**[1] as:

**Metre**   The metre is the length equal to 1,650,763.73 wavelengths in vacuum of the radiation corresponding to the transition between the levels $2p_{10}$ and $5d_5$ of the krypton 86 atom.

**Kilogram**   The kilogram is the unit of mass; it is equal to the mass of the international prototype of the kilogram.

**Second**   The second is the duration of 9,192,631,770 periods of the radiation corresponding to the transition between the two hyperfine levels of the ground state of the cesium 133 atom.

**Ampere**   The ampere is that constant current which, if maintained in two straight parallel conductors of infinite length, of negligible cross section, and placed 1 metre apart in vacuum, would produce between these conductors a force equal to $2 \times 10^{-7}$ newton per metre of length.

**Kelvin**   The kelvin, unit of thermodynamic temperature, is the fraction 1/273.16 of the thermodynamic temperature of the triple point of water.

**Mole**   The mole is the amount of substance of a system which contains as many elementary entities as there are atoms

[1]ASTM E380-74.

in 0.012 kilogram of carbon 12. (When the mole is used, the elementary entities must be specified and may be atoms, molecules, ions, electrons, other particles, or specified groups of such particles.)

**Candela** The candela is the luminous intensity, in the perpendicular direction, of a surface of 1/600,000 square metre of a blackbody at the temperature of freezing platinum under a pressure of 101,325 newtons per square metre.

**Radian** The unit of measure of a plane angle with its vertex at the center of a circle and subtended by an arc equal in length to the radius.

**Steradian** The unit of measure of a solid angle with its

## SI Units

| Quantity | Unit | SI symbol | Formula |
|---|---|---|---|
| **Base Units*** | | | |
| length | metre | m | |
| mass | kilogram | kg | |
| time | second | s | |
| electric current | ampere | A | |
| thermodynamic temperature | kelvin | K | |
| amount of substance | mole | mol | |
| luminous intensity | candela | cd | |
| **Supplementary units*** | | | |
| plane angle | radian | rad | |
| solid angle | steradian | sr | |
| **Derived units*** | | | |
| acceleration | metre per second squared | | m/s² |
| activity (of a radioactive source) | disintegration per second | | (disintegration)/s |
| angular acceleration | radian per second squared | | rad/s² |
| angular velocity | radian per second | | rad/s |
| area | square metre | | m² |
| density | kilogram per cubic metre | | kg/m³ |
| electric capacitance | farad | F | A·s/V |
| electrical conductance | siemens | S | A/V |
| electric field strength | volt per metre | | V/m |
| electric inductance | henry | H | V·s/A |
| electric potential difference | volt | V | W/A |
| electric resistance | ohm | Ω | V/A |
| electromotive force | volt | V | W/A |
| energy | joule | J | N·m |
| entropy | joule per kelvin | | J/K |
| force | newton | N | kg·m/s² |
| frequency | hertz | Hz | 1/s |
| illuminance | lux | lx | lm/m² |
| luminance | candela per square metre | | cd/m² |
| luminous flux | lumen | lm | cd·sr |
| magnetic field strength | ampere per metre | | A/m |
| magnetic flux | weber | Wb | V·s |
| magnetic flux density | tesla | T | Wb/m² |
| magnetomotive force | ampere | A | |
| power | watt | W | J/s |
| pressure | pascal | Pa | N/m² |
| quantity of electricity | coulomb | C | A·s |
| quantity of heat | joule | J | N·m |
| radiant intensity | watt per steradian | | W/sr |
| specific heat | joule per kilogram-kelvin | | J/kg·K |
| stress | pascal | Pa | N/m² |
| thermal conductivity | watt per metre-kelvin | | W/m·K |
| velocity | metre per second | | m/s |
| viscosity, dynamic | pascal-second | | Pa·s |
| viscosity, kinematic | square metre per second | | m²/s |
| voltage | volt | V | W/A |
| volume | cubic metre | | m³ |
| wavenumber | reciprocal metre | | (wave)/m |
| work | joule | J | N·m |

*ASTM E380-74.

vertex at the center of a sphere and enclosing an area of the spherical surface equal to that of a square with sides equal in length to the radius.

### SI Prefixes*

| Multiplication factors | Prefix | SI symbol |
|---|---|---|
| $1\ 000\ 000\ 000\ 000 = 10^{12}$ | tera | T |
| $1\ 000\ 000\ 000 = 10^{9}$ | giga | G |
| $1\ 000\ 000 = 10^{6}$ | mega | M |
| $1\ 000 = 10^{3}$ | kilo | k |
| $100 = 10^{2}$ | hecto† | h |
| $10 = 10^{1}$ | deka† | da |
| $0.1 = 10^{-1}$ | deci† | d |
| $0.01 = 10^{-2}$ | centi† | c |
| $0.001 = 10^{-3}$ | milli | m |
| $0.000\ 001 = 10^{-6}$ | micro | $\mu$ |
| $0.000\ 000\ 001 = 10^{-9}$ | nano | n |
| $0.000\ 000\ 000\ 001 = 10^{-12}$ | pico | p |
| $0.000\ 000\ 000\ 000\ 001 = 10^{-15}$ | femto | f |
| $0.000\ 000\ 000\ 000\ 000\ 001 = 10^{-18}$ | atto | a |

*ASTM E380-74.
†To be avoided where possible.

**SI Conversion factors** are listed here alphabetically (adapted from E. A. Mechtly, "The International System of Units: Physical Constants and Conversion Factors," revised, National Aeronautics and Space Administration, 1969). Conversion factors are written as a number greater than one and less than ten with six or fewer decimal places. This number is followed by the letter E (for exponent), a plus or minus symbol, and two digits which indicate the power of 10 by which the number must be multiplied to obtain the correct value. For example:

3.523 907 E − 02 is $3.523\ 907 \times 10^{-2}$ or 0.035 239 07

An asterisk (*) after the sixth decimal place indicates that the conversion factor is exact and that all subsequent digits are zero. All other conversion factors have been rounded off.

### SYSTEMS OF UNITS

The principal units of interest to mechanical engineers can all be derived from the three units of **force, length,** and **time.** These three units may be chosen at pleasure; each such choice gives rise to a "system" of units. The table on p. 1-40 gives the units of the four "systems" most often met with in the literature.

| To convert from | to | Multiply by |
|---|---|---|
| abampere | ampere (A) | 1.000 000*E+01 |
| abcoulomb | coulomb (C) | 1.000 000*E+01 |
| abfarad | farad (F) | 1.000 000*E+09 |
| abhenry | henry (H) | 1.000 000*E−09 |
| abmho | siemens (S) | 1.000 000*E+09 |
| abohm | ohm ($\Omega$) | 1.000 000*E−09 |
| abvolt | volt (V) | 1.000 000*E−08 |
| acre-foot (U.S. survey)[a] | metre³ (m³) | 1.233 489 E+03 |
| acre (U.S. survey)[a] | metre² (m²) | 4.046 873 E+03 |
| ampere, international U.S. ($A_{\text{INT−US}}$)[b] | ampere (A) | 9.998 43  E−01 |
| ampere, U.S. legal 1948 ($A_{\text{US−48}}$) | ampere (A) | 1.000 008 E+00 |
| ampere-hour | coulomb (C) | 3.600 000*E+03 |
| angstrom | metre (m) | 1.000 000*E−10 |
| are | metre² (m²) | 1.000 000*E+02 |
| astronomical unit | metre (m) | 1.495 98  E+11 |
| atmosphere (normal) | pascal (Pa) | 1.013 25  E+05 |
| atmosphere (technical = 1 $kg_f$/cm²) | pascal (Pa) | 9.806 650*E+04 |
| bar | pascal (Pa) | 1.000 000*E+05 |
| barn | metre² (m²) | 1.000 000*E−28 |
| barrel (for crude petroleum, 42 gal) | metre³ (m³) | 1.589 873 E−01 |
| board foot | metre³ (m³) | 2.359 737 E−03 |
| British thermal unit (International Table)[c] | joule (J) | 1.055 056 E+03 |
| British thermal unit (mean) | joule (J) | 1.055 87  E+03 |
| British thermal unit (thermochemical) | joule (J) | 1.054 350 E+03 |
| British thermal unit (39°F) | joule (J) | 1.059 67  E+03 |
| British thermal unit (59°F) | joule (J) | 1.054 80  E+03 |
| British thermal unit (60°F) | joule (J) | 1.054 68  E+03 |
| Btu (thermochemical)/foot²-second | watt/metre² (W/m²) | 1.134 893 E+04 |
| Btu (thermochemical)/foot²-minute | watt/metre² (W/m²) | 1.891 489 E+02 |
| Btu (thermochemical)/foot²-hour | watt/metre² (W/m²) | 3.152 481 E+00 |
| Btu (thermochemical)/inch²-second | watt/metre² (W/m²) | 1.634 246 E+06 |
| Btu (thermochemical)·in/s·ft²·°F ($k$, thermal conductivity) | watt/metre-kelvin (W/m·K) | 5.188 732 E+02 |
| Btu (International Table)·in/s·ft²·°F ($k$, thermal conductivity) | watt/metre-kelvin (W/m·K) | 5.192 204 E+02 |
| Btu (thermochemical)·in/h·ft²·°F ($k$, thermal conductivity) | watt/metre-kelvin (W/m·K) | 1.441 314 E−01 |

| To convert from | to | Multiply by |
|---|---|---|
| Btu (International Table)·in/h·ft²·°F ($k$, thermal conductivity) | watt/metre-kelvin (W/m·K) | 1.442 279 E−01 |
| Btu (International Table)/ft² | joule/metre² (J/m²) | 1.135 653 E+04 |
| Btu (thermochemical)/ft² | joule/metre² (J/m²) | 1.134 893 E+04 |
| Btu (International Table)/h·ft²·°F ($C$, thermal conductance) | watt/metre²-kelvin (W/m²·K) | 5.678 263 E+00 |
| Btu (thermochemical)/h·ft²·°F ($C$, thermal conductance) | watt/metre²-kelvin (W/m²·K) | 5.674 466 E+00 |
| Btu (International Table)/pound-mass | joule/kilogram (J/kg) | 2.326 000*E+03 |
| Btu (thermochemical)/pound-mass | joule/kilogram (J/kg) | 2.324 444 E+03 |
| Btu (International Table)/lbm·°F ($c$, heat capacity) | joule/kilogram-kelvin (J/kg·K) | 4.186 800*E+03 |
| Btu (thermochemical)/lbm·°F ($c$, heat capacity) | joule/kilogram-kelvin (J/kg·K) | 4.184 000 E+03 |
| Btu (International Table)/s·ft²·°F | watt/metre²-kelvin (W/m²·K) | 2.044 175 E+04 |
| Btu (thermochemical)/s·ft²·°F | watt/metre²-kelvin (W/m²·K) | 2.042 808 E+04 |
| Btu (International Table)/hour | watt (W ) | 2.930 711 E−01 |
| Btu (thermochemical)/second | watt (W) | 1.054 350 E+03 |
| Btu (thermochemical)/minute | watt (W) | 1.757 250 E+01 |
| Btu (thermochemical)/hour | watt (W) | 2.928 751 E−01 |
| bushel (U.S.) | metre (m³) | 3.523 907 E−02 |
| calorie (International Table) | joule (J) | 4.186 800*E+00 |
| calorie (mean) | joule (J) | 4.190 02 E+00 |
| calorie (thermochemical) | joule (J) | 4.184 000*E+00 |
| calorie (15°C) | joule (J) | 4.185 80 E+00 |
| calorie (20°C) | joule (J) | 4.181 90 E+00 |
| calorie (kilogram, International Table) | joule (J) | 4.186 800*E+03 |
| calorie (kilogram, mean) | joule (J) | 4.190 02 E+03 |
| calorie (kilogram, thermochemical) | joule (J) | 4.184 000*E+03 |
| calorie (thermochemical)/centimetre²-minute | watt/metre² (W/m²) | 6.973 333 E+02 |
| cal (thermochemical)/cm² | joule/metre² (J/m²) | 4.184 000*E+04 |
| cal (thermochemical)/cm²·s | watt/metre² (W/m²) | 4.184 000*E+04 |
| cal (thermochemical)/cm·s·°C | watt/metre-kelvin (W/m·K) | 4.184 000*E+02 |
| cal (International Table)/g | joule/kilogram (J/kg) | 4.186 800*E+03 |
| cal (International Table)/g·°C | joule/kilogram-kelvin (J/kg·K) | 4.186 800*E+03 |
| cal (thermochemical)/g | joule/kilogram (J/kg) | 4.184 000*E+03 |
| cal (thermochemical)/g·°C | joule/kilogram-kelvin (J/kg·K) | 4.184 000*E+03 |
| calorie (thermochemical)/second | watt (W) | 4.184 000*E+00 |
| calorie (thermochemical)/minute | watt (W) | 6.973 333 E−02 |
| carat (metric) | kilogram (kg) | 2.000 000*E−04 |
| centimetre of mercury (0°C) | pascal (Pa) | 1.333 22 E+03 |
| centimetre of water (4°C) | pascal (Pa) | 9.806 38 E+01 |
| centipoise | pascal-second (Pa·s) | 1.000 000*E−03 |
| centistokes | metre²/second (m²/s) | 1.000 000*E−06 |
| chain (engineer or ramden) | meter (m) | 3.048* E+01 |
| chain (surveyor or gunter) | meter (m) | 2.011 68* E+01 |
| circular mil | metre² (m²) | 5.067 075 E−10 |
| cord | metre³ (m³) | 3.624 556 E+00 |
| coulomb, international U.S. ($C_{INT-US}$)[b] | coulomb (C) | 9.998 43 E−01 |
| coulomb, U.S. legal 1948 ($C_{US-48}$) | coulomb (C) | 1.000 008 E+00 |
| cup | metre³ (m³) | 2.365 882 E−04 |
| curie | bequerel (Bq) | 3.700 000*E+10 |
| day (mean solar) | second (s) | 8.640 000 E+04 |
| day (sidereal) | second (s) | 8.616 409 E+04 |
| degree (angle) | radian (rad) | 1.745 329 E−02 |
| degree Celsius | kelvin (K) | $t_K = t°_C + 273.15$ |
| degree centigrade | kelvin (K) | $t_K = t°_C + 273.15$ |
| degree Fahrenheit | degree Celsius | $t°_C = (t°_F - 32)/1.8$ |
| degree Fahrenheit | kelvin (K) | $t_K = (t°_F + 459.67)/1.8$ |
| deg F· h· ft²/Btu (thermochemical) ($R$, thermal resistance) | kelvin-metre²/watt (K·m²/W) | 1.762 280 E−01 |
| deg F· hft²/Btu (thermochemical) ($R$, thermal resistance) | kelvin-metre²/watt (K·m²/W) | 1.761 102 E−01 |

| To convert from | to | Multiply by |
|---|---|---|
| degree Rankine | kelvin (K) | $t_K = t°_R/1.8$ |
| dram (avoirdupois) | kilogram (kg) | 1.771 845 E−03 |
| dram (troy or apothecary) | kilogram (kg) | 3.887 934 E−03 |
| dram (U.S. fluid) | kilogram (kg) | 3.696 691 E−06 |
| dyne | newton (N) | 1.000 000*E−05 |
| dyne-centimetre | newton-metre (N·m) | 1.000 000*E−07 |
| dyne-centimetre² | pascal (Pa) | 1.000 000*E−01 |
| electron volt | joule (J) | 1.602 19  E−19 |
| EMU of capacitance | farad (F) | 1.000 000*E+09 |
| EMU of current | ampere (A) | 1.000 000*E+01 |
| EMU of electric potential | volt (V) | 1.000 000*E−08 |
| EMU of inductance | henry (H) | 1.000 000*E−09 |
| EMU of resistance | ohm (Ω) | 1.000 000*E−09 |
| ESU of capacitance | farad (F) | 1.112 650 E−12 |
| ESU of current | ampere (A) | 3.335 6   E−10 |
| ESU of electric potential | volt (V) | 2.997 9   E+02 |
| ESU of inductance | henry (H) | 8.987 554 E+11 |
| ESU of resistance | ohm (Ω) | 8.987 554 E+11 |
| erg | joule (J) | 1.000 000*E−07 |
| erg/centimetre²-second | watt/metre² (W/m²) | 1.000 000*E−03 |
| erg/second | watt (W) | 1.000 000*E−07 |
| farad, international U.S. ($F_{INT-US}$) | farad (F) | 9.995 05  E−01 |
| faraday (based on carbon 12) | coulomb (C) | 9.648 70  E+04 |
| faraday (chemical) | coulomb (C) | 9.649 57  E+04 |
| faraday (physical) | coulomb (C) | 9.652 19  E+04 |
| fathom (U.S. survey)[a] | metre (m) | 1.828 804 E+00 |
| fermi (femtometer) | metre (m) | 1.000 000*E−15 |
| fluid ounce (U.S.) | metre³ (m³) | 2.957 353 E−05 |
| foot | metre (m) | 3.048 000*E−01 |
| foot (U.S. survey)[a] | metre (m) | 3.048 006 E−01 |
| foot³/minute | metre³/second (m³/s) | 4.719 474 E−04 |
| foot³/second | metre³/second (m³/s) | 2.831 685 E−02 |
| foot³ (volume and section modulus) | metre³ (m³) | 2.831 685 E−02 |
| foot² | metre² (m²) | 9.290 304*E−02 |
| foot⁴ (moment of section)[d] | metre⁴ (m⁴) | 8.630 975 E−03 |
| foot/hour | metre/second (m/s) | 8.466 667 E−05 |
| foot/minute | metre/second (m/s) | 5.080 000*E−03 |
| foot/second | metre/second (m/s) | 3.048 000*E−01 |
| foot²/second | metre²/second (m²/s) | 9.290 304*E−02 |
| foot of water (39.2°F) | pascal (Pa) | 2.988 98  E+03 |
| footcandle | lumen/metre² (lm/m²) | 1.076 391 E+01 |
| footcandle | lux (lx) | 1.076 391 E+01 |
| footlambert | candela/metre² (cd/m²) | 3.426 259 E+00 |
| foot-pound-force | joule (J) | 1.355 818 E+00 |
| foot-pound-force/hour | watt (W) | 3.766 161 E−04 |
| foot-pound-force/minute | watt (W) | 2.259 697 E−02 |
| foot-pound-force/second | watt (W) | 1.355 818 E+00 |
| foot-poundal | joule (J) | 4.214 011 E−02 |
| ft²/h (thermal diffusivity) | metre²/second (m²/s) | 2.580 640*E−05 |
| foot/second² | metre/second² (m/s²) | 3.048 000*E−01 |
| free fall, standard | metre/second² (m/s²) | 9.806 650*E+00 |
| furlong | metre (m) | 2.011 68 *E+02 |
| gal | metre/second² (m/s²) | 1.000 000*E−02 |
| gallon (Canadian liquid) | metre³ (m³) | 4.546 090 E−03 |
| gallon (U.K. liquid) | metre³ (m³) | 4.546 092 E−03 |
| gallon (U.S. dry) | metre³ (m³) | 4.404 884 E−03 |
| gallon (U.S. liquid) | metre³ (m³) | 3.785 412 E−03 |
| gallon (U.S. liquid)/day | metre³/second (m³/s) | 4.381 264 E−08 |
| gallon (U.S. liquid)/minute | metre³/second (m³/s) | 6.309 020 E−05 |
| gamma | tesla (T) | 1.000 000*E−09 |
| gauss | tesla (T) | 1.000 000*E−04 |
| gilbert | ampere-turn | 7.957 747 E−01 |
| gill (U.K.) | metre³ (m³) | 1.420 654 E−04 |
| gill (U.S.) | metre³ (m³) | 1.182 941 E−04 |
| grad | degree (angular) | 9.000 000*E−01 |
| grad | radian (rad) | 1.570 796 E−02 |

| To convert from | to | Multiply by |
|---|---|---|
| grain (1/7,000 lbm avoirdupois) | kilogram (kg) | 6.479 891*E−05 |
| gram | kilogram (kg) | 1.000 000*E−03 |
| gram/centimetre$^3$ | kilogram/metre$^3$ (kg/m$^3$) | 1.000 000*E+03 |
| gram-force/centimetre$^2$ | pascal (Pa) | 9.806 650*E+01 |
| hectare | metre$^2$ (m$^2$) | 1.000 000*E+04 |
| henry, international U.S. (H$_{INT-US}$) | henry (H) | 1.000 495 E+00 |
| hogshead (U.S.) | metre$^3$ (m$^3$) | 2.384 809 E−01 |
| horsepower (550 ft · lbf/s) | watt (W) | 7.456 999 E+02 |
| horsepower (boiler) | watt (W) | 9.809 50 E+03 |
| horsepower (electric) | watt (W) | 7.460 000*E+02 |
| horsepower (metric) | watt (W) | 7.354 99 E+02 |
| horsepower (water) | watt (W) | 7.460 43 E+02 |
| horsepower (U.K.) | watt (W) | 7.457 0 E+02 |
| hour (mean solar) | second (s) | 3.600 000 E+03 |
| hour (sidereal) | second (s) | 3.590 170 E+03 |
| hundredweight (long) | kilogram (kg) | 5.080 235 E+01 |
| hundredweight (short) | kilogram (kg) | 4.535 924 E+01 |
| inch | metre (m) | 2.540 000*E−02 |
| inch$^2$ | metre$^2$ (m$^2$) | 6.451 600*E−04 |
| inch$^3$ (volume and section modulus) | metre$^3$ (m$^3$) | 1.638 706 E−05 |
| inch$^3$/minute | metre$^3$/second (m$^3$/s) | 2.731 177 E−07 |
| inch$^4$ (moment of section)$^d$ | metre$^4$ (m$^4$) | 4.162 314 E−07 |
| inch/second | metre/second (m/s) | 2.540 000*E−02 |
| inch of mercury (32°F) | pascal (Pa) | 3.386 389 E+03 |
| inch of mercury (60°F) | pascal (Pa) | 3.376 85 E+03 |
| inch of water (39.2°F) | pascal (Pa) | 2.490 82 E+02 |
| inch of water (60°F) | pascal (Pa) | 2.488 4 E+02 |
| inch/second$^2$ | metre/second$^2$ (m/s$^2$) | 2.540 000*E−02 |
| joule, international U.S. (J$_{INT-US}$)$^b$ | joule (J) | 1.000 182 E+00 |
| joule, U.S. legal 1948 (J$_{US-48}$) | joule (J) | 1.000 017 E+00 |
| kayser | 1/metre (1/m) | 1.000 000*E+02 |
| kelvin | degree Celsius | $t_C = t_K - 273.15$ |
| kilocalorie (thermochemical)/minute | watt (W) | 6.973 333 E+01 |
| kilocalorie (thermochemical)/second | watt (W) | 4.184 000*E+03 |
| kilogram-force (kgf) | newton (N) | 9.806 650*E+00 |
| kilogram-force-metre | newton-metre (N·m) | 9.806 650*E+00 |
| kilogram-force-second$^2$/metre (mass) | kilogram (kg) | 9.806 650*E+00 |
| kilogram-force/centimetre$^2$ | pascal (Pa) | 9.806 650*E+04 |
| kilogram-force/metre$^2$ | pascal (Pa) | 9.806 650*E+00 |
| kilogram-force/millimetre$^2$ | pascal (Pa) | 9.806 650*E+06 |
| kilogram-mass | kilogram (kg) | 1.000 000*E+00 |
| kilometre/hour | metre/second (m/s) | 2.777 778 E−01 |
| kilopond | newton (N) | 9.806 650*E+00 |
| kilowatt hour | joule (J) | 3.600 000*E+06 |
| kilowatt hour, international U.S. (kWh$_{INT-US}$)$^b$ | joule (J) | 3.600 655 E+06 |
| kilowatt hour, U.S. legal 1948 (kWh$_{US-48}$) | joule (J) | 3.600 061 E+06 |
| kip (1,000 lbf) | newton (N) | 4.448 222 E+03 |
| kip/inch$^2$ (ksi) | pascal (Pa) | 6.894 757 E+06 |
| knot (international) | metre/second (m/s) | 5.144 444 E−01 |
| lambert | candela/metre$^2$ (cd/m$^2$) | 3.183 099 E+03 |
| langley | joule/metre$^2$ (J/m$^2$) | 4.184 000*E+04 |
| league, nautical (international and U.S.) | metre (m) | 5.556 000*E+03 |
| league (U.S. survey)$^a$ | metre (m) | 4.828 042 E+03 |
| league, nautical (U.K.) | metre (m) | 5.559 552*E+03 |
| light year | metre (m) | 9.460 55 E+15 |
| link (engineer or ramden) | metre (m) | 3.048* E−01 |
| link (surveyor or gunter) | metre (m) | 2.011 68* E−01 |
| litre$^e$ | metre$^3$ (m$^3$) | 1.000 000*E−03 |
| lux | lumen/metre$^2$ (lm/m$^2$) | 1.000 000*E+00 |
| maxwell | weber (Wb) | 1.000 000*E−08 |
| mho | siemens (S) | 1.000 000*E+00 |
| microinch | metre (m) | 2.540 000*E−08 |
| micron | metre (m) | 1.000 000*E−06 |

| To convert from | to | Multiply by |
|---|---|---|
| mil | metre (m) | 2.540 000*E−05 |
| mile, nautical (international and U.S.) | metre (m) | 1.852 000*E+03 |
| mile, nautical (U.K.) | metre (m) | 1.853 184*E+03 |
| mile (international) | metre (m) | 1.609 344*E+03 |
| mile (U.S. survey)[a] | metre (m) | 1.609 347 E+03 |
| mile$^2$ (international) | metre$^2$ (m$^2$) | 2.589 988 E+06 |
| mile$^2$ (U.S. survey)[a] | metre$^2$ (m$^2$) | 2.589 998 E+06 |
| mile/hour (international) | metre/second (m/s) | 4.470 400*E−01 |
| mile/hour (international) | kilometre/hour | 1.609 344*E+00 |
| millimetre of mercury (0°C) | pascal (Pa) | 1.333 224 E+02 |
| minute (angle) | radian (rad) | 2.908 882 E−04 |
| minute (mean solar) | second (s) | 6.000 000 E+01 |
| minute (sidereal) | second (s) | 5.983 617 E+01 |
| month (mean calendar) | second (s) | 2.268 000 E+06 |
| oersted | ampere/metre (A/m) | 7.957 747 E+01 |
| ohm, international U.S. ($\Omega_{INT-US}$) | ohm ($\Omega$) | 1.000 495 E+00 |
| ohm-centimetre | ohm-metre ($\Omega \cdot$m) | 1.000 000*E−02 |
| ounce-force (avoirdupois) | newton (N) | 2.780 139 E−01 |
| ounce-force-inch | newton-metre (N·m) | 7.061 552 E−03 |
| ounce-mass (avoirdupois) | kilogram (kg) | 2.834 952 E−02 |
| ounce-mass (troy or apothecary) | kilogram (kg) | 3.110 348 E−02 |
| ounce-mass/yard$^2$ | kilogram/metre$^2$ (kg/m$^2$) | 3.390 575 E−02 |
| ounce (avoirdupois) (mass)/inch$^3$ | kilogram/metre$^3$ (kg/m$^3$) | 1.729 994 E+03 |
| ounce (U.K. fluid) | metre$^3$ (m$^3$) | 2.841 307 E−05 |
| ounce (U.S. fluid) | metre$^3$ (m$^3$) | 2.957 353 E−05 |
| parsec | metre (m) | 3.083 74 E+16 |
| peck (U.S.) | metre$^3$ (m$^3$) | 8.809 768 E−03 |
| pennyweight | kilogram (kg) | 1.555 174 E−03 |
| perm (0°C) | kilogram/pascal-second-metre$^2$ (kg/Pa·s·m$^2$) | 5.721 35 E−11 |
| perm (23°C) | kilogram/pascal-second-metre$^2$ (kg/Pa·s·m$^2$) | 5.745 25 E−11 |
| perm-inch (0°C) | kilogram/pascal-second-metre (kg/Pa·s·m) | 1.453 22 E−12 |
| perm-inch (23°C) | kilogram/pascal-second-metre (kg/Pa·s·m) | 1.459 29 E−12 |
| phot | lumen/metre$^2$ (lm/m$^2$) | 1.000 000*E+04 |
| pica (printer's) | metre (m) | 4.217 518 E−03 |
| pint (U.S. dry) | metre$^3$ (m$^3$) | 5.506 105 E−04 |
| pint (U.S. liquid) | metre$^3$ (m$^3$) | 4.731 765 E−04 |
| point (printer's) | metre | 3.514 598*E−04 |
| poise (absolute viscosity) | pascal-second (Pa·s) | 1.000 000*E−01 |
| poundal | newton (N) | 1.382 550 E−01 |
| poundal-foot$^2$ | pascal (Pa) | 1.488 164 E+00 |
| poundal-second/foot$^2$ | pascal-second (Pa·s) | 1.488 164 E+00 |
| pound-force (lbf avoirdupois) | newton (N) | 4.448 222 E+00 |
| pound-force-inch | newton-metre (N·m) | 1.129 848 E−01 |
| pound-force-foot | newton-metre (N·m) | 1.355 818 E+00 |
| pound-force-foot/inch | newton-metre/metre (N·m/m) | 5.337 866 E+01 |
| pound-force-inch/inch | newton-metre/metre (N·m/m) | 4.448 222 E+00 |
| pound-force/inch | newton/metre (N/m) | 1.751 268 E+02 |
| pound-force/foot | newton/metre (N/m) | 1.459 390 E+01 |
| pound-force/foot$^2$ | pascal (Pa) | 4.788 026 E+01 |
| pound-force/inch$^2$ (psi) | pascal (Pa) | 6.894 757 E+03 |
| pound-force-second/foot$^2$ | pascal-second (Pa·s) | 4.788 026 E+01 |
| pound-mass (lbm avoirdupois) | kilogram (kg) | 4.535 924 E−01 |
| pound-mass (troy or apothecary) | kilogram (kg) | 3.732 417 E−01 |
| pound-mass-foot$^2$ (moment of inertia) | kilogram-metre$^2$ (kg·m$^2$) | 4.214 011 E−02 |
| pound-mass-inch$^2$ (moment of inertia) | kilogram-metre$^2$ (kg·m$^2$) | 2.926 397 E−04 |
| pound-mass-foot$^2$ | kilogram/metre$^2$ (kg/m$^2$) | 4.882 428 E+00 |
| pound-mass/second | kilogram/second (kg/s) | 4.535 924 E−01 |
| pound-mass/minute | kilogram/second (kg/s) | 7.559 873 E−03 |
| pound-mass/foot$^3$ | kilogram/metre$^3$ (kg/m$^3$) | 1.601 846 E+01 |
| pound-mass/inch$^3$ | kilogram/metre$^3$ (kg/m$^3$) | 2.767 990 E+04 |
| pound-mass/gallon (U.K. liquid) | kilogram/metre$^3$ (kg/m$^3$) | 9.977 633 E+01 |
| pound-mass/gallon (U.S. liquid) | kilogram/metre$^3$ (kg/m$^3$) | 1.198 264 E+02 |

| To convert from | to | Multiply by |
|---|---|---|
| pound-mass/foot-second | pascal-second (Pa·s) | 1.488 164 E+00 |
| quart (U.S. dry) | metre³ (m³) | 1.101 221 E−03 |
| quart (U.S. liquid) | metre³ (m³) | 9.463 529 E−04 |
| rad (radiation dose absorbed) | joule/kilogram (J/kg) | 1.000 000*E−02 |
| rhe | metre²/newton-second (m²/N·s) | 1.000 000*E+01 |
| rod (U.S. survey)[a] | metre (m) | 5.029 210 E+00 |
| roentgen | coulomb/kilogram (C/kg) | 2.579 760*E−04 |
| second (angle) | radian (rad) | 4.848 137 E−06 |
| second (sidereal) | second (s) | 9.972 696 E−01 |
| section (U.S. survey)[a] | metre² (m²) | 2.589 998 E+06 |
| shake | second (s) | 1.000 000*E−08 |
| slug | kilogram (kg) | 1.459 390 E+01 |
| slug/foot³ | kilogram/metre³ (kg/m³) | 5.153 788 E+02 |
| slug/foot-second | pascal-second (Pa·s) | 4.788 026 E+01 |
| statampere | ampere (A) | 3.335 640 E−10 |
| statcoulomb | coulomb (C) | 3.335 640 E−10 |
| statfarad | farad (F) | 1.112 650 E−12 |
| stathenry | henry (H) | 8.987 554 E+11 |
| statmho | siemens (S) | 1.112 650 E−12 |
| statohm | ohm (Ω) | 8.987 554 E+11 |
| statvolt | volt (V) | 2.997 925 E+02 |
| stere | metre³ (m³) | 1.000 000*E+00 |
| stilb | candela/metre² (cd/m²) | 1.000 000*E+04 |
| stokes (kinematic viscosity) | metre²/second (m²/s) | 1.000 000*E−04 |
| tablespoon | metre³ (m³) | 1.478 676 E−05 |
| teaspoon | metre³ (m³) | 4.928 922 E−06 |
| ton (assay) | kilogram (kg) | 2.916 667 E−02 |
| ton (long, 2,240 lbm) | kilogram (kg) | 1.016 047 E+03 |
| ton (metric) | kilogram (kg) | 1.000 000*E+03 |
| ton (nuclear equivalent of TNT) | joule (J) | 4.20      E+09 |
| ton (register) | metre³ (m³) | 2.831 685 E+00 |
| ton (short, 2,000 lbm) | kilogram (kg) | 9.071 847 E+02 |
| ton (short, mass)/hour | kilogram/second (kg/s) | 2.519 958 E−01 |
| ton (long, mass)/yard³ | kilogram/metre³ (kg/m³) | 1.328 939 E+03 |
| tonne | kilogram (kg) | 1.000 000*E+03 |
| torr (mm Hg, 0°C) | pascal (Pa) | 1.333 22  E+02 |
| township (U.S. survey)[a] | metre² (m²) | 9.323 994 E+07 |
| unit pole | weber (Wb) | 1.256 637 E−07 |
| volt, international U.S. ($V_{INT-US}$)[b] | volt (V) | 1.000 338 E+00 |
| volt, U.S. legal 1948 ($V_{US-48}$) | volt (V) | 1.000 008 E+00 |
| watt, international U.S. ($W_{INT-US}$)[b] | watt (W) | 1.000 182 E+00 |
| watt, U.S. legal 1948 ($W_{US-48}$) | watt (W) | 1.000 017 E+00 |
| watt/centimetre² | watt/metre² (W/m²) | 1.000 000*E+04 |
| watt-hour | joule (J) | 3.600 000*E+03 |
| watt-second | joule (J) | 1.000 000*E+00 |
| yard | metre (m) | 9.144 000*E−01 |
| yard² | metre² (m²) | 8.361 274 E−01 |
| yard³ | metre³ (m³) | 7.645 549 E−01 |
| yard³/minute | metre³/second(m³/s) | 1.274 258 E−02 |
| year (calendar) | second (s) | 3.153 600 E+07 |
| year (sidereal) | second (s) | 3.155 815 E+07 |
| year (tropical) | second (s) | 3.155 693 E+07 |

[a]Based on the U.S. survey foot (1 ft = 1,200/3,937 m).

[b]In 1948 a new international agreement was reached on absolute electrical units, which changed the value of the volt used in this country by about 300 parts per million. Again in 1969 a new base of reference was internationally adopted making a further change of 8.4 parts per million. These changes (and also changes in ampere, joule, watt, coulomb) require careful terminology and conversion factors for exact use of old information. Terms used in this guide are:

Volt as used prior to January 1948—volt, international U.S. ($V_{INT-US}$)

Volt as used between January 1948 and January 1969—volt, U.S. legal 1948 ($V_{US-48}$)

Volt as used since January 1969—volt (V)

Identical treatment is given the ampere, coulomb, watt, and joule.

[c]This value was adopted in 1956. Some of the older International Tables use the value 1.055 04 E+03. The exact conversion factor is 1.055 055 852 62*E+03.

[d]Moment of inertia of a plane section about a specified axis.

[e]In 1964, the General Conference on Weights and Measures adopted the name litre as a special name for the cubic decimetre. Prior to this decision the litre differed slightly (previous value, 1.000028 dm³) and in expression of precision volume measurement this fact must be kept in mind.

## Systems of Units

| Name of unit | Dimensions of units in terms of $F$, $L$, $T$ | British "gravitational" system, or "foot-pound-second" system | Metric "gravitational" system, or "kilogram-meter-second" system | Metric "absolute" system, or "cgs" system | SI newton, metre, second |
|---|---|---|---|---|---|
| Force | $F$ | 1 lb | 1 kg | 1 dyne | 1 newton |
| Length | $L$ | 1 ft | 1 m | 1 cm | 1 m |
| Time | $T$ | 1 sec | 1 sec | 1 sec | 1 sec |

In these systems the "standard pound body" and the "standard kilogram body" refer to two material standards of mass, carefully preserved at London and Paris, respectively (the U.S. pound is derived from the kilogram); the "standard locality" means sea level, 45 deg latitude, or more strictly any locality in which the acceleration due to gravity has the value 980.665 cm per sec per sec = 32.1740 ft per sec per sec, which may be called the **standard acceleration.**

The **pound force** is the force required to support the standard pound body against gravity, *in vacuo*, in the standard locality; or, it is the force which, if applied to the standard pound body, supposed free to move, would give that body the "standard acceleration." The word "pound" is used for the unit of both force and mass and consequently is ambiguous. To avoid uncertainty, it is desirable to call the units "pound force" and "pound mass," respectively.

The **kilogram force** is the force required to support the standard kilogram against gravity, *in vacuo*, in the standard locality; or, it is the force which, if applied to the standard kilogram body, supposed free to move, would give that body the "standard acceleration." The word "kilogram" is used for the unit of both force and mass and consequently is ambiguous. To avoid uncertainty, it is desirable to call the units "kilogram force" and "kilogram mass," respectively.

The **dyne** is the force which, if applied to the standard gram body, would give that body an acceleration of 1 cm per sec per sec; i.e., 1 dyne = 1/980.665 of a gram force.

The **newton** is that force which will impart to a 1-kilogram mass an acceleration of 1 metre per sec per sec.

### TEMPERATURE

The SI unit for thermodynamic temperature is the **kelvin, K,** which is the fraction 1/273.16 of the thermodynamic temperature of the triple point of water. Thus 273.16 K is the **fixed (base) point** on the **kelvin scale.**

Another unit used for the measurement of temperature is degrees **Celsius** (formerly **centigrade**), °C. The relation between a thermodynamic temperature $T$ and a Celsius temperature $t$ is

$$t = T - 273.16 \text{ K}$$

Thus the unit Celsius degree is equal to the unit kelvin, and a difference of temperature would be the same on either scale.

In the USCS temperature is measured in degrees **Fahrenheit, F.** The relation between the Celsius and the Fahrenheit scales is

$$t_{°C} = (t_{°F} - 32)/1.8$$

(For temperature-conversion tables, see Sec. 4.)

### TERRESTRIAL GRAVITY

**Standard acceleration of gravity** is $g^0 = 9.80665$ m per sec per sec, or 32.1740 ft per sec per sec. This value $g^0$ is assumed to be the value of $g$ at sea level and latitude 45 deg.

### MOHS SCALE OF HARDNESS

This scale is an arbitrary one which is used to describe the hardness of several mineral substances on a scale of 1 through 10. The given number indicates a higher relative hardness compared with that of substances below it; and a lower relative hardness than those above it. For example, an unknown substance is scratched by quartz, but it, in turn, scratches feldspar. The unknown has a hardness of between 6 and 7 on the Mohs scale.

### Mohs Scale of Hardness

| | | |
|---|---|---|
| 1. Talc | 5. Apatite | 8. Topaz |
| 2. Gypsum | 6. Feldspar | 9. Sapphire |
| 3. Calc-spar | 7. Quartz | 10. Diamond |
| 4. Fluorspar | | |

### Acceleration of Gravity
(U.S. Coast and Geodetic Survey, 1912)

| Latitude, deg | $g$ m/s² | $g$ ft/sec² | $g/g^0$ | Latitude, deg | $g$ m/s² | $g$ ft/sec² | $g/g^0$ |
|---|---|---|---|---|---|---|---|
| 0 | 9.780 | 32.088 | 0.9973 | 50 | 9.811 | 32.187 | 1.0004 |
| 10 | 9.782 | 32.093 | 0.9975 | 60 | 9.819 | 32.215 | 1.0013 |
| 20 | 9.786 | 32.108 | 0.9979 | 70 | 9.826 | 32.238 | 1.0020 |
| 30 | 9.793 | 32.130 | 0.9986 | 80 | 9.831 | 32.253 | 1.0024 |
| 40 | 9.802 | 32.158 | 0.9995 | 90 | 9.832 | 32.258 | 1.0026 |

Correction for altitude above sea level: − 3mm per sec² for each 1,000 m; − 0.003 ft per sec² for each 1,000 ft.

## TIME

**Kinds of Time**  Three kinds of time are recognized by astronomers: sidereal, apparent solar, and mean solar time. The **sidereal day** is the interval between two consecutive transits of some fixed celestial object across any given meridian, or it is the interval required by the earth to make one complete revolution on its axis. This interval is constant, but it is inconvenient as a time unit because the noon of the sidereal day occurs at all hours of the day and night. The **apparent solar day** is the interval between two consecutive transits of the sun across any given meridian. On account of the variable distance between the sun and earth, the variable speed of the earth in its orbit, the effect of the moon, etc., this interval is not constant and consequently cannot be kept by any simple mechanism, such as clocks or watches. To overcome the objection noted above, the **mean solar day** was devised. The mean solar day is the length of the average apparent solar day. Like the sidereal day it is constant, and like the apparent solar day its noon always occurs at approximately the same time of day. By international agreement, beginning Jan. 1, 1925, the astronomical day, like the civil day, is from midnight to midnight. The hours of the astronomical day run from 0 to 24, and the hours of the civil day usually run from 0 to 12 A.M. and 0 to 12 P.M. In some countries the hours of the civil day also run from 0 to 24.

**The Year**  Three different kinds of year are used: the sidereal, the tropical, and the anomalistic. The **sidereal year** is the time taken by the earth to complete one revolution around the sun from a given star to the same star again. Its length is 365 days, 6 hours, 9 minutes, and 9 seconds. The **tropical year** is the time included between two successive passages of the vernal equinox by the sun, and since the equinox moves westward 50.2 seconds of arc a year, the tropical year is shorter by 20 minutes 23 seconds in time than the sidereal year. As the seasons depend upon the earth's position with respect to the equinox, the tropical year is the year of civil reckoning. The **anomalistic year** is the interval between two successive passages of the perihelion, viz., the time of the earth's nearest approach to the sun. The anomalistic year is used only in special calculations in astronomy.

**The Second**  Although the second is ordinarily defined as 1/86,400 of the mean solar day, this is not sufficiently precise for many scientific purposes. Scientists have adopted more precise definitions for specific purposes: one in terms of the length of the tropical year 1900 and one in terms of a specific atomic frequency.

**Frequency** is the reciprocal of time; the unit of frequency is the **hertz** (Hz), defined as 1 cycle/s.

**The Calendar**  The **Gregorian calendar,** now used in most of the civilized world, was adopted in Catholic countries of Europe in 1582 and in Great Britain and her colonies Jan. 1, 1752. The average length of the Gregorian calendar year is $365\frac{1}{4} - \frac{3}{400}$ days, or 365.2425 days. This is equivalent to 365 days, 5 hours, 49 minutes, 12 seconds. The length of the tropical year is 365.2422 days, or 365 days, 5 hours, 48 minutes, 46 seconds. Thus the Gregorian calendar year is longer than the tropical year by 0.0003 day, or 26 seconds. This difference amounts to 1 day in slightly more than 3,300 years and can properly be neglected.

**Standard Time**  Prior to 1883, each city of the United States had its own time, which was determined by the time of passage of the sun across the local meridian. A system of standard time had been used since its first adoption by the railroads in 1883 but was first legalized on Mar. 19, 1918, when Congress directed the Interstate Commerce Commission to establish limits of the standard time zones. Congress took no further steps until the **Uniform Time Act of 1966** was enacted, followed with an amendment in 1972. This legislation, referred to as "the Act," transferred the regulation and enforcement of the law to the Department of Transportation.

By the legislation of 1918, with some modifications by the Act, the contiguous United States is divided into four **time zones,** each of which, theoretically, was to span 15 degrees of longitude. The first, the **Eastern zone,** extends from the Atlantic westward to include most of Michigan and Indiana, the eastern parts of Kentucky and Tennessee, Georgia, and Florida, except the west half of the panhandle. **Eastern standard time** is based upon the mean solar time of the 75th meridian west of Greenwich, and is 5 hours slower than **Greenwich Mean Time (GMT).** (See also discussion of UTC below.) The second or **Central zone** extends westward to include most of North Dakota, about half of South Dakota and Nebraska, most of Kansas, Oklahoma, and all but the two most westerly counties of Texas. **Central standard time** is based upon the mean solar time of the 90th meridian west of Greenwich, and is 6 hours slower than GMT. The third or **Mountain zone** extends westward to include Montana, most of Idaho, one county of Oregon, Utah, and Arizona. **Mountain standard time** is based upon the mean solar time of the 105th meridian west of Greenwich, and is 7 hours slower than GMT. The fourth or **Pacific zone** includes all of the remaining 48 contiguous states. **Pacific standard time** is based on the mean solar time 120th meridian west of Greenwich, and is 8 hours slower than GMT. Exact locations of boundaries may be obtained from the Department of Transportation.

In addition to the above four zones there are four others that apply to the noncontiguous states and islands. The most easterly is the **Atlantic zone,** which includes Puerto Rico and the Virgin Islands, where the time is 4 hours slower than GMT. Eastern standard time is used in the Panama Canal strip. To the west of the Pacific time zone there are the **Yukon,** the **Alaska-Hawaii,** and **Bering zones** where the times are, respectively, 9, 10, and 11 hours slower than GMT. The system of standard time had been adopted in all civilized countries and is used by ships on the high seas.

The 1966 Act directs that from the last Sunday in April to the last Sunday in October, the time in each zone is to be advanced one hour for advanced time or daylight saving time (DST). However, any state-by-state enactment may exempt the entire state from using advanced time. By this provision Arizona and Hawaii do not observe advanced time (as of 1973). By the 1972 amendment to the Act, a state split by a time-zone boundary may exempt from using advanced time all that part which is in one zone without affecting the rest of the state. By this amendment, 80 counties of Indiana in the Eastern zone are exempt from using advanced time, while 6 counties in the northwest corner and 6 counties in the southwest, which are in the Central zone, do observe advanced time.

Pursuant to its assignment of carrying out the Act, the Department of Transportation has stipulated that municipalities located on the boundary between the Eastern and Central zones are in the Central zone; those on the boundary between the Central and Mountain zones are in the Mountain zone (except that Murdo, SD, is in the Central zone); those on the boundary between Mountain and Pacific time zones are in the Mountain zone. In such places, when the time is given, it should be specified as Central, Mountain, etc.

**Standard Time Signals** The National Bureau of Standards broadcasts time signals from **station WWV**, Ft. Collins, CO, and from **station WWVH**, near Kekaha, Kaui, HI. The broadcasts by WWV are on radio carrier frequencies of 2.5, 5, 10, 15, 20, and 25 MHz, while those by WWVH are on radio carrier frequencies of 2.5, 5, 10, 15, and 20 MHz. Effective Jan. 1, 1975, time announcements by both WWV and WWVH are referred to **Coordinated Universal Time, UTC,** the international coordinated time scale used around the world for most timekeeping purposes. UTC is generated by reference to atomic standards. Time (i.e., clock time) is given in terms of 0 to 24 hours a day, starting with 0000 at midnight at Greenwich zero longitude. The beginning of each 0.8-second-long audio tone marks the end of an announced time interval. For example, at 2.15 P.M., UTC, the voice announcement would be: "At the tone fourteen hours fifteen minutes Coordinated Universal Time." From WWV this is given during the last 7.5 seconds of each minute. From WWVH the voice announcement of UTC is given between 45 and 52.5 seconds *after* the minute. The tone markers from both stations are given simultaneously, but owing to propagation interferences may not be received simultaneously.

Beginning 1 minute after the hour from WWVH and 2 minutes after the hour from WWV, the **standard musical pitch** of 440 Hz is broadcast for about 45 seconds. In addition to providing the musical pitch, this tone signal may be of use as a marker for automated recorders and other such devices.

## DENSITY AND SPECIFIC GRAVITY

**Density** of a body is its mass per unit volume. With SI units densities are in kilograms per cubic meter. However, giving densities in grams per cubic centimeter has been common, and $kg/m^3 = 1,000 \ g/cm^3$. With the USCS, densities are given in pounds-mass per cubic foot.

**Specific gravity** is the ratio of the density of one substance to that of a second (or reference) substance, both at some specified temperature. For solids and liquids water is almost universally used as the reference substance. Physicists use a reference temperature of 4°C (= 39.2°F); U.S. engineers commonly use 60°F. With the introduction of SI units, it may be found desirable to use 59°F, since 59°F and 15°C are equivalents.

For gases, specific gravity is generally the ratio of the density of the gas to that of air, both at the same temperature, pressure, and dryness (as regards water vapor). Instead of using density, the ratio of the molecular weight of the gas to that of air may be used as the specific gravity of the gas. When this is done, the molecular weight of air may be taken as 28.9644.

The specific gravity of liquids is usually measured by means of a **hydrometer**. In addition to a scale reading in specific gravity, as defined above, other arbitrary scales for hydrometers are used in various trades and industries. The most common of these are the **API** and **Baumé**. The API (American Petroleum Institute) scale is approved by the American Petroleum Institute, the ASTM, the U.S. Bureau of Mines, and the National Bureau of Standards and is recommended for exclusive use in the United States petroleum industry, superseding the Baumé scale for liquids lighter than water. The relation between **API degrees** and specific gravity (see table below) is expressed by the following equation:

### Specific Gravities at $\frac{60°}{60°}$ F Corresponding to Degrees API and Weights per U.S. Gallon at 60°F

$$\left( \text{Calculated from the formula, specific gravity} = \frac{141.5}{131.5 + \text{deg API}} \right)$$

| Degrees API | Specific gravity | Lb per U.S. gallon | Degrees API | Specific gravity | Lb per U.S. gallon | Degrees API | Specific gravity | Lb per U.S. gallon | Degrees API | Specific gravity | Lb per U.S. gallon |
|---|---|---|---|---|---|---|---|---|---|---|---|
| 10 | 1.0000 | 8.328 | 33 | 0.8602 | 7.163 | 56 | 0.7547 | 6.283 | 79 | 0.6722 | 5.595 |
| 11 | 0.9930 | 8.270 | 34 | 0.8550 | 7.119 | 57 | 0.7507 | 6.249 | 80 | 0.6690 | 5.568 |
| 12 | 0.9861 | 8.212 | 35 | 0.8498 | 7.076 | 58 | 0.7467 | 6.216 | 81 | 0.6659 | 5.542 |
| 13 | 0.9792 | 8.155 | 36 | 0.8448 | 7.034 | 59 | 0.7428 | 6.184 | 82 | 0.6628 | 5.516 |
| 14 | 0.9725 | 8.099 | 37 | 0.8398 | 6.993 | 60 | 0.7389 | 6.151 | 83 | 0.6597 | 5.491 |
| 15 | 0.9659 | 8.044 | 38 | 0.8348 | 6.951 | 61 | 0.7351 | 6.119 | 84 | 0.6566 | 5.465 |
| 16 | 0.9593 | 7.989 | 39 | 0.8299 | 6.910 | 62 | 0.7313 | 6.087 | 85 | 0.6536 | 5.440 |
| 17 | 0.9529 | 7.935 | 40 | 0.8251 | 6.870 | 63 | 0.7275 | 6.056 | 86 | 0.6506 | 5.415 |
| 18 | 0.9465 | 7.882 | 41 | 0.8203 | 6.830 | 64 | 0.7238 | 6.025 | 87 | 0.6476 | 5.390 |
| 19 | 0.9402 | 7.830 | 42 | 0.8155 | 6.790 | 65 | 0.7201 | 5.994 | 88 | 0.6446 | 5.365 |
| 20 | 0.9340 | 7.778 | 43 | 0.8109 | 6.752 | 66 | 0.7165 | 5.964 | 89 | 0.6417 | 5.341 |
| 21 | 0.9279 | 7.727 | 44 | 0.8063 | 6.713 | 67 | 0.7128 | 5.934 | 90 | 0.6388 | 5.316 |
| 22 | 0.9218 | 7.676 | 45 | 0.8017 | 6.675 | 68 | 0.7093 | 5.904 | 91 | 0.6360 | 5.293 |
| 23 | 0.9159 | 7.627 | 46 | 0.7972 | 6.637 | 69 | 0.7057 | 5.874 | 92 | 0.6331 | 5.269 |
| 24 | 0.9100 | 7.578 | 47 | 0.7927 | 6.600 | 70 | 0.7022 | 5.845 | 93 | 0.6303 | 5.246 |
| 25 | 0.9042 | 7.529 | 48 | 0.7883 | 6.563 | 71 | 0.6988 | 5.817 | 94 | 0.6275 | 5.222 |
| 26 | 0.8984 | 7.481 | 49 | 0.7839 | 6.526 | 72 | 0.6953 | 5.788 | 95 | 0.6247 | 5.199 |
| 27 | 0.8927 | 7.434 | 50 | 0.7796 | 6.490 | 73 | 0.6919 | 5.759 | 96 | 0.6220 | 5.176 |
| 28 | 0.8871 | 7.387 | 51 | 0.7753 | 6.455 | 74 | 0.6886 | 5.731 | 97 | 0.6193 | 5.154 |
| 29 | 0.8816 | 7.341 | 52 | 0.7711 | 6.420 | 75 | 0.6852 | 5.703 | 98 | 0.6166 | 5.131 |
| 30 | 0.8762 | 7.296 | 53 | 0.7669 | 6.385 | 76 | 0.6819 | 5.676 | 99 | 0.6139 | 5.109 |
| 31 | 0.8708 | 7.251 | 54 | 0.7628 | 6.350 | 77 | 0.6787 | 5.649 | 100 | 0.6112 | 5.086 |
| 32 | 0.8654 | 7.206 | 55 | 0.7587 | 6.316 | 78 | 0.6754 | 5.622 | | | |

The weights in this table are weights in air at 60°F with humidity 50 percent and pressure 760 mm.

$$\text{Degrees API} = \frac{141.5}{\text{sp gr } 60°/60°\text{F}} - 131.5$$

The specific gravities corresponding to the indications of the **Baumse hydrometer** are given in the two tables below.

## CONVERSION AND EQUIVALENCY TABLES

### Note for Use of Conversion Tables

In many tables both equivalents and their logarithms (to the base 10) are given. The equivalents are shown in heavier type.

The logarithms are immediately below. In some cases the equivalents have been rounded off, while the logarithm corresponds to the equivalent carried to a greater number of decimal places. The logarithm may have a minus sign over the whole-number part. This indicates that the characteristic is negative. For example:

$$\log 0.03281 = \bar{2}.51598 = 0.51598 - 2 = 8.51598 - 10$$

Subscripts after any figure, $0_8$, $9_8$, etc., mean that that figure is to be repeated the indicated number of times.

**Specific Gravities at $\frac{60°}{60°}$F Corresponding to Degrees Baumé for Liquids Lighter than Water and Weights per U.S. Gallon at 60°F**

$$\left(\text{Calculated from the formula, specific gravity} \frac{60°}{60°}\text{ F} = \frac{140}{130 + \text{deg Baumé}}\right)$$

| Degrees Baumé | Specific gravity | Lb per gallon | Degrees Baumé | Specific gravity | Lb per gallon | Degrees Baumé | Specific gravity | Lb per gallon | Degrees Baumé | Specific gravity | Lb per gallon |
|---|---|---|---|---|---|---|---|---|---|---|---|
| 10.0 | 1.0000 | 8.328 | 33.0 | 0.8589 | 7.152 | 55.0 | 0.7568 | 6.300 | 78.0 | 0.6731 | 5.602 |
| 11.0 | 0.9929 | 8.269 | 34.0 | 0.8537 | 7.108 | 56.0 | 0.7527 | 6.266 | 79.0 | 0.6699 | 5.576 |
| 12.0 | 0.9859 | 8.211 | 35.0 | 0.8485 | 7.065 | 57.0 | 0.7487 | 6.233 | 80.0 | 0.6667 | 5.549 |
| 13.0 | 0.9790 | 8.153 | 36.0 | 0.8434 | 7.022 | 58.0 | 0.7447 | 6.199 | 81.0 | 0.6635 | 5.522 |
| 14.0 | 0.9722 | 8.096 | 37.0 | 0.8383 | 6.980 | 59.0 | 0.7407 | 6.166 | 82.0 | 0.6604 | 5.497 |
| 15.0 | 0.9655 | 8.041 | 38.0 | 0.8333 | 6.939 | 60.0 | 0.7368 | 6.134 | 83.0 | 0.6573 | 5.471 |
| 16.0 | 0.9589 | 7.986 | 39.0 | 0.8284 | 6.898 | 61.0 | 0.7330 | 6.102 | 84.0 | 0.6542 | 5.445 |
| 17.0 | 0.9524 | 7.931 | 40.0 | 0.8235 | 6.857 | 62.0 | 0.7292 | 6.070 | 85.0 | 0.6512 | 5.420 |
| 18.0 | 0.9459 | 7.877 | 41.0 | 0.8187 | 6.817 | 63.0 | 0.7254 | 6.038 | 86.0 | 0.6482 | 5.395 |
| 19.0 | 0.9396 | 7.825 | 42.0 | 0.8140 | 6.777 | 64.0 | 0.7216 | 6.007 | 87.0 | 0.6452 | 5.370 |
| 20.0 | 0.9333 | 7.772 | 43.0 | 0.8092 | 6.738 | 65.0 | 0.7179 | 5.976 | 88.0 | 0.6422 | 5.345 |
| 21.0 | 0.9272 | 7.721 | 44.0 | 0.8046 | 6.699 | 66.0 | 0.7143 | 5.946 | 89.0 | 0.6393 | 5.320 |
| 22.0 | 0.9211 | 7.670 | 45.0 | 0.8000 | 6.661 | 67.0 | 0.7107 | 5.916 | 90.0 | 0.6364 | 5.296 |
| 23.0 | 0.9150 | 7.620 | 46.0 | 0.7955 | 6.623 | 68.0 | 0.7071 | 5.886 | 91.0 | 0.6335 | 5.272 |
| 24.0 | 0.9091 | 7.570 | 47.0 | 0.7910 | 6.586 | 69.0 | 0.7035 | 5.856 | 92.0 | 0.6306 | 5.248 |
| 25.0 | 0.9032 | 7.522 | 48.0 | 0.7865 | 6.548 | 70.0 | 0.7000 | 5.827 | 93.0 | 0.6278 | 5.225 |
| 26.0 | 0.8974 | 7.473 | 49.0 | 0.7821 | 6.511 | 71.0 | 0.6965 | 5.798 | 94.0 | 0.6250 | 5.201 |
| 27.0 | 0.8917 | 7.425 | 50.0 | 0.7778 | 6.476 | 72.0 | 0.6931 | 5.769 | 95.0 | 0.6222 | 5.178 |
| 28.0 | 0.8861 | 7.378 | 51.0 | 0.7735 | 6.440 | 73.0 | 0.6897 | 5.741 | 96.0 | 0.6195 | 5.155 |
| 29.0 | 0.8805 | 7.332 | 52.0 | 0.7692 | 6.404 | 74.0 | 0.6863 | 5.712 | 97.0 | 0.6167 | 5.132 |
| 30.0 | 0.8750 | 7.286 | 53.0 | 0.7650 | 6.369 | 75.0 | 0.6829 | 5.685 | 98.0 | 0.6140 | 5.110 |
| 31.0 | 0.8696 | 7.241 | 54.0 | 0.7609 | 6.334 | 76.0 | 0.6796 | 5.657 | 99.0 | 0.6114 | 5.088 |
| 32.0 | 0.8642 | 7.196 | | | | 77.0 | 0.6763 | 5.629 | 100.0 | 0.6087 | 5.066 |

**Specific Gravities at $\frac{60°}{60°}$ F Corresponding to Degrees Baumé for Liquids Heavier than Water**

$$\left(\text{Calculated from the formula, specific gravity} \frac{60°}{60°}\text{ F} = \frac{145}{145 - \text{deg Baumé}}\right)$$

| Degrees Baumé | Specific gravity | Degrees Baumé | Specific gravity | Degrees Baumé | Specific gravity | Degrees Baumé | Specific gravity | Degrees Baumé | Specific gravity | Degrees Baumé | Specific gravity |
|---|---|---|---|---|---|---|---|---|---|---|---|
| 0 | 1.0000 | 12 | 1.0902 | 24 | 1.1983 | 36 | 1.3303 | 48 | 1.4948 | 60 | 1.7059 |
| 1 | 1.0069 | 13 | 1.0985 | 25 | 1.2083 | 37 | 1.3426 | 49 | 1.5104 | 61 | 1.7262 |
| 2 | 1.0140 | 14 | 1.1069 | 26 | 1.2185 | 38 | 1.3551 | 50 | 1.5263 | 62 | 1.7470 |
| 3 | 1.0211 | 15 | 1.1154 | 27 | 1.2288 | 39 | 1.3679 | 51 | 1.5426 | 63 | 1.7683 |
| 4 | 1.0284 | 16 | 1.1240 | 28 | 1.2393 | 40 | 1.3810 | 52 | 1.5591 | 64 | 1.7901 |
| 5 | 1.0357 | 17 | 1.1328 | 29 | 1.2500 | 41 | 1.3942 | 53 | 1.5761 | 65 | 1.8125 |
| 6 | 1.0432 | 18 | 1.1417 | 30 | 1.2609 | 42 | 1.4078 | 54 | 1.5934 | 66 | 1.8354 |
| 7 | 1.0507 | 19 | 1.1508 | 31 | 1.2719 | 43 | 1.4216 | 55 | 1.6111 | 67 | 1.8590 |
| 8 | 1.0584 | 20 | 1.1600 | 32 | 1.2832 | 44 | 1.4356 | 56 | 1.6292 | 68 | 1.8831 |
| 9 | 1.0662 | 21 | 1.1694 | 33 | 1.2946 | 45 | 1.4500 | 57 | 1.6477 | 69 | 1.9079 |
| 10 | 1.0741 | 22 | 1.1789 | 34 | 1.3063 | 46 | 1.4646 | 58 | 1.6667 | 70 | 1.9333 |
| 11 | 1.0821 | 23 | 1.1885 | 35 | 1.3182 | 47 | 1.4796 | 59 | 1.6860 | ...... | ........ |

## Length Equivalents

| Centimetres | Inches | Feet | Yards | Metres | Chains | Kilometres | Miles |
|---|---|---|---|---|---|---|---|
| 1 | 0.3937 <br> $\overline{1}$.59517 | 0.03281 <br> $\overline{2}$.51598 | 0.01094 <br> $\overline{2}$.03886 | 0.01 <br> $\overline{2}$.00000 | $0.0_3 4971$ <br> $\overline{4}$.69644 | $10^{-5}$ <br> $\overline{5}$.00000 | $0.0_6 6214$ <br> $\overline{6}$.79335 |
| 2.540 <br> 0.40483 | 1 | 0.08333 <br> $\overline{2}$.92082 | 0.02778 <br> $\overline{2}$.44370 | 0.0254 <br> $\overline{2}$.40483 | 0.001263 <br> $\overline{3}$.10127 | $0.0_4 254$ <br> $\overline{5}$.40483 | $0.0_4 1578$ <br> $\overline{5}$.19818 |
| 30.48 <br> 1.48401 | 12 <br> 1.07918 | 1 | 0.3333 <br> $\overline{1}$.52288 | 0.3048 <br> $\overline{1}$.48401 | 0.01515 <br> $\overline{2}$.18046 | $0.0_3 3048$ <br> $\overline{4}$.48401 | $0.0_3 1894$ <br> $\overline{4}$.27736 |
| 91.44 <br> 1.96114 | 36 <br> 1.55630 | 3 <br> 0.47712 | 1 | 0.9144 <br> $\overline{1}$.96114 | 0.04545 <br> $\overline{2}$.65758 | $0.0_3 9144$ <br> $\overline{4}$.96114 | $0.0_3 5682$ <br> $\overline{4}$.75449 |
| 100 <br> 2.00000 | 39.37 <br> 1.59517 | 3.281 <br> 0.51598 | 1.0936 <br> 0.03886 | 1 | 0.04971 <br> $\overline{2}$.69644 | 0.001 <br> $\overline{3}$.00000 | $0.0_6 6214$ <br> $\overline{4}$.79335 |
| 2012 <br> 3.30356 | 792 <br> 2.89873 | 66 <br> 1.81954 | 22 <br> 1.34242 | 20.12 <br> 1.30356 | 1 | 0.02012 <br> $\overline{2}$.30356 | 0.0125 <br> $\overline{2}$.09691 |
| 100000 <br> 5.00000 | 39370 <br> 4.59517 | 3281 <br> 3.51598 | 1093.6 <br> 3.03886 | 1000 <br> 3.00000 | 49.71 <br> 1.69644 | 1 | 0.6214 <br> $\overline{1}$.79335 |
| 160934 <br> 5.20665 | 63360 <br> 4.80182 | 5280 <br> 3.72263 | 1760 <br> 3.24551 | 1609 <br> 3.20665 | 80 <br> 1.90309 | 1.609 <br> 0.20665 | 1 |

(As used by metrology laboratories for precise measurements, including measurements of surface texture)*

| Angstrom units Å | Surface texture, (U.S.), microinch μin | Light bands,† monochromatic helium light count‡ | Surface texture foreign, μm | Precision measurements,§ 0.0001 in | Close-tolerance measurements, 0.001 in (mils) | Metric unit, mm | USCS unit, in |
|---|---|---|---|---|---|---|---|
| 1 | 0.003937 <br> $\overline{3}$.59517 | 0.0003404 <br> $\overline{4}$.53199 | 0.0001 <br> $\overline{4}$.0000 | $0.0_4 3937$ <br> $\overline{5}$.59517 | $0.0_5 3937$ <br> $\overline{6}$.59517 | $0.0_6 1$ <br> $\overline{7}$.00000 | $0.0_8 3937$ <br> $\overline{9}$.59517 |
| 254 <br> 2.40483 | 1 | 0.086 <br> $\overline{2}$.93682 | 0.0254 <br> $\overline{2}$.40483 | 0.01 <br> $\overline{2}$.00000 | 0.001 <br> $\overline{3}$.00000 | $0.0_4 254$ <br> $\overline{5}$.40483 | $0.0_5 1$ <br> $\overline{6}$.00000 |
| 2937.5 <br> 3.46797 | 11.566 <br> 1.06318 | 1 | 0.29375 <br> $\overline{1}$.46797 | 0.11566 <br> $\overline{1}$.06318 | 0.011566 <br> $\overline{2}$.06318 | $0.0_3 29375$ <br> $\overline{4}$.46797 | $0.0_4 11566$ <br> $\overline{5}$.06318 |
| 10,000 <br> 4.00000 | 39.37 <br> 1.59517 | 3.404 <br> 0.53199 | 1 | 0.3937 <br> $\overline{1}$.59517 | 0.03937 <br> $\overline{2}$.59517 | 0.001 <br> $\overline{3}$.00000 | $0.0_4 3937$ <br> $\overline{5}$.59517 |
| 25,400 <br> 4.40483 | 100 <br> 2.00000 | 8.646 <br> 0.93682 | 2.54 <br> 0.40483 | 1 | 0.1 <br> $\overline{1}$.00000 | 0.00254 <br> $\overline{3}$.40483 | 0.0001 <br> $\overline{4}$.00000 |
| 254,000 <br> 5.40483 | 1000 <br> 3.00000 | 86.46 <br> 1.93682 | 25.4 <br> 1.40483 | 10 <br> 1.00000 | 1 | 0.0254 <br> $\overline{2}$.40483 | 0.001 <br> $\overline{3}$.00000 |
| 10,000,000 <br> 7.00000 | 39,370 <br> 4.59517 | 3404 <br> 3.53199 | 1000 <br> 3.00000 | 393.7 <br> 2.59517 | 39.37 <br> 1.59517 | 1 | 0.03937 <br> $\overline{2}$.59517 |
| 254,000,000 <br> 8.40483 | 1,000,000 <br> 6.00000 | 86,460 <br> 4.93682 | 25,400 <br> 4.40483 | 10,000 <br> 4.00000 | 1000 <br> 3.00000 | 25.4 <br> 1.40483 | 1 |

*Computed by J. A. Broadston.

†One light band equals one-half corresponding wavelength. Visible-light wavelengths range from red at 6500Å to violet at 4100Å.

‡One helium light band = 0.000011661 in = 2937.5Å; one krypton 86 light band = 0.0000119 in = 3022.5Å; one mercury 198 light band = 0.00001075 in = 2730Å.

§The designations "precision measurements," etc., are not necessarily used in all metrology laboratories.

**Conversion of Lengths***

| | Inches to milli- metres | Milli- metres to inches | Feet to metres | Metres to feet | Yards to metres | Metres to yards | Miles to kilo- metres | Kilo metres to miles |
|---|---|---|---|---|---|---|---|---|
| 1 | 25.40 | 0.03937 | 0.3048 | 3.281 | 0.9144 | 1.094 | 1.609 | 0.6214 |
| 2 | 50.80 | 0.07874 | 0.6096 | 6.562 | 1.829 | 2.187 | 3.219 | 1.243 |
| 3 | 76.20 | 0.1181 | 0.9144 | 9.843 | 2.743 | 3.281 | 4.828 | 1.864 |
| 4 | 101.60 | 0.1575 | 1.219 | 13.12 | 3.658 | 4.374 | 6.437 | 2.485 |
| 5 | 127.00 | 0.1969 | 1.524 | 16.40 | 4.572 | 5.468 | 6.047 | 3.107 |
| 6 | 152.40 | 0.2362 | 1.829 | 19.69 | 5.486 | 6.562 | 9.656 | 3.728 |
| 7 | 177.80 | 0.2756 | 2.134 | 22.97 | 6.401 | 7.655 | 11.27 | 4.350 |
| 8 | 203.20 | 0.3150 | 2.438 | 26.25 | 7.315 | 8.749 | 12.87 | 4.971 |
| 9 | 228.60 | 0.3543 | 2.743 | 29.53 | 8.230 | 9.843 | 14.48 | 5.592 |

*EXAMPLE: 1 in = 25.40 mm.

Common fractions of an inch to millimetres (from 1/64 to 1 in)

| 64ths | Milli- metres | 64ths | Milli- metres | 64ths | Milli- metres | 64ths | Milli- metres | 64ths | Milli- metres | 64ths | Milli- metres |
|---|---|---|---|---|---|---|---|---|---|---|---|
| 1 | 0.397 | 13 | 5.159 | 25 | 9.922 | 37 | 14.684 | 49 | 19.447 | 57 | 22.622 |
| 2 | 0.794 | 14 | 5.556 | 26 | 10.319 | 38 | 15.081 | 50 | 19.844 | 58 | 23.019 |
| 3 | 1.191 | 15 | 5.953 | 27 | 10.716 | 39 | 15.478 | 51 | 20.241 | 59 | 23.416 |
| 4 | 1.588 | 16 | 6.350 | 28 | 11.112 | 40 | 15.875 | 52 | 20.638 | 60 | 23.812 |
| 5 | 1.984 | 17 | 6.747 | 29 | 11.509 | 41 | 16.272 | 53 | 21.034 | 61 | 24.209 |
| 6 | 2.381 | 18 | 7.144 | 30 | 11.906 | 42 | 16.669 | 54 | 21.431 | 62 | 24.606 |
| 7 | 2.778 | 19 | 7.541 | 31 | 12.303 | 43 | 17.066 | 55 | 21.828 | 63 | 25.003 |
| 8 | 3.175 | 20 | 7.938 | 32 | 12.700 | 44 | 17.462 | 56 | 22.225 | 64 | 25.400 |
| 9 | 3.572 | 21 | 8.334 | 33 | 13.097 | 45 | 17.859 | | | | |
| 10 | 3.969 | 22 | 8.731 | 34 | 13.494 | 46 | 18.256 | | | | |
| 11 | 4.366 | 23 | 9.128 | 35 | 13.891 | 47 | 18.653 | | | | |
| 12 | 4.762 | 24 | 9.525 | 36 | 14.288 | 48 | 19.050 | | | | |

Decimals of an inch to millimetres (from 0.01 to 0.99 in)

| | 0 | 1 | 2 | 3 | 4 | 5 | 6 | 7 | 8 | 9 |
|---|---|---|---|---|---|---|---|---|---|---|
| .0 | | 0.254 | 0.508 | 0.762 | 1.016 | 1.270 | 1.524 | 1.778 | 2.032 | 2.286 |
| .1 | 2.540 | 2.794 | 3.048 | 3.302 | 3.556 | 3.810 | 4.064 | 4.318 | 4.572 | 4.826 |
| .2 | 5.080 | 5.334 | 5.588 | 5.842 | 6.096 | 6.350 | 6.604 | 6.858 | 7.112 | 7.366 |
| .3 | 7.620 | 7.874 | 8.128 | 8.382 | 8.636 | 8.890 | 9.144 | 9.398 | 9.652 | 9.906 |
| .4 | 10.160 | 10.414 | 10.668 | 10.922 | 11.176 | 11.430 | 11.684 | 11.938 | 12.192 | 12.446 |
| .5 | 12.700 | 12.954 | 13.208 | 13.462 | 13.716 | 13.970 | 14.224 | 14.478 | 14.732 | 14.986 |
| .6 | 15.240 | 15.494 | 15.748 | 16.002 | 16.256 | 16.510 | 16.764 | 17.018 | 17.272 | 17.526 |
| .7 | 17.780 | 18.034 | 18.288 | 18.542 | 18.796 | 19.050 | 19.304 | 19.558 | 19.812 | 20.066 |
| .8 | 20.320 | 20.574 | 20.828 | 21.082 | 21.336 | 21.590 | 21.844 | 22.098 | 22.352 | 22.606 |
| .9 | 22.860 | 23.114 | 23.368 | 23.622 | 23.876 | 24.130 | 24.384 | 24.638 | 24.892 | 25.146 |

Millimetres to decimals of an inch (from 1 to 99 mm)

| | 0. | 1. | 2. | 3. | 4. | 5. | 6. | 7. | 8. | 9. |
|---|---|---|---|---|---|---|---|---|---|---|
| 0 | | 0.0394 | 0.0787 | 0.1181 | 0.1575 | 0.1969 | 0.2362 | 0.2756 | 0.3150 | 0.3543 |
| 1 | 0.3937 | 0.4331 | 0.4724 | 0.5118 | 0.5512 | 0.5906 | 0.6299 | 0.6693 | 0.7087 | 0.7480 |
| 2 | 0.7874 | 0.8268 | 0.8661 | 0.9055 | 0.9449 | 0.9843 | 1.0236 | 1.0630 | 1.1024 | 1.1417 |
| 3 | 1.1811 | 1.2205 | 1.2598 | 1.2992 | 1.3386 | 1.3780 | 1.4173 | 1.4567 | 1.4961 | 1.5354 |
| 4 | 1.5748 | 1.6142 | 1.6535 | 1.6929 | 1.7323 | 1.7717 | 1.8110 | 1.8504 | 1.8898 | 1.9291 |
| 5 | 1.9685 | 2.0079 | 2.0472 | 2.0866 | 2.1260 | 2.1654 | 2.2047 | 2.2441 | 2.2835 | 2.3228 |
| 6 | 2.3622 | 2.4016 | 2.4409 | 2.4803 | 2.5197 | 2.5591 | 2.5984 | 2.6378 | 2.6772 | 2.7165 |
| 7 | 2.7559 | 2.7953 | 2.8346 | 2.8740 | 2.9134 | 2.9528 | 2.9921 | 3.0315 | 3.0709 | 3.1102 |
| 8 | 3.1496 | 3.1890 | 3.2283 | 3.2677 | 3.3071 | 3.3465 | 3.3858 | 3.4252 | 3.4646 | 3.5039 |
| 9 | 3.5433 | 3.5827 | 3.6220 | 3.6614 | 3.7008 | 3.7402 | 3.7795 | 3.8189 | 3.8583 | 3.8976 |

## Area Equivalents
(1 hectare = 100 ares = 10,000 centiares or square metres)

| Square metres | Square inches | Square feet | Square yards | Square rods | Square chains | Roods | Acres | Square miles or sections |
|---|---|---|---|---|---|---|---|---|
| **1** | 1550<br>3.19033 | 10.76<br>1.03197 | 1.196<br>0.07773 | 0.0395<br>$\bar{2}$.59700 | 0.002471<br>$\bar{3}$.39288 | $0.0_39884$<br>$\bar{3}$.99495 | $0.0_32471$<br>$\bar{4}$.39288 | $0.0_63861$<br>$\bar{7}$.58670 |
| $0.0_36452$<br>$\bar{4}$.80967 | **1** | 0.006944<br>$\bar{3}$.84164 | $0.0_37716$<br>$\bar{4}$.88740 | $0.0_42551$<br>$\bar{5}$.40667 | $0.0_51594$<br>$\bar{6}$.20255 | $0.0_66377$<br>$\bar{7}$.80461 | $0.0_61594$<br>$\bar{7}$.20255 | $0.0_92491$<br>$\bar{10}$.39637 |
| 0.09290<br>$\bar{2}$.96803 | 144<br>2.15836 | **1** | 0.1111<br>$\bar{1}$.04576 | 0.003673<br>$\bar{3}$.56503 | $0.0_32296$<br>$\bar{4}$.36091 | $0.0_49183$<br>$\bar{5}$.96297 | $0.0_42296$<br>$\bar{4}$.36091 | $0.0_73587$<br>$\bar{8}$.55473 |
| 0.8361<br>$\bar{1}$.92227 | 1296<br>3.11260 | 9<br>0.95424 | **1** | 0.03306<br>$\bar{2}$.51927 | 0.002066<br>$\bar{3}$.31515 | $0.0_38264$<br>$\bar{4}$.91721 | 0.0002066<br>$\bar{4}$.31515 | $0.0_63228$<br>$\bar{7}$.50898 |
| 25.29<br>1.40300 | 39204<br>4.59333 | 272.25<br>2.43497 | 30.25<br>1.48072 | **1** | 0.0625<br>$\bar{2}$.79588 | 0.02500<br>$\bar{2}$.39794 | 0.00625<br>$\bar{3}$.79588 | $0.0_59766$<br>$\bar{6}$.98970 |
| 404.7<br>2.60712 | 627264<br>5.79745 | 4356<br>3.63909 | 484<br>2.68484 | 16<br>1.20412 | **1** | 0.4<br>$\bar{1}$.60206 | 0.1<br>$\bar{1}$.00000 | 0.0001562<br>$\bar{4}$.19382 |
| 1012<br>3.00506 | 1568160<br>6.19539 | 10890<br>4.03703 | 1210<br>3.08278 | 40<br>1.60206 | 2.5<br>0.39794 | **1** | 0.25<br>$\bar{1}$.39794 | $0.0_33906$<br>$\bar{4}$.59176 |
| 4047<br>3.60712 | 6272640<br>6.79745 | 43560<br>4.63909 | 4840<br>3.68484 | 160<br>2.20412 | 10<br>1.00000 | 4<br>0.60206 | **1** | 0.001562<br>$\bar{3}$.19382 |
| 2589988<br>6.41330 | | 27878400<br>7.44527 | 3097600<br>6.49102 | 102400<br>5.01030 | 6400<br>3.80618 | 2560<br>3.40824 | 640<br>2.80618 | **1** |

## Conversion of Areas*

| | Sq in to sq cm | Sq cm to sq in | Sq ft to sq m | Sq m to sq ft | Sq yd to sq m | Sq m to sq yd | Acres to hectares | Hectares to acres | Sq mi to sq km | Sq km to sq mi |
|---|---|---|---|---|---|---|---|---|---|---|
| 1 | 6.452 | 0.1550 | 0.0929 | 10.76 | 0.8361 | 1.196 | 0.4047 | 2.471 | 2.590 | 0.3861 |
| 2 | 12.90 | 0.3100 | 0.1858 | 21.53 | 1.672 | 2.392 | 0.8094 | 4.942 | 5.180 | 0.7722 |
| 3 | 19.35 | 0.4650 | 0.2787 | 32.29 | 2.508 | 3.588 | 1.214 | 7.413 | 7.770 | 1.158 |
| 4 | 25.81 | 0.6200 | 0.3716 | 43.06 | 3.345 | 4.784 | 1.619 | 9.884 | 10.360 | 1.544 |
| 5 | 32.26 | 0.7750 | 0.4645 | 53.82 | 4.181 | 5.980 | 2.023 | 12.355 | 12.950 | 1.931 |
| 6 | 38.71 | 0.9300 | 0.5574 | 64.58 | 5.017 | 7.176 | 2.428 | 14.826 | 15.540 | 2.317 |
| 7 | 45.16 | 1.085 | 0.6503 | 75.35 | 5.853 | 8.372 | 2.833 | 17.297 | 18.130 | 2.703 |
| 8 | 51.61 | 1.240 | 0.7432 | 86.11 | 6.689 | 9.568 | 3.237 | 19.768 | 20.720 | 3.089 |
| 9 | 58.06 | 1.395 | 0.8361 | 96.88 | 7.525 | 10.764 | 3.642 | 22.239 | 23.310 | 3.475 |

*EXAMPLE: 1 sq in = 6.452 sq cm.

## Volume and Capacity Equivalents

| Cubic inches | Cubic feet | Cubic yards | U.S. Apothecary fluid ounces | U.S. quarts Liquid | U.S. quarts Dry | U.S. gallons | U.S. bushels | Cubic decimetres or litres |
|---|---|---|---|---|---|---|---|---|
| 1 | 0.0$_3$5787 $\bar{4}$.76246 | 0.0$_4$2143 $\bar{5}$.33109 | 0.5541 $\bar{1}$.74360 | 0.01732 $\bar{2}$.23845 | 0.01488 $\bar{2}$.17263 | 0.0$_2$4329 $\bar{3}$.63639 | 0.0$_3$4650 $\bar{4}$.66748 | 0.01639 $\bar{2}$.21450 |
| 1728 3.23754 | 1 | 0.03704 $\bar{2}$.56864 | 957.5 2.98114 | 29.92 1.47599 | 25.71 1.41017 | 7.481 0.87393 | 0.8036 $\bar{1}$.90502 | 28.32 1.45205 |
| 46656 4.66891 | 27 1.43136 | 1 | 25853 4.41251 | 807.9 2.90736 | 694.3 2.84153 | 202.2 2.30530 | 21.70 1.33638 | 764.6 2.88341 |
| 1.805 0.25640 | 0.001044 $\bar{3}$.01886 | 0.0$_4$3868 $\bar{5}$.58749 | 1 | 0.03125 $\bar{2}$.49485 | 0.02686 $\bar{2}$.42903 | 0.007812 $\bar{3}$.89279 | 0.0$_3$8392 $\bar{4}$.92388 | 0.02957 $\bar{2}$.47091 |
| 57.75 1.76155 | 0.03342 $\bar{2}$.52401 | 0.001238 $\bar{3}$.09264 | 32 1.50515 | 1 | 0.8594 $\bar{1}$.93418 | 0.25 $\bar{1}$.39794 | 0.02686 $\bar{2}$.42903 | 0.9464 $\bar{1}$.97606 |
| 67.20 1.82737 | 0.03889 $\bar{2}$.58983 | 0.001440 $\bar{3}$.15847 | 37.24 1.57097 | 1.164 0.06582 | 1 | 0.2909 $\bar{1}$.46376 | 0.03125 $\bar{2}$.49485 | 1.101 0.04187 |
| 231 2.36361 | 0.1337 $\bar{1}$.12607 | 0.004951 $\bar{3}$.69470 | 128 2.10721 | 4 0.60206 | 3.437 0.53624 | 1 | 0.1074 $\bar{1}$.03109 | 3.785 0.57812 |
| 2150 3.33252 | 1.244 0.09498 | 0.04609 $\bar{2}$.66362 | 1192 3.07612 | 37.24 1.57097 | 32 1.50515 | 9.309 0.96891 | 1 | 35.24 1.54696 |
| 61.02 1.78550 | 0.03531 $\bar{2}$.54795 | 0.001308 $\bar{3}$.11659 | 33.81 1.52909 | 1.057 0.02394 | 0.9081 $\bar{1}$.95812 | 0.2642 $\bar{1}$.42188 | 0.02838 $\bar{2}$.45297 | 1 |

## Conversion of Volumes or Cubic Measure*

| | Cu in to ml | ml to cu in | Cu ft to cu m | Cu m to cu ft | Cu yd to cu m | Cu m to cu yd | Gallons to cu ft | Cu ft to gallons |
|---|---|---|---|---|---|---|---|---|
| 1 | 16.39 | 0.06102 | 0.02832 | 35.31 | 0.7646 | 1.308 | 0.1337 | 7.481 |
| 2 | 32.77 | 0.1220 | 0.05663 | 70.63 | 1.529 | 2.616 | 0.2674 | 14.96 |
| 3 | 49.16 | 0.1831 | 0.08495 | 105.9 | 2.294 | 3.924 | 0.4010 | 22.44 |
| 4 | 65.55 | 0.2441 | 0.1133 | 141.3 | 3.058 | 5.232 | 0.5347 | 29.92 |
| 5 | 81.94 | 0.3051 | 0.1416 | 176.6 | 3.823 | 6.540 | 0.6684 | 37.40 |
| 6 | 98.32 | 0.3661 | 0.1699 | 211.9 | 4.587 | 7.848 | 0.8021 | 44.88 |
| 7 | 114.7 | 0.4272 | 0.1982 | 247.2 | 5.352 | 9.156 | 0.9358 | 52.36 |
| 8 | 131.1 | 0.4882 | 0.2265 | 282.5 | 6.116 | 10.46 | 1.069 | 59.84 |
| 9 | 147.5 | 0.5492 | 0.2549 | 317.8 | 6.881 | 11.77 | 1.203 | 67.32 |

*EXAMPLE: 1 cu in = 16.39 ml.

## Conversion of Volumes or Capacities*

| | Fluid ounces to ml | ml to fluid ounces | Liquid pints to litres | Litres to liquid pints | Liquid quarts to litres | Litres to liquid quarts | Gallons to litres | Litres to gallons | Bushels to hecto-litres | Hecto-litres to bushels |
|---|---|---|---|---|---|---|---|---|---|---|
| 1 | 29.57 | 0.03381 | 0.4732 | 2.113 | 0.9463 | 1.057 | 3.785 | 0.2642 | 0.3524 | 2.838 |
| 2 | 59.15 | 0.06763 | 0.9463 | 4.227 | 1.893 | 2.113 | 7.571 | 0.5284 | 0.7048 | 5.676 |
| 3 | 88.72 | 0.1014 | 1.420 | 6.340 | 2.839 | 3.170 | 11.36 | 0.7925 | 1.057 | 8.513 |
| 4 | 118.3 | 0.1353 | 1.893 | 8.454 | 3.785 | 4.227 | 15.14 | 1.057 | 1.410 | 11.35 |
| 5 | 147.9 | 0.1691 | 2.366 | 10.57 | 4.732 | 5.284 | 18.93 | 1.321 | 1.762 | 14.19 |
| 6 | 177.4 | 0.2092 | 2.839 | 12.68 | 5.678 | 6.340 | 22.71 | 1.585 | 2.114 | 17.03 |
| 7 | 207.0 | 0.2367 | 3.312 | 14.79 | 6.624 | 7.397 | 26.50 | 1.849 | 2.467 | 19.86 |
| 8 | 236.6 | 0.2705 | 3.785 | 16.91 | 7.571 | 8.454 | 30.28 | 2.113 | 2.819 | 22.70 |
| 9 | 266.2 | 2.3043 | 4.259 | 19.02 | 8.517 | 9.510 | 34.07 | 2.378 | 3.171 | 25.54 |

*EXAMPLE; 1 fluid oz = 29.57 ml.

## Mass Equivalents

| Kilograms | Grains | Ounces | | Pounds | | Tons | | |
|---|---|---|---|---|---|---|---|---|
| | | Troy and apoth | Avoir-dupois | Troy and apoth | Avoir-dupois | Short | Long | Metric |
| **1** | 15432 | 32.15 | 35.27 | 2.6792 | 2.205 | $0.0_21102$ | $0.0_39842$ | 0.001 |
| | 4.18843 | 1.50719 | 1.54745 | 0.42801 | 0.34333 | $\bar{3}.04230$ | $\bar{4}.99309$ | $\bar{3}.00000$ |
| 0.0₆6480 | **1** | $0.0_22083$ | $0.0_22286$ | $0.0_31736$ | $0.0_31429$ | $0.0_77143$ | $0.0_76378$ | $0.0_76480$ |
| $\bar{5}.81157$ | | $\bar{3}.31876$ | $\bar{3}.35902$ | $\bar{4}.23958$ | $\bar{4}.15490$ | $\bar{8}.85387$ | $\bar{8}.80465$ | $\bar{8}.81157$ |
| 0.03110 | 480 | **1** | 1.09714 | 0.08333 | 0.06857 | $0.0_43429$ | $0.0_43061$ | $0.0_43110$ |
| $\bar{2}.49281$ | 2.68124 | | 0.04026 | $\bar{2}.92082$ | $\bar{2}.83614$ | $\bar{5}.53511$ | $\bar{5}.48590$ | $\bar{5}.49281$ |
| 0.02835 | 437.5 | 0.9115 | **1** | 0.07595 | 0.0625 | $0.0_43125$ | $0.0_42790$ | $0.0_42835$ |
| $\bar{2}.45255$ | 2.64098 | $\bar{1}.95974$ | | $\bar{2}.88056$ | $\bar{2}.79588$ | $\bar{5}.49485$ | $\bar{5}.44563$ | $\bar{5}.45255$ |
| 0.3732 | 5760 | 12 | 13.17 | **1** | 0.8229 | $0.0_34114$ | $0.0_33673$ | $0.0_33732$ |
| $\bar{1}.57199$ | 3.76042 | 1.07918 | 1.11944 | | $\bar{1}.91532$ | $\bar{4}.61429$ | $\bar{4}.56508$ | $\bar{4}.57199$ |
| 0.4536 | 7000 | 14.58 | 16 | 1.215 | **1** | 0.0005 | $0.0_34464$ | $0.0_34536$ |
| $\bar{1}.65667$ | 3.84510 | 1.16386 | 1.20412 | 0.08468 | | $\bar{4}.69897$ | $\bar{4}.64975$ | $\bar{4}.65667$ |
| 907.2 | 140₆ | 29167 | 320₂ | 2431 | 2000 | **1** | 0.8929 | 0.9072 |
| 2.95770 | 7.14613 | 4.46489 | 4.50515 | 3.38571 | 3.30103 | | $\bar{1}.95078$ | $\bar{1}.95770$ |
| 1016 | 15680₄ | 32667 | 35840 | 2722 | 2240 | 1.12 | **1** | 1.016 |
| 3.00691 | 7.19535 | 4.51411 | 4.55437 | 3.43492 | 3.35025 | 0.04922 | | 0.00691 |
| 1000 | 15432356 | 32151 | 35274 | 2679 | 2205 | 1.102 | 0.9842 | **1** |
| 3.00000 | 7.18843 | 4.50719 | 4.54745 | 3.42801 | 3.34333 | 0.04230 | $\bar{1}.99309$ | |

## Conversion of Masses*

| | Grains to grams | Grams to grains | Ounces (avdp) to grams | Grams to ounces (avdp) | Pounds (avdp) to kilo-grams | Kilo-grams to pounds (avdp) | Short tons (2000 lb) to metric tons | Metric tons (1000 kg) to short tons | Long tons (2240 lb) to metric tons | Metric tons to long tons |
|---|---|---|---|---|---|---|---|---|---|---|
| 1 | 0.06480 | 15.43 | 28.35 | 0.03527 | 0.4536 | 2.205 | 0.907 | 1.102 | 1.016 | 0.984 |
| 2 | 0.1296 | 30.86 | 56.70 | 0.07055 | 0.9072 | 4.409 | 1.814 | 2.205 | 2.032 | 1.968 |
| 3 | 0.1944 | 46.30 | 85.05 | 0.1058 | 1.361 | 6.614 | 2.722 | 3.307 | 3.048 | 2.953 |
| 4 | 0.2592 | 61.73 | 113.40 | 0.1411 | 1.814 | 8.818 | 3.629 | 4.409 | 4.064 | 3.937 |
| 5 | 0.3240 | 77.16 | 141.75 | 0.1764 | 2.268 | 11.02 | 4.536 | 5.512 | 5.080 | 4.921 |
| 6 | 0.3888 | 92.59 | 170.10 | 0.2116 | 2.722 | 13.23 | 5.443 | 6.614 | 6.096 | 5.905 |
| 7 | 0.4536 | 108.03 | 198.45 | 0.2469 | 3.175 | 15.43 | 6.350 | 7.716 | 7.112 | 6.889 |
| 8 | 0.5184 | 123.46 | 226.80 | 0.2822 | 3.629 | 17.64 | 7.257 | 8.818 | 8.128 | 7.874 |
| 9 | 0.5832 | 138.89 | 255.15 | 0.3175 | 4.082 | 19.84 | 8.165 | 9.921 | 9.144 | 8.858 |

*EXAMPLE: 1 grain = 0.06480 grams.

**Pressure Equivalents**

| Pascals N/m² | Bars 10⁵ N/m² | Pounds_f per in² | Atmospheres | Columns of mercury at temperature 0°C and $g = 9.80665$ m/s² | | Columns of water at temperature 15°C and $g = 9.80665$ m/s² | |
|---|---|---|---|---|---|---|---|
| | | | | cm | in | cm | in |
| **1** | **10⁻⁵** | **0.000145** | **0.00001** | **0.00075** | **0.000295** | **0.01021** | **0.00402** |
| | $\bar{5}$.00000 | $\bar{4}$.16148 | $\bar{6}$.99427 | $\bar{4}$.87510 | $\bar{4}$.47025 | $\bar{2}$.00884 | $\bar{3}$.60408 |
| **100000** | **1** | **14.504** | **0.9869** | **75.01** | **29.53** | **1020.7** | **401.8** |
| 5.00000 | | 1.16148 | $\bar{1}$.99427 | 1.87510 | 1.47025 | 3.00891 | 2.60408 |
| **6894.8** | **0.068948** | **1** | **0.06805** | **5.171** | **2.036** | **70.37** | **27.703** |
| 3.83852 | $\bar{2}$.83852 | | $\bar{2}$.83280 | 0.71360 | 0.30876 | 1.84738 | 1.44253 |
| **101326** | **1.0132** | **14.696** | **1** | **76.000** | **29.92** | **1034** | **407.1** |
| 5.00572 | 0.00572 | 1.16720 | | 1.88081 | 1.47598 | 3.01459 | 2.60975 |
| **1333** | **0.0133** | **0.1934** | **0.01316** | **1** | **0.3937** | **13.61** | **5.357** |
| 3.12498 | $\bar{2}$.12498 | $\bar{1}$.28640 | $\bar{2}$.11919 | | $\bar{1}$.59517 | 1.13378 | 0.72894 |
| **3386** | **0.03386** | **0.4912** | **0.03342** | **2.540** | **1** | **34.56** | **13.61** |
| 3.52975 | $\bar{2}$.52975 | $\bar{1}$.69124 | $\bar{2}$.52402 | 0.40484 | | 1.53861 | 1.13378 |
| **97.98** | **0.0009798** | **0.01421** | **0.000967** | **0.07349** | **0.02893** | **1** | **0.3937** |
| 1.99114 | $\bar{4}$.99114 | $\bar{2}$.15202 | $\bar{4}$.98541 | $\bar{2}$.86622 | $\bar{2}$.46139 | | $\bar{1}$.59517 |
| **248.9** | **0.002489** | **0.03609** | **0.002456** | **0.1867** | **0.07349** | **2.540** | **1** |
| 2.39598 | $\bar{3}$.39598 | $\bar{2}$.55745 | $\bar{3}$.39024 | $\bar{1}$.27106 | $\bar{2}$.86622 | 0.40484 | |

**Conversion of Pressures***

| | Lb/in² to bars | Bars to lb/in² | Lb/in² to atmospheres | Atmospheres to lb/in² | Bars to atmospheres | Atmospheres to bars |
|---|---|---|---|---|---|---|
| 1 | 0.06895 | 14.504 | 0.06805 | 14.696 | 0.98692 | 1.01325 |
| 2 | 0.13790 | 29.008 | 0.13609 | 29.392 | 1.9738 | 2.0265 |
| 3 | 0.20684 | 43.511 | 0.20414 | 44.098 | 2.9607 | 3.0397 |
| 4 | 0.27579 | 58.015 | 0.27218 | 58.784 | 3.9477 | 4.0530 |
| 5 | 0.34474 | 72.519 | 0.34823 | 73.480 | 4.9346 | 5.0663 |
| 6 | 0.41368 | 87.023 | 0.40826 | 88.176 | 5.9215 | 6.0795 |
| 7 | 0.48263 | 101.53 | 0.47632 | 102.87 | 6.9085 | 7.0927 |
| 8 | 0.55158 | 116.03 | 0.54436 | 117.57 | 7.8954 | 8.1060 |
| 9 | 0.62053 | 130.53 | 0.61241 | 132.26 | 8.8823 | 9.1192 |

*Example: 1 lb/in² = 0.06895 bar.

**Velocity Equivalents**

| Centi-metres per sec | Metres per sec | Metres per min | Kilo-metres per hour | Feet per sec | Feet per min | Miles per hour | Knots |
|---|---|---|---|---|---|---|---|
| 1 | 0.01 | 0.6<br>$\bar{1}$.77815 | 0.036<br>$\bar{2}$.55630 | 0.03281<br>$\bar{2}$.51598 | 1.9685<br>0.29414 | 0.02237<br>$\bar{2}$.34965 | 0.01944<br>$\bar{2}$.28866 |
| 100<br>2.00000 | 1 | 60<br>1.77815 | 3.6<br>0.55630 | 3.281<br>0.51598 | 196.85<br>2.29414 | 2.237<br>0.34965 | 1.944<br>0.28866 |
| 1.667<br>0.22185 | 0.01667<br>$\bar{2}$.22185 | 1 | 0.06<br>$\bar{2}$.77815 | 0.05468<br>$\bar{2}$.73783 | 3.281<br>0.51598 | 0.03728<br>$\bar{2}$.57150 | 0.03240<br>$\bar{2}$.51050 |
| 27.78<br>1.44370 | 0.2778<br>$\bar{1}$.44370 | 16.67<br>1.22185 | 1 | 0.9113<br>$\bar{1}$.95968 | 54.68<br>1.73783 | 0.6214<br>$\bar{1}$.79335 | 0.53996<br>$\bar{1}$.73236 |
| 30.48<br>1.48401 | 0.3048<br>$\bar{1}$.48401 | 18.29<br>1.26217 | 1.097<br>0.04032 | 1 | 60<br>1.77815 | 0.6818<br>$\bar{1}$.83367 | 0.59248<br>$\bar{1}$.77268 |
| 0.5080<br>$\bar{1}$.70586 | 0.005080<br>$\bar{3}$.70586 | 0.3048<br>$\bar{1}$.48401 | 0.01829<br>$\bar{2}$.26217 | 0.01667<br>$\bar{2}$.22185 | 1 | 0.01136<br>$\bar{2}$.05553 | 0.00987<br>$\bar{3}$.99453 |
| 44.70<br>1.65035 | 0.4470<br>$\bar{1}$.65035 | 26.82<br>1.42850 | 1.609<br>0.20670 | 1.467<br>0.16633 | 88<br>1.94448 | 1 | 0.86898<br>$\bar{1}$.93901 |
| 51.44<br>1.71133 | 0.5144<br>$\bar{1}$.71133 | 30.87<br>1.48949 | 1.852<br>0.26764 | 1.688<br>0.22732 | 101.3<br>2.00547 | 1.151<br>1.06100 | 1 |

**Conversion of Linear and Angular Velocities***

| | Cm per sec to feet per min | Feet per min to cm per sec | Cm per sec to miles per hour | Miles per hour to cm per sec | Feet per sec to miles per hour | Miles per hour to feet per sec | Radians per sec to rev per min | Rev per min to radians per sec |
|---|---|---|---|---|---|---|---|---|
| 1 | 1.97 | 0.508 | 0.0224 | 44.70 | 0.682 | 1.47 | 9.55 | 0.1047 |
| 2 | 3.94 | 1.016 | 0.0447 | 89.41 | 1.364 | 2.93 | 19.10 | 0.2094 |
| 3 | 5.91 | 1.524 | 0.0671 | 134.1 | 2.045 | 4.40 | 28.65 | 0.3142 |
| 4 | 7.87 | 2.032 | 0.0895 | 178.8 | 2.727 | 5.87 | 38.20 | 0.4189 |
| 5 | 9.84 | 2.540 | 0.1118 | 223.5 | 3.409 | 7.33 | 47.75 | 0.5236 |
| 6 | 11.81 | 3.048 | 0.1342 | 268.2 | 4.091 | 8.80 | 57.30 | 0.6283 |
| 7 | 13.78 | 3.556 | 0.1566 | 312.9 | 4.773 | 10.27 | 66.84 | 0.7330 |
| 8 | 15.75 | 4.064 | 0.1790 | 357.6 | 5.455 | 11.73 | 76.39 | 0.8378 |
| 9 | 17.72 | 4.572 | 0.2013 | 402.3 | 6.136 | 13.20 | 85.94 | 0.9425 |

*EXAMPLE: 1cm/s = 1.97 ft/min.

**Acceleration Equivalents**

| Centimetres per sec per sec | Metres per sec per sec | Metres per hr per sec | Kilometres per hr per sec | Feet per hr per sec | Feet per sec per sec | Feet per min per min | Miles per hr per sec | Knots per sec |
|---|---|---|---|---|---|---|---|---|
| **1** | 0.01 / $\overline{1}$.00000 | 36.00 / 1.55630 | 0.036 / $\overline{2}$.55630 | 118.1 / 2.07225 | 0.03281 / $\overline{2}$.51599 | 118.1 / 2.07225 | 0.02237 / $\overline{2}$.34965 | 0.01944 / $\overline{2}$.29865 |
| 100 / 2.00000 | **1** | 3600 / 3.55630 | 3.6 / 0.55630 | 11811 / 4.07225 | 3.281 / 0.51599 | 11811 / 4.07225 | 2.237 / 0.34965 | 1.944 / 0.29865 |
| 0.02778 / $\overline{2}$.44370 | 0.0002778 / $\overline{4}$.44370 | **1** | 0.001 / $\overline{3}$.00000 | 3.281 / 0.51599 | 0.0009113 / $\overline{4}$.95968 | 3.281 / 0.51599 | 0.0006214 / $\overline{4}$.79325 | 0.0005400 / $\overline{4}$.73235 |
| 27.78 / 1.44370 | 0.2778 / $\overline{1}$.44370 | 1000 / 3.00000 | **1** | 3281 / 3.51599 | 0.9113 / $\overline{1}$.95968 | 3281 / 3.51599 | 0.6214 / $\overline{1}$.79335 | 0.5400 / $\overline{1}$.73235 |
| 0.008467 / $\overline{3}$.92771 | 0.00008467 / $\overline{5}$.92771 | 0.3048 / $\overline{1}$.48401 | 0.0003048 / $\overline{4}$.48401 | **1** | 0.0002778 / $\overline{4}$.44370 | **1** | 0.0001894 / $\overline{4}$.27737 | 0.0001646 / $\overline{4}$.21640 |
| 30.48 / 1.48401 | 0.3048 / $\overline{1}$.48401 | 1097 / 3.04030 | 1.097 / 0.04030 | 3600 / 3.55630 | **1** | 3600 / 3.55630 | 0.6818 / $\overline{1}$.83366 | 0.4572 / $\overline{1}$.66008 |
| 0.008467 / $\overline{3}$.92771 | 0.00008467 / $\overline{5}$.92771 | 0.3048 / $\overline{1}$.48401 | 0.0003048 / $\overline{4}$.48401 | **1** | 0.0002778 / $\overline{4}$.44370 | **1** | 0.0001894 / $\overline{4}$.27737 | 0.0001646 / $\overline{4}$.21640 |
| 44.70 / 1.65035 | 0.4470 / $\overline{1}$.65035 | 1609 / 3.20665 | 1.609 / 0.20665 | 5280 / 3.72263 | 1.467 / 0.13636 | 5280 / 3.72263 | **1** | 0.8690 / $\overline{1}$.93901 |
| 51.44 / 1.71134 | 0.5144 / $\overline{1}$.71134 | 1852 / 3.26764 | 1.852 / 0.26764 | 6076 / 3.78362 | 1.688 / 0.22732 | 6076 / 3.78362 | 1.151 / 0.06099 | **1** |

**Conversion of Accelerations\***

| | Centimetres per sec per sec to ft per min per min | Kilometres per hr per sec to miles per hr per sec | Kilometres per hr per sec to knots per sec | Feet per sec per sec to miles per hr per sec | Feet per sec per sec to knots per sec | Feet per min per min to cm per sec per sec | Miles per hr per sec to kilometres per hr per sec | Miles per hr per sec to knots per sec | Knots per sec to miles per hr per sec | Knots per sec to kilometres per hr per sec |
|---|---|---|---|---|---|---|---|---|---|---|
| 1 | 118.1 | 0.6214 | 0.5400 | 0.6818 | 0.4572 | 0.008467 | 1.609 | 0.8690 | 1.151 | 1.852 |
| 2 | 236.2 | 1.243 | 1.080 | 1.364 | 0.9144 | 0.01693 | 3.219 | 1.738 | 2.302 | 3.704 |
| 3 | 354.3 | 1.864 | 1.620 | 2.045 | 1.372 | 0.02540 | 4.828 | 2.607 | 3.452 | 5.556 |
| 4 | 472.4 | 2.485 | 2.160 | 2.727 | 1.829 | 0.03387 | 6.437 | 3.476 | 4.603 | 7.408 |
| 5 | 590.6 | 3.107 | 2.700 | 3.409 | 2.286 | 0.04233 | 8.046 | 4.345 | 5.754 | 9.260 |
| 6 | 708.7 | 3.728 | 3.240 | 4.091 | 2.743 | 0.05080 | 9.656 | 5.214 | 6.905 | 11.11 |
| 7 | 826.8 | 4.350 | 3.780 | 4.772 | 3.200 | 0.05927 | 11.27 | 6.083 | 8.056 | 12.96 |
| 8 | 944.9 | 4.971 | 4.320 | 5.454 | 3.658 | 0.06774 | 12.87 | 6.952 | 9.206 | 14.82 |
| 9 | 1063 | 5.592 | 4.860 | 6.136 | 4.115 | 0.07620 | 14.48 | 7.821 | 10.36 | 16.67 |

\*EXAMPLE: 1 cm/s/s = 118.1 ft/min/min.

## Energy or Work Equivalents

| Joules or Newton·metre | Kilogramf· metres | Foot· poundsf | Kilo- watt- hours | Metric horse- power- hours | Horse- power· hours | Litre- atmos- pheres | Kilo- calories | British thermal units |
|---|---|---|---|---|---|---|---|---|
| **1** | 0.10197 $\bar{1}$.00848 | 0.7376 $\bar{1}$.86780 | $0.0_6$2778 $\bar{7}$.44370 | $0.0_6$3777 $\bar{7}$.57711 | $0.0_6$3725 $\bar{7}$.57113 | 0.009869 $\bar{3}$.99427 | $0.0_2$2388 $\bar{4}$.37809 | $0.0_2$9478 $\bar{4}$.97670 |
| 9.80665 0.9915207 | **1** | 7.233 0.85932 | $0.0_5$2724 $\bar{6}$.43521 | $0.0_5$37037 $\bar{6}$.56863 | $0.0_5$3653 $\bar{6}$.56265 | 0.09678 $\bar{2}$.98579 | 0.002342 $\bar{3}$.36961 | 0.009295 $\bar{3}$.96825 |
| 1.356 0.13220 | 0.1383 $\bar{1}$.14068 | **1** | $0.0_6$3766 $\bar{7}$.57590 | $0.0_6$51206 $\bar{7}$.70932 | $0.0_6$50505 $\bar{7}$.70333 | 0.01338 $\bar{2}$.12647 | $0.0_2$3238 $\bar{4}$.51029 | 0.001285 $\bar{3}$.10890 |
| $3.600 \times 10^6$ 6.55630 | $3.671 \times 10^5$ 5.56478 | $2.655 \times 10^6$ 6.42410 | **1** | 1.3596 0.13342 | 1.341 0.12743 | 35528 4.55057 | 859.9 2.93443 | 3412 3.53303 |
| $2.648 \times 10^6$ 6.42288 | 270000 5.43136 | $1.9529 \times 10^6$ 6.29068 | 0.7355 $\bar{1}$.86658 | **1** | 0.9863 $\bar{1}$.99401 | 26131 4.41715 | 632.4 2.80098 | 2510 3.39961 |
| $2.6845 \times 10^6$ 6.42887 | $2.7375 \times 10^5$ 5.43735 | $1.98 \times 10^6$ 6.29667 | 0.7457 $\bar{1}$.87356 | 1.0139 0.00598 | **1** | 26493 4.42314 | 641.2 2.80699 | 2544 3.40557 |
| 101.33 2.00573 | 10.333 1.01421 | 74.74 1.87353 | $0.0_4$2815 $\bar{5}$.44952 | $0.0_4$3827 $\bar{5}$.58284 | $0.0_4$3775 $\bar{5}$.57686 | **1** | 0.02420 $\bar{2}$.38382 | 0.09604 $\bar{2}$.98246 |
| 4186.8 3.62191 | 426.9 2.63036 | 3088 3.48971 | 0.001163 $\bar{3}$.06558 | 0.001581 $\bar{3}$.19902 | 0.001560 $\bar{3}$.19304 | 41.32 1.61618 | **1** | 3.968 0.59861 |
| 1055 3.02300 | 107.6 2.03178 | 778.2 2.89110 | $0.0_3$2931 $\bar{4}$.46697 | $0.0_3$3985 $\bar{4}$.60042 | $0.0_3$3930 $\bar{4}$.59444 | 10.41 1.01757 | 0.25200 $\bar{1}$.40139 | **1** |

## Conversion of Energy, Work, Heat*

| | Ft·lbf to joules | Joules to ft·lbf | Ft·lbf to Btu | Btu to ft·lbf | Kilogramf· metres to kilocalories | Kilocalories to kilogramf· metres | Joules to calories | Calories to joules |
|---|---|---|---|---|---|---|---|---|
| 1 | 1.3558 | 0.7376 | 0.001285 | 778.2 | 0.002342 | 426.9 | 0.2388 | 4.187 |
| 2 | 2.7116 | 1.4751 | 0.002570 | 1,556 | 0.004685 | 853.9 | 0.4777 | 8.374 |
| 3 | 4.0674 | 2.2127 | 0.003855 | 2,334 | 0.007027 | 1,281 | 0.7165 | 12.56 |
| 4 | 5.4232 | 2.9503 | 0.005140 | 3,113 | 0.009369 | 1,708 | 0.9554 | 16.75 |
| 5 | 6.7790 | 3.6879 | 0.006425 | 3,891 | 0.01172 | 2,135 | 1.194 | 20.93 |
| 6 | 8.1348 | 4.4254 | 0.007710 | 4,669 | 0.01405 | 2,562 | 1.433 | 25.12 |
| 7 | 9.4906 | 5.1630 | 0.008995 | 5,447 | 0.01640 | 2,989 | 1.672 | 29.31 |
| 8 | 10.8464 | 5.9006 | 0.01028 | 6,225 | 0.01874 | 3,415 | 1.911 | 33.49 |
| 9 | 12.2022 | 6.6381 | 0.01156 | 7,003 | 0.02108 | 3,842 | 2.150 | 37.68 |

*Example: 1 ft·lbf = 1.3558 J.

**Power Equivalents**

| Horse-power | Kilo-watts | Metric horse-power | Kgf·m per sec | Ft·lbf per sec | Kilo-calories per sec | Btu per sec |
|---|---|---|---|---|---|---|
| **1** | 0.7457 $\overline{1}$.87256 | 1.014 0.00599 | 76.04 1.88105 | 550 2.74036 | 0.1781 $\overline{1}$.25066 | 0.7068 $\overline{1}$.84936 |
| 1.341 0.12743 | **1** | 1.360 0.13343 | 102.0 2.00848 | 737.6 2.86780 | 0.2388 $\overline{1}$.37813 | 0.9478 $\overline{1}$.97673 |
| 0.9863 $\overline{1}$.99402 | 0.7355 $\overline{1}$.86658 | **1** | 75 1.87506 | 542.5 2.73438 | 0.1757 $\overline{1}$.24467 | 0.6971 $\overline{1}$.84328 |
| 0.01315 $\overline{2}$.11896 | 0.009807 $\overline{3}$.99152 | 0.01333 $\overline{2}$.12493 | **1** | 7.233 0.85932 | 0.002342 $\overline{3}$.36961 | 0.009295 $\overline{3}$.96825 |
| 0.00182 $\overline{3}$.25946 | 0.001356 $\overline{3}$.13220 | 0.00184 $\overline{3}$.26562 | 0.1383 $\overline{1}$.14067 | **1** | $0.0_33238$ $\overline{4}$.51029 | 0.001285 $\overline{3}$.10890 |
| 5.615 0.74934 | 4.187 0.62187 | 5.692 0.75530 | 426.9 2.63036 | 3088 3.48971 | **1** | 3.968 0.59861 |
| 1.415 0.15074 | 1.055 0.02320 | 1.434 0.15668 | 107.6 2.03178 | 778.2 2.89110 | 0.2520 $\overline{1}$.40138 | **1** |

**Conversion of Power***

| | Horsepower to kilowatts | Kilowatts to horsepower | Metric horsepower to kilowatts | Kilowatts to metric horsepower | Horsepower to metric horsepower | Metric horsepower to horsepower |
|---|---|---|---|---|---|---|
| 1 | 0.7457 | 1.341 | 0.7355 | 1.360 | 1.014 | 0.9863 |
| 2 | 1.491 | 2.682 | 1.471 | 2.719 | 2.028 | 1.973 |
| 3 | 2.237 | 4.023 | 2.206 | 4.079 | 3.042 | 2.959 |
| 4 | 2.983 | 5.364 | 2.942 | 5.438 | 4.055 | 3.945 |
| 5 | 3.729 | 6.705 | 3.677 | 6.798 | 5.069 | 4.932 |
| 6 | 4.474 | 8.046 | 4.412 | 8.158 | 6.083 | 5.918 |
| 7 | 5.220 | 9.387 | 5.147 | 9.520 | 7.097 | 6.904 |
| 8 | 5.966 | 10.73 | 5.883 | 10.88 | 8.111 | 7.891 |
| 9 | 6.711 | 12.07 | 6.618 | 12.24 | 9.125 | 8.877 |

*EXAMPLE: 1 hp = 0.7457 kW.

**Density Equivalents and Conversion Factors***

| Equivalents | | | | | Conversion factors | | | | |
|---|---|---|---|---|---|---|---|---|---|
| Grams per ml | Lb per cu in | Lb per cu ft | Short tons (2,000 lb) per cu yd | Lb per U.S. gal | | Grams per ml to lb per cu ft | Lb per cu ft to grams per ml | Grams per ml to short tons per cu yd | Short tons per cu yd to grams per ml |
| **1** | 0.03613 | 62.43 | 0.8428 | 8.345 | 1 | 62.43 | 0.01602 | 0.8428 | 1.187 |
| | $\overline{2}$.55787 | 1.79539 | $\overline{1}$.92572 | 0.92143 | 2 | 124.86 | 0.03204 | 1.6856 | 2.373 |
| 27.68 | **1** | 1728 | 23.33 | 231 | 3 | 187.28 | 0.04805 | 2.5283 | 3.560 |
| 1.44217 | | 3.23754 | 1.36792 | 2.36361 | 4 | 249.71 | 0.06407 | 3.3711 | 4.746 |
| 0.01602 | 0.0$_3$5787 | **1** | 0.0135 | 0.1337 | 5 | 312.14 | 0.08009 | 4.2139 | 5.933 |
| $\overline{2}$.20466 | $\overline{4}$.76245 | | $\overline{2}$.13033 | $\overline{1}$.12613 | 6 | 374.57 | 0.09611 | 5.0567 | 7.119 |
| 1.187 | 0.04287 | 74.07 | **1** | 9.902 | 7 | 437.00 | 0.11213 | 5.8995 | 8.306 |
| 0.07428 | $\overline{2}$.63212 | 1.86964 | | 0.99572 | 8 | 499.43 | 0.12814 | 6.7423 | 9.492 |
| 0.1198 | 0.004329 | 7.481 | 0.1010 | **1** | 9 | 561.85 | 0.14416 | 7.5850 | 10.679 |
| $\overline{1}$.07855 | $\overline{3}$.63639 | 0.87396 | $\overline{1}$.00432 | | 10 | 624.28 | 0.16018 | 8.4278 | 11.866 |

*EXAMPLE: 1 g per ml = 62.43 lb per cu ft.

**Thermal Conductivity**

| Calories per sec per sq cm per cm per deg C | Watts per sq cm per cm per deg C | Calories per hr per sq cm per cm per deg C | Btu per hr per sq ft per ft per deg F | Btu per day per sq ft per in. per deg F |
|---|---|---|---|---|
| 1 | 4.1868 | 3,600 | 241.9 | 69,670 |
| 0.2388 | 1 | 860 | 57.79 | 16,641 |
| 0.0002778 | 0.001163 | 1 | 0.0672 | 19.35 |
| 0.004134 | 0.01731 | 14.88 | 1 | 288 |
| 0.00001435 | 0.00006009 | 0.05167 | 0.00347 | 1 |

**Thermal Conductance**

| Calories per sec per sq cm per deg C | Watts per sq cm per deg C | Calories per hr per sq cm per deg C | Btu per hr per sq ft per deg F | Btu per day per sq ft per deg F |
|---|---|---|---|---|
| 1 | 4.1868 | 3,600 | 7,373 | 176,962 |
| 0.2388 | 1 | 860 | 1,761 | 42,267 |
| 0.0002778 | 0.001163 | 1 | 2.048 | 49.16 |
| 0.0001356 | 0.0005678 | 0.4882 | 1 | 24 |
| 0.000005651 | 0.00002366 | 0.02034 | 0.04167 | 1 |

**Heat Flow**

| Calories per sec per sq cm | Watts per sq cm | Calories per hr per sq cm | Btu per hr per sq ft | Btu per day per sq ft |
|---|---|---|---|---|
| 1 | 4.1868 | 3,600 | 13,272 | 318,531 |
| 0.2388 | 1 | 860 | 3,170 | 76,081 |
| 0.0002778 | 0.001163 | 1 | 3.687 | 88.48 |
| 0.00007535 | 0.0003154 | 0.2712 | 1 | 24 |
| 0.000003139 | 0.00001314 | 0.01130 | 0.04167 | 1 |

Section **2**

# Mathematics

**BY**

**PHILIP FRANKLIN**  *Late Professor of Mathematics, Massachusetts Institute of Technology.*
**GEORGE J. MOSHOS**  *Chairman, Dept. of Computer and Information Science, New Jersey Institute of Technology*

## ARITHMETIC
### By Philip Franklin

Numerical Computation ............................... 2-2
Logarithms ......................................... 2-3
The Slide Rule ..................................... 2-5
Financial Arithmetic ................................ 2-7

## GEOMETRY AND MENSURATION
### By Philip Franklin

Geometrical Theorems ............................... 2-7
Geometrical Constructions .......................... 2-9
Lengths and Areas of Plane Figures ................. 2-11
Surfaces and Volumes of Solids ..................... 2-12

## ALGEBRA
### By Philip Franklin

Formal Algebra .................................... 2-15
Solution of Equations in One Unknown Quantity ......... 2-18
Solution of Simultaneous Equations ................. 2-20
Determinants ...................................... 2-22
Imaginary or Complex Quantities .................... 2-22

## TRIGONOMETRY
### By Philip Franklin

Formal Trigonometry ................................ 2-25
Solution of Plane Triangles ........................ 2-28
Solution of Spherical Triangles .................... 2-29
Hyperbolic Functions ............................... 2-29

## ANALYTICAL GEOMETRY
### By Philip Franklin

The Point and the Straight Line ...................... 2-30
The Circle ........................................ 2-31
The Parabola ...................................... 2-31
The Ellipse ....................................... 2-32
The Hyperbola ..................................... 2-34
The Catenary ...................................... 2-36
Other Useful Curves ............................... 2-39

## DIFFERENTIAL AND INTEGRAL CALCULUS
### By Philip Franklin

Derivatives and Differentials ........................ 2-42
Maxima and Minima ................................. 2-43
Expansion in Series ................................ 2-44
Indeterminate Forms ............................... 2-45
Curvature ......................................... 2-45
Indefinite Integrals ............................... 2-46
Definite Integrals ................................. 2-48
Differential Equations ............................. 2-49
Linear Equations .................................. 2-51
The Laplace Transformation ......................... 2-53

## GRAPHICAL REPRESENTATION OF FUNCTIONS
### By Philip Franklin

Equations Involving Two Variables .................... 2-54
Equations Involving Three Variables ................. 2-57
Alignment Charts .................................. 2-58

## VECTOR ANALYSIS
### By Philip Franklin

Newtonian Mechanics ............................... 2-63
Analytic Representation ............................ 2-64
Differential Operators ............................. 2-64

## COMPUTERS
### By George J. Moshos

Computer Types .................................... 2-64
Digital Computers ................................. 2-65
Analog Computers .................................. 2-73
Comparison of Analog and Digital Computers .......... 2-74
Calculators ....................................... 2-75

REFERENCES Brenke, "Plane and Spherical Trigonometry," Dryden. Fine and Thompson, "Coordinate Geometry," Macmillan. Fine, "College Algebra," Ginn. Franklin, "Differential and Integral Calculus," McGraw-Hill. Franklin, "Methods of Advanced Calculus," McGraw-Hill. Hildebrand, "Introduction to Numerical Analysis," McGraw-Hill. Sokolnikoff and Sokolnikoff, "Higher Mathematics for Engineers and Physicists," McGraw-Hill. "Handbook of Mathematical Functions," NBS.

# ARITHMETIC

## by Philip Franklin

### NUMERICAL COMPUTATION

EDITOR'S NOTE: *The techniques described in this section are for use when no calculator is available.*

**Number of Significant Figures** In any engineering computations, the data are ordinarily the result of measurement and are correct only to a limited number of significant figures. Each of the numbers 3.840 and 0.003840 is said to be given "correct to four figures"; the true value lies in the first case between 3.8395 and 3.8405; in the second case between 0.0038395 and 0.0038405. The **absolute error** is less than 0.001 in the first case, and less than 0.000001 in the second; but the **relative error** is the same in both cases, namely, an error of less than "one part in 3,840."

If a number is written as 384,000, the reader is left in doubt whether the number of correct significant figures is 3, 4, 5, or 6. This doubt can be removed by writing the number as 3.84 × $10^5$, or 3.840 × $10^5$, or 3.8400 × $10^5$, or 3.84000 × $10^5$.

In any numerical computation, the possible or desirable degree of accuracy should be decided on and the computation should then be so arranged that the required number of significant figures, and no more, is secured. Carrying out the work to a larger number of places than is justified by the data is to be avoided, (1) because the form of the results leads to an erroneous impression of their accuracy and (2) because time and labor are wasted in superfluous computation. The labor of working with six-place tables is nearly three times as great as that with four-place tables. In computations involving several steps, it is desirable to retain one extra figure until just before the final result is reached, in order to protect the last figure against the possible cumulative effect of small tabular errors. In **discarding superfluous figures,** if the first discarded figure is 5 or more, increase the preceding figure by 1. Thus, 3.14159, written correct to four figures, is 3.142; correct to three figures, 3.14. Again, 6.1297, correct to four figures, is 6.130.

**Addition** In adding numbers, note that a doubtful final figure in any one number will render doubtful the whole column in which the figure lies; hence all figures to the right of that column are superfluous and contribute nothing to the accuracy of the result.

```
  0.2056x
  2.572xx
 14.25xxx
576.1xxxx
---------
593.1
```

**Subtraction** The Austrian or "shop" method is recommended. The mental process is as follows, the figures here printed in boldface type being the only ones written down:

| | |
|---|---|
| (3 plus how many is 12?) 3 plus **9** is 12; 1 to carry. | 14752 |
| (7 plus how many is 15?) 7 plus **8** is 15; 1 to carry. | 8463 |
| 5 plus **2** is 7. 8 plus **6** is 14. | **6289** |

This method is especially useful when it is desired to subtract from a given number the sum of several other numbers.

| | |
|---|---|
| 7 plus 1 is 8; plus 5 is 13; plus **9** is 22; 2 to carry. | 14752 |
| 5 plus 0 is 5; plus 2 is 7; plus **8** is 15; 1 to carry. | 3125 |
| 3 plus 1 is 4; plus 1 is 5; plus **2** is 7. | 101 |
| 5 plus 3 is 8; plus **6** is 14. | 5237 |
| | **6289** |

The use of a wavy line to indicate subtraction is also recommended, as it will minimize the danger of adding when subtraction is intended.

```
  4956
  8372
------
 39648
 1486|8
  346|92
    9|912
------
41492|xxx
```

**Multiplication** In long examples in multiplication, the arrangement of work here illustrated is recommended, since it facilitates the abbreviation of the work by the omission, in practice, of all the figures on the right of the vertical line.

The **position of the decimal point** should be determined by reference to the first, or left-hand, figures of the numbers, rather than by "pointing off" so-and-so many places from the right-hand end. For the right-hand figures of a number are the least important ones, and in many cases are entirely unknown (especially when the slide rule or a desk calculator is used). The mental process for determining the decimal point is as follows:

1. If the multiplier is a number like 3.1416, with only one figure preceding the decimal point, think of this number as "a little over 3"; then the product must be "a little over three times the number which is being multiplied"; and this gives the position of the decimal point at once, by inspection.

2. If the multiplier is a number like 3,141.6 (or 0.000 003 141 6), think of this number as "about 3, with the point moved three places to the right" (or "about 3, with the point moved six places to the left"); then think what the answer would be if the multiplier were simply "about 3," and shift the decimal point accordingly.

**Multiplication Tables** Crelle's large volume (Reimer, Berlin) gives the product of every three-figure number by every three-figure number; Peters (Reimer, Berlin), of every four-figure number by every two-figure number.

```
23026)31416(1
      23026
 2303)  8390(3
        6909
  230)  1481(6
        1380
   23)   101(4
         92
    2)     9(4
```

**Division** In long division, where the numbers are given only approximately, the work can be much abbreviated without loss of accuracy by "cutting off" one figure of the divisor at each step, instead of "bringing down" a doubtful zero in the dividend. Thus, 3.1416 ÷ 2.3026 = 1.3644.

To determine the **position of the decimal point** in a problem of fractional division, shift the point (mentally) in both numerator and denominator (the same number of places in each) until the denominator is a number in the "standard form," i.e., a number with only one figure preceding the decimal point. (This will not change the value of the fraction.) Then estimate the approximate magnitude of the quotient by inspection. Thus:

$$\frac{0.2718}{3141.6} = \frac{0.000\ 2718}{3.1416} = \text{"about } 0.000\ 09\text{"} = 0.000\ 08652$$

$$\frac{31.416}{0.002718} = \frac{31,416}{2.718} = \text{"about } 10,000\text{"} = 11,558$$

**Reciprocals**  The reciprocal of $N$ is $1/N$. Instead of dividing by a long number $N$, it is often better to multiply by the reciprocal of $N$. The Mathematical Tables (Sec. 1) give the reciprocal of any number, correct to four figures. Barlow's table (Spon & Chamberlain, New York) gives the reciprocal of every four-figure number correct to seven figures (but without facilities for interpolation). The reciprocals of numbers having more than four figures are best found by the use of a large table of logarithms or a calculator.

**Reciprocals of $1 \pm x$ When $x$ Is Small**

$1/(1 + x) = 1 - x + (\text{error} < x^2, \text{ if } x \text{ is between } 0 \text{ and } 1),$
$\qquad\qquad = 1 - x + x^2 - (\text{error} < x^3, \text{ if } x \text{ is between } 0 \text{ and } 1).$
$1/(1 - x) = 1 + x + (\text{error} < x^2 + 2x^3, \text{ if } x \text{ is between } 0 \text{ and } \frac{1}{2}),$
$\qquad\qquad = 1 + x + x^2 + (\text{error} < x^3 + 2x^4, \text{ if } x \text{ is between } 0 \text{ and } \frac{1}{2}).$

Note  $1/(a \pm b) = (1/a)[1/(1 \pm x)]$, where $x = b/a$.

**Notation by Powers of 10**  All questions concerning the position of the decimal point are readily answered if each number is expressed in the "standard form," i.e., as the product of two factors, one of which is a number with only one figure preceding the decimal point, while the other is a positive or negative power of 10. Thus, $3.1416 \times 10^3$ means 3.1416 with the point moved three places to the right, that is, 3141.6. Again, $3.1416 \times 10^{-6}$ means 3.1416 with the point moved six places to the left, i.e., 0.000 003 1416. This notation by powers of 10 should always be used in dealing with very large or very small numbers.

**Square Root**  (1) If four figures of the root are sufficient, take the answer directly from the table of square roots, Sec. 1, Mathematical Tables. (2) To obtain a root of six or seven figures, use the formula: $\sqrt{N} = a + [(N - a^2)/2a]$ (approx), where $a$ is the nearest value of $\sqrt{N}$ obtainable from the table, with three or four ciphers annexed. Here $a^2$ must be found exactly, by direct multiplication, so that at least three significant figures of the difference $N - a^2$ shall be known correctly; but this done, the division of $N - a^2$ by $2a$ should be carried to only three figures (logarithms or slide rule may be used).

**Square Roots of $1 \pm x$ When $x$ Is Small**

$(1 + x)^{1/2} = 1 + \frac{1}{2}x - (\text{error less than } \frac{1}{8}x^2 \text{ if } 0 < x < 1),$
$\qquad\qquad = 1 + \frac{1}{2}x - \frac{1}{8}x^2 + (\text{error} < \frac{1}{16}x^3 \text{ if } 0 < x < 1).$
$(1 - x)^{1/2} = 1 - \frac{1}{2}x - (\text{error} < \frac{1}{8}x^2 + \frac{1}{10}x^3 \text{ if } 0 < x < \frac{1}{2}),$
$\qquad\qquad = 1 - \frac{1}{2}x - \frac{1}{8}x^2 - (\text{error} < \frac{1}{16}x^3 + \frac{1}{16}x^4 \text{ if } 0 < x < \frac{1}{2}).$

Note.  $\sqrt{a + b} = \sqrt{a}\,(1 + x)^{1/2}$, where $x = b/a$.

**Cube Root**  (1) If four figures of the root are sufficient, take the answer directly from the table of cube roots, Sec. 1, Mathematical Tables. (2) To obtain a root of six or seven

figures, use the formula: $\sqrt[3]{N} = a + [(N - a^3)/3a^2]$ (approx), where $a$ is the nearest value of $\sqrt[3]{N}$ obtainable from the table, with three or four ciphers annexed. Here $a^3$ must be found correct to seven or eight figures, by direct multiplication, so that at least three significant figures of the difference $N - a^3$ shall be known; but this done, the division of $N - a^3$ by $3a^2$ should be carried to only three or four figures (logarithms or the slide rule may be used).

**Cube Roots of $1 \pm x$ When $x$ Is Small**

$(1 + x)^{1/3} = 1 + \frac{1}{3}x - (\text{error} < \frac{1}{9}x^2 \text{ if } 0 < x < 1),$
$\qquad\qquad = 1 + \frac{1}{3}x - \frac{1}{9}x^2 + (\text{error} < \frac{1}{16}x^3 \text{ if } 0 < x < 1).$
$(1 - x)^{1/3} = 1 - \frac{1}{3}x - (\text{error} < \frac{1}{9}x^2 + 1/10x^3 \text{ if } 0 < x < \frac{1}{2}),$
$\qquad\qquad = 1 - \frac{1}{3}x - \frac{1}{9}x^2 - (\text{error} < \frac{1}{16}x^3 + \frac{1}{15}x^4 \text{ if } 0 < x < \frac{1}{2}).$

Note.  $\sqrt[3]{a + b} = \sqrt[3]{a}\,(1 + x)^{1/3}$, where $x = b/a$.

## LOGARITHMS

**Tables of Logarithms**  The use of a table of logarithms greatly reduces the labor of multiplication, division, raising to powers, and extracting roots. The table in Sec. 1 is carried out to four significant figures, and the following explanations should be sufficient to permit the use of the table readily, even by one without previous experience. For algebraic theory, see Sec. 2, Algebra.

If more than four-figure accuracy is required, recourse must be had to a larger table. Five-place tables are available in great variety. If more than five figures are required, use a seven-place table: Schrön (Vieweg & Sohn, Braunschweig); Bruhns; Vega-Bremiker. If extreme accuracy is required, use the eight-place table by Bauschinger and Peters (Engelmann, Leipzig). For logarithmic paper, see Sec. 2, Graphical Representation of Functions.

**To Find the Logarithm of Any Given (Positive) Number**

1. When the given number is between 1 and 10.  An inspection of the table in Section 1 shows that as the number increases from 1 to 9.99 . . . the logarithm of that number increases continuously from 0 to 0.999 . . . For example, $\log 2.97 = 0.4728$; $\log 2.98 = 0.4742$.

If the given number contains four significant figures, it is necessary to interpolate between the tabulated values, as follows:

To find $\log 2.973$, notice that this number is 3/10 of the way from 2.97 to 2.98; hence its logarithm will be (approx) 3/10 of the way from 0.4728 to 0.4742. The difference here is 14 units, and 3/10 of this difference is 4 (to the nearest unit); hence, by adding this 4 to 4728, $\log 2.973 = 0.4732$. This process of interpolating should be performed mentally; the step of finding the tabular difference will be facilitated by a glance at the last column on the right, which gives, for each line of the table, the average of the differences along that line.

Again, to find $\log 4.098$: From table, $\log 4.09 = 0.6117$; adding 8/10 of the difference (11), or about 9, gives $\log 4.098 = 0.6126$. Or better, since 8/10 of the way forward is equal to 2/10 of the way back, find in table $\log 4.10 = 0.6128$, and subtract 2/10 of 11, or 2, giving $\log 4.098 = 0.6126$. It should be noted that any interpolated value may be in error by 1 in the last place.

If the given number contains more than four significant figures, it should be cut down to four figures, since the later figures will not affect the result in four-place computations.

2. WHEN THE GIVEN NUMBER IS LESS THAN 1 OR MORE THAN 10, it is simply necessary to notice that every such number can be regarded as obtainable from some number between 1 and 10 by merely shifting the decimal point, and that according to the rule at the foot of the table, moving the decimal point $n$ places to the right (or left) in the number-column is equivalent to adding $n$ (or $-n$) to the logarithm in the body of the table.

For example, to find log 2,973. Here $2,973 = 2.973 \times 10^3$ (i.e., 2.973 with the decimal point moved 3 places to the right). From the table, log $2.973 = 0.4732$. Hence, log $2,973 = 0.4732 + 3$, which may be written as 3.4732.

Again, to find log 0.0002973. Here $0.0002973 = 2.973 \times 10^{-4}$ (i.e., 2.973 with the decimal point moved 4 places to the left). From the table, log $2.973 = 0.4732$. Hence, log $0.0002973 = 0.4732 - 4$. (This may be written as $\overline{4}.4732$, if desired, and is equal, of course, to $-3.5268$; this latter form, however, is not convenient in practice.)

It is thus evident that the logarithm of every positive number may be regarded as consisting of two parts: a decimal fraction, which is always positive (or zero); and a whole number, which may be positive, negative, or zero. The fractional part is called the **mantissa,** and is found from the table; the whole-number part is called the **characteristic,** and is determined by inspection.

### To Find the Number Corresponding to a Given Logarithm

1. WHEN THE GIVEN LOGARITHM IS A POSITIVE DECIMAL FRACTION (CHARACTERISTIC ZERO), simply reverse the process for finding the logarithm of a number between 1 and 10.

For example, given log $N = 0.4732$; to find $N$. In the body of the table it is seen that 0.4732 lies a little beyond 0.4728; hence $N$ must lie a little beyond 2.97. By taking differences it is found that 0.4732 is in fact 4/14 of the way from 0.4728 to the next higher logarithm; therefore, $N$ must be 4/14 of the way from 2.97 to the next higher number. But 4/14 of 1 is 0.3 (to the nearest tenth), hence $N = 2.973$.

Again, given log $N = 0.6126$; to find $N$. Here, 0.6126 is 9/11 of the way from 0.6117 to the next higher logarithm; therefore, $N$ must be 9/11 of the way from 4.09 to the next higher number. But 9/11 of 1 is 0.8 (to the nearest tenth), hence $N = 4.098$.

2. WHEN THE GIVEN LOGARITHM HAS ANY GIVEN VALUE (CHARACTERISTIC NOT ZERO), proceed as follows: First, be sure the given logarithm is in the "standard form," i.e., a positive decimal fraction (mantissa) plus a positive or negative whole number (characteristic). For example, if log $N$ is originally given in the form log $N = -3.5268$, this must first be reduced to the (equivalent) form log $N = 0.4732 - 4$ (or $\overline{4}.4732$), before entering the table. Having the logarithm given in the standard form, suppose for the moment that the characteristic is zero, and find in the table the number corresponding to the given mantissa; then move the decimal point to the right or left according as the value of the characteristic is positive or negative.

For example, given log $N = 0.4732 + 3$; to find $N$. From the table, the number corresponding to 0.4732 is 2.973. The characteristic ($+3$) directs that the decimal point be moved 3 places to the right; hence $N = 2.973 \times 10^3 = 2,973$.

Again, given log $N = 0.4732 - 4$; to find $N$. From the table, the number corresponding to 0.4732 is 2.973. The characteristic ($-4$) indicates that the decimal point is to be moved 4 places to the left; hence $N = 2.973 \times 10^{-4} = 0.0002973$.

The number corresponding to a given logarithm is called its **antilogarithm.** Thus, if log $2,973 = 0.4732 + 3$, then $2973 =$ antilog $(0.4732 + 3)$.

NOTE 1. In most tables of logarithms the decimal point is omitted, the tables being in fact not tables of logarithms, but tables of mantissas. This omission is of no consequence to the experienced computer but is often perplexing to one who makes only occasional use of such tables.

NOTE 2. Many computers prefer to write negative characteristics in the form of some positive number minus some multiple of 10; thus, $0.4732 - 4 = 6.4732 - 10$; $0.4732 - 13 = 7.4732 - 20$; etc.

**Fundamental Properties of Logarithms** The usefulness of logarithms in computation depends on the following properties:

$$\log (ab) = \log a + \log b$$
$$\log (a/b) = \log a - \log b$$
$$\log (a^n) = n \log a$$

$$\log \sqrt[n]{a} = (1/n) \log a$$
$$\log 10^n = n$$

It is to be noted also that log $1 = 0$, log $10 = 1$, and log $(1/a) = -\log a$.

**To Multiply by Logarithms** Find from the table the log of each factor, and add; the result will be the log of the product. Then find the product itself from the table.

EXAMPLE. Find $x = (4.098) (0.0002973) (72.1)$
*Answer:* $x = 8.784 \times 10^{-2} = 0.08784$

*Method:*
| | |
|---|---|
| log 4.098 | $= 0.6126$ |
| log 0.0002973 | $= 0.4732 - 4$ |
| log 72.1 | $= 0.8579 + 1$ |
| log $x$ | $= 1.9437 - 3 = 0.9437 - 2.$ |

**To Divide by Logarithms. Method 1** Find from the table the log of the numerator and the log of the denominator, and subtract the second from the first; the result will be the logarithm of the quotient. Then find the quotient itself from the table.

EXAMPLE. Find $x = \dfrac{4.098}{0.0002973}$

*Answer:* $x = 1.378 \times 10^4 = 13780$

*Method:*
| | |
|---|---|
| log 4.098 | |
| log 0.0002973 | $= 0.4732 - 4$ |
| log $x$ | $= 0.1394 + 4$ |

In order **to avoid negative mantissas** in cases where a larger mantissa would have to be subtracted from a smaller, modify the upper logarithm by adding and subtracting 1.

EXAMPLE. Find $x = \dfrac{0.0291}{63.4}$

*Answer:* $x = 4.590 \times 10^{-4} = 0.0004590$

*Method:*
| | | |
|---|---|---|
| log 0.0291 | $= 0.4639 - 2$ | $= 1.4639 - 3$ |
| log 63.4 | $= 0.8021 + 1$ | $= 0.8021 + 1$ |
| log $x$ | | $= 0.6618 - 4$ |

But if the logarithms are written with the characteristics in front, and the "shop method" of subtraction is used (see p. 2-2), then no such special device is here required. Thus:

| | |
|---|---|
| log 0.0291 | $= \overline{2}.4639$ |
| log 63.4 | $= 1.8021$ |
| log $x$ | $= \overline{4}.6618$ |

**To Divide by Logarithms. Method 2** Instead of subtracting the log of a number, it is often convenient to add the **cologarithm** of that number; the colog of $N$ being defined by: colog $N = \log (1/N) = -\log N$.

To find the colog of a number, write the log of the number in the standard form, and subtract it from $1.000 - 1$, as in the following examples:

$$1.000 - 1$$
$$\log 69.5 = \underline{0.8420 + 1}$$
$$\text{colog } 69.5 = 0.1580 - 2$$

$$1.0000 - 1$$
$$\log 0.0002973 = \underline{0.4732 - 4}$$
$$\text{colog } 0.0002973 = 0.5268 + 3$$

This subtraction should be performed mentally. Thus, to subtract the mantissa, subtract each digit from 9 until the last nonzero digit is arrived at, and subtract this from 10; to subtract the characteristic, follow the regular rule of algebra ("reverse the sign and add"). Hence, if the logarithm itself is already written down, or can be read off from the table without interpolation, the cologarithm can be written down at once, by inspection. The use of cologarithms is not essential in logarithmic computation, but it often facilitates a compact arrangement of the work, especially in cases where the denominator of a fraction is itself the product of two or more factors.

**To Find the $n$th Power of a Number by Logarithms**  Find from the table the log of the number, and multiply it by $n$; the result will be the logarithm of the $n$th power of that number. Then find the power itself from the tables.

EXAMPLE 1.    Find $x = (0.0291)^3$
*Answer:*        $x = 2.464 \times 10^{-5} = 0.00002464$

$$\overset{3}{\phantom{x}}$$
$$\log x \quad = 1.3917 - 6 = 0.3917 - 5$$
EXAMPLE 2.    Find $x = (0.0291)^{1.41}$
*Answer:*        $x = 6.825 \times 10^{-3} = 0.006825$

*Method:* $\log 0.0291 = 0.4639 - 2 = -1.5361$
$$\underline{\phantom{xxxxxxxx}1.41}$$
$$15361$$
$$61444$$
$$\underline{15361}$$
$$\log x = \quad -2.1659$$
$$= 0.8341 - 3$$

**To Find the $n$th Root of a Number by Logarithms**  Find from the table the log of the number, and divide it by $n$; the result will be the log of the $n$th root of that number. Then find the root itself from the table.

EXAMPLE.    Find $x = \sqrt[3]{4.098}$
*Answer:*        $x = 1.600$

*Method:* $\log 4.098 = 0.6126$
$$\log x = 0.2042$$

In order **to avoid fractional characteristics**, if the characteristic is not divisible by $n$, make it so divisible by adding and subtracting a suitable number before dividing.

EXAMPLE.    Find $x = \sqrt[3]{0.0004590}$
*Answer:*        $x = 7.714 \times 10^{-2} = 0.07714$

*Method:* $\log 0.0004590 = 0.6618 - 4$
$$3)\underline{2.6618 - 6}$$
$$\log x = 0.8873 - 2$$

But if the characteristic is positive, it is simpler to write it in front of the mantissa, and then divide directly.

## THE SLIDE RULE

The slide rule is an indispensable aid in all problems in multiplication, division, proportion, square roots, etc., in which a limited degree of accuracy is sufficient. The ordinary 10-in Mannheim rule (see below) gives three significant figures correctly; the 20-in rule gives from three to four figures. For many problems the slide rule gives results more rapidly than a table of logarithms; it requires, however, more care in placing the decimal point in the answer. In all work with the slide rule, the position of the decimal point should be determined by inspection, and only the sequence of digits should be obtained from the instrument itself. Rapidity in the use of the instrument depends mainly on the skill with which the eye can estimate the values of the various divisions on the scale; expertness in this respect comes only with practice. The following explanations should be sufficient to permit the use of the ordinary slide rule successfully without previous experience and without knowledge of logarithms.

**Multiplication and Division with a (Theoretical) Complete Logarithmic Scale**  Consider a *complete* logarithmic scale ($D$, Fig. 1), assumed to extend indefinitely in both directions, only the main section, from 1 to 10, however, being usually available. Note that the divisions within the several sections are identical, except that the numeral attached to each division of any one section is ten times the numeral attached to the corresponding division in the preceding section. (The distances laid off from 1 are proportional to the logarithms of the corresponding numbers, the distance from 1 to 10 being taken as unity.) Consider also a duplicate scale $C$, numbered from 1 to 10, and arranged to slide along the fixed scale $D$ as in the figures. By means of such a scale $D$ and slide $C$ any two numbers between 1 and 10 (and hence any two numbers whatever, with proper attention to the decimal point) can be multiplied or divided, as in the following examples.

To MULTIPLY 4 BY 6.  In Fig. 1, starting with point 1 of the fixed scale, run the eye along from 1 to 4; then set the 1 of the slide opposite this point 4, and *run the eye forward along the slide from 1 to 6*; the point thus reached on the fixed scale is 24, which is equal to $4 \times 6$. This process gives the distance from 1 to 4 *plus* the distance from 1 to 6, and is, in fact, a mechanical method of adding the logarithms of these numbers; hence the result is the product of the numbers.

To DIVIDE 4 BY 6.  In Fig. 2, starting with the point 1 of the fixed scale, run the eye along from 1 to 4; then set the 6 of the

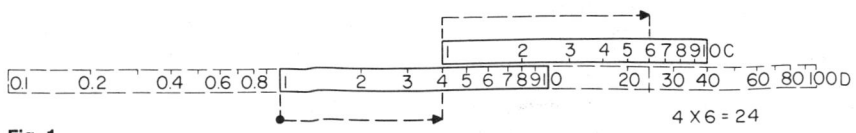

**Fig. 1**

slide opposite the point 4, and *run the eye backward along the slide from 6 to* 1; the point thus reached on the fixed scale is 0.667, which is equal to 4 ÷ 6. This process gives the distance from 1 to 4 *minus* the distance from 1 to 6 and is, in fact, a mechanical method of subtracting the logarithms of these numbers; hence the result is their quotient.

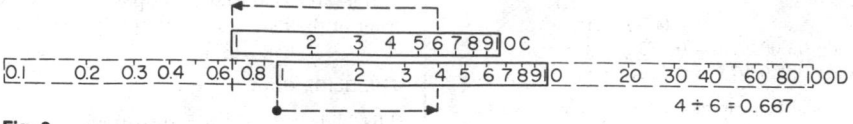

**Fig. 2**

**Multiplication and Division, Using Only a Single Section of the Scale**  If only the main section of scale *D* is available (as is usually the case in practice), the result of multiplication may fall beyond the scale, as it does in Fig. 1. In such cases *divide the first factor by 10 before beginning to multiply;* this will bring the result within the scale, without affecting the sequence of digits.

For example, to multiply 4 by 6. Having found that the setting shown in Fig. 1 is not successful, *reset the slide as in Fig. 3, with 10 instead of 1 opposite* 4; run the eye backward along the slide from 10 to 1, thus reaching the (unrecorded) point corresponding to 4 ÷ 10; then, continuing from this point, run the eye forward along the slide from 1 to 6, as before; the point finally reached on the main scale is 2.4, which has the same sequence of digits as the required value 24. After a little practice, this preliminary step of dividing by 10 will be performed almost intuitively. Whether or not this step is necessary in any given case can be determined by trial.

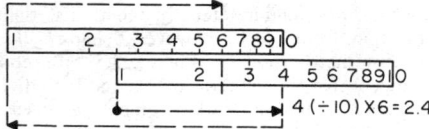

**Fig. 3**

The **general rule of multiplication** may be stated as follows: To find the product of two factors, find one factor on the fixed scale; opposite this, set (tentatively) point 1 of the slide; on the slide find the second factor, and opposite this read the product on the main scale, if possible. If the product falls beyond the scale, begin over again, using point 10 of the slide instead of point 1.

In division also, the result may fall beyond the main section of the scale, as it does in Fig. 2. In such cases, it suffices merely to *multiply the result by* 10 in order to bring it within the scale; this will not affect the sequence of digits.

For example, to divide 4 by 6, set the slide as in Fig. 4, and follow out mentally the steps indicated by the arrows. It will be noticed that the supplementary step of multiplying by 10 is performed by simply running the eye along the slide from 1 to 10 without resetting the slide; for this reason, division on the slide rule is slightly easier than multiplication.

The ordinary **Mannheim** slide rule has four scales, *A, B, C, D*, as shown in Fig. 5. Scales *C* and *D* are essentially the same as the *C* and *D* scales described above, and the principle just explained shows how they are used in multiplication and division. The fact that the *D* scale covers only the main section from 1 to 10 (all decimal points being omitted) is practically no

restriction on the scope of the scale, as is seen in the preceding examples. A runner is provided, so that intermediate positions reached in the course of an extended computation may be indicated temporarily on the scale without the necessity of reading off their numerical values. The best runners are those which have no side frame to obscure the numerals.

In problems involving successive multiplications and divisions, arrange the work so that multiplication and division are performed alternately.

For example, to calculate $\dfrac{a \times b \times c}{d \times e}$, divide the product $a \times b$ by $d$; multiply this quotient by $c$; and divide this product by $e$. Each operation will require only one shifting either of the slide (for multiplication) or of the runner (for division).

To multiply a number of different quantities by a *constant multiplier x*, set the point 1 of slide opposite *x*, and read, by aid of the runner, the products of *x* by all the quantities which do not fall beyond the scale; then reset the slide, setting 10 instead of 1 opposite *x*, and read the products of *x* by all the remaining quantities.

To divide a number of different quantities by a *constant divisor y*, first find (by the slide rule) the quotient $1 \div y$, and then use this as a constant multiplier.

Scales *A* and *B* are exactly like scales *C* and *D*, except that they cover two sections of the complete logarithmic scale, the graduations being only half as fine. Either pair of scales may be used for multiplication and division; *C* and *D* give more accurate readings, but have the disadvantage that in the case of multiplication the slide must often be shifted to the other end in order to keep the result on the scale—an inconvenience which is not present when the less accurate scales *A* and *B* are employed.

By the use of both pairs of scales, problems in squares and square roots may be readily solved; for every number on *A*, except for the decimal point, is the square of the number directly below it on *D* (use the runner).

A scale of sines, tangents, and logarithms is often printed on the back of the slide. For further details concerning the use of the slide rule in various problems, see the instruction books furnished with each instrument: Cox, "Manual of the Mannheim Slide Rule"; Halsey, "Manual of the Slide Rule"; etc.

**Other Types of Slide Rules**  The **duplex slide rule** has two faces providing space for additional fixed as well as movable scales. The slide can be set and points read, on either side, by means of a runner encircling the whole rule with hairlines on

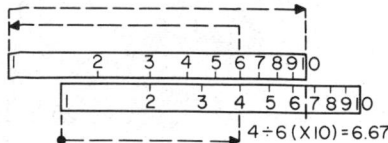

**Fig. 4**

each face. The additional scales frequently include inverted scales *CI* like *C* but numbered in the reverse order, and folded scales *CF*, *DF*, like *C* and *D* but starting with $\pi = 3.1416$, bringing the mark for 1 or 10 near the middle of the scale. These scales make possible the solution of more complicated problems with fewer settings of the slide. The **polyphase rule** is constructed like a Mannheim rule with the addition of a *CI*, or inverted *C* scale, on the slide and a fixed *K* scale. The *K* scale is like the *D* scale but covers three sections of the complete logarithmic scale, and so may be used with the *D* scale to find cube roots. The **log log slide rule** is of duplex type and carries on the fixed part a log in scale in several parts. This may be used

reading the figures in oblique positions, or else continually turning the instrument as a whole in the hand. There are also many **special slide rules,** adapted to various special types of computation, such as calculating discharge of water through pipes, horsepower of engines, dimensions of lumber, stadia measurements, and electric circuits. For a description of a large variety of slide rules and other calculating apparatus, see the Catalogue of the Collection in the Science Museum, South Kensington, or Horsburgh, "Napier Tercentenary Celebration" (Royal Society of Edinburgh).

Mass produced **portable solid state electronic calculators,** sold at reasonable cost, have displaced the slide rule to a large extent.

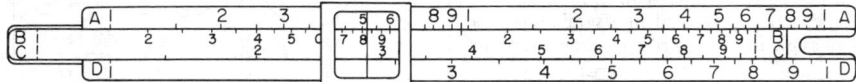

**Fig. 5**

directly with the *D* scale (or *B* scale for some parts) to read the values of exponentials, $e^x$ or natural logarithms, ln *x*. And in conjunction with the *C* (or *A* scale for some parts) it may be used to find the value of $a^b$ from a single setting, where *b* may have any (e.g., fractional or negative) value. Scales for reading the hyperbolic sine, sinh *x*, and hyperbolic tangent, tanh *x*, are sometimes added. A number of **circular slide rules** are on the market, the best of which are operated by a milled thumbnut, like the stem wind of a watch. The advantage of the circular rule, aside from its compact size (some models are scarcely larger than a watch), lies in the fact that the scale is endless, so that the slide never has to be reset in order to bring the result within the scale. A disadvantage is found in the necessity of

Sophistication of the instruments will rise and costs will fall further in the years ahead. Accordingly, it is expected that, by and large, slide rules per se will become obsolete. See Sect. 2, Computers.

### FINANCIAL ARITHMETIC

For the facts that are commonly required in regard to compound interest, sinking funds, etc., see the headings of the appropriate tables in Sec. 1, Mathematical Tables. More extended tables may be found in Glover, "Tables of Applied Mathematics" (Ann Arbor, Mich., 1923).

# GEOMETRY AND MENSURATION

## by Philip Franklin

### GEOMETRICAL THEOREMS

**Right Triangles**   $a^2 + b^2 = c^2$. (See Fig. 1.) $< A\ + < B = 90°$ $p^2 = mn.\ a^2 = mc.\ b^2 = nc.$
**Oblique Triangles**   Sum of angles = 180°. An exterior angle = sum of the two opposite interior angles (Fig. 1).

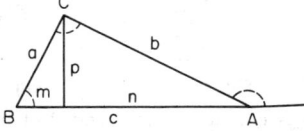

**Fig. 1**

The medians, joining each vertex with the middle point of the opposite side, meet in the center of gravity *G* (Fig. 2), which trisects each median.

The altitudes meet in a point called the **orthocenter,** *O*.

The perpendiculars erected at the midpoints of the sides meet in a point *C*, the center of the circumscribed circle. (In

any triangle *G*, *O*, and *C* lie in line, and *G* is two-thirds of the way from *O* to *C*.)

The bisectors of the angles meet in the center of the inscribed circle (Fig. 3).

The largest side of a triangle is opposite the largest angle; it is less than the sum of the other two sides, and greater than their difference.

**Similar Figures**   Any two similar figures, in a plane or in space, can be placed in "perspective," i.e., so that straight lines joining corresponding points of the two figures will pass through a common point (Fig. 4). That is, of two similar

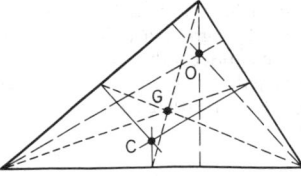

**Fig. 2**

figures, one is merely an enlargement of the other. Assume that each length in one figure is $k$ times the corresponding length in the other; then each area in the first figure is $k^2$ times the corresponding area in the second, and each volume in the

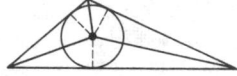

**Fig. 3**

first figure is $k^3$ times the corresponding volume in the second. If two lines are cut by a set of parallel lines (or parallel planes), the corresponding segments are proportional.

**The Circle**  An angle that is inscribed in a semicircle is a right angle (Fig. 5). An angle that is inscribed in a circle, or an

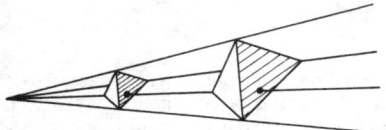

**Fig. 4**

angle between a chord and a tangent, is measured by half the intercepted arc (Fig. 6). An angle formed by any two lines which meet a circle is measured by half the sum or half the difference of the intercepted arcs, according as the point of intersection of the lines lies inside (Fig. 7) or outside the circle (Fig. 8).

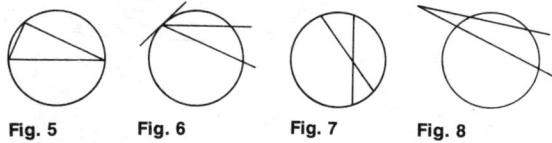

**Fig. 5**          **Fig. 6**          **Fig. 7**          **Fig. 8**

A tangent is perpendicular to the radius drawn to the point of contact.

If a variable line through $A$ (Figs. 9 and 10) cuts a circle in $P$ and $Q$, then $\overline{AP} \times \overline{AQ}$ is constant; in particular, if $A$ is an external point, $\overline{AP} \times \overline{AQ} = \overline{AT}^2$, where $AT$ is the tangent from $A$.

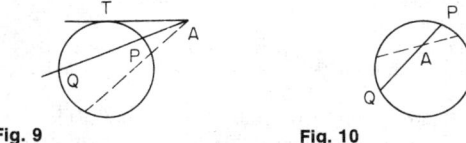

**Fig. 9**                    **Fig. 10**

The radical axis (Fig. 11) of two circles is a straight line such that the tangents drawn from any point of this line to the two circles are of equal length. If the two circles intersect, the radical axis passes through their points of intersection. In any case, the radical axis bisects the common tangents of the two circles. The three radical axes of a set of three circles meet in a common point. (For equations, see Sec. 2, Analytic geometry.)

**Dihedral Angles**  The dihedral angle between two planes is measured by a plane angle formed by two lines, one in each

plane, perpendicular to the edge (Fig. 12). (For solid angles, see Surfaces and Volumes of Solids below)

In a **tetrahedron,** or triangular pyramid, the four medians, joining each vertex with the center of gravity of the opposite face, meet in a point, the center of gravity of the tetrahedron; this point is ¾ of the way from any vertex to the center of

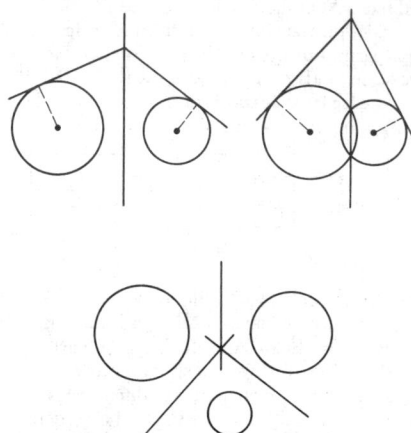

**Fig. 11**

gravity of the opposite face. The four perpendiculars erected at the circumcenters of the four faces meet in a point, the center of the circumscribed sphere. The four altitudes meet in a point called the orthocenter of the tetrahedron. The planes bisecting the six dihedral angles meet in a point, the center of the inscribed sphere.

**Regular Polyhedra** (see also Surfaces and Volumes of Solids below) Regular tetrahedron (Fig. 13), bounded by four equilateral triangles; cube (Fig. 14), bounded by six squares;

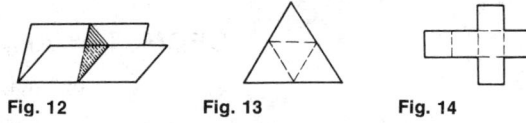

**Fig. 12**          **Fig. 13**          **Fig. 14**

octahedron (Fig. 15), bounded by eight equilateral triangles; dodecahedron (Fig. 16), bounded by twelve regular pentagons; icosahedron (Fig. 17), bounded by twenty equilateral triangles. Figures 13 to 17 show how these solids can be made by cutting the surface out of paper and folding it together.

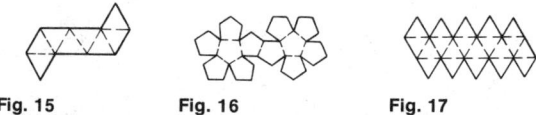

**Fig. 15**          **Fig. 16**          **Fig. 17**

**The Sphere** (see also Surfaces and Volumes of Solids below) If $AB$ is a diameter, any plane perpendicular to $AB$ cuts the sphere in a circle, of which $A$ and $B$ are called the poles. A great circle on the sphere is formed by a plane passing through the center. A spherical triangle is bounded by arcs of great circles. In two polar triangles, each angle in one is the supplement of the corresponding side in the other. In two symmetrical triangles, the sides and angles of one are equal to the corresponding sides and angles of the other, but arranged in the reverse order (like right-handed and left-handed gloves).

## GEOMETRICAL CONSTRUCTIONS

**To Bisect a Line** *AB* (Fig. 18)   (1) From *A* and *B* as centers, and with equal radii, describe arcs intersecting at *P* and *Q*, and draw *PQ*, which will bisect *AB* in *M*. (2) Lay off *AC* = *BD* = approximately half of *AB*, and then bisect *CD*.

**To Draw a Parallel to a Given Line** *l* **through a Given Point** *A* (Fig. 19)   With point *A* as center draw an arc just touching the line *l*; with any point *O* of the line as center, draw an arc

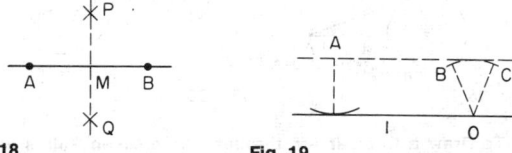

**Fig. 18**          **Fig. 19**

*BC* with the same radius. Then a line through *A* touching this arc will be the required parallel. Or, use a straightedge and triangle. Or, use a sheet of celluloid with a set of lines parallel to one edge and about ¼ in apart ruled upon it.

**To Draw a Perpendicular to a Given Line from a Given Point** *A* **outside the Line** (Fig. 20)   (1) With *A* as center, describe an

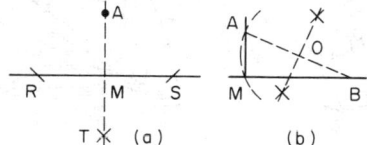

**Fig. 20**

arc cutting the line at R and S, and bisect *RS* at *M*. Then *M* is the foot of the perpendicular. (2) If *A* is nearly opposite one end of the line, take any point *B* of the line and bisect *AB* in *O*; then with *O* as center, and *OA* or *OB* as radius, draw an arc cutting the line in *M*. Or, (3) use a straightedge and triangle.

**To Erect a Perpendicular to a Given Line at a Given Point** *P*   (1) Lay off *PR* = *PS* (Fig. 21), and with *R* and *S* as centers draw arcs intersecting at *A*. Then *PA* is the required perpendicular. (2) If *P* is near the end of the line, take any convenient point *O* (Fig. 22) above the line as center, and with radius *OP* draw an arc cutting the line at *Q*. Produce *QO* to meet the arc at *A*; then *PA* is the required perpendicular. (3) Lay off *PB* = 4 units of any scale (Fig. 23); from *P* and *B* as centers lay off *PA* = 3 and *BA* = 5; then *APB* is a right angle.

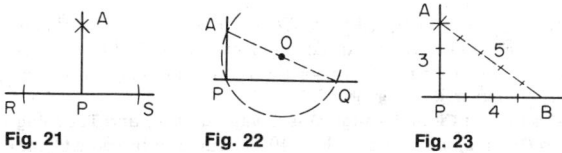

**Fig. 21**          **Fig. 22**          **Fig. 23**

**To Divide a Line** *AB* **into** *n* **Equal Parts** (Fig. 24)   Through *A* draw a line *AX* at any angle, and lay off *n* equal steps along this line. Connect the last of these divisions with *B*, and draw

parallels through the other divisions. These parallels will divide the given line into *n* equal parts. A similar method may be used to divide a line into parts which shall be proportional to any given numbers.

**To Construct a Mean Proportional (or Geometric Mean) between Two Lengths,** *m* **and** *n* (Fig. 25)   Lay off *AB* = *m* and *BC* = *n* and construct a semicircle on *AC* as diameter. Let the

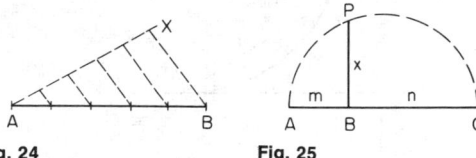

**Fig. 24**          **Fig. 25**

perpendicular erected at *B* meet the circumference at *P*. Then *BP* = $\sqrt{mn}$.)

**To Divide a Line** *AB* **in Extreme and Mean Ratio** (the "golden section")   At one end, *B*, of the given line (Fig. 26), erect a perpendicular, *BO*, equal to half *AB*, and join *OA*. Along *OA* lay off *OP* = *OB*, and along *AB* lay off *AX* = *AP*. Then *X* is the required point of division; i.e., $\overline{AX}^2 = AB \times BX$.

Numerically, $AX = \frac{1}{2}(\sqrt{5} - 1)(AB) = 0.618(AB)$.

**To Bisect an Angle** *AOB* (Fig. 27)   Lay off *OA* = *OB*. From *A* and *B* as centers, with any convenient radius, draw arcs meeting at *M*; then *OM* is the required bisector.

To draw the bisector of an angle when the vertex of the angle is not accessible (Fig. 28). Parallel to the given lines *a*, *b*, and equidistant from them, draw two lines *a'*, *b'* which intersect; then bisect the angle between *a'* and *b'*.

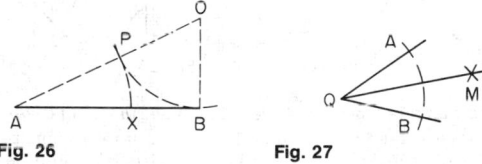

**Fig. 26**          **Fig. 27**

**To Draw a Line through a Given Point** *A* **and in the Direction of the Point of Intersection of Two Given Lines,** when this point of intersection is inaccessible (Fig. 29)   Draw any two parallel lines *PQ* and *P'Q'* as in the figure; through *P'* draw a line parallel to *PA*, and through *Q'* draw a line parallel to *QA*;

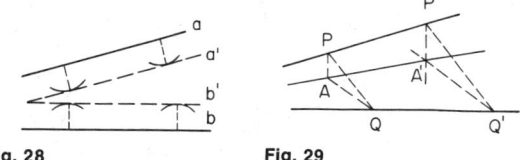

**Fig. 28**          **Fig. 29**

let these lines intersect at *A'*, and draw the line *AA'*. This line *AA'* will (if produced) pass through the intersection of the two given lines.

**To Construct, Approximately, the Length of a Circular Arc** (Rankine)   In Fig. 30 draw a tangent at *A*. Prolong the chord *BA* to *C*, making *AC* = ½*AB*. With *C* as center, and radius *CB*, draw an arc cutting the tangent at *D*. Then *AD* = arc

*AB*, approximately (error about 4 min in an arc of 60°). Conversely, to find an arc *AB* on a given circle to equal a given length *AD*, take *E* one-fourth of the way from *A* to *D*, and with *E* as center and radius *ED* draw an arc cutting the circumference in *B*. Then arc *AB* = *AD*, approximately.

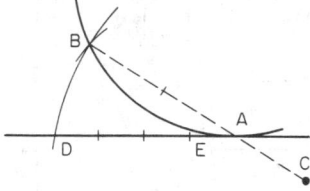

**Fig. 30**

**To Inscribe a Hexagon in a Circle** (Fig. 31)  Step around the circumference with a chord equal to the radius. Or, use a 60° triangle.

**To Circumscribe a Hexagon about a Circle** (Fig. 32)  Draw a chord *AB* equal to the radius. Bisect the arc *AB* at *T*. Draw the tangent at *T* (parallel to *AB*), meeting *OA* and *OB* at *P* and *Q*. Then draw a circle with radius *OP* or *OQ* and inscribe in it a hexagon, one side being *PQ*.

**To Inscribe an Octagon in a Square** (Fig. 33)  From the corners as centers, and with radius equal to half the diagonal, draw four arcs, cutting the sides in eight points. The points will be the vertices of the octagon.

 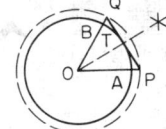 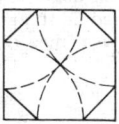

**Fig. 31**          **Fig. 32**          **Fig. 33**

**To Construct a Polygon of *n* Sides, One Side *AB* Being Given** (Fig. 34)  With *A* as center and *AB* as radius, draw a semicircle, and divide it into *n* parts, of which *n* − 2 parts (counting from *B*) are to be used. Draw rays from *A* through these points of division, and complete the construction as in the figure (in which *n* = 7). Note that the center of the polygon must lie in the perpendicular bisector of each side.

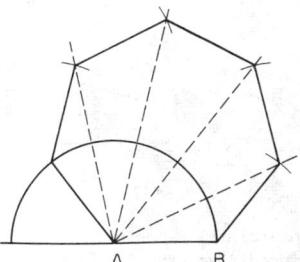

**Fig. 34**

**To Draw a Tangent to a Circle** from an external point *A* (Fig. 35)  Bisect *AC* in *M*; with *M* as center and radius *MC*, draw arc cutting circle at *P*; then *P* is the required point of tangency.

**To Draw a Common Tangent to Two Given Circles** (Fig. 36)  Let *C* and *c* be the centers and *R* and *r* the radii (*R* > *r*). From *C* as center, draw two concentric circles with radii *R* + *r* and *R* − *r*; draw tangents to these circles from *c*; then draw parallels to these lines at distance *r*. These parallels will be the required common tangents.

**To Draw a Circle through Three Given Points** *A*, *B*, *C*, or to find the center of a given circular arc (Fig. 37)  Draw the perpendicular bisectors of *AB* and *BC*; these will meet at the center, *O*.

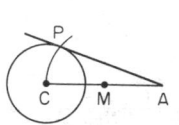 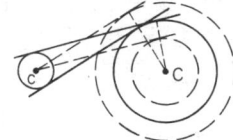

**Fig. 35**          **Fig. 36**

**To Draw a Circular Arc through Three Given Points When the Center Is Not Available** (Fig. 38)  With *A* and *B* as centers, and chord *AB* as radius, draw arcs, cut by *BC* at *R* and by *AC* at *S*. Divide *RA* into *n* equal parts, 1, 2, 3,.... On *BS* produced take points 1′, 2′, 3′, . . . to give arcs *S*1′, 1′2′, etc., all equal to *R*1. . . . Connect *A* with 1′, 2′, 3′, . . .

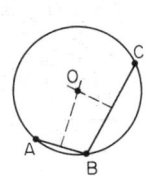 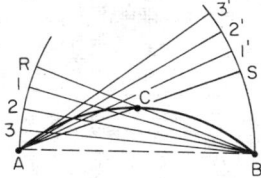

**Fig. 37**          **Fig. 38**

and *B* with 1, 2, 3,. . . . Then the points of intersection of corresponding lines will be points of the required arc.

**To Draw a Circle through Two Given Points,** *A*, *B*, **and Touching a Given Line,** *l* (Fig. 39)  Let *AB* meet line *l* at *C*. Draw any circle through *A* and *B*, and let *CT* be tangent to this

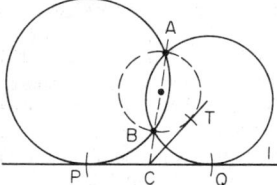

**Fig. 39**

circle from *C*. Along *l*, lay off *CP* and *CQ* equal to *CT*. Then either *P* or *Q* is the required point of tangency. (Two solutions.) Note that the center of the required circle lies on the perpendicular bisector of *AB*.

**To Draw a Circle through One Given Point,** *A*, **and Touching Two Given Lines,** *l* **and** *m* (Fig. 40)  Draw the bisector of the angle between *l* and *m*, and let *B* be the reflection of *A* in this line. Then draw a circle through *A* and *B* and touching *l* (or *m*), as in preceding construction. (Two solutions.)

**To Draw a Circle Touching Three Given Lines** (Fig. 41)  Draw the bisectors of the three angles; these will meet in the center $O$. (Four solutions.) The perpendiculars from $O$ to the three lines give the points of tangency.

**To Draw a Circle through Two Given Points** $A$, $B$, **and Touching a Given Circle** (Fig. 42)   Draw any circle through $A$ and $B$, cutting the given circle at $C$ and $D$. Let $AB$ and $CD$ meet at

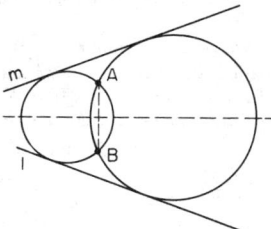

**Fig. 40**

$E$, and let $ET$ be tangent from $E$ to the circle just drawn. With $E$ as center, and radius $ET$, draw an arc cutting the given circle at $P$ and $Q$. Either $P$ or $Q$ is the required point of contact. (Two solutions.)

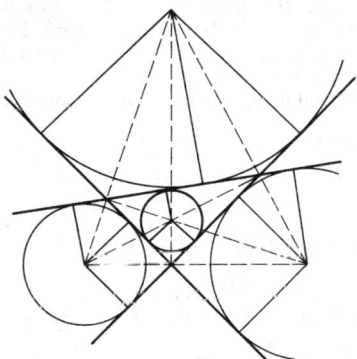

**Fig. 41**

**To Draw a Circle through One Given Point,** $A$, **and Touching Two Given Circles** (Fig. 43)   Let $S$ be a center of similitude for the two given circles, i.e., the point of intersection of two external (or internal) common tangents. Through $S$ draw any line cutting one circle at two points, the nearer of which shall be called $P$, and the other at two points, the more remote of which shall be called $Q$. Through $A$, $P$, $Q$ draw a circle cutting $SA$ at $B$. Then draw a circle through $A$ and $B$ and touching one of the given circles (see preceding construction). This circle will touch the other given circle also. (Four solutions.)

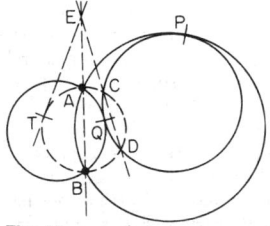

**Fig. 42**

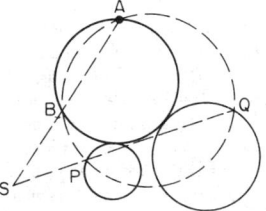

**Fig. 43**

**To Draw an Annulus Which Shall Contain a Given Number of Equal Contiguous Circles** (Fig. 44)   (An annulus is a ring-shaped area enclosed between two concentric circles.) Let $R + r$ and $R - r$ be the inner and outer radii of the annulus, $r$ being the radius of each of the $n$ circles. Then the required relation between these quantities is given by $r = R \sin (180°/n)$, or $r = (R + r)[\sin (180°/n)]/[1 + \sin (180°/n)]$.

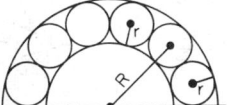

**Fig. 44**

For **methods of constructing ellipses and other curves,** see Sec. 2, Analytic Geometry.

## LENGTHS AND AREAS OF PLANE FIGURES

**Right Triangle** (Fig. 45)   $a^2 + b^2 = c^2$.
Area $= \frac{1}{2}ab = \frac{1}{2}a^2 \cot A = \frac{1}{2}b^2 \tan A = \frac{1}{4}c^2 \sin 2A$.

**Equilateral Triangle** (Fig. 46)   Area $= \frac{1}{4}a^2\sqrt{3} = 0.43301a^2$.
**Any Triangle** (Fig. 47).

$s = \frac{1}{2}(a + b + c)$, $t = \frac{1}{2}(m_1 + m_2 + m_3)$.

$r = \sqrt{(s - a)(s - b)(s - c)/s}$ = radius inscribed circle.

$R = \frac{1}{2}a/\sin A = \frac{1}{2}b/\sin B = \frac{1}{2}c/\sin C$ = radius circumscribed circle.

Area $= \frac{1}{2}$ base $\times$ altitude $= \frac{1}{2}ab = \frac{1}{2}ab \sin C = rs = abc/4R$,

$= \sqrt{s(s - a)(s - b)(s - c)} = \frac{4}{3}\sqrt{t(t - m_1)(t - m_2)(t - m_3)}$, $= r^2 \cot \frac{1}{2}A \cot \frac{1}{2}B \cot \frac{1}{2}C = 2R^2 \sin A \sin B \sin C$, $= \pm\frac{1}{2}\{(x_1y_2 - x_2y_1) + (x_2y_3 - x_3y_2) + (x_3y_1 - x_1y_3)\}$, where $(x_1,y_1)$, $(x_2,y_2)$, $(x_3,y_3)$, are coordinates of vertices. See also Sec. 2, Trigonometry.

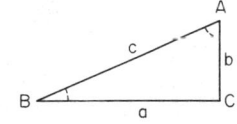
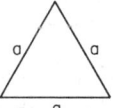

**Fig. 45**          **Fig. 46**

**Rectangle** (Fig. 48)   Area $= ab = \frac{1}{2}D^2 \sin u$, where $u =$ angle between diagonals $D$, $D$.
**Rhombus** (Fig. 49)   Area $= a^2 \sin C = \frac{1}{2}D_1D_2$, where $C =$ angle between two adjacent sides; $D_1$, $D_2 =$ diagonals.
**Parallelogram** (Fig. 50)   Area $= bh = ab \sin C = \frac{1}{2}D_1D_2 \sin u$, where $u =$ angle between diagonals $D_1$ and $D_2$; $D_1^2 + D_2^2 = 2(a^2 + b^2)$.
**Trapezoid** (Fig. 51)   Area $= \frac{1}{2}(a + b)h = \frac{1}{2}D_1D_2 \sin u$, where bases $a$ and $b$ are parallel; $u =$ angle between diagonals $D_1$ and $D_2$.

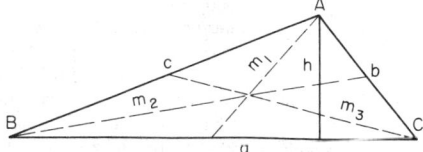

**Fig. 47**

**Quadrilateral Inscribed in a Circle** (Fig. 52)   Area $= \frac{1}{2}D_1D_2$ sin $u = \sqrt{(s - a)(s - b)(s - c)(s - d)} = \frac{1}{2}(ac + bd)$ sin $u$, where $s = \frac{1}{2}(a + b + c + d)$.

**Any Quadrilateral** (Fig. 53)   Area $= \frac{1}{2}D_1D_2$ sin $u$.

NOTE.   $a^2 + b^2 + c^2 + d^2 = D_1^2 + D_2^2 + 4m^2$, where $m =$ distance between midpoints of $D_1$ and $D_2$.

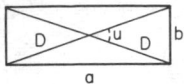
**Fig. 48**

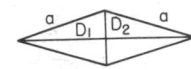
**Fig. 49**

**Polygons**   See table, Sec. 1, Mathematical Tables.

**Circle**   Area $= \pi r^2 = \frac{1}{2}Cr = \frac{1}{4}Cd = \frac{1}{4}\pi d^2 = 0.785398d^2$, where $r =$ radius, $d =$ diameter, $C =$ circumference $= 2\pi r = \pi d$ (see also Sec. 1, Mathematical Tables).

**Annulus** (Fig. 54)   Area $= \pi(R^2 - r^2) = \pi(D^2 - d^2)/4 = 2\pi R'b$, where $R' =$ mean radius $= \frac{1}{2}(R + r,)$ and $b = R - r$.

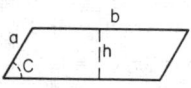
**Fig. 50**

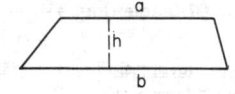
**Fig. 51**

**Sector** (Fig. 55)   Area $= \frac{1}{2}rs = \pi r^2 A/360° = \frac{1}{2} r^2$ rad $A$, where rad $A =$ radian measure of angle $A$, and $s =$ length of arc $= r$ rad $A$.

**Segment** (Fig. 56)   Area $= \frac{1}{2}r^2$(rad $A -$ sin $A) = \frac{1}{2}[r(s - c) + cb]$, where rad $A =$ radian measure of angle $A$ (see also

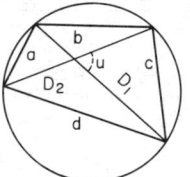
**Fig. 52**

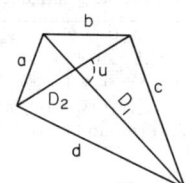
**Fig. 53**

Sec. 1, Mathematical Tables). For small arcs, $s = \frac{1}{3}(8c' - c)$, where $c' =$ chord of half of the arc (Huygen's approximation).

NOTE.   $c = 2\sqrt{b(d - b)}$; $c' = \sqrt{db}$ or $d = c'^2/b$, where $d =$ diameter of circle; $b = r(1 - \cos \frac{1}{2}A)$, $s = 2r$ rad $\frac{1}{2}A$.

**Ribbon** bounded by two parallel curves (Fig. 57). If a straight line $AB$ moves so that it is always perpendicular to the path traced by its middle point $G$, then the area of the ribbon or strip thus generated is equal to the length of $AB$ times the length of the path traced by $G$. (It is assumed that the radius of curvature of $G$'s path is never less than $\frac{1}{2}AB$, so that successive positions of the generating line will not intersect.)

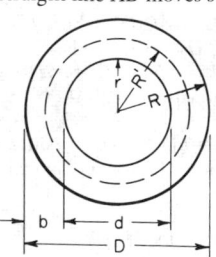
**Fig. 54**

**Simpson's Rule** (Fig. 58)   Divide the given area into $n$ panels (where $n$ is some even number) by means of $n$ + 1 parallel lines, called ordinates, drawn at constant distance $b$ apart; and denote the lengths of these ordinates by $y_0$, $y_1, y_2, \ldots, y_n$. (Note that $y_0$ or $y_n$ may be zero.) Area $= \frac{1}{3}b[(y_0 + y_n) + 1(y_1 + y_3 + y_5 \ldots) + 2(y_2 + y_4 + y_6 \ldots)]$, approx.

The greater the number of divisions, the more accurate the result.

NOTE.   Taking $y = f(x)$, where $x$ varies from $x = a$ to $x = b$, and $b = (b - a)/n$, then the error $= -\frac{1}{180}\frac{(b - a)^5}{n^4}f''''(X)$, where $f''''(X)$ is the value of the fourth derivative of $f(x)$ for some (unknown) value $x = X$, between $a$ and $b$.

**Ellipse** (Fig. 59; see Sec. 2, Trigonometry)   Area of ellipse $= \pi ab$. Area of shaded segment $= xy + ab \sin^{-1}(x/a)$. Length of perimeter of ellipse $= \pi(a + b)K$, where $K = (1 + \frac{1}{4}m^2 + \frac{1}{64}m^4 + \frac{1}{256}m^6 + \ldots)$, $m = (a - b)/(a + b)$.

| For $m =$ | 0.1 | 0.2 | 0.3 | 0.4 | 0.5 |
|---|---|---|---|---|---|
| $K =$ | 1.002 | 1.010 | 1.023 | 1.040 | 1.064 |
| For $m =$ | 0.6 | 0.7 | 0.8 | 0.9 | 1.0 |
| $K =$ | 1.092 | 1.127 | 1.168 | 1.216 | 1.273 |

**Hyperbola** (Fig. 60)   In any hyperbola, shaded area $A = ab$ $\ln\left(\frac{x}{a} + \frac{y}{b}\right)$. In an equilateral hyperbola $(a = b)$, area $A = a^2$ $\sinh^{-1}(y/a) = a^2 \cosh^{-1}(x/a)$. For tables of hyperbolic functions, see Sec. 1, Mathematical Tables. Here $x$ and $y$ are coordinates of point $P$.

**Parabola** (Fig. 61)   Shaded area $A = \frac{2}{3}ch$. In Fig. 62, length of arc $OP = s = \frac{1}{2}PT + \frac{1}{2}p \ln \cot \frac{1}{2}u$. Here $c =$ any chord; $p =$ semilatus rectum; $PT =$ tangent at $P$.

NOTE.   $OT = OM = x$.

For lengths and areas of **other curves** see Sec. 2, Analytical Geometry.

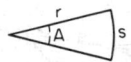
**Fig. 55**

**SURFACES AND VOLUMES OF SOLIDS**

**Regular Prism** (Fig. 63)   Volume $= \frac{1}{2}nrah = Bh$. Lateral area $= nah = Ph$. Here $n =$ number of sides; $B =$ area of base; $P =$ perimeter of base.

**Right Circular Cylinder** (Fig. 64)   Volume $= \pi r^2 h = Bh$. Lateral area $= 2\pi rh = Ph$. Here $B =$ area of base; $P =$ perimeter of base.

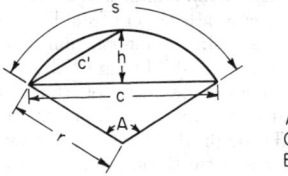
**Fig. 56**

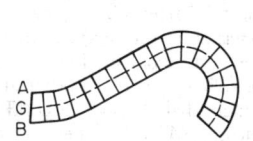
**Fig. 57**

**Truncated Right Circular Cylinder** (Fig. 65)   Volume $= \pi r^2 h = Bh$. Lateral area $= 2\pi rh = Ph$. Here $h =$ mean height $= \frac{1}{2}(h_1 + h_2)$; $B =$ area of base; $P =$ perimeter of base.

**Any Prism or Cylinder** (Fig. 66)   Volume $= Bh = Nl$. Lateral area $= Ql$. Here $l =$ length of an element or lateral edge; $B =$ area of base; $N =$ area of normal section; $Q =$ perimeter of normal section.

**Any Truncated Prism or Cylinder** (Fig. 67)   Volume $= Nl$. Lateral area $= Qk$. Here $l =$ distance between centers of gravity of areas of the two bases; $k =$ distance between centers of gravity of perimeters of the two bases; $N =$ area of normal section; $Q =$ perimeter of normal section. For a truncated triangular prism with lateral edges $a,b,c$, $l = k = \frac{1}{3}(a + b + c)$. Note that $l$ and $k$ will always be parallel to the elements.

**Special Ungula of a Right Circular Cylinder** (Fig. 68) Volume $= \frac{2}{3}r^2H$. Lateral area $= 2rH$. $r =$ radius. (Upper surface is a semiellipse.)

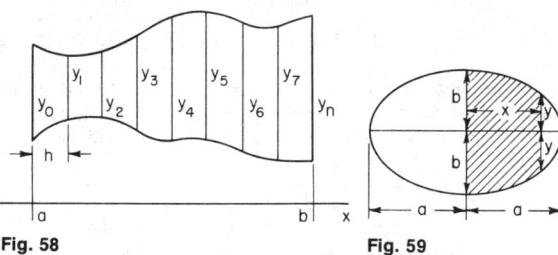

**Fig. 58**          **Fig. 59**

**Any Ungula** of a right circular cylinder (Figs. 69 and 70) Volume $= H(\frac{2}{3}a^3 \pm cB)/(r \pm c) = H[a(r^2 - \frac{1}{3}a^2) \pm r^2c$ rad $u]/(r \pm c)$. Lateral area $= H(2ra \pm cs)/(r \pm c) = 2rH(a \pm c$ rad $u)/(r \pm c)$. If base is greater (less) than a semicircle, use $+$ $(-)$ sign. $r =$ radius of base; $B =$ area of base; $s =$ arc of base; $u =$ half the angle subtended by arc $s$ at center; rad $u =$ radian measure of angle $u$.

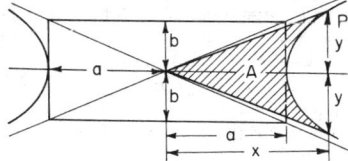

**Fig. 60**

**Hollow Cylinder** (right and circular) Volume $= \pi b(R^2 - r^2) = \pi b b(D - b) = \pi b b(d + b) = \pi b b D' = \pi b b(R + r)$. Here $b =$ altitude; $r,R(d,D) =$ inner and outer radii (diameters); $b =$ thickness $= R - r$; $D' =$ mean diam $- \frac{1}{2}(d + D) = D - b$ $= d + b$.

**Regular Pyramid** (Fig. 71) Volume $= \frac{1}{3}$ altitude $\times$ area of base $= \frac{1}{6}bran$. Lateral area $= \frac{1}{2}$ slant height $\times$ perimeter of base $= \frac{1}{2}san$. Here $r =$ radius of inscribed circle; $a =$ side (of regular polygon); $n =$ number of sides; $s = \sqrt{r^2 + b^2}$. Vertex of pyramid directly above center of base.

**Right Circular Cone** Volume $= \frac{1}{3}\pi r^2 b$. Lateral area $= \pi rs$. Here $r =$ radius of base; $b =$ altitude; $s =$ slant height $= \sqrt{r^2 + b^2}$.

**Frustum of Regular Pyramid** (Fig. 72)
Volume $= \frac{1}{6}bran[1 + (a'/a) + (a'/a)^2]$.
Lateral area $=$ slant height $\times$ half sum of perimeters of

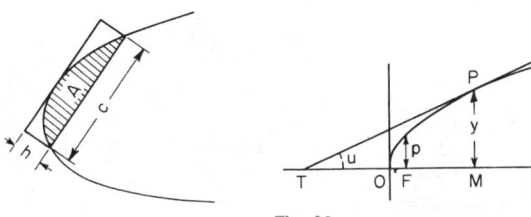

**Fig. 62**

bases $=$ slant height $\times$ perimeter of mid-section $= \frac{1}{2}sn(r + r')$. Here $r,r' =$ radii of inscribed circles; $s = \sqrt{(r - r')^2 + b^2}$; $a,a' =$ sides of lower and upper bases; $n =$ number of sides.

**Frustum of Right Circular Cone** (Fig. 73) Volume $= \frac{1}{3}\pi r^2b[1 + (r'/r) + (r'/r)^2] = \frac{1}{3}\pi b(r^2 + rr' + r'^2) = \frac{1}{4}\pi b[(r + r')^2 + \frac{1}{3}(r - r')^2]$. Lateral area $= \pi s(r + r')$; $s = \sqrt{(r - r')^2 + b^2}$.

**Any Pyramid or Cone** Volume $= \frac{1}{3}Bb$. $B =$ area of base; $b =$ perpendicular distance from vertex to plane in which base lies.

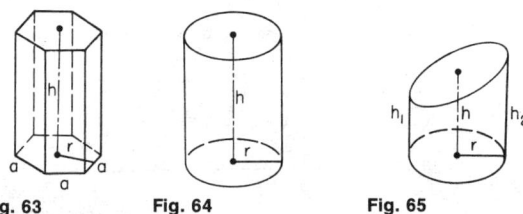

**Fig. 63**          **Fig. 64**          **Fig. 65**

**Any Pyramidal or Conical Frustum** (Fig. 74) Volume $= \frac{1}{3}b(B + \sqrt{BB'} + B') = \frac{1}{3}bB[1 + (P'/P) + (P'/P)^2]$. Here $B$, $B' =$ areas of lower and upper bases; $P, P' =$ perimeters of lower and upper bases.

**Obelisk** (frustum of a rectangular pyramid in Fig. 75) Volume $= \frac{1}{6}b[(2a + a_1)b + (2a_1 + a)b_1] = \frac{1}{6}b[ab + (a + a_1)(b + b_1) + a_1b_1]$.

**Wedge** (rectangular base; $a_1$ parallel to $a,a$ and at distance $b$ above base in Fig. 76) Volume $= \frac{1}{6}bb(2a + a_1)$.

**Sphere** Volume $= V = \frac{4}{3}\pi r^3 = 4.188790r^3 = \frac{1}{6}\pi d^3 = 0.523599d^3$ (table, Sec. 1, Mathematical Tables) $= \frac{2}{3}$ volume

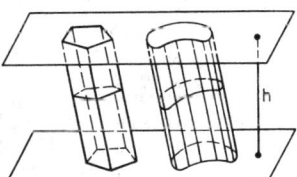

**Fig. 66**

of circumscribed cylinder. Area $= A = 4\pi r^2 =$ four great circles $= \pi d^2 = 3.14159d^2 =$ lateral area of circumscribed cylinder. Here $r =$ radius; $d = 2r =$ diameter $= \sqrt[3]{6V/\pi} = 1.24070 \sqrt[3]{V} = \sqrt{A/\pi} = 0.56419 \sqrt{A}$.

**Hollow Sphere,** or spherical shell. Volume $= \frac{4}{3}\pi(R^3 - r^3) = \frac{1}{6}\pi(D^3 - d^3) = 4\pi R_1^2t + \frac{1}{3}\pi t^3$. Here $R,r =$ outer and inner radii; $D,d =$ outer and inner diameters; $t =$ thickness $= R - r$; $R_1 =$ mean radius $= \frac{1}{2}(R + r)$.

**Spherical Segment of One Base. Zone** (spherical "cap" of Fig. 78) Volume $= \frac{1}{6}\pi b(3a^2 + b^2) = \frac{1}{3}\pi b^2(3r - b)$ (table, Sec. 1, Mathematical Tables). Lateral area (of zone) $= 2\pi rb = \pi(a^2 + b^2)$.

NOTE. $a^2 = b(2r - b)$, where $r =$ radius of sphere.

**Any Spherical Segment. Zone** (Fig. 77) Volume $= \frac{1}{6}\pi b(3a^2 + 3a_1^2 + b^2)$. Lateral area (zone) $= 2\pi rb$. Here $r =$ radius of sphere. If the inscribed frustum of a cone be removed from the spherical segment, the volume remaining is $\frac{1}{6}\pi bc^2$, where $c =$ slant height of frustum $= \sqrt{b^2 + (a - a_1)^2}$.

**Spherical Sector** (Fig. 78)   Volume = $\frac{1}{3}r$ × area of cap = $\frac{2}{3}\pi r^2 h$. Total area = area of cap + area of cone = $2\pi rh + \pi ra$.
  Note.   $a^2 = h(2r - h)$.

  **Spherical wedge** bounded by two plane semicircles and a **lune** (Fig. 79). Volume of wedge ÷ volume of sphere = $u/360°$. Area of lune ÷ area of sphere = $u/360°$. $u$ = dihedral angle of the wedge.

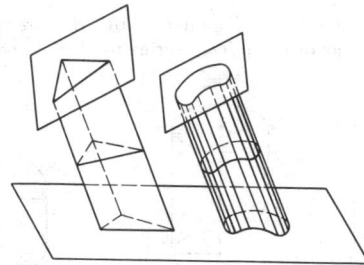

**Fig. 67**

  **Spherical triangle** bounded by arcs of three great circles (Fig. 80). Area of triangle = $\pi r^2 E/180°$ = area of octant × $E/90°$. $E$ = spherical excess = $180° - (A + B + C)$, where $A$, $B$, and $C$ are angles of the triangle. See also Sec. 2, Trigonometry.

  **Solid Angles**   Any portion of a spherical surface subtends what is called a **solid angle** at the center of the sphere. If the area

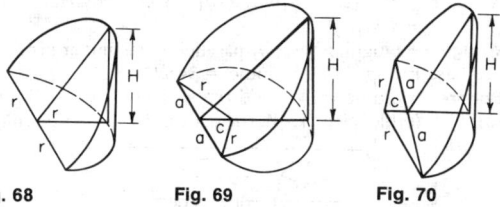

**Fig. 68**        **Fig. 69**        **Fig. 70**

of the given portion of spherical surface is equal to the square of the radius, the subtended solid angle is called a **steradian**, and this is commonly taken as the unit. The entire solid angle about the center is called a **steregon**, so that $4\pi$ steradians = 1 steregon. A so-called "solid right angle" is the solid angle

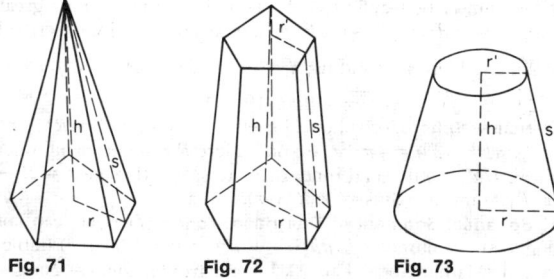

**Fig. 71**        **Fig. 72**        **Fig. 73**

subtended by a quadrantal (or trirectangular) spherical triangle, and a "spherical degree" (now little used) is a solid angle equal to $\frac{1}{90}$ of a solid right angle. Hence 720 spherical degrees = 1 steregon, or $\pi$ steradians = 180 spherical degrees. If $u$ = the angle which an element of a cone makes with its axis, then the solid angle of the cone contains $2\pi(1 - \cos u)$ steradians.
  **Regular Polyhedra**   $A$ = area of surface; $V$ = volume; $a$ = edge.

| Name of solid | Bounded by | $A/a^2$ | $V/a^3$ |
|---|---|---|---|
| Tetrahedron | 4 triangles | 1.7321 | 0.1179 |
| Cube | 6 squares | 6.0000 | 1.0000 |
| Octahedron | 8 triangles | 3.4641 | 0.4714 |
| Dodecahedron | 12 pentagons | 20.6457 | 7.6631 |
| Icosahedron | 20 triangles | 8.6603 | 2.1817 |

  **Ellipsoid** (Fig. 81)   Volume = $\frac{4}{3}\pi abc$, where $a$, $b$, $c$ = semiaxes.
  **Spheroid** (or ellipsoid of revolution)   The volume of any segment made by two planes perpendicular to the axis of revolution may be found accurately by the prismoidal formula below.

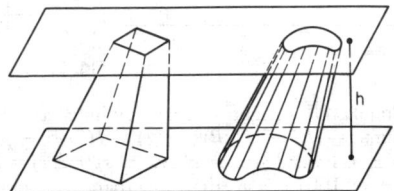

**Fig. 74**

  **Paraboloid of Revolution** (Fig. 82)   Volume = $\frac{1}{2}\pi r^2 h = \frac{1}{2}$ volume of circumscribed cylinder.
  **Segment of Paraboloid of Revolution** (bases perpendicular to axis in Fig. 83)   Volume of segment = $\frac{1}{2}\pi(R^2 + r^2)h$.
  **Barrels or Casks** (Fig. 84)   Volume = $\frac{1}{12}\pi h(2D^2 + d^2)$ approx for circular staves. Volume = $\frac{1}{15}\pi h(2D^2 + Dd + \frac{3}{4}d^2)$ exactly for parabolic staves. For a standing cask, partially full, compute contents by the prismoidal formula below. Roughly,

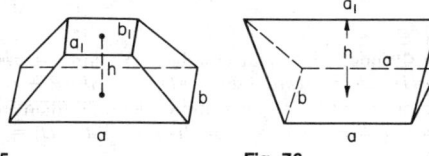

**Fig. 75**        **Fig. 76**

the number of gallons, $G$, in a cask is given by $G = 0.0034n^2h$, where $n$ = number of inches in the mean diameter, or $\frac{1}{2}(D + d)$, and $h$ = number of inches in the height.
  **Torus, or Anchor Ring** (Fig. 85)   Volume = $2\pi^2 cr^2$. Area = $4\pi^2 cr$ (proof by theorems of Pappus).
  **Theorems of Pappus**   1. Assume that a plane figure, area $A$, revolves about an axis in its plane but not cutting it; and let $s$ = length of circular arc traced by its center of gravity. Then volume of the solid generated by $A$ is $V = As$. For a complete revolution, $V = 2\pi rA$, where $r$ = distance from axis to center of gravity of $A$.

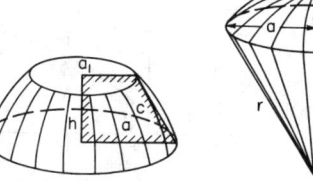

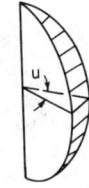

**Fig. 77**        **Fig. 78**        **Fig. 79**

2. Assume that a plane curve, length $l$, revolves about an axis in its plane but not cutting it; and let $s$ = length of circular arc traced by its center of gravity. Then area of the surface generated by $l$ is $S = ls$. For a complete revolution, $S = 2\pi rl$, where $r$ = distance from axis to center of gravity of $l$.

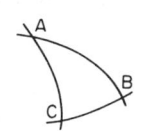

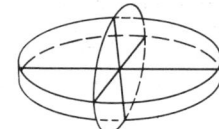

**Fig. 80**          **Fig. 81**

NOTE.  If $V_1$ or $S_1$ about any axis is known, then $V_2$ or $S_2$ about any parallel axis can be readily computed when the distance between the axes is known.

**Generalized Theorems of Pappus**  Consider any curved path of length $s$. If (1) a plane figure, area $A$ [or (2) a plane curve, length $l$] moves so that its center of gravity slides along

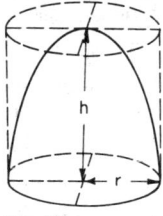

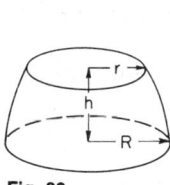

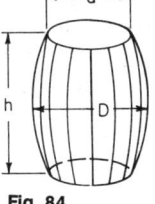

**Fig. 82**          **Fig. 83**          **Fig. 84**

this curved path (Fig. 86), while the plane of $A$ (or $l$) remains always perpendicular to the path, then (1) the volume generated by $A$ is $V = As$ [and (2) the area generated by $l$ is $S = ls$]. The path is assumed to curve so gradually that successive positions of $A$ (or $l$) will not intersect.

**Prismoidal Formula** (Fig. 87)  Volume = $\frac{1}{6}h(A + B + 4M)$, where $h$ = altitude, $A$ and $B$ = areas of bases and $M$ = area of a plane section midway between the bases. This formula is exactly true for any solid lying between two parallel planes and such that the area of a section at distance $x$ from one of these planes is expressible as a polynomial of not higher than the third degree in $x$. It is approximately true for many other solids.

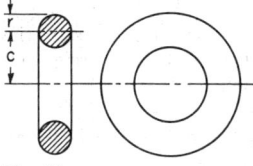

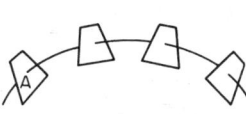

**Fig. 85**          **Fig. 86**

**Simpson's rule** may be applied to finding volumes, if the ordinates $y_1$, $y_2$, are interpreted as the areas of plane sections, at constant distance $h$ apart.

**Cavalieri's Theorem**  Assume two solids to have their bases in the same plane. If the plane section of one solid at every

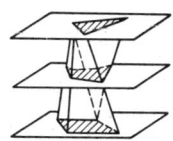

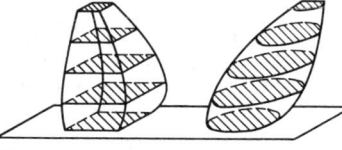

**Fig. 87**          **Fig. 88**

distance $x$ above the base is equal in area to the plane section of the other solid at the same distance $x$ above the base, then the volumes of the two solids will be equal. See Fig. 88.

# ALGEBRA

## by Philip Franklin

### FORMAL ALGEBRA

**Notation**  The main points of separation in a simple algebraic expression are the + and − signs. Thus, $a + b \times c - d \div x + y$ is to be interpreted as $a + (b \times c) - (d \div x) + y$. In other words, the range of operation of the symbols $\times$ and $\div$ extends only so far as the next + or − sign. As between the signs $\times$ and $\div$ themselves, $a \div b \times c$ means, properly speaking, $a \div (b \times c)$; i.e., the $\div$ sign is the stronger separative; but this rule is not always strictly followed, and in order to avoid ambiguity it is better to use the parentheses.

The range of influence of exponents and radical signs extends only over the next adjacent quantity. Thus, $2ax^3$

means $2a(x^3)$, and $\sqrt{2}\,ax$ means $(\sqrt{2})\,(ax)$. Instead of $\sqrt{2}\,ax$, it is safer, however, to write $\sqrt{2} \cdot ax$, or, better, $ax\sqrt{2}$.

Any expression within parentheses is to be treated as a single quantity. A horizontal bar serves the same purpose as parentheses.

The notation $a \cdot b$, or simply $ab$, means $a \times b$; and $a:b$, or $a/b$, means $a \div b$.

The symbol $|a|$ means the "absolute value of $a$," or numerical value regardless of sign; thus, $|-2| = |+2| = 2$.

The symbol $n!$ (where $n$ is a whole number) is read: "$n$ factorial," and means the product of the natural numbers from

1 to $n$, inclusive. Thus $1! = 1$; $2! = 1 \times 2$; $3! = 1 \times 2 \times 3$; $4!$ $= 1 \times 2 \times 3 \times 4$; etc.

The symbol $\neq$ or $\mp$ means "not equal to"; $\pm$ means "plus or minus."

The symbol $\approx$ is sometimes used for "approximately equal to."

**Addition and Subtraction**   $a + b = b + a$.
$(a + b) + c = a + (b + c)$. $a - (-b) = a + b$. $a - a = 0$.
$a + (x - y + z) = a + x - y + z$. $a - (x - y + z) = a - x + y - z$.
A minus sign preceding a parenthesis operates to reverse the sign of every term within, when the parentheses are removed.

**Multiplication and Simple Factoring**   $ab = ba$. $(ab)c = a(bc)$. $a(b + c) = ab + ac$. $a(b - c) = ab - ac$. Also, $a \times (-b) = -ab$, and $(-a) \times (-b) = ab$; "unlike signs give minus; like signs give plus."

$(a + b)(a - b) = a^2 - b^2$.
$(a + b)^2 = a^2 + 2ab + b^2$, $(a - b)^2 = a^2 - 2ab + b^2$.
$(a + b)^3 = a^3 + 3a^2b + 3ab^2 + b^3$, $(a - b)^3 = a^3 - 3a^2b + 3ab^2 - b^3$, etc.
(See table of binomial coefficients, Sec. 1, Mathematical Tables, and Binomial Theorem, below.)
$a^2 - b^2 = (a - b)(a + b)$, $a^3 - b^3 = (a - b)(a^2 + ab + b^2)$.
$a^n - b^n = (a - b)(a^{n-1} + a^{n-2}b + a^{n-3}b^2 + \cdots + ab^{n-2} + b^{n-1})$.
$a^n + b^n$ is factorable by $a + b$ only when $n$ is odd; thus,
$a^3 + b^3 = (a + b)(a^2 - ab + b^2)$.
$a^5 + b^5 = (a + b)(a^4 - a^3b + a^2b^2 - ab^3 + b^4)$, etc.
The following transformation is sometimes useful:

$$ax^2 + bx + c = a\left[\left(x + \frac{b}{2a}\right)^2 - \left(\frac{\sqrt{b^2 - 4ac}}{2a}\right)^2\right]$$

**Fractions**   If $m$ is not zero, $\dfrac{ma + mb + mc}{mx + my} = \dfrac{a + b + c}{x + y}$; i.e., both numerator and denominator of a fraction may be multiplied or divided by any quantity different from zero, without altering the value of the fraction.

To add two fractions, reduce each to a common denominator and add the numerators: $\dfrac{a}{b} + \dfrac{x}{y} = \dfrac{ay}{by} + \dfrac{bx}{by} = \dfrac{ay + bx}{by}$.

To multiply two fractions: $\dfrac{a}{b} \times \dfrac{x}{y} = \dfrac{ax}{by}$; $\dfrac{a}{b} \times x = \dfrac{a}{b} \times \dfrac{x}{1} = \dfrac{ax}{b}$.

To divide one fraction by another, invert the divisor and multiply:

$$\frac{a}{b} \div \frac{x}{y} = \frac{a}{b} \times \frac{y}{x} = \frac{ay}{bx} \qquad \frac{a}{b} \div x = \frac{a}{b} \times \frac{1}{x} = \frac{a}{bx}$$

**Ratio and Proportion**   The notation $a:b::c:d$, or the preferable form $a:b = c:d$, is read: "$a$ is to $b$ as $c$ is to $d$," and means simply $(a/b) = (c/d)$, or $ad = bc$. $a$ and $d$ are called the "extremes," $b$ and $c$ the "means," and $d$ the "fourth proportional" to $a$, $b$, and $c$. The "mean proportional" between two numbers is the square root of their product; also called the "geometric mean" of the numbers. If $a/b = c/d$, then $(a + b)/b = (c + d)/d$, and $(a - b)/b = (c - d)/d$; whence also, $(a + b)/(a - b) = (c + d)/(c - d)$. If $a/x = b/y = c/z = \cdots r$, then

$$(a + b + c + \cdots)/(x + y + \cdots) = r$$

See also Geometric Progression, below.

**Variation**   The notation $x \propto y$ is read: "$x$ varies directly as $y$," or "$x$ is directly proportional to $y$," and means $x = ky$, where $k$ is some constant. To determine the constant $k$, it is sufficient to know any pair of values, as $x_1$ and $y_1$, which belong together; then $x_1 = ky_1$, and hence $x/x_1 = y/y_1$, or $x = (x_1/y_1)y$. The expression "$x$ varies inversely as $y$," or "$x$ is inversely proportional to $y$," means that $x$ is proportional to $1/y$, or $x = k/y$.

**Exponents**   $a^{m+n} = a^m a^n$. $a^{m-n} = a^m/a^n$. $a^0 = 1$ (if $a \neq 0$). $a^{-m} = 1/a^m$. $(a^m)^n = a^{mn}$. $a^{1/n} = \sqrt[n]{a}$. Thus: $a^{1/2} = \sqrt{a}$, and $a^{1/3} = \sqrt[3]{a}$. $a^{m/n} = \sqrt[n]{a^m}$. Thus: $a^{2/3} = \sqrt[3]{a^2}$ and $a^{3/2} = \sqrt{a^3}$. $(\sqrt[n]{a})^n = a$. $(ab)^n = a^n b^n$. $(a/b)^n = a^n/b^n$. $(-a)^n = a^n$ if $n$ is even. $(-a)^n = -a^n$ if $n$ is odd. If $n$ is positive and increases indefinitely, $a^n$ becomes infinite if $a > 1$, and approaches 0 if $a < 1$ ($a$ being always positive). Graphs, Sec. 2, Graphical Representation of Functions; series, Sec. 2, Differential and Integral Calculus.

**Radicals**   Except in the simple cases of square root and cube root, radical signs should always be replaced by fractional exponents: $\sqrt[n]{a} = a^{1/n}$. $(\sqrt[n]{a})^n = (a^{1/n})^n = a$. If $n$ is odd, $\sqrt[n]{-a} = -\sqrt[n]{a}$; but if $n$ is even, $\sqrt[n]{-a}$ is imaginary. Every positive number $a$ has two square roots, one positive and the other negative; but the notation $\sqrt{a}$ always means the positive root; thus, $\sqrt{9} = 3$; $-\sqrt{9} = -3$. If the denominator of a fraction is of the form $\sqrt{a} \pm \sqrt{b}$, it is possible to "rationalize the denominator" by multiplying both numerator and denominator by $\sqrt{a} \mp \sqrt{b}$. Thus:

$$\frac{\sqrt{a} + \sqrt{b}}{\sqrt{a} - \sqrt{b}} = \frac{(\sqrt{a} + \sqrt{b})(\sqrt{a} + \sqrt{b})}{(\sqrt{a} - \sqrt{b})(\sqrt{a} + \sqrt{b})} = \frac{a + b + 2\sqrt{ab}}{a - b}$$

**Logarithms**   (For the use of logarithms in numerical computation, see Sec. 2, Arithmetic.) The logarithm of a (positive) number $N$ is the exponent of that power to which the base (10 or $e$) must be raised to produce $N$. Thus, $x = \log_{10} N$ means that $10^x = N$, and $x = \log_e N$ or $\ln N$ means that $e^x = N$. Logarithms to base 10 are called **common, denary,** or **Briggsian** logarithms. For table of four-place common logarithms see Sec. 1, Mathematical Tables.

Logarithms to base $e$ are called **natural** or **Napierian** logarithms. Here $e = 1 + 1 + 1/2! + 1/3! + 1/4! + \cdots = 2.718281828459\ldots$. For table of four-place natural logarithms see Sec. 1, Mathematical Tables.

If the subscript 10 or $e$ is omitted, the base must be inferred from the context, the base 10 being used in numerical computation, and the base $e$ in theoretical work. In either system,

$$\log (ab) = \log a + \log b$$
$$\log (a/b) = \log a - \log b$$
$$\log (1/n) = -\log n$$
$$\log (a^n) = n \log a$$

$$\log (\sqrt[n]{a}) = (1/n) \log a$$
$$\log (\text{base}) = 1$$
$$\log 0 = -\infty$$
$$\log 1 = 0$$

The two systems are related as follows:

$\log_{10} e = M = 0.4342944819 \ldots$
$\log_{10} x = 0.4343 \log_e x$
$\ln 10 = \log_e 10 = 1/M = 2.3025850930 \ldots$
$\ln x = \log_e x = 2.3026 \log_{10} x$

For tables of multiples of $M$ and $1/M$, see Sec. 1, Mathematical Tables. For graphs of the logarithmic and exponential functions, see Sec. 2, Graphical Representation of Functions; series, Sec. 2, Differential and Integral Calculus.

**Binomial Theorem**   (For table of binomial coefficients, see below and Sec. 1, Mathematical Tables.)

Let

$$(n)_1 = n,$$
$$(n)_2 = \frac{n(n-1)}{1 \times 2}$$
$$(n)_3 = \frac{n(n-1)(n-2)}{1 \times 2 \times 3}$$
$$(n)_4 = \frac{n(n-1)(n-2)(n-3)}{1 \times 2 \times 3 \times 4}$$

etc.

Then, for any value of $n$, provided $|x| < 1$,

$$(1+x)^n = 1 + (n)_1 x + (n)_2 x^2 + (n)_3 x^3 + (n)_4 x^4 + \cdots$$

(If $n$ is a positive integer, the series breaks off with the term in $x^n$, and is valid without restrictions on $x$.)

The most useful **special cases** are the following:

$$\sqrt{1+x} = (1+x)^{1/2}$$
$$= 1 + \frac{1}{2}x - \frac{1}{8}x^2 + \frac{1}{16}x^3 - \frac{5}{128}x^4 + \cdots \qquad [|x|<1]$$

$$\sqrt[3]{1+x} = (1+x)^{1/3}$$
$$= 1 + \frac{1}{3}x - \frac{1}{9}x^2 + \frac{5}{81}x^3 - \frac{10}{243}x^4 + \cdots \qquad [|x|<1]$$

$$\frac{1}{1+x} = (1+x)^{-1} = 1 - x + x^2 - x^3 + x^4 - \cdots \qquad [|x|<1]$$

$$\frac{1}{\sqrt{1+x}} = (1+x)^{-1/2}$$
$$= 1 - \frac{1}{2}x + \frac{3}{8}x^2 - \frac{5}{16}x^3 + \frac{35}{128}x^4 - \cdots \qquad [|x|<1]$$

$$\frac{1}{\sqrt[3]{1+x}} = (1+x)^{-1/3}$$
$$= 1 - \frac{1}{3}x + \frac{2}{9}x^2 - \frac{14}{81}x^3 + \frac{35}{243}x^4 - \cdots \qquad [|x|<1]$$

$$\sqrt{(1+x)^3} = (1+x)^{3/2}$$
$$= 1 - \frac{3}{2}x + \frac{3}{8}x^2 - \frac{1}{16}x^3 + \frac{3}{128}x^4 - \cdots \qquad [|x|<1]$$

$$\frac{1}{\sqrt{(1+x)^3}} = (1+x)^{-3/2}$$
$$= 1 - \frac{3}{2}x + \frac{15}{8}x^2 - \frac{35}{16}x^3 + \frac{315}{128}x^4 - \cdots \qquad [|x|<1]$$

with corresponding formulas for $\sqrt{1-x}$, etc., obtained by

reversing the signs of the odd powers of $x$. Also, provided $|b| < |a|$:

$$(a+b)^n = a^n \left(1 + \frac{b}{a}\right)^n$$
$$= a^n + (n)_1 a^{n-1}b + (n)_2 a^{n-2}b^2 + (n)_3 a^{n-3}b^3 + \cdots$$

where $(n)_1$, $(n)_2$, etc., have the values given above.

**Arithmetical Progression**   In an arithmetical progression, $a; a+d; a+2d; a+3d; \ldots$, each term is obtained from the preceding term by adding a constant, called the constant difference, $d$. If $n$ is the number of terms, the last term is $l = a + (n-1)d$; the "average" term is $\frac{1}{2}(a+l)$; and the sum of the $n$ terms is $n$ times the average term, or $S = \frac{1}{2}n(a+l)$. The **arithmetical mean** between $a$ and $b$ is $(a+b)/2$.

**Geometrical Progression**   In a geometrical progression, $a; ar; ar^2; ar^3; \ldots$, each term is obtained from the preceding term by multiplying by a constant, called the constant ratio, $r$. the $n$th term is $ar^{n-1}$. The sum of the first $n$ terms is $S = a(r^n - 1)/(r-1) = a(1 - r^n)/(1 - r)$. If $r$ is a positive or negative fraction, that is, if $-1 < r < +1$, then $r^n$ will approach zero as $n$ increases, and the sum of $n$ terms will approach $a/(1-r)$ as a limit. The **geometric mean** between $a$ and $b$ is $\sqrt{ab}$; also called the **mean proportional** between $a$ and $b$.

The **harmonic mean** between $a$ and $b$ is $2ab/(a+b)$.

**Summation of Certain Series by Second and Third Differences**   Let $a_1, a_2, a_3, \ldots a_n$ be any series of $n$ numbers, as in the first column of the adjoining scheme. By subtracting each number from the next following, form the column of "first differences," and by repeating this process, form the columns of second, third, etc., differences. If the $k$th differences are all equal, so that subsequent differences are all zero, the original series is called an arithmetical series of the $k$th order. In this special case the series can be summed as follows: Denote the numbers which stand at the head of the successive columns of differences by $D'$, $D''$, $D'''$, $\ldots$. Then the $n$th term of the series is $a_n$, and the sum of the first $n$ terms is $S_n$, where

| Numbers | 1st diff | 2nd diff | 3rd diff |
|---|---|---|---|
| $-64$ | $37$ | $-18$ | $6$ |
| $-27$ | $19$ | $-12$ | $6$ |
| $-8$ | $7$ | $-6$ | $6$ |
| $-1$ | $1$ | $0$ | $6$ |
| $0$ | $1$ | $6$ | |
| $1$ | $7$ | | |
| $8$ | | | |

$$a_n = a_1 + (n-1)D' + \frac{(n-1)(n-2)}{1 \times 2}D''$$
$$+ \frac{(n-1)(n-2)(n-3)}{1 \times 2 \times 3}D''' + \cdots$$
$$S_n = na_1 + \frac{n(n-1)}{1 \times 2}D' + \frac{n(n-1)(n-2)}{1 \times 2 \times 3}D''$$
$$+ \frac{n(n-1)(n-2)(n-3)}{1 \times 2 \times 3 \times 4}D''' + \cdots$$

If the series is, for example, of the third order, each of these formulas will stop with the term involving $D'''$; and only a few terms of the series are required for the computation of the $D$'s.

**Sum of the Squares or Cubes of the First $n$ Natural Numbers**

$$1 + 2 + 3 + \cdots + (n-1) + n = \frac{1}{2}n(n+1)$$
$$1^2 + 2^2 + 3^2 + \cdots + (n-1)^2 + n^2 = \frac{1}{6}n(n+1)(2n+1)$$
$$1^3 + 2^3 + 3^3 + \cdots + (n-1)^3 + n^3 = [\frac{1}{2}n(n+1)]^2$$

**Formula for Interpolation by Second Differences**   In any ordinary table giving a quantity $y$ as a function of a variable $x$, let it be required to find the value of $y$ corresponding to a value of $x$ which is not given directly in the table, but which lies between two tabulated values, as $x_1$ and $x_2$. If $x = x_1 + md$, where $d = x_2 - x_1 =$ the constant interval between two successive $x$'s, and $m$ is some proper fraction, then the corresponding value of $y$ will be given by the formula

$$y = y_1 + mD' + \frac{m(m-1)}{1 \times 2} D'' + \frac{m(m-1)(m-2)}{1 \times 2 \times 3} D''' + \cdots$$

where $D'$, $D''$, $D'''$, $\ldots$ are the first, second, third, $\ldots$ differences in the series of $y$'s which begins with $y_1$ (see above), provided the function is of such a nature that the differences of higher orders become negligibly small.

The coefficients of $D'$, $D''$, $D'''$, $\ldots$ in the formula are the binomial coefficients for fractional values of $m$ (see following table). The several terms of the formula (with careful attention to sign) are the successive corrections which must be added to $y_1$; the sum of these corrections should be rounded out to the nearest unit of the last significant place before adding. If $D' < 4$, the term involving $D''$, and later terms, can be neglected; the formula then reduces to $y = y_1 + mD'$, which is the familiar formula for ordinary, or "linear," interpolation. If $D''' < 8$ (or $D'''' < 12$, or $D'''''' < 16$), the term involving $D'''$ (or $D''''$, or $D''''''$) can be neglected.

**Permutations**   The number of possible permutations or arrangements of $n$ different elements is $1 \times 2 \times 3 \times \cdots \times n = n!$ (read: "$n$ factorial").

If among the $n$ elements there are $p$ equal ones of one sort, $q$ equal ones of another sort, $r$ equal ones of a third sort, etc., then the number of possible permutations is $(n!)/(p! \times q! \times r! \times \cdots)$, where $p + q + r + \cdots = n$.

**Combinations**   The number of possible combinations or groups of $n$ elements taken $r$ at a time (without repetition of any element within any one group) is $[n(n-1)(n-2)(n-3) \cdots (n-r+1)]/(r!) = (n)_r$. (See table of binomial coefficients, Sec. 1, Mathematical Tables.) If repetitions are allowed, so that a group, for example, may contain as many as $r$ equal elements, then the number of combinations of $n$ elements taken $r$ at a time is $(m)_r$, where $m_r = n + r - 1$.

NOTE.   $(n)_1 + (n)_2 + \cdots + (n)_n = 2^n - 1$.

## SOLUTION OF EQUATIONS IN ONE UNKNOWN QUANTITY

**Roots of an Equation**   An equation containing a single variable $x$ will in general be true for some values of $x$ and false for

other values. Any value of $x$ for which the equation is true is called a **root** of the equation. To "solve" an equation means to find all its roots. Any root of an equation, when substituted therein for $x$, will "satisfy" the equation. An equation which is true for all values of $x$, like $(x + 1)^2 = x^2 + 2x + 1$, is called an **identity** (often written $(x + 1)^2 \equiv x^2 + 2x + 1$).

### Types of Equations

1. ALGEBRAIC EQUATIONS
Of the first degree (linear), e.g., $2x + 6 = 0$ (root: $x = -3$).
Of the second degree (quadratic), e.g., $x^2 - 2x - 3 = 0$ (roots: $-1$, $3$).
Of the third degree (cubic), e.g., $x^3 - 6x^2 + 5x + 12 = 0$ (roots: $-1$, $3$, $4$).

2. TRANSCENDENTAL EQUATIONS
Exponential equations, e.g., $2^x = 32$ (root: $x = 5$); $2^x = -32$ (no real root).
Trigonometric equations, e.g., $10 \sin x - \sin 3x = 4$ (roots: $30°$, $150°$).

**Legitimate Operations on Equations**   An equation which is true for a particular value of $x$ will remain true for that value of $x$ after any one of the following operations is performed:

Adding any quantity to both sides; subtracting any quantity from both sides; transposing any term from one side to the other, provided its sign be changed; multiplying or dividing both sides by any quantity which is not zero; changing the signs of all the terms; raising both sides to any positive integral power; extracting any odd root of both sides; extracting any even root of both sides, provided the $\pm$ sign is used; taking the logarithms of both sides (both sides being positive); taking the sin, cos, tan, etc., of both sides.

Notice, however, that the new equation obtained by some of these operations may possess "additional roots" which did not belong to the original equation. This occurs especially when both sides are squared; thus, $x = -2$ has only one root, namely, $-2$; but $x^2 = 4$, obtained by squaring, has not only the root $-2$ but also another root, $+2$.

**Equations of the First Degree** (Linear Equations)   *Solution:* Collect all the terms involving $x$ on one side of the equation, thus: $ax = b$, where $a$ and $b$ are known numbers. Then divide through by the coefficient of $x$, obtaining $x = b/a$ as the root.

**Equations of the Second Degree** (Quadratic Equations)   *Solution:* Throw the equation into the standard form $ax^2 + bx + c = 0$. Then the two roots are

$$x_1 = \frac{-b + \sqrt{b^2 - 4ac}}{2a} \qquad x_2 = \frac{-b - \sqrt{b^2 - 4ac}}{2a}$$

**Binomial Coefficients for Fractional Values of $m$**

| $m$ | $(m)_2$ | $(m)_3$ | $(m)_4$ | $(m)_5$ |
|-----|---------|---------|---------|---------|
| 0.0 | −0.0000 | 0.0000 | −0.0000 | 0.0000 |
| 0.1 | −0.0450 | 0.0285 | −0.0207 | 0.0161 |
| 0.2 | −0.0800 | 0.0480 | −0.0336 | 0.0255 |
| 0.3 | −0.1050 | 0.0595 | −0.0402 | 0.0297 |
| 0.4 | −0.1200 | 0.0640 | −0.0416 | 0.0300 |
| 0.5 | −0.1250 | 0.0625 | −0.0391 | 0.0273 |
| 0.6 | −0.1200 | 0.0560 | −0.0336 | 0.0228 |
| 0.7 | −0.1050 | 0.0455 | −0.0262 | 0.0173 |
| 0.8 | −0.0800 | 0.0320 | −0.0176 | 0.0113 |
| 0.9 | −0.0450 | 0.0165 | −0.0087 | 0.0054 |

Here $(m)_2 = \dfrac{m(m-1)}{1 \times 2}$, $(m)_3 = \dfrac{m(m-1)(m-2)}{1 \times 2 \times 3}$, $(m)_4 = \dfrac{m(m-1)(m-2)(m-3)}{1 \times 2 \times 3 \times 4}$, etc.

The roots are real and distinct, coincident, or imaginary, according as $b^2 - 4ac$ is positive, zero, or negative. The sum of the roots is $x_1 + x_2 = -b/a$; the product of the roots is $x_1 x_2 = c/a$.

*Graphical Solution:* Write the equation in the form $x^2 = px + q$, and plot the parabola $y_1 = x^2$, and the straight line $y_2 = px + q$. The abscissas of the points of intersection will be the roots of the equation. If the line does not cut the parabola, the roots are imaginary.

**Equations of the Third Degree with Term in $x^2$ Absent** *Solution:* After dividing through by the coefficient of $x^3$, any equation of this type can be written $x^3 = Ax + B$. Let $p = A/3$ and $q = B/2$. The general solution is as follows:

CASE 1.  $q^2 - p^3$ positive. One root is real, viz.,

$$x_1 = \sqrt[3]{q + \sqrt{q^2 - p^3}} + \sqrt[3]{q - \sqrt{q^2 - p^3}}$$

the other two roots are imaginary. If $\cosh u = q/p\sqrt{p}$, $x_1 = 2\sqrt{p} \cosh(u/3)$.

CASE 2.  $q^2 - p^3 =$ zero. Three roots real, but two of them equal.

$$x_1 = 2\sqrt[3]{q} \qquad x_2 = -\sqrt[3]{q} \qquad x_3 = -\sqrt[3]{q}$$

CASE 3.  $q^2 - p^3$ negative. All three roots real and distinct. Determine an angle $u$ between 0 and 180°, such that $\cos u = q/(p\sqrt{p})$. Then

$$x_1 = 2\sqrt{p} \cos(u/3)$$
$$x_2 = 2\sqrt{p} \cos(u/3 + 120°)$$
$$x_3 = 2\sqrt{p} \cos(u/3 + 240°)$$

*Graphical Solution:* Plot the curve $y_1 = x^3$, and the straight line $y_2 = Ax + B$. The abscissas of the points of intersection will be the roots of the equation.

**Equations of the Third Degree (General Case)** *Solution:* The general cubic equation, after dividing through by the coefficient of the highest power, may be written $x^3 + ax^2 + bx + c = 0$. To get rid of the term in $x^2$, let $x = x_1 - a/3$. The equation then becomes $x_1^3 = Ax_1 + B$, where $A = 3(a/3)^2 - b$, and $B = -2(a/3)^3 + b(a/3) - c$. Solve this equation for $x_1$, by the method above, and then find $x$ itself from $x = x_1 - (a/3)$.

*Graphical Solution:* Without getting rid of the term in $x^2$, write the equation in the form $x^3 = -a[x + (b/2a)]^2 + [a(b/2a)^2 - c]$, and solve by the graphical method.

**General Properties of Algebraic Equations** An algebraic equation of the $n$th degree in $x$ is an equation of the type

$$a_0 x^n + a_1 x^{n-1} + a_2 x^{n-2} + \cdots + a_{n-1}x + a_n = 0$$

where the $a$'s are any given numbers ($a_0$ not zero), the expression on the left being called a **polynomial** of the $n$th degree in $x$. Such an equation will, in general, have $n$ roots; but some of these $n$ roots may be equal, and some may be imaginary. **Imaginary roots** always occur in pairs if the $a$'s are all real numbers.

If the equation is written in the form: (a polynomial in $x$) = 0, then (1) if $a$ is a root of the equation, $x - a$ is a factor of the polynomial; (2) if the polynomial can be factored in the form $(x$

$- p)(x - q)(x - r)\cdots = 0$, each of the quantities $p, q, r, \ldots$ is a root of the equation; (3) if $x$ is very large (either positive or negative), the higher powers of $x$ are the most important; (4) if $x$ is very small, the higher powers may be neglected.

**Short Method of Substitution in a Polynomial** To find the value of $4x^4 - 14x^3 + 23x - 26$ when $x = 3$, for example, first arrange the terms in order of descending powers of $x$, and write the detached coefficients, with their signs, in a row, taking care to supply a zero coefficient for any missing term, including the constant term. Then, beginning at the left, bring down the first coefficient; multiply this by 3, and add to the second coefficient; multiply this result by 3 again, and add to the third coefficient; and so on. The final result, $-11$, is the value of the polynomial when $x = 3$.

$$\begin{array}{rrrrrr}
4 & -14 & 0 & 23 & -26 & (3 \\
  & 12 & -6 & -18 & 15 & \\ \hline
4 & -2 & -6 & 5 & -11 &
\end{array}$$

**Short Method of Dividing a Polynomial by** $x - a$ The device just explained gives not only the value of the polynomial when $x = 3$, but also the result of dividing the polynomial by $x - 3$. Thus, in the case illustrated, the quotient is $4x^3 - 2x^2 - 6x + 5$ and the remainder is $-11$. That is, $4x^4 - 14x^3 + 0x^2 + 23x - 26 = (x - 3)(4x^3 - 2x^2 - 6x + 5) - 11$.

**Exponential Equation** To solve an equation of the form $a^x = b$, take the logarithms of both sides: $x \log a = \log b$, whence $x = (\log b/\log a)$. For example, if $3^x = 0.4$, $x = \log 0.4/\log 3 = (0.6021 - 1)/0.4771 = -0.3979/0.4771 = -0.8340$. Notice that the complete logarithm must be taken, not merely the mantissa.

**Trigonometric Equations** (1) To solve $a \cos x + b \sin x = c$, where $a$ and $b$ are positive: Find the acute angle $u$ for which $\tan u = b/a$, and the angle $v$ (between 0 and 180°) for which $\cos v = c/\sqrt{a^2 + b^2}$. Then $x_1 = u + v$ and $x_2 = u - v$ are roots of the equation. (2) To solve $a \cos x - b \sin x = c$, where $a$ and $b$ are positive: Find $u$ and $v$ as above. Then $x_1 = -(u + v)$ and $x_2 = (u - v)$ are roots of the equation.

**General Method of Solution by Trial and Error** This method is applicable to a numerical equation of any form, and can be carried out to any desired degree of approximation. It is especially useful when a first approximation to a root is already known. Write the equation in the form $f(x) = 0$, where $f(x)$ means any function of $x$, and plot the curve $y = f(x)$ for a sufficient number of values of $x$ to obtain a general idea of the shape of the curve. Then pick out the regions in which the curve appears to cross the axis of $x$, and plot the curve more accurately in each of these regions. Thus, by successive approximations, plotting the important parts of the curve on a larger and larger scale, determine as accurately as necessary the points where the curve crosses the axis, i.e., the values of $x$ which make $f(x)$ equal to zero.

Thus, suppose that $f(x) = 3.0$ when $x = 2.6$ and $-5.0$ when $x = 2.7$ (see Fig. 1). Then the curve must cross the axis somewhere between $x = 2.6$ and $x = 2.7$; and since it will not vary greatly from a straight line between those points, it is seen that it must cross near 2.64. Suppose the value of $f(x)$, when computed for $x = 2.64$, is $-0.2$, and when computed for $x = 2.63$ is $+0.7$; then the root lies between $x = 2.63$ and 2.64. Plotting this section on the larger scale, it is seen that the next guess should be about 2.638; and so on.

Instead of writing the original equation with all the terms on the left-hand side, it is often better to divide the expression into two parts, say $f_1(x)$ and $f_2(x)$, writing the equation in the form $f_1(x) = f_2(x)$. If then the two curves $y_1 = f_1(x)$ and $y_2 = f_2(x)$ be plotted separately, on the same diagram, the value of $x$ corresponding to their point of intersection will be the desired root.

## SOLUTION OF SIMULTANEOUS EQUATIONS

**Meaning of a System of Simultaneous Equations**  To solve a system of $n$ simultaneous equations in $n$ unknowns means to find all the sets of values of the unknowns (if any) which, when substituted in the given equations, will satisfy all the equations at the same time. If a system of equations has no solution, the equations are "inconsistent"; if it has an infinite number of solutions, the equations are "not all independent."

**Simultaneous Equations of the First Degree in Two Unknowns**

$$\begin{array}{ll} \text{Factors} \\ \textbf{(1)}\ a_1x + b_1y = c_1 & \begin{vmatrix} -b_2 \end{vmatrix} \begin{vmatrix} -a_2 \end{vmatrix} \\ \textbf{(2)}\ a_2x + b_2y = c_2 & \begin{vmatrix} -b_1 \end{vmatrix} \begin{vmatrix} a_1 \end{vmatrix} \end{array}$$

$$(a_1b_2 - a_2b_1)x = b_2c_1 - b_1c_2$$
$$\therefore x = (b_2c_1 - b_1c_2)/(a_1b_2 - a_2b_1)$$

$$(a_1b_2 - a_2b_1)y = a_1c_2 - a_2c_1$$
$$\therefore y = (a_1c_2 - a_2c_1)/(a_1b_2 - a_2b_1)$$

Here **(1)** is multiplied by $b_2$, **(2)** by $-b_1$, and the products added so as to eliminate $y$; again, **(1)** is multiplied by $-a_2$, **(2)** by $a_1$, and the products added so as to eliminate $x$. (The process is most conveniently performed as follows: Write the multipliers, as $b_2$ and $-b_1$, at the right of the equations; multiply the first term of each equation by its proper multiplier and add; then multiply the second term of each equation by its proper multiplier, and add; and so on. This is simpler than the common practice of multiplying out each equation separately before adding.) If $a_1b_2 - a_2b_1 = 0$, the equations have no solution when $c_1 \neq c_2$, and an infinite number of solutions when $c_1 = c_2$. The following **special solution** is possible when the sum and difference of the two unknowns are given:

$$\begin{array}{lll} \text{Let} & x + y = m & \textbf{(1)} \\ \text{and} & x - y = n & \textbf{(2)} \end{array}$$

$$\textbf{(1)} + \textbf{(2)}: \quad 2x = m + n \quad \therefore x = \tfrac{1}{2}(m + n)$$
$$\textbf{(1)} - \textbf{(2)}: \quad 2y = m - n \quad \therefore y = \tfrac{1}{2}(m - n)$$

**Simultaneous Equations of the Second Degree in Two Unknowns**

CASE *a.*  When the product of the unknowns, and their sum or difference, are given:

$$\begin{array}{ll} x + y = 5 & \textbf{(1)} \\ xy = 4 & \textbf{(2)} \end{array}$$

Squaring **(1)**,
From **(2)**,
$$\begin{array}{rr} x^2 + 2xy + y^2 = & 25 \\ -4xy = & -16 \end{array}$$

Adding,
$$x^2 - 2xy + y^2 = 9$$

Hence,
$$x - y = 3 \text{ or } -3$$

But
$$x + y = 5 \text{ or } 5$$

Therefore,
$$\left. \begin{array}{l} x = 4 \\ y = 1 \end{array} \right| \text{ or } \left| \begin{array}{l} x = 1 \\ y = 4 \end{array} \right.$$

$$\begin{array}{ll} x - y = 3 & \textbf{(1)} \\ xy = 4 & \textbf{(2)} \end{array}$$

$$\begin{array}{rr} x^2 - 2xy + y^2 = & 9 \\ 4xy = & 16 \end{array}$$

$$x^2 + 2xy + y^2 = 25$$

$$x + y = 5 \text{ or } -5$$
$$x - y = 3 \text{ or } 3$$

$$\left. \begin{array}{l} x = 4 \\ y = 1 \end{array} \right| \text{ or } \left| \begin{array}{l} x = -1 \\ y = -4 \end{array} \right.$$

CASE *b.*  When the product and the sum of the squares are given:

$$\begin{array}{ll} xy = 5 & \textbf{(1)} \\ x^2 + y^2 = 26 & \textbf{(2)} \\ 2xy = 10 & \textbf{(3)} \end{array}$$

From **(1)**,

$$\textbf{(2)} + \textbf{(3)}: x^2 + 2xy + y^2 = 36 \quad \textbf{(4)}$$
$$\textbf{(2)} - \textbf{(3)}: x^2 - 2xy + y^2 = 16 \quad \textbf{(5)}$$
$$\sqrt{\textbf{(4)}}: x + y = 6 \text{ or } 6 \text{ or } -6 \text{ or } -6$$
$$\sqrt{\textbf{(5)}}: x - y = 4 \text{ or } -4 \text{ or } 4 \text{ or } -4$$

$$\begin{array}{l} \therefore x = 5 \\ \therefore y = 1 \end{array} \bigg| \text{ or } \begin{array}{l} 1 \\ 5 \end{array} \bigg| \text{ or } \begin{array}{l} -1 \\ -5 \end{array} \bigg| \text{ or } \begin{array}{l} -5 \\ -1 \end{array}$$

CASE *c.*  When the sum or difference, and the sum of the squares, are given:

$$\begin{array}{lll} x + y & = 5 & \textbf{(1)} \\ x^2 + y^2 & = 17 & \textbf{(2)} \end{array}$$

$$\begin{array}{lll} \textbf{(1)}^2: & x^2 + 2xy + y^2 = 25 \\ \textbf{(2)}: & x^2 \qquad\ + y^2 = 17 \end{array}$$

$$\textbf{(1)}^2 - \textbf{(2)}: \quad 2xy = 8$$
$$xy = 4$$

Then proceed as in Case *a*, above.

$$\begin{array}{lll} x - y & = 3 & \textbf{(1)} \\ x^2 + y^2 & = 7 & \textbf{(2)} \end{array}$$

$$\begin{array}{lll} \textbf{(1)}^2: & x^2 - 2xy + y^2 = 9 \\ \textbf{(2)}: & x^2 \qquad\ + y^2 = 17 \end{array}$$

$$\textbf{(1)}^2 - \textbf{(2)}: \quad -2xy = -8$$
$$xy = 4$$

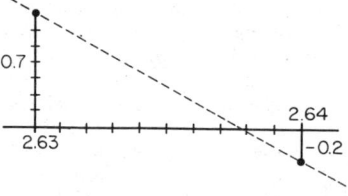

**Fig. 1**

Then proceed as in Case $a$, above.

CASE $d$.   When one equation is of the first degree and the other of the second, as $ax + by = c$, and $Ax^2 + Bxy + Cy^2 + Dx + Ey + F = 0$: Solve the first equation for $y$ in terms of $x$, and substitute in the second. This will give a quadratic equation in $x$. Solve this quadratic for the two values of $x$, and for each of these values of $x$ find the corresponding value of $y$ by substituting in the equation of the first degree.

**Simultaneous Equation of the First Degree in $n$ Unknowns**
For example:

|  |  |  | Factors |  |  |
|---|---|---|---|---|---|
| ($a$) | $2x - y + 3z + 5w = 29$ | $3$ | $1$ | $2$ |  |
| ($b$) | $5x + 2y - 2z + 3w = 15$ | $-5$ |  |  |  |
| ($c$) | $3x - 4y + 7z - w = 12$ | $5$ |  |  |  |
| ($d$) | $4x + 3y - 5z + 2w = 3$ |  |  | $-5$ |  |
| ($e$) | $-19x - 13y + 19z = 12$ | $-2$ | $-31$ |  |  |
| ($f$) | $17x - 21y + 38z = 89$ | $1$ |  |  |  |
| ($g$) | $-16x - 17y + 31z = 43$ |  | $19$ |  |  |
| ($h$) | $55x + 5y = 65$ | $16$ |  |  |  |
| ($i$) | $285x + 80y = 445$ | $-1$ |  |  |  |

($j$)   $595x = 595$;
   $\therefore x = 1$

   $5y = 65 - 55x = 65 - 55 = 10$;
   $\therefore y = 2$

   $19z = 12 + 19x + 13y = 12 + 19 + 26 = 57$;
   $\therefore z = 3$

   $2w = 3 - 4x - 3y + 5z = 3 - 4 - 6 + 15 = 8$;
   $\therefore w = 4$

Here $w$ is eliminated from ($a$) and ($b$), obtaining ($e$); from ($a$) and ($c$), obtaining ($f$); and from ($a$) and ($d$), obtaining ($g$). Then $z$ is eliminated from ($e$) and ($f$), obtaining ($h$), and from ($e$) and ($g$), obtaining ($i$). Then $y$ is eliminated from ($h$) and ($i$), obtaining ($j$), which contains only the single variable $x$. Hence $x = 1$. Now substituting this value of $x$ in either ($h$) or ($i$), $y$ is found; substituting these values of $x$ and $y$ in either ($e$), ($f$), or ($g$), $z$ is found; and so on. (Solution by determinants, see below, Determinants.)

**Approximate Solution of a Set of Simultaneous Equations of the First Degree When the Number of Equations is Greater Than the Number of Unknowns** (Method of Least Squares)

CASE 1. SINGLE UNKNOWN QUANTITY.  Given $n$ equations in one unknown $x$; e.g., $n$ equally careful, independent measurements of some physical quantity:

$$x = x_1 \qquad x = x_2 \qquad \cdots x = x_n$$

As the "best" value of $x$, take the arithmetic mean, $x_0$, of the several determinations, namely, $x_0 = (x_1 + x_2 + \cdots + x_n)/n$. The quantities $v_1 = x_0 - x_1$, $v_2 = x_0 - x_2$, $\ldots v_n = x_0 - x_n$ are called the **residuals** of the observed values with respect to $x_0$, and their absolute values (i.e., their numerical values without regard to sign) are denoted by $|v_1|, |v_2|, \ldots |v_n|$. (It can be shown that the sum of the squares of the residuals with respect to $x_0$ is smaller than the sum of the squares of the residuals with respect to any other value $x_0'$; hence the name of the method: "least squares.")

The quantities $r$ and $r_0$, defined exactly by **Bessel's** formulas:

$$r = \frac{0.6745}{\sqrt{n-1}} \sqrt{v_1^2 + v_2^2 + \cdots + v_n^2}$$

$$r_0 = \frac{0.6745}{\sqrt{n(n-1)}} \sqrt{v_1^2 + v_2^2 + \cdots + v_n^2}$$

or given approximately by the simpler formulas of **Peters:**

$$r = \frac{0.8453}{\sqrt{n(n-1)}} (|v_1| + |v_2| + \cdots + |v_n|)$$

$$r_0 = \frac{0.8453}{n\sqrt{n-1}} (|v_1| + |v_2| + \cdots + |v_n|)$$

are called the **probable error of a single observation** ($r$), and the **probable error of the mean** ($r_0$), for the given series of observations. Note that $r_0 = r/\sqrt{n}$. For tables of the coefficients, see Sec. 1, Mathematical Tables. This quantity $r$ (or $r_0$) is best regarded as merely a conventional means of recording the relative precision of different sets of observations. If $r$ is small, it may be inferred that most errors of the "accidental" class have been eliminated; but it should be especially noted that the smallness of $r$ gives no information in regard to "constant" or "systematic" errors.

A statement like "$x$ is equal to 2.36 with a probable error of 0.02," is written: $x = 2.36 \pm 0.02$, and is usually understood to mean that **the true value** of $x$, as far as it can be told, **is just as likely to lie inside as outside the interval** from 2.34 to 2.38.

To test the **distribution of residuals,** arrange the residuals in order of magnitude, without regard to sign, and count the number, $y$, of residuals which are numerically less than some assigned value $a$; divide $y$ by $n$, the total number of observations, and divide $a$ by $r$, the probable error of a single observation. Do this for various values of $a$, and compare the results with the table in Sec. 1, which gives the standard distribution of residuals, as found from experience from a large number of different series of observations. In particular, the number of residuals numerically less than $r$ should be about equal to the number numerically greater than $r$ (if $n$ is large). If any large discrepancy appears, the series of observations should be regarded as unsatisfactory.

NOTE.  The "mean square error" sometimes met with is equal to the probable error divided by 0.6745.

CASE 2. SEVERAL UNKNOWN QUANTITIES.  Assume that there have been obtained by measurement or observation $n$ different equations of the first degree involving, say, three unknown quantities, $x$, $y$, $z$. There are then $n$ simultaneous equations in three unknowns, and if $n > 3$ there will be, in general, no set of values of $x$, $y$, $z$ which will satisfy all these $n$ equations exactly. In such a case, the "best" set of values, $x_0$, $y_0$, $z_0$, may be found by the method of least squares as follows. (The process usually involves a large amount of labor; the use of a desk calculator is advisable.)

Given Equations
$$a_1x + b_1y + c_1z = p_1$$
$$a_2x + b_2y + c_2z = p_2$$
$$\cdot$$
$$a_nx + b_ny + c_nz + p_n$$

First, arrange the $n$ given equations in the form indicated, being careful not to modify any of them by multiplication or division. (Any of the coefficients may of course be zero.)

Next, form the three "normal equations" as follows: (1) Multiply each of the given equations by the coefficient of $x$ in that equation, and add; the result will be the first normal equation. (2) Multiply each of the given equations by the coefficient of $y$ in that equation, and add; the re-

Normal Equations
$$[aa]x_0 + [ab]y_0 + [ac]z_0 = [ap]$$
$$[ba]x_0 + [bb]y_0 + [bc]z_0 = [bp]$$
$$[ca]x_0 + [cb]y_0 + [cc]z_0 = [cp]$$

sult will be the second normal equation. (3)
Similarly for the third. {Notation: $[aa] = a_1^2 + a_2^2 + \cdots + a_n^2$; $[ab] = a_1b_1 + a_2b_2 + \cdots + a_nb_n$; $[ap] = a_1p_1 + a_2p_2 + \cdots + a_np_n$; etc.}

Finally, solve the three normal equations for the three unknowns in the usual way.

The quantities $v_1 = a_1x_0 + b_1y_0 + c_1z_0 - p_1$, etc., are called the **residuals** with respect to $x_0, y_0, z_0$. (It can be shown that the sum of the squares of the residuals with respect to $x_0, y_0, z_0$ is smaller than the corresponding quantity with respect to any other set of values, $x_0', y_0', z_0'$; this relation is taken as the criterion for the "best" set of values of $x, y, z$.)

One application of the above least-squares process is to find the constants defining a curve of assumed form best fitting a set of observed points. We illustrate for a straight line $y = mx + b$ best fitting a set of points $x_i, y_i$. We assume that the $x_i$ are exact and only the $y_i$ are subject to error. Here the normal equations are $m(xx) + b(x1) = (xy)$, $m(x1) + b(11) = (y1)$, where $(x1) = x_1 + x_2 + \cdots + x_n$, $(11) = 1 + 1 + \cdots + 1 = n$. Thus the second normal equation expresses the fact that the center of gravity of the points $(x_i/n, y_i/n)$ lies on the line of best fit.

The **probable error of a single observation** is

$$r = \frac{0.6745}{\sqrt{n - m}} \sqrt{v_1^2 + v_2^2 + \cdots + v_n^2}, \text{ or approx.}$$

$$r = \frac{0.8453}{\sqrt{n(n - m)}} (|v_1| + |v_2| + \cdots + |v_n|)$$

where $m = $ the number of unknown quantities (here $m = 3$).

## DETERMINANTS

Determinants are used chiefly in formulating theoretical results; they are seldom of use in numerical computation.

### Evaluation of Determinants

Of the second order:

$$\begin{vmatrix} a_1 & b_1 \\ a_2 & b_2 \end{vmatrix} = a_1b_2 - a_2b_1$$

Of the third order:

$$\begin{vmatrix} a_1 & b_1 & c_1 \\ a_2 & b_2 & c_2 \\ a_3 & b_3 & c_3 \end{vmatrix} = a_1 \begin{vmatrix} b_2 & c_2 \\ b_3 & c_3 \end{vmatrix} - a_2 \begin{vmatrix} b_1 & c_1 \\ b_3 & c_3 \end{vmatrix} + a_3 \begin{vmatrix} b_1 & c_1 \\ b_2 & c_2 \end{vmatrix}$$

$$= a_1(b_2c_3 - b_3c_2) - a_2(b_1c_3 - b_3c_1) + a_3(b_1c_2 - b_2c_1)$$

Of the fourth order:

$$\begin{vmatrix} a_1 & b_1 & c_1 & d_1 \\ a_2 & b_2 & c_2 & d_2 \\ a_3 & b_3 & c_3 & d_3 \\ a_4 & b_4 & c_4 & d_4 \end{vmatrix}$$

$$= a_1 \begin{vmatrix} b_2 & c_2 & d_2 \\ b_3 & c_3 & d_3 \\ b_4 & c_4 & d_4 \end{vmatrix} - a_2 \begin{vmatrix} b_1 & c_1 & d_1 \\ b_3 & c_3 & d_3 \\ b_4 & c_4 & d_4 \end{vmatrix} + a_3 \begin{vmatrix} b_1 & c_1 & d_1 \\ b_2 & c_2 & d_2 \\ b_4 & c_4 & d_4 \end{vmatrix} - a_4 \begin{vmatrix} b_1 & c_1 & d_1 \\ b_2 & c_2 & d_2 \\ b_3 & c_3 & d_3 \end{vmatrix}$$

etc. In general, to evaluate a determinant of the $n$th order, take the elements of the first column with signs alternately plus and minus, and form the sum of the products obtained by multiplying each of these elements by its corresponding **minor**. The minor corresponding to any element $a_1$ is the determinant (of next lower order) obtained by striking out from the given determinant the row and column containing $a_1$.

### Properties of Determinants

**1.** The columns may be changed to rows and the rows to columns:

$$\begin{vmatrix} a_1 & b_1 & c_1 \\ a_2 & b_2 & c_2 \\ a_3 & b_3 & c_3 \end{vmatrix} = \begin{vmatrix} a_1 & a_2 & a_3 \\ b_1 & b_2 & b_3 \\ c_1 & c_2 & c_3 \end{vmatrix}$$

**2.** Interchanging two adjacent columns changes the sign of the result.

**3.** If two columns are equal, the determinant is zero.

**4.** If the elements of one column are $m$ times the elements of another column, the determinant is zero.

**5.** To multiply a determinant by any number $m$, multiply all the elements of any one column by $m$.

**6.**
$$\begin{vmatrix} a_1 + p_1 + q_1, & b_1 & c_1 \\ a_2 + p_2 + q_2, & b_2 & c_2 \\ a_3 + p_3 + q_3, & b_3 & c_3 \end{vmatrix} = \begin{vmatrix} a_1 & b_1 & c_1 \\ a_2 & b_2 & c_2 \\ a_3 & b_3 & c_3 \end{vmatrix} + \begin{vmatrix} p_1 & b_1 & c_1 \\ p_2 & b_2 & c_2 \\ p_3 & b_3 & c_3 \end{vmatrix} + \begin{vmatrix} q_1 & b_1 & c_1 \\ q_2 & b_2 & c_2 \\ q_3 & b_3 & c_3 \end{vmatrix}$$

**7.**
$$\begin{vmatrix} a_1 & b_1 & c_1 \\ a_2 & b_2 & c_2 \\ a_3 & b_3 & c_3 \end{vmatrix} = \begin{vmatrix} a_1 + mb_1, & b_1 & c_1 \\ a_2 + mb_2, & b_2 & c_2 \\ a_3 + mb_3, & b_3 & c_3 \end{vmatrix}$$

### Solution of Simultaneous Equations by Determinants

If
$$a_1x + b_1y + c_1z = p_1$$
$$a_2x + b_2y + c_2z = p_2 \text{ where } D = \begin{vmatrix} a_1 & b_1 & c_1 \\ a_2 & b_2 & c_2 \\ a_3 & b_3 & c_3 \end{vmatrix} \neq 0,$$
$$a_3x + b_3y + c_3z = p_3$$

then
$$x = D_1/D,$$
$$y = D_2/D,$$
$$z = D_3/D,$$

where $D_1 = \begin{vmatrix} p_1 & b_1 & c_1 \\ p_2 & b_2 & c_2 \\ p_3 & b_3 & c_3 \end{vmatrix}$, $D_2 = \begin{vmatrix} a_1 & p_1 & c_1 \\ a_2 & p_2 & c_2 \\ a_3 & p_3 & c_3 \end{vmatrix}$, $D_3 = \begin{vmatrix} a_1 & b_1 & p_1 \\ a_2 & b_2 & p_2 \\ a_3 & b_3 & p_3 \end{vmatrix}$

Similarly for a larger (or smaller) number of equations.

## IMAGINARY OR COMPLEX QUANTITIES

In the algebra of imaginary or complex quantities, the objects on which the operations of the algebra are performed are not numbers in any ordinary sense of the word, but are best thought of as **points in a plane** (or as **vectors** drawn from a fixed origin to these points). The **"complex plane"** is determined by three fundamental points, $O$, $U$, $i$, arranged as in Fig. 2 and

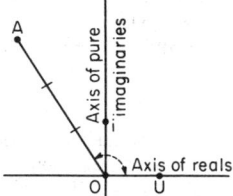

**Fig. 2**

called the **zero point**, the **unit point**, and the **imaginary unit point**, respectively. All points on the line through $O$ and $U$ are called **real points** —positive if on the right of $O$, negative if on the left. All the remaining points in the plane are called **imaginary points** —those on the line through $O$ and $i$ being called the **pure imaginary points**.

The position of any point $A$ in the plane may be determined by the *distance* from the origin $O$, measured in terms of $OU$ as the unit length, and the *angle* $\varphi$ which $OA$ makes with the positive direction of the axis of reals. The distance $r$ is sometimes called the modulus or absolute value of the point; the angle $\varphi$ is sometimes called the amplitude or argument of the point. The notation $A = (3/120°)$ means the point whose *distance*, $r$, is 3 times $OU$, and whose *angle*, $\varphi$, is 120°. The development of the algebra depends wholly on the definitions of three fundamental operations denoted by $A + B$, $A \times B$, and $e_A$, as follows.

**Addition and Subtraction**   The sum, $A + B$, of two points $A$ and $B$ is defined as the point reached by starting from $A$ and performing a journey equal in length and direction to the journey from $O$ to $B$. That is, the vector from $O$ to $A + B$ is the vector sum of the vectors $OA$ and $OB$. In case $A$ and $B$ are not in line with $O$, the point $A + B$ is the fourth vertex of a

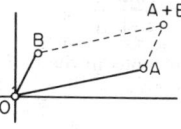

**Fig. 3**

parallelogram of which $OA$ and $OB$ are the sides (Fig. 3). Conversely, if any two points $A$ and $B$ are given, there is a definite point $X$ such that $A = B + X$; this point $X$ is called the **remainder**, $A$ minus $B$, and is denoted by $A - B$. The point $O - B$ is denoted for brevity by $-B$. With these definitions of $A + B$ and $A - B$, all the ordinary laws of addition and subtraction that hold in the algebra of real numbers hold also in the algebra of complex quantities. In particular, the zero point $O$ has all the formal properties of the number zero, and is denoted by 0.

NOTE.   If $A$ and $B$ are "real" points, $A + B$ and $A - B$ will also be real.

**Repeated Addition. Multiples and Submultiples**   The point $A + A + A + \cdots + A$ to $n$ terms is called the **nth multiple of $A$** and is denoted by $nA$. The points $U$, $2U$, $3U$, ... are denoted, for brevity, by $1, 2, 3, \ldots$. Conversely, if any point $A$ and any positive integer $n$ are given, there is a definite point $X$ such that $nX = A$; this point $X$ is called the **$n$ th submultiple of $A$**, and is denoted by $A/n$. The points $U/2$, $U/3$, ... are denoted, for brevity, by $\frac{1}{2}$, $\frac{1}{3}$, ....

**Multiplication and Division**   The **product**, $A \times B$, or $A \cdot B$, or $AB$, of two points $A$ and $B$ is defined as the point whose angle is the sum of the angles of the given points, and whose distance is the product of the distances. (See Fig. 4.) Thus, if $A = (5/120°)$ and $B = (2/270°)$, then $AB = (10/30°)$. Conversely, if any two points $A$ and $B$ are given, provided $B$ is not zero, there is a definite point $X$ such that $A = BX$. This point $X$ is called the **quotient**, $A$ divided by $B$, and is denoted by $A/B$ (where $B \neq 0$). Thus, the point $A/B$ is a point whose angle is

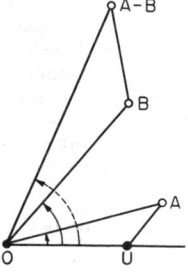

**Fig. 4**

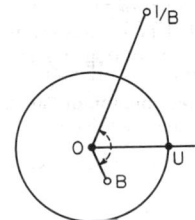

**Fig. 5**

the angle of $A$ minus the angle of $B$, and whose distance is the distance of $A$ divided by the distance of $B$. The point $U/B$ ($B \neq 0$) is called the **reciprocal** of the point $B$, and is denoted by $1/B$ (see Fig. 5). With these definitions of $AB$ and $A/B$, the elementary laws of multiplication and division that hold in the algebra of real numbers hold also in the algebra of complex quantities. In particular, the point $U$ has all the formal properties of the number unity, and is denoted by 1.

NOTE.   If $A$ and $B$ are real, $AB$ and $A/B$ will also be real.

**Repeated Multiplication. Powers and Roots**   The point $A \times A \times A \times \cdots \times A$ to $n$ factors is called the $n$th power of $A$ and is denoted by $A^n$ (Fig. 6). Conversely, if any point $A$ (not 0) and any positive integer $n$ are given, there will be $n$ distinct points $X$ such that $X^n = A$; each of these points is called an **nth root of $A$**, some one of them, usually the one with the smallest positive angle, being denoted by $\sqrt[n]{A}$ or $A^{1/n}$. Thus, the point $\sqrt[n]{A}$ is a point whose distance is the $n$th root of the distance of $A$, and whose angle is $1/n$th of the angle of $A$. All the $n$th roots of $A$ will lie on the circumference of a circle

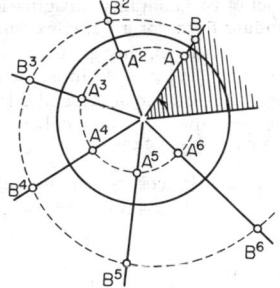

**Fig. 6**

about $O$ as center, and will divide that circumference into $n$ equal parts (Fig. 7). Every point $A$ (not 0) has two square roots, three cube roots, etc. Hence the theorem "If $A^n = B^n$ then $A = B$" does not hold in this algebra, and the ordinary rules for radical signs must be applied with caution. For example, if $A$ and $B$ are positive reals, $\sqrt{-A} \cdot \sqrt{-B} = -\sqrt{AB}$ and not $\sqrt{(-A)(-B)}$, which would give $+ \sqrt{AB}$.

NOTE.   If $A$ is real and positive, $\sqrt[n]{A}$ will be real and positive; if $A$ is real and negative, $\sqrt[n]{A}$ will be real if $n$ is odd and imaginary if $n$ is even.

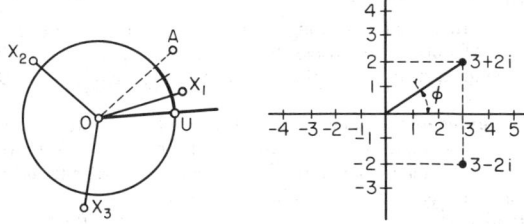

**Fig. 7**          **Fig. 8**

**Properties of $i$**   The point $i$ is the point whose distance is 1 and whose angle is 90°. It follows from the definition above that **multiplying any point $A$ by $i$ has the effect of rotating the point through an angle of $+90°$** without changing its distance from $O$. In particular, $i^2 = -1$, $i^3 = -i$, $i^4 = 1$, $i^5 = i$, etc.; $i - \sqrt{-1}$, $-i = -\sqrt{-1}$; where "1" denotes not the *number* one, but the *point U*.

Similarly, multiplying any point $A$ by $-1$ has the effect of rotating the point through $180°$.

**First Standard Form for a Complex Quantity** (Fig. 8)   Any point $A$ can be expressed in the form $x + iy$, where $x$ and $y$ are real points. For example, the three cube roots of 1 are 1, $-\frac{1}{2} + \frac{1}{2}i\sqrt{3}$, and $-\frac{1}{2} - \frac{1}{2}i\sqrt{3}$. In general, $(x_1 + iy_1) + (x_2 + iy_2) = (x_1 + x_2) + i(y_1 + y_2)$

$$(x_1 + iy_1)(x_2 + iy_2) = (x_1x_2 - y_1y_2) + i(x_2y_1 + x_1y_2)$$
$$\frac{x_1 + iy_1}{x_2 + iy_2} = \frac{x_1x_2 + y_1y_2}{x_2^2 + y_2^2} + i\,\frac{x_2y_1 - x_1y_2}{x_2^2 + y_2^2}$$

If two complex quantities are equal, their real parts must be equal, and the coefficients of their pure imaginary parts must also be equal; i.e., if $x_1 + iy_1 = x_2 + iy_2$, then $x_1 = x_2$ and $y_1 = y_2$. Thus a single equation between complex quantities is equivalent to two equations between real quantities.

**Conjugate Imaginaries**   Two points $A = x + iy$ and $B = x - iy$ are called conjugate imaginaries. Two such points are symmetrically situated with regard to the axis of reals. The sum and product of two conjugate imaginaries will be real.

**Second Standard Form for a Complex Quantity**   Since $x = r \cos \varphi$ and $y = r \sin \varphi$, any point $A = x + iy$ can be expressed as $A = r (\cos \varphi + i \sin \varphi)$, where $r$ is real and positive (namely, the distance of $A$), and $\varphi$ is real (namely, the angle of $A$). For example, the three cube roots of 1 are 1, $\cos 120° + i \sin 120°$, and $\cos 240° + i \sin 240°$. In general,

$$[r_1(\cos \varphi_1 + i \sin \varphi_1)] [r_2(\cos \varphi_2 + i \sin \varphi_2)]$$
$$= r_1r_2[\cos (\varphi_1 + \varphi_2) + i \sin (\varphi_1 + \varphi_2)]$$

and

$$[r(\cos \varphi + i \sin \varphi)]^n$$
$$= r^n[\cos (n\varphi) + i \sin (n\varphi)] \quad \textbf{(De Moivre's theorem)}$$

The **exponential function,** $e^A$, or exp $A$, of any point $A = x + iy$ is defined as the point whose distance is $e^x$ and whose angle (measured in radians) is $y$; i.e., $e^{x+iy} = e^x(\cos y + i \sin y)$. Here $e^x$ means the ordinary exponential function of the real quantity $x$, where $e = 2.718$.

From this definition, the usual formal laws of exponents can be deduced: $e^A e^B = e^{A+B}$, $(e^A)^n = e^{nA}$, $e^{-A} = 1/e^A$; $e^1 = e$, $e^0 = 1$.

The function $e^A$ is a periodic function with a pure imaginary period $2\pi i$; that is, $e^{A\pm k2\pi i} = e^A$, where $k$ is any positive integer.

If $A$ is made to move along a line parallel to the axis of reals (or axis of pure imaginaries), the corresponding point $e^A$ will move along a straight line through $O$ (or along a circle about $O$ as center).

**Properties of** $e^{i\varphi}$   The point $e^{i\varphi}$ is a point whose distance is 1 and whose angle is $\varphi$. It follows from the definitions above that **multiplying any point** $A$ **by** $e^{i\varphi}$ **has the effect of rotating the point through an angle** $\varphi$, without changing its distance from $O$. In particular, $e^{i\pi} = -1$, $e^{-i\pi} = -1$; $e^{i\pi/2} = i$; $e^{-i\pi/2} = -i$; $e^{2\pi i} = 1$.

**Third Standard Form for a Complex Quantity**   Any point $A$ can be expressed in the form $A = re^{i\varphi}$, where $r$ is the distance and $\varphi$ the angle of the point. For example, the three cube roots of 1 are 1, $e^{(1/3)2\pi i}$, $e^{(2/3)2\pi i}$. In general,

$$(r_1e^{i\varphi_1})(r_2e^{i\varphi_2}) = (r_1r_2)e^{i(\varphi_1+\varphi_2)} \qquad (re^{i\varphi})^n = (r^n)e^{in\varphi}$$

If $x + iy = re^{i\varphi}$, then $r = \sqrt{x^2 + y^2}$, $\sin \varphi = y/r$, $\cos \varphi = x/r$, $\tan \varphi = y/x$.

If two complex quantities are equal, their distances will be equal, and their angles will differ at most by some multiple of $2\pi$. Thus, if $r_1e^{i\varphi_1} = r_2e^{i\varphi_2}$ then $r_1 = r_2$ and $\varphi_1 = \varphi_2$ or $\varphi_2 \pm k2\pi$. Here again a single equation between complex quantities is equivalent to two equations between real quantities.

**Definition of** $A^B$   Let $A = re^{i\varphi}$; then $A^B = \exp [(\ln r + i\varphi)B]$.

For example, $i^i = e^{-\pi/2}$ where $i = \sqrt{-1}$

If $a$ is a positive real, $a^{x+iy} = a^x [\cos (y \ln a) + i \sin (y \ln a)]$.

**Trigonometric and Hyperbolic Functions of a Complex Variable**   If $A$ is any point, then, by definition,

$$\sin A = \frac{e^{iA} - e^{-iA}}{2i} \qquad \sinh A = \frac{e^A - e^{-A}}{2}$$
$$\cos A = \frac{e^{iA} + e^{-iA}}{2} \qquad \cosh A = \frac{e^A + e^{-A}}{2}$$
$$\tan A = \frac{\sin A}{\cos A} \qquad \tanh A = \frac{\sinh A}{\cosh A}$$

Hence the formulas that hold for these functions in the real case hold also for the complex case. Further;

$$\sin (x + iy) = \sin x \cosh y + i \cos x \sinh y$$
$$\cos (x + iy) = \cos x \cosh y - i \sin x \sinh y$$
$$\sinh (x + iy) = \sinh x \cos y + i \cosh x \sin y$$
$$\cosh (x + iy) = \cosh x \cos y + i \sinh x \sin y$$
$$\sin iy \;\; = i \sinh y$$
$$\cos iy \;\; = \cosh y$$
$$\sinh iy \;\; = i \sin y$$
$$\cosh iy \;\; = \cos y$$

where $\sin x$, $\sinh x$, etc., are the ordinary trigonometric and hyperbolic functions of the real variables $x$ and $y$. The functions $\sin A$ and $\cos A$ are periodic with a real period $2\pi$. The functions $\sinh A$ and $\cosh A$ are periodic with a pure imaginary period $2\pi i$.

**Logarithmic and Other Inverse Functions of a Complex Variable.** If any point $A$ is given, there will be an infinite number of points $X$ such that $e^x = A$; any one of these points may be called a logarithm of $A$, and be denoted by $\ln A$. All the values of the logarithm of $A$ may be obtained from any one value by adding multiples of $2\pi i$.

If $x + iy = re^{i\varphi}$, then $\ln (x + iy) = \ln r + i\varphi \pm k \cdot 2\pi i$.

If any point $A$ is given, there will be an infinite number of points $X$ such that $\sin X = A$; any one of these may be denoted by $\sin^{-1} A$. The functions $\cos^{-1} A$, $\sinh^- A$, etc., are defined in a similar way.

The elementary laws of operation which hold for these functions in the algebra of reals hold also, in a general way, in the algebra of complex quantities; but caution must be used, on account of the ambiguity in the symbols $\ln A$, $\sin^{-1}A$, etc., which denote many-valued functions. For a method of numerically computing possible correct determinations, see Franklin, "Fourier Methods," pp. 19–26, McGraw-Hill.

**Differentiation of Functions of a Complex Variable**   If $w = f(z)$, the derivative of $w$ with respect to $z$ is defined as

$$dw/dz = \lim \{[f(z + \Delta z) - f(z)]/\Delta z\}$$

when $\Delta z$ approaches 0

It can be shown that $\lim [(\exp \Delta z - 1)/\Delta z] = 1$; hence $d(e^z)$

= $e^z \, dz$, $d(\sin z) = \cos z \, dz$, etc., so that the formulas for differentiation here are the same as in the case of a real variable.

NOTE. For the algebra of vector analysis, which differs in important respects from the algebra of complex quantities, see Sec. 2, Vector Analysis.

# TRIGONOMETRY

## by Philip Franklin

## FORMAL TRIGONOMETRY

**Angles or Rotations** An **angle** is generated by the rotation of a ray, as $Ox$, about a fixed point $O$ in the plane. Every angle has an **initial line** ($OA$) from which the rotation started (Fig. 1), and a **terminal line** ($OB$) where it stopped; and the counterclockwise direction of rotation is taken as positive. Since the rotating ray may revolve as often as desired, angles of any magnitude, positive or negative, may be obtained. Two angles are **congruent** if they may be superposed so that their initial lines coincide and their terminal lines coincide; i.e., two congruent angles are either equal or differ by some multiple of 360°. Two angles are **complementary** if their sum is 90°; **supplementary** if their sum is 180°. (The acute angles of a right-angled triangle are complementary.) If the initial line is placed so that it runs horizontally to the right, as in Fig. 2, then the angle is said to be an angle in the 1st, 2nd, 3rd, or 4th **quadrant** according as

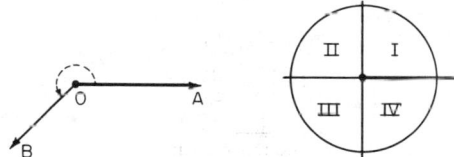

**Fig. 1**          **Fig. 2**

the terminal line lies across the region marked, I, II, III, or IV. The angles 0, 90, 180, 270° are called the quadrantal angles.

### Units of Angular Measurement

1. SEXAGESIMAL MEASURE. (360 degrees = 1 revolution.) 1 degree = 1° = $\frac{1}{90}$ of a right angle. The degree is usually divided into 60 equal parts called minutes ('), and each minute into 60 equal parts called seconds("); while the second is subdivided decimally. But for many purposes it is more convenient to divide the degree itself into decimal parts, thus avoiding the use of minutes and seconds.

2. CENTESIMAL MEASURE, used chiefly in France. (400 grades = 1 revolution.) 1 grade = $\frac{1}{100}$ of a right angle. The grade is always divided decimally, the following terms being sometimes used: 1 "centesimal minute" = $\frac{1}{100}$ of a grade; 1 "centesimal second" = $\frac{1}{100}$ of a centesimal minute. In reading Continental books it is important to notice carefully which system is employed.

3. RADIAN, OR CIRCULAR, MEASURE. ($\pi$ radians = 180 degrees.) 1 radian = the angle subtended by an arc whose length is equal to the length of the radius. The radian is constantly used in higher mathematics and in mechanics, and is always divided decimally.

1 radian = 57°.30 − = 57°.2957795131 = 57° 17′ 44″ .806247 = 180°/$\pi$.

1° = 0.01745 . . . radian = 0.01745 32925 radian.

1′ = 0.00029 08882 radian.

1″ = 0.00000 48481 radian.

(For ten-place conversion tables, see the Smithsonian Tables of Hyperbolic Functions, Washington, D.C.)

**Definitions of the Trigonometric Functions** Let $x$ be any angle whose initial line is $OA$ and terminal line $OP$ (see Fig. 3). Drop a perpendicular from $P$ on $OA$ or $OA$ produced. In the right triangle $OMP$, the three sides are $MP$ − "side opposite" $O$ (positive if running upward); $OM$ = "side adjacent" to $O$ (positive if running to the right); $OP$ = "hypotenuse" or "radius" (may always be taken as positive); and the six ratios between these sides are the principal trigonometric functions of the angle $x$; thus:

$$\text{sine of } x = \sin x = \text{opp/hyp} = MP/OP$$
$$\text{cosine of } x = \cos x = \text{adj/hyp} = OM/OP$$
$$\text{tangent of } x = \tan x = \text{opp/adj} = MP/OM$$
$$\text{cotangent of } x = \cot x = \text{adj/opp} = OM/MP$$
$$\text{secant of } x = \sec x = \text{hyp/adj} = OP/OM$$
$$\text{cosecant of } x = \csc x = \text{hyp/opp} = OP/MP$$

The last three are best remembered as the reciprocals of the first three:

$$\cot x = 1/\tan x \quad \sec x = 1/\cos x \quad \csc x = 1/\sin x$$

### Signs of the Trigonometric Functions

| If $x$ is in quadrant | I | II | III | IV |
|---|---|---|---|---|
| sin $x$ and csc $x$ are.......................... | + | + | − | − |
| cos $x$ and sec $x$ are.......................... | + | − | − | + |
| tan $x$ and cot $x$ are.......................... | + | − | + | − |

Vers $x$ and covers $x$ are always positive.

| $x$ | 0° to 90° | 90° to 180° | 180° to 270° | 270° to 360° | Values at | | |
|---|---|---|---|---|---|---|---|
| | | | | | 30° | 45° | 60° |
| sin $x$ | +0 to +1 | +1 to +0 | −0 to −1 | −1 to −0 | ½ | ½$\sqrt{2}$ | ½$\sqrt{3}$ |
| csc $x$ | +∞ to +1 | +1 to +∞ | −∞ to −1 | −1 to −∞ | 2 | $\sqrt{2}$ | ⅔$\sqrt{3}$ |
| cos $x$ | +1 to +0 | −0 to −1 | −1 to −0 | +0 to +1 | ½$\sqrt{3}$ | ½$\sqrt{2}$ | ½ |
| sec $x$ | +1 to +∞ | −∞ to −1 | −1 to −∞ | +∞ to +1 | ⅔$\sqrt{3}$ | $\sqrt{2}$ | 2 |
| tan $x$ | +0 to +∞ | −∞ to −0 | +0 to +∞ | −∞ to −0 | ⅓$\sqrt{3}$ | 1 | $\sqrt{3}$ |
| cot $x$ | +∞ to +0 | −0 to −∞ | +∞ to +0 | −0 to −∞ | $\sqrt{3}$ | 1 | ⅓$\sqrt{3}$ |
| vers $x$ | +0 to +1 | +1 to +2 | +2 to +1 | +1 to +0 | | | |
| covers $x$ | +1 to +0 | +0 to +1 | +1 to +2 | +2 to +1 | | | |

$$\sqrt{2} = 1.4142; \; \tfrac{1}{2}\sqrt{2} = 0.7071; \; \sqrt{3} = 1.7321; \; \tfrac{1}{2}\sqrt{3} = 0.8660; \; \tfrac{1}{3}\sqrt{3} = 0.5774; \; \tfrac{2}{3}\sqrt{3} = 1.1547.$$

Other functions in use are the versed sine, the coversed sine, and the exterior secant:

$$\text{vers } x = 1 - \cos x \qquad \text{covers } x = 1 - \sin x$$
$$\text{exsec } x = \sec x - 1$$

**Variations in the functions as $x$ varies from 0 to 360°** are shown in the table above. The variations in the sine and cosine are best remembered by noting the changes in the lines $MP$ and $OM$ (Fig. 4) in the "unit circle" (i.e., a circle with radius = $OP = 1$), as $P$ moves around the circumference.

**Trigonometrical Tables**    The tables shown in Sec. 1 give the values of the principal trigonometric functions and of their logarithms, correct to four places of decimals, the angle advancing either by tenths of a degree intervals or by 10 min intervals. These tables will be found adequate for most computations in which an accuracy of 1 part in 1,000 is sufficient. If much computing is to be done, it is advisable to use a separate volume of tables, containing more facilities for interpolation, and printed in larger type, such as the four-place tables of Huntington (Houghton Mifflin), with convenient marginal tabs; the five-place tables published by Macmillan or many others; the six-place tables of Bremiker; the standard seven-place tables of Schrön, Vega, or Bruhns (angles advancing by 10 sec); or the great eight-place of Bauschinger and Peters (angles advancing at intervals of 1 sec from 0 to 90°). The larger tables give only the logarithms of the functions, not the natural values.

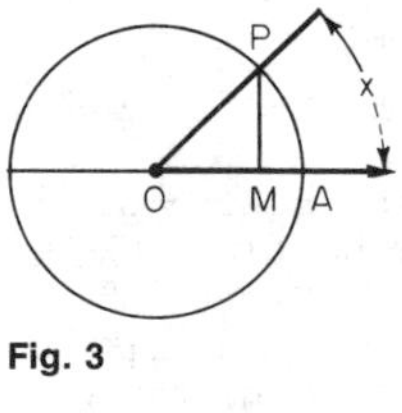

**To Find Any Function of a Given Angle**    (Reduction to the first quadrant.) It is often required to find the functions of any angle $x$ from a table that includes only angles between 0 and 90°. If $x$ is not already between 0 and 360°, first "reduce to the first revolution" by simply adding or subtracting the proper multiple of 360° [for any function of $(x)$ = the same function of $(x \pm n \times 360°)$]. Next **reduce to first quadrant** per table below.

The "reduced angle" ($x - 90°$, or $x - 180°$, or $x - 270°$) will in each case be an angle between 0 and 90°, whose functions can then be found in the table.

NOTE.   The formulas for sine and cosine are best remembered by aid of the unit circle.

**To Find the Angle When One of Its Functions Is Given**    In general, there will be two angles between 0 and 360° corresponding to any given function. The rules showing how to find these angles are tabulated at the top of the next page.

**Relations between the Functions of a Single Angle** (See Fig. 5)

$$\sin^2 x + \cos^2 x = 1$$
$$\tan x = \frac{\sin x}{\cos x}$$
$$\cot x = \frac{1}{\tan x} = \frac{\cos x}{\sin x}$$
$$1 + \tan^2 x = \sec^2 x = \frac{1}{\cos^2 x}$$
$$1 + \cot^2 x = \csc^2 x = \frac{1}{\sin^2 x}$$
$$\sin x = \sqrt{1 - \cos^2 x} = \frac{\tan x}{\sqrt{1 + \tan^2 x}} = \frac{1}{\sqrt{1 + \cot^2 x}}$$
$$\cos x = \sqrt{1 - \sin^2 x} = \frac{1}{\sqrt{1 + \tan^2 x}} = \frac{\cot x}{\sqrt{1 + \cot^2 x}}$$

**Fig. 3**

| If $x$ is between | 90° and 180° | 180° and 270° | 270° and 360° |
|---|---|---|---|
| Subtract | 90° from $x$ | 180° from $x$ | 270° from $x$ |
| Then sin $x$ .......... | = + cos $(x - 90°)$ | = − sin $(x - 180°)$ | = − cos $(x - 270°)$ |
| csc $x$ .......... | = + sec $(x - 90°)$ | = − csc $(x - 180°)$ | = − sec $(x - 270°)$ |
| cos $x$ .......... | = − sin $(x - 90°)$ | = − cos $(x - 180°)$ | = + sin $(x - 270°)$ |
| sec $x$ .......... | = − csc $(x - 90°)$ | = − sec $(x - 180°)$ | = + csc $(x - 270°)$ |
| tan $x$ .......... | = − cot $(x - 90°)$ | = + tan $(x - 180°)$ | = − cot $(x - 270°)$ |
| cot $x$ .......... | = − tan $(x - 90°)$ | = + cot $(x - 180°)$ | = − tan $(x - 270°)$ |
| vers $x$ ......... | = 1 + sin $(x - 90°)$ | = 1 + cos $(x - 180°)$ | = 1 − sin $(x - 270°)$ |
| covers $x$ ....... | = 1 − cos $(x - 90°)$ | = 1 + sin $(x - 180°)$ | = 1 + cos $(x - 270°)$ |

| Given | First find from the tables an *acute* angle $x_0$ such that | Then the required angles $x_1$ and $x_2$ will be |
|---|---|---|
| $\sin x = +a$ | $\sin x_0 = a$ | $x_0$ and $180° - x_0$ |
| $\cos x = +a$ | $\cos x_0 = a$ | $x_0$ and $[360° - x_0]$ |
| $\tan x = +a$ | $\tan x_0 = a$ | $x_0$ and $[180° + x_0]$ |
| $\cot x = +a$ | $\cot x_0 = a$ | $x_0$ and $[180° + x_0]$ |
| $\sin x = -a$ | $\sin x_0 = a$ | $[180° + x_0]$ and $[360° - x_0]$ |
| $\cos x = -a$ | $\cos x_0 = a$ | $180° - x_0$ and $[180° + x_0]$ |
| $\tan x = -a$ | $\tan x_0 = a$ | $180° - x_0$ and $[360° - x_0]$ |
| $\cot x = -a$ | $\cot x_0 = a$ | $180° - x_0$ and $[360° - x_0]$ |

The angles enclosed in brackets lie outside the range 0 to 180 deg and hence cannot occur as angles in a triangle.

**Functions of Negative Angles**   $\sin(-x) = -\sin x$; $\cos(-x) = \cos x$; $\tan(-x) = -\tan x$.

**Functions of the Sum and Difference of Two Angles**

$\sin(x + y) = \sin x \cos y + \cos x \sin y$.
$\cos(x + y) = \cos x \cos y - \sin x \sin y$.
$\tan(x + y) = (\tan x + \tan y)/(1 - \tan x \tan y)$.
$\cot(x + y) = (\cot x \cot y - 1)/(\cot x + \cot y)$.
$\sin(x - y) = \sin x \cos y - \cos x \sin y$.
$\cos(x - y) = \cos x \cos y + \sin x \sin y$.
$\tan(x - y) = (\tan x - \tan y)/(1 + \tan x \tan y)$.
$\cot(x - y) = (\cot x \cot y + 1)/(\cot y - \cot x)$.
$\sin x + \sin y = 2 \sin \frac{1}{2}(x + y) \cos \frac{1}{2}(x - y)$.
$\sin x - \sin y = 2 \cos \frac{1}{2}(x + y) \cos \frac{1}{2}(x - y)$.
$\cos x + \cos y = 2 \cos \frac{1}{2}(x + y) \cos \frac{1}{2}(x - y)$.
$\cos x - \cos y = -2 \sin \frac{1}{2}(x + y) \sin \frac{1}{2}(x - y)$.

$$\tan x + \tan y = \frac{\sin(x + y)}{\cos x \cos y}; \quad \cot x + \cot y = \frac{\sin(x + y)}{\sin x \sin y}.$$

$$\tan x - \tan y = \frac{\sin(x - y)}{\cos x \cos y}; \quad \cot x - \cot y = \frac{\sin(y - x)}{\sin x \sin y}.$$

$\sin^2 x - \sin^2 y = \cos^2 y - \cos^2 x = \sin(x + y) \sin(x - y)$.
$\cos^2 x - \sin^2 y = \cos^2 y - \sin^2 x = \cos(x + y) \cos(x - y)$.
$\sin(45° + x) = \cos(45° - x)$; $\tan(45° + x) = \cot(45° - x)$.
$\sin(45° - x) = \cos(45° + x)$; $\tan(45° - x) = \cot(45° + x)$.

In the following transformations, $a$ and $b$ are supposed to be positive, $c = \sqrt{a^2 + b^2}$, $A =$ the positive acute angle for which $\tan A = a/b$, and $B =$ the positive acute angle for which $\tan B = b/a$:

$a \cos x + b \sin x = c \sin(A + x) = c \cos(B - x)$.
$a \cos x - b \sin x = c \sin(A - x) = c \cos(B + x)$.

**Functions of Multiple Angles and Half Angles**

$\sin 2x = 2 \sin x \cos x$; $\sin x = 2 \sin \frac{1}{2}x \cos \frac{1}{2}x$.
$\cos 2x = \cos^2 x - \sin^2 x = 1 - 2 \sin^2 x = 2 \cos^2 x - 1$.

$$\tan 2x = \frac{2 \tan x}{1 - \tan^2 x}; \quad \cot 2x = \frac{\cot^2 x - 1}{2 \cot x}.$$

$\sin 3x = 3 \sin x - 4 \sin^3 x$; $\tan 3x = \dfrac{3 \tan x - \tan^3 x}{1 - 3 \tan^2 x}$.

$\cos 3x = 4 \cos^3 x - 3 \cos x$.
$\sin(nx) = n \sin x \cos^{n-1} x - (n)_3 \sin^3 x \cos^{n-3} x + (n)_5 \sin^5 x \cos^{n-5} x - \cdots$
$\cos(nx) = \cos^n x - (n)_2 \sin^2 x \cos^{n-2} x + (n)_4 \sin^4 x \cos^{n-4} x - \cdots$,

where $(n)_2$, $(n)_3$, . . . are the binomial coefficients.

$\sin \frac{1}{2}x = \pm \sqrt{\frac{1}{2}(1 - \cos x)}$. $1 - \cos x = 2 \sin^2 \frac{1}{2}x$.

$\cos \frac{1}{2}x = \pm \sqrt{\frac{1}{2}(1 + \cos x)}$. $1 + \cos x = 2 \cos^2 \frac{1}{2}x$.

$$\tan \frac{1}{2}x = \pm \sqrt{\frac{1 - \cos x}{1 + \cos x}} = \frac{\sin x}{1 + \cos x} = \frac{1 - \cos x}{\sin x}$$

$$\tan\left(\frac{x}{2} + 45°\right) = \pm \sqrt{\frac{1 + \sin x}{1 - \sin x}}.$$

Here the $+$ or $-$ sign is to be used according to the sign of the left-hand side of the equation.

**Relations between Three Angles Whose Sum Is 180°**

$\sin A + \sin B + \sin C = 4 \cos \frac{1}{2}A \cos \frac{1}{2}B \cos \frac{1}{2}C$.
$\cos A + \cos B + \cos C = 4 \sin \frac{1}{2}A \sin \frac{1}{2}B \sin \frac{1}{2}C + 1$.
$\sin A + \sin B - \sin C = 4 \sin \frac{1}{2}A \sin \frac{1}{2}B \cos \frac{1}{2}C$.
$\cos A + \cos B - \cos C = 4 \cos \frac{1}{2}A \cos \frac{1}{2}B \sin \frac{1}{2}C - 1$.
$\sin^2 A + \sin^2 B + \sin^2 C = 2 \cos A \cos B \cos C + 2$.
$\sin^2 A + \sin^2 B - \sin^2 C = 2 \sin A \sin B \cos C$.
$\tan A + \tan B + \tan C = \tan A \tan B \tan C$.
$\cot \frac{1}{2}A + \cot \frac{1}{2}B + \cot \frac{1}{2}C = \cot \frac{1}{2}A \cot \frac{1}{2}B \cot \frac{1}{2}C$.
$\cot A \cot B + \cot A \cot C + \cot B \cot C = 1$.
$\sin 2A + \sin 2B + \sin 2C = 4 \sin A \sin B \sin C$.
$\sin 2A + \sin 2B - \sin 2C = 4 \cos A \cos B \sin C$.

**Inverse Trigonometric Functions**   The notation $\sin^{-1} x$ (read: antisine of $x$, or inverse sine of $x$; sometimes written arc $\sin x$) means the principal angle whose sine is $x$. Similarly for $\cos^{-1} x$, $\tan^{-1} x$, etc. (The principal angle means an angle between $-90$ and $+90°$ in case of $\sin^{-1}$ and $\tan^{-1}$, and between 0 and $180°$ in the case of $\cos^{-1}$.)

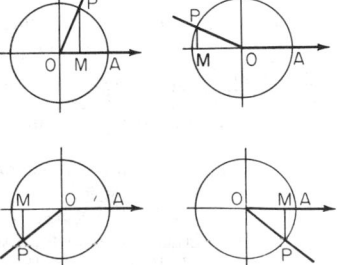

**Fig. 4**

## SOLUTION OF PLANE TRIANGLES

The "parts" of a plane triangle are its three sides, $a$, $b$, $c$, and its three angles $A$, $B$, $C$ ($A$ being opposite $a$, $B$ opposite $b$, $C$ opposite $c$, and $A + B + C = 180°$). A triangle is, in general, determined by any three parts (not all angles). To "solve" a

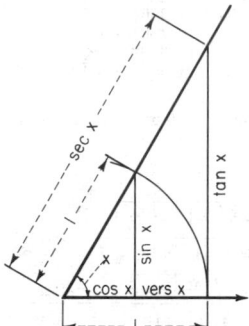

**Fig. 5**

triangle means to find the unknown parts from the known. The fundamental formulas are

$$\text{Law of sines: } \frac{a}{b} = \frac{\sin A}{\sin B}$$

$$\text{Law of cosines: } c^2 = a^2 + b^2 - 2ab \cos C$$

**Right Triangles** Use the definitions of the trigonometric functions, selecting for each unknown part a relation which connects that unknown with known quantities; then solve the resulting equations. Thus, in Fig. 6, if $C = 90°$, then $A + B = 90°$, $c^2 = a^2 + b^2$,

$$\sin A = a/c \qquad \cos A = b/c \qquad \tan A = a/b \qquad \cot A = b/a$$

If $A$ is very small, use $\tan \frac{1}{2}A = \sqrt{(c - b)/(c + b)}$

**Oblique Triangles** There are four cases. It is highly desirable in all these cases to draw a sketch of the triangle approximately to scale before commencing the computation, so that any large numerical error may be readily detected.

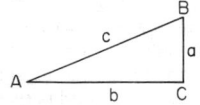

**Fig. 6**    **Fig. 7**

CASE 1. GIVEN TWO ANGLES (provided their sum is $< 180°$) AND ONE SIDE (say $a$, Fig. 7). The third angle is known, since $A + B + C = 180°$. To find the remaining sides, use $b = \dfrac{a \sin B}{\sin A}$, $c = \dfrac{a \sin C}{\sin A}$. Or, drop a perpendicular from either $B$ or $C$ on the opposite side, and solve by right triangles.
*Check:* $c \cos B + b \cos C = a$.

CASE 2. GIVEN TWO SIDES (say $a$ and $b$) AND THE INCLUDED ANGLE ($C$); AND SUPPOSE $a > b$ (Fig. 8).
*Method 1:* Find $c$ from $c^2 = a^2 + b^2 - 2ab \cos C$ [or $c^2 = (a - b)^2 + 2ab$ vers $C$]; then find the smaller angle, $B$, from $\sin B = (b/c) \sin C$; and finally, find $A$ from $A = 180° - (B + C)$. *Check:* $a \cos B + b \cos A = c$.

*Method 2:* Find $\frac{1}{2}(A - B)$ from the law of tangents:

$$\tan \tfrac{1}{2}(A - B) = [(a - b)/(a + b)] \cot \tfrac{1}{2}C$$

and $\frac{1}{2}(A + B)$ from $\frac{1}{2}(A + B) = 90° - C/2$; hence $A = \frac{1}{2}(A + B) + \frac{1}{2}(A - B)$ and $B = \frac{1}{2}(A + B) - \frac{1}{2}(A - B)$. Then find $c$ from $c = a \sin C/\sin A$ or $c = b \sin C/\sin B$. *Check:* $a \cos B + b \cos A = c$.

*Method 3:* Drop a perpendicular from $A$ to the opposite side, and solve by right triangles.

CASE 3. GIVEN THE THREE SIDES (provided the largest is less than the sum of the other two) (Fig. 9).

*Method 1:* Find the largest angle $A$ (which may be acute or obtuse) from $\cos A = (b^2 + c^2 - a^2)/2bc$ {or vers $A = [a^2 - (b - c)^2]/2bc$} and then find $B$ and $C$ (which will always be acute) from $\sin B = b \sin A/a$ and $\sin C = c \sin A/a$. *Check:* $A + B + C = 180°$.

*Method 2:* Find $A$, $B$, and $C$ from $\tan \frac{1}{2}A = r/(s - a)$, $\tan \frac{1}{2}B = r/(s - b)$, $\tan \frac{1}{2}C = r/(s - c)$, where $s = \frac{1}{2}(a + b + c)$, and $r = \sqrt{(s - a)(s - b)(s - c)/s}$. *Check:* $A + B + C = 180°$.

*Method 3:* If only one angle, say $A$, is required, use

$$\sin \tfrac{1}{2}A = \sqrt{(s - b)(s - c)/bc}$$

or

$$\cos \tfrac{1}{2}A = \sqrt{s(s - a)/bc}$$

according as $\frac{1}{2}A$ is nearer $0°$ or nearer $90°$.

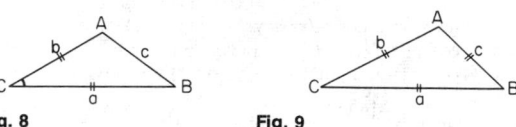

**Fig. 8**    **Fig. 9**

CASE 4. GIVEN TWO SIDES (say $b$ and $c$) AND THE ANGLE OPPOSITE ONE OF THEM ($B$). This is the "ambiguous case" in which there may be two solutions, or one, or none (see Fig. 10).

First, try to find $C = c \sin B/b$. If $\sin C > 1$, there is no solution. If $\sin C = 1$, $C = 90°$ and the triangle is a right triangle. If $\sin C < 1$, this determines two angles $C$, namely, an acute angle $C_1$, and an obtuse angle $C_2 = 180° - C_1$. Then $C_1$ will yield a solution when and only when $C_1 + B < 180°$ (see Case 1); and similarly $C_2$ will yield a solution when and only when $C_2 + B < 180°$ (see Case 1).

**Other Properties of Triangles** (See also Sec. 2, Geometry and Mensuration)

Area $= \frac{1}{2}ab \sin C = \sqrt{s(s - a)(s - b)(s - c)} = rs$, where $s = \frac{1}{2}(a + b + c)$, and $r$ = radius of inscribed circle $= \sqrt{(s - a)(s - b)(s - c)/s}$.

Radius of circumscribed circle = $R$, where

$$2R = a/\sin A = b/\sin B = c/\sin C$$

$$r = 4R \sin \frac{A}{2} \sin \frac{B}{2} \sin \frac{C}{2} = \frac{abc}{4Rs}$$

The length of the bisector of the angle $C$ is

$$z = \frac{2\sqrt{abs(s - c)}}{a + b} = \frac{\sqrt{ab[(a + b)^2 - c^2]}}{a + b}$$

The median from $C$ to the middle point of $c$ is $m = \frac{1}{2}\sqrt{2(a^2 + b^2) - c^2}$.

## SOLUTION OF SPHERICAL TRIANGLES

For the occasional solution of a spherical triangle the following formulas will be sufficient. For a detailed discussion, see any textbook on spherical trigonometry.

Let $a$, $b$, $c$ be the sides of the spherical triangle, i.e., portions of arcs of great circles of the sphere; and let $A$, $B$, $C$ be the angles of the triangle, i.e., the angles made by tangents drawn to the sides at their points of intersection on the sphere. The sum of the angles will always be greater than two right angles, and may be nearly six right angles. The angle $E = A + B + C - 180°$ is called the **spherical excess** of the triangle.

$$\frac{\sin a}{\sin A} = \frac{\sin b}{\sin B}$$
$$\frac{\sin b}{\sin B} = \frac{\sin c}{\sin C}$$
$$\frac{\sin c}{\sin C} = \frac{\sin a}{\sin A}$$

$$\cos a = \cos b \cos c + \sin b \sin c \cos A$$

with similar formulas for $\cos b$ and $\cos c$.

$$\cos A = - \cos B \cos C + \sin B \sin C \cos a$$

with similar formulas for $\cos B$ and $\cos C$. Other sample formulas are

$$\sin a \cos B = \cos b \sin c - \sin b \cos c \cos A$$
$$\sin A \cos b = \cos B \sin C + \sin B \cos C \cos a$$

If $\quad s = \frac{1}{2}(a + b + c)$ and $S = \frac{1}{2}(A + B + C)$

and if $\quad \tan r = \sqrt{\sin (s - a) \sin (s - b) \sin (s - c)/\sin s}$

and $\tan R = \sqrt{(- \cos S)/[\cos (S - A) \cos (S - B) \cos (S - C)]}$
then $\qquad \tan \frac{1}{2}A = (\tan r)/\sin (s - a)$
and $\qquad \tan \frac{1}{2}a = \tan R \cos (S - A)$

with similar formulas for $\tan \frac{1}{2}B$, $\tan \frac{1}{2}C$, and for $\tan \frac{1}{2}b$, $\tan \frac{1}{2}c$. (Here $r$ and $R$ are the radii, on the spherical surface, of the inscribed and circumscribed circles, respectively.) If $E = A + B + C - 180°$,

$$\tan \frac{1}{4}E = \sqrt{\tan \frac{1}{2}s \tan \frac{1}{2}(s - a) \tan \frac{1}{2}(s - b) \tan \frac{1}{2}(s - c)}$$

If $E$ is small, then approximately, $\sin E = F/K^2$, where $K =$ radius of sphere, and $F =$ area of triangle, regarded as a plane triangle. In any case,

$$\frac{\text{Area of a spherical triangle}}{\text{Area of a great circle}} = \frac{\text{spherical excess}}{180°}$$

In the special case of a right spherical triangle, in which $C = 90°$,

$$\cos c = \cos a \cos b = \cot A \cot B$$
$$\cos a = \cos A/\sin B$$
$$\cos b = \cos B/\sin A$$
$$\sin A = \sin a/\sin c$$
$$\cos A = \tan b/\tan c$$
$$\tan A = \tan a/\sin b$$

## HYPERBOLIC FUNCTIONS

The **hyperbolic sine, hyperbolic cosine**, etc., of any number $x$, are functions of $x$ which are closely related to the exponential $e^x$, and which have formal properties very similar to those of the trigonometric functions, sine, cosine, etc. Their definitions and fundamental properties are as follows:

$$\sinh x = \frac{1}{2}(e^x - e^{-x})$$
$$\cosh x = \frac{1}{2}(e^x + e^{-x})$$
$$\tanh x = \sinh x/\cosh x$$

$$\operatorname{csch} x = 1/\sinh x$$
$$\operatorname{sech} x = 1/\cosh x$$
$$\coth x = 1/\tanh x$$

$$\cosh^2 x - \sinh^2 x = 1$$
$$1 - \tanh^2 x = \operatorname{sech}^2 x$$
$$1 - \coth^2 x = -\operatorname{csch}^2 x$$

$$\sinh (-x) = -\sinh x$$
$$\cosh (-x) = \cosh x$$
$$\tanh (-x) = -\tanh x$$

$$\sinh (x \pm y) = \sinh x \cosh y \pm \cosh x \sinh y$$
$$\cosh (x \pm y) = \cosh x \cosh y \pm \sinh x \sinh y$$
$$\tanh (x \pm y) = (\tanh x \pm \tanh y)/(1 \pm \tanh x \tanh y)$$
$$\sinh 2x = 2 \sinh x \cosh x.$$
$$\cosh 2x = \cosh^2 x + \sinh^2 x$$
$$\tanh 2x = (2 \tanh x)/(1 + \tanh^2 x)$$

$$\sinh \frac{1}{2}x = \sqrt{\frac{1}{2}(\cosh x - 1)}$$
$$\cosh \frac{1}{2}x = \sqrt{\frac{1}{2}(\cosh x + 1)}$$
$$\tanh \frac{1}{2}x = (\cosh x - 1)/(\sinh x) = (\sinh x)/(\cosh x + 1)$$

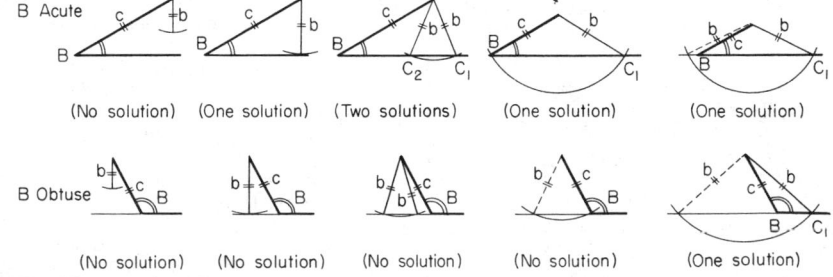

B Acute

(No solution)   (One solution)   (Two solutions)   (One solution)   (One solution)

B Obtuse

(No solution)   (No solution)   (No solution)   (No solution)   (One solution)

**Fig. 10**

The hyperbolic functions are related to the rectangular hyperbola, $x^2 - y^2 = a^2$ (Fig. 12), in much the same way that the trigonometric functions are related to the circle $x^2 + y^2 = a^2$ (Fig. 11); the analogy, however, concerns not angles but areas. Thus, in either figure, let $A$ represent the shaded area,

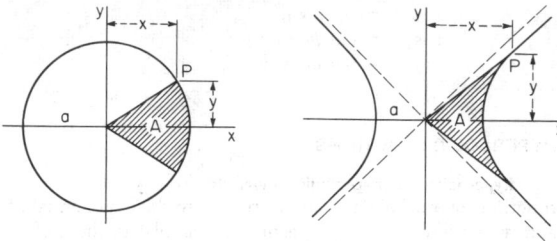

**Fig. 11**              **Fig. 12**

and let $u = A/a^2$ (a pure number). Then for the coordinates of the point $P$ we have, in Fig. 11, $x = a \cos u$, $y = a \sin u$; and in Fig. 12, $x = a \cosh u$, $y = a \sinh u$.

The **inverse hyperbolic sine** of $y$, denoted by $\sinh^{-1}y$, is the number whose hyperbolic sine is $y$; that is, the notation $x = \sinh^{-1}y$ means $\sinh x = y$. Similarly for $\cosh^{-1}y$, $\tanh^{-1}y$, etc. These functions are closely related to the logarithmic function, and are especially valuable in the integral calculus.

$$\sinh^{-1}(y/a) = \ln(y + \sqrt{y^2 + a^2}) - \ln a.$$
$$\cosh^{-1}(y/a) = \ln(y + \sqrt{y^2 - a^2}) - \ln a.$$
$$\tanh^{-1}\frac{y}{a} = \tfrac{1}{2}\ln\frac{a+y}{a-y};$$
$$\coth^{-1}\frac{y}{a} = \tfrac{1}{2}\ln\frac{y+a}{y-a}.$$

The **Gudermannian** of $x$ (written gd $x$) is an angle $u$ such that $x = \ln \tan(\tfrac{1}{4}\pi + \tfrac{1}{2}u)$. See Smithsonian Tables of the Hyperbolic Functions.

The **antigudermannian** of an angle $u$, denoted by $\mathrm{gd}^{-1} u$, is a number defined by $\mathrm{gd}^{-1}u = \ln \tan(\tfrac{1}{4}\pi + \tfrac{1}{2}u) = \int \sec u \, du = \sinh^{-1}(\tan u) = \cosh^{-1}(\sec u) = \tanh^{-1}(\sin u) = 2\tanh^{-1}(\tan \tfrac{1}{2}u)$. When $u$ is small, $\mathrm{gd}^{-1}u = u + \tfrac{1}{6}u^3 + \tfrac{1}{24}u^5 + \tfrac{61}{5040}u^7 + \cdots$.

# ANALYTICAL GEOMETRY

## by Philip Franklin

### THE POINT AND THE STRAIGHT LINE

**Rectangular Coordinates**   (Fig. 1) Let $P_1 = (x_1, y_1)$, $P_2 = (x_2, y_2)$. Then, distance $P_1P_2 = \sqrt{(x_2 - x_1)^2 + (y_2 - y_1)^2}$; slope of $P_1P_2 = m = \tan u = (y_2 - y_1)/(x_2 - x_1)$; coordinates of midpoint are $x = \tfrac{1}{2}(x_1 + x_2)$, $y = \tfrac{1}{2}(y_1 + y_2)$; coordinates of point 1/nth of the way from $P_1$ to $P_2$ are $x = x_1 + (1/n)(x_2 - x_1)$, $y = y_1 + (1/n)(y_2 - y_1)$.

Let $m_1$, $m_2$ be the slopes of two lines; then, if the lines are parallel, $m_1 = m_2$; if the lines are perpendicular to each other, $m_1 = -1/m_2$.

### Equations of a Straight Line

1. Intercept Form (Fig. 2): $\dfrac{x}{a} + \dfrac{y}{b} = 1$. ($a$, $b$ = intercepts of the line on the axes.)

2. Slope Form (Fig. 3): $y = mx + b$. ($m = \tan u$ = slope; $b$ = intercept on the $y$ axis.

3. Normal Form (Fig. 4): $x \cos v + y \sin v = p$. ($p$ = perpendicular from origin to line; $v$ = angle from the $x$ axis to $p$.)

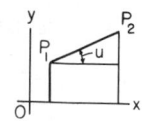 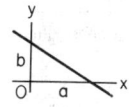 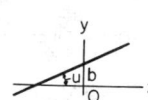

**Fig. 1**              **Fig. 2**              **Fig. 3**

4. Parallel-Intercept Form (Fig. 5): $\dfrac{y-b}{c-b} = \dfrac{x}{k}$. ($b$, $c$ = intercepts on two parallels at distance $k$ apart.)

5. General Form: $Ax + By + C = 0$. [Here $a = -C/A$, $b = -C/B$, $m = -A/B$, $\cos v = A/R$, $\sin v = B/R$, $p = -C/R$, where $R = \pm\sqrt{A^2 + B^2}$ (sign to be so chosen that $p$ is positive).]

6. Line Through $(x_1, y_1)$ with Slope $m$: $y - y_1 = m(x - x_1)$.

7. Line Through $(x_1, y_1)$ and $(x_2, y_2)$: $y - y_1 = \dfrac{y_2 - y_1}{x_2 - x_1}(x - x_1)$.

8. Line Parallel to $x$ Axis: $y = a$; to $y$ axis: $x = b$.

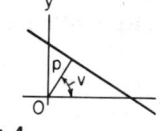 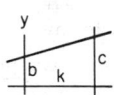

**Fig. 4**              **Fig. 5**

**Angles and Distances**   If $u$ = angle from the line with slope $m_1$ to the line with slope $m_2$, then

$$\tan u = \frac{m_2 - m_1}{1 + m_2 m_1}.$$

If parallel, $m_1 = m_2$.
If perpendicular, $m_1 m_2 = -1$.

If $u$ = angle between the lines $Ax + By + C = 0$ and $A'x + B'y + C' = 0$, then

$$\cos u = \frac{AA' + BB'}{\pm\sqrt{(A^2 + B^2)(A'^2 + B'^2)}}.$$

If parallel, $A/A' = B/B'$.

If perpendicular, $AA' + BB' = 0$.

The equations of the bisectors of the angles between the two lines just mentioned are

$$\frac{Ax + By + C}{\sqrt{A^2 + B^2}} \mp \frac{A'x + B'y + C'}{\sqrt{A'^2 + B'^2}} = 0$$

The equation of a line through $(x_1, y_1)$ and meeting a given line $y = mx + b$ at an angle $u$, is

$$y - y_1 = \frac{m + \tan u}{1 - m \tan u}(x - x_1)$$

The distance from $(x_0, y_0)$ to the line $Ax + By + C = 0$ is

$$D = \left| \frac{Ax_0 + By_0 + C}{\sqrt{A^2 + B^2}} \right|$$

where the vertical bars mean "the absolute value of."

The distance from $(x_0, y_0)$ to a line which passes through $(x_1, y_1)$ and makes an angle $u$ with the $x$ axis is

$$D = (x_0 - x_1) \sin u - (y_0 - y_1) \cos u$$

**Polar Coordinates** (Fig. 6) Let $(x, y)$ be the rectangular and $(r, \theta)$ the polar coordinates of a given point $P$. Then $x = r \cos \theta$; $y = r \sin \theta$; $x^2 + y^2 = r^2$.

**Transformation of Coordinates** If origin is moved to point $(x_0, y_0)$, the new axes being parallel to the old, $x = x_0 + x'$, $y = y_0 + y'$.

If axes are turned through the angle $u$, without change of origin,

$$x = x' \cos u - y' \sin u \qquad y = x' \sin u + y' \cos u$$

## THE CIRCLE

The **equation of a circle** with center $(a,b)$ and radius $r$ is

$$(x - a)^2 + (y - b)^2 = r^2$$

If center is at the origin, the equation becomes $x^2 + y^2 = r^2$. If circle goes through the origin and center is on the $x$ axis at point $(r, 0)$, equation becomes $x^2 + y^2 = 2rx$. The **general equation** of a circle is

$$x^2 + y^2 + Dx + Ey + F = 0$$

It has center at $(-D/2, -E/2)$, and radius $= \sqrt{(D/2)^2 + (E/2)^2 - F}$ (which may be real, null, or imaginary).

The **equation of the radical axis** of two circles, $x^2 + y^2 + Dx + Ey + F = 0$ and $x^2 + y^2 + D'x + E'y + F' = 0$, is $(D - D')x + (E - E')y + (F - F') = 0$. The tangents drawn to two circles from any point of their radical axis are of equal length. If circles intersect, the radical axis passes through their points of intersection.

The **equation of the tangent** to $x^2 + y^2 = r^2$ at $(x_1, y_1)$ is $x_1x + y_1y = r^2$. The tangent to $x^2 + y^2 + Dx + Ey + F = 0$ at $(x_1, y_1)$ is $x_1x + y_1y + \frac{1}{2}D(x + x_1) + \frac{1}{2}E(y + y_1) + F = 0$. The line $y = mx + b$ will be tangent to the circle $x^2 + y^2 = r^2$ if $b = r \sqrt{1 + m^2}$.

**Equations of Circle in Parametric Form** It is sometimes convenient to express the coordinates $x$ and $y$ of the moving point $P$ (Fig. 7) in terms of an auxiliary variable, called a **parameter**. Thus, if the parameter be taken as the angle $u$ from

the $x$ axis to the radius vector $OP$, then the equations of the circle in parametric form will be $x = a \cos u$; $y = a \sin u$. For every value of the parameter $u$, there corresponds a point $(x, y)$ on the circle. The ordinary equation $x^2 + y^2 = a^2$ can be obtained from the parametric equations by eliminating $u$.

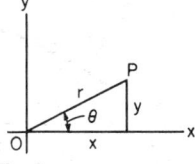

**Fig. 6**

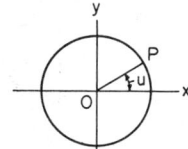

**Fig. 7**

## THE PARABOLA

The **parabola** is the locus of a point which moves so that its distance from a fixed line (called the **directrix**) is always equal to its distance from a fixed point $F$ (called the **focus**). See Fig. 8. The point halfway from focus to directrix is the **vertex**, $O$. The line through the focus, perpendicular to the directrix, is the **principal axis**. The breadth of the curve at the focus is called the **latus rectum**, or **parameter**, $= 2p$, where $p$ is the distance from focus to directrix.

NOTE. Any section of a right circular cone made by a plane parallel to a tangent plane of the cone will be a parabola.

**Equation of parabola**, origin at vertex (Fig. 8): $y^2 = 2px$.

**Polar equation of parabola**, referred to $F$ as origin and $Fx$ as axis (Fig. 9): $r = p/(1 - \cos \theta)$.

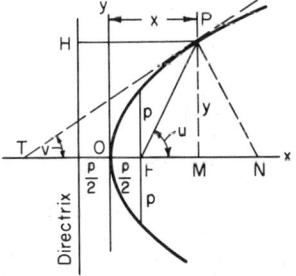

**Fig. 8**

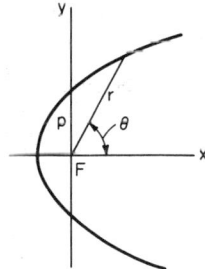

**Fig. 9**

**Equation referred to the tangents** at the ends of the latus rectum as axes (Fig. 10): $x^{1/2} + y^{1/2} = a^{1/2}$, where $a = p \sqrt{2}$.

**Equation of tangent** to $y^2 = 2px$ at $(x_1, y_1)$: $y_1y = p(x + x_1)$. The line $y = mx + b$ will be tangent to $y^2 = 2px$ if $b = p/(2m)$.

The **tangent** $PT$ at any point $P$ bisects the angle between $PF$ and $PH$ (Fig. 8). A ray of light from $F$, reflected at $P$, will move off parallel to the principal axis. The **subtangent**, $TM$, is bisected at $O$. The **subnormal**, $MN$, is constant and equal to $p$. The locus of the foot of the perpendicular from the focus on a moving tangent is the tangent at the vertex (Fig. 11). The locus of the point of intersection of perpendicular tangents is the directrix (Fig. 12). The locus of the midpoints of a set of parallel chords whose slope is $m$ is a

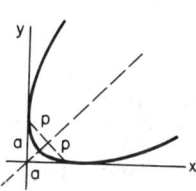

**Fig. 10**

straight line parallel to the principal axis at a distance $p/m$ and is called a **diameter** (Fig. 13). If $M$ is the midpoint of a chord $PQ$ and if $T$ is the point of intersection of the tangents at $P$ and $Q$, then $TM$ is parallel to the principal axis and is bisected by the curve (Fig. 13).

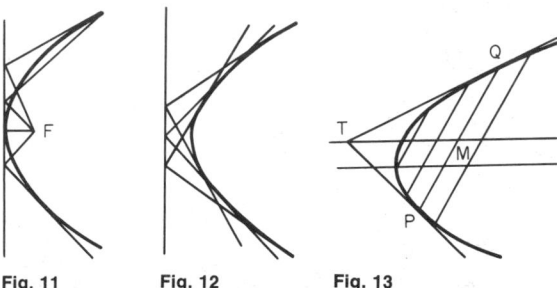

**Fig. 11**          **Fig. 12**          **Fig. 13**

**To Construct a Tangent to a Given Parabola** (1) At a given point of contact, $P$ (Fig. 14): Find $T$ so that $OT = OM$, or $FT = FP$. Then $TP$ is the tangent at $P$. Or, make $MN = p = 2(OF)$; then $PN$ is the normal at $P$. (2) From a given external point, $Q$ (Fig. 15): With $Q$ as center and radius $QF$ draw circle

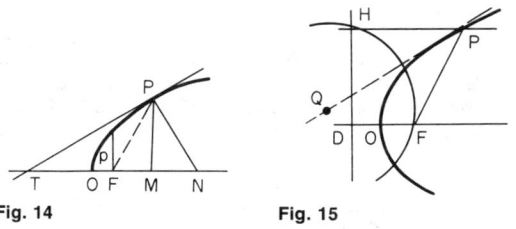

**Fig. 14**          **Fig. 15**

cutting the directrix in $H$; draw $HP$ parallel to principal axis; then $P$ is required point of contact. As check, note that $QP$ is the perpendicular bisector of $FH$.

**To Construct a Parabola** 1. GIVEN ANY TWO POINTS, $P$ AND $Q$, THE TANGENT $PT$ AT ONE OF THEM, AND THE DIRECTION OF THE PRINCIPAL AXIS $OX$. In Fig. 16, let $K$ be a variable point on a line through $Q$ parallel to $OX$. Draw $KR$ parallel to $PT$ (meeting $PQ$ in $R$), and draw $RS$ parallel to $OX$ (meeting $PK$ in $S$); then $S$ is a point of the curve. Note that a line through $P$ parallel to the principal axis bisects all chords parallel to the tangent $PT$.

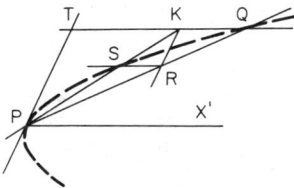

**Fig. 16**

2. GIVEN THE VERTEX $O$ AND FOCUS $F$. (*a*) In Fig. 17 draw $Oy$ perpendicular to $OF$, and slide the vertex of a right angle along $Oy$ so that one side always passes through $F$; then the other side will always be a tangent to the parabola. (*b*) Take a piece of paper (Fig. 18) with a straightedge, $d$, and mark a point $F$ near the edge. Let $K$ be a variable point of the edge, and fold the paper so that $K$ coincides with $F$. The crease will be a

tangent to the parabola which has focus $F$ and directrix $d$. (*c*) In Fig. 19, let $M$ be a variable point of the principal axis, and lay off $MN = 2(OF) = p$. With $F$ as center and radius $FN$ draw a circle, cutting the perpendicular at $M$ in $P$. Then $P$ is a point of the curve, and $PT$ and $PN$ are the tangent and normal at $P$.

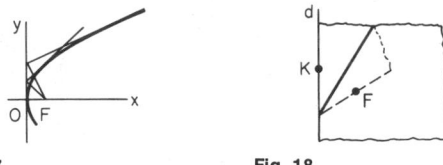

**Fig. 17**          **Fig. 18**

3. GIVEN TWO TANGENTS AND THEIR POINTS OF CONTACT, $P$ AND $Q$ (Fig. 20). Divide $TP$ and $QT$ into any number of equal parts (here 4). Then the lines 11, 22, 33, . . . will be tangents to the parabola. This method is especially advantageous in drawing rather flat arcs.

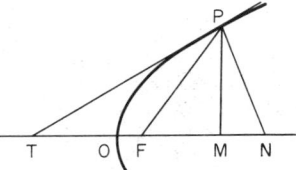

**Fig. 19**

The **radius of curvature** of $y^2 = 2px$ at a point $P = (x,y)$ is $R = (p + 2x)^{3/2}/\sqrt{p}$, or $R = p/\sin^3 v$, where $v =$ the angle which the tangent at $P$ makes with $PF$ (Fig. 21). At the vertex, $R = p$. To construct the radius of curvature at any point $P$, lay off $PR = 2(PF)$ parallel to the principal axis, and draw $RC$ perpendic-

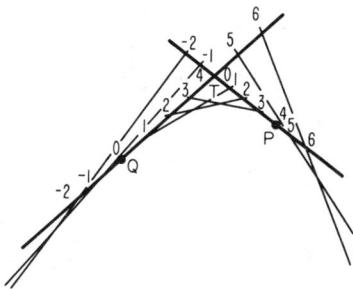

**Fig. 20**

ular to the axis, meeting the normal, $PN$, in $C$. Then $C$ is the center of curvature for the point $P$, and a circle about $C$ with radius $CP$ will closely approximate the parabola in the neighborhood of $P$.

## THE ELLIPSE

The **ellipse** (as shown in Figure 22), has two **foci**, $F$ and $F'$ and two **directrices**, $DH$ and $D'H'$. If $P$ is any point on the curve, $PF + PF'$ is constant, $= 2a$; and $PF/PH$ (or $PF'/PH'$) is also constant, $= e$, where $e$ is the **eccentricity** ($e < 1$). Either of these properties may be taken as the definition of the curve. The relations between $e$ and the semiaxes $a$ and $b$ are as shown in Fig. 23. Thus, $b^2 = a^2(1 - e^2)$, $ae = \sqrt{a^2 - b^2}$, $e^2 = 1 - (b/a)^2$. The **semilatus rectum** $= p = a(1 - e^2) = b^2/a$. Note that $b$ is always less than $a$, except in the special case of the circle, in which $b = a$ and $e = 0$.

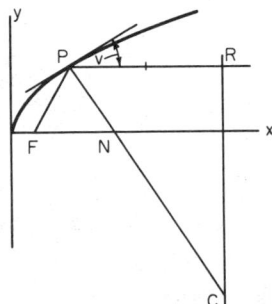

**Fig. 21**

Any section of a right circular cone made by a plane which cuts all the elements of one nappe of the cone will be an ellipse; if the plane is perpendicular to the axis of the cone, the ellipse becomes a circle.

**Equation of ellipse,** center as origin:

$$\frac{x^2}{a^2} + \frac{y^2}{b^2} = 1 \qquad \text{or} \qquad y = \pm \frac{b}{a}\sqrt{a^2 - x^2}$$

If $P = (x,y)$ is any point of the curve, $PF = a + ex$, $PF' = a - ex$.

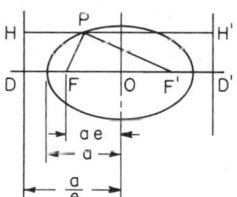

**Fig. 22**

**Equations of the ellipse in parametric form** $x = a \cos u$, $y = b \sin u$, where $u$ is the eccentric angle of the point $P = (x,y)$. See Fig. 28.

**Polar equation,** focus as origin, axes as in Fig. 24: $r = p/(1 - e \cos \theta)$.

**Equation of the tangent** at $(x_1,y_1)$: $b^2x_1x + a^2y_1y = a^2b^2$.

The line $y = mx + k$ will be a tangent if $k = \pm\sqrt{a^2m^2 + b^2}$. The normal at any point $P$ bisects the angle between $PF$ and $PF'$ (Fig. 25). The locus of the foot of the perpendicular from the focus on a moving tangent is the circle on the major axis as diameter (Fig. 26). The locus of the point of intersection of

perpendicular tangents is a circle with radius $\sqrt{a^2 + b^2}$ (Fig. 27).

**Ellipse as a Flattened Circle. Eccentric Angle** If the ordinates in a circle are diminished in a constant ratio, the resulting points will lie on an ellipse (Fig. 28). If $Q$ traces the circle

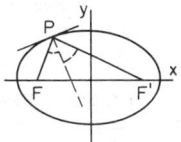

**Fig. 25**          **Fig. 26**

with uniform velocity, the corresponding point $P$ will trace the ellipse, with varying velocity. The angle $u$ in the figure is called the eccentric angle of the point $P$.

**Conjugate diameters** are lines through the center, each of which bisects all the chords parallel to the other (Fig. 29). If

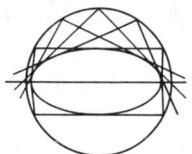 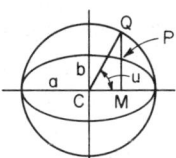

**Fig. 27**          **Fig. 28**

$m_1$ and $m_2$ are the slopes, then $m_1m_2 = -b^2/a^2$. One pair of conjugate diameters are the diagonals of the rectangle circumscribing the ellipse. The eccentric angles of the ends of two conjugate diameters differ by 90°. Thus (Fig. 30), if $CQ$ and $CQ'$ are perpendicular radii in the circle, $CP$ and $CP'$ will be conjugate semidiameters in the ellipse. A parallelogram formed by tangents drawn parallel to a pair of conjugate diameters has a constant area, $= 4ab$ (Fig. 31). Also, if $a'$, $b'$ are conjugate semidiameters, and $w$ the angle between them, then $a'^2 + b'^2 = a^2 + b^2$ and $a'b = ab/\sin w$.

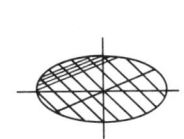 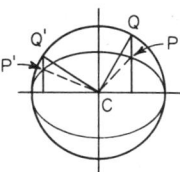

**Fig. 29**          **Fig. 30**

**To Construct a Tangent to a Given Ellipse** 1. AT A GIVEN POINT OF CONTACT, $P$. Bisect the angle between the focal radii $PF$ and $PF'$ (Fig. 25).

2. FROM A GIVEN EXTERNAL POINT, $R$. (a) Through $R$ draw any two lines cutting the ellipse, one in $A$ and $B$, the other in $C$ and $D$ (Fig. 32). Through the point of intersection of $AD$

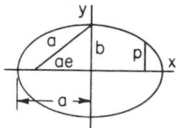

**Fig. 23**        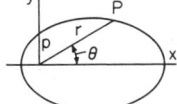 **Fig. 24**

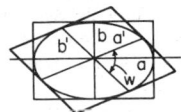

**Fig. 31**          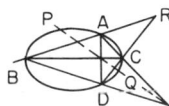 **Fig. 32**

and *BC* and the point of intersection of *AC* and *BD*, draw a line cutting the ellipse in *P* and *Q*. Then *P* and *Q* are the required points of contact. (*b*) With *R*

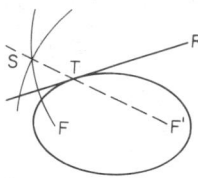

**Fig. 33**

as a center and radius *RF*, draw an arc; with *F'* as center and radius *2a* draw an arc, intersecting the first in *S;* and let *SF'* meet the curve in *T*. Then *T* is the point of contact (Fig. 33).

**To Construct an Ellipse, Given** *a* **and** *b*    1. In Fig. 34, with *O* as center, draw circles with radii *a* and *b* (and also a third circle with radius *a* + *b*). Let a variable ray through *O* cut these circles in *J*, *K* (and *S*); through *J* and *K* draw parallels to the axes, meeting in *P*. Then *P* is a point of the ellipse (and *SP* is the normal at *P*).

2. In Fig. 35, let *P* divide a line *AB* so that *PA* = *a* and *PB* = *b*. Then if *A* and *B* slide on the axes, *P* will describe an ellipse.

3. In Fig. 36, let *PBA* be a straight line such that *PA* = *a* and *PB* = *b*. Then if *A* and *B* slide on the axes, *P* will trace an

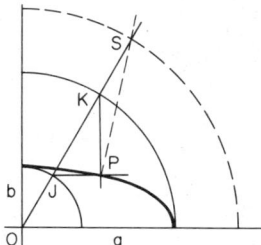

**Fig. 34**

ellipse. (Use a strip of paper, with the points *P*, *B*, and *A* marked on it.)

4. Find the foci, *F* and *F'*, by striking an arc of radius *a* with center at *B* (Fig. 37). Drive pins at *F*, *F'*, and *B*, and adjust a loop of thread around them. Then remove the pin at

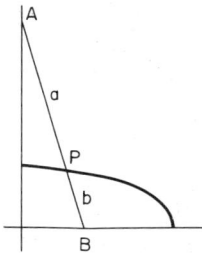

**Fig. 35**

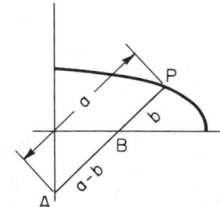

**Fig. 36**

*B*, and replace it by a pencil point; by moving the pencil so as to keep the string taut, the complete ellipse can be drawn at one sweep. Or, use a mechanical ellipsograph.

5. Apply the method of the following paragraph to the special case in whch *OP* and *OQ* are perpendicular semiaxes.

**To Construct an Ellipse, Given a Pair of Conjugate Semidiameters,** *OP* **and** *OQ*    Complete the parallelogram, as in Fig. 38. Divide *QD* and *QO* into *n* equal parts, 1, 2, 3, . . . and 1', 2', 3', . . . Connect *P* with 1, 2, 3, . . . and *P'* with 1', 2', 3'. . . . The points of intersection of corresponding lines will be points of the ellipse.

**To Construct an Ellipse Approximately by Two Circular Arcs**    In Fig. 39, lay off *OL* = *OA* and *BS* = *BL* = *a* − *b*. Bisect *SA* in *T*, and draw *THK* perpendicular to *BA*. Then *H* is one center, with radius *HA*, and *K* is the other center, with radius *KB*. The junction point *Q* of the two arcs will fall a little outside the true ellipse.

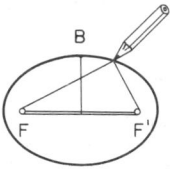

**Fig. 37**

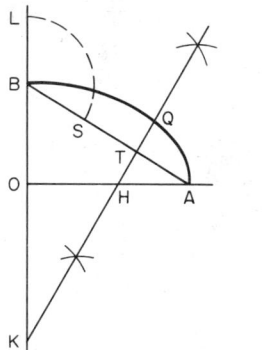

**Fig. 38**

The **radius of curvature of an ellipse at any point** *P* = (*x*, *y*) is

$$R = a^2b^2(x^2/a^4 + y^2/b^4)^{3/2} = p/\sin^3 v$$

where *v* is the angle which the tangent at *P* makes with *PF* or *PF'*. At end of major axis, *R* = $b^2/a$ = *MA*; at end of minor axis, *R* = $a^2/b$ = *NB* (see Fig. 40). To construct the radius of curvature at any other point *P* (Fig. 41), draw the normal at *P* (by bisecting the angle between *PF* and *PF'*) and let it meet the major axis in *N*. At *N* draw a perpendicular to *PN* meeting *PF* in *H*. At *H* draw a perpendicular to *PH* meeting *PN* in *C*. Then *C* is the center of curvature for the point *P*, and a circle

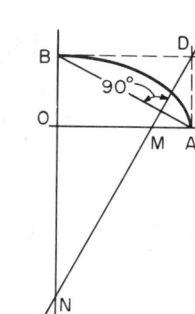

**Fig. 39**                    **Fig. 40**

about *C* with radius *CP* will closely approximate the ellipse in the neighborhood of *P*. Note that, if the circle of curvature for *P* cuts the ellipse again at *Q*, then the tangent at *P* and the line *PQ* are equally inclined to the axis.

**THE HYPERBOLA**

The **hyperbola** has two **foci**, *F* and *F'*, at distances ±*ae* from the center, and two **directrices**, *DH* and *D'H'*, at distances ±*a/e* from the center (Fig. 42). If *P* is any point of the curve, | *PF* −

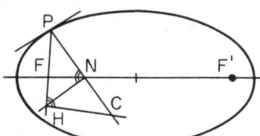

**Fig. 41**

$PF'$ | is constant, $= 2a$; and $PF/PH$ (or $PF'/PH'$) is also constant, $= e$ (called the **eccentricity**), where $e > 1$. Either of these properties may be taken as the definition of the curve. The curve has two branches which approach more and more nearly two straight lines called the **asymptotes.** Each asymptote makes

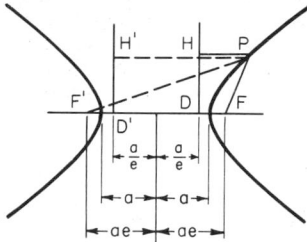

**Fig. 42**

with the principal axis an angle whose tangent is $b/a$. The relations between $e$, $a$, and $b$ are shown in Fig. 43: $b^2 = a^2(e^2 - 1)$, $ae = \sqrt{a^2 + b^2}$, $e^2 = 1 + (b/a)^2$. The semilatus rectum, or ordinate at the focus, is $p = a(e^2 - 1) = b^2/a$.

Any section of a right circular cone made by a plane which cuts both nappes of the cone will be a hyperbola.

**Equation of the hyperbola,** center as origin:

$$\frac{x^2}{a^2} - \frac{y^2}{b^2} = 1 \qquad \text{or} \qquad y = \pm \frac{b}{a}\sqrt{x^2 - a^2}$$

If $P = (x,y)$ is on the right-hand branch, $PF = ex - a$, $PF' = ex + a$. If $P$ is on the left-hand branch, $PF = -ex + a$, $PF' = -ex - a$.

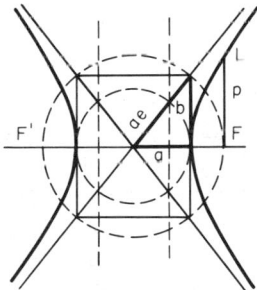

**Fig. 43**

**Equations of Hyperbola in Parametric Form**   (1) $x = a \cosh u$, $y = b \sinh u$. Here $u$ may be interpreted as $A/ab$, where $A$ is the area shaded in Fig. 44. (2) $x = a \sec v$, $y = b \tan v$, where $v$ is an auxiliary angle of no special geometric interest.

**Polar equation,** referred to focus as origin, axes as in Fig. 45:

$$r = p/(1 - e \cos \theta)$$

**Equation of the tangent** at $(x_1,y_1)$: $b^2 x_1 x - a^2 y_1 y = a^2 b^2$. The line $y = mx + k$ will be a tangent if $k = \pm\sqrt{a^2 m^2 - b^2}$. The tangent at any point $P$ (Fig. 46) bisects the angle between $PF$ and $PF'$. The locus of the foot of the perpendicular from the focus on a moving tangent is the circle on the principal axis as diameter (Fig. 47). The locus of the point of intersection of perpendicular tangents is a circle with radius $\sqrt{a^2 - b^2}$, which will be imaginary if $b > a$ (Fig. 48).

**Properties of the Asymptotes** (Fig. 49)   If $P$ is any point of the curve, the product of the perpendicular distances from $P$ to the two asymptotes is constant, $= a^2 b^2/(a^2 + b^2)$. Also, the product of the oblique distances (the distance to each asymptote being measured parallel to the other) is constant and equal to $\frac{1}{4}(a^2 + b^2)$. If a line cuts the hyperbola and its asymptotes, the parts of the line intercepted between the curve and the asymptotes are equal. The part of a tangent intercepted between the asymptotes is bisected by the point of contact.

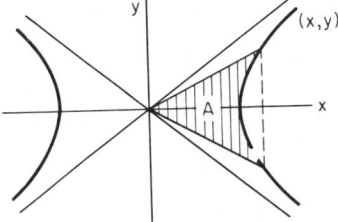

**Fig. 44**

The triangle bounded by the asymptotes and a variable tangent is of constant area, $= ab$. If a line through $Q$ perpendicular to the principal axis meets the asymptotes in $R$ and $S$ (see Fig. 49), then $\overline{QR} \times \overline{QS} = b^2$. If a line through $Q$ parallel to the principal axis meets the asymptotes in $U$ and $V$, then $\overline{QU} \times \overline{QV} = a^2$.

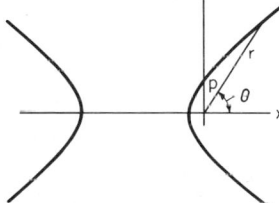

**Fig. 45**

**Conjugate hyperbolas** are two hyperbolas having the same asymptotes with semiaxes interchanged (Fig. 50). The equation of the hyperbola conjugate to $\frac{x^2}{a^2} - \frac{y^2}{b^2} = 1$ is $\frac{x^2}{a^2} - \frac{y^2}{b^2} = -1$.

**Conjugate diameters** are lines through the center, each of which bisects all the chords parallel to the other—a chord which does not meet the given hyperbola being understood to be terminated by the conjugate hyperbola (Fig. 50). If $m_1$ and $m_2$ are the slopes, then $m_1 m_2 = b^2/a^2$. Each asymptote,

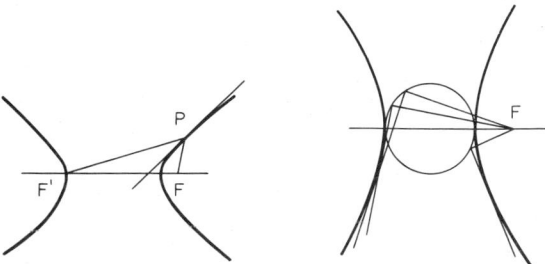

**Fig. 46**                                **Fig. 47**

regarded as a diameter, is its own conjugate. If a parallelogram is formed by tangents drawn parallel to a pair of conjugate diameters, its vertices will lie on the asymptotes, and its area will be constant $= 4ab$. If $a'$, $b'$ are conjugate semidiameters, and $w$ the angle between them, then $a'^2 - b'^2 = a^2 - b^2$, and $a'b' = ab/\sin w$.

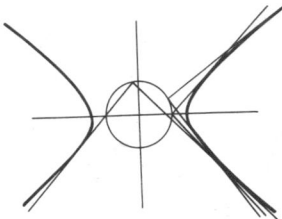

**Fig. 48**

**Equilateral Hyperbola** ($a = b$)   Equation referred to principal axes (Fig. 51): $x^2 - y^2 = a^2$.

NOTE:   $p = a$ (Fig. 51). Equation referred to asymptotes as axes (Fig. 52): $xy = a^2/2$.

Asymptotes are perpendicular. Eccentricity $= \sqrt{2}$. Any diameter is equal in length to its conjugate diameter.

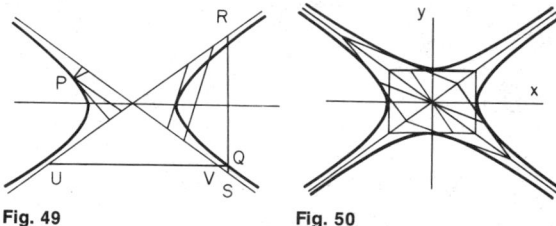

**Fig. 49**          **Fig. 50**

**To construct a tangent** at any given point $P$ of a hyperbola. In Fig. 53, draw $PA$ and $PB$ parallel to the asymptotes, and take $OS = 2(OA)$ and $OT = 2(OB)$. Then $ST$ is the tangent at $P$.

**To construct a hyperbola**, given the asymptotes and any point $P$.

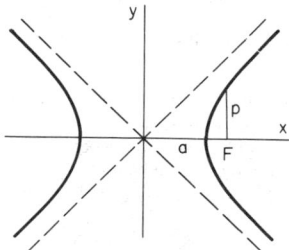

**Fig. 51**

1. In Fig. 54 let $TPT'$ be a variable line through $P$, and lay off $T'P' = TP$; then $P'$ is a point of the curve.

2. In Fig. 55, draw $PA$ and $PB$ parallel to the asymptotes. Lay off $OA' = n(OA)$ and $OB' = (1/n)(OB)$, where $n$ is any number; and through $A'$ and $B'$ draw parallels to the axes; these will meet in a point $P'$ of the curve.

3. (Fig. 56). Take any point $K$ in the ordinate $PM$, and draw $OK$ meeting the line through $P$ parallel to the $x$ axis in $R$.

Draw a parallel to the $x$ axis through $K$ and a parallel to the $y$ axis through $R$, meeting in $Q$. Then $Q$ is a point of the curve.

## THE CATENARY

The **catenary** is the curve in which a flexible chain or cord of uniform density will hang when supported by the two ends. Let $w$ = weight of the chain per unit length; $T$ = the tension at any point $P$; and $T_h$, $T_v$ = the horizontal and vertical components of $T$. The horizontal component $T_h$ is the same at all points of the curve.

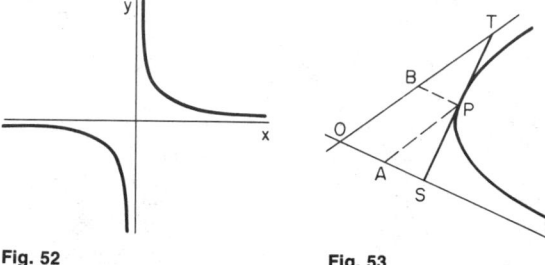

**Fig. 52**          **Fig. 53**

The length $a = T_h/w$ is called the **parameter** of the catenary, or the distance from the lowest point $O$ to the **directrix** $DQ$ (Fig. 57). When $a$ is very large, the curve is very flat. For methods of finding $a$ in any given case, see problems 1 to 6 below.

The rectangular **equation**, referred to the lowest point as origin, is $y = a \, [\cosh (x/a) - 1]$. In case of very flat arcs ($a$ large), $y = \dfrac{x^2}{2a} + \cdots$; $s = x + \frac{1}{6}\dfrac{x^3}{a^2} + \cdots$, approx, so that in such a case the catenary closely resembles a parabola.

If the perpendicular from $O$ to the tangent at $P$ meets the directrix in $Q$, then $DQ = $ arc $OP = s$ and $OQ = y + a$. The **radius of curvature** at $P$ is $R = (y + a)^2/a$, which is equal in length to the portion of the normal intercepted between $P$ and the directrix.

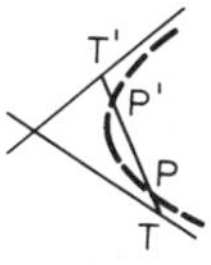

**Problems on the Catenary** (Fig. 57)   When any two of the four quantities, $x$, $y$, $s$, $T/w$ are known, the remaining two, and also the parameter $a$, can be found, as follows:

1. GIVEN $x$ AND $y$. Compute $y/x$, and find from Table 1 the value of the auxiliary variable $z$. Then compute $a = x/z$, $s = a \sinh z$, and $T = wa \cosh z$. Or, having $z$, find $s/x$ and $wx/T$ by using Tables 3 and 2 inversely, and hence (since $x$ is known) compute $s$ and $T/w$ without the use of $a$.

**Fig. 54**

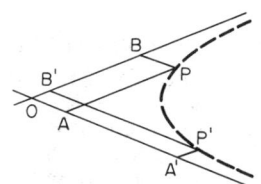

**Fig. 55**

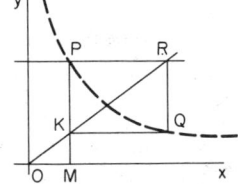

**Fig. 56**

2. Given $x$ and $T/w$.  Compute $wx/T$, and find from Table 2 the value of the auxiliary variable $z$.  Then compute $a = x/z$, $y = a(\cosh z - 1)$ and $s = a \sinh z$.  Or, having $z$, find $y/x$ and $s/x$ by using Tables 1 and 3 inversely, and hence (since $x$ is known) compute $y$ and $s$ without the use of $a$.

3. Given $x$ and $s$.  Compute $s/x$, and find from Table 3 the value of the auxiliary variable $z$.  Then compute $a = x/z$, $y = a(\cosh z - 1)$, and $T = wa \cosh z$.  Or, having $z$, find $y/x$ and

$wx/T$ by using Table 1 and 2 inversely, and hence (since $x$ is known) compute $y$ and $T/w$ without the use of $a$.

4. Given $y$ and $s$.  Then $\dfrac{T}{w} = \dfrac{s^2}{2y} + \dfrac{y}{2}$, $x = \left(\dfrac{s^2}{y} - y\right) \tanh^{-1}\left(\dfrac{y}{s}\right)$, $a = \dfrac{s^2}{2y} - \dfrac{y}{2}$.  Or, if $y/s$ is small, $x = s\left[1 - \dfrac{2}{3}\left(\dfrac{y}{s}\right)^2 - \dfrac{2}{15}\left(\dfrac{y}{s}\right)^4 - \cdots\right]$.

**Table 1. Giving $z$ When $y/x$ Is Known. Then $a = x/z$**

| $y/x$ | 0 | 1 | 2 | 3 | 4 | 5 | 6 | 7 | 8 | 9 |
|---|---|---|---|---|---|---|---|---|---|---|
| 0.0 | 0.0000 | 0.0200 | 0.0400 | 0.0600 | 0.0800 | 0.0999 | 0.1199 | 0.1398 | 0.1597 | 0.1795 |
| 0.1 | 0.1993 | 0.2191 | 0.2389 | 0.2586 | 0.2782 | 0.2978 | 0.3173 | 0.3368 | 0.3562 | 0.3756 |
| 0.2 | 0.3948 | 0.4140 | 0.4332 | 0.4522 | 0.4712 | 0.4901 | 0.5089 | 0.5276 | 0.5463 | 0.5648 |
| 0.3 | 0.5833 | 0.6016 | 0.6199 | 0.6381 | 0.6561 | 0.6741 | 0.6919 | 0.7097 | 0.7274 | 0.7449 |
| 0.4 | 0.7623 | 0.7797 | 0.7969 | 0.8140 | 0.8311 | 0.8480 | 0.8647 | 0.8814 | 0.8980 | 0.9145 |
| 0.5 | 0.9308 | 0.9471 | 0.9632 | 0.9792 | 0.9951 | 1.0109 | 1.0266 | 1.0422 | 1.0576 | 1.0730 |
| 0.6 | 1.0883 | 1.1034 | 1.1184 | 1.1334 | 1.1482 | 1.1629 | 1.1775 | 1.1920 | 1.2064 | 1.2207 |

NOTE.  $y/x = (\cosh z - 1)/z$.

**Table 2. Giving $z$ When $wx/T$ Is Known. Then $a = x/z$**

| $wx/T$ | 0 | 1 | 2 | 3 | 4 | 5 | 6 | 7 | 8 | 9 |
|---|---|---|---|---|---|---|---|---|---|---|
| 0.0 | 0.0000 | 0.0100 | 0.0200 | 0.0300 | 0.0400 | 0.0501 | 0.0601 | 0.0702 | 0.0803 | 0.0904 |
| 0.1 | 0.1005 | 0.1107 | 0.1209 | 0.1311 | 0.1414 | 0.1517 | 0.1621 | 0.1725 | 0.1830 | 0.1936 |
| 0.2 | 0.2042 | 0.2149 | 0.2256 | 0.2365 | 0.2474 | 0.2584 | 0.2695 | 0.2807 | 0.2920 | 0.3035 |
| 0.3 | 0.3150 | 0.3267 | 0.3385 | 0.3505 | 0.3626 | 0.3749 | 0.3874 | 0.4000 | 0.4129 | 0.4259 |
| 0.4 | 0.4392 | 0.4528 | 0.4666 | 0.4806 | 0.4950 | 0.5097 | 0.5248 | 0.5403 | 0.5562 | 0.5726 |
| 0.5 | 0.5894 | 0.6068 | 0.6249 | 0.6436 | 0.6632 | 0.6836 | 0.7051 | 0.7277 | 0.7517 | 0.7775 |
| 0.6 | 0.8053 | 0.8357 | 0.8695 | 0.9082 | 0.9541 | 1.0132 | 1.1110 | ...... | ...... | ...... |

NOTE.  $wx/T = z/\cosh z$.  If $wx/T$ is less than 0.6627, there are two values of $z$, one less than 1.200 and one greater than 1.200; only the smaller of these values is tabulated.  If $wx/T$ is greater than 0.6627, the problem is impossible.

**Table 3. Giving $z$ When $s/x$ is Known. Then $a = x/z$**

| $s/x$ | 0 | 1 | 2 | 3 | 4 | 5 | 6 | 7 | 8 | 9 |
|---|---|---|---|---|---|---|---|---|---|---|
| 1.000 | ...... | 0.0245 | 0.0346 | 0.0424 | 0.0490 | 0.0548 | 0.0600 | 0.0648 | 0.0693 | 0.0735 |
| 1 | 0.0774 | 0.0812 | 0.0848 | 0.0883 | 0.0916 | 0.0948 | 0.0980 | 0.1010 | 0.1039 | 0.1067 |
| 2 | 0.1095 | 0.1122 | 0.1149 | 0.1174 | 0.1200 | 0.1224 | 0.1249 | 0.1272 | 0.1296 | 0.1319 |
| 3 | 0.1341 | 0.1363 | 0.1385 | 0.1407 | 0.1428 | 0.1448 | 0.1469 | 0.1489 | 0.1509 | 0.1529 |
| 4 | 0.1548 | 0.1567 | 0.1586 | 0.1605 | 0.1623 | 0.1642 | 0.1660 | 0.1678 | 0.1696 | 0.1713 |
| 1.005 | 0.1731 | 0.1748 | 0.1765 | 0.1782 | 0.1799 | 0.1815 | 0.1831 | 0.1848 | 0.1864 | 0.1880 |
| 6 | 0.1896 | 0.1911 | 0.1927 | 0.1942 | 0.1958 | 0.1973 | 0.1988 | 0.2003 | 0.2018 | 0.2033 |
| 7 | 0.2047 | 0.2062 | 0.2076 | 0.2091 | 0.2105 | 0.2119 | 0.2133 | 0.2147 | 0.2161 | 0.2175 |
| 8 | 0.2188 | 0.2202 | 0.2215 | 0.2229 | 0.2242 | 0.2255 | 0.2269 | 0.2282 | 0.2295 | 0.2308 |
| 9 | 0.2321 | 0.2334 | 0.2346 | 0.2359 | 0.2372 | 0.2384 | 0.2397 | 0.2409 | 0.2421 | 0.2434 |
| 1.01 | 0.2446 | 0.2565 | 0.2678 | 0.2787 | 0.2892 | 0.2993 | 0.3091 | 0.3186 | 0.3278 | 0.3367 |
| 2 | 0.3454 | 0.3539 | 0.3621 | 0.3702 | 0.3781 | 0.3859 | 0.3934 | 0.4009 | 0.4082 | 0.4153 |
| 3 | 0.4224 | 0.4293 | 0.4361 | 0.4428 | 0.4494 | 0.4559 | 0.4623 | 0.4686 | 0.4748 | 0.4809 |
| 4 | 0.4870 | 0.4930 | 0.4989 | 0.5047 | 0.5105 | 0.5162 | 0.5218 | 0.5274 | 0.5329 | 0.5383 |
| 1.05 | 0.5437 | 0.5490 | 0.5543 | 0.5595 | 0.5647 | 0.5698 | 0.5749 | 0.5799 | 0.5849 | 0.5898 |
| 6 | 0.5947 | 0.5996 | 0.6044 | 0.6091 | 0.6139 | 0.6186 | 0.6232 | 0.6278 | 0.6324 | 0.6369 |
| 7 | 0.6414 | 0.6459 | 0.6504 | 0.6548 | 0.6591 | 0.6635 | 0.6678 | 0.6721 | 0.6763 | 0.6806 |
| 8 | 0.6848 | 0.6889 | 0.6931 | 0.6972 | 0.7013 | 0.7053 | 0.7094 | 0.7134 | 0.7174 | 0.7213 |
| 9 | 0.7253 | 0.7292 | 0.7331 | 0.7369 | 0.7408 | 0.7446 | 0.7484 | 0.7522 | 0.7559 | 0.7597 |
| 1.10 | 0.7634 | ...... | ...... | ...... | ...... | ...... | | | | |

NOTE.  $s/x = \sinh z/z$

5. GIVEN $y$ AND $T/w$. Then $a = \dfrac{T}{w} - y$, $x = \left(\dfrac{T}{w} - y\right) \cosh^{-1} \dfrac{T/w}{(T/w) - y}$.

$s = \sqrt{2y(T/w) - y^2}$. Or, if $y(T/w)$ is small,

$$x = \sqrt{\frac{2yT}{w}}\left[1 - \frac{7}{12}\frac{wy}{T} - \cdots\right], \quad \frac{s-x}{s} = \frac{1}{3}\frac{wy}{T}, \text{ approx,}$$

$$s = \sqrt{\frac{2yT}{w}}\left[1 - \frac{1}{4}\frac{wy}{T} - \frac{1}{32}\left(\frac{wy}{T}\right)^2 - \frac{1}{128}\left(\frac{wy}{T}\right)^3 - \cdots\right].$$

6. GIVEN $s$ AND $T/w$. Then

$$x = \frac{T}{w}\sqrt{1 - \left(\frac{ws}{T}\right)^2}\,\tanh^{-1}\left(\frac{ws}{T}\right),$$

$$y = \frac{T}{w} - \frac{T}{w}\sqrt{1 - \left(\frac{ws}{T}\right)^2},$$

$$a = \frac{T}{w}\sqrt{1 - \left(\frac{ws}{T}\right)^2}. \text{ Or, if } ws/T \text{ is small,}$$

$$x = s\left[1 - \frac{1}{6}\left(\frac{ws}{T}\right)^2 - \frac{11}{120}\left(\frac{ws}{T}\right)^4 - \cdots\right],$$

$$y = s\left[\frac{1}{2}\left(\frac{ws}{T}\right) + \frac{1}{8}\left(\frac{ws}{T}\right)^3 + \cdots\right],$$

$$a = \frac{T}{w}\left[1 - \frac{1}{2}\left(\frac{ws}{T}\right)^2 - \frac{1}{8}\left(\frac{ws}{T}\right)^4 - \cdots\right].$$

**Given the Length $2L$ of a Chain Supported at Two Points** $A$ **and** $B$ **not in the Same Level, to Find** $a$ (See Fig. 58; $b$ and $c$ are supposed known.) Let $(\sqrt{L^2 - b^2}\,)/c = s/x$; enter Table 3 with this value of $s/x$, and find the corresponding value of the auxiliary variable $z$. Then $a = c/z$.

NOTE. The coordinates of the midpoint $M$ of $AB$ (see Fig. 58) are $x_0 = a \tanh^{-1}(b/L)$, $y_0 = (L/\tanh z) - a$, so that the position of the lowest point is determined.

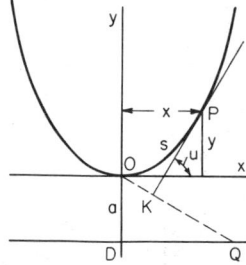

**Fig. 57**          **Fig. 58**

**Correction for Sag in Chaining Uphill** (Fig. 59) Let $l =$ length of tape (corrected for stretch and temperature), $w =$ weight per unit length of tape, $A =$ angle between the chord $AB$ and the horizontal.

If the tension $P$ at the upper end is known, compute $wl/P$ and find $k$ from Table 4. If the tension $Q$ at the lower end is known, compute $wl/Q$ and find $k$ from Table 5. In either case, chord $AB = l - kl$.

NOTE. $k = 1 - [(1 - \sqrt{1 - 2m\sin u + m^2})/(m\sin A)]$, where $m = wl/P$ and $u$ is given by

$(1 - \sqrt{1 - 2m\sin u + m^2})\sec u = [\sinh^{-1}(\tan u) - \sinh^{-1}(\tan u - m\sec u)]\tan A$.

**Table 4. Giving $k$**

| $\dfrac{wl}{P}$ | $A = 0°$ | 10° | 20° | 30° | 40° | 50° | 60° | 70° | 80° |
|---|---|---|---|---|---|---|---|---|---|
| .01 | .00000 | 000 | 000 | 000 | 000 | 000 | 000 | 000 | 000 |
| .02 | | 002 | 002 | 001 | 001 | 001 | 001 | 000 | 000 | 000 |
| .03 | | 004 | 004 | 003 | 003 | 002 | 002 | 001 | 000 | 000 |
| .04 | | 007 | 006 | 006 | 005 | 004 | 003 | 002 | 001 | 000 |
| .05 | | 011 | 010 | 009 | 008 | 006 | 004 | 003 | 001 | 000 |
| .06 | .00015 | 015 | 013 | 012 | 009 | 006 | 004 | 002 | 000 |
| .07 | | 020 | 020 | 018 | 016 | 012 | 009 | 005 | 003 | 001 |
| .08 | | 027 | 026 | 024 | 021 | 016 | 012 | 007 | 003 | 001 |
| .09 | | 034 | 033 | 031 | 026 | 021 | 015 | 009 | 004 | 001 |
| .10 | | 042 | 041 | 038 | 033 | 026 | 019 | 011 | 005 | 001 |
| .11 | .00051 | 050 | 046 | 040 | 032 | 023 | 014 | 007 | 002 |
| .12 | | 060 | 060 | 055 | 048 | 038 | 027 | 017 | 008 | 002 |
| .13 | | 070 | 070 | 065 | 057 | 045 | 032 | 020 | 009 | 002 |
| .14 | | 082 | 081 | 076 | 066 | 053 | 038 | 023 | 011 | 003 |
| .15 | | 094 | 094 | 087 | 076 | 061 | 044 | 027 | 013 | 003 |
| .16 | .00107 | 107 | 100 | 087 | 070 | 050 | 031 | 015 | 004 |
| .17 | | 121 | 121 | 113 | 099 | 079 | 057 | 035 | 017 | 004 |
| .18 | | 136 | 136 | 128 | 112 | 090 | 065 | 040 | 019 | 005 |
| .19 | | 151 | 152 | 143 | 125 | 101 | 073 | 045 | 021 | 006 |
| .20 | | 168 | 168 | 159 | 140 | 113 | 082 | 050 | 024 | 006 |

Also, $Q = P - wl(1 - k)\sin A$, where $k$ is the value in Table 4 corresponding to the given values of $P$ and $A$.

**Correction for Stretch in Chaining Uphill** Let $L =$ unstretched length of tape at working temperature, $w =$ weight per unit length of tape, $A =$ angle between chord $AB$ and the horizontal, $F =$ area of cross section, $E =$ Young's modulus of elasticity (for steel, $E = 29{,}000{,}000$ lb/in²), $l =$ stretched length (along curve).

If the tension $P$ at the upper end is known, compute $wL/P$ and find $m$ from Table 6. Then $l = L + (LP/FE)(1 - m)$.

If the tension $Q$ at the lower end is known, compute $wL/Q$ and find $n$ from Table 7. Then $l = L + (LQ/FE)(1 + n)$.

**Table 5. Giving $k$**

| $\dfrac{wl}{Q}$ | $A = 0°$ | 10° | 20° | 30° | 40° | 50° | 60° | 70° | 80° |
|---|---|---|---|---|---|---|---|---|---|
| .01 | .00000 | 000 | 000 | 000 | 000 | 000 | 000 | 000 | 000 |
| .02 | | 002 | 002 | 001 | 001 | 001 | 001 | 000 | 000 | 000 |
| .03 | | 004 | 004 | 003 | 003 | 002 | 002 | 001 | 001 | 000 |
| .04 | | 007 | 006 | 006 | 005 | 004 | 003 | 002 | 001 | 000 |
| .05 | | 011 | 010 | 009 | 008 | 006 | 004 | 002 | 001 | 000 |
| .06 | .00015 | 014 | 013 | 011 | 008 | 006 | 004 | 002 | 000 |
| .07 | | 020 | 020 | 018 | 015 | 011 | 008 | 005 | 002 | 001 |
| .08 | | 027 | 026 | 023 | 019 | 015 | 011 | 006 | 003 | 001 |
| .09 | | 034 | 032 | 029 | 024 | 019 | 013 | 008 | 004 | 001 |
| .10 | | 042 | 040 | 036 | 030 | 023 | 016 | 010 | 004 | 001 |
| .11 | .00051 | 048 | 043 | 036 | 028 | 019 | 011 | 005 | 001 |
| .12 | | 060 | 057 | 051 | 043 | 033 | 023 | 014 | 006 | 002 |
| .13 | | 070 | 067 | 060 | 050 | 038 | 026 | 016 | 007 | 002 |
| .14 | | 082 | 078 | 069 | 057 | 044 | 030 | 018 | 008 | 002 |
| .15 | | 094 | 089 | 079 | 066 | 050 | 035 | 021 | 010 | 002 |
| .16 | .00107 | 101 | 090 | 074 | 057 | 039 | 022 | 011 | 003 |
| .17 | | 121 | 114 | 101 | 084 | 064 | 044 | 026 | 012 | 003 |
| .18 | | 136 | 128 | 113 | 092 | 071 | 049 | 029 | 013 | 003 |
| .19 | | 151 | 142 | 125 | 103 | 079 | 054 | 032 | 015 | 004 |
| .20 | | 168 | 157 | 138 | 114 | 087 | 060 | 035 | 016 | 004 |

## OTHER USEFUL CURVES

The **cycloid** is traced by a point on the circumference of a circle which rolls without slipping along a straight line. Equations of cycloid, in parametric form (axes as in Fig. 60): $x = a$ (rad $u - \sin u$), $y = a(1 - \cos u)$, where $a$ is the radius of the rolling

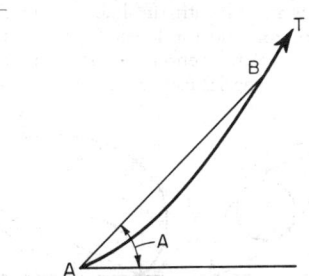

**Fig. 59**

circle, and rad $u$ is the radian measure of the angle $u$ through which it has rolled. The tangent and normal at any point pass through the highest and lowest points of the corresponding position of the generating circle. The **radius of curvature** at any point $P$ is $PC = 4a \sin (u/2) = 2\sqrt{2ay}$ = twice the length of the normal, $PN$. The **evolute**, or locus of centers of curvature,

**Table 6. Giving** $m$

| $\dfrac{wL}{P}$ | $A =$ 0° | 10° | 20° | 30° | 40° | 50° | 60° | 70° | 80° | 90° |
|---|---|---|---|---|---|---|---|---|---|---|
| .00 | .000 | .000 | .000 | .000 | .000 | .000 | .000 | .000 | .000 | .000 |
| .10 | .001 | .010 | .018 | .026 | .033 | .039 | .044 | .047 | .049 | .050 |
| .20 | .003 | .021 | .038 | .053 | .067 | .078 | .088 | .094 | .099 | .100 |

**Table 7. Giving** $n$

| $\dfrac{wL}{Q}$ | $A =$ 10° | 20° | 30° | 40° | 50° | 60° | 70° | 80° | 90° |
|---|---|---|---|---|---|---|---|---|---|
| .00 | .000 | .000 | .000 | .000 | .000 | .000 | .000 | .000 | .000 |
| .10 | .008 | .016 | .024 | .032 | .038 | .043 | .047 | .049 | .050 |
| .20 | .014 | .031 | .047 | .062 | .075 | .086 | .094 | .099 | .100 |

is an equal cycloid. **To construct** a cycloid (Fig. 61), divide the semicircumference of the generating circle into $n$ equal parts (here 4) and lay off these arcs along the base (from $O$ to $4'$). Describe arcs with centers at $1'$, $2'$, . . . and radii equal to the chords $O1$, $O2$, . . . , and sketch the cycloid as a curve tangent to all of these arcs. Or, on horizontal lines through 1, 2, . . . lay off distances equal to $O1'$, $O2'$, etc. The points thus reached will lie on the cycloid.

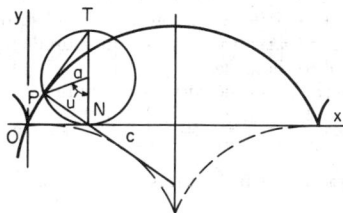

**Fig. 60**   Cycloid.

The area of one arch $= 3\pi a^2$, length of arc of one arch $= 8a$. Area bounded by the ordinate of the point $P$ corresponding to any value of $u$ is $a^2(\tfrac{3}{2}$ rad $u - 2 \sin u + \tfrac{1}{4} \sin 2u) = \tfrac{3}{2}ax - \tfrac{1}{2} y \sqrt{(2a - y)y}$. Length of arc $OP = 4a(1 - \cos \tfrac{1}{2}u) = 4a - 2\sqrt{2a(2a - y)}$.

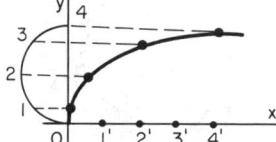

**Fig. 61**

The **trochoid** is a more general curve, traced by any point on a radius of the rolling circle, at distance $b$ from the center (Fig. 62). It is a prolate trochoid if $b < a$, and a curtate or looped trochoid if $b > a$. The equations in either case are $x = a$ rad $u - b \sin u$, $y = a - b \cos u$.

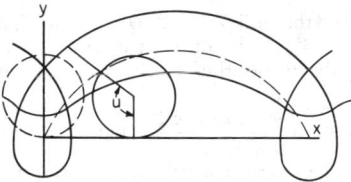

**Fig. 62**   Trochoid.

The **epicycloid** (or **hypocycloid**) is a curve generated by a point on the circumference of a circle of radius $a$ which rolls without slipping on the outside (or inside) of a fixed circle of radius $c$. For the **equations**, put $b = a$ in the equations of the epi- or hypotrochoid, below. The normal at any point $P$ passes through the point of contact $N$ of the corresponding position of the rolling circle. To construct the curve (Figs. 63 and 64), divide the semicircumference of the rolling circle into $n$ equal parts, by points 1, 2, 3 . . . , and lay off these arcs ($A1$, $A2$, $A3$) along the circumference of the base circle, as $A1'$, $A2'$,

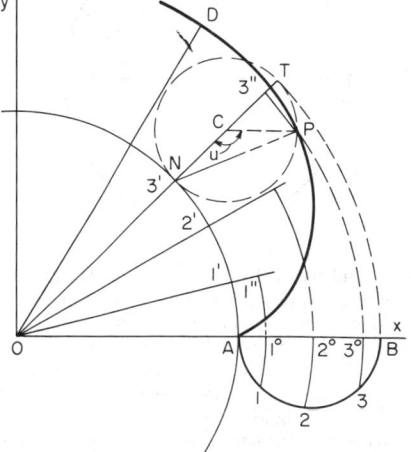

**Fig. 63**   Epicycloid.

$A3', \ldots$ Describe circles with centers at $1', 2', 3', \ldots$ and radii equal to the chords $A1, A2, A3, \ldots$; then the required curve will be tangent to all these circles. Or, with $O$ as center, draw arcs through $1, 2, 3, \ldots$, meeting the radius $OA$ in $1^0$,

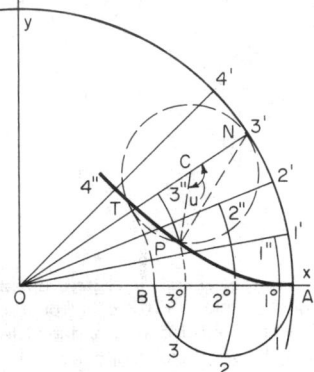

**Fig. 64**  Hypocycloid.

$2^0, 3^0, \ldots$, and the radii $O1', O2', O3', \ldots$ in $1'', 2'', 3'', \ldots$; then from $1'', 2'', 3'', \ldots$ lay off arcs equal to $1^01, 2^02, 3^03, \ldots$, respectively; the points thus reached will be points of the curve.

The area $OAP = \dfrac{a(c \pm a)(c \pm 2a)}{2c}(\text{rad } u - \sin u)$, where the upper sign applies to the epicycloid, the lower to the hypocycloid, and rad $u$ = the radian measure of the angle $u$ shown in Figs. 63 and 64. Arc $AP = (4a/c)(c \pm a)(1 - \cos \frac{1}{2}u)$; arc $AD = (4a/c)(c \pm a)$. (In Fig. 64, $D = 4''$.)

Radius of curvature at any point $P$ is $R = \dfrac{4a(c \pm a)}{c \pm 2a} \sin \frac{1}{2}u$;

at $A$, $R = 0$; at $D$, $R = \dfrac{4a(c \pm a)}{c \pm 2a}$.

**Special Cases**  If $a = \frac{1}{2}c$, the hypocycloid becomes a straight line, diameter of the fixed circle (Fig. 65). In this case

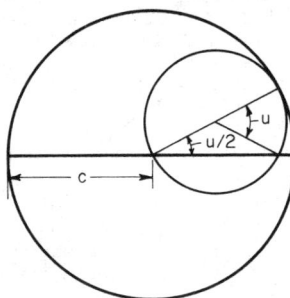

**Fig. 65**

the hypotrochoid traced by any point rigidly connected with the rolling circle (not necessarily on the circumference) will be an ellipse. If $a = \frac{1}{4}c$, the curve generated will be the four-cusped hypocycloid, or **astroid** (Fig. 66), whose equation is $x^{2/3} + y^{2/3} = c^{2/3}$. If $a = c$, the epicycloid is the **cardioid,** whose equation in polar coordinates (axes as in Fig. 67) is $r = 2c(1 + \cos \theta)$. Length of cardioid = $16c$.

The **epitrochoid** (or **hypotrochoid** ) is a curve traced by any point rigidly attached to a circle of radius $a$, at distance $b$ from the center, when this circle rolls without slipping on the

outside (or inside) of a fixed circle of radius $c$. The equations are $x = (c \pm a) \cos \left( \dfrac{a}{c} u \right) \mp b \cos \left[ \left( 1 \pm \dfrac{a}{c} \right) u \right]$, $y = (c \pm a)$

$\sin \left( \dfrac{a}{c} u \right) - b \sin \left[ \left( 1 \pm \dfrac{a}{c} \right) u \right]$, where $u$ = the angle which the moving radius makes with the line of centers; take the upper sign for the epi- and the lower for the hypotrochoid. The curve is called prolate or curtate according as $b < a$ or $b > a$. When $b = a$, the special case of the epi- or hypocycloid arises.

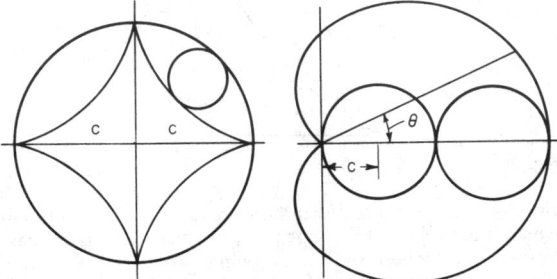

**Fig. 66**  Astroid.          **Fig. 67**  Cardioid.

The **involute of a circle** is the curve traced by the end of a taut string which is unwound from the circumference of a fixed circle, of radius $c$. If $QP$ is the free portion of the string at any instant (Fig. 68), $QP$ will be tangent to the circle at $Q$, and the length of $QP$ = length of arc $QA$; hence the construction of the curve. The equations of the curve in parametric form (axes as in figure) are $x = c(\cos u + \text{rad } u \sin u)$, $y = c(\sin u - \text{rad } u \cos u)$, where rad $u$ is the radian measure of the angle $u$ which $OQ$ makes with the $x$ axis. Length of arc $AP = \frac{1}{2}c(\text{rad } u)^2$; radius of curvature at $P$ is $QP$. Polar equations, in terms of parameter $v( = \text{angle } POQ)$, are $r = c \sec v$, rad $\theta = \tan v - \text{rad } v$. Here, $r = OP$, and rad $\theta$ = radian measure of angle, $AOP$ (Fig. 68).

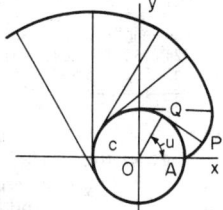

**Fig. 68**  Involute of circle.

The **spiral of Archimedes** (Fig. 69) is traced by a point $P$ which, starting from $O$, moves with uniform velocity along a ray $OP$, while the ray itself revolves with uniform angular velocity about $O$. Polar equation: $r = k$ rad $\theta$, or $r = a$ $(\theta°/360°)$. Here $a = 2\pi k$ = the distance, measured along a radius, from each coil to the next.

In order to construct the curve, draw radii $O1, O2, O3, \ldots$ making angles $\dfrac{1}{n}(360°)$, $\dfrac{2}{n}(360°)$, $\dfrac{3}{n}(360°)$, $\ldots$ with $Ox$, and along these radii lay off distances equal to $\dfrac{1}{n}a, \dfrac{2}{n}a, \dfrac{3}{n}a, \ldots$; the points thus reached will lie on the spiral. The figure shows one-half of the curve, corresponding to positive values of $\phi$.

Construction for tangent and normal: Let $PT$ and $PN$ be the tangent and normal at any point $P$, the line $TON$ being perpendicular to $OP$. Then $OT = r^2/k$, and $ON = k$, where $k = a/(2\pi)$. Hence the construction.

The radius of curvature at $P$ is $R = (k^2 + r^2)^{3/2}/(2k^2 + r^2)$. To construct the center of curvature, $C$, draw $NQ$ perpendicular to $PN$ and $PQ$ perpendicular to $OP$; then $OQ$ will meet $PN$ in $C$. Length of arc $OP = \frac{1}{2}k$ [rad $\theta \sqrt{1 + (\text{rad } \theta)^2} + \sinh^{-1}(\text{rad } \theta)$]. After many windings, arc $OP = \frac{1}{2}r^2/k$, approx.

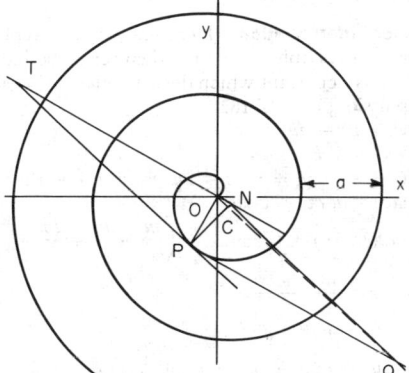

**Fig. 69**  Spiral of Archimedes.

The **logarithmic spiral** (Fig. 70) is a curve which cuts the radii from $O$ at a constant angle $v$, whose cotangent is $m$. Polar equation: $r = ae^{m \text{ rad } \theta}$. Here $a$ is the value of $r$ when $\theta = 0$. For large negative values of $\theta$, the curve winds around $O$ as an asymptotic point. If $PT$ and $PN$ are the tangent and normal at $P$, the line $TON$ being perpendicular to $OP$ (not shown in

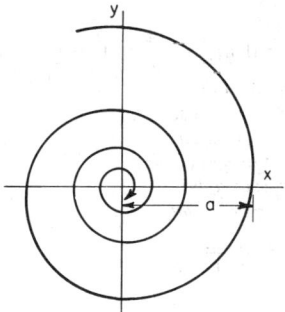

**Fig. 70**  Logarithmic spiral.

figure), then $ON = rm$, and $PN = r \sqrt{1 + m^2} = r/\sin v$. Radius of curvature at $P$ is $PN$. The evolute of the spiral is an equal spiral whose axis makes an angle $\frac{1}{2}\pi - (\ln m)/m$ with the axis of the given spiral. Area swept out by the radius $r$ from $r = 0$ (where $\theta = -\infty$) to $r = r$ is $A = r^2/4m =$ half the triangle $OPT$. Length of arc from $O$ to $P = s = r/\cos v = PT$.

The **tractrix**, or Schiele's antifriction curve (Fig. 71), is a curve such that the portion $PT$ of the tangent between the point of contact and the $x$ axis is constant $= a$. Its equation is $x = \pm a \left[ \cosh^{-1} \dfrac{a}{y} - \sqrt{1 - \left(\dfrac{y}{a}\right)^2} \right]$, or, in parametric form, $x$

$= \pm a(t - \tanh t)$, $y = a/\cosh t$. The $x$ axis is an asymptote of the curve. Length of arc $BP = a \log_e (a/y)$. The evolute (locus of centers of curvature) is the catenary whose lowest point is at $B$, and whose directrix is $Ox$.

The **lemniscate** (Fig. 72) is the locus of a point $P$ the product of whose distances from two fixed points $F$, $F'$ is constant,

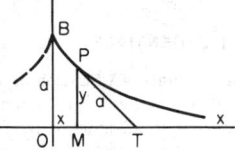

**Fig. 71**  Tractrix.

equal to $\frac{1}{2}a^2$. The distance $FF' = a\sqrt{2}$. Polar equation is $r = a\sqrt{\cos 2\theta}$. Angle between $OP$ and the normal at $P$ is $2\theta$. The two branches of the curve cross at right angles at $O$. Maximum $y$ occurs when $\theta = 30°$ and $r = a/\sqrt{2}$, and is equal to $\frac{1}{4}a\sqrt{2}$. Area of one loop $= a^2/2$.

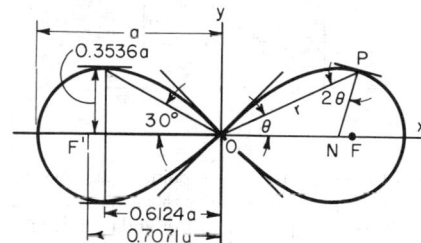

**Fig. 72**  Lemniscate.

The **helix** (Fig. 73) is the curve of a screw thread on a cylinder of radius $r$. The curve crosses the elements of the cylinder at a constant angle, $v$. The pitch, $h$, is the distance between two coils of the helix, measured along an element of the cylinder; hence $h = 2\pi r \tan v$. Length of one coil $= \sqrt{(2\pi r)^2 + h^2} = 2\pi r/\cos v$. To construct the projection of a helix on a plane containing the axis of the cylinder, draw a rectangle, breadth $2r$ and height $h$, to represent the plane, with a semicircle below it, as in the figure, to represent the base of the cylinder. Divide $h$ into equal parts (here 8), numbered from 1 to 8; think of the circumference as also divided into 8 equal parts, represented on the semicircle by numbers from 1' to 4' and back again from 4' to 8'. Then the point of intersection of a horizontal line through 1, 2, . . . with a vertical line through 1', 2', . . . will be a point of the required projection. If the cylinder is rolled out on a plane, the development of the helix will be a straight line, with slope equal to $\tan v$.

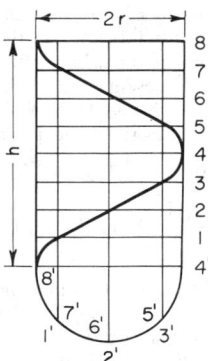

**Fig. 73**  Helix.

# DIFFERENTIAL AND INTEGRAL CALCULUS

## by Philip Franklin

### DERIVATIVES AND DIFFERENTIALS

**Derivatives and Differentials** A **function** of a single variable $x$ may be denoted by $f(x)$, $F(x)$, etc. The value of the function when $x$ has the value $x_0$ is then denoted by $f(x_0)$, $F(x_0)$, etc. The **derivative** of a function $y = f(x)$ may be denoted by $f'(x)$, or by $dy/dx$. The value of the derivative at a given point $x = x_0$ is the **rate of change** of the function at that point; or, if the function is represented by a curve in the usual way (Fig. 1),

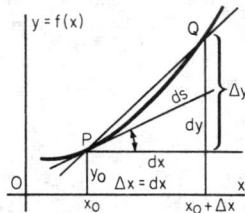

**Fig. 1**

the value of the derivative at any point shows the **slope of the curve** (i.e., the slope of the tangent to the curve) at that point (positive if the tangent points upward, and negative if it points downward, moving to the right).

The **increment** $\Delta y$ (read: "delta $y$"), in $y$ is the change produced in $y$ by increasing $x$ from $x_0$ to $x_0 + \Delta x$; i.e., $\Delta y = f(x_0 + \Delta x) - f(x_0)$. The **differential**, $dy$, of $y$ is the value which $\Delta y$ would have if the curve coincided with its tangent. (The differential, $dx$, of $x$ is the same as $\Delta x$ when $x$ is the independent variable.) Note that the derivative depends only on the value of $x_0$, while $\Delta y$ and $dy$ depend not only on $x_0$ but on the value of $\Delta x$ as well. The ratio $\Delta y/\Delta x$ represents the secant slope, and $dy/dx$ the slope of the tangent (see Fig. 1). If $\Delta x$ is made to approach zero, the secant approaches the tangent as a limiting position, so that the derivative $= f'(x) = \dfrac{dy}{dx} =$

$$\lim_{\Delta x \to 0} \left[ \frac{\Delta y}{\Delta x} \right] = \lim_{\Delta x \to 0} \left[ \frac{f(x_0 + \Delta x) - f(x_0)}{\Delta x} \right].$$ Also, $dy = f'(x)\,dx$.

The symbol "lim" in connection with $\Delta x \to 0$ means "the limit, as $\Delta x$ approaches 0, of . . ." (A constant $c$ is said to be the **limit** of a variable $u$ if, whenever any quantity $m$ has been assigned, there is a stage in the variation process beyond which $|c - u|$ is always less than $m$; or, briefly, $c$ is the limit of $u$ if the difference between $c$ and $u$ can be made to become and remain as small as we please.)

**To find the derivative** of a given function at a given point: (1) If the function is given only by a curve, measure graphically the slope of the tangent at the point in question; (2) if the function is given by a mathematical expression, use the following rules for differentiation. These rules give, directly, the differential,

$dy$, in terms of $dx$; to find the derivative, $dy/dx$, divide through by $dx$.

**Rules for Differentiation** (Here $u$, $v$, $w$, . . . represent any functions of a variable $x$, or may themselves be independent variables. $a$ is a constant which does not change in value in the same discussion; $e = 2.71828$.)

1. $d(a + u) = du$.
2. $d(au) = a\,du$.
3. $d(u + v + w + \cdots) = du + dv + dw + \cdots$.
4. $d(uv) = u\,dv + v\,du$.
5. $d(uvw\ldots) = (uvw\ldots)\left( \dfrac{du}{u} + \dfrac{dv}{v} + \dfrac{dw}{w} + \cdots \right)$.
6. $d\dfrac{u}{v} = \dfrac{v\,du - u\,dv}{v^2}$.
7. $d(u^m) = mu^{m-1}\,du$.

Thus, $d(u^2) = 2u\,du$; $d(u^3) = 3u^2\,du$; etc.

8. $d\sqrt{u} = \dfrac{du}{2\sqrt{u}}$.
9. $d\left(\dfrac{1}{u}\right) = -\dfrac{du}{u^2}$.
10. $d(e^u) = e^u\,du$.
11. $d(a^u) = (\ln a)a^u\,du$.
12. $d\ln u = \dfrac{du}{u}$.
13. $d\log_{10} u = (\log_{10} e)\,\dfrac{du}{u} = (0.4343\ldots)\,\dfrac{du}{u}$.
14. $d\sin u = \cos u\,du$.
15. $d\csc u = -\cot u\csc u\,du$.
16. $d\cos u = -\sin u\,du$.
17. $d\sec u = \tan u\sec u\,du$.
18. $d\tan u = \sec^2 u\,du$.
19. $d\cot u = -\csc^2 u\,du$.
20. $d\sin^{-1} u = \dfrac{du}{\sqrt{1 - u^2}}$.
21. $d\csc^{-1} u = -\dfrac{du}{u\sqrt{u^2 - 1}}$.
22. $d\cos^{-1} u = -\dfrac{du}{\sqrt{1 - u^2}}$.
23. $d\sec^{-1} u = \dfrac{du}{u\sqrt{u^2 - 1}}$.
24. $d\tan^{-1} u = \dfrac{du}{1 + u^2}$.
25. $d\cot^{-1} u = -\dfrac{du}{1 + u^2}$.
26. $d\ln\sin u = \cot u\,du$.
27. $d\ln\tan u = \dfrac{2\,du}{\sin 2u}$.
28. $d\ln\cos u = -\tan u\,du$.

29. $d \ln \cot u = -\dfrac{2du}{\sin 2u}$.

30. $d \sinh u = \cosh u \, du$.

31. $d \operatorname{csch} u = -\operatorname{csch} u \coth u \, du$.

32. $d \cosh u = \sinh u \, du$.

33. $d \operatorname{sech} u = -\operatorname{sech} u \tanh u \, du$.

34. $d \tanh u = \operatorname{sech}^2 u \, du$.

35. $d \coth u = -\operatorname{csch}^2 u \, du$.

36. $d \sinh^{-1} u = \dfrac{du}{\sqrt{u^2 + 1}}$.

37. $d \operatorname{csch}^{-1} u = -\dfrac{du}{u\sqrt{u^2 + 1}}$.

38. $d \cosh^{-1} u = \dfrac{du}{\sqrt{u^2 - 1}}$.

39. $d \operatorname{sech}^{-1} u = -\dfrac{du}{u\sqrt{1 - u^2}}$.

40. $d \tanh^{-1} u = \dfrac{du}{1 - u^2}$.

41. $d \coth^{-1} u = \dfrac{du}{1 - u^2}$.

42. $d(u^v) = (u^{v-1})(u \ln u \, dv + v \, du)$.

**Derivatives of Higher Orders**  The derivative of the derivative is called the second derivative; the derivative of this, the third derivative; and so on. If $y = f(x)$,

$$f'(x) = D_x y = \frac{dy}{dx}$$

$$f''(x) = D_x^2 y = \frac{d^2 y}{dx^2}$$

$$f'''(x) = D_x^3 y = \frac{d^3 y}{dx^3} \quad \text{etc.}$$

NOTE.  If the notation $d^2y/dx^2$ is used, this must not be treated as a fraction, like $dy/dx$, but as an inseparable symbol, made up of a symbol of operation $d^2/dx^2$, and an operand $y$.

The geometric meaning of the second derivative is this: if the original function $y = f(x)$ is represented by a curve in the usual way, then at any point where $f''(x)$ is *positive*, the curve is *concave upward*, and at any point where $f''(x)$ is *negative*, the curve is *concave downward* (Fig. 2). When $f''(x) = 0$, the curve usually has a **point of inflection.**

**Differentials of Higher Orders**  The differential of the differential is called the second differential; the differential of this, the third differential; etc. These quantities are of little importance except in the case where $dx$ = a constant. In this case

$$dy = f'(x) \, dx$$

$$d^2 y = f''(x) \cdot (dx)^2$$

$$d^3 y = f'''(x) \cdot (dx)^3 \quad \dots$$

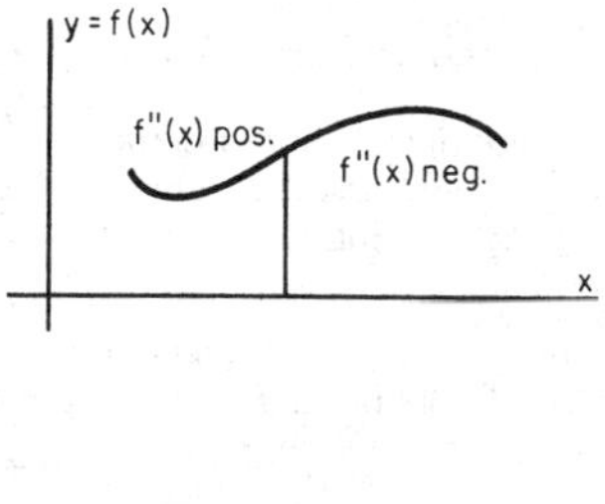

Fig. 2

The first, second, third, etc., differentials are close approximations to the first, second, third, etc., differences and are, therefore, sometimes useful in constructing tables. Thus, denoting the first, second, third, etc., differences by $D'$, $D''$, $D'''$, etc., and, assuming always that $dx$ = a constant,

$$D' = dy + \tfrac{1}{2}d^2 y + \tfrac{1}{6}d^3 y + \tfrac{1}{24}d^4 y + \cdots$$
$$d^3 y = D''' - \tfrac{3}{2}D'''' + \cdots$$
$$D'' = d^2 y + d^3 y + \tfrac{7}{12}d^4 y + \cdots$$
$$d^2 y = D'' - D''' + \tfrac{11}{12}D'''' + \cdots$$
$$D''' = d^3 y + \tfrac{3}{2}d^4 y + \cdots$$
$$dy = D' - \tfrac{1}{2}D'' + \tfrac{1}{3}D''' - \tfrac{1}{4}D'''' + \cdots$$

**Functions of two or more variables** may be denoted by $f(x, y, \dots)$, $F(x, y, \dots)$, etc. The derivative of such a function $u = f(x, y, \dots)$ formed on the assumption that $x$ is the only variable ($y, \dots$ being regarded for the moment as constants) is called the **partial derivative of $u$ with respect to** $x$, and is denoted by $f_x(x,y)$, or $D_x u$, or $d_x u/dx$, or $\partial u/\partial x$. Similarly, the partial derivative of $u$ with respect to $y$ is $f_y(x,y)$, or $D_y u$, or $d_y u/dy$, or $\partial u/\partial y$.

NOTE.  In the third notation, $d_x u$ denotes the differential of $u$ formed on the assumption that $x$ is the only variable. If the fourth notation, $\partial u/\partial x$, is used, this must not be treated as a fraction like $du/dx$; the $\partial/\partial x$ is a symbol of operation, operating on $u$, and the "$\partial x$" must not be separated.

Partial derivatives of the second order are denoted by $f_{xx}$, $f_{xy}$, $f_{yy}$, or by $Du$, $D_x(D_y u)$, $D_y^2 u$, or by $\partial^2 u/\partial x^2$, $\partial^2 u/\partial x \, \partial y$, $\partial^2 u/\partial y^2$, the last symbols being "inseparable." Similarly for higher derivatives. Note that $f_{xy} = f_{yx}$.

If increments $\Delta x$, $\Delta y$ (or $dx$, $dy$) are assigned to the independent variables $x$, $y$, the increment, $\Delta u$, produced in $u = f(x,y)$ is

$$\Delta u = f(x + \Delta x, y + \Delta y) - f(x,y)$$

while the **differential,** $du$, i.e., the value which $\Delta u$ would have if the partial derivatives of $u$ with respect to $x$ and $y$ were constant, is given by

$$du = (f_x) \cdot dx + (f_y) \cdot dy$$

Here the coefficients of $dx$ and $dy$ are the values of the partial derivatives of $u$ at the point in question.

If $x$ and $y$ are functions of a third variable $t$, then the equation

$$\frac{du}{dt} = (f_x)\frac{dx}{dt} + (f_y)\frac{dy}{dt}$$

expresses the rate of change of $u$ with respect to $t$, in terms of the separate rate of change of $x$ and $y$ with respect to $t$.

**Implicit Functions**  If $f(x,y) = 0$, either of the variables $x$ and $y$ is said to be an implicit function of the other. To find $dy/dx$, either (1) solve for $y$ in terms of $x$, and then find $dy/dx$ directly; or (2) differentiate the equation through as it stands, remembering that both $x$ and $y$ are variables, and then divide by $dx$; or (3) use the formula $dy/dx = -(f_x/f_y)$, where $f_x$ and $f_y$ are the partial derivatives of $f(x,y)$ at the point in question.

## MAXIMA AND MINIMA

**A function of one variable,** as $y = f(x)$, is said to have a **maximum** at a point $x = x_0$, if at that point the slope of the curve is zero and the concavity downward (see Fig. 3); a sufficient condition for a maximum is $f'(x_0) = 0$ and $f''(x_0)$ negative. Similarly, $f(x)$

has a **minimum** if the slope is zero and the concavity upward; a sufficient condition for a minimum is $f'(x_0) = 0$ and $f''(x_0)$ positive. If $f'(x_0)$ and $f'(x_0) = 0$ and $f''(x_0) \neq 0$, the point $x_0$ will be a **point of inflection.** If $f'(x_0) = 0$ and $f''(x_0) = 0$ and $f'''(x_0) = 0$, the point $x_0$ will be a maximum if $f''''(x_0) < 0$, and a minimum if $f''''(x_0) > 0$. It is usually sufficient, however, in any practical case, to find the values of $x$ which make $f'(x) = 0$, and then decide, from a general knowledge of the curve or the sign of $f'(x)$ to the right and left of $x_0$, which of these values (if any) give maxima or minima, without investigating the higher derivatives.

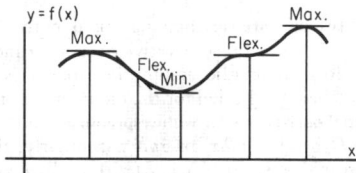

**Fig. 3**

A **function of two variables,** as $u = f(x,y)$, will have a **maximum** at a point $(x_0,y_0)$ if at that point $f_x = 0$, $f_y = 0$, and $f_{xx} < 0$, $f_{yy} < 0$; and a **minimum** if at that point $f_x = 0$, $f_y = 0$, and $f_{xx} > 0$, $f_{yy} > 0$; provided, in each case, $(f_{xx})(f_{yy}) - (f_{xy})^2$ is positive. If $f_x = 0$ and $f_y = 0$, and $f_{xx}$ and $f_{yy}$ have opposite signs, the point $(x_0,y_0)$ will be a "saddle point" of the surface representing the function.

## EXPANSION IN SERIES

The range of values of $x$ for which each of the series is convergent is stated at the right of the series.

**Arithmetical and Geometrical Series, and the Binomial Theorem** See Sec. 2, Algebra.

**Exponential and Logarithmic Series**

$$e^x = 1 + \frac{x}{1!} + \frac{x^2}{2!} + \frac{x^3}{3!} + \frac{x^4}{4!} + \cdots \qquad [-\infty < x < +\infty]$$

$$a^x = e^{mx} = 1 + \frac{m}{1!}x + \frac{m^2}{2!}x^2 + \frac{m^3}{3!}x^3 + \cdots$$
$$[a > 0, -\infty < x < +\infty]$$

where $m = \ln a = (2.3026)(\log_{10} a)$.

$$\ln(1 + x) = x - \frac{x^2}{2} + \frac{x^3}{3} - \frac{x^4}{4} + \frac{x^5}{5} \cdots \qquad [-1 < x < +1]$$

$$\ln(1 - x) = -x - \frac{x^2}{2} - \frac{x^3}{3} - \frac{x^4}{4} - \frac{x^5}{5} - \cdots \quad [-1 < x < +1]$$

$$\ln\left(\frac{1+x}{1-x}\right) = 2\left(x + \frac{x^3}{3} + \frac{x^5}{5} + \frac{x^7}{7} + \cdots\right) \quad [-1 < x < +1]$$

$$\ln\left(\frac{x+1}{x-1}\right) = 2\left(\frac{1}{x} + \frac{1}{3x^3} + \frac{1}{5x^5} + \frac{1}{7x^7} + \cdots\right)$$
$$[x < -1 \text{ or } +1 < x]$$

$$\ln x = 2\left[\frac{x-1}{x+1} + \frac{1}{3}\left(\frac{x-1}{x+1}\right)^3 + \frac{1}{5}\left(\frac{x-1}{x+1}\right)^5 + \cdots\right]$$
$$[0 < x < \infty]$$

$$\ln(a + x) = \ln a + 2\left[\frac{x}{2a+x} + \frac{1}{3}\left(\frac{x}{2a+x}\right)^3\right.$$
$$\left. + \frac{1}{5}\left(\frac{x}{2a+x}\right)^5 + \cdots\right]$$
$$[0 < a < +\infty, -a < x < +\infty]$$

**Series for the Trigonometric Functions** In the following formulas, *all angles must be expressed in radians.* If $D$ = the number of degrees in the angle, and $x$ = its radian measure, then $x = 0.017453D$.

$$\sin x = x - \frac{x^3}{3!} + \frac{x^5}{5!} - \frac{x^7}{7!} + \cdots \qquad [-\infty < x < +\infty]$$

$$\cos x = 1 - \frac{x^2}{2!} + \frac{x^4}{4!} - \frac{x^6}{6!} + \frac{x^8}{8!} - \cdots \qquad [-\infty < x < +\infty]$$

$$\tan x = x + \frac{x^3}{3} + \frac{2x^5}{15} + \frac{17x^7}{315} + \frac{62x^9}{2835} + \cdots$$
$$[-\pi/2 < x < +\pi/2]$$

$$\cot x = \frac{1}{x} - \frac{x}{3} - \frac{x^3}{45} - \frac{2x^5}{945} - \frac{x^7}{4725} - \cdots \quad [-\pi < x < +\pi]$$

$$\sin^{-1} y = y + \frac{y^3}{6} + \frac{3y^5}{40} + \frac{5y^7}{112} + \cdots \qquad [-1 \leq y \leq +1]$$

$$\tan^{-1} y = y - \frac{y^3}{3} + \frac{y^5}{5} - \frac{y^7}{7} + \cdots \qquad [-1 \leq y \leq +1]$$

$$\cos^{-1} y = \tfrac{1}{2}\pi - \sin^{-1} y; \qquad \cot^{-1} y = \tfrac{1}{2}\pi - \tan^{-1} y.$$

**Series for the Hyperbolic Functions** ($x$ a pure number)

$$\sinh x = x + \frac{x^3}{3!} + \frac{x^5}{5!} + \frac{x^7}{7!} + \cdots \qquad [-\infty < x < \infty]$$

$$\cosh x = 1 + \frac{x^2}{2!} + \frac{x^4}{4!} + \frac{x^6}{6!} + \cdots \qquad [-\infty < x < \infty]$$

$$\sinh^{-1} y = y - \frac{y^3}{6} + \frac{3y^5}{40} - \frac{5y^7}{112} + \cdots \qquad [-1 < y < +1]$$

$$\tanh^{-1} y = y + \frac{y^3}{3} + \frac{y^5}{5} + \frac{y^7}{7} + \cdots \qquad [-1 < y < +1]$$

**General Formulas of Maclaurin and Taylor** If $f(x)$ and all its derivatives are continuous in the neighborhood of the point $x = 0$ (or $x = a$), then, for any value of $x$ in this neighborhood, the function $f(x)$ may be expressed as a power series arranged according to ascending powers of $x$ (or of $x - a$), as follows:

$$f(x) = f(0) + \frac{f'(0)}{1!}x + \frac{f''(0)}{2!}x^2 + \frac{f'''(0)}{3!}x^3 + \cdots$$
$$+ \frac{f^{(n-1)}(0)}{(n-1)!}x^{n-1} + (P_n)x^n \text{ (Maclaurin)}$$

$$f(x) = f(a) + \frac{f'(a)}{1!}(x-a) + \frac{f''(a)}{2!}(x-a)^2 + \frac{f'''(a)}{3!}(x-a)^3$$
$$+ \cdots + \frac{f^{(n-1)}(a)}{(n-1)!}(x-a)^{n-1} + (Q_n)(x-a)^n \text{ (Taylor)}$$

Here $(P_n)x^n$, or $(Q_n)(x-a)^n$, is called the **remainder term;** the values of the coefficients $P_n$ and $Q_n$ may be expressed as follows:

$$P_n = [f^{(n)}(sx)]/n! = [(1-t)^{n-1}f^{(n)}(tx)]/(n-1)!$$

$$Q_n = \{f^{(n)}[a + s(x-a)]\}/n!$$
$$= \{(1-t)^{n-1}f^{(n)}[a + t(x-a)]\}/(n-1)!$$

where $s$ and $t$ are certain unknown numbers between 0 and 1; the $s$ form is due to Lagrange, the $t$ form to Cauchy.

The error due to neglecting the remainder term is less than $(\overline{P}_n)x^n$, or $(\overline{Q}_n)(x-a)^n$, where $\overline{P}_n$, or $\overline{Q}_n$, is the largest value taken on by $P_n$, or $Q_n$, when $s$ or $t$ ranges from 0 to 1. If this error, which depends on both $n$ and $x$, approaches 0 as $n$ increases (for any given value of $x$), then the general expression with remainder becomes (for that value of $x$) a convergent infinite series.

The sum of the first few terms of Maclaurin's series gives a good approximation to $f(x)$ for values of $x$ near $x = 0$; Taylor's series gives a similar approximation for values near $x = a$.

**Reversing a Series**   If $y = x + bx^2 + cx^3 + dx^4 + ex^5 + \cdots$, then $x = y - by^2 + (2b^2 - c)y^3 - (5b^3 - 5bc + d)y^4 + (14b^4 - 21b^2c + 6bd + 3c^2 - e)y^5 + \cdots$, provided the latter series is convergent.

**Fourier's Series**   Let $f(x)$ be a function which is finite in the interval from $x = -c$ to $x = +c$ and whose graph has finite arc length in that interval (see note below). Then, for any value of $x$ between $-c$ and $c$,

$$f(x) = \tfrac{1}{2}a_0 + a_1 \cos\frac{\pi x}{c} + a_2 \cos\frac{2\pi x}{c} + a_3 \cos\frac{3\pi x}{c} + \cdots$$

$$+ b_1 \sin\frac{\pi x}{c} + b_2 \sin\frac{2\pi x}{c} + b_3 \sin\frac{3\pi x}{c} + \cdots$$

where the constant coefficients are determined as follows:

$$a_n = \frac{1}{c}\int_{-c}^{c} f(t)\cos\frac{n\pi t}{c}\,dt \qquad b_n = \frac{1}{c}\int_{-c}^{c} f(t)\sin\frac{n\pi t}{c}\,dt$$

In case the curve $y = f(x)$ is symmetrical with respect to the origin, the $a$'s are all zero, and the series is a sine series. In case the curve is symmetrical with respect to the $y$ axis, the $b$'s are all zero, and a cosine series results. (In this case, the series will

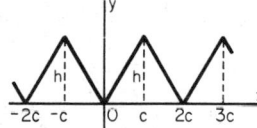

**Fig. 4**

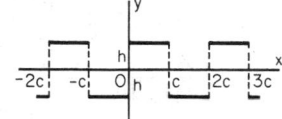

**Fig. 5**

be valid not only for values of $x$ between $-c$ and $c$, but also for $x = -c$ and $x = c$.) A Fourier series can always be integrated term by term; but the result of differentiating term by term may not be a convergent series.

NOTE.   If $x = x_0$ is a point of discontinuity, $f(x_0)$ is to be defined as $\tfrac{1}{2}[f_1(x_0) + f_2(x_0)]$, where $f_1(x_0)$ is the limit of $f(x)$ when $x$ approaches $x_0$ from below, and $f_2(x_0)$ is the limit of $f(x)$ when $x$ approaches $x_0$ from above.

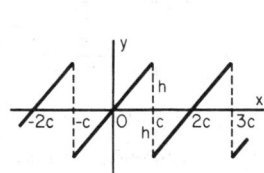

**Fig. 6**

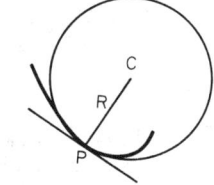

**Fig. 7**

**Examples of Fourier's Series**   If $y = f(x)$ is the curve in Figs. 4 to 6, then

In Fig. 4, $\quad y = \dfrac{b}{2} - \dfrac{4b}{\pi^2}\left(\cos\dfrac{\pi x}{c} + \dfrac{1}{9}\cos\dfrac{3\pi x}{c}\right.$
$$\left. + \dfrac{1}{25}\cos\dfrac{5\pi x}{c} + \cdots\right)$$

In Fig. 5, $\quad y = \dfrac{4b}{\pi}\left(\sin\dfrac{\pi x}{c} + \dfrac{1}{3}\sin\dfrac{3\pi x}{c}\right.$
$$\left. + \dfrac{1}{5}\sin\dfrac{5\pi x}{c} + \cdots\right)$$

In Fig. 6, $\quad y = \dfrac{2b}{\pi}\left(\sin\dfrac{\pi x}{c} - \dfrac{1}{2}\sin\dfrac{2\pi x}{c}\right.$
$$\left. + \dfrac{1}{3}\sin\dfrac{3\pi x}{c} - \cdots\right)$$

## INDETERMINATE FORMS

In the following paragraphs, $f(x)$, $g(x)$ denote functions which approach 0; $F(x)$, $G(x)$ functions which increase indefinitely; and $U(x)$ a function which approaches 1, when $x$ approaches a definite quantity $a$. The problem in each case is to find the limit approached by certain combinations of these functions when $x$ approaches $a$. The symbol $\to$ is to be read "approaches" or "tends to."

CASE 1.   "$\dfrac{0}{0}$." To find the limit of $f(x)/g(x)$ when $f(x)\to 0$ and $g(x)\to 0$, use the theorem that $\lim \dfrac{f(x)}{g(x)} = \lim \dfrac{f'(x)}{g'(x)}$, where $f'(x)$ and $g'(x)$ are the derivatives of $f(x)$ and $g(x)$. This second limit may be easier to find than the first. If $f'(x)\to 0$ and $g'(x)\to 0$, apply the same theorem a second time: $\lim \dfrac{f'(x)}{g'(x)} = \lim \dfrac{f''(x)}{g''(x)}$, and so on.

CASE 2.   "$\dfrac{\infty}{\infty}$." If $F(x)\to \infty$ and $G(x)\to \infty$, then $\lim \dfrac{F(x)}{G(x)} = \lim \dfrac{F'(x)}{G'(x)}$, precisely as in Case 1.

CASE 3.   "$0 \cdot \infty$." To find the limit of $f(x)\cdot F(x)$ when $f(x)\to 0$ and $F(x)\to \infty$, write $\lim [f(x)\cdot F(x)] = \lim \dfrac{f(x)}{1/F(x)}$, or $= \lim \dfrac{F(x)}{1/f(x)}$; then proceed as in Case 1 or Case 2.

CASE 4.   The limit of combinations "$0^0$" or $[f(x)]^{g(x)}$; "$1^\infty$" or $[U(x)]^{F(x)}$, "$\infty^0$" or $[F(x)]^{b(x)}$ may be found since their logarithms are limits of the type evaluated in Case 3.

CASE 5.   "$\infty - \infty$." If $F(x)\to \infty$ and $G(x)\to \infty$, write $\lim [F(x) - G(x)] = \lim \dfrac{\dfrac{1}{G(x)} - \dfrac{1}{F(x)}}{\dfrac{1}{F(x)\cdot G(x)}}$; then proceed as in Case 1. Sometimes it is shorter to expand the functions in series. It should be carefully noticed that expressions like $0/0$, $\infty/\infty$, etc., do not represent mathematical quantities.

## CURVATURE

The **radius of curvature** $R$ of a plane curve at any point $P$ (Fig. 7) is the distance, measured along the normal, on the concave side of the curve, to the **center of curvature**, $C$, this point being the limiting position of the point of intersection of the normals at $P$ and a neighboring point $Q$, as $Q$ is made to approach $P$ along the curve. If the equation of the curve is $y = f(x)$,

$$R = \frac{ds}{du} = \frac{[1 + (y')^2]^{3/2}}{y''}$$

where $ds = \sqrt{dx^2 + dy^2}$ = the differential of arc, $u = \tan^{-1}[f'(x)]$ = the angle which the tangent at $P$ makes with the $x$ axis, and $y' = f'(x)$ and $y'' = f''(x)$ are the first and second derivatives of $f(x)$ at the point $P$. Note that $dx = ds \cos u$ and $dy = ds \sin u$. The **curvature**, $K$, at the point $P$, is $K = 1/R = du/ds$; i.e., the curvature is the rate at which the angle $u$ is changing with respect to the length of arc $s$. If the slope of the curve is small, $K \approx f''(x)$.

If the equation of the curve in polar coordinates is $r = f(\theta)$, where $r$ = radius vector and $\theta$ = polar angle, then

$$R = \frac{[r^2 + (r')^2]^{3/2}}{r^2 - rr'' + 2(r')^2}$$

where $r' = f'(\theta)$ and $r'' = f''(\theta)$.

The **evolute** of a curve is the locus of its centers of curvature. If one curve is the evolute of another, the second is called the **involute** of the first.

## INDEFINITE INTEGRALS

An **integral** of $f(x)\,dx$ is any function whose differential is $f(x)\,dx$, and is denoted by $\int f(x)\,dx$. All the integrals of $f(x)\,dx$ are included in the expression $\int f(x)\,dx + C$, where $\int f(x)\,dx$ is any particular integral, and $C$ is an arbitrary constant. The process of finding (when possible) an integral of a given function consists in recognizing by inspection a function which, when differentiated, will produce the given function; or in transforming the given function into a form in which such recognition is easy. The most common integrable forms are collected in the following brief table; for a more extended list, see Peirce, "Table of Integrals," Ginn, or Dwight, "Table of Integrals and other Mathematical Data," Macmillan.

### GENERAL FORMULAS

1. $\int a\,du = a \int du = au + C.$

2. $\int (u + v)\,dx = \int u\,dx + \int v\,dx.$

3. $\int u\,dv = uv - \int v\,du.$

4. $\int f(x)\,dx = \int f[F(y)]F'(y)\,dy, \; x = F(y).$

5. $\int dy \int f(x,y)\,dx = \int dx \int f(x,y)\,dy.$

### FUNDAMENTAL INTEGRALS

6. $\int x^n\,dx = \dfrac{x^{n+1}}{n + 1} + C$, when $n \neq -1.$

7. $\int \dfrac{dx}{x} = \ln x + C = \ln cx.$

8. $\int e^x\,dx = e^x + C.$

9. $\int \sin x\,dx = -\cos x + C.$

10. $\int \cos x\,dx = \sin x + C.$

11. $\int \dfrac{dx}{\sin^2 x} = -\cot x + C.$

12. $\int \dfrac{dx}{\cos^2 x} = \tan x + C.$

13. $\int \dfrac{dx}{\sqrt{1 - x^2}} = \sin^{-1} x + C = -\cos^{-1} x + c.$

14. $\int \dfrac{dx}{1 + x^2} = \tan^{-1} x + C = -\cot^{-1} x + c.$

### RATIONAL FUNCTIONS

15. $\int (a + bx)^n\,dx = \dfrac{(a + bx)^{n+1}}{(n + 1)b} + C.$

16. $\int \dfrac{dx}{a + bx} = \dfrac{1}{b} \ln (a + bx) + C = \dfrac{1}{b} \ln c(a + bx).$

17. $\int \dfrac{1}{x^2}\,dx = -\dfrac{1}{x} + C.$

18. $\int \dfrac{dx}{(a + bx)^2} = -\dfrac{1}{b(a + bx)} + C.$

19. $\int \dfrac{dx}{1 - x^2} = \frac{1}{2} \ln \dfrac{1 + x}{1 - x} + C$
$$= \tanh^{-1} x + C, \text{ when } x < 1.$$

20. $\int \dfrac{dx}{x^2 - 1} = \frac{1}{2} \ln \dfrac{x - 1}{x + 1} + C$
$$= -\coth^{-1} x + C, \text{ when } x > 1.$$

21. $\int \dfrac{dx}{a + bx^2} = \dfrac{1}{\sqrt{ab}} \tan^{-1}\left(\sqrt{\dfrac{b}{a}}\,x\right) + C$

22. $\int \dfrac{dx}{a - bx^2} = \dfrac{1}{2\sqrt{ab}} \ln \dfrac{\sqrt{ab} + bx}{\sqrt{ab} - bx} + C$ $\quad \left. \right\} [a > 0, b > 0].$
$$= \dfrac{1}{\sqrt{ab}} \tanh^{-1}\left(\sqrt{\dfrac{b}{a}}\,x\right) + C$$

23. $\int \dfrac{dx}{a + 2bx + cx^2} = $
$$\dfrac{1}{\sqrt{ac - b^2}} \tan^{-1}\dfrac{b + cx}{\sqrt{ac - b^2}} + C \quad \left.\right\} [ac - b^2 > 0].$$
$$= \dfrac{1}{2\sqrt{b^2 - ac}} \ln \dfrac{\sqrt{b^2 - ac} - b - cx}{\sqrt{b^2 - ac} + b + cx} + C \quad \left.\right\} [b^2 - ac > 0].$$
$$= -\dfrac{1}{\sqrt{b^2 - ac}} \tanh^{-1}\dfrac{b + cx}{\sqrt{b^2 - ac}} + C$$

24. $\int \dfrac{dx}{a + 2bx + cx^2} = -\dfrac{1}{b + cx} + C$, when $b^2 = ac.$

25. $\int \dfrac{(m + nx)\,dx}{a + 2bx + cx^2} = \dfrac{n}{2c} \ln (a + 2bx + cx^2)$
$$+ \dfrac{mc - nb}{c} \int \dfrac{dx}{a + 2bx + cx^2}.$$

26. In $\int \dfrac{f(x)\,dx}{a + 2bx + cx^2}$, if $f(x)$ is a polynomial of higher than the first degree, divide by the denominator before integrating.

27. $\int \dfrac{dx}{(a + 2bx + cx^2)^p} = \dfrac{1}{2(ac - b^2)(p - 1)}$
$$\times \dfrac{b + cx}{(a + 2bx + cx^2)^{p-1}} +$$
$$\dfrac{(2p - 3)c}{2(ac - b^2)(p - 1)} \int \dfrac{dx}{(a + 2bx + cx^2)^{p-1}}$$

28. $\int \dfrac{(m + nx)\,dx}{(a + 2bx + cx^2)^p} = -\dfrac{n}{2c(p - 1)}$
$$\times \dfrac{1}{(a + 2bx + cx^2)^{p-1}} + \dfrac{mc - nb}{c} \int \dfrac{dx}{(a + 2bx + cx^2)^{p-1}}$$

29. $\int x^{m-1}(a + bx)^n \ dx = \dfrac{x^{m-1}(a + bx)^{n+1}}{(m + n)b}$

$\qquad - \dfrac{(m - 1)a}{(m + n)b} \int x^{m-2}(a + bx)^n dx,$

$\qquad = \dfrac{x^m(a + bx)^n}{m + n} + \dfrac{na}{m + n} \int x^{m-1}(a + bx)^{n-1} \ dx.$

## IRRATIONAL FUNCTIONS

30. $\int \sqrt{a + bx} \ dx = \dfrac{2}{3b} (\sqrt{a + bx})^3 + C.$

31. $\int \dfrac{dx}{\sqrt{a + bx}} = \dfrac{2}{b} \sqrt{a + bx} + C.$

32. $\int \dfrac{(m + nx) \ dx}{\sqrt{a + bx}} = \dfrac{2}{3b^2} (3mb - 2an$

$\qquad\qquad\qquad\qquad + nbx) \sqrt{a + bx} + C.$

33. $\int \dfrac{dx}{(m + nx) \sqrt{a + bx}};$ substitute $y$

$\qquad\qquad = \sqrt{a + bx},$ and use 21 and 22.

34. $\int \dfrac{f(x, \sqrt[n]{a + bx})}{F(x, \sqrt[n]{a + bx})} \ dx;$ substitute $\sqrt[n]{a + bx} = y.$

35. $\int \dfrac{dx}{\sqrt{a^2 - x^2}} = \sin^{-1} \dfrac{x}{a} + C = -\cos^{-1} \dfrac{x}{a} + c.$

36. $\int \dfrac{dx}{\sqrt{a^2 + x^2}} = \ln (x + \sqrt{a^2 + x^2}) + C = \sinh^{-1} \dfrac{x}{a} + c.$

37. $\int \dfrac{dx}{\sqrt{x^2 - a^2}} = \ln (x + \sqrt{x^2 - a^2}) + C = \cosh^{-1} \dfrac{x}{a} + c.$

38. $\int \dfrac{dx}{\sqrt{a + 2bx + cx^2}} = \dfrac{1}{\sqrt{c}} \ln (b + cx$

$\qquad + \sqrt{c} \sqrt{a + 2bx + cx^2}) + C,$ when $c > 0.$

$\qquad = \dfrac{1}{\sqrt{c}} \sinh^{-1} \dfrac{b + cx}{\sqrt{ac - b^2}} + C,$ when $ac - b^2 > 0.$

$\qquad = \dfrac{1}{\sqrt{c}} \cosh^{-1} \dfrac{b + cx}{\sqrt{b^2 - ac}} + C,$ when $b^2 - ac > 0.$

$\qquad = \dfrac{-1}{\sqrt{-c}} \sin^{-1} \dfrac{b + cx}{\sqrt{b^2 - ac}} + C,$ when $c < 0.$

39. $\int \dfrac{(m + nx) \ dx}{\sqrt{a + 2bx + cx^2}}$

$\qquad = \dfrac{n}{c} \sqrt{a + 2bx + cx^2} + \dfrac{mc - nb}{c} \int \dfrac{dx}{\sqrt{a + 2bx + cx^2}}.$

40. $\int \dfrac{x^m \ dx}{\sqrt{a + 2bx + cx^2}} = \dfrac{x^{m-1}X}{mc} - \dfrac{(m - 1)a}{mc} \int \dfrac{x^{m-2} \ dx}{X}$

$\qquad - \dfrac{(2m - 1)b}{mc} \int \dfrac{x^{m-1} \ dx}{X},$ when $X = \sqrt{a + 2bx + cx^2}.$

41. $\int \sqrt{a^2 + x^2} \ dx = \dfrac{x}{2} \sqrt{a^2 + x^2} + \dfrac{a^2}{2} \ln (x + \sqrt{a^2 + x^2})$

$\qquad + C = \dfrac{x}{2} \sqrt{a^2 + x^2} + \dfrac{a^2}{2} \sinh^{-1} \dfrac{x}{a} + c.$

42. $\int \sqrt{a^2 - x^2} \ dx = \dfrac{x}{2} \sqrt{a^2 - x^2} + \dfrac{a^2}{2} \sin^{-1} \dfrac{x}{a} + C.$

43. $\int \sqrt{x^2 - a^2} \ dx = \dfrac{x}{2} \sqrt{x^2 - a^2} - \dfrac{a^2}{2} \ln (x + \sqrt{x^2 - a^2})$

$\qquad + C = \dfrac{x}{2} \sqrt{x^2 - a^2} - \dfrac{a^2}{2} \cosh^{-1} \dfrac{x}{a} + c.$

44. $\int \sqrt{a + 2bx + cx^2} \ dx = \dfrac{b + cx}{2c} \sqrt{a + 2bx + cx^2}$

$\qquad + \dfrac{ac - b^2}{2c} \int \dfrac{dx}{\sqrt{a + 2bx + cx^2}} + C.$

## TRANSCENDENTAL FUNCTIONS

45. $\int a^x \ dx = \dfrac{a^x}{\ln a} + C.$

46. $\int x^n e^{ax} \ dx = \dfrac{x^n e^{ax}}{a} \left[ 1 - \dfrac{n}{ax} \right.$

$\qquad \left. + \dfrac{n(n - 1)}{a^2 x^2} - \cdots \pm \dfrac{n!}{a^n x^n} \right] + C.$

47. $\int \ln x \ dx = x \ln x - x + C.$

48. $\int \dfrac{\ln x}{x^2} \ dx = - \dfrac{\ln x}{x} - \dfrac{1}{x} + C.$

49. $\int \dfrac{(\ln x)^n}{x} \ dx = \dfrac{1}{n + 1} (\ln x)^{n+1} + C.$

50. $\int \sin^2 x \ dx = -\frac{1}{4} \sin 2x + \frac{1}{2}x + C$

$\qquad\qquad = -\frac{1}{2} \sin x \cos x + \frac{1}{2}x + C.$

51. $\int \cos^2 x \ dx = \frac{1}{4} \sin 2x + \frac{1}{2}x + C$

$\qquad\qquad = \frac{1}{2} \sin x \cos x + \frac{1}{2}x + C.$

52. $\int \sin mx \ dx = - \dfrac{\cos mx}{m} + C.$

53. $\int \cos mx \ dx = \dfrac{\sin mx}{m} + C.$

54. $\int \sin mx \cos nx \ dx = - \dfrac{\cos (m + n)x}{2(m + n)}$

$\qquad\qquad\qquad - \dfrac{\cos (m - n)x}{2(m - n)} + C.$

55. $\int \sin mx \sin nx \ dx$

$\qquad = \dfrac{\sin (m - n)x}{2(m - n)} - \dfrac{\sin (m + n)x}{2(m + n)} + C.$

56. $\int \cos mx \cos nx \ dx$

$\qquad = \dfrac{\sin (m - n)x}{2(m - n)} + \dfrac{\sin (m + n)x}{2(m + n)} + C.$

57. $\int \tan x \ dx = -\ln \cos x + C.$

58. $\int \cot x \ dx = \ln \sin x + C.$

59. $\int \dfrac{dx}{\sin x} = \ln \tan \dfrac{x}{2} + C.$

60. $\int \dfrac{dx}{\cos x} = \ln \tan \left( \dfrac{\pi}{4} + \dfrac{x}{2} \right) + C.$

61. $\int \dfrac{dx}{1 + \cos x} = \tan \dfrac{x}{2} + C.$

62. $\int \dfrac{dx}{1 - \cos x} = -\cot \dfrac{x}{2} + C.$

63. $\displaystyle\int \sin x \cos x \, dx = \frac{1}{2} \sin^2 x + C.$

64. $\displaystyle\int \frac{dx}{\sin x \cos x} = \ln \tan x + C.$

65.* $\displaystyle\int \sin^n x \, dx = -\frac{\cos x \sin^{n-1} x}{n}$
$$+ \frac{n-1}{n} \int \sin^{n-2} x \, dx.$$

66.* $\displaystyle\int \cos^n x \, dx = \frac{\sin x \cos^{n-1} x}{n} + \frac{n-1}{n} \int \cos^{n-2} x \, dx.$

67. $\displaystyle\int \tan^n x \, dx = \frac{\tan^{n-1} x}{n-1} - \int \tan^{n-2} x \, dx.$

68. $\displaystyle\int \cot^n x \, dx = -\frac{\cot^{n-1} x}{n-1} - \int \cot^{n-2} x \, dx.$

69. $\displaystyle\int \frac{dx}{\sin^n x} = -\frac{\cos x}{(n-1)\sin^{n-1} x} + \frac{n-2}{n-1} \int \frac{dx}{\sin^{n-2} x}$

70. $\displaystyle\int \frac{dx}{\cos^n x} = \frac{\sin x}{(n-1)\cos^{n-1} x} + \frac{n-2}{n-1} \int \frac{dx}{\cos^{n-2} x}.$

71.† $\displaystyle\int \sin^p x \cos^q x \, dx = \frac{\sin^{p+1} x \cos^{q-1} x}{p+q}$
$$+ \frac{q-1}{p+q} \int \sin^p x \cos^{q-2} x \, dx,$$
$$= -\frac{\sin^{p-1} x \cos^{q+1} x}{p+q}$$
$$+ \frac{p-1}{p+q} \int \sin^{p-2} x \cos^q x \, dx.$$

72.† $\displaystyle\int \sin^{-p} x \cos^q x \, dx = -\frac{\sin^{-p+1} x \cos^{q+1} x}{p-1}$
$$+ \frac{p-q-2}{p-1} \int \sin^{-p+2} x \cos^q x \, dx.$$

73.† $\displaystyle\int \sin^p x \cos^{-q} x \, dx = \frac{\sin^{p+1} x \cos^{-q+1} x}{q-1}$
$$+ \frac{q-p-2}{q-1} \int \sin^p x \cos^{-q+2} x \, dx.$$

74. $\displaystyle\int \frac{dx}{a + b \cos x} = \frac{2}{\sqrt{a^2 - b^2}} \tan^{-1} \left( \sqrt{\frac{a-b}{a+b}} \tan \tfrac{1}{2}x \right)$
$$+ C, \text{ when } a^2 > b^2,$$

$$= \frac{1}{\sqrt{b^2 - a^2}} \ln \frac{b + a \cos x + \sin x \sqrt{b^2 - a^2}}{a + b \cos x}$$
$$+ C, \text{ when } a^2 < b^2,$$

$$= \frac{2}{\sqrt{b^2 - a^2}} \tanh^{-1} \left( \sqrt{\frac{b-a}{b+a}} \tan \tfrac{1}{2}x \right)$$
$$+ C, \text{ when } a^2 < b^2$$

*If $n$ is an odd number, substitute $\cos x = z$ or $\sin x = z$.
†If $p$ or $q$ is an odd number, substitute $\cos x = z$ or $\sin x = z$.

75. $\displaystyle\int \frac{\cos x \, dx}{a + b \cos x} = \frac{x}{b} - \frac{a}{b} \int \frac{dx}{a + b \cos x} + C.$

76. $\displaystyle\int \frac{\sin x \, dx}{a + b \cos x} = -\frac{1}{b} \ln (a + b \cos x) + C.$

77. $\displaystyle\int \frac{A + B \cos x + C \sin x}{a + b \cos x + c \sin x} \, dx = A \int \frac{dy}{a + p \cos y}$
$$+ (B \cos u + C \sin u) \int \frac{\cos y \, dy}{a + p \cos y} - (B \sin u -$$
$$C \cos u) \int \frac{\sin y \, dy}{a + p \cos y}, \text{ where } b = p \cos u, c = p \sin u \text{ and}$$
$$x - u = y.$$

78. $\displaystyle\int e^{ax} \sin bx \, dx = \frac{a \sin bx - b \cos bx}{a^2 + b^2} e^{ax} + C.$

79. $\displaystyle\int e^{ax} \cos bx \, dx = \frac{a \cos bx + b \sin bx}{a^2 + b^2} e^{ax} + C.$

80. $\displaystyle\int \sin^{-1} x \, dx = x \sin^{-1} x + \sqrt{1 - x^2} + C.$

81. $\displaystyle\int \cos^{-1} x \, dx = x \cos^{-1} x - \sqrt{1 - x^2} + C.$

82. $\displaystyle\int \tan^{-1} x \, dx = x \tan^{-1} x - \frac{1}{2} \ln (1 + x^2) + C.$

83. $\displaystyle\int \cot^{-1} x \, dx = x \cot^{-1} x + \frac{1}{2} \ln (1 + x^2) + C.$

84. $\displaystyle\int \sinh x \, dx = \cosh x + C.$

85. $\displaystyle\int \tanh x \, dx = \ln \cosh x + C.$

86. $\displaystyle\int \cosh x \, dx = \sinh x + C.$

87. $\displaystyle\int \coth x \, dx = \ln \sinh x + C.$

88. $\displaystyle\int \operatorname{sech} x \, dx = 2 \tan^{-1} (e^x) + C.$

89. $\displaystyle\int \operatorname{csch} x \, dx = \ln \tanh (x/2) + C.$

90. $\displaystyle\int \sinh^2 x \, dx = \frac{1}{2} \sinh x \cosh x - \frac{1}{2}x + C.$

91. $\displaystyle\int \cosh^2 x \, dx = \frac{1}{2} \sinh x \cosh x + \frac{1}{2}x + C.$

92. $\displaystyle\int \operatorname{sech}^2 x \, dx = \tanh x + C.$

93. $\displaystyle\int \operatorname{csch}^2 x \, dx = -\coth x + C.$

## DEFINITE INTEGRALS

The definite integral of $f(x) \, dx$ from $x = a$ to $x = b$, denoted by $\displaystyle\int_a^b f(x) \, dx$, is the limit (as $n$ increases indefinitely) of a sum of $n$ terms:

$$\int_a^b f(x) \, dx = \lim_{n = \infty} [f(x_1) \, \Delta x + f(x_2) \, \Delta x$$
$$+ f(x_3) \, \Delta x + \cdots + f(x_n) \, \Delta x]$$

built up as follows: Divide the interval from $a$ to $b$ into $n$ equal parts, and call each part $\Delta x, = (b - a)/n$; in each of these intervals take a value of $x$ (say $x_1, x_2, \ldots x_n$), find the value of the function $f(x)$ at each of these points, and multiply it by $\Delta x$, the width of the interval; then take the limit of the sum of the terms thus formed, when the number of terms increases indefinitely, while each individual term approaches zero.

Geometrically, $\int_a^b f(x)\,dx$ is the area bounded by the curve $y = f(x)$, the $x$ axis, and the ordinates $x = a$ and $x = b$ (Fig. 8); i.e., briefly, the "area under the curve, from $a$ to $b$." The

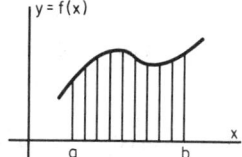

**Fig. 8**

**fundamental theorem** for the evaluation of a definite integral is the following:

$$\int_a^b f(x)\,dx = \left[ \int f(x)\,dx \right]_{x=b} - \left[ \int f(x)\,dx \right]_{x=a}$$

i.e., the definite integral is equal to the difference between two values of any one of the indefinite integrals of the function in question. In other words, the limit of a sum can be found whenever the function can be integrated.

**Properties of Definite Integrals**

$$\int_a^b = -\int_b^a; \qquad \int_a^c + \int_c^b = \int_a^b$$

MEAN-VALUE THEOREM FOR INTEGRALS.

$$\int_a^b F(x)\,f(x)\,dx = F(X)\int_a^b f(x)\,dx$$

provided $f(x)$ does not change sign from $x = a$ to $x = b$; here $X$ is some (unknown) value of $x$ intermediate between $a$ and $b$.

MEAN VALUE. The **mean value** of $f(x)$ with respect to $x$, between $a$ and $b$, is

$$\bar{f} = \frac{1}{b - a}\int_a^b f(x)\,dx$$

THEOREM ON CHANGE OF VARIABLE. In evaluating $\int_{x=a}^{x=b} f(x)\,dx$, $f(x)\,dx$ may be replaced by its value in terms of a new variable $t$ and $dt$, and $x = a$ and $x = b$ by the corresponding values of $t$, provided that throughout the interval the relation between $x$ and $t$ is a one-to-one correspondence (i.e., to each value of $x$ there corresponds one and only one value of $t$, and to each value of $t$ there corresponds one and only one value of $x$).

DIFFERENTIATION WITH RESPECT TO THE UPPER LIMIT. If $b$ is variable, then $\int_a^b f(x)\,dx$ is a function of $b$, whose derivative is

$$\frac{d}{db}\int_a^b f(x)\,dx = f(b)$$

DIFFERENTIATION WITH RESPECT TO A PARAMETER

$$\frac{\partial}{\partial c}\int_a^b f(x,c)\,dx = \int_a^b \frac{\partial f(x,c)}{\partial c}\,dx$$

**Functions Defined by Definite Integrals** The following definite integrals have received special names, and their values

have been tabulated. (See, for example, Peirce, "Table of Integrals.")

1. Elliptic integral of the first kind $= F(u,\ k) = \int_0^u \dfrac{dx}{\sqrt{1 - k^2 \sin^2 x}}$ when $k^2 < 1$.

2. Elliptic integral of the second kind $= E(u,\ k) = \int_0^u \sqrt{1 - k^2 \sin^2 x}\,dx$, when $k^2 < 1$.

3, 4. Complete elliptic integrals of the first and second kinds; put $u = \pi/2$ in (1) and (2).

5. The probability integral $= \dfrac{2}{\sqrt{\pi}}\displaystyle\int_0^x e^{-x^2}\,dx$.

6. The gamma function $= \Gamma(n) = \displaystyle\int_0^\infty x^{n-1}e^{-x}\,dx$.

**Approximate Methods of Integration. Mechanical Quadrature** 1. Use Simpson's rule (see also Scarborough, "Numerical Mathematical Analyses," Johns Hopkins Press).

2. Expand the function in a converging power series, and integrate term by term.

3. Plot the area under the curve $y = f(x)$ from $x = a$ to $x = b$ on squared paper, and measure this area roughly by "counting squares," or more accurately, by the use of a planimeter, or by graphical means (see Franklin, "Methods of Advanced Calculus," McGraw-Hill).

4. Coradi's Mechanical Integraph provides a means of drawing on paper the curve $y = \int f(x)\,dx$, when the curve $y = f(x)$ is given, and can be used to facilitate the solution of certain differential equations.

**Double Integrals** The notation $\int \int f(x,y)\,dy\,dx$ means $\int[\int f(x,y)\,dy]\,dx$, the limits of integration in the inner, or first, integral being functions of $x$ (or constants).

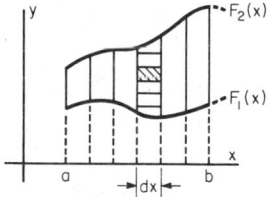

**Fig. 9**

EXAMPLE. To find the weight of a plane area whose density, $w$, is variable, say $w = f(x,y)$. The weight of a typical element, $dx\,dy$, is $f(x,y)\,dx\,dy$. Keeping $x$ and $dx$ constant and summing these elements from, say, $y = F_1(x)$ to $y = F_2(x)$, as determined by the shape of the boundary, the weight of a typical strip perpendicular to the $x$ axis is

$$dx \int_{y=F_1(x)}^{y=F_2(x)} f(x,y)\,dy$$

Finally, summing these strips from, say, $x = a$ to $x = b$, the weight of the whole area is

$$\int_{x=a}^{x=b}\left[ dx \int_{y=F_1(x)}^{y=F_2(x)} f(x,y)\,dy \right] \qquad \text{or, briefly,} \quad \int \int f(x,y)\,dy\,dx$$

## DIFFERENTIAL EQUATIONS

An **ordinary differential equation** is one which contains a single independent variable, or argument, and a single dependent variable, or function, with its derivatives of various orders. **A**

**partial differential equation** is one which contains a function of several independent variables, and its partial derivatives of various orders. The order of a differential equation is the order of the highest derivative which occurs in it. A solution of a differential equation is any relation between the variables, which, when substituted in the given equation, will satisfy it. The general solution of an ordinary differential equation of the $n$th order will contain $n$ arbitrary constants. A differential equation is usually said to be solved when the problem is reduced to simple quadratures, i.e., integrations of the form $y = \iint f(x)\,dx$.

### Methods of Solving Ordinary Differential Equations

DIFFERENTIAL EQUATIONS OF THE FIRST ORDER

1. If possible, separate the variables; i.e., collect all the $x$'s and $dx$ on one side, and all the $y$'s and $dy$ on the other side; then integrate both sides, and add the constant of integration.

2. If the equation is homogeneous in $x$ and $y$, the value of $dy/dx$ in terms of $x$ and $y$ will be of the form $dy/dx = f(y/x)$. Substituting $y = xt$ will enable the variables to be separated.

*Solution:* $\log_e x = \int \dfrac{dt}{f(t) - t} + C$.

3. The expression $f(x,y)\,dx + F(x,y)\,dy$ is an *exact differential* if $\dfrac{\partial f(x,y)}{\partial y} = \dfrac{\partial F(x,y)}{\partial x} (= P$, say). In this case the solution of $f(x,y)$ $dx + F(x,y)\,dy = 0$ is

$$\int f(x,y)\,dx + \int [F(x,y) - \int P\,dx]\,dy = C$$

or

$$\int F(x,y)\,dy + \int [f(x,y) - \int P\,dy]\,dx = C$$

4. Linear differential equation of the first order: $\dfrac{dy}{dx} + f(x) \cdot y = F(x)$. *Solution:* $y = e^{-P}[\int e^P F(x)\,dx + C]$, where $P = \int f(x)\,dx$.

5. Bernoulli's equation: $\dfrac{dy}{dx} + f(x) \cdot y = F(x) \cdot y^n$. Substituting $y^{1-n} = v$ gives $(dv/dx) + (1 - n)f(x) \cdot v = (1 - n)F(x)$, which is linear in $v$ and $x$.

6. Clairaut's equation: $y = xp + f(p)$, where $p = dy/dx$. The solution consists of the family of lines given by $y = Cx + f(C)$, where $C$ is any constant, together with the curve obtained by eliminating $p$ between the equations $y = xp + f(p)$ and $x + f'(p) = 0$, where $f'(p)$ is the derivative of $f(p)$.

DIFFERENTIAL EQUATIONS OF THE SECOND ORDER

7. $\dfrac{d^2y}{dx^2} = -n^2y$. *Solution:* $y = C_1 \sin (nx + C_2)$,

or

$$y = C_3 \sin nx + C_4 \cos nx.$$

8. $\dfrac{d^2y}{dx^2} = +n^2y$. *Solution:* $y = C_1 \sinh (nx + C_2)$,

or

$$y = C_3 e^{nx} + C_4 e^{-nx}$$

9. $\dfrac{d^2y}{dx^2} = f(y)$. *Solution:* $x = \int \dfrac{dy}{\sqrt{C_1 + 2P}}$

$+ C_2$, where $P = \int f(y)\,dy$.

10. $\dfrac{d^2y}{dx^2} = f(x)$. *Solution:* $y \int P\,dx + C_1x + C_2$, where $P = \int f(x)\,dx$,

or

$$y = xP - \int xf(x)\,dx + C_1x + C_2.$$

11. $\dfrac{d^2y}{dx^2} = f\left(\dfrac{dy}{dx}\right)$. Putting $\dfrac{dy}{dx} = z$, $\dfrac{d^2y}{dx^2} = \dfrac{dz}{dx}$, $x = \int \dfrac{dz}{f(z)} + C_1$ and $y = \int \dfrac{z\,dz}{f(z)} + C_2$; then eliminate $z$ from these two equations.

12. The equation for damped vibration: $\dfrac{d^2y}{dx^2} + 2b\dfrac{dy}{dx} + a^2y = 0$.

CASE 1. If $a^2 - b^2 > 0$, let $m = \sqrt{a^2 - b^2}$. *Solution:*

$y = C_1 e^{-bx} \sin (mx + C_2)$ or $y = e^{-bx}[C_3 \sin (mx) + C_4 \cos (mx)]$

CASE 2. If $a^2 - b^2 = 0$, solution is $y = e^{-bx}(C_1 + C_2x)$.

CASE 3. If $a^2 - b^2 < 0$, let $n = \sqrt{b^2 - a^2}$ *Solution:*

$y = C_1 e^{-bx} \sinh (nx + C_2)$ or $y = C_3 e^{-(b+n)x} + C_4 e^{-(b-n)x}$

13. $\dfrac{d^2y}{dx^2} + 2b\dfrac{dy}{dx} + a^2y = c$. *Solution:* $y = \dfrac{c}{a^2} + y_1$, where $y_1 =$ the solution of the corresponding equation with second member zero [see (12) above].

14. $\dfrac{d^2y}{dx^2} + 2b\dfrac{dy}{dx} + a^2y = c \sin (kx)$. *Solution:*

$$y = R \sin (kx - S) + y_1$$

where $R = c/\sqrt{(a^2 - k^2)^2 + 4b^2k^2}$, $\tan S = 2bk/(a^2 - k^2)$, and $y_1 =$ the solution of the corresponding equation with second member zero [see (12) above].

15. $\dfrac{d^2y}{dx^2} + 2b\dfrac{dy}{dx} + a^2y = f(x)$. *Solution:* $y = y_0 + y_1$, where $y_0 =$ any particular solution of the given equation, and $y_1 =$ the general solution of the corresponding equation with second member zero [see (12) above].

If $b^2 < a^2$, $y_0 = \dfrac{1}{2\sqrt{b^2 - a^2}}\left[e^{m_1 x} \int e^{-m_1 x} f(x)\,dx - e^{m_2 x} \int e^{-m_2 x} f(x)\,dx\right]$,

where $m_1 = -b + \sqrt{b^2 - a^2}$ and $m_2 = -b - \sqrt{b^2 - a^2}$.

If $b^2 < a^2$, let $m = \sqrt{a^2 - b^2}$; then $y_0 = \dfrac{1}{m} e^{-bx}\left[\sin (mx) \int e^{bx} \cos (mx) \cdot f(x)\,dx - \cos (mx) \int e^{bx} \sin (mx) \cdot f(x)\,dx\right]$.

If $b^2 = a^2$, $y_0 = e^{-bx}[x \int e^{bx} f(x)\,dx - \int x \cdot e^{bx} f(x)\,dx]$.

Types 12 to 15 are examples of linear differential equations with constant coefficients. The solutions of such equations are often found most simply by the use of Laplace transforms. (See Franklin, "Fourier Methods," pp. 198–229, McGraw-Hill.)

## LINEAR EQUATIONS

For the linear equation of the $n$th order

$$A_n(x)d^ny/dx^n + A_{n-1}(x)d^{n-1}y/dx^{n-1} + \cdots + A_1(x)dy/dx + A_0(x)y = E(x)$$

the general solution is $y = u + c_1u_1 + c_2u_2 + \cdots + c_nu_n$. Here $u$, the particular integral, is any solution of the given equation, and $u_1, u_2, \cdots, u_n$ form a fundamental system of solutions of the homogeneous equation obtained by replacing $E(x)$ by zero. A set of solutions is fundamental, or independent, if its Wronskian determinant $W(x)$ is not zero, where

$$W(x) = \begin{vmatrix} u_1 & u_2 & \cdots & u_n \\ u_1' & u_2' & \cdots & u_n' \\ \cdot & \cdot & \cdots & \cdot \\ \cdot & \cdot & \cdots & \cdot \\ \cdot & \cdot & \cdots & \cdot \\ u_1^{(n-1)} & u_2^{(n-1)} & \cdots & u_n^{(n-1)} \end{vmatrix}$$

For any $n$ functions, $W(x) = 0$ if some one $u_i$ is linearly dependent on the others, as $u_n = k_1u_1 + k_2u_2 + \cdots + k_{n-1}u_{n-1}$ with the coefficients $k_i$ constant. And for $n$ solutions of a linear differential equation of the $n$th order, if $W(x) \neq 0$, the solutions are linearly independent.

### Constant Coefficients

To solve the homogeneous equation of the $n$th order $A_nd^ny/dx^n + A_{n-1}d^{n-1}y/dx^{n-1} + \ldots + A_1dy/dx + A_0y = 0$, $A_n \neq 0$, where $A_n, A_{n-1}, \cdots, A_0$ are constants, find the roots of the auxiliary equation

$$A_np^n + A_{n-1}p^{n-1} + \cdots + A_1p + A_0 = 0.$$

For each simple real root $r$, there is a term $ce^{rx}$ in the solution. The terms of the solution are to be added together. When $r$ occurs twice among the $n$ roots of the auxiliary equation, the corresponding term is $e^{rx}(c_1 + c_2x)$. When $r$ occurs three times, the corresponding term is $e^{rx}(c_1 + c_2x + c_3x^2)$, and so forth. When there is a pair of conjugate complex roots $a + bi$ and $a - bi$, the real form of the terms in the solution is $e^{ax}(c_1 \cos bx + d_1 \sin bx)$. When the same pair occurs twice, the corresponding term is $e^{ax}[(c_1 + c_2x) \cos bx + (d_1 + d_2x) \sin bx]$, and so forth.

Consider next the general nonhomogeneous linear differential equation of order $n$, with constant coefficients, or

$$A_nd^ny/dx^n + A_{n-1}d^{n-1}y/dx^{n-1} + \cdots + A_1 dy/dx + A_0y = E(x).$$

We may solve this by adding any particular integral to the complementary function, or general solution, of the homogeneous equation obtained by replacing $E(x)$ by zero. The complementary function may be found from the rules just given. And the particular integral may be found by the methods of the following paragraphs.

### Undetermined Coefficients

In the last equation, let the right member $E(x)$ be a sum of terms each of which is of the type $k$, $k \cos bx$, $k \sin bx$, $ke^{ax}$, $kx$, or more generally, $kx^me^{ax}$, $kx^me^{ax} \cos bx$, or $kx^me^{ax} \sin bx$. Here $m$ is zero or a positive integer, and $a$ and $b$ are any real numbers. Then the form of the particular integral $I$ may be predicted by the following rules.

**CASE 1.** $E(x)$ **is a single term** $T$. Let $D$ be written for $d/dx$, so that the given equation is $P(D)y = E(x)$, where $P(D) = A_nD^n + A_{n-1}D^{n-1} + \cdots + A_1D + A_0y$. With the term $T$ associate the simplest polynomial $Q(D)$ such that $Q(D)T = 0$. For the particular types $k$, etc., $Q(D)$ will be $D$, $D^2 + b^2$, $D^2 + b^2$, $D - a$, $D^2$; and for the general types $kx^me^{ax}$, etc., $Q(D)$ will be $(D - a)^{m+1}$, $(D^2 - 2aD + a^2 + b^2)^{m+1}$, $(D^2 - 2aD + a^2 + b^2)^{m+1}$. Thus $Q(D)$ will always be some power of a first- or second-degree factor, $Q(D) = F^v$, $F = D - a$, or $F = D^2 - 2aD + a^2 + b^2$.

Use the method described under **Constant Coefficients** to find the terms in the solution of $P(D)y = 0$ and also the terms in the solution of $Q(D)P(D)y = 0$. Then assume that the particular integral $I$ is a linear combination with unknown coefficients of those terms in the solution of $Q(D)P(D)y = 0$ which are not in the solution of $P(D)y = 0$. Thus if $Q(D) = F^q$ and $F$ is *not* a factor of $P(D)$, assume $I = (Ax^{q-1} + Bx^{q-2} + \cdots + L)e^{ax}$ when $F = D - a$, and assume $I = (Ax^{q-1} + Bx^{q-2} + \cdots + L)e^{ax} \cos bx + (Mx^{q-1} + Nx^{q-2} + \cdots + R)e^{ax} \sin bx$ when $F = D^2 - 2aD + a^2 + b^2$. When $F$ is a factor of $P(D)$ and the highest power of $F$ which is a divisor of $P(D)$ is $F^k$, try the $I$ above multiplied by $x^k$.

**CASE 2.** $E(x)$ **is a sum of terms.** With each term in $E(x)$, associate a polynomial $Q(D) = F^q$ as before. Arrange in one group all the terms that have the same $F$. The particular integral of the given equation will be the sum of solutions of equations each of which has one group on the right. For any one such equation, the form of the particular integral is given as for Case 1, with $q$ the highest power of $F$ associated with any term of the group on the right.

After the form has been found in Case 1 or 2, the unknown coefficients follow when we substitute back in the given differential equation, equate coefficients of like terms, and solve the resulting system of simultaneous equations.

### Variation of Parameters

Whenever a fundamental system of solutions $u_1, u_2, \ldots, u_n$ for the homogeneous equation is known, a particular integral of

$$A_n(x)d^ny/dx^n + A_{n-1}(x)d^{n-1}y/dx^{n-1} + \cdots + A_1(x)\,dy/dx + A_0(x)y = E(x)$$

may be found in the form $y = \Sigma v_ku_k$. In this and the next few summations, $k$ runs from 1 to $n$. The $v_k$ are functions of $x$, found by integrating their derivatives $v_k'$, and these derivatives are the solutions of the $n$ simultaneous equations $\Sigma v_k'u_k = 0$, $\Sigma v_k'u_k' = 0$, $\Sigma v_k'u_k'' = 0$, $\cdots$, $\Sigma v_k'u_k^{(n-2)} = 0$, $A_n(x)\Sigma v_k'u_k^{(n-1)} = E(x)$. To find the $v_k$ from $v_k = \int v_k' \, dx + c_k$, any choice of constants will lead to a particular integral. The special choice $v_k = \int_0^x v_k' \, dx$ leads to the particular integral having $y, y', y'', \ldots, y^{(n-1)}$ each equal to zero when $x = 0$.

**The Cauchy-Euler Equidimensional Equation** This has the form

$$k_nx^nd^ny/dx^n + k_{n-1}x^{n-1}d^{n-1}y/dx^{n-1} + \cdots + k_1 x \, dy/dx + k_0y = F(x)$$

The substitution $x = e^t$, which makes

$$x \, dy/dx = dy/dt$$
$$x^k \, d^ky/dx^k = (d/dt - k + 1) \cdots (d/dt - 2)(d/dt - 1) \, dy/dt$$

### Table 1. Laplace Transforms
*(Thaler)*

| $f(t)$ | $F(s) = \mathcal{L}[f(t)]$ |
|---|---|
| 1. $A$ | $A/s$ |
| 2. $1 = u(t)$ | $1/s$ |
| 3. $e^{-\alpha t}$ | $\dfrac{1}{s + \alpha}$ |
| 4. $\dfrac{1}{\tau} e^{-t/\tau}$ | $\dfrac{1}{\tau s + 1}$ |
| 5. $Ae^{-\alpha t}$ | $\dfrac{A}{s + \alpha}$ |
| 6. $\sin \beta t$ | $\dfrac{\beta}{s^2 + \beta^2}$ |
| 7. $\cos \beta t$ | $\dfrac{s}{s^2 + \beta^2}$ |
| 8. $\dfrac{1}{\beta} e^{-\alpha t} \sin \beta t$ | $\dfrac{1}{s^2 + 2\alpha s + \alpha^2 + \beta^2}$ |
| 9. $\dfrac{e^{-\alpha t}}{\beta - \alpha} - \dfrac{e^{-\beta t}}{\beta - \alpha}$ | $\dfrac{1}{(s + \alpha)(s + \beta)}$ |
| 10. $\dfrac{Ae^{-\alpha t} - Be^{-\beta t}}{C}$ where $A = a - \alpha$, $B = a - \beta$, $C = \beta - \alpha$ | $\dfrac{s + a}{(s + \alpha)(s + \beta)}$ |
| 11. $\dfrac{e^{-\alpha t}}{A} + \dfrac{e^{-\beta t}}{B} + \dfrac{e^{-\delta t}}{C}$ where $A = (\beta - \alpha)(\delta - \alpha)$ $B = (\alpha - \beta)(\delta - \beta)$ $C = (\alpha - \delta)(\beta - \delta)$ | $\dfrac{1}{(s + \alpha)(s + \beta)(s + \delta)}$ |
| 12. $t$ | $\dfrac{1}{s^2}$ |
| 13. $t^2$ | $2/s^3$ |
| 14. $t^n$ | $\dfrac{n!}{s^{n+1}}$ |
| 15. $d/dt[f(t)]$ | $sF(s) - f(0^+)$ |
| 16. $d^2/dt^2[f(t)]$ | $s^2F(s) - sf(0^+) - \dfrac{df}{dt}(0^+)$ |
| 17. $d^3/dt^3[f(t)]$ | $s^3F(s) - s^2f(0^+) - s\dfrac{df(0^+)}{dt} - \dfrac{d^2f(0^+)}{dt^2}$ |
| 18. $\int f(t)dt$ | $\dfrac{1}{s}[F(s) + \int f(t)dt|_{0^+}]$ |
| 19. $\dfrac{1}{\alpha} \sinh \alpha t$ | $\dfrac{1}{s^2 - \alpha^2}$ |
| 20. $\cosh \alpha t$ | $\dfrac{s}{s^2 - \alpha^2}$ |

transforms this into a linear differential equation with constant coefficients. Its solution $y = g(t)$ leads to $y = g(\ln x)$ as the solution of the given Cauchy-Euler equation.

## THE LAPLACE TRANSFORMATION

One form of operational calculus, the Laplace transformation, has as its basis the Laplace integral. When a differential equation expressed in terms of time $t$ is operated upon by the Laplace integral, a new equation results which is expressed in terms of a complex variable of the form $\sigma + j\omega$. The transformed equation is in purely algebraic terms and may be manipulated algebraically to solve for the desired quantity as an explicit function of the complex variable. It is necessary to perform an inverse process to return to the time domain. There are essentially three reasons for the use of the Laplace transformation. They are (1) the solution of high-order differential equations may be performed by purely algebraic manipulation of the transformed equation; (2) boundary conditions are easily handled; (3) the Laplace-transform method is suited to the complex-variable theory associated with the Nyquist stability criterion.

In Laplace-transformation mathematics the following symbols and equations are used:

$f(t)$ = a function of time
$s$ = a complex variable of the form $(\sigma + j\omega)$
$F(s)$ = an equation expressed in the transform variable $s$, resulting from operating on a function of time with the Laplace integral
$\mathcal{L}$ = an operational symbol indicating that the quantity which it prefixes is to be transformed into the frequency domain

Therefore, $F(s) = \mathcal{L}[f(t)]$. The Laplace integral is defined as $\mathcal{L} = \int_0^\infty e^{-st}\, dt$. Therefore, $\mathcal{L}[f(t)] = \int_0^\infty e^{-st}f(t)\, dt$.

### Direct Transforms

EXAMPLE:

$$f(t) = \sin \beta t$$

$$\mathcal{L}[f(t)] = \mathcal{L}(\sin \beta t) = \int_0^\infty \sin \beta t\; e^{-st}\, dt$$

but $\sin \beta t = \dfrac{e^{j\beta t} - e^{-j\beta t}}{2j}$

$$\mathcal{L}(\sin \beta t) = \frac{1}{2j}\int_0^\infty (e^{j\beta t} - e^{-j\beta t})e^{-st}\, dt$$

$$= \frac{1}{2j}\left(\frac{-1}{s - j\beta}\right)e^{(-s+j\beta)t}\Big|_0^\infty - \frac{1}{2j}\left(\frac{-1}{s + j\beta}\right)e^{(-s-j\beta)t}\Big|_0^\infty$$

$$= \frac{\beta}{s^2 + \beta^2}$$

Table 1 lists the transforms of common time-variable expressions normally encountered.

**Transformation Calculus**  By applying the preceding technique, the Laplace transform of derivative or integral functions can be derived.

The transform of a first derivative of a function of time is

$$\mathcal{L}\left[\frac{d}{dt}\, f(t)\right] = sF(s) - f(0^+)$$

where $F(s)$ = Laplace transform of $f(t)$
$s$ = transform variable
$f(0^+)$ = initial value of $f(t)$, evaluated as $t$ approaches zero from positive values

For a step function of amplitude $A$ at $t = 0$, $f(0^+) = A$ and $f(0^-) = 0$. The transform of a second derivative of a function of time

is $\dfrac{d}{dt}\left[\dfrac{d}{dt}\, f(t)\right] = f''(t)$

$$\mathcal{L}[f''(t)] = s^2F(s) - sf(0^+) - f'(0^+)$$

The transform of $\int\!\int f(t)\, dt$ is

$$\mathcal{L}\left[\int f(t)\, dt\right] = \frac{f^{-1}(0^+)}{s} + \frac{F(s)}{s}$$

where $f^{-1}(0^+) = \int\!\int f(t)\, dt$, evaluated as $t$ approaches zero from positive values.

**Inversion**  When an equation has been transformed, an explicit solution for the unknown may be directly determined through algebraic manipulation. In automatic-control design, the equation is usually the differential equation describing the system, and the unknown is either the output quantity or the error. The solution gained from the transformed equation is expressed in terms of the complex variable $s$. For many design or analysis purposes, the solution in $s$ is sufficient, but in some cases it is necessary to retransform the solution in terms of time. The process of passing from the complex-variable (frequency domain) expression to that of time (time domain) is called an **inverse transformation.** It is represented symbolically as

$$\mathcal{L}^{-1}F(s) = f(t)$$

For any $f(t)$ there is only one direct transform, $F(s)$. For any given $F(s)$ there is only one inverse transform $f(t)$. Therefore, tables are generally used for determining inverse transforms. Very complete tables of inverse transforms may be found in Gardner and Barnes, "Transients in Linear Systems." As an example of the inversion procedure consider an equation of the form

$$K = \alpha x(t) + \int \frac{x(t)}{\beta}\, dt$$

It is desired to obtain an expression for $x(t)$ resulting from an instantaneous change in the quantity $K$. Transforming the last equation yields

$$\frac{K}{s} = X(s)\alpha + \frac{X(s)}{s\beta} + \frac{f^{-1}(0^+)}{s}$$

If $f^{-1}(0^+)/s = 0$

then $X(s) = \dfrac{K/\alpha}{s + 1/\alpha\beta}$

$x(t) = \mathcal{L}^{-1}[X(s)] = \mathcal{L}^{-1}\dfrac{K/\alpha}{s + 1/\alpha\beta}$

From Table 1, $x(t) = \dfrac{K}{\alpha}\, e^{-t/\alpha\beta}$

# GRAPHICAL REPRESENTATION OF FUNCTIONS

## by Philip Franklin

### EQUATIONS INVOLVING TWO VARIABLES

**Curve** $y = f(x)$  To represent graphically any function, $y$, of a single variable, $x$, lay off the values of $x$ as **abscissas** along a uniformly graduated horizontal axis, whose positive direction (as usually chosen) runs to the right, and at each point on this $x$ axis erect a perpendicular (called an **ordinate**) whose length represents the value of $y$ at that point. The unit of measurement for the $y$ scale, whose positive direction (as usually chosen) runs upward, need not be the same as the unit for the $x$ scale. Draw a smooth curve through the extremities of the ordinates; this is the **graph** of the given function in rectangular coordinates, or the curve of the function.

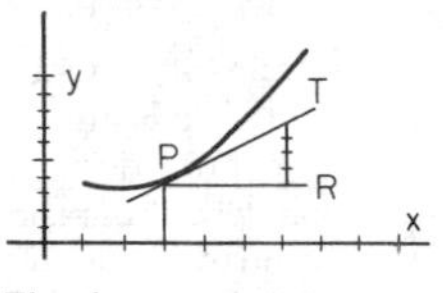

To measure graphically the rate of change of the function at any point $P$ (Fig. 1), draw the tangent at $P$; then **rate of change** at $P = RT/PR$, where $RT$ and $PR$ are measured in units of the $y$ axis and $x$ axis, respectively. This ratio, which is positive if $RT$ runs upward, negative if $RT$ runs downward, is equal to the derivative of the function at the point $P$.

**Fig. 1**

**Graphs of Important Functions**  Figures 2 to 9 show the graphs (in rectangular coordinates) of the most important elementary functions, namely:

The **linear function**, $y = mx + b$ (Fig. 2).

The **power functions**, $y = x^n$ [$n$ positive (parabolic type); $n$ negative (hyperbolic type)] (Fig. 3).

The **exponential function**, $y = 10^x$ or $y = e^x$, and the **logarithmic function**, $y = \log_{10} x$ or $y = \ln x$ (Fig. 4).

The **trigonometric functions** (Fig. 5), and the inverse trigonometric functions (Fig. 6).

The **hyperbolic functions** (Figs. 7 and 8) and the inverse hyperbolic functions (Fig. 9).

Various **special functions** (Figs. 10 to 12).

By a slight modification, each of these diagrams may be made to represent a somewhat more general function than that for which it is primarily intended. For, if $x$ is replaced by $x - a$ in the equation, this merely requires renumbering the $x$ axis so that each number is moved $a$ units to the left; and similarly, if $y$ is replaced by $y - b$ in the equation, this merely requires renumbering the $y$ axis so that each number is moved $b$ units downward. (Such a change is called a translation of the curve to the right, or upward.) Further, if $x$ is replaced by $x/c$ (or $y$ by $y/c$) in the equation, it is merely necessary to multiply each of the numbers written along the $x$ axis (or $y$ axis) by $c$, in order to adapt the graph to the new equation. (Such a change is called a "stretching" of the curve along one of the axes.)

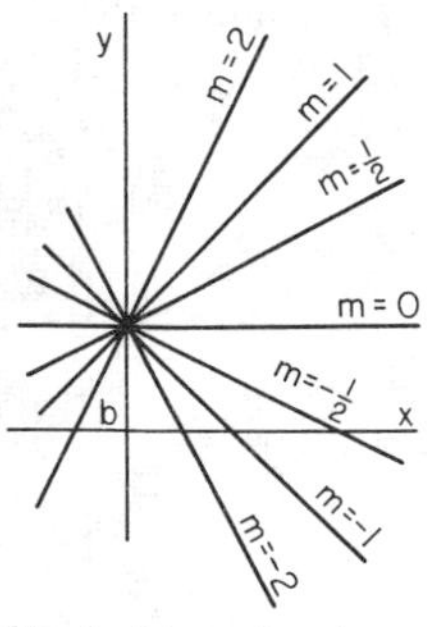

**Fig. 2**  Linear function, $y = mx + b$.

**Empirical Curves**  Any set of values of two variables $x$ and $y$ can be represented by plotting the points $(x,y)$ on rectangular coordinate paper, and drawing a smooth curve through these points. The points which correspond to actual data should be clearly indicated by small circles or crosses, intermediate points being spoken of as interpolated points. While this process of graphically interpolating a continuous series of points *between* given values is usually fairly safe, the process of extrapolation—i.e., extending the curve *beyond* the range of the given values—is dangerous.

**To Find a Mathematical Equation to Fit a Given Empirical Curve**  This problem is one which in general requires much patience and ingenuity. Only the simplest cases can be mentioned here.

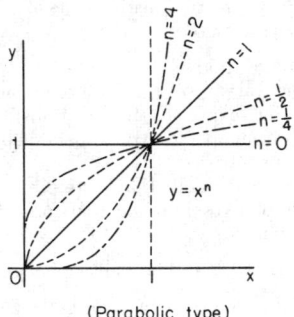

(Parabolic type)

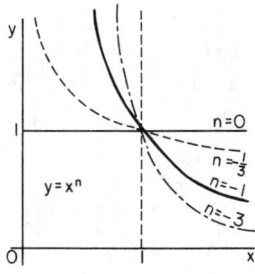

(Hyperbolic type)

**Fig. 3**  Power function, $y = x^n$.

CASE 1.  If the given empirical curve is a straight line, then the law connecting the given values of $x$ and $y$ is $y = mx + b$, where $m$ = the slope of the line, and $b$ = the value of $y$ at the point where the line crosses the $y$ axis. If the points lie only approximately on a straight line, the best position for this line can usually be found by stretching a black thread among the points; or, assume a law of the form $y = mx + b$, and by substituting in this formula $n$ pairs of values of $x$ and $y$, obtain $n$ equations connecting the coefficients $m$ and $b$; various pairs of these equations may then be solved for $m$ and $b$, and the average of the results taken. Or, if great accuracy is required, all $n$ of the equations may be solved for $m$ and $b$ by the method of least squares.

If any law of the form $f(x,y) = m \cdot F(x,y) + b$ is suspected, where $f(x,y)$ and $F(x,y)$ are any expressions involving either $x$ or $y$ or both $x$ and $y$, such a law may be tested by plotting $F(x,y)$ instead of $x$, and $f(x,y)$ instead of $y$, on rectangular cross-section paper, and seeing whether or not the points lie

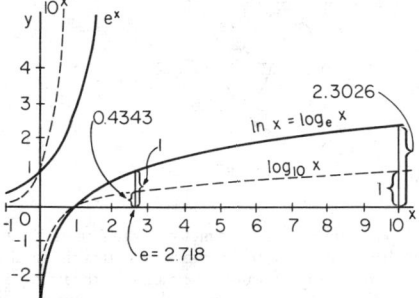

**Fig. 4**  Exponential function, $10^x$ or $e^x$. Logarithmic function, $\log_{10} x$ or $\ln x$.

on a straight line. If they do, the form of the law is verified, and the values of $m$ and $b$ can be read from the figure as before. For example, if $y^2 = mxy + b$, a straight line will be obtained by plotting $y^2$ against $xy$. Again, if $xy = bx + my$, a straight line will be obtained by plotting $y$ against $y/x$, since the equation may be written $y = b + m(y/x)$.

CASE 2.  If a law of the form $y = cx^n$ is suspected, plot the points $(x,y)$ on logarithmic paper (see below).

CASE 3.  If a law of the form $y = c \cdot 10^{mx}$ (or $y = c \cdot e^{mx}$) is suspected, plot the points $(x,y)$ on semilogarithmic paper (see below).

CASE 4.  If the given curve resembles the logarithmic curve, $y = \log x$, interchange $x$ and $y$ and proceed as in Case 3.

CASE 5.  If the given curve is a wavy line, resembling a sine or cosine curve, try an equation of the form $y = a \sin bx$ or $y = a \cos bx$. If the heights of the waves diminish as $x$ increases, try an equation of the form $y = ae^{-nx} \sin bx$.

NOTE.  Any periodic function (satisfying certain simple conditions) can be expressed by a Fourier series.

CASE 6.  A great variety of functions can be represented approximately by a polynomial of the form $y = a + bx + cx^2 + dx^3 + ex^4 + \ldots$, the first three or four terms being usually sufficient. To determine the coefficients $a, b, c, \ldots$ most

accurately, substitute in the formula all the given pairs of values of $x$ and $y$, and solve the resulting equations for $a, b, c, \ldots$ by the method of least squares.

CASE 7.  Many simple curves can be represented approximately by an equation of the hyperbolic form, $xy = c + bx + ay$, where $a$, $b$, and $c$ are determined by substituting the coordinates of three conspicuous points of the curve. The lines $x = a$ and $y = b$ are the asymptotes of the hyperbola. The equation may also be written $(x - a)(y - b) = k$, where $k = ab + c$.

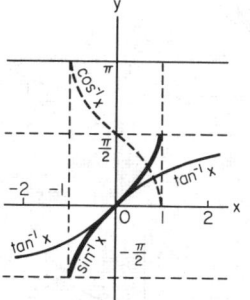

**Fig. 6**  Inverse trigonometric functions.

**Logarithmic Cross-Section Paper**  In this form of cross-section paper (Fig. 13), the distance from the origin to any point on the $x$ or $y$ axis is equal to the logarithm of the number written against that point. Thus, in Fig. 13 the distances (shown for clearness on two auxiliary scales $X$ and $Y$) are the logarithms of the numbers written along $x$ and $y$.

Accurately made logarithmic paper can be obtained from the principal dealers in draftsmen's supplies. Logarithmic paper can be easily constructed, in case of need, by copying the logarithmic scale from any ordinary slide rule. The actual figures along the $x$ and $y$ axes are usually left for the user to insert; in so doing, notice that the numbers . . . , 0.01, 0.1, 1, 10, 100, . . . , or such of them as may be needed to cover any given range of values, must be placed at the points of division which separate the main squares. It is often convenient, however, to omit the decimal point, numbering each square independently from 1 to 10. The length of the side of one square is called the *unit* or *base* of the logarithmic paper; the larger the unit, the finer the possible subdivisions of the scale.

To plot a point $(x,y)$ on logarithmic paper, e.g., the point $(3,5)$, means to find the point of intersection of the vertical line marked $x = 3$ and the horizontal line marked $y = 5$. In

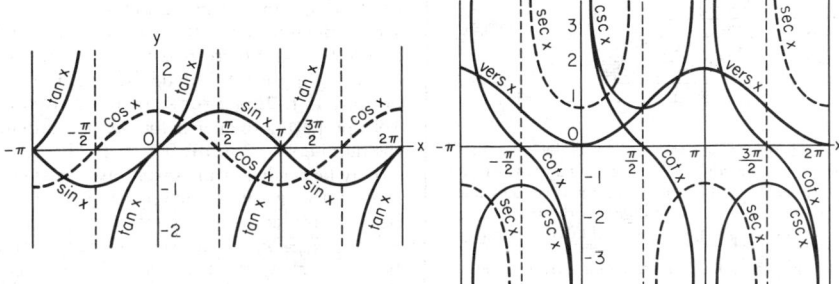

**Fig. 5**  Trigonometric functions.

interpolating between two lines, account should be taken of the fact that the divisions are not of uniform length.

Any equation of the form $y = cx^n$ when plotted on logarithmic paper will be represented by a straight line whose slope is $n$. For, if $y_1 = cx_1^n$ and $y_2 = cx_2^n$, then $y_1/y_2 = (x_1/x_2)^n$, or $(\log y_1 - \log y_2)/(\log x_1 - \log x_2) = n$. The slope must be measured by aid of an auxiliary *uniform* scale.

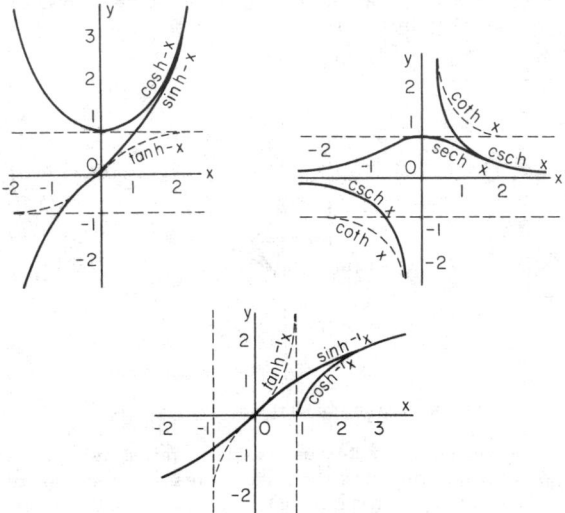

**Figs. 7, 8, and 9**  Hyperbolic functions and inverse hyperbolic functions.

EXAMPLE.   Let $y = x^{3/2}$. When $x = 1$, $y = 1$; plot this point $A$ on the logarithmic paper, and draw the straight line $AE$ with a slope equal to $\frac{3}{2}$ (Fig. 13). By the aid of this line, the value of $y$ for any value of $x$ between 1 and 100 can be read off directly; for example, if $x = 2.50$, $y = 3.95$, as shown by dotted lines, so that $(2.50)^{3/2} = 3.95$. To find the value of $y$ for any value of $x$ outside this range, note that moving the decimal point 2 places in $x$ is equivalent to moving it 3 places in $y$. The line shown in Fig. 13 is thus equivalent to a complete table of three-halves powers.

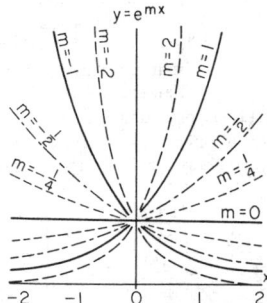

**Fig. 10**

It will be noticed that this line crosses four squares of the logarithmic paper. By superposing these four squares the whole diagram may be condensed into a single square (Fig. 14), in which, however, the scales for $x$ and $y$ now give only the sequence of digits in the answer, the position of the decimal point having to be determined by inspection.

**To determine whether a given set of values, $x$ and $y$, satisfies a law of the form** $y = cx^n$, plot the values on logarithmic paper, and see whether they lie on a straight line; if they do, then the

given values satisfy a law of this form; moreover, the slope of the line gives the value of $n$, and the value of $y$ when $x = 1$ gives the value of $c$.

If the plotted points fail to lie exactly in line but form a curve slightly concave upward, try subtracting some constant $b$ from all the $y$'s, i.e.,

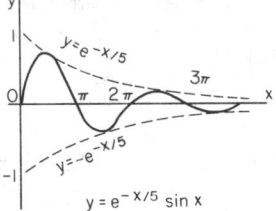

**Fig. 11**

move each point downward a distance equal to $b$ units of the $y$ scale at that point. If it proves possible to choose $b$ so that the resulting points lie in line, then the original values obey a law of the form $y - b = cx^n$, where $n$ is again the slope of the line, and $c$ is the value of $y - b$ when $x = 1$. (Conversely if the curve is concave *downward*, try *adding* $b$ to all the $y$'s; i.e., move each point *upward*; if the new points lie in line, the original values obey a law of the form $y + b = cx^n$.) Another method of "straightening" the curve consists of adding some constant, $\pm a$, to all the values of $x$, which has the effect of shifting all the points to the right or left (by varying amounts); if this method succeeds, the original values obey a law of the form $y = c(x + a)^n$.

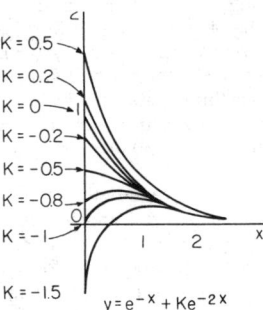

**Fig. 12**

**Semilogarithmic Cross-Section Paper**  This form of paper (Fig. 15) has a logarithmic scale along $y$ and a uniform scale along $x$. The **scale value**, $k$, of the paper is the number which stands, on the $x$ axis, at a distance from the origin equal to the width of one of the main horizontal strips. Thus, in Fig. 15, each number shown along the auxiliary scale $Y$ is the logarithm of the corresponding number along $y$, and each number shown along the auxiliary scale $X$ is $1/k$th of the corresponding number along $x$ (here $k = 5$). The number $k$, which may be chosen at pleasure, should be taken equal to some simple integer, as 1, 2, or 5, or some integral power of 10.

In preparing the paper for use it is important to notice that the numbers . . . , 0.01, 0.1, 1, 10, 100, . . . (or such of them as may be needed in any given case) must be placed along the $y$ axis at the points which mark the main lines of division between the horizontal strips; while the numbers . . . , $-2k$, $-k$, 0, $+k$, $+2k$, . . . (or such of them as may be needed) must be placed along the $x$ axis at uniform intervals, each interval (from 0 to $k$, from $k$ to $2k$, etc.) being equal to the width of one of the main horizontal strips. The width of one of these strips is called the *unit* or *base* of the semilogarithmic paper; the larger the unit, the finer the possible subdivisions of the scale.

To plot a point $(x,y)$, as $x = 3$, $y = 5$, on semilogarithmic paper means to find the point of intersection of the vertical line marked $x = 3$ with the horizontal line marked $y = 5$.

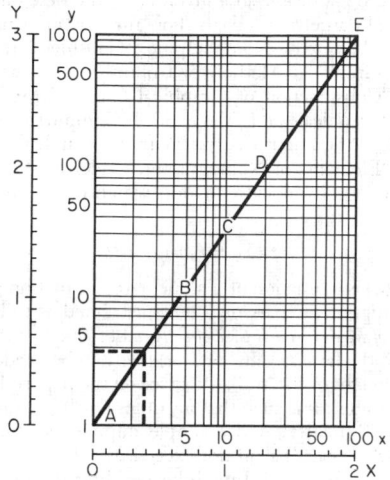

**Fig. 13**

Any equation of the form $y = c \cdot 10^{mx}$ (or $y = c \cdot e^{mx}$), when plotted on semilogarithmic paper with scale value $k$, will be represented by a straight line whose slope is $km$ (or $0.4343$ $km$). By a suitable choice of the scale value $k$, any given range of values of $x$ can be brought within the size of the paper. Note that $e = 10^{0.4343}$.

EXAMPLE.  Given $y = 4 \cdot 10^{-0.1x}$ (or $y = 4 \cdot e^{-0.1x}$). In Fig. 15, when $x = 0$, $y = 4$. By plotting this point ($A$) on the semilogarithmic paper, with scale value 5, and drawing through it a straight line with slope equal to $-0.5$ (or $-0.217$) a graphical representation is obtained from which, for any value of $x$, the corresponding value of $y$ can be read off. If it is desired to condense the figure, several horizontal strips may be superposed on a single strip; this of course renders the decimal point in the $y$ scale undetermined (unless a separate $y$ scale is provided for each section of the graph).

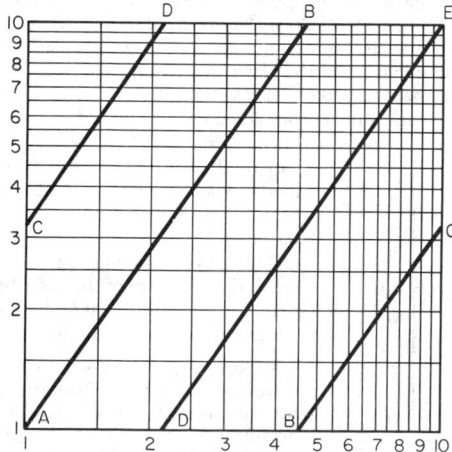

**Fig. 14**

In order to determine whether a given set of values of $x$ and $y$ satisfies a law of the form $y = c \cdot 10^{mx}$ (or $y = c \cdot e^{mx}$), plot the values of $x$ and $y$ on semilogarithmic paper, with a suitable scale value $k$, and see whether they lie on a straight line; if they do so, the law is satisfied, and the values of $m$ and $c$ may be found as follows: $m$ = the slope of the line divided by $k$ (or the slope of the line divided by $0.4343k$), and $c$ = the value of $y$ when $x = 0$.

If the plotted points fail to lie exactly in line, but form a curve slightly concave upward, try subtracting some constant $b$ from all the $y$'s, and plot the values thus modified; if $b$ can be so chosen that the revised points lie in line, then the original values obey a law of the form $y - b = c \cdot 10^{mx}$ (or $y - b = c \cdot e^{mx}$), where $m$ and $c$ are to found as before. If the curve is concave downward, add $b$, instead of subtracting; and replace $y - b$ by $y + b$ in the law.

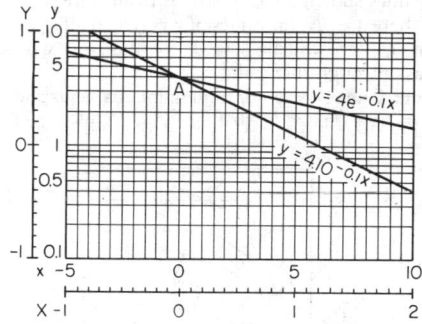

**Fig. 15**

**Curves in Polar Coordinates**  Any function, $r$, of a single variable, $\theta$, can be represented by a curve in polar coordinates. First, lay off the given values of $\theta$ as angles, the initial line $0x$ running toward the right, and the counterclockwise direction about the origin being taken as positive. Along the terminal side of each angle $\theta$, lay off the corresponding value of $r$, forward if $r$ is positive, backward if $r$ is negative; and pass a smooth curve through the points thus determined.

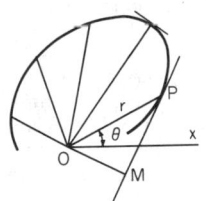

**Fig. 16**

The rate of change of $r$ with respect to $\theta$ at a given point $P$ is represented graphically as follows (Fig. 16): On the tangent at $P$ drop a perpendicular $OM$ from the origin: then $r(MP/OM)$ represents the rate of change, $dr/d\theta$, provided $\theta$ is measured in radians. Specially ruled polar-coordinate paper is supplied by dealers in drafting supplies.

## EQUATIONS INVOLVING THREE VARIABLES

**The Surface** $z = f(x,y)$  Any function, $z$, of two variables, $x$ and $y$, may be represented by a surface, as follows: Plot the given pairs of values of $x$ and $y$ as points in a horizontal $x,y$ plane, called the base plane; at each of these points erect an ordinate, parallel to a vertical axis $z$, and representing by its length the value of $z$ at that point. Then conceive a smooth surface passed through the extremities of these ordinates: this surface is said to represent the function. In practice, the ordinates may be made by implanting stiff vertical rods in a horizontal board of soft wood which serves as the base plane; the surface may then be constructed by filling in the spaces with plaster of paris. Or, more simply, pieces of cardboard may be cut out to represent parallel plane sections of the surface, and then stood on edge in slots cut in the board to receive them. The units employed along $x$, $y$, and $z$ need not be equal to each other.

**Contour-Line Charts** All the points of a surface $z = f(x,y)$ which are at any given height above the base plane form a curve on the surface, called a contour line of the surface. If each of these contour lines be projected on the base plane, and each labeled with the value of $z$ to which it corresponds, a complete representation of the function $z = f(x,y)$ is obtained, all in one plane. A topographical map, with contour lines showing elevations above the sea, and a weather map, with contour lines showing barometric pressure, are familiar examples. If there are several values of $z$ corresponding to any given point $(x,y)$, there will be several contour lines whose projections pass through that point.

**Contour-Line Charts for Simultaneous Equations** [of the form $z = f(x,y)$, $w = F(x,y)$] In Fig. 17, plot the function $z$

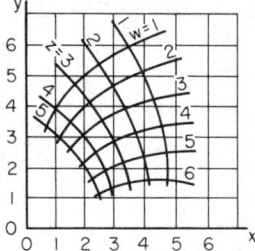

**Fig. 17**

$= f(x,y)$ by contour lines on an $xy$ plane, and plot the function $w = F(x,y)$ by contour lines on the same $xy$ plane. Then every point on the diagram (either directly or by interpolation) is the intersection of four curves—an $x$ curve, a $y$ curve, a $z$ curve, and a $w$ curve. Here, by "curve" is meant any line, straight or curved. By the aid of such a diagram, when the values of any two of these four variables are given, the values of the other two can be found. The method of use consists simply in entering the diagram along the two given curves (or lines), tracing them to their point of intersection, and then coming out again along the two curves (or lines) whose values are required. The best manner of numbering the curves is indicated in the figure.

## ALIGNMENT CHARTS

**Alignment Charts for Three Variables,** $t$, $u$, $v$ Any relation between three variables, $t$, $u$, $v$, which can be thrown into one of the forms listed in later paragraphs, can be represented graphically by a very convenient form of diagram called an alignment chart. In the simplest form of an alignment chart for three variables there are three scales (straight or curved), along which the values of the three variables, $t$, $u$, $v$, are marked in such a way that any three values of $t$, $u$, $v$ which satisfy the given equation are represented by three points which lie in line. Hence, if the values of any two of the variables are given,

the corresponding value of the third can be found by simply drawing a straight line through the two given points and reading the value of the point where it crosses the third scale.

The most important methods of constructing alignment charts for three variables are described below. Where several methods are applicable in a given case, the best one must be determined largely by trial. For further information, see d'Ocagne, "Traité de Nomographie," Gauthier-Villars, Paris; Runge, "Graphical Methods," Columbia University Press; Peddle, "Construction of Graphical Charts," McGraw-Hill; Lipka, "Graphical and Mechanical Computation," Wiley; Kraitchik, "Alignment Charts," Van Nostrand; Douglass and Adams, "Elements of Nomography," McGraw-Hill.

METHOD 1. Given, an equation which can be thrown into the form

$$f_1(u) + f_2(v) = f_3(t)$$

where $f_1(u)$ is a function of $u$ alone, $f_2(v)$ a function of $v$ alone, etc. An alignment chart may be constructed as follows:

*Choice of Moduli to Fit Size of Paper.* Let $f_1(u')$ be the smallest and $f_1(u'')$ the largest value of $f_1(u)$ likely to be needed, and let $h$ be the height of the available space on the paper. Then find a simple number, $m_1$, such that $m_1$ times $f_1(u'') - f_1(u')$ shall not exceed $h$. Similarly, find a simple number $m_2$ such that $m_2$ times $f_1(v'') - f_1(v')$ shall not exceed $h$.

Also, compute a third modulus, $m_3$ by the formula

$$m_3 = (m_1 m_2)/(m_1 + m_2)$$

*Construction of the First Two Scales.* Draw two parallel vertical axes, at any distance, $k$, apart. On the first axis, marked $u$, starting with any convenient origin, lay off the distances $x = m_1 f_1(u)$ for successive values of $u$, labeling each point thus plotted with the corresponding value of $u$. Similarly, on the second axis, marked $v$, starting with any convenient origin, lay off $y = m_2 f_2(v)$ for successive values of $v$, labeling each point with the corresponding value of $v$. The $u$ scale and the $v$ scale are thus completed.

*Construction of the Third Scale.* Draw a third line, $t$, parallel to the first two lines, dividing the distance $k$ in the ratio $m_1/m_2$; that is, the distance from $u$ to $t$ is $m_1 k/(m_1 + m_2)$. Compute the value $t_0$ corresponding to any convenient values $u_0$ and $v_0$, and label with this value, $t_0$, the point where the $t$ axis is cut by a straight line joining the points $u_0$ and $v_0$. Using this point $t_0$ as an anchorage, lay off along the $t$ line the scale determined by $z = m_3 f_3(t)$ where $m_3 = (m_1 m_2)/(m_1 + m_2)$. The third scale is thus completed, and the chart is ready for use.

Note that the units of measurement for $x$, $y$, and $z$ (which do not appear on the completed chart) must of course be the same. Note also that to ensure accuracy on the third scale, especially if the modulus $m_3$ is small, it is well to compute more than one anchorage point, $t_0$.

The construction is greatly facilitated by the use of previously constructed uniform and logarithmic scales with various moduli.

EXAMPLE (Fig. 18). Let $uv^{1.41} = t$, for a range of values of $u$ and $v$ between 1 and 10. By taking the logarithm of both sides, reduce the equation to the form $\log u + 1.41 \log v = \log t$. Here $f_1(u) = \log u$, $f_2(v) = 1.41 \log$

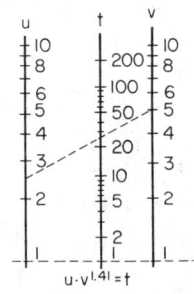

**Fig. 18**

$v$, $f_3(t) = \log t$. For a height of paper $b = 10$, and a width $k = 5$, we may take $m_1 = 10$ and $m_2 = 10/1.41 = 7.00$; whence $m_3 = 4.15$ and $m_3 k(m_1 + m_2) = 2.92$. Hence the chart is readily constructed, as shown.

METHOD 1a.   Method 1 may be readily extended to equations of the form

$$f_1(u) + f_2(v) + f_3(w) = f_4(t)$$

involving four variables, $t$, $u$, $v$, $w$.

Let $f_1(u) + f_2(v) = q$ and chart this equation by Method 1. Then chart the equation $q + f_3(w) = f_4(t)$ by the same method, using as one of the scales the $q$ scale already drawn. (The $q$ scale need not be graduated; the position of the $q$ axis is all that is important.) In reading the completed chart, we use two index lines, one joining points $u$ and $v$, and cutting the $q$ axis in an (unlabeled) point $q$; the other joining points $w$ and $t$, and cutting the $q$ axis in the same (unlabeled) point $q$. Thus when any three of the four variables are given, the fourth can be found.

A further extension to equations of the same form involving five or more variables is obvious.

EXAMPLE (Fig. 19).   Let $t = w \sqrt{uv}$ whence $\frac{1}{2} \log u + \frac{1}{2} \log v + \log w = \log t$. Here $f_1(u) = \frac{1}{2} \log u$, $f_2(v) = \frac{1}{2} \log v$, $f_3(w) = \log w$, and $f_4(t) = \log t$.

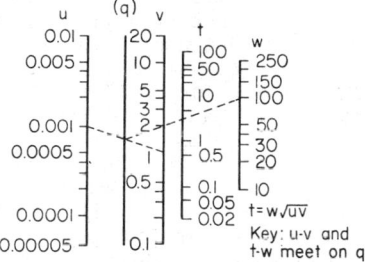

**Fig. 19**

METHOD 2.   Given, an equation which can be thrown into the form

$$f_1(u) = f_2(v) \cdot f_3(t)$$

where $f_1(u)$ is a function of $u$ alone, $f_2(v)$ a function of $v$ alone, etc. First, choose two "moduli" $m_1$ and $m_2$ (to fit size of paper) exactly as in Method 1.

Second, draw two parallel vertical axes, $AX$ and $BY$, oppositely directed, the diagonal line $AB$ being of any convenient length $k$. With $A$ and $B$ as origins, lay off along these axes the distances

$$x = m_1 f_1(u) \qquad \text{and} \qquad y = m_2 f_2(v)$$

for successive values of $u$ and $v$, respectively, and label each point thus plotted with the corresponding value of $u$ (or $v$). The $u$ and $v$ scales are thus completed.

Third, on $BY$ select a point $F$ at any convenient distance, $l$, from $B$; compute an auxiliary modulus, $n$, by the formula $n = l m_1 / m_2$; and lay off along $AX$ an auxiliary scale, $x' = n f_3(t)$ for successive values of $t$, marking each point (temporarily) with the corresponding value of $t$. Then transfer this auxiliary scale to the axis $AB$ by means of projecting lines drawn through the

point $F$, marking each point (permanently) with the corresponding value of $t$. The $t$ scale, along the axis $AB$, is thus completed, and the chart is ready for use.

As a check, note that along the $t$ axis, the distance $z$ from $A$ to any point labeled $t$ should be given by

$$v = k m_1 f_3(t)/[m_1 f_3(t) + m_2]$$

Indeed the points of the $t$ scale may be laid down independently by the use of this formula, if desired, instead of by the graphical method above described.

This type of chart is known as a $Z$ chart.

EXAMPLE (Fig. 20).   Let $u = 0.196\, t^3 v$, where $u$ is to range from 0 to 150,000 and $v$ from 0 to 15,000. The equation may be written $u = (10v) (0.0196 t^3)$.

Here $f_1(u) = u$, $f_2(v) = 10v$, $f_3(t) = 0.0196 t^3$.

The theory underlying Methods 1 and 2 depends only on simple properties of similar triangles. The following methods

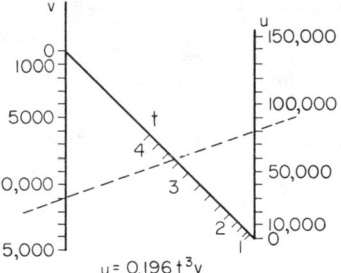

**Fig. 20**

are based on certain standard equations of the straight line in analytical geometry, and the notation in what follows has been suggested by the form of these equations.

**Notation**   In each of the equations which follow, $U$ stands for any function of $u$ alone, $V$ for any function of $v$ alone, and $F_1(t)$, $F_2(t)$ for any functions of $t$ alone. Any of these functions may reduce to a constant. The axes of $x$, $y$, and $y'$ which are mentioned are of merely temporary use in constructing the diagram, and the letters $x$, $y$, $y'$ should not be written on the chart. It is not necessary that the axes be at right angles, provided the $x$ of a point is always measured parallel to the $x$ axis, and its $y$ parallel to the $y$ axis.

METHOD 3.   Given, an equation which can be thrown into the form

$$U \cdot F_1(t) + V \cdot F_2(t) = 1$$

where, for the given range of values of $u$ and $v$, the largest variations in $U$ and $V$ are less than a certain number $m$.

Draw a pair of (temporary) $x,y$ axes (Fig. 21), and through the point $x = 1$ draw a third axis, which may be called the axis of $y'$, parallel to the axis of $y$. In ordinary cases, the unit of measurement along $x$ should be nearly equal to the full width of the paper. Now choose a unit for $y$ and $y'$ such that $m$ times this unit will about equal the height of the paper, and plot, in the usual way, the points $(x,y)$ given by

$$x = \frac{F_2(t)}{F_1(t) + F_2(t)} \qquad y = \frac{1}{F_1(t) + F_2(t)}$$

labeling each point with the value of $t$ to which it corresponds.

Connect these points by a smooth curve, which gives the $t$ scale of the diagram. [If $F_1(t)/F_2(t) =$ a constant, the $t$ scale will prove to be a straight line parallel to the $y$ axis.]

Then, using the same units as above, plot along $y$ the points given by $y = U$, labeling each point with the corresponding value of $u$; and plot along $y'$ the points given by $y' = V$, labeling each of these points with the corresponding value of $v$. This gives the $u$ and $v$ scales of the diagram. The three scales being thus constructed, the $x$ axis may now be erased, and the diagram is ready for use. Any three points $t$, $u$, $v$ which lie in line correspond to three values of $t$, $u$, $v$, which

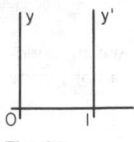

**Fig. 21**

satisfy the given equation. The numbering on each scale should be shown at sufficiently frequent intervals to permit of easy interpolation.

EXAMPLE (Fig. 22). Let $v = ut + 16t^2$, which reduces to the form $(-u/16)(1/t) + (v/16)(1/t^2) = 1$. Here $U = -u/16$, $V = v/16$, $F_1(t) = 1/t$, $F_2(t) = 1/t^2$ and $x = 1/(1 + t)$. $y = t^2/(1 + t)$.

NOTE. If $m = \infty$, values of $u$ and $v$ which give large values of $U$ and $V$ cannot be shown within the limits of the paper. In such cases, the chart may be supplemented by a second chart, made according to Method 4, below.

METHOD 4. Given, an equation which can be thrown into the form

$$\frac{F_1(t)}{U} + \frac{F_2(t)}{V} = 1$$

where, for the given range of values of $u$ and $v$, the largest variation in $U$ is less than a certain number $m$, and the largest variation in $V$ is less than a certain number $n$.

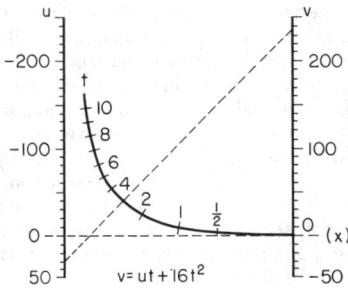

**Fig. 22**

Draw a pair of temporary $x,y$ axes, and having chosen a unit for the $x$ axis equal to about $1/m$th of the width of the paper, and a unit for the $y$ axis equal to about $1/n$th of the height, plot the points $(x,y)$ given by

$$x = F_1(t) \qquad y = F_2(t)$$

labeling each point of this curve with the value of $t$ to which it corresponds. Connect these points by a smooth curve, which gives the $t$ scale of the diagram. [If $F_1(t)/F_2(t) =$ a constant, the $t$ scale will be a straight line through the origin.]

Then, using the same units as above, plot along $x$ the values of $U$, labeling each point with the corresponding value of $u$; and plot along $y$ the values of $V$, labeling each point with the corresponding value of $v$. This gives the $u$ and $v$ scales of the diagram. On the chart as thus completed, any three points $t$,

$u$, $v$ which lie in line correspond to three values of $t$, $u$, $v$ which satisfy the given equation.

EXAMPLE (Fig. 23). Let $t = (uv)/(u + v)$, which may be written in the form $t/v + t/u = 1$. Here $U = u$, $V = v$, $F_1(t) = t$, $F_2(t) = t$.

NOTE. If $m = \infty$ and $n = \infty$, values of $u$ and $v$ which give large values of $U$ and $V$ cannot be shown within the limits of the paper. In such cases the chart may be supplemented by a second chart, made according to Method 3, above.

METHOD 5. Given, an equation which can conveniently be thrown into the form

$$F_2(t) = V \cdot F_1(t) + U$$

where, for the given range of values of $t$, the largest variation in $F_1(t)$ is less than a certain number $m$, and the largest variation in $F_2(t)$ is less than a certain number $n$.

Draw a pair of temporary $x,y$ axes, and, having chosen a unit for $x$ equal to about $1/m$th of the width of the paper and a unit for $y$ equal to about $1/n$th of the height, plot the points $(x,y)$ given by

$$x = F_1(t) \qquad y = F_2(t)$$

labeling each point of the curve with the value of $t$ to which it

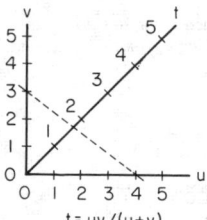

**Fig. 23**

corresponds. Connect these points by a smooth curve, which forms the $t$ scale. Next using the same unit for $y$ as above, plot along the $y$ axis the values of $U$, labeling each point with the corresponding value of $u$. This gives the $u$ scale. Finally, with the origin as center, and any convenient radius, draw a circle cutting the $x$ axis in $A$. Along this circular arc, starting from $A$ in the counterclockwise direction, lay off the angles whose slopes are equal to $V$, labeling each point of the arc with the value of $v$ to which it corresponds. This gives the $v$ scale, which in this case, however, plays a peculiar role, since, in using this form of chart, two straight lines are required instead of one. Thus:

In order to determine whether three values, $t$, $u$, $v$, satisfy the given equation, lay one straight line through the points $t$ and $u$, and another straight line through the point $v$ and the origin; if these lines are parallel, the three values of $t$, $u$, $v$ satisfy the equation. It will be noticed that the function of the $v$ scale here is to measure, in a certain sense, the slope of the line joining $t$ and $u$. A chart of this type may be called "an alignment chart with a sliding scale for one of the variables."

EXAMPLE. Let $\sin u = \sin 60° \sin t - \cos 60° \cos t \cos v$ (Fig. 24), which may be put in the form $(\sin 60° \sin t) = \cos v (\cos 60° \cos t) + \sin u$. Here $F_1(t) = \cos 60° \cos t$, $F_2(t) = \sin 60° \sin t$, $U = \sin u$, $V = \cos v$.

**Alignment Charts for Four Variables** The extension of Methods 3, 4, and 5 to the case of four variables, say $r$, $s$, $u$, $v$ consists essentially in replacing the $t$ scale of the earlier diagram by a network of two scales, one for $r$ and one for $s$. The point where a curve $r = r_1$ and a curve $s = s_1$ intersect may be spoken of as the point $(r_1,s_1)$. In the following equations, $U$ denotes as before any function of $u$ alone $V$ any function of $v$ alone; while $F_1(r,s)$ and $F_2(r,s)$ represent any functions of $r$ and $s$.

METHOD 3a.   Given, an equation of the form

$$U \cdot F_1(r,s) + V \cdot F_2(r,s) = 1$$

Draw axes $x$, $y$, and $y'$ as in Method 3, and plot the network of curves given by the equations

$$x = \frac{F_2(r,s)}{F_1(r,s) + F_2(r,s)} \qquad y = \frac{1}{F_1(r,s) + F_2(r,s)}$$

To do this (Fig. 25), find the point $(x,y)$ that corresponds to each given pair of values of $r$ and $s$, by direct substitution in

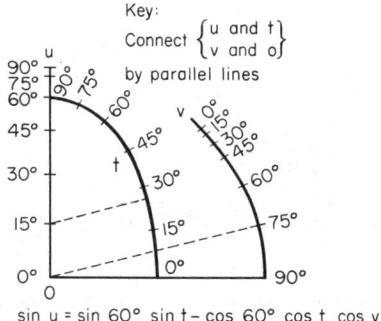

Key:

Connect $\begin{Bmatrix} u \text{ and } t \\ v \text{ and } o \end{Bmatrix}$ by parallel lines

$\sin u = \sin 60° \sin t - \cos 60° \cos t \cos v$

**Fig. 24**

the equations for $x$ and $y$. Connect all the points for which $r = 1$ by a curve, and label it $r = 1$; connect all the points for which $r = 2$ by another curve, and label it $r = 2$; etc. This gives the family of $r$ curves. Similarly, through all the points for which $s = 1$ draw a curve labeled $s = 1$; through all the points for which $s = 2$ draw a curve labeled $s = 2$; etc. This gives the family of $s$ curves, intersecting the family of $r$ curves. Note, however, that if it is possible to eliminate $s$ (or $r$) from the equations that give $x$ and $y$, the resulting equation in $x$, $y$, and $r$ (or $x$, $y$, and $s$) can often be plotted directly for each given value of $r$ (or of $s$).

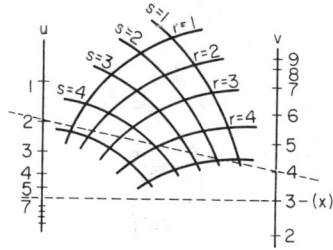

**Fig. 25**

Next, construct the $u$ and $v$ scales along the axes of $y$ and $y'$ as in Method 3. (The letters $x$, $y$, and $y'$, and the units used in plotting along these axes, should be omitted from the finished diagram, as should also the axis of $x$.)

In the chart, as thus completed, any three points, $(r,s)$, $u$, and $v$ which lie in a straight line, correspond to values of $r$, $s$, $u$, $v$ which satisfy the given equation. Hence, when any three of these four values are given, the fourth can be found from the chart.

METHOD 4a.   Given, an equation of the form

$$\frac{F_1(r,s)}{U} + \frac{F_2(r,s)}{V} = 1$$

Draw axes of $x$ and $y$ as in Method 4, and plot the network of curves given by

$$x = F_1(r,s) \qquad y = F_2(r,s)$$

To do this, follow the plan outlined for a similar case under Method 3a, labeling each curve of the $r$ family (Fig. 26) with the corresponding value of $r$, and each curve of the $s$ family

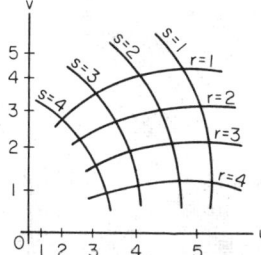

**Fig. 26**

with the corresponding value of $s$. Next, construct the $u$ and $v$ scales along the $x$ and $y$ axes, precisely as in Method 4. Then any three points, $(r,s)$, $u$, and $v$, which lie in a straight line correspond to values of $r$, $s$, $u$, $v$ which satisfy the given equation.

METHOD 5a.   Given, an equation of the form

$$F_2(r,s) = V \cdot F_1(r,s) + U$$

Draw axes of $x$ and $y$, as in Method 5, and plot the network of curves given by $x = F_1(r,s)$, $y = F_2(r,s)$, following the plan outlined for a similar case under Method 3a, and labeling each curve of the $r$ family (or $s$ family) with the value of $r$ (or $s$) to which it corresponds. Next, construct the $u$ scale along the $y$ axis, and the $v$ scale along a circular arc, precisely as in Method 5. Then any three points, $(r,s)$, $u$, and $v$, which are so related that the line through $(r,s)$ and $u$ is parallel to the line joining $v$ with the origin, will correspond to values of $r$, $s$, $u$, $v$ which satisfy the given equation.

EXAMPLE.   Method 5a (Fig. 27). Let $\cot v = \cot r \cos s + \csc r \sin s \cot u$, which may be written $(\cos r \cot s) = \cot v (\sin r \csc s) - \cot u$. Here $U = -\cot u$, $V = \cot v$. $F_1(r,s) = \sin r \csc s$, $F_2(r,s) = \cos r \cot s$, whence

$$\frac{x^2}{\csc^2 s} + \frac{y^2}{\cot^2 s} = 1 \qquad \frac{x^2}{\sin^2 r} - \frac{y^2}{\cos^2 r} = 1$$

so that the $s$ curves are ellipses and the $r$ curves hyperbolas.

**Parallel Charts, or Proportional Charts, for Four Variables**
In the following methods of representation there are four scales, one for each of the four variables, and the method of using the diagram consists in connecting two pairs of points by parallel lines.

METHOD 1.   Given, an equation of the form

$$R - S = U - V$$

where $R$, $S$, $U$, $V$ are any functions of the variables $r$, $s$, $u$, $v$, respectively. (It will be noted that any proportion $R/S = U/V$ can at once be thrown into this form by taking the logarithm of both sides.)

In Fig. 28, draw four vertical axes, $y_1$, $y_2$, $y_1'$, $y_2'$, such that the distance between $y_1$ and $y_1'$ (which may be zero) is equal to the distance between $y_2$ and $y_2'$, and so that the four zero points lie in line. Along these axes, using the same unit for all, plot the points given by $y_1 = R$, $y_1' = S$, $y_2 = U$, $y_2' = V$, and label each point with the value of $r$, $s$, $u$, or $v$ to which it corresponds. (The letters $y_1$, $y_2$, $y_1'$, $y_2'$ are temporary, and

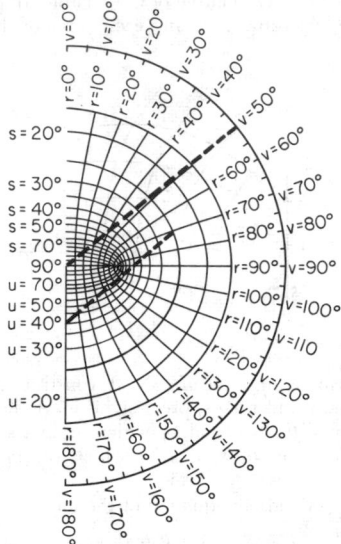

cot v = cot r cos s + csc r sin s cot u

Key: connect $\begin{cases}(r,s) \text{ and } u \\ v \text{ and } o\end{cases}$ by parallel lines

**Fig. 27**

should not appear on the diagram.) Then if the line joining two points $r$ and $u$ is parallel to the line joining two points $s$ and $v$, the four values of $r$, $s$, $u$, $v$ will satisfy the given equation. In this and the following methods, a parallel ruler, or a pair of draftsman's triangles, will be useful in reading the chart. A "key" stating which points are to be joined with which should be clearly given on the diagram.

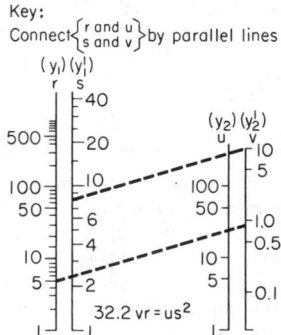

Key:
Connect $\begin{cases}r \text{ and } u \\ s \text{ and } v\end{cases}$ by parallel lines

32.2 vr = us²

**Fig. 28**

EXAMPLE (Fig. 28).    Let $32.2vr = us^2$, or $\log r - 2 \log s = \log u - \log(32.2v)$. Here $R = \log r$, $S = 2 \log s$, $U = \log u$, $V = \log(32.2v)$.

METHOD 2.    Given, an equation of the form

$$\frac{R}{S} = \frac{U}{V}$$

In Fig. 29, draw a pair of axes, $x,y$, and parallel to them (or coinciding with them) a second pair of axes, $x_1, y_1$. Using any convenient horizontal unit, plot along $x$ and $x_1$ the points given by $x = R$, $x_1 = U$, and using any convenient vertical unit, plot along $y$ and $y_1$ the points given by $y = S$, $y_1 = V$. Label each point with the value of $r$, $s$, $u$, $v$, to which it

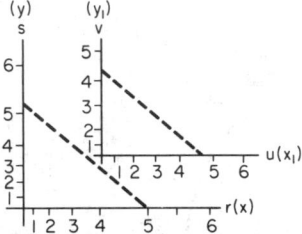

Key:
Connect $\begin{cases}r \text{ and } s \\ u \text{ and } v\end{cases}$ by parallel lines

**Fig. 29**

corresponds. (The letters $x$, $y$, $x_1$, $y_1$ should not appear on the diagram.) Then if the line joining two points $r$ and $s$ is parallel to the line joining two points $u$ and $v$, the four values $r$, $s$, $u$, $v$ will satisfy the given equation.

METHOD 3.    Given, an equation of the form

$$R - S = \frac{V}{U}$$

In Fig. 30, take a pair of axes, $x,y$, and through the point $x = 1$ draw a third axis, $y'$, parallel to $y$. Also, take a second pair of axes, $x_2, y_2$, parallel to (or coinciding with) the axes of $x$ and

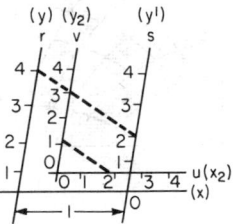

Key:
Connect $\begin{cases}r \text{ and } s \\ u \text{ and } v\end{cases}$ by parallel lines

**Fig. 30**

$y$. Having chosen a suitable unit for $x$ and $x_2$, and a suitable unit for $y$, $y'$, and $y_2$, lay off the values of $R$ and $S$ along $y$ and $y'$, respectively, labeling each point with the value of $r$ or $s$ to which it corresponds; and lay off the values of $U$ and $V$ along $x_2$ and $y_2$, labeling each point with the value of $u$ or $v$ to which it corresponds. Then if the line joining two points $r$ and $s$ is parallel to the line joining two points $u$ and $v$, the four values $r$, $s$, $u$, $v$ will satisfy the given equation.

For further examples, see Strachan, Nomographic Solutions for Formulas of Various Types, *Trans. ASCE*, **78**, 1915.

# VECTOR ANALYSIS

## by Philip Franklin

Many problems involving directed magnitudes can be advantageously treated by the methods of vector analysis. The following is a brief summary of the principal definitions and formulas.

A set of arrows, each arrow having a given *length* and pointing in a given *direction*, is called a set of **vectors,** provided they combine by addition according to the parallelogram law (see below.) Notation: **a** for a vector; $a$ or $|a|$ for its length. Two "free" vectors are equal if they have the same length and point in the same direction; two "sliding" vectors are equal if they have the same length and direction, and also lie in the same line.

A **scalar** is any real number, positive, negative, or zero.

**Addition of Vectors** If an arrow **a** is immediately followed, tip to tail, by a second arrow **b**, then the arrow which runs from the beginning of **a** to the end of **b** is called the **sum** of **a** and **b**, denoted by **a** + **b**. Conversely, if **a** + **x** = **b**, then **x** = **b** − **a**. The laws of operation for + and − are the same as in ordinary algebra. If $m$ is a scalar, then $m$**a** means a vector having the same direction as **a**, and $m$ times its length.

**Multiplication of vectors** is of two kinds, as follows:

The **scalar product,** or dot product, of two vectors **a** and **b**, denoted by **a·b**—or by (**ab**) in round parentheses—is defined as the scalar quantity $ab \cos \theta$, where $\theta$ is the angle between **a** and **b**.

EXAMPLE. If **F** is a force whose point of application moves along a vector distance **x**, then **F·x** − *work* done by **F** this during displacement.

PECULIARITIES OF SCALAR PRODUCTS. (1) Since **a·b** is not a vector, expressions like (**a·b**)·**c** will not occur; (2) from **a·x** = **a·y** we cannot infer that **x** = **y**; hence, quotients will not occur; (3) from **a·b** 0, it follows that **a** is **perpendicular** to **b** (unless **a** or **b** is zero).

On the other hand, scalar products are like ordinary products in the following respects: **a·b** = **b·a**, and (**a** + **b**)·(**c** + **d**) = **a·c** + **a·d** + **b·c** + **b·d**; also, $m$(**a·b**) = ($m$**a·b**) = **a·** ($m$**b**), where $m$ is any scalar.

The **vector product,** or cross product, of two vectors **a** and **b**, denoted by **a** × **b** or by [**ab**] in square brackets—is defined as the *vector* whose length is $ab \sin \theta$, where $\theta$ is the angle between **a** and **b,** and whose direction is perpendicular to the plane of **a** and **b** (in such a sense that a right-handed screw advancing along **a** × **b** would turn **a** toward **b**). The angle $\theta$ must be between 0 and 180°.

EXAMPLE. If **F** is a force acting on a particle whose radius vector is **r**, then **r** × **F** = the *torque* of **F** about the origin.

PECULIARITIES OF VECTOR PRODUCTS. (1) **a** × **b** = −**b** × **a**, so that the order of the factors is always important; (2) **a** × **a** = 0; (3) it is *not* true that **a** × (**b** × **c**) = (**a** × **b**) × **c**; (4) from **a** × **x** = **a** × **y** it does *not* follow that **x** = **y**; hence, quotients will not occur; (5) from **a** × **b** = 0, it follows that **a** and **b** are **parallel** (unless **a** or **b** is zero).

On the other hand, as in ordinary algebra,

$$(a + b) \times (c + d) = a \times c + a \times d + b \times c + b \times d$$

provided the order of factors in each product is preserved; also, $m$(**a** × **b**) = ($m$**a**) × **b** = **a** × ($m$**b**), where $m$ is any scalar. Further laws are

$$a \cdot (b \times c) = b \cdot (c \times a) = c \cdot (a \times b)$$

and

$$a \times (b \times c) = (a \cdot c)b - (a \cdot b)c$$

**Vector Differentiation** If **r** = **f**($t$) gives a vector **r** as a function of a scalar $t$, then $d\mathbf{r}/dt = \lim \{[\mathbf{f}(t + \Delta t) - \mathbf{f}(t)]/\Delta t\}$ as $\Delta t$ approaches zero.

$$d(a + b) = da + db \qquad d(ma) = m(da) + (dm)a$$
$$d(a \cdot b) = (da) \cdot b + a \cdot (db) \qquad d(a \times b) = (da) \times b + a \times (db)$$

EXAMPLE. If **r** = **f**($t$) gives the position vector of a moving particle as a function of the time $t$, then $d\mathbf{r}/dt$ = its vector velocity, **v**, and $d\mathbf{v}/dt$ = its vector acceleration, **a.** If **m** and **n** are unit vectors in the direction of the tangent and normal to the path at the time $t$, then **v** = $v$**m**, where $v = ds/dt$ = the (scalar) path velocity, and $d\mathbf{m} = [(ds/R)]\mathbf{n}$, where $R$ = the (scalar) radius of curvature of the path. Then

$$a = \frac{d(v\mathbf{m})}{dt} = \frac{dv}{dt}\mathbf{m} + v\frac{d\mathbf{m}}{dt} = \frac{dv}{dt}\mathbf{m} + \frac{v^2}{R}\mathbf{n}$$

Here $dv/dt$ and $v^2/R$ are the familiar expressions for the components of acceleration along the tangent and normal.

## NEWTONIAN MECHANICS

In many linkages and machine elements it is often convenient to choose several sets of cartesian coordinate systems for the determination of the acceleration, stress, and other pertinent properties of the several parts. Formulas have been derived that facilitate the simultaneous use of these several systems which are most often rendered in vector notation.

The first set of axes, $X_o$, $Y_o$, $Z_o$, has its origin at $O$. This set is considered fixed in space. The second set of axes, $X'$, $Y'$, $Z'$, has its origin at $Q$. The position of $Q$, measured in the first system, is $\mathbf{r}_q$. The second set is always parallel to the first, i.e., $X'\|X_o$, $Y'\|Y_o$, $Z'\|Z_o$; and is in general translating in the first system with a velocity $\mathbf{V}_q$, i.e., the velocity of the origin $Q$ measured in the first system. The third set of axes, $X$, $Y$, $Z$, also has its origin at $Q$ and therefore is not translating in the second system but is in general rotating in the second system with an angular velocity of $\boldsymbol{\omega}$. Neither $\mathbf{V}_q$ nor $\boldsymbol{\omega}$ is constant, but they change at a rate of $\mathbf{a}_q$ and $\dot{\boldsymbol{\omega}}$, respectively.

The point in the mechanism whose dynamic properties are to be considered usually possesses a position, **r**; a velocity, **V**; and an acceleration, **a**; all measured in the $X$, $Y$, $Z$ system. The purpose is to find the position, velocity, and acceleration of this point in fixed space, i.e., in the first system, the $X_o$, $Y_o$, $Z_o$ set, which are $\mathbf{r}_o$, $\mathbf{V}_o$, and $\mathbf{a}_o$, respectively:

$$\mathbf{r}_o = \mathbf{r}_q + \mathbf{r} \qquad \mathbf{V}_o = \mathbf{V}_q + \boldsymbol{\omega} \times \mathbf{r} + \mathbf{V}$$
$$\mathbf{a}_o = \mathbf{a}_q + \dot{\boldsymbol{\omega}} \times \mathbf{r} + \boldsymbol{\omega} \times (\boldsymbol{\omega} \times \mathbf{r}) + 2(\boldsymbol{\omega} \times \mathbf{V}) + \mathbf{a}$$

## ANALYTIC REPRESENTATION

Let $\mathbf{i}, \mathbf{j}, \mathbf{k}$ represent three vectors of unit magnitude along the three mutually perpendicular lines $OX$, $OY$, $OZ$, respectively, forming a right-hand system. Then the vector $\mathbf{r}$ may be represented by $\mathbf{r} = a\mathbf{i} + b\mathbf{j} + c\mathbf{k}$. Also, for any scalar $m$, $m\mathbf{r} = ma\ \mathbf{i} + mb\ \mathbf{j} + mc\ \mathbf{k}$.

Let $\mathbf{r}_1 = a_1\mathbf{i} + b_1\mathbf{j} + c_1\mathbf{k}$ and $\mathbf{r}_2 = a_2\mathbf{i} + b_2\mathbf{j} + c_2\mathbf{k}$. Then the sum $\mathbf{r}_1 + \mathbf{r}_2 = (a_1 + a_2)\mathbf{i} + (b_1 + b_2)\mathbf{j} + (c_1 + c_2)\mathbf{k}$. The scalar product $\mathbf{r}_1 \cdot \mathbf{r}_2 = a_1 a_2 + b_1 b_2 + c_1 c_2$. The vector product

$$\mathbf{r}_1 \times \mathbf{r}_2 = (b_1 c_2 - b_2 c_1)\mathbf{i} + (c_1 a_2 - c_2 a_1)\mathbf{j} + (a_1 b_2 - a_2 b_1)\mathbf{k}$$

$$\mathbf{r}_1 \times \mathbf{r}_2 = \begin{vmatrix} \mathbf{i} & \mathbf{j} & \mathbf{k} \\ a_1 & b_1 & c_1 \\ a_2 & b_2 & c_2 \end{vmatrix} \quad \text{and} \quad \mathbf{r}_1 \cdot (\mathbf{r}_2 \times \mathbf{r}_3) = \begin{vmatrix} a_1 & b_1 & c_1 \\ a_2 & b_2 & c_2 \\ a_3 & b_3 & c_3 \end{vmatrix}$$

## DIFFERENTIAL OPERATORS

By definition, $\nabla = \mathrm{del} = \mathbf{i}\dfrac{\partial}{\partial x} + \mathbf{j}\dfrac{\partial}{\partial y} + \mathbf{k}\dfrac{\partial}{\partial z}$ and $\nabla^2 = $ Laplacian $= \nabla \cdot \nabla = \dfrac{\partial^2}{\partial x^2} + \dfrac{\partial^2}{\partial y^2} + \dfrac{\partial^2}{\partial z^2}$ $\nabla S = \mathrm{grad}\ S = \dfrac{\partial S}{\partial x}\mathbf{i} + \dfrac{\partial S}{\partial y}\mathbf{j} + \dfrac{\partial S}{\partial z}\mathbf{k}$, the gradient of a scalar function $S(x,y,z)$. For a vector function $\mathbf{V}(x,y,z) = P\mathbf{i} + Q\mathbf{j} + R\mathbf{k}$, the divergence of $\mathbf{V}$ is $\nabla \cdot \mathbf{V} = \dfrac{\partial P}{\partial x} + \dfrac{\partial Q}{\partial y} + \dfrac{\partial R}{\partial z}$. And the curl of $\mathbf{V}$ is

$$\nabla \times \mathbf{V} = \mathrm{curl}\ \mathbf{V} = \mathrm{rot}\ \mathbf{V} = \begin{vmatrix} \mathbf{i} & \mathbf{j} & \mathbf{k} \\ \dfrac{\partial}{\partial x} & \dfrac{\partial}{\partial y} & \dfrac{\partial}{\partial z} \\ P & Q & R \end{vmatrix}$$

The **divergence theorem** states that if $\mathbf{F}$ is a vector function and $V$ is a volume bounded by a surface $S$, then

$$\iiint_V \mathrm{div}\ \mathbf{F}\ dv = \iiint_V \nabla \cdot \mathbf{F}\ dv = \iint_S \mathbf{F} \cdot d\mathbf{S}$$

The integrations are to be carried out over the volume $V$ and the surface $S$. And if $\mathbf{n}$ is the unit outward normal and $dS = |d\mathbf{S}|$ is the scalar element of surface, $d\mathbf{S} = \mathbf{n}\ dS$, $\mathbf{F} \cdot d\mathbf{S} = \mathbf{F} \cdot \mathbf{n}\ dS$.

**Stokes' theorem** states that if $\mathbf{F}$ is a vector function and $\mathbf{S}$ is a surface bounded by a simple closed curve $\mathbf{C}$, then

$$\int_C \mathbf{F}\,d\mathbf{r} = \iint_S \mathrm{curl}\ \mathbf{F} \cdot d\mathbf{S} = \iint_S (\nabla \times \mathbf{F})\,d\mathbf{S}$$

Here, $d\mathbf{r} = dx\ \mathbf{i} + dy\ \mathbf{j} + dz\ \mathbf{k}$. The theorem implies that the surface integral of $(\nabla \times \mathbf{F})$ over any surface $S$ which is bounded by $C$ is equal to the line integral of $\mathbf{F}$ over the contour $C$, taken in the direction related to that of $\mathbf{n}$ (in $d\mathbf{S} = \mathbf{n}\ dS$) by the right-hand rule.

# COMPUTERS

## by George J. Moshos

REFERENCES: Manuals of Computer Manufacturers. *Datamation* Magazine. *Modern Data Magazine*. Knuth, "The Art of Computer Programming," vols. 1, 2, and 3, Addison-Wesley. Madnick and Donovan, "Operating Systems," McGraw-Hill. Martin, "System Analysis for Data Transmission," Prentice-Hall. Bowers and Sedore, "SCEPTRE: A Computer Program for Circuit and System Analysis," Prentice-Hall. Hamming, "Numerical Methods for Scientists and Engineers," McGraw-Hill. *Communications, Journal,* and *Computing Surveys*, ACM. *Computer* Magazine, IEEE.

## COMPUTER TYPES

### Machine Organization

Computers are machines used for automatically processing information represented by mechanical or electrical means. They may be classified as analog or digital according to the techniques used to represent and process the information. Information in an **analog computer** is represented as continuous quantities which are physically measurable. This information is processed by components which are interconnected to form an analogous model of the problem to be solved. **Digital computers,** on the other hand, represent information as discrete physical states which are encoded into symbolic formats. Digital information is processed by sequences of operational steps which are preplanned to solve the given problem.

Information for digital computers may be represented in two fundamentally different ways, which leads to different machine organizations. Digital information may be represented (1) as a series of small incremental steps or (2) encoded into symbolic characters such as the digits or alphabet, or as whole words such as arithmetic numbers.

Data presented as incremental steps may be manipulated with mechanical versions of the operators of the difference calculus or may be used to position control devices using stepping motors. Incremental digital computers are designed to handle problems similar to those solved on the analog computer. However, greater precision and accuracy can be expected from an incremental computer than from an equivalent analog computer.

Digital data presented as characters or whole words may be manipulated by a variety of operators. Machines organized in this manner can handle both numerical and nonnumerical information and so can be applied to more general classes of problems. They are by far the most common type of digital computer. In fact, the term computer has become synonymous with this machine organization and is often used in this restricted way.

An important characteristic of a digital computer is that the sequence of steps which are processed to solve a problem operate automatically without human intervention between the operational steps. When the computational process is controlled directly by a human operator, the machine is called a calculator. This distinction between computer and calculator, however, is arbitrary and vague with modern machines. Modern calculators offer the opportunity to program a number of operations before human intervention is required. On the other hand, problem solutions on a computer are often programmed to interrogate the human operator for a response before continuing with the solution.

More complex computer organizations may be formed by combining several individual computers. When an analog and digital computer are combined, the resulting machine is called a **hybrid computer.** In such an organization, the bulk of the computations are usually assigned to the analog part of the machine. The digital part of the computer combination is used to perform calculations that require higher accuracy than can be measured or to set up parameters for different cases of a problem. Communication between these two different computer types is handled through special electronic translating equipment.

Several digital computers may also be combined to form a more complex machine organization. Such a machine is called a **multiprocessor.** Greater effectiveness and speed is usually possible on a multiprocessor, since the individual machines can be assigned those parts of the problem solution for which they are especially designed, e.g., input-output operations or computational operations.

### Distinction from Other Machines

Computers differ from other kinds of electrical and mechanical machines in a basic way. The work is performed on information rather than upon forces and displacements. A common form of information is numbers. Numbers can be represented in some coded form and processed by the four rules of arithmetic. But numbers are only one kind of information that can be manipulated by the computer. Given an encoded alphabet, words and languages can be formed and the computer can be used to perform such functions as information storage and retrieval, translation, and editing. Given an encoded representation for points and lines, the computer can be used to perform such functions as editing and displaying graphs, recognizing patterns, and organizing and connecting objects.

Because computers have been easily accessible, scientists from every discipline have reformulated their professional activities to mechanize those aspects which do not require human thought and decision. In this way, mechanical processes can be viewed as augmenting man's physical skills and strength, and information processes can be viewed as augmenting human mental skills and intelligence.

## DIGITAL COMPUTERS

### Binary Notation

Information in a digital computer is represented by strings of digits which assume one of the two values 0 or 1. These units of information are called **bits.** The term bit is a contraction for binary digit. Bits are used internally in the computer to represent both numerical and nonnumerical information.

In order to achieve efficiency in dealing with information, a fixed number of bits are grouped together and referenced as a discrete unit. These units are used to encode and format the information that can be processed by the computer. Units of 8 bits are common and are called a **byte.** Bytes are used to encode the basic symbolic characters which provide the computer with input-output information such as the alphabet, decimal digits, punctuation marks, and special characters. The byte size is not universally adopted on all machines but is perhaps the most popular basic unit of information.

Larger bit groups are organized into units called **words** (and sometimes into smaller units called half words or larger units called double words). These units are used to encode the basic instruction repertoire of the machine and format the numerical data. Common word sizes used on various commercially available computers are 16, 24, 32, 36, 48, and 60 bits.

Numerical information processed by a computer is represented in the binary numbering system. The binary numbering system uses a position notation analogous to that used in decimal numbers. For example, the decimal number 596.37 represents the value $5 \times 10^2 + 9 \times 10^1 + 6 \times 10^0 + 3 \times 10^{-1} + 7 \times 10^{-2}$. The value assigned to any of the 10 possible digits 0 through 9 depends on its position relative to the decimal point; i.e., zero and positive exponents of 10 are assigned to digits appearing to the left of the decimal point and negative exponents of 10 to the right. In a similar manner, the **binary** number 1011.011 represents $1 \times 2^3 + 0 \times 2^2 + 1 \times 2^1 + 1 \times 2^0 + 0 \times 2^{-1} + 1 \times 2^{-2} + 1 \times 2^{-3}$. The radix (base) of this system is 2. Each position of a binary number is occupied by one of the two possible digits 0 or 1. The radix point in this system serves a purpose similar to that of the decimal point in the decimal system.

The functional operators available in the computer for setting up the solution of a problem are encoded into the words of the machine. To handle numerical calculations, the instruction repertoire usually includes the four rules of arithmetic, i.e., $+$, $-$, $\times$, and $\div$. These instructions operate on data encoded in the binary system. This, however, is not a serious operational problem, since the user specifies information for the problem in the decimal system, or through mnemonics, and the computer is used to convert these formats into its own internal representations.

### Formats for Numerical Data

Three different formats are used to represent numerical information internal to the computer, fixed-point, floating-point, and encoded-decimal. These formats are described in this section.

A word in **fixed-point** format is given as a string of 0's and 1's representing a binary number. The radix point is not explicitly given but is implied by the program processing the data to be at a fixed position in the word, e.g., immediately to the right of the word so that the number represented would be an integer. To represent algebraic numbers, several alternate forms are used. These are listed in Table 1. Most often, 1's or 2's complement forms are adopted because they lead to a reduction and simplification of the electronics needed to perform the arithmetic operations.

**Floating-point** format is a mechanized version of the scientific notation, i.e., $\pm M \times 10^{\pm E}$, where $\pm M$ and $\pm E$ represent the signed mantissa and signed exponent of the number, respectively. Using a machine word to represent numbers in this format, a large range of numbers is possible. The signed mantissa and signed exponent are explicitly given in each word. The exponent, however, is implied as a power of 2 or 16 rather than 10. The radix point is implied to the left of the mantissa. After each operation the exponent is adjusted so that

**Table 1. Alternate Forms of Signed Fixed-Point Numbers**

| Format | Negative representation* | Representation of zero† | Range of integers‡ |
|---|---|---|---|
| 1's complement | Each bit is the complement of the positive number and vice versa | 1111 0000 | $-2^{n-1} + 1 \leq N \leq 2^{n-1} - 1$ |
| 2's complement | Same as 1's complement followed by adding 1 to the resulting number in binary | 0000 | $-2^{n-1} \leq N \leq 2^{n-1} - 1$ |
| Sign-magnitude | Leftmost bit is used to encode the sign, 0 for + and 1 for − | 0000 1000 | $-2^{n-1} + 1 \leq N \leq 2^{n-1} - 1$ |

*To complement a digit, 1 is changed to 0 and 0 is changed to 1.

†In the examples, representation of zero is given for four binary digits. Zero has a unique representation in 2's complement notation, while there are two different representations in both 1's complement and sign-magnitude notations.

‡Assume radix point is at extreme right so each number is an integer. $N$ is the value of the integer. $n$ is the total number of bits in the word.

the most significant digit is other than zero; i.e., the mantissa $M$ is maintained so that its value is in the range $0.1 \leq M < 1$, where these bounds are in the radix of the system. This is termed **normalizing.** Zero is treated as a special case in this notation, e.g., zero mantissa and zero exponent.

Often greater precision is needed in the calculations than is possible with one word. In this case two words are used to represent the number. Since the exponent need not be defined again, the added word is appended to the mantissa to extend the precision of the number. Such a representation is called **double precision,** although extended precision would be a more accurate term, since the precision is more than doubled.

The **range** of possible numbers in floating-point notation depends on the number of bits designated to the exponent and implied base of the system. For example, if 7 bits are used for the signed exponent and if 16 is implied as the base, then $16^{-64} \leq$ range $\leq 16^{63}$.

The **precision** of a floating-point number depends on the number of bits used for the unsigned mantissa and also depends on the implied base. If $m$ is the number of bits in the unsigned mantissa, the precision expressed as equivalent number of decimal digits $d$ is

$$0.301(m - 1) \leq d \leq 0.301m \text{ when implied base is 2}$$
$$0.301(m - 4) \leq d \leq 0.301m \text{ when implied base is 16}$$

For example, if 24 bits of a 32-bit word are used for the unsigned mantissa, 6.02 to 7.22 decimal digits can be represented when the implied base of the system is 16. The fractional component of 6.02 indicates that a fraction of 7-digit decimal numbers cannot be represented with a 24-bit mantissa. In a similar manner 7.55 indicates that some 8-digit decimal numbers cannot be represented. A double-precision number in this system, on the other hand, can represent numbers which can be expressed with 15.65 to 16.85 equivalent decimal digits.

**Encoded-decimal** representation is usually not used in scientific calculations. However, it is supplied in many computers as a convenience in commercial applications. Table 2 gives some typical schemes used to encode the decimal digits in which each decade is represented with 4 bits. A large selection of other schemes is possible and is also used in computers.

### Formats for Nonnumerical Data

A large variety of codes is used for representing the alphabet, digits, punctuation marks, and special symbols internal to the computer. The most popular ones are the 7-bit ASCII code and the 8-bit EBCDIC code. These are available in the literature in many different sources and are not presented here. Associated with these codes is also a coded representation for input and output devices. For example, EBCDIC is used to represent information on punch cards where each character is encoded as a combination of punches in columns of the card.

**Table 2. Schemes for Encoding Decimal Digits**

| Decimal digit | BCD | Excess-3 | 4221 code |
|---|---|---|---|
| 0 | 0000 | 0011 | 0000 |
| 1 | 0001 | 0100 | 0001 |
| 2 | 0010 | 0101 | 0010 |
| 3 | 0011 | 0110 | 0011 |
| 4 | 0100 | 0111 | 0110 |
| 5 | 0101 | 1000 | 1001 |
| 6 | 0110 | 1001 | 1100 |
| 7 | 0111 | 1010 | 1101 |
| 8 | 1000 | 1011 | 1110 |
| 9 | 1001 | 1100 | 1111 |

### Digital-Computer Components

Figure 1 shows a schematic which can be used to discuss the hardware configuration of a digital computer system.

The principal part of the system is composed of the central processing unit (CPU) and the memory of the system. The memory is organized into words (bytes, half words, and/or double words) which can be located by an address. The CPU makes available a repertoire of instructions which are used to set up the problem solutions. The general format of each instruction is as follows:

name: operator, operand(s)

The name designates an address which contains the operator and one or more operands. The operations permitted by the hardware of the CPU are encoded into the operator. The operand(s) refer to other named words which are used in the operation and may refer to either data or other instructions by their address.

The instructions and data are organized into sequences which, when executed in order, will produce the problem

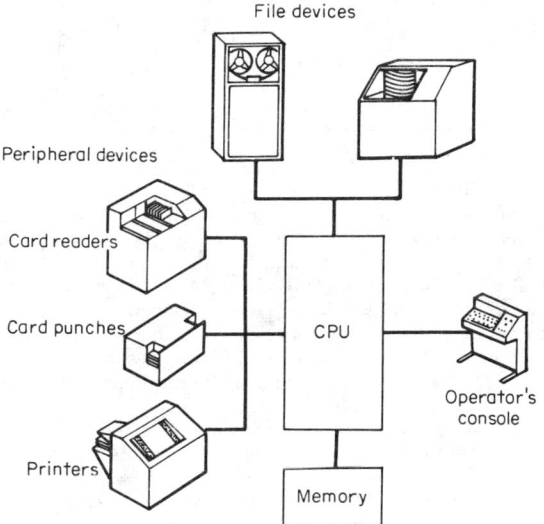

File devices

Peripheral devices

Card readers

Card punches

CPU

Operator's console

Printers

Memory

**Fig. 1**  Elementary computer-system organization.

solution. These sequences are variously called programs, subprograms, routines, subroutines, functions, etc. They are inserted into the memory prior to machine execution. The functions of the CPU are to monitor the flow of new data and instructions into and out of memory during the execution of the problem solution, control the order in which the instructions are executed, decode the operation and locate the operand(s) of each instruction, and perform the operation specified.

The size of the CPU and memory subsystems varies from small computers called **minicomputers** to very large computers which serve as a computing utility to many users. The speed of these two subsystems depends on the speed of the electronics used and on the number of activities which can occur concurrently. For a family of machines, that is, machines that have the same internal organization and instruction sets but vary in speed, it usually suffices to compare them on the basis of their memory-cycle time. This parameter is the time required to read or write a unit of information between the memory and the CPU.

To compare the speeds of machines with different organizations, more complex techniques are used which are based on the intended use of the machine. The most popular method is to calculate the average execution time of an instruction. Instruction mixes such as that shown in Table 3 are used to calculate the average execution time of an instruction. The instruction mixes used are compiled from typical programs and so are dependent on the types of problems used in deriving the mix, e.g., scientific, commercial, or communication. The speed of each instruction is weighted according to how frequently it is executed in such problems. The average

instruction execution time $\bar{t}$ is calculated using the instruction mix as follows:

$$\bar{t} = \sum_i w_i t_i$$

where $t_i$ is the execution time for instruction of type $i$.

A computing system can be viewed as a library of assembled instructions which have been packaged to solve specific problems or perform utility functions for the users. This collection of programs and routines, called the software of the system, is stored on the file devices of the system until they are needed by a user. Typically, the file devices utilize magnetic-recording technology. These devices are categorized as sequential- or direct-access depending on how selected information is located on the file. Magnetic-tape devices are the primary **sequential-access** devices; and magnetic cards, disks, or drums are the primary direct-access devices.

**Table 3. Instruction Mix**

| $i$ | Instruction type | Weight $w_i$ |
|---|---|---|
| 1 | Add: Floating-point | 0.07 |
| 2 | Fixed-point | 0.16 |
| 3 | Multiply: floating-point | 0.06 |
| 4 | Load/store register | 0.12 |
| 5 | Shift: one character | 0.11 |
| 6 | Branch: Conditional | 0.21 |
| 7 | Unconditional | 0.17 |
| 8 | Move: 3 words in memory | 0.10 |
| | Total | 1.00 |

The most common form magnetic-recording media consist of reels of **tape** ½ in wide, 0.0015 in thick, and 2,400 ft long. Information is recorded across this tape in either 7- or 9-bit frames. One of these bits is used for checking purposes only and is not transferred to the memory of the machine. The eight information bits are recorded using EBCDIC, a modified 8-bit ASCII code, or some other standard code. Lengthwise, the information can be recorded in various densities; e.g., 200, 556, 800, and 1,600 frames per inch are most often used. Besides the disadvantage that information must be passed over sequentially in order to locate the needed information, magnetic-tape recording does not permit information to be changed in situ. It must be recopied onto another tape.

**Direct-access** devices offer a wide choice of characteristics. The recording surface of a magnetic **drum** consists of a drum coated with magnetic material which revolves at high speeds. A number of read-write heads are located at fixed positions above the surface of the drum. The magnetic-**disk** recording surface consists of platters of magnetic material on a common spindle which revolves at high speed. The read-write heads may be permanently positioned or may be mounted on a common arm which locates the recording track by mechanical positioning. Magnetic-card devices consist of strips of magnetic material mounted in a magazine. A card is selected by mechanical means and automatically wound about a revolving drum. The time required to access information on any of these devices depends on the relative position of the information on the circumference of the recording surface when a read-write instruction is given. On the average this time corresponds to one-half a revolution of the device. The time required for this

**Table 4. Typical Characteristics of Direct-Access Devices**

| Characteristic | Magnetic cards | Movable-head disk | Fixed-head disk | Magnetic drum |
|---|---|---|---|---|
| Capacity, $10^6$ bytes | 5–500 | 7–200 | 2–5 | 1.5–4 |
| Transfer rate, $10^3$ bytes/s | 40–80 | 150–800 | 300–3,000 | 100–1,400 |
| Positioning time, $10^{-3}$ s | 200–500 | 75–250 | | |
| Latency time, $10^{-3}$ s | 20–30 | 8–12 | 2.5–20 | 4–20 |

half revolution is called the **latency** time. In addition to the latency time, magnetic-card and movable-head disk devices involve mechanical positioning. Typical values for these accessing times are given in Table 4. Also tabulated in this table is the amount of information that can be stored on these devices and the rate at which that information can be transferred to and from the memory unit.

The operator's **console** of a computing system is a relatively slow speed input-output keyboard device which is used by the computing-system staff to control and monitor the operations of the machine. The operators of the system may use this device to interrogate the status of the computing tasks in the system and may also receive instructions from the computer for performing clerical tasks such as mounting a particular magnetic tape on one of the file devices.

The number of different peripheral devices that are attached to the computing system for input and output have increased dramatically over the last decade. **Input devices** may be defined as any devices that put information into a machine-readable form. For engineering work the most common techniques for input are punched cards, punched tape, magnetic tape, keyboard in conjunction with a printing mechanism, or video scope. Other methods of input which up to now have had only limited acceptance to engineering work include handwritten- or printed-character recognition, magnetic ink, audio headset, touch-tone dialing, and mark sensing.

Contrasted to input devices, **output devices** have several roles in a computing system. The primary role is that of producing displays for the interpretation of the results. A large variety of printers, graphical plotters, and video displays have been developed for this purpose. At times, intermediate results are produced by the computer which are later reentered into the computer for further computations or transported to another machine for processing. Punched cards, punched tapes, and magnetic tape are usually used for this purpose. Output from a computer can also be used to drive mechanisms directly for **automatic control.** The number of variations of devices that serve these purposes are too specialized to list.

Figure 2 shows a schematic of a computing system that is located remotely from the area of application. In this case communication lines are used to transmit the information between stations. In order to interface the lines, special devices are needed to put the information signals in a form suitable for transmission or receiving. At a minimum, these interface units consist of circuits for modulation and demodulation of the signal, but they may consist of more sophisticated equipment such as a **multiplexer** which permits several devices to share a line or a minicomputer which preprocesses the information prior to transmission.

The amount of information that can be transmitted over a line depends on the quality and the frequency bandwidth of the line. The theoretical maximum number of bits per second $B$ that can be transmitted over a line with bandwidth $\Delta f$ and signal-to-noise power ratio $P_s/P_n$ is given by

$$B = \Delta f \, \mathrm{dB}/3$$

where

$$\mathrm{dB} = 10[\log(P_s/P_n)]$$

**Software Systems**

Computer software is organized into systems which provide the operating and functional facilities for computer users. Three systems generally provided are (1) **operating systems,** which consist of components for the control and operation of the computer hardware and software; (2) **program-preparation systems,** which consist of components for preparing and modifying programs for computer execution; and (3) **data-management systems,** which consist of components for generating, storing, updating, accessing, retrieving, editing, revising, and maintaining the information on the computer files. Data-management systems are of major concern to data processing rather than scientific or engineering applications. The components which support file management and input-output control, which are important parts of data-management systems, are usually included in the design of operating and data-preparation systems.

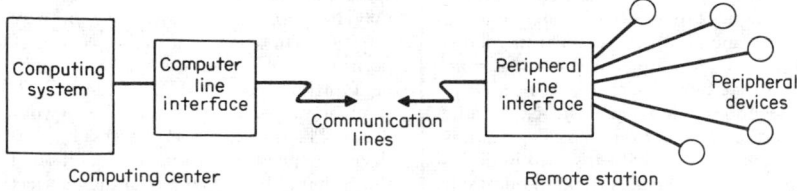

**Fig. 2**  Configuration for remote computing.

## Operating Systems

Prior to the development of operating systems each user was assigned a reserved time period for using the computer on a dedicated basis. The program setup, operation, and cleanup were done manually at a cost of valuable computer time. The current speed of computers makes such a mode of operation intolerable except for some isolated uses of minicomputer systems.

The purpose of computer operating systems is to allocate and manage the hardware and software resources of the computer, set up and schedule users' tasks and jobs so as to resolve conflicts and attempt to optimize production, provide control of the system input-output devices, protect the system and users' programs from faults and illegal use, and account for system use. The well-designed operating system makes itself and the computer operating staff transparent to the user. While the number and types of operating systems are too numerous to classify concisely, it is instructive to single out three basic types for further discussion: batch, multiprogramming, and timesharing.

**Batch** operating systems process jobs in sequential order. Jobs are collected in batches and are entered into the computer. Setup instructions, resources needed, processing instructions, etc., are specified with each job through a specialized language called the **job-control language**, which is interpreted by the operating system. Each job is then read into the memory of the computer and is processed, and the results are reported in sequence. In case of a malfunction or fault, the job is terminated in an orderly fashion before the next job in sequence is initiated. Hardware changes in system configurations can be used to improve performance. For example, input and output information may be recorded onto magnetic tape, disk, or drum devices rather than using slow-speed peripheral devices.

**Multiprogramming** operating systems process jobs in a concurrent fashion. Jobs may be entered into the computer any time that memory and other resources needed by the job are available. Many jobs are simultaneously active in such a system and are in a partial state of completion. The order of execution is usually based on some priority system. The highest-priority job is executed until it is completed or until pending requests from peripheral devices are serviced. While a job is waiting for service, control is passed to the highest-priority job that can be processed. It is important to note that only one program is being executed at any time, but operation with peripheral devices can occur simultaneously. The objective of such a system then is to balance the input, output, and processing aspects of several jobs so as to achieve greater productivity of all the equipment. Multiprogramming systems may vary in sophistication. For example, one or more batch systems may be integrated in such a system or several machines may be used to share in the work load.

**Timesharing** operating systems permit on-line computing by several users simultaneously. Each user is connected directly to the computer through an input-output terminal which permits him to work interactively with the computer on his program. The resources of the computer, such as memory and central processing time, are shared by all the users. Typically, each timesharing-system user is given a short slice of the available computer time to complete his task. If the task is not completed within this time slice or if the program requests service of some device, the user is put on a wait status and control is passed to the next user in a round-robin fashion. Both multiprogramming and batch-type processing may be incorporated into the more sophisticated timesharing systems. The attempt to treat each user in an equitable fashion makes the scheduling of these multifacet systems more complex.

An important component in any operating system is called the **loader.** Its function is to locate and enter into the memory of the machine the programs and routines which are to be executed. The most basic type of loader is called an absolute loader. Loaders of this type work with programs which are organized to execute only at specific locations in memory. A **relocatable loader** is more popular, especially with multiprogramming operating system. These loaders work with programs that can be executed in any area of memory, and it is the function of the loader to adjust the addresses of the program accordingly. The most sophisticated loaders, called **dynamic-link loaders,** are usually found as part of a timesharing operating system. Such loaders permit programs whose execution has already been started, and which are in partial state of completion, to be moved in memory. The operating system resolves the location of key addresses in such programs only when these addresses are referenced during execution.

It is difficult to compare timesharing systems with batch or multiprogramming systems because their objectives are different. Simply stated, the objectives of a pure batch or multiprogramming system are to achieve greater utilization of the equipment while those of timesharing are to achieve greater utilization of the people using the equipment. Parameters which are useful in comparing batch or multiprogramming systems are turnaround time and throughput. **Turnaround time** is defined as the average elapsed time between the time a job is submitted for computer processing to the time the processing is completed. **Throughput** is the ratio of the number of jobs processed by the system to the maximum number of jobs the system can process. System performance can be given as a trade off between these two parameters. Contrasted to these two parameters, the performance of a timesharing system can be given as a trade-off between the number of users which are active concurrently on the systems to the average response time each user is given. **Response time** is defined as the elapsed time from the time a user has completed a request for service by the computing system to the time the system has completed the report back to the user. More complex systems may be studied in terms of these parameters using queuing theory or through physical measurements of the system.

### Program-Preparation Systems

Most of the facilities provided for program preparation can be classified under two broad categories; (1) libraries of prepared program modules and (2) facilities for preparing new modules or programs. These modules may be combined to produce programs which may be inserted into the memory of the system for execution by the system loader. The system program which is used to combine these modules into a format acceptable to the system loader is called the **linkage editor.** The functions of the linkage editor are to locate and assemble the modules which are needed, and to position these modules relative to each other to produce a program which can be inserted in memory by the system loader for execution.

If a program is too large to be contained completely in memory, it can be divided into pieces called **segments.** The segments may also be assembled by the linkage editor. During execution, the segments which comprise the program are overlaid by the loader one at a time into the same area of memory and executed accordingly. Some modern computers provide hardware facilities for handling segments during program execution rather than by the linkage editor. These hardware facilities permit the program to be divided into units of fixed size, called **pages,** or units of variable size, called **segments,** or combination of these two. The advantage of this hardware is that it permits the user to develop programs which are independent of the size of the physical memory. Systems which use such facilities are known as **virtual memory** systems and are used with timesharing operating systems.

Functions which are commonly needed by the users are preprepared modules; they are stored in libraries on the file devices of the system for easy access. A computer **library** is faced with problems analogous to those of any library, that is, how to catalog, list, and advertise the items which are available so that a user may identify and select those items of possible interest. To aid the computer user, the preprepared modules are organized into packages which can be categorized by topic. Typically, a scientific computer library contains packages of statistics routines, higher mathematical functions, matrix routines, linear programming, etc.

Individual items stored in computer libraries may be selected and assembled into users' programs by the linkage editor. However, computer mathematical routines are only approximations to mathematical functions, and so each routine is designed as a trade-off of speed, accuracy, reliability, maintainability, flexibility, and memory-space economy. In cases where there is a major conflict in these design criteria, alternate routines are provided. The documentation of these routines should be consulted to select the best versions for building blocks in developing customized programs.

Program-preparation systems which are used to prepare new programs or modules are known categorically as language systems. A language system is composed of three parts: (1) the language, (2) a translator or compiler, and (3) a run-time system. The **language** is the symbols and the rules that are used to compose acceptable statements. Computer languages are more precise and have a simpler structure than natural languages. A clear separation is made between what is written to compose an acceptable statement, i.e., the syntax, and what is the meaning of the statement, i.e., the semantics. For a program to be executed on the computer, it must ultimately be composed of a sequence of the basic instructions of the computer. The statements written by the user are put on some input media such as punched cards or are typed on a keyboard device and entered into the machine. The translator, or **compiler,** is a computer program that accepts these statements as input and translates (or compiles) them into a sequence of basic computer instructions which form an executable version of the program. The **run-time program** is present in the memory of the machine when the program is being executed. The purpose of the run-time program is to perform services that may be required by the user's program during execution. For example, in case of a program fault the run-time system will identify the error and terminate the program in an orderly fashion.

In using a language system, the user should be aware of the distinction between translating and executing a program. However, for the user, the crucial part of a language system is the language, i.e., the part that is used to compose acceptable computer statements. Languages may be classified as procedure-oriented or problem-oriented. The difference is in the types of statements permitted in composing a computer program. With **procedure-oriented** languages, all the detailed steps must be specified by the user. These languages are usually characterized as being more verbose than problem-oriented languages but more flexible in dealing with a wider range of problems. **Problem-oriented** languages are more specialized and deal with a restricted class of problems. The elements of these languages are usually familiar to a knowledgeable user and so are easier to learn and use than procedure-oriented languages.

### Procedure-oriented Languages

The most elementary form of a procedure-oriented language is called an **assembler.** This language permits a computer program to be written directly as basic computer instructions using mnemonic operators and symbolic operands. These instructions, and any data used by the program, are converted by the assembler translator into a machine-representable form.

A further refinement of an assembler is to permit the use of macros. A **macro** is a sequence of computer instructions (or other macros) written in assembler format which are identified by an assigned name. A program written in the language of a macroassembler may include macro names as statements as well as basic assembler instructions. The sequence of instructions designated by the macro name is substituted in place of the macro name each time the macro name is encountered during the translation of the program.

Besides these basic language systems there exists a large spectrum of language systems which are used to write computer programs as a sequence of detailed steps. These languages are called higher-level languages since they permit more complex statements than are permitted in a macroassembler. An added advantage of higher-level languages is that they can be used on machines produced by different manufacturers or machines with different instruction repertoires. This feature permits programs to be written independent of the machine used for translation. However, a high degree of standardization in the language is needed in order to achieve this objective.

Higher-level languages may be further classified according to the types of applications for which they are especially well adapted. The range of languages, however, is limitless. The following examples are typical of some of the languages which have been developed.

In the commercial field the most popular language is **COBOL** (for COmmon Business-Oriented Language). This language is designed to permit the handling of complex information files which are found in business and data-processing problems. Its popularity stems from the fact that it has been standardized by a joint committee of users and manufacturers.

A series of languages have been developed for handling problems involving strings of text. These languages are called string-manipulation or list-processing languages. SNOBOL, $L^6$, and LISP are examples of such languages. Applications of these languages vary from problems of generating concordances to sophisticated symbolic-formula manipulation.

The procedure-oriented languages of most interest to engineers are those designed for problems involving complex or repetitive scientific calculations. PL/I (also used for business problems), ALGOL 60, ALGOL 68, BASIC, APL, and FORTRAN (FORmula TRANslator) are examples of such languages. **FORTRAN** is the most widely accepted language for problems of this type. This language has been standarized by ANSI (American National Standards Institute) under the sponsorship of BEMA (Business Equipment Manufacturers Association). Unfortunately, this standard is not always adhered to in writing FORTRAN source programs, and so some minor modifications are often required to translate programs written for another computer. Some dialect of FORTRAN exists for every commercially available machine.

### Problem-oriented Languages

Problem-oriented languages have been developed for every discipline. Such languages are too numerous to attempt to classify. Typically they deal with specialized applications such as the production of the control tapes for automatic control of machine tools (APT) or a whole gamut of civil engineering applications such as ICES, which includes many specialized languages (COGO, STRESS, etc.).

A class of problem-oriented languages that deserve special mention are those designed for solving problems in discrete simulation. GPSS, Simscript, and SIMULA are three of the most popular languages of this type. These languages are used to set up and solve problems dealing with discrete events such as problems in machine scheduling and queuing theory.

Problem-oriented languages make the computer directly accessible to the specialist with little training. This is achieved by permitting the user to describe problems to the computer in terms that are familiar in the discipline for which the language is designed. Table 5 lists a few of the many problem-oriented languages which have been developed for engineering problems.

**Table 5. Typical Engineering Problem-oriented Languages**

| Language | EE | CE | ChE | IE | ME |
|----------|----|----|----|----|----|
| MIMIC    | X  | X  | X  | X  | X  |
| CSSLIII  | X  | X  | X  | X  | X  |
| ECAP     | X  |    |    |    |    |
| SCEPTRE  | X  |    |    |    | X  |
| STRUDL   |    | X  |    |    | X  |
| BRIDGE   |    | X  |    |    | X  |
| COGO     |    | X  |    |    | X  |
| STRESS   |    | X  |    |    | X  |
| CHESS    |    |    | X  |    |    |

### Application of Procedure-oriented Languages

Many of the following applications can also be effectively handled using problem-oriented languages. The distinction is primarily that the solution techniques are built into the translator or run-time system of problem-oriented language systems while the solution techniques must be explicitly specified with procedure-oriented languages. Nevertheless, the following applications are typical for procedure-oriented languages.

A **simulation** (another word for model) of a system is used whenever it is desirable to watch the succession of many interrelated events or when there is interplay between the

system itself and outside forces. Examples are problems in man-machine interaction and in the modeling of business systems. Typical man-machine problems are the servicing of automatic equipment by a crew of operators and the repair by a maintenance crew of equipment subject to unpredictable breakdown. A business-system model consists of the interactions between businesses and the rest of the economy, e.g., competitive buyers in a raw-materials market or competitive manufacturers marketing their products. Business models often involve transportation and warehousing of commodities.

Physical or chemical systems may also be modeled; e.g., for the application of automatic control valves in pipelines, a model is made with portions which represent the attributes of the control system, the valve, the piping system, and the fluid. Such a model, when tested, can indicate whether fluid hammer will occur or whether valve action is fast enough, or it can be used to demonstrate pressure and temperature conditions in the fluid when subject to control by the valve.

One of the major uses of the digital computer is **data reduction.** In installations where large amounts of data are recorded and kept, it is often advisable to reduce the amount of these data by ganging the data together, by averaging the data before they are stored, or by converting the data to a more useful form, e.g., a form more appropriate to storage or future processing.

A technique much used by engineers is the solving of equations by **trial and error.** A computer lends itself well to such solutions because of its speed and accuracy. It can try thousands of solutions in a very short time, thus freeing technical manpower for evaluation, rather than calculation, of these trials. The computer itself can often be programmed to evaluate these trials.

Many standard mathematical techniques such as **relaxation** can be practicably applied only through the use of a computer. A typical relaxation problem may require the setting up of a grid of hundreds of points. Such a problem lends itself ideally to the digital computer because similar calculations are done at many points within the grid.

Problems that can be formulated into **matrix** form can be readily solved on the computer. Good matrix routines are available as preprepared modules. The user's task is to select those routines that can best be applied to the specific problem and to prepare the input for these routines.

**Optimization** problems find wide applications in engineering, and in recent years many excellent procedures have been developed for solving these problems on computers. Mathematically these problems are often formulated as a problem in finding the maximum or minimum of a nonlinear function. This results in the need to calculate the value of complex functions repetitively many times in order to find a reasonable solution. The speed of the computer is a great advantage in such calculations.

There is a class of problems which is characterized by values or cost functions which plot as families of parallel straight lines. These are functions of many variables, perhaps hundreds. The fact that the value function is linear enables a particular method of solution known as **linear programming** to be considered. The area of solution to a linear-programming problem is circumscribed by limits on the magnitudes of the variables. These limits are imposed by the real world and must also plot as straight lines, usually in many dimensions. The statements of these limits describe a space bounded by them which is called the **solution space.** All linear-programming algorithms search about in the corners of this solution space to find the point at which the cost function is minimized or the value function maximized. All modern computer-service centers have available large, efficient, high-speed, standard programs for solving linear-programming problems.

Also available with most digital computers are programs for the solution of **statistical problems.** The most widely used and the cheapest and simplest both to formulate and to solve are programs of linear statistics, which involve a technique known as **regression,** or **least squares.** In some areas, however, linear statistics will not suffice. For these problems, there exist more specialized programs which involve **nonlinear** statistical techniques. This type of problem is more difficult and expensive both to state and to solve, and in certain cases no solution can be obtained.

A class of problems susceptible to **Monte Carlo** solution became practical with the advent of digital computers. The Monte Carlo technique is used when the list of possible conditions in which the activity under investigation can find itself is too large or too complex to be easily stated. As its name implies, the Monte Carlo technique uses random numbers (which are easily made available to the computer) to determine statistically what conditions exist or what changes will take place. A large number of solutions is then run, and statistical inferences are drawn.

Computers have been widely used by engineering organizations (with individuals of widely varying skills) in such diverse areas as process control, economic evaluation, industrial-engineering decisions, cost estimating, and curve and graph plotting.

### Application of Problem-oriented Languages

For the engineering type of application, two basic approaches are used by problem-oriented languages in problem solving. These are illustrated in Figs. 3 and 4.

**Fig. 3**  Illustration of MIMIC program.

| MIMIC statements (punched onto successive cards) | Explanation |
|---|---|
| $\cdot DY2 = (1 - Z * DY1 - K * Y)/M$ | Differential equation to be solved. "$*$" is used for multiplication and $DY2$, $DY1$ and $Y$ are defined mnemonics for $\ddot{y}$, $\dot{y}$, and $y$ |
| $DY1 = INT(DY2,0.)$<br>$Y = INT(DY1,0.)$ | $INT(A,B)$ is used to perform integration. It forms successive values of $B + \int A dt$. |
| $FIN(T,10.)$ | $T$ is a reserved name representing the independent variable. This statement will terminate execution when $T \geqslant 10$. |
| $CON(M,K,Z)$ | Values must be furnished for $M$, $K$, and $Z$. A card with these values must appear after the END card. |
| $PLO(T,DY2)$<br>$PLO(T,DY1)$<br>$PLO(T,Y)$<br>END | Three point plots are produced on the line printer; $\ddot{y}$, $y$, and $y$ vs. $t$. |
| | Necessary last card of program. |

The approach illustrated in Fig. 3 is to set up the computer program directly from mathematical equations. In this example the **MIMIC** language is used to write the computer program for the solution of the initial-value problem

$$M\ddot{y} + Z\dot{y} + Ky = 1 \qquad \dot{y}(0) = y(0) = 0$$

MIMIC is a digital simulation language used to solve systems of ordinary differential equations. The solution technique is similar to that used on analog computers. The key step in setting up the solution is to isolate the highest-order derivative on the left-hand side of the equation and equate it to an expression composed of the remaining terms. In the problem illustrated in Fig. 3, this results in the equation

$$\ddot{y} = (1 - Z\dot{y} - Ky)/M$$

The solution function and the lower-order derivatives are generated successively by integrating the highest-order derivative. The MIMIC language permits the user to write this solution setup in a natural manner.

The alternate approach used in problem-oriented languages is to permit the setup to be described to the computer directly from the block diagram of the problem to be solved. **SCEPTRE,** the language illustrated in Fig. 4, uses this approach. SCEPTRE statements are written under headings and subheadings which identify the type of component being described. This language may be applied to network problems composed of electrical digital-logic elements, mechanical-translation or rotational elements, or transfer-function blocks.

**Fig. 4**  Illustration of SCEPTRE program. (a) Problem to be solved; (b) SCEPTRE program.

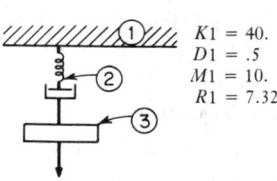

| | |
|---|---|
| $K1 = 40.$<br>$D1 = .5$<br>$M1 = 10.$<br>$R1 = 7.32$ | A node is assigned to:<br>• ground<br>• any mass<br>• point between two elements<br><br>The prefix of element name specifies its type; i.e., $M$ for mass, $K$ for spring, $D$ for damper, and $R$ for force. |

*(a)*

| SCEPTRE statements (punched onto successive cards) | Explanation |
|---|---|
| MECHANICAL<br> DESCRIPTION<br>ELEMENTS<br>$M1, 1 - 3 = 10.$<br>$K1, 1 - 2 = 40.$<br>$D1, 2 - 3 = .5$<br>$R1, 1 - 3 = 7.32$ | Specifies the elements and their position in the diagram using the node numbers. |
| OUTPUT<br>$SM1, VM1$ | Results are listed on the line printer. Prefix on the element specifies the quantity to be listed; $S$ for displacement, $V$ for velocity. |
| RUN CONTROL<br>STOPTIME = 10. | TIME is reserved name for independent variable. Statement will terminate execution of program when TIME is equal to or greater than 10. |
| END | Necessary last card. |

*(b)*

The translator for this language system develops and sets up the equations directly from the network diagram, and so relieves the user of the mathematical aspects of the problem.

## ANALOG COMPUTERS

### Components

Present-day analog computers are generally electrical. Most computers of this type are built into other equipment for performing specialized and dedicated computations. However, one form of the analog computer, called a differential analyzer, is used for setting up and solving complex differential equations. The basic components of a differential analyzer are shown as functional elements in Fig. 5. Typically, all problem variables are represented as dc voltages in the range $\pm 100$ V.

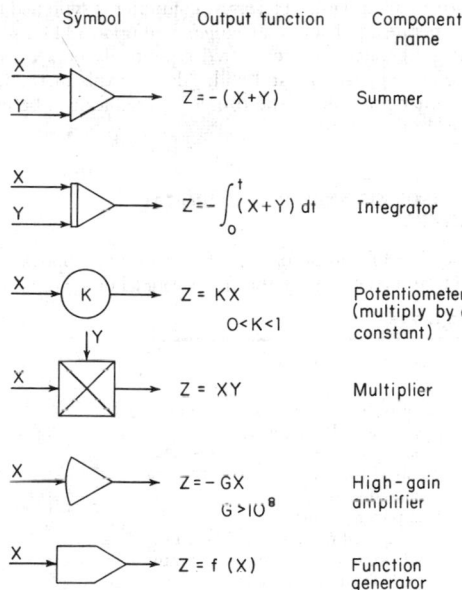

Fig. 5   Analog-computer components.

A **summer** is used to perform addition; the sign of the answer is opposite to that of the input components. To perform subtraction, the sign of the subtrahend must be changed (e.g., run through another summer with nothing added to it). An **integrator,** which generally integrates with respect to time, is the component which gives the analog computer its ability to solve problems stated in terms of differential equations. Also, the integrator requires a mechanism to reset it to its initial conditions. A **potentiometer** (also called a **pot)** is used to multiply an incoming variable (a voltage) by a constant factor which is less than 1. A **multiplier** multiplies two variables together.

A **function generator** is a specially prepared device to provide an arbitrary function $f(X)$ for a given input signal $X$. For Monte Carlo solutions, a signal generator providing random output is used.

A **high-gain** amplifier produces an output which is over $10^8$ times the sum of the input voltages (with the sign reversed). A high-gain amplifier is a part of both the summer and the integrator, which differ only in the way the high-gain ampli-

fier is connected and in some additional components. Often the hardware representing an amplifier can be used either as a summer or as an integrator. It is cheaper to buy high-gain amplifiers that do not have this multiple use; so there are summers and integrators that are not interchangeable merely by using different connections. The use of the high-gain amplifier for division is illustrated by Fig. 6. Voltage $X$ goes

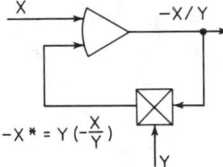

Fig. 6   High-gain amplifier used for division.

in. Assume that $(-X/Y)$ comes out. The $(-X/Y)$ goes to the multiplier; $(-X/Y) \cdot (Y) = -X^*$ comes out of the multiplier and goes into the amplifier to be added to $X$. $(-X/Y)$ is the only output value for which $X^*$ will balance $X$, and no current will flow. Any unbalance between $X$ and $-X^*$ will be raised by $10^8$ by the amplifier and fed back into the multiplier, where it goes back in with reversed sign.

### Using Analog Computers

Consider the equation

$$a_1 \frac{d^3x}{dt^3} + a_2 \frac{d^2x}{dt^2} + a_3 \frac{dx}{dt} + a_4 x = K \tag{1}$$

This can be converted to

$$\frac{d^3x}{dt^3} = -\frac{a_2}{a_1}\frac{d^2x}{dt^2} - \frac{a_3}{a_1}\frac{dx}{dt} - \frac{a_4}{a_1}x + \frac{K}{a_1} \tag{2}$$

To integrate, run the signal into integrators as often as necessary and get a setup as shown in Fig. 7. To set up Eq. (2), a circuit as shown in Fig. 8a is required. Figure 8a shows the use of potentiometers to multiply the derivatives by their respective coefficients, of summers 2 and 3 to change the sign of variables, and of summer 1 to sum up the component parts of $d^3x/dt^3$.

The translations of algebraic and differential equations into an analog-computer diagram appear straightforward. This is, however, not the case. In the simple example above, all problems associated with scaling have been neglected. **Scaling** means that all voltages must be between $+100$ and $-100$ V. Most voltmeters used for readout of answers are precise to only 0.1 V. Therefore, all answers that must be read out for the user must be scaled to avoid becoming meaningless. In addition, most of the components are precise to only 1 part in $10^4$, so that if at any point in the network a variable becomes less than that part of the total voltage, it can either be completely lost or be considered at twice its correct value. This error can then be propagated to the point that the so-called *answers* represent errors or noise and not the problem under investigation. The solution to this problem, particularly, requires skillful analog programming.

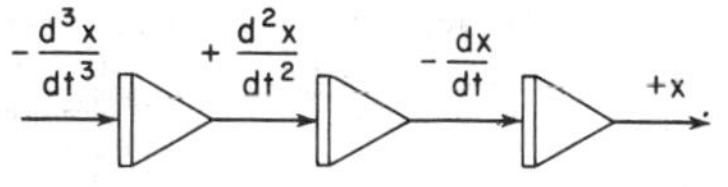

Fig. 7   Integration in an analog computer.

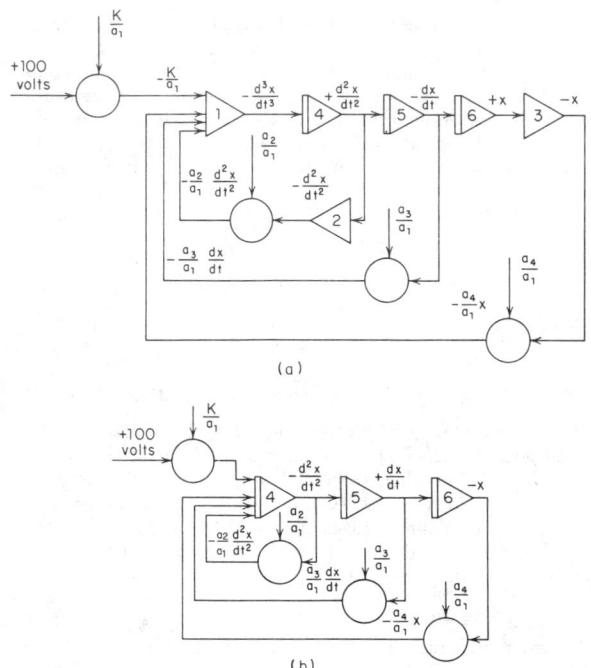

**Fig. 8**  Diagram representing Eq. (2). (*a*) Direct solution; (*b*) economical solution.

Another problem that usually confronts the analog user is that each symbol on his flow chart represents some electronic hardware in the racks of the computer. In many problems, there are not nearly enough pieces. Much of the cost of obtaining answers may well be spent in overcoming this limitation.

A typical way to save components is to remove excess summers and do the summing at the integrators. For example, in Fig 8*a* removing summer 1 and summing at integrator 4 changes the signs of all quantities involved, permitting removal of summers 2 and 3 but putting one in as No. 7 (see Fig. 8*b*), with a net gain of two summers.

The results of an analog simulation may be recorded in several ways. The most common method is with an **X-Y plotter,** which allows the plotting of one variable vs. another as a graph. Another common form of graphic display is on a **cathode-ray tube,** which requires the use of the repetitive-operation feature called **rep-op.** Here, the computer is switched from reset to operate and back at a frequency of up to 50 Hz, so that the voltages created can be plotted repetitively on a cathode-ray tube and appear to stand still. Also, by this method, the effect of changing one of the problem variables can be easily and quickly determined.

## COMPARISON OF ANALOG AND DIGITAL COMPUTERS

The selection of computer type for a specific application is often determined solely by the application. For a large class of

**Table 6. Contrasting Factors of Analog and Digital Computers**

| Factors | Digital | Analog |
|---|---|---|
| Solution setup | Requires accurate and detailed specification of the sequence of operations needed for problem solutions | Intuitively easy for engineers and scientists. Requires designing analogous circuit of problem to be solved from a few different components |
| Machine setup | After sequence of operations are proved correct, they may be stored in the computer library and reused on demand to solve additional problems | Components must be physically connected through patchboard which is inserted into computer to execute the solution |
| Versatility | Can be programmed to solve a large range of scientific and commercial problems involving numerical and nonnumerical information | Usually limited to network problems and differential equations |
| Speed | Operate sequentially, usually slower than analogs | Operates in real time, usually faster than digitals |
| Cost | More expensive unit and system costs than analogs | Low production costs and will often be used as special-purpose computers |
| Modularity | Well suited for composing problem solutions from pretested modules | Well suited for production circuits integrated into real-time applications |
| Interfacing | Requires complex analog-to-digital and digital-to-analog conversion to interface mechanisms | Offers a convenient interface to other equipment in the configuration |
| Data formats | May represent both numerical and nonnumerical information | Limited to numerical data |
| Accuracy | Many different accuracies are possible. Accuracy equivalents of 7 to 18 significant decimal digits are common for commercially available computers | Limited to measurement accuracy. Accuracy of 1 to 0.1% of full scale is commonly quoted for commercially available equipment |

engineering problems, however, either type of computer may be used. The selection, in this case, of the computer type that is most appropriate is based on a trade-off of a number of interdependent factors. These factors are listed in Table 6 together with how they differ for digital and analog computers.

## CALCULATORS

Desk and hand calculators are available with almost every conceivable combination of attributes. They come desk-top or hand-held, with or without paper print tape for recording answers, and will perform a wide variety of functions. Some are programmable so that they can remember calculation steps once performed. On others, the program can be stored on plastic magnetic strips which can be read in later to reload the program. Many complicated programs are available from the calculator manufacturer. Many varieties of calculators are available, some scientific and some business. They have different collections of function keys appropriate to their intended use.

Function keys include addition, subtraction, multiplication, division, square roots, squares, factorials, reciprocals, conversions to and from SI (metric) units, conversion between degrees and radians, logarithms and exponentials (base 10 and base $e$), raising to powers, elementary statistical regression, percentages, sign reversals, and generation of SI constants (e.g., 2.54 cm/in). Intermediate answers can be stored for later recall. Trigonometric and inverse trigonometric functions are available for both degree and radian measure.

All the calculators have a readout consisting of a series of LEDs (light emitting diodes) which display the digits, a decimal point, and a sign. Some units use scientific notation for very large or very small numbers (e.g., 3.1216114 −12 for 0.0000000000031216114). Some displays are hard to see in bright light or in light surroundings; others can be easily seen only at a very narrow viewing angle. The number of digits stored and displayed varies widely among models.

Nearly all units come with a battery charger and can be supplied with rechargeable batteries. These usually suffice for several hours of use between rechargings. Calculator prices have been reduced drastically during the past several years despite vastly increased capability, and are expected to continue to decrease in the forseeable future.

# Mechanics of Solids and Fluids

BY

**ROBERT F. STEIDEL, JR.** *Professor of Mechanical Engineering, University of California, Berkeley.*
**DUDLEY D. FULLER** *Professor of Mechanical Engineering, Columbia University.*
**J. W. MURDOCK** *Associate Professor, Mechanical Engineering & Mechanics Dept., Drexel University.*

## MECHANICS OF SOLIDS
### by Robert F. Steidel, Jr.

Physical Mechanics ............................................. 3-2
Systems and Units of Measurements ...................... 3-2
Statics of Rigid Bodies ...................................... 3-3
Center of Gravity ............................................. 3-7
Moment of Inertia ........................................... 3-8
Kinematics ..................................................... 3-11
Dynamics of Particles ....................................... 3-16
Work and Energy ............................................ 3-20
Impulse and Momentum ................................... 3-21
Gyroscopic Motion and the Gyroscope ................. 3-22

## FRICTION
### by Dudley D. Fuller

Static and Sliding Coefficients of Friction ............... 3-24
Rolling Friction .............................................. 3-28
Friction of Machine Elements ............................. 3-28

## MECHANICS OF FLUIDS
### by J. W. Murdock

Fluids and Other Substances ............................... 3-33
Fluid Properties .............................................. 3-34
Fluid Statics .................................................. 3-37
Fluid Kinematics ............................................. 3-40
Fluid Dynamics ............................................... 3-41
Dimensionless Parameters .................................. 3-46
Dynamic Similarity ........................................... 3-47
Dimensional Analysis ........................................ 3-49
Forces on Immersed Objects ............................... 3-52
Flow in Pipes ................................................. 3-53
Piping Systems ............................................... 3-57
ASME Pipeline Flowmeters ................................ 3-60
Pitot Tubes ................................................... 3-64
ASME Weirs .................................................. 3-65
Open-Channel Flow .......................................... 3-66
Flow of Liquids from Tank Openings ..................... 3-68
Water Hammer ............................................... 3-69

# MECHANICS OF SOLIDS
## by Robert F. Steidel, Jr.

REFERENCES: Beer and Johnston, "Mechanics for Engineers," McGraw-Hill. Halfman, "Dynamics," Addison-Wesley. Ham and Crane, "Mechanics of Machinery," McGraw-Hill. Higdon and Stiles, "Engineering Mechanics," Prentice-Hall. Holowenko, "Dynamics of Machinery," Wiley. Housner and Hudson, "Applied Mechanics," Van Nostrand. Meriam, "Mechanics," Wiley. Synge and Griffith, "Principles of Mechanics," McGraw-Hill. Timoshenko and Young, "Advanced Dynamics," McGraw-Hill. Timoshenko and Young, "Engineering Mechanics," McGraw-Hill.

## PHYSICAL MECHANICS

### Definitions

**Force** is the action of one body on another which will cause acceleration of the second body unless acted on by an equal and opposite action counteracting the effect of the first body. It is a **vector** quantity.

**Time** is a measure of the sequence of events. In newtonian mechanics it is an absolute quantity. In relativistic mechanics it is relative to the *frames of reference* in which the sequence of events is observed. The common unit of time is the second.

**Inertia** is that property of matter which causes a resistance to any change in the motion of a body.

**Mass** is a quantitative measure of *inertia*.

**Acceleration of Gravity** Every object which falls in a vacuum at a given position on the earth's surface will have the same acceleration $g$. Accurate values of the acceleration of gravity as measured *relative* to the earth's surface include the effect of the earth's rotation and flattening at the poles. The international gravity formula for the acceleration of gravity at the earth's surface is $g = 32.0881(1 + 0.005305 \sin^2 \phi - 0.0000059 \sin^2 2\phi)$ ft/s², where $\phi$ is latitude in degrees. For extreme accuracy, the local acceleration of gravity must also be corrected for the presence of large water or land masses and for height above sea level. The absolute acceleration of gravity for a nonrotating earth discounts the effect of the earth's rotation and is rarely used, except outside the earth's atmosphere. If $g_0$ represents the absolute acceleration at sea level, the absolute value at an altitude $h$ is $g = g_0 R^2/(R + h)^2$, where $R$ is the radius of the earth, approximately 3,960 mi (6,373 km).

**Weight** is the resultant force of attraction on the mass of a body due to a gravitational field. On the earth, units of weight are based upon an acceleration of gravity of 32.1740 ft/s² (9.80665 m/s²).

**Linear momentum** is the product of mass and the linear velocity of a particle and is a vector. The moment of the linear-momentum vector about a fixed axis is the **angular momentum** of the particle about that fixed axis. For a rigid body rotating about a fixed axis, angular momentum is defined as the product of moment of inertia and angular velocity, each measured about the fixed axis.

An increment of **work** is defined as the product of an incremental displacement and the component of the force vector in the direction of the displacement or the component of the displacement vector in the direction of the force. The increment of work done by a couple acting on a body during a rotation of $d\theta$ in the plane of the couple is $dU = M\, d\theta$.

**Energy** is defined as the capacity of a body to do work by reason of its motion or configuration (see **Work and Energy**).

A **vector** is a directed line segment that has both magnitude and direction. In script or text, a vector is distinguished from a scalar $V$ by a boldface-type **v**. The magnitude of the scalar is the magnitude of the vector, $V = |\mathbf{V}|$.

A **frame of reference** is a specified set of geometric conditions to which other locations, motion, and time are referred. In newtonian mechanics, the fixed stars are referred to as the **primary (inertial) frame of reference**. Relativistic mechanics denies the existence of primary reference frame and holds that all reference frames must be described relative to each other.

## SYSTEMS AND UNITS OF MEASUREMENTS

In *absolute systems,* the units of **length, mass,** and **time** are considered fundamental quantities, and all other units including that of **force** are derived.

In *gravitational systems,* the units of **length, force,** and **time** are considered fundamental qualities, and all other units including that of **mass** are derived.

In the SI system of units, the unit of mass is the kilogram (kg) and the unit of length is the metre (m). A force of one newton (N) is derived as the force that will give 1 kilogram an acceleration of 1 m/s².

In the English engineering system of units, the unit of mass is the pound mass (lbm) and the unit of length is the foot (ft). A force of one pound (1 lbf) is the force that gives a pound mass (1 lbm) an acceleration equal to the standard acceleration of gravity on the earth, 32.1740 ft/s². A slug is the mass that will be accelerated 1 ft/s² by a force of 1 lbf. Therefore, 1 slug = 32.1740 lbm. When described in the gravitational system, mass is a derived unit, being the constant of proportionality between force and acceleration, as determined by Newton's second law.

### General Laws

NEWTON'S LAWS

**I.** If a balanced force system acts on a particle at rest, it will remain at rest. If a balanced force system acts on a particle in motion, it will remain in motion in a straight line without acceleration.

**II.** If an unbalanced force system acts on a particle, it will accelerate in proportion to the magnitude and in the direction of the resultant force.

**III.** When two particles exert forces on each other, these forces are equal in magnitude, opposite in direction, and collinear.

**Fundamental Equation** The basic relation between mass, acceleration, and force is contained in Newton's second law of motion. As applied to a particle of mass, $\mathbf{F} = m\mathbf{a}$, force = mass × acceleration. This equation is a vector equation, since the direction of $\mathbf{F}$ must be the direction of $\mathbf{a}$, as well as having $\mathbf{F}$ equal in magnitude to $m\mathbf{a}$. An alternative form of Newton's second law states that the resultant force is equal to the time rate of change of momentum, $\mathbf{F} = d(m\mathbf{v})/dt$.

**Law of the Conservation of Mass**  The mass of a body remains unchanged by any ordinary physical or chemical change to which it may be subjected.

**Law of the Conservation of Energy**  The principle of conservation of energy requires that the total mechanical energy of a system remain unchanged if it is subjected only to forces which depend on position or configuration.

**Law of the Conservation of Momentum**  The linear momentum of a system of bodies is unchanged if there is no resultant external force on the system. The angular momentum of a system of bodies about a fixed axis is unchanged if there is no resultant external moment about this axis.

**Law of Mutual Attraction (Gravitation)**  Two particles attract each other with a force $F$ proportional to their masses $m_1$ and $m_2$ and inversely proportional to the square of the distance $r$ between them, or $F = km_1m_2/r^2$, in which $k$ is the gravitational constant. The value of the gravitational constant is $k = 6.673 \times 10^{-11}$ m³/kg·s² in SI or absolute units, or $k = 3.44 \times 10^{-8}$ ft⁴ lb⁻¹ s⁻⁴ in engineering gravitational units. It should be pointed out that the unit of force $F$ in the SI system is the **newton** and is derived, while the unit force in the gravitational system is the **pound-force** and is a fundamental quantity.

EXAMPLE.  Each of two solid steel spheres 6 in diam will weigh 32.0 lb on the earth's surface. This is the force of attraction between the earth and the steel sphere. The force of mutual attraction between the spheres if they are just touching is 0.000000136 lb.

## STATICS OF RIGID BODIES

### General Considerations

If the forces acting on a rigid body do not produce any acceleration, they must neutralize each other, i.e., form a **system of forces in equilibrium.** Equilibrium is said to be **stable** when the body with the forces acting upon it returns to its original position after being displaced a very small amount from that position; **unstable** when the body tends to move still farther from its original position than the very small displacement; and **neutral** when the forces retain their equilibrium when the body is in its new position.

**External and Internal Forces**  The forces by which the individual particles of a body act on each other are known as internal forces. All other forces are called external forces. If a body is supported by other bodies while subject to the action of forces, deformations and forces will be produced at the points of support or contact and these internal forces will be distributed throughout the body until equilibrium exists and the body is said to be in a state of tension, compression, or shear. The forces exerted by the body on the supports are known as **reactions.** They are equal in magnitude and opposite in direction to the forces with which the supports act on the body, known as **supporting forces.** The supporting forces are external forces applied to the body.

In considering a body at a definite section, it will be found that all the internal forces act in pairs, the two forces being equal and opposite. The external forces act singly.

**General Law**  When a body is at rest, the forces acting externally to it must form an equilibrium system. This law will hold for any part of the body, in which case the forces acting at any section of the body become external forces when the part on either side of the section is considered alone. In the case of a **rigid body,** any two forces of the same magnitude, but acting in opposite directions in any straight line, may be added or removed without change in the action of the forces acting on the body, provided the strength of the body is not affected.

### Composition, Resolution, and Equilibrium of Forces

The **resultant** of several forces acting at a point is a force which will produce the same effect as all the individual forces acting together.

**Forces Acting on a Body at the Same Point**  The resultant $R$ of two forces $F_1$ and $F_2$ applied to a rigid body at the same point is represented in magnitude and direction by the diagonal of the parallelogram formed by $F_1$ and $F_2$ (see Figs. 1 and 2).

$$R = \sqrt{F_1^2 + F_2^2 + 2F_1F_2 \cos a}$$

$$\sin a_1 = (F_2 \sin a)/R \qquad \sin a_2 = (F_1 \sin a)/R$$

When $a = 90°$, $R = \sqrt{F_1^2 + F_2^2}$, $\sin a_1 = F_2/R$, and $\sin a_2 = F_1/R$.

When $a = 0°$, $R = F_1 + F_2$  } Forces act in same
When $a = 180°$, $R = F_1 - F_2$ } straight line.

A **force $R$ may be resolved into two component forces** intersecting anywhere on $R$ and acting in the same plane as $R$, by the reverse of the operation shown by Figs. 1 and 2; and by repeating the operation with the components, $R$ may be resolved into any number of component forces intersecting $R$ at the same point and in the same plane.

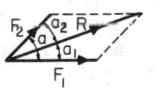 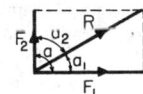

**Resultant of Any Number of Forces Applied to a Rigid Body at the Same Point**  Resolve each of the given forces $F$ into components along three rectangular coordinate axes. If $A$, $B$, and $C$ are the angles made with $XX$, $YY$, and $ZZ$, respectively, by any force $F$, the components will be $F \cos A$ along $XX$, $F \cos B$ along $YY$, $F \cos C$ along $ZZ$; add the components of all the forces along each axis algebraically and obtain $\Sigma F \cos A = \Sigma X$ along $XX$, $\Sigma F \cos B = \Sigma Y$ along $YY$, and $\Sigma F \cos C = \Sigma Z$ along $ZZ$.

The resultant $R = \sqrt{(\Sigma X)^2 + (\Sigma Y)^2 + (\Sigma Z)^2}$. The angles made by the resultant with the three axes are $A_r$ with $XX$, $B_r$ with $YY$, $C_r$ with $ZZ$, where

$$\cos A_r = \Sigma X/R \qquad \cos B_r = \Sigma Y/R \qquad \cos C_r = \Sigma Z/R$$

The **direction** of the **resultant** can be determined by plotting the algebraic sums of the components.

**If the forces are all in the same plane,** the components of each of the forces along one of the three axes (say, $ZZ$) will be 0; i.e., angle $C_r = 90°$ and $R = \sqrt{(\Sigma X)^2 + (\Sigma Y)^2}$, $\cos A_r = \Sigma X/R$, and $\cos B_r = \Sigma Y/R$.

**For equilibrium,** it is necessary that $R = 0$; i.e., $\Sigma X$, $\Sigma Y$, and $\Sigma Z$ must each be equal to zero.

**General Law**   In order that a number of forces acting at the same point shall be in equilibrium, the algebraic sum of their components along any *three* coordinate axes must each be equal to zero. When the forces all act in the same plane, the algebraic sum of their components along any *two* coordinate axes must each equal zero.

**When the Forces Form a System in Equilibrium**   Three **unknown forces** can be determined if the lines of action of the forces are *all* known and are in different planes. If the forces are all in the same plane, the lines of action being known, only *two* unknown forces can be determined. If the lines of action of the unknown forces are *not* known, only *one* unknown force can be determined in either case.

### Couples and Moments

**Couple**   Two parallel forces of equal magnitude (Fig. 3) which act in opposite directions and are not collinear form a couple. **A couple cannot be reduced to a single force.**

**Displacement and Change of a Couple**   The forces forming a couple may be moved about and their magnitude and direction changed, provided they always remain parallel to each other and remain in either the original plane or one parallel to it, and provided the product of one of the forces and the perpendicular distance between the two is constant and the direction of rotation remains the same.

**Moment of a Couple**   The moment of a couple is the product of the magnitude of one of the forces and the perpendicular distance between the lines of action of the forces. $Fa =$ moment of couple; $a =$ arm of couple. If the forces are measured in pounds and the distance $a$ in feet, the **unit of rotation moment** is the foot-pound. If the force is measured in kilograms and the distance in metres, the unit is the metre-kilogram. In the cgs system the unit of rotation moment is 1 cm-dyne.

**Rotation moments of couples** acting in the same plane are conventionally considered to be positive for counterclockwise moments and negative for clockwise moments, although it is only necessary to be consistent within a given problem. The magnitude, direction, and sense of rotation of a couple are completely determined by its moment axis, or moment vector, which is a line drawn perpendicular to the plane in which the couple acts, with an arrow indicating the direction from which the couple will appear to have right-handed rotation; the length of the line represents the magnitude of the moment of the couple. See Fig. 4, in which $AB$ represents the magnitude of the moment of the couple. Looking along the line in the direction of the arrow, the couple will have right-handed rotation in any plane perpendicular to the line.

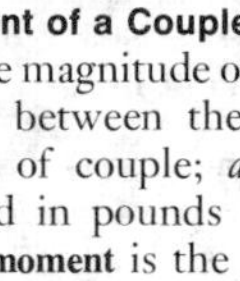

**Fig. 3** 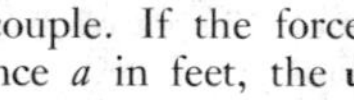          **Fig. 4**

**Composition of Couples**   Couples may be combined by adding their moment vectors geometrically, in accordance with the parallelogram rule, in the same manner in which forces are combined.

**Couples lying in the same or parallel planes are added algebraically.** Let $+28$ lbf·ft ($+38$ N·m), $-42$ lbf·ft ($-57$ N·m), and $+70$ lbf·ft ($95$ N·m) be the moments of three couples in the same or parallel planes; their resultant is a single couple lying in the same or in a parallel plane, whose moment is $\Sigma M = +28 - 42 + 70 = +56$ lbf·ft ($\Sigma M = +38 - 57 + 95 = 76$ N·m).

**If the polygon formed by the moment vectors of several couples closes itself, the couples form an equilibrium system. Two couples will balance each other** when they lie in the same or parallel planes and have the same moment in magnitude, but opposite in sign.

**Combination of a Couple and a Single Force in the Same Plane** (Fig. 5)   Given a force $F = 18$ lbf ($80$ N) acting as shown at distance $x$ from $YY$, and a couple whose moment is $-180$ lbf·ft ($244$ N·m) in the same or parallel plane, to find the resultant. A couple may be changed to any other couple in the same or a parallel plane having the same moment and same sign. Let the couple consist of two forces of 18 lbf (80 N) each and let the arm be 10 ft (3.05 m). Place the couple in such a manner that one of its forces is opposed to the given force at $p$. This force of the couple and the given force being of the same magnitude and opposite in direction will neutralize each other, leaving the other force of the couple acting at a distance of 10 ft (3.05 m) from $p$ and parallel and equal to the given force 18 lbf (80 N).

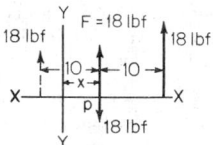

**Fig. 5**

**General Rule**   The resultant of a couple and a single force lying in the same or parallel planes is a single force, equal in magnitude, in the same direction and parallel to the single force, and acting at a distance from the line of action of the single force equal to the moment of the couple divided by the single force. The moment of the resultant force about any point on the line of action of the given single force must be of the same sense as that of the couple, positive if the moment of the couple is positive, and negative if the moment of the couple is negative. If the moment of the couple in Fig. 5 had been $+$ instead of $-$, the resultant would have been a force of 18 lbf (80 N) acting in the same direction and parallel to $F$, but at a distance of 10 ft (3.05 m) to the left of it (shown dotted), making the moment of the resultant about any point on $F$ positive.

**To effect a parallel displacement of a single force** $F$ over a distance $a$, a couple whose moment is $Fa$ must be added to the system. The sense of the couple will depend upon which way it is desired to displace force $F$.

**The moment of a force with respect to a point** is the product of the force $F$ and the perpendicular distance from the point to the line of action of the force.

**The Moment of a Force with Respect to a Straight Line**   If the force is resolved into components parallel and perpendicular to the given line, the moment of the force with respect to the line is the product of the magnitude of the perpendicular component and the distance from its line of action to the given line.

### Forces with Different Points of Application

**Composition of Forces**  If each force **F** is resolved into components parallel to three rectangular coordinate axes $XX$, $YY$, and $ZZ$, the magnitude of the resultant is $R = \sqrt{(\Sigma X)^2 + (\Sigma Y)^2 + (\Sigma Z)^2}$, and its line of action makes angles $A_r$, $B_r$, and $C_r$ with axes $XX$, $YY$, and $ZZ$, where $\cos A_r = \Sigma X/R$, $\cos B_r = \Sigma Y/R$, and $\cos C_r = \Sigma Z/R$; and there are three couples which may be combined by their moment vectors into a single resultant couple having the moment $M_r = \sqrt{(M_x)^2 + (M_y)^2 + (M_z)^2}$, whose moment vector makes angles of $A_m$, $B_m$, and $C_m$ with axes $XX$, $YY$, and $ZZ$, such that $\cos A_m = M_x/M_r$, $\cos B_m = M_y/M_r$, and $\cos C_m = M_z/M_r$. If this single resulting couple is in the same plane as the single resulting force at the origin or a plane parallel to it, the system may be reduced to a single force **R** acting at a distance from **R** equal to $M_r/R$. If the couple and force are not in the same or parallel planes, it is impossible to reduce the system to a single force. If **R** = 0, i.e., if $\Sigma X$, $\Sigma Y$, and $\Sigma Z$ all equal zero, the system will reduce to a single couple whose moment is **M**$_r$. If **M**$_r$ = 0, i.e., if $M_x$, $M_y$, and $M_z$ all equal zero, the resultant will be a single force **R**.

**When the forces are all in the same plane,** cosine of one of the angles $A_r$, $B_r$, or $C_r$, = 0, say, $C_r = 90°$. Then $R = \sqrt{(\Sigma X)^2 + (\Sigma Y)^2}$, $M_r = \sqrt{M_x^2 + M_y^2}$, and the final resultant is a force equal and parallel to $R$, acting at a distance from $R$ equal to $M_r/R$.

A system of forces in the same plane can always be replaced by either a couple or a single force. If **R** = 0 and **M**$_r$ > 0, the resultant is a couple. If **M**$_r$ = 0 and **R** > 0, the resultant is a single force.

**A rigid body is in equilibrium** when acted upon by a system of forces whenever R = 0 and $M_r$ = 0, i.e., when the following six conditions hold true: $\Sigma X = 0$, $\Sigma Y = 0$, $\Sigma Z = 0$, $M_x = 0$, $M_y = 0$, and $M_z = 0$. When the system of forces is in the same plane, equilibrium prevails when the following three conditions hold true: $\Sigma X = 0$, $\Sigma Y = 0$, $\Sigma M = 0$.

### Forces Applied to Support Rigid Bodies

The external forces in equilibrium acting upon a body may be statically determinate or indeterminate according to the number of unknown forces existing. When the forces are all in the same plane and act at a common point, two unknown forces may be determined if their lines of action are known, one if unknown.

When the forces are all in the same plane and are parallel, two unknown forces may be determined if the lines of action are known, one if unknown.

When the forces are anywhere in the same plane, three unknown forces may be determined if their lines of action are known, if they are not parallel or do not pass through a common point; if the lines of action are unknown, only one unknown force can be determined.

If the forces all act at a common point but are in different planes, three unknown forces can be determined if the lines of action are known, one if unknown.

If the forces act in different planes but are parallel, three unknown forces can be determined if their lines of action are known, one if unknown.

The first step in the solution of problems in statics is the determination of the supporting forces. The following data are required for the complete knowledge of supporting forces: magnitude, direction, and point of application. According to the nature of the problem, none, one, or two of these quantities are known.

**One Fixed Support**  The point of application, direction, and magnitude of the load are known. See Fig. 6. As the body on which the forces act is in equilibrium, the supporting force $P$ must be equal in magnitude and opposite in direction to the resultant of the loads $L$.

In the case of a **rolling surface,** the point of application of the support is obtained from the center of the connecting bolt $A$ (Fig. 7), both the direction and magnitude being unknown. The point of application and line of action of the support at $B$ are known, being determined by the rollers.

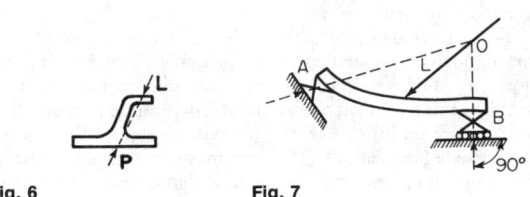

**Fig. 6**            **Fig. 7**

When three forces acting in the same plane on the same rigid body are in equilibrium, their lines of action must pass through the same point $O$. The load $L$ is known in magnitude and direction. The line of action of the support at $B$ is known on account of the rollers. The point of application of the support at $A$ is known. The three forces are in equilibrium and are in the same plane; therefore, the lines of action must meet at the point $O$.

In the case of the rolling surfaces shown in Fig. 8, the direction of the support at $A$ is known, the magnitude and point of application unknown. The line of action and point of application of the supporting force at $B$ are known, its magnitude unknown. The lines of action of the three forces must meet in a point, and the supporting force at $A$ must be perpendicular to the plane $XX$. In the case shown in Fig. 9, the directions and points of application of the supporting forces are known, and the magnitudes unknown. The lines of action of resultant of supports $A$ and $B$, the support at $C$ and load $L$ must meet at a point. Resolve the resultant of supports at $A$ and $B$ into components at $A$ and $B$, their direction being determined by the rollers.

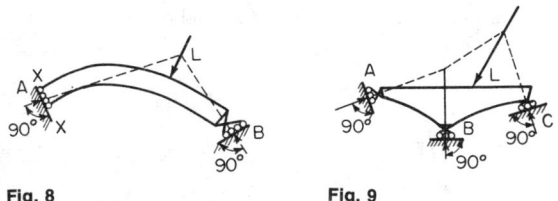

**Fig. 8**            **Fig. 9**

If a member of a truss or frame in equilibrium is pinned at two points and loaded at these two points only, the line of action of the forces exerted on the member or by the member at these two points must be along a line connecting the pins.

If the external forces acting upon a rigid body in equilibrium are all in the same plane, the equations $\Sigma X = 0$, $\Sigma Y = 0$, and $\Sigma M = 0$ must be satisfied. When trusses, frames, and other structures are under discussion, these equations are usually used as $\Sigma V = 0$, $\Sigma H = 0$, $\Sigma M = 0$, where $V$ and $H$ represent vertical and horizontal components, respectively.

The **supports** are said to be **statically determinate** when the laws of equilibrium are sufficient for their determination. When the conditions are not sufficient for the determination of the supports or other forces, the structure is said to be **statically indeterminate**; the unknown forces can then be determined from considerations involving the deformation of the material.

When several bodies are so connected to one another as to make up a rigid structure, the forces at the points of connection must be considered as internal forces and are not taken into consideration in the determination of the supporting forces for the structure as a whole.

The distortion of any practically rigid structure under its working loads is so small as to be negligible when determining supporting forces. When the forces acting at the different joints in a built-up structure cannot be determined by dividing the structure up into parts, the structure is said to be **statically indeterminate internally**. A structure may be statically indeterminate internally and still be statically determinate externally.

### Fundamental Problems in Graphical Statics

A force may be represented by a straight line in a determined position, and its magnitude by the length of the straight line. The direction in which it acts may be indicated by an arrow.

**Polygon of Forces**   The parallelogram of two forces intersecting each other (see Fig. 45) leads directly to the graphic composition by means of the triangle of forces. In Fig. 10, $R$ is called the **closing side**, and represents the resultant of the forces $F_1$ and $F_2$ in magnitude and direction. Its position is given by

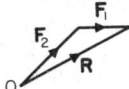

**Fig. 10**

the point of application $O$. By means of repeated use of the triangle of forces and by omitting the closing sides of the individual triangles, the magnitude and direction of the resultant $R$ of any number of forces in the same plane and intersecting at a single point can be found. In Fig. 11 the lines representing the forces start from point $O$, and in the force polygon (Fig. 12) they are joined in any order, the arrows showing their directions following around the polygon in the

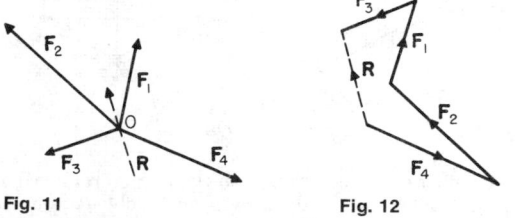

**Fig. 11**          **Fig. 12**

same direction. The magnitude of the resultant at the point of application of the forces is represented by the closing side $R$ of

the force polygon; its direction, as shown by the arrow, is counter to that in the other sides of the polygon.

**If the forces are in equilibrium, R must equal zero,** i.e., **the force polygon must close.**

If in a closed polygon one of the forces is reversed in direction, this force becomes the resultant of all the others.

If the **forces** do **not** all lie **in the same plane,** the diagram becomes a polygon in space. The resultant **R** of this system may be obtained by adding the forces in space. The resultant is the vector which closes the space polygon. The space polygon may be projected onto three coordinate planes, giving three related plane polygons. Any two of these projections will involve all static equilibrium conditions and will be sufficient for a full description of the force system (see Fig. 13).

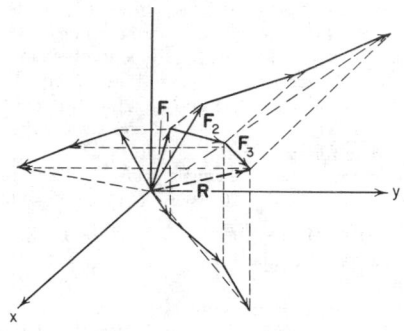

**Fig. 13**

### Determination of Stresses in Members of a Statically Determinate Plane Structure with Loads at Rest

It will be assumed that the loads are applied at the joints of the structure, i.e., at the points where the different members are connected, and that the connections are pins with no friction. The stresses in the members must then be along lines connecting the pins, unless any member is loaded at more than two points by pin connections. If the members are straight, the forces exerted on them or by them must coincide with the axes of the members. In other words, there shall be no bending stresses in any of the members of the structure.

**Equilibrium**   In order that the whole structure should be in equilibrium, it is necessary that the external forces (loads and supports) shall form a balanced system. Graphical and analytical methods are both of service.

**Supporting Forces**   When the supporting forces are to be determined, it is not necessary to pay any attention to the makeup of the structure under consideration so long as it is practically rigid; the loads may be taken as they occur, or the resultant of the loads may be used instead. When the stresses in the members of the structure are being determined, the loads *must* be distributed at the joints where they belong.

**Method of Joints**   When all the external forces have been determined, any joint at which there are not more than two unknown forces may be taken and these unknown forces determined by the methods of the stress polygon, resolution or moments. In Fig. 14, let $O$ be the joint of a structure and **F** be the only known force; but let $O1$ and $O2$ be two members of the structure joined at $O$. Then the lines of action of the unknown forces are known and their magnitude may be determined (1) by a **stress polygon** which, for equilibrium, must close; (2) by resolution into $H$ and $V$ components, using the condition of equilibrium $\Sigma H = 0$, $\Sigma V = 0$; or (3) by moments, using any convenient point on the line of action of $O1$ and $O2$ and the condition of equilibrium $\Sigma M = 0$. No more than *two* unknown forces can be determined. In this manner, proceeding from joint to joint, the stresses in all the members of the truss can usually be determined if the structure is statically determinate internally.

**Method of Sections**   The structure may be divided into parts by passing a section through it cutting some of its members; one part may then be treated as a rigid body and the external forces acting upon it determined. Some of these forces will be the stresses in the members themselves. For example, let $xx$ (Fig. 15) be a section taken through a truss loaded at $P_1$, $P_2$, and $P_3$, and supported on rollers at $S$. As the whole truss is in equilibrium, any part of it must be also, and

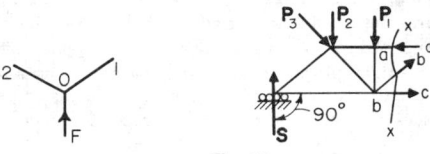

**Fig. 14**          **Fig. 15**

consequently the part shown to the left of $xx$ must be in equilibrium under the action of the forces acting externally to it. Three of these forces are the stresses in the members $aa$, $bb$, and $bc$, and are the unknown forces to be determined. They can be determined by applying the condition of equilibrium of forces acting in the same plane but not at the same point. $\Sigma H = 0$, $\Sigma V = 0$, $\Sigma M = 0$. The three unknown forces can be determined only if they are not parallel or do not pass through the same point; if, however, the forces are parallel or meet in a point, two unknown forces only can be determined. Sections may be passed through a structure cutting members in any convenient manner, as a rule, however, cutting not more than three members, unless members are unloaded.

For the determination of stresses in framed structures, see Sec. 12.

## CENTER OF GRAVITY

Consider a three-dimensional body of any size, shape, and weight. If it is suspended as in Fig. 16 by a cord from any point $A$, it will be in equilibrium under the action of the tension in the cord and the resultant of the gravity or body forces $W$. If the experiment is repeated by suspending the

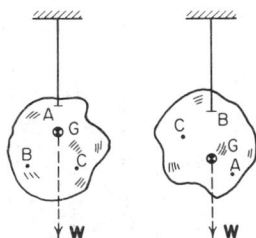

**Fig. 16**

body from point $B$, it will again be in equilibrium. If the lines of action of the resultant of the body forces were marked in each case, they would be concurrent at a point $G$ known as the **center of gravity** or **center of mass.** Whenever the density of the body is uniform, it will be a constant factor and like geometric shapes of different densities will have the same center of gravity. The term **centroid** is used in this case since the location of the center of gravity is of geometric concern only. If densities are nonuniform, like geometric shapes will have the same centroid but different centers of gravity.

**Centroids of Technically Important Lines, Areas, and Solids**

### Centroids of Lines

**Straight Line**   The centroid is at its middle point.

**Circular Arc** $AB$ (Fig. 17a)    $x_0 = r \sin c / \text{rad } c$; $y_0 = 2r \sin^2 \frac{1}{2}c / \text{rad } c$. (rad $c$ = angle $c$ measured in radians.)

**Circular Arc** $AC$ (Fig. 17b)    $x_0 = r \sin c / \text{rad } c$; $y_0 = 0$.

**Quadrant,** $AB$ (Fig. 18)    $x_0 = y_0 = 2r/\pi = 0.6366r$.

**Semicircumference,** $AC$ (Fig. 18)    $y_0 = 2r/\pi = 0.6366r$; $x_0 = 0$.

**Combination of Arcs and Straight Line** (Fig. 19)    $AD$ and $BC$ are two quadrants of radius $r$. $y_0 = \{(AB)r + 2[0.5\pi r(r - 0.6366r)]\} \div [AB + 2(0.5\pi r)]$.

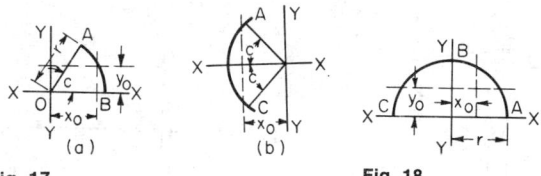

**Fig. 17**              **Fig. 18**

### Centroids of Plane Areas

**Triangle**   Centroid lies at the intersection of the lines joining the vertices with the midpoints of the sides, and at a distance from any side equal to one-third of the corresponding altitude.

**Parallelogram**   Centroid lies at the point of intersection of the diagonals.

**Trapezoid** (Fig. 20)   Centroid lies on the line joining the middle points $m$ and $n$ of the parallel sides. The distances $h_a$ and $h_b$ are

$$h_a = h(a + 2b)/3(a + b) \qquad h_b = h(2a + b)/3(a + b)$$

Draw $BE = a$ and $CF = b$; $EF$ will then intersect $mn$ at centroid.

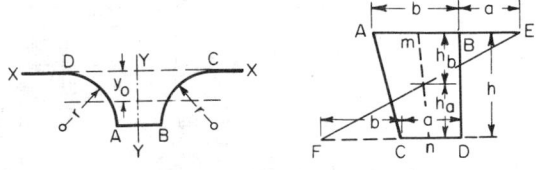

**Fig. 19**              **Fig. 20**

**Any Quadrilateral**   The centroid of any quadrilateral may be determined by the general rule for areas, or graphically by dividing it into two sets of triangles by means of the diagonals.

Find the centroid of each of the four triangles and connect the centroids of the triangles belonging to the same set. The intersection of these lines will be centroid of area. Thus, in Fig. 21, $O$, $O_1$, $O_2$, and $O_3$ are, respectively, the centroids of the triangles $ABD$, $ABC$, $BDC$, and $ACD$. The intersection of $O_1O_3$ with $OO_2$ gives the centroids.

**Segment of a Circle** (Fig. 22)   $x_0 = \frac{2}{3}r \sin^3 c/(\text{rad } c - \cos c \sin c)$. A segment may be considered to be a sector from which a triangle is subtracted, and the general rule applied.

**Sector of a Circle** (Fig. 23)   $x_0 = \frac{2}{3}r \sin c/\text{rad } c$; $y_0 = \frac{1}{3}r \sin^2 \frac{1}{2}c/\text{rad } c$.

**Semicircle**   $x_0 = \frac{4}{3}r/\pi = 0.4244r$; $y_0 = 0$.

**Quadrant** (90° sector)   $x_0 = y_0 = \frac{4}{3}r/\pi = 0.4244r$.

**Parabolic Half Segment** (Fig. 24)   Area $ABO$: $x_0 = \frac{3}{5}x_1$; $y_0 = \frac{3}{8}y_1$.

**Parabolic Spandrel** (Fig. 24)   Area $AOC$: $x_0' = \frac{3}{10}x_1$; $y_0' = \frac{3}{4}y_1$.

**Quadrant of an Ellipse** (Fig. 25)   Area $OAB$: $x_0 = \frac{4}{3}(a/\pi)$; $y_0 = \frac{4}{3}(b/\pi)$.

The centroid of a figure such as that shown in Fig. 26 may be determined as follows: Divide the area $OABC$ into a number of parts by lines drawn perpendicular to the axis $XX$, e.g., 11, 22, 33, etc. These parts will be approximately either triangles, rectangles, or trapezoids. The area of each division

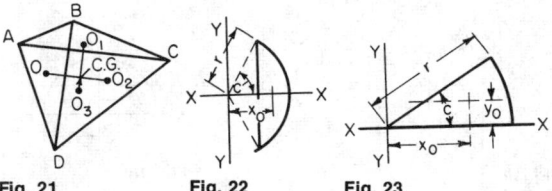

**Fig. 21          Fig. 22          Fig. 23**

may be obtained by taking the product of its mean height and its base. The centroid of each area may be obtained as previously shown. The sum of the moments of all the areas about $XX$ and $YY$, respectively, divided by the sum of the areas will give approximately the distances from the center of gravity of the whole area to the axes $XX$ and $YY$. The greater the number of areas taken the more nearly exact the result.

Centroids of Solids

**Prism or Cylinder with Parallel Bases**   The centroid lies in the center of the line connecting the centers of gravity of the bases.

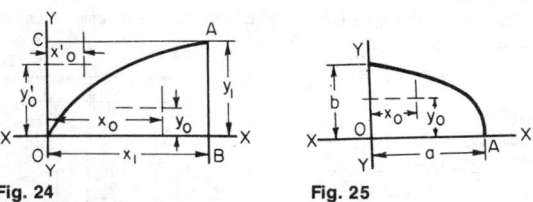

**Fig. 24                    Fig. 25**

**Oblique Frustum of a Right Circular Cylinder** (Fig. 27)   Let 1 2 3 4 be the plane of symmetry. The distance from the base to the centroid is $\frac{1}{2}h + (r^2 \tan^2 c)/8h$, where $c$ is the angle of inclination of the oblique section to the base. The distance of the centroid from the axis of the cylinder is $r^2 \tan c/4h$.

**Pyramid or Cone**   The centroid lies in the line connecting

the centroid of the base with the vertex and at a distance of one-fourth of the altitude above the base.

**Truncated Pyramid**   If $h$ is the height of the truncated pyramid and $A$ and $B$ the areas of its bases, the distance of its centroid from the surface of $A$ is

$$h(A + 2\sqrt{AB} + 3B)/4(A + \sqrt{AB} + B)$$

**Truncated Circular Cone**   If $h$ is the height of the frustum and $R$ and $r$ the radii of the bases, the distance from the surface of the base whose radius is $R$ to the centroid is $h(R^2 + 2Rr + 3r^2)/4(R^2 + Rr + r^2)$.

**Segment of a Sphere** (Fig. 28), Volume $ABC$:   $x_0 = 3(2r - h)^2/4(3r - h)$.

**Hemisphere**   $x_0 = 3r/8$.

**Hollow Hemisphere**   $x_0 = 3(R^4 - r^4)/8(R^3 - r^3)$, where $R$ and $r$ are, respectively, the outer and inner radii.

**Sector of a Sphere** (Fig. 28), Volume $OABCO$:   $x_0' = \frac{3}{8}(2r - h)$.

**Ellipsoid, with semiaxes a, b, and c**   For each octant, distance from center of gravity to each of the bounding planes = $\frac{3}{8} \times$ length of semiaxis perpendicular to the plane considered.

The formulas given for the determination of the centroid of lines and areas can be used to determine the areas and volumes of surfaces and solids of revolution, respectively, by employing the theorems of Pappus, Sec. 2.

**Determination of Center of Gravity of a Body by Experiment**   The center of gravity may be determined by hanging the body up from different points and plumbing down; the point of intersection of the plumb lines will give the center of gravity. It may also be determined as shown in Fig. 29. The body is placed on knife-edges which rest on platform scales. The sum of the weights registered on the two scales ($w_1 + w_2$) must equal the weight ($w$) of the body. Taking a moment axis at either end (say, $O$), $w_2A/w = x_0 = $ distance from $O$ to plane containing the center of gravity.

**Graphical Determination of the Centroids of Plane Areas**   See Fig. 40.

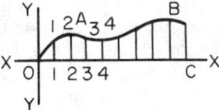

**Fig. 26**

## MOMENT OF INERTIA

The **moment of inertia of a solid body** with respect to a given axis is the limit of the sum of the products of the masses of each of

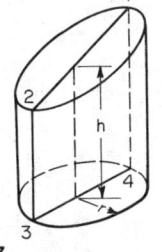

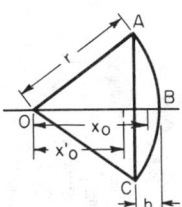

**Fig. 27**                    **Fig. 28**

the elementary particles into which the body may be conceived to be divided and the square of their distance from the given axis.

If $dm = dw/g$ represents the mass of an elementary particle and $y$ its distance from an axis, the moment of inertia $I$ of the body about this axis will be $I = \int y^2 \, dm = \int y^2 \, dw/g$.

The moment of inertia may be expressed in weight units ($I_w = \int y^2 \, dw$), in which case the moment of inertia in weight units, $I_w$, is equal to the moment of inertia in mass units, $I$, multiplied by $g$.

If $I = k^2 m$, the quantity $k$ is called the **radius of gyration** or the **radius of inertia**.

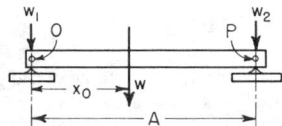

**Fig. 29**

If a body is considered to be composed of a number of parts, its moment of inertia about an axis is equal to the sum of the moments of inertia of the several parts about the same axis, or $I = I_1 + I_2 + I_3 + \cdots + I_n$.

The **moment of inertia of an area** with respect to a given axis is the limit of the sum of the products of the elementary areas into which the area may be conceived to be divided and the square of their distance ($y$) from the axis in question. $I = \int y^2 \, dA = k^2 A$, where $k$ = **radius of gyration.**

The quantity $\int y^2 \, dA$ is more properly referred to as the *second moment of area* since it is not a measure of *inertia* in a true sense.

Formulas for moments of inertia and radii of gyration of various areas follow later in this section.

**Relation between the Moments of Inertia of an Area and a Solid** The moment of inertia of a solid of elementary thickness about an axis is equal to the moment of inertia of the area of one face of the solid about the same axis multiplied by the mass per unit volume of the solid times the elementary thickness of the solid.

**Moments of Inertia about Parallel Axes** The moment of inertia of an area or solid about any given axis is equal to the moment of inertia about a parallel axis through the center of gravity plus the square of the distance between the two axes times the area or mass.

In Fig. 30a, the moment of inertia of the area $ABCD$ about axis $YY$ is equal to $I_0$ (or the moment of inertia about $Y_0Y_0$ through the center of gravity of the area and parallel to $YY$) plus $x_0^2 A$, where $A$ = area of $ABCD$. In Fig. 30b, the moment of inertia of the mass $m$ about $YY = I_0 + x_0^2 m$. $Y_0Y_0$ passes through the centroid of the mass and is parallel to $YY$.

**Polar Moment of Inertia** The polar moment of inertia (Fig. 31) is taken about an axis perpendicular to the plane of the area. Referring to Fig. 31, if $I_y$ and $I_x$ are the moments of inertia of the area $A$ about $YY$ and $XX$, respectively, then the polar moment of inertia $I_p = I_x + I_y$, or the polar moment of inertia is equal to the sum of the moments of inertia about any two axes at right angles to each other in the plane of the area and intersecting at the pole.

**Product of Inertia** This quantity will be represented by $I_{xy}$, and is $\int\int xy \, dy \, dx'$ where $x$ and $y$ are the coordinates of any

elementary part into which the area may be conceived to be divided. $I_{xy}$ may be positive or negative, depending upon the position of the area with respect to the coordinate axes $XX$ and $XY$.

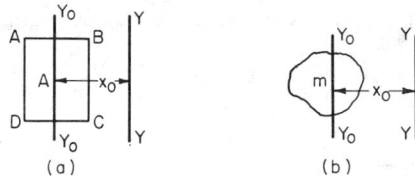

**Fig. 30**

**Relation between Moments of Inertia about Axes Inclined to Each Other** Referring to Fig. 32, let $I_y$ and $I_x$ be the moments of inertia of the area $A$ about $YY$ and $XX$, respectively, $I_y'$ and $I_x'$ the moments about $Y'Y'$ and $X'X'$, and $I_{xy}$ and $I_{xy}'$ the products of inertia for $XX$ and $YY$, and $X'X'$ and $Y'Y'$, respectively. Also, let $c$ be the angle between the respective pairs of axes, as shown. Then,

$$I_y' = I_y \cos^2 c + I_x \sin^2 c + I_{xy} \sin 2c$$
$$I_x' = I_x \cos^2 c + I_y \sin^2 c - I_{xy} \sin 2c$$
$$I_{xy}' = \frac{I_x - I_y}{2} \sin 2c + I_{xy} \cos 2c$$

**Principal Moments of Inertia** In every plane area, a given point being taken as the origin, there is at least one pair of rectangular axes in the plane of the area about one of which the moment of inertia is a maximum, and a minimum about the other. These moments of inertia are called the **principal moments of inertia,** and the axes about which they are taken are the **principal axes of inertia.** One of the conditions for principal moments of inertia is that the product of inertia $I_{xy}$ shall equal zero. **Axes of symmetry** of an area are always principal axes of inertia.

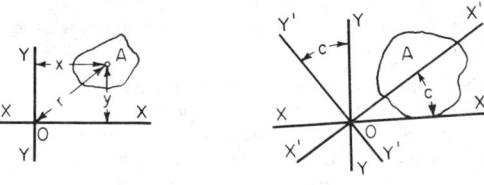

**Fig. 31**                **Fig. 32**

**Relation between Products of Inertia and Parallel Axes** In Fig. 33, $X_0X_0$ and $Y_0Y_0$ pass through the center of gravity of the area parallel to the given axes $XX$ and $YY$. If $I_{xy}$ is the product of inertia for $XX$ and $YY$, and $I_{x_0 y_0}$ that for $X_0X_0$ and $Y_0Y_0$, then $I_{xy} = I_{x_0 y_0} + abA$.

**Mohr's Circle** The **principal moments of inertia** and the location of the **principal axes of inertia** for any point of a plane area may be established graphically as follows.

Given at any point $A$ of a plane area (Fig. 34), the moments of inertia $I_x$ and $I_y$ about axes $X$ and $Y$, and the product of inertia $I_{xy}$ relative to $X$ and $Y$. The graph shown in Fig. 34b is plotted on rectangular coordinates with moments of inertia as abscissas and products of inertia as ordinates. Lay out $Oa = I_x$ and $ab = I_{xy}$ (upward for positive products of inertia, downward for negative). Lay out $Oc = I_y$ and $cd = $ *negative of* $I_{xy}$.

Draw circle with $bd$ as diameter. This is **Mohr's circle.** The *maximum* moment of inertia is $I_x' = Of$; the *minimum* moment of inertia is $I_y' = Og$. The principal axes of inertia are located as follows. From axis $AX$ (Fig. 34$a$) lay out angular distance $\theta = {}^1/_2 \angle bef$. This locates axis $AX'$, one principal axis $(I_x' = Of)$. The other principal axis of inertia is $AY'$, perpendicular to $AX'$ $(I_x' = Og)$.

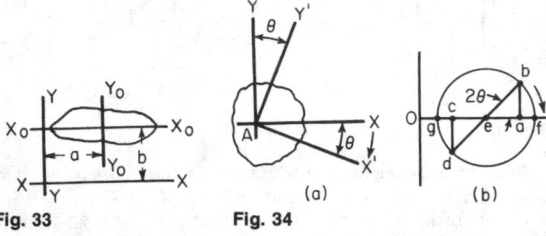

**Fig. 33**          **Fig. 34**

The **moment of inertia of any area** may be considered to be made up of the sum or difference of the known moments of inertia of simple figures. For example, the dimensioned figure shown in Fig. 35 represents the section of a rolled shape with hole $oprs$ and may be divided into the semicircle $abc$, rectangle $edkg$, and triangles $mfg$ and $hkl$, from which the rectangle $oprs$ is to be subtracted. Referring to axis $XX$,

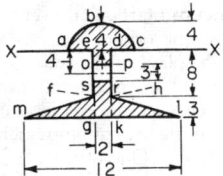

**Fig. 35**

$I_{xx} = \pi 4^4/8$ for semicircle $abc$ $= (2 \times 11^3)/3$ for rectangle $edkg$ $= 2[(5 \times 3^3)/36 + 10^2(5 \times 3)/2]$ for the two triangles $mfg$ and $hkl$

From the sum of these there is to be subtracted $I_{xx} = [(2 \times 3^2)/12 + 4^2(2 \times 3)]$ for the rectangle $oprs$.

If the moment of inertia of the whole area is required about an axis parallel to $XX$, but passing through the center of gravity of the whole area, $I_0 = I_{xx} - x_0^2 A$, where $x_0 =$ distance from $XX$ to center of gravity. The **moments of inertia of built-up sections** used in structural work may be found in the same manner, the moments of inertia of the different **rolled sections** being given in Sec. 12.

**Moments of Inertia of Solids**  For moments of inertia of solids about parallel axes, $I_x = I_0 + x_0^2 m$.

**Moment of Inertia with Reference to Any Axis**  Let a mass particle $dm$ of a body have $x$, $y$, and $z$ as coordinates, $XX$, $YY$, and $ZZ$ being the coordinate axes and $O$ the origin. Let $X'X'$ be any axis passing through the origin and making angles of $A$, $B$, and $C$ with $XX$, $YY$, and $ZZ$, respectively. The moment of inertia with respect to this axis then becomes equal to

$$I_x' = \cos^2 A \int (y^2 + z^2)\, dm + \cos^2 B \int (z^2 + x^2)\, dm$$
$$+ \cos^2 C \int (x^2 + y^2)\, dm - 2 \cos B \cos C \int yz\, dm$$
$$- 2 \cos C \cos A \int zx\, dm - 2 \cos A \cos B \int xy\, dm$$

Let the moment of inertia about $XX = I_x = \int (y^2 + z^2)\, dm$,

about $YY = Iy = \int (z^2 + x^2)\, dm$, and about $ZZ = I_z = \int (x^2 + y^2)\, dm$. Let the products of inertia about the three coordinates axes be

$$I_{yz} = \int yz\, dm \qquad I_{zx} = \int zx\, dm \qquad I_{xy} = \int xy\, dm$$

Then the moment of inertia $I_x'$ becomes equal to

$$I_x \cos^2 A + I_y \cos^2 B + I_z \cos^2 C - 2I_{yz} \cos B \cos C$$
$$- 2I_{zx} \cos C \cos A - 2I_{xy} \cos A \cos B$$

The moment of inertia of any solid may be considered to be made up of the sum or difference of the moments of inertia of simple solids of which the moments of inertia are known.

**Moments of Inertia of Important Solids: (Homogeneous)**

$m = w/g =$ mass per unit of volume of the body
$M = W/g =$ total mass of body
$r =$ radius
$I =$ moment of inertia (mass units)
$I_w = I \times g =$ moment of inertia (weight units)

**Solid circular cylinder** about its axis: $I = \pi r^4 ma/2 = Mr^2/2$. ($a =$ length of axis of cylinder.)

**Solid circular cylinder** about an axis through the center of gravity and perpendicular to axis of cylinder: $I = M[r^2 + (a^2/3)]/4$.

**Hollow circular cylinder** about its axis: $I = \pi ma(r_1^4 - r_2^4)/2$. ($r_1$ and $r_2 =$ outer and inner radii; $a =$ length.)

**Thin hollow circular cylinder** about its axis: $I = Mr^2$.

**Solid sphere** about a diameter: $I = 8m\pi r^5/15 = 2Mr^2/5$.

**Thin hollow sphere** about a diameter: $I = 2Mr^2/3$.

**Thick hollow sphere** about a diameter: $I = 8m\pi(r_1^5 - r_2^5)/15$. ($r_1$ and $r_2$ are outer and inner radii.)

**Rectangular prism** about an axis through center of gravity and perpendicular to a face whose dimensions are $a$ and $b$: $I = M(a^2 + b^2)/12$.

**Solid right circular cone** about an axis through its apex and perpendicular to its axis: $I = 3M[(r^2/4) + h^2]/5$. ($h =$ altitude of cone, $r =$ radius of base.)

**Solid right circular cone** about its axis of revolution: $I = 3Mr^2/10$.

**Ellipsoid** with semiaxes $a$, $b$, and $c$: $I$ about diameter $2c$ ($z$ axis) $= 4m\pi abc (a^2 + b^2)/15$. [Equation of ellipsoid: $(x^2/a^2) + (y^2/b^2) + (z^2/c^2) = 1$.]

**Ring with Circular Section** (Fig. 36)  $I_{yy} = {}^1/_2 m\pi^2 Ra^2(4R^2 + 3a^2)$; $I_{xx} = m\pi^2 Ra^2[R^2 + (5a^2/4)]$.

**Approximate Moments of Inertia of Solids**  In order to determine the moment of inertia of a solid, it is necessary to know all its dimensions. In the case of a rod of mass $M$ (Fig. 37) and length $l$, with shape and size of the cross section unknown, making the approximation that the weight is all concentrated along the axis of the rod, the moment of inertia about $YY$ will be $I_{yy} = \int_0^l (M/l)x^2\, dx = Ml^2/3$.

A **thin plate** may be treated in the same way (Fig. 38): $I_{yy} = \int_0^l (M/l)x^2\, dx$. Here the mass of the plate is assumed concentrated at its middle layer.

**Thin Ring, or Cylinder** (Fig. 39)  Assume the mass $M$ of the ring or cylinder to be concentrated at a distance $r$ from $O$. The moment of inertia about an axis through $O$ perpendicular to plane of ring or along the axis of the cylinder will be $I = Mr^2$; this will be greater than the exact moment of inertia, and $r$ is sometimes taken as the distance from $O$ to the center of gravity of the cross section of the rim.

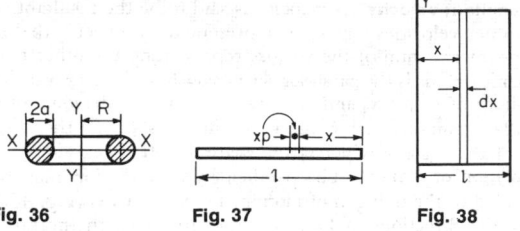

**Fig. 36**     **Fig. 37**     **Fig. 38**

**Flywheel Effect**  The moment of inertia of a solid is often called flywheel effect in the solution of problems dealing with rotating bodies, and is usually expressed in lb·ft² ($I_w$).

**Graphical Determination of the Centroids and Moments of Inertia of Plane Areas**  Required to find the center of gravity of the area $MNP$ (Fig. 40) and its moment of inertia about any axis $XX$.

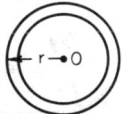

**Fig. 39**

Draw any line $SS$ parallel to $XX$ and at a distance $d$ from it. Draw a number of lines such as $AB$ and $EF$ across the figure parallel to $XX$. From $E$ and $F$ draw $ER$ and $FT$ perpendicular to $SS$. Select as a pole any point on $XX$, preferably the point nearest the area, and draw $OR$ and $OT$, cutting $EF$ at $E'$ and $F'$. If the same construction is repeated, using other lines parallel to $XX$, a number of points will be obtained, which, if connected by a smooth curve, will give the area $M'N'P'$. Project $E'$ and $F'$ onto $SS$ by lines $E'R'$ and $F'T'$. Join $F'$ and $T'$ with $O$, obtaining $E''$ and $F''$; connect the points obtained using other lines parallel to $XX$ and obtain an area $M''N''P''$. The area $M'N'P' \times d =$ moment of area $MNP$ about the line $XX$, and the distance from $XX$ to the centroid $MNP =$ area $M'N'P' \times d/$area $MNP$. Also, area $M''N''P'' \times d^2 =$ moment of inertia of $MNP$ about $XX$. The areas $M'N'P'$ and $M''N''P''$ can best be obtained by use of a planimeter.

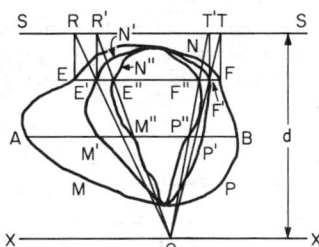

**Fig. 40**

## KINEMATICS

**Kinematics** is the study of the motion of bodies without refer-

ence to the forces causing that motion or the mass of the bodies.

The **displacement** of a point is the directed distance that a point has moved on a geometric path from a convenient origin. It is a **vector,** having both magnitude and direction, and is subject to all the laws and characteristics attributed to vectors. In Fig. 41, the displacement of the point $A$ from the origin $O$ is the directed distance $O$ to $A$, symbolized by the vector **s.**

The **velocity** of a point is the time rate of change of displacement, or $\mathbf{v} = ds/dt$.

The **acceleration** of a point is the time rate of change of velocity, or $\mathbf{a} = dv/dt$.

The kinematic definitions of velocity and acceleration involve the four variables, *displacement, velocity, acceleration,* and *time.* If we eliminate the variable of time, a third equation of motion is obtained, $ds/v = dt = dv/a$. This differential equation, together with the definitions of velocity and acceleration, make up the *three kinematic equations* of motion, $v = ds/dt$, $a = dv/dt$, and $a\,ds = v\,dv$. These differential equations are usually limited to the scalar form when expressed together, since the last can only be properly expressed in terms of the scalar $dt$. The first two, since they are definitions for velocity and acceleration, are **vector equations.**

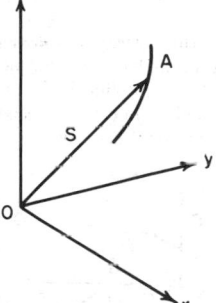

**Fig. 41**

A **space-time curve** offers a convenient means for the study of the motion of a point. The **slope of the curve** at any point will represent the **velocity** at that time. In Fig. 42$a$ the slope is constant, as the graph is a straight line; the velocity is therefore uniform. In Fig. 42$b$ the slope of the curve varies from point to point, and the velocity must also vary. At $p$ and $q$ the slope is zero; therefore, the velocity of the point at the corresponding times must also be zero.

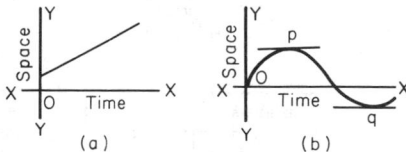

**Fig. 42**

A **velocity-time curve** offers a convenient means for the study of acceleration. The **slope of the curve** at any point will represent the **acceleration** at that time. In Fig. 43$a$ the slope is constant; so the acceleration must be constant. In the case represented by the full line, the acceleration is positive; so the velocity is increasing. The dotted line shows a negative acceleration and therefore a decreasing velocity. In Fig. 43$b$ the slope of the curve varies from point to point; so the acceleration must also vary. At $p$ and $q$ the slope is zero; therefore, the acceleration of the point at the corresponding times must also be zero. The area under the velocity-time curve between any two ordinates such as $NL$ and $HT$ will represent the distance moved in time interval $LT$. In the case of the uniformly accelerated motion shown by the full line in Fig. 43$a$, the area $LNHT$ is $\frac{1}{2}(NL + HT) \times (OT - OL)$ = mean velocity multiplied by the time interval = space passed over during this time interval. In Fig. 43$b$ the mean velocity can be obtained from the equation of the curve by means of the calculus, or graphically by approximation of the area.

An **acceleration-time curve** (Fig. 44) may be constructed by plotting accelerations as ordinates, and times as abscissas. The area under this curve between any two ordinates will represent the total increase in velocity during the time interval. The area $ABCD$ represents the total increase in velocity between time $t_1$ and time $t_2$.

### General Expressions Showing the Relations between Space, Time, Velocity, and Acceleration for Rectilinear Motion

SPECIAL MOTIONS

**Uniform Motion**  If the **velocity** is **constant**, the **acceleration** must be **zero**, and the point has uniform motion. The space-time curve becomes a straight line inclined toward the time axis (Fig. 42$a$). The velocity-time curve becomes a straight line parallel to the time axis. For this motion $a = 0$, $v$ = constant, and $s = s_0 + vt$.

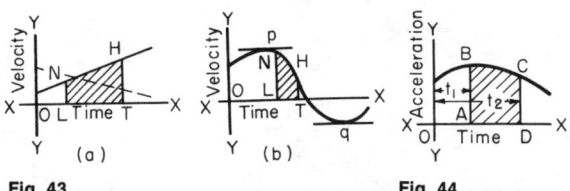

**Fig. 43**                **Fig. 44**

**Uniformly Accelerated or Retarded Motion**  If the **velocity** is **not uniform** but the **acceleration** is **constant**, the point has uniformly accelerated motion; the acceleration may be either positive or negative. The space-time curve becomes a parabola and the velocity-time curve becomes a straight line inclined toward the time axis (Fig. 43 a). The acceleration-time curve becomes a straight line parallel to the time axis. For this motion $a$ = constant, $v = v_0 + at$, $s = s_0 + v_0 t + \frac{1}{2}at^2$.

If the point starts from rest, $v_0 = 0$. Care should be taken concerning the sign $+$ or $-$ for acceleration.

### Composition and Resolution of Velocities and Acceleration

**Resultant Velocity**  A velocity is said to be the resultant of two other velocities when it is represented by a vector that is the geometric sum of the vectors representing the other two velocities. This is the **parallelogram of motion**. In Fig. 45, **v** is the resultant of $v_1$ and $v_2$ and is represented by the diagonal of a parallelogram of which $v_1$ and $v_2$ are the sides; or it is the third side of a triangle of which $v_1$ and $v_2$ are the other two sides.

**Polygon of Motion**  The parallelogram of motion may be extended to the polygon of motion. Let $v_1$, $v_2$, $v_3$, $v_4$ (Fig. 46$a$) show the directions of four velocities imparted in the same plane to point $O$. If the lines $v_1$, $v_2$, $v_3$, $v_4$ (Fig. 46$b$) are drawn parallel to and proportional to the velocities imparted to point $O$, $v$ will represent the resultant velocity imparted to $O$. It will make no difference in what order the velocities are taken in constructing the motion polygon. As long as the arrows showing the direction of the motion follow each other in order about the polygon, the resultant velocity of the point will be represented in magnitude by the closing side of the polygon, but opposite in direction.

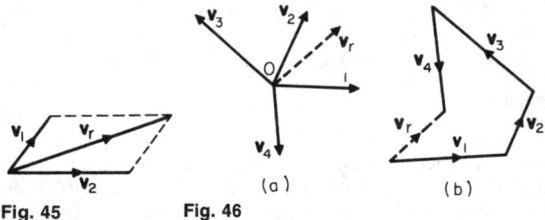

**Fig. 45**          **Fig. 46**

**Resolution of Velocities**  Velocities may be resolved into **component velocities** in the same plane, as shown by Fig. 47. Let the velocity of point $O$ be $v_r$. In Fig. 47$a$ this velocity is resolved into two components in the same plane as $v_r$ and at right angles to each other.

$$v_r = \sqrt{(v_1)^2 + (v_2)^2}$$

In Fig. 47$b$ the components are in same plane as $v_r$, but are not at right angles to each other. In this case,

$$v_r = \sqrt{(v_1)^2 + (v_2)^2 + 2v_1 v_2 \cos B}$$

If the components $v_1$ and $v_2$ and angle $B$ are known, the direction of $v_r$ can be determined. $\sin bOc = (v_1/v_r) \sin B$. $\sin cOa = (v_2/v_r) \sin B$. Where $v_1$ and $v_2$ are at right angles to each other, $\sin B = 1$.

**Table 1**

| Variables | $s = f(t)$ | $v = f(t)$ | $a = f(t)$ | $a = f(s,v)$ |
|---|---|---|---|---|
| Displacement.... | ......... | $s = s_0 + \int_{t_0}^{t} v\,dt$ | $s = s_0 + \int_{t_0}^{t}\int_{t_0}^{t} a\,dt\,dt$ | $s = s_0 + \int_{v_0}^{v} (v/a)\,dv$ |
| Velocity......... | $v = ds/dt$ | ............. | $v = v_0 + \int_{t_0}^{t} a\,dt$ | $\int_{v_0}^{v} v\,dv = \int_{s}^{s_0} a\,ds$ |
| Acceleration..... | $a = d^2s/dt^2$ | $a = dv/dt$ | .................. | $a = v\,dv/ds$ |

**Resultant Acceleration**  Accelerations may be combined and resolved in the same manner as velocities, but in this case the lines or vectors represent accelerations instead of velocities. If the acceleration had components of magnitude $a_1$ and $a_2$, the magnitude of the resultant acceleration would be $a = \sqrt{(a_1)^2 + (a_2)^2 + 2a_1a_2 \cos B}$, where $B$ is the angle between the vectors $a_1$ and $a_2$.

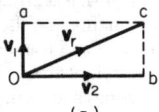 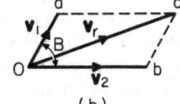

(a)                    (b)

**Fig. 47**

### Curvilinear Motion in a Plane

The linear velocity $v = ds/dt$ of a point in curvilinear motion is the same as for rectilinear motion. Its direction is tangent to the path of the point. In Fig. 48a, let $p_1 p_2 p_3$ be the path of a moving point and $v_1$, $v_2$, $v_3$ represent its velocity at points $p_1$, $p_2$, $p_3$, respectively. If $O$ is taken as a pole (Fig. 48b) and vectors $v_1$, $v_2$, $v_3$ representing the velocities of the point at $p_1$, $p_2$, and $p_3$ are drawn, the curve connecting the terminal points of these vectors is known as the **hodograph** of the motion. This velocity diagram is applicable only to motions all in the same plane.

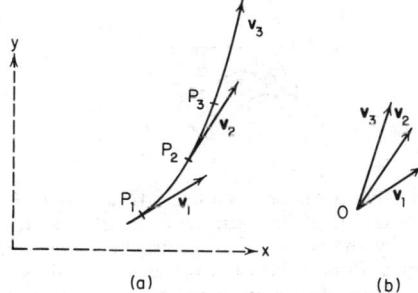

(a)                    (b)

**Fig. 48**

**Acceleration**  Tangents to the curve (Fig. 48b) indicate the directions of the **instantaneous velocities**. The direction of the tangents does not, as a rule, coincide with the direction of the accelerations as represented by tangents to the path. If the

acceleration $a$ at some point in the path is resolved by means of a parallelogram into components tangent and normal to the path, the normal acceleration $a_n = v^2/\rho$, where $\rho$ = radius of curvature of the path at the point in question, and the tangential acceleration $a_t = dv/dt$, where $v$ = velocity tangent to the path at the same point. $a = \sqrt{a_n^2 + a_t^2}$. The normal acceleration is constantly directed toward the center of the path.

EXAMPLE.  Figure 49 shows a point moving in a curvilinear path. At $p_1$ the velocity is $\mathbf{v}_1$; at $p_2$ the velocity is $\mathbf{v}_2$. If these velocities are drawn from pole $O$ (Fig. 49b), $\Delta \mathbf{v}$ will be the difference between $\mathbf{v}_2$ and $\mathbf{v}_1$. The acceleration during travel $p_1 p_2$ will be $\Delta \mathbf{v}/\Delta t$, where $\Delta t$ is the time interval. The approximation becomes closer to instantaneous acceleration as shorter intervals $\Delta t$ are employed. The acceleration $\Delta \mathbf{v}/\Delta t$ can be resolved into normal and tangential components leading to $\mathbf{a}_n = \Delta \mathbf{v}_n/\Delta t$, normal to the path, and $\mathbf{a}_r = \Delta \mathbf{v}_p/\Delta t$, tangential to the path.

*Velocity* and *acceleration* may be expressed in **polar coordinates** such that $v = \sqrt{v_r^2 + v_\theta^2}$ and $a = \sqrt{a_r^2 + a_\theta^2}$. Figure 50 may be used to explain the $r$ and $\theta$ coordinates.

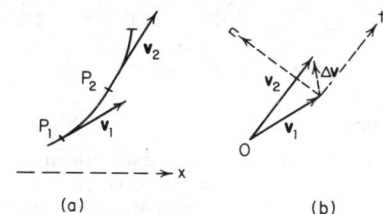

(a)                    (b)

**Fig. 49**

EXAMPLE.  At $p_1$ the velocity is $\mathbf{v}_1$, with components $\mathbf{v}_{1r}$ in the $r$ direction and $\mathbf{v}_{1\theta}$ in the $\theta$ direction. At $p_2$ the velocity is $\mathbf{v}_2$, with components $\mathbf{v}_{2r}$ in the $r$ direction and $\mathbf{v}_{2\theta}$ in the $\theta$ direction. It is evident that the difference in velocities $\mathbf{v}_2 - \mathbf{v}_1 = \Delta \mathbf{v}$ will have components $\Delta \mathbf{v}_r$ and $\Delta \mathbf{v}_\theta$, giving rise to accelerations $\mathbf{a}_r$ and $\mathbf{a}_\theta$ in a time interval $\Delta t$.

*In polar coordinates*, $v_r = dr/dt$, $a_r = d^2r/dt^2 - r(d\theta/dt)^2$, $v_\theta = r(d\theta/dt)$, and $a_\theta = r(d^2\theta/dt^2) + 2(dr/dt)(d\theta/dt)$.

If a point $P$ moves on a circular path of radius $r$ with an angular velocity of $\omega$ and an angular acceleration of $\alpha$, the linear velocity of the point $P$ is $v = \omega r$ and the two components of the linear acceleration are $a_n = v^2/r = \omega^2 r = v\omega$ and $a_t = \alpha r$.

If the angular velocity is constant, the point $P$ travels equal circular paths in equal intervals of time. The projected dis-

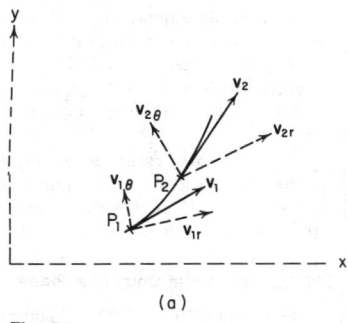

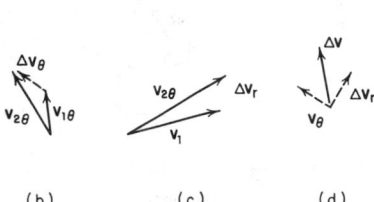

(a)            (b)        (c)        (d)

**Fig. 50**

placement, velocity, and acceleration of the point $P$ on the $x$ and $y$ axes are sinusoidal functions of time, and the motion is said to be **harmonic motion.** Angular velocity is usually expressed in radians per second, and when the number $(N)$ of revolutions traversed per minute (r/min) by the point $P$ is known, the angular velocity of the radius $r$ is $\omega = 2\pi N/60 = 0.10472N$. In Fig. 51, let the angular velocity of the line $OP$ be a constant $\omega$. Let the point $P$ start at $X'$ and move to $P$ in time $t$. Then the angle $\theta = \omega t$. If $OP = r$, $X'A = r - OA = r - r \cos \omega t = s$. The velocity $V$ of the point $A$ on the $x$ axis will equal $ds/dt = \omega r \sin \omega t$, and the acceleration $a = dv/dt = -\omega^2 r \cos \omega t$. The period $r$ is the time necessary for the point $P$ to complete one cycle of motion $r = 2\pi/\omega$, and it is also equal to the time necessary for $A$ to complete a full cycle on the $x$ axis from $X'$ to $X$ and return.

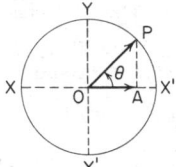

**Fig. 51**

### Curvilinear Motion in Space

If **three dimensions** are used, velocities and accelerations may be resolved into components not in the same plane by what is known as the **parallelepiped of motion.** Three coordinate systems are widely used, cartesian, cylindrical, and spherical. In **cartesian coordinates,** $v = \sqrt{v_x^2 + v_y^2 + v_z^2}$ and $a = \sqrt{a_x^2 + a_y^2 + a_z^2}$. In **cylindrical coordinates,** the radius vector **R** of displacement lies in the $rz$ plane, which is at an angle with the $xz$ plane. Referring to $(a)$ of Fig. 52, the $\theta$ coordinate is perpendicular to the $rz$ plane. In this system, $v = \sqrt{v_r^2 + v_\theta^2 + v_z^2}$ and $a = \sqrt{a_r^2 + a_\theta^2 + a_z^2}$, where $v_r = dr/dt$, $a_r = d^2r/dt^2 - r(d\theta/dt)^2$, $\mathbf{v}_\theta = r(d\theta/dt)$, and $a_\theta = r(d^2\theta/dt^2) + 2(dr/dt)(d\theta/dt)$. In **spherical coordinates,** the three coordinates are the $R$ coordinate, the $\theta$ coordinate, and the $\phi$ coordinate as in $(b)$ of Fig. 52. The velocity and acceleration are $v = \sqrt{v_R^2 + v_\theta^2 + v_\phi^2}$ and $a = \sqrt{a_R^2 + a_\theta^2 + a_\phi^2}$, where $v_R = dR/dt$, $v_\phi = R(d\phi/dt)$, $v_\theta = R \cos \phi(d\theta/dt)$, $a_R = d^2R/dt^2 - R(d\phi/dt)^2 - R \cos^2 \phi(d\theta/dt)^2$, $a_\phi = R(d^2\phi/dt^2) + R \cos \phi \sin \phi (d\theta/dt)^2 + 2(dR/dt)(d\phi/dt)$, and $a_\theta = R \cos \phi (d^2\theta/dt^2) + 2[(dR/dt) \cos \phi - R \sin \phi (d\phi/dt)] \, d\theta/dt$.

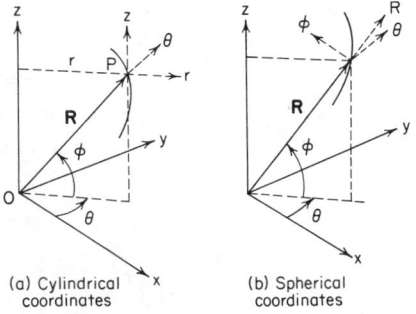

(a) Cylindrical coordinates

(b) Spherical coordinates

**Fig. 52**

### Motion of Rigid Bodies

A body is said to be **rigid** when the distances between all its particles are invariable. Theoretically, rigid bodies do not exist, but materials used in engineering are rigid under most practical working conditions. The motion of a rigid body can be completely described by knowing the **angular motion** of a line on the rigid body and the **linear motion** of a point on this line and relating the motion of all other parts of the rigid body to these motions. If a rigid body moves so that a straight line connecting any two of its particles remains parallel to its original position at all times, it is said to have **translation.** In **rectilinear translation,** all points move in straight lines. In **curvilinear translation,** all points move on congruent curves but without rotation. **Rotation** is defined as angular motion about an axis, which may or may not be fixed. Rigid body motion in which the paths of all particles lie on parallel planes is called **plane motion.**

### Angular Motion

**Angular displacement** is the change in angular position of a given line as measured from a convenient reference line. In Fig. 53, consider the motion of the line $AB$ as it moves from its original

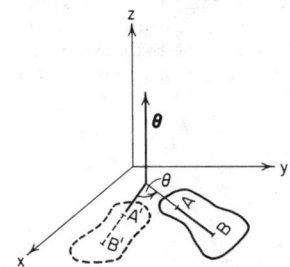

**Fig. 53**

position $A'B'$. The angle between lines $AB$ and $A'B'$ is the angular displacement of line **AB,** symbolized as $\theta$. It is a directed quantity and is a vector. The usual notation used to designate angular displacement is a vector normal to the plan in which the angular displacement occurs. The length of the vector is proportional to the magnitude of the angular displacement. For a rigid body moving in three dimensions, the line $AB$ may have angular motion about any three orthogonal axes. For example, the angular displacement can be described in cartesian coordinates as $\boldsymbol{\theta} = \boldsymbol{\theta}_x + \boldsymbol{\theta}_y + \boldsymbol{\theta}_z$, where $\theta = \sqrt{\theta_x^2 + \theta_y^2 + \theta_z^2}$.

*Angular velocity* is defined as the time rate of change of angular displacement, $\boldsymbol{\omega} = d\boldsymbol{\theta}/dt$. Angular velocity may also have components about any three orthogonal axes.

*Angular acceleration* is defined as the time rate of change of angular acceleration, $\boldsymbol{\alpha} = d\boldsymbol{\omega}/dt = d^2\boldsymbol{\theta}dt^2$. Angular acceleration may also have components about any three orthogonal axes.

The *kinematic equations* of angular motion of a line are analogous to those for the motion of a point. In referring to Table 1, $\omega = d\theta/dt$ $\alpha = d\omega/dt$, and $\alpha \, d\theta = \omega \, d\omega$. Substitute $\theta$ for $s$, $\omega$ for $v$, and $\alpha$ for $a$.

### Motion of a Rigid Body in a Plane

**Plane motion** is the motion of a rigid body such that the paths of all particles of that rigid body lie on parallel planes.

**Instantaneous Axis**   When the axis about which any body may be considered to rotate changes its position, any one position is known as an instantaneous axis, and the line through all positions of the instantaneous axis as the **centrode.**

When the velocity of two points in the same plane of a rigid body having plane motion is known, the instantaneous axis for the body will be at the intersection of the lines drawn from each point and perpendicular to its velocity. See Fig. 54, in which $A$ and $B$ are two points on the rod $AB$, $\mathbf{v}_1$ and $\mathbf{v}_2$ representing their velocities. $O$ is the instantaneous axis for $AB$; therefore point $C$ will have velocity shown in a line perpendicular to $OC$.

**Linear velocities** of points in a body rotating about an instantaneous axis are proportional to their distances from this axis.

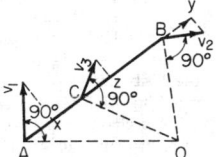

**Fig. 54**

In Fig. 54, $\mathbf{v}_1 : \mathbf{v}_2 : \mathbf{v}_3 = AO : OB : OC$. If the velocities of $A$ and $B$ were parallel, the lines $OA$ and $OB$ would also be parallel and there would be no instantaneous axis. The motion of the rod would be translation, and all points would be moving with the same velocity in parallel straight lines.

**If a body has plane motion, the components of the velocities of any two points in the body along the straight line joining them must be equal.** $Ax$ must be equal to $By$ and $Cz$ in Fig. 54.

EXAMPLE.   In Fig. 55a, the velocities of points $A$ and $B$ are known—they are $\mathbf{v}_1$ and $\mathbf{v}_2$, respectively. To find the instantaneous axis of the body, perpendiculars $AO$ and $BO$ are drawn. $O$, at the intersection of the perpendiculars, is the **instantaneous axis** of the body. To find the velocity of any other point, like $C$, line $OC$ is drawn and $\mathbf{v}_3$ erected perpendicular to $OC$ with magnitude equal to $v_1 (CO/AO)$. The **angular velocity** of the body will be $\omega = v_1/AO$ or $v_2/BO$ or $v_3/CO$. The instantaneous axis of a wheel rolling on a rack without slipping (Fig. 55b) lies at the point of contact $O$, which has zero linear velocity. All points of the wheel will have velocities perpendicular to radii to $O$ and proportional in magnitudes to their respective distances from $O$.

Another way to describe the plane motion of a rigid body is with the use of **relative motion.** In Fig. 56 the velocity of point $A$ is $\mathbf{v}_1$. The angular velocity of the line $AB$ is $v_1/r_{AB}$, the velocity of $B$ relative to $A$ is $\omega_{AB} \times r_{AB}$. Point $B$ is considered to be moving on a circular path around $A$ as a center. The

direction of relative velocity of $B$ to $A$ would be tangent to the circular path in the direction that $\omega_{AB}$ would make $B$ move. The velocity of $B$ is the vector sum of the velocity $A$ added to the velocity of $B$ relative to $A$, $\mathbf{v}_B = \mathbf{v}_A + \mathbf{v}_{B/A}$.

The **acceleration** of $B$ is the vector sum of the acceleration of $A$ added to the acceleration of $B$ relative to $A$, $\mathbf{a}_B = \mathbf{a}_A + \mathbf{a}_{B/A}$. Care must be taken to include the complete relative acceleration of $B$ to $A$. If $B$ is considered to move on a circular path about $A$, with a velocity relative to $A$, it will have an acceleration relative to $A$ that has both normal and tangential components: $\mathbf{a}_{B/A} = \mathbf{a}_B{}^n/A + \mathbf{a}_B{}^t/A$.

If $B$ is a point on a path which lies on the same rigid body as the line $AB$, a **particle** $P$ **traveling on the path** will have a velocity $\mathbf{v}_P$ at the instant $P$ passes over point $B$ such that $\mathbf{v}_P = \mathbf{v}_A + \mathbf{v}_{B/A} + \mathbf{v}_{P/B}$, where the velocity $\mathbf{v}_{P/B}$ is the velocity of $P$ relative to path $B$.

The particle $P$ will have an acceleration $\mathbf{a}_P$ at the instant $P$ passes over the point $B$ such that $\mathbf{a}_P = \mathbf{a}_A + \mathbf{a}_{B/A} + \mathbf{a}_{P/B} + 2\omega_{AB} \times \mathbf{v}_{P/B}$. The term $\mathbf{a}_{P/B}$ is the acceleration of $P$ relative to the

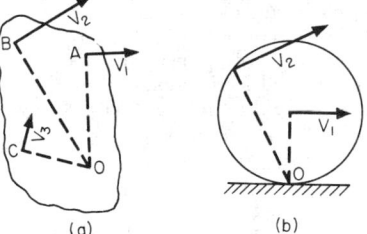

**Fig. 55**

(a)            (b)

path at point $B$. The last term, $2\omega_{AB} \mathbf{v}_{P/B}$ is frequently referred to as the **coriolis acceleration.** The direction is always normal to the path in a sense which would rotate the head of the vector $\mathbf{v}_{P/B}$ about its tail in the direction of the angular velocity of the rigid body $\omega_{AB}$.

EXAMPLE.   In Fig. 57, arm $AB$ is rotating counterclockwise about $A$ with a constant angular velocity of 38 r/min or 4 rad/s, and the slider moves outward with a velocity of 10 ft/s (3.05 m/s). At an instant when the slider $P$ is 30 in (0.76 m) from the center $A$, the acceleration of the slider will have two components. One component is the normal acceleration directed toward the center $A$. Its magnitude is $\omega^2 r = 4^2 \left(\dfrac{30}{12}\right) = 40$ ft/s² [$\omega^2 r = 4^2 (0.76) = 12.2$ m/s²]. The second is the coriolis acceleration directed normal to the arm $AB$, upward and to the left. Its magnitude is $2\omega v = 2(4)(10) = 80$ ft/s² [$2\omega v = 2(4)(3.05) = 24.4$ m/s²].

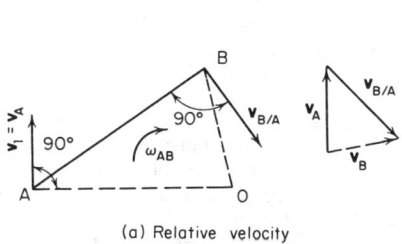

(a) Relative velocity

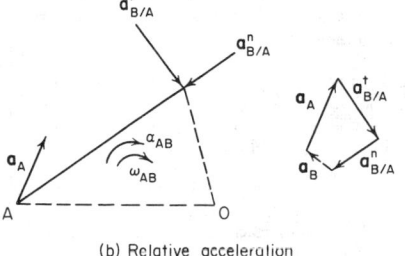

(b) Relative acceleration

**Fig. 56**

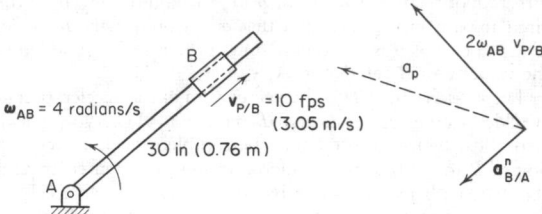

**Fig. 57**

### General Motion of a Rigid Body

The general motion of a point moving in a coordinate system which is itself in motion is complicated and can best be summarized by using vector notation. Referring to Fig. 58, let the point $P$ be displaced a vector distance $\mathbf{R}$ from the origin $O$ of a moving reference frame $x$, $y$, $z$ which has a velocity $\mathbf{v}_o$ and an acceleration $\mathbf{a}_o$. If point $P$ has a velocity and an acceleration relative to the moving reference plane, let these be $\mathbf{v}_r$ and $\mathbf{a}_r$. The angular velocity of the moving reference frame is $\omega$, and the origin of the moving reference frame is displaced a vector distance $\mathbf{R}_1$ from the origin of a primary (fixed) reference frame $X$, $Y$, $Z$. The velocity and acceleration of $P$ are $\mathbf{v}_P = \mathbf{v}_o + \boldsymbol{\omega} \times \mathbf{R} + \mathbf{v}_r$ and $\mathbf{a}_P = \mathbf{a}_o + (d\boldsymbol{\omega}/dt) \times \mathbf{R} + \boldsymbol{\omega} \times (\boldsymbol{\omega} \times \mathbf{R}) + 2\boldsymbol{\omega} \times \mathbf{v}_r + \mathbf{a}_r$.

### DYNAMICS OF PARTICLES

Consider a particle of mass $m$ subjected to the action of forces $\mathbf{F}_1$, $\mathbf{F}_2$, $\mathbf{F}_3$, . . . , whose vector resultant is $\mathbf{R} = \Sigma\mathbf{F}$. According to

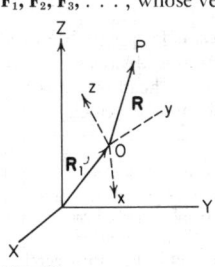

**Fig. 58**

Newton's first law of motion, if $\mathbf{R} = 0$, the body is acted on by a balanced force system, and it will either remain at rest or move uniformly in a straight line. If $\mathbf{R} \neq 0$, Newton's second law of motion states that the body will accelerate in the direction of and proportional to the magnitude of the resultant $R$. This may be expressed as $\Sigma\mathbf{F} = m\mathbf{a}$. If the resultant of the force system has components in the $x$, $y$, and $z$ directions, the resultant acceleration will have proportional components in the $x$, $y$, and $z$ direction so that $F_x = ma_x$, $F_y = ma_y$, and $F_z = ma_z$. If the resultant of the force system varies with time, the acceleration will also vary with time.

In **rectilinear motion,** the acceleration and the direction of the unbalanced force must be in the direction of motion. **Forces must be in balance and the acceleration equal to zero in any direction other than the direction of motion.**

EXAMPLE 1.    The body in Fig. 59 has a mass of 90 lbm (40.8 kg) and is subjected to an external horizontal force of 36 lbf (160 N) applied in the direction shown. The coefficient of friction between the body and the inclined plane is 0.1. Required, the velocity of the body at the end of 5 s, if it starts from rest.

First determine all the forces acting externally on the body. These are the applied force $F = 36$ lbf (106 N), the weight $W = 90$ lbf (400 N), and the force with which the plane reacts on the body. The latter force can be resolved into component forces, one normal and one parallel to the surface of the plane. Motion will be downward along the plane since a static analysis will show that the body will slide downward unless the static coefficient of friction is greater than 0.269. In the

direction normal to the surface of the plane, the forces must be balanced. The normal force is $(3/5)(36) + (4/5)(90) = 93.6$ lbf (416 N). The frictional force is $93.6 \times 0.1 = 9.36$ lbf (41.6 N). The unbalanced force acting on the body along the plane is $(3/5)(90) - (4/5)(36) - 9.36 = 15.84$ lbf (70.46 N) downward. $F = (W/9)\ a = (90/g)\ a$; therefore, $a = 0.176\ g = 56.6$ ft/s² (1.725 m/s²). In SI units, $F = ma = 70.46 = 40.8a$; and $a = 1.725$ m/s². The body is acted upon by constant forces and starts from rest; therefore, $v = \int_0^5 a\ dt$, and at the end of 5 s, the velocity would be 28.35 ft/s (8.91 m/s).

EXAMPLE 2.    The force with which a rope acts on a body is equal and opposite to the force with which the body acts on the rope, and each is equal to the tension in the rope. In Fig. 60a, neglecting the weight of the pulley and the rope, the tension in the cord must be the force of 27 lbf. For the 18-lb mass, the unbalanced force is $27 - 18 = 9$ lbf in the upward direction, i.e., $27 - 18 = (18/g)a$, and $a = 16.1$ ft/s² upward. In Fig. 60b the 27-lb force is replaced by a 27-lb mass. The unbalanced force is still $27 - 18 = 9$ lbf, but it now acts on two masses so that $27 - 18 = (45/g)$ and $a = 6.44$ ft/s². The 18-lb mass is accelerated upward, and the 27-lb mass is accelerated downward. The tension in the rope is equal to 18 lbf plus the unbalanced force necessary to give it an upward acceleration of $g/5$ or $T = 18 + (18/g)(g/5) = 21.6$ lbf. The tension is also equal to 27 lbf less the unbalanced force necessary to give it a downward acceleration of $g/5$ or $T = 27 - (27/g) \times (g/5) = 21.6$ lbf.

In SI units, in Fig. 60a, the unbalanced force is $120 - 80 = 40$ N, in the upward direction, i.e., $120 - 80 = 8.16a$, and $a = 4.9$ m/s² (16.1 ft/s²). In Fig. 60b the unbalanced force is still 40 N, but it now acts on the two masses so that $120 - 80 = 20.4a$ and $a = 1.96$ m/s² (6.44 ft/s²). The tension in the rope is the weight of the 8.16-kg mass in newtons plus the unbalanced force necessary to give it an upward acceleration of 1.96 m/s², $T = 9.807(8.16) + (8.16)(1.96) = 96$ N (21.6 lbf).

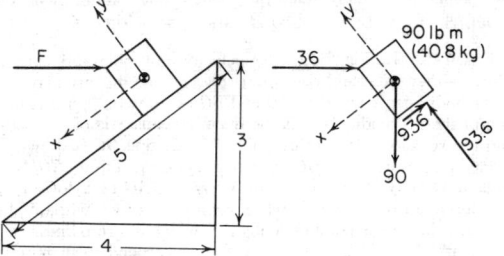

**Fig. 59**

### General Formulas for the Motion of a Body under the Action of a Constant Unbalanced Force

Let $s$ = space, ft; $a$ = acceleration, ft/s²; $v$ = velocity, ft/s; $v_0$ = initial velocity, ft/s; $b$ = height, ft; $F$ = force; $m$ = mass; $w$ = weight; $g$ = acceleration due to gravity.

$$\text{Initial velocity} = 0$$
$$F = ma = (w/g)a$$
$$v = at$$
$$s = \tfrac{1}{2}at^2 = \tfrac{1}{2}vt$$
$$v = \sqrt{2as}$$
$$= \sqrt{2gb} \text{ (falling freely from rest)}$$
$$\text{Initial velocity} = v$$
$$F = ma = (w/g)a$$
$$v = v_0 + at$$
$$s = v_0 t + \tfrac{1}{2}at^2 = \tfrac{1}{2}v_0 t + \tfrac{1}{2}vt$$

If a body is to be moved in a straight line by a force, the line of action of this force must pass through its center of gravity.

**General Rule for the Solution of Problems When the Forces Are Constant in Magnitude and Direction** Resolve all the forces acting on the body into two components, one in the direction of the body's motion and one at right angles to it. Add the components in the direction of the body's motion algebraically and find the **unbalanced force,** if any exists.

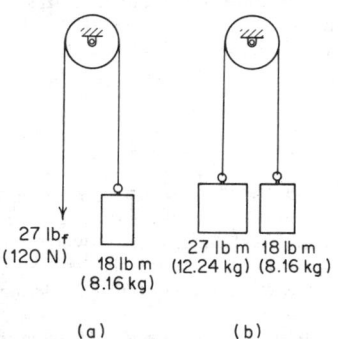

27 lb$_f$
(120 N)    18 lb m      27 lb m  18 lb m
           (8.16 kg)    (12.24 kg)  (8.16 kg)

(a)          (b)

**Fig. 60**

In **curvilinear motion,** a particle moves along a curved path, and the resultant of the unbalanced force system may have components in directions other than the direction of motion. **The acceleration in any given direction is proportional to the component of the resultant in that direction.** It is common to utilize orthogonal coordinate systems such as **cartesian coordinates, polar coordinates,** and **normal and tangential coordinates** in analyzing forces and accelerations.

EXAMPLE. A conical pendulum consists of a weight suspended from a cord or light rod and made to rotate in a horizontal circle about a vertical axis with a constant angular velocity of $N$ r/min. For any given constant speed of rotation, the angle $\theta$, the radius $r$, and the height $h$ will have fixed values. Looking at Fig. 61, we see that the forces in the vertical direction must be balanced, $T \cos \theta = w$. The forces in the direction normal to the circular path of rotation are unbalanced such that $T \sin \theta = (w/g)a_n = (w/g)\omega^2 r$. Substituting $r - l \sin \theta$ in this last equation gives the value of the tension in the cord $T = (w/g)l\omega^2$. Dividing the second equation by the first and substituting $\tan \theta = r/h$ yields the additional relation that $h = g/\omega^2$.

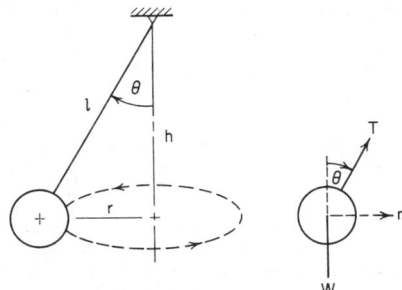

**Fig. 61**

An unresisted projectile has a motion compounded of the vertical motion of a falling body, and of the horizontal motion due to the horizontal component of the velocity of projection. In Fig. 62 the only force acting after the projectile starts is gravity, which causes an acceleration downward. The horizontal component of the original velocity $v_0$ is not changed by

gravity. The projectile will rise until the velocity given to it by gravity is equal to the vertical component of the starting velocity $v_0$, and the equation $v_0 \sin \theta = gt$ gives the time $t$ required to reach the highest point in the curve. The same time will be taken in falling if the surface $XX$ is level, and the projectile will therefore be in flight $2t$ s. The distance $s = v_0 \cos \theta \times 2t$, and the maximum height of ascent $h = (v_0 \sin \theta)^2/2g$. The expressions for the coordinates of any point on the path of the projectile are: $x = (v_0 \cos \theta)t$, and $y = (v_0 \sin \theta)t - \frac{1}{2}gt^2$, giving $y = x \tan \theta - (gx^2/2v_0^2 \cos^2 \theta)$ as the equation for the curve of the path. The radius of curvature of the highest point may be found by using the general expression $v^2 = gr$ and solving for $r$, $v$ being taken equal to $v_0 \cos \theta$.

**Simple Pendulum** The period of oscillation $= \tau = 2\pi \sqrt{l/g}$, where $l$ is the length of the pendulum and the length of the swing is not great compared to $l$.

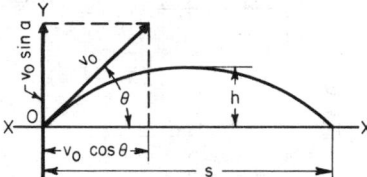

**Fig. 62**

**Centrifugal and Centripetal Forces** When a body revolves about an axis, some connection must exist capable of applying force enough to the body to constantly deviate it toward the axis. This deviating force is known as **centripetal force.** The equal and opposite resistance offered by the body to the connection is called a **centrifugal force** . The acceleration toward the axis necessary to keep a particle moving in a circle about that axis is $v^2/r$; therefore, the force necessary is $ma = mv^2/r = wv^2/gr = w\pi^2N^2r/900g$, where $N$ = r/min. This force is constantly directed toward the axis.

**The centrifugal force of a solid body revolving about an axis is the same as if the whole mass of the body were concentrated at its center of gravity.** Centrifugal force $= wv^2/gr = mv^2/r = w\omega^2r/g$, where $w$ and $m$ are the weight and mass of the whole body, $r$ is the distance from the axis about which the body is rotating to the center of gravity of the body, $\omega$ the angular velocity of the body about the axis in radians, and $v$ the linear velocity of the center of gravity of the body.

**Balancing**

A rotating body is said to be in **standing balance** when its center of gravity coincides with the axis upon which it revolves. Standing balance may be obtained by resting the axis carrying the body upon two horizontal plane surfaces, as in Fig. 63. If the center of gravity of the wheel $A$ coincides with the center of the shaft $B$, there will be no movement, but if the center of gravity does not coincide with the center of the shaft, the shaft will roll until the center of gravity of the wheel comes directly under the center of the shaft. The center of gravity may be brought to the center of the shaft by adding or taking away

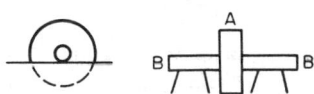

**Fig. 63**

weight at proper points on the diameter passing through the center of gravity and the center of the shaft. Weights may be added to or subtracted from any part of the wheel so long as its center of gravity is brought to the center of the shaft.

A rotating body may be in standing balance and not in **dynamic balance**. In Fig. 64, $AA$ and $BB$ are two disks whose centers of gravity are at $o$ and $p$, respectively. The shaft and the disks are in standing balance if the disks are of the same weight and the distances of $o$ and $p$ from the center of the shaft are equal, and $o$ and $p$ lie in the same axial plane but on opposite sides of the shaft. Let the weight of each disk be $w$ and the distances of $o$ and $p$ from the center of the shaft each

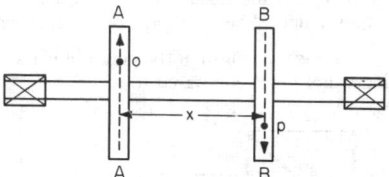

**Fig. 64**

be equal to $r$. The force exerted on the shaft by $AA$ is equal to $w\omega^2 r/g$, where $\omega$ is the angular velocity of shaft. Also, the force exerted on the shaft by $BB = w\omega^2 r/g$. These two equal and opposite parallel forces act at a distance $x$ apart and constitute a couple with a moment tending to rotate the shaft, as shown by the arrows, of $(w\omega^2 r/g)x$. A couple cannot be balanced by a single force; so two forces at least must be added to or subtracted from the system to get dynamic balance.

**Systems of Particles**   The principles of motion for a single particle can be extended to cover a **system of particles**. In this case, **the vector resultant of all external forces acting on the system of particles must equal the total mass of the system times the acceleration of the mass center, and the direction of the resultant must be the direction of the acceleration of the mass center.** This is the **principle of motion of the mass center.**

### Rotation of Solid Bodies in a Plane about Fixed Axes

For a rigid body revolving **in a plane** about a fixed axis, **the resultant moment about that axis must be equal to the product of the moment of inertia (about that axis) and the angular acceleration,** $\Sigma M_0 = I_0\alpha$. This is a general statement which includes the particular case of rotation about an axis that passes through the center of gravity.

**Rotation about an Axis Passing through the Center of Gravity**   The rotation of a body about its center of gravity can only be caused or changed by a **couple**. See Fig. 65. If a single force $F$ is applied to the wheel, the axis immediately acts on the

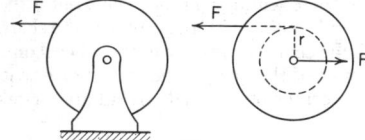

**Fig. 65**

wheel with an equal force to prevent translation, and the result is a couple (moment $Fr$) acting on the body and causing rotation about its center of gravity.

**General formulas for rotation of a body** about a fixed axis through the center of gravity, if a constant unbalanced moment is applied (Fig. 65).

Let $\theta$ = angular displacement, rad; $\omega$ = angular velocity, rad/s; $\alpha$ = angular acceleration, rad/s²; $M$ = unbalanced moment, ft·lb; $I$ = moment of inertia (mass); $g$ = acceleration due to gravity; $t$ = time of application of $M$.

Initial angular velocity = 0     Initial angular velocity = $\omega_0$

$$M = I\alpha \qquad\qquad M = I\alpha$$
$$\theta = 1/2\alpha t^2 \qquad\qquad \theta = \omega_0 t + 1/2\alpha t^2$$
$$\omega = \sqrt{2\alpha\theta} \qquad\qquad \omega = \sqrt{\omega_0^2 + 2\alpha\theta}$$

**General Rule for Rotating Bodies**   Determine all the external forces acting and their moments about the axis of rotation. If these moments are balanced, there will be no change of motion. If the moments are unbalanced, this unbalanced moment, or **torque**, will cause an angular acceleration about the axis.

**Rotation about an Axis Not Passing through the Center of Gravity**   **The resultant force acting on the body must be proportional to the acceleration of the center of gravity and directed along its line of action.** If the axis of rotation does not pass through the center of gravity, the center of gravity will have a resultant acceleration with a component $a_n = \omega^2 r$ directed toward the axis of rotation and a component $a_t = \alpha r$ tangential to its circular path. The resultant force acting on the body must also have two components, one directed normal and one directed tangential to the path of the center of gravity. The line of action of this resultant does not pass through the center of gravity because of the unbalanced moment $M_0 = I_0\alpha$ but at a point $Q$, as in Fig. 66. The point of application of this resultant is known as the **center of percussion** and may be **defined** as the point of application of the resultant of all the forces tending to cause a body to rotate about a certain axis. It is the point at which a suspended body may be struck without causing any pressure on the axis passing through the point of suspension.

**Center of Percussion**   The distance from the axis of suspension to the center of percussion is $q_0 = I/mx_0$, where $I$ = moment of inertia of the body about its axis of suspension to the center of gravity of the body.

EXAMPLES.   1. Find the center of percussion of the homogeneous rod (Fig. 67) of length $L$ and mass $m$, suspended at $XX$.

$$q_0 = \frac{I}{mx_0}$$

$$I \text{ (approx)} = \frac{m}{L}\int_0^L x^2\, dx \qquad r_0 = \frac{L}{2} \qquad \therefore q_0 = \frac{2}{L^2}\int_0^L x^2\, dx = 2L/3$$

2. Find the center of percussion of a solid cylinder, of mass $m$, resting on a horizontal plane. In Fig. 68, the instantaneous center of the cylinder is at $A$. The center of percussion will therefore be a height above the plane equal to $q_0 = I/mx_0$. Since $I = (mr^2/2) + mr^2$ and $x_0 = r$, $q_0 = 3r/2$.

**Wheel or Cylinder Rolling down a Plane**   In this case the component of the weight along the plane tends to make it roll down and is treated as a force causing rotation. The forces acting on the body should be resolved into components along the line of motion and perpendicular to it. If the forces are all known, their resultant is at the center of percussion. If one force is to be determined (the exact conditions as regards slipping or not slipping must be known), the center of percussion can be determined and the unknown force found.

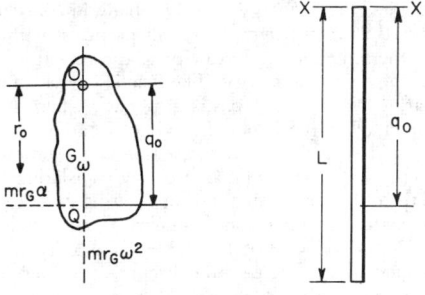

**Fig. 66**                    **Fig. 67**

**Relation between the Center of Percussion and Radius of Gyration** $q_0 = I-mx_0 = k^2/x_0$ ∴ $k^2 = x_0 q_0$ where $k =$ radius of gyration. Therefore, the radius of gyration is a mean proportional between the distance from the axis of oscillation to the center of percussion and the distance from the same axis to the center of gravity.

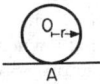

**Interchangeability of Center of Percussion and Axis of Oscillation** If a body is suspended from an axis, the center of percussion for that axis can be found. If the body is suspended from this center of percussion as an axis, the original axis of suspension will then become the center of percussion. The center of percussion is sometimes known as the **center of oscillation.**

**Period of Oscillation of a Compound Pendulum** The length of an equivalent simple pendulum is the distance from the axis of suspension to the center of percussion of the body in question. To find the **period of oscillation** of a body about a given axis, find the distance $q_0 = I/mx_0$ from that axis to the center of percussion of the swinging body. The length of the simple pendulum that will oscillate in the same time is this distance $q_0$. The period of oscillation for the equivalent single pendulum is $\tau = 2\pi \sqrt{q_0/g}$.

**Determination of Moment of Inertia by Experiment** To find the moment of inertia of a body, suspend it from some axis not passing through the center of gravity and, by swinging it, determine the period of a single oscillation in seconds. The known values will then be $\tau =$ time of single oscillation, $r_0 =$ distance from axis to center of gravity, and $m =$ mass of rod. The length of the equivalent simple pendulum is $q_0 = I/mx_0$.

Substituting this value of $q_0$ in $\tau = 2\pi \sqrt{q_0/g}$ gives $\tau = 2\pi \sqrt{I/mx_0 g}$, from which $\tau^2 = 4\pi^2 I/mx_0 g$, or $I = mx_0 g \tau^2 / 4\pi^2$.

### Plane Motion of a Rigid Body

**Plane motion** may be considered to be a combination of translation and rotation (see Kinematics). For translation, Newton's second law of motion must always be satisfied, and the resultant of the external force system must be equal to the product of the mass times the acceleration of the center of gravity in any system of coordinates. In rotation, the body moving in plane motion will not have a fixed axis. When the methods of relative motion are being used, any point on the body may be used as a reference axis to which the motion of all other points is referred.

**The sum of the moments of all external forces about the reference axis must be equal to the vector sum of the centroidal moment of inertia times the angular acceleration and the moment of the resultant force about the reference axis.**

EXAMPLE. Determine the forces acting on the piston pin $A$ and the crankpin $B$ of the connecting rod of a reciprocating engine shown in Fig. 69 for a position of 30° from TDC. The crankshaft speed is constant at 2,000 r/min. Assume that the pressure of expanding gases on the 4-lbm (1.81-kg) piston at this point is 145 lb/in² (10⁶ N/m²). The connecting rod has a mass of 5 lbm (2.27 kg) and has a centroidal radius of gyration of 3 in (0.076 m).

The kinematics of the problem are such that the angular velocity of the crank is $\omega_{OB} = 209.4$ rad/s clockwise, the angular velocity of the connecting rod is $\omega_{AB} = 45.7$ rad/s counterclockwise, and the angular acceleration is $\alpha_{AB} = 5{,}263$ rad/s² clockwise. The linear acceleration of the piston is 7,274 ft/s² in the direction of the crank. From the free-body diagram of the piston, the horizontal component of the piston-pin force is $145 \times (\pi/4)(5^2) - P = (4/32.2)(7{,}274)$, $P = 1{,}943$ lbf. The acceleration of the center of gravity $G$ is the vector sum of the component accelerations $a_G = a_B + a_{G/B}^n + a_{G/B}^t$ where $a_{G/B}^n = \omega_{GB}^2 \cdot r_{GB} = 3/12(45.7)^2 = 522$ ft/s² and $a_{G/B}^t = \alpha_{GB} \cdot r_{GB} = 3/12(5{,}263) = 1{,}316$ ft/s². The resultant acceleration of the center of gravity is 6,685 ft/s² in the x direction and 2,284 ft/s² in the negative y direction. The resultant of the external force system will have corresponding components such that $ma_{Gx} = (5/32.2)(6{,}685) = 1{,}039$ lbf and $ma_{Gy} = (5/32.2)(2{,}284) = 355$ lbf. The three remaining unknown forces can be found from the three equations of motion for the connecting rod.

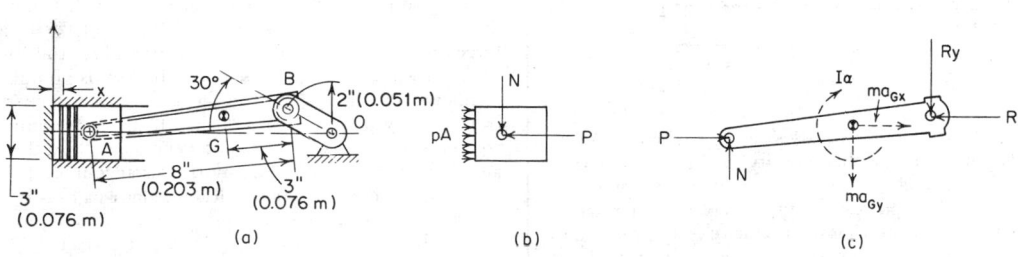

**Fig. 69**

Taking the sum of the forces in the $x$ direction, $\epsilon\ F = ma_{G_x}$; $P - R_x = ma_{G_x}$, and $R_x = 905.4$ lbf. In the $y$ direction, $\Sigma F = ma_{G_y}$; $R_y - N = ma_{gy}$; this has two unknowns, $R_y$ and $N$. Taking the sum of the movements of the external forces about the center of mass $g$, $\Sigma M_G = I_G \alpha AB$; $(N)(5) \cos (7.18°) - (P)(5) \sin (7.18°) + (R_y)(3) \cos (7.18°) - R_x (3) \sin (7.18°) - (5/386.4)(3)^2(5,263)$. Solving for $R_y$ and $N$ simultaneously, $R_y = 494.7$ lbf and $N = 140$ lbf. We could have avoided the solution of two simultaneous algebraic equations by taking the moment summation about end $A$, which would determine $R_y$ independently, or about end $B$, which would determine $N$ independently.

In SI units, the kinematics would be identical, the linear acceleration of the piston being 2,217 m/s² (7,274 ft/s²). From the free-body diagram of the piston, the horizontal component of the piston-pin force is $(10^6) \times (\pi/4)(0.127)^2 - P = (1.81)(2,217)$, and $P = 8,640$ N. The components of the acceleration of the center of gravity $G$ are $a_{G/B}^N = 522$ ft/s² and $a_{G/B}^T = 1,315$ ft/s². The resultant acceleration of the center of gravity is 2,037.5 m/s² (6,685 ft/s²) in the $x$ direction and 696.3 m/s² (2,284 ft/s²) in the negative $y$ direction. The resultant of the external force system will have the corresponding components; $ma_{Gx} = (2.27)(2,037.5) = 4,620$ N; $ma_{Gy} = (2.27)(696.3) = 1,579$ N. $R_x = 4,027$ N, $R_y = 2,201$ N, force $N = 623$ newtons.

## WORK AND ENERGY

**Work**   When a body is displaced against resistance or accelerated, work must be done upon it. An increment of work is defined as the product of an incremental displacement and the component of the force vector in the direction of the displacement or the product of the component of the incremental displacement and the force in the direction of the force. $dU = F \cdot ds \cos \alpha$, where $\alpha$ is the angle between the vector displacement and the vector force. The increment of work done by a couple $M$ acting in a body during an increment of angular rotation $d\theta$ in the plane of the couple is $dU = M\ d\theta$. In a force-displacement or moment-angle diagram, called a **work diagram** (Fig. 70), force is plotted as a function of displacement. The area under the curve represents the work done, which is equal to $\int_{s_1}^{s_2} F\ ds \cos \alpha$ or $\int_{\theta_1}^{\theta_2} M\ d\theta$.

**Units of Work.**   When the force of 1 lb acts through the distance of 1 ft, 1 lb·ft of work is done. In SI units, a force of 1 newton acting through 1 metre is 1 joule of work. 1.356 N·m = 1 lb·ft.

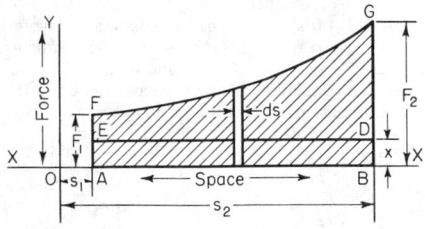

**Fig. 70**

**Energy**   A body is said to possess energy when it can do work. A body may possess this capacity through its **position** or **condition.** When a body is so held that it can do work, if released, it is said to possess energy of position or **potential energy.** When a body is moving with some velocity, it is said to possess energy of motion or **kinetic energy.** An example of potential energy is a body held suspended by a rope; the

position of the body is such that if the rope is removed work can be done by the body.

Energy is expressed in the same units as work. The kinetic energy of a particle is expressed by the formula $E = \frac{1}{2}mv^2 = \frac{1}{2}(w/g)v^2$. The kinetic energy of a rigid body in translation is also expressed as $E = \frac{1}{2}mv^2$. Since all particles of the rigid body have the same identical velocity $v$, the velocity $v$ is the velocity of the center of gravity. The kinetic energy of a rigid body, rotating about a fixed axis is $E = I_o \omega^2$, where $I_o$ is the mass moment of inertia about the axis of rotation. In plane motion, a rigid body has both translation and rotation. The kinetic energy is the algebraic sum of the translating kinetic energy of the center of gravity and the rotating kinetic energy about the center of gravity, $E = \frac{1}{2}mv^2 + \frac{1}{2}I\omega^2$. Here the velocity $v$ is the velocity of the center of gravity, and the moment of inertia $I$ is the centroidal moment of inertia.

If a force which varies acts through a space on a body of mass $m$, the work done is $\int_{s_1}^{s} F\ ds$, and if the work is all used in giving kinetic energy to the body it is equal to $\frac{1}{2}m(v_2^2 - v_1^2)$ = **change in kinetic energy,** where $v_2$ and $v_1$ are the velocities at distances $s_2$ and $s_1$, respectively. This is a specific statement of the law of conservation of energy. **The principle of conservation of energy requires that the mechanical energy of a system remain unchanged if it is subjected only to forces which depend on position or configuration.**

Certain problems in which the velocity of a body at any point in its straight-line path when acted upon by varying forces is required can be easily solved by the use of a **work diagram.**

In Fig. 70, let a body start from rest at $A$ and be acted upon by a force that varies in accordance with the diagram $AFGBA$. Let the resistance to motion be a constant force $= x$. Find the velocity of the body at point $B$. The area $AFGBA$ represents the work done upon the body and the area $AEDBA$ (= force $x$ × distance $AB$) represents the work that must be done to overcome resistance. The difference of these areas, or $EFGDE$, will represent work done in excess of that required to overcome resistance, and consequently is equal to the increase in kinetic energy. Equating the work represented by the area $EFGDE$ to $\frac{1}{2}wv^2/g$ and solving for $v$ will give the required velocity at $B$. If the body did not start from rest, this area would represent the change in kinetic energy, and the velocity could be obtained by the formula: Work $= \frac{1}{2}(w/g)(v_1^2 - v_0^2)$, $v_1$ being the required velocity.

**General Rule for Rectilinear Motion**   Resolve each force acting on the body into components, one of which acts along the line of motion of the body and the other at right angles to the line of motion. Take the sum of all the components acting in the direction of the motion and multiply this sum by the distance moved through for constant forces. (Take the average force times distance for forces that vary.) This product will be the total work done upon the body. If there is no unbalanced component, there will be no change in kinetic energy and consequently no change in velocity. If there is an unbalanced component, the change in kinetic energy will be this unbalanced component multiplied by the distance moved through.

The **work done by a system of forces acting on a body** is equal to the algebraic sum of the work done by each force taken separately.

Power is the rate at which work is performed, or the number of units of work performed in unit time. In the

English engineering system, the units of power are the horsepower, or 33,000 lb·ft/min = 550 lb·ft/s, and the kilowatt = 1.341 hp = 737.55 lb·ft/s. In SI units, the unit of power is the watt, which is 1 newton-metre per second or 1 joule per second.

**Friction Brake**  In Fig. 71 a pulley revolves under the band and in the direction of the arrow, exerting a pull of $T$ on the spring. The friction of the band on the rim of the pulley is ($T - w$), where $w$ is the weight attached to one end of the band. Let the pulley make $N$ r/min; then the work done per minute against friction by the rim of the pulley is $2\pi RN(T - w)$, and the horsepower absorbed by brake = $2\pi RN(T - w)/33,000$.

## IMPULSE AND MOMENTUM

**The product of force and time is defined as linear impulse.** The impulse of a constant force over a time interval $t_2 - t_1$ is $F(t_2 - t_1)$. If the force is not constant in magnitude but is constant in direction, the impulse is $\int_{t_1}^{t_2} F\,dt$. The dimensions of linear impulse are (force) × (time) in pound-seconds, or newton-seconds.

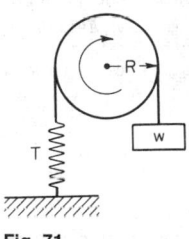

**Fig. 71**

Impulse is a **vector** quantity which has the direction of the resultant force. Impulses may be added vectorially by means of a vector polygon, or they may be resolved into components by means of a parallelogram. The **moment** of a **linear impulse** may be found in the same manner as the moment of a force. The linear impulse is represented by a directed line segment, and the moment of the impulse is the product of the magnitude of the impulse and the perpendicular distance from the line segment to the point about which the moment is taken. **Angular impulse** over a time interval $t_2 - t_1$ is a product of the sum of applied moments on a rigid body about a reference axis and time. The dimensions for angular impulse are (force) × (time) × (displacement) in foot-pound-seconds or newton-metre-seconds. **Angular impulse and linear impulse cannot be added.**

**Momentum** is also a vector quantity and can be added and resolved in the same manner as force and impulse. The dimensions of linear momentum are (force) × (time) in pound-seconds or newton-seconds, and are identical to linear impulse. An alternate statement of Newton's second law of motion is that the resultant of an unbalanced force system must be equal to the time rate of change of linear momentum, $\Sigma \mathbf{F} = d(mv)/dt$.

If a variable force acts for a certain time on a body of mass $m$, the quantity $\int_{t_1}^{t_2} F\,dt = m(v_1 - v_2)$ = the change of momentum of the body.

The **moment of momentum** can be determined by the same methods as those used for the moment of a force or moment of an impulse. The dimensions of the moment of momentum are (force) × (time) × (displacement) in foot-pound-seconds, or newton-metre-seconds.

In **plane motion** the angular momentum of a rigid body about a reference axis perpendicular to the plane of motion is the sum of the moments of linear momenta of all particles in the body about the reference axes. Specifically, **the angular momentum of a rigid body in plane motion is the vector sum of the angular momentum about the reference axis and the moment of the linear momentum of the center of gravity about the reference axis,** $\mathbf{H}_0 = I_0\omega + \mathbf{d} \times m\mathbf{v}$.

In three-dimensional rotation about a fixed axis, the angular momentum of a rigid body has components along three coordinate axes, which involve both the moments of inertia about the $x$, $y$, and $z$ axes, $I_{0_{xx}}$, $I_{0_{yy}}$, and $I_{0_{zz}}$, and the products of inertia, $I_{0_{xy}}$, $I_{0_{xz}}$, and $I_{0_{yz}}$; $H_{0_x} = I_{0_{xx}} \cdot \omega_x - I_{0_{xy}} \cdot \omega_y - I_{0_{xz}} \omega_z$, $H_{0_y} = -I_{0_{xy}} \cdot \omega_x + I_{0_{yy}} \cdot \omega_y - I_{0_{yz}} \cdot \omega_z$, and $H_{0_z} = -I_{0_{xz}}\omega_x - I_{0_{zy}} \cdot \omega_y + I_{0_{zz}} \cdot \omega_z$, where $\mathbf{H}_0 = \mathbf{H}_{0x} + \mathbf{H}_{0y} + \mathbf{H}_{0z}$.

These same equations may express the angular momentum of a rigid body in general three-dimensional motion. In this case the reference axis $O$ must be the center of gravity.

### Impact

The collision between two bodies, where relatively large forces result over a comparatively short interval of time, is called **impact**. A straight line perpendicular to the plane of contact of two colliding bodies is called the **line of impact**. If the centers of gravity of the two bodies lie on the line of contact, the impact is called **central impact,** in any other case, **eccentric impact.** If the linear momenta of the centers of gravity are also directed along the line of impact, the impact is **collinear or direct central impact.** In any other case impact is said to be **oblique.**

**Collinear Impact**  When two masses $m_1$ and $m_2$, having respective velocities $u_1$ and $u_2$, move in the same line, they will collide if $u_2 > u_1$ (Fig. 72a). During collision (Fig. 72b), kinetic energy is absorbed in the deformation of the bodies. There follows a period of restoration which may or may not be complete. If complete restoration of the energy of deformation occurs, the impact is **elastic.** If the restoration of energy is incomplete, the impact is referred to as **inelastic.** After collision (Fig. 72c), the bodies continue to move with changed velocities of $v_1$ and $v_2$. Since the contact forces on one body are equal to and opposite the contact forces on the other, the sum of the linear momenta of the two bodies is conserved; $m_1u_1 + m_2u_2 = m_1v_1 + m_2v_2$.

**The law of conservation of momentum states that the linear momentum of a system of bodies is unchanged if there is no resultant external force on the system.**

**Coefficient of Restitution**  The ratio of the velocity of separation $v_1 - v_2$ to the velocity of approach $u_2 - u_1$ is called the **coefficient of restitution** $e$. $e = (v_1 - v_2)/(u_2 - u_1)$.

The value of $e$ will depend on the shape and material properties of the colliding bodies. In elastic impact, the coefficient of restitution is unity and there is no energy loss. A coefficient of restitution of zero indicates perfectly inelastic or plastic impact, where there is no separation of the bodies after collision and the energy loss is a maximum. In **oblique impact,** the coefficient of restitution applies only to those components of velocity along the line of impact or normal to the plane of impact. The coefficient of restitution between two materials can be measured by making one body many times larger than the other so that $m_2$ is infinitely large in comparison to $m_1$. The velocity of $m_2$ is unchanged for all practical purposes during impact and $e = v_1/u_1$. For a small ball dropped from a height $H$ upon an extensive horizontal surface and rebounding to a height $h$, $e = \sqrt{h/H}$.

**Impact of Jet Water on Flat Plate**   When a jet of water strikes a flat plate perpendicularly to its surface, the force exerted by the water on the plate is $wv/g$, where $w$ is the weight of water striking the plate in a unit of time and $v$ is the velocity. When the jet is inclined to the surface by an angle, $A$, the pressure is $(wv/g) \cos A$.

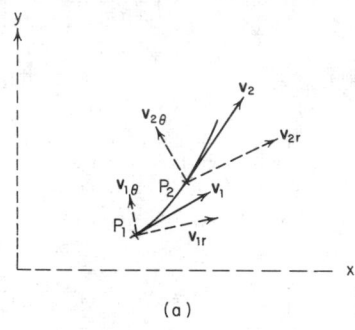

(a)

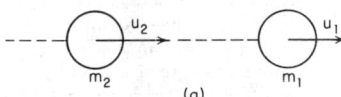

(a)

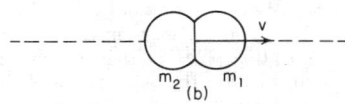

(b)

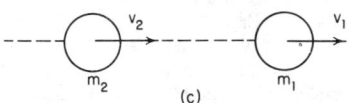

(c)

**Fig. 72**

**Variable Mass**

If the mass of a body is variable such that mass is being either added or ejected, an alternate form of Newton's second law of motion must be used which accounts for changes in mass:

$$\mathbf{F} = m\frac{d\mathbf{v}}{dt} + \frac{dm}{dt}\mathbf{u}$$

The mass $m$ is the instantaneous mass of the body, and $dv/dt$ is the time rate of change of the absolute of velocity of mass $m$. The velocity $u$ is the velocity of the mass $m$ relative to the added or ejected mass, and $dm/dt$ is the time rate of change of mass. In this case, care must be exercised in the choice of coordinates and expressions of sign. If mass is being added, $dm/dt$ is plus, and if mass is ejected, $dm/dt$ is minus.

**Fields of Force—Attraction**

The space within which the action of a physical force comes into play on bodies lying within its boundaries is called the **field of the force.**

The **strength** or **intensity of the field** at any given point is the relation between a force $F$ acting on a mass $m$ at that point and the mass. Intensity of field $= i = F/m$; $F = mi$.

The **unit of field intensity** is the same as the unit of acceleration, i.e., 1 ft/s² or 1 m/s². The intensity of a field of force may be represented by a line (or **vector** ).

A field of force is said to be **homogeneous** when the intensity of all points is uniform and in the same direction.

A field of force is called a **central field of force** with a center $O$, if the direction of the force acting on the mass particle $m$ in every point of the field passes through $O$ and its magnitude is a function only of the distance $r$ from $O$ to $m$. A line so drawn through the field of force that its direction coincides at every point with that of the force prevailing at that point is called a **line of force.**

**Rotation of Solid Bodies about Any Axis**

The general moment equations for three-dimensional motion are usually expressed in terms of the angular momentum. For a reference axis $O$, which is either a fixed axis of the center of gravity, $M_{0_x} = (dH_{0_x}/dt) - H_{0_y} \cdot \omega_z + H_{0_z} \cdot \omega_y$, $M_{0_y} = (dH_{0_y}/dt) - H_{0_z} \cdot \omega_x + H_{0_x} \cdot \omega_z$, and $M_{0_z} = (dH_{0_z}/dt) - H_{0_x}\omega_y + H_{0_y}\omega_x$. If the coordinate axes are oriented to coincide with the principal axes of inertia, $I_{0_{xx}}$, $I_{0_{yy}}$, and $I_{0_{zz}}$, a similar set of three differential equations results, involving moments, angular velocity, and angular acceleration; $M_{0_x} = I_{0_{xx}}(d\omega_x/dt) + (I_{0_{zz}} - I_{0_{yy}})\omega_y \cdot \omega_z$, $M_{0_y} = I_{0_{yy}}(d\omega_y/dt) + (I_{0_{xx}} - I_{0_{zz}})\omega_z \cdot \omega_x$, and $M_{0_z} = I_{0_{zz}}(d\omega_z/dt) + (I_{0_{yy}} - I_{0_{xx}})\omega_x\omega_y$. These equations are known as **Euler's equations of motion** and may apply to any rigid body.

**GYROSCOPIC MOTION AND THE GYROSCOPE**

**Gyroscopic motion** can be explained in terms of Euler's equations. Let $I_1$, $I_2$, and $I_3$ represent the principal moments of inertia of a gyroscope spinning with a constant angular velocity $\omega$, about axis 1, the subscripts 1, 2, and 3 representing a right-hand set of reference axes (Figs. 73 and 74). If the gyroscope is precessed about the third axis, a vector moment results along the second axis such that

$$M_2 = I_2 \, (d\omega_2/dt) + (I_1 - I_3)\omega_3\omega_1$$

Where the precession and spin axes are at right angles, the term $(d\omega_2/dt)$ equals the component of $\omega_3 \times \omega_1$ along axis 2. Because of this, in the simple case for a body of symmetry,

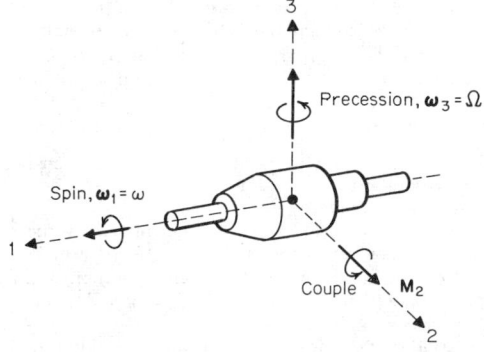

**Fig. 73**

where $I_2 = I_3$, the gyroscopic moment can be reduced to the common expression $M = I\omega\Omega$, where $\Omega$ is the rate of precession, $\omega$ the rate of spin, and $I$ the moment of inertia about the spin axis. It is important to realize that these are equations of motion and relate the applied or resulting gyroscopic moment due to forces which act *on* the rotor, as disclosed by a free-body diagram, to the resulting motion of the rotor.

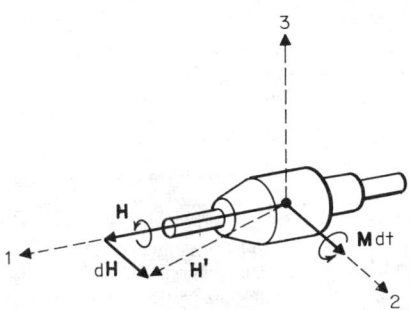

**Fig. 74**

Physical insight into the behavior of a steady precessing gyro with mutually perpendicular moment, spin, and precession axes is gained by recognizing from Fig. 74 that the change $dH$ in angular momentum $H$ is equal to the angular impulse $M$ $dt$. In time $dt$, the angular-momentum vector swings from $H$ to $H'$, owing to the velocity of precession $\omega_3$. The vector change $dH$ in angular momentum is in the direction of the applied moment $M$. This fact is inherent in the basic moment-momentum equation and can always be used to establish the correct spatial relationships between the moment, precessional, and spin vectors. It is seen, therefore, from Fig. 74 that the spin axis always turns toward the moment axis. Just as the change in direction of the mass-center velocity is in the same direction as the resultant force, so does the change in angular momentum follow the direction of the applied moment.

For example, suppose an airplane is driven by a right-handed propeller (turning like a right-handed screw when moving forward). If a gust of wind or other force turns the machine to the *left*, the gyroscopic action of the propeller will make the forward end of the shaft strive to *rise*; if the wing surface is large, this motion will be practically prevented by the resistance of the air, and the gyroscopic forces become effective merely as internal stresses, whose maximum value can be computed by the formula above. Similarly, if the airplane is dipped *downward*, the gyroscopic action will make the forward end of the shaft strive to turn to the *left*.

Modern **applications** of the gyroscope are based on one of the following properties: (1) a gyroscope mounted in three gimbal rings so as to be entirely free angularly in all directions will retain its direction *in space* in the absence of outside couples; (2) if the axis of rotation of a gyroscope turns or precesses in space, a couple or torque acts on the gyroscope (and conversely on its frame).

Devices operating on the first principle are satisfactory only for short durations, say less than half an hour, because no gyroscope is entirely without outside couple. The friction couples at the various gimbal bearings, although small, will precess the axis of rotation so that after a while the axis of rotation will have changed its direction in space. The chief device based on the first principle is the **airplane compass,** which is a freely mounted gyro, keeping its direction in space during fast maneuvers of a fighting airplane. No magnetic compass will indicate correctly during such maneuvers. After the plane is back on an even keel in steady flight, the magnetic compass once more reads the true magnetic north, and the gyro compass has to be reset to point north again.

An example of a device operating on the second principle is the **automatic pilot** for keeping a vehicle on a given course. This device has been installed on torpedoes, ships, airplanes. When the ship or plane turns from the chosen course, a couple is exerted on the gyro axis, which makes it precess and this operates electric contacts or hydraulic or pneumatic valves. These again operate on the rudders, through relays, and bring the ship back to its course.

Another application is the ship **antirolling** gyroscope. This very large gyroscope spins about a vertical axis and is mounted in a ship so that the axis can be tipped fore and aft by means of an electric motor, the precession motor. The gyro can exert a large torque on the ship about the fore-and-aft axis, which is along the "rolling" axis. The sign of the torque is determined by the direction of rotation of the precession motor, which in turn is controlled by electric contacts operated by a small pilot gyroscope on the ship, which feels which way the ship rolls and gives the signals to apply a countertorque.

The **turn indicator** for airplanes is a gyro, the frame of which is held by springs. When the airplane turns, it makes the gyro axis turn with it, and the resultant couple is delivered by the springs. Thus the elongation of the springs is a measure of the rate of turn, which is suitably indicated by a pointer.

The most complicated and ingenious application of the gyroscope is the **marine compass.** This is a pendulously suspended gyroscope which is affected by gravity and also by the earth's rotation so that the gyro axis is in equilibrium only when it points north, i.e., when it lies in the plane formed by the local vertical and by the earth's north-south axis. If the compass is disturbed so that it points away from north, the action of the earth's rotation will restore it to the correct north position in a few hours.

# FRICTION

## by Dudley D. Fuller

REFERENCES: Bowden and Tabor, "The Friction and Lubrication of Solids," Oxford. Fuller, "Theory and Practice of Lubrication for Engineers," Wiley. Ham and Crane, "Mechanics of Machinery," McGraw-Hill. Bevan, "Theory of Machines," Longmans. Shigley, "Mechanical Design," McGraw-Hill. Rabinowicz, "Friction and Wear of Materials," Wiley. Ling, Klaus, and Fein, "Boundary Lubrication—An Appraisal of World Literature," ASME, 1969.

**Friction** is the resistance that is encountered when two solid surfaces slide or tend to slide over each other. The surfaces may be either dry or lubricated. In the first case, when the surfaces are free from contaminating fluids, or films, the resistance is called **dry friction.** The friction of brake shoes on the rim of a railroad wheel is an example of dry friction.

When the rubbing surfaces are separated from each other by a very thin film of lubricant, the friction is that of **boundary** (or **greasy**) **lubrication.** The lubrication depends in this case on the strong adhesion of the lubricant to the material of the rubbing surfaces; the layers of lubricant slip over each other instead of the dry surfaces. A journal when starting, reversing, or turning at very low speed under a heavy load is an example of the condition that will cause boundary lubrication. Other examples are gear teeth (especially hypoid gears), cutting tools, wire-drawing dies, power screws, bridge trunnions, and the running-in process of most lubricated surfaces.

When the lubrication is arranged so that the rubbing surfaces are separated by a fluid film, and the load on the surfaces is carried entirely by the hydrostatic or hydrodynamic pressure in the film, the friction is that of **complete** (or **viscous**) **lubrication.** In this case, the frictional losses are due solely to the internal fluid friction in the film. Oil ring bearings, bearings with forced feed of oil, pivoted shoe-type thrust and journal bearings, bearings operating in an oil bath, hydrostatic oil pads, oil lifts, and step bearings are instances of complete lubrication.

**Incomplete lubrication** or mixed lubrication takes place when the load on the rubbing surfaces is carried partly by a fluid viscous film and partly by areas of boundary lubrication. The friction is intermediate between that of fluid and boundary lubrication. Incomplete lubrication exists in bearings with drop-feed, waste-packed, or wick-fed lubrication, or on parallel-surface bearings.

## STATIC AND SLIDING COEFFICIENTS OF FRICTION

In the absence of friction, the resultant of the forces between the surfaces of two bodies pressing upon each other is normal to the surface of contact. With friction, the resultant deviates from the normal.

If one body is pressed against another by a force $P$, as in Fig. 1, the first body will not move, provided the angle $a_0$ included between the line of action of the force and a normal to the surfaces in contact does not exceed a certain value which depends upon the nature of the surfaces. The resultant force $R$ has the same magnitude and line of action as the force $P$. In

Fig. 1, $R$ is resolved into two components: a force $N$ normal to the surfaces in contact and a force $F_r$ parallel to the surfaces in contact. From the above statement it follows that

$$F_r \leqslant N \tan a_0 \leqslant N f_0$$

where $f_0 = \tan a_0$ is called the **coefficient of friction of rest** (or of **static friction**) and $a_0$ is the **angle of friction of rest** (or angle of **repose**).

If the normal force $N$ between the surfaces is kept constant, and the tangential force $F_r$ is gradually increased, there will be no motion while $F_r < N f_0$. A state of *impending motion* is reached when $F_r$ nears the value of $N f_0$. If one surface slides over the other, being pressed together by a normal force $N$, a frictional force $F$ resisting the motion must be overcome. This force is usually smaller than $F_r$.

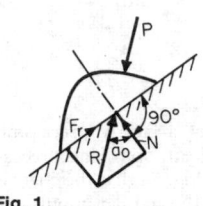

**Fig. 1**

The force $F$ is commonly expressed as $F = fN$, where $f$ is the **coefficient of sliding friction,** or **kinetic friction.** In the range of practical velocities of sliding, the coefficients of sliding friction are smaller than the coefficients of static friction. With small velocities of sliding and very clean surfaces, the two coefficients do not differ appreciably.

Under moderate pressures, the frictional force is proportional to the normal load on the rubbing surfaces. It is independent of the pressure per unit area of the surfaces. The coefficient of friction is approximately independent of the rubbing speed, when the speed is sufficiently low so as not to affect the temperature of the surface; at higher velocities, the coefficient of friction decreases as the velocity increases.

The coefficients of friction for dry surfaces (dry friction) depend on the materials sliding over each other and on the finished condition of the surfaces. With greasy (boundary) lubrication, the coefficients depend both on the materials and conditions of the surfaces and on the lubricants employed.

Coefficients of friction are sensitive to atmospheric dust and humidity, oxide films, surface finish, velocity of sliding, temperature, vibration, and the extent of contamination. In many instances the degree of contamination is perhaps the most important single variable. For example, in the table below, values for the static coefficient of friction of steel on steel are listed, and, depending upon the degree of contamination of the specimens, the coefficient of friction varies effectively from ∞ (infinity) to 0.013.

The most effective lubricants for non fluid lubrication are generally those which react chemically with the solid surface and form an adhering film that is attached to the surface with a chemical bond. This action depends upon the nature of the lubricant and upon the reactivity of the solid surface. The table below indicates that a fatty acid, such as found in animal, vegetable, and marine oils, reduces the coefficient of friction markedly only if it can react effectively with the solid surface. Paraffin oil is almost completely nonreactive.

## Coefficients of Static Friction for Steel on Steel

| Test condition | $f_0$ | Ref. |
|---|---|---|
| Degassed at elevated temp in high vacuum ∞ (weld on contact) | | 1 |
| Grease-free in vacuum | 0.78 | 2 |
| Grease-free in air | 0.39 | 3 |
| Clean and coated with oleic acid | 0.11 | 2 |
| Clean and coated with solution of stearic acid | 0.013 | 4 |

(1) Bowden and Young, *Proc. Roy. Soc.*, 1951. (2) Campbell, *Trans. ASME*, 1939. (3) Tomlinson, *Phil. Mag.*, 1929. (4) Hardy and Doubleday, *Proc. Roy. Soc.*, 1923.

Values in Table 1 of sliding and static coefficients have been selected largely from investigations where these variables have been very carefully controlled. They are representative values for smooth surfaces. It has been generally observed that sliding friction between hard materials is smaller than that between softer surfaces.

## Coefficients of Static Friction at Room Temperature*

| Surfaces | Clean | Paraffin oil | | Degree of reactivity of solid |
|---|---|---|---|---|
| Nickel | 0.7 | 0.3 | 0.28 | Low |
| Chromium | 0.4 | 0.3 | 0.3 | Low |
| Platinum | 1.2 | 0.28 | 0.25 | Low |
| Silver | 1.4 | 0.8 | 0.7 | Low |
| Glass | 0.9 | | 0.4 | Low |
| Copper | 1.4 | 0.3 | 0.08 | High |
| Cadmium | 0.5 | 0.45 | 0.05 | High |
| Zinc | 0.6 | 0.2 | 0.04 | High |
| Magnesium | 0.6 | 0.5 | 0.08 | High |
| Iron | 1.0 | 0.3 | 0.2 | Mild |
| Aluminum | 1.4 | 0.7 | 0.3 | Mild |

*From Bowden and Tabor, "The Friction and Lubrication of Solids," Oxford.

**Effect of Surface Films**  Campbell[1] observed a lowering of the coefficient of friction when oxide or sulfide films were present on metal surfaces. The reductions listed in the following table were obtained with oxide films formed by heating in air at temperatures from 100 to 500 C, and sulfide films produced by immersion in a 0.02 percent sodium sulfide solution.

## Static Coefficients of Friction $f_0$

| | Clean and dry | Oxide film | Sulfide film |
|---|---|---|---|
| Steel-steel | 0.78 | 0.27 | 0.39 |
| Brass-brass | 0.88 | | 0.57 |
| Copper-copper | 1.21 | 0.76 | 0.74 |

**Effect of Sliding Velocity**  It has generally been observed that coefficients of friction reduce on dry surfaces as sliding velocity increases. (See results of railway brake-shoe tests below.) Dokos[2] measured this reduction in friction for mild steel on medium steel. Values are for the average of four tests with high contact pressures.

[1]*Trans. ASME*, 1939; see footnotes to Table 1.
[2]*Trans. ASME*, 1946; see footnotes to Table 1.

| Sliding velocity, in/s | 0.0001 | 0.001 | 0.01 | 0.1 | 1 | 10 | 100 |
|---|---|---|---|---|---|---|---|
| $f$ | 0.53 | 0.48 | 0.39 | 0.31 | 0.23 | 0.19 | 0.18 |

**Effect of Surface Finish**  The degree of surface roughness has been found to influence the coefficient of friction. Burwell[3] evaluated this effect for conditions of boundary or greasy friction. The values listed in the table, top of page 3-27, are for sliding coefficients of friction, hard steel on hard steel.

**Solid Lubricants**  In certain applications solid lubricants are used successfully. Boyd and Robertson[4] with pressures ranging from 50,000 to 400,000 lb/in² (344,700 to 2,757,000 kN/m²) found sliding coefficients of friction $f$ for hard steel on hard steel as follows: powdered mica, 0.305; powdered soapstone, 0.306; lead iodide, 0.071; silver sulfate, 0.054; graphite, 0.058; molybdenum disulfide, 0.033; tungsten disulfide, 0.037; stearic acid, 0.029.

## Coefficients of Static Friction for Special Cases

**Masonry and Earth**  Dry masonry on brickwork, 0.6–0.7; timber on polished stone, 0.40; iron on stone, 0.3 to 0.7; masonry on dry clay, 0.51; masonry on moist clay, 0.33.

**Earth on Earth**  Dry sand, clay, mixed earth, 0.4 to 0.7; damp clay, 1.0; wet clay, 0.31; shingle and gravel, 0.8 to 1.1.

**Natural Cork**  On cork, 0.59; on pine with grain, 0.49; on glass, 0.52; on dry steel, 0.45; on wet steel, 0.69; on hot steel, 0.64; on oiled steel, 0.45; water-soaked cork on steel, 0.56; oil-soaked cork on steel, 0.42.

## Coefficients of Sliding Friction for Special Cases

**Soapy Wood**  Lesley gives for wood on wood, copiously lubricated with tallow, stearine, and soft soap (as used in launching practice), a starting coefficient of friction equal to 0.036, diminishing to an average value of 0.019 for the first 50 ft of motion of the ship. Rennic gives 0.0385 for wood on wood, lubricated with soft soap, under a load of 56 lb/in².

**Asbestos-Fabric Brake Material**  The coefficient of sliding friction $f$ of asbestos fabric against a cast-iron brake drum, according to Taylor and Holt (*NBS*, 1940) is 0.35 to 0.40 when at normal temperature. It drops somewhat with rise in brake temperature up to 300°F. (149°C) With a further increase in brake temperature from 300 to 500°F (149 to 260°C) the value of $f$ may show an increase caused by disruption of the brake surface.

**Steel Tires on Steel Rails** (Galton)

| Speed mi/h | Start | 6.8 | 13.5 | 27.3 | 40.9 | 54.4 | 60 |
|---|---|---|---|---|---|---|---|
| Values of $f$ | 0.242 | 0.088 | 0.072 | 0.07 | 0.057 | 0.038 | 0.027 |

**Railway Brake Shoes on Steel Tires**  Galton and Westinghouse give, for cast-iron brakes, the following values for $f$, which decrease rapidly with the speed of the rim; the coefficient $f$ decreases also with time, as the temperature of the shoe increases.

| Speed, mi/h | 10 | 20 | 30 | 40 | 50 | 60 |
|---|---|---|---|---|---|---|
| $f$, when brakes were applied | 0.32 | 0.21 | 0.18 | 0.13 | 0.10 | 0.06 |
| $f$, after 5 s | 0.21 | 0.17 | 0.11 | 0.10 | 0.07 | 0.05 |
| $f$, after 12 s | | 0.13 | 0.10 | 0.08 | 0.06 | 0.05 |

[3]*Jour. SAE*, 1942; see footnotes to Table 1.
[4]*Trans. ASME*, 1945; see footnotes to Table 1.

## Table 1. Coefficients of Static and Sliding Friction

(Reference letters indicate the lubricant used; numbers in parentheses give the sources. See footnote)

| Materials | Static Dry | Static Greasy | Sliding Dry | Sliding Greasy |
|---|---|---|---|---|
| Hard steel on hard steel | 0.78 (1) | 0.11 (1, a) | 0.42 (2) | 0.029 (5, h) |
|  |  | 0.23 (1, b) |  | 0.081 (5, c) |
|  |  | 0.15 (1, c) |  | 0.080 (5, i) |
|  |  | 0.11 (1, d) |  | 0.058 (5, j) |
|  |  | 0.0075 (18, p) |  | 0.084 (5, d) |
|  |  | 0.0052 (18, h) |  | 0.105 (5, k) |
|  |  |  |  | 0.096 (5, l) |
|  |  |  |  | 0.108 (5, m) |
|  |  |  |  | 0.12 (5, a) |
| Mild steel on mild steel | 0.74 (19) |  | 0.57 (3) | 0.09 (3, a) |
|  |  |  |  | 0.19 (3, u) |
| Hard steel on graphite | 0.21 (1) | 0.09 (1, a) |  |  |
| Hard steel on babbitt (ASTM No. 1) | 0.70 (11) | 0.23 (1, b) | 0.33 (6) | 0.16 (1, b) |
|  |  | 0.15 (1, c) |  | 0.06 (1, c) |
|  |  | 0.08 (1, d) |  | 0.11 (1, d) |
|  |  | 0.085 (1, e) |  |  |
| Hard steel on babbitt (ASTM No. 8) | 0.42 (11) | 0.17 (1, b) | 0.35 (11) | 0.14 (1, b) |
|  |  | 0.11 (1, c) |  | 0.065 (1, c) |
|  |  | 0.09 (1, d) |  | 0.07 (1, d) |
|  |  | 0.08 (1, e) |  | 0.08 (11, h) |
| Hard steel on babbitt (ASTM No. 10) |  | 0.25 (1, b) |  | 0.13 (1, b) |
|  |  | 0.12 (1, c) |  | 0.06 (1, c) |
|  |  | 0.10 (1, d) |  | 0.055 (1, d) |
|  |  | 0.11 (1, e) |  |  |
| Mild steel on cadmium silver |  |  |  | 0.097 (2, f) |
| Mild steel on phosphor bronze |  |  | 0.34 (3) | 0.173 (2, f) |
| Mild steel on copper lead |  |  |  | 0.145 (2, f) |
| Mild steel on cast iron |  | 0.183 (15, c) | 0.23 (6) | 0.133 (2, f) |
| Mild steel on lead | 0.95 (11) | 0.5 (1, f) | 0.95 (11) | 0.3 (11, f) |
| Nickel on mild steel |  |  | 0.64 (3) | 0.178 (3, x) |
| Aluminum on mild steel | 0.61 (8) |  | 0.47 (3) |  |
| Magnesium on mild steel |  |  | 0.42 (3) |  |
| Magnesium on magnesium | 0.6 (22) | 0.08 (22, y) |  |  |
| Teflon on Teflon | 0.04 (22) |  |  | 0.04 (22, f) |
| Teflon on steel | 0.04 (22) |  |  | 0.04 (22, f) |
| Tungsten carbide on tungsten carbide | 0.2 (22) | 0.12 (22, a) |  |  |
| Tungsten carbide on steel | 0.5 (22) | 0.08 (22, a) |  |  |
| Tungsten carbide on copper | 0.35 (23) |  |  |  |
| Tungsten carbide on iron | 0.8 (23) |  |  |  |
| Bonded carbide on copper | 0.35 (23) |  |  |  |
| Bonded carbide on iron | 0.8 (23) |  |  |  |
| Cadmium on mild steel |  |  | 0.46 (3) |  |
| Copper on mild steel | 0.53 (8) |  | 0.36 (3) | 0.18 (17, a) |
| Nickel on nickel | 1.10 (16) |  | 0.53 (3) | 0.12 (3, w) |
| Brass on mild steel | 0.51 (8) |  | 0.44 (6) |  |
| Brass on cast iron |  |  | 0.30 (6) |  |
| Zinc on cast iron | 0.85 (16) |  | 0.21 (7) |  |
| Magnesium on cast iron |  |  | 0.25 (7) |  |
| Copper on cast iron | 1.05 (16) |  | 0.29 (7) |  |
| Tin on cast iron |  |  | 0.32 (7) |  |
| Lead on cast iron |  |  | 0.43 (7) |  |
| Aluminum on aluminum | 1.05 (16) |  | 1.4 (3) |  |
| Glass on glass | 0.94 (8) | 0.01 (10, p) | 0.40 (3) | 0.09 (3, a) |
|  |  | 0.005 (10, q) |  | 0.116 (3, v) |
| Carbon on glass |  |  | 0.18 (3) |  |
| Garnet on mild steel |  |  | 0.39 (3) |  |
| Glass on nickel | 0.78 (8) |  | 0.56 (3) |  |
| Copper on glass | 0.68 (8) |  | 0.53 (3) |  |
| Cast iron on cast iron | 1.10 (16) |  | 0.15 (9) | 0.070 (9, d) |
|  |  |  |  | 0.064 (9, n) |
| Bronze on cast iron |  |  | 0.22 (9) | 0.077 (9, n) |
| Oak on oak (parallel to grain) | 0.62 (9) |  | 0.48 (9) | 0.164 (9, r) |
|  |  |  |  | 0.067 (9, s) |
| Oak on oak (perpendicular) | 0.54 (9) |  | 0.32 (9) | 0.072 (9, s) |
| Leather on oak (parallel) | 0.61 (9) |  | 0.52 (9) |  |
| Cast iron on oak |  |  | 0.49 (9) | 0.075 (9, n) |
| Leather on cast iron |  |  | 0.56 (9) | 0.36 (9, t) |
|  |  |  |  | 0.13 (9, n) |
| Laminated plastic on steel |  |  | 0.35 (12) | 0.05 (12, t) |
| Fluted rubber bearing on steel |  |  |  | 0.05 (13, t) |

(1) Campbell, *Trans. ASME*, 1939; (2) Clarke, Lincoln, and Sterrett, *Proc. API*, 1935; (3) Beare and Bowden, *Phil. Trans. Roy. Soc.*, 1935; (4) Dokos, *Trans. ASME*, 1946; (5) Boyd and Robertson, *Trans. ASME*, 1945; (6) Sachs, *Zeit. f. angew. Math. und Mech.*, 1924; (7) Honda and Yama la, *Jour. I of M*, 1925; (8) Tomlinson, *Phil. Mag.*, 1929; (9) Morin, *Acad. Roy. des Sciences*, 1838; (10) Claypoole, *Trans. ASME*, 1943; (11) Tabor, *Jour. Applied Phys.*, 1945; (12) Eyssen, General Discussion on Lubrication, *ASME*, 1937; (13) Brazier and Holland-Bowyer, General Discussion on Lubrication, *ASME*, 1937; (14) Burwell, *Jour. SAE*, 1942; (15) Stanton, "Friction," Longmans; (16) Ernst and Merchant, Conference on Friction and Surface Finish, M.I.T., 1940; (17) Gongwer, Conference on Friction and Surface Finish, M.I.T., 1940; (18) Hardy and Bircumshaw, *Proc. Roy. Soc.*, 1925; (19) Hardy and Hardy, *Phil. Mag.*, 1919; (20) Bowden and Young, *Proc. Roy. Soc.*, 1951; (21) Hardy and Doubleday, *Proc. Roy. Soc.*, 1923; (22) Bowden and Tabor, "The Friction and Lubrication of Solids," Oxford; (23) Shooter, *Research*, 4, 1951.

(a) Oleic acid; (b) Atlantic spindle oil (light mineral); (c) castor oil; (d) lard oil; (e) Atlantic spindle oil plus 2 percent oleic acid; (f) medium mineral oil; (g) medium mineral oil plus ½ percent oleic acid; (h) stearic acid; (i) grease (zinc oxide base); (j) graphite; (k) turbine oil plus 1 percent graphite; (l) turbine oil plus 1 percent stearic acid; (m) turbine oil (medium mineral); (n) olive oil; (p) palmitic acid; (q) ricinoleic acid; (r) dry soap; (s) lard; (t) water; (u) rape oil; (v) 3-in-1 oil; (w) octyl alcohol; (x) triolein; (y) 1 percent lauric acid in paraffin oil.

| | Surface | | | | | |
|---|---|---|---|---|---|---|
| | Super-finished | Ground | Ground | Ground | Ground | Grit-blasted |
| Roughness, microinches | 2 | 7 | 20 | 50 | 65 | 55 |
| Mineral oil | 0.128 | 0.189 | 0.360 | 0.372 | 0.378 | 0.212 |
| Mineral oil + 2% oleic acid | 0.116 | 0.170 | 0.249 | 0.261 | 0.230 | 0.164 |
| Oleic acid | 0.099 | 0.163 | 0.195 | 0.222 | 0.238 | 0.195 |
| Mineral oil + 2% sulfonated sperm oil | 0.095 | 0.137 | 0.175 | 0.251 | 0.197 | 0.165 |

Schmidt and Schrader confirm the marked decrease in the coefficient of friction with the increase of rim speed. They also show an irregular slight decrease in the value of $f$ with higher shoe pressure on the wheel, but they did not find the drop in friction after a prolonged application of the brakes. Their observations are as follows:

| Speed, mi/h | 20 | 30 | 40 | 50 | 60 |
|---|---|---|---|---|---|
| Coefficient of friction | 0.25 | 0.23 | 0.19 | 0.17 | 0.16 |

**Wood Brake Blocks**  According to Klein, $f$ is practically constant for velocities from 200 to 4,000 ft/min (61 to 1219 m/min) and for pressures from 7 to 142 lb/in² (48.3 to 979 kN/m²). The following values of $f$ are for wood on lengthwise fiber brake blocks carefully machined:

| | Beech | Oak | Poplar |
|---|---|---|---|
| Cast iron | 0.29–0.37 | 0.30–0.34 | 0.35–0.40 |
| Wrought iron | 0.54 | 0.51–0.40 | 0.65–0.60 |

| | Elm | Willow |
|---|---|---|
| Cast iron | 0.36–0.37 | 0.46–0.47 |
| Wrought iron | 0.60–0.49 | 0.63–0.60 |

The higher values apply for cast iron when the brake wheel is cleaned with gasoline, the lower when it is only wiped clean; the reverse holds for wrought iron.

**Hydraulic Hoists**  According to Lang, for bronze or lignum vitae sliding surfaces on bronze, $f$ is constant for slow reversing motion and for pressures of 30 to 1,500 lb/in² (20.7 to 10,340 kN/m²). For surfaces continuously lubricated, $f = 0.06$; for surfaces wet with water through numerous slots, 0.10; for surfaces running dry and creaking, up to 0.30.

**Stuffing Boxes Packed with Hemp, Cotton, or Leather**  $f$ is constant for hydraulic pressures between 15 and 750 lb/in². For **hemp or cotton packing**, loose or woven, soaked in hot tallow, smooth rod, box not set up too tightly so that the packing is still elastic, usual dimensions—even after months of use, $f = 0.06$ to $0.11$ (see also *Am. Mach.*, Feb. 3, 1898). When the packing is rendered difficult by unfavorable conditions, $f$ = up to 0.25. For well-made **leather packing rings** of soft leather $f = 0.03$ to $0.07$; if of hard, stiff tanned leather, 0.10 to 0.13; under unfavorable conditions such as rough piston and dirty water, up to 0.20. The coefficient of friction is found to change inversely to the diameter of the cylinder. The depth of the leather does not influence the friction.

**Grindstones**  The coefficient of friction between coarse-grained sandstone and cast iron is $f = 0.21$ to $0.24$; for steel, 0.29; for wrought iron, 0.41 to 0.46, according as the stone is freshly trued or dull; for fine-grained sandstone (wet grinding) $f = 0.72$ for cast iron, 0.94 for steel, and 1.0 for wrought iron.

Honda and Yamada give $f = 0.28$ to $0.50$ for carbon steel on emery, depending on the roughness of the wheel.

**Rubber Tires on Pavement**  Arnoux gives $f = 0.67$ for dry macadam, 0.71 for dry asphalt, and 0.17 to 0.06 for soft, slippery roads. For a cord tire on a sand-filled brick surface in fair condition, Agg (*Bull.* 88, *Iowa State College Engineering Experiment Station*, 1928) gives the following values of $f$ depending on the inflation of the tire:

| Inflation pressure, lb/in² | Dry pavement | | Wet pavement | |
|---|---|---|---|---|
| | Static $f_0$ | Sliding $f$ | Static $f_0$ | Sliding $f$ |
| 40 | 0.90 | 0.85 | 0.74 | 0.69 |
| 50 | 0.88 | 0.84 | 0.64 | 0.58 |
| 60 | 0.80 | 0.76 | 0.63 | 0.56 |

Tests of the Goodrich Company on wet brick pavement with balloon tires of different treads gave the following values of $f$:

| | Coefficients of friction | | | |
|---|---|---|---|---|
| | Static (before slipping) | | Sliding (after slipping) | |
| Speed, mi/h | 5 | 30 | 5 | 30 |
| Smooth tire | 0.49 | 0.28 | 0.43 | 0.26 |
| Circumferential grooves | 0.58 | 0.42 | 0.52 | 0.36 |
| Angular grooves at 60° | 0.75 | 0.55 | 0.70 | 0.39 |
| Angular grooves at 45° | 0.77 | 0.55 | 0.68 | 0.44 |

Currently, much development work is underway using various manufacturing techniques (bias ply, belted, radial, studs), tread patterns, and rubber compounds, so that it is not possible to present average values applicable to present conditions.

**Sleds**  For **unshod wooden runners** on smooth wood or stone surfaces, $f = 0.07$ (0.15) when tallow (dry soap) is used as a lubricant ( = 0.38 when not lubricated); on snow and ice, $f = 0.035$. For **runners with metal shoes** on snow and ice, $f = 0.02$. Rennie found for steel on ice, $f = 0.014$. However, as the temperature falls, the coefficient of friction will get larger. Bowden cites the following data for brass on ice:

| Temp, C | $f$ |
|---|---|
| 0 | 0.025 |
| −20 | 0.085 |
| −40 | 0.115 |
| −60 | 0.14 |

**Compound Sliding**  A body sliding across another may be

deflected crosswise from its original direction by a small force. This explains the ease with which an automobile may skid on the road or with which a plug gage can be inserted into a hole if it is rotated while being pushed in.

## ROLLING FRICTION

Rolling is substituted frequently for sliding friction, as in the case of wheels under vehicles, balls or rollers in bearings, rollers under skids when moving loads; frictional resistance to the rolling motion is substantially smaller than to sliding motion. The coefficient of rolling friction $f_r = P/L$ where $L$ is the load and $P$ is the frictional resistance.

The frictional resistance $P$ to the rolling of a cylinder under a load $L$ applied at the center of the roller (Fig. 2) is inversely proportional to the radius $r$ of the roller; $P = (k/r)L$. If $r$ is in inches, values of $k$ are as follows: hardwood on hardwood, 0.02; iron on iron, steel on steel, 0.002; hard polished steel on hard polished steel, 0.0002 to 0.0004.

Data on rolling friction are scarce. Noonan and Strange give, for steel rollers on steel plates and for loads varying from light to those causing a permanent set of the material, the following values of $k$: surfaces well finished and clean, 0.0005

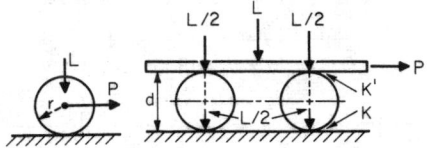

**Fig. 2**

to 0.001; surfaces well oiled, 0.001 to 0.002; surfaces covered with silt, 0.003 to 0.005; surfaces rusty, 0.005 to 0.01.

If a load $L$ is moved on rollers (Fig. 3) and if $k$ and $k'$ are the respective coefficients of friction for the lower and upper surfaces, the frictional force $P = (k + k')L/d$.

McKibben and Davidson (*Agri. Eng.*, 1939) give the data in Table 2 on the rolling resistance of various types of wheels for typical road and field conditions.

Moyer found the following average values of $f_r$ for pneumatic rubber tires properly inflated and loaded: hard road,

0.008; dry, firm, and well-packed gravel, 0.012; wet loose gravel, 0.06.

## FRICTION OF MACHINE ELEMENTS

**Work of Friction—Efficiency**  In a simple machine or assemblage of two elements, the work done by an applied force $P$ acting through the distance $s$ is measured by the product $Ps$. The useful work done is less and is measured by the product $Ll$ of the resistance $L$ by the distance $l$ through which it acts. The **efficiency** $e$ of the machine is the ratio of the useful work performed to the total work received, or $e = Ll/Ps$. The **work expended in friction** $W_f$ is the difference between the total work received and the useful work, or $W_f = Ps - Ll$. The **lost-work ratio** $= V = W_f/Ll$, and $e = 1/(1 + V)$.

If a machine consists of a train of mechanisms having the respective efficiencies $e_1, e_2, e_3 \ldots e_n$, the combined efficiency of the machine is equal to the product of these efficiencies.

**Efficiencies of Machines and Machine Elements**  The values for machine elements shown in the table at the top of p. 3-29 are from "Elements of Machine Design," by Kimball and Barr. Those for machines are from Goodman's "Mechanics Applied to Engineering." The quantities given are percentage efficiencies.

### Wedges

**Sliding in V Guides**  If a wedge-shaped slide having an angle $2b$ is pressed into a $V$ guide by a force $P$ (Fig. 4), the force normal to the wedge faces will be $N = P/\sin b$, and the frictional force opposing motion along the axis of the wedge is $F = fN = fP/\sin b = f'P$, where $f' = f/\sin b$ is improperly called the coefficient of friction. In these formulas, the fact that the elasticity of the materials permits an advance of the wedge into the guide under the load $P$ has been neglected. The common efficiency for $V$ guides is $e = 0.88$ to $0.90$.

**Taper Keys**  In Fig. 5 if the key is moved in the direction of the force $P$, the force $H$ must be overcome. The supporting reactions $K_1$, $K_2$, and $K_3$ together with the required force $P$ may be obtained by drawing the force polygon (Fig. 6). The friction angles of these faces are $a_1$, $a_2$, and $a_3$, respectively. In Fig. 6, draw $AB$ parallel to $H$ in Fig. 5, and lay it off to scale to

**Table 2. Coefficients of Rolling Friction $f_r$ for Wheels with Steel and Pneumatic Tires**

| Wheel | Inflation press, lb/in² | Load, lb | Concrete | Blue-grass sod | Tilled loam | Loose sand | Loose snow 10–14 in deep |
|---|---|---|---|---|---|---|---|
| 2.5 × 36 steel............ | .. | 1,000 | 0.010 | 0.087 | 0.384 | 0.431 | 0.106 |
| 4 × 24 steel............ | .. | 500 | 0.034 | 0.082 | 0.468 | 0.504 | 0.282 |
| 4.00–18 4-ply............ | 20 | 500 | 0.034 | 0.058 | 0.366 | 0.392 | 0.210 |
| 4 × 36 steel............ | .. | 1,000 | 0.019 | 0.074 | 0.367 | 0.413 | |
| 4.00–30 4-ply............ | 36 | 1,000 | 0.018 | 0.057 | 0.322 | 0.319 | |
| 4.00–36 4-ply............ | 36 | 1,000 | 0.017 | 0.050 | 0.294 | 0.277 | |
| 5.00–16 4-ply............ | 32 | 1,000 | 0.031 | 0.062 | 0.388 | 0.460 | |
| 6 × 28 steel............ | .. | 1,000 | 0.023 | 0.094 | 0.368 | 0.477 | 0.156 |
| 6.00–16 4-ply............ | 20 | 1,000 | 0.027 | 0.060 | 0.319 | 0.338 | 0.146 |
| 6.00–16 4-ply*............ | 30 | 1,000 | 0.031 | 0.070 | 0.401 | 0.387 | |
| 7.50–10 4-ply†............ | 20 | 1,000 | 0.029 | 0.061 | 0.379 | 0.429 | |
| 7.50–16 4-ply............ | 20 | 1,500 | 0.023 | 0.055 | 0.280 | 0.322 | |
| 7.50–28 4-ply............ | 16 | 1,500 | 0.026 | 0.052 | 0.197 | 0.205 | |
| 8 × 48 steel............ | .. | 1,500 | 0.013 | 0.065 | 0.236 | 0.264 | 0.118 |
| 7.50–36 4-ply............ | 16 | 1,500 | 0.018 | 0.046 | 0.185 | 0.177 | 0.0753 |
| 9.00–10 4-ply†............ | 20 | 1,000 | 0.031 | 0.060 | 0.331 | 0.388 | |
| 9.00–16 6-ply............ | 16 | 1,500 | 0.042 | 0.054 | 0.249 | 0.272 | 0.099 |

* Skid-ring tractor tire.
† Ribbed tread tractor tire.
All other pneumatic tires with implement-type tread.

represent $H$. From the point $A$, draw $AC$ parallel to $K_1$, i.e., making the angle $b + a_1$ with $AB$; from the other extremity of $AB$, draw $BC$ parallel to $K_2$ in Fig. 5. $AC$ and $CB$ then give the magnitudes of $K_1$ and $K_2$, respectively. Now through $C$ draw $CD$ parallel to $K_3$ to its intersection with $AD$ which has been drawn through $A$ parallel to $P$. The magnitudes of $K_3$ and $P$ are then given by the lengths of $CD$ and $DA$.

By calculation,

$$K_1/H = \cos a_2/\cos (b + a_1 + a_2)$$
$$P/K_1 = \sin (b + a_1 + a_3)/\cos a_3$$
$$P/H = \cos a_2 \sin (b + a_1 + a_3)/\cos a_3 \cos (b + a_1 + a_2)$$

If $a_1 = a_2 = a_3 = a$, then $P = H \tan (b + 2a)$, and efficiency $e$

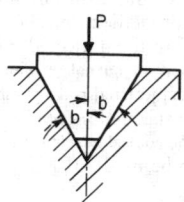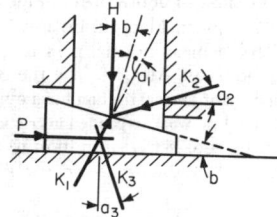

**Fig. 4**

$= \tan b/\tan (b + 2a)$. Force required to loosen the key $= P_1 = H \tan (2a - b)$. In order for the key not to slide out when force $P$ is removed, it is necessary that $b < (a_1 + a_3)$, or $b < 2a$.

The forces acting upon the taper key of Fig. 7 may be found in a similar way (see Fig. 8).

$$\begin{aligned} P &= 2H \cos a \sin (b + a)/\cos (b + 2a) \\ &= 2H \tan (b + a)/[1 - \tan a \tan (b + a)] \\ &= 2H \tan (b + a) \text{ approx} \end{aligned}$$

The force to loosen the key is $P_1 = 2H \tan (a - b)$ approx, and the efficiency $e = \tan b/\tan (b + a)$. The key will be self-locking when $b < a$, or, more generally, when $2b < (a_1 + a_3)$.

**Screws**

**Screws with Square Threads** (Fig. 9) Let $r =$ mean radius of the thread $= \frac{1}{2}$ (radius at root + outside radius), and $l =$ pitch (or lead of a single-threaded screw), both in inches; $b =$ angle of inclination of thread to a plane at right angles to the axis of screw ($\tan b = l/2\pi r$); and $f =$ coefficient of sliding friction $= \tan a$. Then for a screw in uniform motion (friction of the root and outside surfaces being neglected) there is required a force $P$ acting at right angles to the axis at the distance $r$. $P = L \tan (b \pm a) = L(l \pm 2\pi rf)/(2\pi r \pm fl)$, where the upper signs are for motion in a direction opposed to that of $L$ and the lower for motion in the same direction as that of $L$. When $b \leq a$, the screw will not "overhaul" (or move under the action of the load $L$).

The **efficiency** for motion opposed to direction in which $L$

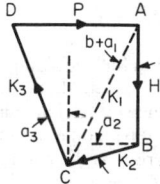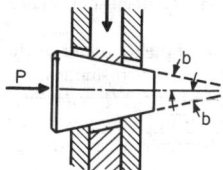

**Fig. 6** **Fig. 7**

Common bearing (singly)................ 96–98
Common bearing, long lines of shafting..... 95
Roller bearings......................... 98
Ball bearings........................... 99
Spur gear, including bearings
    Cast teeth......................... 93
    Cut teeth.......................... 96
Bevel gear, including bearings
    Cast teeth......................... 92
    Cut teeth.......................... 95
Worm gear
    Thread angle, 30°.................. 85–95
    Thread angle, 15°.................. 75–90
Belting................................ 96–98
Pin-connected chains (bicycle)........... 95–97
High-grade transmission chains........... 97–99
Weston pulley block (½ ton).......... 30–47
Epicycloidal pulley block................ 40–45
1-ton steam hoist or windlass............ 50–70
Hydraulic windlass...................... 60–80
Hydraulic jack......................... 80–90
Cranes (steam)........................ 60–70
Overhead traveling cranes.............. 30–50
Locomotives (drawbar hp/ihp).......... 65–75
Hydraulic couplings, max............... 98

acts $= e = \tan b/\tan (b + a)$; for motion in the same direction in which $L$ acts, $e = \tan (b - a)/\tan b$.

The value of $e$ is a maximum when $b = 45° - \frac{1}{2}a$; e.g., $e_{max} = 0.81$ for $b = 42°$ and $f = 0.1$. Since $e$ increases rapidly for values of $b$ up to 20°, this angle is generally not exceeded; for $b = 20°$, and $f_1 = 0.10$, $e = 0.74$. In presses, where the mechanical advantage is required to be great, $b$ is taken down to 3°, for which value $e = 0.34$ with $f = 0.10$.

Kingsbury found for square-threaded screws running in loose-fitting nuts, the following coefficients of friction: lard oil, 0.09 to 0.25; heavy mineral oil, 0.11 to 0.19; heavy oil with graphite, 0.03 to 0.15.

Ham and Ryan give for screws the following values of coefficients of friction, with medium mineral oil: high-grade materials and workmanship, 0.10; average quality materials and workmanship, 0.12; poor workmanship, 0.15.

The use of castor oil as a lubricant lowered $f$ from 0.10 to 0.066.

The coefficients of friction of a **plain collar thrust bearing** used with a **power screw** with medium mineral oil were: soft steel on cast iron, 0.12; hardened steel on cast iron, 0.09; soft steel on bronze, 0.08; hardened steel on bronze, 0.06.

The coefficients of static friction (at starting) were 30 percent higher.

**Screws with V Threads** Let $c =$ half the angle between the faces of a thread. Then, using the same notation as for square-

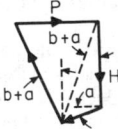

**Fig. 8**

threaded screws, for a screw in motion (neglecting friction of root and outside surfaces),

$$P = L(l \pm 2\pi rf \sec d)/(2\pi r \mp lf \sec d)$$

$d$ is the angle between a plane normal to the axis of the screw

| Rubbing speed of worm, ft/min (m/min[9]) | 100 (30.5) | 200 (61) | 300 (91.5) | 500 (152) | 800 (244) | 1200 (366) |
|---|---|---|---|---|---|---|
| Phosphor-bronze wheel, polished-steel worm | 0.054 | 0.045 | 0.039 | 0.030 | 0.024 | 0.020 |
| Single-threaded cast-iron worm and gear | 0.060 | 0.051 | 0.047 | 0.034 | 0.025 | |

through the point of the resultant thread friction, and a plane which is tangent to the surface of the thread at the same point (see Groat, *Proc. Engs. Soc. West. Penn,* **34**). Sec $d$ = sec $c$ $\sqrt{1 - (\sin b \sin c)^2}$. For small values of $b$ this reduces practically to sec $d$ = sec $c$, and, for all cases the approximation, $P = L(l \pm 2\pi rf \sec c)/(2\pi r \pm lf \sec c)$ is within the limits of probable error in estimating values to be used for $f$.

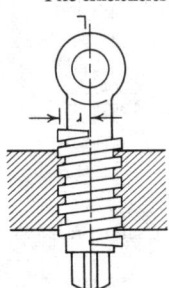

**Fig. 9**

The **efficiencies** are: $e = \tan b(1 - f \tan b \sec d)/(\tan b + f \sec d)$ for motion opposed to $L$, and $e = (\tan b - f \sec d)/\tan b(1 + f \tan b \sec d)$ for motion with $L$. If we let $\tan d' = f \sec d$, these equations reduce, respectively, to $e = \tan b/\tan (b + d')$ and $e = \tan (b - d')/\tan b$. Negative values in the latter case merely mean that the thread will not overhaul. Subtract the values from unity for actual efficiency, considering the external moment and not the load $L$ as being the driver. The efficiency of a $V$ thread is lower than that of a square thread of the same helix angle, since $d' > a$.

For a **V-threaded screw and nut,** let $r_1$ = outside radius of thread, $r_2$ = radius at root of thread, $r = (r_1 + r_2)/2$, $\tan d' = f \sec d$, $r_0$ = mean radius of nut seat = $1.5r$ (approx) and $f'$ = coefficient of friction between nut and seat.

**To tighten up the nut** the turning moment required is $M = Pr + Lr_0 f = Lr[\tan (d' + b) + 1.5f']$. **To loosen** $M = Lr[\tan (d' - b) + 1.5f']$.

The **total tension in a bolt** due to tightening up with a moment $M$ is $T = 2\pi M/(l + fl \sec b \sec d \csc b + f' 3\pi r)$. $T \div$ area at root gives unit pure tensile stress induced, $S_t$. There is also a unit torsional stress: $S_s = 2(M - 1.5rf'T)/\pi r_2^3$. The equivalent combined stress is $S = 0.35S_t + 0.65 \sqrt{S_t^2 + 4S_s^2}$.

Kingsbury, from tests on U.S. standard bolts, finds efficiencies for tightening up nuts from 0.06 to 0.12, depending upon the roughness of the contact surfaces and the character of the lubrication.

### Toothed and Worm Gearing

The efficiency of **spur and bevel gearing** depends on the material and the workmanship of the gears and on the lubricant employed. For high-speed gears of good quality the efficiency of the gear transmission is 99 percent; with slow-speed gears of average workmanship the efficiency of 96 percent is common. On the average, efficiencies of 97 to 98 percent can be considered normal.

In **helical gears,** where considerable transverse sliding of the meshing teeth on each other takes place, the friction is much greater. If $b$ and $c$ are, respectively, the spiral angles of the teeth of the driving and driven helical gears (i.e., the complements of their angles of inclination), $b + c$ is the shaft angle of the two gears, and $f = \tan a$ is the coefficient of sliding friction of the teeth, the efficiency of the gear transmission is $e = [\cos b \cos (c + a)]/[\cos c \cos (b - a)]$.

In the case of **worm gearing** when the shafts are normal to each other ($b + c = 90$), the efficiency is $e = \tan c/\tan (c + a) = (1 - pf/2\pi r)/(1 + 2\pi rf/p)$, where $c$ is the spiral angle of the worm wheel, or the lead angle of the worm; $p$ the lead, or pitch of the worm thread; and $r$ the mean radius of the worm. Typical values of $f$ are shown at the top of this page:

### Journals and Bearings

**Friction of Journal Bearings** If $P$ = total load on journal, lb; $l$ = journal length, in; and $2r$ = journal diam, in, then $p = P/2rl$ = mean normal pressure, lb/in$^2$ of the projected area of the journal. Also, if $f_1$ is the **coefficient of journal friction,** the **moment of journal friction** for a cylindrical journal is $M = f_1Pr$ in in·lb. The **work expended in friction** at a speed of $n$ rpm is $W_f = 2\pi Mn = 6.283f_1Prn$ in·lb/min. For the **conical bearing** (Fig. 10) the mean radius $r_m = (r + R)/2$ is to be used.

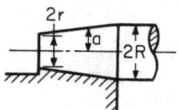

**Fig. 10**

**Values of Coefficient of Friction** For very low velocities of rotation (e.g., below 10 r/min), high loads, and with good lubrication, the coefficient of friction approaches the value of greasy friction, 0.07 to 0.15 (see Table 1 above). This is also the "pullout" coefficient of friction *on starting* the journal. With higher velocities, a fluid film is established between the journal and bearing, and the values of the coefficient of friction depend on the speed of rotation, the pressure on the bearing, and the viscosity of the oil. For journals running in complete bearing bushings, with a small clearance, i.e., with the diameter of the bushing slightly larger than the diameter of the journal, the experimental data of McKee give approximate values of the coefficient of friction as in Fig. 11.

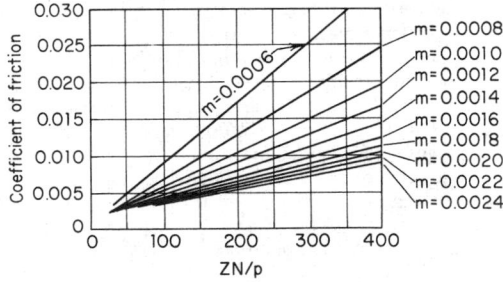

**Fig. 11** Coefficient of friction of journal.

If $d_1$ is the diameter of the bushing in inches, $d$ the diameter of the journal in inches, then $(d_1 - d)$ is the diametral clearance and $m = (d_1 - d)/d$ is the clearance ratio. The diagram of McKee (Fig. 11) gives the coefficient of friction as a function of the characteristic number $ZN/p$, where $N$ is the r/min, $p = P/$

$dl$ the average pressure, lb/in², on the projected area of the bearing, where $P$ is the load, lb; $l$ the length of bushing, in; $Z$ the absolute viscosity of the oil in centipoises (see p. 3-35). Approximate values of $Z$ at 100 (130)°F are as follows: light machine oil, 30 (16); medium machine oil, 60 (25); medium-heavy machine oil, 120 (40); heavy machine oil, 160 (60).

For purposes of design of ordinary machinery with bearing pressures from 50 to 300 lb/in² (344.7 to 2,068 kN/m²) and speeds of 100 to 3,000 rpm, values for the coefficient of journal friction can be taken from 0.008 to 0.020.

### Thrust Bearings

**Frictional Resistance**   Step bearings or pivots may be used to resist the end thrust of shafts. Let $L$ = total load, lb in the direction of the shaft axis; $dA$ = an elementary area of the thrust-bearing surface, in²; $y$ = distance of the area $dA$ from its axis of revolution, in; $p$ = pressure on $dA$ due to load $L$, lb/in²; and $f$ = coefficient of sliding friction. Then, **moment of thrust friction** = $M = fp \int y \, dA$ in·lb; and the **work expended in friction** per min at $n$ r/min = $W_f = 2\pi M n$ in·lb.

For a **ring-shaped flat step bearing** such as that shown in Fig. 12 (or a **collar bearing**), $M = \frac{1}{3}fL(D^3 - d^3)/(D^2 - d^2)$, where $D$ and $d$ are in in. For a **flat circular step bearing**, $d = 0$, and $M = \frac{1}{3}fLD$.

The value of the coefficient of sliding friction is 0.08 to 0.15 when the speed of rotation is very slow. At higher velocities when a collar or step bearing is used, $f = 0.04$ to 0.06. If the design provides for the formation of a load carrying oil film, as in the case of the Kingsbury thrust bearing, the coefficient of friction has values $f = 0.001$ to 0.0025.

Where oil is supplied from an external pump with such pressure as to separate the surfaces and provide an oil film of thickness $b$ (Fig. 12), the frictional moment is

$$M = \frac{Z \times n(D^4 - d^4)}{67 \times 10^7 \times b}$$

where $b$ is the film thickness, in, $Z$ is viscosity of lubricant in centipoises, and $n$ is rotation speed, r/min. With this kind of lubrication the frictional moment depends upon the speed of rotation of the shaft and actually approaches zero for zero shaft

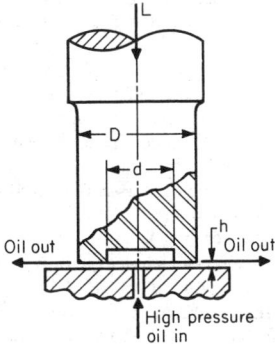

**Fig. 12**

speeds. The thrust load will be carried on a film of oil regardless of shaft rotation for as long as the pump continues to supply the required volume and pressure (see also Sec. 8).

EXAMPLE.   A hydrostatic thrust bearing carries 101,000 lb, $D$ is 16

in, $d$ is 10 in, oil-film thickness $b$ is 0.006 in, oil viscosity $Z$, 30 centipoises at operating temperature, and $n$ is 750 r/min. Substituting these values, the frictional torque $M$ is 310 in·lb (358 cm·kg). The oil supply pressure was 82.5 lb/in² (569 kN/m²); the oil flow, 12.2 gal/min (46.2 l/min).

### Frictional Forces in Pin Joints of Mechanisms

In the absence of friction, or when the effect of friction is negligible, the force transmitted by the link $b$ from the driver $a$ to the driven link $c$ (Figs. 13 and 14) acts through the centerline $OO$ of the pins connecting the link $b$ with links $a$ and $c$. With friction, this line of action shifts to the line $AA$, tangent to small circles of diameter $d$. The diameter $d$ of the circle, called the **friction circle**, for each individual joint, is equal to $fD$, where $D$ is the diameter of the pin and $f$ is the coefficient of friction between the pin and the link. The choice of the proper disposition of the tangent $AA$ with respect to the two friction circles is dictated by the consideration that friction always opposes the action of the linkage. The force $f$ opposes the motion of $a$; therefore, with friction it acts on a longer lever than without friction (Figs. 13 and 14). On the other hand, the force $F$ drives the link $c$; friction hinders its

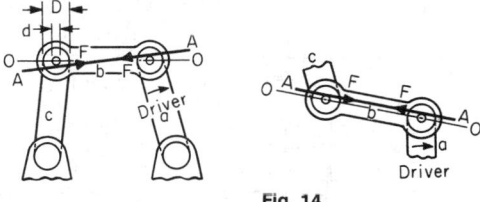

**Fig. 14**

action, and the equivalent lever is shorter with friction than without friction; the friction throws the line of action toward the center of rotation of link $c$.

EXAMPLE.   An engine eccentric (Fig. 15) is a joint where the friction loss may be large. For the dimensions shown and with a torque of 250 in·lb applied to the rotating shaft, the resultant horizontal force, with no friction, will act through the center of the eccentric and be 250/(2.5 sin 60) or 115.5 lb. With friction coefficient 0.1, the resultant force (which for a long rod remains approximately horizontal) will be tangent to the friction circle of radius 0.1 × 5, or 0.5 in, and have a magnitude of 250/(2.5 sin 60 + 0.5), or 93.8 lb (42.6 kg).

### Tension Elements

**Frictional Resistance**   In Fig. 16, let $T_1$ and $T_2$ be the tensions with which a rope, belt, chain, or brake band is strained over a drum, pulley, or sheave, and let the rope or

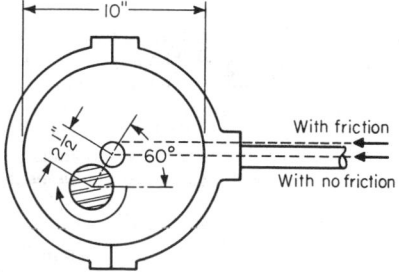

**Fig. 15**

**Table 3. Values of $e^{fa}$**

| $\dfrac{a°}{360°}$ | $f$ | | | | | | | | |
|---|---|---|---|---|---|---|---|---|---|
| | 0.1 | 0.15 | 0.2 | 0.25 | 0.3 | 0.35 | 0.4 | 0.45 | 0.5 |
| 0.1 | 1.06 | 1.1 | 1.13 | 1.17 | 1.21 | 1.25 | 1.29 | 1.33 | 1.37 |
| 0.2 | 1.13 | 1.21 | 1.29 | 1.37 | 1.46 | 1.55 | 1.65 | 1.76 | 1.87 |
| 0.3 | 1.21 | 1.32 | 1.45 | 1.60 | 1.76 | 1.93 | 2.13 | 2.34 | 2.57 |
| 0.4 | 1.29 | 1.46 | 1.65 | 1.87 | 2.12 | 2.41 | 2.73 | 3.10 | 3.51 |
| 0.425 | 1.31 | 1.49 | 1.70 | 1.95 | 2.23 | 2.55 | 2.91 | 3.33 | 3.80 |
| 0.45 | 1.33 | 1.53 | 1.76 | 2.03 | 2.34 | 2.69 | 3.10 | 3.57 | 4.11 |
| 0.475 | 1.35 | 1.56 | 1.82 | 2.11 | 2.45 | 2.84 | 3.30 | 3.83 | 4.45 |
| 0.5 | 1.37 | 1.60 | 1.87 | 2.19 | 2.57 | 3.00 | 3.51 | 4.11 | 4.81 |
| 0.525 | 1.39 | 1.64 | 1.93 | 2.28 | 2.69 | 3.17 | 3.74 | 4.41 | 5.20 |
| 0.55 | 1.41 | 1.68 | 2.00 | 2.37 | 2.82 | 3.35 | 3.98 | 4.74 | 5.63 |
| 0.6 | 1.46 | 1.76 | 2.13 | 2.57 | 3.10 | 3.74 | 4.52 | 5.45 | 6.59 |
| 0.7 | 1.55 | 1.93 | 2.41 | 3.00 | 3.74 | 4.66 | 5.81 | 7.24 | 9.02 |
| 0.8 | 1.65 | 2.13 | 2.73 | 3.51 | 4.52 | 5.81 | 7.47 | 9.60 | 12.35 |
| 0.9 | 1.76 | 2.34 | 3.10 | 4.11 | 5.45 | 7.24 | 9.60 | 12.74 | 16.90 |
| 1.0 | 1.87 | 2.57 | 3.51 | 4.81 | 6.59 | 9.02 | 12.35 | 16.90 | 23.14 |
| 1.5 | 2.57 | 4.11 | 6.59 | 10.55 | 16.90 | 27.08 | 43.38 | 69.49 | 111.32 |
| 2.0 | 3.51 | 6.59 | 12.35 | 23.14 | 43.38 | 81.31 | 152.40 | 285.68 | 535.49 |
| 2.5 | 4.81 | 10.55 | 23.14 | 50.75 | 111.32 | 244.15 | 535.49 | 1,174.5 | 2,575.9 |
| 3.0 | 6.59 | 16.90 | 43.38 | 111.32 | 285.68 | 733.14 | 1,881.5 | 4,828.5 | 12,391 |
| 3.5 | 9.02 | 27.08 | 81.31 | 244.15 | 733.14 | 2,199.90 | 6,610.7 | 19,851 | 59,608 |
| 4.0 | 12.35 | 43.38 | 152.40 | 535.49 | 1,881.5 | 6,610.7 | 23,227 | 81,610 | 286,744 |

$e^{\pi} = 23.1407.$　$\log e^{\pi} = 1.3643764.$

belt be on the point of slipping from $T_2$ toward $T_1$ by reason of the difference of tension $T_1 - T_2$. Then $T_1 - T_2 =$ circumferential force $P$ transferred by friction must be equal to the frictional resistance $W$ of the belt, rope, or band on the drum or pulley. Also, let $a$ = angle subtending the arc of contact between the drum and tension element, measured in radians. Then, disregarding centrifugal forces,

$$T_1 = T_2 e^{fa} \quad \text{and} \quad P = (e^{fa} - 1)T_1/e^{fa} = (e^{fa} - 1)T_2 = W$$

where $e$ = base of the napierian system of logarithms = 2.718+.

$f$ is the coefficient of friction of repose ($f_0$) when there is no slip of the belt or band on the drum and the coefficient of sliding friction ($f$) when slip takes place.

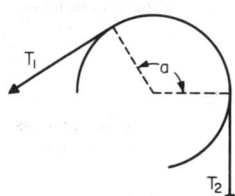

Fig. 16

Average values of $f_0$ for belts, ropes, and brake bands are as follows: For leather belt on slightly greasy wood pulley, 0.47. For leather belt on cast-iron pulley, very greasy, 0.12; slightly greasy, 0.28; moist, 0.38. For hemp rope on cast-iron drum, 0.25; on wooden drum, 0.40; on rough wood, 0.50; on polished wood, 0.33. For iron brake bands on cast-iron pulleys, 0.18. For wire ropes, Tichvinsky reports coefficients of static friction, $f_0$, for a ⅝ rope (8 × 19) on a worn-in cast-iron groove: 0.113 (dry); 0.124 (with fuller's earth); and 0.104 (when lubricated with Texaco Crater compound No. 1).

**Effect of Friction and Stiffness in Tension Elements**　Let $d$ = the rope diameter, the diameter of the chain stock, or the diameter of the pin in link chains, and $R$ = the radius of the pitch circle in which the tension element travels, both in inches. The internal friction rigidity of the tension element causes a shortening at the driving end $T_1$ of the lever arm $R$ an amount $b_1$ in inches, and at the following end a lengthening of the lever arm $R$ an amount = $b_2$ in. Then, for simultaneous winding on and off, $T_1(R - b_1) = T_2(R + b_2)$. Approximately, $b_1 = b_2 = b$, whence $b_1 + b_2 = 2b$ (approx) and $T_1 = [1 + (2b/R)]T_2$.

This results in a frictional loss, and the efficiency of transmission is $e = 1 - (2b/R)$. If the tension element is only wound on the drum, $b_1 = 0$ and $T_1 = [1 + (b/R)]T_2$, with the corresponding efficiency $e = 1 - b/R$.

For **chains** the coefficient of friction between the link faces or pivots is $f = 0.2$ to $0.3$; when $d$ = the diam of the link pin, $b = fd/2$.

For **hemp ropes**, $b = 0.03d^2$ to $0.09d^2$ according to the construction, material, and condition of the rope.

The **elastic rigidity** of the material is not a factor in simultaneous winding on and off, since the lever arm $R$ is increased equally at the points of winding on and off; the work expended in bending the tension element as it is wound on is recovered as it straightens out in unwinding. But if there is only winding on, the effect of the bending is to be taken into account.

**Efficiency of rope and chain sheaves** at low speeds, including journal friction (180° contact):

For fixed sheaves, chain, and wire rope, $e = 0.94$ to $0.96$.
For floating sheaves, chain, and wire rope, $e = 0.97$.
Hemp rope sheaves:

| Rope diam., in (cm) | ⅝ (1.59) | 1 (2.54) | 1½ (3.81) | 1¾ (4.44) | 2 (3.08) |
|---|---|---|---|---|---|
| Fixed sheaves: $e =$ | 0.95–0.96 | 0.91–0.96 | 0.89–0.93 | 0.84–0.92 | 0.85–0.91 |
| Floating sheaves: $e =$ | 0.97 | 0.96 | 0.95 | 0.94 | 0.93 |

# MECHANICS OF FLUIDS

## by J. W. Murdock

REFERENCES: *Specific.* "Handbook of Chemistry and Physics," Chemical Rubber Company. "Smithsonian Physical Tables," Smithsonian Institution. "Petroleum Measurement Tables," ASTM. "Steam Tables," ASME. "American Institute of Physics Handbook," McGraw-Hill. "International Critical Tables," McGraw-Hill. "Tables of Thermal Properties of Gases," *NBS Circular 564*. Murdock, "Fluid Mechanics and its Applications," Houghton Mifflin, 1976. "Pipe Friction Manual," Hydraulic Institute. "Flow of Fluids," ASME, 1971. "Fluid Meters," 6th ed., ASME, 1971. Murdock, ASME 64-WA/FM-6. Horton, *Engineering News*, **75**, 373, 1916. Belvins, ASME 72/WA/FE-39. Staley and Graven, ASME 72PET/30. "Temperature Measurement," PTC 19.3, ASME. Moody, *Trans. ASME*, 1944, pp. 671–684. *General.* Binder, "Fluid Mechanics," Prentice-Hall. Langhaar, "Dimensional Analysis and Theory of Models," Wiley. Murdock, "Fluid Mechanics," Drexel University Press. Rouse, "Elementary Mechanics of Fluids," Wiley. Shames, "Mechanics of Fluids," McGraw-Hill. Streeter, "Fluid Mechanics," McGraw-Hill.

## Notation

$a$ = acceleration, area, exponent
$A$ = area
$c$ = velocity of sound
$C$ = coefficient
$\mathbf{C}$ = Cauchy number
$C_p$ = pressure coefficient
$d$ = diameter, distance
$E$ = bulk modulus of elasticity, velocity of approach factor, specific energy
$\mathbf{E}$ = Euler number
$f$ = frequency, friction factor
$F$ = dimension of force, force
$\mathbf{F}$ = Froude number
$g$ = acceleration due to gravity
$g_c$ = proportionality constant = 32.1740 lbm/(lbf) (ft/s²)
$G$ = mass velocity
$h$ = head, vertical distance below a liquid surface
$H$ = geopotential altitude
$i$ = ideal
$I$ = moment of inertia
$J$ = mechanical equivalent of heat, 778.169 ft·lbf
$k$ = isentropic exponent, ratio of specific heats
$K$ = constant, resistant coefficient, weir coefficient
$\mathbf{K}$ = flow coefficient
$L$ = dimension of length, length
$m$ = mass, lbm
$\dot{m}$ = mass rate of flow, lbm/s
$M$ = dimension of mass, mass (slugs)
$\dot{M}$ = mass rate of flow, slugs/s
$\mathbf{M}$ = Mach number
$n$ = exponent for a polytropic process, roughness factor
$N$ = dimensionless number
$p$ = pressure
$P$ = perimeter, power
$q$ = heat added
$\mathbf{q}$ = flow rate per unit width

$Q$ = volumetric flow rate
$r$ = pressure ratio, radius
$R$ = gas constant, reactive force
$\mathbf{R}$ = Reynolds number
$R_h$ = hydraulic radius
$s$ = distance, second
sp. gr. = specific gravity
$S$ = scale reading, slope of a channel
$\mathbf{S}$ = Strouhal number
$t$ = time
$T$ = dimension of time, absolute temperature
$u$ = internal energy
$U$ = stream-tube velocity
$v$ = specific volume
$V$ = one-dimensional velocity, volume
$\mathbf{V}$ = velocity ratio
$W$ = work done by fluid
$\mathbf{W}$ = Weber number
$x$ = abscissa
$y$ = ordinate
$Y$ = expansion factor
$z$ = height above a datum
$Z$ = compressibility factor, crest height
$\alpha$ = angle, kinetic energy correction factor
$\beta$ = ratio of primary element diameter to pipe diameter
$\gamma$ = specific weight
$\delta$ = boundary-layer thickness
$\epsilon$ = absolute surface roughness
$\theta$ = angle
$\mu$ = dynamic viscosity
$\nu$ = kinematic viscosity
$\pi$ = 3.14159 . . ., dimensionless ratio
$\rho$ = density
$\sigma$ = surface tension
$\tau$ = unit shear stress
$\omega$ = rotational speed

## FLUIDS AND OTHER SUBSTANCES

**Substances** may be classified by their response when at rest to the imposition of a shear force. Consider the two very large plates, one moving, the other stationary, separated by a small distance $y$ as shown in Fig. 1. The space between these plates is filled with a substance whose surfaces adhere to these plates in such a manner that its upper surface moves at the same velocity as the upper plate and the lower surface is stationary. The upper surface of the substance attains a velocity of $U$ as the result of the application of shear force $F_s$. As $y$ approaches $dy$, $U$ approaches $dU$, and the rate of deformation of the substance becomes $dU/dy$. The **unit shear stress** is defined by $\tau = F_s/A_s$, where $A_s$ is the shear or surface area. The deformation characteristics of various substances are shown in Fig. 2.

An ideal or **elastic solid** will resist the shear force, and its rate of deformation will be zero regardless of loading and hence is coincident with the ordinate of Fig. 2. A **plastic** will resist the shear until its yield stress is attained, and the application of additional loading will cause it to deform continuously, or flow. If the deformation rate is directly proportional to the flow, it is called an **ideal plastic**.

If the **substance** is unable to resist even the slightest amount

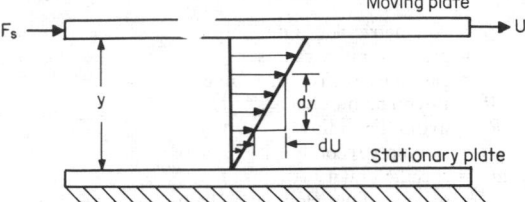

**Fig. 1**  Flow of a substance between parallel plates.

of shear without flowing, it is a **fluid**. An **ideal fluid** has no internal friction, and hence its deformation rate coincides with the abscissa of Fig. 2. All real fluids have internal friction so that their rate of deformation is proportional to the applied shear stress. If it is directly proportional, it is called a **newtonian fluid**; if not, a **nonnewtonian fluid.**

Two kinds of **fluids** are considered in this section, incompressible and compressible. A **liquid** except at very high pressures and/or temperatures may be considered incompressible. **Gases** and **vapors** are compressible fluids, but only **ideal gases** (those that follow the ideal-gas laws) are considered in this section. All others are covered in Sec. 4.

## FLUID PROPERTIES

The **density** $\rho$ of a fluid is its mass per unit volume. Its dimensions are $M/L^3$. In **fluid mechanics,** the units are slugs/ft³ and lbf·s²/ft⁴ (515.3788 kg/m³), but in **thermodynamics** (Sec. 4), the units are lbm/ft³ (16.01846 kg/m³). Numerical values of densities for selected liquids are shown in Table 1. The temperature change at 68°F (20°C) required to produce a 1

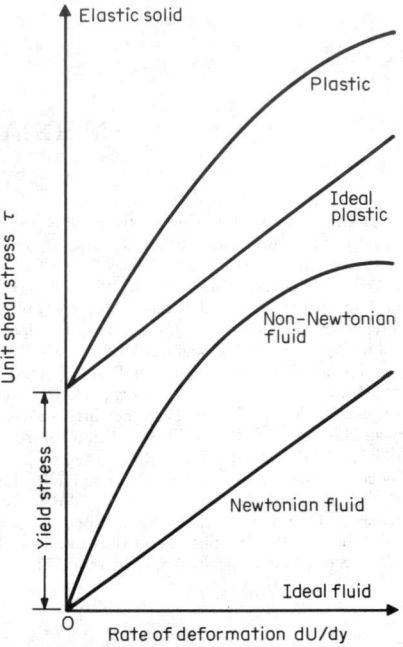

**Fig. 2**  Deformation characteristics of substances.

percent change in density varies from 12°F (6.7°C) for kerosene to 99°F (55°C) for mercury.

The **specific volume** $v$ of a fluid is its volume per unit mass. Its dimensions are $L^3/M$. The units are ft³/lbm. Specific volume is related to density by $v = 1/\rho g_c$, where $g_c$ is the proportionality constant [32.1740 (lbm/lbf)(ft/s²)]. **Specific volumes** of ideal gases may be computed from the equation of state: $v = RT/p$, where $R$ is the gas constant in ft·lbf/(lbm) (°R) (see Sec. 4), $T$ is the temperature in degrees Rankine (°F + 459.67), and $p$ is the pressure in lbf/ft² abs.

**Table 1. Density of Liquids at Atmospheric Pressure**

| Temp: °C | 0 | 20 | 40 | 60 | 80 | 100 |
|---|---|---|---|---|---|---|
| °F | 32 | 68 | 104 | 140 | 176 | 212 |
| Liquid | $\rho$, slugs/ft³ (515.4 kg/m³) | | | | | |
| Alcohol, ethyl[f] | 1.564 | 1.532 | 1.498 | 1.463 | | |
| Benzene[a,b] | 1.746 | 1.705 | 1.663 | 1.621 | 1.579 | |
| Carbon tetrachloride[a,b] | 3.168 | 3.093 | 3.017 | 2.940 | 2.857 | |
| Gasoline,[c] sp. gr. 0.68 | 1.345 | 1.310 | 1.275 | 1.239 | | |
| Glycerin[a,b] | 2.472 | 2.447 | 2.423 | 2.398 | 2.372 | 2.346 |
| Kerosene,[c] sp. gr. 0.81 | 1.630 | 1.564 | 1.536 | 1.508 | 1.480 | |
| Mercury[b] | 26.379 | 26.283 | 26.188 | 26.094 | 26.000 | 25.906 |
| Oil, machine,[c] sp. gr. 0.907 | 1.778 | 1.752 | 1.727 | 1.702 | 1.677 | 1.651 |
| Water, fresh[d] | 1.940 | 1.937 | 1.925 | 1.908 | 1.885 | 1.859 |
| Water, salt[e] | 1.995 | 1.988 | 1.975 | | | |

Computed from data given in:
[a]"Handbook of Chemistry and Physics," 52d ed., Chemical Rubber Company, 1971–1972.
[b]"Smithsonian Physical Tables," 9th rev. ed., 1954.
[c]ASTM-IP, "Petroleum Measurement Tables."
[d]"Steam Tables," ASME, 1967.
[e]"American Institute of Physics Handbook," 3d ed., McGraw-Hill, 1972.
[f]"International Critical Tables," McGraw-Hill.

**Table 2. Bulk Modulus of Elasticity, Ratio of Specific Heats of Liquids, and Velocity of Sound at One Atmosphere and 68°F (20°C)**

| Liquid | $E$ in lbf/in$^2$ (6,895 N/m$^2$) | | $k = c_p/c_v$ | $c$ in ft/s (0.3048 m/s) |
| | Isothermal $E_T$ | Isentropic $E_s$ | | |
|---|---|---|---|---|
| Alcohol, ethyl[a],[e] | 130,000 | 155,000 | 1.19 | 3,810 |
| Benzene[a],[f] | 154,000 | 223,000 | 1.45 | 4,340 |
| Carbon tetrachloride[a],[b] | 139,000 | 204,000 | 1.47 | 3,080 |
| Glycerin[f] | 654,000 | 719,000 | 1.10 | 6,510 |
| Kerosene,[a],[e] sp. gr. 0.81 | 188,000 | 209,000 | 1.11 | 4,390 |
| Mercury[e] | 3,590,000 | 4,150,000 | 1.16 | 4,770 |
| Oil, machine,[f] sp. gr. 0.907 | 189,000 | 219,000 | 1.13 | 4,240 |
| Water, fresh[a] | 316,000 | 319,000 | 1.01 | 4,860 |
| Water, salt[a],[e] | 339,000 | 344,000 | 1.01 | 4,990 |

Computed from data given in:
[a]"Handbook of Chemistry and Physics," 52d ed., Chemical Rubber Company, 1971–1972.
[b]"Smithsonian Physical Tables," 9th rev. ed., 1954.
[c]ASTM-IP, "Petroleum Measurement Tables."
[d]"Steam Tables," ASME, 1967.
[e]"American Institute of Physics Handbook," 3d ed., McGraw-Hill, 1972.
[f]"International Critical Tables," McGraw-Hill.

The **specific weight** $\gamma$ of a fluid is its weight per unit volume and has dimensions of $F/L^3$ or $M/(L^2)(T^2)$. The units are lbf/ft$^3$ or slugs/(ft$^2$) (s$^2$) (157.087 N/m$^3$). Specific weight is related to density by $\gamma = \rho g$, where $g$ is the acceleration of gravity.

The **specific gravity** (sp. gr.) of a substance is a dimensionless ratio of the density of a fluid to that of a reference fluid. Water is used as the reference fluid for solids and liquids, and air is used for gases. Since the density of liquids changes with temperature for a precise definition of **specific gravity**, the temperature of the fluid and the reference fluid should be stated, for example, 60/60°F, where the upper temperature pertains to the liquid and the lower to water. If no temperatures are stated, reference is made to water at its maximum density, which occurs at 3.98°C and atmospheric pressure. The maximum density of water is 1.9403 slugs/ft$^3$ (999.973 kg/m$^3$). See Sec. 1 for conversion factors for API and Baumé hydrometers. For gases, it is common practice to use the ratio of the molecular weight of the gas to that of air (28.9644), thus eliminating the necessity of stating the pressure and temperature for ideal gases.

The **bulk modulus of elasticity** $E$ of a fluid is the ratio of the pressure stress to the volumetric strain. Its dimensions are $F/L^2$. The units are lbf/in$^2$ or lbf/ft$^2$. $E$ depends upon the thermodynamic process causing the change of state so that $E_x = -v(\partial p/\partial v)_x$, where $x$ is the process. For ideal gases, $E_T = p$ for an isothermal process and $E_s = kp$ for an isentropic process where $k$ is the ratio of specific heats. Values of $E_T$ and $E_S$ for liquids are given in Table 2. For liquids, a mean value is used by integrating the equation over a finite interval, or $E_{xm} = -v_1 (\Delta p/\Delta v)_x = v_1 (p_2 - p_1)/(v_1 - v_2)_x$.

EXAMPLE.  What pressure must be applied to ethyl alcohol at 68°F (20°C) to produce a 1 percent decrease in volume at constant temperature?

$$\Delta p = -E_T (\Delta v/v) = -(130,000) (-0.01)$$
$$= 1,300 \text{ lbf/in}^2 (9 \times 10^6 \text{ N/m}^2)$$

In a like manner, the pressure required to produce a 1 percent decrease in the volume of mercury is found to be 35,900 lbf/in$^2$ (248 × 10$^6$ N/m$^2$). For most engineering purposes, liquids may be considered as incompressible fluids.

The **acoustic velocity**, or velocity of sound in a fluid, is given by $c = \sqrt{E_s/\rho}$. For an ideal gas $c = \sqrt{kp/\rho} = \sqrt{kg_c pv} = \sqrt{kg_c RT}$. Values of the speed of sound in liquids are given in Table 2.

EXAMPLE.  Check the value of the velocity of sound in benzene at 68°F (20°C) given in Table 2 using the isentropic bulk modulus. $c = \sqrt{E_s/\rho} = \sqrt{144 \times 223,000/1.705} = 4,340$ ft/s (1,320 m/s). Additional information on the velocity of sound is given in Secs. 4, 11, and 12.

Application of shear stress to a fluid results in the continual and permanent distortion known as flow. **Viscosity** is the resistance of a fluid to shear motion—its internal friction. This resistance is due to two phenomena: (1) cohesion of the molecules and (2) molecular transfer from one layer to another, setting up a tangential or shear stress. In liquids, cohesion predominates, and since cohesion decreases with increasing temperature, the viscosity of liquids does likewise. Cohesion is relatively weak in gases; hence increased molecular activity with increasing temperature causes an increase in molecular transfer with corresponding increase in viscosity.

The **dynamic viscosity** $\mu$ of a fluid is the ratio of the shearing stress to the rate of deformation from Fig. 1, $\mu = \tau/(dU/dy)$. Its dimensions are $(F)(T)/L^2$ or $M/(L)(T)$. The units are lbf·s/ft$^2$ or slugs/(ft) (s) [47.88164 N·s/m$^2$].

In the cgs system, the unit of dynamic viscosity is the **poise**, 2,089 × 10$^{-6}$ (lbf·s)/ft$^2$) [0.1 (N·s)/m$^2$)], but for convenience the **centipoise** (1/100 poise) is widely used. The dynamic viscosity of water at 68°F (20°C) is approximately 1 centipoise.

Table 3 gives values of dynamic viscosity for selected liquids at atmospheric pressure. Values of viscosity for fuels and lubricants are given in Sec. 6. The effect of pressure on liquid viscosity is generally unimportant in fluid mechanics except in lubricants (Sec. 6). The viscosity of water changes little at pressures up to 15,000 lbf/in$^2$, but for animal and vegetable oils it increases about 350 percent and for mineral oils about 1,600 percent at 15,000 lbf/in$^2$ pressure.

The dynamic viscosity of gases is primarily a temperature function and essentially independent of pressure. Table 4 gives values of dynamic viscosity of selected gases.

The **kinematic viscosity** $\nu$ of a fluid is its dynamic viscosity

**Table 3. Dynamic Viscosity of Liquids at Atmospheric Pressure**

| Temp:<br>°C<br>°F | 0<br>32 | 20<br>68 | 40<br>104 | 60<br>140 | 80<br>176 | 100<br>212 |
|---|---|---|---|---|---|---|
| Liquid | $\mu$, (lbf·s)/(ft²)[47.88 (N·s)/(m²)] × 10⁶ | | | | | |
| Alcohol, ethyl[a,e] | 37.02 | 25.06 | 17.42 | 12.36 | 9.028 | |
| Benzene[a] | 19.05 | 13.62 | 10.51 | 8.187 | 6.871 | |
| Carbon tetrachloride[e] | 28.12 | 20.28 | 15.41 | 12.17 | 9.884 | |
| Gasoline,[b] sp. gr. 0.68 | 7.28 | 5.98 | 4.93 | 4.28 | | |
| Glycerin[d] | 252,000 | 29,500 | 5,931 | 1,695 | 666.2 | 309.1 |
| Kerosene,[b] sp. gr. 0.81 | 61.8 | 38.1 | 26.8 | 20.3 | 16.3 | |
| Mercury[a] | 35.19 | 32.46 | 30.28 | 28.55 | 27.11 | 25.90 |
| Oil, machine,[a] sp. gr. 0.907 | | | | | | |
| "Light" | 7,380 | 1,810 | 647 | 299 | 164 | 102 |
| "Heavy" | 66,100 | 9,470 | 2,320 | 812 | 371 | 200 |
| Water, fresh[c] | 36.61 | 20.92 | 13.61 | 9.672 | 7.331 | 5.827 |
| Water, salt[d] | 39.40 | 22.61 | 18.20 | | | |

Computed from data given in:
[a]"Handbook of Chemistry and Physics," 52d ed., Chemical Rubber Company, 1971–1972.
[b]"Smithsonian Physical Tables," 9th rev. ed., 1954.
[c]"Steam Tables," ASME, 1967.
[d]"American Institute of Physics Handbook, " 3d ed., McGraw-Hill, 1972.
[e]"International Critical Tables," McGraw-Hill.

divided by its density, or $\nu = \mu/\rho$. Its dimensions are $L^2/T$. The units are ft²/s ($9.290304 \times 10^{-2}$ m²/s).

In the cgs system, the unit of kinematic viscosity is the **stoke** ($1 \times 10^{-4}$ m²/s²), but for convenience, the **centistoke** (1/100 stoke) is widely used. The kinematic viscosity of water at 68°F (20°C) is approximately 1 centistoke.

The standard device for **experimental determination of kinematic viscosity** in the United States is the **Saybolt Universal** viscometer. It consists essentially of a metal tube and an orifice built to rigid specifications and calibrated. The time required for a gravity flow of 60 cubic centimeters is called the SSU (Saybolt seconds Universal). Approximate conversions of SSU to stokes may be made as follows:

$32 < $ SSU $< 100$ seconds, stokes
$$= 0.00226\,(SSU) - 1.95/(SSU)$$
SSU $> 100$ seconds, stokes $= 0.00220\,(SSU) - 1.35/(SSU)$

For viscous oils, the **Saybolt Furol** viscometer is used. Approximate conversions of SSF (Saybolt seconds Furol) may be made as follows:

$25 < $ SSF $< 40$ seconds, stokes $= 0.0224\,(SSF) - 1.84/\,(SSF)$
SSF $> 40$ seconds, stokes $= 0.0216\,(SSF) - 0.60/\,(SSF)$

For exact conversions of Saybolt viscosities, see ASTM D445-71 and Sec. 6.

The **surface tension** $\sigma$ of a fluid is the work done in extending the surface of a liquid one unit of area or work per unit area. Its dimensions are F/L. The units are lbf/ft (14.5930 N/m).

Values of $\sigma$ for various interfaces are given in Table 5. Surface tension decreases with increasing temperature. Surface tension is of importance in the formation of bubbles and in problems involving atomization.

**Capillary** action is due to surface tension, **cohesion** of the

**Table 4. Viscosity of Gases at One Atmosphere**

| Temp:<br>°C<br>°F | 0<br>32 | 20<br>68 | 60<br>140 | 100<br>212 | 200<br>392 | 400<br>752 | 600<br>1112 | 800<br>1472 | 1000<br>1832 |
|---|---|---|---|---|---|---|---|---|---|
| Gas | $\mu$, (lbf·s)/ft²[47.88(N·s)/(m²)] × 10⁸ | | | | | | | | |
| Air* | 35.67 | 39.16 | 41.79 | 45.95 | 53.15 | 70.42 | 80.72 | 91.75 | 100.8 |
| Carbon dioxide* | 29.03 | 30.91 | 35.00 | 38.99 | 47.77 | 62.92 | 74.96 | 87.56 | 97.71 |
| Carbon monoxide† | 34.60 | 36.97 | 41.57 | 45.96 | 52.39 | 66.92 | 79.68 | 91.49 | 102.2 |
| Helium* | 38.85 | 40.54 | 44.23 | 47.64 | 55.80 | 71.27 | 84.97 | 97.43 | |
| Hydrogen*·† | 17.43 | 18.27 | 20.95 | 21.57 | 25.29 | 32.02 | 38.17 | 43.92 | 49.20 |
| Methane* | 21.42 | 22.70 | 26.50 | 27.80 | 33.49 | 43.21 | | | |
| Nitrogen*·† | 34.67 | 36.51 | 40.14 | 43.55 | 51.47 | 65.02 | 76.47 | 86.38 | 95.40 |
| Oxygen† | 40.08 | 42.33 | 46.66 | 50.74 | 60.16 | 76.60 | 90.87 | 104.3 | 116.7 |
| Steam‡ | | 18.49 | 21.89 | 25.29 | 33.79 | 50.79 | 67.79 | 84.79 | |

Computed from data given in:
*"Handbook of Chemistry and Physics," 52d ed., Chemical Rubber Company, 1971–1972.
†"Tables of Thermal Properties of Gases," *NBS Circular* 564, 1955.
‡"Steam Tables," ASME, 1967.

liquid molecules, and the **adhesion** of the molecules on the surface of a solid. This action is of importance in fluid mechanics because of the formation of a meniscus (curved section) in a tube. When the adhesion is greater than the cohesion, a liquid "wets" the solid surface, and the liquid will rise in the tube and conversely will fall if the reverse. Figure 3 illustrates this effect on manometer tubes. In the reading of a manometer, all data should be taken at the center of the meniscus.

**Table 5. Surface Tension of Liquids at One Atmosphere and 68°F (20°C)**

| Liquid | $\delta$, lbf/ft (14.59 N/m) $\times 10^3$ | | |
| --- | --- | --- | --- |
| | In vapor | In air | In water |
| Alcohol, ethyl* | 1.56 | 1.53 | |
| Benzene* | 2.00 | 1.98 | 2.40 |
| Carbon tetrachloride* | 1.85 | 1.83 | 3.08 |
| Gasoline,* sp. gr. 0.68 | 1.3–1.6 | | 2.7–3.6 |
| Glycerin* | 4.30 | 4.35 | |
| Kerosene,* sp. gr. 0.81 | 1.6–2.2 | | |
| Mercury* | 32.6§ | 32.8 | 25.7 |
| Oil, machine,† sp. gr. 0.907 | 2.5 | 2.6 | 2.3–3.7 |
| Water, fresh‡ | | 4.99 | |
| Water, salt‡ | | 5.04 | |

Computed from data given in:
*"International Critical Tables," McGraw-Hill.
†ASTM-IP, "Petroleum Measurement Tables."
‡"American Institute of Physics Handbook," 3d ed., McGraw-Hill, 1972.
§In vacuum.

The **vapor pressure** $p_v$ of a fluid is the pressure at which its liquid and vapor are in equilibrium at a given temperature. See Sec. 4 for further definitions and values.

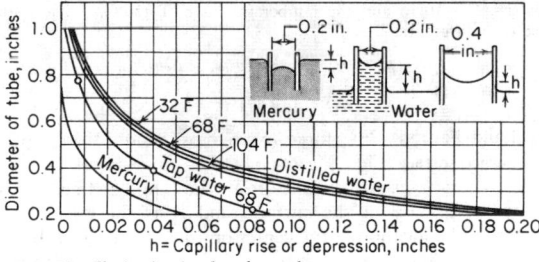

**Fig. 3** Capillarity in circular glass tubes.

## FLUID STATICS

**Pressure** $p$ is the force per unit area exerted on or by a fluid and has dimensions of $F/L^2$. In fluid mechanics and in thermodynamic equations, the units are lbf/ft$^2$ (47.88026 N/m$^2$), but engineering practice is to use units of lbf/in$^2$ (6,894.757 N/m$^2$).

The relationship between **absolute pressure, gage pressure,** and **vacuum** is shown in Fig. 4. Most fluid-mechanics equations and **all** thermodynamic equations require the use of **absolute pressure,** and unless otherwise designated, a pressure should be understood to be **absolute pressure.** Common practice is to denote absolute pressure as lbf/ft$^2$ abs, or psfa, lbf/in$^2$ abs or psia; and in a like manner for gage pressure lbf/ft$^2$ g, lbf/in$^2$ g, and psig.

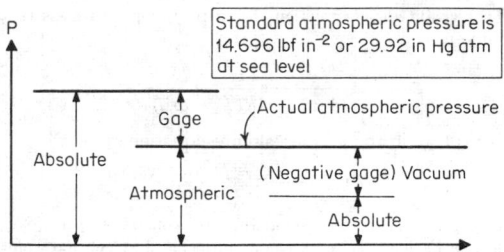

**Fig. 4** Pressure relations.

According to **Pascal's** principle, the pressure in a static fluid is the same in all directions.

The **basic equation of fluid statics** is obtained by consideration of a fluid particle at rest with respect to other fluid particles, all being subjected to body-force accelerations of $a_x$, $a_y$, and $a_z$ opposite the directions of $x$, $y$, and $z$, respectively, and the acceleration of gravity in the $z$ direction, resulting in the following:

$$dp = -\rho \left[ a_x \, dx + a_y \, dy + (a_z + g) \, dz \right]$$

**Pressure-Height Relations** For a fluid at rest and subject only to the gravitational force, $a_x$, $a_y$, and $a_z$ are zero and the **basic equation for fluid statics reduces to** $dp = -\rho g \, dz = \gamma \, dz$.

**Liquids** (Incompressible Fluids) The pressure-height equation integrates to $(p_1 - p_2) = \rho g (z_2 - z_1) = \gamma (z_2 - z_1) = \Delta p = \gamma h$, where $h$ is measured from the liquid surface (Fig. 5).

EXAMPLE. A large closed tank is partly filled with 68°F (20°C) benzene. If the pressure on the surface is 10 lb/in$^2$, what is the pressure in the benzene at a depth of 11 ft below the liquid surface?

$$p_1 = \rho g h + p_2 = \frac{1.705 \times 32.17 \times 11}{144} + 10$$
$$= 14.19 \text{ lbf/in}^2 \ (9,784 \times 10^4 \text{ N/m}^2)$$

**Ideal Gases** (Compressible Fluids) For problems involving the upper atmosphere, it is necessary to take into account the variation of gravity with altitude. For this purpose, the

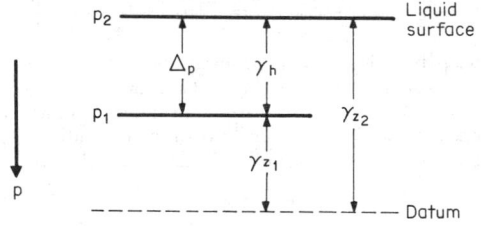

**Fig. 5** Pressure equivalence.

**geopotential altitude** $H$ is used, defined by $H = Z/(1 + z/r)$, where $r$ is the radius of the earth ($\approx 21 \times 10^6$ ft $\approx 6.4 \times 10^6$ m) and $z$ is the height above sea level. The integration of the pressure-height equation depends upon the thermodynamic process. For an **isothermal process** $p_2/p_1 = e^{-(H_2 - H_1)/RT}$ and for a **polytropic process** ($n \neq 1$)

$$\frac{p_2}{p_1} = \left[ 1 - \frac{(n - 1)(H_2 - H_1)}{nRT_1} \right]^{n/(n-1)}$$

**Temperature-height relations** for a polytropic process ($n \neq 1$) are given by

$$\frac{n}{1-n} = \frac{H_2 - \dot{H}_1}{R(T_2 - T_1)}$$

Substituting in the **pressure-altitude** equation,

$$p_2/p_1 = (T_2/T_1)^{(H_2 - H_1)/R(T_1 - T_2)}$$

EXAMPLE. The U.S. Standard Atmosphere 1962 (Sec. 11) is defined as having a sea-level temperature of 59°F (15°C) and a pressure of 2,116.22 lbf/ft². From sea level to a geopotential altitude of 36,089 ft (11,000 m) the temperature decreases linearly with altitude to −69.70°F (−56.5°C). Check the value of pressure ratio at this altitude given in the standard table.

Noting that $T_1 = 59 + 459.67 = 518.67$, $T_2 = -69.70 + 459.67 = 389.97$, and $R = 53.34$ ft·lbf/(lbm) (°R),

$$\begin{aligned}
p_2/p_1 &= (T_2/T_1)^{(H_2 - H_1)/R(T_1 - T_2)}\\
&= (389.97/518.67)^{(36,089-0)/53.34(518.67-389.97)}\\
&= 0.2233 \text{ vs. tabulated value of } 0.2234
\end{aligned}$$

**Pressure-sensing Devices**  The two principal devices using liquids are the **barometer** and the **manometer.** The barometer senses absolute pressure and the manometer senses pressure differential. For discussion of the barometer and other pressure-sensing devices, refer to Sec. 16.

**Manometers** are a direct application of the basic equation of fluid statics and serve as a pressure standard in the range of $\frac{1}{10}$ in of water to 100 lbf/in². The most familiar type of manometer is the **U tube** shown in Fig. 6a. Because of the necessity of

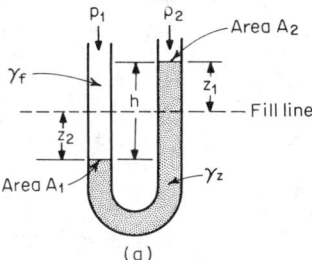

(a)

**Fig. 6**  (a) U-tube manometer;

observing both legs simultaneously, the **well** or **cistern** type (Fig. 6b) is sometimes used. The **inclined manometer** (Fig. 6c) is a special form of the well-type manometer designed to enhance the readability of small pressure differentials. Application of the basic equation of fluid statics to each of the types

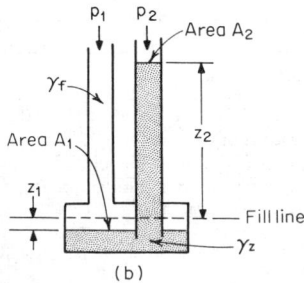

(b)

**Fig. 6**  (b) Well or cistern type of manometer.

results in the following equations. For the **U tube**, $p_1 - p_2 = (\gamma_m - \gamma_f)h$, where $\gamma_m$ and $\gamma_f$ are the specific weights of the manometer and sensed fluids, respectively, and $h$ is the verti-

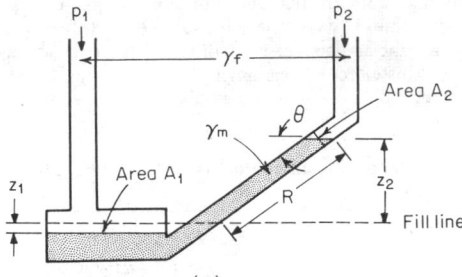

(c)

**Fig. 6**  (c) Inclined manometer.

cal distance between the liquid interfaces. For the **well type**, $p_1 - p_2 = (\gamma_m - \gamma_f)(z_2)(1 + A_2/A_1)$, where $A_1$ and $A_2$ are as shown in Fig. 6b and $z_2$ is the vertical distance from the fill line to the upper interface. Commercial manufacturers of well-type manometers correct for the area ratios so that $p_1 - p_2 = (\gamma_m - \gamma_f)S$, where $S$ is the scale reading and is equal to $z_2(1 + A_2/A_1)$. For this reason, scales should not be interchanged between U type or well type or between well types without consulting the manufacturer. For inclined manometers,

$$p_1 - p_2 = (\gamma_m - \gamma_f)(A_2/A_1 + \sin\theta)R$$

where $R$ is the distance along the inclined tube. Commercial inclined manometers also have special scales so that $p_1 - p_2 = (\gamma_m - \gamma_f)S$, where $S = (A_2/A_1 + \sin\theta)R$.

EXAMPLE. A U-tube manometer containing mercury is used to sense the difference in water pressure. If the height between the interfaces is 10 in and the temperature is 68°F (20°C), what is the pressure differential?

$$\begin{aligned}
p_1 - p_2 &= (\gamma_m - \gamma_f)h = g(\rho_m - \rho_f)h\\
&= 32.17(26.283 - 1.937)(10/12)\\
&= 652.7 \text{ lbf/ft}^2 \ (3.152 \times 10^4 \text{ N/m}^2)
\end{aligned}$$

**Liquid Forces**  The force exerted by a liquid on a **plane submerged surface** (Fig. 7) is given by $F = \int p\, dA = \gamma\int h\, dA = \gamma h_c A$, where $h_c$ is the distance from the liquid surface to the center of gravity of the surface, and $A$ is the area of the surface. The location of the center of this force is given by

$$s_F = s_c + I_G/s_c A$$

where $s_F$ is the inclined distance from the liquid surface to the center of force, $s_c$ the inclined distance to the center of gravity of the surface, and $I_G$ the moment of inertia around its center of gravity. Values of $I_G$ are given in Sec. 5. From Fig. 7, $h = R\sin\theta$, so that the vertical center of force becomes

$$h_F = h_c + I_G(\sin\theta)^2/h_c A$$

EXAMPLE. Determine the force and its location acting on a rectangular gate 3 ft wide and 5 ft high at the bottom of a tank containing 68°F (20°C) water, 12 ft deep, (1) if the gate is vertical, and (2) if it is inclined 30° from horizontal.

1. *Vertical Gate*

$$\begin{aligned}
F &= \gamma g h_c A = \rho g h_c A\\
&= 1.937 \times 32.17(12 - 5/2)(5 \times 3)\\
&= 8,800 \text{ lbf} \ (3.914 \times 10^4 \text{ N})
\end{aligned}$$

$h_F = h_c + I_G (\sin \theta)^2/h_c A$, from Sec. 5, $I_G$ for a rectangle = (width) (height)$^3$/12

$h_F = (12 - 5/2) + (3 \times 5^3/12) (\sin 90°)^2/(12 - 5/2) (3 \times 5)$

$h_F = 9.719$ ft (2.962 m)

   2. *Inclined Gate*

$F = \gamma h_c A = \rho g h_c A$
   $= 1.937 \times 32.17 (12 - 5/2 \sin 30°) (5 \times 3)$
   $= 10,048$ lbf ($4.470 \times 10^4$ N)

$h_F = h_c + I_G (\sin \theta)^2/h_c A$
   $= (12 - 5/2 \sin 30°) + (3 \times 5^3/12) (\sin 30°)^2/ (12 - 5/2 \sin 30°) (3 \times 5)$
   $= 10.80$ ft (3.291 m)

**Forces on irregular surfaces** may be obtained by considering their horizontal and vertical components. The vertical component $F_z$ equals the weight of liquid above the surface and acts through the centroid of the volume of the liquid above the surface. The horizontal component $F_x$ equals the force on a vertical projection of the irregular surface. This force may be calculated by $F_x = \gamma h_{cx} A_x$, where $h_{cx}$ is the distance from the surface center of gravity of the horizontal projection, and $A_x$ is the projected area. The forces may be combined by $F = \sqrt{F_z^2 + F_x^2}$.

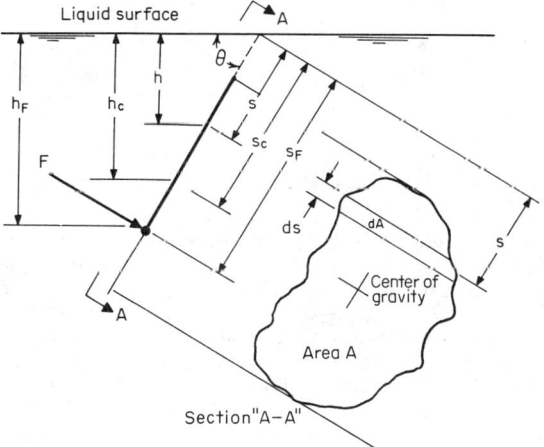

**Fig. 7**   Notation for liquid force on submerged surfaces.

When fluid masses are **accelerated** without relative motion between fluid particles, the basic equation of fluid statics may be applied. For **translation** of a **liquid** mass due to uniform acceleration, the basic equation integrates to

$p_2 - p_1 = -\rho[(x_2 - x_1)a_x$
$\qquad\qquad + (y_2 - y_1)a_y + (z_2 - z_1) (a_z + g)]$

**EXAMPLE.**   An open tank partly filled with a liquid is being accelerated up an inclined plane as shown in Fig. 8. The uniform acceleration is 20 ft/s² and the angle of the incline is 30°. What is the angle of the free surface of the liquid? Noting that on the free surface $p_2 = p_1$ and that the acceleration in the $y$ direction is zero, the basic equation reduces to

$$(x_2 - x_1)a_x + (z_2 - z_1) (a_z + g) = 0$$

Solving for tan $\theta$,

$$\tan \theta = \frac{z_1 - z_2}{x_2 - x_1} = \frac{a_x}{a_z + g} = \frac{a \cos \alpha}{a \sin \alpha + g}$$
$$= (20 \cos 30°)/(20 \sin 30° + 32.17) = 0.4107$$
$$\theta = 22°20'$$

For **rotation** of liquid masses with uniform rotational acceleration, the basic equation integrates to

$$p_2 - p_1 = \rho \left[ \frac{\omega^2}{2} (x_2^2 - x_1^2) - g(z_2 - z_1) \right]$$

where $\omega$ is the rotational speed in rad/s and $x$ is the radial distance from the axis of rotation.

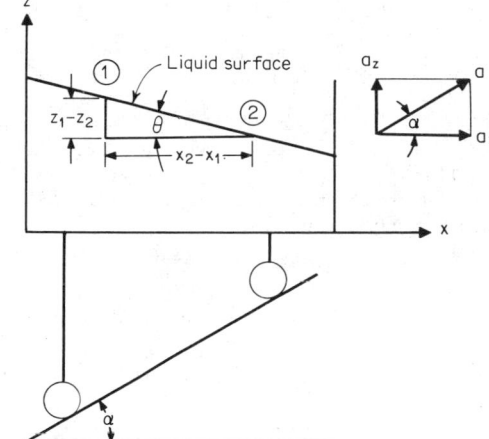

**Fig. 8**   Notation for translation example.

**EXAMPLE.**   The closed cylindrical tank shown in Fig. 9 is 4 ft in diameter and 10 ft high and is filled with 104°F (40°C) benzene. The tank is rotated at 250 r/min about an axis 3 ft from its centerline. Compute the maximum pressure differential in the tank. Analysis of the rotation equation indicates that the maximum pressure will occur at the maximum rotational radius and the minimum elevation and, con-

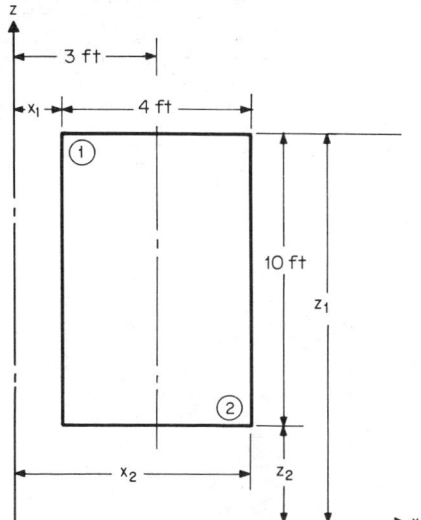

**Fig. 9**   Notation for rotation example.

versely, the minimum at the minimum rotational radius and maximum elevation. From Fig. 9, $x_1 = 3 - 4/2 = 1$ ft, $x_2 = 1 + 4 = 5$ ft, $z_2 - z_1 = -10$ ft, and the rotational speed $\omega = 2\pi N/60 = 2\pi(250)/60 = 26.18$ rad/s. Substituting into the rotational equation,

$$p_2 - p_1 = \rho \left[ \frac{\omega^2}{2} (x_2^2 - x_1^2) - g(z_2 - z_1) \right]$$
$$= \frac{1.663}{144} \left[ \frac{(26.18)^2}{2} (5^2 - 1^2) - 32.17(-10) \right]$$
$$p_2 - p_1 = 98.70 \text{ lbf/in}^2 \ (6.805 \times 10^5 \text{ N/m}^2)$$

**Buoyancy** Archimedes' principle states that a body immersed in a fluid is buoyed up by a force equal to the weight of the fluid displaced. If an object immersed in a fluid is heavier than the fluid displaced, it will sink to the bottom, and if lighter, it will rise. From the free-body diagram of Fig. 10, it is seen that for vertical equilibrium,

$$\Sigma F_z = 0 = F_B - F_g - F_D$$

where $F_B$ is the buoyant force, $F_g$ the gravity force (weight of body), and $F_D$ the force required to prevent the body from rising. The buoyant force being the weight of the displaced liquid, the equilibrium equation may be written as

$$F_D = F_B - F_g = \gamma_f V - \gamma_0 V = (\gamma_f - \gamma_0) \ V$$

where $\gamma_f$ is the specific weight of the fluid, $\gamma_0$ is the specific weight of the object, and $V$ is the volume of the object.

EXAMPLE. An airship has a volume of 3,700,000 ft³ and is filled with hydrogen. What is its gross lift in air at 59°F (15°C) and 14.696 psia? Noting that $\gamma = p/RT$,

$$F_D = (\gamma_f - \gamma_0)V = \left( \frac{p}{R_a T} - \frac{p}{R_{H_2} T} \right) V$$
$$= \frac{pV}{T} \left( \frac{1}{R_a} - \frac{1}{R_{H_2}} \right)$$
$$= \frac{144 \times 14.696 \times 3,700,000}{59 + 459.7} \left( \frac{1}{53.34} - \frac{1}{766.8} \right)$$
$$= 263,300 \text{ lbf} \ (1.171 \times 10^6 \text{ N})$$

**Flotation** is a special case of buoyancy where $F_D = 0$, and hence $F_B = F_g$.

EXAMPLE. A crude **hydrometer** consists of a cylinder of ½ in diameter and 2 in length surmounted by a cylinder of ⅛ in diameter and 10 in long. Lead shot is added to the hydrometer until its total weight is 0.32 oz. To what depth would this hydrometer float in 104°F (40°C) glycerin? For flotation, $F_B = F_g = \gamma_f V = \rho_f g V$ or $V = F_B/\rho_f g = (0.32/16)/(2.423 \times 32.17) = 2.566 \times 10^{-4}$ ft³. Volume of cylindrical

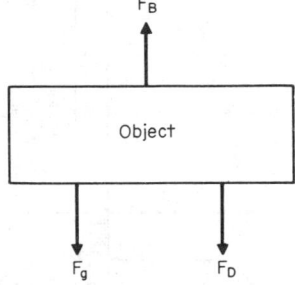

**Fig. 10**  Free-body diagram of an immersed object.

portion of hydrometer $= V_c = \pi D^2 L/4 = \pi(0.5/12)^2(2/12)/4 = 2.273 \times 10^{-4}$ ft³. Volume of stem immersed $= V_S = V - V_c = 2.566 \times 10^{-4} - 2.273 \times 10^{-4} = 2.930 \times 10^{-5}$ ft³. Length of immersed stem $= L_S = 4 V_S/\pi D^2 = (4 \times 2.930 \times 10^{-5})/\pi(0.125/12)^2 = 0.3438$ ft $= 0.3438 \times 12 = 4.126$ in. Total immersion $= L + L_S = 2 + 4.126 = 6.126$ in (0.156 m).

**Static Stability** A body is in static equilibrium when the imposition of a small displacement brings into action forces that tend to restore the body to its original position. For **completely submerged bodies,** the center of buoyancy and the center of gravity must lie on the same vertical line and the center of buoyancy must be located above the center of gravity. Figure 11a shows a balloon and its basket in its normal position with the center of buoyancy $B$ above and on the same vertical line as the center of gravity $G$. Figure 11b shows the balloon displaced from its normal position. In this position,

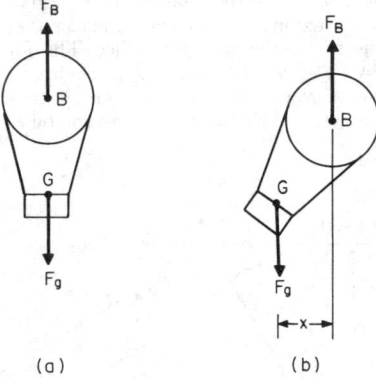

**Fig. 11**  Stability of an immersed body.

there is a couple $F_g x$ which tends to restore the balloon and its basket to its original position. For **floating bodies,** the center of gravity and the center of buoyancy must lie on the same vertical line, but the center of buoyancy may be below the center of gravity, as is common practice in surface-ship design. It is required that when displaced, the line of action of the buoyant force intersect the centerline above the center of gravity. Figure 12a shows a floating body in its normal position with its center of gravity $G$ on the same vertical line and above the center of buoyancy $B$. Figure 12b shows the object displaced. The intersection of the line of action of the buoyant force with the centerline of the body at $M$ is called the metacenter. As shown, this is above the center of buoyancy and sets up a restoring couple. When the metacenter is below the center of gravity, the object will capsize (see Sec. 11).

## FLUID KINEMATICS

**Steady and Unsteady Flow** If at every point in the fluid stream, none of the local fluid properties changes with time, the flow is said to be steady. The mathematical conditions for steady flow are met when $\partial(\text{fluid properties})/\partial t = 0$. While flow is generally unsteady by nature, many real cases of unsteady flow may be treated as steady flow by using average properties or by changing the space reference. The amount of error produced by the averaging technique depends upon the nature of the unsteady flow, but the latter technique is error-free when it can be applied.

**Streamlines and Stream Tubes**  Velocity has both magnitude and direction and hence is a vector. A **streamline** is a line which gives the *direction* of the velocity of a fluid particle at each point in the flow stream. When streamlines are connected by a closed curve in steady flow, they will form a boundary

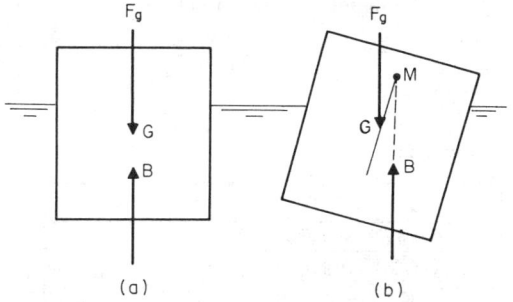

*(a)*                              *(b)*

**Fig. 12**  Stability of a floating body.

through which the fluid particles cannot pass. The space between the streamlines becomes a **stream tube**. The stream-tube concept broadens the application of fluid-flow principles; for example, it allows treating the flow inside a pipe and the flow around an object with the same laws. A stream tube of decreasing size approaches its own axis, a central streamline; thus equations· developed for a stream tube may also be applied to a streamline.

**Velocity and Acceleration**  In the most general case of fluid motion, the resultant **velocity** $U$ along a streamline is a function of both distance $s$ and time $t$, or $U = f(s,t)$. In differential form,

$$dU = \frac{\partial U}{\partial s} ds + \frac{\partial U}{\partial t} dt$$

An expression for **acceleration** may be obtained by dividing the velocity equation by $dt$, resulting in

$$\frac{dU}{dt} = \frac{\partial U}{\partial s}\frac{ds}{dt} + \frac{\partial U}{\partial t}$$

for steady flow $\partial U/\partial t = 0$.

**Velocity Profile**  In the flow of real fluids, the individual streamlines will have different velocities past a section. Figure 13 shows the steady flow of a fluid past a section (*A-A*) of a circular pipe. The **velocity profile** is obtained by plotting the velocity $U$ of each streamline as it passes *A-A*. The stream tube that is formed by the space between the streamlines is the annulus whose area is $dA$, as shown in Fig. 13 for the stream tube whose velocity is $U$. The volumetric rate of flow $Q$ for the flow past section *A-A* is $Q = \int U\, dA$. All flows take place between boundaries that are **three-dimensional**. The terms **one-dimensional, two-dimensional,** and **three-dimensional** flow refer to the number of dimensions required to describe the velocity profile. For **three-dimensional** flow, a **volume** ($L^3$) is required, for example, the flow of a fluid in a circular pipe. For **two-dimensional** flow, an **area** ($L^2$) is necessary, for example, the flow between two parallel plates. For **one-dimensional** flow, a **line** ($L$) describes the profile. In cases of two- or three-dimensional flow, $\int U\, dA$ can be integrated either mathematically if the equations are known or graphically if velocity-measurement data are available. In many engineering applications, the

**average velocity** $V$ may be used where $V = Q/A = (1/A)\int U\, dA$.

The **continuity equation** is a special case of the general physical law of the conservation of mass. It may be simply stated for a control volume:

Mass rate entering = mass rate of storage + mass rate leaving

This may be expressed mathematically as

$$\rho U\, dA = \left[\frac{\partial}{\partial t}(\rho\, dA\, ds)\right] + \left[\rho U\, dA + \frac{\partial}{\partial s}(\rho U\, dA)ds\right]$$

where $ds$ is an incremental distance along the control volume. For steady flow, $\partial/\partial t\,(\rho\, dA\,\, ds) = 0$, the general equation reduces to $d(\rho U\, dA) = 0$. Integrating the steady-flow **continuity equation** for the average velocity along a flow passage:

$$\rho VA = \text{a constant} = \rho_1 V_1 A_1 = \\ \rho_2 V_2 A_2 = \ldots = \rho_n V_n A_n = \dot{M}$$

where $\dot{M}$ is the mass flow rate in slugs/s (14.5939 kg/s). In many engineering applications, the flow rate in pounds mass per second is desired, so that

$$\dot{m} = \frac{V_1 A_1}{v_1} = \frac{V_2 A_2}{v_2} = \ldots = \frac{V_n A_n}{v_n}$$

where $\dot{m}$ is the flow rate in lbm/s (0.4535924 kg/s).

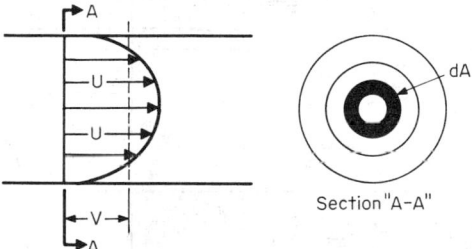

Section "A-A"

**Fig. 13**  Velocity profile.

EXAMPLE.  Air discharges from a 12-in-diameter duct through a 4-in-diameter nozzle into the atmosphere. The pressure in the duct is 20 lbf/in², and atmospheric pressure is 14.7 lbf/in². The temperature of the air in the duct just upstream of the nozzle is 150°F, and the temperature in the jet is 147°F. If the velocity in the duct is 18 ft/s, compute (1) the mass flow rate in lbm/s and (2) the velocity in the nozzle jet. From the equation of state

$v = RT/p$
$v_D = RT_D/p_D = 53.34\,(150 + 459.7)/(144 \times 20)$
$\quad = 11.29$ ft³/lbm
$v_J = RT_J/p_J = 53.34\,(147 + 459.7)/(144 \times 14.7)$
$\quad = 15.29$ ft³/lbm
(1) $\dot{m} = V_D A_D/v_D = 18\,[(\pi/4)\,(12/12)^2]/11.29$
$\quad \dot{m} = 1.252$ lbm/s (0.5680 kg/s)
(2) $V_J = m\,v_J/A_J = (1.252)\,(15.29)/[(\pi/4)\,(4/12)^2]$
$\quad v_J = 219.2$ ft/s (66.82 m/s)

## FLUID DYNAMICS

**Equation of Motion**  For steady one-dimensional flow, consideration of forces acting on a fluid element of length $dL$, flow area $dA$, boundary perimeter in fluid contact $dP$, and change in elevation $dz$ with a unit shear stress $\tau$ moving at a velocity of $V$ results in

$$v\, dp + \frac{V\, dV}{g_c} + \frac{g}{g_c}\, dz + v\tau \left(\frac{dP}{dA}\right) dL = 0$$

Substituting $v = g/g_c\gamma$ and simplifying,

$$\frac{dp}{\gamma} + \frac{V\, dV}{g} + dz + db_f = 0$$

where $db_f = (\tau/\gamma)(dP/dA)dL = \tau\, dL/\gamma R_h$

The expression $1/(dP/dA)$ is the **hydraulic radius** $R_h$ and

equals the flow area divided by the perimeter of the solid boundary in contact with the fluid. This perimeter is usually called the "wetted" perimeter. The hydraulic radius of a pipe flowing full is $(\pi D^2/4)/\pi D = D/4$. Values for other configurations are given in Table 6. Integration of the equation of motion for an incompressible fluid results in

$$\frac{p_1}{\gamma} + \frac{V_1^2}{2g} + z_1 = \frac{p_2}{\gamma} + \frac{V_2^2}{2g} + z_2 + b_{1,2}$$

**Table 6. Values of Flow Area $A$ and Hydraulic Radius $R_h$ for Various Cross Sections**

| Cross section | Condition | | | Equations |
|---|---|---|---|---|
| | Flowing full | $h/D=1$ | | $A = \pi D^2/4 \quad R_h = D/4$ |
| | Upper half partly full | $0.5 < h/D < 1$ | | $\cos^{-1}(\theta/2) = (2h/D - 1)$ $A = [\pi(360-\theta) + 180\sin\theta](D^2/1{,}440)$ $R_h = [1 + (180\sin\theta)/(\pi\theta)](D/4)$ |
| | | $h/D = 0.8128$ | | $A = 0.6839\,D^2 \quad R_h\text{ max} = 0.3043D$ |
| | Lower half partly full | $h/D = 0.5$ | | $A = \pi D^2/8 \quad R_h\text{ max} = h/2$ |
| | | $0 < h/D < 0.5$ | | $\cos^{-1}(\theta/2 = (1 - 2h/D)$ $A = (\pi\theta - 180\sin\theta)(D^2/1{,}440)$ $R_h = [1 - (180\sin\theta)/(\pi\theta)](D/4)$ |
| | Flowing full | $h/D=1$ | | $A = bD \quad R_h = bD/2(b+D)$ |
| | | Square $b = D$ | | $A = D^2 \quad R_h = D/4$ |
| | Partly full | $h/D < 1$ | | $A = bh \quad R_h = bh/(2h+b)$ |
| | | $h/b = 0.5$ | | $A = b^2/2 \quad R_h\text{ max} = h/2$ |
| | | $b \to \infty, h \to 0$ | | $R_h \to h$ (wide shallow stream) |
| | $\alpha \neq \beta$ | | | $R_h\text{ max} = h/2$ $A = [b + 1/2\,h(\cot\alpha + \cot\beta)]h$ $R_h = A/[b + h(\csc\alpha + \csc\beta)]$ |
| | $\alpha = \beta$ | $\dfrac{h}{a} = \dfrac{1}{2}$ | $\alpha = 26°34'$ | $A = (b + 2h)h$ $R_h = (b + 2h)h/(b + 4.472h)$ |
| | | $\dfrac{h}{a} = \dfrac{\sqrt{3}}{3}$ | $\alpha = 30°$ | $A = (b + 1.732h)h$ $R_h = (b + 1.732h)h/(b + 4h)$ |
| | | $\dfrac{h}{a} = \dfrac{2}{3}$ | $\alpha = 33°41'$ | $A = (b + 1.5h)h$ $R_h = (b + 1.5h)h/(b + 3.606h)$ |
| | | $\dfrac{h}{a} = 1$ | $\alpha = 45°$ | $A = (b + h)h$ $R_h = (b + h)h/(b + 2.828h)$ |
| | | $\dfrac{h}{a} = \dfrac{3}{2}$ | $\alpha = 56°19'$ | $A = (b + 0.6667h)h$ $R_h = (b + 0.6667h)h/(b + 2.404h)$ |
| | | $\dfrac{h}{a} = \sqrt{3}$ | $\alpha = 60°$ | $A = (b + 0.5774h)h$ $R_h = (b + 0.5774h)h/(b + 2.309h)$ |
| | $\theta$ = any angle | | | $A = \tan(\theta/2)h^2 \quad R_h = \sin(\theta/2)h/2$ |
| | $\theta = 30$ | | | $A = 0.2679\,h^2 \quad R_h = 0.1294h$ |
| | $\theta = 45$ | | | $A = 0.4142\,h^2 \quad R_h = 0.1913h$ |
| | $\theta = 60$ | | | $A = 0.5774\,h^2 \quad R_h = 0.2500h$ |
| | $\theta = 90$ | | | $A = h^2 \quad R_h = 0.3536h$ |

Each term of the equation is in feet and is equivalent to the height the fluid would rise in a tube if its energy were converted into potential energy. For this reason, in hydraulic practice, each type of energy is referred to as a **head**. The **static pressure head** is $p/\gamma$. The **velocity head** is $V^2/2g$, and the **potential head** is $z$. The energy loss between sections $b_{f_2}$ is called the **lost head** or **friction head**. The **energy grade line** at any point is $\Sigma(p/\gamma + V^2/2g + z)$, and the **hydraulic grade line** is $\Sigma(p/\gamma + z)$ as shown in Fig. 14.

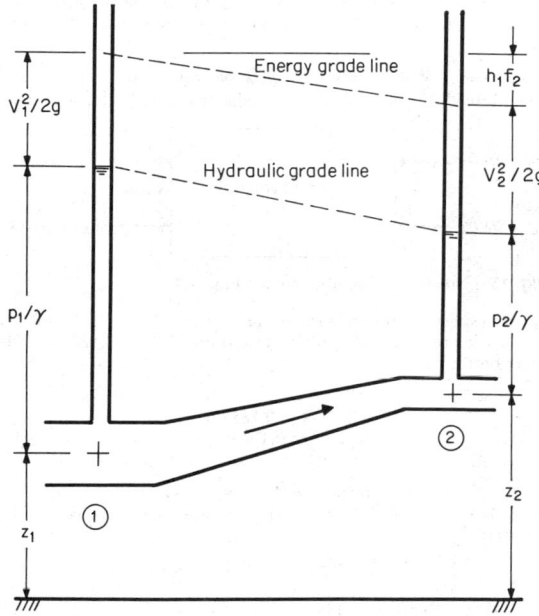

**Fig. 14** Energy relations.

EXAMPLE.   A 12-in pipe (11.938 in inside diameter) reduces to a 6-in pipe (6.065 in inside diameter). Benzene at 68°F (20°C) flows steadily through this system. At section 1, the 12-in pipe centerline is 10 ft above the datum, and at section 2, the 6-in pipe centerline is 15 ft above the datum. The pressure at section 1 is 20 lbf/in$^2$ and the velocity is 4 ft/s. If the head loss due to friction is 0.05 $V_2^2/2g$, compute the pressure at section 2. Assume $g = g_c$, $\gamma = \rho g = 1.705 \times 32.17 = 54.85$ lbf/ft$^3$. From the continuity equation,

$$\dot{M} = \rho_1 A_1 V_1 = \rho_2 A_2 V_2 \qquad (\rho_1 = \rho_2)$$
$$V_2 = V_1 (A_1/A_2) = V_1 (\pi D_1^2/4)/(\pi D_2^2/4) = V_1 (D_1/D_2)^2$$
$$V_2 = 4(11.938/6.065)^2 = 15.50 \text{ ft/s}$$

From the equation of motion,

$$\frac{p_2}{\gamma} = \frac{p_1}{\gamma} + \frac{V_1^2}{2g} + z_1 - \left( \frac{V_2^2}{2g} + z_2 + b_{f_2} \right)$$

$$\frac{p_2}{\gamma} = \frac{p_1}{\gamma} + \frac{V_1^2 - V_2^2 - 0.05 V_2^2}{2g} + z_1 - z_2$$

$$\frac{p_2}{\gamma} = \frac{144 \times 20}{54.85} + \frac{4^2 - 1.05 (15.50)^2}{2 \times 32.17} + 10 - 15$$

$$\frac{p_2}{\gamma} = 43.83 \text{ ft}$$

$$p_2 = \frac{54.88 \times 43.83}{144} = 16.70 \text{ lbf/in}^2 \ (1.151 \times 10^5 \text{ N/m}^2)$$

**Energy Equation**   Application of the principle of conservation of energy to a control volume for one-dimensional flow results in the following for steady flow:

$$J \, dq = dW + \frac{V \, dV}{g_c} + \frac{g}{g_c} \, dz + J \, du + d(pv)$$

where $J$ is the mechanical equivalent of heat, 778.169 ft · lbf/Btu; $q$ is the heat added, Btu/lbm (2,326 J/kg); $W$ is the steady-flow shaft work done *by* the fluid; and $u$ is the internal energy, Btu/lbm (2,326 J/kg). If the energy equation is integrated for an incompressible fluid,

$$J_1 q_2 = {_1}W_2 + \frac{V_2^2 - V_1^2}{2g_c} + \frac{g}{g_c} (z_2 - z_1)$$
$$+ J(u_2 - u_1) + v(p_2 - p_1)$$

The equation of motion does not consider thermal energy or steady-flow work; the energy equation has no terms for friction. Subtracting the differential equation of motion from the energy equation and solving for friction results in

$$db_f = (dW + J \, du + p \, dv - J \, dq)(g_c/g)$$

Integrating for an incompressible fluid ($dv = 0$),

$$b_{f_2} = [{_1}W_2 + J(u_2 - u_1) - J_1 q_2](g_c/g)$$

In the absence of steady-flow work in the system, the effect of friction is to increase the internal energy and/or to transfer heat from the system.

For steady frictionless, incompressible flow, both the equation of motion and the energy equation reduce to

$$\frac{p_1}{\gamma} + \frac{V_1^2}{2g} + z_1 = \frac{p_2}{\gamma} + \frac{V_2^2}{2g} + z_2$$

which is known as the **Bernoulli equation.**

**Area-Velocity Relations**   The continuity equation may be written as $\log_e \dot{M} = \log_e V + \log_e A + \log_e \rho$, which when differentiated becomes

$$\frac{dA}{A} = -\frac{dV}{V} - \frac{d\rho}{\rho}$$

For incompressible fluids, $d\rho = 0$; so

$$\frac{dA}{A} = -\frac{dV}{V}$$

Examination of this equation indicates

1. If the area increases, the velocity decreases.
2. If the area is constant, the velocity is constant.
3. There are no critical values.

For the frictionless flow of compressible fluids, it can be demonstrated that

$$\frac{dA}{A} = -\frac{dV}{V} \left[ 1 - \left( \frac{V}{c} \right)^2 \right]$$

Analysis of the above equation indicates:

1. **Subsonic velocity** $V < c$. If the area increases, the velocity decreases. Same as for incompressible flow.
2. **Sonic velocity** $V = c$. Sonic velocity can exist only where the change in area is zero, i.e., at the *end* of a convergent passage or at the *exit* of a constant-area duct.
3. **Supersonic velocity** $V > c$. If area increases, the velocity increases, the reverse of incompressible flow. Also, supersonic

velocity can exist only in the **expanding** portion of a passage **after** a constriction where **sonic** velocity existed.

**Frictionless adiabatic compressible flow** of an ideal gas in a horizontal passage must satisfy the following requirements:

1. **Conservation of mass** as expressed by the continuity equation $\dot{M} = \rho_1 A_1 V_1 = \rho_2 A_2 V_2$.

2. **Conservation of energy.** As expressed by the energy equation

$$\frac{V^2}{2g_c} + Ju + pv = \frac{V_1^2}{2g_c} + Ju_1 + p_1v_1 = \frac{V_2^2}{2g_c} + Ju_2 + p_2v_2$$

3. **Process relationship.** For an ideal gas undergoing a frictionless adiabatic (isentropic) process,

$$pv^k = p_1v_1^k = p_2v_2^k$$

4. **Ideal-gas law.** The equation of state for an ideal gas

$$pv = RT$$

In an expanding supersonic flow, a **compression shock wave** will be formed if the requirements for the conservation of mass and energy are not satisfied. This type of wave is associated with large and sudden rises in pressure, density, temperature, and entropy. The shock wave is so thin that for computation purposes it may be considered as a single line. For compressible flow of gases and vapors in passages, refer to Sec. 4; for steam-turbine passages, Sec. 9; for compressible flow around immersed objects, see Sec. 11.

The **impulse-momentum equation** is an application of the principle of conservation of momentum and is derived from Newton's second law. It is used to calculate the forces exerted on a solid boundary by a moving stream. Because velocity and force have both magnitude and direction, they are vectors. The impulse-momentum equation may be written for all three directions:

$$\Sigma F_x = \dot{M}(V_{x_2} - V_{x_1})$$
$$\Sigma F_y = \dot{M}(V_{y_2} - V_{y_1})$$
$$\Sigma F_z = \dot{M}(V_{z_2} - V_{z_1})$$

Figure 15 shows a free-body diagram of a control volume. The pressure forces shown are those imposed by the boundaries on the fluid and on the atmosphere. The reactive force $R$ is that imposed by the downstream boundary on the fluid for equilibrium. Application of the impulse-momentum equation yields

$$\Sigma F = (F_{p_1} + F_{a_2}) - (F_{a_1} + F_{p_2} + R) = \dot{M}(V_2 - V_1)$$

Solving for $R$,

$$R = (p_1 - p_a)A_1 - (p_2 - p_a)A_2 = \dot{M}(V_2 - V_1)$$

The impulse-momentum equation is often used in conjunction with the continuity and energy equations to solve engineering problems. Because of the wide variety of possible applications, some examples are given to illustrate the methods of attack.

EXAMPLE. **Compressible Fluid in a Duct.** Nitrogen flows steadily through a 6-in (5.761 in inside diameter) straight, horizontal pipe at a mass rate of 25 lbm/s. At section 1, the pressure is 120 lbf/in² and the temperature is 100°F. At section 2, the pressure is 80 lbf/in² and the temperature is 110°F. Find the friction force opposing the motion. From the equation of state,

$$v = RT/p$$
$$v_1 = 55.16(459.7 + 100)/(144 \times 120) = 1.787 \text{ ft}^3/\text{lbm}$$
$$v_2 = 55.16(459.7 + 110)/(144 \times 80) = 2.728 \text{ ft}^3/\text{lbm}$$

Flow area of pipe = $\pi D^2/4 = \pi(5.761/12)^2/4 = 0.1810 \text{ ft}^2$

From the continuity equation,

$$v = \dot{m}V/A$$
$$V_1 = (25 \times 1.787)/0.1810 = 246.8 \text{ ft/s}$$
$$V_2 = (25 \times 2.728)/0.1810 = 376.8 \text{ ft/s}$$

Applying the free-body equation for impulse momentum ($A = A_1 = A_2$),

$$R = (p_1 - p_a)A_1 - (p_2 - p_a)A_2 - \dot{M}(V_2 - V_1)$$
$$= (p_1 - p_2)A - \dot{M}(V_2 - V_1) = 144(120 - 80)0.1810$$
$$-(25/32.17)(376.8 - 246.8) = 941.5 \text{ lbf } (4.188 \times 10^3 \text{ N})$$

EXAMPLE. **Water Flow through a Nozzle.** Water at 68°F (20°C) flows through a horizontal 12- by 6-in-diameter nozzle discharging into

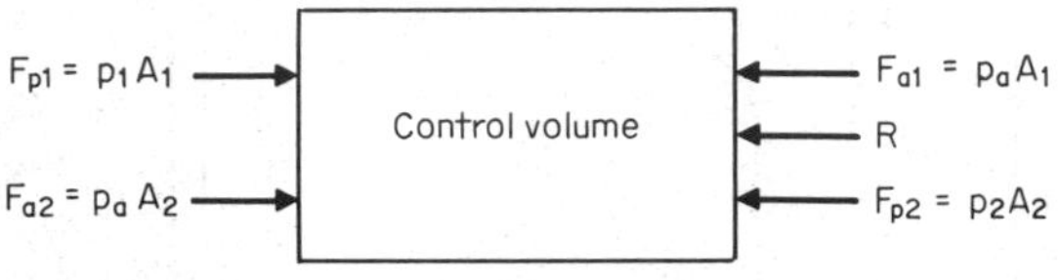

**Fig. 15** Notation of impulse momentum.

the atmosphere. The pressure at the nozzle inlet is 65 lbf/in² and barometric pressure is 14.7 lbf/in². Determine the force exerted by the water on the nozzle.

$$A = \pi D^2/4$$
$$A_1 = \pi(12/12)^2/4 = 0.7854 \text{ ft}^2$$
$$A_2 = \pi(6/12)^2/4 = 0.1963 \text{ ft}^2$$
$$\gamma = \rho g = 1.937 \times 32.17 = 62.31 \text{ lbf/ft}^3$$

From the continuity equation $\rho_1 A_1 V_1 = \rho_2 A_2 V_2$ for $\rho_1 = \rho_2$, $V_2 = V_1 A_1/A_2 = (0.7854/0.1963)V_1 = 4 V_1$. From Bernoulli's equation ($z_1 = z_2$),

$$p_1/\gamma + V_1^2/2g = p_2/\gamma + V_2^2/2g = p_2/\gamma + (4V_1)^2/2g$$

or

$$V_1 = \sqrt{2g(p_1 - p_2)/15\gamma}$$
$$= \sqrt{2 \times 32.17 \times 144(65 - 14.7)/15 \times 62.31} = 22.33 \text{ ft/s}$$
$$V_2 = 4 \times 22.33 = 89.32 \text{ ft/s}$$

Again from the equation of continuity,

$$\dot{M} = \rho_1 A_1 V_1 = 1.937 \times 0.7854 \times 22.33 = 33.97 \text{ slugs/s}$$

Applying the free-body equation for impulse momentum,

$$R = (p_1 - p_a)A_1 - (p_2 - p_a)A_2 - \dot{M}(V_2 - V_1)$$
$$= 144(65 - 14.7)0.7854 - 144(14.7 - 14.7)0.1963$$
$$= (33.97)(89.32 - 22.33)$$
$$= 3,413 \text{ lbf } (1.518 \times 10^4 \text{ N})$$

EXAMPLE. **Incompressible Flow through a Reducing Bend.** Carbon tetrachloride flows steadily without friction at 68°F (20°C) through a 90° horizontal reducing bend. The mass flow rate is 4 slugs/s, the inlet diameter is 6 in, and the outlet is 3 in. The inlet pressure is 50 lbf/in² and the barometric pressure is 14.7 lbf/in². Compute the magnitude and direction of the force required to "anchor" this bend.

$$A = \pi D^2/4$$
$$A_1 = (\pi/4)(6/12)^2 = 0.1963 \text{ ft}^2$$
$$A_2 = (\pi/4)(3/12)^2 = 0.04909 \text{ ft}^2$$

From continuity,

$$V = M/\rho A$$
$$V_1 = 4/(3.093)(0.1963) = 6.588 \text{ ft/s}$$
$$V_2 = 4/(3.093)(0.04909) = 26.35 \text{ ft/s}$$

From the Bernoulli equation ($z_1 = z_2$),

$$\frac{p_2}{\gamma} = \frac{p_1}{\gamma} + \frac{V_1^2}{2g} - \frac{V_2^2}{2g} = \frac{144 \times 50}{3.093 \times 32.17} + \frac{(6.588)^2 - (26.35)^2}{2 \times 32.17} = 62.24 \text{ ft}$$

$$p_2 = \frac{(3.093 \times 32.17)(62.24)}{144} = 43.01 \text{ lbf/in}^2$$

From Fig. 16,

$$\Sigma F_x = (p_1 - p_a)A_1 - (p_2 - p_a)A_2 \cos \alpha - R_x$$
$$= \dot{M}(V_2 \cos \alpha - V_1)$$

or $\quad R_x = (p_1 - p_a)A_1 - (p_2 - p_a)A_2 \cos \alpha - \dot{M}(V_2 \cos \alpha - V_1)$

$$\Sigma F_y = 0 - (p_2 - p_a)A_2 \sin \alpha + R_y = \dot{M}(V_2 \sin \alpha - 0)$$

or $\quad R_y = (p_2 - p_a)A_2 \sin \alpha + \dot{M} V_2 \sin \alpha$

$$R_x = 144 (50 - 14.7)0.1963 - 144 (43.01 - 14.7)$$
$$(\cos 90°) - 4 (26.35 \cos 90° - 6.588)$$
$$= 1,024 \text{ lbf}$$
$$R_y = 144 (43.01 - 14.7)(0.04909) \sin 90° + 4(26.35)(\sin 90°)$$
$$= 305.5 \text{ lbf}$$
$$R = \sqrt{R_x^2 + R_y^2} = \sqrt{(1,024)^2 + (305.5)^2}$$
$$R = 1,068 \text{ lbf } (4.753 \times 10^3 \text{ N})$$
$$\theta = \tan^{-1}(F_y/F_x) = \tan^{-1}(305.5/1,024)$$
$$\theta = 16°37'$$

**Forces on Blades and Deflectors** The forces imposed on a fluid jet whose velocity is $V_J$ by a blade moving at a speed of $V_b$ away from the jet are shown in Fig. 17. The following equations were developed from the application of the impulse-momentum equation for an open jet $(p_2 = p_1)$ and for **frictionless** flow:

$$F_x = \rho A_J (V_J - V_b)^2 (1 - \cos \alpha)$$
$$F_y = \rho A_J (V_J - V_b)^2 \sin \alpha$$
$$F = 2\rho A_J (V_J - V_b)^2 \sin (\alpha/2)$$

EXAMPLE. In the nozzle-blade system of Fig. 17, water at 68°F (20°C) enters a 3- by $1\frac{1}{2}$-in-diameter horizontal nozzle with a pressure 23 lbf/in² and discharges at 14.7 lbf/in² (atmospheric pressure). The blade moves away from the nozzle at a velocity of 10 ft/s and deflects the stream through an angle of 80°. For frictionless flow, calculate the total force exerted by the jet on the blade. Assume $g = g_c$; then $\gamma = \rho g$. From the continuity equation $(\rho_I = \rho_J)$, $\rho_I A_I V_I = \rho_J A_J V_J$, $V_I = (A_J/A_I)V_J$,

$$V_I = \frac{\pi D_J^2/4}{\pi D_I^2/4} V_J = \left(\frac{D_J}{D_I}\right)^2 V_J$$
$$V_I = (1.5/3)^2 V_J = V_J/4$$

From the Bernoulli equation $(z_2 = z_1)$,

$$\frac{V_J^2}{2g} = \frac{V_I^2}{2g} + \frac{p_I - p_J}{\rho g}$$
$$\frac{V_J^2 - (V_J/4)^2}{2g} = \frac{p_I - p_J}{\rho g}$$
$$V_J = \sqrt{\frac{2(16/15)(p_I - p_J)}{\rho}} = \sqrt{\frac{2 \times (16/15) \, 144 \, (23 - 14.7)}{1.937}} = 36.28 \text{ ft/s}$$

The total force $F = 2\rho A_J (V_J - V_b)^2 \sin (\alpha/2)$

$$F = 2 \times 1.937 \, (\pi/4)(1.5/12)^2(36.28 - 10)^2 \sin (80/2)$$
$$F = 21.11 \text{ lbf } (93.90 \text{ N})$$

**Impulse Turbine** In a turbine, the total of the separate forces acting simultaneously on each blade equals that caused by the combined mass flow rate $\dot{M}$ discharged by the nozzle or

$$\Sigma P = \Sigma F_x V_b = \dot{M}(V_J - V_b)(1 - \cos \alpha)V_b$$

The maximum value of power $P$ is found by differentiating $P$ with respect to $V_b$ and setting the result equal to zero. Solving for $V_b$ yields $V_b = V_J/2$ or that maximum power which occurs when the velocity of the jet is equal to twice the velocity of the

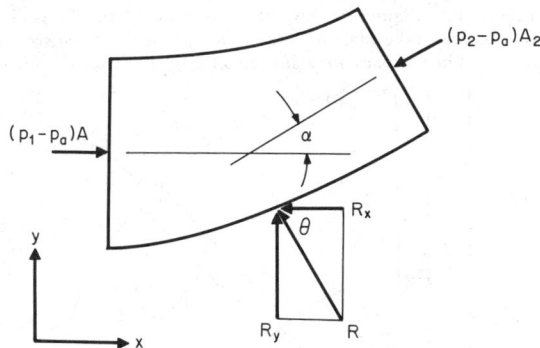

**Fig. 16** Forces on a bend.

blade. Examination of the power equation also indicates that the angle $\alpha$ for a maximum power results when $\cos \theta = -1$ or $\alpha = 180°$. For theoretical maximum power of a blade, $2V_b = V_J$ and $\alpha = 180°$. It should be noted that in any practical impulse-turbine application, $\alpha$ cannot be 180° because the discharge interferes with the next set of blades. Substituting $V_b = V_J/2$, $\alpha = 180°$ in the power equation,

$$\Sigma P_{max} = \dot{M}(V_J - V_J/2) [1 - (-1)] \, V_J/2$$
$$= \dot{M} \, V_J^2/2 = \dot{m} \, V_J^2/2g_c$$

or the maximum power per unit mass is equal to the total power of the jet. Application of the Bernoulli equation between the surface of a reservoir and the discharge of the turbine shows that $\Sigma P_{max} = \dot{M} \sqrt{2g \, (z_2 - z_1)}$. For design details, see Sec. 9.

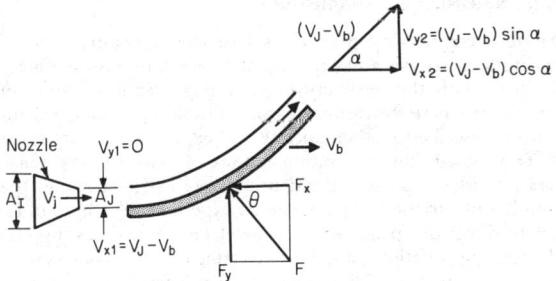

**Fig. 17** Notation for blade study.

**Flow in a Curved Path** When a fluid flows through a bend, it is also rotated around an axis and the energy required to produce rotation must be supplied from the energy already in the fluid mass. This fluid rotation is called a **free vortex** because it is free of outside energy. Consider the fluid mass $\rho(r_o - r_i)dA$ of Fig. 18 being rotated as it flows through a bend of outer radius $r_o$, inner radius $r_i$, with a velocity of $V$. Application of Newton's second law to this mass results in

$$dF = p_o \, dA - p_i \, dA = [\rho(r_o - r_i)dA][V^2/(r_o + r_i)/2]$$

which reduces to

$$p_o - p_i = 2(r_o - r_i) \rho V^2/(r_o + r_i)$$

Because of the difference in fluid pressure between the inner and outer walls of the bend, secondary flows are set up, and this is the primary cause of friction loss of bends. These secondary flows set up turbulence that requires 50 or more straight pipe diameters downstream to dissipate. Thus this loss does not take place in the bend, but in the downstream system. These losses may be reduced by the use of splitter

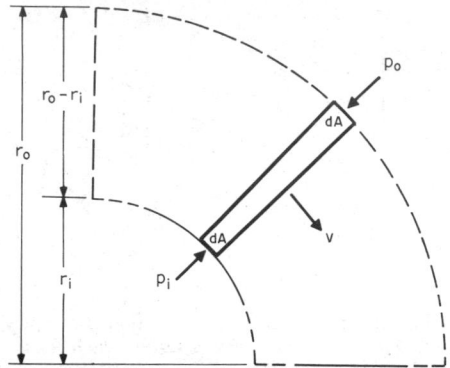

**Fig. 18** Notation for flow in a curved path.

plates which help minimize the secondary flows by reducing $r_o - r_i$ and hence $p_o - p_i$.

EXAMPLE. 104°F (40°C) benzene flows at a rate of 8 ft³/s in a square horizontal duct. This duct makes a 90° turn with an inner radius of 1 ft and an outer radius of 2 ft. Calculate the difference between the walls of this bend. The area of this duct is $(r_o - r_i)^2 = (2 - 1)^2 = 1$ ft². From the continuity equation $V = Q/A = 8/1 = 8$ ft/s. The pressure difference

$$p_o - p_i = 2(r_o - r_i)\rho V^2/(r_o + r_i)$$
$$= 2(2 - 1) 1.663 (8)^2/(2 + 1) = 70.95 \text{ lbf/ft}^2$$
$$= 70.95/144 = 0.4927 \text{ lbf in}^2 (3.397 \times 10^3 \text{ N/m}^2)$$

## DIMENSIONLESS PARAMETERS

Modern engineering practice is based on a combination of theoretical analysis and experimental data. Often the engineer is faced with the necessity of obtaining practical results in situations where for various reasons, physical phenomena cannot be described mathematically and experimental data must be considered. The generation and use of **dimensionless parameters** provides a powerful and useful tool in (1) reducing the number of variables required for an experimental program, (2) establishing the principles of model design and testing, (3) developing equations, and (4) converting data from one system of units to another. Dimensionless parameters may be generated from (1) **physical equations,** (2) the principles of **similarity,** and (3) **dimensional analysis.** All **physical equations** must be dimensionally correct so that a dimensionless parameter may be generated by simply dividing one side of the equation by the other. A minimum of two dimensionless parameters will be formed, one being the inverse of the other.

EXAMPLE. It is desired to generate a series of dimensionless parameters to describe the ratios of static pressure head, velocity head, and potential head to total head for frictionless incompressible flow. From the Bernoulli equation,

$$\frac{p}{\gamma} + \frac{V^2}{2g} + z = \Sigma h = \text{total head}$$

$$N_1 = \frac{p/\gamma + V^2/2g + z}{\Sigma h} = \frac{p/\gamma}{\Sigma h} + \frac{V^2/2g}{\Sigma h} + \frac{z}{\Sigma h} = N_p + N_V + N_z$$

or
$$N_2 = \frac{\Sigma h}{p_2/\gamma + V^2/2g + z} = N_1^{-1}$$

$N_1$ and $N_2$ are total energy ratios and $N_p$, $N_V$, and $N_z$ are the ratios of the static pressure head, velocity head, and potential head, respectively, to the total head.

**Models vs. Prototypes** There are times when for economic or other reasons it is desirable to determine the performance of a structure or machine by testing another structure or machine. This type of testing is called model testing. The equipment being tested is called a **model,** and the equipment whose performance is to be predicted is called a **prototype.** A **model** may be smaller than, the same size as, or larger than the **prototype.** Model experiments on aircraft, rockets, missiles, pipes, ships, canals, turbines, pumps, and other structures and machines have resulted in savings that more than justified the expenditure of funds for the design, construction, and testing of the model. In some situations, the model and the prototype may be the same piece of equipment, for example, the laboratory calibration of a flowmeter with water to predict its performance with other fluids. Many manufacturers of fluid machinery have test facilities that are limited to one or two fluids and are forced to test with what they have available in order to predict performance with nonavailable fluids. For towing-tank testing of ship models and for wind-tunnel testing of aircraft and aircraft-component models, see Sec. 11. For wind-tunnel testing of model buildings and structures, see Sec. 12.

**Similarity Requirements** For complete similarity between a model and its prototype, it is necessary to have **geometric, kinematic,** and **dynamic** similarity. **Geometric similarity** exists between model and prototype when the **ratios** of all corresponding dimensions of the model and prototype are equal. These ratios may be written as follows:

Length:
$$L_{model}/L_{prototype} = L_{ratio}$$
$$= L_m/L_p = L_r$$

Area:
$$L_{model}^2/L_{prototype}^2 = L_{ratio}^2$$
$$= L_m^2/L_p^2 = L_r^2$$

Volume:
$$L_{model}^3/L_{prototype}^3 = L_{ratio}^3$$
$$= L_m^3/L_p^3 = L_r^3$$

**Kinematic similarity** exists between model and prototype when their streamlines are geometrically similar. The kinematic ratios resulting from this condition are

Acceleration:
$$a_r = a_m/a_p = L_m T_m^{-2}/L_p T_p^{-2}$$
$$= L_r T_r^{-2}$$

Velocity:
$$V_r = V_m/V_p = L_m T_m^{-1}/L_p T_p^{-1}$$
$$= L_r T_r^{-1}$$

Volume flow rate:
$$Q_r = Q_m/Q_p = L_m^3 T_m^{-1}/L_p^3 T_p^{-1}$$
$$= L_r^3 T_r^{-1}$$

**Dynamic similarity** exists between model and prototype having geometric and kinematic similarity when the ratios of all forces are the same. Consider the model/prototype relations for the flow around the object shown in Fig. 19. For **geometric similarity** $D_m/D_p = L_m/L_p = L_r$ and for **kinematic similarity** $U_{Am}/U_{Ap} = U_{Bm}/U_{Bp} = V_r = L_r T_r^{-1}$. Next consider the three forces acting on point $C$ of Fig. 19 without specifying their nature. From the geometric similarity of their vector polygons and Newton's law, for **dynamic similarity** $F_{1m}/F_{1p} = F_{2m}/F_{2p} = F_{3m}/F_{3p} = M_m a_{Cm}/M_p a_{Cp} = F_r$. For dynamic similarity, these force ratios must be maintained on all corresponding fluid particles throughout the flow pattern. From the force polygon of Fig. 19, it is evident that $F_1 +\!\!\rightarrow F_2 +\!\!\rightarrow F_3 = Ma_C$. For total model/prototype force ratio, comparisons of force polygons yield

$$F_r = \frac{F_{1m} +\!\!\rightarrow F_{2m} +\!\!\rightarrow F_{3m}}{F_{1p} +\!\!\rightarrow F_{2p} +\!\!\rightarrow F_{3p}} = \frac{M_m a_{Cm}}{M_p a_{Cp}}$$

**Fluid Forces**   The fluid forces that are considered here are those acting on a fluid element whose mass $= \rho L^3$, area $= L^2$, length $= L$, and velocity $= L/T$.

**Inertia force**

$$\begin{aligned}F_i &= \text{(mass)(acceleration)} = (\rho L^3)(L/T^2) \\ &= \rho L(L^2/T^2) = \rho L^2 V^2\end{aligned}$$

**Viscous force**

$$\begin{aligned}F_\mu &= \text{(viscous shear stress)(shear area)} \\ &= \tau L^2 = \mu(dU/dy)L^2 = \mu(V/L)L^2 \\ &= \mu LV\end{aligned}$$

**Gravity force**

$$\begin{aligned}F_g &= \text{(mass)(acceleration due to gravity)} \\ &= (\rho L^3)(g) = \rho L^3 g\end{aligned}$$

**Pressure force**

$$F_p = \text{(pressure)(area)} = pL^2$$

**Centrifugal force**

$$\begin{aligned}F_\omega &= \text{(mass)(acceleration)} = (\rho L^3)(L/T^2) \\ &= (\rho L^3)(L\omega^2) = \rho L^4 \omega^2\end{aligned}$$

**Elastic force**

$$F_E = \text{(modulus of elasticity)(area)} = EL^2$$

**Surface-tension force**

$$F_\sigma = \text{(surface tension)(length)} = \sigma L$$

**Vibratory force**

$$\begin{aligned}F_f &= \text{(mass)(acceleration)} = (\rho L^3)(L/T^2) \\ &= (\rho L^4)(T^{-2}) = \rho L^4 f^2\end{aligned}$$

If all fluid forces were acting on a fluid element,

$$\begin{aligned}F_r &= \frac{F_{\mu m} +\!\!\rightarrow F_{gm} +\!\!\rightarrow F_{pm} +\!\!\rightarrow F_{\omega m} +\!\!\rightarrow F_{Em} +\!\!\rightarrow F_{\sigma m} +\!\!\rightarrow F_{fm}}{F_{\mu p} +\!\!\rightarrow F_{gp} +\!\!\rightarrow F_{pp} +\!\!\rightarrow F_{\omega p} +\!\!\rightarrow F_{Ep} +\!\!\rightarrow F_{\sigma p} +\!\!\rightarrow F_{fp}} \\ &= \frac{F_{im}}{F_{ip}}\end{aligned}$$

Examination of the above equation and the force polygon of Fig. 19 lead to the conclusion that dynamic similarity can be characterized by an equality of force ratios one less than the total number involved. Any force ratio may be eliminated, depending upon the quantities which are desired. Fortunately, in most practical engineering problems, not all of the eight forces are involved because some may not be acting, may be of negligible magnitude, or may be in opposition to each

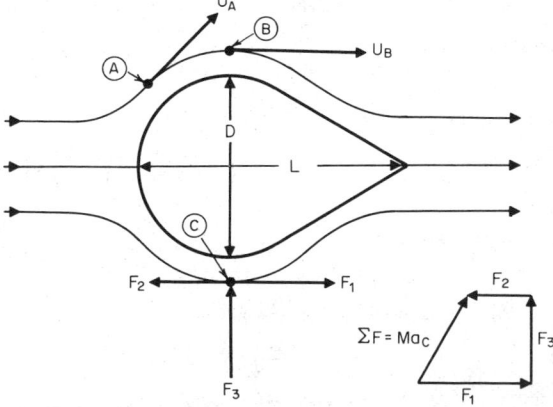

**Fig. 19**   Notation for dynamic similarity.

other in such a way as to compensate. In each application of similarity, a good understanding of the fluid phenomena involved is necessary to eliminate the irrelevant, trivial, or compensating forces. When the flow phenomenon is too complex to be readily analyzed, or is not known, only experimental verification with the prototype of results from a model test will determine what forces should be considered in future model testing.

**Standard Numbers**   With eight fluid forces that can act in flow situations, the number of dimensionless parameters that can be formed from their ratios is 56. However, conventional practice is to ratio the inertia force to the other fluid forces, usually by division because the inertia force is the vector sum of all the other forces involved in a given flow situation. Results obtained by dividing the inertia force by each of the other forces are shown in Table 7 compared with the standard numbers that are used in conventional practice.

## DYNAMIC SIMILARITY

**Vibration**   In the flow of fluids around objects and in the motion of bodies immersed in fluids, **vibration** may occur because of the formation of a wake caused by alternate shedding of eddies in a periodic fashion or by the vibration of the object or the body. The **Strouhal number S** is the ratio of the velocity of vibration $Lf$ to the velocity of the fluid $V$. Since the vibration may be fluid-induced or structure-induced, two frequencies must be considered, the wake frequency $f\omega$ and the natural frequency of the structure $f_n$. Fluid-induced forces are usually of small magnitude, but as the wake frequency approaches the natural frequency of the structure, the vibratory forces increase very rapidly. When $f\omega = f_n$, the structure will go into resonance and fail. This imposes on the model designer the requirement of matching to scale the natural-frequency characteristics of the prototype. This subject is treated later under Wake Frequency. All further discussions of model/prototype relations are made under the assumption that either vibratory forces are absent or they are taken care of in the design of the model or in the test program.

**Incompressible Flow**   Considered in this category are the flow of fluids around an object, motion of bodies immersed in incompressible fluids, and the flow of incompressible fluids in conduits. It includes, for example, a submarine traveling under water but not partly submerged, and liquids flowing in pipes and passages when the liquid completely fills them, but not when partly full as in open-channel flow. It also includes aircraft moving in atmospheres that may be considered incompressible. Incompressible flow in rotating machinery is considered separately.

In these situations the gravity force, although acting on all fluid particles, does not affect the flow pattern. Excluding rotating machinery, centrifugal forces are absent. By definition of an incompressible fluid, elastic forces are zero, and since there is no liquid-gas interface, surface-tension forces are absent.

The only forces now remaining for consideration are the inertia, viscous, and pressure. Using standard numbers, the parameters are **Reynolds number** and **pressure coefficient.** The Reynolds number may be converted into a kinematic ratio by noting that by definition $v = \mu/\rho$ and substituting in $\mathbf{R} = \rho LV/\mu = LV/v$. In this form, Reynolds number is the ratio of the fluid velocity $V$ and the "shear velocity" $v/L$. For this reason,

**Table 7. Standard Numbers.**

| Force ratio | Equations | Result | Conventional practice | | |
|---|---|---|---|---|---|
| | | | Form | Symbol | Name |
| $\dfrac{\text{Inertia}}{\text{Viscous}}$ | $\dfrac{F_i}{F_\mu} = \dfrac{\rho L^2 V^2}{\mu LV}$ | $\dfrac{\rho LV}{\mu}$ | $\dfrac{\rho LV}{\mu}$ | **R** | Reynolds |
| $\dfrac{\text{Inertia}}{\text{Gravity}}$ | $\dfrac{F_i}{F_g} = \dfrac{\rho L^2 V^2}{\rho L^3 g}$ | $\dfrac{V^2}{Lg}$ | $\dfrac{V}{\sqrt{Lg}}$ | **F** | Froude |
| $\dfrac{\text{Inertia}}{\text{Pressure}}$ | $\dfrac{F_i}{F_p} = \dfrac{\rho L^2 V^2}{\rho L^2}$ | $\dfrac{\rho V^2}{p}$ | $\dfrac{\rho V^2}{p}$ | **E** | Euler |
| | | | $\dfrac{2\Delta p}{\rho V^2}$ | **C$_p$** | Pressure coefficient |
| $\dfrac{\text{Inertia}}{\text{Centrifugal}}$ | $\dfrac{F_i}{F_\omega} = \dfrac{\rho L^2 V^2}{\rho L^4 \omega^2}$ | $\dfrac{V^2}{L^2\omega^2}$ | $\dfrac{V}{DN}$ | **V** | Velocity ratio |
| $\dfrac{\text{Inertia}}{\text{Elastic}}$ | $\dfrac{F_i}{F_E} = \dfrac{\rho L^2 V^2}{EL^2}$ | $\dfrac{\rho V^2}{E}$ | $\dfrac{\rho V^2}{E}$ | **C** | Cauchy |
| | | | $\dfrac{V}{\sqrt{E/\rho}}$ | **M** | Mach |
| $\dfrac{\text{Inertia}}{\text{Surface tension}}$ | $\dfrac{F_i}{F_\sigma} = \dfrac{\rho L^2 V^2}{\sigma^L}$ | $\dfrac{\rho LV^2}{\sigma}$ | $\dfrac{\rho LV^2}{\sigma}$ | **W** | Weber |
| $\dfrac{\text{Inertia}}{\text{Vibration}}$ | $\dfrac{F_i}{F_f} = \dfrac{\rho L^2 V^2}{\rho L^4 f^2}$ | $\dfrac{V^2}{L^2 f^2}$ | $\dfrac{Lf}{V}$ | **S** | Strouhal |

Computed from data given in J. W. Murdock, "Fluid Mechanics and Its Applications," Houghton Mifflin, 1976.

Reynolds number is used to characterize the velocity profile. Forces and pressure losses are then determined by the pressure coefficient.

EXAMPLE. A submarine is to move submerged through 32°F (0°C) seawater at a speed of 10 knots. (1) At what speed should a 1:20 model be towed in 68°F (20°C) fresh water? (2) If the thrust on the model is found to be 42,500 lbf, what horsepower will be required to propel the submarine?

1. Speed of model for Reynolds-number similarity

$$R_m = R_p = \left(\frac{\rho VL}{\mu}\right)_m = \left(\frac{\rho VL}{\mu}\right)_p$$
$$V_m = V_p \, (\rho_p/\rho_m)(L_p/L_m)(\mu_m/\mu_p)$$
$$V_m = (10)(1.995/1.937)(20/1)(20.92 \times 10^{-6}/39.40 \times 10^{-6})$$
$$= 109.4 \text{ knots } (56.27 \text{ m/s})$$

2. Prototype horsepower

$$\mathbf{C}_{p_p} = \mathbf{C}_{p_m} = \left(\frac{2\Delta p}{\rho V^2}\right)_p = \left(\frac{2\Delta p}{\rho V^2}\right)_m$$
$$F = \Delta pL^2, \quad \Delta p = \frac{F}{L^2}, \text{ so that } \left(\frac{2F}{\rho V^2 L^2}\right)_p = \left(\frac{2F}{\rho V^2 L^2}\right)_m$$
$$F_p = F_m(\rho_p/\rho_m)(V_p/V_m)^2(L_p/L_m)^2$$
$$= 42,500 \, (1.995/1.937)(10/109.4)^2(20/1)^2 = 146,300 \text{ lbf}$$
$$P = FV = \left(\frac{146,300}{550}\right)\left(\frac{10 \times 6,076}{3,600}\right) = 4,490 \text{ hp } (3.35 \times 10^6 \text{ W})$$

**Compressible Flow** Considered in this category are the flow of compressible fluids under the conditions specified for incompressible flow in the preceding paragraphs. In addition to the forces involved in incompressible flow, the elastic force must be added. Conventional practice is to use the square root of the inertia/elastic force ratio or **Mach number.**

**Mach number** is the ratio of the fluid velocity to its speed of sound and may be written $\mathbf{M} = V/c = V\sqrt{E/\rho}$. For an ideal gas, $\mathbf{M} = V/\sqrt{kg_cRT}$. In compressible-flow problems, practice is to use the Mach number to characterize the velocity or kinematic similarity, the Reynolds number for dynamic similarity, and the pressure coefficient for force or pressure-loss determination.

EXAMPLE. An airplane is to fly at 500 mi/h in an atmosphere whose temperature is 32°F (0°C) and pressure is 12 lbf/in². A 1:20 model is tested in a wind tunnel where a supply of air at 392°F (200°C) and variable pressure is available. At (1) what speed and (2) what pressure should the model be tested for dynamic similarity?

1. Speed for Mach-number similarity.

$$\mathbf{M}_m = \mathbf{M}_p = \left(\frac{V}{\sqrt{E/\rho}}\right)_m = \left(\frac{V}{\sqrt{E/\rho}}\right)_p = \left(\frac{V}{\sqrt{kg_cRT}}\right)_m = \left(\frac{V}{\sqrt{kg_cRT}}\right)_p$$
$$V_m = V_p \,(k_m/k_p)^{1/2}(R_m/R_p)^{1/2}(T_m/T_p)^{1/2}$$

For the same gas $k_m = k_p$, $R_m = R_p$, and

$$V_m = V_p \sqrt{T_m/T_p} = 500 \sqrt{(851.7/491.7)} = 658.1 \text{ mi/h}$$

2. Pressure for Reynolds-number similarity

$$\mathbf{R}_m = \mathbf{R}_p = \left(\frac{\rho VL}{\mu}\right)_m = \left(\frac{\rho VL}{\mu}\right)_p$$
$$\rho_m = \rho_p \,(V_p/V_m)(L_p/L_m)(\mu_m/\mu_p)$$

Since $\rho = p/g_cRT$

$$\left(\frac{p}{g_cRT}\right)_m = \left(\frac{p}{g_cRT}\right)_p (V_p/V_m)(L_p/L_m)(\mu_m/\mu_p)$$

$p_m = p_p (T_m/T_p)(V_p/V_m)(L_p/L_m)(\mu_m/\mu_p)$
$p_m = 12(851.7/491.7)(500/658.1)(20/1)(53.15 \times 10^{-6}/35.67 \times 10^{-6})$
$p_m = 470.6$ lbf/in² ($3.245 \times 10^6$ N/m²)

For information about wind-tunnel testing and its limitations, refer to Sec. 11.

**Centrifugal Machinery**   This category includes the flow of fluids in such centrifugal machinery as compressors, fans, and pumps. In addition to the inertia, pressure, viscous, and elastic forces, centrifugal forces must now be considered. Since centrifugal force is really a special case of the inertia force, their ratio as shown in Table 7 is **velocity ratio** and is the ratio of the fluid velocity to the machine tangential velocity. In model/prototype relations for centrifugal machinery, $DN$ ($D$ = diameter, ft, $N$ = rotational speed) is substituted for the velocity $V$, and $D$ for $L$, which results in the following:

$$\mathbf{M} = DN/\sqrt{kg_cRT} \qquad \mathbf{R} = \rho D^2 N/\mu \qquad \mathbf{C}_p = 2\Delta p/\rho D^2 N^2$$

EXAMPLE.   A centrifugal compressor operating at 100 r/min is to compress methane delivered to it at 50 lbf/in² and 68°F (20°C). It is proposed to test this compressor with air from a source at 140°F (60°C) and 100 lbf/in². Determine compressor speed and inlet-air pressure required for dynamic similarity:
Find speed for Mach-number similarity:

$$\mathbf{M}_m = \mathbf{M}_p = (DN/\sqrt{kg_cRT})_m = (DN/\sqrt{kg_cRT})_p$$
$$N_m = N_p (D_p/D_m)\sqrt{(k_m/k_p)(R_m/R_p)(T_m/T_p)}$$
$$= 100 (1) \sqrt{(1.40/1.32)(53.34/96.33)(599.7/527.7)}$$
$$= 81.70 \text{ r/min}$$

Find pressure for Reynolds-number similarity:

$$\mathbf{R}_m = \mathbf{R}_p = \left(\frac{\rho D^2 N}{\mu}\right)_m = \left(\frac{\rho D^2 N}{\mu}\right)_p$$

For an ideal gas $\rho = p/g_cRT$, so that

$(\rho D^2 N/g_cRT\mu)_m = (\rho D^2 N/g_cRT\mu)_p$
$p_m = p_p (D_p/D_m)^2(N_p/N_m)(R_m/R_p)(T_m/T_p)(\mu_m/\mu_p)$
$p_m = 50(1)^2(100/81.70)(53.34/96.33)(599.7/527.7) \times$
    $(41.79 \times 10^{-8}/22.70 \times 10^{-8}) = 70.90$ lbf/in² ($4.888 \times 10^5$ N/m²)

See Sec. 14 for specific information on pump and compressor similarity.

**Liquid Surfaces**   Considered in this category are ships, seaplanes during takeoff, submarines partly submerged, piers, dams, rivers, open-channel flow, spillways, harbors, etc. Resistance at liquid surfaces is due to surface tension and wave action. Since wave action is due to gravity, the gravity force and surface-tension force are now added to the forces that were considered in the last paragraph. These are expressed as the square root of the inertia/gravity force ratio or **Froude number** $\mathbf{F} = V/\sqrt{Lg}$ and as the inertia/surface tension force ratio or **Weber number** $\mathbf{W} = \rho LV^2/\sigma$. On the other hand, elastic and pressure forces are now absent. Surface tension is a minor property in fluid mechanics and it normally exerts a negligible effect on wave formation *except* when the waves are small, say less than 1 in. Thus the effects of surface tension on the model might be considerable, but negligible on the prototype. This type of "scale effect" must be avoided. For accurate results, the inertia/surface tension force ratio or Weber number should

be considered. It is never possible to have complete dynamic similarity of liquid surfaces unless the model and prototype are the same size, as shown in the following example.

EXAMPLE.   An ocean vessel 500 ft long is to travel at a speed of 15 knots. A 1:25 model of this ship is to be tested in a towing tank using seawater at design temperature. Determine the model speed required for (1) wave-resistance similarity, (2) viscous or skin-friction similarity, (3) surface-tension similarity, and (4) the model size required for complete dynamic similarity.

1. Speed for **Froude-number** similarity

or
$$\mathbf{F}_m = \mathbf{F}_p = (V/\sqrt{Lg})_m = (V/\sqrt{Lg})_p$$
$$V_m = V_p \sqrt{L_m/L_p} = 15\sqrt{1/25} = 3 \text{ knots}$$

2. Speed for **Reynolds-number** similarity

$\mathbf{R}_m = \mathbf{R}_p = (\rho LV/\mu)_m = (\rho LV/\mu)_p$
$V_m = V_p(\rho_p/\rho_m)(L_p/L_m)(\mu_m/\mu_p)$
$V_m = 15(1) (25/1) (1) = 375 \text{ knots}$

3. Speed for **Weber-number** similarity

$$\mathbf{W}_m = \mathbf{W}_p = (\rho LV^2/\sigma)_m = (\rho LV^2/\sigma)_p$$
$$V_m = V_p \sqrt{(\rho_p/\rho_m)(L_p/L_m)(\sigma_m/\sigma_p)}$$
$$V_m = 15 \sqrt{(1)(25)(1)} = 75 \text{ knots}$$

4. Model size for complete similarity. First try Reynolds and Froude similarity; let

$$V_m = V_p(\rho_p/\rho_m)(L_p/L_m)(\mu_m/\mu_p) = V_p \sqrt{L_m/L_p}$$

which reduces to

$$L_m/L_p = (\rho_p/\rho_m)^{2/3} (\mu_m/\mu_p)^{2/3}$$

Next try Weber and Froude similarity; let

$$V_m = V_p \sqrt{(\rho_p/\rho_m)(L_p/L_m)(\sigma_m/\sigma_p)} = V_p \sqrt{L_m/L_p}$$

which reduces to

$$L_m/L_p = (\rho_p/\rho_m)^{1/2} (\sigma_m/\sigma_p)^{1/2}$$

For the same fluid at the same temperature, either of the above solves for $L_m = L_p$, or the model must be the same size as the prototype. For use of different fluids and/or the same fluid at different temperatures,

$$L_m/L_p = (\rho_p/\rho_m)^{2/3} (\mu_m/\mu_p)^{2/3} = (\rho_p/\rho_m)^{1/2} (\sigma_m/\sigma_p)^{1/2}$$

which reduces to

$$(\mu^4/\rho\sigma^3)_m = (\mu^4/\rho\sigma^3)_p$$

No practical way has been found to model for complete similarity. Marine engineering practice is to model for wave resistance and correct for skin-friction resistance. See Sec. 11.

## DIMENSIONAL ANALYSIS

**Dimensional Analysis** is the mathematics of dimensions and quantities and provides procedural techniques whereby the variables that are assumed to be significant in a problem can be formed into dimensionless parameters, the number of parameters being less than the number of variables. This is a great advantage, because fewer experimental runs are then required to establish a relationship between the parameters than between the variables. While the user is not presumed to have any knowledge of the fundamental physical equations, the more knowledgeable the user, the better the results. If any significant variable or variables are omitted, the relationship obtained from dimensional analysis will not apply to the physical problem. On the other hand, inclusion of all possible variables will result in losing the principal advantage of dimensional analysis, i.e., the reduction of the amount of experimental data required to establish relationships. Two formal methods of dimensional analysis are used, the **method of Lord Rayleigh** and **Buckingham's Π theorem.**

**Dimensions** used in mechanics are mass $M$, length $L$, time $T$, and force $F$. Corresponding **units** for these dimensions are the slug (kilogram), the foot (metre), the second (second), and the pound force (newton). Any system in mechanics can be defined by three fundamental dimensions. Two systems are used, the force $(FLT)$ and the mass $(MLT)$. In the force system, mass is a derived quantity and in the mass system, force is a derived quantity. Force and mass are related by Newton's law: $F = MLT^{-2}$ and $M = FL^{-1}T^2$. Table 8 shows common variables and their dimensions and units.

**Lord Rayleigh's method** uses algebra to determine interrelationships among variables. While this method may be used for any number of variables, it becomes relatively complex and is not generally used for more than four. This method is most easily described by example.

EXAMPLE. In laminar flow, the unit shear stress $\tau$ is some function of the fluid dynamic viscosity $\mu$, the velocity difference $dU$ between adjacent laminae separated by the distance $dy$. Develop a relationship.

1. Write a functional relationship of the variables:

$$\tau = f(\mu, dU, \mathbf{dy})$$

Assume $\tau = K(\mu^a dU^b dy^c)$

2. Write a dimensional equation in either $FLT$ or $MLT$ system:

$$(FL^{-2}) = K(FL^{-2}T)^a(LT^{-1})^b(L)^c$$

3. Solve the dimensional equation for exponents:

|  |  | $\tau$ | $\mu$ | $dU$ | $dy$ |
|---|---|---|---|---|---|
| Force | $F$ | $1 =$ | $a +$ | $0 +$ | $0$ |
| Length | $L$ | $-2 =$ | $-2a +$ | $b +$ | $c$ |
| Time | $T$ | $0 =$ | $a -$ | $b +$ | $0$ |

*Solution:* $a = 1, b = 1, c = -1$

4. Insert exponents in the functional equation:

$\tau = K(\mu^a dU^b dy^c) = K(\mu^1 du^1 dy^{-1})$, or $K = (\mu dU/\tau dy)$. This was based on the assumption of $\tau = K (\mu^o dU^b dy^c)$. The general relationship is $K = f(\mu dU/\tau dy)$. The functional relationship cannot be obtained from dimensional analysis. Only physical analysis and/or experiments can determine this. From both physical analysis and experimental data,

$$\tau = \mu \, dU/dy$$

The **Buckingham Π theorem** serves the same purpose as the method of Lord Rayleigh for deriving equations expressing one variable in terms of its dependent variables. The Π theorem is preferred when the number of variables exceeds four. Application of the Π theorem results in the formation of dimensionless parameters called $\pi$ ratios. These $\pi$ ratios have no relation to $3.14159. . . .$ The Π theorem will be illustrated in the following example.

EXAMPLE. Experiments are to be conducted with gas bubbles rising in a still liquid. Consider a gas bubble of diameter $D$ rising in a liquid whose density is $\rho$, surface tension $\sigma$, viscosity $\mu$, rising with a velocity of $V$ in a gravitational field of $g$. Find a set of parameters for organizing experimental results.

1. List all the physical variables considered according to type: geometric, kinematic, or dynamic.

2. Choose either the $FLT$ or $MLT$ system of dimensions.
3. Select a "basic group" of variables characteristic of the flow as follows:
   - *a.* $B_G$ A geometric variable
   - *b.* $B_K$ A kinematic variable
   - *c.* $B_D$ A dynamic variable (if three dimensions are used)
4. Assign A numbers to the remaining variables starting with $A_1$.

| Type | Symbol | Description | Dimensions | Number |
|---|---|---|---|---|
| Geometric | $D$ | Bubble diameter | $L$ | $B_G$ |
| Kinematic | $V$ | Bubble velocity | $L T^{-1}$ | $B_K$ |
|  | $g$ | Acceleration of Gravity | $L T^{-2}$ | $A_1$ |
| Dynamic | $\rho$ | Liquid density | $M L^{-3}$ | $B_D$ |
|  | $\sigma$ | Surface tension | $M T^{-2}$ | $A_2$ |
|  | $\mu$ | Liquid viscosity | $M L^{-1}T^{-1}$ | $A_3$ |

5. Write the basic equation for each $\pi$ ratio as follows:

$$\pi_1 = (B_G)^{x_1}(B_K)^{y_1}(B_D)^{z_1}(A_1) \quad \pi_2 = (B_G)^{x_2}(B_K)^{y_2}(B_D)^{z_2}(A_2) \quad \cdots$$
$$\pi_n = (B_G)^{x_n}(B_K)^{y_n}(B_D)^{z_n}(A_n)$$

Note that the number of $\pi$ ratios is equal to the number of $A$ numbers and thus equal to the number of variables less the number of fundamental dimensions in a problem.

6. Write the dimensional equations and use the algebraic method to determine the value of exponents $x$, $y$, and $z$ for each $\pi$ ratio. Note that for all $\pi$ ratios, the sum of the exponents of a given dimension is zero.

$$\pi_1 = (B_G)^{x_1}(B_K)^{y_1}(B_D)^{z_1}(A_1) = (D)^{x_1}(V)^{y_1}(\rho)^{z_1}(g)$$
$$(M^0L^0T^0) = (L^{x_1})(L^{y_1}T^{-y_1})(M^{z_1}L^{-3z_1})(LT^{-2})$$

*Solution:* $x_1 = 1, y_1 = -2, z_1 = 0$

$$\pi_1 = D^1V^{-2}\rho^0g = Dg/V^2$$
$$\pi_2 = (B_G)^{x_2}(B_K)^{y_2}(B_D)^{z_2}(A_2) = (D)^{x_2}(V)^{y_2}(\rho)^{z_2}(\sigma)$$
$$(M^0L^0T^0) = (L^{x_2})(L^{y_2}T^{-y_2})(M^{z_2}L^{-3z_2})(MT^{-2})$$

*Solution:* $x_2 = -1, y_2 = -2, z_2 = -1$

$$\pi_2 = D^{-1}V^{-2}\rho^{-1}\sigma = \sigma/DV^2\rho$$
$$\pi_3 = (B_G)^{x_3}(B_K)^{y_3}(B_D)^{z_3}(A_3) = (D)^{x_3}(V)^{y_3}(\rho)^{z_3}(\mu)$$
$$(M^0L^0T^0) = (L^{x_3})(L^{y_3}T^{-y_3})(M^{z_3}L^{-3z_3})(ML^{-1}T^{-1})$$

*Solution:* $x_3 = -1, y_3 = -1, z_3 = -1$

$$\pi_3 = D^{-1}V^{-1}\rho^{-1}\mu = \mu/DV\rho$$

7. Convert $\pi$ ratios to conventional practice. One statement of the Buckingham Π theorem is that any $\pi$ ratio may be taken as a function of all the others, or $f(\pi_1, \pi_2, \pi_3, . . . , \pi_n) = 0$. This equation is mathematical shorthand for a functional statement. It could be written, for example, as $\pi_2 = f(\pi_1, \pi_3, . . . , \pi_n)$. This equation states that $\pi_2$ is some function of $\pi_1$ and $\pi_3$ through $\pi_n$ but is not a statement of *what* function $\pi_2$ is of the other $\pi$ ratios. This can be determined only by physical and/or experimental analysis. Thus we are free to substitute *any* function in the equation; for example, $\pi_1$ may be replaced with $2\pi_1^{-1}$ or $\pi_n$ with $a\pi_n^b$.

The procedures set forth in this example are designed to produce $\pi$ ratios containing the same terms as those resulting from the application of the principles of similarity so that the physical significance may be understood. However, any other combinations might have been used. The only real requirement for a "basic group" is that it contain the same number of terms as there are dimensions in a problem and that each of these dimensions be represented in it.

The $\pi$ ratios derived for this example may be converted into conventional practice as follows:

$$\pi_1 = Dg/V^2$$

**Table 8. Dimensions and Units of Common Variables**

| Symbol | Variable | Dimensions MLT | FLT | Units USCU* | SI |
|--------|----------|------|-----|--------|-----|
| | | Geometric | | | |
| $L$ | Length | $L$ | | ft | m |
| $A$ | Area | $L^2$ | | ft$^2$ | m$^2$ |
| $V$ | Volume | $L^3$ | | ft$^3$ | m$^3$ |
| | | Kinematic | | | |
| $t$ | Time | $T$ | | s | s |
| $\omega$ $f$ | Angular velocity Frequency | $T^{-1}$ | | s$^{-1}$ | s$^{-1}$ |
| $V$ | Velocity | $LT^{-1}$ | | ft/s | m/s |
| $v$ | Kinematic viscosity | $L^2T^{-1}$ | | ft$^2$/s | m$^2$/s |
| $Q$ | Volume flow rate | $L^3T^{-1}$ | | ft$^3$/s | m$^3$/s |
| $\alpha$ | Angular acceleration | $T^{-2}$ | | s$^{-2}$ | s$^{-2}$ |
| $a$ | Acceleration | $LT^{-2}$ | | ft/s$^{-2}$ | m/s$^2$ |
| | | Dynamic | | | |
| $\rho$ | Density | $ML^{-3}$ | $FL^{-4}T^2$ | slug/ft$^3$ | kg/m$^3$ |
| $M$ | Mass | $M$ | $FL^{-1}T^2$ | slugs | kg |
| $I$ | Moment of inertia | $ML^2$ | $FLT^2$ | slug·ft$^2$ | kg·m$^2$ |
| $\mu$ | Dynamic viscosity | $ML^{-1}T^{-1}$ | $FL^{-2}T$ | slug/ft/s | kg/m/s |
| $M$ | Mass flow rate | $MT^{-1}$ | $FL^{-1}T^{-1}$ | slug/s | kg/s |
| $MV$ $Ft$ | Momentum Impulse | $MLT^{-1}$ | $FT$ | lbf·s | N·s |
| $M\omega$ | Angular momentum | $ML^2T^{-1}$ | $FLT$ | slug·ft$^2$/s | kg·m$^2$/s |
| $\gamma$ | Specific weight | $ML^{-2}T^{-2}$ | $FL^{-3}$ | lbf/ft$^3$ | N/m$^3$ |
| $p$ $\tau$ $E$ | Pressure Unit shear stress Modulus of elasticity | $ML^{-1}T^{-2}$ | $FL^{-2}$ | lbf/ft$^2$ | N/m$^2$ |
| $\sigma$ | Surface tension | $MT^{-2}$ | $FL^{-1}$ | lbf/ft | N/m |
| F | Force | $MLT^{-2}$ | $F$ | lbf | N |
| $E$ $W$ $FL$ | Energy Work Torque | $ML^2T^{-2}$ | $FL$ | lbf·ft | J |
| $P$ | Power | $ML^2T^{-3}$ | $FLT^{-1}$ | lbf·ft/s | W |
| $v$ | Specific volume | $M^{-1}L^3$ | $F^{-1}L^4T^{-2}$ | ft$^3$/lbm | m$^3$/kg |

*United States customary units.

is recognized as the inverse of the square root of the Froude number $\mathbf{F}$

$$\pi_2 = \sigma/DV^2\rho$$

is the inverse of the Weber number $\mathbf{W}$

$$\pi = \mu/DV\rho$$

is the inverse of the Reynolds number $\mathbf{R}$

| Let | $\pi_1 = f(\pi_2, \pi_3)$ |
| Then | $V = K(Dg)^{1/2}$ |
| where | $K = f(\mathbf{W}, \mathbf{R})$ |

This agrees with the results of the dynamic-similarity analysis of liquid surfaces. This also permits a reduction in the experimental program from variations of six variables to three dimensionless parameters.

## FORCES ON IMMERSED OBJECTS

**Drag and Lift** When a fluid impinges on an object as shown in Fig. 20, the undisturbed fluid pressure $p$ and the velocity $V$ change. Writing Bernoulli's equation for two points on the surface of the object, the point $S$ being the most forward point

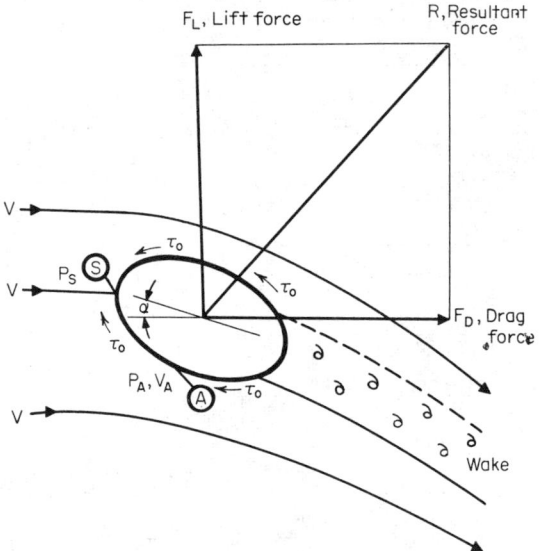

**Fig. 20** Notation for drag and lift.

and point $A$ being any other point, we have, for horizontal flow,

$$p + \rho V^2/2 = p_S + \rho V_S^2/2 = p_A + \rho V_A^2/2$$

At point $S$, $V_S = 0$, so that $p_S = p + \rho V^2/2$. This is called the **stagnation** point, and $p_S$ is the **stagnation pressure**. Since point $A$ is any other point, the result of the fluid impingement is to create a pressure $p_A = p + \rho(V^2 - V_A^2)/2$ acting normal to every point on the surface of the object. In addition, a frictional force $F_f = \tau_0 A_s$ tangential to the surface area $A_s$ opposes the motion. The sum of these forces gives the resultant force $R$ acting on the body. The resultant force $R$ is resolved into the **drag** component $F_D$ parallel to the flow and **lift** component $F_L$ perpendicular to the fluid motion. Depending upon the shape

of the object, a wake may be formed which sheds eddies with a frequency of $f$. The angle $\alpha$ is called the angle of attack. (See Sec. 11.)

From dimensional analysis or dynamic similarity,

$$f(\mathbf{C}_p, \mathbf{R}, \mathbf{M}, \mathbf{S}) = 0$$

The formation of a wake depends upon the Reynolds number, or $\mathbf{S} = f(\mathbf{R})$. This reduces the functional relation to $f(\mathbf{C}_p, \mathbf{R}, \mathbf{M}) = 0$. Since the drag and lift forces may be considered independently,

$$F_D = C_D \rho V^2(A)/2$$

where $C_D = f(\mathbf{R}, \mathbf{M})$
    $A$ = characteristic area

$$F_L = C_L \rho V^2(A)/2$$

where $C_L = f(\mathbf{R}, \mathbf{M})$

It is evident from Fig. 20 that $C_D$ and $C_L$ are also functions of the angle of attack. Since the drag force arises from two sources, the pressure or shape drag $F_p$ and the skin-friction drag $F_f$ due to wall shear stress $\tau_0$, the drag coefficient is made up of two parts:

$$F_D = F_p + F_f = C_p \rho A V^2/2 + C_f \rho A_s V^2/2$$
or $\quad C_D = C_p + C_f A_s/A$

where $C_p$ is the coefficient of pressure, $C_f$ the skin-friction coefficient, and $A_s$ the characteristic area for shear.

**Skin-Friction Drag** Figure 21 shows a fluid approaching a smooth flat plate with a uniform velocity profile of $V$. As the fluid passes over the plate, the velocity at the plate surface is zero and increases to $V$ at some distance $\delta$ from the surface. The region in which the velocity varies from 0 to $V$ is called the **boundary layer.** For some distance along the plate, the flow within the boundary layer is laminar, with viscous forces predominating, but in the transition zone as the inertia forces become larger, a turbulent layer begins to form and increases as the laminar layer decreases.

Boundary-layer thickness and skin-friction drag for incompressible flow over smooth flat plates may be calculated from the following equations, where $\mathbf{R}_X = \rho V X/\mu$:

**Laminar**

| | |
|---|---|
| $\delta/X = 5.20\ \mathbf{R}_X^{-1/2}$ | $0 < \mathbf{R}_X < 5 \times 10^5$ |
| $C_f = 1.328\ \mathbf{R}_X^{-1/2}$ | $0 < \mathbf{R}_X < 5 \times 10^5$ |

**Turbulent**

| | |
|---|---|
| $\delta/X = 0.377\ \mathbf{R}_X^{-1/5}$ | $5 \times 10^4 < \mathbf{R}_X < 10^6$ |
| $\delta/X = 0.220\ \mathbf{R}_X^{-1/6}$ | $10^6 < \mathbf{R}_X < 5 \times 10^8$ |
| $C_f = 0.0735\ \mathbf{R}_X^{-1/5}$ | $2 \times 10^5 < \mathbf{R}_X < 10^7$ |
| $C_f = 0.455\ (\log_{10}\mathbf{R}_X)^{-2.58}$ | $10^7 < \mathbf{R}_X < 10^8$ |
| $C_f = 0.05863\ (\log_{10}C_f\mathbf{R}_X)^{-2}$ | $10^8 < \mathbf{R}_X < 10^9$ |

**Transition** The Reynolds number at which the boundary layer changes depends upon the roughness of the plate and degree of turbulence. The generally accepted number is 500,000, but the transition can take place at Reynolds numbers higher or lower. (Refer to Sec. 11.) For transition at any Reynolds number $\mathbf{R}_X$,

$$C_f = 0.455\ (\log_{10}\mathbf{R}_X)^{-2.58} - (0.0735\mathbf{R}_t^{4/5} - 1.328\ \mathbf{R}^{1/2})\ \mathbf{R}^{-1}$$

For $\mathbf{R}_t = 5 \times 10^5$, $C_f = 0.455\ (\log_{10}\mathbf{R}_X)^{-2.58} - 1{,}725\ \mathbf{R}_X^{-1}$.

**Pressure Drag** Experiments with sharp-edged objects

placed perpendicular to the flow stream indicate that their drag coefficients are essentially constant at Reynolds numbers over 1,000. This means that the drag for $\mathbf{R}_x > 10^3$ is pressure drag. Values of $C_D$ for various shapes are given in Sec. 11 along with the effects of Mach number.

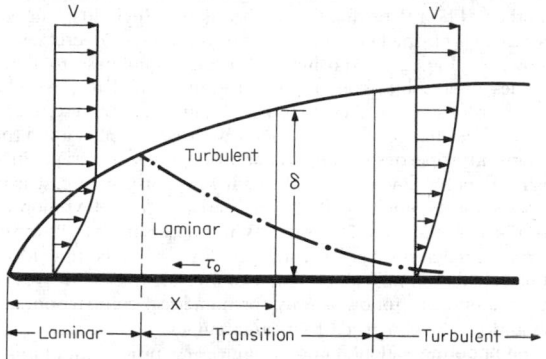

**Fig. 21**   Boundary layer along a smooth flat plate.

**Wake Frequency**   An object in a fluid stream may be subject to the downstream periodic shedding of vortices from first one side and then the other. The frequency of the resulting transverse (lift) force is a function of the stream Strouhal number. As the wake frequency approaches the natural frequency of the structure, the periodic lift force increases asymptotically in magnitude, and when resonance occurs, the structure fails. Neglecting to take this phenomenon into account in design has been responsible for failures of electric transmission lines, submarine periscopes, smokestacks, bridges, and thermometer wells. The wake-frequency characteristics of **cylinders** are shown in Fig. 22. At a Reynolds number of about 20, vortices begin to shed alternately. Behind the cylinder is a staggered stable arrangement of vortices known as the "Kármán vortex trail." At a Reynolds number of about $10^5$, the flow changes from laminar to turbulent. At the end of the transition zone ($\mathbf{R} \approx 3.5 \times 10^5$), the flow becomes turbulent, the alternate shedding stops, and the wake is aperiodic. At the end of the supercritical zone ($\mathbf{R} \approx 3.5 \times 10^6$), the wake continues to be turbulent, but the shedding again becomes alternate and periodic.

The alternating **lift force** is given by

$$F_L(t) = C_L \rho V^2 A \sin (2\pi f t)/2$$

where $t$ is the time. For an analysis of this force in the subcritical zone, see R. Belvins.[1] For design of steel stacks, C. Staley and G. Graven[2] recommend $C_L = 0.8$ for $10^4 < \mathbf{R} < 10^5$, $C_L = 2.8 - 0.4 \log_{10} \mathbf{R}$ for $\mathbf{R} = 10^5$ to $10^6$, and $C_L = 0.4$ for $10^6 < \mathbf{R} < 10^7$.

The **Strouhal number** is nearly constant to $\mathbf{R} = 10^5$, and a nominal design value of 0.2 is generally used. Above $\mathbf{R} = 10^5$, data from different experimenters vary widely, as indicated by the crosshatched zone of Fig. 22. This wide zone is due to experimental and/or measurement difficulties and the dependence on surface roughness to "trigger" the boundary layer. Examination of Fig. 22 indicates an inverse relation of Strouhal number to drag coefficient.

[1] In Murdock, "Fluid Mechanics and Its Applications," Houghton Mifflin, 1976.
[2] ASME 72PET/30.

Observation of actual structures shows that they vibrate at their natural frequency and with a mode shape associated with their fundamental (first) mode during vortex excitation. Based on observations of actual stacks and wind-tunnel tests, Staley and Graven recommend a constant Strouhal number of 0.2 for all ranges of Reynolds number. The ASME[3] recommends $\mathbf{S} = 0.22$ for thermowell design. Until such time as the value of the Strouhal number above $\mathbf{R} = 10^5$ has been firmly established, designers of structures in this area should proceed with caution. See Secs. 11 and 12 for more design detail.

## FLOW IN PIPES

**Parameters for Pipe Flow**   The forces acting on a fluid flowing through and completely filling a horizontal pipe are inertia, viscous, pressure, and elastic. If the surface roughness of the pipe is $\epsilon$, either similarity or dimensional analysis leads to $\mathbf{C}_p = f(\mathbf{R}, \mathbf{M}, L/D, \epsilon/D)$, which may be written for incompressible fluids as $\Delta p = \mathbf{C}_p V^2/2 = \bar{K} \rho V^2/2$, where $K$ is the **resistance coefficient** and $\epsilon/D$ the **relative roughness** of the pipe surface, and the resistance coefficient $K = f(\mathbf{R}, L/D, \epsilon/D)$. The pressure loss may be converted to the terms of **lost head**: $b_f = \Delta p/\gamma = KV^2/2g$. Conventional practice is to use the **friction factor** $f$, defined as $f = KD/L$ or $b_f = KV^2/2g = (fL/D)V^2/2g$, where $f = f(\mathbf{R}, \epsilon/D)$. When a fluid flows into a pipe, the boundary layer starts at the entrance, as shown in Fig. 23, and grows continuously until it fills the pipe. From the equation of motion $db_f = \tau \, dL/\gamma R_h$ and for circular ducts $R_h = D/4$. Comparing wall shear stress $\tau_0$ with friction factor results in the following: $\tau_0 = f\rho V^2/8$.

[3] "Temperature Measurement," PTC 19.3.

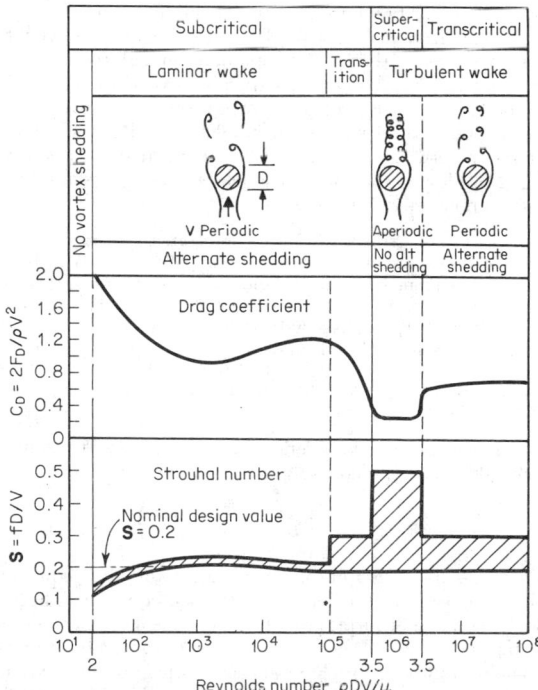

**Fig. 22**   Flow around a cylinder. (*From J. W. Murdock, "Fluid Mechanics and Its Applications," Houghton Mifflin, 1976.*)

**Laminar Flow**   In this type of flow, the resistance is due to viscous forces only so that it is independent of the pipe surface roughness, or $\tau_0 = \mu \ dU/dy$. Application of this equation to the equation of motion and the friction factor yields $f = 64/\mathbf{R}$. Experiments show that it is possible to maintain laminar flow to very high Reynolds numbers if care is taken to increase the

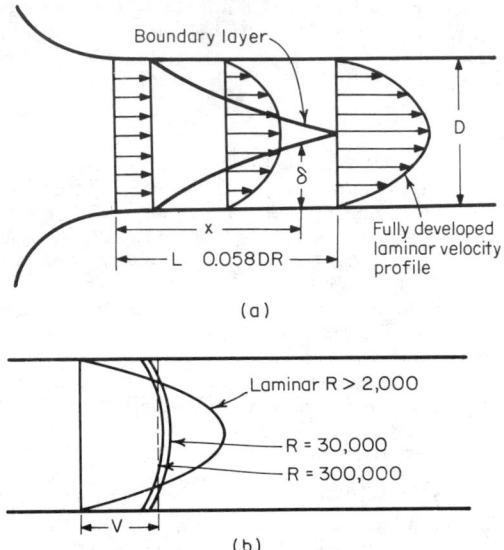

**Fig. 23**   Velocity profiles in pipes.

flow gradually, but normally the slightest disturbance will destroy the laminar boundary layer if the value of Reynolds number is greater than 4,000. In a like manner, flow initially turbulent can be maintained with care to very low Reynolds numbers, but the slightest upset will result in laminar flow if the Reynolds number is less than 2,000. The Reynolds-number range between 2,000 and 4,000 is called the **critical zone** (Fig. 24). Flow in the zone is **unstable,** and designers of piping systems must take this into account.

EXAMPLE.   Glycerin at 68°F (20°C) flows through a horizontal pipe 1 in in diameter and 20 ft long at a rate of 0.090 lbm/s. What is the pressure loss? From the continuity equation $V = Q/A = (\dot{m}/\rho g)/(\pi D^2/4)$ $= [0.090/(2.447 \times 32.17)]/[(\pi/4)(1/12)^2] = 0.2096$ ft/s. The Reynolds number $\mathbf{R} = \rho V D/\mu = (2.447)(0.2096)(1/12)/(29,500 \times 10^{-6}) = 1.449$. $\mathbf{R} < 2,000$; therefore, flow is laminar and $f = 64/\mathbf{R} = 64/1.449 = 44.17$. $K = fL/D = 44.17 \times 20/(1/12) = 10,600$. $\Delta p = K\rho V^2/2 = 10,600 \times 2.447 \ (0.2096)^2/2 = 569.8$ lbf/ft$^2$ = 569.8/144 = 3.957 lbf/in$^2$ (2.728 $\times$ 10$^4$ N/m$^2$).

**Turbulent Flow**   The friction factor for Reynolds number over 4,000 is computed using the Colebrook equation:

$$\frac{1}{\sqrt{f}} = -\ 2\ \log_{10}\left(\frac{\epsilon/D}{3.7} + \frac{2.51}{\mathbf{R}\ \sqrt{f}}\right)$$

Figure 24 is a graphical presentation of this equation (Moody[1]). Examination of the Colebrook equation indicates that if the value of surface roughness $\epsilon$ is small compared with the pipe diameter ($\epsilon/D \to 0$), the friction factor is a function of Reynolds number only. A **smooth pipe** is one in which the ratio

[1]*Trans. ASME*, 1944, pp. 671–684.

$(\epsilon/D)/3.7$ is small compared with $2.51/\ \mathbf{R}\ \sqrt{f}$. On the other hand, as the Reynolds number increases so that $2.51/\ \mathbf{R}\sqrt{f} \to 0$, the friction factor becomes a function of relative roughness only and the pipe is called a **rough pipe**. Thus the same pipe may be smooth under one flow condition, and rough under another. The reason for this is that as the Reynolds number increases, the thickness of the laminar sublayer decreases as shown in Fig. 21, exposing the surface roughness to flow. Values of absolute roughness $\epsilon$ are given in Table 9. The variation of friction factor shown in Fig. 9 is for new, clean pipes. The change of friction factor with **age** depends upon the chemical properties of the fluid and the piping material. Published data for flow of water through wrought-iron or cast-iron pipes show as much as 20 percent increase after a few months to 500 percent after 20 years. When necessary to allow for service life, a study of specific conditions is recommended. The calculation of friction factor to four significant figures in the examples to follow is only for numerical comparison and should not be construed to mean accuracy.

**Engineering Calculations**   Engineering pipe computations usually fall into one of the following classes:

1. Determine pressure loss $\Delta p$ when $Q$, $L$, and $D$ are known.

2. Determine flow rate $Q$ when $L$, $D$, and $\Delta p$ are known.

3. Determine pipe diameter $D$ when $Q$, $L$, and $\Delta p$ are known.

Pressure-loss computations may be made to engineering accuracy using an expanded version of Fig. 24. Greater precision may be obtained by using a combination of Table 9 and the Colebrook equation, as will be shown in the example to follow. Flow rate may be determined by direct solution of the Colebrook equation. Computation of pipe diameter necessitates the trial-and-error method of solution.

EXAMPLE.   *Case 1:* 2,000 gal/min of 68°F (20°C) water flow through 500 ft of cast-iron pipe having an internal diameter of 10 in. At point 1 the pressure is 10 lbf/in$^2$ and the elevation 150 ft, and at point 2 the elevation is 100 ft. Find $p_2$.

From continuity $V = Q/A = [2,000 \times (231/1,728)/60]/[(\pi/4)(10/12)^2]$ $= 8.170$ ft/s. Reynolds number $\mathbf{R} = \rho V D/\mu = (1.937)(8.170)(10/12)(20.92 \times 10^{-6}) = 6.304 \times 10^5$. $\mathbf{R} > 4,000 \therefore$ flow is turbulent. $\epsilon/D = (850 \times 10^{-6})/(10/12) = 1.020 \times 10^{-3}$.

Determine $f$: from Fig. 24 by interpolation $f = 0.02$. Substituting this value on the right-hand side of the Colebrook equation,

$$\frac{1}{\sqrt{f}} = -2\ \log_{10}\left(\frac{\epsilon/D}{3.7} + \frac{2.51}{\mathbf{R}\sqrt{f}}\right)$$

$$= -2\ \log_{10}\left[\frac{1.020 \times 10^{-3}}{3.7} + \frac{2.51}{(6.305 \times 10^5)\ \sqrt{0.02}}\right]$$

$$\frac{1}{\sqrt{f}} = 7.035 \qquad f = 0.02021$$

Resistance coefficient $K = \dfrac{fL}{D} = \dfrac{0.02021 \times 500}{10/12}$

$$K = 12.13$$
$$h_{1f_2} = KV^2/2g = 12.13 \times (8.170)^2/2 \times 32.17$$
$$h_{1f_2} = 12.58 \text{ ft}$$

Equation of motion: $p_1/\gamma + V_1^2/2g + z_1 = p_2/\gamma + V_2^2/2g + z_2 + h_{1f_2}$. Noting that $V_1 = V_2 = V$ and solving for $p_2$,

$$p_2 = p_1 + \gamma(z_1 - z_2 - h_{1f_2})$$
$$= 144 \times 10 + (1.937 \times 32.17)(150 - 100 - 12.58)$$
$$p_2 = 3,772 \text{ lbf/ft}^2 = 3,772/144 = 26.20 \text{ lbf/in}^2 \ (1.806 \times 10^5 \text{ N/m}^2)$$

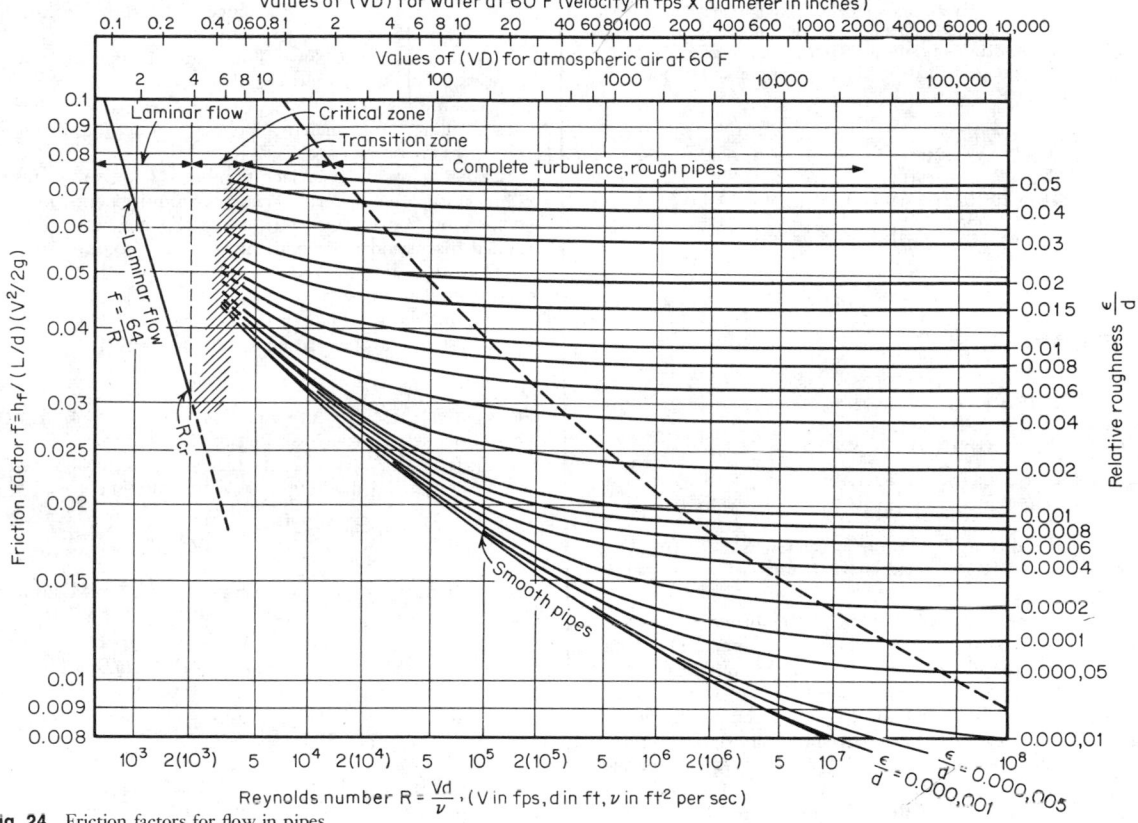

Values of (VD) for water at 60 F (velocity in fps X diameter in inches)

Values of (VD) for atmospheric air at 60 F

**Fig. 24** Friction factors for flow in pipes.

EXAMPLE. *Case 2:* Gasoline (sp. gr. 0.68) at 68°F (20°C) flows through a 6-in schedule 40 (ID = 0.5054 ft) welded steel pipe with a head loss of 10 ft in 500 ft. Determine the flow. This problem may be solved directly by deriving equations that do not contain the flow rate $Q$.

From $h_f = \left(\dfrac{fL}{D}\right)\dfrac{V^2}{2g}$, $V = (2g\,h_f D)^{1/2}/(fL)^{1/2}$

From $\mathbf{R} = \rho VD/\mu$, $V = \mathbf{R}\mu/\rho D$

Equating the above and solving,

$$\mathbf{R}\sqrt{f} = (\rho D/\mu)(2g\,h_f D/L)^{1/2}$$
$$\mathbf{R}\sqrt{f} = (1.310 \times 0.5054/5.98 \times 10^{-6})$$
$$\times (2 \times 32.17 \times 10 \times 0.5054/500)^{1/2} = 89{,}285$$

which is now in a form that may be used directly in the Colebrook equation:

$$\epsilon/D = 150 \times 10^{-6}/0.5054 = 2.968 \times 10^{-4}$$

From the Colebrook equation,

$$\frac{1}{\sqrt{f}} = -2\log_{10}\left(\frac{\epsilon/D}{3.7} + \frac{2.51}{\mathbf{R}\sqrt{f}}\right) = -2\log_{10}\left(\frac{2.968 \times 10^{-4}}{3.7} + \frac{2.51}{89{,}285}\right)$$
$$\frac{1}{\sqrt{f}} = 7.931 \qquad f = 0.01590$$

$\mathbf{R} = 89{,}285/\sqrt{f} = 89{,}285 \times 7.93 = 7\,08 \times 10^5$
$\mathbf{R} > 4{,}000$ ∴ flow is turbulent
$V = \mathbf{R}\mu/\rho D = (7.08 \times 10^5 \times 5.98$
$\qquad \times 10^{-6})/(1.310 \times 0.5054) = 6.396$ ft/s
$Q = AV = (\pi/4)(0.5054)^2(6.396)$
$Q = 1.283$ ft³/s (3.633 × 10⁻² m³/s¹)

### Table 9. Values of Absolute Roughness, New Clean Commercial Pipes

| Type of pipe or tubing | $\epsilon$ ft (0.3048 m) × 10⁶ | | Probable max variation of $f$ from design, % |
|---|---|---|---|
| | Range | Design | |
| Asphalted cast iron | 400 | 400 | −5 to +5 |
| Brass and copper | 5 | 5 | −5 to +5 |
| Concrete | 1,000   10,000 | 4,000 | −35 to 50 |
| Cast iron | 850 | 850 | −10 to +15 |
| Galvanized iron | 500 | 500 | 0 to +10 |
| Wrought iron | 150 | 150 | −5 to 10 |
| Steel | 150 | 150 | −5 to 10 |
| Riveted steel | 3,000   30,000 | 6,000 | −25 to 75 |
| Wood stave | 600   3,000 | 2,000 | −35 to 20 |

Compiled from data given in "Pipe Friction Manual," Hydraulic Institute, 3d ed., 1961.

EXAMPLE. *Case 3:* Water at 68°F (20°C) is to flow at a rate of 500 ft³/s through a concrete pipe 5,000 ft long with a head loss not to exceed 50 ft. Determine the diameter of the pipe. This problem may be solved by trial and error using methods of the preceding example. First trial: Assume any diameter (say 1 ft).

$$\mathbf{R}\sqrt{f} = (\rho D/\mu)(2gb_f D/L)^{1/2}$$
$$= (1.937D/20.92 \times 10^{-6}) \times (32.17 \times 50D/5,000)^{1/2}$$
$$= 74,269D^{3/2} = 74,269(1)^{3/2} = 74,269$$
$$\epsilon/D_1 = 4,000 \times 10^{-6}/D = 4,000 \times 10^{-6}/(1) = 4,000 \times 10^{-6}$$
$$\frac{1}{\sqrt{f_1}} = -2\log_{10}\left(\frac{\epsilon/D_1}{3.7} + \frac{2.51}{\mathbf{R}\sqrt{f_1}}\right)$$
$$= -2\log_{10}\left(\frac{4,000 \times 10^{-6}}{3.7} + \frac{2.51}{74,269}\right)$$
$$\frac{1}{\sqrt{f_1}} = 5.906 \qquad f_1 = 0.02867$$

$$\mathbf{R}_1 = 74,269/\sqrt{f_1} = 74,269 \times 5.906 = 438,600$$
$$V_1 = \mathbf{R}\mu/\rho D_1 = (438,600 \times 20.92 \times 10^{-6})/(1.937 \times 1)$$
$$V_1 = 4.737 \text{ ft/s}$$
$$Q_1 = A_1 V_1 = [\pi(1)^2/4]4.737 = 3.720 \text{ ft}^3/\text{s}$$

For the same loss and friction factor,

$$D_2 = D_1 (Q/Q_1)^{2/5} = (1)(500/3.720)^{2/5} = 7.102 \text{ ft}$$

For the second trial use $D_2 = 7.102$, which results in $Q = 502.2$ ft³/s. Since the nearest standard size would be used, additional trials are unnecessary.

**Velocity Profile** Figure 23a shows the formation of a laminar velocity profile. As the fluid enters the pipe, the boundary layer starts at the entrance and grows continuously until it fills the pipe. The flow while the boundary is growing is called **generating flow.** When the boundary layer completely fills the pipe, the flow is called **established flow.** The distance required for establishing **laminar flow** is $L/D \approx 0.058 \mathbf{R}$. For **turbulent flow,** the distance is much shorter because of the turbulence and not dependent upon Reynolds number, $L/D$ being from 25 to 50.

Examination of Fig. 23b indicates that as the Reynolds number increases, the velocity distribution becomes "flatter" and the flow approaches **one-dimensional.** The velocity profile for laminar flow is parabolic, $U/V = 2[1 - (r/r_o)^2]$ and for **turbulent flow, logarithmic** (except for the very thin laminar boundary layer), $U/V = 1 + 1.43\sqrt{f} + 2.15\sqrt{f}\log_{10}(1 - r/r_o)$. The use of the average velocity produces an error in the computation of kinetic energy. If $\alpha$ is the **kinetic-energy correction factor,** the true kinetic-energy change per unit mass between two points on a flow system $\Delta KE = \alpha_1 V_1^2/2g_c - \alpha_2 V_2^2/2g_c$, where $\alpha = (1/AV^3)\int U^3 dA$. For **laminar flow,** $\alpha = 2$ and for **turbulent flow,** $\alpha \approx 1 + 2.7f$. Of interest is the **pipe factor** $V/U_{max}$; for **laminar flow,** $V/U_{max} = 1/2$ and for **turbulent flow,** $V/U_{max} = 1 + 1.43\sqrt{f}$. The location at which the local velocity equals the average velocity for **laminar flow** is $U = V$ at $r/r_o = 0.7071$ and for **turbulent flow** is $U = V$ at $r/r_o = 0.7838$.

**Compressible Flow** At the present time, there are no true analytical solutions for the computation of actual characteristics of compressible fluids flowing in pipes. In the real flow of a compressible fluid in a pipe, the amount of heat transferred and its direction are dependent upon the amount of insulation, the temperature gradient between the fluid and ambient temperatures, and the heat-transfer coefficient. Each condition requires an individual application of the principles of thermodynamics and heat transfer for its solution.

Conventional engineering practice is to use one of the following methods for flow computation.

1. Assume **adiabatic** flow. This approximates the flow of compressible fluids in short, insulated pipelines.

2. Assume **isothermal** flow. This approximates the flow of gases in long, uninsulated pipelines where the fluid and ambient temperatures are nearly equal.

**Adiabatic Flow** If the Mach number is less than ¼, results within normal engineering-accuracy requirements may be obtained by considering the fluid to be incompressible.* A detailed discussion of and methods for the solution of compressible adiabatic flow are beyond the scope of this section, and any standard gas-dynamics text should be consulted.

**Isothermal Flow** The equation of motion for a horizontal piping system may be written as follows:

$$dp + \rho V \, dV + \gamma \, db_f = 0$$

noting, from the continuity equation, that $\rho V = \dot{M}/A = G$, where $G$ is the **mass velocity** in slugs/(ft²)(s), and that $\gamma \, db_f = [(f/D)\rho V^2/2]dL = [(f/D)GV/2]dL$. Substituting in the above equation of motion and dividing by $GV/2$ results in

$$\frac{2\rho \, dp}{G^2} + \frac{2dV}{V} + \left(\frac{f}{D}\right) dL = 0$$

Integrating for an isothermal process ($p/\rho = C$) and assuming $f$ is a constant,

$$\frac{\rho_1 p_1}{G^2}\left[\left(\frac{p_2}{p_1}\right)^2 - 1\right] + 2\log_e V_2/V_1 + fL/D = 0$$

Noting that $A_1 = A_2$, $V_2/V_1 = \rho_1/\rho_2 = p_1/p_2$, and solving for $G$,

$$G = \left\{\frac{\rho_1 p_1 [1 - (p_2/p_1)^2]}{2\log_e p_1/p_2 + fL/D}\right\}^{1/2}$$

The Reynolds number may be written as

$$\mathbf{R} = \frac{\rho VD}{\mu} = \frac{GD}{\mu} \qquad \text{and} \qquad G = \mathbf{R}\frac{\mu}{D}$$

The value of $\mathbf{R}\sqrt{f}$ may be obtained from the simultaneous solution of the two equations for $G$, assuming that $2\log_e p_1/p_2$ is small compared with $fL/D$.

$$\mathbf{R}\sqrt{f} \approx \left\{\frac{D^3 \rho_1 p_1}{\mu^2 L}\left[1 - \left(\frac{p_2}{p_1}\right)\right]\right\}^{1/2}$$

EXAMPLE. Air at 68°F (20°C) is flowing isothermally through a horizontal straight standard 1-in steel pipe (inside diameter = 1.049 in). The pipe is 200 ft long, the pressure at the pipe inlet is 74.7 lbf/in², and the pressure drop through the pipe is 5 lbf/in². Find the flow rate in lbm/s. From the equation of state $\rho_1 = p/g_c RT = (144 \times 74.7)/(32.17 \times 53.34 \times 527.7) = 0.01188$ slugs/ft³.

$$\mathbf{R}\sqrt{f} = \{[(D^3\rho_1 p_1/\mu^2 L)][1 - (p_2/p_1)^2]\}^{1/2} = \{[(1.049/12)^3(0.01188)$$
$$\times (144 \times 74.7)/(39.16 \times 10^{-8})^2(200)][1 - (69.7/74.7)^2]\}^{1/2}$$
$$= 18,977$$

For steel pipe $\epsilon = 150 \times 10^{-6}$ ft, $\epsilon/D = (150 \times 10^{-6})/(1.049/12) = 1.716 \times 10^{-3}$. From the Colebrook equation,

$$\frac{1}{\sqrt{f}} = -2\log_{10}\left(\frac{\epsilon/D}{3.7} + \frac{2.51}{\mathbf{R}\sqrt{f}}\right)$$
$$= 2\log_{10}[(1.716 \times 10^{-3}/3.7) + (2.51)/(18,977)] = 6.449$$
$$f = 0.02404$$

$R = (R \, \bar{y})(1/\sqrt{f}) = (18,953)(6.449) = 122,200$

$R > 4,000 \; \therefore$ flow is turbulent

$$G = \left\{ \frac{\rho_1 p_1 \, [1 - (p_2/p_1)^2]}{2 \log_e p_1/p_2 + fL/D} \right\}^{1/2}$$

$$= \left\{ \frac{(0.01188)(144 \times 74.7)[1 - (69.7/74.7)^2]}{2 \log_e (74.7/69.7) + (0.02404)(200)/(1.049/12)} \right\}^{1/2}$$

$= 0.5476 \text{ slug}/(\text{ft}^2)(\text{s})$

$\dot{m} = g_c AG = (32.17)(\pi/4)(1.049/12)^2(0.5476)$

$\dot{m} = 0.1057 \text{ lbm/s } (47.94 \times 10^{-3} \text{ kg/s})$

**Noncircular Pipes**  For the flow of fluids in noncircular pipes, the **hydraulic diameter** $D_h$ is used. From the definition of hydraulic radius, the diameter of a circular pipe was shown to be four times its **hydraulic radius**; thus $D_h = 4R_h$. The Reynolds number thus may be written as $R = \rho V D_h/\mu = GD_h/\mu$, the relative roughness as $\epsilon/D_h$, and the resistance coefficient $K = fL/D_h$. With the above modifications, flows through noncircular pipes may be computed in the same manner as for circular pipes.

EXAMPLE.  Air at 68°F (20°C) and 100 lbf/in² enters a rectangular duct 1 by 3 ft at a rate of 720 lbm/s. The duct is horizontal, 100 ft long, and made of galvanized iron. Assuming isothermal flow, estimate the pressure loss due to friction in this line. From the equation of state, $\rho_1 = p_1/g_c RT_1 = (144 \times 100)/(32.17)(53.34)(527.7) = 0.01590 \text{ slug/ft}^3$. From Table 6, $R_h = bD/2(b + D) = 3 \times 1/2(3 + 1) = 0.375$ ft, and $D_h = 4R_h = 4 \times 0.375 = 1.5$ ft. For galvanized iron, $\epsilon/D_h = 500 \times 10^{-6}/1.5 = 3.333 \times 10^{-4}$

$G = (\dot{m}/g_c)/A = (720/32.17)/(1 \times 3) = 7.460 \text{ slugs}/(\text{ft}^2)(\text{s})$

$R = GD_h/\mu = (7.460)(1.5)/(39.16 \times 10^{-8})$

$= 28,580,000 > 4,000 \; \therefore$ flow is turbulent

From Fig. 24, $f \approx 0.015$.

$$\frac{1}{\sqrt{f}} = -2 \log_{10} \left( \frac{\epsilon/D_h}{3.7} + \frac{2.51}{R \sqrt{f}} \right)$$

$$= -2 \log_{10} \left( \frac{3.333 \times 10^{-4}}{3.7} + \frac{2.51}{28,580,000 \sqrt{0.015}} \right)$$

$$f = 0.01530$$

Solving the isothermal equation for $p_2/p_1$,

$$\frac{p_2}{p_1} = [1 - (G^2/\rho_1 p_1)(2 \log_e p_1/p_2 + fL/D_h)]^{1/2}$$

For first trial, assume $2 \log_e p_1/p_2$ is small compared with $fL/D$:

$p_2/p_1 = \{1 - [(7.460)^2/(0.01590)(144$

$\times 100)][0 + (0.01530)(100)/1.5]\}^{1/2}$

$p_2/p_1 = 0.8672$

Second trial using first-trial values results in 0.8263. Subsequent trials result in a balance at $p_2/p_1 = 0.8036$, $p_2 = 100 \times 0.8036 = 80.36$ lbf/in² $(5.541 \times 10^5 \text{ N/m}^2)$.

## PIPING SYSTEMS

**Resistance Parameters**  The resistance to flow of a piping system is similar to the resistance of an object immersed in a flow stream and is made up of pressure (inertia) or shape drag and skin-friction (viscous) drag. For long, straight pipes, the pressure drag is characterized by the **relative roughness** $\epsilon/D$ and the skin friction by the **Reynolds number R.** For other piping components, two parameters are used to describe the resistance to flow, the **resistance coefficient** $K = fL/D$ and the **equivalent length** $L/D = K/f$. The resistance-coefficient method assumes that the component loss is all due to pressure drag and that the flow through the component is completely turbulent

and independent of Reynolds number. The equivalent-length method assumes that resistance of the component varies in the same manner as does a straight pipe. The basic assumption then is that its pressure drag is the same as that for the relative roughness $\epsilon/D$ of the pipe and that the friction drag varies with the Reynolds number **R** in the same manner as the straight pipe. Both methods have the inherent advantage of simplicity in application, but neither is correct except in the fully developed turbulent region. Two excellent sources of information on the resistance of piping-system components are the Hydraulic Institute "Pipe Friction Manual," which uses the resistance-coefficient method, and the Crane Company Technical Paper 410,[1] which uses the equivalent-length concept.

For valves, branch flow through tees, and the type of components listed in Table 10, the pressure drag is predominant, is "rougher" than the pipe to which it is attached, and will extend the completely turbulent region to lower values of Reynolds number. For bends and elbows, the loss is made up of pressure drag due to the change of direction and the consequent secondary flows which are dissipated in 50 diameters or more downstream piping. For this reason, loss through adjacent bends will not be twice that of a single bend.

In long pipelines, the effect of bends, valves, and fittings is usually negligible, but in systems where there is little straight pipe, they are the controlling factor. Under-design will result in the failure of the system to deliver the required capacity. Over-design will result in inefficient operation because it will be necessary to "throttle" one or more of the valves. For estimating purposes, Tables 10 and 11 may be used as shown in the examples. When available, the manufacturers' data should be used, particularly for valves, because of the wide variety of designs for the same type.

**Series Systems**  In a single piping system made of various

[1] "Fluid Meters," 6th ed., ASME, 1971.

### Table 10.  Representative Values of Resistance Coefficient $K$

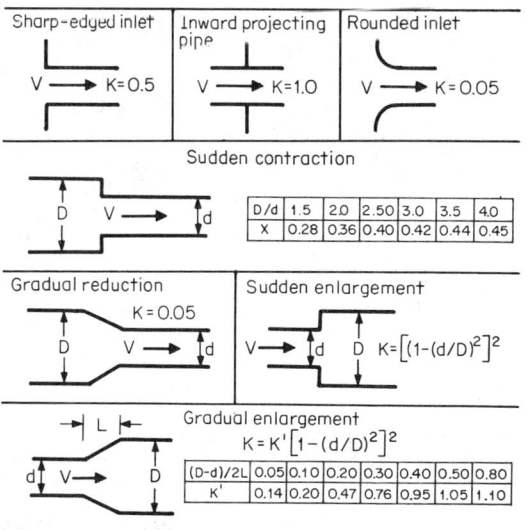

Exit loss = (sharp edged, projecting, Rounded ), K = 1.0

Compiled from data given in "Pipe Friction Manual," 3d ed., Hydraulic Institute, 1961.

**Table 11. Representative Equivalent Length in Pipe Diameters (L/D) of Various Valves and Fittings**

| | |
|---|---|
| Globe valves, fully open | 450 |
| Angle valves, fully open | 200 |
| Gate valves, fully open | 13 |
| ¾ open | 35 |
| ½ open | 160 |
| ¼ open | 900 |
| Swing check valves, fully open | 135 |
| In line, ball check valves, fully open | 150 |
| Butterfly valves, 6 in and larger, fully open | 20 |
| 90° standard elbow | 30 |
| 45° standard elbow | 16 |
| 90° long-radius elbow | 20 |
| 90° street elbow | 50 |
| 45° street elbow | 26 |
| Standard tee: | |
| Flow through run | 20 |
| Flow through branch | 60 |

Compiled from data given in "Flow of Fluids," Crane Company Technical Paper 410, ASME, 1971.

sizes, the practice is to group all of one size together and apply the continuity equation, as shown in the following example.

EXAMPLE.   Water at 68°F (20°C) leaves an open tank whose surface elevation is 180 ft and enters a 2-in schedule 40 steel pipe via a sharp-edged entrance. After 50 ft of straight 2-in pipe that contains a 2-in globe valve, the line enlarges suddenly to an 8-in schedule 40 steel pipe which consists of 100 ft of straight 8-in pipe, two standard 90° elbows, and one 8-in angle valve. The 8-in line discharges below the surface of another open tank whose surface elevation is 100 ft. Determine the volumetric flow rate.

$D_1 = 2.067/12 = 0.1723$    and    $D_2 = 7.981/12 = 0.6651$
$\epsilon/D_1 = 150 \times 10^{-6}/0.1723 = 8.706 \times 10^{-4}$
$\epsilon/D_2 = 150 \times 10^{-6}/0.6651 = 2.255 \times 10^{-4}$

For turbulent flow,

$$\frac{1}{\sqrt{f_1}} = -2 \log_{10} \left( \frac{\epsilon/D}{3.7} \right)$$
$$\frac{1}{\sqrt{f_1}} = -2 \log_{10} \left( \frac{8.706 \times 10^{-4}}{3.7} \right) \qquad f_1 = 0.01899$$
$$\frac{1}{\sqrt{f_2}} = -2 \log_{10} \left( \frac{2.255 \times 10^{-4}}{3.7} \right) \qquad f_2 = 0.01407$$

1. 2-in components                                                **K**
   Entrance loss, sharp-edged entrance          $= 0.5$
   50 ft straight pipe $= f_1 (50/0.1723)$         $= 290.2 f_1$
   Globe valve $= f_1 (L/D)$                             $= 450.0 f_1$
   Sudden enlargement $K = [1 - (D_1/D_2)^2]^2$
   $= [1 - (2.067/7.981)^2]^2$                           $= 0.87$
   $\qquad\qquad\qquad\qquad\qquad \Sigma K_1 = 1.37 + 740.2 f_1$

2. 8-in components
   100 ft of straight pipe $f_2 (100/0.6651)$      $= 150.4 f_2$
   2 standard 90° elbows $2 \times 30 f_2$         $= 60 \ f_2$
   1 angle valve $200 f_2$                               $= 200 \ f_2$
   Exit loss                                                  $= 1$
   $\qquad\qquad\qquad\qquad\qquad \Sigma K_2 = 1 + 410.4 f_2$

3. Apply equation of motion

$$h_{1f_2} = z_1 - z_2 = (\Sigma K_1) \frac{V_1^2}{2g} + (\Sigma K_2) \frac{V_2^2}{2g}$$

From continuity, $\rho_1 A_1 V_1 = \rho_2 A_2 V_2$     for $\rho_1 = \rho_2$

$V_2 = V_1(A_1/A_2) = V_1(D_1/D_2)^2$
$h_{1f_2} = z_1 - z_2 = [\Sigma K_1 + \Sigma K_2 (D_1/D_2)^4] V_1^2/2g$
$V_1 = \{[2g(z_1 - z_2)]/[\Sigma K_1 + \Sigma K_2(D_1/D_2)^4]\}^{1/2}$

$$V_1 = \left[ \frac{2 \times 32.17 \times (180 - 100)}{(1.37 + 740.2 f_1) + (1 + 410.4 f_2)(2.067/7.981)^4} \right]^{1/2}$$

$$V_1 = \frac{71.74}{(1.374 + 740.2 f_1 + 1.846 f_2)^{1/2}}$$

4. For first trial assume $f_1$ and $f_2$ for complete turbulence

$$V_1 = \frac{71.74}{(1.374 + 740.2 \times 0.01899 + 1.846 \times 0.01407)^{1/2}}$$
$V_1 = 18.25$ ft/s
$V_2 = 18.25 (2.067/7.981)^2 = 1.224$ ft/s
$\mathbf{R}_1 = \rho_1 V_1 D_1/\mu = (1.937)(18.25)(0.1723)/(20.92 \times 10^{-6})$
$\mathbf{R}_1 = 291,100 > 4,000 \ \therefore$ flow is turbulent
$\mathbf{R}_2 = \rho_2 V_2 D_2/\mu_2 = (1.937)(1.224)(0.6651)/(20.92 \times 10^{-6})$    *1 pt.*
$\mathbf{R}_2 = 75,420 > 4,000 \ \therefore$ flow is turbulent

5. For second trial use first trial $V_1$ and $V_2$. From Fig. 24 and the Colebrook equation,

$$\frac{1}{\sqrt{f_1}} = -2 \log_{10} \left( \frac{8.706 \times 10^{-4}}{3.7} + \frac{2.51}{291,100 \sqrt{0.020}} \right)$$
$f_1 = 0.02008$
$$\frac{1}{\sqrt{f_2}} = -2 \log_{10} \left( \frac{2.255 \times 10^{-4}}{3.7} + \frac{2.51}{75,420 \sqrt{0.020}} \right)$$
$f_2 = 0.02008$
$$V_1 = \frac{71.74}{(1.374 + 740.2 \times 0.02008 + 1.864 \times 0.02008)^{1/2}}$$
$V_1 = 17.78$

A third trial results in $V = 17.77$ ft/s or $Q = A_1 V_1 = (\pi/4)$ $(0.1723)^2(17.77) = 0.4143$ ft³/s $(1.173 \times 10^{-2}$ m³/s ).

**Parallel Systems**   In solution of problems involving two or more parallel pipes, the head loss for each of the pipes is the same as shown in the following example:

EXAMPLE.   Benzene at 68°F (20°C) flows at a rate of 0.5 ft³/s through two parallel straight, horizontal pipes connecting two pressurized tanks. The pipes are both schedule 40 steel, one being 1 in, the other 2 in. They both are 100 ft long and have connections that project inwardly in the supply tank. If the pressure in the supply tank is maintained at 100 lbf/in², what pressure should be maintained on the receiving tank?

$D_1 = 1.049/12 = 0.08742$ ft    and    $D_2 = 2.067/12 = 0.1723$ ft
$\epsilon/D_1 = 150 \times 10^{-6}/0.08742 = 1.716 \times 10^{-3}$
$\epsilon/D_2 = 150 \times 10^{-6}/0.1723 = 8.706 \times 10^{-4}$

For turbulent flow,

$$\frac{1}{\sqrt{f}} = -2 \log_{10} \left( \frac{\epsilon/D}{3.7} \right)$$
$$\frac{1}{\sqrt{f_1}} = -2 \log_{10} \left( \frac{1.716 \times 10^{-3}}{3.7} \right) \qquad f_1 = 0.02249$$
$$\frac{1}{\sqrt{f_2}} = -2 \log_{10} \left( \frac{8.706 \times 10^{-4}}{3.7} \right) \qquad f_2 = 0.01899$$

1. 1-in components                                              **K**
   Entrance loss, inward projection                $= 1.0$
   100 ft straight pipe $f_1 (100/0.08742)$        $= 1,144 f_1$
   Exit loss                                                  $= 1.0$
   $\qquad\qquad\qquad\qquad\qquad \Sigma K_1 = 2.0 + 1,144 f_1$

2. 2-in components

Entrance loss, inward projection = 1.0
100 ft straight pipe $f_2$ (100/0.1723) = 580.4 $f_2$
Exit loss = 1.0

$$\Sigma K_2 = 2.0 + 580.4 f_2$$

$b_f = \Sigma K_1 V_1^2/2g = \Sigma K_2 V_2^2/2g$
From the continuity equation, $Q = AV$

$$\Sigma K_1 \frac{Q_1^2}{2gA_1^2} = \Sigma K_2 \frac{Q_2^2}{2gA_2^2}$$

Solving for $Q_1/Q_2$,

$$\frac{Q_1}{Q_2} = \frac{A_1}{A_2}\sqrt{\frac{\Sigma K_2}{\Sigma K_1}} = \left(\frac{D_1}{D_2}\right)^2 \sqrt{\frac{\Sigma K_2}{\Sigma K_1}} = \left(\frac{D_1}{D_2}\right)^2 \sqrt{\frac{2.0 + 580.4 f_2}{2.0 + 1,144 f_1}}$$

For first trial assume flow is completely turbulent,

$$\frac{Q_1}{Q_2} = \left(\frac{0.08742}{0.1723}\right)^2 \sqrt{\frac{2.0 + 580.4 \times 0.01899}{2.0 + 1,144 \times 0.02249}}$$

$$\frac{Q_1}{Q_2} = 0.1764 \qquad Q = Q_1 + Q_2 = 0.1764 Q_2 + Q_2$$

$0.5000 = 1.1764 Q_2 \qquad Q_2 = 0.4250$
$Q_1 = 0.5000 - 0.4250 = 0.0750$

For second trial use first-trial values,

$V_1 = Q_1/A_1 = 0.0750/(\pi/4)(0.08742)^2 = 12.50$
$V_2 = Q_2/A_2 = 0.4250/(\pi/4)(0.1723)^2 = 18.23$
$R_1 = \rho_1 V_1 D_1/\mu_1 = (1.705)(12.50)(0.08742)/(13.62 \times 10^{-6})$
$R_1 = 136,800 > 4,000$ ∴ flow is turbulent
$R_2 = \rho_2 V_2 D_2/\mu_2 = (1.705)(18.23)(0.1723)/(13.62 \times 10^{-6})$
$R_2 = 393,200 > 4,000$ ∴ flow is turbulent

Using the Colebrook equation and Fig. 24,

$$\frac{1}{\sqrt{f_1}} = -2 \log_{10}\left(\frac{1.716 \times 10^{-3}}{3.7} + \frac{2.51}{136,800 \sqrt{0.024}}\right)$$
$f_1 = 0.02389$
$$\frac{1}{\sqrt{f_2}} = -2 \log_{10}\left(\frac{8.706 \times 10^{-4}}{3.7} + \frac{2.51}{393,200 \sqrt{0.020}}\right)$$
$f_2 = 0.01981$
$$b_f = \Sigma K_1 \frac{V_1^2}{2g} = \Sigma K_2 \frac{V_2^2}{2g}$$
$\Sigma K_1 V_1^2/2g$
$= (2.0 + 1,144 \times 0.02389)(12.50)^2/(2 \times 32.17) = 71.23$
$\Sigma K_2 V_2^2/2g$
$= (2.0 + 580.4 \times 0.01981)(18.23)^2/(2 \times 32.17) = 69.80$

71.23 = 69.80; further trials not justifiable because of accuracy of $f$, $K$, $L/D$. Use average or 70.52, so that $\Delta p = \rho g \, b_f = (1.705 \times 32.17 \times 70.52)/144 = 26.86$ lbf/in² $= p_1 - p_2 = 100 - p_2$, $p_2 = 100 - 26.86 = 73.40$ lbf/in² ($5.061 \times 10^5$ N/m²).

**Branch Flow** Problems of a single line feeding several points may be solved as shown in the following example.

EXAMPLE. Ethyl alcohol at 68°F (20°C) flows from tank $A$, which is maintained at a constant pressure of 100 lb/in² through 200 ft of 2-in cast-iron schedule 40 pipe to a Y branch connection ($K = 0.5$) where 100 ft of 2-in pipe goes to tank $B$, which is maintained at 80 lbf/in² and 50 ft of 2-in pipe to tank $C$, which is also maintained at 80 lbf/in². All tank connections are flush and sharp-edged and are at the same elevation. Estimate the flow rate to each tank.

$$D = 2.067/12 = 0.1723 \text{ ft}$$
$$\epsilon/D = 850 \times 10^{-6}/0.1723 = 4.933 \times 10^{-3}$$

For turbulent flow, $\dfrac{1}{\sqrt{f}} = -2 \log_{10}\left(\dfrac{\epsilon/D}{3.7}\right)$

$$\frac{1}{\sqrt{f}} = -2 \log_{10}\left(\frac{4.933 \times 10^{-3}}{3.7}\right) \qquad f = 0.03025$$
$b_{A f_B} = (p_A - p_B)/\rho g = 144(100 - 80)/(1.532 \times 32.17) = 58.44$
$b_{A f_C} = (p_A - p_C)/\rho g = b_{A f_B} = 58.44$

Let point $X$ be just before the Y; then

1. From tank $A$ to Y $\qquad\qquad\qquad\qquad\qquad$ **K**
   Entrance loss, sharp-edged $\qquad\qquad\qquad = 0.5$
   200 ft straight pipe $= f_{AX}$ (200/0.1723) $= 1,161 f_{AX}$
   $\qquad\qquad\qquad\qquad \Sigma K_{AX} = 0.5 + 1,161 f_{AX}$

2. From Y to tank $B$
   Y branch $\qquad\qquad\qquad\qquad\qquad\qquad = 0.5$
   100 ft straight pipe $= f_{XB}$ (100/0.1723) $= 580.4 f_{XB}$
   Exit loss $\qquad\qquad\qquad\qquad\qquad\qquad = 1.0$
   $\qquad\qquad\qquad\qquad \Sigma K_{XB} = 1.5 + 580.4 f_{XB}$

3. From Y to tank $C$
   Y branch $\qquad\qquad\qquad\qquad\qquad\qquad = 0.5$
   50 ft straight pipe $= f_{XC}$ (50/0.1723) $= 290.2 f_{XC}$
   Exit loss $\qquad\qquad\qquad\qquad\qquad\qquad = 1.0$
   $\qquad\qquad\qquad\qquad \Sigma K_{XC} = 1.5 + 290.2 f_{XC}$

Balance of flows:

$$Q_{AX} = Q_{XB} + Q_{XC}$$

and from continuity, $(A_{AX} = A_{XB} = A_{XC})$, $V_{AX} = V_{XB} + V_{XC}$; then

$$b_{A f_B} = \Sigma K_{AX}\frac{V_{AX}^2}{2g} + \Sigma K_{XB}\frac{V_{XB}^2}{2g}$$
$$b_{A f_C} = \Sigma K_{AX}\frac{V_{AX}^2}{2g} + \Sigma K_{XC}\frac{V_{XC}^2}{2g}$$

For first trial assume completely turbulent flow

$$b_{A f_B} = \frac{(0.5 + 1,161 f_{AX})V_{AX}^2}{2g} + \frac{(1.5 + 580.4 f_{XB})V_{XB}^2}{2g}$$
$$58.44 = \frac{(0.5 + 1,161 \times 0.03025)V_{AX}^2}{2 \times 32.17} + \frac{(1.5 + 580.4 \times 0.03025)V_{XB}^2}{2 \times 32.17}$$
$$58.44 = 0.5536 V_{AX}^2 + 0.2962 V_{XB}^2$$

and in a like manner

$$b_{A f_C} = 58.44 = 0.5536 V_{AX}^2 + 0.1598 V_{XC}^2$$

Equating $b_{A f_B} = b_{A f_C}$,

$$0.5536 V_{AX}^2 + 0.2962 V_{XB}^2 = 0.5536 V_{AX}^2 + 0.1598 V_{XC}^2$$

or $V_{XC} = 1.3615 V_{XB}$ and since $V_{AX} = V_{XB} + V_{XC}$
$V_{AX} = V_{XB} + 1.3615 V_{XB} = 2.3615 V_{XB}$

so that

$$b_{A f_B} = 58.44 = 0.5536(2.3615 V_{XB}^2) + 0.2962 V_{XB}^2$$
$$V_{XB} = 4.156$$
$$V_{XC} = 1.3615(4.156) = 5.658$$
$$V_{AX} = 4.156 + 5.658 = 9.814$$

Second trial,

$$R_{AX} = \frac{\rho V_{AX} D}{\mu} = \frac{1.532 \times 9.814 \times 0.1723}{25.06 \times 10^{-6}}$$
$$R_{AX} = 103,400 > 4,000 \therefore \text{ flow is turbulent}$$

In a like manner,

$$R_{XB} = 43,780 \qquad R_{XC} = 59,600$$

Using the Colebrook equation and Fig. 24,

$$\frac{1}{\sqrt{f_{AX}}} = -2 \log_{10} \left( \frac{4.933 \times 10^{-3}}{3.7} + \frac{2.51}{103,400 \sqrt{0.031}} \right)$$
$$f_{AX} = 0.03116$$

In a like manner,

$$f_{XB} = 0.03231 \qquad f_{XC} = 0.03179$$

$$b_{A f_B} = \frac{(0.5 + 1,161 \times 0.03116) V_{AX}^2}{2 \times 32.17} + \frac{(1.5 + 580.4 \times 0.03231) V_{XB}^2}{2 \times 32.17}$$
$$b_{A f_B} = 0.5700 \, V_{AX}^2 + 0.3148 \, V_{XB}^2$$
$$b_{A f_C} = 0.5700 \, V_{AX}^2 + \frac{(1.5 + 290.2 \times 0.03179) V_{XC}^2}{2 \times 32.17}$$
$$b_{A f_C} + 0.5700 \, V_{AX}^2 + 0.1667 \, V_{XC}^2$$
$$0.3148 \, V_{XB}^2 = 0.1667 \, V_{XC}^2$$
$$V_{XC} = 1.374 \, V_{XB}$$
$$V_{AX} = V_{XB} + 1.374 \, V_{XB} = 2.374 \, V_{XB}$$

so that

$$b_{A f_B} = 58.44 = 0.5700 \, (2.374 \, V_{XB})^2 + 0.3148 \, V_{XB}^2$$
$$V_{XB} = 4.070 \qquad V_{XC} = 5.592 \qquad V_{AX} = 9.663$$

Further trials are not justified.

$$A = \pi D^2/4 = (\pi/4)(0.1723)^2 = 0.02332 \text{ ft}^2$$
$$Q_{XB} = V_{XB} A = 4.070 \times 0.02332 = 0.09491 \text{ ft}^3/\text{s} \ (2.686 \times 10^{-3} \text{ m}^3/\text{s})$$
$$Q_{XC} = V_{XC} A = 5.592 \times 0.02332 = 0.1304 \text{ ft}^3/\text{s} \ (3.693 \times 10^{-3} \text{ m}^3/\text{s})$$

**Siphons** are arrangements of hose or pipe which cause liquids to flow from one level $A$ in Fig. 25 to a lower level $C$ over an intermediate summit $B$. Performance of siphons may be evalu-

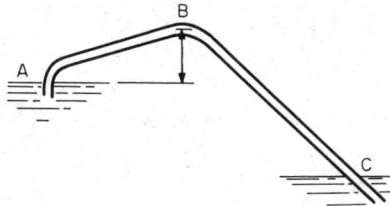

**Fig. 25**  Siphon.

ated from the equation of motion between points $A$ and $B$:

$$\frac{p_A}{\gamma} + \frac{V_A^2}{2g} + z_A = \frac{p_B}{\gamma} + \frac{V_B^2}{2g} + z_B + b_{A f_B}$$

Noting that on the surface $V_A = 0$ and the minimum pressure that can exist at point $B$ is the vapor pressure $p_v$, the maximum elevation of point $B$ is

$$z_B - z_A = \frac{p_A}{\gamma} - \left( \frac{p_v}{\gamma} + \frac{V_B^2}{2g} + b_{A f_B} \right)$$

The friction loss $b_f = \Sigma K_{AB} V_B^2/2g$, and let $V_B = V$; then

$$z_B - z_A = \frac{p_A - p_v}{\rho g} - (1 + \Sigma K_{AB}) \frac{V^2}{2g}$$

Flow under this maximum condition will be uncertain. The air pump or ejector used for priming the pipe (flow will not take place unless the siphon is full of water) might have to be operated occasionally to remove accumulated air and vapor. Values of $z_B - z_A$ less than those calculated by the above equation should be used.

EXAMPLE.  The siphon shown in Fig. 25 is composed of 2,000 ft of

6-in schedule 40 cast-iron pipe. Reservoir $A$ is at elevation 800 ft and $C$ at 600 ft. Estimate the maximum height for $z_B - z_A$ if the water temperature may reach 104°F (40°C), and the amount of straight pipe from $A$ to $B$ is 100 ft. For the first bend $L/D = 25$ and the second (at $B$) $L/D = 50$. Atmospheric pressure is 14.70 lbf/in². For 6-in schedule 40 pipe $D = 6.065/12 = 0.5054$ ft, $\epsilon/D = 850 \times 10^{-6}/0.5054 = 1.682 \times 10^{-3}$. Turbulent friction factor $1/\sqrt{f} = -2 \log_{10} \left( \frac{\epsilon/D}{3.7} \right) = -2 \log_{10} (1.682 \times 10^{-3}/3.7) = 0.02238$.

1. Components from $A$ to $B$ $\qquad\qquad\qquad\qquad K$
   (Note loss in second bend takes place
   in downstream piping)

| | |
|---|---|
| Entrance (inward projection) | = 1.0 |
| 100 ft straight pipe $f(100/0.5054)$ | = 197.9 $f$ |
| First bend | = 25 $f$ |
| | $\Sigma K_{AB} = 1.0 + 227.9 f$ |

2. Components from $A$ to $C$

| | |
|---|---|
| $\Sigma K_{AB}$ | = 1.0 + 2,229 $f$ |
| 1,900 ft of straight pipe $f(1,900/0.5054)$ | = 3,759.4 $f$ |
| Second bend | = 50 $f$ |
| Exit loss | = 1 |
| | $\Sigma K_{AC} = 2.0 + 4,032 f$ |

First trial assume complete turbulence. Writing the equation of motion between $A$ and $C$,

$$\frac{p_A}{\gamma} + \frac{V_A^2}{2g} + z_A = \frac{p_C}{\gamma} + \frac{V_C^2}{2g} + z_C + \Sigma K_{AC} \frac{V^2}{2g}$$

Noting $V_A = V_C = 0$, and $p_A = p_C = 14.7$ lbf/in²,

$$V = \sqrt{\frac{2g(z_A - z_C)}{\Sigma K_{AC}}} = \sqrt{\frac{2g(z_A - z_C)}{2.0 + 4,032 f}} = \sqrt{\frac{2 \times 32.17 (800 - 600)}{2.0 + 4,032 f}}$$
$$= \frac{113.44}{\sqrt{2.0 + 4,032 \times 0.02238}} = 11.81$$

Second trial, use first trial values,

$$\mathbf{R} = \frac{\rho V D}{\mu} = (1.925)(11.81)(0.5054)/(13.61 \times 10^{-6})$$
$$\mathbf{R} = 846,200 > 4,000 \therefore \text{ flow is turbulent}$$

From Fig. 24 and the Colebrook equation,

$$\frac{1}{\sqrt{f}} = -2 \log_{10} \left( \frac{1.682 \times 10^{-3}}{3.7} + \frac{2.51}{844,200 \sqrt{0.023}} \right)$$
$$f = 0.02263$$
$$V = \frac{113.44}{\sqrt{2.0 + 4,032 \times 0.02263}} = 11.75 \qquad \text{(close check)}$$

From Sec. 4 steam tables at 104°F, $p_v = 1.070$ lbf/in², the maximum height

$$z_B - z_A = \frac{p_A - p_v}{\rho g} - (1 + \Sigma K_{AB}) \frac{V^2}{2g}$$
$$z_B - z_A = \frac{144(14.70 - 1.070)}{1.925 \times 32.17} - (1 + 1 + 227.9)$$
$$\times 0.02262 \frac{(11.75)^2}{2 \times 32.17} = 16.58 \text{ ft } (5.053 \text{ m})$$

Note that if a ±10 percent error exists in calculation of pressure loss, maximum height should be limited to ≈ 15 ft (5 m).

## ASME PIPELINE FLOWMETERS

**Parameters**  Dimensional analysis of the flow of an incompressible fluid flowing in a pipe of diameter $D$, surface roughness $\epsilon$, through a primary element (venturi, nozzle or orifice) whose diameter is $d$ with a velocity of $V$, producing a pressure drop of $\Delta p$ sensed by pressure taps located a distance $L$ apart results in $f(\mathbf{C}_p, \mathbf{R}_d, \epsilon/D, d/D) = 0$, which may be written as $\Delta p = \mathbf{C}_p \, \rho V^2/2$. Conventional practice is to express the relations as

$V = \mathbf{K}\sqrt{2\Delta p/\rho}$, where $\mathbf{K}$ is the **flow coefficient**, $\mathbf{K} = 1/\sqrt{\mathbf{C}_p}$, and $\mathbf{K} = f(\mathbf{R}_d, L/D, \epsilon/d, d/D)$. The ratio of the diameter of the primary element to meter tube (pipe) diameter $D$ is known as the **beta ratio**, where $\beta = d/D$. Application of the continuity

equation leads to $Q = \mathbf{K}A_2\sqrt{2\Delta p/\rho}$, where $A_2$ is the area of the primary element.

Conventional practice is to base flowmeter computations on the assumption of one-dimensional frictionless flow of an incompressible fluid in a horizontal meter tube and to correct for actual conditions by the use of a coefficient for viscous effects and a factor for elastic effects. Application of the Bernoulli equation for horizontal flow from section 1 (inlet tap) to section 2 (outlet tap) results in $p_1/\rho g + V_1^2/2g = p_2/\rho g + V_2^2/2g$ or $(p_1 - p_2)/\rho = V_2^2 - V_1^2 = \Delta p/\rho$. From the equation of continuity, $Q_i = A_1 V_1 = A_2 V_2$, where $Q_i$ is the **ideal flow rate**. Substituting, $2\Delta p/\rho = Q_i^2/A_1^2 - Q_i^2/A_2^2$, and solving for $Q_i$, $Q_i = A_2\sqrt{2\Delta p/\rho} / \sqrt{1 - (A_2/A_1)^2}$, noting that $A_2/A_1 = (d/D)^2 = \beta^2$, $Q_i = A_2\sqrt{2\Delta p/\rho} / \sqrt{1 - \beta^4}$. The **discharge coefficient** $C$ is defined as the ratio of the actual flow $Q$ to the ideal flow $Q_i$, or $C = Q/Q_i$, so that $Q = CQ_i = CA_2\sqrt{2\Delta p/\rho} / \sqrt{1 - \beta^4}$. It is customary to write the volumetric-flow equation as $Q = CEA_2\sqrt{2\Delta p/\rho}$, where $E = 1/\sqrt{1 - \beta^4}$. $E$ is called the **velocity-of-approach factor** because it accounts for the one-dimensional kinetic energy at the upstream tap. Comparing the equation from dimensional analysis with the modified Bernoulli equation, $Q = \mathbf{K}A_2\sqrt{2\Delta p/\rho} = CEA_2\sqrt{2\Delta p/\rho}$, or $\mathbf{K} = CE$ and $C = f(\mathbf{R}_d, L/D, \beta)$.

For compressible fluids, the incompressible equation is modified by the **expansion factor** $Y$, where $Y$ is defined as the ratio of the flow of a compressible fluid to that of an incompressible fluid at the same value of Reynolds number. Calculations are then based on inlet-tap-fluid properties, and the compressible equation becomes

$$Q_1 = \mathbf{K}YA_2\sqrt{2\Delta p/\rho_1} = CEYA_2\sqrt{2\Delta p/\rho_1}$$

where $\qquad Y = f(L/D, \epsilon/D, \beta, \mathbf{M})$

Reynolds number $\mathbf{R}_d$ is also based on inlet-fluid properties, but on the primary-element diameter or

$$\mathbf{R}_d = \rho_1 V_2 d/\mu_1 = \rho_1(Q_1/A_2)d/\mu_1 = 4\rho_1 Q_1/\pi d \mu_1$$

**Caution** The numerical values of coefficients for flowmeters given in the paragraphs to follow are based on experimental data obtained with long, straight pipes where the velocity profile approaching the primary element was fully developed. The presence of valves, bends, and fittings upstream of the primary element can cause serious errors. For approach and discharge, straight-pipe requirements, "Fluid Meters,"[1] should be consulted.

**Venturi Tubes** Figure 26 shows a typical venturi tube consisting of a cylindrical inlet, convergent cone, throat, and

[1]6th ed., ASME, 1971.

divergent cone. The convergent entrance has an included angle of about 21° and the divergent cone 7 to 8°. The purpose of the divergent cone is to reduce the overall pressure loss of the meter; its removal will have no effect on the coefficient of discharge. Pressure is sensed through a series of holes in the inlet and throat. These holes lead to an annular chamber, and

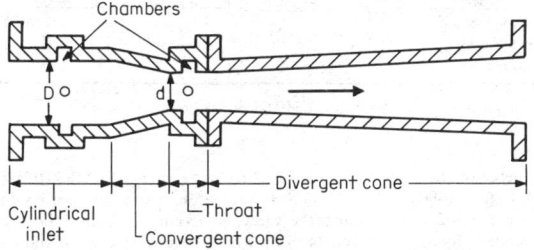

**Fig. 26** Venturi tube.

the two chambers are connected to a pressure-differential sensor. Discharge coefficients for venturi tubes as established by the American Society of Mechanical Engineers are given in Table 12. Coefficients of discharge outside the tabulated limits must be determined by individual calibrations.

EXAMPLE. Benzene at 68°F (20°C) flows through a machined-inlet venturi tube whose inlet diameter is 8 in and whose throat diameter is 3.5 in. The differential pressure is sensed by a U-tube manometer. The manometer contains mercury under the benzene, and the level of the mercury in the throat leg is 4 in. Compute the volumetric flow rate. Noting that $D = 8$ in (0.6667 ft) and $\beta = 3.5/8 = 0.4375$ are within the limits of Table 12, assume $C = 0.995$, and then check $\mathbf{R}_d$ to verify if it is within limits. For a U-tube manometer (Fig. 6a), $p_2 - p_1 = (\gamma_m - \gamma_f)b = \Delta p$ and $\Delta p/\rho_1 = (\rho_m g - \rho_f g)b/\rho_f = g(\rho_m/\rho_f - 1)b = 32.17(26.283/1.705 - 1)(4/12) = 154.6$. For a liquid, $Y = 1$ (incompressible fluid), $E = 1/\sqrt{1 - \beta^4} = 1/\sqrt{1 - (0.4375)^4} = 1.019$.

$$Q_1 = CEY A_d\sqrt{2\Delta p/\rho_1} = (0.995)(1.019)(\pi/4)(3.5/12)^2 \sqrt{2 \times 154.6}$$
$$= 1.192 \text{ ft}^3/\text{s } (3.373 \times 10^{-3} \text{ m}^3/\text{s})$$

$\mathbf{R}_d = 4\rho_1 Q_1/\pi \, d\mu_1 = 4(1.705)(1.192)/\pi(3.5/12)(13.62 \times 10^{-6})$
$\mathbf{R}_d = 651,400$, which lies between 200,000 and 1,000,000 of Table 12
$\therefore$ solution is valid.

**Flow Nozzles** Figure 27 shows an ASME flow nozzle. This nozzle is built to rigid specifications, and pressure differential may be sensed by either throat taps or pipe-wall taps. Taps are located one pipe diameter upstream and one-half diameter downstream from the nozzle inlet. Discharge coefficients for ASME flow nozzles may be computed from $C = 0.9975 - 0.00653 \, (10^6/\mathbf{R}_d)^a$, where $a = 1/2$ for $\mathbf{R}_d < 10^6$ and $a = 1/5$ for $\mathbf{R}_d > 10^6$. Most of the data were obtained for $D$ between 2 and 15.75 in, $\mathbf{R}_d$ between $10^4$ and $10^6$, and beta between 0.15 and 0.75. For values of $C$ within these ranges, a tolerance of 2 percent may be anticipated, and outside these limits, the tolerance may be greater than 2 percent. Because slight variations in form or dimension of either pipe or nozzle may affect the observed pressures, and thus cause the exponent $a$ and the slope term ($-0.00653$) to vary considerably, nozzles should be individually calibrated.

EXAMPLE. An ASME flow nozzle is to be designed to measure the flow of 400 gal/min of 68°F (20°C) water in a 6-in schedule 40 (inside diameter = 6.065 in) steel pipe. The pressure differential across the nozzle is not to exceed 75 in of water. What should be the throat

**Table 12. ASME Coefficients for Venturi Tubes**

| Type of inlet cone | Reynolds number $R_d$ | | Inlet diam D in $(2.54 \times 10^{-2}\text{m})$ | | $\beta$ | | C | Tolerance, % |
|---|---|---|---|---|---|---|---|---|
| | Min | Max | Min | Max | Min | Max | | |
| Machined | | $1 \times 10^6$ | 2 | 10 | 0.4 | 0.75 | 0.995 | ±1.0 |
| Rough welded sheet metal | $5 \times 10^5$ | $2 \times 10^6$ | 8 | 48 | | 0.70 | 0.985 | ±1.5 |
| Rough cast | | | 4 | 32 | 0.3 | 0.75 | 0.984 | ±0.7 |

Compiled from data given in "Fluid Meters," 6th ed., ASME, 1971.

diameter of the nozzle? $\Delta p = h\rho_1 g$, $\Delta p/\rho_1 = hg = (75/12)(32.17) = 201.1$, $Q = (400/60)(231/1,728) = 0.8912$ ft³/s. A trial-and-error solution is necessary to establish the values of $C$ and $E$ because they are dependent upon $\beta$ and $R_d$, both of which require that $d$ be known. Since $K = CE \approx 1$, assume for first trial that $CE = 1$. Since a liquid is involved, $Y = 1$, $A_2 = Q_1/(CE)(Y)\sqrt{2\Delta p/\rho_1} = (0.8912)/(1)(1)$ $\sqrt{2 \times 201.1} = 0.04444$ ft², $d = \sqrt{4A_2/\pi} = \sqrt{4(0.04444)/\pi} = 0.2379$ ft or $d = 0.2379 \times 12 = 2.854$ in, $\beta = d/D = 2.854/6.065 = 0.4706$.
For second trial use first-trial value:

$$E = 1/\sqrt{1 - \beta^4} = 1/\sqrt{1 - (0.4706)^4} = 1.025.$$

$R_d = 4\rho_1 Q_1/\pi d\mu_1 = 4(1.937)(0.8912)/\pi(0.2379)(20.92 \times 10^{-6})$ $= 442,600 < 10^6 \therefore a = 1/2$ and $C = 0.9975 - 0.00653(10^6/R_d)^{1/2}$. $C = 0.9975 - 0.00653(10^6/442,600)^{1/2} = 0.9877$. $A_2 = (0.8912)/(0.9877 \times 1.025) \sqrt{2 \times 201.1} = 0.04389$, $d_2 = \sqrt{4 \times (0.04389/\pi)} = 0.2364$, $d_2 = 0.2364 \times 12 = 2.837$ in $(7.205 \times 10^{-2}$ m). Further trials are not necessary in view of the ±2 percent tolerance of $C$.

**Compressible Flow—Venturi Tubes and Flow Nozzles** The expansion factor $Y$ is computed based on the assumption of a frictionless adiabatic (isentropic) expansion of an ideal gas

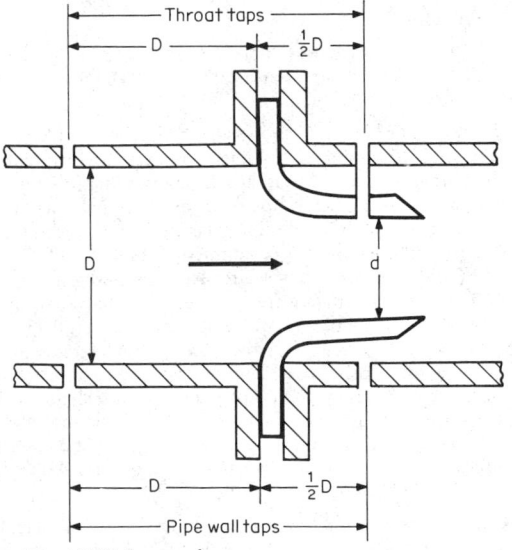

**Fig. 27** ASME flow nozzle.

from the inlet to the throat of the primary element, resulting in (see Sec. 4)

$$Y = \left[ \frac{kr^{2/k}(1 - r^{(k-1)/k})(1 - \beta^4)}{(1 - r)(k - 1)(1 - \beta^4 r^{2/k})} \right]^{1/2}$$

Maximum flow is obtained when the critical pressure ratio is reached. The critical pressure ratio $r_c$ may be calculated from

$$r^{(1-k)/k} + \frac{k - 1}{2} \beta^4 r^{2/k} = \frac{k + 1}{2}$$

Table 13 gives selected values of $Y$ and $r$.

EXAMPLE. A piping system consists of a compressor, a horizontal straight length of 2-in-inside-diameter pipe, and a 1-in-throat-diameter ASME flow nozzle attached to the end of the pipe, discharging into the atmosphere. The compressor is operated to maintain a flow of air with 115 lbf/in² and 140°F (60°C) conditions in the pipe just one pipe diameter before the nozzle inlet. Barometric pressure is 14.7 lbf/in². Estimate the flow rate of the air in lbm/s.
From the equation of state, $\rho_1 = p_1/g_c RT_1 = (144 \times 115)/(32.17)(53.34)(140 + 459.7) = 0.01609$ slug/ft³, $\beta = d/D = 1/2 = 0.5$, $E = 1/\sqrt{1 - \beta^4} = 1/\sqrt{1 - (0.5)^4} = 1.033$, $r = p_2/p_1 = 14.7/115 = 0.1278$, but from Table 13 at $\beta = 0.5$, $k = 1.4$, $r_c = 0.5362$, and $Y_c = 0.6973$, so that because of critical flow the throat pressure $p_c = 115 \times 0.5362 = 61.66$ lbf/in². $\Delta p_c/\rho_1 = 144(115 - 61.66)/0.01609 = 477,375$. A trial-and-error solution is necessary to obtain $C$. For the first trial assume $10^6/R_d = 0$ or $C = 0.9975$. Then $Q_1 = CEY_c A_2 \sqrt{2\Delta p_c/\rho_1} = (0.9975)(1.033)(0.6973)(\pi/4)(1/12)^2 \sqrt{2 \times 477,375} = 3.829$ ft³/s, $R_d = 4\rho_1 Q/\pi d\mu_1 = (4)(0.01609)(3.828)/\pi(1/12)(41.79 \times 10^{-8}) = 2,252,000$.
Second trial, use first-trial values:

$R > 10^6$, $a = 115$, $C = 0.9975 - (0.00653)(10^6/2,252,000)^{1/5}$
$C = 0.9919$, $Q_1 = 3.828(0.9919/0.9975) = 3.806$ ft³/s.

Further trials are not necessary in view of ±2 percent tolerance on $C$.

$\dot{m} = Q_1\rho_1 g = 3.806 \times 0.01609 \times 32.17$
$\qquad\qquad\qquad\qquad = 1.970$ lbm/s (0.8935 kg/s).

**Orifice Meters** When a fluid flows through a square-edged thin-plate orifice, the minimum-flow area is found to occur downstream from the orifice plate. This minimum area is called the vena contracta, and its location is a function of beta ratio. Figure 28 shows the relative pressure difference due to the presence of the orifice plate. Because the location of the pressure taps is vital, it is necessary to specify the exact position of the downstream pressure tap. The jet contraction amounts to about 60 percent of the orifice area; so orifice

**Table 13. Expansion Factors and Critical Pressure Ratios for Venturi Tubes and Flow Nozzles**

| $\beta$ | $k$ | Critical Values | | Expansion | Factor $Y$ | | |
| | | $r_c$ | $Y_c$ | $r = 0.60$ | $r = 0.70$ | $r = 0.80$ | $r = 0.90$ |
|---|---|---|---|---|---|---|---|
| 0 | 1.10 | 0.5846 | 0.6894 | 0.7021 | 0.7820 | 0.8579 | 0.9304 |
| | 1.20 | 0.5644 | 0.6948 | 0.7228 | 0.7981 | 0.8689 | 0.9360 |
| | 1.30 | 0.5457 | 0.7000 | 0.7409 | 0.8119 | 0.8783 | 0.9408 |
| | 1.40 | 0.5282 | 0.7049 | 0.7568 | 0.8240 | 0.8864 | 0.9449 |
| 0.20 | 1.10 | 0.5848 | 0.6892 | 0.7017 | 0.7817 | 0.8577 | 0.9303 |
| | 1.20 | 0.5546 | 0.6946 | 0.7225 | 0.7978 | 0.8687 | 0.9359 |
| | 1.30 | 0.5459 | 0.6998 | 0.7406 | 0.8117 | 0.8781 | 0.9407 |
| | 1.40 | 0.5284 | 0.7047 | 0.7576 | 0.8237 | 0.8862 | 0.9448 |
| 0.50 | 1.10 | 0.5921 | 0.6817 | 0.6883 | 0.7699 | 0.8485 | 0.9250 |
| | 1.20 | 0.5721 | 0.6872 | 0.7094 | 0.7864 | 0.8600 | 0.9310 |
| | 1.30 | 0.5535 | 0.6923 | 0.7278 | 0.8007 | 0.8699 | 0.9361 |
| | 1.40 | 0.5362 | 0.6973 | 0.7440 | 0.8133 | 0.8785 | 0.9405 |
| 0.60 | 1.10 | 0.6006 | 0.6729 | | 0.7556 | 0.8374 | 0.9186 |
| | 1.20 | 0.5808 | 0.6784 | 0.6939 | 0.7727 | 0.8495 | 0.9250 |
| | 1.30 | 0.5625 | 0.6836 | 0.7126 | 0.7875 | 0.8599 | 0.9305 |
| | 1.40 | 0.5454 | 0.6885 | 0.7292 | 0.8006 | 0.8689 | 0.9352 |
| 0.70 | 1.10 | 0.6160 | 0.6570 | | 0.7290 | 0.8160 | 0.9058 |
| | 1.20 | 0.5967 | 0.6624 | 0.6651 | 0.7469 | 0.8292 | 0.9131 |
| | 1.30 | 0.5788 | 0.6676 | 0.6844 | 0.7626 | 0.8405 | 0.9193 |
| | 1.40 | 0.5621 | 0.6726 | 0.7015 | 0.7765 | 0.8505 | 0.9247 |
| 0.80 | 1.10 | 0.6441 | 0.6277 | | 0.6778 | 0.7731 | 0.8788 |
| | 1.20 | 0.6238 | 0.6331 | | 0.6970 | 0.7881 | 0.8877 |
| | 1.30 | 0.6087 | 0.6383 | | 0.7140 | 0.8012 | 0.8954 |
| | 1.40 | 0.5926 | 0.6433 | 0.6491 | 0.7292 | 0.8182 | 0.9021 |

SOURCE: J. W. Murdock, "Fluid Mechanics and Its Applications," Houghton Mifflin, 1976.

coefficients are in the order of 0.6 compared with the nearly unity obtained with venturi tubes and flow nozzles.

Three pressure-differential-measuring tap locations are specified by the ASME. These are the flange, vena contracta, and the 1 D and 1/2 D. In the flange tap, the location is always 1 in from either face of the orifice plate regardless of the size of the pipe. In the vena contracta tap, the upstream tap is located one pipe diameter from the inlet face of the orifice plate and the downstream tap at the location of the vena contracta. In the 1 D and 1/2 D tap, the upstream tap is located one pipe diameter from the inlet face of the orifice plate and down-stream one-half pipe diameter from the inlet face of the orifice plate.

Flange taps are used because they can be prefabricated, and flanges with holes drilled at the correct locations may be purchased as off-the-shelf items, thus saving the cost of field fabrication. The disadvantage of flange taps is that they are not symmetrical with respect to pipe size. Because of this, coefficients of discharge for flange taps vary greatly with pipe size.

Vena contracta taps are used because they give the maximum differential for any given flow. The disadvantage of the vena contracta tap is that if the orifice size is changed, a new downstream tap must be drilled. The 1 D and 1/2 D taps incorporate the best features of the vena contracta taps and are symmetrical with respect to pipe size.

Discharge coefficients for orifices may be calculated from

$$C = C_o + \Delta C \mathbf{R}_d^{-a} \qquad (\mathbf{R}_d > 10^4)$$

where $C_o$, $\Delta C$, and $a$ are obtained from Table 14.

Tolerances for uncalibrated orifice meters are in the order of $\pm 1$ to $\pm 2$ percent depending upon $\beta$, $D$, and $\mathbf{R}_d$.

**Compressible Flow through ASME Orifices** As shown in Fig. 28, the minimum flow area for an orifice is at the vena contracta located downstream of the orifice. The stream of compressible fluid is not restrained as it leaves the orifice throat and is free to expand transversely and longitudinally to

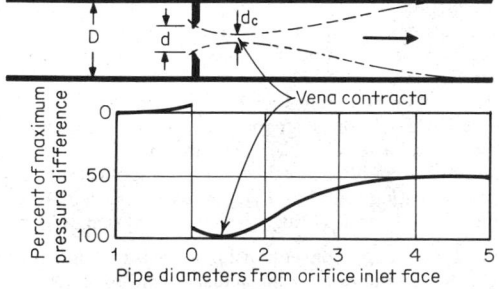

**Fig. 28** Relative-pressure changes due to flow through an orifice.

**Table 14. Values of $C_o$, $\Delta C$, and $a$ for Use in Orifice-Coefficient Equation**

| | $D = 2$ in | | $D = 4$ in | | $D = 8$ in | | $D = 16$ in | |
|---|---|---|---|---|---|---|---|---|
| $\beta$ | $C_o$ | $\Delta C$ | $C_o$ | $\Delta C$ | $C_o$ | $\Delta C$ | $C_o$ | $\Delta C$ |
| | | | | Flange taps, $a = 1$ | | | | |
| 0.20 | 0.5972 | 127 | 0.5946 | 200 | 0.5951 | 327 | 0.5955 | 551 |
| 0.30 | 0.5978 | 144 | 0.5977 | 209 | 0.5978 | 307 | 0.5980 | 457 |
| 0.40 | 0.6014 | 181 | 0.6005 | 256 | 0.6002 | 362 | 0.6001 | 514 |
| 0.50 | 0.6050 | 260 | 0.6034 | 386 | 0.6026 | 584 | 0.6022 | 903 |
| 0.60 | 0.6078 | 392 | 0.6055 | 622 | 0.6040 | 1,015 | 0.6032 | 1710 |
| 0.70 | 0.6068 | 573 | 0.6030 | 953 | 0.6006 | 1,637 | 0.5991 | 2898 |
| | | | | Vena contracta taps, $a = \frac{1}{2}$ | | | | |
| 0.20 | 0.5938 | 1.61 | 0.5928 | 1.61 | 0.5925 | 1.61 | 0.5924 | 1.61 |
| 0.30 | 0.5938 | 1.78 | 0.5934 | 1.78 | 0.5933 | 1.78 | 0.5932 | 1.78 |
| 0.40 | 0.5970 | 2.01 | 0.5954 | 2.01 | 0.5953 | 2.01 | 0.5953 | 2.01 |
| 0.50 | 0.5994 | 2.29 | 0.5992 | 2.29 | 0.5992 | 2.29 | 0.5991 | 2.29 |
| 0.60 | 0.6042 | 2.68 | 0.6041 | 2.68 | 0.6041 | 2.69 | 0.6041 | 2.70 |
| 0.70 | 0.6069 | 3.34 | 0.6068 | 3.37 | 0.6067 | 3.44 | 0.6068 | 3.57 |
| | | | | 1D and $\frac{1}{2}$D taps, $a = \frac{1}{2}$ | | | | |
| 0.20 | 0.5909 | 2.03 | 0.5922 | 1.41 | 0.5936 | 1.10 | 0.5948 | 0.94 |
| 0.30 | 0.5915 | 2.02 | 0.5930 | 1.50 | 0.5944 | 1.24 | 0.5956 | 1.12 |
| 0.40 | 0.5936 | 2.17 | 0.5951 | 1.72 | 0.5963 | 1.49 | 0.5974 | 1.38 |
| 0.50 | 0.5979 | 2.40 | 0.5978 | 1.99 | 0.5999 | 1.79 | 0.6007 | 1.69 |
| 0.60 | 0.6036 | 2.67 | 0.6040 | 2.31 | 0.6044 | 2.12 | 0.6048 | 2.11 |
| 0.70 | 0.6078 | 3.19 | 0.6072 | 2.98 | 0.6068 | 3.07 | 0.6064 | 3.51 |

Compiled from data given in Murdock, ASME 64-WA/FM-6.

the point of minimum-flow area. Thus the contraction of the jet will be less for a compressible fluid than for a liquid. Because of this, the theoretical-expansion-factor equation may not be used with orifices. Neither may the critical-pressure-ratio equation be used, as the phenomenon of critical flow has not been observed during testing of orifice meters.

For orifice meters, the following equation, which is based on experimental data, is used:

$$Y = 1 - (0.41 + 0.35\beta^4)(\Delta p/p_1)/k$$

EXAMPLE. Air at 68°F (20°C) and 150 lbf/in² flows in a 2-in schedule 40 pipe (inside diameter = 2.067 in) at a volumetric rate of 15 ft³/min. A 0.5500-in ASME orifice equipped with flange taps is used to meter this flow. What deflection in inches could be expected on a U-tube manometer filled with 60°F water? From the equation of state, $\rho_1$ = $p_1/g_cRT_1$ = $(144 \times 150)/(32.17)(53.34)(68 + 459.7)$ = 0.02385 slug/ft³, $\beta$ = 0.5500/2.067 = 0.2661. $Q_1$ = 15/60 = 0.25 ft³/s, $A_2$ = $(\pi/4)(0.5500/12)^2$ = 1.650 × 10⁻³ ft². $E$ = $1/\sqrt{1 - \beta^4}$ = $1/\sqrt{1 - (0.2661)^4}$ = 1.003. $\mathbf{R}_d$ = $4\rho_1Q_1/\pi \ d\mu_1$ = $4(0.02385)(0.25)/\pi(0.5500/12)(39.16 \times 10^{-8})$. $\mathbf{R}_d$ = 423,000.

From Table 14 at $\beta$ = 0.2661, $D$ = 2.067-in flange taps, $a$ = 1, by interpolation, $C_o$ = 0.5976, $\Delta C$ = 140, from orifice-coefficient equation $C$ = $C_o$ + $\Delta C \mathbf{R}_d^{-a}$. $C$ = 0.5976 + $(140)(423,000)^{-1}$ = 0.5979. A trial-and-error solution is required because the pressure loss is needed in order to compute $Y$. For the first trial, assume $Y$ = 1, $\Delta p$ = $(Q_1/CEYA_2)^2(\rho_1/2)$ = $[(0.25)/(0.5979)(1.003)(Y)(1.650 \times 10^{-3})]^2 \ (0.02385/2)$ = 761.2/$Y^2$ = 761.2/(1)² = 761.2 lbf/ft².

For the second trial we use first-trial values.

$Y$ = 1 − (0.41 − 0.35 $\beta^4)(\Delta p/p_1)/k$
$Y$ = 1 − [0.41 − 0.35(0.2661)⁴](761.2/144 × 150)/(1.4) = 0.9897
$\Delta p$ = 761.2/$Y^2$ = 761.2/(0.9897)² = 777.1

For the third trial, use second-trial values.

$Y$ = 1 − [0.41 − 0.35(0.2661)⁴](777.1/144 × 150)/(1.4)
$Y$ = 0.9895, $\Delta p$ = 777.1/(0.9895)² = 793.7

resubstitution does not produce any further change in $Y$. From the U-tube-manometer equation,

$h$ = $\Delta p/(\gamma_m - \gamma_f)$ = $\Delta p/g(\rho_m - \rho_f)$ = 793.7/(32.17)(1.937 − 0.02385)
$h$ = 12.90 ft = 12.90 × 12 = 154.8 in (3.932 m).

## PITOT TUBES

**Definition** A Pitot tube is a device that is shaped in such a manner that it senses stagnation pressure. The name "Pitot tube" has been applied to two general classifications of instruments, the first being a tube that measures the impact or stagnation pressures only, and the second a combined tube that measures both impact and static pressures with a single primary instrument. The combined sensor is also called a Pitot-static tube.

**Tube Coefficient** From Fig. 29, it is evident that the Pitot tube can sense only the stagnation pressure resulting from the local stream-tube velocity $U$. The local ideal velocity $U_i$ for an incompressible fluid is obtained by the application of the Bernoulli equation $(z_S = z)$, $U_i^2/2g + p/\rho g = U_S^2/2g + p_S/\rho g$. Solving for $U_i$ and noting that by definition $U_S = 0$,

$U_i = \sqrt{2(p_S - p)/\rho}$. Conventional practice is to define the **tube coefficient** $C_T$ as the ratio of the actual stream-tube velocity to the ideal stream-tube velocity, or $C_T = U/U_i$ and $U$ =

$C_TU_i = C_T\sqrt{2\Delta p/\rho}$. The numerical value of $C_T$ depends

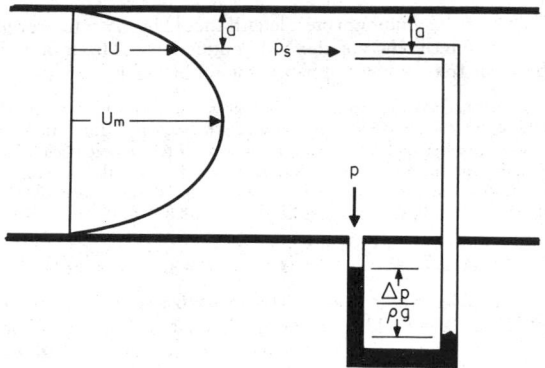

**Fig. 29**   Notation for Pitot-tube study.

primarily upon its geometry. The value of $C_T$ may be established (1) by calibration with a uniform velocity, (2) from published data for similar geometry, or (3) in the absence of other information, may be assumed to be unity.

**Pipe Coefficient**   For the calculation of volumetric flow rate, it is necessary to integrate the continuity equation, $Q = \int U\,da = AV$. The **pipe coefficient** $C_P$ is defined as the ratio of the average velocity to the stream-tube velocity, or $C_P = V/U$, and $Q = C_P A_1 V = C_P C_T A_1 \sqrt{2\Delta p/\rho}$. The numerical value of $C_P$ is dependent upon the location of the tube and the **velocity profile**. The values of $C_P$ may be established by (1) making a "traverse" by taking data at various points in the flow stream and determining the velocity profile experimentally (see "Fluid Meters," 6th ed., ASME, 1971, for locations of traverse points), (2) using standard velocity profiles, (3) locating the Pitot tube at a point where $U = V$, and (4) assuming one-dimensional flow of $C_P = 1$ **only** in the absence of other data.

**Compressible Flow**   For compressible flow, the **compression factor** $Z$ is based on the assumption of a frictionless adiabatic (isentropic) compression of an ideal gas from the moving stream tube to the stagnation point (see Sec. 4), which results in

$$Z = \left[\frac{k}{k-1}\,\frac{(p_S/p)^{(k-1)/k} - 1}{(p_S/p) - 1}\right]^{1/2}$$

and the volumetric flow rate becomes

$$Q = C_P C_T Z A_1 \sqrt{2\Delta p/\rho}$$

EXAMPLE.   Carbon dioxide flows at 68°F (20°C) and 20 lbf/in² in an 8-in schedule 40 galvanized-iron pipe. A Pitot tube located on the pipe centerline indicates a pressure differential of 6.986 lbf/in². Estimate the mass flow rate. For 8-in schedule 40 pipe $D = 7.981/12 = 0.6651$, $\epsilon/D = 500 \times 10^{-6}/0.6651 = 7.518 \times 10^{-4}$, $A_1 = \pi D^2/4 = (\pi/4)(0.6651)^2 = 0.3474$ ft², $p_S = p + \Delta p = 20 + 6.986 = 26.986$ lbf/in². From the equation of state, $\rho = p/g_c RT_o = (20 \times 144)/(32.17)(35.11)(68 + 459.7) = 0.004832$,

$$Z = \left[\frac{k}{k-1}\,\frac{(p_S/p)^{(k-1)/k} - 1}{(p_S/p) - 1}\right]^{1/2} = \left\{[1.3/(1.3 - 1)]\right.$$
$$\left. \times \frac{(26.986/20)^{(1.3 - 1)/1.3} - 1}{(26.986/20) - 1}\right\}^{1/2} = 0.9423$$

In the absence of other data, $C_T$ may be assumed to be unity. A trial and error solution is necessary to determine $C_P$, since $f$ requires flow rate. For the first trial assume complete turbulence.

$$1/\sqrt{f} = -2\,\log_{10}\,(7.518 \times 10^{-4}/3.7) \qquad \sqrt{f} = 0.1354$$
$$C_P = V/U = V/U_{max} = 1/(1 + 1.43\,\sqrt{f})$$
$$= 1/(1 + 1.43 \times 0.1354) = 0.8378$$
$$V = C_P C_T Z \sqrt{2\Delta p/\rho} = (0.8378)(1)(0.9423)\sqrt{2 \times 144(6.987)/(0.004832)}$$
$$V = 509.4 \text{ ft/s}$$
$$\mathbf{R} = \rho VD/\mu = (0.004832)(509.4)(0.6651)/(30.91 \times 10^{-8})$$
$$\mathbf{R} = 5,296,000 > 4,000 \;\therefore\; \text{flow is turbulent}$$

From the Colebrook equation and Fig. 24,

$$\frac{1}{\sqrt{f}} = -2\,\log_{10}\left(\frac{7.518 \times 10^{-4}}{3.7} + \frac{2.51}{5,296,000\,\sqrt{0.018}}\right)$$

$$\sqrt{f} = 0.1357$$
$$C_P = 1/(1 + 1.43 \times 0.1357) = 0.8375 \qquad \text{(close check)}$$
$$V = 509.4(0.8375/0.8378) = 509.2 \text{ ft/s}$$

From the continuity equation, $m = \rho A_1 V g_c = (0.004832)(0.3474)(509.2)(32.17) = 27.50$ lbm/s (12.47 kg/s).

## ASME WEIRS

**Definitions**   A weir is a dam over which liquids are forced to flow. Weirs are used to measure the flow of liquids in open channels or in conduits which do not flow full; i.e., there is a free liquid surface. Weirs are almost exclusively used for measuring water flow, although small ones have been used for metering other liquids. Weirs are classified according to their notch or opening as follows: (1) **rectangular notch** (original form); (2) V or **triangular notch**; (3) **trapezoidal notch**, which when designed with end slopes one horizontal to four vertical is called the **Cipolletti weir**; (4) the **hyperbolic weir** designed to give a constant coefficient of discharge; and (5) the **parabolic weir** designed to give a linear relationship of head to flow. As shown in Fig. 30, the top of the weir is the **crest** and the distance from the liquid surface to the crest $h$ is called the **head**.

The sheet of liquid flowing over the weir crest is called the **nappe**. When the nappe falls downstream of the weir plate, it is said to be free, or **aerated**. When the width of the approach channel $L_c$ is greater than the crest length $L_w$, the nappe will contract so that it will have a minimum width less than the crest length. For this reason, the weir is known as a **contracted weir**. For the special case where $L_w = L_c$, the contractions do not take place, and such weirs are known as **suppressed weirs**.

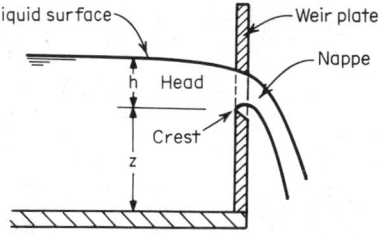

**Fig. 30**   Notation for weir study.

**Parameters**   The forces acting on a liquid flowing over a weir are inertia, viscous, surface tension, and gravity. If the weir head produced by the flow is $h$, the characteristic length of the weir is $L_w$, and the channel width is $L_c$, either similarity or dimensional analysis leads to $f(\mathbf{F}, \mathbf{W}, \mathbf{R}, L_w/L_c) = 0$, which may be written as $V = K\sqrt{2gh}$, where $K$ is the **weir coefficient**

and $K = f(\mathbf{W}, \mathbf{R}, L_w/L_c)$. Since the weir has been almost exclusively used for metering water flow over limited temperature ranges, the effects of surface tension and viscosity have not been adequately established by experiment.

**Caution** The numerical values of coefficients for weirs are based on experimental data obtained from calibration of weirs with long approaches of straight channels. Head measurement should be made at a distance at least three to four times the expected maximum head $h$. Screens and baffles should be used as necessary to ensure steady uniform flow without waves or local eddy currents. The approach channel should be relatively wide and deep.

**Rectangular Weirs** Figure 31 shows a rectangular weir whose crest width is $L_w$. The volumetric flow rate may be computed from the continuity equation: $Q = AV = (L_w h)(K\sqrt{2gh}) = KL_w\sqrt{2g}\;h^{3/2}$. The ASME "Fluid Meters"

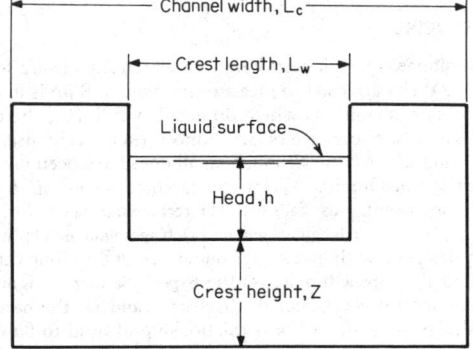

**Fig. 31** Rectangular weir.

report recommends the following equation for rectangular weirs: $Q = (2/3)CL_a\sqrt{2g}\;h_a^{3/2}$, where $C$ is the **coefficient of discharge** $C = f(L_w/L_c, h/Z)$, $L_a$ is the adjusted crest length $L_a = L_w + \Delta L$, and $h_a$ is the adjusted weir head $h_a = h + 0.003$ ft. Values of $C$ and $\Delta L$ may be obtained from Table 15. To avoid the possibility that the liquid drag along the sides of the channel will affect side contractions, $L_c - L_w$ should be at

least $4h$. The minimum crest length should be 0.5 ft to prevent mutual interference of the end contractions. The minimum head for free flow of the nappe should be 0.1 ft.

EXAMPLE. Water flows in a channel whose width is 40 ft. At the end of the channel is a rectangular weir whose crest width is 10 ft and whose crest height is 4 ft. The water flows over the weir at a height of 3 ft above the crest of the weir. Estimate the volumetric flow rate. $L_w/L_c = 10/40 = 0.25$, $h/Z = 3/4 = 0.75$, from Table 15 (interpolated), $C = 0.589$, $\Delta L = 0.008$, $L_a = L_w + \Delta L = 10 + 0.008 = 10.008$ ft, $h_a = h + 0.003 = 3 + 0.003 = 3.003$ ft, $Q = (2/3)\;CL_a\sqrt{2g}\;h^{3/2}$, $Q = (2/3)(0.589)(10.008)(2 \times 32.17)^{1/2}(3.003)^{3/2}$, $Q = 164.0$ ft³/s (4.644 m³/s).

**Triangular Weirs** Figure 32 shows a triangular weir whose notch angle is $\theta$. The volumetric flow rate may be computed from the continuity equation $Q = AV = (h^2 \tan \theta/2)(K\sqrt{2gh}) = K\tan(\theta/2)\sqrt{2g}\;h^{5/2}$. The ASME "Fluid Meters" report recommends the following for triangular weirs: $Q = (8/15)\;C\tan(\theta/2)\sqrt{2g}\;(h + \Delta h)^{5/2}$, where $C$ is the coefficient of discharge $C = f(\theta)$ and $\Delta h$ is the correction for head/crest ratio $\Delta h = f(\theta)$. Values of $C$ and $\Delta h$ may be obtained from Table 16.

EXAMPLE. It is desired to maintain a flow of 167 ft³/s in an open channel whose width is 20 ft at a height of 7 ft by locating a triangular weir at the end of the channel. The weir has a crest height of 2 ft. What notch angle is required to maintain these conditions? A trial-and-error solution is required. For the first trial assume $\theta = 60°$ (mean value 20 to 100°); then $C = 0.576$ and $\Delta h = 0.004$.

$$h + Z = 7 = h + 2 \therefore h = 5$$
$$Q = (8/15)\;C\tan(\theta/2)\sqrt{2g}\;(h + \Delta h)^{5/2}$$

$167 = (8/15)(0.576)\tan(\theta/2)\sqrt{2 \times 32.17}\;(5 + 0.004)^{5/2}$, $\tan^{-1}(\theta/2) = 1.20993$, $\theta = 100°51'$.

Second trial, using $\theta = 100$, $C = 0.581$, $\Delta h = 0.003$, $167 = (8/15)(0.581)\tan(\theta/2)\sqrt{2 \times 32.17}\;(5 + 0.003)^{5/2}$, $\tan^{-1}(\theta/2) = 1.20012$, $\theta = 100°39'$ (close check).

### OPEN-CHANNEL FLOW

**Definitions** An **open channel** is a conduit in which a liquid flows with a free surface subjected to a constant pressure. Flows of water in natural streams, artificial canals, irrigation

**Table 15. Values of $C$ and $\Delta L$ for Use in Rectangular-Weir Equation**

| h/Z | 0 | 0.2 | 0.4 | 0.6 | 0.7 | 0.8 | 0.9 | 1.0 |
|-----|-----|-----|-----|-----|-----|-----|-----|-----|
| | \multicolumn{8}{c}{Crest length/channel width $= L_w/L_c$} | | | | | | | |
| | \multicolumn{8}{c}{Coefficient of discharge $C$} | | | | | | | |
| 0 | 0.587 | 0.589 | 0.591 | 0.593 | 0.595 | 0.597 | 0.599 | 0.603 |
| 0.5 | 0.586 | 0.588 | 0.594 | 0.602 | 0.610 | 0.620 | 0.631 | 0.640 |
| 1.0 | 0.586 | 0.587 | 0.597 | 0.611 | 0.625 | 0.642 | 0.663 | 0.676 |
| 1.5 | 0.584 | 0.586 | 0.600 | 0.620 | 0.640 | 0.664 | 0.695 | 0.715 |
| 2.0 | 0.583 | 0.586 | 0.603 | 0.629 | 0.655 | 0.687 | 0.726 | 0.753 |
| 2.5 | 0.582 | 0.585 | 0.608 | 0.637 | 0.671 | 0.710 | 0.760 | 0.790 |
| 3.0 | 0.580 | 0.584 | 0.610 | 0.647 | 0.687 | 0.733 | 0.793 | 0.827 |
| | \multicolumn{8}{c}{Adjustment for crest length $\Delta L$, ft} | | | | | | | |
| Any | 0.007 | 0.008 | 0.009 | 0.012 | 0.013 | 0.014 | 0.013 | −0.005 |

Compiled from data given in "Fluid Meters," ASME, 1971.

ditches, sewers, and flumes are examples where the water surface is subjected to atmospheric pressure. The flow of any liquid in a pipe where there is a free liquid surface is an example of open-channel flow where the liquid surface will be

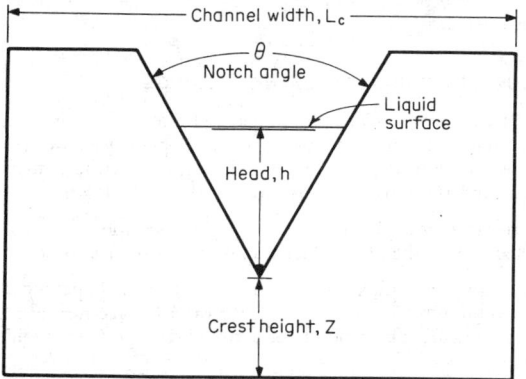

**Fig. 32**   Triangular weir.

subjected to the pressure existing in the pipe. The **slope** $S$ of a channel is the change in elevation per unit of horizontal distance. For small slopes, this is equivalent to dividing the change in elevation by the distance $L$ measured along the channel bottom between two sections. For steady uniform flow, the velocity distribution is the same at all sections of the channel, so that the energy grade line has the same angle as the bottom of the channel, thus:

$$S = h_f/L$$

The distance between the liquid surface and the bottom of the channel is sometimes called the **stage** and is denoted by the symbol $y$ in Fig. 33. When the stages between the sections are not uniform, that is, $y_1 \neq y_2$ or the cross section of the channel changes, or both, the flow is said to be **varied**. When a liquid flows in a channel of uniform cross section and the slope of the surface is the same as the slope of the bottom of the channel ($y_1 = y = y_2$), the flow is said to be **uniform**.

**Parameters**   The forces acting on a liquid flowing in an open channel are inertia, viscous, surface tension, and gravity. If the channel has a surface roughness of $\epsilon$, a hydraulic radius of $R_h$, and a slope of $S$, either similarity or dimensional analysis leads to $f(\mathbf{F}, \mathbf{W}, \mathbf{R}, \epsilon/4R_h) = 0$, which may be written as $V = C\sqrt{R_h S}$, where $C = f(\mathbf{W}, \mathbf{R}, \epsilon/4R_h)$ and is known as the **Chézy coefficient**. The relationship between the Chézy coefficient $C$ and the friction factor may be determined by equating

$$V = \sqrt{8 R_h h_f g/fL} = C\sqrt{R_h S} = C\sqrt{(R_h h_f)/L}$$

or $C = (8g/f)^{1/2}$. Although this establishes a relationship between the Chézy coefficient and the friction factor, it should be noted that $f = f(\mathbf{R}, \epsilon/4R_h)$ and $C = f(\mathbf{W}, \mathbf{R}, \epsilon/4R_h)$, because in open-channel flow, pressure forces are absent and in pipe flow, surface-tension and gravity forces are absent. For these reasons, data obtained in pipe flow should not be applied to open-channel flow.

**Roughness Factors**   For open-channel flow, the Chézy coefficient is calculated by the Manning equation, which was developed from examination of experimental results of water tests. The **Manning relation** is stated as

$$C = \frac{1.486}{n} R_h^{1/6}$$

where $n$ is a roughness factor and should be a function of Reynolds number, Weber number, and relative roughness. Since only water-test data obtained at ordinary temperatures support these values, it must be assumed that $n$ is the value for turbulent flow only. Since surface tension is a weak property, the effects of Weber-number variation are negligible, leaving $n$ to be some function of surface roughness. Design values of $n$ are given in Table 17. Maximum flow for a given slope will take place when $R_h$ is a maximum, and values of $R_{hmax}$ are given in Table 6.

EXAMPLE.   It is necessary to carry 150 ft³/s of water in a rectangular unplaned timber flume whose width is to be twice the depth of water. What are the required dimensions for various slopes of the flume? From Table 6, $A = b^2/2$ and $R_h = b/2 = b/4$. From Table 17, $n = 0.013$ for unplaned wood. From Manning's equation, $C = 1.486/n$, $R_h^{1/6} = (1.486/0.013) (b)^{1/6}/(4)^{1/6} = 90.73 \, b^{1/6}$. From the continuity equation, $V = Q/A = 150/(b^2/2)$, $V = 300/b^2$. From the Chézy equation, $V = C\sqrt{R_h S} = 300/b^2 = 90.73 b^{1/6} \sqrt{(b/4)S}$; solving for $b$, $b = 2.0308/S^{3/16}$.

| Assumed $S$: | $1 \times 10^{-1}$ | $1 \times 10^{-2}$ | $1 \times 10^{-3}$ | $1 \times 10^{-4}$ | $1 \times 10^{-5}$ | $1 \times 10^{-6}$ | ft/ft |
|---|---|---|---|---|---|---|---|
| Required $b$: | 3.127 | 4.816 | 7.416 | 11.42 | 17.59 | 27.08 | ft |

EXAMPLE.   A rubble-lined trapezoidal canal with 45° sides is to carry 360 ft³/s of water at a depth of 4 ft. If the slope is $9 \times 10^{-4}$ ft/ft, what should be the dimensions of the canal? From Table 17, $n = 0.025$ for rubble. From Table 6 for $\alpha = 45°$, $A = (b + h)h = 4(b + 4)$, and $R_h = (b + h)h/(b + 2.828h) = 4(b + 4)/(b + 11.312)$. From the Manning relation, $C = (1.486/n) (R_h^{1/6}) = (1.486/0.025)R_h^{1/6} = 59.44 R_h^{1/6}$. For the first trial, assume $R_h = R_{hmax} = b/2 = 4/2 = 2$; then $C = 59.44(2)^{1/6} = 66.72$ and $V = C\sqrt{R_h S} = 66.72 \sqrt{2 \times 9 \times 10^{-4}} = 2.831$. From the continuity equation, $A = Q/V = 360/2.831 = 127.2 = 4(b + 4)$, $b = 27.79$ ft. Second trial, use the first trial, $R_h = 4(27.79 + 4)/(27.79 + 11.312)$, $R_h = 3.252$, $V = 59.44(3.252)^{1/6} \sqrt{3.252 \times 9 \times 10^{-4}} = 3.914$.

**Table 16.   Values of $C$ and $\Delta h$ for Use in Triangular-Weir Equation**

| Item | Weir notch angle $\theta$, deg | | | | | | |
|---|---|---|---|---|---|---|---|
| | 20 | 30 | 45 | 60 | 75 | 90 | 100 |
| $C$ | 0.592 | 0.586 | 0.580 | 0.576 | 0.576 | 0.579 | 0.581 |
| $\Delta h$, ft | 0.010 | 0.007 | 0.005 | 0.004 | 0.003 | 0.003 | 0.003 |

Compiled from data given in "Fluid Meters," ASME, 1971.

From the equation of continuity, $Q/V = 360/3.914 = 91.97 = 4(b + 4)$, $b = 18.99$. Subsequent trial-and-error solutions result in a balance at $b = 19.93$ ft (6.075 m).

**Specific Energy**    Specific energy is defined as the energy of the fluid referred to the bottom of the channel as the datum. Thus the specific energy $E$ at any section is given by $E = y + V^2/2g$; from the continuity equation $V = Q/A$ or $E = y + (Q/$

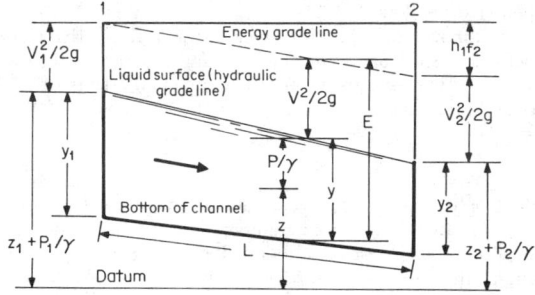

**Fig. 33**    Notation for open-channel flow.

$A)^2/2g$. For a rectangular channel whose width is $b$, $A = by$; and if **q** is defined as the flow rate per unit width, $\mathbf{q} = Q/b$ and $E = y + (\mathbf{q}b/by)^2/2g = y + (\mathbf{q}/y)^2/2g$.

**Critical Values**    For rectangular channels, if the specific-energy equation is differentiated and set equal to zero, critical values are obtained; thus $dE/dy = d/dy\,[y + (\mathbf{q}/y)^2/2g] = 0 = 1 - \mathbf{q}^2/y^3g$ or $y_c^3 = \mathbf{q}^2/g$. Substituting in the specific-energy equation, $E = y_c + y_c^3g/2gy_c^2 = 3/2\,y_c$. Figure 34 shows the relation between depth and specific energy for a constant flow rate. If the depth is greater than critical, the flow is *subcritical;* at critical depth it is *critical* and at depths below critical the flow is *supercritical.* For a given specific energy, there is a maximum unit flow rate that can exist.

The **Froude number** $\mathbf{F} = V/\sqrt{gy}$, when substituted in the specific-energy equation, yields $E = y + (\mathbf{F}^2gy)/2g = y(1 + \mathbf{F}^2/2)$ or $E/y = 1 + \mathbf{F}^2/2$. For critical flow, $E_c/y_c = 3/2$. Substituting $E_c/y_c = 3/2 = 1 + \mathbf{F}_c^2/2$, or $\mathbf{F} = 1$,

| | |
|---|---|
| $\mathbf{F} < 1$ | Flow is subcritical |
| $\mathbf{F} = 1$ | Flow is critical |
| $\mathbf{F} > 1$ | Flow is supercritical |

It is seen that for open-channel flow the Froude number determines the type of flow in the same manner as Mach number for compressible flow.

EXAMPLE    Water flows at a rate of 600 ft³/s in a rectangular channel 10 ft wide at a depth of 4 ft. Determine (1) specific energy and (2) type of flow.

1. From the continuity equation,
$$V = Q/A = 600/(10 \times 4) = 15 \text{ ft/s}$$
$$E = y + V^2/2g = 4 + (15)^2/2(2 \times 32.17) = 7.497 \text{ ft}$$

2. $\mathbf{F} = V/\sqrt{gy} = 15/\sqrt{32.17 \times 4} = 1.322$; $\mathbf{F} > 1$ ∴ flow is supercritical.

### FLOW OF LIQUIDS FROM TANK OPENINGS

**Steady State**    Consider the jet whose velocity is $V$ discharging from an open tank through an opening whose area is $a$, as shown in Fig. 35. The liquid height above the centerline is $b$, and the cross-sectional area of the tank at $h$ is $A$. The ideal velocity of the jet is $V_i = \sqrt{2gh}$. The ratio of the actual velocity $V$ to the ideal velocity $V_i$ is the **coefficient of velocity** $C_v$, or $V = C_vV_i = C_v\sqrt{2gh}$. The ratio of the actual opening $a$ to the minimum area of the jet $a_c$ is the **coefficient of contraction** $C_c$, or $a = C_ca_c$. The ratio of the actual discharge $Q$ to the ideal

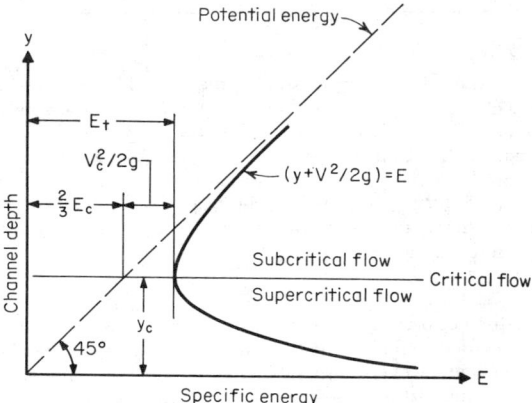

**Fig. 34**    Specific-energy diagram, constant flow rate.

discharge $Q_i$ is the **coefficient of discharge** $C$, or $Q = CQ_i = C_aV_i = C_cC_va\sqrt{2gh}$, and $C = C_cC_v$. Nominal values of coefficients for various openings are given in Fig. 36.

**Unsteady State**    If the rate of liquid entering the tank $Q_{in}$ is different from that leaving, the level $h$ in the tank will change because of the change in storage. For liquids, the conservation-of-mass equation may be written as $Q_{in} - Q_{out} = Q_{stored}$; for a time interval $dt$, $(Q_{in} - Q_{out})dt = A\,dh$, neglecting fluid acceleration,

**Table 17. Values of Roughness Factor $n$ for Use in Manning Equation**

| Surface | $n$ | Surface | $n$ |
|---|---|---|---|
| Brick ................ | 0.015 | Earth, with stones and weeds  .. | 0.035 |
| Cast iron   ............ | 0.015 | Gravel ..................... | 0.029 |
| Concrete, finished ..... | 0.012 | Riveted steel ................ | 0.017 |
| Concrete, unfinished  .. | 0.015 | Rubble  .................... | 0.025 |
| Brass pipe  .......... | 0.010 | Wood, planed ............... | 0.012 |
| Earth ............... | 0.025 | Wood, unplaned  ............. | 0.013 |

Compiled from data given in R. Horton, *Engineering News*, **75**, 373, 1916.

$Q_{out}\, dt = Ca\sqrt{2gh}\, dt$, or $(Q_{in} - Ca\sqrt{2gh})\, dt$

$$= A\, dh,\text{ or } \int_{t_1}^{t_2} dt = \int_{h_1}^{h_2} \frac{A\, dh}{Q_{in} - Q_{out}} = \int_{h_1}^{h_2} \frac{A\, dh}{Q_{in} - Ca\sqrt{2gh}}$$

EXAMPLE.   An open cylindrical tank is 6 ft in diameter and is filled with water to a depth of 10 ft. A 4-in-diameter sharp-edged orifice is installed on the bottom of the tank. A pipe on the top of the tank

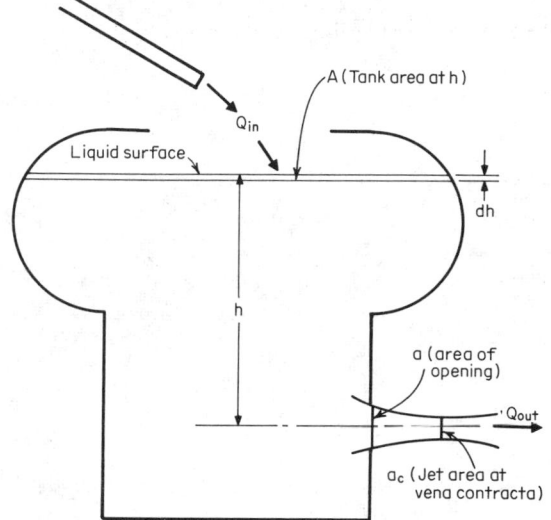

**Fig. 35**   Notation for tank flow

supplies water at the rate of 1 ft³/s. Estimate (1) the steady-state level of this tank, (2) the time required to reduce the tank level by 2 ft.

1. Steady-state level. From Fig. 36, $C = 0.61$ for a sharp-edged orifice, $a = (\pi/4)d^2 = (\pi/4)(4/12)^2 = 0.08727$ ft². For steady state, $Q_{in} = Q_{out} = Ca\sqrt{2gh} = 1 = (0.61)(0.08727)(2 \times 32.17\,h)^{1/2}$; $h = 5.484$ ft.

2. Time required to lower level 2 ft, $A = (\pi/4)D^2 = (\pi/4)(6)^2 = 28.27$ ft²

$$t_2 - t_1 = \int_{h_1}^{h_2} \frac{A\, dh}{Q_{in} - Ca\sqrt{2gh}}$$

This equation may be integrated by letting $Q = Ca\sqrt{2g}\, h^{1/2}$; then $dh = 2Q\, dQ/(Ca\sqrt{2g})^2$; then

$$t_2 - t_1 = \frac{2A}{(Ca\sqrt{2g})^2}\left[Q_{in}\log_e\left(\frac{Q_{in} - Q_1}{Q_{in} - Q_2}\right) + Q_2 - Q_1\right]$$

At $t_1$: $Q_1 = 0.61 \times 0.08727\sqrt{2 \times 32.17 \times 10} = 1.350$ ft³/s

At $t_2$: $Q_2 = 0.61 \times 0.08727\sqrt{2 \times 32.17 \times 8} = 1.208$ ft³/s

$$t_2 - t_1 = \frac{2 \times 28.27}{(0.61 \times 0.08727\sqrt{2 \times 32.17})^2}$$
$$\times \left[(1)\log_e\left(\frac{1 - 1.350}{1 - 1.208}\right) + 1.208 - 1.350\right]$$

$t_2 - t_1 = 117.3$ s

## WATER HAMMER

**Equations**   Water hammer is the series of shocks, sounding like hammer blows, produced **by suddenly reducing the flow** of a

fluid in a pipe. Consider a fluid flowing frictionlessly in a rigid pipe of uniform area $A$ with a velocity $V$. The pipe has a length $L$, and inlet pressure $p_1$ and a pressure $p_2$ at $L$. At length $L$, there is a valve which can suddenly reduce the velocity at $L$ to $V - \Delta V$. The equivalent mass rate of flow of a pressure wave traveling at sonic velocity $c$, $\dot{M} = \rho A c$. From the impulse-momentum equation, $\dot{M}(V_2 - V_1) = p_2 A_2 - p_1 A_1$; for this application, $(\rho A c)(V - \Delta V - V) = p_2 A - p_1 A$, or the increase in pressure $\Delta p = -\rho c\,\Delta V$. When the liquid is flowing in an elastic pipe, the equation for pressure rise must be modified to account for the expansion of the pipe; thus

$$c = \sqrt{\frac{E_s}{\rho[1 + (E_s/E_p)(D_o + D_i)/(D_o - D_i)]}}$$

where $E_p$ is the bulk modulus of elasticity of the pipe material, $D_o$ the outside diameter of the pipe, and $D_i$ the inside diameter.

**Time of Closure**   The time for a pressure wave to travel the length of pipe $L$ and return is $t = 2L/c$. If the time of closure $t_c \leq t$, the approximate pressure rise $\Delta p \approx -2\,\rho V(L/t_c)$. When it is not feasible to close the valve slowly, **air chambers** or **surge tanks** may be used to absorb all or most of the pressure rise.

EXAMPLE.   Water flows at 68°F (20°C) in a 3-in steel schedule 40 pipe at a velocity of 10 ft/s. A valve located 200 ft downstream is suddenly closed. Determine (1) the increase in pressure considering pipe to be rigid, (2) the increase considering pipe to be elastic, and (3) the maximum time of valve closure to be considered "sudden." For

| Type | Coefficient | | |
|---|---|---|---|
| | $c$ | $c_c$ | $c_v$ |
| Sharp-edged orifice | 0.61 | 0.62 | 0.98 |
| Rounded-edged orifice | 0.98 | 1.00 | 0.98 |
| Short tube | 0.80 | 1.00 | 0.80 |
| Borda | 0.51 | 0.52 | 0.98 |

**Fig. 36**   Nominal coefficients of orifices.

water $\rho = 1.937$, $E_s = 319,000$, $c = 4,860$; for steel (Sec. 5) $E_p = 28.5 \times 10^6$, for schedule 40 pipe $D_o = 3.500$ in, $D_i = 3.068$ in.

1. Inelastic pipe

$$\Delta p = -\rho c\, \Delta V = -(1.937)(4,860)(-10) = 94,138 \text{ lbf/ft}^2$$
$$= 94,138/144 = 653.8 \text{ lbf/in}^2 \ (4.507 \times 10^6 \text{ N/m}^2)$$

2. Elastic pipe

$$c = \sqrt{\dfrac{E_s}{\rho[1 + (E_s/E_p)(D_o + D_i)/(D_o - D_i)]}}$$

$$= \sqrt{\dfrac{319,000 \times 144}{1.937[1 + (319,000/28.6 \times 10^6)(3.500 + 3.067)/(3.500 - 3.067)]}}$$

$$= 4,504 \qquad \Delta p = -(1.937)(4,504)(-10)$$
$$= 87,242 \text{ lbf/ft}^2 = 87,242/144 = 605.9 \text{ lbf/in}^2 \ (4.177 \times 10^6 \text{ N/m}^2)$$

3. Maximum time for closure

$$t = 2L/c = 2 \times 200/4,860 = 0.08230 \text{ s or less than } 1/10 \text{ s}$$

Water hammer can be dangerous. See Sec. 9.

# Section 4

# Heat

BY

**G. A. HAWKINS**  *Vice President Emeritus for Academic Affairs, Purdue University.*
**G. C. WILLIAMS**  *Professor of Chemical Engineering, Massachusetts Institute of Technology.*
**K. A. SMITH**  *Professor of Chemical Engineering, Massachusetts Institute of Technology.*
**HOYT C. HOTTEL**  *Professor Emeritus, Massachusetts Institute of Technology.*
**ADEL F. SAROFIM**  *Professor of Chemical Engineering, Massachusetts Institute of Technology.*

## THERMAL PROPERTIES OF SUBSTANCES AND THERMODYNAMICS
### by G. A. Hawkins

Thermal Properties of Substances ........................... 4-2
Temperature Measurement, Thermometers ................ 4-2
Expansion of Solids and Liquids by Heat .................... 4-3
Melting Points of Solids .................................. 4-5
Freezing and Boiling Points of Liquids .................... 4-5
Units of Force and Mass ................................. 4-8
Specific Heats of Solids and Liquids ..................... 4-8
Specific Heats of Gases ................................. 4-11
Heats of Solution, Fusion, and Vaporization ............... 4-11
General Principles of Thermodynamics .................... 4-12
Entropy .................................................. 4-14
Availability .............................................. 4-15
Maxwell Relations ....................................... 4-15
Ideal Gas Laws ......................................... 4-16
Ideal Gas Mixtures ...................................... 4-18
Special Changes of State for Ideal Gases ................. 4-18
Graphical Representation of Changes of State ............. 4-19
Ideal Cycles with Perfect Gases ......................... 4-20
Air Compression ........................................ 4-21
Vapors .................................................. 4-22
Thermal Properties of Saturated Vapors and of Vapor and Liquid Mixtures ......................................... 4-22
Charts for Saturated and Superheated Vapors ............. 4-23
Changes of State. Superheated Vapors and Mixtures of Liquid and Vapor ............................................... 4-24
Properties of Ammonia and Other Refrigerants ............ 4-26
Steam Cycles ........................................... 4-27
Mixtures of Air and Water Vapor ......................... 4-28
Steam Tables ........................................... 4-29

Humidity Measurements ................................. 4-31
Psychrometric Charts .................................... 4-40
Air-Conditioning Processes .............................. 4-41
Refrigeration ........................................... 4-45
Thermodynamics of Flow of Compressible Fluids .......... 4-46
Flow of Fluids in Circular Pipes ......................... 4-49
Throttling ............................................... 4-50
Combustion ............................................. 4-51
Internal Energy and Enthalpy of Common Gases .......... 4-53
Temperature Attained by Combustion .................... 4-53
Dissociation ............................................ 4-54
Combustion of Solid Fuels ............................... 4-56
Surface Combustion ..................................... 4-58

## TRANSMISSION OF HEAT BY CONDUCTION AND CONVECTION
### by G. C. Williams and K. A. Smith

Thermal Conductivities .................................. 4-60
Conduction and Convection .............................. 4-61
Film Coefficients ........................................ 4-63
Laminar Flow ........................................... 4-67
Heat Transmission through Tubes ........................ 4-67
Combined Convection and Radiation Coefficients .......... 4-69
Heat Transmission through Pipe Insulation ............... 4-69

## RADIANT-HEAT TRANSFER
### by Hoyt C. Hottel and Adel F. Sarofim

Black-Body Radiation .................................... 4-71
Radiative Exchange between Surfaces of Solids ........... 4-71
Flames, Combustion Products, and Particle Clouds ........ 4-77
Enclosures—Combustion Chambers ...................... 4-80

# THERMAL PROPERTIES OF SUBSTANCES AND THERMODYNAMICS
## by G. A. Hawkins

REFERENCES: "ASHRAE Guide and Data Book," Applications Volume (1971), American Society of Heating, Refrigerating, and Air-Conditioning Engineers. "ASHRAE Guide and Data Book," Equipment Volume, 1972, American Society of Heating, Refrigerating, and Air-Conditioning Engineers. "ASHRAE Guide and Data Book," Handbook of Fundamentals, 1972, American Society of Heating, Refrigerating, and Air-Conditioning Engineers. "ASHRAE Guide and Data Book," Systems Volume, 1970, American Society of Heating, Refrigerating, and Air-Conditioning Engineers. DeGroot, "Thermodynamics of Irreversible Processes," North-Holland Publishing Co., 1961. Editors of Power, "Power-Generation Systems," McGraw-Hill, 1967. Fryling, "Combustion Engineering," Combustion Engineering, Inc., 1966. Glasstone, "Thermodynamics for Chemists," Van Nostrand, 1958. Guggenheim, "Thermodynamics," 3d ed., Interscience, 1957. Harris, "Modern Air-Conditioning Practice," McGraw-Hill, 1959. Hatsopoulos and Keenan, "Principles of General Thermodynamics," Wiley, 1965. Jones and Hawkins, "Engineering Thermodynamics," Wiley, 1960. Keenan, "Thermodynamics," Wiley, 1953. Keenan and Kaye, "Gas Tables," Wiley, 1948. Keenan, Keyes, Hill, and Moore, "Steam Tables," Wiley, 1969. Kestin, "A Course in Thermodynamics," vol. I, Blaisdell Publishing Co., 1966. Kestin, "A Course in Thermodynamics," vol. II, Blaisdell Publishing Co., 1968. Lang, "Principles of Air-Conditioning," Delmar Publishers, 1961. Laub, "Air-Conditioning and Heating Practice," Holt, 1963. Lee and Sears, "Thermodynamics," Addison-Wesley, 1955. Lewis and Randall, "Thermodynamics," McGraw-Hill, 1961. Prigogine, "Introduction to Thermodynamics of Irreversible Processes," 3d ed., Interscience, 1967. Reynolds, "Thermodynamics," McGraw-Hill, 1965. Rocard, "Thermodynamique," Masson et Cie., Editeurs, Paris, 1952. Rossini, "Chemical Thermodynamics," Wiley, 1950. Rossini, "Thermodynamics and Physics of Matter," Princeton University Press, 1955. Ruhemann, "Separation of Gases," Clarendon, 1945. Schlichting, "Boundary Layer Theory," McGraw-Hill, 1960. Stoecker, "Design of Thermal Systems," McGraw-Hill, 1971. Stoecker, "Refrigeration and Air-Conditioning," McGraw-Hill, 1958. "Tables of Thermal Properties of Gases," National Bureau of Standards, circ. 564, 1955. Van Wylen and Sonntag, "Fundamentals of Classical Thermodynamics," Wiley, 1965. Wrangham, "The Theory and Practice of Heat Engines," 2d ed., Cambridge University Press, 1951. Zemansky, "Heat and Thermodynamics," 4th ed., McGraw-Hill, 1957.

## THERMAL PROPERTIES OF SUBSTANCES

### Measurement of Temperature

**Thermometers** The basis for establishing a temperature scale to serve as a primary standard is the fact that at constant temperature, the pressure-volume product for a pure gas approaches a constant finite value as the pressure approaches zero. The scale of the **hydrogen thermometer** is based on this fact. Because of the complexity and difficulty of using such a thermometer precisely the Seventh General Conference of Weights and Measures, representing 31 nations, adopted in 1927 the more readily usable **International Temperature Scale** based on basic fixed points. These are: temperature of equilibrium between liquid and gaseous oxygen at the pressure of 1 standard atmosphere (oxygen point) $-182.97°C$; between ice and air-saturated water at normal atmospheric pressure (ice point) $0.000°C$ (changed to $0.01°C$ in 1960); between liquid water and its vapor at the pressure of 1 standard atmosphere (steam point) $100.000°C$; between liquid sulphur and its vapor at the pressure of 1 standard atmosphere (sulphur point) $444.00°C$; between solid silver and liquid silver at normal atmospheric pressure (silver point) $960.5°C$; between solid gold and liquid gold at normal atmospheric pressure (gold point) $1063°C$. For ordinary use, the mercury thermometer agrees closely with the International Temperature scale, but above $500°F$ the divergence between the two scales may be appreciable.

The usual **mercury thermometer** can be used to about $600°F$; this limit may be extended to $1000°F$ if the capillary tube above the mercury is filled with nitrogen or carbon dioxide under high pressure. The lower temperature limit for the mercury thermometer is $-39°F$. For lower temperatures, alcohol, pentane, or petroleum ether may be used as the thermometric substance. Temperatures are also measured by means of thermocouples, resistance thermometers, and various forms of **pyrometers.**

At the 1948 meeting of the General Conference of Weights and Measures, the word **Celsius** was adopted to replace the adjective **centigrade.** The use of the term Celsius avoids confusion in the French language. All other temperature scales are designated by the first letter of the inventor's name: F for Fahrenheit, R for Reaumur, K for Kelvin, and R for Rankine. It was logical to designate the international scale by C, in recognition of its inventor (in 1742), Anders Celsius, a Swedish astronomer. In 1960, the General Conference changed the defining fixed point from the ice point to the triple point of water, $0.01°C$, which is more easily reproduced than is the ice point. The triple point is the condition under which three phases of matter can coexist in equilibrium. Pure water can exist in three phases in a system in equilibrium. As long as the three phases coexist, the pressure and temperature have fixed values which are characteristic of the chemical nature of the substance. The temperature or pressure can be changed only if one phase is allowed to disappear. The triple point temperature and pressure for water are $0.01°C$ and $0.006028$ atm. For carbon dioxide the triple-point temperature and pressure are $-56.60°C$ and $5.112$ atm, respectively. By this action, the interval between the triple point and steam became $99.99$ deg instead of $100$ and the international scale ceased to be a "centigrade scale." The word *Celsius* is accordingly used in the material that follows. (See Celsius versus Centigrade, the Nomenclature of the Temperature Scale of Science, *Science,* **136,** Apr. 20, 1962, pp. 254–255.)

**Thermometer Scales** Let $F$ and $C$ denote the readings on the Fahrenheit and Celsius (or centigrade) scales, respectively, for the same temperature. Then

$$C = 5/9(F - 32) \qquad F = 9/5C + 32$$

Table 1 gives corresponding readings on the two scales.

If the pressure readings of a constant-volume hydrogen thermometer are extrapolated to zero pressure, it is found that the corresponding temperatures is $-273.15°C$, or $-459.67°F$. It is convenient to have a so-called **absolute temperature scale** on which zero corresponds with zero pressure on the hydrogen thermometer. Such a scale has fundamental thermodynamic significance and agrees very closely with the thermodynamic temperature scale. The absolute scales in use are:

Degrees Kelvin (K) = degrees Celsius + 273.15
Degrees Rankine (R) = degrees Fahrenheit + 459.67

**Fixed Temperatures**  Standard samples for calibration of temperature-measuring instruments at certain fixed points are available from the National Bureau of Standards with certificates giving the exact freezing point of each lot of metal. These are: tin 449.6°F (232°C); lead, 621.5 (327.5); zinc, 787.2 (419.6); aluminum, 1220.7 (660.4); copper, 1984.1 (1084.5).

Convenient standards for use in less precise calibrations are listed in Table 2. Additional information which may be found useful in estimating temperatures is presented in Tables 3 to 9.

## Expansion of Bodies by Heat

**Coefficients of Expansion**  The **coefficient of linear expansion** of a solid is defined as the increment of length in a unit of length for a rise in temperature of 1°. Likewise, the **coefficient of cubical expansion** of a solid, liquid, or gas is the increment of volume of a unit volume for a rise of temperature of 1 deg. Denoting these coefficients by $a'$ and $a'''$, respectively,

$$a' = \frac{1}{l}\frac{dl}{dt} \qquad a''' = \frac{1}{V}\frac{dV}{dt}$$

in which $l$ denotes length, $V$ volume, and $t$ temperature. For homogenous solids $a''' = 3a'$ and the **coefficient of area expansion** $a'' = 2a'$.

The coefficients of expansion are, in general, dependent upon the temperature, but for ordinary ranges of temperature, constant mean values may be taken. If lengths, areas, and volumes at 32°F (0°C) be taken as standard, then these magnitudes at other temperatures $t_1$ and $t_2$ are related as follows:

$$\frac{l_1}{l_2} = \frac{1 + a't_1}{1 + a't_2} \qquad \frac{A_1}{A_2} = \frac{1 + a''t_1}{1 + a''t_2} \qquad \frac{V_1}{V_2} = \frac{1 + a'''t_1}{1 + a'''t_2}$$

### Table 1. Conversion of Thermometer Readings
Degrees Celsius to degrees Fahrenheit

| C | F | C | F | C | F | C | F | C | F | C | F |
|---|---|---|---|---|---|---|---|---|---|---|---|
| −40 | −40.0 | +5 | +41.0 | +40 | +104.0 | +175 | +347 | +350 | +662 | +750 | +1382 |
| −38 | −36.4 | 6 | 42.8 | 41 | 105.8 | 180 | 356 | 355 | 671 | 800 | 1472 |
| −36 | −32.8 | 7 | 44.6 | 42 | 107.6 | 185 | 365 | 360 | 680 | 850 | 1562 |
| −34 | −29.2 | 8 | 46.4 | 43 | 109.4 | 190 | 374 | 365 | 689 | 900 | 1652 |
| −32 | −25.6 | 9 | 48.2 | 44 | 111.2 | 195 | 383 | 370 | 698 | 950 | 1742 |
| −30 | −22.0 | 10 | 50.0 | 45 | 113.0 | 200 | 392 | 375 | 707 | 1000 | 1832 |
| −28 | −18.4 | 11 | 51.8 | 46 | 114.8 | 205 | 401 | 380 | 716 | 1050 | 1922 |
| −26 | −14.8 | 12 | 53.6 | 47 | 116.6 | 210 | 410 | 385 | 725 | 1100 | 2012 |
| −24 | −11.2 | 13 | 55.4 | 48 | 118.4 | 215 | 419 | 390 | 734 | 1150 | 2102 |
| −22 | − 7.6 | 14 | 57.2 | 49 | 120.2 | 220 | 428 | 395 | 743 | 1200 | 2192 |
| −20 | − 4.0 | 15 | 59.0 | 50 | 122.0 | 225 | 437 | 400 | 752 | 1250 | 2282 |
| −19 | − 2.2 | 16 | 60.8 | 55 | 131.0 | 230 | 446 | 405 | 761 | 1300 | 2372 |
| −18 | − 0.4 | 17 | 62.6 | 60 | 140.0 | 235 | 455 | 410 | 770 | 1350 | 2462 |
| −17 | + 1.4 | 18 | 64.4 | 65 | 149.0 | 240 | 464 | 415 | 779 | 1400 | 2552 |
| −16 | 3.2 | 19 | 66.2 | 70 | 158.0 | 245 | 473 | 420 | 788 | 1450 | 2642 |
| −15 | 5.0 | 20 | 68.0 | 75 | 167.0 | 250 | 482 | 425 | 797 | 1500 | 2732 |
| −14 | 6.8 | 21 | 69.8 | 80 | 176.0 | 255 | 491 | 430 | 806 | 1550 | 2822 |
| −13 | 8.6 | 22 | 71.6 | 85 | 185.0 | 260 | 500 | 435 | 815 | 1600 | 2912 |
| −12 | 10.4 | 23 | 73.4 | 90 | 194.0 | 265 | 509 | 440 | 824 | 1650 | 3002 |
| −11 | 12.2 | 24 | 75.2 | 95 | 203.0 | 270 | 518 | 445 | 833 | 1700 | 3092 |
| −10 | 14.0 | 25 | 77.0 | 100 | 212.0 | 275 | 527 | 450 | 842 | 1750 | 3182 |
| − 9 | 15.8 | 26 | 78.8 | 105 | 221.0 | 280 | 536 | 455 | 851 | 1800 | 3272 |
| − 8 | 17.6 | 27 | 80.6 | 110 | 230.0 | 285 | 545 | 460 | 860 | 1850 | 3362 |
| − 7 | 19.4 | 28 | 82.4 | 115 | 239.0 | 290 | 554 | 465 | 869 | 1900 | 3452 |
| − 6 | 21.2 | 29 | 84.2 | 120 | 248.0 | 295 | 563 | 470 | 878 | 1950 | 3542 |
| − 5 | 23.0 | 30 | 86.0 | 125 | 257.0 | 300 | 572 | 475 | 887 | 2000 | 3632 |
| − 4 | 24.8 | 31 | 87.8 | 130 | 266.0 | 305 | 581 | 480 | 896 | 2050 | 3722 |
| − 3 | 26.6 | 32 | 89.6 | 135 | 275.0 | 310 | 590 | 485 | 905 | 2100 | 3812 |
| − 2 | 28.4 | 33 | 91.4 | 140 | 284.0 | 315 | 599 | 490 | 914 | 2150 | 3902 |
| − 1 | 30.2 | 34 | 93.2 | 145 | 293.0 | 320 | 608 | 495 | 923 | 2200 | 3992 |
| 0 | 32.0 | 35 | 95.0 | 150 | 302.0 | 325 | 617 | 500 | 932 | 2250 | 4082 |
| + 1 | 33.8 | 36 | 96.8 | 155 | 311.0 | 330 | 626 | 550 | 1022 | 2300 | 4172 |
| 2 | 35.6 | 37 | 98.6 | 160 | 320.0 | 335 | 635 | 600 | 1112 | 2350 | 4262 |
| 3 | 37.4 | 38 | 100.4 | 165 | 329.0 | 340 | 644 | 650 | 1202 | 2400 | 4352 |
| 4 | 39.2 | 39 | 102.2 | 170 | 338.0 | 345 | 653 | 700 | 1292 | 2450 | 4442 |

TABLES OF VALUES FOR INTERPOLATION IN THE ABOVE TABLE

| Degrees Celsius | 1 | 2 | 3 | 4 | 5 | 6 | 7 | 8 | 9 |
|---|---|---|---|---|---|---|---|---|---|
| Degrees Fahrenheit | 1.8 | 3.6 | 5.4 | 7.2 | 9.0 | 10.8 | 12.6 | 14.4 | 16.2 |

**Table 1. Conversion of Thermometer Readings (*Continued*)**
Degrees Fahrenheit to degrees Celsius*

| F | C | F | C | F | C | F | C | F | C | F | C |
|---|---|---|---|---|---|---|---|---|---|---|---|
| −40 | −40.00 | +30 | −1.11 | +80 | +26.67 | +250 | +121.11 | +500 | +260.00 | +900 | +482.22 |
| −38 | −38.89 | 31 | −0.56 | 81 | 27.22 | 255 | 123.89 | 505 | 262.78 | 910 | 487.78 |
| −36 | −37.78 | 32 | 0.00 | 82 | 27.78 | 260 | 126.67 | 510 | 265.56 | 920 | 493.33 |
| −34 | −36.67 | 33 | +0.56 | 83 | 28.33 | 265 | 129.44 | 515 | 268.33 | 930 | 498.89 |
| −32 | −35.56 | 34 | 1.11 | 84 | 28.89 | 270 | 132.22 | 520 | 271.11 | 940 | 504.44 |
| −30 | −34.44 | 35 | 1.67 | 85 | 29.44 | 275 | 135.00 | 525 | 273.89 | 950 | 510.00 |
| −28 | −33.33 | 36 | 2.22 | 86 | 30.00 | 280 | 137.78 | 530 | 276.67 | 960 | 515.56 |
| −26 | −32.22 | 37 | 2.78 | 87 | 30.56 | 285 | 140.55 | 535 | 279.44 | 970 | 521.11 |
| −24 | −31.11 | 38 | 3.33 | 88 | 31.11 | 290 | 143.33 | 540 | 282.22 | 980 | 526.67 |
| −22 | −30.00 | 39 | 3.89 | 89 | 31.67 | 295 | 146.11 | 545 | 285.00 | 990 | 532.22 |
| −20 | −28.89 | 40 | 4.44 | 90 | 32.22 | 300 | 148.89 | 550 | 287.78 | 1000 | 537.78 |
| −18 | −27.78 | 41 | 5.00 | 91 | 32.78 | 305 | 151.67 | 555 | 290.55 | 1050 | 565.56 |
| −16 | −26.67 | 42 | 5.56 | 92 | 33.33 | 310 | 154.44 | 560 | 293.33 | 1100 | 593.33 |
| −14 | −25.56 | 43 | 6.11 | 93 | 33.89 | 315 | 157.22 | 565 | 296.11 | 1150 | 621.11 |
| −12 | −24.44 | 44 | 6.67 | 94 | 34.44 | 320 | 160.00 | 570 | 298.89 | 1200 | 648.89 |
| −10 | −23.33 | 45 | 7.22 | 95 | 35.00 | 325 | 162.78 | 575 | 301.67 | 1250 | 676.67 |
| − 8 | −22.22 | 46 | 7.78 | 96 | 35.56 | 330 | 165.56 | 580 | 304.44 | 1300 | 704.44 |
| − 6 | −21.11 | 47 | 8.33 | 97 | 36.11 | 335 | 168.33 | 585 | 307.22 | 1350 | 732.22 |
| − 4 | −20.00 | 48 | 8.89 | 98 | 36.67 | 340 | 171.11 | 590 | 310.00 | 1400 | 760.00 |
| − 2 | −18.89 | 49 | 9.44 | 99 | 37.22 | 345 | 173.89 | 595 | 312.78 | 1450 | 787.78 |
| 0 | −17.78 | 50 | 10.00 | 100 | 37.78 | 350 | 176.67 | 600 | 315.56 | 1500 | 815.56 |
| + 1 | −17.22 | 51 | 10.56 | 105 | 40.55 | 355 | 179.44 | 610 | 321.11 | 1550 | 843.33 |
| 2 | −16.67 | 52 | 11.11 | 110 | 43.33 | 360 | 182.22 | 620 | 326.67 | 1600 | 871.11 |
| 3 | −16.11 | 53 | 11.67 | 115 | 46.11 | 365 | 185.00 | 630 | 332.22 | 1650 | 898.89 |
| 4 | −15.56 | 54 | 12.22 | 120 | 48.89 | 370 | 187.78 | 640 | 337.78 | 1700 | 926.67 |
| 5 | −15.00 | 55 | 12.78 | 125 | 51.67 | 375 | 190.55 | 650 | 343.33 | 1750 | 954.44 |
| 6 | −14.44 | 56 | 13.33 | 130 | 54.44 | 380 | 193.33 | 660 | 348.89 | 1800 | 982.22 |
| 7 | −13.89 | 57 | 13.89 | 135 | 57.22 | 385 | 196.11 | 670 | 354.44 | 1850 | 1010.00 |
| 8 | −13.33 | 58 | 14.44 | 140 | 60.00 | 390 | 198.89 | 680 | 360.00 | 1900 | 1037.78 |
| 9 | −12.78 | 59 | 15.00 | 145 | 62.78 | 395 | 201.67 | 690 | 365.56 | 1950 | 1065.56 |
| 10 | −12.22 | 60 | 15.56 | 150 | 65.56 | 400 | 204.44 | 700 | 371.11 | 2000 | 1093.33 |
| 11 | −11.67 | 61 | 16.11 | 155 | 68.33 | 405 | 207.22 | 710 | 376.67 | 2050 | 1121.11 |
| 12 | −11.11 | 62 | 16.67 | 160 | 71.11 | 410 | 210.00 | 720 | 382.22 | 2100 | 1148.89 |
| 13 | −10.56 | 63 | 17.22 | 165 | 73.89 | 415 | 212.78 | 730 | 387.78 | 2150 | 1176.67 |
| 14 | −10.00 | 64 | 17.78 | 170 | 76.67 | 420 | 215.56 | 740 | 393.33 | 2200 | 1204.44 |
| 15 | − 9.44 | 65 | 18.33 | 175 | 79.44 | 425 | 218.33 | 750 | 398.89 | 2250 | 1232.22 |
| 16 | − 8.89 | 66 | 18.89 | 180 | 82.22 | 430 | 221.11 | 760 | 404.44 | 2300 | 1260.00 |
| 17 | − 8.33 | 67 | 19.44 | 185 | 85.00 | 435 | 223.89 | 770 | 410.00 | 2350 | 1287.78 |
| 18 | − 7.78 | 68 | 20.00 | 190 | 87.78 | 440 | 226.67 | 780 | 415.56 | 2400 | 1315.56 |
| 19 | − 7.22 | 69 | 20.56 | 195 | 90.55 | 445 | 229.44 | 790 | 421.11 | 2450 | 1343.33 |
| 20 | − 6.67 | 70 | 21.11 | 200 | 93.33 | 450 | 232.22 | 800 | 426.67 | 2500 | 1371.11 |
| 21 | − 6.11 | 71 | 21.67 | 205 | 96.11 | 455 | 235.00 | 810 | 432.22 | 2550 | 1398.89 |
| 22 | − 5.56 | 72 | 22.22 | 210 | 98.89 | 460 | 237.78 | 820 | 437.78 | 2600 | 1426.67 |
| 23 | − 5.00 | 73 | 22.78 | 215 | 101.67 | 465 | 240.55 | 830 | 443.33 | 2650 | 1454.44 |
| 24 | − 4.44 | 74 | 23.33 | 220 | 104.44 | 470 | 243.33 | 840 | 448.89 | 2700 | 1482.22 |
| 25 | − 3.89 | 75 | 23.89 | 225 | 107.22 | 475 | 246.11 | 850 | 454.44 | 2750 | 1510.00 |
| 26 | − 3.33 | 76 | 24.44 | 230 | 110.00 | 480 | 248.89 | 860 | 460.00 | 2800 | 1537.78 |
| 27 | − 2.78 | 77 | 25.00 | 235 | 112.78 | 485 | 251.67 | 870 | 465.56 | 2850 | 1565.59 |
| 28 | − 2.22 | 78 | 25.56 | 240 | 115.56 | 490 | 254.44 | 880 | 471.11 | 2900 | 1593.33 |
| 29 | − 1.67 | 79 | 26.11 | 245 | 118.33 | 495 | 257.22 | 890 | 476.67 | 2950 | 1621.11 |

TABLE OF VALUES FOR INTERPOLATION IN THE ABOVE TABLE

| Degrees Fahrenheit.... | 1 | 2 | 3 | 4 | 5 | 6 | 7 | 8 | 9 |
|---|---|---|---|---|---|---|---|---|---|
| Degrees Celsius*....... | 0.56 | 1.11 | 1.67 | 2.22 | 2.78 | 3.33 | 3.89 | 4.44 | 5.00 |

*All decimals in the table are repeating decimals; 37.78 is really 37.777. . . .

### Table 2. Standards for Less Precise Calibrations

| Substance | °F | °C | Substance | °F | °C |
|---|---|---|---|---|---|
| Liquid naphthalene boils at | 424.33 | 217.9 | Liquid copper-silver eutectic | | |
| Liquid tin solidifies at | 449.4 | 231.9 | solidifies at | 1434 | 779 |
| Liquid benzophenone boils at | 582.6 | 305.9 | Solid silver melts at | 1761 | 961 |
| Liquid lead solidifies at | 621.2 | 327.3 | Solid gold melts at | 1945 | 1063 |
| Liquid zinc solidifies at | 787.1 | 419.5 | Liquid copper solidifies at | 1981 | 1083 |
| Liquid sulphur boils at | 832.28 | 444.6 | Solid nickel melts at | 2646 | 1452 |
| Liquid antimony solidifies at | 1166.9 | 630.7 | Solid palladium melts at | 2831 | 1555 |
| Liquid aluminum (97.7 percent | | | Solid platinum melts at | 3223 | 1773 |
| pure) solidifies at | 1218 | 660.4 | Solid alumina melts at | 3722 | 2050 |
| | | | Solid tungsten melts at | 6134 | 3390 |

### Table 3. Melting Points of Nonmetallic Elements, °F

| | | | | | |
|---|---|---|---|---|---|
| Helium | −456* | Argon | −309 | Phosphorus | 111 |
| Hydrogen | −434 | Krypton | −272 | Iodine | 236 |
| Neon | −416 | Xenon | −220 | Sulphur | 235 |
| Fluorine | −367 | Chlorine | −151 | Silicon | 2588 |
| Oxygen | −362 | Bromine | +19 | Carbon | >6500 |
| Nitrogen | −346 | | | | |

*At 25 atm.

### Table 4. Melting Points of Various Solids, °F

For pure metals and refractories, see Sec. 6

| Alloys: | | Fusible alloys: | | Enamel colors | 1760 |
|---|---|---|---|---|---|
| Bismuth solder | 200–262 | 33 Bi + 33 Pb + 33 Sn | 250 | India rubber | 257 |
| Brass and bronze | | 18 Bi + 36 Pb + 46 Sn | 305 | Paraffin | 129 |
| (about) | 1650 | 10 Bi + 40 Pb + 50 Sn | 324 | Phosphorus (white or yellow) | 111 |
| 80 Cu + 20 Zn | 1845 | | | Porcelain | 2820 |
| 50 Cu + 50 Zn | 1615 | Tin solder | 275–350 | Potassium | 147 |
| 20 Cu + 80 Zn | 1300 | Blast-furnace slag | 2370–2600 | Sodium | 208 |
| Delta metal | 1742 | Borax | 1040 | Spermaceti | 120 |
| 20 Sn + 80 Pb | 530 | Cast iron, gray | 2460–2550 | Stearine | 122 |
| 50 Sn + 50 Pb | 400 | Cast iron, white | 1920–2010 | Steel | 2370–2550 |
| 80 Sn + 20 Pb | 388 | Chlorides: | | Wrought iron | 2460–2640 |
| | | Calcium | 1422 | | |
| | | Potassium | 1454 | | |
| | | Sodium | 1479 | | |
| | | Zinc | 541 | | |

### Table 5. Melting Points of Reactor Materials, °F

| | | | |
|---|---|---|---|
| Niobium carbide, NbC | 6330 | Beryllium carbide, Be$_2$C | 3812 |
| Graphite, C | 6700 | Aluminum oxide, Al$_2$O$_3$ | 3722 |
| Zirconium carbide, ZrC | 6400 | Beryllium silicide, 2BeO·SiO$_2$ | 3630 |
| Zirconium nitride, ZrN | 5400 | Beryllium oxide-aluminum, BeO·Al$_2$O$_3$ | 3398 |
| Zirconium oxide, ZrO$_2$ | 4900 | Zirconium beryllide, ZrBe$_2$ | 3180 |
| Beryllium oxide, BeO | 4568 | Zirconium disilicide, ZrSi$_2$ | 3090 |
| Zirconium silicides, Zr$_3$Si$_2$, Zr$_4$Si$_3$, | | Titanium, Ti | 3547 |
| Zr$_6$Si$_5$ | 4010–4080 | Zirconium aluminide, ZrAl$_2$ | 3000 |
| Aluminum nitride, AlN | 4060 | | |
| Beryllium nitride, Be$_3$N$_4$ | 4000 | | |

SOURCE: Power's Data Sheet No. 317, *Power*, Jan. 1959.

### Table 6. Freezing Points of Liquids at Atmospheric Pressure, °F

| | | | | | |
|---|---|---|---|---|---|
| Ammonia | −107.8 | Calcium chloride (sat sol) | −40 | Methyl alcohol | −144.2 |
| Aniline | 20.8 | Ether | −180 | Rapeseed oil | 25.7 |
| Benzol | 41.9 | Ethyl alcohol | −174.6 | Turpentine | <−75 |
| Carbon bisulphide | −168.1 | Glycerin | 64 | Sulphuric acid | −105 |
| Carbon dioxide | −110.2 | Naphthalene | 176 | Salt (NaCl) sol, sat | −0.4 |
| Chloroform | −82.3 | Linseed oil | −4 | Sea water | 27.5 |
| | | Mercury | −38.8 | Toluene | −149 |

**Table 6. Freezing Points of Liquids at Atmospheric Pressure, °F (Continued)**

| Mixtures of glycerin and water (Bolley) | | | Mixtures of ethyl alcohol and water (F. Beilstein) | | | |
|---|---|---|---|---|---|---|
| Percent by weight of glycerin | Specific gravity | Freezing point, deg F | Percent by weight of alcohol | Freezing point, deg F | Percent by weight of alcohol | Freezing point, deg F |
| 10 | 1.0245 | 30.2 | 2.58 | 30.2 | 21.7 | 10.4 |
| 20 | 1.0498 | 27.5 | 5.22 | 28.4 | 23.8 | 6.8 |
| 30 | 1.0771 | 20.8 | 7.36 | 26.6 | 26.0 | 3.2 |
| 40 | 1.1045 | 1.0 | 9.58 | 24.8 | 28.0 | − 0.4 |
| 45 | 1.1183 | −15.2 | 11.50 | 23.0 | 30.0 | − 4.0 |
| 50 | 1.1320 | −25.6 | 13.27 | 21.2 | 33.5 | −11.2 |
| 60 | 1.1582 | {Below −31.0 | 16.53 | 17.6 | 37.3 | −18.4 |
| | | | 19.09 | 14.0 | 41.2 | −25.6 |

**Table 7. Boiling Points at Atmospheric Pressure, °F**

| | | | |
|---|---|---|---|
| Zinc | 1665 | Glycerin | 554 | Turpentine | 320.0 |
| Sulphur | 832 | Phosphorus | 536 | Toluene | 231.0 |
| Mercury | 675 | Naphthalene | 424 | Sodium chloride (sat sol) | 226.4 |
| Linseed oil | 549 | Aniline | 364 | | |
| Paraffin | 572 | Calcium chloride (sat sol) | 356 | Helium | −452.0 |

**Table 8. Boiling Points of Hydrocarbons, °F**

| Name | Chemical formula | Deg F |
|---|---|---|
| **Paraffins or Alkanes: $C_nH_{2n+2}$** | | |
| Methane | $CH_4$ | −258.9 |
| Ethane | $C_2H_6$ | −127.5 |
| Propane | $C_3H_8$ | −44.2 |
| Butane | $C_4H_{10}$ | 31.3 |
| Pentane | $C_5H_{12}$ | 96.9 |
| Hexane | $C_6H_{14}$ | 155.6 |
| Heptane | $C_7H_{16}$ | 209.1 |
| Octane | $C_8H_{18}$ | 258.2 |
| Nonane | $C_9H_{20}$ | 303.2 |
| Decane | $C_{10}H_{22}$ | 345.2 |
| **Olefins or Alkenes: $C_nH_{2n}$** | | |
| Ethylene | $C_2H_4$ | −155.0 |
| Propylene | $C_3H_6$ | −53.7 |
| 1-Butene | $C_4H_8$ | 20.8 |
| 1-Pentene | $C_5H_{10}$ | 86.0 |
| **Diolefins or Dienes: $C_nH_{2n-2}$** | | |
| 1,3-Butadiene | $C_4H_6$ | 24.1 |
| Isoprene | $C_5H_8$ | 93.3 |
| **Acetylenes or Alkines: $C_nH_n$** | | |
| Acetylene | $C_2H_2$ | −119.2 |
| **Aromatic Series** | | |
| Benzene | $C_6H_6$ | 176.1 |
| Toluene | $C_7H_8$ | 231.2 |
| Ethylbenzene | $C_8H_{10}$ | 277.1 |
| Naphthalene | $C_{10}H_8$ | 424.2 |

SOURCE: Arthur and Elizabeth Rose, "The Condensed Chemical Dictionary," 6th ed., Reinhold, 1961.

**Table 9. Relation of Color to Temperature of Iron or Steel**

| | Deg F | | Deg F |
|---|---|---|---|
| Dark blood red, black red | 990 | Orange, free scaling heat | 1650 |
| Dark red, blood red, low red | 1050 | Light orange | 1725 |
| Dark cherry red | 1175 | Yellow | 1825 |
| Medium cherry red | 1250 | Light yellow | 1975 |
| Cherry, full red | 1375 | White | 2200 |
| Light cherry, light red | 1550 | | |

Since for solids and liquids the expansion is small, the preceding formulas for these bodies become approximately

$$l_2 - l_1 = a'l_1(t_2 - t_1)$$
$$A_2 - A_1 = a''A_1(t_2 - t_1)$$
$$V_2 - V_1 = a'''V_1(t_2 - t_1)$$

For certain metals, the variation of the coefficient of expansion with temperature is given by an equation in which, denoting by $l_0$ the length at 32°F, and by $l$ the length at temperature $t$, the following relation is obtained:

$$l = l_0\left[1 + a\left(\frac{t - 32}{1000}\right) + b\left(\frac{t - 32}{1000}\right)^2\right]$$

Information relative to the constants in this equation is presented in Table 10.

Gruneisen finds that $a'$ varies directly as the specific heat.

Additional information regarding the compressibility of water, coefficients of expansion, and linear shrinkage of castings is presented in Tables 11 to 14.

### Table 10. Constants for the Equation

| Metal | 1000 a | 1000 b | Temperature range, deg F |
|---|---|---|---|
| Aluminum | 12.58 | 3.0 | 32–1130 |
| Cast iron | 5.441 | 1.747 | 32–1160 |
| Ingot iron | 6.375 | 1.636 | 32–1380 |
| Malleable iron | 6.503 | 1.622 | 32– 930 |
| Ingot steel | 6.212 | 1.623 | 32–1380 |
| Copper | 9.278 | 1.244 | 32–1160 |
| Nickel | 7.652 | 1.023 | 32–1830 |

### Table 11. Compressibility of Water $(v_f - v) \times 10^5$

| Pressure, psia | Temperature, °F | | | | | |
|---|---|---|---|---|---|---|
|  | 32 | 100 | 200 | 300 | 400 | 500 |
| Saturated liquid { $p$, psia | 0.08859 | 0.9503 | 11.529 | 66.98 | 247.1 | 680.0 |
| { $100v_f$, ft³/lb | 1.6022 | 1.6130 | 1.6635 | 1.7453 | 1.8668 | 2.060 |
| 1,000 | 5.5 | 4.8 | 5.5 | 7.4 | 11.8 | 24.0 |
| 1,500 | 8.3 | 7.2 | 8.1 | 11.0 | 17.5 | 36.0 |
| 2,000 | 11.0 | 9.6 | 10.8 | 14.5 | 22.9 | 46.0 |
| 2,500 | 13.7 | 12.0 | 13.4 | 17.9 | 28.2 | 56.0 |
| 3,000 | 16.3 | 14.3 | 15.9 | 21.3 | 33.4 | 66.0 |
| 3,204 | 17.4 | 15.2 | 17.0 | 22.7 | 35.5 | 69.0 |
| 3,500 | 18.9 | 16.6 | 18.5 | 24.7 | 38.4 | 75.0 |
| 4,000 | 21.5 | 18.8 | 21.0 | 27.9 | 43.3 | 82.0 |
| 5,000 | 26.7 | 23.3 | 25.9 | 34.3 | 52.7 | 100.0 |
| 6,000 | 31.7 | 27.7 | 30.7 | 40.5 | 61.7 | 115.0 |

SOURCE: Abstracted from Keenan, Keyes, Hill, and Moore, "Steam Tables, Thermodynamic Properties of Water, Including Vapor, Liquid, and Solid Phases—English Units," Wiley, 1969.

### Table 12. Coefficients of Expansion
For pure metals, see Sec. 6

(COEFFICIENTS OF LINEAR EXPANSION)
(Mean values of 10,000a' between 32 and 212°F; 0 and 100°C)

| METALS | | OTHER MATERIALS | | | |
|---|---|---|---|---|---|
| Aluminum bronze | 0.094 | Type metal | 0.108 | Masonry | 0.025–0.050 |
| Brass, cast | 0.104 | Bakelite, bleached | 0.122 | Paraffin: | |
| Brass, wire | 0.107 | Brick | 0.053 | 32 F–61 F | 0.592 |
| Bronze | 0.100 | Caoutchouc | 0.372 | 61 F–100 F | 0.724 |
| Constantan (60 Cu, 40 Ni) | 0.095 | Carbon—coke | 0.030 | 100 F–120 F | 2.612 |
| German silver | 0.102 | Cement, neat | 0.060 | Porcelain | 0.02 |
| Iron: | | Concrete | 0.080 | Quartz: | |
| Cast | 0.059 | Ebonite | 0.468 | Parallel to axis | 0.044 |
| Soft forged | 0.063 | Glass: | | Perpend. to axis | 0.074 |
| Wire | 0.080 | Thermometer | 0.045 | Quartz, fused | 0.0028 |
| Magnalium (85 Al, 15 Mg) | 0.133 | Hard | 0.033 | Rubber | 0.428 |
| Phosphor bronze | 0.094 | Plate and crown | 0.050 | Vulcanite | 0.400 |
| Solder | 0.134 | Flint | 0.044 | Wood (‖ to fiber): | |
| Speculum metal | 0.107 | Pyrex | 0.018 | Ash | 0.053 |
| Steel: | | Granite | 0.04–0.05 | Chestnut and maple | 0.036 |
| Bessemer, rolled hard | 0.056 | Graphite | 0.044 | Oak | 0.027 |
| Bessemer, rolled soft | 0.063 | Gutta percha | 0.875 | Pine | 0.030 |
| Nickel (10% Ni) | 0.073 | Ice | 0.283 | Across the fiber: | |
|  |  | Limestone | 0.023–0.05 | Chestnut and pine | 0.019 |
|  |  | Marble | 0.02–0.09 | Maple | 0.027 |
|  |  |  |  | Oak | 0.030 |

**Table 13. Coefficients of Cubical Expansion**
Mean values of $1,000a'''$ at ordinary room temperatures

| LIQUIDS | | | | | |
|---|---|---|---|---|---|
| Acetic acid | 0.60 | Hydrochloric acid, 50 % solution | 0.52 | Sulphuric acid, 50 % solution | 0.45 |
| Alcohol (ethyl) | 0.61 | Mercury | 0.10 | Turpentine | 0.54 |
| Alcohol (methyl) | 0.80 | Olive oil | 0.41 | Water | 0.115 |
| Benzene | 0.77 | Petroleum, Pennsylvania | 0.50 | **SOLIDS** | |
| Benzol | 0.70 | Petroleum, California | 0.43 | Fluorspar | 0.035 |
| Calcium chloride ($CaCl_2$), 5 to 50 % solution | 0.28 | Petroleum, Texas | 0.42 | Ice (4 to 30 F) | 0.62 |
| | | Phenol ($C_6H_6O$) | 0.50 | Paraffin wax | 0.61 |
| Chloroform | 0.77 | Rapeseed oil | 0.50 | Rock salt | 0.67 |
| Ether | 0.92 | Salt, 1.6 % solution | 0.60 | Sulphur | 0.40 |
| Glycerin | 0.28 | Salt, 26 % solution | 0.24 | Wood (beech) | 0.016 |
| Hydrochloric acid | 0.27 | Sulphuric acid | 0.31 | Wood (pine) | 0.028 |

**Table 14. Linear Shrinkage of Castings**

| | | | | | |
|---|---|---|---|---|---|
| Bar iron, rolled | 1:55 | Cast iron | 1:96 | Steel, puddled | 1:72 |
| Bell metal | 1:65 | Gun metal | 1:134 | Steel, wrought | 1:64 |
| Bismuth | 1:265 | Iron, fine grained | 1:72 | Tin | 1:128 |
| Brass | 1:65 | Lead | 1:92 | Zinc, cast | 1:624 |
| Bronze | 1:63 | Steel castings | 1:50 | 8 Cu + 1 Sn (by wt) | 1:13 |

The coefficients of cubical expansion for different gases at ordinary temperatures are about the same. From 0 to 212°F and at atmospheric pressure, the values multiplied by 1,000 are as follows: for $NH_3$, 2.11; CO, 2.04; $CO_2$, 2.07; $H_2$, 2.03; NO, 2.07.

### Units of Force and Mass

Force mass, length, and time are related by Newton's second law of motion, which may be expressed as

$$F \sim ma$$

In order to write this as an equality, a constant must be introduced which has magnitude and dimensions. For convenience, the constant may be designated as $1/g_c$. Thus,

$$F = \frac{ma}{g_c}$$

Since this equation must be homogeneous insofar as the dimensions are concerned, the units for $g_c$ are $mL/t^2F$. Consider a 1-lb mass, lbm, in the earth's gravitational field, where the acceleration is 32.1740 ft/s². The force exerted on the pound mass will be defined as the pound force, lbf. This system of units gives for $g_c$ the following magnitude and dimensions:

$$1 \text{ lbf} = \frac{(1 \text{ lbm})(32.174 \text{ ft/s}^2)}{g_c}$$

hence

$$g_c = 32.174 \text{ lbm·ft/lbf·s}^2$$

$g_c$ may be used with other units, in which case the numerical value changes. The numerical value of $g_c$ for four systems of units is

$$g_c = 32.174 \frac{\text{lbm·ft}}{\text{lbf·s}^2} = 1 \frac{\text{slug·ft}}{\text{lbf·s}^2} = 1 \frac{\text{lbm·ft}}{\text{pdl·s}^2} = 1 \frac{\text{g cm}}{\text{dyne·s}^2}$$

Consider now the relationship which involves weight, a gravitational force, and mass by applying the basic equation for a body of fixed mass acted upon by a gravitational force $g$ and no other forces. The acceleration of the mass caused by the gravitational force is the acceleration due to gravity $g$.

Substituting gives the relationship between weight and mass

$$w = \frac{mg}{g_c}$$

If the gravitational acceleration is constant, the weight and mass are in a fixed proportion to each other; hence for accounting purposes in mass balances they can be used interchangeably. This is not possible if $g$ is a variable.

We may now write the relation between mass $m$ and weight $w$ as

$$w = m \frac{g}{g_c}$$

The constant $g_c$ is used throughout the following paragraphs. (An extensive table of conversion factors from customary units to SI units is found in Sec. 1.)

### Measurement of Heat

**Units of Heat** Many units of heat have been dependent on the experimentally determined properties of some substance. To eliminate experimental variations, the unit of heat may be defined in terms of fundamental units. The International Steam Table Conference (London, 1929) defines the Steam Table (IT) calorie as $\frac{1}{860}$ of a watthour. One British thermal unit (Btu) is defined as 251.996 IT cal, or 778.26 ft·lb.

Previously, the Btu was defined as the heat necessary to raise one pound of water one degree Fahrenheit at some arbitrarily chosen temperature level. Similarly, the calorie was defined as the heat required to heat one gram of water one degree Celsius at 15°C (or at 17.5°C). These units are roughly the same in value as those mentioned above.

**Heat Capacity and Specific Heat** The heat capacity of a material is the amount of heat transferred to raise unit mass of a material 1 deg in temperature. The ratio of the amount of heat transferred to raise unit mass of a material 1 deg to that required to raise unit mass of water 1 deg at some specified temperature is the **specific heat** of the material. For most engineering purposes, heat capacities may be assumed numerically equal to specific heats. Two heat capacities are generally used, that at constant pressure $c_p$ and that at constant volume $c_v$. For unit mass, the instantaneous heat capacities are defined as

$$\left(\frac{\partial h}{\partial t}\right)_p = c_p \qquad \left(\frac{\partial u}{\partial t}\right)_v = c_v$$

Over a range in temperature, the mean heat capacities are given by

$$c_{pm} = \frac{1}{t_2 - t_1}\int_{t_1}^{t_2} c_p\, dt \qquad c_{vm} = \frac{1}{t_2 - t_1}\int_{t_1}^{t_2} c_v\, dt$$

Denoting by $c$ the heat capacity, the heat required to raise the temperature of $w$ lb of a substance from $t_1$ to $t_2$ is $Q = mc(t_2 - t_1)$, provided $c$ is a constant.

In general, $c$ varies with the temperature, though for moderate temperature ranges a constant mean value may be taken.

If, however, $c$ is taken as variable, then $Q = m\int_{t_1}^{t_2} c\, dt$. The

*mean* heat capacity from 0 to $t$ deg is given by $c_m = \dfrac{1}{t}\int_0^t c\, dt$.

If $c = a_1 + a_2 t + a_3 t^2 + \cdots$

$$c_m = a_1 + \tfrac{1}{2}a_2 t + \tfrac{1}{3}a_3 t^2 + \cdots$$

Data relative to the specific heat of water, mean specific heats for solids and liquids, the specific heat of gases at 1 atm, and the mean specific heat of iron are presented in Tables 15 to 20.

### Table 15. Specific Heat of Water at Constant Pressure

Interpolated from Keenan and Keyes tables

| Temp, °F | Pressure, psia | | | |
|---|---|---|---|---|
| | 1,000 | 2,000 | 4,000 | 6,000 |
| 200 | 1.002 | 0.998 | 0.992 | 0.986 |
| 400 | 1.072 | 1.063 | 1.047 | 1.034 |
| 600 | 0.888 | 1.453 | 1.295 | 1.211 |
| 680 | 0.707 | 1.365 | 1.789 | 1.450 |
| 800 | 0.605 | 0.800 | 1.967 | 2.872 |
| 1000 | 0.566 | 0.633 | 0.829 | 1.110 |

### Table 16. Specific Heat of Liquid Water

| Temp, °F | $C_p$, Btu/(lb·°F) |
|---|---|
| 32 | 1.007 |
| 50 | 1.003 |
| 100 | 0.998 |
| 150 | 1.000 |
| 200 | 1.006 |
| 212 | 1.008 |
| 250 | 1.015 |
| 300 | 1.029 |
| 350 | 1.060 |
| 400 | 1.085 |
| 450 | 1.116 |
| 500 | 1.180 |

SOURCE: ASHRAE, "Handbook of Fundamentals," New York, 1972. The full tables, for 43 temperatures, are contained in a report ASHRAE, "Thermophysical Properties of Refrigerants," 227 pp., submitted to ASHRAE by Purdue University in 1972.

### Table 17. Mean Specific Heats of Various Solids and Liquids between 32 and 212°F, Btu/(lb·°F)

| SOLIDS | | | | | |
|---|---|---|---|---|---|
| Alloys: | | Glass: | | Tufa | 0.33 |
| Bismuth-tin | 0.040–0.045 | Normal | 0.199 | Vulcanite | 0.331 |
| Bell metal | 0.086 | Crown | 0.16 | Wood: | |
| Brass, yellow | 0.0883 | Flint | 0.12 | Fir | 0.65 |
| Brass, red | 0.090 | Gneiss | 0.18 | Oak | 0.57 |
| Bronze | 0.104 | Granite | 0.195 | Pine | 0.67 |
| Constantan | 0.098 | Graphite | 0.201 | | |
| D'Arcet's metal | 0.050 | Gypsum | 0.259 | LIQUIDS | |
| German silver | 0.095 | Hornblende | 0.195 | Acetic acid | 0.51 |
| Lipowitz's metal | 0.040 | Humus (soil) | 0.44 | Acetone | 0.51 |
| Nickel steel | 0.109 | Ice: | | Alcohol (absolute) | 0.58 |
| Rose's metal | 0.050 | −4 F | 0.465 | Aniline | 0.51 |
| Solders (Pb and Sn) | | 32 F | 0.487 | Benzol | 0.40 |
| | 0.040–0.045 | India rubber (Para) | 0.27–0.48 | Chloroform | 0.23 |
| Type metal | 0.0388 | Kaolin | 0.224 | Ether | 0.54 |
| Wood's metal | 0.040 | Limestone | 0.217 | Ethyl acetate | 0.47 |
| 40 Pb + 60 Bi | 0.0317 | Marble | 0.210 | Ethylene glycol | 0.60 |
| 25 Pb + 75 Bi | 0.030 | Oxides: | | Fusel oil | 0.56 |
| Asbestos | 0.20 | Alumina ($Al_2O_3$) | 0.183 | Gasoline | 0.50 |
| Ashes | 0.20 | $Cu_2O$ | 0.111 | Glycerin | 0.58 |
| Bakelite | 0.3–0.4 | Lead oxide (PbO) | 0.055 | Hydrochloric acid | 0.60 |
| Basalt (lava) | 0.20 | Lodestone | 0.156 | Kerosene | 0.50 |
| Borax | 0.229 | Magnesia | 0.222 | Naphthalene | 0.31 |
| Brick | 0.22 | Magnetite ($Fe_3O_4$) | 0.168 | Machine oil | 0.40 |
| Carbon-coke | 0.203 | Silica | 0.191 | Mercury | 0.03 |
| Chalk | 0.215 | Soda | 0.231 | Olive oil | 0.40 |
| Charcoal | 0.20 | Zinc oxide (ZnO) | 0.125 | Paraffin oil | 0.52 |
| Cinders | 0.18 | Paraffin wax | 0.69 | Petroleum | 0.50 |
| Coal | 0.3 | Porcelain | 0.22 | Sulphuric acid | 0.33 |
| Concrete | 0.156 | Quartz | 0.17–0.28 | Sea water | 0.94 |
| Cork | 0.485 | Quicklime | 0.217 | Toluene | 0.44 |
| Corundum | 0.198 | Salt, rock | 0.21 | Turpentine | 0.42 |
| Dolomite | 0.222 | Sand | 0.195 | Molten metals: | |
| Ebonite | 0.33 | Sandstone | 0.22 | Bismuth (535–725 F) | 0.036 |
| | | Serpentine | 0.25 | Lead (590–680 F) | 0.041 |
| | | Sulphur | 0.180 | Sulphur (246–297 F) | 0.235 |
| | | Talc | 0.209 | Tin (460–660 F) | 0.058 |

### Table 18. Constants for Kopp's Law

| Elements | Specific heats of the atoms | Elements | Specific heats of the atoms |
|---|---|---|---|
| Heavy elements.................. | 6.4 | Oxygen........................ | 4.0 |
| Boron........................... | 2.7 | Phosphorus.................... | 5.4 |
| Carbon.......................... | 1.8 | Silicon........................ | 3.5 |
| Fluorine........................ | 5.0 | Sulphur........ ............... | 5.4 |
| Hydrogen....................... | 2.3 | | |

### Table 19. Specific Heats of Gases at 1 Atm

| Gas | Symbol | Equation for $C_p$ in Btu per mol | Temp range, deg R | Source |
|---|---|---|---|---|
| Oxygen........... | $O_2$ | $11.515 - \left(\dfrac{172}{\sqrt{T}}\right) + \left(\dfrac{1530}{T}\right)$ | 540–5000 | $a$ |
| | | $11.515 - \left(\dfrac{172}{\sqrt{T}}\right) + \left(\dfrac{1530}{T}\right)$ $+ \left(\dfrac{0.05(T - 4000)}{1000}\right)$ | 5000–9000 | $a$ |
| Nitrogen........... | $N_2$ | $9.47 - \left(\dfrac{3.47 \times 10^3}{T}\right) + \left(\dfrac{1.16 \times 10^6}{T^2}\right)$ | 540–5000 | $a$ |
| Carbon monoxide.... | $CO$ | $9.46 - \left(\dfrac{3.29 \times 10^3}{T}\right) + \left(\dfrac{1.07 \times 10^6}{T^2}\right)$ | 540–5000 | $a$ |
| Hydrogen.......... | $H_2$ | $5.76 + \left(\dfrac{0.578T}{1000}\right) + \left(\dfrac{20}{\sqrt{T}}\right)$ | 540–4000 | $a$ |
| | | $5.76 + \left(\dfrac{0.578T}{1000}\right) + \left(\dfrac{20}{\sqrt{T}}\right)$ $- \left(\dfrac{0.33(T - 4000)}{1000}\right)$ | 4000–9000 | $a$ |
| Water............. | $H_2O$ | $19.86 - \left(\dfrac{597}{\sqrt{T}}\right) + \left(\dfrac{7500}{T}\right)$ | 540–5000 | $a*$ |
| Carbon dioxide..... | $CO_2$ | $16.2 - \left(\dfrac{6.53 \times 10^3}{T}\right) + \left(\dfrac{1.41 \times 10^6}{T^2}\right)$ | 540–6300 | $a$ |
| Methane.......... | $CH_4$ | $4.22 + 8.211 \times 10^{-3}T$ | 492–1800 | $b$ |
| | | $27.0 - \dfrac{14,400}{T}$ | 1800–5940 | $b$ |
| Ethylene.......... | $C_2H_4$ | $6.0 + 8.33 \times 10^{-3}T$ | 720–1400 | $c$ |
| Ethane............ | $C_2H_6$ | $6.6 + 13.33 \times 10^{-3}T$ | 720–1440 | $c$ |
| Ethyl alcohol....... | $C_2H_6O$ | $4.5 + 21.1 \times 10^{-3}T$ | 680–1120 | $c$ |
| Methyl alcohol..... | $CH_4O$ | $2.0 + 16.67 \times 10^{-3}T$ | 680–1100 | $c$ |
| Benzene........... | $C_6H_6$ | $6.5 + 28.9 \times 10^{-3}T$ | 520–1120 | $c$ |
| Octane............ | $C_8H_{18}$ | $14.4 + 53.3 \times 10^{-3}T$ | 720–1440 | $c$ |
| Dodecane.......... | $C_{12}H_{26}$ | $19.6 + 80.0 \times 10^{-3}T$ | 720–1440 | $c$ |

[a]Sweigert and Beardsley, Empirical Specific Heat Equations Based upon Spectroscopic Data, *Ga. School Tech., State Eng. Expt. Sta. Bull.* 2, 1938.

[b]Schwarz, Die Spezifischen Wärmen der Gase als Hiltswerte zur Berechnung von Gleichgewichten, *Arch. Eisenbüttenw.*, 9, 1936, p. 389.

[c]Parks and Huffman, *ACS, Mon.* 60, 1932.

*Approximate. An equation based on the most recent data is given by Keyes in *J. Chem. Phys.*, 15, Aug. 1947, p. 602.

### Table 20. Mean Specific Heat of Iron ($c_m$) between 32 and $t°F$ (Oberhoffer)

| $t$...... | 600 | 800 | 1000 | 1200 |
|---|---|---|---|---|
| $c_m$...... | 0.127 | 0.133 | 0.139 | 0.148 |

**Specific Heat of Solids**  For elements near room temperature, the specific heat may be approximated by the rule of Dulong and Petit, that the specific heat at constant volume for one atomic weight of any solid element is 6.4. For solid compounds at about room temperature, Kopp's approximation is often useful. This states that the specific heat of a solid compound at room temperature is equal to the sum of the specific heats of the atoms forming the compound. The values given in Table 18 are to be used in connection with Kopp's law.

EXAMPLE.   Specific heat of the atoms of $Na_2SO_4$ = 2(6.4) + 5.4 + 4(4) = 34.2. Molecular weight $Na_2SO_4$ = 142. Specific heat = 34.2/142 = 0.24: (I.C.T. value = 0.20). Errors of 20 percent are not uncommon using Kopp's law.

**Specific Heats of Gases**   For monatomic gases, the specific heats do not vary with temperature, and $k$, the value of $c_p/c_v$, is 1.66. For diatomic gases (oxygen, nitrogen, etc.), the specific heats vary with temperature but for many purposes may be assumed constant over considerable ranges of temperature. For diatomic gases, $k$ is approximately 1.40. For more complex gases, generalizations are not possible. Specific heat increases with molecular complexity, and the value of $k$ decreases.

Properties of gases are, usually, most readily correlated on the mol basis. A **pound mol** is the mass in pounds equal to the molecular weight. Thus 1 pound mol of oxygen is 32 lb. At the same pressure and temperature, the volume of one mol is the same for all perfect gases, i.e., following the gas laws. Experimental findings led Avogadro (1776–1856) to formulate the microscopic hypothesis now known as **Avogadro's principle,** which states that one mol of any perfect gas contains the same number of molecules. The number is known as the **Avogadro number** and is equal to

$$N = 6.02486 \times 10^{26} \text{ molecules/(k·mol)}$$
$$= 2.73283 \times 10^{26} \text{ molecules/(lb·mol)}$$

For perfect gases, $Mc_p - Mc_v = AMR = 1.987$.

$$c_v = AR/(k - 1) \qquad c_p = ARk/(k - 1)$$

On a molal basis, the average values of specific heat for some of the more common gases are given by Fig. 1. These values

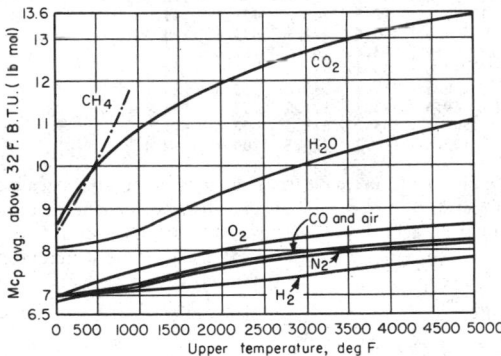

**Fig. 1**   Values of the mean molal specific heat at constant pressure, above 32°F.

have been corrected to zero pressure but they are practically the same at 1 atm. Table 19 gives equations for specific heats for several gases at a constant pressure of 1 atm.

**Specific Heat of Mixtures**   If $w_1$ lb of a substance at temperature $t_1$ and with specific heat $c_1$ is mixed with $w_2$ lb of a second substance at temperature $t_2$ and with specific heat $c_2$, provided chemical reaction, heat evolution, or heat absorption do not occur, the specific heat of the mixture is

$$c_m = (w_1c_1 + w_2c_2)/(w_1 + w_2)$$

and the temperature of the mixture is

$$t_m = (w_1c_1t_1 + w_2c_2t_2)/(w_1c_1 + w_2c_2)$$

In general, $t_m = \Sigma wct/\Sigma wc$.

To raise the temperature of $w_1$ lb of a substance having a specific heat $c_1$ from $t_1$ to $t_m$, the weight $w_2$ of a second substance required is

$$w_2 = w_1c_1(t_m - t_1)/c_2(t_2 - t_m)$$

For mixing two bodies of the same (perfect) gas at constant pressure,

$$t_m = [(V_1 + V_2)/(V_1/T_1 + V_2/T_2)] - 459.69$$

**Specific Heat of Solutions**   For aqueous solutions of salts, the specific heat may be estimated by assuming the specific heat of the solution equal to that of the water alone. Thus, for a 20 percent by weight solution of sodium chloride in water, the specific heat would be approximately 0.8.

**Latent Heats**   For pure substances, the heat effects accompanying changes in state at constant pressure are known as latent effects, because no temperature change is evident. Heat of fusion, vaporization, sublimation, and change in crystal form are examples. Heats of vaporization at low pressures for pure liquids of similar chemical characteristics are well correlated by the methods proposed by Hildebrand. Such a correlation is given in Fig. 2.

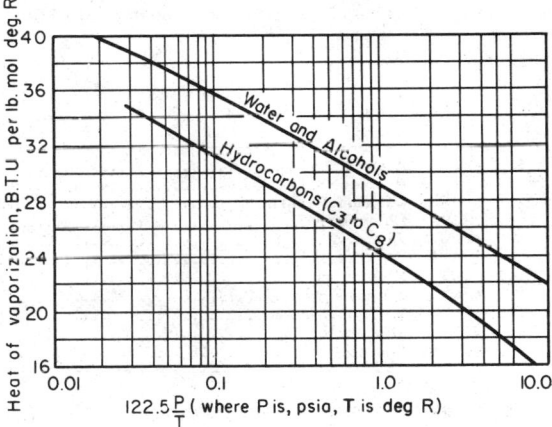

**Fig. 2**   Hildebrand function for enthalpy of vaporization.

EXAMPLE.   For water at 25 psia and 240°F, the heat of vaporization is 952 Btu. Referring to Fig. 2, $122.5 \dfrac{25}{240 + 460} = 4.4$, and the corresponding value of the molal heat of vaporization is 24.6(240 + 460) = 17,200 Btu per lb·mol or 956 Btu per lb.

The values for the heat of fusion and latent heat of vaporization are presented in Tables 21 and 22.

**Vapor Pressures**   At a specified temperature, a pure liquid can exist in equilibrium contact with its vapor at but one pressure, its vapor pressure. A plot of these pressures against the corresponding temperatures is known as a vapor-pressure curve.

Various values for the evaporation temperatures of certain liquids at a pressure of 1 atm are presented in Table 23.

### Table 21. Heat of Fusion,* Btu/lb

| | | | | | |
|---|---|---|---|---|---|
| Aluminum | 171.0 | Potassium | 26.2 | Wood's metal, | |
| Bismuth | 22.0 | Silver | 44.0 | 25.8 Pb + 14.7 Sn | |
| Cadmium | 23.4 | Sodium | 49.5 | + 52.4 Bi + 7.1 Cd.. | 15.1 |
| Chromium | 136.0 | Sulphur | 16.7 | Acetic acid, $C_2H_4O_2$.... | 88.4 |
| Cobalt | 115.2 | Tin | 25.4 | Benzene, $C_6H_6$ | 54.5 |
| Copper | 88.2 | Zinc | 47.0 | Diphenyl, $C_{12}H_{10}$ | 47.0 |
| Gold | 28.7 | Alloys: | | Ethyl alcohol, $C_2H_6O$... | 45.0 |
| Iron | 117.0 | 30.5 Pb + 69.5 Sn | 30.6 | Glycerol, $C_3H_8O_3$ | 85.0 |
| Lead | 10.0 | 36.9 Pb + 63.1 Sn | 28.0 | Methyl alcohol, $CH_4O$. | 29.5 |
| Mercury | 5.0 | 64.7 Pb + 36.3 Sn | 21.0 | Napnthalene, $C_{10}H_8$.... | 64.0 |
| Nickel | 130.0 | 77.8 Pb + 22.2 Sn | 17.0 | Phenol, $C_6H_6O$ | 52.2 |
| Palladium | 65.0 | Rose's metal, | | Ice | 144.0 |
| Phosphorus | 9.0 | 24.0 Pb + 27.3 Sn | | | |
| Platinum | 49.0 | + 48.7 Bi | 12.3 | | |

*Data compiled from "Handbook of Chemistry and Physics," 43d ed., Chemical Rubber Publishing Co.; "Chemical Engineers' Handbook," 3d ed., McGraw-Hill; Kelley, *U.S. Bureau of Mines Bull.* 476, 1949; "Smithsonian Physical Tables," 8th ed. (rev.), 1st reprint, 1934.

### Table 22. Latent Heat of Vaporization at Atmospheric Pressure, Btu/lb

| | | | | | |
|---|---|---|---|---|---|
| Ethyl alcohol | 367 | Hexane | 156 | Hydrogen | 194 |
| Methyl alcohol | 482 | Heptane | 133 | Nitrogen | 86 |
| Aniline | 198 | Octane | 128 | Oxygen | 92 |
| Benzene | 172 | Decane | 110 | Chlorine | 121 |
| Toluene | 151 | Gasoline | 133–145 | Sulphur | 120 |
| Chloroform | 110 | Kerosene | 105–110 | Acetone | 239 |
| Ether | 162 | Turpentine | 126 | Carbon bisulphide | 152 |
| Diphenyl | 134 | | | Carbon tetrachloride | 83.5 |

### Table 23. Evaporation Temperatures of Certain Liquids at Pressures up to 1 Atm

| Substance | Pressures, mm of mercury | | | | | |
|---|---|---|---|---|---|---|
| | 10 | 100 | 200 | 400 | 600 | 760 |
| | Temperature, deg F | | | | | |
| Acetone | − 25.6 | 45.1 | 71.2 | 101.7 | 121.1 | 133.1 |
| Benzene | − 9.5 | 71.2 | 109.4 | 141.9 | 163.7 | 176.2 |
| Carbon tetrachloride | − 4.0 | 72.2 | 102.3 | 134.6 | 156.1 | 170.2 |
| Chloroform | − 22.0 | 49.1 | 77.3 | 108 | 128.8 | 142.0 |
| Diphenyl | 246.0 | 357.5 | 400.5 | 441.1 | 471.8 | 493.0 |
| Ethyl alcohol | 26.6 | 94.8 | 118.8 | 145.5 | 162.4 | 173.0 |
| Ethyl ether | − 56.2 | 17.6 | 35.4 | 63.5 | 83.1 | 94.2 |
| Ethylene glycol | 190.5 | 280.5 | 313.0 | 348.8 | 372.0 | 384.7 |
| Methyl alcohol | 3.74 | 69.6 | 93.7 | 121.0 | 138.2 | 148.4 |
| Mercury | 363 | 501 | 555 | 613.5 | 650.5 | 672.5 |
| Water | 52.3 | 123.8 | 141.2 | 176.5 | 200.4 | 212.0 |

## GENERAL PRINCIPLES OF THERMODYNAMICS

Thermodynamics is the study which deals with energy, the various concepts and laws describing the conversion of one form of energy to another, and the various systems employed to effect the conversions. Thermodynamics deals in general with systems in equilibrium. By means of its fundamental concepts and basic laws the behavior of an engineering system may be described when the various variables are altered. Thermodynamics covers a very broad field and includes many systems, for example, those dealing with chemical, thermal, mechanical, and electrical force fields and potentials. The quantity of matter under consideration is called the **system,** and everything else is spoken of as the **surroundings.** With a **closed system** there is no interchange of matter between system and surroundings; with an **open system** there is such an interchange. Any change that the system may undergo is known as a **process.** Any process or series of processes in which the system returns to its original condition or state is called a **cycle.**

**Heat** is energy **in transit** from one mass to another because of a temperature difference between the two. Whenever a force of any kind acts through a distance, **work** is done. Like heat, work is also energy in transit. Work is to be differentiated from the capacity of a quantity of energy to do work.

### Notation

$A = 1/J = 1/778$ Btu/(ft·lb)

$B$ = availability (by definition, $B = H - T_o S$)

$c_p$ = specific heat at constant pressure

$c_v$ = specific heat at constant volume

$E, e$ = total energy associated with a system

$g$ = local acceleration of gravity, ft/s²

$g_c$ = a dimensional constant

$H, h$ = enthalpy, Btu (by definition $h = u + Apv$)

$J$ = mechanical equivalent of heat = 778.26 ft·lb/Btu = 4.1861 J/cal

$k = c_p/c_v$

$m$ = mass of substance under consideration, lb$m$

$M$ = molecular weight

$p$ = absolute pressure, lb/ft²

$Q, q$ = quantity of heat absorbed by the system from the surroundings, Btu

$R$ = ideal gas constant

$R_u$ = universal gas constant

$S, s$ = entropy

$t$ = temperature, °F

$T = t + 459.69$ = absolute temperature = °R

$T_0$ = sink or discard temperature

$U, u$ = internal energy

$V$ = linear velocity, ft/s (or total volume)

$v$ = volume

$w$ = weight of substance under consideration, lb

$W$ = external work performed on surroundings during change of state, ft·lb

$Y = \left(\dfrac{p_1}{p_2}\right)^{(k-1)/k} - 1$

$z$ = distance above or below chosen datum

$g$ = free energy (by definition, $g = h - Ts$)

$f$ = Helmholtz free energy (by definition, $f = u - Ts$)

In this notation, small letters usually denote magnitudes referred to unit mass of the substance, capital letters corresponding magnitudes referred to $m$ units of mass. Thus, $v$ denotes the volume of 1 lb, $V = mv$, the volume of $m$ lb. Similarly, $U = mu$, $S = ms$, etc. Subscripts are used to indicate different states; thus, $p_1$, $v_1$, $T_1$, $u_1$, $s_1$ refer to state 1, $p_2$, $v_2$, $T_2$, $u_2$, $s_2$ refer to state 2. $Q_{12}$ is used to denote the heat transferred during the change from state 1 to state 2, and $W_{12}$ denotes the external work done during the same change.

The two fundamental and general laws of thermodynamics are: (1) Energy may be neither created nor destroyed. (2) It is impossible to bring about any change or series of changes the sole net result of which is transfer of energy as heat from a low to a high temperature; in other words, heat will not of itself flow from low to high temperatures.

The **first law of thermodynamics,** one of the very important laws of nature, is the law of conservation of energy. Although the law has been stated in a variety of ways, all have essentially the same meaning. The following are examples of typical statements: whenever energy is transformed from one form to another, energy is always conserved; energy can neither be created nor destroyed; the sum total of all energy remains constant. The energy conservation hypothesis was stated by a number of investigators; however, experimental evidence was not available until the famous work of J. P. Joule.

It has long been the custom to designate the law of conservation of energy, the first law of thermodynamics, when it is used in the analysis of engineering systems involving heat transfer and work. Statements of the first law may be written as follows: heat and work are mutually convertible; or, since energy can neither be created nor destroyed, the total energy associated with an energy conversion remains constant.

Before the first law may be applied to the analysis of engineering systems it is necessary to express it in some form of expression. Thus it may be stated for an **open system** as

$$\begin{bmatrix}\text{Net amount of}\\\text{energy added to}\\\text{system as heat}\\\text{and all forms}\\\text{of work}\end{bmatrix} + \begin{bmatrix}\text{stored}\\\text{energy}\\\text{of mass}\\\text{entering}\\\text{system}\end{bmatrix} - \begin{bmatrix}\text{stored}\\\text{energy}\\\text{of mass}\\\text{leaving}\\\text{system}\end{bmatrix} = \begin{bmatrix}\text{net increase}\\\text{in stored}\\\text{energy of}\\\text{system}\end{bmatrix}$$

For an open system with fluid entering only at section 1 and leaving only at section 2 and with no electrical, magnetic, or surface-tension effects, this equation may be written as

$$JQ - W + \int \left(Jh_1 + \frac{V_1^2}{2g_c} + \frac{gz_1}{g_c}\right)\delta m_1$$
$$- \int \left(Jh_2 + \frac{V_2^2}{2g_c} + \frac{gz_2}{g_c}\right)\delta m_2 = JU_f - JU_i$$
$$+ \frac{m_f V_f^2 - m_i V_i^2}{2g_c} + \frac{g}{g_c}(m_f z_f - m_i z_i)$$

The subscripts $i$ and $f$ refer to entire systems before and after the process occurs. $\delta m$ refers to a differential quantity of matter.

It must be remembered that all terms in the first-law equation must be expressed in the same units.

For a **closed stationary system,** the first-law expression reduces to

$$JQ - W = J(U_2 - U_1)$$

For an **open system** fixed in position but undergoing **steady flow,** e.g., a turbine or reciprocating steam engine, for a mass flow rate of $m$ is

$$JQ - W = m\left[J(h_2 - h_1) + \frac{V_2^2 - V_1^2}{2g_c} + \frac{g}{g_c}(z_2 - z_1)\right]$$

In a **steady-flow** process, the mass rate of flow into the apparatus is equal to the mass rate of flow out and, in addition, at any point in the apparatus, the conditions are unchanging with time.

Since for many processes the last two terms are often negligible they will be omitted for simplicity except when such omission would introduce appreciable error.

Work done in overcoming a fluid pressure is measured by $W = \int p \, dv$, where $p$ is the pressure *effectively* applied to the surroundings for doing work and $dv$ represents the change in volume of the system.

**Reversible and Irreversible Processes**  A reversible process is one in which both the system and the surroundings may be returned to their original states. After an irreversible process, this is not possible. No process involving friction or an unbalanced potential can be reversible. No loss in ability to do work is suffered because of a reversible process but there is always a loss in ability to do work because of an irreversible process. All actual processes are irreversible. Any series of reversible processes that starts and finishes with the system in the same state is called a **reversible cycle.**

**Steady-flow Processes**  With **steady-flow,** the conditions at any point in an apparatus through which a fluid is flowing do not change progressively with time. Steady-flow processes involving only mechanical effects are equivalent to similar nonflow processes occurring between two weightless frictionless diaphragms or pistons moving at constant pressure with the system as a whole in motion. Under these circumstances, the total work done by or on a unit amount of fluid is made up of that done on the two diaphragms $p_2 v_2 - p_1 v_1$ and that done on the rest of the surroundings $\int p \, dv - p_2 v_2 + p_1 v_1$. Differentiating, $p \, dv - d(pv) = -v \, dp$. The net, useful flow work done on the surroundings is $-\int v \, dp$. This is often called the shaft work. The net, useful or shaft work differs from the total work by $p_2 v_2 - p_1 v_1$. The first-law equation may be written to indicate this result for a unit mass flow rate as

$$Jq - W_{\text{net}} = J(u_2 - u_1)$$
$$+ p_2 v_2 - p_1 v_1 + \frac{1}{2g_c}(V_2^2 - V_1^2) + \frac{g}{g_c}(z_2 - z_1)$$

or since by definition

$$Ju + pv = Jb$$

$$Jq - W_{\text{net}} = J(b_2 - b_1) + \frac{1}{2g_c}(V_2^2 - V_1^2) + \frac{g}{g_c}(z_2 - z_1)$$

If all net work effects are mechanical,

$$Jq + \int v\, dp = J(b_2 - b_1) + \frac{1}{2g_c}(V_2^2 - V_1^2) + \frac{g}{g_c}(z_2 - z_1)$$

Since in evaluating $\int v\, dp$ the pressure is that *effectively* applied to the surroundings, the integration cannot usually be performed except for reversible processes.

If a fluid is passed adiabatically through a conduit (i.e., without heat exchange with the conduit), without doing any net or useful work, and if velocity and potential effects are negligible, $b_2 = b_1$. A process of the kind indicated is the Joule-Thomson flow and the ratio $(\partial T/\partial p)$ for such a flow is the Joule-Thomson coefficient.

If a fluid is passed through a nonadiabatic conduit without doing any net or useful work and if velocity and potential effects are negligible, $q = b_2 - b_1$. This equation is important in the calculation of heat balances on flow apparatus, e.g., condensers, heat exchangers, and coolers.

In many engineering processes the movement of materials is not independent of time; hence the steady-flow equations do not apply. For example, the process of oxygen discharging from a storage bottle represents a transient condition. The pressure within the bottle changes as the amount of oxygen in the tank decreases. The analysis of some transient processes is very complex; however, in order to show the general approach, a simple case will be considered.

The quantity of material flowing into and out of the engineering system shown in Fig. 3 varies with time. The amount

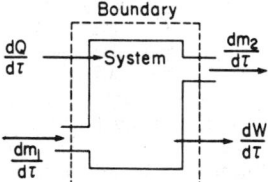

**Fig. 3**  Variable-flow system.

of work and the heat transfer crossing the system boundary are likewise dependent upon time. According to the law of conservation of mass, the rate of change of mass within the system is equal to the net rate of mass flow into and out of the system. Hence, in terms of mass flow rates,

$$\frac{dm_s}{d\tau} = \frac{dm_1}{d\tau} - \frac{dm_2}{d\tau}$$

For a finite period of time, this relation may be expressed as

$$\Delta m_s = \Delta m_1 - \Delta m_2$$

The first law may be written as follows:

$$J\frac{dU_s}{d\tau} = J\frac{dQ}{d\tau} + \frac{dW}{d\tau} + \left(Jb_1 + \frac{V_1^2}{2g_c} + \frac{g}{g_c}z_1\right)\frac{dm_1}{d\tau}$$
$$- \left(Jb_2 + \frac{V_2^2}{2g_c} + \frac{g}{g_c}z_2\right)\frac{dm_2}{d\tau}$$

Under non-steady-flow conditions the variables $b$, $V$, $z$ may change with time as well as flow rate, in which case the solution is very involved.

If steady-flow conditions prevail, then $\Delta U_s$ is equal to 0 and the integrands are independent of time, in which case the above equation reduces to the familiar steady-flow relation.

The **second law of thermodynamics** is a statement that conversion of heat to work is limited by the temperature at which conversion occurs. It may be shown that:

1. No cycle can be more efficient than a reversible cycle operating between given temperature limits.

2. The efficiency of all reversible cycles absorbing heat only at a single constant higher temperature $T_1$ and rejecting heat only at a single constant lower temperature $T_2$ must be the same.

3. For all such cycles, the efficiency is

$$e = \frac{W}{Q_1} = \frac{T_1 - T_2}{T_1}$$

This is usually called the **Carnot cycle efficiency**. By the first law $W = Q_1 + Q_2$,

$$(Q_1 + Q_2)/Q_1 = (T_1 - T_2)/T_1$$

By algebraic rearrangement,

$$(Q_1/T_1) + (Q_2/T_2) = 0$$

**Clapeyron Equation**

$$\frac{dp}{dT} = \frac{Q}{ATV_{12}}$$

This important relation is useful in calculations relating to constant-pressure evaporation of pure substances. In that case the equation may be written

$$v_{fg} = \frac{b_{fg}}{AT}\frac{1}{(dp/dT)}$$

**Entropy**   For reversible cyclical processes in which the temperature varies during heat absorption and rejection, i.e., for any *reversible cycle*, $\int \dfrac{dQ}{T} = 0$. Consequently, for any *reversible process*, $\int \dfrac{dQ}{T}$ is not a function of the particular *reversible path* followed. This integral is called the entropy change, or $\int_1^2 \dfrac{dQ_{\text{rev}}}{T} = S_2 - S_1 = S_{12}$. The entropy of a substance is dependent only on its state or condition. Mathematically, $dS$ is a complete or perfect differential and $S$ is a point function in contrast with $Q$ and $W$ which are path functions. For any reversible process, the change in entropy of the system and surroundings is zero, whereas for any irreversible process, the net entropy change is positive.

All actual processes are irreversible and therefore occur with a decrease in the amount of energy available for doing work, i.e., with an increase in unavailable energy. The increase in unavailable energy is the product of two factors, $T_0$

the lowest available temperature for heat discard (practically always the temperature of the atmosphere) and the net change in entropy. The increase in unavailable energy is $T_0 \, \Delta S_{net}$. Any process that occurs of itself (any spontaneous process) will proceed in such a direction as to result in a net increase in entropy. This is an important concept in the application of thermodynamics to chemical processes.

Three important potentials used in the Maxwell relations are:

1. The familiar potential, known as enthalpy,

$$h = u + pv$$

2. The free energy or the Helmholtz function is defined by the following relation:

$$f = u - Ts$$

3. The free enthalpy or the Gibbs function is defined by

$$g = h - Ts = f + pv = u + pv - Ts$$

The names used for these potentials have not gained universal acceptance. In particular, the name **free energy** is used for $g$ in many textbooks on chemical thermodynamics. One should be very cautious when referring to different books or technical papers and should verify by definition, rather than rely on the name of the potential.

**Availability** of a system or quantity of energy is defined as $g = h - T_0 s$. In this equation, all quantities except $T_0$ refer to the system irrespective of the state of the surroundings. $T_0$ is the lowest temperature available for heat discard. The preceding definition assumes the absence of velocity, potential, and similar effects. When these are not negligible, proper allowance must be made, e.g., $g = h - T_0 s + \dfrac{V^2}{2g_c} + \dfrac{g}{g_c} z$. By substitution of $Q = T_0(S_2 - S_1)$ in the appropriate first law expressions, it may be shown that for any steady-flow process, or for any constant-pressure nonflow process, decrease in availability is equal to the maximum possible (reversible) net work effect with sink for heat discard at $T_0$.

The availability function $g$ is of particular value in the thermodynamic analysis of changes occurring in the stages of a turbine and is of general utility in determining thermodynamic efficiencies, i.e., the ratio of actual work performed during a process to that which theoretically should have been performed.

The Gibbs function is of particular importance in processes where chemical changes occur. For reversible isothermal steady-flow processes, or for reversible constant-pressure isothermal nonflow processes, change in free energy is equal to net work.

**Helmholtz free energy,** $f = u - Ts$, is equal to the work during a constant-volume isothermal reversible nonflow process.

All of these functions $g$ and $f$ are point functions, and like $E$, $h$, and $s$ their differentials are complete or perfect.

**Perfect Differentials. Maxwell Relations**   If $z$ is some function of $x$ and $y$, in general

$$dz = \left(\frac{\partial z}{\partial x}\right)_y dx + \left(\frac{\partial z}{\partial y}\right)_x dy$$

Substituting $M$ for $\left(\dfrac{\partial z}{\partial x}\right)_y$ and $N$ for $\left(\dfrac{\partial z}{\partial y}\right)_x$

$$dz = M \, dx + N \, dy$$

But $\dfrac{\partial}{\partial y}\left(\dfrac{\partial z}{\partial x}\right) = \dfrac{\partial}{\partial x}\left(\dfrac{\partial z}{\partial y}\right)$ or $\dfrac{\partial M}{\partial y} = \dfrac{\partial N}{\partial x}$. This is Euler's criterion for integrability. A perfect differential has the characteristics of $dz$ stated above. Many important thermodynamic relations may be derived from the appropriate point function by the use of this relation. Table 24 gives some of them.

From the third column of the bottom half of the table, by equating various of the terms which are equal, one may obtain

$$\left(\frac{\partial u}{\partial s}\right)_v = \left(\frac{\partial h}{\partial s}\right)_p \qquad \left(\frac{\partial u}{\partial v}\right)_s = \left(\frac{\partial f}{\partial v}\right)_T$$

$$\left(\frac{\partial h}{\partial p}\right)_s = \left(\frac{\partial g}{\partial p}\right)_T \qquad \left(\frac{\partial g}{\partial T}\right)_p = \left(\frac{\partial f}{\partial T}\right)_v$$

By mathematical manipulation of equations previously given, the following important relations may be formulated:

$$c_v = \left(\frac{\partial q}{\partial T}\right)_v = T\left(\frac{\partial s}{\partial T}\right)_v = \left(\frac{\partial u}{\partial T}\right)_v$$

$$c_p = \left(\frac{\partial q}{\partial T}\right)_p = T\left(\frac{\partial s}{\partial T}\right)_p = \left(\frac{\partial h}{\partial T}\right)_p$$

$$c_p - c_v = AT\left(\frac{\partial v}{\partial T}\right)_p\left(\frac{\partial p}{\partial T}\right)_v$$

$$\left(\frac{\partial c_v}{\partial v}\right)_T = AT\left(\frac{\partial^2 p}{\partial T^2}\right)_v \qquad \left(\frac{\partial c_p}{\partial p}\right)_T = -AT\left(\frac{\partial^2 v}{\partial T^2}\right)_p$$

Relations involving $q$, $u$, $h$, and $s$:

$$dq = c_v \, dT + AT\left(\frac{\partial p}{\partial T}\right)_v dv = c_p \, dT - AT\left(\frac{\partial v}{\partial T}\right)_p dp$$

$$du = c_v \, dt + A\left[T\left(\frac{\partial p}{\partial T}\right)_v - p\right] dv$$

$$dh = c_p \, dT - A\left[T\left(\frac{\partial v}{\partial T}\right)_p - v\right] dp$$

$$ds = c_v \frac{dT}{T} + A\left(\frac{\partial p}{\partial T}\right)_v dv = c_p \frac{dT}{T} - A\left(\frac{\partial v}{\partial T}\right)_p dp$$

Since $q - AW = u$ and $h = u + Apv$, for reversible processes,

$$du = T \, ds - Ap \, dv \qquad \text{and} \qquad dh = du + Ap \, dv + Av \, dp$$

it follows that

$$Av = -T\left(\frac{\partial s}{\partial p}\right)_T + \left(\frac{\partial h}{\partial p}\right)_T$$

But from Table 24,

$$A\left(\frac{\partial v}{\partial T}\right)_p = -\left(\frac{\partial s}{\partial p}\right)_T$$

Therefore,

$$\left(\frac{\partial h}{\partial p}\right)_T = A\left[v - T\left(\frac{\partial v}{\partial T}\right)_p\right]$$

Similarly,

$$\left(\frac{\partial u}{\partial v}\right)_T = -A\left[p - T\left(\frac{\partial p}{\partial T}\right)_v\right]$$

These last two equations give in terms of $p$, $v$, and $T$ the necessary relations that must hold for any system, however complex. An equation in $p$, $v$, and $T$ for the properties of a substance is called an equation of state. These two equations

**Table 24. Maxwell Relations**

$A$ is a conversion factor, $1/778$ Btu/(ft·lb)

| Function | Differential | Maxwell relation |
|---|---|---|
| $\Delta u = q - W$ | $du = T\,ds - Ap\,dv$ | $\left(\dfrac{\partial T}{\partial v}\right)_s = -A\left(\dfrac{\partial p}{\partial s}\right)_v$ |
| $h = u + Apv$ | $dh = T\,ds + Av\,dp$ | $\left(\dfrac{\partial T}{\partial p}\right)_s = A\left(\dfrac{\partial v}{\partial s}\right)_p$ |
| $f = u - Ts$ | $df = -s\,dT - Ap\,dv$ | $\left(\dfrac{\partial s}{\partial v}\right)_T = A\left(\dfrac{\partial p}{\partial T}\right)_v$ |
| $g = h - Ts$ | $dg = -s\,dT + Av\,dp$ | $\left(\dfrac{\partial s}{\partial p}\right)_T = -A\left(\dfrac{\partial v}{\partial T}\right)_p$ |

By holding certain variables constant, a second set of relations is obtained:

| Differential | Independent variable held constant | Relation |
|---|---|---|
| $du = T\,ds - Ap\,dv$ | $s$ | $\left(\dfrac{\partial u}{\partial v}\right)_s = -Ap$ |
| | $v$ | $\left(\dfrac{\partial u}{\partial s}\right)_v = T$ |
| $dh = T\,ds + Av\,dp$ | $s$ | $\left(\dfrac{\partial h}{\partial p}\right)_s = Av$ |
| | $p$ | $\left(\dfrac{\partial h}{\partial s}\right)_p = T$ |
| $df = -s\,dT - Ap\,dv$ | $T$ | $\left(\dfrac{\partial f}{\partial v}\right)_T = -Ap$ |
| | $v$ | $\left(\dfrac{\partial f}{\partial T}\right)_v = -s$ |
| $dg = -s\,dT + Av\,dp$ | $T$ | $\left(\dfrac{\partial g}{\partial p}\right)_T = Av$ |
| | $p$ | $\left(\dfrac{\partial g}{\partial T}\right)_p = -s$ |

applicable to any substance or system are known as **thermodynamic equations of state.**

**Presentation of Thermal Properties**  Before the laws of thermodynamics can be applied and quantitative results obtained in the analysis of an engineering system, it is necessary to have available the properties of the system, some of which are temperature, pressure, internal energy, entropy, and enthalpy. In general, the property of a pure substance under equilibrium conditions may be expressed as a function of two other properties. This is based on the assumption that certain effects, such as gravitational and magnetic, are not important for the condition under investigation. The various properties of a pure substance under equilibrium conditions may be expressed by an equation of state, which in general form follows:

$$p = f(T,v)$$

In this relation the pressure is shown to be a function of both the temperature and the specific volume. Many special forms of equations of state are used in the analysis of engineering systems. Plots of the properties of various pure substances are very useful in studies dealing with thermodynamics. Two-dimensional plots, such as $p - v$, $p - h$, $p - T$, $T - s$, etc., show phase relations and are important in the analysis of cycles.

The constants in the equations of state are usually based on experimental data. The properties may be presented in many different ways, some of which are:

1. As equations of state, e.g., the perfect gas laws and the van der Waals equation.
2. As charts or graphs.
3. As tables.
4. As approximations which may be useful when more reliable data are not available.

**Ideal Gas Laws**  At low pressures and high enough temperatures, in the absence of chemical reaction, all gases approach a condition such that their $P$-$V$-$T$ properties may be expressed by the simple relation

$$pv = RT$$

If $v$ is expressed as volume per unit weight, the value of the constant $R$ will be different for different gases. If $v$ is expressed as the volume of one molecular weight of gas, then $R_u$ is the same for all gases in any chosen system of units. Hence $R = R_u/M$.

In general, for any amount of gas, the ideal gas equation

becomes

$$pV = NMRT$$

where $V$ is now the total gas volume, $N$ is the number of moles of gas in the volume $V$, $M$ is the molecular weight, and $R_u = MR$ the universal gas constant.

For all ideal gases, $R_u = MR$ in lb·ft is 1,546. One pound mol of any perfect gas occupies a volume of 359 ft³ at 32°F and 1 atm.

For many engineering purposes, use of the gas laws is permissible up to pressures of 100 to 200 psi if the absolute temperatures are at least twice the critical temperatures. Below the critical temperature, errors introduced by use of the gas laws may usually be neglected up to 15 psi pressure although errors of 5 percent are often met when dealing with saturated vapors.

The **van der Waals equation of state,** $p = [BT/(v - b)] - a/v^2$, is a modification of the ideal gas law which is sometimes useful at high pressures. The quantities $B$, $a$, and $b$ are constants.

**Approximate *P-V-T* Relations** For most gases, suitable P-V-T data are not available. An approximation useful under such circumstances is based on the observation of van der Waals that in terms of reduced properties most gases approxi-

mate a common **reduced equation of state.** The reduced quantities are the actual ones divided by the corresponding critical quantities, e.g., the reduced temperature $T_R = T_{actual}/T_{critical}$, the reduced volume $v_R = v_{actual}/v_{critical}$, the reduced pressure $p_R = p_{actual}/p_{critical}$. The gas laws may be made to apply to any nonperfect gas by the introduction of a correction factor $\mu$

$$pV = \mu NR_u T$$

When the gas laws apply, $\mu = 1$ and on a molal basis $\mu = pV/R_u T$. If on a plot of $\mu$ versus $p_R$ lines of constant $T_R$ are drawn, for different substances these are found to fall in narrow bands. Single $T_R$ lines may be drawn to represent approximately the various bands. This has been done in Fig. 4. To use the chart, only the critical pressure and temperature of the gas need be known.

EXAMPLE. Find the volume of 1 lb of steam at 5,500 psia and 1200°F (by steam tables, $v = 0.1516$ ft³/lb).

For water, critical temperature = 705.4°F; critical pressure = 3,206.4 psia; reduced temp = 1660/1165 = 1.43; reduced pressure 5,500/3206.4 = 1.72; $\mu$ (see Fig. 4) = 0.83. $v = 0.83\dfrac{(1546)(1660)}{(18)(5500)(144)} = 0.149$ ft³. Error = 100(0.152 − 0.149)/0.152 = 1.7 percent. If the gas laws had been used, the error would have been 17 percent.

**Table 25. Properties of Various Gases**
Approximate values[a] at 68°F and 14.7 psia; 20°C and 760 mm

| Gas | Symbol | Mol wt | Sp. gr, air = 1.00 | Sp vol, ft³/lb | Gas constant R, ft·lb per lb, deg F | Sp heat, $c_p$, Btu per lb,deg F | $c_p/c_v = k$ | Boiling point at atm press, deg F | Critical temp, deg F | Critical press., psia |
|---|---|---|---|---|---|---|---|---|---|---|
| Acetylene (ethyne) | $C_2H_2$ | 26.0 | 0.907 | 14.75 | 59.4 | 0.350 | 1.30 | −118 | 96 | 910 |
| Air | . . . . | 29.0 | 1.000 | 13.26 | 53.3 | 0.241 | 1.40 | −318 | −221 | 546 |
| Ammonia | $NH_3$ | 17.0 | 0.596 | 22.51 | 91.0 | 0.523 | 1.32 | −28 | 271 | 1,657 |
| Argon | A | 39.9 | 1.379 | 9.51 | 38.7 | 0.124 | 1.67 | −302 | −187 | 705 |
| N-Butane | $C_4H_{10}$ | 58.1 | 2.067 | 6.61 | 26.5 | 0.395 | 1.11 | 31 | 307 | 550 |
| Butene-1 (Butylene α) | $C_4H_8$ | 56.1 | 1.935 | 6.85 | 27.5 | 0.360 | 1.11 | 21 | 291 | 621 |
| Carbon dioxide | $CO_2$ | 44.0 | 1.529 | 8.72 | 35.1 | 0.205 | 1.30 | −109 | 88 | 1,072 |
| Carbon monoxide | CO | 28.0 | 0.967 | 13.70 | 55.2 | 0.243 | 1.40 | −313 | −218 | 514 |
| Chlorine | $Cl_2$ | 70.9 | 2.486 | 5.41 | 21.8 | 0.115 | 1.33 | −30 | 291 | 118 |
| Ethane | $C_2H_6$ | 30.0 | 1.049 | 12.77 | 51.5 | 0.386 | 1.22 | −127 | 90 | 717 |
| Ethyl chloride | $C_2H_5Cl$ | 64.5 | 2.365 | 5.96 | 24.0 | 0.275 | 1.13 | 54 | 370 | 764 |
| Ethylene | $C_2H_4$ | 28.0 | 0.975 | 13.69 | 55.1 | 0.400 | 1.22 | −155 | 50 | 732 |
| Freon (F-12) | $CCl_2F_2$ | 120.9 | 4.520 | 3.18 | 12.6 | . . . | 1.13 | −21 | 233 | 597 |
| Helium | He | 4.0 | 0.1381 | 95.9 | 386.3 | 1.250 | 1.66 | −452 | −450 | 33 |
| Hydrogen | $H_2$ | 2.0 | 0.0695 | 191.5 | 766.8 | 3.420 | 1.41 | −423 | −400 | 191 |
| Hydrogen chloride | HCl | 36.5 | 1.268 | 10.53 | 42.4 | 0.191 | 1.41 | −121 | 124 | 1,198 |
| Hydrogen sulphide | $H_2S$ | 34.1 | 1.190 | 11.26 | 45.2 | 0.243 | 1.30 | −75 | 212 | 1,306 |
| Isobutane (2-Methyl propane) | $C_4H_{10}$ | 58.1 | 2.018 | 6.61 | 26.5 | 0.392 | 1.11 | 10 | 273 | 543 |
| Methane | $CH_4$ | 16.0 | 0.554 | 24.0 | 96.4 | 0.593 | 1.32 | −258 | −116 | 672 |
| Methyl chloride | $CH_3Cl$ | 50.5 | 1.785 | 7.60 | 30.6 | 0.240 | 1.20 | −11 | 289 | 966 |
| Natural gas[b] | . . . | 19.5 | 0.667 | 19.75 | 79.1 | 0.560 | 1.27 | −240 | −80 | 670 |
| Neon | Ne | 20.2 | 0.696 | 19.06 | 76.4 | 0.248 | 1.64 | −410 | −380 | 389 |
| Nitric oxide | NO | 30.0 | 1.037 | 12.80 | 51.5 | 0.231 | 1.40 | −240 | −137 | 954 |
| Nitrogen | $N_2$ | 28.0 | 0.967 | 13.66 | 55.2 | 0.247 | 1.41 | −320 | −232 | 492 |
| Nitrous oxide | $N_2O$ | 44.0 | 1.530 | 8.72 | 35.1 | 0.221 | 1.31 | −129 | 98 | 1,053 |
| Oxygen | $O_2$ | 32.0 | 1.105 | 12.00 | 48.3 | 0.217 | 1.40 | −297 | −182 | 730 |
| Pentane | $C_5H_{12}$ | 72.1 | 2.471 | 5.33 | 21.3 | 0.040 | 1.06 | 97 | 387 | 485 |
| Propane | $C_3H_8$ | 44.1 | 1.562 | 8.72 | 35.0 | 0.393 | 1.15 | −48 | 204 | 662 |
| Propene (propylene) | $C_3H_6$ | 42.1 | 1.451 | 9.14 | 36.8 | 0.358 | 1.14 | −52 | 198 | 661 |
| Sulphur dioxide | $SO_2$ | 64.1 | 2.264 | 5.99 | 24.0 | 0.154 | 1.26 | 14 | 315 | 1,141 |

[a]Approximate values listed are adapted from the following sources: "CAGI Handbook." Rose and Rose, "Condensed Chemical Dictionary," Reinhold. "Handbook of Chemistry and Physics," Chemical Rubber Publishing Co. Selected Values of Physical and Thermodynamic Properties of Hydrocarbons and Related Compounds, *API Research Project* 44, 1953. Selected Values of Chemical Thermodynamic Properties, *NBS Circ.* 500, 1952.

[b]Representative value; exact characteristics require knowledge of exact constituents.

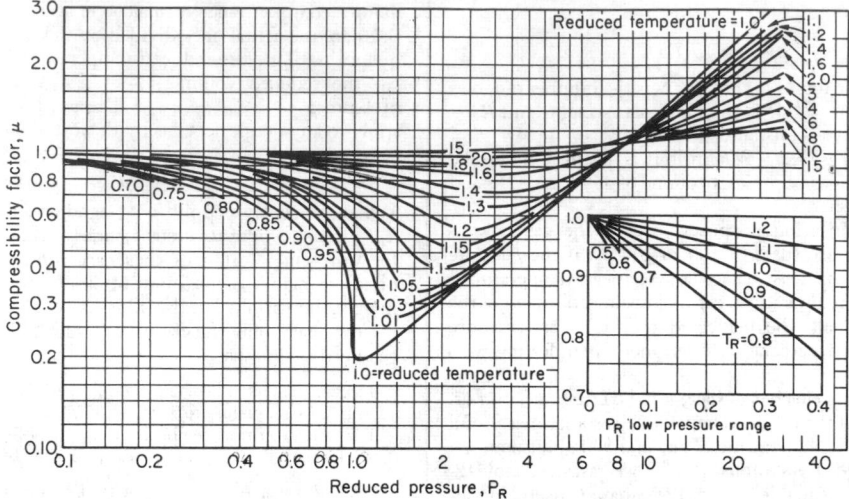

**Fig. 4**   Compressibility factors of gases and vapors. *(From Hougan and Watson, "Chemical Process Principles," Wiley.)*

No entirely satisfactory method for calculation for gaseous mixtures has been developed, but the use of average critical constants as proposed by Kay (*Ind. Eng., Chem.*, **28**, 1936, p. 1014) is easy and gives satisfactory results under conditions considerably removed from the critical. He assumes the gaseous mixture can be treated as if it were a single pure gas with a pseudocritical pressure and temperature estimated by a method of molar averaging.

$$(T_c)_{\text{mixture}} = (T_c)_a y_a + (T_c)_b y_b + (T_c)_c y_c + \ldots$$
$$(p_c)_{\text{mixture}} = (p_c)_a y_a + (p_c)_b y_b + (p_c)_c y_c + \ldots$$

where $(T_c)_a$ is the critical temperature of pure $a$, etc.; $(p_c)_a$ is the critical pressure of pure $a$, etc.; and $y_a$ is the mol fraction of $a$, etc. For a gaseous mixture made up of gases $a$, $b$, $c$, etc., the pseudocritical constants having been determined, the gaseous mixture is handled on the $\mu$ charts as if it were a single pure gas.

**Ideal Gas Mixtures**   Many of the fluids involved in engineering systems are physical mixtures of the permanent gases or one or more of these with superheated or saturated vapors. For example, normal atmospheric air is a mixture of oxygen and nitrogen with traces of other gases, plus superheated or saturated water vapor, or at times saturated vapor and liquid. If the properties of each constituent of a mixture would have to be considered individually during an analysis of a system, the procedures would be very complex. Experience has demonstrated that a mixture of gases may be regarded as an equivalent gas, the properties of which depend upon the kind and proportion of each of the constituents. The general relations applicable to a mixture of perfect gases will be presented. Let $V$ denote the total volume of the mixture, $m_1$, $m_2$, $m_3$, ... the masses of the constituent gases, $R_1$, $R_2$, $R_3$, ... the corresponding gas constants, and $R_m$ the constant for the mixture. The **partial pressures** of the constituents, i.e., the pressures that the constituents would have if occupying the total volume $V$, are $p_1 = m_1 R_1 T/V$, $p_2 = m_2 R_2 T/V$, etc.

According to Dalton's law, the **total pressure** $p$ **of the mixture** is the sum of the partial pressures; i.e., $p = p_1 + p_2 + p_3$

$+ \cdots$ Let $m = m_1 + m_2 + m_3 + \cdots$ denote the total mass of the mixture; then $pV = mR_mT$ and $R_m = \Sigma(m_i R_i)/m$. Also $p_1/p = m_1 R_1/mR_m$, $p_2/p = m_2 R_2/mR_m$, etc.

Let $V_1$, $V_2$, $V_3$ ... denote the volumes that would be occupied by the constituents at pressure $p$ and temperature $T$ (these are given by the volume composition of the gas). Then $V = V_1 + V_2 + V_3 + \cdots$ and the apparent molecular weight $m_m$ of the mixture is $m_m = \Sigma(m_i V_i)/V$. Then $R_m = 1546/m_m$. The subscript $i$ denotes an individual constituent.

**Volume of 1 lb** at 32°F and atm pressure $= 359/m_m$.

**Mass of 1 ft³** at 32°F and atm pressure $= 0.002788m_m$.

The specific **heats of the mixture** are, respectively,

$$c_p = \Sigma(m_i c_{pi})/m \qquad c_v = \Sigma(m_i c_{vi})/m$$

**Internal Energy, Enthalpy, and Entropy of an Ideal Gas.**   If an ideal gas with constant specific heats changes from an initial state $p_1$, $V_1$, $T_1$ to a final state $p_2$, $V_2$, $T_2$, the following equations hold:

$$u_2 - u_1 = mc_v(T_2 - T_1) = A(p_2 v_2 - p_1 v_1)/(k - 1)$$
$$h_2 - h_1 = mc_p(T_2 - T_1) = Ak(p_2 v_2 - p_1 v_1)/(k - 1)$$
$$s_2 - s_1 = m\left(c_v \ln \frac{T_2}{T_1} + AR \ln \frac{v_2}{v_1}\right)$$
$$= m\left(c_p \ln \frac{T_2}{T_1} - AR \ln \frac{p_2}{p_1}\right) = m\left(c_p \ln \frac{v_2}{v_1} + c_v \ln \frac{p_2}{p_1}\right)$$

In general, the energy per unit mass is $u = c_v T + u_0$, the enthalpy is $h = cpT + h_0$, and the entropy is $s = c_v \ln T + AR$ $\ln v + s_0 = c_p \ln T - AR \ln p + s'_0 = c_p \ln v + c_p \ln p = s''_0$.

The two fundamental equations for ideal gases are

$$dq = c_v \, dT + Ap \, dv \qquad dq = c_p \, dT - Av \, dp$$

**Special Changes of State for Ideal Gases**
(Specific heats assumed constant)

In the following formulas, the subscripts 1 and 2 refer to the initial and final states, respectively.

**1. Constant Volume:** $p_2/p_1 = T_2/T_1$.

$$Q_{12} = U_2 - U_1 = mc_v(t_2 - t_1) = AV(p_2 - p_1)/(k - 1)$$
$$W_{12} = 0 \qquad s_2 - s_1 = mc_v \ln (T_2/T_1)$$

**2. Constant Pressure:** $V_2/V_1 = T_2/T_1$.

$$W_{12} = p(V_2 - V_1) = mR(t_2 - t_1)$$
$$Q_{12} = mc_p(t_2 - t_1) = AkW_{12}/(k - 1)$$
$$s_2 - s_1 = mc_p \ln (T_2/T_1)$$

**3. Isothermal (Constant Temperature):** $p_2/p_1 = V_1/V_2$.

$$U_2 - U_1 = 0 \qquad W_{12} = mRT \ln (V_2/V_1) = p_1V_1 \ln (V_2/V_1)$$
$$Q_{12} = AW_{12} \qquad s_2 - s_1 = Q_{12}/T = mAR \ln (V_2/V_1)$$

**4. Reversible Adiabatic. Isentropic:** $p_1V_1^k = p_2V_2^k$.

$$T_2/T_1 = (V_1/V_2)^{k-1} = (p_2/p_1)^{(k-1)/k}$$
$$AW_{12} = U_1 - U_2 = mc_v(t_1 - t_2)$$
$$Q_{12} = 0 \qquad s_2 - s_1 = 0$$
$$W_{12} = (p_1V_1 - p_2V_2)/(k - 1)$$
$$= p_1V_1[1 - (p_2/p_1)^{(k-1)/k}]/(k - 1)$$

**5. Polytropic** This name is given to the change of state which is represented by the equation $pV^n = $ const. A polytropic curve usually represents actual expansion and compression curves in motors and air compressors for pressures up to a few hundred pounds. By giving $n$ different values and assuming specific heats constant, the preceding changes may be made special cases of the polytropic change, thus,

For $n = 1$,     $pv = $ const     isothermal
$n = k$,     $pv^k = $ const     isentropic
$n = 0$,     $p = $ const     constant pressure
$n = \infty$,     $v = $ const     constant volume

For a polytropic change of an ideal gas (for which $c_v$ is constant), the specific heat is given by the relation $c_n = c_v(n - k)/(n - 1)$; hence for $1 < n < k$, $c_n$ is negative. This is approximately the case in air compression up to a few hundred pounds pressure. The following are the principal formulas:

$$p_1V_1^n = p_2V_2^n$$
$$T_2/T_1 = (V_1/V_2)^{n-1} = (p_2/p_1)^{(n-1)/n}$$
$$W_{12} = (p_1V_1 - p_2V_2)/(n - 1)$$
$$= p_1V_1[1 - (p_2/p_1)^{(n-1)/n}]/(n - 1)$$
$$Q_{12} = mc_n(t_2 - t_1)$$
$$AW_{12}:U_2 - U_1:Q_{12} = k - 1:1 - n:k - n$$

The quantity $(p_1/p_2)^{(k-1)/k} - 1$ occurs frequently in calculations for perfect gases.

**Determination of Exponent $n$** Lay off successive values of $p$ and $V$, measured at chosen points on the curve under investigation, on logarithmic cross-section paper; or, lay off values of $\log p$ and $\log V$ on ordinary cross-section paper. If $n$ is a constant, the points will lie in a straight line, and the slope of the line gives the value of $n$.

If two representative points ($p_1$, $V_1$ and $p_2$, $V_2$) be chosen, then

$$n = (\log p_1 - \log p_2)/(\log V_2 - \log V_1)$$

Several pairs of points should be used to test the constancy of $n$.

**Changes of State with Variable Specific Heat** In case of a considerable range of temperature, the assumption of constant specific heat is not permissible, and the equations referring to changes of state must be suitably modified. Experiments on the specific heat of various gases show that the specific heat may sometimes be taken as a linear function of the temperature: thus, $c_v = a + bT$; $c_p = a' + b'T$. In that case, the following expressions apply for the change of internal energy and entropy, respectively:

$$U_2 - U_1 = m[a(T_2 - T_1) + 0.5b(T_2^2 - T_1^2)]$$
$$S_2 - S_1 = m[a \ln (T_2/T_1) + b(T_2 - T_1) + AR \ln (V_2/V_1)]$$

and for an isentropic change,

$$W_{12} = J(U_1 - U_2)$$
$$AR \ln (V_1/V_2) = a \ln (T_2/T_1) + b(T_2 - T_1)$$

For more complex specific-heat relations (see Table 19), integration of the equations between specified limits must be undertaken.

**Graphical Representation** The change of state of a substance may be shown graphically by taking any two of the six variables $p$, $V$, $T$, $S$, $U$, $H$ as independent coordinates and drawing a curve to represent the successive values of these two variables as the change proceeds. While any pair may be chosen, there are three systems of graphical representation that are specially useful.

**1. $p$ and $V$** The curve (Fig. 5) represents the simultaneous values of $p$ and $V$ during the change (reversible) from state 1 to state 2. The area between the curve and the axis $OV$ is given by the integral $\int_{V_1}^{V_2} p \, dV$ and therefore represents the external work $W_{12}$ done by the gas during the change. The area included by a closed cycle represents the work of the cycle (as in the indicator diagram of the steam engine).

**2. $T$ and $S$** (Fig. 6) The absolute temperature $T$ is taken as the ordinate, the entropy $S$ as the abscissa. The area between the curve of change of state and the $S$ axis is given by the integral $\int_{S_1}^{S_2} T \, dS$, and it therefore represents the heat $Q_{12}$ absorbed by the substance from external sources provided there are no irreversible effects. On the $T$-$S$ diagram, an isothermal is a straight line, as $AB$, parallel to the $S$ axis; a *reversible* adiabatic is a straight line, as $CD$, parallel to the $T$ axis.

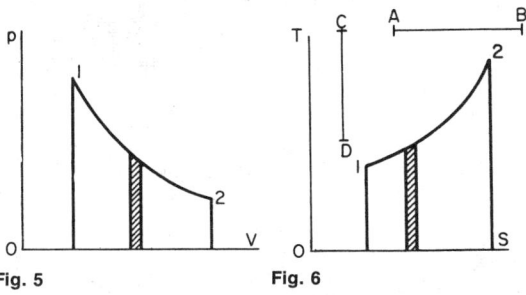

**Fig. 5**          **Fig. 6**

In the case of internal generation of heat through friction, as in steam turbines, the increase of entropy is given by $\int_{T_1}^{T_2} \dfrac{dQ'}{T}$ and the area under the curve represents the heat $Q'$ thus generated. In this case, an adiabatic is *not* a straight line parallel to the $T$ axis.

**3. H and S** In the system of representation devised by Dr. Mollier, the enthalpy $H$ is taken as the ordinate and the entropy $S$ as the abscissa. If on this diagram (Fig. 7) a line of constant pressure, as 12, be drawn, the heat absorbed during the change at constant pressure is given by $Q_{12} = H_2 - H_1$, and this is represented by the line segment 23. The **Mollier diagram** is specially useful in problems that involve the flow of fluids, throttling, and the action of steam in turbines. A Mollier diagram for steam is presented in Fig. 22.

### Ideal Cycles with Perfect Gases

Gases are used as heat mediums in several important types of machines. In air compressors, air engines, and air refrigerating machines, atmospheric air is the medium. In the internal-combustion engine, the medium is a mixture of products of combustion. Engines using gases are operated in certain well-defined cycles, which are described below. In the analyses given, ideal conditions that cannot be attained by actual motors are assumed. However, conclusions derived from such analyses are usually approximately valid for the modified actual cycle.

In the following, the subscripts 1, 2, 3, etc., refer to corresponding points shown in the figures. The work of the cycle is denoted by $(W)$ and the net heat absorbed by $(Q)$.

**Carnot Cycle** The Carnot cycle (Fig. 8) is of historic interest. It consists of two isothermals and two isentropics. The heat absorbed along the upper isothermal 12 is $Q_{12} = mART \ln (V_2/V_1)$, and the heat transformed into work, represented by the cycle area, is $A(W) = Q_{12}[1 - (T_0/T)]$.

Hence,

$$(W) = mR(T - T_0) \ln (V_2/V_1)$$

If the cycle is traversed in the reverse sense, $Q_{43} = mART_0 \ln (V_3/V_4)$ is the heat absorbed from the cold body (brine), and the ratio $Q_{43}{:}A(W) = T_0{:}(T - T_0)$ is the **coefficient of performance** of the refrigerating machine.

**Otto and Diesel Cycles** The ideal cycles usually employed for internal-combustion engines may be classified in two groups: (1) explosive—Otto, (2) nonexplosive—Diesel, Joule.

**Otto Cycle** (Fig. 9 for pressure-volume plane, Fig. 10 for temperature-entropy plane) Isentropic compression 12 is followed by ignition and rapid heating at constant volume 23. This is followed by isentropic expansion, 34. Assuming constant specific heats the following relations hold:

$$\frac{T_2}{T_1} = \frac{T_2}{T_4} = \left(\frac{p_2}{p_1}\right)^{k-1/k} = \left(\frac{p_3}{p_4}\right)^{k-1/k} = \left(\frac{V_1}{V_2}\right)^{k-1}$$
$$Q_{23} = mc_v(T_3 - T_2)$$
$$(W) = JQ_{23}[1 - (T_1/T_2)] = Jmc_v(T_3 - T_4 - T_2 + T_1)$$
$$\text{Efficiency} = 1 - \frac{T_1}{T_2} = 1 - \left(\frac{V_2}{V_1}\right)^{k-1} = 1 - \left(\frac{p_1}{p_2}\right)^{k-1/k}$$

If the compression and expansion curves are polytropics with the same value of $n$, replace $k$ by $n$ in the first relation above. In this case,

$$(W) = [(p_3V_3 - p_4V_4) - (p_2V_2 - p_1V_1)]/(n - 1)$$
$$= mR(T_3 - T_4 - T_2 + T_1)/(n - 1)$$

The **mean effective pressure** of the diagram is given by

$$p_m = ap_1[(p_3/p_2) - 1]$$

where $a$ has the values given in the following table.

| | $p_2/p_1 = 3$ | 4 | 5 | 6 | 8 | 10 | 12 | 14 | 16 |
|---|---|---|---|---|---|---|---|---|---|
| $(n = 1.4)$ | $a = 1.70$ | 1.94 | 2.13 | 2.31 | 2.62 | 2.88 | 3.10 | 3.31 | 3.50 |
| $(n = 1.3)$ | $a = 1.69$ | 1.92 | 2.11 | 2.28 | 2.57 | 2.81 | 3.03 | 3.22 | 3.39 |
| $(n = 1.2)$ | $a = 1.68$ | 1.90 | 2.08 | 2.25 | 2.51 | 2.74 | 2.94 | 3.12 | 3.27 |

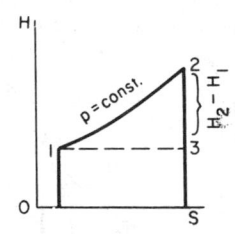

**Fig. 7**

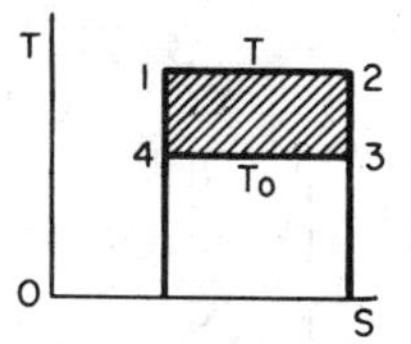

**Fig. 8** Carnot cycle.

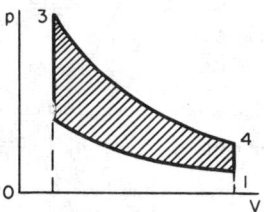

**Fig. 9** Otto cycle.

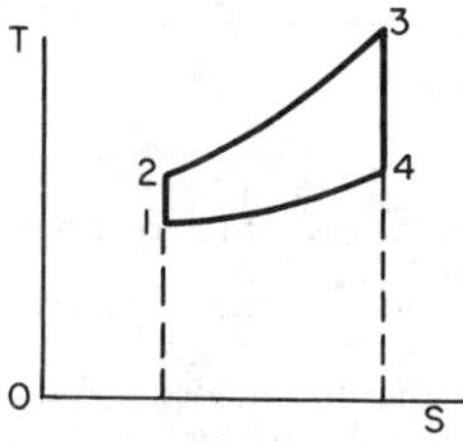

**Fig. 10** Otto cycle.

**Diesel Cycle** In the Diesel oil engine, air is compressed to a high pressure. Fuel is then injected into the air, which is at a temperature above the ignition point, and it burns at nearly constant pressure (23, in Fig. 11). Isentropic expansion of the products of combustion is followed by exhaust and suction of fresh air, as in the Otto cycle.

The work obtained is

$$(W) = Jm[c_p(T_3 - T_2) - c_v(T_4 - T_1)]$$

and the efficiency of the ideal cycle is

$$1 - [(T_4 - T_1)/k(T_3 - T_2)]$$

The **Joule cycle,** also called the **Brayton cycle** (Fig. 12), consists of two isentropics and two constant-pressure lines. The following relations hold:

$$V_3/V_2 = V_4/V_1 = T_3/T_2 = T_4/T_1$$
$$\frac{T_2}{T_1} = \frac{T_3}{T_4} = \left(\frac{V_1}{V_2}\right)^{k-1} = \left(\frac{V_4}{V_3}\right)^{k-1} = \left(\frac{p_2}{p_1}\right)^{k-1/k}$$
$$(W) = Jmc_p(T_3 - T_2 - T_4 + T_1)$$
$$\text{Efficiency} = (W)/JQ_{23} = 1 - (T_1/T_2)$$

The Joule cycle has assumed renewed importance as a basis for analysis of gas turbine operation.

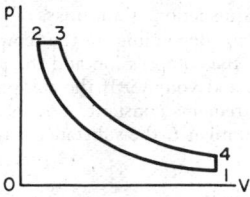

**Fig. 11**  Diesel cycle.

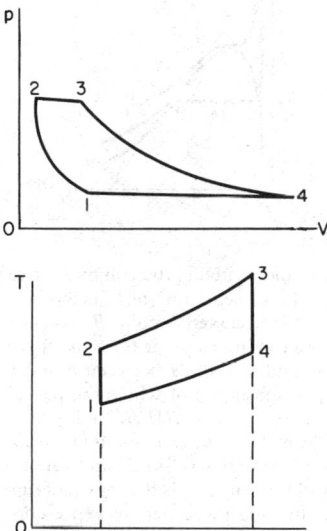

**Fig. 12**  Joule or Brayton cycle.

**Stirling Cycle**  The Stirling engine may be visualized as a cylinder with a piston at each end. Between the pistons is a regenerator. The cylinder is assumed to be insulated except for a contact with a hot reservoir at one end and a contact with a cold reservoir at the other end.

Starting with state 1, Fig. 13, heat from the hot reservoir is added to the gas at $T_H$ (or $T_H - dT$). During the reversible isothermal process, the left piston moves outward, doing work

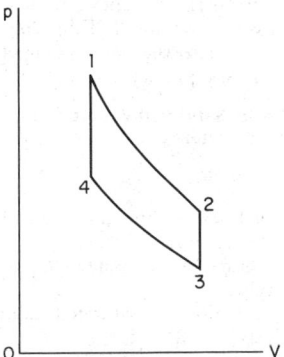

**Fig. 13**  Stirling cycle.

as the system volume increases and the pressure falls. Both pistons are then moved to the right at the same rate to keep the system volume constant (process 2–3). No heat transfer occurs with either reservoir. As the gas passes through the regenerator, heat is transferred from the gas to the regenerator, causing the gas temperature to fall to $T_L$ by the time the gas leaves the right end of the regenerator. For this heat-transfer process to be reversible, the temperature of the regenerator at each point must equal the gas temperature at that point. Hence there is a temperature gradient through the regenerator from $T_H$ at the left end to $T_L$ at the right end. No work is accomplished during this process. During the path 3–4, heat is removed from the gas at $T_L$ (or $T_L + dT$) to the reservoir at $T_L$. To hold the gas temperature constant, the right piston is moved inward—doing work on the gas, with a resulting increase in pressure. During process 4–1, both pistons are moved to the left at the same rate to keep the system volume constant. The pistons are closer together during this process than they were during process 2–3, since $V_4 = V_1 < V_2 = V_3$. No heat is transferred to either reservoir. As the gas passes back through the regenerator, the energy stored in the regenerator during 2–3 is returned to the gas. The gas emerges from the left end of the regenerator at the temperature $T_H$. No work is performed during this process since the volume remains constant. Thus the cycle is completed and is externally reversible. The system exchanges a net amount of heat with only the two energy reservoirs $T_H$ and $T_L$. Two types of Stirling engines are shown in Fig. 14. Extensive research-and-development effort has been devoted to the Stirling engines for future use as prime movers in space power systems operating on solar energy. (See also Sec. 9.)

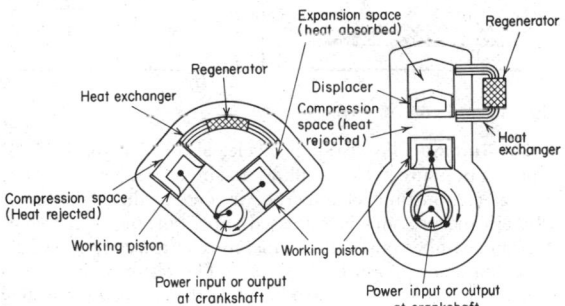

**Fig. 14**  Two main types of Stirling engine: (1) left, double-cylinder, two-piston; (2) right, single-cylinder, piston-plus-displacer. Each has two variable-volume working spaces filled with the working fluid—one for expansion and one for compression of the gas. Spaces are at different temperatures—the extreme temperatures of the working cycle—and are connected by a duct, which holds the regenerator and heat exchangers. (*International Science & Technology, May, 1962.*)

**Air Compression**  It is assumed that the compressor works under ideal reversible conditions without clearance and without friction losses and that the changes are over ranges where the gas laws are applicable. Where the gas laws cannot be used, analysis in terms of $\mu$ charts is convenient. If the compression from $p_1$ to $p_2$ (Fig. 15) follows the law $pV^n = $ const, the work represented by the indicator diagram is

$$-W = n(p_2 V_2 - p_1 V_1)/(n - 1)$$
$$= n p_1 V_1 [(p_2/p_1)^{(n-1)/n} - 1]/(n - 1)$$

The temperature at the end of compression is given by $T_2/T_1 = (p_2/p_1)^{(n-1)/n}$ The work $W$ is smaller the smaller the value of $n$, and the purpose of the water jacket is to reduce $n$ from the isentropic value 1.4. Under usual working conditions, $n$ is about 1.3.

When the pressure $p_2$ is high, it is advantageous to divide the process into two or more stages and cool the air between the cylinders. The saving effected is best shown on the $T$-$S$ plane (Fig. 16). With single-stage compression, 12 represents the compression from $p_1$ to $p_2$, and if the constant-pressure line 23 is drawn cutting the isothermal through point 1 in point 3, the area 1'1233' represents the work $W$. When two stages are used, 14 represents the compression from $p_1$ to an intermediate pressure $p'$, 45 cooling at constant pressure in the intercooler between the cylinders, and 56 the compression in the second stage. The area under 14563 represents the work of the two stages and the area 2456 the saving effected by compounding. This saving is a maximum when $T_4 = T_6$, and this is the case when the intermediate pressure $p'$ is given by

$$p' = \sqrt{p_1 p_2} \quad \text{(see Sec. 14).}$$

The total work in two-stage compression is

$$-n p_1 V_1 [(p'/p_1)^{(n-1)/n} + (p_2/p')^{(n-1)/n} - 2]/(n-1)$$

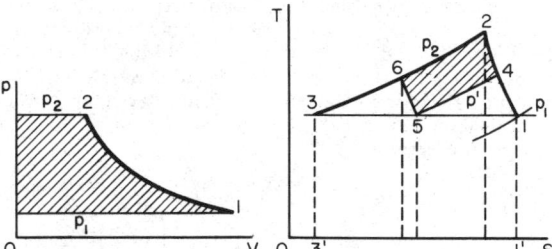

**Figs. 15 and 16**  Air-compressor cycle.

#### Gas Turbine
The Brayton cycle, also called the Joule or constant-pressure cycle, employs an air engine, a compressor, and a combustion chamber. Air enters the compressor wherein the pressure is increased. Fuel burning in the combustion chamber raises the temperature of the compressed air under constant-pressure conditions. The resulting high-temperature gases are then introduced to the engine where they expand and perform work. The excess work of the engine over that required to compress the air is available for operating other devices, such as a generator.

Basically, the simple gas-turbine cycle is the same as the Brayton cycle, except that the air compressor and engine are replaced by an axial flow compressor and gas turbine. Air is compressed in the compressor, after which it enters a combustion chamber where the temperature is increased while the pressure remains constant. The resulting high-temperature air then enters the turbine, thereby performing work. For more detailed information regarding the actual gas-turbine cycles the reader is referred to Sec. 9.

#### Vapors

**General Characteristics of Vapors**  Let a gas be compressed at constant temperature; then, provided this temperature does not exceed a certain critical value, the gas begins to liquefy at a definite pressure, which depends upon the temperature. At the beginning of liquefaction, a unit mass of gas will also have a definite volume $v_g$, depending on the temperature. In Fig. 17, $AB$ represents the compression and the point $B$ gives the **saturation** pressure and volume. If the compression is continued, the pressure remains constant with the temperature, as indicated by $BC$, until at $C$ the substance is in the liquid state with the volume $v_f$.

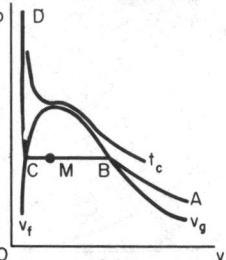

**Fig. 17**

The curves $v_f$ and $v_g$ giving the volumes for various temperatures at the end and beginning of liquefaction, respectively, may be called the **limit curves**. A point $B$ on curve $v_g$ represents the state of **saturated vapor**; a point $C$ on the curve $v_f$ represents the liquid state; and a point $M$ between $B$ and $C$ represents a mixture of vapor and liquid of which the part $x = MC/BC$ is vapor and the part $1 - x = BM/BC$ is liquid. The ratio $x$ is called the **quality of the mixture**. The region between the curves $v_f$ and $v_g$ is thus the region of liquid and vapor mixtures. The region to the right of curve $v_g$ is the region of **superheated vapor**. The curve $v_g$ dividing these regions represents the so-called **saturated vapor**.

For saturated vapor or a mixture of vapor and liquid, the pressure is a function of the temperature only, and the volume of the mixture depends upon the temperature and quality $x$. That is, $p = f(t)$, $v = F(t, x)$.

For the vapor in the superheated state, the volume depends on pressure and temperature [$v = F_1(p, t)$], and these may be varied independently.

**Critical State**  If the temperature of the gas lies above a definite temperature $t_c$ called the **critical temperature,** the gas cannot be liquefied by compression alone. The saturation pressure corresponding to $t_c$ is the **critical pressure** and is denoted by $p_c$. At the critical state, the limit curves $v_f$ and $v_g$ merge; hence for temperatures above $t_c$, it is impossible to have a mixture of vapor and liquid. Table 26 gives the critical data for various gases; also the boiling temperature $t_b$ corresponding to atmospheric pressure.

#### Thermal Properties of Saturated Vapors and of Vapor and Liquid Mixtures

**Notation**

$v_f$, $v_g$ = volume of 1 lb of saturated liquid and vapor, respectively, ft³

$c_f$, $c_g$ = specific heat of saturated liquid and vapor, respectively

$h_f$, $h_g$ = specific enthalpy of saturated liquid and vapor, respectively

$u_f$, $u_g$ = specific internal energy of saturated liquid and vapor, respectively

**Table 26. Critical Data for Various Gases**

| Substance | $t_b$ deg F | $t_c$ deg F | $p_c$ atm | $v_c$ cu ft per lb | Substance | $t_b$ deg F | $t_c$ deg F | $p_c$ atm | $v_c$ cu ft per lb |
|---|---|---|---|---|---|---|---|---|---|
| Air | −317.6 | −220.3 | 37.2 | 0.0457 | $C_2H_6$....... | −127.5 | 90.0 | 48.2 | 0.079 |
| Helium | −452.0 | −450.2 | 2.26 | 0.231 | $C_3H_8$....... | − 48.1 | 206.26 | 42.01 | 0.071 |
| Hydrogen | −422.9 | −399.8 | 12.8 | 0.516 | $C_4H_{10}$...... | 31.5 | 307.4 | 37.48 | 0.071 |
| Argon | −302.3 | −187.7 | 48.0 | 0.03 | $C_5H_{12}$...... | 97.0 | 387.0 | 33 | 0.069 |
| Nitrogen | −320.4 | −232.8 | 33.5 | 0.053 | $C_6H_{14}$...... | 156.1 | 454.6 | 29.5 | 0.0685 |
| Oxygen | −297.2 | −181.8 | 49.7 | 0.037 | $C_7H_{16}$...... | 209.2 | 517.1 | 27.65 | 0.0685 |
| Bromine | 137.8 | 575.6 | ........ | ........ | $C_8H_{18}$...... | 259.2 | 565.7 | 24.8 | 0.0685 |
| Chlorine | − 30.3 | 291.3 | 76.1 | 0.028 | $C_2H_2$...... | −118.5 | 96.3 | 62 | 0.0693 |
| HCl | −121 | 124.5 | 81.6 | 0.038 | $C_2H_4$....... | −155.0 | 49.3 | 50.9 | 0.073 |
| $H_2S$ | − 74.9 | 212.7 | 88.9 | ........ | $C_6H_6$...... | 176 | 551.4 | 47.7 | 0.0526 |
| NO | −239.8 | −136.7 | 65.0 | 0.031 | $C_7H_8$...... | 231.3 | 609.1 | 41.6 | 0.055 |
| $N_2O$ | −129.1 | 97.7 | 71.7 | 0.036 | $CH_4O$...... | 147.3 | 464.0 | 78.7 | 0.059 |
| $NH_3$ | − 28 | 270.3 | 111.5 | 0.068 | $C_2H_6O$...... | 173.0 | 469.6 | 63.1 | 0.058 |
| $H_2O$ | 212 | 705.45 | 218.53 | 0.0503 | $C_3H_6O$...... | 133 | 455 | 47 | 0.060 |
| $SO_2$ | 14 | 315.0 | 77.7 | 0.031 | $C_4H_{10}O$..... | 101.8 | 380.8 | 35.5 | 0.061 |
| CO | −313.6 | −220.33 | 34.53 | 0.053 | $CHCl_3$...... | 142.2 | 505.4 | ........ | 0.031 |
| $CO_2$ | −109.3 | 88.0 | 73.0 | 0.035 | $CH_3Cl$..... | − 10.25 | 289.6 | 65.8 | 0.043 |
| $CS_2$ | 115.3 | 523.4 | 76.0 | ........ | $C_2H_5Cl$..... | 54.9 | 369.0 | 52 | 0.0485 |
| $CH_4$ | −258.5 | −116.5 | 45.8 | 0.099 | | | | | |

SOURCE: Condensed from "International Critical Tables."

$s_f, s_g$ = specific entropy of saturated liquid and vapor, respectively

$v_{fg} = v_g − v_f$ = increase of volume during vaporization

$h_{fg} = h_g − h_f$ = heat of vaporization, or heat required to vaporize 1 lb of liquid at constant pressure and temperature

$r$ may be used for $h_{fg}$ when several heats of vaporization (as $r_1, r_2, r_3$, etc.) are under consideration.

$u_{fg} = u_g − u_f$ = increase of internal energy during vaporization

$s_{fg} = s_g − s_f = h_{fg}/T$ = increase of entropy during vaporization

$Apv_{fg}$ = work performed during vaporization

The energy equation applied to the vaporization process is

$$h_{fg} = u_{fg} + Apv_{fg}$$

The properties of a unit mass of a mixture of liquid and vapor of quality $x$ are given by the following expressions:

$$v = v_f + xv_{fg}$$
$$h = h_f + xh_{fg}$$
$$u = u_f + xu_{fg}$$
$$s = s_f + xs_{fg}$$

Tables of superheated vapor usually give values of $v$, $h$, and $s$ per unit mass. The internal energy $u$ per unit mass can be found from the equation

$$u = h − Apv$$

or with $p$ in psi,

$$u = h − 0.1852pv$$

**Charts for Saturated and Superheated Vapors**

Certain properties of vapor mixtures and superheated vapors may be shown graphically by means of charts. Such charts show the behavior of vapors and have a practical application in the solution of certain problems.

**Temperature-Entropy Chart** Figure 18 shows the temperature-entropy chart for water vapor. The liquid curve is obtained by plotting corresponding values of $T$ and $s_f$, and the saturation curve by plotting values of $T$ and $s_g$. The values are taken from Tables 28 and 29. The two curves merge into each other at the critical temperature $T = 1165.4°R$. Between these two curves, constant pressure lines are also lines of constant temperature; but at the saturation curve the constant pressure lines show a sharp break with rising temperature. The constant quality lines $x = 0.2, 0.4$, etc., are equally spaced between the liquid and saturation curves.

Figure 19 is a temperature-entropy chart for air.

Figure 20 shows a temperature-entropy chart for isobutane. The form of the saturation curve is worthy of note. In the case of water vapor, this curve has a negative slope throughout; but in the case of isobutane, the curve has a positive slope except near the critical temperature. In the case of water vapor, therefore, isentropic expansion of dry vapor ($s$ = const) will be accompanied by condensation, isentropic compression by superheating. For isobutane, in the region where the saturation curve has the positive slope, the conditions are reversed; isentropic expansion is accompanied by superheating, isentropic compression by condensation.

**Enthalpy-Entropy Chart (Mollier Chart)** In this chart, the enthalpy $h$ is taken as the ordinate and entropy as the abscissa. Figure 21 shows a Mollier chart for water vapor. A large-scale chart, covering only the region near the saturation curve, is shown in Fig. 22.

EXAMPLES. The following illustrate the use of the Mollier chart:

1. Steam enters a superheater at a pressure of 240 psia containing 2 percent water. It leaves the superheater at a temperature of 580°F. Required: the heat per pound of steam to effect this change.

The initial and final points are located on the Mollier chart. From the enthalpy scale, $h_1 = 1184.2$ and $h_2 = 1308.5$; therefore $q_{12} = 1308.5 − 1184.2 = 124.3$ Btu.

2. Steam at $p = 240$ psia, $t = 580°F$ expands isentropically to a pressure of 60 psia. Required: the final condition of the steam and the decrease in enthalpy.

Following a constant entropy line from the point 240 psia, 580°F, it is found that this line intersects the line $p = 60$ psia on the saturation curve and that the value of $h$ at the point of intersection is 1177.6 Btu. Hence the steam in the second state is just saturated, and the decrease in enthalpy is 1308.5 − 1177.6 = 130.9 Btu.

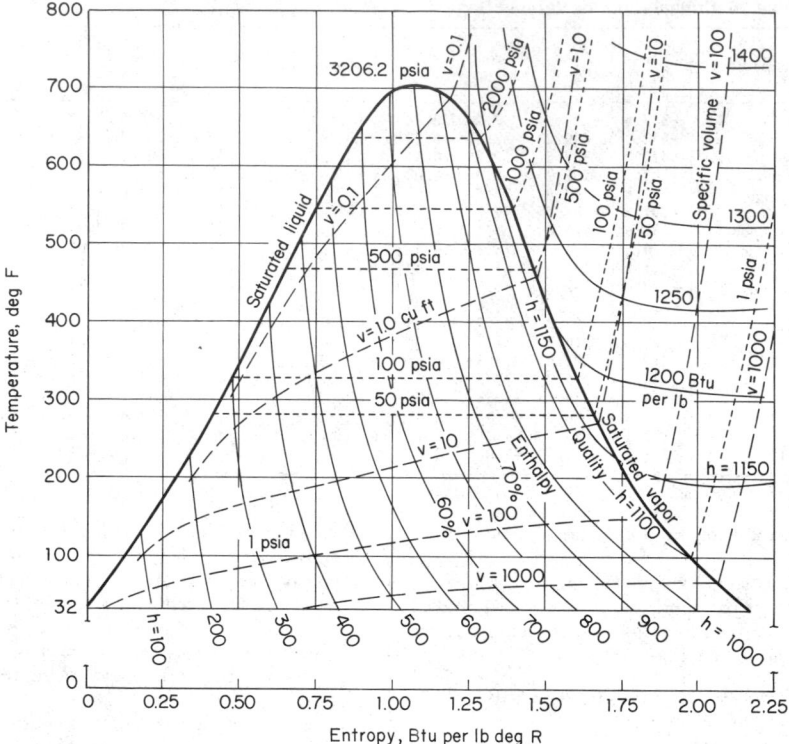

**Fig. 18**  Temperature-entropy diagram for steam. (*Data from Keenan and Keyes, "Thermodynamic Properties of Steam," Wiley.*)

A Mollier diagram for sodium vapor is shown in Fig. 23.

**Pressure-Enthalpy Chart**  For refrigeration media it is convenient to use pressure and enthalpy as the variables to be plotted. Such a chart for ammonia is shown in Fig. 24.

### Changes of State. Superheated Vapors and Mixtures of Liquid and Vapor

**Isothermal**  In the only important cases, the fluid is a mixture of liquid and vapor in both initial and final states.

$$t = \text{const} \quad p = \text{const} \quad x_1, x_2 = \text{initial and final qualities}$$
$$Q_{12} = m h_{fg}(x_2 - x_1)$$
$$W_{12} = m p v_{fg}(x_2 - x_1)$$
$$U_2 - U_1 = m u_{fg}(x_2 - x_1)$$
$$S_2 - S_1 = Q_{12}/T$$

**Constant Pressure**  If the fluid is a mixture at the beginning and end of the change, the constant pressure change is also isothermal. If the initial state is in the mixture region and the final state is that of a superheated vapor, the following are the equations for $Q_{12}$, etc. Let $h_2$, $u_2$, $v_2$, and $s_2$ be the properties of 1 lb of superheated vapor in the final state 2: then

$$Q_{12} = m(h_2 - h_1)$$
$$U_2 - U_1 = m(u_2 - u_1)$$
$$S_2 - S_1 = m(s_2 - s_1)$$
$$W_{12} = m p(v_2 - v_1)$$

$$h_1 = h_{f1} + x_1 h_{fg1}$$
$$u_1 = u_{f1} + x_1 u_{fg1}$$
$$s_1 = s_{f1} + x_1 s_{fg1}$$
$$v_1 = v_{f1} + x_1 v_{fg1}$$

**Constant Volume**  Since $v_f$ the liquid volume is nearly constant,

$$x_1 v_{fg1} = x_2 v_{fg2}$$
$$x_2 = x_1 v_{fg1}/v_{fg2} \quad \text{or} \quad x_2 = x_1 v_{g1}/v_{g2} \text{ approx}$$
$$Q_{12} = U_2 - U_1 = m(u_2 - u_1) \qquad W_{12} = 0$$

**Isentropic**  $s = $ const.

If the fluid is a mixture in the initial and final states,

$$s_{f1} + x_1 s_{fg1} = s_{f2} + x_2 s_{fg2}$$

If the initial state is that of superheated vapor,

$$s_1 = s_{f2} + x_2 s_{fg2}$$

in which $s_1$ is read from the table of superheated vapor. The final value $x_2$ is determined from one of these equations, and the final internal energy $u_2$ is then

$$u_{f2} + x_2 u_{fg2} \qquad Q_{12} = 0$$
$$W_{12} = J(U_1 - U_2) = Jm(u_1 - u_2)$$

For water vapor, the relation between $p$ and $v$ during an isentropic change may be represented approximately by the equation $p v^n = $ constant. The exponent $n$ is not constant, but

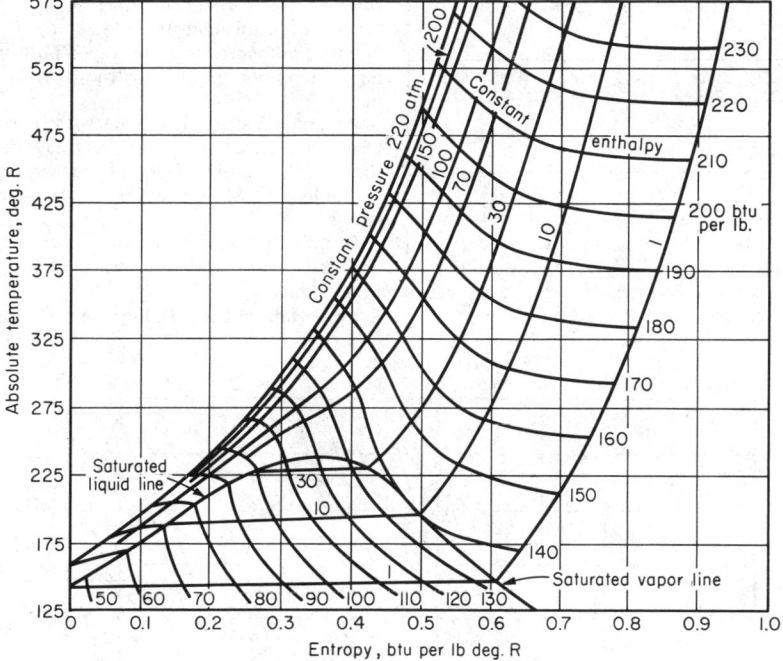

**Fig. 19**  Temperature-entropy diagram for air. *(From Williams, Thermodynamic Properties of Air at Low Temperatures, Trans. AICE, 39, Feb. 1943.)*

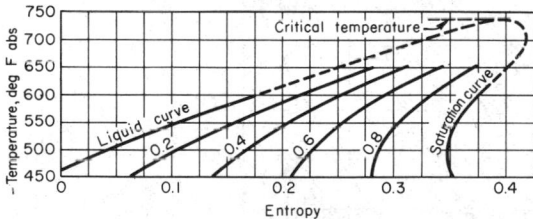

**Fig. 20**  Temperature-entropy chart for isobutane.

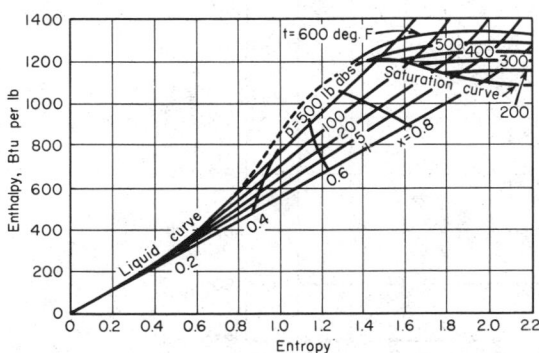

**Fig. 21**  Enthalpy-entropy chart for steam.

varies with the initial quality and initial pressure, as shown in Table 27.

The isentropic expansion of superheated steam is fairly represented by $pv^n$ = const, with $n$ = 1.315.

The volume at the end of expansion (or compression) is $V_2$ = $V_1(p_1/p_2)^{1/n}$, and the external work is

$$W_{12} = (p_1V_1 - p_2V_2)/(n - 1)$$
$$= p_1V_1[1 - (p_2/p_1)^{(n-1)/n}]/(n - 1)$$

If the initial state is in the region of superheat and the final state in the mixture region, two values of $n$ must be used: $n$ = 1.315 for the expansion to the state of saturation, and the appropriate value from the first row of Table 27 for the expansion of the mixture.

**Tables of Thermal Properties of Vapors**

**Steam**  Tables 28 to 30, abstracted from Keenan, Keyes, Hill, and Moore, "Steam Tables," give internationally accepted values which are believed to be correct within very narrow tolerance. They are a representation of the results obtained from an organized body of research and agreed to in a series of international conferences.

**Mercury Vapor**  Properties of mercury vapor are presented in Table 31.

**Diphenyl** $(C_6H_5)_2$ has the following properties (Chipman and Peltier, *Ind. Eng. Chem.*, Nov., 1929, p. 1106): boiling point, 491.5°F; density of the liquid, 53 lb/ft³; density of saturated vapor, 0.242 lb/ft³; heat of vaporization, 134 Btu/lb: all measured at the boiling point. The specific heat at constant pressure is given by the relation $c_p$ = 0.279 + 0.000667t. The vapor pressures (*p*, psia) at various temperatures (*t*, deg F) are as follows:

| *t* | 200 | 250 | 300 | 350 | 400 | 450 | 500 |
|-----|-----|-----|-----|-----|-----|-----|-----|
| *p* | 0.060 | 0.227 | 0.701 | 1.832 | 4.117 | 8.638 | 16.29 |

| *t* | 550 | 600 | 650 | 700 | 750 | 800 |
|-----|-----|-----|-----|-----|-----|-----|
| *p* | 28.54 | 47.01 | 73.55 | 110.1 | 158.6 | 221.0 |

**Dowtherm A** is the eutectic mixture of diphenyl oxide and diphenyl containing 73.5 percent of diphenyl oxide and 26.5 percent of diphenyl and melting at 53.6°F. It is used as a liquid heating medium at elevated temperatures. Its low vapor pressure permits high temperature without attendant high pres-

sures. Table 32 (Badger, *Ind. Eng. Chem.*, Sept., 1937) gives properties of this eutectic.

**Pure Hydrocarbons** The vapor pressures of various commercially important pure hydrocarbons are shown graphically in Fig. 25.

**Ammonia Vapor** The properties of saturated and superheated ammonia vapor have been determined accurately by the NBS (*Circ.* 142, 1923). The principal properties are given in Tables 33 and 34 and Fig. 24. Properties of **aqua-ammonia** are given in Fig. 26.

In these tables, the entropy $s_f$ and the heat of the liquid $h_f$ are taken as zero at −40°F instead of at 32°F, as is customary in most tables.

**Properties of Other Refrigerants** Complete and consistent

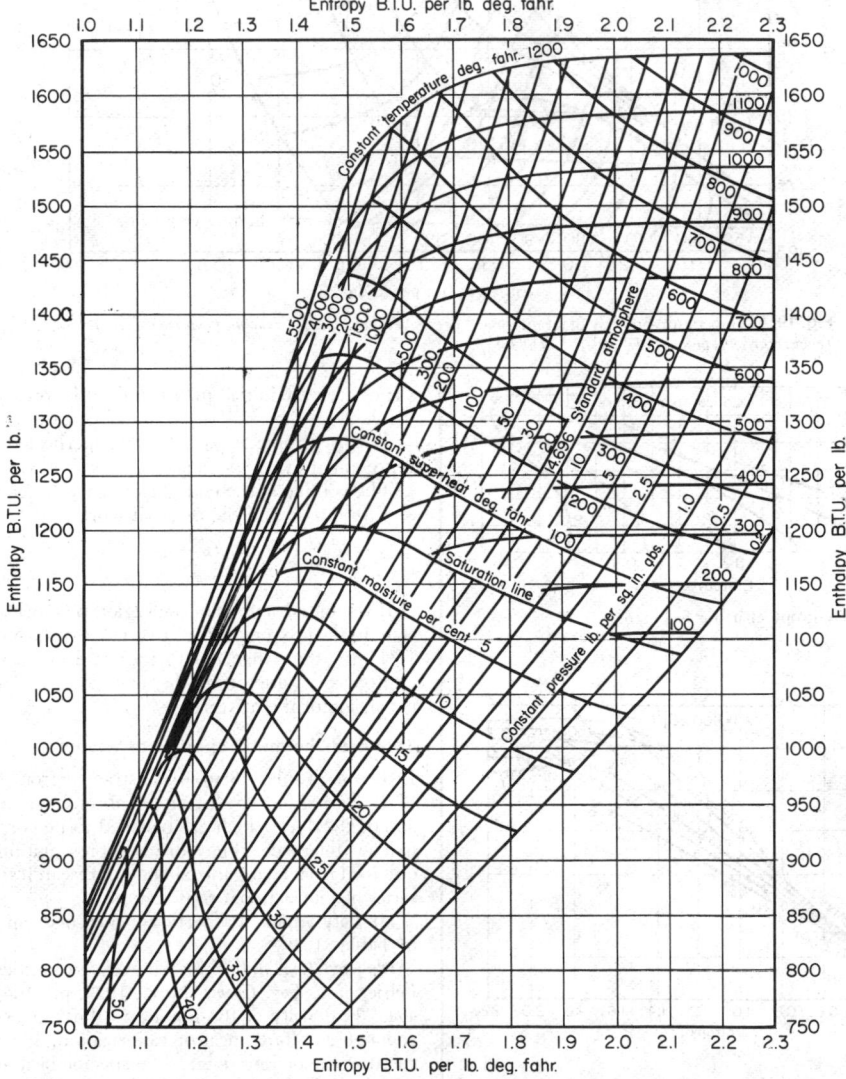

**Fig. 22**   Enthalpy-entropy (Mollier) chart for steam. (*From "Steam, Its Generation and Use," The Babcock & Wilcox Co., 1963.*)

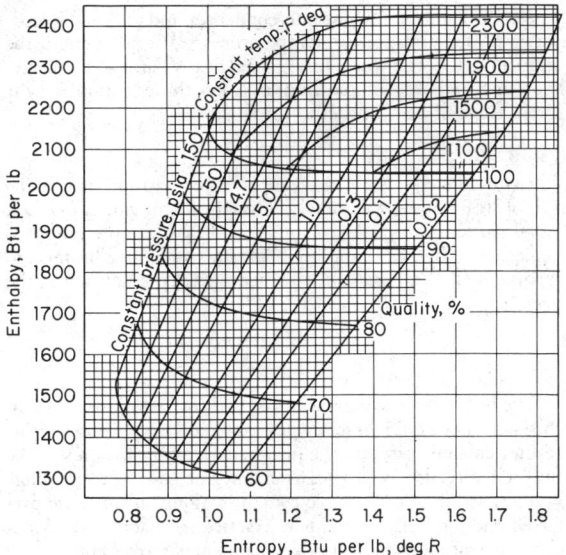

**Fig. 23**  Mollier diagram for sodium vapor. Datum is saturated liquid, 208°F and $1.68 \times 10^{-9}$ psia. *(Power, July 1961.)*

data are not available on most of the working fluids for refrigeration. The data presented in this section on refrigerating fluids other than ammonia are, in many cases, not of as high order of accuracy as the values given for steam (see also Sec. 19).

### Steam Cycles

**Rankine Cycle**  The ideal Rankine cycle is generally employed by engineers as a standard of reference for comparing the performance of actual steam engines and steam turbines. Figure 27 shows this cycle on the *T-S* and *p-V* planes. *AB* represents the heating of the water in the boiler, *BC* represents evaporation (and superheating if there is any), *CD* the assumed isentropic expansion in the engine cylinder, and *DA* condensation in the condenser.

Let $h_a$, $h_b$, $h_c$, $h_d$ represent the enthalpy per pound of steam in the four states *A*, *B*, *C*, and *D*, respectively. Then the energy transformed into work, represented by the area *ABCD*, is $h_c - h_d$.

The energy expended on the fluid is $h_c - h_a$; hence the **Rankine cycle efficiency** is $e_t = (h_c - h_d)/(h_c - h_a)$.

The **steam consumption** of the ideal Rankine engine in pounds per horsepower hour is $N_r = 2{,}544/(h_c - h_d)$. Expressed in pounds per kilowatthour, the steam consumption of the ideal Rankine cycle is $3{,}412.7/(h_c - h_d)$.

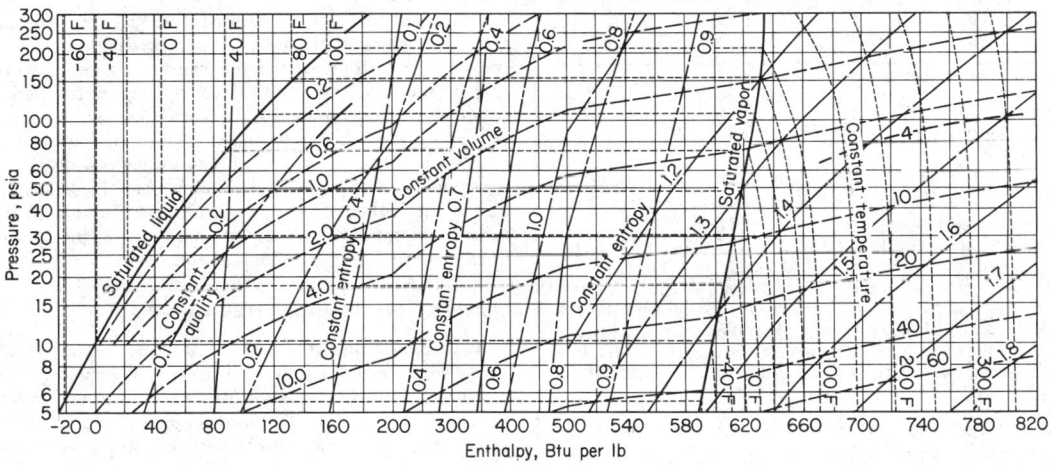

**Fig. 24**  Pressure-enthalpy diagram for ammonia. *(Based on NBS Circ. 142.)*

The performance of an engine is frequently stated in terms of the heat used per horsepower-hour. For the ideal Rankine engine, this is

$$Q_r = 2544/e_t = 2544(h_c - h_a)/(h_c - h_d)$$

**Efficiency of the Actual Engine**  Let $Q$ denote the heat transformed into work per pound of steam by the actual engine; then if $Q_1$ is the heat furnished by the boiler per pound of steam, the **thermal efficiency of the engine** is $e_t = Q/Q_1$.

### Table 27. Value of $n$ (Water Vapor)

| Initial quality | Initial pressure, psia | | | | | | | | | | | |
|---|---|---|---|---|---|---|---|---|---|---|---|---|
| | 20 | 40 | 60 | 80 | 100 | 120 | 140 | 160 | 180 | 200 | 220 | 240 |
| 1.00 | 1.131 | 1.132 | 1.133 | 1.134 | 1.136 | 1.137 | 1.138 | 1.139 | 1.141 | 1.142 | 1.143 | 1.145 |
| 0.95 | 1.127 | 1.128 | 1.129 | 1.130 | 1.131 | 1.131 | 1.132 | 1.133 | 1.134 | 1.135 | 1.136 | 1.137 |
| 0.90 | 1.123 | 1.123 | 1.124 | 1.124 | 1.125 | 1.125 | 1.126 | 1.126 | 1.127 | 1.127 | 1.128 | 1.129 |
| 0.85 | 1.119 | 1.119 | 1.119 | 1.119 | 1.120 | 1.120 | 1.120 | 1.120 | 1.120 | 1.120 | 1.120 | 1.121 |
| 0.80 | 1.115 | 1.115 | 1.114 | 1.114 | 1.114 | 1.114 | 1.113 | 1.113 | 1.113 | 1.113 | 1.112 | 1.112 |
| 0.75 | 1.111 | 1.110 | 1.110 | 1.109 | 1.109 | 1.108 | 1.107 | 1.106 | 1.106 | 1.105 | 1.104 | 1.104 |

The efficiency thus defined is misleading, as it takes no account of the conditions of boiler and condenser pressure, superheat, or quality of steam. It is customary therefore to define the efficiency as the ratio $Q/Q_a$, where $Q_a$ is the **available heat,** or the heat that could be transformed under ideal conditions. For steam engines and turbines, the Rankine cycle is usually taken as the ideal, and the quantity $Q/Q_a = Q(b_c - b_d)$ is called the **engine efficiency.** For engines and turbines, this efficiency ranges from 0.50 to 0.85. The engine efficiency $e$ may also be expressed in terms of steam consumed; thus, if $N_a$ is the steam consumption of the actual engine and $N_r$ is the steam consumption of the ideal Rankine engine under similar conditions, then $e = N_r/N_a$.

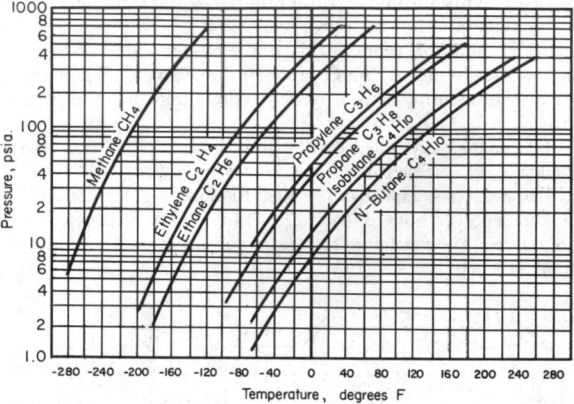

**Fig. 25**  Vapor pressures of pure hydrocarbons.

EXAMPLE.  Suppose the boiler pressure to be 180 psia, superheat 150°F, and the condenser pressure 3 in of mercury. From the steam tables or diagram, the following values are found: $b_c = 1283.3$, $b_d = 942$, $b_a = 82.99$. The available heat is $Q_a = 1283.3 - 942 = 341.3$ Btu, and the thermodynamic efficiency of the cycle is $341.3/(1283.3 - 82.99) = 0.284$. The steam consumption per horsepower-hour is 2,544/341.3 = 7.46 lb, and the heat used per horsepower-hour is 2544/0.284 = 8960 Btu. If an actual engine working under the same conditions has a steam consumption of 11.4 lb/(hp·h), its efficiency is 7.46/11.4 = 0.655, and its heat consumption per horsepower-hour is 8960/0.655 = 13,680 Btu.

**Reheating Cycle**  Let the steam after expansion from $p_1$ to an intermediate pressure $p_2$ (cd, Fig. 28) be reheated at constant pressure $p_2$, as indicated by de. Then follows the isentropic expansion to pressure $p_3$, represented by ef.

The energy absorbed by 1 lb of steam is $b_c - b_a$ from the boiler, and $b_e - b_d$ from the reheating. The work done, neglecting the energy required to operate the boiler feed pump, etc., is $b_c - b_d + b_e - b_f$. Hence the efficiency of the cycle is

$$e_t = \frac{b_c - b_d + b_e - b_f}{b_c - b_a + b_e - b_d}$$

**Bleeding Cycle**  In the **regenerative** or bleeding cycle, steam is drawn from the turbine at one or more stages and used to heat the feed water. Figure 29 shows a diagrammatic arrangement for bleeding at one stage. Entering the turbine is $1 + w$ lb, of steam at $p_1$, $t_1$, and enthalpy $b_1$. At the bleeding point $w$ lb at $p_2$, $t_2$, $b_2$ enters the feed-water heater. The remaining 1 lb

passes through the turbine and condenser and enters the feed-water heater as water at temperature $t_3$. Let $t'$ denote the temperature of the water leaving the heater, and $b'$ the corresponding enthalpy of the liquid. Then the equation for the interchange of heat in the heater is

$$w(b_2 - b') = b' - b_{f3}$$

The work done by the bled steam is $w(b_1 - b_2)$ and that by the 1 lb of steam going completely through the turbine is $b_1 - b_3$. Total work = $w(b_1 - b_2) + (b_1 - b_3)$ if work to the pumps is neglected. The heat supplied between feed-water heater and turbine is $(1 + w) (b_1 - b')$. Hence the ideal efficiency of the cycle is

$$e_t = \frac{w(b_1 - b_2) + b_1 - b_3}{(1 + w)(b - b')}$$

The heat from **nuclear reactors** is used for heating services or, through thermodynamic cycles, for power purposes. The reactor coolant transfers the heat generated by fission so as to be used directly, or through an intermediate heat-exchange system, avoiding radioactive contamination. Steam is the preferred thermodynamic fluid in practice so that the Rankine-cycle performance standards with regenerative and reheat variations, prevail. Adaptation of gas-turbine cycles, using various gases, can be expected as allowable reactor temperatures are raised.

Many engineers and scientists are actively engaged in research dealing with the location, production, utilization, transmission, storage, and distribution of new forms of energy. Examples are the study of the energy released in the fusion of hydrogen nuclei and research in solar energy. Considerable effort is being expended in studying the feasibility of combining the gas turbine with a steam-generating plant. The possibility of efficiency improvement over the steam cycle is due to the higher inlet temperatures associated with the gas turbine. Significant advances are predicted in the near future in expanding our energy sources and reserves.

## MIXTURES OF AIR AND WATER VAPOR

**Atmospheric Humidity**  The atmosphere is a mixture of air and water vapor. Dalton's law of partial pressures (for the mixture) and the ideal gas law (for each constituent) may safely be assumed to apply. The **total pressure** $p_t$ (barometric pressure) is the sum of the **vapor pressure** $p_v$ and the **air pressure** $p_a$.

The temperature of the atmosphere, as indicated by an ordinary thermometer, is the **dry-bulb temperature** $t_d$. If the atmosphere is cooled under constant total pressure, the partial pressures remain constant until a temperature is reached at which condensation of vapor begins. This temperature is the **dew point** $t_c$ (condensation temperature) and is the saturation temperature, or boiling point, corresponding to the actual vapor pressure $p_v$. If a thermometer bulb is covered with absorbent material, e.g., linen, wet with distilled water and exposed to the atmosphere, evaporation will cool the water and the thermometer bulb to the **wet-bulb temperature** $t_w$. This is the temperature given by a psychrometer. The wet-bulb temperature lies between the dry-bulb temperature and the dew point. These three temperatures are distinct except for a saturated atmosphere, for which they are identical. For each of these temperatures, there is a corresponding vapor pressure. The actual vapor pressure $p_v$ corresponds with the dew

**Table 28. Properties of Saturated Steam**

| Abs press, lb/in² | Temp., °F | Specific volume | | Enthalpy | | | Entropy | | | Internal energy, Evap. |
|---|---|---|---|---|---|---|---|---|---|---|
| | | Liquid | Vapor | Liquid | Evap. | Vapor | Liquid | Evap. | Vapor | |
| 0.08866 | 32.02 | 0.016022 | 3302 | 0.01 | 1075.4 | 1075.4 | 0.00000 | 2.1869 | 2.1869 | 1021.2 |
| 1.0 | 101.70 | 0.016136 | 333.6 | 69.74 | 1036.0 | 1105.8 | 0.13266 | 1.8453 | 1.9779 | 974.3 |
| 1.5 | 115.65 | 0.016187 | 227.7 | 83.65 | 1028.0 | 1111.7 | 0.15714 | 1.7867 | 1.9438 | 964.8 |
| 2 | 126.04 | 0.016230 | 173.75 | 94.02 | 1022.1 | 1116.1 | 0.17499 | 1.7448 | 1.9198 | 957.8 |
| 3 | 141.43 | 0.016300 | 118.72 | 109.39 | 1013.1 | 1122.5 | 0.20089 | 1.6852 | 1.8861 | 947.2 |
| 4 | 152.93 | 0.016358 | 90.64 | 120.89 | 1006.4 | 1127.3 | 0.21983 | 1.6426 | 1.8624 | 939.3 |
| 5 | 162.21 | 0.016407 | 73.53 | 130.17 | 1000.9 | 1131.0 | 0.23486 | 1.6093 | 1.8441 | 932.9 |
| 10 | 193.19 | 0.016590 | 38.42 | 161.23 | 982.1 | 1143.3 | 0.28358 | 1.5041 | 1.7877 | 911.0 |
| 14.696 | 211.99 | 0.016715 | 26.80 | 180.15 | 970.4 | 1150.5 | 0.31212 | 1.4446 | 1.7567 | 897.5 |
| 15 | 213.03 | 0.016723 | 26.29 | 181.19 | 969.7 | 1150.9 | 0.31367 | 1.4414 | 1.7551 | 896.8 |
| 20 | 227.96 | 0.016830 | 20.09 | 196.26 | 960.1 | 1156.4 | 0.33580 | 1.3962 | 1.7320 | 885.8 |
| 25 | 240.08 | 0.016922 | 16.306 | 208.52 | 952.2 | 1160.7 | 0.35345 | 1.3607 | 1.7142 | 876.9 |
| 30 | 250.34 | 0.017004 | 13.748 | 218.93 | 945.4 | 1164.3 | 0.36821 | 1.3314 | 1.6996 | 869.2 |
| 35 | 259.30 | 0.017078 | 11.900 | 228.04 | 939.3 | 1167.4 | 0.38093 | 1.3064 | 1.6873 | 862.4 |
| 40 | 267.26 | 0.017146 | 10.501 | 236.16 | 933.8 | 1170.0 | 0.39214 | 1.2845 | 1.6767 | 856.2 |
| 45 | 274.46 | 0.017209 | 9.403 | 243.51 | 928.8 | 1172.3 | 0.40218 | 1.2651 | 1.6673 | 850.7 |
| 50 | 281.03 | 0.017269 | 8.518 | 250.24 | 924.2 | 1174.4 | 0.41129 | 1.2476 | 1.6589 | 845.5 |
| 55 | 287.10 | 0.017325 | 7.789 | 256.46 | 919.9 | 1176.3 | 0.41963 | 1.2317 | 1.6513 | 840.8 |
| 60 | 292.73 | 0.017378 | 7.177 | 262.25 | 915.8 | 1178.0 | 0.42733 | 1.2170 | 1.6444 | 836.3 |
| 65 | 298.00 | 0.017429 | 6.657 | 267.67 | 911.9 | 1179.6 | 0.43450 | 1.2035 | 1.6380 | 832.1 |
| 70 | 302.96 | 0.017478 | 6.209 | 272.79 | 908.3 | 1181.0 | 0.44120 | 1.1909 | 1.6321 | 828.1 |
| 75 | 307.63 | 0.017524 | 5.818 | 277.61 | 904.8 | 1182.4 | 0.44749 | 1.1790 | 1.6265 | 824.3 |
| 80 | 312.07 | 0.017570 | 5.474 | 282.21 | 901.4 | 1183.6 | 0.45344 | 1.1679 | 1.6214 | 820.6 |
| 85 | 316.29 | 0.017613 | 5.170 | 286.58 | 898.2 | 1184.8 | 0.45907 | 1.1574 | 1.6165 | 817.1 |
| 90 | 320.31 | 0.017655 | 4.898 | 290.76 | 895.1 | 1185.9 | 0.46442 | 1.1475 | 1.6119 | 813.8 |
| 95 | 324.16 | 0.017696 | 4.654 | 294.76 | 892.1 | 1186.9 | 0.46952 | 1.1380 | 1.6076 | 810.6 |
| 100 | 327.86 | 0.017736 | 4.434 | 298.61 | 889.2 | 1187.8 | 0.47439 | 1.1290 | 1.6034 | 807.5 |
| 150 | 358.48 | 0.018089 | 3.016 | 330.75 | 864.2 | 1194.9 | 0.51422 | 1.0562 | 1.5704 | 781.0 |
| 200 | 381.86 | 0.018387 | 2.289 | 355.6 | 843.7 | 1199.3 | 0.5440 | 1.0025 | 1.5464 | 759.6 |
| 250 | 401.04 | 0.018653 | 1.8448 | 376.2 | 825.8 | 1202.1 | 0.5680 | 0.9594 | 1.5274 | 741.4 |
| 300 | 417.43 | 0.018896 | 1.5442 | 394.1 | 809.8 | 1203.9 | 0.5883 | 0.9232 | 1.5115 | 725.1 |
| 350 | 431.82 | 0.019124 | 1.3267 | 409.9 | 795.0 | 1204.9 | 0.6060 | 0.8917 | 1.4978 | 710.3 |
| 400 | 444.70 | 0.019340 | 1.1620 | 424.2 | 781.2 | 1205.5 | 0.6218 | 0.8638 | 1.4856 | 696.7 |
| 450 | 456.39 | 0.019547 | 1.0326 | 437.4 | 768.2 | 1205.6 | 0.6360 | 0.8385 | 1.4746 | 683.9 |
| 500 | 467.13 | 0.019748 | 0.9283 | 449.5 | 755.8 | 1205.3 | 0.6490 | 0.8154 | 1.4645 | 671.7 |
| 550 | 477.07 | 0.019943 | 0.8423 | 460.9 | 743.9 | 1204.8 | 0.6611 | 0.7941 | 1.4551 | 660.2 |
| 600 | 486.33 | 0.02013 | 0.7702 | 471.7 | 732.4 | 1204.1 | 0.6723 | 0.7742 | 1.4464 | 649.1 |
| 700 | 503.23 | 0.02051 | 0.6558 | 491.5 | 710.5 | 1202.0 | 0.6927 | 0.7378 | 1.4305 | 628.2 |
| 800 | 518.36 | 0.02087 | 0.5691 | 509.7 | 689.6 | 1199.3 | 0.7110 | 0.7050 | 1.4160 | 608.4 |
| 900 | 532.12 | 0.02123 | 0.5009 | 526.6 | 669.5 | 1196.0 | 0.7277 | 0.6750 | 1.4027 | 589.6 |
| 1000 | 544.75 | 0.02159 | 0.4459 | 542.4 | 650.0 | 1192.4 | 0.7432 | 0.6471 | 1.3903 | 571.5 |
| 1500 | 596.39 | 0.02346 | 0.2769 | 611.5 | 557.2 | 1168.7 | 0.8082 | 0.5276 | 1.3359 | 486.9 |
| 2000 | 636.00 | 0.02565 | 0.18813 | 671.9 | 464.4 | 1136.3 | 0.8623 | 0.4238 | 1.2861 | 404.2 |
| 2500 | 668.31 | 0.02860 | 0.13059 | 730.9 | 360.5 | 1091.4 | 0.9131 | 0.3196 | 1.2327 | 313.4 |
| 3000 | 695.52 | 0.03431 | 0.08404 | 802.5 | 213.0 | 1015.5 | 0.9732 | 0.1843 | 1.1575 | 185.4 |
| 3203.6 | 705.44 | 0.05053 | 0.05053 | 902.5 | 0 | 902.5 | 1.0580 | 0 | 1.0580 | 0 |

SOURCE: Abstracted from Keenan, Keyes, Hill, and Moore, "Steam Tables," 1969.

**Table 29. Superheated Steam Tables**
($v$ = specific volume, ft$^3$/lb; $h$ = enthalpy, Btu/lb; $s$ = entropy)

| Pressure, psia (saturation temp., °F) | | Temperature of steam, °F | | | | | | | |
|---|---|---|---|---|---|---|---|---|---|
| | | 200 | 300 | 400 | 500 | 600 | 800 | 1000 | 1200 |
| 10 | $v$ | 38.85 | 44.99 | 51.03 | 57.04 | 63.03 | 74.98 | 86.91 | 98.84 |
| (193.19) | $h$ | 1146.6 | 1193.7 | 1240.5 | 1287.7 | 1335.5 | 1433.3 | 1534.6 | 1639.4 |
| | $s$ | 1.7927 | 1.8592 | 1.9171 | 1.9690 | 2.0164 | 2.1009 | 2.1755 | 2.2428 |
| 50 | $v$ | | 8.772 | 10.061 | 11.305 | 12.529 | 14.949 | 17.352 | 19.747 |
| (281.03) | $h$ | | 1184.4 | 1235.0 | 1284.0 | 1332.8 | 1431.7 | 1533.5 | 1638.7 |
| | $s$ | | 1.6722 | 1.7348 | 1.7887 | 1.8371 | 1.9225 | 1.9975 | 2.0650 |
| 100 | $v$ | | | 4.934 | 5.587 | 6.216 | 7.445 | 8.657 | 9.861 |
| (327.86) | $h$ | | | 1227.5 | 1279.1 | 1329.3 | 1429.6 | 1532.1 | 1637.7 |
| | $s$ | | | 1.6517 | 1.7085 | 1.7582 | 1.8449 | 1.9204 | 1.9882 |
| 150 | $v$ | | | 3.221 | 3.679 | 4.111 | 4.944 | 5.759 | 6.566 |
| (358.48) | $h$ | | | 1219.5 | 1274.1 | 1325.7 | 1427.5 | 1530.7 | 1454.5 |
| | $s$ | | | 1.5997 | 1.6598 | 1.7110 | 1.7989 | 1.8750 | 1636.7 |
| 200 | $v$ | | | 2.361 | 2.724 | 3.058 | 3.693 | 4.310 | 4.918 |
| (381.86) | $h$ | | | 1210.8 | 1268.8 | 1322.1 | 1425.3 | 1529.3 | 1635.7 |
| | $s$ | | | 1.5600 | 1.6239 | 1.6767 | 1.7660 | 1.8425 | 1.9109 |
| 300 | $v$ | | | | 1.7662 | 2.004 | 2.442 | 2.860 | 3.270 |
| (417.43) | $h$ | | | | 1257.5 | 1314.5 | 1421.0 | 1526.5 | 1633.8 |
| | $s$ | | | | 1.5701 | 1.6266 | 1.7187 | 1.7964 | 1.8653 |
| 400 | $v$ | | | | 1.2843 | 1.4760 | 1.8163 | 2.136 | 2.446 |
| (444.70) | $h$ | | | | 1245.2 | 1306.6 | 1416.6 | 1523.6 | 1631.8 |
| | $s$ | | | | 1.5282 | 1.5892 | 1.6844 | 1.7632 | 1.8327 |
| 500 | $v$ | | | | .9924 | 1.1583 | 1.4407 | 1.7008 | 1.9518 |
| (467.13) | $h$ | | | | 1231.5 | 1298.3 | 1412.1 | 1520.7 | 1629.8 |
| | $s$ | | | | 1.4923 | 1.5585 | 1.6571 | 1.7371 | 1.8072 |
| 600 | $v$ | | | | .7947 | .9456 | 1.1900 | 1.4108 | 1.6222 |
| (486.33) | $h$ | | | | 1216.2 | 1289.5 | 1407.6 | 1517.8 | 1627.8 |
| | $s$ | | | | 1.4592 | 1.5320 | 1.6343 | 1.7155 | 1.7861 |
| 700 | $v$ | | | | | .7929 | 1.0109 | 1.2036 | 1.3868 |
| (503.23) | $h$ | | | | | 1280.2 | 1402.9 | 1514.9 | 1625.8 |
| | $s$ | | | | | 1.5081 | 1.6145 | 1.6970 | 1.7682 |
| 800 | $v$ | | | | | .6776 | .8764 | 1.0482 | 1.2102 |
| (518.36) | $h$ | | | | | 1270.4 | 1398.2 | 1511.9 | 1623.8 |
| | $s$ | | | | | 1.4861 | 1.5969 | 1.6807 | 1.7526 |
| 900 | $v$ | | | | | .5871 | .7717 | .9273 | 1.0729 |
| (532.14) | $h$ | | | | | 1260.0 | 1393.4 | 1508.9 | 1621.7 |
| | $s$ | | | | | 1.4652 | 1.5810 | 1.6662 | 1.7386 |
| 1000 | $v$ | | | | | 0.5140 | 0.6878 | 0.8305 | 0.9630 |
| (544.75) | $h$ | | | | | 1248.8 | 1388.5 | 1505.9 | 1619.7 |
| | $s$ | | | | | 1.4450 | 1.5664 | 1.6530 | 1.7261 |
| 1500 | $v$ | | | | | 0.2816 | 0.4350 | 0.5400 | 0.6334 |
| (596.39) | $h$ | | | | | 1174.8 | 1362.5 | 1490.3 | 1609.3 |
| | $s$ | | | | | 1.3416 | 1.5058 | 1.6001 | 1.6765 |
| 2000 | $v$ | | | | | | 0.3071 | 0.3945 | 0.4685 |
| (636.00) | $h$ | | | | | | 1333.8 | 1474.1 | 1598.6 |
| | $s$ | | | | | | 1.4562 | 1.5598 | 1.6398 |
| 2500 | $v$ | | | | | | 0.2291 | 0.3069 | 0.3696 |
| (668.31) | $h$ | | | | | | 1301.7 | 1457.2 | 1587.7 |
| | $s$ | | | | | | 1.4112 | 1.5262 | 1.6101 |
| 3000 | $v$ | | | | | | 0.17572 | 0.2485 | 0.3036 |
| (695.52) | $h$ | | | | | | 1265.2 | 1439.6 | 1576.6 |
| | $s$ | | | | | | 1.3675 | 1.4967 | 1.5848 |

SOURCE: Abstracted from Keenan, Keyes, Hill, and Moore, "Steam Tables," 1969.

point $t_c$. The vapor pressures $p_d$ and $p_w$, corresponding with $t_d$ and $t_w$, do not represent pressures actually appearing in the atmosphere but are used in computations.

**Relative humidity** $r$ is the ratio of the actual vapor pressure to the pressure of saturated vapor at the prevailing dry-bulb temperature $r = p_v/p_d$. Within the limits of usual accuracy, this equals the ratio of actual vapor density to the density of saturated vapor at dry-bulb temperature, $r = \rho_v/\rho_d$. It is to be noted that relative humidity is a property of the vapor alone; it has nothing to do with the fact that the vapor is mixed with air. It is a method of expressing the departure of the vapor from saturation.

**Table 30. Properties of Saturated Steam**

| Temp, °F, $t$ | Abs. pressure lb/in² $p$ | Abs. pressure InHg | Specific volume Sat. vapor, $v_g$ | Enthalpy Sat. liquid, $h_f$ | Enthalpy Evap., $h_{fg}$ | Enthalpy Sat. vapor, $h_g$ | Entropy Sat. liquid, $s_f$ | Entropy Sat. vapor $s_g$ |
|---|---|---|---|---|---|---|---|---|
| 50  | 0.17803 | 0.363 | 1704.2 | 18.06  | 1065.2 | 1083.3 | 0.03607 | 2.1259 |
| 55  | 0.2140  | 0.436 | 1431.4 | 23.07  | 1062.4 | 1085.5 | 0.04586 | 2.1099 |
| 60  | 0.2563  | 0.522 | 1206.9 | 28.08  | 1059.6 | 1087.7 | 0.05555 | 2.0943 |
| 65  | 0.3057  | 0.623 | 1021.5 | 33.09  | 1056.8 | 1089.9 | 0.06514 | 2.0791 |
| 70  | 0.3632  | 0.739 | 867.7  | 38.09  | 1054.0 | 1092.0 | 0.07463 | 2.0642 |
| 75  | 0.4300  | 0.876 | 739.7  | 43.09  | 1051.1 | 1094.2 | 0.08402 | 2.0497 |
| 80  | 0.5073  | 1.032 | 632.8  | 48.09  | 1048.3 | 1096.4 | 0.09332 | 2.0356 |
| 85  | 0.5964  | 1.215 | 543.1  | 53.08  | 1045.5 | 1098.6 | 0.10252 | 2.0218 |
| 90  | 0.6988  | 1.422 | 467.7  | 58.07  | 1042.7 | 1100.7 | 0.11165 | 2.0083 |
| 95  | 0.8162  | 1.663 | 404.0  | 63.06  | 1039.8 | 1102.9 | 0.12068 | 1.9951 |
| 100 | 0.9503  | 1.933 | 350.0  | 68.05  | 1037.0 | 1105.0 | 0.12963 | 1.9822 |
| 105 | 1.1029  | 2.144 | 304.2  | 73.03  | 1034.1 | 1107.2 | 0.13850 | 1.9697 |
| 110 | 1.2763  | 2.596 | 265.1  | 78.02  | 1031.3 | 1109.3 | 0.14730 | 1.9574 |
| 115 | 1.4727  | 2.998 | 231.7  | 83.01  | 1028.4 | 1111.4 | 0.15601 | 1.9454 |
| 120 | 1.6945  | 3.446 | 203.0  | 88.00  | 1025.5 | 1113.5 | 0.16465 | 1.9336 |
| 125 | 1.9444  | 3.958 | 178.41 | 92.99  | 1022.7 | 1115.7 | 0.17322 | 1.9221 |
| 130 | 2.225   | 4.525 | 157.17 | 97.98  | 1019.8 | 1117.8 | 0.18172 | 1.9109 |
| 135 | 2.540   | 5.174 | 138.81 | 102.97 | 1016.9 | 1119.8 | 0.19015 | 1.8999 |
| 140 | 2.892   | 5.881 | 122.88 | 107.96 | 1014.0 | 1121.9 | 0.19851 | 1.8892 |
| 145 | 3.285   | 6.692 | 109.04 | 112.96 | 1011.0 | 1124.0 | 0.20681 | 1.8787 |
| 150 | 3.277   | 7.569 | 96.99  | 117.96 | 1008.1 | 1126.1 | 0.21503 | 1.8684 |
| 155 | 4.207   | 8.569 | 86.45  | 122.96 | 1005.1 | 1128.1 | 0.22320 | 1.8583 |
| 160 | 4.745   | 9.672 | 77.23  | 127.96 | 1002.2 | 1130.1 | 0.23130 | 1.8484 |
| 165 | 5.340   | 10.87 | 69.14  | 132.96 | 999.2  | 1132.2 | 0.23934 | 1.8387 |
| 170 | 5.996   | 12.21 | 62.02  | 137.97 | 996.2  | 1134.2 | 0.24732 | 1.8293 |

SOURCE: Abstracted from Keenan, Keyes, Hill, and Moore, "Steam Tables," 1969.

**Molal humidity** $f$ is the mass of water vapor in mols per one mol of air. The laws of Dalton and Avogadro state that the molal composition of a mixture is proportional to the distribution of partial pressures, or $f = p_v/p_a = p_v/(p_t - p_v)$.

**Specific humidity** (humidity ratio) $W$ is the mass of water vapor (pounds or grains) per pound of air. Mass in pounds equals mass in mols multiplied by the molecular weight. The molecular weight of water is 18, and the equivalent molecular weight of air is 28.97. The ratio $28.97/18 = 1.608$, or 1.61 with ample accuracy. Thus $W = f/1.61$.

**Air density** $\rho_a$ is the pounds of air in one cubic foot. **Vapor density** $\rho_v$ is the pounds of vapor in one cubic foot. **Mixture density** $\rho_m$ is the sum of these, i.e., the pounds of air plus vapor in one cubic foot.

**Notation** The subscripts $a$, $v$, $m$, and $f$ apply to air, vapor, mixture, and liquid water, respectively. The subscripts $d$ and $w$ apply to conditions pertaining to the dry- and wet-bulb temperature, respectively.

**Humidity Measurements** Many methods are in use: (1) The **dew-point** method measures the temperature at which condensation begins; water-vapor pressure can then be found from steam tables. Dew-point apparatus can either cool a surface or compress and expand moist air. (2) **Hygrometers** measure relative humidity, often by using the change in dimensions of a hygroscopic material such as human hair, wood, or paper; these instruments are simple and inexpensive but require frequent calibration. The electrical resistance of an electrolytic film can also be used as an indication of relative humidity. (3) The wet- and dry-bulb **psychrometer** is widely used. Humidity measurements of air flowing in ducts can be made with psychrometers that use mercury-in-glass thermometers, thermocouples, or resistance thermometers. Humidity measurements of still air can be made with sling psychrometers or aspiration psychrometers. Psychrometric wet bulb temperatures must be corrected to obtain thermodynamic wet-bulb temperatures, or there must be adequate air motion past the wet-bulb thermometer, 800 to 900 fpm (with duct walls at air temperature), to ensure a proper balance between radiation and convection. (Carrier and Mackey, *Trans. ASME*, 1937, p. 33.) (4) **Chemical analysis** by the use of dessicants such as sulphuric acid, phosphorus pentoxide, lithium chloride, or silica gel can be used as primary standards of humidity measurement.

Table 42 gives the pressure-temperature relations for saturated water vapor over water or ice.

The following equations give various properties in terms of pressure in inches Hg and temperature in degrees F.

Relative humidity:   $r = p_v/p_a$
Specific humidity: $W = p_v/1.61 (p_t - p_v)$    lb/lb dry air

Volume of mixture per pound of dry air:

$$v_a = \frac{1}{\rho_a} = 0.754(t_d + 460)/(p_t - rp_a) \quad \text{ft}^3$$

Volume of mixture per pound of mixture:

$$v_m = \frac{1}{\rho_m} = v_a/(1 + W) \quad \text{ft}^3$$

**Table 31. Properties of Mercury Vapor**

$h_f$ and $s_f$ are measured from 32°F

By L. A. Sheldon, General Electric Co.

| Pressure, psia, $p$ | Temp, deg F, $t$ | Specific vol, cu ft, per lb, $v_g$ | Enthalpy, Btu | | | Entropy | | |
|---|---|---|---|---|---|---|---|---|
| | | | Sat liquid, $h_f$ | Vapor- ization, $h_{fg}$ | Sat vapor, $h_g$ | Sat liquid, $s_f$ | Vapor- ization, $s_{fg}$ | Sat vapor, $s_g$ |
| 0.4 | 402.3 | 114.5 | 13.81 | 128.1 | 141.9 | 0.02094 | 0.1486 | 0.1696 |
| 0.6 | 426.1 | 78.23 | 14.70 | 127.6 | 142.3 | 0.02195 | 0.1441 | 0.1660 |
| 0.8 | 443.8 | 59.71 | 15.36 | 127.2 | 142.6 | 0.02269 | 0.1408 | 0.1635 |
| 1.0 | 458.1 | 48.45 | 15.89 | 126.9 | 142.8 | 0.02328 | 0.1382 | 0.1615 |
| 1.5 | 485.1 | 33.14 | 16.90 | 126.3 | 143.2 | 0.02436 | 0.1337 | 0.1580 |
| 2 | 505.2 | 25.31 | 17.65 | 125.8 | 143.5 | 0.02514 | 0.1304 | 0.1556 |
| 3 | 535.4 | 17.34 | 18.78 | 125.2 | 144.0 | 0.02629 | 0.1258 | 0.1521 |
| 4 | 558.0 | 13.26 | 19.62 | 124.7 | 144.3 | 0.02714 | 0.1225 | 0.1497 |
| 5 | 576.2 | 10.77 | 20.30 | 124.3 | 144.6 | 0.02780 | 0.1200 | 0.1478 |
| 6 | 591.4 | 9.096 | 20.87 | 123.9 | 144.8 | 0.02834 | 0.1179 | 0.1462 |
| 7 | 605.0 | 7.882 | 21.37 | 123.6 | 145.0 | 0.02882 | 0.1161 | 0.1450 |
| 8 | 616.8 | 6.963 | 21.81 | 123.4 | 145.2 | 0.02923 | 1.1146 | 0.1439 |
| 9 | 627.5 | 6.244 | 22.21 | 123.2 | 145.4 | 0.02960 | 0.1133 | 0.1429 |
| 10 | 637.3 | 5.661 | 22.58 | 122.9 | 145.5 | 0.02993 | 0.1121 | 0.1420 |
| 15 | 676.5 | 3.892 | 24.04 | 122.1 | 146.1 | 0.03124 | 0.1074 | 0.1387 |
| 20 | 706.2 | 2.983 | 25.15 | 121.4 | 146.6 | 0.03220 | 0.1041 | 0.1363 |
| 25 | 730.4 | 2.429 | 26.05 | 120.9 | 146.9 | 0.03297 | 0.1016 | 0.1345 |
| 30 | 750.9 | 2.053 | 26.81 | 120.4 | 147.2 | 0.03360 | 0.09953 | 0.1331 |
| 35 | 769.0 | 1.781 | 27.49 | 120.0 | 147.5 | 0.03416 | 0.09774 | 0.1319 |
| 40 | 784.8 | 1.576 | 28.08 | 119.7 | 147.8 | 0.03464 | 0.09621 | 0.1308 |
| 45 | 799.3 | 1.414 | 28.62 | 119.4 | 148.0 | 0.03507 | 0.09486 | 0.1299 |
| 50 | 812.5 | 1.284 | 29.11 | 119.1 | 148.2 | 0.03546 | 0.09364 | 0.1291 |
| 60 | 836.1 | 1.086 | 29.99 | 118.6 | 148.6 | 0.03614 | 0.09154 | 0.1276 |
| 70 | 856.6 | 0.9436 | 30.75 | 118.1 | 148.9 | 0.03672 | 0.08976 | 0.1264 |
| 80 | 874.8 | 0.8349 | 31.43 | 117.7 | 149.1 | 0.03725 | 0.08824 | 0.1254 |
| 90 | 891.6 | 0.7497 | 32.06 | 117.3 | 149.4 | 0.03771 | 0.08687 | 0.1245 |
| 100 | 906.9 | 0.6811 | 32.63 | 117.0 | 149.6 | 0.03813 | 0.08565 | 0.1237 |
| 120 | 934.4 | 0.5767 | 33.60 | 116.4 | 150.1 | 0.03887 | 0.08353 | 0.1224 |
| 140 | 958.3 | 0.5012 | 34.55 | 115.9 | 150.4 | 0.03951 | 0.08175 | 0.1212 |
| 160 | 979.9 | 0.4438 | 35.35 | 115.4 | 150.8 | 0.04007 | 0.08019 | 0.1202 |
| 180 | 999.6 | 0.3990 | 36.09 | 115.0 | 151.1 | 0.04058 | 0.07881 | 0.1193 |

The **enthalpy** of a mixture of dry air and stream, when each constituent is assumed to be an ideal gas, in Btu per pound of dry air, is the sum of the enthalpy of 1 lb of dry air and the enthalpy of the $W$ lb of steam mixed with that air. The specific enthalpy of dry air (above 0°F) is $h_a = 0.240 t_d$ (up to 130°F, the specific heat of dry air is 0.240; at higher temperatures, it is larger). The specific enthalpy of low-pressure steam (saturated or superheated) is nearly independent of the vapor pressure and depends only on $t_d$. An empirical equation for the specific enthalpy of low-pressure steam for the range of temperatures from −40 to 250°F is

$$h_v = 1,062 + 0.44 t_d \quad \text{Btu/lb}$$

The enthalpy of a mixture of air and steam is

$$h_m = 0.240 t_d + W(1,062 + 0.44 t_d)$$

The specific heat of a mixture of dry air and steam per pound of dry air may be called **humid specific heat** and is 0.240 + 0.44W Btu/lb dry air. For a steady-flow process without change of specific humidity, heat transfers per pound of dry air may be computed as the product of humid specific heat and change in dry-bulb temperature.

**Thermodynamic Wet-bulb Temperature (Temperature of Adiabatic Saturation** The thermodynamic wet-bulb temperature $t^*$ is an important property of state of mixtures of dry air and superheated steam; it is the temperature at which water (or ice), by evaporating into a mixture of air and steam, will bring the mixture to saturation at the same temperature in a steady-flow process in the absence of external heat transfer. For a mixture of dry air and saturated steam only, $t^* = t_d$; where $r < 1$, $t^* < t_d$. By writing energy and mass balances for the process of adiabatic saturation with water supplied at $t^*$, the following equation may be derived:

$$W = W^* - \frac{(0.240 + 0.44 W^*)(t_d - t^*)}{1,094 + 0.44 t_d - t^*}$$

where $W^*$ = specific humidity for saturation at $t^*$ and total pressure of $p_t$.

The enthalpy of a mixture of dry air and **saturated** steam at the total pressure $p_t$ and thermodynamic wet-bulb temperature $t^*$ exceeds the enthalpy of a mixture of dry air and **superheated** steam at the same $p_t$ and $t^*$, for

$$h_m^* = h_m + (W^* - W) h_f^*$$

A property of the mixture that remains constant for constant $p_t$ and $t^*$ has been called the $\Sigma$ **function**, for

$$\Sigma^* = h_m^* - W^* h_f^* = \Sigma = h_m - W h_f^*$$

### Table 32. Properties of Saturated Dowtherm A*

| Temp, deg F | Pressure, psia | Enthalpy, Btu per lb, above 53.6 F | | | Specific heat, liquid | Density, lb per cu ft | |
|---|---|---|---|---|---|---|---|
| | | Sat liquid, $h_f$ | Vapori-zation, $h_{fg}$ | Sat vapor, $h_g$ | | Liquid | Vapor |
| 500.0 | 14.7 | 222.0 | 123 | 345 | 0.63 | 54.1 | 0.28 |
| 510.0 | 18.1 | 228.0 | 121 | 349 | 0.63 | 53.7 | 0.32 |
| 520.0 | 20.4 | 234.0 | 120 | 354 | 0.64 | 53.2 | 0.36 |
| 530.0 | 22.7 | 240.0 | 119 | 359 | 0.64 | 53.0 | 0.40 |
| 540.0 | 25.1 | 247.0 | 118 | 365 | 0.65 | 52.7 | 0.44 |
| 550.0 | 27.0 | 253.0 | 117 | 370 | 0.65 | 52.3 | 0.48 |
| 560.0 | 30.8 | 260.0 | 115 | 375 | 0.65 | 51.9 | 0.54 |
| 570.0 | 34.6 | 267.0 | 114 | 381 | 0.66 | 51.6 | 0.60 |
| 580.0 | 36.6 | 274.0 | 112 | 386 | 0.66 | 51.2 | 0.67 |
| 590.0 | 41.4 | 281.0 | 111 | 392 | 0.66 | 50.8 | 0.75 |
| 600.0 | 44.3 | 288.0 | 110 | 398 | 0.66 | 50.4 | 0.88 |
| 610.0 | 46.2 | 295.0 | 109 | 404 | 0.67 | 50.1 | 1.00 |
| 620.0 | 53.0 | 302.0 | 107 | 409 | 0.67 | 49.8 | 1.10 |
| 630.0 | 57.6 | 309.0 | 106 | 415 | 0.67 | 49.3 | 1.17 |
| 640.0 | 63.6 | 316.0 | 105 | 421 | 0.67 | 49.1 | 1.24 |
| 650.0 | 68.4 | 323.0 | 104 | 427 | 0.67 | 48.6 | 1.29 |
| 660.0 | 74.2 | 330.0 | 102 | 432 | 0.68 | 48.4 | 1.34 |
| 670.0 | 80.8 | 337.0 | 101 | 438 | 0.68 | 47.9 | 1.40 |
| 680.0 | 87.7 | 344.0 | 99 | 443 | 0.68 | 47.5 | 1.5 |
| 690.0 | 95.4 | 351.0 | 98 | 449 | 0.68 | 47.2 | 1.6 |
| 700.0 | 104.0 | 358.0 | 97 | 455 | 0.68 | 46.9 | 1.7 |
| 710.0 | 113.0 | 365.0 | 95 | 460 | 0.68 | 46.3 | 1.8 |
| 720.0 | 119.0 | 372.0 | 93 | 465 | 0.68 | 45.9 | 1.9 |
| 730.0 | 131.0 | 379.0 | 92 | 471 | 0.68 | 45.5 | 2.1 |
| 740.0 | 142.0 | 386.0 | 90 | 476 | 0.68 | 44.9 | 2.3 |
| 750.0 | 150.0 | 393.0 | 89 | 482 | 0.68 | 44.4 | 2.5 |

*Dowtherm F boils at 350°F and freezes at 0°F. Additional information on this fluid is available from the "Dowtherm Handbook," published by the Dow Chemical Company, Midland, Michigan.

### Table 33. Properties of Saturated Ammonia

$h_f$ and $s_f$ are measured from −40°F

| Temp, deg F, $t$ | Pres-sure, psia, $p$ | Specific volume, cu ft per lb | | Enthalpy, Btu | | | Entropy | | |
|---|---|---|---|---|---|---|---|---|---|
| | | Sat liquid, $v_f$ | Sat vapor, $v_g$ | Sat liquid, $h_f$ | Vapor-ization, $h_{fg}$ | Sat vapor, $h_g$ | Sat liquid, $s_f$ | Vapor-ization, $s_{fg}$ | Sat vapor, $s_g$ |
| −40 | 10.41 | 0.02322 | 24.86 | 0.0 | 597.6 | 597.6 | 0.000 | 1.4242 | 1.4242 |
| −38 | 11.04 | 0.02326 | 23.53 | 2.1 | 596.2 | 598.3 | 0.0051 | 1.4142 | 1.4193 |
| −36 | 11.71 | 0.02331 | 22.27 | 4.3 | 594.8 | 599.1 | 0.0101 | 1.4043 | 1.4144 |
| −34 | 12.41 | 0.02335 | 21.10 | 6.4 | 593.5 | 599.9 | 0.0151 | 1.3945 | 1.4096 |
| −32 | 13.14 | 0.02340 | 20.00 | 8.5 | 592.1 | 600.6 | 0.0201 | 1.3847 | 1.4048 |
| −30 | 13.90 | 0.02345 | 18.97 | 10.7 | 590.7 | 601.4 | 0.0250 | 1.3751 | 1.4001 |
| −28 | 14.71 | 0.02349 | 18.00 | 12.8 | 589.3 | 602.1 | 0.0300 | 1.3655 | 1.3955 |
| −26 | 15.55 | 0.02354 | 17.09 | 14.9 | 587.9 | 602.8 | 0.0350 | 1.3559 | 1.3909 |
| −24 | 16.42 | 0.02359 | 16.24 | 17.1 | 586.5 | 603.6 | 0.0399 | 1.3464 | 1.3863 |
| −22 | 17.34 | 0.02364 | 15.43 | 19.2 | 585.1 | 604.3 | 0.0448 | 1.3370 | 1.3818 |
| −20 | 18.30 | 0.02369 | 14.68 | 21.4 | 583.6 | 605.0 | 0.0497 | 1.3277 | 1.3774 |
| −18 | 19.30 | 0.02374 | 13.97 | 23.5 | 582.2 | 605.7 | 0.0545 | 1.3184 | 1.3729 |
| −16 | 20.34 | 0.02378 | 13.29 | 25.6 | 580.8 | 606.4 | 0.0594 | 1.3092 | 1.3686 |
| −14 | 21.43 | 0.02383 | 12.66 | 27.8 | 579.3 | 607.1 | 0.0642 | 1.3001 | 1.3643 |
| −12 | 22.56 | 0.02384 | 12.06 | 30.0 | 577.8 | 607.8 | 0.0690 | 1.2910 | 1.3600 |
| −10 | 23.74 | 0.02393 | 11.50 | 32.1 | 576.4 | 608.5 | 0.0738 | 1.2820 | 1.3558 |
| − 8 | 24.97 | 0.02399 | 10.97 | 34.3 | 574.9 | 609.2 | 0.0786 | 1.2730 | 1.3516 |
| − 6 | 26.26 | 0.02404 | 10.47 | 36.4 | 573.4 | 609.8 | 0.0833 | 1.2641 | 1.3474 |
| − 4 | 27.59 | 0.02409 | 9.991 | 38.6 | 571.9 | 610.5 | 0.0880 | 1.2553 | 1.3433 |
| − 2 | 28.98 | 0.02414 | 9.541 | 40.7 | 570.4 | 611.1 | 0.0928 | 1.2465 | 1.3393 |

## Table 33. Properties of Saturated Ammonia—(*Continued*)

| Temp, deg F, | Pressure, psia, | Specific volume, cu ft per lb | | Enthalpy, Btu | | | Entropy | | |
|---|---|---|---|---|---|---|---|---|---|
| | | Sat liquid, $v_f$ | Sat vapor, $v_g$ | Sat liquid, $h_f$ | Vaporization, $h_{fg}$ | Sat vapor, $h_g$ | Sat liquid, $s_f$ | Vaporization, $s_{fg}$ | Sat vapor, $s_g$ |
| $t$ | $p$ | | | | | | | | |
| 0 | 30.42 | 0.02419 | 9.116 | 42.9 | 568.9 | 611.8 | 0.0975 | 1.2377 | 1.3352 |
| 2 | 31.92 | 0.02424 | 8.714 | 45.1 | 567.3 | 612.4 | 0.1022 | 1.2290 | 1.3312 |
| 4 | 33.47 | 0.02430 | 8.333 | 47.2 | 565.8 | 613.0 | 0.1069 | 1.2204 | 1.3273 |
| 6 | 35.09 | 0.02435 | 7.971 | 49.4 | 564.2 | 613.6 | 0.1115 | 1.2119 | 1.3234 |
| 8 | 36.77 | 0.02440 | 7.629 | 51.6 | 562.7 | 614.3 | 0.1162 | 1.2033 | 1.3195 |
| 10 | 38.51 | 0.02446 | 7.304 | 53.8 | 561.1 | 614.9 | 0.1208 | 1.1949 | 1.3157 |
| 12 | 40.31 | 0.02451 | 6.996 | 56.0 | 559.5 | 615.5 | 0.1254 | 1.1864 | 1.3118 |
| 14 | 42.18 | 0.02457 | 6.703 | 58.2 | 557.9 | 616.1 | 0.1300 | 1.1781 | 1.3081 |
| 16 | 44.12 | 0.02462 | 6.425 | 60.3 | 556.3 | 616.6 | 0.1346 | 1.1697 | 1.3043 |
| 18 | 46.13 | 0.02468 | 6.161 | 62.5 | 554.7 | 617.2 | 0.1392 | 1.1614 | 1.3006 |
| 20 | 48.21 | 0.02474 | 5.910 | 64.7 | 553.1 | 617.8 | 0.1437 | 1.1532 | 1.2969 |
| 22 | 50.36 | 0.02479 | 5.671 | 66.9 | 551.4 | 618.3 | 0.1483 | 1.1450 | 1.2933 |
| 24 | 52.59 | 0.02485 | 5.443 | 69.1 | 549.8 | 618.9 | 0.1528 | 1.1369 | 1.2897 |
| 26 | 54.90 | 0.02491 | 5.227 | 71.3 | 548.1 | 619.4 | 0.1573 | 1.1288 | 1.2861 |
| 28 | 57.28 | 0.02497 | 5.021 | 73.5 | 546.4 | 619.9 | 0.1618 | 1.1207 | 1.2825 |
| 30 | 59.74 | 0.02503 | 4.825 | 75.7 | 544.8 | 620.5 | 0.1663 | 1.1127 | 1.2790 |
| 32 | 62.29 | 0.02508 | 4.637 | 77.9 | 543.1 | 621.0 | 0.1708 | 1.1047 | 1.2755 |
| 34 | 64.91 | 0.02514 | 4.459 | 80.1 | 541.4 | 621.5 | 0.1753 | 1.0968 | 1.2721 |
| 36 | 67.63 | 0.02521 | 4.289 | 82.3 | 539.7 | 622.0 | 0.1797 | 1.0889 | 1.2686 |
| 38 | 70.43 | 0.02527 | 4.126 | 84.6 | 537.9 | 622.5 | 0.1841 | 1.0811 | 1.2652 |
| 40 | 73.32 | 0.02533 | 3.971 | 86.8 | 536.2 | 623.0 | 0.1885 | 1.0733 | 1.2618 |
| 42 | 76.31 | 0.02539 | 3.823 | 89.0 | 534.4 | 623.4 | 0.1930 | 1.0655 | 1.2585 |
| 44 | 79.38 | 0.02545 | 3.682 | 91.2 | 532.7 | 623.9 | 0.1974 | 1.0578 | 1.2552 |
| 46 | 82.55 | 0.02551 | 3.547 | 93.5 | 530.9 | 624.4 | 0.2018 | 1.0501 | 1.2519 |
| 48 | 85.82 | 0.02558 | 3.418 | 95.7 | 529.1 | 624.8 | 0.2062 | 1.0424 | 1.2486 |
| 50 | 89.19 | 0.02564 | 3.294 | 97.9 | 527.3 | 625.2 | 0.2105 | 1.0348 | 1.2453 |
| 52 | 92.66 | 0.02571 | 3.176 | 100.2 | 525.5 | 6.257 | 0.2149 | 1.0272 | 1.2421 |
| 54 | 96.23 | 0.02577 | 3.063 | 102.4 | 523.7 | 626.1 | 0.2192 | 1.0197 | 1.2389 |
| 56 | 99.91 | 0.02584 | 2.954 | 104.7 | 521.8 | 625.5 | 0.2236 | 1.0121 | 1.2357 |
| 58 | 103.7 | 0.02590 | 2.851 | 106.9 | 520.0 | 626.9 | 0.2279 | 1.0046 | 1.2325 |
| 60 | 107.6 | 0.02597 | 2.751 | 109.2 | 518.1 | 627.3 | 0.2322 | 0.9972 | 1.2294 |
| 62 | 111.6 | 0.02604 | 2.656 | 111.5 | 516.2 | 627.7 | 0.2365 | 0.9897 | 1.2262 |
| 64 | 115.7 | 0.02611 | 2.565 | 113.7 | 514.3 | 628.0 | 0.2408 | 0.9823 | 1.2231 |
| 66 | 120.0 | 0.02618 | 2.477 | 116.0 | 512.4 | 628.4 | 0.2451 | 0.9750 | 1.2201 |
| 68 | 124.3 | 0.02625 | 2.393 | 118.3 | 510.5 | 628.8 | 0.2494 | 0.9676 | 1.2170 |
| 70 | 128.8 | 0.02632 | 2.312 | 120.5 | 508.6 | 629.1 | 0.2537 | 0.9603 | 1.2140 |
| 72 | 133.4 | 0.02639 | 2.235 | 122.8 | 506.6 | 629.4 | 0.2579 | 0.9531 | 1.2110 |
| 74 | 138.1 | 0.02646 | 2.161 | 125.1 | 504.7 | 629.8 | 0.2622 | 0.9458 | 1.2080 |
| 76 | 143.0 | 0.02653 | 2.089 | 127.4 | 502.7 | 630.1 | 0.2664 | 0.9386 | 1.2050 |
| 78 | 147.9 | 0.02661 | 2.021 | 129.7 | 500.7 | 630.4 | 0.2706 | 0.9314 | 1.2020 |
| 80 | 153.0 | 0.02668 | 1.955 | 132.0 | 498.7 | 630.7 | 0.2749 | 0.9242 | 1.1991 |
| 82 | 158.3 | 0.02675 | 1.892 | 134.3 | 496.7 | 631.0 | 0.2791 | 0.9171 | 1.1962 |
| 84 | 163.7 | 0.02684 | 1.831 | 136.6 | 494.7 | 631.3 | 0.2833 | 0.9100 | 1.1933 |
| 86 | 169.2 | 0.02691 | 1.772 | 138.9 | 492.6 | 631.5 | 0.2875 | 0.9029 | 1.1904 |
| 88 | 174.8 | 0.02699 | 1.716 | 141.2 | 490.6 | 631.8 | 0.2917 | 0.8958 | 1.1875 |
| 90 | 180.6 | 0.02707 | 1.661 | 143.5 | 488.5 | 632.0 | 0.2958 | 0.8888 | 1.1846 |
| 92 | 186.6 | 0.02715 | 1.609 | 145.8 | 486.4 | 632.2 | 0.3000 | 0.8818 | 1.1818 |
| 94 | 192.7 | 0.02723 | 1.559 | 148.2 | 484.3 | 632.5 | 0.3041 | 0.8748 | 1.1789 |
| 96 | 198.9 | 0.02731 | 1.510 | 150.5 | 482.1 | 632.6 | 0.3083 | 0.8678 | 1.1761 |
| 98 | 205.3 | 0.02739 | 1.464 | 152.9 | 480.0 | 632.9 | 0.3125 | 0.8608 | 1.1733 |
| 100 | 211.9 | 0.02747 | 1.419 | 155.2 | 477.8 | 633.0 | 0.3166 | 0.8539 | 1.1705 |
| 105 | 228.9 | 0.02769 | 1.313 | 161.1 | 472.3 | 633.4 | 0.3269 | 0.8366 | 1.1635 |
| 110 | 247.0 | 0.02790 | 1.217 | 167.0 | 466.7 | 633.7 | 0.3372 | 0.8194 | 1.1566 |
| 115 | 266.2 | 0.02813 | 1.128 | 173.0 | 460.9 | 633.9 | 0.3474 | 0.8023 | 1.1497 |
| 120 | 286.4 | 0.02836 | 1.047 | 179.0 | 455.0 | 634.0 | 0.3576 | 0.7851 | 1.1427 |

## Table 34. Properties of Superheated Ammonia

$v$ = specific volume in ft$^3$/lb; $h$ = enthalpy in Btu/lb; $s$ = entropy. $h_f$ and $s_f$ are measured from $-40°$F

| Pressure, psia | Temp of saturated vapor, deg F | | Temperature of superheated vapor, deg F | | | | | | | | |
|---|---|---|---|---|---|---|---|---|---|---|---|
| | | | −30 | −20 | −10 | 0 | 10 | 20 | 30 | 40 | 50 |
| 10 | −41.34 | $v$ | 26.58 | 27.26 | 27.92 | 28.58 | 29.24 | 29.90 | 30.55 | 31.20 | 31.85 |
| | | $h$ | 603.2 | 608.5 | 613.7 | 618.9 | 624.0 | 629.1 | 634.2 | 639.3 | 644.4 |
| | | $s$ | 1.4420 | 1.4542 | 1.4659 | 1.4773 | 1.4884 | 1.4992 | 1.5097 | 1.5200 | 1.5301 |
| 20 | −16.64 | $v$ | ........ | ........ | 13.74 | 14.09 | 14.44 | 14.78 | 15.11 | 15.45 | 15.78 |
| | | $h$ | ........ | ........ | 610.0 | 615.5 | 621.0 | 626.4 | 631.7 | 637.0 | 642.3 |
| | | $s$ | ........ | ........ | 1.3784 | 1.3907 | 1.4025 | 1.4138 | 1.4248 | 1.4356 | 1.4460 |
| 30 | −0.57 | $v$ | ........ | ........ | ........ | 9.250 | 9.492 | 9.731 | 9.966 | 10.20 | 10.43 |
| | | $h$ | ........ | ........ | ........ | 611.9 | 617.8 | 623.5 | 629.1 | 634.6 | 640.1 |
| | | $s$ | ........ | ........ | ........ | 1.3371 | 1.3497 | 1.3618 | 1.3733 | 1.3845 | 1.3953 |
| 40 | 11.66 | $v$ | ........ | ........ | ........ | ........ | ........ | 7.203 | 7.387 | 7.568 | 7.746 |
| | | $h$ | ........ | ........ | ........ | ........ | ........ | 620.4 | 626.3 | 632.1 | 637.8 |
| | | $s$ | ........ | ........ | ........ | ........ | ........ | 1.3231 | 1.3353 | 1.3470 | 1.3583 |
| 50 | 21.67 | $v$ | ........ | ........ | ........ | ........ | ........ | ........ | 5.838 | 5.988 | 6.135 |
| | | $h$ | ........ | ........ | ........ | ........ | ........ | ........ | 623.4 | 629.5 | 635.4 |
| | | $s$ | ........ | ........ | ........ | ........ | ........ | ........ | 1.3046 | 1.3169 | 1.3286 |

| Pressure, psia | Temp of saturated vapor, deg F | | 100 | 120 | 140 | 160 | 180 | 200 | 240 | 280 | 320 |
|---|---|---|---|---|---|---|---|---|---|---|---|
| 80 | 44.40 | $v$ | 4.190 | 4.371 | 4.548 | 4.722 | 4.893 | 5.063 | 5.398 | 5.730 | |
| | | $h$ | 658.7 | 670.4 | 681.8 | 693.2 | 704.4 | 715.6 | 738.1 | 760.7 | |
| | | $s$ | 1.3199 | 1.3404 | 1.3598 | 1.3784 | 1.3963 | 1.4136 | 1.4467 | 1.4781 | |
| 100 | 56.05 | $v$ | 3.304 | 3.454 | 3.600 | 3.743 | 3.883 | 4.021 | 4.294 | 4.562 | |
| | | $h$ | 655.2 | 667.3 | 679.2 | 690.8 | 702.3 | 713.7 | 736.5 | 759.4 | |
| | | $s$ | 1.2891 | 1.3104 | 1.3305 | 1.3495 | 1.3678 | 1.3854 | 1.4190 | 1.4507 | |
| 120 | 66.02 | $v$ | 2.712 | 2.842 | 2.967 | 3.089 | 3.190 | 3.326 | 3.557 | 3.783 | |
| | | $h$ | 651.6 | 664.2 | 676.5 | 688.5 | 700.2 | 711.8 | 734.9 | 758.0 | |
| | | $s$ | 1.2628 | 1.2850 | 1.3058 | 1.3254 | 1.3441 | 1.3620 | 1.3960 | 1.4281 | |
| 140 | 74.79 | $v$ | 2.288 | 2.404 | 2.515 | 2.622 | 2.727 | 2.830 | 3.030 | 3.227 | 3.420 |
| | | $h$ | 647.8 | 661.1 | 673.7 | 686.0 | 698.0 | 709.9 | 733.3 | 756.7 | 780.0 |
| | | $s$ | 1.2396 | 1.2628 | 1.2843 | 1.3045 | 1.3236 | 1.3418 | 1.3763 | 1.4088 | 1.4395 |
| 160 | 82.64 | $v$ | 1.969 | 2.075 | 2.175 | 2.272 | 2.365 | 2.457 | 2.635 | 2.809 | 2.980 |
| | | $h$ | 643.9 | 657.8 | 670.9 | 683.5 | 695.8 | 707.9 | 731.7 | 755.3 | 778.9 |
| | | $s$ | 1.2186 | 1.2429 | 1.2652 | 1.2859 | 1.3054 | 1.3240 | 1.3591 | 1.3919 | 1.4229 |
| 180 | 89.78 | $v$ | 1.720 | 1.818 | 1.910 | 1.999 | 2.084 | 2.167 | 2.328 | 2.484 | 2.637 |
| | | $h$ | 639.9 | 654.4 | 668.0 | 681.0 | 693.6 | 705.9 | 730.1 | 753.9 | 777.7 |
| | | $s$ | 1.1992 | 1.2247 | 1.2477 | 1.2691 | 1.2891 | 1.3081 | 1.3436 | 1.3768 | 1.4081 |
| 200 | 96.34 | $v$ | 1.520 | 1.612 | 1.698 | 1.780 | 1.859 | 1.935 | 2.082 | 2.225 | 2.364 |
| | | $h$ | 635.6 | 650.9 | 665.0 | 678.4 | 691.3 | 703.9 | 728.4 | 752.5 | 776.5 |
| | | $s$ | 1.1809 | 1.2077 | 1.2317 | 1.2537 | 1.2742 | 1.2935 | 1.3296 | 1.3631 | 1.3947 |
| 220 | 102.42 | $v$ | ........ | 1.443 | 1.525 | 1.601 | 1.675 | 1.745 | 1.881 | 2.012 | 2.140 |
| | | $h$ | ........ | 647.3 | 662.0 | 675.8 | 689.1 | 701.9 | 726.8 | 751.1 | 775.3 |
| | | $s$ | ........ | 1.1917 | 1.2167 | 1.2394 | 1.2604 | 1.2801 | 1.3168 | 1.3507 | 1.3825 |
| 240 | 108.09 | $v$ | ........ | 1.302 | 1.380 | 1.452 | 1.521 | 1.587 | 1.714 | 1.835 | 1.954 |
| | | $h$ | ........ | 643.5 | 658.8 | 673.1 | 686.7 | 699.8 | 725.1 | 749.8 | 774.1 |
| | | $s$ | ........ | 1.1764 | 1.2025 | 1.2259 | 1.2475 | 1.2677 | 1.3049 | 1.3392 | 1.3712 |
| 260 | 113.42 | $v$ | ........ | 1.182 | 1.257 | 1.326 | 1.391 | 1.453 | 1.572 | 1.686 | 1.796 |
| | | $h$ | ........ | 639.5 | 655.6 | 670.4 | 684.4 | 697.7 | 723.4 | 748.4 | 772.9 |
| | | $s$ | ........ | 1.1617 | 1.1889 | 1.2132 | 1.2354 | 1.2560 | 1.2938 | 1.3258 | 1.3608 |

SOURCE: Condensed from *NBS Circ.* 142, 1923.

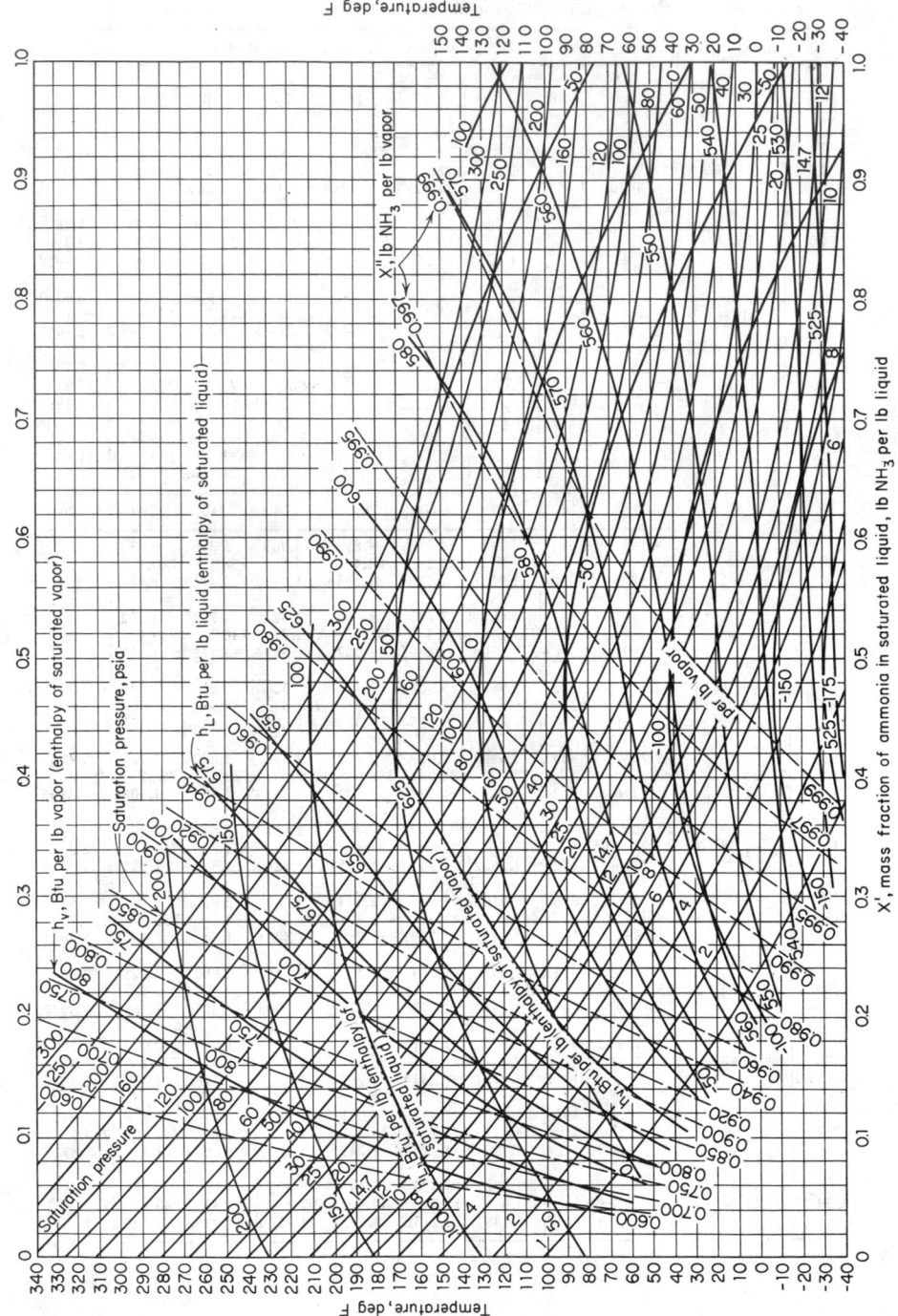

**Fig. 26**  Properties of aqua-ammonia. *(Kobloss and Scott, Refrig. Eng., Oct. 1950, reproduced by permission of ASHRAE.)*

### Table 35. Properties of Saturated Sulphur Dioxide

$h_f$ and $s_f$ are measured from $-40°F$

| Temp, deg F, | Pressure, psia, | Specific volume, cu ft per lb | | Enthalpy, Btu | | | Entropy | | |
|---|---|---|---|---|---|---|---|---|---|
| | | Sat liquid, | Sat vapor, | Sat liquid, | Vaporization, | Sat vapor, | Sat liquid, | Vaporization, | Sat vapor, |
| $t$ | $p$ | $v_f$ | $v_g$ | $h_f$ | $h_{fg}$ | $h_g$ | $s_f$ | $s_{fg}$ | $s_g$ |
| $-40$ | 3.136 | 0.01044 | 22.42 | 0.00 | 178.6 | 178.6 | 0.0000 | 0.4256 | 0.4256 |
| $-30$ | 4.331 | 0.01053 | 16.56 | 2.93 | 177.0 | 179.9 | 0.00674 | 0.4119 | 0.4186 |
| $-20$ | 5.883 | 0.01062 | 12.42 | 5.98 | 175.1 | 181.1 | 0.01366 | 0.3983 | 0.4119 |
| $-10$ | 7.863 | 0.01072 | 9.44 | 9.16 | 173.0 | 182.1 | 0.02075 | 0.3847 | 0.4054 |
| 0 | 10.35 | 0.01082 | 7.28 | 12.44 | 170.6 | 183.1 | 0.02795 | 0.3712 | 0.3992 |
| 10 | 13.42 | 0.01092 | 5.682 | 15.80 | 168.1 | 183.9 | 0.03519 | 0.3579 | 0.3931 |
| 20 | 17.18 | 0.01103 | 4.487 | 19.20 | 165.3 | 184.5 | 0.04241 | 0.3447 | 0.3871 |
| 30 | 21.70 | 0.01114 | 3.581 | 22.64 | 162.4 | 185.0 | 0.04956 | 0.3316 | 0.3812 |
| 40 | 27.10 | 0.01126 | 2.887 | 26.12 | 159.3 | 185.4 | 0.05668 | 0.3187 | 0.3754 |
| 50 | 33.45 | 0.01138 | 2.348 | 29.61 | 156.0 | 185.6 | 0.06370 | 0.3060 | 0.3697 |
| 60 | 40.90 | 0.01150 | 1.926 | 33.10 | 152.5 | 185.6 | 0.07060 | 0.2935 | 0.3641 |
| 70 | 49.62 | 0.01163 | 1.590 | 36.58 | 148.9 | 185.5 | 0.07736 | 0.2811 | 0.3585 |
| 80 | 59.68 | 0.01176 | 1.321 | 40.05 | 145.1 | 185.2 | 0.08399 | 0.2690 | 0.3529 |
| 90 | 71.25 | 0.01190 | 1.104 | 43.50 | 141.2 | 184.7 | 0.09038 | 0.2569 | 0.3473 |
| 100 | 84.52 | 0.01204 | 0.9262 | 46.90 | 137.2 | 184.1 | 0.09657 | 0.2452 | 0.3417 |
| 110 | 99.76 | 0.01219 | 0.7804 | 50.26 | 133.1 | 183.3 | 0.1025 | 0.2336 | 0.3361 |
| 120 | 120.9 | 0.01236 | 0.6598 | 53.58 | 128.8 | 182.4 | 0.1083 | 0.2222 | 0.3305 |
| 130 | 136.5 | 0.01253 | 0.5595 | 56.85 | 124.4 | 181.2 | 0.1138 | 0.2110 | 0.3247 |
| 140 | 158.6 | 0.01272 | 0.4758 | 60.04 | 119.9 | 179.9 | 0.1189 | 0.1999 | 0.3189 |

### Table 36. Properties of Superheated Sulphur Dioxide

$v$ = specific volume in ft³/lb; $h$ = enthalpy in Btu/lb; $s$ = entropy. $h_f$ amd $s_f$ are measured from $-40°F$

| Pressure, psia | Temp of saturated vapor, deg F | | Temperature of superheated vapor, deg F | | | | | | | | |
|---|---|---|---|---|---|---|---|---|---|---|---|
| | | | 0 | 20 | 40 | 60 | 80 | 100 | 120 | 140 | 160 |
| 6 | $-19.37$ | $v$ | 12.75 | 13.34 | 13.93 | 14.52 | 15.11 | 15.69 | 16.26 | 16.82 | 17.35 |
| | | $h$ | 184.3 | 187.5 | 190.7 | 193.9 | 197.2 | 200.5 | 203.8 | 207.1 | 210.4 |
| | | $s$ | 0.4185 | 0.4254 | 0.4320 | 0.4383 | 0.4444 | 0.4504 | 0.4561 | 0.4618 | 0.4672 |
| 10 | $-1.34$ | $v$ | 7.545 | 7.939 | 8.316 | 8.681 | 9.038 | 9.389 | 9.736 | 10.08 | 10.42 |
| | | $h$ | 183.2 | 186.7 | 190.1 | 193.5 | 196.9 | 200.3 | 203.7 | 207.1 | 210.5 |
| | | $s$ | 0.4005 | 0.4080 | 0.4151 | 0.4216 | 0.4280 | 0.4341 | 0.4400 | 0.4457 | 0.4512 |
| 15 | 14.43 | $v$ | ........ | 5.192 | 5.470 | 5.734 | 5.988 | 6.233 | 6.471 | 6.705 | 6.937 |
| | | $h$ | ........ | 185.4 | 189.2 | 192.8 | 196.4 | 199.9 | 203.3 | 206.7 | 210.1 |
| | | $s$ | ........ | 0.3927 | 0.4005 | 0.4078 | 0.4144 | 0.4208 | 0.4268 | 0.4326 | 0.4383 |
| 20 | 26.44 | $v$ | ........ | ........ | 4.035 | 4.251 | 4.454 | 4.648 | 4.834 | 5.015 | 5.193 |
| | | $h$ | ........ | ........ | 187.8 | 191.8 | 195.6 | 199.3 | 202.9 | 206.5 | 209.9 |
| | | $s$ | ........ | ........ | 0.3896 | 0.3972 | 0.4043 | 0.4109 | 0.4173 | 0.4232 | 0.4290 |
| 25 | 36.33 | $v$ | ........ | ........ | 3.181 | 3.363 | 3.536 | 3.696 | 3.848 | 3.998 | 4.145 |
| | | $h$ | ........ | ........ | 186.1 | 190.6 | 194.7 | 198.6 | 202.4 | 206.0 | 209.6 |
| | | $s$ | ........ | ........ | 0.3793 | 0.3880 | 0.3958 | 0.4029 | 0.4095 | 0.4157 | 0.4216 |
| | | | 60 | 80 | 100 | 120 | 140 | 160 | 180 | 200 | 220 |
| 30 | 44.76 | $v$ | 2.747 | 2.907 | 3.052 | 3.189 | 3.318 | 3.443 | 3.565 | 3.685 | 3.803 |
| | | $h$ | 189.3 | 193.8 | 197.9 | 201.8 | 205.6 | 209.3 | 212.9 | 216.5 | 220.1 |
| | | $s$ | 0.3797 | 0.3885 | 0.3960 | 0.4029 | 0.4094 | 0.4154 | 0.4211 | 0.4266 | 0.4318 |
| 40 | 58.83 | $v$ | 1.980 | 2.121 | 2.246 | 2.360 | 2.465 | 2.565 | 2.662 | 2.755 | 2.845 |
| | | $h$ | 185.9 | 191.3 | 196.1 | 200.4 | 204.6 | 208.5 | 212.3 | 216.0 | 219.7 |
| | | $s$ | 0.3654 | 0.3754 | 0.3842 | 0.3918 | 0.3988 | 0.4053 | 0.4113 | 0.4169 | 0.4223 |
| 60 | 80.29 | $v$ | ........ | ........ | 1.288 | 1.403 | 1.514 | 1.608 | 1.689 | 1.751 | 1.819 |
| | | $h$ | ........ | ........ | 191.4 | 197.0 | 201.9 | 206.5 | 210.7 | 214.8 | 218.7 |
| | | $s$ | ........ | ........ | 0.3640 | 0.3738 | 0.3822 | 0.3896 | 0.3964 | 0.4026 | 0.4084 |
| 80 | 96.88 | $v$ | ........ | ........ | 0.993 | 1.084 | 1.163 | 1.232 | 1.292 | 1.347 | 1.400 |
| | | $h$ | ........ | ........ | 185.6 | 192.5 | 198.6 | 203.9 | 208.7 | 213.3 | 217.5 |
| | | $s$ | ........ | ........ | 0.3457 | 0.3580 | 0.3682 | 0.3769 | 0.3846 | 0.3915 | 0.3978 |

### Table 37. Properties of Carbon Dioxide

$h_f$ and $s_f$ are measured from 32°F

| Temp, deg F, | Pressure, psia, | Density, lb per cu ft | | Enthalpy, Btu | | | Entropy | |
|---|---|---|---|---|---|---|---|---|
| | | Sat liquid | Sat vapor | Sat liquid, $h_f$ | Vapor-ization, $h_{fg}$ | Sat vapor, $h_g$ | Sat liquid, $s_f$ | Sat vapor, $s_g$ |
| $t$ | $p$ | | | | | | | |
| −40 | 145.87 | 69.8 | 1.64 | −38.5 | 136.5 | 98.0 | −0.0850 | 0.2400 |
| −35 | 161.33 | 69.1 | 1.83 | −35.8 | 134.3 | 98.5 | −0.0793 | 0.2367 |
| −30 | 177.97 | 68.3 | 2.02 | −33.1 | 132.1 | 99.0 | −0.0735 | 0.2336 |
| −25 | 195.85 | 67.6 | 2.23 | −30.4 | 129.8 | 99.4 | −0.0676 | 0.2306 |
| −20 | 215.02 | 66.9 | 2.44 | −27.7 | 127.5 | 99.8 | −0.0619 | 0.2277 |
| −15 | 235.53 | 66.1 | 2.66 | −24.9 | 125.0 | 100.1 | −0.0560 | 0.2250 |
| −10 | 257.46 | 65.3 | 2.91 | −22.1 | 122.4 | 100.3 | −0.0500 | 0.2220 |
| − 5 | 280.85 | 64.5 | 3.17 | −19.4 | 120.0 | 100.6 | −0.0440 | 0.2198 |
| 0 | 305.76 | 63.6 | 3.46 | −16.7 | 117.5 | 100.8 | −0.0381 | 0.2173 |
| 5 | 332.2 | 62.8 | 3.77 | −14.0 | 115.0 | 101.0 | −0.0322 | 0.2151 |
| 10 | 360.4 | 61.9 | 4.12 | −11.2 | 112.2 | 101.0 | −0.0264 | 0.2124 |
| 15 | 390.2 | 61.0 | 4.49 | − 8.4 | 109.4 | 101.0 | −0.0204 | 0.2100 |
| 20 | 421.8 | 60.0 | 4.89 | − 5.5 | 106.3 | 100.8 | −0.0144 | 0.2071 |
| 25 | 455.3 | 59.0 | 5.33 | − 2.5 | 103.1 | 100.6 | −0.0083 | 0.2043 |
| 30 | 490.6 | 58.0 | 5.81 | + 0.4 | 99.7 | 100.1 | −0.0021 | 0.2012 |
| 35 | 528.0 | 57.0 | 6.35 | 3.5 | 95.8 | 99.3 | +0.0039 | 0.1975 |
| 40 | 567.3 | 55.9 | 6.91 | 6.6 | 91.8 | 98.4 | 0.0099 | 0.1934 |
| 45 | 608.9 | 54.7 | 7.60 | 9.8 | 87.5 | 97.3 | 0.0160 | 0.1892 |
| 50 | 652.7 | 53.4 | 8.37 | 12.9 | 83.2 | 96.1 | 0.0220 | 0.1852 |
| 55 | 698.8 | 52.1 | 9.27 | 16.1 | 78.7 | 94.8 | 0.0282 | 0.1809 |
| 60 | 747.4 | 50.7 | 10.2 | 19.4 | 74.0 | 93.4 | 0.0345 | 0.1767 |
| 65 | 798.6 | 49.1 | 11.3 | 22.9 | 68.9 | 91.8 | 0.0412 | 0.1724 |
| 70 | 852.4 | 47.3 | 12.6 | 26.6 | 62.7 | 89.3 | 0.0482 | 0.1665 |
| 75 | 909.3 | 45.1 | 14.2 | 30.9 | 54.8 | 85.7 | 0.0562 | 0.1587 |
| 80 | 969.3 | 42.4 | 16.2 | 35.6 | 44.0 | 79.6 | 0.0649 | 0.1464 |
| 85 | 1032.7 | 38.2 | 19.1 | 41.7 | 27.5 | 69.2 | 0.0761 | 0.1265 |
| 88 | 1072.1 | 32.9 | 25.4 | Critical point at 88.43 F | | | | |

### Table 38. Properties of Ethyl Chloride and of Methyl Chloride

| Temp, deg F, | Ethyl chloride | | | | | Methyl chloride | | | | |
|---|---|---|---|---|---|---|---|---|---|---|
| | Pressure, psia, | Specific vol of vapor, cu ft per lb, | Enthalpy, Btu from 32 F | | | Pressure, psia, | Specific vol of vapor, cu ft per lb, | Enthalpy, Btu from 32 F | | |
| | | | Sat liquid, $h_f$ | Vapor-ization, $h_{fg}$ | Sat vapor, $h_g$ | | | Sat liquid, $h_f$ | Vapor-ization, $h_{fg}$ | Sat vapor, $h_g$ |
| $t$ | $p$ | $v_g$ | | | | $p$ | $v_g$ | | | |
| −20 | 2.16 | 29.54 | −19.0 | 177.6 | 158.6 | 11.75 | 8.09 | −19.0 | 186.4 | 167.4 |
| −15 | 2.53 | 26.07 | −17.2 | 176.8 | 159.6 | 13.43 | 7.22 | −17.19 | 185.3 | 168.1 |
| −10 | 2.94 | 22.95 | −15.4 | 175.9 | 160.5 | 15.00 | 6.46 | −15.38 | 184.2 | 168.8 |
| − 5 | 3.41 | 20.28 | −13.5 | 175.1 | 161.6 | 16.79 | 5.80 | −13.58 | 183.1 | 169.5 |
| 0 | 3.93 | 18.04 | −11.7 | 174.2 | 162.5 | 18.80 | 5.18 | −11.75 | 182.0 | 170.2 |
| + 5 | 4.50 | 16.10 | − 9.8 | 173.4 | 163.6 | 21.00 | 4.65 | − 9.93 | 180.8 | 170.9 |
| 10 | 5.13 | 14.36 | − 8.1 | 172.6 | 164.5 | 23.30 | 4.18 | − 8.06 | 179.6 | 171.6 |
| 15 | 5.86 | 12.86 | − 6.3 | 171.8 | 165.5 | 25.92 | 3.78 | − 6.24 | 178.5 | 172.2 |
| 20 | 6.65 | 11.56 | − 4.4 | 170.9 | 166.5 | 28.8 | 3.41 | − 4.32 | 177.3 | 172.9 |
| 25 | 7.48 | 10.35 | − 2.5 | 170.1 | 167.6 | 31.9 | 3.09 | − 2.48 | 176.1 | 173.6 |
| 30 | 8.40 | 9.22 | − 0.6 | 169.2 | 168.6 | 35.2 | 2.81 | − 0.62 | 174.9 | 174.3 |
| 35 | 9.42 | 8.27 | + 1.3 | 168.4 | 169.6 | 38.7 | 2.54 | + 1.25 | 173.7 | 174.9 |
| 40 | 10.53 | 7.43 | 3.2 | 167.5 | 170.7 | 42.6 | 2.31 | 3.15 | 172.4 | 175.6 |
| 45 | 11.77 | 6.70 | 5.1 | 166.6 | 171.7 | 47.0 | 2.10 | 5.04 | 171.2 | 176.2 |
| 50 | 13.20 | 6.04 | 6.9 | 165.6 | 172.5 | 51.5 | 1.93 | 6.88 | 169.9 | 176.8 |
| 55 | 14.67 | 5.42 | 8.9 | 164.6 | 173.5 | 56.4 | 1.76 | 8.80 | 168.6 | 177.4 |
| 60 | 16.45 | 4.84 | 10.8 | 163.6 | 174.4 | 61.6 | 1.61 | 10.70 | 167.3 | 178.0 |
| 65 | 18.17 | 4.38 | 12.8 | 162.6 | 175.4 | 67.3 | 1.48 | 12.63 | 166.0 | 178.6 |
| 70 | 20.03 | 4.00 | 14.7 | 161.6 | 176.3 | 73.3 | 1.34 | 14.52 | 164.6 | 179.2 |
| 75 | 22.11 | 3.65 | 16.6 | 160.6 | 177.2 | 79.2 | 1.24 | 16.46 | 163.3 | 179.7 |
| 80 | 24.33 | 3.39 | 18.6 | 159.5 | 178.1 | 85.3 | 1.14 | 18.36 | 161.9 | 180.2 |
| 85 | 26.67 | 3.17 | 20.6 | 158.5 | 179.1 | 94.1 | 1.05 | 20.26 | 160.5 | 180.7 |
| 90 | 29.24 | 2.99 | 22.6 | 157.3 | 179.9 | 102.1 | 0.98 | 22.13 | 159.1 | 181.2 |
| 95 | 32.03 | 2.83 | 24.6 | 156.1 | 180.6 | 110.3 | 0.906 | 24.07 | 157.7 | 181.8 |
| 100 | 34.93 | 2.70 | 26.6 | 154.9 | 181.5 | 118.8 | 0.85 | 26.06 | 156.3 | 182.3 |
| 105 | 37.97 | 2.61 | 28.6 | 153.8 | 182.4 | 128.1 | 0.804 | 28.02 | 154.9 | 182.9 |
| 110 | 41.10 | 2.51 | 30.5 | 152.6 | 183.1 | 137.6 | 0.765 | 30.03 | 153.4 | 183.5 |

### Table 39. Properties of Propane and Butane

| Temp, deg F | Propane ($C_3H_8$) (Heat measurements are from 0 F) | | | | | | Butane ($C_4H_{10}$) (Heat measurements are from 0 F) | | | | | |
| --- | --- | --- | --- | --- | --- | --- | --- | --- | --- | --- | --- | --- |
| | Pressure, psia | Specific volume of vapor, cu ft per lb | Enthalpy, Btu per lb Liquid $h_f$ | Vapor $h_g$ | Entropy Liquid $S_f$ | Vapor $S_g$ | Pressure, psia | Specific volume of vapor, cu ft per lb | Enthalpy, Btu per lb Liquid $h_f$ | Vapor $h_g$ | Entropy Liquid $S_f$ | Vapor $S_g$ |
| −70 | 7.37 | 12.9 | −37.0 | 152.5 | −0.086 | 0.400 | | | | | | |
| −60 | 9.72 | 9.93 | −32.0 | 155.0 | −0.074 | 0.393 | | | | | | |
| −50 | 12.6 | 7.74 | −26.5 | 158.0 | −0.061 | 0.389 | | | | | | |
| −40 | 16.2 | 6.13 | −21.5 | 160.0 | −0.049 | 0.384 | | | | | | |
| −30 | 20.3 | 4.93 | −16.0 | 163.0 | −0.036 | 0.380 | | | | | | |
| −20 | 25.4 | 4.00 | −11.0 | 165.0 | −0.024 | 0.377 | | | | | | |
| −10 | 31.4 | 3.26 | −5.5 | 168.0 | −0.012 | 0.374 | | | | | | |
| 0 | 38.2 | 2.71 | 0 | 170.5 | 0.000 | 0.371 | 7.3 | 11.10 | 0 | 170.5 | 0.000 | 0.371 |
| +10 | 46.0 | 2.27 | 5.5 | 173.5 | 0.012 | 0.370 | 9.2 | 8.95 | 5.5 | 174.0 | 0.011 | 0.370 |
| 20 | 55.5 | 1.90 | 11.0 | 176.0 | 0.024 | 0.368 | 11.6 | 7.23 | 10.5 | 177.5 | 0.022 | 0.370 |
| 30 | 66.3 | 1.60 | 17.0 | 179.0 | 0.035 | 0.366 | 14.4 | 5.90 | 16.0 | 181.5 | 0.033 | 0.371 |
| 40 | 78.0 | 1.37 | 23.0 | 182.0 | 0.047 | 0.366 | 17.7 | 4.88 | 21.5 | 185.0 | 0.044 | 0.371 |
| 50 | 91.8 | 1.18 | 29.0 | 185.0 | 0.059 | 0.365 | 21.6 | 4.07 | 27.0 | 188.5 | 0.056 | 0.373 |
| 60 | 107.1 | 1.01 | 35.0 | 188.0 | 0.070 | 0.364 | 26.3 | 3.40 | 33.0 | 192.5 | 0.067 | 0.374 |
| 70 | 124.0 | 0.883 | 41.0 | 190.5 | 0.082 | 0.364 | 31.6 | 2.88 | 38.5 | 196.0 | 0.078 | 0.375 |
| 80 | 142.8 | 0.770 | 47.5 | 193.5 | 0.093 | 0.364 | 37.6 | 2.46 | 44.5 | 199.5 | 0.089 | 0.376 |
| 90 | 164.0 | 0.673 | 54.0 | 196.5 | 0.105 | 0.364 | 44.5 | 2.10 | 51.0 | 203.0 | 0.100 | 0.377 |
| 100 | 187.0 | 0.591 | 60.5 | 199.0 | 0.116 | 0.363 | 52.2 | 1.81 | 57.0 | 206.5 | 0.111 | 0.378 |
| 110 | 212.0 | 0.521 | 67.0 | 201.0 | 0.128 | 0.363 | 60.8 | 1.58 | 63.5 | 210.5 | 0.122 | 0.380 |
| 120 | 240.0 | 0.459 | 73.5 | 202.5 | 0.140 | 0.363 | 70.8 | 1.38 | 70.0 | 213.5 | 0.134 | 0.382 |
| 130 | ...... | ...... | ...... | ...... | ...... | ...... | 81.4 | 1.21 | 76.5 | 217.0 | 0.145 | 0.384 |
| 140 | ...... | ...... | ...... | ...... | ...... | ...... | 92.6 | 1.07 | 83.5 | 221.0 | 0.157 | 0.386 |

### Table 40. Properties of Freon 11 and Freon 12

| Temp, deg F | Freon 11 ($CCl_3F$) (Heat measurements are from −40 F) | | | | | | Freon 12 ($CCl_2F_2$) (Heat measurements are from −40 F) | | | | | |
| --- | --- | --- | --- | --- | --- | --- | --- | --- | --- | --- | --- | --- |
| | Pressure, psia | Specific volume of vapor, cu ft per lb | Enthalpy, Btu per lb Liquid $h_f$ | Vapor $h_g$ | Entropy Liquid $S_f$ | Vapor $S_g$ | Pressure, psia | Specific volume of vapor, cu ft per lb | Enthalpy, Btu per lb Liquid $h_f$ | Vapor $h_g$ | Entropy Liquid $S_f$ | Vapor $S_g$ |
| −40 | 0.739 | 44.2 | 0.00 | 87.48 | 0.0000 | 0.2085 | 9.3 | 3.91 | 0.00 | 73.50 | 0.0000 | 0.1752 |
| −30 | 1.03 | 32.3 | 1.97 | 88.67 | 0.0046 | 0.2064 | 12.0 | 3.09 | 2.03 | 74.70 | 0.00471 | 0.1739 |
| −20 | 1.42 | 24.1 | 3.94 | 89.87 | 0.0091 | 0.2046 | 15.3 | 2.47 | 4.07 | 75.87 | 0.00940 | 0.1727 |
| −10 | 1.92 | 18.2 | 5.91 | 91.07 | 0.0136 | 0.2030 | 19.2 | 2.00 | 6.14 | 77.05 | 0.01403 | 0.1717 |
| 0 | 2.55 | 13.9 | 7.89 | 92.27 | 0.0179 | 0.2015 | 23.9 | 1.64 | 8.25 | 78.21 | 0.01869 | 0.1709 |
| 10 | 3.35 | 10.8 | 9.88 | 93.48 | 0.0222 | 0.2003 | 29.3 | 1.35 | 10.39 | 79.36 | 0.02328 | 0.1701 |
| 15 | 3.82 | 9.59 | 10.88 | 94.09 | 0.0244 | 0.1997 | 32.4 | 1.23 | 11.48 | 79.94 | 0.02556 | 0.1698 |
| 20 | 4.34 | 8.52 | 11.87 | 94.69 | 0.0264 | 0.1991 | 35.7 | 1.12 | 12.55 | 80.49 | 0.02783 | 0.1695 |
| 25 | 4.92 | 7.58 | 12.88 | 95.30 | 0.0285 | 0.1986 | 39.3 | 1.02 | 13.66 | 81.06 | 0.03008 | 0.1692 |
| 30 | 5.56 | 6.75 | 13.88 | 95.91 | 0.0306 | 0.1981 | 43.2 | 0.939 | 14.76 | 81.61 | 0.03233 | 0.1689 |
| 35 | 6.26 | 6.07 | 14.88 | 96.51 | 0.0326 | 0.1976 | 47.3 | 0.862 | 15.87 | 82.16 | 0.03458 | 0.1686 |
| 40 | 7.03 | 5.45 | 15.89 | 97.11 | 0.0346 | 0.1972 | 51.7 | 0.792 | 17.00 | 82.71 | 0.03680 | 0.1683 |
| 45 | 7.88 | 4.90 | 16.91 | 97.72 | 0.0366 | 0.1968 | 56.4 | 0.730 | 18.14 | 83.26 | 0.03903 | 0.1681 |
| 50 | 8.80 | 4.42 | 17.92 | 98.32 | 0.0386 | 0.1964 | 61.4 | 0.673 | 19.27 | 83.78 | 0.04126 | 0.1678 |
| 55 | 9.81 | 4.00 | 18.95 | 98.93 | 0.0406 | 0.1960 | 66.7 | 0.622 | 20.41 | 84.31 | 0.04348 | 0.1676 |
| 60 | 10.9 | 3.63 | 19.96 | 99.53 | 0.0426 | 0.1958 | 72.4 | 0.575 | 21.57 | 84.82 | 0.04568 | 0.1674 |
| 70 | 13.4 | 2.99 | 22.02 | 100.73 | 0.0465 | 0.1951 | 84.8 | 0.493 | 23.90 | 85.82 | 0.05009 | 0.1670 |
| 80 | 16.3 | 2.49 | 24.09 | 101.93 | 0.0504 | 0.1947 | 98.8 | 0.425 | 26.28 | 86.80 | 0.05446 | 0.1666 |
| 90 | 19.7 | 2.09 | 26.18 | 103.12 | 0.0542 | 0.1942 | 114.3 | 0.368 | 28.70 | 87.74 | 0.05882 | 0.1662 |
| 100 | 23.6 | 1.76 | 28.27 | 104.30 | 0.0580 | 0.1938 | 131.6 | 0.319 | 31.16 | 88.62 | 0.06316 | 0.1658 |
| 110 | 28.1 | 1.50 | 30.40 | 105.47 | 0.0617 | 0.1935 | 150.7 | 0.277 | 33.65 | 89.43 | 0.06749 | 0.1654 |
| 120 | 33.2 | 1.28 | 32.53 | 106.63 | 0.0654 | 0.1933 | 171.8 | 0.240 | 36.16 | 90.15 | 0.07180 | 0.1649 |
| 130 | 39.0 | 1.10 | 34.67 | 107.78 | 0.0691 | 0.1931 | 194.9 | 0.208 | 38.69 | 90.76 | 0.07607 | 0.1644 |

### Table 41. Properties of Freon 21 and Freon 22

| Temp, deg F | Freon 22 (CHClF₂) (Heat measurements are from −40 F) | | | | | | Freon 21 (CHCl₂F) (Heat measurements are from −40 F) | | | | | |
|---|---|---|---|---|---|---|---|---|---|---|---|---|
| | Pressure, psia | Specific volume of vapor, cu ft per lb | Enthalpy, Btu per lb Liquid $h_f$ | Enthalpy, Btu per lb Vapor $h_g$ | Entropy Liquid $S_f$ | Entropy Vapor $S_g$ | Pressure, psia | Specific volume of vapor, cu ft per lb | Enthalpy, Btu per lb Liquid $h_f$ | Enthalpy, Btu per lb Vapor $h_g$ | Entropy Liquid $S_f$ | Entropy Vapor $S_g$ |
| −40 | 15.31 | 3.279 | 0.00 | 100.46 | 0.0000 | 0.2394 | 1.36 | 32.1 | 0.00 | 114.6 | 0.0000 | 0.2730 |
| −30 | 19.72 | 2.590 | 2.62 | 101.63 | 0.0062 | 0.2367 | 1.89 | 23.6 | 2.36 | 115.8 | 0.0055 | 0.2695 |
| −20 | 25.01 | 2.074 | 5.28 | 102.79 | 0.0123 | 0.2341 | 2.58 | 17.7 | 4.71 | 117.0 | 0.0109 | 0.2663 |
| −10 | 31.29 | 1.681 | 7.96 | 103.92 | 0.0182 | 0.2316 | 3.46 | 13.4 | 7.07 | 118.2 | 0.0162 | 0.2633 |
| 0 | 38.79 | 1.373 | 10.63 | 105.02 | 0.0240 | 0.2293 | 4.58 | 10.3 | 9.44 | 119.4 | 0.0214 | 0.2606 |
| 10 | 47.63 | 1.130 | 13.29 | 106.08 | 0.0296 | 0.2272 | 5.98 | 8.08 | 11.81 | 120.6 | 0.0265 | 0.2581 |
| 15 | 52.63 | 1.028 | 14.63 | 106.61 | 0.0313 | 0.2262 | 6.80 | 7.18 | 13.01 | 121.2 | 0.0291 | 0.2569 |
| 20 | 57.98 | 0.9369 | 15.98 | 107.13 | 0.0352 | 0.2253 | 7.70 | 6.39 | 14.21 | 121.8 | 0.0316 | 0.2559 |
| 25 | 63.75 | 0.8552 | 17.34 | 107.63 | 0.0377 | 0.2244 | 8.70 | 5.71 | 15.40 | 122.4 | 0.0341 | 0.2548 |
| 30 | 69.93 | 0.7816 | 18.74 | 108.13 | 0.0409 | 0.2235 | 9.79 | 5.11 | 16.61 | 123.0 | 0.0365 | 0.2538 |
| 35 | 76.59 | 0.7157 | 20.20 | 108.62 | 0.0439 | 0.2226 | 11.0 | 4.59 | 17.83 | 123.6 | 0.0390 | 0.2528 |
| 40 | 83.72 | 0.6559 | 21.70 | 109.09 | 0.0469 | 0.2218 | 12.3 | 4.13 | 19.04 | 124.2 | 0.0414 | 0.2519 |
| 45 | 91.31 | 0.6024 | 23.20 | 109.54 | 0.0499 | 0.2209 | 13.8 | 3.73 | 20.27 | 124.8 | 0.0439 | 0.2510 |
| 50 | 99.4 | 0.5537 | 24.73 | 109.98 | 0.0528 | 0.2201 | 15.3 | 3.37 | 21.49 | 125.4 | 0.0463 | 0.2502 |
| 55 | 108.0 | 0.5099 | 26.27 | 110.39 | 0.0558 | 0.2193 | 17.0 | 3.10 | 22.73 | 126.0 | 0.0487 | 0.2493 |
| 60 | 117.2 | 0.4695 | 27.83 | 110.78 | 0.0588 | 0.2185 | 18.9 | 2.77 | 23.98 | 126.6 | 0.0511 | 0.2486 |
| 70 | 137.2 | 0.4000 | 30.99 | 111.49 | 0.0648 | 0.2168 | 23.1 | 2.30 | 26.49 | 127.8 | 0.0559 | 0.2471 |
| 80 | 159.7 | 0.3417 | 34.27 | 112.13 | 0.0708 | 0.2151 | 28.0 | 1.92 | 29.03 | 129.0 | 0.0606 | 0.2458 |
| 90 | 184.8 | 0.2928 | 37.61 | 112.67 | 0.0768 | 0.2133 | 33.6 | 1.62 | 31.59 | 130.1 | 0.0652 | 0.2446 |
| 100 | 212.6 | 0.2517 | 40.98 | 113.06 | 0.0827 | 0.2115 | 40.0 | 1.37 | 34.18 | 131.3 | 0.0699 | 0.2434 |
| 110 | 243.4 | 0.2167 | 44.35 | 113.23 | 0.0886 | 0.2096 | 47.4 | 1.17 | 36.79 | 132.4 | 0.0745 | 0.2424 |
| 120 | 277.3 | 0.1871 | 47.85 | 113.52 | 0.0945 | 0.2078 | 55.7 | 1.00 | 39.46 | 133.5 | 0.0791 | 0.2414 |
| 130 | 313.5 | 0.1629 | 51.5 | 113.71 | | | 65.1 | 0.862 | 42.13 | 134.6 | 0.0837 | 0.2405 |

EXAMPLE. A mixture of dry air and **saturated** steam; $p_t = 24$ inHg; $t_d = 76°F$. Partial pressure of water vapor: from Table 42,

$$p_v = p_d = 0.905 \text{ in Hg}$$

Partial pressure of dry air: $p_a = p_t - p_v = 23.095$ in Hg.
Specific humidity:

$$W = 0.905/1.61(23.095) = 0.0243 \text{ lb/lb dry air}$$

Volume of mixture per pound of dry air:

$$v_a = 0.754(536)/23.095 = 17.5 \text{ ft}^3$$

Volume of mixture per pound of mixture:

$$v_m = 17.5/1.0243 = 17.1 \text{ ft}^3$$

Enthalpy of mixture:

$$b_m = 0.240(76) + 0.0243(1,095) = 44.85 \text{ Btu/lb dry air}$$

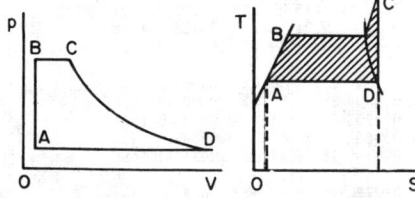

**Fig. 27** Rankine cycle.

EXAMPLE. A mixture of dry air and **superheated** steam; $p_t = 24$ inHg; $t_d = 76$ F; $t_w = t^* = 62°F$.
Pressure of saturated steam at $t^* = 0.560$ inHg (from Table 42):

$$W^* = 0.560/1.61(23.44) = 0.01484 \text{ lb/lb dry air}$$

Specific humidity:

$$W = 0.01484 - \frac{0.2465(14)}{1,065.4} = 0.0116 \text{ lb/lb dry air}$$

Partial pressure of water vapor:

and

$$0.0116 = p_v/1.61(24 - p_v)$$
$$p_v = 0.44 \text{ inHg}$$

Relative humidity: $r = 0.44/0.905 = 0.486$.
Volume of mixture per pound of dry air:

$$v_a = 0.754(536)/23.56 = 17.2 \text{ ft}^3$$

Volume of mixture per pound of mixture:

$$v_m = 17.2/1.0116 = 17.0 \text{ ft}^3$$

Enthalpy of mixture:

$$b_m = 0.240(76) + 0.0116(1,095) = 30.95 \text{ Btu/lb dry air}$$

**Psychrometric Charts** For occasional use, algebraic equations are less confusing and more reliable; for frequent use, a **psychrometric chart** may be preferable. A disadvantage of charts is that each applies for only one value of barometric pressure, usually 760 mm or 30 inHg. Correction to other barometric readings is not simple. The equations have the advantage that the actual barometric pressure is taken into account. The equations are often more convenient for equal accuracy or more accurate for equal convenience.

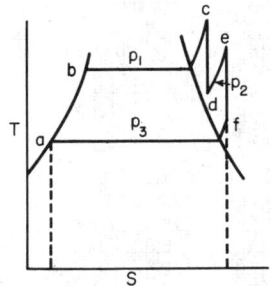

**Fig. 28**  Reheating cycle.

Psychrometric charts are usually plotted, as indicated by Fig. 30, with dry-bulb temperature as abscissa and specific humidity as ordinate. Since the specific humidity is determined by the vapor pressure and the barometric pressure (which is constant for a given chart), and is nearly proportional to the vapor pressure, a second ordinate scale, departing slightly from uniform graduations, will give the vapor pressure. The saturation curve ($r = 1.0$) gives the specific humidity and vapor pressure for a mixture of air and saturated vapor. Similar curves below it give results for various values of relative humidity. Inclined lines of one set carry fixed values of the wet-bulb temperature, and those of another set carry fixed values of $v_a$, cubic feet per pound of air. Many charts carry additional scales of enthalpy or $\Sigma$ function.

Any two values will locate the point representing the state of the atmosphere, and the desired values can be read directly.

Figures 31 and 32 are psychrometric charts from the General Electric Company and Ellenwood and Mackey, "Thermodynamic Charts," covering a dry-bulb temperature range from

32 to 300°F. They are accurate only for a barometric pressure of 29.92 inHg.

**Air-conditioning processes** alter the temperature and specific humidity of the atmosphere. The weight of dry air remains constant and consequently computations are best based upon 1 lb of dry air.

Liquid water may enter or leave the apparatus. Its weight $m_f$ lb per lb of air is often merely the difference between the specific humidities of the entering and leaving atmospheres.

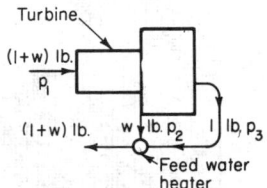

**Fig. 29**  Regenerative feedwater heating.

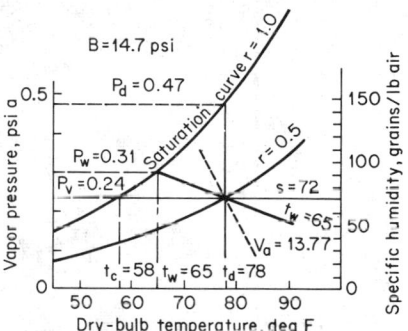

**Fig. 30**  Skeleton humidity chart.

**Table 42. Vapor Pressure of Water in inHg at 32°F**
Ice below 32°F

| t | 0 | 1 | 2 | 3 | 4 | 5 | 6 | 7 | 8 | 9 |
|---|---|---|---|---|---|---|---|---|---|---|
| −40 | 0.0039 | | | | | | | | | |
| −30 | 0.0071 | 0.0067 | 0.0062 | 0.0058 | 0.0055 | 0.0051 | 0.0048 | 0.0045 | 0.0043 | 0.0041 |
| −20 | 0.0126 | 0.0119 | 0.0112 | 0.0106 | 0.0100 | 0.0094 | 0.0089 | 0.0084 | 0.0079 | 0.0075 |
| −10 | 0.0220 | 0.0208 | 0.0197 | 0.0187 | 0.0176 | 0.0167 | 0.0158 | 0.0149 | 0.0141 | 0.0133 |
| − 0 | ...... | 0.0358 | 0.0339 | 0.0322 | 0.0305 | 0.0289 | 0.0274 | 0.0259 | 0.0246 | 0.0233 |
| + 0 | 0.0377 | 0.0397 | 0.0419 | 0.0441 | 0.0465 | 0.0489 | 0.0514 | 0.0541 | 0.0569 | 0.0598 |
| 10 | 0.0629 | 0.0661 | 0.0695 | 0.0730 | 0.0767 | 0.0806 | 0.0847 | 0.0889 | 0.0933 | 0.0980 |
| 20 | 0.103 | 0.108 | 0.113 | 0.119 | 0.124 | 0.130 | 0.137 | 0.143 | 0.150 | 0.157 |
| 30 | 0.165 | 0.172 | 0.180 | 0.188 | 0.196 | 0.204 | 0.212 | 0.220 | 0.229 | 0.238 |
| 40 | 0.248 | 0.258 | 0.268 | 0.278 | 0.289 | 0.300 | 0.312 | 0.324 | 0.336 | 0.349 |
| 50 | 0.363 | 0.376 | 0.391 | 0.405 | 0.420 | 0.436 | 0.452 | 0.469 | 0.486 | 0.504 |
| 60 | 0.522 | 0.541 | 0.560 | 0.580 | 0.601 | 0.622 | 0.644 | 0.667 | 0.690 | 0.714 |
| 70 | 0.739 | 0.765 | 0.791 | 0.818 | 0.846 | 0.875 | 0.905 | 0.935 | 0.967 | 0.999 |
| 80 | 1.032 | 1.066 | 1.102 | 1.138 | 1.175 | 1.213 | 1.253 | 1.293 | 1.335 | 1.378 |
| 90 | 1.422 | 1.467 | 1.513 | 1.561 | 1.610 | 1.660 | 1.712 | 1.765 | 1.819 | 1.875 |
| 100 | 1.933 | 1.992 | 2.052 | 2.114 | 2.178 | 2.243 | 2.310 | 2.379 | 2.449 | 2.521 |

SOURCE: Keenan and Keyes, "Thermodynamic Properties of Steam," Wiley, 1936.

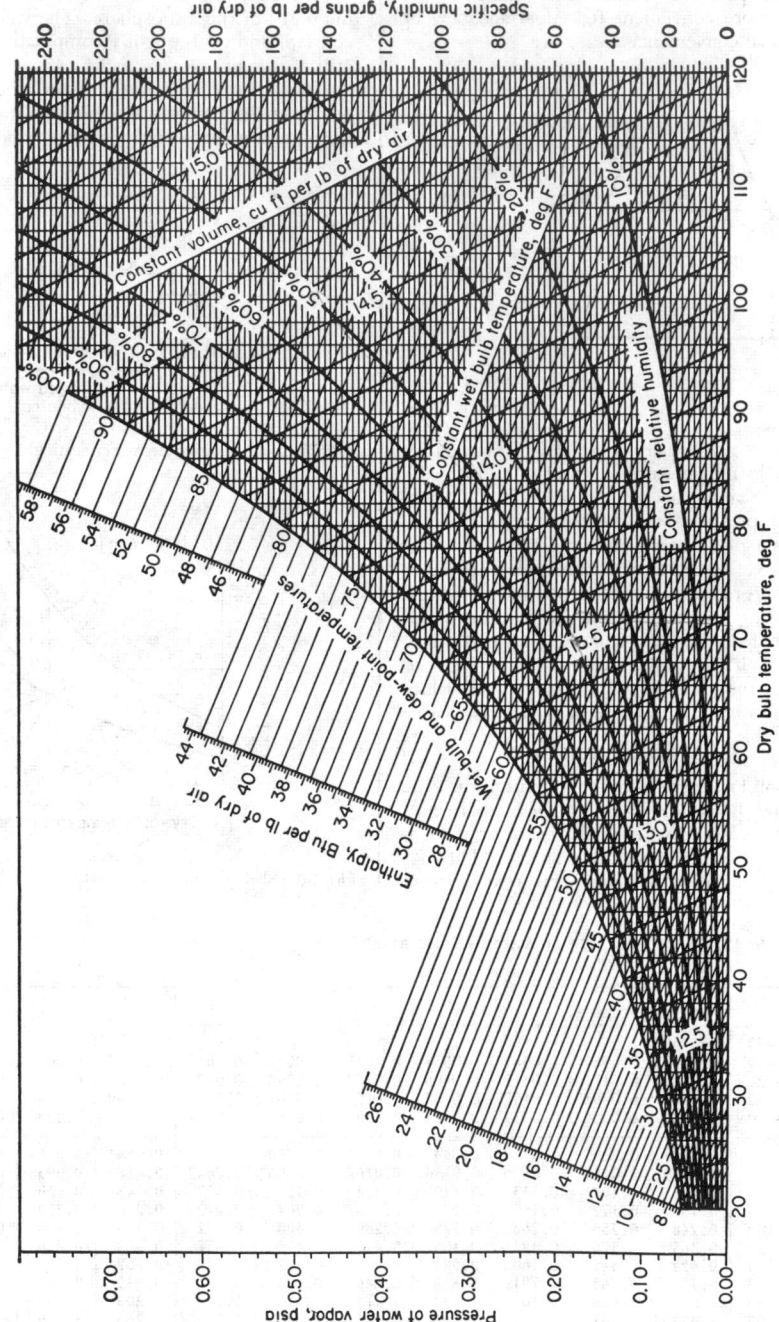

**Fig. 31** Humidity chart for low temperatures. Barometric pressure 14.696 psia. *(Copyright by General Electric Co. Adapted by permission.)*

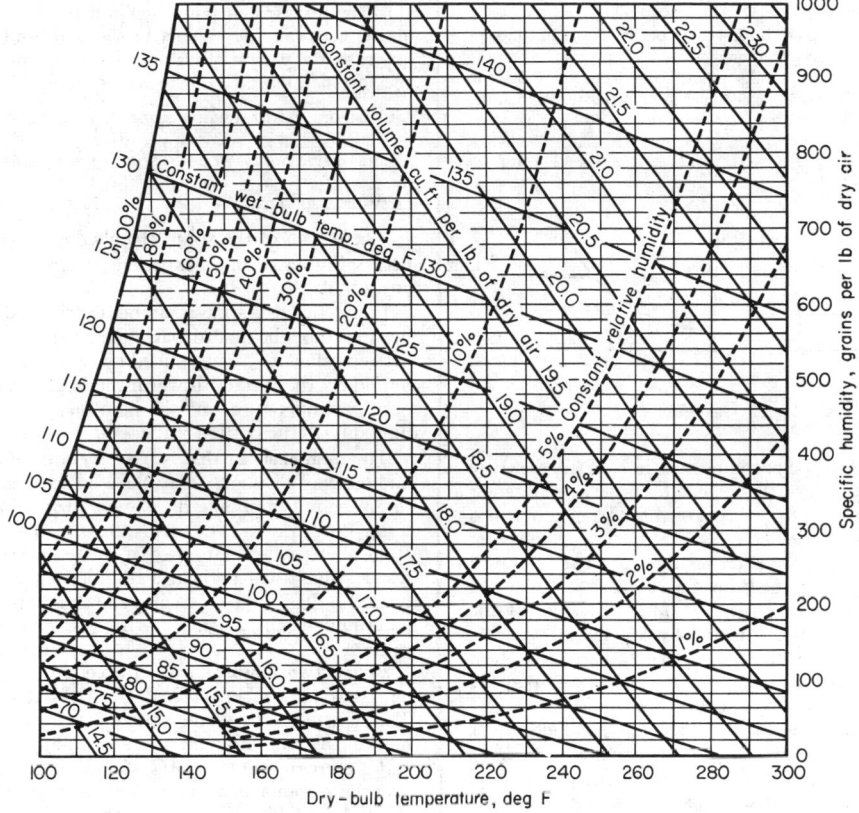

**Fig. 32** Humidity chart for medium temperatures. *(From Ellenwood and Mackey, "Vapor Charts," Wiley.)*

Its specific enthalpy at the observed or assumed temperature of supply or removal $t_f$ is

$$h_f = t_f - 32 \text{ Btu per lb of liquid}$$

Because most air conditioning involves steady-flow processes, thermal results are computed by the steady flow equation, written for 1 lb of air. Using subscript 1 for entering atmosphere and liquid water, and for heat supplied; and 2 for departing atmosphere and water, and for heat abstracted; the equation becomes (in the absence of work)

$$h_{m1} + m_{f1}h_{f1} + q_1 = h_{m2} + m_{f2}h_{f2} + q_2 \text{ Btu per lb air}$$

Either or both values of $m_f$ or $q$ may be zero.

In terms of the sigma function, the steady-flow equation becomes

$$\Sigma_1 + W_1(t_{w1} - 32) + m_{f1}h_{f1} + q_1 = \Sigma_2 + W_2(t_{w2} - 32) + m_{f2}h_{f2} + q_2 \text{ Btu per lb air}$$

Unit processes involved in air conditioning include heating and cooling an atmosphere above its dew point, cooling below the dew point, adiabatic saturation, and mixing of two atmospheres. These, in various sequences, make it possible to start with any given atmosphere and produce an atmosphere of any required characteristics.

**Heating and cooling above the dew point** entail no condensation of vapor. Barometric pressure and composition being unaltered, partial pressures remain constant. The process is represented in Fig. 33.

EXAMPLE. Initial conditions: $p_t = 28$ inHg; $t_d = 60°F$; $t_w = 50°F$; $p_v = 0.26$ inHg; $V = 1,200$ ft³.
Final conditions: $t_d = 82°F$.
Initial computed values: $r = 0.50$; $W = 0.0058$ lb vapor per lb air; $\rho_a = 0.0707$ lb air per ft³; $m_a = V \times \rho_a = 1,200 \times 0.0707 = 84.9$ lb air; $h_m = 20.7$ Btu per lb air.
Final computed values: $p_v$, $W$, and $m_a$ are unaltered; $r = 0.24$; $\rho_a = 0.0679$ lb air per ft³; $V = m_a/\rho_a = 84.9/0.0679 = 1,250$ ft³; $h_m = 26.1$ Btu per lb air.
Heat added: $q = h_{m2} - h_{m1} = 26.1 - 20.7 = 5.4$ Btu per lb air; $Q = q \times m_a = 5.4 \times 84.9 = 458$ Btu.

**Cooling below the dew point,** or **dehumidification,** entails condensation of vapor; the final atmosphere will be saturated, liquid will appear (see Fig. 34).

EXAMPLE. Initial conditions: $p_t = 29$ inHg; $t_d = 75°F$; $t_w = 65°F$; $V = 1,500$ ft³.
Final conditions: $t_d = 45°F$.
Initial computed values: $W = 0.0113$ lb vapor per lb air; $\rho_a = 0.0706$ lb air per ft³; $m_a = 1,500 \times 0.0706 = 106.0$ lb air; $h_m = 30.4$ Btu per lb air; $t_c = 60°F$.

Final computed values: $t_d = 45°F$; $p_v = 0.30$ inHg; $r = 1.0$; $W = 0.0065$ lb vapor per lb air; $\rho_a = 0.0754$ lb air per ft³; $V = 106.0/0.0754 = 1,406$ ft³; $b_m = 17.8$ Btu per lb air.

Liquid formed: $m_f = W_1 - W_2 = 0.0113 - 0.0065 = 0.0048$ lb liquid per lb air; $b_f = 50 - 32 = 18$ Btu per lb liquid (assuming that the liquid is drained out at an average temperature $t_f = 50°F$).

Heat abstracted: $q = b_{m1} - b_{m2} - m_f b_f = 30.4 - 17.8 - 0.0048 \times 18 = 12.5$ Btu per lb air; $Q = q \times m_a = 12.5 \times 106.0 = 1325$ Btu.

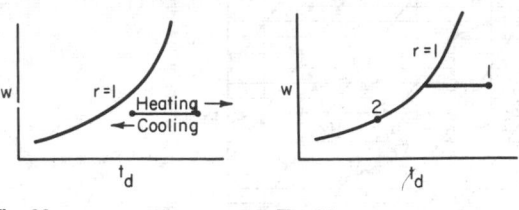

**Fig. 33**          **Fig. 34**

**Dehumidification** may be accomplished in a **surface cooler,** in which the air passes over tubes cooled by brine or refrigerant flowing through them. The solution of this type of problem is most easily handled on the chart (see Fig. 35). Locate the point representing the state of the entering atmosphere, and draw a straight line to a point on the saturation curve ($r = 1.0$) at the temperature of the cooling surface. The final state of the issuing atmosphere is approximated by a point on this line whose position on the line is determined by the heat abstracted by the cooling medium. This depends upon the extent of surface and the coefficient of heat transfer.

**Adiabatic saturation (humidification)** may be conducted in a spray chamber through which atmosphere flows. A large excess of water is recirculated through spray nozzles, and evaporation is made up by a suitable water supply. After the process has been operating for some time, the water in the spray chamber will have been cooled to the temperature of adiabatic saturation, which differs from the wet-bulb temperature only because of radiation and velocity errors that affect the wet-bulb thermometer. No heat is added or abstracted; the process is adiabatic. The heat of vaporization for the water that is evaporated is supplied by the cooling of the air passing through the chamber. The wet-bulb temperature of the atmosphere is constant throughout the chamber (Fig. 36). If the chamber is sufficiently large, the issuing atmosphere will be

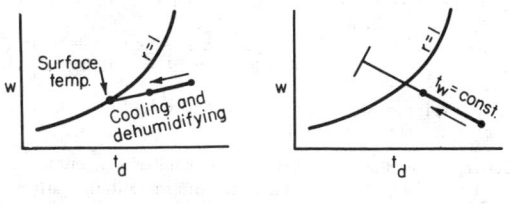

**Fig. 35**          **Fig. 36**

saturated at the wet-bulb temperature of the entering atmosphere; i.e., as the atmosphere passes through the chamber, $t_w$ remains constant, $t_d$ is reduced from its initial value to $t_w$, and $t_c$ is increased from its initial value to $t_w$. In a chamber of commercial size, the action may terminate somewhat short of

this, the precise end point being determined by the duration and effectiveness of contact between air and spray water. In any case, the weight of water evaporated equals the increase in the specific humidity of the atmosphere.

EXAMPLE. Initial conditions $p_t = 30$ inHg; $t_d = 78°F$; $t_w = 55°F$; $r = 0.20$; $W = 28$ grains vapor per lb air.

Final condition: $t_d = t_w = 55°F$; $r = 1.0$; $W = 64$ grains vapor per lb air.

Water evaporated: $W_2 - W_1 = 64 - 28 = 36$ grains water per lb air.

The design of the spray chamber to produce this result is necessarily based upon experience with like apparatus previously built.

In practice, the spray chamber is preceded and followed by heating coils, the first to warm the entering atmosphere to the desired value of $t_w$, determined by the prescribed final specific humidity, the second to warm the issuing atmosphere to the desired temperature, and simultaneously to reduce its relative humidity to the desired value.

The **spray chamber** that is used for adiabatic saturation (humidification) in winter may be used for dehumidification in summer by supplying the spray nozzles with refrigerated water instead of recirculated water. In this case, the issuing atmosphere will be saturated at the temperature of the spray water, which will be held at the desired dew point. Subsequent heating of the atmosphere to an acceptable temperature will simultaneously reduce the relative humidity to the desired value.

**Mixing Two Atmospheres** In recirculating ventilation systems, two atmospheres (1 and 2) are mixed to form a third (3). The state of the final atmosphere is readily found graphically on the psychrometric chart (see Fig. 37). Locate the points 1 and 2 representing the states of the initial atmospheres. Connect these points by a straight line. Locate a point that divides this line into segments inversely proportional to the weights of air in the respective atmospheres. The division point represents the state of the final mixture, so long as it falls below the saturation curve ($r = 1$). If the final point falls above the saturation curve, as in Fig. 38, condensation will ensue, and the true final point 4 is found by drawing a line from the apparent point 3, parallel to the lines of constant wet-bulb temperature, to its intersection with the saturation curve. From all the points involved, readings of specific humidity may be taken, including point 3 when it falls above the saturation curve, and in this case the difference between $W_3$ and $W_4$ will be the weight of condensate, pounds per pound air.

If the chart is sectional and the two points do not fall in the same section, or in any case in which it is preferred, the same method may be carried out arithmetically.

For adiabatic mixing in a steady-flow process of two masses of "moist" air, each at the total pressure of $p_t$,

$$m_{a3} = m_{a1} + m_{a2}$$

In the absence of condensation,

and
$$m_{a3}W_3 = m_{a1}W_1 + m_{a2}W_2$$
$$m_{a3}b_{m3} = m_{a1}b_{m1} + m_{a2}b_{m2}$$

When condensation occurs, assume that the condensate is removed at the final temperature $t_4$ and that the final mixture consists of dry air and saturated water vapor at this same temperature. The weight of condensate is

$$m_c = m_{a1}W_1 + m_{a2}W_2 - m_{a3}W_4$$

where $W_4$ is the specific humidity for saturation at temperature $t_4$ and total pressure $p_t$. Also

$$m_{a1}b_{m1} + m_{a2}b_{m2} = m_{a2}b_{m4} + m_c b_{f4}$$

In the case of condensation, a trial solution is necessary to find the temperature $t_4$ that will satisfy these relations.

EXAMPLE. Two thousand ft³ of air per min at $t_{d1} = 80°F$ and $t_{w1} = 65°F$ are mixed in an adiabatic, steady-flow process with 1,000 ft³ of air per min at $t_{d2} = 95°F$ and $t_{w2} = 75°F$; the total pressure of each mixture is 29 inHg.
By computation, $m_{a1} = 140$ lb dry air/min; $W_1 = 0.010$ lb/lb dry air; $m_{a2} = 67.6$ lb dry air/min; $W_2 = 0.0146$ lb/lb dry air.
   $m_{a3} = 207.6$ lb dry air/min.
   $W_3 = 0.0116$ lb/lb dry air.
   $b_{m1} = 30.3$ Btu/lb dry air and $b_{m2} = 38.9$ Btu/lb dry air.
   $b_{m3} = 33.1$ Btu/lb dry air.
   $t_{d3} = 84.9°F$.

EXAMPLE. Fifteen hundred ft³ of air per min at $t_{d1} = 0°F$ and $r_1 = 0.8$ are mixed in an adiabatic, steady-flow process with 1,000 ft³ of air per min at $t_{d2} = 100°F$ and $r_2 = 0.9$; the total pressure of each mixture is 30 inHg.
By computation, $m_{a1} = 129.6$ lb of dry air/min; $W_1 = 0.000626$ lb/lb dry air; $m_{a2} = 66.9$ lb of dry air/min; $W_2 = 0.03824$ lb/lb dry air; $b_{m1} = 0.90$ Btu/lb dry air; $b_{m2} = 66.29$ Btu/lb dry air.
The three equations that must be satisfied by a choice of the terminal temperature, $t_4 = t_{d4} = t_{w4}$, are

$$m_c = 2.64 - 196.5W_4$$
$$4551 = 196.5b_{m4} + m_c b_{f4}$$
$$W_4 = p_{v4}/1.61(30 - p_{v4}) \quad \text{for } r_4 = 1$$

The value of $t_4$ that satisfies these equations is 55°F; condensation amounts to 0.84 lb per min.

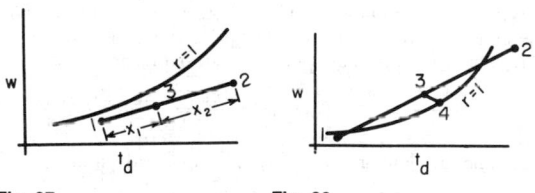

**Fig. 37**        **Fig. 38**

The **cooling tower** is a chamber in which outdoor atmosphere flows through a spray of entering hot water, which is to be cooled. The temperature of the water is reduced in part by the warming of the air, and in greater part by the evaporation of a portion of the water. The atmosphere enters at given conditions and emerges at a higher temperature and usually saturated ($r = 1$). It is commonly possible to cool the water below the temperature of the entering air, often to about halfway between $t_d$ and $t_w$. The volume of atmosphere per pound of entering water and the weight of water evaporated are to be computed.

EXAMPLE. A cooling tower is to receive water at 120°F and atmosphere at $t_d = 90$, $t_w = 80$, whence $p_v = 0.92$, $W = 0.0196$ lb vapor per lb air, $\rho_a = 0.0702$ lb air per ft³, and $b_m = 43.2$. The water is to be cooled to 85°F. What volume of atmosphere must be passed through the tower, and what weight of water will be lost by evaporation?
The issuing atmosphere will be assumed to be saturated at 115°F. Then $t_d = 115°F$, $p_v = 3.0$ inHg, $W = 0.0690$ lb vapor per lb air, $\rho_a = 0.0623$ lb air per ft³, and $b_m = 104.4$ Btu per lb air.

The two unknowns are the weight of air to be passed through the tower and the weight of water to be evaporated. The two equations are the water-weight balance and the enthalpy balance (the steady-flow equation for zero heat transfer to or from outside). Assume that 1 lb water enters, of which $x$ lb are evaporated. The water-weight balance $1 + m_aW_1 = 1 - x + m_aW_2$ becomes $x = m_a(W_2 - W_1) = m_a(0.0690 - 0.0196) = 0.0494m_a$. The enthalpy balance $1 \times (120 - 32) + m_a b_{m1} = (1 - x)(85 - 32) + m_a b_{m2}$ becomes $88 + 43.2m_a = 53(1 - x) + 104.4m_a$; whence $53x = 53 - 88 + m_a(104.4 - 43.2) = -35 + 61.2m_a$.
Solving these simultaneous equations, $x = 0.0295$ lb water evaporated per pound of water entering and $m_a = 0.597$ lb air per pound water entering.

In an **evaporative condenser** vapor is condensed within tubes that are cooled by the evaporation of water flowing over the outside of the tubes; the water evaporates into the atmosphere. The computation of results is similar to that for the cooling tower.

## REFRIGERATION

**Vapor-Compression Machines** The essential parts of a vapor-compression system are the same as in the system using air, except that the expansion cylinder is replaced by an expansion value through which the liquified medium flows from the high-pressure condensing coils to the low-pressure brine coils. The cycle of operation is best shown on the $T$-$S$ plane (Fig. 39). The point $B$ represents the state of the refrigerating medium leaving the brine coils and entering the compressor. Usually in this state the fluid is nearly dry saturated vapor; i.e., point $B$ is near the saturation curve $S_g$. $BC$ represents the assumed reversible adiabatic compression, during which the fluid is usually superheated. In the state $C$, the superheated vapor passes into the cooling coils and is cooled at constant pressure, as indicated by $CD$, and then condensed at temperature $T_2$, as shown by $DE$. The liquid now flows through the expansion valve to the brine coils. This is a throttling process, and the final-state point $A$ is located on the $T_1$-line in such a position as to make the enthalpy for state $A$ (= area $OHGAA_1$) equal to the enthalpy at $E$ (= area $OHEE_1$). The mixture of liquid and vapor now absorbs heat from the brine and vaporizes, as indicated by $AB$.

The **heat absorbed from the brine**, represented by area $A_1ABC_1$, is

$$Q_1 = b_b - b_a = b_b - b_e$$

The **heat rejected to the cooling water**, represented by area $C_1CDEE_1$, is

$$Q_2 = b_c - b_e = c_p(T_c - T_d) + r_2 \quad \text{approx}$$

where $r_2$ denotes the enthalpy of vaporization at the upper temperature $T_2$, and $c_p$ the specific heat of the superheated vapor. The **work** that must be **supplied per pound of fluid circulated** is $W = J(Q_2 - Q_1) = J(b_c - b_b)$. The ratio $JQ_1/W = (b_b - b_c)/(b_c - b_b)$ is sometimes called the **coefficient of performance**.
If $Q$ denotes the heat to be absorbed from the brine per hour, then the **quantity of fluid circulated per hour** is $m = Q/(b_b - b_a)$; or, if $B$ is taken on the saturation curve, $m = Q/(b_{g1} - b_{f2})$.
The work per hour is $W = Jm(b_c - b_{g1}) = JQ(b_c - b_{g1})/(b_{g1} - b_{f2})$ ft·lb, and the **horsepower required** is $H = Q(b_c - b_{g1})/2544(b_{g1} - b_{f2})$.
If $v_{g1}$ is the volume of the saturated vapor at the temperature $T_1$ in the brine coils, and $n$ the number of working strokes

per minute, the **displacement volume** of the **compressor cylinder** is
$V = m v_{g1}/60n$.

The values of $h_{g1}$, $h_{f2}$ in the preceding formulas are found in the tables of saturated vapors. The enthalpy $h_c$ of the superheated vapor may be determined in the case of ammonia from Table 34 for superheated ammonia.

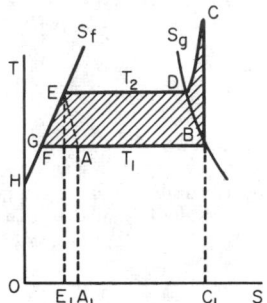

**Fig. 39**  Vapor-compression refrigerating cycle.

The work necessary for operating a refrigerator, although usually supplied through the compressor, may be supplied in other ways. Thus in **absorption refrigerators** (see Secs. 12, and 19) an absorbent, usually water, absorbs the refrigerant, usually ammonia. The water, by its affinity for the ammonia, has, in a thermodynamic sense, ability to do work. Having absorbed the ammonia and thereby lost its ability to do work, the water may have its work capacity restored by passing the ammonia-water solution through a rectifying column from which water and ammonia emerge. With operation under a suitable pressure, the ammonia is condensed to a liquid. This, in turn, may be evaporated, yielding refrigeration, the ammonia vapors being once again absorbed in the water. Thermodynamically the analysis for these absorption cycles is similar to that for compression cycles. See Fig. 26 for properties of aqua-ammonia.

### THERMODYNAMICS OF FLOW OF COMPRESSIBLE FLUIDS

Important examples of the flow of compressible fluids are the following: (1) the flow of air and steam through orifices and short tubes or nozzles, as in the steam turbine, (2) the flow of compressed air, steam, and illuminating gas in long mains, (3) the flow of low-pressure gases, as furnace gases in ducts and chimneys or air in ventilating ducts, and (4) the flow of gases in moving channels, as in the centrifugal fan.

### Notation

Let $A$ = area of section, $\text{ft}^2$

$C$ = empirically determined coefficient of discharge

$D$ = inside diameter of pipe, ft

$d = 12D$ = inside diameter of pipe, in

$F_{12}$ = energy expended in overcoming internal and external friction between sections $A_1$ and $A_2$

$F'$ = energy used in overcoming friction, ft·lb per lb of fluid flowing

$f$ = friction factor = $4f'$

$g = 32.2$ = local acceleration of gravity, $\text{ft}^2$

$g_c$ = a dimensional constant

$h$ = enthalpy, Btu per lb

$J = 778.3$

$k = c_p/c_v$

$L$ = equivalent length of pipe, ft

$m$ = mass of fluid flowing past a given section per s, lb

$\mu$ = viscosity, centipoises

$P$ = pressure, psia

$\Delta P$ = differential pressure across nozzle, psi

$p$ = pressure of fluid at given section, $\text{lb/ft}^2$ abs

$p_m$ = critical flow pressure

$Q_{12}$ = heat entering the flowing fluid between sections $A_1$ and $A_2$

$q$ = volume of fluid flowing past section, cfm

$R$ = ideal gas constant

$\rho$ = density, $\text{lb/ft}^3$

$T$ = temperature $R$

$V$ = mean velocity at the given section, fps

$v$ = specific volume

$w$ = weight of fluid flowing past a given section per s, lb

$z$ = height from center of gravity of flow to a fixed base level, ft

The cross sections of the tube or channel are denoted by $A_1$, $A_2$, etc. (Fig. 40), and the various magnitudes pertaining to these sections are denoted by corresponding subscripts. Thus, at section $A_1$, the velocity, specific volume, and pressure are, respectively, $V_1$, $v_1$, $p_1$; at section $A_2$, they are $V_2$, $v_2$, $p_2$.

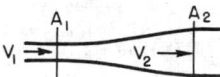

**Fig. 40**

**Fundamental Equations**  In the interpretation of fluidflow phenomena, three fundamental equations are of importance.

1. The continuity equation, or material balance,

$$\frac{A_1 V_1}{v_1} = \frac{A_2 V_2}{v_2} \quad \text{or} \quad \frac{dv}{v} = \frac{dA}{A} + \frac{dV}{V}$$

2. The first law or energy balance for steady flow,

$$Jq = J(h_2 - h_1) + \frac{V_2^2 - V_1^2}{2g_c} + \frac{g}{g_c}(z_2 - z_1)$$

3. The available energy balance for a steady-flow process, based on unit weight, is

$$v \, dp + \frac{V \, dV}{g} + dF + dz = 0$$

In the processes here discussed, no net external or shaft work is performed.

For most actual processes, the third equation cannot be integrated because the actual path is not known. Usually, adiabatic flow is assumed, but occasionally the assumption of isothermal conditions may be more nearly correct.

**Flow through Orifices and Nozzles**  As a compressible fluid passes through a nozzle, drop in pressure and simultaneous increase in velocity result. By assuming the type of flow, e.g., adiabatic, it is possible to calculate from the properties of the fluid the required area for the cross section of the nozzle at any point in order that the flowing fluid may just fill the provided space. From this calculation, it is found that for all compressible fluids the nozzle form must first be converging but eventually, if the pressure drops sufficiently, a place is reached where to accommodate the increased volume due to the expansion the nozzle must become diverging in form. The smallest cross section of the nozzle is called the throat, and the pressure at the throat is the **critical flow pressure** (not to be confused with the critical pressure). If the nozzle is cut off at the throat with no diverging section and the pressure at the discharge end is progressively decreased, with fixed inlet pressure, the amount of fluid passing increases until the discharge pressure equals the critical, but further decrease in discharge pressure does not result in increased flow. This is not true for thin plate orifices.* For any particular gas, the ratio of critical to inlet pressure is approximately constant. For gases, $p_m/p_1 = 0.53$ approx; for saturated steam the ratio is about 0.575; and for moderately superheated steam it is about 0.55.

**Formulas for Orifice Computations**  The general fundamental relation is given by the energy balance ($V_2^2 - V_1^2)/2g_c = -Jh_{12}$. Referring to Fig. 41, let section 2 be taken at the orifice,

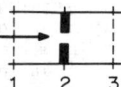

**Fig. 41**

section 3 is somewhat beyond the orifice on the downstream side, and section 1 is before the orifice on the upstream side. Then

$$V_2 = C\sqrt{2g_cJ(h_1 - h_2)}\Big/\sqrt{1 - \left(\frac{A_2}{A_1}\right)^2\left(\frac{v_1}{v_2}\right)^2}$$

The coefficient of discharge factor $C$ is discussed in Sec. 16.

The volume of gas passing is $V_2A_2$ ft³/s, and the quantity is $V_2A_2\rho$. For ideal gases, assuming reversible adiabatic expansion through the orifice,

$$V_2 = C\sqrt{2g_cp_1v_1\,\frac{k}{k-1}\left[1 - \left(\frac{p_2}{p_1}\right)^{k-1/k}\right]}\Big/$$

$$\sqrt{1 - \left(\frac{A_2}{A_1}\right)^2\left(\frac{p_2}{p_1}\right)^{2/k}}$$

$$m = CA_2p_2\sqrt{\frac{2g_c}{RT_1}\frac{k}{k-1}\left(\frac{p_1}{p_2}\right)^{k-1/k}\left[\left(\frac{p_1}{p_2}\right)^{k-1/k} - 1\right]}\Big/$$

$$\sqrt{1 - \left(\frac{A_2}{A_1}\right)^2\left(\frac{p_2}{p_1}\right)^{2/k}}$$

$V_1$ is often small compared with $V_2$, and under these conditions the denominators in the preceding equations become approximately equal to unity. For air, assuming $R = 53.3$, $k = 1.3937$, and $V_1$ negligible,

$$m = 2.05\,CA_2p_2\sqrt{(1/T_1)(p_1/p_2)^{0.283}[(p_1/p_2)^{0.283} - 1]}$$

*Cf. Perry, "Critical Flow through Sharp-edged Orifices," ASME Paper 48-A-146.

Although the preceding formulas are generally applicable under the assumed conditions, it must be remembered that irrespective of the value of $p_3$, $p_2$ cannot become less than $p_m$. When $p_3$ is less than $p_m$, the flow rate becomes independent of the downstream pressure; for ideal gases,

$$m = CA_2p_1\sqrt{\frac{g_c}{RT_1}k\left(\frac{2}{k+1}\right)^{(k+1)/(k-1)}}$$

or for air

$$m = 0.53\,Cp_1\frac{A_2}{\sqrt{T_1}}$$

The following formula (see ASME Power Test Code No. 10 for 1949) is useful for calculating the flow rate, in cubic feet per minute, of any gas (provided no condensation occurs) through a nozzle for pressure drops less than the critical range:

$$q_1 = \frac{31.5\,Cd_n^2Y'}{\rho_1}\sqrt{\rho_1\Delta P}$$

In this equation,

$$Y' = \left(\frac{k}{k-1}\right)^{1/2}\left(\frac{P_2}{P_1}\right)^{1/k}$$

$$\times \sqrt{\left[1 - \left(\frac{P_2}{P_1}\right)^{k-1/k}\right]\Big/\left(1 - \frac{P_2}{P_1}\right)\left[1 - \left(\frac{d_n}{d_1}\right)^4\left(\frac{P_2}{P_1}\right)^{2/k}\right]}$$

where $P_1$ = static pressure on upstream side of nozzle, psia; $P_2$ = static pressure on downstream side of nozzle, psia; $d_1$ = diameter of pipe upstream of nozzle, in; $d_n$ = nozzle throat diameter, in; $\rho_1$ = specific weight of gas at upstream side of nozzle, lb/ft³. Values of $Y'$ are given in Table 43.

Where the pressure drop through the orifice is small, the hydraulic formulas applicable to incompressible fluids may be employed for gases and other compressible fluids.

In general, the formulas of the preceding section are applicable to nozzles. When so used, however, the proper value of the discharge coefficient must be employed. For steam nozzles, this may be as high as 0.94 to 0.96, although for many orifice installations it is as low as 0.50 to 0.60. Steam nozzles constitute a most important type, and calculations for these are best carried out with the aid of a Mollier or similar chart.

**Formulas for Discharge of Steam**  When the back pressure $p_3$ is less than the critical pressure $p_m$, the discharge depends upon the area of orifice $A_2$ and reservoir pressure $p_1$. There are three formulas widely used to express, approximately, the discharge $m$ of **saturated** steam in terms of $A_2$ and $p_1$ as follows:

1. Napier's equation, $m = A_2p_1/70$.
2. Grashof's formula, $m = 0.0165A_2p_1^{0.97}$.
3. Rateau's formula, $m = A_2p_1(16.367 - 0.96\log p_1)/1{,}000$.

In these formulas, $A_2$ is to be taken in square inches, $p_1$ in pounds per square inch. Napier's formula is merely convenient as a rough check. Formulas 2 and 3 are applicable to well-rounded convergent orifices, in which case the coefficient of discharge may be taken as 1; i.e., no correction is required.

When the back pressure $p_2$ is greater than the critical flow pressure $p_m$, the velocity and discharge are found most conveniently from the general formulas of flow. From the steam tables or from the **Mollier chart,** find the initial enthalpy $h_1$ and

**Table 43. Values for $Y'$**

| $P_2/P_1$ | $k = 1.40$ $d_n/d_1$ | | | | | $k = 1.35$ $d_n/d_1$ | | | | | $k = 1.30$ $d_n/d_1$ | | | | |
|---|---|---|---|---|---|---|---|---|---|---|---|---|---|---|---|
| | 0 | 0.2 | 0.3 | 0.4 | 0.5 | 0 | 0.2 | 0.3 | 0.4 | 0.5 | 0 | 0.2 | 0.3 | 0.4 | 0.5 |
| 1.00 | 1.000 | 1.001 | 1.004 | 1.013 | 1.033 | 1.00 | 1.001 | 1.004 | 1.013 | 1.033 | 1.000 | 1.001 | 1.004 | 1.013 | 1.033 |
| 0.99 | 0.995 | 0.995 | 0.999 | 1.007 | 1.027 | 0.994 | 0.995 | 0.999 | 1.007 | 1.027 | 0.994 | 0.995 | 0.998 | 1.007 | 1.026 |
| 0.98 | 0.989 | 0.990 | 0.993 | 1.002 | 1.021 | 0.989 | 0.990 | 0.993 | 1.001 | 1.020 | 0.988 | 0.989 | 0.992 | 1.001 | 1.020 |
| 0.97 | 0.984 | 0.985 | 0.988 | 0.996 | 1.015 | 0.983 | 0.984 | 0.987 | 0.995 | 1.014 | 0.983 | 0.983 | 0.986 | 0.995 | 1.013 |
| 0.96 | 0.978 | 0.979 | 0.982 | 0.990 | 1.009 | 0.978 | 0.978 | 0.981 | 0.990 | 1.008 | 0.977 | 0.977 | 0.980 | 0.989 | 1.007 |
| 0.95 | 0.973 | 0.974 | 0.977 | 0.985 | 1.002 | 0.972 | 0.973 | 0.976 | 0.984 | 1.001 | 0.971 | 0.972 | 0.974 | 0.982 | 1.000 |
| 0.94 | 0.967 | 0.968 | 0.971 | 0.979 | 0.996 | 0.966 | 0.967 | 0.970 | 0.978 | 0.995 | 0.965 | 0.966 | 0.968 | 0.976 | 0.993 |
| 0.93 | 0.962 | 0.963 | 0.965 | 0.973 | 0.990 | 0.961 | 0.961 | 0.964 | 0.972 | 0.989 | 0.959 | 0.960 | 0.962 | 0.970 | 0.987 |
| 0.92 | 0.956 | 0.957 | 0.960 | 0.967 | 0.984 | 0.955 | 0.955 | 0.958 | 0.966 | 0.982 | 0.953 | 0.954 | 0.956 | 0.964 | 0.980 |
| 0.91 | 0.951 | 0.951 | 0.954 | 0.961 | 0.978 | 0.949 | 0.950 | 0.952 | 0.960 | 0.976 | 0.947 | 0.948 | 0.950 | 0.957 | 0.973 |
| 0.90 | 0.945 | 0.946 | 0.948 | 0.956 | 0.971 | 0.943 | 0.944 | 0.946 | 0.953 | 0.969 | 0.941 | 0.942 | 0.944 | 0.951 | 0.966 |
| 0.89 | 0.939 | 0.940 | 0.943 | 0.950 | 0.965 | 0.937 | 0.938 | 0.940 | 0.947 | 0.963 | 0.935 | 0.935 | 0.938 | 0.945 | 0.959 |
| 0.88 | 0.934 | 0.934 | 0.937 | 0.944 | 0.959 | 0.931 | 0.932 | 0.934 | 0.941 | 0.956 | 0.929 | 0.929 | 0.932 | 0.938 | 0.953 |
| 0.87 | 0.928 | 0.928 | 0.931 | 0.938 | 0.953 | 0.925 | 0.926 | 0.926 | 0.935 | 0.950 | 0.922 | 0.923 | 0.926 | 0.932 | 0.946 |
| 0.86 | 0.922 | 0.923 | 0.925 | 0.932 | 0.946 | 0.919 | 0.920 | 0.922 | 0.929 | 0.943 | 0.916 | 0.917 | 0.919 | 0.926 | 0.939 |
| 0.85 | 0.916 | 0.917 | 0.919 | 0.926 | 0.940 | 0.913 | 0.914 | 0.916 | 0.923 | 0.936 | 0.910 | 0.911 | 0.913 | 0.923 | 0.932 |
| 0.84 | 0.910 | 0.911 | 0.913 | 0.920 | 0.933 | 0.907 | 0.908 | 0.910 | 0.916 | 0.930 | 0.904 | 0.904 | 0.907 | 0.919 | 0.925 |
| 0.83 | 0.904 | 0.905 | 0.907 | 0.913 | 0.927 | 0.901 | 0.902 | 0.904 | 0.910 | 0.923 | 0.897 | 0.898 | 0.900 | 0.916 | 0.918 |
| 0.82 | 0.898 | 0.899 | 0.901 | 0.907 | 0.920 | 0.895 | 0.895 | 0.898 | 0.904 | 0.917 | 0.891 | 0.891 | 0.894 | 0.900 | 0.911 |
| 0.81 | 0.892 | 0.893 | 0.895 | 0.901 | 0.914 | 0.889 | 0.889 | 0.891 | 0.897 | 0.910 | 0.885 | 0.885 | 0.887 | 0.893 | 0.904 |
| 0.80 | 0.886 | 0.887 | 0.889 | 0.895 | 0.907 | 0.883 | 0.883 | 0.885 | 0.891 | 0.903 | 0.878 | 0.879 | 0.880 | 0.886 | 0.897 |
| 0.79 | 0.880 | 0.881 | 0.883 | 0.889 | 0.901 | 0.876 | 0.877 | 0.879 | 0.834 | 0.896 | 0.872 | 0.872 | 0.874 | 0.880 | 0.890 |
| 0.78 | 0.874 | 0.875 | 0.877 | 0.882 | 0.894 | 0.870 | 0.870 | 0.872 | 0.878 | 0.889 | 0.865 | 0.865 | 0.868 | 0.873 | 0.883 |
| 0.77 | 0.868 | 0.869 | 0.871 | 0.876 | 0.887 | 0.864 | 0.864 | 0.866 | 0.871 | 0.882 | 0.859 | 0.859 | 0.861 | 0.866 | 0.876 |
| 0.76 | 0.862 | 0.862 | 0.864 | 0.869 | 0.881 | 0.857 | 0.858 | 0.859 | 0.865 | 0.876 | 0.852 | 0.852 | 0.854 | 0.859 | 0.869 |
| 0.75 | 0.856 | 0.856 | 0.858 | 0.863 | 0.874 | 0.851 | 0.851 | 0.853 | 0.858 | 0.869 | 0.845 | 0.846 | 0.848 | 0.852 | 0.862 |
| 0.74 | 0.849 | 0.850 | 0.852 | 0.857 | 0.867 | 0.844 | 0.845 | 0.846 | 0.851 | 0.862 | 0.839 | 0.839 | 0.841 | 0.845 | 0.855 |
| 0.73 | 0.843 | 0.844 | 0.845 | 0.850 | 0.860 | 0.838 | 0.838 | 0.840 | 0.845 | 0.855 | 0.832 | 0.832 | 0.834 | 0.838 | 0.848 |
| 0.72 | 0.837 | 0,837 | 0.839 | 0.844 | 0.854 | 0.831 | 0.831 | 0.833 | 0.838 | 0.848 | 0.825 | 0.825 | 0.827 | 0.831 | 0.841 |
| 0.71 | 0.820 | 0.831 | 0.832 | 0.837 | 0.847 | 0.825 | 0.825 | 0.827 | 0.831 | 0.840 | 0.818 | 0.819 | 0.820 | 0.824 | 0.834 |
| 0.70 | 0.824 | 0.824 | 0.826 | 0.830 | 0.840 | 0.818 | 0.818 | 0.820 | 0.824 | 0.833 | 0.811 | 0.812 | 0.813 | 0.817 | 0.826 |

If the velocity of approach is zero (as with a nozzle taking in air from the outside), $d_1$ is infinite and $d_n/d_1$ is zero.

the enthalpy $h_2$ after isentropic expansion; also the specific volume $v_2$ (see Fig. 42). Then

$$V_2 = 223.7\sqrt{h_1 - h_2} \quad \text{and} \quad m = A_2 V_2/v_2$$

The same method is used in the case of steam initially superheated.

EXAMPLE. Required the discharge through an orifice ½ in diam of steam at 140 psia superheated 110°F, back pressure, 90 psia.

From the Mollier chart and the steam tables, $h_1 = 1255.7$, $h_2 = 1214$, $v_2 = 5.30$ ft³.

$$V_2 = 233.7\sqrt{1,255.7 - 1,214} = 1,455$$
$$A_2 = 0.1964 \text{ in}^2 = (0.1964/144) \text{ ft}^2$$
$$m = A_2 V_2/v_2 = (0.1964/144) \times (1,455/5.30) = 0.372 \text{ lb/s}$$

This calculation assumes ideal conditions, and the results must be multiplied by the correct coefficient of discharge to get actual results.

**Flow through Converging-Diverging Nozzles**    At the throat, or smallest cross section of the nozzle (Fig. 43), the pressure of saturated steam takes the value $p_m = 0.57p_1$. The quantity discharged is fixed by the area $A_2$ of the throat and the initial pressure $p_1$. For saturated steam, Grashof's or Rateau's formula (see above) may be used. The diverging part of the nozzle permits further expansion to the break pressure $p_3$, the velocity of the jet meanwhile increasing from $V_m(= V_2)$, the

critical velocity at the throat, to $V_3$ given by the fundamental equation $V_3 = 223.7\sqrt{h_1 - h_3}$.

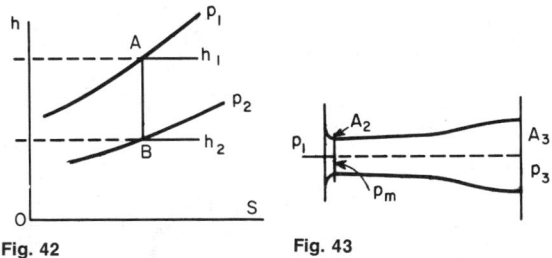

Fig. 42        Fig. 43

The frictional resistances in the nozzle have the effect of decreasing the jet energy $V_3^2/2g_c$ and correspondingly increasing the enthalpy of the flowing fluid. Thus, if $h_3$ is the enthalpy in the final state with frictionless expansion, $h_3'$ ($>h_3$) is the enthalpy when friction is taken into account; hence $(V_3')^2/2g_c = J(h_1 - h_3')$ is less than $V_3^2/2g_c = J(h_1 - h_3)$. The loss of kinetic energy, in Btu, is $h_3' - h_3$, and the ratio of this loss to the available kinetic energy, i.e. $(h_3' - h_3)/(h_1 - h_3)$, is denoted by $y$.

The **design of a nozzle for a given discharge** $m$ with pressures $p_1$ and $p_3$ is most conveniently effected with the aid of the Mollier chart. Determine $p_m$, the critical pressure, and $h_1$, $h_m$, $h_3$, assuming frictionless flow. Then

$$V_m = 223.7\sqrt{h_1 - h_m}$$

and

$$V_3' = 223.7\sqrt{(1 - y)(h_1 - h_3)}$$

Next find $v_m$ and $v_3'$. Then, from the equation of continuity,

$$A_m = mv_m/V_m$$

and

$$A_3' = mv_3'/V_3'$$

The following example illustrates the method.

EXAMPLE.   Required the throat and end sections of a nozzle to deliver 0.7 lb of steam per second. The initial pressure is 160 psia, the back pressure 15 psia, and the steam is initially superheated 100°F; $y = 0.15$.

The critical pressure is $160 \times 0.55 = 88$ lb. On the Mollier chart (Fig. 44), the point $A$ representing the initial state is located, and line of constant entropy (frictionless adiabatic) is drawn from $A$. This cuts the curves $p = 88$ and $p = 15$ in the points $B$ and $C$, respectively. The three values of $h$ are found to be $h_1 = 1253$, $h_m = 1199$, $h_3 = 1067$. Of the available drop in enthalpy, $h_1 - h_3 = 185.5$ Btu, 15 percent or 27.9 Btu is lost through friction. Hence, $CD = 27.9$ is laid off and $D$ is projected horizontally to point $C'$ on the curve $p = 15$. Then $C'$ represents the final state of the steam, and the quality is found to be $x = 0.943$. The specific volume in the state $C'$ is $26.29 \times 0.943 = 24.8$ ft³. Likewise, the specific volume for the state $B$ is found to be 5.29 ft³/lb.

For the velocities at throat and end sections,

$$V_m = 223.7\sqrt{1,253 - 1,199} = 1,643 \text{ fps}$$
$$V_3 = 223.7\sqrt{185.5 - 27.9} = 2,813 \text{ fps}$$
$$A_m = (0.7 \times 5.29)/1,643 = 0.00225 \text{ ft}^2 = 0.324 \text{ in}^2$$
$$A_3 = (0.7 \times 24.8)/2,813 = 0.00617 \text{ ft}^2 = 0.89 \text{ in}^2$$

The diameters are $d_m = 0.613$ in and $d_3 = 1.064$ in

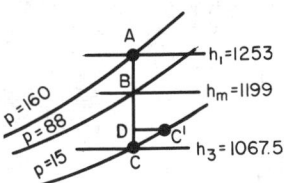

**Fig. 44**

**Divergence of Nozzles**   Figure 45 gives, for various ratios of expansion, the required "divergence" of the nozzle, i.e., the ratio of the area of any section to the throat area. Thus in the case of saturated steam, if the final pressure is $\frac{1}{15}$ of the initial pressure the ratio of the areas is 3.25. The curves apply to frictionless flow; the effect of friction is to increase the divergence.

**Theory of Supersaturation**   Certain discrepancies between the discharge of saturated steam through an orifice as calculated from the preceding theory and the discharge actually observed are explained by a hypothesis first advanced by Martin, viz., that steam when expanded rapidly, as in turbine nozzles, becomes supersaturated; in other words, the condensation required by the ordinary theory of adiabatic expansion does not occur on account of the rapidity of the expansion (see Goodenough, *Power*, Sept. 27, Oct. 4, 1927).

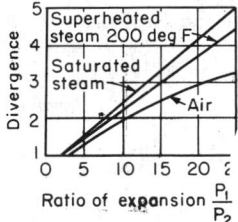

**Fig. 45**

The effect of supersaturation in turbines is a loss of energy, the amount of which may be 1.5 to 3 percent of the available energy of the steam.

**Flow of Wet Steam**   When the steam entering a nozzle is wet, the speed of the water particles at exit is not the same as the speed of the steam. Denoting by $V$ the speed of the steam, the speed of the water drops is $fV$, and $f$ may vary perhaps 0.20 to 0.05 or less, depending on the pressure. The actual velocity $V$ of the steam is greater than the velocity $V_0$ calculated on the usual assumption that steam and water have the same velocity. If $x$ is the quality of the steam, the ratio of these velocities is

$$V/V_0 = 1/\sqrt{x + f^2(1 - x)}$$

Thus with $x = 0.92$, $f = 0.15$, $V/V_0 = 1.036$. Since the discharge is practically proportional to the steam velocity, the actual discharge in this case is 3.6 percent greater than the discharge computed on the usual assumptions.

**Velocity Coefficients. Loss of Energy** $y$   On account of friction losses, the actual velocity $V$ attained by the jet is less than the velocity $V_0$ calculated under ideal conditions. That is, $V = xV_0$, where $x$ ($< 1$) is a velocity coefficient. The coefficient $x$ is connected with the coefficient $y$, giving the loss of energy, by the relation, $y = 1 - x^2$.

The elaborate and accurate experiments of the General Electric Co. on turbine nozzles (Warren and Keenan, *Trans. ASME*, **48**, p. 33) give for convergent nozzles values of $x$ in excess of 0.98, with a corresponding loss of energy $y = 0.025$ to 0.04. For similar nozzles, the experiments of the Steam Nozzles Research Committee (of England) by a different method give values of $x$ around 0.96, or $y = 0.08$. In the case of divergent nozzles, the velocity coefficient may be somewhat lower.

## Flow of Fluids in Circular Pipes

The fundamental equation as previously given on a unit weight basis, assuming the pipe horizontal, is

$$(V \, dV/g) + v \, dp + dF = 0$$

The friction term $dF$ includes not only losses due to frictional flow along the pipe but also those due to fittings, valves, etc., as well as losses occasioned by any enlargement or contraction of the pipe as, for instance, the loss occurring when a fluid passes from a pipe into a tank. For long straight pipes of uniform diameter, $dF$ is approximately equal to $2f'(V^2\,dL/gD)$. It is usual to express friction due to fittings, etc., in terms of additional length of pipe, adding this to the actual pipe length to get the equivalent pipe length.

Integration of the fundamental equation leads to two sets of formulas.

1. For pressure drops, small relative to the initial pressure, the specific volume $v$ and the velocity $V$ may be assumed constant. Then approximately

$$p_1 - p_2 = 2f'V^2L/vgD$$

Expressing pressure in pounds per square inch, $p'$, the diameter in inches, and $V$ as a function of $wv/d^2$, this equation becomes

$$p_1' - p_2' = 174.2f'w^2vL/d^5$$

2. For considerable pressure drops, when dealing with approximately isothermal flow of gases and vapors to which the gas laws are applicable, the fundamental equation on a weight basis may be integrated to give

$$p_1^2 - p_2^2 = \frac{2w^2RT}{gA^2}\ln\frac{v_2}{v_1} + \frac{4f'RTw^2L}{gA^2D}$$

**Coefficients of Friction**   The coefficient of friction $f$ is not a constant but is a function of the dimensionless expression $\mu/\rho Vd$ or $\mu v/Vd$, which is the reciprocal of the Reynolds number. McAdams and Sherwood (*Mech. Eng.*, Oct., 1926) formulate the expression

$$f' = 0.0054 + 0.375\,(\mu v/Vd)$$

This formula is applicable to water and other fluids. For high-pressure steam, the second term in the expression is small and $f'$ is approximately equal to 0.0054. Babcock has suggested the approximation $f' = 0.0027(1 + 3.6/d)$ for steam.

Values of $f = 4f'$ as a function of pipe surface are given in Sec. 3.

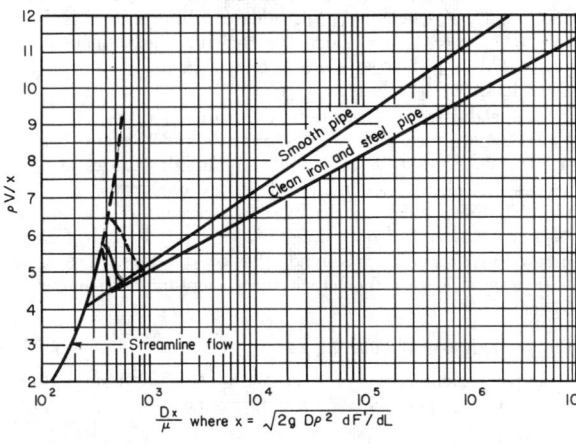

**Fig. 46**   Chart for estimating rate of flow from the pressure gradient.

For predicting the capacity of a given pipe operating on a chosen fluid with fixed pressure drop, the use of Fig. 46 eliminates the trial-and-error methods usually involved. Fig-

ure 47 gives the viscosity of various gases. For conversion factors, see Hawkins, Solberg, and Sibbitt, Units and Conversion Factors for Absolute Viscosity, *Power Plant Eng.*, Nov., 1941; see also Sec. 3.

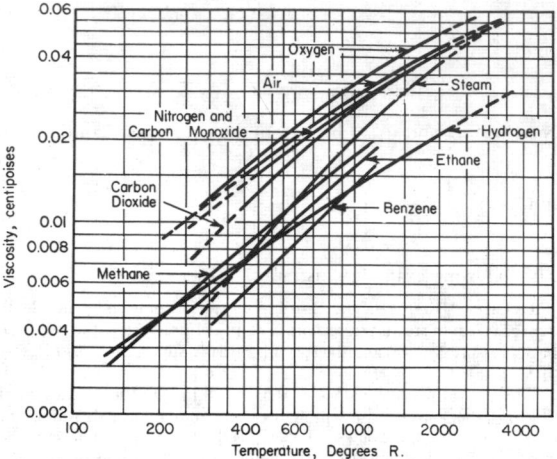

**Fig. 47**   Viscosity of gases.

**EXAMPLE.**   Air at 68°F is flowing isothermally through a straight standard 1-in pipe (ID = 1.049 in). The pipe is 200 ft long, the pressure at the pipe inlet is 60 psig, and the pressure drop through the pipe is 5 lb/in². The average density of the air in the pipe is

$$\rho_{avg} = 0.0753[(60 + 14.7) + (55 + 14.7)]/(2 \times 14.7) = 0.370\text{ lb/ft}^3$$

$$F' = \frac{(5.0)(144)}{0.370} = 1945\text{ ft·lbf per lbf}$$

$$D = \frac{1.049}{12} = 0.0875\text{ ft}$$

$$x = \sqrt{\frac{2(32.2)(0.0875)(0.370)^2(1,945)}{200}} = 2.74$$

$$\mu = 0.018\text{ centipoise (Fig. 47), or }(0.018)(0.000672)$$
$$= 0.0000121\text{ in lb·ft·s units}$$

$$\frac{Dx}{\mu} = \frac{(0.0875)(2.74)}{0.0000121} = 19,800$$

and from Fig. 37 the corresponding value of $\dfrac{\rho V}{x}$ is 7.05

$$V = \frac{7.05(2.74)}{0.370} = 52.2\text{ ft/s}$$

$$m = AV\rho = \frac{(3.1416)(0.0875)^2}{4}(52.2)(0.370) = 0.116\text{ lb/s}$$

Resistances due to fittings, expressed in terms of $L/D$, are as follows: 90-deg elbows, $1-2\frac{1}{2}$ (3–6) [7–10] in, 30 (40) [50]; 90-deg curves, radius of center line of curve 2–8 pipe diameters, 10; globe valves, $1-2\frac{1}{2}$ (3–6) [7–10] in, 45 (60) [75]; tees, 1–4 in, 60. The resistance in energy units, due to sudden enlargement in a pipe, is approximately $(V_1 - V_2)^2/2g$. For sudden contraction it is $1.5(1 - r)V_2^2/2g(3 - r)$, where $r = A_2/A_1$.

**Throttling**

**Throttling or Wiredrawing**   When a fluid flows from a region of higher pressure into a region of lower pressure through a valve or constricted passage, it is said to be throttled or wiredrawn. Examples are seen in the passage of steam through pressure-reducing valves, in the flow through ports and passages in the steam engine, and in the expansion valve of the refrigerating machine.

The **general equation applicable to throttling processes** is

$$(V_2^2 - V_1^2)/2g_c = (h_1 - h_2)J$$

The velocities $V_2$ and $V_1$ are practically equal, and it follows that $h_1 = h_2$; i.e., in a throttling process there is no change in enthalpy.

For a mixture of liquid and vapor, $h = h_f + x h_{fg}$; hence the equation of throttling is $h_{f1} + x_1 h_{fg1} = h_{f2} + x_2 h_{fg2}$. In the case of a perfect gas, $h = c_p T + b_0$; hence the equation of throttling is $c_p T_1 + b_0 = c_p T_2 + b_0$, or $T_1 = T_2$.

**Joule-Thomson Effect** The investigations of Joule and Lord Kelvin showed that a gas drops in temperature when throttled. This is not universally true. For some gases, notably hydrogen, the temperature rises for throttling processes over ordinary ranges of temperature and pressure. Whether there is a rise or fall in temperature depends on the particular range of pressure and temperature over which the change occurs. For every gas, there is one temperature at which no temperature change occurs during a Joule-Thomson expansion; this is called the **inversion temperature.** Below this temperature, a gas cools on throttling; above this temperature, its temperature rises. The ratio of the observed drop in temperature to the drop in pressure, i.e., $dT/dp$, is the **Joule-Thomson coefficient.**

The variations of the Joule-Thomson coefficient, for **air,** with both temperature and pressure, can be determined from the constant enthalpy curves of Fig. 19. Values of the coefficient for **steam** vary from 0.465 at 300 deg to 0.165 at 530°F, the unit being degrees per pound per square inch.

The cooling **effect produced by throttling** has been applied to the liquefaction of gases.

**Loss Due to Throttling** A throttling process in a cycle of operations always introduces a loss of efficiency. If $T_0$ is the temperature corresponding to the back pressure, the loss of available energy is the product of $T_0$ and the increase of entropy during the throttling process. The following example illustrates the calculation in the case of ammonia passing through the expansion valve of a refrigerating machine.

EXAMPLE.    The liquid ammonia at a temperature of 70°F passes through the valve into the brine coil in which the temperature is 20 deg and the pressure is 48.21 psia. The initial enthalpy of the liquid ammonia (see Table 33) is $h_{f1} = 120.5$, and therefore the final enthalpy is $h_{f2} + x_2 h_{fg2} = 64.7 + 553.1x_2 = 120.5$, whence $x_2 = 0.101$. From the table, the initial entropy is $s_{f1} = 0.254$. The final entropy is $s_{f2} + (x_2 h_{fg2}/T_2) = 0.144 + 0.101 \times 1.153 = 0.260$. $T_0 = 20 + 460 = 480$; hence the loss of refrigerating effect is $480 \times (0.260 - 0.254) = 2.9$ Btu.

## COMBUSTION

**Fuels** For special properties of various fuels, see Sec. 7. In general, fuels may be classed under three heads: (1) gaseous fuels, (2) liquid fuels, and (3) solid fuels.

The combustible elements that characterize fuels are carbon, hydrogen, and, in some cases, sulphur. The complete combustion of carbon gives, as a product, carbon dioxide, $CO_2$; the combustion of hydrogen gives water, $H_2O$.

### Combustion of Gaseous and Liquid Fuels

**Combustion Equations** The approximate molecular weights of the important elements and compounds entering into combustion calculations are:

| Gas | H$_2$ | O$_2$ | N$_2$ | CO | CO$_2$ | H$_2$O | CH$_4$ | C$_2$H$_4$ | C$_2$H$_6$O |
|---|---|---|---|---|---|---|---|---|---|
| Molecular weight | 2 | 32 | 28 | 28 | 44 | 18 | 16 | 28 | 46 |

For the elements C and H, the equations of complete combustion are

$$\begin{cases} C + O_2 = CO_2 \\ 12\ lb + 32\ lb = 44\ lb \end{cases} \quad \begin{array}{l} H_2 + \frac{1}{2}O_2 = H_2O \\ 2\ lb + 16\ lb = 18\ lb \end{array}$$

For a combustible compound, as $CH_4$, the equation may be written

$$CH_4 + x \cdot O_2 = y \cdot CO_2 + z \cdot H_2O$$

Taking, as a basis, 1 molecule of $CH_4$ and making a balance of the atoms on the two sides of the equation, it is seen that

$$y = 1 \qquad z = 2 \qquad 2x = 2y + z \qquad \text{or} \qquad x = 2$$

Hence,

$$\begin{cases} CH_4 + 2O_2 = CO_2 + 2H_2O \\ 16\ lb + 64\ lb = 44\ lb + 36\ lb \end{cases}$$

The coefficients in the combustion equation give the combining volumes of the gaseous components. Thus, in the last equation 1 ft$^3$ of $CH_4$ requires for combustion 2 ft$^3$ of oxygen and the resulting gaseous products of combustion are 1 ft$^3$ of $CO_2$ and 2 ft$^3$ of $H_2O$. The coefficients multiplied by the corresponding molecular weights give the combining weights. These are conveniently referred to 1 lb of the fuel. In the combustion of $CH_4$, for example, 1 lb of $CH_4$ requires 64/16 =

4 lb of oxygen for complete combustion and the products are 44/16 = 2.75 lb of $CO_2$ and 36/16 = 2.25 lb of $H_2O$.

**Air Required for Combustion** The composition of air is approximately 0.232 $O_2$ and 0.768 $N_2$ on a pound basis, or 0.21 $O_2$ and 0.79 $N_2$ by volume. For exact analyses, it may be necessary sometimes to take account of the water vapor mixed with the air, but ordinarily this may be neglected.

The minimum amount of air required for the combustion of 1 lb of a fuel is the quantity of oxygen required, as found from the combustion equation, divided by 0.232. Likewise, the minimum volume of air required for the combustion of 1 ft$^3$ of a fuel gas is the volume of oxygen divided by 0.21. For example, in the combustion of $CH_4$ the air required per pound of $CH_4$ is 4/0.232 = 17.24 lb and the volume of air per cubic foot of $CH_4$ is 2/0.21 = 9.52 ft$^3$. Ordinarily, more air is provided than is required for complete combustion. Let $a$ denote the minimum amount required and $xa$ the quantity of air admitted; then $x - 1$ is the **excess coefficient.**

**Products of Combustion** The products arising from the complete combustion of a fuel are $CO_2$, $H_2O$, and, if sulphur is present, $SO_2$. Accompanying these are the nitrogen brought in with the air and the oxygen in the excess of air. Hence the products of complete combustion are principally $CO_2$, $H_2O$, $N_2$, and $O_2$. The **presence of CO indicates incomplete combustion.** The composition of the products of combustion is readily calculated from the combustion equations, as shown by the following illustrative example:

EXAMPLE.  A producer gas having the volume composition given is burned with 20 percent excess of air; required the volume composition of the exhaust gases.

|  | $V$ |
|---|---|
| $H_2$ ......... | 0.08 |
| CO ........ | 0.22 |
| $CH_4$ ....... | 0.024 |
| $CO_2$ ....... | 0.066 |
| $N_2$ ......... | 0.61 |
|  | 1.0 |

| Coefficients in reaction equations | | | Coefficients multiplied by $V$ | | |
|---|---|---|---|---|---|
| $O_2$ | $CO_2$ | $H_2O$ | $O_2$ | $CO_2$ | $H_2O$ |
| 0.5 | 0 | 1 | 0.04 | 0 | 0.08 |
| 0.5 | 1 | 0 | 0.11 | 0.22 | 0 |
| 2 | 1 | 2 | 0.048 | 0.024 | 0.048 |
| 0 | 1 | 0 | 0 | 0.066 | 0 |
| 0 | 0 | 0 | 0 | 0 | 0 |
|  |  |  | 0.198 | 0.31 | 0.128 |

For 1 ft³ of the producer gas, 0.198 ft³ of $O_2$ is required for complete combustion. The minimum volume of air required is 0.198/0.21 = 0.943 ft³ and with 20 percent excess the air supplied is 0.943 × 1.2 = 1.132 ft³. Of this, 0.238 ft³ is oxygen and 0.894 ft³ is $N_2$. Consequently, for 1 ft³ of the fuel gas, the exhaust gas contains

| $CO_2$ ........... |  | 0.31 ft³ |
|---|---|---|
| $H_2O$ ........... |  | 0.128 ft³ |
| $N_2$ ............. | 0.61 + 0.894 = | 1.504 ft³ |
| $O_2$ (excess) ...... | 0.238 − 0.198 = | 0.040 ft³ |
|  |  | 1.982 ft³ |

or

| $CO_2$ ........... | 15.7 percent |
|---|---|
| $H_2O$ ........... | 6.5 percent |
| $N_2$ ............. | 75.8 percent |
| $O_2$ ............. | 2.0 percent |
|  | 100.0 percent |

**Volume Contraction**   As a result of chemical action, there is often a change of volume; for example, in the reaction $2H_2 + O_2 = 2H_2O$, three volumes (two of $H_2$ and one of $O_2$) contract to two volumes of water vapor. In the example just given, the volume of producer gas and air supplied is 1 ft³ gas + 1.132 ft³ air = 2.132 ft³, and the corresponding volume of the exhaust gas is 1.982 ft³, showing a contraction of about 7 percent. For a hydrocarbon having the composition $C_mH_n$, the relative volume contraction is $[1 - (n/4)]$; thus for $CH_4$ and $C_2H_4$ there is no change of volume, for $C_2H_2$ the contraction is half the volume, and for $C_2H_6$ there is an increase of one-half in volume.

The change of volume accompanying a chemical reaction, such as a combustion, causes a corresponding change in the gas constant $R$. Let $R'$ denote the constant for the mixture of gas and air (1 lb of gas and $xa$ lb of air) before combustion, and $R''$ the constant of the mixture of resulting products of combustion. Then, if $y$ is the resulting contraction of volume, $R''/R' = (1 + xa - y)/(1 + xa)$.

**Heat of Combustion**   Usually, a chemical change is accompanied by the generation or absorption of heat. The union of a combustible with oxygen produces heat, and the heat thus generated when 1 lb of combustible is completely burned is called the **heat of combustion** or the **heat value** of the combustible. Heat values are determined experimentally by calorimeters in which the products of combustion are cooled to the initial temperature and the heat absorbed by the cooling medium is measured. This is called the **high heat value.**

The heat transferred (heat of combustion) during a combustion reaction is computed on either a constant-pressure or a constant-volume basis. The first law is used in the analysis of either process.

1. The **heat value at constant volume** ($H_v$). Consider a constant-volume combustion process where several reactants combine under proper conditions to form one or more products. The heat of combustion under constant-volume conditions ($H_v$) according to the first law may be expressed as follows:

$$H_v = \Sigma(Nu)_P - \Sigma(Nu)_R$$

The term $N$ refers to the amount of material and the symbol $u$ signifies the internal energy per unit quantity of material. The subscripts $P$ and $R$ refer to the products and reactants respectively. Hence, it may be concluded that $H_v$ is equal to the change in internal energy. The heat of combustion under constant-volume conditions may also be described as the quantity of heat transferred from a calorimeter to the external surroundings when the temperature and volume of the combustion products are brought to the temperature and volume, respectively, of the gaseous mixture before burning.

2. The **heat value at constant pressure** ($H_p$). For a constant-pressure process the first law may be expressed as

$$H_p = \Sigma(Nu)_P - \Sigma(Nu)_R + pV_P - pV_R$$

Here the symbols $p$ and $V$ refer to the pressure and total volume, respectively. Usually in combustion reactions that part of the change in internal energy resulting from a volume change is small in comparison to the total change; hence it may usually be neglected. Assuming therefore that the internal energy change for a constant-volume reaction is approximately equal to that for a constant-pressure change, the following equation results:

$$H_p = H_v + p(V_P - V_R)$$

Since $H_v$ is equal to the change in internal energy, this relation may be changed to enthalpy values, from which it may be concluded that $H_p$ is equal to the change in enthalpy. The heat of combustion under constant-pressure conditions may also be described as the heat transferred from a calorimeter when the pressure and temperature of the products are brought back to the pressure and temperature, respectively, of the gaseous mixture before burning.

If the reactants and products are assumed to be ideal gases, then the relation for ($H_p$) may be expressed as follows, where $\Delta N$ represents the change in number of moles and $R$ the universal gas constant:

$$H_p = H_v + \Delta NRT$$

From this relation the heat transferred (heat of combustion) at constant pressure may be found from the heat of reaction at constant volume, or vice versa, if the temperature and molar-volume change are known.

If there is no change of volume due to the combustion, the heat values $H_p$ and $H_v$ are the same. When there is a contraction of volume, $H_p$ exceeds $H_v$ by the heat equivalent of the work done on the gas during the contraction. For example, in the burning of CO according to the equation $CO + \frac{1}{2}O_2 = CO_2$, there is a contraction of $\frac{1}{2}$ volume. Taking 62°F as the temperature, the volume of 1 lb CO at atmospheric pressure is 13.6 ft$^3$; hence the equivalent of the work done at atmospheric pressure is $\frac{1}{2} \times 13.6 \times 2116/778 = 18.5$ Btu, which is about 0.4 percent of the heat value of CO. Since the difference between $H_p$ and $H_v$ is small in most fuels, it is usually neglected.

It is also to be noted that heat values vary with the initial temperature (which is also the final temperature), but the variation is usually negligible.

**Heat Value per Unit Volume**   Since the consumption of a fuel gas is more easily measured by volume than by mass, it is convenient to express heat values in terms of volumes. For this purpose, a standard temperature and pressure must be assumed. It is customary to take atmospheric pressure (14.70 psi) as standard, but there is diversity of practice in the matter of a **standard temperature.** The temperature of 68°F (20°C) is generally accepted in metric countries and has been recommended by the American delegates to the meeting of the International Committee of Weights and Measures and also by the ASME Power Test Codes Committee. The American Gas Assoc. uses 60°F as the standard temperature of reference. Conversion of density and heat values from 68 to 60°F of dry (saturated) gas is obtained by multiplying by the factor 1.0154 (1.0212). Conversion of specific volumes of dry (saturated) gas is obtained by multiplying by the factor 0.9848 (0.9792).

If the gas is at some other pressure and temperature, say $p_1$ psia and $T_1$ deg R, the heat value per cubic foot is found by multiplying the heat value per cubic foot under standard conditions by $35.9 p_1/T_1$.

The heat values of a few of the more common fuels per pound and per cubic foot are given in Table 44.

**Heat Value per Unit Volume of Mixture**   Let $a$ denote the volume of air required for the combustion of 1 ft$^3$ of fuel gas and $xa$ the value of air actually admitted, $(x - 1)a$ being therefore the excess. Then the volume of the mixture of fuel gas and air is $1 + xa$, and the quotient $H/(1 + xa)$ may be called the heat value per cubic foot of mixture. This magnitude is useful in comparing the relative volumes of mixture required with different fuel gases. Thus a lean gas, as blast-furnace gas or producer gas, has a low heat value $H$, but the value of $a$ is correspondingly low. On the other hand, a rich gas, like natural gas, has a high heat value but requires a large volume of air for combustion.

**Low and High Heat Values**   Any fuel containing hydrogen yields water as one product of combustion. At atmospheric pressure, the partial pressure of the water vapor in the resulting combustion gas mixture will usually be sufficiently high to cause water to condense out if the temperature is allowed to fall below 120 to 140°F. This causes liberation of the heat of vaporization of any water condensed. The low heat value is evaluated assuming no water vapor condensed, whereas the high heat value is calculated assuming all water vapor condensed.

To facilitate calculations of the temperature attained by combustion, it is desirable to make use of the **low heat value.** The necessity of taking into account the heat of vaporization of the water vapor and the difference between the specific heats of liquid water and of water vapor is thus avoided. The high heat of combustion exceeds the low heat of combustion by the difference between the heat actually given up on cooling the products to the initial temperature and that which would have been given up if the products had remained in the gaseous state. A bomb calorimeter (constant volume) gives practically correct values of the high heat value; a gas calorimeter (constant pressure) gives values which, for the usual fuels, may be incorrect by a fraction of 1 percent. The quantity to be substracted from the high heat value to obtain the low heat value will vary with the composition of the fuel; an approximate value is 1050 $m$, where $m$ is the number of pounds of $H_2O$ formed per pound of fuel burned.

In Germany, the low heat value of the fuel is used in calculating efficiencies of internal-combustion engines. In the United States, the high value is specified by the ASME Power Test Codes.

**Heat of Formation**   The change in enthalpy resulting when a compound is formed from its elements isothermally and at constant pressure is numerically equal to, but of opposite sign to, the heat of formation, $\Delta H_f = -Q_f$. It is equal to the difference between the heats of combustion of the constituents forming the compound and the heat of combustion of the compound itself. The following values for heats of formation are in Btu per pound of the compound. The elements before the change and the compounds formed are assumed in their ordinary stable states at 65°F and 1 atm. A plus sign indicates heat evolved on forming the compound, a minus sign heat absorbed from the surroundings.

**Fuels**   Methane, $CH_4$ (gas), 2001.4; ethane, $C_2H_6$ (gas), 1206.1; propane, $C_3H_8$ (vapor), 1008.5; acetylene, $C_2H_2$ (gas), 3747, ethylene, $C_2H_4$ (gas), $-805.3$; benzene, $C_6H_6$ (vapor), $-459$; toluene, $C_7H_8$ (vapor), $-234.9$; methyl alcohol, $CH_3OH$ (liquid), 3227.3; ethyl alcohol, $C_2H_5OH$ (liquid), 2623.3.

**Inorganic Compounds**   $Al_2O_3$, 6710; CaO, 4869; $CaCO_3$, 5206; FeO, 1611; $Fe_2O_3$, 2238; $Fe_3O_4$, 2075; $FeS_2$, 532.7; HCL (gas), 1089; $HNO_3$ (liquid), 1190; $H_2O$ (liquid), 6827; $H_2S$ (gas), 279.9; $H_2SO_4$ (liquid), 3555.8; $K_2O$, 164.7; MgO, 6522; MnO, 2449; NO, $-1296$; $N_2O$, $-803.5$; $Na_2O$, 2888; $NH_3$, 1163; $NH_4CL$, 1480; NiO, 1407; $P_2O_5$, 5394; PbO (red), 423.0; $PbO_2$, 489.1; $SO_2$, 1933; $SO_3$, 2112; SnO, 904.7; ZnO, 1847.

**Internal Energy and Enthalpy of Gases**   Table 46 gives the internal energy of various common gases in Btu per lb mol measured above 520°R (60°F). The corresponding values of the enthalpy are obtained by adding the value of $APv$ from the last column.

**Temperature Attained by Combustion**   Excluding the effect of dissociation, the temperature attained at the end of combustion may be calculated by a simple energy balance. The heat of combustion less the heat lost by conduction and radiation during the process is equal to the increase in internal energy of the products mixture if the combustion is at constant volume;

**Table 44. Heats of Combustion**

| Fuel | Chemical symbol | High heat value, Btu | | Low heat value, Btu | |
|---|---|---|---|---|---|
| | | Per lb | Per cu ft[a] | Per lb | Per cu ft[a] |
| Carbon to $CO_2$........................ | C | 14,096 | | | |
| Carbon to CO......................... | C | 3,960 | | | |
| CO to $CO_2$.....,.................... | CO | 4,346 | 316.0 | | |
| Sulphur to $SO_2$...................... | S | 3,984 | | | |
| Hydrogen............................ | $H_2$ | 61,031 | 319.4 | 51,593 | 270.0 |
| Methane............................. | $CH_4$ | 23,890 | 994.7 | 21,518 | 896.0 |
| Ethane.............................. | $C_2H_6$ | 22,329 | 1742.6 | 20,431 | 1594.5 |
| Propane............................. | $C_3H_8$ | 21,670 | 2480.1 | 19,944 | 2282.6 |
| Butane.............................. | $C_4H_{10}$ | 21,316 | 3215.6 | 19,679 | 2968.7 |
| Pentane............................. | $C_5H_{12}$ | 21,095 | 3950.2 | 19,513 | 3654.0 |
| Hexane (liquid)...................... | $C_6H_{14}$ | 20,675 | ...... | 19,130 | |
| Octane (liquid)...................... | $C_8H_{18}$ | 20,529 | ...... | 19,029 | |
| $n$-Decane (liquid)................... | $C_{10}H_{22}$ | 20,371 | ...... | 19,175 | |
| Ethylene............................ | $C_2H_4$ | 21,646 | 1576.1 | 20,276 | 1477.4 |
| Propene (propylene).................. | $C_3H_6$ | 21,053 | 2299.4 | 19,683 | 2151.3 |
| Acetylene (ethyne).................... | $C_2H_2$ | 21,477 | 1451.4 | 20,734 | 1402.0 |
| Benzene............................. | $C_6H_6$ | 18,188 | 3687.5 | 17,446 | 3539.3 |
| Toluene (methyl benzene).............. | $C_7H_8$ | 18,441 | 4410.1 | 17,601 | 4212.6 |
| Methanol (methyl alcohol, liquid)....... | $CH_4O$ | 9,758 | ...... | 8,570 | |
| Ethanol (ethyl alcohol, liquid).......... | $C_2H_6O$ | 12,770 | ...... | 11,531 | |
| Naphthalene (solid)................... | $C_{10}H_8$ | 17,310 | ...... | 13,110 | |

[a]Measured as a gas at 68°F and 14.70 psia. Multiply by 1.0154 for 60°F and 14.70 psia.

or, if the combustion is at constant pressure, the difference is equal to the increase in enthalpy of the products mixture.

EXAMPLE. To calculate the temperature of combustion of a fuel gas having the composition $H_2$ = 0.50, CO = 0.46, $CO_2$ = 0.04. The gas is burned with 15 percent excess air at constant volume, and the initial temperature is 62 F; i.e., $T$ = 522°R.

The volume compositions of the initial mixture of fuel gas and air and of the mixture of products are, respectively:

Initial: $H_2$, 0.50; CO, 0.46; $CO_2$, 0.04; $O_2$, 0.552; $N_2$, 2.098
Products: $H_2O$, 0.50; $CO_2$, 0.50; $O_2$, 0.072; $N_2$, 2.098

Since a volume composition is also a mol composition, the products mixture may be regarded as made up of 0.5 mol each of $H_2O$ and $CO_2$, 0.072 mol of $O_2$, and 2.098 mols of $N_2$. If values are taken from Table 45, the heat generated by combustion of the fuel mixture is 0.50 × 51,593 + 0.46 × 28 × 4346 = 107,569 Btu. The internal energy $u$ of the products mixture at $T$ = 522 is now calculated. For 0.5 mol $H_2O$ + 0.5 mol $CO_2$ + 0.072 mol $O_2$ + 2.098 mols $N_2$ this is 6.1 + 6.95 + 0.72 + 20.34 = 34.11 Btu.

The energy $u$ of the mixture is next calculated for various assumed temperatures, the proper values being taken from Table 46:

If the heat of combustion, 107,569 Btu, is entirely used in the increase of energy, the temperature attained lies somewhere between 5000 and 5100; by interpolation, the value 5073° is obtained.

Loss of heat during combustion may readily be taken into account; thus if 10 percent of the heat of combustion is lost, the amount available for increasing the energy of the products is 107,569 × 0.90 = 96,812 Btu, and this increase gives $T_2$ = 4671°. If the fuel is burned at constant pressure, $H_p$ is used instead of $H_v$ and values of $h$ are determined from Table 45 instead of values of $u$.

**Effect of Dissociation** The maximum temperature that can be obtained by the combustion of any fuel is limited by the dissociation of the products formed. The dissociation and equilibriums involved in high-temperature combustion are exceedingly complex, involving such chemical species as $CO_2$, CO, $H_2O$, $H_2$, H, OH, $N_2$, NO, N, $O_2$, and O. Equilibrium constants for the reactions involved are given in Lewis and von Elbe, "Combustion, Flames, and Explosions of Gases," pp. 382–383, Macmillan, 1938. Calculations of this type are facilitated by the use of charts. See Hottel, Williams, and Satterfield, "Thermodynamic Charts for Combustion Processes," Wiley.

| $T_2$ assumed | 4700 | 4800 | 4900 | 5000 | 5100 |
|---|---|---|---|---|---|
| Energy 0.5 mol $H_2O$ | 18,308 | 18,851 | 19,396 | 19,943 | 20,492 |
| Energy 0.5 mol $CO_2$ | 23,742 | 24,388 | 25,035 | 25,683 | 26,331 |
| Energy 0.072 mol $O_2$ | 1,973 | 2,026 | 2,079 | 2,132 | 2,185 |
| Energy 2.098 mols $N_2$ | 54,596 | 55,018 | 56,447 | 57,882 | 59,321 |
| $u_2$ = | 97,619 | 100,283 | 102,957 | 105,640 | 108,329 |
| $u_1$ = | 34 | 34 | 34 | 34 | 34 |
| | 97,585 | 100,249 | 102,923 | 105,606 | 108,295 |

Calculated flame temperatures, allowing for dissociation, for gaseous fuels with stated amounts of air present are given in Table 47. The combustion is assumed to be adiabatic and at 14.7 psia.

In the case of **explosion** in the **internal-combustion engine,** the figures in Table 47 will be somewhat changed. The effect of compression is to increase both the initial temperature and the initial pressure. The resulting increase in the explosion tem-

**Table 45. Products of Combustion**

| Fuel | Chemical formula | Molecular weight $O_2 = 32$ | Specific weight, lb per cu ft at 68 F and 14.70 lb per sq in. | Volume of air necessary for combustion of unit volume of fuel at same temperature and pressure | Products of combustion of 1 cu ft of fuel in theoretical amount of air, cu ft | | | Weight of air necessary for combustion of unit weight of fuel | Products of combustion of 1 lb of fuel in theoretical amount of air, lb | | |
|------|------|------|------|------|------|------|------|------|------|------|------|
| | | | | | $CO_2$ | $H_2O$ | $N_2$ | | $CO_2$ | $H_2O$ | $N_2$ |
| Oxygen..... | $O_2$ | 32 | 0.0831 | | | | | | | | |
| Nitrogen.... | $N_2$ | 28.08 | 0.0727 | | | | | | | | |
| Air......... | ........ | ........ | 0.0753 | | | | | | | | |
| Hydrogen... | $H_2$ | 2.016 | 0.0052 | 2.39 | 0 | 1 | 1.89 | 34.2 | 0.0 | 8.94 | 26.28 |
| Steam....... | $H_2O$ | 18.016 | | | | | | | | | |
| Carbon monoxide.. | CO | 28.00 | 0.0727 | 2.39 | | | | | | | |
| Carbon dioxide.... | $CO_2$ | 44.00 | 0.1142 | | | | | | | | |
| Methane.... | $CH_4$ | 16.03 | 0.0416 | 9.55 | 1 | 2 | 7.55 | 17.21 | 2.75 | 2.248 | 13.22 |
| Ethane...... | $C_2H_6$ | 30.05 | 0.0779 | 16.71 | 2 | 3 | 13.21 | 16.07 | 2.93 | 1.799 | 12.34 |
| Propane..... | $C_3H_8$ | 44.06 | 0.1142 | 23.87 | 3 | 4 | 18.87 | 15.65 | 3.00 | 1.635 | 12.02 |
| Butane...... | $C_4H_{10}$ | 58.1 | 0.1506 | 30.94 | 4 | 5 | 24.53 | 15.44 | 3.03 | 1.551 | 11.86 |
| Pentane..... | $C_5H_{12}$ | 72.1 | 0.1869 | 38.08 | 5 | 6 | 30.2 | 15.31 | 3.05 | 1.499 | 11.76 |
| Hexane..... | $C_6H_{14}$ | 86.1 | 0.2232 | 45.3 | 6 | 7 | 35.8 | 15.22 | 3.07 | 1.465 | 11.69 |
| Heptane..... | $C_7H_{16}$ | 100.1 | 0.2596 | 52.5 | 7 | 8 | 41.5 | 15.15 | 3.08 | 1.439 | 11.64 |
| Octane...... | $C_8H_{18}$ | 114.1 | 0.2959 | 59.7 | 8 | 9 | 47.2 | 15.11 | 3.08 | 1.421 | 11.60 |
| Nonane..... | $C_9H_{20}$ | 128.2 | 0.3323 | 66.8 | 9 | 10 | 52.8 | 15.07 | 3.09 | 1.406 | 11.57 |
| Benzene.... | $C_6H_6$ | 78.0 | 0.2025 | 35.8 | 6 | 3 | 28.3 | 13.26 | 3.38 | 0.693 | 10.18 |
| Toluene..... | $C_7H_8$ | 92.1 | 0.2388 | 42.9 | 7 | 4 | 34.0 | 13.50 | 3.35 | 0.783 | 10.36 |
| Xylene...... | $C_8H_{10}$ | 106.2 | 0.2752 | 50.1 | 8 | 5 | 39.6 | 13.57 | 3.31 | 0.845 | 10.42 |
| Cyclohexane.... | $C_6H_{12}$ | 84.0 | 0.2180 | 43.0 | 6 | 6 | 34.0 | 14.76 | 3.14 | 1.285 | 11.34 |
| Ethylene.... | $C_2H_4$ | 28.03 | 0.0728 | 14.32 | 2 | 2 | 11.32 | 14.76 | 3.14 | 1.285 | 11.34 |
| Propylene... | $C_3H_6$ | 42.0 | 0.1090 | 21.48 | 3 | 3 | 16.98 | 14.76 | 3.14 | 1.285 | 11.34 |
| Butylene.... | $C_4H_8$ | 64.1 | 0.1454 | 28.64 | 4 | 4 | 22.64 | 14.76 | 3.14 | 1.285 | 11.34 |
| Acetylene.... | $C_2H_2$ | 26.02 | 0.0675 | 11.93 | 2 | 1 | 9.43 | 13.26 | 3.38 | 0.693 | 10.18 |
| Allylene..... | $C_3H_4$ | 40.0 | 0.1038 | 19.09 | 3 | 2 | 15.09 | 13.78 | 3.30 | 0.900 | 10.59 |
| Napthalene.. | $C_{10}H_8$ | 128.1 | 0.3322 | 57.3 | 10 | 4 | 45.28 | 12.93 | 3.44 | 0.563 | 9.93 |
| Methyl alcohol.... | $CH_4O$ | 32.0 | 0.0830 | 7.16 | 1 | 2 | 5.66 | 6.46 | 1.37 | 1.125 | 4.96 |
| Ethyl alcohol.... | $C_2H_6O$ | 46.0 | 0.1194 | 14.32 | 2 | 3 | 11.32 | 8.99 | 1.91 | 1.174 | 6.90 |

SOURCE: Marks, "The Airplane Engine."

perature will tend to increase the dissociation, the increase of pressure will tend to reduce it. The net effect will be a small reduction.

**Combustion of Liquid Fuels**  For properties of fuel oils, heat values, etc., see Sec. 7. Calculations for the burning of liquid fuels are fundamentally the same as for gaseous fuels. Liquid fuels are almost always gasified before or during actual combustion.

Tizard and Pye (*Automobile Engr.*, Feb., 1921) give the following result on a theoretically correct mixture of benzene vapor and air at 212°F and atmospheric pressure. This was compressed to ⅕ of its volume and exploded. It was assumed that no heat was lost during explosion. With no dissociation, the maximum temperature reached is 5473°F and the maximum pressure 658 psi. With dissociation, the maximum tem-

perature is 4811°F and the maximum pressure 600.7 psi. Similar results are found for other liquid fuels, as alcohol, kerosene, etc.

As the temperature falls after reaching the maximum, the uncombined CO and $H_2$ burn with the oxygen present, and when the temperature has reached about 3000°F the combustion is practically complete. Thus in the internal-combustion engine dissociation has the effect of reducing the explosion temperature and pressure and afterward of raising the expansion curve. The effect on the efficiency is not large. For an engine using benzene according to the conditions just noted, the ideal efficiency with no heat loss and without dissociation was found by Tizard and Pye to be 35.9 percent; with dissociation and subsequent recombination, it was found to be 33.8 percent.

**Table 46. Internal Energy of Gases**
Btu per lb mol above 520°R

| Temp, R | $O_2$ | $N_2$ | Air | $CO_2$ | $H_2O$ | $H_2$ | CO | $Apv$ |
|---|---|---|---|---|---|---|---|---|
| 520 | 0 | 0 | 0 | 0 | 0 | 0 | 0 | 1,033 |
| 540 | 100 | 97 | 97 | 139 | 122 | 96 | 97 | 1,072 |
| 560 | 200 | 196 | 196 | 280 | 244 | 193 | 196 | 1,112 |
| 580 | 301 | 295 | 295 | 424 | 357 | 291 | 295 | 1,152 |
| 600 | 402 | 395 | 395 | 570 | 490 | 390 | 396 | 1,192 |
| 700 | 920 | 896 | 897 | 1,320 | 1,110 | 887 | 896 | 1,390 |
| 800 | 1,449 | 1,399 | 1,403 | 2,120 | 1,734 | 1,386 | 1,402 | 1,589 |
| 900 | 1,989 | 1,905 | 1,915 | 2,965 | 2,366 | 1,886 | 1,913 | 1,787 |
| 1000 | 2,539 | 2,416 | 2,431 | 3,852 | 3,009 | 2,387 | 2,430 | 1,986 |
| 1100 | 3,101 | 2,934 | 2,957 | 4,778 | 3,666 | 2,889 | 2,954 | 2,185 |
| 1200 | 3,675 | 3,461 | 3,492 | 5,736 | 4,399 | 3,393 | 3,485 | 2,383 |
| 1300 | 4,262 | 3,996 | 4,036 | 6,721 | 5,030 | 3,899 | 4,026 | 2,582 |
| 1400 | 4,861 | 4,539 | 4,587 | 7,731 | 5,740 | 4,406 | 4,580 | 2,780 |
| 1500 | 5,472 | 5,091 | 5,149 | 8,764 | 6,468 | 4,916 | 5,145 | 2,979 |
| 1600 | 6,092 | 5,652 | 5,720 | 9,819 | 7,212 | 5,429 | 5,720 | 3,178 |
| 1700 | 6,718 | 6,224 | 6,301 | 10,896 | 7,970 | 5,945 | 6,305 | 3,376 |
| 1800 | 7,349 | 6,805 | 6,889 | 11,993 | 8,741 | 6,464 | 6,899 | 3,575 |
| 1900 | 7,985 | 7,393 | 7,485 | 13,105 | 9,526 | 6,988 | 7,501 | 3,773 |
| 2000 | 8,629 | 7,989 | 8,087 | 14,230 | 10,327 | 7,517 | 8,109 | 3,972 |
| 2100 | 9,279 | 8,592 | 8,698 | 15,368 | 11,146 | 8,053 | 8,722 | 4,171 |
| 2200 | 9,934 | 9,203 | 9,314 | 16,518 | 11,983 | 8,597 | 9,339 | 4,369 |
| 2300 | 10,592 | 9,817 | 9,934 | 17,680 | 12,835 | 9,147 | 9,961 | 4,568 |
| 2400 | 11,252 | 10,435 | 10,558 | 18,852 | 13,700 | 9,703 | 10,588 | 4,766 |
| 2500 | 11,916 | 11,056 | 11,185 | 20,033 | 14,578 | 10,263 | 11,220 | 4,965 |
| 2600 | 12,584 | 11,682 | 11,817 | 21,222 | 15,469 | 10,827 | 11,857 | 5,164 |
| 2700 | 13,257 | 12,313 | 12,453 | 22,419 | 16,372 | 11,396 | 12,499 | 5,362 |
| 2800 | 13,937 | 12,949 | 13,095 | 23,624 | 17,288 | 11,970 | 13,144 | 5,561 |
| 2900 | 14,622 | 13,590 | 13,742 | 24,836 | 18,217 | 12,549 | 13,792 | 5,759 |
| 3000 | 15,309 | 14,236 | 14,394 | 26,055 | 19,160 | 13,133 | 14,443 | 5,958 |
| 3100 | 16,001 | 14,888 | 15,051 | 27,281 | 20,117 | 13,723 | 15,097 | 6,157 |
| 3200 | 16,693 | 15,543 | 15,710 | 28,513 | 21,086 | 14,319 | 15,754 | 6,355 |
| 3300 | 17,386 | 16,199 | 16,369 | 29,750 | 22,066 | 14,921 | 16,414 | 6,554 |
| 3400 | 18,080 | 16,855 | 17,030 | 30,991 | 23,057 | 15,529 | 17,078 | 6,752 |
| 3500 | 18,776 | 17,512 | 17,692 | 32,237 | 24,057 | 16,143 | 17,744 | 6,951 |
| 3600 | 19,475 | 18,171 | 18,356 | 33,487 | 25,067 | 16,762 | 18,412 | 7,150 |
| 3700 | 20,179 | 18,833 | 19,022 | 34,741 | 26,085 | 17,385 | 19,082 | 7,348 |
| 3800 | 20,887 | 19,496 | 19,691 | 35,998 | 27,110 | 18,011 | 19,755 | 7,547 |
| 3900 | 21,598 | 20,162 | 20,363 | 37,258 | 28,141 | 18,641 | 20,430 | 7,745 |
| 4000 | 22,314 | 20,830 | 21,037 | 38,522 | 29,178 | 19,274 | 21,107 | 7,944 |
| 4100 | 23,034 | 21,500 | 21,714 | 39,791 | 30,221 | 19,911 | 21,784 | 8,143 |
| 4200 | 23,757 | 22,172 | 22,393 | 41,064 | 31,270 | 20,552 | 22,462 | 8,341 |
| 4300 | 24,482 | 22,845 | 23,073 | 42,341 | 32,326 | 21,197 | 23,140 | 8,540 |
| 4400 | 25,209 | 23,519 | 23,755 | 43,622 | 33,389 | 21,845 | 23,819 | 8,738 |
| 4500 | 25,938 | 24,194 | 24,437 | 44,906 | 34,459 | 22,497 | 24,499 | 8,937 |
| 4600 | 26,668 | 24,869 | 25,120 | 46,193 | 35,535 | 23,154 | 25,179 | 9,136 |
| 4700 | 27,401 | 25,546 | 25,805 | 47,483 | 36,616 | 23,816 | 25,860 | 9,334 |
| 4800 | 28,136 | 26,224 | 26,491 | 48,775 | 37,701 | 24,480 | 26,542 | 9,533 |
| 4900 | 28,874 | 26,905 | 27,180 | 50,069 | 38,791 | 25,418 | 27,226 | 9,731 |
| 5000 | 29,616 | 27,589 | 27,872 | 51,365 | 39,885 | 25,819 | 27,912 | 9,930 |
| 5100 | 30,361 | 28,275 | 28,566 | 52,663 | 40,983 | 26,492 | 28,600 | 10,129 |
| 5200 | 31,108 | 28,961 | 29,262 | 53,963 | 42,084 | 27,166 | 29,289 | 10,327 |
| 5300 | 31,857 | 29,648 | 29,958 | 55,265 | 43,187 | 27,842 | 29,980 | 10,526 |
| 5400 | 32,607 | 30,337 | 30,655 | 56,569 | 44,293 | 28,519 | 30,674 | 10,724 |

SOURCE: L. C. Lichty, "Internal Combustion Engines," 1939, p. 582, derived from data given by Hershey, Eberhardt, and Hottel, *Trans. SAE*, **31**, 1936, p. 409.

## Combustion of Solid Fuels

For **properties of solid fuels,** heat values, etc., see Sec. 7.

**Air Required for Combustion** Let $c$, $h$, and $o$, denote, respectively, the parts of carbon, hydrogen, and oxygen in 1 lb of the fuel. Then the **minimum amount of oxygen** required for complete combustion is $2.67c + 8h - o$ lb, and the **minimum quantity of air** required is $a = (2.67c + 8h - o)/0.23 = 11.6[c + 3(h - o/8)]$ lb.

With air at 62°F and at atmospheric pressure, the minimum volume of air required is $v_m = 147[c + 3(h - o/8)]$ ft$^3$. In

**Table 47. Flame Temperatures, Deg R, at 14.7 psia, Allowing for Dissociation**

| Fuel | Percent of theoretical air | | | | |
|---|---|---|---|---|---|
| | 80 | 90 | 100 | 120 | 140 |
| Hydrogen..................................... | 4,210 | 4,330 | 4,390 | 4,000 | 3,670 |
| Carbon monoxide............................. | 4,280 | 4,370 | 4,320 | 4,140 | 3,850 |
| Methane..................................... | 4,050 | .... | 4,010 | 3,660 | 3,330 |
| Carbureted water gas......................... | 3,940 | .... | 4,150 | 3,820 | 3,510 |
| Coal gas.................................... | 3,920 | .... | 4,050 | 3,780 | 3,440 |
| Natural gas................................. | 4,010 | .... | 4,180 | 3,840 | 3,520 |
| Producer gas................................ | 3,040 | .... | 3,330 | 3,130 | 2,970 |
| Blast-furnace gas............................ | 2,810 | .... | 3,060 | 2,920 | 2,750 |

SOURCE: Satterfield, "Generalized Thermodynamics of High-temperature Combustion," Sc.D. thesis, M.I.T., 1946.

The volumetric compositions of the fuels of Table 47 are given below:

| Fuels | CO | $H_2$ | $CH_4$ | $C_2H_6$ | Illuminants (assumed $C_2H_4$) | $CO_2$ | $O_2$ | $N_2$ |
|---|---|---|---|---|---|---|---|---|
| $H_2$.......................... | ..... | 100.0 | | | | | | |
| CO.......................... | 100.0 | | | | | | | |
| $CH_4$.......................... | ..... | ..'.. | 100.0 | | | | | |
| Carbureted water gas......... | 24.1 | 32.5 | 9.0 | 2.2 | 10.3 | 4.6 | 0.6 | 16.7 |
| Coal gas.................... | 5.9 | 53.2 | 29.6 | .... | 2.7 | 1.4 | 0.7 | 6.5 |
| Natural gas................. | ..... | ..... | 78.8 | 14.0 | .... | 0.4 | ... | 6.8 |
| Producer gas................ | 26.0 | 3.0 | 0.5 | .... | .... | 2.50 | ... | 56.0 |
| Blast-furnace gas............ | 26.5 | 3.5 | 0.2 | .... | .... | 12.8 | 0.1 | 56.9 |

practice, an excess of air over that required for combustion is admitted to the furnace. The actual quantity admitted per pound of fuel may be denoted by $xa$. Then $x$ = amount admitted ÷ minimum amount.

**Combustion Products**  If $v_m$ is the minimum volume of air required for complete combustion and $xv_m$ the actual volume supplied, then the products will contain per pound of fuel, $O_2$ = $0.21v_m(x - 1)$ ft³, $N_2$ = $0.79 xv_m$ ft³.

From the reaction equation $C + O_2 - CO_2$, the volume of $CO_2$ formed is equal to the volume of oxygen required for the carbon constituent alone; hence volume of $CO_2 = 0.21 v_m c/[c + 3(b - 0.125 o)]$.

Of the *dry* gaseous products (i.e., without water), the $CO_2$ content by volume is therefore given by the expression

$$CO_2 = 0.21c/[xc + (x - 0.21)3(b - 0.125 o)]$$

The combined $CO_2$ and $O_2$ content is

$$CO_2 + O_2 = 0.21\left\{1 - 0.79 \Big/ \left[\frac{x + cx}{3(b - 0.125 o) - 0.21}\right]\right\}$$

If the fuel is all carbon, the combined $CO_2$ and $O_2$ is by volume 21 percent of the gaseous products. The more hydrogen contained in the fuel, the smaller is the $CO_2 + O_2$ content. The $CO_2$ content depends in the first instance on the excess of air. Thus, for pure carbon, it is $CO_2 = 0.21/x$.

The excess of air may be calculated from the composition of the gases and that of the fuel. Thus

$$x = 0.21\left[\frac{c}{[CO_2]} + 3(b - 0.125 o)\right] \Big/ [c + 3(b - 0.125 o)]$$

in which $[CO_2]$ denotes the percent by volume of the $CO_2$ in the dry gas.

The temperature of combustion is calculated by the same method as for gaseous fuels.

**Loss Due to Incomplete Combustion**  The loss due to incomplete combustion of the carbon in the fuel, in Btu per lb of fuel, is

$$L = 10,136C \times CO/(CO + CO_2)$$

where $10,136$ − difference in heat evolved in burning 1 lb of carbon to $CO_2$ and to CO; CO and $CO_2$ = percentages by volume of carbon monoxide and carbon dioxide as found by analysis; and C = fraction of quantity of carbon in the fuel which is actually burned and passes up the stack, either as CO or $CO_2$. The presence of 1 percent of CO in the flue gases will represent a decrease in the boiler efficiency of 4.5 percent. An additional loss is caused by passage through the grate to the ashpit of any unburned or partly burned fuel.

It is generally assumed that high $CO_2$ readings are indicative of good combustion and, hence, of high efficiencies. Such readings are not satisfactory when considered apart from the CO determination. The best percentage of $CO_2$ to maintain varies with different fuels and is lower for those with a high hydrogen content than for fuel mainly composed of carbon.

Hydrogen in a fuel increases the nitrogen content of the flue gases. This is due to the fact that the water vapor formed by the combustion of hydrogen will condense at the temperature at which the analysis is made, while the nitrogen which accompanied the oxygen maintains its gaseous form and passes in that form into the sampling apparatus. For this reason, where highly volatile coals containing considerable hydrogen are burned, the flue gas contains an apparently increased amount of nitrogen. The effect is even more pronounced when burning gaseous or liquid hydrocarbon fuels.

The amount of flue gases per pound of fuel, including moisture formed by the hydrogen component, is approximately = $3.02[N/(CO_2 + CO)]C + (1 - A)$, where A = percent of ash found in test. The quantity of dry flue gases per pound of fuel may be approximated from the formula: $W_2$ =

$C[11CO_2 + 8O + 7(CO + N)]/3(CO_2 + CO)$. In these formulas, the amount of gas is per pound of dry or moist fuel as the percentage of C is referred to a dry or moist basis.

The ratio of air supplied per pound of fuel to the air theoretically required is

$$\frac{W_1}{W} = \frac{3.02C\left(\dfrac{N}{CO_2 + CO}\right)}{34.56\left(\dfrac{C}{3} + H - \dfrac{O}{8}\right)}.$$

The ratio of air supplied per pound of combustible to that theoretically required is $N/[N - 3.782(O - \frac{1}{2}CO)]$, on the assumption that all the nitrogen in the flue gas comes from the air supplied. Figure 48 gives the value of this ratio for varying flue-gas analyses where there is no CO present.

For petroleum fuels with hydrogen content from 9 to 16 percent, the excess air can be determined from the $CO_2$ content of the flue gases (with no CO present) by the use of Fig. 49. The curves are based on the assumption of 0.4 percent sulphur in the oil.

**Surface Combustion**   If, after forcing combustible gas through a porous mass of refractory material, igniting, and allowing it to burn in the air with the usual flame, air is forced in with the gas so that the resulting explosive mixture flows with too great velocity for backfiring to occur, the flame becomes non-luminous and, as the surface of the refractory mass is heated up, gradually disappears from the surface of the material, which glows under the action of the surface combustion. The combustion, which is supported by the oxygen admitted with the gas and is not influenced by the oxygen in front of the refractory mass, may be perfect with a minimum of excess air. In the case of a porous slab or diaphragm, the combustion takes place within a distance of from $\frac{1}{8}$ to $\frac{1}{4}$ in from the surface. This method of heating has been applied to

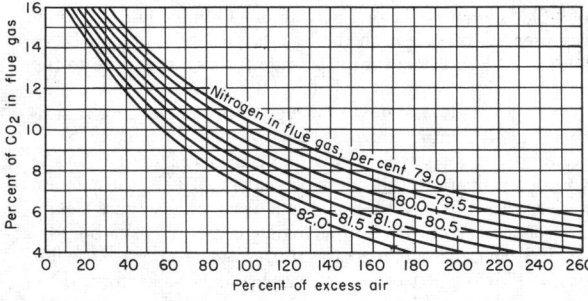

**Fig. 48**   Ratio of air supplied per pound of combustible to that theoretically required.

muffles and crucibles, where the porous refractory material, in the form of small lumps, is closely packed in the interspaces. Muffles can be kept steadily at a temperature of 2700°F (3600°F attainable) with the use of only one-half the amount of gas required in the older method of firing. The advantages of this method of combustion, according to Prof. W. A. Bone (*Engineering*, Apr. 14, 1911), are that combustion is accelerated by the incandescent surface, the heat developed can be

concentrated just where it is required, very high temperatures are attainable without the use of regenerators, and the energy is converted into radiant form which is transmitted very rapidly to the object exposed to it.

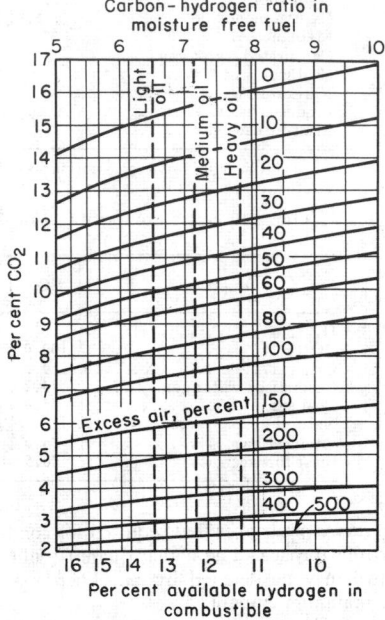

**Fig. 49**   Relation of $CO_2$ to excess air for fuels.

Surface combustion has been applied to boilers, the tubes being packed throughout their length with pieces of firebrick, the air-gas mixture entering each tube through a $\frac{3}{4}$-in hole in a fireclay plug or nozzle (which serves to protect the joints in the tube sheet) at the front end. For gas firing, 3- to 6-in tubes have been used, and 9- to 12-in for oil firing, the oil being first vaporized in a chamber at the firing end of the tube. The combustion is confined to about 4 in next to the admission nozzles. In one boiler, it was found that 70 (22) [8] percent of the total evaporation took place over the first (2d) [3d] third of the length of the tubes (*Engineering*, May 10, 1912). The length of the boiler is not increased for greater capacities, the increased capacity being obtained by an increase in the diameter of the boiler and the use of a greater number of tubes.

In another type of surface-combustion furnace, the air-fuel mixture impinges directly on a bed of refractory material and a zone of combustion forms at that surface in the bed at which the velocity of the mixture equals the rate of flame propagation in the mixture. With this type, the delivery tube is accessible for cleaning.

The term surface combustion is also applied to a furnace in which the combustible mixture discharges into a refractory-lined tube or tunnel of small dimensions opening out into a furnace. As the tube or tunnel surface heats to incandescence, the combustion is accelerated and is completed before the mixture leaves it (see Blake, *Proc. Engrs'. Soc. West. Penna.*, 1920).

# TRANSMISSION OF HEAT BY CONDUCTION AND CONVECTION

## by G. C. Williams and K. A. Smith

REFERENCES: McAdams, "Heat Transmission," McGraw-Hill. Eckert and Drake, "Heat and Mass Transfer," McGraw-Hill. Carslaw and Jaeger, "Conduction of Heat in Solids," Oxford. Jakob, "Heat Transfer," vols. I and II, Wiley. Grober, Erk, and Grigull, "Fundamentals of Heat Transfer," translated by J. Moszynski, McGraw-Hill. Wilkes, "Heat Insulation," Wiley. Kays and London, "Compact Heat Exchangers," McGraw-Hill. "Thermophysical Properties Data Book," Purdue University.

## Notation and Units

The units are based on feet, pounds, hours, degrees Fahrenheit, and Btu. Any other consistent set may be used in the dimensionless relations given, but for the dimensional equations the units of this table must be used.

$A$ = area of heat-transfer surface, ft$^2$

$A_i$ = inside area

$A_o$ = outside area

$A_m$ = average value of $A$, ft$^2$

$a$ = an empirical constant

$C_p$ = specific heat at constant pressure, Btu/lb/°F

$D$ = diameter, ft

$D_o$ = outside diam, ft

$D_i$ = inside diam, ft

$D'$ = diameter, in

$D_o'$ = outside diam, in

$D_i'$ = inside diam, in

$e$ = base of natural logarithms, 2.718

$G$ = mass velocity, equals $w/S$, lb/h/ft$^2$ of cross section occupied by fluid

$G_{max}$ = mass velocity through minimum free area in a row of pipes normal to fluid stream, lb/h/ft$^2$

$g_c$ = conversion factor, equal to $4.18 \times 10^8$ (mass lb)(ft)/(force lb)(h)$^2$

$g_L$ = local acceleration due to gravity, $4.18 \times 10^8$ ft/h/h at sea level

$h$ = local individual coefficient of heat transfer, equals $dq/dA \, \Delta t$, Btu/h(ft$^2$)(°F)diff

$h_c + h_r$ = combined coefficient by conduction, convection, and radiation between surface and surroundings

$h_m$ = mean value of $h$ for entire surface, based on $(\Delta t)_m$

$h_{a.m.}$ = average $h$, arbitrarily based on arithmetic-mean temperature difference

$h_s$ = heat-transfer coefficient through scale deposits

$J$ = mechanical equivalent of heat, 778 ft·lb/Btu

$k$ = thermal conductivity, Btu/h/ft$^2$/unit temperature gradient °F/ft

$$k_m = -\frac{1}{t_1 - t_2} \int_1^2 k \, dt$$

$k_f$ = $k$ at the "film" temperature, $t_f = (t + t_w)/2$

$l$ = thickness of material normal to heat flow, ft

$L$ = length of heat-transfer surface, heated length, ft

$N$ = number of rows of tubes

$N_{Gr}$ = Grashof number, $L_c^3 \rho_f^2 g_L \beta_f (\Delta t)_s / \mu_f^2$

$Q$ = quantity of heat, Btu

$q$ = total rate of heat flow, Btu/h

$\dot{\mathbf{q}}$ = heat-flux vector, Btu/h/ft$^2$

$\dot{\mathbf{q}}_x$ = $x$-component of heat-flux vector

$R$ = thermal resistances, $1/UA$, $1/hA$, $1/(h_c + h_r)A_0$

$R_g$ = gas constant, 1,546/mol wt, 53.35 for air

$\Re$ = recovery factor

$r$ = radius, ft

$S$ = cross section, filled by fluid, in plane normal to direction of fluid flow, ft$^2$

$T$ = temperature, °R = $t + 460$

$T_1, T_2$ = inlet and outlet bulk temperatures, respectively, of warmer fluid, °F

$t$ = bulk temperature (based on heat balance), °F

$t_w$ = wall temperature, °F

$t_1, t_2$ = inlet and outlet bulk temperatures of colder fluid °F

$t_i, t_o$ = temperatures of fluid inside and outside, °F

$t_f = (t + t_w)/2$

$t_{sat}$ = saturation temperature, °F

$U$ = overall coefficient of heat transfer, Btu/h/ft$^2$/°F; $U_i$, $U_o$ based on inside and outside surface, respectively

$V$ = mean velocity, ft/h

$V_s$ = average velocity, volumetric rate divided by cross section filled by fluid, ft/s

$V_{sm}$ = maximum velocity, through minimum cross section ft/s

$x$ = one of the axes of a Cartesian reference frame, ft

$X = (t_2 - t_1)/(T_1 - t_1)$

$w$ = mass rate of flow per tube, lb/h/tube

$Z = (T_1 - T_2)/(t_2 - t_1)$

$\alpha$ = a dimensional constant

$\beta$ = volumetric coefficient of thermal expansion, having units of reciprocal of Fahrenheit temperature

$\Gamma$ = mass rate of flow, lb/(h) (ft of wetted periphery measured on a plane normal to direction of fluid flow); $= w/\pi D$ for a vertical and $w/2L$ for a horizontal tube

$\gamma$ = ratio of specific heats, $c_p/c_v$ 1.4 for air

$\nabla$ = gradient operator

$\Delta t$ = temperature difference, °F

$(\Delta t)_{ave}, (\Delta t)_{l.m.}$ = arithmetic and logarithmic means of terminal temperature differences, respectively, °F

$(\Delta t)_m$ = true mean value of the terminal temperature differences, °F

$(\Delta t)_o$ = overall temperature difference, °F

$(\Delta t)_s$ = temperature difference between surface and surroundings, °F

$\lambda$ = latent heat (enthalpy) of vaporization, Btu/lb

$\mu$ = viscosity at bulk temperature, lb per h per ft; equals 2.42 times centipoises; equals 116,000 times viscosity in (lb force) (s)/ft²

$\mu_f$ = viscosity, lb/h/ft, at arithmetic mean of wall and fluid temperatures

$\mu_w$ = viscosity at wall temperature, lb/h/ft

$\rho$ = density, lb/ft³

$\rho_\infty$ = density of stream of great depth, lb/ft³

$\sigma$ = surface tension, (lb force)/ft

**Subscripts:**

$l$ = liquid

$v$ = vapor

**Preliminary Statements**  The transfer of heat is usually considered to occur by three processes:

1. **Conduction** is the transfer of heat from one part of a body to another part or to another body by short-range interaction of molecules and/or electrons.

2. **Convection** is the transfer of heat by the combined mechanisms of fluid mixing and conduction.

3. **Radiation** is the emission of energy in the form of electromagnetic waves. All bodies above absolute zero temperature radiate. Radiation incident on a body may be absorbed, reflected, and transmitted. (See Sec. 4.)

## CONDUCTION

The **basic Fourier conduction law** for an isotropic material is

$$\dot{\mathbf{q}} = -k\nabla t \qquad (1)$$

In cartesian coordinates, the $x$ component of this equation is $\dot{\mathbf{q}}_x = -k(\partial t/\partial x)$, and if the heat flow is unidimensional, $q = \dot{\mathbf{q}}A(x) = -kA(x)(dt/dx)$. This states that the steady-state rate of heat conduction $q$ is proportional to the cross-sectional $A(x)$ normal to the direction of flow and to the temperature gradient $\partial t/\partial x$ along the conduction path. The proportionality constant $k$ is called the **"true" thermal conductivity** of the material.

The thermal conductivity of a given material varies with temperature, and the mean thermal conductivity is defined by

$$k_m = \frac{1}{t_0' - t_i'} \int_{t_i'}^{t_0'} k\,dt$$

Over moderate range, $k$ varies linearly with $t$, and hence $k_m$ is the value of $k$ at the arithmetic mean of $t_i'$ and $t_0'$.

For unidimensional heat flow through a material of thickness $l$

$$q \int_0^l \frac{dx}{A(x)} = \int_{t_i'}^{t_0'} k\,dt = k_m(t_i' - t_0') \qquad (2)$$

with an obvious definition for the mean area:

$$\frac{q}{A_m} = k_m(t_i' - t_0')$$

For flat plates, $A_m = A_i = A_0$; for hollow cylinders, $A_m = (A_0 - A_i)/\ln (A_0/A_i)$; for hollow spheres, $A_m = \sqrt{A'A_0}$. For

## Table 1. Thermal Conductivities of Metals*

$k$ = Btu/h/ft²/°F/ft

$kt = kt_0 - a(t - t_0)$

| Substance | Temp range, deg F | $k_{t0}$ | $a$ | Substance | Temp range, deg F | $k_{t0}$ | $a$ |
|---|---|---|---|---|---|---|---|
| Metals: | | | | Uranium.............. | 70–770 | 14 | −0.007 |
| Aluminum...... | 70–700 | 130 | 0.03 | Vanadium............. | 70 | 20 | ...... |
| Antimony...... | 70–212 | 10.6 | 0.006 | Zinc................... | 60–212 | 65 | 0.007 |
| Beryllium...... | 70–700 | 80 | 0.027 | Zirconium ............ | 32 | 11 | — |
| Cadmium ...... | 60–212 | 53.7 | 0.01 | Alloys: | | | |
| Cobalt......... | 70 | 28 | ...... | Admiralty metal....... | 68–460 | 58.1 | −0.054 |
| Copper......... | 70–700 | 232 | 0.032 | Brass.................. | −265–360 | 61.0 | −0.066 |
| Germanium..... | 70 | 34 | ...... | (70 % Cu, 30 % Zn)... | 360–810 | 84.6 | 0 |
| Gold........... | 60–212 | 196 | ...... | Bronze, 7.5 % Sn..... | 130–460 | 34.4 | −0.042 |
| Iron, pure...... | 70–700 | 41.5 | 0.025 | 7.7 % Al........ | 68–392 | 39.1 | −0.038 |
| Iron, wrought... | 60–212 | 34.9 | 0.002 | Constantan............ | −350–212 | 12.7 | −0.0076 |
| Steel (1 % C)... | 60–212 | 26.2 | 0.002 | (60 % Cu, 40 % Ni)... | 212–950 | 10.1 | −0.019 |
| Lead........... | 32–500 | 20.3 | 0.006 | Dural 24S (93.6 % Al, | | | |
| Magnesium..... | 32–370 | 99 | 0.015 | 4.4 % Cu, 1.5 % Mg, | −321–550 | 63.8 | −0.083 |
| Mercury........ | 32 | 4.8 | ...... | 0.5 % Mn)........... | 550–800 | 130. | +0.038 |
| Molybdenum... | 32–800 | 79 | 0.016 | Inconel X (73 % Ni | | | |
| Nickel.......... | 70–560 | 36 | 0.0175 | 15 % Cr, 7 % Fe, | | | |
| Palladium...... | 70 | 39 | ...... | 2.5 % Ti) | 27–1070 | 7.62 | −0.0068 |
| Platinum....... | 70–800 | 41 | 0.0014 | Manganin (84 % Cu, | 1070–1650 | 3.35 | −0.0111 |
| Plutonium...... | 70 | 5 | ...... | 12 % Mn, 4 % Ni).... | −256–212 | 11.5 | −0.015 |
| Rhodium....... | 70 | 88 | ...... | Monel (67.1 % Ni, | | | |
| Silver.......... | 70–600 | 242 | 0.058 | 29.2 % Cu, 1.7 % Fe, | | | |
| Tantalum...... | 212 | 32 | ...... | 1.0 % Mn).......... | −415–1470 | 12.0 | −0.008 |
| Thallium....... | 32 | 29 | ...... | Nickel silver | | | |
| Thorium....... | 70–570 | 17 | −0.0045 | (64 % Cu, 17 % Zn, | | | |
| Tin............ | 60–212 | 36 | 0.0135 | 18 % Ni)............ | 68–390 | 18.1 | −0.0156 |
| Titanium....... | 70–570 | 9 | 0.001 | | | | |
| Tungsten....... | 70–570 | 92 | 0.02 | | | | |

*For refractories see Sec. 6; for pipe coverings, Sec. 8; for building materials, Sec. 4. Conversion factors for various units are given in Sec. 1. Tables 3–7 were revised by G. B. Wilkes.

**Thermal Conductivity of Nickel-Chromium Alloys with Iron**

$k_t = k_{t0} - a(t - t_0)$

| ANSI Number | Temp, range, deg F | $k_{t0}$ | $a$ |
|---|---|---|---|
| 301, 302, 303, 304 (303 Se, 304 L)....... | 95–1650 | 8.08 | − 0.0052 |
| 310 (310S)......................... | 32–1650 | 6.85 | − 0.0072 |
| 314.............................. | 80–572 | 10.01 | − 0.00124 |
|  | 572–1650 | 8.20 | − 0.0045 |
| 316 (316 L)........................ | − 60–1750 | 7.50 | − 0.0042 |
| 321, 347 (348)..................... | −100–1650 | 8.22 | − 0.0050 |
| 403, 410 (416, 416 Se, 420)............. | −100–1850 | 15.0 | 0 |
| 430 [430 F, 430 F (Se)]............... | 122–1650 | 12.60 | − 0.0012 |
| 440 C............................. | 212–932 | 12.77 | − 0.0043 |
| 446.............................. | 32–1850 | 12.96 | − 0.0050 |
| 501, 502.......................... | 80–1520 | 21.4 | +0.0037 |

**Properties of Molten Metals***

| Metal and melting point | Temperature, deg F | $k$, Btu (hr)(ft)(deg F) | $\rho$, lb cu ft | $c_p$, Btu (lb)(deg F) | $\mu$, lb (ft)(hr) |
|---|---|---|---|---|---|
| Bismuth................... | 600 | 9.5 | 625 | 0.0345 | 3.92 |
| (520F) | 1000 | 9.0 | 608 | 0.0369 | 2.66 |
|  | 1400 | 9.0 | 591 | 0.0393 | 1.91 |
| Lead...................... | 700 | 10.5 | 658 | 0.038 | 5.80 |
| (621F) | 900 | 11.4 | 650 | 0.037 | 4.65 |
|  | 1300 | ..... | 633 | ..... | 3.31 |
| Mercury.................. | 50 | 4.7 | 847 | 0.033 | 3.85 |
| (−38F) | 300 | 6.7 | 826 | 0.033 | 2.66 |
|  | 600 | 8.1 | 802 | 0.032 | 2.09 |
| Potassium............... | 300 | 26.0 | 50.4 | 0.19 | 0.90 |
| (147F) | 800 | 22.8 | 46.3 | 0.18 | 0.43 |
|  | 1300 | 19.1 | 42.1 | 0.18 | 0.31 |
| Sodium................... | 200 | 49.8 | 58.0 | 0.33 | 1.69 |
| (208F) | 700 | 41.8 | 53.7 | 0.31 | 0.68 |
|  | 1300 | 34.5 | 48.6 | 0.30 | 0.43 |
| Na, 56 wt %.............. | 200 | 14.8 | 55.4 | 0.270 | 1.40 |
| K, 44 wt % | 700 | 15.9 | 51.3 | 0.252 | 0.570 |
| (66.2F) | 1300 | 16.7 | 46.2 | 0.249 | 0.389 |
| Na, 22 wt %.............. | 200 | 14.1 | 53.0 | 0.226 | 1.19 |
| K, 78 wt % | 750 | 15.4 | 48.4 | 0.210 | 0.500 |
| (12F) | 1400 | ..... | 43.1 | 0.211 | 0.353 |
| Pb, 44.5 wt %............ | 300 | 5.23 | 657 | 0.035 |  |
| Bi, 55.5 wt % | 700 | 6.85 | 639 | 0.035 | 3.71 |
| (257F) | 1200 | ..... | 614 | ..... | 2.78 |

*Based largely on the "Liquid-Metals Handbook," 2d ed., United States Government Printing Office, Washington, D.C., 1952.

more complex shapes, Eq. (1) must be employed. Mean areas may then be evaluated by a graphical procedure (Awbery and Schofield, *Proc. 5th Intern. Congr. Refrig.*, **3**, 1929, pp. 591–610) or by the relaxation procedure (Emmons, *Trans. ASME*, **65**, 1943, pp. 607–612).

## CONDUCTION AND CONVECTION

**Phenomena of Heat Transmission**   In many practical cases of heat transmission—e.g., boilers, condensers, the cooling of engine cylinders—heat is transmitted from one fluid to another through a wall separating the two. The processes occurring in the fluids may be extremely complex. However, to facilitate discussion, it is convenient to imagine that most of the fluid offers no resistance to heat transmission but that a thin film of fluid adjacent to the wall offers considerable resistance. This situation is depicted in Fig. 1. Then, by definition,

$$q = h_i A_i(t_i - t_i') = \frac{k}{l}A_m(t_i' - t_0) = h_0 A_0(t_0' - t_0)$$

The terms $h_i$ and $h_0$ are the **film coefficients**, or **unit conductances,** of the films $f_1$ and $f_2$, respectively, and $k$ is the thermal conductivity of the wall. Since $q$, $A$, $t_i - t_i'$, and $t_0' - t_0$ are susceptible to direct measurement, $h_i$ and $h_0$ are simply defined quantities and the propriety of the above equation does not rest upon the heuristic film concept. Indeed, for laminar flow, the film concept is a gross misrepresentation,

**Table 2. Thermal Conductivities of Liquids and Gases**

| Substance | Temp, deg F | $k$ | Substance | Temp, deg F | $k$ |
|---|---|---|---|---|---|
| LIQUIDS | | | GASES | | |
| Acetone......................... | 68 | 0.103 | Air (see below)............... | 32 | 0.0140 |
| Ammonia...................... | 45 | 0.29 | Ammonia, vapor.............. | 32 | 0.0126 |
| Aniline......................... | 32 | 0.104 | Ammonia.................... | 212 | 0.0192 |
| Benzol......................... | 86 | 0.089 | Argon....................... | 32 | 0.00915 |
| Carbon bisulphide.............. | 68 | 0.0931 | Carbon dioxide.............. | 32 | 0.0084 |
| Ethyl alcohol.................. | 68 | 0.105 | | 212 | 0.0128 |
| Ether.......................... | 68 | 0.0798 | Carbon monoxide............ | 32 | 0.0135 |
| Glycerin, USP, 95%............ | 68 | 0.165 | Chlorine.................... | 32 | 0.0043 |
| Kerosene....................... | 68 | 0.086 | Ethane..................... | 32 | 0.0106 |
| Methyl alcohol................. | 68 | 0.124 | Ethylene.................... | 32 | 0.0101 |
| n-Pentane...................... | 68 | 0.0787 | Helium...................... | 32 | 0.0818 |
| Petroleum ether............... | 68 | 0.0758 | n-Hexane.................... | 32 | 0.0072 |
| Toluene........................ | 86 | 0.086 | Hydrogen.................... | 32 | 0.0966 |
| Water......................... | 32 | 0.343 | | 212 | 0.124 |
| | 140 | 0.377 | Methane..................... | 32 | 0.0175 |
| Oil, castor.................... | 39 | 0.104 | Neon........................ | 32 | 0.0267 |
| Oil, olive..................... | 39 | 0.101 | Nitrogen.................... | 32 | 0.0140 |
| Oil, turpentine................. | 54 | 0.0734 | Nitrous oxide................ | 32 | 0.0088 |
| Vaseline....................... | 59 | 0.106 | | 212 | 0.0090 |
| | | | Nitric oxide................. | 32 | 0.0138 |
| | | | Oxygen..................... | 32 | 0.0142 |
| | | | n-Pentane................... | 32 | 0.0074 |
| | | | Sulphur dioxide.............. | 32 | 0.005 |

**Thermal Conductivities of Air and Steam***

| Temperature, deg F | 32 | 200 | 400 | 600 | 800 | 1000 |
|---|---|---|---|---|---|---|
| Air, 1 atm | 0.0140 | 0.0181 | 0.0225 | 0.0266 | 0.0303 | 0.0337 |
| Steam, 1 psia | ...... | 0.0132 | 0.0184 | 0.0238 | 0.0292 | 0.0345 |

*Values from F. G. Keyes, *Tech. Rept.* 37, Project Squid (Apr. 1, 1952).

and yet the definition of a film coefficient (or heat-transfer coefficient) remains convenient and valid.

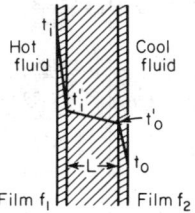

**Fig. 1** Temperature gradients in heat flow through a wall.

If $t_i'$ and $t_0$ are eliminated from the above equation, a relation is obtained for steady flow through several resistances in series:

$$q = \frac{t_i - t_0}{(1/b_i A_i) + (l/kA_m) + (1/b_o A_0)} \qquad (3)$$

Each of the terms in the denominator represents a resistance to heat transfer. There may also be a resistance $1/b_s A_s$, due to the presence of a scale deposit on the surface. Thus, if the over-all heat transfer is given by $q = UA(t_i - t_0)$, then the total thermal resistance is given by

$$1/UA = (1/b_i A_i) + (l/kA_m) + (1/b_o A_o) + (1/b_s A_s) \qquad (4)$$

Coefficients for scale deposits are given in Table 8.

**Mean Temperature Difference** The basic equation for any steadily operated heat exchanger is $dq = U(\Delta t)_o \, dA$, in which $U$ is the overall coefficient [Eq. (4)], $(\Delta t)_o$ is the overall

temperature difference between hot and cold fluids, and $dq/dA$ is the local rate of flow per unit surface. In order to apply this relation to a finite exchanger, it is necessary to integrate it. The assumptions usually made are constant $U$, constant mass rates of flow, no changes in phase, constant specific heats, and negligible heat losses. The resulting equation for parallel or countercurrent flow of fluids is

$$q = UA(\Delta t)_m = UA[(\Delta t)_{01} - (\Delta t)_{02}]/\ln [(\Delta t)_{01}/(\Delta t)_{02}] \qquad (5a)$$

in which $(\Delta t)_m$ is the logarithmic mean of the terminal temperature differences, $(\Delta t)_{01}$ and $(\Delta t)_{02}$, between hot and cold fluid. The value of $UA$ is evaluated from the resistance concept of Eq. (4) and the values of $b$ are obtained from the following pages. For the more complex cases of multipass and cross flow, $(\Delta t)_m$ is given by Fig. 2. The shell-side fluid is assumed to be well mixed by suitable baffles.

EXAMPLE: Assume an exchanger in which the hot fluid enters at 400°F and leaves at 327°F; the cold fluid enters at 100°F and leaves at 283°F. Assuming $U$ independent of temperature, what will be the true mean temperature difference from hot to cold fluid, (1) for counterflow and (2) for a reversed current apparatus with one well-baffled pass in the shell and two equal passes in the tubes?

1. With counterflow, the terminal differences are $400 - 283 = 117$°F and $327 - 100 = 227$°F; the logarithmic mean difference is $110/0.662 = 166$°F.

2. $Z = (400 - 327)/(283 - 100) = 0.4$; $X = (283 - 100)/(400 - 100) = 0.61$; from section A of Fig. 2, $Y = 0.9 = (\Delta t)_m/166$; $(\Delta t)_m = 149$°F.

If one of the temperatures remains constant, as in a condenser or in an evaporative cooler, Eq. (5a) applies for parallel flow, counterflow, reversed current, and cross flow.

### Table 3. Thermal Conductivities of Miscellaneous Solid Substances*

Values of $k$ are to be regarded as rough average values for the temperature range indicated

| Material | Bulk density, lb per cu ft | Temp, deg F | $k$ | Material | Bulk density, lb per cu ft | Temp, deg F | $k$ |
|---|---|---|---|---|---|---|---|
| Asbestos board, compressed asbestos and cement............ | 123. | 86. | 0.225 | Quartz, crystal, parallel to $C$-axis........ | ... | −300. | 25.0 |
| Asbestos millboard.... | 60.5 | 86. | 0.070 | | | 0. | 8.3 |
| Asbestos wool........ | 25. | 212. | 0.058 | | | 300. | 4.2 |
| Ashes, soft wood...... | 12.5 | 68. | 0.018 | Rubber, hard......... | 74.3 | 100. | 0.092 |
| Ashes, volcanic....... | 51. | 300. | 0.123 | Rubber, soft, vulcanized.............. | 68.6 | 86. | 0.08 |
| Carbon black......... | 12. | 133. | 0.012 | Sand, dry........... | 94.8 | 68. | 0.188 |
| Cardboard, corrugated. | ... | ... | 0.037 | Sawdust, dry......... | 13.4 | 68. | 0.042 |
| Celluloid............. | 87.3 | 86. | 0.12 | Silica, fused.......... | ... | 200. | 0.83 |
| Cellulose sponge, du Pont.......... | 3.4 | 82. | 0.033 | Silica gel, powder..... | 32.5 | 131. | 0.049 |
| Concrete, sand. and gravel............ | 142. | 75. | 1.05 | Soil, dry............. | ... | 68. | 0.075 |
| Concrete, cinder...... | 97. | 75. | 0.41 | Soil, dry, including stones............. | 127. | 68. | 0.30 |
| Charcoal, powder..... | 11.5 | 63. | 0.029 | Snow................ | 7–31 | 32. | 0.34–1.3 |
| Cork, granulated...... | 5.4 | 23. | 0.028 | Titanium oxide, finely ground........... | 52. | 1000. | 0.041 |
| Cotton wool.......... | 5.0 | 100. | 0.035 | Wool, pure.......... | 5.6 | 86. | 0.021 |
| Diamond............. | 151. | 70. | 320. | Zirconia grain........ | 113. | 600. | 0.11 |
| Earth plus 42% water. | 108. | 0. | 0.62 | Woods, oven dry, across grain†: | | | |
| Fiber, red............ | 80.5 | 68. | 0.27 | Aspen.............. | 26. | 85. | 0.069 |
| "Flotofoam" (U.S. Rubber Co.)........ | 1.6 | 92. | 0.017 | Bald cypress........ | 24. | 85. | 0.063 |
| Glass, pyrex.......... | 139 | 200. | 0.59 | Balsa.............. | 10. | 85. | 0.034 |
| Glass, soda lime...... | ... | 200. | 0.59 | Basswood.......... | 24. | 85. | 0.058 |
| Graphite, solid....... | 93.5 | 122. | 87. | Douglas Fir........ | 29. | 85. | 0.063 |
| Gravel............... | 116. | 68. | 0.22 | Elm, rock.......... | 48. | 85. | 0.097 |
| Gypsum board....... | 51. | 99. | 0.062 | Fir, white.......... | 26. | 85. | 0.069 |
| Ice................. | 57.5 | ... | 1.26 | Hemlock........... | 29. | 85. | 0.066 |
| Kaolin wool.......... | 10.6 | 800. | 0.059 | Larch, western..... | 36. | 85. | 0.078 |
| Leather, sole........ | 62.4 | ... | 0.092 | Maple, sugar....... | 43. | 85. | 0.094 |
| Mica................ | 122. | ... | 0.25 | Oak, red........... | 42. | 85. | 0.099 |
| Pearlite, Arizona, spherical shell of siliceous material..... | 9.1 | 112. | 0.035 | Pine, southern yellow............. | 35. | 85. | 0.078 |
| Polystyrene, expanded "Styrofoam"....... | 1.7 | ... | 0.021 | Pine, white........ | 25. | 85. | 0.060 |
| Pumice, powdered.... | 49. | 300. | 0.11 | Red cedar, western.. | 21. | 85. | 0.053 |
| Quartz, crystal, perpendicular to $C$-axis. | ... | −300. | 12.5 | Redwood.......... | 25. | 85. | 0.062 |
| | | 0. | 4.3 | Spruce............ | 21. | 85. | 0.052 |
| | | 300. | 2.3 | | | | |

*The thermal conductivity of different materials varies greatly. For metals and alloys $k$ is high, while for certain insulating materials, such as glass wool, cork, and kapok, it is very low. In general, $k$ varies with the temperature, but in the case of metals, the variation is relatively small. With most other substances, $k$ increases with rising temperatures, but in the case of many crystalline materials, the reverse is true.

†With heat flow parallel to the grain, $k$ may be 2 to 3 times that with heat flow perpendicular to the grain; the values for wool are taken chiefly from J. D. MacLean, *Trans. ASHRAE*, **47**, 1941, p. 323.

$Y$ = ordinate

$$= \frac{\text{true mean temp difference}}{\text{logarithmic mean temp difference for counter flow}}$$

For the other symbols see Sec. 4. (*From Trans. ASME*, **62**, 1940, pp. 283–294.)

If $U$ varies considerably with temperature, the apparatus should be considered to be divided into stages, in each of which variation of $U$ with temperature or temperature difference is linear. Then for parallel or counterflow operation, the following relation may be applied to each stage:

$$q = A[U_2(\Delta t)_{01} - U_1(\Delta t)_{02}]/\ln[U_2(\Delta t)_{01}/U_1(\Delta t)_{02}] \qquad \textbf{(5b)}$$

The above discussion focuses on the concepts of an overall coefficient and a mean-temperature difference. An alternative approach focuses on the concepts of effectiveness and the number of transfer units. The alternatives are basically equivalent, but one or the other may enjoy a computational advantage. The latter method is presented in detail by Kays and London and by Mickley and Korchak (*Chem. Eng.*, **69**, 1962, pp. 181–188 and 239–242).

### FILM COEFFICIENTS

The important physical properties which affect film coefficients (see Sec. 4) are thermal conductivity, viscosity, density, and specific heat. Factors within the control of the designer include fluid velocity and shape and arrangement of the heating surface. With forced flow of gases or water, under the conditions usually met in practice, the flow is turbulent (see Sec. 3) and under these conditions the film coefficient can be greatly increased by increasing the velocity of the fluid at the

**Table 4. Thermal Conductivities for Building Insulation**

| Material | Bulk density, lb per cu ft | Temp, deg F | k |
|---|---|---|---|
| Balsam wool, blanket | 3.6 | 70. | 0.021 |
| Cabot's Quilt, eel-grass............ | 15.6 | 86. | 0.027 |
| Glass wool, blanket.. | 3.25 | 100. | 0.022 |
| Hairfelt, blanket.... | 11.0 | 86. | 0.022 |
| Insulating boards, Insulite, Celotex, etc.............. | 12–19 | 100. | 0.027–0.031 |
| Kapok, "DryZero," blanket.......... | 1.6 | 75. | 0.019 |
| Redwood bark, loose, shredded, "Palco Bark"..... | 4.0 | 100. | 0.025 |
| Rock wool, loose.... | 7. | 117. | 0.024 |
| Sil-O-Cel powder.... | 10.6 | 86. | 0.026 |
| Vermiculite, loose, "Zonolite"....... | 8.2 | 60. | 0.038 |

expense of a greater power requirement. For a given velocity and fluid, the film coefficient depends upon the direction of flow of fluid relative to the heating surface. With free or natural convection, for a given arrangement of surface, the film coefficient depends on an additional fluid property, the coefficient of thermal expansion, on the temperature difference between surface and fluid, and on the local gravitational acceleration. With forced convection at low rates of flow, particularly with viscous fluids such as oils, laminar motion may prevail and the film coefficient depends on thermal conductivity, specific heat, mass rate of flow per tube, and length and diameter of the tube. In any event, the film coefficients $h$ are correlated in terms of dimensionless groups of the controlling factors.

**Turbulent Flow inside Clean Tubes (No Change in Phase),** $DG/\mu_f > 7000$:

$$\frac{h_m}{C_p G}\left(\frac{C_p \mu_f}{k_f}\right)^{2/3} = \frac{0.023}{(DG/\mu_f)^{0.2}} \quad \textbf{(6a)}$$

For $L/D$ less than 60, multiply the right-hand side of Eqs. **(6a)**, **(6b)**, and **(6c)** by $[1 + (D/L)^{0.7}]$.

**Table 5. Thermal Conductivities of Material for Refrigeration and Extreme Low Temperatures**

| Material | Bulk density, lb per cu ft | Temp, deg F | k | Material | Bulk density, lb per cu ft | Temp, deg F | k |
|---|---|---|---|---|---|---|---|
| Corkboard........ | 6.9 | 100 | 0.022 | Rubber board, expanded, "Rubatex"..... | 4.9 | 100 | 0.018 |
| | | −100 | 0.018 | | | −100 | 0.015 |
| | | −300 | 0.010 | | | −300 | 0.004 |
| Fiberglas with asphalt coating (board)........ | 11.0 | 100 | 0.023 | Silica aerogel, powder, "Santocel". | 5.3 | 100 | 0.013 |
| | | −100 | 0.014 | | | 0 | 0.012 |
| | | −300 | 0.007 | | | −100 | 0.010 |
| Glass blocks, expanded, "Foamglas".......... | 10.6 | 100 | 0.036 | Vegetable fiberboard, asphalt coating........ | 14.4 | 100 | 0.028 |
| | | −100 | 0.033 | | | −100 | 0.021 |
| | | −300 | 0.018 | | | −300 | 0.013 |
| Mineral wool board, "Rockcork".... | 14.3 | 100 | 0.024 | Foams: Polystyrene[a]... | 2.9 | −100 | 0.015 |
| | | −100 | 0.017 | Polyurethane[b].. | 5.0 | −100 | 0.019 |
| | | −300 | 0.008 | | | | |

[a]Test space pressure, 1.0 atm; $k = 0.0047$ at $10^{-5}$ mmHg.
[b]Test space pressure, 1.0 atm; $k = 0.007$ at $10^{-3}$ mmHg.

**Turbulent Flow of Gases inside Clean Tubes,** $DG/\mu_f > 7000$:

$$h_m = 0.024 C_p G^{0.8}/(D_i')^{0.2} \quad \textbf{(6b)}$$

**Turbulent Flow of Water inside Clean Tubes,** $DG/\mu_f > 7000$:

$$h_m = 160(1 + 0.012\, t_f)V_s^{0.8}/(D_i')^{0.2} \quad \textbf{(6c)}$$

**Turbulent Flow of Liquid Metals inside Clean Tubes,** $(C_p \mu/k < 0.05)$: The equation of Sleicher and Tribus ("Recent Advances in Heat Transfer," p. 281, McGraw-Hill, 1961) is recommended for isothermal tube walls—

$$\frac{h_m D}{k} = 6.3 + 0.016\left(\frac{DGC_p}{k}\right)^{0.91}\left(\frac{C_p \mu}{k}\right)^{0.3} \quad \textbf{(6d)}$$

**Turbulent Flow of Gases or Water in Annuli**  Use Eq. **(6b)** or **(6c)**,with $D'$ taken as the clearance, inches.

**Water in Coiled Pipes**  Multiply $h_m$ for the straight pipe by the term $[1 + (3.5\, D_i/D_c)]$, where $D_i$ is the inside diameter of the pipe and $D_c$ is that of the coil.

**Turbulent Boundary Layer on a Flat Plate,** $V_\infty \rho_f x/\mu_f > 4 \times 10^5$, no pressure gradient:

$$\frac{h}{\rho_f C_p V_\infty}\left(\frac{C_p \mu}{k}\right)^{2/3} = \frac{0.0148}{(\rho_f V_\infty x/\mu_f)^{0.2}} \quad \textbf{(6e)}$$

$$\frac{h_m}{\rho_f C_p V_\infty}\left(\frac{C_p \mu}{k}\right)_f^{2/3} = \frac{0.0185}{(\rho_f V_\infty L/\mu_f)^{0.2}} \quad \textbf{(6f)}$$

**Fluid Flow Normal to a Single Tube,** $D_o G/\mu_f$ from 1000 to 50,000:

$$\frac{h_m D_o}{k_f} = 0.26\left(\frac{D_o G}{\mu_f}\right)^{0.6}\left(\frac{C_p \mu}{k}\right)_f^{0.3} \quad \textbf{(7)}$$

**Gas Flow Normal to a Single Tube,** $D_o G/\mu_f$ from 1000 to 50,000:

**Table 6. Thermal Conductivities of Insulating Materials for High Temperatures**

| Material | Bulk density, lb per cu ft | Max temp, deg F | 100 F | 300 F | 500 F | 1000 F | 1500 F | 2000 F |
|---|---|---|---|---|---|---|---|---|
| Asbestos paper, laminated........ | 22. | 400 | 0.038 | 0.042 | | | | |
| Asbestos paper, corrugated...... | 16. | 300 | 0.031 | 0.042 | | | | |
| Diatomaceous earth, silica, powder.................... | 18.7 | 1500 | 0.037 | 0.045 | 0.053 | 0.074 | | |
| Diatomaceous earth, asbestos and bonding material........ | 18. | 1600 | 0.045 | 0.049 | 0.053 | 0.065 | | |
| Fiberglas block, PF612......... | 2.5 | 500 | 0.023 | 0.039 | | | | |
| Fiberglas block, PF614......... | 4.25 | 500 | 0.021 | 0.033 | | | | |
| Fiberglas block, PF617......... | 9. | 500 | 0.020 | 0.033 | | | | |
| Fiberglas, metal mesh blanket, #900....................... | ...... | 1000 | 0.020 | 0.030 | 0.040 | | | |
| Glass blocks, average values.... | 14–24 | 1600 | ..... | 0.046 | 0.053 | 0.074 | | |
| Hydrous calcium silicate, "Kaylo".................... | 11. | 1200 | 0.032 | 0.038 | 0.045 | | | |
| 85% magnesia................ | 12. | 600 | 0.029 | 0.035 | | | | |
| Micro-quartz fiber, blanket..... | 3. | 3000 | 0.021 | 0.028 | 0.042 | 0.075 | 0.108 | 0.142 |
| Potassium titanate, fibers....... | 71.5 | .... | ..... | 0.022 | 0.024 | 0.030 | | |
| Rock wool, loose............... | 8–12 | .... | 0.027 | 0.038 | 0.049 | 0.078 | | |
| Zirconia grain................ | 113. | 3000 | ..... | ..... | | 0.108 | 0.129 | 0.163 | 0.217 |

**Table 7. Thermal Conductance across Air Spaces**

Btu/(h) (ft²)—Reflective insulation

| Air space, in | Direction of heat flow | Temp diff, deg F | Mean temp, deg F | Aluminum surfaces, $\epsilon = 0.05$ | Ordinary surfaces, non-metallic, $\epsilon = 0.90$ |
|---|---|---|---|---|---|
| Horizontal, ¾–4 across....... | Upward | 20. | 80. | 0.60 | 1.35 |
| Vertical, ¾–4 across......... | Across | 20. | 80. | 0.49 | 1.19 |
| Horizontal, ¾ across........ | Downward | 20. | 75. | 0.30 | 1.08 |
| Horizontal, 4 across......... | Downward | 20. | 80. | 0.19 | 0.93 |

**Values of $K$ for $N$ Rows Deep**

| $N$ | 1 | 2 | 3 | 4 | 5 | 6 | 7 | 10 |
|---|---|---|---|---|---|---|---|---|
| $K$ | 0.24 | 0.25 | 0.27 | 0.29 | 0.30 | 0.31 | 0.32 | 0.33 |

$$b_m = 0.30 C_p G^{0.6}/(D_o')^{0.4} \qquad (7a)$$

**Fluid Flow Normal to a Bank of Staggered Tubes,** $D_o G_{max}/\mu_f$ from 2000 to 40,000:

$$\frac{b_m D_o}{k_f} = K \left(\frac{C_p \mu}{k}\right)_f^{1/3} \left(\frac{D_o G_{max}}{\mu_f}\right)^{0.6} \qquad (8)$$

**Water Flow Normal to a Bank of Staggered Tubes,** $D_o G_{max}/\mu_f$ from 2000 to 40,000

$$b_m = 370(1 + 0.0067 t_f) V_{sm}^{0.6}/(D_o')^{0.4} \qquad (8a)$$

For baffled exchangers, to allow for leakage of fluids around the baffles, use 60 percent of the values of $b_m$ from Eq. (8); for tubes in line, deduct 25 percent from the values of $b_m$ given by Eq. (8).

**Water Flow in Layer Form over Horizontal Tubes,** $4\Gamma/\mu <$ 2100

$$b_{a.m.} = 150(\Gamma/D_o')^{1/3} \qquad (9)$$

for $\Gamma$ ranging from 100 to 1,000 lb of water per h per ft (each side).

**Water Flow in Layer down Vertical Tubes,** $w/\pi D > 500$

$$b_m = 120\Gamma^{1/3} \qquad (9a)$$

**Heat Transfer to Gases Flowing at Very High Velocities** If a nonreactive gas stream is brought to rest adiabatically, as at the true stagnation point of a blunt body, the temperature rise will be

$$t_s - t_\infty = V^2/2g_c J C_p \qquad (9b)$$

where $t_s$ is the stagnation temperature and $t_\infty$ is the temperature of the free stream moving at velocity $V$. At every other point on the body, the gas is brought to rest partly by pressure changes and partly by viscous effects in the boundary layer. In general, this process is not adiabatic, even though the body transfers no heat. The thermal conductivity of the gas will transfer heat from one layer of gas to another. At an insulated surface, the gas temperature will therefore be neither the free-stream temperature nor the stagnation temperature. In general, the rise in gas temperature will be given by the equation

$$t_{aw} - t_\infty = \Re(t_s - t_\infty) = \Re V^2/2g_c J C_p \qquad (9c)$$

where $t_{aw}$ is the gas temperature at the adiabatic wall and $\Re$ is the recovery factor.

If a given point on the surface of a body is not at the

### Table 8. Heat Transfer Coefficients ($h_s$) for Scale Deposits from Water[a]
For use in Eq. (4)

| Temp of heating medium | Up to 240 F | | 240 to 400 F | |
|---|---|---|---|---|
| Temp of water | 125 F or less | | Above 125 F | |
| Water velocity, fps | 3 and less | Over 3 | 3 and less | Over 3 |
| Distilled | 2,000 | 2,000 | 2,000 | 2,000 |
| Sea water | 2,000 | 2,000 | 1,000 | 1,000 |
| Treated boiler feed water | 1,000 | 2,000 | 500 | 1,000 |
| Treated make-up for cooling tower | 1,000 | 1,000 | 500 | 500 |
| City, well, Great Lakes | 1,000 | 1,000 | 500 | 500 |
| Brackish, clean river water | 500 | 1,000 | 330 | 500 |
| River water, muddy, silty[b] | 330 | 500 | 250 | 330 |
| Hard (over 15 grains per gal) | 330 | 330 | 200 | 200 |
| Chicago Sanitary Canal | 130 | 170 | 100 | 130 |

Miscellaneous cases: Refrigerating liquids, brine, clean petroleum distillates, organic vapors, 1,000; refrigerant vapor, 500; vegetable oils, 330; fuel oil (topped crude), 200.
[a]From standards of Tubular Exchanger Manufacturers Assoc., 1952.
[b]Delaware, East River (N.Y.), Mississippi, Schuylkill, and New York Bay.

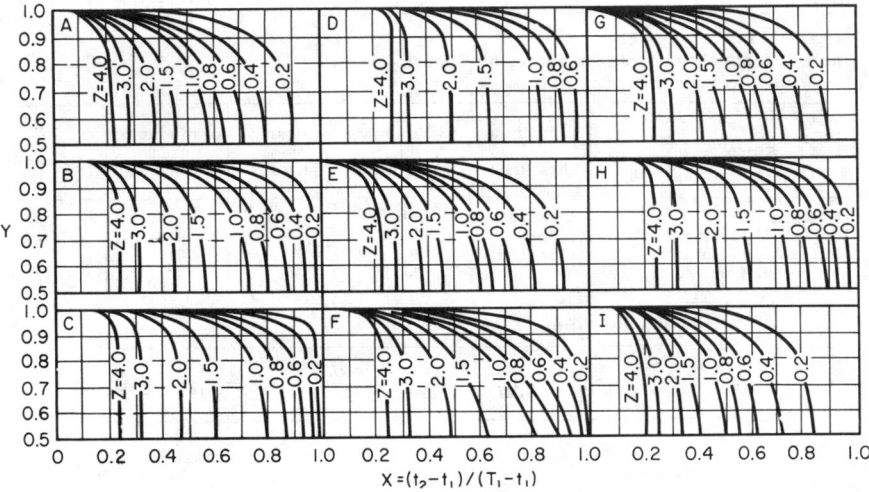

$$X = (t_2 - t_1)/(T_1 - t_1)$$

**Fig. 2** (A) One shell pass and two tube passes; (B) two shell passes and four tube passes; (C) three shell passes and six tube passes; (D) four shell passes and eight tube passes; (E) cross-flow, one shell pass and one tube pass; (F) single-pass cross-flow exchanger, both fluids unmixed; (G) single-pass cross-flow exchanger, one fluid mixed, other unmixed; (H) two-pass cross-flow exchanger, shell fluid mixed, tube fluid unmixed, shell fluid flowing across second and first passes in series; (I) same as (H), but shell fluid crosses first and second passes in series.

### Table 9. Typical Values of $h_m$ for Heating and Cooling, Forced Convection
$D_o' = 1.31$ in, $D_i' = 1.05$ in

| Fluid and arrangement | $t_f$, deg F | Velocity | | Btu per hr per sq ft per deg F | Eq. No. |
|---|---|---|---|---|---|
| | | Fps[a] | Lb per hr per sq ft | | |
| Air inside tubes | ... | $V_s = 31.8$, $G = 8600$ | | 8.0 | 6b |
| Air normal to staggered tubes | 170 | $V_s = 8.92$, $G_m = 2000$ | | 7.5 | 8 |
| Water inside tubes | 100 | $V_s = 5.0$, $G = 1.12 \times 10^6$ | | 1260 | 6c |
| Water normal to staggered tubes | 100 | $V_s = 2.0$, $G_m = 0.448 \times 10^6$ | | 800 | 8a |
| Trickle cooler, water | ... | $\Gamma = 100$ lb per hr per ft | | 640 | 9 |
| Falling water film, vertical tube | ... | $\Gamma = 1,000$ lb per hr per ft | | 1200 | 9a |

[a]Velocity in ft/s at 70°F and 1 atm = $G/3600\rho$.

temperature $t_{aw}$ given by Eq. (9c) with the proper local value of $\Re$ inserted, there will be a transfer of heat to or from the body. This suggests defining the coefficient of heat transfer in the usual way, except that the difference $t_w - t_{aw}$ should be used:

$$q/A = h(t_\infty - t_{aw}) = h[t_w - (t + \Re V^2/2g_c JC_p)] \qquad \text{(9d)}$$

where $t_w$ is the surface temperature of the heated wall. With this modification, it is found that the correlations for $h$ are nearly independent of Mach number; e.g., Eq. (6a) may be used for turbulent, compressible flow in a pipe. Obviously, $\Re = 1.0$ at a forward stagnation point. For flows parallel to surfaces which have little or no curvature in the direction of flow, the following are recommended:

Laminar flow $\qquad \Re = \left(\dfrac{C_p\mu}{k}\right)^{1/2}$

Turbulent flow $\qquad \Re = \left(\dfrac{C_p\mu}{k}\right)^{1/3}$

Very little is presently known about point values of the recovery factor for flow over more complex shapes. Thus, special thermocouples should be used to measure the temperature of high-velocity gas streams (Hottel and Kalitinsky, *Jour. Applied Mechanics*, 1945, pp. A25–A32; and Franz, *Jahrb 1938 deut. Luftfahrt-Forsch* II, pp. 215–218). Eckert (*Trans. ASME*, **78**, 1956, pp. 1273–1283) recommends that all property values be evaluated at a film temperature defined by

$$t_f = (t_\infty + t_w)/2 + 0.22(t_{aw} - t_\infty) \qquad \text{(9e)}$$

Nielsen (*NACA Wartime Rept.* L-179) give graphs for predicting the heat transfer and pressure drop for air flow at Mach numbers up to 1.0, in tubes having a uniform wall temperature.

Heat transfer from a reacting gas to a surface is treated by Lees ("Recent Advances in Heat and Mass Transfer," p. 161, McGraw-Hill).

### Laminar Flow

Pipe Flow, $DG/\mu < 2100$. Use the Sieder-Tate modification of the Graetz equation for isothermal tube walls and $wC_p/kL > 10$:

$$h_{a.m.}D/k = 2.0(wC_p/kL)^{1/3}(\mu/\mu_w)^{0.14} \qquad \text{(10)}$$

or

$$(h_{a.m.}/C_p G)(C_p\mu/k)^{2/3}(\mu_w/\mu)^{0.14} = 1.85(D/L)^{1/3}(DG/\mu)^{-2/3} \qquad \text{(10a)}$$

As shown in Fig. 3, as $DG/\mu$ increases from 2,100 to 7,000,

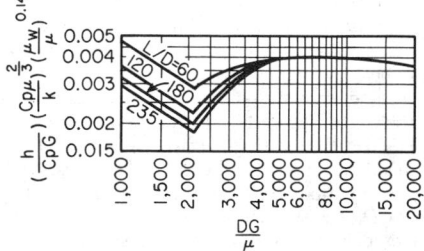

**Fig. 3** Heating and cooling of viscous oils flowing inside tubes. [The curves for $DG/\mu$ below 2,100 are based on Eq. (10).]

the effect of $L/D$ diminishes and finally becomes negligible for $L/D > 60$.

Laminar Boundary Layer on a Flat Plate, $\rho V_\infty x/\mu < 4 \times 10^5$. Isothermal plate, no pressure gradient:

$$\frac{h}{\rho_f C_p V_\infty}\left(\frac{C_p\mu}{k}\right)_f^{2/3} = \frac{0.332}{(\rho_f V_\infty x/\mu_f)^{1/2}} \qquad \text{(10b)}$$

$$\frac{h_m}{\rho_f C_p V_\infty}\left(\frac{C_p\mu}{k}\right)_f^{2/3} = \frac{0.664}{(\rho_f V_\infty L/\mu_f)^{1/2}}$$

**Extended Surfaces** Fin efficiency is defined as the ratio of the mean temperature difference from surface to fluid divided by the temperature difference from fin to fluid at the base or root of the fin. Graphs of fin efficiency for extended surfaces of various types are given by Gardner (*Trans. ASME*, **67**, pp. 621–628, 1945). Heat transfer coefficients for various extended surfaces are given by Kays and London.

**Natural Convection** Heat transfer by natural convection is governed by relations of the form

$$\frac{h_m L_c}{k_f} = f[L_c^3 \rho_f^2 g_L \beta_f (\Delta t)_s/\mu_f^2, (C_p\mu/k)_f] \qquad \text{(11)}$$

where $\beta_f$ is defined by the equation $\rho_f = \rho_\infty[1 - \beta_f(\Delta t)_s)]$. For perfect gases, $\beta_f = 1/T_\infty$. The dimensionless group $L_c^3 \rho_f^2 g_L \beta_f (\Delta t)_s/\mu_f^2 \equiv N_{Gr}$ represents the ratio of the product (inertial force times buoyant force) to (viscous force squared).

If the flow is of the laminar-boundary-layer type and if $(C_p\mu/k)_f > 1$, an effective correlation is

$$\frac{h_m L_c}{k_f} = B_1[N_{Gr}(C_p\mu/k)_f]^{0.25} \qquad \text{(11a)}$$

where $B_1$ is a weak function of $(C_p\mu/k)_f$. Similarly, for $(C_p\mu/k)_f < 1$,

$$\frac{h_m L_c}{k_f} = B_2[N_{Gr}(C_p\mu/k)_f^2]^{0.25} \qquad \text{(11b)}$$

Vertical Flat Plates. For this case, $L = L_c$ and the flow in the laminar-boundary-layer type will be laminar if

$$(C_p\mu/k)_f > 1; \ 10^9 > N_{Gr}(C_p\mu/k)_f > 10^4$$
$$(C_p\mu/k)_f < 1; \ ? > N_{Gr}(C_p\mu/k)_f^2 > 10^4$$

Lefevre (*Rept. Heat* 113, National Engineering Laboratory, Great Britain, Aug. 1956) gives an interpolation formula which contains the proper limiting forms and is in complete agreement with existing numerical results:

$$\frac{h_m L_c}{k_f} = \left[\frac{N_{Gr}(C_p\mu/k)^2}{2.435 + 4.884(C_p\mu/k)_f^{1/2} + 4.953(C_p\mu/k)_f}\right]^{0.25} \qquad \text{(11c)}$$

If $(C_p\mu/k)_f$ is in the vicinity of unity and if $N_{Gr}(C_p\mu/k)_f > 10^9$, the boundary layer will be turbulent and

$$\frac{hL}{k_f} = 0.13[N_{Gr}(C_p\mu/k_f]^{1/3} \qquad \text{(11d)}$$

Horizontal Cylinders. Replace $L$ in the vertical-flat-plate formulas by $\pi D_0/2$.

Heated Horizontal Plates Facing Upward or Cooled Horizontal Plates Facing Downward.

$$2 \times 10^7 > N_{Gr}(C_p\mu/k)_f > 10^5$$
$$\frac{h_m L}{k_f} = 0.54[N_{Gr}(C_p\mu/k)_f]^{0.25} \qquad \text{(11e)}$$
$$N_{Gr}(C_p\mu/k)_f > 2 \times 10^7$$
$$\frac{h_m L}{k_f} = 0.14[N_{Gr}(C_p\mu/k)_f]^{1/3} \qquad \text{(11f)}$$

Heated Horizontal Plates Facing Downward or Cooled Horizontal Plates Facing Upward.

$$3 \times 10^{10} > N_{Gr}(C_p\mu/k)_f > 3 \times 10^5$$

$$\frac{hL}{k_f} = 0.27[N_{Gr}(C_p\mu/k)_f]^{0.25} \tag{11g}$$

Equations (11c) to (11g) should not be considered reliable if $(C_p\mu/k)_f$ differs greatly from unity.

For more complex systems, it is best to consult plots of experimental data (McAdams).

For any particular fluid, the above equations may be greatly simplified. For air which is at room temperature and atmospheric pressure and is subjected to the gravitational attraction at sea level:

Vertical Plates.

$$10^3 > L^3(\Delta t)_s > 10^{-2}$$
$$h_m = 0.28[(\Delta t)_s/L]^{0.25} \tag{12a}$$
$$L^3(\Delta t)_s > 10^3$$
$$h_m = 0.19(\Delta t)_s^{1/3} \tag{12b}$$

Horizontal Cylinders.

$$10^2 > D^3(\Delta t)_s > 10^{-3}$$
$$h_m = 0.25[(\Delta t)_s/D]^{0.25} \tag{12c}$$
$$D^3(\Delta t)_s > 10^2$$
$$h_m = 0.19(\Delta t)_s^{1/3} \tag{12d}$$

Heated Horizontal Plates Facing Upward or Cooled Horizontal Plates Facing Downward.

$$10 > L^3(\Delta t)_s > 0.1$$
$$h_m = 0.27[(\Delta t)_s/L]^{0.25} \tag{12e}$$
$$10^4 > L^3(\Delta t)_s > 10$$
$$h_m = 0.22(\Delta t)_s^{1/3} \tag{12f}$$

Heated Horizontal Plates Facing Downward or Cooled Horizontal Plates Facing Upward.

$$10^4 > L^3(\Delta t)_s > 0.1$$
$$h_m = 0.12[(\Delta t)_s/L]^{0.25} \tag{12g}$$

**Condensing Vapors** If the condensate of a single pure vapor, saturated or supersaturated, wets the surface, film-type condensation is obtained. The rate of heat transfer equals $h_m(\Delta t)_m$, where $(\Delta t)_m$ is the mean difference between the saturation temperature and the temperature of the surface. As long as the condensate flow is laminar ($4\Gamma/\mu_f < 2,100$), the following dimensionless equations may be used:

For **horizontal tubes,**

$$h_mD/k = 0.73[D^3\rho^2\lambda g_L/k\mu_f N(\Delta t)_m]^{0.25} = 0.76(D^3\rho^2g_L/\mu_f\Gamma)^{1/3} \tag{13}$$

For **vertical tubes,**

$$h_mL/k = 0.94[L^3\rho^2\lambda g_L/k\mu_f(\Delta t)_m]^{0.25} = 0.93(L^3\rho^2g_L/\mu_f\Gamma)^{1/3} \tag{13a}$$

The equations show that a tube of given dimensions, for the usual case where $L/ND$ is greater than 2.76, is more effective in a horizontal than in a vertical position. Thus for $L/ND = 100$, a horizontal tube gives an average $h$ which is 2.5 times that for a vertical tube. Since there is but little variation in the thermal conductivity or viscosity of the condensate at the condensing temperature at 1 atm, there is little variation in $h_m$. Thus with horizontal tubes, $h_m$ may be taken as 200 to 400 for the

following vapors condensing at atmospheric pressure: benzene, carbon tetrachloride, dichlormethane, dichlordifluoromethane, diphenyl ethyl alcohol, heptane, hexane, methyl alcohol, octane, toluene, and xylene. Ammonia gives $h_m$ of 1,000, and mixtures of steam and organic vapors, forming immiscible condensates, give $h_m$ ranging from 250 to 750, increasing with increasing proportion of steam. With film-type condensation of clean steam on horizontal tubes, $h_m$ ranges from 1,000 to 3,000 see Eq. (13). With vertical tubes 10 to 20 ft long, ripples form in the film; values of $h_m$ from Eq. (13a) should be increased 20 percent.

For long vertical tubes, $4\Gamma/\mu_f$ may exceed 2100; in that case:

$$h_m(\mu_f^2/k_f^3\rho_f^2g_L)^{1/3} = 0.0077(4\Gamma/\mu_f)^{0.4} \tag{13b}$$

The presence of **noncondensible gas**, such as air, seriously reduces $h$, and consequently all vapor-heated apparatus should be well vented.

With steam, small traces of certain promoters (Nagle, U.S. Patent 1,995,361) such as oleic acid and benzyl mercaptan become adsorbed in a very thin layer on the surface of the tubes, preventing the condensate from wetting the metal and inducing dropwise condensation, which gives much higher values of $h_m$ (7,000 to 70,000) than film-type condensation. However, with dirty or corroded surfaces, it is difficult to maintain dropwise condensation. Figure 4 shows overall coefficients $U_o$ for condensing steam at 1 atm on a vertical 10 ft length of copper tube, ⅝ in OD, 0.049-in wall, at various water velocities.

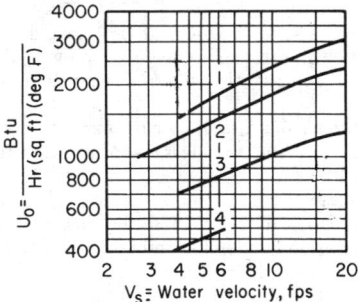

**Fig. 4** Overall coefficients between condensing steam and water. Curve 1, chromium-plated copper, oleic acid; curve 2, copper, benzyl mercaptan; curve 3, copper, oleic acid; curve 4, admiralty metal, no promoter.

**Boiling Liquids** The nature of the heat transfer from a submerged heater to a pool of boiling water is shown in Fig. 5. Other liquids exhibit the same qualitative features. In the range $AB$, heat transfer to the liquid occurs solely by natural convection, and evaporation occurs at the free surface of the pool. In the range $BC$, **nucleate boiling** occurs. Bubbles form at active nuclei on the heating surface, detach, and rise to the pool surface. At point $C$, the heat flux passes through a maximum at a temperature difference called the **critical** $\Delta t$. In the range $CD$, transitional boiling occurs. At point $D$, the transition is complete and the heating surface is completely blanketed by a vapor film. This is the point of minimum heat flux, or the **Leidenfrost point**. In the range $DE$, the heating surface continues to be blanketed by a vapor film.

The range $AB$ is adequately correlated by the usual natural-

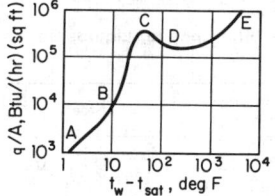

**Fig. 5**  Boiling of water at 212°F on a platinum surface.

convection equations. No truly adequate correlation is available for the range $BC$ because the complex processes of nucleation and interfacial interaction are only partially understood. However, the relation due to Rohsenow (*Trans. ASME*, **74**, 1952, pp. 969–976) is one of the best and can be reliably used for modest extrapolations of existing data.

$$\frac{C_{p,l}(t_w - t_{sat})}{\lambda}$$

$$= C_{fs} \left[ \frac{q/A}{\mu_l \lambda} \sqrt{\frac{g_c \sigma}{g_L(\rho_l - \rho_v)}} \right]^{1/3} \left( \frac{C_{p,l}\mu_l}{k_l} \right)^{1.7} \quad \textbf{(14a)}$$

The value of the constant $C_{fs}$ is intimately dependent on the nature of the particular fluid-solid pair and must be determined by experiment. It usually assumes values in the range $0.003 < C_{fs} < 0.05$ and is not affected by moderate subcooling or the shape of the heating surface.

Zuber (*USAEC Rept.* AECU-4439, June, 1959) has presented a theoretical equation for the maximum heat flux from a flat, horizontal surface. The analysis is based on considerations of hydrodynamic stability. For saturated liquids,

$$(q/A)_{max} = K_1 \rho_v \lambda \left[ \frac{\sigma g_L g_c(\rho_l - \rho_v)}{\rho_v^2} \right]^{1/4} \left( \frac{\rho_l}{\rho_l + \rho_v} \right)^{1/2} \quad \textbf{(14b)}$$
$$0.12 < K_1 < 0.157 \quad \text{(theoretical)}$$

Berenson (Sc.D. thesis, Mechanical Engineering Department, MIT, 1960) used a similar analysis and obtained a relation which is identical for $\rho_l >> \rho_v$, but he found that $K_1 = 0.18$ gives better agreement with the data. The theoretical basis of this equation has been subject to attack, but the correlation appears to be the best available. Zuber also performed an analysis for subcooled liquids and proposed a modification which is also in excellent agreement with experiment:

$$(q/A)_{max} = K_1 \rho_v [\lambda + C_{p,l}(t_{sat} - t_l)] \left[ \frac{\sigma g_L g_c(\rho_l - \rho_v)}{\rho_v^2} \right]^{0.25} \left( \frac{\rho_l}{\rho_l + \rho_v} \right)^{1/2} \left\{ 1 + \frac{5.33(\rho_l C_{p,l} k_l)^{1/2}(t_{sat} - t_l)}{\rho_v[\lambda + C_{p,l}(t_{sat} - t_l)]} \left[ \frac{g_L(\rho_l - \rho_v)\rho_v^2}{\sigma^3 g_c^3} \right]^{1/8} \right\} \textbf{(14c)}$$

Zuber's hydrodynamic analysis of the Leidenfrost point yields

$$(q/A)_{min} = K_2 \lambda \rho_v \left[ \frac{\sigma g_L g_c(\rho_l - \rho_v)}{\rho l^2} \right]^{1/4} \quad 0.144 < K_2 < 0.177 \quad \textbf{(14d)}$$

on any size of pipe, Eqs. **(2)**, **(4)**, and **(15)** combine to give

$$\frac{q_o}{A_o} = \frac{(\Delta t)_o}{\dfrac{r_o}{k_1}\ln\dfrac{r_2}{r_1} + \dfrac{r_o}{k_2}\ln\dfrac{r_3}{r_2} + \cdots + \dfrac{1}{b_c + b_r}} \quad \textbf{(16)}$$

where $q_o/A_o$ is the Btu/h/ft$^2$ of outer surface of the last layer; $(\Delta t)_o$ is the overall temperature difference (°F) between pipe and air; $r_o$ is the radius, feet, of the outer surface; $r_1$ is the

Berenson finds better agreement with the data if $K_2 = 0.09$. For very small wires, the heat flux will exceed that predicted by this flat-plate formula. A reliable prediction of the critical temperature is not available.

For nucleate boiling accompanied by forced convection, the heat flux may be approximated by the sum of the heat flux for pool boiling alone and the heat flux for forced convection alone. This procedure will not be satisfactory at high qualities, and no satisfactory correlation exists for the maximum heat flux.

For a given liquid and boiling pressure, the nature of the surface may substantially influence the flux at a given $(\Delta t)$, Table 10. These data may be used as rough approximations for a bank of submerged tubes. Film coefficients for scale deposits are given in Table 8.

For **forced-circulation evaporators**, vapor binding is also encountered. Thus with liquid benzene entering a 4-pass steam-jacketed pipe at 0.9 fps, up to the point where 60 percent by weight was vaporized, the maximum flux of 60,000 Btu/h/ft$^2$ was obtained at an overall temperature difference of 60°F; beyond this point, the coefficient and flux decreased rapidly, approaching the values obtained in superheating vapor, see Eq. **(6b)**. For comparison, in a natural convection evaporator, a maximum flux of 73,000 Btu/h/ft$^2$ was obtained at $(\Delta t)_o$ of 100°F.

**Combined Convection and Radiation Coefficients**  In some cases of heat loss, such as that from bare and insulated pipes, where loss is by convection to the air and radiation to the walls of the enclosing space it is convenient to use a combined convection and radiation coefficient $(b_c + b_r)$. The rate of heat loss thus becomes

$$q = (b_c + b_r)A(\Delta t)_s \quad \textbf{(15)}$$

where $(\Delta t)_s$ is the temperature difference, deg F, between the surface of the hot body and the walls of the space. In evaluating $(h_c + b_r)$, $b_c$ should be calculated by the appropriate convection formula [see Eqs. **(11c)** to **(11g)**] and $b_r$ from the equation

$$b_r = 0.00685\epsilon(T_{av}/100)^3$$

where $\epsilon$ is the black body coefficient of the radiating surface (see Sec. 4). $T_{av}$ is the average temperature of the surface and the enclosing walls, deg R. For oxidized bare steel pipe, the sum $b_c + b_r$ may be taken directly from Table 11.

**Heat Transmission through Pipe Insulation** (McMillan, *Trans. ASME*, 1915.) For any number of layers of insulation

outside radius, feet, of the pipe, $r_2 = r_1$ + thickness of first layer of insulation, foot; $r_3 = r_2$ plus the thickness of second layer, etc.; and $k_1, k_2, k_3$, etc., are the conductivities of the respective layers. For average indoor conditions, $b_c + b_r$ is often taken as 2 as an approximation, since a substantial error in $b_c + b_r$ will have but little effect on the overall loss of heat. Figure 6 shows the variation in $U_o$ with pipe size and thickness of insulation (for $k = 0.042$) for pipe and air temperatures of 375 and 75°F, respectively.

**Table 10. Maximum Flux and Corresponding Over-all Temperature Difference for Liquids Boiled at 1 atm with a Submerged Horizontal Steam-heated Tube**

| Liquid | Aluminum | | Copper | | Chromium-plated copper | | Steel | |
|---|---|---|---|---|---|---|---|---|
| | $\dfrac{q/A}{1000}$ | $(\Delta t)_o$ | $\dfrac{q/A}{1000}$ | $(\Delta t)_o$ | $\dfrac{q/A}{1000}$ | $(\Delta t)_o$ | $\dfrac{q/A}{1000}$ | $(\Delta t)_o$ |
| Ethyl acetate........ | 41 | 70 | 61 | 55 | 77 | 55 | | |
| Benzene............ | 51 | 80 | 58 | 70 | 73 | 100 | 82 | 100 |
| Ethyl alcohol........ | 55 | 80 | 85 | 65 | 124 | 65 | | |
| Methyl alcohol...... | .. | .. | 100 | 95 | 110 | 110 | 155 | 110 |
| Distilled water...... | .. | .. | 230 | 85 | 350 | 75 | 410 | 150 |

**Table 11. Values of** $(h_c + h_r)$

For horizontal bare or insulated standard steel pipe of various sizes and for flat plates in a room at 80°F

| Nominal pipe diam, in. | $(\Delta t)_s$, temperature difference, deg F, from surface to room | | | | | | | | | | | | | | |
|---|---|---|---|---|---|---|---|---|---|---|---|---|---|---|---|
| | 50 | 100 | 150 | 200 | 250 | 300 | 400 | 500 | 600 | 700 | 800 | 900 | 1000 | 1100 | 1200 |
| ½ | 2.12 | 2.48 | 2.76 | 3.10 | 3.41 | 3.75 | 4.47 | 5.30 | 6.21 | 7.25 | 8.40 | 9.73 | 11.20 | 12.81 | 14.65 |
| 1 | 2.03 | 2.38 | 2.65 | 2.98 | 3.29 | 3.62 | 4.33 | 5.16 | 6.07 | 7.11 | 8.25 | 9.57 | 11.04 | 12.65 | 14.48 |
| 2 | 1.93 | 2.27 | 2.52 | 2.85 | 3.14 | 3.47 | 4.18 | 4.99 | 5.89 | 6.92 | 8.07 | 9.38 | 10.85 | 12.46 | 14.28 |
| 4 | 1.84 | 2.16 | 2.41 | 2.72 | 3.01 | 3.33 | 4.02 | 4.83 | 5.72 | 6.75 | 7.89 | 9.21 | 10.66 | 12.27 | 14.09 |
| 8 | 1.76 | 2.06 | 2.29 | 2.60 | 2.89 | 3.20 | 3.88 | 4.68 | 5.57 | 6.60 | 7.73 | 9.05 | 10.50 | 12.10 | 13.93 |
| 12 | 1.71 | 2.01 | 2.24 | 2.54 | 2.83 | 3.13 | 3.83 | 4.61 | 5.50 | 6.52 | 7.65 | 8.96 | 10.42 | 12.03 | 13.84 |
| 24 | 1.64 | 1.93 | 2.15 | 2.45 | 2.72 | 3.03 | 3.70 | 4.48 | 5.37 | 6.39 | 7.52 | 8.83 | 10.28 | 11.90 | 13.70 |
| FLAT PLATES | | | | | | | | | | | | | | | |
| Vertical............ | 1.82 | 2.13 | 2.40 | 2.70 | 2.99 | 3.30 | 4.00 | 4.79 | 5.70 | 6.72 | 7.86 | 9.18 | 10.64 | 12.25 | 14.06 |
| HFU............... | 2.00 | 2.35 | 2.65 | 2.97 | 3.26 | 3.59 | 4.31 | 5.12 | 6.04 | 7.08 | 8.21 | 9.54 | 11.01 | 12.63 | 14.45 |
| HFD............... | 1.58 | 1.85 | 2.09 | 2.36 | 2.63 | 2.93 | 3.61 | 4.38 | 5.27 | 6.27 | 7.40 | 8.71 | 10.16 | 11.76 | 13.57 |

HFU, horizontal, facing upward; HFD, horizontal, facing downward.

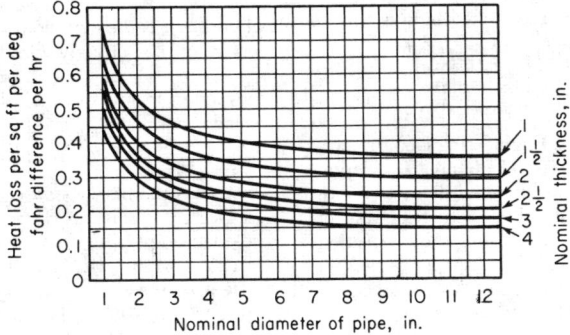

**Fig. 6** Variation with pipe size of overall coefficient $U_o$ for a given thickness of insulation, for $k = 0.042$.

# RADIANT-HEAT TRANSFER

## by Hoyt C. Hottel and Adel F. Sarofim

REFERENCES: Hottel and Sarofim, "Radiative Transfer," McGraw-Hill, 1967. Siegel and Howell, "Thermal Radiation Heat Transfer," McGraw-Hill, 1972. Sparrow and Cess, "Radiation Heat Transfer," Brooks-Cole, 1966.

A heated body loses energy continuously by radiation, at a rate dependent on the shape, the size, and, particularly, the temperature of the body. In contrast to conductive energy transport, such emitted radiation is capable of passage to a distant body, where it may be absorbed, reflected, scattered, or transmitted.

Consider a pencil of radiation, defined as all the rays passing through each of two small, widely separated areas $dA_1$ and $dA_2$. The rays at $dA_1$ will have a solid angle of divergence $d\Omega_1$, equal to the apparent area of $dA_2$ viewed from $dA_1$, divided by the square of the separating distance. Let the normal to $dA_1$ make the angle $\theta_1$ with the pencil. The flux density $q$ [energy/(time)(area normal to beam)] per unit solid angle of divergence is called the **intensity** $I$, and the flux $dQ_1$ (energy/time) through area $dA_1$ (of apparent area $dA_1 \cos \theta_1$ normal to the beam) is therefore given by

$$d\dot{Q}_1 = dA_1 \cos \theta_1 q_1 = I \, dA_1 \cos \theta_1 \, d\Omega_1 \qquad (1)$$

**The intensity $I$ along a pencil, in the absence of absorption or scatter is constant** (unless the beam passes into a medium of different refractive index $n$; $I_1/n_1^2 = I_2/n_2^2$). The **emissive power**\* of a surface is the flux density [energy/(time)(surface area)] due to emission from it throughout a hemisphere. If the intensity $I$ of emission from a surface is independent of the angle of emission, Eq. (1) may be used to show that the surface emissive power is $\pi I$, though the emission is throughout $2\pi$ steradians.

### Black-Body Radiation

Engineering calculations of thermal radiation from surfaces are best keyed to the radiation characteristics of the **black body**, or **ideal radiator**. The characteristic properties of a black body are that it absorbs all the radiation incident on its surface and that the quality and intensity of the radiation it emits are completely determined by its temperature. The total radiative flux throughout a hemisphere from a black surface of area $A$ and absolute temperature $T$ is given by the **Stefan-Boltzmann law**: $\dot{Q} = A\sigma T^4$ or $q = \sigma T^4$. Stefan-Boltzmann constant $\sigma$ has the value $5.67 \times 10^{-8}$ W/m²(K)⁴, $0.1713 \times 10^{-8}$ Btu/(ft)²(h)(°R)⁴ or $1.356 \times 10^{-12}$ cal/(cm)²(s) (K)⁴. From the above definition of emissive power, $\sigma T^4$ is the total emissive power of a black body, called $E$; and the intensity $I_B$ of emission from a black body is $E/\pi$, or $\sigma T^4/\pi$.

The spectral distribution of energy flux from a black body is expressed by Planck's law

$$E_\lambda \, d\lambda = \frac{2\pi hc^2 n^2 \lambda^{-5}}{e^{hc/k\lambda T} - 1} \, d\lambda \equiv \frac{n^2 c_1 \lambda^{-5}}{e^{c_2/\lambda T} - 1} \, d\lambda \qquad (2)$$

wherein $E_\lambda \, d\lambda$ is the hemispherical flux density W/m² lying in the wavelength range $\lambda$ to $\lambda + d\lambda$; $h$ is Planck's constant, $6.6256 \times 10^{-34}$ J·s; $c$ is the velocity of light in vacuo, $2.9979 \times 10^8$ m/s; $k$ is the Boltzmann constant, $1.3805 \times 10^{-23}$ J/(K); $\lambda$ is the wavelength measured in vacuo, $m$; $n$ is the refractive index of the emitter; $c_1$ and $c_2$, the first and second Planck's law constants, are $3.7415 \times 10^{-16}$ W · m² and $1.4388 \times 10^{-2}$ m (K). It may be shown from Planck's law that, of the total energy flux from a black body at any temperature, the fraction $f$ in wavelengths below $\lambda$ is dependent only on $\lambda T$. Values of $f$ versus $\lambda T$ appear in Table 1.

Table 1 indicates that half of black-body radiation lies on either side of a wavelength given by $\lambda T = 4107$ $\mu$m K. It may be shown that the maximum intensity per unit fractional change in wavelength **or** frequency occurs at $\lambda T = 3670(\mu$m)(K), (a more useful displacement law than Wien's),

\*Variously called, in the literature, **emittance, total hemispherical intensity**, or **radiant flux density**.

and that a two-fold range of $\lambda T$ geometrically centered on the latter number spans about half the energy.

### Radiative Exchange between Surfaces of Solids

The ratio of the total radiating power of a real surface to that of a black surface at the same temperature is called the **emittance** of the surface (for a perfectly plane surface, the **emissivity**), designated by $\epsilon$. Subscripts $\lambda$, $\theta$, and $n$ may be assigned to differentiate monochromatic, directional, and surface-normal values, respectively, from the total hemispherical value. If radiation is incident on a surface, the fraction absorbed is called the **absorptance (absorptivity)**, a term in which two subscripts may be appended, the first to identify the temperature of the surface and the second to identify the quality of the incident radiation. According to **Kirchhoff's law**, the emissivity and absorptivity of a surface *in surroundings at its own temperature* are the same, for both monochromatic and total radiation. When the temperatures of the surface and its surroundings differ, the total emissivity and absorptivity of the surface are found often to be different, but because absorptivity is substantially independent of irradiation density, the monochromatic emissivity and absorptivity of surfaces are for all practical purposes the same. The difference between total emissivity and absorptivity depends on the variation, with wavelength, of $\epsilon_\lambda$ and on the difference between the emitter temperature and the effective source temperature.

Consider radiative exchange between a body of area $A_1$ and temperature $T_1$ and its black surroundings at $T_2$. The net interchange is given by

$$\dot{Q}_{1 \rightleftharpoons 2} = A_1 \int_0^\alpha [\epsilon_\lambda E_\lambda(T_1) - \alpha_\lambda E_\lambda(T_2)] \, d\lambda$$
$$= A_1(\epsilon_1 \sigma T_1^4 - \alpha_{12} \sigma T_2^4) \qquad (3)$$

where

$$\epsilon_1 = \int_0^1 \epsilon_\lambda \, df_{\lambda T_1} \qquad \text{and} \qquad \alpha_{12} = \int_0^1 \epsilon_\lambda \, df_{\lambda T_2} \qquad (4)$$

*i.e.*, $\epsilon_1$ (or $\alpha_{12}$) is the area under a curve of $\epsilon_\lambda$ versus $f$, read as a function of $\lambda T$ at $T_1$ (or $T_2$) from Table 1. If $\epsilon_\lambda$ does not change with wavelength, the surface is called **gray**, and $\epsilon_1 = \alpha_{12} = \epsilon_\lambda$. A selective surface is one whose $\epsilon_\lambda$ changes dramatically with wavelength. If this change is monotonic, $\epsilon_1$ and $\alpha_{12}$ are, according to Eqs. (3) and (4), markedly different when the absolute temperature ratio is far from 1; e.g., when $T_1 = 293$ K (ambient temperature) and $T_2 = 5700$ K (effective solar temperature), $\epsilon_1 = 0.9$ and $\alpha_{12} = 0.1 - 0.2$ for a white paint, but $\epsilon_1$ can be as low as $0.12$ and $\alpha_{12}$ above $0.9$ for a thin layer of copper oxide on bright aluminum, or of chromic oxide on bright nickel.

Although values of emittances and absorptances depend in very complex ways on the real and imaginary components of the refractive index and on the geometrical structure of the surface layer, some generalizations are possible.

**Polished Metals** (1) $\epsilon_\lambda$ is quite low in the infrared and, for $\lambda > 8$ $\mu$m, can be adequately approximated by $0.00365 \sqrt{r/\lambda}$, where $r$ is the resistivity in ohm-cm and $\lambda$ is in microns; at shorter wavelengths, $\epsilon_\lambda$ increases and, for many metals, has values of 0.4 to 0.8 in the visible (0.4–0.7 $\mu$m). $\epsilon_\lambda$ is approximately proportional to the square root of the absolute temper-

**Table 1. Fraction $f$ of Black-body Radiation below $\lambda$**
$\lambda T = (\mu m)(K)$

| $\lambda T$ | 1200 | 1600 | 1800 | 2000 | 2200 | 2400 | 2600 | 2800 | 3000 | 3200 |
|---|---|---|---|---|---|---|---|---|---|---|
| $f$ | 0.002 | 0.020 | 0.039 | 0.067 | 0.101 | 0.140 | 0.183 | 0.228 | 0.273 | 0.318 |
| $\lambda T$ | 3400 | 3600 | 3800 | 4000 | 4200 | 4500 | 4800 | 5100 | 5500 | 6000 |
| $f$ | 0.362 | 0.404 | 0.443 | 0.480 | 0.516 | 0.564 | 0.608 | 0.646 | 0.691 | 0.738 |
| $\lambda T$ | 6500 | 7000 | 7600 | 8400 | 10000 | 12000 | 14000 | 20000 | 50000 | |
| $f$ | 0.776 | 0.808 | 0.839 | 0.871 | 0.914 | 0.945 | 0.963 | 0.986 | 0.999 | |

ature ($\epsilon_\lambda \propto \sqrt{r}$ and $r \propto T$) in the far infrared ($\lambda > 8 \mu m$), is temperature insensitive in the near infrared ($0.7-1.5\mu$) and, in the visible, decreases slightly as temperature increases. (2) Total emittance is substantially proportional to absolute temperature; at moderate temperature, $\epsilon_n = 0.58T \sqrt{r_0/T_0}$, where $T$ is in degrees K. (3) Total absorptance of a metal at $T_1$ for radiation from a black or gray source at $T_2$ is equal to the emissivity evaluated at the geometric mean of $T_1$ and $T_2$. (4) The ratio of hemispherical to normal emittance (absorptance) varies from 1.33 at very low $\epsilon$'s ($\alpha$'s) to about 1.03 at an $\epsilon(\alpha)$ of 0.4.

Unless extraordinary pains are taken to prevent oxidation, however, a metallic surface may exhibit several times the emittance or absorptance of a polished specimen. The emittance of iron and steel, for example, varies widely with degree of oxidation and roughness—clean metallic surfaces have an emittance of from 0.05–0.45 at ambient temperatures to 0.4–0.7 at high temperatures; oxidized and/or rough surfaces range from 0.6–0.95 at low temperatures to 0.9–0.95 at high temperatures.

**Refractory Materials** Grain size and concentration of trace impurities are important. (1) Most refractory materials have an $\epsilon_\lambda$ of 0.8 to 1.0 at wavelengths beyond 2 to 4 micrometers; $\epsilon_\lambda$ decreases rapidly toward shorter wavelengths for materials that are white in the visible but retains its high value for black materials such as FeO and $Cr_2O_3$. Small concentrations of FeO and $Cr_2O_3$ or other colored oxides can cause marked increases in the emittance of materials that are normally white. $\epsilon_\lambda$ for refractory materials varies little with temperature. (2) Refractory materials generally have a total emittance which is high (0.7 to 1.0) at ambient temperatures and decreases with increase in temperature; a change from 1000 to 1600°C may cause a decrease in $\epsilon$ of one-fourth to one-third. (3) The emittance and absorptance increase with increase in grain size over a grain-size range of 1–200 $\mu m$. (4) The ratio $\epsilon/\epsilon_n$ of hemispherical to normal emissivity of polished surfaces varies with refractive index from 1 at $n = 0$ to 0.93 at $n = 1.5$ (common glass) and back to 0.96 at $n = 3$. (5) The ratio $\epsilon/\epsilon_n$ for a surface composed of particulate matter which scatters isotropically varies with $\epsilon$ from 1 when $\epsilon = 1$ to 0.8 when $\epsilon = 0.07$. (6) The total absorptance shows a decrease with increase of the temperature of the radiation source similar to the decrease in emittance with increase in the specimen temperature. Figure 1 shows the effect of the temperature of the radiation source on the absorptance of surfaces of various materials at room temperature. It will be noted that polished aluminum (line 15) and anodized aluminum (line 13), representative of metals and nonmetals, respectively, respond oppositely to a change in the temperature of the radiation source. The absorptance of surfaces for sunlight may be read from the right of Fig. 1, assuming sunlight to consist of black-body radiation from a source at 10300 R (5700 K).

When $T_2$ is not too different from $T_1$, $\alpha_{12}$ may be expressed as $\epsilon_1(T_2/T_1)^n$, with $n$ determined from Fig. 1. For this case, Eq. (3) becomes

$$\dot{Q}_{1,net} = \sigma A_1 \epsilon_{AV}(1 + n/4)(T_1^4 - T_2^4) \qquad (5)$$

where $\epsilon_{AV}$ is evaluated at the arithmetic mean of $T_1$ and $T_2$.

Table 2 gives the emittance of various surfaces and emphasizes the variation possible in a single material. The values in

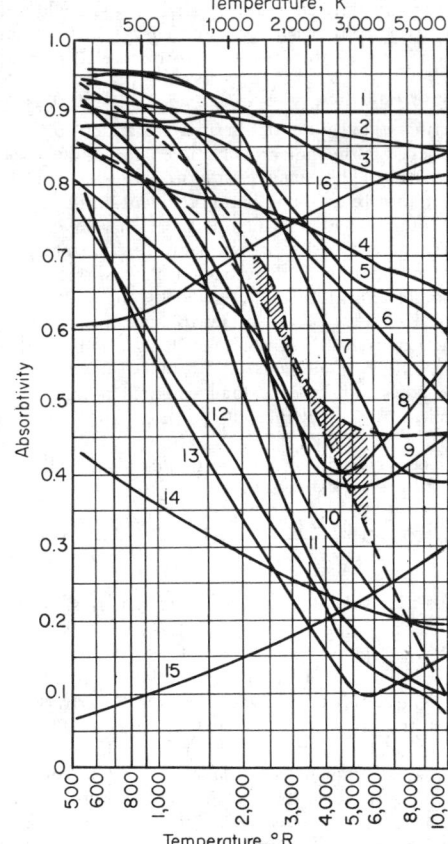

**Fig. 1** Variation of absorptivity with temperature of radiation source. (1) Slate composition roofing; (2) linoleum, red brown; (3) asbestos slate; (4) soft rubber, gray; (5) concrete; (6) porcelain; (7) vitreous enamel, white; (8) red brick; (9) cork; (10) white Dutch tile; (11) white chamotte; (12) MgO, evaporated; (13) anodized aluminum; (14) aluminum paint; (15) polished aluminum; (16) graphite. The two dotted lines bound the limits of data on gray paving brick, asbestos paper, wood, various cloths, plaster of paris, lithopone, and paper.

**Table 2. Emissivity of Surfaces**

| Surface | Temp,* °C | Emissivity* |
|---|---|---|
| **METALS AND THEIR OXIDES** | | |
| Aluminum: | | |
| Highly polished | 230–580 | 0.039–0.057 |
| Polished | 23 | 0.040 |
| Rough plate | 26 | 0.055–0.07 |
| Oxidized at 600°C | 200–600 | 0.11–0.19 |
| Oxide | 280–830 | 0.63–0.26 |
| Alloy 75ST | 24 | 0.10 |
| 75ST, repeated heating | 230–480 | 0.22–0.16 |
| Brass: | | |
| Highly polished | 260–380 | 0.03–0.04 |
| Rolled plate, natural | 22 | 0.06 |
| Rolled, coarse-emeried | 22 | 0.20 |
| Oxidized at 600°C | 200–600 | 0.61–0.59 |
| Chromium | 40–540 | 0.08–0.26 |
| Copper: | | |
| Electrolytic, polished | 80 | 0.02 |
| Comm'l plate, polished | 20 | 0.030 |
| Heated at 600°C | 200–600 | 0.57–0.57 |
| Thick oxide coating | 25 | 0.78 |
| Cuprous oxide | 800–1100 | 0.66–0.54 |
| Molten copper | 1080–1280 | 0.16–0.13 |
| Dow metal, cleaned, heated | 230–400 | 0.24–0.20 |
| Gold, highly polished | 230–630 | 0.02–0.40 |
| Iron and steel: | | |
| Pure Fe, polished | 180–980 | 0.05–0.37 |
| Wrought iron, polished | 40–250 | 0.28 |
| Smooth sheet iron | 700–1040 | 0.55–0.60 |
| Rusted plate | 20 | 0.69 |
| Smooth oxidized iron | 130–530 | 0.78–0.82 |
| Strongly oxidized | 40–250 | 0.95 |
| Molten iron and steel | 1500–1770 | 0.40–0.45 |
| Lead: | | |
| 99.96%, unoxidized | 130–230 | 0.06–0.08 |
| Gray oxidized | 24 | 0.28 |
| Oxidized at 190°C | 190 | 0.63 |
| Mercury, pure clean | 0–100 | 0.09–0.12 |
| Molybdenum filament | 730–2590 | 0.10–0.29 |
| Monel metal, K5700 | | |
| Washed, abrasive soap | 24 | 0.17 |
| Repeated heating | 230–875 | 0.46–0.65 |
| Nickel and alloys: | | |
| Electrolytic, polished | 23 | 0.05 |
| Electroplated, not polished | 20 | 0.11 |
| Wire | 190–1010 | 0.10–0.19 |
| Plate, oxid. at 600°C | 200–600 | 0.37–0.48 |
| Nickel oxide | 650–1250 | 0.59–0.86 |
| Copper-nickel, polished | 100 | 0.06 |
| Nickel-silver, polished | 100 | 0.14 |
| Nickelin, gray oxide | 21 | 0.26 |
| Nichrome wire, bright | 50–1000 | 0.65–0.79 |
| Nichrome wire, oxide | 50–500 | 0.95–0.98 |
| ACI-HW (60Ni, 12Cr); firm black ox. coat | 270–560 | 0.89–0.82 |
| Platinum, polished plate | 230–1630 | 0.05–0.17 |
| Silver, pure polished | 230–630 | 0.02–0.03 |
| Stainless steels: | | |
| Type 316, cleaned | 24 | 0.28 |
| 316, repeated heating | 230–870 | 0.57–0.66 |
| 304, 42 hr at 520°C | 220–530 | 0.62–0.73 |
| 310, furnace service | 220–530 | 0.90–0.97 |
| Allegheny #4, polished | 100 | 0.13 |
| Tantalum filament | 1330–3000 | 0.194–0.33 |
| Thorium oxide | 280–830 | 0.58–0.21 |
| Tin, bright | 24 | 0.04–0.06 |

| Surface | Temp,* °C | Emissivity* |
|---|---|---|
| Tungsten, aged filament | 25–3320 | 0.03–0.35 |
| Zinc, 99.1%, comm'l, polished | 230–330 | 0.05 |
| Galv., iron, bright | 28 | 0.23 |
| Galv. gray oxid | 24 | 0.28 |
| **Refractories, Building Materials, Paints, Misc.** | | |
| Alumina, 50μ grain size | 1010–1570 | 0.39–0.28 |
| Alumina-silica, cont'g | 1010–1570 | |
| 0.4% Fe$_2$O$_3$ | | 0.61–0.43 |
| 1.7% Fe$_2$O$_3$ | | 0.73–0.62 |
| 2.9% Fe$_2$O$_3$ | | 0.78–0.68 |
| Al paints (vary with am't lacquer body, age) | 100 | 0.27–0.67 |
| Asbestos | 40–370 | 0.93–0.95 |
| Candle soot; lampblack-water glass | 20–370 | 0.95 ± 0.01 |
| Carbon plate, heated | 130–630 | 0.81–0.79 |
| Oil layers: | | |
| Lube oil, 0.01 in on pol. Ni | 20 | 0.82 |
| Linseed, 1–2 coats on Al | 20 | 0.56–0.57 |
| Rubber, soft gray reclaimed | 24 | 0.86 |
| Misc. I: shiny black lacquer, planed oak, white enamel, serpentine, gypsum, white enamel paint, roofing paper, lime plaster, black matte shellac | 21 | 0.87–0.91 |
| Misc. II: glazed porcelain, white paper, fused quartz, polished marble, rough red brick, smooth glass, hard glossy rubber, flat black lacquer, water, electrographite | 21 | 0.92–0.96 |

*When two temperatures and two emissivities are given they correspond, first to first and second to second, and linear interpolation is suggested.

the table apply, with a few exceptions, to normal radiation from the surface.

For opaque materials, the **reflectance** $\rho$ is the complement of the absorptance. The directional distribution of the reflected radiation depends on the material, its degree of roughness or grain size, and if a metal, its state of oxidation. Polished surfaces of homogeneous materials reflect specularly. In contrast, the intensity of the radiation reflected from a **perfectly diffuse**, or **Lambert**, surface is independent of direction. The directional distribution of reflectance of many oxidized metals, refractory materials, and natural products approximates that of a perfectly diffuse reflector. A better model, adequate for many calculational purposes, is achieved by assuming that the total reflectance $\rho$ is the sum of diffuse and specular components $\rho_D$ and $\rho_S$.

**Black-Surface Enclosures** When several surfaces are present, the need arises for evaluating a geometrical factor $F$, called the **view factor**. Restriction is temporarily to black surfaces, the intensity from which is independent of angle of emission. Define $F_{12}$ as the fraction of the radiation leaving surface $A_1$ in all directions which is intercepted by surface $A_2$. Since the net interchange between $A_1$ and $A_2$ must be zero when their temperatures are alike, it follows that $A_1F_{12} = A_2F_{21}$. From the definition of $F$ and Eq. (1),

$$A_1F_{12} = \int_{A_1}\int_{A_2} \frac{dA_1 \cos\theta_1 \, d\Omega_1}{\pi} = \int_{A_1}\int_{A_2} \frac{dA_1 \cos\theta_1 \, dA_2 \cos\theta_2}{\pi r^2}$$

The product $A_1F_{12}$, having the dimensions of area, will be called the **direct-interchange area** and be designated by $\overline{s_1s_2}$,

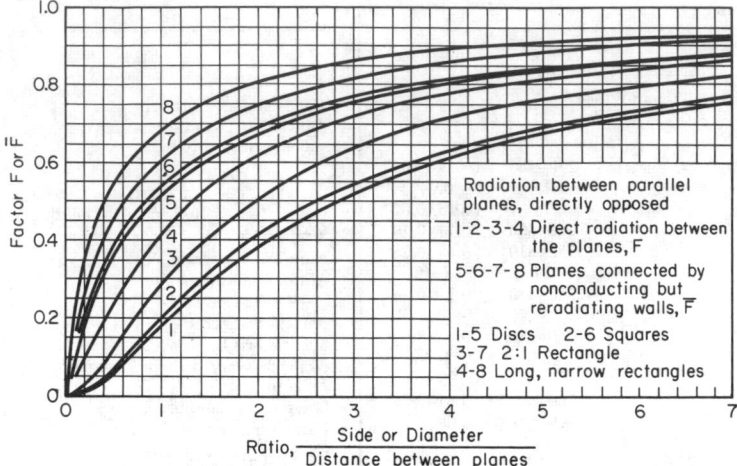

**Fig. 2**   Values of the factor $F$ or $\bar{F}$ for parallel planes directly opposed.

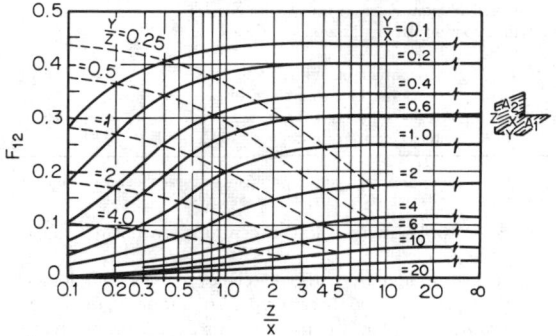

**Fig. 3**   Values of the factor $F$ for perpendicular adjacent rectangles.

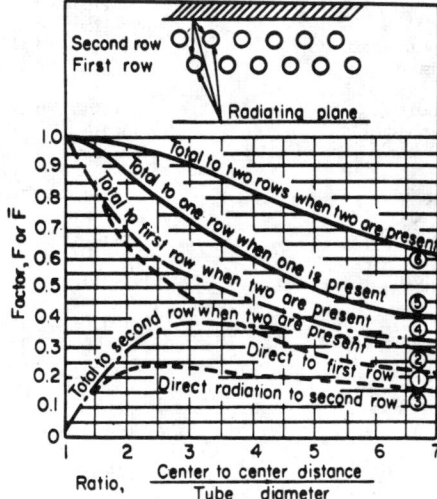

**Fig. 4**   Values of $F$ or $\bar{F}$ for a plane parallel to rows of tubes.

sometimes for brevity by $\overline{12}$ ($\equiv \overline{21}$). Clearly, $\overline{11} + \overline{12} + \overline{13} + \ldots = A_1$; and when $A_1$ cannot "see" itself, $\overline{11} = 0$. Values of $F$ have been calculated for various surface arrangements and appear for opposed parallel rectangles and disks of equal size as lines 1 to 4 of Fig. 2, for perpendicular adjacent rectangles in Fig. 3, and for an infinite plane parallel to a system of rows of parallel tubes as lines 1 and 3 of Fig. 4. Multiple graphical integration to determine $F$ is unnecessary for surfaces in two-dimensional systems (third dimension infinite). In the cross-sectional view, the sum of lengths of crossed strings from the ends of $A_1$ to the ends of $A_2$ less the sum of uncrossed strings from and to the same points, all divided by 2, equals $A_1F_{12}$ ($\equiv \overline{12}$) per unit length normal to the drawing. For other cases, see References.

The view factor $F$ may often be evaluated from that for simpler configurations by the application of three principles: that of reciprocity, $A_iF_{ij} = A_jF_{ji}$; that of conservation, $\Sigma F_{ij} = 1$; and that due to Yamauti, showing that the exchange areas $AF$ between two pairs of surfaces are equal when there is a one-to-one correspondence for all sets of symmetrically placed pairs of elements in the two surface combinations.

EXAMPLE.   The exchange area between the two squares 1 and 4 of Fig. 5 is to be evaluated. The following exchange areas may be obtained from the values, in Fig. 3, of $F$ for commonside rectangles: $\overline{13} = 0.24$, $\overline{24} = 2 \times 0.29 = 0.58$, $\overline{(1 + 2)(3 + 4)} = 3 \times 0.32 = 0.96$. Expression of $\overline{(1 + 2)(3 + 4)}$ in terms of its components yields $\overline{(1 + 2)(3 + 4)} = \overline{13} + \overline{14} + \overline{23} + \overline{24}$. And by the Yamauti principle $\overline{14} = \overline{23}$, since for every pair of elements in 1 and 4, there is a corresponding pair in 2 and 3. Therefore,

$$\overline{14} = [\overline{(1 + 2)(3 + 4)} - \overline{13} - \overline{24}]/2 = 0.07$$

Figure 2 may be used in the same way. Another example is the evaluation of $AF$ for exchange between the outside of the smaller of two coaxial cylinders and the inside of the larger when they are *not* coextensive, given the view factor for coextensive cylinders (Hottel and Sarofim, pp. 54, 59–62).

**Enclosures Containing Gray Sources and Sinks, Refractory Surfaces, and No Absorbing Gas**  The calculation of interchange between a source and a sink under conditions involving successive multiple reflections from other source-sink surfaces in the enclosure, as well as reradiation from refractory surfaces, can become complicated. Let a **zone** of a furnace enclosure be an area small enough to make all elements of itself have substantially equivalent "views" of the rest of the enclosure. (In a furnace containing a symmetry plane, parts of a single zone would lie on either side of the plane.) Zones are of two classes, source-sink surfaces, designated by numerical subscripts and having areas $A_1$, $A_2$, ... and emissivities $\epsilon_1$, $\epsilon_2$, ...; and surfaces at which the net radiant-heat flux is zero (fulfilled by the average refractory wall where difference between internal convection and external loss is minute compared with incident radiation), designated by letter subscripts starting with $r$, and having areas $A_r$, $A_s$, .... It may be shown that the net radiation interchange between source-sink zones $i$ and $j$ is given by

$$\dot{Q}_i \rightleftharpoons_j = \overline{S_iS_j}\sigma T_i^4 - \overline{S_jS_i}\sigma T_j^4 \qquad (6)$$

The term $\overline{S_iS_j}$ is called the **total-interchange area** shared by areas $A_i$ and $A_j$ and depends on the shape of the enclosure and the emissivity and absorptivity of the source and sink zones. It is sometimes called $A_i\mathcal{F}_{ij}$. Restriction here is to gray source-sink zones, for which $\overline{S_iS_j} = \overline{S_jS_i}$ the more general case is treated elsewhere (Hottel and Sarofim, Chaps. 3 and 5).

Evaluation of the $\overline{SS}$'s that characterize an enclosure involves solution of a system of radiation balances on the surfaces. If at a surface the total **leaving-flux density**, emitted plus reflected, is denoted by $W$, radiation balances take the form

For source-sink surface $j$:

$$A_j\epsilon_jE_j + \rho_j\sum_i \overline{(ij)}W_i = A_jW_j \qquad (7)$$

For adiabatic surface $r$:

$$\sum_i \overline{(ir)}W_i = A_rW_r \qquad (8)$$

where $\rho$ is reflectance and the summation is over all surfaces in the enclosure. These equations apply to surfaces which emit and reflect diffusely (i.e., their leaving intensity $W_i/\pi$ is independent of direction—of $j$). Most nonmetallic, tarnished, or rough metal surfaces correspond reasonably well to this restriction (but see p. 4-77). In matrix notation, Eqs. (7) and (8) become

This represents a system of simultaneous equations equal in number to the number of rows of the square matrix. Each equation consists, on the left, of the sum of the products of the members of a row of the square matrix and the corresponding members of the $W$-column matrix, and, on the right, of the member of that row in the third matrix. With the above set of equations solved for $W$, the net flux at any surface $A_i$ is given by

$$\dot{Q}_{i,\text{net}} = \frac{A_i\epsilon_i}{\rho_i}(E_i - W_i) \qquad (10)$$

Refractory temperature is obtained from $W_r = E_r = \sigma T_r^4$.

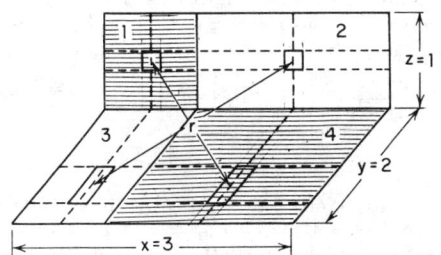

**Fig. 5**   Illustration of the Yamauti principle.

The more general use of Eq. (9) is to obtain the set of total-interchange areas $\overline{SS}$ which constitute a complete description of the effect of shape, size, and emissivity on radiative flux, independent of the presence or absence of other transfer mechanisms. It may be shown that

$$\overline{S_iS_j} \equiv \overline{S_jS_i} \equiv A_i\mathcal{F}_{ij} = \frac{A_i\epsilon_i}{\rho_i}\frac{A_j\epsilon_j}{\rho_j}\left(-\frac{D_{ij}'}{D}\right) \qquad (11)$$

where $D$ is the determinant of the square coefficient matrix in Eq. (9) and $D_{ij}'$ is the cofactor of its $i$th row and $j$th column, or $-1^{i+j}$ times the minor of $D$ formed by crossing out the $i$th row and $j$th column.

As an example, consider radiant interchange between concentric gray spheres of inner and outer radii $r_1$ and $r_2$.

$$\overline{S_iS_j} = \frac{A_1\epsilon_1}{\rho_1}\frac{A_2\epsilon_2}{\rho_2}\begin{array}{c}\overline{12}\\\hline\begin{vmatrix}\overline{11}-\dfrac{A_1}{\rho_1} & \overline{12}\\[2mm] \overline{12} & \overline{22}-\dfrac{A_2}{\rho_2}\end{vmatrix}\end{array}$$

$\overline{12} = A_1$; $\overline{11} = 0$; $\overline{22} = A_2 - \overline{21} = A_2 - \overline{12} = A_2 - A_1$. Substitution gives

$$\begin{bmatrix} \overline{11}-\dfrac{A_1}{\rho_1} & \overline{12} & \cdots & \overline{1r} & \overline{1s} & \cdots \\[2mm] \overline{12} & \overline{22}-\dfrac{A_2}{\rho_2} & \cdots & \overline{2r} & \overline{2s} & \cdots \\ \cdots & \cdots & \cdots & & & \\ \overline{1r} & \overline{2r} & \cdots & \overline{rr}-A_r & \overline{rs} & \cdots \\ \overline{1s} & \overline{2s} & \cdots & \overline{rs} & \overline{ss}-A_s & \cdots \\ \cdots & & & & & \end{bmatrix}\begin{bmatrix}W_1\\[2mm]W_2\\ \cdots \\ W_r \\ W_s \\ \cdots\end{bmatrix} = \begin{bmatrix}-\dfrac{A_1\epsilon_1}{\rho_1}E_1\\[2mm]-\dfrac{A_2\epsilon_2}{\rho_2}E_2\\ \cdots \\ 0 \\ 0 \\ \cdots\end{bmatrix} \qquad (9)$$

$$\overline{S_iS_j} = \frac{A_1\epsilon_1}{\rho_1}\frac{A_2\epsilon_2}{\rho_2}\frac{A_1}{\begin{vmatrix} -A_1/\rho_1 & A_1 \\ A_1 & A_2 - A_1 - A_2/\rho_2 \end{vmatrix}}$$

$$= \frac{1}{\dfrac{1}{A_1\epsilon_1} + \dfrac{1}{A_2}\left(\dfrac{1}{\epsilon_2}-1\right)} \tag{12}$$

This case includes that of infinite parallel planes ($\mathfrak{F} = 1/[1/\epsilon_1 + 1/\epsilon_2 - 1]$), and that of a small body $A_1$ enclosed in a large one ($\overline{S_1S_2} = A_1\mathfrak{F}_{12} = A_1\epsilon_1$).

Many furnace problems are adequately handled by dividing the enclosure into but two source-sink zones $A_1$ and $A_s$, and any number of no-flux zones, $A_r$, $A_s$, .... For this case Eq. (11) yields

$$\frac{1}{\overline{S_1S_2}}\left(\equiv \frac{1}{\overline{S_2S_1}}\right) = \frac{1}{A_1}\left(\frac{1}{\epsilon_1}-1\right)$$
$$+ \frac{1}{A_2}\left(\frac{1}{\epsilon_2}-1\right) + \frac{1}{(\overline{S_1S_2})_B} \tag{13}$$

Here the expression $(\overline{S_1S_2})_B (\equiv \overline{S_2S_1})_B$ represents the total interchange area for the limiting case of a black source and black sink (the refractory emissivity is of no moment). The factor $(\overline{S_1S_2})_B/A_1$, called $\overline{F}_{12}$, is known exactly for a few geometrically simple cases and may be approximated for others. If $A_1$ and $A_2$ are equal parallel disks, squares, or rectangles, connected by non-conducting but reradiating refractory walls, then $\overline{F}$ is given by Fig. 2, lines 5 to 8. If $A_1$ represents an infinite plane and $A_2$ is one or two rows of infinite parallel tubes in a parallel plane, and if the only other surface is a refractory surface behind the tubes, $\overline{F}_{12}$ is given by line 5 or 6 of Fig. 4. If an enclosure may be divided into several radiant-heat sources or sinks $A_1$, $A_2$, etc., and the rest of the enclosure (reradiating refractory surface) may be lumped together as $A_r$ at a uniform temperature $T_r$, then the total-interchange area for zone pairs in the black system is given by

$$(\overline{S_1S_2})_B(\equiv A_1\overline{F}_{12}) = \overline{12} + \frac{(\overline{1r})(\overline{r2})}{A_r - \overline{rr}} \tag{14}$$

For the two-source-sink-zone system to which Eq. (13) applies, Eq. (14) simplifies to $(\overline{S_1S_2})_B = \overline{12} + 1/(1/\overline{1r} + 1/\overline{2r})$; and if $A_1$ and $A_2$ each can see none of itself, there is further simplification to

$$(\overline{S_1S_2})_B = \overline{12} + \frac{1}{1/(A_1 - \overline{12}) + 1/(A_2 - \overline{12})}$$
$$= \frac{A_1A_2 - (\overline{12})^2}{A_1 + A_2 - 2(\overline{12})} \tag{15}$$

which necessitates the evaluation of but one geometrical factor $F$.

Equation (13) covers many problems of radiant-heat interchange between source and sink in furnace enclosures involving no radiating gas. The error due to single zoning of source and sink is small even if the "views" of the enclosure from different parts of each zone are quite different, provided the emissivity is fairly high; the error in $\overline{F}$ is zero if it is obtainable

from Fig. 2 or 4, small if Eq. (14) is used and the variation in temperature over the refractory is small. Approach to any desired accuracy can be made by use of Eq. (11) with division of the surfaces into more zones.

From the definitions of $F$, $\overline{F}$, and $\mathfrak{F}$ or of $\overline{s}$, $(\overline{SS})_B$ and $\overline{SS}$ it is to be noted that

$$\left.\begin{array}{l} F_{11} + F_{12} + \cdots + F_{1r} + F_{1s} + \cdots = 1 \\ \overline{F}_{11} + \overline{F}_{12} + \cdots = 1 \end{array}\right\}$$

$$\mathfrak{F}_{11} + \mathfrak{F}_{12} + \cdots = \epsilon_1$$

$$\left\{\begin{array}{l} \overline{s_1s_2} + \overline{s_1s_2} + \cdots + \overline{s_1s_r} + \overline{s_1s_s} + \cdots = A_1 \\ \text{or}\ \ (\overline{S_1S_1})_B + (\overline{S_1S_2})_B + \cdots = A_1 \\ \overline{S_1S_1} + \overline{S_1S_2} + \cdots = A_1\epsilon_1 \end{array}\right.$$

EXAMPLE. A furnace chamber of rectangular parallelepipedal form is heated by the combustion of gas inside vertical radiant tubes lining the side walls. The tubes are on centers 2.4 diameters apart. The stock forms a continuous plane on the hearth. Roof and end walls are refractory. Dimensions are shown in Fig. 6. The radiant tubes and stock are gray bodies having emissivities 0.8 and 0.9, respectively. What is the net rate of heat transmission to the stock by radiation when the mean temperature of the tube surface is 1500°F (1089 K) and that of the stock is 1200°F (922 K)?

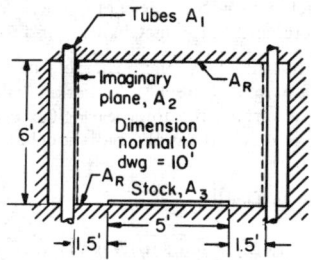

**Fig. 6** Dimensions of a furnace chamber.

This problem must be broken up into two parts, first considering the walls with their refractory-backed tubes. To imaginary planes $A_2$ of area 6 × 10 ft located parallel to and inside the rows of radiant tubes, the tubes emit radiation $\sigma T_1^4 A_1 \overline{F}_{12}$, which equals $\sigma T_1^4 A_2 \mathfrak{F}_{21}$.

To find $\mathfrak{F}_{21}$ use Fig 4, line 5, from which $\overline{F}_{21} = 0.81$. Then from Eq. (13),

$$\mathfrak{F}_{21} = 1/[(1/0.81) + (1/1 - 1) + (2.4/\pi)(1/0.8 - 1)] = 0.702$$

This amounts to saying that the system of refractory-backed tubes is equal in radiating power to a continuous plane $A_2$ replacing the tubes and refractory back of them, having a temperature equal to that of the tubes and an equivalent or effective emissivity of 0.702.

The new simplified furnace now consists of an enclosure formed by two 6 × 10 ft radiating side walls (area $A_2$, of emissivity 0.702), a 5 × 10 ft receiving plane on the floor ($A_3$), and refractory surfaces ($A_R$) to complete the enclosure (ends, roof, and floor side strips); the desired heat transfer is

$$q_2 \rightleftharpoons_3 = \sigma(T_1^4 - T_3^4)A_2\mathfrak{F}_{23}$$

To evaluate $\mathfrak{F}_{23}$, start with the direct interchange factor $F_{23}$. $F_{23} = F$ from ($A_2$) to ($A_3$ + a strip of $A_R$ alongside $A_3$ which has a common edge with $A_2$) minus $F$ from ($A_2$) to (the strip only). These two $F$'s may be evaluated from Fig. 3. For the first $F$, $Y/X = 6/10$, $Z/\overline{X} = 6.5/10$, $F = 0.239$; for the second $F$, $Y/X = 6/10$, $Z/\overline{X} = 1.5/10$, $F = 0.100$. Then

$F_{23} = 0.239 - 0.10 = 0.139$. Now $\bar{F}$ may be evaluated. From Eq. (14) et seq.,

$$A_2 \bar{F}_{23} = \overline{23} + \frac{1}{(1/2\bar{R} + 1/3\bar{R})}; \quad \bar{F}_{23} = F_{23} + \frac{1}{(1/F_{2R}) + (A_2/A_3)(1/F_{3R})}$$

Since $A_2$ "sees" $A_R$, $A_3$, and some of itself (the plane opposite), $F_{2R} = 1 - F_{22} - F_{26}$. $F_{22}$, the direct interchange factor between parallel $6 \times 10$ ft rectangles separated by 8 ft, may be taken as the geometric mean of the factors for 6-ft squares separated by 8 ft, and 10-ft squares separated by 8 ft. These come from Fig. 2, line 2, according to which $F_{22} = \sqrt{0.13 \times 0.255} = 0.182$. Then $F_{2R} = 1 - 0.182 - 0.139 = 0.679$. The other required direct factor is $F_{3R} = 1 - F_{32} = 1 - F_{23}A_2/A_3 = 1 - 0.139 \times {}^{120}\!/_{50} = 0.666$. Then $\bar{F}_{23} = 0.139 \dfrac{1}{(1/0.679) + (120/50)(1/0.666)}$ $= 0.336$. Having $\bar{F}_{23}$, we may now evaluate the factor $\mathcal{F}_{23}$.

$$\mathcal{F}_{23} = \frac{1}{(1/0.336) + [(1/0.702) - 1] + ({}^{120}\!/_{50})[(1/0.9) - 1]} = 0.273$$
$$\dot{Q}_{\text{net}} = \sigma(T_1^4 - T_3^4)A_2 \mathcal{F}_{23} = 0.171(19.6^4 - 16.6^4)(120)(0.273)$$
$$= 402{,}000 \text{ Btu per hr}$$

In S/I units,

$$\dot{Q}_{\text{net}} = 5.67(10.89^4 - 9.22^4)(120 \times .3048^2)(0.273) = 118{,}000 \text{ W}$$

A result of interest is obtained by dividing the term $A_2 \mathcal{F}_{23}(120 \times 0.273$, or 32.7 ft$^2$) by the actual area $A_1$ of the radiating tubes $\left(\dfrac{\pi}{2.4} \times 60 \times 2 = 157 \text{ sq ft}\right)$. This is $32.7/157 = 0.208$, which means that the net radiation from a tube to the stock is 20.8 percent as much as if the tube were black and completely surrounded by black stock.

### Enclosures of Surfaces That Are Not Diffuse Reflectors

The total-interchange-area concept has been generalized to include surfaces the reflectance $\rho$ of which can be divided into a diffuse, or Lambert-reflecting, component $\rho_D$ and a specular component $\rho_S$ independent of angle of incidence, with $\epsilon + \rho_S + \rho_D = 1$. In application to concentric spheres or infinite cylinders, with $A_1$ the inner surface, the method yields (Hottel and Sarofim, p. 181).

$$A_1 \mathcal{F}_{12} \equiv \overline{S_1 S_2} = \frac{1}{\dfrac{1}{A_1 \epsilon_1} + \dfrac{1}{A_2}\left(\dfrac{1}{\epsilon_2} - 1\right) + \dfrac{\rho_{S2}}{1 - \rho_{S2}}\left(\dfrac{1}{A_1} - \dfrac{1}{A_2}\right)}$$

When there is no specular reflectance, the third term in the denominator drops out, in agreement with Eq. (12). When the reflectance is exclusively specular, the denominator becomes $1/A_1 \epsilon_1 + \rho_{S2}/A_1(1 - \rho_{S2})$, easily derivable from first principles.

### Flames, Combustion Products, and Particle Clouds

The radiation from a flame consists of (a) continuum radiation from burning soot particles of microscopic and submicroscopic dimensions, from suspended larger particles of coal, coke, or ash, all contributing to what is spoken of as flame luminosity, (b) infrared radiation, mostly from the water vapor and carbon dioxide in the hot gaseous combustion products, and (c) nonequilibrium radiation associated with the combustion process itself, called chemiluminescence and not a significant contributor to the total radiation. A major problem is the effect of the shape of the emitting volume on the radiative flux; this will be considered first.

**Mean Beam Lengths** Evaluation of radiation from a nonisothermal volume is beyond the scope of this section (see Hottel and Sarofim, Chap. 11). If a volume emitter is isothermal and at a temperature $T$, the ratio of the emission from an element of its volume subtending the solid angle $d\Omega$ at a receiver element $dA$, and making the angle $\theta$ with the normal thereto, to blackbody radiation arriving from within the same solid angle is called the **gas emissivity**. Clearly, $\epsilon$ depends on the path length $L$ through the volume to $dA$. A hemispherical volume radiating to a spot on the center of its base represents the case in which $L$ is independent of direction. Flux at that spot relative to hemispherical black-body flux is thus an alternative way to visualize emissivity. The flux density to an area of interest on the envelope of an emitter volume of any shape can be matched by that at the base of a hemispherical volume of some radius $L$, which will be called the **mean beam length**. It is found that, although the ratio of $L$ to a characteristic dimension $D$ of the shape varies with opacity, the variation is small enough for most engineering purposes to permit use of a constant ratio, $L_M/D$, where $L_M$ is the **average mean beam length**. $L_M$ can be defined to apply either to a spot on the envelope or to any finite portion of its area. An important limiting case is that of opacity approaching zero ($PD \to 0$, where $P = $ partial pressure of the emitter constituent). For this case, $L$ (called $L_0$) equals $4 \times$ (ratio of gas volume to bounding area) when interest is in radiation to the entire envelope. For the range of $PD$ encountered in practice, $L$ (now $L_M$) is always less. For various shapes, 0.8 to 0.95 times $L_0$ has been found optimum (see Table 3); for shapes not reported in Table 3, a factor of 0.88 (or $L_M = 0.88 L_0 = 3.5 V/A$) is recommended.

**Soot Luminosity** is important where combustion occurs under such conditions that the hydrocarbons in the flame are subject to heat in the absence of sufficient air well mixed on a molecular scale. Because soot particles are small relative to the wavelength of radiation of interest (diameters 200 to 1,400$A$), the monochromatic emissivity $\epsilon_\lambda$ depends on the total particle volume per unit volume of space $f_v$, regardless of particle size. It is given by

$$\epsilon_\lambda = 1 - e^{-Kf_v L/\lambda}$$

where $L$ is the path length. Integration of the above over the energy spectrum can be dodged by use of a mean $\lambda$ from the displacement law on p. 4-71 ($\lambda T = 3670 \ \mu\text{m} = c_2/3.92$) to give

$$\epsilon = 1 - e^{-3.92 KTf_v L/c_2} \tag{16a}$$

More nearly rigorous integration gives

$$\epsilon = 1 - (1 + KTf_v L/c_2)^{-4} \tag{16b}$$

Though the latter equation is better, either is adequate for most engineering use. $c_2$ is the second Planck constant, and $K/c_2$ can be obtained from the complex refractive index of soot, in turn dependent on its hydrogen-carbon ratio $H/C$. Based on a study of coals, $K/c_2$ varies from 480 m$^{-1}$ K$^{-1}$ at $H/C = 0$ to 240 at $H/C = 0.4$; some experimental work at Ijmuiden on two oil-flame types leads to 440 and 980. A tentative value of 500 m$^{-1}$ K$^{-1}$ (85 ft$^{-1}$ °R$^{-1}$) is recommended.

There is at present no method of predicting soot concentration of a luminous flame analytically; reliance must be placed on experimental measurement on flames similar to that of

**Table 3. Mean Beam Lengths for Volume Radiation**

| Shape | Characteristic dimension, $D$ | $L_0/D$ | $L_M/D$ |
|---|---|---|---|
| Sphere......... | Diameter | 0.67 | 0.63 |
| Infinite cylinder......... | Diameter | 1 | 0.94 |
| Semi-infinite cylinder, radiating to: | | | |
|   Center of base......... | Diameter | 1 | 0.90 |
|   Entire base......... | Diameter | 0.81 | 0.65 |
| Right-circle cylinder, ht = diam, radiating to: | | | |
|   Center of base......... | Diameter | 0.76 | 0.71 |
|   Whole surface......... | Diameter | 0.67 | 0.60 |
| Right-circle cylinder, ht = 0.5 diam, radiating to: | | | |
|   End......... | Diameter | 0.47 | 0.43 |
|   Side......... | Diameter | 0.52 | 0.46 |
|   Total surface......... | Diameter | 0.50 | 0.45 |
| Right-circle cylinder, ht = 2 × diam, radiating to: | | | |
|   End......... | Diameter | 0.73 | 0.60 |
|   Side......... | Diameter | 0.82 | 0.76 |
|   Total surface......... | Diameter | 0.80 | 0.73 |
| Infinite cylinder, half-circle cross section, radiating to spot on middle of flat side......... | Radius | | 1.26 |
| Rectangular parallelepipeds: | | | |
|   1:1:1 (cube)......... | Edge | 0.67 | 0.60 |
|   1:1:4, radiating to: | | | |
|     1 × 4 face......... | Shortest edge | 0.90 | 0.82 |
|     1 × 1 face......... | Shortest edge | 0.86 | 0.71 |
|     Whole surface......... | Shortest edge | 0.89 | 0.81 |
|   1:2:6, radiating to: | | | |
|     2 × 6 face......... | Shortest edge | 1.18 | |
|     1 × 6 face......... | Shortest edge | 1.24 | |
|     1 × 2 face......... | Shortest edge | 1.18 | |
|     Whole surface......... | Shortest edge | 1.2 | |
| Infinite parallel planes......... | Clearance | 2.00 | 1.76 |
| Space outside infinite bank of tubes, centers on equilateral triangles; tube diam = clearance......... | Clearance | 3.4 | 2.8 |
| Same, except tube diam = 0.5 clearance......... | Clearance | 4.45 | 3.8 |
| Same, except tube centers on squares, diam = clearance......... | Clearance | 4.1 | 3.5 |

interest. Visual observation is misleading; a flame so bright as to hide the wall behind it may be far from a "black" radiator. The International Flame Foundation at Ijmuiden has recorded data on many luminous flames from gas, oil, and coal (see *Jour. Inst. Fuel*, 1956–present). Although the conversion of fuel carbon to soot locally in fuel-rich parts of the flame can be as high as 1 to 2 percent, the soot-carbon in few flames is as high as the smaller of these. One percent conversion corresponds, at 1600 K, to $f_v = 5 \times 10^{-8}$. For flames with $L = 1-10$ meters, this corresponds to a soot emissivity of about 0.15–0.75. A crude allowance for soot in moderately luminous flames is to add 0.1 to the nonluminous gas emissivity; this low number is a consequence of luminosity being so localized in the volume and of radiation from the luminous core's being attenuated by absorption in the colder mantle.

**Clouds of Large Black Particles**   Let the particles have a projected area per unit volume of space a/v, a characteristic mean dimension d, a projected mean particle surface-to-volume ratio b/d, a number-concentration c, and a mean projected area A per particle. (All randomly oriented particles without dimples have a ratio of surface area to mean projected area of 4.) The emissivity of a cloud of particles is

$$\epsilon = 1 - e^{-(a/v)L} \equiv 1 - e^{-bf_v L/d} \equiv 1 - e^{-cAL} \quad \textbf{(17)}$$

As an example, consider heavy fuel oil ($CH_{1.5}$, s.g. 0.95) atomized to a surface-mean particle diameter of $d$ microns, burned with 20 percent excess air to produce coke residue particles having the original drop diameter, and suspended in combustion products at 1500 K. From stoichiometry, $f_v = 1.27 \times 10^{-5}$. For spherical particles $b = \frac{3}{2}$, and the flame emissivity due to the particles along a path $L$ will be $1 - e^{-1.9 \times 10^{-2} L/d}$. With $200\mu$ particles and an $L$ of 3 m, the particle contribution to emissivity will be 0.25. Soot luminosity will increase this; particle burnout will decrease it. The combined emissivity due to several kinds of emitters can be calculated from the separately calculated emissivities provided only one of these (gaseous combustion products) is a selective emitter. If $\epsilon_G, \epsilon_p, \epsilon_S$ are the separate emissivities due to gas, massive particles, and soot, all calculated as though no other emitter were present, the combined emissivity is $1 - (1 - \epsilon_G)(1 - \epsilon_p)(1 - \epsilon_S)$. The emissivity of a cloud of gray particles of surface emissivity $\epsilon_m$ lies between values predicted from the use of exponents of $(a/v)L$ and $\epsilon_m(a/v)L$ in the cloud-emissivity relationship, because of multiple scatter by reflection. Particles with perimeter lying between 0.5 and 5 times the λ of interest are difficult to handle (see Hottel and Sarofim, Chap. 12).

**Gaseous Combustion Products**   Radiation from water vapor and carbon dioxide occurs in spectral bands in the infrared. In magnitude it overshadows convection at furnace temperatures. The emissivity $\epsilon_G$ of a $CO_2$-containing gas volume depends on gas temperature $T_G$, on the $CO_2$ partial-pressure-beam-length product $P_C L$, and to a much lesser extent, on the total pressure. Figure 7 gives $\epsilon_G$ for carbon dioxide at a total pressure of 1 atm. The gas absorptivity $\alpha_G$ equals the emissivity when the absorbing gas and the emitter are at the same temperature. When the emitter surface temperature is $T_s$, $\alpha_G$ is $(T_G/T_S)^{0.65}$ times the $\epsilon_G$ read from Fig. 7 at $T_S$ instead of $T_G$ and at $P_CLT_S/T_G$ instead of $P_CL$. If, at constant $P_CL$, the total pressure $P_T$ is increased above 1 atm, $\epsilon_G$ and $\alpha_G$ increase a small fraction due to line broadening. Over the range of $P_C + P_T$ of 1 to 10 atm, a multiplying factor $(P_C + P_T)^m$ may be used to estimate the effect. $m$ is 0.06, 0.09, or 0.11 (0.008, 0.01, or 0.03) at $T_G = 500$ R (2000 R), when $(P_C + P_T)L$ is 10, 4, or 0.5 ft atm; it is at maximum in the $PL$ range 0.1–1.0; it decreases as temperature rises.

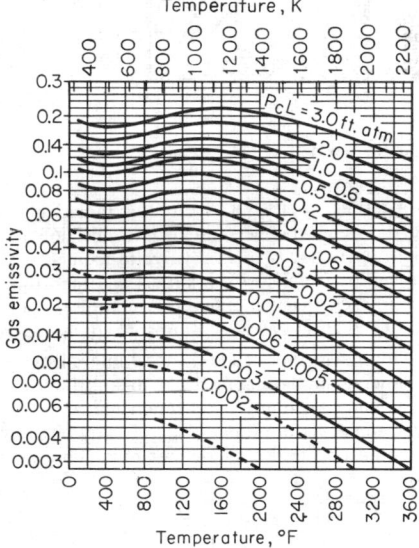

**Fig. 7**  Emissivity of carbon dioxide.

For water vapor, the gas emissivity depends on $T_G$ and $P_wL$ and on total pressure $P_T$ and on the partial pressure of water vapor $P_w$. Emissivity due to water vapor is given in Fig. 8 as a function of $T_G$ and $P_wL$, for the special case of $P_w = 0$ and $P_T = 1$. Allowance for departure from these special conditions is made by multiplying $\epsilon_G$ from Fig. 8 by a factor $C_w$ read from Fig. 9 as a function of $(P_w + P_T)$ and $P_wL$. The absorptivity $\alpha_G$ of water vapor for black-body radiation is $\epsilon_G$ from Fig. 8, read at $T_S$ and at $P_wL(T_S/T_G)$ instead of $P_wL$, then multiplied by $(T_G/T_S)^{0.45}$. The correction factor $C_w$ still applies.

When carbon dioxide and water vapor are present together, the total radiation due to both is somewhat less than the sum of the separately calculated effects because each gas is somewhat opaque to radiation from the other. The amount $\Delta\epsilon$ by which to reduce the sum of $\epsilon_G$ for $CO_2$ and $\epsilon_G$ for $H_2O$ (each evaluated as if the other were absent) to obtain the $\epsilon_G$ due to the two together is read from Fig. 10. The same type of correction applies in calculating $\alpha_G$.

Effective use can sometimes be made of the fact that, at furnace temperatures and for gases containing $H_2O$ and $CO_2$ in a fixed ratio, $\epsilon_G$ decreases with rising temperature in such a way that the product $(\epsilon_G)(T_G)$ depends almost exclusively on $(P_c + P_w)L$. A single design chart, much easier to use than Figs. 7 to 10 in combination, may be constructed on this basis (Hottel and Sarofim, p. 234); but it will have validity over a restricted range of variables.

The final formulation of radiant interchange between a gas

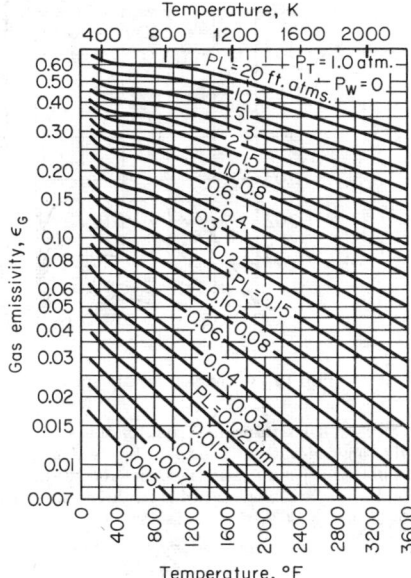

**Fig. 8**  Emissivity of water vapor.

and its bounding surface, when the gas contains $CO_2$ and $H_2O$, is then

$$q = \sigma\epsilon_S'[\epsilon_G T_G^4 - \alpha_G T_S^4] \qquad (18)$$

If the surface is gray, multiplication by $\epsilon_S$ makes proper allowance for reduction in the primary beam from gas to surface and surface to gas, respectively; but some of the gas radiation initially reflected from the surface has further opportunity for absorption at the surface because the gas is but incompletely opaque to the reflected beam. Consequently, the

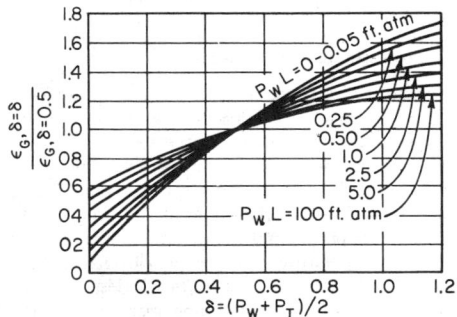

**Fig. 9**  Correction for effects of total pressure and partial pressure on water-vapor radiation.

factor to allow for surface emissivity lies between $\epsilon_S$ and unity, nearer the latter the more transparent the gas (low $PL$) and the more convoluted the surface. In the emissivity range of most industrial surface, 0.7 to 1.0, an adequate approximation consists in use of an effective emissivity $\epsilon_S'$ halfway between the actual value of $\epsilon_S$ and unity.

EXAMPLE.  Flue gas containing 9.5 percent $CO_2$ and 7.1 percent

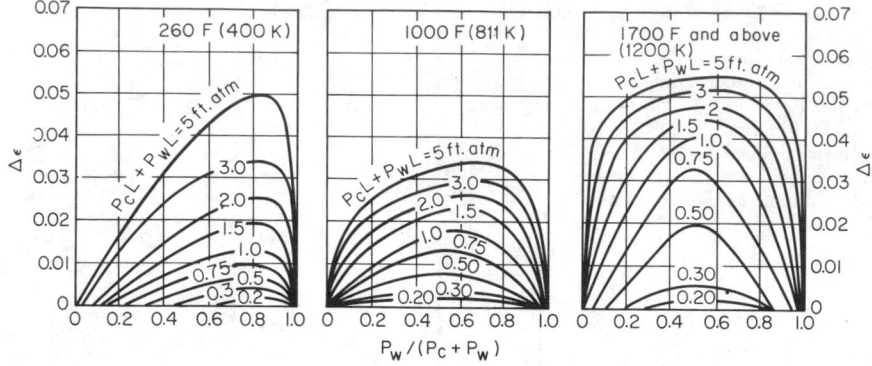

**Fig. 10**  Correction for superimposed radiation from mixtures of carbon dioxide and water vapor.

$H_2O$, wet basis, flows through a bank of tubes of 1.5-in OD on equilateral triangular centers 4.5 in apart. In a section in which the gas and tube surface temperatures are 1700 and 1000°F, what is the heat-transfer rate per square foot of tube area, due to gas radiation only? Tube surface emissivity = 0.8.

$$L_M = 3.8(4.5 - 1.5)/12 = 0.95 \text{ ft}$$
$$P_c L = 0.90 \text{ ft atm}$$
$$P_w L = 0.067 \text{ ft atm}$$

From Fig. 7, for $CO_2$: $\epsilon_G = 0.069$, $\alpha_G = 0.086$.
From Fig. 8, for $H_2O$: $\epsilon_G = 0.036$, $\alpha_G = 0.056$.
From Fig. 9, $c_u = 1.04$.
From Fig. 10, $\Delta\epsilon = 0.001$, $\Delta\alpha = 0.002$.

$$\epsilon_{G,\text{Total}} = 0.069 + 0.036 \times 1.04 - 0.001 = 0.105$$
$$\alpha_{G,\text{Total}} = 0.086 + 0.056 \times 1.04 - 0.002 = 0.142$$

The effective surface-emissivity factor $\epsilon_s = (0.8 + 1)/2 = 0.9$. From Eq. (18),

$$q = 0.9 \times 0.171(0.105 \times 21.6^4 - 0.142 \times 14.6^4) = 2529 \text{ btu/ft}^2\text{h}$$

equivalent to a convection coefficient of $2529/700$, or 3.6 (English units). The emissivity of an "equivalent gray" flame is $(0.105 \times 21.6^4 - 0.142 \times 14.6^4)/(21.6^4 - 14.6^4) = 0.095$.

### Enclosures—Combustion Chambers

The so-called *radiant section* of a furnace presents a heat-transfer problem in which there enters the combined action of direct radiation from the flame to the stock or heat sink and radiation from the flame to refractory surfaces and thence back through the flame (with partial absorption) to the sink, convection, and external losses. Solutions of the problem based on varying degrees of simplification are available, including allowance for temperature variation in both gas and refractory walls (Hottel and Sarofim, Chap. 14). A less rigorous treatment suffices, however, for handling many problems. There are two limiting cases: the long chamber with gas temperature varying only in the direction of gas flow and the compact chamber containing a gas or flame at a uniform temperature. The latter will be considered first.

**The One-Gas-Zone Model**  The following simple mathematical model of a furnace chamber is found to make substantially correct predictions of the relation among the dominant variables for a wide range of furnace types: The gas mass and flame transfer heat as though at a mean temperature $T_G$; the gas is gray; the bounding walls of total area $A_T$ are divisible into a heat-sink surface $A_1$, gray and at a single temperature

$T_1$, and a refractory area $A_r$ treated as radiatively adiabatic (i.e., the difference between interior convection to $A_r$ and conduction through it is small relative to its irradiation). The net radiative flux from gas to sink must be proportional to the difference in their black emissive powers; the proportionality constant, of dimension area, will by analogy to the treatment of surface interchange be called the total-interchange area $\overline{GS_1}$, a term allowing for both direct gas-to-sink interchange and indirect exchange via multiple reflection and refractory action. (The older nomenclature was $A_1\mathcal{F}_{1G}$.) It may be shown [from Eq. (11) with exchange areas redefined to include gas transmittance] that

$$\frac{A_1}{GS_1} = \frac{1}{\epsilon_1} - 1 + \cfrac{1}{\epsilon_G\left[1 + \cfrac{A_r/A_1}{\epsilon_G/(1-\epsilon_G)F_{r1}}\right]} \tag{19}$$

For the commonly encountered case of the sink consisting of a row of tubes mounted on a refractory wall, $A_1$ is the area of the whole plane in which the tubes lie, $T_1$ is tube surface temperature, and $\epsilon_1$ is the effective emissivity of the tube-row-refractory-wall combination, as in the earlier numerical example associated with Fig. 6, where $\epsilon_1 = 0.702$. A further simplification is to replace $A_1/A_T$ by $C$, the "cold" fraction of the wall.

It may be shown that (19) takes one of three limiting forms:

$$\overline{GS_1} = \cfrac{A_T}{\cfrac{1}{C\epsilon_1} + \left(\cfrac{1}{\epsilon_g} - 1\right)\left(1 \text{ or } \cfrac{1-\epsilon_g}{1-C\epsilon_g} \text{ or } \cfrac{1-\epsilon_g(1-C)/C}{1-\epsilon_g(1-C)}\right)} \tag{20}$$

The first corresponds to the "speckled furnace" model with surfaces $A_1$ and $A_r$ so intimately mixed that $F_{r1} = F_{11} = C$ and $F_{1r} = F_{rr} = 1 - C$; the second to complete segregation of heat sink $A_1$ in a single plane so that $F_{1r} = 1$; the third to complete segregation of $A_r$ in a single plane so that $F_{r1} = 1$.

If the convective flux from gas to sink is $A_1 h_1(T_G - T_1)$ (and this is small compared with radiation), convection may be linearized in $T^4$ to give $A_1 h_1(T_G - T_1) = A_1(h_1/4\sigma T_{G1}^3)\sigma(T_G^4 - T_1^4)$. $T_{G1}$ in the dimensionless term in the first parentheses may be approximated as the arithmetic mean of $T_g$ and $T_1$ (guessed the first time around, in calculating). The total gas-to-sink flux is then $\sigma(T_G^4 - T_1^4)$ multiplied by $(\overline{GS_1} + A_1 h_1/4\sigma T_{G1}^3)$; this sum will hereafter be called $A_S$, the effective area of the sink.

Let the loss through the refractory walls be represented by $U_r A_r(T_G - T_0)$, where $T_0$ is the outside ambient (and base) temperature and $1/U_r = 1/b_i + $ (wall thickness)$/k + 1/b_{c+r,o}$. The net total flux $\dot{Q}_G$ from the gas is then

$$\dot{Q}_G = A_S \sigma (T_G^4 - T_1^4) + U_r A_r (T_G - T_0) \qquad (21)$$

An energy balance on the gas yields

$$\dot{Q}_G = \dot{H} - (T_G' - T_0)\dot{m}\overline{C}_p \qquad (22)$$

where $\dot{H}$ is the hourly enthalpy input, above the base temperature $T_0$, in the fuel plus preheated air; $\dot{m}$ is the mass flow rate of combustion products; $\overline{C}_p$ is their mean specific heat; and $T_{G'}$ is the gas temperature leaving the radiant section. If the same mean specific heat is used to define $T_{AF}$—a pseudoadiabatic flame temperature given by $T_{AF} = T_0 + \dot{H}/\dot{m}\overline{C}_p$—Eq. (22) becomes $Q_G = H(T_{AF} - T_{G'})/(T_{AF} - T_0)$. The leaving-enthalpy temperature $T_{G'}$ may be markedly less than the mean gas heat-transfer temperature $T_G$. Let $T_{AF} - T_{G'} = d(T_{AF} - T_G)$, with $d$ always greater than 1 and in the neighborhood of 4/3 (further discussion later). The last equation then becomes

$$\dot{Q}_G = \frac{\dot{H}}{d}\frac{T_{AF} - T_G}{T_{AF} - T_0} \qquad (23)$$

If $d$ is known or capable of estimation, (21) and (23) constitute two equations in two unknowns, $\dot{Q}_G$ and $T_G$. To make the relations dimensionless, let the ratios of $T_1$, $T_G$, and $T_0$ to $T_{AF}$ be $T_1^*$, $T_G^*$ and $T_0^*$, and let

$\eta_G^* \equiv$ reduced furnace efficiency based on gas loss $\equiv$
$$\frac{\dot{Q}_G}{\dot{H}}\frac{1 - T_0^*}{d}$$

$D^* \equiv$ reduced firing density $= \dfrac{\dot{H}}{A_S \sigma T_{AF}^4(1 - T_0^*)}$

$L^* \equiv$ wall-loss group $= \dfrac{U_r A_r}{\sigma T_{AF}^3 A_S}$

Equations (21) and (23) then become

$$\eta_G^*(dD^*) = T_G^{*4} - T_1^{*4} + L^*(T_G^* - T_0^*) \qquad (24)$$
$$\text{and } \eta_G^* = 1 - T_G^* \qquad (25)$$

Elimination of $T_G^*$ between (24) and (25) gives

$$(1 - \eta_G^*)^4 - (\eta_G)^*(L^* + dD^*) = T_1^{*4} - L^*(1 - T_0^*) \qquad (26)$$

The efficiency $\eta_G$ based on the gas loss is $\eta_G^* d/(1 - T_0^*)$; the true furnace efficiency $\eta$ based on the energy gain of the sink is

$$\eta = \eta_G^* \frac{d}{1 - T_0^*} - \frac{L^*(1 - \eta_G^* - T_0^*)}{D^*(1 - T_0^*)} \qquad (27)$$

Solution of (26) for $(\eta_G^*)$ yields an efficiency which, from its derivation, depends on firing density, heat-sink temperature, gas-temperature drop from $T_G$ to $T_{G'}$ (at this point the one arbitrary quantity), and the wall loss; and the firing-density term makes due allowance for such operating variables as fuel type, excess air, or air preheat which affect flame temperature or gas emissivity, for fractional occupancy of the walls by sink surfaces, and for sink emissivity. Figure 11, from Eq. (26) with $L^*$ assumed 0, shows the reduced efficiency $\eta_G^*$ as a function of $dD^*$ (proportional to firing density) for various values of normalized sink temperature $T_1^*$. Shaded areas indicate the

operating regimes of various types of furnaces. Note the significant properties of the function presented: (1) As firing rate $D^*$ goes down, the efficiency rises and approaches $1 - T_1^*$ in the limit. (2) Changes in sink temperature have little effect if $T_1^* < 0.3$. (3) As the furnace walls approach complete

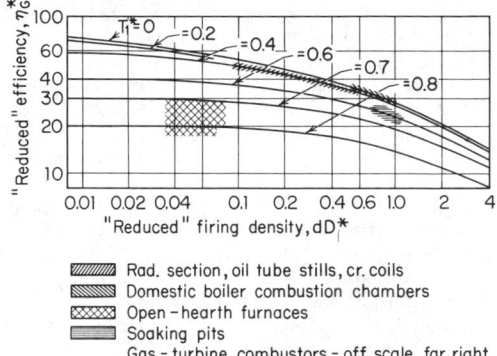

**Fig. 11** The thermal performance of "well-stirred" furnace chambers.

coverage by a black sink ($C = \epsilon_1 = 1$), $\overline{GS_1}$ according to Eq. (20) becomes $\epsilon_g A_T$; and as convection becomes unimportant $A_s \rightarrow A_T \epsilon_G$ and $D^* \propto 1/\epsilon_G$; thus at very high firing rates where $\eta_G^*$ approaches inverse proportionality to $D^*$, the efficiency of heat transfer varies directly as $\epsilon_G$ (gas-turbine chambers), but at low firing rates $\epsilon_G$ has relatively little effect. (4) When $C\epsilon_1 << 1$ because of a nonblack sink or much refractory surface, the effect of changing flame emissivity is to produce a much less than proportional effect on heat flux.

Equation (26) or (27) and Fig. 11 serve as a framework for correlating the performance of furnaces with flow patterns—plug flow, parabolic profile, and recirculatory flow—differing from the well-stirred model (Hottel and Sarofim, Chap. 14). As expected, plug-flow furnaces show somewhat higher efficiency, mild-recirculation types somewhat lower efficiency, and strong-recirculation furnaces performances closely similar to that of the well-stirred model. Equation (26) has also been used to correlate data on the radiant section of a tube still in which $T_1^*$ varied from 0.34 to 0.47. The assumption of a mean $T_1^*$ of 0.4 and a value of $d$ of 4/3 leads, from Eq. (26), to

$$4/3\eta_G^* D^* = (1 - \eta_G^*)^4 - (0.4)^4$$

Though $A_s$, to which $D^*$ is inversely proportional, varied with gas temperature and excess air, its mean value was used. The average deviation of 10 measured performance points from the above relation was only 3.8 percent, even though no constants of the equation were determined empirically from furnace data (Hottel, *Jour. Inst. Fuel*, 1961).

Equation (26) [or (27)] is capable of covering a wide range of furnace types; its structure makes it safer to use than the many empirical relations in the literature. The formulation of gas and flame radiation to yield a value for $\overline{GS_1}$, together with an estimation of $d$, permits a determination, by use of Eqs. (26) and (27), of the efficiency of a furnace of almost any type. An entirely different use of those equations is in fitting existing furnace data to a model equation of performance. Suppose that furnace performance data are available in the form of efficiency $\eta$ versus firing rate $\dot{H}$. The quantities $T_0$, $T_1$, $T_{AF}$,

$U_r A_r$ are known or readily calculable, and a value of $A_S$ sufficiently good for determining the not too important quantity $L^*$ should also be calculable. Let the efficiency $\eta$ be plotted versus $H/\sigma T_{AF}^4 (1 - T_0^*)$ on translucent logarithmic paper, and let the plot be superimposed on a similar figure which has been constructed from Eqs. (26) and (27) for the $T_1^*$ and $L^*$ of interest, with $\eta/d$ plotted versus $dD^*$. Let the data plot be displaced vertically and horizontally until the data fit the theoretical curve. The relative vertical displacement of the two plots yields the value of $d$; their relative horizontal displacement yields the value $D^* d/[H/\sigma T_F^4 (1 - T_0^*)]$, which is $d/A_S$. The two quantities $d$ and $A_S$ are a characterization of the furnace; one is an empirical constant taking account of the difference between $T_G$ and $T_G'$, the other a measure of the many factors affecting radiative exchange. They permit a computation of what will happen under other furnace operating conditions (provided those don't change the general character of the flow pattern or the system's radiation geometry); and when determined for each of several furnaces in a particular class they lead to an empirical determination of the effects of design variables on performance. (For nongray gas effects, sink-temperature variation, reformer furnace shape problems, see "Heat Transfer in Flames," Afgan and Beer, eds., Halsted Press, Washington, D.C., 1974, Chap. 1.)

**The Long Chamber** When the gas-temperature transverse to the flow direction is reasonably uniform and the chamber is long compared with its mean hydraulic radius, the opposed upstream and downstream fluxes through the flow cross section will substantially cancel (e.g., tunnel kilns, billet-reheating furnaces, the numerical example of p. 4-76). Under these conditions, the radiative contribution to local flux may be formulated in terms of local temperatures and of view factors or exchange areas evaluated as for a two-dimensional system. The local flux density at the sink $A_1$ is then

$$q(T_G, T_1) = (\overline{GS_1}/A_1)\sigma(T_G^4 - T_1^4) + b_1(T_G - T_1) \quad (28)$$

where $q$ is written to indicate that it is a function of $T_G$ and $T_1$ and $\overline{GS_1}$ is evaluated as in Eq. (19). If $\dot{m}c_p$ is the hourly heat capacity of the gas stream, the temperature of which changes by $dT_G$ over the sink area increment $dA_1$, then

$$[q(T_G, T_1)] \, dA_1 = -\dot{m}c_p \, dT_G$$

From this it is clear that the reciprocal of the mean height of a curve of $1/q$ versus $\dot{m}c_p T_G$ over the gas-temperature interval of interest is $q_{avg}$. Three points generally suffice.

Section **5**

# Strength of Materials

**CHARLES W. MacGREGOR**   *Formerly Engineering Consultant and Manager of Advanced Technology, Systems Development Division, International Business Machines Corporation, Endicott, N.Y. Presently, Consulting Engineer in Private Practice, Cohasset, MA.*

**JOHN SYMONDS**   *Fellow Engineer, Oceanic Division, Westinghouse Electric Corporation.*

**J. P. VIDOSIC**   *Regents' Professor Emeritus of Mechanical Engineering, Georgia Institute of Technology.*

**HAROLD V. HAWKINS**   *Manager, Product Standards and Services, Columbus McKinnon Corporation, Tonawanda, N.Y.*

**WILLIAM T. THOMSON**   *Emeritus Professor of Engineering, University of California, Santa Barbara.*

**DONALD D. DODGE**   *Principal Staff Engineer, Engineering & Research, Ford Motor Company.*

### MECHANICAL PROPERTIES OF MATERIALS
#### by Charles W. MacGregor, Revised by John Symonds

Stress-Strain Diagrams ............................... 5-2
Fracture at Low Stresses ............................. 5-6
Fatigue ............................................... 5-9
Creep ................................................ 5-10
Hardness ............................................. 5-12
Testing of Materials ................................. 5-14

### MECHANICS OF MATERIALS
#### by J. P. Vidosic

Simple Stresses and Strains .......................... 5-16
Combined Stresses .................................... 5-19
Plastic Design ....................................... 5-21
Design Stresses ...................................... 5-22
Beams ................................................ 5-22
Torsion .............................................. 5-37
Columns .............................................. 5-40
Eccentric Loads ...................................... 5-42
Curved Beams ......................................... 5-45
Impact ............................................... 5-46
Theory of Elasticity ................................. 5-47
Cylinders and Spheres ................................ 5-49
Pressure between Bodies with Curved Surfaces ......... 5-51
Flat Plates .......................................... 5-52
Theories of Failure .................................. 5-53

Plasticity ........................................... 5-54
Rotating Disks ....................................... 5-55
Experimental Stress Analysis ......................... 5-57

### PIPELINE FLEXURE STRESSES CAUSED BY EXPANSION OR MOVEMENT OF SUPPORTS
#### by Harold V. Hawkins

Pipeline Flexure Stresses ............................ 5-61

### VIBRATION
#### by William T. Thomson

Free, Damped, and Forced Vibrations .................. 5-67
Vibration-measuring Instruments ...................... 5-75

### NONDESTRUCTIVE TESTING
#### by Donald D. Dodge

Nondestructive Testing ............................... 5-76
Magnetic-Particle Methods ............................ 5-81
Penetrant Methods .................................... 5-81
Radiographic Methods ................................. 5-81
Ultrasonic Methods ................................... 5-82
Eddy-Current Methods ................................. 5-83
Microwave Methods .................................... 5-83
Infrared Methods ..................................... 5-83
Acoustic-Signature Analysis .......................... 5-84

# MECHANICAL PROPERTIES OF MATERIALS

## by Charles W. MacGregor (Revised by John Symonds)

REFERENCES: Davis et al., "Testing and Inspection of Engineering Materials," McGraw-Hill. Timoshenko, "Strength of Materials," pt. II, Van Nostrand. Richards, "Engineering Materials Science," Wadsworth. Nadai, "Plasticity," McGraw-Hill. Tetelman and McEvily, "Fracture of Structural Materials," Wiley. "Fracture Toughness Testing and Its Applications," ASTM STP-381. McClintock and Argon (eds.), "Mechanical Behavior of Materials," Addison-Wesley. Dieter, "Mechanical Metallurgy," McGraw-Hill. "Creep Data," ASME. ASTM Standards, ASTM.

## STRESS-STRAIN DIAGRAMS

**The Stress-Strain Curve** The engineering tensile stress-strain curve is obtained by **static loading** of a standard specimen, that is, by applying the load slowly enough that all parts of the specimen are in equilibrium at any instant. The curve is usually obtained by controlling the loading rate in the tensile machine. ASTM Standard E8 requires a loading rate not exceeding 100,000 lb/in² (70 kg/mm²)/min. An alternate method of obtaining the curve is to specify the strain rate as the independent variable, in which case the loading rate is continuously adjusted to maintain the required strain rate. A strain rate of 0.05 in/in/min is commonly used. It is measured usually by an extensometer attached to the gage length of the specimen. Figure 1 shows several stress-strain curves.

For most engineering materials, the curve will have an initial linear elastic region (Fig. 2) in which deformation is reversible and time-independent. The slope in this region is **Young's modulus** $E$. The **proportional elastic limit** (PEL) is the point where the curve starts to deviate from a straight line. The **elastic limit** (frequently indistinguishable from PEL) is the

point on the curve beyond which plastic deformation is present after release of the load. If the stress is increased further, the stress-strain curve departs more and more from the straight line. Unloading the specimen at point $X$ (Fig. 2), the portion $XX'$ is linear and is essentially parallel to the original line $OX''$. The horizontal distance $OX'$ is called the **permanent set** corresponding to the stress at $X$. This is the basis for the

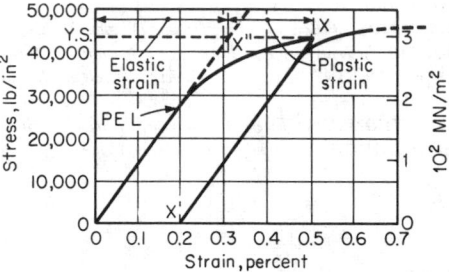

**Fig. 2** General stress-strain diagram.

construction of the arbitrary **yield strength.** To determine the yield strength, a straight line $XX'$ is drawn parallel to the initial elastic line $OX''$ but displaced from it by an arbitrary value of permanent strain. The permanent strain commonly used is 0.20 percent of the original gage length. The intersection of this line with the curve determines the stress value called the yield strength. In reporting the yield strength, the amount of permanent set should be specified. The arbitrary yield strength is used especially for those materials not exhibiting a natural yield point such as nonferrous metals; but it is not limited to these. Plastic behavior is somewhat time-dependent, particularly at high temperatures. Also at high temperatures, a small amount of time-dependent reversible strain may be detectable, indicative of **anelastic** behavior.

The **ultimate tensile strength** (UTS) is the maximum load sustained by the specimen divided by the original specimen cross-sectional area. The percent **elongation at failure** is the plastic extension of the specimen at failure expressed as (the change in original gage length × 100) divided by the original gage length. This extension is the sum of the **uniform** and **nonuniform** elongations. The uniform elongation is that which occurs prior to the UTS. It has an unequivocal significance, being associated with uniaxial stress, whereas the nonuniform elongation which occurs during localized extension (necking) is associated with triaxial stress. The nonuniform elongation will depend on geometry, particularly the ratio of specimen gage length $L_0$ to diameter $D$ or square root of cross-sectional area $A$. ASTM Standard E8 specifies test-specimen geometry for a number of specimen sizes. The ratio $L_0/\sqrt{A}$ is maintained at 4.5 for flat- and round-cross-section specimens. The original gage length should always be stated in reporting elongation values.

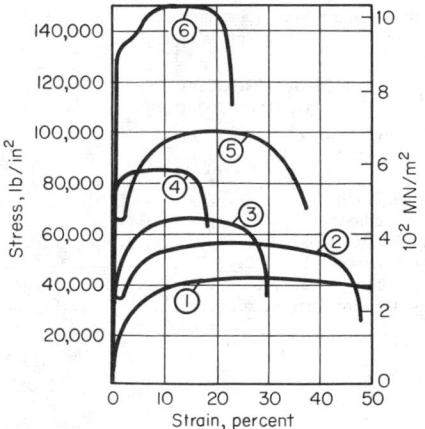

**Fig. 1** Comparative stress-strain diagrams. (1) Soft brass; (2) low-carbon steel; (3) hard bronze; (4) cold-rolled steel; (5) medium-carbon steel, annealed; (6) medium-carbon steel, heat-treated.

**Table 1. Typical Mechanical Properties at Room Temperature**
(Based on ordinary stress-strain values)

| Metal | Tensile strength, 1,000 lb/in² | Yield strength, 1,000 lb/in² | Ultimate elonga-tion, % | Reduc-tion of area, % | Brinell No. |
|---|---|---|---|---|---|
| Cast iron | 18–60 | 8–40 | 0 | 0 | 100–300 |
| Wrought iron | 45–55 | 25–35 | 35–25 | 55–30 | 100 |
| Commercially pure iron, annealed | 42 | 19 | 48 | 85 | 70 |
| Hot rolled | 48 | 30 | 30 | 75 | 90 |
| Cold rolled | 100 | 95 | | | 200 |
| Structural steel, ordinary | 50–65 | 30–40 | 40–30 | | 120 |
| Low alloy, high strength | 65–90 | 40–80 | 30–15 | 70–40 | 150 |
| Steel, SAE 1300, annealed | 70 | 40 | 26 | 70 | 150 |
| Quenched, drawn 1300°F | 100 | 80 | 24 | 65 | 200 |
| Drawn 1000°F | 130 | 110 | 20 | 60 | 260 |
| Drawn 700°F | 200 | 180 | 14 | 45 | 400 |
| Drawn 400°F | 240 | 210 | 10 | 30 | 480 |
| Steel, SAE 4340, annealed | 80 | 45 | 25 | 70 | 170 |
| Quenched, drawn 1300°F | 130 | 110 | 20 | 60 | 270 |
| Drawn 1000°F | 190 | 170 | 14 | 50 | 395 |
| Drawn 700°F | 240 | 215 | 12 | 48 | 480 |
| Drawn 400°F | 290 | 260 | 10 | 44 | 580 |
| Cold-rolled steel, SAE 1112 | 84 | 76 | 18 | 45 | 160 |
| Stainless steel, 18-S | 85–95 | 30–35 | 60–55 | 75–65 | 145–160 |
| Steel castings, heat-treated | 60–125 | 30–90 | 33–14 | 65–20 | 120–250 |
| Aluminum, pure, rolled | 13–24 | 5–21 | 35–5 | | 23–44 |
| Aluminum-copper alloys, cast | 19–23 | 12–16 | 4–0 | | 50–80 |
| Wrought, heat-treated | 30–60 | 10–50 | 33–15 | | 50–120 |
| Aluminum die castings | 30 | | 2 | | |
| Aluminum alloy 17ST | 56 | 34 | 26 | 39 | 100 |
| Aluminum alloy 51ST | 48 | 40 | 20 | 35 | 105 |
| Copper, annealed | 32 | 5 | 58 | 73 | 45 |
| Copper, hard-drawn | 68 | 60 | 4 | 55 | 100 |
| Brasses, various | 40–120 | 8–80 | 60–3 | | 50–170 |
| Phosphor bronze | 40–130 | | 55–5 | | 50–200 |
| Tobin bronze, rolled | 63 | 41 | 40 | 52 | 120 |
| Magnesium alloys, various | 21–45 | 11–30 | 17–0.5 | | 47–78 |
| Monel metal, 70Ni, 30Cu | 100 | 50 | 35 | | 170 |
| Molybdenum, arc-cast | 97 | 91 | 28* | 40 | 260 |
| Zirconium, crystal bar | 24–43 | 8–26 | 54–24 | 75–25 | 70–130 |
| Titanium (99.0 Ti), annealed bar | 95 | 80 | 47 | 27 | |
| Ductile iron, Grade 90–65–02, as cast | 95–105 | 70–75 | 2.5–5.5 | | 225–265 |

Compressive strength of cast iron, 80,000 to 150,000 lb/in².
Compressive yield strength of all metals, except those cold-worked, = tensile yield strength.
Stress, 1,000 lb/in² × 6.894 = stress, MN/m².
*1 in gage length.

The specimen percent **reduction in area** (RA) is the contraction in cross-sectional area at the fracture expressed as a percentage of the original area. It is obtained by measurement of the cross section of the broken specimen at the fracture location. The RA along with the load at fracture can be used to obtain the **fracture stress,** that is, fracture load divided by cross-sectional area at the fracture.

The type of fracture in tension gives some indication of the quality of the material, but this is considerably affected by the testing temperature, speed of testing, the shape and size of the test piece, and other conditions. Contraction is greatest in tough and ductile materials and least in brittle materials. In general, fractures are either of the **shear** or of the **separation** (loss of cohesion) type. Flat tensile specimens of ductile metals often show shear failures if the ratio of width to thickness is greater than 6:1. A completely shear-type failure may termi-

nate in a chisel edge, for a flat specimen, or a point rupture, for a round specimen. Separation failures occur in brittle materials, such as certain cast irons. Combinations of both shear and separation failures are common on round specimens of ductile metal. Failure often starts at the axis in a necked region and produces a relatively flat area which grows until the material shears along a cone-shaped surface at the outside of the specimen, resulting in what is known as the cup-and-cone fracture. Double cup-and-cone and rosette fractures sometimes occur. Several types of tensile fractures are shown in Fig. 3.

Annealed or hot-rolled mild steels generally exhibit a **yield point** (see Fig. 4). Here, in a constant strain-rate test, a large increment of extension occurs under constant load at the elastic limit or at a stress just below the elastic limit. In the latter event the stress drops suddenly from the **upper yield point** to the **lower yield point.** Subsequent to the drop, the yield-point extension occurs at constant stress, followed by a rise to the UTS. Plastic flow during the yield-point extension is discontinuous; successive zones of plastic deformation, known as **Luder's bands** or **stretcher strains,** appear until the entire specimen gage length has been uniformly deformed at the end of the yield-point extension. This behavior causes a banded or stepped appearance on the metal surface. The exact form of the stress-strain curve for this class of material is sensitive to test temperature, test strain rate, and the characteristics of the tensile machine employed.

**Fig. 3**   Typical metal fractures in tension.

The plastic behavior in a uniaxial tensile test can be represented as the **true stress-strain curve.** (See MacGregor, The Tension Test, *Proc. ASTM*, **40**, 1940, pp. 508–534; also The True Stress-Strain Tension Test—Its Role in Modern Materials Testing, *Jour. Franklin Inst.*, **238**, nos. 2, 3, Aug., Sept. 1944, pp. 111–135, 159–176.) The **true stress** $\sigma$ is based on the instantaneous cross section $A$, so that $\sigma = \text{load}/A$. The instantaneous **true strain** increment is $-dA/A$, or $dL/L$ prior to necking. Total true strain $\epsilon$ is

$$\int_{A_0}^{A} \frac{-dA}{A} = \ln\left(\frac{A_0}{A}\right)$$

or $\ln (L/L_0)$ prior to necking. The true stress-strain curve or flow curve obtained has the typical form shown in Fig. 5. In the part of the test subsequent to the maximum load point (UTS), when necking occurs, the true strain of interest is that which occurs in an infinitesimal length at the region of minimum cross section. True strain for this element can still be expressed as $\ln (A_0/A)$, where $A$ refers to the minimum cross section. Methods of constructing the true stress-strain curve are described in the previously cited articles. In the range between initial yielding and the neighborhood of the maximum load point the relationship between plastic strain $\epsilon_p$ and true stress often approximates

$$\sigma = k\epsilon_p^n$$

where $k$ is the **strength coefficient** and $n$ is the **work-hardening exponent.** (See Low, "Properties of Metals in Materials Engineering," ASM, 1949.) For a material which shows a yield point the relationship applies only to the rising part of the curve beyond the lower yield. It can be shown that at the maximum load point the slope of the true stress-strain curve equals the true stress, from which it can be deduced that for a material obeying the above exponential relationship between $\epsilon_p$ and $n$, $\epsilon_p = n$ at the maximum load point. The exponent strongly influences the spread between YS and UTS on the engineering stress-strain curve. Values of $n$ and $k$ for some materials are shown in Table 2. A point on the flow curve identifies the **flow stress** corresponding to a certain strain, that is, the stress required to bring about this amount of plastic deformation. The concept of true strain is useful for accurately describing large amounts of plastic deformation. The

linear strain definition $(L - L_0)/L_0$ fails to correct for the continuously changing gage length, which leads to an increasing error as deformation proceeds.

During extension of a specimen under tension, the change in the specimen cross-sectional area is related to the elongation

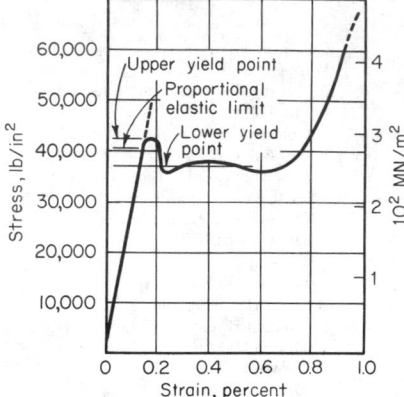

**Fig. 4**   Yielding of annealed steel.

by **Poisson's ratio** $\mu$, which is the ratio of strain in a transverse direction to that in the longitudinal direction. Values of $\mu$ for the elastic region are shown in Table 3. For plastic strain it is approximately 0.5.

The general effect of increased strain rate is to increase the resistance to plastic deformation and thus to raise the flow curve. Decreasing test temperature also raises the flow curve. The effect of strain rate is expressed as **strain-rate sensitivity** $m$.

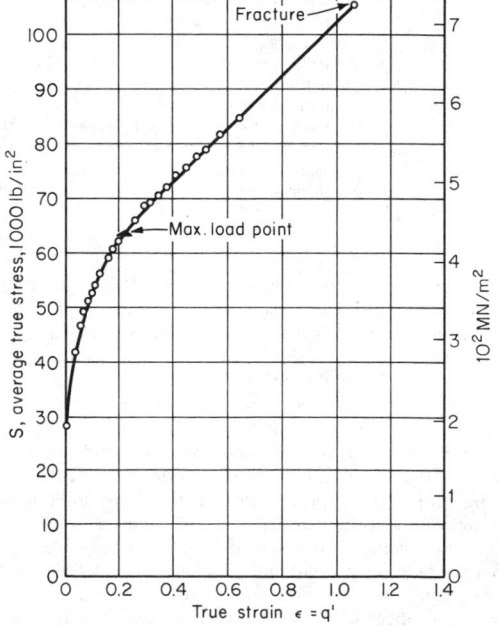

**Fig. 5**   True stress-strain curve for 20°C annealed mild steel.

**Table 2. Room-Temperature Plastic-Flow Constants for a Number of Metals\***

| Material | Condition | $k$, 1,000 in² (MN/m²) | $n$ |
|---|---|---|---|
| 0.40% C steel | Quenched and tempered at 400°F (478°K) | 416(2,860) | 0.088 |
| 0.05% C steel | Annealed and temper-rolled | 72(49.6) | 0.235 |
| 2024 aluminum | Precipitation-hardened | 100(689) | 0.16 |
| 2024 aluminum | Annealed | 49(338) | 0.21 |
| Copper | Annealed | 46.4(319) | 0.54 |
| 70–30 brass | Annealed | 130(895) | 0.49 |

*\*Reproduced by permission from "Properties of Metals in Materials Engineering," ASM, 1949.*

Its value can be measured in the tension test if the strain rate is suddenly increased by a small increment during the plastic extension. The flow stress will then jump to a higher value. The strain-rate sensitivity is the ratio of incremental changes of $\log \sigma$ and $\log \dot{\epsilon}$:

$$m = \left(\frac{\delta \log \sigma}{\delta \log \dot{\epsilon}}\right)_{\dot{\epsilon}}$$

For most engineering materials at room temperature the strain-rate sensitivity is of the order of 0.01. The effect becomes more significant at elevated temperatures, with values ranging to 0.2 and sometimes higher. (See Lubahn, Creep-Tension Relations at Low Temperature, *Proc. ASTM,* **52,** 1952, p. 908.)

**Compression Testing**   The compressive stress-strain curve is similar to the tensile stress-strain curve up to the yield strength. Thereafter, the progressively increasing specimen cross section causes the compressive stress-strain curve to diverge from the tensile curve. Some ductile metals will not fail in the compression test. Complex behavior occurs when the direction of stressing is changed, because of the **Bauschinger effect,** which can be described as follows: If a specimen is first plastically strained in tension, its yield stress in compression is reduced and vice versa.

**Combined Stresses**   This refers to the situation in which stresses are present on each of the faces of a cubic element of the material. For a given cube orientation the applied stresses may include shear stresses over the cube faces as well as stresses normal to them. By a suitable rotation of axes the problem can be simplified: applied stresses on the new cubic element are equivalent to three mutually orthogonal **principal stresses** $\sigma_1$, $\sigma_2$, $\sigma_3$ alone, each acting normal to a cube face. Combined stress behavior in the elastic range is described in Sec. 5, Mechanics of Materials.

Prediction of the conditions under which plastic yielding will occur under combined stresses can be made with the help of several empirical theories. In the **maximum-shear-stress theory** the criterion for yielding is that yielding will occur when

$$\sigma_1 - \sigma_3 = \sigma_{ys}$$

in which $\sigma_1$ and $\sigma_3$ are the largest and smallest principal stresses, respectively, and $\sigma_{ys}$ is the uniaxial tensile yield strength. This is the simplest theory for predicting yielding under combined stresses. A more accurate prediction can be

**Table 3. Elastic Constants of Metals**
(Mostly from tests of R. W. Vose)

| Metal | $E$ Modulus of elasticity (Young's modulus). 1,000,000 lb/in² | $G$ Modulus of rigidity (shearing modulus). 1,000,000 lb/in² | $K$ Bulk modulus. 1,000,000 lb/in² | $\mu$ Poisson's ratio |
|---|---|---|---|---|
| Cast steel | 28.5 | 11.3 | 20.2 | 0.265 |
| Cold-rolled steel | 29.5 | 11.5 | 23.1 | 0.287 |
| Stainless steel 18–8 | 27.6 | 10.6 | 23.6 | 0.305 |
| All other steels, including high-carbon, heat-treated | 28.6–30.0 | 11.0–11.9 | 22.6–24.0 | 0.283–0.292 |
| Cast iron | 13.5–21.0 | 5.2–8.2 | 8.4–15.5 | 0.211–0.299 |
| Malleable iron | 23.6 | 9.3 | 17.2 | 0.271 |
| Copper | 15.6 | 5.8 | 17.9 | 0.355 |
| Brass, 70–30 | 15.9 | 6.0 | 15.7 | 0.331 |
| Cast brass | 14.5 | 5.3 | 16.8 | 0.357 |
| Tobin bronze | 13.8 | 5.1 | 16.3 | 0.359 |
| Phosphor bronze | 15.9 | 5.9 | 17.8 | 0.350 |
| Aluminum alloys, various | 9.9–10.3 | 3.7–3.9 | 9.9–10.2 | 0.330–0.334 |
| Monel metal | 25.0 | 9.5 | 22.5 | 0.315 |
| Inconel | 31 | 11 | | |
| Z-nickel | 30 | 11 | | |
| Beryllium copper | 17 | 7 | | |
| Elektron (magnesium alloy) | 6.3 | 2.5 | 4.8 | 0.281 |
| Titanium (99.0 Ti), annealed bar | 15–16 | | | |
| Zirconium, crystal bar | 11–14 | | | |
| Molybdenum, arc-cast | 48–52 | | | |

made by the **distortion-energy theory,** according to which the criterion is

$$(\sigma_1 - \sigma_2)^2 + (\sigma_2 - \sigma_3)^2 + (\sigma_2 - \sigma_1)^2 = 2(\sigma_{ys})^2$$

Stress-strain curves in the plastic region for combined stress loading can be constructed. However, a particular stress state does not determine a unique strain value. The latter will depend on the stress-state path which is followed.

**Plane strain** is a condition where strain is confined to two dimensions. There is generally stress in the third direction, but because of mechanical constraints, strain in this dimension is prevented. Plane strain occurs in certain metalworking operations. It can also occur in the neighborhood of a crack tip in a tensile loaded member if the member is sufficiently thick. The material at the crack tip is then in triaxial tension, which condition promotes brittle fracture. On the other hand, ductility is enhanced and fracture is suppressed by triaxial compression.

**Stress Concentration** In a structure or machine part having a notch or any abrupt change in cross section, the maximum stress will occur at this location and will be greater than the stress calculated by elementary formulas based upon simplified assumptions as to the stress distribution. The ratio of this maximum stress to the nominal stress (calculated by the elementary formulas) is the stress-concentration factor $K_t$. This is a constant for the particular shape and is independent of the material, provided it is isotropic. The stress-concentration factor may be determined experimentally or, in many cases, theoretically from the mathematical theory of elasticity. The factors shown in Figs. 6 to 13 were determined from both

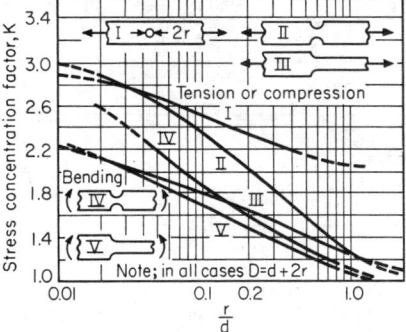

**Fig. 6** Flat plate with semicircular fillets and grooves or with holes, in tension or compression.

photoelastic tests and the theory of elasticity. Stress concentration will cause failure of brittle materials if the concentrated stress is larger than the ultimate strength of the material. In ductile materials, concentrated stresses higher than the yield strength will generally cause local plastic deformation and redistribution of stresses (rendering them more uniform). On the other hand, even with ductile materials areas of stress concentration are possible sites for fatigue if the component is cyclically loaded.

### FRACTURE AT LOW STRESSES

Materials under tension sometimes fail by rapid fracture at stresses much below their strength level as determined in tests

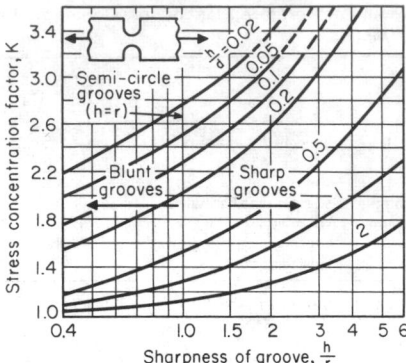

**Fig. 7** Flat plate with grooves, in tension.

on carefully prepared specimens. These **brittle, unstable,** or **catastrophic failures** originate at preexisting stress-concentrating flaws which may be inherent in a material.

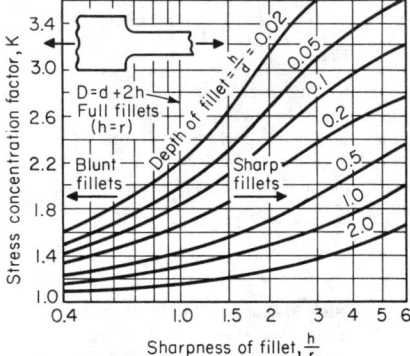

**Fig. 8** Flat plate with fillets, in tension.

The **transition-temperature approach** is often used to ensure fracture-safe design in structural-grade steels. These materials exhibit a characteristic temperature, known as the **ductile brittle transition** (DBT) temperature, below which they are susceptible to brittle fracture. The transition-temperature approach to fracture-safe design ensures that the transition temperature of a material selected for a particular application is suitably

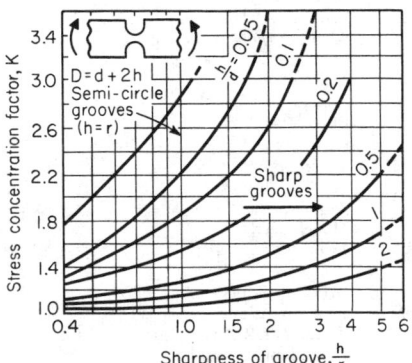

**Fig. 9** Flat plate with grooves, in bending.

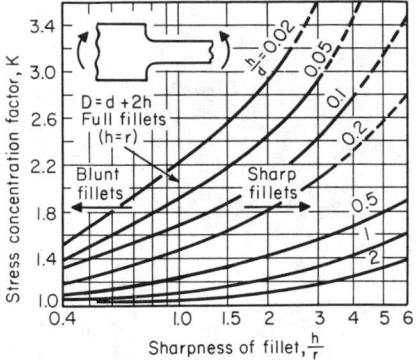

**Fig. 10** Flat plate with fillets, in bending.

matched to its intended use temperature. The DBT can be detected by plotting certain measurements from tensile or impact tests against temperature. Usually the transition to brittle behavior occurs over a transition-temperature range in which behavior is complex, being neither fully ductile nor fully brittle. The range may extend over a 200°F (110 K)

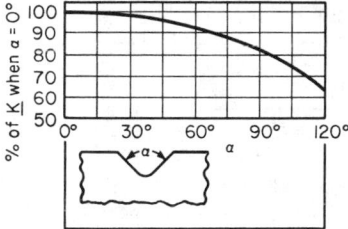

**Fig. 11** Flat plate with angular notch, in tension or bending.

interval. The **nil-ductility temperature** (NDT), determined by the **drop-weight test** (see ASTM F208), is an important reference point in the transition range. When NDT for a particular steel is known, temperature-stress combinations can be specified which define the limiting conditions under which catastrophic fracture can occur. (See W. S. Pellini and P. P. Puzak, *NRL Repts.* 5920 and 6030 1963.)

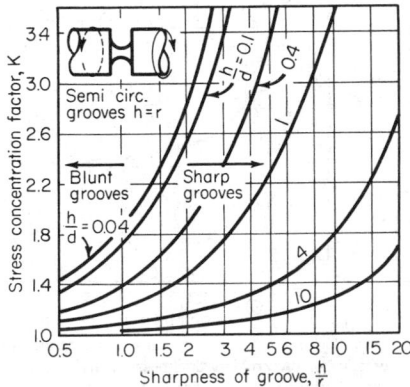

**Fig. 12** Stress-concentration factors for grooved shaft in torsion.

In the **Charpy V-notch** (CVN) impact test, a notched-bar specimen (Fig. 25) is used which is loaded in bending (see ASTM A370-71). The energy absorbed from a swinging pendulum in fracturing the specimen is measured. The pendulum strikes the specimen at 16 to 19 ft (4.88 to 5.80 m)/s so that the specimen deformation associated with fracture occurs at a rapid strain rate. This ensures a conservative measure of toughness, since in some materials, toughness is reduced by high strain rates. A CVN impact energy vs. temperature

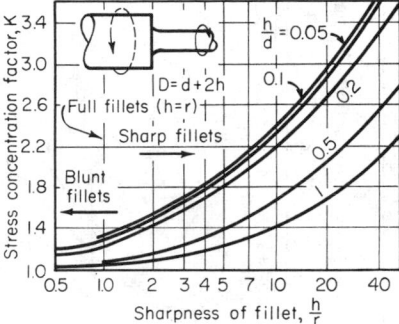

**Fig. 13** Stress-concentration factors for filleted shaft in torsion.

curve is shown in Fig. 14, which also shows the transitions as given by percent brittle fracture and by percent lateral expansion. The CVN energy has no analytical significance. The test is useful mainly as a guide to the fracture behavior of a material for which an empirical correlation has been established between impact energy and some rigorous fracture criterion. For a particular grade of steel the CVN curve can be correlated with NDT. (See ASME Boiler and Pressure Vessel Code, sec. III, NB-2300.)

**Fracture Mechanics** This analytical method is used for ultra-high-strength alloys, transition-temperature materials below the DBT temperature, and some low-strength materials in heavy section thickness.

Fracture-mechanics theory (see Irwin, Dimensional and Geometric Aspects of Fracture, in "Fracture of Engineering Materials," ASM, 1964; also STP-381, ASTM, 1965) deals with cracks of maximum acuity where crack-tip radius is of the order of interatomic dimensions. This ensures that the effect of a crack is conservatively evaluated, since minimum energy is used in initiating unstable propagation. The present discussion is concerned with a through-thickness crack in a tension-loaded plate (Fig. 15) which is large enough so that the crack-tip stress field is not affected by the plate edges. Fracture-mechanics theory states that unstable crack extension occurs when the work required for an increment of crack extension, namely, surface energy and energy consumed in local plastic deformation, is exceeded by the elastic-strain energy released at the crack tip. The elastic-stress field surrounding one of the crack tips in Fig. 15 is characterized by the **stress intensity** $K_I$, which has units of lb/in² $\sqrt{\text{in}}$ or MN $\sqrt{\text{m}}$/m². It is a function of applied nominal stress $\sigma$, crack half length $a$, and a geometry factor $Q$:

$$K_I^2 = Q\sigma^2\pi a \qquad (1)$$

for the situation of Fig. 15. For a particular material it is found

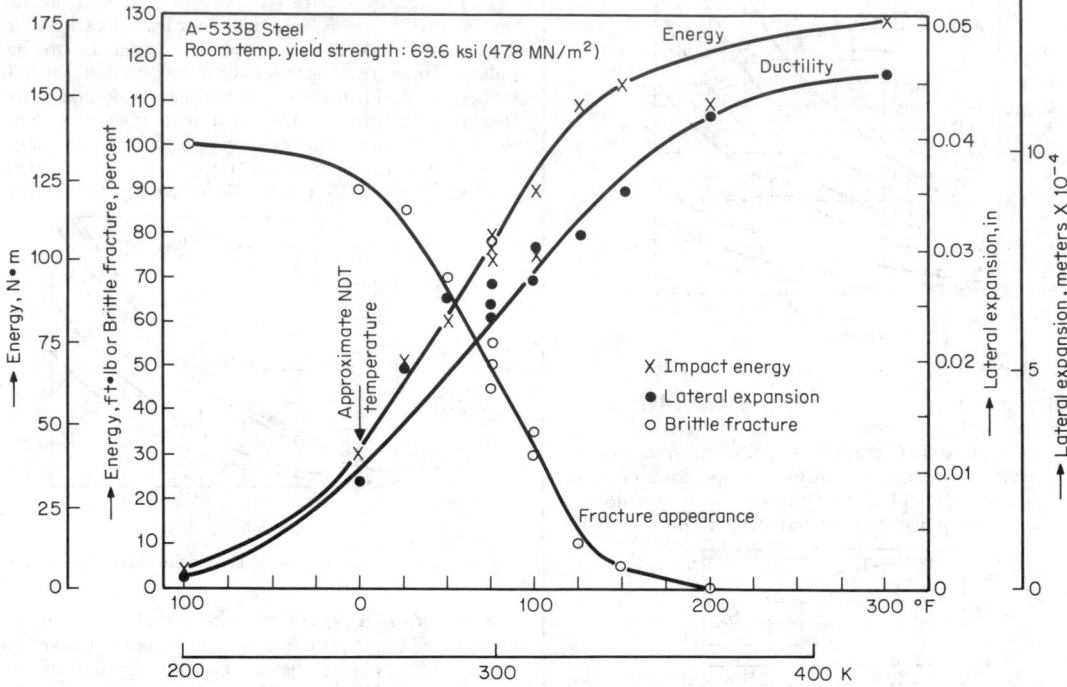

**Fig. 14** CVN transition curves. *(Data from Westinghouse R&D Lab.)*

that as $K_I$ is increased, a value $K_c$ is reached at which unstable crack propagation occurs. $K_c$ depends on plate thickness $B$, as

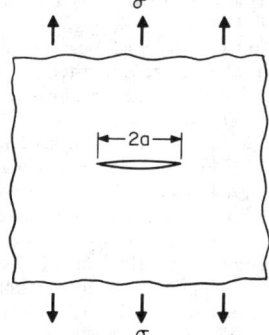

**Fig. 15** Through-thickness crack geometry.

shown in Fig. 16. It attains a constant value when $B$ is great enough to provide plane-strain conditions at the crack tip. The low plateau value of $K_c$ is an important material property known as the **plane-strain critical stress intensity** or **fracture toughness** $K_{Ic}$. Values for a number of materials are shown in Table 4. They are influenced strongly by processing and small changes in composition, so that the values shown are not necessarily typical. $K_{Ic}$ can be used in the critical form of Eq. (1):

$$(K_{Ic})^2 = Q\sigma^2\pi a_{cr} \qquad (2)$$

to predict failure stress when a maximum flaw size in the

material is known or to determine maximum allowable flaw size when the stress is set. The predictions will be accurate so long as plate thickness $B$ satisfies the **plane-strain criterion:** $B \geqslant (2.5)(K_{Ic}/\sigma_{ys})^2$. They will be conservative if a plane-strain condition does not exist. A big advantage of the fracture-mechanics approach is that stress intensity can be calculated by equations analogous to (1) for a wide variety of geometries,

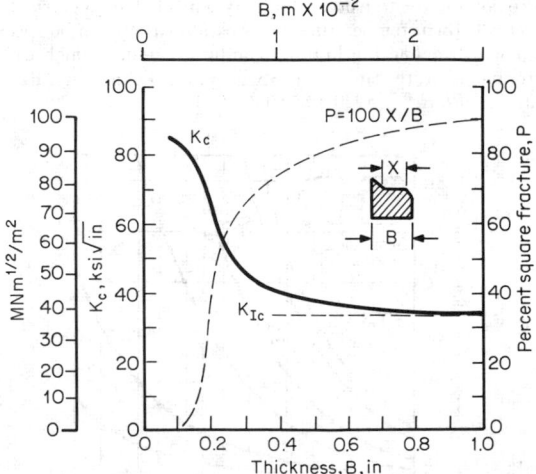

**Fig. 16** Dependence of $K_c$ and fracture appearance (in terms of percentage of square fracture) on thickness of plate specimens. Based on data for aluminum 7075-T6. *(From Srawley and Brown, STP-381, ASTM.)*

**Table 4. Room-Temperature $K_{Ic}$ Values on High-Strength Materials***

| Material | 0.2% YS, 1,000 in² (MN/m²) | $K_{Ic}$, 1,000 in² $\sqrt{\text{in}}$ (MN m$^{1/2}$/m²) |
|---|---|---|
| 18% Ni maraging steel | 300 (2,060) | 46 (50.7) |
| 18% Ni maraging steel | 270 (1,850) | 71 (78) |
| 18% Ni marging steel | 198 (1,360) | 87 (96) |
| Titanium 6-4 alloy | 152 (1,022) | 39 (43) |
| Titanium 6-4 alloy | 140 (960) | 75 (82.5) |
| Aluminum alloy 7075-T6 | 75 (516) | 26 (28.6) |
| Aluminum alloy 7075-T6 | 64 (440) | 30 (33) |

*Determined at Westinghouse Research Laboratories.

types of crack, and loadings (P. C. Paris and G. C. Sih, "Stress Analysis of Cracks," STP-381, ASTM, 1965). Failure occurs in all cases when $K_I$ reaches $K_{Ic}$. Fracture mechanics also provides a framework for predicting the occurrence of **stress-corrosion cracking** by using Eq. **(2)** with $K_{Ic}$ replaced by $K_{Iscc}$, which is the material parameter denoting resistance to stress-corrosion-crack propagation in a particular medium (see B. F. Brown, The Application of Fracture Mechanics to Stress-Corrosion Cracking, *Metall. Rev.*, **13**, 1968, p. 171).

Two standard test specimens for $K_{Ic}$ determination are specified in ASTM E399, which also specifies details of specimen preparation and test procedure.

## FATIGUE

Fatigue is generally understood as the gradual deterioration of a material which is subjected to repeated loads. In fatigue testing, a specimen is subjected to periodically varying constant-amplitude stresses by means of mechanical or magnetic devices. The applied stresses may alternate between equal positive and negative values, from zero to maximum positive or negative values, or between unequal positive and negative values. The most common loading is alternate tension and compression of equal numerical values obtained by rotating a smooth cylindrical specimen while under a bending load. A series of fatigue tests are made on a number of specimens of the material at different stress levels. The stress endured is then plotted against the number of cycles sustained. By choosing lower and lower stresses, a value may be found which will not produce failure, regardless of the number of applied

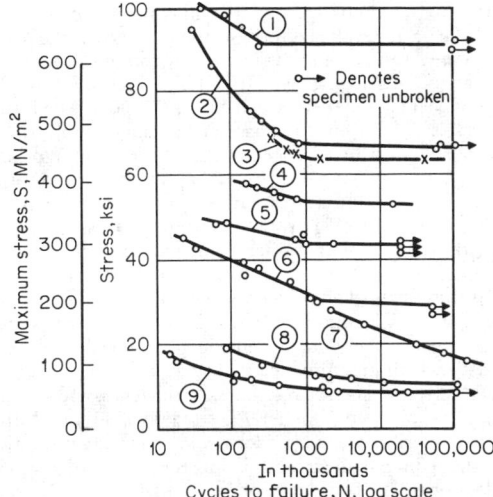

**Fig. 17**  *S-N* diagrams from fatigue tests: (1) 1.20 C steel, quenched, drawn 860°F(733 K); (2) SAE 3420, quenched, drawn 1200°F(923 K); (3) alloy structural steel; (4) SAE 1050, quenched, drawn 1200°F; (5) SAE 4130, normalized, annealed; (6) ordinary structural steel; (7) Duralumin; (8) copper, annealed; (9) cast iron. (Reversed bending.)

cycles. This stress value is called the **fatigue limit.** The diagram is called the stress-cycle diagram or *S-N diagram.* Instead of recording the data on cartesian coordinates, either stress is plotted vs. the logarithm of the number of cycles (Fig. 17) or both stress and cycles are plotted to logarithmic scales. Both diagrams show a relatively sharp bend in the curve near the fatigue limit for ferrous metals. The fatigue limit may be established for most steels between 2 and 10 million cycles. Nonferrous metals usually show no clearly defined fatigue limit. The *S-N* curves in these cases indicate a continuous decrease in stress values to several hundred million cycles, and both the stress value and the number of cycles sustained should be reported.

The mean stress (the average of the maximum and minimum stress values for a cycle) has a pronounced influence on the stress range (the algebraic difference between the maximum and minimum stress values). Several empirical formulas

**Table 5. Typical Approximate Fatigue Limits for Reversed Bending**

| Metal | Tensile strength, 1,000 lb/in² | Fatigue limit, 1,000 lb/in² | Metal | Tensile strength, 1,000 lb/in² | Fatigue limit, 1,000 lb/in² |
|---|---|---|---|---|---|
| Cast iron | 20–50 | 6–18 | Copper | 32–50 | 12–17 |
| Malleable iron | 50 | 24 | Monel | 70–120 | 20–50 |
| Cast steel | 60–80 | 24–32 | Phosphor bronze | 55 | 12 |
| Armco iron | 44 | 24 | Tobin bronze, hard | 65 | 21 |
| Plain carbon steels | 60–150 | 25–75 | Cast aluminum alloys | 18–40 | 6–11 |
| SAE 6150, heat-treated | 200 | 80 | Wrought aluminum alloys | 25–70 | 8–18 |
| Nitralloy | 125 | 80 | Magnesium alloys | 20–45 | 7–17 |
| Brasses, various | 25–75 | 7–20 | Molybdenum, as cast | 98 | 45 |
| Zirconium crystal bar | 52 | 16–18 | Titanium (Ti-75A) | 91 | 45 |

Stress, 1,000 lb/in² × 6.894 = stress, MN/m².

and graphical methods such as the "modified Goodman diagram" have been developed to show the influence of the mean stress on the stress range for failure (see Moore and Kommers, "The Fatigue of Metals," McGraw-Hill; "Prevention of the Failure of Metals under Repeated Stress," Wiley). Much further work is needed on this phase of the fatigue problem. A simple but conservative procedure (see Soderberg, Working Stresses, *Jour. Appl. Mech.*, **2**, Sept. 1935) is to plot the variable stress $S_v$ (one-half the stress range) as ordinate vs. the mean stress $S_m$ as abscissa (Fig. 18). At zero mean stress, the ordinate is the fatigue limit under completely reversed stress. Yielding will occur if the mean stress exceeds the yield stress $S_o$, and this establishes the extreme right-hand point of the diagram. A straight line is drawn between these two points. The coordinates of any other point along this line are values of $S_m$ and $S_v$ which may produce failure.

Surface defects, such as roughness or scratches, and notches or shoulders, all reduce the fatigue strength of a part. With a notch of prescribed geometric form and known concentration factor, the reduction in strength is appreciably less than would be called for by the concentration factor itself, but the various metals differ widely in their susceptibility to the effect of roughness and concentrations, or notch sensitivity. Further, notch sensitivity seems to be higher, and ordinary fatigue strength lower in large specimens, necessitating full-scale tests in many cases (see Peterson, Stress Concentration Phenomena in Fatigue of Metals, *Trans. ASME*, **55**, 1933, p. 157, and Buckwalter and Horger, Investigation of Fatigue Strength of Axles, Press Fits, Surface Rolling and Effect of Size, *Trans. ASM*, **25**, Mar. 1937, p. 229). **Corrosion** and **galling** (due to rubbing of mating surfaces) cause great reduction of fatigue strengths, sometimes amounting to as much as 90 percent of the original endurance limit (see Gough, Contact Corrosion under Pressure, *Natl. Phys. Lab. Rept., Dept. Ind. Sci. Res.*, 1935, p. 150). Although any corroding agent is productive of severe corrosion fatigue, there is so much difference between the effects of "sea water" or "tap water" from different localities that numerical values are not quoted here.

**Overstressing** specimens above the fatigue limit for periods shorter than necessary to produce failure at that stress reduces the fatigue limit in a subsequent test. Similarly, **understressing** below the fatigue limit may increase it. Shot peening, nitriding, and cold work usually improve fatigue properties.

No very good overall correlation exists between fatigue properties and any other mechanical property of a material. The best correlation is between the fatigue limit under completely reversed bending stress and the ordinary tensile strength. For many ferrous metals, the fatigue limit is approximately 0.40 to 0.60 times the tensile strength if the latter is below 200,000 lb/in². Low-alloy high-yield-strength steels often show higher values than this. The fatigue limit for nonferrous metals is approximately 0.20 to 0.50 times the tensile strength. The fatigue limit in reversed shear is approximately 0.57 times that in reversed bending.

In some very important engineering situations components are cyclically stressed into the plastic range. Examples are thermal strains resulting from temperature oscillations and notched regions subjected to secondary stresses. Fatigue life in the plastic or **"low-cycle" fatigue** range has been found to be a function of plastic strain, and low-cycle fatigue testing is done with strain as the controlled variable rather than stress.

Fatigue life $N$ and cyclic plastic strain $\epsilon_p$ tend to follow the relationship

$$N\epsilon_p^2 = C$$

where $C$ is a constant for a material when $N < 10^5$. (See Coffin, A Study of Cyclic-Thermal Stresses in a Ductile Material, *Trans. ASME*, **76**, 1954, p. 947.)

The type of physical change occurring inside a material as it is repeatedly loaded to failure varies as the life is consumed, and a number of stages in fatigue can be distinguished on this basis. The early stages comprise the events causing nucleation of a crack or flaw. This is most likely to appear on the surface of the material: fatigue failures generally originate at a surface. Following nucleation of the crack, it grows during the crack-propagation stage. Eventually the crack becomes large enough for some rapid terminal mode of failure to take over such as ductile rupture or brittle fracture. The rate of crack growth in the crack-propagation stage can be accurately quantified by fracture-mechanics methods. Assuming an initial flaw and a loading situation as shown in Fig. 15, the rate of crack growth per cycle can generally be expressed as

$$da/dN = C_0(\Delta K_1)^n \tag{3}$$

where $C_0$ and $n$ are constants for a particular material and $\Delta K_1$ is the range of stress intensity per cycle. $K_I$ is given by (1). (See Paris, "The Fracture Mechanics Approach to Fatigue—An Interdisciplinary Approach," pp. 107–132, Syracuse University Press, 1963.) Using (3), it is possible to predict the number of cycles for the crack to grow to a size at which some other mode of failure can take over. Values of the constants $C_0$ and $n$ are determined from specimens of the same type as those used for determination of $K_{Ic}$ but instrumented for accurate measurement of slow crack growth.

Constant-amplitude fatigue-test data are relevant to many rotary-machinery situations where constant cyclic loads are encountered. There are important situations where the component undergoes variable loads and where it may be advisable to use **random-load testing**. In this method, the load spectrum which the component will experience in service is determined and is applied to the test specimen artificially. (See Swanson, Random Load Fatigue Testing, *Mater. Res. Stand.*, Apr. 1968).

## CREEP

Experience has shown that, for the design of equipment subjected to sustained loading at elevated temperatures, little reliance can be placed on the usual short-time tensile properties of metals at those temperatures. Under the application of a constant load it has been found that materials, both metallic and nonmetallic, show a gradual flow or **creep** even for stresses below the proportional limit at elevated temperatures (see Kanter and Spring, Long-time or Flow Tests on Carbon Steels at Various Temperatures with Particular Reference to Stresses below the Proportional Limit, *Proc. ASTM*, **28**, Pt. II, 1928, p. 80). Similar effects are present in low-melting metals such as lead at room temperature. The deformation which can be permitted in the satisfactory operation of most high-temperature equipment is limited.

In metals, creep is a plastic deformation caused by slip occurring along crystallographic directions in the individual crystals, together with some flow of the grain-boundary material. After complete release of load, a small fraction of this plastic deformation is recovered with time. Most of the flow is nonrecoverable for metals.

Since the early creep experiments (see Andrade, On the Viscous Flow in Metals and Allied Phenomena, *Proc. Roy. Soc.*, 1910) many different types of tests have come into use. The most common are the **long-time creep test** under constant tensile load and the **stress-rupture** test. Other special forms are the **stress-relaxation test** and the **constant-strain-rate test.**

The **long-time creep test** is conducted by applying a dead weight to one end of a lever system, the other end being attached to the specimen surrounded by a furnace and held at constant temperature. The axial deformation is read periodically throughout the test and a curve is plotted of the strain $\epsilon_0$ as a function of time $t$ (Fig. 19). This is repeated for various loads at the same testing temperature. The portion of the curve $OA$ in Fig. 19 is the region of **primary creep**, $AB$ the region of **secondary creep**, and $BC$ that of **tertiary creep.** The strain rates, or the slopes of the curve, are decreasing, constant, and increasing, respectively, in these three regions. Since the period of the creep test is usually much shorter than the duration of the part in service, various extrapolation procedures are followed (see McVetty, Working Stresses for High Temperature Service, *Mech. Eng.*, **56**, no. 3, Mar. 1939, p.

149, and Creep of Metals at Elevated Temperatures; the Hyperbolic-Sine Relation between Stress and Creep Rate, *Trans. ASME*, **65**, 1943, pp. 761–767).

In practical applications the region of constant-strain rate (secondary creep) is often used to estimate the probable deformation throughout the life of the part. It is thus assumed that

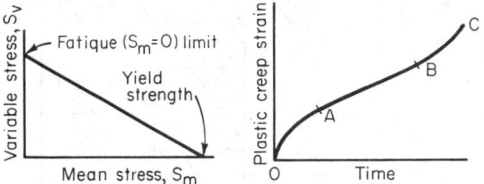

**Fig. 18** Effect of mean stress on the variable stress for failure.  **Fig. 19** Typical creep curve.

this rate will remain constant during periods beyond the range of the test data. The working stress is chosen so that this total deformation will not be excessive. An **arbitrary creep strength,** which is defined as the stress which at a given temperature will result in 1 percent deformation in 100,000 h, has received a certain amount of recognition, but it is advisable to determine

**Table 6. Stresses for Given Creep Rates and Temperatures***

| Material Temp, °F | Creep rate 0.1 % per 1,000 h | | | | | Creep rate 0.01 % per 1,000 h | | | | |
|---|---|---|---|---|---|---|---|---|---|---|
| | 800 | 900 | 1000 | 1100 | 1200 | 800 | 900 | 1000 | 1100 | 1200 |
| Wrought steels: | | | | | | | | | | |
| SAE 1015 | 17–27 | 11–18 | 3–12 | 2–7 | 1 | 10–18 | 6–14 | 3–8 | 1 | |
| 0.20 C, 0 50 Mo | 26–33 | 18–25 | 9–16 | 2–6 | 1–2 | 16–24 | 11–22 | 4–12 | 2 | 1 |
| 0.10–0.25 C, 4–6 Cr + Mo | 22 | 15–18 | 9–11 | 3–6 | 2–3 | 14–17 | 11–15 | 4–7 | 2–3 | 1–2 |
| SAE 4140 | 27–33 | 20–25 | 7–15 | 4–7 | 1–2 | 19–28 | 12–19 | 3–8 | 2–4 | 1 |
| SAE 1030–1045 | 8–25 | 5–15 | 5 | 2 | 1 | 5–15 | 3–7 | 2–4 | 1 | |
| Commercially pure iron | 7 | | 4 | | 3 | 5 | | 2 | | |
| 0.15 C, 1–2.5 Cr, 0.50 Mo | 25–35 | 18–28 | 8–20 | 6 8 | 3–4 | 20–30 | 12–18 | 3–12 | 2–5 | 1–2 |
| SAE 4340 | 20–40 | 15–30 | 2–12 | 1–3 | | 8–20 | | 1–6 | | |
| SAE X3140 | 7–10 | | 5–4 | | | 3–8 | | 1–2 | | |
| 0.20 C, 4–6 Cr | 30 | 10–20 | 7–10 | 1 | | | | 3–5 | | |
| 0.25 C, 4–6 Cr + W | 30 | 10–15 | 4–10 | 2–8 | | | 6–11 | 2–7 | | |
| 0.16 C, 1.2 Cu | | 18 | 10–15 | 3 | 1 | | 10–18 | 7–12 | | |
| 0.20 C, 1 Mo | 35 | 27 | 12 | | | 25 | 12 | 6 | | |
| 0.10–0.40 C, 0.2–0.5 Mo, 1–2 Mn | 30–40 | 12–20 | 4–14 | | | 25–28 | 8–15 | 2–8 | | 0.5 |
| SAE 2340 | 7–12 | 5 | 2 | | | | | | | |
| SAE 6140 | 30 | 12 | 4 | | | 7 | 6 | 1 | | |
| SAE 7240 | 30 | 21 | 6–15 | 2 | | 30 | 11 | 3–9 | 1 | |
| Cr + Va + W, various | 20–70 | 14–30 | 5–15 | | | 18–50 | 8–18 | 2–13 | | |

| Temp, °F | 1100 | 1200 | 1300 | 1400 | 1500 | 1000 | 1100 | 1200 | 1300 | 1400 |
|---|---|---|---|---|---|---|---|---|---|---|
| Wrought chrome-nickel steels: | | | | | | | | | | |
| 18–8† | 10–18 | 5–11 | 3–10 | 2–5 | 2.5 | 11–16 | 5–12 | 2–10 | | 1–2 |
| 10–25 Cr, 10–30 Ni‡ | 10–20 | 5–15 | 3–10 | 2–5 | | | 6–15 | 3–10 | 2–8 | 1–3 |

| Temp, °F | 800 | 900 | 1000 | 1100 | 1200 | 800 | 900 | 1000 | 1100 | 1200 |
|---|---|---|---|---|---|---|---|---|---|---|
| Cast steels: | | | | | | | | | | |
| 0.20–0.40 C | 10–20 | 5–10 | 3 | | | 8–15 | | 1 | | |
| 0.10–0.30 C, 0.5–1 Mo | 28 | 20–30 | 6–12 | 2 | | 20 | 10–15 | 2–5 | | |
| 0.15–0.30 C, 4–6 Cr + Mo | 25–30 | 15–25 | 8–15 | 8 | | 20–25 | 9–15 | 2–7 | 2 | |
| 18–8§ | | | 20–25 | 15 | 10 | | | 20 | 15 | 8 |
| Cast iron | 20 | 8 | 4 | | | 10 | | 2 | | |
| Cr Ni cast iron | | | 9 | | | | | 3 | | |

*Based on 1,000-h tests. Stresses in 1,000 lb/in².
†Additional data. At creep rate 0.1 percent and 1000 (1600°)F the stress is 18–25 (1); at creep rate 0.01 percent at 1500°F, the stress is 0.5.
‡Additional data. At creep rate 0.1 percent and 1000 (1600°)F the stress is 10–30 (1).
§Additional data. At creep rate 0.1 percent and 1600°F the stress is 3; at creep rate 0.01 and 1500°F, the stress is 2–3.

the proper stress for each individual case from diagrams of stress vs. creep rate (Fig. 20) (see "Creep Data," ASTM and ASME, 1938).

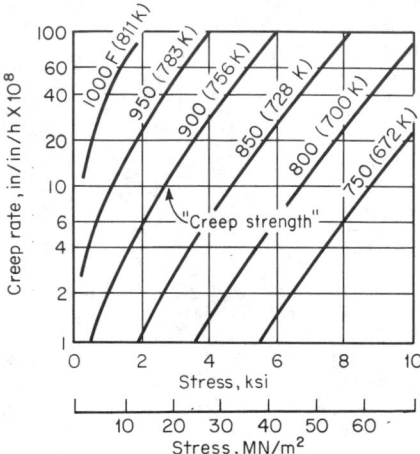

**Fig. 20** Creep rates for 0.35 carbon steel.

Additional temperatures (°F) and stresses (in 1,000 lb/in²) for stated creep rates (percent per 1,000 h) for wrought nonferrous metals are as follows:

60–40 BRASS. Rate 0.1, temp. 350 (400), stress 8 (2); rate 0.01, temp 300 (350) [400], stress 10 (3) [1].

PHOSPHOR BRONZE. Rate 0.1, temp 400 (550) [700] [800], stress 15 (6) [4] [4]; rate 0.01, temp 400 (550) [700], stress 8 (4) [2].

NICKEL. Rate 0.1, temp 800 (1000), stress 20 (10).

70 CU, 30 NI. Rate 0.1, temp 600 (750), stress 28 (13–18); rate 0.01, temp 600 (750), stress 14 (8–9).

ALUMINUM ALLOY 17S *(Duralumin)*. Rate 0.1, temp 300 (500) [600], stress 22 (5) [1.5].

LEAD, PURE (Commercial) (0.03 percent Ca). At 110°F, for rate 0.1 percent the stress range, lb/in², is 150–180 (60–140) [200–220]; for rate of 0.01 percent, 50–90 (10–50) [110–150].

Stress, 1,000 lb/in² × 6.894 = stress, MN/m², $t_k = (\frac{5}{9})(t_F + 459.67)$.

Structural changes may occur during a creep test, thus altering the metallurgical condition of the metal. In some cases, premature rupture appears at a low fracture strain in a normally ductile metal, indicating that the material has become embrittled. This is a very insidious condition and difficult to predict. The **stress-rupture test** (see Thielemann and Parker, Fracture of Steels at Elevated Temperatures after Prolonged Loading, *Metals Technol.*, Apr. 1939) is well adapted to study this effect. It is conducted by applying a constant load to the specimen in the same manner as for the long-time creep test. The nominal stress is then plotted vs. the time for fracture at constant temperature on a log-log scale (Fig. 21).

The stress reaction is measured in the **constant-strain-rate test** while the specimen is deformed at a constant strain rate. In the **relaxation test,** the decrease of stress with time is measured while the total strain (elastic + plastic) is maintained constant. The latter test has direct application to the loosening of turbine bolts and to similar problems. Although some correlation has been indicated between the results of these various types of tests, no general correlation is yet available, and it has been found necessary to make tests under each of these special conditions to obtain satisfactory results.

The interrelationship between strain rate and temperature in the form of a velocity-modified temperature (see MacGregor and Fisher, A Velocity-modified Temperature for the Plastic Flow of Metals, *Jour. Appl. Mech.*, Mar. 1945) simplifies the creep problem in reducing the number of variables.

## HARDNESS

Hardness has been variously defined as resistance to local penetration, to scratching, to machining, to wear or abrasion, and to yielding. The multiplicity of definitions, and corresponding multiplicity of hardness-measuring instruments, together with the lack of a fundamental definition, indicates that hardness may not be a fundamental property of a material but rather a composite one including yield strength, work hardening, true tensile strength, modulus of elasticity, and others.

**Scratch hardness** is measured by **Mohs scale** of minerals (Sec. 1) which is so arranged that each mineral will scratch the mineral of the next lower number. In recent mineralogical work and in certain microscopic metallurgical work, jeweled scratching points either with a set load or else loaded to give a

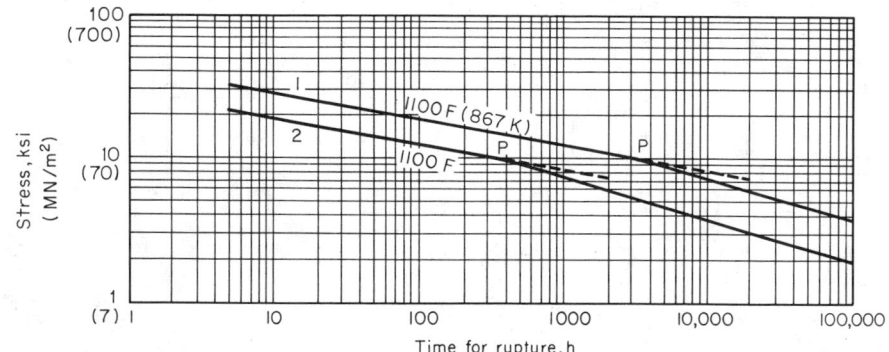

**Fig. 21** Relation between time for failure and stress for a 3 percent chromium steel. (1) Heat-treated 2 h at 1740°F (1223 K) and furnace-cooled; (2) hot-rolled and annealed at 1580°F (1134 K).

set width of scratch have been used (see Talmadge, Quantitative Standards for Hardness of the Ore Minerals, *Econ. Geol.*, **20**, 1925, p. 531; and Bierbaum, The Microcharacter, *Trans. ASST*, **18**, 1930, p. 1009). Hardness in its relation to machinability and to wear and abrasion is generally dealt with in direct machining or wear tests, and little attempt is made to separate hardness itself, as a numerically expressed quantity, from the results of such tests.

The resistance to localized penetration, or **indentation hardness,** is widely used industrially as a measure of hardness, and indirectly as an indicator of other desired properties in a manufactured product. The indentation tests described below are essentially nondestructive, and in most applications may be considered nonmarring, so that they may be applied to each piece produced; and through the empirical relationships of hardness to such properties as tensile strength, fatigue strength, and impact strength, pieces likely to be deficient in the latter properties may be detected and rejected.

**Brinell hardness** is determined by forcing a hardened sphere under a known load into the surface of a material and measuring the diameter of the indentation left after the test. The **Brinell hardness number,** or simply the **Brinell number,** is obtained by dividing the load used, in kilograms, by the actual surface area of the indentation, in square millimeters. The result is a pressure, but the units are rarely stated.

$$BHN = P \bigg/ \frac{\pi D}{2}(D - \sqrt{D^2 - d^2})$$

where *BHN* is the Brinell hardness number; *P* the imposed load, kg; *D* the diameter of the spherical indenter, mm; and *d* the diameter of the resulting impression, mm.

Hardened-steel bearing balls may be used for hardness up to 450, but beyond this hardness specially treated steel balls or jewels should be used to avoid flattening the indenter. The standard-size ball is 10 mm and the standard loads 3,000, 1,500, and 500 kg, with 100, 125, and 250 kg sometimes used for softer materials. If for special reasons any other size of ball is used, the load should be adjusted approximately as follows: for iron and steel, $P = 30D^2$; for brass, bronze, and other soft metals, $P = 5D^2$; for extremely soft metals, $P = D^2$ (see "Methods of Brinell Hardness Testing," ASTM E10-61T). Readings obtained with other than the standard ball and loadings should have the load and ball size appended, as such readings are only approximately equal to those obtained under standard conditions.

The size of the specimen should be sufficient to ensure that no part of the plastic flow around the impression reaches a free surface, and in no case should the thickness be less than 10 times the depth of the impression. The load should be applied steadily and should remain on for at least 15 s in the case of ferrous materials and 30 s in the case of most nonferrous materials. Longer periods may be necessary on certain soft materials that exhibit creep at room temperature. In testing thin materials, it is not permissible to pile up several thicknesses of material under the indenter, as the readings so obtained will invariably be lower than the true readings. With such materials, smaller indenters and loads, or different methods of hardness testing, are necessary.

In the standard Brinell test, the diameter of the impression is measured with a low-power hand microscope, but for production work several testing machines are available which automatically measure the depth of the impression and from

this give readings of hardness. Such machines should be calibrated frequently on test blocks of known hardness.

In the **Rockwell method** of hardness testing, the depth of penetration of an indenter under certain arbitrary conditions of test is determined. The indenter may be either a steel ball of some specified diameter or a spherical-tipped conical diamond of 120° angle and 0.2 mm tip radius, called a "Brale." A *minor load* of 10 kg is first applied which causes an initial penetration and holds the indenter in place. Under this condition, the dial is set to zero and the major load applied. The values of the latter are 60, 100, or 150 kg. Upon removal of the major load, the reading is taken while the minor load is still on. The hardness number may then be read directly from the scale which measures penetration, and this scale is so arranged that soft materials with deep penetration give low hardness numbers.

A variety of combinations of indenter and major load are possible; the most commonly used are $R_B$ using as indenter a $\frac{1}{16}$-in ball and a major load of 100 kg and $R_C$ using a Brale as indenter and a major load of 150 kg (see "Rockwell Hardness and Rockwell Superficial Hardness of Metallic Materials," ASTM E18-61).

As compared with the Brinell test, the Rockwell method makes a smaller indentation, may be used on thinner material, and is much more rapid, since hardness numbers are read directly and need not be calculated. However, the Brinell test may be made without special apparatus and is somewhat more widely recognized for laboratory use. There is also a **Rockwell superficial hardness test** similar to the regular Rockwell, except that the indentation is much shallower.

The **Vickers** method of hardness testing is similar in principle to the Brinell in that it expresses the result in terms of the pressure under the indenter and uses the same units, kilograms per square millimeter. The indenter is a diamond in the form of a square pyramid with an apical angle of 136°, the loads are much lighter, varying between 1 and 120 kg, and the impression is measured by means of a medium-power compound microscope.

$$V = P/0.5393d^2$$

where *V* is the Vickers hardness number, sometimes called the **diamond-pyramid hardness** *(DPH); P* the imposed load, kg; and *d* the diagonal of indentation, mm. The Vickers method is more flexible and is considered to be more accurate than either the Brinell or the Rockwell, but the equipment is more expensive than either of the others and the Rockwell is somewhat faster in production work (ASTM E92-57).

Among the other hardness methods may be mentioned the **Scleroscope,** in which a diamond-tipped "hammer" is dropped on the surface and the rebound taken as an index of hardness. This type of apparatus is seriously affected by the resilience as well as the hardness of the material and has largely been superseded by other methods. In the **Monotron** method, a penetrator is forced into the material to a predetermined depth and the load required is taken as the indirect measure of the hardness. This is the reverse of the Rockwell method in principle, but the loads and indentations are smaller than those of the latter. In the **Herbert pendulum,** a 1-mm steel or jewel ball resting on the surface to be tested acts as the fulcrum for a 4-kg compound pendulum of 10 s period. The swinging of the pendulum causes a rolling indentation in the material, and from the behavior of the pendulum several factors in hardness, such as **work hardenability,** may be determined which are not revealed by other methods. Although the Herbert results are of considerable significance, the instrument is suitable for laboratory use only (see Herbert, The Pendulum Hardness Tester, and Some Recent Developments in Hardness Testing, *Engineer*, **135**, 1923, pp. 390, 686). In the **Herbert cloudburst** test, a shower of steel balls, dropped from a predetermined height, dulls the surface of a hardened part in proportion to its softness and thus reveals defective areas. A variety of **mutual indentation methods** (Cowdrey, Hardness by Mutual Indentation, *Proc. ASTM*, **30**, Pt. II, 1930, p. 559), in which crossed cylinders or prisms of the material to be tested are forced together, give results comparable with the Brinell test. These are particularly useful on wires and on materials at high temperatures.

The relation among the scales of the various hardness methods is not exact, since no two measure exactly the same sort of hardness, and a relationship determined on steels of different hardnesses will be found only approximately true with other materials. The **Vickers-Brinell relation** is nearly linear up to at least 400, with the Vickers approximately 5 percent higher than the Brinell (actual values run from +2 to +11 percent) and nearly independent of the material. Beyond 500, the values become more widely divergent owing to the flattening of the Brinell ball. The **Brinell-Rockwell relation** is fairly satisfactory and is shown in Fig. 22. Approximate relations for the **Shore Scleroscope** are also given on the same plot.

The **hardness of wood** is defined by the ASTM as the load in pounds required to force a ball 0.444 in in diameter into the wood to a depth of 0.222 in, the speed of penetration being ¼ in/min. For a summary of the work in hardness see Williams, "Hardness and Hardness Measurements," ASM, 1942.

## TESTING OF MATERIALS

**Testing Machines**  Machines for the mechanical testing of materials usually contain elements (1) for gripping the specimen, (2) for deforming it, and (3) for measuring the load required in performing the deformation. Some machines (ductility testers) omit the measurement of load and substitute a measurement of deformation, whereas other machines include the measurement of both load and deformation through apparatus either integral with the testing machine (stress-strain recorders) or auxiliary to it (strain gages). In most general-purpose testing machines, the deformation is controlled as the independent variable and the resulting load measured, and in many special-purpose machines, particularly those for light loads, the load is controlled and the resulting deformation is measured. Special features may include those for constant rate of loading (pacing disks), for constant rate of straining, for constant load maintenance, and for cyclical load variation (fatigue).

A nut-and-screw combination, driven through clutches and change gears, was formerly common for **producing deformation,** but this arrangement has been largely superseded by a pump and hydraulic cylinder. The hydraulic arrangement is quieter, more flexible in control, and more durable, but cannot maintain constant deformation for any length of time on account of leakage. In **weighing light loads,** direct weights are accurate, but unless lead shot or water is used small variations in load are inconvenient. Ordinary springs used for this purpose are not sufficiently accurate, but the "Iso-Elastic" spring (John Chatillon & Sons, New York) has negligible errors. For larger loads, a reducing device with a **hydraulic weighing system** has largely displaced the lever system. Accurately lapped pistons may be made nearly frictionless; in some systems, an oscillatory rotation imposed on either the piston or the cylinder is used to remove any remaining friction. In the Emery hydraulic support, the load is taken by liquid pressure and is applied through a piston of considerable clearance sealed to the cylinder wall by a flexible flat metal diaphragm; although only minute displacement is possible, this system possesses great accuracy and sensitivity. For weighing the reduced load of a lever system, there may be used dead weights (lead shot), gravity pendulums, scale-and-rider combinations, or Iso-Elastic springs. In hydraulic systems, the liquid pressure is occasionally balanced by hydrostatic head, but may be transferred to a mechanical force through small pistons or Emery supports and balanced as for lever systems. Specially accurate pressure gages may be used (Emery-Tatnall system, Baldwin-Southwark Corp., Philadelphia); in the Tate-Emery system (Baldwin-Southwark), the pressure gage element is used as a null indicator, being balanced by Iso-Elastic springs, and gives particularly accurate and sensitive readings. Many systems have included means of **recording automatically** the load and deformation of the test piece, but few have been satisfactory until recent years (Emery-Tatnall and Tate-Emery). (For further information, see Gibbons, "Materials Testing Machines," Instruments Publishing Co., also in *Instruments*, **7, 8**, 1934–1935, and *Bull. ASTM*, Oct. 1939.)

**Grips** should not only hold the test specimen against slippage but should also apply the load in the desired manner. Centering of the load is of great importance in compression testing, and should not be neglected in tension testing if the material is brittle. Figure 23 shows the theoretical errors due to off-center loading; the results are directly applicable to compression tests using swivel loading blocks. Swivel (ball-and-socket) holders or compression blocks should be used with all except the most ductile materials, and in compression testing of brittle materials (concrete, stone, brick), any rough faces should be smoothly capped with plaster of paris or, for greater strength, a mixture of two-thirds plaster of paris and one-third portland cement. Serrated grips may be used to hold ductile materials or the shanks of other holders in tension; a taper of 1 in 6 on the wedge faces gives a self-tightening action without excessive jamming. Ropes are ordinarily held by wet

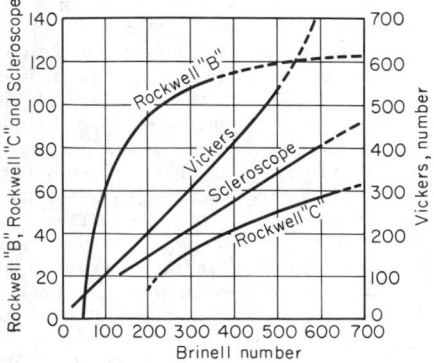

**Fig. 22**  Hardness scales.

eye splices, but braided ropes or small cords may be given several turns over a fixed pin and then clamped. Wire ropes should be zincked into forged sockets (solder and lead have insufficient strength).

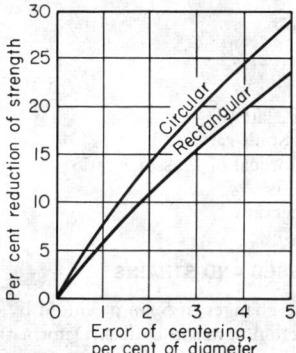

**Fig. 23**  Effect of centering errors on brittle test specimens.

**Accuracy and Calibration**  ASTM E4-61T requires that commercial machines shall have errors of less than 1 percent within the "loading range" when checked against acceptable standards of comparison at at least five suitably spaced loads. The "loading range" may be any range through which the preceding requirements for accuracy are satisfied, except that it shall not extend below 100 times the least load to which the machine will respond or which can be read on the indicator. The use of calibration plots or tables to correct the results of an otherwise inaccurate machine is not permitted under any circumstances. Machines with errors less than 0.1 percent are commercially available (Tate-Emery, and others), and somewhat greater accuracy is possible in the most refined research apparatus.

Dead loads may be used to check machines of low capacity; accurately calibrated proving levers may be used to extend the range of available weights. Various elastic devices (such as the Morehouse proving ring) made of specially treated steel, with sensitive distortion-measuring devices, and calibrated by dead weights at the Bureau of Standards are among the most satisfactory means of checking the higher loads.

Two **standard forms of test specimens** (ASTM) are shown in Figs. 24 and 25. In wrought materials, and particularly in

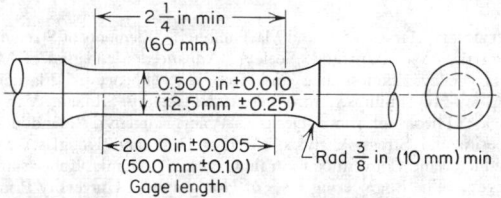

**Fig. 24**  Test specimen, 2-in (50-mm) gage length, ½-in (12.5-mm) diameter. Others available for 0.350-in (8.75-mm) and 0.250-in (6.25-mm) diameter (ASTM E8-69).

those which have been cold-worked, different properties may be expected in different directions with respect to the direction of the applied work, and the test specimen should be cut from the parent material in such a way as to give the strength

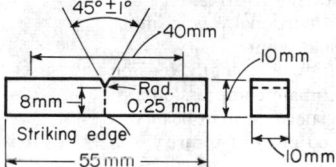

**Fig. 25**  Charpy V-notch impact specimen (ASTM E23-62).

in the desired direction. With the exception of fatigue specimens and specimens of extremely brittle materials, surface finish is of little practical importance, although extreme roughness tends to decrease the ultimate elongation.

# MECHANICS OF MATERIALS

## by J. P. Vidosic

REFERENCES: Timoshenko and MacCullough, "Elements of Strength of Materials," Van Nostrand. Seeley, "Advanced Mechanics of Materials," Wiley. Timoshenko and Goodier, "Theory of Elasticity," McGraw-Hill. Phillips, "Introduction to Plasticity," Ronald. Van Den Broek, "Theory of Limit Design," Wiley. Hetényi, "Handbook of Experimental Stress Analysis," Wiley. Dean and Douglas, "Semi-Conductor and Conventional Strain Gages," Academic. Robertson and Harvey, "The Engineering Uses of Holography," University Printing House, London. Sellers, "Basic Training Guide to the New Metrics and SI Units," National Tool, Die and Precision Machining Association.

### Main Symbols

#### Unit Stress

$S$ = apparent stress
$S_v$ or $S_s$ = pure shearing
$T$ = true (ideal) stress
$S_p$ = proportional elastic limit
$S_y$ = yield point
$S_M$ = ultimate strength, tension
$S_c$ = ultimate compression
$S_v$ = vertical shear in beams
$S_R$ = modulus of rupture

#### Moment

$M$ = bending
$M_t$ = torsion

#### External Action

$P$ = force
$G$ = weight of body
$W$ = weight of load
$V$ = external shear

#### Modulus of Elasticity

$E$ = longitudinal
$G$ = shearing
$K$ = bulk
$U_p$ = modulus of resilience
$U_R$ = ultimate resilience

#### Geometrical

$l$ = length
$A$ = area
$V$ = volume
$v$ = velocity
$r$ = radius of gyration
$I$ = rectangular moment of inertia
$I_P$ or $J$ = polar moment of inertia

#### Deformation

$e$ = gross, longitudinal
$s$ = unit, longitudinal
$d$ or $\alpha$ = unit, angular

$s'$ = unit, lateral
$\mu$ = Poisson's ratio
$n$ = reciprocal of Poisson's ratio
$r$ = radius
$f$ = deflection

## SIMPLE STRESSES AND STRAINS

**Deformations** are changes in form produced by external forces or loads that act on nonrigid bodies. Deformations are **longitudinal**, $e$, a lengthening ($+$) or shortening ($-$) of the body; and **angular**, $d$, a change of angle between the faces.

**Unit deformation** (dimensionless number) is the deformation in unit distance. Unit longitudinal deformation $s = e/l$ (Fig. 1). Unit angular-deformation tan $\alpha$ equals $\alpha$ approx (Fig. 2).

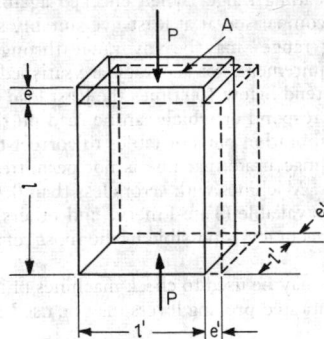

**Fig. 1**

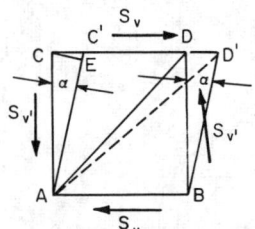

**Fig. 2**

Accompanying a longitudinal deformation $e$ is a **lateral** deformation of $e'$ (Fig. 1). The ratio of $s'/s$ is **Poisson's ratio** $\mu$. Values of $\mu$ are: glass, 0.244; brass, 0.333; copper, 0.333; cast iron, 0.270; wrought iron, 0.278; steel, 0.303; lead, 0.430; concrete, 0.10 to 0.20 at working stresses and 0.25 at higher stresses.

**Stress** is an internal distributed force; it is the internal mechanical reaction of the material accompanying deformation. Stresses always occur in pairs. Stresses are **normal**

[tensile stress (+) and compressive stress (−)]; and **tangential**, or **shearing**.

**Intensity of stress,** or **unit stress,** $S$, lb/in² (kgf/cm²) is the amount of force per unit of area (Fig. 3). $P$ is the load acting through the center of gravity of the area. The uniformly distributed normal stress is

$$S = P/A$$

When the stress is not uniformly distributed, $S = dP/dA$.

A **long rod** will stretch under its own weight $G$ and a terminal load $P$ (see Fig. 4). The total elongation $e$ is that due to the terminal load plus that due to one-half the weight of the rod considered as acting at the end.

$$e = [Pl + (Gl/2)]/AE$$

The maximum stress is at the upper end.

When a **load** is **carried by several paths** to a support, the different paths take portions of the load in proportion to their stiffness, which is controlled by material ($E$) and by design.

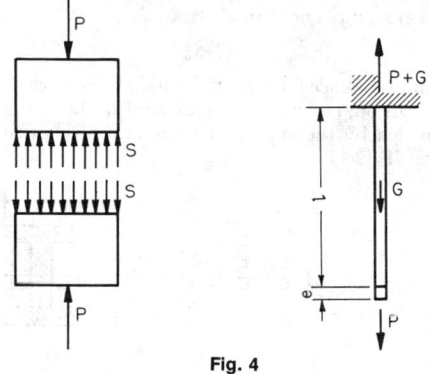

**Fig. 3**          **Fig. 4**

EXAMPLE. Two pairs of bars rigidly connected (with the same elongation) carry a load $P_0$ (Fig. 5). $A_1$, $A_2$ and $E_1$, $E_2$ and $P_1$, $P_2$ and $S_1$, $S_2$ are cross sections, moduli of elasticity, loads, and stresses of the bars, respectively; $e$ = elongation.

$$e = P_1 l/(E_1 A_1) = P_2 l/(E_2 A_2)$$
$$P_0 = 2P_1 + 2P_2$$
$$S_2 = P_2/A_2 = \frac{1}{2}[P_0 E_2/(E_1 A_1 + E_2 A_2)]$$
$$S_1 = \frac{1}{2}[P_0 E_1/(E_1 A_1 + E_2 A_2)]$$

**Temperature Stresses**  When the deformation arising from change of temperature is prevented, temperature stresses arise that are proportional to the amount of deformation that is prevented. Let $a$ = coefficient of expansion per degree of temperature, $l_1$ = length of bar at temperature $t_1$, and $l_2$ = length at temperature $t_2$. Then

$$l_2 = l_1[1 + a(t_2 - t_1)]$$

If, subsequently, the bar is cooled to a temperature $t_1$, the proportionate deformation is $s = a(t_2 - t_1)$ and the corresponding unit stress $S = Ea \times (t_2 - t_1)$. For **coefficients of expansion,** see Sec. 4. In the case of steel, a change of temperature of 12°F (6.7 K, 6.7°C) will cause in general a unit stress of 2,340 lb/in² (164 kgt/cm²).

**Shearing stresses** (Fig. 2) act tangentially to surface of contact and do not change length of sides of elementary volume; they

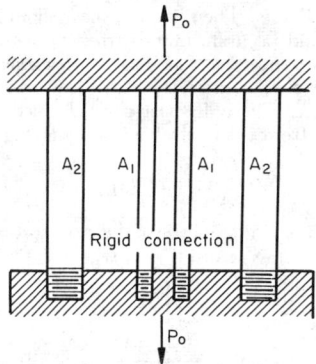

**Fig. 5**

change the angle between faces and the **length of diagonal.** Two pairs of shearing stresses must act together. **Shearing stress intensities are of equal magnitude on all four faces of an element.** $S_v = S_v'$ (Fig. 6).

In the presence of **pure shear** on external faces (Fig. 6), the **resultant stress** $S$ on one diagonal plane at 45° is pure tension and on the other diagonal plane pure compression; $S = S_v = S_v'$. $S$ on diagonal plane is called "diagonal tension" by writers on reinforced concrete. Failure under pure shear is difficult to produce experimentally, except under torsion and in certain special cases. Figure 7 shows an ideal case, and Fig. 8 a common form of test piece that introduces bending stresses.

Let Fig. 9 represent the section of area $A$ on which a

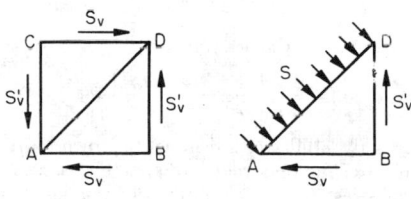

**Fig. 6**

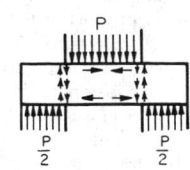

**Fig. 7**          **Fig. 8**

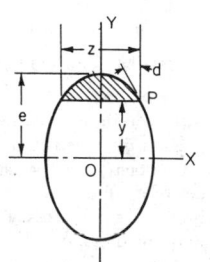

**Fig. 9**

shearing force $V$ acts. Then, if pure shear should exist, $S_v = V/A$. This would be uniformly distributed over the area $A$. When **shear** is **accompanied by bending (transverse shear in beams)**, the unit shear $S_v$ increases from the extreme fiber to the neutral axis $O\,X$. The unit shear parallel to axis $O\,X$ at any point $y$ distant from the neutral axis as at $P$ (Fig. 9) is

$$S_v = \left( V \int_y^e yz \, dy \right) \Big/ Iz$$

where $I$ is the moment of inertia of the cross section about $OX$. For a **rectangular cross section** (Fig. 10a),

$$S_v = \frac{3}{2} \frac{V}{bh} \left[ 1 - \left( \frac{2y}{h} \right)^2 \right]$$

$$S_v \text{ (max)} = \frac{3}{2} \frac{V}{bh} = \frac{3}{2} \frac{V}{A}, \text{ for } y = 0$$

For a **circular cross section** (Fig. 10b),

$$S_v = \frac{4}{3} \frac{V}{\pi r^2} \left[ 1 - \left( \frac{y}{r} \right)^2 \right]$$

$$S_v \text{ (max)} = \frac{4}{3} \frac{V}{\pi r^2} = \frac{4}{3} \frac{V}{A}, \text{ for } y = 0$$

For a **circular ring** (thickness small in comparison with the major diameter), $S_v$ (max) $= 2V/A$, for $y = 0$.

For a **square cross section** (diagonal vertical, Fig. 10c),

$$S_v = \frac{V\sqrt{2}}{a^2} \left[ 1 + \frac{y\sqrt{2}}{a} - 4 \left( \frac{y}{a} \right)^2 \right]$$

$$S_v \text{ (max)} = 1.591 \frac{V}{A}, \text{ for } y = \frac{e}{4}$$

For an **I-shaped cross section** (Fig. 10d),

$$S_v \text{ (max)} = \frac{3}{4} \frac{V}{a} \left[ \frac{be^2 - (b-a)f^2}{be^3 - (b-a)f^3} \right] \text{ for } y = 0$$

**Elasticity** is the ability of a material to return to its original dimensions after the removal of stresses. The **elastic limit** $S_p$ is the limit of stress within which the deformation completely disappears after the removal of stress; i.e., no set remains.

**Hooke's law** states that, within the elastic limit, deformation produced is proportional to the stress. Unless modified, the deduced formulas of mechanics apply only within the elastic limit. Beyond this, they are modified by experimental coefficients, as, for instance, the modulus of rupture.

The **modulus of elasticity,** lb/in² (kgf/cm²), is the ratio of the increment of unit stress to increment of unit deformation within the elastic limit.

The **modulus of elasticity in tension,** or **Young's modulus,**

$$E = \text{unit stress/unit deformation} = Pl/Ae$$

The modulus of elasticity in compression is similarly measured.

The **modulus of elasticity in shear** or **coefficient of rigidity,** $G = S_v/\alpha$ where $\alpha$ is expressed in radians (see Fig. 2).

The **bulk modulus of elasticity** $K$ is the ratio of normal stress, applied to all six faces of a cube, to the change of volume.

**Change of volume** under normal stress. Let $l$, $d$, and $b$ represent length, width, and thickness; $\mu =$ Poisson's ratio; $s =$ unit deformation. Then deformed volume $= (1 + s)l(1 - \mu s)b(1 - \mu s)d \approx (1 + s - 2\mu s)lbd$. Fractional change of volume $\approx (1 -$

$2\mu)s$. When $\mu$ is less than ½, the volume is increased in tension and decreased in compression. For steel ($\mu \approx$ ⅓), change of volume is about 1/3,000 part of the elastic limit.

The following relationships exist between the modulus of elasticity in tension or compression $E$, modulus of elasticity in shear $G$, bulk modulus of elasticity $K$, and Poisson's ratio $\mu$:

$$E = 2G(1 + \mu)$$
$$G = E/2(1 + \mu)$$
$$\mu = (E - 2G)/2G$$
$$K = E/3(1 - 2\mu)$$
$$\mu = (3K - E)/6K$$

**Resilience** $U$ (in·lb) (cm·kgf) is the potential energy stored up in a deformed body. The amount of resilience is equal to the work required to deform the body from zero stress to stress $S$, when $S$ does not exceed the elastic limit. For normal stress, resilience $=$ work of deformation $=$ average force times deformation $= \frac{1}{2}Pe = \frac{1}{2}AS \times Sl/E = \frac{1}{2}S^2V/E$.

**Modulus of resilience** $U_p$ (in·lb/in³), (cm·kgf/cm³), or **unit resilience,** is the elastic energy stored up in a cubic inch of material at the elastic limit. For normal stress,

$$U_p = \frac{1}{2}S_p^2/E$$

The unit resilience for any other kind of stress, as shearing, bending, torsion, is a constant times one-half the square of the stress divided by the appropriate modulus of elasticity. For values, see Table 1.

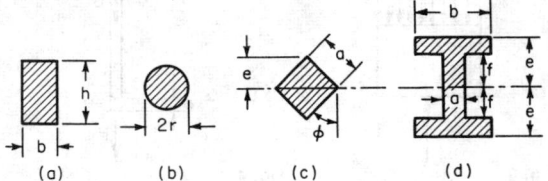

**Fig. 10**

**Unit rupture work** $U_R$, sometimes called **ultimate resilience,** is measured by the area of the stress-deformation diagram to rupture.

$$U_R = \frac{1}{3}e_u(S_y + 2S_M) \quad \text{approx}$$

where $e_u$ is the total deformation at rupture.

For structural steel, $U_R = \frac{1}{3} \times {}^{27}\!/_{100} \times [35,000 + (2 \times 60,000)] = 13,950$ in·lb/in³ (982 cm·kgf/cm³).

**EXAMPLE 1.** A load $P = 40,000$ lb compresses a wooden block of cross-sectional area $A = 10$ in² and length $= 10$ in, an amount $e = \frac{4}{100}$ in. Stress $S = \frac{1}{10} \times 40,000 = 4,000$ lb/in². Unit elongation $s = \frac{4}{100} \div 10 = \frac{1}{250}$. Modulus of elasticity $E = 4,000 \div \frac{1}{250} = 1,000,000$ lb/in². Unit resilience $U_p = \frac{1}{2} \times 4,000 \times 4,000/1,000,000 = 8$ in·lb/in³ (0.563 cm·kgf/cm³).

**EXAMPLE 2.** A weight $G = 5,000$ lb falls through a height $b = 2$ ft; $V =$ number of cubic inches required to absorb the shock without exceeding a stress of 4,000 lb/in². Neglect compression of block. Work done by falling weight $= Gb = 5,000 \times 2 \times 12$ in·lb (2,271 × 61 cm·kgf) Resilience of block $= V \times 8$ in·lb $= 5,000 \times 2 \times 12$. Therefore, $V = 15,000$ in³ (245,850 cm³).

**Thermal Stresses** A bar will change its length when its temperature is raised (or lowered) by the amount $\Delta l_0 = \alpha l_0(t_2 - 32)$. The linear coefficient of thermal expansion $\alpha$ is assumed constant at normal temperatures and $l_0$ is the length

**Table 1. Resilience per Unit of Volume, $U_p$**

($S$ = longitudinal stress; $S_v$ = shearing stress; $E$ = tension modulus of elasticity; $G$ = shearing modulus of elasticity)

| | | | |
|---|---|---|---|
| Tension or compression | $\frac{1}{2}S^2/E$ | Torsion | |
| Shear | $\frac{1}{2}S_v^2/G$ | Solid, circular | $\frac{1}{4}S_v^2/G$ |
| Beams(free ends) | | Hollow, radii $R_1$ and $R_2$ | $\dfrac{R_1^2 + R_2^2}{R_1^2}\dfrac{1}{4}\dfrac{S_v^2}{G}$ |
| Rectangular section, bent in arc of circle; no shear | $\frac{1}{6}S^2/E$ | | |
| Ditto, circular section | $\frac{1}{8}S^2/E$ | Springs | |
| Concentrated center load; rectangular cross section | $\frac{1}{18}S^2/E$ | Carriage | $\frac{1}{6}S^2/E$ |
| Ditto, circular cross section | $\frac{1}{24}S^2/E$ | Flat spiral, rectangular section | $\frac{1}{24}S^2/E$ |
| | | Helical: axial load, circular wire | $\frac{1}{4}S_v^2/G$ |
| Uniform load, rectangular cross section | $\frac{5}{36}S^2/E$ | Helical: axial twist | $\frac{1}{8}S^2/E$ |
| I-beam section, concentrated center load | $\frac{3}{32}S^2/E$ | Helical: axial twist, rectangular section | $\frac{1}{6}S^2/E$ |

at 32°F (273.2K, 0°C). If this expansion (or contraction) is prevented, a **thermal-time stress** is developed, equal to $S = E\alpha(t_2 - t_1)$, as the temperature goes from $t_1$ to $t_2$. In thin flat plates the stress becomes $S = E\alpha(t_2 - t_1)/(1 - \mu)$; $\mu$ is Poisson's ratio. Such stresses can occur in castings containing large and small sections. Similar stresses also occur when heat flows through members because of the difference in temperature between one point and another. The heat flowing across a length $b$ as a result of a linear drop in temperature $\Delta t$ equals $Q = k\,A\Delta t/b$ Btu/h (cal/h). The thermal conductivity $k$ is in Btu/(h)(ft²)(°F) (in of thickness) [cal/(h)(m²)(k)(m)]. The **thermal-flow stress** is then $S = E\alpha Qb/kA$. Note, when $Q$ is substituted, the stress becomes $S = E\alpha\Delta t$ as above, only $t$ is now a function of distance rather than time.

EXAMPLE. A cast-iron plate 3 ft square and 2 in thick is used as a fire wall. The temperature is 330°F on the hot side and 160°F on the other. What is the thermal-flow stress developed across the plate?

$$S = E\alpha\Delta t = 13 \times 10^6 \times 6.5 \times 10^{-6} \times 170$$
$$= 14{,}360 \text{ lb/in}^2 \text{ (1,010 kgf/cm}^2)$$

or $\quad Q = 2.3 \times 9 \times 170/2 = 1{,}760$ Btu/h

and $\quad S = 13 \times 10^6 \times 6.5 \times 10^{-6} \times 1{,}760 \times 2/2.3 \times 9$
$$= 14{,}360 \text{ lb/in}^2 \text{ (1,010 kgf/cm}^2)$$

## COMBINED STRESSES

**Simple stresses,** defined as such by the flexure and torsion theories, lie in planes normal or parallel to the line of action of the forces. Normal, as well as shearing, stresses may, however, exist in other directions. A particle out of a loaded member will contain normal and shearing stresses as shown in Fig. 11. Note that the four shearing stresses must be of the same magnitude, if equilibrium is to be satisfied.

If the particle is "cut" along the plane $\overline{AA}$, equilibrium will reveal that, in general, normal as well as shearing stresses act upon the plane $\overline{AC}$ (Fig. 12). The normal stress on plane $\overline{AC}$ is labeled $S_n$ and shearing $S_s$. The application of equilibrium yields

$$S_n = \frac{S_x + S_y}{2} + \frac{S_x - S_y}{2}\cos 2\theta + S_{xy}\sin 2\theta$$

and $\quad S_s = \dfrac{S_x - S_y}{2}\sin 2\theta - S_{xy}\cos 2\theta$

A *sign convention* must be used. A tensile stress is positive while compression is negative. A shearing stress is positive when directed as on plane $\overline{AB}$ of Fig. 12; i.e., when the shearing stresses on the vertical planes form a clockwise couple, the stress is positive.

The planes defined by $\tan 2\theta = 2S_{xy}/S_x - S_y$, the *principal planes*, contain the **principal stresses** —the maximum and minimum normal stresses. These stresses are

$$S_M,\ S_m = \frac{S_x + S_y}{2} \pm \sqrt{\left(\frac{S_x - S_y}{2}\right)^2 + S_{xy}^2}$$

The radical in the equation yields the **maximum and minimum shearing stresses.**

These act upon the planes, $\tan 2\theta = -\dfrac{S_x - S_y}{2S_{xy}}$.

EXAMPLE. The steam in a boiler subjects a particular particle on the boiler shell to a circumferential stress of 8,000 lb/in² and a longitudinal stress of 4,000 lb/in² as shown in Fig. 13. Find the stresses acting on the plane $\overline{XX}$, making an angle of 60° with the direction of the 8,000 lb/in² stress. Find the principal stresses and locate the principal planes. Also find the maximum-minimum shearing stresses.

$$60°\ S_n = \frac{4{,}000 + 8{,}000}{2} + \frac{4{,}000 - 8{,}000}{2}(-0.5000) + 0$$
$$= 7{,}000 \text{ lb/in}^2$$

$$60°\ S_s = \frac{4{,}000 - 8{,}000}{2}(0.8660) - 0 = -1{,}732 \text{ lb/in}^2$$

$$S_{M,m} = \frac{4{,}000 + 8{,}000}{2} \pm \sqrt{\left(\frac{4{,}000 - 8{,}000}{2}\right)^2 + 0}$$
$$= 6{,}000 \pm 2{,}000$$
$$= 8{,}000 \text{ and } 4{,}000 \text{ lb/in}^2 \text{ (564 and 282 kgf/cm}^2)$$

at $\tan 2\theta = \dfrac{2 \times 0}{4{,}000 - 8{,}000} = 0 \quad$ or $\quad \theta = 90°$ and $0°$

$$S_{s\,M,m} = \pm\sqrt{\left(\frac{4{,}000 - 8{,}000}{2}\right)^2 + 0}$$
$$= \pm 2{,}000 \text{ lb/in}^2 \text{ (}\pm141 \text{ kgf/cm}^3)$$

**Mohr's Stress Circle** The biaxial stress field with its combined stresses can be represented graphically by the Mohr stress circle. For instance, for the particle given in Fig. 11, Mohr's circle is as shown in Fig. 14. The stress *sign convention* previously defined must be adhered to. Furthermore, in order to locate the point (on Mohr's circle) that yields the stresses on a plane $\theta°$ from the vertical side of the particle (such as plane $\overline{AA}$ in Fig. 11), $2\theta°$ must be laid off in the same direction from the radius to $(S_x,S_{xy})$. For the previous example, Mohr's circle becomes Fig. 15.

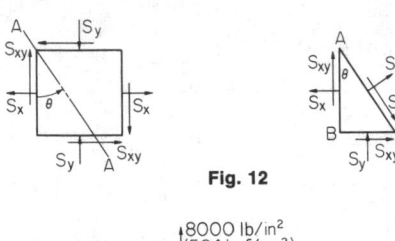

**Fig. 11**  **Fig. 12**

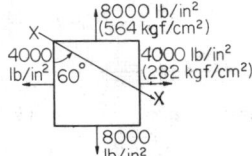

**Fig. 13**

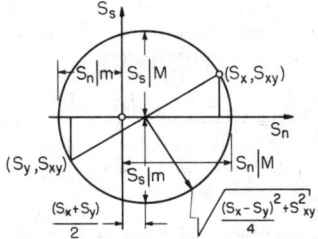

**Fig. 14**

Eight special stress fields are shown in Figs. 16 to 23, along with Mohr's circle for each.

**Combined Loading**  Combined flexure and torsion arise, for instance, when a shaft twisted by a torque $M_t$ is bent by forces produced by belts or gears. An element on the surface, such as *ABCD* on the shaft of Fig. 24, is subjected to a flexure stress $S_x = Mc/I = 8\ Fl/\pi d^3$ and a torsional shearing stress $S_{xy} = M_t c/J = 16\ M_t/\pi d^3$. These stresses will induce combined stresses. The maximum combined stresses will be

and
$$S_n = \frac{1}{2}\ (S_x \pm \sqrt{S_x^2 + 4S_{xy}^2})$$
$$S_s = \pm\ \frac{1}{2}\ \sqrt{S_x^2 + 4S_{xy}^2}$$

The above situation applies to any case of normal stress with shear, as when a bolt is under both tension and shear. A beam particle subjected to both flexure and transverse shear is another case.

**Combined torsion and longitudinal loads** exist on a propeller shaft. A particle on this shaft will contain a tensile stress computed using $S = F/A$ and a torsion shearing stress equal to $S_s = M_t c/J$. The free body of a particle on the surface will be the same as that of Fig. 24. A particle on the surface of a vertical turbine shaft is subjected to direct compression and torsion.

When combined loading results in stresses of the same type and direction, the addition is algebraic. Such a situation exists on an offset link like that of Fig. 25.

**Mohr's Strain Circle**  Strain equations can also be derived for plane-strain fields. Strains $e_x$ and $e_y$ are the extensional strains (tension or compression) occurring at a point in two right-angle directions, and the change of the angle between

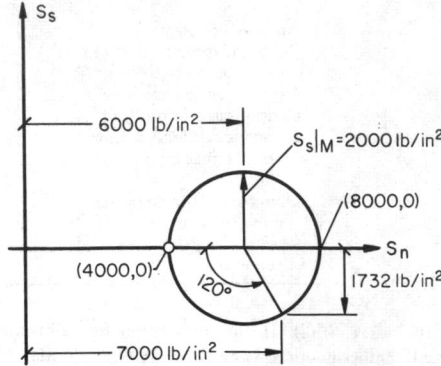

**Fig. 15**

them is $\gamma_{xy}$. The strain $e$ at the point in any direction $a$ at an angle $\theta$ with the $x$ direction derives as

$$e_a = \frac{e_x + e_y}{2} + \frac{e_x - e_y}{2} \cos 2\theta + \frac{\gamma_{xy}}{2} \sin 2\theta$$

Similarly, the shearing strain $\gamma_{ab}$ (change in the original right angle between directions $a$ and $b$) is defined by

$$\gamma_{ab} = (e_x - e_y) \sin 2\theta + \gamma_{xy} \cos 2\theta$$

Inspection easily reveals that the above equations for $e_a$ and $\gamma_{ab}$ are mathematically identical to those for $S_n$ and $S_s$. Thus, once a sign convention is established, a Mohr circle for strain can be constructed and used as the stress circle is used. The

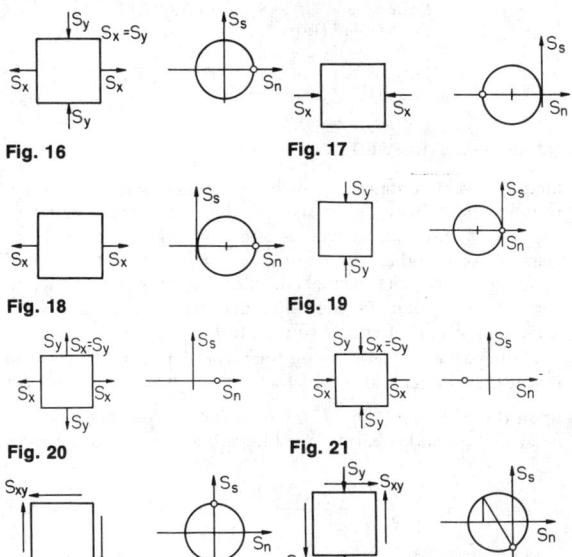

**Fig. 16**  **Fig. 17**

**Fig. 18**  **Fig. 19**

**Fig. 20**  **Fig. 21**

**Fig. 22**  **Fig. 23**

strain $e$ is positive when an extension and negative when a contraction. If the direction associated with the first subscript $a$ rotates counterclockwise during straining with respect to the direction indicated by the second subscript $b$, the shearing strain is positive; if clockwise, it is negative. In constructing the circle, positive extensional strains will be plotted to the right as abscissas and positive **half-shearing** strains will be plotted upward as ordinates.

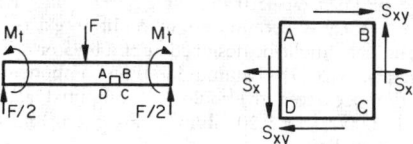

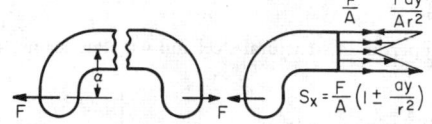

**Fig. 24**

**Fig. 25**

For the strains shown in Fig. 26a, Mohr's strain circle becomes that shown in Fig. 26b. The extensional strain in the direction $a$, making an angle of $\theta_a$ with the $x$ direction, is $e_a$, and the shearing strain is $\gamma_{ab}$ counterclockwise. The strain 90° away is $e_b$. The maximum principal strain is $e_M$ at an angle $\theta_M$ clockwise from the $x$ direction. The other principal or minimum strain is $e_m$ 90° away.

## PLASTIC DESIGN

Early efforts in stress analysis were based on limit loads, that is, loads which stress a member "wholly" to the yield strength. Euler's famous column paper ("Sur la Force des Colonnes," Academie des Sciences de Berlin, 1757) deals with the column problem this way. More recently, the concept of limit loads, referred to as **limit,** or **plastic, design,** has found strong application in the design of certain structures. The theory presupposes a ductile material, absence of stress raisers, and fabrication free of embrittlement. Local load overstress is allowed, provided the structure does not deform appreciably.

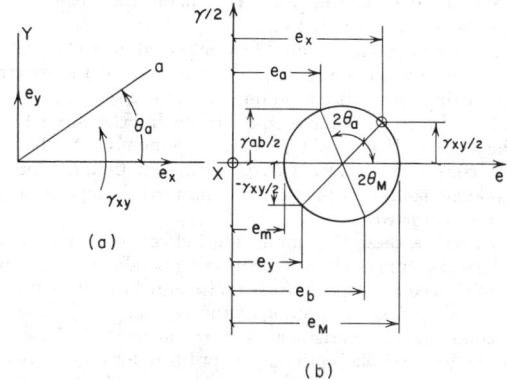

**Fig. 26**

To visualize the limit-load approach, consider a simple beam of uniform section subjected to a concentrated load of midspan, as depicted in Fig. 27a. According to elastic theory, the outermost fiber on each side and at midspan—the section of maximum bending moment—will first reach the yield-strength value. Across the depth of the beam, the stress distribution will, of course, follow the triangular pattern, becoming zero at the neutral axis. If the material is ductile, the stress in the outermost fibers will remain at the yield value until every other fiber reaches the same value as the load increases. Thus the stress distribution assumes the rectangular pattern before the *plastic hinge* forms and failure ensues.

The problem is that of finding the final limit load. Elastic-flexure theory gives the maximum load—triangular distribution—as

$$F_y = \frac{2S_y bh^2}{3l}$$

For the rectangular stress distribution, the limit load becomes

$$F_L = \frac{S_y bh^2}{l}$$

The ratio $F_L/F_y = 1.50$—an increase of 50 percent in load capability. The ratio $F_L/F_y$ has been named **shape factor** (Jenssen, Plastic Design in Welded Structures Promises New Economy and Safety, *Welding Jour.*, Mar. 1959). See Fig. 27b for shape factors for some other sections. The shape factor may also be determined by dividing the first moment of area about the neutral axis by the section modulus.

A constant-section beam with both ends fixed, supporting a uniformly distributed load, illustrates another application of the plastic-load approach. The bending-moment diagram based on the elastic theory drawn in Fig. 28 (broken line) shows a moment at the center equal to one-half the moment at either end. A preferable situation, it might be argued, is one in

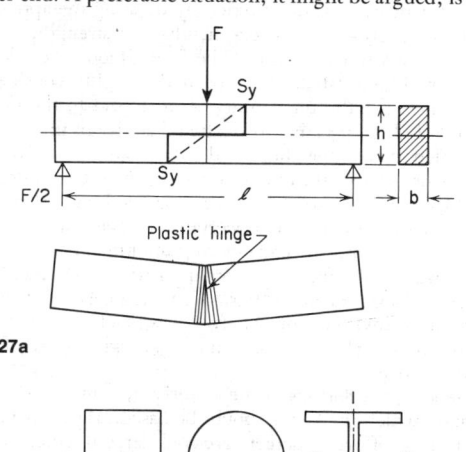

**Fig. 27a**

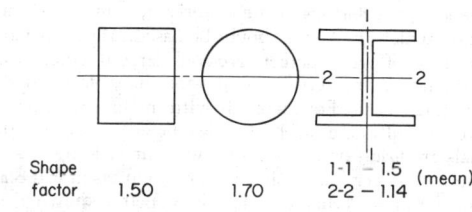

| Shape factor | 1.50 | 1.70 | 1-1 – 1.5  (mean) 2-2 – 1.14 |

**Fig. 27b**

which the moments are the same at the three stations—solid line. Thus, applying equilibrium to, say, the left half of the beam yields a bending moment at each of the three plastic hinges of

$$M_L = \frac{wl^2}{16}$$

## DESIGN STRESSES

If a machine part is safely to transmit loads acting upon it, a permissible maximum stress must be established and used in the design. This is the allowable stress, the working stress, or preferably, the **design stress.** The design stress should not waste material, yet should be large enough to prevent failure in case loads exceed expected values, or other uncertainties react unfavorably.

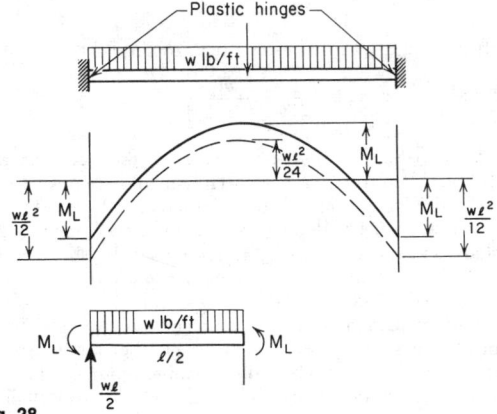

**Fig. 28**

The design stress is determined by dividing the applicable material property—yield strength, ultimate strength, fatigue strength—by a **factor of safety.** The factor should be selected only after all **uncertainties** have been thoroughly considered. Among these are the uncertainty with respect to the magnitude and kind of operating load, the reliability of the material from which the component is made, the assumptions involved in the theories used, the environment in which the equipment might operate, the extent to which localized and fabrication stresses might develop, the uncertainty concerning causes of possible failure, and the endangering of human life in case of failure. Factors of safety vary from industry to industry, being the result of accumulated experience with a class of machines or a kind of environment. Many codes, such as the ASME code for power shafting, recommend design stresses found safe in practice.

In general, the **ductility** of the material determines the property upon which the factor should be based. Materials having an elongation of over 5 percent are considered ductile. In such cases, the factor of safety is based upon the yield strength or the endurance limit. For materials with an elongation under 5 percent, the ultimate strength must be used because these materials are **brittle** and so fracture without yielding.

Factors of safety based on yield are often taken **between 1.5 and 4.0.** For more reliable materials or better-known design and operating conditions, the lower factors are appropriate. In the case of untried materials or otherwise uncertain conditions, the higher factors are safer. The same values can be used when loads vary, but in such cases they are applied to the fatigue or endurance strength. When the ultimate strength determines the design stress (in the case of brittle materials), the factors of safety can be doubled.

Thus, under static loading, the design stress for, say, SAE 1020, which has a yield strength of 45,000 lb/in² (3,170 kgf/cm²) may be taken at 45,000/2, or 22,500 lb/in² (1,585 kgf/cm²), if a reasonably certain design condition exists. A Class 30 cast-iron part might be designed at 30,000/5 or 6,000 lb/in² (423 kgf/cm²). A 2017S-0 aluminum-alloy component (13,000 lb/in² endurance strength) could be computed at a design stress of 13,000/2.5 or 5,200 lb/in² (366 kgf/cm²) in the usual fatigue-load application.

## BEAMS

(For properties of structural steel and wooden beams, see Sec. 12.)

### Notation

$R$ = reaction
$M$ = bending moment
$W$ = total distributed load
$S_s$ or $S_v$ = transverse shearing unit stress
$Z$ = horizontal shearing unit stress
$I$ = rectangular moment of inertia
$w$ = distributed load per longitudinal unit
$P$ = concentrated load
$V$ or $Q$ = total vertical shear
$S$ = unit normal apparent stress
$I_P$ = polar moment of inertia
$r$ = radius of curvature
$i$ = slope
$f$ = deflection
$l$ = distance between supports

A **simple beam** rests on supports at its ends which permit rotation. A **cantilever beam** is fixed (no rotation) at one end. When computing reactions and moments, distributed loads may be replaced by their resultants acting at the center of gravity of the distributed-load area.

**Reactions** are the forces and/or couples acting at the supports and holding the beam in place. In general, the weight of the beam should be accounted for.

The **bending moment** (pound·feet or pound·inches) (kgf·m or kgf·cm) at any section is the algebraic sum of the external forces acting on the beam on one side of the section, $M = \Sigma F_x$. It is also equal to the moment of the internal-stress forces at the section, $M = \int s(dA)y$. A bending moment that bends a beam convex downward (tensile stress on bottom fiber) is considered **positive,** while convex upward (compression on bottom) is **negative.**

The **vertical shear** $V$ (pounds) (kgf) effective on a section is the algebraic sum of all the forces acting parallel to and on one side of the section, $V = \Sigma F$. It is also equal to the sum of the transverse shear stresses acting on the section, $V = \int S_s \, dA$.

**Moment** and **shear diagrams** may be constructed by plotting to scale the particular entity as the ordinate for each section of the beam. Such diagrams show in continuous form the variation along the length of the beam.

**Moment-Shear Relation** The shear $V$ is the first derivative of moment with respect to distance along the beam, $V = dM/dx$. This relationship does not, however, account for any sudden changes in moment.

EXAMPLES. Figure 29 illustrates a simple beam subjected to a uniform load. $M = R_1 x - wx \times \dfrac{x}{2} = \dfrac{wlx}{2} - \dfrac{wx^2}{2}$ and $V = R_1 - wx = \dfrac{wl}{2} - wx$. Note also that $V = \dfrac{d}{dx}\left(\dfrac{wlx}{2} - \dfrac{wx^2}{2}\right) = \dfrac{wl}{2} - wx$.

Figure 30 is a simple beam carrying a uniformly varying load; $M = R_1 x - b\dfrac{x}{l} \times \dfrac{x}{2} \times \dfrac{x}{3} = \dfrac{blx}{6} - \dfrac{bx^3}{6l}$, if $b$ is in pounds per foot and weight of beam is neglected. The vertical shear $V = R_1 - \dfrac{bx}{l} \times \dfrac{x}{2} = \dfrac{bl}{6} - \dfrac{bx^2}{2l}$.

Note again that $V = \dfrac{d}{dx}\left(\dfrac{blx}{6} - \dfrac{bx^3}{6l}\right) = \dfrac{bl}{6} - \dfrac{bx^2}{2l}$.

Table 2 gives the reactions, bending-moment equations, vertical shear equations, and the deflection of some of the more common types of beams.

**Maximum Safe Load on Steel Beams** To obtain max safe load (or max deflection under max safe load) for any of the conditions of loading given in Table 5, multiply the corresponding coefficient in that table by the greatest safe load (or deflection) for distributed load for the particular section under consideration as given in Table 4.

The following factors for reducing the load should be used when beams are long in comparison with their breadth:

Ratio of unsupported (lateral)

| length to flange width or breadth | 20 | 30 | 40 | 50 | 60 | 70 |
|---|---|---|---|---|---|---|
| Ratio of greatest safe load to calculated load | 1 | 0.9 | 0.8 | 0.7 | 0.6 | 0.5 |

**Theory of Flexure** A bent beam is shown in Fig. 31. The concave side is in compression and the convex side in tension. These are divided by the **neutral plane** of zero stress $A'B'BA$. The intersection of the neutral plane with the face of the beam is in the **neutral line or elastic curve** $AB$. The intersection of the neutral plane with the cross section is the **neutral axis** $NN'$.

It is assumed that a beam is prismatic, of a length at least 10 times its depth, and that the external forces are all at right angles to the axis of the beam and in a plane of symmetry, and that flexure is slight. Other assumptions are: (1) That the material is homogeneous, and obeys Hooke's law. (2) That **stresses are within the elastic limit.** (3) That every layer of material is free to expand and contract longitudinally and laterally under stress as if separate from other layers. (4) That the

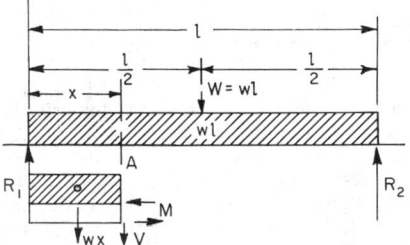

**Fig. 29**

tensile and compressive moduli of elasticity are equal. (5) That the cross section remains a plane surface. (The assumption of plane cross sections is strictly true only when the shear is constant or zero over the cross section, and when the shear is constant throughout the length of the beam.)

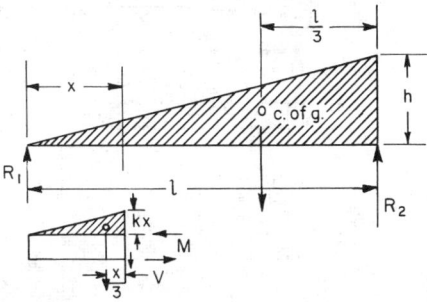

**Fig. 30**

It follows then that: (1) The internal forces are in horizontal balance. (2) The **neutral axis contains the center of gravity** of the cross section, when there is no resultant axial stress. (3) The stress intensity varies directly with the distance from the neutral axis.

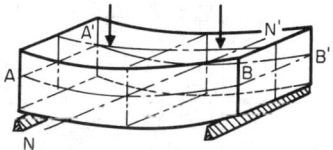

**Fig. 31**

The moment of the elastic forces about the neutral axis, i.e., the **stress moment** or **moment of resistance,** is $M = SI/c$, where $S$ is an elastic unit stress at outer fiber whose distance from the neutral axis is $c$; and $I$ is the rectangular moment of inertia about the neutral axis.

This formula is for the **strength of beams.** For rectangular beams, $M = \frac{1}{6}Sbh^2$, where $b$ = breadth and $h$ = depth; i.e., the elastic **strength of beam sections** varies as follows: (1) for equal width, as the square of the depth; (2) for equal depth, directly as the width; (3) for equal depth and width, directly as the strength of the material; (4) if span varies, then for equal depth, width, and material, inversely as the span.

If a beam is cut in halves horizontally, the two halves laid side by side will carry only one-half as much as the original beam.

The term **section modulus** is given to the value of $I/c$, where $c$ is the distance to the fiber carrying greatest stress. Moment of inertia of cross section = $I$.

Tables 6 to 8 give the properties of various beam cross-sections. For properties of structural-steel shapes, see Sec. 12.

**Oblique Loading** It should be noted that Table 6 includes certain cases for which the horizontal axis is not a neutral axis, assuming the common case of vertical loading. The rectangular section with the diagonal as a horizontal axis (Table 6) is such a case. These cases must be handled by the principles of oblique loading.

Every section of a beam has two principal axes passing

**Table 2. Beams of Uniform Cross Section, Loaded Transversely**

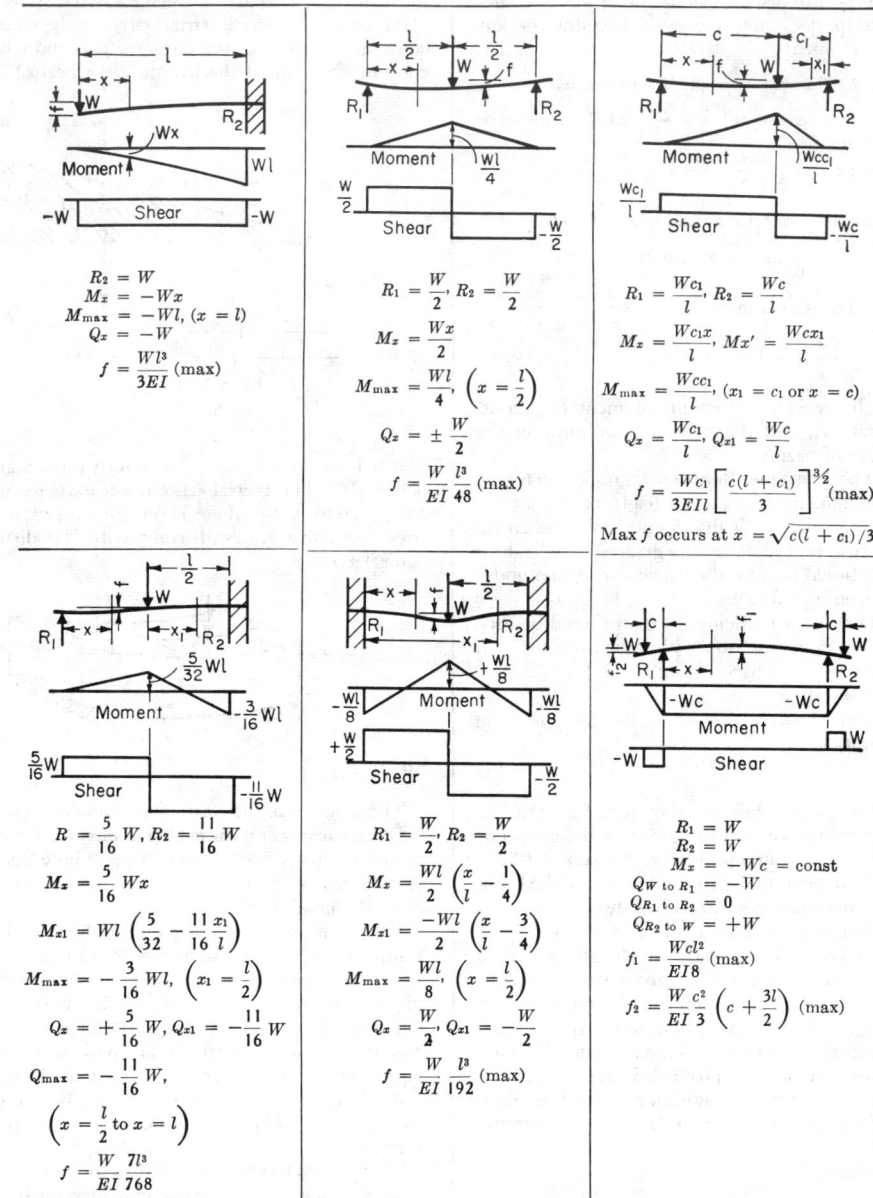

$$R_2 = W$$
$$M_x = -Wx$$
$$M_{max} = -Wl, \ (x = l)$$
$$Q_x = -W$$
$$f = \frac{Wl^3}{3EI} \ \text{(max)}$$

$$R_1 = \frac{W}{2}, \ R_2 = \frac{W}{2}$$
$$M_x = \frac{Wx}{2}$$
$$M_{max} = \frac{Wl}{4}, \ \left(x = \frac{l}{2}\right)$$
$$Q_x = \pm \frac{W}{2}$$
$$f = \frac{W}{EI} \frac{l^3}{48} \ \text{(max)}$$

$$R_1 = \frac{Wc_1}{l}, \ R_2 = \frac{Wc}{l}$$
$$M_x = \frac{Wc_1 x}{l}, \ M_{x'} = \frac{Wcx_1}{l}$$
$$M_{max} = \frac{Wcc_1}{l}, \ (x_1 = c_1 \ \text{or} \ x = c)$$
$$Q_x = \frac{Wc_1}{l}, \ Q_{x1} = \frac{Wc}{l}$$
$$f = \frac{Wc_1}{3EIl} \left[\frac{c(l + c_1)}{3}\right]^{3/2} \text{(max)}$$
Max $f$ occurs at $x = \sqrt{c(l + c_1)/3}$

$$R = \frac{5}{16} W, \ R_2 = \frac{11}{16} W$$
$$M_x = \frac{5}{16} Wx$$
$$M_{x1} = Wl \left(\frac{5}{32} - \frac{11}{16} \frac{x_1}{l}\right)$$
$$M_{max} = -\frac{3}{16} Wl, \ \left(x_1 = \frac{l}{2}\right)$$
$$Q_x = +\frac{5}{16} W, \ Q_{x1} = -\frac{11}{16} W$$
$$Q_{max} = -\frac{11}{16} W,$$
$$\left(x = \frac{l}{2} \text{ to } x = l\right)$$
$$f = \frac{W}{EI} \frac{7l^3}{768}$$

$$R_1 = \frac{W}{2}, \ R_2 = \frac{W}{2}$$
$$M_x = \frac{Wl}{2} \left(\frac{x}{l} - \frac{1}{4}\right)$$
$$M_{x1} = \frac{-Wl}{2} \left(\frac{x}{l} - \frac{3}{4}\right)$$
$$M_{max} = \frac{Wl}{8}, \ \left(x = \frac{l}{2}\right)$$
$$Q_x = \frac{W}{2}, \ Q_{x1} = -\frac{W}{2}$$
$$f = \frac{W}{EI} \frac{l^3}{192} \ \text{(max)}$$

$$R_1 = W$$
$$R_2 = W$$
$$M_x = -Wc = \text{const}$$
$$Q_{W \text{ to } R_1} = -W$$
$$Q_{R_1 \text{ to } R_2} = 0$$
$$Q_{R_2 \text{ to } W} = +W$$
$$f_1 = \frac{Wcl^2}{EI8} \ \text{(max)}$$
$$f_2 = \frac{W}{EI} \frac{c^2}{3} \left(c + \frac{3l}{2}\right) \ \text{(max)}$$

through the center of gravity, and these two axes are always at right angles to each other. The principal axes are axes with respect to which the moment of inertia is, respectively, a maximum and a minimum, and for which the product of inertia is zero. For symmetrical sections, axes of symmetry are always principal axes. For unsymmetrical sections, like a **rolled angle** section (Fig. 32), the inclination of the principal axis with the X axis may be found from the formula $\tan 2\theta = 2I_{xy}/(I_y -$

$I_x)$, in which $\theta$ = angle of inclination of the principal axis to the X axis, $I_{xy}$ = the product of inertia of the section with respect to the X and Y axes, $I_y$ = moment of inertia of the section with respect to the Y axis, $I_x$ = moment of inertia of the section with respect to the X axis. When this principal axis has been found, the other principal axis is at right angles to it.

Calling the moments of inertia with respect to the principal axes $I'_x$ and $I'_y$, the unit stress existing anywhere in the section

**Table 2. Beams of Uniform Cross Section, Loaded Transversely (Continued)**

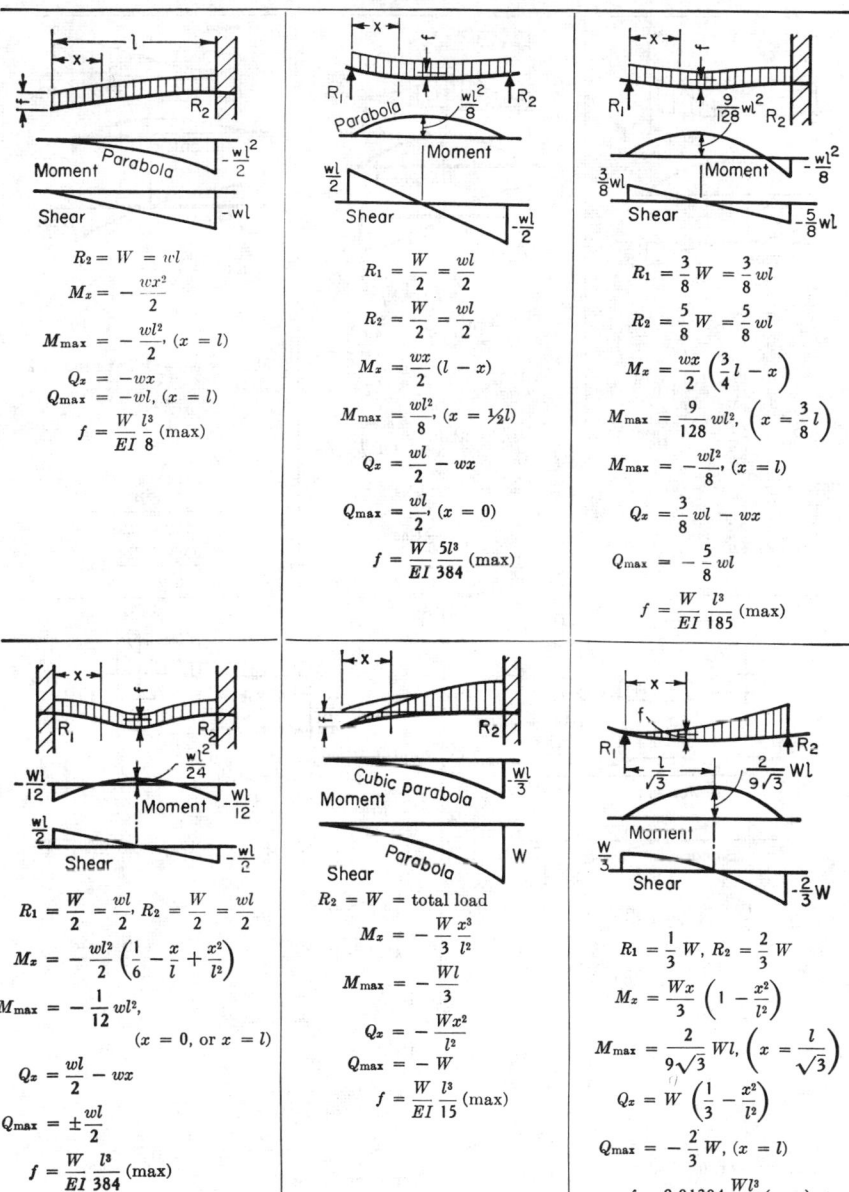

$$R_2 = W = wl$$

$$M_x = -\frac{wx^2}{2}$$

$$M_{max} = -\frac{wl^2}{2}, \; (x = l)$$

$$Q_x = -wx$$
$$Q_{max} = -wl, \; (x = l)$$

$$f = \frac{W}{EI}\frac{l^3}{8} \; (max)$$

$$R_1 = \frac{W}{2} = \frac{wl}{2}$$

$$R_2 = \frac{W}{2} = \frac{wl}{2}$$

$$M_x = \frac{wx}{2}(l - x)$$

$$M_{max} = \frac{wl^2}{8}, \; (x = \tfrac{1}{2}l)$$

$$Q_x = \frac{wl}{2} - wx$$

$$Q_{max} = \frac{wl}{2}, \; (x = 0)$$

$$f = \frac{W}{EI}\frac{5l^3}{384} \; (max)$$

$$R_1 = \frac{3}{8}W = \frac{3}{8}wl$$

$$R_2 = \frac{5}{8}W = \frac{5}{8}wl$$

$$M_x = \frac{wx}{2}\left(\frac{3}{4}l - x\right)$$

$$M_{max} = \frac{9}{128}wl^2, \; \left(x = \frac{3}{8}l\right)$$

$$M_{max} = -\frac{wl^2}{8}, \; (x = l)$$

$$Q_x = \frac{3}{8}wl - wx$$

$$Q_{max} = -\frac{5}{8}wl$$

$$f = \frac{W}{EI}\frac{l^3}{185} \; (max)$$

$$R_1 = \frac{W}{2} = \frac{wl}{2}, \; R_2 = \frac{W}{2} = \frac{wl}{2}$$

$$M_x = -\frac{wl^2}{2}\left(\frac{1}{6} - \frac{x}{l} + \frac{x^2}{l^2}\right)$$

$$M_{max} = -\frac{1}{12}wl^2,$$
$$(x = 0, \text{ or } x = l)$$

$$Q_x = \frac{wl}{2} - wx$$

$$Q_{max} = \pm\frac{wl}{2}$$

$$f = \frac{W}{EI}\frac{l^3}{384} \; (max)$$

$$R_2 = W = \text{total load}$$

$$M_x = -\frac{W}{3}\frac{x^3}{l^2}$$

$$M_{max} = -\frac{Wl}{3}$$

$$Q_x = -\frac{Wx^2}{l^2}$$
$$Q_{max} = -W$$

$$f = \frac{W}{EI}\frac{l^3}{15} \; (max)$$

$$R_1 = \frac{1}{3}W, \; R_2 = \frac{2}{3}W$$

$$M_x = \frac{Wx}{3}\left(1 - \frac{x^2}{l^2}\right)$$

$$M_{max} = \frac{2}{9\sqrt{3}}Wl, \; \left(x = \frac{l}{\sqrt{3}}\right)$$

$$Q_x = W\left(\frac{1}{3} - \frac{x^2}{l^2}\right)$$

$$Q_{max} = -\frac{2}{3}W, \; (x = l)$$

$$f = 0.01304\frac{Wl^3}{EI} \; (max)$$

at a point whose coordinates are $x$ and $y$ (Fig. 33) is $S = (My \cos \alpha/I'_x) + (Mx \sin \alpha/I'_y)$, in which $M$ = bending moment with respect to the section in question, $\alpha$ = the angle which the plane of bending moment or the plane of the loads makes with the $y$ axis, $M \cos \alpha$ = the component of bending moment causing bending about the principal axis which has been designated as the $X$ axis, $M \sin \alpha$ = the component of bending

moment causing bending about the principal axis which has been designated as the $Y$ axis. The sign of the two terms for unit stress may be determined by inspection in the usual way, and the result will be tension or compression as determined by the algebraic sum of the two terms.

In general, it may be stated that when the plane of the bending moment coincides with one of the principal axes, the

**Table 2. Beams of Uniform Cross Section, Loaded Transversely** *(Continued)*

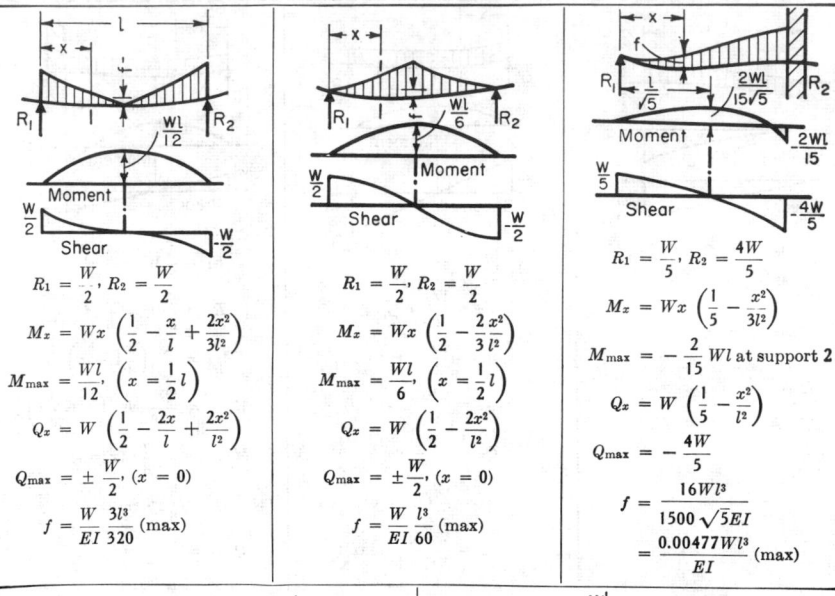

$$R_1 = \frac{W}{2}, \; R_2 = \frac{W}{2}$$

$$M_x = Wx \left( \frac{1}{2} - \frac{x}{l} + \frac{2x^2}{3l^2} \right)$$

$$M_{max} = \frac{Wl}{12}, \; \left( x = \frac{1}{2} l \right)$$

$$Q_x = W \left( \frac{1}{2} - \frac{2x}{l} + \frac{2x^2}{l^2} \right)$$

$$Q_{max} = \pm \frac{W}{2}, \; (x = 0)$$

$$f = \frac{W}{EI} \frac{3l^3}{320} \; (max)$$

$$R_1 = \frac{W}{2}, \; R_2 = \frac{W}{2}$$

$$M_x = Wx \left( \frac{1}{2} - \frac{2}{3} \frac{x^2}{l^2} \right)$$

$$M_{max} = \frac{Wl}{6}, \; \left( x = \frac{1}{2} l \right)$$

$$Q_x = W \left( \frac{1}{2} - \frac{2x^2}{l^2} \right)$$

$$Q_{max} = \pm \frac{W}{2}, \; (x = 0)$$

$$f = \frac{W}{EI} \frac{l^3}{60} \; (max)$$

$$R_1 = \frac{W}{5}, \; R_2 = \frac{4W}{5}$$

$$M_x = Wx \left( \frac{1}{5} - \frac{x^2}{3l^2} \right)$$

$$M_{max} = -\frac{2}{15} Wl \text{ at support 2}$$

$$Q_x = W \left( \frac{1}{5} - \frac{x^2}{l^2} \right)$$

$$Q_{max} = -\frac{4W}{5}$$

$$f = \frac{16Wl^3}{1500 \sqrt{5} EI}$$
$$= \frac{0.00477 Wl^3}{EI} \; (max)$$

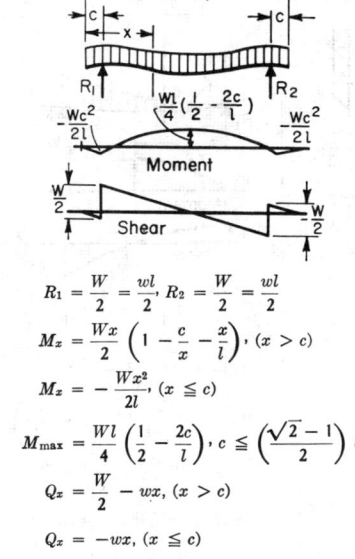

$$R_1 = \frac{W}{2} = \frac{wl}{2}, \; R_2 = \frac{W}{2} = \frac{wl}{2}$$

$$M_x = \frac{Wx}{2} \left( 1 - \frac{c}{x} - \frac{x}{l} \right), \; (x > c)$$

$$M_x = -\frac{Wx^2}{2l}, \; (x \leqq c)$$

$$M_{max} = \frac{Wl}{4} \left( \frac{1}{2} - \frac{2c}{l} \right), \; c \leqq \left( \frac{\sqrt{2}-1}{2} \right) l$$

$$Q_x = \frac{W}{2} - wx, \; (x > c)$$

$$Q_x = -wx, \; (x \leqq c)$$

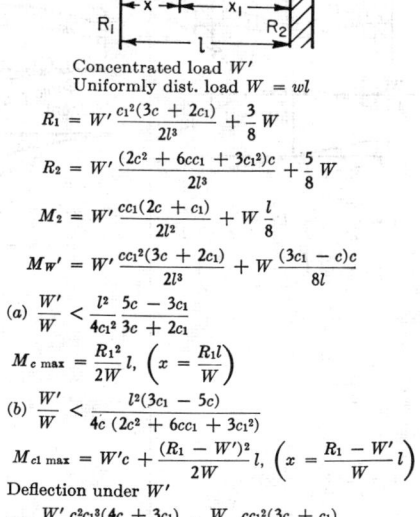

Concentrated load $W'$
Uniformly dist. load $W = wl$

$$R_1 = W' \frac{c_1^2(3c + 2c_1)}{2l^3} + \frac{3}{8} W$$

$$R_2 = W' \frac{(2c^2 + 6cc_1 + 3c_1^2)c}{2l^3} + \frac{5}{8} W$$

$$M_2 = W' \frac{cc_1(2c + c_1)}{2l^2} + W \frac{l}{8}$$

$$M_{w'} = W' \frac{cc_1^2(3c + 2c_1)}{2l^3} + W \frac{(3c_1 - c)c}{8l}$$

$$(a) \; \frac{W'}{W} < \frac{l^2}{4c_1^2} \frac{5c - 3c_1}{3c + 2c_1}$$

$$M_{c \, max} = \frac{R_1^2}{2W} l, \; \left( x = \frac{R_1 l}{W} \right)$$

$$(b) \; \frac{W'}{W} < \frac{l^2(3c_1 - 5c)}{4c(2c^2 + 6cc_1 + 3c_1^2)}$$

$$M_{c1 \, max} = W'c + \frac{(R_1 - W')^2}{2W} l, \; \left( x = \frac{R_1 - W'}{W} l \right)$$

Deflection under $W'$

$$f = \frac{W'}{EI} \frac{c^2 c_1^3(4c + 3c_1)}{12l^3} + \frac{W}{EI} \frac{cc_1^2(3c + c_1)}{48l}$$

other principal axis is the neutral axis. This is the ordinary case, in which the ordinary formula for unit stress may be applied. When the plane of the bending moment does not coincide with one of the principal axes, the above formula for oblique loading may be applied.

### Internal Moment beyond the Elastic Limit

Ordinarily, the expression $M = SI/c$ is used for stresses above the elastic limit, in which case $S$ becomes an experimental coefficient $S_R$, the **modulus of rupture,** and the formula is empirical. The true relation is obtained by applying to the cross

**Table 2. Beams of Uniform Cross Section, Loaded Transversely**  *(Continued)*

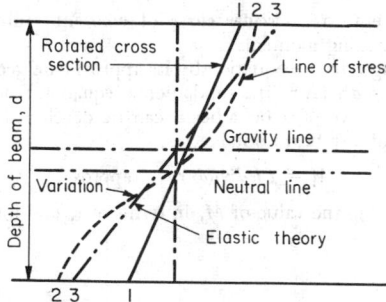

Concentrated load $W'$
Uniformly dist. load $W = wl; c < c_1$

$$R_1 = W'\frac{c_1}{l} + \frac{W}{2}$$

$$R_2 = W'\frac{c}{l} + \frac{W}{2}$$

(a) $\dfrac{W'}{W} < \dfrac{c_1 - c}{2c}$

$$M_{max} = R_2\frac{x_1}{2} = \frac{R_2{}^2 l}{2W}, \left(x_1 = \frac{R_2 l}{W}\right)$$

(b) $\dfrac{W'}{W} > \dfrac{c_1 - c}{2c}$

$$M_{max} = \left(W' + \frac{W}{2}\right)\frac{cc_1}{l}, (x_1 = c_1)$$

Deflection of beam under $W'$:

$$f = \left(W' + \frac{l^2 + cc_1}{8cc_1}W\right)\frac{c^2 c_1{}^2}{3EIl}$$

$c < c_1$

$$R_1 = W'\frac{(3c + c_1)c_1{}^2}{l^3} + \frac{W}{2}$$

$$R_2 = W'\frac{(c + 3c_1)c^2}{l^3} + \frac{W}{2}$$

$$M_{max} = M_1 = W'\frac{cc_1{}^2}{l^2} + \frac{Wl}{12}$$

Deflection under $W'$

$$f = \frac{1}{EI}\left(W'\frac{c^3 c_1{}^3}{3l^3} + W\frac{c^2 c_1{}^2}{24l}\right)$$

section a stress-strain diagram from a tension and compression test, as in Fig. 34. Figure 34 shows the side of a beam of depth $d$ under flexure beyond its elastic limit; line 1–1 shows the distorted cross section; line 3–3, the usual rectilinear relation of stress to strain; and line 2–2, an actual stress-strain diagram, applied to the cross section of the beam, compression above and tension below. The neutral axis is then below the gravity axis. The **outer material** may be expected to develop **greater ultimate strength** than in simple stress, on account of the reinforcing action of material nearer the neutral axis that is not yet overstrained. This leads to an **equalization of stress** over the cross section. $S_R$ exceeds the ultimate strength $S_M$ in tension as follows: for cast iron, $S_R = 2S_M$; for sandstone, $S_R = 3S_M$; for concrete, $S_R = 2.2S_M$; for wood (green), $S_R = 2.3S_M$.

In the case of steel I beams, failure begins practically when the elastic limit in the compression flange is reached.

Because of the support of adjoining material, the **elastic limit in flexure** $S_p$ is also greater than in tension, depending upon the relation of breadth to depth of section. For the same breadth, the difference decreases with increase of height. No difference

will occur in the case of an I beam, or with hard materials. Bauschinger quotes for soft steel plates, 1.27; Considère, 1.37; Hatt, 1.5 (*Railroad Gaz.*, 1899).

Wide plates will not expand and contract freely, and the value of $E$ will be increased on account of side constraint. As a consequence of lateral contraction of the fibers of the tension side of a beam and lateral swelling of fibers at the compression side, the cross section becomes distorted to a trapezoidal shape, and the neutral axis is at the center of gravity of the trapezoid. Strictly, this shape is one with a curved perimeter, the radius being $r/\mu$, where $r$ is the radius of the neutral line of the beam, and $\mu$ is Poisson's ratio.

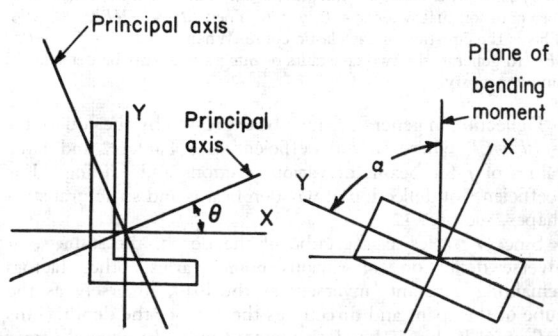

**Fig. 32**          **Fig. 33**

Fig. 34

**Deflection of Beams**

When a beam is subjected to bending, the fibers on one side elongate, while the fibers on the other side shorten (Fig. 35). These changes in length cause the beam to deflect. All points on the beam except those directly over the support fall below their original position, as shown in Figs. 31 and 35.

**Table 3. Uniformly Distributed Loads on Rectangular Beams 1 In Wide***
(Calculated for unit fiber stress at 1,000 lb/in² (70 kgf/cm²): nominal size)
Total load in pounds (kgf)† including weight of beam

| Span , ft (m)‡ | Depth of beam, in (cm)§ | | | | | | | | | | | |
|---|---|---|---|---|---|---|---|---|---|---|---|---|
| | 6 | 7 | 8 | 9 | 10 | 11 | 12 | 13 | 14 | 15 | 16 |
| 5 | 800 | 1090 | 1420 | 1800 | 2220 | 2690 | 3200 | 3750 | 4350 | 5000 | 5690 |
| 6 | 670 | 910 | 1180 | 1500 | 1850 | 2240 | 2670 | 3130 | 3630 | 4170 | 4740 |
| 7 | 570 | 780 | 1010 | 1290 | 1590 | 1920 | 2280 | 2680 | 3110 | 3570 | 4060 |
| 8 | 500 | 680 | 890 | 1120 | 1390 | 1680 | 2000 | 2350 | 2720 | 3130 | 3560 |
| 9 | 440 | 600 | 790 | 1000 | 1230 | 1490 | 1780 | 2090 | 2420 | 2780 | 3160 |
| 10 | 400 | 540 | 710 | 900 | 1110 | 1340 | 1600 | 1880 | 2180 | 2500 | 2840 |
| 11 | 360 | 490 | 650 | 820 | 1010 | 1220 | 1450 | 1710 | 1980 | 2270 | 2590 |
| 12 | 330 | 450 | 590 | 750 | 930 | 1120 | 1330 | 1560 | 1810 | 2080 | 2370 |
| 13 | 310 | 420 | 550 | 690 | 850 | 1030 | 1230 | 1440 | 1680 | 1920 | 2190 |
| 14 | 290 | 390 | 510 | 640 | 790 | 960 | 1140 | 1340 | 1560 | 1790 | 2030 |
| 15 | 270 | 360 | 470 | 600 | 740 | 900 | 1070 | 1250 | 1450 | 1670 | 1900 |
| 16 | 250 | 340 | 440 | 560 | 690 | 840 | 1000 | 1170 | 1360 | 1560 | 1780 |
| 17 | 230 | 320 | 420 | 530 | 650 | 790 | 940 | 1100 | 1280 | 1470 | 1670 |
| 18 | 220 | 300 | 400 | 500 | 620 | 750 | 890 | 1040 | 1210 | 1390 | 1580 |
| 19 | 210 | 290 | 380 | 470 | 590 | 710 | 840 | 990 | 1150 | 1320 | 1500 |
| 20 | 200 | 270 | 360 | 450 | 560 | 670 | 800 | 940 | 1090 | 1250 | 1420 |
| 22 | 180 | 250 | 320 | 410 | 500 | 610 | 730 | 850 | 990 | 1140 | 1290 |
| 24 | 160 | 230 | 290 | 370 | 460 | 560 | 670 | 780 | 910 | 1040 | 1180 |
| 26 | 150 | 210 | 270 | 340 | 420 | 520 | 610 | 720 | 840 | 960 | 1090 |
| 28 | 140 | 190 | 250 | 320 | 390 | 480 | 570 | 670 | 780 | 890 | 1010 |
| 30 | 130 | 180 | 240 | 300 | 370 | 450 | 530 | 630 | 730 | 830 | 950 |

*This table is convenient for wooden beams. For any other fiber stress $S'$, multiply the values in table by $S'/1,000$. See Sec. 12 for properties of wooden beams of commercial sizes.
†To change to kgf, multiply by 0.454.
‡To change to m, multiply by 0.305.
§To change to cm, multiply by 2.54.

The **elastic curve** is the curve taken by the neutral axis. The radius of curvature at any point is

$$r = EI/M$$

A beam bent to a **circular curve** of constant radius has a constant bending moment.

Replacing $r$ in the equation by its approximate geometrical value, $1/r = d^2y/(dx)^2$, the fundamental equation from which the elastic curve of a bent beam can be developed and the deflection of any beam obtained is,

$$M = EI\, d^2y/(dx)^2 \qquad \text{approx}$$

Substituting the value of $M$, in terms of $x$, and integrating

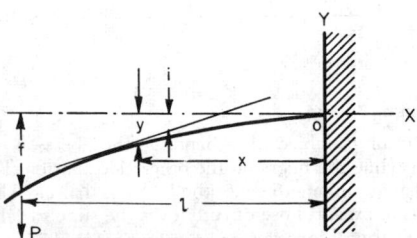

**Fig. 35**

once, gives the slope of the tangent to the elastic curve of the beam at point $x$; $\tan i = dy/dx = \int_0^x M\, dx/EI$. Since $i$ is usually small, $\tan i = i$, expressed in radians. A second integration gives the vertical deflection of any point of the elastic curve from its original position.

EXAMPLE. In the cantilever beam shown in Fig. 35, the bending moment at any section $= -P(l - x) = EI\, d^2y/(dx)^2$. Integrate and determine constant by the condition that when $x = 0$, $dy/dx = 0$. Then $EI\, dy/dx = -Plx + \frac{1}{2}Px^2$. Integrate again, and determine constant by the condition that when $x = 0$, $y = 0$. Then $EIy = -\frac{1}{2}Plx^2 + Px^3/6$. This is the equation of the elastic curve. When $x = l$, $y = f = -Pl^3/3EI$. In general, the two constants of integration must be determined simultaneously.

Deflection in general, $f$, may be expressed by the equation $f = Pl^3/mEI$, where $m$ is a coefficient. See Tables 2 and 4 for values of $f$ for beams of various sections and loadings. For coefficients of deflection of wooden beams and structural steel shapes, see Sec. 12.

Since $I$ varies as the cube of the depth, the **stiffness,** or inverse deflection, of various **beams** varies, other factors remaining constant, inversely as the load, inversely as the cube of the span, and directly as the cube of the depth. This deflection is due to bending moment only. In general, however, the bending of beams involves transverse shearing stresses which cause **shearing strains** and thus **add to the total**

**Table 4. Approximate Safe Loads in Pounds (kgf) on Steel Beams***

Allowable fiber stress for steel, 16,000 lb/in² (1,127 kgf/cm²) (basis of table); for iron, reduce values given in table by one-eighth.

Beams supported at both ends.

$L$ = distance between supports, ft (m)      $a$ = interior area, in² (cm²)
$A$ = sectional area of beam, in² (cm²)      $d$ = interior depth, in (cm)
$D$ = depth of beam, in (cm)                 $w$ = total working load, net tons (kgf)

| Shape of section | Greatest safe load, lb | | Deflection, in | |
|---|---|---|---|---|
| | Load in middle | Load distributed | Load in middle | Load distributed |
| Solid rectangle | $\dfrac{890AD}{L}$ | $\dfrac{1,780AD}{L}$ | $\dfrac{wL^3}{32AD^2}$ | $\dfrac{wL^3}{52AD^2}$ |
| Hollow rectangle | $\dfrac{890(AD-ad)}{L}$ | $\dfrac{1,780(AD-ad)}{L}$ | $\dfrac{wL^3}{32(AD^2-ad^2)}$ | $\dfrac{wL^3}{52(AD^2-ad^2)}$ |
| Solid cylinder | $\dfrac{667AD}{L}$ | $\dfrac{1,333AD}{L}$ | $\dfrac{wL^3}{24AD^2}$ | $\dfrac{wL^3}{38AD^2}$ |
| Hollow cylinder | $\dfrac{667(AD-ad)}{L}$ | $\dfrac{1,333(AD-ad)}{L}$ | $\dfrac{wL^3}{24(AD^2-ad^2)}$ | $\dfrac{wL^3}{38(AD^2-ad^2)}$ |
| Even-legged angle or tee | $\dfrac{885AD}{L}$ | $\dfrac{1,770AD}{L}$ | $\dfrac{wL^3}{32AD^2}$ | $\dfrac{wL^3}{52AD^2}$ |
| Channel or Z bar | $\dfrac{1,525AD}{L}$ | $\dfrac{3,050AD}{L}$ | $\dfrac{wL^3}{53AD^2}$ | $\dfrac{wL^3}{85AD^2}$ |
| Deck beam | $\dfrac{1,380AD}{L}$ | $\dfrac{2,760AD}{L}$ | $\dfrac{wL^3}{50AD^2}$ | $\dfrac{wL^3}{80AD^2}$ |
| I beam | $\dfrac{1,795AD}{L}$ | $\dfrac{3,390AD}{L}$ | $\dfrac{wL^3}{58AD^2}$ | $\dfrac{wL^3}{93AD^2}$ |

**Table 5. Coefficients for Correcting Values in Table 4 for Various Methods of Support and of Loading**

| Conditions of loading | Max relative safe load | Max relative deflection under max relative safe load |
|---|---|---|
| Beam supported at ends: | | |
| Load uniformly distributed over span | 1.0 | 1.0 |
| Load concentrated at center of span | ½ | 0.80 |
| Two equal loads symmetrically concentrated | $l/4c$ | |
| Load increasing uniformly to one end | 0.974 | 0.976 |
| Load increasing uniformly to center | ¾ | 0.96 |
| Load decreasing uniformly to center | ³⁄₂ | 1.08 |
| Beam fixed at one end, cantilever: | | |
| Load uniformly distributed over span | ¼ | 2.40 |
| Load concentrated at end | ⅛ | 3.20 |
| Load increasing uniformly to fixed end | ⅜ | 1.92 |
| Beam continuous over two supports equidistant from ends: | | |
| Load uniformly distributed over span | | |
| 1. If distance $a > 0.2071l$ | $l^2/4a^2$ | |
| 2. If distance $a < 0.2071l$ | $\dfrac{l}{l-4a}$ | |
| 3. If distance $a = 0.2071l$ | 5.83 | |
| Two equal loads concentrated at ends | $l/4a$ | |

$l$ = length of beam; $c$ = distance from support to nearest concentrated load; $a$ = distance from support to end of beam.

## Table 6. Properties of Various Cross Sections

($I$ = moment of inertia; $I/c$ = section modulus; $r = \sqrt{I/A}$ = radius of gyration)

| Section | Moment of inertia | Section modulus | Radius of gyration |
|---|---|---|---|
| $I = \dfrac{bh^3}{12}$ $\dfrac{I}{c} = \dfrac{bh^2}{6}$ $r = \dfrac{h}{\sqrt{12}} = 0.289h$ | $\dfrac{bh^3}{3}$ $\dfrac{bh^2}{3}$ $\dfrac{h}{\sqrt{3}} = 0.577h$ | $\dfrac{b^3h^3}{6(b^2+h^2)}$ $\dfrac{b^2h^2}{6\sqrt{b^2+h^2}}$ $\dfrac{bh}{\sqrt{6(b^2+h^2)}}$ | $\dfrac{bh}{12}(h^2\cos^2 a + b^2\sin^2 a)$ $\dfrac{bh}{6}\left(\dfrac{h^2\cos^2 a + b^2\sin^2 a}{h\cos a + b\sin a}\right)$ $\sqrt{\dfrac{h^2\cos^2 a + b^2\sin^2 a}{12}}$ |
| $I = \dfrac{b}{12}(H^3 - h^3)$ $\dfrac{I}{c} = \dfrac{b}{6}\dfrac{H^3-h^3}{H}$ $r = \sqrt{\dfrac{H^3-h^3}{12(H-h)}}$ | $\dfrac{H^4 - h^4}{12}$ $\dfrac{1}{6}\dfrac{H^4-h^4}{H}$ $\sqrt{\dfrac{H^2+h^2}{12}}$ | $\dfrac{H^4-h^4}{12}$ $\dfrac{\sqrt{2}}{12}\dfrac{H^4-h^4}{H}$ $\sqrt{\dfrac{H^2+h^2}{12}}$ | $\dfrac{bh^3}{36}; c = \dfrac{2}{3}h$ $\dfrac{bh^2}{24}$ $\dfrac{h}{\sqrt{18}}$ |
| $I = \dfrac{bh^3}{12}$ $\dfrac{I}{c} = \dfrac{bh^2}{12}$ $r = \dfrac{h}{\sqrt{6}}$ | $\dfrac{5}{8}R^2$ | $\dfrac{5\sqrt{3}}{16}R^4$ $\dfrac{5\sqrt{3}}{16}R^3$ $\sqrt{\dfrac{5}{24}}R$ | $\dfrac{1+2\sqrt{2}}{6}R^4$ $0.6906R^3$ $0.475R$ |

Square, axis same as first rectangle, side = $b$; $I = b^4/12$; $I/c = b^3/6$; $r = 0.289b$.
Square, diagonal taken as axis: $I = b^4/12$; $I/c = 0.1179b^3$; $r = 0.289b$.

**deflection.** These strains may affect substantially the strength as well as the deflection of beams. When deflection due to transverse shear is to be accounted for, the differential equation of the elastic curve takes the form

$$EI\frac{d^2y}{dx^2} = EI\left(\frac{d^2y_b}{dx^2} + \frac{d^2y_s}{dx^2}\right) = M - \frac{kEI}{AG} \times \frac{d^2M}{dx^2}$$

where $k$ is a factor dependent upon the beam cross section. Sergius Sergev, in "The Effect of Shearing Forces on the Deflection and Strength of Beams" (*Univ. Wash. Eng. Exp. Stn.*

*Bull.* 114) gives $k = 1.2$ for rectangular sections, 10/9 for circular sections, and 2.4 for I beams. He also points out that in the case of a deep, rectangular-section cantilever, carrying a concentrated load at the free end, the deflection due to shear may be up to 3.1 percent of that due to bending moment; if this beam supports a uniformly distributed load, it may be up to 4.1 percent. A deep, simple beam deflection may increase up to 15.6 percent when carrying a uniformly distributed load and up to 12.5 percent when the load is concentrated at midspan. These added deflections can be even larger for I beams.

**Table 6. Properties of Various Cross Sections** *(Continued)*

| Section | Moment of inertia | Section modulus | Radius of gyration |
|---|---|---|---|
| Equilateral Polygon<br>$A$ = area, (see p. 1–39)<br>$R$ = rad circumscribed circle<br>$r$ = rad inscribed circle<br>$n$ = no. sides<br>$a$ = length of side<br>Axis as in preceding section of octagon | $I = \dfrac{A}{24}(6R^2 - a^2)$<br><br>$= \dfrac{A}{48}(12r^2 + a^2)$<br><br>$= \dfrac{AR^2}{4}$ (approx) | $\dfrac{I}{c} = \dfrac{I}{r}$<br><br>$= \dfrac{I}{R\cos\dfrac{180°}{n}}$<br><br>$= \dfrac{AR}{4}$ (approx) | $\sqrt{\dfrac{6R^2 - a^2}{24}} \approx \dfrac{R}{2}$<br><br>$\sqrt{\dfrac{12r^2 + a^2}{48}}$ |

| | $I = \dfrac{6b^2 + 6bb_1 + b_1^2}{36(2b + b_1)}h^3$<br>$c = \dfrac{1}{3}\dfrac{3b + 2b_1}{2b + b_1}h$ | $\dfrac{I}{c} = \dfrac{6b^2 + 6bb_1 + b_1^2}{12(3b + b_1)}h^2$ | $\dfrac{h\sqrt{12b^2 + 12bb_1 + 2b_1^2}}{6(2b + b_1)}$ |

| | $I = \dfrac{BH^3 + bh^3}{12}$<br>$\dfrac{I}{c} = \dfrac{BH^3 + bh^3}{6H}$ | | $\sqrt{\dfrac{BH^3 + bh^3}{12(BH + bh)}}$ |

| | $I = \dfrac{BH^3 - bh^3}{12}$<br>$\dfrac{I}{c} = \dfrac{BH^3 - bh^3}{6H}$ | | $\sqrt{\dfrac{BH^3 - bh^3}{12(BH - bh)}}$ |

| | $I = \frac{1}{3}(Bc_1^3 - B_1h^3 + bc_2^3 - b_1h_1^3)$<br>$c_1 = \dfrac{1}{2}\dfrac{aH^2 + B_1d^2 + b_1d_1(2H - d_1)}{aH + B_1d + b_1d_1}$ | | $\sqrt{\dfrac{I}{(Bd + bd_1) + a(h + h_1)}}$ |

| | $I = \frac{1}{3}(Bc_1^3 - bh^3 + ac_2^3)$<br>$c_1 = \dfrac{1}{2}\dfrac{aH^2 + bd^2}{aH + bd}$<br>$c_2 = H - c_1$<br>$r = \sqrt{\dfrac{I}{[Bd + a(H - d)]}}$ | | |

| | $I = \dfrac{\pi d^4}{64} = \dfrac{\pi r^4}{4} = \dfrac{A}{4}r^2$<br>$= 0.05d^4$ (approx) | $\dfrac{I}{c} = \dfrac{\pi d^3}{32} = \dfrac{\pi r^3}{4} = \dfrac{A}{4}r$<br>$= 0.1d^3$ (approx) | $\dfrac{r}{2} = \dfrac{d}{4}$ |

**Design of beams** may be based on **strength** (stress) or on **stiffness** if deflection must be limited. When more than one beam shares a load, each beam will assume a portion of the load that is proportional to its stiffness. **Superposition** may be used in connection with both stresses and deflections.

EXAMPLE (Fig. 36). Two wooden stringers—one $(A)$ 8 × 16 in in cross section and 20 ft in span, the other $(B)$ 8 in × 8 in × 16 ft— carrying the center load $P_0 = 22{,}000$ lb. Required, the load carried by each stringer. The deflections $f$ of the two stringers must be equal. Load on $A = P_1$, and on $B = P_2$. $f = P_1 l_1^3/48EI_1 = P_2 l_2^3/48EI_2$. Then

### Table 6. Properties of Various Cross Sections   (Continued)

| Section | Moment of inertia | Section modulus | Radius of gyration |
|---|---|---|---|
| $d_m = \tfrac{1}{2}(D + d)$ <br> $s = \tfrac{1}{2}(D - d)$ | $\begin{aligned} I &= \frac{\pi}{64}(D^4 - d^4) \\ &= \frac{\pi}{4}(R^4 - r^4) \\ &= \tfrac{1}{4}A(R^2 + r^2) \\ &= 0.05(D^4 - d^4) \\ & \qquad \text{(approx)} \end{aligned}$ | $\begin{aligned} \frac{I}{c} &= \frac{\pi}{32}\frac{D^4 - d^4}{D} \\ &= \frac{\pi}{4}\frac{R^4 - r^4}{R} \\ &= 0.8d_m{}^2s \ \text{(approx)} \\ & \text{when } \frac{s}{d_m} \text{ is very small} \end{aligned}$ | $\dfrac{\sqrt{R^2 + r^2}}{2} = \dfrac{\sqrt{D^2 + d^2}}{4}$ |
| | $\begin{aligned} I &= r^4\left(\frac{\pi}{8} - \frac{8}{9\pi}\right) \\ &= 0.1098r^4 \end{aligned}$ | $\begin{aligned} \frac{I}{c_2} &= 0.1908r^3 \\ \frac{I}{c_1} &= 0.2587r^3 \\ c_1 &= 0.4244r \end{aligned}$ | $\dfrac{\sqrt{9\pi^2 - 64}}{6\pi}\,r = 0.264r$ |
| | $\begin{aligned} I &= 0.1098(R^4 - r^4) \\ & \quad - \frac{0.283R^2r^2(R - r)}{R + r} \\ &= 0.3tr_1{}^3 \ \text{(approx)} \\ & \text{when } \frac{t}{r_1} \text{ is very small} \end{aligned}$ | $\begin{aligned} c_1 &= \frac{4}{3\pi}\frac{R^2 + Rr + r^2}{R + r} \\ c_2 &= R - c_1 \end{aligned}$ | $\begin{aligned} & \sqrt{\frac{2I}{\pi(R^2 - r^2)}} \\ & = 0.31r_1 \ \text{(approx)} \end{aligned}$ |
| | $I = \dfrac{\pi a^3 b}{4} = 0.7854a^3 b$ | $\dfrac{I}{c} = \dfrac{\pi a^2 b}{4} = 0.7854a^2 b$ | $\dfrac{a}{2}$ |
| | $\begin{aligned} I &= \frac{\pi}{4}(a^3 b - a_1{}^3 b_1) \\ &= \frac{\pi}{4}a^2(a + 3b)t \\ & \qquad \text{(approx)} \end{aligned}$ | $\begin{aligned} \frac{I}{c} &= \frac{\pi}{4}a(a + 3b)t \\ & \qquad \text{(approx)} \end{aligned}$ | $\begin{aligned} & \sqrt{\frac{I}{(\pi ab - a_1 b_1)}} = \\ & \frac{a}{2}\sqrt{\frac{a + 3b}{a + b}} \ \text{(approx)} \end{aligned}$ |
| | $\begin{aligned} I &= \frac{1}{12}\left[\frac{3\pi}{16}d^4 + b(h^3 - d^3) + b^3(h - d)\right] \\ \frac{I}{c} &= \frac{1}{6h}\left[\frac{3\pi}{16}d^4 + b(h^3 + d^3) + b^3(h - d)\right] \end{aligned}$ | | $\begin{aligned} & \sqrt{\frac{I}{\pi\frac{d^2}{4} + 2b(h - d)}} \\ & \qquad \text{(approx)} \end{aligned}$ |
| | $\begin{aligned} I &= \frac{t}{4}\left(\frac{\pi B^3}{16} + B^2 h + \frac{\pi B h^2}{2} + \frac{2}{3}h^3\right) \\ & \qquad h = H - \tfrac{1}{2}B \\ \frac{I}{c} &= \frac{2I}{H + t} \end{aligned}$ | | $\sqrt{\dfrac{I}{2\left(\dfrac{\pi B}{4} + h\right)t}}$ |

**Table 6. Properties of Various Cross Sections**  *(Continued)*

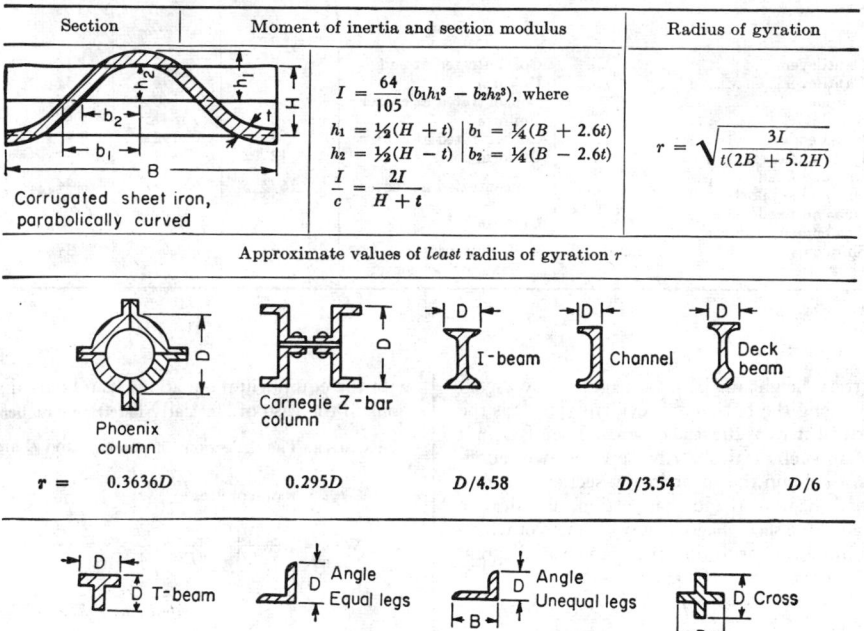

| Section | Moment of inertia and section modulus | Radius of gyration |
|---|---|---|
| Corrugated sheet iron, parabolically curved | $I = \dfrac{64}{105}(b_1 h_1^3 - b_2 h_2^3)$, where $h_1 = \frac{1}{2}(H + t)$   $b_1 = \frac{1}{4}(B + 2.6t)$ $h_2 = \frac{1}{2}(H - t)$   $b_2 = \frac{1}{4}(B - 2.6t)$ $\dfrac{I}{c} = \dfrac{2I}{H + t}$ | $r = \sqrt{\dfrac{3I}{t(2B + 5.2H)}}$ |

Approximate values of *least* radius of gyration $r$

Phoenix column — Carnegie Z-bar column — I-beam — Channel — Deck beam

| $r =$ | $0.3636D$ | $0.295D$ | $D/4.58$ | $D/3.54$ | $D/6$ |

T-beam — Angle Equal legs — Angle Unequal legs — Cross

| $r =$ | $D/4.74$ | $D/5$ | $BD/2.6(B + D)$ | $D/4.74$ |

$P_1/P_2 = l_2^3 I_1 / l_1^3 I_2 = 4$. $P_0 = P_1 + P_2 = 4P_2 + P_2$, whence $P_2 = 22,000/5 = 4,400$ lb (1,998 kgf) and $P_1 = 4 \times 4,400 = 17,600$ lb (7,990 kgf).

### Relation between Deflection and Stress

Combine the formula $M = SI/c = Pl/n$, where $n$ is a constant, $P =$ load, and $l =$ span, with formula $f = Pl^3/mEI$, where $m$ is a constant. Then

$$f = C''Sl^2/Ec$$

where $C''$ is a new constant $= n/m$. Other factors remaining the same, the **deflection varies directly as the stress and inversely as** $E$. If the span is constant, a shallow beam will submit to greater deformations than a deeper beam without exceeding a safe stress. If depth is constant, a beam of double span will attain a given deflection with only one-quarter the stress. Values of $n$, $m$, and $C''$ are given in Table 7 (for other values, see Table 2):

### Graphical Relations

Referring to Fig. 37, the shear $V$ acting at any section is equal to the total load on the right of the section, or

$$V = \int w \, dx$$

Since $w \, dx$ is the product of $w$, a loading intensity (which is

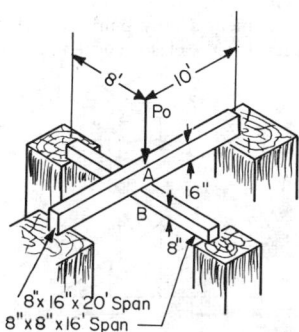

8" × 16" × 20' Span
8" × 8" × 16' Span

**Fig. 36**

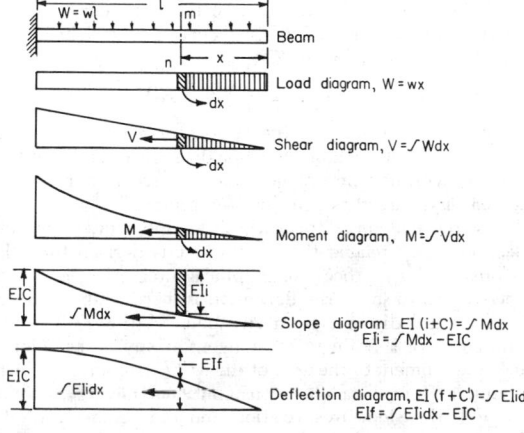

| | |
|---|---|
| $W = wl$ | Beam |
| Load diagram, $W = wx$ | |
| $V$ | Shear diagram, $V = \int W dx$ |
| $M$ | Moment diagram, $M = \int V dx$ |
| $EIC$  $\int M dx$  $EIi$ | Slope diagram $EI(i + C) = \int M dx$  $EIi = \int M dx - EIC$ |
| $EIC$  $\int EIi dx$  $EIf$ | Deflection diagram, $EI(f + C') = \int EIi dx$  $EIf = \int EIi dx - EIC$ |

**Fig. 37**

**Table 7**

| Beam | Load | $n$ | $m$ | $C''$ |
|---|---|---|---|---|
| Cantilever.................... | Concentrated at end | 1 | 3 | ⅓ |
| Cantilever.................... | Uniform | 2 | 8 | ¼ |
| Simple....................... | Concentrated at center | 4 | 48 | ¹⁄₁₂ |
| Simple....................... | Uniform | 8 | 384/5 | ⁵⁄₄₈ |
| Fixed ends................... | Concentrated at center | 8 | 192 | ¹⁄₂₄ |
| Fixed ends................... | Uniform | 12 | 384 | ¹⁄₃₂ |
| One end fixed<br>One end supported } ............. | Concentrated at center | 16/3 | 768/7 | ⁷⁄₁₄₄ |
| One end fixed<br>One end supported } ............. | Uniform | 128/9 | 185 | ¹⁄₁₃ |
| Simple....................... | Uniformly varying,<br>maximum at center | 6 | 60 | ¹⁄₁₀ |

expressed as a vertical height in the load diagram), by $dx$, an elementary length along the horizontal, evidently $w\ dx$ is the area of a small vertical strip of the **load diagram.** Then $\int w\ dx$ is the summation of all such vertical strips between two indefinite points. Thus, to obtain the shear in any section $mn$, find the area of the load diagram up to that section, and draw a second diagram called the **shear diagram,** any ordinate of which is proportional to the shear, or to the area in the load diagram to the right of $mn$. Since $V = dM/dx$,

$$\int V\ dx = M$$

By similar reasoning, a **moment diagram** may be drawn, such that the ordinate at any point is proportional to the area of the shear diagram to the right of that point. Since $M = EI\ d^2f/(dx)^2$,

$$\int M\ dx = EI[(df/dx) + C] = EI(i + C)$$

if $I$ is constant. Here $C$ is a constant of integration. Thus $i$, the slope or grade of the elastic curve at any point, is proportional to the area of the moment diagram $\int M\ dx$ up to that point; and a **slope diagram** may be derived from the moment diagram in the same manner as the moment diagram was derived from the shear diagram.

If $I$ is not constant, draw a new curve whose ordinates are $M/I$ and use these $M/I$ ordinates just as the $M$ ordinates were used in the case where $I$ was constant; that is, $\int (M/I)dx = E(i + C)$. The ordinate at any point of the slope curve is thus proportional to the area of the $M/I$ curve to the right of that point. Again, since $iE = E\ df/dx$,

$$\int iE\ dx = \int E\ df = E(f + C')$$

and thus the ordinate $f$ to the elastic curve at any point is proportional to the area of the slope diagram $\int i\ dx$ up to that point. The equilibrium polygon may be used in drawing the **deflection curve** directly from the $M/I$ diagram.

Thus, the five curves of load, shear, moment, slope, and deflection are so related that each curve is derived from the previous one by a process of graphical integration, and with proper regard to scales the deflection is thereby obtained.

The vertical distance from any point $A$ (Fig. 38) on the elastic curve of a beam to the tangent at any other point $B$ equals the moment of the area of the $M/EI$ diagram from $A$ to $B$ about $A$. This distance, the **tangential deviation** $t_{AB}$, may be used with the slope-area relation and the geometry of the elastic curve to obtain deflections. These theorems, together

with the equilibrium equations, can be used to compute reactions in the case of statically **indeterminate beams.**

EXAMPLE. The deflections of points $B$ and $D$ are

$$y_B = -t_{AB} = \text{moment area } \frac{M}{EI}\Big|_A^B = -\frac{1}{EI} \times \frac{Pl}{4} \times \frac{l}{4} \times \frac{l}{3} = -\frac{Pl^3}{48EI}$$

$$\theta_C = \nabla\theta\Big|_B^C = \text{area } \frac{M}{EI}\Big|_B^C = \frac{1}{EI} \times \frac{Pl}{4} \times \frac{l}{4} = \frac{Pl^2}{16EI}$$

$$y_D = -(\theta_C \times \frac{l}{4} - t_{DC}) = -\frac{Pl^2}{16EI} \times \frac{l}{4} + \frac{1}{EI} \times \frac{Pl}{8} \times \frac{l}{8} \times \frac{l}{12} = -\frac{11Pl^3}{768EI}$$

### Resilience of Beams

The external work of a load gradually applied to a beam, and which increases from zero to $P$, is $\frac{1}{2}Pf$ and equals the **resilience** $U$. But, from the formulas $P = nSI/cl$ and $f = nSl^2/mcE$, where $n$ and $m$ are constants that depend upon loading and supports, $S$ = fiber stress, $c$ = distance from neutral axis to outer fiber, and $l$ = length of span. Substitute for $P$ and $f$, and

$$U = \frac{n^2}{m}\left(\frac{k}{c}\right)^2 \frac{S^2V}{2E}$$

where $k$ is the radius of gyration and $V$ the volume of the beam. For values of $U$, see Table 1.

The resilience of beams of similar cross section at a given stress is proportional to their volumes. The **internal resilience,** or the elastic deformation energy in the material of a beam in a length $x$ is $dU$, and

$$U = \frac{1}{2}\int M^2\ dx/EI = \frac{1}{2}\int M\ di$$

where $M$ is the moment at any point $x$, and $di$ is the angle between the tangents to the elastic curve at the ends of $dx$. The

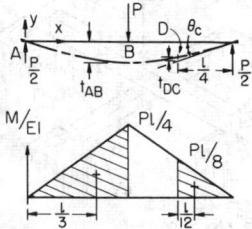

**Fig. 38**

values of resilience and deflection in special cases are easily developed from this equation.

### Rolling Loads

**Rolling** or **moving loads** are those loads which may change their position on a beam. Figure 39 represents a beam with two equal concentrated moving loads, such as two wheels on a crane girder, or the wheels of a truck on a bridge. Since the maximum moment occurs where the shear is zero, it is evident from the shear diagram that the maximum moment will occur under a wheel. $x < a/2$:

$$R_1 = P\left(1 - \frac{2x}{l} + \frac{a}{l}\right)$$

$$M_2 = \frac{Pl}{2}\left(1 - \frac{a}{l} + \frac{2x}{l}\frac{a}{l} - \frac{4x^2}{l^2}\right)$$

$$R_2 = P\left(1 + \frac{2x}{l} - \frac{a}{l}\right)$$

$$M_1 = \frac{Pl}{2}\left(1 - \frac{a}{l} - \frac{2a^2}{l^2} + \frac{2x}{l}\frac{3a}{l} - \frac{4x^2}{l^2}\right)$$

$$M_2 \text{ max when } x = \tfrac{1}{4}a$$

$$M_1 \text{ max when } x = \tfrac{3}{4}a$$

$$M_{\max} = \frac{Pl}{2}\left(1 - \frac{a}{2l}\right)^2 = \frac{P}{2l}\left(l - \frac{a}{2}\right)^2$$

EXAMPLE. Two wheel loads of 3,000 lb each, spaced on 5-ft centers, move on a span of $l = 15$ ft, $x = 1.25$ ft, and $R_2 = 2,500$ lb. $\therefore M_{\max} = M_2 = 2,500 \times 6.25 \, (1,135 \times 1.90) = 15,600$ lb:ft (2,159 kgf:m.)

Figure 40 shows the condition when two equal loads are

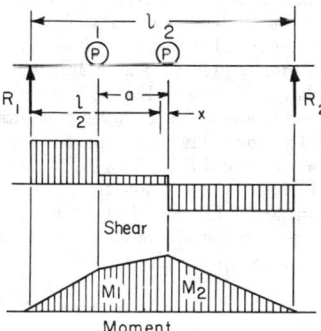

**Fig. 39**

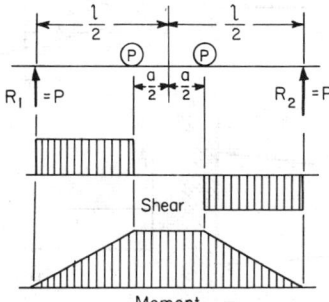

**Fig. 40**

equally distant on opposite sides of the center. The moment is equal under the two loads.

If the **two moving loads** are **of unequal weight,** the condition for **maximum moment** is that the maximum moment will occur under the heavy wheel when the center of the beam bisects the distance between the resultant of the loads and the heavy wheel. Figure 41 shows this position and the shear and moment diagrams.

When **several wheel loads** constituting a system occur, the several suspected wheels must be examined in turn to determine which will cause the greatest moment. The **position for the greatest moment** that can occur under a given wheel is, as stated, when the center of the span bisects the distance between the wheel in question and the resultant of all the loads then on the span. The **position for maximum shear** at the support will be when one wheel is passing off the span.

### Constrained Beams

Constrained beams are those so held or "built in" at one or both ends that the tangent to the elastic curve remains fixed in direction. These beams are held at the ends in such a manner as to allow free horizontal motion, as illustrated by Fig. 42. A constrained beam is stiffer than a simple beam of the same material, on account of the modification of the moment by an end resisting moment. Figure 43 shows the two most common cases of constrained beams. See also Table 2.

### Continuous Beams

A continuous beam is one resting upon several supports which may or may not be in the same horizontal plane. The general discussion for beams holds for continuous beams. $S_vA = V$, $SI/c = M$, and $d^2f/dx^2 = M/EI$. The **shear** at any section is equal to the algebraic sum of the components parallel to the section of all external forces on either side of the section. The

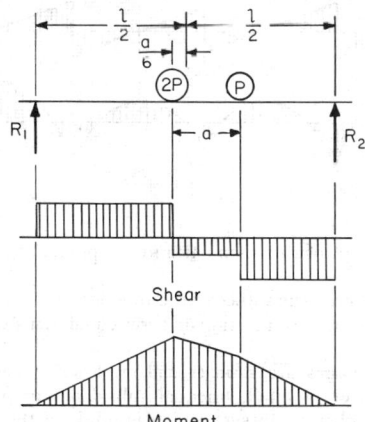

**Fig. 41**

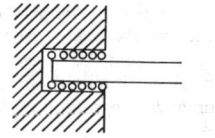 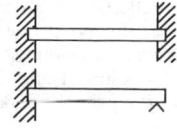

**Fig. 42**          **Fig. 43**

bending moment at any section is equal to the moment of all external forces on either side of the section. The relations stated above between shear and moment diagrams hold true for continuous beams. The bending moment at any section is equal to the bending moment at any other section, plus the shear at that section times its arm, plus the product of all the intervening external forces times their respective arms. To illustrate (Fig. 44):

$$V_x = R_1 + R_2 + R_3 - P_1 - P_2 - P_3$$
$$M_x = R_1(l_1 + l_2 + x) + R_2(l_2 + x) + R_3 x$$
$$\quad - P_1(l_2 + c + x) - P_2(b + x) - P_3 a$$
$$M_x = M_3 + V_3 x - P_3 a$$

Table 8 gives the value of the moment at the various supports of a uniformly loaded continuous beam over equal spans, and it also gives the values of the shears on each side of the supports. Note that the shear is of opposite sign on either side of the supports and that the sum of the two shears is equal to the reaction.

Figure 45 shows the relation between the moment and shear diagrams for a uniformly loaded continuous beam of four equal spans (see Table 8). Table 8 also gives the **maximum bending moment** which will occur **between supports**, and in addi-

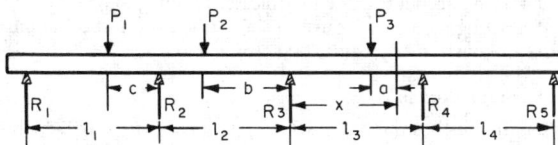

**Fig. 44**

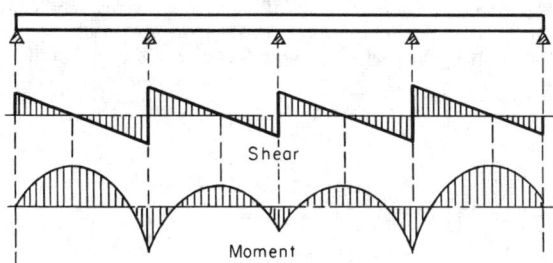

**Fig. 45**

tion the position of this moment and the points of inflection (see Fig. 46).

Figure 46 shows the values of the functions for a uniformly loaded continuous beam resting on three equal spans with four supports.

**Continuous beams** are **stronger and much stiffer than simple beams.** However, a small, unequal subsidence of piers will cause serious changes in sign and magnitude of the bending stresses, reactions, and shears.

**Maxwell's Theorem** When a number of loads rest upon a beam, the deflection at any point is equal to the sum of the deflections at this point due to each of the loads taken separately. Maxwell's theorem states that if unit loads rest upon a beam at two points A and B, the deflection at A due to the unit load at B equals the deflection at B due to the unit load at A.

**Castigliano's theorem** states that the deflection of the point of

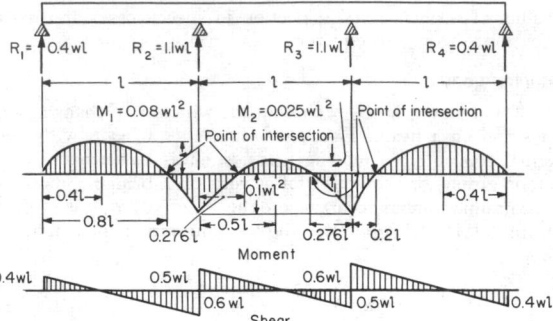

**Fig. 46**

application of an external force acting on a beam is equal to the partial derivative of the work of deformation with respect to this force. Thus, if $P$ be the force, $f$ the deflection, and $U$ the work of deformation, which equals the resilience,

$$dU/dP = f$$

According to the **principle of least work,** the deformation of any structure takes place in such a manner that the work of deformation is a minimum.

**Beams of Uniform Strength**

Beams of uniform strength so vary in section that the unit stress $S$ remains constant, and $I/c$ varies as $M$. For **rectangular beams,** of breadth $b$ and depth $d$, $I/c = bd^2/6$; and $M = Sbd^2/6$. Thus, for a cantilever beam of rectangular cross section, under a load $P$, $Px = Sbd^2/6$. If $b$ is constant, $d^2$ varies with $x$, and the profile of the shape of the beam will be a parabola, as Fig. 47. If $d$ is constant, $b$ will vary as $x$ and the beam will be triangular in plan, as shown in Fig. 48.

**Shear at the end of a beam** necessitates a modification of the forms determined above. The area required to resist shear will be $P/S_v$ in a cantilever and $R/S_v$ in a simple beam. The dotted extensions in Figs. 47 and 48 show the changes necessary to enable these cantilevers to resist shear. The waste in material and extra cost in fabricating, however, make many of the forms impractical, except for cast iron.

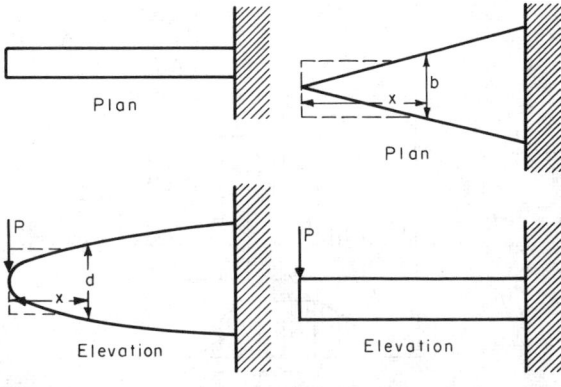

**Fig. 47**          **Fig. 48**

**Table 8. Uniformly Loaded Continuous Beams over Equal Spans**
(Uniform load per unit length = $w$; length of each span = $l$)

| Number of supports | Notation of support of span | Shear on each side of support. $L$ = left, $R$ = right. Reaction at any support is $L + R$ | | Moment over each support | Max moment in each span | Distance to point of max moment, measured to right from support | Distance to point of inflection, measured to right from support |
| --- | --- | --- | --- | --- | --- | --- | --- |
| | | $L$ | $R$ | | | | |
| 2 | 1 or 2 | 0 | ½ | 0 | 0.125 | 0.500 | None |
| 3 | 1 | 0 | ⅜ | 0 | 0.0703 | 0.375 | 0.750 |
| | 2 | ⅝ | ⅝ | ⅛ | 0.0703 | 0.625 | 0.250 |
| 4 | 1 | 0 | ⁴⁄₁₀ | 0 | 0.080 | 0.400 | 0.800 |
| | 2 | ⁶⁄₁₀ | ⁵⁄₁₀ | ¹⁄₁₀ | 0.025 | 0.500 | 0.276, 0.724 |
| 5 | 1 | 0 | ¹¹⁄₂₈ | 0 | 0.0772 | 0.393 | 0.786 |
| | 2 | ¹⁷⁄₂₈ | ¹⁵⁄₂₈ | ³⁄₂₈ | 0.0364 | 0.536 | 0.266, 0.806 |
| | 3 | ¹³⁄₂₈ | ¹³⁄₂₈ | ²⁄₂₈ | 0.0364 | 0.464 | 0.194, 0.734 |
| 6 | 1 | 0 | ¹⁵⁄₃₈ | 0 | 0.0779 | 0.395 | 0.789 |
| | 2 | ²³⁄₃₈ | ²⁰⁄₃₈ | ⁴⁄₃₈ | 0.0332 | 0.526 | 0.268, 0.783 |
| | 3 | ¹⁸⁄₃₈ | ¹⁹⁄₃₈ | ³⁄₃₈ | 0.0461 | 0.500 | 0.196, 0.804 |
| 7 | 1 | 0 | ⁴¹⁄₁₀₄ | 0 | 0.0777 | 0.394 | 0.788 |
| | 2 | ⁶³⁄₁₀₄ | ⁵⁵⁄₁₀₄ | ¹¹⁄₁₀₄ | 0.0340 | 0.533 | 0.268, 0.790 |
| | 3 | ⁴⁹⁄₁₀₄ | ⁵¹⁄₁₀₄ | ⁸⁄₁₀₄ | 0.0433 | 0.490 | 0.196, 0.785 |
| | 4 | ⁵³⁄₁₀₄ | ⁵³⁄₁₀₄ | ⁹⁄₁₀₄ | 0.0433 | 0.510 | 0.215, 0.804 |
| 8 | 1 | 0 | ⁵⁶⁄₁₄₂ | 0 | 0.0778 | 0.394 | 0.789 |
| | 2 | ⁸⁶⁄₁₄₂ | ⁷⁵⁄₁₄₂ | ¹⁵⁄₁₄₂ | 0.0338 | 0.528 | 0.268, 0.788 |
| | 3 | ⁶⁷⁄₁₄₂ | ⁷⁰⁄₁₄₂ | ¹¹⁄₁₄₂ | 0.0440 | 0.493 | 0.196, 0.790 |
| | 4 | ⁷²⁄₁₄₂ | ⁷¹⁄₁₄₂ | ¹²⁄₁₄₂ | 0.0405 | 0.500 | 0.215, 0.785 |
| Values apply to | | $wl$ | $wl$ | $wl^2$ | $wl^2$ | $l$ | $l$ |

The numerical values given are coefficients of the expressions at the foot of each column.

Table 9 shows some of the simple **sections of uniform strength.** In none of these, however, is shear taken into account.

## TORSION

Under torsion, a bar (Fig. 49) is twisted by a couple of the value $Pp$. Elements of the surface become helices of angle $d$, and a radius rotates through an angle $a$ in a length $l$, both $d$ and $a$ being expressed in radians. $S_v$ = shearing unit stress at distance $r$ from center; $I_P$ = polar moment of inertia; $G$ = shearing modulus of elasticity. It is assumed that the cross sections remain plane surfaces. The strain on the cross section is wholly tangential, and is zero at the center of the section. $ld = ra$.

In the case of a **circular cross section**, the stress $S_v$ increases directly as the distance of the strained element from the center.

The polar moment of inertia $I_P$ for any section may be obtained from $I_P = I_1 + I_2$, where $I_1$ and $I_2$ are the rectangular moments of inertia of the section about any two lines at right angles to each other, through the center of gravity.

The **external twisting moment** $M_t$ is balanced by the internal resisting moment.

For strength, $M_t = S_v I_P / r$.

For stiffness, $M_t = aGI_P/l$.

The torsional resilience $U = \frac{1}{2}Ppa = S_v^2 I_P l / 2r^2 G = a^2 GI_P / 2l$.

The state of stress on an element taken from the surface of

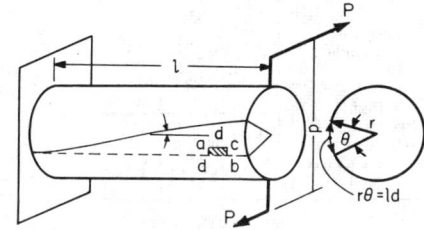

**Fig. 49**

### Table 9. Beams of Uniform Strength (in Bending)

1. FIXED AT ONE END, LOAD $P$ CONCENTRATED AT OTHER END

| Beam | Cross section | Elevation and plan | Formulas |
|---|---|---|---|
| | Rectangle: width ($b$) constant, depth ($y$) variable | Elevation: 1, top, straight line; bottom, parabola. 2, complete parabola.  Plan: rectangle | $y^2 = \dfrac{6P}{bS_s} x$   $h = \sqrt{\dfrac{6Pl}{bS_s}}$   Deflection at $A$: $f = \dfrac{8P}{bE}\left(\dfrac{l}{h}\right)^3$ |
| | Rectangle: width ($y$) variable, depth ($h$) constant | Elevation: rectangle  Plan: triangle | $y = \dfrac{6P}{h^2 S_s} x$   $b = \dfrac{6Pl}{h^2 S_s}$   Deflection at $A$: $f = \dfrac{6P}{bE}\left(\dfrac{l}{h}\right)^3$ |
| | Rectangle: width ($z$) variable, depth ($y$) variable  $\dfrac{z}{y} = k$ (const) | Elevation: cubic parabola  Plan: cubic parabola | $y^3 = \dfrac{6P}{kS_s} x$   $z = ky$   $h = \sqrt[3]{\dfrac{6Pl}{kS_s}}$   $b = kh$ |
| | Circle: diam ($y$) variable | Elevation: cubic parabola  Plan: cubic parabola | $y^3 = \dfrac{32P}{\pi S_s} x$   $d = \sqrt[3]{\dfrac{32Pl}{\pi S_s}}$ |

2. FIXED AT ONE END, LOAD $P$ UNIFORMLY DISTRIBUTED OVER $l$

| Beam | Cross section | Elevation and plan | Formulas |
|---|---|---|---|
| | Rectangle: width ($b$) constant, depth ($y$) variable | Elevation: triangle  Plan: rectangle | $y = x\sqrt{\dfrac{3P}{blS}}$   $h = \sqrt{\dfrac{3Pl}{bS_s}}$   $f = 6\dfrac{P}{bE}\left(\dfrac{l}{h}\right)^3$ |
| | Rectangle: width ($y$) variable, depth ($h$) constant | Elevation: rectangle  Plan: two parabolic curves with vertices at free end | $y = \dfrac{3Px^2}{lS_s h^2}$   $b = \dfrac{3Pl}{S_s h^2}$   Deflection at $A$: $f = \dfrac{3P}{bE}\left(\dfrac{l}{h}\right)^3$ |

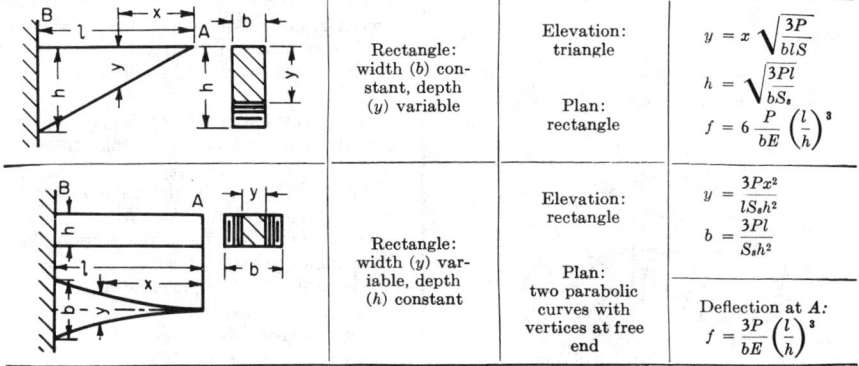

**Table 9. Beams of Uniform Strength (in Bending)**   *(Continued)*

2. FIXED AT ONE END, LOAD $P$ UNIFORMLY DISTRIBUTED OVER $l$

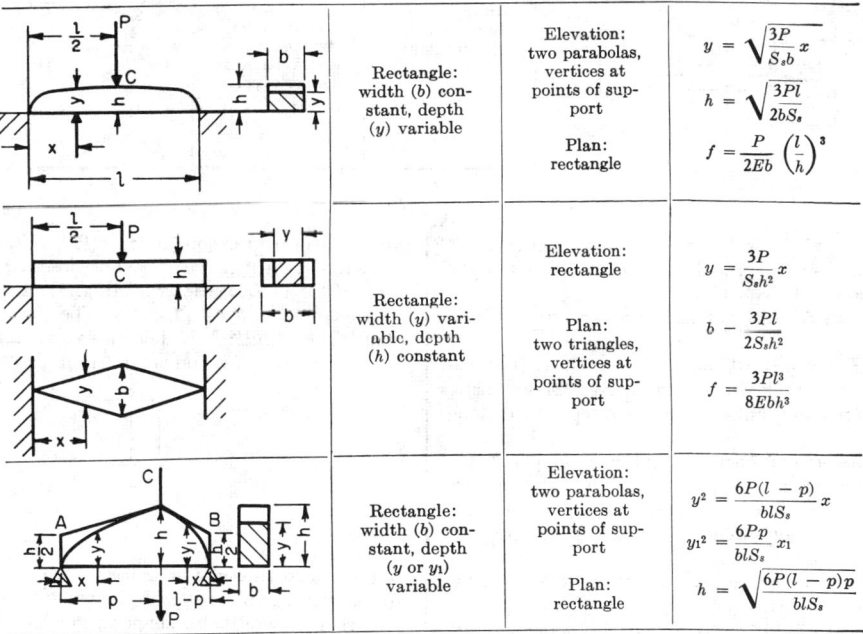

| Beam | Cross section | Elevation and plan | Formulas |
|---|---|---|---|
| | Rectangle: width ($z$) variable, depth ($y$) variable, $\dfrac{z}{y} = k$ | Elevation: semicubic parabola<br><br>Plan: semicubic parabola | $y^3 = \dfrac{3Px^2}{kS_sl}$<br><br>$z = ky$<br><br>$h = \sqrt[3]{\dfrac{3Pl}{kS_s}}$<br><br>$b = kh$ |
| | Circle: diam ($y$) variable | Elevation: semicubic parabola<br><br>Plan: semicubic parabola | $y^3 = \dfrac{16P}{\pi lS_s}x^2$<br><br>$d = \sqrt[3]{\dfrac{16Pl}{\pi S_s}}$ |

3. SUPPORTED AT BOTH ENDS, LOAD $P$ CONCENTRATED AT POINT $C$

| Beam | Cross section | Elevation and plan | Formulas |
|---|---|---|---|
| | Rectangle: width ($b$) constant, depth ($y$) variable | Elevation: two parabolas, vertices at points of support<br><br>Plan: rectangle | $y = \sqrt{\dfrac{3P}{S_sb}}\,x$<br><br>$h = \sqrt{\dfrac{3Pl}{2bS_s}}$<br><br>$f = \dfrac{P}{2Eb}\left(\dfrac{l}{h}\right)^3$ |
| | Rectangle: width ($y$) variable, depth ($h$) constant | Elevation: rectangle<br><br>Plan: two triangles, vertices at points of support | $y = \dfrac{3P}{S_sh^2}x$<br><br>$b = \dfrac{3Pl}{2S_sh^2}$<br><br>$f = \dfrac{3Pl^3}{8Ebh^3}$ |
| | Rectangle: width ($b$) constant, depth ($y$ or $y_1$) variable | Elevation: two parabolas, vertices at points of support<br><br>Plan: rectangle | $y^2 = \dfrac{6P(l-p)}{blS_s}x$<br><br>$y_1^2 = \dfrac{6Pp}{blS_s}x_1$<br><br>$h = \sqrt{\dfrac{6P(l-p)p}{blS_s}}$ |

LOAD $P$ MOVING ACROSS SPAN

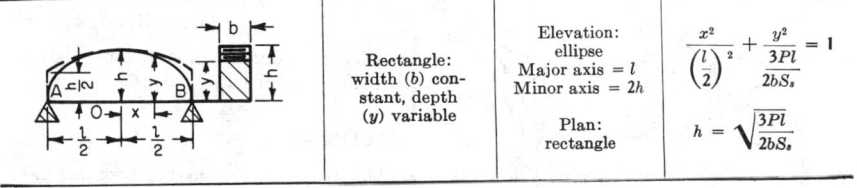

| Beam | Cross section | Elevation and plan | Formulas |
|---|---|---|---|
| | Rectangle: width ($b$) constant, depth ($y$) variable | Elevation: ellipse<br>Major axis $= l$<br>Minor axis $= 2h$<br><br>Plan: rectangle | $\dfrac{x^2}{\left(\dfrac{l}{2}\right)^2} + \dfrac{y^2}{\dfrac{3Pl}{2bS_s}} = 1$<br><br>$h = \sqrt{\dfrac{3Pl}{2bS_s}}$ |

**Table 9. Beams of Uniform Strength (in Bending)** *(Continued)*

4. SUPPORTED AT BOTH ENDS, LOAD $P$ UNIFORMLY DISTRIBUTED OVER $l$

| Beam | Cross section | Elevation and plan | Formulas |
|---|---|---|---|
| | Rectangle: width ($b$) constant, depth ($y$) variable | Elevation: ellipse | $\dfrac{x^2}{\left(\dfrac{l}{2}\right)^2} + \dfrac{y^2}{\dfrac{3Pl}{4bS_s}} = 1$ |
| | | | $h = \sqrt{\dfrac{3Pl}{4bS_s}}$ |
| | | Plan: rectangle | Deflection at $O$: |
| | | | $f = \dfrac{1}{64}\dfrac{Pl^3}{EI}$ |
| | | | $= \dfrac{3}{16}\dfrac{P}{bE}\left(\dfrac{l}{h}\right)^3$ |
| | Rectangle: width ($y$) variable, depth ($h$) constant | Elevation: rectangle | $y = \dfrac{3P}{S_s h^2}\left(x - \dfrac{x^2}{l}\right)$ |
| | | Plan: two parabolas with vertices at center of span | $b = \dfrac{3Pl}{4S_s h^2}$ |

the shaft, as in Fig. 50, is pure shear. Pure tension exists at right angles to one 45° helix and pure compression at right angles to the opposite helix.

Reduced **formulas for shafts of various sections** are given in Table 11.

**Failure under torsion** in brittle materials is a tensile failure at right angles to a helical element on the surface. Plastic materials twist off squarely. Fibrous materials separate in long strips.

**Torsion of Noncircular Sections**   When a section is not circular, the unit stress no longer varies directly as the distance from the center. Cross sections become warped, and the greatest unit stress usually occurs at a point on the perimeter of the cross section *nearest* the axis of twist. There is no stress at the corners of square and rectangular sections, and the analyses become complex.

Assuming the stress distribution from the point of maximum stress to the corner to be parabolic, Bach derived the

approximate expression, $S_s M = 9M_t/2b^2h$ for a rectangular section, $b$ by $b$, where $h > b$. For closer results, the shearing stresses for a **rectangular section** (Fig. 51) may be expressed $S_A = M_t/\alpha_A b^2 h$ and $S_B = M_t/\alpha_B b^2 h$. The angle of twist for these shafts is $\theta = M_t l/\beta G b^3 h$. The factors $\alpha_A$, $\alpha_B$, and $\beta$ are functions of the ratio $h/b$ and are given in Table 10.

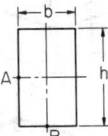

**Fig. 51**

In the case of **composite sections,** such as a tee or angle, the torque that can be resisted is $M_t = G\theta\Sigma\beta b h^3$; the summation applies to each of the rectangles into which the section can be divided. The maximum stress occurs on the component rectangle having the largest $b$ value. It is computed from

$$S_A = M_t\beta_A b_A/\alpha_A\Sigma\beta b h^3$$

Torque, deflection, and work relations for some additional sections are given in Table 11.

## COLUMNS

Members subjected to direct compression can be grouped into three classes. **Compression blocks** are so short (slenderness ratios below 30) that bending of member is not pending. At the other

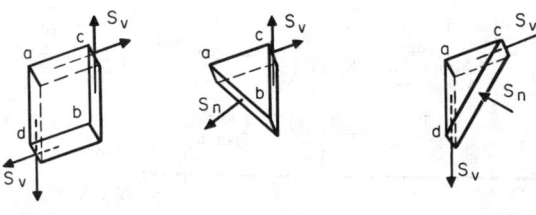

**Fig. 50**

**Table 10. Factors for Torsion of Rectangular Shafts**

| $b/b$ | 1.00 | 1.50 | 1.75 | 2.00 | 2.50 | 3.00 | 4.00 | 5.00 | 6.00 | 8.00 | 10.0 | ∞ |
|---|---|---|---|---|---|---|---|---|---|---|---|---|
| $\alpha_A$ | 0.208 | 0.231 | 0.239 | 0.246 | 0.258 | 0.267 | 0.282 | 0.291 | 0.299 | 0.307 | 0.312 | 0.333 |
| $\alpha_B$ | 0.208 | 0.269 | 0.291 | 0.309 | 0.336 | 0.355 | 0.378 | 0.392 | 0.402 | 0.414 | 0.421 | |
| $\beta$ | 0.141 | 0.196 | 0.214 | 0.229 | 0.249 | 0.263 | 0.281 | 0.291 | 0.299 | 0.307 | 0.312 | 0.333 |

limit, columns so slender that bending is primary, are the **long columns** defined by Euler's theory. The intermediate columns, quite common in practice, are called **short columns.**

Long columns and the more slender short columns usually fail by **buckling** when the **critical load** is reached. This is a matter of **instability;** that is, the column may continue to yield and deflect even though the load is not being increased above critical. The **slenderness ratio** is the unsupported length divided by the least radius of gyration, parallel to which it can bend.

**Long columns** are handled by **Euler's column formula,**

$$P_{cr} = n\pi^2 EI/l^2 = n\pi^2 EA/(l/r)^2$$

The **coefficient** $n$ accounts for **end conditions.** When the column is pivoted at both ends, $n = 1$; when one end is fixed and other rounded, $n = 2$; when both are fixed, $n = 4$; and when one end is fixed with the other free, $n = \frac{1}{4}$. The slenderness ratio that separates long columns from short ones depends upon the modulus of elasticity and the yield strength of the column material. When Euler's formula results in $(P_{cr}/A) > S_y$,

**Table 11. Torsion of Shafts of Various Cross Sections**
(For strength and stiffness of shafts, see Sec. 8)

| Cross section | Torsional resisting moment $M_t$ | Angular twist, $\theta_1$ (length = 1 in., radius = 1 in.) | | Work of torsion ($V$ = volume) |
|---|---|---|---|---|
| | | In terms of torsional moment | In terms of max shear | |
| | $\dfrac{\pi}{16} d^3 S_v$ | $\dfrac{M_t}{GI_P} = \dfrac{32}{\pi d^4}\dfrac{M_t}{G}$ | $2\dfrac{S_{v\max}}{G}\dfrac{1}{d}$ | $\dfrac{1}{4}\dfrac{S_v^2{}_{\max}}{G} V$ (Note 1) |
| | $\dfrac{\pi}{16}\dfrac{D^4 - d^4}{D} S_v$ | $\dfrac{32}{\pi(D^4 - d^4)}\dfrac{M_t}{G}$ | $2\dfrac{S_{v\max}}{G}\dfrac{1}{D}$ | $\dfrac{1}{4}\dfrac{S_v^2{}_{\max}}{G}\dfrac{D^2 + d^2}{D^2} V$ (Note 2) |
| | $\dfrac{\pi}{16} b^2 h S_v$ ($h > b$) | $\dfrac{16}{\pi}\dfrac{b^2 + h^2}{b^3 h^3}\dfrac{M_t}{G}$ | $\dfrac{S_{v\max}}{G}\dfrac{b^2 + h^2}{bh^2}$ | $\dfrac{1}{8}\dfrac{S_v^2{}_{\max}}{G}\dfrac{b^2 + h^2}{h^2} V$ (Note 3) |
| | $\frac{2}{9} b^2 h S_v$ ($h > b$) | $3.6\dfrac{b^2 + h^2}{b^3 h^3}\dfrac{M_t{}^*}{G}$ | $0.8\dfrac{S_{v\max}}{G}\dfrac{b^2 + h^2}{bh^2}{}^*$ | $\dfrac{4}{45}\dfrac{S_v^2{}_{\max}}{G}\dfrac{b^2 + h^2}{h^2} V$ (Note 4) |
| | $\frac{2}{9} h^3 S_v$ | $7.2\dfrac{1}{h^4}\dfrac{M_t}{G}$ | $1.6\dfrac{S_{v\max}}{G}\dfrac{1}{h}$ | $\dfrac{8}{45}\dfrac{S_v^2{}_{\max}}{G} V$ (Note 5) |
| | $\dfrac{b^3}{20} S_v$ | $46.2\dfrac{1}{b^4}\dfrac{M_t}{G}$ | $2.31\dfrac{S_{v\max}}{G}\dfrac{1}{b}$ | |
| | $\dfrac{b^3}{1.09} S_v$ | $0.967\dfrac{1}{b^4}\dfrac{M_t}{G}$ | $0.9\dfrac{S_{v\max}}{G}\dfrac{1}{b}$ | |

*When

| $b/b$ = | 1 | 2 | 4 | 8 |
|---|---|---|---|---|
| Coefficient 3.6 becomes = | 3.56 | 3.50 | 3.35 | 3.21 |
| Coefficient 0.8 becomes = | 0.79 | 0.78 | 0.74 | 0.71 |

NOTES. (1) $S_{v\max}$ at circumference. (2) $S_{v\max}$ at outer circumference. (3) $S_{v\max}$ at $A$; $S_{vB} = 16M_t/\pi bh^2$. (4) $S_{v\max}$ at middle of side $h$; in middle of $b$, $S_v = 9M_t/2bh^2$. (5) $S_{v\max}$ at middle of side.

strength rather than buckling causes failure, and the column ceases to be long. In round numbers, this **critical slenderness ratio** falls between 120 and 150. Table 12 gives additional facts concerning long columns.

**Short Columns**   The stress in a short column may be considered partly due to compression and partly due to bending. A theoretical equation has not been derived. Empirical, though rational, expressions are, in general, based on the assumption that the permissible stress must be reduced below that which could be permitted were it due to compression only. The manner in which this reduction is made determines the type of equation as well as the slenderness ratio beyond which the equation does not apply. Figure 52 illustrates the situation. Some typical formulas are given in Table 13.

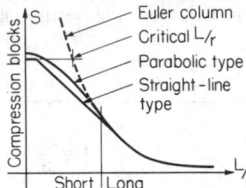

**Fig. 52**

EXAMPLE. A machine member unsupported for a length of 15 in has a square cross section 0.5 in on a side. It is to be subjected to compression. What maximum safe load can be applied centrally, according to the AISC formula? At the computed load, what size section (also square) would be needed, if it were to be designed according to the AREA formula?

$$l/r = 15/.5\sqrt{12} = 104 \therefore \text{ short column}$$
$$P/A = 17,000 - 0.485\,(104)^2 = 11,730$$
or $\quad P = 0.25 \times 11,730 = 2,940 \text{ lb } (1,335 \text{ kgf})$

and $\quad \dfrac{2,940}{a^2} \leqslant 15,000 - 50\,(15/a\sqrt{12}) = 15,000 - \dfrac{2,600}{a}$

thus $\quad a^2 - 0.173a - 0.196 = 0 \quad$ or $\quad a = 0.536$ in (1.36 cm)

**Combined Flexure and Longitudinal Force**   Figure 53

shows a bar under flexure due to transverse and longitudinal loads. The maximum fiber stress $S$ is made up of $S_0$, due to the direct action of load $P$, and $S_b$, due to the entire bending moment $M$. $M$ is the algebraic sum of two bending moments, $M_1$ due to longitudinal load (+ for compression and − for tension), and $M_2$ due to transverse load. $M = M_2 \pm M_1$. Here $M_1 = Pf$ and $f = CS_b l^2/Ec$.

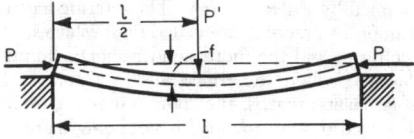

**Fig. 53**

FOR THE CASE OF LONGITUDINAL COMPRESSION.   $S_b I/c = M_2 + CPS_b l^2/Ec$, or $S_b = M_2 c(I - CPl^2/E)$. The maximum stress is $S = S_b + S_0$ compression. The constant $C$ for the case of Fig. 53 is derived from the equations $P'l/4 = S_b I/c$ and $f = P'l^3/48EI$. Solving for $f; f = \frac{1}{12} S_b l^2/Ec$, or $C = \frac{1}{12}$. For a beam supported at the ends and uniformly loaded, $C = \frac{5}{48}$. Other cases can be similarly calculated.

FOR THE CASE OF LONGITUDINAL TENSION.   $M = M_2 - Pf$, and $S_b = M_2 c/(I + CPl^2/E)$. The maximum stress is $S = S_b + S_0$, tension.

## ECCENTRIC LOADS

When **short blocks** are **loaded eccentrically** in compression or in tension, i.e., not through the center of gravity (cg), a combination of axial and bending stress results. The maximum unit stress $S_M$ is the algebraic sum of these two unit stresses.

In Fig. 54 a load $P$ acts in a line of symmetry at the distance $e$ from cg; $r = $ radius of gyration. The unit stresses are (1) $S_c$, due to $P$, as if it acted through cg, and (2) $S_b$, due to the bending moment of $P$ acting with a leverage of $e$ about cg. Thus unit stress $S$ at any point $y$ is

### Table 12. Strength of Round-ended Columns According to Euler's Formula

| Material | Cast iron | Wrought iron | Low-carbon steel | Medium-carbon steel |
|---|---|---|---|---|
| Ultimate compressive strength, lb/in² | 107,000 | 53,400 | 62,600 | 89,000 |
| Allowable compressive stress, lb/in² (maximum) | 7,100 | 15,400 | 17,000 | 20,000 |
| Modulus of elasticity | 14,200,000 | 28,400,000 | 30,600,000 | 31,300,000 |
| Factor of safety | 8 | 5 | 5 | 5 |
| Smallest $I$ allowable at worst section, in⁴ | $\dfrac{Pl^2}{17,500,000}$ | $\dfrac{Pl^2}{56,000,000}$ | $\dfrac{Pl^2}{60,300,000}$ | $\dfrac{Pl^2}{61,700,000}$ |
| Limit of ratio, $l/r >$ | 50.0 | 60.6 | 59.4 | 55.6 |
| Rectangle $(r = b\sqrt{1/12})$, $l/b >$ | 14.4 | 17.5 | 17.2 | 16.0 |
| Circle $(r = \frac{1}{4}d)$, $l/d >$ | 12.5 | 15.2 | 14.9 | 13.9 |
| Circular ring of small thickness $(r = d\sqrt{1/8})$, $l/d >$ | 17.6 | 21.4 | 21.1 | 19.7 |

($P$ = allowable load, lb; $l$ = length of column, in; $b$ = smallest dimension of a rectangular section, in; $d$ = diameter of a circular section, in; $r$ = least radius of gyration of section.)

**Table 13. Typical Short-Column Formulas**

| Formula | Material | Code | Slenderness ratio |
|---|---|---|---|
| $S_w = 17,000 - 0.485\left(\dfrac{l}{r}\right)^2$ | Carbon steels | AISC | $l/r < 120$ |
| $S_w = 16,000 - 70\,(l/r)$ | Carbon steels | Chicago | $l/r < 120$ |
| $S_w = 15,000 - 50\left(\dfrac{l}{r}\right)$ | Carbon steels | AREA | $l/r < 150$ |
| $S_w = 19,000 - 100\,(l/r)$ | Carbon steels | Am. Br. Co. | $60 < \dfrac{l}{r} < 120$ |
| $*\,S_{cr} = 135,000 - \dfrac{15.9}{c}\left(\dfrac{l}{r}\right)^2$ | Alloy-steel tubing | ANC | $\dfrac{l}{\sqrt{c}\,r} < 65$ |
| $S_w = 9,000 - 40\left(\dfrac{l}{r}\right)$ | Cast iron | NYC | $\dfrac{l}{r} < 70$ |
| $*\,S_{cr} = 34,500 - \dfrac{245}{\sqrt{c}}\left(\dfrac{l}{r}\right)$ | 2017ST Aluminum | ANC | $\dfrac{l}{\sqrt{c}\,r} < 94$ |
| $*\,S_{cr} = 5,000 - \dfrac{0.5}{c}\left(\dfrac{l}{r}\right)^2$ | Spruce | ANC | $\dfrac{l}{\sqrt{c}\,r} < 72$ |
| $*\,S_{cr} = S_y\left[1 - \dfrac{S_y}{4n\pi^2 E}\left(\dfrac{l}{r}\right)^2\right]$ | Steels | Johnson | $\dfrac{l}{r} < \sqrt{\dfrac{2n\pi^2 E}{S_y}}$ |
| $\dagger\,S_{cr} = \dfrac{S_y}{1 + \dfrac{ec}{r^2}\sec\left(\dfrac{l}{r}\sqrt{\dfrac{P}{4AE}}\right)}$ | Steels | Secant | $\dfrac{l}{r} < $ critical |

$*S_{cr}$ = theoretical maximum, $c$ = end fixity coefficient,
  $c = 2$, both ends pivoted, $c = 2.86$, one pivoted, other fixed,
  $c = 4$, both ends fixed, $c = 1$ one fixed, one free.
$\dagger e$ is initial eccentricity at which load is applied to center of column cross section.

$$
\begin{aligned}
S &= S_c \pm S_b \\
&= (P/A) \pm Pey/I \\
&= S_c(1 \pm ey/r^2)
\end{aligned}
$$

$y$ is positive for points on the same side of cg as $P$, and negative on the opposite side. For a **rectangular cross section** of width $b$, the maximum stress $S_M = S_c\,(1 + 6e/b)$. When $P$ is outside the middle third of width $b$ and is a compressive load, tensile stresses occur.

For a **circular cross section** of diameter $d$, $S_M = S_c(1 + 8e/d)$. The stress due to the weight of the solid will modify these relations.

NOTE. In these formulas $e$ is measured from the gravity axis, and gives tension when $e$ is greater than one-sixth the width (measured in the same direction as $e$), for rectangular sections; and when greater than one-eighth the diameter for solid circular sections.

If, as in certain classes of masonry construction, the **material cannot withstand tensile stress** and thus no tension can occur, the center of moments (Fig. 55) is taken at the center of stress. For a **rectangular section**, $P$ acts at distance $k$ from the nearest edge. Length under compression = $3k$, and $S_M = \frac{2}{3}P/bk$. For a

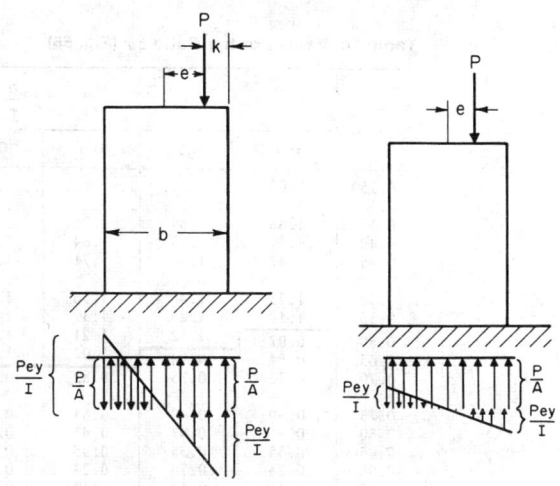

**Fig. 54**

**circular section,** $S_M = [0.372 + 0.056(k/r)]P/k\sqrt{rk}$, where $r = $ radius and $k = $ distance of $P$ from circumference. For a **circular ring,** $S = $ average compressive stress on cross section produced by $P$; $e = $ eccentricity of $P$; $z = $ length of diameter under compression (Fig. 56). Values of $z/r$ and of the ratio of $S_{max}$ to average $S$ are given in Tables 14 and 15.

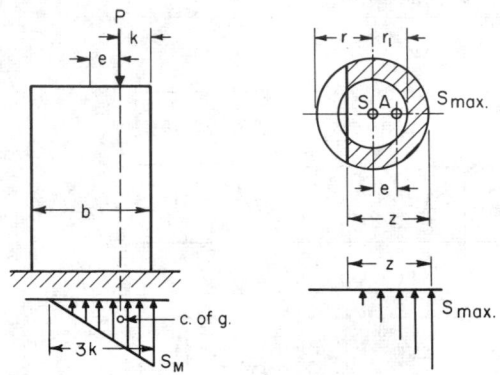

**Fig. 55**          **Fig. 56**

CHIMNEY PROBLEM. Weight of chimney $= 563,000$ lb; $e = 1.56$ ft; OD of chimney $= 10$ ft 8 in; ID $= 6$ ft $6^1/_2$ in. Overturning moment $= Pe = 878,000$ ft·lb, $r_1/r = 0.6$. $e/r = 0.29$. This gives $(z/r) > 2$. Therefore, the entire area of the base is under compression. Area under compression $= 55.8$ ft²; $I = 546$; $S = (563,000/55.8) \pm (878,000 \times 5.33)/546 = 18,700$ (max) and $1,500$ (min) lb compression per ft². From Table 15, by interpolation, $S_{max}/S_{avg} = 1.85$. ∴ $S_{max} = (563,000/55.8) \times 1.85 = 18,685$ lb/ft² (91,313 kgf/m²).

The **kern** is the area around the center of gravity of a cross section within which any load applied will produce stress of only one sign throughout the entire cross section. Outside the kern, a load produces stresses of different sign. Figure 57 shows kerns (shaded) for various sections.

For a **circular ring,** the radius of the kern $r = D[1 + (d/D)^2]/8$.

For a **hollow square** ($H$ and $b = $ lengths of outer and inner sides), the kern is a square similar to Fig. 57a, where

$$r_{min} = \frac{H}{6}\frac{1}{\sqrt{2}}\left[1 + \left(\frac{b}{H}\right)^2\right] = 0.1179H\left[1 = \left(\frac{b}{H}\right)^2\right]$$

For a **hollow octagon** $R_a$ and $R_i = $ radii of circles circumscribing the outer and inner sides; thickness of wall $= 0.9239(R_a - R_i)]$, the kern is an octagon similar to Fig. 57c, where $0.2256R$ becomes $0.2256R_a[1 + (R_i/R_a)^2]$.

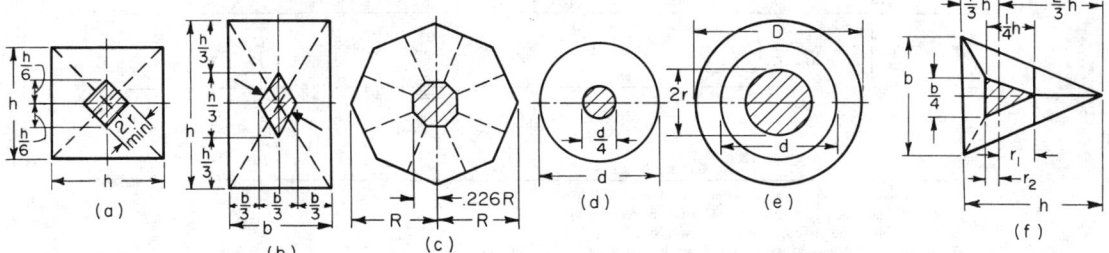

**Fig. 57**

**Table 14. Values of the Ratio** $z/r$ **(Fig. 56)**

| $\dfrac{e}{r}$ | $\dfrac{r_1}{r}$ | | | | | | | $\dfrac{e}{r}$ |
|---|---|---|---|---|---|---|---|---|
| | 0.0 | 0.5 | 0.6 | 0.7 | 0.8 | 0.9 | 1.0 | |
| 0.25 | 2.00 | ..... | ..... | ..... | ..... | ..... | ..... | 0.25 |
| 0.30 | 1.82 | ..... | ..... | ..... | ..... | ..... | ..... | 0.30 |
| 0.35 | 1.66 | 1.89 | 1.98 | ..... | ..... | ..... | ..... | 0.35 |
| 0.40 | 1.51 | 1.75 | 1.84 | 1.93 | ..... | ..... | ..... | 0.40 |
| 0.45 | 1.37 | 1.61 | 1.71 | 1.81 | 1.90 | ..... | ..... | 0.45 |
| 0.50 | 1.23 | 1.46 | 1.56 | 1.66 | 1.78 | 1.89 | 2.00 | 0.50 |
| 0.55 | 1.10 | 1.29 | 1.39 | 1.50 | 1.62 | 1.74 | 1.87 | 0.55 |
| 0.60 | 0.97 | 1.12 | 1.21 | 1.32 | 1.45 | 1.58 | 1.71 | 0.60 |
| 0.65 | 0.84 | 0.94 | 1.02 | 1.13 | 1.25 | 1.40 | 1.54 | 0.65 |
| 0.70 | 0.72 | 0.75 | 0.82 | 0.93 | 1.05 | 1.20 | 1.35 | 0.70 |
| 0.75 | 0.59 | 0.60 | 0.64 | 0.72 | 0.85 | 0.99 | 1.15 | 0.75 |
| 0.80 | 0.47 | 0.47 | 0.48 | 0.52 | 0.61 | 0.77 | 0.94 | 0.80 |
| 0.85 | 0.35 | 0.35 | 0.35 | 0.36 | 0.42 | 0.55 | 0.72 | 0.85 |
| 0.90 | 0.24 | 0.24 | 0.24 | 0.24 | 0.24 | 0.32 | 0.49 | 0.90 |
| 0.95 | 0.12 | 0.12 | 0.12 | 0.12 | 0.12 | 0.12 | 0.25 | 0.95 |

**Table 15. Values of the Ratio $S_{max}/S_{avg}$**
(In determining $S$ average, use load $P$ divided by total area of cross section)

| $\dfrac{e}{r}$ | $\dfrac{r_1}{r}$ | | | | | | | $\dfrac{e}{r}$ |
|---|---|---|---|---|---|---|---|---|
| | 0.0 | 0.5 | 0.6 | 0.7 | 0.8 | 0.9 | 1.0 | |
| 0.00 | 1.00 | 1.00 | 1.00 | 1.00 | 1.00 | 1.00 | 1.00 | 0.00 |
| 0.05 | 1.20 | 1.16 | 1.15 | 1.13 | 1.12 | 1.11 | 1.10 | 0.05 |
| 0.10 | 1.40 | 1.32 | 1.29 | 1.27 | 1.24 | 1.22 | 1.20 | 0.10 |
| 0.15 | 1.60 | 1.48 | 1.44 | 1.40 | 1.37 | 1.33 | 1.30 | 0.15 |
| 0.20 | 1.80 | 1.64 | 1.59 | 1.54 | 1.49 | 1.44 | 1.40 | 0.20 |
| 0.25 | 2.00 | 1.80 | 1.73 | 1.67 | 1.61 | 1.55 | 1.50 | 0.25 |
| 0.30 | 2.23 | 1.96 | 1.88 | 1.81 | 1.73 | 1.66 | 1.60 | 0.30 |
| 0.35 | 2.48 | 2.12 | 2.04 | 1.94 | 1.85 | 1.77 | 1.70 | 0.35 |
| 0.40 | 2.76 | 2.29 | 2.20 | 2.07 | 1.98 | 1.88 | 1.80 | 0.40 |
| 0.45 | 3.11 | 2.51 | 2.39 | 2.23 | 2.10 | 1.99 | 1.90 | 0.45 |
| 0.50 | 3.55 | 2.80 | 2.61 | 2.42 | 2.26 | 2.10 | 2.00 | 0.50 |
| 0.55 | 4.15 | 3.14 | 2.89 | 2.67 | 2.42 | 2.26 | 2.17 | 0.55 |
| 0.60 | 4.96 | 3.58 | 3.24 | 2.92 | 2.64 | 2.42 | 2.26 | 0.60 |
| 0.65 | 6.00 | 4.34 | 3.80 | 3.30 | 2.92 | 2.64 | 2.42 | 0.65 |
| 0.70 | 7.48 | 5.40 | 4.65 | 3.86 | 3.33 | 2.95 | 2.64 | 0.70 |
| 0.75 | 9.93 | 7.26 | 5.97 | 4.81 | 3.93 | 3.33 | 2.89 | 0.75 |
| 0.80 | 13.87 | 10.05 | 8.80 | 6.53 | 4.93 | 3.96 | 3.27 | 0.80 |
| 0.85 | 21.08 | 15.55 | 13.32 | 10.43 | 7.16 | 4.50 | 3.77 | 0.85 |
| 0.90 | 38.25 | 30.80 | 25.80 | 19.85 | 14.60 | 7.13 | 4.71 | 0.90 |
| 0.95 | 96.10 | 72.20 | 62.20 | 50.20 | 34.60 | 19.80 | 6.72 | 0.95 |
| 1.00 | ∞ | ∞ | ∞ | ∞ | ∞ | ∞ | ∞ | 1.00 |

## CURVED BEAMS

The application of the flexure formula for a straight beam to the case of a curved beam results in error. When all "fibers" of a member have the same center of curvature, the **concentric** or common type of curved beam exists (see Fig. 58). Such a beam is defined by the Winkler-Bach theory. The stress at a point $y$ units from the centroidal axis is

$$S = \frac{M}{AR}\left[1 + \frac{y}{Z(R + y)}\right]$$

$M$ is the bending moment, positive when it increases curvature; $Y$ is positive when measured toward the convex side; $A$ is the cross-sectional area; $R$ is the radius of the centroidal axis; $Z$ **is a cross-section property** defined by

$$Z = -\frac{1}{A}\int \frac{y}{R + y}\,dA$$

**Analytical** expressions for $Z$ of certain sections are given in

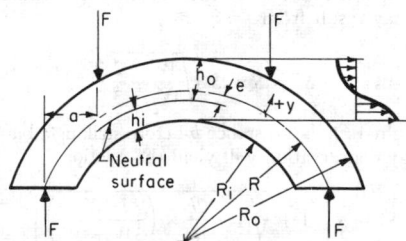

**Fig. 58**

Table 16. $Z$ can also be found by **graphical** integration methods (see any advanced strength book). The **neutral surface** shifts toward the center of curvature, or inside fiber, an amount equal to $e = ZR/(Z + 1)$. The Winkler-Bach theory, though practically satisfactory, disregards radial stresses as well as lateral deformations and assumes pure bending. The **maximum stress** occurring on the inside fiber is $S = Mb_i/AeR_i$, while that on the outside fiber is $S = Mb_0/AeR_0$.

EXAMPLE. A split steel ring of rectangular cross section is subjected to a diametral force of 1,000 lb as shown in Fig. 59a. Compute the stress at the point 0.5 in from the outside fiber on plane $mm$. Also compute the maximum stress.

$$Z = -1 + \frac{R}{b}\left(\ln\frac{R + C}{R - C}\right)$$

$$= -1 + \frac{10}{4}\left(\ln\frac{10 + 2}{10 - 2}\right) = 0.0133$$

$$S_{1.5} = \frac{M}{AR}\left[1 + \frac{y}{Z(R + y)}\right] + \frac{F}{A}$$

$$= \frac{-1,000 \times 10}{8 \times 10}\left[1 + \frac{1.5}{0.0133(10 + 1.5)}\right] + \frac{1,000}{8}$$

$$= -1,250 + 125 = -1,125\ \text{lb/in}^2\ (\text{compr.})\ (79\ \text{kgf/cm}^2)$$

$$S_M = \frac{-1,000}{8}\left[1 + \frac{-2}{0.0133(10 - 2)}\right] + \frac{1,000}{8}$$

$$= 2,230 + 125 = 2,355\ \text{lb/in}^2\ (166\ \text{kgf/cm}^2)$$

or

$$e = \frac{ZR}{Z + 1} = \frac{0.0133 \times 10}{0.0133 + 1} = 0.131$$

and

$$S_M = \frac{Mb_i}{AeR_i} + \frac{P}{A} = \frac{1,000 \times 1.87}{8 \times 0.131 \times 8} + \frac{1,000}{8}$$

$$= 2,355\ \text{lb/in}^2\ (166\ \text{kgf/cm}^2)$$

The **deflection** in curved beams can be computed by means of the moment-area theory. If the origin of axes is taken at the point whose deflection is wanted, it can be shown that the component displacements in the $x$ and $y$ directions are

$$\Delta_x = \int_0^s \frac{Myds}{EI} \quad \text{and} \quad \Delta_y = \int_0^s \frac{Mxds}{EI}$$

The resultant deflection is then equal to $\Delta_0 = \sqrt{\Delta_x^2 + \Delta_y^2}$ in the direction defined by $\tan \theta = \Delta_y/\Delta_x$. Deflections can also be found conveniently by use of **Castigliano's theorem.** It states that in an elastic system the displacement in the direction of a force (or couple) and due to that force (or couple) is the partial derivative of the strain energy with respect to the force (or couple). Stated mathematically, $\Delta_z = \partial U/\partial F_z$. If a force does not exist at the point and/or in the direction desired, a dummy force may be applied. This force must then be eliminated by equating it to zero at the end.

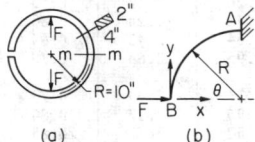

**Fig. 59**

EXAMPLE.  A quadrant of radius $R$ is fixed at one end as shown in Fig. 59$b$. The force $F$ is applied in the radial direction at the free end $B$. Find the deflection of $B$.

By moment area:

$$y = R \sin \theta \qquad x = R(1 - \cos \theta)$$
$$ds = Rd\theta \qquad M = FR \sin \theta$$
$$_B\Delta_x = \frac{FR^3}{EI} \int_0^{\pi/2} \sin^2\theta\, d\theta = \frac{\pi FR^3}{4EI}$$
$$_B\Delta_y = \frac{FR^3}{EI} \int_0^{\pi/2} \sin \theta\, (1 - \cos \theta) d\theta = -\frac{FR^3}{2EI}$$

and

$$\Delta_B = \frac{FR^3}{2EI} \sqrt{1 + \pi^2/4}$$

at

$$\theta_x = \tan^{-1}\left(-\frac{FR^3}{2EI} \times \frac{4EI}{\pi FR^3}\right) = \tan^{-1}\frac{2}{\pi} = 32.5°$$

By Castigliano:

$$_B\Delta_x = \frac{\partial U}{\partial F} = \frac{\partial}{\partial F} \int_0^{\pi/2} \frac{F^2 R^3}{2EI} \sin^2\theta\, d\theta = \frac{\pi FR^3}{4EI}$$
$$_B\Delta_y = \frac{\partial U}{\partial F_y} = \frac{\partial}{\partial F_y} \int_0^{\pi/2} \frac{[FR \sin \theta - F_yR\,(1 - \cos \theta)]^2\, Rd\theta}{2EI}$$
$$= -\frac{FR^3}{2EI}$$

The $F_y$, assumed downward, is equated to zero, after the integration and differentiation are performed to find $_B\Delta_y$. The remainder of the computation is exactly as in the moment-area method.

**Eccentrically Curved Beams**  These beams (Fig. 60) are bounded by arcs having different centers of curvature. In addition, it is possible for either radius to be the larger one. The one in which the section depth shortens as the central section is approached may be called the **arch beam.** When the central section is the largest, the beam is of the crescent type.

**Crescent I** denotes the beam of larger outside radius and **crescent II** of larger inside radius. The stress at the **central section** of such beams may be found from $S = KMC/I$. In the case of rectangular cross section, the equation becomes $S = 6KM/bh^2$ where $M$ is the bending moment, $b$ is the width of the beam section, and $h$ its height. The **stress factors** $K$ for the **inner boundary,** established from photoelastic data, are given in Table 17. The outside radius is denoted by $R_o$ and the inside by $R_i$. The geometry of crescent beams is such that the stress can be larger in **off-center sections.** The stress at the central section determined above must then be multiplied by the **position factor** $k$,

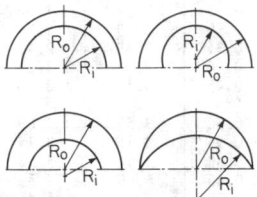

**Fig. 60**

given in Table 18. As in the concentric beam, the **neutral surface** shifts slightly toward the inner boundary (see Vidosic, Curved Beams with Eccentric Boundaries, *Trans. ASME,* **79,** pp. 1317–1321).

## IMPACT

A force or stress is considered **suddenly applied** when the duration of load application is less than one-half the **fundamental natural period** of vibration of the member upon which the force acts. Under impact, a compression wave propagates through the member at a velocity $c = \sqrt{E/\rho}$, where $\rho$ is the mass density. As this compression wave travels back and forth by reflection from one end of the bar to the other, a maximum stress is produced which is many times larger than what it would be statically. An exact determination of this stress is most difficult. However, if conservation of kinetic and strain energies is applied, the **impact stress** is found to be

$$S' = S \sqrt{\frac{W}{W_b}\left(\frac{3W}{3W + W_b}\right)}$$

The weight of the striking mass is here denoted by $W$, that of the struck bar by $W_b$, while $S$ is the static stress, $W/A$ ($A$ is the cross-sectional area of the bar). Above is the case of **sudden impact.** When the ratio $W/W_b$ is small, the stress computed by the above equation may be erroneous. A better solution of this problem may result from

$$S' = S + S \sqrt{\frac{W}{W_b} + \frac{2}{3}}$$

If a weight $W$ falls a distance $h$ before striking a bar of mass $W_b$, energy conservation will yield the relation

$$S' = S\left[ 1 + \sqrt{1 + \frac{2h}{e} \times \left(\frac{3W}{3W + W_b}\right)} \right]$$

The elongation $e = \epsilon l = Sl/E$. When the striking mass $W$ is

**Table 16. Analytical Expressions for $Z$**

| Section | Expression |
|---|---|
| | $$Z = -1 + \frac{R}{h}\left(\ln\frac{R+C}{R-C}\right)$$ |
| | $$Z = -1 + 2\left(\frac{R}{r}\right)\left[\frac{R}{r} - \sqrt{\left(\frac{R}{r}\right)^2 - 1}\right]$$ |
| | $$Z = -1 + \frac{R}{A}\left[t\ln(R + C_1) + (b - t)\right.$$ $$\left.\ln(R - C_3) - b\ln(R - C_2)\right]$$ and $A = tC_1 - (b - t)C_3 + bC_2$ |
| | $$Z = -1 + \frac{R}{A}\left[b\ln\left(\frac{R+C_2}{R-C_2}\right) + (t - b)\ln\left(\frac{R+C_1}{R-C_1}\right)\right]$$ $$A = 2\left[(t - b)C_1 + bC_2\right]$$ |

assumed rigid, the elasticity factor is taken equal to 1. Thus the equation becomes

$$S' = S\left(1 + \sqrt{1 + 2b/e}\right)$$

If, in addition, $b$ is taken equal to zero (**sudden impact**), the radical equals 1, and so the stress becomes $S' = 2S$. Since Hooke's law is applicable, the relations

$$e' = e\left(1 + \sqrt{1 + 2b/e}\right) \quad \text{and} \quad e' = 2e$$

are also true for the same conditions.

The expression may be converted, by using $v^2 = 2gb$, to

$$S' = S\left(1 + \sqrt{1 + v^2/eg}\right)$$

This might be called the **energy impact** form. If the **natural frequency** $f_n$ of the bar is used, the stress equation is

$$S' = S\left(1 + \sqrt{1 + 0.204\, bf_n^2}\right)$$

In general, the **maximum impact stress** in a **beam** and a **shaft** can be approximated from the simplified falling-weight equa-

tion. It is necessary, though, to substitute the maximum **deflection** $y$ for $e$, in the case of beams, and for the **angle of twist** $\theta$ in the case of shafts. Of course $S = MC/I$ and $M_tC/J$, respectively. Thus

$$S' = S\left[1 + \sqrt{1 + \frac{2b}{y\,(\text{or }\theta)}}\right]$$

For a more exact solution, elastic yield in each member must be considered. The theory then yields

$$S' = S\left[1 + \sqrt{1 + 2b/y\left(\frac{35W}{35W + 17W_b}\right)}\right]$$

for a **simply supported beam** struck in the middle by a weight $W$.

## THEORY OF ELASTICITY

Loaded members in which the stress distribution cannot be estimated fail of solution by elementary strength-of-material methods. To such cases, the more advanced mathematical

**Table 17. Stress Factors for Inner Boundary at Central Section (see Fig. 60)**

1. For the arch-type beams

   (a) $K = 0.834 + 1.504 \dfrac{h}{R_o + R_i}$ if $\dfrac{R_o + R_i}{h} < 5$.

   (b) $K = 0.899 + 1.181 \dfrac{h}{R_o + R_i}$ if $5 < \dfrac{R_o + R_i}{h} < 10$.

   (c) In the case of larger section ratios use the equivalent beam solution.

2. For the crescent I-type beams

   (a) $K = 0.570 + 1.536 \dfrac{h}{R_o + R_i}$ if $\dfrac{R_o + R_i}{h} < 2$.

   (b) $K = 0.959 + 0.769 \dfrac{h}{R_o + R_i}$ if $2 < \dfrac{R_o + R_i}{h} < 20$.

   (c) $K = 1.092 \left(\dfrac{h}{R_o + R_i}\right)^{0.0296}$ if $\dfrac{R_o + R_i}{h} > 20$.

3. For the crescent II-type beams

   (a) $K = 0.897 + 1.098 \dfrac{h}{R_o + R_i}$ if $\dfrac{R_o + R_i}{h} < 8$.

   (b) $K = 1.119 \left(\dfrac{h}{R_o + R_i}\right)^{0.0378}$ if $8 < \dfrac{R_o + R_i}{h} < 20$.

   (c) $K = 1.081 \left(\dfrac{h}{R_o + R_i}\right)^{0.0270}$ if $\dfrac{R_o + R_i}{h} > 20$.

orthogonal planes that bound the point element. Thus there are at each point **six stress components**, $\sigma_x$, $\sigma_y$, $\sigma_z$, $\tau_{xy} = \tau_{yx}$, $\tau_{xz} = \tau_{zx}$, and $\tau_{yz} = \tau_{zy}$. Similarly, if the normal unit strain is designated by the letter $\epsilon$ and shearing unit strain by $\gamma$, the **six components of strain** are defined by

$$\epsilon_x = \partial u/\partial x \quad \epsilon_y = \partial v/\partial y \quad \epsilon_z = \partial w/\partial z$$
$$\gamma_{xy} = \partial u/\partial y + \partial v/\partial x \quad \gamma_{yz} = \partial v/\partial z + \partial w/\partial y$$

and $\quad \gamma_{xz} = \partial u/\partial z + \partial w/\partial x$

The elastic **displacements** of particles on the body in the $x$, $y$, and $z$ directions are identified as the $u$, $v$, and $w$ **components**, respectively.

Since metals have the usually assumed elastic as well as isotropic properties, Hooke's law holds. Therefore, the interrelationships between stress and strain can easily be obtained.

$$\epsilon_x = \frac{1}{E}\left[\sigma_x - \mu(\sigma_y + \sigma_z)\right]$$

$$\epsilon_y = \frac{1}{E}\left[\sigma_y - \mu(\sigma_x + \sigma_z)\right]$$

$$\epsilon_z = \frac{1}{E}\left[\sigma_z - \mu(\sigma_x + \sigma_y)\right]$$

$$\gamma_{xy} = \tau_{xy}/G \qquad \gamma_{xz} = \tau_{xz}/G$$

and $\quad \gamma_{yz} = \tau_{yz}/G$

The general case of strain can be obtained by superposing the elongation strains upon the shearing strains.

Problems depending upon theories of elasticity are considerably simplified if the stresses are all parallel to one plane or if all deformations occur in planes perpendicular to the length of the member. The first case is one of **plane stress,** as when a thin plate of uniform thickness is subjected to central, boundary forces parallel to the plane of the plate. The second is a case of **plane strain,** such as a gate subjected to hydrostatic pressure, the intensity of which does not vary along the gate's length. All particles therefore displace at right angles to the length, and so cross sections remain plane. In plane-stress problems, three of the six stress components vanish, thus leaving only $\sigma_x$, $\sigma_y$, and $\tau_{xy}$. Similarly, in plane strain, only $\epsilon_x$, $\epsilon_y$, and $\gamma_{xy}$ will not equal zero; thus the same three stresses $\sigma_x$, $\sigma_y$, and $\tau_{xy}$ remain to be considered. Plane problems can thus be

principles of the **theory of elasticity** must be applied. When this is not possible, experimental stress analysis has to be used. Because of the complexity of solution, only some of the more practical problems have been solved by the theory of elasticity. The more general concepts and methods are presented.

Two kinds of forces may act on a body. **Surface forces** are distributed over the surface as the result of, for instance, the pressure of one body on another. Forces due to gravity, inertia, magnetism, etc., which act over the entire volume of a body, are called **body forces.** Both surface and body forces can be best handled if resolved into three orthogonal components.

Surface forces are thus designated $\bar{X}$, $\bar{Y}$, and $\bar{Z}$, while body forces are labeled $X$, $Y$, and $Z$.

In general, there exists a normal stress $\sigma$ and a shearing stress $\tau$ at each point of a loaded member. It is convenient to deal with components of each of these stresses on each of six

**Table 18. Crescent-Beam Position Stress Factors (see Fig. 60)**

| Angle $\theta$, deg | $k$ | |
|---|---|---|
| | Inner | Outer |
| 10 | $1 + 0.055\,H/h$ | $1 + 0.03\,H/h$ |
| 20 | $1 + 0.164\,H/h$ | $1 + 0.10\,H/h$ |
| 30 | $1 + 0.365\,H/h$ | $1 + 0.25\,H/h$ |
| 40 | $1 + 0.567\,H/h$ | $1 + 0.467\,H/h$ |
| 50 | $1.521 - \dfrac{(0.5171 - 1.382\,H/h)^{1/2}}{1.382}$ | $1 + 0.733\,H/h$ |
| 60 | $1.756 - \dfrac{(0.2416 - 0.6506\,H/h)^{1/2}}{0.6506}$ | $1 + 1.123\,H/h$ |
| 70 | $2.070 - \dfrac{(0.4817 - 1.298\,H/h)^{1/2}}{0.6492}$ | $1 + 1.70\,H/h$ |
| 80 | $2.531 - \dfrac{(0.2939 - 0.7084\,H/h)^{1/2}}{0.3542}$ | $1 + 2.383\,H/h$ |
| 90 | | $1 + 3.933\,H/h$ |

Note: All formulas are valid for $0 < H/b \leq 0.325$. Formulas for the inner boundary, except for 40 deg, may be used to $H/b \leq 0.36$. $H$ = distance between centers.

represented by the element shown in Fig. 61. Equilibrium considerations applied to this particle result in the **differential equations of equilibrium** which reduce to

$$\frac{\partial \sigma_x}{\partial x} + \frac{\partial \tau_{xy}}{\partial y} + X = 0$$

and

$$\frac{\partial \sigma_y}{\partial y} + \frac{\partial \tau_{xy}}{\partial x} + Y = 0$$

Since the two differential equations of equilibrium are insufficient to find the three stresses, a third equation must be used. This is the **compatibility equation** relating the three strain components. It is

$$\frac{\partial^2 \epsilon_x}{\partial y^2} + \frac{\partial^2 \epsilon_y}{\partial x^2} = \frac{\partial^2 \gamma_{xy}}{\partial x \partial y}$$

If strains are expressed in terms of the stresses, the compatibility equation becomes

$$\left( \frac{\partial^2}{\partial x^2} + \frac{\partial^2}{\partial y^2} \right) (\sigma_x + \sigma_y) = 0$$

Now, in any **two-dimensional** problem, the compatibility equation along with the differential equilibrium equations must be simultaneously solved for the three unknown stresses. This is accomplished using **stress functions,** which permit the integration and satisfy boundary conditions in each particular situation.

In **three-dimensional** problems, the third dimension must be considered. This results in three differential equations of equilibrium, as well as three compatibility equations. The six stress components can thus be found. The complexity involved in the solution of these equations is such, however, that only a few special cases have been solved.

In certain problems, such as rotating circular disks, **polar coordinates** become more convenient. In such cases, the stress components in a two-dimensional field are the **radial stress** $\sigma_r$, the **tangential stress** $\sigma_\theta$, and the **shearing stress** $\tau_{r\theta}$. In terms of these stresses the **polar differential equations** become

$$\frac{\partial \sigma_r}{\partial r} + \frac{1}{r} \frac{\partial \tau_{r\theta}}{\partial \theta} + \frac{\sigma_r - \sigma_\theta}{r} + R = 0$$

and

$$\frac{1}{r} \frac{\partial \sigma_\theta}{\partial \theta} + \frac{\partial \tau_{r\theta}}{\partial r} + \frac{2\tau_{r\theta}}{r} = 0$$

The body force per unit volume is represented by $R$. The **compatibility equation** in polar coordinates is

$$\left( \frac{\partial^2}{\partial r^2} + \frac{1}{r} \frac{\partial}{\partial r} + \frac{1}{r^2} \frac{\partial^2}{\partial \theta^2} \right) \left( \frac{\partial^2 \phi}{\partial r^2} + \frac{1}{r} \frac{\partial \phi}{\partial r} + \frac{1}{r^2} \frac{\partial^2 \phi}{\partial \theta^2} \right) = 0$$

$\phi$ is again a stress function of $r$ and $\theta$ that will provide a

solution of the differential equations and satisfy boundary conditions. As an example, the exact solution of a **simply supported beam** carrying a uniformly distributed load $w$ yields

$$\sigma_x = \frac{w}{2I} (l^2 - x^2) y + \frac{w}{2I} \left( \frac{2y^3}{3} - \frac{2c^2 y}{5} \right)$$

The origin of coordinates is at the center of the beam, $2c$ is the beam depth, and $2l$ is the span length. Thus the maximum stress at $x = 0$ and $y = C$ is $\sigma_x = \frac{wl^2 C}{2I} + \frac{9}{30} \frac{wC^3}{30}$. The first term represents the stress as obtained by the elementary flexure theory; the second is a correction. The second term becomes negligible when $c$ is small compared to $l$.

The important case of a **flat plate** of unit width with a **circular hole** of diameter $2a$ at its center, subjected to a uniform tensile load, has been solved using polar coordinates. If $S$ is the uniform stress at some distance from the hole, $r$ is measured from the center of the hole, and $\theta$ is the angle of $r$ with respect to the longitudinal axis of the member, the stresses are

$$\sigma_r = \frac{S}{2} \left( 1 - \frac{a^2}{r^2} \right) + \frac{S}{2} \left( 1 + \frac{3a^4}{r^4} - \frac{4a^2}{r^2} \right) \cos 2\theta$$

$$\sigma_\theta = \frac{S}{2} \left( 1 + \frac{a^2}{r^2} \right) - \frac{S}{2} \left( 1 + \frac{3a^4}{r^4} \right) \cos 2\theta$$

$$\tau_{r\theta} = -\frac{S}{2} \left( 1 - \frac{3a^4}{r^4} + \frac{2a^2}{r^2} \right) \sin 2\theta$$

### CYLINDERS AND SPHERES

**A thin-wall cylinder** has a wall thickness such that the assumption of constant stress across the wall results in negligible error. Cylinders having internal-diameter–to–thickness $(D/t)$ ratios greater than 10 are usually considered thin-walled. Boilers, drums, tanks, and pipes are often treated as such. Equilibrium equations reveal the circumferential, or hoop, stress to be $S = pr/t$ under an internal pressure $p$ (see Fig. 62).

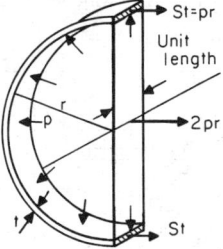

**Fig. 62**

If the cylinder is closed at the ends, a longitudinal stress of $pr/2t$ is developed. The tensile stress developed in a thin hollow sphere subjected to internal pressure is also $pr/2t$.

When thin-walled cylinders, such as vacuum tanks and submarines, are subjected to **external pressure, collapse** becomes the mode of failure. The shell is assumed perfectly round and of uniform thickness, the material obeys Hooke's law, the radial stress is negligible, and the normal stress distribution is linear. Other, lesser assumptions are also made. Using the theory of elasticity, R. G. Sturm (*Univ. Ill. Eng. Exp. Stn.*

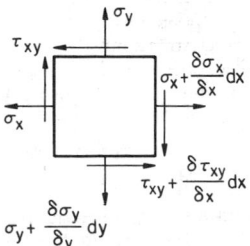

**Fig. 61**

*Bull.*, no. 12, Nov. 11, 1941) derived the **collapsing pressure** as

$$W_c = KE \left(\frac{t}{D}\right)^3 \quad \text{lb/in}^2$$

The factor $K$, a numerical coefficient, depends upon the $L/D$ and $D/t$ ratios ($D$ is outside-shell diameter), the kind of end support, and whether pressure is applied radially only, or at the ends as well. Figures 63 to 66, reproduced from the bulletin, supply the $K$ values. $N$ on these charts indicates the number of lobes into which the shell collapses. These values are for materials having Poisson's ratio of 0.3. It may also be pointed out that in the case of long cylinders (infinitely long, theoretically) the value of $K$ approaches $2/(1 - \mu^2)$.

When the **cylinder** is **stiffened with rings,** the shell may be assumed to be divided into a series of shorter shells, equal in length to the ring spacing. The previous equation can then be applied to a ring-to-ring length of cylinder. However, the flexural rigidity of the combined stiffener and shell $EI_c$ neces-

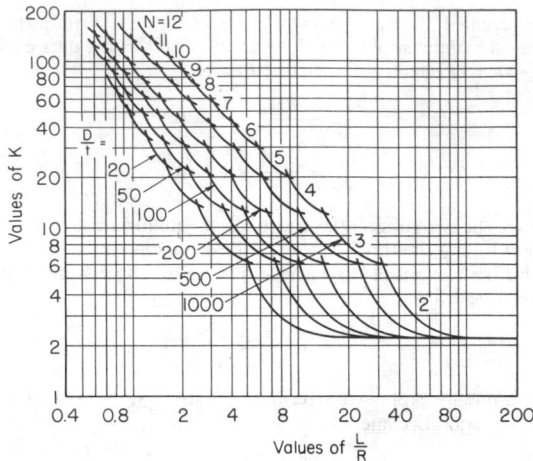

**Fig. 65**  Radial and end external pressure with simply supported edges.

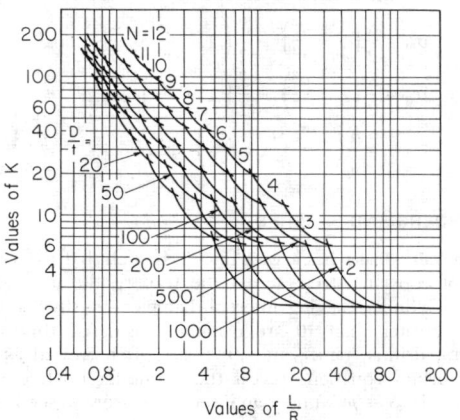

**Fig. 63**  Radial external pressure with simply supported edges.

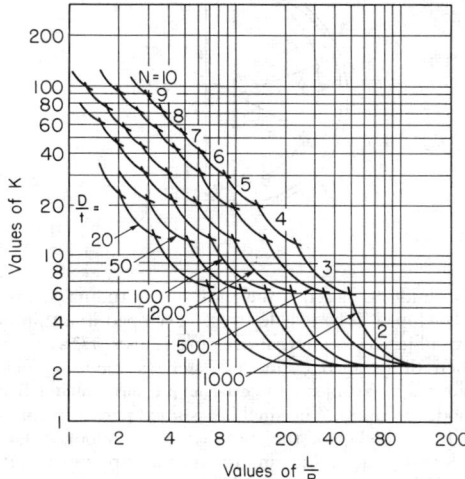

**Fig. 64**  Radial external pressure with fixed edges.

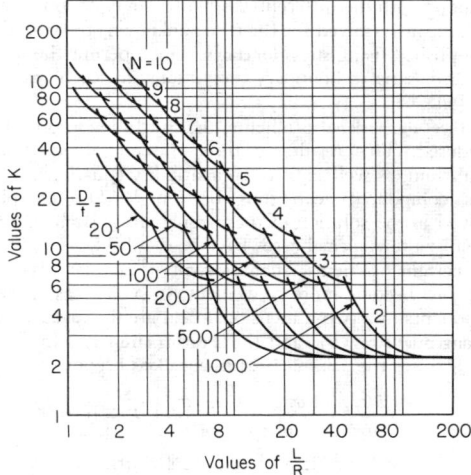

**Fig. 66**  Radial and end external pressure with fixed edges.

sary to withstand the pressure is $EI_c = W_s \, D^3 \, L_s/24$. $W_s$ is the pressure, $L_s$ the length between rings, and $I_c$ the combined moment of inertia of the ring and that portion of the shell assumed acting with the ring.

In some instances, cylinders collapse only after a stress in excess of the elastic limit has been reached; that is, **plastic range** stresses are present. In such cases the same equation applies, but the modulus of elasticity must be modified.

When the average stress $S_a$ is less than the proportional limit $S_p$, and the maximum stress (direct, plus bending) is $S$, the modified modulus

$$E' = E \left[ 1 - \frac{1}{4} \left( \frac{S - S_l}{S_u - S_a} \right)^2 \right]$$

$S_u$ is the modulus of rupture. When the average stress is larger than the proportional limit, the modified modulus is taken as the tangent at the average stress.

In **thick-walled cylinders** (Fig. 67a) the circumferential, hoop, or tangential stress $S_t$ is not uniform. In addition a radial stress $S_r$ is present. When equilibrium is applied to the annulus taken out of Fig. 67a and shown in Fig. 67b and the equation is integrated, the general tangential and radial stress relations, called the **Lamé** equations, are derived.

$$S_t = \frac{(r_1^2 p_1 - r_2^2 p_2) + (p_1 - p_2)\, r_1^2 r_2^2/r^2}{r_2^2 - r_1^2}$$

and

$$S_r = \frac{(r_1^2 p_1 - r_2^2 p_2) - (p_1 - p_2)\, r_1^2 r_2^2/r^2}{r_2^2 - r_1^2}$$

When the external pressure $p_2 = 0$, the equations reduce to

$$S_t = \frac{r_1^2 p_1}{r_2^2 - r_1^2}\,(1 + r_2^2/r^2)$$

and

$$S_r = \frac{r_1^2 p_1}{r_2^2 - r_1^2}\,(1 - r_2^2/r^2)$$

At the inner boundary the tangential elongation $\epsilon_t$ is equal to

$$\epsilon_t = (S_t - \mu S_r)/E.$$

The increase in the bore radius $\Delta r_1$ resulting therefrom is

$$\Delta r_1 = \frac{r_1 p_1}{E_h}\left(\frac{1 + r_1^2/r_2^2}{1 - r_1^2/r_2^2} + \mu\right)$$

Similarly a solid shaft of $r$ radius under external pressure $p_2$ will have its radius decreased by the amount

$$\Delta r = -\frac{r p_2}{E_s}\,(1 - \mu)$$

In the case of a **press or shrink fit**, $p_1 = p_2 = p$. The sum of $\Delta r_1$ and $\Delta r_1$ absolute is the radial interference; twice this sum is the **diametral interference** $\Delta$ or

$$\Delta = 2 r_1 p \left[\frac{1}{E_h}\left(\frac{1 + r_1^2/r_2^2}{1 - r_1^2/r_2^2} + \mu\right) + \frac{1 - \mu}{E_s}\right]$$

If the hub and shaft materials are the same, $E_h = E_s = E$, and

$$\Delta = \frac{4 r_1 r_2^2 p}{E\,(r_2^2 - r_1^2)}$$

If the equation is solved for $p$ and this value is substituted in

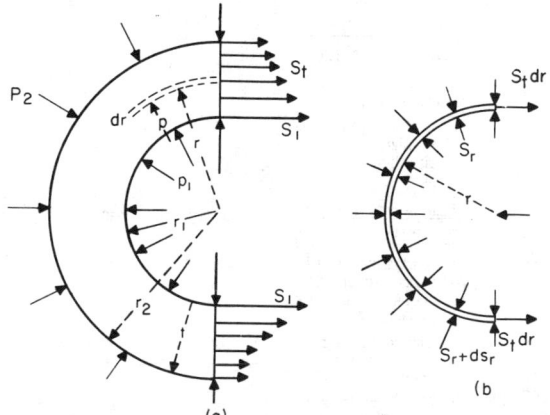

**Fig. 67**

Lamé's equation, the maximum tangential stress on the inner surface of the hub is found to be

$$S_t = \frac{E\Delta}{4 r_1 r_2^2}\,(r_2^2 + r_1^2)$$

EXAMPLE. The barrel of a field gun has an outside diameter of 9 in and a bore of 4.7 in. An internal pressure of 16,000 lb/in² is developed during firing. What maximum stress occurs in the barrel? An investigation of Lamé's equations for internal pressure reveals the maximum stress to be the tangential one on the inner surface. Thus,

$$S_t = p_1 \frac{r_1^2 + r_2^2}{r_2^2 - r_1^2} = \frac{16{,}000\,(2.39^2 + 4.5^2)}{4.5^2 - 2.39^2} = 31{,}900 \text{ lb/in}^2 \,(2{,}246 \text{ kgf/cm}^2)$$

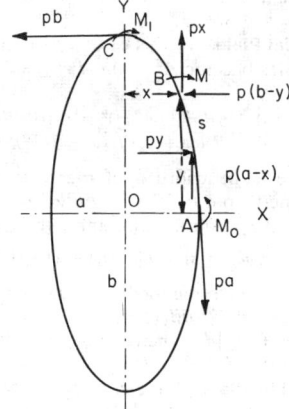

**Fig. 68**

**Oval Hollow Cylinders**  In Fig. 68, let $a$ and $b$ be the semiminor and semimajor axes. The bending moments at $A$ and $C$ will then be

$$M_0 = (pa^2/2) - (pI_x/2S) - (pI_y/2S)$$
$$M_1 = M_0 - p(a^2 - b^2)/2$$

where $I_x$ and $I_y$ are the moments of inertia of the arc $AC$ about the $x$ and $y$ axes, respectively. The **bending moment at any point** will be

$$M = M_0 - (pa^2/2) + (px^2/2) + (py^2/2)$$

**Thick Hollow Spheres**  With an **internal pressure** $p$, where $p < T/0.65$,

$$r_2 = r_1[T + 0.4p)/(T - 0.65p)]^{1/3}$$

The maximum tensile stress is on the inner surface, in the direction of the circumference. With an **external pressure** $p$, where $p < T/1.05$,

$$r_2 = r_1[T/(T - 1.05p)]^{1/3}$$

In both cases $T$ is the true stress.

## PRESSURE BETWEEN BODIES WITH CURVED SURFACES

(See Hertz, "Gesammelte Werke," vol. 1, pp. 159 et seq., Barth)

**Two Spheres**  The radius $A$ of the compressed area is obtained from the formula $A^3 = 0.68P(c_1 + c_2)/[(1/r_1) + (1/r_2)]$, in which $P$ is the compressing force, $c_1$ and $c_2$ ($= 1/E_1$ and $1/E_2$) are reciprocals of the respective moduli of elasticity, and $r_1$

and $r_2$ are the radii. (Reciprocal of Poisson's ratio assumed to be $n = 10/3$.) The **greatest contact pressure** in the middle of the compressed surface will be $S_{max} = 1.5(P/\pi A^2)$, and

$$S_{max}^3 = 0.235P[(1/r_1) + (1/r_2)]^2/(c_1 + c_2)^2$$

The **total deformation** of the two spheres will be $Y$, which is obtained from

$$Y^3 = 0.46P^2(c_1 + c_2)^2[(1/r_1) + (1/r_2)]$$

**For** $c_1 = c_2 = 1/E$, i.e., **two spheres with the same modulus of elasticity,** it follows that $A^3 = 1.36\ P/E[(1/r_1) + (1/r_2)]$, $S_{max}^3 = 0.059PE^2[(1/r_1) + (1/r_2)]^2$, and $Y^3 = 1.84P^2[(1/r_1) + (1/r_2)]/E^2$. If the radii of these spheres are also equal, $A^3 = 0.68Pr/E = 0.34Pd/E$; $S_{max}^3 = 0.235PE^2/r^2 = 0.94PE^2/d^2$; and $Y^3 = 3.68P^2/E^2r = 7.36P^2/E^2d$.

**Sphere and Flat Plate**   In this case $r_1 = r$ and $r_2 = \infty$, and the above formulas become $A^3 = 0.68Pr(c_1 + c_2) = 1.36Pr/E$, and

$$S_{max}^3 = 0.235P/r^2(c_1 + c_2)^2 = 0.059PE^2/r^2$$
$$Y^3 = 0.46P^2(c_1 + c_2)^2/r = 1.84P^2/E^2r$$

**Two Cylinders**   The width $b$ of the rectangular pressure surface is obtained from $(b/4)^2 = 0.29P(c_1 + c_2)/l[(1/r_1) + (1/r_2)]$, where $r_1$ and $r_2$ are the radii, and $l$ the length

$$S_{max}^2 = (4P/\pi bl)^2 = 0.35P[(1/r_1) + (1/r_2)]/l(c_1 + c_2)$$

For cylinders with the same moduli of elasticity, $c_1 = c_2 = 1/E$, and $(b/4)^2 = 0.58P/El[(1/r_1) + (1/r_2)]$; and $S_{max}^2 = 0.175PE[(1/r_1) + (1/r_2)]/l$. When $r_1 = r_2 = r$, $(b/4)^2 = 0.29Pr/El$, and $S_{max}^2 = 0.35PE/lr$.

**Cylinder and Flat Plate**   Here $r_1 = r$, $r_2 = \infty$, and the above formulas reduce to $(b/4)^2 = 0.29Pr(c_1 + c_2)/l = 0.58Pr/El$, and

$$S_{max}^2 = 0.35P/lr(c_1 + c_2) = 0.175PE/lr$$

For application to ball and roller bearings and to gear teeth, see Sec. 8.

## FLAT PLATES

The analysis of flat plates subjected to **lateral loads** is very involved because plates bend in all vertical planes. Strict mathematical derivations have therefore been accomplished only in some special cases. Most of the available formulas contain some amount of rational empiricism. Plates may be classified as (1) **thick plates,** in which transverse shear is important; (2) **average-thickness plates,** in which flexure stress predominates; (3) **thin plates,** which depend in part upon direct tension; and (4) **membranes,** which are subject to direct tension only. However, exact lines of demarcation do not exist.

The flat-plate formulas given apply primarily to symmetrically loaded **average-thickness plates** of constant thickness. In the mathematical analyses, allowance for stress redistribution, because of slight local yielding, is usually not made. Since this yielding, especially in ductile materials, is beneficial, the formulas generally err on the side of safety. Certain cases of symmetrically loaded **circular and rectangular plates** are presented in Figs. 69 and 70. The maximum stresses are calculated from

$$S_M = k\,\frac{wr^2}{t^2} \qquad S_M = k\,\frac{P}{t^2} \qquad \text{or} \qquad S_M = k\,\frac{C}{t^2}$$

The first equation is for a uniformly distributed load $w$, lb/in$^2$; the second supports a concentrated load $P$, pounds; and the third a couple $C$, per unit length. Combinations of these loadings may be treated by superposition. The factors $k$ are given in Tables 19 and 20, $R$ is the radius of circular plates or one side of rectangular plates, and $t$ is the plate thickness.

[In Figs. 69 & 70, $r = R$ for circular plates and $r =$ smaller side for rectangular plates]

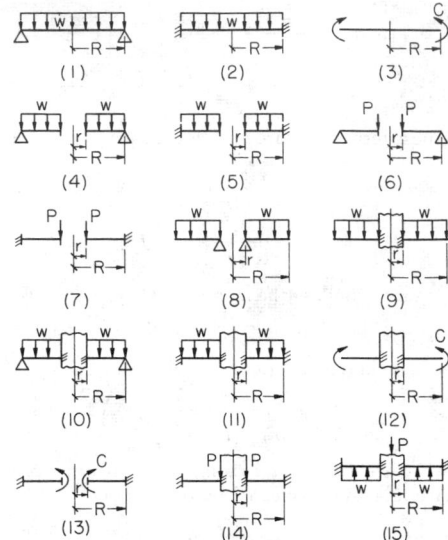

**Fig. 69**   Circular plates. Cases (4), (5), (6), (7), (8), and (13) have a central hole of radius $r$; cases (9), (10), (11), (12), (14), and (15) have a central piston of radius $r$ to which the plate is fixed.

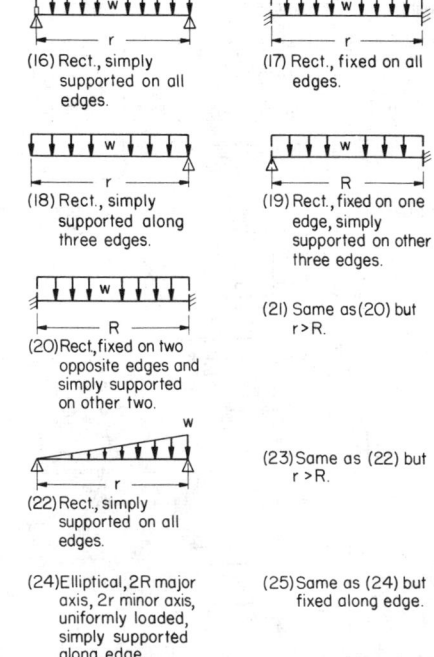

**Fig. 70**   Rectangular and elliptical plates. [R is the longer dimension except in cases (21) and (23).]

**Table 19. Coefficients $k$ and $k_1$ for Circular Plates**
($\mu = 0.3$)

| R/r | 1.25 | | 1.5 | | 2 | | 3 | | 4 | | 5 | |
|---|---|---|---|---|---|---|---|---|---|---|---|---|
| Case | $k$ | $k_1$ | $k$ | $k_1$ | $k$ | $k_1$ | $k$ | $k_1$ | $k$ | $k_1$ | $k$ | $k_1$ |
| 1 | 1.24 | 0.696 | | | | | | | | | | → |
| 2 | 0.75 | 0.171 | | | | | | | | | | → |
| 3 | 6.0 | 4.2 | | | | | | | | | | |
| 4 | 0.592 | 0.184 | 0.976 | 0.414 | 1.440 | 0.664 | 1.880 | 0.824 | 2.08 | 0.830 | 2.19 | 0.813 |
| 5 | 0.105 | 0.0025 | 0.259 | 0.0129 | 0.481 | 0.057 | 0.654 | 0.130 | 0.708 | 0.163 | 0.730 | 0.176 |
| 6 | 1.10 | 0.341 | 1.26 | 0.519 | 1.48 | 0.672 | 1.88 | 0.734 | 2.17 | 0.724 | 2.34 | 0.704 |
| 7 | 0.195 | 0.0036 | 0.320 | 0.024 | 0.455 | 0.081 | 0.670 | 0.171 | 1.00 | 0.218 | 1.30 | 0.238 |
| 8 | 0.660 | 0.202 | 1.19 | 0.491 | 2.04 | 0.902 | 3.34 | 1.220 | 4.30 | 1.300 | 5.10 | 1.310 |
| 9 | 0.135 | 0.0023 | 0.410 | 0.0183 | 1.04 | 0.0938 | 2.15 | 0.293 | 2.99 | 0.448 | 3.69 | 0.564 |
| 10 | 0.122 | 0.00343 | 0.336 | 0.0313 | 0.740 | 0.1250 | 1.21 | 0.291 | 1.45 | 0.417 | 1.59 | 0.492 |
| 11 | 0.072 | 0.00068 | 0.1825 | 0.005 | 0.361 | 0.023 | 0.546 | 0.064 | 0.627 | 0.092 | 0.668 | 0.112 |
| 12 | 6.865 | 0.2323 | 7.448 | 0.6613 | 8.136 | 1.493 | 8.71 | 2.555 | 8.930 | 3.105 | 9.036 | 3.418 |
| 13 | 6.0 | 0.196 | 6.0 | 0.485 | 6.0 | 0.847 | 6.0 | 0.940 | 6.0 | 0.801 | 6.0 | 0.658 |
| 14 | 0.115 | 0.00129 | 0.220 | 0.0064 | 0.405 | 0.0237 | 0.703 | 0.062 | 0.933 | 0.092 | 1.13 | 0.114 |
| 15 | 0.090 | 0.00077 | 0.273 | 0.0062 | 0.710 | 0.0329 | 1.54 | 0.110 | 2.23 | 0.179 | 2.80 | 0.234 |

The **maximum deflection** for the same cases is given by

$$y_M = k_1 \frac{wr^4}{Et^3} \qquad y_M = k_1 \frac{Pr^2}{Et^3} \qquad \text{and} \qquad y_M = k_1 \frac{Cr^2}{Et^3}$$

The factors $k_1$ are also given in the tables. For additional information, including shells, refer to ASME Handbook, "Metals Engineering: Design," McGraw-Hill.

### THEORIES OF FAILURE

Material properties are usually determined from tests in which specimens are subjected to **simple stresses** under static or fluctuating loads. The attempt to apply these data to **bi- or triaxial stress fields** has resulted in the proposal of various theories of failure. Figure 71 shows the principal stresses on a triaxially stressed element. It is assumed, for simplicity, that $S_1 > S_2 > S_3$. Compressive stresses are negative.

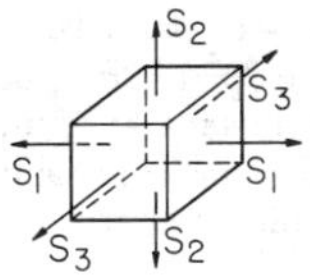

**Fig. 71**

1. **Maximum-stress theory** (Rankine) assumes failure occurs when the largest principal stress reaches the yield stress in a tension (or compression) specimen. That is, $S_1 = \pm S_y$.

2. **Maximum-shear theory** (Coulomb) assumes yielding (failure) occurs when the maximum shearing stress equals that in a simple tension (or compression) specimen at yield. Mathematically, $S_1 - S_3 = \pm S_y$.

3. **Maximum-strain-energy theory** (Beltrami) assumes failure occurs when the energy absorbed per unit volume equals the strain energy per unit volume in a tension (or compression) specimen at yield. Mathematically, $S_1^2 + S_2^2 + S_3^2 - 2\mu(S_1 S_2 + S_2 S_3 + S_3 S_1) = S_y^2$.

4. **Maximum-distortion-energy theory** (Huber, von Mises, Hencky) assumes yielding occurs when the distortion energy equals that in simple tension at yield. The distortion energy, that portion of the total energy which causes distortion rather than volume change, is

$$U_d = \frac{1 + \mu}{3E}(S_1^2 + S_2^2 + S_3^2 - S_1 S_2 - S_2 S_3 - S_3 S_1)$$

Thus failure is defined by

$$S_1^2 + S_2^2 + S_3^2 - (S_1 S_2 + S_2 S_3 + S_3 S_1) = S_y^2$$

5. **Maximum-strain theory** (Saint-Venant) claims failure

**Table 20. Coefficients $k$ and $k_1$ for Rectangular and Elliptical Plates**
($\mu = 0.3$)

| R/r | 1.0 | | 1.5 | | 2.0 | | 3.0 | | 4.0 | |
|---|---|---|---|---|---|---|---|---|---|---|
| Case | $k$ | $k_1$ | $k$ | $k_1$ | $k$ | $k_1$ | $k$ | $k_1$ | $k$ | $k_1$ |
| 16 | 0.287 | 0.0443 | 0.487 | 0.0843 | 0.610 | 0.1106 | 0.713 | 0.1336 | 0.741 | 0.1400 |
| 17 | 0.308 | 0.0138 | 0.454 | 0.0240 | 0.497 | 0.0277 | 0.500 | 0.028 | 0.500 | 0.028 |
| 18 | 0.672 | 0.140 | 0.768 | 0.160 | 0.792 | 0.165 | 0.798 | 0.166 | 0.800 | 0.166 |
| 19 | | 0.030 | | 0.070 | | 0.101 | | | | |
| 20 | | 0.0209 | | 0.0582 | | 0.0987 | | 0.1276 | | |
| 21* | | 0.0216 | | 0.0270 | | 0.0284 | | 0.0284 | | 0.0284 |
| 22 | | 0.0221 | | 0.0421 | | 0.0553 | | 0.0668 | | 0.0700 |
| 23* | | 0.0220 | | 0.0436 | | 0.0592 | | 0.0772 | | 0.0908 |
| 24 | 1.24 | 0.70 | 1.92 | 1.26 | 2.26 | 1.58 | 2.60 | 1.88 | 2.78 | 2.02 |
| 25 | 0.75 | 0.171 | 1.34 | 0.304 | 1.63 | 0.379 | 1.84 | 0.419 | 1.90 | 0.431 |

*Length ratio is $r/R$ in cases 21 and 23.

occurs when the maximum strain equals the strain in simple tension at yield or $S_1 - \mu(S_2 + S_3) = S_y$.

6. **Internal-friction theory** (Mohr). When the ultimate strengths in tension and compression are the same, this theory reduces to that of maximum shear. For principal stresses of opposite sign, failure is defined by $S_1 - (S_{uc}/S_u) S_2 = -S_{uc}$; if the signs are the same $S_1 = S_u$ or $- S_{uc}$, where $S_{uc}$ is the ultimate strength in compression. If the principal stresses are both either tension or compression, then the larger one, say $S_1$, must equal $S_u$ when $S_1$ is tension or $S_{uc}$ when $S_1$ is compression.

A **graphical representation** of the first four theories applied to a biaxial stress field is presented in Fig. 72. Stresses outside the bounding lines in the case of each theory mean failure (yield or fracture). A comparison with experimental data proves the distortion-energy theory (4) best for ductile materials of equal tension-compression properties. When these properties are unequal, the internal-energy theory (6) appears best. In practice, judging by some accepted codes, the maximum-shear theory (2) is generally used for ductile materials, and the maximum-stress theory (1) for brittle materials.

Fatigue failures cannot be related, theoretically, to elastic strength and thus to the theories described. However, experimental results justify this, at least to a limited extent. Therefore, the theory evaluation given above holds for **fluctuating stresses,** provided that principal stresses at the maximum load are used and the **endurance strength** in simple bending is substituted for the yield strength.

EXAMPLE. A steel shaft, 4 in in diameter, is subjected to a bending moment of 120,000 in·lb, as well as a torque. If the yield strength in tension is 40,000 lb/in², what maximum torque can be applied under the (1) maximum-shear theory and (2) the distortion-energy theory?

$$S_x = \frac{M_c}{I} = \frac{120,000 \times 2}{12.55}$$

$$= 19,100 \text{ lb/in}^2 \qquad S_{xy} = \frac{TC}{J} = \frac{T \times 2}{25.1} = 0.0798T$$

and

$$S_{M,m} = \frac{S_x}{2} \pm \sqrt{\left(\frac{S_x}{2}\right)^2 + S_{xy}^2}$$

(1) $S_M - S_m = S_y$ or $2\sqrt{\left(\frac{19,100}{2}\right)^2 + (0.0798T)^2} = (40,000)^2$

or $\qquad\qquad T = 221,000 \text{ in·lb } (254,150 \text{ cm·kgf})$

(2) $\qquad\qquad S_M^2 + S_m^2 - S_M S_m = S_y^2$

substituting and simplifying,

$$(9,550)^2 + 3\sqrt{\left(\frac{19,100}{2}\right)^2 + (0.0798T)^2} = (40,000)^2$$

or $\qquad\qquad T = 255,000 \text{ in·lb } (293,250 \text{ cm·kgf})$

## PLASTICITY

The reaction of materials to stress and strain in the plastic range is not yet fully defined. However, some concepts and theories have been proposed.

Ideally, a **purely elastic material** is one complying explicitly with Hooke's law. In a **viscous material,** the shearing stress is proportional to the shearing strain. The **purely plastic material** yields indefinitely, but only after reaching a certain stress. Combinations of these are the **elastico-viscous** and the **elastico-plastic** materials.

Engineering materials are not ideal, but usually contain some of the elastico-plastic characteristics. The **total strain** $\epsilon_t$ is the sum of the **elastic strain** $\epsilon_o$ plus the **plastic strain** $\epsilon_p$, as shown in Fig. 73, where the stress-strain curve is approximated by two straight lines. The **natural strain,** which is at the same time the total strain, is $\epsilon = \int_{l_o}^{l} dl/l = \ln (l/l_o)$. In this equation, $l$ is the instantaneous length, while $l_o$ is the original length. In

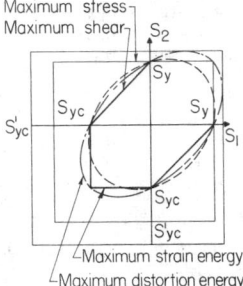

terms of the normal strain, the natural strain becomes $\bar{\epsilon} = \ln(1 + \epsilon_o)$. Since it is assumed that the volume remains constant, $l/l_o = A_o/A$, and so the natural stress becomes $\bar{S} = P/A = (P/A_o) (1 + \epsilon_o)$. $A_o$ is the original cross-sectional area. If the natural stress is plotted against the natural strain on log-log paper, the graph is very nearly a straight line. The plastic-range relation is thus approximated by $\bar{S} = K\bar{\epsilon}^n$, where the proportionality factor $K$ and the **strain-hardening coefficient** $n$ are determined from best fits to experimental data. Values of $K$ and $n$ determined by Low and Garofalo (*Proc. Soc. Exp. Stress Anal.*, vol. IV, no. 2, 1947) are given in Table 21.

The geometry of Fig. 73 can be used to arrive at a second approximate relation

$$\bar{S} = S_o + (\epsilon_p - \epsilon_o) \tan \theta = S_o \left(1 - \frac{H}{E}\right) + \epsilon_p H$$

where $H = \tan \theta$ is a kind of **plastic modulus.**

The **deformation theory of plastic flow** for the general case of combined stress is developed using the above concepts. Certain additional assumptions involved include: principal plastic-strain directions are the same as principal stress directions; the elastic strain is negligible compared to plastic strain; and the ratios of the three principal shearing strains—$(\bar{\epsilon}_1 - \bar{\epsilon}_2)$, $(\bar{\epsilon}_2 - \bar{\epsilon}_3)$, $(\bar{\epsilon}_3 - \bar{\epsilon}_1)$—to the principal shearing stresses—$(\bar{S}_1 - \bar{S}_2)/2$, $(\bar{S}_2 - \bar{S}_3)/2$, $(\bar{S}_3 - \bar{S}_1)/2$—are equal. The relations between the

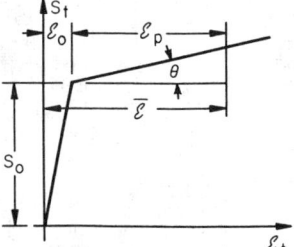

**Fig. 73**

**Fig. 72**

principal strains and stresses in terms of the simple tension quantities become

$$\bar{\epsilon}_1 = \bar{\epsilon}/\bar{S} \, [\bar{S}_1 - (\bar{S}_2 + \bar{S}_3)/2]$$
$$\bar{\epsilon}_2 = \bar{\epsilon}/\bar{S} \, [\bar{S}_2 - (\bar{S}_3 + \bar{S}_1)/2]$$
$$\bar{\epsilon}_3 = \bar{\epsilon}/\bar{S} \, [\bar{S}_3 - (\bar{S}_1 + \bar{S}_2)/2]$$

If these equations are added, the plastic-flow theory is expressed:

$$\frac{\bar{S}}{\bar{\epsilon}} = \sqrt{\frac{[(\bar{S}_1 - \bar{S}_2)^2 + (\bar{S}_2 - \bar{S}_3)^2 + (\bar{S}_3 - \bar{S}_1)^2]/2}{2\,(\bar{\epsilon}_1^2 + \bar{\epsilon}_2^2 + \bar{\epsilon}_3^2)/3}}$$

In the above equation

$$\sqrt{[(\bar{S}_1 - \bar{S}_2)^2 + (\bar{S}_2 - \bar{S}_3)^2 + (\bar{S}_3 - \bar{S}_1)^2]/2} = \bar{S}_e$$

and

$$\sqrt{[\bar{\epsilon}_1^2 + \bar{\epsilon}_2^2 + \bar{\epsilon}_3^2]} = \epsilon_e$$

are the effective, or significant, stress and strain, respectively.

EXAMPLE. An annealed, stainless-steel type 430 tank is 41 in in inside diameter and has a wall 0.375 in thick. The ultimate strength of the stainless steel is 85,000 lb/in². Compute the maximum strain as well as the pressure at fracture.

The tank constitutes a biaxial stress field where $S_1 = pd/2t$, $S_2 = pd/4t$, and $S_3 = 0$. Taking the power stress-strain relation

$$\bar{S}_e = K\bar{\epsilon}_e^n \quad \text{or} \quad \bar{\epsilon}/\bar{S} = \bar{S}_e^{\left(\frac{1-n}{n}\right)} / K^{1/n}$$

thus

$$\bar{\epsilon}_1 = \frac{\bar{S}_e^{\left(\frac{1-n}{n}\right)}}{K^{1/n}} \left(\frac{3}{4}\bar{S}_1\right) \bar{\epsilon}_2 = 0,$$

and

$$\bar{\epsilon}_3 = \frac{\bar{S}_e^{\left(\frac{1-n}{n}\right)}}{K^{1/n}} \left(-\frac{3}{4}\bar{S}_1\right) = -\bar{\epsilon}_1$$

The maximum-shear theory, which is applicable to a ductile material under combined stress, is acceptable here. Thus rupture will occur at $\bar{S}_1 - \bar{S}_3 = \bar{S}_1 = \bar{S}_u$, and

$$\bar{S}_e = \sqrt{\frac{1}{2}\left[\left(\frac{\bar{S}_1}{2}\right)^2 + \left(\frac{\bar{S}_1}{2}\right)^2 + \bar{S}_1^2\right]} = \sqrt{\frac{3}{4}\bar{S}_1^2} = \left(\frac{3}{4}\right)^{1/2}\bar{S}_u$$

So

$$\bar{\epsilon}_1 = \frac{[(3/4)^{1/2}\bar{S}_u]^{\frac{1-n}{n}}}{K^{1/n}}\left(\frac{3}{4}\bar{S}_u\right) = \left[\frac{3}{4}\right]^{\frac{1+0.229}{0.458}}\left(\frac{85,000}{143,000}\right)^{\frac{1}{0.229}}$$
$$= 0.0475 \text{ in/in } (0.0475 \text{ cm/cm})$$

And since $\bar{S}_u = S_1 = \dfrac{pd}{2t}$, then $p = \dfrac{2t\bar{S}_u}{d}$

or

$$p = \frac{2 \times 0.375 \times 85,000}{41} = 1,550 \text{ lb/in}^2 \ (109 \text{ kgf/cm}^2)$$

## ROTATING DISKS

Rotating circular disks may be of various profiles, of constant thickness or variable thickness, with or without centrally and noncentrally located holes, and with radial, tangential, and shearing stresses.

Solution starts with the differential equations of equilibrium and compatibility and the subsequent application of appropriate boundary conditions for the derivation of working-stress equations.

If the disk thickness is small compared with the diameter,

the variation of stress with thickness can be assumed to be negligible, symmetry eliminating the shearing stress. In the rotating case, the disk weight is neglected, but its inertia force becomes the body-force term in the equilibrium equations.

Thus solved, the stress components in a solid disk become

$$\sigma_r = \frac{3+\mu}{8}\rho\omega^2(R^2 - r^2)$$
$$\sigma_\theta = \frac{3+\mu}{8}\rho\omega^2 R^2 - \frac{1+3\mu}{8}\rho\omega^2 r^2$$

where  $\mu$ = Poisson's ratio
$\rho$ = mass density, lb·s²/in⁴
$\omega$ = angular speed, rad/s
$R$ = outside disk radius
$r$ = radius to point in question

The largest stresses occur at the center of the solid disk and are

$$\sigma_r = \sigma_\theta = \frac{3+\mu}{8}\rho\omega^2 R^2$$

**A disk with a central hole** of radius $r_h$ (no external forces) is subjected to the following stresses:

$$\sigma_r = \frac{3+\mu}{8}\rho\omega^2\left(R^2 + r_h^2 - \frac{R^2 r_h^2}{r^2} - r^2\right)$$
$$\sigma_\theta = \frac{3+\mu}{8}\rho\omega^2\left(R^2 + r_h^2 + \frac{R^2 r_h^2}{r^2} - \frac{1+3\mu}{3+\mu}r^2\right)$$

The maximum radial stress $\sigma_r|_M$ occurs at $r = \sqrt{Rr_h}$, and

$$\sigma_r|_M = \frac{3+\mu}{8}\rho\omega^2(R - r_h)^2$$

The largest tangential stress $\sigma_\theta|_M$ exists at the inner boundary, and

$$\sigma_\theta|_M = \frac{3+\mu}{4}\rho\omega^2\left(R^2 + \frac{1-\mu}{3+\mu}r_h^2\right)$$

As the hole radius $r_h$ approaches zero, the tangential stress assumes a value twice that at the center of a rotating solid disk, given above.

**Stresses in Turbine Disks**  Explicit solutions for cases other than those cited are not available; so approximate solutions, such as those proposed by Stodola, Thompson, Hetényi, and Robinson, are necessary. Manson (The Determination of Elastic Stresses in Gas-Turbine Disks, *N A C A Rept.* 871, 1947) uses the calculus of finite differences.

The customary, simplifying assumptions of axial symmetry—no variation of stress in the thickness direction and a completely elastic stress situation—are made. The differential equations of equilibrium and compatibility are rewritten in finite-difference form.

Solution of the finite-difference equations, appreciation of their linear nature, and successive application of them yield the stresses at any station in terms of those at a boundary station such as $r_0$. The equations thus derived are

$$\sigma_{r,n} = A_{r,n}\sigma_{t,r_0} + B_{r,n}$$
$$\sigma_{t,n} = A_{t,n}\sigma_{t,r_0} + B_{t,n} \tag{1}$$

The finite-difference expressions yield Eqs. (2), which permit

**Table 21. Constants $K$ and $n$ for Sheet Materials**

| Material | Treatment | $K$, lb/in² | $n$ |
|---|---|---|---|
| 0.05%C rimmed steel | Annealed | 77,100 | 0.261 |
| 0.05%C killed steel | Annealed and tempered | 73,100 | 0.234 |
| Decarburized 0.05%C steel | Annealed in wet $H_2$ | 75,500 | 0.284 |
| 0.05/0.07% phos. low C | Annealed | 93,330 | 0.156 |
| SAE 4130 | Annealed | 169,400 | 0.118 |
| SAE 4130 | Normalized and tempered | 154,500 | 0.156 |
| Type 430 stainless | Annealed | 143,000 | 0.229 |
| Alcoa 24-S | Annealed | 55,900 | 0.211 |
| Reynolds R-301 | Annealed | 48,450 | 0.211 |

the coefficients at station $n$ to be computed from those at station $n - 1$.

$$
\begin{aligned}
A_{r,n} &= K_n A_{r,n-1} + L_n A_{t,n-1} \\
A_{t,n} &= K_n' A_{r,n-1} + L_n' A_{t,n-1} \\
B_{r,n} &= K_n B_{r,n-1} + L_n B_{t,n-1} + M_n \\
B_{t,n} &= K_n' B_{r,n-1} + L_n' B_{t,n-1} + M_n'
\end{aligned}
\tag{2}
$$

The coefficients at the first station can be established by inspection. For a solid disk, for instance, where both stresses are equal to the tangential stress at the center, the coefficients in Eqs. (1) are $A_{r,n} = A_{t,n} = 1$ and $B_{r,n} = B_{t,n} = 0$. In the case of the disk with a central hole, where $\sigma_{r,r_h} = 0$. $A_{r,r_h} = B_{r,r_h} = B_{t,r_h} = 0$ and $A_{t,r_1} = 1$. Knowing these, all others can be found from Eqs. (2).

At the outer boundary, $\sigma_{r,R} = A_{r,R}\sigma_{t,r_0} + B_{r,R}$ and $\sigma_{t,r_0} = (\sigma_{r,R} - B_{r,R})/A_{r,R}$. The radial and tangential stresses at each station are successively obtained, knowing $\sigma_{t,r_0}$ and all the coefficients, using Eqs. (1).

The remaining coefficients in Eqs. (2), extracted from the finite-difference equations, are defined below, where $E$ is Young's modulus at the temperature of the point in question, $h$ is the profile thickness, $\alpha$ is the thermal coefficient of expansion, $\Delta T$ is the temperature increment above that at which the thermal stress is zero, $\mu$ is Poisson's ratio, $\omega$ is the angular velocity of disk, and $\rho$ is the mass density of disk material.

$$
\begin{aligned}
C_n &= r_n b_n \\
C_n' &= \mu_n/E_n + [(1 + \mu_n)(r_n - r_{n-1})/(2E_n r_n)] \\
D_n &= \tfrac{1}{2}(r_n - r_{n-1})b_n \\
D_n' &= 1/E_n + [(1 + \mu_n)(r_n - r_{n-1})/(2E_n r_n)] \\
F_n &= r_{n-1} b_{n-1} \\
F_n' &= (\mu_{n-1}/E_{n-1}) - [(1 + \mu_{n-1})(r_n - r_{n-1})/(2E_{n-1} r_{n-1})] \\
G_n &= \tfrac{1}{2}(r_n - r_{n-1})b_{n-1} \\
G_n' &= (1/E_{n-1}) - [(1 + \mu_{n-1})(r_n - r_{n-1})/(2E_{n-1} r_{n-1})] \\
H_n &= \tfrac{1}{2}\omega^2(r_n - r_{n-1})(\rho_n b_n r_n^2 + \rho_{n-1} b_{n-1} r_{n-1}^2) \\
H_n' &= \alpha_n \Delta T_n - \alpha_{n-1} \Delta T_{n-1} \\
K_n &= (F_n' D_n - F_n D_n')/(C_n' D_n - C_n D_n') \\
K_n' &= (C_n F_n' - C_n' F_n)/(C_n' D_n - C_n D_n') \\
L_n &= -(G_n' D_n + G_n D_n')/(C_n' D_n - C_n D_n') \\
L_n' &= -(C_n' G_n + C_n G_n')/(C_n' D_n - C_n D_n') \\
M_n &= (H_n' D_n + H_n D_n')/(C_n' D_n - C_n D_n') \\
M_n' &= (C_n' H_n + C_n H_n')/(C_n' D_n - C_n D_n')
\end{aligned}
$$

Stations need not be equally spaced between the two boundaries. It is best to space them more closely where the profile, temperature, or other property is changing rapidly. In cases of sudden or abrupt section changes, it is best to fair in across the change; the material density should, however, be adjusted to give a total mass equal to the actual. Six to ten stations are often sufficient.

The modulus of elasticity has a significant effect, and its exact value at the temperature of each station should be used. The coefficients of thermal expansion are usually averaged for the temperature between the station and that at which no thermal stress occurs.

The first two Eqs. (2) and the last two must be worked simultaneously.

At the outer boundary, loads external to the disk may be imposed, e.g., the radial stress $\sigma_{r,R}$ from the centrifuged pull of a bucket. At the center, the disk may be shrunk on a shaft with the fit pressures causing a radial external push at this boundary.

Numerical solutions are most expeditiously accomplished by use of a table with column-to-column procedures.

**Disks with Noncentral Holes**    This case has not been solved explicitly, but approximations are useful (e.g., Armstrong, Stresses in Rotating Tapered Disks with Noncentral Holes, Ph.D. dissertation, Iowa State University, 1960). The area between the holes is considered removed and replaced by uniform spokes, each one with a cross-sectional area equal to the original minimum spoke area and with a length equal to the diameter of the noncentral holes. The higher stress in such a spoke results in an additional extension, which is then applied to the outer annulus according to thin-ring theory and based on the average radius of the ring. The additional stress is considered constant and is added to the tangential stress which would be present in a disk of the same dimensions but filled (that is, no noncentral holes).

The stress in the substitute spoke is computed by adjusting the stress at the hole-center radius in the solid or filled disk in proportion to the areas, or $S_{sp} = \sigma_{r,h}(A_g/A_{sp})$, where $\sigma_{r,h}$ is the radial stress in the filled disk at the radius of the hole circle, $A_g$ is the gross circumferential area at the same radius of the filled disk, and $A_{sp}$ is the area of the substitute spoke. The increase in total strain is $\delta = \sigma_{r,h}/E[(A_g/A_{sp}) - 1]l_{sp}$, where $l_{sp}$ is the length of the substitute spoke.

The spoke-effect correction to be applied to the tangential stress is therefore $\sigma\theta_c = \delta E/r'$, where $r'$ is the average outer-rim or annulus radius. This is added to the tangential stress found at the corresponding radius in the filled disk. The final step is to adjust the tangential and radial stresses as determined for stress concentrations caused by the holes in the actual disk. The factors for this adjustment are those in an infinite plate of uniform thickness having the same size hole.

The method is claimed to yield stresses within 5 percent of those measured photoelastically at points of highest stress.

## EXPERIMENTAL STRESS ANALYSIS

Analytical methods of stress analysis can reach limits of applicability. Many experimental techniques have been suggested and tried; several have been developed to a state of great usefulness, e.g., photoelasticity, strain-gage measurement, brittle coating, birefringent coating, and holography.

### Photoelasticity

Most transparent materials exhibit temporary double refraction, or **birefringence**, when stressed. Light is resolved into components along the two principal plane directions. The effect is temporary as long as the elastic stress is not exceeded and is in direct proportion to the applied load. The stress magnitude can be established by the amount of component wave retardation, as given in the white and black band field (fringe pattern) obtained when a monochromatic light source is used. The polariscope, consisting of the light source, the polarizer, the model in a loading frame, an analyzer (same as polarizer), and a screen or camera, is used to produce and evaluate the fringe effect. Quarter-wave plates may be placed on either side of the model, making the light components through the model independent of the absolute orientation of polarizer and analyzer. The polarizer is a plane polariscope and yields the directions of principal stresses (the isoclinics); the analyzer is a circular polariscope yielding the fringes (isochromatics) as well.

Figure 74 shows the fringe pattern and the 20° isoclinics of a disk loaded radially at four places.

The isochromatics in the fringe pattern depict the **difference** between principal stresses. At free boundaries where the normal stress is zero, the difference automatically becomes the tangential stress. Starting at such a boundary and proceeding into the interior, the stresses can be separated by numerical calculation.

**The Stress-Optic Law**  In a transparent, isotropic plate subjected to a biaxial stress field within the elastic limit, the relative retardation $R_t$ between the two components produced by temporary double refraction is $R_t = Ct(p - q) = n\lambda$, where $C$ is the stress-optic coefficient, $t$ is the plate thickness, $p$ and $q$ are the principal stresses, $n$ is the fringe order (the number of fringes which have passed the point during application of load), and $\lambda$ is the wavelength of monochromatic light used.

Thus, $(p - q/2 = \tau\big|_M = (n\lambda)/2Ct = nf/t$.

If the **material-fringe value** $f$ is determined with the same light source (generally a mercury-vapor lamp emitting light having a wavelength of 5,461 Å) as used in the model study, the maximum shearing stress, or one-half the difference between the principal stresses, is directly determined. The calibration is a matter of obtaining the material-fringe value in lb/in² per fringe per inch (kgf/cm² per fringe per cm).

**Isoclinics**, or the direction of the principal planes, can be obtained with a plane polariscope. A new isoclinic parameter is observed each time the polarizer and analyzer are rotated simultaneously into a new position. A white-light source reveals a more distinct isoclinic, as the black curve is more distinguishable against a colored background.

**Isostatics**, or stress trajectories, are curves the tangents to which represent the progressive change in principal-plane directions. They are constructed graphically using the isoclinics. Since there are two principal planes at each point, two families of orthogonal curves are drawn. Care must be exercised in the drawing of trajectories for practical accuracy.

**Stress Separation**  If knowledge of each principal stress is required, the photoelastic data must be treated to separate the stresses from the difference given by the data. If the sum of the two stresses is also obtained somehow, a simultaneous solution of the sum and difference values will yield each principal stress. One can also start at a boundary where the normal stress value is zero. There, the photoelastic reading gives the principal stress parallel to the boundary. Starting with the single value, methods have been developed which can be used to proceed with the separation. Typical of the former are lateral-extensometer, iteration, and membrane-analogy techniques; typical of the latter are the slope-equilibrium, shear-difference, graphical-integration, and alternating-summation methods, and oblique incidence. Often, however, the surface stresses are the maximum valued ones. (See Frocht, "Photoelasticity," McGraw-Hill.)

EXAMPLE. The fringe pattern of a Homalite disk 1.31 in in diam, 0.282 in thick, and carrying four radial loads of 155 lb each is shown in Fig. 74.

A closed solution is not known. However, by counting the fringe order at any point, the stress can be determined photoelastically. For

(a) Fringe pattern

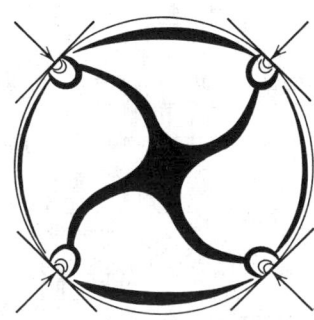

(b) 20° isoclinics

**Fig. 74**

instance, the dark spot at the center marks a fringe of zero order, as do the disk edges except in the immediate vicinity of the concentrated loads. The point at the center, which remained dark throughout the loading, is an isotropic point (zero stress difference and normal stresses are equal in all directions). Counting out from the center toward the load, the first "circular" fringe is of order 3. Therefore, anywhere along it $(p - q/2 = \tau \Big|_M = nf/t = 3 \times 65/0.282 = 692$ lb/in² (49 kgf/cm²). Carefully inspected, fringe 12 can be counted at the point of load application. Therefore, $\tau \Big|_M = 12 \times 65/0.282 = 2{,}770$ lb/in² (195 kgf/cm²).

## THREE-DIMENSIONAL PHOTOELASTICITY

Stress "freezing" and slicing, wherein a plastic model is brought up to its critical temperature, loaded as desired, and while loaded, slowly brought back to room temperature, are techniques which freeze the fringe pattern into the model. The model can be cut into slices without disturbing the "frozen" strains. Two-dimensional models are usually machined from plate stock, and three-dimensional models are cast. The frozen stress model is sliced so that the desired information can be obtained by normal incidence using the previous formulations.

When normal incidence is not possible, **oblique incidence** becomes necessary. Oblique-incidence patterns are usable in two-dimensional as well as three-dimensional stress separation. The measurement of fractional fringes is often required when using oblique incidence. With a crossed, circular, monochromatic polariscope, oriented to the principal stresses at a point, the analyzer is rotated through some angle $\phi$ until extinction occurs. The fringe value $n$ is $n = n_n \pm \phi/180$, where $n_n$ is the order of the last visible fringe. Whether the fractional term is added or subtracted depends upon the direction in which the analyzer is rotated (established by inspection).

Oblique-incidence calculations are based on the stress-optic law: $n_n = R_t = t(p - q)/f = tp/f - tq/f = n_p - n_q$. Also, when polarized light is directed through the slice at an angle $\theta_x$ to a principal plane, either by rotating the slice away from normal to the light ray or by cutting it at the angle $\theta_x$ (see Fig. 75), the fringe order becomes

$$n_{\theta x} = \frac{t'}{f}(p' - q') = \frac{t}{f \cos \theta_x}(p - q \cos^2 \theta_x)$$
$$= (n_p - n_q \cos^2 \theta_x)/\cos \theta_x$$

Solving algebraically,

and
$$n_p = (n_{\theta x} \cos \theta_x - n_n \cos^2 \theta_x)/\sin^2 \theta_x$$
$$n_q = (n_{\theta x} \cos \theta_x - n_n)/\sin^2 \theta_x$$

If orders $n_n$ and $n_{\theta x}$ are thus measured at a point, $n_p$ and $n_q$

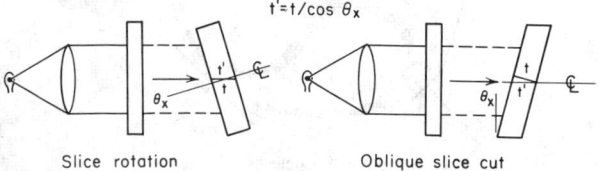

$t' = t/\cos \theta_x$

Slice rotation        Oblique slice cut

**Fig. 75**

can be computed. The principal stresses are then determined from $p = f n_p/t$ and $q = f n_q/t$.

The material-fringe value $f$ in these equations is at the "freezing" temperature (critical temperature). The angle of incidence, as well as the fringe orders, must be accurately measured if errors are to be minimized.

### Bonded Metallic Gages

Strain measurements down to one-millionth inch per inch (one-millionth cm/cm) are possible with electrical-resistance wire gages. Such gages can be used to measure surface strains (stress by Hooke's law) on any shape or size of object. Figure 76 illustrates schematically the gage construction with a grid of fine alloy wire or thin foil, bonded to paper and covered for protection with a felt pad. In use, the gage is cemented rigidly to the surface of the member to be analyzed. The strain relation is $\epsilon = (\Delta R/R)(1/G_f)$ in/in (cm/cm). Thus, if the resistance $R$ and gage factor $G_f$ (given by the gage manufacturer) are known and the change in resistance $\Delta R$ is measured, the strain which caused the resistance change can be determined and Hooke's law can be applied to determine the stress.

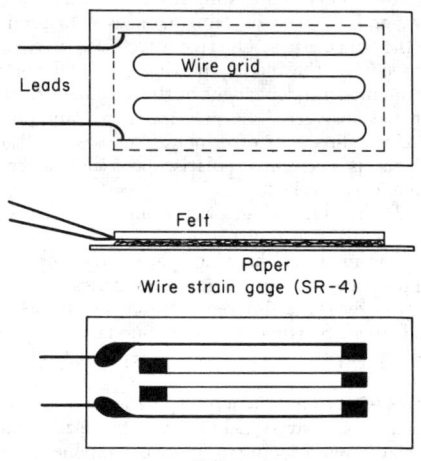

Leads — Wire grid

Felt

Paper
Wire strain gage (SR-4)

Foil strain gage

**Fig. 76**

Gages must be properly selected in accordance with manufacturer's recommendations. The surface to which the gage is applied must be clean, the proper cement must be used, and the gage assembly must be coated for protection against environmental conditions (e.g., moisture).

A gaging unit, usually a Wheatstone bridge or a ballast circuit (see Fig. 77 and Sec. 15), is needed to detect the signal resulting from the change in resistance of the strain gage. The strain and, therefore, the signal are often too small for direct handling, so that amplification is needed, with a metering discriminator for magnitude evaluation.

The signal is read or recorded by a galvanometer, oscilloscope, or other device. Equipment specifically constructed for strain measurement is available to indicate or record the signal directly in strain units.

Static strains are best gaged on a **Wheatstone bridge,** with strain gages wired to it as indicated in Fig. 77a. With the bridge set so that the only unbalance is the change of resistance in the active-strain gage, the potential difference between the output terminals becomes a measurement of strain. Since the gage is sensitive to temperature as well as strain, it will measure the combined effect. However, if a "dummy" gage, cemented to an unstressed piece of the same metal subjected to the same climatic conditions, is wired into the bridge leg adjacent to the one containing the "active" gage, the electric-resistance temperature effect is canceled out. Thus the active gage reports only that which is taking place in the stressed plate. The power supply can be either ac or dc.

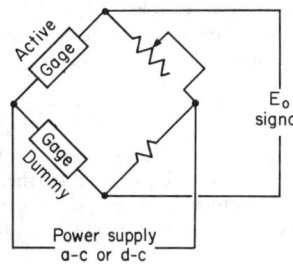

(a) Wheatstone bridge circuit

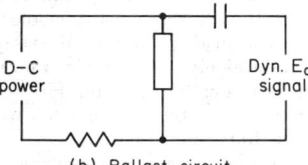

(b) Ballast circuit

**Fig. 77**

It is sometimes useful to make both gages active—e.g., mounted on opposite sides of a beam, with one gage subjected to tension and the other to compression. Temperature effects are still compensated, but the bridge output is doubled. In other instances, it may be desirable to make all four bridge arms active gages. The experimenter must determine the most practical arrangement for the problem at hand and must bear in mind that the bridge unbalances in proportion to the difference in the strains of gages located in adjacent legs and to the sum of strain in gages located in opposite legs.

Dynamic strains can be detected using circuits such as the **ballast** type shown in Fig. 77b. The capacitor coupling passes only rapidly varying or dynamic strains. The capacitor's infinite impedance to a steady voltage filters out any static effects or strains. The circuit is dc powered.

**Transverse Sensitivity** Grid-type gages possess some strain sensitivity in the direction perpendicular to the gage axis. In a uniaxial stress field, this transverse sensitivity is of no concern because the gage factor was obtained in such a field. However, in a biaxial stress field, neglect of transverse sensitivity will give slightly erroneous strains. When accounted for, the true strains in the axial direction of gage, $\epsilon_1$, and at right angles to it, $\epsilon_2$, are $\epsilon_1 = (1 - \mu k)(\epsilon_{a1} - k\epsilon_{a2})/(1 - k^2)$ and $\epsilon_2 = (1 - \mu k)(\epsilon_{a2} - k\epsilon_{a1})/(1 - k^2)$, where the apparent strains are $\epsilon_{a1} = \Delta R_1/RG_f$ and $\epsilon_{a2} = \Delta R_2/RG_f$, measured by cementing a gage in each direction 1 and 2. The factor $\mu$ is Poisson's ratio of the material to which gages are cemented, and $k$ (usually provided by the gage manufacturer) is the coefficient of transverse sensitivity of the gage. The gage is

cemented to the test piece, a uniaxial stress is applied in its axial direction, and the resistance change and strain are measured. The gage factor $G_1 = \Delta R_1/R\epsilon_1$ is computed. A uniaxial stress is next applied transversely to the gage. Again the resistance change and strain are measured and $G_2$ computed. Then $k = (G_2 + \mu G_1)/(G_1 + \mu G_2)$.

**Strain Rosettes** In a general biaxial stress field, the principal plane directions, as well as the stresses, are unknown. Thus, three gages mounted in three differing directions are needed if the three unknowns are to be determined. Three standard gage combinations, called strain rosettes, are commercially available and are best for the purpose. These are the *rectangular strain rosette* (Fig. 78a), which covers a minimum of area and is therefore best where the strain gradient is high; the *equiangular strain rosette* (Fig. 78b), where the gages do not overlap and which can be used where the strain gradient is low; the *T-delta strain rosette* (Fig. 78c), which occupies no more area than the equiangular rosette and which provides an extra check, or "insurance" gage. The wiring and instrumentation of gages in rosettes do not differ from those of individual gages.

The true strains along the gage-length directions are found according to the following equations, in which $R_n = \Delta R_n/RF_1(1 - k^2)$ and $b = 1/k$.

RECTANGULAR ROSETTE (SEE FIG. 78a.)

$$\epsilon_1 = R_1 - R_3/b$$
$$\epsilon_2 = R_2(1 + 1/b) - (1/b)(R_1 + R_3)$$
$$\epsilon_3 = R_3 - R_1/b$$

EQUIANGULAR ROSETTE (SEE FIG. 78b)

$$\epsilon_1 = R_1 - (1/b)(R_2 + R_3)$$
$$\epsilon_2 = R_2 - (1/b)(R_1 + R_3)$$
$$\epsilon_3 = R_3 - (1/b)(R_1 + R_2)$$

T-DELTA ROSETTE (SEE FIG. 78c)

$$\epsilon_1 = R_1(1 + 1/b) - (1/b)(R_3 + R_4)$$
$$\epsilon_2 = R_2(1 + 1/b) - (1/b)(R_3 + R_4)$$
$$\epsilon_3 = R_3 - (1/b)R_4$$
$$\epsilon_4 = R_4 - (1/b)R_3$$

**Foil Gages** Foil gages are produced from thin foil by photoetching techniques and are applied, instrumented, read, and evaluated just like the wire-grid type. Foil gages, being much thinner, may be applied easily to curved surfaces, have lower transverse sensitivity, exhibit negligible hysteresis under cycling loads, creep little under sustained loads, and can be stacked on top of each other.

### Brittle-Coating Analysis

Brittle coatings which adhere to the surface well can reveal the strain in the underlying material. Probably the first such coating used was mill scale, a thin iron oxide which forms on hot-rolled steel stock. Many coatings such as whitewash, portland cement, and shellac have been tried.

The most popular of presently available strain-indicating brittle coatings are the wood-rosin lacquers supplied by the Magnaflux Corporation under the trade name **Stresscoat.** Several Stresscoat compositions are available; the suitability of a particular lacquer depends upon the prevailing temperature and humidity. The lacquer is usually sprayed to a thickness of 0.004 to 0.008 in (0.01 to 0.02 cm) upon the surface, which must be clean and free of grease and loose particles. Calibration bars are sprayed at the same time. Both must be dried at an even temperature for up to 24 h. To facilitate observation of cracks, an undercoating of bright aluminum is often applied.

When the cured test piece is subjected to loads, the lacquer will first begin to crack at its threshold sensitivity in the area of the largest principal stress, with the parallel cracks perpendicular to the principal stress. This information is often sufficient, as it reveals the critical area and the direction of normal stress.

The threshold sensitivity of Stresscoat lacquers is 600 to 800 microinches per inch (1,500 to 2,000 microcentimeters per centimeter) in a uniaxial stress field. Exact control of lacquer selection, thickness, curing, and testing temperatures may reduce the threshold to 400 microinches per inch (1,000 microcentimeters per centimeter). If desired, the approximate strain (probably within 10 percent) may be established using the calibration strip sprayed with the test part. The strip is placed in a loading device and bent as a cantilever beam by means of a cam at the free end, causing the coating to crack on the tension surface. Crack spacing varies with the strain, being close at the fixed end and diminishing toward the free end down to threshold sensitivity values. The strip is placed in a holder containing strain graduations. A visual comparison of cracks on the test-part surface with those on the strip reveals the strain magnitude which caused the cracks.

### Birefringent Coatings

A birefringent coating is one which becomes double refractive when strained. The principle is quite old, but plastics, which adhere to all kinds of materials, which have stable optical-strain constants, and which are sufficiently sensitive to be practical, are of recent development. The trade name applied to this technique is **Photostress**. Photostress plastics can be obtained either as thin sheets (0.040, 0.080, and 0.120 in) or in liquid form. The sheet material can be bonded to a surface with a special adhesive. The liquid can be brushed or sprayed on, or the part can be dipped in the liquid. The layer should be at least 0.004 in (0.010 cm) thick. It is often necessary to apply several successive coatings, with heat curing of each layer in turn. Two sheet types and two liquids are available; these differ in stretching ability and in magnitude of the strain-optical constant. Each of the sheet materials is available metallized on one face, to reflect polarized light even when cemented to a dull surface.

The principles involved are the same as those for conventional photoelasticity. One frequent advantage is the fact that the plastic (sheet or liquid) can be applied directly to the part, which can then be subjected to actual operating loads. A special reflecting polariscope must be used. It contains only one polarizer and quarter-wave disk because the light passes back through the same pair after reflection by the stressed surface-plastic interface. The only limitation rests in the geometry of the structural component to be examined; not only must it be possible to apply the plastic to the surface, but the surface must be accessible to light.

The strain-optic law, since the light passes the plastic thickness twice, becomes

$$p - q = \frac{n}{2t} \frac{E}{K(1 + \mu)}$$

where $n$ is fringe order, $E$ is modulus, $\mu$ is Poisson's ratio of workpiece material, and $K$ (supplied by the manufacturer) is the strain-optic coefficient of the plastic. As in conventional photoelasticity, isoclinics are present as well.

### Holography

A more recently developed technique applicable to stress, or rather, strain analysis as well as to many other purposes is that of holography. It is made possible by the laser, an instrument which produces a highly concentrated, thin beam of light of single wavelength. The helium-neon (He-Ne) laser, emitting at the red end of the visible spectrum at a wavelength of 633 nanometers, has found much favor.

The laser beam is split into two components, one of which is directed upon the object (or specimen) and then onto the photographic plate. It is identified as the object beam. The other component, referred to as the reference beam, propagates directly to the plate. Interference between the beams resulting from retardations caused by displacements or strains forms fringes which in turn provide a measure of the disturbance. Spacing of such fringes depends upon Bragg's law:

$$d = \frac{\lambda}{2 \sin \theta/2}$$

where $d$ is the distance between fringes, $\lambda$ is the wavelength of the light source, and $\theta$ is the angle between the object and reference ray at the plate.

A simple holographic setup consists of the laser source,

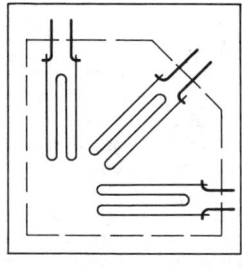
(a) Rectangular

(b) Equiangular

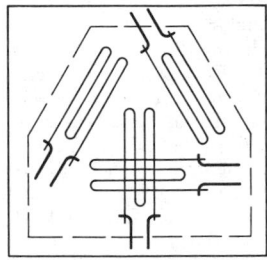
(c) Tee – delta

**Fig. 78**

beam splitter, reflecting surfaces, filters, and the recording plate. A possible arrangement is depicted in Fig. 79. Some arrangement for loading the specimen must also be provided. Additional auxiliary and refining hardware becomes necessary

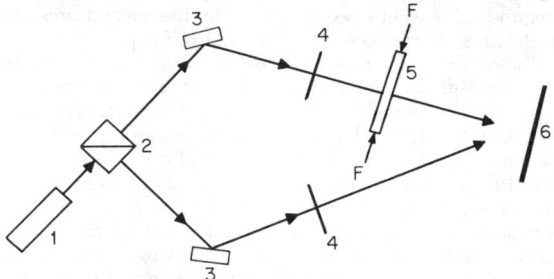

**Fig. 79** Simple holographic setup. (1) Laser source; (2) beam splitter; (3) reflecting surfaces; (4) circular polarizers; (5) loaded specimen (birefringent); (6) photographic plate.

as the analysis assumes greater complexity. Thus the system layout is limited only by test requirements and the experimenter's imagination. However, only a thorough understanding of the laws of optics and interferometry will make possible a reliable investigation and interpretation of results.

Stability of setup must be assured via a rigid optical bench and supporting brackets. Component instruments must be spaced upon the bench so that beam coherence is assured, required coherence depth satisfied, and the object/reference angle $\theta$ consistent with the fringe spacing desired. The film must also possess adequate sensitivity in the spectral range of the laser beam used.

Holography, using pulsed lasers, can be used to measure transient disturbances. Thus vibration studies are possible. Fatigue detection using holographic techniques has also been undertaken. Holography has been used in acoustical studies and in automatic gaging as well. It holds much promise as a versatile engineering tool. However, much additional experimentation, refinement, and simplification must be performed before its potential becomes a practical technique.

# PIPELINE FLEXURE STRESSES CAUSED BY EXPANSION OR MOVEMENT OF SUPPORTS

## by Harold V. Hawkins

References: Shipman, Design of Steam Piping to Care for Expansion, *Trans. ASME*, 1929. Wahl, Stresses and Reactions in Expansion Pipe Bends, *Trans. ASME*, 1927. Hovgaard, The Elastic Deformation of Pipe Bends, *Jour. Math. Phys.*, Nov. 1926, Oct. 1928, and Dec. 1929. M. W. Kellog Co., "The Design of Piping Systems," Wiley.
For details of Pipe and Pipe Fittings see Sec. 8.

**Nomenclature** (see Figs. 1 and 2)

$M_0$ = end moment at origin, in:lb (N:m)

$M$ = max moment, in:lb (N:m)

$F_x$ = end reaction at origin in $x$ direction, lb (N)

$F_y$ = end reaction at origin in $y$ direction, lb (N)

$S_l = (Mr/I)\alpha$ = max unit longitudinal flexure stress, lb/in² (N/m²)

$S_t = (Mr/I)\beta$ = max unit transverse flexure stress, lb/in² (N/m²)

$S_s = (Mr/I)\gamma$ = max unit shearing stress, lb/in² (N/m²)

$\Delta x$ = relative deflection of ends of pipe parallel to $x$ direction caused by either temperature change or support movement, or both, in (m)

$\Delta y$ = same as $\Delta x$ but parallel to $y$ direction, in. Note that $\Delta x$ and $\Delta y$ are positive if under the change in temperature the end opposite the origin tends to move in a positive $x$ or $y$ direction, respectively.

$t$ = wall thickness of pipe, in (m)

$r$ = mean radius of pipe cross section, in (m)

$\lambda$ = constant = $tR/r^2$

$I$ = moment of inertia of pipe cross section about pipe centerline, in⁴ (m⁴)

$E$ = modulus of elasticity of pipe at *actual working* temperature, lb/in² (N/m²)

$K$ = flexibility index of pipe. $K = 1$ for all straight pipe sections, $K = (10 + 12\lambda^2)/(1 + 12\lambda^2)$ for all curved pipe sections where $\lambda > 0.335$ (see Fig. 3)

$\alpha, \beta, \gamma$ = ratios of actual max longitudinal flexure, transverse flexure, and shearing stresses to $Mr/I$ for curved sections of pipe (see Fig. 3)

$A, B, C, F, G, H$ = constants given by Table 2

$\theta$ = angle of intersection between tangents to direction of pipe at reactions

$\Delta\theta$ = change in $\theta$ caused by movements of supports, or by temperature change, or both, radians

$ds$ = an infinitesimal element of length of pipe

$s$ = length of a particular curved section of pipe, in (m)

$R$ = radius of curvature of pipe centerline, in (m)

### General Discussion

Under the effect of changes in temperature of the pipeline, or of movement of support reactions (either translation or rotation), or both, the determination of stress distribution in a pipe becomes a statically indeterminate problem. In general the problem may be solved by a slight modification of the standard arch theory: $\Delta x = -K\int My\, ds/EI$, $\Delta y = K\int Mx\, ds/EI$, and $\Delta\theta = K\int M\, ds/EI$ where the constant $K$ is introduced to correct

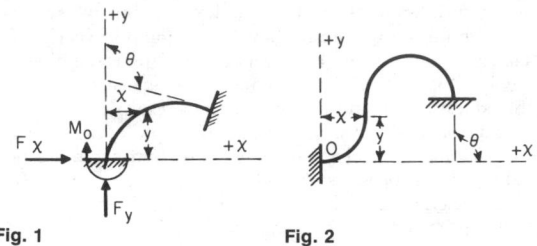

**Fig. 1**          **Fig. 2**

for the increased flexibility of a curved pipe, and where the integration is over the entire length of pipe between supports. In Table 1 are given equations derived by this method for

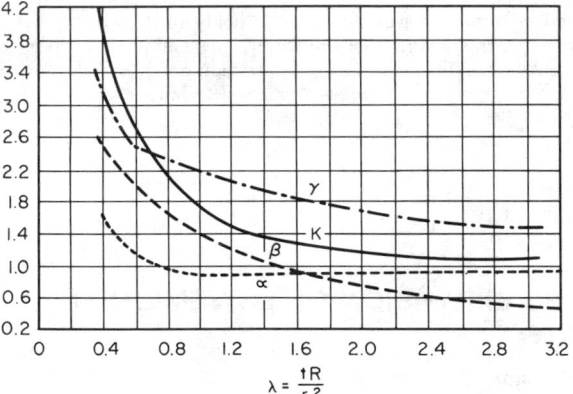

$$\lambda = \frac{tR}{r^2}$$

**Fig. 3**   Flexure constants of initially curved pipes.

moment and thrust at one reaction point for pipes in one plane that are fully fixed, hinged at both ends, hinged at one end and fixed at the other, or partly fixed. If the reactions at one end of the pipe are known, the moment distribution in the entire pipe then can be obtained by simple statics.

Since an initially curved pipe is more flexible than indicated by its moment of inertia, the constant $K$ is introduced. Its value may be taken from Fig. 3, or computed from the equation given below. $K = 1$ for all straight pipe sections, since they act according to the simple flexure theory.

In Fig. 3 are given the flexure constants $K$, $\alpha$, $\beta$, and $\gamma$ for initially curved pipes as functions of the quantity $\lambda = tR/r^2$. The flexure constants are derived from the equations

$$K = (10 + 12\lambda^2)/(1 + 12\lambda^2) \qquad \text{when } \lambda > 0.335$$

$$\alpha = \tfrac{2}{3}K\sqrt{(5 + 6\lambda^2)/18} \qquad \lambda \leq 1.472$$

$$\alpha = K(6\lambda^2 - 1)/(6\lambda^2 + 5) \qquad \lambda > 1.472$$

$$\beta = 18\lambda/(1 + 12\lambda^2)$$

$$\gamma = [8\lambda - 36\lambda^3 + (32\lambda^2 + 20/3)\sqrt{(4/3)\lambda^2 + 5/18}] \div$$
$$(1 + 12\lambda^2) \qquad \text{when } \lambda < 0.58$$
$$= (12\lambda^2 + 18\lambda - 2)/(1 + 12\lambda^2) \qquad \text{when } \lambda > 0.58$$

The increased flexibility of the curved pipe is brought about by the tendency of its cross section to flatten. This flattening causes a transverse flexure stress whose maximum is $S_t$. Because the maximum longitudinal and maximum transverse

stresses do not occur at the same point in the pipe's cross section, the resulting maximum shear is not one-half the difference of $S_l$ and $S_t$; it is $S_s$. In the straight sections of the pipe, $\alpha = 1$, the transverse stress disappears, and $\gamma = \frac{1}{2}$. This discussion of $S_s$ does not include the uniform transverse or longitudinal tension stresses induced by the internal pressure in the pipe; their effects should be added if appreciable.

Table 2 gives values of the constants $A$, $B$, $C$, $F$, $G$, and $H$ for use in equations listed in Table 1. The values may be used (1) for the solution of any pipeline, or (2) for the derivation of equations for standard shapes composed of straight sections and arcs of circles as in Fig. 5. Equations for shapes not given may be obtained by algebraic addition of those given. All measurements are from the left-hand end of the pipeline. Reactions and stresses are greatly influenced by end conditions. Formulas are given to cover the extreme conditions. The following suggestions and comments should be considered when laying out a pipeline:

**Avoid expansion bends,** and design the entire pipeline to take care of its own expansion.

The movement of the equipment to which the ends of the pipeline are attached must be included in the $\Delta x$ and $\Delta y$ of the equations.

Maximum flexibility is obtained by placing **supports** and **anchors** so that they will not interfere with the natural movement of the pipe.

That shape is most efficient in which the **maximum length of pipe** is **working** at the maximum safe stress.

Excessive **bending moment at joints** is more likely to cause trouble than excessive stresses in pipe walls. Hence, keep pipe joints away from points of high moment.

Reactions and stresses are greatly influenced by **flattening of the cross section** of the curved portions of the pipeline.

It is recommended that **cold springing** allowances be discounted in stress calculations.

**Application to Two- and Three-Plane Pipelines**   Pipelines in more than one plane may be solved by the successive application of the preceding data, dividing the pipeline into two or more one-plane lines.

EXAMPLE 1. The unsymmetrical pipeline of Fig. 4 has fully fixed ends. From Table 2 use $K = 1$ for all sections, since only straight segments are involved.

Upon introduction of $a = 120$ in (3.05 m), $b = 60$ in (1.52 m), and $c = 180$ in (4.57 m), into the preceding relations (Table 3) for $A$, $B$, $C$, $F$, $G$, $H$, the equations for the reactions at 0 from Table 1 become

$$M_0 = EI\Delta x(-7.1608 \times 10^{-5}) + EI\Delta y(-8.3681 \times 10^{-5})$$
$$F_x = EI\Delta x(+1.0993 \times 10^{-5}) + EI\Delta y(+3.1488 \times 10^{-6})$$
$$F_y = EI\Delta x(+3.1488 \times 10^{-6}) + EI\Delta y(+1.33717 \times 10^{-6})$$

Also it follows that

$$M_1 = M_0 + F_y a = EI\,\Delta x(+3.0625 \times 10^{-4}) + EI\,\Delta y(+7.6779 \times 10^{-5})$$
$$M_2 = M_1 - F_x b = EI\,\Delta x(+2.4029 \times 10^{-4}) + EI\,\Delta y(-1.1215 \times 10^{-4})$$
$$M_3 = M_2 + F_y c = EI\,\Delta x(+8.0707 \times 10^{-4}) + EI\,\Delta y(+1.2854 \times 10^{-4})$$

Thus the maximum moment $M$ occurs at 3.

The total maximum longitudinal fiber stress ($\alpha = 1$ for straight pipe)

$$= \frac{F_x}{2\pi r t} \pm \frac{M_3 r}{I}$$

There is no transverse flexure stress since all sections are straight. The maximum shearing stress is either (1) one-half of the maximum

**Table 1. General Equations for Pipelines in One Plane (see Figs. 1 and 2)**

| Type of supports | Unsymmetrical | Symmetrical about $y$-axis |
|---|---|---|
| Both ends fully fixed.............. | $M_0 = \dfrac{EI\Delta x(CF - AB) + EI\Delta y(BF - AG)}{2ABF + CGH - B^2H - A^2G - CF^2}$ <br> $F_x = \dfrac{EI\Delta x(CH - A^2) + EI\Delta y(BH - AF)}{2ABF + CGH - B^2H - A^2G - CF^2}$ <br> $F_y = \dfrac{EI\Delta x(BH - AF) + EI\Delta y(GH - F^2)}{2ABF + CGH - B^2H - A^2G - CF^2}$ <br> $\Delta\theta = 0$ | $M_0 = \dfrac{EI\Delta x F}{GH - F^2}$ <br> $F_x = \dfrac{EI\Delta x H}{GH - F^2}$ <br> $F_y = 0$ <br> $\Delta\theta = 0$ |
| Both ends hinged................ | $M_0 = 0$ <br> $F_x = \dfrac{EI\Delta x C + EI\Delta y B}{CG - B^2}$ <br> $F_y = \dfrac{EI\Delta x B + EI\Delta y G}{CG - B^2}$ <br> $\Delta\theta = \dfrac{\Delta x(AB - CF) + \Delta y(AG - BF)}{CG - B^2}$ | $M_0 = 0$ <br> $F_x = \dfrac{EI\Delta x}{G}$ <br> $F_y = 0$ <br> $\Delta\theta = \dfrac{-\Delta x F}{G}$ |
| Origin end only hinged, other end fully fixed..................... | $M_0 = 0$ <br> $F_x = \dfrac{EI\Delta x C + EI\Delta y B}{CG - B^2}$ <br> $F_y = \dfrac{EI\Delta x B + EI\Delta y G}{CG + B^2}$ <br> $\Delta\theta = \dfrac{\Delta x(AB - CF) + \Delta y(AG - BF)}{CG - B^2}$ | |
| In general for any specific rotation $\Delta\theta$ and movement $\Delta x$ and $\Delta y$..... | $M_0 = \dfrac{EI\Delta x(CF - AB) + EI\Delta y(BF - AG) + EI\Delta\theta(CG - B^2)}{2ABF + CGH - A^2G - CF^2 - B^2H}$ <br> $F_x = \dfrac{EI\Delta x(CH - A^2) + EI\Delta y(BH - AF) + EI\Delta\theta(CF - AB)}{2ABF + CGH - A^2G - CF^2 - B^2H}$ <br> $F_y = \dfrac{EI\Delta x(BH - AF) + EI\Delta y(GH - F^2) + EI\Delta\theta(BF - AG)}{2ABF + CGH - A^2G - CF^2 - B^2H}$ | |

longitudinal fiber stress as given above, (2) one-half of the hoop-tension stress caused by any internal radial pressure that might exist in the pipe, or (3) one-half the difference of the maximum longitudinal fiber stress and hoop-tension stress, whichever of these three possibilities is numerically greatest.

EXAMPLE 2. The equations of Table 1 may be employed to develop the solution of generalized types of pipe configurations for which Fig. 5

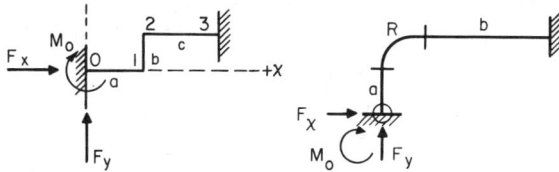

**Fig. 4**

**Fig. 5**  Right-angle pipeline.

is a typical example. If only temperature changes are considered, the reactions for the right-angle pipeline (Fig. 5) may be determined from the following equations:

$$M_0 = C_1 EI\,\Delta x/R^2$$
$$F_x = C_2 EI\,\Delta x/R^3$$
$$F_y = C_3 EI\,\Delta x/R^3$$

In these equations, $\Delta x$ is the $x$ component of the deflection between reaction points caused by temperature change only. The values of $C_1$,

$C_2$, and $C_3$ are given in Fig. 6 for $K = 1$ and $K = 2$. For other values of $K$, interpolation may be employed.

EXAMPLE 3. With $a/R = 20$ and $b/R = 3$, the value of $C_1$ is 0.185 for $K = 1$ and 0.165 for $K = 2$. If $K = 1.75$, the interpolated value of $C_1$ is 0.175.

**Elimination of Flexure Stresses**  Pipeline flexure stresses that normally would result from movement of supports or from the tendency of the pipes to expand under temperature change often may be avoided entirely through the use of expansion joints (Sec. 8). Their use may simplify both the design of the pipeline and the support structure. When using expansion joints, the following suggestions should be considered: (1) select expansion joint carefully for maximum temperature range (and deflection) expected so as to prevent damage to expansion fitting; (2) provide guides to limit movement at expansion joint to direction permitted by joint; (3) provide adequate anchors at one end of each straight section or along their midlength, forcing movement to occur at expansion joint yet providing adequate support for pipeline; (4) mount expansion joints adjacent to an anchor point to prevent sagging of the pipeline under its own weight and do not depend upon the expansion joint for stiffness—it is intended to be flexible; (5) give consideration to effects of corrosion, since corrugated character of expansion joints makes cleaning difficult.

Table 2.  Values of $A$, $B$, $C$, $F$, $G$, and $H$ for Various Piping Elements

| Element | $A = K\int x\,ds$ | $B = K\int xy\,ds$ | $C = K\int x^2\,ds$ | $F = K\int y\,ds$ | $G = K\int y^2\,ds$ | $H = K\int ds$ |
|---|---|---|---|---|---|---|
| $s$ line $(x_1,y)$—$(x_2,y_2)$ | $\dfrac{s}{2}(x_1+x_2)$ | $Ay$ | $s\left(\dfrac{s^2}{3}+x_1x_2\right)$ | $sy$ | $Fy$ | $s$ |
| vertical $(x,y_2)$—$(x,y_1)$ | $sx$ | $\dfrac{A}{2}(y_1+y_2)$ | $Ax$ | $\dfrac{s}{2}(y_1+y_2)$ | $s\left(\dfrac{s^2}{3}+y_1y_2\right)$ | $s$ |
| diagonal $(x_2,y_2)$—$(x_1,y_1)$ | $\dfrac{s}{2}(x_1+x_2)$ | $\dfrac{A}{3}(y_1+y_2)$ $+\dfrac{s}{6}(x_1y_1+x_2y_2)$ | $\dfrac{s}{3}(x_1^2+x_1x_2+x_2^2)$ | $\dfrac{s}{2}(y_1+y_2)$ | $\dfrac{s}{3}(y_1^2+y_1y_2+y_2^2)$ | $s$ |
| semicircle $(x,y)$ | $\pi KRx$ | $A\left(y+\dfrac{2R}{\pi}\right)$ | $A\left(x+\dfrac{R^2}{2x}\right)$ | $(\pi y+2R)KR$ | $Fy+\left(2y+\dfrac{\pi}{2}\,R\right)KR^2$ | $\pi KR$ |
| semicircle $(x,y)$ | | $A\left(y-\dfrac{2R}{\pi}\right)$ | | $(\pi y-2R)KR$ | $Fy-\left(2y-\dfrac{\pi}{2}\,R\right)KR^2$ | |
| quarter circle $(x,y)$ | $\left(\dfrac{\pi x}{2}-R\right)KR$ | $Ay+\left(x-\dfrac{R}{2}\right)KR^2$ | $Ax+\left(\dfrac{\pi R}{4}-x\right)KR^2$ | $\left(\dfrac{\pi y}{2}+R\right)KR$ | $Fy+\left(\dfrac{\pi R}{4}+y\right)KR^2$ | $\dfrac{\pi KR}{2}$ |

| Shape | | | | | | Top |
|---|---|---|---|---|---|---|
| quarter circle, $(x,y)$ | $\left(\dfrac{\pi x}{2} + R\right)KR$ | $Ay + \left(x + \dfrac{R}{2}\right)KR^2$ | $\left(\dfrac{\pi y}{2} + R\right)KR$ | $Ax + \left(\dfrac{\pi R}{4} + x\right)KR^2$ | $Fy + \left(\dfrac{\pi R}{4} + y\right)KR^2$ | $\dfrac{\pi KR}{2}$ |
| quarter circle, $(x,y)$ | $\left(\dfrac{\pi x}{2} - R\right)KR$ | $Ay - \left(x + \dfrac{R}{2}\right)KR^2$ | $\left(\dfrac{\pi y}{2} - R\right)KR$ | $Ax + \left(\dfrac{\pi R}{4} - x\right)KR^2$ | $Fy + \left(\dfrac{\pi R}{4} - y\right)KR^2$ | |
| quarter circle, $(x,y)$ | $\left(\dfrac{\pi x}{2} - R\right)KR$ | $Ay - \left(x - \dfrac{R}{2}\right)KR^2$ | $\left(\dfrac{\pi y}{2} - R\right)KR$ | $Ax + \left(\dfrac{\pi R}{4} - x\right)KR^2$ | $Fy + \left(\dfrac{\pi R}{4} - y\right)KR^2$ | |
| triangle $45°$, $(x,y)$ | $\left(\dfrac{\pi x}{4} - \dfrac{R}{\sqrt{2}}\right)KR$ | $Ay + \left[\left(1 - \dfrac{\sqrt{2}}{2}\right)x - \dfrac{R}{4}\right]KR^2$ | $\left[\dfrac{\pi y}{4} + \left(1 - \dfrac{\sqrt{2}}{2}\right)R\right]KR$ | $Ax + \left[\left(\dfrac{\pi}{8} + \dfrac{1}{4}\right)R - \dfrac{x\sqrt{2}}{2}\right]KR^2$ | $Fy + \left[\left(1 - \dfrac{\sqrt{2}}{2}\right)y + \left(\dfrac{\pi}{8} - \dfrac{1}{4}\right)R\right]KR^2$ | $\dfrac{\pi KR}{4}$ |
| triangle $45°$, $(x,y)$ | $\left(\dfrac{\pi x}{4} - \dfrac{R}{\sqrt{2}}\right)KR$ | $Ay - \left[\left(1 - \dfrac{\sqrt{2}}{2}\right)x - \dfrac{R}{4}\right]KR^2$ | $\left[\dfrac{\pi y}{4} - \left(1 - \dfrac{\sqrt{2}}{2}\right)R\right]KR$ | $Ax + \left[\left(\dfrac{\pi}{8} + \dfrac{1}{4}\right)R - \dfrac{x\sqrt{2}}{2}\right]KR^2$ | $Fy - \left[\left(1 - \dfrac{\sqrt{2}}{2}\right)y - \left(\dfrac{\pi}{8} - \dfrac{1}{4}\right)R\right]KR^2$ | |
| triangle $45°$, $(x,y)$ | $\left(\dfrac{\pi x}{4} + \dfrac{R}{\sqrt{2}}\right)KR$ | $Ay + \left[\left(1 - \dfrac{\sqrt{2}}{2}\right)x + \dfrac{R}{4}\right]KR^2$ | $\left[\dfrac{\pi y}{4} + \left(1 - \dfrac{\sqrt{2}}{2}\right)R\right]KR$ | $Ax + \left[\left(\dfrac{\pi}{8} + \dfrac{1}{4}\right)R + \dfrac{x\sqrt{2}}{2}\right]KR^2$ | $Fy + \left[\left(1 - \dfrac{\sqrt{2}}{2}\right)y + \left(\dfrac{\pi}{8} - \dfrac{1}{4}\right)R\right]KR^2$ | |
| triangle $45°$, $(x,y)$ | $\left(\dfrac{\pi x}{4} + \dfrac{R}{\sqrt{2}}\right)KR$ | $Ay - \left[\left(1 - \dfrac{\sqrt{2}}{2}\right)x + \dfrac{R}{4}\right]KR^2$ | $\left[\dfrac{\pi y}{4} - \left(1 - \dfrac{\sqrt{2}}{2}\right)R\right]KR$ | $Ax + \left[\left(\dfrac{\pi}{8} + \dfrac{1}{4}\right)R + \dfrac{x\sqrt{2}}{2}\right]KR^2$ | $Fy - \left[\left(1 - \dfrac{\sqrt{2}}{2}\right)y - \left(\dfrac{\pi}{8} - \dfrac{1}{4}\right)R\right]KR^2$ | |
| sector, $\theta_1$, $\theta_2$, $(x,y)$ | $[x(\theta_2 - \theta_1) - R(\sin\theta_2 - \sin\theta_1)]KR$ | $Ay - \left[\begin{array}{l}x(\cos\theta_2 - \cos\theta_1) + \\ \dfrac{R}{2}(\sin^2\theta_2 - \sin^2\theta_1)\end{array}\right]KR^2$ | $\left[\begin{array}{l}y(\theta_2 - \theta_1) - \\ R(\cos\theta_2 - \cos\theta_1)\end{array}\right]KR$ | $Ax - \left[\begin{array}{l}x(\sin\theta_2 - \sin\theta_1) - \\ \dfrac{R}{4}(\sin 2\theta_2 - \sin 2\theta_1) - \\ \dfrac{R}{2}(\theta_2 - \theta_1)\end{array}\right]KR^2$ | $Fy - \left[\begin{array}{l}y(\cos\theta_2 - \cos\theta_1) + \\ \dfrac{R}{4}(\sin 2\theta_2 - \sin 2\theta_1) - \\ \dfrac{R}{2}(\theta_2 - \theta_1)\end{array}\right]KR^2$ | $(\theta_2 - \theta_1)KR$ |

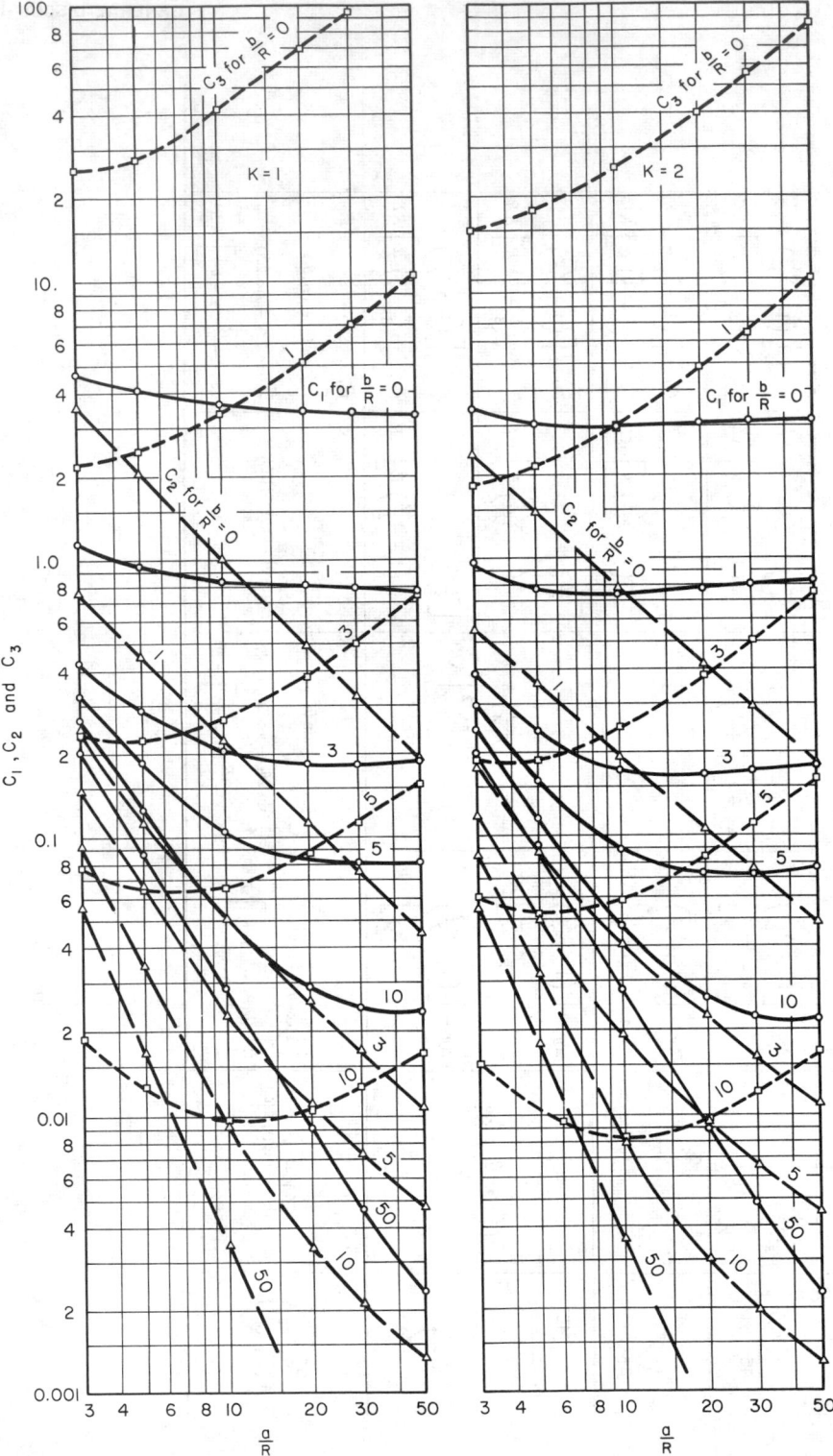

**Fig. 6** Reactions for right-angle pipelines.

**Table 3. Example 1 Showing Determination of Integrals**

| Part of pipe | Values of integrals | | | | | |
|---|---|---|---|---|---|---|
| | $A$ | $B$ | $C$ | $F$ | $G$ | $H$ |
| 0–1 | $\dfrac{a^2}{2}$ | $0$ | $\dfrac{a^3}{3}$ | $0$ | $0$ | $a$ |
| 1–2 | $ab$ | $\dfrac{ab^2}{2}$ | $a^2b$ | $\dfrac{b^2}{2}$ | $\dfrac{b^3}{3}$ | $b$ |
| 2–3 | $\dfrac{c}{2}(2a+c)$ | $\dfrac{bc}{2}(2a+c)$ | $\dfrac{c^3}{3}+ac(a+c)$ | $bc$ | $b^2c$ | $c$ |
| Total 0–3 | $\dfrac{a^2}{2}+ab$ $+\dfrac{c}{2}(2a+c)$ | $\dfrac{ab^2}{2}$ $+\dfrac{bc}{2}(2a+c)$ | $\dfrac{a^3}{3}+a^2b+\dfrac{c^3}{3}$ $+ac(a+c)$ | $\dfrac{b^2}{2}+bc$ | $\dfrac{b^3}{3}+b^2c$ | $a+b+c$ |

# VIBRATION

## by William T. Thomson

REFERENCES: Den Hartog, "Mechanical Vibration," 4th ed., McGraw-Hill, 1956. Timoshenko, "Vibration Problems in Engineering," 3d ed., Van Nostrand, 1954. Thomson, "Theory of Vibration," Prentice-Hall, 1973. Harris and Crede, "Shock and Vibration Handbook," McGraw-Hill, 1961.

### FREE, DAMPED, AND FORCED VIBRATIONS

The simplest vibrating system is one of a single degree of freedom, the motion of which can be described by a single coordinate $x$, as shown in Fig. 1. The mass attached to the spring has freedom to vibrate in the vertical direction, and the mass of the spring will be considered to be small in comparison to it.

**Free Vibrations** When the system previously described is put into motion by an initial displacement or velocity, the oscillations will gradually diminish in amplitude, because of resisting forces which are always present. These damping forces may arise from various sources, such as air or fluid resistance, internal friction of the material of the vibrating body, or friction between sliding surfaces. When a body is vibrating in air or in a liquid and velocities are small, the resisting force is very nearly proportional to the velocity and equal to $c\dot{x}$. The differential equation of motion then becomes

$$\frac{W}{g}\ddot{x} + c\dot{x} + kx = 0 \qquad (1)$$

where $\dot{x}$ and $\ddot{x}$ are the first and second derivatives of $x$ with respect to time.

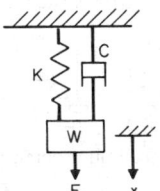

**Fig. 1**

The general solution of this equation in terms of the initial displacement $x_0$ and initial velocity $\dot{x}_0$ is

$$x = e^{-\zeta\omega_n t}\left[ x_0 \cos \omega_d t + \left(\frac{\dot{x}_0 + \zeta\omega_n x_0}{\omega_d}\right)\sin\omega_d t\right] \qquad (2)$$

where the symbols used are related to the original quantities as follows:

$\omega_n = \sqrt{\dfrac{kg}{W}} = \dfrac{2\pi}{\tau} = $ undamped natural frequency, rad/s

$\tau = $ period of free vibration with no damping, $\zeta = 0$

$\omega_d = \omega_n\sqrt{1 - \zeta^2} = \dfrac{2\pi}{\tau_d}$

$\tau_d = $ period of free vibration with damping $\zeta$

$\zeta = c/c_c = $ damping factor

$c_c = 2\sqrt{\dfrac{kW}{g}} = $ critical damping

By substituting $x_0 = A\sin\alpha$ and $\dfrac{\dot{x}_0 + \zeta\omega_n x_0}{\omega_d} = A\cos\alpha$, Eq. (2) can also be written as

$$x = A e^{-\zeta\omega_n t}\sin(\omega_d t + \alpha) \qquad (3)$$

In general, damping, $\zeta$, is small, and the difference between $\omega_d$ and $\omega_n$ is a small quantity of the second order. It can be assumed, therefore, that small damping forces do not affect the period of free vibration, which is

$$\tau = 2\pi\sqrt{\frac{W}{kg}} = 2\pi\sqrt{\frac{\delta_{st}}{g}} \qquad (4)$$

where $\delta_{st}$ is the statical deflection of the spring under the action of the weight $W$.

**Logarithmic Decrement** A convenient way to determine the amount of damping present in an oscillatory system is to measure the rate of decay of free oscillations shown in Fig. 2. This is best expressed by the logarithmic decrement, which is defined as the natural logarithm of the ratio of any two successive amplitudes, which results in the equation

$$\delta = \ln \frac{x_i}{x_{i+1}} = \frac{2\pi\zeta}{\sqrt{1 - \zeta^2}} \cong 2\pi\zeta \qquad (5)$$

The logarithmic decrement can also be obtained from the resonance curve of forced vibration as

$$\delta = \frac{f_2 - f_1}{f_n} \pi \qquad (6)$$

where $f_n$ is the resonant frequency and $f_1$ and $f_2$ are frequencies on each side of resonance, for which the amplitude is 0.707 times the amplitude at resonance. Since $f_n/(f_2 - f_1)$ is a measure of the sharpness of the resonance curve, it is often referred to as the $Q$ of the system.

$$Q = \frac{f_n}{f_2 - f_1} = \frac{1}{2\zeta} \qquad (7)$$

**Forced Vibration**   When a harmonic disturbing force $P \sin \omega t$ acts on the spring-mass system, the differential equation becomes

$$\frac{W}{g}\ddot{x} + c\dot{x} + kx = P \sin \omega t \qquad (8)$$

The resulting vibration will consist of two parts, free damped

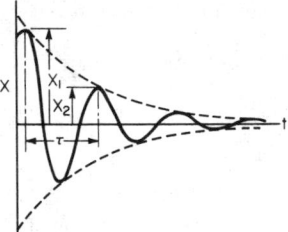

**Fig. 2**

vibration such as that represented by Eq. (2) and forced vibration. The free vibration will be damped out in a short time, after which only the forced vibration will exist.

The expression for the forced vibration has the form

$$x = X \sin (\omega t - \phi) \qquad (9)$$

where the amplitude and phase are given by the following equations:

$$X = \frac{P}{k} \frac{1}{\sqrt{[1 - (\omega/\omega_n)^2]^2 + 4\zeta^2(\omega/\omega_n)^2}} \qquad (10)$$

$$\phi = \tan^{-1} \frac{2\zeta(\omega/\omega_n)}{1 - (\omega/\omega_n)^2} \qquad (11)$$

The first factor on the right side of Eq. (10) is the **zero-frequency deflection,** i.e., the statical deflection which would result if a steady force of magnitude $P$ was applied to the spring of stiffness $k$. The second factor is due to dynamical conditions and is called the magnification factor. The phase angle given by Eq. (11) determines the amount the displacement lags behind the disturbing force, in radians. In Fig. 3, the magnification factor and phase angle are plotted against $\omega/\omega_n$ for various values of the damping factor $\zeta$.

When the disturbing force has a very low frequency as compared with the natural frequency of the system, the ratio $\omega/\omega_n$ is small and the magnification factor approaches unity with phase approaching zero. When the disturbing force has a very high frequency, $\omega/\omega_n$ is large, and the magnification factor approaches zero with phase angle tending to 180°. In either of these extreme cases the effect of damping on the magnification factor is small.

When $\omega$ approaches $\omega_n$, a condition of resonance is encountered, and the amplitude of vibration is limited only by the amount of damping present. The phase angle in this neighborhood also undergoes a large variation and has the value of 90° at $\omega/\omega_n = 1.0$.

**Structural Damping**   In structural analysis, it is found that damping is encountered proportional to displacement but out of phase with the velocity of harmonic oscillation. For such cases the following differential equation applies:

$$\frac{W}{g}\ddot{x} + (1 + i\gamma)kx = P \sin \omega t \qquad (12)$$

where $\gamma$ is the structural damping coefficient (usually between 0 and 0.05). The magnification factor then becomes

$$\frac{1}{\sqrt{[1 - (\omega/\omega_n)^2]^2 + \gamma^2}} \qquad (13)$$

and its resonant value becomes $1/\gamma$, in contrast to $1/2\zeta$ of the viscous damping. Thus, in the neighborhood of resonance, structural damping behaves like viscous damping of value $2\zeta$.

**Energy Method**   Examination of the frequency equation $\omega_d = \omega_n\sqrt{1 - \zeta^2}$ indicates that damping can usually be neglected when determining the resonant frequencies of the system. For an undamped system, the total energy is a constant, and there is a continual interchange between the potential and kinetic energy. Thus the maximum kinetic energy of the system as it moves through the equilibrium position is equal to the maximum potential energy at the extreme displacement, where the velocity is zero.

The energy principle can be used for the determination of the fundamental frequency of multidegree and continuous systems as well as that of single-degree systems. For such systems, it is necessary to assume some form of deflection distribution $y(x)$, from which the kinetic and potential energies to be equated are calculated. This procedure, known as Rayleigh's method, gives surprisingly accurate values of the fundamental frequency.

As an example, the effect of the mass of the spring of the system of Fig. 1 (which was previously neglected) on the natural frequency can be determined by Rayleigh's method as follows: Let $w$ be the total weight of the spring, which will be considered to be small in comparison to the end weight $W$. Assuming the displacement amplitude of any point a distance $y$ from the fixed end to be $x_0(y/l)$, where $x_0$ is the amplitude of $W$, the kinetic energy of the system becomes

$$T_{max} = \frac{1}{2}\frac{W}{g}\dot{x}_0^2 + \frac{1}{2}\frac{w}{g}\int_0^l \left(\frac{\dot{x}_0 y}{l}\right)^2 dy$$
$$= \frac{1}{2g}\left(W + \frac{w}{3}\right)\omega^2 x_0^2 \qquad (14)$$

Equating this to the maximum potential energy

$$U_{max} = 1/2 kx_0^2 \qquad (15)$$

the natural frequency of the system including the mass of the spring becomes

$$\omega_n = \sqrt{\frac{kg}{W + \frac{1}{3}w}} \qquad (16)$$

Equation (16) indicates that one-third of the weight of the spring should be added to the end weight for a more accurate estimate of the natural frequency of the system of Fig. 1.

As a second example, the Rayleigh procedure is applied to a single-span beam of uniform mass and stiffness. In this case, the end fixity $F$ is defined as the moment developed at the ends, divided by the end moment for a perfectly rigid support. For instance, $F = 0$ for pinned supports, while $F = 1$ for rigidly clamped supports. Using statical deflection for any value of $F$, the stiffness and mass of an equivalent spring-mass system can be determined, from which the fundamental frequency can be expressed in the form

$$f_1 = \frac{\alpha}{2\pi} \sqrt{\frac{EIg}{Wl^3}} \qquad (17)$$

where $E$ = modulus of elasticity, lb/in² = $6.899 \times 10^3$ N/m²
$I$ = moment of inertia of cross-sectional area about neutral axis, in⁴ = $41.6 \times 10^{-8}$ m⁴
$W$ = weight of beam, lb = 4.45 N
$l$ = length of beam, in = $2.54 \times 10^{-2}$ m
$g$ = 386 in/s² = 9.806 m/s²
$\alpha$ = nondimensional coefficient depending on $F$

The coefficient $\alpha$ and the equivalent stiffness and mass are plotted in Fig. 4 as functions of $F$.

**Natural Frequencies of Simple Systems**  (Table 1)  It is often possible to reduce a mechanical system to one of the simple systems shown in this section. The equations for the natural frequencies are given in $\omega_n$ rad/s = $2\pi f_n$. Table 2 for the stiffness of elements can be used for the general equation $\omega_n = \sqrt{k/m}$.

**Vibration Isolation**  Vibrations originating from machines or other sources are in general transmitted to the neighboring structure, to the detriment of the environment. To reduce the transmitted vibrations, isolators in the form of springs, rubber mounts, or cork padding are frequently used. As an example, automobile engines are supported on rubber mounts to reduce the transmission of engine vibrations to the remainder of the car.

Of interest here is the ratio of the force transmitted to the disturbing force, this ratio being designated as the **transmissibility.** Assuming that the isolator can be represented by a spring and a dashpot, as shown in Fig. 5, the force transmitted is the vector sum of the spring and dashpot force, which is

$$F_{TR} = \sqrt{(kx)^2 + (c\omega x)^2} = kx\sqrt{1 + 2\zeta(\omega/\omega_n)^2} \qquad (18)$$

The amplitude is equal to that of forced vibration given by Eq. (10), and hence the transmissibility becomes

$$\frac{F_{TR}}{F_0} = \frac{\sqrt{1 + 4\zeta^2(\omega/\omega_n)^2}}{\sqrt{[1 - (\omega/\omega_n)^2]^2 + 4\zeta^2(\omega/\omega_n)^2}} \qquad (19)$$

which is plotted in Fig. 5. The above ratio is less than 1 if $\omega/\omega_n$ is greater than $\sqrt{2}$ and decreases with increasing values of $\omega/\omega_n$. Thus for an isolator to perform its function the natural frequency $\omega_n$ of the supported structure must be small in comparison to the frequency $\omega$ of the disturbing force.

The actual design of an isolator frequently offers difficulties when very low natural frequencies are required. Since $\omega_n \cong \sqrt{g/\delta_{st}}$, the statical deflection $\delta_{st}$ necessary for small $\omega_n$ is often beyond the practical range.

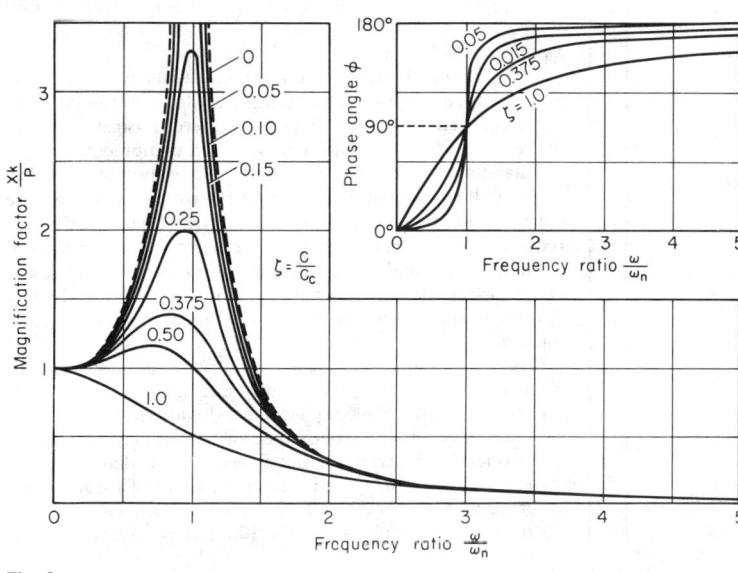

**Fig. 3**

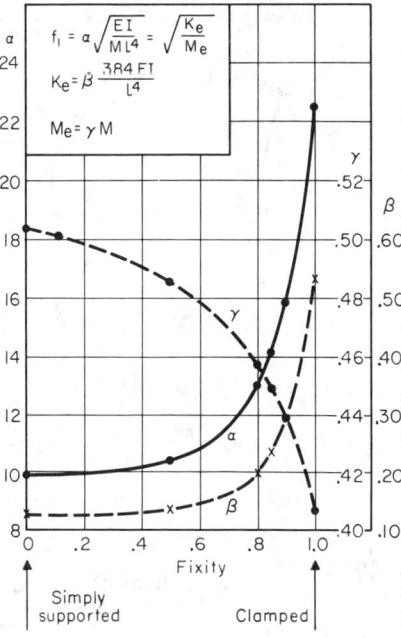

**Fig. 4**

**Table 1. Natural Frequencies**

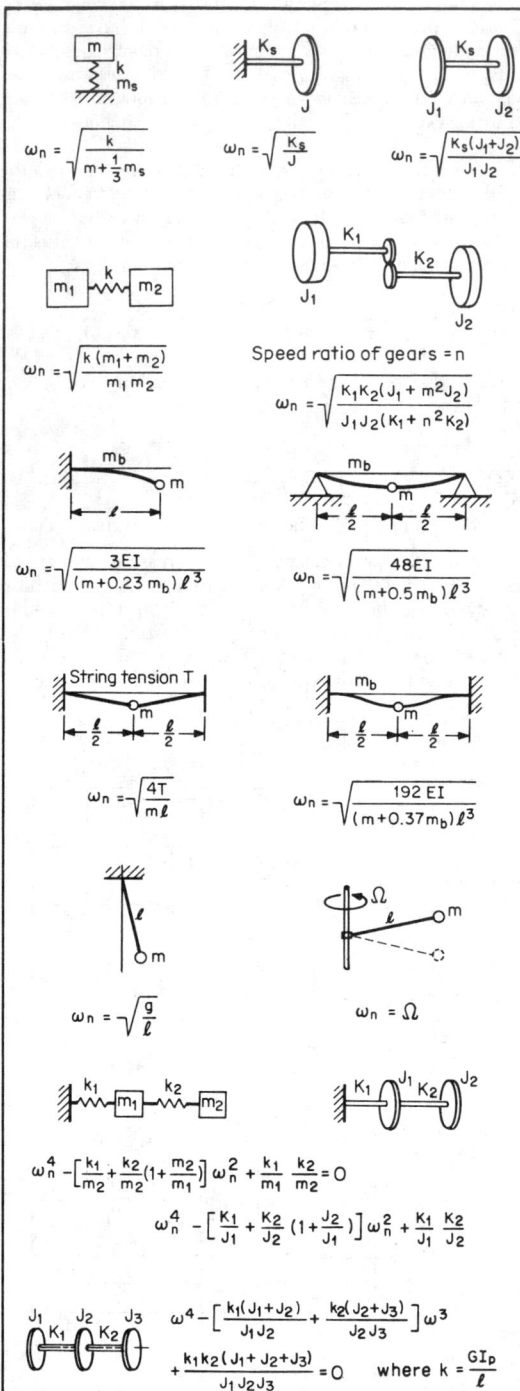

**Table 2. Stiffness of Elements**

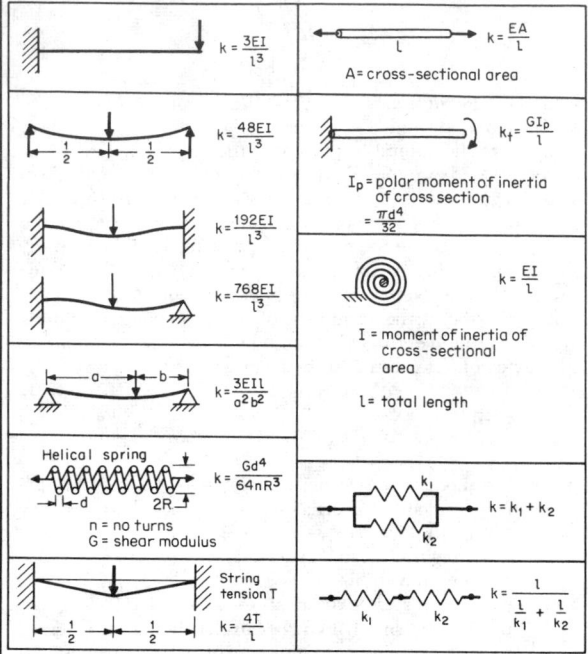

The relationship between the disturbing frequency $f$ in cycles per minute, the statical deflection $\delta$ in inches, and the percent reduction in vibration, defined as $R = (1 - \text{transmissibility})$, is

$$f = 188 \sqrt{\frac{1}{\delta}\left(\frac{2-R}{1-R}\right)} \qquad (20)$$

a plot of which is shown as Fig. 6.

In Fig. 7 are shown two applications of isolators. In (*a*) the isolation is accomplished by springs, while in (*b*) the bearings are supported in rubber rings. Commercial isolators of all shapes and sizes are available as standard equipment.

**Balancing of Rotating Machines**   An important requirement of all rotating machines is that the rotation axis coincide with one of the principal axes of inertia of the body. This requirement is difficult to satisfy exactly in the process of manufacturing, and hence balancing becomes necessary, especially for high-speed machines. This is evident from the fact that the magnitude of any unbalance is equal to the centrifugal force $m\omega^2 r$.

The condition of unbalance of a rotating body may be classified as static or dynamic unbalance. In the case of static unbalance, the unbalance appears in a single axial plane and on the same side of the axis of rotation, as shown in Fig. 8*a*. Consequently this type of unbalance can be detected by a static test, where the rotor is placed on a pair of parallel rails. In practice, however, the effect of the unbalance is magnified by rotation. The unbalance of a thin disk is essentially static unbalance.

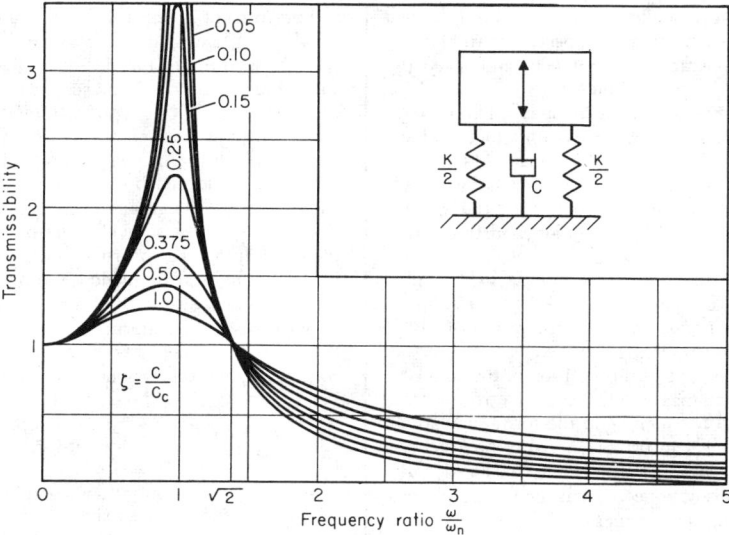

**Fig. 5**

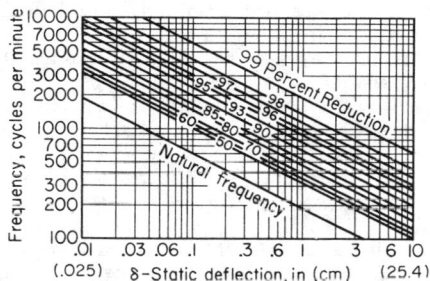

**Fig. 6**

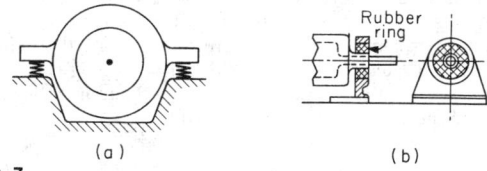

**Fig. 7**

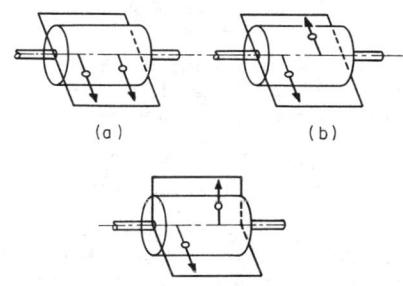

**Fig. 8**

In the case of dynamic unbalance, the unbalance can be in a single axial plane and on opposite sides of the rotation axis, as in Fig. 8b, or in two different axial planes, as shown in Fig. 8c. In (b) it is possible for the center of gravity to lie on the rotation axis, and hence the rotor may be in static balance. However, under rotation, the two unbalance forces form a couple, which has a tendency to rock the axis of rotation. The unbalance indicated in (c) is the most general case, where the rotor is in static and dynamic unbalance.

In all cases, a complete balance can be obtained by adding or removing correcting weights in two arbitrary and separated transverse planes. In general the end planes of the rotor are convenient correction planes; e.g., for the rotor of Fig. 9, with unbalance $w_1$ and $w_2$, the correction weights at the ends are $w_1(a/l)$ and $w_1[(l - a)/l]$ in the axial plane of $w_1$, and $w_2$ may

be resolved similarly. Combining the necessary corrections for each unbalance, a single weight at each end placed at a proper radial distance will completely balance the rotor.

The determination of the magnitude and angular position of the unbalance is the task of the balancing machine. All balancing machines are provided with elastically supported bearings in which the rotor may spin (Fig. 10). Because of unbalance, the bearings will oscillate laterally, and the amplitude and phase of the rotor are indicated with respect to an arbitrary rotor position by electrical pickups and strobo-flash light.

In the case in which the rotor is very long and flexible, the position of the unbalance will depend on the elastic configura-

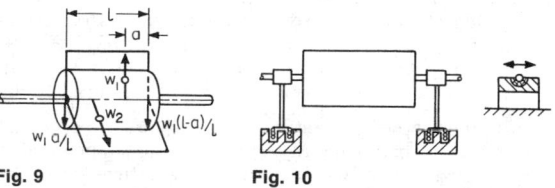

**Fig. 9**          **Fig. 10**

tion of the rotor, which is dependent on the speed of rotation, its temperature, etc. In such cases, it is necessary to balance the rotor in its normal operating environment and speed by means of a portable balancing instrument.

**Inertial Unbalance of Reciprocating Engines** The moving parts of a reciprocating engine produce dynamic forces which may result in undesirable vibrations. Rotating parts such as the crankshaft can be balanced; however, translating parts, such as the piston, and parts, like the connecting rod, with a more complex motion of combined rotation and translation cannot be so easily balanced.

In the calculation for the unbalanced forces in a single-cylinder engine, the moving parts are divided into a recipro-cating weight and a rotating weight. This is accomplished by assigning portions of the connecting-rod weight to the piston end and the crank end. In general this division of the connect-ing rod into two lumped masses will lead to errors in the moment of inertia and consequently in the torque equation; however, the force analysis can be considered to be exact.

Assuming the rotating weight to be counterbalanced, only the reciprocating weight is of concern, and the force equation for the single-cylinder engine becomes

$$\begin{aligned} F &= m_{rec}\, r\omega^2 \left( \cos \omega t + \frac{r}{l}\cos 2\omega t \right) \\ &= \quad X_1 \quad + \quad X_2 \end{aligned} \qquad (21)$$

The component $X_1$, which alternates once per revolution, is referred to as the primary force, while the component $X_2$, which alternates twice per revolution, is called the secondary force.

In addition to the inertia force, there will be an unbalanced torque about the crankshaft axis due to the reciprocating weight. This torque, however, is in general considered together with the torque due to the power stroke, and the torsional oscillations resulting from these excitations are mini-mized by means of the pendulum-type absorber or the tor-sional damper, consisting of a floating disk within a closed hub filled with viscous fluid.

The analysis of the single-cylinder engine can be extended to the multicylinder in-line and V-block engines by superposi-tion. For the in-line engine or one block of the V engine, the force equation (**21**) can be added for each crank position $\varphi_n$ so that the equation becomes

$$F = m_{rec}r\omega^2 \sum_n \left[ \cos(\omega t + \varphi_n) + \frac{r}{l}\cos 2(\omega t + \varphi_n) \right] \qquad (22)$$

It is thus possible by proper angular spacing $\varphi_n$ of the $n$ cylinders to eliminate the vibration force $F$.

It is evident that even if $F = 0$, there may be pitching and yawing moments due to the spacing of the cylinders. Table 3 gives the inertia unbalance and pitching moments of the pri-mary and secondary forces for various crank-angle arrange-ments of $n$-cylinder engines.

**Vibration Absorbers** In certain cases a secondary spring-mass system, called the absorber, can be effectively designed to reduce the vibrations of the system. If a system is forced to vibrate at a frequency $\omega_1$, an absorber $k_2$, $W_2$, tuned to the same frequency $\omega_1 = \sqrt{k_2 g/W_2}$, will introduce an opposing force equal to that of the disturbing force to suppress com-pletely the vibratory motion of the original system. Since the

counterforce of the absorber is equal to the inertial force ($W_2/g$,$) \omega^2 X_2$ of the secondary mass, the size of the absorber system is established by the magnitude of the disturbing force and the allowable amplitude of the secondary mass.

An absorber of this type has limitations in that it is effective only at a single frequency and also in that two other resonant frequencies in the neighborhood of the suppressed frequency will be introduced. Thus, for a variable-frequency distur-bance, such as the automobile engine, the simple spring-mass absorber is useless. For the variable-speed rotational system, the pendulum-type absorber shown in Fig. 11 is ideally suited in that its natural frequency is proportional to the rotational speed and hence such an absorber will always remain in tune with the disturbing torque. With $r$ very much smaller than $R$, the gravitational field $g$ is replaced by the centrifugal field $\omega^2 R$, so that the natural frequency of the pendulum becomes

$$\omega_n = \omega \sqrt{\frac{R}{r}} \qquad (23)$$

In the actual absorber, a U-shaped counterweight, fitted loosely by two pins, as shown in Fig. 12, is used, and the effective pendulum length is here equal to $r = r_1 - r_2$. As an example, in the six-cylinder four-cycle engine, there will be three power strokes per revolution, and to absorb the torsional vibrations at this frequency, $R/r$ must be equal to 9. The countertorque imparted by the centrifugal absorber of weight $W$ is $(W/g)\, \omega^2 R^2\theta$.

**Critical Speed** At certain speeds, known as critical speeds, rotating shafts become dynamically unstable with large lateral amplitudes. This phenomenon is due to the resonance fre-quency when the rotation speed in revolutions per second corresponds to the natural frequencies of lateral vibration of the shaft.

Consider first a vertical shaft with a centrally mounted thin disk of weight $W$, as shown in Fig. 13. We will assume that the center of gravity of the disk is a distance $e$ from the geometric center corresponding to the center of the shaft. When rotating at speed $\omega$ the angle between $x$ and $e$ will depend on whether $\omega$ is less or greater than $\omega_n$, the natural frequency of lateral vibration. With zero damping this angle will be zero for $\omega < \omega_n$ and 180° for $\omega > \omega_n$, and in either case the centrifugal force acting on the shaft is

$$\frac{W}{g} \omega^2(x + e) \qquad (24)$$

where $x$ may be positive or negative. The elastic restoring force of the shaft is $kx$, $k$ being the lateral stiffness of the shaft

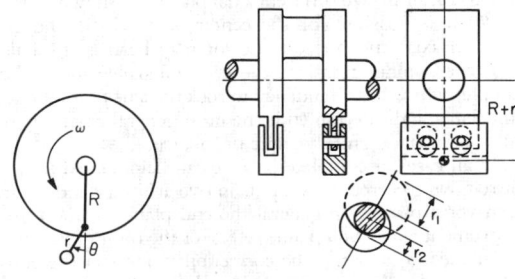

**Fig. 11**                    **Fig. 12**

**Table 3. Inertial Unbalance of Four Stroke per Cycle Engines**

| No. cylinders | Crank arrangement, $\varphi n$ | Unbalanced forces | | Unbalanced pitching moments about 1st cylinder | |
|---|---|---|---|---|---|
| | | Primary | Secondary | Primary | Secondary |
| 1 | | $X_1$ | $X_2$ | — | — |
| 2 | 0–180 | 0 | $2X_2$ | $lX_1$ | $2lX_2$ |
| 4 | 0–180–80–0 | 0 | $4X_2$ | 0 | $6lX_2$ |
| 4 | 0–90–270–180 | 0 | 0 | $lX_1\sqrt{1+3^2}$ | 0 |
| 6 | 0–120–240–240–120–0 | 0 | 0 | 0 | 0 |
| 8 | 0–180–90–270–270–90–180–0 | 0 | 0 | 0 | 0 |
| 90°–V-8 | 0–90–270–180 | 0 | 0 | | |

Rotating primary couple of constant magnitude $\sqrt{10}\ lX_1$ which may be completely counterbalanced

at the disk, or $48\ EI/l^3$, if the ends of the shaft are considered pinned; and by equating these two forces and replacing $kg/W$ by $\omega_n^2$, we arrive at the result

$$x = \frac{e(\omega/\omega_n)^2}{1 - (\omega/\omega_n)^2} \qquad (25)$$

As stated previously, $x$ and $e$ have the same sign for $\omega < \omega_n$ and opposite signs for $\omega > \omega_n$, the two conditions being illustrated by Fig. 13a and b.

When $\omega = \omega_n$ a condition of resonance is encountered and the amplitude $x$ becomes excessively large, which often leads to failure. Thus the critical speed of the system coincides with the natural frequency of lateral vibration of the system, which for the single-disk case is

$$\omega_n = \sqrt{\frac{kg}{W}} \qquad (26)$$

In simple cases, where the mass can be considered to be concentrated at a single point, the critical speed can be found from the equation

$$N_c(\text{rpm}) = \frac{60}{2\pi}\omega_n = \frac{60}{2\pi}\sqrt{\frac{g}{\delta_{st}}} = \frac{187.7}{\sqrt{\delta_{st}}} \qquad (27)$$

where $\delta_{st}$ is the statical deflection, measured in inches, of the shaft at the position of the concentrated mass. The quantity $\delta_{st}$ is always proportional to $Wl^3/EI$ and can be readily determined from $\delta_{st} = W/k$, where $k$ is given for beams with various end conditions in Table 2.

When a single-span shaft is loaded by several weights $W_1$, $W_2$, $W_3$ . . . , Rayleigh's energy method results in the equation

$$N_c(\text{rpm}) = 187.7\sqrt{\frac{W_1y_1 + W_2y_2 + W_3y_3 + \cdots}{W_1y_1^2 + W_2y_2^2 + W_3y_3^2 + \cdots}} \qquad (28)$$

for the first critical speed.

The deflections $y_1$, $y_2$, $y_3$, etc., may be determined from static loads $W_1$, $W_2$, $W_3$, etc., for the first approximation, after which a recalculation of the deflection due to dynamic loads

$W_1y_1$, $W_2y_2$, $W_3y_3$, etc., will improve the accuracy. $N_c$ calculated from the first approximation of static loads is generally within 1 percent of the correct value.

One other procedure for the calculation of the natural frequency of lateral vibration of shafts carrying several weights is given by Dunkerley. This formula is given as

$$\frac{1}{\omega_c^2} = \frac{1}{\omega_0^2} + \frac{1}{\omega_1^2} + \frac{1}{\omega_2^2} + \frac{1}{\omega_3^2} + \cdots \qquad (29)$$

where $\omega_c$ = critical speed of entire shaft assembly, rad/s
$\omega_0$ = critical speed of shaft only
$\omega_1$ = critical speed of shaft carrying only $W_1$
$\omega_2$ = critical speed of shaft carrying only $W_2$, etc.

**Lateral Vibrations of Shafts on Three Supports** (See Timoshenko, "Vibration Problems in Engineering," 3d ed., p. 268, Van Nostrand, 1954.) The critical speed of rotation of a shaft is equal to the natural frequency of the shaft in lateral vibration. For two disks mounted on a uniform shaft and simply supported at three points, the two natural frequencies for the lateral vibration can be computed from the equation (Fig. 14)

$$N_{cpm} = 187.7\sqrt{\frac{1}{2(ac-b^2)}\left[\frac{c}{W_1} + \frac{a}{W_2} \pm \sqrt{\left(\frac{c}{W_1} + \frac{a}{W_2}\right)^2 - \frac{4(ac-b^2)}{W_1W_2}}\right]} \qquad (30)$$

where the three influence coefficients are given as

$$a = \frac{1}{12ll_1^2EI}[4l_1^2(l-c_1)^2c_1^2 - c_1(-c_1^3 + l^2c_1 - l_2^2c_1)(l^2 - l_2^2 - c_1^2)]$$

$$b = \frac{1}{12ll_1l_2EI}[2l_1l_2c_1c_2(l^2 - c_1^2 - c_2^2)$$
$$- c_1c_2(l^2 - l_2^2 - c_1^2)(l^2 - l_1^2 - c_2^2)]$$

$$c = \frac{1}{12ll_2^2EI}[4l_2^2(l-c_2)^2c_2^2 - c_2(-c_2^3 + l^2c_2 - l_1^2c_2)(l^2 - l_1^2 - c_2^2)]$$

**Lateral Vibrations of Uniform Beams** For uniform beams

vibrating in flexure, the natural frequencies are expressible in the form

$$f_n \text{ (c/s)} = c_n \sqrt{\frac{gEI}{\omega l^4}} \qquad (31)$$

where $EI$ = flexural stiffness, lb/in² = 28.70 × 10⁻⁴ N/m²
$w$ = weight per unit length, lb/in = 1.752 × 10² N/m
$l$ = length of beam, in = 2.54 × 10⁻² m
$g$ = 386 in/s² = 9.806 m/s²
$c_n$ = number depending on boundary conditions and mode number

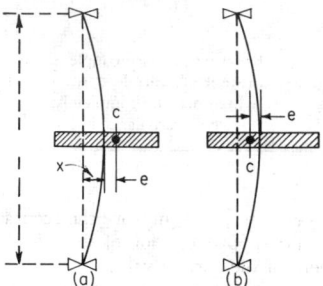

**Fig. 13**

The following table gives values of $c_n$ for the first five natural frequencies of beams with six different boundary conditions.

| Beam Configuration | $c_1$ | $c_2$ | $c_3$ | $c_4$ | $c_5$ |
|---|---|---|---|---|---|
| Simply supported ends | 1.56 | 6.28 | 14.1 | 25.1 | 39.3 |
| Clamped-free | 0.560 | 3.57 | 9.82 | 19.2 | 31.8 |
| Free-free or clamped-clamped | 3.58 | 9.82 | 19.2 | 31.8 | 47.5 |
| Clamped-hinged or hinged-free | 2.45 | 7.96 | 16.6 | 28.4 | 43.3 |

**Longitudinal and Torsional Vibration of Uniform Rods**   The equations for the natural frequencies of slender uniform rods in longitudinal or torsional vibration are identical except for the stiffness factor and can be written as

$$f \text{(c/s)} = c_n \sqrt{\frac{g\,AE}{wl^2}} \qquad \text{(longitudinal)} \qquad (32)$$

$$= c_n \sqrt{\frac{g\,AG}{wl^2}} \qquad \text{(torsional)} \qquad (33)$$

where $E$ = modulus of elasticity, lb/in² = 6.899 × 10³ N/m²
$A$ = cross-sectional area, in² = 6.45 × 10⁻⁴ m²
$w$ = weight per unit length, lb/in = 1.752 × 10² N/m
$l$ = length, in = 2.54 × 10⁻² m
$g$ = 386 in/s² = 9.806 m/s²
$G$ = shear modulus, lb/in² = 6.899 × 10³ N/m²
$c_n$ = nondimensional number depending on boundary conditions and mode number $n$, as shown in accompanying table. The same number $c_n$ applies for the two equations above.

| End Conditions | $c_n, n = 1, 2, 3, \ldots$ |
|---|---|
| Free-free or clamped-clamped | $1/2n$ |
| Clamped-free | $1/4(2n - 1)$ |

**Vibration of Membranes**   Assume that the membrane in a perfectly flexible and infinitely thin sheet of uniform material and thickness is uniformly stretched in all directions in its plane by a tension so large that the fluctuation in this tension due to the small deflections during vibrations can be neglected.

Let $S$ = uniform tension per unit length of boundary
$A$ = area of membrane
$w$ = weight of membrane per unit area

The frequency of the fundamental mode of vibration of the membrane is

$$f = \frac{\alpha}{2\pi} \sqrt{\frac{gS}{wA}} \qquad (34)$$

where $\alpha$ for various shapes of the boundary is as follows:

| Shape | $\alpha$ |
|---|---|
| Circle | 4.261 |
| Square | 4.443 |
| Quadrant of a circle | 4.551 |
| 60° sector of a circle | 4.616 |
| Equilateral traiangle | 4.774 |
| Semicircle | 4.803 |
| Rectangle 3 × 2 | 4.624 |
| Rectangle 2 × 1 | 4.967 |
| Rectangle 3 × 1 | 5.736 |

The frequencies of higher modes of vibrations of a circular membrane are given by the equation

$$f_{n,m} = \frac{\alpha_{nm}}{2\pi a} \sqrt{\frac{gS}{w}} \qquad (35)$$

where $a$ = radius of circular membrane
$\alpha_{nm}$ = constant given in table below and depending on number $n$ of nodal diameters and number $m$ of nodal circles (the boundary circle is included in $m$).

| $m$ \ $n$ | 0 | 1 | 2 | 3 | 4 | 5 |
|---|---|---|---|---|---|---|
| 1 | 2.40 | 3.83 | 5.13 | 6.38 | 7.59 | 8.78 |
| 2 | 5.52 | 7.02 | 8.42 | 9.76 | 11.06 | 12.3 |
| 3 | 8.65 | 10.17 | 11.6 | 13.02 | 14.4 | 15.7 |
| 4 | 11.8 | 13.3 | 14.8 | 16.2 | 17.6 | 19.0 |
| 5 | 14.9 | 16.5 | 18.0 | 19.4 | 20.8 | 22.2 |
| 6 | 18.1 | 19.6 | 21.1 | 22.6 | 24.0 | 25.4 |
| 7 | 21.2 | 22.8 | 24.3 | 25.7 | 27.2 | 28.6 |
| 8 | 24.4 | 25.9 | 27.4 | 28.9 | 30.4 | 31.8 |

**Vibration of Plates**   It is assumed that the plate consists of a perfectly elastic, homogeneous, isotropic material, and that it has a uniform thickness, considered small in comparison with its other dimensions. The deflections are assumed to be small in comparison with the thickness of the plates.

Let $h$ = thickness of plate
$D = Eh^3/12(1 - v^2)$ = flexural rigidity of plate ($v$ is Poisson's ratio)
$d$ = weight per unit volume of material of plate

The frequencies of the consecutive modes of vibration of a rectangular plate with the sides $a$ and $b$ and simply supported along the edges are

$$f_{mn} = \frac{\pi}{2} \sqrt{\frac{gD}{hd}} \left( \frac{m^2}{a^2} + \frac{n^2}{b^2} \right) \qquad (36)$$

where $m = 1, 2, 3 \ldots$
$n = 1, 2, 3 \ldots$

For the fundamental type of vibration of a square plate,

$$f = (\pi/a^2) \sqrt{gD/dh} \qquad (37)$$

Consecutive frequencies of a square plate with free edges are given by the equation

$$f_i = (\alpha_i/2\pi a^2) \sqrt{gD/dh} \qquad (38)$$

in which $\alpha_i$ is a constant depending on the mode of vibration. For the three lowest modes, the values of this constant are

$$\alpha_1 = 14.10 \qquad \alpha_2 = 20.56 \qquad \alpha_3 = 23.91$$

**Circular Plate Clamped at the Boundary** Equation (38) can be used for calculating frequencies, $\alpha$ denoting in this case the

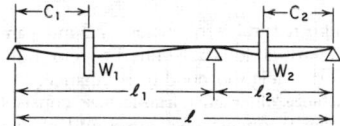

**Fig. 14**

radius of the boundary. Denoting with $m$ the number of nodal circles and with $n$ the number of nodal diameters, the magnitude of the factor $\alpha_i$ in Eq. (38) will be given by the following table. The table also gives values of $\alpha_i$ in Eq. (38) for a circular plate with free boundary.

**Circular Plate Fixed at the Center** The constant $\alpha_i$ for consecutive modes of vibration having $s$ nodal circles is as follows:

| $s =$ | 0 | 1 | 2 | 3 |
|---|---|---|---|---|
| $\alpha_i =$ | 3.75 | 20.91 | 60.68 | 119.7 |

| | Plate clamped at boundary | | | Plate with free boundary | | | |
|---|---|---|---|---|---|---|---|
| $m$ | $n = 0$ | $n = 1$ | $n = 2$ | $n = 0$ | $n = 1$ | $n = 2$ | $n = 3$ |
| 0 | 10.21 | 21.22 | 34.84 | | | 5.251 | 12.23 |
| 1 | 39.78 | | | 9.076 | 20.52 | 35.24 | 52.91 |
| 2 | 88.90 | | | 38.52 | 59.86 | | |

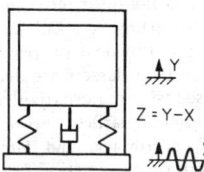

$Z = Y - X$

**Fig. 15**

## VIBRATION-MEASURING INSTRUMENTS

Vibrations to be measured can be classified as either (1) periodic, (2) shock or transient, or (3) random or statistical.

Of these, periodic motion is best understood, and instruments for measuring frequency, amplitude, velocity, acceleration, or wave slope are well developed. In the measurement of shocks the same quantities mentioned above may be of interest; however, in general, peak accelerations are of prime importance. In the case of random motions a frequency spectrum of mean-square values is desirable, and the instrumentation for such measurements is quite complex and somewhat recent in development.

The seismic spring-mass system of Fig. 15 represents the basic transducer element of many vibration-measuring instruments. Depending on the frequency range utilized, displacement, velocity, or acceleration is indicated by the relative motion of the suspended mass with respect to the case. Since vibrations are often too small for mechanical indication, the relative motion is generally converted to electrical voltage by motion of a coil in a magnetic field.

The relationship between the amplitude $X$ of the harmonic motion to be measured and the relative displacement $Z$ of the seismic mass is given by the equations

$$Z = \frac{(\omega/\omega_n)^2 \, X}{\sqrt{[\, 1 - (\omega/\omega_n)^2\,]^2 + [\, 2\zeta \, (\omega/\omega_n)]^2}} \qquad (39)$$

$$\varphi = \tan^{-1} \frac{2\zeta \, (\omega/\omega_n)}{1 - (\omega/\omega_n)^2} \qquad (40)$$

and plotted in Fig. 16. These equations indicate that the parameters involved are the frequency ratio $\omega/\omega_n$ and the damping factor $\zeta$. For minimum amplitude and phase distortion, $\zeta$ is chosen in the neighborhood of 0.7.

**Accelerometers** are high-natural-frequency instruments, and their useful frequency range is below resonance. For small values of $\omega/\omega_n$, Eq. (39) becomes approximately

$$Z = \frac{1}{\omega_n^2} \, (\omega^2 X)$$

and the relative motion becomes proportional to the acceleration.

**Displacement and velocity instruments** are low-natural-frequency devices, and their useful range in a region $\omega/\omega_n \gg 1$, or from Eq. (39),

$$Z \cong X$$

However, both velocity and displacement for harmonic motion can be obtained from accelerometers by means of electronic integrators.

Aside from the seismic-mass type of transducer, there are a

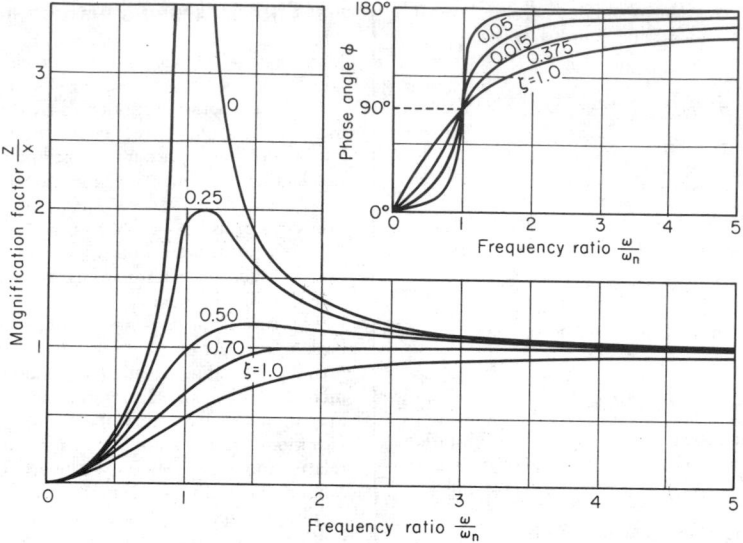

**Fig. 16**

variety of other vibration-measuring devices, which are listed as follows.

**Electrical-resistance strain gage** consists of a grid of fine wire cemented to the vibrating object to measure fluctuating strains. These wires are also used as suspensions for seismic weights in some accelerometers.

**Piezoelectric transducer** utilizes quartz or barium titanate crystals, has high natural frequencies, and must in general be used together with a low-pass filter to cut out frequency components in the neighborhood of resonance.

**Variable-reluctance differential transformer** consists of a primary with carrier-frequency excitation and of opposing secondary coils which indicate a voltage due to displacement of an iron core attached to the vibrating structure.

# NONDESTRUCTIVE TESTING

## by Donald D. Dodge

REFERENCES: McMaster, "Handbook of Nondestructive Testing," Ronald. Frederick, "Nondestructive Testing," Ann Arbor Press. Heuter and Bolt, "Sonics," Wiley. Krautkramer, "Ultrasonic Testing of Materials," Springer-Verlag. Betz, "Principles of Penetrants," Magnaflux Corp. "Radiography in Modern Industry," Eastman Kodak Co. Crowther, "Handbook of Industrial Radiography," Arnold. Wiltshire, "A Further Handbook of Industrial Radiography," Arnold. Commonly Used Specifications and Standards for Nondestructive Testing, *Mater. Eval.*, Mar. 1966. "Standards," Pt. 31, ASTM. Boiler and Pressure Vessel Code, Secs. III, XI, Case Interpretation, ASME. ASME Handbook, "Metals Engineering—Design," McGraw-Hill. "Standards," Secs. J358, J359, J420, J425–J428, SAE. Materials Evaluation, *Jour. Am. Soc. Nondestructive Test.*

**Nondestructive tests** are those tests which determine the usefulness, serviceability, or quality of a part or material without limiting its usefulness. Nondestructive tests are used in machinery maintenance to avoid costly unscheduled loss of service due to fatigue or wear; they are used in manufacturing to assure product quality and minimize warranty costs. Consideration of test requirements early in the design of a product may facilitate testing and minimize testing cost. Nearly every form of energy is used in nondestructive tests, including all wavelengths of the electromagnetic spectrum as well as vibrational mechanical energy. Physical properties, composition, and structure are determined; defects are detected; and thickness is measured. These tests are here divided into the following basic methods: **magnetic-particle, penetrant, radiographic, ultrasonic, eddy-current, microwave, and infrared.** Numerous techniques are utilized in the application of each test method. Table 1 gives a summary of the principal nondestructive test methods.

**Table 1. Nondestructive Test Methods***

| Method | Measures or detects | Applications | Advantages | Limitations |
|---|---|---|---|---|
| Acoustic emission | Crack initiation and growth rate<br>Internal cracking in welds during cooling<br>Boiling or cavitation<br>Friction or wear<br>Plastic deformation<br>Phase transformations | Pressure vessels<br>Stressed structures<br>Turbine or gearboxes<br>Fracture-mechanics research<br>Weldments<br>Sonic-signature analysis | Remote and continuous surveillance<br>Permanent record<br>Dynamic (rather than static) detection of cracks<br>Portable<br>Triangulation techniques to locate flaws | Transducers must be placed on part surface<br>Highly ductile materials yield low-amplitude emissions<br>Part must be stressed or operating<br>Test-system noise needs to be filtered out |
| Acoustic-impact (tapping) | Debonded areas or delaminations in metal or nonmetal composites or laminates<br>Cracks under bolt or fastener heads<br>Cracks in turbine wheels or turbine blades<br>Loose rivets or fastener heads<br>Crushed core | Brazed or adhesive-bonded structures<br>Bolted or riveted assemblies<br>Turbine blades<br>Turbine wheels<br>Composite structures | Portable<br>Easy to operate<br>May be automated<br>Permanent record or positive meter readout<br>No couplant required | Part geometry and mass influences test results<br>Impactor and probe must be repositioned to fit geometry of part<br>Reference standards required |
| Eddy current | Surface and subsurface cracks and seams<br>Alloy content<br>Heat-treatment variations<br>Wall thickness, coating thickness<br>Crack depth<br>Conductivity<br>Permeability | Tubing<br>Wire<br>Ball bearings<br>"Spot checks" on all types of surfaces<br>Proximity gage<br>Metal detector<br>Metal sorting<br>Measure conductivity in % IACS | No special operator skills required<br>High speed, low cost<br>Automation possible for symmetrical parts<br>Permanent-record capability for symmetrical parts<br>No couplant or probe contact required | Conductive materials<br>Shallow depth of penetration (thin walls only)<br>Masked or false indications caused by sensitivity to variations such as part geometry<br>Reference standards required<br>Permeability variations |
| Eddy-sonic | Debonded areas in metal-core or metal-faced honeycomb structures<br>Delaminations in metal laminates or composites<br>Crushed core | Metal-core honeycomb<br>Metal-faced honeycomb<br>Conductive laminates such as boron or graphite-fiber composites<br>Bonded-metal panels | Portable<br>Simple to operate<br>No couplant required<br>Locates far-side debonded areas<br>Access to only one surface required<br>May be automated | Specimen or part must contain conductive materials to establish eddy-current field<br>Reference standards required<br>Part geometry |
| Electric current | Cracks<br>Crack depth<br>Resistivity<br>Wall thickness<br>Corrosion-induced wall thinning | Metallic materials<br>Electrically conductive materials<br>Train rails<br>Nuclear fuel elements | Access to only one surface required<br>Battery or dc source<br>Portable | Edge effect<br>Surface contamination<br>Good surface contact required<br>Difficult to automate<br>Electrode spacing<br>Reference standards required |
| Electrified particle | Surface defects in nonconducting material<br>Through-to-metal pinholes on metal-backed material<br>Tension, compression, cyclic cracks<br>Brittle-coating stress cracks | Glass<br>Porcelain enamel<br>Nonhomogeneous materials such as plastic or asphalt coatings<br>Glass-to-metal seals | Portable<br>Useful on materials not practical for penetrant inspection | Poor resolution on thin coatings<br>False indications from moisture streaks or lint<br>Atmospheric conditions<br>High-voltage discharge |

**Table 1. Nondestructive Test Methods\*** *(Continued)*

| Method | Measures or detects | Applications | Advantages | Limitations |
|---|---|---|---|---|
| Filtered particle | Cracks<br>Porosity<br>Differential absorption | Porous materials such as clay, carbon, powdered metals, concrete<br>Grinding wheels<br>High-tension insulators<br>Sanitary ware | Colored or fluorescent particles<br>Leaves no residue after baking part over 400°F<br>Quickly and easily applied<br>Portable | Size and shape of particles must be selected before use<br>Penetrating power of suspension medium is critical<br>Particle concentration must be controlled<br>Skin irritation |
| Fluoroscopy (cine-fluorography) (kine-fluorography) | Level of fill in containers<br>Foreign objects<br>Internal components<br>Density variations<br>Voids<br>Formation of casting defects | Flow of liquids<br>Presence of cavitation<br>Operation of valves and switches<br>Burning in small solid-propellant rocket motors | High-brightness images<br>Real-time viewing<br>Image magnification<br>Permanent record<br>Moving subject can be observed | Costly equipment<br>Geometric unsharpness<br>Thick specimens<br>Speed of event to be studied<br>Viewing area |
| Holography (acoustical-liquid surface levitation) | Lack of bond<br>Delaminations<br>Voids<br>Porosity<br>Resin-rich or resin-starved areas<br>Inclusions<br>Density variations | Metals<br>Plastics<br>Composites<br>Laminates<br>Honeycomb structures<br>Ceramics<br>Biological specimens | No hologram film development required<br>Real-time imaging provided<br>Liquid surface responds rapidly to ultrasonic energy | Through-transmission techniques only<br>Object and reference beams must superimpose on special liquid surface<br>Immersion test only<br>Laser required |
| Holography (interferometry) | Strain<br>Plastic deformation<br>Cracks<br>Debonded areas<br>Voids and inclusions<br>Vibration | Bonded and composite structures<br>Automotive or aircraft tires<br>Three-dimensional imaging | Surface of test object can be uneven<br>No special surface preparations or coatings required<br>No physical contact with test specimen | Vibration-free environment is required<br>Heavy base to damp vibrations<br>Difficult to identify type of flaw detected |
| Infrared (radiometers) | Lack of bond<br>Hot spots<br>Heat transfer<br>Isotherms<br>Temperature ranges | Brazed joints<br>Adhesive-bonded joints<br>Metallic platings or coatings; debonded areas or thickness<br>Electrical assemblies<br>Temperature monitoring | Sensitive to 1.5°F temperature variation<br>Permanent record or thermal picture<br>Quantitative<br>Remote sensing; need not contact part<br>Portable | Emissivity<br>Liquid-nitrogen-cooled detector<br>Critical time-temperature relationship<br>Poor resolution for thick specimens<br>Reference standards required |
| Leak testing | Leaks:<br>  Helium<br>  Ammonia<br>  Smoke<br>  Water<br>  Air bubbles<br>  Radioactive gas<br>  Halogens | Joints:<br>  Welded<br>  Brazed<br>  Adhesive-bonded<br>Sealed assemblies<br>Pressure or vacuum chambers<br>Fuel or gas tanks | High sensitivity to extremely small, light separations not detectable by other NDT methods<br>Sensitivity related to method selected | Accessibility to both surfaces of part required<br>Smeared metal or contaminants may prevent detection<br>Cost related to sensitivity |
| Magnetic particles | Surface and slightly subsurface defects; cracks, seams, porosity, inclusions<br>Permeability variations<br>Extremely sensitive for locating small tight cracks | Ferromagnetic materials; bar, forgings, weldments, extrusions, etc. | Advantage over penetrant in that it indicates subsurface defects, particularly inclusions<br>Relatively fast and low-cost<br>May be portable | Alignment of magnetic field is critical<br>Demagnetization of parts required after tests<br>Parts must be cleaned before and after inspection<br>Masking by surface coatings |

**Table 1. Nondestructive Test Methods\*** *(Continued)*

| Method | Measures or detects | Applications | Advantages | Limitations |
|---|---|---|---|---|
| Magnetic field | Cracks<br>Wall thickness<br>Hardness<br>Coercive force<br>Magnetic anisotropy<br>Magnetic field<br>Nonmagnetic coating thickness on steel | Ferromagnetic materials<br>Ship degaussing<br>Liquid-level control<br>Treasure hunting<br>Wall thickness of nonmetallic materials<br>Material sorting | Measurement of magnetic material properties<br>May be automated<br>Easily detects magnetic objects in nonmagnetic material<br>Portable | Permeability<br>Reference standards required<br>Edge effect<br>Probe lift-off |
| Microwave (300 MHz–300 GHz) | Cracks, holes, debonded areas, etc., in nonmetallic parts<br>Changes in composition, degree of cure, moisture content<br>Thickness measurement<br>Dielectric constant<br>Loss tangent | Reinforced plastics<br>Chemical products<br>Ceramics<br>Resins<br>Rubber<br>Wood<br>Liquids<br>Polyurethane foam<br>Radomes | Between radio waves and infrared in the electromagnetic spectrum<br>Portable<br>Contact with part surface not normally required<br>Can be automated | Will not penetrate metals<br>Reference standards required<br>Horn-to-part spacing critical<br>Part geometry<br>Wave interference<br>Vibration |
| Mössbauer effect | Nuclear magnetic resonance in materials, most common being iron 57<br>Polarization of magnetic domains in steel | Detect and identify iron in specimen or sample<br>Detect iron films on stainless steel<br>Measure retained austenite (2 – 35%) in steels<br>Determine nitrided surfaces on steel<br>Interaction of domains with dislocation in ferromagnetic materials | Provide unique information about the surroundings of the iron 57 nuclei | Radiation hazard<br>Trained engineers or physicists required<br>Nonportable<br>Precision equipment for vibrating source and spectrum analysis |
| Neutron activation analysis | Radiation emission resulting from neutron activation<br>Oxygen in steel<br>Nitrogen in food products<br>Silicon in metals and ores | Metallurgical<br>Prospecting<br>Well logging<br>Oceanography<br>On-line process control of liquid or solid materials | Automatic systems<br>Accurate (ppm range)<br>Fast<br>No contact with sample<br>Sample preparation minimal | Radiation hazard<br>Fast decay time |
| Penetrants | Defects open to surface of parts; cracks, porosity, seams, laps, etc.<br>Through-wall leaks | All parts with nonabsorbing surfaces (forgings, weldments, castings, etc.). Note: Bleed-out from porous surfaces can mask indications of defects | Low cost<br>Portable<br>Indications may be further examined visually<br>Results easily interpreted | Surface films such as coatings, scale, and smeared metal may prevent detection of defects<br>Parts must be cleaned before and after inspection<br>Defect must be open to surface |
| Radiography (thermal neutron) | Hydrogen contamination of titanium or zirconium alloys<br>Defective or improperly loaded pyrotechnic devices<br>Improper assembly of metal, nonmetal parts | Pyrotechnic devices<br>Metallic, nonmetallic assemblies<br>Biological specimens<br>Nuclear-reactor-fuel elements and control rods | High neutron absorption by hydrogen, boron, lithium, cadmium, uranium, plutonium<br>Low neutron absorption by most metals<br>Complement to X-ray or gamma-ray radiography | Very costly equipment<br>Nuclear reactor or accelerator required<br>Trained physicists required<br>Radiation hazard<br>Nonportable<br>Indium or gadolinium screens required |

**Table 1. Nondestructive Test Methods\*** *(Continued)*

| Method | Measures or detects | Applications | Advantages | Limitations |
|---|---|---|---|---|
| Radiography (gamma rays), cobalt 60, Iridium 192 | Internal defects and variations, porosity, inclusions, cracks, lack of fusion, geometry variations, corrosion | Usually where X-ray machines are not suitable because source cannot be placed in part with small openings and/or power source not available | Low initial cost<br>Permanent records; film<br>Small sources can be placed in parts with small openings<br>Portable<br>Low contrast | One energy level per source<br>Source decay<br>Radiation hazard<br>Trained operators needed<br>Lower image resolution<br>Cost related to energy range |
| Radiography (X-rays—film) | Internal defects and variations; porosity, inclusions, cracks, lack of fusion, geometry variations, corrosion<br>Density variations | Castings<br>Electrical assemblies<br>Weldments<br>Small, thin, complex wrought products<br>Nonmetallics<br>Solid-propellant rocket motors | Permanent records; film<br>Adjustable energy levels (5 kV–25 meV)<br>High sensitivity to density changes<br>No couplant required<br>Geometry variations do not affect direction of X-ray beam | High initial costs<br>Orientation of linear defects in part may not be favorable<br>Radiation hazard<br>Depth of defect not indicated<br>Sensitivity decreases with increase in scattered radiation |
| Radiometry (X-ray, gamma-ray, beta-ray) | Wall thickness<br>Plating thickness<br>Variations in density or composition<br>Fill level in cans or containers<br>Inclusions or voids | Sheet, plate, strip, tubing<br>Nuclear-reactor-fuel rods<br>Cans or containers<br>Plated parts | Fully automatic<br>Fast<br>Extremely accurate<br>In-line process control<br>Portable | Radiation hazard<br>Beta-ray useful for ultrathin coatings only<br>Source decay<br>Reference standards required |
| Sonic (less than 0.1 MHz) | Debonded areas or delaminations in metal or nonmetal composites or laminates<br>Cohesive bond strength under controlled conditions<br>Crushed or fractured core<br>Bond integrity of metal insert fasteners | Metal or nonmetal composite or laminates brazed or adhesive-bonded<br>Plywood<br>Rocket-motor nozzles<br>Honeycomb | Portable<br>Easy to operate<br>Locates far-side debonded areas<br>May be automated<br>Access to only one surface required | Surface geometry influences test results<br>Reference standards required<br>Adhesive or core-thickness variations influence results |
| Thermal (thermochromic paint, liquid crystals) | Lack of bond<br>Hot spots<br>Heat transfer<br>Isotherms<br>Temperature ranges | Brazed joints<br>Adhesive-bonded joints<br>Metallic platings or coatings<br>Electrical assemblies<br>Temperature monitoring | Very low initial cost<br>Can be readily applied to surfaces which may be difficult to inspect by other methods<br>No special operator skills | Thin-walled surfaces only<br>Critical time-temperature relationship<br>Image retentivity affected by humidity<br>Reference standards required |
| Thermoelectric | Thermoelectric potential<br>Coating thickness<br>Physical properties<br>Thompson effect<br>*P-N* junctions in semiconductors | Metal sorting<br>Ceramic coating thickness on metals<br>Semiconductors | Portable<br>Simple to operate<br>Access to only one surface required | Hot probe<br>Difficult to automate<br>Reference standards required<br>Surface contaminants<br>Conductive coatings |
| Ultrasonic (0.1–25 MHz) | Internal defects and variations; cracks, lack of fusion, porosity, inclusions, delaminations, lack of bond, texturing<br>Thickness or velocity<br>Poisson's ratio, elastic modulus | Wrought metals<br>Welds<br>Brazed joints<br>Adhesive-bonded joints<br>Nonmetallics<br>In-service parts | Most sensitive to cracks<br>Test results known immediately<br>Automating and permanent-record capability<br>Portable<br>High penetration capability | Couplant required<br>Small, thin, complex parts may be difficult to check<br>Reference standards required<br>Trained operators for manual inspection |

\*From Donald J. Hagemaier, "Metal Progress Databook," Douglas Aircraft Co., McDonnell-Douglas Corp., Long Beach, CA.

## MAGNETIC-PARTICLE METHODS

**Magnetic-particle** testing is a nondestructive method for detecting discontinuities at or near the surface in **ferromagnetic materials.** The test object is properly magnetized, and then finely divided magnetic particles are applied to its surface. When the object is properly oriented to the induced magnetic field, a discontinuity creates a leakage field which attracts and holds the particles, forming a visible indication. Magnetic-field direction and character are dependent upon how the magnetizing force is applied and upon the type of current used. For best sensitivity, the **magnetizing current** must flow in a direction parallel to the principal direction of the expected defect. Circular fields, produced by passing current through the object, are almost completely contained within the test object. Longitudinal fields, produced by coils or yokes, create external poles and a general-leakage field. Alternating, direct, or half-wave direct current may be used for the location of surface defects. Half-wave direct current is most effective for locating subsurface defects. Magnetic particles may be applied dry or as a wet suspension in a liquid similar to kerosene. Colored **dry powders** are advantageous when testing for subsurface defects and when testing objects which have rough surfaces, such as castings, forgings, and weldments. **Wet particles** are preferred for detection of very fine cracks, such as fatigue or grinding cracks. Fluorescent wet particles are used to inspect objects with the aid of ultraviolet light. Fluorescent inspection is most widely used because of its greater sensitivity. Application of particles while magnetizing current is on (continuous method) produces stronger indications than those obtained if the particles are applied after the current is shut off (residual method). Interpretation of subsurface-defect indications requires experience. Demagnetization of the test object after inspection is advisable.

**Electrified-particle** testing indicates minute cracks in nonconducting materials. Particles of calçium carbonate are positively charged as they are blown through a spray gun at the test object. If the object is metal-backed, such as porcelain enamel, no preparation other than cleaning is necessary. When it is not metal-backed, the object must be dipped in an aqueous penetrant solution and dried. The penetrant remaining in cracks provides a mobile electron supply for the test. A readily visible powder indication forms at a crack owing to the attraction of the positively charged particles.

## PENETRANT METHODS

**Liquid-penetrant** testing is used to locate defects open to the surface of **nonporous materials.** The test object must be thoroughly cleaned before testing. Penetrating liquid is applied to the surface of a test object by a brush, spray, flow, or dip method. A time allowance (1 to 30 min) is required for liquid penetration of surface defects. Excess penetrant is then carefully removed from the surface, and an absorptive coating, known as developer, is applied to the object to draw penetrant out of defects, thus showing their location, shape, and approximate size. The developer is typically a fine powder, such as talc usually in suspension in a liquid. Penetrating-liquid types are (1) for test in visible light, and (2) for test under ultraviolet light (3650 Å). Sensitivity of penetrant testing is greatest when a fluorescent penetrant is used and the object is observed in a

semidarkened location. After testing, the penetrant and developer are removed by washing with water, sometimes aided by an emulsifier, or with a solvent.

In **filtered-particle** testing, cracks in **porous objects** (100 mesh or smaller) are indicated by the difference in absorption between a cracked and a flaw-free surface. A fluid containing suspended particles is sprayed on a test object. If a crack exists, particles are filtered out and concentrate at the surface as liquid flows into the additional absorbent area created by the crack. Fluorescent or colored particles are used to locate defects in unfired dried clay, certain fired ceramics, concrete, some powdered metals, carbon, and partially sintered tungsten and titanium carbides.

## RADIOGRAPHIC METHODS

**Radiographic** test methods employ X-rays, gamma rays, or similar penetrating radiation to reveal flaws, voids, inclusions, thickness, or structure of objects. Electromagnetic-energy wavelengths in the region of 0.01 to 10 Å (1 Å = $10^{-8}$ cm) are used to examine the interior of opaque materials. Penetrating radiation proceeds from its source in straight lines to the test object. Rays are differentially absorbed by the object, depending upon the energy of the radiation and the nature and thickness of the material.

**X-rays** of a variety of wavelengths result when high-speed electrons in a vacuum tube are suddenly stopped. An X-ray tube contains a heated filament (cathode) and a target (anode); radiation intensity is almost directly proportional to filament current (mA); tube voltage (kV) determines the penetration capability of the rays. As tube voltage increases, shorter wavelengths and more intense X-rays are produced. When the energy of penetrating radiation increases, shorter wavelengths and more intense X-rays are produced. Also, when the energy of penetrating radiation increases, the difference in attenuation between materials decreases. Consequently, more film-image contrast is obtained at lower voltage, and a greater range of thickness can be radiographed at one time at higher voltage.

**Gamma rays** of a specific wavelength are emitted from the disintegrating nuclei of natural radioactive elements, such as radium, and from a variety of artificial radioactive isotopes produced in nuclear reactors. Cobalt 60 and iridium 192 are commonly used for industrial radiography. The half-life of an isotope is the time required for half of the radioactive material to decay. This time ranges from a few hours to many years.

A radiograph is a photographic record produced by passage of penetrating radiation onto a film. A void or reduced mass appears as a darker image on the film because of the lesser absorption of energy and the resulting additional exposure of the film. The quantity of X-rays absorbed by a material generally increases as the atomic number increases.

**Radiographic films** vary in speed, contrast, and grain size. Slow films generally have smaller grain size and produce more contrast. Slow films are used where optimum sharpness and maximum contrast are desired. Fast films are used where objects with large differences in thickness are to be radiographed or where sharpness and contrast can be sacrificed to shorten exposure time. Exposure of a radiographic film comes from direct radiation and scattered radiation. **Direct radiation** is desirable, image-forming radiation; **scattered radiation,** which occurs in the object being X-rayed or in neighboring objects, produces undesirable images on the film and loss of contrast. **Intensifying screens** made of 0.005- or 0.010-in-thick lead are often used for radiography at voltages above 100 kV. The lead filters out much of the low-energy scatter radiation. Under action of X-rays or gamma rays above 88 kV, a lead screen also emits electrons which, when in intimate contact with the film, produce additional coherent darkening of the film. Exposure time can be materially reduced by use of intensifying screens above and below the film.

A radiograph is a shadow picture, since X-rays and gamma rays follow the laws of light in shadow formation. Four factors determine the best geometric sharpness of a picture: (1) The effective focal-spot size of the radiation source should be as small as possible. (2) The source-to-object distance should be adequate for proper definition of the area of the object farthest from the film. (3) The film should be as close as possible to the object. (4) The area of interest should be in the center of and perpendicular to the X-ray beams and parallel to the X-ray film.

**Penetrameters** are used to indicate the contrast and definition which exist in a radiograph. The type generally used in the United States is a small rectangular plate of the same material as the object being X-rayed. It is uniform in thickness (usually 2 percent of the object thickness) and has holes drilled through it. ASTM specifies hole diameters one, two, and four times the thickness of the penetrameter. Step, wire, and bead penetrameters are also used. (See ASTM Materials Specification E94.)

Because of the variety of factors which relate to the production and measurements of an X-ray image, operating factors are generally selected from reference tables or graphs which have been prepared from test data obtained for a range of operating conditions.

All materials may be inspected by radiographic means, but there are limitations to the configurations of materials. With optimum techniques, wires 0.0001 in in diameter can be resolved in small electrical components. At the other extreme, welded steel pressure vessels with 20-in wall thickness can be routinely inspected by use of high-energy accelerators as a source of radiation. **Neutron radiation** penetrates extremely dense materials such as lead more readily than X-rays or gamma rays but is attenuated by lighter-atomic-weight materials such as plastics.

**Radiographic** standards are published by ASTM, ASME, AWS, and API, primarily for detecting lack of penetration or lack of fusion in welded objects. Cast-metal objects are radiographed to detect conditions such as shrink, porosity, hot tears, cold shuts, inclusions, coarse structure, and cracks.

The usual method of utilizing penetrating radiation is film. However, Geiger counters, semiconductors, phosphors (fluoroscopy), photoconductors (xeroradiography), scintillation crystals, and vidicon tubes (image intensifiers) are also used.

The **dangers** connected with exposure of the human body to X-rays and gamma rays should be fully understood by any person responsible for the use of radiation equipment. NBS is a prime source of information concerning radiation safety. USAEC specifies maximum permissible exposure to be 1.25 r/¼ year.

## ULTRASONIC METHODS

**Ultrasonic** nondestructive test methods employ high-frequency mechanical vibrational energy to detect and locate structural discontinuities or differences and to measure thickness of a variety of materials. An electric pulse is generated in a test instrument and transmitted to a transducer, which converts the electric pulse into mechanical vibrations. These low-energy-level vibrations are transmitted through a coupling liquid into the test object, where the ultrasonic energy is attenuated, scattered, reflected, or resonated to indicate conditions within material. Reflected, transmitted, or resonant sound energy is reconverted to electrical energy by a transducer and returned to the test instrument, where it is amplified. The received energy is then usually displayed on a cathode-ray tube. The presence, position, and amplitude of echoes indicate conditions of the test-object material.

**Materials** capable of being tested by ultrasonic energy are those which transmit vibrational energy. Metals are tested in dimensions of up to 30 ft (9.14 m). Noncellular plastics, ceramics, glass, new concrete, organic materials, and rubber can be tested. Each material has a characteristic sound velocity, which is a function of its density and modulus (elastic or shear).

**Material characteristics** determinable through ultrasonics include structural discontinuities, such as flaws and unbonds, physical constants and metallurgical differences, and thickness (measured from one side). A common application of ultrasonics is the inspection of welds for inclusions, porosity, lack of penetration, and lack of fusion. Other applications include location of unbond in nuclear-fuel elements, location of fatigue cracks in machinery, and medical applications. Automatic testing is frequently performed in manufacturing applications.

Ultrasonic systems are classified as either **pulse-echo,** in which a single transducer is used, or **through-transmission,** in which separate sending and receiving transducers are used. Pulse-echo systems are more common. In either system, ultrasonic energy must be transmitted into, and received from, the test object through a **coupling medium,** since air will not transmit ultrasound of these frequencies. Water, oil, grease, and glycerin are commonly used couplants. Two types of testing are used: contact and immersion. In contact testing, the transducer is placed directly on the test object. In immersion testing, the transducer and test object are separated from one another in a tank filled with water or by a column of water or by a liquid-filled wheel. Immersion testing eliminates transducer wear and facilitates scanning of the test object. Scanning systems usually have paper-printing equipment for readout of test information.

**Ultrasonic transducers** are piezoelectric units which convert electric energy into acoustic energy and convert acoustic energy into electric energy of the same frequency. Quartz, barium titanate, lithium sulfate, lead metaniobate, and lead zirconate titanate are commonly used transducer crystals, which are generally mounted with a damping backing in a housing. Transducers range in size from 1/16 to 5 in (0.15 to 12.7 cm) and are of circular or rectangular shape. Ultrasonic beams can be focused to improve resolution and definition. Transducer characteristics and beam patterns are dependent upon frequency, size, crystal material, and construction.

**Test frequencies** used range from 40 kHz to 50 MHz. Flaw-detection and thickness-measurement applications use frequencies between 500 kHz and 25 MHz, with 2.25 and 5 MHz being most commonly employed for flaw detection. Low frequencies (40 kHz to 1.0 MHz) are used on materials of low elastic modulus or large grain size. High frequencies (2.25 to 25 MHz) provide better resolution of smaller defects and are used on fine-grain materials and thin sections. Frequencies above 25 MHz are employed for investigation and measurement of physical properties related to acoustic attenuation.

**Wave-vibrational modes** other than longitudinal are effective in detecting flaws that do not present a reflecting surface to the ultrasonic beam, or other characteristics not detectable by the longitudinal mode. Wedges of plastic, water, or other material are inserted between the transducer face and the test object to convert, by refraction, to shear, transverse, surface, or Lamb vibrational modes. As in optics, Snell's law expresses the relationship between incident and refracted beam angles; i.e., the ratio of the sines of the angle from the normal, of the incident and refracted beams in two mediums, is equal to the ratio of the mode acoustic velocities in the two mediums.

**Limiting conditions** for ultrasonic testing may be the test-object shape, surface roughness, grain size, material structure, defect orientation, selectivity of discontinuities, and the skill of the operator. Test sensitivity is less for cast metals than for wrought metals because of grain size and surface differences.

**Standards** for acceptance are established in many government, national society, and company specifications (see references above). Evaluation is made by comparing (visually or by automated electronic means) received signals with signals obtained from reference blocks containing flat bottom holes between $1/64$ and $8/64$ in (0.040 and 0.325 cm) diameter, or from parts containing known defects, drilled holes, or machined notches.

## EDDY-CURRENT METHODS

**Eddy-current** nondestructive tests are based upon correlation between electromagnetic properties and physical or structural properties of a test object. Eddy currents are induced in metals whenever they are brought into an ac magnetic field. These eddy currents create a secondary magnetic field, which opposes the inducing magnetic field. The presence of discontinuities or material variations alters eddy currents, thus changing the apparent impedance of the inducing coil or of a detection coil. Coil impedance indicates the magnitude and phase relationship of the eddy currents to their inducing magnetic-field current. This relationship is dependent upon the mass, conductivity, permeability, and structure of the metal and upon the frequency, intensity, and distribution of the alternating magnetic field. Conditions such as heat treatment, composition, hardness, phase transformation, case depth, cold working, strength, size, thickness, cracks, seams, and inhomogeneities are indicated by eddy-current tests. Correlation data must usually be obtained to determine whether test conditions for desired characteristics of a particular test object can be established. Because of the many factors which cause variation in electromagnetic properties of metals, care must be taken that the instrument response to the condition of interest is not nullified or duplicated by variations due to other conditions.

Alternating-current **frequencies** between 1 and 5,000,000 Hz are used for eddy-current testing. Test frequency determines the depth of current penetration into the test object, owing to the ac phenomenon of "skin effect." Depth of current penetration is the depth at which the eddy currents are equal to 37 percent of their value at the surface. In a plane conductor, depth of penetration varies inversely as the square root of the product of conductivity, permeability, and frequency. High-frequency eddy currents are more sensitive to surface defects or conditions while low-frequency eddy currents are sensitive also to deeper internal defects or conditions.

**Test coils** are of three general types: the **circular coil,** which surrounds an object; the **bobbin coil,** which is inserted within an object; and the **probe coil,** which is placed on the surface of an object. Coils are further classified as **absolute,** when testing is conducted without direct comparison with a reference object in another coil; or **differential,** when comparison is made through use of two coils connected in series opposition. Many variations of these coil types are utilized. Axial length of a circular test coil should not be more than 4 in (10.2 cm), and its shape should correspond closely to the shape of the test object for best results. Coil diameter should be no more than twice the test-object diameter for consistent and useful results. Coils may be of the air-core or magnetic-core type.

**Instrumentation** for the analysis and presentation of electrical signals resulting from eddy-current testing includes a variety of means, ranging from meters to oscilloscopes. Instrument, meter, or alarm circuits are adjusted to be sensitive only to signals of a certain electrical phase, so that selected conditions are indicated while others are ignored. Automatic testing is one of the principal advantages of the method.

**Thickness** measurement of metallic and nonmetallic coatings on metals is performed using eddy-current principles. Coating thicknesses measured typically range from 0.0001 to 0.100 in (0.00025 to 0.25 cm). For measurement to be possible, metallic-coating conductivity must differ from that of the base metal.

## MICROWAVE METHODS

**Microwave** test methods utilize electromagnetic energy to determine characteristics of nonmetallic substances, either solid or liquid. Frequencies used range from 1 to 3,000 GHz. Microwaves generated in a test instrument are transmitted by a waveguide through air to the test object. Analysis of reflected or transmitted energy indicates certain material characteristics, such as moisture content, composition, structure, density, degree of cure, aging, and presence of flaws. Other applications include thickness and displacement measurement in the range of 0.001 in (0.0025 cm) to more than 12 in (30.4 cm). Materials that can be tested include most solid and liquid nonmetals, such as chemicals, minerals, plastics, wood, ceramics, glass, and rubber.

## INFRARED METHODS

Infrared nondestructive tests involve the detection of infrared electromagnetic energy emitted by a test object. Infrared radiation is produced naturally by all matter at all temperatures above absolute zero. Materials emit radiation at varying intensities, depending upon their temperature and surface characteristics. A **passive** infrared system detects the natural radiation of an unheated test object, while an **active** system employs a source to heat the test object, which then radiates infrared energy to a detector. Sensitive indication of temperature or temperature distribution through infrared detection is useful in locating irregularities in materials, in processing, or in the functioning of parts. Emission in the infrared range of

0.8 to 15 $\mu$m is collected optically, filtered, detected, and amplified by a test instrument which is designed around the characteristics of the detector material. Temperature variations on the order of 0.01°F can be indicated by meter or graphic means. Infrared theory and instrumentation are based upon radiation from a blackbody; therefore, **emissivity correction** must be made electrically in the test instrument or arithmetically from instrument readings.

### ACOUSTIC-SIGNATURE ANALYSIS

**Acoustic-signature analysis** involves the analysis of sound energy emitted from an object to determine characteristics of the object. The object may be a simple casting or a complex manufacturing system. A passive test is one in which **sonic** energy is transmitted into the object. In this case, a mode of resonance is usually detected to correlate with cracks or structure variations, which cause a change in effective modulus of the object, such as a nodular iron casting. An active test is one in which the object emits sound as a result of being struck or as a result of being in operation. In this case, characteristics of the object may be correlated to damping time of the sound energy or to the presence or absence of a certain frequency of sound energy. Bearing wear in rotating machinery can often be detected prior to actual failure, for example. More complex analytical systems can monitor and control manufacturing processes, based upon analysis of emitted sound energy.

Acoustic-signature analysis is a technology distinctly separate from **acoustic emission,** where strain produces bursts of energy in an object. These are detected by **ultrasonic** transducers coupled to the object. Growth of microcracks, and incipient failure, is monitored by counting the pulses of energy from the object or recording the time rate of the pulses of energy in the ultrasonic range (usually a discrete frequency between 1 kHz and 1 MHz).

Section **6**

# Materials of Engineering

**BY**

**HOWARD S. BEAN**  *Late Physicist, National Bureau of Standards.*
**HAROLD E. McGANNON**  *Technical Writer and Editor.*
**CHARLES WILLERS BRIGGS**  *Late Consulting Engineer.*
**L. D. KUNSMAN**  *Fellow Engineer, Research Labs, Westinghouse Electric Corp.*
**C. L. CARLSON**  *Fellow Engineer, Research Labs, Westinghouse Electric Corp.*
**HILARY E. BACON**  *Consulting Chemical Engineer and Partner, Sheppard T. Powell Associates.*
**GEORGE G. SWARD**  *Retired Consultant, National Paint and Coatings Association.*
**CARL H. de ZEEUW**  *Professor, Department of Wood Products Engineering, College of Environmental Science and Forestry, State University of New York, Syracuse.*
**ANTONIO F. BALDO**  *Professor of Mechanical Engineering, The City College, The City University of New York.*
**WILLIAM L. GAMBLE**  *Professor of Civil Engineering, University of Illinois at Urbana-Champaign.*
**KENNETH A. ROE**  *Chairman and President, Burns and Roe, Inc.*
**FRED J. VILLFORTH, JR.**  *Technologist, Beacon Research Laboratories, Research and Development Department, Texaco, Inc.*

## GENERAL PROPERTIES OF MATERIALS
### by Howard S. Bean
Chemistry ................................................. 6-3
Specific Gravities and Densities ......................... 6-7
Physical Data ............................................ 6-10

## IRON AND STEEL
### by Harold E. McGannon
Classification of Iron and Steel ......................... 6-12
Commercially Pure Iron ................................... 6-12
Wrought Iron ............................................. 6-12
Ingot Iron ............................................... 6-13
Steel .................................................... 6-13
Effect of Alloying Elements on the Properties of Steel .... 6-20
Principles of Heat Treatment of Iron and Steel ........... 6-21
Casehardening ............................................ 6-23
Thermomechanical Treatment ............................... 6-25
Commercial Steels ........................................ 6-25
Tool Steels .............................................. 6-33
Spring Steel ............................................. 6-35
Stainless Steels ......................................... 6-36
Special-Alloy Steels ..................................... 6-38
Wire, Sheets, and Bars ................................... 6-42

## IRON AND STEEL CASTINGS
### by Charles Willers Briggs
Classification of Castings ............................... 6-47
Cast Iron ................................................ 6-47
Malleable-Iron Castings .................................. 6-51
Steel Castings ........................................... 6-52

## NONFERROUS METALS
### by L. D. Kunsman and C. L. Carlson
Copper and Its Alloys .................................... 6-63
Brasses .................................................. 6-68
Bronzes .................................................. 6-68
Processing of Copper Alloys .............................. 6-69
Copper-Base Alloys for Casting ........................... 6-70
Nickel and Nickel Alloys ................................. 6-70
Aluminum and Its Alloys .................................. 6-75
Magnesium and Magnesium Alloys ........................... 6-81
Zinc and Zinc Alloys ..................................... 6-85
Titanium and Zirconium Alloys ............................ 6-87
Low-Melting Metals and Alloys ............................ 6-87
Bearing Metals ........................................... 6-91
Metals and Alloys for Use at Elevated Temperatures ....... 6-94
Metals and Alloys for Atomic-Energy Applications ......... 6-100
Jewelry Metals ........................................... 6-103
Powder Metallurgy ........................................ 6-104
Cemented Carbides ........................................ 6-104

## CORROSION
### by Hilary E. Bacon
General Considerations ................................... 6-106
Factors in Electrochemical Corrosion ..................... 6-107
Corrosion Resistance of Various Metals ................... 6-108
Corrosion Problems in Steam-Generating Plants ............ 6-109
Corrosion of Equipment in Chemical Plants ................ 6-112
Corrosion of Pipes and Structures ........................ 6-113
Methods for Minimizing Corrosion ......................... 6-114

### PAINTS AND PROTECTIVE COATINGS
#### by George G. Sward

Paint Ingredients ........................................ 6-118
Paints .................................................. 6-119
Varnish ................................................. 6-121

### WOOD
#### by Carl H. de Zeeuw

Composition, Structure, and Nomenclature ................ 6-122
Physical Properties of Wood ............................ 6-122
Strength and Elastic Properties of Wood ................ 6-126
Glued-Laminated Wood .................................. 6-132
Plywood ................................................ 6-135
Preservative Treatment of Wood ........................ 6-136
Commercial Lumber Standards ........................... 6-138

### NONMETALLIC MATERIALS
#### revised by Antonio F. Baldo

Abrasives ............................................... 6-138
Tools ................................................... 6-139
Adhesives (BY ANTONIO F. BALDO) ........................ 6-140
Brick, Block, and Tile ................................. 6-141
Ceramics (BY ROBERT S. FEIGELSON) ..................... 6-149
Cleansing Materials (BY DOUGLAS MELZARD) .............. 6-150
Cordage ................................................. 6-151
Electrical Insulating Materials (BY JOHN T. BLAKE) ......... 6-151
Fibers and Fabrics (BY MILTON M. PLATT AND ROBERT E. SEBRING, REVISED BY ANTONIO F. BALDO) ............... 6-153
Freezing Preventives ................................... 6-156
Glass ................................................... 6-157
Natural Stones (BY DAVID ROSTOKER) .................... 6-158
Paper (BY R. CLAIRE CANTY AND F. G. PERRY, JR., REVISED BY ANTONIO F. BALDO) ................................... 6-159
Plastics (BY HARRIS J. BIXLER) ......................... 6-160

Roofing Materials ....................................... 6-163
Rubber and Rubberlike Materials (BY C. C. DAVIS) ......... 6-165
Solvents (BY C. G. HARFORD) ............................ 6-168
Thermal Insulations (BY PETER E. GLASER) ............... 6-169
Silicones ............................................... 6-170
Refractories (BY F. H. NORTON) ......................... 6-171

### CEMENT, MORTAR, AND CONCRETE
#### by William L. Gamble

Cement ................................................. 6-177
Lime ................................................... 6-179
Aggregates ............................................. 6-179
Water .................................................. 6-180
Admixtures ............................................. 6-180
Mortars ................................................ 6-180
Concrete ............................................... 6-183

### WATER
#### by Kenneth A. Roe

Water Resources ........................................ 6-189
Measurements and Definitions ........................... 6-190
Industrial Water ....................................... 6-191
Water Pollution ........................................ 6-192
Water Desalinization ................................... 6-192

### LUBRICANTS AND LUBRICATION
#### by Fred J. Villforth, Jr.

Lubricants ............................................. 6-196
Liquid Lubricants ...................................... 6-196
Solid Lubricants ....................................... 6-200
Animal, Vegetable, and Fish Oils ....................... 6-202
Greases ................................................ 6-202
Lubrication ............................................ 6-203

# GENERAL PROPERTIES OF MATERIALS

## Howard S. Bean

REFERENCES: "International Critical Tables," McGraw-Hill. "Smithsonian Physical Tables," Smithsonian Institution. Landolt, "Landolt-Börnstein, Zahlenwerte und Funktionen aus Physik, Chemie, Astronomie, Geophysik und Technik," Springer. "Handbook of Chemistry and Physics," Chemical Rubber Co. "Book of ASTM Standards," ASTM. "ASRE Refrigerating Data Book," ASRE. Brady, "Materials Handbook," McGraw-Hill. Mantell, "Engineering Materials Handbook," McGraw-Hill. International Union of Pure and Applied Chemistry, Butterworth Scientific Publications. "U.S. Standard Atmosphere," Government Printing Office, 1962. Tables of Thermodynamic Properties of Gases, *NBS Circ.* 564, 1960. ASME Steam Tables, 1967, ASME.

## CHEMISTRY

Every **elementary substance** is made up of exceedingly small particles called **atoms** which are all alike and which cannot be further subdivided or broken up by **chemical processes**. It will be noted that this statement is virtually a definition of the term elementary substance and a limitation of the term chemical process. There are as many different classes or families of atoms as there are **chemical elements**.

Two or more atoms, either of the same kind or of different kinds, are, in the case of most elements, capable of uniting with one another to form a higher order of distinct particles called **molecules**. If the molecules or atoms of which any given material is composed are all exactly alike, the material is a **pure substance**. If they are not all alike, the material is a **mixture**.

If the atoms which compose the molecules of any pure substances are all of the same kind, the substance is, as already stated, an **elementary substance**. If the atoms which compose the molecules of a pure chemical substance are not all of the same kind, the substance is a **compound substance**. The atoms are to

## Chemical Elements[a]

| Element | Symbol | Atomic No. | Atomic weight[b] | Valence |
|---|---|---|---|---|
| Actinium................... | Ac | 89 | | |
| Aluminum.............. .... | Al | 13 | 26.9815 | 3 |
| Americium.................. | Am | 95 | | |
| Antimony.................. | Sb | 51 | 121.75 | 3, 5 |
| Argon[c]..................... | Ar | 18 | 39.948 | 0 |
| Arsenic[d]................... | As | 33 | 74.9216 | 3, 5 |
| Astatine................... | At | 85 | | |
| Barium................... | Ba | 56 | 137.34 | 2 |
| Berkelium.................. | Bk | 97 | | |
| Beryllium.................. | Be | 4 | 9.0122 | 2 |
| Bismuth................... | Bi | 83 | 208.980 | 3, 5 |
| Boron[d]................... | B | 5 | 10.811[l] | 3 |
| Bromine[e].................. | Br | 35 | 79.904[m] | 1, 3, 5 |
| Cadmium.................. | Cd | 48 | 112.40 | 2 |
| Calcium................... | Ca | 20 | 40.08 | 2 |
| Californium................ | Cf | 98 | | |
| Carbon[d]................... | C | 6 | 12.01115[l] | 2, 4 |
| Cerium.................... | Ce | 58 | 140.12 | 3, 4 |
| Cesium[k]................... | Cs | 55 | 132.905 | 1 |
| Chlorine[f]................. | Cl | 17 | 35.453[m] | 1, 3, 5, 7 |
| Chromium................. | Cr | 24 | 51.996[m] | 2, 3, 6 |
| Cobalt.................... | Co | 27 | 58.9332 | 2, 3 |
| Columbium (see Niobium).... | | | | |
| Copper.................... | Cu | 29 | 63.546[m] | 1, 2 |
| Curium.................... | Cm | 96 | | |
| Dysprosium................ | Dy | 66 | 162.50 | 3 |
| Einsteinium................ | Es | 99 | | |
| Erbium.................... | Er | 68 | 167.26 | 3 |
| Europium.................. | Eu | 63 | 151.96 | 2, 3 |
| Fermium................... | Fm | 100 | | |
| Fluorine[g]................. | F | 9 | 18.9984 | 1 |
| Francium ................. | Fr | 87 | | |
| Gadolinium................ | Gd | 64 | 157.25 | 3 |
| Gallium[k]................. | Ga | 31 | 69.72 | 2, 3 |
| Germanium................ | Ge | 32 | 72.59 | 2, 4 |
| Gold..................... | Au | 79 | 196.967 | 1, 3 |
| Hafnium.................. | Hf | 72 | 178.49 | 4 |
| Helium[c].................. | He | 2 | 4.0026 | 0 |
| Holmium.................. | Ho | 67 | 164.930 | 3 |
| Hydrogen[h]............... | H | 1 | 1.00797[l] | 1 |
| Indium................... | In | 49 | 114.82 | 1, 2, 3 |
| Iodine[d].................. | I | 53 | 126.9044 | 1, 3, 5, 7 |

## Chemical Elements[a]—(Continued)

| Element | Symbol | Atomic No. | Atomic weight[b] | Valence |
|---|---|---|---|---|
| Iridium | Ir | 77 | 192.2 | 2, 3, 4, 6 |
| Iron | Fe | 26 | 55.847[m] | 2, 3 |
| Krypton[c] | Kr | 36 | 83.80 | 0 |
| Lanthanum | La | 57 | 138.91 | 3 |
| Lead | Pb | 82 | 207.19 | 2, 4 |
| Lithium[i] | Li | 3 | 6.939 | 1 |
| Lutetium | Lu | 71 | 174.97 | 3 |
| Magnesium | Mg | 12 | 24.312 | 2 |
| Manganese | Mn | 25 | 54.9380 | 2, 3, 4, 6, 7 |
| Mendelevium | Md | 101 | | |
| Mercury[e] | Hg | 80 | 200.59 | 1, 2 |
| Molybdenum | Mo | 42 | 95.94 | 3, 4, 5, 6 |
| Neodymium | Nd | 60 | 144.24 | 3 |
| Neon[c] | Ne | 10 | 20.183 | 0 |
| Neptunium | Np | 93 | | |
| Nickel | Ni | 28 | 58.71 | 2, 3, 4 |
| Niobium | Nb | 41 | 92.906 | 2, 3, 4, 5 |
| Nitrogen[f] | N | 7 | 14.0067 | 3, 5 |
| Nobelium | No | 102 | | |
| Osmium | Os | 76 | 190.2 | 2, 3, 4, 6, 8 |
| Oxygen[f] | O | 8 | 15.9994[l] | 2 |
| Palladium | Pd | 46 | 106.4 | 2, 4 |
| Phosphorus[d] | P | 15 | 30.9738 | 3, 5 |
| Platinum | Pt | 78 | 195.09 | 2, 4 |
| Plutonium | Pu | 94 | | |
| Polonium | Po | 84 | | 2, 4 |
| Potassium | K | 19 | 39.102 | 1 |
| Praseodymium | Pr | 59 | 140.907 | 3 |
| Promethium | Pm | 61 | | 5 |
| Protactinium | Pa | 91 | | |
| Radium | Ra | 88 | | 2 |
| Radon[j] | Rn | 86 | | 0 |
| Rhenium | Re | 75 | 186.2 | 1, 4, 7 |
| Rhodium | Rh | 45 | 102.905 | 3, 4 |
| Rubidium | Rb | 37 | 85.47 | 1 |
| Ruthenium | Ru | 44 | 101.07 | 3, 4, 6, 8 |
| Samarium | Sm | 62 | 150.35 | 3 |
| Scandium | Sc | 21 | 44.956 | 3 |
| Selenium[d] | Se | 34 | 78.96 | 2, 4, 6 |
| Silicon[d] | Si | 14 | 28.086[l] | 4 |
| Silver | Ag | 47 | 107.868[m] | 1 |
| Sodium | Na | 11 | 22.9898 | 1 |
| Strontium | Sr | 38 | 87.62 | 2 |
| Sulphur[d] | S | 16 | 32.064[l] | 2, 4, 6 |
| Tantalum | Ta | 73 | 180.948 | 4, 5 |
| Technetium | Tc | 43 | | |
| Tellurium[d] | Te | 52 | 127.60 | 2, 4, 6 |
| Terbium | Tb | 65 | 158.924 | 3 |
| Thallium | Tl | 81 | 204.37 | 1, 3 |
| Thorium | Th | 90 | 232.038 | 3 |
| Thulium | Tm | 69 | 168.934 | 3 |
| Tin | Sn | 50 | 118.69 | 2, 4 |
| Titanium | Ti | 22 | 47.90 | 3, 4 |
| Tungsten | W | 74 | 183.85 | 3, 4, 5, 6 |
| Uranium | U | 92 | 238.03 | 4, 6 |
| Vanadium | V | 23 | 50.942 | 1, 2, 3, 4, 5 |
| Xenon[c] | Xe | 54 | 131.30 | 0 |
| Ytterbium | Yb | 70 | 173.04 | 2, 3 |
| Yttrium | Y | 39 | 88.905 | 3 |
| Zinc | Zn | 30 | 65.37 | 2 |
| Zirconium | Zr | 40 | 91.22 | 4 |

Note: Table courtesy IUPAC and Butterworth Scientific Publications.

[a]All the elements for which atomic weights are listed are metals, except as otherwise indicated. No atomic weights are listed for most radioactive elements, as these elements have no fixed value.

[b]The atomic weights are based upon nuclidic mass of $C^{12} = 12$.

[c]Inert gas. [d]Metalloid. [e]Liquid. [f]Gas. [g]Most active gas. [h]Lightest gas. [i]Lightest metal. [j]Not placed. [k]Liquid at 25°C.

[l]The atomic weight varies because of natural variations in the isotopic composition of the element. The observed ranges are boron, ±0.003; carbon, ±0.00005; hydrogen, ±0.00001; oxygen, ±0.0001; silicon, ±0.001; sulfur, ±0.003.

[m]The atomic weight is believed to have an experimental uncertainty of the following magnitude: bromine, ±0.001; chlorine, ±0.001; chromium, ±0.001; copper, ±0.001; iron, ±0.003; silver, ±0.001. For other elements, the last digit given is believed to be reliable to ±0.5.

be considered as the smallest particles which occur separately in the structure of molecules of either compound or elementary substances, so far as can be determined by ordinary chemical analysis. The molecule of an element consists of a definite (usually small) number of its atoms. The molecule of a compound consists of one or more atoms of each of its several elements, the numbers of the various kinds of atoms and their arrangement being definite and fixed and determining the character of the compound. This notion of molecules and their constituent atoms is useful for interpreting the observed fact that chemical reactions—e.g., the analysis of a compound into its elements, the synthesis of a compound from the elements, or the changing of one or more compounds into one or more different compounds—take place so that the masses of the various substances concerned in a given reaction stand in definite and fixed ratios.

It appears from recent researches that some substances which cannot by any available means be decomposed into simpler substances and which must, therefore, be defined as elements, are continually undergoing spontaneous changes or **radioactive transformation** into other substances which can be recognized as physically and chemically different from the original substance. Radium is an element by the definition given and may be considered as made up of atoms. But it is assumed that these atoms, so called because they resist all efforts to break them up and are, therefore, apparently indivisible, nevertheless split up spontaneously, at a rate which scientists have not been able to influence in any way, into other atoms, thus forming other elementary substances of totally different properties.

The view generally accepted at present is that the atoms of all the chemical elements, including those not yet known to be

**Solubility of Inorganic Substances in Water**

(Number of grams of the anhydrous substance soluble in 1,000 g of water. The common name of the substance is given in parentheses)

| | Composition | Temperature, °F (°C) | | |
|---|---|---|---|---|
| | | 32 (0) | 122 (50) | 212 (100) |
| Aluminum sulfate | $Al_2(SO_4)_3$ | 313 | 521 | 891 |
| Aluminum potassium sulfate (potassium alum) | $Al_2K_2(SO_4)_4 \cdot 24H_2O$ | 30 | 170 | 1540 |
| Ammonium bicarbonate | $NH_4HCO_3$ | 119 | | |
| Ammonium chloride (sal ammoniac) | $NH_4Cl$ | 297 | 504 | 760 |
| Ammonium nitrate | $NH_4NO_3$ | 1183 | 3440 | 8710 |
| Ammonium sulfate | $(NH_4)_2SO_4$ | 706 | 847 | 1033 |
| Barium chloride | $BaCl_2 \cdot 2H_2O$ | 317 | 436 | 587 |
| Barium nitrate | $Ba(NO_3)_2$ | 50 | 172 | 345 |
| Calcium carbonate (calcite) | $CaCO_3$ | 0.018* | | 0.88 |
| Calcium chloride | $CaCl_2$ | 594 | | 1576 |
| Calcium hydroxide (hydrated lime) | $Ca(OH)_2$ | 1.77 | | 0.67 |
| Calcium nitrate | $Ca(NO_3)_2 \cdot 4H_2O$ | 931 | 3561 | 3626 |
| Calcium sulfate (gypsum) | $CaSO_4 \cdot 2H_2O$ | 1.76 | 2.06 | 1.69 |
| Copper sulfate (blue vitriol) | $CuSO_4 \cdot 5H_2O$ | 140 | 334 | 753 |
| Ferrous chloride | $FeCl_2 \cdot 4H_2O$ | 644§ | 820 | 1060 |
| Ferrous hydroxide | $Fe(OH)_2$ | 0.0067‡ | | |
| Ferrous sulfate (green vitriol or copperas) | $FeSO_4 \cdot 7H_2O$ | 156 | 482 | |
| Ferric chloride | $FeCl_3$ | 730 | 3160 | 5369 |
| Lead chloride | $PbCl_2$ | 6.73 | 16.7 | 33.3 |
| Lead nitrate | $Pb(NO_3)_2$ | 403 | | 1255 |
| Lead sulfate | $PbSO_4$ | 0.042† | | |
| Magnesium carbonate | $MgCO_3$ | 0.13‡ | | |
| Magnesium chloride | $MgCl_2 \cdot 6H_2O$ | 524 | | 723 |
| Magnesium hydroxide (milk of magnesia) | $Mg(OH)_2$ | 0.009‡ | | |
| Magnesium nitrate | $Mg(NO_3)_2 \cdot 6H_2O$ | 665 | 903 | |
| Magnesium sulfate (Epsom salts) | $MgSO_4 \cdot 7H_2O$ | 269 | 500 | 710 |
| Potassium carbonate (potash) | $K_2CO_3$ | 893 | 1216 | 1562 |
| Potassium chloride | $KCl$ | 284 | 435 | 566 |
| Potassium hydroxide (caustic potash) | $KOH$ | 971 | 1414 | 1773 |
| Potassium nitrate (saltpeter or niter) | $KNO_3$ | 131 | 851 | 2477 |
| Potassium sulfate | $K_2SO_4$ | 74 | 165 | 241 |
| Sodium bicarbonate (baking soda) | $NaHCO_3$ | 69 | 145 | |
| Sodium carbonate (sal soda or soda ash) | $NaCO_3 \cdot 10H_2O$ | 204 | 475 | 452 |
| Sodium chloride (common salt) | $NaCl$ | 357 | 366 | 392 |
| Sodium hydroxide (caustic soda) | $NaOH$ | 420 | 1448 | 3388 |
| Sodium nitrate (Chile saltpeter) | $NaNO_3$ | 733 | 1148 | 1755 |
| Sodium sulfate (Glauber salts) | $Na_2SO_4 \cdot 10H_2O$ | 49 | 466 | 422 |
| Zinc chloride | $ZnCl_2$ | 2044 | 4702 | 6147 |
| Zinc nitrate | $Zn(NO_3)_2 \cdot 6H_2O$ | 947 | | |
| Zinc sulfate | $ZnSO_4 \cdot 7H_2O$ | 419 | 768 | 807 |

*59°F.
†68°F.
‡In cold water.
§50°F.

# Periodic table of the elements

Legend:
- Atomic number
- Symbol
- Element
- Atomic weight based on $C^{12} = 12.00$
- Valence
- ( ) denotes mass number of most stable known isotope

Category groupings shown on the chart: Light metals, Brittle metals, Ductile metals, Low melting, Nonmetallic elements, Inert gases, Rare earth elements, Lanthanide series, Actinide series, Transuranium elements.

| Atomic No. | Symbol | Element | Atomic weight | Valence |
|---|---|---|---|---|
| 1 | H | Hydrogen | 1.00797 | 1 |
| 2 | He | Helium | 4.0026 | 0 |
| 3 | Li | Lithium | 6.939 | 1 |
| 4 | Be | Beryllium | 9.0122 | 2 |
| 5 | B | Boron | 10.811 | 3 |
| 6 | C | Carbon | 12.01115 | 2,4 |
| 7 | N | Nitrogen | 14.0067 | 3,5 |
| 8 | O | Oxygen | 15.9994 | 2 |
| 9 | F | Fluorine | 18.9984 | 1 |
| 10 | Ne | Neon | 20.183 | 0 |
| 11 | Na | Sodium | 22.9898 | 1 |
| 12 | Mg | Magnesium | 24.312 | 2 |
| 13 | Al | Aluminum | 26.9815 | 3 |
| 14 | Si | Silicon | 28.086 | 4 |
| 15 | P | Phosphorus | 30.9738 | 3,5 |
| 16 | S | Sulphur | 32.064 | 2,4,6 |
| 17 | Cl | Chlorine | 35.453 | 1,3,5,7 |
| 18 | Ar | Argon | 39.948 | 0 |
| 19 | K | Potassium | 39.102 | 1 |
| 20 | Ca | Calcium | 40.08 | 2 |
| 21 | Sc | Scandium | 44.956 | 3 |
| 22 | Ti | Titanium | 47.90 | 3,4 |
| 23 | V | Vanadium | 50.942 | 1,2,3,4,5 |
| 24 | Cr | Chromium | 51.996 | 2,3,6 |
| 25 | Mn | Manganese | 54.938 | 2,3,4,6,7 |
| 26 | Fe | Iron | 55.847 | 2,3 |
| 27 | Co | Cobalt | 58.9332 | 2,3 |
| 28 | Ni | Nickel | 58.71 | 2,3,4 |
| 29 | Cu | Copper | 63.546 | 1,2 |
| 30 | Zn | Zinc | 65.37 | 2 |
| 31 | Ga | Gallium | 69.72 | 2,3 |
| 32 | Ge | Germanium | 72.59 | 2,4 |
| 33 | As | Arsenic | 74.9216 | 3,5 |
| 34 | Se | Selenium | 78.96 | 2,4,6 |
| 35 | Br | Bromine | 79.904 | 1,3,5 |
| 36 | Kr | Krypton | 83.80 | 0 |
| 37 | Rb | Rubidium | 85.47 | 1 |
| 38 | Sr | Strontium | 87.62 | 2 |
| 39 | Y | Yttrium | 88.905 | 3 |
| 40 | Zr | Zirconium | 91.22 | 4 |
| 41 | Nb | Niobium | 92.906 | 2,3,4,5 |
| 42 | Mo | Molybdenum | 95.94 | 3,4,5,6 |
| 43 | Tc | Technetium | (99) | |
| 44 | Ru | Ruthenium | 101.07 | 3,4,6,8 |
| 45 | Rh | Rhodium | 103.905 | 3,4 |
| 46 | Pd | Palladium | 106.4 | 2,4 |
| 47 | Ag | Silver | 107.868 | 1 |
| 48 | Cd | Cadmium | 112.40 | 2 |
| 49 | In | Indium | 114.82 | 1,2,3 |
| 50 | Sn | Tin | 118.69 | 2,4 |
| 51 | Sb | Antimony | 121.75 | 3,5 |
| 52 | Te | Tellurium | 127.60 | 2,4,6 |
| 53 | I | Iodine | 126.9044 | 1,3,5,7 |
| 54 | Xe | Xenon | 131.30 | 0 |
| 55 | Cs | Cesium | 132.905 | 1 |
| 56 | Ba | Barium | 137.34 | 2 |
| 57 | La | Lanthanum | 137.91 | 3 |
| 72 | Hf | Hafnium | 178.49 | 4 |
| 73 | Ta | Tantalum | 180.948 | 4,5 |
| 74 | W | Tungsten | 183.85 | 3,4,5,6 |
| 75 | Re | Rhenium | 186.2 | 1,4,7 |
| 76 | Os | Osmium | 190.2 | 2,3,4,6,8 |
| 77 | Ir | Iridium | 192.2 | 2,3,4,6 |
| 78 | Pt | Platinum | 195.09 | 2,4 |
| 79 | Au | Gold | 196.967 | 1,3 |
| 80 | Hg | Mercury | 200.59 | 1,2 |
| 81 | Tl | Thallium | 204.37 | 1,3 |
| 82 | Pb | Lead | 207.19 | 2,4 |
| 83 | Bi | Bismuth | 208.98 | 3,5 |
| 84 | Po | Polonium | (210) | 2,4 |
| 85 | At | Astatine | (210) | |
| 86 | Rn | Radon | (212) | 0 |
| 87 | Fr | Francium | (223) | |
| 88 | Ra | Radium | (227) | 2 |
| 89 | Ac | Actinium | (227) | 3 |

## Lanthanide series

| Atomic No. | Symbol | Element | Atomic weight | Valence |
|---|---|---|---|---|
| 58 | Ce | Cerium | 140.12 | 3,4 |
| 59 | Pr | Praseodymium | 140.907 | 3 |
| 60 | Nd | Neodymium | 144.24 | 3 |
| 61 | Pm | Promethium | (147) | 5 |
| 62 | Sm | Samarium | 150.35 | 3 |
| 63 | Eu | Europium | 151.96 | 2,3 |
| 64 | Gd | Gadolinium | 157.25 | 3 |
| 65 | Tb | Terbium | 158.924 | 3 |
| 66 | Dy | Dysprosium | 162.50 | 3 |
| 67 | Ho | Holmium | 164.93 | 3 |
| 68 | Er | Erbium | 167.26 | 3 |
| 69 | Tm | Thulium | 168.934 | 3 |
| 70 | Yb | Ytterbium | 173.04 | 2,3 |
| 71 | Lu | Lutetium | 174.97 | 3 |

## Actinide series

| Atomic No. | Symbol | Element | Atomic weight | Valence |
|---|---|---|---|---|
| 90 | Th | Thorium | 232.038 | 4 |
| 91 | Pa | Protactinium | (231) | |
| 92 | U | Uranium | 236.03 | |
| 93 | Np | Neptunium | (237) | |
| 94 | Pu | Plutonium | (242) | |
| 95 | Am | Americium | (243) | |
| 96 | Cm | Curium | (245) | |
| 97 | Bk | Berkelium | (249) | |
| 98 | Cf | Californium | (249) | |
| 99 | Es | Einsteinium | (254) | |
| 100 | Fm | Fermium | (252) | |
| 101 | Md | Mendelevium | (256) | |
| 102 | No | Nobelium | (254) | |
| 103 | Lw | Lawrenium | (257) | |

**Solubility of Gases in Water**
(By volume. at atmospheric pressure)

| $t$, °F (°C) | 32 (0) | 68 (20) | 212 (100) | $t$, °F (°C) | 32 (0) | 68 (20) | 212 (100) |
|---|---|---|---|---|---|---|---|
| Air | 0.032 | 0.020 | 0.012 | Hydrogen | 0.023 | 0.020 | 0.018 |
| Acetylene | 1.89 | 1.12 | | Hydrogen sulfide | 5.0 | 2.8 | 0.87 |
| Ammonia | 1250 | 700 | | Hydrochloric acid | 560 | 480 | |
| Carbon dioxide | 1.87 | 0.96 | 0.26 | Nitrogen | 0.026 | 0.017 | 0.0105 |
| Carbon monoxide | 0.039 | 0.025 | | Oxygen | 0.053 | 0.034 | 0.0185 |
| Chlorine | 5.0 | 2.5 | 0.00 | Sulfuric acid | 87 | 43 | |

radioactive, consist of several kinds of still smaller particles, three of which are known as **protons, neutrons,** and **electrons.** The protons are bound together in the atomic nucleus with other particles, including neutrons, and are positively charged. The neutrons are particles having approximately the mass of a proton but are uncharged. The electrons are negatively charged particles, all alike, external to the nucleus, and sufficient in number to neutralize the nuclear charge in an atom. The differences between the atoms of different chemical elements are due to the different numbers of these smaller particles composing them. According to the original Bohr theory, an ordinary atom is conceived as a stable system of such electrons revolving in closed orbits about the nucleus like the planets of the solar system around the sun. In a hydrogen atom, there is 1 proton and 1 electron; in a radium atom, there are 88 electrons surrounding a nucleus 226 times as massive as the hydrogen nucleus. Only a few, in general the outermost or valence electrons of such an atom, are subject to rearrangement within, or ejection from, the atom, thereby enabling it, because of its increased energy, to combine with other atoms to form molecules of either elementary substances or compounds. The **atomic number** of an element is the number of excess positive charges on the nucleus of the atom. The essential feature that distinguishes one element from another is this charge of the nucleus. It also determines the position of the element in the periodic table. Modern researches have shown the existence of **isotopes,** that is, two or more species of atoms having the same atomic number and thus occupying the same place in the periodic system, but differing somewhat in atomic weight. These isotopes are chemically identical and are merely different species of the same chemical element. Most of the ordinary inactive elements have been shown to consist of a mixture of isotopes. This convenient atomic model should be regarded as only a working hypothesis for coordinating a number of phenomena about which much yet remains to be known.

**Calculation of the Percentage Composition of Substances** Add together the atomic weights of the elements in the compound to obtain its molecular weight. Multiply the atomic weight of the element to be calculated by the number of atoms present (indicated in the formula by a subscript number) and by 100, and divide by the molecular weight of the compound. For example, hematite iron ore ($Fe_2O_3$) contains 69.94 percent of iron by weight, determined as follows: Molecular weight of $Fe_2O_3 = (55.84 \times 2) + (16 \times 3) = 159.68$. Percentage of iron in compound $= (55.84 \times 2) \times 100/159.68 = 69.94$.

## SPECIFIC GRAVITIES AND DENSITIES

### Approximate Specific Gravities and Densities
(Water at 39°F and normal atmospheric pressure taken as unity) For more detailed data on any material, see the section dealing with the properties of that material. Data given are for usual room temperatures.

| Substance | Specific gravity | Avg density lb/ft³ | Avg density kg/m³ |
|---|---|---|---|
| **Metals, Alloys, Ores** | | | |
| Aluminum, cast-hammered | 2.55–2.80 | 165 | 2,643 |
| Brass, cast-rolled | 8.4–8.7 | 534 | 8,553 |
| Bronze, aluminum | 7.7 | 481 | 7,702 |
| Bronze, 7.9–14% Sn | 7.4–8.9 | 509 | 8,153 |
| Bronze, phosphor | 8.88 | 554 | 8,874 |
| Copper, cast-rolled | 8.8–8.95 | 556 | 8,906 |
| Copper ore, pyrites | 4.1–4.3 | 262 | 4,197 |
| German silver | 8.58 | 536 | 8,586 |
| Gold, cast-hammered | 19.25–19.35 | 1205 | 19,300 |
| Gold coin (U.S.) | 17.18–17.2 | 1073 | 17,190 |
| Iridium | 21.78–22.42 | 1383 | 22,160 |
| Iron, gray cast | 7.03–7.13 | 442 | 7,079 |
| Iron, cast, pig | 7.2 | 450 | 7,207 |
| Iron, wrought | 7.6–7.9 | 485 | 7,658 |
| Iron, spiegeleisen | 7.5 | 468 | 7,496 |
| Iron, ferrosilicon | 6.7–7.3 | 437 | 6,984 |
| Iron ore, hematite | 5.2 | 325 | 5,206 |
| Iron ore, limonite | 3.6–4.0 | 237 | 3,796 |
| Iron ore, magnetite | 4.9–5.2 | 315 | 5,046 |
| Iron slag | 2.5–3.0 | 172 | 2,755 |
| Lead | 11.34 | 710 | 11,370 |
| Lead ore, galena | 7.3–7.6 | 465 | 7,449 |
| Manganese | 7.42 | 475 | 7,608 |
| Manganese ore, pyrolusite | 3.7–4.6 | 259 | 4,149 |
| Mercury | 13.546 | 847 | 13,570 |
| Monel metal, rolled | 8.97 | 555 | 8,688 |
| Nickel | 8.9 | 537 | 8,602 |
| Platinum, cast-hammered | 21.5 | 1330 | 21,300 |
| Silver, cast-hammered | 10.4–10.6 | 656 | 10,510 |
| Steel, cold-drawn | 7.83 | 489 | 7,832 |
| Steel, machine | 7.80 | 487 | 7,800 |
| Steel, tool | 7.70–7.73 | 481 | 7,703 |
| Tin, cast-hammered | 7.2–7.5 | 459 | 7,352 |
| Tin ore, cassiterite | 6.4–7.0 | 418 | 6,695 |
| Tungsten | 19.22 | 1200 | 18,820 |
| Uranium | 18.7 | 1170 | 18,740 |
| Zinc, cast-rolled | 6.9–7.2 | 440 | 7,049 |
| Zinc, ore, blende | 3.9–4.2 | 253 | 4,052 |
| **Various Solids** | | | |
| Cereals, oats, bulk | 0.41 | 26 | 417 |

| Substance | Specific gravity | Avg density lb/ft³ | kg/m³ |
|---|---|---|---|
| Cereals, barley, bulk | 0.62 | 39 | 625 |
| Cereals, corn, rye, bulk | 0.73 | 45 | 721 |
| Cereals, wheat, bulk | 0.77 | 48 | 769 |
| Cork | 0.22–0.26 | 15 | 240 |
| Cotton, flax, hemp | 1.47–1.50 | 93 | 1,491 |
| Fats | 0.90–0.97 | 58 | 925 |
| Flour, loose | 0.40–0.50 | 28 | 448 |
| Flour, pressed | 0.70–0.80 | 47 | 753 |
| Glass, common | 2.40–2.80 | 162 | 2,595 |
| Glass, plate or crown | 2.45–2.72 | 161 | 2,580 |
| Glass, crystal | 2.90–3.00 | 184 | 1,950 |
| Glass, flint | 3.2–4.7 | 247 | 3,960 |
| Hay and straw, bales | 0.32 | 20 | 320 |
| Leather | 0.86–1.02 | 59 | 945 |
| Paper | 0.70–1.15 | 58 | 929 |
| Potatoes, piled | 0.67 | 44 | 705 |
| Rubber, caoutchouc | 0.92–0.96 | 59 | 946 |
| Rubber goods | 1.0–2.0 | 94 | 1,506 |
| Salt, granulated, piled | 0.77 | 48 | 769 |
| Saltpeter | 2.11 | 132 | 2,115 |
| Starch | 1.53 | 96 | 1,539 |
| Sulfur | 1.93–2.07 | 125 | 2,001 |
| Wool | 1.32 | 82 | 1,315 |
| **Timber, Air-Dry** | | | |
| Apple | 0.66–0.74 | 44 | 705 |
| Ash, black | 0.55 | 34 | 545 |
| Ash, white | 0.64–0.71 | 42 | 973 |
| Birch, sweet, yellow | 0.71–0.72 | 44 | 705 |
| Cedar, white, red | 0.35 | 22 | 352 |
| Cherry, wild red | 0.43 | 27 | 433 |
| Chestnut | 0.48 | 30 | 481 |
| Cypress | 0.45–0.48 | 29 | 465 |
| Fir, Douglas | 0.48–0.55 | 32 | 513 |
| Fir, balsam | 0.40 | 25 | 401 |
| Elm, white | 0.56 | 35 | 561 |
| Hemlock | 0.45–0.50 | 29 | 465 |
| Hickory | 0.74–0.80 | 48 | 769 |
| Locust | 0.67–0.77 | 45 | 722 |
| Mahogany | 0.56–0.85 | 44 | 705 |
| Maple, sugar | 0.68 | 43 | 689 |
| Maple, white | 0.53 | 33 | 529 |
| Oak, chestnut | 0.74 | 46 | 737 |
| Oak, live | 0.87 | 54 | 866 |
| Oak, red, black | 0.64–0.71 | 42 | 673 |
| Oak, white | 0.77 | 48 | 770 |
| Pine, Oregon | 0.51 | 32 | 513 |
| Pine, red | 0.48 | 30 | 481 |
| Pine, white | 0.43 | 27 | 433 |
| Pine, Southern | 0.61–0.67 | 38–42 | 610–673 |
| Pine, Norway | 0.55 | 34 | 541 |
| Poplar | 0.43 | 27 | 433 |
| Redwood, California | 0.42 | 26 | 417 |
| Spruce, white, red | 0.45 | 28 | 449 |
| Teak, African | 0.99 | 62 | 994 |
| Teak, Indian | 0.66–0.88 | 48 | 769 |
| Walnut, black | 0.59 | 37 | 593 |
| Willow | 0.42–0.50 | 28 | 449 |
| **Various Liquids** | | | |
| Alcohol, ethyl (100%) | 0.789 | 49 | 802 |
| Alcohol, methyl (100%) | 0.796 | 50 | 809 |
| Acid, muriatic, 40% | 1.20 | 75 | 1,201 |
| Acid, nitric, 91% | 1.50 | 94 | 1.506 |
| Acid, sulfuric, 87% | 1.80 | 112 | 1,795 |
| Chloroform | 1.500 | 95 | 1,532 |
| Ether | 0.736 | 46 | 738 |

| Substance | Specific gravity | Avg density lb/ft³ | kg/m³ |
|---|---|---|---|
| Lye, soda, 66% | 1.70 | 106 | 1,699 |
| Oils, vegetable | 0.91–0.94 | 58 | 930 |
| Oils, mineral, lubricants | 0.88–0.94 | 57 | 914 |
| Turpentine | 0.861–0.867 | 54 | 866 |
| **Various Liquids** | | | |
| Water, 4°C, max density | 1.0 | 62.426 | 999.97 |
| Water, 100°C | 0.9584 | 59.812 | 958.10 |
| Water, ice | 0.88–0.92 | 56 | 897 |
| Water, snow, fresh fallen | 0.125 | 8 | 128 |
| Water, seawater | 1.02–1.03 | 64 | 1,025 |
| **Gases,** see Sec. 4, Thermal Properties and Thermodynamics | | | |
| **Ashlar Masonry** | | | |
| Granite, syenite, gneiss | 2.4–2.7 | 159 | 2,549 |
| Limestone | 2.1–2.8 | 153 | 2,450 |
| Marble | 2.4–2.8 | 162 | 2,597 |
| Sandstone | 2.0–2.6 | 143 | 2,290 |
| Bluestone | 2.3–2.6 | 153 | 2,451 |
| **Rubble Masonry** | | | |
| Granite, syenite, gneiss | 2.3–2.6 | 153 | 2,451 |
| Limestone | 2.0–2.7 | 147 | 2,355 |
| Sandstone | 1.9–2.5 | 137 | 2,194 |
| Bluestone | 2.2–2.5 | 147 | 2,355 |
| Marble | 2.3–2.7 | 156 | 2,500 |
| **Dry Rubble Masonry** | | | |
| Granite, syenite, gneiss | 1.9–2.3 | 130 | 2,082 |
| Limestone, marble | 1.9–2.1 | 125 | 2,001 |
| Sandstone, bluestone | 1.8–1.9 | 110 | 1,762 |
| **Brick Masonry** | | | |
| Hard brick | 1.8–2.3 | 128 | 2,051 |
| Medium brick | 1.6–2.0 | 112 | 1,794 |
| Soft brick | 1.4–1.9 | 103 | 1,650 |
| Sand-lime brick | 1.4–2.2 | 112 | 1,794 |
| **Concrete Masonry** | | | |
| Cement, stone, sand | 2.2–2.4 | 144 | 2,309 |
| Cement, slag, etc. | 1.9–2.3 | 130 | 2,082 |
| Cement, cinder, etc. | 1.5–1.7 | 100 | 1,602 |
| **Various Building Materials** | | | |
| Ashes, cinders | 0.64–0.72 | 40–45 | 640–721 |
| Cement, portland, loose | 1.5 | 94 | 1,505 |
| Portland cement | 3.1–3.2 | 196 | 3,140 |
| Lime, gypsum, loose | 0.85–1.00 | 53–64 | 849–1,025 |
| Mortar, lime, set | 1.4–1.9 | 103 | 1,650 |
|  |  | 94 | 1,505 |
| Mortar, portland cement | 2.08–2.25 | 135 | 2,163 |
| Slags, bank slag | 1.1–1.2 | 67–72 | 1,074–1,153 |
| Slags, bank screenings | 1.5–1.9 | 98–117 | 1,570–1,874 |
| Slags, machine slag | 1.5 | 96 | 1,538 |
| Slags, slag sand | 0.8–0.9 | 49–55 | 785–849 |
| **Earth, etc., Excavated** | | | |
| Clay, dry | 1.0 | 63 | 1,009 |
| Clay, damp, plastic | 1.76 | 110 | 1,761 |
| Clay and gravel, dry | 1.6 | 100 | 1,602 |
| Earth, dry, loose | 1.2 | 76 | 1,217 |
| Earth, dry, packed | 1.5 | 95 | 1,521 |
| Earth, moist, loose | 1.3 | 78 | 1,250 |
| Earth, moist, packed | 1.6 | 96 | 1,538 |
| Earth, mud, flowing | 1.7 | 108 | 1,730 |
| Earth, mud, packed | 1.8 | 115 | 1,841 |
| Riprap, limestone | 1.3–1.4 | 80–85 | 1,282–1,361 |
| Riprap, sandstone | 1.4 | 90 | 1,441 |
| Riprap, shale | 1.7 | 105 | 1,681 |
| Sand, gravel, dry, loose | 1.4–1.7 | 90–105 | 1,441–1,681 |
| Sand, gravel, dry, packed | 1.6–1.9 | 100–120 | 1,602–1,922 |
| Sand, gravel, wet | 1.89–2.16 | 126 | 2,019 |

| Substance | Specific gravity | Avg density lb/ft³ | Avg density kg/m³ |
|---|---|---|---|
| **Excavations in Water** | | | |
| Sand or gravel | 0.96 | 60 | 951 |
| Sand or gravel and clay | 1.00 | 65 | 1,041 |
| Clay | 1.28 | 80 | 1,281 |
| River mud | 1.44 | 90 | 1,432 |
| Soil | 1.12 | 70 | 1,122 |
| Stone riprap | 1.00 | 65 | 1,041 |
| **Minerals** | | | |
| Asbestos | 2.1–2.8 | 153 | 2,451 |
| Barytes | 4.50 | 281 | 4,504 |
| Basalt | 2.7–3.2 | 184 | 2,950 |
| Bauxite | 2.55 | 159 | 2,549 |
| Bluestone | 2.5–2.6 | 159 | 2,549 |
| Borax | 1.7–1.8 | 109 | 1,746 |
| Chalk | 1.8–2.8 | 143 | 2,291 |
| Clay, marl | 1.8–2.6 | 137 | 2,196 |
| Dolomite | 2.9 | 181 | 2,901 |
| Feldspar, orthoclase | 2.5–2.7 | 162 | 2,596 |
| Gneiss | 2.7–2.9 | 175 | 2,805 |
| Granite | 2.6–2.7 | 165 | 2,644 |
| Greenstone, trap | 2.8–3.2 | 187 | 2,998 |
| Gypsum, alabaster | 2.3–2.8 | 159 | 2,549 |
| Hornblende | 3.0 | 187 | 2,998 |
| Limestone | 2.1–2.86 | 155 | 2,484 |
| Marble | 2.6–2.86 | 170 | 2,725 |
| Magnesite | 3.0 | 187 | 2,998 |
| Phosphate rock, apatite | 3.2 | 200 | 3,204 |
| Porphyry | 2.6–2.9 | 172 | 2,758 |
| Pumice, natural | 0.37–0.90 | 40 | 641 |
| Quartz, flint | 2.5–2.8 | 165 | 2,645 |
| Sandstone | 2.0–2.6 | 143 | 2,291 |
| Serpentine | 2.7–2.8 | 171 | 2,740 |
| Shale, slate | 2.6–2.9 | 172 | 2,758 |
| Soapstone, talc | 2.6 2.8 | 169 | 2,709 |
| Syenite | 2.6–2.7 | 165 | 2,645 |
| **Stone, Quarried, Piled** | | | |
| Basalt, granite, gneiss | 1.5 | 96 | 1,579 |
| Limestone, marble, quartz | 1.5 | 95 | 1,572 |
| Sandstone | 1.3 | 82 | 1,314 |
| Shale | 1.5 | 92 | 1,474 |
| Greenstone, hornblend | 1.7 | 107 | 1,715 |
| **Bituminous Substances** | | | |
| Asphaltum | 1.1–1.5 | 81 | 1,298 |
| Coal, anthracite | 1.4–1.8 | 97 | 1,554 |
| Coal, bituminous | 1.2–1.5 | 84 | 1,346 |

| Substance | Specific gravity | Avg density lb/ft³ | Avg density kg/m³ |
|---|---|---|---|
| Coal, lignite | 1.1–1.4 | 78 | 1,250 |
| Coal, peat, turf, dry | 0.65–0.85 | 47 | 753 |
| Coal, charcoal, pine | 0.28–0.44 | 23 | 369 |
| Coal, charcoal, oak | 0.47–0.57 | 33 | 481 |
| Coal, coke | 1.0–1.4 | 75 | 1,201 |
| Graphite | 1.64–2.7 | 135 | 2,163 |
| Paraffin | 0.87–0.91 | 56 | 898 |
| Petroleum | 0.87 | 54 | 856 |
| Petroleum, refined (kerosene) | 0.78–0.82 | 50 | 801 |
| Petroleum, benzine | 0.73–0.75 | 46 | 737 |
| Petroleum, gasoline | 0.70–0.75 | 45 | 721 |
| Pitch | 1.07–1.15 | 69 | 1,105 |
| Tar, bituminous | 1.20 | 75 | 1,201 |
| **Coal and Coke, Piled** | | | |
| Coal, anthracite | 0.75–0.93 | 47–58 | 753–930 |
| Coal, bituminous, lignite | 0.64–0.87 | 40–54 | 641–866 |
| Coal, peat, turf | 0.32–0.42 | 20–26 | 320–417 |
| Coal, charcoal | 0.16–0.23 | 10–14 | 160–224 |
| Coal, coke | 0.37–0.51 | 23–32 | 369–513 |

## Compressibility of Liquids

If $v_1$ and $v_2$ are the volumes of the liquids at pressures of $p_1$ and $p_2$ atm, respectively, at any temperature, the coefficient of compressibility $b$ is given by the equation

$$b = \frac{1}{v_1} \times \frac{v_1 - v_2}{p_2 - p_1}$$

The value of $b \times 10^6$ for oils at low pressures at about 70°F varies from about 55 to 80; for mercury at 32°F, it is 3.9; for chloroform at 32°F, it is 100 and increases with the temperature to 200 at 140°F; for ethyl alcohol, it increases from about 100 at 32°F and low pressures to 125 at 104°F; for glycerin, it is about 24 at room temperature and low pressure.

**Specific Gravity and Density of Water at Atmospheric Pressure**
(Weights are in vacuo)

| Temp, °C | Specific gravity | Density, lb/ft³ | Density, kg/m³ | Temp, °C | Specific gravity | Density, lb/ft³ | Density, kg/m³ |
|---|---|---|---|---|---|---|---|
| 0 | 0.99987 | 62.4183 | 999.845 | 40 | 0.99224 | 61.9428 | 992.228 |
| 2 | 0.99997 | 62.4246 | 999.946 | 42 | 0.99147 | 61.894 | 991.447 |
| 4 | 1.00000 | 62.4266 | 999.955 | 44 | 0.99066 | 61.844 | 990.647 |
| 6 | 0.99997 | 62.4246 | 999.946 | 46 | 0.98982 | 61.791 | 989.797 |
| 8 | 0.99988 | 62.4189 | 999.854 | 48 | 0.98896 | 61.737 | 988.931 |
| 10 | 0.99973 | 62.4096 | 999.706 | 50 | 0.98807 | 61.682 | 988.050 |
| 12 | 0.99952 | 62.3969 | 999.502 | 52 | 0.98715 | 61.624 | 987.121 |
| 14 | 0.99927 | 62.3811 | 999.272 | 54 | 0.98621 | 61.566 | 986.192 |
| 16 | 0.99897 | 62.3623 | 998.948 | 56 | 0.98524 | 61.505 | 985.215 |
| 18 | 0.99862 | 62.3407 | 998.602 | 58 | 0.98425 | 61.443 | 984.222 |
| 20 | 0.99823 | 62.3164 | 998.213 | 60 | 0.98324 | 61.380 | 983.213 |
| 22 | 0.99780 | 62.2894 | 997.780 | 62 | 0.98220 | 61.315 | 982.172 |
| 24 | 0.99732 | 62.2598 | 997.304 | 64 | 0.98113 | 61.249 | 981.113 |
| 26 | 0.99681 | 62.2278 | 996.793 | 66 | 0.98005 | 61.181 | 980.025 |
| 28 | 0.99626 | 62.1934 | 996.242 | 68 | 0.97894 | 61.112 | 978.920 |
| 30 | 0.99567 | 62.1568 | 995.656 | 70 | 0.97781 | 61.041 | 977.783 |
| 32 | 0.99505 | 62.1179 | 995.033 | 72 | 0.97666 | 60.970 | 976.645 |
| 34 | 0.99440 | 62.0770 | 994.378 | 74 | 0.97548 | 60.896 | 975.460 |
| 36 | 0.99371 | 62.0341 | 993.691 | 76 | 0.97428 | 60.821 | 974.259 |
| 38 | 0.99299 | 61.9893 | 992.973 | 78 | 0.97307 | 60.745 | 973.041 |

## PHYSICAL DATA

### Average Composition of Air between Sea Level and 90 km Altitude and Dry

| Element | Formula | % by vol. | % by mass | Molecular weight |
|---|---|---|---|---|
| Nitrogen | $N_2$ | 78.084 | 75.55 | 28.0134 |
| Oxygen | $O_2$ | 20.948 | 23.15 | 31.9988 |
| Argon | Ar | 0.934 | 1.325 | 39.948 |
| Carbon dioxide | $CO_2$ | 0.0314 | 0.0477 | 44.00995 |
| Neon | Ne | 0.00182 | 0.00127 | 20.183 |
| Helium | He | 0.00052 | 0.000072 | 4.0026 |
| Krypton | Kr | 0.000114 | 0.000409 | 83.80 |
| Methane | $CH_4$ | 0.0002 | 0.000111 | 16.043 |

From 0.0 to 0.00005 percent by volume of 9 other gases.
Average composite molecular weight of air 28.9644.
Data from "U.S. Standard Atmosphere, 1962," Government Printing Office.

### Volume of Water as a Function of Pressure and Temperature
(From "International Critical Tables")

| Temp, °F (°C) | Pressure in atmospheres | | | | | | | | |
|---|---|---|---|---|---|---|---|---|---|
| | 0 | 500 | 1,000 | 2,000 | 3,000 | 4,000 | 5,000 | | |
| 32(0) | 1.0000 | 0.9769 | 0.9566 | 0.9223 | 0.8954 | 0.8739 | 0.8565 | 0.8361 | |
| 68(20) | 1.0016 | 0.9804 | 0.9619 | 0.9312 | 0.9065 | 0.8855 | 0.8675 | 0.8444 | 0.8244 |
| 122(50) | 1.0128 | 0.9915 | 0.9732 | 0.9428 | 0.9183 | 0.8974 | 0.8792 | 0.8562 | 0.8369 |
| 176(80) | 1.0287 | 1.0071 | 0.9884 | 0.9568 | 0.9315 | 0.9097 | 0.8913 | 0.8679 | 0.8481 |

**Basic Properties of Several Metals**
(Staff contribution)*

| Material | Density,† g/cm³ | Coefficient of thermal expansion, in/(in)(°F) × 10⁻⁶ | Thermal conductivity, Btu/(h)(ft)(°F) | Specific heat, Btu/(lb)(°F) | Approx melting temp, °F | Modulus of elasticity lb/in² × 10⁶ | Poisson's ratio | Yield stress, lb/in² × 10³ | Ultimate stress, lb/in² × 10³ | Elongation, % |
|---|---|---|---|---|---|---|---|---|---|---|
| Aluminum, 2024-T3 | 2.77 | 12.6 | 110 | 0.23 | 940 | 10.6 | 0.33 | 50 | 70 | 18 |
| Aluminum, 6061-T6 | 2.70 | 13.5 | 90 | 0.23 | 1080 | 10.6 | 0.33 | 40 | 45 | 17 |
| Aluminum, 7079-T6 | 2.74 | 13.7 | 70 | 0.23 | 900 | 10.4 | 0.33 | 68 | 78 | 14 |
| Beryllium, QMV | 1.85 | 6.4–10.2 | 85 | 0.45 | 2340 | 40–44 | 0.024–0.030 | 27–38 | 33–51 | 1–3.5 |
| Copper, pure | 8.90 | 9.2 | 227 | 0.092 | 1980 | 17.0 | 0.42 | See "Metals Handbook" | 18 | 30 |
| Gold, pure | 19.32 | 9.2 | 172 | 0.031 | 1950 | 10.8 | 0.42 | | | |
| Lead, pure | 11.34 | 29.3 | 21.4 | 0.031 | 620 | 2.0 | 0.40–0.45 | 1.3 | 2.6 | 20–50 |
| Magnesium AZ31B-H24 (sheet) | 1.77 | 14.5 | 55 | 0.25 | 1100 | 6.5 | 0.35 | 22 | 37 | 15 |
| Magnesium, HK31A-H24 | 1.79 | 14.0 | 66 | 0.13 | 1100 | 6.4 | 0.35 | 29 | 37 | 8 |
| Molybdenum, wrought | 10.3 | 3.0 | 83 | 0.07 | 4730 | 40.0 | 0.32 | 80 | 120–200 | Small |
| Nickel, pure | 8.9 | 7.2 | 53 | 0.11 | 2650 | 32.0 | | See "Metals Handbook" | | |
| Platinum | 21.45 | 5.0 | 40 | 0.031 | 3217 | 21.3 | 0.39 | | 20–24 | 35–40 |
| Plutonium, alpha phase | 19.0–19.7 | 30.0 | 4.8 | 0.034 | 1184 | 14.0 | 0.15–0.21 | 40 | 60 | Small |
| Silver, pure | 10.5 | 11.0 | 241 | 0.056 | 1760 | 10–11 | 0.37 | 8 | 18 | 48 |
| Steel, AISI C1020 (hot-worked) | 7.85 | 8.4 | 27 | 0.10 | 2750 | 29–30 | 0.29 | 48 | 65 | 36 |
| Steel, AISI 304 (sheet) | 8.03 | 9.9 | 9.4 | 0.12 | 2600 | 28 | 0.29 | 39 | 87 | 65 |
| Tantalum | 16.6 | 3.6 | 31 | 0.03 | 5425 | 27.0 | 0.35 | | 50–145 | 1–40 |
| Thorium, induction melt | 11.6 | 6.95 | 21.7 | 0.03 | 3200 | 7–10 | 0.27 | 21 | 32 | 34 |
| Titanium, B 120VCA (aged) | 4.85 | 5.2 | 4.3 | 0.13 | 3100 | 14.8 | 0.3 | 190 | 200 | 9 |
| Tungsten | 19.3 | 2.5 | 95 | 0.033 | 6200 | 50 | 0.28 | | 18–600 | 1–3 |
| Uranium D-38 | 18.97 | 4.0–8.0 | 17 | 0.028 | 2100 | 24 | 0.21 | 28 | 56 | 4 |

Room-temperature properties are given. For further information, consult the "Metals Handbook" or a manufacturer's publication.
*Compiled by Anders Lundberg, University of California, and reproduced by permission.
†To obtain the preferred density units, kg/m³, multiply these values by 1,000.

# IRON AND STEEL

## by Harold E. McGannon

REFERENCES: "Metals Handbook," ASM. ASTM Standards, Part 1. SAE Handbook. "Steel Products Manual," AISI.

### CLASSIFICATION OF IRON AND STEEL

Iron (Fe) is not a high-purity metal commercially but contains other chemical elements which have a large effect on its physical and mechanical properties. The amount and distribution of these elements are dependent upon the method of manufacture. The most important commercial forms of iron are listed below.

**Pig iron** is the product of the blast furnace and is made by the reduction of iron ore.

**Cast iron** is an alloy of iron containing so much carbon that, as cast, it is not appreciably malleable at any temperature.

**Gray cast iron** is an iron which, as cast, has combined carbon (in the form of cementite, $Fe_3C$) not in excess of a eutectoid percentage—the balance of the carbon occurring as graphite flakes. The term "gray iron" is derived from the characteristic gray fracture of this metal.

**White cast iron** contains carbon in the combined form. The presence of cementite or iron carbide ($Fe_3C$) makes this metal hard and brittle, and the absence of graphite gives the fracture a white color.

**Malleable cast iron** is an alloy in which all the combined carbon in a special white cast iron has been changed to free or temper carbon by suitable heat treatment.

**Nodular (ductile) cast iron** is produced by adding alloys of magnesium or cerium to molten iron. These additions cause the graphite to form into small nodules, resulting in a higher-strength, ductile iron.

**Ingot iron** is an open-hearth iron very low in carbon, manganese, and other impurities.

**Wrought iron** is a ferrous material aggregated from a solidifying mass of pasty particles of highly refined metallic iron with which is incorporated, without subsequent fusion, a minutely and uniformly distributed quantity of slag.

**Steel** is a malleable alloy of iron and carbon, usually containing substantial quantities of manganese.

**Carbon steel** is steel that owes its distinctive properties chiefly to the carbon it contains.

**Alloy steel** is steel that owes its distinctive properties chiefly to some element or elements other than carbon, or jointly to such other elements and carbon. Some alloy steels necessarily contain an important percentage of carbon, even as much as 1.25 percent. There is no complete agreement about where to draw the line between the alloy steels and the carbon steels.

**Basic oxygen steel, open-hearth steel,** and **electric-furnace steel** are steels made by the basic-oxygen-furnace, open-hearth, and electric-furnace processes, irrespective of carbon content. (**Bessemer steel,** formerly made in furnaces called converters, is no longer produced in the United States.)

Iron ore is reduced in a blast furnace to form **pig iron,** which is the raw material for practically all iron and steel products. Formerly, nearly 90 percent of the iron ore used in the United States came from the Lake Superior district; the ore had the advantages of high quality and the cheapness with which it could be mined and transported by way of the Great Lakes. The higher-grade ores in these deposits have incurred such heavy depletion that the steel industry must depend in the future on concentrates produced from low-grade ores such as the taconites and on foreign ores.

The modern **blast furnace** consists of a vertical shaft 9.1 m or 30 ft or more in diameter and over 30 m (100 ft) high containing a descending column of iron ore, coke, and limestone and a large volume of ascending hot gas. The gas is produced by the burning of coke in the hearth of the furnace and contains about 34 percent carbon monoxide. This gas reduces the iron ore to metallic iron, which melts and picks up considerable quantities of carbon, manganese, phosphorus, sulfur, and silicon. The **gangue** (mostly silica) of the iron ore and the ash in the coke combine with the limestone to form the blast-furnace slag. The pig iron and slag are drawn off at intervals from the hearth through the iron notch and cinder notch, respectively. Some larger blast furnaces produce over 4,000 tons of pig iron per day. The blast furnace produces iron for three general applications: (1) conversion pig irons, which are used to make many varieties of steel; (2) pig iron for use in foundries for making castings; and (3) ferroalloys, which contain a considerable percentage of another metallic element and are used as addition agents in steelmaking. Compositions of commercial pig irons and two ferroalloys (ferromanganese and ferrosilicon) are listed in Table 1.

### COMMERCIALLY PURE IRON

Commercial quantities of **electrolytic iron** have been produced since 1904, but for economic reasons, most of the processes have been short-lived. Electrolytic iron is brittle as deposited so that it can be readily pulverized to produce a high-purity iron powder. Commercially pure iron powders also are produced by direct-reduction processes. Typical compositions of different irons are given in Table 2.

### WROUGHT IRON

**Wrought iron** is the oldest form of iron made by man. It is no longer made commercially in the United States. The normal wrought iron contained a considerable amount of slag, giving it a characteristic fibrous structure. It could readily be worked and welded at temperatures close to its melting point. It could be obtained in the form of plates, sheets, structural shapes, bars, pipe, and tubing. The principal use for wrought iron was in the form of pipe for mildly corrosive conditions.

**Table 1. Types of Pig Iron for Steelmaking and Foundry Use***

| Designation | Chemical composition, %† | | | Principal use |
|---|---|---|---|---|
| | Si | P | Mn | |
| Bessemer pig | 1.00–2.25 | 0.04–0.135 | 0.50–1.00 | Acid Bessemer steel |
| Basic pig, northern | 1.50 max | 0.400 max | 1.01–2.00 | Basic oxygen and basic open-hearth steel |
| In steps of | 0.25 | | 0.50 | |
| Foundry, northern | 3.50 max | 0.301–0.700 | 0.50–1.25 | A wide variety of castings |
| In steps of | 0.25 | | 0.25 | |
| Foundry, southern | 3.50 max | 0.700–0.900 | 0.40–0.75 | Cast-iron pipe |
| In steps of | 0.25 | | 0.25 | |
| Ferromanganese (3 grades) | 1.2 max | 0.35 max | 74–82 | Addition of manganese to steel or cast iron |
| Ferrosilicon (silvery pig) | 5.00–17.00 | 0.300 max | 1.00–2.00 | Addition of silicon to steel or cast iron |

*Carbon content not specified. Sulfur specified 0.05 max, except 0.045 for Bessemer pig and 0.06 max for ferrosilicon.
†Excerpted from "The Making, Shaping and Treating of Steel," U.S. Steel Corp., 1971; further information in "Steel Products Manual," AISI and ASTM Standards, Part I.

Typical compositions of wrought iron are given in Table 2. Table 3 gives the former ASTM mechanical-test requirements for Grade A, double-refined bars, and Grade B, single-refined bars.

## INGOT IRON

**Ingot iron** (Armco iron) is a relatively pure iron which is made in commercial tonnage quantities in a basic open-hearth furnace in a manner similar to that used for plain carbon steel. The refining operation is carried on considerably further by the addition of a very pure grade of iron ore, which oxidizes out the impurities to a low point. A high furnace temperature is required for this operation owing to the high melting point of pure iron. A typical composition of ingot iron is given in Table 2.

**Physical constants:** specific gravity, 7.866; melting point, 1539°C (2802°F); specific heat at 25°C (77°F), 0.108; heat of fusion, 272 kJ/kg (117 Btu/lbm); thermal conductivity at 100°C (212°F), 66.9 W/(m) (°C) [465 Btu/(h) (ft²) (in) (°F)]; thermal coefficient of expansion at 100°C (212°F), 13 millionths per °C (7 millionths per °F); electrical resistivity at 0°C (32°F), 9.50 $\mu\Omega\cdot$cm; temperature coefficient of electrical resistance between 0 and 100°C (32 and 212°F), 0.0056 per °C (0.0031 per °F). Many of these constants are affected considerably by small changes in composition, grain size, or mechanical treatment.

**Mechanical Properties**    The average mechanical properties of ingot iron after various treatments are given in Table 4.

Young's modulus for ingot iron is 202,017 MPa (29,300,000 lb/in²) in both tension and compression, and the shear modulus is 81,358 MPa (11,800,000 lb/in²). Poisson's ratio is 0.28.

The **effect of cold rolling** on the tensile strength, yield strength, elongation, and shape of the stress-strain curve is shown in Fig. 1. Ingot iron welds evenly and easily; it holds paint well and has superior enameling properties; it has a high magnetic permeability at high inductions and low retentivity.

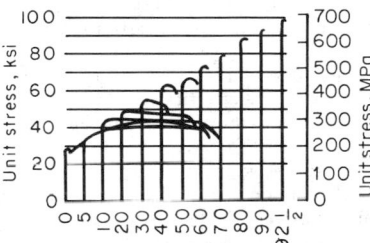

**Fig. 1**    Effect of cold rolling on the stress-strain relations of Armco ingot iron. (*Kenyon and Burns.*)

**Uses**    In galvanized-sheet form for culverts, flumes, roofing and siding; in plate form in oil tanks, water tanks, boilers, gas holders, and large pipe; in enameled form for ranges, refrigerators, tables and kitchen furniture, lighting fixtures, and similar articles. The purity of ingot iron makes it valuable as melting stock in making high-grade tool steel.

## STEEL

### Steel Manufacturing

Steel is produced by the removal of impurities from pig iron in

**Table 2. Typical Compositions of Wrought Iron, Ingot Iron, Electrolytic Iron, and Steel (Percent)**

| Material | C | Mn | P | S | Si | Slag by weight |
|---|---|---|---|---|---|---|
| Wrought iron (hand-puddled) | 0.06 | 0.05 | 0.068 | 0.009 | 0.10 | 2.0 |
| Wrought iron (Byers) | 0.08 | 0.02 | 0.062 | 0.010 | 0.16 | 1.20 |
| Ingot iron (Armco) | 0.015 | 0.020 | 0.005 | 0.025 | Trace | |
| Electrolytic iron | 0.006 | | 0.005 | 0.005 | 0.005 | |
| Low-carbon steel | 0.06 | 0.40 | 0.012 | 0.030 | 0.009 | |

**Table 3. Mechanical Properties of Wrought Iron (Round, Square, and Hexagonal Bars)***

| | Under 41.3 mm (1⅝ in) in diam or thickness | 41.3 mm (1⅝ in) and up to but not incl. 63.5 mm (2½ in) in diam or thickness | 63.5 mm (2½ in) and over in diam or thickness and flat bars |
|---|---|---|---|
| Grade A, Double-Refined | | | |
| Tensile strength, 1,000 lb/in² | 48–54 | 47–54 | 46–54 |
| Tensile strength, MPa | 331–372 | 324–372 | 317–372 |
| Yield point, min | 0.60 T.S. | 0.55 T.S. | 0.50 T.S. |
| Elongation in 200 mm (8 in), min % | 28 | 25 | 22 |
| Reduction of area, min % | 45 | 40 | 35 |
| Grade B, Double-Refined | | | |
| Tensile strength, 1,000 lb/in² | 48 | 47 | 46 |
| Tensile strength, MPa | 331 | 324 | 317 |
| Yield point, min | 0.60 T.S. | 0.55 T.S. | 0.50 T.S. |
| Elongation in 200 mm (8 in), min % | 25 | 22 | 20 |
| Reduction of area, min % | 40 | 35 | 30 |

*ASTM A189-68. Specification discontinued in 1971 because wrought iron is no longer produced.

a basic oxygen furnace, an open-hearth furnace, or an electric furnace.

**Basic Oxygen Steel** This steel is produced by blowing pure (99 percent) oxygen vertically under high pressure (1.2 MPa or 175 lb/in²) onto the surface of molten pig iron. This is an autogenous process that requires no external heat to be supplied. The furnaces are similar in shape to the former Bessemer converters but range in capacity to 275 metric tons (300 net tons) or more. The barrel-shaped furnace or vessel is closed on the bottom, open at the top, and can rotate in a vertical plane about a horizontal axis for charging and for pouring the finished steel. Scrap is dumped into the furnace first (up to 30 percent of the total charge by weight), and hot metal (molten pig iron) is poured on top of the scrap. An oxygen lance is lowered into the furnace, and the flow of oxygen is started. Within seconds after the oxygen is turned on, some iron in the charge is converted to ferrous oxide, which reacts rapidly with the impurities of the charge to remove them from the metal. As soon as reaction starts, limestone is added as a flux. Blowing is continued until the desired degree of purification is attained. The reactions take place very rapidly, and blowing of a heat is completed in about 20 min in a 200 net ton furnace. Because of the speed of the process, a computer is used to calculate the charge required for making a given heat of steel, the rate and duration of oxygen blowing, and to regulate the quantity and timing of additions during the blow and for finishing the steel. Production rates of well over 270 metric tons per furnace hour (300 net tons) can be attained. The comparatively low investment cost and low cost of operation have already made the basic oxygen process

**Table 4. Mechanical Properties of Ingot Iron***

| | Hot-rolled rods or plates | Dead soft | Cold-worked (approx max) | Finished cold | Finished hot | Finished hot, annealed | Finished hot, quenched from 940°C (1725°F) in water |
|---|---|---|---|---|---|---|---|
| Tensile strength: | | | | | | | |
| MPa | 179–221 | 131 | | 186 | 131 | 124 | 207 |
| 1,000 lb/in² | 26–32 | 19 | | 27 | 19 | 18 | 30 |
| Yield strength: | | | | | | | |
| MPa | 290–331 | 262 | 689 | 303 | 290 | 283 | 324 |
| 1,000 lb/in² | 42–48 | 38 | 100 | 44 | 42 | 41 | 47 |
| Elongation, % | 22–28 | 43–48 | | 42 | 45 | 47 | 36 |
| Gage length, in | 8 | 2 | | 2 | 2 | 2 | 2 |
| Reduction of area, % | 65–78 | 70–77 | 65–70 | 76 | 77 | 71 | 70 |
| Hardness, BHN | 82–100 | 67 | 220 | 101 | 90 | 82 | 110 |

*R. L. Kenyon, "Metals Handbook," ASM, 1948.

the largest producer of steel in the world, replacing the basic open-hearth as the major steelmaking process.

**Basic open-hearth steel** is produced in the basic open-hearth furnace. This is a furnace having a capacity up to 550 metric tons (600 net tons) or larger, although capacities in the range of 115 to 270 metric tons (130 to 300 net tons) have been most common in the United States. The hearth has a basic lining of magnesia and dolomite, and the roof generally is made of high-grade basic brick. A practice in wide use for making structural steels is to charge limestone, iron ore, and scrap and then melt the scrap with an oxidizing flame. When melting is well advanced, molten pig iron is added, and a reaction takes place between the iron oxide and the impurities in the pig iron. The rate of this reaction and the initial melting of scrap can be increased by the addition of oxygen to the furnace by roof lances. The acid slag formed by the oxidation of silicon and manganese is run off, and with further heating, the lime comes up through the bath, causing a violent boil and forming a basic slag. This slag makes it possible to remove a large percentage of the phosphorus in the iron. When the percentages of impurities are lowered to the desired extent, the metal is tapped through a hole in the rear of the furnace and poured into a ladle. Various additions are made to the steel, either in the furnace before tapping or in the ladle, to deoxidize the steel and to obtain the desired composition. The time required for the complete operation is 6 to 12 h, depending on the rate of oxygen injection. Production rates for the open hearth vary from 22.5 to 54.5 metric tons (25 to 60 net tons) per furnace hour.

**Acid open-hearth steel** is produced in an open-hearth furnace having an acid or siliceous lining. The process is similar to the basic process except that an acid slag is used, so that no phosphorus can be removed from the steel. This requires a scrap and pig-iron charge of low phosphorus content.

**Bessemer steel** (no longer produced in this country) is made in a pear- or barrel-shaped converter which is mounted on trunnions so as to be tilted easily for charging and pouring. Molten pig iron is poured into the converter, and air is blown through the liquid metal. Heat is liberated by the oxidation of the impurities—silicon, manganese, and carbon. When the carbon has been nearly eliminated, the metal is recarburized by the addition of alloys containing carbon, manganese, and silicon. From 7.5 to 27.0 metric tons (8 to 30 net tons) of steel are made in one blow requiring only 10 to 20 min.

Acid Bessemer steel was made in a converter having a siliceous lining and was the only Bessemer steel made in the United States, for our ores were not suitable for the production of pig iron for **basic Bessemer steel.** The latter (still produced in some foreign countries) is made in a converter having a magnesite lining, and limestone is added during the process to form a basic slag which removes a large proportion of the phosphorus in the pig. The process requires a pig iron very high in phosphorus, whereas the basic open-hearth process can use pig iron containing phosphorus in all but the highest amounts.

**Electric Steel** The electric furnace commonly used in the production of electric steel is the three-phase arc furnace in which electric arcs heat the bath. The furnace has a basic lining and is usually charged with cold steel scrap. When the bath is molten, the impurities are removed by oxidation by the use of iron oxide or by the introduction of gaseous oxygen. The resultant slag is then removed, and a new slag is formed

from lime, sand, and spar. This slag may be made carbidic by the addition of coke dust, and the oxygen content of the steel is reduced to a low level. Alloying elements are then added to meet the specified composition, and the heat is tapped into a ladle. The **induction furnace** is simply a melting furnace to which the various metals are added to make the desired alloy. When steel scrap is used as a charge, it will be a high-grade scrap the composition of which is well known (see also Sec. 7).

**Vacuum Treatment** Hydrogen in amounts as low as 5 ppm may cause internal flakes in large steel sections. By pouring the liquid steel in a vacuum chamber, it is possible to reduce the hydrogen to about 1 ppm. Vacuum casting is used extensively for large forgings such as electrical rotors. Carbon deoxidation is also carried out in a vacuum but is carried out for the purpose of reducing the oxygen content in the steel to less than 0.001 percent by removal of oxygen as carbon monoxide. This treatment substantially decreases the number of nonmetallic inclusions in the steel and is used for bearing steel and other high-quality steels.

**Steel Ingots** After being refined by one of the preceding methods, the steel is tapped into a ladle. Most steel still is poured from the ladle into iron ingot molds. Many defects in rolled-steel products may be introduced by incorrect ingot practice, and the production of high-grade steel requires great care at this stage. Increasing quantities of molten steel are being poured from the ladle into continuous-casting machines to convert the liquid steel directly into solid semifinished forms such as blooms, slabs, and billets.

Steels in which no gas evolution occurs on solidification are called **killed steels.** When the steel cools in the mold, shrinkage of the steel on solidifying causes **piping**—the cavity, or "pipe," being found usually in the upper portion of the ingot. To minimize this condition, a large-end-up mold is used together with a refractory "hot top" which supplies molten steel to the main body of the ingot while solidification proceeds. This minimizes the amount of metal that has to be discarded on account of pipe, but the top discard still amounts to 10 or 15 percent of the total weight of the ingot.

To reduce the cost of hot tops and the large percentage of metal discard when making mild steel for structural purposes, the steel is not fully deoxidized. This results in blowholes in the steel on solidification; the presence of these blowholes minimizes piping by distributing small voids throughout the ingot instead of having one large one in the upper center of the ingot. If not exposed at the surface of the ingot, these blowholes weld together during rolling. Steel deoxidized in this manner is called **semikilled steel.** If the steel is deoxidized still less in the ladle, a reaction takes place during solidification in which the oxygen and the carbon in the steel form carbon monoxide, which is freely evolved from the ingot. The intensity of this reaction affects the ingot structure greatly. If the reaction is allowed to go to completion, the product is called **rimmed steel,** but if the reaction is stopped after a short while by preventing, in a mechanical manner, further evolution of gas from the top of the ingot, the steel is called **capped steel.** The gas evolution results in an outer skin on the ingot which is clean and very low in carbon. In capped steel, the skin is thinner and there is less segregation or concentration of impurities than in rimmed steel. The presence of this nearly pure iron skin enables the production of an excellent surface finish on the rolled product, and therefore sheet and strip are made nearly exclusively from rimmed or capped steel.

Some defects besides the occurrence of pipe in rolled-steel products are segregation, ingot cracks, seams, scabs, laps, and inclusions. **Segregation,** or the concentration of impurities, occurs in all steels upon solidification, but it can be minimized by proper mold design and low pouring temperatures. Rough mold surfaces or molds containing cracks or cavities interfere with the normal contraction of the ingot, and transverse cracks in the ingot skin may result. Cracks thus produced soon have their surfaces oxidized, and when the ingot is rolled out, these defects will be elongated in the direction of rolling and are called **seams.** Any oxidized crack or blowhole at the surface will roll out into a seam. Improper pouring conditions such as splashing of steel in the molds will form **scabs.** When rolling with grooved rolls which are not properly designed or set up, fins are liable to result from the flow of metal between the flat bodies of the rolls. If the fin is thin and wide, it will be folded over when the steel passes through the next set of rolls and will form a **lap. Nonmetallic inclusions** consisting of sulfides, silicates, etc., are found to some extent in all steels and are introduced in the refining and deoxidation of the steel. Steels containing many inclusions are said to be "dirty," but the presence of these inclusions, unless they are large in size, is not necessarily detrimental to the mechanical properties.

Pipe, segregation, and inclusions are defects in the steel which cannot be remedied to any large extent. Surface defects such as seams, laps, and scabs can be removed by **chipping** the surface with air hammers, **scarfing** or **deseaming** with oxyacetylene torches, grinding, or by machining. **Scale,** composed principally of iron oxides, forms on the surface of steel during heating and hot working. To facilitate inspection of the surface of steel products, the scale can be removed by **pickling** in acid, by the use of a torch in **flame scaling,** or by mechanical means.

### Mechanical Treatment of Steel

At present, most of the total steel produced is cast in the form of ingots and then subjected to some form of mechanical treatment. The solid ingot is heated to between 1095 and 1425°C (2000 and 2600°F) depending on the composition of the steel, and then hot-worked by rolling, pressing, or hammering. **Hot work** consists of any mechanical treatment at temperatures above the thermal critical range of the steel. Considerable improvement in mechanical properties is obtained by hot work. It will increase to some extent the yield and tensile strength, but it is especially beneficial to the ductility of the steel, since it breaks up the dendritic structure in the ingot and minimizes the effects of segregation and inclusions. The temperature at which hot work is completed is important and should be above the thermal critical range, but as near to as practicable, the exact temperature depending upon the carbon and alloy content.

The effect of **rolling** is to elongate the inclusions in the direction of rolling, giving the steel excellent properties on samples taken parallel to this direction, although tests on samples taken in a transverse direction will not have as high mechanical properties. **Hammer forging** is more effective than rolling in that working can be done in more than one direction, thus eliminating directional properties to the steel. The slow application of pressure in a **forging press** works the interior of a large forging more effectively than hammering; the press is used to a large extent for large high-quality forgings. Rolling, hammering, or pressing effectively breaks up the coarse crys-

tallization of a cast ingot when a reduction in area of 3 or 4 to 1 has been obtained.

Shaping of steel by rolling is adopted whenever possible because of the rapidity of the operation and its relatively low cost. Rolling operations are carried out in mills which derive their name from the name of the product that they produce. Thus an ingot may be rolled into blooms in a blooming mill and slabs in a slabbing mill. In the blooming mill, the ingot is reduced by several passes to a **bloom,** having dimensions 150 mm (about 6 in) square or larger. The bloom may then be further reduced to a **billet** which is somewhere between 30 and 150 mm (1¼ and 6 in) square. The names blooms and billets still apply if the products are rectangular in form when the widths are less than twice the thickness. However, if the width far exceeds the thickness of the rectangular section, it is called a **slab.** Blooms are rolled on finishing mills into **structural shapes, rails, wheels,** and **skelp** to be used in the manufacture of pipe. Blooms may also be used for forging purposes. Slabs are rolled into plates and coils of hot-rolled sheets. Billets are rolled into rods, bars, bands, hoops, small shapes, and seamless tubes.

As stated earlier, some steel is cast directly into slabs, blooms, and billets, using continuous casting which eliminates the teeming, stripping, reheating, and rolling steps involved in ingot practice. Continuous casting saves time and increases yield.

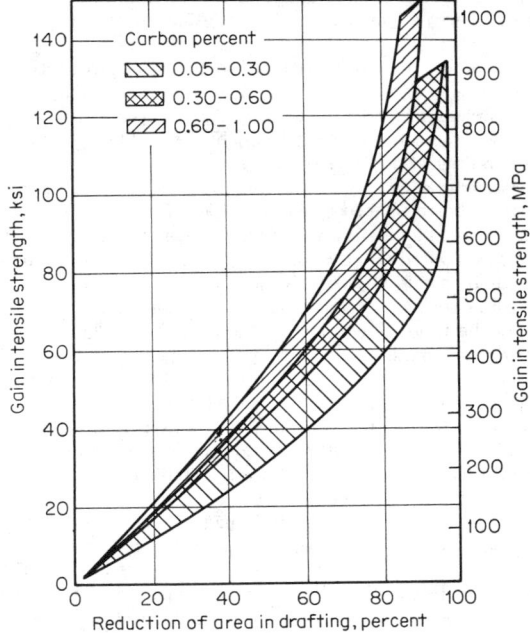

**Fig. 2**  Increase of tensile strength of plain carbon steel with increasing amounts of cold working by drafting through a wire-drawing die.

**Cold work** is work done on the metal below the thermal critical range and usually is done at atmospheric temperature. It greatly increases the yield strength and tensile strength, especially the former, and reduces the ductility (see Figs. 2 and 3). It includes operations such as cold rolling, cold pressing, twisting, and wire drawing. Large tonnages of sheets are cold-rolled, which greatly improves their surface finish as well as increases their strength.

The **extrusion** process shapes metal into the desired form by forcing it through a die under pressure. The advantages of extrusion are that it reduces scrap loss and labor and machin-

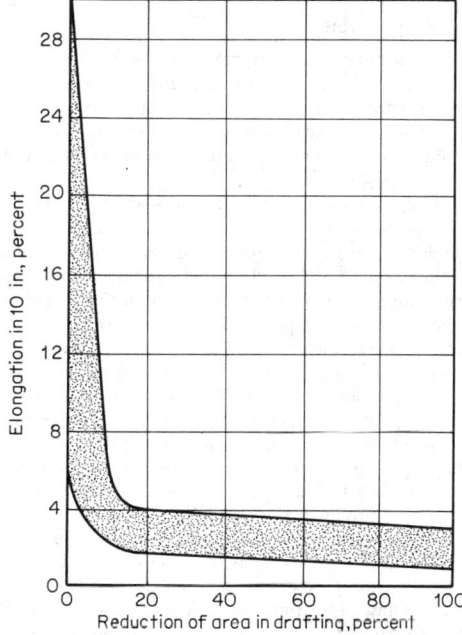

**Fig. 3** Effect of cold working by drafting through a wire-drawing die on the ductility of plain carbon steel.

ing costs and is economically more suitable for both small-unit or large-scale production of odd-shaped pieces. **Cold extrusion** of steel markedly increases the mechanical properties and is therefore used extensively in production of ordnance. A low-carbon, aluminum-killed steel is normally used, and a phosphate coating serves as an excellent lubricant. For higher-alloy and stainless steels **hot extrusion** is necessary. Hot extrusion achieved practicality with the invention of the **Ugine-Sejournet** process, which solved the basic problem of lubrication by using glass. The glass melts at the high temperatures involved and forms a lubricating film over the section to be extruded. The problem of scale-free heating is solved in several ways, including induction heating, heating in molten salt baths, and using controlled heating atmospheres. The process is being improved continuously by changes in dies, presses, and operating speeds, and it is probably the best method for forming alloys with poor hot-working characteristics.

### Constitution and Structure of Steel

As a result of the methods of production, the following elements are always present in steel: carbon, manganese, phosphorus, sulfur, silicon, and traces of oxygen, nitrogen, and aluminum. Various alloying elements are frequently added, such as nickel, chromium, copper, molybdenum, and vanadium. The most important of the above elements in steel is

carbon, and it is necessary to understand the effect of carbon on the internal structure of steel to understand the heat treatment of carbon and low-alloy steels.

In Fig. 4 is the iron-iron carbide equilibrium diagram which shows the phases that are present in steels of various carbon contents over a range of temperatures under equilibrium conditions. Pure iron when heated to 910°C (1670°F) changes its internal crystalline structure from a body-centered cubic arrangement of atoms, **alpha iron,** to a face-centered cubic structure, **gamma iron.** At 1390°C (2535°F), it changes back to the body-centered cubic structure, **delta iron,** and at 1539°C (2802°F) the iron melts. When carbon is added to iron, it is found that it has only slight solid solubility in alpha iron (less than 0.001 percent at room temperature). On the other hand, gamma iron will hold up to 2.0 percent carbon in solution at 1130°C (2066°F). The alpha iron containing carbon or any other element in solid solution is called **ferrite.** and the gamma iron containing elements in solid solution is called **austenite.** Usually when not in solution in the iron, the carbon forms a compound $Fe_3C$ (iron carbide) which is extremely hard and brittle and is known as **cementite.**

The temperatures at which the phase changes occur are called **critical points** (or temperatures) and, in the diagram, represent equilibrium conditions. In practice there is a lag in the attainment of equilibrium, and the critical points are found at lower temperatures on cooling and at higher temperatures on heating than those given, the difference increasing with the rate of cooling or heating.

The various critical points have been designated by the letter A; when obtained on cooling, they are referred to as Ar, on the heating as Ac. The various critical points are distinguished from each other by numbers after the letters, being numbered in the order in which they occur as the temperature increases. $Ac_1$ represents the beginning of transformation of ferrite to austenite on heating; $Ac_3$ the end of transformation of ferrite to austenite on heating, and $Ac_4$ the change from austenite to delta iron on heating. On cooling, the critical

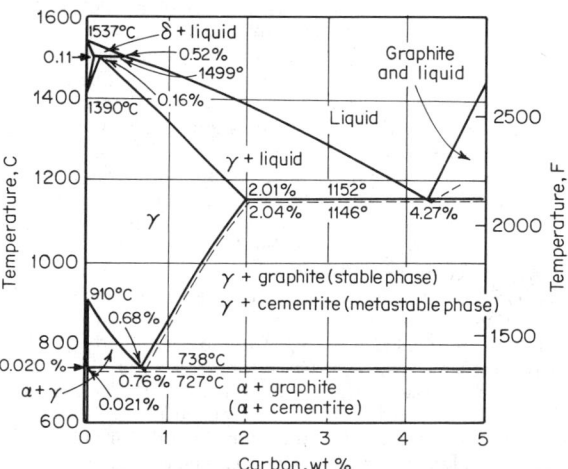

**Fig. 4** Iron–iron carbide equilibrium diagram, for carbon contents up to 5 percent. (Dashed lines represent equilibrium with cementite or iron carbide: adjacent solid lines indicate equilibrium with graphite.)

points would be referred to as $Ar_4$, $Ar_3$, and $Ar_1$, respectively. It must be remembered that the diagram represents the pure iron-iron carbide system at equilibrium. The varying amounts of impurities in commercial steels affect to a considerable extent the position of the curves and especially the lateral position of the eutectoid point.

Carbon steel in equilibrium at room temperature will have present both ferrite and cementite. The physical properties of the ferrite are approximately those of pure iron and are characteristic of the metal. The presence of cementite does not in itself cause steel to be hard, but rather it is the shape and distribution of the carbides in the iron that determines the hardness of the steel. The fact that the carbides can be dissolved in austenite is the basis of the heat treatment of steel, since the steel can be heated above the $A_1$ critical temperature to dissolve all the carbides, and then suitable cooling through the cooling range will produce the desired size and distribution of carbides in the ferrite.

If austenite containing about 0.80 percent carbon ( **eutectoid** composition at 0.76 percent carbon in Fig. 4) is slowly cooled through the critical temperature, ferrite and cementite are rejected simultaneously, forming alternate plates or lamellae. This microstructure is called **pearlite**, since when polished and etched it has a pearly luster. When examined under a microscope, however, the individual plates of cementite can easily be distinguished. If the austenite contains less than 0.80 percent carbon (**hypoeutectoid** composition), free ferrite will first be rejected on slow cooling through the critical until the composition of the remaining austenite reaches 0.80 percent carbon when the simultaneous rejection of both ferrite and carbide will again occur, producing pearlite. So a hypoeutectoid steel at room temperature will be composed of areas of free ferrite and areas of pearlite; the higher the carbon percentage, the greater the amount of pearlite present in the steel. When austenite containing more than 0.80 percent carbon (**hypereutectoid** composition) is slowly cooled, cementite is thrown out at the austenite grain boundaries, forming a cementite network until the austenite contains 0.80 percent carbon, at which time pearlite is again formed. Thus a hypereutectoid steel when slowly cooled will have areas of pearlite surrounded by a thin carbide network.

As the cooling rate is increased, the spacing between the pearlite lamellae becomes smaller; with the resulting greater dispersion of carbide preventing slip in the iron crystals, the steel becomes harder. Also, with an increase in the rate of cooling, there is less time for the separation of excess ferrite or cementite, and the equilibrium amount of these constituents will not be precipitated before the austenite transforms to pearlite. Thus with a fast rate of cooling, pearlite may contain more or less carbon than given by the eutectoid composition. When the cooling rate becomes very rapid (as obtained by quenching), the carbon does not have sufficient time to separate out in the form of carbide, and the austenite transforms to a highly stressed structure supersaturated with carbon called **martensite.** This structure is exceedingly hard but brittle and requires **tempering** to increase the ductility. Tempering consists of heating martensite to some temperature below the critical causing the carbide to precipitate in the form of small spheroids. The higher the tempering temperature, the larger the carbide particle size, the greater the ductility of the steel, and the lower the hardness.

In a carbon steel, it is possible to have a structure consisting either of parallel plates of carbide in a ferrite matrix, the distance between the plates depending upon the rate of cooling, or of carbide spheroids in a ferrite matrix, the size of the spheroids depending upon the temperature to which the hardened steel was heated. (Some spheroidization occurs when pearlite is heated, but only at high temperatures close to the critical temperature range.)

**Heat-Treating Operations**

The following definitions of terms have been adopted by the ASTM, SAE, and ASM in substantially identical form.

**Heat Treatment** An operation, or combination of operations, involving the heating and cooling of a metal or an alloy in the solid state, for the purpose of obtaining certain desirable conditions or properties.

**Quenching** Rapid cooling by immersion in liquids or gases or by contact with metal.

**Hardening** Heating and quenching certain iron-base alloys from a temperature either within or above the critical range for the purpose of producing a hardness superior to that obtained when the alloy is not quenched. Usually restricted to the formation of martensite.

**Annealing** A heating and cooling operation implying usually a relatively slow cooling. The purpose of such a heat treatment may be (1) to remove stresses; (2) to induce softness; (3) to alter ductility, toughness, electrical, magnetic, or other physical properties; (4) to refine the crystalline structure; (5) to remove gases; or (6) to produce a definite microstructure. The temperature of the operation and the rate of cooling depend upon the material being heat-treated and the purpose of the treatment. Certain specific heat treatments coming under the comprehensive term annealing are as follows:

**Full Annealing** Heating iron-base alloys above the critical temperature range, holding above that range for a proper period of time, followed by slow cooling to below that range. The annealing temperature is generally about 100°F above the upper limit of the critical temperature range, and the time of holding is usually not less than 1 h for each inch of section of the heaviest objects being treated. The objects being treated are ordinarily allowed to cool slowly in the furnace. They may, however, be removed from the furnace and cooled in some medium that will prolong the time of cooling as compared with unrestricted cooling in the air.

**Process Annealing** Heating iron-base alloys to a temperature below or close to the lower limit of the critical temperature range followed by cooling as desired. This heat treatment is commonly applied in the sheet and wire industries, and the temperatures generally used are from 540 to 705°C (about 1000 to 1300°F).

**Normalizing** Heating iron-base alloys to approximately 40°C (about 100°F) above the critical temperature range followed by cooling to below that range in still air at ordinary temperature.

**Patenting** Heating iron-base alloys above the critical temperature range followed by cooling below that range in air, in molten lead, or a molten mixture of nitrates or nitrites maintained at a temperature usually between 425 and 565°C (about 800 to 1050°F), depending on the carbon content of the steel and the properties required of the finished product. This treatment is applied in the wire industry to medium- or high-carbon steel as a treatment to precede further wire drawing.

**Spheroidizing** Any process of heating and cooling steel

that produces a rounded or globular form of carbide. The following spheroidizing methods are used: (1) Prolonged heating at a temperature just below the lower critical temperature, usually followed by relatively slow cooling. (2) In the case of small objects of high-carbon steels, the spheroidizing result is achieved more rapidly by prolonged heating to temperatures alternately within and slightly below the critical temperature range. (3) Tool steel is generally spheroidized by heating to a temperature of 750 to 805°C (about 1380 to 1480°F) for carbon steels and higher for many alloy tool steels, holding at heat from 1 to 4 h, and cooling slowly in the furnace.

**Tempering** (also termed **Drawing**) Reheating hardened steel to some temperature below the lower critical temperature, followed by any desired rate of cooling. Although the terms "tempering" and "drawing" are practically synonymous as used in commercial practice, the term "tempering" is preferred.

Figure 5 summarizes the rates of decomposition of a eutectoid carbon steel over a range of temperatures. Various cooling rates are shown diagrammatically, and it will be seen that the faster the rate of cooling, the lower the temperature of transformation, and the harder the product formed. At around 540°C (1000°F), the austenite transforms rapidly to fine pearlite; to form martensite it is necessary to cool very rapidly through this temperature range to avoid the formation of pearlite before the specimen reaches the temperature for the formation of martensite. The minimum rate of cooling that is required to form a fully martensite structure is called the **critical cooling rate.** No matter at what rate the steel is cooled, the only products of transformation of this steel will be pearlite or martensite. However, if the steel is given an interrupted quench in a molten bath at some temperature between 205 and 540°C (about 400 and 1000°F), an acicular structure, called **bainite,** of considerable toughness, combining high strength with high ductility, is obtained, and this heat treatment is known as austempering. A somewhat similar heat treatment called martempering can be utilized to produce a fully martensitic structure of high hardness, but free of the cracking, distortion, and residual stresses often associated with such a structure. Instead of quenching to room temperature, the steel is quenched to just above the martensitic transformation temperature and held for a short time to permit equilization of the temperature gradient throughout the piece. Then the steel may be cooled relatively slowly through the martensitic transformation range without superimposing thermal stresses on those introduced during transformation. The limitation of austempering and martempering for carbon steels is that these two heat treatments can be applied only to articles of small cross section, since the rate of cooling in salt baths is not sufficient to prevent the formation of pearlite in samples or more than ½ in diameter.

The maximum hardness obtainable in a high-carbon steel with a fine pearlite structure is approximately 400 Brinell, although a martensitic structure would have a hardness of approximately 700 Brinell. Besides being able to obtain structures of greater hardness by forming martensite, a spheroidal structure will have considerably higher proof stress (i.e., stress to cause a permanent deformation of 0.01 percent) and ductility than a lamellar structure of the same tensile strength and hardness. It is essential, therefore, to form martensite when optimum properties are desired in the steel. This can be done with a piece of steel having a small cross section by heating the steel above the critical and quenching in water; but when the cross section is large, the cooling rate at the center of the section will not be sufficiently rapid to prevent the formation of pearlite. The characteristic of steel that determines its capacity to harden throughout the section when quenched is called **hardenability.** This term should not be confused with the ability of a steel to attain a certain hardness. The intensity of hardening, i.e., the maximum hardness of the martensite formed, is largely dependent upon the carbon content of the steel.

**Determination of Hardenability** The standard method of determining the capacity of a steel to harden is the **Jominy hardenability test.** This test consists of heating a steel bar about 75 mm (3 in) long and 25 mm (1 in) in diameter to the desired austenitizing temperature and quenching one end in water, allowing the other end to air cool. A continuous change in cooling rate is obtained along the bar varying from 333°C (600°F) per s at about 1.6 mm (¹⁄₁₆ in) from the quenched end to 2.2°C (about 4°F) per s on the other. Rockwell C hardness measurements are then made at about 1.6 mm (¹⁄₁₆ in) intervals along the bar, and a plot is made of hardness versus the distance from the quenched end. Since the cooling rate is known for any given position along the Jominy bar, the relation is obtained between the cooling rate and the hardness developed in the steel. In addition, the cooling rates for many simple geometric shapes such as bars and plates in different quenching mediums have been determined, so that it is possible to predict the hardness that will be developed in a given steel part if the Jominy hardenability curve is known; or if a minimum hardness is required in a given steel article, it is possible to specify the minimum Jominy hardenability needed to develop the required hardness in a given quenching medium. For the detailed procedure of the Jominy test and its application, see the 1955 SAE Handbook.

Three main factors affect the hardenability of steel: (1) austenite composition; (2) austenite grain size; and (3) amount, nature, and distribution of undissolved or insoluble particles in the austenite. The austenite composition will determine the rate of decomposition, in the range of 540°C (about 1000°F). The slower the rate of decomposition, the larger the section that can be hardened throughout, and therefore the greater the

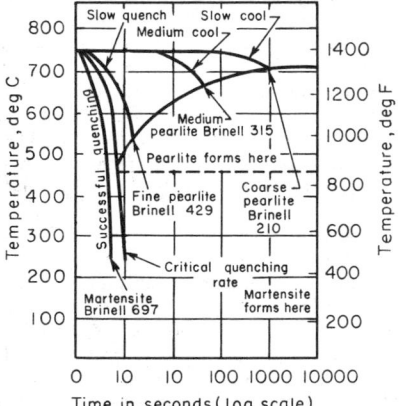

**Fig. 5** Influence of cooling rate on the product of transformation in a eutectoid carbon steel.

hardenability of the steel. Everything else being equal, the higher the carbon content, the greater the hardenability. The question of **austenitic grain size** is of considerable importance in any steel that is to be heat-treated, since it affects the properties of the steel to a considerable extent. When a steel is heated to just above the critical temperature, small polyhedral grains of austenite are formed. With increase in temperature, there is an increase in the size of grains, until at temperatures close to the melting point the grains are very large. The relation between the grain size developed and the heating temperature will vary considerably among steels on account of variations in the deoxidation practice. By suitable deoxidation, steels can be made to be coarse-grained at comparatively low temperatures, and others are fine-grained over a considerable range of temperatures. Since the transformation of austenite usually starts at grain boundaries, a fine-grained steel will transform more rapidly than a coarse-grained steel because the latter has much less surface area bounding the grains than a steel with a fine grain size. Everything else being equal, coarse-grained austenite will have a higher hardenability than fine-grained. Small particles in the austenite will act as nuclei for the beginning of transformation in a manner similar to grain boundaries, and therefore the presence of a large number of small particles (sometimes submicroscopic in size) will result in low hardenability.

**Determination of Austenitic Grain Size** The subject of ausenite grain size is of considerable interest because of the fact that the grain size developed during heat treatment has a large effect on the physical properties of the steel. In steels of similar chemical analysis, the steel developing the finer austenitic grain size will have a lower hardenability but will, in general, have greater toughness, show less tendency to crack or warp on quenching, be less susceptible to grinding cracks, have lower internal stresses, and retain less austenite than coarse-grained steel. It is for these reasons that most alloy steels are fine-grain steels. There are several methods of determining the grain-size characteristics of a steel, the one most commonly used in steel specifications being the **McQuaid-Ehn test.** In this test, a representative sample is carburized for 8 h at 925°C (1700°F) and cooled slowly. The high-carbon case on slow cooling will reject cementite at the austenite grain boundaries, and, by polishing and etching, the grains will be clearly seen under a microscope. There are several ways to report the grain size observed under the microscope, the one used most extensively being the ASTM index numbers. In fps or English units, the numbers are based on the formula: number of grains per square inch at $100x = 2^{N-1}$, in which $N$ is the grain-size index. The usual range in steels will be from 1 to 128 grains/in² at $100x$, and the corresponding ASTM numbers will be 1 to 8. Steels having an ASTM grain size of 1 to 4 are usually considered coarse-grained, and those from 5 to 8 are fine-grained steels. Grain-size relationships in SI units are covered in detail in Designation E112 of ASTM Standards. It should be noted that the McQuaid-Ehn test will give only the grain size developed in the steels when heated to one temperature for a given length of time. To determine fully the grain-size characteristics of a steel, tests should be made over a range of temperatures, but the McQuaid-Ehn test has proved of great help to both producers and consumers of steel, since the test is inexpensive and reproducible. For further information on grain size, reference should be made to the ASM "Metals Handbook."

## EFFECT OF ALLOYING ELEMENTS ON THE PROPERTIES OF STEEL

When relatively large amounts of alloying elements are added to steel, the characteristic behavior of carbon steels is obliterated. Most **alloy steel** is medium- or high-carbon steel to which various elements have been added to modify its properties to an appreciable extent, but it still owes its distinctive characteristics to the carbon that it contains. The percentage of alloy element required for a given purpose ranges from a few hundredths of 1 percent to possibly as high as 5 percent.

When ready for service, these steels will usually contain only two constituents, ferrite and carbide. The only way that an alloying element can affect the properties of the steel is to change the dispersion of carbide in the ferrite, change the properties of the ferrite, or change the properties of the carbide. The effect on the distribution of carbide is the most important factor, since in sections amenable to close control of structure, carbon steel is only moderately inferior to alloy steel. However, in large sections where carbon steels will fail to harden throughout the section even on a water quench, the hardenability of the steel can be increased by the addition of any alloying element (except possibly cobalt). The increase in hardenability permits the hardening of a larger section of alloy steel than of plain carbon steel. The quenching operation does not have to be so drastic. Consequently, there is a smaller difference in temperature between the surface and center during quenching, and cracking and warping resulting from sharp temperature gradients in a steel during hardening can be avoided. The elements most effective in increasing the hardenability of steel are manganese, silicon, and chromium.

Elements such as molybdenum, tungsten, and vanadium are effective in increasing the hardenability when dissolved in the austenite, but are usually present in the austenite in the form of carbides. The main advantage of these carbide-forming elements is that they prevent the agglomeration of carbides in tempered martensite. Tempering relieves the internal stresses in the hardened steel and causes spheroidization of the carbide particles with resultant loss in hardness and strength. The presence of these stable carbide-forming elements enables higher tempering temperatures to be used without sacrificing strength. This permits these alloy steels to have a greater ductility for a given strength, or, conversely, greater strength for a given ductility, than plain carbon steels.

The third factor which contributes to the strength of alloy steel is the presence of the alloying element in the ferrite. Any element in solid solution in a metal will increase the strength of the metal, so that these elements will materially contribute to the strength of hardened and tempered steels. The elements most effective in strengthening the ferrite are phosphorus, silicon, manganese, nickel, molybdenum, tungsten, and chromium.

A final important effect of alloying elements is their influence on the austenitic grain size. Martensite formed from a fine-grained austenite has considerably greater resistance to shock than when formed from a coarse-grained austenite. The oxides formed by the deoxidation of the steel by different elements apparently prevent grain growth above the critical temperature over a considerable temperature range. Aluminum is the most effective element to form grain-growth inhibitors; most killed steels have some aluminum added during deoxidation. The presence of finely scattered carbides in the austenite appears to have a similar effect on the austenite grain

**Table 5. Trends of Influence of the Alloying Elements**

| Element | As dissolved in ferrite, strength | As dissolved in austenite, hardenability | As undissolved carbide in austenite, fine grain, toughness | As dispersed carbide in tempering, high temp. strength and toughness | As fine non-metallic dispersion, fine grain, toughness |
|---------|-----------------------------------|------------------------------------------|------------------------------------------------------------|----------------------------------------------------------------------|--------------------------------------------------------|
| Al | Moderate | Mild | None | None | Very strong |
| Cr | Mild | Strong | Strong | Moderate | Slight |
| Co | Strong | Negative | None | None | None |
| Cb | Little | Strong | Strong | Strong | None |
| Cu | Strong | Moderate | None | None | None |
| Mn | Strong | Moderate | Mild | Mild | Slight |
| Mo | Moderate | Strong | Strong | Strong | None |
| Ni | Mild | Mild | None | None | None |
| P | Strong | Mild | None | None | None |
| Si | Moderate | Moderate | None | None | Moderate |
| Ta | Moderate ? | Strong ? | Strong | Strong | None |
| Ti | Strong | Strong | Very strong | Little ? | Moderate |
| W | Moderate | Strong | Strong | Strong | None |
| V | Mild | Very strong | Very strong | Very strong | Moderate ? |

size; so the elements forming stable carbides will also contribute to the formation of a fine-grained austenite.

In Table 5, a summary of the effects of various alloying elements is given. It must be remembered that this table indicates only the trends of the various elements, and the fact that one element has an important influence on one factor does not prevent it from having an equally strong influence on another one.

## PRINCIPLES OF HEAT TREATMENT OF IRON AND STEEL

When heat-treating a steel for a given part, certain precautions have to be taken to develop optimum mechanical properties in the steel. Some of the major factors that have to be taken into consideration are outlined below.

**Heating**  The first step in the heat treatment of steel is the heating of the material to above the critical to make it fully austenitic. The **heating rate** should be sufficiently slow to avoid injury to the material through excessive thermal and transformational stresses. In general, hardened steel should be heated more slowly and uniformly than is necessary for soft stress-free materials. Large sections should not be placed in a hot furnace, the allowable size depending upon the carbon and alloy content. For high-carbon steels, care should be taken in heating sections as small as 50 mm (2 in) diameter and in medium-carbon steels precautions are required for sizes over 150 mm (6 in) diameter. The **maximum temperature** selected will be determined by the chemical composition of the steel and its grain-size characteristics. In hypoeutectoid steel (below 0.80 percent carbon), a temperature just above the upper critical range is used, and in hypereutectoid steels (above 0.80 percent carbon), a temperature between the lower and the upper critical is generally used so as to avoid heating to high temperatures with consequent grain growth. See Table 6 for heat-treating temperatures for plain carbon steels. The **time** at maximum temperature should be such that a uniform temperature is obtained throughout the cross section of the steel. Care should be taken to avoid undue length of time at temperature, since this will result in undesirable grain growth, scaling, or decarburization of the surface. A practical figure often given for the total time in the hot furnace is 12 min/cm (about $\frac{1}{2}$ h/in) of cross-sectional thickness. When the steel has attained a uniform temperature, the **cooling rate** must be such

**Table 6. Temperatures for Heat Treatments for Carbon Steels***

| AISI steel No. | Normalize °C† | Normalize °F | Anneal °C† | Anneal °F | Quench °C† | Quench °F |
|----------------|---------------|--------------|------------|-----------|------------|-----------|
| 1010 | 900–955 | 1650–1750 | 540–730 | 1000–1350 | 900–925 | 1650–1700 |
| 1020 | 900–955 | 1650–1750 | 540–730 | 1000–1350 | 870–910 | 1600–1675 |
| 1030 | 870–915 | 1600–1675 | 675–745 | 1250–1375‡ | 855–900 | 1575–1650 |
| 1035 | 855–900 | 1575–1650 | 855–900 | 1575–1650‡ | 830–870 | 1525–1600 |
| 1040 | 855–900 | 1575–1650 | 845–885 | 1550–1625‡ | 815–855 | 1500–1575 |
| 1045 | 845–900 | 1550–1650 | 830–870 | 1525–1600 | 800–845 | 1475–1550 |
| 1050 | 845–900 | 1550–1650 | 830–870 | 1525–1600 | 800–845 | 1475–1550 |
| 1060 | 830–885 | 1525–1625 | 815–855 | 1500–1575 | 790–845 | 1450–1550 |
| 1070 | 830–885 | 1525–1625 | 815–855 | 1500–1575 | 790–845 | 1450–1550 |
| 1080 | 830–885 | 1525–1625 | 815–855 | 1500–1575‡ | 790–845 | 1450–1550 |
| 1095 | 830–885 | 1525–1625 | 815–855 | 1500–1575‡ | 790–845 | 1450–1550 |

*Based on ASM "Metals Handbook," 1948.
†Calculated from temperature in °F and rounded to nearest 5°C.
‡For spheroidizing, anneal at 675 to 745°C (1250 to 1375°F).

as to develop the desired structure: slow cooling rates (furnace or air cooling) to develop the softer pearlitic structures and high cooling rates (quenching) to form the hard martensitic structures. In selecting a **quenching medium** (see ASM "Metals Handbook"), it is important to select the quenching medium for a particular job on the basis of size, shape, and allowable distortion before choosing the steel composition. It is convenient to classify steels in two groups on the basis of depth of hardening: shallow hardening and deep hardening. **Shallow-hardening steels** may be defined as those which, in the form of 25-mm-(1-in-)diameter rounds, have, after brine quenching, a completely martensitic shell not deeper than 6.4 mm (¼ in). The shallow-hardening steels are those of low or no alloy content, whereas the deep-hardening steels have a substantial content of those alloying elements that increase penetration of hardening, notably chromium, manganese, and nickel. The high cooling rates required to harden shallow-hardening steel produce severe distortion in all but simple, symmetrical shapes having a low ratio of length to diameter or thickness. Plain carbon steels cannot be used for complicated shapes where distortion must be avoided. In this case, water quenching must be abandoned and a less active quench used which materially reduces the temperature gradient during quenching. Certain oils are satisfactory but are incapable of hardening shallow-hardening steels of substantial size. A change in steel composition is required with a change from water to an oil quench. Quenching in oil does not entirely prevent distortion. When the degree of distortion produced by oil quenching is objectionable, recourse is taken to air hardening. The cooling rate in air is very much slower than in oil or water; so an exceptionally high alloy content is required. This means that a high price is paid for the advantage gained, in terms of both metal cost and loss in machinability, though it may be well justified when applied to expensive tools. In this case, danger of cracking is negligible.

**Liquids for Quenching Shallow-Hardening Steels**   Shallow-hardening steels require extremely rapid surface cooling in the quench, particularly in the temperature range around 550°C (1020°F). A submerged water spray will give the fastest and most reproducible quench practicable. Such a quench is limited in application to simple short objects which are not likely to warp. Because of difficulty in obtaining symmetrical flow of the water relative to the work, the spray quench is conducive to warping. The ideal practical quench is one that will give the required surface cooling without agitation of the bath. The addition of ordinary salt, sodium chloride, greatly improves the performance of water in this respect, the best concentration being around 10 percent. Most inorganic salts are effective in suppressing the formation of vapor at the surface of the steel and thus aid in cooling steel uniformly and eliminating the formation of soft spots. To minimize the formation of vapor, water-base quenching liquids must be kept cold, preferably under 20°C (about 70°F). The addition of some other soluble materials to water such as soap is extremely detrimental because of increased formation of vapor.

**Liquids for Quenching Deep-Hardening Steels**   When oil quenching is required, use a steel of sufficient alloy content to produce a completely martensitic structure at the surface over the heaviest section of the work. To minimize the possibility of cracking, especially when hardening tool steels, keep the quenching oil warm, preferably between 40 and 65°C (about 100 and 150°F). If this expedient is insufficient to prevent

cracking, the work may be removed just before the start of the hardening transformation and cooled in air. Whether or not transformation has started can be determined with a permanent magnet, the work being completely nonmagnetic before transformation if completely hardened by the quench.

The cooling characteristics of quenching oils are difficult to evaluate and have not been satisfactorily correlated with the physical properties of the oils as determined by the usual tests. The standard tests are important with regard to secondary requirements of quenching oils. Low viscosity assures free draining of oil from the work and therefore low oil loss. A high flash and fire point assures a high boiling point and reduces the fire hazard which is increased by keeping the oil warm. A low carbon residue indicates stability of properties with continued use and little sludging. The steam-emulsion number should be low to assure low water content, water being objectionable because of its vapor-film-forming tendency and high cooling power. A low saponification number assures that the oil is of mineral base and not subject to organic deterioration of fatty oils which give rise to offensive odors. Viscosity index is a valuable property for maintenance of composition.

**Effect of the Condition of Surface**   The factors that affect the depth of hardening are the hardenability of the steel, the size of specimen, the quenching medium, and finally the condition of the surface of the steel before quenching. Steel that carries a heavy coating of scale will not cool so rapidly as a steel that is comparatively scale-free, and soft spots may be produced; or, in extreme cases, complete lack of hardening may result. It is therefore essential to minimize scaling as much as possible. Decarburization can also produce undesirable results such as nonuniform hardening and thus lower the resistance of the material to alternating stresses.

**Tempering** of fully hardened steel is carried out for relief of quenching stresses and for the recovery of a limited degree of toughness and ductility. The operation will give almost any desired combination of properties by proper selection of the time-temperature conditions. Table 7 gives the Brinell hardness obtained in several carbon and alloy steels when tempered for 1 h at various temperatures. It should be emphasized that steel should be tempered immediately after hardening to prevent cracking caused by internal stresses.

### The Relation of Design to Heat Treatment

Care must be taken in the design of a machine part to prevent cracking or distortion during heat treatment. With proper design the entire piece may be heated and cooled at approximately the same rate during the heat-treating operation. A light section should never be joined to a heavy section. Sharp reentrant angles should be avoided. Sharp corners and inadequate fillets produce serious stress concentration, causing the actual service stresses to build up to a point where they amount to two to five times the normal working stress calculated by the engineer in the original layout. The use of generous fillets is especially desirable with all high-strength alloy steels.

It is well for the designer to remember that the modulus of elasticity of all commercial steels, either carbon or alloy, is the same so far as practical designing is concerned. The deflection under load of a given part is, therefore, entirely a function of the section of the part and is not affected by the composition or heat treatment of the steel. Consequently if a part deflects excessively, a change in design is necessary; either a heavier

**Table 7. Brinell Hardness of Carbon and Alloy Steels**

| Composition of steel | As quenched | Brinell hardness when hardened and tempered at | | | | | |
|---|---|---|---|---|---|---|---|
| | | 95°C (200°F) | 205°C (400°F) | 315°C (600°F) | 425°C (800°F) | 540°C (1000°F) | 650°C (1200°F) |
| 0.35 carbon | 550 | 534 | 495 | 415 | 352 | 269 | 217 |
| 0.80 carbon | 697 | 712 | 653 | 555 | 461 | 363 | 269 |
| 1.2 carbon | 728 | 755 | 682 | 601 | 477 | 375 | 285 |
| AISI 2340 | 668 | 614 | 534 | 461 | 375 | 307 | 237 |
| AISI 1340 | 627 | 614 | 539 | 469 | 401 | 311 | 241 |
| AISI 5140 | 601 | 589 | 545 | 477 | 408 | 331 | 262 |
| AISI 4140 | 614 | 614 | 545 | 485 | 429 | 363 | 302 |
| AISI 6145 | 601 | 601 | 555 | 485 | 415 | 363 | 302 |
| 0.70 C; 18 W; 4 Cr; 1 V | 720 | 720 | 682 | 663 | 668 | 710 | 653 |

section must be used or the points of support must be increased.

## CASEHARDENING

The production of articles having a soft ductile interior and a very hard surface can be accomplished by carburizing a low-carbon steel at an elevated temperature and then quenching. The process of carburizing followed by hardening is known as casehardening.

**Carburizing methods** can be divided into three groups: pack carburizing where a solid carburizing agent is used, gas carburizing, and liquid carburizing. In **pack carburizing,** the usual compound contains 20 percent metallic carbonates (mostly barium carbonate) bound to a hardwood charcoal by the use of oil, tar, or molasses. Sometimes up to 20 percent coke is added in order to increase the rate of heat transfer. Before using again, used compound should have some new compound added to make up for loss of carbonates, powdering of the compound, and burning of the charcoal and coke. One part new compound is often added to 3 parts used compound, but with care in handling this can be cut down to 1 part new to 5 parts old.

**Carburizing boxes** are usually made from a high-nickel-chromium heat-resisting alloy. Cast-steel or welded-steel plate boxes may be substituted when the boxes are used infrequently or when the work being carburized is large and nonsymmetrical. Work should be packed with its longest dimension vertical, and the minimum amount of compound should be used so as to decrease the heating time. Normal **carburizing temperatures** are around 925°C (1700°F), but for nickel steels a temperature of 900°C (1650°F) is used and for shallow-case depth, temperatures as low as 845°C (1550°F). It is difficult to prevent variations of 0.25 mm (0.010 in) in case depth, and therefore, it is standard practice to specify case depths of at least 0.63 mm (0.025 in) when pack carburizing. By using a minimum of carburizing compound, the overall carburizing time for a case of 1.00 to 1.25 mm (0.040 to 0.050 in) will be about 9 h. The pack method of carburizing is adaptable to both batch and continuous furnace operation; warping is held to a minimum. The main disadvantages are the time consumed in heating the charge and the high labor cost of packing and unpacking.

**Gas carburizing** is usually carried out in heated retorts that can be rotated to tumble the work. Low labor costs and automatic quenching reduce the operating costs to a low

figure, but the high initial investment is justified only by large units operating at high output. The carburizing gases used are carbon monoxide, methane, ethane, and propane. The hydrocarbons will break down, if undiluted, liberating an excessive amount of carbon in the form of soot on all exposed surfaces. By diluting with gases such as nitrogen and hydrogen, the amount of free carbon deposit will be decreased. Cases of 1.00 to 1.25 mm (0.040 to 0.050 in) can be obtained in 4 h at 925°C (1700°F). By suitable adjustment of time, temperature, and gas composition, the surface carbon content and the carbon gradient can be varied to meet almost any requirement.

**Liquid carburizing** is done in activated baths of calcium cyanamid, sodium or potassium cyanide, and controlling chemicals which govern the decomposition of the cyanides. The baths are operated at between 815 and 900°C (1500 and 1650°F), and cases of 0.5 mm (0.020 in) are obtained in 90 min. For case depths of more than about 0.75 mm (0.030 in) it is more economical to use either pack or gas carburizing. The process is extremely flexible, easily controlled, and particularly well adapted to small units. With continuous operation and automatic quenching, labor costs are low.

The type of **steel for carburizing** will depend upon section and distortion limits of the finished part and the stresses to which it will be subjected. Plain carbon steels of between 0.15 and 0.25 percent carbon are used extensively and after heat treatment will develop core strengths as high as 689 MPa (100,000 lb/in²). To develop higher core strengths, it is necessary to have alloying elements such as nickel, chromium, manganese, or molybdenum to give sufficient hardenability. In regard to the correct **heat treatment,** it must be remembered that a carburized steel is a duplex material having a high-carbon case and a low-carbon core with a more or less gradual transition from one to the other. Two critical temperatures are thus involved in the hardening of a carburized part.

**Annealing and Hardening of Carburized Parts**    To minimize distortion, a part should be annealed at a temperature about 30°C (about 50°F) above the carburizing temperature, and the rate of cooling from the annealing temperature should be regulated to provide the required structure for best machinability. For **hardening,** there are three treatments in general use. First, a direct quench from the carburizing temperature into a suitable quenching medium. This method was not used until recently except with parts of regular shape because of the danger of cracking and distortion. With the advent of steels that maintain a fine grain size at carburizing temperatures, distortion is found to be less by direct quenching than by any other method. A second treatment is to cool slowly from the carburizing temperature, reheat to above the critical of the case, and quench. This facilitates handling of large quantities of carburized parts delivered discontinuously from a batch-type furnace and minimizes the retention of austenite in the high-carbon case. The old "double-quench" method in which the part was slowly cooled from the carburizing temperature, reheated and quenched from above the core critical temperature, then reheated and quenched from above the critical of the case is no longer favored.

**Cyaniding**   Where it is necessary to have only a superficial hard wear-resisting surface, a rapid method of casehardening is to use a molten cyanide bath. The bath usually consists of sodium cyanide with sodium chloride and sodium carbonate to retard the decomposition of the cyanide. The cyaniding should be carried out at a temperature just above the critical of the core, and the steel should be quenched directly from the cyanide bath. A uniform case depth of around 0.25 mm (0.010 in) is obtained in 1 h in both carbon and low-allow steels, and besides carbon the case will contain very hard iron nitrides which increase the wear resistance of the surface of the steel part. Cyanide baths are frequently used simply as a heating medium in connection with the hardening of steels to prevent surface decarburization and produce work with a clean surface.

**Nitriding**   The introduction of nitrogen into the outer surface of steel parts in order to give an extremely hard, wear-resisting case is called nitriding. The treatment consists in heating steel to a temperature of 510 to 540°C (950 to 1000°F) inside a chamber through which a controlled stream of ammonia gas is passed. The treatment usually lasts 50 to 90 h depending upon the composition of the steel and the depth to which it is desired to effect nitrification. The nitriding temperature is below the thermal critical range and even below the usual tempering temperature so that the mechanical properties of the previously heat-treated core metal are not affected by nitriding. The big advantage of nitriding is that hardening is accomplished without a quenching operation, so that complicated shapes and articles of uneven cross section can be treated with safety. The steel used is usually heat-treated before machining, rough-machined, given a heat treatment at the nitriding temperature without ammonia in order to produce whatever slight distortion is likely to occur, then the part is finish machined and nitrided. Besides wear resistance, nitriding aids in the retention of hardness at elevated temperatures and to some extent increases the resistance to corrosion and resistance to fatigue. The disadvantages of nitriding include high cost, time of operation, and need for expert attention.

It has been found that chromium and aluminum are desirable in steels for nitriding, and compositions especially adapted for nitriding have been developed. A typical composition is:

Carbon . . . . . .  0.20–0.30 percent
Manganese . .  0.40–0.60 percent
Aluminum . . .  0.90–1.40 percent
Chromium . . .  0.90–1.40 percent
Molybdenum  0.15–0.25 percent

**Flame hardening** is the local heating of steel above the critical temperature so that on subsequent quenching a hardened layer will be produced. The depth of the flame-hardened layer will vary from about 1.5 to 6.5 mm (about $\frac{1}{16}$ to $\frac{1}{4}$ in) depending upon the service requirements. For local hardening

or for hardening the surface of large steel parts, this method has been very useful, especially since distortion of the part is kept to a minimum. The chemical composition of the steel has to be such as to respond readily to heat treatment. For plain carbon steels, the carbon content should be between 0.35 to 0.70 percent, although steels with higher carbon content may be flame-hardened if care is taken to prevent surface cracking. To obtain satisfactory results by flame hardening, the character of the flame, its distance from the surface of the work, its speed of travel, and the timing of the quench must all be under perfect control. After quenching, tempering is essential to relieve the stresses, a temperature of 205°C (about 400°F) usually being sufficient. Flame hardening may be adapted to castings, forgings, or rolled sections irrespective of size. Typical applications are for the hardening of gear teeth, cams, wheel treads, rail ends, and many machine parts.

**Local hardening** can also be accomplished by local heating with electricity. Resistance heating is useful in hardening local sections of some forgings and castings, but in general its principal application is for heating parts having a uniform cross section. Induction heating is well adapted for the surface hardening of cylindrical parts. Crankshaft bearings are hardened by applying a high-frequency current to the bearing section for a few seconds. When the steel is heated to the desired depth, water is sprayed on the heated surface through holes in the inductor blocks surrounding the bearing. The extent of the heated zone can be so closely controlled that the fillets will remain soft while the bearing surface is hardened, thereby reducing the possibility of fatigue failure without sacrificing wear resistance at the bearing surface.

A **hard working surface** also can be obtained on a **ductile backing steel** by use of composite steels or surfacing by welding. **Composite steels** consist of two or more steels of different composition and are used for hacksaw blades, shear blades, knives of all types, lathe tools, drills, dies, vault steels, jail bars, and stoker bars. The usual usual method of making composite steel is to cast molten steel around a steel insert in a mold. For best results the insert is pickled to remove all oxide and may even be plated with a protective coating of iron. **Surfacing by welding** (see Sec. 13) permits the use of hard-facing alloys on steel parts subject to abrasion. In addition to increasing the life of equipment and decreasing costs, hard facing can be used for salvaging or reclaiming worn parts. No single type of hard-facing material is satisfactory for all applications. Various facing alloys of widely different composition are used for different requirements of hardness and shock resistance. These alloys may be divided into four groups: (1) alloy steels containing less than 20 percent of alloying elements such as chromium, tungsten, and manganese which are comparatively economical and possess greatest shock resistance; (2) ferrous alloys containing more than 20 percent alloying elements which are therefore harder and give longer service; (3) nonferrous alloys of cobalt, chromium, and tungsten which are resistant to wear and corrosion and maintain a high hardness at elevated temperatures; and (4) tungsten, boron, or tantalum carbides supported in a matrix of cobalt or nickel. These carbides are the hardest facing alloys and give the maximum resistance to abrasion.

**Clad Steels**   A clad steel plate is a composite plate made of mild steel with a cladding of corrosion- or heat-resistant metal on one or both sides.

The cladding may consist of various grades of stainless

steels, nickel, monel, Inconel, cupronickel, or silver. There are many ways of producing clad steel, but the usual method is to make a "sandwich" of two mild steel plates and two cladding plates separated by an inert material in the center. The sandwich is welded around the edges to prevent oxidation on heating and is rolled at about 1200°C (about 2200°F). During rolling, welding occurs and the plates are reduced to the desired thickness. The edges are sheared and the sandwich separates into two plates, each clad on one face. The thickness of the clad material may vary from 5 to 50 percent of the thickness of the clad plate but normally is held to 10 or 20 percent. The clad steels are available in the form of plate, sheet, and strip and may be obtained as wire.

The clad steels are used in place of solid corrosion- or heat-resistant materials and thus are found in a wide variety of industries. They are also used where corrosion is a minor problem but where freedom from contamination of the materials handled is essential. In addition to savings in material costs, the clad steels are frequently easier to fabricate than solid plates of the cladding material. Their high heat conductivity is another factor in their selection for many applications. For instance, for cooking vessels, in the canning, varnish, and soap industries, the inside of such vessels must be stainless, but the higher heat conductivity of the mild-steel backing prevents hot spots and burning that occur with solid stainless vessels. Clad steels are used for processing equipment in the chemical, food, beverage, drug, paper, textile, oil, and associated industries.

**Chromizing** Chromizing of low-carbon steel is effective in improving corrosion resistance by developing a surface containing up to 40 percent chromium. Some forming operations can be carried out on chromized material. Most chromizing is accomplished by packing the steel to be treated in a powdered mixture of chromium and alumina and then heating to above 1260°C (2300°F) for 3 or 4 h in a reducing atmosphere. Another method is to expose the parts to be treated to gaseous chromium compounds at temperatures above 845°C (about 1550°F). Flat rolled sheets for corrosive applications such as auto mufflers can thus be chromized in open-coil annealing facilities.

## THERMOMECHANICAL TREATMENT

The effects of mechanical treatment and heat treatment on the mechanical properties of steel have been discussed earlier in this section. Thermomechanical treatment consists of combining controlled (sometimes large) amounts of plastic deformation with the heat-treatment cycle to achieve improvements in yield strength beyond those attainable by the usual rolling practices alone or rolling followed by a separate heat treatment. The tensile strength, of course, is increased at the same time as the yield strength (not necessarily to the same degree), and other properties such as ductility, toughness, creep resistance, and fatigue life can be improved. However, the high strength and hardness of thermomechanically treated steels limit their usefulness to the fabrication of components that require very little cold forming or machining, or very simple shapes such as strip and wire that can be used as part of a composite structure. Although the same yield strength may be achieved in a given steel by different thermomechanical treatments, the other mechanical properties (particularly the toughness) are not necessarily the same.

**Table 8. Classification of Thermomechanical Treatments***

Class I. Deformation before austenite transformation
   *a.* Normal hot-working processes
      (hot/cold working)
   *b.* Deformation before transformation to martensite
      (ausforming, austforming, austenrolling, hot-cold working,
      marworking, warm working)
   *c.* Deformation before transformation to ferrite-carbide aggregates
      (austentempering)
Class II. Deformation during austenite transformation
   *a.* Deformation during transformation to martensite (Zerolling and
      Ardeform processes)
   *b.* Deformation during transformation to ferrite-carbide aggregates
      (flow tempering of bainite and isoforming)
Class III. Deformation after austenite transformation
   *a.* Deformation of martensite followed by tempering
   *b.* Deformation of tempered martensite followed by aging
      (flow tempering, marstraining, strain tempering, tempforming,
      warm working)
   *c.* Deformation of isothermal transformation products
      (patenting, flow tempering, warm working)

*From Radcliffe and Kula, Syracuse University Press, 1964.

There are many possible combinations of deformation schedules and time-temperature relationships in heat treatment that can be used for thermomechanical treatment, and individual treatments cannot be discussed here. Table 8 classifies broadly thermomechanical treatments into three principal groups related to the time-temperature dependence of the transformation of austenite discussed earlier under heat treatment. The names in parentheses following the subclasses in the table are those of some types of thermomechanical treatments that have been used commercially or have been discussed in the literature.

## COMMERCIAL STEELS

The variety of applications of steel for engineering purposes is due to the wide range of mechanical properties obtainable by changes in carbon content and heat treatment. Applications of carbon steels are given in Table 9. Carbon steels can be subdivided roughly into three groups: (1) low-carbon steel, 0.05 to 0.25 percent carbon, for use where only moderate strength is required together with considerable plasticity; (2) machinery steels, 0.30 to 0.55 percent carbon, which can be heat-treated to develop high strength; and (3) tool steels, containing from 0.60 to 1.30 percent carbon (this range also includes rail and spring steels).

The chemical compositions and mechanical and physical properties of many of the steels whose uses are listed in Table 9 are covered by specifications adopted by the American Society for Testing and Materials (ASTM), the Society of Automotive Engineers (SAE), the American Society of Mechanical Engineers (ASME), and other groups including governmental agencies engaged in military procurement. Because of the large number of these documents, only a few that are pertinent to the present discussion can be mentioned here. The mechanical properties of a variety of steels excerpted from some ASTM specifications are listed in Tables 11*a* and 11*b*.

**Low-Carbon Steels** Of the many low-carbon-steel prod-

### Table 9. Applications of Carbon Steels

| Percent C | Uses |
|---|---|
| 0.05–0.10 | Sheet, strip, tubing, wire nails |
| 0.10–0.20 | Rivets, screws, parts to be case-hardened |
| 0.20–0.35 | Structural steel, plate, forgings such as camshafts |
| 0.35–0.45 | Machinery steel—shafts, axles, connecting rods, etc. |
| 0.45–0.55 | Large forgings—crankshafts, heavy-duty gears, etc. |
| 0.60–0.70 | Bolt-heading and drop-forging dies, rails, setscrews |
| 0.70–0.80 | Shear blades, cold chisels, hammers, pickaxes, band saws |
| 0.80–0.90 | Cutting and blanking punches and dies, rock drills, hand chisels |
| 0.90–1.00 | Springs, reamers, broaches, small punches, dies |
| 1.00–1.10 | Small springs and lathe, planer, shaper, and slotter tools |
| 1.10–1.20 | Twist drills, small taps, threading dies, cutlery, small lathe tools |
| 1.20–1.30 | Files, ball races, mandrels, drawing dies, razors |

ucts, sheet and strip steels are becoming increasingly important. The consumption of steel in the sheet and tinplate industry has accounted for approximately 50 percent of the total steel production in the United States. This large production has been made possible by the development of the continuous sheet and strip rolling mills. Applications in which large quantities of sheet are employed are tinplate for food containers; black, galvanized, and terne-coated sheets for building purposes; and high-quality sheets for automobiles, furniture, refrigerators, and countless other stamped, formed, and welded products. The difference between **sheet** and **strip** is based on width and is arbitrary. The difference between hot- and cold-rolled products is that in hot rolling the steel is heated before the final rolling and in cold rolling it is not. Cold working produces a better surface finish, improves the mechanical properties, and permits the rolling of thinner-gage material than hot rolling. Approximate mechanical properties of cold-rolled strip steel are given in Table 10.

Sheets for deep-drawing applications must be dead soft so as to have a maximum amount of plasticity. They must also have a relatively fine grain size, since a large grain size will cause a rough finish, an "orange-peel" effect, on the deep-drawn article. The sharp yield point characteristic of low-carbon steel must be eliminated to prevent sudden local elongations in the sheet during forming, which result in strain marks called **stretcher strains** or **Lüders lines**. This can be done by cold working (Fig. 1), a reduction of only 1 percent in thickness usually being sufficient. This cold reduction is

### Table 10. Approximate Mechanical Properties for Various Tempers of Cold-Rolled Carbon Strip Steel*

| Temper | Tensile strength | | Elongation in 50 mm or 2 in for 1.27-mm (0.050 in) thickness of strip, % | Remarks |
|---|---|---|---|---|
| | MPa | 1,000 lb/in² | | |
| No. 1 (hard) | 621 ± 69 | 90 ± 10 | | A very stiff cold-rolled strip intended for flat blanking only, and not requiring ability to withstand cold forming |
| No. 2 (half-hard) | 448 ± 69 | 65 ± 10 | 10 ± 6 | A moderately stiff cold-rolled strip intended for limited bending |
| No. 3 (quarter-hard) | 379 ± 69 | 55 ± 10 | 20 ± 7 | A medium-soft cold-rolled strip intended for limited bending, shallow drawing, and stamping |
| No. 4 (skin-rolled) | 331 ± 41.4 | 48 ± 6 | 32 ± 8 | A soft ductile cold-rolled strip intended for deep drawing where no stretcher strains or fluting are permissible |
| No. 5 (dead-soft) | 303 ± 41.4 | 44 ± 6 | 39 ± 6 | A soft ductile cold-rolled strip intended for deep drawing where stretcher strains or fluting are permissible. Also for extrusions |

* ASTM A109. Complete specification should be consulted in ASTM Standards (latest edition).

**Table 11a. Mechanical Properties of Some Constructional Steels***

| ASTM designation | Thickness range, mm(in) | Yield point, min | | Tensile strength | | Elongation in 200 mm (8 in) min, % | Suit-ability for welding |
|---|---|---|---|---|---|---|---|
| | | MPa | 1,000 lb/in² | MPa | 1,000lb/in² | | |
| Structural carbon-steel plates | | | | | | | |
| ASTM A7† | All thicknesses | 228 | 33 | 414–517 | 60–75 | 21 | No |
| ASTM A373† | To 100 mm (4 in), incl. | 221 | 32 | 400–517 | 58–75 | 21 | Yes |
| ASTM A36 | To 100 mm (4 in), incl. | 248 | 36 | 400–552 | 58–80 | 20 | Yes |
| Low- and intermediate-tensile-strength carbon-steel plates | | | | | | | |
| ASTM A283 | (structural quality) | | | | | | |
| Grade A | All thicknesses | 165 | 24 | 310 | 45 | 28 | Yes |
| Grade B | All thicknesses | 186 | 27 | 345 | 50 | 25 | Yes |
| Grade C | All thicknesses | 207 | 30 | 379 | 55 | 22 | Yes |
| Grade D | All thicknesses | 228 | 33 | 414 | 60 | 20 | Yes |
| Carbon-silicon steel plates for machine parts and general construction | | | | | | | |
| ASTM A284 | | | | | | | |
| Grade A | To 305 mm (12 in) | 172 | 25 | 345 | 50 | 25 | Yes |
| Grade B | To 305 mm (12 in) | 159 | 23 | 379 | 55 | 23 | Yes |
| Grade C | To 305 mm (12 in) | 145 | 21 | 414 | 60 | 21 | Yes |
| Grade D | To 200 mm (8 in) | 145 | 21 | 414 | 60 | 21 | Yes |
| Carbon-steel pressure-vessel plates | | | | | | | |
| ASTM A285 | | | | | | | |
| Grade A | To 50 mm (2 in) | 165 | 24 | 303–379 | 44–55 | 27 | Yes |
| Grade B | To 50 mm (2 in) | 186 | 27 | 345–414 | 50–60 | 25 | Yes |
| Grade C | To 50 mm (2 in) | 207 | 30 | 379–448 | 55–65 | 23 | Yes |
| Structural steel for locomotives and cars | | | | | | | |
| ASTM A113 | | | | | | | |
| Grade A | All thicknesses | 228 | 33 | 414–496 | 60–72 | 21 | No |
| Grade B | All thicknesses | 186 | 27 | 345–427 | 50–62 | 24 | No |
| Grade C | All thicknesses | 179 | 26 | 331–400 | 48–58 | 26 | No |
| Structural steel for ships | | | | | | | |
| ASTM A131 | | | | | | | |
| Grade A | To 13 mm (½ in) | | | | | | |
| Grade B | To 25 mm (1 in) | | | | | | |
| Grade C | To 50 mm (2 in) | 221 | 32 | 400–490 | 58–71 | 21 | No |
| Grade E | To 50 mm(2 in) | | | | | | |
| Grade CS | To 50 mm (2 in) | | | | | | |
| Grade R | To 50 mm (2 in) | | | | | | |
| High-strength low-alloy steel plates | | | | | | | |
| ASTM A242 | To 19 mm (¾ in), incl. | 345 | 50 | 485 | 70 min | 18 | Yes |
| ASTM A440 | Over 19 to 38 mm (¾ to 1½ in), incl. | 315 | 46 | 460 | 67 min | 18 | No |
| ASTM A441 | Over 38 to 102 mm (1½ to 4 in), incl. | 290 | 42 | 435 | 63 min | 18 | Yes |
| ASTM A588 | Up to 102 mm (4 in), incl. | 345 | 50 | 485 | 70 | 18 | Yes |

**Table 11a. Mechanical Properties of Some Constructional Steels (*Continued*)**

| ASTM designation | Thickness range, mm(in) | Yield point, min | | Tensile strength | | Elongation in 200 mm (8 in) min, % | Suitability for welding |
|---|---|---|---|---|---|---|---|
| | | MPa | 1,000 lb/in² | MPa | 1,000lb/in² | | |
| *Heat-treated constructional alloy-steel plates* | | | | | | | |
| ASTM A514 | To 64 mm (2½ in), incl. | 700 | 100 | 800–950 | 115–135 | 18‡ | Yes |
| | Over 64 to 102 mm (2½ to 4 in), incl. | 650 | 90 | 750–950 | 105–135 | 17‡ | Yes |

*See appropriate ASTM documents for properties of other plate steels, shapes, bars, wire, tubing, etc.
†Included for comparison only; no longer used for specification.
‡Elongation in 50 mm (2 in), min.

**Table 11b. Specifications of ASTM and SAE for Some High-Strength Low-Alloy (HSLA) Steels**

| Society | Designation | Min yield point[a] | | Min tensile strength | | Min thickness[b] | |
|---|---|---|---|---|---|---|---|
| | | MPa | 1,000 lb/in² | MPa | 1,000 lb/in² | mm | in |
| SAE | J410b Grade 42X | 290 | 42 | 414 | 60 | 9.5 | ⅜ |
| ASTM | A572 Grade 42 | 290 | 42 | 414 | 60 | 101.6 | 4 |
| SAE | J410b Grade 945X | 310 | 45 | 414 | 60 | 9.5 | ⅜ |
| ASTM | A572 Grade 45 | 310 | 45 | 414 | 60 | 38.1 | 1½ |
| ASTM | A607 Grade 45 | 310 | 45 | 414 | 60 | c | c |
| ASTM | A606 | 310 | 45 | 448 | 65 | c | c |
| SAE | J410b Grades 945A, C[d] | 310 | 45 | 448 | 65[e] | 12.7 | ½[f] |
| SAE | J410b Grade 950X | 345 | 50 | 448 | 65 | 9.5 | ⅜ |
| ASTM | A572 Grade 50 | 345 | 50 | 448 | 65 | 38.1 | 1½ |
| ASTM | A607 Grade 50 | 345 | 50 | 448 | 65 | c | c |
| SAE | J410b Grades 950A, B, C, D[d] | 345 | 50 | 483 | 70 | 38.1 | 1½[f] |
| ASTM | A242 | 345 | 50 | 483 | 70 | 19.1 | ¾[f] |
| ASTM | A440[d] | 345 | 50 | 483 | 70 | 19.1 | ¾[f] |
| ASTM | A441 | 345 | 50 | 483 | 70 | 19.1 | ¾[f] |
| ASTM | A588 | 345 | 50 | 483 | 70 | 101.6 | 4[f] |
| SAE | J410b Grade 955X | 379 | 55 | 483 | 70 | 9.5 | ⅜ |
| ASTM | A572 Grade 55 | 379 | 55 | 483 | 70 | 38.1 | 1½ |
| ASTM | A607 Grade 55 | 378 | 55 | 483 | 70 | c | c |
| SAE | J410b Grade 960X | 414 | 60 | 517 | 75 | 9.5 | ⅜ |
| ASTM | A572 Grade 60 | 414 | 60 | 517 | 75 | 25.4 | 1 |
| ASTM | A607 Grade 60 | 414 | 60 | 517 | 75 | c | c |
| SAE | J410b Grade 965X | 448 | 65 | 552 | 80 | 9.5 | ⅜ |
| ASTM | A572 Grade 65 | 448 | 65 | 552 | 80 | 12.7 | ½ |
| ASTM | A607 Grade 65 | 448 | 65 | 552 | 80 | c | c |
| SAE | J410b Grade 970X | 483 | 70 | 586 | 85 | 9.5 | ⅜ |
| ASTM | A607 Grade 70 | 483 | 70 | 586 | 85 | c | c |
| SAE | J410b Grade 980X | 552 | 80 | 655 | 95 | 9.5 | ⅜ |

[a]SAE steels specify minimum yield strength.
[b]Applies to plates and bars, approximate web thickness for structurals.
[c]ASTM A606 and A607 apply to sheet and strip only.

[d]SAE J410b Grades 945C and 950C and ASTM A440 steels are high-strength carbon-manganese steels rather than HSLA steels.
[e]Reduced 34.5 MPa (5,000 lbf/in²) for sheet and strip.
[f]Available in heavier thickness at reduced strength levels.

usually done by cold rolling, known as **temper rolling,** followed by alternate bending and reverse bending in a roller leveler. Temper rolling must always precede roller leveling because soft-annealed sheets will "break" (yield locally) in the roller leveler. An important phenomenon in these temper-rolled low-carbon sheets is the return of the sharp yield point after a period of time. This is known as **aging** in steel. The return of the yield point is accompanied by an increase in hardness and a loss in ductility. Greater reductions by temper rolling decrease the rate of the aging process, and an increase in temperature greatly increases its rate. It is therefore necessary to fabricate the sheets soon after temper rolling in order to avoid stretcher strains. Nonaging sheets can be supplied from aluminum killed steels, but the higher cost of killed steel usually restricts its use to severe drawing applications.

High-strength hot-rolled, cold-rolled, and galvanized sheets are now available with specified yield strengths. By the use of small alloy additions of columbium or vanadium, it is possible to meet the requirements of ASTM A607-70 for hot-rolled and cold-rolled sheets—345 MPa (50,000 lb/in²) yield point, 483MPa (70,000 lb/in²) tensile strength, and 22 percent elongation in 2 in. By further alloy additions, sheets are produced to a minimum of 448MPa (65,000 lb/in²) yield point, 552 MPa (80,000 lb/in²) tensile strength, and 16 percent elongation in 2 in. The sheets are available in coil form and are used extensively for metallic buildings and for welding into tubes for construction of furniture, etc.

**Structural Carbon Steels** Bridges and buildings are constructed with structural carbon steel meeting the requirements of ASTM A36 (Table 11a). This steel has a minimum yield point of 248 MPa (36,000 lb/in²) and was developed to fill the need for a higher-strength structural carbon steel than the steels formerly covered by ASTM A7 and A373 (Table 11a). The controlled composition of A36 steel provides good weldability and furnishes a significant improvement in the economics of steel construction. The structural carbon steels are available in the forms of plates, shapes, sheet piling, and bars, all in the hot-rolled condition. A uniform strength over a range of section thickness is provided by adjusting the amount of carbon, manganese, and silicon in the A36 steel.

**High-Strength Low-Alloy Steels** These steels have in the past been referred to as "high-tensile steels" and "low-alloy steels," but the name "high-strength low-alloy steels," abbreviated as "HSLA steels," is now the generally accepted designation.

HSLA steels are a group of steels, intended for general structural and miscellaneous applications, that have minimum yield strengths above about 40,000 lb/in². These steels typically contain small amounts of alloying elements to achieve their strength in the hot rolled or normalized condition. Among the elements used in small amounts, singly or in combination, are columbium, vanadium, manganese, copper, and phosphorus. A complete listing of HSLA steels available from producers in the United States and Canada would include over 200 brands or variations, many of which are not covered by ASTM or other specifications. Table 11b lists numerous ASTM and SAE specifications that cover a large number of HSLA steels. These steels generally are available as sheet, strip, plates, bars, and shapes and generally are sold as proprietary grades.

HSLA steels have characteristics and properties that result in economies to the user when the steels are properly applied. They are considerably stronger, and in many instances tougher, than structural carbon steel, yet have sufficient ductility, formability, and weldability to be fabricated successfully by customary shop methods. In addition, many of the steels have improved resistance to corrosion, so that equal service life in a thinner section or longer life in the same section is obtained in comparison with that of a structural-carbon-steel member. Good resistance to repeated loading and good abrasion resistance in service may be other characteristics of some of the steels. While high strength is a common characteristic of all HSLA steels, the other properties mentioned above may or may not be, singly or in combination, exhibited by any particular steel.

HSLA steels have found wide acceptance in many fields, among which are the construction of railroad cars, trucks, trailers, and buses; welded steel bridges; television and power-transmission towers and lighting standards; columns in high-rise buildings; portable liquefied-petroleum-gas containers; ship construction; oil-storage tanks; air-conditioning equipment; agricultural and earthmoving equipment.

**Quenched and Tempered Low-Carbon Constructional Alloy Steels** These steels, having yield strengths at the 689 MPa (100,000 lb/in²) level, are covered by ASTM A514, by military specifications, and for pressure-vessel applications, by ASME Code Case 1204. They are available in plates, shapes, and bars, and are readily welded. Since they are heat-treated to tempered martensitic structure, they retain excellent toughness at temperatures as low as minus 45°C (minus 50°F). Major cost savings have been effected by using these steels in the construction of pressure vessels, in mining and earthmoving equipment, and for major members of large steel structures.

**Ultraservice Low-Carbon Alloy Steels** (Quenched and Tempered) The need for high-performance materials with higher strength-to-weight ratios for critical military needs, for hydrospace explorations, and for aerospace applications has led to the development of quenched and tempered ultraservice alloy steels. Although these steels are similar in many respects to the quenched and tempered low-carbon constructional alloy steels described above, their significantly higher notch toughness at yield strengths up to 965 MPa (140,000 lb/in²) distinguishes the ultraservice alloy steels from the constructional alloy steels. These steels are not included in the AISI-SAE classification of alloy steels. There are numerous proprietary grades of ultraservice steels in addition to those covered by ASTM designations A543 and A579. Ultraservice steels may be used in large welded structures subjected to unusually high loads, and must exhibit excellent weldability and toughness. In some applications, such as hydrospace operations, the steels must have high resistance to fatigue and corrosion as well.

**Maraging Steels** Two groups of age-hardenable martensitic steels containing nominally about 18 and 12 percent nickel, respectively, called maraging steels (not included in the AISI-SAE classification of alloy steels) contain so little carbon that they are referred to as low-carbon martensites. Iron-carbon martensite is hard and brittle in the as-quenched condition and becomes softer and more ductile when tempered. Carbon-free iron-nickel martensite, on the other hand, is relatively soft and ductile and becomes hard, strong, and tough when aged. Thus maraging steels can be fabricated while they are in a comparatively ductile martensitic condition and later strengthened by a simple aging treatment.

The nominally 18 percent nickel steels contain 17 to 19 percent nickel, 7 to 9½ percent cobalt, 3 to 5 percent molybdenum, and 0.1 to 0.8 percent titanium. By adjusting the percentages of these elements, yield strengths in the range of 1,379 to 2,068 MPa (200,000 to 300,000 lb/in²) can be obtained. In addition to their high mechanical properties, the 18 percent nickel steels have a resistance to stress-corrosion cracking that is superior to that of any known alloy steel of comparable strength.

The 12 percent nickel maraging steels contain 10 to 12 percent nickel, 3 to 5 percent chromium, about 3 percent molybdenum, and titanium and aluminum contents adjusted to develop yield strengths in the range of 1,034 to 1,379 MPa (150,000 to 200,000 lb/in²) or higher. At these yield strengths, the 12 percent nickel maraging steels generally exhibit notch toughness significantly higher than that attained for alloy steels quenched and tempered to these strength levels.

**Cryogenic-Service Steels**    For the economical construction of cryogenic vessels operating from room temperature down to the temperature of liquid nitrogen (−195°C or −320°F) a 9 percent nickel alloy steel has been developed. The mechanical properties as specified by ASTM A353 are 517 MPa (75,000 lb/in²) minimum yield strength and 689 to 827 MPa (100,000 to 120,000 lb/in²) minimum tensile strength. The minimum Charpy impact requirement is 20.34 J (15 ft·lbf) at −195°C (−320°F). For lower temperatures, it is necessary to use austenitic stainless steel.

**Machinery Steels**    A large variety of carbon and alloy steels has been used in the automotive and allied industries. Specifications are published by AISI and SAE on all types of steel, and these specifications should be referred to for detailed information.

A numerical index is used to identify the compositions of AISI (and SAE) steels. Most AISI and SAE alloy steels are made by the basic oxygen, basic open-hearth, or basic electric-furnace processes; a few steels that must be made in the electric furnace carry the prefix E before their number, i.e., E52100. A series of four numerals designates the composition of the AISI steels: the first two indicate the steel type, and the last two indicate, as far as feasible, the average carbon content in "points" or hundredths of 1 percent. Thus 1020 is a carbon steel with a carbon range of 0.18 to 0.23 percent, probably made in the basic oxygen or basic open-hearth furnace, and E4340 is a nickel-chromium molybdenum steel with 0.38 to 0.43 percent carbon made in the electric-arc furnace. The compositions for the standard steels are listed in Tables 12a and 12b. A group of steels developed during World War II,

known as **NE steels,** has been discontinued, although the better alloys have been included in the AISI steels. The NE steels that were retained are those with the first two digits 86, 87, or 98. A group of steels known as **H steels,** which are similar to the standard AISI steels, is being produced with a specified Jominy hardenability; these steels are identified by a suffix H added to the conventional series number. In general, these steels have a somewhat greater allowable variation in chemical composition but a smaller variation in hardenability than would be normal for a given grade of steel. This smaller variation in hardenability results in greater reproducibility of the mechanical properties of the steels on heat treatment; therefore, H steels have become increasingly important in machinery steels.

**Boron steels** are designated by the letter B inserted between the second and third digits, e.g., 50B44. The effectiveness of boron in increasing hardenability was a discovery of the late thirties, when it was noticed that heats treated with complex deoxidizers (containing boron) showed exceptionally good hardenability, high strength, and ductility after heat treatment. It was found that as little as 0.0005 percent of boron increased the hardenability of steels with 0.15 to 0.60 carbon, whereas boron contents of over 0.005 percent had an adverse effect on hot workability. Boron steels achieve special importance in times of alloy shortages, for they can replace such critical alloying elements as nickel, molybdenum, chromium, and manganese and, when properly heat-treated, possess physical properties comparable to the alloy grades they replace. Additional advantages for the use of boron in steels are a decrease in susceptibility to flaking, formation of less

### Table 12a. Chemical Composition of AISI Carbon Steels

| AISI grade designation | Chemical composition limits (ladle analyses), % | | | |
|---|---|---|---|---|
| | C | Mn | P | S |
| 1006 | 0.08 max | 0.25–0.40 | 0.04 max | 0.05 max |
| 1008 | 0.10 max | 0.30–0.50 | 0.04 max | 0.05 max |
| 1010 | 0.08–0.13 | 0.30–0.60 | 0.04 max | 0.05 max |
| 1012 | 0.10–0.15 | 0.30–0.60 | 0.04 max | 0.05 max |
| 1015 | 0.13–0.18 | 0.30–0.60 | 0.04 max | 0.05 max |
| 1016 | 0.13–0.18 | 0.60–0.90 | 0.04 max | 0.05 max |
| 1017 | 0.15–0.20 | 0.30–0.60 | 0.04 max | 0.05 max |
| 1018 | 0.15–0.20 | 0.60–0.90 | 0.04 max | 0.05 max |
| 1019 | 0.15–0.20 | 0.70–1.00 | 0.04 max | 0.05 max |
| 1020 | 0.18–0.23 | 0.30–0.60 | 0.04 max | 0.05 max |
| 1021 | 0.18–0.23 | 0.60–0.90 | 0.04 max | 0.05 max |
| 1022 | 0.18–0.23 | 0.70–1.00 | 0.04 max | 0.05 max |
| 1023 | 0.20–0.25 | 0.30–0.60 | 0.04 max | 0.05 max |
| 1025 | 0.22–0.28 | 0.30–0.60 | 0.04 max | 0.05 max |
| 1026 | 0.22–0.28 | 0.60–0.90 | 0.04 max | 0.05 max |
| 1030 | 0.28–0.34 | 0.60–0.90 | 0.04 max | 0.05 max |
| 1035 | 0.32–0.38 | 0.60–0.90 | 0.04 max | 0.05 max |
| 1037 | 0.32–0.38 | 0.70–1.00 | 0.04 max | 0.05 max |
| 1038 | 0.35–0.42 | 0.60–0.90 | 0.04 max | 0.05 max |
| 1039 | 0.37–0.44 | 0.70–1.00 | 0.04 max | 0.05 max |
| 1040 | 0.37–0.44 | 0.60–0.90 | 0.04 max | 0.05 max |
| 1042 | 0.40–0.47 | 0.60–0.90 | 0.04 max | 0.05 max |
| 1043 | 0.40–0.47 | 0.70–1.00 | 0.04 max | 0.05 max |
| 1045 | 0.43–0.50 | 0.60–0.90 | 0.04 max | 0.05 max |
| 1046 | 0.43–0.50 | 0.70–0.90 | 0.04 max | 0.05 max |
| 1049 | 0.46–0.53 | 0.60–0.90 | 0.04 max | 0.05 max |
| 1050 | 0.48–0.55 | 0.60–0.90 | 0.04 max | 0.05 max |
| 1055 | 0.50–0.60 | 0.60–0.90 | 0.04 max | 0.05 max |
| 1060 | 0.55–0.65 | 0.60–0.90 | 0.04 max | 0.05 max |
| 1064 | 0.60–0.70 | 0.50–0.80 | 0.04 max | 0.05 max |
| 1065 | 0.60–0.70 | 0.60–0.90 | 0.04 max | 0.05 max |

| AISI grade designation | Chemical composition limits (ladle analyses), % | | | |
|---|---|---|---|---|
| | C | Mn | P | S |
| 1070 | 0.65–0.75 | 0.60–0.90 | 0.04 max | 0.05 max |
| 1078 | 0.72–0.85 | 0.30–0.60 | 0.04 max | 0.05 max |
| 1080 | 0.75–0.88 | 0.60–0.90 | 0.04 max | 0.05 max |
| 1084 | 0.80–0.93 | 0.60–0.90 | 0.04 max | 0.05 max |
| 1086 | 0.80–0.93 | 0.30–0.50 | 0.04 max | 0.05 max |
| 1090 | 0.85–0.98 | 0.60–0.90 | 0.04 max | 0.05 max |
| 1095 | 0.90–1.03 | 0.30–0.50 | 0.04 max | 0.05 max |
| Resulfurized (free-cutting) steels | | | | |
| 1108 | 0.08–0.13 | 0.50–0.80 | 0.04 max | 0.08–0.13 |
| 1109 | 0.08–0.13 | 0.60–0.90 | 0.04 max | 0.08–0.13 |
| 1117 | 0.14–0.20 | 1.00–1.30 | 0.04 max | 0.08–0.13 |
| 1118 | 0.14–0.20 | 1.30–1.60 | 0.04 max | 0.08–0.13 |
| 1119 | 0.14–0.20 | 1.00–1.30 | 0.04 max | 0.24–0.33 |
| 1132 | 0.27–0.34 | 1.35–1.65 | 0.04 max | 0.08–0.13 |
| 1137 | 0.32–0.39 | 1.35–1.65 | 0.04 max | 0.08–0.13 |
| 1139 | 0.35–0.43 | 1.35–1.65 | 0.04 max | 0.13–0.20 |
| 1140 | 0.37–0.44 | 0.70–1.00 | 0.04 max | 0.08–0.13 |
| 1141 | 0.37–0.45 | 1.35–1.65 | 0.04 max | 0.08–0.13 |
| 1144 | 0.40–0.48 | 1.35–1.65 | 0.04 max | 0.24–0.33 |
| 1145 | 0.42–0.49 | 0.70–1.00 | 0.04 max | 0.04–0.07 |
| 1146 | 0.42–0.49 | 0.70–1.00 | 0.04 max | 0.08–0.13 |
| 1151 | 0.48–0.55 | 0.70–1.00 | 0.04 max | 0.08–0.13 |
| Rephosphorized and resulfurized (free-cutting) steels | | | | |
| 1110 | 0.08–0.13 | 0.30–0.60 | 0.04 max | 0.08–0.13 |
| 1211 | 0.13 max | 0.60–0.90 | 0.07–0.12 | 0.10–0.15 |
| 1212 | 0.13 max | 0.70–1.00 | 0.07–0.12 | 0.16–0.23 |
| 1213 | 0.13 max | 0.70–1.00 | 0.07–0.12 | 0.24–0.33 |
| 1116 | 0.14–0.20 | 1.10–1.40 | 0.04 max | 0.16–0.23 |
| 1215 | 0.09 max | 0.75–1.05 | 0.04–0.09 | 0.26–0.35 |
| 12L14 | 0.15 max | 0.85–1.15 | 0.04–0.09 | 0.26–0.35 |
| High-manganese carbon steels | | | | |
| 1513 | 0.10–0.16 | 1.10–1.40 | 0.04 max | 0.05 max |
| 1518 | 0.15–0.21 | 1.10–1.40 | 0.04 max | 0.05 max |
| 1522 | 0.18–0.24 | 1.10–1.40 | 0.04 max | 0.05 max |
| 1524 | 0.19–0.25 | 1.35–1.65 | 0.04 max | 0.05 max |
| 1525 | 0.23–0.29 | 0.80–1.10 | 0.04 max | 0.05 max |
| 1526 | 0.22–0.29 | 1.10–1.40 | 0.04 max | 0.05 max |
| 1527 | 0.22–0.29 | 1.20–1.50 | 0.04 max | 0.05 max |
| 1536 | 0.30–0.37 | 1.20–1.50 | 0.04 max | 0.05 max |
| 1541 | 0.36–0.44 | 1.35–1.65 | 0.04 max | 0.05 max |
| 1547 | 0.43–0.51 | 1.35–1.65 | 0.04 max | 0.05 max |
| 1548 | 0.44–0.52 | 1.10–1.40 | 0.04 max | 0.05 max |
| 1551 | 0.45–0.56 | 0.85–1.15 | 0.04 max | 0.05 max |
| 1552 | 0.47–0.55 | 1.20–1.50 | 0.04 max | 0.05 max |
| 1561 | 0.55–0.65 | 0.75–1.05 | 0.04 max | 0.05 max |
| 1566 | 0.60–0.71 | 0.85–1.15 | 0.04 max | 0.05 max |
| 1572 | 0.65–0.76 | 1.00–1.30 | 0.04 max | 0.05 max |

adherent scale, greater softness in the unhardened condition, and better machinability.

Specific applications of these steels cannot be given, since the selection of a steel for a given part must depend upon an intimate knowledge of factors such as the availability and cost of the material, the detailed design of the part, and the severity of the service to be imposed. However, the mechanical properties desired in the part to be heat-treated will determine to a large extent the carbon and alloy content of the steel. Table 13 gives a résumé of mechanical properties that can be expected on heat-treating AISI steels, and Table 14 gives an indication of the effect of mass on the mechanical properties of heat-treated steels.

The low-carbon AISI steels are used for carburized parts, cold-headed bolts and rivets, and for similar applications where high quality is required. The AISI 1100 series are low-carbon **free-cutting** steels for high speed screw-machine stock and other machining purposes. These steels have high sulfur present in the steel in the form of manganese sulfide inclusions causing the chips to break short on machining. Manganese and phosphorus harden and embrittle the steel, which also contributes toward free machining. For a further improvement in

**Table 12b. Alloy-Steel Compositions**[a,b,c,d]

| AISI No. | Chemical composition limits (ladle analyses), %[c,d] | | | | | | | | |
| | C | Mn | P, max | S, max | Si | Ni | Cr | Mo | V |
|---|---|---|---|---|---|---|---|---|---|
| 1330 | 0.28–0.33 | 1.60–1.90 | 0.035 | 0.040 | 0.20–0.35 | | | | |
| 1335 | 0.33–0.38 | 1.60–1.90 | 0.035 | 0.040 | 0.20–0.35 | | | | |
| 1340 | 0.38–0.43 | 1.60–1.90 | 0.035 | 0.040 | 0.20–0.35 | | | | |
| 1345 | 0.43–0.48 | 1.60–1.90 | 0.035 | 0.040 | 0.20–0.35 | | | | |
| 4012 | 0.09–0.14 | 0.75–1.00 | 0.035 | 0.040 | 0.20–0.35 | | | 0.15–0.25 | |
| 4023 | 0.20–0.25 | 0.70–0.90 | 0.035 | 0.040 | 0.20–0.35 | | | 0.20–0.30 | |
| 4024 | 0.20–0.25 | 0.70–0.90 | 0.035 | 0.035–0.050 | 0.20–0.35 | | | 0.20–0.30 | |
| 4027 | 0.25–0.30 | 0.70–0.90 | 0.035 | 0.040 | 0.20–0.35 | | | 0.20–0.30 | |
| 4028 | 0.25–0.30 | 0.70–0.90 | 0.035 | 0.035–0.050 | 0.20–0.35 | | | 0.20–0.30 | |
| 4037 | 0.35–0.40 | 0.70–0.90 | 0.035 | 0.040 | 0.20–0.35 | | | 0.20–0.30 | |
| 4047 | 0.45–0.50 | 0.70–0.90 | 0.035 | 0.040 | 0.20–0.35 | | | 0.20–0.30 | |
| 4118 | 0.18–0.23 | 0.70–0.90 | 0.035 | 0.040 | 0.20–0.35 | | 0.40–0.60 | 0.08–0.15 | |
| 4130 | 0.28–0.33 | 0.40–0.60 | 0.035 | 0.040 | 0.20–0.35 | | 0.80–1.10 | 0.15–0.25 | |
| 4137 | 0.35–0.40 | 0.70–0.90 | 0.035 | 0.040 | 0.20–0.35 | | 0.80–1.10 | 0.15–0.25 | |
| 4140 | 0.38–0.43 | 0.75–1.00 | 0.035 | 0.040 | 0.20–0.35 | | 0.80–1.10 | 0.15–0.25 | |
| 4142 | 0.40–0.45 | 0.75–1.00 | 0.035 | 0.040 | 0.20–0.35 | | 0.80–1.10 | 0.15–0.25 | |
| 4145 | 0.43–0.48 | 0.75–1.00 | 0.035 | 0.040 | 0.20–0.35 | | 0.80–1.10 | 0.15–0.25 | |
| 4147 | 0.45–0.50 | 0.75–1.00 | 0.035 | 0.040 | 0.20–0.35 | | 0.80–1.10 | 0.15–0.25 | |
| 4150 | 0.48–0.53 | 0.75–1.00 | 0.035 | 0.040 | 0.20–0.35 | | 0.80–1.10 | 0.15–0.25 | |
| 4320 | 0.17–0.22 | 0.45–0.65 | 0.035 | 0.040 | 0.20–0.35 | 1.65–2.00 | 0.40–0.60 | 0.20–0.30 | |
| 4340 | 0.38–0.43 | 0.60–0.80 | 0.035 | 0.040 | 0.20–0.35 | 1.65–2.00 | 0.70–0.90 | 0.20–0.30 | |
| E4340[e] | 0.38–0.43 | 0.65–0.85 | 0.025 | 0.025 | 0.20–0.35 | 1.65–2.00 | 0.70–0.90 | 0.20–0.30 | |
| 4419 | 0.18–0.23 | 0.45–0.65 | 0.035 | 0.040 | 0.20–0.35 | | | 0.45–0.60 | |
| 4615 | 0.13–0.18 | 0.45–0.65 | 0.035 | 0.040 | 0.20–0.35 | 1.65–2.00 | | 0.20–0.30 | |
| 4620 | 0.17–0.22 | 0.45–0.65 | 0.035 | 0.040 | 0.20–0.35 | 1.65–2.00 | | 0.20–0.30 | |
| 4621 | 0.18–0.23 | 0.70–0.90 | 0.035 | 0.040 | 0.20–0.35 | 1.65–2.00 | | 0.20–0.30 | |
| 4626 | 0.24–0.29 | 0.45–0.65 | 0.035 | 0.040 | 0.20–0.35 | 0.70–1.00 | | 0.15–0.25 | |
| 4718 | 0.16–0.21 | 0.70–0.90 | | | | 0.90–1.20 | 0.35–0.55 | 0.30–0.40 | |
| 4720 | 0.17–0.22 | 0.50–0.70 | 0.035 | 0.040 | 0.20–0.35 | 0.90–1.20 | 0.35–0.55 | 0.15–0.25 | |
| 4815 | 0.13–0.18 | 0.40–0.60 | 0.035 | 0.040 | 0.20–0.35 | 3.25–3.75 | | 0.20–0.30 | |
| 4817 | 0.15–0.20 | 0.40–0.60 | 0.035 | 0.040 | 0.20–0.35 | 3.25–3.75 | | 0.20–0.30 | |
| 4820 | 0.18–0.23 | 0.50–0.70 | 0.035 | 0.040 | 0.20–0.35 | 3.25–3.75 | | 0.20–0.30 | |
| 5015 | 0.12–0.17 | 0.30–0.50 | 0.035 | 0.040 | 0.20–0.35 | | 0.30–0.50 | | |
| 50B44[f] | 0.43–0.48 | 0.75–1.00 | 0.035 | 0.040 | 0.20–0.35 | | 0.40–0.60 | | |
| 50B46[f] | 0.44–0.49 | 0.75–1.00 | 0.035 | 0.040 | 0.20–0.35 | | 0.20–0.35 | | |
| 50B50[f] | 0.48–0.53 | 0.75–1.00 | 0.035 | 0.040 | 0.20–0.35 | | 0.40–0.60 | | |
| 50B60[f] | 0.56–0.64 | 0.75–1.00 | 0.035 | 0.040 | 0.20–0.35 | | 0.40–0.60 | | |
| 5120 | 0.17–0.22 | 0.70–0.90 | 0.035 | 0.040 | 0.20–0.35 | | 0.70–0.90 | | |
| 5130 | 0.28–0.33 | 0.70–0.90 | 0.035 | 0.040 | 0.20–0.35 | | 0.80–1.10 | | |
| 5132 | 0.30–0.35 | 0.60–0.80 | 0.035 | 0.040 | 0.20–0.35 | | 0.75–1.00 | | |
| 5135 | 0.33–0.38 | 0.60–0.80 | 0.035 | 0.040 | 0.20–0.35 | | 0.80–1.05 | | |
| 5145 | 0.43–0.48 | 0.70–0.90 | 0.035 | 0.040 | 0.20–0.35 | | 0.70–0.90 | | |
| 5147 | 0.46–0.51 | 0.70–0.95 | 0.035 | 0.040 | 0.20–0.35 | | 0.85–1.15 | | |
| 5150 | 0.48–0.53 | 0.70–0.90 | 0.035 | 0.040 | 0.20–0.35 | | 0.70–0.90 | | |
| 5155 | 0.51–0.59 | 0.70–0.90 | 0.035 | 0.040 | 0.20–0.35 | | 0.70–0.90 | | |
| 5160 | 0.56–0.64 | 0.75–1.00 | 0.035 | 0.040 | 0.20–0.35 | | 0.70–0.90 | | |
| 51B60[f] | 0.56–0.64 | 0.75–1.00 | 0.035 | 0.040 | 0.20–0.35 | | 0.70–0.90 | | |
| E51100[e] | 0.98–1.10 | 0.25–0.45 | 0.025 | 0.025 | 0.20–0.35 | | 0.90–1.15 | | |
| E52100[e] | 0.98–1.10 | 0.25–0.45 | 0.025 | 0.025 | 0.20–0.35 | | 1.30–1.60 | | |
| 6118 | 0.16–0.21 | 0.50–0.70 | 0.035 | 0.040 | 0.20–0.35 | | 0.50–0.70 | | 0.10–0.15 |
| 6150 | 0.48–0.53 | 0.70–0.90 | 0.035 | 0.040 | 0.20–0.35 | | 0.80–1.10 | | 0.15 |
| 81B45[f] | 0.43–0.48 | 0.75–1.00 | 0.035 | 0.040 | 0.20–0.35 | 0.20–0.40 | 0.35–0.55 | 0.08–0.15 | |
| 8615 | 0.13–0.18 | 0.70–0.90 | 0.035 | 0.040 | 0.20–0.35 | 0.40–0.70 | 0.40–0.60 | 0.15–0.25 | |
| 8617 | 0.15–0.20 | 0.70–0.90 | 0.035 | 0.040 | 0.20–0.35 | 0.40–0.70 | 0.40–0.60 | 0.15–0.25 | |
| 8620 | 0.18–0.23 | 0.70–0.90 | 0.035 | 0.040 | 0.20–0.35 | 0.40–0.70 | 0.40–0.60 | 0.15–0.25 | |
| 8622 | 0.20–0.25 | 0.70–0.90 | 0.035 | 0.040 | 0.20–0.35 | 0.40–0.70 | 0.40–0.60 | 0.15–0.25 | |
| 8625 | 0.23–0.28 | 0.70–0.90 | 0.035 | 0.040 | 0.20–0.35 | 0.40–0.70 | 0.40–0.60 | 0.15–0.25 | |
| 8627 | 0.25–0.30 | 0.70–0.90 | 0.035 | 0.040 | 0.20–0.35 | 0.40–0.70 | 0.40–0.60 | 0.15–0.25 | |
| 8630 | 0.28–0.33 | 0.70–0.90 | 0.035 | 0.040 | 0.20–0.35 | 0.40–0.70 | 0.40–0.60 | 0.15–0.25 | |
| 8637 | 0.35–0.40 | 0.75–1.00 | 0.035 | 0.040 | 0.20–0.35 | 0.40–0.70 | 0.40–0.60 | 0.15–0.25 | |

| AISI No. | Chemical composition limits (ladle analyses), % [c,d] C | Mn | P, max | S, max | Si | Ni | Cr | Mo | V |
|---|---|---|---|---|---|---|---|---|---|
| 8640 | 0.38–0.43 | 0.75–1.00 | 0.035 | 0.040 | 0.20–0.35 | 0.40–0.70 | 0.40–0.60 | 0.15–0.25 | |
| 8642 | 0.40–0.45 | 0.75–1.00 | 0.035 | 0.040 | 0.20–0.35 | 0.40–0.70 | 0.40–0.60 | 0.15–0.25 | |
| 8645 | 0.43–0.48 | 0.75–1.00 | 0.035 | 0.040 | 0.20–0.35 | 0.40–0.70 | 0.40–0.60 | 0.15–0.25 | |
| 8655 | 0.51–0.59 | 0.75–1.00 | 0.035 | 0.040 | 0.20–0.35 | 0.40–0.70 | 0.40–0.60 | 0.15–0.25 | |
| 8720 | 0.18–0.23 | 0.70–0.90 | 0.035 | 0.040 | 0.20–0.35 | 0.40–0.70 | 0.40–0.60 | 0.20–0.30 | |
| 8740 | 0.38–0.43 | 0.75–1.00 | 0.035 | 0.040 | 0.20–0.35 | 0.40–0.70 | 0.40–0.60 | 0.20–0.30 | |
| 8822 | 0.20–0.25 | 0.75–1.00 | 0.035 | 0.040 | 0.20–0.35 | 0.40–0.70 | 0.40–0.60 | 0.30–0.40 | |
| 9255 | 0.51–0.59 | 0.70–0.95 | 0.035 | 0.040 | 1.80–2.20 | | | | |
| 9260 | 0.56–0.64 | 0.75–1.00 | 0.035 | 0.040 | 1.80–2.20 | | | | |
| 94B17[f] | 0.15–0.20 | 0.75–1.00 | 0.035 | 0.040 | 0.20–0.35 | 0.30–0.60 | 0.30–0.50 | 0.08–0.15 | |
| 94B30[f] | 0.28–0.33 | 0.75–1.00 | 0.035 | 0.040 | 0.20–0.35 | 0.30–0.60 | 0.30–0.50 | 0.08–0.15 | |

[a]These tables are subject to change from time to time, with new steels sometimes added, other steels eliminated, and compositions of retained steels occasionally altered. Current publications of AISI and SAE should be consulted for latest information.

[b]Applicable to blooms, billets, slabs, and hot-rolled and cold-rolled bars.

[c]These steels may be produced by the basic oxygen, basic open-hearth, or basic electric steelmaking process.

[d]Small quantities of certain elements which are not specified or required may be found in alloy steels. These elements are considered to be incidental and are acceptable up to the following maximum amounts: copper to 0.35 percent, nickel to 0.25 percent, chromium to 0.20 percent, and molybdenum to 0.06 percent.

[e]Electric-furnace steel.

[f]Boron content is 0.0005 percent minimum.

machinability, **lead** is added to steel. The usual range is from 0.20 to 0.35 percent lead.

**Cold-finished carbon-steel bars** are used for bolts, nuts, typewriter and cash-register parts, motor and transmission power shafting, piston pins, bushings, oil-pump shafts and gears, etc. Average mechanical properties of cold-drawn steel are given in Table 15. Besides improved mechanical properties, cold-finished steel has better machining properties than hot-rolled products. The surface finish and dimensional accuracy are also greatly improved by cold finishing.

**Forging steels,** between 0.30 and 0.40 percent carbon, are used for axles, bolts, pins, connecting rods, and similar applications. These steels are readily forged and, after heat treatment, develop considerably higher mechanical properties than low-carbon steels. For heavy sections where high strength is required, such as in crankshafts and heavy-duty gears, the carbon may be increased to 0.40 to 0.50 percent and sufficient alloy content used to obtain the necessary hardenability.

## TOOL STEELS

The application of tool steels can generally be fitted into one of the following categories or types of operations: cutting, shearing, forming, drawing, extruding, rolling, and battering. Each of these operations requires in the tool steel a particular physical property or a combination of such metallurgical characteristics as hardness, strength, toughness, wear resistance, and resistance to heat softening, before optimum performance can be realized. These considerations are of prime importance in tool selection; but hardenability, permissible distortion, surface decarburization during heat treatment, and machinability of the tool steel are a few of the additional factors to be weighed in reaching a final decision. In actual practice, the final selection of a tool steel represents a compromise of the most desirable physical properties with the best overall economic performance. Tool steels have been identified and classified by the SAE and the AISI into six major groups, based upon quenching methods, applications, special characteristics, and use in specific industries. These six classes are water-hardening, shock-resisting, cold-work, hot-work, high-speed, and special-purpose tool steels. A simplified classification of these six basic types and their subdivisions is given in Table 16.

**Water-hardening tool steels,** containing 0.60 to 1.40 percent carbon, are widely used because of their low cost, good toughness, and excellent machinability. They are shallow-hardening steels, unsuitable for nondeforming applications because of high warpage, and possess poor resistance to softening at elevated temperatures. Water-hardening tool steels have the widest applications of all major groups and are used for files, twist drills, shear knives, chisels, hammers, and forging dies.

**Shock-resisting tool steels,** with chromium-tungsten, silicon-molybdenum, or silicon-manganese as the dominant alloys, combine good hardenability with outstanding toughness. A tendency to distort easily is their greatest disadvantage. However, oil quenching can minimize this characteristic.

**Cold-work tool steels** are divided into three groups: oil-hardening, medium-alloy air-hardening, and high-carbon, high-chromium. In general, this class possesses high wear resistance and hardenability, develops little distortion, but at best is only average in toughness and in resistance to heat softening. Machinability ranges from good in the oil-hardening grade to poor in the high-carbon, high-chromium steels.

**Hot-work tool steels** are either chromium- or tungsten-based alloys possessing fine nondeforming, hardenability, toughness, and resistance to heat-softening characteristics, with fair machinability and wear resistance. Either air or oil hardening can be employed. Applications are blanking, forming, extrusion and casting dies where temperatures may rise to 540°C (1000°F).

**High-speed tool steels,** the best-known tool steels, possess the best combination of all properties excepting toughness, which is not critical for high-speed cutting operations, and are either tungsten- or molybdenum-base types. Cobalt is added in some cases to improve the cutting qualities in roughing operations. They retain considerable hardness at a red heat. Very high heating temperatures are required for the heat treatment of

### Table 13. Mechanical Properties of Certain AISI Steels with Various Heat Treatments
(Sections up to 40 mm or 1½ in diam or thickness)

| Tempering temp | | Tensile strength | | Yield point | | Reduction of area, % | Elongation in 50 mm (2 in), % | Brinell hardness |
|---|---|---|---|---|---|---|---|---|
| °C | °F | MPa | 1,000 lb/in² | MPa | 1,000 lb/in² | | | |
| AISI 1040 quenched in water from 815°C (1500°F) | | | | | | | | |
| 315 | 600 | 862 | 125 | 717 | 104 | 46 | 11 | 260 |
| 425 | 800 | 821 | 119 | 627 | 91 | 53 | 13 | 250 |
| 540 | 1000 | 758 | 110 | 538 | 78 | 58 | 15 | 220 |
| 595 | 1100 | 745 | 108 | 490 | 71 | 60 | 17 | 216 |
| 650 | 1200 | 717 | 104 | 455 | 66 | 62 | 20 | 210 |
| 705 | 1300 | 676 | 98 | 414 | 60 | 64 | 22 | 205 |
| AISI 1340 normalized at 865°C (1585°F), quenched in oil from 845°C (1550°F) | | | | | | | | |
| 315 | 600 | 1,565 | 227 | 1,420 | 206 | 43 | 11 | 448 |
| 425 | 800 | 1,248 | 181 | 1,145 | 166 | 51 | 13 | 372 |
| 540 | 1000 | 966 | 140 | 834 | 121 | 58 | 17.5 | 297 |
| 595 | 1100 | 862 | 125 | 710 | 103 | 62 | 20 | 270 |
| 650 | 1200 | 793 | 115 | 607 | 88 | 65 | 23 | 250 |
| 705 | 1300 | 758 | 110 | 538 | 78 | 68 | 25.5 | 234 |
| AISI 4042 normalized at 870°C (1600°F), quenched in oil from 815°C (1500°F) | | | | | | | | |
| 315 | 600 | 1,593 | 231 | 1,448 | 210 | 41 | 12 | 448 |
| 425 | 800 | 1,207 | 175 | 1,089 | 158 | 50 | 14 | 372 |
| 540 | 1000 | 966 | 140 | 862 | 125 | 58 | 19 | 297 |
| 595 | 1100 | 862 | 125 | 758 | 110 | 62 | 23 | 260 |
| 650 | 1200 | 779 | 113 | 683 | 99 | 65 | 26 | 234 |
| 705 | 1300 | 724 | 105 | 634 | 92 | 68 | 30 | 210 |
| AISI 4140 normalized at 870°C (1600°F), quenched in oil from 815°C (1500°F) | | | | | | | | |
| 315 | 600 | 1,531 | 225 | 1,434 | 208 | 42.5 | 10 | 426 |
| 425 | 800 | 1,241 | 180 | 1,124 | 163 | 49 | 13 | 372 |
| 540 | 1000 | 931 | 135 | 827 | 120 | 57 | 18 | 283 |
| 595 | 1100 | 827 | 120 | 724 | 105 | 61 | 20 | 250 |
| 650 | 1200 | 745 | 108 | 655 | 95 | 62 | 22.5 | 228 |
| 705 | 1300 | 689 | 100 | 607 | 88 | 63 | 25 | 216 |
| AISI 4340 normalized at 870°C (1600°F), quenched in oil from 825°C (1525°F) | | | | | | | | |
| 315 | 600 | 1,724 | 250 | 1,586 | 230 | 40 | 9 | 484 |
| 425 | 800 | 1,455 | 211 | 1,379 | 200 | 44 | 10 | 426 |
| 540 | 1000 | 1,193 | 173 | 1,103 | 160 | 52 | 12.5 | 352 |
| 595 | 1100 | 1,089 | 158 | 966 | 140 | 56 | 15 | 313 |
| 650 | 1200 | 966 | 140 | 848 | 123 | 60 | 18 | 283 |
| 705 | 1300 | 848 | 123 | 745 | 108 | 63 | 22.5 | 250 |
| AISI 5140 normalized at 860°C (1575°F), quenched in oil from 815°C (1500°F) | | | | | | | | |
| 315 | 600 | 1,600 | 232 | 1,455 | 211 | 43 | 11 | 448 |
| 425 | 800 | 1,310 | 190 | 1,117 | 162 | 49 | 12.5 | 372 |
| 540 | 1000 | 966 | 140 | 855 | 124 | 58 | 17.5 | 283 |
| 595 | 1100 | 848 | 123 | 745 | 108 | 62 | 21 | 250 |
| 650 | 1200 | 758 | 110 | 655 | 95 | 65 | 24 | 228 |
| 705 | 1300 | 689 | 100 | 607 | 88 | 68 | 29 | 210 |
| AISI 8640 normalized at 870°C (1600°F), quenched in oil from 825°C (1525°F) | | | | | | | | |
| 315 | 600 | 1,655 | 240 | 1,517 | 220 | 42 | 10 | 472 |
| 425 | 800 | 1,393 | 202 | 1,296 | 188 | 45 | 12.5 | 415 |
| 540 | 1000 | 1,138 | 165 | 1,020 | 148 | 53 | 16 | 332 |
| 595 | 1100 | 1,000 | 145 | 896 | 130 | 57 | 18 | 297 |
| 650 | 1200 | 896 | 130 | 779 | 113 | 61 | 21 | 283 |
| 705 | 1300 | 800 | 116 | 689 | 100 | 63 | 22.5 | 250 |

**Table 14. Effect of Mass of Specimen on the Mechanical Properties of Some AISI Steels**

| Diam of section in mm | | Tensile strength | | Yield point | | Reduction of area, % | Elongation in 50 mm (2 in), % | Brinell hardness |
|---|---|---|---|---|---|---|---|---|
| | | MPa | 1,000 lb/in² | MPa | 1,000 lb/in² | | | |
| AISI 1040, water-quenched, tempered at 540°C (1000°F) | | | | | | | | |
| 1 | 25 | 758 | 110 | 538 | 78 | 58 | 15 | 230 |
| 2 | 50 | 676 | 98 | 448 | 65 | 49 | 20 | 194 |
| 3 | 75 | 641 | 93 | 407 | 59 | 48 | 23 | 185 |
| 4 | 100 | 621 | 90 | 393 | 57 | 47 | 24.5 | 180 |
| 5 | 125 | 614 | 89 | 372 | 54 | 46 | 25 | 180 |
| AISI 4140, oil-quenched, tempered at 540°C (1000°F) | | | | | | | | |
| 1 | 25 | 1,000 | 145 | 883 | 128 | 56 | 18 | 297 |
| 2 | 50 | 986 | 143 | 802 | 125 | 58 | 19 | 297 |
| 3 | 75 | 945 | 137 | 814 | 118 | 59 | 20 | 283 |
| 4 | 100 | 869 | 126 | 758 | 110 | 60 | 18 | 270 |
| 5 | 125 | 841 | 122 | 724 | 105 | 59 | 17 | 260 |
| AISI 8640, oil-quenched, tempered at 540°C (1000°F) | | | | | | | | |
| 1 | 25 | 1,158 | 168 | 1,000 | 145 | 44 | 16 | 332 |
| 2 | 50 | 1,055 | 153 | 910 | 132 | 45 | 20 | 313 |
| 3 | 75 | 951 | 138 | 807 | 117 | 46 | 22 | 283 |
| 4 | 100 | 889 | 129 | 745 | 108 | 46 | 23 | 270 |
| 5 | 125 | 869 | 126 | 724 | 105 | 45 | 23 | 260 |

high-speed steel and, in general, the tungsten-cobalt high-speed steels require higher quenching temperatures than the molybdenum steels. High-speed steel should be tempered at about 595°C (1100°F) to increase the toughness; owing to a secondary hardening effect, the hardness of the tempered steels may be higher than as quenched.

**Special-purpose tool steels** are comprised of the low-carbon, low-alloy, carbon-tungsten, mold, and other miscellaneous types.

## SPRING STEEL

For small springs, steel is often supplied to spring manufacturers in a form that requires no heat treatment except perhaps a low-temperature anneal to relieve forming strains. Types of previously treated steel wire for small helical springs are **music wire** which has been given a special heat treatment called patenting and then cold-drawn to develop a high yield strength, **hard-drawn wire** which is of lower quality than music wire since it is usually made of lower-grade material and is seldom patented, and **oil-tempered wire** which has been quenched and tempered. The wire usually has a Brinell hardness between 352 and 415, although this will depend on the application of the spring and the severity of the forming operation. Steel for small flat springs has either been cold-rolled or quenched and tempered to a similar hardness.

**Table 15. Average Mechanical Properties of Cold-Drawn Steel***

| AISI No. | Tensile strength | | Yield strength | | Elongation in 50 mm (2 in), % | Reduction of area, % | Brinell hardness |
|---|---|---|---|---|---|---|---|
| | MPa | 1,000 lb/in² | MPa | 1,000 lb/in² | | | |
| 1010 | 462 | 67 | 379 | 55.0 | 25.0 | 57 | 137 |
| 1015 | 490 | 71 | 416 | 60.3 | 22.0 | 55 | 149 |
| 1020 | 517 | 75 | 439 | 63.7 | 20.0 | 52 | 156 |
| 1025 | 552 | 80 | 469 | 68.0 | 18.5 | 50 | 163 |
| 1030 | 600 | 87 | 509 | 73.9 | 17.5 | 48 | 179 |
| 1035 | 634 | 92 | 539 | 78.2 | 17.0 | 45 | 187 |
| 1040 | 669 | 97 | 568 | 82.4 | 16.0 | 40 | 197 |
| 1045 | 703 | 102 | 598 | 86.7 | 15.0 | 35 | 207 |
| 1117 | 552 | 80 | 469 | 68.0 | 19.0 | 51 | 163 |
| 1118 | 569 | 82.5 | 483 | 70.1 | 18.5 | 50 | 167 |
| 1137 | 724 | 105 | 615 | 89.2 | 16.0 | 35 | 217 |
| 1141 | 772 | 112 | 656 | 95.2 | 14.0 | 30 | 223 |

Sizes 16 to 50 mm (⅝ to 2 in) diam, test specimens 50 × 13 mm (2 × 0.505 in).
* ASM "Metals Handbook," 1948.

**Table 16. Simplified Tool-Steel Classification**

| Major grouping | Symbol | Types |
|---|---|---|
| Water-hardening tool steels | W | |
| Shock-resisting tool steels | S | |
| Cold-work tool steels | O | Oil hardening |
| | A | Medium-alloy air hardening |
| | D | High-carbon, high-chromium |
| Hot-work tool steels | H | H10–H19 chromium base |
| | | H20–H39 tungsten base |
| | | H40–H59 molybdenum base |
| High-speed tool steels | T | Tungsten base |
| | M | Molybdenum base |
| Special-purpose tool steels | F | Carbon-tungsten |
| | L | Low-alloy |
| | P | Mold steels |
| | | P1–P19 low carbon |
| | | P20–P39 other types |

*Each subdivision is further identified as to type by a suffix number which follows the letter symbol.

Steel for both helical and flat springs which is hardened and tempered after forming is usually supplied in an annealed condition. Plain carbon steel is satisfactory for small springs; for large springs it is necessary to use alloy steels such as chrome-vanadium or silicon-manganese steel in order to obtain a uniform structure throughout the cross section. Table 17 gives the chemical composition and heat treatment of several spring steels. It is especially important for springs that the surface of the steel be free from all defects.

## STAINLESS STEELS

**Corrosion- and Heat-Resisting Steels**   Certain alloys of iron and chromium are highly resistant to corrosion and oxidation at high temperatures and maintain considerable strength at these temperatures. These alloys sometimes contain nickel and small percentages of silicon, molybdenum, tungsten, copper, and other elements. This large and complex group of alloys is known as **stainless steels,** and they are normally classified in one of three groups: (A) **austenitic steels,** containing both nickel and chromium; (B) **martensitic steels,** which are hardenable alloys containing up to 18 percent chromium and which are martensitic when quenched; and (C) **ferritic steels,** which are low-carbon, nonhardenable alloys containing up to 27 percent chromium. (See Table 18a.)

**Group A (Austenitic)**   The addition of substantial quantities of Ni to high-chromium alloys stabilizes the austenite to such

an extent that the alloys are austenitic at room temperature. The most common composition is 18 Cr and 8 Ni (known as **18-8**), and many modifications have been developed for special applications (Table 18). They cannot be hardened except by cold work, although excellent properties in the lower nickel grades are obtained in this manner (see Table 19). These alloys are highly resistant to many acids, including hot or cold nitric acid. They have excellent toughness at temperatures as low as liquid helium ($-269°C$ or $-452°F$) and are useful for parts subjected to severe stresses at elevated temperatures. The 25 Cr alloys can be used up to 1095°C (2000°F) without excessive scaling.

The austenitic stainless steels are not highly resistant to hot sulfurous gases and are sometimes subject to intergranular corrosion if chromium carbides are present in the grain boundaries. These carbides form during prolonged exposure in the 425 to 870°C (800 to 1600°F) temperature range. Normal corrosion resistance can be restored by heating the steel above 925°C (1700°F) and cooling rapidly. When titanium or columbium is added to a low-carbon 18-8 (types 321 and 347, respectively), the steel is relatively immune to this intergranular attack. Two grades, 304L and 316L, with a carbon content below 0.03 percent have been developed to minimize the amount of carbide precipitation. They find excellent application where welding is involved and postannealing is impractical.

**Table 17. Type of Steel and Heat Treatment for Large Hot-Formed Flat, Leaf, and Helical Springs**

| AISI steel no. | Normalizing temp* | | Quenching temp† | | Tempering temp | |
|---|---|---|---|---|---|---|
| | °C | °F | °C | °F | C | F |
| 1095 | 860–885 | 1575–1625 | 800–830 | 1475–1525 | 455–565 | 850–1050 |
| 6150 | 870–900 | 1600–1650 | 870–900 | 1600–1650 | 455–565 | 850–1050 |
| 9260 | 870–900 | 1600–1650 | 870–900 | 1600–1650 | 455–565 | 850–1050 |
| 5150 | 870–900 | 1600–1650 | 800–830 | 1475–1525 | 455–565 | 850–1050 |
| 8650 | 870–900 | 1600–1650 | 870–900 | 1600–1650 | 455–565 | 850–1050 |

*These normalizing temperatures should be used as the forming temperature whenever feasible.
†Quench in oil at 45 to 60°C (110 to 140°F).

**Table 18a. Composition of AISI Grades of Wrought Stainless and Heat-Resisting Steels**

| AISI type no. | Nominal composition, % | | | | | |
|---|---|---|---|---|---|---|
| | C | Mn, max | Si, max | Cr | Ni | Other[a] |
| | | | | *Austenitic steels* | | |
| 201 | 0,15 max | 7.50[b] | 1.00 | 16.00–18.00 | 3.50–5.50 | 0.25 max N |
| 202 | 0.15 max | 10.00[c] | 1.00 | 17.00–19.00 | 4.00–6.00 | 0.25 max N |
| 301 | 0.15 max | 2.00 | 1.00 | 16.00–18.00 | 6.00–8.00 | |
| 302 | 0.15 max | 2.00 | 1.00 | 17.00–19.00 | 8.00–10.00 | |
| 302B | 0.15 max | 2.00 | 3.00[d] | 17.00–19.00 | 8.00–10.00 | |
| 303 | 0.15 max | 2.00 | 1.00 | 17.00–19.00 | 8.00–10.00 | 0.15 min S[f] |
| 303 (Se) | 0.15 max | 2.00 | 1.00 | 17.00–19.00 | 8.00–10.00 | 0.15 min Se |
| 304 | 0.08 max | 2.00 | 1.00 | 18.00–20.00 | 8.00–10.50 | |
| 304L | 0.03 max | 2.00 | 1.00 | 18.00–20.00 | 8.00–12.00 | |
| 305 | 0.12 max | 2.00 | 1.00 | 17.00–19.00 | 10.50–13.00 | |
| 308 | 0.08 max | 2.00 | 1.00 | 19.00–21.00 | 10.00–12.00 | |
| 309 | 0.20 max | 2.00 | 1.00 | 22.00–24.00 | 12.00–15.00 | |
| 309S | 0.08 max | 2.00 | 1.00 | 22.00–24.00 | 12.00–15.00 | |
| 310 | 0.25 max | 2.00 | 1.50 | 24.00–26.00 | 19.00–22.00 | |
| 310S | 0.08 max | 2.00 | 1.50 | 24.00–26.00 | 19.00–22.00 | |
| 314 | 0.25 max | 2.00 | 3.00[e] | 23.00–26.00 | 19.00–22.00 | |
| 316 | 0.08 max | 2.00 | 1.00 | 16.00–18.00 | 10.00–14.00 | 2.00–3.00 Mo |
| 316L | 0.03 max | 2.00 | 1.00 | 16.00–18.00 | 10.00–14.00 | 2.00–3.00 Mo |
| 317 | 0.08 max | 2.00 | 1.00 | 18.00–20.00 | 11.00–15.00 | 3.00–4.00 Mo |
| 321 | 0.08 max | 2.00 | 1.00 | 17.00–19.00 | 9.00–12.00 | 5 × C min Ti |
| 347 | 0.08 max | 2.00 | 1.00 | 17.00–19.00 | 9.00–12.00 | 10 × C min Cb-Ta |
| 348 | 0.08 max | 2.00 | 1.00 | 17.00–19.00 | 9.00–13.00 | 10 × C min Cb-Ta (0.10 max Ta), 0.20 Co |
| | | | | *Martensitic steels* | | |
| 403 | 0.15 max | 1.00 | 0.50 | 11.50–13.00 | | |
| 410 | 0.15 max | 1.00 | 1.00 | 11.50–13.50 | | |
| 414 | 0.15 max | 1.00 | 1.00 | 11.50–13.50 | 1.25–2.50 | |
| 416 | 0.15 max | 1.25 | 1.00 | 12.00–14.00 | | 0.15 min S[f] |
| 416 (Se) | 0.15 max | 1.25 | 1.00 | 12.00–14.00 | | 0.15 min Se |
| 420 | 0.15 min | 1.00 | 1.00 | 12.00–14.00 | | |
| 420F | 0.15 min | 1.25 | 1.00 | 12.00–14.00 | | 0.15 min S[f] |
| 431 | 0.20 max | 1.00 | 1.00 | 15.00–17.00 | 1.25–2.50 | |
| 440A | 0.60–0.75 | 1.00 | 1.00 | 16.00–18.00 | | 0.75 max Mo |
| 440B | 0.75–0.95 | 1.00 | 1.00 | 16.00–18.00 | | 0.75 max Mo |
| 440C | 0.95–1.20 | 1.00 | 1.00 | 16.00–18.00 | | 0.75 max Mo |
| 501 | 0.10 min | 1.00 | 1.00 | 4.00–6.00 | | 0.40–0.65 Mo |
| | | | | *Ferritic steels* | | |
| 405 | 0.08 max | 1.00 | 1.00 | 11.50–14.50 | | 0.10–0.30 Al |
| 429 | 0.12 max | 1.00 | 1.00 | 14.00–16.00 | | |
| 430 | 0.12 max | 1.00 | 1.00 | 16.00–18.00 | | |
| 430F | 0.12 max | 1.25 | 1.00 | 16.00–18.00 | | 0.15 min S |
| 430F (Se) | 0.12 max | 1.25 | 1.00 | 16.00–18.00 | | 0.15 min Se |
| 434, 436 | 0.12 max | 1.00 | 1.00 | 16.00–18.00 | | 0.75–1.25 Mo[g] |
| 442 | 0.20 max | 1.00 | 1.00 | 18.00–23.00 | | |
| 446 | 0.20 max | 1.50 | 1.00 | 23.00–27.00 | | 0.25 max N |
| 502 | 0.10 max | 1.00 | 1.00 | 4.00–6.00 | | 0.40–0.65 Mo |

[a]Other elements in addition to those shown below are as follows: phosphorus is 0.20 percent max in types 303 and 303 (Se); 0.06 percent max in type 201, 202, 416, 416 (Se), 420F, 430F, 430F (Se); 0.045 max in types 301, 302, 302B, 304, 304L, 305, 308, 309, 309S, 310, 310S, 314, 316, 316L, 317, 321, 347, 348; 0.040 percent max in types 403, 405, 410, 414, 420, 420F, 429, 430, 431, 434, 436, 440A, 440B, 440C, 442, 446. Sulfur is 0.030 percent max in types 201, 202, 301, 302, 302B, 304, 304L, 305, 308, 309, 309S, 310, 310S, 314, 316, 316L, 317, 321, 347, 348, 403, 405, 410, 414, 420, 429, 430, 431, 434, 440A, 440B, 440C, 442, 446, 501, and 502; 0.060 percent max in 303 (Se), 416 (Se), 430F, and 430F (Se); 0.15 percent min in type 303, 416, 420F, and 430F.
[b]Mn range, 5.50–7.50.
[c]Mn range 7.50–10.00.
[d]Si range 2.00–3.00.
[e]Si range 1.50–3.00.
[f]Mo 0.60 (optional).
[g]For 436, 5 × C min Cb-Ta, 0.70 max.

**Table 18b. Typical Compositions of Some Nonstandard Grades of Wrought Stainless Steels**

| Tentative designation | Composition, % | | | | | | |
|---|---|---|---|---|---|---|---|
| | C | Mn | Si | Cr | Ni | Mo | Other |
| 308L | 0.025 | 1.75 | 0.40 | 21.00 | 10.00 | | |
| 316F | 0.06 | 1.50 | 0.50 | 18.00 | 13.00 | 2.25 | 0.13 P, 0.15 S |
| 317L | 0.025 | 1.75 | 0.50 | 18.50 | 13.50 | 3.25 | |
| 329 | 0.07 | 0.60 | 0.50 | 27.50 | 4.50 | 2.25 | |
| 18-18-2 | 0.07 | 2.00 | 2.20 | 19.00 | 18.50 | | |
| 418 | 0.17 | 0.40 | 0.30 | 12.75 | 2.00 | | 3.00 W |
| 420F | 0.38 | 0.45 | 0.35 | 13.50 | | | 0.21 Se or 0.18 S |
| 422 | 0.20 | 0.65 | 0.50 | 12.00 | 0.75 | 1.00 | 1.00 W, 0.30 V |
| 440F | 1.00 | 0.40 | 0.40 | 17.00 | | | 0.18 Se or 0.08 S |
| 442 | 0.06 | 0.50 | 0.50 | 21.00 | | | |
| 443 | 0.06 | 0.50 | 0.50 | 21.00 | | | 1.00 Cu |
| Stainless W | 0.07 | 0.50 | 0.50 | 16.75 | 6.75 | | 0.80 Ti, 0.20 Al |
| 17-4 PH | 0.04 | 0.40 | 0.50 | 16.50 | 4.25 | | 0.25 Cb, 3.60 Cu |
| 17-7 PH | 0.07 | 0.70 | 0.40 | 17.00 | 7.00 | | 1.15 Al |
| PH 15-7 Mo | 0.07 | 0.70 | 0.40 | 15.00 | 7.00 | 2.25 | 1.15 Al |
| AM-350 | 0.10 | 0.75 | 0.35 | 16.50 | 4.25 | 2.75 | 0.10 N |
| AM-355 | 0.13 | 0.85 | 0.35 | 15.50 | 4.25 | 2.75 | 0.12 N |
| 16-18 | 0.05 | 0.50 | 0.40 | 16.00 | 19.00 | | |
| 20-29 Cu-Mo | 0.05 | 0.75 | 1.00 | 20.00 | 29.00 | 2.20 | 3.20 Cu |
| 17-10 P | 0.12 | 0.75 | 0.50 | 17.00 | 10.50 | | 0.28 P |
| HMN | 0.30 | 3.50 | 0.50 | 18.50 | 9.50 | | 0.25 P |
| Tenelon | 0.08 | 14.50 | 0.50 | 17.00 | | | 0.40 N |

**Group B (Martensitic)** The hardenable alloys can be heat-treated to a high hardness and because of their oxidation resistance are used extensively for cutlery, razor blades, surgical and dental instruments, springs for high-temperature operation, ball valves and seats, and similar applications. Compositions and obtainable properties are given in Table 18 and 19, respectively. The hardening-temperature range depends on composition, but in general, the higher the quenching temperature, the harder the article. Oil quenching is preferable, but with thin and intricate shapes, hardening should be obtained by cooling in air. Tempering at 425°C (800°F) does not lower the hardness of the part, and in this condition these steels show remarkable resistance to fruit and vegetable acids, lye, ammonia, and other corrosive agents to which cutlery may be subjected.

**Group C (Ferritic)** This group is frequently called **stainless iron** because of its low carbon content. The alloys possess considerable ductility, ability to be worked hot or cold, and excellent corrosion resistance and are relatively inexpensive. Although these low-carbon chromium alloys cannot be hardened by heat treatment, they can be hardened to a considerable extent by cold working. Alloys containing 16 to 18 percent Cr are probably the most useful of the straight chromium steels because of their forming and medium-deep-drawing properties. They are used extensively for kitchen equipment, dairy machinery, interior decorative work, automobile trimmings, and chemical equipment (to resist nitric acid corrosion).

For resisting oxidizing conditions at high temperatures, the Cr content is increased to between 25 and 30 percent. These alloys are useful for all types of furnace parts not subjected to high stress. Since the oxidation resistance is independent of carbon content, soft, forgeable alloys low in carbon can be rolled into plates, shapes, and sheets, and hard and wear-resistant castings can be made from higher-carbon, nonforgeable alloys.

The compositions and mechanical properties of the ferritic alloys are given in Tables 18 and 19, respectively.

There are many nonstandard grades of stainless steels developed for specific applications. An important group is the **precipitation-hardening alloys,** which consist essentially of the 18-8 composition to which age-hardening elements such as titanium, columbium, aluminum, copper, and molybdenum have been added. Typical compositions are **stainless W,** with 17 Cr, 7 Ni, 0.70 Ti, and 0.20 Al; **17-7 PH,** with 17 Cr, 7 Ni, and 1.00 Al; and **AM 350,** with 17 Cr, 4 Ni, 3 Mo, and 0.10 N. Typical properties obtained in the 17-7 PH by solution annealing at 760°C (1400°F) and cooling and aging at 565°C (1050°F) are 1,276 MPa (185,000 lb/in²) yield strength, 1,379 MPa (200,000 lb/in²) tensile strength, and 9 percent elongation. These precipitation-hardening stainless steels are available in coils, sheets, strip, plate, forging billets, bars, rods, and wire. Because strength does not vary with bar size and can be developed in any thickness, many new applications are feasible. Final-machining operations can be performed before heat treatment if allowance is made for the slight growth that occurs. Wide application is made of these steels by the aircraft industry because of their high strength-weight ratio and strength at elevated temperatures.

## SPECIAL-ALLOY STEELS

**Iron-Silicon Alloys** The principal iron-silicon alloy products of the steel industry are **electrical sheet steels,** which are alloys of iron and silicon with C, Mn, P, and S kept as low as possible. The silicon increases the electrical resistivity of iron and greatly decreases the hysteresis loss; silicon-alloy sheets are used in almost all magnetic circuits where alternating current is used. For transformers, the silicon content is around 5 percent, but in structures subjected to vibration, such as motor armatures, the silicon is usually kept below 4 percent because of the brittleness of high-silicon sheets.

**Table 19. Nominal Mechanical Properties of AISI Stainless Steels**

| AISI No. | Form tested[e] | Condition | Tensile strength MPa | Tensile strength 1,000 lb/in² | 0.2% yield strength MPa | 0.2% yield strength 1,000 lb/in² | Elongation in 50 mm (2 in), % | Reduction of area, % | Hardness Rockwell | Hardness BHN |
|---|---|---|---|---|---|---|---|---|---|---|
| *Austenitic steels* | | | | | | | | | | |
| 201 | S | Annealed | 793 | 115 | 379 | 55 | 55 | | B90 | |
| | S | ¼ hard | 862[a] | 125[a] | 517[a] | 75[a] | 20[a] | | C25 | |
| | S | ½ hard | 1,034[a] | 150[a] | 758[a] | 110[a] | 10[a] | | C32 | |
| | S | ¾ hard | 1,207[a] | 175[a] | 931[a] | 135[a] | 5[a] | | C37 | |
| | S | Full hard | 1,276[a] | 183[a] | 966[a] | 140[a] | 4[a] | | C41 | |
| 202 | S | Annealed | 724 | 105 | 379[a] | 55 | 55 | | B90 | |
| | S | ¼ hard | 862[a] | 125[a] | 517[a] | 75[a] | 12[a] | | C27 | |
| 301 | S | Annealed | 758 | 110 | 276 | 40 | 60 | | B85 | |
| | S | ¼ hard | 862[a] | 125[a] | 517[a] | 75[a] | 25[a] | | C25 | |
| | S | ½ hard | 1,034[a] | 150[a] | 758[a] | 110[a] | 15[a] | | C32 | |
| | S | ¾ hard | 1,207[a] | 175[a] | 931[a] | 135[a] | 12[a] | | C37 | |
| | S | Full hard | 1,276[a] | 185[a] | 966[a] | 140[a] | 8[a] | | C41 | |
| 302 | S | Annealed | 621 | 90 | 276 | 40 | 50 | | B85 | |
| | S | ¼ hard | 862[a] | 125[a] | 517[a] | 75[a] | 12[a] | | C25 | |
| 302B | B,W | Cold-drawn[b] To | 2,413 | 350 | | | | | | |
| 303, 303 (Se) | S | Annealed | 655 | 95 | 276 | 40 | 55 | 55 | B85 | 145 |
| 304 | B | Annealed | 586 | 85 | 241 | 35 | 60 | 65 | B76 | 150 |
| 304L | B | Annealed | 545 | 79 | 228 | 33 | 60 | 65 | B80 | 143 |
| 305 | P | Annealed | 586 | 85 | 262 | 38 | 50 | | B79 | |
| 308 | S | Annealed | 586 | 85 | 241 | 35 | 50 | | B82 | |
| 309, 309S | S | Annealed | 621 | 90 | 310 | 45 | 45 | | B80 | |
| 310, 310S | S | Annealed | 655 | 95 | 310 | 45 | 45 | | B85 | |
| 314 | B,P | Annealed | 689 | 100 | 345 | 50 | 45 | 60 | B85 | 170 |
| 316 | B,W | Annealed | 552 | 80 | 207 | 30 | 60 | 65 | B87 | 142 |
| | | Cold-drawn[b] To | 2,413 | 350 | | | | | B78 | |
| 316L | S | Annealed | 558 | 81 | 290 | 42 | 50 | 60 | B79 | 160 |
| 317 | B,P | Annealed | 586 | 85 | 276 | 40 | 50 | 65 | B84 | 165 |
| 321 | P | Annealed | 586 | 85 | 207 | 30 | 55 | 65 | B85 | 160 |
| 347, 348 | B,P | Annealed | 621 | 90 | 241 | 35 | 50 | | B84 | |
| *Martensitic steels* | | | | | | | | | | |
| 403, 410, 416, 416 (Se) | B | Annealed | 517 | 75 | 276 | 40 | 30 | 65 | B82 | 155 |
| | | Hardened[c] | | | | | | | C43 | 410 |
| | | Tempered at: | | | | | | | | |
| | | 205°C (400°F) | 1,310 | 190 | 1,000 | 145 | 15 | 55 | C41 | 390 |
| | | 315°C (600°F) | 1,241 | 180 | 966 | 140 | 15 | 55 | C39 | 375 |
| | | 425°C (800°F) | 1,344 | 195 | 1,034 | 150 | 17 | 55 | C41 | 390 |
| | | 540°C (1000°F) | 1,000 | 145 | 793 | 115 | 20 | 65 | C31 | 300 |
| | | 650°C (1200°F) | 758 | 110 | 586 | 85 | 23 | 65 | B97 | 225 |
| | | 760°C (1400°F) | 621 | 90 | 414 | 60 | 30 | 70 | B89 | 180 |
| 414 | B | Annealed | 827 | 120 | 655 | 95 | 17 | 55 | C22 | 235 |
| | | Hardened[c] | | | | | | | C44 | 426 |

Table 19. Nominal Mechanical Properties of AISI Stainless Steels  (Continued)

| AISI No. | Form tested[e] | Condition | Tensile strength MPa | Tensile strength 1,000 lb/in² | 0.2% yield strength MPa | 0.2% yield strength 1,000 lb/in² | Elongation in 50 mm (2 in), % | Reduction of area, % | Rockwell | BHN |
|---|---|---|---|---|---|---|---|---|---|---|
|  |  | Tempered at: |  |  |  |  |  |  |  |  |
|  |  | 205°C (400°F) | 1,379 | 200 | 1,034 | 150 | 15 | 55 | C43 | 415 |
|  |  | 315°C (600°F) | 1,310 | 190 | 1,000 | 145 | 15 | 55 | C41 | 460 |
|  |  | 425°C (800°F) | 1,379 | 200 | 1,034 | 150 | 16 | 58 | C43 | 415 |
|  |  | 540°C (1000°F) | 1,000 | 145 | 827 | 120 | 20 | 60 | C34 | 325 |
|  |  | 650°C (1200°F) | 827 | 120 | 724 | 105 | 20 | 65 | C24 | 260 |
| 420, 420F | B | Annealed | 655 | 95 | 345 | 50 | 25 | 55 | B92 | 195 |
|  |  | Hardened[d] |  |  |  |  |  |  | C54 | 540 |
|  |  | Tempered at: |  |  |  |  |  |  |  |  |
| 431 | B | 315°C (600°F) | 1,586 | 230 | 1,344 | 195 | 8 | 25 | C50 | 500 |
|  |  | Annealed | 862 | 125 | 655 | 95 | 20 | 60 | C24 | 260 |
|  |  | Hardened[d] |  |  |  |  |  |  | C45 | 440 |
|  |  | Tempered at: |  |  |  |  |  |  |  |  |
|  |  | 205°C (400°F) | 1,413 | 205 | 1,069 | 155 | 15 | 55 | C43 | 415 |
|  |  | 315°C (600°F) | 1,344 | 195 | 1,034 | 150 | 15 | 55 | C41 | 400 |
|  |  | 425°C (800°F) | 1,413 | 205 | 1,069 | 155 | 15 | 60 | C43 | 415 |
|  |  | 540°C (1000°F) | 1,034 | 150 | 896 | 130 | 18 | 60 | C34 | 325 |
|  |  | 650°C (1200°F) | 862 | 125 | 655 | 95 | 20 | 60 | C24 | 260 |
| 440A | B | Annealed | 724 | 105 | 414 | 60 | 20 | 45 | B95 | 215 |
|  |  | Hardened[d] |  |  |  |  |  |  | C56 | 570 |
|  |  | Tempered at: |  |  |  |  |  |  |  |  |
| 440B | B | 315°C (600°F) | 1,931 | 280 | 1,862 | 270 | 3 | 15 | C55 | 555 |
|  |  | Annealed | 738 | 107 | 427 | 62 | 18 | 35 | B96 | 220 |
|  |  | Hardened[d] |  |  |  |  |  |  | C58 | 590 |
|  |  | Tempered at: |  |  |  |  |  |  |  |  |
| 440C | B | 315°C (600°F) | 1,931 | 280 | 1,862 | 270 | 3 | 15 | C55 | 555 |
|  |  | Annealed | 758 | 110 | 448 | 65 | 13 | 25 | B97 | 230 |
|  |  | Hardened[d] |  |  |  |  |  |  | C60 | 610 |
|  |  | Tempered at: |  |  |  |  |  |  |  |  |
| 501 | B | 315°C (600°F) | 1,965 | 285 | 1,896 | 275 | 2 | 10 | C57 | 500 |
|  |  | Annealed | 483 | 70 | 207 | 30 | 28 | 65 | B84 | 160 |
| Ferritic steels |  |  |  |  |  |  |  |  |  |  |
| 405 | B | Annealed | 483 | 70 | 276 | 40 | 30 | 60 | B80 | 150 |
| 429 | P | Annealed | 483 | 70 | 276 | 40 | 30 |  | B83 | 163 |
| 430 | P | Annealed | 517 | 75 | 276 | 40 | 30 | 60 | B82 | 160 |
| 430F, 430F (Se) | B | Annealed | 552 | 80 | 379 | 55 | 25 | 60 | B87 | 170 |
| 434 | S | Annealed | 531 | 77 | 365 | 53 | 23 |  | B83 |  |
| 436 | S | Annealed | 531 | 77 | 365 | 53 | 23 |  | B83 |  |
| 442 | B | Annealed | 552 | 80 | 310 | 45 | 20 |  | B90 |  |
| 446 | B | Annealed | 552 | 80 | 345 | 50 | 25 | 50 | B87 | 170 |
| 502 | B | Annealed | 483 | 70 | 207 | 30 | 30 | 75 | B80 | 150 |

[a]Minimum.
[b]Depending upon size and amount of reduction.
[c]Hardening temperature 980°C (1800°F), 25 mm (1 in) diam bars.
[d]Hardening temperature 1040°C (1900°F), 25 mm (1 in) diam bars.
[e]S = sheet and strip, B = bars, P = plates, W = wire.

**Iron-nickel** alloys are used extensively in the electrical industry owing to their exceptional magnetic properties. Alloys containing 20 to 30 Ni are nonmagnetic and are used to some extent for nonmagnetic parts in electrical machinery. Alloys having a high permeability and low hysteresis loss have a composition between 45 and 80 Ni., two of the better known being **Permalloy** with 78.5 Ni and **Hipernik** with 50 Ni. **Perminvar** (45 Ni, 25 Co) has a constant permeability over a range of flux densities. The magnetic properties of various alloys are given in Table 20.

Another important group of iron-nickel base alloys are those with low coefficients of expansion. **Invar**, containing 36 Ni, has an exceedingly low coefficient of linear expansion. Within limits of atmospheric-temperature change, its expansion is proportional to the temperature, and it is therefore used for secondary standards of length. **Elinvar** (32 Ni with small percentages of Cr, W, Mn, Si, and C) not only has a low coefficient of expansion but also has a constant modulus of elasticity over the temperature range of 0 to 100°F and is thus useful in hairsprings for watches and springs for other precision instruments. **Platinite**, a 46 Ni alloy, has the same thermal coefficient of expansion as platinum; and **Dumet wire**, a 42 Ni alloy covered with copper to prevent gassing at the seal, is used to replace platinum as the "seal-in" wire in incandescent lamps and vacuum tubes.

**Iron-cobalt alloys** containing various percentages of cobalt (up to 65 percent) have magnetic-saturation values higher than that of pure iron, the highest values being obtained at 34.5

### Table 20. Magnetic and Physical Properties of Magnetically Soft Materials*

| Material | Permeability Initial | Permeability Max | Induction, G at max permeability | Hysteresis loss, ergs/cm³/Hz | Residual induction, G | Coercive force, O | Saturation value, G | Resistivity, $\mu\Omega\cdot$cm |
|---|---|---|---|---|---|---|---|---|
| Thermenol (16 Al, 3.5 Mo) | 6,000 | 60,000 | 1,500 | | 2,070 | 0.018 | 6,100 | 162 |
| 16 Alfenol (16 Al) | 4,000 | 80,000 | 3,500 | 76.4 | 4,000 | 0.044 | 8,000 | 153 |
| 12 Alfenol (12 Al) | 2,370 | 20,600 | | | | 0.10 | 13,320 | |
| Sinimax (43 Ni, 3 Si) | 2,200 | 50,000 | 5,400 | 400 | 5,500 | 0.06 | 11,000 | 90 |
| Monimax (48 Ni, 3 Mo) | 3,000 | 60,000 | 6,200 | 800 | 8,900 | 0.06 | 14,500 | 80 |
| Supermalloy (79 Ni, 5 Mo) | 55,000 min | 300,000 min | 4,000 | 20 | 4,000–5,500 | 0.006 | 6,800–7,800 | 65 |
| 4-79 Moly Permalloy, Hymu 80 (79 Ni, 4 Mo) | 20,000 min | 90,000 min | 4,000 | 200 | 4,000–5,500 | 0.003 | 7,000–7,800 | 58 |
| Mumetal (77 Ni, 5 Cu, 1.5 Cr) | 20,000 min | 100,000 min | 2,000 | | 2,300 | 0.30 | 6.500 | 60 |
| 1040 alloy (72 Ni, 14 Cu, 11 Fe, 3 Mo) | 20,000 min | 100,000 min | 2,000 | 200 | 2,400 | 0.20 | 6,000 | 56 |
| High Permalloy 49, A-L 4750, Armco 48, Hipernik (47–50 Ni) | 5,000 | 70,000 | 4,500 | 300 | 10,000 | 0.50 | 16,000 | 48 |
| 45 Permalloy (45 Ni) | 2,500 | 25,000 | | | | 0.25 | 16,000 | 45 |
| Supermendur (49 Co, 2 V) | 800 | 70,000 | 20,000 | 1,500 | 21,400 | 0.23 | 24,000 | 40 |
| 2V Permendur (49 Co, 2 V) | 800 | 6,000 | 13,000 | | 10,000 | 2.20 | 23,500 | 40 |
| 35% Co, 1% Cr | 650 | 10,000 | 10,000 | 2,900 | 11,000 | 0.63 | 24,200 | 20 |
| Ingot iron | 150 | 5,000 | 8,000 | 2,700 | 7,700 | 1.00 | 21,400 | 10 |
| 0.5% Si steel | 280 | 3,000 | 8,000 | 2,300 | | 0.90 | 20,500 | 28 |
| 1.75% Si steel | 280 | 5,000 | 6,000 | 2,100 | | 0.80 | 20,000 | 37 |
| 3.0% Si steel | 290 | 8,000 | 8,000 | 1,600 | | 0.70 | 20,100 | 47 |
| Grain-oriented 3% Si steel | 1,400 | 50,000 | 10,000 | 400 | 12,000 | 0.09 | 20,100 | 50 |
| Grain-oriented 50% Ni iron | 500 | 150,000 | 13,000 | 450 | 14,500 | 0.09 | 16,000 | 45 |
| 50% Ni iron | 3,500 | 100,000 | 5,500 | 250 | 9,000 | 0,05 | 15,500 | 50 |

*ASM Handbook, 1961.

**Galvanized-Sheet Gage**

| Galva-nized-sheet gage No. | Gage weights | | | Mean thick-ness in§ | Galva-nized-sheet gage No. | Gage weights | | | Mean thick-ness in§ |
|---|---|---|---|---|---|---|---|---|---|
| | oz/ft²* | lb/ft²† | lb/in²‡ | | | oz/ft²* | lb/ft²† | lb/in²‡ | |
| 8 | 112.5 | 7.0312 | 0.048828 | 0.1681 | 20 | 26.5 | 1.6562 | 0.011502 | 0.0396 |
| 9 | 102.5 | 6.4062 | 0.044488 | 0.1532 | 21 | 24.5 | 1.5312 | 0.010634 | 0.0366 |
| 10 | 92.5 | 5.7812 | 0.040148 | 0.1382 | 22 | 22.5 | 1.4062 | 0.0097656 | 0.0336 |
| 11 | 82.5 | 5.1562 | 0.035807 | 0.1233 | 23 | 20.5 | 1.2812 | 0.0088976 | 0.0306 |
| 12 | 72.5 | 4.5312 | 0.031467 | 0.1084 | 24 | 18.5 | 1.1562 | 0.0080295 | 0.0276 |
| 13 | 62.5 | 3.9062 | 0.027127 | 0.0934 | 25 | 16.5 | 1.0312 | 0.0071615 | 0.0247 |
| 14 | 52.5 | 3.2812 | 0.022786 | 0.0785 | 26 | 14.5 | 0.90625 | 0.0062934 | 0.0217 |
| 15 | 47.5 | 2.9688 | 0.020616 | 0.0710 | 27 | 13.5 | 0.84375 | 0.0058594 | 0.0202 |
| 16 | 42.5 | 2.6562 | 0.018446 | 0.0635 | 28 | 12.5 | 0.78125 | 0.0054253 | 0.0187 |
| 17 | 38.5 | 2.4062 | 0.016710 | 0.0575 | 29 | 11.5 | 0.71875 | 0.0049913 | 0.0172 |
| 18 | 34.5 | 2.1562 | 0.014974 | 0.0516 | 30 | 10.5 | 0.65625 | 0.0045573 | 0.0157 |
| 19 | 30.5 | 1.9062 | 0.013238 | 0.0456 | 31 | 9.5 | 0.59375 | 0.0041233 | 0.0142 |
| | | | | | 32 | 9.0 | 0.56250 | 0.0039062 | 0.0134 |

*1 ozm/ft² = 0.30515 kg/m²
†1 lbm/ft² = 4.8824 kg/m²
‡1 lbm/m² = 0.03391 kg/m²
§1 in = 25.4 mm.

percent cobalt. Use of alloys containing between 25 and 50 percent cobalt is limited by their low resistivity and high hysteresis loss, the brittleness of alloys containing more than 30 percent cobalt, and the cost of cobalt. However, small additions of vanadium and chromium make it possible, with special processing, to produce strip containing 35 percent cobalt with magnetic properties suitable for both alternating-current and direct-current applications that is ductile enough to be punched and sheared. High-saturation alloys containing up to 50 percent cobalt also are used (Table 20).

**Iron-Aluminum Alloys** Aluminum and silicon (see Iron-Silicon Alloys above) have similar effects on the electrical resistivity and some magnetic properties of iron, but simple binary alloys of iron and aluminum can be difficult to fabricate. Ternary alloys of iron, silicon, and aluminum have high resistivity and good permeability at low flux densities, the saturation value being decreased by increasing amounts of silicon and aluminum. Magnetic properties of the ternary alloys can approach those of iron-nickel alloys (see above) at low flux densities.

**Austenitic manganese steel** (Hadfield's manganese steel) is a nonmagnetic alloy containing around 12 Mn and 1 C. It is relatively soft but work-hardens on the surface when subjected to severe abrasion, so that it is extremely useful in crushing machinery, for railroad crossings and frogs, tractor shoes, etc. As cast, this alloy is partly martensitic and therefore hard and brittle. By quenching from a high temperature (1040°C or about 1900°F), a homogeneous austenite is retained and the alloy has the high toughness, strength, and ductility characteristic of austenitic steels.

## WIRE, SHEETS, AND BARS

**Wire and Sheet-Metal Gages** In the metal industries, the word gage has been used in various systems, or scales, for expressing the thickness or weight per unit area of thin plates, sheet, and strip, or the diameters of rods and wire. Specific diameters, thicknesses, or weights per square foot have been or are denoted in gage systems by certain numerals followed by the word gage; for example, No. 12 gage, or simply 12 gage. Gage numbers for flat-rolled products have been used only in connection with thin materials; that is, usually when the thickness was not more than $\frac{1}{4}$ in or the weight per square foot was not more than 10 lb, although most gage tables began at about $\frac{1}{2}$ in or 20 lb/ft², and one table began at double these quantities. Heavier and thicker flat-rolled materials always have been designated by weight per unit area or length, or by thickness in English or metric units.

There is considerable danger of confusion in the use of gage numbers in both foreign and domestic trade, which can be avoided by specifying thickness or diameter in inches or millimeters or in weights per square foot or per square meter, or by giving other equivalents in absolute units. The latter method of specification in absolute units has become general for most coated and uncoated flat-rolled products, and is gaining favor in specifications for wire sizes.

Wire and black and galvanized sheet metal of the smaller thicknesses have been made to various gages. Steel wire has usually been made to the Washburn and Moen (W & M) or Roebling gage. The U.S. standard for sheet metal was based upon weight per square foot; the tabulated values are the corresponding thicknesses for wrought iron weighing 480 and for steel and open-hearth iron weighing 489.6 lb/ft³. Stubs' steel-wire gage has also been used for numbered twist drill sizes (see Sec. 13). The Birmingham wire gage has been used in the United States for brass wire. The Brown and Sharpe (B & S) gage is a uniform geometrical progression, each gage being equal to 0.89053 times the preceding gage. See Sec. 15 for more data on the Brown and Sharpe gage.

## Weights of Rolled Sheet Steel

| Gage No. | Weight per sq ft, lb* BWG | USSG | Gage No. | Weight per sq ft, lb* BWG | USSG | Gage No. | Weight per sq ft, lb* BWG | USSG | Gage No. | Weight per sq ft, lb* BWG | USSG |
|---|---|---|---|---|---|---|---|---|---|---|---|
| 7-0's | | 20.00 | 6 | 8.2824 | 8.125 | 18 | 1.9992 | 2 | 29 | 0.5304 | 0.5625 |
| 6-0's | | 18.75 | 7 | 7.344 | 7.5 | 19 | 1.7126 | 1.75 | 30 | 0.4896 | 0.5 |
| 5-0's | | 17.50 | 8 | 6.732 | 6.875 | 20 | 1.428 | 1.50 | 31 | 0.408 | 0.4375 |
| 0000 | 18.5232 | 16.25 | 9 | 6.0384 | 6.25 | 21 | 1.3056 | 1.375 | 32 | 0.3672 | 0.4063 |
| 000 | 17.34 | 15 | 10 | 5.4672 | 5.625 | 22 | 1.1424 | 1.25 | 33 | 0.3264 | 0.375 |
| 00 | 15.504 | 13.75 | 11 | 4.896 | 5 | 23 | 1.02 | 1.125 | 34 | 0.2856 | 0.3438 |
| 0 | 13.872 | 12.50 | 12 | 4.4972 | 4.375 | 24 | 0.8976 | 1 | 35 | 0.2040 | 0.3125 |
| 1 | 12.24 | 11.25 | 13 | 3.876 | 3.75 | 25 | 0.816 | 0.875 | 36 | 0.1632 | 0.2813 |
| 2 | 11.5872 | 10.625 | 14 | 3.8864 | 3.125 | 26 | 0.7344 | 0.75 | 37 | | 0.2656 |
| 3 | 10.5672 | 10 | 15 | 2.9376 | 2.813 | 27 | 0.6528 | 0.6875 | 38 | | 0.25 |
| 4 | 9.7104 | 9.375 | 16 | 2.651 | 2.5 | 28 | 0.5712 | 0.625 | | | |
| 5 | 8.976 | 8.75 | 17 | 2.3664 | 2.25 | | | | | | |

*To convert to kg per m², multiply by 4.8824.

## Properties of Steel Wire
(Breaking stress = 689 MPa or 100,000 lb/in²)

| No., Roebling gage | Breaking load, lbf* | Weight, lb/1,000 ft† | No., Roebling gage | Breaking load, lbf* | Weight, lb/1,000 ft† | No., Roebling gage | Breaking load, lbf* | Weight, lb/1,000 ft† | No., Roebling gage | Breaking load, lbf* | Weight, lb/1,000 ft† |
|---|---|---|---|---|---|---|---|---|---|---|---|
| 6-0's | 16,619 | 558.4 | 6 | 2,895 | 97.3 | 17 | 229 | 7.70 | 27 | 23.0 | 0.763 |
| 5-0's | 14,522 | 487.9 | 7 | 2,461 | 82.7 | 18 | 174 | 5.83 | 28 | 20.0 | 0.676 |
| 0000 | 12,130 | 407.6 | 8 | 2,061 | 69.3 | 19 | 132 | 4.44 | 29 | 18.0 | 0.594 |
| 000 | 10,292 | 345.8 | 9 | 1,720 | 57.8 | 20 | 96 | 3.23 | 30 | 15.0 | 0.517 |
| 00 | 8,605 | 289.1 | 10 | 1,431 | 48.1 | 21 | 80 | 2.70 | 31 | 14.0 | 0.481 |
| 0 | 7,402 | 248.7 | 11 | 1,131 | 38.0 | 22 | 62 | 2.07 | 32 | 13.0 | 0.446 |
| 1 | 6,290 | 211.4 | 12 | 866 | 29.1 | 23 | 49 | 1.65 | 33 | 9.5 | 0.319 |
| 2 | 5,433 | 182.5 | 13 | 665 | 22.3 | 24 | 42 | 1.40 | 34 | 7.9 | 0.264 |
| 3 | 4,676 | 157.1 | 14 | 503 | 16.9 | 25 | 31 | 1.06 | 35 | 7.1 | 0.238 |
| 4 | 3,976 | 133.6 | 15 | 407 | 13.7 | 26 | 25 | 0.855 | 36 | 6.4 | 0.214 |
| 5 | 3,365 | 113.1 | 16 | 312 | 10.5 | | | | | | |

*1 lbf = 4.448222 N.
†1 lbm/1.000 ft = 0.14886 kg/100 m.

### Weights of Flat Rolled Steel, Pounds per Linear Foot*

(The last line of the table gives weights per square feet)
(For wrought iron, subtract 2 percent)

| Width, in.† | Thickness, in.† | | | | | | | | | | | | | | | |
|---|---|---|---|---|---|---|---|---|---|---|---|---|---|---|---|---|
| | $\frac{1}{16}$ | $\frac{1}{8}$ | $\frac{3}{16}$ | $\frac{1}{4}$ | $\frac{5}{16}$ | $\frac{3}{8}$ | $\frac{7}{16}$ | $\frac{1}{2}$ | $\frac{9}{16}$ | $\frac{5}{8}$ | $\frac{11}{16}$ | $\frac{3}{4}$ | $\frac{13}{16}$ | $\frac{7}{8}$ | $\frac{15}{16}$ | 1 |
| $\frac{1}{4}$ | 0.053 | 0.106 | 0.159 | 0.213 | 0.27 | 0.32 | 0.37 | 0.43 | 0.48 | 0.53 | 0.58 | 0.64 | 0.69 | 0.74 | 0.80 | 0.85 |
| $\frac{1}{2}$ | 0.106 | 0.213 | 0.319 | 0.425 | 0.53 | 0.64 | 0.74 | 0.85 | 0.96 | 1.06 | 1.17 | 1.28 | 1.38 | 1.49 | 1.59 | 1.70 |
| $\frac{3}{4}$ | 0.159 | 0.319 | 0.478 | 0.638 | 0.80 | 0.96 | 1.12 | 1.28 | 1.43 | 1.59 | 1.75 | 1.91 | 2.07 | 2.23 | 2.39 | 2.55 |
| 1 | 0.213 | 0.425 | 0.638 | 0.850 | 1.06 | 1.23 | 1.49 | 1.70 | 1.91 | 2.13 | 2.34 | 2.55 | 2.76 | 2.98 | 3.19 | 3.40 |
| 2 | 0.425 | 0.850 | 1.275 | 1.700 | 2.13 | 2.55 | 2.98 | 3.40 | 3.83 | 4.25 | 4.68 | 5.10 | 5.53 | 5.95 | 6.38 | 6.80 |
| 3 | 0.638 | 1.275 | 1.913 | 2.550 | 3.19 | 3.83 | 4.46 | 5.10 | 5.74 | 6.38 | 7.01 | 7.65 | 8.29 | 8.93 | 9.56 | 10.20 |
| 4 | 0.850 | 1.700 | 2.550 | 3.400 | 4.25 | 5.10 | 5.95 | 6.80 | 7.65 | 8.50 | 9.35 | 10.20 | 11.05 | 11.90 | 12.75 | 13.60 |
| 5 | 1.063 | 2.125 | 3.188 | 4.250 | 5.31 | 6.38 | 7.44 | 8.50 | 9.56 | 10.63 | 11.69 | 12.75 | 13.81 | 14.88 | 15.94 | 17.00 |
| 6 | 1.275 | 2.550 | 3.825 | 5.100 | 6.38 | 7.65 | 8.93 | 10.20 | 11.48 | 12.75 | 14.03 | 15.30 | 16.58 | 17.85 | 19.13 | 20.40 |
| 7 | 1.488 | 2.975 | 4.463 | 5.950 | 7.44 | 8.93 | 10.41 | 11.90 | 13.39 | 14.88 | 16.36 | 17.85 | 19.34 | 20.83 | 22.31 | 23.80 |
| 8 | 1.700 | 3.400 | 5.100 | 6.800 | 8.50 | 10.20 | 11.90 | 13.60 | 15.30 | 17.00 | 18.70 | 20.40 | 22.10 | 23.80 | 25.50 | 27.20 |
| 9 | 1.913 | 3.825 | 5.738 | 7.650 | 9.56 | 11.48 | 13.39 | 15.30 | 17.21 | 19.13 | 21.04 | 22.95 | 24.86 | 26.78 | 28.69 | 30.60 |
| 10 | 2.125 | 4.250 | 6.375 | 8.500 | 10.63 | 12.75 | 14.88 | 17.00 | 19.13 | 21.25 | 23.38 | 25.50 | 27.63 | 29.75 | 31.88 | 34.00 |
| 20 | 4.25 | 8.50 | 12.75 | 17.00 | 21.25 | 25.50 | 29.75 | 34.00 | 38.25 | 42.50 | 46.80 | 51.0 | 55.30 | 59.50 | 63.8 | 68.00 |
| 30 | 6.38 | 12.75 | 19.13 | 25.50 | 31.88 | 38.25 | 44.63 | 51.00 | 57.38 | 63.75 | 70.10 | 76.5 | 82.90 | 89.30 | 95.6 | 102.00 |
| 40 | 8.50 | 17.00 | 25.50 | 34.00 | 42.50 | 51.00 | 59.50 | 68.00 | 76.50 | 85.00 | 93.50 | 102.0 | 110.50 | 119.00 | 127.5 | 136.00 |
| 12 | 2.55 | 5.10 | 7.65 | 10.20 | 12.75 | 15.30 | 17.85 | 20.40 | 22.95 | 25.50 | 28.05 | 30.60 | 33.15 | 35.70 | 38.25 | 40.80 |

NOTE: For other widths the weights are obtainable by addition; for example, $5\frac{1}{4} \times \frac{3}{4}$ in = $[(10 \times 5) + 4] \times \frac{3}{4}$ in, and weight = $(10 \times 12.75) + 10.20 = 137.7$ lbm. Similarly, for greater thicknesses, the weights are obtainable by addition. Note also that this table is based on steel having a density of 489.6 lbm/ft³; densities of different varieties of steel and iron are given below.

† 1 lbm/1.000 ft = 0.14886 kg/lin m.

† 1 in = 25.4 mm.

### Approximate Densities of Different Varieties of Iron and Steel

| Material (in wrought form) | Density at 15.5°C (60°F) | | |
|---|---|---|---|
| | g/cm³ | lbm/in³ | lbm/ft³ |
| Pure iron (99.9% Fe) | 7.86 | 0.284 | 491 |
| Soft steel (0.06% C) | 7.87 | 0.284 | 491 |
| Carbon steel (0.40% C) | 7.84 | 0.283 | 489 |
| Carbon tool steel (0.90% C) | 7.82 | 0.282 | 487 |
| Wrought iron | 7.40–7.90 | 0.267–0.285 | 461–493 |
| Stainless steel (18% Cr, 8% Ni) | 8.03 | 0.29 | 501 |
| Stainless steel (17% Cr, 0.12% C) | 7.75 | 0.28 | 484 |
| Stainless steel (27% Cr, 0.35% C) | 7.47 | 0.27 | 467 |
| High-speed tool steel (18% W) | 8.75 | 0.316 | 546 |

**Comparison of Standard Gages***

(Thickness or diam in decimals of an inch)†

| Gage No. | BWG; Stubs Iron Wire | AWG; B&S | U.S. Steel Wire; Am. Steel & Wire; Washburn & Moen; Steel Wire | U.S. Standard (old) | SWG | Manufacturers' standard |
|---|---|---|---|---|---|---|
| 0000000 | . . . . . | . . . . . . . . | 0.4900 | 0.5000 | 0.500 | |
| 000000 | . . . . . | 0.580000 | 0.4615 | 0.4687 | 0.464 | |
| 00000 | . . . . . | 0.516500 | 0.4305 | 0.4375 | 0.432 | |
| 0000 | 0.454 | 0.460000 | 0.3938 | 0.4062 | 0.400 | |
| 000 | 0.425 | 0.409642 | 0.3625 | 0.3750 | 0.372 | |
| 00 | 0.380 | 0.364796 | 0.3310 | 0.3437 | 0.348 | |
| 0 | 0.340 | 0.324861 | 0.3065 | 0.3125 | 0.324 | |
| 1 | 0.300 | 0.289297 | 0.2830 | 0.2812 | 0.300 | |
| 2 | 0.284 | 0.257627 | 0.2625 | 0.2656 | 0.276 | |
| 3 | 0.259 | 0.229423 | 0.2437 | 0.2500 | 0.252 | 0.2391 |
| 4 | 0.238 | 0.204307 | 0.2253 | 0.2344 | 0.232 | 0.2242 |
| 5 | 0.220 | 0.181940 | 0.2070 | 0.2187 | 0.212 | 0.2092 |
| 6 | 0.203 | 0.162023 | 0.1920 | 0.2031 | 0.192 | 0.1943 |
| 7 | 0.180 | 0.144285 | 0.1770 | 0.1875 | 0.176 | 0.1793 |
| 8 | 0.165 | 0.128490 | 0.1620 | 0.1719 | 0.160 | 0.1644 |
| 9 | 0.148 | 0.114423 | 0.1483 | 0.1562 | 0.144 | 0.1495 |
| 10 | 0.134 | 0.101897 | 0.1350 | 0.1406 | 0.128 | 0.1345 |
| 11 | 0.120 | 0.090742 | 0.1205 | 0.1250 | 0.116 | 0.1196 |
| 12 | 0.109 | 0.080808 | 0.1055 | 0.1094 | 0.104 | 0.1046 |
| 13 | 0.095 | 0.071962 | 0.0915 | 0.0937 | 0.092 | 0.0897 |
| 14 | 0.083 | 0.064084 | 0.0800 | 0.0781 | 0.080 | 0.0747 |
| 15 | 0.072 | 0.057068 | 0.0720 | 0.0703 | 0.072 | 0.0673 |
| 16 | 0.065 | 0.050821 | 0.0625 | 0.0625 | 0.064 | 0.0598 |
| 17 | 0.058 | 0.045257 | 0.0540 | 0.0562 | 0.056 | 0.0538 |
| 18 | 0.049 | 0.040303 | 0.0475 | 0.0500 | 0.048 | 0.0478 |
| 19 | 0.042 | 0.035890 | 0.0410 | 0.0437 | 0.040 | 0.0418 |
| 20 | 0.035 | 0.031961 | 0.0348 | 0.0375 | 0.036 | 0.0359 |
| 21 | 0.032 | 0.028462 | 0.03175 | 0.0344 | 0.032 | 0.0329 |
| 22 | 0.028 | 0.025346 | 0.0286 | 0.0312 | 0.028 | 0.0299 |
| 23 | 0.025 | 0.022572 | 0.0258 | 0.0281 | 0.024 | 0.0269 |
| 24 | 0.022 | 0.020101 | 0.0230 | 0.0250 | 0.022 | 0.0239 |
| 25 | 0.020 | 0.017900 | 0.0204 | 0.0219 | 0.020 | 0.0209 |
| 26 | 0.018 | 0.015941 | 0.0181 | 0.0187 | 0.018 | 0.0179 |
| 27 | 0.016 | 0.014195 | 0.0173 | 0.0172 | 0.0164 | 0.0164 |
| 28 | 0.014 | 0.012641 | 0.0162 | 0.0156 | 0.0148 | 0.0149 |
| 29 | 0.013 | 0.011257 | 0.0150 | 0.0141 | 0.0136 | 0.0135 |
| 30 | 0.012 | 0.010025 | 0.0140 | 0.0125 | 0.0124 | 0.0120 |
| 31 | 0.010 | 0.008928 | 0.0132 | 0.0109 | 0.0116 | 0.0105 |
| 32 | 0.009 | 0.007950 | 0.0128 | 0.0102 | 0.0108 | 0.0097 |
| 33 | 0.008 | 0.007080 | 0.0118 | 0.0094 | 0.0100 | 0.0090 |
| 34 | 0.007 | 0.006305 | 0.0104 | 0.0086 | 0.0092 | 0.0082 |
| 35 | 0.005 | 0.005615 | 0.0095 | 0.0078 | 0.0084 | 0.0075 |
| 36 | 0.004 | 0.005000 | 0.0090 | 0.0070 | 0.0076 | 0.0067 |
| 37 | . . . . . | 0.004453 | 0.0085 | 0.0066 | 0.0068 | 0.0064 |
| 38 | . . . . . | 0.003965 | 0.0080 | 0.0062 | 0.0060 | 0.0060 |
| 39 | . . . . . | 0.003531 | 0.0075 | . . . . . . | 0.0052 | |
| 40 | . . . . . | 0.003144 | 0.0070 | . . . . . . | 0.0048 | |

*Principal uses—BWG: strips, bands, hoops, and wire; AWG or B&S: nonferrous sheets, rod, and wire; U.S. Steel Wire: steel wire except music wire; U.S. Standard (old): stainless-steel sheets; SWG: English legal standard wire gage; manufacturers' standard : uncoated steel sheets.

†1 in = 25.4 mm.

## Weights of Square and Round Steel Bars*

(For wrought iron, subtract 2 percent)

| Size, in | Weight, lb /lin ft‡ Square | Round | Size, in† | Weight, lb /lin ft‡ Square | Round | Size, in† | Weight, lb /lin ft‡ Square | Round | Size, in† | Weight, lb /lin ft‡ Square | Round |
|---|---|---|---|---|---|---|---|---|---|---|---|
| 0 | ....... | ....... | 3 | 30.60 | 24.03 | 6 | 122.4 | 96.1 | 9 | 275.4 | 216.3 |
| 1/16 | 0.013 | 0.010 | 1/16 | 31.89 | 25.05 | 1/16 | 125.0 | 98.2 | 1/16 | 279.2 | 219.3 |
| 1/8 | 0.053 | 0.042 | 1/8 | 33.20 | 26.08 | 1/8 | 127.6 | 100.2 | 1/8 | 283.1 | 222.4 |
| 3/16 | 0.120 | 0.094 | 3/16 | 34.54 | 27.13 | 3/16 | 130.2 | 102.2 | 3/16 | 287.0 | 225.4 |
| 1/4 | 0.213 | 0.167 | 1/4 | 35.91 | 28.21 | 1/4 | 132.8 | 104.3 | 1/4 | 290.9 | 228.5 |
| 5/16 | 0.332 | 0.261 | 5/16 | 37.31 | 29.30 | 5/16 | 135.5 | 106.4 | 5/16 | 294.9 | 231.6 |
| 3/8 | 0.478 | 0.376 | 3/8 | 38.73 | 30.42 | 3/8 | 138.2 | 108.5 | 3/8 | 298.8 | 234.7 |
| 7/16 | 0.651 | 0.511 | 7/16 | 40.18 | 31.55 | 7/16 | 140.9 | 110.7 | 7/16 | 302.8 | 237.8 |
| 1/2 | 0.850 | 0.668 | 1/2 | 41.65 | 32.71 | 1/2 | 143.7 | 112.8 | 1/2 | 306.9 | 241.0 |
| 9/16 | 1.076 | 0.845 | 9/16 | 43.15 | 33.89 | 9/16 | 146.4 | 115.0 | 9/16 | 310.9 | 244.2 |
| 5/8 | 1.328 | 1.043 | 5/8 | 44.68 | 35.09 | 5/8 | 149.2 | 117.2 | 5/8 | 315.0 | 247.4 |
| 11/16 | 1.607 | 1.262 | 11/16 | 46.23 | 36.31 | 11/16 | 152.1 | 119.4 | 11/16 | 319.1 | 250.6 |
| 3/4 | 1.913 | 1.502 | 3/4 | 47.81 | 37.55 | 3/4 | 154.9 | 121.7 | 3/4 | 323.2 | 253.9 |
| 13/16 | 2.245 | 1.763 | 13/16 | 49.42 | 38.81 | 13/16 | 157.8 | 123.9 | 13/16 | 327.4 | 257.1 |
| 7/8 | 2.603 | 2.044 | 7/8 | 51.05 | 40.10 | 7/8 | 160.7 | 126.2 | 7/8 | 331.6 | 260.4 |
| 15/16 | 2.988 | 2.347 | 15/16 | 52.71 | 41.40 | 15/16 | 163.6 | 128.5 | 15/16 | 335.8 | 263.7 |
| 1 | 3.400 | 2.670 | 4 | 54.40 | 42.73 | 7 | 166.6 | 130.9 | 10 | 340.0 | 267.0 |
| 1/16 | 3.838 | 3.015 | 1/16 | 56.11 | 44.07 | 1/16 | 169.6 | 133.2 | 1/16 | 344.3 | 270.4 |
| 1/8 | 4.303 | 3.380 | 1/8 | 57.85 | 45.44 | 1/8 | 172.6 | 135.6 | 1/8 | 348.6 | 273.8 |
| 3/16 | 4.795 | 3.766 | 3/16 | 59.62 | 46.83 | 3/16 | 175.6 | 137.9 | 3/16 | 352.9 | 277.1 |
| 1/4 | 5.313 | 4.172 | 1/4 | 61.41 | 48.23 | 1/4 | 178.7 | 140.4 | 1/4 | 357.2 | 280.6 |
| 5/16 | 5.857 | 4.600 | 5/16 | 63.23 | 49.66 | 5/16 | 181.8 | 142.8 | 5/16 | 361.6 | 284.0 |
| 3/8 | 6.428 | 5.049 | 3/8 | 65.08 | 51.11 | 3/8 | 184.9 | 145.2 | 3/8 | 366.0 | 287.4 |
| 7/16 | 7.026 | 5.518 | 7/16 | 66.95 | 52.58 | 7/16 | 188.1 | 147.7 | 7/16 | 370.4 | 290.9 |
| 1/2 | 7.650 | 6.008 | 1/2 | 68.85 | 54.07 | 1/2 | 191.3 | 150.2 | 1/2 | 374.9 | 294.4 |
| 9/16 | 8.301 | 6.519 | 9/16 | 70.78 | 55.59 | 9/16 | 194.5 | 152.7 | 9/16 | 379.3 | 297.9 |
| 5/8 | 8.978 | 7.051 | 5/8 | 72.73 | 57.12 | 5/8 | 197.7 | 155.3 | 5/8 | 383.8 | 301.5 |
| 11/16 | 9.682 | 7.604 | 11/16 | 74.71 | 58.67 | 11/16 | 200.9 | 157.8 | 11/16 | 388.4 | 305.0 |
| 3/4 | 10.413 | 8.178 | 3/4 | 76.71 | 60.25 | 3/4 | 204.2 | 160.4 | 3/4 | 392.9 | 308.6 |
| 13/16 | 11.170 | 8.773 | 13/16 | 78.74 | 61.85 | 13/16 | 207.5 | 163.0 | 13/16 | 397.5 | 312.2 |
| 7/8 | 11.953 | 9.388 | 7/8 | 80.80 | 63.46 | 7/8 | 210.9 | 165.6 | 7/8 | 402.1 | 315.8 |
| 15/16 | 12.763 | 10.024 | 15/16 | 82.89 | 65.10 | 15/16 | 214.2 | 168.2 | 15/16 | 406.7 | 319.5 |
| 2 | 13.600 | 10.681 | 5 | 85.00 | 66.76 | 8 | 217.6 | 170.9 | 11 | 411.4 | 323.1 |
| 1/16 | 14.463 | 11.359 | 1/16 | 87.14 | 68.44 | 1/16 | 221.0 | 173.6 | 1/16 | 416.1 | 326.8 |
| 1/8 | 15.353 | 12.058 | 1/8 | 89.30 | 70.14 | 1/8 | 224.5 | 176.3 | 1/8 | 420.8 | 330.5 |
| 3/16 | 16.270 | 12.778 | 3/16 | 91.49 | 71.85 | 3/16 | 227.9 | 179.0 | 3/16 | 425.5 | 334.2 |
| 1/4 | 17.213 | 13.519 | 1/4 | 93.71 | 73.60 | 1/4 | 231.4 | 181.8 | 1/4 | 430.3 | 338.0 |
| 5/16 | 18.182 | 14.280 | 5/16 | 95.96 | 75.36 | 5/16 | 234.9 | 184.5 | 5/16 | 435.1 | 341.7 |
| 3/8 | 19.178 | 15.062 | 3/8 | 98.23 | 77.15 | 3/8 | 238.5 | 187.3 | 3/8 | 439.9 | 345.5 |
| 7/16 | 20.201 | 15.866 | 7/16 | 100.53 | 78.95 | 7/16 | 242.1 | 190.1 | 7/16 | 444.8 | 349.3 |
| 1/2 | 21.250 | 16.690 | 1/2 | 102.85 | 80.78 | 1/2 | 245.7 | 192.9 | 1/2 | 449.7 | 353.2 |
| 9/16 | 22.326 | 17.534 | 9/16 | 105.20 | 82.62 | 9/16 | 249.3 | 195.8 | 9/16 | 454.6 | 357.0 |
| 5/8 | 23.428 | 18.400 | 5/8 | 107.58 | 84.49 | 5/8 | 252.9 | 198.7 | 5/8 | 459.5 | 360.9 |
| 11/16 | 24.557 | 19.287 | 11/16 | 109.98 | 86.38 | 11/16 | 256.6 | 201.5 | 11/16 | 464.4 | 364.8 |
| 3/4 | 25.713 | 20.195 | 3/4 | 112.41 | 88.29 | 3/4 | 260.3 | 204.5 | 3/4 | 469.4 | 368.7 |
| 13/16 | 26.895 | 21.123 | 13/16 | 114.87 | 90.22 | 13/16 | 264.0 | 207.4 | 13/16 | 474.4 | 372.6 |
| 7/8 | 28.103 | 22.072 | 7/8 | 117.35 | 92.17 | 7/8 | 267.8 | 210.3 | 7/8 | 479.5 | 376.6 |
| 15/16 | 29.338 | 23.042 | 15/16 | 119.86 | 94.14 | 15/16 | 271.6 | 213.3 | 15/16 | 484.5 | 380.5 |

*Based on steel having a density of 489.6 lbm/ft³.

†1 in = 25.4 mm.

‡1 lbm/lin ft = 1.48815 kgm/lin m.

# IRON AND STEEL CASTINGS

## by Charles Willers Briggs

REFERENCES: "Metals Handbook," ASM. "Cast Metals Handbook," AFS. "Gray Iron Castings Handbook," Gray Iron Founders' Society. "Malleable Iron Castings," Malleable Founders' Society. "Steel Castings Handbook," Steel Founders' Society. Briggs, "The Metallurgy of Steel Castings," McGraw-Hill.

## CLASSIFICATION OF CASTINGS

**Cast-Iron Castings** The term **cast iron** covers a wide range of iron-carbon-silicon alloys containing from 2.0 to 4.0 percent carbon and 0.25 to 3.00 percent silicon in combination with varying percentages of manganese, sulfur, and phosphorus, and sometimes one or more alloying elements, such as nickel, chromium, molybdenum, copper, vanadium, and titanium. Cast irons may be grouped broadly into two classes.

GRAY CAST IRON. A cast iron that contains a relatively large percentage of its carbon in the form of graphite. It has a gray fracture. It may be a plain or an alloy cast iron varying from 20,000 to 60,000 lb/in² (140 to 400 MPa) minimum tensile strength. The soft irons are readily machined and are used for the ordinary run of machine construction work. Strong iron castings are adapted for medium and heavy sections where strength properties are desirable.

CHILLED-IRON CASTINGS. Cast iron with some section purposely cooled by chills so fast that the carbon is retained in the combined form (white iron) while other sections are allowed to cool more slowly, retaining the carbon in the form found in gray iron. Such castings as crusher jaws, chilled rolls, or car wheels, requiring hard surfaces for wear resistance and soft bodies, are made from low-silicon irons.

**Special Processed Irons** Cast iron produced by licensed or patented controlled processes that permit the achievement of specific cast irons for specific purposes. **Meehanite** is the name of numerous cast irons each having a different combination of mechanical and engineering properties. Four general classification types are produced: (1) general engineering, (2) heat resisting, (3) wear resisting, and (4) corrosion resisting. Tensile strengths vary from 25,000 to 55,000 lb/in² (170 to 380 MPa) and, when oil-quenched and tempered a strength of 75,000 lb/in² (500 MPa) can be obtained. **Nodular iron,** or ductile iron, is cast iron with the graphite substantially spherulitic or in nodular shape and substantially free of flake graphite. There are primarily two grades: an as-cast grade and a graphitizing annealed grade. Tensile strengths vary from 60,-000 to 120,000 lb/in² (400 to 800 MPa).

**Malleable-Iron Castings** Malleable iron is a mixture of iron and carbon including small amounts of silicon, manganese, phosphorus, and sulfur, which, after being cast, is converted, by heat treatment, into a matrix of ferrite containing nodules of temper carbon. The ordinary malleable-iron grade has a tensile strength from 50,000 to 55,000 lb/in² (380 MPa). The pearlite malleable grade is 60,000 lb/in² (400 MPa) and the heat-treated pearlite malleable irons develop tensile strengths

of upwards of 85,000 lb/in² (600 MPa). Most malleable-iron castings weigh from a few ounces to 100 lb (50 kg). Machinery and automotive castings are outstanding applications.

**Steel Castings** There are two main classes of steel castings: carbon and alloy. There are five classes of commercial steel castings: (1) **low-carbon steels** (carbon content below 0.20 percent), (2) **medium-carbon steels** (carbon between 0.20 and 0.50 percent), (3) **high-carbon steels** (carbon content above 0.50 percent), (4) **low-alloy steels** (alloy content totaling less than 8 percent), (5) **high-alloy steels** (alloying content totaling greater than 8 percent). The tensile strength of cast steel varies from 60,000 to 250,000 lb/in² (400 to 1,700 MPa) depending on composition and heat treatment. Steel castings are produced weighing from a few ounces to over 200 tons. They are used in the transportation, machinery, and allied fields. Steel castings should be selected where strength, toughness, and reliability are essential.

## CAST IRON

**Composition** The properties of cast iron are regulated by the control of the amount, type, size, and distribution of the various carbon formations. The important factors are (1) casting design, (2) chemical composition, (3) type of melting scrap, (4) melting process, (5) rate of cooling in the mold, and (6) subsequent heat treatment.

It is not advisable to specify cast irons by chemical composition. Different foundries use different ranges of composition to secure the desired properties, owing to differences in raw materials, melting practices, etc. In general, the engineer who must be familiar with many metals need not know the intimate details of the production of each. Therefore, the method of producing the irons and their composition should be left to the discretion of the foundryman and his metallurgist, who are familiar with the many composition variables to obtain the properties which the engineer desires. Since cast irons are specified on the basis of tensile strength, long tables of compositions for various classes of cast iron are not given, because they are of little value.

The carbon in cast iron is of two types: (1) **combined carbon** as iron carbide and (2) **graphite,** present as a mechanical admixture. The graphite is in the form of dispersed flakes occupying from 6 to 10 percent of the volume of the typical gray irons. These flakes impair the continuity of the matrix to such an extent that they exert a very pronounced effect upon the mechanical properties of the metal. An increase in the amount of graphite present in cast iron, such as an increase in the flake size or an unfavorable distribution of the graphite, affects adversely the strength of the metal.

**Silicon** has a powerful softening effect. Its presence in cast iron reduces the ability of the iron to retain carbon in chemical combination. With very little silicon, the iron retains all its carbon in combination and produces white iron. With about 3 percent silicon, almost no carbon can be held in chemical combination. Manganese, chromium, molybdenum, titanium, and vanadium promote the retention of carbon in the combined form (carbide stabilizers) and counteract silicon. Nickel and copper improve the matrix and increase the strength of the iron, but they do not lessen the amount of graphite present and keep the iron readily machinable.

### Mechanical Properties

The **tensile strength** of cast iron, including the special irons, varies from 20,000 to 80,000 lb/in² (140 to 550 MPa). Eight classes of increasing strength are recognized by ASTM A48 for gray iron castings. Some of the property values that may be obtained are listed in Tables 1 to 3 and Fig. 1. The tensile test is the standard strength test for cast iron. The dimensions of the test bar employed depend on the thickness of the walls of the controlling section of the casting, as indicated in Table 1. Test bars are cast to shape and machined to size. The length of the test bar varies, depending on the bar diameter. Tensile strengths range from 20,000 to 60,000 lb/in² (140 to 400 MPa) for the eight grades.

**Table 1. Cast-Iron-Test-Bar Thicknesses**

| Casting-wall thickness, in* | Machined test-bar diam, in* | As-cast test-bar diam, in* |
|---|---|---|
| 0.25–0.50 | 0.50 | 0.88 |
| 0.51–1.00 | 0.75 | 1.20 |
| 1.01–2.00 | 1.25 | 2.00 |
| Over 2 in | To be agreed upon | |

*× 25.4 = mm.

The **elastic limit** of cast iron is close to its ultimate breaking strength. Gray iron can sustain indefinitely a static load just short of the tensile strength without distortion or breakage. Gray iron has low ductility and breaks without perceptible distortion. Since gray iron does not distort prior to breaking, it

is essential that service stresses be known or that a conservative safety factor be employed.

With static loading the ultimate strength of cast iron in tension is less than that shown in compression; the impact strength of most cast irons is low.

The tensile strength of gray iron is reduced by temperatures over 700°F, and gray iron is limited to a maximum temperature of 450°F when used under the ASME code for unfired pressure vessels. Within this range there is no perceptible

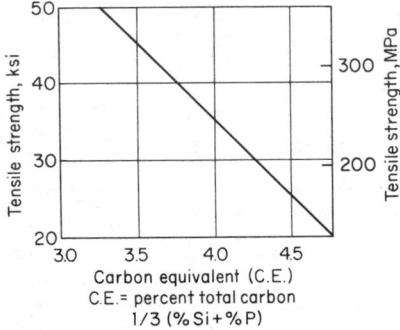

**Fig. 1** General relation of tensile strength to carbon equivalent for gray cast iron in 1.2-in-diameter cast test bars. (*Reprinted by permission from "Gray Iron Castings Handbook," Gray Iron Founders' Society, Inc., 1957.*)

change in the tensile strength of gray irons. The **damping capacity**, or the ability to absorb vibration, is high. High-strength pearlite irons, such as **meehanite**, or nodular iron, have the highest damping capacities of all engineering metals; this accounts for the common selection of cast iron for machinery bases and housings. The endurance limit of gray cast iron is from 12,000 to 24,000 lb/in² (80 to 164 MPa). Meehanite metal of the GA type is reported at 28,000 lb/in² (193 MPa) in the quenched and tempered condition.

**Table 2. Gray Cast Iron**

| Carbon equivalent | Tensile strength, lb/in²* | Modulus of elasticity in tension at ½ load, lb/in² × 10⁶† | Modulus of rupture, lb/in²* | Deflection in 18-in span, in‡ | Shear strength, lb/in²* | Endurance limit, lb/in²* | Compressive strength, lb/in²* | Brinell hardness, 3,000-kg load | Izod impact, unnotched, ft·lb |
|---|---|---|---|---|---|---|---|---|---|
| 4.8 | 20,400 | 8.0 | 48,300 | 0.370 | 29,600 | 10,000 | 72,800 | 146 | 3.6 |
| 4.6 | 22,400 | 8.7 | 49,200 | 0.251 | 33,000 | 11,400 | 91,000 | 163 | 3.6 |
| 4.5 | 25,000 | 9.7 | 58,700 | 0.341 | 35,500 | 11,800 | 95,000 | 163 | 4.9 |
| 4.3 | 29,300 | 10.5 | 63,300 | 0.141 | 37,000 | 12,300 | 90,900 | 179 | 2.2 |
| 4.1 | 32,500 | 13.6 | 73,200 | 0.301 | 44,600 | 16,500 | 128,800 | 192 | 4.2 |
| 4.0 | 35,100 | 13.3 | 77,000 | 0.326 | 47,600 | 17,400 | 120,800 | 196 | 4.4 |
| 3.7 | 40,900 | 14.8 | 84,200 | 0.308 | 47,300 | 19,600 | 119,100 | 215 | 3.9 |
| 3.3 | 47,700 | 20.0 | 92,000 | 0.230 | 60,800 | 25,200 | 159,000 | 266 | 4.4 |

*× 6.89 = kPa.
†× 0.00689 = MPa.
‡× 25.4 = mm.

**Table 3. Alloy and Special Alloy Cast Irons**

| Commercial name or type | Average composition, percent | | | | | | Mechanical properties | | Resistant to |
|---|---|---|---|---|---|---|---|---|---|
| | Total carbon | Ni | Cr | Si | Mo | Other elements | BHN | Tensile strength, lb/in²* | |
| Ni................. | 2.61 | 1.08 | ..... | 2.38 | .... | .... | 269 | 51,500 | Wear |
| Ni-Cr.............. | 2.79 | 0.50 | 0.20 | 2.44 | .... | .... | 250 | 56,700 | Wear |
| Ni-Cr-No........... | 2.72 | 0.85 | 0.15 | 2.35 | 1.05 | .... | 306 | 70,000 | Wear |
| Ni-Hard............ | 3.50 | 4.50 | 1.50 | 0.60 | .... | .... | 675 | 35,000 | Wear |
| Nitrided cast iron.... | 3.00 | ..... | 1.25 | 1.25 | 0.15 | Al 1.00 | ... | ...... | Wear |
| Cr.................. | 3.70 | ..... | 0.75 | 1.10 | .... | Mn 0.50 | 250 | 35,000 | Wear |
| High Cr............. | 2.60 | ..... | 26.70 | 0.33 | .... | .... | 477 | ...... | Wear |
| Oil quenched........ | 3.00 | 1.20 | 0.40 | 1.50 | .... | .... | 480 | ...... | Wear |
| Ni-Resist........... | 3.00 | 13.50 | 3.20 | 1.60 | .... | Cu 6.48 | 140 | 24,000 | Heat and corrosion |
| Ni-Mo.............. | 3.25 | 1.00 | ..... | 2.00 | 0.40 | .... | 156 | 26,000 | Heat |
| Cr.................. | 3.00 | ..... | 1.00 | 2.00 | .... | .... | ... | 40,000 | Heat |
| 17% Cr............. | 2.00 | ..... | 17.00 | 1.75 | .... | .... | 420 | 70,000 | Oxidation |

* × 6.89 = kPa.

The ease of **machining** gray irons usually is inversely proportional to the strength of the casting. Chilling, heat treatment, and alloy additions reduce the machinability.

**Corrosion Resistance** Ordinary grades of cast iron offer considerable resistance to underground corrosion; water pipes have given constant service for over 100 years. High-alloy cast irons of the chromium, nickel, and silicon types are especially resistant to sulfur and acid corrosion. Chromium (15 to 30 percent) imparts a protective coating to cast iron which is especially resistant to sulfur and acid corrosion. High-nickel (18 to 22 percent Ni) cast irons (Ni-Resist, Mechanite CRS) are austenitic in structure and extremely corrosion resistant to many acids and alkalies. High-silicon (11 to 17 percent Si) cast irons (Duriron, Tantiron, etc.) are remarkably good for withstanding all acids except hydrofluoric and hot concentrated hydrochloric acid.

**General Information on Properties** Cast iron is widely used in engineering and allied industries because of the ease with which it may be cast, its moderate cost, and its wide range of useful properties. The engineer should realize that the term "cast iron" is a general term used to designate a class of materials including soft, weak irons; hard, brittle irons; and strong cast irons. Modern engineering practice classifies gray irons according to minimum-tensile-strength properties. The bulk of commercial gray irons is found in the classes from 25,000 to 50,000 lb/in² (170 to 345 MPa). However, irons upward to 80,000 lb/in² (550 MPa) are obtainable by special processes, compositions, and treatments (Meehanite GA60, SP80).

**Gray iron** breaks without perceptible distortion because of its low ductility. However, a gray iron of 40,000 lb/in² (275 MPa) tensile strength could indefinitely sustain a static load equivalent to a stress of about 37,000 lb/in² (255 MPa) without distortion or breakage, whereas a bronze of 40,000 lb/in² (275 MPa) tensile strength and 18,000 lb/in² (124 MPa) yield point would be grossly distorted by a load of 37,000 lb/in² (255 MPa); in fact, it would continue to distort and soon break. The ultimate strength of gray iron in tension under conditions of static loading is less than that shown in compression, transverse loading, and shear. There is no constant relationship

between the Brinell hardness and the tensile strength of cast iron; while strong irons are somewhat harder, tensile strength cannot be consistently predicted from hardness tests. The modulus of elasticity of cast iron is proportioned to the tensile strength. The effective modulus of elasticity at 25 percent of the ultimate strength ranges from 12 million lb/in² (80 million kPa) for the weaker irons to 18 million (124 million) for the strong gray irons. Cast iron has little or no impact or shock-resistance properties, and is difficult to weld.

### Iron-Foundry Practice

Cast iron is produced from pig iron, cast-iron scrap, and steel scrap. There are eight classes of pig iron based on increasing phosphorus content. Each class has a number of grades; ASTM A43 recognizes 243 grades. The steel used is medium to light in character, such as clippings and punchings from structural-steel plate. Melting is done in cupolas, air furnaces, and electric induction and arc furnaces. Uniformity of composition and mechanical properties can be held within fairly close limits. The electric melting units have an advantage that adjustments to the composition can be made to the bath. The molten iron is poured into sand molds. Castings from a few ounces to upwards of 100 tons are produced. (See also Sec. 13.)

### Purchase Specification for Cast-Iron Castings

Engineers should purchase cast-iron castings to the ASTM specifications. They are known and understood by cast-iron founders. The most generally used specification is ASTM A48 Gray Iron Castings, summarized as follows:

Gray-iron castings are classified according to **tensile strength.** The following minimum values are specified:

| Class No. | 20 | 25 | 30 | 35 | 40 | 50 | 60 |
|---|---|---|---|---|---|---|---|
| Tensile strength, min. klb/in²* | 20 | 25 | 30 | 35 | 40 | 50 | 60 |

*× 6.89 = MPa.

**Transverse tests** are optional. The following minimum breaking loads are specified:

| Thickness of critical casting section, in* | Test bar | Diam of test bar, in* | Span length, in* | Class No. | | | | | | |
|---|---|---|---|---|---|---|---|---|---|---|
| | | | | 20 | 25 | 30 | 35 | 40 | 50 | 60 |
| | | | | Min breaking load at center, lb† | | | | | | |
| 0.50 and under | A | 0.875 | 12 | 900 | 1,025 | 1,150 | 1,275 | 1,400 | 1,675 | 1,925 |
| 0.51–1.00 | B | 1.20 | 18 | 1,800 | 2,000 | 2,200 | 2,400 | 2,600 | 3,000 | 3,400 |
| 1.01 and up | C | 2.00 | 24 | 6,000 | 6,800 | 7,600 | 8,300 | 9,100 | 10,300 | 12,500 |

\*×25.4 = mm.
†×4.448 = N.

The **test bars** are cast separately from the casting but made under the same sand conditions as the casting and receive the same thermal treatment.

Other ASTM specifications for cast iron are as follows: A126 Gray Iron Castings for Valves, Flanges and Pipe Fittings; A159 Gray Iron Castings, Automotive; A278 Gray Iron Castings for Pressure-Containing Parts for Temperatures up to 650 deg F; A319 Gray Iron Castings for Elevated Temperatures for Non-Pressure-Containing Parts; A74 Cast Iron Soil Pipe and Fittings; A142 Cast Iron Culvert Pipe; A377 Cast Iron Pressure Pipe; A436 Gray Iron Castings, Austenitic. The methods of testing cast-iron castings are as follows: A256 Compression Testing of Cast Iron; A327 Impact Testing of Cast Iron; A438 Transverse Testing of Gray Cast Iron.

### Tolerances for Gray-Iron Castings

**Pattern shrinkage tolerances** compensate for the shrinkage of gray iron on cooling from solidification to room temperature. The value of 1 percent, or approximately $\frac{1}{8}$ in/ft, is the one commonly used. The amount of shrinkage varies somewhat with the size of the casting and the resistance offered by the mold as follows:

| Pattern dimension, in* | Type of construction | Shrinkage allowance | |
|---|---|---|---|
| | | in/ft | cm/cm |
| Up to 24 | Open | $\frac{1}{8}$ | 0.01 |
| 25–48 | Open | $\frac{1}{10}$ | 0.008 |
| Over 48 | Open | $\frac{1}{12}$ | 0.007 |
| Up to 24 | Cored | $\frac{1}{8}$ | 0.01 |
| 25–36 | Cored | $\frac{1}{10}$ | 0.008 |
| Over 36 | Cored | $\frac{1}{12}$ | 0.007 |
| High-alloy irons up to 48 | | $\frac{5}{32}$ | 0.005 |

\*× 2.54 = cm.

**Casting finish tolerances** (unmachined) are related to the pattern tolerances. The purchaser should allow, for miscellaneous castings, a dimensional tolerance approximately one-half of the maximum shrinkage tolerance or $\pm\frac{1}{16}$ in (0.06 in or 1.6 mm) for castings approaching a foot in length. Certain small castings, such as cams, camshafts, etc., are being held to $\pm\frac{1}{32}$ in (0.03 in) (0.8 mm). Medium-sized castings (3 ft (1m) in length) are produced to a value of $\pm\frac{3}{20}$ in or 4 mm. Tolerances for heavy-cored areas are $\pm\frac{1}{10}$ in/ft (0.008 cm/cm). Tolerances for highly repetitive jobs, such as automotive cylinder blocks, can be held to $\pm\frac{1}{64}$ in (0.015 in or 0.4 mm) for length up to an 8-in (21-cm) span. Over 8 in (21 cm) a tolerance of $\pm\frac{1}{32}$ in (0.8 mm) is considered normal.

**Machine finish tolerances** are normally $\frac{3}{32}$ to $\frac{1}{8}$ in (2.4 to 3.2 mm). In the machine-tool industry, where casting of 100 to 2,000 lb (50 to 1,000 kg) is common, machining allowances of $\frac{3}{16}$ to $\frac{1}{4}$ in (4.7 to 6.4 mm) usually are specified. The following table shows some common machining tolerances or allowances:

| Pattern size, in* | Bore, in† | Finish, in† |
|---|---|---|
| Up to 12 | $\frac{1}{8}$ | $\frac{3}{32}$ |
| 13–24 | $\frac{3}{16}$ | $\frac{1}{8}$ |
| 25–42 | $\frac{1}{4}$ | $\frac{3}{16}$ |
| 43–60 | $\frac{5}{16}$ | $\frac{1}{4}$ |
| 61–80 | $\frac{3}{8}$ | $\frac{5}{16}$ |
| 81–120 | $\frac{7}{16}$ | $\frac{3}{8}$ |
| Over 120 | Special instructions required | |

Surfaces to be machined should be located in the lower, or "drag," sections of the mold for minimum machine tolerances. Additional finish allowances of $\frac{1}{32}$ to $\frac{1}{16}$ in greater than shown may be desirable if certain upper, or "cope," surfaces are to be machined.
\*× 2.54 = cm.
†× 25.4 = mm.

### Nodular (Ductile) Cast Iron

Nodular iron is a cast iron produced by adding to it graphite spherulitic alloys, such as magnesium and cerium. The addition causes the graphite to form as small nodules or spheroids instead of as the normal angular flakes. Quality-control processes are required to produce cast iron of high strengths with reasonable amounts of ductility, such as Meehanite Ductliron Type S.

As-cast grades of ductile iron have a tensile strength between 60,000 and 105,000 lb/in² (415 and 720 MPa) with an elongation of 10.0 to 1.0 percent, depending on composition. Annealing increases the elongation to 10 to 25 percent with a tensile strength of 60,000 to 75,000 lb/in² (415 to 520 MPa). Normalized and tempered specimens have a tensile strength of 90,000 to 120,000 lb/in² (620 to 830 MPa) with an elongation of 2.0 to 5.0 percent. The varying composition of the base metal, the methods of nodularizing, and the control measures account for the wide range of tensile values in these test bars.

The notched impact value as measured by the standard Charpy specimen is very low, being less than 5 ft·lb (7 Nm) approx at room temperatures. The endurance limit varies from 35,000 to 40,000 lb/in² (240 to 275 MPa) for tensile strengths of 70,000 to 95,000 lb/in² (480 to 650 MPa). The material appears to have the advantages of cast iron (fluidity, low melting point, good machinability) with the tensile strength of carbon cast steel.

Nodular iron can be produced by any of the furnaces used for the melting of cast iron, although the electric-arc and induction furnaces produce more closely controlled nodular iron. **Specifications for nodular iron** have been prepared by ASTM: A339 Nodular-iron Castings; A395 Cast Nodular Iron for Pressure Containing Parts for Use at Elevated Temperatures; A396 High Strength Nodular Iron Castings; A436 Austenitic Gray Iron Castings.

The property requirements for the low- and high-strength grades are as follows:

to 3.30, silicon 1.10 to 0.60, and manganese 0.40 to 0.65 percent. Most of the commercial malleable is produced in the standard grade.

**Mechanical Properties**  The tensile strength of malleable irons vary from 50,000 to 110,000 lb/in² (345 to 760 MPa) approx. The tensile strength of the standard grade is from 50,000 to 60,000 lb/in² (345 to 414 MPa) approx. Specification minimum values for malleable irons are given in Table 4.

Standard malleable iron has an average endurance limit of 54,000 lb/in² (372 MPa) and an endurance ratio of 0.57. The

|  | A339 | | A396 | |
|---|---|---|---|---|
|  | 60–45–10 | 80–60–03 | 100–70–03 | 120–90–02 |
| Tensile strength, min lb/in²* | 60,000 | 80,000 | 100,000 | 120,000 |
| Yield strength, min lb/in²* | 45,000 | 60,000 | 70,000 | 90,000 |
| Elongation in 2 in, min % | 10 | 3† | 3 | 2 |

*× 6.89 = kPa.

†Where strength is the prime requirement, the elongation requirement may be waived by agreement.

Tensile test bars (0.505 in diam) are machined from coupons of three different section thicknesses: ½, 1, and 3 in. The coupons are to be cast at the same time the castings are cast and are to be of a comparable section thickness.

The application of nodular iron in brief summary is as follows:

notch-fatigue ratio is 0.33. The ultimate strength of malleable iron in shear is roughly 90 percent of the ultimate tensile strength. The modulus of rigidity is about 11 million lb/in² (76,000 MPa). The impact resistance of standard-grade malleable iron, measured by the Charpy test, keyhole notch, is 6.5 to 8.0 ft·lb (8.9 to 11 N·m).

| Type | BHN | Characteristics | Application |
|---|---|---|---|
| 60–45–10 | 140–190 | Annealed ferritic matrix, good machinability and ductility | Valves, pumps, and pressure parts |
| 80–60–03 | 200–270 | Pearlitic matrix, as-cast high strength | Heavy-duty machinery gears, rolls, statically loaded castings |
| 100–70–03 | 240–300 | Normalized and tempered or quenched and tempered | Wear-resistant parts, pinions, gears, cams, guides, rollers |
| 120–90–02 | 270–350 | | |

The engineer should employ factors of safety similar to those used for cast iron. Materials with less than 5 percent ductility are not usually considered as ductile materials. Notched-impact properties are low, and shock-loading applications need confirming service testing. Welding is difficult and not recommended.

**MALLEABLE-IRON CASTINGS**

**Composition**  Malleable-iron castings may be broadly divided as follows: (1) standard malleable irons, (2) pearlitic malleable irons, (3) special malleable irons, and (4) cupola malleable irons. The chemical composition of the white iron from which the standard malleable iron is produced falls generally within the following limits: carbon 2.00 to 2.70, silicon 1.20 to 0.80, manganese less than 0.55, phosphorus less than 0.20, and sulfur less than 0.18 percent. Pearlitic malleable irons are made from a chemical composition similar to that of standard malleable iron together with additional alloys or so heat-treated that some of the carbon in the resultant material is in the combined form. Special malleable irons consist principally of those with a high silicon content, those alloyed with copper, and those alloyed with copper and molybdenum. High-silicon malleable permits short annealing times. Copper is used as an alloy to increase strength and endurance limit. The copper-molybdenum malleable iron is used for extra-high strengths. Cupola malleable-iron composition is carbon 2.80

Malleable iron has excellent machinability. With SAE 1112 steel rated at the base value of 100, the following ratings have been given:

| Malleable irons | Machinability value | BHN |
|---|---|---|
| Standard............ | 120 | 110–145 |
| Pearlitic ............ | 90 | 180–200 |
| Pearlitic ............ | 80 | 200–240 |

Malleable iron has a high resistance to atmospheric corrosion and finds wide use as pole-line hardware, bridge railings, panels, street signs, etc. Malleable-iron castings are used extensively for automobile, agricultural-implement, conveyor, and handling equipment, and in electrical, power, and railroad industries.

**Specifications and Use**  ASTM specification A47 Malleable Iron Castings leads the field in production volume and is the standard grade of malleable, consisting largely of ferrite interspersed by nodules of free carbon; A220 Pearlitic Malleable Iron Castings, which contains some carbon in the combined form through special heat treatment and by alloying to make it pearlitic, has rapidly increased in use during the past decade; A197 Cupola Malleable Iron has been almost completely supplanted by the other classes. Specification values are given in Table 4. Another standard specification, A338

**Table 4. Specification Minimum Properties of Malleable Irons**

| Type | ASTM specification | Tensile strength, min lb/in²* | Yield point, min lb/in²* | Elongation, min % | BHN, typical range |
|---|---|---|---|---|---|
| Cupola....... | A197–47 | 40,000 | 30,000 | 5 | |
| Standard..... | A47–61(32510) | 50,000 | 32,500 | 10 | 110–130 |
| Standard..... | A47–61(35018) | 53,000 | 35,000 | 18 | 120–145 |
| Pearlitic...... | A220–61T(45010) | 65,000 | 45,000 | 10 | 163–207 |
| Pearlitic...... | A220–61T(45007) | 68,000 | 45,000 | 7 | 163–217 |
| Pearlitic...... | A220–61T(48004) | 70,000 | 48,000 | 4 | 163–229 |
| Pearlitic...... | A220–61T(50007) | 75,000 | 50,000 | 7 | 179–229 |
| Pearlitic...... | A220–61T(53004) | 80,000 | 53,000 | 4 | 197–241 |
| Pearlitic...... | A220–61T(60003) | 80,000 | 60,000 | 3 | 197–255 |
| Pearlitic...... | A220–61T(80003) | 100,000 | 80,000 | 2 | 241–269 |

\* $\times$ 6.89 kPa.

covers Malleable Iron Flanges, Pipe Fillings, and Valve Parts for Railroad, Marine and Other Heavy Duty Service at Temperatures up to 650°F (350°C).

**Processing Methods** The production of malleable-iron castings consists of two steps: the manufacture of the white-iron casting and its subsequent conversion into the tough malleable product. This conversion is accomplished by heating the white iron to 1500 to 1600°F (815 to 870°C). If this temperature is maintained, the combined carbon ($Fe_3C$) will be dissolved and will tend to precipitate out as a "temper" carbon. The resultant matrix consists of iron interspersed with nodules of tempered graphite.

Castings having sections over 2 in (5 cm) do not lend themselves to production in malleable iron. There are very few malleable castings that weigh over 500 lb (230 kg). Most castings weigh from a few ounces to less than 100 lb (45 kg). Fusion welding is not recommended.

**Tolerances for Malleable-Iron Castings**

**Pattern shrinkage tolerances** for malleable-iron castings are approximately ⅛ in/ft (0.0096 cm/cm). Draft allowances are ¹⁄₆₄ in/in for production patterns; ¼ in/in for loose patterns.

**Casting-finish tolerances** (unmachined) for green-sand molding on the basis of outside dimensions are: up to 4 in, ±¹⁄₃₂ (10 cm, ±0.8 mm); 4 to 8 in, ±³⁄₆₄ (10 to 20 cm, ±1.2 mm); 8 to 12 in, ±¹⁄₁₆ (20 to 30 cm, ±1.6 mm); 12 to 24 in, ±⅛ in (30 to 60 cm, ±3.2 mm). Shell-molding tolerances are about two-thirds of these values.

**Machine-finish allowances** recommended for malleable iron castings are: (1) milling—¹⁄₁₆ to ³⁄₃₂ (1.5 to 2.4 mm) for small castings, ⅛ to ³⁄₁₆ (3.2 to 4.8 mm) for medium castings, and somewhat greater for large castings weighing 100 lb (50 kg) or more; (2) reaming—³⁄₃₂ in (2.4 mm) on the diameter for cored holes under 1 in (2.5 cm), ⅛ to ³⁄₁₆ in (3.2 to 4.8 mm) on the diameter for medium holes, and more for large holes; (3) turning or boring—diameters larger than 5 in (13 cm), ¼ to ⅜ in (6.4 to 9.5 mm) on the diameter.

## STEEL CASTINGS

**Composition** There are five classes of commercial steel castings:

1. Low-carbon steels (carbon below 0.20 percent)
2. Medium-carbon steels (carbon between 0.20 and 0.50 percent)

3. High-carbon steels (carbon above 0.50 percent)
4. Low-alloy steels (total alloy less than 8 percent)
5. High-alloy steels (total alloy more than 8 percent)

**Carbon-Steel Castings** Carbon-steel castings contain less than 1.70 percent C, along with other elements normally present. These elements may be present in percentages ranging as follows: Mn 0.50 to 1.00, Si 0.20 to 0.70, max P 0.05, and max S 0.06. In addition, carbon-steel castings contain small percentages of other elements, which were not added but were residual in the scrap steel used as part of the melting charge.

The medium-carbon class, constituting the bulk of the steel casting output, is the regular grade product. Low- and high-carbon grades have been developed for specialized products and uses.

**Alloy-Steel Castings** A steel casting is considered to be an alloy-steel casting if the alloying elements, either residual or added, are present in percentages greater than the following: Mn, 1.00; Si, 0.70; Cu, 0.50; Cr, 0.25; Mo, 0.10; V, 0.05; W, 0.05; Al, 0.05; Ti, 0.05. Limitations on phosphorus and sulfur contents apply to cast alloy steels as well as to cast carbon steels unless they are specified for the purpose of producing an alloying effect. The low-alloy casting class represents a considerable portion of the total steel-casting production in this country. High-alloy steel castings of the heat- and corrosion-resistant type are similar to the wrought stainless steels.

**Weight Range** Steel castings range from a few ounces to many tons. The largest on record had a finished weight of 230 tons (210 tonnes). Steel castings may be made of any thickness down to about ¼ in (6 mm).

**Mechanical Properties** The outstanding mechanical properties of cast steel are strength, ductility, and resistance to impact. Steel castings possess high rigidity and are capable of withstanding both high and low temperatures, are weldable and have excellent endurance properties.

The mechanical properties of carbon steel are shown in Figs. 2 and 3. The properties of low-alloy cast steels are given in Figs. 4 and 5. In the heat- and corrosion-resistant class of alloy castings there are 46 class designations. Table 7 lists the ASTM requirements for heat and corrosion alloy castings; the strength values normally expected are from 10 to 25 percent above the specification minimum values.

**Tensile and Yield Strength** Ferritic steels of a given hardness or hardenability have the same tensile strength whether cast, rolled, wrought, or welded, regardless of the alloy content. For design purposes involving tensile and yield properties, rolled, wrought, cast, and welded steels can be interchanged with the fullest confidence.

**Ductility** If the ductility properties of steels are compared with their hardness values, cast, wrought, rolled steels, and welds are almost identical. The longitudinal properties of the forged and rolled steels are slightly higher than those of cast steel or weld metal. The transverse properties are lower, by an

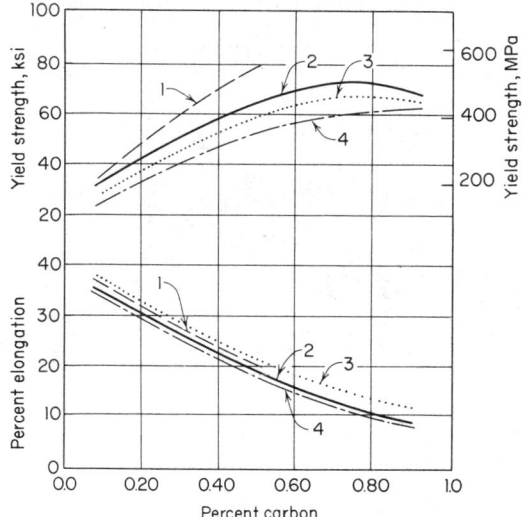

**Fig. 2** Yield strength and elongation vs. carbon content for carbon cast steels. (1) Water quenched and tempered—1200°F. (2) Normalized. (3) Normalized and tempered—1200°F. (4) Annealed.

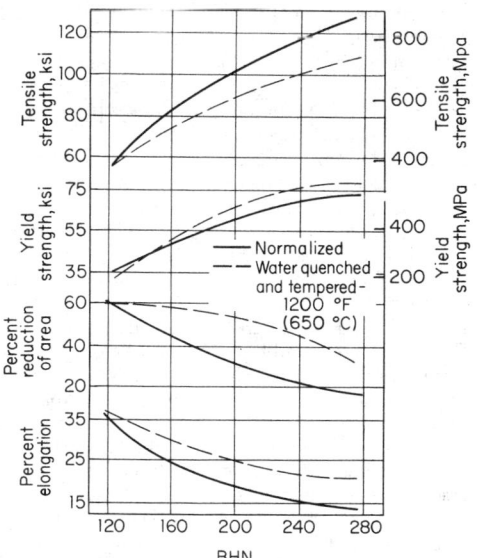

**Fig. 3** Tensile properties of carbon cast steel as a function of hardness.

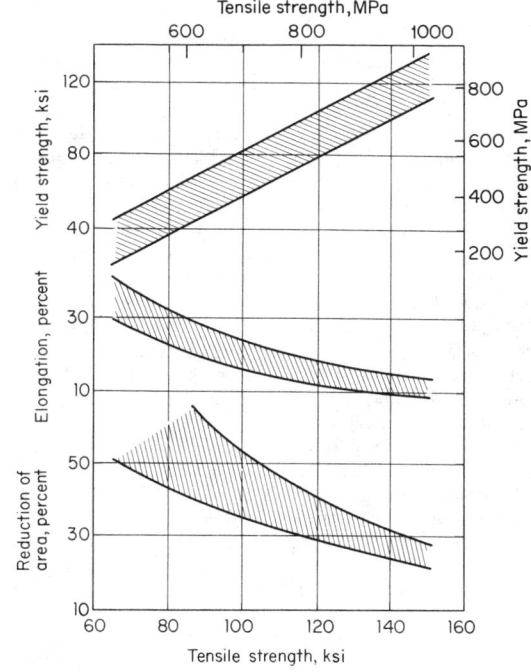

**Fig. 4** Properties of normalized and tempered low-alloy cast steels.

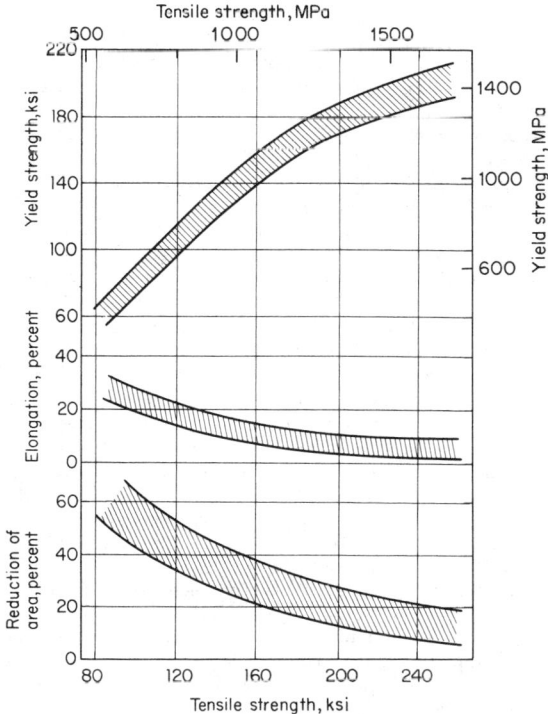

**Fig. 5** Properties of quenched and tempered low-alloy cast steels.

amount that depends on the degree of working. Since most service conditions involve several directions of loading, the securing of uniform directional properties of cast steel is sometimes particularly advantageous.

**Impact**   The notched-bar impact test is often used as a measurement of the toughness of materials. Cast steels have excellent impact resistance at normal and low temperatures. Generally, wrought steels are tested in the direction of rolling, and show higher impact values than cast steels of similar composition. Transverse impact values will be 50 to 70 percent of these values. Cast steels do not show directional properties. If the directional properties are averaged for wrought steels, the values obtained are comparable to the values obtained for cast steels of similar composition. The **hardenability** of cast steels is influenced by composition and other variables in the same manner as the hardenability of wrought steels. The ratio of the **endurance limit** to the tensile strength for cast steel varies from 0.42 to 0.50, depending somewhat upon the composition and heat treatment of the steel. The notch-fatigue ratio varies from 0.28 to 0.32 for cast steels and is the same for wrought steels (Fig. 6).

In **wear-resistance testing,** cast steels react similarly to rolled steels and give corresponding values, depending on composition, structure, and hardness. Carbon cast steels of approximately 0.50 percent C and low-alloy cast steels of the chromium, chromium-molybdenum, nickel-chromium, chromium-vanadium, and medium-manganese types, all of which contain more than 0.40 percent C, have given excellent resistance to wear in service.

**Corrosion Resistance**   Cast steel and wrought steel of similar composition and heat treatment appear to be equally resistant to corrosion in the same environments. Small amounts of copper in cast steel increase the resistance of steel to atmospheric corrosion. High-alloy cast steels of chromium and chromium-nickel types are normally used for corrosion resistance.

**Heat Resistance**   Although not comparable to the high-alloy steels of the nickel-chromium type especially designed for heat resistance, the 4.0 to 6.5 percent Cr cast steels, particularly with additions of 0.75 to 1.25 W, or 0.40 to 0.70 Mo and 0.75 to 1.00 Ti, show good strength and considerable resistance to scaling at 1000°F (550°C) and below.

The **machinability** of carbon and alloy cast steels is comparable to that of wrought steels having equivalent strength, ductility, and hardness, and similar microstructure. Factors influencing the machinability of cast steel are as follows: (1) Microstructure has a definite effect on the machinability of cast steels. In some cases it is possible to improve machining characteristics as much as 100 to 200 percent through heat treatments, which alter the microstructure. (2) Generally speaking, hardness alone cannot be taken as the criterion for predicting tool life in the cutting of cast steels. (3) In general, for a given structure, the plain carbon steels possess better machining properties than the alloy steels. (4) The tool life of carbon (1040) cast steel, when machined with carbides, varies as the ratio of ferrite to pearlite in its microstructure, the $^{60}/_{40}$ ratio machining best. (5) To obtain equivalent tool life, the skin of a cast steel should be machined at approximately one-half of the cutting speed recommended for the base metal. The machinability of various carbon and low-alloy cast steels is given in Table 5.

The **welding** of steel castings presents the same problems as the welding of wrought steels.

**Purchase Specifications for Steel Castings**   Most steel castings are purchased according to mechanical-property specifications rather than according to the SAE or AISI numbering system for ranges of chemical composition. These composition ranges are used as the basis of purchase requirements when the purchaser is considering certain definite engineering properties, such as wear resistance, weldability, high-temperature service, and corrosion resistance. Engineers should pur-

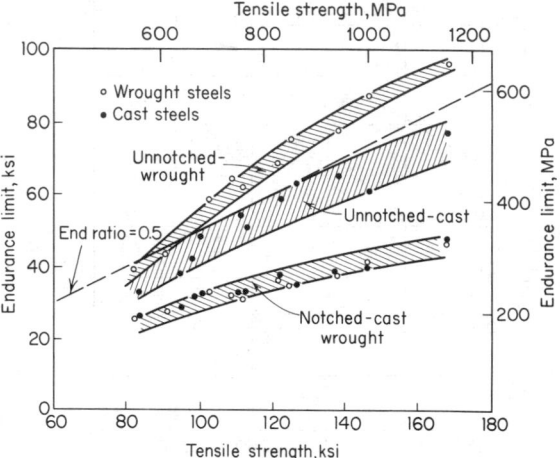

**Fig. 6**   Variation of endurance limit with tensile strength for comparable carbon and low-alloy and wrought steels.

chase steel castings to the ASTM specifications. They are jointly prepared by purchasers and manufacturers of steel castings and are well known and understood by steel foundrymen. The most generally used specifications are ASTM A27 Mild- to Medium-Strength Carbon-Steel Castings for General Application and A148 High-Strength Steel Castings for Structural Purposes. A summary of these specifications is given in Table 6. The values specified in the table are minimum values used as rejection limits. The normal expected values of cast steel meeting these requirements are from 10 to 40 percent above the specification minimums.

Other ASTM specifications for steel castings are as follows: A128, Austenitic-Manganese-Steel Castings; A95, Carbon-steel Castings for Valves, Flanges, and Fittings for High-Temperature Service; A216, Carbon-steel Castings Suitable for Fusion Welding for High-Temperature Service; A217, Alloy-Steel Castings Suitable for Fusion Welding for High-Temperature Service; A296, Corrosion-Resistant Iron-Chromium and Iron-Chromium-Nickel-Alloy Castings for General Applications; A297, Heat-Resistant Iron-Chromium and Iron-Chromium-Nickel-Alloy Castings for General Application; A351, Ferritic and Austenitic Steel Castings for High-Temperature Service; A352, Ferritic Steel Castings for Pressure-Containing Parts Suitable for Low-Temperature Service; A356, Heavy-Walled Carbon and Low-Alloy Steel Castings for Steam Turbines; A362, Iron-Chromium and Iron-Chromium-Nickel-Alloy Tubular Centrifugal Castings for General

**Table 5. Machinability Index for Cast Steels**

| Steel | BHN | Conventional | | Metcut* | |
|---|---|---|---|---|---|
| | | Carbide | HSS | HSS | Carbide |
| B1112 Free machining steel (wrought)............... | 179 | .. | 100 | | |
| 1020 Annealed........................ | 122 | 10 | 90 | 160 | 400 |
| 1020 Normalized........ ................. | 134 | 6 | 75 | 135 | 230 |
| 1040 Double normalized................... | 185 | 11 | 70 | 130 | 400 |
| 1040 Normalized and annealed..................... | 175 | 10 | 75 | 135 | 380 |
| 1040 Normalized........................ | 190 | 6 | 65 | 120 | 325 |
| 1040 Normalized and oil quenched.................. | 225 | 6 | 45 | 80 | 310 |
| 1330 Normalized........................ | 187 | 2 | 40 | 75 | 140 |
| 1330 Normalized and tempered...................... | 160 | 3 | 65 | 120 | 230 |
| 4130 Annealed......................... | 175 | 4 | 55 | 95 | 260 |
| 4130 Normalized and spheroidized................... | 175 | 3 | 50 | 90 | 200 |
| 4340 Normalized and annealed..................... | 200 | 3 | 35 | 60 | 210 |
| 4340 Normalized and spheroidized.................. | 210 | 6 | 55 | 95 | 290 |
| 4340 Quenched and tempered...................... | 300 | 2 | 25 | 45 | 200 |
| 4340 Quenched and tempered...................... | 400 | ½ | 20 | 35 | 180 |
| 8430 Normalized and tempered at 1200°F (660°C)........ | 200 | 3 | 50 | 90 | 200 |
| 8430 Normalized and tempered at 1275°F (702°C)........ | 180 | 4 | 60 | 110 | 240 |
| 8630 Normalized............................. | 240 | 2 | 40 | 75 | 180 |
| 8630 Annealed............................ ..... | 175 | 5 | 65 | 120 | 290 |

*The metcut speed index number is the actual cutting speed (surface ft/min) which will give 1 h tool life in turning.

Applications; A389, Alloy-Steel Castings Normalized and Drawn for High-Pressure and Elevated-Temperature Service; A447, Chromium-Nickel-Iron-Alloy Castings (25–12 Class) for High-Temperature Service; A448, Nickel-Chromium-Iron-Alloy Castings (35–15 Class) for High-Temperature Service.

A summary of the requirements for the heat and corrosion specifications A296 and A297 is given in Table 7 to indicate the types of stainless steels cast in steel foundries.

**Steel-Melting Practice**  Steel for steel castings is produced commercially by almost all the currently used processes of making steel: open-hearth (acid and basic), electric-arc (acid and basic), and electric-induction.

The melting methods employed are similar to those for ingot production, except the acid electric-arc and high-fre-

quency furnace practices which apply directly to steel-casting production. The primary differences between the production of steel for castings and for ingots are that the steel for castings must always be a dead-killed steel having a high degree of fluidity. Approximately 75 grades of carbon and alloy cast steels and 40 grades of heat- and corrosion-resistant alloy-steel castings are being made by the industry.

**Tolerances for Steel Castings**

**Pattern shrinkage tolerances** for steel castings vary from 9/32 to 1/16 in/ft (0.03 to 0.005 cm/cm). The figure often used is 3/16 in/ft (0.016 cm/cm), but its universal adoption would lead to trouble and errors in steel casting dimensions and tolerances. The best policy is to discuss pattern shrinkage tolerances with the foundry that is to make the castings.

**Table 6. ASTM Requirements for Steel Castings, Mechanical Properties**

| Grade | Tensile strength, min | | Yield point, min | | Elongation in 2 in (5 cm) min, % | Reduction of area, min, % |
|---|---|---|---|---|---|---|
| | 1,000 lb/in² | MPa | 1,000 lb/in² | MPa | | |
| | | | ASTM A27 | | | |
| 60–30 | 60 | 414 | 30 | 207 | 24 | 35 |
| 65–35 | 65 | 448 | 35 | 241 | 24 | 35 |
| 70–36 | 70 | 482 | 36 | 248 | 22 | 30 |
| | | | ASTM A148 | | | |
| 80–40 | 80 | 552 | 40 | 276 | 18 | 30 |
| 80–50 | 80 | 552 | 50 | 345 | 22 | 35 |
| 90–60 | 90 | 621 | 60 | 414 | 20 | 40 |
| 105–85 | 105 | 724 | 85 | 586 | 17 | 35 |
| 120–95 | 120 | 827 | 95 | 655 | 14 | 30 |
| 150–125 | 150 | 1,034 | 125 | 853 | 9 | 22 |
| 175–145 | 175 | 1,210 | 145 | 1,000 | 6 | 12 |

**Table 7. ASTM Requirements for Corrosion and Heat-Resistant Alloy-Steel Castings**

| Specification | Class | Tensile strength, lb/in²* | Yield point, lb/in²* | Elongation in 2 in, % | C | Mn | Si | Ni | Cr | Mo and other |
|---|---|---|---|---|---|---|---|---|---|---|
| ASTM A296 | CA-15 | 90,000 | 65,000 | 18 | 0.15 | 1.00 | 1.50 | 1.00 | 11.5–14 | 0.50 |
| | CB-30 | 65,000 | 30,000 | | 0.30 | 1.00 | 1.50 | 2.00 | 18–21 | |
| | CC-50 | 55,000 | | | 0.50 | 1.00 | 1.50 | 4.00 | 26–30 | |
| | CE-30 | 80,000 | 40,000 | 10 | 0.30 | 1.50 | 2.00 | 8–11 | 26–30 | |
| | CF-8 | 65,000 | 28,000 | 35 | 0.08 | 1.50 | 2.00 | 8–11 | 18–21 | |
| | CF-8C | 70,000 | 30,000 | 30 | 0.08 | 1.50 | 2.00 | 9–12 | 18–21 | |
| | CF-8M | 70,000 | 30,000 | 30 | 0.08 | 1.50 | 2.00 | 9–12 | 18–21 | 2.00–3.00 |
| | CF-16F | 70,000 | 30,000 | 25 | 0.16 | 1.50 | 2.00 | 9–12 | 18–21 | |
| | CF-20 | 70,000 | 30,000 | 30 | 0.20 | 1.50 | 2.00 | 8–11 | 18–21 | |
| | CG-12 | 70,000 | 28,000 | 35 | 0.12 | 1.50 | 2.00 | 10–13 | 20–23 | |
| | CH-20 | 70,000 | 30,000 | 30 | 0.20 | 1.50 | 2.00 | 12–15 | 22–26 | |
| | CK-20 | 65,000 | 28,000 | 30 | 0.20 | 2.00 | 2.00 | 19–22 | 23–27 | |
| | CA-40 | 90,000 | 65,000 | 18 | 0.20–0.40 | 1.00 | 1.50 | 1 | 11.5–14 | 0.5 |
| | CF-3 | 65,000 | 28,000 | 35 | 0.03 | 1.50 | 2.00 | 8–12 | 17–21 | |
| | CF-3M | 70,000 | 30,000 | 30 | 0.03 | 1.50 | 1.50 | 9–13 | 17–21 | 2–3 |
| | CG-8M | 75,000 | 35,000 | 25 | 0.08 | 1.50 | 1.50 | 9–13 | 18–21 | 3–4 |
| | CN-7M | 62,500 | 25,000 | 35 | 0.07 | 1.50 | 1.50 | 28–30 | 19–22 | Cu 3–4, Co 2.5, W 5.25 |
| | CW-12M | 72,000 | 46,000 | 4 | 0.12 | 1.00 | 1.50 | Bal. | 16–20 | Mo 16–20, V 0.4, Fe 7.50 |
| | CY-40 | 70,000 | 28,000 | 30 | 0.40 | 1.50 | 3.00 | Bal. | 14–17 | Fe 1.10 |
| | CZ-100 | 50,000 | 18,000 | 10 | 1.00 | 1.50 | 2.00 | 95 | 0 | Fe 3.0 + Cu 1.25 |
| | M-35 | 65,000 | 30,000 | 25 | 0.35 | 1.50 | 2.00 | Bal. | | Cu 26–33, Fe 3.5 |
| | N-12M | 72,000 | 46,000 | 6 | 0.12 | 1.00 | 1.00 | Bal. | | Mo 26–33, Co 2.5, V 0.6, Fe 6.0 |
| ASTM A297 | CA-6NM | 110,000 | 80,000 | 15 | 0.06 | 1.00 | 1.00 | 3.5–4.5 | 11.5–14 | Mo 0.4–1.0 |
| | HC | 55,000 | | | 0.50 | 1.00 | 2.00 | 4.00 | 26–30 | 0.50 |
| | HE | 85,000 | 40,000 | 9 | 0.20–0.50 | 2.00 | 2.00 | 8–11 | 26–30 | 0.50 |
| | HF | 70,000 | 35,000 | 25 | 0.20–0.40 | 2.00 | 2.00 | 8–12 | 18–23 | 0.50 |
| | HH | 75,000 | 35,000 | 10 | 0.20–0.50 | 2.00 | 2.00 | 11–14 | 24–28 | 0.50 |
| | HI | 70,000 | 35,000 | 10 | 0.20–0.50 | 2.00 | 2.00 | 14–18 | 26–30 | 0.50 |
| | HK | 65,000 | 35,000 | 10 | 0.20–0.60 | 2.00 | 2.00 | 18–22 | 24–28 | 0.50 |
| | HT | 65,000 | | 4 | 0.35–0.75 | 2.00 | 2.50 | 33–37 | 13–17 | 0.50 |
| | HU | 65,000 | | 4 | 0.35–0.75 | 2.00 | 2.50 | 37–41 | 17–21 | 0.50 |
| | HW | 60,000 | | | 0.35–0.75 | 2.00 | 2.50 | 58–62 | 10–14 | 0.50 |
| | HX | 60,000 | | | 0.35–0.75 | 2.00 | 2.50 | 64–68 | 15–19 | 0.50 |
| | HD | 75,000 | 35,000 | 8 | 0.50 | 1.50 | 2.00 | 4–7 | 26–30 | 0.50 |
| | HL | 65,000 | 35,000 | 10 | 0.20–0.60 | 2.00 | 2.00 | 18–22 | 28–32 | 0.50 |
| | HN | 63,000 | | 8 | 0.20–0.50 | 2.00 | 2.00 | 23–27 | 19–23 | 0.50 |

Mechanical properties, min. Chemical composition % max.

\* × 6.89 = kPa.

**Table 8. Dimensional Tolerances for Steel Castings***
(Deviation from the design dimension)

| Pattern type | Blueprint dimension, in† | | | |
|---|---|---|---|---|
| | 0–3.0 | 3.1–7.0 | 7.1–20.0 | 20.1–100.0 |
| Metal match plate.................. | +1/32, −1/16 | +3/32, −1/16 | +1/8, −1/16 | +1/8, −1/8 |
| Metal pattern mounted on cope and drag boards...................... | +1/16, −1/16 | +3/32, −3/32 | +1/8, −3/32 | +7/32, −1/8 |
| Hardwood pattern mounted on cope and drag boards...................... | +3/32, −1/16 | +1/8, −3/32 | +1/8, −3/32 | +1/4, −5/32 |

*Surfaces taht are not to be machined.
† × 2.54 = cm.

**Minimum Section Thickness** The fluidity of steel in comparison with other metals is known to be low. In order that sections may be completely run, it is necessary that a minimum value of section thickness be adopted as a function of the largest dimension of the casting. Values suggested for design use are as follows:

| Min section thickness | | Max length of section | |
|---|---|---|---|
| in | mm | in | cm |
| 1/4 | 6 | 12 | 30 |
| 1/2 | 13 | 50 | 125 |

**Casting finish tolerances** (unmachined) are based on the longest dimension of the casting, and the values vary, depending on the type of pattern equipment employed. The average tolerances are given in Table 8.

**Machine finish tolerances** to be added to the casting section for machining purposes will depend entirely on the casting design. Definite values cannot be established for all casting designs, but guides can be suggested to designers. Table 9 presents a guide to machine allowances on gears, wheels, and circular-shaped and flat castings.

### Precision-Casting Processes

**Investment Process** This is the *lost wax* method used by ancient craftsmen, readapted after the First World War for jewelry and denture manufacture, and used extensively in the Second World War for aircraft engine parts. Patterns are formed by pressure injection of wax or plastic into a precision metallic die. Patterns, either singly or in groups, are fitted with wax gates and risers for casting metals and precoated with a fine silica mixture by spraying or dipping. This is followed by a coarser "stucco" coat. The assembly is then

**Table 9. Machine Allowances for Steel Castings**

| Specific dimension or reference line (plane) distance in inches | | Greatest dimension of the casting | | | | | | | |
|---|---|---|---|---|---|---|---|---|---|
| | | 10 in | | 10–20 in | | 20–100 in | | Over 100 in* | |
| | | 0.xx in* | X/32 in* | 0.xx in* | X/32 in* | 0.xx in* | X/32 in* | 0.xx in* | X/32 in |
| Greater than | But not exceeding | + | + | + | + | + | + | + | + |
| 0 | 2 | 0.187 | 6 | 0.218 | 7 | 0.25 | 8 | 0.312 | 10 |
| 2 | 5 | 0.25 | 8 | 0.281 | 9 | 0.312 | 10 | 0.437 | 14 |
| 5 | 10 | 0.312 | 10 | 0.344 | 11 | 0.375 | 12 | 0.531 | 17 |
| 10 | 20 | | | 0.406 | 13 | 0.468 | 15 | 0.625 | 20 |
| 20 | 50 | | | | | 0.562 | 18 | 0.75 | 24 |
| 50 | 75 | | | | | 0.656 | 21 | 0.875 | 28 |
| 75 | 100 | | | | | 0.75 | 24 | 1.00 | 32 |
| 100 | 500 | | | | | | | 1.25 | 40 |

| | | Machine tolerances on section thickness | |
|---|---|---|---|
| | | 0.XXX in* | X/32 in* |
| | | + | + |
| 0 | 1/2 | 0.062 | 2 |
| 1/2 | 1 1/2 | 0.092 | 3 |
| 1 1/2 | 4 | 0.187 | 6 |
| 4 | 7 | 0.25 | 8 |
| 7 | 10 | 0.344 | 11 |
| 10 | | 0.50 | 16 |

* × 2.54 = cm.

placed in a flask and covered (invested) with a slurry of refractory material which is chemically bonded. Entrapped air is removed by vibrating the filled flask or by application of vacuum. After the mold has set, the pattern is melted out, and the mold is heated slowly to 1900°F (1040°C). Metal is poured into the hot mold at proper casting temperatures. Silica, zircon, zirconia, or sillimanite in powder form are used as investments for casting steels and irons.

A modification of the above is the use of frozen mercury as the pattern. The frozen pattern is dipped several times in a refractory mixture to build upon investment around the frozen mercury pattern. The mercury is melted out and the mold fired to a hard ceramic shell.

Carbon steels, low-alloy steels, stainless steels, and high-alloy base alloys are produced normally in weights from a few ounces to 20 lb (10 kg). A limited number of patterns for castings in the 20 to 100 lb (10 to 50 kg) weight class are being produced.

The preferred carbon and low-alloy steel investment castings are AlSl 1020, 4140, and 8620 compositions. The preferred corrosion-resistant investment castings are AlSl 302, 304, 310, 347, 410, 440, and 17-4 PH. However, all the steels of Table 7 are being produced by the investment process. Tool and die steel alloys made as castings by the investment process are AlSl A2, D-2, H-11, M-2, O-1, S-1, and E52100.

**Ceramic Process**   The mold is formed of a mixture of finely graded refractory fillers, hydrolyzed ethyl silicate, and a liquid catalyst that are blended to a slurry consistency. This is poured over the wood, plastic, or metal pattern, and sets within a few minutes—to form first a gel and finally a rigid mold. During the gel stage, the mold is moderately flexible, which facilitates stripping from the pattern at this time. The mold is then heated to a high temperature, as is normally done in making ceramic materials. Molds can be poured with steel when they are hot, or after cooling.

The process has limited application but is particularly adaptable to intricate castings, such as jet-engine manifolds, brake backing plates, blades and vanes for gas turbines, and intricate pumps and profile-type dies for metal and plastic production, many of which are most precise and would otherwise take die sinkers many hours to reproduce. Castings from ounces up to several tons are produced in ceramic molds.

**Shell Molding**   A thermosetting sand and resin mixture is applied mechanically to a hot pattern plate for a controlled time to form a plastic layer adjacent to the pattern. Excess sand mixture is dumped off the pattern. The shell is cured on the pattern and then stripped from the pattern. Matching shells, with shell cores inserted for complex parts, are pasted or clamped together for pouring. Shells may be backed with sand or shot or may receive metal unsupported. The resin burns out of the sand, allowing the mold to crumble away from the solidified casting. Precision and tolerances are closer

than those obtained by the usual green-sand molding but not as close as attained by the ceramic and investment processes.

All the ferrous metals are cast in the shell-molding process—carbon- and low-alloy steel castings, gray-iron, nodular-iron, and malleable-iron castings, as well as stainless steels. Current commercial production is largely castings ranging from ½ to 20 lb (0.2 to 10 kg) in weight. However, castings up to 72 in (1.8 m) in length are being made.

**True Centrifugal Castings**   Cast-iron steel, or sand molds are used in a horizontal-spindle machine for the casting of tubular castings such as liners and tubes. Weighed molten metal is poured into the metal mold while spinning. Centrifugal force permits uniformly thick wall sections to form. A wide range of sizes is produced commercially in cast iron and steel.

**Graphite-Mold Process**   This is a process for the manufacture of permanent molds for casting steel car wheels. The mold cavity is machined in a graphite block. Matching mold halves are clamped together and the metal is poured under pressure. Quick chilling of graphite molds improves strength of castings. Surface finishes are superior to those produced by sand casting. Production sizes range up to 200 lb (100 kg) in steel.

### Use of Steel Castings

All major construction industries employ steel castings. They constitute the major truck construction for railroad cars and are employed in large numbers in road-building equipment, trucks, trailers, agricultural equipment, tractors, hoists, and power shovels. Steel castings are also employed as valves, fittings, and other parts for refinery and chemical industries. Large castings find service in rolling mills, ships, and marine, mining, logging, and machine-tool industries. The military requires ordnance, naval, and aircraft castings. If service requirements include dynamic loading and impact and fatigue-stress conditions, steel castings are employed.

### Casting Design

Maximum service and properties can be obtained from castings only when they are properly designed. Design rules for castings have been prepared in detail to aid design engineers in preparing efficient designs. Engineers are referred to the following publications, which are procurable: Handbooks on Steel Castings, Gray Iron Castings, and Malleable Iron Castings of the Cast Metals Federation, 20611 Center Ridge Road, Rocky River, OH 44116; "Casting Design Handbook," American Society for Metals, Metals Park, Novelty, OH 44073; Meehanite Casting Handbooks, Meehanite Worldwide, New King St., White Plains, NY 10604; Design of Ferrous Castings, American Foundrymen's Society, Des Plaines, IL 60016; "Investment Casting Handbook," Investment Casting Institute, Chicago, IL 60645.

# NONFERROUS METALS

## by L. D. Kunsman and C. L. Carlson

REFERENCES: "Metals Handbook," 8th ed., vols. 1–6, incl., American Society for Metals. "Aluminum Standards and Data," 4th ed., The Aluminum Association. "Standards Handbook, Part 2—Wrought Alloy Data," Copper Development Association, 1973. "Standards Handbook, Part 7—Cast Alloy Data," Copper Development Association, 1970. "1975 Annual Book of ASTM Standards," Parts 6, 7, and 8, American Society for Testing and Materials. "Materials Selector, 76," vol. 82, no. 4, 1975, Materials Engineering, Reinhold Publishing. "Data Book 1975," 3d ed., Metal Progress, American Society for Metals. Mantell, "Engineering Materials Handbook," McGraw-Hill, 1958. Publications of various metals-producing companies.

**Foreword**  In the interest of preserving familiarity with recognized contemporary units of measurement, the English system is largely used here. However, in view of a world-wide movement toward adoption of an international system (SI) of metrication, the reader is encouraged to become familiar with this new system. This can be done best by referring to "Metric Practice Guide (A Guide to the Use of SI—the International System of Units)," ASTM E380, 1974, American Society for Testing and Materials (see Sec. 1).

Seven **nonferrous metals** are of primary commercial importance: copper, zinc, lead, tin, aluminum, nickel, and magnesium. Some 40 other elements are frequently alloyed with these to make the commercially important alloys. There are also about 15 minor metals that have important specific uses. The properties of these elements are given in Table 1. (See also Sec. 4, Thermal Properties and Thermodynamics; Sec. 5. Mechanical Properties of Materials; Sec. 6, General Properties of Materials.)

**Metallic Properties**  Metals are substances that characteristically are opaque crystalline solids of high reflectivity having good electrical and thermal conductivities, a positive chemical valence, and, usually, the important combination of considerable strength and the ability to flow before fracture. These characteristics are exhibited by the metallic elements (e.g., iron, copper, aluminum), or **pure metals,** and by combinations of elements (e.g., steel, brass, dural), or **alloys.** Metals are composed of many small **crystals,** which have grown individually until they have filled the intervening spaces by abutting neighboring crystals. Although the external shape of these crystals and their orientation with respect to each other are usually random, within each such **grain,** or crystal, the atoms are arranged on a regular three-dimensional lattice. Most metals are arranged according to one of the three common types of **lattice,** or **crystal structure:** face-centered cubic, body-centered cubic, or hexagonal close-packed. (See Table 1.)

The several properties of metals are influenced to varying degrees by the testing or service **environment** (temperature, surrounding medium) and the **internal structure** of the metal, which is a result of its chemical composition and previous history such as casting, hot rolling, cold extrusion, annealing, and heat treatment. These relationships, and the discussions below, are best understood in the framework of the several phenomena that may occur in metals processing and service and their general effect on metallic structure and properties.

**Ores, Extractions, and Refining**  All metals begin with the mining of ores and are successively brought through suitable physical and chemical processes to arrive at commercially useful degrees of purity. At the higher-purity end of this process sequence, scrap metal is frequently combined with that derived from ore. Availability of suitable ores and the extracting and refining processes used are specific for each metal and largely determine the price of the metal and the impurities that are usually present in commercial metals.

**Melting**  Once a metal arrives at a useful degree of purity, it is brought to the desired combination of shape and properties by a series of physical processes, each of which influences the internal structure of the metal. The first of these is usually melting, during which several elements can be combined to produce an alloy of the desired composition. Depending upon the metal, its container, the surrounding atmosphere, the addition of alloy-forming materials, or exposure to vacuum, various chemical reactions may be utilized in the melting stage to achieve optimum results.

**Casting**  The molten metal of desired composition is poured into some type of mold in which the heat of fusion is dissipated and the melt becomes a solid of suitable shape for the next stage of manufacture. Such castings are used directly (e.g., sand casting, die casting, permanent-mold casting) or transferred to subsequent operations (e.g., ingots to rolling, billets to forging or extrusion). The principal advantage of castings is their design flexibility and low cost. The typical casting has, to a greater or lesser degree, a large crystal size (**grain size**), extraneous **inclusions** from slag or mold, and **porosity** caused by gas evolution and/or shrinking during solidification. The foundryman's art lies in minimizing the possible defects in castings while maximizing the economies of the process.

**Metalworking**  (See also Sec. 13.)  The major portion of all metals is subjected to additional shape and size changes in the solid state. These metalworking operations substantially alter the internal structure and eliminate many of the defects typical of castings. The usual sequence involves first **hot working** and then, quite often, **cold working.** Some metals are only hot-worked, some only cold-worked; most are both hot- and cold-worked. These terms have a special meaning in metallurgical usage.

A cast metal consists of an aggregate of variously oriented grains, each one a single crystal. Upon deformation, the grains flow by a process involving the slip of blocks of atoms over each other, along definite crystallographic planes. The metal is hardened, strengthened, and rendered less ductile, and further deformation becomes more difficult. This is an important method of increasing the strength of nonferrous metals. The effect of progressive **cold rolling** on brass (Fig. 1) is typical. Terms such as "soft," "quarter hard," "half hard," and "full hard" are frequently used to indicate the degree of hardening produced by such working. For most metals, the hardness resulting from cold working is stable at room temperature, but with lead, zinc, or tin, it will decrease with time.

## Table 1. Physical Constants of the Principal Alloy-Forming Elements*

| Element | Atomic No. | Atomic weight | Density, lb/in³ | Melting point, °F | Boiling point, °F | Specific heat | Latent heat of fusion, Btu/lb | Linear coef of thermal exp, per °F × 10⁻⁶ | Thermal conductivity (near 68°F), Btu/(ft²)(h)(in)(°F) | Electrical resistivity, $\mu\Omega\cdot$cm | Modulus of elasticity (tension), lb/in² × 10⁶ | Crystal structure† | Transition temp, °F | Symbol |
|---|---|---|---|---|---|---|---|---|---|---|---|---|---|---|
| Aluminum | 13 | 26.97 | 0.09751 | 1220.4 | 4520 | 0.215 | 170 | 13.3 | 1,540 | 2.655 | 10 | FCC | | Al |
| Antimony | 51 | 121.76 | 0.239 | 1166.9 | 2620 | 0.049 | 68.9 | 4.7–7.0 | 131 | 39.0 | 11.3 | Rhom | | Sb |
| Arsenic | 33 | 74.91 | 0.207 | 1497 | 1130 | 0.082 | 159 | 2.6 | | 35 | 11 | Rhom | | As |
| Barium | 56 | 137.36 | 0.13 | 1300 | 2980 | 0.068 | | (10) | | 50 | 1.8 | BCC | | Ba |
| Beryllium | 4 | 9.02 | 0.0658 | 2340 | 5020 | 0.52 | 470 | 6.9 | 1,100 | 5.9 | 37 | HCP | | Be |
| Bismuth | 83 | 209.00 | 0.354 | 520.3 | 2590 | 0.029 | 22.5 | 7.4 | 58 | 106.8 | 4.6 | Rhom | | Bi |
| Boron | 5 | 10.82 | 0.083 | 3812 | 4620 | 0.309 | | 4.6 | | $1.8 \times 10^{12}$ | | O | | B |
| Cadmium | 48 | 112.41 | 0.313 | 609.6 | 1409 | 0.055 | 23.8 | 16.6 | 639 | 6.83 | 8 | HCP | | Cd |
| Calcium | 20 | 40.08 | 0.056 | 1560 | 2625 | 0.149 | 100 | 12 | 871 | 3.43 | 3 | FCC/BCC | 867 | Ca |
| Carbon | 6 | 12.010 | 0.0802 | 6700 | 8730 | 0.165 | | 0.3–2.4 | 165 | 1,375 | 0.7 | Hex/D | | C |
| Cerium | 58 | 140.13 | 0.25 | 1460 | 4500 | 0.042 | 27.2 | 3.4 | | 78 | | HCP/FCC | 572/1328 | Ce |
| Chromium | 24 | 52.01 | 0.260 | 3350 | 4380 | 0.11 | 146 | 6.8 | 464 | 13 | 36 | BCC/FCC | 3344 | Cr |
| Cobalt | 27 | 58.94 | 0.32 | 2723 | 6420 | 0.099 | 112 | 4.0 | 479 | 6.24 | 30 | HCP/FCC/HCP | 783/2048 | Co |
| Columbium (niobium) | 41 | 92.91 | 0.310 | 4380 | 5970 | 0.065 | 91.1 | 9.2 | | 13.1 | 15 | BCC | | Cb |
| Copper | 29 | 63.54 | 0.324 | 1981.4 | 4700 | 0.092 | | | 2,730 | 1.673 | 16 | FCC | | Cu |
| Gadolinium | 64 | 156.9 | 0.287 | | | | | | | | | HCP | | Gd |
| Gallium | 31 | 69.72 | 0.216 | 85.5 | 3600 | 0.0977 | 34.5 | 10.1 | 232 | 56.8 | 1 | O | | Ga |
| Germanium | 32 | 72.60 | 0.192 | 1756 | 4890 | 0.086 | 205.7 | (3.3) | | $60 \times 10^{6}$ | 11.4 | D | | Ge |
| Gold | 79 | 197.2 | 0.698 | 1945.4 | 5380 | 0.031 | 29.0 | 7.9 | 2,060 | 2.19 | 12 | FCC | | Au |
| Hafnium | 72 | 178.6 | 0.473 | 3865 | 9700 | 0.0351 | | (3.3) | | 32.4 | 20 | HCP/BCC | 3540 | Hf |
| Hydrogen | 1 | 1.0080 | $3.026 \times 10^{-6}$ | -434.6 | -422.9 | 3.45 | 27.0 | | 1.18 | | | Hex | | H |
| Indium | 49 | 114.76 | 0.264 | 313.5 | 3630 | 0.057 | 12.2 | 18 | 175 | 8.37 | 1.57 | FCT | | In |
| Iridium | 77 | 193.1 | 0.813 | 4449 | 9600 | 0.031 | 47 | 3.8 | 406 | 5.3 | 75 | FCC | | Ir |
| Iron | 26 | 55.85 | 0.284 | 2802 | 4960 | 0.108 | 117 | 6.50 | 523 | 9.71 | 28.5 | BCC/FCC/BCC | 1663/2554 | Fe |
| Lanthanum | 57 | 138.92 | 0.223 | 1535 | 8000 | 0.0448 | | | | 59 | 5 | HCP/FCC/? | 662/1427 | La |
| Lead | 82 | 207.21 | 0.4097 | 621.3 | 3160 | 0.031 | 11.3 | 16.3 | 241 | 20.65 | 2.6 | FCC | | Pb |
| Lithium | 3 | 6.940 | 0.019 | 367 | 2500 | 0.79 | 286 | 31 | 494 | 11.7 | 1.7 | BCC | | Li |
| Magnesium | 12 | 24.32 | 0.0628 | 1202 | 2030 | 0.25 | 160 | 14 | 1,100 | 4.46 | 6.5 | HCP | | Mg |
| Manganese | 25 | 54.93 | 0.268 | 2273 | 3900 | 0.115 | 115 | 12 | | 185 | 23 | CCX/CCX/FCT | 1340/2010/2080 | Mn |

|  |  |  |  |  |  |  |  |  |  |  |  |  |  |  |
|---|---|---|---|---|---|---|---|---|---|---|---|---|---|---|
| Mercury | 80 | 200.61 | 0.4896 | −37.97 | 675 | 0.033 | 4.9 | 58 | 33.8 | 94.1 | 50 | Rhom |  | Hg |
| Molybdenum | 42 | 95.95 | 0.369 | 4750 | 8670 | 0.061 | 126 | 1,020 | 3.0 | 5.17 | 30 | BCC |  | Mo |
| Nickel | 28 | 58.69 | 0.322 | 2651 | 4950 | 0.105 | 133 | 639 | 7.4 | 6.84 |  | FCC |  | Ni |
| Niobium (see Columbium) |  |  |  |  |  |  |  |  |  |  |  |  |  | Nb |
| Nitrogen | 7 | 14.008 | $0.042 \times 10^{-3}$ | −346 | −320.4 | 0.247 | 11.2 | 0.147 |  |  |  | Hex |  | N |
| Osmium | 76 | 190.2 | 0.813 | 4900 | 9900 | 0.031 |  |  | 2.6 | 9.5 | 80 | HCP |  | Os |
| Oxygen | 8 | 16.000 | $0.048 \times 10^{-3}$ | −361.8 | −297.4 | 0.218 | 5.9 | 0.171 | 6.6 |  |  | C |  | O |
| Palladium | 46 | 106.7 | 0.434 | 2829 | 7200 | 0.058 | 69.5 | 494 | 70 | 10.8 | 17 | FCC |  | Pd |
| Phosphorus | 15 | 30.98 | 0.0658 | 111.4 | 536 | 0.177 | 9.0 |  |  | $10^{17}$ |  | C |  | P |
| Platinum | 78 | 195.23 | 0.7750 | 3224.3 | 7970 | 0.032 | 49 | 494 | 4.9 | 9.83 | 21 | FCC |  | Pt |
| Plutonium | 94 | 239 | 0.686 | 1225 |  |  |  |  | 50–65 | 150 |  | MC | 6 forms | Pu |
| Potassium | 19 | 39.096 | 0.031 | 145 | 1420 | 0.177 | 26.1 | 697 | 46 | 6.15 | 0.5 | BCC |  | K |
| Radium | 88 | 226.05 | 0.18 | 1300 |  |  |  |  |  |  |  | ? |  | Ra |
| Rhenium | 75 | 186.31 | 0.765 | 5733 | 10,700 | 0.0326 | 76 |  | 4.6 | 21 | 75 | HCP |  | Re |
| Rhodium | 45 | 102.91 | 0.4495 | 3571 | 8100 | 0.059 |  | 610 | 21 | 4.5 | 54 | FCC |  | Rh |
| Selenium | 34 | 78.96 | 0.174 | 428 | 1260 | 0.084 | 29.6 | 3 |  | 12 | 8.4 | MC/Hex | 248 | Se |
| Silicon | 14 | 28.06 | 0.084 | 2605 | 4200 | 0.162 | 607 | 581 | 1.6–4.1 | $10^{5}$ | 16 | D |  | Si |
| Silver | 47 | 107.88 | 0.379 | 1760.9 | 4010 | 0.056 | 45.0 | 2,900 | 10.9 | 1.59 | 11 | FCC |  | Ag |
| Sodium | 11 | 22.997 | 0.035 | 207.9 | 1638 | 0.295 | 49.5 | 929 | 39 | 4.2 | 1.3 | BCC |  | Na |
| Sulfur | 16 | 32.066 | 0.0748 | 246.2 | 832.5 | 0.175 | 16.7 | 1.83 | 36 | $2 \times 10^{23}$ |  | Rhom/FCC | 204 | S |
| Tantalum | 73 | 180.88 | 0.600 | 5420 | 9570 | 0.036 |  | 377 | 3.6 | 12.4 | 27 | BCC |  | Ta |
| Tellurium | 52 | 127.61 | 0.225 | 840 | 2530 | 0.047 | 13.1 | 41 | 9.3 | $2 \times 10^{5}$ | 6 | Hex |  | Te |
| Thallium | 81 | 204.39 | 0.428 | 577 | 2655 | 0.031 | 9.1 | 2,700 | 16.6 | 18 | 1.2 | HCP/BCC | 446 | Tl |
| Thorium | 90 | 232.12 | 0.422 | 3348 | 8100 | 0.0355 | 35.6 |  | 6.2 | 18.6 | 11.4 | FCC |  | Th |
| Tin | 50 | 118.70 | 0.264 | 449.4 | 4120 | 0.054 | 26.1 | 464 | 13 | 11.5 | 6 | D/BCT | 55 | Sn |
| Titanium | 22 | 47.90 | 0.164 | 3074 | 6395 | 0.139 | 187 | 1,190 | 4.7 | 47.8 | 16.8 | HCP/BCC | 1650 | Ti |
| Tungsten | 74 | 183.92 | 0.697 | 6150 | 10,700 | 0.032 | 79 | 900 | 2.4 | 5.5 | 50 | BCC |  | W |
| Uranium | 92 | 238.07 | 0.687 | 2065 | 7100 | 0.028 | 19.8 | 186 | 11.4 | 29 | 29.7 | O/Tet/BCC | 1229/1427 | U |
| Vanadium | 23 | 50.95 | 0.217 | 3452 | 5430 | 0.120 |  | 215 | 4.3 | 26 | 18.4 | BCC | 2822 | V |
| Zinc | 30 | 65.38 | 0.258 | 787 | 1663 | 0.092 | 43.3 | 784 | 9.4–22 | 5.92 | 12 | HCP |  | Zn |
| Zirconium | 40 | 91.22 | 0.23 | 3326 | 9030 | 0.066 |  | 116 | 3.1 | 41.0 | 11 | HCP/BCC | 1585 | Zr |

NOTE: See Sec. 1 for conversion factors to SI units.

*From ASM "Metals Handbook," revised and supplemented where necessary from Hampel's "Rare Metals Handbook" and elsewhere.

†Cal/g°C at room temperature equals Btu/lb°F at room temperature.

‡FCC = face-centered cubic; BCC = body-centered cubic; C = cubic; HCP = hexagonal closest packing; Rhom = rhombohedral; Hex = hexagonal; FCT = face-centered tetragonal; O = orthorhombic; FCO = face-centered orthorhombic; CCX = cubic complex; D = diamond cubic; BCT = body-centered tetragonal; MC = monoclinic.

Upon **annealing, or heating the cold-worked metal,** the first effect is to relieve macrostresses in the object without loss of strength; indeed, the strength is often increased slightly. Above a certain temperature, softening commences and proceeds rapidly with increase in temperature (Fig. 2). The cold-worked distorted metal undergoes a change called **recrystallization.** New grains form and grow until they have consumed the old, distorted ones. The temperature at which this occurs in a

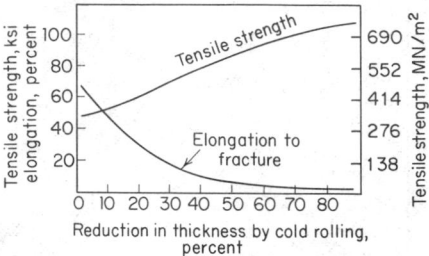

**Fig. 1**   Effect of cold rolling on annealed brass (Cu 72, Zn 28).

given time, called the **recrystallization temperature,** is lower and the resulting grain size finer, the more severe the working of the original piece. If the original working is carried out above this temperature, it does not harden the piece, and the operation is termed **hot working.** If the annealing temperature is increased beyond the recrystallization temperature or if the piece is held at temperature for a long time, the average grain size increases. This generally softens and decreases the strength of the piece still further.

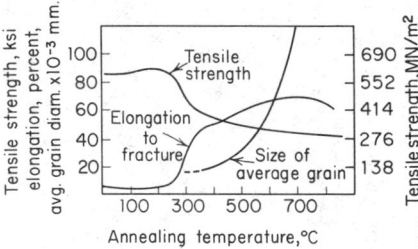

**Fig. 2**   Effect of annealing on cold-rolled brass (Cu 72, Zn 28).

**Hot working** occurs when deformation and annealing proceed simultaneously, so that the resulting piece of new size and shape emerges in a soft condition roughly similar to the annealed condition.

**Alloys**   The above remarks concerning structure and property control through casting, hot working, cold working, and annealing generally apply both to pure metals and to alloys. The addition of alloying elements makes possible other means for controlling properties. In some cases, the addition to a metal A of a second element B simply results in the appearance of some new crystals of B as a **mixture** with crystals of A; the resulting properties tend to be an average of A and B. In other cases, an entirely new substance will form—the **intermetallic compound** AB, having its own set of distinctive properties (usually hard and brittle). In still other cases, element B will

simply dissolve in element A to form the **solid solution** A(B). Such solid solutions have the characteristics of the **solvent** A modified by the presence of the **solute** B, usually causing increased hardness, strength, electrical resistance, and recrystallization temperature. The most interesting case involves the combination of solid solution A(B) and the precipitation of a second constituent, either B or AB, brought about by the precipitation-hardening heat treatment, which is particularly important in the major nonferrous alloys.

**Precipitation Hardening**   Many alloys, especially those of aluminum but also some alloys of copper, nickel, magnesium, and other metals, can be hardened and strengthened by heat treatment. The heat treatment is usually a two-step process which involves (1) a solution heat treatment followed by rapid quenching and (2) a precipitation or aging treatment to cause separation of a second phase from solid solution and thereby cause hardening. These alloys after a solution treatment are comparatively soft and consist of homogeneous grains of solid solution generally indistinguishable microscopically from a pure metal. If very slowly cooled from the solution-treatment temperature, the alloy will deposit crystals of a second constituent, the amount of which increases as the temperature decreases. Rapid cooling after a solution treatment will retain the supersaturated solution at room temperature, but if the alloy is subsequently reheated to a suitable temperature, fine particles of a new phase will form and in time will grow to a microscopically resolvable size. At some stage in this precipitation process the hardness, the tensile strength, and particularly the yield strength of the alloy will be considerably increased. If the reheating treatment is carried out for too long a time, the alloy will overage and soften. The temperature and time for both solution and precipitation heat treatments must be closely controlled to obtain the best results. To some extent, precipitation hardening may be superimposed upon hardening resulting from cold working. Precipitation-hardened alloys have an unusually high ratio of proportional limit to tensile strength, but the endurance limit in fatigue is not increased to nearly the same extent.

**Effect of Environment**   The properties of metals under the "normal" conditions of 70°F and 50 percent relative humidity in clean air can be markedly changed under other conditions, with the various metals differing greatly in their degree of response to such changing conditions. The topic of **corrosion** is both important and highly specific (see Sec. 6). **Oxidation** is a chemical reaction specific to each metal and, wherever important, is so treated. The **effect of temperature,** however, can be profitably considered in general terms. The primary consequence of increase in temperature is increased atom movement, or **diffusion.** In the discussion above, the effects of recrystallization, hot working, solution treatment, and precipitation were all made possible by increased diffusion at elevated temperatures. The temperature at which such atom movements become appreciable is roughly proportional to the melting point of the metal. If their melting points are expressed on an absolute-temperature scale, then various metals can be expected to exhibit comparable effects at about the same fraction of their melting points. Thus the recrystallization temperatures of lead, zinc, aluminum, copper, nickel, molybdenum, and tungsten are successively higher. The consequent rule of thumb is that alloys do not have useful structural strength at temperatures above about 0.5 of their absolute melting temperatures.

## COPPER AND ITS ALLOYS

**Commercial Coppers** Most copper is refined electrolytically from anodes of crude **blister copper. Lake copper** originates in the Great Lakes district; it is not electrolytically refined but is of high purity and conductivity. **Secondary copper,** if electrolytically refined, is equal to primary copper, but if fire-refined, it may be less pure and is commonly used as casting copper, produced for brass-foundry use in making alloys.

The electrolytic refining operation produces **cathodes** that may measure about 3 ft square and weigh 180 to 300 lb. Cathodes may be used directly as melting stock for making alloys, but if copper is to be rolled to fabricated forms, it is melted and cast into wire bars, cakes, or billets. The remelting, or so-called "refining operation," involves melting in a large fuel-fired reverberatory furnace, oxidizing to eliminate sulfur and gas absorbed from the fuel, and "poling" to reduce the oxygen to about 0.04 percent. When correctly refined, castings solidify with an approximately level surface, the gas evolved during solidification balancing the shrinkage that would otherwise occur. This is known as **tough-pitch copper.** It has a density of 0.321 to 0.323 lb/in³ (8.4 to 8.7 g/cm³) when worked and annealed.

The presence of **oxygen** is desirable in making the copper slightly harder and, before electrolytic refining was adopted, was useful in neutralizing the effect of certain impurities. Oxygen is harmful if the copper is to be heated in a reducing gas, as it causes an embrittlement of the copper and renders it useless. **Deoxidized copper** is usually made by adding phosphorus, but this decreases the conductivity. Most copper tubing is made of phosphorus-deoxidized copper. **Oxygen-free copper** of high conductivity and density is available, made by casting without contact with air. Like deoxidized copper, this is more ductile than tough-pitch copper and is immune to embrittlement by hot reducing gases encountered in some service conditions and frequently in brazing or heat-treating operations.

Copper may be hot-worked extensively at 1400 to 1600°F and has excellent ductility in cold work. Sheet, rod, and tube are employed after cold rolling or drawing to increase strength and hardness according to demands of application but may be softened by annealing at 700 to 1400°F. **Copper wire** about 0.04 in diameter is commonly made by drawing from a hot-rolled rod without annealing, but smaller sizes may involve intermediate anneals. Copper shapes for switch parts are made by **extrusion,** brushes and commutator sections by rolling and **drawing.** Copper for **electrical purposes** must be very pure; the presence of even traces of certain impurities (particularly phosphorus, arsenic, iron, titanium, and silicon) decreases the conductivity considerably. Such copper should have a minimum copper content of 99.90 percent (silver being counted as copper) and a resistance, measured on a drawn and annealed wire, not exceeding 1.7241 $\mu\Omega \cdot$cm at 20°C (ASTM specification B3–74). Selected values of the strength of copper wire from ASTM specifications B1–70, B2–70, and B3–74 are shown in Table 2.

Copper containing small amounts of **silver** retains the effect of cold working to a higher temperature than pure copper (about 610°F compared with about 400°F). This is useful where comparatively high temperatures are to be withstood, as in soldering operations or for stressed conductors designed to operate at moderately elevated temperatures. **Tellurium** and **sulfur** are added to enhance the machinability of copper with minimum sacrifice in conductivity. Similarly, **chromium** and **zirconium** are added to increase elevated-temperature strength with little decrease in conductivity. **Beryllium copper** is a precipitation-hardening alloy that combines moderate conductivity with very high strength. [Typical heat treatment, 1450°F (788°C) for 1 h; water quench, 600°F (316°C) for 3 h.] See Table 3.

**Copper alloys** are useful because of their good thermal or electrical conductivity, good cold- or hot-working properties, machinability, or corrosion resistance. For high thermal or electrical conductivity, commercially pure copper should be used; if greater strength combined with high conductivity is required, alloys containing zirconium or other elements are used. The cheapest copper alloy is **brass** of high zinc content and is used unless high corrosion resistance under stress or the special mechanical properties of other alloys are required.

**Table 2. ASTM Specification Properties of Copper Wire**

| Diam | | Hard-drawn wire | | Medium-drawn wire | | | Soft-annealed wire | |
| | | | | | | | | |
| in | mm | Tensile strength, ksi,† min | Elongation, % in 60 in, min | Tensile strength, ksi† Min | Max | Elongation, % in 60 in, min | Tensile strength, ksi,† max | Elongation, % in 10 in, min |
|---|---|---|---|---|---|---|---|---|
| 0.460 | 11.7 | 49 | 3.75* | 42 | 49 | 3.75* | 36 | 35 |
| 0.325 | 8.3 | 54.5 | 2.40* | 45 | 52 | 3.0* | 36 | 35 |
| 0.229 | 5.8 | 59 | 1.79* | 48 | 55 | 2.25* | 37 | 30 |
| 0.162 | 4.1 | 62.1 | 1.14 | 49 | 56 | 1.15 | 37 | 30 |
| 0.114 | 2.9 | 64.3 | 1.02 | 50 | 57 | 1.06 | 37 | 30 |
| 0.081 | 2.06 | 65.7 | 0.95 | 51 | 58 | 1.00 | 38.5 | 25 |
| 0.057 | 1.45 | 66.4 | 0.89 | 52 | 59 | 0.94 | 38.5 | 25 |
| 0.040 | 1.02 | 67 | 0.85 | 53 | 60 | 0.88 | 38.5 | 25 |
| Electrical resistivity, $\mu\Omega$/cm | | 1.7930 | | 1.7837‡ | | | 1.7241 | |

*Elongation in 10 in. gage length.
†ksi × 6.895 = MN/m².
‡On wire below 0.324 in diam, 1.7654 for larger wires.

**Table 3. Composition and Properties of Selected Wrought-Copper-Base Alloys***

| Material | Copper alloy No. | Form tested | Nominal composition, % | | | | | Machinability index | Tensile strength ksi | |
|---|---|---|---|---|---|---|---|---|---|---|
| | | | Cu | Zn | Pb | Sn | Others | | Hard | Soft |
| **Coppers** | | | | | | | | | | |
| Electrolytic tough-pitch copper† | 110 | Sheet | 99.92 | | | | 0.04 O | 20 | 50 | 32 |
| | | Wire | | | | | | | 55 | 35 |
| | | Rod | | | | | | | 44 | 32 |
| Oxygen free copper‡ | 102 | | 99.95 | | | | | 20 | 50 | 32 |
| Phosphorus deoxidized copper | 122 | | 99.9+ | | | | 0.02 P | 20 | 48 | 32 |
| Tellurium copper | 145 | | 99.5 | | | | 0.50 Te | 85 | 50 | 33 |
| Sulfur copper | 147 | | 99.7 | | | | 0.4 S | 85 | 50 | 34 |
| Zirconium copper | 150 | | 99.8+ | | | | 0.15 Zr | 20 | 70 | 37 |
| Beryllium copper | 172 | | 98.1 | | | | 2.15 Be | 20 | 118 | 70 |
| Beryllium copper, H.T. | 172 | | 98.1 | | | | 2.15 Be | 20 | 180 | 70 |
| Chromium copper | 182 | | 99.1 | | | | 0.85 Cr | 20 | 67 | 34 |
| **Plain Brasses** | | | | | | | | | | |
| Gilding, 95% | 210 | Sheet | 95.0 | 5.0 | | | | 20 | 56 | 34 |
| Commercial bronze, 90% | 220 | Sheet | 90.0 | 10.0 | | | | 20 | 61 | 37 |
| Red brass, 85% | 230 | Sheet | 85.0 | 15.0 | | | | 30 | 70 | 40 |
| | | Tube | 85.0 | 15.0 | | | | | 70 | 40 |
| Low brass, 80% | 240 | Sheet | 80.0 | 20.0 | | | | 30 | 74 | 44 |
| | | Wire | 80.0 | 20.0 | | | | | 107 | 44 |
| Cartridge brass, 70% | 260 | Sheet | 70.0 | 30.0 | | | | 30 | 76 | 47 |
| Yellow brass | 270 | Wire | 65.0 | 35.0 | | | | 30 | 74 | 62 |
| | | Rod | 65.0 | 35.0 | | | | | 110 | 50 |
| Muntz metal | 280 | Sheet | 60.0 | 40.0 | | | | 40 | 70 | 48 |
| Leaded commercial bronze | 314 | Rod | 89.0 | 9.25 | 1.75 | | | 80 | 52 | 54 |
| Low-leaded brass | 335 | Sheet | 67.0 | 32.5 | 0.5 | | | 60 | 75 | 37 |
| Medium-leaded brass | 340 | Sheet | 64.5 | 34.5 | 2.0 | | | 70 | 74 | 47 |
| High-leaded brass | 342 | Sheet | 65.0 | 33.0 | 2.0 | | | 90 | 75 | 47 |
| Extra-high-leaded brass | 356 | Sheet | 62.5 | 35.0 | 2.5 | | | 100 | 74 | 52 |
| Free-cutting brass | 360 | Rod | 61.5 | 35.5 | 3.0 | | | 100 | 58 | 49 |
| Leaded Muntz metal | 365 | Plate | 60.0 | 39.5 | 0.5 | | | 60 | 70 | 49 |
| Free-cutting Muntz metal | 370 | Tube | 60.5 | 38.4 | 1.1 | | | 70 | 80 | 54 |
| Forging brass | 377 | Rod | 60.0 | 38.0 | 2.0 | | | 80 | | 52 |
| Architectural bronze | 385 | Rod | 57.0 | 40.0 | 3.0 | | | 90 | 60 | 60 |

Tin and aluminum brasses

| Name | No. | Form | | | | | | | | |
|---|---|---|---|---|---|---|---|---|---|---|
| Admiralty, inhibited | 443 | Tube | 71.0 | 28.0 | 1.0 | | | 30 | | 53 |
| Naval brass | 464 | Rod | 60.0 | 39.25 | 0.75 | | | 30 | 75 | 57 |
| | | Sheet | 50.0 | 39.25 | 0.75 | | | 30 | 70 | 62 |
| Leaded naval brass | 485 | Rod | 60.0 | 37.5 | 0.75 | 1.75 | | 70 | 75 | 57 |
| Manganese bronze | 675 | Rod | 58.5 | 39.2 | 1.0 | | 1.0 Fe; 0.3 Mn | 30 | 84 | 65 |
| Aluminum brass | 687 | Tube | 77.5 | 20.5 | | | 2.0 Al | 30 | | 60 |

Phosphor bronzes (tin bronzes)

| Name | No. | Form | | | | | | | | |
|---|---|---|---|---|---|---|---|---|---|---|
| Phosphor bronze, 5% | 510 | Sheet | 95.0 | | 5.0 | | | 20 | 81 | 47 |
| Phosphor bronze, 8% | 521 | Sheet | 92.0 | | 8.0 | | | 20 | 93 | 55 |
| Phosphor bronze, 10% | 524 | Sheet | 90.0 | | 10.0 | | | 20 | 100 | 66 |
| Phosphor bronze, 1.25% | 502 | Sheet | 98.75 | | 1.25 | | | 20 | 65 | 40 |

Aluminum bronzes

| Name | No. | Form | | | | | | | | |
|---|---|---|---|---|---|---|---|---|---|---|
| Aluminum bronze, 5% | 608 | Tube | 95.0 | | | | 5.0 Al | 20 | | 60 |
| Aluminum bronze, 8% | 610 | Rod | 92.0 | | | | 8.0 Al | 20 | 80 | 70 |
| Aluminum bronze | 614 | Sheet | 90.0 | | | | 7.0 Al; 2.0 Fe | 20 | 89 | 82 |

Cupronickel and nickel silver

| Name | No. | Form | | | | | | | | |
|---|---|---|---|---|---|---|---|---|---|---|
| Cupronickel, 30% | 715 | Tube | 70.0 | | | | 30.0 Ni | 20 | | 60 |
| Cupronickel, 10% | 706 | Tube | 88.6 | | | | 10 Ni; 1.4 Fe | 20 | 60 | 44 |
| Nickel-silver, 18% | 752 | Sheet | 65.0 | 17.0 | | | 18.0 Ni | 20 | 85 | 58 |
| Nickel-silver, 18% | 770 | Sheet | 55.0 | 27.0 | | | 18.0 Ni | 30 | 100 | 60 |

Silicon bronzes

| Name | No. | Form | | | | | | | | |
|---|---|---|---|---|---|---|---|---|---|---|
| High-silicon bronze | 655 | Sheet | 97.0 | | | | 3.0 Si | 30 | 94 | 60 |
| | | Rod | | | | | | | 92 | 58 |
| Low-silicon bronze | 651 | Rod | 98.5 | | | | 1.5 Si | | 70 | 40 |

This table is largely prepared from "Alloy Data" of the Copper Association, New York. Values shown are typical and should not be used for specification purposes since they vary with the size and shape tested and with manufacturing variables. Development tests on sheet 0.040 in thick; rod 1 in diam, wire 0.080 in diam, tube 1 in OD by 0.065 in wall thickness. "Hard" corresponds to commercial hard-drawn temper. Higher values of yield and tensile strength and hardness are obtained by a greater amount of cold working, but this is limited to relatively narrow sheets and to rod, wire, and tube of small diameter.

H.T. = precipitation heat treatment on form shown; F = Rockwell F hardness.
*See Sec. 1 for conversion factors to SI units.
†Tough pitch and oxygen-free coppers containing 0.027 to 0.085 percent Ag for softening resistance at elevated temperature are also available.
‡Copper 101, 99.99 Cu, is available for electronic applications.

**Table 3. Composition and Properties of Selected Wrought-Copper-Base Alloys (Continued)**

| | Elongation, % in 2 in | | Yield strength (0.5% extension under load), ksi | | Rockwell B hardness | | Melting point, °F | Density, lb/in³ | Coef. of expansion (avg 77–572 °F) per °F × 10⁻⁶ | Elec conductivity (annealed), % IACS | Thermal conductivity, Btu/(h)(ft²)(ft/°F) |
|---|---|---|---|---|---|---|---|---|---|---|---|
| | Hard | Soft | Hard | Soft | Hard | Soft | | | | | |
| **Coppers** | | | | | | | | | | | |
| Electrolytic tough-pitch copper† | 12 | 50 | 45 | 10 | 50 | F40 | 1981 | 0.322 | 9.8 | 101 | 226 |
| | 1.5 | 35 | | | | | | | | | |
| Oxygenfree copper‡ | 16 | 55 | 44 | 10 | 47 | F40 | 1981 | 0.323 | 9.8 | 101 | 226 |
| Phosphorus deoxidized copper | 16 | 55 | 45 | 10 | 47 | F40 | 1981 | 0.323 | 9.8 | 85 | 196 |
| Tellurium copper | 15 | 50 | 40 | 11 | 50 | F45 | 1967 | 0.323 | 9.9 | 95 | 205 |
| Sulfur copper | 10 | 46 | 49 | 10 | 50 | F40 | 1969 | 0.323 | 9.8 | 93 | 216 |
| Zirconium copper | 12 | 42 | 48 | 11 | 45 | F40 | 1976 | 0.321 | 11.2 | 93 | 212 |
| Beryllium copper | 5 | 50 | 65 / 85 | 30 | 75 / 102 | F40 | 1955 | 0.298 | 9.9 | 17 | 62 |
| Beryllium copper, H.T. | 17 | 45 | 160 | | 114 | 60 | | | | 22 | 75 |
| Chromium copper | 14 | 40 | 59 | 19 | 79 | 16 | 1967 | 0.321 | 9.8 | 85 | 187 |
| **Plain brasses** | | | | | | | | | | | |
| Gilding, 95% | 5 | 45 | 50 | 10 | 64 | F46 | 1950 | 0.320 | 10.0 | 56 | 135 |
| Commercial bronze, 90% | 5 | 45 | 54 | 10 | 70 | F53 | 1910 | 0.318 | 10.2 | 44 | 109 |
| Red brass, 85% | 5 | 47 | 57 | 12 | 77 | F59 | 1880 | 0.316 | 10.4 | 37 | 92 |
| Low brass, 80% | 8 | 55 | 58 | 12 | 77 | F60 | 1830 | 0.313 | 10.6 | 32 | 81 |
| Cartridge brass, 70% | 8 | 62 | 63 | 15 | 82 | F64 | 1750 | 0.308 | 11.1 | 28 | 70 |
| Yellow brass | 8 | 62 | 60 | 15 | 80 | F64 | 1710 | 0.306 | 11.3 | 27 | 67 |
| Muntz metal | 10 | 45 | 50 | 16 | 75 | F80 | 1660 | 0.303 | 11.6 | 28 | 71 |
| Leaded commercial bronze | 18 | 45 | 45 | 21 | 58 | F55 | 1900 | 0.319 | 10.2 | 42 | 104 |
| Low-leaded brass | 7 | 60 | 60 | 15 | 80 | F64 | 1700 | 0.306 | 11.2 | 26 | 67 |
| Medium-leaded brass | 7 | 60 | 60 | 15 | 80 | F64 | 1700 | 0.306 | 11.3 | 26 | 67 |

Copper alloys — mechanical and physical properties

| Material | | | | | | | | | | | |
|---|---|---|---|---|---|---|---|---|---|---|---|
| High-leaded brass | 7 | 50 | 60 | 20 | 80 | F75 | 1670 | 0.306 | 11.3 | 26 | 67 |
| Extra-high-leaded brass | 7 | 50 | 68 | 17 | 80 | F68 | 1660 | 0.307 | 11.4 | 26 | 67 |
| Free-cutting brass | 25 | 53 | 45 | 18 | 78 | F68 | 1650 | 0.307 | 11.4 | 26 | 67 |
| Leaded Muntz metal | | 45 | | 20 | | F80 | 1650 | 0.304 | 11.6 | 28 | 71 |
| Free-cutting Muntz metal | 6 | 40 | 60 | 20 | 85 | F80 | 1650 | 0.304 | 11.6 | 27 | 69 |
| Forging brass | | 45 | | 20 | | F78 | 1640 | 0.305 | 11.5 | 27 | 69 |
| Architectural bronze | | 30 | | 20 | | 65 | 1630 | 0.306 | 11.6 | 28 | 71 |
| **Tin and aluminum brasses** | | | | | | | | | | | |
| Admiralty, inhibited | 20 | 65 | 53 | 22 | 82 | F75 | 1720 | 0.308 | 11.2 | 25 | 64 |
| Naval brass | 17 | 47 | 58 | 25 | 75 | 55 | 1650 | 0.304 | 11.8 | 26 | 67 |
| Leaded naval brass | 15 | 40 | 53 | 30 | 82 | 60 | 1650 | 0.305 | 11.8 | 26 | 67 |
| Manganese bronze | 19 | 33 | 60 | 25 | 90 | 55 | 1630 | 0.302 | 11.8 | 24 | 61 |
| Aluminum brass | | 55 | | 27 | | F77 | 1780 | 0.301 | 10.3 | 23 | 58 |
| **Phosphor bronzes (tin bronzes)** | | | | | | | | | | | |
| Phosphor bronze, 5% | 10 | 64 | 75 | 19 | 87 | 26 | 1920 | 0.320 | 9.9 | 15 | 40 |
| Phosphor bronze, 8% | 10 | 70 | 72 | | 93 | F75 | 1880 | 0.318 | 10.1 | 13 | 36 |
| Phosphor bronze, 10% | 13 | 68 | | 28 | 97 | 55 | 1830 | 0.317 | 10.2 | 11 | 29 |
| Phosphor bronze, 1.25% | 8 | 48 | 50 | 14 | 75 | F68 | 1970 | 0.320 | 9.9 | 48 | 120 |
| **Aluminum bronzes** | | | | | | | | | | | |
| Aluminum bronze, 5% | 30 | 55 | 50 | 27 | 85 | F77 | 1945 | 0.295 | 10.0 | 17 | 46 |
| Aluminum bronze, 8% | 32 | 65 | 60 | 25 | 87 | 60 | 1905 | 0.281 | 9.9 | 15 | 40 |
| Aluminum bronze | | 40 | | 45 | | 84 | 1915 | 0.285 | 9.0 | 14 | 39 |
| **Cupronickel and nickel silver** | | | | | | | | | | | |
| Cupronickel, 30% | 10 | 45 | 57 | 25 | 22 | 45 | 2260 | 0.323 | 9.0 | 4.6 | 17 |
| Cupronickel, 10% | 3 | 42 | 74 | 16 | 87 | 25 | 2100 | 0.323 | 9.3 | 9.0 | 26 |
| Nickel-silver, 18% | 3 | 40 | 85 | 25 | 91 | 40 | 2030 | 0.316 | 9.0 | 6.0 | 19 |
| Nickel-silver, 18% | | 40 | | 27 | | 55 | 1930 | 0.314 | 9.3 | 5.5 | 17 |
| **Silicon bronzes** | | | | | | | | | | | |
| High-silicon bronze | 8 | 60 | 58 | 25 | 93 | 62 | 1880 | 0.308 | 10.0 | 7.0 | 21 |
| Low-silicon bronze | 15 | 50 | 55 | 15 | 80 | F55 | 1940 | 0.316 | 9.9 | 12 | 33 |

When good cold-working properties are desired, as in deep-drawing or forming operations, a brass with 30 to 35 percent zinc is used. **Leaded brass** is used when much machining must be done, particularly for automatic-screw-machine work. For high elastic strength, the **tin bronzes** are used. The **alloys of** copper with **aluminum** or **silicon** or **nickel** are good for corrosion resistance. Several hundred different copper alloys are available, but requirements can generally be met satisfactorily by one or more of the alloys listed in Table 3.

## BRASSES

The useful copper-zinc alloys contain up to 40 percent zinc. Those with 30 to 35 percent find the greatest application as they are cheap, very ductile, and readily worked. With decreasing zinc content, the alloys approach copper more and more in their properties and improve in corrosion resistance. **Season cracking** may occur with high-zinc brasses but rarely with 15 percent zinc or below; this is spontaneous cracking, occurring on exposure to atmospheric corrosion, in brass objects with high residual tensile stresses at the surface. It may be prevented by avoiding the production of internal macrostresses or by removing such stresses by relief annealing at 475 to 530°F (246 to 276°C) without softening the work. It should be noted that alloys susceptible to spontaneous season cracking, even if they are free from internal strains, will crack when exposed to corrosive conditions under high service stresses.

The 5 to 20 percent zinc alloys find application because of freedom from season cracking, because of their red color, and because their high melting point is desirable in brazing operations. The properties of these alloys are included in Table 3. Figures 1 and 2 show the effect of progressive cold rolling and progressive annealing on a brass containing 30 percent zinc. Cold working increases the hardness and tensile strength and decreases the ductility as measured by elongation or reduction in area. Annealing below a certain temperature has practically no effect, but in the recrystallization range a rapid decrease in strength and increase in ductility occurs. At this point, the effect of cold working is almost entirely removed. Heating beyond this point results in the growth of the grains with comparatively little further increase in ductility. Figures 3 and 4 show the variation of properties of brass with composition after annealing at the temperatures indicated.

Table 4 summarizes the ASTM specification requirements for 65/35 brass of various rolled and annealed tempers. Practically all wrought-copper alloys are used in the cold-worked condition to gain additional strength. Articles are often made from annealed stock and depend on the work of the forming operation to shape and harden them. When the work involved is too small to do this, brass rolled with a degree of temper depending on requirements should be used.

Brass for springs should be rolled as hard as consistent with the subsequent forming operations. For articles requiring sharp bends, or for deep-drawing operations, annealed brass must be used. In general, the smaller grain sizes are to be preferred.

The addition of **lead** to brass renders it free-cutting and remarkably machinable. Additions of 0.75 to 1.25 percent of **tin** improve the corrosion resistance. **Aluminum** is added to brass to improve the corrosion resistance particularly in condenser-tube applications. **Manganese bronze** is a complex brass with hot-working properties, high strength, and abrasion resistance. **Naval brass** is used as boat shafting.

All the brasses may be hot-worked if they are free from lead, particularly those alloys containing 60 percent or less of copper, for these contain a constituent, beta, which is extremely plastic at high temperatures even in the presence of lead. Alloys for extrusion and hot pressing are all in this range.

**Extruded** sections of many copper alloys are made in a wide variety of shapes. In addition to those used architecturally for moldings, etc., extrusion is important to the engineer since many objects such as pinions, hinges, brackets, and lock barrels can be made directly from extruded bars. Special extruded shapes used as stock for automatic screw-machine operations frequently reduce scrap considerably.

## BRONZES

The three most common *tin bronzes* contain about 5, 8, and 10 percent tin and are known as alloys A, C, and D, respectively. They usually contain phosphorus, from a trace up to 0.4 percent, which improves the casting qualities, hardens them somewhat, and has given rise to the misleading name **phosphor bronze.** The bronzes are characterized by excellent elastic properties.

The **aluminum bronzes** with 5 and 8 percent aluminum find application because of their high strength and corrosion resistance, and sometimes because of their golden color. Those with 10 percent aluminum and other alloys with even higher amounts are very plastic when hot and have exceptionally high strength, particularly after heat treatment.

**Silicon Bronzes** A number of alloys are made (and sold under trade names) in which silicon is the primary alloying agent but which also contain appreciable amounts of zinc, iron, tin, or manganese. These alloys are as corrosion-resistant

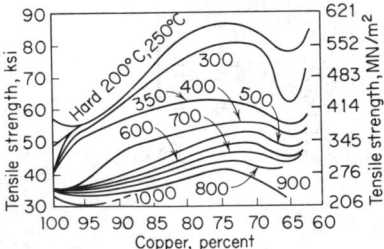

**Fig. 3**  Tensile strengths of copper-zinc alloys.

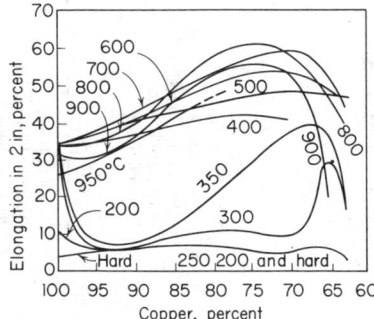

**Fig. 4**  Percent elongation in 2 in of copper-zinc alloys.

**Table 4. Properties of Rolled Yellow Brass**
(ASTM specification B36–74)

| Temper | Nominal grain size, mm | Annealing temp,°F,* (approx)† | Tensile strength, ksi‡ | | Rockwell hardness |
|---|---|---|---|---|---|
| | | | Min | Max | |
| Annealed A | 0.120 | 1300 | | | F50–62 |
| B | 0.070 | 1180 | | | 52–67 |
| C | 0.050 | 1070 | | | 61–73 |
| D | 0.035 | 930 | | | 65–76 |
| E | 0.025 | 800 | | | 67–79 |
| | 0.015 | 700 | | | 72–85 |

| Temper | Nominal reduction, B. & S. gage No. § | Percentage reduction of thickness | Tensile strength, ksi‡ | | Rockwell hardness |
|---|---|---|---|---|---|
| | | | Min | Max | |
| Quarter hard | 1 | 10.9 | 49 | 59 | B40–65 |
| Half hard | 2 | 20.7 | 55 | 65 | 57–74 |
| Three-quarter hard | 3 | 29.4 | 62 | 72 | 70–80 |
| Hard | 4 | 37.1 | 68 | 78 | 76–84 |
| Extra hard | 6 | 50.0 | 79 | 89 | 83–89 |
| Spring | 8 | 60.5 | 86 | 95 | 87–92 |
| Extra spring | 10 | 68.7 | 90 | 99 | 88–93 |

*°C = (°F −32)/1.8.

†The grain size obtained after a given annealing treatment is dependent on the prior grain size and on the amount of final rolling.

‡ksi × 6.895 = MN/m².

§Since the B & S scale is a uniform geometric progression (see Sec. 6) the actual reduction in thickness of one number will decrease as the piece is progressively cold-rolled, but since the actual thickness also decreases, the effect of cold work remains approximately the same.

as copper (slightly more so in some solutions) and possess excellent hot workability with high strength. Their outstanding characteristic is that of ready weldability by all methods. The alloys are extensively fabricated by arc or acetylene welding into tanks and vessels for hot-water storage and chemical processing.

The **cupronickels** and so-called **nickel silvers** are white in color and find application on this account, and because of their comparative freedom from tarnishing under atmospheric conditions. Nickel silver is the base for most silver-plated ware. The cupronickels are extremely malleable and may be worked extensively without annealing. Because of their excellent corrosion resistance, they are used for condenser tubes for most severe service. Alloys containing nickel have the best high-temperature properties of any copper alloy.

## PROCESSING OF COPPER ALLOYS

**Fabrication** Practically any of the copper alloys listed in Table 3 can be obtained in sheet, rod, and wire form and many in the form of tubes. Most of them, in the annealed condition, will withstand extensive amounts of cold work and may be shaped to the desired form by deep drawing, flanging, forming, bending, and similar operations. If extensive cold work is planned, the material should be purchased in the annealed condition. Very extensive operations need intermediate annealing either to avoid failure of the metal or to minimize the power consumption. This is done at 900 to 1300°F (482 to 704°C), depending on the alloy, and is usually followed by air cooling. Because of the ready workability of brass, it is often cheaper to use than steel. Brass may be drawn

at higher speeds than ferrous metals and with less wear on the tools. In cupping operations, a take-in of 45 percent is usual and on some jobs, may be larger. Brass hardened by cold working is softened by annealing at about 1100°F (593°C).

**Welding** Deoxidized copper will weld satisfactorily by the oxyacetylene method. Sufficient heat input to overcome its high heat conductivity must be maintained by the use of torches considerably more powerful than those customary for steel, and preferably by preheating the work in addition. The filler rod must be deoxidized. Gas-shielded arc welding is preferred, however. Tough-pitch copper will not give high-strength welds because of embrittlement due to the oxygen content. Copper may be arc-welded, using shielded metal arc, gas metal arc (MIG), or gas tungsten arc (TIG) welding with experienced operators. Filler rods of phosphor bronze or silicon bronze will give strong welds more consistently and are used where the presence of a weld of different composition and corrosion-resistance characteristics is not harmful. Brass may be welded by the oxyacetylene process but not with the arc. A filler rod of about the same composition is used, although silicon is frequently added to prevent zinc fumes. (See also Sec. 13.)

The copper-silicon alloys are characterized by remarkable ease of welding by all methods. The conductivity is not too high, and the alloy is to a large extent self-fluxing.

All copper alloys, except those containing large amounts of aluminum, can be readily soft-**soldered** or silver-soldered. **Brazing** is possible only with those alloys whose melting points are sufficiently greater than that of the solder used.

(See ASM "Metals Handbook, vol. 6, for information on metals joining.)

**Machining** Free-cutting brass rod (alloy 360) is the standard material for automatic screw-machine work where the very highest machinability is necessary. The use of this material will often result in considerable savings over steel at a lower base price. Most copper alloys are readily machined by usual methods using standard tools designed for steel but at higher speeds. Consideration of the wide range of characteristics presented by various types of copper alloys and the adaptation of the machining practice to the particular material concerned will give greatly improved results. For purposes of machining, copper alloys can be divided into three groups depending on their structure and related characteristics:

GROUP A is composed of alloys of homogeneous structure; copper, the wrought bronzes up to 10 percent tin, brasses and nickel silvers up to 37 percent zinc, aluminum bronzes up to 8 percent aluminum, the silicon bronzes, and cupronickel. These alloys are all tough and ductile and form a long continuous chip. When they are severely cold-worked, they approach the second classification in their characteristics.

GROUP B includes lead-free alloys of duplex structure. Some cast bronzes and most of the high-strength copper alloys in the wrought condition belong in this group. They form a continuous but brittle chip by a process of intermittent shearing against the tool edge which causes chatter unless work and tool are rigid.

GROUP C. Many of the basic brasses and bronzes are rendered particularly fit for machining operations by the addition of 0.5 to 3.0 percent of lead. This exists in the structure as minute uniformly distributed droplets which apparently serve the function of breaking up the chip and lubricating the tool. Chips are fine, almost needlelike, and readily removed. Very little heat is evolved, but the tendency to chatter is greater than in Type A alloy. Lead additions may be made to most copper alloys, but its low melting point makes hot working impossible. Tellurium and sulfur additions are used in place of lead when the combination of hot workability and good machinability is desired.

For lathe turning, the tough alloys of Group A need a sharp top rake angle (20 to 30° for copper and cupronickel; 12 to 16° for the brasses, bronzes, and silicon bronzes with high-speed tools; 8 to 12° with carbide tools except for copper for which 16 is recommended). Type C (leaded) materials should have a much smaller rake to minimize chatter, a maximum of 8° with high-speed steels and 3 to 6° with carbide. Type B materials are intermediate, working best with 6 to 12° with high-speed tools, 3 to 8° with carbide, the higher angle being used with the tougher materials. Side clearance angle should be 5 to 7° except for tough "sticky" materials like copper and cupronickel, where a side rake of 10 to 15° will give better service.

Many copper alloys will drill satisfactorily with the standard helix angle of 30°. Straight fluted tools (helix angle 0°) are preferable for the leaded Group C alloys. An angle of 10° is to be preferred for Group B and 40° for copper and cupronickel. Feeds may generally be 2 to 3 times those used for steel. With Type B alloy, a fairly coarse feed helps breaking up the chip, and with Type A a fine feed and high speed gives best results provided that sufficient feed be used to prevent rubbing and work hardening. (See ASM "Metals Handbook," vol. 3, for information on machining.)

**Corrosion Resistance** All copper alloys are highly resistant to atmospheric attack, but for outdoor exposure those containing over 80 percent of copper (or copper itself) are preferred because of their resistance to **season cracking.**

Water pipes are commonly made of deoxidized copper, red brass, and yellow brass, the last being least resistant to acid waters. Alloys with over 80 percent copper, such as red brass, the silicon-copper alloys, cupronickel, or aluminum bronze, are generally more resistant to corrosion than the alloys with low copper content. Deoxidized copper, arsenical copper, or admiralty brass have proved satisfactory for condenser tubes operating with fresh water. In seawater, copper is less suitable because of its inability to form protective films. Admiralty brass was for years the standard alloy for seawater installations; as now made it contains minute additions of arsenic, antimony, or phosphorus to retard dezincification, which is in effect selective corrosion of the alloy with redeposition of copper. Aluminum brass and the cupronickels are preferred for the most severe seawater service.

Copper and its alloys are not suitable for use in oxidizing acidic solutions or in the presence of moist free ammonia, or mercury. They are also subject to velocity effects in solutions.

**Effect of Temperature** Copper and all its alloys increase in strength slightly and uniformly as temperature decreases from room temperature. No low-temperature brittleness is encountered. Copper is useless for prolonged stressed service much above 400°F (204°C), but some of its alloys may be used up to 550°F (287°C). For service above this, the cupronickel and copper-aluminum alloys alone have satisfactory properties. Aluminum bronzes, particularly those containing 10 percent aluminum, find application for valve seats in internal-combustion engines. For specific data, see "Elevated-Temperature Properties of Copper Base Alloys," ASTM STP No. 181, 1956; "Low-Temperature Properties of Copper and Selected Copper Alloys," NBS Monograph 101, 1967; and R. A. Wilkens and E. S. Bunn, "Copper and Copper Base Alloys," McGraw-Hill, 1943.

## COPPER-BASE ALLOYS FOR CASTING

(See Copper Development Association "Standard Handbook, Part 7, Cast Alloy Data")

Table 5 shows the composition of the important copper-base casting alloys together with typical properties to be expected from sand castings. Test results obtained on standard test pieces (either attached to the casting or separately poured) indicate the quality of the metal used but not the specific properties of the casting itself because of variations of thickness, soundness, and other factors. The ideal casting is one with a fairly uniform metal section with ample fillets and a gradual transition from thin to thick parts.

By far the greatest tonnage of copper castings is made by the **sand-casting** process, and alloys 836 and 937 are the most popular for general-purpose work. **Brass die castings** are made when greater accuracy of dimensions or a better surface finish is desired. While inferior in properties to hot-pressed parts, die castings are adaptable to a wider range of design, for they may be made with intricate coring and with considerable variation in section thickness. (See "Copper Alloy Pressure Die Casting," International Copper Research Association.)

## NICKEL AND NICKEL ALLOYS

The majority of the world's production of nickel comes from the sulfide deposits in Ontario. After intermediate processing, the pure metal is achieved by electrolysis or by decomposition of nickel carbonyl. In either case, the metal is then melted and cast into ingots for further working. **Monel,** an important alloy

**Table 5. Composition and Properties of Cast Copper-Base Alloys***

| Common name | Copper alloy No. | Nominal composition, % Cu | Other | Tensile strength, ksi† | Yield strength, ksi† | Elongation in 2 in, % | Casting types‡ | Conductivity, % IACS | Machinability rating | Weldability |
|---|---|---|---|---|---|---|---|---|---|---|
| Copper | 801 | 99.95 | | 25 | 9 | 40 | C,T,I,M,P,S | 100 | 10 | Fair |
| Leaded red brass | 836 | 80 | 5 Sn, 5 Pb, 5 Zn | 37 | 17 | 30 | C,T,I,S | 15 | 84 | Poor |
| Commercial red brass | 838 | 83 | 4 Sn, 4 Pb, 7 Zn | 35 | 16 | 25 | C,T,S | 15 | 90 | Poor |
| Leaded yellow brass | 852 | 71 | 1 Sn, 3 Pb, 24 Zn | 3 | 1 | 5 | C,T | 18 | 80 | Poor |
| Yellow die-casting brass | 858 | 58 | 1 Sn, 1 Pb, 40 Zn | 55 | 30 | 15 | D | 20 | 80 | Poor |
| High-strength manganese bronze | 863 | 63 | 25 Zn, 3 Fe, 6 Al, 3 Mn | 119 | 83 | 18 | C,I,P,S | 8 | 8 | Poor |
| Tin bronze | 903 | 88 | 8 Sn, 4 Zn | 45 | 21 | 30 | C,T,I,P,S | 12 | 30 | Fair |
| Tin bronze | 905 | 88 | 10 Sn, 2 Zn | 45 | 22 | 25 | C,T,I,S | 11 | 30 | Poor |
| High leaded tin bronze | 937 | 80 | 10 Sn, 10 Pb | 35 | 18 | 20 | C,T,M,S | 10 | 80 | Poor |
| Aluminum bronze | 955 | 81 | 4 Ni, 4 Fe, 11 Al | 100 | 44 | 12 | C,T,M,P,S | 8.5 | 50 | Good |
| Incramet 800 | 993 | 72 | 15 Ni, 0.7 Fe, 11 Al, 1.5 Co | 95 | 55 | 2 | T,S | 9 | 20 | Good |

*Properties are for sand castings, except for alloy 858.
†ksi × 6.895 = MN/m².
‡C, centrifugal; T, continuous; D, die; I, investment; M, permanent mold; P, plaster; S, sand.

**Table 6. Composition of Selected Nickel Alloys**

| Alloy | Nominal composition, % Ni | Cu | Fe | Cr | Mo | W | Al | Si | Mn | C | S | Nb | Ti |
|---|---|---|---|---|---|---|---|---|---|---|---|---|---|
| Nickel 270 | 99.98* min | | | | | | | | | | | | |
| Nickel 200 | 99.4* | 0.1 | 0.15 | | | | | 0.05 | 0.25 | 0.05 | 0.005 | | |
| Duranickel 301 | 93.7* | 0.05 | 0.35 | | | | 4.4 | 0.5 | 0.3 | 0.17 | 0.005 | | |
| 80 Ni | 77.0 min | | | 19.5 | | | | 1.5 max | 2.5 max | 0.25 max | | | |
| Monel 400 | 67 | 30 | 1.4 | | | | | 0.1 | 1.0 | 0.15 | 0.01 | | |
| Monel K-500 | 66 | 29 | 0.9 | | | | 2.7 | 0.5 | 0.7 | 0.15 | 0.005 | | |
| Hastelloy C | 51 | | 6 | 17 | 19 | 5 | | 0.1 | 1 | | | | |
| Incoloy 825 | 42 | 2.2 | 30 | 21.5 | 3 | | | 0.25 | 0.5 | 0.03 | 0.015 | | 0.9 |
| Inconel 600 | 76* | 0.25 | 8 | 15.5 | | | 0.1 | 0.25 | 0.5 | 0.08 | 0.007 | | |
| Inconel 718 | 70* min | 0.50 | 5.9 | 15.5 | | | 0.7 | 0.7 | 1 | 0.08 | 0.01 | | 2.3 |
| TD Ni† | Bal. | 0.15 max | 0.05 max | | | | | | | 0.02 max | | | |

*Including cobalt.
†Co 0.2 max, Th added.

of nickel, is made by reducing an intermediate smelting product, copper-nickel matte, without separating the metals from each other. Other alloys are made in the usual way by incorporation of the elements in the molten state.

The nominal composition and typical properties of selected nickel and various nickel-rich alloys are summarized in Tables 6 to 8.

Commercially pure nickel, known as **nickel 200,** is available as sheet, rod, wire, tubing, and other fabricated forms. It is used where the thermal or electrical properties of nickel are required and where corrosion resistance is needed in parts that have to be worked extensively.

Commercial nickel may be forged or rolled at 1600 to

2300°F (871 to 1260°C). It becomes increasingly harder below 1600°F (871°C) but has no brittle range. The recrystallization temperature of cold-worked pure nickel is about 660°F (349°C), but commercial nickel recrystallizes at about 1100°F (538°C) and is usually annealed at temperatures between 1100 and 1750°F (538 to 954°C).

The addition of certain elements to nickel renders it susceptible to a precipitation or aging treatment to increase its strength and hardness. In the unhardened or quenched state, Duranickel 301 fabricates almost as easily as pure nickel and when finished may be hardened by heating for about 8 h at 1000 to 1100°F (538 to 593°C). Intermediate anneals during fabrication are at 1650 to 1750°F (900 to 954°C). The increase

### Table 7. Physical Properties of Selected Nickel Alloys

| Alloy | Density, lb/in³ | Modulus of elasticity 10⁶ lb/in² | Specific heat (70°F) Btu/(lb)(°F) | Thermal expansion (70–200°F) 10⁻⁶in/in/°F | Thermal conductivity (70°F), Btu/ (ft²)(h)(ft)(°F) | Electrical resistivity, $\mu\Omega \cdot$ cm (annealed) |
|---|---|---|---|---|---|---|
| Nickel 270 | 0.321 | 30 | 0.11 | 8.6 | 7.48 | 7.48 |
| Nickel 200 | 0.321 | 29.6 | 0.109 | 8.5 | 43.3 | 7.6 |
| Duranickel 301 | 0.298 | 30 | 0.14 | 8.2 | 13.7 | 42.4 |
| 80 Ni | 0.300 | 31 | 0.107 | 6.0 | 15.0 | 108 |
| Monel 400 | 0.319 | 26 | 0.102 | 7.7 | 12.6 | 48.2 |
| Monel K-500 | 0.306 | 26 | 0.100 | 7.6 | 10.1 | 57.8 |
| Hastelloy C | 0.323 | 29.8 | 0.092 | 6.3 | 7.5 | 130 |
| Incoloy 825 | 0.294 | 28 | | 7.8 | 6.4 | |
| Inconel 600 | 0.301 | 31.4 | 0.106 | 7.4 | 8.6 | |
| Inconel 718 | 0.296 | 31 | 0.104 | 7.1 | 6.5 | |
| TD Ni | 0.322 | 18.3 | 0.106 | 8.6 | 27 | |

### Table 8. Mechanical Properties of Selected Nickel Alloys

| Alloy | Yield strength, 1,000 lb/in² | Tensile strength, 1,000 lb/in² | Elongation, % in 2 in | Brinell hardness | Remarks |
|---|---|---|---|---|---|
| 270 | 16–90 | 50–95 | 50–4 | 80–210 | Typical soft-hard values |
| 201 | 10–100 | 50–110 | 60–3 | 75–230 | Typical soft-hard values |
| Duranickel 301 | 30–175 | 90–210 | 55–15 | 135–375 | Soft age-hardened values |
| 80 Ni | | 95–200 | 35–0 | 210–270 | Soft-hard ribbon values |
| Monel 400 | 25–100 | 70–100 | 60–22 | 110–241 | Typical soft-hard values |
| Monel K-500 | 40–160 | 90–190 | 45–13 | 140–346 | Soft age-hardened values |
| Hastelloy C | 58 | 116 | 56 | 195 | As-cast values |
| Incoloy 825 | 36 | 92 | 50 | 180 | As-annealed values |
| Inconel 600 | 35–100 | 90–125 | 50–20 | 145–240 | Typical soft-hard values |
| Inconel 718 | 140–171 | 185–201 | 32–26 | 393 | Age-hardened values |
| TD Ni | 40 | 60 | 15 | 144 | Stress-relieved plate |

in strength due to aging may be superimposed on that due to cold work.

Nickel castings are made in dry sand molds but need special technique because of the high temperatures involved. The addition of 1½ percent silicon and lesser amounts of carbon and manganese is necessary to obtain good casting properties.

Alloys of nickel with 2 to 5 percent manganese are used generally for spark-plug electrodes, an application that requires resistance to hot corrosive gases. They are slightly stronger than nickel but have, in general, the same properties.

**Copper-Nickel Alloys**  Alloys containing less than 50 percent nickel are discussed under Copper Alloys. An important nickel-rich alloy is known by the trade name **Monel 400** (see Tables 6 to 8). It combines high strength, high ductility, and excellent resistance to corrosion. It is a homogeneous solid solution alloy; hence its strength can be increased by cold working alone. In the annealed state, its tensile strength is about 70,000 lb/in², and this may be increased to 170,000 lb/in² in the hardest drawn wires. It is available in practically all fabricated forms. Monel 400 is hot-worked in the range 1600 to 2150°F (871 to 1177°C) after rapid heating in a reducing but sulfur-free atmosphere. It must be cold-worked in the same manner as mild steel, but requires more power. Very heavily cold-worked Monel 400 may commence to recrystallize at 800°F (427°C), but in normal practice no softening will occur below 1200°F (649°C). Annealing is done in boxes for 2 to 6 h at about 1400°F (760°C) or open for 2 to 5 min at about 1725°F (940°C). Nonscaling atmospheres are desirable and sulfur-free atmospheres essential.

Because of its toughness, Monel 400 must be machined with slower cutting speed and lighter cuts than mild steel. High-speed tools are necessary. The special grade of Monel containing sulfur (Monel R405) should be used where high cutting speeds must be maintained. This alloy is essentially the same as the sulfur-free alloy in mechanical properties and corrosion resistance, and it can be hot-forged.

The short-time tensile strengths of Monel at elevated temperatures are summarized in Table 9.

The fatigue endurance limit of Monel 400 is about 35,000 (47,000) lb/in² when annealed (hard-drawn). The action of corrosion during fatigue is much less drastic on Monel 400 than on steels of equal or higher endurance limit. McAdam found a limit of 30,000 lb/in² in brackish water.

Monel 400 may be welded by the usual electric and gas methods but needs special fluxes. The gas flame should be slightly reducing and the work done rapidly without rewelding. Flux-coated electrodes should be used for arc welding. Spot and seam welding can be used on thin sheet. Soft soldering, brazing, and silver soldering are readily applied.

Monel 400 is highly resistant to atmospheric action, seawater, steam, foodstuffs, and many industrial chemicals. It deteriorates rapidly in the presence of moist chlorine and ferric, stannic, or mercuric salts in acid solutions. It must not be exposed when hot to molten metals, sulfur, or gaseous products of combustion containing sulfur.

If about 2.75 percent aluminum is added to the base 70/30 nickel-copper alloy, it becomes susceptible to precipitation hardening. This alloy (trade name Monel K500) is available in rod, strip, tubing, sheet, and wire form. It is sufficiently ductile in the annealed state to permit drawing, forming, bending, or other cold-working operations but work-hardens rapidly and requires more power than mild steel. It is hot-worked at 1700 to 2150°F (927 to 1177°C) and should be cooled and quenched from about 1600°F if the metal is to be further worked or to be hardened. Heat treatment consists of quenching from 1600°F (871°C), cold working if desired, and reheating for 10 to 16 h at 1000 to 1100°F (538 to 593°C). If no cold working is intended, the quench may be omitted on sections less than 2 in thick, and the alloy hardened at 1100°F (593°C). The properties of the heat-treated alloy remain quite stable, at least up to 1000°F (538°C) for several months. It is nonmagnetic down to −150°F (−101°C).

**Table 9. Short-Time High-Temperature Properties of Hot-Rolled Nickel and Its Alloys***

| | Temp, °F | | | | | | | |
|---|---|---|---|---|---|---|---|---|
| | 70 | 600 | 800 | 1000 | 1200 | 1500 | 1800 | 2000 |
| Nickel 200† | | | | | | | | |
| Tensile strength, ksi | 73 | 83 | 76 | 46 | 34 | 25 | 8 | |
| Yield strength, 0.2% offset, ksi | 24 | 22 | 21 | 17 | 15 | | | |
| Elongation, % in 2 in | 49 | 50 | 52 | 55 | 57 | 65 | 91 | |
| Monel 400 | | | | | | | | |
| Tensile strength, ksi | 81 | 78 | 71 | 51 | 30 | 16 | 8 | |
| Yield strength, 0.2% offset, ksi | 32 | 28 | 29 | 23 | 18 | 9 | | |
| Elongation, % in 2 in | 46 | 51 | 52 | 29 | 34 | 58 | 45 | |
| Inconel 600 | | | | | | | | |
| Tensile strength, ksi | 85 | 79 | 83 | 79 | 71 | 32 | 15 | 11 |
| Yield strength, 0.2% offset, ksi | 36 | 27 | 28 | 22 | 22 | | | |
| Elongation, % in 2 in | 50 | 51 | 50 | 21 | 5 | 23 | 51 | 67 |

*See also data on metals and alloys for use at elevated temperature.

†Nickel 200 is not recommended for use above 600°F; low-carbon nickel 201 is the preferred substitute.

**Table 10. Limiting Temperatures or Usual Service Temperatures for Nickel Alloys, °F***

| Service or environment | Nickel 200 | Monel alloy 400 | Inconel alloy 600 | Incoloy alloy 800 | Other alloys |
|---|---|---|---|---|---|
| Ammonia synthesis | | | 750–1200 | | |
| Ammonia dissociation | | | 1700–1800 | | |
| Carburizing | | | 1550–1750 | 1550–1750 | |
| Carbonitriding | | | 1550–1750 | | |
| Fused salts: | | | | | |
|   Neutral heat-treating | | | 1550–1750 | | |
|   Sodium hydroxide | | | 1500† | | |
|   Sodium nitrate | | | | | Alloy 201–1000† |
| Gases: | | | | | |
|   Air | 1900† | 1000† | 2150† | 2000† | Alloy 702–2400† |
|   Chlorine | 1000† | 800† | 1000† | 650† | |
|   Fluorine | 1000† | 900† | 750† | | |
|   Hydrogen chloride | 950† | 450† | 900† | 800† | |
|   Hydrogen fluoride | 1400† | 1200† | 1200† | | |
|   Hydrogen | 2300† | 2000† | 2300† | 2300† | |
| Glass | | | 2300† | | Alloys 301, 400, K-500‡ |
| Liquid metals: | | | | | |
|   Sodium, potassium, sodium-potassium | | | | | |
|     Static | 1650† | 1100† | 1650† | 1600† | |
|     Dynamic | | | 1300† | | |
|   Mercury, static | 600† | | 900† | | |
| Nitriding | | | 900–1050 | | |
| Steam | 800† | 800† | 1800† | 1800† | |

°C = (°F − 32)/1.8.
*"Handbook of Huntington Alloys," International Nickel Co.
†Limiting temperatures.
‡Water-cooled molds for fluoride glass.

**Inconel 600** is a high-strength nonmagnetic (−40°F or C) nickel-chromium-iron alloy which is used widely for corrosion- and heat-resisting purposes at temperatures up to 2200°F (1204°C) in nonsulfidizing atmospheres. In sulfidizing atmospheres, the maximum recommended temperature is 1500°F (816°C). Inconel X 750 is an age-hardenable modification suitable for stressed applications in the range 1200 to 1500°F (649 to 816°C).

**High-Temperature Properties of Nickel Alloys** The combination of useful strength and oxidation resistance makes nickel alloys frequent choices for high-temperature service. Table 9 indicates the changes in mechanical properties with temperature, and Table 10 expresses the limiting service temperatures for major nickel alloys in various environments.

**Low-Temperature Properties of Nickel Alloys** Several nickel-base alloys have very good properties at low temperatures in contrast to ferrous alloys where impact strength (the index of brittleness) falls off very rapidly with decreasing temperature. The impact strength remains nearly constant with most nickel-rich alloys, while the tensile and yield strengths increase as they do in the other alloys. Specific data on low-temperature properties may be found in Mudge, *Inco Mag.*, vol. 19, no. 1; White and Siebert, "Literature Survey of Low-Temperature Properties of Metals," Edwards; and "Mechanical Properties of Metals at Low Temperatures," *NBS Circ.* 520, 1952.

A series of nickel-rich alloys with molybdenum and some other elements are sold under the trade name of **Hastelloy** (see Table 6). Hastelloy A is a nickel-manganese alloy. These materials are exceptionally resistant to corrosion by hot concentrated acids such as hydrochloric, sulfuric, and nitric. Hastelloy C has good strength at elevated temperatures; its properties are given in the Section on high-temperature metals. Nickel-chromium electrical-resistance alloys are represented by 80 Ni in Tables 6, 7, and 8.

**Magnetic Properties of Nickel Alloys** Nickel is slightly ferromagnetic but loses its magnetism at a temperature of 695°F (368°C) when pure. In commercial nickel, this temperature is about 650°F (343°C). Monel is feebly magnetic and loses all ferromagnetism above 200°F (93°C). Monel K500 is nonmagnetic down to at least −110°F (−79°C). The degree of ferromagnetism and the temperature sensitivity are very susceptible to variations in composition and mechanical and thermal treatment.

An important group of magnetic alloys are the nickel-irons, and their modifications, of exceptionally high permeability. Other nickel-iron alloys are used for their special thermal expansion and thermoelastic properties.

**Nickel Electroplates** Electroplating accounts for about 10 percent of the total annual consumption of nickel. Normal nickel electroplate has properties approximating those of wrought nickel, but special baths and techniques can give very much harder plates. Bonds between nickel and base metal are usually strong. The largest use of nickel plate is for corrosion protection of iron and steel parts and zinc-base die castings for automotive use. A 1.5- to 3-mil nickel plate is covered with a chromium plate only about 1/100 as thick to give a bright, tarnish-resistant, hard surface. Nickel electrodeposits are also used to facilitate brazing of chromium containing alloys, to reclaim worn parts, and for electroformed parts, such as dies for phonograph records.

## ALUMINUM AND ITS ALLOYS

REFERENCES: Edwards, Frary, and Jeffries, "The Aluminum Industry," McGraw-Hill. Fink et al., "Physical Metallurgy of Aluminum Alloys," ASM. von Zeerleder, "Technology of Light Metals," Elsevier. K. Van Horn et al., "Aluminum," vols. I, II, III, ASM.

Aluminum owes most of its applications to its low density and to the relatively high strength of its alloys, although other uses depend upon its comparatively good corrosion resistance, good working properties, or electrical or thermal conductivity and reflectivity. Commercial aluminum is a soft and ductile metal and is used for many applications where high strength is not necessary. It is available in extruded or rolled forms and can be **hardened** by cold working but not by heat treatment. The alloys of aluminum possess better casting and machining characteristics and better mechanical properties and, therefore, are used more extensively than the pure metal.

**Aluminum Alloys for Sand Castings** The compositions and typical properties of aluminum alloys used for foundry work are listed in Table 11. Most of these are based on either aluminum-copper or aluminum-silicon systems with additions to improve the casting or service characteristics. Among **aluminum-copper** alloys, the one containing 8 percent copper has been used longest as a general-purpose alloy, although the additions of silicon and iron to this improve the casting characteristics, particularly in rendering the alloy less hot-short. Additions of zinc to this alloy, 213.0, are made to improve the machinability. The 4 percent alloys do not have as good casting properties as those containing 8 percent copper.

Alloys containing 12 percent copper are slightly stronger than the 8 percent alloy but considerably less tough. They owe their employment to the fact that it is easy to produce castings free from leaks in this alloy, although they have now been largely replaced by aluminum-silicon alloys.

Alloy 222.0, which has a small magnesium content, retains a larger percentage of its strength and hardness up to comparatively high temperatures. It was developed primarily for pistons but is used in other high-temperature applications. It age-hardens spontaneously after casting, and further hardening may be produced by a precipitation heat treatment at a moderate temperature, particularly if this follows a solution heat treatment.

**Aluminum-silicon** alloys have come into considerable use because of their excellent casting qualities and resistance to corrosion. The alloys are not hot-short and are easy to cast sound in thin or thick sections. They are rather difficult to machine. The most commonly used aluminum-silicon alloy is that containing 5 percent silicon B443.0. Alloy B443.0 solidifies normally with a coarse hypereutectic structure, but this is "modified" before casting by the addition of a small amount of sodium to give a fine eutectic structure of greater strength and toughness. With all alloys containing substantial amounts of silicon, the iron content must be low to avoid brittleness.

**Aluminum-magnesium** alloys are superior to practically all other aluminum casting alloys with respect to resistance to corrosion and machinability. In addition, these alloys exhibit good to exceptional combinations of mechanical strength and

ductility. Two important alloys of this group are ASTM 514.0 and 520.0. In general, however, the alloys are difficult to cast into intricate or pressure-tight castings, and carefully controlled foundry practices are required to minimize a marked tendency for oxidation of the molten alloys.

**Permanent-Mold Casting Alloys** Alloys for casting in permanent molds must be free from hot-shortness. The silicon alloys mentioned above are extensively used, also those containing copper, and alloys with both silicon and copper.

In the United States, the greater use of permanent-mold castings is for internal-combustion-engine pistons for which light weight, low thermal expansion, and good properties at high temperatures are desirable. Most of these are cast in alloy 222.0. Most cast alloys can be heat-treated to improve their properties or to render them less liable to dimensional change during use.

**Die-Casting Alloys** Aluminum alloys for pressure die casting must possess considerable fluidity and be free from hot-shortness. The physical properties are usually of less importance than the casting qualities. Absorption of iron is difficult to avoid under operating conditions but should be kept low.

**Aluminum-Alloy Designations** Wrought alloys are designated by a four-digit number following the system adopted by the Aluminum Association in 1954. The first digit indicates the alloy type, according to the major alloying element. The second digit indicates specific alloy modifications. The last two digits identify the specific aluminum alloy or indicate the aluminum purity. In most cases, the last two digits correspond to the alloy number in the former alloy-designation system (as 2024 was formerly 24S). The first digit is assigned according to the principal alloying element: 1 = 99 + percent aluminum, 2 = copper, 3 = manganese, 4 = silicon, 5 = magnesium, 6 = magnesium and silicon, 7 = zinc.

The **temper designation** indicates the treatment the alloy has received in arriving at its present condition and associated properties. The temper of an alloy is indicated by the letters O, F, H, or T followed by one or more numbers. In wrought alloys, O indicates annealed; F, as-fabricated; T, heat-treated; H, cold-worked. Specific heat treatments and degrees of cold working are indicated by numbers, e.g., T7 or H14. When an alloy is customarily used only in the as-cast condition, the temper designation in omitted. If the alloy is used in both the as-cast and the heat-treated conditions, the letter F is used to indicate the as-cast condition. T2 indicates an annealed casting.

**Wrought-aluminum alloys** (Table 12) are divided into two classes: those hardened and strengthened by cold working alone and those which owe their improved properties to heat treatment. Some of the latter alloys age-harden spontaneously at room temperature, although other wrought alloys of both types need to be heated moderately.

The most important work-hardening alloys are commercially pure aluminum (1100) or the 1.25 percent manganese alloy (3003). Both of these are available in a wide range of sheet, rod, tube, and wire sizes and extruded shapes. The 3003 alloy is used extensively for the manufacture of cooking utensils, conduit pipe, and other products where strength and hardness somewhat greater than those of pure aluminum are desired. These alloys can be hardened by cold work but are not heat treated except when annealed to remove hardness resulting from cold work.

**Table 11a. Compositions of Aluminum Casting Alloys***

| AA† No. | Former designation | Product | Silicon | Iron | Copper | Manganese | Magnesium | Chromium | Nickel | Zinc | Tin | Titanium | Others Each | Others Total |
|---|---|---|---|---|---|---|---|---|---|---|---|---|---|---|
| 208.0 | 108 | S | 2.5–3.5 | 1.2 | 3.5–4.5 | 0.50 | 0.10 | | 0.35 | 1.0 | | 0.25 | | 0.50 |
| 213.0 | C113 | S&P | 1.0–3.0 | 1.2 | 6.0–8.0 | 0.6 | 0.10 | | 0.35 | 2.5 | | 0.25 | | 0.50 |
| 222.0 | 122 | S&P | 2.0 | 1.5 | 9.2–10.7 | 0.50 | 0.15–0.35 | | 0.50 | 0.8 | | 0.25 | | 0.35 |
| 242.0 | 142 | S&P | 0.7 | 1.0 | 3.5–4.5 | 0.35 | 1.2–1.8 | 0.25 | 1.7–2.3 | 0.35 | | 0.25 | 0.05 | 0.15 |
| 295.0 | 195 | S | 0.7–1.5 | 1.0 | 4.0–5.0 | 0.35 | 0.03 | | | 0.35 | | 0.25 | 0.05 | 0.15 |
| B295.0 | B195 | P | 2.0–3.0 | 1.2 | 4.0–5.0 | 0.35 | 0.05 | | 0.35 | 0.50 | | 0.25 | | 0.35 |
| 308.0 | A108 | S&P | 5.0–6.0 | 1.0 | 4.0–5.0 | 0.50 | 0.10 | | | 1.0 | | 0.25 | | 0.50 |
| 319.0 | 319, Allcast | S&P | 5.5–6.5 | 1.0 | 3.0–4.0 | 0.50 | 0.10 | | 0.35 | 1.0 | | 0.25 | | 0.50 |
| 328.0 | Red X-8 | S | 7.5–8.5 | 1.0 | 1.0–2.0 | 0.20–1.6 | 0.20–0.6 | 0.35 | 0.25 | 1.5 | | 0.25 | | 0.50 |
| A332.0 | A132 | P | 11.0–13.0 | 1.2 | 0.50–1.5 | 0.35 | 0.7–1.3 | | 2.0–3.0 | 0.35 | | 0.25 | 0.05 | |
| F332.0 | F132 | P | 8.5–10.5 | 1.2 | 2.0–4.0 | 0.50 | 0.50–1.5 | | 0.50 | 1.0 | | 0.25 | | 0.50 |
| 333.0 | 333 | P | 8.0–10.0 | 1.0 | 3.0–4.0 | 0.50 | 0.05–0.50 | | 0.50 | 1.0 | | 0.25 | | 0.50 |
| 355.0 | 355 | S&P | 4.5–5.5 | 0.6 | 1.0–1.5 | 0.50 | 0.40–0.6 | 0.25 | | 0.35 | | 0.25 | 0.05 | 0.15 |
| C355.0 | C355 | S&P | 4.5–5.5 | 0.20 | 1.0–1.5 | 0.10 | 0.40–0.6 | | | 0.10 | | 0.20 | 0.05 | 0.15 |
| 356.0 | 356 | S&P | 6.5–7.5 | 0.6 | 0.25 | 0.35 | 0.20–0.40 | | | 0.35 | | 0.25 | 0.05 | 0.15 |
| A356.0 | A356 | S&P | 6.5–7.5 | 0.20 | 0.20 | 0.10 | 0.20–0.40 | | | 0.10 | | 0.20 | 0.05 | 0.15 |
| 357.0 | 357 | S&P | 6.5–7.5 | 0.15 | 0.05 | 0.03 | 0.45–0.6 | | | 0.05 | | 0.20 | 0.05 | 0.15 |
| 360.0 | 360 | D | 9.0–10.0 | 2.0 | 0.6 | 0.35 | 0.40–0.6 | | 0.50 | 0.50 | 0.15 | | | 0.25 |
| A360.0 | A360 | D | 9.0–10.0 | 1.3 | 0.6 | 0.35 | 0.40–0.6 | | 0.50 | 0.50 | 0.15 | | | 0.25 |
| 380.0 | 380 | D | 7.5–9.5 | 2.0 | 3.0–4.0 | 0.50 | 0.10 | | 0.50 | 3.0 | 0.35 | | | 0.50 |
| A380.0 | A380 | D | 7.5–9.5 | 1.3 | 3.0–4.0 | 0.50 | 0.10 | | 0.50 | 3.0 | 0.35 | | | 0.50 |
| 383.0 | 383 | D | 9.5–11.5 | 1.3 | 2.0–3.0 | 0.50 | 0.10 | | 0.30 | 3.0 | 0.15 | | | 0.50 |
| 384.0 | 384 | D | 10.5–12.0 | 1.3 | 3.0–4.5 | 0.50 | 0.10 | | 0.50 | 1.0 | 0.35 | | | 0.50 |
| 413.0 | 13 | D | 11.0–13.0 | 2.0 | 1.0 | 0.35 | 0.10 | | 0.50 | 0.50 | 0.15 | | | 0.25 |
| A413.0 | A13 | D | 11.0–13.0 | 1.3 | 1.0 | 0.35 | 0.10 | | 0.50 | 0.50 | 0.15 | | | 0.25 |
| B443.0 | 43 (0.15 max Cu) | S&P | 4.5–6.0 | 0.8 | 0.15 | 0.35 | 0.05 | | | 0.35 | | | | 0.15 |
| C443.0 | A43 | D | 4.5–6.0 | 2.0 | 0.6 | 0.35 | 0.10 | | 0.50 | 0.50 | 0.15 | | | 0.25 |
| 514.0 | 214 | S | 0.35 | 0.50 | 0.15 | 0.35 | 3.5–4.5 | | | 0.15 | | 0.25 | 0.05 | 0.15 |
| A514.0 | A214 | P | 0.30 | 0.40 | 0.10 | 0.30 | 3.5–4.5 | | | 1.4–2.2 | | 0.20 | 0.05 | 0.15 |
| B514.0 | B214 | S | 1.4–2.2 | 0.6 | 0.35 | 0.8 | 3.5–4.5 | 0.25 | | 0.35 | | 0.25 | 0.05 | 0.15 |
| 518.0 | 218 | D | 0.35 | 1.8 | 0.25 | 0.35 | 7.5–8.5 | | | 0.15 | 0.15 | | | 0.25 |
| 520.0 | 220 | S | 0.25 | 0.30 | 0.25 | 0.15 | 9.5–10.6 | | | 0.15 | | 0.25 | 0.05 | 0.15 |
| 535.0 | Almag 35 | S | 0.15 | 0.15 | 0.05 | 0.10–0.25 | 6.2–7.5 | | | | | 0.10–0.25 | 0.05 | 0.15 |
| 705.0 | 603, Ternalloy 5 | S&P | 0.20 | 0.8 | 0.20 | 0.40–0.6 | 1.4–1.8 | 0.20–0.40 | | 2.7–3.3 | | 0.25 | 0.05 | 0.15 |
| 707.0 | 607, Ternalloy 7 | S&P | 0.20 | 0.8 | 0.20 | 0.40–0.6 | 1.8–2.4 | 0.20–0.40 | | 4.0–4.5 | | 0.25 | 0.05 | 0.15 |
| A712.0 | A612 | S | 0.15 | 0.50 | 0.35–0.65 | 0.05 | 0.6–0.8 | | | 6.0–7.0 | | 0.25 | 0.05 | 0.15 |
| D712.0 | D612, 40E | S | 0.30 | 0.50 | 0.25 | 0.10 | 0.50–0.65 | 0.40–0.6 | | 5.0–6.5 | | 0.15–0.25 | 0.05 | 0.20 |
| 713.0 | 613, Tenzaloy | S&P | 0.25 | 1.1 | 0.40–1.0 | 0.6 | 0.20–0.50 | 0.35 | 0.15 | 7.0–8.0 | | 0.25 | 0.10 | 0.25 |
| 771.0 | Precedent 71A | S | 0.15 | 0.15 | 0.10 | 0.10 | 0.8–1.0 | 0.06–0.20 | | 6.5–7.5 | | 0.10–0.20 | 0.05 | 0.15 |
| 850.0 | 750 | S&P | 0.7 | 0.7 | 0.7–1.3 | 0.10 | 0.10 | | 0.7–1.3 | | 5.5–7.0 | 0.20 | | 0.30 |
| A850.0 | A750 | S&P | 2.0–3.0 | 0.7 | 0.7–1.3 | 0.10 | 0.10 | | 0.30–0.7 | | 5.5–7.0 | 0.20 | | 0.30 |
| B850.0 | B750 | S&P | 0.40 | 0.7 | 1.7–2.3 | 0.10 | 0.6–0.9 | | 0.9–1.5 | | 5.5–7.0 | 0.20 | | 0.30 |

*From "Aluminum Standards and Data," Aluminum Association. ASTM Specifications B26–75, B108–75, B85–73, and B179–75 apply to sand castings (S), permanent-mold castings (P), die castings (D), and ingot, respectively. Composition is percent maximum unless shown as range; aluminum is the balance.
†Aluminum Association.

**Table 11b. Mechanical Properties of Aluminum Castings***

| AA No. | Former designation | Product* | Temper | Tensile | Yield | Elongation in 2 in, % min |
|---|---|---|---|---|---|---|
| 208.0 | 108 | S | F | 19.0 | | 1.5 |
| 208.0 | 108 | S | T55 | 21.0 | | |
| 213.0 | C113 | S | F | 19.0 | | |
| 213.0 | C113 | P | F | 23.0 | | |
| 222.0 | 122 | S | T2 | 23.0 | | |
| 222.0 | 122 | S | T61 | 30.0 | | |
| 222.0 | 122 | P | T551 | 30.0 | | |
| 222.0 | 122 | P | T65 | 40.0 | | |
| 242.0 | 142 | S | T21 | 23.0 | | |
| 242.0 | 142 | S | T571 | 29.0 | | |
| 242.0 | 142 | P | T571 | 34.0 | | |
| 242.0 | 142 | P | T61 | 40.0 | | |
| 295.0 | 195 | S | T4 | 29.0 | | 6.0 |
| 295.0 | 195 | S | T6 | 32.0 | 20.0 | 3.0 |
| 295.0 | 195 | S | T62 | 36.0 | | |
| 295.0 | 195 | S | T7 | 29.0 | | 3.0 |
| B295.0 | B195 | P | T4 | 33.0 | | 4.5 |
| B295.0 | B195 | P | T6 | 35.0 | | 2.0 |
| B295.0 | B195 | P | T7 | 33.0 | | 3.0 |
| 308.0 | A108 | P | F | 24.0 | | |
| 319.0 | 319, Allcast | S | F | 23.0 | | |
| 319.0 | 319, Allcast | S | T5 | 25.0 | | |
| 319.0 | 319, Allcast | S | T6 | 31.0 | | 1.5 |
| 319.0 | 319, Allcast | P | F | 28.0 | | 1.5 |
| 319.0 | 319, Allcast | P | T6 | 34.0 | | 2.0 |
| 328.0 | Red X-8 | S | F | 25.0 | | 1.0 |
| 328.0 | Red X-8 | S | T6 | 34.0 | | 1.0 |
| A332.0 | A132 | P | T551 | 31.0 | | |
| A332.0 | A132 | P | T65 | 40.0 | | |
| F332.0 | F132 | P | T5 | 31.0 | | |
| 333.0 | 333 | P | F | 28.0 | | |
| 333.0 | 333 | P | T5 | 30.0 | | |
| 333.0 | 333 | P | T6 | 35.0 | | |
| 333.0 | 333 | P | T7 | 31.0 | | |
| 355.0 | 355 | S | T51 | 25.0 | | |
| 355.0 | 355 | S | T6 | 32.0 | 20.0 | 2.0 |
| 355.0 | 355 | S | T7 | 35.0 | | |
| 355.0 | 355 | S | T71 | 30.0 | | |
| 355.0 | 355 | P | T51 | 27.0 | | |
| 355.0 | 355 | P | T6 | 37.0 | | |
| 355.0 | 355 | P | T62 | 42.0 | | |
| 355.0 | 355 | P | T71 | 34.0 | | |
| C355.0 | C355 | P | T61 | 40.0 | | 3.0 |
| 356.0 | 356 | S | T51 | 23.0 | | |
| 356.0 | 356 | S | T6 | 30.0 | 20.0 | 3.0 |
| 356.0 | 356 | S | T7 | 31.0 | 29.0 | |
| 356.0 | 356 | P | T51 | 25.0 | | |
| 356.0 | 356 | P | T6 | 33.0 | | 3.0 |
| 356.0 | 356 | P | T7 | 29.0 | | 4.0 |
| A356.0 | A356 | P | T61 | 37.0 | | 5.0 |
| 357.0 | 357 | P | T6 | 45.0 | | 3.0 |
| B443.0 | 43 (0.15 max Cu) | S | F | 17.0 | | 3.0 |
| B443.0 | 43 (0.15 max Cu) | P | F | 21.0 | | 5.0 |
| 514.0 | 214 | S | F | 22.0 | | 6.0 |
| A514.0 | A214 | P | F | 22.0 | | 2.5 |
| B514.0 | B214 | S | F | 17.0 | 10.0 | |
| 520.0 | 220 | S | T4 | 42.0 | 22.0 | 12.0 |
| 535.0 | Almag 35 | S | F | 35.0 | 18.0 | 9.0 |
| 535.0 | Almag 35 | S | T2 | 35.0 | 18.0 | 9.0 |
| 705.0 | 603, Ternalloy 5 | S | F or T5 | 30.0 | 17.0 | 5.0 |
| 705.0 | 603, Ternalloy 5 | P | T5 | 37.0 | | 10.0 |
| 707.0 | 607, Ternalloy 7 | S | F or T5 | 33.0 | 22.0 | 2.0 |
| 707.0 | 607, Ternalloy 7 | P | T5 | 42.0 | | 4.0 |
| 707.0 | 607, Ternalloy 7 | P | T7 | 45.0 | | 3.0 |
| A712.0 | A612 | S | T5 | 32.0 | 20.0 | 2.0 |
| D712.0 | D612, 40E | S | F or T5 | 34.0 | 25.0 | 4.0 |
| 713.0 | 613, Tenzaloy | S | F or T5 | 32.0 | 22.0 | 3.0 |
| 713.0 | 613, Tenzaloy | P | T5 | 32.0 | | 4.0 |
| 771.0 | Precedent 71A | S | T6 | 45.0 | 37.0 | 5.0 |
| 850.0 | 750 | S | T5 | 16.0 | | 5.0 |
| 850.0 | 750 | P | T5 | 18.0 | | 8.0 |
| A850.0 | A750 | S | T5 | 17.0 | | 3.0 |
| A850.0 | A750 | P | T5 | 17.0 | | 3.0 |
| B850.0 | B750 | S | T5 | 24.0 | 18.0 | |
| B850.0 | B750 | P | T5 | 27.0 | | 3.0 |

*From "Aluminum Standards and Data," Aluminum Association. ASTM Specifications B26-75, B108-75, B85-73, and B179-75 apply to sand castings (S), permanent-mold castings (P), die castings (D), and ingot, respectively. Values represent properties obtained from separately cast test bars.

†ksi = 6.895 = MN/m².

**Table 11c. Mechanical Properties of Aluminum Die Castings**

| Alloy New | Alloy Former | Typical tensile strength, ksi | Typical yield strength (0.2% offset), ksi | Typical elongation in 2 in, % |
|---|---|---|---|---|
| 360.0 | SG100B | 44 | 25 | 2.5 |
| A360.0 | SG100A | 46 | 24 | 3.5 |
| 380.0 | SC84B | 46 | 23 | 2.5 |
| A380.0 | SC84A | 47 | 23 | 3.5 |
| 383.0 | SC102A | 45 | 22 | 3.5 |
| 384.0 | SC114A | 48 | 24 | 2.5 |
| 413.0 | S12B | 43 | 21 | 2.5 |
| A413.0 | S12A | 42 | 19 | 3.5 |
| C443.0 | S5C | 33 | 14 | 9.0 |
| 518.0 | G8A | 45 | 28 | 5 |

ksi = 6.895 = MN/m².

Where high strength is necessary, it is usual to use an alloy of the **duralumin** type. The most common is 2024. This alloy is readily hot-worked. It hardens spontaneously at room temperature after solution heat treatment. Extensive cold working must be done within a few hours after quenching. Cold working following age hardening is less easy but gives the highest strength obtainable with the alloys. For aircraft construction, 2017 has been largely superseded by 2024 because this alloy has a higher strength. 6061 and 6063 have better fabricating qualities in the quenched condition. Alloy 6053 is used because of its combination of good physical properties and corrosion resistance. It is available in various rolled structural shapes; most architectural extruded sections are made from it. Alloy 2219 is by far the most weldable of the high-strength heat-treatable alloys. Alloy 7178 is the highest-strength heat-treatable alloy commercially available today and is readily welded.

Most of the heat-treatable alloys are less resistant to corrosion than is pure aluminum or aluminum-manganese alloy; alloy 6053 is an exception. The corrosion resistance of the duralumin-type alloys is greatest in the quenched condition. Oil quenching and artificial aging decrease the resistance still further. Many of the heat-treatable alloys are available in sheet form with an integral coating of high-purity aluminum or corrosion-resistant alloy on each side; the thickness of each layer is approximately 2½ to 10 percent of the total, depending upon the alloy and gage. These products, known as **alclad alloys**, possess an excellent resistance to corrosion. Because of electrolytic action, exposed cut edges of the base metal are protected, and ordinary bare rivets can be used. The strength is slightly less than that of bare sheet of similar gage.

A new type of wrought-aluminum commodity, designated **APM**, has elevated-temperature strength, stability, and resistance to creep superior to any known aluminum alloy. The process consists of making a compact of fine, high-purity aluminum powder containing large but controlled amounts of aluminum oxide, followed by extruding, rolling, or forging into a final product. Alloy XAP001 contains 6 percent oxide and has 47 percent IACS conductivity, 37,000 lb/in² tensile strength, 27,000 lb/in² yield strength, and 13 percent elongation at normal temperature. As high as 800°F (427°C), the tensile strength is 11,000 lb/in² with 10,000 lb/in² yield strength and 4 percent elongation. (See Table 13.)

**Heat Treatment** Intermediate annealing to relieve cold work is done at a temperature of about 650°F (343°C) for pure aluminum and 5052 or about 750°F (399°C) for 3003. The rate of cooling is unimportant. The heat-treatable alloys are best cold-worked when in the quenched condition after heat treatment. They may be fully annealed only by heating to 750 to 800°F (399 to 427°C) and cooling slowly to 500°F (260°C). A partial softening of heat-treated material can be obtained by an ordinary anneal at 650°F (343°C).

The heat-treatable alloys must have a double heat treatment: one at a high temperature to dissolve the alloy constituents later responsible for hardening and the other at a low temperature to permit them to cause precipitation hardening of the alloy. The second action may take place spontaneously at room temperatures on some alloys and is then known as **natural aging,** but on other alloys it has to be carried out at a somewhat elevated temperature referred to as *artificial aging* or precipitation treatment. The correct treatment for the various alloys is given in Table 14. The solution treatment is usually done in a nitrate bath or in a furnace with forced air circulation. The temperature must be controlled closely. The solution heat treatment of duralumin should be followed by a rapid quench, preferably in cold water since slower quenches in hot water or oil, although they minimize distortion, render the alloy susceptible to intergranular corrosion. In the alloys that age spontaneously at room temperature, hardening starts immediately after quenching and is practically complete in 4 days. Severe cold-forming operations must be done within less than an hour after quenching. If it is desired to hold the alloy for later cold working, aging may be retarded by storing the quenching material at low temperatures, e.g., in ice, which will permit working up to 24 h, or in "dry ice" (solid $CO_2$) which will retard aging almost indefinitely.

**Machining** (see Sec. 13) Many aluminum alloys are easily machined without special technique. Pure aluminum and the aluminum-manganese alloys are hard to machine unless special tools are used with greater rake than is customary for steel. In general, such tools are similar to those used for working wood but should be harder. Cemented hard carbide tools are almost essential for aluminum-silicon alloys. The casting alloys containing copper and all the wrought heat-treated alloys possess good machinability. For products in which physical properties are subordinate to high machinability, as in automatic-screw-machine work, alloy 2011 is used. The additions of lead and bismuth render this alloy free-cutting.

**Table 12. Composition and Typical Room-Temperature Properties of Wrought-Aluminum Alloys***

In the table below, columns *Cu*, *Si*, *Mn*, *Mg*, and *Other elements* fall under the heading "Nominal composition, % (balance aluminum)"; columns *Yield strength*, *Tensile strength*, and *Elong* fall under the heading "Mechanical properties."

| Aluminum Assoc. alloy designation† | Cu | Si | Mn | Mg | Other elements | Temper‡ | Density, lb/in³ | Electrical conductivity, % IACS | Brinell hardness, 500-kg load, 10-mm ball | Yield strength,§ 1,000 lb/in² | Tensile strength, 1,000 lb/in² | Elong, % in 2 in | Endurance limit, 1,000 lb/in²¶ |
|---|---|---|---|---|---|---|---|---|---|---|---|---|---|
| **Work-hardenable alloys** | | | | | | | | | | | | | |
| 1100 | | | | | | 0 | 0.098 | 59 | 23 | 5 | 13 | 35 | 5 |
| | | | | | | H14 | | | 32 | 17 | 18 | 9 | 7 |
| | | | | | | H18 | | 57 | 44 | 22 | 24 | 5 | 9 |
| 3003 | | | 1.2 | | | 0 | 0.099 | 50 | 28 | 6 | 16 | 30 | 7 |
| | | | | | | H14 | | 41 | 40 | 21 | 22 | 8 | 9 |
| | | | | | | H18 | | 40 | 55 | 27 | 29 | 4 | 10 |
| 5052 | | | | 2.5 | 0.25 Cr | 0 | 0.096 | 40 | 47 | 13 | 28 | 25 | 16 |
| | | | | | | H34 | | | 68 | 31 | 38 | 10 | 18 |
| | | | | | | H38 | | 40 | 77 | 37 | 42 | 7 | 19 |
| 5056 | | | 0.1 | 5.2 | 0.1 Cr | 0 | 0.095 | 29 | 65 | 22 | 42 | 35 | 20 |
| | | | | | | H18 | | 27 | 105 | 59 | 63 | 10 | 22 |
| **Heat-treatable alloys** | | | | | | | | | | | | | |
| 2011 | 5.5 | | | | 0.5 Pb, 0.5 Bi | T3 | 0.102 | 40 | 95 | 43 | 55 | 15 | 18 |
| | | | | | | T8 | | | 100 | 45 | 59 | 12 | 18 |
| 2014 | 4.4 | 0.8 | 0.8 | 0.4 | | 0 | 0.101 | 40 | 45 | 14 | 27 | 18 | 13 |
| | | | | | | T4, 451 | | | 105 | 42 | 62 | 20 | 20 |
| | | | | | | T6, 651 | | | 135 | 60 | 70 | 13 | 18 |
| Alclad 2014 | | | | | | 0 | 0.101 | 40 | | 10 | 25 | 21 | 18 |
| | | | | | | T4, 451 | | | | 37 | 61 | 22 | 37 |
| | | | | | | T6, 651 | | | | 60 | 68 | 10 | 41 |
| 2017 | 4.0 | 0.5 | 0.5 | 0.5 | | 0 | 0.101 | 45 | 45 | 10 | 26 | 22 | 13 |
| | | | | | | T4, 451 | | | 105 | 40 | 62 | 22 | 18 |
| 2024 | 4.4 | 0.5 | 0.5 | 1.5 | | 0 | 0.100 | 50 | 47 | 11 | 27 | 20 | 13 |
| | | | | | | T3 | | 30 | 120 | 50 | 70 | 18 | 20 |
| | | | | | | T4, 351 | | | 120 | 47 | 68 | 20 | 20 |
| Alclad 2024 | | | | | | 0 | 0.100 | 30 | | 11 | 26 | 20 | |
| | | | | | | T3 | | | | 45 | 65 | 18 | |
| | | | | | | T4, 351 | | | | 42 | 64 | 19 | |
| 2025 | 4.5 | 0.8 | 0.8 | | 0.15 Zr | T6 | 0.101 | 40 | 110 | 37 | 58 | 19 | 18 |
| 2219 | 6.1 | | 0.3 | | 0.10 V, 0.06 Ti | 0 | | | | 11 | 25 | 18 | |
| | | | | | | T351 | | | | 36 | 52 | 17 | |
| | | | | | | T851 | | | 130 | 51 | 66 | 10 | |
| 6053 | | 0.7 | | 1.3 | 0.25 Cr | 0 | 0.097 | 45 | 26 | 8 | 16 | 35 | 8 |
| | | | | | | T6 | | | 80 | 32 | 37 | 13 | 13 |
| 6061 | 0.25 | 0.6 | | 1.0 | 0.25 Cr | 0 | 0.098 | 40 | 30 | 8 | 18 | 25 | 9 |
| | | | | | | T6, 651 | | 45 | 95 | 40 | 45 | 12 | 14 |
| 6063 | | 0.4 | | 0.7 | | 0 | 0.098 | 50 | 25 | 7 | 13 | 25 | 8 |
| | | | | | | T6 | | 55 | 73 | 31 | 35 | 12 | 10 |
| 7075 | 1.6 | | | 2.5 | 5.6 Zn, 0.3 Cr | 0 | 0.101 | 30 | 60 | 15 | 33 | 17 | |
| | | | | | | T6, 651 | | | 150 | 73 | 83 | 11 | 23 |
| Alclad 7075 | | | | | 0.3 Cr | 0 | 0.101 | | | 14 | 32 | 17 | |
| | | | | | | T6, 651 | | | | 67 | 76 | 11 | |
| 7178 | 2.0 | | | 2.7 | 0.3 Cr, 6.8 Zn | T6, 651 | | | | 78 | 88 | 10 | |

NOTE: See Sec. 1 for conversion factors for SI units.

*Typical values for 0.064-in-thick sheet.

†Aluminum Association Standardized System of Alloy Designation adopted October 1954.

‡Standard temper designations: 0 = fully annealed; H14 and H34 correspond to half-hard and H18 and H38 to hard strain-hardened tempers. T3 = solution heat-treated and then cold-worked; T4 = solution heat-treated; T6 = solution treated and then artificially aged; T42 = heat-treated by user. T × 51 = stretcher stress-relieved. T8 = solution treated, and cold-worked, and artificially aged.

§At 0.2% offset.

¶500,000,000 cycles of completely reversed stress; rotating beam type specimen.

**Table 13. Typical Tensile Properties of Aluminum Alloys at Elevated Temperatures**

**Wrought**

| Alloy and temper | Property* | 75 | 300 | 400 | 500 | 700 |
|---|---|---|---|---|---|---|
| 1100-H18 | T.S. | 24 | 18 | 6 | 4 | 2 |
|  | Y.S. | 22 | 14 | 4 | 3 | 1.5 |
|  | El. | 15 | 20 | 65 | 75 | 85 |
| 3003-H18 | T.S. | 29 | 23 | 14 | 8 | 3 |
|  | Y.S. | 27 | 16 | 9 | 4 | 2 |
|  | El. | 10 | 11 | 8 | 60 | 70 |
| 3004-H18 | T.S. | 26 | 22 | 14 | 10 | 5 |
|  | Y.S. | 10 | 10 | 10 | 8 | 3 |
|  | El. | 25 | 35 | 55 | 70 | 90 |
| 2017-T4 | T.S. | 62 | 40 | 16 | 9 | 4 |
|  | Y.S. | 40 | 30 | 13 | 8 | 3.5 |
|  | El. | 22 | 15 | 35 | 45 | 70 |
| 2024-T4 | T.S. | 68 | 45 | 26 | 11 | 5 |
|  | Y.S. | 47 | 36 | 19 | 9 | 4 |
|  | El. | 19 | 17 | 27 | 35 | 100 |
| 5052-H34 | T.S. | 38 | 30 | 24 | 12 | 5 |
|  | Y.S. | 31 | 27 | 15 | 8 | 3 |
|  | El. | 36 | 27 | 45 | 80 | 130 |
| 6053-T6 | T.S. | 37 | 25 | 13 | 6 | 3 |
|  | Y.S. | 32 | 24 | 12 | 4 | 2 |
|  | El. | 13 | 13 | 25 | 70 | 90 |
| 6061-T6 | T.S. | 45 | 34 | 19 | 8 | 2 |
|  | Y.S. | 40 | 31 | 15 | 5 | 2 |
|  | El. | 17 | 20 | 28 | 60 | 95 |
| 7075-T6 | T.S. | 83 | 31 | 16 | 11 | 6 |
|  | Y.S. | 73 | 27 | 13 | 9 | 5 |
|  | El. | 11 | 30 | 55 | 65 | 70 |
| XAP001 | T.S. | 37 |  | 23 | 20 | 14 |
|  | Y.S. | 27 |  | 19 | 16 | 12 |
|  | El. | 13 |  | 13 | 12 | 8 |

**Sand castings**

| Alloy and temper | Property* | 75 | 300 | 400 | 500 | 600 |
|---|---|---|---|---|---|---|
| 122-T2 | T.S. | 27 | 25 | 22 | 17 | 8 |
|  | Y.S. | 20 | 17 | 14 | 11 | 5 |
|  | El. | 1 | 1 | 2 | 3 | 14 |
| 122-T61 | T.S. | 41 | 36 | 24 | 17 | 9 |
|  | Y.S. | 40 | 35 | 17 | 11 | 5 |
|  | El. | 0.5 | 1 | 2 | 6 | 14 |
| 142-T571 | T.S. | 32 | 30 | 26 | 13 | 8 |
|  | Y.S. | 30 | 28 | 21 | 8 | 4 |
|  | El. | 0.5 | 0.5 | 1 | 8 | 20 |
| 195-T4 | T.S. | 32 | 28 | 15 | 9 | 4 |
|  | Y.S. | 16 | 15 | 9 | 6 | 3 |
|  | El. | 9 | 5 | 15 | 25 | 75 |
| 355-T51 | T.S. | 28 | 24 | 14 | 10 | 6 |
|  | Y.S. | 23 | 19 | 10 | 5 | 3 |
|  | El. | 2 | 3 | 8 | 16 | 36 |
| 356-T6 | T.S. | 33 | 23 | 12 | 8 | 4 |
|  | Y.S. | 24 | 20 | 9 | 5 | 3 |
|  | El. | 4 | 6 | 18 | 35 | 60 |

**Permanent-mold castings**

| Alloy and temper | Property* | 75 | 300 | 400 | 500 | 600 |
|---|---|---|---|---|---|---|
| 356-T6 | T.S. | 38 | 21 | 12 | 8 | 4 |
|  | Y.S. | 27 | 17 | 9 | 5 | 3 |
|  | El. | 5 | 10 | 30 | 55 | 70 |
| A132-T551 | T.S. | 36 | 31 | 26 | 18 | 10 |
|  | Y.S. | 28 | 22 | 15 | 10 | 4 |
|  | El. | 0.5 | 1 | 2 | 2 | 10 |
| 142-T571 | T.S. | 40 | 37 | 28 | 13 | 8 |
|  | Y.S. | 34 | 33 | 22 | 8 | 4 |
|  | El. | 1 | 1 | 2 | 15 | 35 |
| B195-T4 | T.S. | 36 | 30 | 15 | 8 | 4 |
|  | Y.S. | 22 | 20 | 9 |  | 3 |
|  | El. | 8 | 8 | 12 | 25 | 65 |
| 355-T51 | T.S. | 30 | 23 | 15 | 10 | 6 |
|  | Y.S. | 24 | 20 | 10 | 5 | 3 |
|  | El. | 2 | 4 | 9 | 33 | 98 |

*T.S., tensile strength, ksi. Y.S., yield strength, 0.2 percent offset, ksi. El., elongation in 2 in. percent.
ksi × 6.895 = MN/m². °C = (°F − 32)/1.8.
Tensile tests made on ASTM standard test pieces, 0.5-in. diam. maintained at elevated temperatures for 10,000 h.
Speed of straining 0.1 in. per min per in. of gage length.

**Table 14. Conditions for Heat Treatment of Aluminum Alloys**

| Alloy | Solution heat treatment* | | Precipitation heat treatment | | |
| | Temp °F | Temper designation | Temp °F | Time of aging | Temper designation |
|---|---|---|---|---|---|
| 2014 | 925–945 | T4 | 315–325 | 18 h | T6 |
| 2017 | 925–945 | T4 | Room | 4 days | |
| 2024 | 910–930 | T4 | Room | 4 days | |
| 6053 | 960–980 | T4 | 315–325 | 18 h | T6 |
| 6061 | 975–995 | T4 | 315–325 | 18 h | T6 |
| 7075 | 860–880 | W | 215–235 ⎱† | 6–8 h | T6 |
| | | | 340–360 ⎰ | 8–10 h | T73 |

°C = (°F − 32)/1.8.

*In a molten nitrate bath, the time varies from 10 to 60 min depending upon the size of the load and the thickness of the material. In an air furnace, proper allowance must be made for a slower rate of bringing the load up to temperature. For heavy material, a longer time at temperature may be necessary. All quenching is performed in cold water.

†This is a two-stage treatment with air cooling to room temperature between the low-temperature and the high-temperature stage for extrusions.

**Riveting** is the most commonly used method of joining aluminum alloys, especially in structures of the heat-treatable alloys that cannot be welded without loss of strength. In general, rivets of similar composition to the base metal are used. When heat-treatable rivets are driven cold, it is important that they be used in the freshly quenched condition prior to aging, but they may be kept for long times in cold storage. Large rivets can sometimes be driven hot from their solution treatment temperature, depending on contact with tools and surrounding metal to produce an effective quench.

**Welding** (see Sec. 13) The wrought-aluminum alloys are readily welded by experienced operators by either the fusion or resistance method. Fusion welding of the strong alloys is not recommended unless subsequent heat treatment is possible, but spot and seam welding can be done if automatically controlled. Most casting alloys may be welded, but experience is necessary to overcome the danger of strains and cracks resulting from thermal contraction. Welding should be done prior to heat treatment. The rod used should generally be of the same composition as the alloy. Aluminum alloys can be **soldered,** but the resultant joints are rarely satisfactory and are not recommended for highly stressed parts or for joints that cannot be given adequate protection against corrosion action.

**Corrosion Resistance** Although aluminum is chemically active, the presence of a firmly adherent self-healing oxide coat on the surface prevents action except under conditions that tend to remove this surface film. Concentrated nitric and acetic acids are handed in aluminum not only because of its resistance to attack but also because any resulting corrosion products are colorless. For the same reason, aluminum is employed in the preparation of foods and beverages. Hydrochloric acid and most alkalies dissolve the protective film at the surface and permit fairly rapid attack. Moderately alkaline soaps and the like can be used with aluminum if a small amount of sodium silicate is added. Aluminum is very resistant to sulfur and most of its gaseous compounds.

Ordinary atmospheric corrosion is resisted by aluminum and most of its alloys, and they may be used without any protective coating. The pure metal is most resistant to attack, and additions of alloying elements usually decrease resistance, particularly after heat treatment. Under severe conditions of exposure such as may prevail on shipboard or where the metal is continually in contact with wood or other absorbent material in the presence of moisture, a protective coat of paint is desirable as an added precaution.

The resistance to corrosion of aluminum alloys may be augmented by coating the material with a surface layer of high-purity aluminum, or in some cases an alloy, which is rolled as an integral part of the sheet. The corrosion resistance of any of the alloys may be improved by giving an **anodizing treatment,** which comprises making the parts to be treated the anode in an electrolytic bath (chromic, sulfuric, or oxalic acid). This produces a tough, adherent coating of aluminum oxide. The film will be colorless on pure aluminum and tends to be gray or colored on alloys containing silicon, copper, or other constituents. This film is very adherent and cannot be readily detached by bending or ordinary fabricating processes. If a colored finish is desired, the electrolytically oxidized article may be treated with a dye solution. Chemical methods are available for producing a similar but thinner film without electrolytic action by mere dipping in hot alkaline oxidizing solution or applying a paste.

When painting or lacquering aluminum it is important that the surface be properly prepared, prior to the application of the paint. A thin anodic film makes an excellent paint base, or the metal may be chemically treated with a dilute phosphoric acid solution. Where corrosive conditions are to be met, zinc chromate may be used as the pigment in the primer coat and aluminum paint for the top.

**Aluminum Conductors** On a weight basis, aluminum has twice the electrical conductance of copper; on a volume basis, the conductivity of aluminum is about 62 percent that of copper. For electrical applications, aluminum is used as a special, high-purity grade (EC) or as relatively dilute alloys designed to improve strength with minimum sacrifice in conductivity. Table 15 lists common conductor alloys and gives their strengths and conductivities in various forms and treatments. In power-transmission lines, the necessary strength for long spans is obtained by stranding aluminum wires about a core wire of steel (**ACSR**).

## MAGNESIUM AND MAGNESIUM ALLOYS

REFERENCES: Publications of the Dow Chemical Co. "Metals Handbook," ASM. Beck, "Technology of Magnesium and Its Alloys," Hughes. 1975 Annual Standards, Part 7, ASTM.

**Table 15. Aluminum Electrical Conductors**

| Form, alloy,* treatment† | Tensile strength, ksi | Min electrical conductivity, % IACS |
|---|---|---|
| Redraw rod (0.375 in diam) (9.53 mm): | | |
| EC-O | 8.5–14 | 61.8 |
| EC-H12 | 12–17 | 61.5 |
| EC-H14 | 15–20 | 61.4 |
| EC-H16 | 17–22 | 61.3 |
| 5005-O | 14–20 | 54.3 |
| 5005-H12 | 17–23 | 54.0 |
| 5005-H14 | 20–26 | 53.9 |
| 5005-H16 | 24–30 | 53.8 |
| Wire (0.0801 in diam) (2.03 mm): | | |
| EC-H19 | 26 min | 61.0 |
| 5005-H19 | 37 | 53.5 |
| Bus conductor: | | |
| EC-H111 (extruded) | 8.5 min | 61.0 |
| EC-H17 (cold-finished) | 12 | 61.0 |
| 6101-T6 | 29 | 55.0 |
| 6061-T6 (extruded) | 38 | 43.0 |
| 6063-T6 (extruded) | 30 | 53.0 |

ksi × 6.895 = MN/m².
*Nominal compositions—EC: 99.6 min Al; 6101: 0.4 Si, 0.6 Mg; 5005: 0.8 Mg; 6061: 0.6
Si, 1.3 Mg, 0.25 Cu, 0.25 Cr; 6063: 0.4 Si, 0.7 Mg.
†Treatments: O = annealed; H-12, 14, 16, 19 = cold-worked; T-6 = solution-treated, aged.

Magnesium is the lightest metal of structural importance (108 lb/ft³) (1,730 kg/m³). Its principal applications are in aircraft, in portable and manually operated tools, and in moving and reciprocating parts of machinery (particularly textile and printing machinery and lawn mowers). Because of its chemical activity, it is used in pyrotechnical materials and for sacrificial galvanic protection of other metals exposed to corrosive mediums.

Commercially pure magnesium contains a minimum of 99.8 percent magnesium. Aluminum, iron, silicon, and manganese are the chief impurities. The major part of the magnesium produced is used in the form of magnesium-rich alloys. Aluminum is the chief addition, with smaller amounts of zinc and manganese being used in nearly all cases. These alloys can be heat-treated and aged to increase their strength. Designs employing magnesium should take into account the low value of the modulus of elasticity ($6.5 \times 10^6$ lb/in²) and the high thermal coefficient of expansion [$14 \times 10^{-6}$ per °F at 32°F (0°C) and $16 \times 10^{-6}$ for the range 68 to 752°F (20 to 400°C)]. See Tables 16 and 17 for compositions and properties.

Alloys of the Mg-Al-Zn system are most used for sand and permanent-mold castings. Alloy AZ63A is the most difficult to use. Pressure tightness is most easily obtained in castings of alloy EZ33A. Most permanent-mold castings are made with AM100A and AZ92A alloys. The casting alloy K1A was developed for applications requiring high damping capacity; its damping capacity exceeds that of cast iron. The normal unhindered shrinkage factor can be reduced to as low as ⅛ in/ft on large castings or where shrinkage is restricted by cores.

Magnesium-alloy **forgings** are used for applications requiring higher properties than are obtainable in castings. They are generally press-forged. Alloy AZ61 is a general-purpose alloy, while alloy AZ80 is used for the highest-strength press forgings of simple design. This alloy may be aged to increase its strength when required.

A wide range of **extruded shapes** is available in a number of alloy compositions. Alloys AZ31, AZ61, and AZ80 increase in cost and strength in the order named. Impact extrusion is now used for small symmetrical tubular parts.

**Sheet** is available in several alloys (see Table 17), in both the soft-annealed and the hard-rolled form. Magnesium-alloy sheet is usually hot-formed at temperatures between 400 and 650°F, although simple bends of large radius can be made cold.

The development of jet engines and high-velocity aircraft, missiles, and spacecraft has accelerated the development of new magnesium-base alloys with improved elevated-temperature properties. These alloys have been obtained by the addition of some combination of **rare earths**, manganese, zirconium, and thorium. Such alloys have extended the temperatures at which magnesium can be used in structural applications to as high as 700 to 800°F (370 to 425°C).

**Joining** Magnesium alloys may be joined by riveting or welding. Riveting is most widely employed. Aluminum-alloy rivets are used; 5052 is preferred since contact corrosion is minimized, although other alloys can be used (with some danger from contact corrosion). All rivets should be anodized to prevent such attack. Adhesive bonding is now an accepted method for joining magnesium. It offers advantages in weight saving, fatigue strength, and corrosion resistance.

**Arc welding** with inert-gas (helium or argon) shielding of the molten metal produces satisfactory joints. Butt joints are preferred, but any type of joint permissible for mild steel can be used. After welding, a stress-relief anneal is necessary to relieve the welding stresses. Typical times are 15 min at 500°F (260°C) for soft-annealed alloys and 1 h at 400°F (204°C) for hard-rolled alloys. (See also Sec. 13.)

**Machining** Magnesium in all its forms is a free-machining metal. Standard tools such as those used for brass and steel can be used with slight modification. Relief angles should be from

**Table 16. Typical Mechanical Properties of Magnesium Casting Alloys**

| Alloy | Condition or temper* | Al | Zn | Mn min | Zr | Other | Tensile strength, ksi | Tensile yield strength, ksi | Elongation, in 2 in | Shear strength, ksi | Strength, ksi — Tensile | Strength, ksi — Yield | Hardness BHN | Electrical conductivity, IACS‡ |
|---|---|---|---|---|---|---|---|---|---|---|---|---|---|---|
| **Sand-casting alloys:** | | | | | | | | | | | | | | |
| AM100A | −T6 | 10.0 | | 0.10 | | | 40 | 22 | 1 | 22 | 60 | 40 | 70 | 14 |
| AZ63A | −F | 6.0 | 3.0 | | | | 29 | 14 | 6 | 18 | 60 | 44 | 50 | 14 |
| | −T4 | | | | | | 40 | 13 | 12 | 17 | 60 | 40 | 55 | 12 |
| | −T5 | | | | | | 30 | 14 | 4 | 17 | 75 | 52 | 55 | |
| | −T6 | | | | | | 40 | 19 | 5 | 20 | 60 | 44 | 73 | |
| AZ81A | −T4 | 7.5 | 0.7 | 0.13 | | | 40 | 12 | 15 | 17 | 60 | 40 | 55 | 15 |
| AZ91C | −F | 8.7 | 0.7 | 0.13 | | | 24 | 14 | 2 | 18 | 60 | 44 | 52 | 12 |
| | −T4 | | | | | | 40 | 12 | 14 | 17 | 75 | 52 | 53 | 13 |
| | −T5 | | | | | | 26 | 17 | 3 | 20 | 50 | 46 | | 11 |
| | −T6 | | | | | | 40 | 19 | 5 | 18 | 68 | 46 | | |
| AZ92A | −F | 9.0 | 2.0 | 0.10 | | | 24 | 14 | 2 | 20 | 50 | 46 | 66 | 13 |
| | −T4 | | | | | | 40 | 14 | 9 | 19 | 80 | 65 | 65 | 12 |
| | −T5 | | | | | | 26 | 16 | 2 | 22 | 57 | 40 | 63 | 10 |
| | −T6 | | | | | | 40 | 21 | 2 | 22 | 61 | 40 | 84 | |
| EZ33A | −T5 | | 2.5 | | 0.8 | 3.5 RE† | 23 | 15 | 3 | 22 | 60 | 37 | 50 | 25 |
| HK31A | −T6 | | | | 0.8 | 3.5 Th | 32 | 15 | 8 | 22 | | | 55 | 22 |
| HZ32A | −T5 | | 2.0 | | 0.8 | 3.5 Th | 30 | 14 | 7 | | | | 57 | 27 |
| K1A | −F | | | | 0.7 | | 25 | 7 | 19 | 8 | | | | 31 |
| QE22A | −T6 | | | | 0.7 | 2.0 RE, 2.5 Ag | 40 | 30 | 4 | 23 | 70 | 51 | 62 | 25 |
| ZE41A | −T5 | | 4.0 | | 0.7 | 1.3 RE | 30 | 20 | 3.5 | 22 | 72 | 49 | 70 | 31 |
| ZH62A | −T5 | | 5.5 | | 0.8 | 1.8 Th | 40 | 25 | 6 | 23 | 72 | 47 | 65 | 27 |
| ZK51A | −T5 | | 4.5 | | 0.8 | | 40 | 24 | 8 | 22 | | | 65 | 27 |
| ZK61A | −T6 | | 6.0 | | 0.8 | | 45 | 28 | 10 | 26 | | | 70 | 27 |
| **Permanent-mold-casting alloys:** | | | | | | | | | | | | | | |
| AM100A | −F | 10.0 | | 0.10 | | | 22 | 12 | 2 | 18 | | | 53 | 12 |
| | −T4 | | | | | | 40 | 13 | 10 | 20 | | | 52 | 10 |
| | −T6 | | | | | | 40 | 16 | 4 | 21 | | | 60 | 14 |
| | −T61 | | | | | | 40 | 22 | 1 | 22 | | | 70 | 15 |
| AZ81A | −T4 | 7.5 | 0.7 | 0.13 | | | 40 | 14 | 2 | 18 | | | 55 | 12 |
| AZ91C | −F | 8.7 | 0.7 | 0.13 | | | 24 | 12 | 14 | 17 | | | 52 | 13 |
| | −T4 | | | | | | 40 | 17 | 3 | | | | 53 | 11 |
| | −T5 | | | | | | 26 | 19 | 5 | | | | 66 | |
| | −T6 | | | | | | 40 | 14 | 2 | | | | 65 | |
| AZ92A | −F | 9.0 | 2.0 | 0.10 | | | 24 | 14 | 10 | 20 | | | 63 | 13 |
| | −T4 | | | | | | 40 | 16 | 1 | 18 | | | 69 | 12 |
| | −T5 | | | | | | 25 | 21 | 2 | 20 | | | 80 | 10 |
| | −T6 | | | | | | 40 | | 3 | 19 | | | 50 | |
| EZ33A | −T5 | | 2.5 | | 0.8 | 3.5 RE | 23 | 15 | 3 | 22 | | | 66 | 14 |
| HK31A | −T6 | | | | 0.7 | 3.0 Th | 32 | 15 | 8 | | | | | 25 |
| QE22A | −T6 | 78 | | | 0.7 | 2.0 RE | 38 | 28 | 3 | | | | 78 | 22 |
| **Die-casting alloys:** | | | | | | | | | | | | | | |
| AZ91A | −F | | | | | | 33 | 22 | 3 | 22 | | | 63 | |
| AZ91B | −F | | | | | | 33 | 22 | 3 | 22 | | | 63 | |

ksi × 6.895 = kN/m².

*−F = as cast; −T4 = artificially aged; −T5 = solution heat-treated; −T6 = solution heat-treated

†RE = rare-earth mixture.

‡Percent electrical conductivity/100 approximately equals thermal conductivity in cgs units.

6-83

**Table 17. Properties of Wrought-Magnesium Alloys**

| Alloy and temper* | Al | Mn | Th | Zn | Zr | Density, lb/in³ | Thermal conductivity, cgs units, 68°F (20°C) | Electrical resistivity, μΩ·cm, 68°F (20°C) | Tensile strength, ksi | Tensile yield strength, ksi | Elongation in 2 in, % | Compressive yield strength, ksi | Shear strength, ksi | Strength, ksi — Tensile | Strength, ksi — Yield | Hardness, BHN |
|---|---|---|---|---|---|---|---|---|---|---|---|---|---|---|---|---|
| **Extruded bars, rods, shapes** | | | | | | | | | | | | | | | | |
| AZ31B-F | 3.0 | | | 1.0 | | 0.0639 | 0.18 | 9.2 | 38 | 29 | 15 | 14 | 19 | 56 | 33 | 49 |
| AZ61A-F | 6.5 | | | 1.0 | | 0.0647 | 0.14 | 12.5 | 45 | 33 | 16 | 19 | 22 | 68 | 40 | 60 |
| AZ80A-T5 | 8.5 | | | 0.5 | | 0.0649 | 0.12 | 14.5 | 55 | 40 | 7 | 35 | 24 | 60 | 58 | 82 |
| ZK60A-F | | | | 5.7 | 0.5 | 0.0659 | 0.29 | 6.0 | 49 | 38 | 14 | 33 | 24 | 76 | 56 | 75 |
| ZK60A-T5 | | | | 5.7 | 0.5 | | | | 53 | 44 | 11 | 36 | 26 | 79 | 59 | 82 |
| LA141A-T7 (Li 14) | | | | | | 0.049 | 0.105 | 15.2 | 21 | 17 | 25 | | | | | |
| ZE10A-O (RE 0.2) | | | | | | 0.0635 | 0.33 | 5.0 | 32 | 15 | 12 | | | | | |
| ZE10A-H24 | | | | | | | | 5.2 | 34 | 23 | 4 | | | | | |
| **Extruded tube** | | | | | | | | | | | | | | | | |
| AZ31B-F | 3.0 | | | 1.0 | | 0.0639 | 0.18 | 9.2 | 36 | 24 | 16 | 12 | | | | 46 |
| AZ61A-F | 6.5 | | | 1.0 | | 0.0647 | 0.14 | 12.5 | 41 | 24 | 14 | 16 | | | | 50 |
| ZK60A-F | | | | 5.7 | 0.5 | 0.0659 | 0.28 | 6.0 | 47 | 35 | 13 | 25 | | | | 75 |
| ZK60A-T5 | | | | 5.7 | 0.5 | 0.0659 | 0.29 | 5.7 | 50 | 40 | 11 | 30 | | | | 82 |
| M1A | | 1.2 | | | | 0.0635 | 0.31 | 5.4 | 37 | 26 | 11 | 12 | | | | |
| **Sheet and plate** | | | | | | | | | | | | | | | | |
| AZ31B-H24 | 3.0 | | | 1.0 | | 0.0639 | 0.18 | 9.2 | 42, 40, 39 | 32, 29, 27 | 15, 17, 19 | 26, 23, 19 | 29, 28, 27 | 77, 72, 70 | 47, 45, 40 | 73 |
| AZ31B-O | 3.0 | | | 1.0 | | 0.0639 | 0.18 | 9.2 | 37 | 22 | 21 | 16 | 26 | 66 | 37 | |
| HK31A-H24 | | | 3.0 | | 0.7 | 0.0647 | 0.27 | 6.1 | 38, 37, 39 | 30, 28, 31 | 9, 10, 14 | 23, 23, 25 | 26, 26, 27 | 67, 65, 68 | 41, 41, 44 | 56, 57, 57 |
| HK31A-O | | | 3.0 | | 0.7 | 0.0647 | 0.25 | 6.6 | 33 | 20 | 23 | 14 | 24 | 58 | 28 | |
| HM21A-T8 | | 0.6 | 2.0 | | | 0.0640 | 0.33 | 5.0 | 35, 37 | 23, 27 | 11, 12 | 19, 23 | 19 | 63, 67 | 37, 41 | 56 |
| M1A | | 1.2 | | | | 0.0635 | 0.31 | 5.4 | 37 | 26 | 11 | 12 | | | | |
| **Tooling plate** | | | | | | | | | | | | | | | | |
| AZ31-B | 3.0 | | | 1.0 | | 0.0639 | 0.18 | 9.2 | 35 | 19 | 12 | 10 | | | | |
| **Tread plate** | | | | | | | | | | | | | | | | |
| AZ31B | 3.0 | | | 1.0 | | 0.0639 | 0.18 | 9.2 | 35 | 19 | 14 | 11 | | | | 52 |

See Sec. 1 for conversion factors to SI units.

NOTE: For all above alloys: coefficient of thermal expansion = 0.0000145; modulus of elasticity = 6,500,000 lb/in²; modulus of rigidity = 2,400,000 lb/in²; Poisson's ratio = 0.35.

*Temper: -F = as fabricated; -O = fully annealed; -H24 = strain-hardened, then partially annealed; -T5 = artificially aged; -T8 = solution heat-treated, cold-worked, then artificially aged.

7 to 12°, and rake angles from 0 to 15°. High-speed steel is satisfactory and is used for most drills, taps, and reamers. The hard grades of cemented carbides are better for production work and should be used where the tool design permits it. Finely divided magnesium constitutes a fire hazard, and good housekeeping in the machine shop is essential. (See also Sec. 13.)

**Corrosion Resistance and Surface Protection** Magnesium alloys display good resistance to ordinary inland atmospheric exposure, to most alkalies, and to many organic chemicals. Marine atmospheric exposure and most acids and salts attack them rapidly. Galvanic couples formed by contact with other metals, or by impregnation of the surface with other metals during fabrication, can cause rapid attack of the magnesium under conditions of wet corrosion. Protective treatments are available. Prior to shipment, most magnesium parts are pickled in a solution of sodium dichromate and nitric acid (chrome pickle); this forms a protective film on the surface and makes an excellent base for painting. Other protective and decorative finishes are also available.

## ZINC AND ZINC ALLOYS

REFERENCES: Publications of the New Jersey Zinc Co., Metals Handbook, ASM, 1975 Annual Standards, Parts 7, 8, and 9, ASTM.

Zinc, one of the least expensive nonferrous metals, is produced from sulfide, silicate, or carbonate ores by a process involving concentration and roasting followed either by reduction of the zinc ore by carbon and simultaneous distillation of the zinc in batch or continuous retorts or by leaching out the oxide with sulfuric acid and electrolyzing the solution after purification. Distilled zinc contains impurities (principally Pb, Cd, and Fe) that may be eliminated by fractional redistillation to produce zinc of 99.99+ percent purity. Metal of equal purity can be produced by the electrolytic process. Zinc reaches the market in the form of slabs, 1 to 1½ in (25.4 to 38.1 mm) thick, 8½ to 10 in (210 to 254 mm) wide, 18 to 20 in (457 to 508 mm) long. In this form, it is frequently called **spelter.**

The important grades of zinc available in the United States are covered by ASTM specification B6, Table 18.

The special high-grade zinc is used in the manufacture of die castings, where impurities have a marked harmful effect on corrosion resistance and dimensional stability. **Galvanizing** (which consumes by far the largest proportion of zinc) utilizes principally prime western zinc, but large tonnages of high-grade zinc are used in continuous galvanizing mills. All grades are used for rolled-zinc products, as the presence of impurities is often desirable for their strengthening effect. For brass manufacture and other alloys, there is an increasing tendency toward the use of high grades of zinc.

**Wrought Zinc** Zinc rolled in the form of sheet, strip, or plate of various thicknesses is used extensively. It is usually made from commercial slab zinc. It is produced by hot rolling unless some stiffness and temper are required, in which case one or more of the finishing passes are done cold. The softer, purer grades of zinc are used for deep drawing or forming operations, and the less pure metal is used for weather strip, roofing, and other applications where some stiffness is necessary, or where specific chemical properties are desired, as in photoengravers' plates.

Nearly all rolled zinc is custom-rolled to meet the customer's requirements (see ASTM B69). Duplication of characteristics is obtained by careful control of composition and rolling treatment and by application of control tests such as dynamic ductility, temper, hardness, and dynamic bend. Fundamental data for structural design may be obtained from creep tests at room temperature. Zinc should not be used in applications where continuous high stresses are involved.

**Alloys of zinc** containing 0.65 to 1.25 percent copper are significantly stronger than unalloyed zinc and possess good ductility and working properties. They can be work-hardened and may be employed for parts that must withstand loads somewhat higher than would be permissible in unalloyed zinc. The addition of about 0.01 percent magnesium (*Trans. AIME*, 1930, p. 481) to this alloy increases the creep resistance considerably, and the alloy finds some application for roofing and the like with design stresses up to 10,000 lb/in². Magnesium additions, however, decrease the ductility and general fabricating characteristics.

The usual tensile test is practically meaningless with zinc because the creep that occurs even at quite small loads causes the breaking load to vary with testing speed; the results are unrelated to service conditions. The speed in tensile testing is usually controlled at 0.25 in (6.3 mm)/min, under which conditions the soft grades of rolled zinc will have a tensile strength of 16,000 to 19,000 lb/in², hard-rolled impure zinc 19,000 to 26,000 lb/in², and the zinc-copper-magnesium alloy in the cold-worked condition as much as 50,000 lb/in². The elongation will vary between 5 and 65 percent but bears no direct relation to formability because of the different speed of testing. The properties of zinc vary with the direction of testing, and the across-grain tensile values will be approximately 20 percent higher than those obtained with the direction of the grain, although the elongations are correspondingly lower. A comparatively new series of zinc alloys containing titanium has been developed for applications requiring increased strength and creep resistance and/or low thermal expansion. A typical analysis is copper 0.5 to 0.8, titanium 0.08 to 0.16, and as maximum values 0.20 lead, 0.015 iron, 0.01 cadmium, 0.01 manganese, 0.02 chromium.

Zinc strip or sheet can be fabricated by the usual methods,

**Table 18. ASTM Specification B6-70 for Slab Zinc**

| Grade | Max % | | | Zinc, min % |
|-------|-------|------|---------|-------------|
| | Lead | Iron | Cadmium | |
| Special high grade | 0.003 | 0.003 | 0.003 | 99.990 |
| High grade | 0.07 | 0.02 | 0.03 | 99.9 |
| Intermediate | 0.20 | 0.03 | 0.40 | 99.5 |
| Brass special | 0.6 | 0.03 | 0.50 | 99.0 |
| Prime western | 1.6 | 0.05 | 0.50 | 98.0 |

**cupping, forming, etc.,** provided it is not at too low a temperature. A take-in of 40 percent on the first cupping operation is usual. Warm soapy water is widely used as a lubricant. The soft grades are self-annealing at room temperature, and only the harder alloys need intermediate annealing between operations as most other metals do. When necessary, the hard zincs are annealed at 212°F (100°C) and the zinc-copper alloys at about 440°F (227°C). Welding is possible, and soldering is exceptionally easy. Simple extrusion of rods, molding, and tubing is possible but expensive because of the slow speeds necessary. The impact extrusion process, however, is being more and more widely used for producing battery cups and similar articles.

Zinc is resistant to atmospheric **corrosion** but is attacked by acids and alkalies. Soap tends to inhibit the action of water. Surface finishes for corrosion resistance or improving the appearance are readily applied. These include electroplating with copper, nickel, and chromium, lacquering, enameling, or chemically coating.

**Zinc Die Castings**  Zinc alloys are particularly suited for making die castings since the melting point is reasonably low, resulting in long die life even with ordinary steels, and a high accuracy and good surface finish are possible.

The alloys at present used for die castings in the United States are practically limited to those covered by the ASTM specification B86. Nominal compositions and typical properties of die-case test pieces are given in Table 19. The low limits of impurities are necessary to avoid disintegration of the castings by intergranular corrosion under moist atmospheric conditions. The presence of magnesium prevents this effect if the impurities are not higher than the specification values. The mechanical properties in Table 19 are average figures for die-cast tensile-test pieces of 0.25 in diam or impact specimens 0.25 in square. Specification values for these properties, if used, would naturally be lower than the typical ones quoted, and in the case of the actual castings, considerable variations must be expected.

The zinc die-casting alloys are somewhat similar in general properties. Alloy AG40A is distinguished by excellent retention of impact strength and dimension. Alloy AC41A has somewhat greater impact strength and some growth in dimension when used at temperatures near 212°F. For normal-temperature service, there is little choice between these two alloys.

A measurement of the expansion of the die casting after exposure to water vapor at 203°F (95°C) for 10 days is a suitable index of stability and freedom from susceptibility to intergranular corrosion, but analysis for impurity content is more widely used.

**Aging of Die Castings**  Because of changes occurring in the structure of zinc die castings, they commence to shrink immediately after removal from the mold, the change being about two-thirds complete in 5 weeks. The maximum extent of this is about 0.001 in/in. Alloy AG40A (copper-free) is unaffected, and alloy AC41A, with 1 percent copper, is not greatly affected at room temperature. AC40A may be partly stabilized with respect to shrinkage by heating for 3 to 6 h at 212°F (100°C). The castings should be at temperature for this period of time and may be cooled normally in air to room temperature.

**Effect of Temperature on Zinc and Zinc Alloys**  The properties of zinc and zinc alloys are very sensitive to temperature. Creep resistance decreases rapidly with increasing temperature, and this must be considered in designing articles to withstand continuous loads.

Ductility and general fabricating characteristics increase with temperature. Forming and drawing operations on strip or sheet zinc should not be attempted below 70°F (21°C) and the more severe operations can be performed more readily at somewhat higher temperatures [up to 125°F (52°C)].

Zinc and zinc alloys become somewhat brittle below the range 0 to 32°F (−18 to 0°C), depending on the particular composition, but recover their normal properties on reaching room temperature again. Even at low temperature, the die-

### Table 19. Properties of Zinc-Base Die-Casting Alloys

| | Alloy AG40A | Alloy AC41A |
|---|---|---|
| Composition, %: | | |
|   Copper | 0.25, max | 0.75–1.25 |
|   Aluminum | 3.5–4.3 | 3.5–4.3 |
|   Magnesium | 0.020–0.05 | 0.03–0.08 |
|   Iron, max | 0.100 | 0.100 |
|   Lead, max | 0.005 | 0.007 |
|   Cadmium, max | 0.004 | 0.005 |
|   Tin, max | 0.003 | 0.005 |
|   Zinc | Remainder | Remainder |
| Typical properties: | | |
|   Tensile strength, 1,000 lb/in$^2$ | 35 | 40 |
|   Elongation, % in 2 in | 10 | 5 |
|   Brinell Hardness, 500 kg/10-mm ball | 65 | 80 |
|   Charpy impact, ft·lb | 35 | 35 |
|   Electrical resistivity, 77°F, $\mu\Omega$·cm | 64 | 67 |
|   Electrical conductivity, 77°F $\times$ 10$^3$ mho·cm$^3$ | 155 | 151 |
|   Thermal conductivity, Btu/(ft)(ft$^2$)(h)(°F) | 66 | 62 |
|   Thermal expansion $\times$ 10$^{-6}$/°F | 15.2 | 15.2 |
|   Density, lb/ft$^3$ | 410 | 415 |
|   Melting point, °F | 728 | 727 |

See Sec. 1 for conversion factors to SI units

casting alloys have residual impact strength superior to ordinary cast iron.

## TITANIUM AND ZIRCONIUM ALLOYS

REFERENCES: "Handbook on Titanium Metal," Titanium Metals Corp. of America, New York. Abkowitz, Burke, and Hiltz, "Titanium in Industry," Van Nostrand. McQuillan and McQuillan, "Titanium," Butterworth.

Although **titanium** and titanium alloys have been commercially available for less than 25 years, they have already become important structural metals because of their unusual combination of properties. These alloys have strengths comparable with alloy steels while the weight is only 60 percent that of steel. In addition, the corrosion resistance of titanium alloys is far superior to aluminum and even exceeds that of stainless steel under most conditions, particularly those involving saltwater spray. Titanium's low magnetic permeability is also notable.

Titanium is prepared by calcium or magnesium reduction of the chlorinated oxide to yield "sponge," which is then melted in consumable-electrode arc furnaces to ingots. Electrolytic methods of recovery have been studied experimentally. Material of special purity may be prepared by powder-metallurgy techniques or by thermal decomposition of the iodide. Titanium is one of the most abundant metals in the earth's crust, and as costs are reduced by improvements in production methods, consumption should rise rapidly.

The allotropic transformation in titanium from HCP to BCC at about 1560°F (849°C) affords opportunity for property variation by heat treatment comparable with that for steels. The various titanium alloys are usually classified in terms of the crystal structure: commercially pure, all-alpha (HCP) weldable, alpha-beta (two-phase) weldable, alpha-beta nonweldable, and all-beta (BCC). The mechanical properties of titanium and its alloys, particularly commercially pure titanium, depend markedly on the content of fractional percentages of C, O, N, H, and Fe. (See Tables 20 and 21.)

**Zirconium** metal's most important application was formerly as a getter (i.e., absorber) for gases in electronic tubes. The chief consumption of zirconium at present is in nuclear energy where its combination of high corrosion resistance and low neutron-absorption cross section offers special advantages in reactor tubing (Zircoloy). For getter applications parts of the tube are built of zirconium, and the construction is such that they will be heated during operation of the tube and thus be able to absorb gas continuously. Zirconium is also used in chemical equipment, superconductors, surgical materials (skull plates, screws, pegs), and as an alloying element (zirconium copper). It has a chemical inertness similar to that of tantalum. The metallurgical characteristics of zirconium are similar to those of titanium in many respects.

## LOW-MELTING METALS AND ALLOYS

REFERENCES: Metals Handbook, ASM; Mantell, "Engineering Materials Handbook," McGraw-Hill, 1958

Metals with low melting temperatures offer a diversity of industrial applications. In this field, much use is made of the **eutectic**-type alloy, in which two or more elements are combined in proper proportion so as to have a minimum melting temperature. Such alloys melt at a specific temperature, as does a pure metal, rather than over a range of temperature, as with most alloys.

**Liquid Metals**   A few metals are used in their liquid state. **Mercury** [mp, $-39.37°F (-40°C)$] is the only metal that is liquid below room temperature. In addition to its use in thermometers, scientific instruments, and electrical contacts, it is a constituent of some very low-melting alloys. Its application in dental amalgams is unique and familiar to all. It has been used as a heat-exchange fluid, as have **sodium** and the sodium-potassium alloy **NaK** (see Sec. 6).

**Tin**   Moving up the temperature scale, there is tin [mp, 449.4°F (232°C)], the largest single use of which is for coating steel to make **tinplate; solders** and **bronze** are the next largest uses. Tin may be applied to steel, copper, or cast iron either by hot dipping in a molten bath or by electrodeposition. Normally thinner tin coatings are achieved by hot tinning. Electrodeposited tin is a matte coating, but it can be bright-

**Table 20. Typical Compositions of Titanium Alloys**

| Alloy designation | C | Nominal composition, weight % | | | | | | | |
|---|---|---|---|---|---|---|---|---|---|
| | | N | O | Al | Fe | Mn | Mo | V | Other |
| 1 Iodide titanium | 0.01 | 0.002 | 0.005 | 0.02 | 0.01 | 0.01 | 0.001 | | |
| 2 A-40 | 0.20 | | | | | | | | |
| 3 Ti-65A | 0.10 | 0.05 | 0.015 | | 0.12 | | | | |
| 4 Ti-75A | 0.10 | 0.08 | | | 0.20 | | | | |
| 5 Ti-100A | 0.07 | | | | 0.30 | | | | |
| 6 A110 At (α, W) | 0.20 | | | 5 | | | | | 2.5 Sn |
| 7 Ti-6Al-4V* (αβ, W) | 0.10 | 0.05 | | 6 | 0.25 | 0.10 | | 4 | |
| 8 Ti-8Al-1Mo-1V (αβ, W) | | | | 8 | | | 1 | 1 | |
| 9 Ti-4Al-3Mo-1V* (αβ) | | | | 4 | | | 3 | 1 | |
| 10 Ti-6Al-6V-2Sn* (αβ) | | | | 6 | | | | 6 | 2.0 Sn |
| 11 Ti-8-Mn (αβ) | 0.20 | | | | | 8 | | | |
| 12 C 130 AM (αβ) | 0.20 | | | 4 | | 4 | | | |
| 13 RS-140 (αβ) | 0.20 | | | 5 | 1.25 | | | | 2.75 Cr |
| 14 Ti-155A (αβ) | 0.10 | 0.08 | | 5 | 1.5 | | 1.2 | | 1.4 Cr |
| 15 Ti-13V-11Cr-3Al* (β, W) | | | | 3 | | | | 13 | 11.0 Cr |

NOTE: α = all alpha; β = all beta; αβ = both α and β; W = weldable.
*Also available in higher-strength (20,000 to 40,000 lb/in²) heat-treated condition.

**Table 21. Mechanical Properties of Titanium Alloys (Annealed Condition)**

| Alloy No. (see Table 20) | Room temp, 68°F (20°C) | | | | | 400°F (204°C) | | | 800°F (427°C) | | | Max recom. temp. for 1,000 hr service | |
|---|---|---|---|---|---|---|---|---|---|---|---|---|---|
| | Yield strength, ksi | Tensile strength, ksi | Elong., % in 2 in | Hardness† | V-notch Charpy impact, ft:lb | Yield strength, ksi | Tensile strength, ksi | Elong., % in 2 in | Yield strength, ksi | Tensile strength, ksi | Elong., % in 2 in | °F | °C |
| 1 | 15 | 35 | 55 | VPN 60 | 150 | 21 | 37 | 40 | 12 | 25 | 22 | 1000 | 538 |
| 2 | 45 | 50 | 22 | | 90 | 36 | 53 | 29 | 19 | 31 | 21 | 1000 | 538 |
| 3 | 55 | 65 | 20 | R_c 30 | | | | | | | | 1000 | 538 |
| 4 | 70 | 80 | 15 | R_b 98 | 25 | | | | | | | | |
| 5 | 90 | 100 | 15 | R_c 30 | | | | | | | | | |
| 6 | 100 | 115 | 10 | | 20 | 82 | 100 | 17 | 63 | 82 | 16 | 1200 | 649 |
| 7* | 120 | 130 | 10 | R_c 34 | 18 | 102 | 110 | 15 | 79 | 96 | 15 | 850 | 454 |
| 8 | 125 | 135 | 10 | | | 94 | 101 | 20 | 75 | 81 | 18 | | |
| 9 | 155 | 180 | 5 | R_c 40 | | 136 | 157 | 6 | 111 | 140 | 8 | | |
| 10 | 170 | 170 | 8 | | | 144 | 153 | 15 | | | | | |
| 11 | 110 | 120 | 10 | | | 85 | 110 | 18 | 55 | 85 | 20 | | |
| 12 | 130 | 140 | 10 | | 15 | 110 | 120 | 17 | 88 | 105 | 22 | 800 | 427 |
| 13 | 140 | 150 | 10 | | 15 | 120 | 140 | 17 | 93 | 117 | 19 | 700 | 371 |
| 14 | 140 | 150 | 12 | R_c 38 | 15 | 110 | 130 | 17 | 91 | 109 | 21 | 650 | 343 |
| 15 | 160 | 170 | 4 | | | 142 | 163 | 8 | 128 | 137 | 10 | | |

See Sec. 1 for conversion factors for SI units.
*Also available in higher-strength (20,000 to 40,000 lb/in²) heat-treated condition.
†VPN = Vickers pyramid hardness scale; R_c = Rockwell C hardness scale.

**Table 22. Composition Specifications for Lead, Percent**

| | Corroding lead | Chemical lead* | Common desilverized lead A | Common desilverized lead B | Copper lead |
|---|---|---|---|---|---|
| Silver, max........................ | 0.0015 | 0.020 | 0.002 | 0.002 | 0.020 |
| Silver, min........................ | ...... | 0.002 | | | |
| Copper, max....................... | 0.0015 | 0.080 | 0.0025 | 0.0025 | 0.080 |
| Copper, min....................... | ...... | 0.040 | ...... | ...... | 0.040 |
| Silver, plus copper, max.............. | 0.0025 | | | | |
| Arsenic, max....................... | 0.0015 | | | | |
| Antimony plus tin, max.............. | 0.0095 | | | | |
| Arsenic, antimony, and tin together, max................................ | ...... | 0.002 | 0.015 | 0.015 | 0.015 |
| Zinc, max.......................... | 0.0015 | 0.001 | 0.002 | 0.002 | 0.002 |
| Iron, max.......................... | 0.002 | 0.002 | 0.002 | 0.002 | 0.002 |
| Bismuth, max....................... | 0.05 | 0.005 | 0.15 | 0.25 | 0.010 |
| Lead (by difference) min............. | 99.94 | 99.90 | 99.85 | 99.73 | 99.85 |

*Chemical lead designates undesilverized lead from southeast Missouri ores.

ened to the brilliance of hot-dipped coatings by reflowing in hot oils or, in the case of tinplate, by induction, convection, or radiation heating. Alloys of 12 to 25 percent tin, balance lead, are applied to steel by hot dipping and are known as **terneplate.**

Modern **pewter** is a tarnish-resistant alloy used only for ornamental ware and is composed of 91 to 93 percent tin, 6 to 7 percent antimony, and 1 to 2 percent copper. Old pewter, first used about 200 years ago, contained sufficient lead to cause the surface to darken with age.

**Lead** Several varieties of pig lead are recognized and are shown in Table 22. **Corroding lead** is the highest-purity commercial lead and is used for making white lead. **Chemical lead** is extensively employed in chemical plants for withstanding corrosion, particularly in sulfuric acid. Its copper content confers added stiffness. **Common lead** is the usual grade for alloying.

**Antimonal lead** is used in places where greater strength is needed. For storage-battery plates, lead containing 6 to 7 percent antimony is used. In the cast condition, this has a tensile strength of about 7,000 lb/in² with an elongation of about 22 percent, density 677 lb/ft³ (1,080 kg/m³). Lead for sheathing telephone and electric-power cables is usually made of an alloy containing about 1 percent antimony. This alloy, when extruded as cable sheath and aged 1 month at room temperature, has a tensile strength of 2,750 to 3,050 lb/in² at a testing speed of 0.25 in/min/in of free length, elongation 30 to 40 percent, and endurance limit 800 lb/in² (50 million cycles at 700 per min).

**Cast lead-antimony alloys** containing 6 to 14 percent antimony have a tensile strength of 7,000 to 8,000 lb/in², with elongation decreasing from 24 to 10 percent. The lead-antimony alloys, particularly in the range 2 to 8 percent antimony, are susceptible to heat treatment, which considerably increases their strength. This treatment is rarely employed in practice. An alloy also used for cable sheathing is ordinary chemical lead with about 0.06 percent copper. Alloys with 0.1 percent or less tellurium or 0.01 to 0.10 percent calcium have been proposed for special purposes where higher creep and fatigue resistance are needed. Alloys with larger amounts of calcium (0.8 percent) and smaller quantities of alkali metals are used to a limited extent as bearing metals. Lead for coating steel (terneplate) and copper contains 5 to 25 percent tin to aid adhesion to the base metal (see Tin, above). Lead is an important constituent of alloys with tin and copper and of type metals.

**Type Metals** The principal type metals are listed in Table 23; composition variations are encountered in practice. Electrotype metal serves only as a backing to the shell and does not need to be hard. In the linotype machine, the metal must be fluid and capable of rapid solidification; hence metal of very nearly eutectic composition is used. It is rarely used as the actual printing surface and, therefore, need not be so hard as stereotype and monotype metals.

**Fusible Alloys** (See Table 24.) These alloys are used typically as fusible links in sprinkler heads, as electric cutouts, as fire-door links, for making castings, for patterns in making match plates, for making electroforming molds, for setting punches in multiple dies, and for dyeing cloth. Some fusible alloys can be cast or sprayed on wood, paper, and other materials without damaging the base materials, and many of these alloys can be used for making hermetic seals. Since some of these alloys melt below the boiling point of water, they can be used in bending tubing. The properly prepared tubing is filled with the molten alloy and allowed to solidify, and after bending, the alloy is melted out by immersion of the tube in

**Table 23. Composition and Properties of Type Metal***

| Service | Composition, % | | | Melting point | | Brinell hardness |
|---|---|---|---|---|---|---|
| | Sn | Sb | Pb | °F | °C | |
| Electrotype | 4 | 4 | 92 | 570 | 299 | 14 |
| Linotype | 5 | 16 | 79 | 475 | 246 | 22 |
| Stereotype | 4 | 11 | 83.7 | 500 | 260 | 24 |
| Monotype | 8 | 16 | 76 | 515 | 268 | 26 |

* "Metals Handbook", vol. 1, ASM.

boiling water. The volume changes during the solidification of a fusible alloy are, to a large extent, governed by the bismuth content of the alloy. As a general rule, alloys containing more than about 55 percent bismuth expand and those containing less than about 48 percent bismuth contract during solidification; those containing 48 to 55 percent bismuth exhibit little change in volume. The change in volume due to cooling of the solid metal is a simple linear shrinkage, but some of the fusible alloys owe much of their industrial importance to other volume changes, caused by change in structure of the solid alloy, which permit the production of castings having dimensions equal to, or greater than, those of the mold in which the metal was cast.

For fire-sprinkler heads with a rating of 160°F (71°C) **Wood's metal** is used for the fusible-solder-alloy link. Wood's metal gives the most suitable degree of sensitivity at this tempera-ture, but in tropical countries and in situations where industrial processes create a hot atmosphere (e.g., baking ovens, foundries), solders having a higher melting point must be used. Alloys of eutectic compositions are used since they melt sharply at a specific temperature.

Fusible alloys are also used as molds for thermoplastics, for the production of artificial jewelry in pastes and plastic materials, in foundry patterns, chucking glass lenses, as hold-down bolts, and inserts in plastics and wood.

**Solders** are nonferrous filler metals used in a joining process wherein coalescence between metal parts is produced by heating to suitable temperatures below those of the base metals and generally below 800°F (427°C). (See Table 25.) The eutectic composition (63 percent tin) has the lowest melting point [361°F (183°C)] of the binary tin-lead solders. For joints in copper pipe and cables, a wide melting range is needed; a 50

**Table 24. Fusible Alloys**

| Alloy composition, percent | | | | | Melting range, deg F | | |
|---|---|---|---|---|---|---|---|
| Sn | Bi | Pb | Cd | Others | Solidus | Eutectic | Liquidus |
| 12.0 | .... | ..... | .... | 82.0 Ga, 6.0 Zn | ... | 63† | |
| 8.0 | .... | ..... | .... | 92.0 Ga | ... | 68† | |
| 8.3 | 44.7 | 22.6 | 5.3 | 19.1 In | ... | 117† | |
| 12.0 | 49.0 | 18.0 | .... | 21.0 In | ... | 136† | |
| 12.77 | 48.0 | 25.63 | 9.6 | 4.0 In | 142 | .... | 149 |
| 13.2 | 49.3 | 26.3 | 9.8 | 1.4 Ga | 149 | .... | 151 |
| 13.1 | 49.5 | 27.3 | 10.1 | ................. | ... | 158† | |
| 12.5 | 50.0 | 25.0 | 12.5 | (Wood's) | 158 | .... | 162 |
| 13.3 | 50.0 | 26.7 | 10.0 | (Lipowitz's) | 158 | .... | 163 |
| 13.0 | 42.0 | 35.0 | 10.0 | ................. | 158 | .... | 176 |
| 13.0 | 40.0 | 37.0 | 10.0 | ................. | 158 | .... | 185 |
| 24.5 | 45.3 | 17.9 | 12.3 | ................. | 158 | .... | 190 |
| 11.3 | 42.5 | 37.7 | 8.5 | ................. | 158 | .... | 194 |
| 15.4 | 38.4 | 30.8 | 15.4 | ................. | 158 | .... | 207 |
| .... | 51.7 | 40.2 | 8.1 | ................. | ... | 198† | |
| 15.5 | 52.5 | 32.0 | .... | ................. | ... | 205† | |
| 20.0 | 50.0 | 30.0 | .... | (Onion's or Lichtenberg's) | 205 | .... | 212 |
| 18.8 | 50.0 | 31.2 | .... | (Newton's) | 205 | .... | 207 |
| 25.0 | 50.0 | 25.0 | .... | (D'Arcet's) | 205 | .... | 208 |
| 22.0 | 50.0 | 28.0 | .... | (Rose's) | 205 | .... | 230 |
| 34.2 | 46.1 | 19.7 | .... | (Malotte's) | 205 | .... | 253 |
| 33.0 | 34.0 | 33.0 | .... | ................. | 205 | .... | 289 |
| 25.9 | 53.9 | ..... | 20.2 | ................. | ... | 217† | |
| 25.0 | 50.0 | ..... | 25.0 | ................. | 217 | ... | 235 |
| 34.5 | 44.5 | ..... | 21.0 | ................. | 217 | ... | 248 |
| 14.5 | 48.0 | 28.5 | .... | 9.0 Sb | 217 | ... | 440 |
| 1.0 | 55.0 | 44.0 | .... | ................. | 243 | ... | 248 |
| 50.0 | .... | ..... | .... | 50.0 In | 243 | ... | 260 |
| 48.0 | .... | ..... | .... | 52.0 In | ... | 243† | |
| .... | 55.5 | 44.5 | .... | ................. | ... | 255† | |
| 46.0 | .... | ..... | 17.0 | 37.0 Tl | ... | 262† | |
| 40.0 | 56.0 | ..... | .... | 4.0 Zn | ... | 266† | |
| 41.6 | 57.4 | 1.0 | .... | ................. | 273 | .... | 275 |
| 43.0 | 57.0 | ..... | .... | ................. | ... | 281† | |
| 48.8 | 10.2 | 41.0 | .... | ................. | 288 | .... | 331 |
| .... | 60.0 | ..... | 40.0 | ................. | ... | 291† | |
| 51.2 | .... | 30.6 | 18.2 | ................. | ... | 293† | |
| 40.0 | .... | 42.0 | 18.0 | ................. | 293 | .... | 320 |
| 56.5 | .... | ..... | .... | 43.5 Tl | ... | 338† | |
| 67.0 | .... | ..... | 33.0 | ................. | ... | 349† | |
| 63.0 | .... | 37.0 | .... | ................. | ... | 361† | |
| .... | 47.5 | ..... | .... | 52.5 Tl | ... | 370† | |
| 92.0 | .... | ..... | .... | 8.0 Zn | ... | 390† | |
| .... | .... | ..... | 17.0 | 83.0 Tl | ... | 397† | |
| 96.5 | .... | ..... | .... | 3.5 Ag | ... | 430† | |
| 99.25 | .... | ..... | .... | 0.75 Cu | ... | 441† | |

°C = (°F − 32)/1.8.

*From "Fusible Alloys Containing Tin," Booklet 175, Tin Research Inst., Inc., May 1963.

†Indicates eutectic alloy.

**Table 25. Composition, Uses, and Melting Ranges for Selected Solder Alloys**

| Nominal composition, percent | | | | Melting range, deg F | | Uses |
|---|---|---|---|---|---|---|
| Sn | Pb | Sb | Ag | Solidus | Liquidus | |
| 70 | 30 | ... | ... | 361 | 378 | Coating metals |
| 63 | 37 | ... | ... | 361 | 361 | Eutectic solder alloy, electronics |
| 60 | 40 | ... | ... | 361 | 374 | General-purpose, electronics |
| 50 | 50 | ... | ... | 361 | 421 | General-purpose, plumbing |
| 45 | 55 | ... | ... | 361 | 441 | Radiator cores, roofing seams |
| 40 | 60 | ... | ... | 361 | 460 | Wiping solder, general-purpose |
| 35 | 65 | ... | ... | 361 | 477 | Machine and torch soldering |
| 30 | 70 | ... | ... | 361 | 491 | Machine and torch soldering |
| 25 | 75 | ... | ... | 361 | 511 | Machine and torch soldering |
| 20 | 80 | ... | ... | 361 | 531 | Auto-body repair |
| 15 | 85 | ... | ... | 440 | 550 | Radiator solder |
| 10 | 90 | ... | ... | 514 | 570 | Coating metals |
| 5 | 95 | ... | ... | 572 | 596 | Wiping solder |
| 2 | 98 | ... | ... | 601 | 611 | Soldering-can sideseams |
| 40 | 58 | 2 | ... | 365 | 448 | General-purpose, not recommended for use on zinc-containing materials |
| 35 | 63.2 | 1.8 | ... | 365 | 470 | |
| 30 | 68.4 | 1.6 | ... | 364 | 482 | Torch or machine soldering, except on zinc-containing materials |
| 25 | 73.7 | 1.3 | ... | 364 | 504 | |
| 20 | 79 | 1 | ... | 363 | 517 | Machine soldering and coating of metals, except zinc-containing materials |
| 95 | .... | 5 | ... | 450 | 464 | Joints on copper: electrical, plumbing, heating; not recommended for zinc-containing materials |
| ... | 97.5 | ... | 2.5 | 580 | 580 | For use on copper, brass with torch heating; not recommended in humid environments due to its known susceptibility to corrosion |
| 1 | 97.5 | ... | 1.5 | 588 | 588 | For use on copper, brass with torch heating |

°C = (°F − 32)1.8.

percent tin, 50 percent lead solder alloy is used most advantageously. An alloy of 95 percent tin, 5 percent antimony also is used for joining copper water pipe, and this alloy has higher strength at hot-water temperatures than the tin-lead solders. Antimonial solders should not be used to join basis metals containing zinc (e.g., galvanized iron, brass). Tin-antimony, tin-silver, and cadmium-silver solder alloys are used for higher-temperature applications than are permissible with tin-lead alloys.

**Brazing filler metals** are defined by the AWS as metals to be added when making a braze. A filler metal has a melting temperature above 800°F (425°C), but below that of the base metals being joined. The AWS-ASTM have classified brazing filler metals according to their nominal composition (see Table 26); composition and melting points are given in Table 27. REFERENCE: AWS, "Brazing Manual," Reinhold.

## BEARING METALS

**Babbitt metal** is a general term used for soft tin and lead-base alloys which are cast as bearing surfaces on steel, bronze, or cast-iron shells. Babbitts have excellent **embedability** (ability to embed foreign particles in itself) and **conformability** (ability to deform plastically to compensate for irregularities in bearing assembly) characteristics. These alloys may be run satisfactorily against a soft-steel shaft. The limitations of Babbitt alloys

**Table 26. Brazing Filler Metals**

| AWS-ASTM filler-metal classification | Base metals joined |
|---|---|
| BAlSi (aluminum-silicon)........... | Aluminum and aluminum alloys |
| BCuP (copper-phosphorus)......... | Copper and copper alloys; limited use on tungsten and molybdenum; should not be used on ferrous or nickel-base metals |
| BAg (silver)..................... | Ferrous and non-ferrous metals except aluminum and magnesium; iron, nickel, cobalt-base alloys; thin-base metals |
| BCu (copper)................... | Ferrous and non-ferrous metals except aluminum and magnesium |
| RBCuZn (copper-zinc)............. | Ferrous and non-ferrous metals except aluminum and magnesium; corrosion resistance generally inadequate for joining copper, silicon, bronze, copper, nickel, or stainless steel |
| BMg (magnesium)................ | Magnesium-base metals |
| BNi (nickel)..................... | AISI 300 and 400 stainless steels; nickel- and cobalt-base alloys; also carbon steel, low-alloy steels, and copper where specific properties are desired |

**Table 27. Compositions and Melting Ranges of Brazing Alloys***

| AWS designation | Nominal composition, % | Melting temp, °F |
|---|---|---|
| BAg-1 | 45 Ag, 15 Cu, 16 Zn, 24 Cd | 1125–1145 |
| BNI-7 | 13 Cr, 10 P, bal. Ni | 1630 |
| BAu-4 | 81.5 Au, bal. Ni | 1740 |
| BAlSi-2 | 7.5 Si, bal. Al | 1070–1135 |
| BAlSi-5 | 10 Si, 4 Cu, 10 Zn, bal. Al | 960–1040 |
| BCuP-5 | 80 Cu, 15 Ag, 5 P | 1190–1475 |
| BCuP-2 | 93 Cu, 7 P | 1310–1460 |
| BAg-1a | 50 Ag, 15.5 Cu, 16.5 Zn, 18 Cd | 1160–1175 |
| BAg-7 | 56 Ag, 22 Cu, 17 Zn, 5 Sn | 1145–1205 |
| RBCuZn-A | 59 Cu, 40 Zn, 0.6 Sn | 1630–1650 |
| RBCuZn-D | 48 Cu, 41 Zn, 10 Ni, 0.15 Si, 0.25 P | 1690–1715 |
| BCu-1 | 99.90 Cu, min | 1980 |
| BCu-1a | 99.9 Cu, min | 1980 |
| BCu-2 | 86.5 Cu, min | 1980 |

$°C = (°F - 32)/1.8$.
*From "Metals Handbook," vol. 6, ASM, 1971.

are liability to spreading under high, steady loads and to fatigue under high, fluctuating loads. These limitations apply more particularly at higher temperatures, for increase in temperature between 68 and 212°F (20 and 100°C) about halves the metal's strength. By suitable design of the bearing assembly, a properly chosen Babbitt alloy can be made to give satisfactory service in all but the most stringent conditions.

The important tin- and lead-base (Babbitt) bearing alloys are listed in Table 28. Alloy No. 1 is used in internal-combustion engines. Numbers 2 and 3, containing more antimony, are harder and less likely to pound out. Alloy Nos. 7 and 8 are lead-base Babbitts which will function satisfactorily under moderate conditions of load and speed. Alloy No. 15 is an arsenical alloy and, with its better high-temperature hardness, finds the largest-volume use of the three lead-base alloys because of its ability to withstand higher loads and to provide longer fatigue life.

**Cadmium-base bearing alloys** are cadmium-nickel (containing about 1.5 percent nickel and 0.4 to 0.75 percent copper) and cadmium-silver containing about 0.5 to 2 percent silver). These alloys are harder than Babbitts and have less conformability than white-metal alloys. Although they possess a higher fatigue strength (particularly at elevated temperatures) than

Babbitts, they are more liable to corrosion by acidic lubricants.

**Silver bearings** have an excellent record in heavy-duty applications in aircraft engines and diesels. For reciprocating engines, silver bearings normally consist of electrodeposited silver on a steel backing with an overlay of 0.001 to 0.005 in of lead. An indium flash on top of the lead overlay is used to increase corrosion resistance of the material.

**Aluminum alloys** are used in high load-carrying applications but have not replaced Babbitts for equipment operating under a steady, undirectional load. Three major alloys are in use: (1) 6.25 percent tin, 1 percent nickel, and 1 percent copper; (2) 1 to 3 percent cadmium with varying amounts of silicon, copper, and nickel; and (3) 20 to 30 percent tin alloy containing up to 3 percent copper. The first two alloy types may be used as either solid-cast or steel-backed bearings, but the third type is backed by steel for support.

Mention should be made of **cast iron** as a bearing material. The flake graphite in cast iron develops a glazed surface which is useful at surface speeds up to 130 ft/min and at loads up to 150 lb/in² approx. Because of the poor conformability of cast iron, good alignment and freedom from dirt are essential.

**Copper-base bearing alloys** have a wide range of bearing prop-

**Table 28. Composition and Properties of Some Babbitt Alloys (ASTM B23)**

| ASTM grade | Nominal composition, % | | | | Yield point,* ksi | | Compressive strength,† ksi | | Brinell hardness | |
|---|---|---|---|---|---|---|---|---|---|---|
| | Sn | Sb | Pb | Cu | 68°F (20°C) | 212°F (100°C) | 68°F (20°C) | 212°F (100°C) | 68°F (20°C) | 212°F (100°C) |
| 1 | 91 | 4.5 | | 4.5 | 4.4 | 2.7 | 12.9 | 7.0 | 17.0 | 8.0 |
| 2 | 89 | 7.5 | | 3.5 | 6.1 | 3.0 | 14.9 | 8.7 | 24.5 | 12.0 |
| 3 | 84 | 8.0 | | 8.0 | 6.6 | 3.2 | 17.6 | 9.9 | 27.0 | 14.5 |
| 7 | 10 | 15 | 75 | | 3.6 | 1.6 | 15.7 | 6.2 | 22.5 | 10.5 |
| 8 | 5 | 15 | 80 | | 3.4 | 1.8 | 15.7 | 6.2 | 20.0 | 9.5 |
| | | | | Other | | | | | | |
| 15 | 1.0 | 16 | Bal. | 1.4 As | | | | | 21.0 | 13.0 |

ksi × 6.895 = MN/m².
*Based on 0.125 percent change in gage length.
†Compression test specimens were cylinder 1.5 in long and 0.5 in diam machined from chill castings. Compressive-strength values based on a deformation of 25 percent of specimen length.

erties that fit them for many applications. Used alone or in combination with steel, Babbitt (white metal), and graphite, the bronzes and copper-leads meet the conditions of load and speed given in Table 29. **Copper-lead alloys** are cast onto steel backing strips in very thin layers (0.02 in) to provide bearing surfaces.

**Aluminum bronzes** and **silicon bronzes** are used for high-strength and oxidation-resistant bearings. These materials must have the best possible lubrication, or heating, with subsequent failure, will result.

**Porous bearing** materials are used in light- and medium-duty applications as small-sized bearings and bushings. Since they can operate for long periods without an additional supply of lubricant, such bearings are useful in inaccessible or inconvenient places where lubrication would be difficult. Porous bearings are made by pressing mixtures of copper and tin (bronze), and often graphite or Teflon or iron and graphite, and sintering these in a reducing atmosphere without melting. By controlling the conditions under which the bearings are made, porosity may be adjusted so that interconnecting voids of up to 35 percent of the total volume may be available for impregnation by lubricants. ASTM specifications for these bearings are given in Table 30.

**Miscellaneous** A great variety of materials, e.g., rubber, wood, phenolic, carbon-graphite, cermets, ceramics, and plastics, are in use for special applications. Carbon-graphite is used where contamination by oil or grease lubricants is undesirable (e.g., textile machinery, pharmaceutical equipment, milk and food processing) and for elevated-temperature applications. Notable among plastic materials are Teflon and

**Table 29. Copper-Base Bearing Metals**

| Specification designations | | Nominal composition, % | | | | | | Min tensile strength, ksi | Uses |
|---|---|---|---|---|---|---|---|---|---|
| Copper alloy No. | Other | Cu | Sn | Pb | Zn | P | Other | | |
| 961 | SAE 64 | 80 | 10 | 10 | | | | 25 | Heavy loads and high speeds; backing for Babbitt-lined bearings, wristpin bushings, valve rocker-arm bushings, crankshaft bearings; electric-motor bushings; lathe, railroad-car, rolling-mill, and trunnion bearings |
| 955 | SAE 660 | 83 | 7 | 7 | 3 | | | 30 | Medium loads and speeds; general utility; automotive generators, distributors, guide bushings, starters, main bearings for presses, spindle bushings for trucks, sleeve bushings |
| 952 | | 85 | 5 | 9 | 1 | | | 25 | Medium loads and speeds; similar to alloy 955 |
| 964 | SAE 67 | 78 | 7 | 15 | | | | 25 | Medium loads and high speeds; pump bushings, gas-engine bearings, and drum bushings for cranes, hydraulic glands, etc. |
| 972 | AMS 4840 | 70 | 5 | 25 | | | | 21 | Light-to-moderate loads and high speeds; unsuitable for extremely heavy compressive and shock loads; pump bushings, compressor bearings, underwater service |
| 954 | ASTM B148 | 85 | | | | | 4 Fe, 11 Al | 75 | Bushings for power shovels, roll-neck bearings, turntable bushings, machine-tool bearings, guidepost bushings, boring-bar bushings |
| 913 | ASTM B22 | 81 | 19 | | | | | 24* | Bridge turntable plates for contact with hardened-steel disks at low speeds under pressures not over 3,000 lb/in$^2$ |

**Table 29. Copper-Base Bearing Metals (*Continued*)**

| Specification designations | | Nominal composition, % | | | | | | Min tensile strength, ksi | Uses |
|---|---|---|---|---|---|---|---|---|---|
| Copper alloy No. | Other | Cu | Sn | Pb | Zn | P | Other | | |
| 911 | ASTM B22 | 84 | 16 | | | | | 18* | As with alloy 913 above, low speeds with pressure not over 2,500 lb/in²/(17.2 kN/m²); with hard steel, 1,500 lb/in²/(10.3 kN/m²); with trunnions of movable bridges, 2,500 lb/in²/(17.2 kN/m²); bearing and expansion plates |
| 937 | ASTM B22 | 80 | 10 | 10 | | | | | Machinery bearings, bearing and expansion plates under pressures not to exceed 1,000 lb/in²/(6.9 kN/m²) |
| 905 | ASTM B22 | 88 | 10 | | 2.0 | | | 40 | Gears, worm gears, and similar parts which are subject to other than compressive stresses |
| 863 | ASTM B22 | 63 | | | 25 | | 3 Fe, 6 Al, 3 Mn | 223 BHN | Bushings for bridge pins and similar applications where angular movement is slight and compressive stresses may attain 8,000 lb/in²/(55.2 kN/m²) |
| 930 | SAE 40 | 85 | 5 | 5 | 5 | | | 35 | Light loads and low-to-medium speeds; bearing shells, automotive-transmission thrust washers, manifold bearings, pump sleeves, spring bushings |
| 905 | SAE 62 | 88 | 10 | | 2 | | | 45 | Heavy loads and low speeds; piston-pin bushings, valve guides, worm bearings, linkage bushings for machine tools |
| 906 | SAE 620 | 88 | 8 | | 4 | | | 40 | Heavy loads and low speeds; bushings for aircraft landing gear, bridge bearings, trunnions, machine-tool bearings |
| 918 | SAE 63 | 88 | 10 | | 2 | | | 40 | Heavy loads, low speeds, and severe working conditions; earthmoving machinery, locomotive bearings, gear bushings, automotive spindle bushings, packaging machinery |

ksi × 6.895 = MN/m².

*Deformation limit (min lb/in²)—compressive stress producing permanent set of 0.001 in on machined sand-cast specimens 1 in square area by 1 in high.

nylon, the polycarbonate Lexan, and the acetal Delrin. Since Lexan and Delrin can be injection-molded easily, bearings can be formed quite economically from these materials.

## METALS AND ALLOYS FOR USE AT ELEVATED TEMPERATURES

REFERENCES: "High Temperature High Strength Alloys," AISI. Simmons and Krivobok, Compilation of Chemical Compositions and Rupture Strength of Super-strength Alloys, *ASTM Tech. Pub.* 170-A. "Metals Handbook," vol. 1, ASM. Smith, "Properties of Metals at Elevated Temperatures," McGraw-Hill. Clark, "High-temperature Alloys," Pittman. Cross, Materials for Gas Turbine Engines, *Metal Progress*, March 1965

Metals are used for an increasing variety of applications at

**Table 30. Metal-powder-sintered Bearings (Oil-impregnated)**
(ASTM specification B202)

| | Grade I<br>Copper base | | Grade II—Iron base | | | |
|---|---|---|---|---|---|---|
| | | | Class A | | | |
| | Class A | Class B | A₁ | A₂ | A₃ | Class B |
| Copper, %.............. | 87.5–90.5 | 82.6–88.5 | ........ | ........ | ....... | 7.0–11.0<br>18.0–22.0 |
| Iron, %............... | 1.0<br>max | 1.0<br>max | 96.25<br>min | 95.9<br>min | 95.5<br>min | Rem.* |
| Tin, %............... | 9.5–10.5 | 9.5–10.5 | | | | |
| Lead, %............... | ........ | 2.0–4.0 | | | | |
| Zinc, max %.......... | ........ | 0.75 | | | | |
| Nickel, max %.......... | ........ | 0.35 | | | | |
| Antimony, max %....... | ........ | 0.25 | | | | |
| Carbon, max %.......... | 1.75† | 1.75† | | | | |
| Silicon, max %.......... | ........ | ........ | 0.3 | 0.3 | 0.3 | |
| Aluminum, max %....... | ........ | ........ | 0.2 | 0.2 | 0.2 | |
| Other elements, %....... | 0.5 | 0.5 | 3.0 | 3.0 | 3.0 | 3.0 |
| Combined carbon,‡ %..... | ........ | ........ | 0.25–0.60 | 0.60–1.0 | | |
| Density (g per cm³)...... | 6.4–6.8 | 6.5–6.9 | 5.7 –6.1 | 5.7 –6.1 | 5.7–6.1 | 5.8–6.2 |

*Total of iron plus copper shall be 97 percent, min.
†Commonly graphite. A maximum of 1.5 percent of another type of solid lubricant may be substituted.
‡The combined carbon may be a metallographic estimate of the carbon in the iron.

elevated temperatures, "elevated" being a relative term that depends upon the specific metal and the specific service environment. Elevated-temperature properties of the common metals and alloys are cited in the several preceding subsections. This subsection deals with metals and alloys whose prime use is in high-temperature applications. [Typical maximum temperatures are compressors, 750°F (399°C); steam turbines, 1100°F (593°C); gas turbines, 2000°F (1093°C); resistance-heating elements, 2400°F (1316°C); electronic vacuum tubes, 3500°F (1926°C); and lamps, 4500°F (2482°C).] In general, alloys for high-temperature service must have melting points above the operating temperature, low vapor pressures at that temperature, resistance to attack (oxidation, sulfidation, corrosion) by the environment, and sufficient strength to withstand the applied load for the service life without deforming beyond permissible limits. At high temperatures, atomic diffusion becomes appreciable, so that time is an important factor with respect to surface chemical reactions, to **creep**, or slow deformation, under constant load, and to internal changes within the alloy during service. The effects of time and temperature are conveniently combined by the empirical **Larson-Miller parameter** $P = T(C + \log t) \times 10^{-3}$, where $T$ = test temp in °R (°F + 460) and $t$ = test time, h. The constant $C$ depends upon the material but is frequently taken to be 20.

A great many alloys have been developed specifically for such applications. The selection of an alloy for a specific high-temperature application is strongly influenced by service conditions (stress, stress fluctuations, temperature, heat shock, atmosphere, service life), and there are hundreds of alloys from which to choose. The following illustrations should be regarded as examples. Vendor literature and more extensive reference volumes should be consulted.

Figure 5 indicates the general stress-temperature range in which various alloy types find application in elevated-temperature service. Figure 6 indicates the important effect of time on the strength of alloys at high temperatures, comparing a familiar stainless steel (type 304) with **superalloy** (M252).

**Common Heat-Resisting Alloys** A number of alloys containing large amounts of **chromium and nickel** are available.

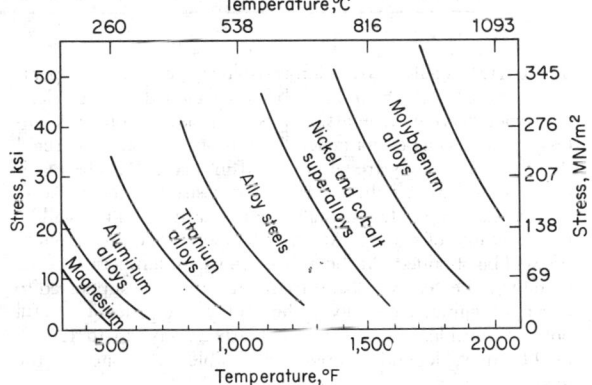

**Fig. 5** Stress-temperature application ranges for various alloy types. (Stress to produce rupture in 1,000 h.)

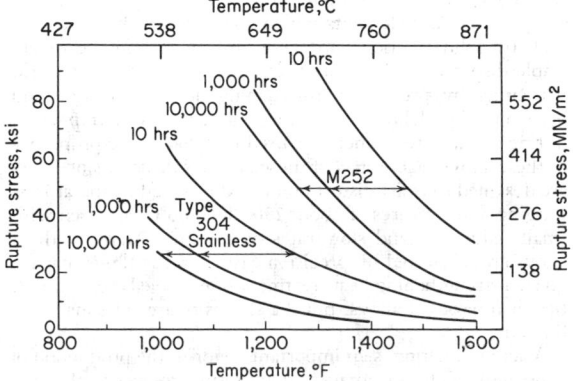

**Fig. 6** Effect of time on the rupture strength of type 304 stainless steel and alloy M252.

### Table 31. Properties of Common Heat-resisting Alloys

| Name or SAE type | Max temp for oxidation resistance °F | Max temp for oxidation resistance °C | Chemical composition, % C | Cr | Ni | Fe | Other | Specific gravity | Coef of thermal expansion per °F ×10⁻⁶ (0–1200°F) | Stress to rupture in 1,000 h, ksi 1200°F (649°C) | 1500°F (816°C) | 1800°F (982°C) | 0.2 percent offset yield strength, ksi Room | 538°C | 1200°F (649°C) |
|---|---|---|---|---|---|---|---|---|---|---|---|---|---|---|---|
| 62 Ni, 15 Cr | 1700 | 926 | | 15 | 62 | Bal | | 8.19 | 9.35 | | | | | | |
| 80 Ni, 20 Cr | 2100 | 1148 | | 20 | 80 | | | 8.4 | 9.8 | | | | 63 | | |
| Kanthal | 2450 | 1371 | | 25 | | Bal | 3 Co, 5 Al | 7.15 | | | | | | | |
| Alloy 10 | 2450 | 1371 | | 37 | | Bal | 7.5 Al | 6.9 | | | | | | | |
| Inconel | 2000 | 1093 | | 13 | 79 | Bal | | 8.4 | | 14.5 | 3.7 | | 36 | 22 | 22 |
| 1015 | 1000 | 537 | 0.15 | | | Bal | | 7.8 | 8.36 | 2.7 | | | 42 | 20 | 10 |
| 502 | 1150 | 621 | 0.12 | 5 | | Bal | 0.5 Mo | 7.8 | 7.31 | 6 | 1.5 | | 27 | 18 | 12 |
| 446 | 2000 | 1093 | 0.12 | 26 | 0.3 | Bal | | 7.6 | 6.67 | 4 | 1.2 | | | | |
| 304 | 1650 | 871 | 0.06 | 18 | 9 | Bal | | 7.9 | 10.4 | 11 | 3.5 | | 32 | 14 | 11 |
| 347 | 1650 | 871 | 0.08 | 18.5 | 11.5 | Bal | 0.8 Cb | 8.0 | 10.7 | 20 | | | 41 | 31 | 26 |
| 316 | 1650 | 871 | 0.07 | 18 | 13 | Bal | 2.5 Mo | 8.0 | 10.3 | 25 | 7 | | 41 | 22 | 21 |
| 310 | 2000 | 1093 | 0.12 | 25 | 20 | Bal | | 7.9 | 9.8 | 13.2 | 3.0 | 2.7 | 34 | 28 | 25 |
| 321 | 1650 | 871 | 0.06 | 18 | 10 | Bal | 0.5 Ti | 8.0 | 10.7 | 17.5 | 3.7 | | 39 | 27 | 25 |
| NA 22 H | 2200 | 1204 | 0.5 | 28 | 48 | Bal | 5W | 8.6 | | 30 | 18 | 3.6 | | | |

ksi × 6.895 = MN/m².

These have excellent oxidation resistance at elevated temperatures. Several of them have been developed as electrical-resistance heating elements; others are modifications of stainless steels, developed for general corrosion resistance. Selected data on these alloys are summarized in Table 31. The maximum temperature value given is for resistance to oxidation with a reasonable life. At higher temperatures, failure will be rapid because of scaling. At lower temperatures, much longer life will be obtained. At the maximum useful temperature, the metal may be very weak and frequently must be supported to prevent sagging. Under load, these alloys are generally useful only at considerably lower temperatures, say up to 1200°F (649°C) max, depending upon permissible creep rate and the load.

**Superalloys** were developed largely to meet the needs of aircraft gas turbines, but they have also been used in other applications demanding high strength at high temperatures. These alloys are based on nickel and/or cobalt, to which are added (typically) chromium for oxidation resistance and a complex of other elements which contribute to hot strength, both by solid solution hardening and by forming relatively stable dispersions of fine particles. Hardening by cold work, hardening by precipitation-hardening heat treatments, and hardening by deliberately arranging for slow precipitation during service are all methods used to enhance the properties of these alloys. **Fabrication** of these alloys is difficult since they are designed to resist distortion even at elevated temperatures. Forging temperatures of about 2300°F (1260°C) are used with small reductions and slow rates of working. Many of these alloys are fabricated by precision casting. Cast alloys that are given a strengthening heat treatment often have better properties than wrought alloys, but the shapes that can be made are limited.

Vacuum melting is an important factor in the production of superalloys. The advantages which result from its application include the ability to melt higher percentages of reactive metals, improved mechanical properties (particularly fatigue strength), decreased scatter in mechanical properties, and improved billet-to-bar stock-conversion ratios in wrought alloys.

Tables 32 to 35 list compositions and properties of some superalloys, and Fig. 7 indicates the temperature dependence of the rupture strength of a number of such alloys.

**Cast tool alloys** are another important group of materials having high-temperature strength and wear resistance. They are principally alloys of cobalt, chromium, and tungsten. They are hard and brittle, and they must be cast and ground to shape. Their most important application is for hard facing, but they compete with high-speed steels and cemented carbides for many applications, in certain instances being superior to both. (See Cemented Carbides, below.) Typical compositions are given in Table 36; typical properties are elastic modulus = 35 × 10⁶ lb/in²; density = 8.8; hardness, BHN at room temp = 660; 1500°F (815°C), 435, and 2000°F (1093°C) 340.

For still higher-temperature applications, the possible choices are limited to a few metals with high melting points, all characterized by limited availability and by difficulty of extraction and fabrication. Alloys of metals such as chromium, columbium, molybdenum, tantalum, and tungsten are still in relatively early stages of development. Table 37 cites properties of some of these new alloys.

**Chromium** with room-temperature ductility has been prepared experimentally as a pure metal and as strong high-temperature alloys. It is not generally available on a commercial basis.

**Columbium** (also known as **niobium**) is available as a pure metal and in several alloys. Alloys of columbium are used in nuclear applications at temperatures above 1600°F (871°C) approx because of their low thermal neutron cross section, high strength, and good corrosion resistance in liquid or gaseous alkali-metal atmospheres. Their oxidation resistance at temperatures above 1800°F approx is so poor that columbium-base alloys are not useful in such an environment.

**Table 32. Wrought Superalloys, Composition‡**

| Common designation | Chemical composition, percent | | | | | | | | | | | | |
|---|---|---|---|---|---|---|---|---|---|---|---|---|---|
| | C | Mn | Si | Cr | Ni | Co | Mo | W | Cb | Ti | Al | Fe | Other |
| 19-9D L............. | 0.2 | 0.5 | 0.6 | 19 | 9 | ... | 1.2 | 1.25 | 0.3 | 0.3 | ... | Bal | |
| Timken.............. | 0.1 | 0.5 | 0.5 | 16 | 25 | ... | 6 | .. | .. | ... | ... | Bal | 0.15N |
| L.C. N-155.......... | 0.1 | 0.5 | 0.5 | 20 | 20 | 20 | 3 | 2 | 1 | ... | ... | Bal | 0.15N |
| S-590............... | 0.4 | 0.5 | 0.5 | 20 | 20 | 20 | 4 | 4 | 4 | ... | ... | Bal | |
| S-816............... | 0.4 | 0.5 | 0.5 | 20 | 20 | Bal | 4 | 4 | 4 | ... | ... | 4 | |
| Inconel-X........... | 0.05 | 0.5 | 0.4 | 15 | 73 | ... | .. | ... | 1 | 2.5 | 0.9 | 7 | |
| Nimonic-80.......... | 0.05 | 0.7 | 0.5 | 20 | 76 | ... | .. | ... | .. | 2.3 | 1.0 | 0.5 | |
| K-42-B.............. | 0.05 | 0.7 | 0.7 | 18 | 42 | 22 | .. | ... | .. | 2.0 | 0.2 | 14 | |
| Hastelloy B......... | 0.1 | 0.5 | 0.5 | .. | 65 | ... | 28 | ... | .. | ... | ... | 6 | 0.4V |
| M 252†.............. | 0.10 | 1.0 | 0.7 | 19 | 53.5 | 10 | 10 | ... | .. | 2.5 | 0.75 | 2 | |
| Inco 700............ | 0.12 | 2.0* | 1.0* | 15 | 48 | 28 | 3 | ... | .. | 2.2 | 2.8 | 5 | |
| J 1570.............. | 0.20 | 0.1* | 0.2* | 20 | 29 | 37.5 | .. | 7 | .. | 4.1 | ... | 2 | |
| HS-R235............. | 0.10 | ... | ... | 16 | Bal | 1.5 | 6 | ... | .. | 3.0 | 1.8 | 8 | |
| Hastelloy X......... | 0.10 | ... | ... | 21 | 48 | 2.0 | 9 | 1.0 | .. | ... | ... | 18 | |
| HS 25 (L605)........ | 0.05 | 1.5 | 1.0* | 20 | 10 | 53 | .. | 15 | .. | ... | ... | 1.0 | |
| A-286.............. | 0.05 | ... | ... | 15 | 26 | ... | 1.3 | ... | .. | 2.0 | 0.35 | 55 | 0.3V |

*Max.

†Waspaloy has a similar composition (except for lower Mo content), and similar properties.

‡For a complete list of superalloys, both wrought and cast, experimental and under development, see *ASTM Spec. Pub.* 170, "Compilation of Chemical Compositions and Rupture Strengths of Superstrength Alloys."

Alloys of columbium are also regarded as the best **superconducting materials** available. This is a rapidly changing field. Currently, the most important high-field superconductors are the alloys of columbium and zirconium and the compound $Cb_3Sn$. At a magnetic field of 50,000 G, the Cb-25 Zr alloy has a critical field of about 90,000 G. Wire of this alloy is ductile and can easily be wound into a coil. The columbium-tin compound has a critical field of over 200,000 G and a critical current at 100,000 G in the range of $10^4$ to $10^5$ amp/cm². This compound is brittle and must either be used as a very thin coating on a ductile substrate or be prepared as a coil in such a way that the compound is formed by high-temperature reaction of columbium and tin after winding the coil.

**Molybdenum** is similar to tungsten in most of its properties. It can be prepared in the massive form by powder metallurgy or by inert-atmosphere or vacuum-arc melting. Its most serious limitation is its ready formation of a volatile oxide at temperatures of 1400°F approx. In the worked form, it is

**Table 33. Wrought Superalloys, Properties**

| Common designation | °F | °C | Coef of thermal expansion in/in/°F × 10⁻⁶ (70–1200°F) | Specific gravity | Stress to rupture, ksi | | | Short-time tensile properties, ksi | | | | | | | |
|---|---|---|---|---|---|---|---|---|---|---|---|---|---|---|---|
| | Max temp under load | | | | At 1200°F | At 1500°F | | Yield Strength 0.2% offset | | | | Tensile strength | | | |
| | | | | | 1,000 h | 100 h | 1,000 h | Room | 1200°F | 1500°F | 1800°F | Room | 1200°F | 1500°F | 1800°F |
| 19-9 DL | 1200 | 649 | 9.7 | 7.75 | 38 | 17 | 10 | 115 | 39 | 30 | | 140 | 75 | 33 | 13 |
| Timken | 1350 | 732 | 9.25 | 8.06 | 36 | 13.5 | 9 | 96* | 70* | 16* | 12 | 134 | 90 | 40 | 18 |
| L.C. N-155 | 1400 | 760 | 9.4 | 8.2 | 48 | 20.0 | 15 | 53 | 40 | 33 | 12 | 115 | 53 | 35 | 19 |
| S-590 | 1450 | 788 | 8.0 | 8.34 | 40 | 19.0 | 15 | 78* | 70 | 47 | 20 | 140 | 82 | 60 | 22 |
| S-816 | 1500 | 816 | 8.3 | 8.66 | 53 | 29.0 | 18 | 63* | 45 | 41 | | 140 | 112 | 73 | 25 |
| Inconel X | 1500 | 816 | 8.4 | 8.33 | 68 | 30.0 | 17 | 100 | 79 | 50 | 6 | 160 | 120 | 60 | 9 |
| Nimonic-80A | 1450 | 788 | 7.56 | | 56 | 24 | 15 | 80* | 73 | 47 | | 150 | 101 | 69 | |
| K-42-B | 1400 | 760 | 8.5 | 8.23 | 38 | 22 | 15 | 105 | 84 | 52 | | 158 | 117 | 54 | |
| Hastelloy B | 1450 | 788 | 6.9 | 9.24 | 36 | 17 | 10 | 58 | 42 | | | 135 | 94 | 66 | 24 |
| M 252 | 1500 | 816 | 7.55 | 8.25 | 70 | 26 | 18 | 90 | 75 | 65 | | 160 | 140 | 80 | 20 |
| Inco 700 | 1600 | 871 | 8.32 | 8.16 | 85 | 42 | 27 | 105 | 93 | 82 | | 170 | 140 | 95 | |
| J 1570 | 1600 | 871 | 8.42 | 8.66 | 84 | 34 | 24 | 81 | 70 | 71 | 17 | 152 | 135 | 82 | 20 |
| HS-R235 | 1600 | 871 | 8.34 | 7.88 | 70 | 35 | 23 | 100 | 90 | 80 | 22 | 170 | 145 | 83 | 25 |
| Hastelloy X | 1350 | 732 | | | 30 | 14 | 10 | 56 | 41 | 37 | | 113 | 83 | 52 | 15 |
| HS 25 (L605) | 1500 | 816 | | | 54 | 22 | 18 | 70 | 35 | | | | 75 | 50 | 23 |
| A-286 | 1350 | 732 | | | 45 | 13.8 | 7.7 | | | | | | | | |

ksi × 6.895 = MN/m².

*0.02 percent offset.

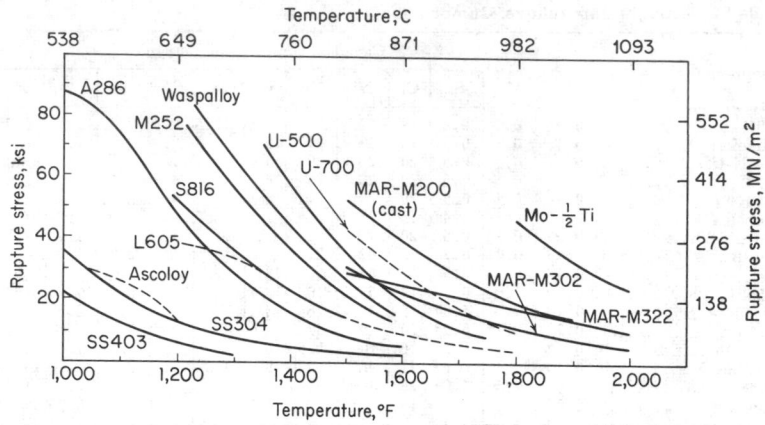

**Fig. 7**  Temperature dependence of the rupture strengths of some high-temperature alloys. (Stress to produce rupture in 1,000 h.)

inferior to tungsten in melting point, tensile strength, vapor pressure, and hardness, but in the recrystallized condition, the ultimate strength and elongation are higher. Tensile strengths up to 350,000 lb/in² have been reported for hard-drawn wire, and to 170,000 lb/in² for soft wire. In the hard-drawn condition, molybdenum has an elongation of 2 to 5 percent, but after recrystallizing, this increases to 10 to 25 percent. Young's modulus is 50 million lb/in². It costs about the same as tungsten, per pound, but its density is much less. It has considerably better forming properties than tungsten and is extensively used for anodes, grids, and supports in vacuum tubes, lamps, and X-ray tubes. Molybdenum is generally used for winding electric furnaces for temperatures up to 3000°F (1649°C). As it must be protected against oxidation, such furnaces are usually operated in hydrogen. It is the common material for contacts in mercury switches. Its principal use is still as an alloying addition to steels, especially tool steels and high-temperature steels.

**Molybdenum alloys** have found increased use in aerospace and commercial structural applications. They are generally stronger than columbium alloys but are not ductile in the welded condition. Molybdenum alloys are resistant to alkali-metal corrosion.

**Tantalum** has a melting point that is surpassed only by tungsten. Its early use was as an electric-lamp filament material. It is more ductile than molybdenum or tungsten; the elongation for annealed material may be as high as 40 percent. The tensile strength of annealed sheet is about 50,000 lb/in². It is used for grids and plates in electronic power tubes. Its most important uses are in chemical-processing equipment, where its high rate of heat transfer (compared with glass or ceramics) is particularly important, although it is equivalent to glass in

**Table 34.  Cast Superalloys, Composition**

| Common designation | Chemical composition, percent | | | | | | | | | | |
|---|---|---|---|---|---|---|---|---|---|---|---|
|  | C | Mn | Si | Cr | Ni | Co | Mo | W | Cb | Fe | Other |
| Vitallium (Haynes 21)............. | 0.25 | 0.6 | 0.6 | 27 | 2 | Bal | 6 | .. | .. | 1 | |
| 61 (Haynes 23).................... | 0.4 | 0.6 | 0.6 | 26 | 1.5 | Bal | .. | 5 | .. | 1 | |
| 422-19 (Haynes 30)................ | 0.4 | 0.6 | 0.6 | 26 | 16 | Bal | 6 | .. | .. | 1 | |
| X-40 (Haynes 31)................. | 0.4 | 0.6 | 0.6 | 25 | 10 | Bal | .. | 7 | .. | 1 | |
| S-816............................ | 0.4 | 0.6 | 0.6 | 20 | 20 | Bal | 4 | 4 | 4 | 5 | |
| HE 1049......................... | 0.45 | 0.7 | 0.7 | 25 | 10 | 45 | .. | 15 | .. | 1.5 | 0.4B |
| Hastelloy C...................... | 0.10 | 0.8 | 0.7 | 16 | 56 | 1 | 17 | 4 | .. | 5.0 | 0.3V |

### Table 35. Cast Superalloys, Properties

| Common designation | Max temp under load °F | °C | Coef of thermal expansion, in/in/°F × 10⁻⁶ (70–1200 °F) | Specific gravity | Stress to rupture in 1,000 h, ksi 1200°F | 1500°F | 1800°F | Yield strength, 0.2% offset Room 68°F (20°C) | 1200°F (650°C) | 1500°F (816°C) | 1800°F (982°C) | Tensile strength Room 68°F (20°C) | 1200°F (650°C) | 1500°F (816°C) | 1800°F (982°C) |
|---|---|---|---|---|---|---|---|---|---|---|---|---|---|---|---|
| Vitallium (Haynes 21) | 1500 | 816 | 8.35 | 8.3 | 44 | 14 | 7 | 82 | 71 | 49 | | 101 | 89 | 59 | 33 |
| 61 (Haynes 23) | 1500 | 816 | 8.5 | 8.53 | 47 | 22 | 5.5 | 58 | 74 | 40 | 33 | 105 | 97 | 58 | 45 |
| 422-19 (Haynes 30) | 1500 | 816 | 8.07 | 8.31 | | 21 | 7 | 55 | 37* | 48 | | 98 | 59* | 64 | 37 |
| X-40 (Haynes 31) | 1500 | 816 | 8.18 | 8.60 | 46 | 23 | 9.8 | 74 | 37* | 44 | | 101 | 77* | 59 | 29 |
| S-816 | 1500 | 816 | 8.27 | 8.66 | 46 | 21 | 9.8 | | | | | 112 | | | |
| HE 1049 | 1650 | 899 | | 8.9 | 75 | 35 | 7 | 80 | 72 | 62 | 36 | 90 | 82 | 81 | 52 |
| Hastelloy C | | | 7.73 | 8.91 | 42.5 | 14.5 | 1.4 | 54 | 50 | | | 130 | 87 | 51 | 19 |

ksi × 6.895 = MN/m².
*Specimen not aged.

### Table 36. Typical Compositions of Cast Tool Alloys

| Name | C | Cr | W | Co | Fe | V | Ta | B | Other |
|---|---|---|---|---|---|---|---|---|---|
| Rexalloy.................................. | 3 | 32 | 20 | 45 | | | | | |
| Stellite 98 M2........................... | 3 | 28 | 18 | 35 | 9 | 4 | 0.1 | 0.1 | 3 Ni |
| Tantung G-2............................. | 3 | 15 | 21 | 40 | ... | ... | 19 | 0.2 | |
| Borcoloy No. 6.......................... | 0.7 | 5 | 18 | 20 | Bal | 1.3 | ... | 0.7 | 6 Mo |
| Colmonoy WCR 100...................... | ... | 10 | 15 | .. | Bal | ... | ... | 3 | |

### Table 37. Properties of Selected Refractory Metals

| Alloy | Melting temp, °F | Density, lb/in³ at 75°F | Elastic modulus at 75°F | Ductile-to-brittle transition temp, °F | Tensile strength, ksi (recrystallized condition) 75°F | 1800°F | 2200°F | 3000°F | 3500°F |
|---|---|---|---|---|---|---|---|---|---|
| Chromium | 3450 | 0.760 | 42 | 625 | | 12 | 8 | | |
| Columbium | 4474 | 0.310 | 16 | −185 | 45 | 12 | 10 | | |
|   F48 (15 W, 5 Mo, 1 Zr, 0.05 C) | | | 25 | | 121 | 75 | 50 | | |
|   Cb74 or Cb752 (10 W, 2 Zr) | | 0.326 | | | 84 | 60 | 36 | | |
| Molybdenum | 4730 | 0.369 | 47 | 85 | 75 | 34 | 22 | 6 | 3 |
|   Mo (½ Ti) | | | | | 110 | 70 | 48 | 8 | 4 |
|   TZM (0.5 Ti, 0.08 Zr) | | | | | 130 | 85 | 70 | 14 | 5 |
| Tantalum | 5425 | 0.600 | 27 | < −320 | 30 | 22 | 15 | 8 | 4 |
|   Ta (10 W) | | | | | 65 | 35 | 12 | 8 |
| Rhenium | 5460 | 0.759 | 68 | <75 | 170 | 85 | 60 | 25 | 11 |
| Tungsten | 6170 | 0.697 | 58 | 645 | 85 | 36 | 32 | 19 | 10 |
|   W (10 Mo) | | | | | | | | 28 | 10 |
|   W (2 ThO₂) | | | | | | | 40 | 30 | 25 |

ksi × 6.895 = MN/m².
°C = (°F × 32)/1.8.

**Table 38. Requirements of Materials for Nuclear-Reactor Components**

| Component | Neutron-absorp-tion cross section | Effect in slowing neu-trons | Strength | Resistance to radia-tion damage | Ther-mal con-duc-tivity | Cor-rosion resist-ance | Cost | Other | Typical material |
|---|---|---|---|---|---|---|---|---|---|
| Moderator and reflector | Low | High | Adequate | ........ | .... | High | Low | Low atomic wt | $H_2O$, graph-ite, Be |
| Fuel*...... | Low | .... | Adequate | High | High | High | Low | ........... | U, Th, Pu |
| Control rod | High | .... | Adequate | Adequate | High | .... | .... | ........... | Cd, $B_4C$ |
| Shield..... | High | High | High | ...... | .... | .... | .... | High $\gamma$ radi-ation ab-sorption | Concrete |
| Cladding... | Low | .... | Adequate | ........ | High | High | Low | ........... | Al, Zr, stainless steel |
| Structural.. | Low | .... | High | Adequate | .... | High | .... | ........... | Zr, stain-less steel |
| Coolant.... | Low | .... | ......... | ........ | High | .... | .... | Low corro-sion rate, high heat capacity | $H_2O$, Na, NaK, $CO_2$, He |

*U, Th, and Pu are used as fuels in the forms of metals, oxides, and carbides.

corrosion resistance. Its corrosion resistance also makes it attractive for surgical implants. Like tungsten and molybdenum, it is prepared by powder metallurgy; so the size of the piece that can be fabricated is limited by the size of the original pressed compact. Tantalum carbide is used in cemented-carbide tools, where it decreases the tendency to seizure and cratering. The stability of the anodic oxide film on tantalum leads to rectifier and capacitor applications. Tantalum and its **alloys** become competitive with columbium at temperatures above 2700°F (1482°C). As in the case of columbium, these materials are resistant to liquid or vapor metal corrosion and have excellent ductility even in the welded condition.

**Tungsten** has the highest melting point of all metals. The massive metal is usually prepared by powder metallurgy from hydrogen-reduced powder. As a metal, its chief use is as filaments in incandescent lamps and electronic tubes, since its vapor pressure is low at high temperatures. Tensile strengths over 600,000 lb/in² have been reported for fine tungsten wires; in larger sizes (0.040 in), tensile strength is only 200,000 lb/in². The hard-drawn wire has an elongation of 2 to 4 percent, but the recrystallized wire is brittle. Young's modulus is 60 million lb/in². It can be sealed directly to hard glass and so is used for lead-in wires. A considerable proportion of tungsten rod and sheet is used for electrical contacts in the form of disks cut from rod or sheet and brazed to supporting elements. The major part of the tungsten used is made into ferroalloys for addition to steels or into tungsten carbide for cutting tools. Other applications include elements of electronic tubes, X-ray tube anodes, and arc-welding electrodes. **Alloys** of tungsten that are commercially available are W-3 Re, W-25 Re, and thoriated tungsten. The rhenium-bearing alloys are more ductile than unalloyed tungsten at room temperature. Thoriated tungsten is stronger than unalloyed tungsten at temperatures up to the recrystallization temperature.

**METALS AND ALLOYS FOR ATOMIC-ENERGY APPLICATIONS**

(See also Sec. 9.)

REFERENCES: Glasstone, "Principles of Nuclear Reactor Engineering," Van Nostrand. Hausner and Roboff, "Materials for Nuclear Power Reactors," Reinhold.

The advent of atomic energy has not only created a demand for new metals and alloys but has also focused attention on certain properties and combinations of properties which theretofore had been of little consequence. Reactor technology requires special materials for fuels, fuel cladding, moderators, reflectors, controls, heat-transfer mediums, operating mechanisms, and auxiliary structures. Some of the properties pertinent to such applications are given in a general way in Table 38. In general, outside the reflector, only normal engineering requirements need be considered.

**Nuclear Properties** A most important consideration in the design of a nuclear reactor is the control of the number and speed of the neutrons resulting from fission of the fuel. It is thus apparent that the designer must have knowledge of the effectiveness of various materials in slowing down neutrons or in capturing them. The slowing-down power depends not only on the relative energy loss per atomic collision but also on the number of collisions per second per unit volume. The former will be larger, the lower the atomic weight, and the latter larger, the greater the atomic density and the higher the probability of a scattering collision. The effectiveness of a moderator is frequently expressed in terms of the moderating ratio, the ratio of the slowing-down power to the capture cross section. The capturing and scattering tendencies are measured in terms of nuclear cross section in barns ($10^{-24}$ cm²). Data for some of the materials of interest are given in Tables 39 to 41. The absorption data apply to slow, or thermal, neutrons; entirely different cross sections obtain for fast neutrons, for which few materials have significantly high capture affinity. Fast neutrons have energies of about $10^6$ eV, while slow, or thermal, neutrons have energies of about $2.5 \times 10^{-2}$ eV. Special consideration must frequently be given to the presence of small amounts of high cross-section elements such as Co or W which are either normal incidental impurities in nickel alloys and steels or important components of high-temperature alloys.

### Table 39. Moderating Properties of Materials*

| Moderator | Slowing-down power, cm$^{-1}$ | Moderating ratio |
|---|---|---|
| H$_2$O | 1.53 | 70 |
| D$_2$O | 0.177 | 21,000 |
| He | 1.6 $\times$ 10$^{-5}$ | 83 |
| Be | 0.16 | 150 |
| BeO | 0.11 | 180 |
| C (graphite) | 0.063 | 170 |

*Adapted from Glasstone, "Principles of Nuclear Reactor Engineering," Van Nostrand.

**Effects of Radiation** Irradiation affects the properties of solids in a number of ways: dimensional changes; decrease in density; increase in hardness, yield, and tensile strengths; decrease in ductility; decrease in electrical conductivity or change in magnetic susceptibility. Another consideration is the activation of certain alloying elements by irradiation. Tantalum[181] and Co[60] have moderate radioactivity but long half-lives; isotopes having short lives but high-activity levels include Cr[51], Mn[56], and Fe[59].

**Metallic Coolants** The need for the efficient transfer of large quantities of heat in a reactor has led to widespread use of several metallic coolants. These have raised new problems of pumping, valving, and corrosion. In addition to their thermal and flow properties, consideration must also be given the nuclear properties of prospective coolants. Extensive thermal, flow, and corrosion data on metallic coolants are given in the "Liquid Metals Handbook," published by USAEC. The resistances of common materials to liquid sodium and NaK are given qualitatively in Table 42. Water, however, remains the most-used coolant; it is used under pressure as a single-phase liquid, as a boiling two-phase coolant, or as steam in a superheat reactor.

### Table 40. Slow-Neutron Absorption by Structural Materials*

| Material | Relative neutron absorption per cm$^3$ $\times$ 10$^3$ | Relative neutron absorption for pipes of equal strength, 68°F (20°C) | Melting point | |
|---|---|---|---|---|
| | | | °F | °C |
| Magnesium | 3.5 | 10 | 1200 | 649 |
| Aluminum | 13 | 102 | 1230 | 666 |
| Stainless steel | 226 | 234 | 2730 | 1499 |
| Zirconium | 12.6 | 16 | 3330 | 1816 |

*Leeser, *Materials & Methods*, **41**, 1955, p. 98.

### Table 41. Slow-Neutron Absorption Cross Sections*

| Low | | Intermediate | | High | |
|---|---|---|---|---|---|
| Element | Cross section, barns | Element | Cross section, barns | Element | Cross section, barns |
| Oxygen | 0.0016 | Zinc | 1.0 | Manganese | 12 |
| Carbon | 0.0045 | Columbium | 1.2 | Tungsten | 18 |
| Beryllium | 0.009 | Barium | 1.2 | Tantalum | 21 |
| Fluorine | 0.01 | Strontium | 1.3 | Chlorine | 32 |
| Bismuth | 0.015 | Nitrogen | 1.7 | Cobalt | 35 |
| Magnesium | 0.07 | Potassium | 2.0 | Silver | 60 |
| Silicon | 0.1 | Germanium | 2.3 | Lithium | 67 |
| Phosphorus | 0.15 | Iron | 2.4 | Gold | 95 |
| Zirconium | 0.18 | Molybdenum | 2.4 | Hafnium | 100 |
| Lead | 0.18 | Gallium | 2.8 | Mercury | 340 |
| Aluminum | 0.22 | Chromium | 2.9 | Iridium | 470 |
| Hydrogen | 0.32 | Thallium | 3.3 | Boron | 715 |
| Calcium | 0.42 | Copper | 3.6 | Cadmium | 3,000 |
| Sodium | 0.48 | Nickel | 4.5 | Samarium | 8,000 |
| Sulphur | 0.49 | Tellurium | 4.5 | Gadolinium | 36,000 |
| Tin | 0.6 | Vanadium | 4.8 | | |
| | | Antimony | 5.3 | | |
| | | Titanium | 5.8 | | |

*Leeser, *Materials & Methods*, **41**, 1955, p. 98.

**Table 42. Resistance of Materials to Liquid Sodium and NaK\***

|  | Good | Limited | Poor |
|---|---|---|---|
| <1000°F | Carbon steels, low-alloy steels, alloy steels, stainless steels, nickel alloys, cobalt alloys, refractory metals, beryllium, aluminum oxide, magnesium oxide, aluminum bronze | Gray cast iron, copper, aluminum alloys, magnesium alloys, glasses | Sb, Bi, Cd, Ca, Au, Pb, Se, Ag, S, Sn, Teflon |
| 1000– 1600°F | Armco iron, stainless steels, nickel alloys,† cobalt alloys, refractory metals‡ | Carbon steels, alloy steels, Monel, titanium, zirconium, beryllium, aluminum oxide, magnesium oxide | Gray cast iron, copper alloys, Teflon, Sb, Bi, Cd, Ca, Au, Pb, Se, Ag, S, Sn, Pt, Si, Magnesium alloys |

\*For more complete details, see "Liquid Metals Handbook."
†Except Monel.
‡Except titanium and zirconium.

**Fuels** There are, at present, only three fissionable materials, $U^{233}$, $U^{235}$, and $Pu^{239}$. Of these, only $U^{235}$ occurs naturally as an isotopic "impurity" with natural uranium $U^{238}$. Uranium$^{233}$ may be prepared from natural thorium, and $Pu^{239}$ from $U^{238}$ by neutron bombardment. Both uranium and thorium are prepared by conversion of the oxide to the tetrafluoride and subsequent reduction to the metal.

The properties of **uranium** are considerably affected by the three allotropic changes which it undergoes, the low-temperature forms being highly anisotropic. The strength of the metal is low (see Table 43) and decreases rapidly with increasing temperature. The corrosion resistance is also poor. Aluminum-base alloys containing uranium-aluminum intermetallic compounds have been used to achieve improved properties. **Thorium** is even softer than uranium but is very ductile. Like

uranium it corrodes readily, particularly at elevated temperatures. Owing to its crystal structure, its properties are isotropic. **Plutonium** is unusual in possessing six allotropic forms, many of which have anomalous physical properties. The electrical resistivity of the alpha phase is greater than that for any other metallic element. Because of the poor corrosion resistance of nuclear-fuel metals, most fuels in power reactors today are in the form of oxide $UO_2$, $ThO_2$, or $PuO_2$, or mixtures of these. The carbides UC and $UC_2$ are also of interest in reactors using liquid-metal coolants.

**Control-Rod Materials** Considerations of neutron absorption cross sections and melting points limit the possible control-rod materials to a very small group of elements, of which only four—boron, cadmium, hafnium, and gadolinium— have thus far been prominent. **Boron** is a very light metal of

**Table 43. Mechanical Properties of Metals for Nuclear Reactors**

| Material | Room temp | | | | | | Elevated temp\* | | |
|---|---|---|---|---|---|---|---|---|---|
|  | Longitudinal | | | Transverse | | | | | |
|  | Yield strength, ksi | Ult strength, ksi | Elong,† % | Ult str, ksi | Elong,† % | Charpy impact ft·lb | °F | Ult str, ksi | Elong,% |
| Beryllium: |  |  |  |  |  |  |  |  |  |
| Cast, extruded and annealed |  | 40 | 1.82 | 16.6 | 0.18 |  | 392 | 62 | 23.5 |
| Flake, extruded and annealed |  | 63.7 | 5.0 | 25.5 | 0.30 |  | 752 | 43 | 29 |
| Powder, hot-extruded | 39.5 | 81.8 | 15.8 | 45.2 | 2.3 | 4.1 | 1112 | 23 | 8.5 |
| Powder, vacuum hot pressed | 32.1 | 45.2 | 2.3 | 45.2 | 2.3 | 0.8 | 1472 | 5.2 | 10.5 |
| Zirconium: |  |  |  |  |  |  |  |  |  |
| Kroll-50% CW |  | 82.6 |  |  |  | 14.8 | 250 | 32 |  |
| Kroll-annealed |  | 49.0 |  |  |  |  | 500 | 23 |  |
| Iodide | 15.9 | 35.9 | 31 |  |  | 2.5-6.0 | 700 | 17 |  |
|  |  |  |  |  |  |  | 900 | 12 |  |
|  |  |  |  |  |  |  | 1500 | 3 |  |
| Uranium | 25 | 53 | <10 |  |  | 15 | 302 | 27 |  |
|  |  |  |  |  |  |  | 1112 | 12 |  |
| Thorium | 27 | 37.5 | 40 |  |  |  | 570 | 22 |  |
|  |  |  |  |  |  |  | 930 | 17.5 |  |

°C = (°F −32)/1.8. ksi × 6.895 = MN/m².
\*All elevated-temperature data on beryllium for hot-extruded powder; on zirconium for iodide material.
†Beryllium is extremely notch-sensitive. The tabulated data have been obtained under very carefully controlled conditions, but ductility values in practice will be found in general to be much lower and essentially zero in the transverse direction.

high hardness (~3000 Knoop) prepared by thermal or hydrogen reduction of $BCl_3$. Because of its high melting point, solid shapes of boron are prepared by powder-metallurgy techniques. Boron has a very high electrical resistivity at room temperature but becomes conductive at high temperatures. The metal in bulk form is oxidation-resistant below 1800°F (982°C) but reacts readily with most halogens at only moderate temperatures. Rather than the elemental form, boron is generally used as boron steel or as the carbide, oxide, or nitride. **Cadmium** is a highly ductile metal of moderate hardness which is recovered as a by-product in zinc smelting. Its properties greatly resemble those of zinc. The relatively low melting point renders it least attractive as a control rod of the four metals cited. **Hafnium** metal is reduced from hafnium tetrachloride by sodium and subsequently purified by the iodide hot-wire process. It is harder and less readily worked than zirconium, to which it is otherwise very similar in both chemical and physical properties. Hafnium reacts easily with oxygen, and its properties are sensitive to traces of most gases. The very high absorption cross section of **gadolinium** renders it advantageous for fast-acting control rods. This metal is one of the rare earths and as yet is of very limited availability. It is most frequently employed as the oxide. Certain of the other rare-earth metals may eventually find similar application.

**Beryllium** Great interest attaches to beryllium because it is unique among the metals with respect to its very low neutron-absorption cross section and high neutron-slowing power. It may also serve as a source of neutrons when subjected to alpha-particle bombardment. Beryllium is currently prepared almost entirely by magnesium reduction of the fluoride, although fused-salt electrolysis is also practicable and has been used. The high affinity of the metal for oxygen and nitrogen renders its processing and fabrication especially difficult. Prior to the intensive effort applied to the problem during World War II, beryllium was regarded as almost hopelessly brittle. Now special techniques, such as vacuum hot pressing, or vacuum casting followed by hot extrusion of clad slugs, have led to material having marginally acceptable, though highly directional, mechanical properties. The extreme toxicity of beryllium powder necessitates special precautions in all operations.

**Zirconium** Zirconium's importance in nuclear technology derives from its low neutron-absorption cross section, excellent corrosion resistance, and high strength at moderate temperatures. The metal is produced by magnesium reduction of the tetrachloride (Kroll process). A subsequent refining by the iodide hot-wire process was formerly customary but is now no longer required. The Kroll product is usually converted to ingot form by consumable-electrode-arc melting. An important step in the processing for many applications is the difficult chemical separation of the 1½ to 3 percent hafnium with which zirconium is contaminated. Unless removed, this small hafnium content results in a prohibitive increase in absorption cross section from 0.18 to 3.5 barns. The mechanical properties of zirconium are particularly sensitive to the impurity content and the fabrication technique. In spite of the high melting point, the mechanical properties are poor at high temperatures, principally because of the allotropic transformation at 1585°F (863°C). A satisfactory annealing temperature is 1100°F (593°C). The low-temperature corrosion behavior is excellent but can be seriously affected by impurities. The oxidation resistance at high temperatures is poor. Special alloys (Zircoloys) have been developed having greatly improved oxidation resistance in the intermediate-temperature range. These alloys have a nominal content of 1.5 percent Sn and minor additions of iron-group elements.

**Stainless Steel** Although stainless steel has a higher thermal neutron-absorption cross section than do the zirconium alloys, its good corrosion resistance, high strength, low cost, and ease of fabrication make it a strong competitor with zirconium alloys as a fuel-cladding material for water-cooled power-reactor applications. Types 348 and 304 stainless have been used in major reactors.

## JEWELRY METALS

**Gold** is used primarily as a monetary standard; the small amount put to metallurgical use is for jewelry or decorative purposes, in dental work, for fountain-pen nibs, and as an electrodeposited protective coating. An alloy with palladium has been used as a platinum substitute for laboratory vessels, but its present price is so high that it does not compete with platinum.

**Silver** has the highest electrical conductivity of any metal, and it found some use in bus bars during World War II, when copper was in short supply. Since its density is higher than that of copper and since government silver frequently has deleterious impurities, it offers no advantage over copper as an electrical conductor. Heavy-duty electrical contacts are usually made of silver. It is used in aircraft bearings and solders. Its largest commercial use is in tableware as sterling silver, which contains 92.5 percent silver (the remainder is usually copper). United States coinage used to contain 90 percent silver, 10 percent copper.

**Platinum** has many uses because of its high melting point, chemical inertness, and catalytic activity. It is the standard catalyst for the oxidation of sulfur dioxide in the manufacture of sulfuric acid. Because it is inert toward most chemicals, even at elevated temperatures, large amounts are used for laboratory apparatus. It is the only metal that can be used for an electric heating element above 2300°F without a protective atmosphere. Thermocouples of platinum with platinum-rhodium alloy are standard for high temperatures. Platinum and platinum alloys are used in large amounts in feeding mechanisms of glass-working equipment to ensure constancy of the orifice dimensions that fix the size of glass products. They are also used for electrical contacts, in dental work, in aircraft spark-plug electrodes, and as jewelry.

**Palladium** follows platinum in importance and abundance among the platinum metals and resembles platinum in most of its properties. Its density and melting point are the lowest of the platinum metals, and it forms an oxide coating at a dull-red heat so that it cannot be heated in air above 800°F (426°C) approx. In the finely divided form, it is an excellent hydrogenation catalyst. It is as ductile as gold and is beaten into leaf as thin as gold leaf. Its hardened alloys find some use in dentistry, jewelry, and electrical contacts.

**Iridium** is one of the platinum metals. Its chief uses are as a hardener for platinum jewelry alloys and as platinum contacts. Its alloys with osmium are used for tipping fountain-pen nibs. It is the most corrosion-resistant element known.

**Rhodium** is used mainly as an alloying addition to platinum. It is a component of many of the pen-tipping alloys. Because of its high reflectivity and freedom from oxidation films, it is frequently used as an electroplate for jewelry and for reflectors for motion-picture projectors, aircraft searchlights, and the like.

**Table 44. Mechanical Properties of Sintered Copper and Iron**
(Goetzal in Wulf, "Powder metallurgy," ASM, 1942)

| Material and treatment | Density lb/in³ | Brinell hardness | Tensile strength, ksi | Elongation, % in 2 in (5 cm) | Yield strength, ksi |
|---|---|---|---|---|---|
| Copper: | | | | | |
| Compacted at 50 tons/in² | 0.283 | 73 | 1 | 0 | |
| Sintered 8 h at 800°C | 0.284 | 34 | 16 | 9.5 | |
| Repressed at 50 tons/in² | 0.302 | 70 | 22 | 4.0 | |
| Repressed and resintered at 800°C | 0.301 | 39 | 25 | 17 | |
| After 25% cold reduction, annealed | 0.300 | 39 | 17 | 16.5 | |
| After 50% cold reduction, annealed | 0.309 | 41 | 25 | 22 | |
| After 75% cold reduction, annealed | 0.318 | 44 | 33 | 27.5 | |
| Cast and wrought copper, annealed | 0.323 | | 32 | 50 | |
| Iron: | | | | | |
| Compacted at 50 tons/in² | 0.224 | 70 | 0.5 | 0 | 0 |
| Sintered 8 h above 1000°C | 0.240 | 47 | 27 | 10 | 0 |
| Repressed at 50 tons/in² | 0.262 | 67 | 43 | 5 | 0 |
| Repressed and resintered above 1000°C | 0.260 | 63 | 35 | 21 | 19 |
| 25% cold reduction, annealed | 0.266 | 63.5 | 30 | 15 | |
| 50% cold reduction, annealed | 0.277 | 68.5 | 32.8 | 21.5 | |
| 75% cold reduction, annealed | 0.279 | 68.5 | 33.8 | 26.0 | |
| Electrolytic iron, cast and annealed | 0.283 | 67 | 37 | 50 | 15 |
| Hot-pressed brass: | | | | | |
| 90 Cu, 10 Zn, pressed at 900°C | | | 30 | 22 | 17 |
| 90 Cu, 10 Zn, cast and wrought | | | 37 | 45 | 10 |
| 80 Cu, 20 Zn, pressed at 900°C | | | 37 | 34 | 18 |
| 80 Cu, 20 Zn, cast and wrought | | | 44 | 50 | 14 |
| 70 Cu, 30 Zn, pressed at 800°C | | | 38 | 16 | 12 |
| 70 Cu, 30 Zn, cast and wrought | | | 47 | 62 | 15 |
| 50 Cu, 50 Zn, pressed at 700°C | | | 21 | 0 | 21 |

ksi × 6.895 = MN/m².

## POWDER METALLURGY

REFERENCES: Schwartzkopf, "Powder Metallurgy," Macmillan. Goetzal, "Treatise on Powder Metallurgy," Interscience.

Powder metallurgy (PM) involves (1) the production of metal powders of suitable characteristics and (2) consolidation of the powder to the desired density and cohesion by pressing and a subsequent or simultaneous heating operation known as sintering. Sintering may involve the formation of a liquid phase or may be carried out below the melting point of all constituents. Reasons for the use of powder metallurgy include (1) the ability to fabricate refractory or reactive metals, (2) the ability homogeneously to combine dissimilar materials, (3) the ability to produce metal of controlled porosity or permeability, (4) the ability to produce large numbers of certain small parts more cheaply than by competitive conventional techniques. Hot hydrostatic pressing of powdered compacts has increased size capability for PM parts, although typically such parts are relatively small (about 1 to 2 lb).

The strength, and more particularly the ductility, of powder-metallurgy products are almost always inferior to those obtained by casting or forging, although by special means, such as additional working and annealing, better properties can be obtained. Typical properties of sintered metals are given in Tables 44 and 45 and are there compared with those for typical cast and wrought metals. An exception to the general trend shown in the tables is the comparatively recent development of sintered aluminum powder. This material,

which incorporates a controlled amount of oxide, has elevated-temperature properties superior to those of any conventional cast- or wrought-aluminum alloy.

A considerable body of empirical knowledge is now at hand with respect to powder characteristics, die design, pressing techniques, sintering atmospheres, etc. (see particularly Goetzal). In general, these problems become more acute the larger and more complex the piece and the more reactive the metal.

## CEMENTED CARBIDES

REFERENCE: Schwartzkopf and Kieffer, "Refractory Hard Metals," Macmillan. Also see powder-metallurgy references above.

Cemented hard carbides constitute a highly important class of materials in modern technology. They are most generally produced by powder-metallurgy techniques. However, some special parts are produced by casting, and flame spraying and arc deposition have recently become prominent, particularly in hard-facing applications. The bulk of the cemented carbides are still based on tungsten monocarbide, WC, although new types have been introduced in recent years based on TiC and on $Cr_3C_2$. The most important application of the cemented carbides is as cutting tools, but a considerable volume is also used for dies and wear-resistant applications.

Carbides are produced by adding carbon to the powdered metal or oxide (or to mixtures of two or more metals or oxides) and then heating in a reducing atmosphere to a temperature in

**Table 45. Comparison of Mechanical Properties of SAE and Sintered Steels**
(Stern, *Trans. AIME*, **166**, 1946, p. 556)

| Treatment | SAE 1020 | | SAE 1040 | | SAE 1060 | | SAE 1080 | |
|---|---|---|---|---|---|---|---|---|
| | SAE | Sin. | SAE | Sin. | SAE | Sin. | SAE | Sin. |
| **Furnace-cooled:** | | | | | | | | |
| Yield point, ksi | 43 | 37 | 52 | 40 | 59 | 42 | 62 | 57 |
| Tensile strength, ksi | 66 | 55 | 88 | 61 | 95 | 69 | 102 | 79 |
| Elongation, % | 35 | 25 | 29 | 21 | 24 | 13.5 | 20 | 7.5 |
| Reduction in area, % | 55 | 27 | 60 | 22 | 51 | 12.5 | 48 | 5.5 |
| **Oil-quenched and drawn at 1300°F (704°C):** | | | | | | | | |
| Yield point, 1,000 lb/in² | 38 | 35 | 55 | 39 | 72 | 44 | 80 | 50 |
| Tensile strength, 1,000 lb/in² | 60 | 52 | 82 | 55 | 98 | 63 | 100 | 69 |
| Elongation, % | 35 | 35 | 30 | 30 | 26 | 18.5 | 23 | 10.5 |
| Reduction in area, % | 65 | 40 | 60 | 35 | 61 | 22 | 56 | 10.5 |
| **Oil-quenched and drawn at 800°F (427°C):** | | | | | | | | |
| Yield point, ksi | 57 | 50 | 76 | 55 | 105 | 65 | 125 | 80 |
| Tensile strength, ksi | 77 | 65 | 110 | 76 | 144 | 91 | 175 | 106 |
| Elongation, % | 30 | 25 | 19 | 18 | 17 | 13.5 | 12 | 8 |
| Reduction in area, % | 58 | 25 | 48 | 19 | 48 | 12 | 40 | 6.5 |

ksi × 6.895 = MN/m².

The sintered specimens were prepared from electrolytic iron powder and graphite. All specimens were pressed at 50 tons/in², presintered 15 min at 2000°F, repressed at 50 tons/in², and sintered 1 h at 2000°F. The sintered alloys do not contain the manganese and silicon of the SAE steels.

the neighborhood of 2600°F (1427°C). The time and temperature must be controlled to give the proper carbon content of the carbides and appropriate particle size. Special properties are claimed for WC-TiC compositions made by reaction in molten nickel alloy, the so-called "menstruum process." The finely granular carbide powder, however made, is intimately mixed with the powdered-metal binder (cobalt for the WC grades) by prolonged ball milling, and then pressed and sintered or hot pressed. In some instances, parts are presintered at low temperatures [~1500°F (816°C)] to give them sufficient strength to be ground or cut to more complex shapes than can be formed by pressing. Shrinkage of the sintered material must be allowed for. Final sintering is then carried out at a much higher temperature, specific for each composition. While the final sintering is usually done below the melting point of the binder metal, a liquid ternary phase is usually formed which is largely responsible for the densification and

bond strength. Tungsten carbide grades are usually sintered in hydrogen, but the newer TiC grades are vacuum sintered. (See Table 46.)

The strength and hardness of the cemented carbides are controlled by varying the amount of the binder metal; the lower the binder content, the harder and more brittle the composite. The strength and toughness may also be improved by reduction in the particle size of the carbides. Admixtures of other carbides to a given base are usually made to control oxidation resistance, thermal conductivity, or elastic modulus.

The modulus of elasticity of tungsten carbide is about 102 × 10⁶ lb/in²—higher than any other known material. The compressive strength of cemented tungsten carbide varies between 900,000 and 500,000 lb/in², depending on cobalt content. Thermal expansion is about 3.3 × 10⁻⁶/°F. Thermal conductivity is about 500 Btu/(h)(ft²)(in)(°F) for cemented tungsten carbide, decreasing to 185 for the double tungsten-

**Table 46. Composition and Properties of Cemented Carbides**

| Composition, percent | Specific gravity | Hardness, Rockwell A | Transverse rupture strength, ksi | Young's modulus, ksi × 10³ |
|---|---|---|---|---|
| WC + 3 Co | 15.25 | 92.7 | 170 | 97.5 |
| WC + 4.5 Co | 15.05 | 92.3 | 200 | 90.5 |
| WC + 6 Co | 14.85 | 90–92 | 225 | 88 |
| WC + 9 Co | 14.6 | 90 | 275 | |
| WC + 13 Co | 14.15 | 89 | 300 | 80 |
| WC + 10 TaC + 6 Co | 14.7 | 91 | 220 | |
| WC + 12 TiC + 6 Co | 11.2 | 92 | 160 | |
| WC + 10 TaC + 15 TiC + 8 Co | 11.7 | 92 | 165 | 72 |
| WC + 15 TaC + 15 TiC + 15 Co | 11.4 | 90 | 190 | 67 |
| WC + 2 TaC + 1 CbC + 20 Co* | 13.4 | 86 | 380 | |
| WC + 6 TaC + 11 TiC + 2 CbC + 9 Co* | 11.1 | 92 | 250 | |
| TiC + 10–40 Ni | 5.5–6.5 | 83–90 | 160 | 55 |
| Cr₃C₂ + 2 WC + 15 Ni | 7.0 | 88 | 100 | |

ksi × 6.895 = MN/m².

*Approximate values for a typical grade; compositions and properties for others in this grouping may vary considerably.

titanium carbides. A low thermal conductivity seems to favor resistance to cratering in machining steels, but in machining cast iron and other materials with discontinuous chip the high thermal conductivity and superior hardness of tungsten carbide are desirable. The lower moduli of the carbide compositions containing tantalum or titanium permit greater deflection before fracture than the harder tungsten carbides. The chromium carbide and particularly the titanium carbide bases have superior oxidation resistance at high temperatures. Their critical materials content is very low. The titanium carbide grades have the further advantages of low density and excellent high-temperature strength. The chromium carbides have a unique feature for gaging applications in that they have about the same expansion coefficient as steels so that results are not affected by ambient-temperature variations when working with steel.

Shaped tips for lathe and other cutting tools are copper-welded, brazed, or silver-soldered to a steel shank both for cheapness and for support of the brittle tip. In designing tips, it is important to avoid large changes of thickness and to design the backer so that stresses due to thermal contraction after brazing cannot crack the tip. In designing tools, it is important to use the lowest rake angle possible so that the tool may have the maximum support. Other applications of hard carbides include lathe centers, gage tips, guides of various kinds, Brinell balls, and the like. The piece should be shaped as closely as possible to final size before sintering, but a limited amount of forming can be done by grinding with special silicon carbide or diamond-impregnated wheels. Numerous other uses are also found for the cemented carbides, e.g., wear-resistant tips for plug and snap gages, scribers, files, wiredrawing dies, thread guides in the manufacture of synthetic fibers, die linings, and wear plates of many kinds. (See also Sec. 13.)

# CORROSION

## by Hilary E. Bacon

REFERENCES: Burns and Bradley, "Protective Coatings for Metals," ACS Monograph 129, Reinhold. Uhlig, "The Corrosion Handbook," Wiley. Evans, "The Corrosion of Metals," Edward Arnold. Friend, "The Corrosion of Iron and Steel," Longmans. Speller, "Corrosion—Cause and Prevention," McGraw-Hill. DePaul, "Corrosion and Wear Handbook," McGraw-Hill. Polar, "A Guide to Corrosion Resistance," Climax Molybdenum Co. Zapffe, "Stainless Steels," ASM. LaQue and Copson, "Corrosion Resistance of Metals and Alloys," ACS. Monograph 158, Reinhold. Bregman, "Corrosion Inhibitors," Macmillan. *Trans. ASTM*, Annual Reports of Committee A-5—Corrosion of Iron and Steel. Fontana and Greene, "Corrosion Engineering," McGraw-Hill. Purcell and Whirl, "Protection against Caustic Embrittlement by Coordinated Phosphate-pH Control," *Proc. 3d Annual Water Conf.*, *Eng. Soc. Pennsylvania*, 45–60. Klein, Use of Coordinated Phosphate Treatment to Prevent Caustic Corrosion in High Pressure Boilers, **10**, no. 4, Oct. 1962, p. 45.

## GENERAL CONSIDERATIONS

Corrosion is a destructive attack on metals which may be chemical or electrochemical in nature. Direct chemical corrosion is limited to unusual conditions involving highly corrosive environments or high temperature or both. Examples are metals in contact with strong acids or alkalies and the formation of iron oxide by dissociation of water in·contact with overheated boiler tubes. However, most of the phenomena involving corrosion of metals containing or submerged in water, or atmospheric corrosion by films of moisture, are electrochemical in nature.

The **mechanism** of electrochemical corrosion is most obvious in the case of electrically coupled dissimilar metals such as zinc and copper submerged in water, so that zinc forms the anode and copper the cathode of a galvanic cell. The reaction proceeds in two parts: (1) the **anodic reaction,** in which the metal dissolves in the electrolyte in the form of positively charged ions, and (2) the **cathodic reaction,** in which positively charged hydrogen ions plate out as atomic hydrogen on the cathodic surface. The electrons released by the anodic reaction flow through the metallic circuit to the cathode, where they neutralize an exactly equivalent number of hydrogen ions.

Anodic reaction: $M \rightarrow M^+ + e$
Cathodic reaction: $H^+ + e \rightarrow H$ (atomic)

The hydrogen film will eventually cover and **polarize** the cathodic surface, stopping the flow of electrons. The positive metal ions released near the surface of the anode combine with negative hydroxide ions from the water to form a neutral metal hydroxide, which frequently coats the anodic surface.

Overall reaction: $M^+ + OH^- \rightarrow MOH$

Thus under favorable conditions the electrochemical reaction will stifle itself at the cathode or anode or both.

The commoner case is a single metal in which the surface is heterogeneous in polarity, composed of small discrete anodic and cathodic areas so close together as to be indistinguishable. Small potential differences arise from minor variations in composition, surface finish, stress, deposits or inclusions, or concentration differences of electrolytic or gaseous solutes in the adjacent liquid phase. Penetration proceeds at the anodic areas resulting in a wide variety of pitting, roughening, or wastage. The cathodic hydrogen film and anodic insoluble precipitates are equally important stifling factors and can be utilized to retard corrosion.

An index of the driving force of the electrochemical reaction is the potential or emf of the galvanic cell, which is the sum of the potentials of the half-cells formed by the anode and cathode and the water surrounding them. These potentials are characteristic of the metal (or other electrode material) and vary with ionic content of the electrolyte, temperature, and other factors. Specific **electrode potentials** are qualified by fixing the equilibrium concentration of metal ions, conventionally, at 1 molal (one gram-molecular weight per liter); they can be determined only by currentless measurement, in the absence of electrode reaction. Otherwise, they are altered by polarization by an increment known as the **overvoltage,** related to the energy changes at the electrodes. The **hydrogen overvoltage** is associated with the release of hydrogen at the cathode.

These basic facts concerning mechanism are fundamental to explanation of electrochemical corrosion phenomena and methods of prevention.

## FACTORS IN ELECTROCHEMICAL CORROSION

**Metal**  The tendency of a metal to dissolve in water, which is known as **solution pressure,** can be measured by the (current-less) electrical potential which must be applied in order to prevent any action when the metal is immersed in a solution of one of its salts at standard (1 molal) concentration. From such information, the metals may be arranged in a series in the order of their solution pressures:

### Electromotive-Force Series of Metals

| | | |
|---|---|---|
| Magnesium | Iron | Copper |
| Beryllium | Cadmium | Mercury |
| Aluminum | Nickel | Silver |
| Manganese | Tin | Palladium |
| Zinc | Lead | Platinum |
| Chromium | Hydrogen (zero) | Gold |

Although the standard concentration of metal salts and complete absence of polarization currents are never encountered in practice, this series nevertheless lists the metals in order of decreasing corrodibility. When two metals in contact are immersed in water, that which is above the other in this series becomes anodic, suffers corrosion, and protects the other metal by rendering it cathodic.

The **electrolyte** is the controllable dominant factor. Normally the water in contact with metals contains impurities such as salts, gases, and vapors. Its **electrolytic activity** is a function of ions which increase its conductivity; thus high salinity promotes electrochemical corrosion. More important, water is rarely **neutral** (pH = 7.0) but is either **acidic** (pH below 7.0) or **basic** (pH above 7.0). (For definition and explanation of pH, see Sec. 9.) The excess of hydrogen ions accompanying a decreasing pH increases the driving force and the reaction rate of electrochemical corrosion; hence acids and acidic salts produce a corrosive environment. Conversely, an overbalance of hydroxyl ions (high pH) depresses electrochemical corrosion and may provide excellent protection even without film formation—as when caustic, ammonia, or amines are added to condensate or demineralized or completely softened water.

**Oxygen**  Gaseous oxygen dissolved in water reacts with the protective atomic hydrogen on cathodic areas of metallic surfaces, destroys the film by **depolarization,** and permits corrosion to continue. The rate of corrosion is roughly limited by the rate of diffusion of dissolved oxygen to the metal surface, and thus extensive oxygen attack occurs at or near the water line. Dissolved oxygen is dominant in the vast majority of corrosion problems, many of which can be completely solved by thorough mechanical, thermal, or chemical deaeration of water.

Figures 1 and 2 give the solubility of oxygen in parts per million in water in contact with saturated air at various pressures from 10 lb/in² gage to a vacuum of 28 in Hg and at temperatures up to 230°F. An entirely secondary action of dissolved oxygen is to oxidize the ferrous hydroxide, formed by the union of the metal and hydroxide ions, to the insoluble ferric state.

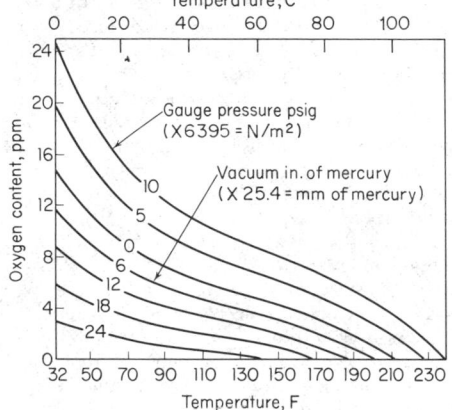

**Fig. 1**  Solubility of oxygen in water under saturated air at pressures up to 10 lb/in². (To convert ppm to cm³ per liter multiply by 0.698.)

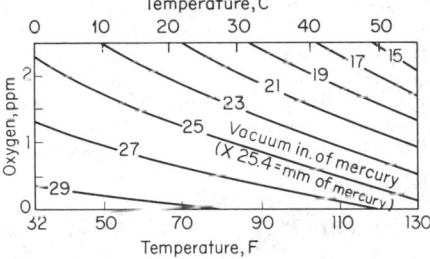

**Fig. 2**  Solubility of oxygen in water under saturated air at pressures down to 29 in vacuum.

**Factors Stimulating Corrosion**  In practice, conditions which promote corrosion may be inferred from what has been said about metals, electrolytes, and dissolved oxygen. Any condition of nonuniformity within the metal, such as may arise from improper annealing or cold working, may increase the heterogeneity and intensity of polarity differences, and obviously the use of **dissimilar metals** may cause one of them to corrode, as in the case of aluminum heat-exchanger tubes adjacent to iron tie rods in the tube bundle. Nonuniformity of concentrations in the electrolyte produces **concentration cells** and is thus favorable to corrosive attack. It is well known that nonuniformity in distribution of dissolved oxygen renders the areas exposed to low oxygen concentration anodic with respect to those in contact with higher oxygen concentration. Thus, **turbulence** and entrained air cause corrosion at the inlet ends of condenser tubes and on the tube sheet. The depolarization of the hydrogen film by dissolved oxygen (and hence corrosion) occurs more rapidly on some metals than on others, zinc and aluminum being less active than iron, and is catalyzed by the presence of **mill scale. Atmospheric corrosion** is stimulated by a damp atmosphere, since this maintains a film of water on the metal, providing the essential electrolyte. Other factors are **acid gases** in the atmosphere or sulfur compounds from cinders, coke, coal dust, etc.; salts that dissociate to produce an acid reaction; oxygen dissolved in the water film. All these factors, although sometimes in other form, apply to **immersed corrosion** and to **underground corrosion.**

**Factors inhibiting corrosion** are likewise deduced from the foregoing relations of metals, electrolytes, and dissolved oxygen to electrochemical corrosion. Protective measures include the use of **sacrificial anodes** such as zinc or magnesium, the impression of a counter emf by various **cathodic protection** devices, and **passivation** of metals (so treating them as to reduce their solubility in acids and the rate of precipitation from other metal ions from solutions). Contact with the electrolyte is prevented by various **protective coatings,** either applied or precipitated from solution, as in the case of calcium carbonate, hydrated chromium, and iron oxides.

**Electrolysis,** which is not to be confused with the fact that corrosion is fundamentally an electrochemical phenomenon, is generally caused by the leakage of current from electric circuits and may take place at a point far removed from where the leakage occurs. Electric-railway track feeder and return systems and improperly grounded welding equipment have caused many cases of corrosion due to electrolysis. Undue emphasis is often given to this phase of the corrosion problem. Stray currents of extremely feeble intensity and voltage may serve to accelerate corrosion, even when they have not initiated it. Corrosion due to electrolysis can be minimized by providing thorough insulation, grounding all metallic conduits, avoiding combinations of dissimilar metals in a circuit, and maintaining apparatus in an electronegative state with reference to possible sources of current, either by a "drainage" system or by "typing up" with some sources of current of a higher potential.

## CORROSION RESISTANCE OF VARIOUS METALS

**Iron and Steel** Under similar conditions, iron and steel corrode at practically the same rate, but the distribution of the corrosion may be different for the two. The slag interspersed through wrought iron may result in a generally distributed attack under atmospheric corrosion rather than a severely localized (pitting) attack. In underground corrosion and in continuous immersion in water, tests have not shown much difference between the two materials.

**Polished surfaces** resist corrosion much better than rougher surfaces. Variations in surface finish may have a greater influence than ordinary variations in chemical composition other than pronounced segregation. The presence of mill scale on the surface favors localized or irregularly distributed corrosion. Frequently a polished surface will withstand exposure for a considerable length of time before showing signs of corrosion.

External conditions overbalance composition of the metal in determining rates of corrosion. Oxygen determines the commencement of corrosion of iron and steel under ordinary conditions; it not only acts as a depolarizer but also unites with ferrous iron at the corrosion anodes. In the general absence of oxygen, corrosion drops to a negligible rate. In local absence of oxygen, a differential oxygen cell may be set up which serves to accelerate corrosion in the oxygen-poor portion. In

salt solutions, corrosion depends both upon the amount of oxygen present and upon the salt in solution.

**Rust** may accelerate corrosion and cause pitting. The probable explanation (Evans) is that surface accumulations of rust shield the underlying metal from free access to oxygen, thus rendering such portions anodic (corrodible) with respect to unshielded areas to which oxygen has freer access (cathodic areas). Under certain conditions of exposure, especially atmospheric, rust may form so continuous and adherent a coating that it protects the underlying metal from further corrosive attack. This is especially true of copper-bearing steel and iron under atmospheric exposure. Rust adheres much better to cast iron than to rolled iron or steel; the superior corrosion resistance of cast iron is attributable in large measure to this fact.

**Cold working** of metals results in an increased rate of attack by **acids**; it also increases heterogeneity in metals, which may lead to increased rate of corrosion. Local cold working, as in the rolling of boiler tubes, builds up internal stresses and greatly increases corrosion rates of such parts; thorough annealing after cold working is the remedy.

**Stress corrosion** is brittle fracture of a normally ductile material. Four factors are involved: (1) the material must be ductile, before attack; (2) a specific environment must be present, under which stress corrosion of the particular metals may occur; (3) duration of exposure must be sufficient, and (4) tensile stress must be present. Examples are ammonia and copper alloys, as in condensers and heat exchangers; fused caustic soda and nickel; chloride solutions and stainless steels, as $MgCl_2$ from seawater in tubular exchangers; concentrated caustic soda solution and ordinary steel if under stress, as in steam boilers.

ASTM has monitored extensive **corrosion tests** by atmospheric exposure and immersion of panels and hardware of black iron, alloys, and metallic coatings. For results see annual reports of Committee A-5 "Corrosion of Iron and Steel," 1918 to date. **Alloys** of commercially pure iron with either **cobalt, nickel,** or **copper** in small amounts (0.25 to 0.30 percent) are resistant to atmospheric corrosion because they form protective rust coatings. Copper or nickel alloyed in small concentrations have no certain effect one way or another on corrosion when alloyed with iron or steel. The presence of a small amount of copper, e.g., 0.20 percent, in black sheets exposed to the atmosphere is very advantageous. Although such materials rust readily when exposed to the atmosphere, the rusting is not progressive. Unlike the rust coating that forms on ordinary steel, the coating on copper-bearing steel is smoother, more adherent, and relatively impervious, and serves to protect the underlying metal. As with ordinary steel, the "life" of copper-bearing steel likewise varies with the conditions to which it is subjected. In ASTM tests of 16 gage sheet steel and iron, in a marine atmosphere at Annapolis, MD, the first failure occurred after 26 years of continuous exposure, whereas, in the atmosphere of Pittsburgh, PA, all the specimens failed within 5 years.

According to other ASTM tests, the presence of copper in the same amounts as in the materials used in atmospheric-exposure tests does not materially improve the life of steel immersed continuously in water, either in treated city water (Washington), brackish river water (Annapolis), or acid mine water (Pittsburgh). Data leading to similar conclusions concerning the life of underground structures were obtained in the soil-corrosion tests carried out by the NBS.

Certain low-alloy low-carbon steels with an alloy content

below 5 percent (see Sec. 6, Iron and Steel) have an improved corrosion resistance. These steels have a relatively high yield strength and tensile strength, and permit thin sections which would not be admissable without the improved corrosion resistance.

**Stainless steels** depend on the presence of chromium for their resistance to corrosion and heat to heat. Such steels fall into four groups: martensitic, ferritic, austenitic, and precipitation-hardened or age-hardened (Fontana and Greene). The martensitic and ferritic types are magnetic, hardenable, and useful for cutlery, machine parts, and resistance to atmospheric corrosion; austenitic types, e.g., type 304 (18 Cr, 8 Ni) are not hardenable but are more widely useful for resistance to corrosive liquids. Precipitation or age-hardened steels which can attain tensile strength of 200,000 lb/in$^2$ (ca. 1,400 MN/m$^2$) are used in missiles and aircraft.

**Passivity** is due to a surface condition which, in the austenitic alloys, is a self-repairing oxide film preserved in an oxidizing environment, e.g., nitric acid. Passivity may be lost if oxygen is withdrawn, as in crevices or under a bit of foreign matter in stagnant water. Corrosion resistance can be seriously impaired by welding or improper heat treatment, whereby the metal becomes susceptible to intercrystalline corrosion. This can be overcome by proper alloy additions stabilizing the material. Paul and Moran (*ACS Monograph* 158, pp. 375 et seq.) note the effectiveness of **halide ions** in penetrating the passivating films, causing severe pitting or stress-corrosion cracking. Thus, when moisture condenses within idle steam turbines, it can form droplets of chloride solution from an imperceptible film of boiler-water salts and cause pinhole penetrations. **Stress-corrosion cracking** has become more important because of the use of austenitic stainless-steel pipe and tubing in high-pressure power stations of both the conventional and the nuclear type. Alloys containing molybdenum (such as types 316 and 329) are more resistant to halides than type 304.

**Cast Iron** Corrosion occurs in buried cast iron in certain types of soil. The product of this **graphitic corrosion** plugs the hole so that a pipe, if undisturbed, is still serviceable in low-pressure systems. Cast iron can be greatly improved in corrosion resistance by alloying elements, such as chromium and nickel. **Silicon cast iron** (13 to 14 percent Si) is resistant to most acids and many other chemicals, though not to alkalies; as the iron is weak and brittle, its use is limited to shapes cast to size. Trade names for this type of alloy are Duriron, Corrosiron, and Tantiron. **Nickel cast iron** (13 to 15 percent Ni) has a high corrosion resistance toward many chemicals and to dilute acids and has the advantage of superior strength and toughness.

**Metals in Brine** According to the work of N. B. Smith (*Ice and Refrig.*, **12**) with neutral calcium chloride brine (sp gr 1.2), copper alloys are most corrosion-resistant, wrought iron, steel, and cast iron constitute an intermediate group, and lead, solder, and zinc constitute a group having lower corrosion resistance. The durability of any metal in brine, however, varies with the condition of the brine. For general purposes, a pH of 8.5 (pH 7.0 being the neutral point) is recommended. Brine with a high pH is corrosive to aluminum and zinc. The corrosiveness of a brine is increased two or three times by saturating it with air. Damage from corrosion fatigue in salt water which is deaerated is less than in fresh water containing air. Chromium is the most effective alloying element for increasing corrosion-fatigue strength in the absence of hydrogen sulfide, and nickel is best in the presence of hydrogen sulfide.

## CORROSION PROBLEMS IN STEAM-GENERATING PLANTS

(See also Sec. 9)

### Corrosion of Boilers

Central steam-electric power-station equipment containing or submerged in water at moderate temperatures and pressures suffers corrosion because of an electrochemical mechanism (see Section 9), and it is affected by similar accelerating and retarding factors and corrective measures. At saturation temperatures above moderately low pressures, a second mechanism predominates, in which iron removes oxygen from water or steam, forming iron oxide and releasing hydrogen:

$$3Fe + 4H_2O \rightarrow Fe_3O_4 + 8H$$

It is noteworthy that this mechanism does not require the intervention of dissolved gaseous oxygen in the water, which is often the rate-limiting factor in the electrochemical corrosion discussed earlier in this subsection.

The stable oxide at boiler temperatures in a nonoxidizing environment is **magnetite**, $Fe_3O_4$ (ferrous ferrite). A normal protective skin of magnetite is formed from the underlying steel. One theory (Potter) holds that loosely adherent magnetite is formed at the water-oxide interface by ferrous ions diffusing through the oxide film and that dense, hard magnetite is formed at the oxide-steel interface by OH$^-$ ions diffusing inward through the film. Both reactions release hydrogen. The magnetite formed at the oxide-steel interface is dense and adherent, that at the water-oxide interface less so. This surface also receives increments of oxide precipitated from iron brought in from the preboiler cycle, forming ferrous hydroxide and ultimately magnetite:

$$3Fe(OH)_2 \rightarrow Fe_3O_4 + 2H_2O + H_2$$

Under favorable conditions all boiler and superheater surfaces are covered with a thin film of black magnetic oxide formed by the above mechanism, but it is constantly being broken down, and repaired with the release of hydrogen. Consequently, the concentration of **hydrogen in steam** (usually of the order of 1 to 2 parts per billion) is an index to the rate of corrosion, and in critical cases a hydrogen recorder will serve as a corrosion alarm. (Hydrogen release by the second reaction above is of a smaller order of magnitude.)

Although the control of internal corrosion in boilers operated at low pressures and high pressures is similar, the problem becomes more complex as the temperature and pressure are elevated. The modern trend toward very large boilers, with operating pressures over 2,000 lb/in$^2$ (14 MN/m$^2$) becoming commonplace, has introduced new forms of corrosion which were not encountered in low-pressure boilers.

The special corrosion problems affecting modern boilers are described in four reports on A Research Study on Internal Corrosion of High Pressure Boilers, *Trans. ASME, Journal of Engineering for Power*, **88**, 232, July 1966; **89**, 378, July 1967; **90**, 21, Jan. 1968; **91**, 75, April 1969.

### Factors in Corrosion of Boilers

**Scale** In earlier times, the conditioning of feedwater was less exacting than now required, and it was common practice to permit accumulations of scale or sludge on the tube surfaces. Such deposits there formed a physical dam between the water and the metal, and corrosion was prevented. In recent years, scale and sludge formations have been minimized or eliminated by modern feedwater conditioning, so that control of corrosion has become essential for the protection of equipment.

**Noncondensible gases** released from water in boilers are probably the greatest single cause of corrosion in boiler tubes, drums, economizers, and superheaters. Oxygen corrosion produces rapid pitting, and it is recognized that the rate of corrosion is directly proportional to the dissolved oxygen concentration of the feed water. Other dissolved gases, such as carbon dioxide, hydrogen sulfide, and compounds which release gaseous products also accelerate the rate of attack at high pressures and temperatures.

The removal of gases from feedwater is effected in modern steam stations by mechanical (vacuum) and thermal deaeration, by the addition of chemicals, or by a combination of those processes. The basic method is to heat the water by direct contact with steam in either open heaters or in the more efficient type known as **deaerators**. In general the water is sprayed, atomized, or trickled over a stack of metal trays, to present the largest possible surface, while steam is admitted and directed so as to sweep out the noncondensible gases as they flash out of solution. By the use of steam at not less than 5 lb/in² [227°F (108°C)], efficient distribution, and liberal venting through a vent condenser, residual oxygen can be reduced to 0.005 parts per million. Refined testing methods have recorded values as low as 0.001 or below. The removal of other dissolved gases is likewise effected by contact heaters or by deaerators, but gases which form compounds with water (ammonia, carbon dioxide) may not be reduced to as low a level as oxygen.

Removal of residual oxygen by chemicals generally follows thermal deaeration as insurance against impaired efficiency. Several chemicals can be added to water which will remove oxygen; these include, in chronological order of use, **tannins**, **sodium sulfite**, and **hydrazine**. This type of treatment is known as scavenging. Gallo-tannins behave like alkaline pyrogallol (as used in gas analysis) in scrubbing oxygen from water at elevated temperatures, but are limited to low-pressure boilers. Sodium sulfite is currently the chemical most widely used at pressures below 900 psig (6.3 MN/m²) for removing traces of oxygen remaining in water after thermal deaeration.

The rate of reaction of trace concentrations of sulfite and oxygen as found in feedwater is slow, but can be greatly speeded up by the addition of **catalysts**. Cobalt nitrate is most widely used for this purpose, although copper sulfate has also been employed. A few years ago excess sulfite in concentrated boiler waters was maintained between 20 and 40 ppm, or higher. Such dosages are now known to be unnecessary and may actually cause trouble in high-pressure boilers. An excess of 10 to 20 ppm is generally satisfactory up to 900 lb/in² (6.3 MN/m²), 3 to 7 ppm up to 1,500 lb/in² (10 MN/m²), while above this level, sulfite may cause difficulty by decomposing.

**Hydrazine hydrate** ($N_2H_4.H_2O$) is a powerful reducing agent and will effectively prevent oxygen corrosion if properly applied and controlled. There is evidence that it functions on the surface of metal and suspended matter, and can coexist with dissolved oxygen. The residual quantities maintained in feedwater vary from 0.02 to 0.10 ppm. Rate of feed required depends on oxygen input, number of concentrations, and other factors. At high temperature, hydrazine partially breaks down to form ammonia, and the rate of input should be held to a level permitting a detectable residual, but just below the

point of ammonia increase. Experience has demonstrated that hydrazine is very effective in preventing corrosion and for limiting the quantity of black iron oxide ($Fe_2O_4$) in boilers.

**The effect of pH** on corrosion at feedwater and boiler-water temperatures is similar to its effect on electrochemical corrosion at ordinary temperatures, i.e., the solution pressure of iron is sharply depressed as the pH is elevated. In the absence of oxygen, iron pickup becomes negligible in most feedwater circuits as the pH is elevated to 8.8 to 9.2. Industrial boilers operating at moderate pressures [below 1,000 lb/in² (7 MN/m²)] may be treated with caustic soda, if it is not already present as a result of makeup treatment, to maintain an alkalinity which will buffer the water within the desired range. Exceptions are central power-station boilers operated at higher pressures and low makeup, in which caustic may attack the protective iron oxide film with the production of a sodium-iron compound:

$$FeO + 2NaOH \rightarrow Fe(ONa)_2 + H_2O$$

The necessary concentrations do not occur in the main body of the boiler water but can develop in the concentrating film next to the tube wall or under porous deposits of metal oxides. In the first case, the result is a **broad, smooth groove** which reduces the wall thickness to the point of failure. In the second case, jagged pits are filled with magnetic iron oxide produced by reaction of the exposed steel. The failure occurs as a pinhole leak. To prevent these types of corrosion, some boilers are operated with "zero caustic" treatment, by maintaining all alkalinity in the form of trisodium phosphate (**coordinated phosphate treatment**) by using di- or monosodium phosphate or a polyphosphate to control pH.

This condition exists when the pH is at or below specific levels in relation to phosphate as $PO_4$, as shown by a graph published by Purcell and Whirl and revised by H. A. Klein. Typical values are as follows:

| $PO_4$ | 10 | 15 | 20 | 25 | 30 | 40 | 60 |
|---|---|---|---|---|---|---|---|
| pH | 10.03 | 10.20 | 10.33 | 10.43 | 10.50 | 10.60 | 10.80 |

If localized caustic attack persists, congruent phosphate control is recommended in which the mol ratio of Na to $PO_4$ is lowered to 2.65:1 or 2.85:1. Volatile or "zero solids" treatment must be used in once-through boilers in which evaporation is complete at the point of steam release and in many high-pressure boilers which will not otherwise produce steam of satisfactory quality. In such cases, only ammonia and hydrazine are added to maintain a pH of about 8.5 to 9.0.

**Copper, nickel,** and **zinc** are often found in boiler sludge and deposits, their sources being in nonferrous-alloy tubes of condensers, evaporators, and heat exchangers. Much of the copper is metallic, in the form of either a spongy mass or very thin sheets. If thin films develop on the boiler tubes, they may provide a mechanism for concentration of solids beneath them, leading to the type of damage described above where caustic is present in the boiler water. Masses of spongy copper bonded to the metal in boiler drums are often found covering corroded areas. Whether the formation of metallic copper occurs by reduction of the oxide or salts by hydrogen or occurs by galvanic action is uncertain.

Histories of tube failures in high-pressure boilers almost always report the presence of excessive amounts of metal oxides, dominated by **black iron oxide**, frequently in the form of a loose, fluffy sludge that accumulates in crevices and at the ends of boiler drums. This may be harmless even when present in considerable quantities. However, on the surface of the boiler tubes the layer of magnetic iron oxide may far exceed the desired thin protective film, the excess being derived from the underlying metal or the preboiler circuits, or both. Mention has been made of the diffusion of ferrous ions from the underlying metal and of hydroxyl ions from the interface with water—forming, respectively, dense or loose magnetic iron oxide. When the thickness exceeds several mils (5 g/ft²) (ca. 50 g/m²) **chemical cleaning** may be necessary to prevent damage to the metal.

**Hydrogen damage** is caused by the diffusion of hydrogen through steel, reacting with carbon to form methane, which builds up high local stresses at the interfaces between grains, forming voids which may ultimately produce failure. Decarburization is visible to varying degrees in the microstructure (Partridge et al.). Hydrogen damage is found only in the steel underlying an area where visible corrosion has occurred and is a common problem in the operation of high-pressure boilers.

There are several mechanisms by which the film on the surface of the boiler metal may be destroyed at an abnormally rapid rate, followed by the re-formation in situ of magnetic iron oxide in such a manner that failure develops. These mechanisms occur in areas subjected to rates of heat input above the average for the boiler furnace, which may be as high as 200,000 Btu/(ft²)(h) (0.63 MW/m²). Under these conditions it is difficult to maintain a thin protective film on the tube surfaces, and water of the highest quality, with very adequate rates of circulation, must be furnished to these sections of the boiler. Deposit-producing impurities such as silica, iron, or copper cannot be tolerated because precipitation is accelerated by high temperatures, and, in turn, the temperature is raised by even a very slight addition to the normal film, or by very thin scale. The deterioration thus feeds on itself and progresses to point of failure unless the boiler is effectively acid-cleaned. At excessively high temperatures the dissociation of steam which is catalyzed by the oxide surface becomes so rapid that the underlying metal cannot be protected. The cause of abnormal heat input and temperature conditions may be flame impingement or the impact of radiant heat; film boiling; or steam blanketing, caused by deficient circulation of water, which is usually avoided by boiler design.

The several corrosion phenomena described above and other types of destructive attack on boiler steel can often be identified by cutting a section of metal from the damaged area, mounting it in plastic, polishing and etching the surface, and taking photomicrographs which will show fine details of grain structure. Boetcher described identifying characteristics of several types of failure which can be classified by this technique. In **corrosion fatigue,** caused by alternating stresses set up by vibration or other periodic mechanical or thermal changes, the attack develops along a concentration of stress and the cracks are predominantly transcrystalline. In severe cases, cracking is produced by periodic stress alone without the interaction of a corrosive medium, but where both mechanisms occur, corrosion penetrates into the metal with the crack. By contrast, high-temperature **creep cracks** resulting from slow plastic flow under sufficient stress and temperature are predominatly intercrystalline and not related to corrosion. Study of properly prepared photomicrographs by a competent metallurgist may reveal the thermal history of the specimen in terms of maximum temperature and duration.

**Caustic embrittlement,** although not strictly classifiable as corrosion, is caused by a selective attack at the grain boundaries, believed to be due to the protective action of silica on the crystal faces. Necessary conditions are high stress and slow leakage accompanied by evaporation which will permit the caustic to build up to many times the concentration maintained in the boiler water. A third condition is the "embrittling character" of the boiler water.

Embrittlement of boiler steel was a dominant factor in boiler-water conditioning for many years. Statistical studies revealed an association of embrittlement failures with high caustic concentrations in the boiler water while relative immunity appeared to be conferred by the presence of sodium sulfate in certain ratios to the caustic. The ASME "Suggested Rules for Care of Power Boilers" recommended specific sulfate-alkalinity ratios in the 1926 and subsequent editions, but research and field experience have failed to confirm the merits of sulfates and have developed more effective inhibitors. These include tannins and related organic substances (suitable only for low-pressure boilers) and sodium nitrate. The "coordinated phosphate" or "zero caustic," conditioning described above is also a successful corrective measure. With the advent of forged and welded drums and the disappearance of riveted joints, opportunities for embrittlement are limited to the rolled joints between boiler tubes and the tube sheets or headers, where embrittlement has occurred in a few cases. Zapffe contends that penetration of hydrogen into the crystalline structure, its reactions and formation of gaseous products, explains cracking ascribed by others to caustic embrittlement.

The **boiler economizer** is particularly susceptible to pitting attack since it is the first high-temperature surface to be exposed to oxygen dissolved in the feedwater. The water passing through it has a relatively low pH value and contains no inhibiting chemicals. Effective treatment consists of deaeration of feedwater and treatment with alkaline and reducing chemicals. In some cases, a portion of the boiler blowdown has been recirculated through the economizer.

**Superheater** corrosion is due largely to the reaction between metal and steam at high temperatures. A secondary cause is carry-over of boiler-water salts. In this case, high concentrations of sodium hydroxide may form on the tube walls with direct chemical attack or may bake on the walls and cause blistering.

### Corrosion in Preboiler Equipment

The feedwater circuits of a modern steam plant usually begin with the surface condensers of turbogenerators or steam-condensing process equipment and include direct-contact or tubular feedwater heaters, surge tanks, pumps, and piping. The metals in this equipment present a large surface area to the circulating feedwater, and appreciable iron, copper, and alloying components go into solution and enter the boilers to become constituents of boiler sludge. This metal pickup is kept within tolerable limits by deaerating in condenser hot wells or feedwater heaters at the earliest point in the cycle where the pressure is adequate, and by maintaining the pH at as high a level as feasible. Where the makeup is softened by hot-process treatment and is an appreciable percentage of the feedwater, the pH is usually 9.5 or higher and presents no control problem. Where ion-exchange or evaporation processes are used to produce makeup, or large amounts of condensate are recovered, the normal alkalinity and pH of the feedwater may be too low to give the desired protection. The usual control measure is continuous feed of caustic soda, ammonia, or a neutralizing amine. To minimize input of chemicals, boiler blowdown may be recirculated to the surge tank of the deaerating heater or some other low-pressure point. Measures for specific control of feedwater pH should aim at a level not below 8.5, and marked reduction in metal pickup is accomplished by raising the pH to 9.0 or a little higher. Ammonia may be present in the water supply in sufficient amounts for pH control or may be added in the form of ammonium hydroxide, but its concentration in contact with copper-alloy tubes or other parts must be limited, as it forms highly soluble ammonium complexes which wash away readily. This is accelerated by dissolved oxygen. Ammonia initiates stress corrosion of yellow-brass alloys in some cases.

In studies of **ammonia, cyclohexylamine,** and **morpholine,** Bigger and others found that these alkalizing materials differed greatly in their distribution between the vapor phase and the liquid phase. For a given concentration in the steam leaving the boiler, morpholine provides the highest fraction in the first-formed film of condensate in turbines, low-pressure heaters, etc., and thus is the most effective of the three for this treatment. Cyclohexylamine has been used to protect condensate return lines of building heating systems but because of its volatility is lost at the deaerating heater and at vents in the system. Roughly twice as much morpholine or cyclohexylamine must be used, in parts per million, as the free $CO_2$ which must be neutralized. Where **hydrazine** is used to prevent oxygen corrosion, partial breakdown contributes ammonia which must be considered in balancing the pH control treatment.

### Corrosion of Other Steam-Plant Equipment

**Condenser tubes** are attacked by circulating cooling water, which may be relatively nonaggressive river or lake water, highly polluted, brackish or seawater, or water recirculated through spray ponds or cooling towers. Tube materials must be carefully selected for resistance to the specific water supply. The metals include admiralty, Muntz, aluminum brass and aluminum bronze, cupronickel, red brass, and copper. Resistance depends upon building up, by initial corrosion, a uniform protective film. Admiralty brass (Cu 70 percent, Zn 29 percent, Sn 1 percent), the most widely used condenser-tube material, is normally satisfactory for seawater or polluted fresh-water service over a wide range and is usually inhibited with a fraction of 1 percent of arsenic, antimony, or phosphorus. For very severe conditions it may be replaced by aluminum brass (Cu 76 percent, Al 2 percent, Zn 22 percent) and 70–30 or 80–20 cupronickel, which are both resistant to impingement pitting and dezincification. Where the latter is persistent, either cupronickel or aluminum bronze (Cu 93.5 percent, Al balance) is used to eliminate zinc. Red brass and copper are not suitable for salt or brackish water, but deoxidized copper has special uses in process and refrigeration equipment.

Even when the material is resistant to corrosive constituents dissolved in the water, attack may occur through several mechanisms: (1) **deposit attack** on areas shielded by adhering foreign matter, producing a local oxygen deficiency and making the underlying metal anodic; (2) **dezincification,** in which both components of the metal are dissolved and the copper is redeposited as a porous plug or layer; (3) **impingement** of air bubbles, usually only near the inlet end, preventing protective film formation or causing undercutting; and (4) **corrosion fatigue** due to unusual stresses or vibration. **Season cracking** due to the latter can be eliminated by proper annealing to remove stress.

**Pumps** may be corroded by the solvent action of the liquid passing through them (see Sec. 14). When dissimilar metals in contact are used, e.g., a cast-iron body with brass glands, seats, etc., autoelectrolysis is almost certain to result when brines or other electrolytes are pumped. If it is not possible to confine the construction to one metal, the less-resistant metal should be used only for easily renewable parts.

Corrosion may result from the leakage of current from the driving motor. This can be prevented by using an insulating coupling between the pump and motor shafts.

Air in the pump, either free or in solution in the liquid pumped, accelerates corrosion. Pumps should be vented so that they may be completely filled with the liquid they are handling.

Manufacturers' design and selection of materials for **boiler feed pumps** are based on a questionnaire covering the mineral analysis, pH, and chemical treatment of the feedwater, as well as temperature and pressure conditions. Corrosion-erosion attack is aggravated by water of high purity, which is frequently required, because of absence of alkaline salts to buffer the pH at a high enough level. For such service, alloys of high chromium content may be necessary, while ordinary ferrous metals of proper physical characteristics are acceptable for feedwater containing appreciable alkalinity with the pH controlled at 8.5 or above. Bronze parts may be attacked by highly alkaline water from hot-process softening systems.

## CORROSION OF EQUIPMENT IN CHEMICAL PLANTS

Chemical plants as a rule are subject to severe damages caused by corrosion. Not only must consideration be given to providing for continuity of operation in order to meet production schedules, but the products of corrosion may detract from the quality of manufactured goods. Maintenance expense is often an important item in manufacturing costs, and an economic balance between the selection of more expensive special materials and equipment upkeep is usually an important operating problem, requiring study and judgment for its satisfactory solution.

Removal of hydrogen from cathodic sufaces by oxygen is often the controlling reaction. When nonoxidizing acids are present, depolarization by hydrogen evolution may overshadow hydrogen oxidation. When an oxidizing acid such as nitric is present, oxidation of the hydrogen film occurs, and is independent of the presence of dissolved oxygen. Strong **nitric acid** passivates iron or steel and reduces the corrosion rate.

**Caustic alkalinity** attacks silica and corrodes high-silicon alloys. Lead, zinc, aluminum, and tin are attacked by caustic solutions. Hydroxide alkalinity is more effective than carbonate or bicarbonate alkalinity in covering iron or steel surfaces with a protective film. The hydroxide film is more dense and impermeable.

**Temperature** plays an important role in corrosion by chemicals; as in other types of corrosion, the rate is sharply increased by higher temperatures.

When strong **sulfuric acid** is stored in steel tanks, and the level fluctuates through comparatively wide limits, air is drawn in or expelled. Moisture in the air is absorbed by the strong acid with which unsubmerged surfaces are wetted, thus reducing its strength. This occurs mostly along the sidewalls, and the diluted acid corrodes the steel. Damage to tanks can be eliminated by installing a desiccator unit on the breather, thus drying all air entering the acid storage tank and consequently preventing corrosion due to formation of weak acid.

In chemical plants where **corrosive vapors or gases** are discharged to the atmosphere, serious corrosion of steel throughout the plant may take place. In many instances it is impracticable to prevent the escape of these gases or to provide special metals which would resist corrosion from this source. Often the most economical solution to the problem is to design steel structures with extra thickness of metal so that even after considerable corrosion, sufficient metal remains to provide the structural strength required. Under such atmospheric conditions, some plants, in designing reinforced concrete, place reinforcing steel further from the surface than would be done if only structural strength were considered. In this case, economy in design is sacrificed to gain better protection for the reinforcing steel.

Tanks or basins used for **storage of acids or alkalies** can be protected by special brick and cement linings. Such linings are usually bonded to the steel by means of an impervious membrane which is designed to withstand the special conditions existing.

**Solvents** liberated when enamel is baked on wire may be burned by means of catalytic combustion which, because of the catalyst, ignite at relatively low temperatures. This method of disposal prevents damage to painted surfaces when the solvent vapors are discharged through stacks to the atmosphere.

In the **contact process for manufacturing sulfuric acid,** it is customary to use cast iron for constructing coolers serving the final absorption towers. When oleum is produced, it has been found necessary to substitute steel pipe with cast-steel bands in preference to cast iron. Sulfur trioxide reacts with silicon in the cast iron, causing rupture of the metal. Coolers of this type are designed for very low velocities of the acid passing through the pipes. This prevents removal of the iron sulfate film which forms and protects the cast iron or steel used in this equipment. Heat transfer is sacrificed in order to obtain long life of the equipment.

**Chemical Equipment** The development of corrosion-resistant steels, of high-nickel alloys, and of ferrosilicon alloys has extended greatly the uses of the ferrous materials in the chemical field. As examples of this extension, **stainless-clad** steel sheet is available for applications that require good thermal conductivity in addition to the corrosion-resistance of stainless steel. **Enameled steel** or iron ware is used widely, but most enamels are generally attacked by alkalies. **Rubber-lined tanks** and other chemical equipment, together with hard-rubber piping and connectors, now serve a very useful purpose. The number of nonferrous metals and alloys useful for chemical equipment is exceedingly large. **Lead** is widely used, and the development of lead-lined steel pipe and tanks has greatly extended its sphere of usefulness. **Nickel,** nickel-clad steel, and alloys high in nickel are widely used. Special **aluminum bronzes** that have been modified by the addition of some other element such as iron are now coming into use for handling acids. For certain uses, especially when natural flavors and odors must be retained, **silver** is used.

## CORROSION OF PIPES AND STRUCTURES

**Underground Pipes** Steel, wrought iron, cast iron, and lead often corrode severely in certain natural soils. Soils containing organic or carbonaceous matter, such as coke, coal, or cinders, or impregnated with acid wastes from manufacturing plants are highly corrosive in their action. The autoelectrolysis of iron buried in soils is recurrent, since oxygen is always present in sufficient amount to act as a depolarizer.

The NBS investigation of soil corrosion, summarized in *NBS Circ.* C450, 1945, has established the following facts:

1. **Ferrous Metals** Serious corrosion underground occurs in the absence of stray currents, although electric currents, which did not originate in power plants, have been detected on pipelines. Soil conditions play a major role. Wrought iron and steel corroded at approximately the same rate, and the presence of copper in the steel did not improve its corrosion resistance. In certain soils, the corrosion rate of cast iron was greater than that of steel. The removal of mill and foundry scale from the surface did not greatly affect the rate of corrosion. Depth of pitting was directly proportional to duration of exposure in certain soils, but in others the pit depth increased very slowly after the soil conditions had become stabilized. Distribution of corrosion tended to become more uniform as the exposure period increased.

2. **Nonferrous Metals** No metal was found outstandingly superior under all conditions. The corrosion rate (loss of weight and depth of penetration) was, with but few exceptions, greater for ferrous than for nonferrous metals. The presence of chlorides, bicarbonates, and sulfates in the soil retarded the rate of corrosion of **lead,** although in certain soils, as in tidal marshes, the rate of attack did not decrease with time. **Copper** and high-copper alloys corroded very slowly in most soils, the presence of sulfides being the principal accelerator of corrosion, with the rate of attack being particularly high in cinders. High-zinc brass became weak and brittle by dezincification in some soils. **Zinc** corroded rapidly in a few soils, the rate being proportional to the duration of exposure. This limits the effective life of galvanized coatings on ferrous metals. The useful service is proportional to the thickness of the zinc coating; a coating of 2.8 oz/ft² (854 g/m²) prevented the formation of pits for 10 years, except in one very corrosive soil. Over a 10-year period, the loss of weight of galvanized steel was one-half, or less, that of the companion bare steel. Three **aluminum** alloys corroded rapidly under most of the soil conditions to which they were exposed.

3. Pipes buried in **mixtures of two dissimilar soils** or in two unlike soils in contact (not mixed) generally corrode more rapidly than in either of the two soils separately. Lead in a mixture or simple contact of dissimilar soils will often corrode markedly. Pipes buried in trenches fulfill these conditions of soils in contact or mixed. Metals corrode at the junction line of dissimilar soils. Cast iron in soils takes on a coating of rust and soil, pits being filled with carbon and black iron oxide. Lead shows both the gray and brown oxides when corroded in soils. Potentials up to 1 V may be generated readily by placing unlike metals in a given soil or by using one metal in two dissimilar soils.

4. **Bituminous coatings** are widely used, especially natural asphalt and blown asphalt made from oil residuum and coal-tar pitch. Coatings of this last substance were somewhat inferior to others in the NBS investigation in that they were

more susceptible to soil stresses, temperature changes, softening, and brittleness, although they were more resistant to water absorption. The method of application of all such coatings is important. Dipping, brushing, and spraying methods are all used. For severe soil conditions, the coating should be reinforced by wrapping the coated pipe spirally with a strong fabric of some kind which has been impregnated with a waterproof bituminous mixture. For exceptionally severe corrosive soil conditions, pipes may be encased in concrete.

**Transite** (asbestos cement) pipes were not deteriorated by 4 years' exposure in a variety of soils.

Pipes are often destroyed by the action of **dissimilar metals** to which they are connected, such as brass valves. Short connections, readily replaced, should be used on either side of the brass fittings.

Application of a negative-potential booster to diminish potential differences in electrical rails has proved to be an effective means of reducing electrolysis of underground iron and steel, such as gas and water mains.

**Bridges, Roofs, Stacks, etc.** Iron and steel structures are corroded by the presence in the atmosphere of moisture and waste products from manufacturing and metallurgical plants such as carbon dioxide, sulfur dioxide, chlorine, ammonia, zinc and acid fumes, and soot. As a rule, they are not corroded in a dry atmosphere. In order to minimize the corrosive effects of gases, protective coatings should be applied are used. Metal work exposed to the action of sulfur fumes should be covered with brick or with a paint highly resistant to sulfur dioxide. Bridges are now protected by means of so-called **drip floors** placed over the floor proper. The failure of **metal smokestacks** by corrosion occurs only during idle periods and particularly in the summer. Deposits of soot and fly ash assist very materially in this by absorbing the corrosive gases in the products of combustion. These combine with atmospheric water vapor, and the corrosive attack is essentially that of a weak acid. The ordinary materials used for the construction of smokestacks do not differ very much among themselves in their resistance to the corrosive attack, although there is some indication that some of the newly developed low-alloy steels are superior.

**Concrete Structures** The alkaline nature of cement assures great permanence to structural steel embedded in concrete. This has been repeatedly confirmed by the examination of the steel in reinforced-concrete buildings which have been demolished. The following information on deterioration by corrosion resulting from electric current is from *NBS Tech. Paper* 18.

When current passes from an iron anode into concrete, oxides of iron form on the anode, occupying about 2.2 times the volume of the equivalent iron, and giving rise to mechanical pressure [sometimes as high as 4,700 lb/in² (32 MN/m²)] which may result in cracking the concrete. At temperatures below 113°F (45°C), and with a potential gradient less than 60 V/ft (200 V/m), this action is very slight, even in wet concrete.

Concrete near the cathode, or metal through which the current leaves the concrete, becomes softened and remains brittle and friable after drying, destroying the bond between the iron and concrete. This effect is noted at all voltages, high or low, and is due to the concentration of sodium or potassium near the cathode. The content of these elements in the cement should therefore be kept low.

Salt (NaCl), or calcium chloride, should never be added to concrete used in structures that will be subject to electrolytic action. Concrete in contact with salt water is very susceptible to electrolysis. In structures exposed to the action of salts, pickling solutions, etc., the potential gradient must be kept low. In the absence of metallic electrodes, the action of the current is similar to slow seepage, the water-soluble elements in the concrete migrating toward the cathode. Grounding of electric conductors in such a structure is equivalent to installing electrodes.

**Waterproofing compounds** when mixed with concrete have but little effect in preventing electrolysis. Waterproofing membranes, properly applied, are fully efficacious in preventing the entry of earth currents. Painting or otherwise coating the reinforcing metal may minimize the danger from electrolysis, but it prevents proper bonding between the metal and the concrete.

All dc circuits within a building should be kept free from grounds. All pipelines entering a building should be installed with insulating joints outside the building; if passing through, they should have insulating joints on both sides of the building. If the potential drop around the isolated section is 8 V or more, the isolated section should be shunted by a copper cable. Lead-covered cable should be kept out of contact with the building.

All metallic structures within a building should be interconnected, provided that all lines entering the building are installed with insulating joints, but they should not be grounded to any ground plate lying outside the insulating joints. Maintaining the reinforcing metal negative is worse than no protection at all.

**Cinder concrete** is frequently used in certain types of building construction. When the concrete is properly made and poured so as to eliminate voids adjacent to the reinforcing steel, very satisfactory service can be relied upon. If it is porous, pipes buried in it sometimes suffer severe corrosive attack by the acid-forming constituents that may be present. Proper precautions should be taken to neutralize any acidity in the cinder, especially that which is in immediate contact with the pipes.

An important use of cement in preventing corrosion is the use of **cement linings for pipes** carrying water. They have proved very useful for corrosive waters such as mine drainage as well as for less severe service, as in water mains. The process is now being applied regularly to cast-iron pipe and also to steel pipe even as small as that used in ordinary water service. A special type of coupling for cement-lined pipe is required. Lead-lined couplings are very useful.

Lead embedded in concrete is sometimes severely corroded as a result of free lime present. Lead shower-bath pans are a common example. Protection of the surface of the lead by a bituminous coating is always recommended, and the use of alumina cement, as an added precaution, can also be recommended. Aluminum and aluminum alloys in intimate contact with plaster or cement may suffer corrosion for the same reasons. Similar precautionary measures are advisable.

## METHODS FOR MINIMIZING CORROSION

Corrosion may be minimized by (1) the use of a coating of protective metal such as zinc, tin, lead, nickel, or copper; (2) the production of oxide, phosphate, or similar coatings on iron and steel surfaces; (3) the application of protective paints; and (4) rendering the surface of the metal passive.

**Coating Iron with Other Metals. Galvanizing** Zinc is applied to metal surfaces by the Sherardizing process, by dipping into a bath of molten zinc, by electrodeposition, or by metal spraying.

In the **Sherardizing process,** the articles, after being thoroughly cleaned by pickling and sandblasting, are placed in a metal drum together with zinc dust and heated to a temperature of from 500 to 600°F (260 to 315°C), depending on their size and shape, the drum being rotated so as to promote "rumbling" of the contents. The coating that results is not pure zinc, but an alloy of about 90 percent zinc and 10 percent iron [melting point, 1260°F (680°C) approx], and is highly resistant to corrosion. The process is especially suited for screws, bolts and nuts, chains, pipe fittings, nails, small castings, and such other articles as may conveniently be placed within the drum. The cost varies with the character of the articles coated. The term **electro-Sherardizing,** which is often used, merely connotes that the Sherardizing furnace is heated electrically. By a suitable annealing or heat treatment, the zinc coating produced on sheet and wire by hot dipping can be converted into a similar alloy coating. A cheap method used for nails, etc., consists in rumbling them in zinc dust. The coating is inferior in quality.

In the **hot process,** the articles after being thoroughly cleaned are dipped into a bath of molten zinc. The bath must be maintained at a temperature somewhat higher than the melting point of zinc, which necessitates a large fuel consumption and also results in a considerable loss of zinc (approx 10 percent). A greater source of loss is the iron-zinc alloy (dross) which forms as a heavy sediment in the zinc bath. That portion of the zinc surface through which the material to be coated enters the zinc bath must be kept covered with a flux; ammonium chloride and zinc chloride are widely used for this. The process is used almost exclusively for sheet and pipe, and, until recently, for wire also. Exposed structural-steel work, such as towers, is generally zinc-coated by this means. Sheet and wire are coated by mechanical means, pipe and structural shapes by hand. Many irregular shapes (pots, vats, tubs, etc.) and small shapes (bolts, nuts, nails, screws) are coated by hand. For the latter, some means for removing excess zinc, such as centrifuges or shaking devices, is generally used. A very small percentage of aluminum renders the zinc bath very fluid and is favored by many in coating irregular shapes. One or two percent tin is often added in the coating of sheets in order to obtain a very uniform coating and to improve the surface appearance. A coating applied by hot dipping never consists of a simple layer of zinc. It is always of a composite nature, the layer adjacent to the base metal consisting of zinc-iron alloys. This layer is relatively brittle and, thereby, imposes some limitations on hot-dipped galvanized sheet and wire for certain uses. By a recent process, formation of the alloy can be practically eliminated. A coating of 1 oz/ft$^2$ (305 g/m$^2$) of exposed surface is considered very suitable for most conditions of service.

The **electrolytic or cold process** consists in setting up the articles to be coated as cathodes in an electrolytic bath of soluble zinc salts, the anode being metallic zinc. Small articles are placed in metallic baskets in contact with the basket and with each other, the basket being attached to the cathode of the system. Both the acid sulfate and the cyanide bath are used, the latter particularly for small articles and, with suitable addition agents, for the recently developed "bright" coatings. The use of zinc plating on a large commercial scale has been advanced by its application to the coating or steel wire. The high ductility of the pure zinc coating obtained is the outstanding feature of such a coating. The ease of control of the uniformity and thickness is also advantageous. The coating of sheet steel by plating appears to be imminent.

In the **metal spraying or metallizing process,** metal wire or powder is fed at a controlled rate into the flame of an oxygas or oxyacetylene torch. The impingement of this atomized metal on a prepared surface produces a layer of flattened and interlocked particles which are mechanically bonded to the surface being coated. To ensure good adhesion, the surface of the base material must be clean and roughened, for example, by sandblasting. The process is especially useful for applying thick coatings to limited areas, for rebuilding or reclamation of worn or undersized parts (such as cylinders, shafts, or pistons), and is also used to supply protective and decorative coatings. The deposited metal is less dense than cast metal; porosity in the coating may be overcome by cold work, heat treatment, increasing the thickness of the coating, or by using two or more metals in the same coating. Sprayed coatings of zinc, aluminum, and cadmium need not be entirely free from pores as these metals are anodic to iron and steel. On the other hand, coatings of metals which are cathodic to iron and steel—for example, nickel, stainless steel, copper, tin, or lead—should be as free from pores as possible since the base metal would be attacked through the pores of the coating. Sprayed coatings of zinc or aluminum range from 0.001 to perhaps 0.015 in (0.0254 mm to 0.38 mm) in thickness, depending upon the severity of exposure. The porous characteristics of the sprayed coats furnish an excellent base for the application of paints, varnishes, and other corrosion-inhibiting or protecting mediums. The process is readily adapted to production coating of small articles of metal or of nonmetallic materials such as wood, leather, or plastics, and is particularly useful for coating large assembled structures such as bridges, towers, storage tanks, and canal gates. Sprayed coatings of aluminum and of chromium-nickel alloys are used for protection against a combination of high temperature and corrosive action, for example, in the oil industry and in furnace equipment.

**Cadmium** behaves similarly to zinc as a coating metal for iron in affording electrochemical protection from corrosion. Cadmium coatings are applied commercially by electroplating and are preferred by many, the claim being made that a thin coating of cadmium gives the same degree of protection as a thicker one of zinc. This is true to some extent under marine conditions, but in an industrial atmosphere, where sulfur compounds are present, the useful life of cadmium coatings is shorter than that of comparable zinc coatings.

**Coating with Tin, Lead, Nickel, Copper, or Chromium** Coatings of tin or of a tin-lead alloy are used principally on thin sheets of iron, being applied as a rule in a manner similar to that of the hot process of galvanizing. When so-called **terneplate** is made, a mixture of tin (25 percent) and lead (75 percent) is generally used. Such coatings, if free from pinholes, are highly resistant to corrosion. The excellent "paint-holding" properties of terneplate fit it for many uses in building construction. Its "lubricating properties," in drawing and stamping processes, assist greatly in the manufacture of containers and fuel tanks.

Lead coatings on steel are most efficient in a polluted atmosphere as in industrial centers. In rural atmosphere, pinhole corrosion is soon evident. A "bonding" agent, either as an alloy in the lead or as an undercoat, is necessary for the process commonly used in applying lead coating, i.e., the hot-dip process.

In coating or electroplating with **nickel or copper,** the object to be coated is made the cathode, the anode consisting of a block of the metal to be deposited, and the electrolyte, a solution of the metal to be deposited. In nickel plating, a copper coating is generally applied before the nickel to render the latter more adherent and more corrosion-resistant by making the entire coating more impervious. **Chromium** coatings which are now widely used are also produced by electroplating from chromic acid solutions. They are nearly always applied as a very thin finish on a nickel coating and are almost perfect in their tarnish resistance. Coatings of these three metals protect the underlying or base metal from corrosion only insofar as they exclude air and moisture. Hence the imperviousness of such coatings as determined by the conditions of deposition is of very great importance. Chromium plating, on account of its great hardness, is also used to some extent as a protection against wear and abrasion.

**Copper** is sometimes applied in coats integral with the base metal. Copper is cast around a steel billet which is afterward worked down to the required size. Most of the steel coated in this way is used for wire and gives a combination of high electrical conductivity and high tensile properties. It is also used for purposes requiring the combination of high corrosion resistance and strength as in concrete revetment mats used in river control. Steel clad with nickel or stainless steel is also available commercially.

**Aluminum coatings** cannot be made by electrodeposition in the ordinary way, although aluminum has been deposited from complex organic liquids in the form of coatings on steel. Aluminum coatings are usually produced by mechanical means. Such a coating has proved very useful on duralumin, but on iron or steel its usefulness is limited by the brittleness of the intermediate alloy layer which generally forms.

Steel coated by immersion in molten aluminum (see below) resists atmospheric corrosion admirably. Its rough unattractive appearance restricts its use.

**Calorizing** is a process by which a coating of aluminum and aluminum-iron alloys is produced on iron and steel (brass, copper, or nickel may also be calorized) which protects the metal against high temperatures because of the formation of aluminum oxide on the surface. It does not protect against ordinary corrosion in the atmosphere or in liquids. Sulfurous acid and carbon monoxide have no appreciable effect on calorized metal. Bending or working of calorized metal should be done at a bright-red heat; threading must be done before calorizing, followed by chasing of the threads to make them fit smoothly without breaking the coating; dimensions and weights are increased by calorizing. Thermal and electrical conductivities are not appreciably changed. The base metal is soft-annealed by the process. A few uses are soot-blowing apparatus, carbonizing boxes, furnace parts, and condenser and economizer tubes. Good results are regularly obtainable up to a temperature of 1700°F (927°C) and in some cases to 1830°F (1000°C). Calorizing is carried out by either a powder or a dip process. In the powder process, the parts to be treated are placed in a tight receptacle partly filled with a mixture of finely divided metallic aluminum and aluminum oxide. The air is replaced by hydrogen, and the receptacle is subjected to a high temperature for a time that depends on the depth of penetration of aluminum desired. In the dip process, the parts are fluxed, then immersed in molten aluminum and then heated to promote alloying. This method gives a thinner coating of aluminum alloy but is much more expeditious. A calorized surface resists continued temperatures up to 1800°F (982°C) but begins to burn at approximately 2000°F (1093°C), whereas ordinary steel begins at about 930°F (500°C). It is not affected by ordinary oxidizing furnace conditions.

**Calite,** which may be considered as representative of a large class of heat-resistant alloys, is an alloy of iron, nickel, and aluminum and resists oxidization up to 2200°F (1204°C) for an indefinite time and, for short periods, up to 2370°F (1300°C). The protective oxide formed does not snap off on quenching from extremely high temperatures. It is practically noncorro-

dible under ordinary conditions of exposure. Many other combinations, containing chromium and nickel as essential alloying constituents, are commercially available.

A process termed **chromizing,** similar in its operation to calorizing, uses powdered chromium or ferrochromium, the articles to be coated being heated while embedded in the powdered metal within a hydrogen or other nonoxidizing atmosphere. It has only a very limited commercial application.

**Magnetic Oxide on Iron Surfaces**  In the **Bower-Barff process,** the iron or steel articles to be coated are heated in a closed retort to a temperature of 1600°F, after which superheated steam is admitted. This results in the formation of red oxide ($Fe_2O_3$) and magnetic oxide ($Fe_3O_4$). Carbon monoxide is then admitted to the retort to reduce the red oxide to magnetic oxide, which is highly resistant to corrosion. Each operation takes about 20 min. The glossy black coating of magnetic oxide on **Russia iron** is produced by laying up sheets of iron with powdered charcoal between, the whole mass being then heated and hammered.

Iron and steel may also be **oxide-coated** by electrolytic means, the object to be coated being made the anode (anodic oxidation) in an alkaline solution. Such coatings are primarily for appearance such as for cast-iron stove parts. Though experimentally successful, the commercial application of the process is limited. The **Chemag process,** a German development, is of this kind.

**Phosphate Coatings for Rustproofing Iron and Steel** (Eckelmann, *Chem. & Met. Eng.,* Dec. 24, 1919.) In the **Coslett process,** iron or steel articles immersed for 3 or 4 h in a boiling solution, made by mixing iron fillings with concentrated $H_3PO_4$ (sufficient to form a paste) and then adding to weak phosphoric acid, become coated with a rust-resisting deposit of basic ferrous phosphate. This process was improved by the addition of an oxidizing agent, the **Parkerizing** process, and later by the addition of other accelerators (**Bonderizing).** The phosphate coating, in itself, affords only a very slight degree of protection against corrosion. Oiling the coating improves the corrosion resistance greatly and imparts an attractive lustrous black appearance. Coatings of this kind are not suitable for severe outdoor service. Phosphating a steel surface is an excellent method of priming prior to subsequent painting or lacquering. The phosphate is applied by spraying. It finds extensive application for automobile bodies and is known commercially as **Bonderizing** and **Granodizing.** Some of the phosphate treatments are electrolytically applied. The phosphate treatment is also applicable to zinc surfaces. The advantages of this latter process over the Bower-Barff and similar process are in greater cheapness and simplicity and the use of low temperatures.

Treatments somewhat analogous to the Parkerizing treatment are widely used for the treatment of magnesium alloys prior to the application of other types of coatings, such as aluminum-pigmented spar varnish and pyroxalin coatings of the Duco type.

**Protection of Aluminum Alloys**  Wrought-aluminum alloys are largely used for aircraft purposes. Alloys containing copper as an essential alloy constituent, in sheet form, are susceptible to intercrystalline corrosive attack which results in the material becoming very brittle with little or no surface evidence of the change. Aluminum alloys containing magnesium or magnesium and silicon, and the newer alloys containing zinc and chromium as the essential alloying constituents, are very stable under prolonged weathering conditions. The protective coatings used on aluminum alloys depend on the severity of the service. A preliminary anodizing treatment to produce a film of oxide on the surface (formed by making the article the anode of a cell with chromic, sulfuric, or oxalic acid as the electrolyte) is now common practice. Such a surface oxide coating forms an excellent basis for the application of other coatings. Aluminum-pigmented spar varnish is excellent for this. For very severe marine conditions, the only coatings that give permanent protection are those of aluminum.

**Alclad** products are made by casting a shell of corrosion-resisting aluminum (2S or 72S) around a billet of aluminum alloy. Fabrication of the coated or clad billet produces sheet, rod, wire, and various rolled and extruded shapes which possess the physical properties of the alloy and the corrosion resistance of the aluminum cladding. Tubing clad only on the inside is also available. Aluminum coating applied by the metal-spraying process is useful, especially for heavy pieces and assembled structures.

The **passivating of iron surfaces** may be accomplished in several ways, the most common consisting of immersion of the metal in nitric acid (sp gr 1.4) after it has been highly polished. Other methods consist in immersing the metal in fuming sulfuric acid, potassium ferrocyanide, or potassium chromate solution or in chromic acid; coating with a manganese dioxide paint; cathodic pickling in a weak acid solution, the metal being made the cathode in a circuit of low voltage; treatment with arsenic-sodium nitrite, etc. This condition of passivity is temporary and, thus far, passivating has been of doubtful value, except for stainless steel for which it is a regular practice. A chromate treatment now used commercially on zinc surfaces is somewhat analogous to some of the above treatments.

Treatment of **cooling water** to form protective films is widely applied in recirculating cooling-tower systems as well as once-through systems serving refrigeration or process condensers, heat exchanges, etc. Deposition of **calcium carbonate scale** may be controlled to avoid interference with heat transfer while still protecting the metal, by adjusting the methyl-orange alkalinity, calcium hardness, and pH of the water so that the Langelier **calcium carbonate saturation index** is slightly positive (0.1 to 0.3) for the existing temperature and total solids. Powell has published a graph which may be used for calculating the index from the above variables. The control may require reduction of calcium hardness by lime or zeolite softening; reduction of alkalinity by lime softening or feeding sulfuric acid, which is widely practiced; adjustment of pH downward or upward with alkalies or acids. Total solids, reflecting ionic strength as a factor in scale formation, must be controlled by blowdown, preferably under 1,500 to 2,000 ppm. Chlorination is usually needed to destroy organic slimes which inhibit crystalline scale formation.

Addition of enough **sodium dichromate** and caustic to maintain 200 to 500 ppm $Na_2CrO_4$ oxidizes ferrous iron to form a tough, self-healing film of mixed oxides of the two metals, covering anodic areas. Optimum pH is about 8.5. The chromate residual requirements rise with temperature and salinity; if enough is not maintained, pitting will result, and the rapid formation of ferrous ions will consume more chromate than would be needed to give more complete protection. To avoid prohibitive chemical cost and waste-disposal problems, the system must be operated at maximum concentration and minimum blowdown. Because of toxicity hazard, exposure of personnel to tower-drift spray or water discharging into funnels or sumps must be avoided. Chromate treatment is excellent for entirely closed bearing-cooling systems, circulating through copper-alloy heat exchangers cooled by a secondary flow of corrosive characteristics, e.g., seawater.

Various **polyphosphates** are added to cooling water to form protective films and also to sequester scale-forming components by **"threshold" treatment**. Effectiveness approaching complete protection may require 10 to 30 ppm in terms of orthophosphate ($PO_4$), and the pH is controlled at 6.5 or lower to avoid precipitation of orthophosphates and control scale. In the **dianodic** treatment, both phosphate and chromate are maintained at a total concentration of the order of 60 ppm in either a 1:2 or 2:1 ratio.

Another modification is the use of chromated organic compounds, such as glucosates, to extend the covering power of the film with lower chromium concentrations.

In 1970, new applications had to be prepared for permits for all discharges into surface waters as part of a complete reclassification of such receiving bodies and establishment of standards for quality. Sudden attention was given to **nutrients** (phosphorus, organic carbon, etc.) and **toxic chemicals** (chromates) which had been present for a long time in cooling-tower blowdown and which had to be abandoned. Prolific improvisation of substitutes followed, with emphasis on organic **polymers**. Some of these are **tannins** and **carbohydrates** which form a protective film by coprecipitation with weaker inhibitors: sodium silicate, small amounts of chromates or zinc salts, chelated iron, etc. Dosages are recommended by the manufacturer, who may also stipulate pH and cycles of concentration in the circulating water. Effectiveness of these treatments varies.

**Slushing oils** are usually nondrying oils or greases which remain soft for prolonged periods, are strongly adhesive on metals, but can readily be removed when desired. The best protection to metals is afforded by acid-free semisolid oils applied in a melted condition. Types of commercial slushing oils are petroleum residues, mixtures of lithopone and iron oxide with heavy petroleum residues, petrolatumlike compounds emulsified with chromate solutions, blown vegetable oils, soft asphalts, and thinned, rosin-base materials. A specification proposed several years ago is as follows: Coating shall adhere firmly at all temperatures to which it will be exposed; coating shall be readily removable with cotton waste wet with kerosine; polished iron, steel, copper, or brass shall show no staining when exposed to weather at any temperature below 212°F (100°C) for not less than 5 days; in the salt-spray test, no rust shall be formed in 24 h, practically none in 5 days, and no appreciable rust in 60 days. Typical formulas are: (1) 20 g rosin of "H" grade + 100 g petrolatum (USP) + 10 cm³ kerosine. The rosin is melted and mixed with the hot petrolatum, after which the kerosine is stirred in. Rosin greatly increases the adhesiveness of the petrolatum. Wax may be added to raise the melting point of the petrolatum if necessary. (2) 3 parts candelilla wax, 6 parts "H" rosin, 50 parts petrolatum (USP). (3) 2 parts carnauba wax, 5 parts "H" rosin, 50 parts petrolatum (USP). In each case, melt the ingredients together at 255°F, stir, and cool. Flow an excess over the metal surface (*NBS Tech. Paper* 176 and *Circ.* 200 and 214). Lanolin is the best grease to use as a basis of slushing oils. A small amount of sodium chromate is desirable in slushing greases unless all traces of water have been eliminated.

# PAINTS AND PROTECTIVE COATINGS

## by George G. Sward

REFERENCES: Keane (ed.), "Good Painting Practice," Steel Structures Painting Council. Levinson and Spindel, "Recent Developments in Architectural and Maintenance Painting." Levinson, "Electrocoat, Powder Coat, Radiate." Reprints, *Journal of Paint Technology*, 121 S. Broad St., Philadelphia. Roberts, "Organic Coatings—Their Properties, Selection and Use," Government Printing Office.

**Paint** is a mixture of filmogen (film-forming material, binder) and pigment. The pigment imparts color, and the filmogen, continuity; together, they create opacity. Most paints require volatile thinner to reduce their consistencies to a level suitable for application. An important exception is the powder paints made with fusible resin and pigment.

In **conventional oil-base paint,** the filmogen is a vegetable oil. Driers are added to shorten drying time. The thinner is usually petroleum spirits.

In **water-thinned paints,** the filmogen may be a material dispersible in water, such as solubilized linseed oil or casein, an emulsified polymer, such as butadiene-styrene, a cementitious material, such as portland cement, or a soluble silicate.

**Varnish** is a blend of resin and drying oil, or other combination of filmogens, in volatile thinner. A solution of resin alone is a **spirit varnish,** e.g., shellac varnish. A blend of resin and drying oil is an **oleoresinous varnish,** e.g., spar varnish. A blend of nonresinous, nonoleaginous filmogens requiring a catalyst to promote the chemical reaction necessary to produce a solid film is a **catalytic coating.**

**Enamel** is paint that dries relatively harder, smoother, and glossier than the ordinary type. These changes come from the use of varnish instead of oil as the liquid portion. The varnish may be oleoresinous, spirit, or catalytic.

**Lacquer** is a term that has been used to designate several types of painting materials; it now generally means a spirit varnish or enamel, based usually on cellulose nitrate or cellulose acetate butyrate.

## PAINT INGREDIENTS

The ingredients of paints are drying and semidrying vegetable oils, resins, plasticizers, thinners (solvents), driers or other catalysts, portland cement, and pigments.

**Drying oils** dry (become solid) when exposed in thin films to air. The drying starts with a chemical reaction of the oil with oxygen. Subsequent or simultaneous polymerization completes the change. The most important drying oil is linseed oil. **Raw linseed oil** requires 3 or 4 days to dry. Addition of small percentages of driers shortens the time to 5 to 10 h.

Heat-treating, or cooking, bodies (thickens) oil. If driers are added during the cooking, the product is **boiled oil.** Boiled oil is made with a minimum of cooking to avoid discoloration, and also by adding concentrated driers to raw oil. **Blown oil** is made by bubbling air (oxygen) through the oil.

**Soybean oil,** with poorer drying properties than linseed oil, is classed as a **semidrying oil. Tung oil** dries much faster than linseed oil, but the film wrinkles so much that it is reminiscent of frost; heat treatment eliminates this tendency. Tung oil gels after a few minutes cooking at high temperatures. It is more resistant to alkali than linseed oil. Its chief use is in oleoresinous varnishes. **Oiticica oil** resembles tung oil in many of its properties. **Castor oil** is nondrying, but heating chemically "dehydrates" it and converts it to a drying oil. The dehydrated oil resembles tung oil but is slower drying and less alkaliresistant. **Fish oil,** extracted from the menhaden and the sardine, has considerable use in paints and varnishes. **Nondrying oils,** like coconut oil, are used in some baking enamels. An oxidizing hydrocarbon oil, made from petroleum, contains no esters or fatty acids.

The drying properties of oils are linked to the amount and nature of unsaturated compounds. By molecular distillation or solvent extraction the better drying portions of an oil can be separated. Drying is also improved by changing the structure of the unsaturated compounds (isomerization) and by modifying the oils with small amounts of such compounds as phthalic acid and maleic acid. The latter process borders on varnish making.

**Driers** are oil-soluble compounds of certain metals, mainly lead, manganese, and cobalt. They accelerate the drying of coatings made with oil. The metals are introduced into the coatings by cooking their oxides or salts in the varnish or oil, or by addition of separately prepared compounds. Certain types of synthetic nonoil coatings (catalytic coatings) dry by baking or by the action of a **catalyst** other than conventional drier. Drier action begins only after the coating has been applied. Catalyst action usually begins immediately upon its addition to the coating; hence addition is delayed until the coating is about to be used.

**Resins** Both natural and synthetic resins are used. **Natural resins** include fossil types from trees now extinct, recent types (rosin, Manila, and dammar), lac (secretion of an insect), and asphalts (gilsonite). **Synthetic resins** include ester gum, phenolic, alkyd, urea, melamine, amide, epoxy, urethane, vinyl, styrene, rubber, petroleum, terpene, cellulose nitrate, cellulose acetate, and ethyl cellulose. In general, the natural resins are compatible with drying oils, or may be made so, and are used in both oleoresinous and spirit varnishes; many are suitable for spirit varnishes only.

**Nonresinous Filmogens** Some filmogens are neither oil nor resin. Among these are casein, soybean protein, glue, starch, portland cement, and water-soluble silicates, all of which are used in water-thinned paint. The latexes might also be included.

**Plasticizers** In an oleoresinous varnish the oil acts as a softening agent or plasticizer for the resin. There are other natural compounds and substances, and many synthetic ones, for plasticizing film formers such as the cellulosics. Plasticizers include blown vegetable oils, camphor, dibutyl phthalate, tricresyl phosphate, dibutyl sebacate, dibutyl tartrate, tributyl citrate, methyl abietate, and chlorinated biphenyl.

**Thinners** As the consistency of the film-forming portion of most paints and varnishes is too high for easy application, thinners (**solvents**) are needed to reduce it. If the nonvolatile portion is already liquid, the term thinner is used; if solid, the term solvent is used. **Petroleum spirits** has largely replaced turpentine for thinning paints in the factory. **Turpentine** is sometimes preferred for on-the-job thinning of architectural paints. The coal-tar products—**toluene, xylene,** and **solvent naphtha**—are used where solvency better than that given by petroleum spirits is needed. Esters, alcohols, and ketones are standard for cellulosic lacquers. **Chlorinated solvents** are required for some of the synthetic resins. Finally, water is used to thin emulsion and cement-base paints, to be thinned at point of use.

**Pigments** may be natural or synthetic, organic or inorganic, opaque or nonopaque, white or colored, chemically active or inert. Factors entering into the selection of a pigment are color, opacity, particle size, compatibility with other ingredients, resistance to light, heat, alkali, and acid, and cost. The most important pigments are: **white**—zinc oxide, leaded zinc oxide, titanium dioxide, lithopone; **red**—iron oxides, red lead (for rust prevention rather than color), chrome orange, molybdate orange, toluidine red, para red; **yellow**—iron oxide, chrome yellow, zinc yellow (for rust prevention rater than color), Hansa; **green**—chrome green, chrome oxide; **blue**—iron, ultramarine, phthalocyanine; **extenders** (nonopaque), i.e., magnesium silicate, calcium silicate, calcium carbonate, barium sulfate, aluminum silicate.

**Miscellany** Paint contains many other ingredients, minor in amount but important in function, such as emulsifiers, antifoamers, and thickeners.

**Ecological Considerations** Many jurisdictions restrict the amounts of solvents such as benzene, ketones, and chlorinated compounds that may be discharged into the atmosphere during coating operations. Users should examine container labels for applicable conditions.

## PAINTS

**Paste paint** contains only a part of the oil and thinner needed for the complete paint; the balance is added by the user. Multiple-pigment pastes contain from 65 to 80 percent of pigment.

**Aluminum paint** is a mixture of aluminum pigment and varnish; from 1 to 2 lb of pigment per gal of varnish. The aluminum is in the form of thin flakes. In the paint film, the flakes "leaf"; i.e., they overlap like leaves fallen from trees. Leafing gives aluminum paint its metallic appearance and its impermeability to moisture. Aluminum paint ranks high as a reflector of the sun's radiation and as a retainer of heat in hot-air or hot-water pipes or tanks.

**Bituminous Paint** Hard asphalts, like gilsonite, cooked with drying oils, and soft asphalts and coal tar, cut back with thinner, may be used to protect metal and masonry wherever their black color is not objectionable.

**Calcimine,** a mixture of glue and pigment, is sold as dry powder to be mixed with water. **Casein paint** is a mixture of casein, solubilizing agent, and pigment; it is sold both as dry powder and as paste. **Emulsion paints** are emulsions in water of oil or varnish base mixed with pigment. The **latex paints** (acrylic, butadiene-styrene, polyvinylacetate, etc.) belong to

this class. Calcimine and casein paints are used mainly on interior plaster and masonry surfaces, and may be used on wood that has been primed. Emulsion paints are designed for both exterior and interior masonry and are more washable than casein paints. Calcimine is not washable.

**Electrical Paint** Most paints, especially those not water-thinned, are electrically insulating to a high degree. Electrically conductive paint contains large percentages of metal powder or carbon pigment.

**Strippable coatings** are intended to give temporary protection to articles during storage or shipment.

**Luminous paints** are of two kinds: **phosphorescent,** which continue to glow after removal of the activating energy; and **fluorescent,** which glow only during activation. Activation may be by sunlight, artifical light, ultraviolet radiation, or by radioactive compound added to the paint. The last method makes paint self-luminous by eliminating the need for activation by light. The luminous properties are imparted by zinc sulfide, calcium sulfide, and strontium sulfide pigments, or by fluorescent dyes.

**Ship-Bottom Paint** Ocean-going steel ships require antifouling paint over the anticorrosive primer. Red lead or zinc yellow is used in the primer. **Antifouling** paints contain ingredients, such as cuprous oxide and mercuric oxide, that are toxic to barnacles and other marine organisms. For rapid drying, so that the time in dry dock is kept short, the vehicle is usually a spirit varnish, such as vinyl, and rosin–coal tar solutions. During World War II, the U.S. Navy developed **hot plastic** ship-bottom paint. It is applied by spray gun in a relatively thick film and dries or sets as it cools. **Copper powder paint** is suitable for small wooden craft.

**Fire-Retardant Paint** Most paint films contain less combustible matter than does wood. To this extent, they are fire-retardant. In addition they cover splinters and fill in cracks. However, flames and intense heat will eventually ignite the wood, painted or not. The most that ordinary paint can do is to delay ignition. Special compositions fuse and give off flame-smothering fumes, or convert to spongy heat-insulating masses, when heated. The relatively thick film and low resistance to abrasion and cleaning make these coatings unsuited for general use as paint, unless conventional decorative paints are applied over them. However, steady improvement in paint properties is being made.

**Traffic paint** is quick-drying, so that traffic is inconvenienced as little as possible. Rough texture and absence of gloss increase visibility. Tiny glass beads may be added for greater night visibility. Bituminous paints are sometimes used on light-colored pavements.

**Heat-Reflecting Paint** Light colors reflect more of the sun's radiation than do dark colors. White is somewhat better than aluminum, but under some conditions, aluminum may retain its reflecting power longer. On heating radiators, flat paint (any color) radiates more heat than aluminum- or copper-bronze paint.

**Heat-Resistant Paint** **Silicone** paint is the most heat-resistant paint yet developed. Its first use was as an insulation varnish for electric motors, but types for other uses, such as on stoves and heaters, have been developed. Since it must be cured by baking, it is primarily a product finish.

**Fungicides** should be added to paint that is to be used in bakeries, breweries, sugar refineries, dairies, and other places where fungi flourish. Until ecological considerations forbade their use, mercury compounds were important fungicides. Their place has been filled by chlorinated phenols and other effective chemical compounds. Fungi-infected surfaces should be sterilized before they are painted. Scrub them with mild alkali; if infection is severe, add a disinfectant.

**Wood preservatives** of the paintable type are used extensively to treat wood windows, doors, and cabinets. They reduce rotting that may be promoted by water that enters at joints. They comprise solutions of compounds, such as chlorinated phenols, and copper and zinc naphthenates, in petroleum thinner. Water repellents, such as paraffin wax, are sometimes added but may interfere with satisfactory painting.

Clear **water repellents** are treatments for masonry to prevent wetting by water. Older types are solutions of waxes, drying oils, or metallic soaps. Silicone solutions, recently developed, are superior, both in repelling water and in effect on appearance.

### Painting

**Exterior Architectural Painting**  Surfaces must be clean and dry, except that wood to be painted with emulsion paint will tolerate a small amount of moisture and masonry to be painted with portland cement–base paint should be damp. Scrape or melt the resin from the knots in wood, or scrub it off with paint thinner or alcohol. As an extra precaution, seal the knots with shellac varnish, aluminum paint, or proprietary knot sealer. When painting wood for the first time, fill nail holes and cracks with putty, after the priming coat has dried. If the wood has been painted before, remove loose paint with a scraper, wire brush, or sandpaper. Prime all bare wood. In bad cases of cracking and scaling, remove all the paint by dry scraping or with paint and varnish remover or with a torch.

Paint on new wood should be from 4 to 5 mils (0.10 to 0.13 mm) thick. This usually requires three coats. A system consisting of nonpenetrating primer and special finish coat may permit the minimum to be reached with two coats. Repainting should be frequent enough to preserve a satisfactory thickness of the finish coat. Nonpenetrating primers stick on some types of wood better than regular house paint does.

**Interior Architectural Painting**  Interior paints are used primarily for decoration, illumination, and sanitation. Enamels are used for kitchen and bathroom walls, where water resistance and easy cleaning are needed; semiglossy and flat types are used on walls and ceilings where it is desired to avoid glare. Wet plaster will eventually destroy oil-base paints. Fresh plaster must be allowed to dry out. Water from leaks and condensation must be kept away from aged plaster. Several coats of paint with low permeability to water vapor make an effective vapor barrier.

**Painting Concrete and Masonry**  Moisture in concrete and masonry brings alkali to the surface where it can destroy oil paint. Allow 2 to 6 months for these materials to dry out before painting; a year or more, if they are massive. Emulsion paints may be used on damp concrete or masonry if there is reason to expect the drying to continue. Portland-cement paint, sometimes preferred for masonry, is especially suitable for first painting of porous surfaces such as cinder block. Before applying this paint, wet the surface so that capillarity will not extract the water from the paint. Scrubbing the paint into the surface with a stiff brush of vegetable fiber gives the best job, but good jobs are also obtained with usual brushes or by spraying. Keep the paint damp for 2 or 3 days so that it will set properly. Chlorinated rubber, chlorinated paraffin, syn-thetic rubber, and vinyl paints are much used on concrete, especially for swimming pools, because of their high resistance to alkali.

**Painting Steel**  Preparation of steel for painting, type of paint, and condition of exposure are closely related. Methods of preparation, arranged in order of increasing thoroughness, include (1) removal of oil with solvent; (2) removal of dirt, loose rust, and loose mill scale with scraper or wire brush; (3) flame cleaning; (4) sandblasting; (5) pickling; (6) phosphating. Exposure environments, arranged in order of increasing severity, include (a) dry interiors, or arid regions; (b) rural or light industrial areas, normally dry; (c) frequently wet; (d) continuously wet; (e) corrosive chemical. Paint systems for condition (a) often consist of a single coat of low-cost paint. For other conditions, the systems comprise one or two coats of rust-inhibitive primer, and one or more finish coats, selected according to severity of conditions.

The primer contains one or more rust-inhibitive pigments, selected mainly from red lead, zinc yellow, and zinc dust. It may also contain zinc oxide, iron oxide, and extender pigments. Of equal importance is the binder, especially for the top coats. For above-water surfaces, linseed oil and alkyd varnish give good service; for underwater surfaces, other binders, like phenolic and vinyl resin, are better. Chemical-resistant binders include epoxy, synthetic rubber, chlorinated rubber, vinyl, and neoprene.

Paint for structural steel is normally air-drying. Large percentages for factory-finished steel products are catalytic-cured, or are baked.

**Galvanized Iron**  Allow new galvanized iron to weather for 6 months before painting. If there is not enough time, treat it with a proprietary etching solution. For the priming coat, without pretreatment, a paint containing a substantial amount of zinc dust should be used. If the galvanizing has weathered, the usual primers for steel are also good.

**Tinplate** is an excellent base for paint. The grease should be washed off with paint thinner and the same primer used as for steel. Much tin roofing is painted with iron oxide paints.

**Painting Copper**  The only preparation needed is washing off any grease and roughening the surface, if it is a polished one. Special primers are not needed. Paint or varnish all copper to prevent corrosion products from staining the adjoining paint.

**Painting Aluminum**  The surface must be clean and free from grease. Highly polished sheet should be etched with phosphoric acid or chromic acid. Zinc-yellow primers give the best protection against corrosion. The only preparation needed for interior aluminum is to have the surface clean; anticorrosive primers are not needed.

**Magnesium** and its alloys corrode readily, especially in marine atmospheres. Red-lead primers must not be used. For factory finishing, it is customary to chromatize the metal and then apply a zinc-yellow primer.

**Water-Tank Interiors**  Among the best paints for these are asphaltic compositions. For drinking-water tanks, select one that imparts no taste. Zinc-dust paints made with phenolic varnish are also good.

**Wood Products**  Finishes may be lacquer or varnish, or their corresponding enamels. High-quality clear finishes may require 10 to 15 operations, such as sanding, staining, filling, sealing, and finishing. Furniture finishing is done mostly by spray. Small articles are often finished by tumbling; shapes like broom handles, by squeegeeing.

**Plastics** Carefully balanced formulations are necessary for satisfactory adhesion, to avoid crazing and to prevent migration of plasticizer.

**Paint-Destroying Agencies** Heavy dew, hot sun, and marine atmospheres shorten the life of paint. Industrial zones where the atmosphere is contaminated with hygroscopic and acidic substances make special attention to painting programs necessary. Dampness within masonry and plaster walls brings alkali to the surface where it can destroy oil-base paints. Interior paints on dry surfaces endure indefinitely; they need renewal to give new color schemes or when it becomes impractical to wash them. Dry temperate climates are favorable to long life of exterior paints.

**Application** Most industrial finishing is done with spray guns. In electrostatic spraying, the spray is charged and attracted to the grounded target. Overspray is largely eliminated. Other methods of application include dipping, flowing, tumbling, doctor blading, rolling, fluid bed, and screen stenciling.

An increasing proportion of maintenance painting is being done with spray and hand roller. The spray requires up to 25 percent more paint than the brush, but the advantage of speed is offsetting. The roller requires about the same amount of paint as the brush.

**Dry finishing** is done by flame-spraying powdered pigment-filmogen compositions or by immersing heated articles in a **fluid bed** of the powdered composition.

**Spreading Rates** When applied by brush, approximate spreading rates for paints on various surfaces are as shown in the table below:

## VARNISH

Oleoresinous varnishes are classed according to oil length, i.e., the number of gallons of oil per 100 lb of resin. Short oil varnishes contain up to 10 gal of oil per 100 lb of resin; medium, from 15 to 25 gal; long, over 30 gal.

**Floor varnish** is of medium length and is often made with modified phenolic resins, tung, and linseed oils. It should dry overnight to a tough hard film. Some floor varnishes are rather thin and penetrating so that they leave no surface film. They show scratches less than the orthodox type.

**Spar varnish** is of long oil length, made usually with phenolic or modified-phenolic resins, tung or dehydrated castor oil, and linseed oil. Other spar varnishes are of the alkyd and urethane types. Spar varnishes dry to a medium-hard glossy film that is resistant to water, actinic rays of the sun, and moderate concentrations of chemicals.

**Chemical-resistant varnishes** are designed to withstand acid, alkali, and other chemicals. They are usually made of synthetic resins, such as chlorinated rubber, cyclized rubber, phenolic resin, melamine, urea-aldehyde, vinyl and epoxy resin. Some of these must be dried by baking. **Catalytic varnish** is made with a nonoxidizing film former, and is cured with a catalyst, such as hydrochloric acid or certain amines. **Flat varnish** is made by adding materials, such as finely divided silica or metallic soap, to glossy varnish. Synthetic latex or other emulsion, or aqueous dispersion, such as glue, is sometimes called **water varnish.**

**Lacquer** The word lacquer has been used for (1) spirit varnishes used especially for coating brass and other metals, (2) Japanese or Chinese lacquer, (3) coatings in which cellulose nitrate (**pyroxylin**), cellulose acetate, or cellulose acetate butyrate is the dominant ingredient, (4) oleoresinous baking varnishes for interior of food cans.

Present-day lacquer primarily refers to cellulosic coatings, clear or pigmented. These lacquers dry by evaporation. By proper choice of solvents, they are made to dry rapidly. Besides cellulosic compounds, they contain resins, plasticizers, and solvent. Cellulose acetate, cellulose acetate butyrate, and cellulose acetate propionate lacquers are nonflammable, and the clear forms have better exterior durability than the cellulose nitrate type. Cellulose acetate and cellulose acetate butyrate are superior to cellulose nitrate for **airplane dopes** because they are nonflammable; the cellulose acetate butyrate is superior to the acetate because it remains taut in humid weather. Although cellulosic and acrylic derivatives dominate the lacquer field, compositions containing vinyls, chlorinated hydrocarbons, or other synthetic thermoplastic polymers are important.

**Lac** is a resinous material secreted by an insect that lives on the sap of certain trees. Most of it comes from India. After removal of dirt, it is marketed in the form of grains, called seed-lac; cakes, called button lac; or flakes, called shellac. Lac contains up to 7 percent of wax, which is removed to make the refined grade. A bleached, or white, grade is also available.

**Shellac varnish** is made by "cutting" the resin in alcohol; the cut is designated by the pounds of lac per gallon of alcohol, generally 4 lb. Shellac varnish should always be used within 6 to 12 months of manufacture, as some of the lac combines with the alcohol to form a soft, sticky material.

| Surface | Type | First coat | | Second and third coats | |
|---------|------|-----------|-------|-----------|-------|
| | | ft²/gal | m²/l | ft²/gal | m²/l |
| Wood | Oil emulsion | 400–500 | 10–12 | 500–600 | 12–15 |
| Wood, primed | Emulsion | 500–600 | 12–15 | 500–600 | 12–15 |
| Structural steel | Oil | 450–600 | 11–15 | 650–900 | 16–22 |
| Sheet metal | Oil | 500–600 | 12–15 | 550–650 | 13–16 |
| Brick, concrete | Oil | 200–300 | 5–7 | 400–500 | 10–12 |
| Brick, concrete | Cement | 100–125 | 2.5–3 | 125–175 | 3–4.5 |
| Brick, concrete | Emulsion | 200–300 | 5–7 | 400–500 | 10–12 |
| Smooth plaster | Oil | 550–650 | 13–16 | 550–650 | 13–16 |
| Smooth plaster | Casein | 500–600 | 12–15 | 500–600 | 12–15 |
| Smooth plaster | Emulsion | 400–500 | 10–12 | 500–600 | 12–15 |
| Concrete floor | Enamel | 400–500 | 10–12 | 550–650 | 13–16 |
| Wood floor | Varnish | 550–650 | 13–16 | 550–650 | 13–16 |

# WOOD

## by Carl H. de Zeeuw

REFERENCES: Brown, Panshin, and de Zeeuw, "Textbook of Wood Technology," McGraw-Hill. "Design of Wood Aircraft Structures," ANC-18, Munitions Board Aircraft Committee, 1951. Freas and Selbo, Fabrication and Design of Glued Laminated Wood Structural Members, *U.S. Dept. Agr. Tech. Bull.* no. 1069, 1954. Hunt and Garratt, "Wood Preservation," McGraw-Hill. "National Design Specification for Stress-Grade Lumber and Its Fastenings," National Lumber Manufacturers Assoc., 1973. Stamm and Harris, "Chemical Processing of Wood," Chemical Publishing. "Technical Data on Plywood," Douglas Fir Plywood Assoc. "Wood Handbook," *U.S. Dept. Agr. Handbook* no. 72, 1974. Scofield and O'Brien, "Modern Timber Engineering," Southern Pine Assoc.

## COMPOSITION, STRUCTURE, AND NOMENCLATURE

**Wood** is a naturally formed organic material consisting essentially of elongated tubular elements called **cells** arranged in a parallel manner for the most part. These cells vary in dimensions and wall thickness with position in the tree, age, conditions of growth, and kind of tree. The walls of the cells are formed principally of chain molecules of cellulose, polymerized from glucose residues and oriented as a partly crystalline material. These chains are aggregated in the cell wall at a variable angle, roughly parallel to the axis of the cell. The cells are cemented by an amorphous material called lignin. The complex structure of the gross wood approximates a rhombic system. The direction parallel to the grain and the axis of the stem is longitudinal (**L**), the two axes across the grain are radial (**R**) and tangential (**T**) with respect to the cylinder of the tree stem. This anisotropy and the molecular orientation accounts for the major differences in physical properties with respect to direction which are present in wood.

**Natural variability** of any given physical measurement in wood approximates the normal probability curve. It is traceable to the differences in the growth of individual samples and at present cannot be controlled. For engineering purposes, statistical evaluation is employed for determination of safe working limits.

Timber is classified broadly as **hardwood,** which is produced by the broad-leaved trees *(angiosperms),* such as oak, maple, ash; and **softwood,** the product of coniferous trees *(gymnosperms),* such as pines, larch, spruce, hemlock. The terms "hard" and "soft" have no relation to actual hardness of the wood, as can be seen by reference to Table 2. **Sapwood** is the living wood of pale color on the outside of the stem. **Heartwood** is the inner core of physiologically inactive wood in a tree and is usually darker in color, somewhat heavier, due to infiltrated material, and more decay-resistant than the sapwood. Other terms relating to wood, veneer, and plywood are defined in ASTM D9–30, D1038–52, and the "Wood Handbook."

**Standard nomenclature** of timber is based on commercial practice which groups woods of similar technical qualities but separate botanical identities under a single name. For listings of domestic hardwoods and softwoods see ASTM D1165–52 and the "Wood Handbook."

The **chemical composition** of woody cell walls is generally about 40 to 50 percent cellulose, 15 to 35 percent lignin, less than 1 percent mineral, 20 to 35 percent hemicellulose, and the remainder extractable matter of a variety of sorts. Softwoods and hardwoods have about the same cellulose content.

## PHYSICAL PROPERTIES OF WOOD

### Moisture Relations

Wood is a hygroscopic material which contains water in varying amounts, depending upon the relative humidity and temperature of the surrounding atmosphere. **Equilibrium conditions** are established as shown in Fig. 1. The standard reference condition for wood is **ovendry weight,** which is determined by drying at 100 to 105°C to constant weight.

**Moisture content of wood** is the weight of water expressed as a percentage of the ovendry weight. The water is absorbed solely in the intermolecular regions of the cell wall up to 31 percent depending on the kind of wood and the temperature. The maximum value for this type of absorption is called the **fiber saturation point** (fsp) and is usually taken as 28 percent for room temperature. Additional water is taken into the cell cavities as "free water" and may be in excess of 700 percent. **Air-dry wood** has 12 to 15 percent moisture content. **Green wood** contains from 40 to 100 percent water for ordinary ranges of specific gravities.

**Strength properties** remain constant as long as the wood is above the fiber saturation point. Changes in moisture content in the cell wall cause the strength to vary inversely in a manner similar to compound interest. Computations for adjustment of strength data with moisture changes below the fiber saturation point can be made using the values from Table 1 as $x$ in the

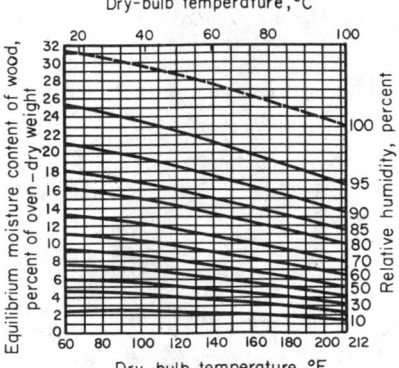

**Fig. 1**  Relation of dry-bulb temperature, relative humidity, and equilibrium moisture content of wood. (*From USDA Tech. Bull. 1069.*)

**Table 1. Strength Changes in Wood as Related to Specific Gravity and Moisture Content**
(From the Wood Handbook)

| Strength property | Specific gravity* | | Change in % for 1 % change in moisture content† |
| --- | --- | --- | --- |
| | Green wood | Air-dry wood, 12% m.c. | |
| Static bending: | | | |
| Fiber stress at prop. limit, lb/in² | $10,200\ G^{1.25}$ | $16,700\ G^{1.25}$ | 5 |
| Modulus of rupture, lb/in² | $17,600\ G^{1.25}$ | $25,700\ G^{1.25}$ | 4 |
| Modulus of elasticity, 1,000 lb/in² | $2,360\ G$ | $2,800\ G$ | 2 |
| Impact bending: | | | |
| Height of drop to failure, in | $114\ G^{1.75}$ | $94.6\ G^{1.75}$ | 0.5 |
| Compression parallel to grain: | | | |
| Fiber stress at prop. limit, lb/in² | $5,250\ G$ | $8,750\ G$ | 5 |
| Maximum crushing strength, lb/in² | $6,730\ G$ | $12,200\ G$ | 6 |
| Modulus of elasticity, 1,000 lb/in² | $2,910\ G$ | $3,380\ G$ | |
| Compression perpendicular to grain: | | | |
| Fiber stress at prop. limit, lb/in² | $3,000\ G^{2.25}$ | $4,630\ G^{2.25}$ | 5.5 |
| Hardness: | | | |
| End, lb | $3,740\ G^{2.25}$ | $4,800\ G^{2.25}$ | 4 |
| Side, lb | $3,420\ G^{2.25}$ | $3,770\ G^{2.25}$ | 2.5 |

*The properties and values should be read as equations; for example, modulus of rupture for green wood = 17,600 $G^{1.25}$, where $G$ represents the specific gravity of ovendry wood based on volume at the moisture condition indicated.
†The percentages are to be applied successively, as in compound-interest calculations.

annual-compound-interest tables (Sec. 1) and the change in moisture-content percentage in place of the number of years.

## Dimensional Changes

**Shrinking or swelling** of wood is the result of change in water content within the cell wall; because of the anisotropic nature of the wall the changes will be unequal for the several axes. Shrinkage is expressed as percentage of dimensional change based on the green or fully swollen size. **Longitudinal shrinkage** from green to ovendry ranges from 0.1 to 0.3 percent and is usually neglected. Across the grain, total shrinkage averages about 50 percent greater tangentially than radially. Average shrinkage values for a number of commercial woods are shown in Table 2. Shrinkage to any moisture condition can be estimated by assuming that the change is linear from green to ovendry and that about half occurs in drying to 12 percent.

**Swelling in polar liquids other than water** is inversely related to the size of the molecule of the liquid. It has been shown that the tendency to hydrogen bonding on the dielectric constant is a close, direct indicator of the swelling power of water-free organic liquids. In general, the strength values for wood swollen in any polar liquid are similar when there is equal swelling of the wood.

**Swelling in aqueous solutions** of sulfuric and phosphoric acids, zinc chloride, and sodium hydroxide above pH 8 may be as much as 25 percent greater in the transverse direction than in water. The transverse swelling may be accompanied by longitudinal shrinkage up to 5 percent. The swelling reflects a chemical change in the cell walls, and the accompanying strength changes are related to the degradation of the cellulose.

**Dimensional stabilization of wood** cannot be completely attained. Two or three coats of varnish, enamel, or synthetic lacquer may be 50 to 85 percent efficient in preventing short-term dimensional changes. Metal foil embedded in multiple coats of varnish may be 90 to 95 percent efficient in short-term cycling. The best long-term stabilization results from internal bulking of the cell wall by the use of materials such as phenolic resins polymerized in situ or water solutions of polyethylene glycol (PEG) on green wood. The presence of the bulking agents alters the physical properties of the treated wood. Phenol increases electrical resistance, hardness, compression strength, weight, and decay resistance but lowers the impact strength. Polyethylene glycol maintains strength values at the green-wood level, reduces electrical resistance, and can be finished only with polyurethane resins. (See Stamm for further discussion.)

## Specific Gravity and Density

**Specific gravity,** $G_m$, of wood at a given moisture condition, $m$, is the ratio of the weight of the ovendry wood $W_o$ to the weight of water displaced by the sample at the given moisture condition $w_m$.

$$G_m = W_o/w_m$$

This definition is required because volume and weight are constant only under special conditions. The **weight density of wood** $D$ (unit weight) at any given moisture content is the weight of ovendry wood and the contained water divided by the volume of the piece at that same moisture content. Average values for specific gravity ovendry and weight density at 12 percent moisture content are given in Table 2. Specific gravity of solid, dry wood substance based on helium displacement is 1.46, or about 91 lb/ft³.

**Conversion of weight density** from one moisture condition to another can be accomplished by the following equation (Skaar):

$$D_2 = D_1 \frac{100 + M_2}{100 + M_1 + 0.0135\ D_1(M_2 - M_1)}$$

**Table 2. Strength and Related Properties of Wood at 12 Percent Moisture Content (Average Values from Tests on Clear Pieces 2 × 2 in in Section per ASTM D143)***

| Kind of wood | Specific gravity, ovendry volume | Density at 12% m.c., lb/ft³ | Shrinkage, % from green to ovendry condition based on dimension when green | | Static bending | | Max crushing strength parallel to grain, lb/in² | Compression perpendicular to grain at proportional limit, lb/in² | Tensile strength perpendicular to grain, lb/in²† | Impact bending, height of drop in inches for failure with 50-lb hammer | Shear strength parallel to grain, lb/in² | Hardness perpendicular to grain, avg of R and T |
|---|---|---|---|---|---|---|---|---|---|---|---|---|
| | | | Rad. | Tan. | Modulus of rupture, lb/in² | Modulus of elasticity, ksi | | | | | | |
| **Hardwoods** | | | | | | | | | | | | |
| Ash, white | 0.64 | 42 | 4.9 | 7.9 | 15,400 | 1,770 | 7,410 | 1,160 | 940 | 43 | 1,950 | 1,320 |
| Basswood | 0.40 | 26 | 6.6 | 9.3 | 8,700 | 1,460 | 4,730 | 370 | 350 | 16 | 990 | 410 |
| Beech | 0.67 | 45 | 5.1 | 11.0 | 14,900 | 1,720 | 7,300 | 1,010 | 1,010 | 41 | 2,010 | 1,300 |
| Birch, yellow | 0.66 | 43 | 7.2 | 9.2 | 16,600 | 2,010 | 8,170 | 970 | 920 | 55 | 1,880 | 1,260 |
| Cherry, black | 0.53 | 35 | 3.7 | 7.1 | 12,300 | 1,490 | 7,110 | 690 | 560 | 29 | 1,700 | 950 |
| Cottonwood, eastern | 0.43 | 28 | 3.9 | 9.2 | 8,500 | 1,370 | 4,910 | 370 | 580 | 20 | 930 | 430 |
| Elm, American | 0.55 | 35 | 4.2 | 9.5 | 11,800 | 1,340 | 5,520 | 690 | 660 | 39 | 1,510 | 830 |
| Elm, rock | 0.66 | 44 | 4.8 | 8.1 | 14,800 | 1,540 | 7,050 | 1,230 | | 56 | 1,920 | 1,320 |
| Sweetgum | 0.55 | 36 | 5.4 | 10.2 | 12,500 | 1,640 | 6,320 | 620 | 760 | 32 | 1,600 | 850 |
| Hickory, shagbark | 0.77 | 50 | 7.0 | 10.5 | 20,200 | 2,160 | 9,210 | 1,760 | | 67 | 2,430 | |
| Mahogany‡ (Swietenia spp) | 0.51 | 34 | 3.5 | 4.8 | 11,460 | 1,500 | 6,800 | 1,100 | 750 | 39 | 1,230 | 800 |
| Maple, sugar | 0.68 | 44 | 4.9 | 9.5 | 15,800 | 1,830 | 7,830 | 1,470 | 800 | 43 | 2,330 | 1,450 |
| Oak, red, northern | 0.66 | 44 | 4.0 | 8.2 | 14,300 | 1,820 | 6,760 | 1,010 | 800 | 43 | 1,780 | 1,290 |
| Oak, white | 0.71 | 48 | 5.3 | 9.0 | 15,200 | 1,780 | 7,440 | 1,070 | 800 | 37 | 2,000 | 1,360 |
| Poplar, yellow | 0.45 | 29 | 4.2 | 7.6 | 10,100 | 1,580 | 5,540 | 500 | 540 | 24 | 1,190 | 540 |
| Tupelo, black | 0.55 | 35 | 4.4 | 7.7 | 9,600 | 1,200 | 5,520 | 930 | 500 | 22 | 1,340 | 810 |
| Walnut, black | 0.56 | 38 | 5.2 | 7.1 | 14,600 | 1,680 | 7,580 | 1,010 | 690 | 34 | 1,370 | 1,010 |
| **Softwoods** | | | | | | | | | | | | |
| Cedar, western red | 0.34 | 23 | 2.4 | 5.0 | 7,500 | 1,110 | 4,560 | 460 | 220 | 17 | 860 | 350 |
| Cypress | 0.48 | 32 | 3.8 | 6.2 | 10,600 | 1,440 | 6,360 | 780 | 270 | 24 | 1,000 | 510 |
| Douglas fir, coast | 0.51 | 34 | 4.8 | 7.6 | 12,400 | 1,950 | 7,240 | 800 | 340 | 31 | 1,160 | 710 |
| Hemlock, eastern | 0.43 | 28 | 3.0 | 6.8 | 8,900 | 1,200 | 5,410 | 650 | | 21 | 1,060 | 500 |
| Hemlock, western | 0.44 | 29 | 4.3 | 7.9 | 11,300 | 1,640 | 7,110 | 550 | 340 | 26 | 1,250 | 540 |
| Larch, western | 0.59 | 38 | 4.5 | 9.1 | 13,100 | 1,870 | 7,640 | 980 | 430 | 35 | 1,360 | 830 |
| Pine, red | 0.47 | 31 | 3.8 | 7.2 | 11,000 | 1,630 | 6,070 | 600 | 460 | 26 | 1,210 | 560 |
| Pine, ponderosa | 0.42 | 31 | 3.9 | 6.3 | 9,400 | 1,290 | 5,320 | 580 | 420 | 19 | 1,130 | 460 |
| Pine, eastern white | 0.37 | 24 | 2.1 | 6.1 | 8,600 | 1,240 | 4,800 | 440 | 310 | 18 | 900 | 380 |
| Pine, western white | 0.42 | 27 | 2.6 | 5.3 | 9,700 | 1,460 | 5,040 | 470 | | 23 | 1,040 | 420 |
| Pine, shortleaf | 0.54 | 36 | 4.4 | 7.7 | 13,100 | 1,760 | 7,270 | 820 | 470 | 33 | 1,390 | 690 |
| Redwood | 0.42 | 28 | 2.6 | 4.4 | 10,000 | 1,340 | 6,150 | 700 | 240 | 19 | 940 | 480 |
| Spruce, sitka | 0.42 | 28 | 4.3 | 7.5 | 10,200 | 1,570 | 5,610 | 580 | 370 | 25 | 1,150 | 510 |
| Spruce, white | 0.45 | 28 | 4.7 | 8.2 | 9,800 | 1,340 | 5,470 | 460 | 360 | 20 | 1,080 | 480 |

*Tabulated from "Wood Handbook," *Tropical Woods* no. 95, and unpublished data from the U.S. Forest Products Laboratory.
†Tensile strength parallel to grain to be taken as equal to modulus of rupture in bending.
‡Central American *Swietenia* spp.

$D_1$ is the weight density, lb/ft³, which is known for some moisture condition $M_1$. $D_2$ is desired weight density at a moisture content $M_2$. Moisture contents $M_1$ and $M_2$ are expressed in percent.

**Specific gravity and strength properties** vary directly in an exponential relationship $S = KG^n$. Table 1 gives values of $K$ and the exponent $n$ for various strength properties. The equation is based on more than 160 kinds of wood and yields estimated average values for wood in general. This relationship is the best general index to wood quality.

### Thermal Properties

The **coefficients of thermal expansion** in wood vary with the structural axes. According to Weatherwax and Stamm (*Trans. ASME*, **69**, 1947, p. 421), the longitudinal coefficient for the temperature range $+ 50°C$ to $- 50 °C$ averages $3.39 \times 10^{-6}/°C$ and is independent of specific gravity. Across the grain, for an average specific gravity ovendry of 0.46, the radial coefficient $\alpha_r$ is $25.7 \times 10^{-6}/°C$ and the tangential $\alpha_t$ is $34.8 \times 10^{-6}/°C$. Both $\alpha_r$ and $\alpha_t$ vary with specific gravity approximately to the first power. Thermal expansions are usually overshadowed by the larger dimensional changes due to moisture.

**Thermal conductivity of wood** varies principally with the direction of heat flow with respect to the grain. **Transverse conductivity** can be calculated by MacLean's equation (*Heating, Piping, Air Conditioning*, **13**, 1941):

$$K = G(1.39 + CM) + 0.165$$

where $G$ is specific gravity, based on ovendry weight and volume at given moisture content $M$, percent. The constant $C$ has two values: for moisture contents under 40 percent, $C = 0.028$; for moisture contents of 40 percent or more, $C = 0.038$. **Longitudinal conductivity** is 2.25 to 2.75 times greater than transverse.

**Fuel value of wood** depends primarily upon the weight of dry cell-wall material plus tannin and resins. Moisture in wood decreases fuel value because of heat loss in vaporizing the water. An approximate relation is

$$\text{Btu/lb} = H \frac{100 - \text{m.c.}/7}{100 + \text{m.c.}}$$

where $H$ is heat of combustion, averaging 8,500 Btu for hardwoods and 9,000 Btu for conifers, and m.c. is moisture content of the wood in percent. (See Sec. 7 for fuel values and Sec. 4 for combustion.)

**Ignition of wood** depends on the exothermic-reaction temperature, which is 273°C approx, the presence of inflammable extractives, and the shape and size of the cross section. Thin, outstanding edges ignite readily, while chamfered, heavy timbers are slow to catch fire. Long-continued heating below the exothermic-reaction temperature can cause ignition.

The **immediate effect of temperature on strength properties** is given by an inverse linear relationship; changes in specific gravity cause direct displacement of the curves, while moisture-content changes result in direct variation of the slopes of the curves. These effects are reversible if the heating time is short. A predicting equation for strength-temperature changes based on Kollmann's experiments is

$$S_2 = S_1 - (37.35) \, G_o(T_2 - T_1)\left(1 + 0.0756 \frac{\text{m.c.}}{100}\right)$$

where $S_2$ is a strength value desired, $S_1$ is a known strength, $G_o$ is specific gravity on an ovendry basis for the wood, $T_2$ and $T_1$ are temperatures, °F, corresponding to the strength values, and m.c. is moisture content, percent.

**Permanent effects of heat on strength of wood** show a direct dependence on time, temperature, and moisture content. Figure 2 illustrates the effect of time on heating on the bending strengths when the other factors are constant. A rough estimate of the effect of moisture is that any given value shown will be less than half as great for ovendry wood. The effect of steam or temperatures above 280°F on wet wood is to cause accelerated rate of strength loss.

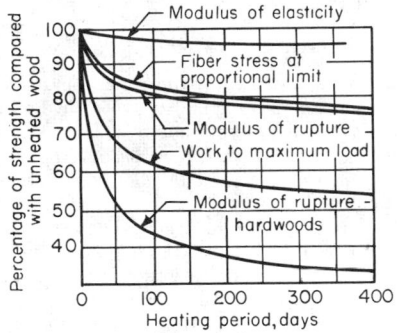

**Fig. 2** Effect of heating in water at 200°F (93.3°C) on bending strength. Wood tested after cooling to room temperature and drying to 12 percent moisture content. All curves for softwoods except as noted. (*From "Wood Handbook."*)

**Fire-retardant treatment** will provide wood which will flame only as long as an external source of heat exists. No treatment can make wood "fireproof" or prevent charring. The best chemicals for fire-retardant treatment are the mono- and diammonium salts. Zinc chloride is also widely used but can cause serious strength losses. Neither of these salts cause corrosive action on metal in contact with the treated lumber. Commercial formulations employ the above salts in combination with borax, boric acid, phosphoric acid, and magnesium chloride. The range of retentions of dry salt in pounds per cubic foot for these fire-retardant materials is from 5 to 6 for thicknesses up to 2 in, from 2 to 3 for lumber up to 4 in. Minimum penetration is specified as ½ in.

### Electrical Properties

The **direct-current electrical resistance** of wood is dependent principally on the moisture content and secondarily on density, grain direction, temperature, minerals, and extractives. According to Clark and Williams (*Jour. Phys. Chem.* **37**, 1933, p. 119), the specific resistance for ovendry wood ranges from $3 \times 10^{17}$ to $3 \times 10^{18}$ Ω·cm and at 16 percent moisture content decreases to $10^8$ Ω·cm. In general, the logarithm of the conductivity (reciprocal of the resistance) will increase linearly with moisture content of the wood up to the fiber saturation point. At this level the specific resistance approximates that of water ($10^5$ to $10^6$ Ω·cm) and increases only slightly up to maximum water content.

**Electrical-resistance moisture meters** measure the resistance between two pins driven into the wood and, when calibrated for species and temperature range, are accurate to within 2 percent for moisture contents between 7 and 25 percent in the surface layers of the wood. Resistance meters are unreliable when used with salt-treated wood.

The **alternating-current characteristics of wood** have been summarized by Skaar (*N.Y. State Coll. Forestry Tech. Publ.* 69, 1948) as follows: The **dielectric constant of wood** without regard to kind is directly proportional to the density at a given moisture content, increases with the latter factor, decreases slightly with increases in frequency of the alternating current, and is 30 to 50 percent greater in the longitudinal than in the perpendicular-to-grain directions.

### Wood in Relation to Sound

The **transmission of sound in wood** for a given direction with respect to the grain is described by the expression $v = \sqrt{E/\rho}$, in which $v$ is the velocity of sound in wood, in/s, $E$ is the dynamic Young's modulus, lb/in$^2$, and $\rho$ is the density of the wood, slugs/in$^3$. Kitazawa (*Jour. Forest Products Research Soc.*, **2**, no. 5, 1952, p. 228) has shown that the directional variation of dynamic modulus with respect to the axes of the wood can be described by Hankinson's formula below. The dynamic modulus is about 10 percent higher than the static value and varies inversely with moisture changes by approximately 1.3 percent for each percent change in moisture content.

**Nondestructive testing of wood** for the determination of elastic moduli is carried out by resonance techniques analogous to those used with steel and concrete. For details, see Kitazawa, *loc. cit.* These methods are particularly useful for a variable material such as wood, since a series of tests can be carried out on a single sample.

### Durability of Wood

**Decay resistance** of timber varies widely between different kinds of trees and between individuals of the same kind. Heartwood is more resistant than sapwood owing to lower permeability and presence of toxic extractives. Experience of the U.S. Forest Products Laboratory shows **high natural decay resistance** of the heartwood for the following important commercial timbers: northern white cedar, southern cypress, western red cedar, redwood, chestnut, black locust, and black walnut. **Moderate decay resistance** is associated with the heartwood of Douglas fir, eastern white pine, southern yellow pine, western larch, white oak, and red gum. Other kinds of native timbers have little or no resistance in an untreated condition.

**Wood-destroying fungi** are primitive forms of plant life which utilize either the lignin portion (white rots) or the cellulose (brown rots) for food. Control of moisture, air, or temperature limits or halts the metabolic process. Optimum conditions for growth are 20 to 25 percent moisture content, 20 percent of wood volume as air, and temperatures from 75 to 95°F. Wood which is saturated with water because of complete submersion has too little air for fungus growth and will remain unattacked. Freezing of wood will not kill the fungus but will cause dormancy. However, elevated temperature from 120 to 145°F will kill even the most resistant fungi. Normal conditions for kiln drying and steaming for preservation treatment fall within the temperature range for sterilization. Service conditions usually preclude control of the growth factors, and fungus growth is inhibited by chemical impregnation of the wood.

**Resistance of wood to insect activity** is poor in sapwood, while heartwood, particularly if resinous, is less subject to attack. **Termites** of the subterranean type are the most serious wood-destroying insect group in the United States. These insects remove the wood internally, leaving few signs of their activity. The cellulose of the wood is acted on by protozoa in the insect intestine and converted to sugars for food. Control of termites is best effected by construction features which prevent the insects from gaining entry to a structure (see *U.S. Dept. Agr. Farmers' Bull.*, 1949, p. 1911). In addition, the soil may be poisoned or the wood pressure-treated with toxic chemicals. **Wood-boring beetles** can also cause extensive damage. Their activity can be noted by the presence of small surface holes and evicted sawdust. Heat sterilization and thorough surface coatings of paint or varnish are the best control measures.

**Marine organisms** will attack and destroy any known kind of wood under the proper conditions in tropical waters. The heartwood of imported timbers such as green heart (*Ocotea rodiaei* Mez.); certain *Eucalyptus* spp., such as turpentine and jarrah; azobe (*Lophira procera* Chev.); and manbarklak (*Eschweilera* spp.) will be durable in coastal waters of the United States. None of the native timbers can be classed as durable under these same conditions unless pressure-treated with creosote to refusal.

**Resistance of wood to chemical action** is best in the less permeable heartwood of the conifers. Aqueous solutions of sodium hydroxide, sulfuric, hydrochloric, and nitric acids cause most damage, since swelling and degradation are simultaneous. Temperature changes for a given solution act roughly with the logarithm of time.

**Submersion of timber in water** will prevent fungus action as noted but will eventually cause deterioration of the wood due to hydrolysis of the cellulose. This change correlates with strength tests in bending and compression. Excavated piles of about 260 years of age have shown shear failures in the early wood zones.

**Long exposure of wood to the atmosphere** also causes changes in the cellulose. A study by Kohara and Okamoto of sound old timbers of a softwood and hardwood of known ages from temple roof beams shows that the percentage of cellulose decreases steadily over a period up to 1,400 years while the lignin remains almost constant. These changes are reflected in strength losses (Fig. 3). Impact properties approximate a loss that is nearly linear with the logarithm of time.

### STRENGTH AND ELASTIC PROPERTIES OF WOOD

#### Primary Strength Data

Primary data on the strength of small pieces of air-dry wood, free from defects, are shown in Table 2. Additional data and tests on green material can be found in the "Wood Handbook" and *U.S. Dept. Agr. Bull.* 479. Test methods are given in ASTM D143 and are similar to those used in other English-speaking countries. Data on foreign timbers imported into the United States can be found in Lavers, The Strength Properties of Timber, *Ministry of Technology, Forest Products Research Bull.* 50, London, 1967.

#### Rheological Properties of Wood

Wood exhibits pronounced viscoelastic characteristics. Under suddenly applied loads there is an immediate deformation approximating the classical pattern, followed by a logarithmic increase with time. Because of this time-dependent relation, the rate of loading is an important factor to consider in the testing and use of wood. Impact and dynamic measures of elasticity for small samples are about 10 percent higher than for static. The strength values are even more influenced by rate (see below). Impact strengths are also affected by this relationship, but present methods of calculating have little meaning, except for height of drop as shown in Table 2, and this for comparative purposes only.

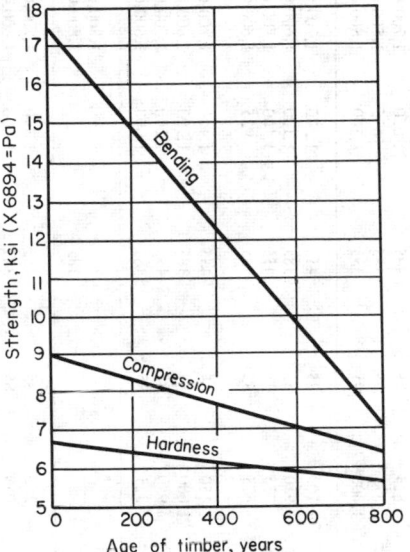

**Fig. 3**  Strength loss with age in a hardwood (*Zelkowa serrata*). (*From Sci. Rpts. Saikyo Univ., no. 7, 1955.*)

**Damping capacity of wood** is superior to most structural materials. Expressed in terms of the logarithmic decrement δ, which is one-half the damping capacity, Greenhill (*Jour. Council Sci. Ind. Research*, **9**, 4, 1936, p. 265) determined a value δ = 0.035 in flexure for sitka spruce. Kimball (*Jour. Applied Mechanics*, **8**, 1941) quoted δ = 0.0049 for mild steel and 0.0034 for aluminum as comparative values.

The **fatigue properties of woods** have not been intensively investigated because they are less sensitive to repeated loads than is the case for more highly crystalline materials. Below the fatigue limit, wood will dissipate an unlimited amount of energy as heat without damage. The U.S. Forest Products Laboratory has found that small cantilever bending specimens subjected to fully reversed stresses at 12 percent moisture content and 75°F have an apparent endurance load of about 30 percent of the modulus of rupture of standard static tests after 30 million cycles.

### Load Direction and Relation to Grain of Wood

All strength properties vary with the orthotropic axes of the wood in a manner which is approximated by an expression first given in *Air Serv. Info. Circ.*, **III,** No. 259, 1921, and is commonly known as **Hankinson's formula:**

$$N = \frac{PQ}{P \sin^2\theta + Q \cos^2\theta}$$

where $N$ = allowable stress induced by a load acting at an angle to the grain direction, lb/in²; $P$ = allowable stress parallel to the grain, lb/in²; $Q$ = allowable stress perpendicular to the grain, lb/in²; and $\theta$ = angle between direction of load and direction of grain.

The deviation of the grain from the long axis of the member to which the load is applied is known as **slope of grain** and is determined by measuring the length of run in inches along the axis for a 1-in deviation of the grain from the axis. Koehler (*U.S. Forest Products Laboratory Rept.* 1585, 1943) gives complete methods and discussions on slope of grain determination. The effect of grain slope on the important strength properties is shown by Table 3.

**Stress-graded construction lumber** is produced under two systems: visual grading rules originating with lumber associations, and machine-stress grading. **The visual grading system** is based on engineering studies made by the U.S. Forest Products Laboratory (see "Wood Handbook" and ASTM 245). These rules take into account natural variability within a species, time-dependent stress behavior, and overload factors. The actual size and location of knots, checks, splits, wane, and slope of grain are used as criteria for assignment of grade. Table 4 shows the grades, and allowable unit stresses established for three of the principal groupings of structural timber. Stresses for other kinds of wood are to be found in the National Design Specification.

Grading rules vary according to the type of end use. For example, lumber 2 to 4 in thick and beams and stringers are graded as simple beams on the extent and placement of defects in the middle third of their length. Any change in the dimen-

**Table 3.  Strength of Wood Members with Various Grain Slopes as Percentages of Straight-Grained Members**
(From Wood Handbook)

| Maximum slope of grain in member | Static bending | | Impact bending; drop height to failure (50 lb hammer), percent | Maximum crushing strength parallel to grain, percent |
| --- | --- | --- | --- | --- |
| | Modulus of rupture, percent | Modulus of elasticity, percent | | |
| Straight-grained................................ | 100 | 100 | 100 | 100 |
| 1 in 25........................................ | 96 | 97 | 95 | 100 |
| 1 in 20........................................ | 93 | 96 | 90 | 100 |
| 1 in 15........................................ | 89 | 94 | 81 | 100 |
| 1 in 10........................................ | 81 | 89 | 62 | 99 |
| 1 in 5......................................... | 55 | 67 | 36 | 93 |

**Table 4. Allowable Unit Stresses for Visually Graded Stress-Grade Lumber**
(Condensed from National Design Specification, 1973 ed.) (The allowable unit stresses below are for normal loading conditions. See text for service modifications.)

Allowable unit stresses, lb/in²

| Species and commercial grade | Size classification | | Extreme fiber in bending $F_b$ Single-member uses | Extreme fiber in bending $F_b$ Repetitive-member uses | Tension parallel to grain $F_t$ | Horizontal shear $F_v$ | Compression perpendicular to grain $F_{c\perp}$ | Compression parallel to grain $F_c$ | Modulus of elasticity $E$ | Grading-rules agency |
|---|---|---|---|---|---|---|---|---|---|---|
| | | | Douglas fir-larch (surfaced dry or surfaced green. Used at 19% max m.c.) | | | | | | | |
| Dense select structural | | 2–4 in thick, 2–4 in wide | 2,450 | 2,800 | 1,400 | 95 | 455 | 1,850 | 1,900,000 | |
| Select structural | | | 2,100 | 2,400 | 1,200 | 95 | 385 | 1,600 | 1,800,000 | |
| Dense No. 1 | | | 2,050 | 2,400 | 1,200 | 95 | 455 | 1,450 | 1,900,000 | West Coast |
| No. 1 | | | 1,750 | 2,050 | 1,050 | 95 | 385 | 1,250 | 1,800,000 | Lumber |
| Dense No. 2 | | | 1,700 | 1,950 | 1,000 | 95 | 455 | 1,150 | 1,700,000 | Inspection |
| No. 2 | | | 1,450 | 1,650 | 850 | 95 | 385 | 1,000 | 1,700,000 | Bureau |
| No. 3 | | | 800 | 925 | 475 | 95 | 385 | 600 | 1,500,000 | and |
| Appearance | | | 1,750 | 2,050 | 1,050 | 95 | 385 | 1,500 | 1,800,000 | Western |
| Stud | | | 800 | 925 | 475 | 95 | 385 | 600 | 1,500,000 | Wood |
| Construction | | 2–4 in thick, 4 in wide | 1,050 | 1,200 | 625 | 95 | 385 | 1,150 | 1,500,000 | Products |
| Standard | | | 600 | 675 | 350 | 95 | 385 | 925 | 1,500,000 | Association |
| Utility | | | 275 | 325 | 175 | 95 | 385 | 600 | 1,500,000 | (see footnotes) |
| Dense select structural | | 2–4 in thick, 6 in and wider | 2,100 | 2,400 | 1,400 | 95 | 455 | 1,650 | 1,900,000 | |
| Select structural | | | 1,800 | 2,050 | 1,200 | 95 | 385 | 1,400 | 1,800,000 | |
| Dense No. 1 | | | 1,800 | 2,050 | 1,200 | 95 | 455 | 1,450 | 1,900,000 | |
| No. 1 | | | 1,500 | 1,750 | 1,000 | 95 | 385 | 1,250 | 1,800,000 | |
| Dense No. 2 | | | 1,450 | 1,700 | 950 | 95 | 455 | 1,250 | 1,700,000 | |
| No. 2 | | | 1,250 | 1,450 | 825 | 95 | 385 | 1,050 | 1,700,000 | |
| No. 3 | | | 725 | 850 | 475 | 95 | 385 | 675 | 1,500,000 | |
| Appearance | | | 1,500 | 1,750 | 1,000 | 95 | 385 | 1,500 | 1,800,000 | |
| Dense select structural | Beams and stringers | | 1,900 | | 1,100 | 85 | 455 | 1,300 | 1,700,000 | |
| Select structural | | | 1,600 | | 950 | 85 | 385 | 1,100 | 1,600,000 | West Coast |
| Dense No. 1 | | | 1,550 | | 775 | 85 | 455 | 1,100 | 1,700,000 | Lumber |
| No. 1 | | | 1,300 | | 675 | 85 | 385 | 925 | 1,600,000 | Inspection |
| Dense select structural | Posts and timbers | | 1,750 | | 1,150 | 85 | 455 | 1,400 | 1,700,000 | Bureau |
| Select structural | | | 1,500 | | 1,000 | 85 | 385 | 1,200 | 1,600,000 | (see footnotes) |
| Dense No. 1 | | | 1,400 | | 950 | 85 | 455 | 1,250 | 1,700,000 | |
| No. 1 | | | 1,200 | | 825 | 85 | 385 | 1,000 | 1,600,000 | |
| Select Dex | Decking | | 1,750 | 2,000 | | | 385 | | 1,800,000 | |
| Commercial Dex | | | 1,450 | 1,650 | | | 385 | | 1,800,000 | |
| Dense select structural | Beams and stringers | | 1,900 | | 1,250 | 85 | 455 | 1,300 | 1,700,000 | |
| Select structural | | | 1,600 | | 1,050 | 85 | 385 | 1,100 | 1,600,000 | |
| Dense No. 1 | | | 1,550 | | 1,050 | 85 | 455 | 1,100 | 1,700,000 | |
| No. 1 | | | 1,350 | | 900 | 85 | 385 | 925 | 1,600,000 | |

**Continued species — West Coast Lumber Inspection Bureau and Western Wood Products Association (see footnotes)**

| Grade | Size | Fb | Ft | Fv | Fc⊥ | Fc∥ | E |
|---|---|---|---|---|---|---|---|
| Dense select structural | | 1,750 | 1,150 | 85 | 455 | 1,350 | 1,700,000 |
| Select structural | Posts and timbers | 1,500 | 1,000 | 85 | 385 | 1,150 | 1,600,000 |
| Dense No. 1 | Posts and timbers | 1,400 | 950 | 85 | 455 | 1,200 | 1,700,000 |
| No. 1 | Posts and timbers | 1,200 | 825 | 85 | 385 | 1,000 | 1,700,000 |
| Selected decking | Decking | 2,000 | | | | | 1,800,000 |
| Commercial decking | Decking | 1,650 | | | | | 1,700,000 |
| Selected decking | Decking | 2,150 | | | | | 1,900,000 |
| Commercial decking | Decking | 1,800 | | | | | 1,700,000 |

(Surfaced at 15% max m.c. and used at 15% max m.c.)

### Hem-fir (surfaced dry or surfaced green. Used at 19% max m.c.)

**West Coast Lumber Inspection Bureau and Western Wood Products Association (see footnotes)**

| Grade | Size | Fb | Ft | Fv | Fc⊥ | Fc∥ | E |
|---|---|---|---|---|---|---|---|
| Select structural | 2–4 in thick, 2–4 in wide | 1,650 | 975 | 75 | 245 | 1,300 | 1,500,000 |
| No. 1 | | 1,400 | 825 | 75 | 245 | 1,000 | 1,500,000 |
| No. 2 | | 1,150 | 675 | 75 | 245 | 800 | 1,400,000 |
| No. 3 | | 625 | 375 | 75 | 245 | 500 | 1,200,000 |
| Appearance | | 1,400 | 825 | 75 | 245 | 1,200 | 1,500,000 |
| Stud | | 625 | 375 | 75 | 245 | 500 | 1,200,000 |
| Construction | 2–4 in thick, 4 in wide | 825 | 475 | 75 | 245 | 925 | 1,200,000 |
| Standard | | 450 | 275 | 75 | 245 | 750 | 1,200,000 |
| Utility | | 225 | 125 | 75 | 245 | 500 | 1,200,000 |
| Select structural | 2–4 in thick, 6 in and wider | 1,400 | 950 | 75 | 245 | 1,150 | 1,500,000 |
| No. 1 | | 1,200 | 800 | 75 | 245 | 1,000 | 1,500,000 |
| No. 2 | | 1,000 | 650 | 75 | 245 | 850 | 1,400,000 |
| No. 3 | | 575 | 375 | 75 | 245 | 550 | 1,200,000 |
| Appearance | | 1,200 | 800 | 75 | 245 | 1,200 | 1,500,000 |

**West Coast Lumber Inspection Bureau**

| Grade | Size | Fb | Ft | Fv | Fc⊥ | Fc∥ | E |
|---|---|---|---|---|---|---|---|
| Select structural | Beams and stringers | 1,250 | 750 | 70 | 245 | 900 | 1,300,000 |
| No. 1 | Beams and stringers | 1,000 | 525 | 70 | 245 | 750 | 1,300,000 |
| Select structural | Posts and timbers | 1,200 | 800 | 70 | 245 | 950 | 1,300,000 |
| No. 1 | Posts and timbers | 975 | 650 | 70 | 245 | 850 | 1,300,000 |
| Select Dex | Decking | 1,400 | | | 245 | | 1,500,000 |
| Commercial Dex | Decking | 1,150 | | | 245 | | 1,400,000 |

**Western Wood Products Association (see footnotes)**

| Grade | Size | Fb | Ft | Fv | Fc⊥ | Fc∥ | E |
|---|---|---|---|---|---|---|---|
| Select structural | Beams and stringers | 1,250 | 850 | 70 | 245 | 900 | 1,300,000 |
| No. 1 | Beams and stringers | 1,050 | 700 | 70 | 245 | 775 | 1,300,000 |
| Select structural | Posts and timbers | 1,200 | 800 | 70 | 245 | 950 | 1,300,000 |
| No. 1 | Posts and timbers | 975 | 650 | 70 | 245 | 850 | 1,300,000 |
| Selected decking | Decking | 1,600 | | | | | 1,500,000 |
| Commercial decking | Decking | 1,300 | | | | | 1,400,000 |
| Selected decking | Decking | 1,750 | | | | | 1,600,000 |
| Commercial decking | Decking | 1,450 | | | | | 1,500,000 |

(Surfaced at 15% max m.c. and used at 15% max m.c.)

### Southern pine (surfaced green. Used any condition)

| Grade | Size | Fb | Fb | Ft | Fv | Fc⊥ | Fc∥ | E |
|---|---|---|---|---|---|---|---|---|
| Select structural | 2½–4 in thick, | 1,700 | 1,950 | 975 | 80 | 270 | 1,050 | 1,600,000 |
| Dense select structural | | 1,950 | 2,250 | 1,150 | 80 | 315 | 1,250 | 1,600,000 |
| No. 1 | | 1,400 | 1,600 | 825 | 80 | 270 | 850 | 1,600,000 |
| No. 1 dense | | 1,650 | 1,900 | 975 | 80 | 315 | 975 | 1,600,000 |
| No. 2 | | 1,000 | | 575 | 70 | 230 | 550 | 1,300,000 |

**Table 4. Allowable Unit Stresses for Visually Graded Stress-Grade Lumber (Continued)**

Allowable unit stresses, lb/in²

| Species and commercial grade | Size classification | Extreme fiber in bending $F_b$ Single-member uses | Extreme fiber in bending $F_b$ Repetitive-member uses | Tension parallel to grain $F_t$ | Horizontal shear $F_v$ | Compression perpendicular to grain $F_{c\perp}$ | Compression parallel to grain $F_c$ | Modulus of elasticity $E$ | Grading-rules agency |
|---|---|---|---|---|---|---|---|---|---|
| | | | | Southern pine (surfaced green. Used any condition) | | | | | |
| No. 2 medium grain | 2½–4 in wide | 1,150 | 1,300 | 675 | 80 | 270 | 650 | 1,400,000 | |
| No. 2 dense | | 1,350 | 1,500 | 800 | 80 | 315 | 775 | 1,500,000 | |
| No. 3 | | 650 | 750 | 375 | 70 | 230 | 400 | 1,300,000 | |
| No. 3 dense | | 775 | 900 | 450 | 80 | 315 | 475 | 1,300,000 | |
| Stud | | 650 | 750 | 375 | 70 | 230 | 400 | 1,300,000 | |
| Construction | 2½–4 in thick, 4 in wide | 850 | 975 | 490 | 70 | 230 | 750 | 1,300,000 | |
| Standard | | 475 | 550 | 275 | 70 | 230 | 620 | 1,300,000 | |
| Utility | | 225 | 250 | 125 | 70 | 230 | 400 | 1,300,000 | |
| Select structural | 2½–4 in thick, 6 in and wider | 1,450 | 1,650 | 950 | 80 | 270 | 925 | 1,600,000 | Southern Pine Inspection Bureau (see footnotes 2 and 8) |
| Dense select structural | | 1,650 | 1,900 | 1,100 | 80 | 315 | 1,100 | 1,600,000 | |
| No. 1 | | 1,200 | 1,400 | 800 | 80 | 270 | 825 | 1,600,000 | |
| No. 1 dense | | 1,400 | 1,600 | 950 | 80 | 315 | 975 | 1,600,000 | |
| No. 2 | | 850 | 975 | 550 | 70 | 230 | 600 | 1,300,000 | |
| No. 2 medium grain | | 1,000 | 1,150 | 650 | 80 | 270 | 700 | 1,400,000 | |
| No. 2 dense | | 1,150 | 1,300 | 775 | 80 | 315 | 825 | 1,500,000 | |
| No. 3 | | 575 | 650 | 375 | 70 | 230 | 425 | 1,300,000 | |
| No. 3 dense | | 675 | 775 | 450 | 80 | 315 | 500 | 1,300,000 | |
| Dense standard factory | 2½–4 in thick, | 1,600 | | 950 | 80 | 315 | 975 | 1,600,000 | |
| No. 1 factory | 2½–4 in wide | 1,150 | | 650 | 80 | 270 | 650 | 1,400,000 | |
| No. 1 dense factory | | 1,300 | | 775 | 80 | 315 | 775 | 1,500,000 | |
| No. 2 factory | | 1,150 | | 650 | 80 | 270 | 650 | 1,400,000 | |
| No. 2 dense factory | | 1,300 | | 775 | 80 | 315 | 775 | 1,500,000 | |
| Dense standard factory | 2½–4 in thick, 6 in and wider | 1,400 | | 950 | 80 | 315 | 975 | 1,600,000 | |
| No. 1 factory | | 1,000 | | 650 | 80 | 270 | 700 | 1,400,000 | |
| No. 1 dense factory | | 1,150 | | 775 | 80 | 315 | 825 | 1,500,000 | |
| No. 2 factory | | 1,000 | | 650 | 80 | 270 | 700 | 1,400,000 | |
| No. 2 dense factory | | 1,150 | | 775 | 80 | 315 | 825 | 1,500,000 | |
| No. 1 SR | 5 in and thicker | 1,300 | | 850 | 110 | 270 | 925 | 1,600,000 | |
| No. 1 dense SR | | 1,500 | | 1,000 | 110 | 315 | 1,050 | 1,600,000 | |
| No. 2 SR | | 1,100 | | 725 | 95 | 270 | 675 | 1,400,000 | |
| No. 2 dense SR | | 1,300 | | 850 | 95 | 315 | 775 | 1,500,000 | |
| Dense structural 65 | | 1,650 | | 1,100 | 105 | 315 | 1,000 | 1,600,000 | |
| Dense structural 86 | 2½ in and thicker | 2,200 | | 1,450 | 140 | 315 | 1,350 | 1,600,000 | |
| Dense structural 72 | | 1,850 | | 1,250 | 120 | 315 | 1,150 | 1,600,000 | |

1. The recommended design values shown are applicable to lumber that will be used under dry conditions such as in most covered structures. For 2- to 4-in-thick lumber the *dry* surfaced size should be used. In calculating design values, the natural gain in strength and stiffness that occurs as lumber dries has been taken into consideration as well as the reduction in size that occurs when unseasoned lumber shrinks. The gain in load-carrying capacity due to increased strength and stiffness resulting from drying more than offsets the design effect of size reductions due to shrinkage. For 5-in and thicker lumber, the surfaced sizes also may be used because design values have been adjusted to compensate for any loss in size by shrinkage which may occur.

2. Values for $F_b$, $F_t$, and $F_c$ for the grades of Construction, Standard, and Utility apply only to 4-in widths. Design values for 2- and 3-in widths of these grades are available from Redwood Inspection Service, Southern Pine Inspection Bureau, West Coast Lumber Inspection Bureau, Western Wood Products Association, and National Lumber Grades Authority.

3. The values are based on edgewise use. For dimension 2 to 4 is thickness, when used flatwise

4. When 2- to 4-in-thick lumber is manufactured at a maximum moisture content of 15 percent and used in a condition where the moisture content does not exceed 15 percent, the design values for surfaced dry or surfaced green lumber shown may be multiplied by the following factors:

| Extreme fiber in bending $F_b$ | Tension parallel to grain $F_t$ | Horizontal shear $F_v$ | Width, in | Compression perpendicular to grain $F_{c\perp}$ — Thickness, in 2 | 3 | 4 | Compression* parallel to grain $F_c$ | Modulus* of elasticity $E$ |
|---|---|---|---|---|---|---|---|---|
| 1.08 | 1.08 | 1.05 | 2–4 | 1.10 | 1.04 | 1.00 | 1.17 | 1.05 |
| | | | 6 and wider | 1.22 | 1.16 | 1.11 | | |
| *For redwood use | | | | | | | 1.15 | 1.04 |

5. When 2- to 4-in-thick lumber is designed for use where the moisture content will exceed 19 percent for an extended period of time, the values shown should be multiplied by the following factors:

| Extreme fiber in bending $F_b$ | Tension parallel to grain $F_t$ | Horizontal shear $F_v$ | Compression perpendicular to grain $F_{c\perp}$ | Compression parallel to grain $F_c$ | Modulus of elasticity $E$ |
|---|---|---|---|---|---|
| 0.86 | 0.84 | 0.97 | 0.67 | 0.70 | 0.97 |

6. When lumber 5 in and thicker is designed for use where the moisture content will exceed 19 percent for an extended period of time, the values shown should be multiplied by the following factors:

| Extreme fiber in bending $F_b$ | Tension parallel to grain $F_t$ | Horizontal shear $F_v$ | Compression perpendicular to grain $F_{c\perp}$ | Compression parallel to grain $F_c$ | Modulus of elasticity $E$ |
|---|---|---|---|---|---|
| 1.00 | 1.00 | 1.00 | 0.67 | 0.91 | 1.00 |

7. The tabulated horizontal shear values shown are based on the conservative assumption of the most severe checks, shakes, or splits possible, as if a piece were split full length. When lumber 4 in and thinner is manufactured unseasoned, the tabulated values should be multiplied by a factor of 0.92. Specific horizontal shear values for any grade and species of lumber may be established by use of the following tables when the length of split or check is known:

| When length of split on wide face is | Multiply tabulated $F_v$ value by (nominal 2-in lumber) |
|---|---|
| No split . . . . . . . . . . . . . | 2.00 |
| ½ × wide face . . . . . . . . . . . | 1.67 |
| ¾ × wide face . . . . . . . . . . . | 1.50 |
| 1 × wide face . . . . . . . . . . . | 1.33 |
| 1½ × wide face or more . . . . . . . | 1.00 |

| When length of split on wide face is | Multiply tabulated $F_v$ value by (3-in and thicker lumber) |
|---|---|
| No split . . . . . . . . . . . . . | 2.00 |
| ½ × narrow face . . . . . . . . . . | 1.67 |
| 1 × narrow face . . . . . . . . . . | 1.33 |
| 1½ × narrow face or more . . . . . . | 1.00 |

8. Stress-rated boards of nominal 1-, 1¼-, and 1½-in thickness, 2 in and wider, are permitted the recommended design values shown for Select Structural, No. 1, Appearance, No. 1, No. 2, and No. 3 grades as shown in 2 to 4 in thick, 2 to 4 in wide, and 2 to 4 in thick, 6 in and wider categories when graded in accordance with those grade requirements.

sions of the piece requires regrading of the member. It should be noted that similarly named grades produced under different Association Rules may not be identical in allowable stresses.

**Machine-stress-rated lumber** is now produced both in the Pacific Northwest and in the southern pine region. The rating is given by passing the lumber flatwise through a series of rolls. The machine continuously measures either the deflection under constant load or the load to constant deflection and feeds the data to a self-contained computer. A program in the computer based on known correlations between modulus of elasticity and breaking strength evaluates the series of readings received, assigns a stress grade, and activates grade-marking stamps for each piece. The allowable unit stresses for machine-stress-rated lumber are shown as Table 5.

This system grades only nominal 2-in-thick material in any width. The advantage over visual grading is that each piece is marked, the stresses assigned are far more reliable, and a wider range of stress grades is now available.

### Adjustments for Working Stresses

Modifications of the unit stresses in Tables 4 and 5 should be made, if applicable, before employing these values in design as working stresses. The principal adjustment is for **duration of load.** Unit allowable stresses have been assigned on the basis of a 10-year loading period, and since the strength of wood varies with the log of time (see Rheology), other load periods can safely use modified stresses. Unit values from Table 4, excluding $E$ for column computations, can be increased as follows: 15 percent for 2 months duration; as for snow; 25 percent for 7 days duration; $33\frac{1}{3}$ percent for wind or earthquake; 100 percent for impact. When a member is fully stressed to maximum design load for more than 10 years either continuously or cumulatively, use working stresses 90 percent of those in Table 4.

**Moisture conditions** which will cause the lumber in use to remain either dry or permanently above the fiber saturation point will employ the same values of stress as shown in Table 4, except that where lumber is fully submerged, compression parallel to the grain is reduced 10 percent, compression perpendicular to the grain is reduced one-third, and Young's modulus is decreased by one-eleventh.

### GLUED-LAMINATED WOOD

The term **glued-laminated** wood, or **glulam,** refers to construction in which a wooden member, either straight or curved, is formed by assembling a set of boards or plank with glue so that the grain of all laminations is essentially parallel to the length of the member. Such construction offers wide latitude for formation of high-strength complex shapes, supplies large members far in excess of sawn solid sections, as, for example, dredge spuds 30 by 30 in and 85 ft long; also, it allows the production of large, fully seasoned members without the long drying period required in the past. The cost of the glue and a normal one-third loss in material from rough to finished form make glued-laminated shapes more expensive than comparable sawn-solid sizes and at present restricts this construction to the special classes noted above.

### Glues and Gluing Practice

Glues for structural laminating are classified for interior or exterior use on the basis of resistance to the moisture conditions to be encountered in service. **Casein** is the most widely used adhesive and is employed where the moisture content of the timbers is lower than 20 percent, as for most indoor uses. This glue is inexpensive and will set at temperatures under 70°F. **Synthetic resins** are required for glued members which will be above 20 percent moisture content in use. All resins used in laminating are of the thermosetting types since joints made with the thermoplastic glues will creep under sustained loads. The hot-setting resins, such as straight phenol-formaldehyde, are generally avoided because the large size of the timbers makes it difficult to attain the proper setting temperatures at the glue line. For information on glues, see adhesives, and Freas and Selbo, *U.S. Dept. Agr. Bull.* 1069.

**Preparation of wood for gluing** requires conditioning to moisture contents of 5 to 6 percent for interior uses and 10 to 12 percent for material to be used outdoors. Just prior to glue application the lumber should be machined, preferably on a double surfacer, using cuts no greater than $\frac{1}{16}$ in, and at a feed speed to produce 20 to 30 knife marks to the inch. Uniformity of thickness is important, and variation should not exceed 0.010 to 0.016 in for good results.

**Gluing practice** should conform closely to the manufacturer's recommendations for use of a given glue. The process of forming a glue joint is essentially a complex reaction in physical chemistry. Quality of the bond formed depends on the mixing and the pot life of the adhesive, assembly time, pressure, curing temperature, and surface preparation.

### Preservative Treatment of Glued-laminated Members

In those cases in which decay hazard requires protection of the finished structure, either this can be supplied by treatment of the member after gluing and framing or the lumber can be treated prior to the laminating. When treatment is applied to a complete member, the glue line must be waterproof and capable of resisting the high temperature and chemical action of the preservative without delamination. For this application the thermosetting phenol derivatives have given best service. When treatment prior to lamination is necessary, creosote has proved to be compatible for both Douglas fir and southern pine with the exterior-use resin adhesives. Good joints are possible using impregnated lumber, if retention is moderate and a final light surfacing is carried out after treatment and before gluing.

### Allowable Unit Stresses for Structural Glued-Laminated Lumber

Allowable unit stresses for Douglas fir, larch, southern pine, and California redwood have been developed by the American Institute of Timber Construction. Since the tables cannot be abbreviated with safety, reference is made to the National Design Specification for allowable unit stresses.

Modifications of allowable stresses for moisture conditions in service and duration of load apply as for sawn-solid members. In addition beam stresses must be modified for slenderness factor and size factor. For the curved parts of bent members the allowable unit stress in bending is multiplied by a **curvature factor:** $1 - 2{,}000\,(t/R)^2$, in which $t$ is the lamination thickness, in, $R$ is the radius of curvature of a lamina, in, and the ratio $t/R$ is not to exceed $\frac{1}{125}$ for soft woods and $\frac{1}{100}$ for hardwoods. The **radial stresses** induced in curved members by bending moments may be taken into account by methods given in paragraph 903-D, National Design Specification.

**Table 5. Allowable Unit Stresses for Machine-Stress-Rated Lumber (From National Design Specification)**

(The allowable unit stresses are for normal loading. See text for service modifications.)

| Grading rules agency, grade designation | Size classification | Extreme fiber in bending $F_b$* Single-member uses | Repetitive-member uses | Tension parallel to grain $F_t$ | Compression parallel to grain $F_c$ | Compression perpendicular to grain $F_{c\perp}$ (dry)† Douglas fir-larch | Hem-fir | Pine‡ | Engelmann spruce | Cedar§ | Modulus of elasticity $E$ |
|---|---|---|---|---|---|---|---|---|---|---|---|
| | | | | | | Western Wood Products Association | | | | | |
| 1,200f-1.2E | Machine-rated lumber 2 in thick or less, all widths | 1,200 | 1,400 | 600 | 950 | 385 | 245 | 240 | 195 | 295 | 1,200,000 |
| 1,500f-1.4E | | 1,500 | 1,750 | 900 | 1,200 | 385 | 245 | 240 | 195 | 295 | 1,400,000 |
| 1,650f-1.5E | | 1,650 | 1,900 | 1,020 | 1,320 | 385 | 245 | 240 | 195 | 295 | 1,500,000 |
| 1,800f-1.6E | | 1,800 | 2,050 | 1,175 | 1,450 | 385 | 245 | 240 | 195 | 295 | 1,600,000 |
| 2,100f-1.8E | | 2,100 | 2,400 | 1,575 | 1,700 | 385 | 245 | 240 | 195 | 295 | 1,800,000 |
| 2,400f-2.0E | | 2,400 | 2,750 | 1,925 | 1,925 | 385 | 245 | 240 | 195 | 295 | 2,000,000 |
| 2,700f-2.2E | | 2,700 | 3,100 | 2,150 | 2,150 | 385 | 245 | 240 | 195 | 295 | 2,200,000 |
| 3,000f-2.4E | | 3,000 | 3,450 | 2,400 | 2,400 | 385 | 245 | 240 | 195 | 295 | 2,400,000 |
| 3,300f-2.6E | | 3,300 | 3,800 | 2,650 | 2,650 | 385 | 245 | 240 | 195 | 295 | 2,600,000 |
| 900f-1.0E | Machine-rated joists 2 in thick or less, all widths | 900 | 1,050 | 350 | 725 | 385 | 245 | 240 | 195 | 295 | 1,000,000 |
| 900f-1.2E | | 900 | 1,050 | 350 | 725 | 385 | 245 | 240 | 195 | 295 | 1,200,000 |
| 1,200f-1.5E | | 1,200 | 1,400 | 600 | 950 | 385 | 245 | 240 | 195 | 295 | 1,500,000 |
| 1,350f-1.8E | | 1,350 | 1,550 | 750 | 1,075 | 385 | 245 | 240 | 195 | 295 | 1,800,000 |
| 1,800f-2.1E | | 1,800 | 2,050 | 1,175 | 1,450 | 385 | 245 | 240 | 195 | 295 | 2,100,000 |
| | | | | | | West Coast Lumber Inspection Bureau | | | | | |
| 900f-1.0E | Machine-rated lumber 2 in thick or less, all widths | 900 | 1,050 | 350 | 725 | 385 | 245 | | | | 1,000,000 |
| 1,200f-1.2E | | 1,200 | 1,400 | 600 | 950 | 385 | 245 | | | | 1,200,000 |
| 1,500f-1.4E | | 1,500 | 1,750 | 900 | 1,200 | 385 | 245 | | | | 1,400,000 |
| 1,650f-1.5E | | 1,650 | 1,900 | 1,020 | 1,320 | 385 | 245 | | | | 1,500,000 |
| 1,800f-1.6E | | 1,800 | 2,050 | 1,175 | 1,450 | 385 | 245 | | | | 1,600,000 |
| 2,100f-1.8E | | 2,100 | 2,400 | 1,575 | 1,700 | 385 | 245 | | | | 1,800,000 |
| 2,400f-2.0E | | 2,400 | 2,750 | 1,925 | 1,925 | 385 | 245 | | | | 2,000,000 |
| 2,700f-2.2E | | 2,700 | 3,100 | 2,150 | 2,150 | 385 | 245 | | | | 2,200,000 |
| 900f-1.0E | Machine-rated joists 2 in thick or less, 6 in and wider | 900 | 1,050 | 350 | 725 | 385 | 245 | | | | 1,000,000 |
| 900f-1.2E | | 900 | 1,050 | 350 | 725 | 385 | 245 | | | | 1,200,000 |
| 1,200f-1.5E | | 1,200 | 1,400 | 600 | 950 | 385 | 245 | | | | 1,500,000 |
| 1,500f-1.8E | | 1,500 | 1,750 | 900 | 1,200 | 385 | 245 | | | | 1,800,000 |
| 1,800f-2.1E | | 1,800 | 2,050 | 1,175 | 1,450 | 385 | 245 | | | | 2,100,000 |

**Table 5. Allowable Unit Stresses for Machine-Stress-Stress-Rated Lumber (From National Design Specification) (Continued)**

| Grading rules agency, grade designation | Size classification | Extreme fiber in bending $F_b$* | | Tension parallel to grain $F_t$ | Compression parallel to grain $F_c$ | Compression perpendicular to grain $F_{c\perp}$ (dry)† | | | | | Modulus of elasticity $E$ |
|---|---|---|---|---|---|---|---|---|---|---|---|
| | | Single-member uses | Repetitive-member uses | | | Douglas fir-larch | Hem-fir | Pine‡ | Engelmann spruce | Cedar§ | |
| | | | | | **Southern Pine Inspection Bureau** | | | | | | |
| | | | | | Southern pine | | | | | | |
| 1,200f-1.2E | Machine-rated lumber 2 in thick or less, 2–4 in wide | 1,200 | 1,375 | 600 | 950 | 405 | | | | | 1,200,000 |
| 1,500f-1.4E | | 1,500 | 1,725 | 900 | 1,200 | 405 | | | | | 1,400,000 |
| 1,650f-1.5E | | 1,650 | 1,900 | 1,020 | 1,320 | 405 | | | | | 1,500,000 |
| 1,800f-1.6E | | 1,800 | 2,100 | 1,175 | 1,450 | 405 | | | | | 1,600,000 |
| 2,100f-1.8E | | 2,100 | 2,400 | 1,575 | 1,700 | 405 | | | | | 1,800,000 |
| 2,400f-2.0E | | 2,400 | 2,750 | 1,925 | 1,925 | 405 | | | | | 2,000,000 |
| 2,700f-2.2E | | 2,700 | 3,100 | 2,150 | 2,150 | 405 | | | | | 2,200,000 |
| 3,000f-2.4E | | 3,000 | 3,450 | 2,400 | 2,400 | 405 | | | | | 2,400,000 |
| 3,300f-2.6E | | 3,300 | 3,800 | 2,650 | 2,650 | 405 | | | | | 2,600,000 |
| 900f-1.0E | Machine-rated lumber 2 in thick or less, 6 in and wider | 900 | 1,025 | 350 | 725 | 405 | | | | | 1,000,000 |
| 900f-1.2E | | 900 | 1,025 | 350 | 725 | 405 | | | | | 1,200,000 |
| 1,200f-1.5E | | 1,200 | 1,375 | 600 | 950 | 405 | | | | | 1,500,000 |
| 1,350f-1.8E | | 1,350 | 1,550 | 750 | 1,075 | 405 | | | | | 1,800,000 |
| 1,800f-2.1E | | 1,800 | 2,050 | 1,175 | 1,450 | 405 | | | | | 2,100,000 |

Allowable unit stresses, lb/in²

*Tabulated extreme fiber in sending values $F_b$ are applicable to lumber loaded on edge. When loaded flatwise, these values should be multiplied by the following factors:

| Nominal width, in | 4 | 6 | 8 | 10 | 12 | 14 |
|---|---|---|---|---|---|---|
| Factor | 1.10 | 1.15 | 1.19 | 1.22 | 1.25 | 1.28 |

†Allowable unit stresses for horizontal shear $F_v$ (dry) for all grade designations are as follows:

| Douglas fir-larch | Hem-fir | Pine | Engelmann spruce | Cedar | Southern pine |
|---|---|---|---|---|---|
| 95 | 75 | 65 | 70 | 75 | 90 |

For southern pine KD: 95

‡Pine includes Idaho white, lodgepole, ponderosa, or sugar pine.
§Cedar includes incense or western red cedar.

**Table 6. Allowable Unit Stresses for Plywood (Douglas Fir and Western Larch)**
(For grades and thicknesses listed in USCS45–60) (Values given are for permanent loads; for normal 10-year loading, increase by 10 percent.)

Dry Location

| Type of stress | Ext A-A | Ext A-B Ext A-C | Ext concrete form B-B Ext B-C Ext C-C Int concrete form B-B Int sheathing C-D Int sheathing C-D (ext glue) | All other grades conforming to USCS45–60, with Coast Douglas fir faces and backs and with inner plies of Douglas fir or any Western softwood species listed in Groups 1, 2, or 3 of USCS122–60 (apply the following percentages to stresses for corresponding Ext grade) |
|---|---|---|---|---|
| **Extreme fiber in bending** | | | | |
| Face grain ∥ to span | 2,188 | 2,000 | 1,875 | 100 % |
| Face grain ⊥ to span | 1,875 | 1,875 | 1,875 | 65 |
| **Tension** | | | | |
| ∥ to face grain (3 ply only*) | 2,188 | 2,000 | 1,875 | 100 %‡ |
| ⊥ to face grain | 1,875 | 1,875 | 1,875 | 65 % |
| ± 45° to face grain | 337 | 320 | 310 | 75 % |
| **Compression** | | | | |
| ∥ to face grain (3 ply only*) | 1,605 | 1,460 | 1,375 | 100 %‡ |
| ⊥ to face grain | 1,375 | 1,375 | 1,375 | 60 % |
| ± 45° to face grain | 496 | 472 | 460 | 70 % |
| Bearing (on face) | 405 | 405 | 405 | 100 % |
| **Shear, rolling, in plane of plies†** | | | | |
| ∥ or ⊥ to face grain | 79 | 72 | 68 | 75 % |
| ± 45° | 105 | 96 | 90 | 75 % |
| **Shear in plane ⊥ to plies** | | | | |
| ∥ or ⊥ to face grain | 260 | 240 | 225 | 80 % |
| ± 45° | 520 | 480 | 450 | 80 % |
| **Modulus of elasticity in bending** | | | | |
| Face grain ∥ to span | 1,600,000 | 1,600,000 | 1,600,000 | 100 % |
| Face grain ⊥ to span | 1,600,000 | 1,600,000 | 1,600,000 | 60 % |

*For tension or compression, ∥ to grain, in 5 ply or thicker, use values for 3 ply, but in next lower grade.

†For the flange-web joints of beams having plywood webs and for joints between the skin and framing members located at the edges of stressed skin plywood panels, these values shall be reduced by 50 percent.

‡For 5 or more plies, use 80 percent.

*Example:* The working stress in compression ∥ for A-D 5-ply (1,100 lb/in²) is found by multiplying the value for Exterior A-C 5-ply, 1,375 lb/in², by 80%, the reduction factor shown in the last column and footnotes * and ‡.

Wet or Damp Location

Where moisture content will exceed 16%, decrease by 20% values shown for dry location for following properties:
Extreme fiber in bending, tension, and compression both ∥ and ⊥ to the grain and at 45°, and bearing. (No change in values for rolling shear or modulus of elasticity.)

## PLYWOOD

**Plywood** is the name given to a wood panel glued under pressure from an odd number of layers or plies of veneer or veneer and lumber, in which alternate plies are laid with the grain at right angles. The alternation of grain direction in adjacent plies equalizes the sheet in the two face directions in strength, stiffness, and dimensional changes.

**Balanced construction** in plywood results in minimum warpage in the sheet. Balance is determined by considering veneer thickness, shrinkage, elasticity, moisture content, and grain angle as factors in the moments of pairs of plies about the neutral axis of the sheet. Balance is especially critical in thin sheets.

The two outside veneer sheets in a plywood panel are known as the **faces** or **face and back.** The interior ply or plies whose grain direction parallels the face is a **core**, if single, or **centers**, if more than one is used. The interior plies whose grain is perpendicular to that in the faces are the **cross bands.** Plywood stock sheets range from three to seven plies and ⅛ to 1³⁄₁₆

in thick and are usually 4 by 8 ft in surface. Special smaller sizes are produced to order, particularly in hardwoods. Lengths up to 30 ft can be made at the mills by splicing. Commercial thicknesses and ply construction of Douglas fir plywood are shown in Table 7.

Plywood can be made of veneers cut from any kind of wood; however, technical difficulties of glue compatibility and veneer cutting restrict production plywood to a few kinds of wood. Depending on the botanical identity of the wood in the sheets, plywood is classed as hardwood or softwood. For both classes the sheets are made in two types that reflect the moisture resistance of the glue joints. Douglas fir plywood, which is the principal constructional plywood in the United States, is made as (1) **interior,** a type that will maintain its shape and most of its strength when occasionally subjected to a thorough wetting and subsequent normal drying; and (2) **exterior,** a type of plywood which must retain its shape and strength when repeatedly wet and dried under adverse conditions and be suitable for permanent outdoor exposure. The latter is also called **marine plywood.** Interior plywood is usually bonded with soybean glue or an extended resin adhesive, while exterior panels are made with hot-pressed phenol resins. The **grade of plywood** is based on the quality of the veneers on the faces of the sheet. Veneer grades for Douglas fir and most other western softwoods are A, B, C, and D, in decreasing order of quality. Hardwood veneers are graded as 1, 2, 3, 4, with number 1 as highest quality. The **grade designation** of plywood will include the type and the veneer grades, as, for example, Ext-DFPA·Plyshield·(A-C), in which the face is grade A and the back C. The inner plies of commercial Douglas fir panels are made up with grade C for exterior type and grade D for the interior-type plywood. The various grades of Douglas fir plywood are shown in Table 6.

### Strength of Plywood

The allowable stress values for solid wood cannot be transposed directly to plywood because of the alternation of ply directions and the fact that knots and other major defects are confined to single plies. Precise equations for strength, based on the composite section and involving all plies, are given in Design of Wood Aircraft Structures, *ANC Bull*. 18. These calculations are involved and yield too refined answers for most applications.

### Recommended Working Stresses for Douglas Fir Plywood

Approximate methods for the calculation of plywood strength have been developed and tested by the USFPL (R-1630, Apr. 1956). These methods are suitable for estimating the strength and stiffness of plywood and are sufficiently accurate for general design use, provided that the plywood is not stressed to buckling. Working stresses based on this system for various grades of Douglas fir are shown in Table 6. These stresses should be used only with the restrictions in the following paragraphs.

**Tension and compression** approximations take into account only those plies with grain parallel to load direction. Consequently, the choice of stress in the table must be made on basis of face-grain direction with respect to load direction. The values for parallel ( $\parallel$ ) to face grain, perpendicular ( $\perp$ ) to face grain, and 3 ply as against 5 or more plies differ because the inner plies are poorer grade veneer which carry lower stresses. Stress direction at 45° to face grain requires the full cross-sectional area to be used in computations because none of the plies in the sheet are parallel to the load direction. Calculations will be simplified for standard plywood constructions by the use of the section properties given in Table 7.

**Bending load for plywood,** based on *Bull*. 1630, is calculated by a modified flexure formula $M = KSI/c$, in which $M$ is the bending moment; $S$ is the unit stress for extreme fiber in bending of those plies in the parallel-to-span direction; $I$ is the moment of inertia computed on basis of plies parallel to span only; $c$ is the distance from neutral axis to outermost fiber of the outermost ply having its grain in the direction of the span; $K = 1.50$ for 3-ply plywood having the grain of the outer plies perpendicular to the span; $K = 0.85$ for all other plywood.

**Deflection in bending** is approximated by the usual equations, except that the moment of inertia $I$ is a composite term made up of the sum of the moments of inertia for the plies parallel to span, plus one-twentieth that of the plies perpendicular to span. For thin sheets of plywood ($\frac{1}{4}$ in and under), the perpendicular plies contribute less than 10 percent to the total

and may be neglected in calculations. However, as sheet thickness increases, the $I$ perpendicular approaches the value of $I$ parallel, and must be included.

**Shear in plywood** is of two types. The stress listed in Table 6 as shear in plane perpendicular ( $\perp$ ) to plies is that developed in the plywood web of a built-up I, or box beam. The second kind of shear is set up in the plane of the plies as normal horizontal shear in a beam. Because the cells of the wood in a direction normal to their length are deformed at low stress levels, the shear is limited in this case by the plies perpendicular to stress direction. Such shear action is called **rolling shear** in plywood. Stress concentrations as exist between flanges and webs of I and box beams and framing members at edges of panels should be computed using 50 percent of tabulated rolling-shear stress.

### Shrinkage in Plywood

The range of possibilities in kinds of wood, combinations within a sheet, thickness, and numbers of plies makes the tabulation of shrinkage data impossibly cumbersome. Calculations can be made using the methods given by C. B. Norris in "Techniques of Plywood." Studies have been made by the U.S. Forest Products Laboratory for 3-ply panels of one kind of wood, and of equal ply thicknesses, dried from soaked to ovendry condition, and ranging in thickness from $\frac{1}{10}$ to $\frac{1}{2}$ in. Shrinkage in such panels averages 0.45 percent parallel to the face grain and 0.67 percent perpendicular to the face grain, and varies from 0.2 to 1 percent and 0.3 to 1.2 percent, respectively.

### PRESERVATIVE TREATMENT OF WOOD

Since few woods are resistant to deterioration under adverse conditions, and the sterilization provided by steam treating and kiln drying is not permanent, the principal method for protection of wood in service is the introduction of toxic chemicals into the wood.

### Materials for Protection against Biological Action

**Oils and oil-borne preservatives** are the chief materials in use today. **Coal-tar creosote,** which is the most widely used, is a by-product distilled from the coal tar produced by high-temperature carbonization of bituminous coal. It consists of a heterogeneous mixture of liquid and solid hydrocarbons having a continuous boiling range from about 200 to 325°C. Analysis methods, as given by the American Wood Preservers Assoc., are important since the chemical composition and related toxicity of coal-tar creosote is variable. The fraction distilling in the 200 to 275°C range appears to be the most toxic. As a means of reducing cost or increasing penetration, **creosote-coal tar solutions** and **creosote-petroleum solutions** have been used extensively. Such solutions are used in proportions as high as 50 percent. Toxicity of the petroleum solutions is reduced in a ratio greater than indicated by the percentage of diluent added. **Pentachlorophenol** in volatile petroleum carries is nearly as important a preservative as creosote and is favored where cleanliness and paintability are necessary. Usual concentrations of pentachlorophenol for wood treatment are not less than 5 percent by weight. This produces a solution in the same range of cost as coal-tar creosote and can be used in roughly the same range of retentions as the latter.

**Table 7. Moments of Inertia, Section Moduli, and Veneer Areas for Selected Plywood Constructions (12-in widths)**
(From "Technical Data on Plywood," Douglas Fir Plywood Assoc.)

| Net plywood thickness | No. of plies | Veneer thickness (nominal), in. | | | Parallel‡ plies only | | | Perpendicular‡ plies only | | |
|---|---|---|---|---|---|---|---|---|---|---|
| | | Faces§ | Centers | Cross-band | Area, sq. in. | Moment of inertia I, in⁴ | Section modulus S, in³ | Area, sq. in | Moment of inertia I, in⁴ | Section modulus S, in³ |
| 1/8"—S† | 3 | 1/16 | 1/16 | ..... | 0.75 | 0.0017 | 0.027 | 0.75 | 0.0002 | 0.0077 |
| 3/16"—S | 3 | 1/12 | 1/12 | ..... | 1.25 | 0.0060 | 0.064 | 1.00 | 0.0006 | 0.0139 |
| 1/4"—R* | 3 | 1/12 | 1/12 | ..... | 2.00 | 0.0150 | 0.120 | 1.00 | 0.0006 | 0.0139 |
| 1/4"—S | 3 | 1/8 | 1/8 | ..... | 1.67 | 0.0143 | 0.114 | 1.33 | 0.0014 | 0.0247 |
| 5/16"—R | 3 | 1/10+ | 1/10+ | ..... | 2.50 | 0.0294 | 0.188 | 1.25 | 0.0011 | 0.0215 |
| 5/16"—S | 3 | 1/8 | 1/8 | ..... | 2.25 | 0.0286 | 0.183 | 1.50 | 0.0020 | 0.0312 |
| 3/8"—R | 3 | 1/8 | 1/8 | ..... | 3.00 | 0.0509 | 0.271 | 1.50 | 0.0020 | 0.0312 |
| 3/8"—S | 3 | 1/8 | 3/16 | ..... | 2.25 | 0.0461 | 0.246 | 2.25 | 0.0066 | 0.0704 |
| 3/8"—S | 5 | 1/10 | 1/12 | 2@1/12 | 2.50 | 0.0377 | 0.201 | 2.00 | 0.0150 | 0.120 |
| 7/16"—S | 5 | 1/10 | 1/10 | 2@1/10 | 2.85 | 0.0575 | 0.263 | 2.40 | 0.0260 | 0.1735 |
| 1/2"—R | 5 | 1/10 | 1/10 | 2@1/10 | 3.60 | 0.0990 | 0.396 | 2.40 | 0.0260 | 0.1735 |
| 1/2"—S | 5 | 1/8 | 1/8 | 2@1/10 | 3.60 | 0.0926 | 0.370 | 2.40 | 0.0324 | 0.1995 |
| 9/16"—S | 5 | 1/8 | 1/8 | 2@1/8 | 3.75 | 0.1273 | 0.452 | 3.00 | 0.0507 | 0.271 |
| 5/8"—R | 5 | 1/8 | 1/8 | 2@1/8 | 4.50 | 0.1934 | 0.619 | 3.00 | 0.0507 | 0.271 |
| 5/8"—S | 5 | 1/8 | 3/16 | 2@1/8 | 4.50 | 0.1670 | 0.534 | 3.00 | 0.0771 | 0.352 |
| 11/16"—S | 5 | 1/8 | 1/8 | 2@3/16 | 3.75 | 0.202 | 0.588 | 4.50 | 0.123 | 0.492 |
| 3/4"—R | 5 | 1/8 | 1/8 | 2@3/16 | 4.50 | 0.299 | 0.798 | 4.50 | 0.123 | 0.492 |
| 3/4"—S | 5 | 1/8 | 3/16 | 2@3/16 | 4.50 | 0.251 | 0.670 | 4.50 | 0.171 | 0.608 |
| 3/4"—S | 7 | 1/8 | 2@1/12 | 3@1/8 | 4.50 | 0.286 | 0.763 | 4.50 | 0.136 | 0.503 |
| 13/16"—S | 7 | 1/8 | 2@1/8 | 3@1/8 | 5.25 | 0.343 | 0.845 | 4.50 | 0.193 | 0.617 |
| 7/8"—S | 7 | 1/8 | 2@5/32 | 3@1/8 | 6.00 | 0.427 | 0.976 | 4.50 | 0.243 | 0.707 |
| 15/16"—S | 7 | 1/8 | 2@3/16 | 3@1/8 | 6.75 | 0.525 | 1.120 | 4.50 | 0.299 | 0.797 |
| 1"—S | 7 | 1/8 | 2@1/8 | 3@3/16 | 5.25 | 0.540 | 1.080 | 6.75 | 0.460 | 1.131 |
| 1 1/16"—S | 7 | 1/8 | 2@1/8 | 3@3/16 | 6.00 | 0.615 | 1.157 | 6.75 | 0.585 | 1.305 |
| 1 1/8"—R¶ | 7 | 1/10 | 2@3/16 | 3@3/16 | 6.90 | 0.798 | 1.419 | 6.66 | 0.627 | 1.115 |
| 1 1/8"—S | 7 | 1/8 | 2@3/16 | 3@3/16 | 6.75 | 0.771 | 1.371 | 6.75 | 0.653 | 1.395 |
| 1 3/16"—S | 7 | 1/8 | 2@7/32 | 3@3/16 | 7.50 | 0.912 | 1.538 | 6.75 | 0.763 | 1.526 |

*Rough.
†Sanded.
‡Refers to direction of face grain.
§For sanded panels, thickness is before sanding.
¶Interior floor panel 2·4·1 grade.

**Water-borne solutions of inorganic salts** have advantages over the oils in greater ease of penetration and freedom from fire hazards and odor; however, they do cause swelling and some react with metal. The principal preservative of this class is chromated zinc chloride. Some other salts used are: acid cupric chromate (Celcure), ammoniacal copper arsenate (Chemonite), chromated copper arsenate (Green salt or Erdalith), chromated zinc arsenate (Boliden salt), copperized chromated zinc chloride. Retention for salts used in wood preserving is stated in pounds of dry salt per cubic foot of material.

### Preparation of Material

Wood which is to be pressure-treated must be peeled to hasten drying and permit maximum preservative absorption. Barking can be done by hand shaving or in a peeling machine which smoothes and straightens the poles. Machine peeling results in lower pole strengths and less penetration for many kinds of wood.

Seasoning ordinarily is done by air drying to about 15 percent moisture content. When green material must be used for treating, two seasoning methods are common. **Steam conditioning** is usually used with southern pine. The wood in the treating cylinder is steamed for several hours at about 30 lb/in², following which a vacuum is drawn on the charge for several hours. MacLean (*U.S. Dept. Agr. Misc. Publ.* 224, rev. 1953) has worked out charts showing cycles required and temperatures attained for various sizes of material. **Boiling**

under vacuum (Boulton process) is extensively used for Douglas fir and red oak. The charge in a cylinder is submerged in hot oil, and a vacuum drawn so that the water boils out of the wood.

Adzing, boring, framing, and all other cutting should be completed prior to treatment, since this will ensure all surfaces are adequately penetrated.

**Incising** is a process by which chisel-pointed teeth are forced into the side grain of the wood to about half-inch depths. Primary purpose is to expose more end grain and give greater penetration in woods such as western red cedar and Douglas fir.

### Methods for Preservative Treatment

Wood can be treated with preservatives under pressure in closed vessels, by dipping, hot and cold soak, and diffusion. On the basis of volume of material treated for use in engineered structures, the pressure treatments are most important.

Three pressure systems are standard today: **full-cell** (Bethell), **Lowry**, and **Rueping**. All employ a combination of pressure and vacuum in sequence on material contained in a pressure vessel. The essential difference is the use of an initial vacuum or pressure. The full-cell process draws a vacuum on the charge, then introduces the preservative without breaking the vacuum, so that a maximum of preservative can be forced into the wood. The Lowry system uses atmospheric pressure at the start, and the Rueping process employs a positive initial pressure. The latter two systems are also called **empty-cell processes** because the air under pressure in the wood will expand at the end of the cycle to force out excess preservative from cell cavities, leaving a minimum retention for maximum penetration. The Rueping process is the most efficient in this respect and is now the principal creosoting process in use in this country. Typical pressure–time cycles for the three processes are shown in Fig. 4. The temperature during the treatment depends mostly on the preservative. Creosote oils are usually used in the range of 180 to 200°F. Salt solutions are used at temperatures from atmospheric to 200°F. Higher temperatures with salts may cause difficulties in reactions with the wood and embrittlement.

**Fire-retardant treatments** of wood use the same pressure methods and equipment as that used for preservation. Salts similar to those used to protect against fungus attack are employed (see above), but in much greater quantities, up to refusal. This high proportion of inorganic material can have a dulling effect on machine tools. The three principal materials used are chromated zinc chloride (FR), Minalith, and Pyresote. (See also Sec. 12, Fire Protection.)

**Allowable working stresses for preservative-treated lumber** should be reduced to account for damage due to the treating process. Tests made by the U.S. Forest Products Laboratory of preservative-treated lumber when undergoing bending and compression perpendicular to grain show reductions in stress in extreme fiber from a few percent up to 25 percent, depending on the treating conditions. Compression parallel to grain is affected less and modulus of elasticity very little. The effect on horizontal shear can be estimated by inspection for increase in shakes and checks after treatment. Keeping temperatures, heating periods, and pressures to a minimum for required penetration and retention will hold the loss in working stress to a reasonable figure.

### COMMERCIAL LUMBER STANDARDS

**Standard abbreviations** for lumber description and **size standards** for yard lumber are given in the "Wood Handbook."

Cross-sectional **dimensions** and **section properties** for beams, stringers, joists, planks, and posts are given in Sec. 12., Structural Design of Buildings.

**Standard patterns** for finish lumber are shown in publications of the grading and dressing rules for the various lumber associations.

Information and **specifications for commercial plywood** are given in Commercial Standards: CS45–55 for Douglas-fir Plywood and CS35–49 for Hardwood Plywood.

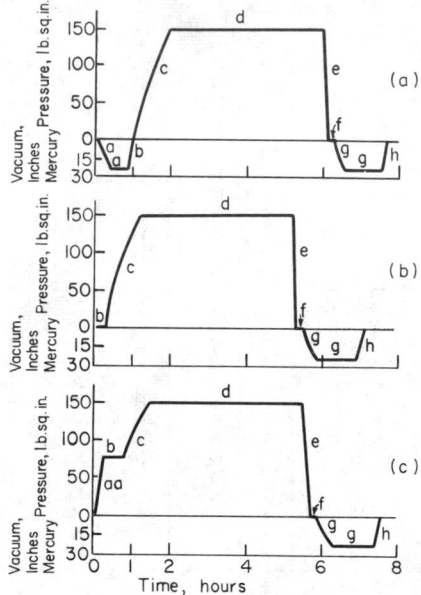

**Fig. 4** Typical pressure diagrams for pressure preservative treatment of wood. (*From Hunt and Garratt, "Wood preservation," McGraw-Hill.*) (*a*) *Full-cell process: a*, preliminary vacuum period; *b*, filling cylinder with preservative; *c*, pressure rising to maximum; *d*, maximum pressure maintained; *e*, pressure released; *f*, preservative withdrawn; *g*, final vacuum period; *h*, vacuum released. (*b*) *Lowry process: b*, filling cylinder with preservative at atmospheric pressure; *c* to *h*, pressure period and final vacuum, as in (*a*). (*c*) *Rueping process: aa*, preliminary air pressure applied; *b*, filling cylinder, while pressure is maintained; *c* to *h*, pressure period and final vacuum, as in (*a*). lb/in² × 6,894 = Pa; in Hg × 3,376 = Pa.

# NONMETALLIC MATERIALS
### by Rienzi B. Parker, Jr.
### revised by Antonio F. Baldo

## ABRASIVES

REFERENCES: "Abrasives: Their History and Development," The Norton Co. Searle, "Manufacture and Use of Abrasive Materials," Pitman. Heywood, "Grinding Wheels and Their Uses," Penton. "Boron Carbide," The Norton Co. "Abrasive Materials," annual review in "Minerals Yearbook," U.S. Bureau of Mines. "Abrasive Engineering," Hitchcock Publishing Co. Coated Abrasives Manfacturer's Institute, "Coated Abrasives—Modern Tool of Industry," McGraw-Hill. Wick, Abrasives; Where They Stand Today, *Manufacturing Engineering and Management*, vol. 69, no. 4, Oct. 1972. Burls, "Diamond Grinding; Recent Research and Development," Mills and Boon, Ltd., London. Coes, Jr., "Abrasives," Springer-Verlag, New York. *Proceedings of the American Society for Abrasives Methods*, 1971.

### Manufactured (Artificial) Abrasives

Manufactured abrasives dominate the scene for commercial and industrial use, because of the greater control over their chemical composition and crystal structure, and their greater uniformity in size, hardness, and cutting qualities, as compared with natural abrasives. Technical advances have resulted in abrasive "machining tools" that are economically competitive with many traditional machining methods, and in some cases replace and surpass them in terms of productivity.

## Grains

Manufactured (industrial or synthetic) **diamonds** are produced from graphite at pressures 29.5 to 66.5 $\times$ $10^6$ N/m² (8 to 18 $\times$ $10^5$ lb/in²) and temperatures from 1090 to 2420°C (2000 to 4400°F), with the aid of metal catalysts. The shape of the crystal is temperature-controllable, with cubes (black) predominating at lower temperatures and octahedra (yellow to white) at higher. General Electric Co. produces synthetics reaching 0.01 carat sizes and of quality comparable to natural diamond powders. Grade MBG-11 (a blocky powder) is harder and tougher and is used for cutting wheels. Diamond hardness (placed at 10 on the Mohs scale) ranges from 5,500 to 7,000 on the Knoop scale. Specific gravity is 3.521.

**Crystalline alumina,** as chemically purified $Al_2O_3$, is very adaptable in operations such as precision grinding of sensitive steels. However, a tougher grain is produced by the addition of $TiO_2$, $Fe_2O_3$, $SiO_2$, $ZrO_2$. The percentage of such additions and the method of cooling the pig greatly influence grain properties. Advantage is taken of this phenomenon to "custom" make the most desirable grain for particular machining needs. Alumina crystals have conchoidal fracture, and the grains when crushed or broken reveal sharp cutting edges and points. Average properties are: density ~ 3.8, coefficient of expansion ~ 0.81 $\times$ $10^{-5}$/°C (0.45 $\times$ $10^{-5}$/°F), hardness ~ 9 (Mohs scale), and melting temperature ~ 2040°C (3700°F). Applications cover the grinding of high-tensile-strength materials such as soft and hard steels, and annealed malleable iron.

**Silicon carbide (SiC)** corresponds to the mineral moissanite, and has a hardness of about 9.5 (Mohs scale) or 2,500 (Knoop), and specific gravity 3.2. It is insoluble in acid and is infusible but decomposes above 2230°C (4060°F). It is manufactured by fusing together coke and sand in an electric furnace of the resistance type. Sawdust is used also in the batch and burns away, leaving passages for the carbon monoxide to escape. The grains are characterized by great brittleness. Abrasives of silicon carbide are best adapted to the grinding of low-tensile-strength materials such as cast iron, brass, bronze, marble, concrete, stone, and glass. It is available under several trade names such as **Carborundum, Carbolon, Crystolon.**

**Boron carbide ($B_4C$),** a black crystal, has a hardness of about 9.32 (Mohs scale) or about 2,800 (Knoop), melts at about 2460°C (4478°F) but reacts with oxygen above 983°C (1800°F), and is not resistant to fused alkalies. Boron carbide powder for grinding and lapping is obtainable in standard mesh sizes to 240, and to 800 in special finer sizes. It is being used in loose grain form for the lapping of cemented-carbide tools. In the form of molded shapes, it is used for pressure blast nozzles, wire-drawing dies, bearing surfaces for gages, etc.

**Boron nitride (BN)** when produced at extremely high pressures and temperatures forms tiny reddish to black grains of cubic crystal structure having hardness equal to diamond, and moreover is stable to 1930°C (3500°F). Abrasive powders, such as Borazon, are used extensively for coated-abrasive applications as for grinding tool and die steels and high-alloy steels, particularly where chemical reactivity of diamonds is a problem. Ease of penetration and free-cutting action minimize heat generation, producing superior surface integrity.

**Crushed steel** is made by heating high-grade crucible steel to white heat and quenching in a bath of cold water. The frag-

ments are then crushed to sizes ranging from fine powder to $^1/_{16}$ in diam. They are classified as diamond crushed steel, diamond steel, emery, and steelite, used chiefly in the stone, brick, glass, and metal trades.

**Rouge and crocus** are finely powdered oxide of iron used for buffing and polishing. Rouge is the red oxide; crocus is purple.

## Natural Abrasives

**Diamond** of the of the **bort** variety, crushed and graded into usable sizes and bonded with synthetic resin, metal powder, or vitrified-type bond, is used extensively for grinding tungsten- and tantalum-carbide cutting tools, and also glass, stone, and ceramics.

**Corundum** is a mineral composed chiefly of crystallized alumina (93 to 97 percent $Al_2O_3$). It has been largely replaced by the manufactured variety.

**Emery,** a cheap and impure form of natural corundum which has been used for centuries as an abrasive, has been largely superseded by manufactured aluminum oxide for grinding. It is still used to some extent in the metal- and glass-polishing trades.

**Garnet** Certain deposits of garnet having a hardness between quartz and corundum are used in the manufacture of abrasive paper. **Quartz** is also used for this purpose. Garnet costs about twice as much as quartz and generally lasts proportionately longer.

**Buhrstones and millstones** are generally made from cellular quartz. Chasers (or stones running on edge) are also made from the same mineral.

**Natural oilstones,** the majority being those quarried in Arkansas, are of either the hard or the soft variety. The hard variety is used for tools requiring an extremely fine edge like those of surgeons, engravers, and dentists. The soft variety, more porous and coarser, is used for less exacting applications.

**Pumice,** of volcanic origin, is extensively used in leather, felt, and woolen industries and in the manufacture of polish for wood, metal, and stone. An artificial pumice is made from sand and clay in five grades of hardness, grain, and fineness.

**Infusorial earth or tripoli** resembles chalk or clay in physical properties. It can be distinguished by absence of effervescence with acid, is generally white or gray in color, but may be brown or even black. Owing to its porosity, it is very absorptive. It is used extensively in polishing powders, scouring soaps, etc., and, on account of its porous structure, in the manufacture of dynamite as a holder of nitroglycerin, also as a nonconductor for steam pipes and as a filtering medium. It is also known as **diatomaceous silica.**

# TOOLS

### Grinding Wheels

**Vitrified Process** In wheels, segments, and other abrasive shapes of this type, the abrasive grains are bonded with a glass or porcelain obtained by mixing the grains with such materials as clays and feldspars in various proportions, molding the wheel, drying, and firing at a temperature of 1370°C (2500°F) approx. It is possible to manufacture wheels as large as 60 in diam by this process, and even larger wheels may be obtained by building up with segments. Most of the grinding wheels and shapes (segments, cylinders, bricks, etc.) now manufactured are of the vitrified type and are very satisfactory for general grinding operations.

**Silicate Process**    Wheels and shapes of the silicate type are manufactured by mixing the abrasive grain with sodium silicate (water glass) and fillers that are more or less inert, molding the wheel by tamping, and baking at a moderate temperature. Silicate bonded wheels are considered relatively "mild acting" and, in the form of large wheels, are still used to some extent for grinding-edge tools in place of the old-fashioned sandstone wheels.

**Organic Bonded Wheels**    Organic bonds are used for high-speed wheels, and are equally well adapted to the manufacture of very thin wheels because of their flexibility compared with vitrified wheels. There are three distinct types in the group. The **shellac process** consists of mixing abrasive grains with shellac, heating the mass until the shellac is viscous, stirring, cooling, crushing, forming in molds, and reheating sufficiently to permit the shellac to set firmly upon cooling. Wheels made by this process are used for saw gumming, roll grinding, ball-race and cam grinding, and in the cutlery trade. In the **rubber process** the bond is either natural or synthetic rubber. The initial mixture of grain, rubber, and sulfur (and such special ingredients as accelerators, fillers, and softeners) may be obtained by rolling or other methods. Having formed the wheel, the desired hardness is then developed by vulcanization. Wheels can be made in a wide variety of grain combinations and grades and have a high factor of safety as regards resistance to breakage in service. Wheels made by the rubber process are used for cutoff service on wet-style machines, ball-race punchings, feed wheels and centerless grinders, and for grinding stainless-steel billets and welds, which usually require a high-quality finish. With the **resinoid process,** the practice is to form the wheel by the cold-press process using a synthetic resin. After heating, the resultant bond is an insoluble, infusible product of notable strength and resiliency. Resinoid bond is used for the majority of high-speed wheels in foundries, welding shops, and billet shops, and also for cutoff wheels. The rate of stock removal is generally in direct proportion to the peripheral speed. Resinoid-bonded wheels are capable of being operated at speeds as high as 2,900 m/min (9,500 surface ft/min), as contrasted with 1,980 m/min (6,500 surface ft/min) for most vitrified bonds.

The **grain size** or **grit** of a wheel is determined by the size or combination of sizes of abrasive grain used. The Grinding Wheel Institute has standardized sizes 8, 10, 12, 14, 16, 20, 24, 30, 36, 46, 54, 60, 70, 80, 90, 100, 120, 150, 180, 200, 220, and 240. The finer sizes, known as flours, are designated as 280, 320, 400, 500, 600, 800, 900, or as F, FF, FFF, and XF.

The **grade** is the hardness or relative strength of bonding of a grinding wheel. The wheel from which grain particles are easily broken away, causing it to wear rapidly, is called soft, and one that is able to retain its particles longer is called hard. The complete range of grade letters used for the order of increasing hardness is

| Soft | Hard |
|---|---|
| A, B, C, D, E, F, G, H, I, J, K, L, M, N, O, P, Q, R, S, T, U, V, W, X, Y, Z | |

### Coated Abrasives

Coated abrasives are "tools" consisting of an abrasive grit, a backing, and an adhesive bond. Grits are generally one of the manufactured variety listed above, and are available in mesh sizes ranging from the coarsest at 12 to the finest at 600. Backings can be cloth, paper, fiber, or combinations and are made in the form of belts and disks for power-operated tools and in cut sheets for both manual and power useage. Adhesive bonds consist of two layers, a "make coat" and a "size coat." Both natural and synthetic adhesives serve for bond materials. Abrasive coatings can be closed-coat, in which the abrasive grains are adjacent to one another without voids, or open-coat, in which the grains are set at a predetermined distance from one another. The flex of the backing is obtained by a controlled, directional, spaced backing of the adhesive bond.

### Industrial Sharpening Stones

Sharpening stones come in several forms such as bench stones, files, rubbing bricks, slip stones, and specialties, and are made chiefly from aluminum oxide, such as Alundum, or silicon carbide, such as Crystolon. Grit sizes of fine, medium, and coarse are available.

# ADHESIVES
### by Antonio F. Baldo

REFERENCES: Cagle (ed.), "Handbook of Adhesive Bonding," McGraw-Hill, 1972. Bikerman, "The Science of Adhesive Joints," Academic, 1968. Shields, "Adhesives Handbook," CRC Press (Division of The Chemical Rubber Co.), 1970. Cook, "Construction Sealants and Adhesives," Wiley, 1970. Patrick, "Treatise on Adhesives," Marcel Dekker, Inc., 1973. NASA SP-5961 (01) Technology Utilization, "Chemistry Technology: Adhesives and Plastics," National Technical Information Services, Virginia. Simonds and Church, "A Concise Guide to Plastics," Reinhold, 1963. Lerner, Kotsher, and Sheckman, "Adhesive Red Book," Palmerton Publishing Co., New York, 1968.

Adhesives are substances capable of holding materials together in a useful manner by surface attachment. Some of the advantages and disadvantages of adhesive bonding are as follows:

**Advantages**    Ability to bond similar or dissimilar materials of different thicknesses; fabrication of complex shapes not feasible by other fastening means; smooth external joint surface; economic and rapid assembly; uniform distribution of stresses; weight reduction; vibration damping; prevention or reduction of galvanic corrosion; insulating properties.

**Disadvantages**    Surface preparation; long cure times; optimum bond strength not realized instantaneously, service-temperature limitations; service deterioration; assembly fire or toxicity; tendency to creep under sustained load.

A broad scheme of classification is given in the table at the top of p. 9-141.

**Thermoplastic adhesives** are a general class of adhesives based upon long-chained polymeric structure and are capable of being softened by the application of heat.

**Thermosetting adhesives** are a general class of adhesives based upon *cross-linked* polymeric structure, and are incapable of being softened once solidified.

Thermoplastic and thermosetting adhesives are cured (set, polymerized, solidified) by heat, catalysis, chemical reaction, free-radical activity, radiation, loss of solvent, etc., as governed by the particular adhesive's chemical nature.

**Elastomers** are a special class of thermoplastic adhesive possessing the common quality of substantial flexibility or elasticity.

| Origin and basic type | | Adhesive material |
|---|---|---|
| Natural | Animal | Albumen, animal glue (inc. fish), casein, shellac, beeswax |
| | Vegetable | Natural resins (gum arabic, tragacanth, colophony, Canada balsam, etc.); oils and waxes (carnauba wax, linseed oils); proteins (soybean); carbohydrates (starch, dextrines) |
| | Mineral | Inorganic materials (silicates, magnesia, phosphates, litharge, sulfur, etc.); mineral waxes (paraffin); mineral resins (copal, amber); bitumen (inc. asphalt) |
| Synthetic | Elastomers | Natural rubber (and derivatives, chlorinated rubber, cyclized rubber, rubber hydrochloride) |
| | | Synthetic rubbers and derivatives (butyl, polyisobutylene, polybutadiene blends (inc. styrene and acylonitrile), polyisoprenes, polychloroprene, polyurethane, silicone, polysulfide, polyolefins (ethylene vinyl chloride, ethylene polypropylene)) |
| | | Reclaim rubbers |
| | Thermoplastic | Cellulose derivatives (acetate, acetate-butyrate, caprate, nitrate, methyl cellulose, hydroxy ethyl cellulose, ethyl cellulose, carboxy methyl cellulose) |
| | | Vinyl polymers and copolymers (polyvinyl acetate, alcohol, acetal, chloride, polyvinylidene chloride, polyvinyl alkyl ethers) |
| | | Polyesters (saturated) (polystyrene, polyamides (nylons and modifications)) |
| | | Polyacrylates (methacrylate and acrylate polymers, cyano-acrylates, acrylamide) |
| | | Polyethers (polyhydroxy ether, polyphenolic ethers) |
| | | Polysulfones |
| | Thermosetting | Amino plastics (urea and melamine formaldehydes and modifications) |
| | | Epoxides and modifications (epoxy polyamide, epoxy bitumen, epoxy polysulfide, epoxy nylon) |
| | | Phenolic resins and modifications (phenol and resorcinol formaldehydes, phenolic-nitrile, phenolic-neoprene, phenolic-epoxy) |
| | | Polyesters (unsaturated) |
| | | Polyaromatics (polyimide, polybenzimidazole, polybenzothiazole, polyphenylene) |
| | | Furanes (phenol furfural) |

Reproduced from Shields, "Adhesives Handbook," CRC Press (Division of the Chemical Rubber Co.), with permission of the publisher.

**Anaerobic adhesives** are a special class of thermoplastic adhesive (polyacrylates) that set *only* in the *absence* of air (oxygen). The two basic types are: (1) *machinery*—possessing shear strength only, and (2) *structural*—possessing both tensile and shear strength.

**Pressure-sensitive adhesives** are permanently (and aggressively) tacky (sticky) solids which form immediate bonds when two parts are brought together under pressure. They are available as films and tapes as well as hot-melt solids.

Table 2 presents a sample of a number of adhesives (with practical information) that are available from various sources. The table is adapted from the rather extensive one found in J. Shields, "Adhesives Handbook," CRC Press (Division of The Chemical Rubber Co.), 1970, by permission of the publisher. Domestic and foreign trade sources are listed there (pages 332–340) and appear coded (in parentheses) in the second column. For other extensive lists of trade sources, the reader is referred to Charles V. Cagle (ed.), "Handbook of Adhesive Bonding," McGraw-Hill, 1973, and "Adhesives Red Book," Palmerton Publishing Co., New York.

The relative performance of a number of adhesives is given in Table 1.

# BRICK, BLOCK, AND TILE

REFERENCES: Plummer and Reardon, "Principles of Brick Engineering, Handbook of Design," Structural Clay Products Institute. Stang, Parsons, and McBurney, Compressive Strength of Clay Brick Walls, *B. of S. Research Paper* 108. ASTM specifications covering paving brick, building brick, sewer brick, and firebrick. Hunting, "Building Construction," Wiley. Amrhein, "Reinforced Masonry Engineering Handbook," Masonry Institute of America, 1972. Simpson and Horrbin (eds.), "The Weathering and Performance of Building Materials," Wiley. SVCE and Jeffers (eds.) "Modern Masonry Panel Construction Systems," Cahners Books (Divison of Cahners Publishing Co.), 1971.

The Committee on Nomenclature and Definitions (E8) of the ASTM has published the following definition of brick:

**Brick**—In the case of structural and road building material, a small unit, solid or practically so, commonly in the form of a rectangular prism, formed from inorganic, non-metallic substances and hardened in its finished shape by heat or chemical action. NOTE—The term is also used collectively for a number of such units, as "a carload of brick." In the present state of the art, the term brick, when used without a qualifying adjective, should be understood to mean such a unit, or a

**Table 1. Performance of Adhesive Resins***
(Rating 1 = poorest or lowest, 10 = best or highest)

| Adhesive resin | Adherence to | | | | | Resistance to | | | | Relative cost |
|---|---|---|---|---|---|---|---|---|---|---|
| | Paper | Wood | Metal | Ceramics | Rubbers | Water | Solvents | Alkali | Acids | |
| Alkyd | 6 | 7 | 5 | 6 | 7 | 7 | 2 | 2 | 5 | 3 |
| Cellulose acetate | 4 | 3 | 1 | 3 | 5 | 2 | 3 | 1 | 3 | 6 |
| Cellulose acetate butyrate | 3 | 3 | 1 | 4 | 5 | 2 | 3 | 1 | 3 | 7 |
| Cellulose nitrate | 5 | 5 | 1 | 5 | 5 | 3 | 2 | 2 | 4 | 5 |
| Ethyl cellulose | 3 | 3 | 1 | 3 | 5 | 2 | 3 | 3 | 3 | 6 |
| Methyl cellulose | 5 | 1 | 1 | 3 | 3 | 1 | 6 | 3 | 3 | 6 |
| Carboxy methyl cellulose | 6 | 1 | 2 | 3 | 2 | 1 | 6 | 1 | 4 | 7 |
| Epoxy resin | 10 | 10 | 8 | 8 | 8 | 8 | 9 | 9 | 8 | 7 |
| Furane resin | 8 | 7 | 1 | 8 | 7 | 8 | 10 | 10 | 8 | 5 |
| Melamine resin | 10 | 10 | 2 | 2 | 2 | 7 | 9 | 5 | 5 | 5 |
| Phenolic resins | 9 | 8 | 2 | 6 | 7 | 8 | 10 | 7 | 8 | 4 |
| Polyester, unsat'd | 6 | 8 | 2 | 5 | 7 | 7 | 6 | 1 | 6 | 4 |
| Polyethylacrylate | 3 | 4 | 3 | 5 | 6 | 8 | 2 | 5 | 7 | 5 |
| Polymethylmethacrylate | 2 | 3 | 2 | 3 | 6 | 8 | 3 | 8 | 7 | 5 |
| Polystyrene | 1 | 3 | 2 | 2 | 5 | 8 | 1 | 10 | 8 | 3 |
| Polyvinylacetate | 8 | 7 | 7 | 7 | 3 | 3 | 3 | 4 | 6 | 3 |
| Polyvinyl alcohol | 6 | 2 | 2 | 4 | 6 | 1 | 7 | 1 | 3 | 6 |
| Polyvinyl acetal | 5 | 7 | 8 | 7 | 7 | 8 | 5 | 3 | 5 | 7 |
| Polyvinyl chloride | 5 | 7 | 6 | 7 | 6 | 8 | 6 | 10 | 9 | 4 |
| Polyvinly acetate chloride | 6 | 8 | 6 | 7 | 5 | 8 | 5 | 9 | 9 | 4 |
| Polyvinylidene copolymer | 4 | 7 | 6 | 7 | 7 | 8 | 7 | 8 | 9 | 4 |
| Silicone T.S. | 4 | 6 | 7 | 7 | 8 | 10 | 7 | 6 | 6 | 6 |
| Urethane T.S. | 8 | 10 | 10 | 9 | 10 | 7 | 8 | 4 | 4 | 9 |
| Acrylonitrile rubber | 3 | 6 | 8 | 6 | 9 | 7 | 5 | 8 | 8 | 5 |
| Polybutene rubber | 3 | 3 | 6 | 2 | 8 | 8 | 3 | 10 | 9 | 5 |
| Chlorinated rubber | 3 | 5 | 7 | 4 | 7 | 6 | 3 | 10 | 9 | 5 |
| Styrene rubber | 5 | 7 | 8 | 5 | 8 | 7 | 3 | 10 | 9 | 4 |

*From Herbert R. Simonds and James M. Church, "A Concise Guide to Plastics," 2 ed., Reinhold, 1963, with permission of the publisher.

**Table 2. Properties and Uses of Various Adhesives***

| Basic type | Trade name or designation | No. of components | Curing cycle, time at temp | Service temp range, °C | Adherends | Main uses | Remarks |
|---|---|---|---|---|---|---|---|
| Animal | | | | | | | |
| Animal (hide) | (64) Jelly Glue | 1 | Melted at 70–75°C. Sets on cooling | | Paper, wood, textiles | Woodworking, carpet materials, paper, bookbinding | May be thinned with water |
| Animal (hide) + plasticizers | (2) Flexhide | 1 | Applied as a melt at 60°C | Less than 60 | Paper, cellulosic materials | Bookbinding, stationery applications | Cures to permanent flexible film |
| Fish glue | (120) Samsom liquid fish glue | 1 | 1 h at 20°C | 60 | Wood, chipboard, paper | General-purpose for porous materials | Rapid setting. Good flexibility. Moderate resistance to water. High tack. |
| Casein | (34) Casco casein glue 1562 | 1 | Cold setting after 20 min standing period on mixing | | Timber with moisture content | Laminated timber arches and beams, plybox beams, and engineering timber work | Full bond strength developed after seasoning period of 48 h |
| Casein + 60% latex | (34) Casco 1060 | 2 | Cold setting after 20 min standing period on mixing | | Aluminum, wood, phenolic formaldehyde (rigid), leather, rubber | Bonding of disimilar materials to give flexible, water-resistant bond | Flexible |
| Vegetable | | | | | | | |
| Dextrine | (134) Lion Glue | 1 | Air drying | | Paper, cardboard, leather, wood, pottery | General-purpose glue for absorbent materials | Medium drying period of 2–3 h |
| Dextrine-starch blend | (211) Swift K1552 | 1 | Applied above 15°C air drying | 48 | Cellulosic materials, cardboard, paper | Labeling, carton sealing, spiral-tube winding | Fast setting. May be diluted with water |
| Gum arabic | (196A) Sello-Bond gum | 1 | Cold setting | | Paper, cardboard | Stationery uses | Fast drying |
| Mineral | | | | | | | |
| Silicate | (15) Fibrous adhesive 81-27 | 1 | 8 h at 20°C | 10–430 | Asbestos, magnesia | Lagging asbestos cloth on high-temperature insulation | Unsuitable where moisture: not recommended for glass or painted surfaces |
| Silicate with china-clay filler | (48) Autostic | 1 | Dried at 80°C before exposure to heat | −183 to 1500 | Asbestos, ceramics, brickwork, glass, silver, aluminum, steel | General-purpose cement for bonding refractory materials and metals. Furnace repairs and gastight jointing of pipe work. Heat-insulating materials | Resistant to oil, gasoline, and weak acids |

**Table 2. Properties and Uses of Various Adhesives (Continued)**

| Basic type | Trade name or designation | No. of components | Curing cycle, time at temp | Service temp range, °C | Adherends | Main uses | Remarks |
|---|---|---|---|---|---|---|---|
| | | | | *Animal* | | | |
| Sodium silicate | (67) Pyramid No. 1 | 1 | Dried at 20–80°C before exposure to heat | 0–850 | Aluminum (foil), paper, wood-wood | Fabrication of corrugated fiber board. Wood bonding, metal foil to paper lamination | Suitable for glass-to-stone bonding |
| Aluminum phosphate + silica filler | (97) Brimor U527 | 1 | Dried ½ h at 20°C, then ½ h at 70°C + 1 h at 100°C + 1 h at 200°C + 1 h at 250°C. Repeat for 2 overcoatings and finally cure 1 h at 350°C | 750 | Steels (low-alloy), iron, brass, titanium, copper, aluminum | Strain-gage attachment to heat-resistant metals. Heater-element bonding | Particularly suited to heat-resistant steels where surface oxidation of metal at high temperatures is less detrimental to adhesion |
| Bitumen/Latex emulsion | (29) Colset 185 | 1 | Dried in air to a tacky state | 0–66 | Cork, Polystyrene (foam), polyvinyl chloride, concrete, asbestos | Lightweight thermal-insulation boards, and preformed sections to porous and nonporous surfaces. Building applications | Not recommended for constructions operated below 0°C |
| | | | | *Elastomers* | | | |
| Natural rubber | (40) BTR Cement 41872/3 | 1 | Air-dried 20 min at 20°C and heat-cured 5 min at 140°C | | Rubber (styrene butadiene, rubber (latex), aluminum, cardboard, leather, cotton | Vulcanizing cement for rubber bonding to textiles and rubbers | May be thinned with toluene |
| Natural rubber in hydrocarbon solvent | (40) Silverlock 5054 | 1 | Air-dried 10 min at 20°C and heat-cured for 20 min at 150°C | 100 | Hair (keratin), bristle, polyamide fiber | Brush-setting cement for natural- and synthetic-fiber materials | Resistant to solvents employed in oil, paint and varnish industries. Can be nailed without splitting |
| Rubber latex | (77) Dunlop A 1020 | 1 | Air drying within 15 min | | Canvas, paper, fabrics, cellulosic materials | Bonding textiles, papers, packaging materials. Carpet bonding | Resistant to heat. Should be protected from frosts, oils |
| Chlorinated rubber in hydrocarbon solvents | (43) Bexol 1528 | 1 | Air-dried 10 min at 20°C and contact bonded | −20 to 60 | Polyvinyl chloride, acrylonitrile butadiene styrene, polystyrene, rubber, wood | General-purpose contact adhesive | Resistant to aging, water, oils, petroleum |
| Styrene-butadiene rubber lattices | (77) Dunlop SF | 1 | Air drying | | Polystyrene (foam), wood, hardboard, asbestos, brickwork | Bonding polystyrene foams to porous surface | |
| Neoprene/nitrile rubbers in solvent | (40) Cement 24876 | 1 | Dried 30 min in air and bonded under pressure while tacky | | Wood, linoleum, leather, paper, metals, nitrile rubbers, glass, fabrics | Cement for bonding synthetic rubbers to metals, woods, fabrics | May be thinned with ketones |

| Base | Tradename | No. | Curing conditions | Temperature range, °C | Materials bonded | Uses | Remarks |
|---|---|---|---|---|---|---|---|
| Acrylonitrile rubber + phenolic resin | (150) Scotchweld AF 30 with EC 1459 primer (10% w/w solids) | 1 | Primer air-dried 60 min at 20°C film cured 60 min at 175°C under pressure. Pressure released on cooling at 50°C | −40 to 130 | Aluminum (alloy)–aluminum to DTD 746 | Metal bonding for structural applications at elevated temperatures | Subject to creep at 150°C for sustained loading |
| Polysulfide rubber in ketone solvent and catalyst | (35) Bosco-frene 2115-5 | 2 | 3 days at 25°C | −50 to 130, withstands higher temps. for short periods | Metals | Sealant for fuel tanks and pressurized cabins in aircraft, where good weatherproof and waterproof properties are required | Resistant to gasoline, oil, hydraulic fluids, ester lubricants. Moderate resistance to acids and alkalies |
| Silicone rubber | (76) Silastic 735 RTV | 1 | 24 h at 20°C (20% R.H.). Full cure in 5 days | −65 to 260 | Aluminum, titanium, steel (stainless), glass, cork, silicone rubber, cured rubber-aluminum, cured rubber-titanium, cured rubber-steel (stainless), aluminum-aluminum (2024 Alclad), cork-cork (phenolic bonded) | General-purpose bonding and sealing applications. Adhesive/sealant for situations where material is expected to support considerable suspended weight. High-pressure exposure conditions | Resistant to weathering and moisture |
| Reclaim Rubber | (40) Cement 21401 | 1 | Contact bonded when tacky | | Fabric, leather, wood, glass, metals (primed) | General industrial adhesive for rubber, fabric, leather, porous materials | May be thinned with toluene |
| Polychloroprene | (77) Dunlop RF | 1 | Air-dried 10–20 min at 20°C | | Rubber, steel, wood, concrete | Bonding all types of rubber flooring to metals, woods, and masonry | Good heat resistance |
| Modified polyurethane | (35) Foaming adhesive 2065 | 2 | 3 h at 18°C to 16 h at −15°C | −80 to 110 | Concrete, plaster, ceramics, glass, hardboards, wood, polyurethane (foam), phenol formaldehyde (foam), polystyrene (foam), copper, lead, steel, aluminum | Bonding rigid and semi-rigid panels to irregular wall surfaces. Wall cladding and floor laying. Building industry applications | Foam remains flexible on aging even at elevated temperatures. Will withstand a 12% movement |
| | | | | Thermoplastic | | | |
| Nitrocellulose in ester solvent | (120) Samsom C110 | 1 | Heat set 1 h at 60°C after wet bonding | 60 | Paper, leather, textiles, silicon carbide, metals | Labeling, general bonding of inorganic materials including metals | Good resistance to mineral oils |
| Modified methyl cellulose | (110) Methofas SA | 1 | Dries in air | | Vinyl-coated paper, polystyrene foam | Heavy-duty adhesive. Decorating paper and plastics | Contains fungicide to prevent biodeterioration |
| Ethylene vinyl acetate copolymer + resins | (83) Evo-Stik Therma-flow RDA 6876 | 1 | Film transfer at 70–80°C followed by bonding at 150–160°C | 60 or 1 h at 90 | Cotton (duck)-cotton, resin rubber-leather, melamine laminate—plywood, steel (mild)-steel, acrylic (sheet) acrylic | Metals, laminated plastics, and textiles. Fabrication of leather goods. Lamination work | Good electrical insulation |

**Table 2. Properties and Uses of Various Adhesives (Continued)**

| Basic type | Trade name or designation | No. of components | Curing cycle, time at temp | Service temp range, °C | Adherends | Main uses | Remarks |
|---|---|---|---|---|---|---|---|
| **Animal** | | | | | | | |
| Polyvinyl acetate | (120) Samson 8100 Resimul | | Rapid setting | | Paper, Cardboard | Carton sealing in packaging industry | Resistant to water |
| Synthetic polymer blend | (80) Eastobond M3 | 1 | Applied as a melt at 177°C | 71 | Paper, cardboard, polythene (coated materials) | Carton and paper-bag sealing. Packaging | |
| Polychloroprene/ resin blend in solvent | (63) Cow PA74 | 2 | Air dried 10 min at 20°C and cured 4 days at 20°C to 7 h at 75°C | | Chlorosulfonated polythene, Polychloroprene fabrics, polyamide fabrics, leather, wood, textiles | Bonding synthetic rubbers and porous materials. Primer for polyamide-coated fabrics such as nylon, terylene | |
| Polychloroprene | (77) Dunlop RF | 1 | Air dried 10–20 min at 20°C | | Rubber, steel, wood, concrete | Bonding all types of rubber flooring to metals, woods, and masonry | Good heat resistance |
| Saturated polyester + isocyanate catalyst in ethyl acetate | (45) Herberts EP adhesive lacquer | 2 | Solvent evaporation and press cured at 40–80°C when tacky | | Cellulose, cellulose acetate, polyolefins (treated film), polyvinyl chloride (rigid), paper, aluminum (foil), copper (foil) | Lamination of plastic films to themselves and metal foils for packaging industry, printed circuits | Resistant to heat, moisture, and many solvents |
| Cyanoacrylate (*anaerobic*) | (121) I.S.-12 | 1 | 15 s to 10 min at 20°C substrate-dependent | Melts at 165 | Steel-steel, steel-aluminum, aluminum-aluminum, butyl rubber-phenolic | Rapid assembly of metal, glass, plastics, rubber components | *Anaerobic* adhesive. Curing action is based on the rapid polymerization of the monomer under the influence of basic catalysts. Absorbed water layer on most surfaces suffices to initiate polymerization and brings about bonding |
| Polyacrylate resin (*anaerobic*) | (121) Loctite 308 impact | 1 | 3 min at 120°C to 45 min at 65°C or 7 days at 20°C | −55 to 95 | Aluminum-aluminum | Assembly requirements requiring high resistance to impact or shock loading. Metals, glass, and thermosetting plastics | *Anaerobic* adhesive |
| **Thermosetting** | | | | | | | |
| Urea formaldehyde | (34) Cascamite "one shot" | 1 | 9 h at 10°C to 1 h at 21°C after mixing powder with water (22%) | | Wood, phenolic laminate | Wood gluing and bonding on plastic laminates to wood. Plywood, chipboard manufacture. Boat building and timber engineering | Excess glue may be removed with soapy water |

| Type | Trade name | No. of parts | Curing conditions | Service temp (°C) | Materials bonded | Applications | Remarks |
|---|---|---|---|---|---|---|---|
| Phenolic formaldehyde + catalyst PX-2Z | (34) Cascophen PC-1 | 2 | Cold setting | | Wood | Timber and similar porous materials for outdoor-exposure conditions. Shop fascia panels | Good resistance to weathering and biodeterioration |
| Resorcinol formaldehyde + catalyst RXS-8 | (34) Cascophen RS-216-M | 2 | Cured at 16°C to 80°C under pressure | | Wood, asbestos, aluminum, phenolic laminate, polystyrene (foam), polyvinyl chloride, polyamide (rigid) | Constructional laminates for marine craft. Building and timber applications. Aluminum-plywood bonding. Laminated plastics | Recommended for severe outdoor-exposure conditions |
| Epoxy resin + catalyst | (150) Scotchweld EC1614B/A | 2 | 24–48 h at 20°C to 20 min at 120°C | 100 | Steel, glass, polyester-glass fiber composite, aluminum-aluminum | General-purpose structural adhesive | |
| Epoxy resin + catalyst | (150) Scotchweld EC1838 B/A | 2 | 8 h at 24°C to 2 h at 66°C to 45 min at 121°C | 65 | Steel, copper, zinc, silicon carbide, wood, masonry, polyester-glass fiber composite, aluminum-aluminum | Bonding of metals, glass, ceramics, and plastic composites | Cures to strong, durable bond |
| Epoxy + steel filler (80% w/w) | (73) Devcon A | 2 | 1–2 h at 21°C | 120 | Iron, steel, aluminum, wood, concrete, ceramics, aluminum-aluminum | Industrial maintenance repairs. Metallic tanks, pipes, valves, engine casings, castings | Good resistance to chemicals, oils, water |
| Epoxy + amine catalyst (ancamine LT) | (198) Epikote 828 | 2 | 2–7 days at 20°C for 33% w/w catalyst content | −5 to 60 | Concrete, stonework | Repair of concrete roads and stone surfaces | Excellent pigment-wetting properties. Effective under water and suited to applications under adverse wet or cold conditions |
| Epoxy resin (modified) | (165) Bondmaster M620 | 1 | 4–5 h at 149°C to 20 min at 230°C to 7 min at 280°C | 150 | Aluminum, steel, ceramics | One-part structural adhesive for high-temperature applications | Good gap-filling properties for poorly fitting joints. Resistant to weather, galvanic action |
| Epoxy | (61) Mereco X-305 | 2 | 45 s at 20°C | | Gem stones, glass, steel, aluminum-aluminum | Rapid assembly of electronic components, instrument parts, printed circuits. Stone setting in jewelry, and as an alternative to soldering | |
| Epoxy resin in solvent + catalyst | (225D) BR600 | 3 | 8 h at 52°C to ½ h at 121°C | −270 to 371 | Aluminum and magnesium alloys for elevated-temperature service | Strain gages for cryogenic and elevated-temperature use. Micro measurement strain gages | Cured material resists outgassing in high vacuum |

**Table 2. Properties and Uses of Various Adhesives (Continued)**

| Basic type | Trade name or designation | No. of components | Curing cycle, time at temp | Service temp range, °C | Adherends | Main uses | Remarks |
|---|---|---|---|---|---|---|---|
| *Animal* | | | | | | | |
| Epoxy polyamide | (13) DEK epoxy | 2 | 8 h at 20°C to 15 min at 100°C | 100 | Copper, lead, concrete, glass, wood, fiberglass, steel-steel, aluminum-aluminum | Metals, ceramics, and plastics bonding. Building and civil engineering applications | Resists water, acids, oils, greases |
| Epoxy/polysulfide | (57) Araldite AY111 + HY111 | 2 | 24 h at 20°C to 3 h at 60°C to 20 min at 100°C | | Asbestos (rigid), ceramics, glass-fiber composites, carbon, polytetrafluoroethylene (treated), polyester (film), polystyrene (treated), rubber (treated), copper (treated), tungsten carbide, magnesium alloys, aluminum-aluminum, steel (stainless)-steel | Cold-setting adhesive especially suitable for bonding materials with differing expansion properties | Cures to flexible material. Resistant to water, petroleum, alkalies, and mild acids |
| Phenol furfural + acid catalyst | (34) Cascote 1376 | 2 | 2 days at 21°C | | Alumina, carbon (graphite) | Formulation of chemically resistant cements. Bedding and joining chemically resistant ceramic tiles | Extremely resistant to abrasion and heat |
| *Pressure-sensitive* | | | | | | | |
| | Fluoro-Plastics, Inc. Adhesive 80 | | Heated by air drying for several hours or 15–30 min at 210°F | | Teflon-Teflon, Teflon-metal | | Good resistance to acids and alkalies. Excellent electrical properties |
| *Miscellaneous* | | | | | | | |
| Ceramic-based | (62A) Denex-2 ceramic cement | 1 | Dried for ½ h at 77°C and cured ½ h at 100°C + 1 h at 200°C + 1 h at 250°C. Postcured, 1 h at 350°C | 816 | Metals | Strain gages, temperature sensors for elevated-temperature work | |

*Adapted from J. Shields, "Adhesives Handbook," CRC Press (Division of The Chemical Rubber Co., 1970), with the permission of the publisher.

**Building Brick Made from Clay or Shale**
(Standard Specifications, ASTM, C62-49)

| Designation* | Min compressive strength (brick flatwise), psi gross area | | Max water absorption by 5 hr boil, percent | | Max saturation coefficient† | |
|---|---|---|---|---|---|---|
| | Avg of 5 | Indiv. | Avg of 5 | Indiv. | Avg of 5 | Indiv. |
| Grade SW.............................. | 3,000 | 2,500 | 17.0 | 20.0 | 0.78 | 0.80 |
| Grade MW............................. | 2,500 | 2,200 | 22.0 | 25.0 | 0.88 | 0.90 |
| Grade NW............................. | 1,500 | 1,250 | No limit | | No limit | |

*Grade SW includes brick intended for use where a high degree of resistance to frost action is desired and the exposure is such that the brick may be frozen when permeated with water.

Grade MW includes brick intended for use where exposed to temperatures below freezing but unlikely to be permeated with water or where a moderate and somewhat nonuniform degree of resistance to frost action is permissible.

Grade NW includes brick intended for use as backup or interior masonry, or if exposed, for use where no frost action occurs; or if frost action occurs, where the average annual precipitation is less than 20 in.

†The saturation coefficient is the ratio of absorption by 24 h submersion in cold water to that after 5 h submersion in boiling water.

collection of such units, made from clay or shale hardened by heat. When other substances are used, the term brick should be suitably qualified unless specifically indicated by the context.

It is, therefore, officially recognized that unless suitably qualified, a brick is a unit of burned clay or shale.

**Brick (Common)**  Any brick made primarily for building purposes and not especially treated for texture or color, but including clinker and oven-burnt brick.

**Brick (Facing)**  A brick made especially for facing purposes, usually treated to produce surface texture or made of selected clays or otherwise treated to produce the desired color.

Brick are manufactured by the dry-press, the stiff-mud, or the soft-mud process. The **dry-press brick** are made in molds under high pressure and from relatively dry clay mixes. Usually all six surfaces are smooth and even, with geometrical uniformity. The **stiff-mud brick** are made from mixes of clay or shale with more moisture than in the dry-press process, but less moisture than used in the soft-mud process. The clay is extruded from an auger machine in a ribbon and cut by wires into the required lengths. These brick may be side-cut or end-cut, depending on the cross section of the ribbon and the length of the section cut off. The two faces cut by wires are rough in texture; the other faces may be smooth or artifically textured. The **soft-mud process** uses a wet mix of clay which is placed in molds under slight pressure.

Brick are highly resistant to freezing and thawing, to attacks of acids and alkalies, and to fire. They furnish good thermal insulation and good insulation against sound transference.

**Paving brick** are made of clay or shale, usually by the stiff-mud or dry-press process. Brick for use as paving brick are burned to vitrification. The common requirements are as follows: size, $8\frac{1}{2} \times 2\frac{1}{2} \times 4$, $8\frac{1}{2} \times 3 \times 3\frac{1}{2}$, $8\frac{1}{2} \times 3 \times 4$, with permissible variations of $\frac{1}{8}$ in in either transverse dimension, and $\frac{1}{4}$ in in length.

Although brick have always been used in construction, their use has been limited, until recently, to resisting compressive-type loadings. By adding steel in the mortar joints to take care of tensile stresses, **reinforced brick masonry** extends the use of brick masonry to additional types of building construction such as floor slabs.

**Sand-lime brick** are made from a mixture of sand and lime, molded under pressure and cured under steam at 200°F. They are usually a light gray in color and are used primarily for backing brick and for interior facing.

**Cement brick** are made from a mixture of cement and sand, manufactured in the same manner as sand-lime brick. In addition to their use as backing brick, they are used where there is no danger of attack from acid or alkaline conditions.

**Firebrick** (see this article, Refractories).

**Specialty Brick**  A number of types of specialty brick are available for important uses, particularly where refractory characteristics are needed. These include **alumina** brick, **silicon carbide** brick, and **boron carbide** brick.

**Other Structural Blocks**  Important structural units which fall outside the classic definition of brick are as follows: **Concrete blocks** are made with portland cement as the basic binder. The choice of aggregate ranges from relatively dense aggregate such as sand to light aggregate such as coal cinders. **Gypsum blocks** are generally used for fire protection in non-load-bearing situations. **Glass blocks** are available where transparency is desired. **Structural clay tile** is widely used in both load-bearing and non-load-bearing situations; variety of complex shapes is available for hollow-wall construction. For classification of all these products with respect to dimensions and specifications for usage, the references should be consulted.

Brick and block can come with porcelain glazed finish for appearance, ease of cleaning, resistance to weathering, corrosion, etc.

**Brick panels** (prefabricated brick walls) are economic and find extensive use in modern high-rise construction. Such panels can be constructed on site or in factories.

New tile units requiring no mortar joint, only a thin epoxy line, form walls which resemble brick and can be installed three times as fast as conventional masonry.

## CERAMICS
### by Robert S. Feigelson
*Stanford University*

REFERENCES: Kingery, "Introduction to Ceramics," Wiley. Norton, "Elements of Ceramics," Addison-Wesley. "Carbon Encyclopedia of Chemical Technology," Interscience. Humenik, "High-Temperature Inorganic Coatings," Reinhold. Kingery, "Ceramic Fabrication Processes," MIT. McCreight, Rauch, Sr., Sutton, "Ceramic and Graphite

Fibers and Whiskers; A Survey of the Technology," Academic, 1965. Rauch, Sr., Sutton, McCreight, "Ceramic Fibers and Fibrous Composite Materials," Academic, 1968. Hague, Lynch, Rudnick, Holden, Duckworth (compilers and editors), "Refractory Ceramics for Aerospace; A Material Selection Handbook," The American Ceramic Society, 1964. Waye, "Introduction to Technical Ceramics," Maclaren and Sons, Ltd., London, 1967. Hove and Riley "Ceramics for Advanced Technologies," Wiley, 1965. McMillan, "Glass-Ceramics," Academic, 1964.

Ceramic materials are a diverse group of nonmetallic, inorganic solids with a wide range of compositions and properties. Their structure may be either crystalline or glassy. The desired properties are often achieved by high-temperature treatment (firing or burning).

**Traditional ceramics** are products based on the silicate industries, where the chief raw materials are naturally occurring minerals such as the clays, silica, feldspar, and talc. While silicate ceramics dominate the industry, newer ceramics, sometimes referred to as **electronic** or **technical** ceramics, are playing a major role in many applications. **Glass ceramics** are important for electrical, electronic, and laboratory-ware uses. Glass ceramics are melted and formed as glasses, then converted, by controlled nucleation and crystal growth, to polycrystalline ceramic materials.

**Manufacture** Typically, the manufacture of traditional-ceramic products involves blending of the finely divided starting materials with water to form a plastic mass which can be formed into the desired shape. The plasticity of clay constituents in water leads to excellent forming properties. Formation processes include extrusion, pressing, and ramming. Unsymmetrical articles can be formed by "slip-casting" techniques, where much of the water is taken up by a porous mold. After the water content of formed articles has been reduced by drying, the ware is **fired** at high temperature for fusion and/or reaction of the components and for attainment of the desired properties. Firing temperatures can usually be considerably below the fusion point of the pure components through the use of a **flux**, often the mineral feldspar. Following burning, a vitreous ceramic coating, or **glaze**, may be applied to render the surface smooth and impermeable.

Quite different is the **fusion casting** of some refractories and refractory blocks and most glasses, where formation of the shaped article is carried out after fusion of the starting materials.

**Properties** The physical properties of ceramic materials are strongly dependent on composition, microstructure (phases present and their distribution), and the history of manufacture. Volume pore concentration can vary widely (0 to 30 percent) and can influence shock resistance, strength, and permeability. Most traditional ceramics have a glassy phase, a crystalline phase, and some porosity. The last can be eliminated at the surface by glazing. Most ceramic materials are resistant to large compressive stresses but fail readily in tension. Resistance to abrasion, heat, and stains, chemical stability, rigidity, good weatherability, and brittleness characterize many common ceramic materials.

**Products** Many traditional-ceramic products are referred to as **whiteware** and include pottery, semivitreous wares, electrical porcelains, sanitary ware, and dental porcelains. Building products which are ceramic include brick and structural tile and conduit, while refractory blocks, as well as many abrasives, are also ceramic in nature. Porcelain **enamels** for metals are opacified, complex glasses which are designed to match the thermal-expansion properties of the substrate. Important **technical ceramics** include magnetic ceramics, with magnetic properties but relatively high electrical resistance; nuclear ceramics, including uranium dioxide fuel elements; barium titanate as a material with very high dielectric constant. Several of the pure oxide ceramics with superior physical properties are being used in electrical and missile applications where high melting and deformation temperatures and stability in oxygen are important. Fibrous ceramic composed of zirconium oxide fibers, such as Zircar, provide optimum combination of strength, low thermal conductivity, and high temperature resistance to about 2490°C (4500°F).

# CLEANSING MATERIALS
## by Douglas Melzard
*Arthur D. Little, Inc.*

REFERENCES: McCutcheon, "Detergents and Emulsifiers," McCutcheon, Inc. Schwartz, Perry, and Berch, "Surface Active Agents and Detergents," Interscience. *ASTM Special Tech. Pub.* 197. Niven, "Industrial Detergency," Reinhold. McLaughlin, "The Cleaning, Hygiene, and Maintenance Handbook," Prentice-Hall, 1973. Hackett, "Maintenance Chemical Specialties," Chemical Publishing Co., 1972. Bennett (ed.), "Cold Cleaning with Halogenated Solvents," *ASTM Special Technical Publication* 403, 1966.

Cleansing is the removal of dirt, soil, and impurities from surfaces of all kinds. Means of soil attachment to the surface include simple entrapment in interstices, electrostatically held dirt, wetting of the surface with liquid soils, and soil-surface chemical reaction. A variety of cleansing systems has been developed which are difficult to classify since several soil-removal mechanisms are often involved. A liquid suspending medium, an active cleansing agent, and mechanical action are usually combined. The last may involve mechanical scrubbing, bath agitation, spray impingement, or ultrasonic energy. Cleansing involves detachment from the surface, suspension of solids or emulsification of liquids, or dissolution, either physical or by chemical reaction.

### Organic Solvents

Both **petroleum solvents** (mineral spirits or naphtha) and **chlorinated hydrocarbons** (trichloroethylene and perchloroethylene) are used to remove solvent-soluble oils, fats, waxes, and greases as well as to flush away insoluble particles. Petroleum solvents are used in both soak-tank and spray equipment. Chlorinated hydrocarbons are widely used in vapor degreasing (where the metal stock or part is bathed in the condensing vapor), their high vapor density and nonflammability being advantageous. Solvent recovery by distillation can be used if the operation is of sufficient size. Both solvent types are used in garment **dry cleaning** with solvent-soluble detergents.

**Fluorocarbons** such as UCON solvent 113-LRI (trichlorotrifluoroethane) are nonflammable and are ideal for critical cleaning of mechanical electrical and electronic equipment, especially for white-room conditions.

**Emulsifiable solvent** cleaners contain a penetrating solvent and dissolved emulsifying agent. Following soaking, the surface is flushed with hot water, the resulting emulsion carrying away both soil and solvent. These cleaners are usually extended with kerosinelike solvents.

## Alkali Cleansers

Alkali cleansers are water-soluble inorganic compounds, often strong cleansers. Carbonates, phosphates, pyrophosphates, and caustic soda are common, with numerous applications in plant maintenance, material processing, and process-water treatment. Cleansing mechanisms vary from beneficial water softening and suspension of solids to chemical reaction in the solubilization of fats and oils. Certain of these compounds are combined with **detergents** to improve efficiency; these are referred to as **builders.** Boiler and process-equipment scales can sometimes be controlled or removed with selected alkali cleansers.

## Synthetic Detergents

Detergents concentrate strongly at a solid-liquid or liquid-liquid interface and are thus characterized as **surface active.** In contrast to soaps, they can be tailored to perform over a wide range of conditions of temperature, acidity, and presence of dissolved impurities with little or no foaming. Detergents promote **wetting** of the surface by the suspending medium (usually water), **emulsification** of oils and greases, and **suspension** of solids without redeposition, the last function being the prime criterion of a good detergent. Detergents are classified as **anionic** (negatively charged in solution), **cationic** (positively charged), and **nonionic.** Germicidal properties may influence detergent choice.

For specific applications, the supplier should be consulted since there is a large variety of available formulations and since many detergent systems contain auxiliary compounds which may be diluents, foam promotors, or alkali chemicals. Additives are designed for pH control, water softening, and enhanced suspending power. Formulations have been developed for cleaning food and dairy-process equipment, metals processing, metal cleaning prior to electroplating, textile fiber and fabric processing, and industrial building maintenance. Strong, improperly selected detergent systems can cause deterioration of masonry or marble floors, aluminum window frames, water-based paints, and floor tiles, whereas detergents matched to the job at hand can result in increased plant efficiency.

**Soaps,** the oldest surface-active cleansers, lack the versatility of synthetic detergents but are widely used in the home and in the laundry industry. Properties depend on the fat or oil and alkali used in their preparation and include solvent-soluble soaps for dry cleaning.

**Chemical cleaners** which attack specific soils include dilute acid for metal oxide removal, for the cleaning of soldered or brazed joints, and for the removal of carbonate scale in process equipment. Oxalic acid (usually with a detergent) is effective on rust.

**Chelating agents** are organic compounds which complex with several metal ions and can aid in removal of common boiler scales and metal oxides from metal surfaces.

**Steam cleaning** in conjunction with a detergent is effective on grease-laden machinery.

There are numerous cleansers for the **hands,** including Boraxo, soap jelly and sawdust, lard for loosening oil grime, and linseed oil for paints; repeated use of solvents can be hazardous.

**Alcohol ethoxylates** are nonionic surfactants suitable for use in the production of maintenance and institutional cleaners. Products such as Tergitol and Neodol 23-6.5 serve in the formulation of liquid detergents for household and industrial uses.

**Alcohol ethoxysulfates** are anionic surfactants which are suited to the formulation of high-foaming liquid detergents, as for manual dishwashing. They offer the advantages of excellent solubility and biodegradability. Neodol 25-3S40 also exhibits low sensitivity to water hardness.

## CORDAGE

REFERENCES: Himmelfarb, "The Technology of Cordage Fibers and Rope," Textile Book Publishers, Inc. (division of Interscience Publishers, Inc., 1957.

The term **cordage** denotes any flexible string or line. Usage includes wrapping, baling, hauling, and power transmission in portable equipment. **Twine** and **cord** generally imply lines of ⅜ in diam or less, with larger sizes being referred to as **rope.**

Natural fibers used in cordage are abaca, sisal, hemp, cotton, and jute. For heavy cordage abaca and manila predominate. Hemp is used for small, tarred lines, and henequen for agricultural binder twine.

**Rope** is made by twisting yarns into strands, with the strands (usually three) **twisted** (laid) into a line. The line twist may be S or Z (see Sec. 10), generally opposite to the twist of the strands, which, in turn, is opposite to the twist of the yarns. The term **lay** designates the number of turns of the strands per unit length of rope but may also characterize the rope properties, a function of the degree of twist of each component. Grades range from **soft lay** (high ultimate strength) to **hard lay** (high abrasion resistance). Cable-laid rope results from twisting together conventional, three-strand rope.

**Synthetic fibers** are gaining usage in cordage because of resistance to rot, high strength, and other special properties. These fibers include nylon for strength, polyester (Dacron) for strength and dimensional stability, vinyls for chemical-plant use, fiber glass for electrical and chemical properties, and polypropylene for strength and flotation (see Sec. 8).

**Braided cordage** has been used largely for small-diameter lines such as sash cord and clothesline. However, braided lines are now available in larger diameters, with lines up to 3 in diam under development. The strength of braided rope is slightly superior, and the line has less tendency to elongate in tension and cannot rotate or unlay under load. These properties are balanced against a somewhat higher cost than that of twisted rope.

A No. 1 common-lay rope will conform to the strength and weight table of the Federal Specification TR601A listed below.

For comparison, a **nylon rope** of a given diameter will have about 3 times the breaking strength given above and a **polyester rope** about 2½ times the strength of manila. Both have substantially greater flex and abrasion resistance.

## ELECTRICAL INSULATING MATERIALS
### by John T. Blake
*Simplex Wire and Cable Co.*

REFERENCES: Plastics Compositions for Dielectrics, *Ind. Eng. Chem.*,

**38,** 1946, p. 1090. High Dielectric Ceramics, *Ind. Eng. Chem.*, **38,** 1946, p. 1097. Polystyrene Plastics as High Frequency Dielectrics, *Ind. Eng. Chem.*, **38,** 1946, p. 1121. Paper Capacitors Containing Chlorinated Impregnants, *Ind. Eng. Chem.*, **38,** 1946, p. 1110. "Contributions of the Chemist to Dielectrics," National Research Council, 1947. National Research Council, Conference on Electrical Insulation, (1) Annual Report, (2) Annual Digest of Literature on Dielectrics. Von Hippel, "Dielectric Materials and Applications," Wiley. Birks (ed.), "Modern Dielectric Materials," Academic, 1960. Saums and Pendelton, "Materials for Electrical Insulating and Dielectric Functions," Hayden Book, Inc., 1973. Licari, "Plastic Coatings for Electronics," McGraw-Hill, 1970. Clark, "Insulating Materials for Design and Engineering Practice," Wiley, 1962. Mayofis, "Plastic Insulating Materials," Illiffe, Ltd., London, 1966. Bruins (ed.), "Plastic for Electrical Insulation," Interscience Publishers, 1968. Swiss Electrochemical Committee, "Encyclopedia of Electrical Insulating Materials," *Bulletin of the Swiss Association of Electrical Engineers*, vol. 48, 1958.

The insulating properties of any material are dependent upon **dielectric strength,** or the ability to withstand high voltages without breakdown; **ohmic resistance,** or the ability to prevent leakage of small currents; and **power loss,** or the absorption of electrical energy that is transformed into heat. Power loss depends upon a number of influences, particularly the molecular symmetry of the insulation and frequency of the voltage, and is the basis of power factor, an important consideration whenever efficient handling of alternating currents is concerned, and a dominating consideration when high frequencies are used, as in radio circuits. Materials may have one of these qualities to a far greater extent than the other; e.g., air has a very high specific resistance but very little dielectric strength and no power loss at any frequency; glass has great dielectric strength yet much lower resistance than air. The **ideal insulator** is one having the maximum dielectric strength and resistance, minimum power loss, and also mechanical strength and chemical stability. Moisture is by far the greatest enemy of insulation; consequently the absence of hygroscopic quality is desirable.

The common insulating materials are described below. For their electrical properties, see Sec. 15.

**Rubber** See Rubber and Rubberlike Materials.

**Mica and Mica Compounds** Mica is a natural mineral varying widely in color and composition, and occurs in sheets that can be subdivided down to a thickness of 0.00025 in. White mica is best for electrical purposes. The green shades are the softest varieties, and the white amber from Canada is the most flexible. Mica has high insulating qualities, the best grades having a dielectric strength of 12,000 volts per tenth of a millimeter. Its lack of flexibility, its nonuniformity, and its surface leakage are disadvantages. To offset these, several mica products have been developed, in which small pieces of mica are built up into finished shapes by means of binders such as shellac, gum, and phenolic resins.

**Micanite** consists of thin sheets of mica built into finished forms with insulating cement. It can be bent when hot and machined when cold, and is obtainable in thicknesses of 0.01 to 0.12 in. Flexible micanite plates, cloth, and paper are also obtainable in various thicknesses. **Megohmit** is similar to micanite except that it is claimed not to contain adhesive matter. It can be obtained in plates, paper, linen, and finished shapes. **Megotalc,** built up from mica and shellac, is similar to the above-named products and is obtainable in similar forms.

**Insulating Varnishes** Two general types of insulating varnish are used: (1) asphalt, bitumen, or wax, in petroleum solvent, and (2) drying-oil varnishes based on natural oils compounded with resins from natural or synthetic sources. Varnishes have changed greatly in the last few years, since new oils have become available, and particularly since phenolic and alkyd resins have been employed in their manufacture.

**Silicone varnishes** harden by baking and have electrical properties similar to those of the phenol-aldehyde resins. They are stable at temperatures up to 300°F. They can be used as wire coatings, and such wire is used in the manufacture of motors that can operate at high temperatures.

**Impregnating Compounds** Bitumens and waxes are used to impregnate motor and transformer coils, the melted mix being forced into the coil in a vacuum tank, forming a solid insulation when cooled. Brittle compounds, which gradually pulverize owing to vibration in service, and soft compounds, which melt and run out under service temperatures, should be avoided as far as possible.

**Oil** Refined grades of petroleum oils are extensively used

## Weight and Strength of Different Sizes of Manila Rope, Specification Values

(From U.S. government specification TR601A, dated Nov. 26, 1935, and formulated jointly by cordage manufacturers and government representatives)

| Approx diam, in * | Circumference, in * | Max net weight, lb/ft | Min breaking strength, lb † | Approx diam, in * | Circumference, in * | Max net weight, lb/ft | Min breaking strength, lb | Approx diam, in | Circumference, in | Max net weight, lb/ft | Min breaking strength, lb |
|---|---|---|---|---|---|---|---|---|---|---|---|
| 3/16 | 5/8 | 0.015 | 450 | 13/16 | 2 1/2 | 0.195 | 6,500 | 1 13/16 | 5 1/2 | 0.895 | 26,500 |
| 1/4 | 3/4 | 0.020 | 600 | 15/16 | 2 3/4 | 0.225 | 7,700 | 2 | 6 | 1.08 | 31,000 |
| 5/16 | 1 | 0.029 | 1,000 | 1 | 3 | 0.270 | 9,000 | 2 1/4 | 7 | 1.46 | 41,000 |
| 3/8 | 1 1/8 | 0.041 | 1,350 | 1 1/16 | 3 1/4 | 0.313 | 10,500 | 2 5/8 | 8 | 1.91 | 52,000 |
| 7/16 | 1 1/4 | 0.053 | 1,750 | 1 3/16 | 3 1/2 | 0.360 | 12,000 | 3 | 9 | 2.42 | 64,000 |
| 1/2 | 1 1/2 | 0.075 | 2,650 | 1 1/4 | 3 3/4 | 0.418 | 13,500 | 3 1/4 | 10 | 2.99 | 77,000 |
| 9/16 | 1 3/4 | 0.104 | 3,450 | 1 5/16 | 4 | 0.480 | 15,000 | 3 5/8 | 11 | 3.67 | 91,000 |
| 11/16 | 2 | 0.133 | 4,400 | 1 1/2 | 4 1/2 | 0.600 | 18,500 | 4 | 12 | 4.36 | 105,000 |
| 3/4 | 2 1/4 | 0.167 | 5,400 | 1 5/8 | 5 | 0.744 | 22,500 | | | | |

The approximate length of coil is 1,200 ft for diam $7/16$ in and larger. For smaller sizes it is longer, up to 3,000 ft for $3/16$ in diam.

\*1 in = 0.0254 m; 1 ft = 0.3048 m.

†1 lbf = 4.448N.

for the insulation of transformers, switches, and lightning arresters. The following specification covers the essential points:

Specific gravity, 0.860; flash test, not less than 335°F; cold test, not more than −10°C (14°F); viscosity (Saybolt) at 37.8°C (100°F), not more than 120 s; loss on evaporation (8 h at 200°F), not more than 0.5 percent; dielectric strength, not less than 35,000 V; freedom from water, acids, alkalies, saponifiable matter, mineral matter, or free sulfur. Moisture is particularly dangerous in oil.

Petroleum oils are used for the impregnation of kraft or manila paper, after wrapping on copper conductors, to form high-voltage power cables for services up to 300,000 V. Oil-impregnated paper insulation is sensitive to moisture, and such cables must be lead-sheathed. The transmission of power at high voltages in underground systems is universally accomplished by such cables.

**Chlorinated Hydrocarbons** Chlorinated hydrocarbons have the advantage of being nonflammable and are used as filling compounds for transformers and condensers where this property is important. Chlorinated naphthalene and chlorinated diphenyl are typical of this class of material. They vary from viscous oils to solids, with a wide range of melting points.

**Impregnated Fabrics** Fabrics serve as a framework to hold a film of insulating material and must therefore be of proper thickness, texture, and mechanical strength, and free from nap and acidity. A wide variety of drying varnishes is used for the impregnation, and the dipping is followed by baking in high-temperature towers. Varnished cambric is used for the wrapping of coils and for the insulation of conductors. These cables have high power factor and must be kept free of moisture, but they are desirable for resisting electrical surges.

**Thermosetting substances** of the phenol-aldehyde type and of the urea-formaldehyde type first soften and then undergo a chemical reaction which converts them quickly to a strong infusible product. Good properties are available in the phenolics, while numerous special types have been developed for high heat resistance, low-power-factor arc resistance, and other specialized properties. A wide variety of resins is available. The urea plastics are lacking in heat resistance but are suitable for general-purpose molding and have fair arc resistance.

**Thermoplastic resins** (see Plastics, below) are used for molding and extruding electrical insulations. They differ from the thermosetting resins in that they do not become infusible. **Polyethylene** softens between 99 and 116°C (210 and 240°F). Its dielectric strength and resistivity are high, its power factor is only 0.0003, and its dielectric constant is 2.28. It is used extensively in high-frequency and radar applications and as insulation on some power and communication cables. **Teflon** is a fluorocarbon resin which has electrical properties similar to those of polyethylene. Its softening point is 750°F approx, and it extrudes and molds with difficulty; however, more tractable grades of Teflon are now available. It is resistant to nearly all chemicals and solvents.

**Nylon** is a synthetic plastic with interesting mechanical and electrical properties. It has only fair water resistance. Its melting point is 198 to 249°C (390 to 480°F), and it is used for the molding of coil forms. It can be extruded onto wire in thin layers. Such wires are used in place of the conventional varnish-coated magnet wires in coils and motors. Operation may be at temperatures up to 127°C (260°F).

**Paper** Except in lead-sheathed telephone cables, the present tendency is to use paper only as a backing or framework for an insulating film or compound, owing to its hygroscopic qualities. **Manila** and **kraft** papers possess the best dielectric and mechanical strength and, when coated with good insulating varnish, are excellent insulators. Various types of paraffined paper are used in condensers. (See also Paper, below.)

**Silicone rubber** (see Silicones) is a rubberlike material of good physical properties [tensile strength, 400 to 700 lb/in² (2.78 to $4.85 \times 10^6$ N/m²) elongation, 200 percent]. It can be operated for long periods of time at temperatures up to 138°C (300°F) or intermittently up to 249°C (480°F).

**Ceramics** and **glasses** find wide usage as insulating materials where brittleness and lack of flexibility are unimportant.

**Polymeric (plastic) films,** particularly the polyester and fluorocarbon types, are being used increasingly where fabrication of the electrical component permits either wrapping or insertion of film chips.

## FIBERS AND FABRICS
### by Milton M. Platt and Robert E. Sebring
*Fabric Research Laboratories, Inc.*

(See also Cordage)

REFERENCES: Matthews-Mauersberger, "The Textile Fibers," Wiley. Von Bergen and Krauss, "Textile Fiber Atlas," Textile Book Publishers, Inc. Hess, "Textile Fibers and Their Use," Lippincott. Sherman and Sherman, "The New Fibers," Van Nostrand. Kaswell, "Textile Fibers, Yarns and Fabrics," Reinhold. "Harris' Handbook of Textile Fibers," Waverly House. Kaswell, "Wellington Sears Handbook of Industrial Textiles," Wellington Sears Co. Fiber Charts, *Textile World*, McGraw-Hill. ASTM Standards. C. Z. Carroll-Porcynski, "Advanced Materials; Refractory Fibers, Fibrous Metals, Composites," Chemical Publishing Co., 1969. Marks, Atlas, and Cernia, "Man-Made Fibers, Fibrous Metals, Composites," Chemical Publishing Co., 1969. Marks, Atlas, and Cernia, "Man-Made Fibers, Science and Technology," Inter-science Publishers, 1968. Frazer, High Temperature Resistant Fibers, *Jour. Polymer Sci.*, Part C, Polymer Symposia, no. 19, 1966. Moncrieff, "Man-Made Fibers," Wiley, 1970. Preston and Economy, "High Temperature and Flame Resistant Fibers," Wiley, 1973.

**Fibers** are threadlike structural materials, adaptable for spinning, weaving, felting, and similar applications. They may be of natural or synthetic origin and of inorganic or organic composition. **Fabrics** are defined by the ASTM as "planar structures produced by interlacing yarns, fibers, or filaments." A **bonded fabric** (or nonwoven fabric) consists of a web of fibers held together with a cementing medium which does not form a continuous sheet of adhesive material. A **braided fabric** is produced by interlacing several ends of yarns such that the paths of the yarns are not parallel to the fabric axis. A **knitted fabric** is produced by interlooping one or more ends of yarn. A **woven fabric** is produced by interlacing two or more sets of yarns, fibers, or filaments such that the elements pass each other essentially at right angles and one set of elements is parallel to the fabric axis. A woven narrow fabric is 12 in or less in width and has a selvage on either side.

### Inorganic Fibers

**Asbestos** is the only mineral fiber of natural origin. Synthetic mineral fibers are **spun glass, rock wool,** and **slag wool.** These fibers can endure high temperatures without substantial loss of strength. Glass fibers possess a higher strength-to-weight ratio at the elastic limit than do other common engineering materials.

**Metal filaments** (wires) are used as textile fibers where their particular material properties are important.

**Carbon (graphite) fibers,** such as Thornel, are high-modulus, highly oriented structures characterized by the presence of carbon crystallites (polycrystalline graphite) preferentially aligned parallel to the fiber axis. Depending on the particular grade, tensile strengths can range from 10.3 to $241 \times 10^7$ N/m$^2$ (15,000 to 350,000 lb/in$^2$). Hybrid composites of carbon and glass fibers find aerospace and industrial applications.

**Zirconia fibers** ($ZrO_2 + HFO_2 +$ stabilizers), such as Zircar, have excellent temperature resistance to about 2200°C (4000°F), and fabrics woven from the fibers serve as thermal insulators.

### Natural Organic Fibers

The **animal fibers** include **wool** from sheep, **mohair** from goats, **camel's hair,** and **silk.** Wool is the most important of these, and it may be processed to reduce its susceptibility to moth damage and shrinkage. Silk is no longer of particular economic importance, having been supplanted by one or another of the synthetic fibers for most applications.

The **vegetable fibers** of greatest utility consist mainly of cellulose and may be classified as follows: seed hairs, such as cotton; bast fibers, such as flax, hemp, jute, and ramie; and vascular fibers. Those containing the most cellulose are the most flexible and elastic and may be bleached white most easily. Those which are more lignified tend to be stiff, brittle, and hard to bleach. Vegetable fibers are much less hygroscopic than wool or silk.

**Mercerized cotton** is cotton fiber that has been treated with strong caustic soda while under tension. The fiber becomes more lustrous, stronger, and more readily dyeable.

### Synthetic Organic Fibers

In recent years, a large and increasing portion of commercial fiber production has been of synthetic or semisynthetic origin. These fibers provide generally superior mechanical properties and greater resistance to degradation than do natural organic fibers and are available in a variety of forms and compositions for particular end-use applications. Generic categories of manmade fibers have been established by the Textile Fiber Products Identification Act, 15 USC 70, 72 Stat. 1717.

Among the important fibers are **rayons,** made from regenerated cellulose, plain or acetylated, which has been put into a viscous solution and extruded through the holes of a spinneret into a setting bath. The types most common at present are **viscose** and **acetate.** Cuprammonium rayon, saponified acetate rayon, and high-wet-modulus rayon are also manufactured and have properties which make them suitable for particular applications. Rayon is generally less expensive than other synthetic fibers.

In contrast to these regenerated fibers are a variety of polymer fibers which are chemically synthesized. The most important is **nylon,** including nylon 6,6, a condensation polymer of hexamethylene diamine-adipic acid, and nylon 6, a polymer of caprolactam. Nylon possesses outstanding mechanical properties and is widely used in industrial fabrics. Nomex, a high-temperature-resistant nylon retains its most important properties at continuous operating temperatures up to 260°C (500°F). Other polymer fibers possess mechanical or

chemical properties which make them a specific material of choice for specialized applications. These include **polyester** fibers, made from polyethylene terephthalate; **acrylic** and **modacrylic** fibers, made from copolymers of acrylonitrile and other chemicals; **Saran,** a polymer composed essentially of vinylidene chloride; and the **olefins,** including polyethylenes and polypropylenes.

**Teflon,** a fluorocarbon resin, has excellent chemical resistance to acids, bases, or solvents. Its useful physical properties range from cryogenic temperatures to 260°C (500°F), high dielectric strength, low dissipation factor, and high resistivity, along with the lowest coefficient of friction of any solid. It is nonflammable and is inert to weather and sunlight.

As part of their manufacture, substantially all synthetic fibers are **drawn** (hot- or cold-stretched) after extrusion to achieve desirable changes in properties. Generally, increases in draw increase breaking strength and modulus and decrease ultimate elongation.

**Proximate Identification of Fibers** Fibers are most accurately distinguished under the microscope, with the aid of chemical reagents and stains. A useful rough test is burning, in which the odor of burned meat distinguishes animal fibers from vegetable and synthetic fibers. Animal fibers, cellulose acetate, and nylon melt before burning and fuse to hard rounded beads. Cellulose fibers burn off sharply. Cellulose acetate dissolves in either acetone or chloroform containing some alcohol.

**Heat Endurance** Fibers of organic origin lose strength when heated over long periods of time above certain temperatures: cellulose, 149°C (300°F); cellulose acetate, 93.5°C (200°F); nylon, 224°C (435°F); casein, 100°C (212°F); and glass, 316°C (600°F).

**Creep** Textile fibers exhibit the phenomenon of creep at relatively low loads. When a textile fiber is subjected to load, it suffers three kinds of distortion: (1) an elastic deformation, closely proportioned to load and fully and instantly recoverable upon load removal; (2) a primary creep, which increases at a decreasing rate with time and which is fully, but not instantaneously, recoverable upon load removal; and (3) a secondary creep, which varies obscurely with time and load and is completely nonrecoverable upon load removal. The relative amounts of these three components, acting to produce the total deformation, vary with the different fibers. The two inelastic components give rise to mechanical hysteresis on loading and unloading.

### Felts and Fabrics

A **felt** is a compacted formation of randomly entangled fibers. Wool felt is cohesive because the scaly structure of the wool fibers promotes mechanical interlocking of the tangled fibers. Felts can be made with blends of natural or synthetic fibers, and they may be impregnated with resins, waxes, or lubricants for specific mechanical uses. Felts are available in sheets or cut into washers or shaped gaskets in a wide range of thicknesses and densities for packing, for vibration absorption, for heat insulation, or as holders of lubricant for bearings.

**Fabrics** are woven or knitted from **yarns.** Continuous-filament synthetic fibers can be made into monofilament or multifilament yarns with little or no twist. Natural fibers, of relatively short length, and synthetic **staple** fibers, which are purposely cut into short lengths, must be twisted together to form yarns. The amount of twist in yarns and the tension and

### Table 1. Fiber Properties*

| Kind | Source | Length of fiber, in. | Width or diam of cells, microns | Specific gravity | Moisture regain,‡ percent | Chemical description | Principal uses |
|---|---|---|---|---|---|---|---|
| Cotton....... | Plant seed hair | 5/8–2 | 8–27 | 1.52 | 8.5 | Cellulose | Industrial, household, apparel |
| Jute†........ | Plant bast | 50–80 | 15–20 | 1.48 | 13.7 | Lignocellulose | Bagging, twine, carpet backing |
| Wool......... | Animal | 2–16 | 10–50 | 1.32 | 17 | Protein | Apparel, household, industrial |
| Viscose....... | Manufactured | Any | 8–43 | 1.52 | 11 | Regenerated cellulose | Apparel, industrial, household |
| Cellulose acetate | Manufactured | Any | 12–46 | 1.33 | 6 | Cellulose ester | Apparel, industrial, household |
| Nylon........ | Manufactured | Any | 8 | 1.14 | 4.2 | Polyamide | Apparel, industrial, household |
| Casein....... | Manufactured | Any | 11–28 | 1.3 | 4.1 | Protein | Apparel |
| Flax†........ | Plant bast | 12–36 | 15–17 | 1.5 | 12 | Cellulose | Household, apparel, industrial |
| Hemp†....... | Plant bast | ..... | 18–23 | 1.48 | 12 | Cellulose | Twine, halyards, rigging |
| Sisal†........ | Plant leaf | 30–48 | 10–30 | .... | .... | Lignocellulose | Twine, cordage |
| Manila†...... | Plant leaf | 60–140 | 10–30 | .... | .... | Lignocellulose | Rope, twine, cordage |
| Ramie†...... | Plant bast | 3–10 | 24–70 | 1.52 | .... | Cellulose | Household, apparel, seines |
| Silk......... | Silkworm | Any | 5–23 | 1.35 | 11 | Protein | Apparel, household, industrial |
| Glass........ | Manufactured | Any | 3 | 2.5 | 0 | Fused metal oxides | Industrial, household |
| Dacron....... | Manufactured | Any | 8 | 1.38 | 0.4 | Polyester | Apparel, industrial, household |

1 in = 0.0254 m; $1\mu = 10^{-6}$ m. The more up-to-date term for the micron ($\mu$) is the micrometer ($\mu$m).

*Adapted from Smith, Textile Fibers, *Proc. ASTM*, 1944; Appel, A Survey of the Synthetic Fibers, *Am. Dyestuff Reporter*, **34**, 1945. pp. 21–26; and other sources.

†These fibers are commercially used as bundles of cells. They vary greatly in width. Width figures given are for the individual cells.

‡In air at 70°F and 65 percent relative humidity.

### Table 2. Tensile Properties of Single Fibers*

| Fiber | Breaking tenacity, gpd | Extension at break, percent | Elastic recovery at corresponding strain, percent | Elastic modulus,† gpd |
|---|---|---|---|---|
| Glass......................... | 6.0–7.3 | 3.0–4.0 | 100 at 2.9 | 200–300 |
| Fortisan (rayon)............... | 6.0–7.0 | 6 | 100 at 1.2 | 150–200 |
|  |  |  | 60 at 2.4 |  |
| Flax.......................... | 2.6–7.7 | 2.7–3.3 | 65 at 2 |  |
| Nylon 6, 6................... | 4.6–9.2 | 16–32 | 100 at 8 | 25–50 |
| Nylon 6...................... | 4.5–8.6 | 16–40 | 100 at 8 | 25–50 |
| Silk......................... | 2.4–5.1 | 10–25 | 92 at 2 | 75–125 |
| Saran........................ | 1.1–2.3 | 15–25 | 95 at 10 |  |
| Cotton....................... | 3.0–4.9 | 3–7 | 74 at 2 | 50–100 |
| Steel (90,000 psi T.S.).......... | 0.9 | 28 | ........ | 300 |
| Steel (music wire).............. | 3.5 | 8 | ........ | 300 |
| Viscose rayon................. | 1.5–5.0 | 15–30 | 82 at 2 | 50–150 |
| Wool........................ | 1.0–1.7 | 25–35 | 99 at 2 | 25–40 |
| Acetate rayon................. | 1.3–1.5 | 23–34 | 100 at 1 | 25–40 |
| Polyester..................... | 4.4–7.8 | 10–25 | 100 at 2 | 50–80 |
| Polypropylene................. | 4.0–7.0 | 15–25 | 95 at 7 | 15–50 |
| Polytetrafluoroethylene......... | 1.7 | 13 | ........ |  |

*From Kaswell, "Wellington Sears Handbook of Industrial Textiles," Wellington Sears Co., Inc.

†From Kaswell, "Textile Fibers, Yarns, and Fabrics." Reinhold.

arrangement of the weaving determine the appearance and mechanical properties of fabrics. Staple-fiber fabrics retain less than 50 percent of the intrinsic fiber strength, but values approaching 100 percent are retained with continuous-filament yarns. Industrial fabrics can be modified by mechanical and chemical treatments, as well as by coatings and impregnations, to meet special demands for strength and other mechanical, chemical, and electrical properties or to resist insect, fungus, and bacterial action and flammability.

**Nomenclature**  There are literally dozens of different numbering systems for expressing the relationship between yarn weight and length, all differing and each used in connection with particular fiber types or in different countries. The most common currently used unit is the **denier**, which is the weight in grams of 9,000 m of yarn. A universal system, based on the **tex** (the weight in grams of 1,000 m of yarn), has been approved by the International Standards Organization and by the ASTM, and this is expected eventually to supersede the present multisystem usage. Relative strengths of fibers and yarns are expressed as the **tenacity**, which reflects the specific gravity and the average cross section of the yarn. Units are grams per denier (gpd) or grams per tex (gpt).

Many textiles can sustain high-energy impact loads because of their considerable elongation before rupture. The total work done per unit length on a fiber or yarn which is extended to the point of rupture can be approximated by multiplying the specific strength by one-half the final extension of that length.

Yarn **twist** direction is expressed as **S** or **Z** twist, with the near-side helical paths of a twisted yarn held in a vertical position comparable in direction of slope to the center portion of one of these letters. Amount of twist is expressed in turns per inch (tpi).

Fabrics are characterized by the composition of the fiber material, the type of weave or construction, the count (the number of yarns per inch in the warp and the filling directions), and the weight of the fabric, usually expressed in ounces per running yard. **Cover factor** is the ratio of fabric surface covered by yarn to the total fabric surface. **Packing factor** is the ratio of fiber volume to total fabric volume.

Tables 1 to 3 give physical data and other information about commercial fibers and yarns. The tabulated quantities involving the denier should be regarded as approximate; they are not absolute values such as are used in engineering calculations.

## FREEZING PREVENTIVES

**Common salt** is sometimes used to prevent the freezing of water; it does not, however, lower the freezing point sufficiently to be of use in very cold weather, and in concentrated solution tends to "creep" and to crystallize all over the receptacle. It also tends to corrode metals. For freezing temperatures, see Sec. 18.

**Calcium chloride** ($CaCl_2$) is a white solid substance widely used for preventing freezing of solutions and (owing to its great hygroscopic power) for keeping sizing materials and other similar substances moist. It does not "creep" as in the case of salt. It does not rust metal but attacks solder.

Calcium chloride solutions are much less corrosive on metal if made alkaline by the use of lime, and also if a trace of sodium chromate is present. They are not suitable for use in automobile radiators, because of corrosive action while hot, and because of tendency of any spray therefrom to ruin the insulation of spark plugs and high-tension cables. For freezing temperatures, see Sec. 18.

**Glycerol** is a colorless, viscid liquid without odor and miscible with water in all proportions. It should have a specific gravity of approximately 1.25. It has no effect on metals but disintegrates rubber and loosens up iron rust.

**Denatured alcohol** is free from the disadvantages of calcium chloride, salt, and glycerin solutions, but is volatile from water mixtures which run hot. A solution containing 50 percent alcohol becomes inflammable, but it is rarely necessary to use more than 30 percent.

**Methyl alcohol,** variously trade-named, is sold widely for automobile antifreeze. It is a desirable and effective antifreeze, but care must be taken not to breathe its fumes, which are poisonous.

**Ethylene glycol (Prestone)** is used as a freezing preventive and also permits the use of high jacket temperatures in aircraft and other engines. Sp gr 1.125 (1.098) at 32 (77)°F; boiling point, 387°F; specific heat, 0.575 (0.675) at 68 (212)°F. Miscible with water in all proportions.

## Table 3. Temperature and Chemical Effects on Textiles*

| Fiber | Temperature limit, deg F | Resistance to chemicals | Resistance to mildew |
|---|---|---|---|
| Cotton | Yellows 250; decomposes 300 | Poor resistance to acids | Attacked |
| Flax | 275 | Poor resistance to acids | Attacked |
| Silk | Decomposes 300 | Attacked | Attacked |
| Glass | Softens 1350 | Resists | Resists |
| Nylon 6 | Sticky 400; melts 420–430 | Generally good | Resists |
| Nylon 6, 6 | Sticky 455; melts 482 | Generally good | Resists |
| Viscose rayon | Decomposes 350–400 | Poor resistance to acids | Attacked |
| Acetate | Sticky 350–400; melts 500 | Poor resistance to acids | Resists |
| Wool | Decomposes 275 | Poor resistance to alkalies | Attacked |
| Asbestos | 1490 | Resists | Resists |
| Polyester | Sticky 455; melts 480 | Generally good | Resists |
| Polypropylene | Softens 300–310; melts 325–335 | Generally good | Resists |
| Polyethylene | Softens 225–235; melts 230–250 | Generally good | Resists |
| Jute | 275 | Poor resistance to acids | Attacked |

Temperature conversion: $t_C = (^5/_9)(t_F - 32)$.
*From Fiber Chart, *Textile World*, McGraw-Hill, 1962.

**Nonfreezing Percentages by Volume in Solution**

| Temperature, °C (°F) | −6.7 (20) | −12.2 (10) | −17.8 (0) | −23.4 (−10) |
|---|---|---|---|---|
| Methyl alcohol | 13 | 20 | 25 | 30 |
| Prestone | 17 | 25 | 32 | 38 |
| Alcohol | 17 | 26 | 34 | 42 |
| Glycerine | 22 | 33 | 40 | 47 |

# GLASS

REFERENCES: *Journal of American Ceramic Society*, Columbus, Ohio. *Journal of the Society of Glass Technology*, Sheffield, England. Morey, "Properties of Glass," Reinhold. Scholes, "Handbook of the Glass Industry," Ogden-Watney. "Non-Silica Glasses," *Chem. & Met. Eng.*, Mar. 1946. Phillips, "Glass the Miracle-Maker," Pitman. Long, "Propriétés physiques et fusion du verre," Dunod. Eitel-Pirani-Scheel, "Glastechnische Tabellen," Springer. Jebsen-Marwedel, "Glastechnische Fabrikationsfehler," Springer. Shand, "Glass Engineering Handbook," McGraw-Hill, 1958. Phillips, "Glass: Its Industrial Applications," Reinhold, 1960. Persson, "Flat Glass Technology," Plenum Press, 1969. McMillan, "Glass-Ceramics," Academic, 1964. Pye, Stevens, and La Course (eds.), "Introduction to Glass Science," Plenum Press, 1972. Jones, "Glass," Chapman & Hall, 1971. Doremus, "Glass Science," Wiley, 1973. Technical Staffs Division, "Glass," Corning Glass Works.

**Glass** is an inorganic product of fusion which has cooled to a rigid condition without crystallizing. It is obtained by melting together silica, alkali, and stabilizing ingredients, such as lime, alumina, lead, and barium. Bottle, plate, and window glass usually contain $SiO_2$, $Al_2O_3$, $CaO$, and $Na_2O$. Small amounts of the oxides of manganese and selenium are added to obtain colorless glass.

Special glasses, such as fiber glass, laboratory ware, thermometer glass, and optical glass, require different manufacturing methods and different compositions. The following oxides are either substituted for or added to the above base glass: $B_2O_3$, $ZnO$, $K_2O$, $As_2O_3$, $PbO$, etc., to secure the requisite properties. Colored glasses are obtained by adding the oxides of iron, manganese, copper, selenium, cobalt, chromium, etc., or colloidal gold.

Molten glass possesses the ability to be fabricated in a variety of ways and to be cooled down to room temperature rapidly enough to prevent crystallization of the constituents. It is a rigid material at ordinary temperatures but may be remelted and molded any number of times by the application of heat. Ordinary glass is melted at about 1430°C (2600°F) and will soften enough to lose its shape at about 594°C (1100°F).

**Window glass** is a soda-lime-silica glass, fabricated in continuous sheets up to a width of 6 ft. The sheets are made in two thicknesses, SS and DS, which are, respectively, $\frac{1}{16}$ and $\frac{1}{8}$ in. Both thicknesses are made in A, B, C, and D grades.

**Reflective-coated glass**, such as Vari-Tran, insulates by reflecting hot sun in summer. Various colors are available, with transmissions ranging from 8 to 50 percent.

**Borosilicate glass** is the oldest type of glass to have significant heat resistance. It withstands higher operating temperatures than either limed or lead glasses and is also more resistant to chemical attack. It is used for piping, sight glasses, boiler-gage glasses, sealed-beam lamps, laboratory beakers, and oven cooking ware. This type of glass was the first to carry the Pyrex trademark.

**Aluminosilicate glass** is similar to borosilicate glass in behavior but is able to withstand higher operating temperatures. It is used for top-of-stove cooking ware, lamp parts, and when coated, as resistors for electronic circuitry.

**Plate glass** and float glass are similar in composition to window glass. They are fabricated in continuous sheets up to a width of 15 ft* and are polished on both sides. They may be obtained in various thicknesses and grades, under names like Parallel-O-Plate and Parallel-O-Float.

**Safety glass** consists of two layers of plate glass firmly held together by an intermediate layer of organic material, such as polyvinyl butyral. Safety glass is ordinarily $\frac{1}{4}$ in* thick but can be obtained in various thicknesses. This plate is shatterproof and is used for windshields, bank cashier's windows, etc.

**Tempered glass** is made from sheet glass in thicknesses up to 1 in.* It possesses great mechanical strength, which is obtained by rapidly chilling the surfaces while the glass is still hot. This process sets up a high compression on the glass surfaces, which have the capacity of withstanding very high tensile forces.

**Wire glass** is a glass having an iron wire screen thoroughly embedded in it. It offers about $1\frac{1}{2}$ times the resistance to bending that plain glass does; even thin sheets may be walked on. If properly made, it does not fall apart when cracked by shocks or heat, and is consequently fireproof. It is used for flooring, skylights, fireproof doors, fire walls, etc.

**Pressed glass** is made by forming heat-softened glass in molds under pressure. Such articles as tableware, lenses, insulators, and glass blocks are made by this process.

**Glass blocks** find wide application for building purposes, and are made $3\frac{7}{8}$ in thick in the following sizes: $6 \times 6$, $8 \times 8$, $12 \times 12$ in.* The thermal conductivity of a glass block panel, for a thickness of $3\frac{7}{8}$ in,* is 0.49 Btu/(ft²)(h)(°F). The solar-heat transmission can be varied within wide limits by using different-colored glasses and by changing the reflection by means of surface configurations. The light transmission can be varied from 25 to 80 percent by changing the surface configurations of the blocks. Vandal-resistant bricks, such as Vistabrik, virtually eliminate breakage problems.

**Fiber glass** is a term used to designate articles that consist of a multitude of tiny glass filaments ranging in size from 0.0001 to 0.01 in in diam. The larger fibers are used in air filters; those 0.0005 in in diam, for thermal insulation; and the 0.0001- to 0.0002-in-diam fibers, for glass fabrics, which are stronger than ordinary textiles of the same size. Insulating tapes made from glass fabric have found wide application in electrical equipment, such as motors and generators and for mechanical uses.

**Glass** is used for many structural purposes, such as store fronts and table tops, and is available in thicknesses of $\frac{5}{16}$ and $\frac{7}{16}$ in,* and $72 \times 130$ in.*

**Cellular glass** is a puffed variety with about 5 million cells per cubic foot.* It is a good heat insulator and makes a durable marine float.

**Vycor** is a 96 percent silica glass having extreme heat-shock resistance to temperatures up to 900°C (1652°F), and is used as furnace sight glasses, drying trays, and space-vehicle outer windows.

*Conversion factors: 1 ft = 0.3048 m; 1 in = 0.0254 m.

**Fused silica** (silicon dioxide) in the noncrystalline or amorphous state shows the maximum resistance to heat shock and the highest permissible operating temperature, 900°C (1652°F) for extended periods, or 1200°C (2192°F) for short periods. It has maximum ultraviolet transmission and the highest chemical-attack resistance. It is used for astronomical telescopes, ultrasonic delay lines, and crucibles for growing single metal crystals.

**Vitreous silica,** also called **fused quartz** is made by melting rock crystal or purest quartz sand in the electric furnace. It is unaffected by changes of temperature, is fireproof and acid-resistant, does not conduct electricity, and has practically no expansion under heat. It is used considerably for high-temperature laboratory apparatus. See Plastics, below, for glass substitutes.

**Glass ceramics,** such as Pyroceram, are melted and formed as glasses, then converted by controlled nucleation and crystal growth to polycrystalline ceramic materials. Most are opaque and stronger than glass, any may also be chemically strengthened.

**Glass beads** are used extensively for reflective paints.

**Properties of Glass** Glass is a brittle material and can be considered perfectly elastic to the fracture point. The range of Young's modulus is 4 to $14 \times 10^3$ kg/mm$^2$ (6 to $20 \times 10^6$ lb/in$^2$), with most commercial glasses falling between 5.5 and 9.0 $\times 10^3$ kg/mm$^2$ (8 and $13 \times 10^6$ lb/in$^2$). Theory predicts glass strength as high as 3,500 kg/mm$^2$ ($5 \times 10^6$ lb/in$^2$); fine glass fibers have shown strengths around 700 kg/mm$^2$ ($10^6$ lb/in$^2$); however, glass products realize but a fraction of such figures owing to surface-imperfection stress-concentration effects. Often design strengths run from 500 to 1,000 times lower than theoretical. Glass strength also deteriorates when held under stress in atmospheric air (static fatigue), an apparent result of reaction of water with glass; high-strength glasses suffer the greater penalty. Glass also exhibits a time-load effect, breaking at lower stresses for sustained loads. At room temperatures the thermal conductivity of glasses ranges from $1.6 \times 10^{-3}$ to $2.9 \times 10^{-3}$ cal/(s)(cm$^2$)(°C/cm) [4.65 to 8.43 Btu/(h)(ft$^2$)(°F/in)].

# NATURAL STONES
### by David Rostoker
*Corning Glass Works*

REFERENCES: Kessler, Insley, and Sligh, *Jour. Res. NBS,* **25,** pp. 161–206. Birch, Schairer, and Spicer, Handbook of Physical Constants, *Geol. Soc. Am. Special Paper* 36. Currier, Geological Appraisal of Dimension Stone Deposits, *USGS Bull.* 1109.

A **stone** or a **rock** is a naturally occurring composite of minerals. Stone has been used for thousands of years as a major construction material because it possesses qualities of strength, durability, architectural adaptibility, and aesthetic satisfaction. There are two principal branches of the natural-stone industry— **dimension** stone and **crushed** or broken stone. The uses of the latter vary from aggregate to riprap, in which stones in a broad range of sizes are used as structural support in a matrix or to provide weathering resistance. Dimension stones are blocks or slabs of stone processed to specifications of size, shape, and surface finish. The largest volume today lies in the use of slabs varying from 1 to 4 in in thickness that are mounted on a structure as a protective and aesthetic veneer.

*Conversion factors: 1 ft = 0.3048 m; 1 in = 0.0254 m.

There are two major types of natural stone: **igneous** and **metamorphic** stones, composed of tightly interlocking crystals of one or more minerals, and **sedimentary** rocks, composed of cemented mineral grains in which the cement may or may not be of the same composition as the grains. The major groups of natural stone used commercially are:

**Granite,** a visibly crystalline rock made of silicate minerals, primarily feldspar and quartz. Commercially, "granite" refers to all stones geologically defined as plutonic, igneous, and gneissic.

**Marble,** generally a visibly crystalline carbonate rock; however, microcrystalline rocks, such as onyx, travertine, and serpentine, are usually included by the trade as long as they can take a polish.

**Limestone,** a sedimentary rock composed of calcium or magnesium carbonate grains in a carbonate matrix.

**Sandstone,** a sedimentary rock composed chiefly of cemented, sand-sized quartz grains. In the trade, quartzites are usually grouped with sandstones, although these rocks tend to fracture through, rather than around, the grains. **Conglomerate** is a term used for a sandstone containing aggregate in sizes from the gravel range up.

The above stones can be used almost interchangeably as dimension stone for architectural or structural purposes.

**Slate,** a fine-grained rock characterized by marked cleavages by which the rock can be split easily into relatively thin slabs. Because of this characteristic, slate was at one time widely used for roofing tiles. It is used today for electrical boards and blackboards, and it has recently become popular for steps, risers, spandrels, flagstones, and other building uses.

Miscellaneous stones, such as traprock (fine-grained black volcanic rock), greenstone, or argillite, are commonly used as crushed or broken stone but rarely as dimension stone.

The table of physical properties contains the range of values that can be obtained from stones in various orientations relative to their textural and structural anisotropy. For a particular application, where one property must be exactly determined, the value must be obtained along a specified axis.

### Selected Terms Applying to the Use of Dimension Stone

**Anchor,** a metal tie or rod used to fasten stone to backup units.

**Arris,** the meeting of two surfaces producing an angle, corner, or edge.

**Ashlar,** a facing of square or rectangular stones having sawed, dressed, or squared beds.

**Bond stones,** stones projecting a minimum of 4 in laterally into the backup wall; used to tie the wall together.

**Cut stone,** finished dimension stone—ready to set in place. The finish may be polished, honed, grooved (for foot traffic), or broken face.

**Bearing wall,** a wall supporting a vertical load in addition to its own weight.

**Cavity wall,** a wall in which the inner and outer parts are separated by an air space but are tied together with cross members.

**Composite wall,** a wall in which the facing and backing are of different materials and are united with bond stones to exert a common reaction under load. It is considered preferable, however, not to require the facing to support a load; thus the bond stones merely tie the facing to the supporting wall, as in the case of a veneer.

**Veneer** or **faced wall,** a wall in which a thin facing and the

**Physical and Thermal Properties of Common Stones**

| Type of stone | Density, lb per cu ft | Compressive strength × $10^{-3}$, psi | Rupture modulus × $10^{-3}$, psi (ASTM C99–52) | Shearing strength × $10^{-3}$, psi | Young's modulus × $10^{-6}$, psi | Modulus of rigidity × $10^{-6}$, psi | Poisson's ratio | Abrasion-hardness index (ASTM C241–51) | Porosity, volume percent | 48-hr water absorption (ASTM C97–47) | Thermal conductivity, Btu per ft per hr per deg F | Coefficient of thermal expansion × $10^{-6}$, per deg F |
|---|---|---|---|---|---|---|---|---|---|---|---|---|
| Granite......... | 160–190 | 13–55 | 1.4–5.5 | 3.5–6.5 | 4–16 | 2–6 | 0.05–0.2 | 37–88 | 0.6–3.8 | 0.02–0.58 | 20–35 | 3.6–4.6 |
| Marble......... | 165–179 | 8–27 | 0.6–4.0 | 1.3–6.5 | 5–11.5 | 2–4.5 | 0.1–0.2 | 8–42 | 0.4–2.1 | 0.02–0.45 | 8–36 | 3.0–8.5 |
| Slate............ | 168–180 | 9–10 | 6–15 | 2.0–3.6 | 6–16 | 2.5–6 | 0.1–0.3 | 6–12 | 0.1–1.7 | 0.01–0.6 | 12–26 | 3.3–5.6 |
| Sandstone...... | 119–168 | 5–20 | 0.7–2.3 | 0.3–3.0 | 0.7–10 | 0.3–4 | 0.1–0.3 | 2–26 | 1.9–27.3 | 2.0–12.0 | 4–40 | 3.9–6.7 |
| Limestone...... | 117–175 | 2.5–28 | 0.5–2.0 | 0.8–3.6 | 3–9 | 1–4 | 0.1–0.3 | 1–24 | 1.1–31.0 | 1.0–10.0 | 20–32 | 2.8–4.5 |

Conversion factors: 1 lbm/ft³ = 16,018 kg/m³. 1 psi (lb/in²) = 6,894.8 N/m². 1 Btu/(ft)(hr)(°F) = 623 W/(m) (°C).
$t_C = (^5/_9)(t_F - 32)$.

backing are of different materials but are not so bonded as to exert a common reaction under load.

# PAPER

**by R. Claire Canty and F. G. Perry, Jr.**
**revised by Antonio F. Baldo**

*Arthur D. Little, Inc.*

REFERENCES: Calkin, "Modern Pulp and Papermaking," Reinhold. Griffin and Little, "Manufacture of Pulp and Paper," McGraw-Hill. Guthrie, "The Economics of Pulp and Paper," State College of Washington Press. "Dictionary of Paper," American Pulp and Paper Assoc. Sutermeister, "Chemistry of Pulp and Papermaking," Wiley. Casey, "Pulp and Paper—Chemistry and Technology," Interscience. TAPPI Technical Information Sheets. TAPPI Monograph Series. "Index of Federal Specifications, Standards and Handbooks," GSA. Lockwoods Directory of the Paper and Allied Trades. Casey, "Pulp and Paper; Chemistry and Chemical Technology," 3 vols., Interscience, 1960–1961. Libby (ed.), "Pulp and Paper Science and Technology," 2 vols., McGraw-Hill, 1962. ASTM Committee D6 on Paper and Paper Products, "Paper and Paperboard Characteristics, Nomenclature, and Significance of Tests," ASTM Special Technical Publication 60-B, 3d ed., 1963. Britt, "Handbook of Pulp and Paper Technology," Van Nostrand Reinhold, 1970. Johnson, "Synthetic Paper from Synthetic Fibers," Noyes Data Corp., New Jersey, 1971.

## Paper Grades

Specific paper **qualities** are achieved in a number of ways: (1) By selecting the composition of the furnish for the paper machine. Usually more than one pulp (prepared by different pulping conditions or processes) is required. The ratio of long-fibered pulp (softwood) to short-fibered pulp (hardwood or mechanical type), the reused-fiber content, and the use of nonfibrous fillers and chemical additives are important factors. (2) By varying the paper-machine operation. Fourdrinier wire machines are most common, although multicylinder machines and high-speed tissue machines with Yankee driers are also used. (3) By using various finishing operations (e.g., calendering, supercalendering, coating, and laminating).

There is a tremendous number of paper grades, which are, in turn, used in a wide variety of converted products. The following broad classifications are included as a useful guide:

**Sanitary papers** are tissue products characterized by bulk, opacity, softness, and water absorbency.

**Glassine, greaseproof,** and **waxing papers**—in glassine, high transparency and density, low grease penetration, and uniformity of formation are important requirements.

**Food-board** products require a good brightness and should be odorless and have good tear, tensile, and fold-endurance properties, opacity, and printability.

**Boxboard** and a variety of other board products are made on multicylinder machines, where layers of fiber are built up to the desired thickness. Interior plys are often made from wastepaper furnishes, while surface plys are from bleached, virgin fiber. Boards are often coated for high brightness and good printing qualities.

**Printing papers** include publication papers (magazines), book papers, bond and ledger papers, newsprint, and catalog papers. In all cases, printability, opacity, and dimensional stability are important.

**Linerboard** and **bag paper** are principally unbleached, long-fiber, kraft products of various weights. Their principal property requirements are high tensile and bursting strengths.

**Corrugating medium,** in combined fiberboard, serves to hold the two linerboards apart in rigid, parallel separation. Stiffness is the most important property, together with good water absorbency for ease of corrugation.

## Pulps

The most important source of fiber for paper pulps is wood, although numerous other vegetable substances are used. Reused fiber (wastepaper) constitutes about 30 percent of the total furnish used in paper, principally in boxboard and other packaging or printing papers.

There are three basic processes used to convert wood to papermaking fibers: mechanical, chemical, and "chemimechanical."

**Mechanical Pulping (Groundwood)** Here, the entire log is reduced to fibers by grinding against a stone cylinder, the simplest route to papermaking fiber. Wood fibers are mingled with extraneous materials which can cause weakening and discoloration in the finished paper. Nonetheless, the resulting pulp imparts good bulk and opacity to a printing sheet. Some long-fibered chemical pulp is usually added. Groundwood is used extensively in low-cost, short-service, and throwaway papers, e.g., newsprint and catalog paper.

**Chemical Pulping** Fiber separation can also be accomplished by chemical treatment of wood chips to dissolve the lignin that cements the fibers together. From 40 to 50 percent of the log is extracted, resulting in relatively pure cellulosic fiber. There are two major chemical-pulping processes, which differ both in chemical treatment and in the nature of the pulp produced. In **sulfate pulping,** also referred to as the **kraft** or **alkaline** process, the pulping chemicals must be recovered and reused for economic reasons. Sulfate pulps result in papers of high physical strength, bulk, and opacity; low unbleached brightness; slow beating rates; and relatively poor sheet-formation properties. Both bleached and unbleached sulfate pulps are used in packaging papers, container board, and a variety of printing and bond papers. In **sulfite pulping,** the delignifying agents are sulfurous acid and an alkali, with several variations of the exact chemical conditions in commercial use. In general, sulfite pulps have lower physical strength properties and lower bulk and opacity than kraft pulps, with higher unbleached brightness and better sheet-formation properties. The pulps are blended with groundwood for newsprint and are used in printing and bond papers, tissues, and glassine.

"**Chemimechanical**" **processes** combine both chemical and mechanical methods of defibration, the most important commercial process being the NSSC (Neutral Sulfite Semi Chemical). A wide range of yields and properties can be obtained.

### Bleaching

Chemical pulps may be bleached to varying degrees of brightness, depending on the end use. During some bleaching operations, the remaining lignin is removed and residual coloring matter destroyed. Alternatively, a nondelignifying bleach lightens the color of high-lignin pulps.

### Refining

The final character of the pulp is developed in the **refining** (beating) operation. Pulp fibers are fibrillated, hydrated, and cut. The fibers are roughened and frayed, and a gelatinous substance is produced. This results in greater fiber coherence in the finished paper.

### Sizing, Loading, and Coating

**Sizing** is used, principally in book papers, to make the paper water-repellent and to enhance interfiber bonding and surface characteristics. Sizing materials may be premixed with the pulp or applied after sheet formation.

**Loading materials,** or **fillers,** are used by the papermaker to smooth the surface, to provide ink affinity, to brighten color, and to increase opacity. The most widely used filler is clay.

**Coatings** consisting of pigment and binder are often applied to the base stock to create better printing surfaces. A variety of particulate, inorganic materials are combined with binders such as starch or casein in paper coatings. Coating is generally followed by calendering.

### Converting and Packaging

A host of products is made from paper; the converting industry represents a substantial portion of the total paper industry. Principal products are in the fine-paper and book-paper fields. Slush pulps are used to make molded pulp products such as egg cartons and paper plates.

Combinations of paper laminated with other materials such as plastic film and metal foil have found wide use in the packaging market.

### Plastic-Fiber Paper

A tough, durable product, such as Tyvek, can be made from 100 percent high-density polyethylene fibers by spinning very fine fibers and then bonding them together with heat and pressure. Binders, sizes, or fillers are not required. Such sheets combine some of the best properties of the fabrics, films, and papers with excellent puncture resistance. They can readily be printed by conventional processes, dyed to pastel colors, embossed for decorative effects, coated with a range of materials, and can be folded, sheeted, die-cut, sewn, hot-melt-sealed, glued, and pasted. Nylon paper, such as Nomex type 410, is produced from short fibers (floc) and smaller binder particles (fibrids) of a high-temperature-resistant polyamide polymer, formed into a sheet product without additional binders, fillers, or sizes, and calendered with heat and pressure to form a nonporous structure. It possesses excellent electrical, thermal, and mechanical properties, and finds use in the electrical industry. Fiber impregnation of ordinary paper, using glass or plastic fibers, produces a highly tear-resistant product.

## PLASTICS
### by Harris J. Bixler
*Amicon Corporation*

REFERENCES: "Modern Plastics Encyclopedia," McGraw-Hill. Reinhold Plastics Applications Series, Reinhold. Roff, "Fibres, Plastics and Rubbers," Butterworth. Lee and Neville, "Epoxy Resins," McGraw-Hill. Billmeyer, "Textbook of Polymer Science," Interscience. Schmidt and Marlies, "Principles of High Polymer Theory and Practice," McGraw-Hill. Beadle (ed.), "Plastics," Morgan-Grampian, Ltd., 2 vols. 1970. Tobolsky and Mark (eds.), "Polymer Science and Materials," Wiley, 1971. Glanville, "The Plastics Engineer's Data Book," Industrial Press, Inc., 1971. Mohr (editor and senior author), Oleesky, Shook, and Meyers, "SPI Handbook of Technology and Engineering of Reinforced Plastics Composites," Van Nostrand Reinhold 1973.

Plastics are materials which are wholly or in part composed of long, chainlike molecules called **high polymers.** While carbon is the element common to all commercial high polymers, hydrogen, oxygen, nitrogen, sulfur, halogens, and silicon can be present in varying proportions. High polymers may be divided into two classes: **thermoplastic** and **thermosetting.** The former reversibly melt to become highly viscous liquids and, upon cooling, solidify to give, depending on their structure, elastic, ductile, tough, or brittle solids. Melting temperatures range from 100 to 300°C. The thermosetting polymers are infusible without thermal or mechanical degradation. Many thermosetting polymers are prepared by curing thermoplastic **prepolymers** or "**B-stage**" resins. **Curing** (a chain-linking chemical reaction) is usually initiated with heat. The more highly cured the polymer, the higher its heat-distortion temperature and the harder and more brittle it becomes.

### Thermoplastics

Thermoplastic polymers are often used with only minor additions (colorants, stabilizers, lubricants). Important characteristics are low density, low cost, toughness, optical clarity, ease of forming complex shapes, low thermal conductivity, chemical resistance, flexibility, and useful electrical properties. Fabrication into plastic objects is achieved by compression, injection, blow, and slush molding; vacuum forming; melt extrusion; solvent casting; stamping; calendering; and powder-metallurgy processes (see also Sec. 13). The starting material in these processes is bulk polymer in the form of small parti-

**Table 1. Thermoplastics—Properties and Uses**

| Item | High-density polyethylene | Low-density polyethylene | Polypropylene |
|---|---|---|---|
| Key properties | Chemical resistance, strength, low cost | Electrical properties, flexibility, low cost | Toughness, gloss, light weight, heat resistance |
| Major uses | Film, bleach, bottles, pipe | Film, containers, wire coating | Housewares, appliance parts, tank liners |
| Density, g per cc | 0.94–0.96 | 0.91–0.93 | 0.90–0.92 |
| Tensile strength, psi | 3,100–5,500 | 1,000–2,300 | 3,500–6,000 |
| Tensile modulus, psi $\times$ $10^{-5}$ | 0.4–1.5 | 0.17–0.35 | 1–2 |
| Maximum continuous-use temp, deg F | 250 | 180–212 | 200–320 |
| Molding temp, deg F | 300–800 | 275–700 | 350–600 |
| Dielectric strength, volts per mil | 440–600 | 420–700 | 450–650 |
| Dielectric constant, 1,000 cps | 2.30–2.35 | 2.25–2.35 | 2.3 |
| Dissipation factor, 1,000 cps | Less than 0.0002 | Less than 0.0005 | 0.0002–0.0008 |
| Water absorption, 24 hr, percent | Less than 0.01 | Less than 0.015 | Less than 0.02 |

| Item | Cellulose acetate | Rigid vinyl | Plasticized vinyl |
|---|---|---|---|
| Key properties | Low cost, toughness, clarity, colorability | Compounding flexibility, low cost, chemical resistance | Compounding flexibility, low cost, pliability |
| Major uses | Door knobs, blister packs, appliance housings | Conduit, pipe, building panels | Tubing, gaskets, upholstery, footwear |
| Density, g per cc | 1.26–1.34 | 1.35–1.45 | 1.16–1.35 |
| Tensile strength, psi | 1,900–8,300 | 5,000–9,000 | 1,500–3,500 |
| Tensile modulus, psi $\times$ $10^{-5}$ | 1.0–4.0 | 4–6 | |
| Maximum continuous-use temp, deg F | 140–220 | 120–160 | 150–175 |
| Molding temp, deg F | 260–490 | 285–400 | 285–385 |
| Dielectric strength, volts per mil | 200–300 | 375–750 | 275–900 |
| Dielectric constant, 1,000 cps | 3.5–7.0 | 3.0–3.3 | 4.0–8.0 |
| Dissipation factor, 1,000 cps | 0.01–0.06 | 0.01–0.02 | 0.07–0.16 |
| Water absorption, 24 hr, percent | 1.9–6.5 | 0.07–0.4 | 0.15–0.75 |

| Item | Cellulose acetate butyrate | Nylon 6, 6 | Polymethyl methacrylate (Lucite or Plexiglas) | Polytetrafluoroethylene (Teflon) |
|---|---|---|---|---|
| Key properties | Impact resistance, clarity, weatherability | Toughness, low coefficient of friction, machinability | Crystal clarity, strength, weatherability, colorability | Thermal stability, low coefficient of friction, chemical resistance |
| Major uses | Outdoor signs, lenses | Gears, bearings, hinges, fibers | Automotive lenses, shaped glazing, signs | Gaskets, non-stick surfaces, insulation, electrical applications |
| Density, g per cc | 1.15–1.22 | 1.09–1.14 | 1.17–1.20 | 2.13–2.22 |
| Tensile strength, psi | 2,600–6,900 | 7,000–12,000 | 7,000–11,000 | 2,000–4,500 |
| Tensile modulus, psi $\times$ $10^{-5}$ | 0.5–2.0 | 2.6–4.0 | 4.5 | 0.58 |
| Maximum continuous-use temp, deg F | 140–220 | 270–300 | 140–190 | 550 |
| Molding temp, deg F | 265–480 | 470–720 | 300–500 | |
| Dielectric strength, volts per mil | 250–350 | 340–410 | 350–400 | 430 |

**Table 1. Thermoplastics—Properties and Uses (Continued)**

| Item | Cellulose acetate butyrate | Nylon 6, 6 | Polymethyl methacrylate (Lucite or Plexiglas) | Polytetrafluoro-ethylene (Teflon) |
|---|---|---|---|---|
| Dielectric constant, 1,000 cps | 3.3–6.3 | 4.0–4.5 | 3.0–3.5 | 2.0 |
| Dissipation factor, 1,000 cps | 0.01–0.04 | 0.02–0.04 | 0.03–0.05 | Less than 0.0002 |
| Water absorption, 24 hr, percent | 0.9–2.2 | 0.4–1.5 | 0.3–0.4 | 0.00 |

| Item | Polystyrene (general-purpose) | High-impact polystyrene | ABS acrylonitrile butadiene-styrene |
|---|---|---|---|
| Key properties | Low cost, crystal clarity | Low cost, impact strength, craze resistance | Toughness, strength, processability |
| Major uses | Food containers, lighting panels, film | Appliance housing, luggage, pipe and fittings | Automotive accessories, refrigerator liners, telephone sets |
| Density, g per cc | 1.05–1.07 | 0.98–1.10 | 1.01–1.15 |
| Tensile strength, psi | 5,000–9,000 | 2,000–6,800 | 2,400–9,000 |
| Tensile modulus, psi $\times 10^{-5}$ | 4–5 | 2–4.5 | 1–4 |
| Maximum continuous-use temp, deg F | 150–170 | 140–175 | 140–250 |
| Molding temp, deg F | 265–600 | 250–600 | 300–600 |
| Dielectric strength, volts per mil | 400–600 | 300–600 | 310–400 |
| Dielectric constant, 1,000 cps | 2.4–2.65 | 2.4–4.5 | 2.4–4.75 |
| Dissipation factor, 1,000 cps | Less than 0.0003 | 0.0004–0.002 | 0.002–0.012 |
| Water absorption, 24 hr, percent | 0.03–0.05 | 0.1–0.3 | 0.1–0.3 |

Conversion factors: 1 psi (lb/in²) = 689.48 N/m². $t_C$ = N/,². $t_C$ = ⁵/₉ ($t_F$ − 32). Cps = Hz.

cles called **molding powders,** extruded sheet, or **plastisol.** Plastisols are liquid or semisolid dispersions of finely powdered polymer in a nonvolatile, nonmigrating liquid, which result in a homogeneous, tough mass upon heating. Expanded, or **foamed,** thermoplastics are finding widespread use as thermal insulation and cushioning material. Hollow beads (or, in some cases, a plastisol) containing a heat-triggerable blowing agent are charged into a heated mold where the polymer softens, is expanded by the blowing agent, and upon cooling, solidifies to a low-density (as low as 2 lb/ft³) open- or closed-cell plastic foam with low thermal conductivity.

Filled thermoplastics (primarily vinyls) containing up to 50 percent by weight of fillers such as asbestos, wood flour, clay, titania, or carbon black are used in flooring tile, electrical-wire insulation, and pipe or conduit.

Table 1 summarizes the **physical properties** and **major end uses** of a number of thermoplastics of industrial importance.

## Thermosetting Plastics

Most **thermosetting plastics** are composite structures, being either multicomponent laminates or blends of polymer and particulate or fibrous fillers. The selection and blending or laminating together (called **compounding**) of suitable polymers and fillers are the heart of thermosetting-plastics technology. These plastics are noted for thermal and dimensional stability, chemical resistance, strength, durability, and good electrical

properties. The costs of thermosetting plastics vary considerably—from very cheap phenolics for container caps, through dinnerware-quality melamine-formaldehydes, to highly engineered epoxy and phenolic composites for space vehicles.

**Compounding** is done in Banbury and dough mixers, ribbon blenders, rubber mills, extruders, or for fibrous fillers like cellulose and asbestos, a papermaker's beater (beater addition products).

**Fillers** include wood flour, silica, mica, clay, glass fiber, cellulose, asbestos, and natural and synthetic textile fibers. To promote adhesion between the filler and the polymeric binder, the former is frequently precoated with sizing agents.

**Fabrication** (shaping and/or curing) is done by injection, compression, and transfer molding; press laminating; casting; tunnel-oven curing (sheet stock and extrudates); lay-up; and bag molding. For lay-up, bag molding, and laminating, prepolymer impregnation of woven and nonwoven felts may be the sole compounding step. Although heat is usually employed to accelerate curing, many room-temperature-curing plastics are now in use. (See also Sec. 13.)

**Polyurethane foams** are a unique class of thermosetting plastics. By mixing two relatively low-viscosity liquids (prepolymer and blowing, plus curing agents), an expanded foam can be formed and cured in minutes at room temperature. These have found widespread use as poured-in-place insulation.

Tables 2 to 4 summarize the **physical properties** and **major end uses** of a number of industrially important thermosets.

**Table 2. Properties and Uses of Phenol-Formaldehyde Plastics**

| | No filler | Mica-filled |
|---|---|---|
| Key properties........................... | Low cost, chemical and heat resistance | Good strength, electrical properties, temperature stability |
| Major uses.............................. | Impregnating resins, buttons, electrical connectors | Electrical components, abrasive products, brake shoes |
| Density, g per cc......................... | 1.25–1.3 | 1.65–1.92 |
| Tensile strength, psi...................... | 7,000–8,000 | 6,500–7,000 |
| Tensile modulus, psi $\times$ $10^{-5}$............... | 7.5–10 | 30–50 |
| Maximum continuous-use temp, deg F....... | 250 | 250–300 |
| Molding temp, deg F...................... | 270–320 | 280–350 |
| Dielectric strength, volts per mil............ | 250–400 | 250–400 |
| Dielectric constant, 1,000 cps............. | 4.5–6.0 | 4.4–5.5 |
| Dissipation factor, 1,000 cps................ | 0.03–0.08 | 0.03–0.04 |
| Water absorption, 24 hr, percent............ | 0.1–0.2 | 0.01–0.05 |

Conversion factors: 1 psi (lb/in²) = 689.48 N/m². $t_1$ = ⅝ $t_1$ − 32). Cps = HZ.

**Table 3. Properties and Uses of Melamine-Formaldehyde and Polyester Plastics**

| | Melamine-formaldehyde | | Polyester woven cloth—reinforced |
|---|---|---|---|
| | No filler | Alpha cellulose-filled | |
| Key properties.............. | Heat resistance, moisture resistance, toughness, scratch resistance | Same as for no filler plus added strength and toughness | Low cost, high strength-to-weight ratio, versatility, abrasion resistance |
| Major uses................. | Buttons, decorative laminates, adhesives, textile finishes | Dinnerware, utensil handles, housewares, electrical moldings | Boats, auto bodies, large lay-ups, wall siding, pipe, filament-wound pressure tanks |
| Density, g per cc............. | 1.48 | 1.47–1.52 | 1.5–2.1 |
| Tensile strength, psi......... | ....... | 7,000–13,000 | 30,000–50,000 |
| Tensile modulus, psi $\times$ $10^{-5}$... | ....... | 12–14 | 15–45 |
| Maximum continuous-use temp, deg F......... | 210 | 210 | 300–350 |
| Molding temp, deg F......... | 300–350 | 280–370 | 70–250 |
| Dielectric strength, volts per mil.................... | ....... | 250–400 | 350–500 |
| Dielectric constant, 1,000 cps.. | ....... | 7.8–9.2 | 4.2–6.0 |
| Dissipation factor, 1,000 cps... | ...... | 0.015–0.036 | 0.01–0.06 |
| Water absorption, 24 hr, percent.................... | 0.3–0.5 | 0.1–0.6 | 0.05–0.5 |

Conversion factors: 1 psi (lb/in²) = 689.48 N/m². $t_1$ = ⅝ $t_1$ − 32). Cps = HZ.

# ROOFING MATERIALS
## Bird & Son, East Walpole, Mass.

REFERENCES: Abraham, "Asphalts and Allied Substances," Van Nostrand. Grondal, "Certigrade Handbook of Red Cedar Shingles," Red Cedar Shingle Bureau, Seattle, Wash. ASTM *Special Technical Publication* 409, "Engineering Properties of Roofing Systems," ASTM, 1967. Griffin, Jr., "Manual of Built-up Roof Systems," McGraw-Hill, 1970.

**Asphalt** Asphalts are bitumens, and the one most commonly seen in roofing and paving is obtained from petroleum residuals. These are obtained by the refining of petroleum. The qualities of asphalt are affected by the nature of the crude and the process of refining. When the flux asphalts obtained from the oil refineries are treated by blowing air through them while the asphalt is maintained at a high temperature, a material is produced which is very stable and has good weathering properties.

**Coal Tar** Coal tar is more susceptible to temperature change than asphalt; therefore, for roofing purposes its use is usually confined to flat decks.

**Asphalt prepared roofing** is manufactured by impregnating a dry roofing felt with a hot asphaltic saturant. A coating consisting of a harder asphalt compounded with a fine mineral **filler** is applied to the weather side of the saturated felt. Into this coating is embedded mineral **surfacing** such as mineral granules, powdered talc, mica, or soap-stone. The reverse side of the roofing has a very thin coating of the same asphalt which is usually covered with powdered talc or mica to prevent the roofing from sticking in the package. The surfacing used on **smooth-surfaced roll roofing** is usually powdered talc or mica. The surfacing used on **mineral-** or **slate-surfaced roll roofing** is roofing granules either in natural colors prepared from slate or artifical colors usually made by applying a coating to a rock granule base. **Asphalt shingles** usually have a granular surfacing. They are made in strips and as individual shingles. The different shapes and sizes of these shingles provide single, double, and triple coverage of the roof deck.

**Table 4. Properties and Uses of Epoxy and Silicone Thermosetting Plastics**

| | Epoxy, mineral-filled | Flexibilized epoxy, no filler | Silicone, mineral-filled |
|---|---|---|---|
| Key properties.............. | Inertness, low shrinkage upon curing, high strength | Adhesion, low shrinkage upon curing, chemical resistance, flexibility | Thermal and oxidative stability, low-temperature flex, excellent electrical properties |
| Major uses................. | Potting of electrical components, adhesives | Adhesives, caulking compounds, surface coatings | Potting compounds, electronic components, sealants, gaskets |
| Density, g per cc............ | 1.6–2.0 | 1.15–1.25 | 1.81–2.82 |
| Tensile strength, psi.......... | 7,000–13,000 | 2,000–10,000 | 3,000–3,500 |
| Tensile modulus, psi $\times 10^{-5}$... | 3.5 | 0.01–3.5 | |
| Maximum continuous-use temp, deg F..................... | 250–550 | 250–300 | Greater than 600 |
| Molding temp, deg F........ | ............ | ............ | 310–370 |
| Dielectric strength, volts per mil...................... | 400–550 | 235–400 | 200–400 |
| Dielectric constant, 1,000 cps.. | 3.2–4.0 | 3–5 | 3.8–6.3 |
| Dissipation factor, 1,000 cps... | 0.008–0.03 | 0.012–0.05 | 0.002–0.005 |
| Water absorption, 24 hr, percent................... | 0.04–0.1 | 0.27–0.25 | 0.13 |

Conversion factors: 1 psi (lb/in²) = 689.48 N/m². $t_C = \frac{5}{9}(T_F - 32)$. Cps = Hz.

**Materials Used in Asphalt Prepared Roofing**  The **felt** is usually composed of a continuous sheet of felted fibers of selected rag, specially prepared wood, and high-quality waste papers. The constituents may be varied to give a felt with the desired qualities of strength, absorbency, and flexibility. (See above, Fibers and Fabrics.)

The most satisfactory roofing **asphalts** are obtained by air-blowing a steam- or vacuum-refined petroleum residual. Saturating asphalts must possess a low viscosity in order for the felt to become thoroughly impregnated. Coating asphalts must have good weather-resisting qualities and possess a high fusion temperature in order that there will be no flowing of the asphalt after the application to the roof.

**Asphalt built-up roof coverings** usually consist of several layers of asphalt-saturated felt with a continuous layer of hot-mopped asphalt between the layers of felt. The top layer of such a roof covering may consist of a top mopping of asphalt only, a top pouring of hot asphalt with slag or gravel embedded therein, or a mineral-surfaced cap sheet embedded in a hot mopping of asphalt.

**Wood shingles** are usually manufactured in three different lengths: 16, 18, and 24 in. There are three grades in each length; the No. 1 is the best, and the No. 3 is intended for purposes where the presence of defects is not objectionable. Red-cedar shingles of good quality are obtainable from the Pacific Coast; in the South, red cypress from the Gulf states is preferable. Redwood shingles come 5½ butts to 2 in; lesser thicknesses are more liable to crack and have shorter life. Shingles 8 in wide or over should be split before laying. Dimension shingles of uniform width are obtainable. Various stains are available for improved weathering resistance and altered appearance.

**Asbestos shingles** are composed of portland cement reinforced with asbestos fiber and are formed under pressure. They resist the destructive effects of time, weather, and fire. Asbestos shingles (American method) weigh about 500 lb per square (roofing to cover 100 ft²) and carry Underwriter's Class A label. Asbestos shingles are made in a variety of colors and shapes. Asbestos roofing shingles have either a smooth surface or a textured surface which represents wood graining.

**Slate** should be hard and tough and have a well-defined vein that is not too coarse. Approximate weight varies from 650 to 850 lb per square. Slate roofing is available in a variety of sizes and colors and has good fire resistance, but it is regarded as expensive. (See above, Natural Stones.)

**Metallic roofings** are usually laid in large panels, often strengthened by corrugating, but they are sometimes cut into small sizes bent into interlocking shapes and laid to interlock with adjacent sheets or shingles. Metallic-roofing panels of both aluminum and steel are available with a variety of prefinished surface treatments to enhance weatherability. Metal tile and metal shingles are usually made of copper, copper-bearing galvanized steel, tinplate, zinc, or aluminum. The lightest metal shingle is the one made from aluminum, which weighs approximately 40 lb per square. The metal radiates solar heat, resulting in lower temperatures beneath than with most other types of uninsulated roofs. Terne-coated stainless steel is unusually resistant to weathering and corrosion.

**Roofing cements** and **coatings** are usually made from asphalt, asbestos fiber, and solvents to make the cement workable. The cements are used for flashings and repairs and contain slow-drying oils so that they will remain plastic on long exposure. Roof coatings are used to renew old asphalt roofings. Asphalt-base aluminum roof coatings are used to renew old asphalt roofs, and to prolong the service life of smooth-surfaced roofs, new or old.

**Tile**  Hard-burned clay tiles with overlapping or interlocking edges cost about the same as slate. They should have a durable glaze and be well made. Unvitrified tiles with slip glaze are satisfactory in warm climates, but only vitrified tiles should be used in the colder regions. Tile roofs weigh from 750 to 1,200 lb per square. Properly made, tile does not deteriorate, is a poor conductor of heat and cold, and is not so brittle as slate.

**Elastomers**

The trend toward irregular roof surfaces—folded plates, hyperbolic paraboloids, domes, barrel shells—has brought the increased use of plastics or synthetic rubber elastomers (applied as fluid or sheets) as roofing membranes. Such membranes offer light weight, adaptability to any roof slope, good heat reflectivity, and high elasticity at moderate temperatures. Negatively, elastomeric membranes have a more limited range of satisfactory substrate materials than conventional ones. The table below presents several such membranes, and the method of use.

## RUBBER AND RUBBERLIKE MATERIALS
### by C. C. Davis
*Boston Woven Hose and Rubber Co.*

REFERENCES: Dawson and Porritt, "Rubber: Physical and Chemical Properties," Research Association of British Rubber Manufacturers. Davis and Blake, "The Chemistry and Technology of Rubber," Reinhold. ASTM Standards on Rubber Products. "Rubber Red Book. Directory of the Rubber Industry," *The Rubber Age.* Flint, "The Chemistry and Technology of Rubber," Van Nostrand. "The Vanderbilt Handbook," R. T. Vanderbilt Co. Whitby, Davis, and Dunbrook, "Synthetic Rubber," Wiley. "Rubber Bibliography," Rubber Division,

American Chemical Society. Noble, "Latex in Industry," *The Rubber Age.* "Annual Report on the Progress of Rubber Technology," Institution of the Rubber Industry. Morton (ed.), "Rubber Technology," Van Nostrand Reinhold, 1973.

To avoid confusion by the use of the word **rubber** for a variety of natural and synthetic products, the term **elastomer** has come into use, particularly in scientific and technical literature, as a name for both natural and synthetic materials which are elastic or resilient and in general resemble natural rubber in feeling and appearance.

The utility of rubber and synthetic elastomers is increased by compounding. In the raw state, elastomers are soft and sticky when hot, and hard or brittle when cold. **Vulcanization** extends the temperature range within which they are flexible and elastic. In addition to vulcanizing agents, ingredients are added to make elastomers stronger, tougher, and harder, to make them age better, to color them, and in general to modify them to meet the needs of service conditions. Few rubber products today are made from rubber or other elastomers alone.

The elastomers of greatest commercial and technical importance today are natural rubber, GR-S, Neoprene, nitrile rubbers, and butyl.

**Natural rubber** of the best quality is prepared by coagulating

**Elastomeric Membranes**

|  | Material | Method of application | Number of coats or sheets |
|---|---|---|---|
| Fluid-applied | Neoprene-Hypalon † | Roller, brush, or spray | 2 + 2 |
|  | Silicone | Roller, brush, or spray | 2 |
|  | Polyurethane foam, Hypalon † coating | Spray | ? |
|  | Clay-type asphalt emulsion reinforced with chopped glass fibers* | Spray | 1 |
| Sheet-applied | Chlorinated polyethylene on foam | Adhesive | 1 |
|  | Hypalon† on asbestos felt | Adhesive | 1 |
|  | Neoprene-Hypalon† | Adhesive | 1 + surface paint |
|  | Tedlar‡ on asbestos felt | Adhesive | 1 |
|  | Butyl rubber | Adhesive | 1 |
| Traffic decks | Silicone plus sand | Trowel | 1 + surface coat |
|  | Neoprene with aggregate§ | Trowel | 1 + surface coat |

From C. W. Griffin, Jr., "Manual of Built-up Roof Systems," McGraw-Hill, 1970, with permission of the publisher.
*Frequently used with coated base sheet.
†Registered trademark of E. I. du Pont de Nemours & Co. for chlorosulfonated polyethylene.
‡Registered trademark of E. I. du Pont de Nemours & Co. for polyvinyl fluoride.
§Aggregate may be flint, sand, or crushed walnut shells.

the latex of the *Hevea brasiliensis* tree, cultivated chiefly in the Far East. This represents nearly all of the natural rubber on the market today.

Unloaded vulcanized rubber will stretch to approximately ten times its length and at this point will bear a **load** of 13.8 × $10^6$ N/m² (10 tons/in²). It can be compressed to one-third its thickness thousands of times without injury. When most types of vulcanized rubber are stretched, their resistance increases in greater proportion than the extension. Even when stretched almost to the point of rupture, they recover very nearly their original dimensions on being released, and then gradually recover a part of the residual distortion.

Freshly cut or torn raw rubber possesses the power of **self-adhesion** which is practically absent in vulcanized rubber. Cold water preserves rubber, but if exposed to the air, particularly to the sun, rubber goods tend to become hard and brittle. Dry heat up to 49°C (120°F) has little deteriorating effect; at temperatures of 181 to 204°C (360 to 400°F) rubber begins to melt and becomes sticky; at higher temperatures, it becomes entirely carbonized. Unvulcanized rubber is **soluble** in gasoline, naphtha, carbon bisulfide, benzene, petroleum ether, turpentine, and other liquids.

Most rubber is **vulcanized**, i.e., made to combine with sulfur or sulfur-bearing organic compounds or with other chemical cross-linking agents. Vulcanization, if properly carried out, improves mechanical properties, eliminates tackiness, renders the rubber less susceptible to temperature changes, and makes it insoluble in all known solvents. It is impossible to dissolve vulcanized rubber unless it is first decomposed. Other ingredients are added for general effects as follows:

To increase tensile strength and resistance to abrasion: carbon black, precipitated pigments, as well as organic vulcanization accelerators.

To cheapen and stiffen: whiting, barytes, talc, silica, silicates, clays, fibrous materials.

To soften (for purposes of processing or for final properties): bituminous substances, coal tar and its products, vegetable and mineral oils, paraffin, petrolatum, petroleum oils, asphalt.

Vulcanization accessories, dispersion and wetting mediums, etc.: magnesium oxide, zinc oxide, litharge, lime, stearic and other organic acids, degras, pine tar.

Protective agents (natural aging, sunlight, heat, flexing): condensation amines, waxes.

Coloring pigments: iron oxides, especially the red grades, lithopone, titanium oxide, chromium oxide, ultramarine blue, carbon and lampblacks, and organic pigments of various shades.

Specifications should state suitable physical tests. Tensile strength and extensibility tests are of importance and differ widely with different compounds.

**GR-S** is an outgrowth and improvement of German Buna S. The quantity now produced far exceeds all other synthetic elastomers. It is made from butadiene and styrene, which are produced from petroleum. These two materials are copolymerized directly to GR-S, which is known as a butadiene-styrene copolymer. GR-S has recently been improved, and now gives excellent results in tires.

**Neoprene** is made from acetylene, which is converted to vinylacetylene, which in turn combines with hydrogen chloride to form chloroprene. The latter is then polymerized to Neoprene.

**Nitrile rubbers,** an outgrowth of German Buna N or Per-bunan, are made by a process similar to that for GR-S, except that acrylonitrile is used instead of styrene. This type of elastomer is a butadiene-acrylonitrile copolymer.

**Butyl,** one of the most important of the synthetic elastomers, is made from petroleum raw materials, the final process being the copolymerization of isobutylene with a very small proportion of butadiene or isoprene.

**Polysulfide rubbers** having unique resistance to oxidation and to softening by solvents are commercially available and are sold under the trademark "Thiokol."

**Ethylene-propylene** rubbers are notable in their oxidation resistance. **Polyurethane** elastomers can have a tensile strength up to twice that of conventional rubber, and solid articles as well as foamed shapes can be cast into the desired form using prepolymer shapes as starting materials. **Silicone** rubbers have the advantages of a wide range of service temperatures and room-temperature curing. **Fluorocarbon** elastomers are available for high-temperature service. Polyester elastomers have excellent impact and abrasion resistance.

No one of these elastomers is satisfactory for all kinds of service conditions, but rubber products can be made to meet a large variety of service conditions.

The following examples show some of the important properties required of rubber products and some typical services where these properties are of major importance:

*Resistance to abrasive wear:* auto-tire treads, conveyor-belt covers, soles and heels, cables, hose covers, V belts.*

*Resistance to tearing:* auto inner tubes, tire treads, footwear, hot-water bags, hose covers, belt covers, V belts.*

*Resistance to flexing:* auto tires, transmission belts, V belts,* mountings, footwear.

*Resistance to high temperatures:* auto tires, auto inner tubes, belts conveying hot materials, steam hose, steam packing.

*Resistance to cold:* airplane parts, automotive parts, auto tires, refrigeration hose.

*Minimum heat buildup:* auto tires, transmission belts, V belts,* mountings.

*High resilience:* auto inner tubes, sponge rubber, mountings, elastic bands, thread, sandblast hose, jar rings, V belts.*

*High rigidity:* packing, soles and heels, valve cups, suction hose, battery boxes.

*Long life:* fire hose, transmission belts, tubing, V belts.*

*Electrical resistivity:* electricians' tape, switchboard mats, electricians' gloves.

*Electrical conductivity:* hospital flooring, nonstatic hose, matting.

*Impermeability to gases:* balloons, life rafts, gasoline hose, special diaphragms.

*Resistance to ozone:* ignition distributor gaskets, ignition cables, windshield wipers.

*Resistance to sunlight:* wearing apparel, hose covers, bathing caps.

*Resistance to chemicals:* tank linings, hose for chemicals.

*Resistance to oils:* gasoline hose, oil-suction hose, paint hose, creamery hose, packinghouse hose, special belts, tank linings, special footwear.

---

*For information concerning types, sizes, strengths, etc., of V belts, see Sec. 8.

*Conversion factors: 1 in = 0.0254 m; $t_C$ = 5/9 $(t_F - 32)$; 1 psi (lb/in²) = 689.48 N/m²; 1 lb$_f$ = 4.448 N; 1 Btu/(ft²)(h)(°F)(in) = 225 W/(m²)(°C)(m).

*Stickiness:* cements, electricians' tapes, adhesive tapes, pressure-sensitive tapes.

*Low specific gravity:* airplane parts, forestry hose, balloons.

*No odor or taste:* milk tubing, brewery and wine hose, nipples, jar rings.

*Special colors:* ponchos, life rafts, welding hose.

The following table gives a comparison of some important characteristics of the most important elastomers when vulcanized. The lower part of the table indicates, for a few representative rubber products, preferences in the use of different elastomers for different service conditions without consideration of cost.

**Gutta-percha** and **balata,** also natural products, are akin to rubber chemically but more leathery and thermoplastic, and are used for some special purposes, principally for submarine cables, golf balls, and various minor products.

### Rubber Derivatives

Rubber derivatives are chemical compounds and modifications of rubber, some of which have become of commercial importance.

### Comparative Properties of Elastomers

| | Natural rubber | GR-S | Neoprene | Nitrile rubbers | Butyl | Thiokol |
|---|---|---|---|---|---|---|
| Tensile properties............ | Excellent | Good | Very good | Good | Good | Fair |
| Resistance to abrasive wear.... | Excellent | Good | Very good | Good | Good | Poor |
| Resistance to tearing......... | Very good | Poor | Good | Fair | Very good | Poor |
| Resilience................... | Excellent | Good | Good | Fair | Poor | Poor |
| Resistance to heat........... | Good | Fair | Good | Excellent | Good | Poor |
| Resistance to cold............ | Excellent | Good | Good | Good | Excellent | Poor |
| Resistance to flexing......... | Excellent | Good | Very good | Good | Excellent | Poor |
| Aging properties............. | Excellent | Excellent | Good | Good | Excellent | Good |
| Cold flow (creep)............ | Very low | Low | Low | Very low | Fairly low | High |
| Resistance to sunlight........ | Fair | Fair | Excellent | Good | Excellent | Excellent |
| Resistance to oils and solvents.. | Poor | Poor | Good | Excellent | Fair | Excellent |
| Permeability to gases......... | Fairly low | Fairly low | Low | Fairly low | Very low | Very low |
| Electrical insulation.......... | Fair | Excellent | Fair | Poor | Excellent | Good |
| Flame resistance............. | Poor | Poor | Good | Poor | Poor | Poor |
| Auto-tire tread.............. | Preferred | Alternate | | | | |
| Inner tube.................. | Alternate | ........ | | ........ | Preferred | |
| Conveyor-belt cover.......... | Preferred | ........ | Alternate | | | |
| Tire sidewall................ | Alternate | Preferred | | | | |
| Transmission belting......... | Preferred | ........ | Alternate | | | |
| Druggist sundries............ | Preferred | | | | | |
| Gasoline and oil hose.......... | ........ | ........ | ........ | Preferred | | |
| Lacquer and paint hose........ | ........ | ........ | ........ | ........ | ........ | Preferred |
| Oil-resistant footwear......... | ........ | ........ | Preferred | | | |
| Balloons.................... | Alternate | ........ | Preferred | | | |
| Jar rings................... | Alternate | Preferred | | | | |
| Wire and cable insulation...... | Alternate | Preferred | | | | |

Specifications for rubber goods may cover the chemical, physical, and mechanical properties, such as elongation, tensile strength, permanent set, and oven tests, minimum rubber content, exclusion of reclaimed rubber, maximum free and combined sulfur contents, maximum acetone and chloroform extracts, ash content, and many construction requirements. It is preferable, however, to specify properties such as resilience, hysteresis, static or dynamic shear and compression modulus, flex fatigue and cracking, creep, electrical properties, stiffening, heat generation, compression set, resistance to oils and chemicals, permeability, brittle point, etc., in the temperature range prevailing in service, and to leave the selection of the elastomer to a competent manufacturer.

**Latex,** imported in stable form from the Far East, is used for various rubber products. In the manufacture of such products, the latex must be compounded for vulcanizing and otherwise modifying properties of the rubber itself. Important products made directly from compounded latex include surgeons' and household gloves, thread, bathing caps, rubberized textiles, balloons, and sponge. A recent important use of latex is for "foam sponge," which may be several inches thick and used for cushions, mattresses, etc.

**Chlorinated rubber,** produced by the action of chlorine on rubber in solution, is nonrubbery, incombustible, and extremely resistant to many chemicals. As commercial **Parlon,** it finds use in corrosion-resistant paints and varnishes, in inks, and in adhesives.

**Rubber hydrochloride,** produced by the action of hydrogen chloride on rubber in solution, is a strong, extensible, tear-resistant, moisture-resistant, oil-resistant material, marketed as **Pliofilm** in the form of tough transparent films for wrappers, packaging material, etc.

**Cyclized rubber** is formed by the action of certain agents, e.g., sulfonic acids and chlorostannic acid, on rubber, and is a thermoplastic, nonrubbery, tough or hard product. One form, **Thermoprene,** is used in the **Vulcalock process** for adhering rubber to metal, wood, and concrete, and in chemical-resistant paints. **Pliolite,** which has high resistance to many chemicals and has low permeability, is used in special paints, paper, and fabric coatings. **Marbon-B** has exceptional electrical properties and is valuable for insulation. **Hypalon** (chlorosulphonated polyethylene) is highly resistant to many important chemicals, notably ozone and concentrated sulfuric acid, for which other rubbers are unsuitable.

## SOLVENTS
### by C. G. Harford
*Arthur D. Little, Inc.*

REFERENCES: Sax, "Dangerous Properties of Industrial Materials," Reinhold, Perry, "Chemical Engineers' Handbook," McGraw-Hill. Doolittle, "The Technology of Solvents and Plasticizers," Wiley, 1954. Riddick and Bunger, "Organic Solvents," Wiley, 1970. Mellan, "Industrial Solvents Handbook," Noges Data Corp., New Jersey, 1970.

The use of solvents has become widespread throughout industry. The health of personnel and the fire hazards involved should always be considered. Generally, solvents are **organic liquids** which vary greatly in solvent power, flammability, volatility, and toxicity.

### Solvents for Polymeric Materials

A wide choice of solvents and solvent combinations is available for use with organic polymers in the manufacture of polymer-coated products and unsupported films. For a given polymer, the choice of solvent system is often critical in terms of solvent power, cost, safety, and evaporation rate. In such instances, the supplier of the base polymer should be consulted.

### Alcohols

**Methyl alcohol (methanol),** is now made synthetically. It is completely miscible with water and most organic liquids. It evaporates rapidly and is a good solvent for dyes, gums, shellac, nitrocellulose, and some vegetable waxes. It is widely used as an antifreeze for automobiles, in shellac solution, spirit varnishes, stain and paint removers. It is toxic; imbibition or prolonged breathing of the vapors can cause blindness. It should be used only in well-ventilated spaces. Flash point 11°C (52°F).

**Ethyl alcohol (ethanol)** is produced by fermentation and synthetically. For industrial use it is generally denatured and sold under various trade names. There are numerous formulations of specially denatured alcohols which can legally be used for specified purposes.

This compound is miscible with water and most organic solvents. It evaporates rapidly and, because of its solvent power, low cost, and agreeable odor, finds a wide range of uses. The common uses are antifreeze (see Freezing Preventives, above), shellac solvent, in mixed solvents, spirit varnishes, and solvent for dyes, oils, and animal greases.

Denatured alcohols are toxic when taken orally. The effect of ethyl alcohol vapors when breathed in high concentration can produce the physiological effects of alcoholic liquors. It should be used in well-ventilated areas. Flash point 15.3 to 16.7°C (60 to 62°F).

**Isopropyl alcohol (isopropanol)** is derived mainly from petroleum gases. It is not as good a solvent as denatured alcohol, although it can be used as a substitute for ethyl alcohol in some instances. It is used as a rubbing alcohol and in lacquer thinners. Flash point is 11°C (52°F).

**Butyl alcohol (normal butanol)** is used extensively in lacquer and synthetic resin compositions and also in penetrating oils, metal cleaners, insect sprays, and paints for application over asphalt. It is an excellent blending agent for otherwise incompatible materials. Flash point is 29°C (84°F).

### Esters

**Ethyl acetate** dissolves a large variety of materials, such as nitrocellulose, oils, fats, gums, and resins. It is used extensively in nitrocellulose lacquers, candy coatings, food flavorings, and in chemical synthesis. On account of its high rate of evaporation, it finds a use in paper, leather, and cloth coatings and cements. Flash point −4.5°C (24°F).

**Butyl acetate** is the acetic-acid ester of normal butanol. This ester is used extensively for dissolving various cellulose esters, mineral and vegetable oils, and many synthetic resins, such as the vinylites, polystyrene, methyl methacrylate, and chlorinated rubber. It is also a good solvent for natural resins. It is the most important solvent used in lacquer manufacture. It is useful in the preparation of perfumes and synthetic flavors. Flash point 22.3°C (72°F).

**Amyl acetate,** sometimes known as banana oil, is used mainly in lacquers. Its properties are somewhat like those of butyl acetate. Flash point 21.1°C (70°F).

### Hydrocarbons

The **aromatic hydrocarbons** are derived from coal-tar distillates, the most common of which are benzene, toluene, xylene (also known as benzol, toluol, and xylol), and high-flash solvent naphtha.

**Benzene** is an excellent solvent for fats, vegetable and mineral oils, rubber, chlorinated rubber; it is also used as a solvent in paints, lacquers, inks, paint removers, asphalt, coal tar. This substance should be used with caution. Flash point 12°F.

**Toluene** General uses are about the same as benzol, in paints, lacquers, rubber solutions, and solvent extractions. Flash point 40°F.

**Xylene** is used in the manufacture of dyestuffs and other synthetic chemicals and as a solvent for paints, rubber, lacquer, and varnishes. Flash point 63°F.

**Hi-flash naphtha** or **coal-tar naphtha** is used mainly as a diluent in lacquers, synthetic enamels, paints, and asphaltic coatings. Flash point 100°F.

### Petroleum

These hydrocarbons, derived from petroleum, are, next to water, the cheapest and most universally used solvents.

**V. M. & P. naphtha,** sometimes called **benzine,** is used by paint and varnish makers as a solvent or diluent. It finds wide use as a solvent for fats, oils, greases and is used as a diluent in paints and lacquers. It is also used as an extractive agent as well as in some specialized fields of cleansing (fat removal). It is used to compound rubber cement, inks, varnish removers. It is relatively nontoxic. Flash point 20 to 45°F.

**Mineral spirits,** also called **Stoddard solvent,** is extensively used in dry cleaning because of its high flash point and clean evaporation. It is also widely used in turpentine substitutes for oil paints. Flash point 100 to 110°F.

**Kerosine,** a No. 1 fuel oil, is a good solvent for petroleum greases, oils, and fats. Flash point 100 to 165°F.

### Chlorinated Solvents

**Carbon tetrachloride** is a colorless nonflammable liquid with a chloroformlike odor. It is an excellent solvent for fats, oils, greases, waxes, and resins and is used in dry cleaning and in degreasing of wool, cotton waste, and glue. It is also used in rubber cements and adhesives, as an extracting agent, and in fire extinguishers. It should be used only in well-ventilated spaces; prolonged inhalation is extremely dangerous.

**Trichlorethylene** is somewhat similar to carbon tetrachloride but is slower in evaporation rate. It is the solvent most commonly used for vapor degreasing of metal parts. It is also used in the manufacture of dyestuffs and other chemicals. It is an excellent solvent and is used in some types of paints, varnishes, and leather coatings.

**Tetrachlorethylene,** also called **perchlorethylene,** is nonflammable and has uses similar to those of carbon tetrachloride and trichlorethylene. Its chief use is in dry cleaning; it is also used somewhat in metal degreasing.

### Ketones

**Acetone** is an exceptionally active solvent for a wide variety of organic materials, gases, liquids, and solids. It is completely miscible with water and also with most of the organic liquids. It can also be used as a blending agent for otherwise immiscible liquids. It is used in the manufacture of pharmaceuticals, dyestuffs, lubricating compounds, and pyroxylin compositions. It is a good solvent for cellulose acetate, ethyl cellulose, vinyl and methacrylate resins, chlorinated rubber, asphalt, camphor, and various esters of cellulose, including smokeless powder, cordite, etc. Some of its more common uses are in paint and varnish removers, the storing of acetylene, and the dewaxing of lubricating oils. It is the basic material for the manufacture of iodoform and chloroform. It is also used as a denature for ethyl alcohol. Flash point 0°F.

**Methylethylketone (MEK)** can be used in many cases where acetone is used as a general solvent, e.g., in the formulation of pyroxylin cements and in compositions containing the various esters of cellulose. Flash point 30°F.

**Glycol ethers** are useful as solvents for cellulose esters, lacquers, varnishes, enamels, wood stains, dyestuffs, and pharmaceuticals.

## THERMAL INSULATIONS
### by Peter E. Glaser
*Arthur D. Little, Inc.*

REFERENCES: "Guide and Data Book," ASHRAE. Glaser, "Aerodynamically Heated Structures," Prentice-Hall. Timmerhaus, "Advances in Cryogenic Engineering," Plenum. Wilkes, "Heat Insulation," Wiley. Wilson, "Industrial Thermal Insulation," McGraw-Hill, 1959. Technical Documentary Report ML-TDR-64-5, "Thermophysical Properties of Thermal Insulating Materials," Air Force Materials Laboratory, Research and Technology Division, Air Force Systems Command; Prepared by Midwest Research Institute, Kansas City, Mo., 1964. Probert and Hub (eds.), "Thermal Insulation," Elsevier, 1968. Malloy, "Thermal Insulation," Van Nostrand Reinhold, 1969.

Thermal insulations consisting of a single material, a mixture of materials, or a composite structure are chosen to reduce heat flow. **Insulating effectiveness** is judged on the basis of thermal conductivity and depends on the physical and chemical structure of the material. The heat transferred through an insulation results from **solid conduction, gas conduction,** and **radiation.** Solid conduction is reduced by small-sized particles or fibers in loose-fill insulation and by thin cell walls in foams. Gas conduction is reduced by providing large numbers of small pores (either interconnected or closed off from each other) of the order of the mean free paths of the gas molecules,

by substituting gases of low thermal conductivity, or by evacuating the pores to a low pressure. Radiation is reduced by adding materials which absorb, reflect, or scatter radiant energy. (See also Sec. 4, Transmission of Heat by Conduction and Convection.)

The **performance** of insulations depends on the temperature of the bounding surfaces and their emittance, the insulation density, the type and pressure of gas within the pores, the moisture content, the thermal shock resistance, and the action of mechanical loads and vibrations. In **transient** applications, the heat capacity of the insulation (affecting the rate of heating or cooling) has to be considered.

The **form** of the **insulations** can be loose fill (bubbles, fibers, flakes, granules, powders), flexible (batting, blanket, felt, multilayer sheets and tubular), rigid (block, board, brick, custom-molded, sheet and pipe covering), cemented, foamed-in-place, or sprayed.

The choice of **insulations** is dictated by the service-temperature range as well as by design criteria and economic considerations.

### Cryogenic Temperatures [below −102°C (−150°F)]
(See also Sec. 19.)

At the low temperatures experienced with cryogenic liquids, **evacuated multilayer insulations,** consisting of a series of radiation shields of high reflectivity separated by low-conductivity spacers, are effective materials. Radiation-shield materials are aluminum foils or aluminized polyester films used in combination with spacers of thin polyester fiber or glass-fiber papers; radiation shields of crinkled, aluminized polyester film without spacers are also used. To be effective, multilayer insulations require a vacuum of at least $10^{-1}$ mmHg. **Evacuated powder** and **fiber insulations** can be effective at gas pressures up to 0.1 mmHg over a wide temperature range. **Powders** include colloidal silica ($8 \times 10^{-7}$ in* particle diam), silica aerogel ($1 \times 10^{-6}$ in*), synthetic calcium silicate (0.001 in), and perlite (an expanded form of glassy volcanic lava particles, 0.05 in* diam). Powder insulations can be **opacified** with copper, aluminum, or carbon particles to reduce radiant-energy transmission. **Fiber insulations** consist of mats of fibers arranged in ordered parallel layers either without binders or with a minimum of binders. Glass fibers ($10^{-5}$ in* diam) are used most frequently. For large process installations and cold boxes, unevacuated perlite powder or mineral fibers are useful.

**Foamed organic plastics,** using either fluorinated hydrocarbons or other gases as expanding agents, are partially evacuated when gases within the closed cells condense when exposed to low temperatures. Polystyrene and polyurethane foams are used frequently. **Gastight barriers** are required to prevent a rise in thermal conductivity with aging due to diffusion of air and moisture into the foam insulation. Gas barriers are made of aluminum foil, polyester film, and polyester film laminated with aluminum foil.

### Refrigeration, Heating, and Air Conditioning, up to 120°C (250°F)

At temperatures associated with commercial refrigeration practice and building insulation, **vapor barriers** resistant to the diffusion of water vapor should be installed on the warm side

*Conversion Factors: 1 in = 0.0254 m; $t_C = 5/9 (t_F - 32)$.

of most types of insulations if the temperature within the insulation is expected to fall below the dew point (this condition would lead to condensation of water vapor within the insulation and result in a substantial decrease in insulating effectiveness). Vapor barriers include oil- or tar-impregnated paper, paper laminated with aluminum foil, and polyester films. Insulations which have an impervious outer skin or structure require a **vaportight sealant** at exposed joints to prevent collection of moisture or ice underneath the insulation. (See also Secs. 12 and 19.)

**Loose-fill insulations** include powders and granules such as perlite, vermiculite (an expanded form of mica), silica aerogel, calcium silicate, expanded organic plastic beads, granulated cork (bark of the cork tree), granulated charcoal, redwood wool (fiberized bark of the redwood tree), and synthetic fibers. The most widely used fibers are those made of glass, rock, or slag produced by centrifugal attenuation or attenuation by hot gases.

**Flexible** or **blanket insulations** include those made from organically bonded glass fibers; rock wool, slag wool, macerated paper, or hair felt placed between or bonded to paper laminate (including vapor-barrier material) or burlap; foamed organic plastics in sheet and pad form (polyurethane, polyethylene); and elastomeric closed-cell foam in sheet, pad, or tube form.

**Rigid** or **board insulations** (obtainable in a wide range of densities and structural properties) include foamed organic plastics such as polystyrene (extruded or molded beads); polyurethane, polyvinyl chloride, phenolics, and ureas; balsa wood, foamed glass, and corkboard (compressed mass of baked-cork particles).

### Moderate Temperatures [up to 650°C (1200°F)]

The widest use of a large variety of insulations is in the temperature range associated with power plants and industrial equipment. Inorganic insulations are available for this temperature range, with several capable of operating over a wider temperature range.

**Loose-fill insulations** include diatomaceous silica (fossilized skeletons of microscopic organisms), perlite, vermiculite, and fibers of glass, rock, or slag. Board and blanket insulations of various shapes and degrees of flexibility and density include glass and mineral fibers, asbestos paper, and millboard [asbestos is a heat-resistant fibrous mineral obtained from Canadian (chrysotile) or South African (amosite) deposits]; asbestos fibers bonded with sodium silicate, 85 percent basic magnesium carbonate, expanded perlite bonded with calcium silicate, calcium silicate reinforced with asbestos fibers, expanded perlite bonded with cellulose fiber and asphalt, organic bonded mineral fibers, and cellulose fiberboard.

**Sprayed insulation** (macerated paper or fibers and adhesive or frothed plastic foam), **insulating concrete** (concrete mixed with expanded perlite or vermiculite), and **foamed-in-place plastic insulation** (prepared by mixing polyurethane components, pouring the liquid mix into the void, and relying on action of generated gas or vaporization of a low-boiling fluorocarbon to foam the liquid and fill the space to be insulated) are useful in special applications.

**Reflective insulations** form air spaces bounded by surfaces of high reflectivity to reduce the flow of radiant energy. Surfaces need not be mirror-bright to reflect long-wavelength radiation emitted by objects below 500°F. Materials for reflective insulation include aluminum foil cemented to one or both sides of kraft paper and aluminum particles applied to the paper with adhesive. Where several reflective surfaces are used, they have to be separated during the installation to form air spaces.

### High Temperatures [above 820°C (1500°F)]

At the high temperatures associated with furnaces and process applications, physical and chemical stability of the insulation in an oxidizing, reducing, or neutral atmosphere or vacuum may be required. **Loose-fill insulations** include glass fibers 538°C (1000°F) useful temperature limit, asbestos fibers 650°C (1200°F), fibrous potassium titanate 1040°C (1900°F), alumina-silica fibers 1260°C (2300°F), microquartz fibers 1370°C (2500°F), opacified colloidal alumina 1310°C (2400°F in vacuum), zirconia fibers 1640°C (3000°F), alumina bubbles 1810°C (3300°F), zirconia bubbles 2360°C (4300°F), and carbon and graphite fibers 2480°C (4500°F) in vacuum or an inert atmosphere.

**Rigid insulations** include reinforced and bonded colloidal silica 1090°C (2000°F), bonded diatomaceous earth brick 1370°C (2500°F), insulating firebrick (see Refractories, below), and anisotropic, pyrolitic graphite (100:1 ratio of thermal conductivity parallel to surface and across thickness).

**Reflective insulations,** forming either an air space or an evacuated chamber between spaced surfaces, include stainless steel, molybdenum, tantalum, or tungsten foils and sheets.

**Insulating cements** are based on asbestos, mineral, or refractory fibers bonded with mixtures of clay or sodium silicate. Lightweight, castable insulating materials consisting of mineral or refractory fibers in a calcium aluminate cement are useful up to 2500°F.

**Ablators** are composite materials capable of withstanding high temperatures and high gas velocities for limited periods with minimum erosion by subliming and charring at controlled rates. Materials include asbestos, carbon, graphite, silica, nylon or glass fibers in a high-temperature resin matrix (epoxy or phenolic resin), and cork compositions.

## SILICONES

Silicones are organosilicon oxide polymers characterized by remarkable temperature stability, chemical inertness, waterproofness, and excellent dielectric properties. The investigations of Prof. Kipping in England for over forty years established a basis for the recent developments of the numerous industrially important products now being made by the Dow-Corning Corp. of Midland, Mich., and by the General Electric Co. Among these products are the following:

**Water repellents** in the form of extremely thin films which can be formed on paper, cloth, ceramics, glass, plastics, leather, powders, or other surfaces. These have great value for the protection of delicate electrical equipment in moist atmospheres.

**Oils** with high flash points [above 315°C (600°F)], low pour points, −84.3°C (−120°F), and with a constancy of viscosity notably superior to petroleum products in the range from 260 to −73.3°C (500 to −100°F). These oil products are practically incombustible. They are in use for hydraulic servomotor fluids, damping fluids, dielectric liquids for transformers, heat-transfer mediums, etc., and are of special value in aircraft because of the rapid and extreme temperature variations to which aircraft are exposed. At present they are not available as lubricants except under light loads.

**Greases and compounds** for plug cocks, spark plugs, and ball bearings which must operate at extreme temperatures and speeds.

**Varnishes and resins** for use in electrical insulation where temperatures are high. Layers of glass cloth impregnated with or bounded by silicone resins withstand prolonged exposure to temperatures up to 260°C (500°F). They form paint finishes of great resistance to chemical agents and to moisture. They have many other industrial uses.

**Silicone rubbers** which retain their resiliency for −45 to 270°C (−50 to 520°F) but with much lower strength than some of the synthetic rubbers. They are used for shaft seals, oven gaskets, refrigerator gaskets, and vacuum gaskets.

# REFRACTORIES
## by F. H. Norton

REFERENCES: Buell, "The Open-Hearth Furnace," 3 vols., Penton. *Bull.* R-2-E, The Babcock & Wilcox Co. "Refractories," General Refractories Co. Chesters, "Steel Plant Refractories," United Steel Companies, Ltd. "Modern Refractory Practice," Harbison Walker Refractories Co. Green and Stewart, "Ceramics: A Symposium," British Ceramic Soc. ASTM Standards on Refractory Materials (Committee C-8). Trinks, "Industrial Furnaces," vol. I, Wiley. Campbell, "High Temperature Technology," Wiley. Budnikov, "The Technology of Ceramics and Refractories," The MIT Press, 1964. Campbell and Sherwood (eds.), "High-Temperature Materials and Technology," Wiley, 1967. Norton, "Refractories," 4th ed., McGraw-Hill, 1969. Clauss, "Engineer's Guide to High-Temperature Materials," Addison-Wesley, 1969. Shaw, "Refractories and Their Uses," Wiley, 1972. Chesters, "Refractories; Production and Properties," The Iron and Steel Institute, London, 1973.

### Types of Refractories

**Fire-Clay Refractories** Fire-clay brick are made from fire clays, which comprise all refractory clays that are not white burning. Fire clays can be divided into plastic clays and hard flint clays; they may also be classified as to alumina content.

Firebricks are usually made of a blended mixture of flint clays and plastic clays which is then formed, after mixing with water, to the required shape. Some or all of the flint clay may be replaced by highly burned or calcined clay, called **grog**. A large proportion of the modern bricks is molded by the dry-press or power-press process where the forming is carried out under high pressure and with a low water content. Some extruded and hand-molded brick are still made.

The dried bricks are burned in either periodic or tunnel kilns at temperatures varying between 1200 and 1480°C (2200 and 2700°F). Tunnel kilns give continuous production and a uniform temperature of burning.

Fire-clay bricks are used for boiler settings, kilns, malleable-iron furnaces, incinerators, and many portions of steel and nonferrous metal furnaces. They are resistant to spalling and stand up well under many slag conditions, but are not generally suitable for use with high-lime slags, fluid-coal-ash slags, or under severe load conditions.

**High-alumina brick** are manufactured from raw materials rich in alumina, such as diaspore and bauxite. They are graded into groups with 50, 60, 70, 80, and 90 percent alumina content. When well fired, these brick contain a large amount of mullite and less of the glassy phase than is present in the firebricks. Corundum is also present in many of these bricks. High-alumina brick are generally used for unusually severe temperature or load conditions. They are employed extensively in limekilns and rotary cement kilns, the ports and regenerators of glass tanks, and for slag resistance in some metallurgical furnaces; their price is higher than that for firebrick.

**Silica brick** are manufactured from crushed ganister rock containing about 97 to 98 percent silica. A bond consisting of 2 percent lime is used, and the bricks are fired in periodic kilns at temperatures of 1480 to 1540°C (2700 to 2800°F) for several days until a stable volume is obtained. They are especially valuable where a good strength is required at high temperatures. Recently, superduty silica brick are finding some use in the steel industry. They have a lowered alumina content, and often a lowered porosity.

Silica brick are used extensively in coke ovens, the roofs and walls of open-hearth furnaces, in the roofs and sidewalls of glass tanks, and as linings of acid electric steel furnaces. Although silica brick is readily **spalled** (cracked by a temperature change) below red heat, it is very stable if the temperature is kept above this range, and for this reason stands up well in regenerative furnaces. Any structure of silica brick should be heated up slowly to the working temperature; a large structure often requires 2 weeks or more to bring up.

**Magnesite brick** are made from crushed magnesium oxide which is produced by calcining raw magnesite rock to high temperatures. A rock containing several percent of iron oxide is preferable, as this permits the rock to be fired at a lower temperature than if pure materials were used. Magnesite brick are generally fired at a comparatively high temperature in periodic or tunnel kilns, though large tonnages of unburned brick are now produced. The latter are made with special grain sizing and a bond such as an oxychloride. A large proportion of magnesite brick made in this country uses raw material extracted from seawater.

Magnesite brick are basic and are used whenever it is necessary to resist high-lime slags such as in the basic open-hearth furnace. They also find use in furnaces for the lead- and copper-refining industry. The highly pressed unburned brick find extensive use as linings for cement kilns. Magnesite brick are not so resistant to spalling as fire-clay brick.

**Dolomite** This rock contains a mixture of $Mg(OH)_2$ and $Ca(OH)_2$, is calcined, and is used in granulated form for furnace bottoms.

**Chrome brick** are manufactured in much the same way as magnesite brick but are made from natural chromite ore. Commercial ores always contain magnesia and alumina. Unburned hydraulically pressed chrome brick are also made.

Chrome bricks are very resistant to all types of slag. They are used as separators between acid and basic refractories, also in soaking pits and floors of forging furnaces. The unburned hydraulically pressed brick now find extensive use in the walls of the open-hearth furnace and are often enclosed in a metal case. Chrome bricks are used in sulfite-recovery furnaces and to some extent in the refining of nonferrous metals. Basic bricks combining various proportions of magnesite and chromite are now made in large quantities and have advantages over either material alone for some purposes.

The **insulating firebrick** is a class of brick which consists of a highly porous fire clay or kaolin. They are light in weight (about ½ to ⅙ that of fireclay), low in thermal conductivity, and yet sufficiently resistant to temperature to be used successfully on the hot side of the furnace wall, thus permitting thin walls of low thermal conductivity and low heat content. The low heat content is particularly valuable in saving fuel and time on heating up, allows rapid changes in temperature to be made, and permits rapid cooling. These bricks are made in a variety of ways, such as mixing organic matter with the clay and later burning it out to form pores; or a bubble structure can be incorporated in the clay-water mixture which is later preserved in the fired brick. The insulating firebricks are classified into several groups according to the maximum use limit; the ranges are up to 872, 1090, 1260, 1420, and above 1540°C (1600, 2000, 2300, 2600, and above 2800°F).

Insulating refractories are used mainly in the heat-treating industry for furnaces of the periodic type; the low heat content permits noteworthy fuel savings as compared with firebrick. They are also used extensively in stress-relieving furnaces, chemical-process furnaces, oil stills or heaters, and in the combustion chambers of domestic-oilburner furnaces. They usually have a life equal to the heavy brick that they replace. They are particularly suitable for constructing experimental or laboratory furnaces because they can be cut or machined readily to any shape. However, they are not resistant to fluid slag.

There are a number of types of **special brick**, obtainable from individual manufactories. High-burned kaolin refractories are particularly valuable under conditions of severe temperature and heavy load, or severe spalling conditions, as in the case of high-temperature oil-fired boiler settings, or piers under enameling furnaces. Another brick for the same uses is a high-fired brick of Missouri aluminous clay.

There are a number of bricks on the market made from electrically fused materials, such as fused mullite, fused alumina, and fused zircon. These bricks, although high in cost, are particularly suitable for certain severe conditions, such as bottoms and walls of glass-melting furnaces.

Bricks of **silicon carbon**, either nitride or clay-bonded, have a high thermal conductivity and find use in muffle walls and as a slag-resisting material.

Other types of refractory that find certain limited use are **forsterite**, and **zirconia**. Acid-resisting brick consisting of a dense body like stoneware are used for lining tanks and conduits in the chemical industry. Carbon blocks are used as linings for the crucibles of blast furnaces.

The **chemical composition** of some of the refractories is given in Table 1. The **physical properties** are given in Table 2. Reference should be made to ASTM standards for details of standard tests.

**Standard and Special Shapes**

There are a large number of **standard refractory shapes** carried in stock by most manufacturers. Their catalogs should be consulted in selecting these shapes, but the common ones are shown in Table 3. These shapes have been standardized by the American Refractories Institute and by the Bureau of Simplification of the U.S. Department of Commerce.

**Regenerator tile** sizes, $a \times b \times c$ are: $18 \times 6$ or $9 \times 3$; $18 \times 9$ or $12 \times 4$; $22\frac{1}{2} \times 6$ or $9 \times 3$; $22\frac{1}{2} \times 9$ or $12 \times 4$; $27 \times 9 \times 3$;

### Table 1. Chemical Composition of Typical Refractories[a]

| No. | Refractory type | SiO$_2$ | Al$_2$O$_3$ | Fe$_2$O$_3$ | TiO$_2$ | CaO | MgO | Cr$_2$O$_3$ | SiC | Alkalies | Resistance to ||||
|---|---|---|---|---|---|---|---|---|---|---|---|---|---|---|
| | | | | | | | | | | | Siliceous steel-slag | High-lime steel-slag | Fused mill-scale | Coal-ash slag |
| 1 | Alumina (fused) | 8–10 | 85–90 | 1–1.5 | 1.5–2.2 | .... | .... | .. | .... | 0.8–1.3[c] | E | G | F | G |
| 2 | Chrome | 6 | 23 | 15[b] | | .... | 17 | 38 | .... | .... | G | E | E | G |
| 3 | Chrome (unburned) | 5 | 18 | 12[b] | | ... | 32 | 30 | .... | .... | G | E | E | G |
| 4 | Fire clay (high-heat duty) | 50–57 | 36–42 | 1.5–2.5 | 1.5–2.5 | .... | .... | .. | .... | 1–3.5[c] | F | P | P | F |
| 5 | Fire clay (super-duty) | 52 | 43 | 1 | 2 | ... | .... | .. | .... | 2[c] | F | P | F | F |
| 6 | Forsterite | 34.6 | 0.9 | 7.0 | | 1.3 | 55.4 | | | | | | | |
| 7 | High-alumina | 22–26 | 68–72 | 1–1.5 | 3.5 | ... | .... | .. | .... | 1–1.5[c] | G | F | F | F |
| 8 | Kaolin | 52 | 45.4 | 0.6 | 1.7 | 0.1 | 0.2 | | | | F | P | G[d] | F |
| 9 | Magnesite | 3 | 2 | 6 | | 3 | 86 | | | | P | E | E | E |
| 10 | Magnesite (unburned) | 5 | 7.5 | 8.5 | | 2 | 64 | 10 | | | P | E | E | E |
| 11 | Magnesite (fused) | .... | | | | | | | | | F | E | E | E |
| 12 | Refractory porcelain | 25–70 | 25–60 | | | ... | .. | .. | .... | 1–5 | G | F | F | F |
| 13 | Silica | 96 | 1 | 1 | | 2 | | | | | E | P | F | P |
| 14 | Silicon carbide (clay bonded) | 7–9 | 2–4 | 0.3–1 | 1 | ... | .. | .. | 85–90 | .... | E | G | F | E |
| 15 | Sillimanite (mullite) | 35 | 62 | 0.5 | 1.5 | ... | .. | .. | .... | 0.5[c] | G | F | F | F |
| 16 | Insulating fire-brick (2600 F) | 57.7 | 36.8 | 2.4 | 1.5 | 0.6 | 0.5 | .. | .... | .... | P | P | G[e] | F |

[a]Many of these data have been taken from a table prepared by Trostel, *Chem. Met. Eng.*, Nov. 1938.
[b]As FeO.
[c]Includes lime and magnesia.
[d]Excellent if left above 1200°F.
[e]Oxidizing atmosphere.
E = Excellent. G = Good. F = Fair. P = Poor.

**Table 2. Physical Properties of Typical Refractories**[a]
(Refractory numbers refer to Table 1)

| Refractory No. | Fusion point Deg F | Fusion point Pyrometric cone | Deformation under load, percent at deg F and lb per sq in. | Spalling resistance | Reheat shrinkage after 5 hr, percent at (deg F) | Wt. of straight 9 in. brick, lb |
|---|---|---|---|---|---|---|
| 1 | 3390+ | 39+ | 1 at 2730 and 50 | Good | +0.5 (2910) | 9–10.6 |
| 2 | 3580+ | 41+ | Shears 2740 and 28 | Poor | −0.5 to 1.0 (3000) | 11.0 |
| 3 | 3580+ | 41+ | Shears 2955 and 28 | Fair | −0.5 to 1.0 (3000) | 11.3 |
| 4 | 3060–3170 | 31–33 | 2.5–10 at 2460 and 25 | Good | ±0 to 1.5 (2550) | 7.5 |
| 5 | 3170–3200 | 33–34 | 2–4 at 2640 and 25 | Excellent | ±0 to 1.5 (2910) | 8.5 |
| 6 | 3430 | 40 | 10 at 2950 | Fair | ............. | 9.0 |
| 7 | 3290 | 36 | 1–4 at 2640 and 25 | Excellent | −2 to 4 (2910) | 7.5 |
| 8 | 3200 | 34 | 0.5 at 2640 and 25 | Excellent | −0.7 to 1.0 (2910) | 7.7 |
| 9 | 3580+ | 41+ | Shears 2765 and 28 | Poor | −1 to 2 (3000) | 10.0 |
| 10 | 3580+ | 41+ | Shears 2940 and 28 | Fair | −0.5 to 1.5 (3000) | 10.7 |
| 11 | 3580+ | 41+ | | Fair | ............. | 10.5 |
| 12 | 2640–3000 | 16–30 | | Good | | 6.5 |
| 13 | 3060–3090 | 31–32 | Shears 2900 and 25 | Poor[b] | +0.5 to 0.8 (2640) | 6.5 |
| 14 | 3390 | 39 | 0–1 at 2730 and 50 | Excellent | +2[c] (2910) | 8–9.3 |
| 15 | 3310–3340 | 37–38 | 0–0.5 at 2640 and 25 | Excellent | −0 to 0.8 (2910) | 8.5 |
| 16 | 2980–3000 | 29–30 | 0.3 at 2200 and 10 | Good | −0.2 (2600) | 2.25 |

| Refractory No. | Porosity | Specific heat 60–1200 F | Mean coefficient of thermal expansion from 60 F shrinkage point $\times 10^5$ | Mean thermal conductivity, Btu per sq ft per hr per deg F per in. thickness — Mean temperatures between the hot and cold face, deg F 200 | 400 | 800 | 1200 | 1600 | 2000 | 2400 |
|---|---|---|---|---|---|---|---|---|---|---|
| 1 | 20–26 | 0.20 | 0.43 | .. | 20 | 22 | 24 | 27 | 30 | 32 |
| 2 | 20–26 | 0.20 | 0.56 | .. | 8 | 9 | 10 | 11 | 12 | 12 |
| 3 | 10–12 | 0.21 | | | | | | | | |
| 4 | 15–25 | 0.23 | 0.25–0.30 | 5 | 6 | 7 | 8 | 10 | 11 | 12 |
| 5 | 12–15 | 0.23 | 0.25–0.30 | 6 | 7 | 8 | 9 | 10 | 12 | 13 |
| 6 | 23–26 | 0.25 | | | | | | | | |
| 7 | 28–36 | 0.23 | 0.24 | 6 | 7 | 8 | 9 | 10 | 12 | 13 |
| 8 | 18 | 0.22 | 0.23 | .. | .... | 11 | 12 | 13 | 13 | 14 |
| 9 | 20–26 | 0.27 | 0.56–0.83 | .. | 40 | 35 | 30 | 27 | 26 | 25 |
| 10 | 10–12 | 0.26 | | | | | | | | |
| 11 | 20–30 | 0.27 | 0.56–0.80 | | | | | | | |
| 12 | .... | 0.23 | 0.30 | .. | 14 | 15 | 17 | 18 | 19 | 20 |
| 13 | 20–30 | 0.23 | 0.46[d] | .. | 8 | 10 | 12 | 13 | 14 | 15 |
| 14 | 13–28 | 0.20 | 0.24 | .. | .... | 100 | 80 | 65 | 55 | 50 |
| 15 | 20–25 | 0.23 | 0.30 | .. | 10 | 11 | 12 | 13 | 14 | 15 |
| 16 | 75 | 0.22 | 0.25 | .. | 1.6 | 2.0 | 2.6 | 3.2 | 3.8 | |

[a] Many of these data have been taken from a table prepared by Trostel, *Chem. Met. Eng.*, Nov. 1938.
[b] Excellent if left above 1200°F.
[c] Oxidizing atmosphere.
[d] Up to 0.56 at red heat.

27 × 9 or 12 × 4; 31½ × 12 × 4; 36 × 12 × 4.

The following **arch, wedge,** and **key bricks** have maximum dimensions, $a \times b \times c$ of $9 \times 4\frac{1}{2} \times 2\frac{1}{2}$ in.* The minimum dimensions $a'$, $b'$, $c'$, are as noted: No. 1 arch, $c' = 2\frac{1}{8}$; No. 2 arch, $c' = 1\frac{3}{4}$; No. 3 arch, $c' = 1$; No. 1 wedge, $c' = 1\frac{7}{8}$; No. 2 wedge, $c' = 1\frac{1}{2}$; No. 3 wedge, $c' = 2$; No. 1 key, $b' = 4$; No. 2 key, $b' = 3\frac{1}{2}$; No. 3 key, $b' = 3$; No. 4 key, $b' = 2\frac{1}{4}$; edge skew, $b' = 1\frac{1}{2}$; feather edge, $c' = \frac{1}{8}$; No. 1 neck, $a' = 3\frac{1}{2}$, $c' = \frac{5}{8}$; No. 2 neck, $a' = 2\frac{1}{2}$, $c = \frac{5}{8}$; No. 3 neck, $a' = 0$, $c' = \frac{5}{8}$; end skew, $a' = 6\frac{3}{4}$; side skew, $b' = 2\frac{1}{4}$; jamb brick, 9 × 2½; bung arch, $c' = 2\frac{3}{8}$.

**Special shapes** are more expensive than the standard refractories, and, as they are usually hand-molded, will not be so dense or uniform in structure as the regular brick. When special shapes are necessary, they should be laid out as simply as possible and the maximum size should be kept down below 30 in if possible. It is also desirable to make all special shapes with the vertical dimension as an even multiple of 2½ in* plus one joint so that they will bond in with the rest of the brickwork.

### Refractory Mortars, Coatings, Plastics, Castables, and Ramming Mixtures

Practically all brickwork is laid up with some type of jointing material to give a more stable structure and to seal the joints. This material may be ground fire clay or a specially prepared mortar containing grog to reduce the shrinkage. The bonding mortars may be divided into three general classes. The first are **air-setting mortars** which often contain chemical or organic binder to give a strong bond when dried or fired at compara-

**Table 3. Shapes of Firebricks**

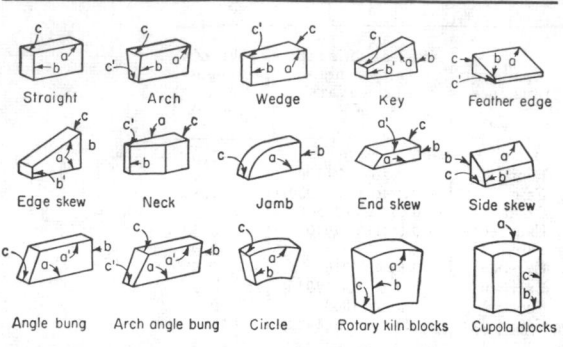

Straight    Arch    Wedge    Key    Feather edge

Edge skew    Neck    Jamb    End skew    Side skew

Angle bung    Arch angle bung    Circle    Rotary kiln blocks    Cupola blocks

tively low temperatures. Many of the air-setting mortars should not be used at extremely high temperatures because the fluxing action of the air-setting ingredient reduces the fusion point. The second class is called **heat-setting mortar** and requires temperatures of over 1090°C (2000°F) to produce a good bond. These mortars vary in vitrifying point, some producing a strong bond in the lower temperature ranges, and the others requiring very high temperatures to give good strength. The third classification comprises **special-base mortars** such as silica, magnesite, silicon carbide, or chrome, which are specially blended for use with their respective bricks. The chrome-base mortar may be satisfactorily used with fire-clay bricks in many cases.

The refractory bonding mortars should preferably be selected on the advice of the manufacturer of the refractory to obtain good service, although there are a considerable number of independent manufacturers of mortars who supply an excellent product. From 1,330 to 1,778 N (300 to 400 lb) of dry mortar per thousand brick is required for thin joints, which are desirable in most furnace construction. For thicker trowel joints, up to 2,220 N (500 lb) per thousand brick is required. In the case of chrome-base mortars, 2,660 N (600 lb) per thousand brick should be allowed, and for magnesite cement 3,550 N (800 lb).

The working properties of the bonding mortar are important. Mortars for insulating refractories should be carefully selected, as many of the commercial products do not retain water sufficiently long to enable a good joint to be made. There are special mortars for this purpose which are entirely satisfactory.

**Coatings** are used to protect the hot surface of the refractories, especially when they are exposed to dust-laden gases or slags. These coatings usually consist of ground grog and fire clay of a somewhat coarser texture than the mortar. There are also chrome-base coatings which are quite resistant to slags, and in a few cases natural clays containing silica and feldspar are satisfactory.

The coatings can be applied to the surface of the brickwork with a brush in thin layers about $\frac{1}{16}$ in thick, or they may be sprayed on with a cement gun, the latter method generally giving the best results. Some types of coating can be put on in much thicker layers, but care should be taken to assure that the coating selected will fit the particular brick used; otherwise it is apt to peel off in service. The coating seals the pores and openings in the brickwork and presents a more continuous and impervious service to the action of the furnace gases and slag. It is not a cure-all for refractory troubles.

**Plastics and ramming mixtures** are generally a mixture of fire clay and coarse grog of somewhat the same composition as the original fire-clay brick. They are used in repairing furnace walls which have been damaged by spalling or slag erosion, and also for making complete furnace walls in certain installations such as small boiler furnaces. They are also used to form special or irregular shapes, in temporary wooden forms, in the actual furnace construction.

Some of the plastics and ramming mixtures contain silicate of soda and are air-setting, so that a strong structure is produced as soon as the material is dry. Others have as a base chrome ore or silicon carbide, which make a mixture having a high thermal conductivity and a good resistance to slag erosion. These mixtures are often used in the water walls of large boiler furnaces; they are rammed around the tubes and held in place by small studs welded to the tube walls. The chrome plastic has been used with good success for heating-furnace floors and subhearths of open-hearth furnaces.

**Castable** mixes are a refractory concrete usually containing high-alumina cement to give the setting properites. These find considerable use in forming intricate furnace parts in wooden molds; large structures have been satisfactorily cast by this method. This type of mixture is much used for baffles in boilers where it can be cast in place around the tubes. Lightweight castables with good insulating properties are used to line furnace doors.

### Furnace Walls

The modern tendency in furnace construction is to make a comparatively thin wall, anchored and supported at frequent intervals by castings or heat-resisting alloys which, in turn, are held by a structural framework so that the weight of the refractory is carried by the framework and does not rest on the base. The wall may be made of heavy refractories backed up with insulating material, or of insulating refractory. Table 4 gives heat losses and heat contents of a number of wall combinations and may enable designers to pick out a wall section to suit their purpose. Solid walls built with standard 9 in* brick are made up in various ways, but the hot face has usually four header courses and one stretcher course alternating.

Many modern furnaces are constructed with air-cooled walls, with refractory blocks held in place against a casing by alloy steel holders. Sectional walls made up with steel panels having lightweight insulating refractories attached to the inner surface are also used and are especially valuable for use in the upper parts of large boiler furnaces, oil stills, and similar types of construction. The sections can be made up at the plant and shipped as a unit. They have the advantage of low cost because of the light ironwork required to support them.

Many failures in furnace construction result from improper expansion joints. **Expansion joints** should usually be installed at least every 10 ft, although in some low-temperature structures the spacing may be greater. For high-temperature construction, the expansion joint allowance per foot in inches should be as follows: fire clay, $\frac{1}{16}$ to $\frac{3}{32}$; high alumina, $\frac{3}{32}$ to $\frac{1}{8}$; silica, $\frac{1}{8}$ to $\frac{3}{16}$; magnesite, $\frac{1}{4}$; chrome, $\frac{5}{32}$; forsterite, $\frac{1}{4}$. Corrugated cardboard is often used in the joints.

*Conversion factors: 1 in = 0.0254 m.

**Table 4. Transmitted Heat Losses and Heat-Storage Capacities of Wall Structures under Equilibrium Conditions**
(Based on still air at 80°F) (Condensed from "B & W Insulating Firebrick" Bulletin of The Babcock & Wilcox Co.)

| Wall | Thickness, in. Of insulating refractory and firebrick | 1200 HL | 1200 HS | 1600 HL | 1600 HS | 2000 HL | 2000 HS | 2400 HL | 2400 HS | 2800 HL | 2800 HS |
|---|---|---|---|---|---|---|---|---|---|---|---|
| 4½ | 4½ 20 | 355 | 1,600 | 537 | 2,300 | 755 | 2,900 | | | | |
| | 4½ 28 | 441 | 2,200 | 658 | 3,100 | 932 | 4,000 | 1,241 | 4,900 | 1,589 | 5,900 |
| | 4½ FB | 1,180 | 8,400 | 1,870 | 11,700 | 2,660 | 14,800 | 3,600 | 18,100 | 4,640 | 21,600 |
| 7 | 4½ 28 + 2½ 20 | 265 | 3,500 | 408 | 4,900 | 567 | 6,500 | 751 | 8,100 | 970 | 9,800 |
| | 4½ FB | 423 | 12,500 | 660 | 17,700 | 917 | 23,000 | 1,248 | 28,200 | | |
| 9 | 4½ 28 + 4½ 20 | 203 | 4,100 | 311 | 5,900 | 432 | 7,900 | 573 | 9,900 | 738 | 12,200 |
| | 4½ FB + 4½ 20 | 285 | 13,700 | 437 | 19,200 | 615 | 24,800 | | | | |
| | 9 20 | 181 | 3,100 | 280 | 4,300 | 395 | 5,500 | | | | |
| | 9 28 | 233 | 4,100 | 349 | 5,800 | 480 | 7,500 | 642 | 9,300 | 818 | 11,100 |
| | 9 FB | 658 | 15,800 | 1,015 | 21,600 | 1,430 | 27,600 | 1,900 | 34,000 | 2,480 | 40,300 |
| 11½ | 9 28 + 2½ 20 | 169 | 5,700 | 260 | 8,000 | 364 | 10,500 | 484 | 13,100 | 623 | 15,800 |
| | 9 FB + 2½ 20 | 335 | 22,300 | 514 | 31,400 | 718 | 40,600 | 962 | 50,400 | 1,233 | 60,300 |
| | 9 28 + 4½ 20 | 143 | 6,500 | 217 | 9,300 | 305 | 12,300 | 404 | 15,300 | 514 | 18,700 |
| | 9 FB + 4½ 20 | 241 | 24,100 | 367 | 34,500 | 516 | 44,800 | 690 | 55,100 | | |
| 13½ | 9 20 + 4½ FB | 165 | 5,300 | 255 | 7,300 | 348 | 9,900 | | | | |
| | 9 28 + 4½ FB | 200 | 6,900 | 302 | 9,700 | 415 | 12,600 | 556 | 15,700 | 710 | 19,100 |
| | 13½ FB | 452 | 22,300 | 700 | 31,000 | 980 | 39,900 | 1,310 | 49,100 | 1,683 | 58,300 |
| 16 | 13½ FB + 2½ 20 | 275 | 31,200 | 423 | 43,300 | 588 | 56,300 | 780 | 70,000 | 994 | 84,200 |
| 18 | 9 20 + 9 FB | 147 | 8,500 | 225 | 11,900 | 319 | 15,700 | | | | |
| | 9 28 + 9 FB | 175 | 10,700 | 266 | 15,100 | 375 | 19,700 | 493 | 24,600 | 635 | 29,800 |
| | 13½ FB + 4½ 20 | 210 | 34,100 | 318 | 48,400 | 440 | 62,600 | 587 | 77,500 | 753 | 92,600 |
| | 18 FB | 355 | 28,800 | 532 | 40,300 | 745 | 52,200 | 1,000 | 64,200 | 1,283 | 76,500 |
| 20½ | 18 FB + 2½ 20 | 234 | 39,000 | 356 | 55,400 | 500 | 72,000 | 665 | 89,200 | 847 | 107,000 |
| 22½ | 18 FB + 4½ 20 | 182 | 43,200 | 281 | 61,000 | 392 | 79,200 | 519 | 97,700 | 667 | 117,600 |
| | 22½ FB | 287 | 36,000 | 435 | 49,500 | 612 | 64,100 | 814 | 78,800 | 1,040 | 93,400 |

Conversion factors: $t_c = \frac{5}{9}(t_F - 32)$; 1 in = 0.0254 m.
HL = heat loss in Btu/(ft²)/(h). HS = heat storage capacity in Btu/ft². 20 = 2000°F insulating refractory. 28 = 2800°F insulating refractory. FB = fire-clay brick.

The **roof** of the furnace is usually either a sprung arch or a suspended arch. A **sprung arch** is generally made of standard shapes using an inside radius equal to the total span. In most cases, it is necessary to build a form on which the arch is sprung.

In the case of arches with a considerable rise, it has been found that an inverted catenary shape is better than a circular shape for stability, and it is possible to run the sidewalls of the furnace right down to the floor in one continuous arch with almost complete elimination of the ironwork. The catenary can be readily laid out by hanging a flexible chain from two points of a vertical wall.

The **suspended arch** is used when it is desirable to have a flat roof (curved suspended arches are also made); it presents certain advantages in construction and repair but is more difficult to insulate than the sprung arch. Special suspended arch shapes are commercially available. The insulating refractory is suited to this type of construction because the steel supports are light and the heat loss is low.

### Selection of Refractories

The selection of the most suitable refractory for a given purpose demands experience in furnace construction. A brick that costs twice as much as another brand and gives twice the life is preferable since the total cost includes the laying cost. Furthermore, a brick that gives longer service reduces the shutdown period of the furnace. Where slag or abrasion is severe, brick with a dense structure is desirable. If spalling conditions are important, a brick with a more flexible structure is better, although there are cases where a very dense structure gives better spalling resistance than a more open one.

High-lime slag can be taken care of with magnesite, chrome, or high-alumina brick, but if severe temperature fluctuations are encountered also, no brick will give long life. For coal-ash slag, dense fire-clay bricks give fairly good service if the temperature is not high. At the higher temperatures, a chrome-plastic or silicon-carbide refractory often proves successful. When the conditions are unusually severe, air- or water-cooled walls must be resorted to; the water-cooled stud-tube wall has been very successful in boiler furnaces.

With a general freedom from slag, it is often most economical to use an insulating refractory. Although this brick may cost more per unit, it allows thinner walls, so that the total construction cost may be no greater than the regular brick.

# Table 5. Physical Properties of Some Dense,* Pure Refractories

| Material | Density, per cm³ | Modulus of rupture, 10³ lb/in² at 70°F | Modulus of rupture, 10³ lb/in² at 1800°F | Modulus of elasticity, 10⁶ lb/in² at 70°F | Fusion point, °F | Linear coef of expansion, $10^{-6}$ in/in °F between 65 and 1800°F | Thermal conductivity, Btu(in)/(ft²)(h)(°F) at 212°F | Thermal conductivity, Btu(in)/(ft²)(h)(°F) at 1800°F | Specific heat, Btu/(lbm)(°F) | Thermal stress resistance | Electrical resistivity, Ω·cm at 70°F | Electrical resistivity, Ω·cm at 1800°F |
|---|---|---|---|---|---|---|---|---|---|---|---|---|
| $Al_2O_3$ | 3.97 | 100 | 60 | 53 | 3690 | 5.0 | 210 | 55 | 0.26 | Good | $>10^{14}$ | $10^7$ |
| $BeO$ | 3.03 | 20 | 10 | 45 | 4660 | 4.9 | 1450 | 130 | 0.50 | Very good | $>10^{14}$ | $10^8$ |
| $MgO$ | 3.58 | 14 | 12 | 31 | 5070 | 7.5 | 240 | 47 | 0.25 | Poor | $>10^{14}$ | $10^7$ |
| $ThO_2$ | 10.00 | 12 | 7 | 21 | 5520 | 5.0 | 62 | 20 | 0.06 | Fair | $>10^{14}$ | $10^5$ |
| $ZrO_2$ | 5.60 | 20 | 15 | 22 | 4600 | 5.5 | 15 | 15 | 0.14 | Fair | $10^8$ | 500 |
| $UO_2$ | 10.96 | 12 |  | 25 | 5070 | 5.6 | 58 | 20 | 0.06 |  |  |  |
| $SiC$ | 3.22 | 24 | 24 | 68 | 5000+ | 2.2 | 390 | 145 | 0.20 | Excel. | 10 |  |
| $BC$ | 2.52 | 50 | 40 | 42 | 4440 | 2.5 | 200 | 145 | 0.36 | Good | 0.5 |  |
| $BN$ | 2.25 | 7 | 1 | 12 | 5000 | 2.6 | 150 | 130 | 0.39 | Good | $10^{10}$ |  |
| $MoSi_2$ | 6.20 | 100 | 40 | 50 | 3890 | 5.1 | 220 | 100 | 0.11 | Good | $10^{-6}$ | 104 |
| $C$ | 2.22 | 3 | 4 | 2 | 7000 | 2.2 | 870 | 290 | 0.34 | Good | $10^{-3}$ | $10^{-2}$ |

Conversion factors: 1 lb/in² = 689.476 N/m². $t_C = \frac{5}{9}(t_F - 32)$. 1 Btu/(ft²)(h)(°F)(in) = 225 W/(m²)(°C)(m). 1 Btu/(lbm)(°F) = 4,190 J/(kg)(°C).

*Porosity, 0 to 5 percent.

†Stabilized.

References: Norton, "Refractories," 3d ed.; Green and Stewart, "Ceramics: A Symposium;" Ryschkewitsch, "Oxydekramik der Einstuffysstemme," Springer; Campbell, "High Temperature Technology;" Kingery, "Property Measurements at High Temperatures," Wiley.

The substitution of insulating refractory for heavy brick in periodic furnaces has sometimes halved the fuel consumption.

The stability of a refractory installation depends largely on the bricklaying. The total cost, in addition to the bricks, of laying brick varies with the type of construction, locality, and refractory.

### Recent Developments in Refractories

**Pure-oxide refractories** have been developed to permit fabrication of parts such as tubes, crucibles, and special shapes. Alumina ($Al_2O_3$) is the most readily formed into nonporous pieces and, up to its softening point of 2040°C (3690°F), is most useful. Mullite ($2SiO_2:3Al_2O_3$), softening at 1820°C (3290°F), is used for thermocouple-protection tubes, crucibles, and other small pieces. Magnesium oxide (MgO), fusing at 2800°C (5070°F), is resistant to metals and slags. Zirconia ($ZrO_2$), softening at 4600°F, is very sensitive to temperature changes but can be stabilized with a few percent of lime. Beryllium oxide (BeO), softening at 2570°C (4660°F), has a very high thermal conductivity but must be fabricated with great care because of health hazards. Thoria ($ThO_2$) softens at 3040°C (5520°F) and has been used in crucibles for melting active metals and as a potential nuclear fuel.

**Refractory carbides, sulfides, borides, silicides, and nitrides** have been developed for special uses. Many have high softening points, but all have limited stability in an oxidizing atmosphere. Silicon carbide (SiC) is the most used because of its high thermal and electrical conductivity, its resistance against certain slags, and its relatively good stability in air. Molybdenum silicide ($MoSi_2$) also has considerable resistance to oxida-

tion and, like SiC, can be used for metal-melting crucibles. Cerium sulfide ($CeS_2$) is a metallic-appearing material of high softening point but no resistance to oxidation. Zirconium nitride (ZrN) and titanium nitride (TiN) are also metalliclike but are not stable when heated in air. Graphite has valuable and well-known refractory properties but is not resistant to oxidation.

**Refractory fibers** are coming into use quite extensively. Fibers of silica-alumina glass have a use limit of about 1090°C (2000°F). They are used for insulating blankets, expansion joints, and other high-temperature insulation. Development of higher-temperature fibers is being carried out on a small scale for use as high-temperature insulation or mechanical reenforcement.

**Nuclear fuels** of uranium, thorium, and plutonium oxides or carbides are now extensively used in high-temperature reactors.

**Space vehicles** are using nozzles of refractories of various kinds to withstand the high temperatures and erosion. Nose cones of sintered alumina are now used extensively because of their excellent refractory and electrical properties. Heat shields to protect space vehicles upon reentry are an important use of special refractories.

### Physical Properties of High-Purity Refractories

In Table 5 are shown the properties of some of the more important pure refractory materials. It should be realized that, as purer materials become available and testing methods become more refined, some of these values will be changed.

# CEMENT, MORTAR, AND CONCRETE

### by William L. Gamble

REFERENCES: Neville, "Properties of Concrete," Wiley. Sahlin, "Structural Masonry," Prentice-Hall. Troxell and Davis, "Composition and Properties of Concrete," McGraw-Hill. Blanks and Kennedy, "The Technology of Cement and Concrete," Wiley. Taylor, Thompson, and Smulski, "Concrete, Plain and Reinforced," Wiley. Baker, "Treatise on Masonry Construction," Wiley. Comm. 318, "Building Code Requirements for Reinforced Concrete," Comm. 613, "Recommended Practice for Selecting Proportions for Concrete," and Comm. 614, "Recommended Practice for Measuring, Mixing, and Placing Concrete," American Concrete Institute.

## CEMENT

**Normal portland cement** is used for concrete, for reinforced concrete, and either with or without lime, for mortar and stucco. It is made from a mixture of about 80 percent carbonate of lime (limestone, chalk, or marl) and about 20 percent clay (in the form of clay, shale, or slag). After being intimately mixed, the materials are finely ground by a wet or dry process and then calcined in kilns to a clinker. When cool, this clinker is ground to a fine powder. During the grinding, a small amount of gypsum is usually added to regulate the setting of the cement. The chemical analysis of 32 American type I

cements gives the following average percentage composition: silica ($SiO_2$), 21.92; alumina ($Al_2O_3$), 6.91; iron oxide ($Fe_2O_3$), 2.91; calcium oxide (CaO), 62.92; magnesium oxide (MgO), 2.54; sulfuric oxide ($SO_3$), 1.72; alkalies ($R_2O_3$), 0.82; loss on ignition, 1.50, insoluble residue 0.20.

**Types and Kinds of Cements** Five types of portland cements are covered by ASTM specification C150.

**Normal portland cement,** type I, is used for purposes for which another type having special properties is not required. Most structures, pavements, and reservoirs are built with type I cement.

**Modified portland cement,** type II, generates less heat from its hydration and is more resistant to sulfate attacks than type I. This cement is used in structures having large cross sections, such as large abutments and heavy retaining walls. It may also be used in drainage where a moderate sulfate concentration exists.

**High-early-strength portland cement,** type III, is used when high strengths are required in a few days. Use of high-early-strength will allow earlier removal of forms and shorter periods of curing.

**Low-heat portland cement,** type IV, generates less heat during hydration than type II and is used for mass concrete constructions such as large dams where high temperature rises would create special problems. Type IV cement gains strength more slowly than type I. The tricalcium aluminate content is limited to 7 percent.

**Sulfate-resisting portland cement,** type V, is a special cement, not readily available, to be used when concrete is exposed to severe sulfate attack. Type V cements gain strengh more slowly than type I cement. The tricalcium aluminate content is limited to a maximum of 5 percent.

**Air-entraining portland cements** purposely cause air, in minute, closely spaced bubbles, to occur in concrete. Entrained air makes the concrete more resistant to the effects of repeated freezing and thawing and of the deicing agents used on pavements. To obtain such cements, air-entraining agents are interground with the cement clinker during manufacture. Types I to III can be obtained as air-entraining cements and are then designated as types IA, IIA, and IIIA, under ASTM C175.

**Portland blast-furnace slag cements** are made by grinding granulated high-quality slag with portland-cement clinker. Portland blast-furnace slag cement type IS and air-entraining portland blast-furnace slag cement type IS-A are covered by ASTM specification C205. Provisions are also made for moderate-heat-of-hydration cements (MH) and moderate-sulfate-resistance cements (MS), or both (MH-MS). Type IS cements initially gain strength more slowly but have about the same 28-day strength as type I cements.

**White portland cement** is used for architectural and ornamental work because of its white color. It is high in alumina and contains less than 0.5 percent iron. The best brands are true portlands in composition.

**Portland-pozzolan cement** is a blended cement made by intergrinding portland cement and pozzolanic materials. Two types, type IP (portland-pozzolan cement) and type IP-A (air-entraining portland-pozzolan cement), are covered in ASTM specification C340.

**Masonry cement,** ASTM specification C91, is a blended cement used in place of job cement-lime mixtures to reduce the number of materials handled and to improve the uniformity of the mortar. These cements are made by combining either natural or portland cements with fattening materials such as hydrated lime and, sometimes, with air-entraining admixtures.

**Waterproofed cement** is sometimes used where a waterproof or water-repellent concrete or mortar is particularly desirable. It is cement ground with certain soaps and oils. The effectiveness is limited to 3 or 4 ft of water pressure.

**Pozzolana cement** has been used under certain conditions for concrete not exposed to the air. It is made by mixing and grinding together slaked lime and granulated blast-furnace slag or other material similar to natural lava, without burning.

The ancient cements were mixtures of lime and volcanic material.

**Natural cement** is sometimes used, either with or without lime, for common mortar for brick or stonework. It is manufactured from limestone containing clay. The chemical constituents are similar to those of portland cement. Formerly, it was made from the rock just as it came from the quarry, so that its chemical composition varied with the composition of the rock, but now greater care is used in maintaining uniformity. The unground rock is fed directly to the kilns, calcined at a temperature much lower than that required for portland cement, and then crushed and ground.

**Shrinkage-compensated cements** are special portland cements which expand slightly during the moist curing period, compensating for the shrinkage accompanying later drying. They are used primarily to aid in producing crack-free concrete floor slabs and other members. An ASTM specification is being developed to cover these cements.

**Regulated-set cements** are special portland cements formulated to set in very short times, producing usable concrete strengths in regulated times of as little as 1 h or less. Such concretes are obviously well suited to repair work done when it is important to minimize downtime.

**Portland-Cement Tests**  Cement should be tested for all but unimportant work. Tests should be made in accordance with the standard specifications of the ASTM or with the federal specifications where they apply. Samples should be taken at the mill, and tests completed before shipments are made. When this is not possible, samples should be taken at random from sound packages, one from every 10 bbl or 40 bags, and mixed. The total sample should weigh about 6 lb. ASTM requirements for standard portland cements are given in ASTM specification C150.

The **autoclave soundness test** consists of determining the expansion of a 1-in-sq neat cement bar 10 in long which, after 24-h storage in 90 percent or greater humidity, is placed in an autoclave, where the pressure is raised to 295 lb/in² in about 1 h, maintained for 3 h, and then brought back to normal in 1½ h. Cements that show over 1 percent expansion may show unsoundness after some years of service; the ASTM allows a maximum of 0.80 percent.

**Time of Setting**  Initial set should not be less than 45 min when Vicat needle is used or 60 min when Gilmore needle is used. Final set should be within 10 h. Cement paste must remain plastic long enough to be properly placed and yet submit to finishing operations in a reasonable time.

**Compressive Strength**  Minimum requirements for average compressive strength of not less than three 2-in cubes composed of 1 part (by weight) cement and 2.75 parts standard graded mortar sand, tested in accordance with ASTM method C109, are as follows:

| Age of test, days | Storage of test pieces | Compressive strength, lb/in² | | | | |
|---|---|---|---|---|---|---|
| | | Normal | Moderate heat | High early strength | Low heat | Sulphate resistant |
| 1 | 1 day moist air | ..... | ..... | 1,700 | | |
| 3 | 1 day moist air, 2 days water | 1,200 | 1,000 | 3,000 | | |
| 7 | 1 day moist air, 6 days water | 2,100 | 1,800 | ..... | 800 | 1,500 |
| 28 | 1 day moist air, 27 days water | 3,500 | 3,500 | ..... | 2,000 | 3,000 |

## LIME

**Common lime,** or **quicklime,** when slaked or hydrated, is used for interior plastering and for lime mortar. Mixed with cement, it is used for lime and cement mortar and for stucco. Mortars made with lime alone are not satisfactory for thick walls because of slow-setting qualities. They must never be used under water. Lime is made by burning pure limestone in kilns about 40 ft high by 10 ft diam at temperatures ranging from 1400 to 2000°F. A typical percentage chemical composition is calcium oxide (CaO), 97; iron oxide ($Fe_2O_3$), 1.3; silica ($SiO_2$), 1.0; magnesium oxide (MgO), 0.7. Quicklime slakes rapidly with water with much heat evolution, forming calcium hydrate ($CaH_2O_2$). With proper addition of water, it becomes plastic, and the volume of putty obtained is 2 or 3 times the loose volume of the lime before slaking, and its weight is about $2\frac{1}{2}$ times the weight of the lime. Plastic lime sets by drying, by crystallization of calcic hydrate, and by absorbing carbonic acid from the air. The process of hardening is very slow. Popping is likely to occur in plaster unless the lime is sound, as indicated by an autoclave test at 120 lb/in² pressure for 2 h.

**Magnesium lime,** used for the same purposes as common or high-calcium lime, contains more than 20 percent magnesium oxide. It slakes more slowly, evolves less heat, expands less, sets more rapidly, and produces higher-strength mortars than does high-calcium quicklime.

**Pulverized and granulated limes** slake completely much more quickly than ordinary lump lime. They are sometimes waterproofed by the addition of stearates and other compounds similar to those used in cement for the same purpose. The waterproofing treatment retards the slaking.

**Hydrated lime** is a finely divided white powder manufactured by slaking quicklime with the requisite amount of water. It has the advantage over lime slaked on the job of giving a more uniform product, free from unslaked lime. It does not have plasticity or water retention equal to freshly slaked quicklime.

**Hydraulic hydrated lime** is used for blending with portland cement and as a masonry cement. It is the hydrated product of calcined impure limestone which contains enough silica and alumina to permit the formation of calcium silicates.

## AGGREGATES

### Sand

Sand to be used for mortar, plaster, and concrete should consist of clean, hard, uncoated grains free from organic matter, vegetable loam, alkali, or other deleterious substances. Vegetable or organic impurities are particularly harmful. A quantity of vegetable matter so small that it cannot be detected by the eye may render a sand absolutely unfit for use with cement. Stone screenings, slab, or other hard inert material may be substituted for or mixed with sand. Sand for concrete should range in size from fine to coarse, with not less than 95 percent (preferably not more than 30 percent) passing a No. 4 sieve, not less than 10 percent retained on a No. 50 sieve, and not more than 5 percent (or 8 percent, if screenings) passing a No. 100 sieve. A straight-line gradation on a graph, with percentages passing plotted as ordinates to normal scale and sieve openings as abscissas to logarithmic scale, gives excellent results.

The grading of sand for mortar depends upon the width of joint, but normally not less than 95 percent should pass a No. 8 sieve, and it should grade uniformly from coarse to fine without more than 8 percent passing a No. 100 sieve.

Sand for plaster should have at least 90 percent passing a No. 8 sieve and not more than 5 percent passing the No. 100 sieve.

Silt or clayey material passing a No. 200 sieve in excess of 2 percent is objectionable.

**Test of Sand**   Sand for use in important concrete structures should always be tested. The strength of concrete and mortar depends to a large degree upon the quality of the sand and the coarseness and relative coarseness of the grains. Sand or other fine aggregate when made into a mortar of 1 part portland cement to 3 parts fine aggregate by weight should show a tensile strength at least equal to the strength of 1:3 mortar of the same consistency made with the same cement and standard sand. If the aggregate is of poor quality, the proportion of cement in the mortar or concrete should be increased to secure the desired strength. If the strength is less than 90 percent that of Ottawa sand mortar, the aggregate should be rejected unless compression tests of concrete made with the selected aggregates pass the requirements. The standard Ottawa sand gradation is described in ASTM specification C109. This sand is supplied by the Ottawa Silica Co., Ottawa, Ill. The compressive strength of 2-in cubes made from a cement and sand mixture with a 0.9 water-cement ratio and a flow of 100 percent should equal 90 percent of the strength of similar cubes made with graded Ottawa sand.

The ASTM standard test (C40) for the presence of injurious organic compounds in natural sands for cement mortar or concrete is as follows: A 12-oz graduated glass prescription bottle is filled to the $4\frac{1}{2}$-oz mark with the sand to be tested. A 3 percent solution of sodium hydroxide (NaOH) in water is then added until the volume of sand and liquid, after shaking, gives a total volume of 7 liquid oz. The bottle is stoppered, shaken thoroughly, and then allowed to stand for 24 h. A standard-reference-color solution of potassium dichromate in sulfuric acid is prepared as directed in ASTM D154. The color of the clear liquid above the sand is then compared with the standard-color solution; if the liquid is darker than the standard color, further tests of the sand should be made before it is used in mortar or concrete. The standard color is similar to light amber.

### Coarse Aggregate

**Broken Stone and Gravel**   Coarse aggregate for concrete may consist of broken stone, gravel, slab, or other hard inert material with similar characteristics. The particles should be clean, hard, durable, and free from vegetable or organic matter, alkali, or other deleterious matter and should range in size from material retained on the No. 4 sieve to the coarsest size permissible for the structure. For reinforced concrete and small masses of unreinforced concrete, the maximum size should be that which will readily pass around the reinforcement and fill all parts of the forms. Either 1 or $1\frac{1}{2}$ in diam is apt to be the maximum. For heavy mass work, the maximum size may run up to 3 in or larger.

**Lightweight aggregates** are usually pumice, lava, slag, burned clay or shale, or cinders from coal and coke. It is recommended that lightweight fine aggregate not be used in conjunction with lightweight coarse aggregate unless it can be demonstrated, from either previous performance or suitable tests, that the particular combination of aggregates results in concrete that is free from soundness and durability problems.

In case of doubt, the concrete mix should be designed using sand, fine aggregate, and lightweight coarse aggregate. Their application is largely for concrete units and floor slabs where saving in weight is important and where special thermal insulation or acoustical properties are desired.

**Heavyweight aggregates** are generally iron or other metal punchings, ferrophosphate, hematite, magnetite, barite, limenite, and similar heavy stones and rocks. They are used in concrete for counterweights, dry docks, and shielding against rays from nuclear reactions.

**Fineness Modulus** The fineness modulus, which is used in the Abrams method as an index of the characteristics of the aggregates, is the sum of the cumulative percentages (divided by 100) which would be retained by all the sieves in a special sieve analysis. The sieves used in this method are Nos. 100, 50, 30, 16, 8, and 4 for fine aggregates and these plus the $\frac{3}{8}$-, $\frac{3}{4}$-, $1\frac{1}{2}$-, and 3-in sizes for coarse aggregates. A high fineness modulus indicates a relatively low surface area because the particles are relatively large, which means less water required and, therefore, a higher concrete strength. Aggregates of widely different gradation may have the same fineness modulus.

ASTM standard sieves for analysis of aggregates for concrete are of the following sizes of opening and wire:

on properties other than the one which is being improved. Most important is consideration of possible changes in the basic mix which might make the admixture unnecessary. Particular care must be used when using two or more admixtures in the same concrete, such as a retarding agent plus an air-entraining agent, to ensure that the materials are compatible with each other when mixed in concrete.

**Air-entraining agents** constitute one of the most important groups of admixtures. They entrain air in small, closely spaced, separated bubbles in the concrete, greatly improving resistance to freezing and thawing and to deicing agents.

**Accelerators** are used to decrease the setting time and increase early strength. They permit shorter curing periods, earlier form removal, and placing at lower temperatures. Calcium chloride is the most frequently used accelerator and can be used in amounts up to 2 percent of the weight of the cement.

**Retarders** increase the setting time. They are particularly useful in hot weather and in grouting operations.

Other admixtures may be classed as gas-forming agents, pozzolanic materials, curing aids, water-repelling agents, and coloring agents.

## MORTARS

Properties desirable in a mortar include (1) good plasticity or

| Sieve No | 100 | 50 | 30 | 16 | 8 | 4 |
|---|---|---|---|---|---|---|
| Sieve opening, in | 0.0059 | 0.0117 | 0.0234 | 0.0469 | 0.0937 | 0.187 |
| Wire diam, in | 0.0043 | 0.0085 | 0.0154 | 0.0256 | 0.0394 | 0.0606 |
| Sieve size, in | $\frac{3}{8}$ | $\frac{3}{4}$ | 1 | $1\frac{1}{2}$ | 2 | 3 |
| Sieve opening, in | 0.375 | 0.750 | 1.00 | 1.50 | 2.00 | 3.00 |
| Wire diam, in | 0.0894 | 0.1299 | 0.1496 | 0.1807 | 0.1988 | 0.2283 |

## WATER

Water for concrete or mortar should be clean and free from oil, acid, alkali, organic matter, or other deleterious substance. Cubes or briquettes made with it should show strength equal to those made with distilled water. Water fit for drinking is normally satisfactory for use with cement. However, many waters not suitable for drinking may be suitable for concrete. Water with less than 2,000 ppm of total dissolved solids can usually be used safely for making concrete. (See Sec. 6, Water.)

**Seawater** can be used as mixing water for plain concrete, although 28-day strength may be lower than for normal concrete. If seawater is used in reinforced concrete, care must be taken to provide adequate cover with a dense air-entrained concrete to minimize risks of corrosion. Seawater should not be used with prestressed concrete.

## ADMIXTURES

Admixtures are substances, other than the normal ingredients, added to mortars or concrete for altering the normal properties so as to improve them for a particular purpose. Admixtures are frequently used to entrain air, increase workability, accelerate or retard setting, provide a pozzolanic reaction with lime, reduce shrinkage, and reduce bleeding. However, before using an admixture, consideration must be given to its effect

workability, (2) low volume change or volume change of the same character as the units bonded, (3) low absorption, (4) low solubility and thus freedom from efflorescence, (5) good strength in bond and ample strength to withstand applied loads, (6) high resistance to weathering.

**Lime mortar** consists of lime paste and sand and is made by slaking lime and mixing it with sand or by making up lime putty in advance to be mixed with sand when needed.

The quality of the lime putty is greatly affected by the procedure used in slaking the quicklime. The method given by ASTM specification C5 may be followed.

**Lime mortar,** because of its slow hardening, is not widely used at the present time; it is suitable for interior non-load-bearing walls. Proportions customarily stated are 2.5 to 3 volumes dry and loose sand to 1 volume lime putty. In practice, unless carefully supervised, the proportion of sand is apt to be greater than this, as the mortar man determines the proportion by the way the mortar falls from his hoe in mixing. Actually, the proportions of mortar should be determined by the characteristics of the masonry, particularly as to absorption. Quicklime is now sold in 80-lb bags, although previously in 200-lb casks. The weight of quicklime and sand required to produce 1 yd³ of lime mortar (as given in Table 1) is based on work by Clair and on the use of well-graded mortar sand with a specific gravity of 2.65.

**Lime-and-cement mortar** is made of mortar cement, lime

**Table 1. Quantities Required per Cubic Yard of Lime-Sand Mortar for 50 Percent Flow**

| Volume mixtures desired | | Weights of constituents, lb | | Water addition, gal |
|---|---|---|---|---|
| Lime putty | Dry sand | Quicklime | Mortar sand, damp | |
| 1 | 3 | 195 | 2,640 | 69 |
| 1 | 4 | 154 | 2,780 | 63 |
| 1 | 5 | 127 | 2,870 | 60 |
| 1 | 6 | 107 | 2,900 | 60 |
| 1 | 7 | 93 | 2,940 | 59 |
| 1 | 9 | 74 | 3,020 | 54 |

**Table 2. Quantities Required per Cubic Yard of Lime-Cement-Sand Mortar for 50 Percent Flow**

| Volume mixtures desired | | | Weight of constituents, lb | | | Water addition, gal |
|---|---|---|---|---|---|---|
| Cement | Lime putty | Dry sand | Cement | Quick-lime | Mortar sand, damp | |
| 1 | 0.1 | 3 | 805 | 20 | 2,680 | 41 |
| 1 | 1 | 4 | 555 | 140 | 2,470 | 57 |
| 1 | 1 | 5 | 475 | 120 | 2,630 | 54 |
| 1 | 1 | 6 | 405 | 100 | 2,690 | 52 |
| 1 | 2 | 5 | 420 | 210 | 2,320 | 67 |
| 1 | 2 | 7 | 325 | 160 | 2,530 | 61 |
| 1 | 2 | 9 | 270 | 135 | 2,700 | 56 |

putty or hydrated lime, and sand in proportions, by volume, normally of 1 cement, 1 or 2 lime, and 5 or 6 sand. This type of mortar is particularly suited for masonry of all kinds. Mortar with unslaked lime should be made up several days before cement is added. When wanted for use, the cement is mixed into the mass and is immediately used. Only small quantities of cement should be mixed with the lime and sand at a time so that there will be no danger of the cement attaining a set before the mortar is used. The time between mixing in of cement and the use of mortar should be limited to 2 h. Prepared hydrated lime and mortar cements are increasingly being used in place of lime putty for the type of work where lime-and-cement mortars apply. Table 2 gives the quantities required to produce 1 yd³ of lime-and-cement mortar and is based on work done by Clair.

**Cement mortar** is composed of cement and sand or screenings mixed with water. Proportions in practice are ordinarily, by dry and loose volumes, 1 part cement to 2 parts sand (1:2) or 1 part cement to 3 parts sand (1:3). Cement mortars are used for masonry where high strength is required or where the work is to be subjected to the action of water. Table 3 gives the quantities of sand and cement required for 1 yd³ of cement mortar. If the specific gravity of the sand and cement being used is known and also the actual water requirements, the quantities for 1 yd³ of mortar can be determined accurately by the sum of the absolute volumes of the solids. The strength of portland-cement mortar (1) increases with the proportion of cement, (2) in general, increases with the coarseness of the sand, (3) increases with the density of the mortar, and (4) decreases with increase in water-cement ratio. With the same aggregate, the strongest and most impermeable mortar is that containing the largest percentage of cement in a given volume of mortar. With the same percentage of cement, the strongest mortar is that which has the greatest density. A small addition of hydrated lime or lime putty, not exceeding 10 percent by weight of the cement, is sometimes used to increase the workability or watertightness of the mortar, but the same results can usually be obtained by increasing the proportion of cement. Unslaked or imperfectly slaked lime must never be used, as it will expand and disintegrate the mortar. Table 4 shows the effect of different gradings of sand upon the strength of mortar.

**Mortar for stonework** is generally in practice 1 cement, 1 lime putty, and 6 sand. Lime mortar should not be used for thick walls, as it hardens very slowly, nor should it be used for masonry under water or in wet soil or for structures requiring great strength or which are subject to shock. Lime-and-cement mortar may be used for masonry above water. The amount of mortar required for 1 yd³ of masonry is given in Table 5.

**Brick mortar** may be made with portland cement and sand, natural cement and sand, lime and sand, or mortar cement and sand. An easy-working mixture suitable for laying brick exposed to severe weathering, for chimneys, and in load-bearing walls is obtained by mixing hydrated lime with water to a creamy consistency, putting in the sand (5 or 6 parts to 1 of hydrated lime), and letting the mixture stand from 1 to 3 days. One part of cement, based on the volume of hydrated lime, is added when the mortar is wanted for use. A good mixture for non-load-bearing walls and mild exposure is 1 cement, 2 lime putty, 7 to 9 sand. On important work where freedom from efflorescence, cracking, and leakage of walls is desired, the mortar is subjected to tests for soluble salts, volume change, absorption, and water retention. The absorption rate of the masonry units to be bonded is determined, and the mortar is adjusted accordingly. High absorption rates during the first 10 min require highly water-retentive mortars and thus higher lime contents, such as 1 cement, 1½ lime, 5 sand. Bond and actual water-penetration tests may be made on wall assemblies. The amount of mortar required in laying brickwork is given in Table 6.

**Mortars for Plastering** Common or lime mortar for interior plastering consists of lime, clean coarse sand, and hair or fiber.

**Table 3. Volume of Compacted Plastic Portland Cement Mortar and Quantities of Materials per Cubic Yard***

(Taylor, Thompson, and Smulski)

| Relative proportions by volume | | Volume | | Materials for 1 cu yd, based on barrel of 4 bags or 4 cu ft | | Relative proportions by volume | | Volume | | Materials for 1 cu yd, based on barrel of 4 bags or 4 cu ft | |
|---|---|---|---|---|---|---|---|---|---|---|---|
| Cement† | Sand‡ | Cubic feet from 1 bag of cement weighing 94 lb | Cubic feet from 1 bbl (4 bags) cement containing 4 cu ft | Packed cement, bbl | Loose sand, cu yd | Cement† | Sand‡ | Cubic feet from 1 bag of cement weighing 94 lb | Cubic feet from 1 bbl (4 bags) cement containing 4 cu ft | Packed cement, bbl | Loose sand, cu yd |
| 1 | 0 | 0.80 | 3.2 | 8.31 | 0.0 | 1 | 4½ | 3.91 | 15.6 | 1.72 | 1.15 |
| 1 | ½ | 1.02 | 4.1 | 6.61 | 0.49 | 1 | 5 | 4.28 | 17.1 | 1.58 | 1.17 |
| 1 | 1 | 1.38 | 5.5 | 4.88 | 0.72 | 1 | 5½ | 4.64 | 18.5 | 1.46 | 1.19 |
| 1 | 1½ | 1.74 | 7.0 | 3.87 | 0.86 | 1 | 6 | 5.00 | 20.0 | 1.35 | 1.20 |
| 1 | 2 | 2.11 | 8.4 | 3.21 | 0.95 | 1 | 6½ | 5.36 | 21.4 | 1.26 | 1.21 |
| 1 | 2½ | 2.47 | 9.9 | 2.74 | 1.01 | 1 | 7 | 5.72 | 22.9 | 1.18 | 1.22 |
| 1 | 3 | 2.83 | 11.3 | 2.39 | 1.06 | 1 | 7½ | 6.08 | 24.3 | 1.11 | 1.23 |
| 1 | 3½ | 3.19 | 12.8 | 2.12 | 1.10 | 1 | 8 | 6.44 | 25.8 | 1.05 | 1.24 |
| 1 | 4 | 3.55 | 14.2 | 1.90 | 1.13 | | | | | | |

*Variations in the fineness of the sand and cement and in the consistency of the mortar may affect the values by 10 percent in either direction.

†Cement as packed by manufacturer (4 bags = 1 bbl).

‡Coarse bank and measured loose.

ASTM (C5) recommendations for white and base coats are as follows:

After slaking, the putty shall be prepared for use as follows:

1. **White Coat.** After the action has ceased, run off the putty through a No. 10 sieve and store for a minimum of 2 weeks unless the manufacturer provides different direction.

2. **Base Coats.** After the action has ceased, run off the putty through a No. 8 sieve. Add sand up to equal parts by weight, all of the hair required, and store for a minimum of 2 weeks.

The mortar should be protected from freezing. One hundred lb, or 1.25 cu ft, of quicklime will make about 3½ cu ft of lime putty or paste. The volume of a lime barrel is about 3 cu ft. **Back plaster** is a scratch coat applied to the back of laths fastened to the boarding between the studs. **Scratch coat** consists of mortar lime, sand, and considerable hair—about 2 to 2½ bu to 200 lb of quicklime. It is put directly on the laths. **Brown coat** consists of mortar with less hair, about 1 to 1½ bu per 200 lb of quicklime. When three-coat work is used, this is troweled directly on to the scratch coat. **Skim coat** is a finish coat composed of lime putty and fine white sand. It is placed in two layers and troweled to a hard finish. **Gaged skim coat** is skimming mixed with a certain

amount of plaster of paris, which makes it practically a hard finish. **Hard finish** consists of 1 part lime putty to 1 or 2 parts plaster of paris.

**Keene's cement,** which is an anhydrous calcined gypsum with an accelerator, is much used as a hard-finish plaster.

**Gypsum plaster** (ASTM C28) lacks the plasticity and sand-carrying capacity of lime plaster but is widely used because of its more rapid hardening and drying and because of the uniformity obtainable as the result of its being put up in bags ready-mixed for use.

Gypsum ready-mixed plaster should contain not more than 3 cu ft of mineral aggregate per 100 lb of calcined gypsum plaster, to which may be added fiber and material to control setting time and workability. Gypsum neat plaster used in place of sanded plaster for second coat should contain at least 66 percent $CaSO_4 \cdot \frac{1}{2}H_2O$; the remainder may be fiber and retarders. Calcined gypsum for finishing coat may be white or gray. If it contains no retarder, it should set between 20 and 40 min; if retarded, it should set between 40 min and 6 hr.

**Cement plaster** is used where a very hard or strong plaster is required, e.g., for thin metal-lath partitions or as a fire protection. It should contain not more than 2 parts sand, by dry and loose volume, to 1 part portland cement. Lime putty or hydrated lime is added up to 15 percent by volume of the cement.

**Table 4. Tests by New York Board of Water Supply of 1:3 Mortar Made with Sands of Different Mechanical Analyses**

(Taylor, Thompson, and Smulski)

| Percentages passing sieves | | | | Tensile test, lb/in² | | Compression test, lb/in² | |
|---|---|---|---|---|---|---|---|
| No. 4 | No. 8 | No. 50 | No. 100 | 7 days | 90 days | 7 days | 90 days |
| 100 | 70 | 12 | 5 | 213 | 613 | 2690 | 5640 |
| 100 | 86 | 21 | 6 | 263 | 412 | 1915 | 4660 |
| 100 | 99 | 26 | 2 | 177 | 325 | 905 | 2170 |
| 100 | 97 | 28 | 6 | 178 | 282 | 1070 | 1500 |
| 100 | 94 | 44 | 12 | 139 | 228 | 905 | 1130 |
| 100 | 100 | 52 | 14 | 122 | 170 | 275 | 810 |
| 100 | 100 | 94 | 48 | 80 | 149 | 330 | 490 |

**Table 5. Amount of Mortar Required for a Cubic Yard of Stone Masonry**

| Description of masonry | Mortar, cu yd | |
| --- | --- | --- |
| | Min | Max |
| Ashlar, 18 in courses and ¼ in joints | 0.03 | 0.04 |
| Ashlar, 12 in courses and ¼ in joints | 0.06 | 0.08 |
| Rubble, small, rough stones | 0.33 | 0.40 |
| Rubble, large stones, rough hammer-dressed | 0.20 | 0.30 |
| Squared stone masonry, 18 in courses and ¾ in joints | 0.12 | 0.15 |
| Squared stone masonry, 12 in courses and ¾ in joints | 0.20 | 0.25 |

**Two-coat work** is used for the cheaper class of houses for plastering on wood laths. For the first coat, the mortar is usually spread on in one layer on the walls and two layers on the ceiling, is smoothed with a **darby** (a board with two handles), and when nearly dry, is floated with a wooden hand float. The finish coat may be of skim or hard finish. The total thickness will be ½ to ⅝ in.

**Three-coat work** has a first or scratch coat of mortar spread on and scratched when nearly dry. The second coat or brown coat consists of mortar, which is smoothed off, darbied, and floated with a wooden hand float. This is then covered, for the third coat, with skim or hard finish. The first coat on metal or wire lathing for three-coat work requires a fatter mortar than for laths; ¾-in. grounds or rails are generally used. The total thickness for three-coat work is, therefore, from ¾ to ⅞ in.

Under moisture conditions where the plaster will not dry rapidly, **curing** at temperatures above 60 and below 90°F is absolutely necessary if cracking is to be avoided.

**Stucco** is used for exterior plastering and is applied to brick or stone or is plastered onto wood or metal lath. For covering wooden buildings, the stucco is plastered either on wood lath or on metal lath in three coats, using mortar similar to that for brick or stone. Concrete in northern climates exposed to frost should never be plastered but should be finished by rubbing down with carborundum brick or similar tool when the surface is comparatively green. It may also be tooled in various ways. Whenever stucco is used, extreme care must be taken to get a good bond to the supporting surface.

For three-coat work on masonry or wood lath, the first or scratch coat should average ¼ in thick outside the lath or surface of the brick. The thickness of the second coat should be ⅜ to ½ in, while the finish coat should be thin, i.e., ⅛ in or not more than ¼ in. The second coat should generally be applied 24 h after the first or scratch coat. The finish coat

should not be applied in less than 1 week after the second coat. Proportions of mix for all coats may be ⅕ part hydrated lime, 1 part portland cement, 3 parts fairly coarse sand, measured by volume. Stucco work should not be put on in freezing weather and must be kept moist for at least 7 days after application of the mortar.

**Coloring of Stucco** Mineral colors, if used, should be of such composition that they will not be affected by cement, lime, or the weather. The best method is to use colored sands when possible. The most satisfactory results with colored mortar are obtained by using white portland cement. Prepared patented stuccos which are combinations of cement, sand, plasticizers, waterproofing agents, and pigment are widely used.

## CONCRETE

Concrete is made by mixing cement and an aggregate composed of hard inert particles of varying size, such as a combination of sand or broken-stone screenings, with gravel, broken stone, lightweight aggregate, or other material. Portland cement should always be used for reinforced concrete, for mass concrete subjected to stress, and for all concrete laid under water.

**Proportioning Concrete** Compressive strength is generally accepted as the principal measure of the quality of concrete, and although this is not entirely true, there is an approximate relation to compressive strength to the other mechanical properties. Methods of proportioning generally aim to give concrete of a predetermined compressive strength.

The concrete mixture is proportioned or designed for a particular condition in various ways: (1) arbitrary selection based on experience and common practice, such as 1 part cement, 2 parts sand, 4 parts stone (written **1:2:4**); (2) proportioning on the basis of the water-cement ratio, either assumed from experience or determined by trial mixtures with the given materials and conditions; (3) combining materials on the basis of either the voids in the aggregates or mechanical-analysis curves so as to obtain the least voids and thus concrete of maximum density for a given cement content.

The method of arbitrary selection, formerly in most common use, has been largely displaced by the water-cement-ratio method. The arbitrary-proportioning method by volume or weight is simple to use but is not economical, nor does it give the best possible results with a given aggregate. In a modifica-

**Table 6. Amount of Mortar Required for Brickwork**
(From investigations of Taylor and Thompson)

| Mortar for common brick (20 brick per cu ft,* size 8 × 3¾ × 2¼ in, with ⅜ in joints) | | | Mortar for selected face brick† (8.8 brick per sq ft, size 7¾ × 3½ × 2¼ in, with ⁵⁄₁₆ in joints) | | | Mortar for pressed face brick‡ (7½ brick per sq ft, size 8¼ × 4 × 2¼ in, with ¼ in joints) | | |
| --- | --- | --- | --- | --- | --- | --- | --- | --- |
| Thickness of joints, in | Cu ft per cu ft of masonry | Cu ft per 1,000 brick | Thickness of joints, in | Cu ft per sq ft of surface | Cu ft per 1,000 brick | Thickness of joints, in | Cu ft per sq ft of surface | Cu ft per 1,000 brick |
| ½ | 0.37 | 19.4 | ⅜ | 0.07 | 8.4 | ¼ | 0.05 | 6.8 |
| ⅜ | 0.28 | 14.0 | ⁵⁄₁₆ | 0.06 | 6.8 | ³⁄₁₆ | 0.035 | 4.7 |
| ¼ | 0.19 | 9.0 | ¼ | 0.05 | 5.6 | ⅛ | 0.02 | 2.6 |

*Actual brick laid in wall. For figuring quantities use 22½ brick per ft³.
†Laid with headers every sixth course.
‡Laid running bond (all stretchers).

**Table 7. Maximum Permissible Water-Cement Ratios (Gal per Bag)[e] for Different Types of Structures and Degrees of Exposure**
(From Recommended Practice for Selecting Proportions for Concrete, *ACI Std.* 613.)

| Type of structure | Exposure conditions[a] | | | | | |
|---|---|---|---|---|---|---|
| | Severe wide range in temperature or frequent alternations of freezing and thawing (air-entrained concrete only) | | | Mild temperature, rarely below freezing, or rainy or arid | | |
| | In air | At the waterline or within the range of fluctuating water level or spray | | In air | At the waterline or within the range of fluctuating water level or spray | |
| | | In fresh water | In seawater or in contact with sulphates[b] | | In fresh water | In seawater or in contact with sulphates[b] |
| Thin sections, such as railings, curbs, sills, ledges, ornamental or architectural concrete, reinforced piles, and pipe and all sections with less than 1 in. concrete cover over reinforcing | 5.5 | 5.0 | 4.5[c] | 6 | 5.5 | 4.5[c] |
| Moderate sections, such as retaining walls, abutments, piers, girders, beams | 6.0 | 5.5 | 5.0[c] | d | 6.0 | 5.0[c] |
| Exterior portions of heavy (mass) sections | 6.5 | 5.5 | 5.0[c] | d | 6.0 | 5.0[c] |
| Concrete deposited by tremie under water | ... | 5.0 | 5.0 | ... | 5.0 | 5.0 |
| Concrete slabs laid on the ground | 6.0 | ... | ... | d | | |
| Concrete protected from the weather, interiors of buildings, concrete below ground | d | ... | ... | d | | |
| Concrete which will later be protected by enclosure or backfill but which may be exposed to freezing and thawing for several years before such protection is offered | 6.0 | ... | ... | d | | |

[a] Air-entrained concrete should be used under all conditions involving severe exposure and may be used under mild-exposure conditions to improve workability of the mixture.
[b] Soil or ground water containing sulphate concentrations of more than 0.2 percent.
[c] When sulphate-resisting cement is used, maximum water-cement ratio may be increased by 0.5 gal per bag.
[d] Water-cement ratio should be selected on basis of strength and workability requirements.
[e] U.S. gallons (8.33 lb/gal) and 94-lb bags of cement.

tion of this method, the minimum cement content is specified along with a maximum water content per bag of cement. The water-cement-ratio method depends on the principle that the strength of the concrete with given aggregates and cement bears a direct relation to the ratio of the water to the cement present. The smaller the water-cement ratio, as long as the mix is workable, the higher the strength of the resulting concrete. Methods devised by Abrams and based on the use of the fineness modulus of the aggregates make it possible to compute the proportions of cement with the fine and coarse aggregates so as to give the most economical mixtures. As the water-cement-ratio–strength relation changes for different materials, it is necessary to make preliminary tests of the materials to determine their exact water-cement-ratio–strength relation.

**Steps in proportioning by the water-cement-ratio method:**

1. Select the proper water-cement ratio for the job; see Tables 7 and 11.
2. Select the proper slump, see Table 9.
3. Estimate the quantity of water needed to provide the required slump. Table 8 gives the approximate quantities needed for a 3-in slump. To raise or decrease the slump 1 in, raise or lower the amount of water 3 percent.
4. Select the maximum size of aggregate that will be suitable for the job. The maximum size is determined by the aggregate available, spacing of reinforcement, and thickness of section.
5. Determine the percentage of the total aggregate that is fine aggregate from Table 8.
6. Compute the starting-mix proportions. Adjustment can be made on the job, but it is better to make trial mixes in the laboratory since only by actually mixing and testing the concrete can its properties be known. The specific gravity of the cement, specific gravity of the aggregates, and moisture content of the aggregates are needed to compute the mix quantities. The quantities for 1 yd³ can be computed as follows:

*a.* Determine the cement content from Table 8 or by dividing the water content, step 3, by the water-cement ratio, step 1.

*b.* Add the absolute volume of water to the absolute volume of the cement and subtract this volume from 1 yd³. If the concrete is air-entrained, the volume of air must also be

**Table 8. Quantities of Material for 1 Cubic Yard of Concrete**

(For coarse sand, fineness modulus of about 2.90, and 3 to 4 in of slump.) (Adapted from "Design of Concrete Mixtures," ST100, Portland Cement Assoc.)

| Water-cement ratio, gal per sack | Entrained air,* percent | Water, gal | Cement,† sack per yd | Fine aggregate, lb | Coarse aggregate, lb | Fine aggregate, percent of total aggregate |
|---|---|---|---|---|---|---|
| NON-AIR-ENTRAINED | | | | | | |
| For ¾-in -max-size aggregate: | | | | | | |
| 5 | 2 | 41 | 8.2 | 1,180 | 1,650 | 42 |
| 6 | 2 | 41 | 6.9 | 1,280 | 1,650 | 44 |
| 7 | 2 | 41 | 5.9 | 1,360 | 1,650 | 45 |
| 8 | 2 | 41 | 5.2 | 1,410 | 1,650 | 46 |
| For 1-in -max-size aggregate: | | | | | | |
| 5 | 1.5 | 39 | 7.8 | 1,100 | 1,820 | 38 |
| 6 | 1.5 | 39 | 6.5 | 1,210 | 1,820 | 40 |
| 7 | 1.5 | 39 | 5.6 | 1,280 | 1,820 | 41 |
| 8 | 1.5 | 39 | 4.9 | 1,330 | 1,820 | 42 |
| For 1½-in -max-size aggregate: | | | | | | |
| 5 | 1 | 36 | 7.2 | 1,030 | 2,030 | 34 |
| 6 | 1 | 36 | 6.0 | 1,120 | 2,030 | 36 |
| 7 | 1 | 36 | 5.2 | 1,190 | 2,030 | 37 |
| 8 | 1 | 36 | 4.5 | 1,240 | 2,030 | 38 |
| AIR-ENTRAINED | | | | | | |
| For ¾-in -max-size aggregate: | | | | | | |
| 5 | 6 | 36 | 7.2 | 1,190 | 1,650 | 42 |
| 6 | 6 | 36 | 6.0 | 1,280 | 1,650 | 44 |
| 7 | 6 | 36 | 5.2 | 1,350 | 1,650 | 45 |
| 8 | 6 | 36 | 4.5 | 1,400 | 1,650 | 46 |
| For 1-in -max-size aggregate: | | | | | | |
| 5 | 6 | 34 | 6.8 | 1,090 | 1,820 | 37 |
| 6 | 6 | 34 | 5.7 | 1,180 | 1,820 | 39 |
| 7 | 6 | 34 | 4.9 | 1,240 | 1,820 | 41 |
| 8 | 6 | 34 | 4.3 | 1,290 | 1,820 | 42 |
| For 1½-in -max-size aggregate: | | | | | | |
| 5 | 5 | 32 | 6.4 | 1,000 | 2,030 | 33 |
| 6 | 5 | 32 | 5.4 | 1,080 | 2,030 | 35 |
| 7 | 5 | 32 | 4.6 | 1,140 | 2,030 | 36 |
| 8 | 5 | 32 | 4.0 | 1,190 | 2,030 | 37 |

*For non-air-entrained concrete, this is the expected amount of entrapped air.

†94 lb per sack.

subtracted at this point. The volume remaining is the amount of aggregates needed.

*c.* Compute the quantities of fine and coarse aggregates needed from the percentage of the total that is fine aggregate, given in Table 8. The computed quantities of aggregates may be different from those shown in Table 8 because of different water content and different specific gravities of aggregates.

The trial batches or first job batches may indicate adjustments needed for workability, slump, or strength.

If the above computations are not made, the suggested proportions in Table 8 may be used for trial batches.

Table 7 is given by ACI as a guide for selection of the maximum water-cement ratio from the standpoint of the exposure of the concrete.

In Table 11, the maximum water content includes the free moisture contained in the aggregates. The water absorbed into the pores of the aggregate is not available for hydration of the cement and is therefore not included in the water content. The water of absorption is usually less than 1 percent by weight. The coarser the aggregate (high fineness modulus), the less free water it will contain. Average allowances for free moisture are as follows in percentage of the dry weight of the material:

|  | Dry | Damp | Wet |
|---|---|---|---|
| Gravel | 0.2 | 1.0 | 2.0 |
| Sand | 2.0 | 4.0 | 7.0 |

The **dry rodded weight** is determined by filling a container, of a diameter approximately equal to the depth, with the aggregate in three equal layers, each layer being rodded 25 times using a bullet-pointed rod ⅝ in diam and 24 in long.

**Measurement of materials** is usually done by weight. The bulking effect of moisture, particularly on the fine aggregate, makes it difficult to keep proportions uniform when volume

**Table 9. Suggested Slump**

| Class of concrete | Limiting values of slump, in | |
|---|---|---|
|  | Min | Max |
| Caissons, heavy foundations, massive walls......... | 1 | 4 |
| Road slabs, floor slabs on ground, pavements....... | 1 | 4 |
| Ordinary reinforced building walls, ............... | 4 | 6 |
| Slabs, beams, girders......................... | 4 | 7 |
| Thin walls, columns........................... | 4 | 7 |

measurement is used. Water is batched by volume or weight.

**Mixing** In order to get good concrete, the cement and aggregates must be throughly mixed so as to obtain a homogeneous mass and coat all particles with the cement paste. Mixing may be done either by hand or by machine, although hand mixing is rare today. **Hand mixing** must be done thoroughly on a watertight mixing platform. A good method of hand mixing is as follows: The sand is measured and spread on the platform; then a proper amount of cement is placed on the top and the two materials are mixed until the color is uniform. In the meantime, screened gravel or stone is placed in the measuring box on the platform. After the box is removed, the gravel is hollowed out slightly in the center and the sand and cement are shoveled on top, covering the gravel with a layer of even thickness. The water is poured on top of these layers and the whole mass is mixed by shovels, each shovelful of material being turned and spread on the platform about 2 ft from its original position. The operation is repeated several times. As a rule, four turnings are sufficient.

**Quality Control of Concrete** Control methods include measuring materials by weight, allowing for the water content of the aggregates; careful limitation of the total water quantity to that designed; frequent tests of the aggregates and changing the proportions as found necessary to maintain yield and workability; constant checks on the consistency by the slump test, careful attention to the placing of steel and the filling of forms; layout of the concrete distribution system so as to eliminate segregation; check on the quality of the concrete as placed by means of specimens made from it as it is placed in the forms; and careful attention to proper curing of the concrete. The field specimens of concrete are usually 6 in diam by 12 in high for aggregates up to $1\frac{1}{2}$ in max size and 8 by 16 in for 3-in aggregate. They are made by rodding the concrete in three layers, each layer being rodded 25 times using a $\frac{5}{8}$-in diam bullet-pointed rod 24 in long.

**Machine-mixed concrete** is employed almost universally. The mixing time for the usual batch mixer of 1 yd³ or less capacity should not be less than 1 min from the time all materials are in the mixer until the time of discharge. Larger mixers require 25 percent increase in mixing time per $\frac{1}{2}$-yd³ increase in capacity. Increased mixing times up to 5 min increase the workability and the strength of the concrete. Mixing should always continue until the mass is homogeneous.

**Concrete mixers** of the batch type give more uniform results; few continuous mixers are used. Batch mixers are either (1) rotating mixers, consisting of a revolving drum or a square box revolving about its diagonal axis and usually provided with deflectors and blades to improve the mixing; or (2) paddle mixers, consisting of a stationary box with movable paddles which perform the mixing. Paddle mixers work better with relatively dry, high-sand, small-size aggregate mixtures and mortars and are less widely used than rotating mixers.

**Ready-mixed concrete** is (1) proportioned and mixed at a central plant (central-mixed concrete) and transported to the job in plain trucks or agitator trucks, or (2) proportioned at a central plant and mixed in a mixer truck (truck-mixed concrete) equipped with water tanks, during transportation. Ready-mixed concrete is largely displacing job-mixed concrete in metropolitan areas. The truck agitators and mixers are essentially rotary mixers mounted on trucks. There is no deleterious effect on the concrete if it is used within 1 h after the cement has been added to the aggregates.

Materials for concrete are also **centrally batched**, particularly for road construction, and transported to the site in batcher trucks with compartments to keep aggregates separated from the cement. The truck discharges into the charging hopper of the job mixer. Road mixers sometimes are arranged in series—the first mixer partly mixes the materials and discharges into a second mixer, which completes the mixing.

**Consistency of Concrete** The consistency to be used depends upon the character of the structure. The proportion of water in the mix is of vital importance. A very wet mixture of the same cement content is much weaker than a dry or mushy mixture. Dry concrete can be employed in dry locations for mass foundations provided that it is carefully spread in layers not over 6 in thick and is thoroughly rammed. Medium, or quaking, concrete is adapted for ordinary mass-concrete uses, such as foundations, heavy walls, large arches, piers, and abutments. Mushy concrete is suitable as rubble concrete and reinforced concrete, for such applications as thin building walls, columns, floors, conduits, and tanks. A medium, or quaking, mixture has a tenacious, jellylike consistency which shakes on ramming; a mushy mixture will settle to a level surface when dumped in a pile and will flow very sluggishly into the forms or around the reinforcing bars; a dry mixture has the consistency of damp earth.

The two methods in common use for measuring the consistency or workability of concrete are the slump test and the flow test. In the **slump test,** which is the more widely used of the two, a form shaped as a frustum of a cone is filled with the concrete and immediately removed. The slump is the subsidence of the mass below its height when in the cone. The form has a base of 8 in diam, a top of 4 in diam, and a height of 12 in. It is filled in three 4-in layers of concrete, each layer being rodded by 25 strokes of a $\frac{5}{8}$-in rod, 24 in long and bullet-pointed at the lower end.

A test using the penetration of a half sphere, called the **Kelly Ball test,** is sometimes used for field control purposes. A 1-in penetration by the Kelly ball corresponds to about 2 in of slump.

Usual limitations on the consistency of concrete as measured by the slump test are given in Table 9.

The **consistency** of concrete has some relation to its **workability,** but a lean mix may be unworkable with a given slump and a rich mix may be very workable. Certain admixtures tend to lubricate the mix and, therefore, increase the workability at certain slumps. The slump at all times should be as small as possible consistent with the requirements of handling and placing. A slump over 7 in is usually accompanied by segregation and low strength of concrete.

**Forms** for concrete should maintain the lines required and prevent leakage of mortar. The pressure on forms is equivalent to that of a liquid with the same density as the concrete, and of the depth placed within 2 h. Dressed lumber or plywood is used for exposed surfaces, and rough lumber for unexposed areas. Wood or steel forms should be oiled before placing concrete.

**Placement** of concrete for most structures is by chutes or buggies. Chutes should have a slope not less than one vertical to two horizontal; the use of flatter slopes encourages the use of excess water, leading to segregation and low strengths. Buggies are preferable to chutes because they handle drier concrete and allow better placement control. Drop-bottom buckets are desirable for large projects and dry concrete. Concrete pumped through pipelines by mechanically applied pressure is sometimes economical for construction spread over large areas. Concrete for tunnels is placed by pneumatic pumps. Underwater concrete is deposited by drop-bottom bucket or by a tremie or pipe. Such concrete should have a cement-content increase of 15 percent to allow for loss of cement in placement.

**Compaction** of concrete (working it into place) is accomplished by manual spading, by walking in, or "booting," the concrete, and by tampers or vibrators. Vibrators are applied to the outside of the forms and to the surface or interior of the concrete; they should be used with care to avoid producing segregation. The frequency of vibrators is usually between 3,000 and 10,000 pulsations per min.

**Curing** of concrete is necessary to ensure proper hydration. Concrete should be kept moist for a period of at least 7 days, and the temperature should not be allowed to fall below 50°F for at least 3 days. Sprayed-on membrane curing compounds may be used to retain moisture. Special precautions must be taken in cold weather and in hot weather.

**Weight of Concrete**   The following are average weights, lb/ft³, of portland-cement concrete:

| | | | |
|---|---|---|---|
| Sand-cinder concrete | 112 | Limestone concrete | 148 |
| Burned-clay or shale concrete | 105 | Sandstone concrete | 143 |
| Gravel concrete | 148 | Traprock concrete | 155 |

**Watertightness**   Concrete can be made practically impervious to water by proper proportioning, mixing, and placing. Leakage through concrete walls is usually due to poor workmanship or occurs at the joints between 2 days' work or through cracks formed by contraction. New concrete can be bonded to old by wetting the old surface, plastering it with neat cement, and then placing the concrete before the neat cement has set. It is almost impossible to prevent contraction cracks entirely, although a sufficient amount of reinforcement may reduce their width so as to permit only seepage of water. For best results, a low-volume-change cement should be used with a concrete of a quaking consistency; the concrete should be placed carefully so as to leave no visible stone pockets, and the entire structure should be made without joints and preferably in one continuous operation. The best waterproofing agent is an additional proportion of cement in the mix. The concrete should contain not less than 6 bags of cement per yd³.

For maximum watertightness, mortar and concrete may require more fine material than would be used for maximum strength. Gravel produces a more watertight concrete than broken stone under similar conditions. Patented compounds are on the market for producing watertight concrete, but under most conditions, equally good results can be obtained for less cost by increasing the percentage of cement in the mix. Membrane waterproofing, consisting of asphalt or tar with layers of felt or tarred paper, or plastic or rubber sheeting in extreme cases, is advisable where it is expected that cracks will occur. Mortar troweled on very hard may produce watertight work.

Concrete to be placed through water should contain at least 7 bags of cement per yd³ and should be of a quaking consistency.

According to Fuller and Thompson (*Trans. ASCE*, **59**, p. 67), watertightness increases (1) as the percentage of cement is increased and in a very much larger ratio; (2) as the maximum size of stone is increased, provided the mixture is homogeneous; (3) materially with age; and (4) with thickness of the concrete, but in a much larger ratio. It decreases uniformly with increase in pressure and rapidly with increase in the water-cement ratio.

**Air-Entrained Concrete**   The entrainment of from 3 to 6 percent by volume of air in concrete by means of vinsol resin or other air-bubble-forming compounds has, under certain conditions, improved the resistance of concrete in roads to frost and salt attack. The air entrainment increases the workability and reduces the compressive strength and the weight of the concrete. As the amount of air entrainment produced by a given percentage of vinsol resin varies with the cement, mixture, aggregates, slump, and mixing time, good results are dependent on very careful control.

**Concrete for Masonry Units**   Mixtures for concrete masonry units, which are widely used for walls and partitions, employ aggregates of a maximum size of ½ in and are proportioned either for casting or for machine manufacture. Cast units are made in steel or wooden forms and employ concrete slumps of from 2 to 4 in. The proportions used are 1 part cement and 3 to 6 parts aggregate by dry and loose volumes. The forms are stripped after 24 h, and the blocks piled for curing. Most **blocks** are made by machine, using very dry mixtures with only enough water present to enable the concrete to hold together when formed into a ball. The proportions used are 1 part cement and 4 to 8 parts aggregate by dry and loose volumes. The utilization of such a lean and dry mixture is possible because the blocks are automatically tamped and vibrated in steel molds. The blocks are stripped from the molds at once, placed on racks on trucks, and cured either in air on in steam. High-pressure-steam (50 to 125 lb/in²) curing develops the needed strength of the blocks in less than 24 h. Most concrete blocks are made with lightweight aggregates such as cinders and burned clay or shale in order to reduce the weight and improve their acoustical and thermal insulating properties. These aggregates not only must satisfy the usual requirements for gradation and soundness but must be limited in the amount of coal, iron, sulfur and phosphorus present because of their effect on durability, discoloration and staining, fire resistance, and the formation of "pops." Pops form on the surface of concrete blocks using cinders or burned clay and shale as a result of the increase in volume of particles of iron, sulfur, and phosphorus when acted on by water, oxygen, and the alkalies of the cement.

### Strength of Concrete

The strength of concrete increases (1) with the quantity of cement in a unit volume, (2) with the decrease in the quantity of mixing water relative to the cement content, and (3) with the density of the concrete. Strength is decreased by an excess of sand over that required to fill the voids in the stone and give sufficient workability. The volume of fine aggregate should not exceed 60 percent of that of coarse aggregate 1½-in max size or larger.

**Compressive Strength**   Table 11 give the results obtainable with first-class materials and under first-class conditions. Growth in strength with age depends in a large measure upon the consistency characteristics of the cement and upon the curing conditions. Table 10 gives the change in relative strength with age for several water-cement ratios and a wide range of consistencies for a cement with a good age-strength gain relation. Many normal portland cements today show very little gain in strength after 28 days.

**Tensile Strength**   The tensile strength of concrete is of less importance than the crushing strength, as it is seldom relied

**Table 10. Variation of Compressive Strength with Age**
(Strength at 28 days taken as 100)

| Water-cement ratio by volume, gal per bag of cement | 3 days | 7 days | 28 days | 3 months | 1 year |
|---|---|---|---|---|---|
| 5 | 40 | 75 | 100 | 125 | 145 |
| 7 | 30 | 65 | 100 | 135 | 155 |
| 9 | 25 | 50 | 100 | 145 | 165 |

upon. The true tensile strength is about 8 percent of the compression strength and must not be confused with the tensile fiber stress in a concrete beam, which is greater.

Recently, a method of obtaining the tensile of concrete by loading a cylinder in compression along two diametrically opposite generators has come into use. Nearly all the concrete in the plane of the two load strips is in tension. The tensile strength may be computed as $T = 2P/\pi ld$, where $T$ = splitting tensile strength, lb/in²; $P$ = maximum load, lb; $l$ = length of cylinder, in; $d$ = diameter of cylinder, in. Average values of tensile splitting strength are given in Table 11.

**Transverse Strength**  There is an approximate relationship between the tensile fiber stress of plain concrete beams and their compressive strength. The modulus of rupture is greatly affected by the size of the coarse aggregate and its bond and transverse strength. Quartzite generally gives low-modulus-of-rupture concrete. Table 11 indicates that the relation of the modulus of rupture of plain concrete beams at 28 days to the water-cement ratio is similar to that of the compressive strength of concrete to the water-cement ratio.

The transverse or beam test is generally used for checking the quality of concrete used for roads. The standard beam is 6 by 6 by 20 in, tested on an 18-in span and loaded at the one-third points.

The strength of concrete in direct **shear** is about 20 percent of the compressive strength.

**Deformation Properties**  Young's modulus for concrete varies with the aggregates used and the concrete strength but will usually be 3 to 5 × 10⁶ lb/in² in short-term tests. According to the ACI Code, the modulus may be expressed as $E_c = 33w^{3/2} \sqrt{f_c}$, where $w$ = unit weight of concrete in lb/ft³ and both $f_c$ and $\sqrt{f_c}$ have lb/in² units. For normal-weight concretes, $E_c = 57{,}000 \sqrt{f_c}$ lb/in².

In addition to the instantaneous deformations, concrete shrinks with drying and is subjected to creep deformations developing with time. Shrinkage strains, which are independent of external stress, may reach 0.05 percent at 50 percent RH, and when restrained may lead to cracking. Creep strains typically reach twice the elastic strains. Both creep and shrinkage depend strongly on the particular aggregates used in the concrete, with limestone usually resulting in the lowest values and river gravel or sandstone in the largest. In all cases, the longer the period of wet curing of the concrete, the lower the final creep and shrinkage values.

**Deleterious Actions and Materials**

**Freezing** retards the setting and hardening of portland-cement concrete and is likely to lower its strength permanently. On exposed surfaces such as walls and sidewalks placed in freezing weather, a thin scale may crack from the surface. Natural cement is completely ruined by freezing.

Concrete laid in freezing weather or when the temperature is likely to drop to freezing should have the materials heated and should be protected from the frost, after laying, by suitable covering or artificial heat. The use of calcium chloride, salt, or other ingredients in sufficient quantities to lower the temperature of freezing significantly is not permitted, since the concrete would be adversely affected.

**Mica and Clay in Sand**  Mica in sand, if over 2 percent, reduces the density of mortar and consequently its strength, sometimes to a very large extent. In crushed-stone screenings, the effect of the same percentage of mica in the natural state is less marked. Black mica, which has a different crystalline form, is not injurious to mortar. Clay in sand may be injurious because it may introduce too much fine material or form balls in the concrete. When not excessive in quantity, it may increase the strength and watertightness of a mortar of proportions 1:3 or leaner.

**Mineral oils** which have not been disintegrated by use do not injure concrete when applied externally. Animal fats and vegetable oils tend to disintegrate concrete unless it has thoroughly hardened. Concrete after it has thoroughly hardened resists the attack of diluted organic acids but is disintegrated by even dilute inorganic acids; protective treatments are magnesium fluosilicate, sodium silicate, or linseed oil. Green concrete is injured by manure but is not affected after it has thoroughly hardened.

**Electrolysis** injures concrete under certain conditions, and electric current should be prevented from passing through it. (See also Sec. 6, Corrosion.)

**Seawater** attacks cement and may disintegrate concrete. Deleterious action is greatly accelerated by frost. To prevent serious damage, the concrete must be made with a sulfate-resisting cement, a rich mix (not leaner than 1:2:4), and exceptionally good aggregates, including a coarse sand, and must be

**Table 11. Strength of Plain Concrete at 28 Days**

| | | | | | |
|---|---|---|---|---|---|
| Max water content, gal per bag of cement | 5 | 5.5 | 6 | 6.5 | 7.0 |
| Compressive strength, lb/in² | 4,000 | 3,700 | 3,350 | 3,000 | 2,650 |
| Modulus of rupture, lb/in² | 650 | 625 | 600 | 550 | 500 |
| Tensile strength (split cyl. method), lb/in² | 350 | 325 | 300 | 275 | 250 |

allowed to harden thoroughly, at least 7 days, before it is touched by the seawater. Although tests indicate that there is no essential difference in the strengths of mortars gaged with fresh water and with seawater, the latter tends to retard the setting and may increase the tendency of the reinforcement to rust.

# WATER

## Kenneth A. Roe

REFERENCES: AID Desalination Manuals, Oct. 1972, Department of State. Ellis, "Fresh Water from the Ocean." Ronald. Saline Water Conversion Report, *OSW Annual Reports*, Department of the Interior. "Water Quality and Treatment." AWWA. Water, *Chem. Eng.*, June 10, 1963; *Chem. Eng. Progr.*, SS107, 1971. "Drinking Water Standards," U.S. Public Health Service, 1962. Arnold, Thermal Pollution of Surface Supplies, *Jour. AWWA*, Nov. 1962. "A B Seas of Desalting," OSW publication, 1968.

## WATER RESOURCES

Oceans cover 70 percent of the earth's surface and are the basic source of all water. Ocean waters contain about 3½ percent by weight of dissolved materials, generally varying from 32,000 to 36,000 ppm and as high as 42,000 ppm in the Persian Gulf.

About 50 percent of the sun's energy falling on the ocean causes evaporation. The vapors form clouds which precipitate pure water as rain. While most rain falls on the sea, land rainfall returns to the sea in rivers or percolates into the ground and back to the sea or is reevaporated. This is known as the **hydrological cycle**. It is a closed distillation cycle, without additions or losses from outer space or from the interior of the earth.

Water supply to the United States depends on an annual rainfall averaging 30 in (76 cm) and equal to 1,664,800 billion gal (6,301 Gm³) per year, or 4,560 billion gal (17.3 Gm³) per day. About 72 percent returns to the atmosphere by direct evaporation and transpiration from trees and plants. The remaining 28 percent, or 1,277 billion gal (4.8 Gm³) daily, is the maximum supply available. This is commonly called runoff and properly includes both surface and underground flows. Between 33 and 40 percent of runoff appears as a groundwater.

Two-thirds of the runoff passes into the ocean as flood flow in one-third of the year. By increased capture, it would be possible to retain about one-half of the 1,277 billion gal (4.8 Gm³), or 638 billion gal (2.4 Gm³), per day as **maximum usable water.**

**Withdrawal use** is the quantity of water removed from the ground or diverted from a body of surface water. **Consumptive use** is the portion of such water that is discharged to the atmosphere or incorporated in growing vegetation or in industrial or food products.

The estimated withdrawal use of water in the United States in 1970 was 372 billion gal per day, including some saline waters (see Table 1).

Fresh-water withdrawal is about 25 percent of runoff, and consumptive use is about 7 percent of runoff. The consumptive use of water for irrigation is about 64 percent, but another 15 percent is allowed for transmission and distribution losses.

Dividing total water withdrawal by the population of the United States shows a 1970 water use, the **Water Index,** of 1,814 gal (6,866 l) per capita day (gpcd). It reflects the great industrial and agricultural uses of water, because man can survive on a theoretical minimum of 1 qt (0.946 l) of water per day. The average **household use** of water in American urban areas is 30 to 60 gpcd (114 to 227 l). But since municipal waterworks also supply industrial customers, the average **production of water utilities** is 166 gpcd (628 l), ranging from 40 to 400 gpcd (150 to 1,500 l). These figures include 15 percent distribution losses.

**Additions to water resources** for the future can be made by (1) increase in storage reservoirs, (2) injection of used water or flood water into underground strata called **aquifers,** (3) covering reservoirs with films to reduce evaporation, (4) rainmaking, (5) saline-water conversion.

It is equally important to improve the efficient use of water supplies by (1) multiple use of cooling water, (2) use of air cooling instead of water cooling, (3) use of cooling towers, (4) reclamation of waste waters, both industrial and sewage, (5)

**Table 1. Water Use in the United States, 1970***
[All quantities in billion gallons/day ($\times$ 0.00378 = Gm³/day)]

|  | Total withdrawal | Fresh water | Saline water | Consumptive use |
|---|---|---|---|---|
| Irrigation | 130 | 130 |  | 73 |
| Public water utilities | 27 | 27 |  | 5.4 |
| Rural domestic | 4.5 | 4.5 |  | 1.69 ⎫ 3.4 |
|  |  |  |  | 1.71 ⎭ |
| Industrial and miscellaneous | 46.2 | 31.8 | 14.4 | 9 |
| Steam-electric plants | 163.8 | 124.2 | 39.6 | 0.5 |
| Total | 371.5 | 317.5 | 54 | 91.3 |

*Compiled from various sources.

abatement of pollution by treatment rather than by dilution, which requires additional fresh water.

## MEASUREMENTS AND DEFINITIONS

Water **quantities** in this country are measured by U.S. gallons, the larger unit being 1,000 U.S. gal. For agriculture and irrigation, water use is measured in acre-feet, i.e., the amount of water covering 1 acre of surface to a depth of 1 ft. In the British-standard area, the imperial gallon is used. In metric-system areas, water is measured in kilograms and the larger unit is the metric ton, which equals a cubic meter of water, expressed as $m^3$.

### Conversion Table

| |
|---|
| 1 acre-ft = 325,850 U.S. gal or 326,000 approx |
| 1 acre-ft = 1,233 metric tons or $m^3$ |
| 1 acre-ft = 43,560 $ft^3$ |
| 1 imperial gal = 1.20 U.S. gal |
| 1 metric ton = 1,000 kg |
| = 2,204 lb |
| = (264.2 U.S. gal) |
| = 220 imperial gal |
| 1 U.S. ton = 240 U.S. gal |
| 1,000,000 U.S. gal = 3.07 acre-ft |

For stream flow and hydraulic purposes, water is measured in cubic feet per second.

$$1,000,000 \text{ U.S. gal per day} = 1.55 \text{ ft}^3/\text{s} \ (0.044 \text{ m}^3/\text{s})$$
$$= 1,120 \text{ acre-ft} \ (1,380,000 \text{ m}^3)$$
$$\text{per year}$$

Water **costs** are expressed in terms of price per 1,000 gal, per acre-foot, per cubic foot, or per metric ton.

$$10¢ \text{ per 1,000 gal} = \$32.59 \text{ per acre-ft}$$
$$= 0.075¢ \text{ per ft}^3$$
$$= 2.64¢ \text{ per metric ton}$$

Water **quality** is measured in terms of the solids, of any character, which are dissolved in the water. The solids are usually expressed in parts per million or in grains per gallon. One grain equals 1/7,000 lb (64.8 mg). Therefore, 17.1 ppm = 1 grain per U.S. gal. In the metric system, 1 ppm = 1 $g/m^3$ = 1 mg/liter. Standards for drinking water, or **potable water,** have been established by the U.S. Public Health Service. The recommended limit is 500 ppm of total dissolved solids (see Table 2). Potable water must also be bacteriologically safe and relatively free of odor, turbidity, and radioactivity. For temporary use, the California State Board of Public Health has permitted usage of water with up to 1,500 ppm total solids but containing not more than 600 ppm sulfate, 600 ppm chloride, and 150 ppm magnesium.

For **agricultural water,** mineral content up to 700 ppm is considered excellent to good. However, certain elements are undesirable, particularly sodium and boron. The California State Water Resources Board limits Class I irrigation water to:

| | Max ppm |
|---|---|
| Sodium, as % of total sodium, potassium, magnesium, and calcium equivalents . | 60 |
| Boron .......................... | 0.5 |
| Chloride .......................... | 177 |
| Sulfate .......................... | 960 |

Class II irrigation water may run as high as 2,100 ppm total dissolved solids, with higher limits on the specific elements, but whether such water is satisfactory or injurious depends on character of soil, climate, agricultural practice, and type of crop.

Waters containing dissolved salts are called **saline waters,** and the lower concentrations are commonly called **brackish;** these waters are defined, in ppm, as follows:

| | |
|---|---|
| Saline........... | All concentrations up to 42,000 |
| Slightly brackish................. | 1,000–3,000 |
| Brackish....................... | 3,000–10,000 |
| Seawaters, average............... | 32,000–36,000 |
| Brine.......................... | Over 42,000 |

**Hardness** of water refers to the content of calcium and magnesium salts, which may be bicarbonates, carbonates, sulfates, chlorides, or nitrates. Bicarbonate content is called **temporary hardness,** as it may be removed by boiling. The salts in "hard water" increase the amount of soap needed to form a lather and also form deposits or "scale" as water is heated or evaporated.

Hardness is a measure of calcium and magnesium salts expressed as equivalent calcium carbonate content and is usually stated in ppm (or in grains per gal) as follows: very soft water, less than 15 ppm; soft water, 15 to 50 ppm; slightly hard water, 50 to 100 ppm; hard water, 100 to 220 ppm; very hard water, over 220 ppm.

### Table 2.  Potable Water Constituent Limitations

| Inorganic maximum levels | | Some other maximum levels | |
|---|---|---|---|
| Constituent | mg per litre | | Units |
| Arsenic | 0.05 | *Chlorinated organics:* | *mg per litre:* |
| Barium | 1.0 | Endrin | 0.002 |
| Cadmium | 0.010 | Lindane | 0.004 |
| Chromium | 0.05 | Methoxychlor | 0.1 |
| Fluoride | (See USPHS) | Toxaphene | 0.005 |
| Lead | 0.05 | 2, 4-D | 0.1 |
| Mercury | 0.002 | 2, 4, 5-TP Silvex | 0.01 |
| Nitrate (as N) | 10 | *Radionuclides:* | *pCi per litre:* |
| Selenium | 0.01 | Tritium | 20,000 |
| Silver | 0.05 | Strontium 90 | 8 |
| Turbidity (monthly average) | 1 TU | Others (see EPA 40CFR141) | |

SOURCE: EPA Interim Drinking Water Regulations (40CFR141, July 9, 1976).

**Table 3. Variance in Industrial Water Withdrawals***

| Product or user and unit | Withdrawal in gallons ($\times$ 3.78 = litres) | | |
| --- | --- | --- | --- |
| | Maximum | Typical | Minimum |
| Steam-electric power per kWh | 170 | 80 | 1.32 |
| Petroleum refining, per gal of crude oil | 44.5 | 18.3 | 1.73 |
| Steel, per finished ton | 65,000 | 40,000 | 1,400 |
| Soaps, edible oils, per lb | 7.5 | | 1.57 |
| Glass containers, per ton | 667 | | 118 |
| Automobiles, per car | 16,000 | | 12,000 |
| Newsprint, per ton | 26,000 | | 6,000 |
| Cannery, per ton | 2,500 | | 1,200 |

*From data by Wolman, AAAS.

## INDUSTRIAL WATER

The use of water within a given industry varies widely because of conditions of price, availability, and process technology (see Table 3).

When a sufficient water supply of suitable quality is available at low cost, plants tend to use maximum volumes. When water is scarce and costly at an otherwise desirable plant site, improved processes and careful water management can reduce water usage to the minimum. Industrial water may be purchased from local public utilities or self-supplied. Figure 1 shows the distribution of water **sources** among 3,000 typical plants. Small industries usually buy water from local suppliers

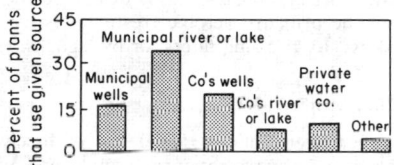

**Fig. 1** Distribution of water sources among 3,000 typical plants. (*Chem. Eng., June 10, 1963.*)

at costs ranging from 12 to 28¢ per 1,000 gal (3.2 to 7.4¢ per m³). Larger industries may provide their own water from sources available to the site. Costs run from 1 to 11¢ per 1,000 gal and include collection, pumping, distribution, storage tanks, and fire-protection system (see Sec. 12). Treatment, where needed, may add materially to these costs. Resistance by the Environmental Protection Agency (EPA) and others to thermal pollution continually causes wider use of more complex cooling systems, and withdrawals will be materially affected in the future.

About 94 percent of industrial water is used for cooling, mostly on a once-through system. An open recirculation system with cooling tower or spray pond (Fig. 2) reduces withdrawal use of water by over 90 percent but increases consump-

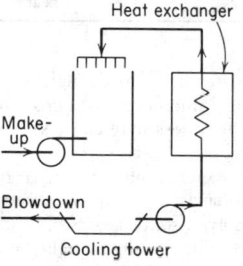

**Fig. 2** Open recirculating system. (*Chem. Eng., June 10, 1963.*)

tive use by 3 to 8 percent because of evaporation loss (see Sec. 9, Power Plant Heat Exchangers). Even more effective reduction in water demand can be achieved by multiple reuse. An example is shown in Fig. 3.

**Quality requirements** for general plant use (nonprocess) are that the water be low in suspended solids to prevent clogging, low in total dissolved solids to prevent depositions, free of organic growth and color, and free of iron and manganese salts. Where the water is also used for drinking, quality must meet the Public Health Service standards.

**Cooling service** requires that water be nonclogging. Reduction in suspended solids is made by settling or by using a coagulating agent such as alum and then settling. For recirculating-type cooling systems, corrosion inhibitors such as poly-

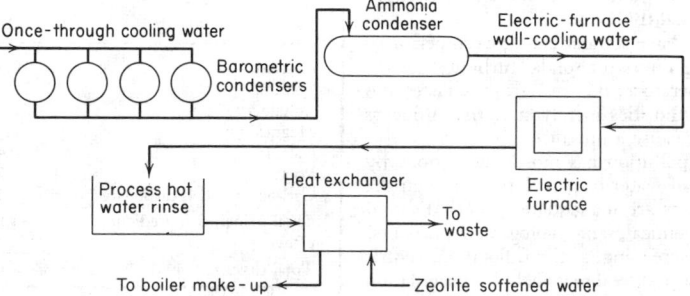

**Fig. 3** Stepwise or cascade cooling system. (*Chem. Eng., June 10, 1963.*)

**Table 4. Water Treatment**

| Treatment | Method | Objective |
|---|---|---|
| Clarification | Presedimentation<br>Coagulation<br>Settling<br>Filtering | Remove suspended solids and reduce turbidity, color, organic matter |
| Disinfection | Add chlorine, 5 to 6 ppm, or continuous-feed to maintain 0.2 to 0.3 ppm residual-free $Cl_2$ | Prevent algae and slime growth |
| Softening | Cold-lime process | Reduce temporary hardness to 85 ppm; also reduce iron and manganese |
|  | Hot-lime-soda process | Reduce total hardness to 25 ppm |
|  | Zeolite | Reduce total hardness to 5 ppm. |
| Membrane processes | Reverse-osmosis electrodialysis | Partial removal of ions; can reduce 10,000 ppm brackish water to 500 ppm or less |
| Demineralization | Ion exchange: two-stage or mixed-bed | Remove both positive and negative ions (cations and anions) to provide very pure water |
| Distillation | Evaporation using steam heat | Produce very pure water—10 ppm or less total solids |

phosphates and chromates are added; algicides and biocides may be needed to control microorganism growths; for cooling jackets on equipment, hardness may cause scaling and should be reduced by softening.

**Process-water quality requirements** are often more exacting than potable-water standards; e.g., boiler feedwater must have less than 1 ppm of dissolved solids (see Sec. 6, Corrosion, and Sec. 9, Steam Boilers). The required quality may be met by the general plant water as available or must be provided by treatment.

Table 4 shows methods and objectives for industrial-water treatment.

## WATER POLLUTION

Pollution may be defined as the return to a natural water supply of waste waters containing ingredients that significantly degrade the quality of the water supply to subsequent users. Most states have regulations to control pollution, and where waters are common to several states, interstate and international agencies have been established. Under EPA guidelines, state regulations are now being standardized.

For the support of fish and aquatic life, water must contain a supply of **dissolved oxygen** (DO). Organic wastes consume oxygen by microbiological action, and this effect is measured by the **biochemical oxygen-demand** (BOD) test.

State **pollution ordinances** have standards to govern permissible plant effluents, setting limits for solids, turbidity, BOD, toxicity, color, pH, temperature, oils and greases, taste, and odor. Many regulatory authorities merely state that effluents must not be obnoxious or cause a nuisance.

**Waste disposal** to avoid pollution has mostly been done by dilution into large bodies of water or percolation through the ground. Where these means are inadequate, proper disposal may require physical, chemical, and biological treatments. **Physical treatments** include screening, settling, flotation, centrifuging, and filtration. **Chemical treatment** includes coagulation or neutralization of acids with soda ash, caustic soda, or lime. Alkali wastes are treated with sulfuric acid or inexpensive

waste acids. Chemical oxidation is effective for certain waste. **Biological treatment** is accomplished by the action of two types of microorganisms: aerobic, which act in the presence of dissolved oxygen, and anaerobic, which act in the absence of oxygen. Most organic wastes can be destroyed by biological treatment. The principal aerobic treatment is by activated-sludge process, by trickling filters, or by lagoons.

## WATER DESALINIZATION

An average seawater contains 35,000 ppm of dissolved solids, equal to $3\frac{1}{2}$ percent by weight of such solids, or 3.5 lb per 100 lb; in 1,000 gal, there are 300 lb of dissolved chemicals in 8,271 lb of pure water. The principal ingredient is sodium chloride (common salt), which accounts for about 80 percent of the total. Other salts are calcium sulfate (gypsum), calcium bicarbonate, magnesium sulfate, magnesium chloride, potassium chloride, and more complex salts. Because these dissolved chemicals are dissociated in solution, the composition of seawater is best expressed by the concentration of the major ions (see Table 5). Even when the content of total solids in seawater varies because of dilution or concentration, the proportion of the ions remains almost constant.

**Table 5. Ions in Seawater\***

| Ions | ppm | lb/1,000 gal | mg/l |
|---|---|---|---|
| Chloride | 19,350 | 165.6 | 19,830 |
| Sodium | 10,600 | 91.2 | 10,863 |
| Sulfate | 2,710 | 23.2 | 2,777 |
| Magnesium | 1,300 | 11.1 | 1,332 |
| Calcium | 405 | 3.48 | 415 |
| Potassium | 385 | 3.30 | 395 |
| Carbonate and bicarbonate | 122 | 1.05 | 125 |
| Total principal ingredients | 34,872 | 298.9 | 35,737 |
| Others | 128 | 1.1 | 131 |
| Total dissolved solids | 35,000 | 300.0 | 35,868 |

\*Compiled from Spiegler, "Salt Water Purification," and Ellis, "Fresh Water from the Ocean."

The **composition of saline waters** varies so widely that no average analysis can be given. Common impurities in land waters are usually the calcium, magnesium, and bicarbonate ions. Some saline waters contain sodium chlorides or sulfates.

**Purification of seawater** requires reduction of 35,000 ppm of solids to less than 1,000 ppm, or a reduction of 35 to 1. For potable water, the sodium content must be further reduced; for agricultural water, the boron content may also have to be lowered. The oldest method of purification of seawater is **distillation.** This technique has been practiced for over a century on oceangoing steamships. Distillation is used in over 95 percent of all land-based conversion plants and is principally accomplished by submerged-surface evaporation, flash distillation, and vapor-compression distillation.

In **submerged-surface distillation,** a heat-transfer surface, such as tubes or coils, is submerged in salt water in a suitable vessel. Steam is passed through the tubes, causing the brine to boil, and some water is evaporated. The vapors pass to a condenser cooled by incoming seawater and so yield a distillate of practically pure water. This is known as **single-effect** distillation. (See Fig. 4 for heat-transfer data.) **Multiple-effect** distillation (Fig. 5) uses several effects in series. While, in a single effect, 1 lb of steam will produce nearly 0.9 lb of distilled water, a double effect will yield 1.75 lb, and so on. The Freeport, Texas, plant uses 11 effects, producing nearly 10 lb of water per lb of steam. The evaporators in this plant employ enhanced vertical tubes in which the seawater is heated and vaporized by steam surrounding the tubes. This is a variation of the submerged-coil evaporator and is commonly known as the VTE, or vertical-tube evaporator.

In **flash distillation,** the salt water is heated in a tubular heater and then passed to a separate chamber where a pressure lower than that in the heating tubes prevails. This causes some of the

hot salt water to vaporize, or "flash," such vapor then being condensed by cooler incoming salt water to produce pure distilled water. A **single-stage flash evaporator** can produce nearly the same amount of distillate per heat unit as the submerged-surface type. Similarly, flash distillation can be

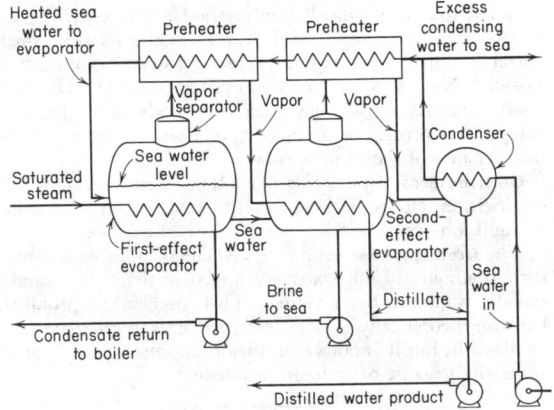

**Fig. 5**   Multiple-(two-) effect submerged-surface distillation process.

carried on in a number of successive stages (Fig. 6) wherein the heated salt water flashes to vapor in a series of chambers, each at a lower pressure than the preceding one. The higher the number of stages in such a **multistage flash** (MSF) system, the better the overall yield per heat unit.

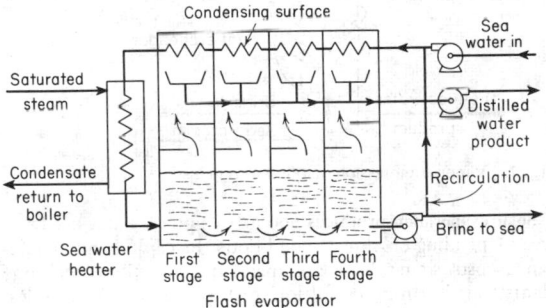

**Fig. 6**   Multiple-(four-) stage flash-distillation process.

Flash distillation offers an advantage over submerged-surface distillation in that the separate heating of salt water without boiling causes fewer scale deposits. No scale occurs in the flash chambers as they contain no heated surfaces, the increased concentration of salts remaining in the seawater. Recent designs of MSF and VTE evaporators provide a great number of stages in one vessel by simple divisions within the vessel. This construction is much less costly than the separate evaporators formerly used in flash distillation and still used in submerged-surface distillation. Therefore, in large modern plants, the MSF and VTE systems have practically replaced the heated-surface evaporators. The Chula Vista, California, OSW demonstration plant uses 68 stages of multiflash evaporation to produce 1 million gal (3,780 m³) of fresh water per day. It has produced yields as high as 19.3 mass units of fresh water per mass unit of saturated steam at the brine heater.

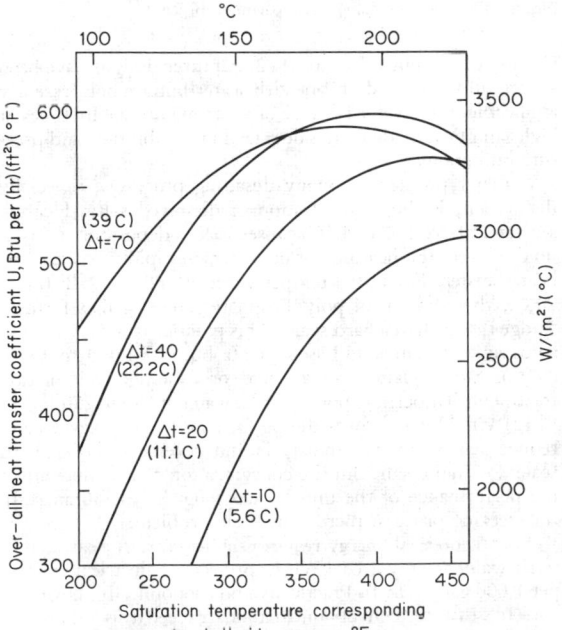

**Fig. 4**   Some heat-transfer rates of submerged-tube, single-effect evaporators. These rates are approximately 65 percent of clean-tube rates.

In **vapor-compression distillation** (Fig. 7), the energy is supplied by a compressor which takes the vapor from boiling salt water and compresses it to a higher pressure and temperature to furnish the heat for vaporization of more seawater. In so doing, the vapor is condensed to yield distilled water. Vapor compression is theoretically a more efficient method of desalinization than other distillation methods. The principle was widely applied during World War II in the form of small, portable units. The largest permanent installations are at Roswell, New Mexcio (1 million gpd) (0.044 m³/s); Dhahran, Saudi Arabia; and Kindley Field, Bermuda. The disadvantages of this process lie in the cost, mechanical operation, and maintenance of the compressors.

**Other methods** of purifying seawater include freezing, solar evaporation, electrodialysis, solvent extraction, reverse osmosis, and ion exchange.

The **freezing process** yields ice crystals of pure water, but a certain amount of salt water is trapped in the crystals and is mostly removed by washing. This process is promising because theoretically it takes less energy to freeze water than to distill it; but it is not yet in major commercial application, primarily because of scale-up problems.

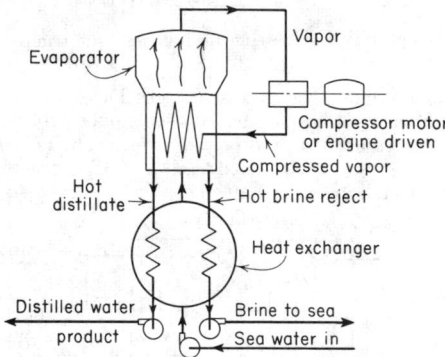

**Fig. 7**  Vapor-compression distillation process.

**Solar evaporation** produces about 0.1 gpd of water per ft² of area, depending on climate conditions. Its capital and maintenance costs so far have not made it competitive for large plants, but it can be useful for small plants in low latitudes. (See also Sec. 9, Power Miscellany.)

**Electrodialysis** (Fig. 8) is a proved method of desalinization, but where used on ocean water, it is not competitive. The purification of brackish or low-saline waters lends itself most advantageously to the process. The ions of dissolved salts are pulled out of saline water by electric forces and pass from the salt-water compartment into adjacent compartments through membranes which are alternately permeable to positively charged ions or to negatively charged ions.

Electrodialysis equipment contains from 10 to 100 or more compartments between one set of electrodes. The number of cells and the amount of electric current required increase with the amount of purification to be done. A brackish water of 5,000 ppm max dissolved salts can be reduced to potable water with less than 500 ppm. A large plant in South Africa reduces 3 million gpd (0.132 m³/s) of a water having a salinity of 2,800 ppm.

Reverse osmosis (Figs. 9 and 10), whereby a membrane permits fresh water to be forced through it but holds back

dissolved solids, has now emerged as the most feasible process for desalting brackish waters containing up to 10,000 ppm of total dissolved solids (tds). Plants of 0.1 to 3.0 mgd (0.0044 to 0.132 m³/s) capacity are now commercially available in three general configurations: (1) nylon hollow fine-filament cartridges; (2) spiral-wrap cellulose acetate sheet modules; (3) Cast tubular cellulose acetate modules.

Operating plants, in capacities up to 0.8 mgd, now serve a variety of municipal and industrial needs, principally in the electronics industry. A total capacity of 12 mgd was on order

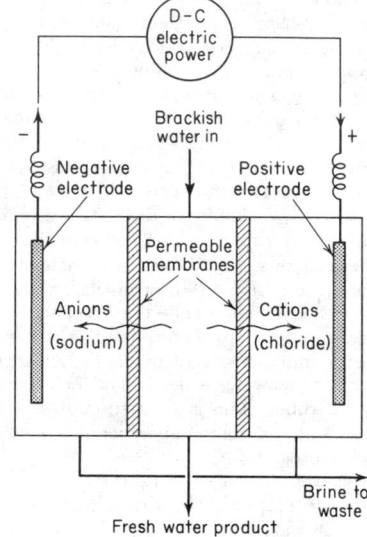

**Fig. 8**  Electrodialysis single-compartment process.

or had been installed by late 1972. All three designs have been successfully operated on brackish water but are not presently applicable to seawater because of short membrane lifetimes at high salinities, inadequate salt-rejection capabilities, and pressure limitations.

A major problem in many desalting processes, especially distillation, is the concentration of dissolved salts. Heating seawater above 150°F (65°C) causes **scale,** a deposition of insoluble salts on the heating surfaces, which rapidly reduces the heat transfer. For initial temperatures of 160 to 195°F (71 to 90°C), the addition of polyphosphates and organics forms a sludge rather than a hard scale. This practice is widely used in ships and present land-based plants but is limited to 195°F (90°C). Some plants use an acid or a sludge-recirculation treatment that permits boiling of seawater at 240 to 250°F (115 to 121°C). This improves the performance of the process and reduces costs. In electrodialysis and reverse osmosis, hard scale does not occur, but the concentration of salts does affect the performance of the unit by clogging the membranes. In the freezing process, there is no scale problem.

The theoretical **energy requirement** to convert seawater to fresh water is only 2.65 kWh (expressed as electrical energy) per 1,000 gal (3,785 l). Practically and economically, however, current seawater-conversion methods exceed this minimum-energy requirement by a factor of not less than 15 times, and usually by about 50 times.

The **costs** of saline water conversion are composed of energy and chemical costs, capital charges, and operation and maintenance costs. Over the past decade (1967 to 1977), because of fuel shortages and more sophisticated environmental requirements, these costs have risen 2½ to 5 times, while the ENR Building Cost Index has risen 2.23 times since 1967. Relative **production costs** in 1967 were about $1 per 1000 gal (26.4¢ per m³) for multistage flash evaporators at the 1 million gpd level, 70¢ per 1000 gal at the 10 million gpd level, and 50¢ per 1000 gal at the 50 million gpd level. These converge to the OSW goal of 38¢ per 1000 gal at the 250 million gpd level but come nowhere near the goal of 12¢ per 1000 for irrigation water. These costs were calculated on the basis of 3.5 percent munici-

pal financing over a 30 year lifetime at a capital cost per daily gallon of $1/(mgd)$^{0.2}$ (mgd being millions of gallons per day), with sulfuric acid selling at $20 per ton, and by using the power credit method of costing steam in a dual-purpose plant. By comparison, cost of municipal fresh water supplies rarely exceeded 25¢ per 1000 gallons (6.6¢ per m³) and ranged mostly around 15¢. All these figures are production costs. **Distribution** and **environmental protection costs** add 20¢ to 40¢ per 1000 gallons. Current **production costs** of a typical 2 mgd dual-purpose power/desalination facility in the United States in 1977 are about $4 per 1000 gal ($1.06 per m³) and capital costs run $2.50 to $5 per gallon per day capacity.

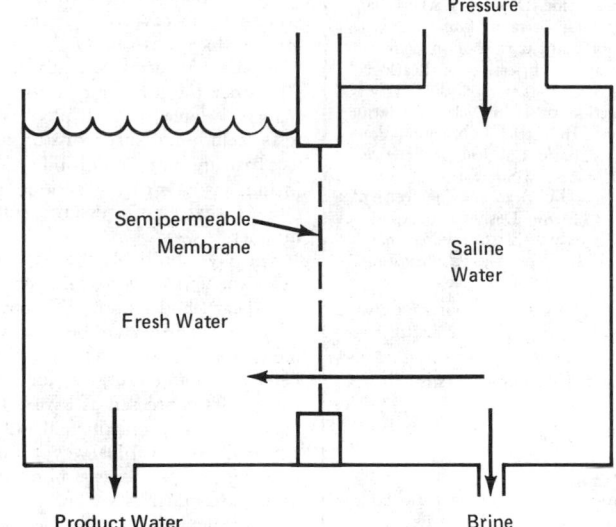

**Fig. 9**  Reverse-osmosis cell. (*Southwest Research Institute.*)

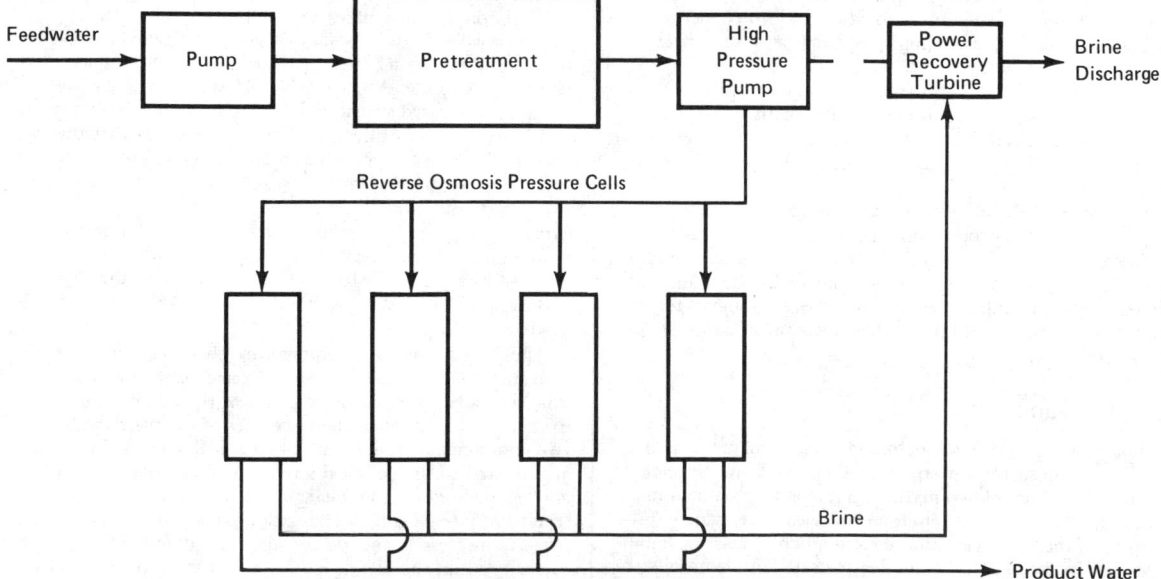

**Fig. 10**  Modular arrangement of reverse-osmosis cell. (*Southwest Research Institute.*)

# LUBRICANTS AND LUBRICATION

## by Fred J. Villforth, Jr.

(For general discussions of friction, viscosity, bearings, and coefficients of friction, see Sec. 3 and 8.)

REFERENCES: ASLE, "Physical Properties of Lubricants," 2d ed. SAE Handbook. Georgi, "Motor Oils and Engine Lubrication," Reinhold. Bondi, "Physical Chemistry of Lubricating Oils," Reinhold. Bauman, "Properties of Lubricating Oil and Engine Deposits," Macmillan. Boner, "Manufacture and Application of Lubricating Greases," Reinhold. ASLE, "Fundamentals of Friction and Lubrication in Engineering." Shaw and Macks, "Analysis and Lubrication of Bearings," McGraw-Hill. Slaymaker, "Bearing Lubrication Analysis," Wiley. Forbes, Pope, and Everitt, "Lubrication of Industrial and Marine Machinery," 2d ed., Wiley. Hobson, "Industrial Lubrication Practice," The Industrial Press. Brewer, "Basic Lubrication Practice," Reinhold. ASTM, 1972 Annual Book of ASTM Standards. O'Connor and Boyd, "Standard Handbook of Lubrication Engineering," McGraw-Hill. Wilcock and Booser, "Bearing Design and Application," McGraw-Hill. Dowson and Higginson, "Elasto-Hydrodynamic Lubrication, The Fundamentals of Roller and Gear Lubrication," Pergamon.

**Lubrication** is concerned with modifying the frictional characteristics and reducing damage and wear at the surface of two solids when one is moved relative to another. Anything introduced between two such surfaces to accomplish this is called a **lubricant.**

## LUBRICANTS

**Types and Properties** Although the substances most frequently used as lubricants have been oils or greases, many other materials, differing widely, may be suitable. Solids and fluids (air or other gases as well as liquids) are employed as lubricants. The lubricant frequently serves a multiple function; it may be a heat-transfer medium, a protection against rust and corrosion, a sealing medium, or a scavenger for contaminants.

The particular application, in its manifold aspects, determines the lubricant selected. Lubricants are manufactured and modified to have certain specific characteristics, which may be defined in terms of physical or chemical properties or by performance.

The concept of the lubricant as a design consideration has helped to place the proper emphasis on the lubrication aspects of the operation of mechanisms and has resulted in more satisfactory performance. The equipment builders and suppliers select the lubricants to suit the equipment for the intended operating conditions; their recommendations should therefore be followed.

## LIQUID LUBRICANTS

Although many liquids, including water, may be used as lubricants, those most frequently employed are petroleum-based or based on refined petroleum fractions or on man-made synthetic fluids. **Liquid-petroleum lubricants** are widely used because of their general suitability to much of existing equipment and/or availability at moderate cost. Petroleum oils are variously prepared by many available refining processes from naturally occurring hydrocarbons. The three main crude-oil types are paraffin, mixed, and naphthene bases. Important crude-oil sources are scattered worldwide in onshore and offshore locations.

**Physical tests** are frequently used to characterize petroleum oils, since the lubricant performance often depends upon or may be related to such physical properties. Usual physical tests include measurements of viscosity, density, pour, gravity, flash and fire, demulsibility, odor, and color. **Chemical tests** include tests for carbon residue, oxidation, corrosion, acidity, oiliness, extreme pressure, sulfur, ash, and precipitation number.

**Viscosity** is probably the single most important property of a lubricant and may be influenced by temperature, pressure, and shear (fluid motion). (See Sec. 3.)

The **Saybolt Standard Universal viscometer** is a standard instrument for determining viscosity of petroleum lubricants between 70 and 210°F (21 and 99°C) (ASTM D88–56), with the results expressed as **Saybolt Universal seconds** (SUS). **Kinematic viscosity,** a fundamental and preferred determination, can be obtained in capillary-type instruments under Newtonian flow conditions. A large number of designs of commercially available capillaries are acceptable (ASTM D445–72).

The calculation of **absolute viscosity** (dynamic) can be made from the determination of the kinematic viscosity and the density. A method of converting kinematic viscosity to Saybolt Universal or Saybolt Furol viscosity is available in the form of convenient tables (ASTM D2161–66).

**The variation of viscosity with temperature of petroleum oils** can be determined when the viscosities are known at any two temperatures. The ASTM D341–43 standard viscosity-temperature charts are available for this purpose for both Saybolt and kinematic viscosities. The procedure states that the two known viscosity-temperature points are plotted on the appropriate chart; then with accuracy, a straight line is drawn through these points. Any point on the line, within the explicit range, indicates the viscosity corresponding to the temperature, or vice versa. A typical chart is shown in Fig. 1. The multigrade SAE 10W-40 oil conforms to the viscosity specification of a 10W grade oil at 0°F (−18°C) and of a 40 grade oil at 210°F (99°C).

The ASTM viscosity-temperature charts can be used very conveniently **to predict the required composition of a two-component lubricating-oil blend** or to estimate the composition of an existing oil blend when the viscosities of the blend and of the two components are known. The vertical scale is almost linearly related to the composition of a two-component oil blend and can be used without change. The horizontal scale between 0 and 100°F (−18 and 38°C) is relabeled 0 to 100 percent and is used to represent the percentage by volume of the high-viscosity oil in the blend. To use the chart, plot the viscosity

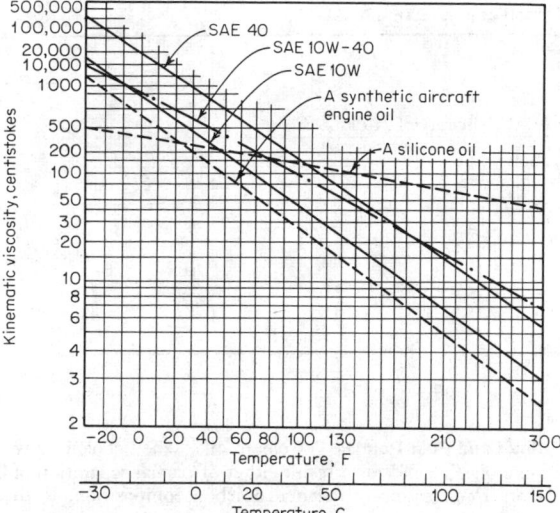

**Fig. 1** Viscosity-temperature relationships (ASTM standard viscosity-temperature chart). Solid lines represent SAE grades of engine oils, the dashed lines two synthetic fluids.

of the low-viscosity oil on the zero percent line and the viscosity of the high-viscosity component on the 100 percent line. If the two points are connected by a straight line, the volumetric composition of any blend of intermediate viscosity can be read. Conversely, the composition of a blend can be estimated when the viscosities of the two components and of the blend are known.

**SAE Viscosity Number** The SAE viscosity numbers for crankcase oils constitute a classification in terms of viscosity only. They are given in Table 1.

Some oils with flat viscosity-temperature curves lie within the limits of two SAE grades; these are called **multigrade oils.** For example, an oil with a viscosity of 15 centistokes at 210°F (99°C) and 1,300 centistokes at 0°F (−18°C) would be a 10W-40 grade. Selected base oils or the addition of viscosity-index (VI) improvers can be effective in blending these multigrade lubricants.

The selection of a winter crankcase oil should be based upon the lowest anticipated atmospheric temperature, whereas a summer crankcase oil should be based on the highest temperature expected. It is generally recommended that the following temperatures be used in selecting the atmospheric-temperature limits for the various SAE viscosity numbers: −10, 0, 10, 32, and 90°F (−23, −18, −12, 0, and 32°C).

The SAE viscosity numbers for gear and transmission lubricants constitute a classification in terms of viscosity only. They are given in Table 2.

The classification is based on the lubricant viscosity measured at both high and low temperatures. The high-temperature viscosities are determined according to ASTM D455 Method. The low-temperature viscosities are determined according to ASTM D2983−71T Method using the Brookfield viscometer. The two viscosity units are related as follows:

$$\frac{cP}{\text{Density, g/ml}} = cSt$$

ASTM D2422−68 provides a common basis for designating, specifying, or selecting the viscosity of industrial fluid lubricants. Sixteen viscosity grades based on a mathematical

**Table 1. SAE Viscosity Numbers for Crankcase Oils (J300c)**

| SAE Viscosity No. | Viscosity units* | Viscosity range† | | | |
| | | At 0°F (−18°C) | | At 210°F (99°C) | |
| | | Min | Max | Min | Max |
| --- | --- | --- | --- | --- | --- |
| 5W | Centipoises | | Less than 1,200 | | |
| | Centistokes | | 1,300 | | |
| | SUS | | 6,000 | | |
| 10W | Centipoises | 1,200‡ | Less than 2,400 | | |
| | Centistokes | 1,300 | 2,600 | | |
| | SUS | 6,000 | 12,000 | | |
| 20W^e | Centipoises | 2,400§ | Less than 9,600 | | |
| | Centistokes | 2,600 | 10,500 | | |
| | SUS | 12,000 | 48,000 | | |
| 20 | Centistokes | | | 5.7 | Less than 9.6 |
| | SUS | | | 45 | 58 |
| 30 | Centistokes | | | 9.6 | Less than 12.9 |
| | SUS | | | 58 | 70 |
| 40 | Centistokes | | | 12.9 | Less than 16.8 |
| | SUS | | | 70 | 85 |
| 50 | Centistokes | | | 16.8 | Less than 22.7 |
| | SUS | | | 85 | 110 |

*The official values in this classification are based upon 210°F (99°C) viscosity in centistokes (ASTM D445) and 0°F (−18°C) viscosities in centipoises (ASTM D2602). Approximate values in other units of viscosity are given for information only. The approximate values of 0°F (−18°C) were calculated using an assumed oil density of 0.9 g/cm³ at that temperature.
†The viscosity of all oils included in this classification shall be not less than 3.9 cSt at 210°F (99°C) (39 SUS).
‡Minimum viscosity at 0°F (−18°C) may be waived provided viscosity at 210°F (99°C) is not below 4.2 cSt (40 SUS).
§Minimum viscosity of 0°F (−18°C) may be waived provided viscosity at 210°F (99°C) is not below 5.7 cSt (45 SUS).
^eSAE 15W may be used to identify SAE 20W oils which have a maximum viscosity at 0°F (−18°C) of 4800 cP. *Note:* 1 cP = 10⁻³ Pₐ·S; 1 cSt = 10⁻⁶ m²/s.

**Table 2. Axle and Manual Transmission Lubricant Viscosity Classification (SAE J306a)**

| SAE viscosity No. | Max temp for viscosity of 150,000 cP (150 Pa/s) | | Viscosity of 210°F (99°C) | | | | | |
|---|---|---|---|---|---|---|---|---|
| | °F | °C | Min | | | Max | | |
| | | | cSt | SUS* | mm²/s | cSt | SUS* | mm²/s |
| 75W | −40 | −40 | 4.2 | 40 | 4.2 | | | |
| 80W | −15 | −26 | 7.0 | 49 | 7.0 | | | |
| 85W | +10 | −12 | 11.0 | 63 | 11.0 | | | |
| 90 | | | 14.0 | 74 | 14.0 | < 25 | 120 | 25 |
| 140 | | | 25.0 | 120 | 25.0 | < 43 | 200 | 43 |
| 350 | | | 43.0 | 200 | 43.0 | | | |

*Approximate.

series of numbers are specified by the nominal viscosity at 100°F (38°C) in both the centistoke and SUS systems. These are applicable to fluids in the range from 2 cSt (32 SUS) to 1,500 cSt (7,000 SUS) at 100°F (38°C), and cover fluids from kerosine to heavy cylinder oils.

**Viscosity index (VI)** is an empirical system for expressing the rate of change of viscosity of an oil with change in temperatures. It is based upon comparison of viscosity measurements of fractions from crude oils L and H, which were chosen because they seemed to possess the maximum and minimum limits of viscosity-temperature sensitivity and were accordingly assigned viscosity indexes of 0 and 100 as the presumed end points of a 100-point viscosity-index scale. While all other oils were expected to fall between these limits, the subsequent advent of solvent refining, the use of modifiers such as the polymers, and the manufacture of synthetics have all produced lubricants that are far outside the viscosity-index scale in both directions.

The **procedure for calculating the VI of an oil** is to determine its viscosity at 100 and 210°F (38 and 99°C). The Standard Method (ASTM D2270–64) and ASTM viscosity tables (ASTM Data Series, DS39a) are used for viscosity-index calculations.

Oils with viscosity indexes above 100 can be made from a wide variety of crude oils by solvent refining, by selective blending of paraffin-base oils, by the addition of relatively small amounts of high-molecular-weight polymeric additives to base oils, or by combinations of these methods. Lubricants with good temperature-viscosity curves (high viscosity index) are desirable where service temperatures vary greatly.

When lubricating oils are subjected to high pressures, several thousand lb/in², their viscosity increases. When oil-film pressures are in this order of magnitude, the influence on viscosity should be considered. A number of empirical equations to relate viscosity to pressure have been derived. In heavily loaded bearings, gears and other machine elements, the high film pressures will influence viscosity with accompanying increase in frictional forces and load-carrying capacity.

The **compressibility** of lubricating oils, especially as they are used in hydraulic systems, is becoming increasingly important. The compressibility should be as low as possible to avoid sluggish operation of these systems. Compressibility is usually expressed as percent reduction in volume at a given pressure or as the ratios of the volumes compressed and uncompressed.

**Cloud and Pour Points**  Petroleum oils, when cooled, may become plastic solids as a result either of partial separation of wax or of congealing of the hydrocarbon composition. With some oils, the separation of wax becomes visible at temperatures slightly above the solidification point, and when that temperature is reached under prescribed conditions, it is known as the **cloud point** (ASTM D2500–66). With oils in which wax does not seperate prior to solidification, or in which the separation is invisible, the cloud point cannot be determined. That temperature at which the oil will just flow under prescribed conditions is known as the **pour point** (ASTM D97–67).

The pour point indicates the lowest temperature at which an oil will flow to the pump, bearings, or cylinder walls. It is particularly important for immediate oil circulation in connection with cold starting of engines, or with gravity lubricating systems, as the fluidity is a factor of pour point and viscosity of the cold oil. Pour-point depressants may be added to wax-containing oils to lower the pour points instead of dewaxing the oils.

**Density and Gravity**  The Standard Method (ASTM D1297–67) may be used for determining density (mass), specific gravity, and API gravity of lubricating oils.

Low-viscosity oils have higher API gravities than the higher-viscosity oils of the same crude-oil series. Paraffinic oils are the lightest or highest API gravities, naphthenic are intermediate, and animal and vegetable oils are the heaviest or low API gravity. The specific gravity of an oil is the ratio of its weight to that of an equal volume of water, both measured at 60°F (16°C). The gravity of lubricating oils is of no value in predicting quality, although it gives a clue to the source of the crude-oil base.

**Flash and Fire Points**  The flash point of an oil is the temperature to which an oil has to be heated until sufficient flammable vapor is driven off to flash when brought into contact with a flame. The fire point is the higher temperature at which the oil vapors will continue to burn when ignited. The ASTM 92–72 standard method for flash and fire points by means of open-cup tester is used for lubricating oils. In general, the open flash point is 30°F (17°C) higher than the closed flash, and the fire point is some 50 to 70°F (28 to 39°C) above the open flash point.

Flash and fire points may vary with the nature of the original crude oil, the viscosity, and the method of refining.

For the same viscosities and degree of refinement, the paraffinic oils are higher than the naphthenic flash and fire points.

**Steam Emulsion Test** ASTM D1401–67 Emulsion Characteristics of Petroleum and Synthetic Oils and ASTM D2711–69 Dimulsibility Characteristics of Lubricating Oils are used for measuring the ability of oil and water to separate. Although the former is specifically intended for steam-turbine oils, this test may be used for other oils. When the oil is emulsified and separated under prescribed conditions, the time required for the emulsion to be reduced to 3 ml or less is recorded (at 5-min intervals).

**Color** The color of a lubricating oil is obtained by reference to transmitted light; the color by reflected light is referred to as **bloom**. The color of an oil indicates the uniformity of a particular grade or brand and not its quality. ASTM D1500–64 is for the visual determination of color of lubricating oils, heating oils, diesel fuels, and petroleum waxes using a standardized colorimeter. The method compares the samples with glass color standards and reports color in terms of the ASTM Color Scale; it also provides for comparison with the former ASTM Union Color. The color scale ranges from 0.5 to 8; oils darker than the 8 color are diluted with kerosine as prescribed by the test method and then observed in the same way as the lighter oils. For determining the color of petroleum products lighter than 0.5, ASTM D156–64 Test for Saybolt Color of Petroleum Products can be used.

**Carbon Residue** The **Conradson carbon-residue test** (ASTM D189–65) is a means of determining the amount of carbon residue left on evaporating an oil under specified conditions and is intended to throw some light on the relative carbon-forming properties of an oil. The carbon-residue test was originally developed for comparison of the carbon-forming properties of lubricating oils for internal-combustion engines. The deposits vary with the type and mechanical condition of the engine, the service conditions, the time of continuous operation, the viscosity of the oil, and the character of the fuel. Modern refining methods and use of detergency additives tend to make this test less useful for predicting the deposit-forming tendency of a lubricating oil. The carbon-residue results are higher with additive-type oils because of the ash content of the additives.

**The Ramsbottom carbon-residue test** (ASTM D524–64) provides an indication of the coke-forming properties of petroleum products. Products containing ash-forming constituents will have comparatively high carbon-residue values. Results from the two methods are not comparable.

**Ash** The ash determination (ASTM D482–63) of lubricating oil is used to measure impurities or contaminants and is limited to products which are free from ash-forming additives. Under prescribed conditions, the sample is ignited and allowed to burn until only ash and carbon remain; then the carbon residue is reduced to ash by heating.

**Sulfated Ash from Lubricating Oils and Additives** (ASTM D874–72) This test indicates the concentration of known metal-containing additives in unused oils and additive concentrates. Usually the additives contain one or more of the following metals: barium, calcium, magnesium, zinc, potassium, sodium, and tin. These may be in combination with one or more of the elements sulfur, phosphorus, and chlorine. Chemical analysis for metals in lubricating oils (ASTM D811–48) determines the barium, calcium, zinc, magnesium, tin, silica, sodium, aluminum, and potassium in new and used detergent-additive-type lubricating oils. Sulfur, phoshorus, and chlorine do not interfere in this method. These methods do not cover the determination of lead, copper, and iron.

**Oxidation Testing Methods** Lubricating oils may be subjected to relatively high temperatures in the presence of air and catalytically active metals or metallic compounds. The process of oxidation becomes critical when oil is operating above 150°F (66°C). It is not uncommon to find lubricating oil sump temperatures in excess of 250°F (121°C). The rate of oxidation doubles for each 18°F (10°C) rise in temperature of the oil above 150°F (66°C). The resultant oxidation of the oil develops increased viscosity, acids, carbon residue, sludge, and asphaltenes. There are numerous oxidation tests. The Navy work factor (Federal Test Method Std. No. 791B Methods 3451.3 to 3452.1) measures the stability of lubricating oils. The oxidation characteristics of inhibited steam-turbine oils are tested by method ASTM D943–54.

**Precipitation Number** Steam cylinder oils and black oils are checked by the (ASTM D91–61) precipitation number. This method is also applied to used crankcase oils of internal-combustion engines to check the oxidized products, carbonaceous matter, and asphaltenes formed through use.

**Acid and Base Number and Corrosion** ASTM D664–58 method determines the acidic or basic constituents in lubricating oils. It also indicates the relative changes caused by oxidation during use. As an oil ages in service its acid number increases.

The acid number of used oil may or may not indicate the corrosive action of the used oil in service. In some cases, an acid number of 1.0 may not attack the alloyed bearing metals; yet in other service tests of a different oil, 0.2 number will show high corrosive tendencies in short periods of operation.

Certain detergent additives used to counteract acidic bodies which cause deposits and corrosive wear are basic or alkaline. ASTM D664–58 is suitable to measure the basic constituents of many basic additive-type lubricating oils.

Certain phosphorus and sulfur additives are used to check corrosion in lubricating oils. The use of oiliness additives such as fatty oils in some instances may result in bearing corrosion especially where cadmium and certain alloy lead-base bearings are used.

Products such as rustproofing oils and similar compounds which are used to protect metal surfaces from rusting or corrosion are generally required to pass special "weathering" tests. The tests measure the degree of protection afforded by the specimen under specified conditions such as salt-spray humidity cabinets, or immersion in water. Rust-preventing characteristics of steam-turbine oil in the presence of water is covered by ASTM D665–60.

**Antifoam** The foam characteristics in crankcase, turbine, or circulating oils is checked by foaming test apparatus (ASTM D892–72). Certain additive oils tend to foam excessively in service. Only a minute quantity of antifoam inhibitor is required to break the foam occurring in oil in service.

**Air Release** Air entrainment differs from foaming, which is a surface effect, in that the former consists of a dispersion of small air bubbles throughout the bulk liquid. Air may be entrained by mechanical means and/or by the release of dissolved air accompanying a change in environment such as a drop in pressure. Some antifoam agents may slow down the release of air from the bulk liquid. The presence of entrained air is apparent from the opaque appearance of the liquid.

Several laboratory tests may be used to measure air-release times.

**Oiliness**   The property of lubrication known as oiliness is of considerable importance. It is a phenomenon that becomes strongly evident only when the oil film separating rubbing surfaces is exceedingly thin. In films of molecular dimensions, viscosity effects are negligible, although oiliness has a marked bearing. Oiliness depends on both the lubricant and the surface to which it is attached. Oiliness is the property that causes a difference in the friction when lubricants of the same viscosity at the same temperature and pressure of the film are used with the same bearings.

**Extreme-Pressure, Antiwear**   The high tooth pressures and high rubbing velocities often encountered in hypoid and spur-type gearing have developed a class of lubricants called extreme-pressure (EP) or **hypoid-gear lubricants.** When mineral oils alone are used, metal-to-metal contact occurs which results in scoring, galling, and local seizure of the gear teeth. Most extreme-pressure lubricants are mineral oils containing active or passive held sulfur, phosphorus, and/or chlorine or some highly reactive material.

EP lubricants inhibit welding, scuffing, and tearing of the metal at the contacts. The additives should be stable at ordinary temperatures but provide the necessary protection by becoming active in the high-friction, high-temperature contact areas.

A variety of methods have been devised to evaluate the load-carrying capacity of lubricants under heavy-duty conditions. Each apparatus tends to emphasize a particular characteristic, and a given lubricant will not necessarily show the same **extreme-pressure** characteristic when tested on different machines. The ASTM D2509–68 and D2782–72 procedures use the **Timken EP Lubricant Tester** to characterize industrial-type gear lubricants. The **SAE Extreme Pressure Lubricant Tester** is designed to determine the load-carrying capacity of hypoid-gear lubricants. The **Falex Tester** is used for evaluating the load-carrying capacity of oils and gear lubricants. Wear tests can also be conducted on this apparatus by maintaining constant load. ASTM D2596–69 and D2783–71 procedures use the **Four-Ball EP Lubricant Tester** for comparing the load-carrying capacity of oils and greases. Numerous procedures are run. The different testers employ different contact configurations and rubbing velocities and employ specimens manufactured from differing metals.

**Synthetic Fluids**   Synthetic lubricants are man-made, as opposed to naturally occurring petroleum fluids. Generally, the synthetics are organic chemicals. Certain groups have been found to have characteristics which make them suitable as lubricants. They often have outstanding properties, such as high viscosity index or thermal stability. The favorable characteristic is usually accompanied by one or more undesirable properties, such as low viscosity, high pour point, or low stability toward water. Some classes of synthetics are polyolefins, polyalkylene glycols, esters, silicones, polyphenol ethers, and halogenated hydrocarbons.

Synthetic lubricants, because of high cost, are used only where the particular property is essential, as in instruments, hydraulic systems, heat-transfer systems, and numerous speciality applications. The development of synthetics was initiated in the aircraft industry especially for the high speeds and high operating temperatures of jet aircraft. As with other lubricants, the basic synthetic fluid is often compounded with modifiers to improve or to impart special properties to the complete lubricant.

Table 3 compares some of the synthetics as to the properties important in considering them as lubricants.

### SOLID LUBRICANTS

A **solid lubricant** is a thin film composed of a single solid or a combination of solids introduced between two rubbing surfaces for the purpose of modifying friction and wear. The operation of mechanisms under severe temperatures, pressures, and environments in which organic fluids are not suitable has fostered the development of solid lubricants. Solid-film lubricants include numerous varieties and types of materials which can have different properties, operating ranges, and methods of application or attachment to the surfaces to be lubricated.

Although solid-lubricant technology developed through the years as an "art," largely through experience, recently a more vigorous scientific approach has been applied to the subject. Systematic studies of the materials used and their application began with the introduction of solid lubrication in the aircraft industry. There are many ways of classifying solid lubricants; one is relative to their manner of attachment to the bearing material. Surface preparation is extremely important in achieving solid-lubricant performance.

**Unbonded solid lubricants,** the simplest type, are in granular or powdered form. These adhere in some degree to the surface by mechanical or molecular action, although there is no purposeful physical or chemical bonding. The most common materials in this class are graphite, molybdenum disulfide, polytetrafluoroethylene, and other polymers, talc, metals, metal oxides, and salts. The properties of the solids generally determine the operating limits as well as influence the coefficient of friction and wear.

Unbonded solid films may be applied in numerous ways; powders may be brushed, dipped, or sprayed. If the solids are applied in carriers, liquid or gas, the fluid does not generally contribute to the adhesive or the lubricating properties but serves mainly to facilitate application. Burnishing surfaces is often beneficial.

**Bonded Solid Lubricants**   The durability of solid films, that is, their ability to sustain and maintain adequate lubrication over an extended life, has been somewhat limiting. Improved lubricant life has been realized through the use of adhesives, of which there are several types suitable for differing operating conditions and applications.

The adhesives, also called binders, are mixed with the solid lubricants (usually those previously mentioned as unbonded lubricants), and the mixture is applied to the bearing surfaces. Two types of organic materials are used as binders—those requiring air curing and those requiring thermal setting. The latter are more durable and superior but require baking at elevated temperatures. Air-cured films are generally limited to operating temperatures below 300°F (260°C), while some thermoset films may be satisfactory to 700°F (371°C).

**Inorganic adhesives** permit bonded solid films to be used at temperatures in excess of 1200°F (649°C). Ceramic binders in combination with powdered metals or solids more stable than graphite and molybdenum disulfide are an example.

In addition to the properties of the solids and the prevailing operating conditions, other factors which may influence the

**Table 3. Comparison of Synthetic Lubricants**

| Fluid (application) | Viscosity index | Pour point | | Property | | | | | |
| | | °F | °C | Oxidative stability | Thermal stability | Fire resistance | Volatility | Lubricity | Hydrolytic stability |
|---|---|---|---|---|---|---|---|---|---|
| Synthetic hydrocarbons (arctic oils, air-cooled 2-cycle engines) | 120 to 150 | −80 to −30 | −62 to −34 | G | G | P | F | E | E |
| Polyglycols (hydraulic brake fluid) | 100 to 200 | −70 to −10 | −57 to −23 | F–G | G | P | P | G | G |
| Phosphate esters (fire-resistant hydraulic fluids) | −18 to 150 | −70 to 30 | −57 to −1 | VG | G | E | F | G | P |
| Dibasic acid esters (aircraft-turbine oils) | 140 to 175 | −70 to −30 | −57 to −34 | G | G | E | F | G | F |
| Silicones (lubricants for rubber and plastics) | 175 | −100 to −10 | −73 to −23 | G | E | P | P | P | G |
| Polyphenyl ethers (speciality, very expensive) | 140 | 10 to 70 | −12 to 21 | E | E | P | F | G | E |
| Fluorocarbon polymers (equipment handling strong oxidizers) | −130 | −20 to 50 | −29 to 10 | E | G | E | F | F | F |

Ratings: E = excellent, VG = very good, G = good, F = fair, P = poor.

Table 4. Properties of Some Animal and Vegetable Oils

| | Free fatty acid or oleic, % | Specific gravity at 60°F (16°C) | Saponification No. | ASTM pour point, °F (°C) | Flash, open-cup, °F (°C) | Viscosity SUS at 100°F (38°C) |
|---|---|---|---|---|---|---|
| Castor, hydrogenated | 0.15–2 | 0.963 | 178–181 | 15 (−9) | 505 (263) | 1,485 |
| Cottonseed | 1–3 | 0.922 | 187–197 | 40 (4) | 580 (304) | 180–195 |
| Lard oil | 15–18 | 0.915 | 192–198 | 50 (10) | 440 (227) | 200–210 |
| Tallow, acidless | 0.2–0.5 | 0.927 | 193–198 | 80 (27) | 610 (321) | 205–215 |

solid-lubricant-film performance include method of application, bearing-surface material finish and hardness, binder–solid lubricant mix, film and surface pretreatment, and cleanliness.

### ANIMAL, VEGETABLE, AND FISH OILS

**Properties of Various Lubricating Fats and Fatty Oils**  Fatty or fixed oils, such as animal, vegetable, and fish oils, are distinguished from mineral oils by being saponifiable with caustic alkalies. These organic oils oxidize, becoming rancid and forming free fatty acids. Oxidation also causes gumming, particularly with cottonseed and corn oil. Subjected to high temperatures, they tend to decompose to corrosive acids.

**Saponification Number**  The saponification number (ASTM D94–71) is used to determine the percentage of fatty oil or fat in a compounded petroleum lubricant. When the sample contains appreciable amounts of sulfur, phosphorus, and halogens, the saponification number obtained will be greater than would the amount of fatty matter actually present. The percentage of fatty oil (or fat) in a compounded petroleum product can be calculated from the saponification number when the fatty oil is known (see Table 4).

The addition of fatty oils or fat to a petroleum mineral oil increases the adsorbed film on the working surface, which raises the load-carrying ability, or "oiliness," of the oil film, a property needed in special applications where boundary lubrication conditions are likely to be encountered. All animal, vegetable, and fish oils will support bacteria; care should be taken to see that such oils are sterile.

The naturally occurring oils are being replaced by synthetics. Sperm oil, one of the more important ones, is not available, since the source, the sperm whale, has been put on the endangered-species list.

### GREASES

Lubricating greases consist primarily of a thickener, a fluid lubricant, and often additional materials to impart specific properties. Although the most common thickener for greases is a soap or mixture of soaps, thickeners may be one of a variety of solid materials including clays or pigments. Soaps are the reaction products of a fatty material, derived of animal or vegetable sources, with alkalies, e.g., aluminum, calcium, sodium, or lithium hydroxide. This chemical reaction to form soap is known as saponification, which also produces water and alcohols or glycerin as secondary-reaction products.

A variety of fats and alkalies are available; thus there is a wide choice of soaps for greases. Usually the fluid employed is a petroleum oil, but many fluids are available, including synthetic esters and silicones to provide different physical and chemical properties. Additional materials may be used to impart certain characteristics to the grease such as oxidation stability, tackiness, extreme-pressure properties, and rust inhibition. Solid materials may also be used to impart special properties; the solids used may be graphite, molybdenum disulfide, talc, metal powders, or polymers.

The ASTM **dropping point** of grease is the temperature at which it changes from a semisolid to a liquid state when the determination is made according to the prescribed ASTM D566–64. Calcium- or lime-soap greases have melting points around 200°F (93°C); sodium- and lithium-soap greases, 300°F (149°C) and higher. Clay-thickened greases may have dropping points higher than 500°F (260°C).

The ASTM method D217–68 is used in measuring the worked or the unworked **consistency** of lubricating greases which have a worked consistency less than 400. In this test, a standardized double-pitch cone is allowed to drop in the product at a definite temperature. The depth of penetration is measured. The unworked (original) consistency of lubricating greases is affected by the soap content, the kind of fat used, the method of manufacture, the final water content, the rate of cooling, and the basic metallic constituent of the soap. It is impractical to control the consistency of a grease to narrow limits. Any working or remelting of a grease after it is in the container will change the consistency. Although many tests are based on the unworked consistency, this property bears no definite relationship to worked values. Final tests are usually based on worked consistency (ASTM).

The texture of a grease refers to its structure such as smooth, fibrous, spongy, or rubbery. Calcium-base greases are smooth, soda-soap greases are fibrous or spongy, lithium-base greases are smooth and buttery, and aluminum-soap greases are stringy or rubbery.

**Grease Numbers**  Greases falling within certain consistency readings are classified in accordance with numbers of National Lubricating Grease Institute (NLGI) as follows:

| Consistency No. | 000 | 00 | 0 | 1 | 2 | 3 | 4 | 5 | 6 |
|---|---|---|---|---|---|---|---|---|---|
| Appearance | Semifluid | Semifluid | Semifluid | Soft | Medium | Medium hard | Hard | Very hard | Block type |
| Worked penetration | 445–475 | 400–430 | 355–385 | 310–340 | 265–295 | 220–250 | 175–205 | 130–160 | 85–115 |

**Table 5. General Characteristics of Greases**

| Base | Texture | Typical dropping point, °F | Max usable temp, °F* | Typical water resistance | Primary use |
|------|---------|---------------------------|---------------------|-------------------------|-------------|
| Sodium tallowate | Fibrous | 390 (199°C) | 250 (121°C) | Poor | Generally restricted to older, slow-moving bearings where water contamination is not significant |
| Calcium 12-hydroxystearate | Smooth | 290 (143°C) | 250 (121°C) | Excellent | Used where high temperature does not occur and maximum water resistance is needed |
| Lithium 12-hydroxystearate | Smooth | 380 (193°C) | 250–325 (121–163°C) | Good | Widely used in general bearing lubrication. Some versions capable of very long life at around 300–325°F (149–163°C) |
| Polyurea | Smooth | 460 (238°C) | 250–325 (121–163°C) | Excellent | Often used for greases having very long life at 300–325°F (149–163°C) |
| Calcium complex | Smooth | 570 (299°C) | 250–325 (121–163°C) | Excellent | Generally used where occasional exposure to very high temperatures is expected. With frequent relubrication, may be used for extended service at temperatures above 400°F (204°C) |
| Lithium complex | Smooth | 550 (288°C) | 250–325 (121–163°C) | Good | |
| Aluminum complex | Smooth | 500 (260°C) | 250–325 (121–163°C) | Excellent | |
| Modified bentonite clay | Smooth | 600 (316°C) | 250–325 (121–163°C) | Excellent | |

*Continuous operation with relatively infrequent relubrication. Depending upon specific oils and other components used to formulate a particular product, the upper service limits may vary over the range shown. The lower value represents the most common service limit.

The ASTM methods (D128–64) of **analysis** permit determination of the constituents of grease likely to be covered by specifications. Such constituents are soap base and content, fat, water, fillers, ash, excess alkali or acid, unsaponifiable matter, and lubricating-oil content. Two greases showing the same analysis may show marked differences in lubricating performance and storage-stability properties. The sodium and lithium soap greases are more stable over much longer periods of service than calcium-base greases. The calcium- and lithium-base greases are water-resistant. The ASTM D942–70 method is used to determine the oxidation stability of lubrication greases (see Table 5).

## LUBRICATION

In engineering practice it is usually desirable to design a mechanism to operate with a fluid film separating the surfaces. This **hydrodynamic lubrication** occurs when the pressures developed in a converging fluid film are sufficient to support the bearing load. When the bearing loads are high, the film thicknesses may be reduced to the size of the surface asperities. The lubricant viscosity itself may be influenced by such operating conditions as film pressure, temperature, and rate of shear. Any of these factors may influence the hydrodynamic system. For a hydrodynamic state to exist, the combination of speed, load, and lubricant viscosity must be in a rather broad but definite range. Increasing speed or viscosity results in a thicker film, and increasing the load reduces film thickness in a given bearing. Bearings are designed to permit maintaining a fluid film under most of the encountered operating conditions. However, when the motion between the bearing surfaces approaches zero (during start and stop), when severe shock loads are applied, or when viscosity is reduced, the complete lubricant film cannot be maintained between the journal and bearing, and metal-to-metal contact occurs. Under these boundary lubrication conditions, mild extreme-pressure (EP) additives are used to provide a measure of safety.

Classical treatment of hydrodynamic lubrication makes the assumption that bearing surfaces are geometrically smooth and that the lubricants and surfaces are unaffected by applied stress. On the other extreme, in "boundary lubrication" the solid surfaces are not in continuous contact and completely supporting the load. In actual liquid-lubrication practice it is unlikely that either extreme prevails, and surface interaction cannot be ignored.

Often, in practice, the regimes of operation are under the conditions in which hydrodynamic films are sufficiently thin so that surface irregularities interact. In the lubrication of rolling elements (in ball and roller bearings), cam and cam-follower mechanisms, and gear tooth mesh, loads tend to deform the metal surfaces in the contact zone. Thus the contact area depends on the deformation of the load-bearing surfaces. Even though the loads may be high, there is usually a thin film between the "contacting" surfaces. This film is called an **elastohydrodynamic (EHD) film.** The formation of such a film depends upon two mutually dependent factors—the hydrodynamic properties of the fluid and the deformation of the bearing contacts.

**Lubrication Systems**   There are many positive methods of applying products to ensure proper lubrication. Bath and circulating systems may be automatic to provide steady or intermittent but positive application as required. There are constant-level lubricators, bottle oilers, gravity-feed oilers, multiple-sight feeds, grease cups, forced-feed lubricators, centralized lubrication systems, and air-mist lubricators (to name a few). The selection of the method of application of lubricant is as important as the lubricant itself. The choice of device and the complexity of the system depend upon many factors, including the type and quantity of the lubricant, the reliability and value of the machine elements, maintenance schedules, accessibility of the lubrication points, and labor costs and other economic considerations, as well as the operating conditions.

**Filtration systems** for the removal of foreign matter in the circulating lubricant may be necessary, particularly where devices having accurately machined and closely fitting parts are used.

**Internal-combustion-engine oils** are required to carry out numerous functions in order to provide adequate lubrication. Crankcase oils, in addition to reducing friction and wear, must keep the engine clean and free from rust and corrosion, must act as a coolant and sealant, and must serve as a hydraulic oil in an engine with hydraulic valve lifters. The lubricant may function under high temperatures and in the presence of dust, water, and other adverse atmospheric conditions as well as with materials formed as a result of incomplete combustion; it must be resistant to oxidation and sludge formation.

Therefore, crankcase oils are compounded and may contain one or more necessary agents and additives, such as **dispersants** to keep insoluble materials suspended; **detergents** to clean mildly or deter deposit formation; **viscosity-index improvers** to increase the viscosity so that it is more, in proportion, at 210°F (99°C) than at 100°F (38°C); **oxidation inhibitors** to decompose peroxides, inhibit free radical formation, and passivate exposed metal; **corrosion inhibitors** to neutralize acid materials and form protective films on metal surfaces; **metal deactivators** to form inactive protective films; **antiwear, extreme-pressure, oiliness, film-strength agents** to form film of lower shear strength than the base metals to reduce friction and prevent welding and seizure if and when the oil film is ruptured; **rust inhibitors** to provide protective, water-repellant films; **pour-point dispersants** to prevent or inhibit growth of wax crystals at reduced temperatures; **foam inhibitors** to reduce surface tension and allow air bubbles to separate more readily.

**Gasoline-engine** operation depends greatly on the engine design and the service. Engines vary in size, speed, displacement, and power output; the operating conditions vary from prolonged high speed and high load under high-temperature conditions to long-time idling at low load and low temperatures with any conceivable intermediate combinations and variations. Engines are used in stationary applications and for propelling vehicles of every size and kind on land, on water, and in the air.

**Diesel-engine** service severity depends first upon power requirements. High continuous power output and overload at high temperature and intermittent power demands at low temperature represent severe conditions. Continuous or intermittent operation at rated load and normal temperatures is considered normal diesel operation. Diesel engines normally operate at lower speeds but higher temperatures than gasoline engines; hence conditions are more conducive to oil oxidation, deposit formation, and corrosion of bearing metals. Engine design, including degree of supercharging, influences susceptibility to deposits, ring sticking, wear, and bearing corrosion. Harmful products of incomplete combustion promote wear and deposit formation; the situation is greatly magnified by low-temperature operation. The oils which perform most satisfactorily in a given engine under each service classification can be selected with certainty only after a series of performance tests.

Laboratory-engine procedures have been established which vehicle and engine manufacturers and the military use in evaluating the performance of crankcase lubricants and in setting specifications. Many engine tests have been standardized, under the auspices of ASTM, SAE, and API, to classify lubricants for various types of engines operating at different severity levels. These tests provide for determining the oils' resistance to oxidation and thickening as well as the oils' influence on engine deposits, corrosion, rust, and wear. Based upon the lubricants' performance, engine manufacturers and suppliers specify the lubricant to be used for given service conditions. These recommendations should be observed by the user.

**Aircraft-engine oils,** because of the necessity of starting at low temperatures, require a low-viscosity, low-pour-point oil. However, the same oil must also provide satisfactory lubrication at warmed-up engine conditions. Oils for aircraft **reciprocating engines** usually have a minimum viscosity index of 95 and viscosities of 90 to 100 SUS at 210°F (99°C). For operation at extremely low temperatures, oils having viscosities down to 65 SUS at 210°F (99°C) are used. Aircraft **turbine engines**, because of their high operating temperatures and low-temperature starting requirements, employ synthetic-type fluids, usually considered to be operable at bulk temperatures of 300°F (149°C), or even as high as 450°F (232°C). The fluids must be pumpable at −65°F (−54°C), have low volatility, good oxidation, and thermal stability at elevated temperatures, be noncorrosive to engine metals, and have adequate gear-load-carrying ability. Specifications require these lubricants to pass stringent static turbine-engine tests and also in-flight tests prior to approval.

**Marine internal-combustion engines** are generally run at rated speed for sustained periods and are exposed to salt-water spray which introduces rust-corrosion problems more severe than those encountered in land vehicles. Heavy-duty oils having detergent, dispersant, and rust-protective qualities are desirable.

**Steam turbines** are usually lubricated with oils of 150 to 250 SUS viscosity at 100°F (38°C). In high-speed turbines with reduction gearing, because of the gearing, the entire system—turbine, reduction gears, and generator—is lubricated with oil having viscosity in the range of 250 to 350 SUS.

Usually turbine and generator bearings, the governor, and other auxiliaries are lubricated with the same oil which serves as the fluid in the control system. To accomplish adequately its numerous functions over an extended period, the lubricant usually contains additives which provide oxidation stability, rust and corrosion protection, and foam resistance. Equipment may also be provided to maintain the lubricant in a clean condition by removing water, sludge, and foreign materials. This purification may be accomplished with settling tanks, centrifuging, or filtering. Where filters are employed, they should be of such a nature as not to remove oil additives. The lubricating system should be designed to reduce all possible contaminations, especially water.

**Gears** are of so many different designs and are used in such a variety of applications with so many different operating conditions that many diverse lubricants have been developed. The gear lubricant may serve a multiple function, such as removing heat and lubricating other machine elements. Selection of the gear lubricant is influenced by the gear type and materials, speed and corresponding sliding velocities, tooth contact and load, environment, temperatures, method of lubricant application, seals, and type of service.

Under moderate-speed and light-load conditions, straight mineral oils may be satisfactory. As operating severity in particular loads increases, extreme-pressure additives are required. Also, one or more modifiers may be employed to improve oxidation stability and service life or to impart other particular qualities. (See also Sec. 8.)

Lubricants for open gears will have special adhesive properties to keep the lubricant on the tooth contact surfaces. In such service, the lubricant may be applied by brush or paddle or by spraying.

Where fluid lubricants would be vaporized or where the heat-transport properties are not essential, dry lubricants may be effective. Where leakage from the gear case could be a factor or with a self-contained sealed unit, greases may be employed.

Aircraft applications often impose severe restrictions, especially where a wide range of operating temperatures may be involved. Aircraft turbine oils of low viscosity because of starting or power requirements at low temperatures may also be required to lubricate high-speed, heavily loaded gears.

Enclosed gears use splash or pressure systems and must be kept free of grit or dust. The viscosities usually used range from 300 to 2,500 SUS at 100°F (38°C). The oil must flow at the lowest temperatures encountered.

**Refrigeration-machinery** lubricant selection must take into consideration the effect on parts not requiring lubrication as well as on the bearing surfaces. Oil, when miscible with the refrigerant, will coat the heat-transfer surfaces, restrict refrigerant flow, and impair performance. It is important that the lubricant remain fluid at the lowest operating temperatures; a low pour point, usually −35°F (−37°C) max, is indicated. Wax separation, referred to as **floc,** must be avoided. In any system operating below 32°F (0°C), the presence of water is objectionable because of ice formation in critical areas. Presence of water may also contribute to corrosion with certain refrigerants. High dielectric constant of the oil is indicative of low water content; care in handling and storage and delivery in sealed containers reduce the possibilities of moisture. The presence of unstable materials in the lubricant leads to sludge formation. To prevent undue carry-over into the refrigeration circuit, the lubricant should neither vaporize excessively nor foam unduly. Where refrigeration systems operate below 5°F (−15°C), a viscosity of 150 SUS at 100°F (38°C) is recommended; for systems operating above 5°F (−15°C), 200 to 300 SUS. Compressor design, refrigerant type, and other factors influence lubricant choice. (See also Sec. 19.)

**Compressors,** particularly air compressors, are essential in many industries, and numerous types and special designs are used. All compressors have bearings—journal, sleeve, or rolling-element—which may be operated at high speeds and high temperatures in the presence of the compressor gas. Most frequently, oil is used as the bearing lubricant, but where atmospheric contaminants can reach the bearings, greases may be more effective in sealing the bearing elements.

In **reciprocating compressors,** valving and porting are exposed to hot gases. Excessive cylinder lubrication may lead to carbon formation at the valves or ports. Accumulation of oil in intercoolers, lines, and receivers where oxygen-containing gases are being compressed enhances the opportunity for explosion and severe damage.

Viscosity at the compression conditions is probably the most important property for the lubricants of reciprocating-compressor piston rings and cylinders. Effective air-filtration systems are essential in keeping contaminants and moisture from causing excessive wear and disrupted lubrication.

Lubricating oils for compressors include straight mineral oils, turbine oils with oxidation and corrosion additives, and heavy-duty engine oils containing oxidation and detergent-dispersant additives. Where gearing is involved, compounded gear oils may be required. (See also Sec. 14.)

**Textile machinery** has a wide variety of lubrication requirements—even within the same unit, which may have many different mechanisms. A unit may operate with plain bearings (e.g., metal or plastic), rolling-element bearings, cams, gears, worms, spindles, sleeves, and chains. Often these parts operate at high speeds and high temperatures, with different methods of application, in dusty or moist atmospheres, with seals, and with intermittent or rapidly changing loads and motions. Some of the elements may require low-viscosity lubricants for low power consumption and small clearances. Other parts of the same machine may require high-viscosity lubricants with special properties such as adhesive and load-carrying abilities. Precision parts, very accurately finished, may require special corrosion and oxidation resistance. One lubricant property unique to the textile industry is that where the possibility of the lubricant's contacting the fabric exists, the lubricant must not stain or should at least be readily removable.

**Electric-motor and generator bearings,** plain or rolling-element type, are either oil- or grease-lubricated. The oils used either are straight mineral or have antirust, antioxidant inhibitors, with viscosities in the range from 150 to 360 SUS at 100°F (38°C). Grease-lubricated units use either sodium or lithium of No. 1 or No. 2 grade. The trend in lubricating electric motors is to grease-lubricate the bearing at manufacture for the life of the motor. The prelubrication is often done by the bearing manufacturer. Relubrication, if required, is done by disassembling the motor, washing the bearings, and repacking. Lubricating these bearings with a grease gun is not good practice, since the quantity applied to the bearing, which is often shielded, is not controllable and overlubrication may result.

**Ball and Roller Bearings** (See also Sec. 8) Rolling-element bearings include tapered and spherical roller bearings, needle bearings, multiple-row roller and ball bearings, and deep-groove bearings. Sizes range from miniature, 0.059 in (0.150 cm) OD, to diameters measured in feet. The wide acceptance of these bearings is partially attributable to their low friction; and friction is little affected by speed, load, and temperature. Their low static friction recommends their selection where low starting torque is sought. They are well suited for heavy loads at low speeds.

Although the primary motion is rolling, sliding does occur between the rolling elements (balls and rollers) and races or thrust collars, particularly in thrust-loaded bearings. Also in assemblies employing separators, sliding occurs between (1) the rolling elements and separators and (2) the separator and races. The lubrication mechanism of rolling-contact bearings cannot be satisfactorily explained in terms of conventional hydrodynamic or boundary location; elastohydrodynamic effects are involved. The lubricant serves several functions: (1) to lubricate the rolling and sliding contacts, (2) to provide cooling, (3) to prevent rusting of highly finished surfaces, (4) to prevent scuffing and wear, and (5) to serve as a seal against moisture and dirt.

The choice of grease or oil as a lubricant is influenced by (1) temperature, (2) speed, (3) load, (4) method of applications, (5) sealing, (6) bearing size, and (7) type of service.

Oil lubrication is generally recommended where high-temperature, high-speed operation requires cooling and circulation. Greases permit the use of simplified lubrication and "sealed-for-life" designs, with accompanying reduced maintenance. Grease applications are used successfully in many places where oil previously had been considered essential.

Oil for ball- and roller-bearing lubrication must be resistant to degradation. If conditions indicate, additives may be used for rust protection, oxidation inhibition, extreme pressure, detergency, and other special properties. The viscosity of the oil depends upon the seals, the speed, and the operating temperatures. Methods of oil supply include spray or mist, oil

**Table 6. Industrial Applications and Their Additive Requirements**

| Additives | Turbines | Hydraulics | Gear boxes | | | General purpose no drip | Electric motor | Spindles Machine tools | Ways | Internal-combustion engines | Steam engines and pumps | Circulating systems | Air compressors |
| | | | Worm | Hypoid | Other | | | | | | | | |
|---|---|---|---|---|---|---|---|---|---|---|---|---|---|
| Antirust............. | x | x | | x | x | | x | x | | x | | x | x |
| Anticorrosion.......... | x | x | x | x | x | | x | x | x | x | x | x | x |
| Antioxidant........ | x | x | | | | | x | | | x | | x | |
| Pour depressant..... | | x | x | x | x | | x | | | x | | | |
| Antifoam.......... | x | x | x | x | x | | x | x | | x | | x | |
| Detergent............ | | | | | | | | | | x | | | |
| Fatty oils............ | | | x | | | | | | | | x | | |
| Extreme pressure..... | | | x | x | | | | | x | | | | |
| Tackiness............ | | | | | | x | | | | | | | |

jet, or more commonly, bath, drip, or pressure feed. The oil level in a bearing should be maintained as low as possible to reduce lubricant churning.

The selection of a suitable grease depends upon load, speed, temperature, method of application, presence of moisture, and bearing-housing design. The mineral-oil, soap-type greases most frequently used are sodium, calcium, sodium and calcium mixed, and lithium.

**Industrial Applications**   Many industrial oils are enhanced or modified with additives, usually chemical compounds which will improve some inherent property or impart new characteristics. In general, lubricant modifiers fall in two general classes: (1) those that affect a physical characteristic, such as viscosity index or pour point; and (2) those whose ultimate influence is chemical in nature, such as oxidation and corrosion inhibitors or detergents. Additives do not transform poor-quality lubricants into high-quality products; it is necessary to start with quality lubricants. Additives in industrial oils take into account different levels of temperature and operating conditions. Table 6 lists some of the common industrial applications and the additives used in the formulation of lubricants.

**Metal Forming**   The functions of fluids in machining operations are (1) to cool and (2) to lubricate. Fluids carry the heat generated by the tool and chip rubbing contact and/or the heat resulting from the plastic deformation of the work. Cooling aids tool life, preserves tool hardness, and helps to maintain the dimensions of the machined parts. Fluids lubricate the chip-tool interface to reduce tool wear, frictional heat, and power consumption. Lubricants aid in the reduction of metal welding and adhesion to improve surface finish. The fluids may also serve to carry away chips and debris from the work as well as to protect machined surfaces, tools, and equipment from rust and corrosion.

Many types of fluids are used. Most frequently they are (1) mineral oils, (2) soluble oil emulsions, (3) chemicals, or (4) synthetics. The mineral oils and synthetics are often compounded with additives to impart specific properties. Some metalworking operations are conducted in a controlled gaseous atmosphere (air, nitrogen, carbon dioxide).

The choice of a metalworking fluid is very complicated. Factors to be considered are (1) the metal to be machined, (2) the tools, and (3) the type of operation. Tools are usually steel,

carbide, or ceramic. In operations where chips are formed, the relative motion between the tool and chip is high speed under high load and often at elevated temperature. In chipless metal forming—drawing, rolling, stamping, extruding, spinning—the function of the fluid is to (1) lubricate and cool the die and work material and (2) reduce adhesion and welding on dies. In addition to fluids, solids such as talc, clay, and soft metals may be used in drawing operations.

In addition to the primary function to lubricate and cool, the cutting fluids should not (1) corrode, discolor, or form deposits in the work; (2) produce undesirable fumes, smoke, or odors; (3) have detrimental physical effects on operators. Fluids should also be stable, resist bacteria growth, and be foam-resistant.

In the machining operations, many combinations of tools, workpieces, and operating conditions may be encountered. In some instances, straight mineral or lightly compounded oils will suffice; while in more severe conditions, highly compounded oils are required. The effectiveness of the compounds depends upon the chemical activity and the newly formed, highly reactive surfaces; these combined with the high temperatures and pressures at the contact points are ideal for chemical reactions. Compounding agents include fatty oils, sulfur, sulfurized fatty oils, and sulfurized, chlorinated and phosphorus additives. These agents react with the metals to form compounds which have a lower shear strength and may posses EP properties. The compounded oils, either dark or transparent, are most widely used in the industry.

Water is probably the most effective coolant available but can seldom be used as an effective cutting fluid. It has little value as a lubricant and will promote rusting. One way of combining the cooling properties of water with the lubricating properties of oil is through the use of soluble oils. These oils are compounded so they form a stable emulsion with water. The main component, water, provides effective cooling while the oil and compounds impart desirable lubricating, EP, and corrosion-resistance properties.

The ratio of water and oil will influence the relative lubricating and cooling properties of the emulsion. For these non-miscible liquids to form a stable emulsion, a substance called an emulsifier must be added. Water hardness is an important consideration in the forming of an emulsion, and protection against bacteria growth should be provided. Care in preparing and handling the emulsion will ensure more satisfactory performance and longer life. Overheating, freezing, water evaporation, contamination, and excessive air mixing will adversely affect the emulsion.

**Fluid-power systems,** frequently referred to as hydraulic systems, are based on the principle that "pressure exerted on a confined liquid is transmitted undiminished in all directions and acts with equal force on equal areas." Some features of fluid over other forms of power transmission are (1) design simplicity, (2) flexibility of location, (3) the possibility of automation, (4) control accuracy, (5) speed and pressure variation, (6) reduction of wear on moving parts, and (7) operational economy. The development of automatic machines and labor-saving devices, the need for remote control, and the desire for compact and accurately controlled power sources have influenced a rapid development of fluid power.

A system usually involves (1) power-input pumps, (2) transmission system (pipes, tubing, or hose), (3) controls (flow, pressure, and relief valves), and (4) power output (cylinders, fluid motors). Very simple and very complex systems have been developed and are in use.

The fluid is primarily used to transmit a force applied at one point in a system to some other location and reproduce quickly any change in direction or in magnitude. The fluid must flow readily and be relatively incompressible. It must also function satisfactorily in other respects, such as (1) providing adequate seal and lubrication between moving parts, (2) undergoing minimum physical or chemical changes, (3) protecting against rust and corrosion, (4) minimizing wear, and (5) permitting rapid settling or separation of insoluble contaminants in the system.

Many liquids fulfill the main requirements for a fluid for a system. From the point of view of availability and cost, water would be an obvious choice. On the other hand, its disadvantages for most applications are obvious. Petroleum oils are suitable for many applications, and selecting the most satisfactory one depends upon its initial suitability and quality. Initial suitability usually involves viscosity and viscosity index, and sometimes involves pour point (petroleum fluids are relatively incompressible). Properties that reflect quality include oxidation stability, rust prevention, foam resistance, demulsibility, and lubricity. Additives are incorporated to improve certain properties.

In pressurized systems, leakage and line rupture are always possible. Where there is a source of ignition for the fluid, a fire hazard exists with hydraulic fluids. For such applications **fire-resistant fluids** have been developed. They are generally two basic types: (1) aqueous-base (water-oil or water-glycol) fluids or (2) nonaqueous or synthetic fluids. The successful application of these fluids depends upon their inherent properties. In changing from a petroleum-base to a fire-resistant fluid, advice and recommendations from suppliers should be sought relative to the fluids' operating temperatures, viscosity and VI, and its effect on paint, seals, and corrosivity.

**Used-Lubricant Disposal** The nonpolluting disposal of used lubricants is becoming increasingly important and requires continual attention. More and more legislation and control is being enacted, at local, state, and national levels, to regulate the disposition of wastes. Depending upon the specific materials, alternative methods are recommended, such as reprocessing, mixing with fuels and burning, and collecting agencies. If wastes are dumped on the ground or directly into sewers, they may eventually be washed into streams and water supplies and become water pollutants. Improper burning may contribute to air pollution. Wastes must be handled in such a way as to assure nonpolluting disposal.

Section **7**

# Fuels and Furnaces

**BY**

**M. D. SCHLESINGER,** *Deputy Director, Energy Research and Development Administration, Pittsburgh, Pennsylvania.*

**J. F. FARNSWORTH,** *Manager Gasification, Engineering and Construction Division, Koppers Company, Inc.*

**M. H. MAWHINNEY,** *Consulting Engineer, Salem, Ohio.*

**CHARLES R VELZY,** *President, Charles R Velzy Associates, Inc.*

**CHARLES O. VELZY,** *Secretary and Treasurer, Charles R Velzy Associates, Inc.*

**WILLIAM E. LEWIS,** *Retired Manager, Field Sales, Lectromelt Furnace Division, McGraw-Edison Co.*

### FUELS
#### by M. D. Schlesinger and Staff at the Bureau of Mines

Coal (BY J. T. MCCARTNEY) ............................ 7-2
Coke (BY D. E. WOLFSON) ............................. 7-11
Peat, Wood, and Miscellaneous Solid Fuels (BY M. D. SCHLESINGER) ......................................... 7-12
Petroleum and Other Liquid Fuels (BY C. C. WARD) .......... 7-14
Gaseous Fuels (BY C. C. WARD) ......................... 7-21
Explosives (BY ROBERT W. VAN DOLAH) .................. 7-24
Dust Explosions (BY HARRY C. VERAKIS AND JOHN NAGY) ...... 7-75
Rocket Fuels (BY GLENN H. DAMON) ..................... 7-33

### CARBONIZATION OF COAL AND GAS MAKING
#### by J. F. Farnsworth

Carbonization of Coal ................................. 7-37
Carbonizing Apparatus ................................ 7-39
Gas Making .......................................... 7-41

### COMBUSTION FURNACES
#### by M. H. Mawhinney

Fuels ............................................... 7-47
Types of Industrial Heating Furnaces .................... 7-48
Size and Economy of Furnaces ......................... 7-48

Furnace Construction ................................. 7-50
Heat-Saving Methods ................................. 7-50
Special Atmospheres ................................. 7-51

### INCINERATION
#### by Charles R Velzy and Charles O. Velzy

Nature of the Fuel ................................... 7-53
Types of Furnaces ................................... 7-53
Plant Design ........................................ 7-53
Furnace Design ..................................... 7-54
Flue-Gas Tempering ................................. 7-56
Recovery ........................................... 7-57

### ELECTRIC FURNACES AND OVENS
#### by William E. Lewis

Classification and Service ............................. 7-60
Resistor Furnaces ................................... 7-60
Dielectric Heating ................................... 7-64
Induction Heating ................................... 7-64
Arc Furnaces ....................................... 7-64
Induction Furnaces .................................. 7-66
Power Requirements for Electric Furnaces .............. 7-68
Submerged-Arc and Resistance Furnaces .............. 7-69

# FUELS

by Staff, Bureau of Mines, U.S. Department of the Interior. Prepared under the direction of M. D. Schlesinger

## COAL
### by J. T. McCartney

REFERENCES: Bibliography of Bureau of Mines Investigations of Coal and Its Products, 1910–1960, *BuMines Inf. Circ.* 8049. ASTM, "Standards on Gaseous Fuels; Coal and Coke." Lowry, "Chemistry of Coal Utilization," Wiley. *Coal Age*, 1972 Mining Guidebook.

**Coal** is a black or brownish-black combustible solid formed by the decomposition of ancient vegetation in the absence of air, under the influence of biochemical action, moisture, pressure, and heat. It is composed primarily of carbon, hydrogen, oxygen, and small amounts of nitrogen and sulfur. Coal is a heterogeneous material, varying in type of parent-plant components and in degree of metamorphic change. Associated with the organic matrix are water and a large variety of inorganic materials (containing as many as 65 chemical elements). Coal is widely used as a fuel and, to a small extent, as a source of organic chemicals.

### Classification and Description

Coal may be classified in various ways: by rank, by variety, by size, and sometimes by use.

Coals are classified by **rank,** i.e., according to their degree of metamorphism, or progressive alteration, in the natural series from lignite to anthracite. Table 1 shows the classification of coals by rank adopted as standard by the ASTM. The basic scheme of classification by this system is according to fixed carbon and heating value calculated to the mineral-matter-free (mmf) basis. The higher-rank coals are classified according to fixed carbon on the dry basis, and the lower-rank coals according to calorific value (Btu) on the moist basis. **Agglomerating** character is used to differentiate between certain adjacent groups. Coals are considered agglomerating if, in a test to determine the amount of volatile matter, they produce either a coherent button that will support a 500-g weight without pulverizing or a button that shows swelling or cell structure.

For classifying coals according to rank, fixed carbon and Btu are calculated to the mmf basis by using either the Parr formulas, Eqs. **(1)** to **(3)**, or the approximation formulas, Eqs. **(4)** to **(6)**. In case of litigation, the appropriate Parr formula is used.

**Parr formulas:**

$$\text{Dry, mmf F.C.} = \frac{\text{F.C.} - 0.15S}{100 - (M + 1.08A + 0.55S)} \times 100 \quad (1)$$

$$\text{Dry, mmf V.M.} = 100 - \text{dry, mmf F.C.} \quad (2)$$

$$\text{Moist, mmf Btu} = \frac{\text{Btu} - 50S}{100 - (1.08A + 0.55S)} \times 100 \quad (3)$$

**Approximation formulas:**

$$\text{Dry, mmf F.C.} = \frac{\text{F.C.}}{100 - (M + 1.1A + 0.1S)} \times 100 \quad (4)$$

$$\text{Dry, mmf V.M.} = 100 - \text{dry, mmf F.C.} \quad (5)$$

$$\text{Moist, mmf Btu} = \frac{\text{Btu}}{100 - (1.1A + 0.1S)} \times 100 \quad (6)$$

where F.C. = percentage of fixed carbon, V.M. = percentage of volatile matter, $M$ = percentage of moisture, $A$ = percentage of ash, $S$ = percentage of sulfur, all on a "moist" basis. "Moist" coal is coal that contains natural bed moisture but does not have any visible water on its surface.

Figure 1 shows **representative proximate analyses and heating values** of various ranks of coal in the United States. The

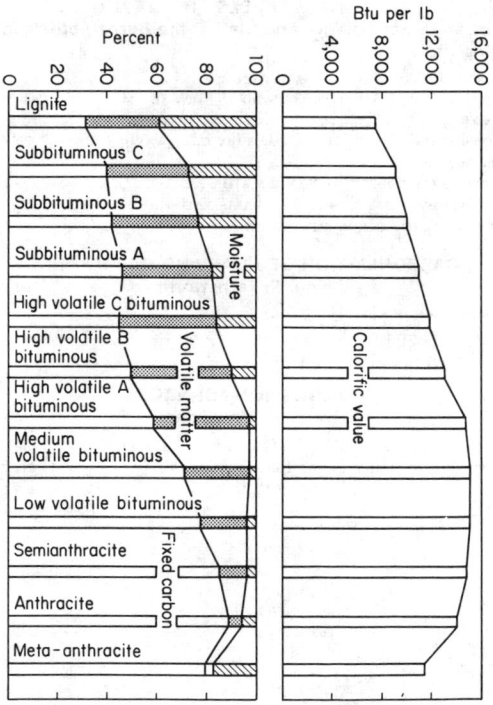

**Fig. 1** Proximate analysis and heating values of various ranks of coal (ash-free basis).

**Table 1. Classification of Coals by Rank (ASTM D388)[a]**

| Class | Group | Fixed-carbon limits, percent (dry, mineral-matter-free basis) | | Volatile-matter limits, percent (dry, mineral-matter-free basis) | | Calorific value limits, Btu per pound (moist,[b] mineral-matter-free basis) | | Agglomerating character |
|---|---|---|---|---|---|---|---|---|
| | | Equal or greater than | Less than | Greater than | Equal or less than | Equal or greater than | Less than | |
| I. Anthracitic | 1. Meta-anthracite | 98 | ... | ... | 2 | ...... | ...... | Non-agglomerating[e] |
| | 2. Anthracite | 92 | 98 | 2 | 8 | ...... | ...... | |
| | 3. Semianthracite | 86 | 92 | 8 | 14 | ...... | ...... | |
| II. Bituminous | 1. Low-volatile bituminous coal | 78 | 86 | 14 | 22 | ...... | ...... | |
| | 2. Medium volatile bituminous coal | 69 | 78 | 22 | 31 | ...... | ...... | Commonly agglomerating[c] |
| | 3. High-volatile A bituminous coal | ... | 69 | 31 | ... | 14,000[d] | ...... | |
| | 4. High-volatile B bituminous coal | ... | ... | ... | ... | 13,000[d] | 14,000 | |
| | 5. High-volatile C bituminous coal | ... | ... | ... | ... | 11,500 | 13,000 | |
| | | | | | | 10,500 | 11,500 | Agglomerating |
| III. Subbituminous | 1. Subbituminous A coal | ... | ... | ... | ... | 10,500 | 11,500 | Non-agglomerating |
| | 2. Subbituminous B coal | ... | ... | ... | ... | 9,500 | 10,500 | |
| | 3. Subbituminous C coal | ... | ... | ... | ... | 8,300 | 9,500 | |
| IV. Lignitic | 1. Lignite A | ... | ... | ... | ... | 6,300 | 8,300 | |
| | 2. Lignite B | ... | ... | ... | ... | ...... | 6,300 | |

[a] This classification does not include a few coals, principally nonbanded varieties, which have unusual physical and chemical properties and which come within the limits of fixed-carbon or calorific value of the high-volatile bituminous and subbituminous ranks. All of these coals either contain less than 48 percent dry, mineral-matter-free fixed carbon or have more than 15,500 moist, mineral-matter-free British thermal units per pound. Btu/lb $\times$ 2,328 = kJ/lb.

[b] Moist refers to coal containing its natural inherent moisture but not including visible water or on the surface of the coal.

[c] If agglomerating, classify in low-volatile group of the bituminous class.

[d] Coals having 69 percent or more fixed carbon on the dry, mineral-matter-free basis are classified according to fixed carbon, regardless of calorific value.

[e] It is recognized that there may be nonagglomerating varieties in these groups of the bituminous class, and there are notable exceptions in the high-volatile C bituminous group.

Table 2. Sources and Analyses of Various Ranks of Coal as Received

| Classification by rank | State | County | Bed | Proximate, % | | | | Ultimate, % | | | | | Calorific value, Btu/lb* |
|---|---|---|---|---|---|---|---|---|---|---|---|---|---|
| | | | | Moisture | Volatile matter | Fixed carbon | Ash† | Sulfur | Hydrogen | Carbon | Nitrogen | Oxygen | |
| Meta-anthracite | Rhode Island | Newport | Middle | 13.2 | 2.6 | 65.3 | 18.9 | 0.3 | 1.9 | 64.2 | 0.2 | 14.5 | 9,310 |
| Anthracite | Pennsylvania | Lackawanna | Clark | 4.3 | 5.1 | 81.0 | 9.6 | 0.8 | 2.9 | 79.7 | 0.9 | 6.1 | 12,880 |
| Semianthracite | Arkansas | Johnson | Lower Hartshorne | 2.6 | 10.6 | 79.3 | 7.5 | 1.7 | 3.8 | 81.4 | 1.6 | 4.0 | 13,880 |
| Low-volatile bituminous coal | West Virginia | Wyoming | Pocahontas No. 3 | 2.9 | 17.7 | 74.0 | 5.4 | 0.8 | 4.6 | 83.2 | 1.3 | 4.7 | 14,400 |
| Medium-volatile bituminous coal | Pennsylvania | Clearfield | Upper Kittanning | 2.1 | 24.4 | 67.4 | 6.1 | 1.0 | 5.0 | 81.6 | 1.4 | 4.9 | 14,310 |
| High-volatile A bituminous coal | West Virginia | Marion | Pittsburgh | 2.3 | 36.5 | 56.0 | 5.2 | 0.8 | 5.5 | 78.4 | 1.6 | 8.5 | 14,040 |
| High-volatile B bituminous coal | Kentucky, western field | Muhlenburg | No. 9 | 8.5 | 36.4 | 44.3 | 10.8 | 2.8 | 5.4 | 65.1 | 1.3 | 14.6 | 11,680 |
| High-volatile C bituminous coal | Illinois | Sangamon | No. 5 | 14.4 | 35.4 | 40.6 | 9.6 | 3.8 | 5.8 | 59.7 | 1.0 | 20.1 | 10,810 |
| Subbituminous A coal | Wyoming | Sweetwater | No. 3 | 16.9 | 34.8 | 44.7 | 3.6 | 1.4 | 6.0 | 60.4 | 1.2 | 27.4 | 10,650 |
| Subbituminous B coal | Wyoming | Sheridan | Monarch | 22.2 | 33.2 | 40.3 | 4.3 | 0.5 | 6.9 | 53.9 | 1.0 | 33.4 | 9,610 |
| Subbituminous C coal | Colorado | El Paso | Fox Hill | 25.1 | 30.4 | 37.7 | 6.8 | 0.3 | 6.2 | 50.5 | 0.7 | 35.5 | 8,560 |
| Lignite | North Dakota | McLean | Unnamed | 36.8 | 27.8 | 29.5 | 5.9 | 0.9 | 6.9 | 40.6 | 0.6 | 45.1 | 7,000 |

*Btu/lb × 2.328 = kJ/kg.
†Ash is part of both the proximate and ultimate analyses.

sources and analyses of the coals selected to represent the various ranks are given in Table 2. The analyses selected for the chart are calculated to an ash-free basis because ash in coal varies without regard to rank. Figure 1 shows that the percentages of fixed carbon and the heating values, except for anthracite, increase from the lowest to the highest rank of coal as the percentages of volatile matter and bed moisture decrease.

**Meta-anthracite** is a high-carbon coal that approaches graphite in structure and composition. It usually is slow to ignite and difficult to burn. At the present time it has little commercial importance.

**Anthracite,** sometimes called hard coal, is hard, compact, and shiny black, with a generally conchoidal fracture. It ignites with some difficulty and burns with a short, smokeless, blue flame. Anthracite is used primarily for space heating and as a source of carbon. It is also used in electric-power-generating plants in or close to the anthracite-producing area. The iron and steel industry uses some anthracite in blends with bituminous coal to make coke, for sintering iron-ore fines, for lining pots and molds, for heating, and as a substitute for coke in foundries.

**Semianthracite** is dense, but softer than anthracite. It burns with a short, clean, bluish flame and is somewhat more easily ignited than anthracite. The uses are about the same as for anthracite.

**Low-volatile bituminous coal** is grayish black, granular in structure and friable on handling. It cakes in a fire and burns with a short flame that is usually considered smokeless under all burning conditions. It is used for space heating and steam raising and as a constituent of blends for improving the coke strength of higher-volatile bituminous coals. Low-volatile bituminous coals cannot be carbonized alone in slot-type ovens because they expand on coking and damage the walls of the ovens.

**Medium-volatile bituminous coal** is an intermediate stage between high-volatile and low-volatile bituminous coal and therefore has some of the characteristics of both. Some are fairly soft and friable, but others are hard and do not disintegrate on handling. They cake in a fuel bed and smoke when improperly fired. These coals make cokes of excellent strength and are either carbonized alone or blended with other bituminous coals. When carbonized alone, only those coals that do not expand appreciably can be used without damaging oven walls.

**High-volatile A bituminous coal** has distinct bands of varying luster. It is hard and handles well with little breakage. It includes some of the best steam and coking coal. On burning in a fuel bed, it cakes and gives off smoke if improperly fired. The coking property is often improved by blending with more strongly coking medium- and low-volatile bituminous coal.

**High-volatile B bituminous coal** is similar to high-volatile A bituminous coal but has slightly higher bed moisture and oxygen content and is less strongly coking. It is good coal for steam raising and space heating. Some of it is blended with more strongly coking coals for making metallurgical coke.

**High-volatile C bituminous coal** is a stage lower in rank than the B bituminous coal and therefore has a progressively higher bed moisture and oxygen content. It is used primarily for steam raising and space heating.

**Subbituminous coals** usually show less evidence of banding than bituminous coals. They have a high moisture content, and on exposure to air, they disintegrate or "slack" because of shrinkage from loss of moisture. They are noncaking and noncoking, and their primary use is for steam raising and space heating.

**Lignites** are brown to black in color and have a bed moisture content of 30 to 45 percent with a resulting lower heating value than higher-rank coals. Like subbituminous coals, they have a tendency to "slack" or disintegrate during air drying. They are noncaking and noncoking. Lignite can be burned on traveling or spreader stokers and in pulverized form.

The principal ranks of coal mined in the major coal-producing states are shown in Table 3. Their analyses depend on several factors, e.g., source, size of coal, and method of preparation. The Reports of Investigations published annually by the Bureau of Mines (Analyses of Tipple and Delivered Samples of Coal Collected during the Fiscal Year 19__) are an excellent source of information on proximate and ultimate analyses, heating values, free-swelling indexes, grindability, and ash-softening temperatures of various coals mined in the United States.

Commercial varieties of coal, as defined by the ASTM (see ASTM "Standards on Coal and Coke"), are as follows:

**Common banded coal**—the common variety of bituminous and subbituminous coal. It consists of a sequence of irregularly alternating layers or lenses of (1) homogeneous black material having a brilliant vitreous luster; (2) grayish-black, less brilliant, striated material, usually of silky luster; and (3) generally thinner bands or lenses of soft, powdery, and fibrous particles of mineral charcoal. The difference in luster of the bands is greater in bituminous than in subbituminous coal.

**Splint coal**—a variety of bituminous or subbituminous coal, commonly having a dull luster and grayish-black color, of compact structure, often containing a few thin, irregular bands with vitreous luster. When struck, it is resonant. It is hard and tough and breaks with an irregular, rough, sometimes splintery fracture. It is free-burning and does not swell on heating.

**Cannel coal**—a variety of bituminous or subbituminous coal of uniform and compact, fine-grained texture with a general absence of banded structure. It is dark gray to black in color, has a greasy luster, and is noticeably of conchoidal or shell-like fracture. It is noncaking, yields a high percentage of volatile matter, ignites easily, and burns with a luminous, smoky flame.

**Boghead coal**—a variety of bituminous or subbituminous coal resembling cannel coal in appearance and behavior during combustion. It is characterized by a high percentage of algal remains and volatile matter. Upon distillation, it gives exceptionally high yields of tar and oil.

### Composition and Characteristics

Proximate analysis, sulfur content, and calorific values are the analytical determinations most commonly used for industrial characterization of coal. The **proximate analysis** is the simplest means for determining the distribution of products obtained during heating. It separates the products into four groups: (1) water or moisture, (2) volatile matter consisting of gases and vapors, (3) fixed carbon consisting of the carbonized residue less ash, and (4) ash derived from the mineral impurities in the coal. For standardized laboratory methods of making those determinations, refer to the latest edition of ASTM "Standards on Coal and Coke."

**Moisture** is the loss in weight obtained by drying the coal at a

Table 3. Principal Ranks of Coal Mined in Various States*

| State | Anthracite | Semianthracite | Low-volatile bituminous | Medium-vol. bituminous | High-vol. A bituminous | High-vol. B bituminous | High-vol. C bituminous | Subbituminous A | Subbituminous B | Subbituminous C | Lignite |
|---|---|---|---|---|---|---|---|---|---|---|---|
| Alabama.......... | | | | x | x | | | | | | |
| Alaska............ | | | | | | x | x | x | x | x | x |
| Arkansas.......... | | x | x | x | x | | | | | x | x |
| Colorado.......... | | | x | | x | x | x | x | x | x | x |
| Illinois............ | | | | | x | x | x | | | | |
| Indiana........... | | | | | | x | x | | | | |
| Iowa.............. | | | | | | | x | | | | |
| Kansas........... | | | | | x | x | | | | | |
| Kentucky | | | | | | | | | | | |
|   Eastern......... | | | | | x | x | | | | | |
|   Western........ | | | | | x | x | x | | | | |
| Maryland......... | | | x | x | x | | | | | | |
| Missouri.......... | | | | | | | x | | | | |
| Montana.......... | | | | | | x | x | x | x | x | x |
| New Mexico....... | | | | | x | x | x | | x | | |
| North Dakota..... | | | | | | | | | | | x |
| Ohio.............. | | | | | x | x | x | | | | |
| Oklahoma........ | | | x | x | x | x | | | | | |
| Pennsylvania...... | x | x | x | x | x | | | | | | |
| South Dakota..... | | | | | | | | | | | x |
| Tennessee........ | | | | x | x | | | | | | |
| Texas............. | | | | | | | x | x | | | x |
| Utah............. | | | | | x | x | x | x | x | | |
| Virginia.......... | | x | x | x | x | | | | | | |
| Washington....... | | | | | x | x | | | x | | x |
| West Virginia..... | | | x | x | x | | | | | | |
| Wyoming......... | | | | | | x | x | x | x | x | |

*Compiled largely from Typical Analyses of Coals of the United States, *BuMines Bull*. 446, and Coal Reserves of the United States, *Geol. Survey Bull*. 1136.

temperature between 104 and 110°C (220 and 230°F) under prescribed conditions. Further heating at higher temperatures may remove more water, but this moisture usually is considered part of the coal substance. The moisture obtained by the standard method consists of (1) surface or extraneous moisture that may come from external sources such as percolating waters in the mine, rain, condensation from the air, or water from a coal washery; (2) inherent moisture, sometimes called **bed moisture,** which is so closely held by the coal substance that it does not produce wetness. The laboratory determination does not separate these two types of moisture. A coal may be air-dried at room temperature or somewhat above, thereby determining an "air-drying loss," but this result is not the extraneous moisture because part of the inherent moisture also vaporizes during air drying.

Mine samples taken at freshly exposed faces in the mine, which are free from visible surface moisture, give the best information as to inherent or bed moisture content. Such moisture content ranges in value from 2 to 4 percent for anthracite and for bituminous coals of the eastern Appalachian field, such as the Pocahontas, Sewell, Pittsburgh, Freeport, and Kittanning beds. In the western part of this field, especially in Ohio, the inherent moisture ranges from 4 to 10 percent. In the interior fields of Indiana, Illinois, western Kentucky, Iowa, and Missouri, the range is from 8 to 17 percent. In subbituminous coals the inherent moisture ranges from 15 to 30 percent, and in lignites from 30 to 45 percent. The total amount of moisture in commercial coal may be greater or less than that of the coal in the mine. Freshly mined subbituminous coal and lignite lose moisture rapidly when exposed to the air. The extraneous or surface moisture in coal is a function of the surface exposed, each surface being able to hold a film of moisture. Fine sizes hold more moisture than lump. Coal which in the mine does not contain more than 4 percent moisture may in slack sizes hold as much as 15 percent; the same coal in lump sizes, even after underwater storage, may contain little more moisture than originally in the mine.

In the standard method of analysis, the **volatile matter** is taken as the loss in weight, less moisture, obtained by heating the coal for 7 min in a covered crucible at about 950°C (1742°F) under specified conditions. Volatile matter does not exist in coal as such but is produced by decomposition of the coal when heated. It consists chiefly of the combustible gases, hydrogen, carbon monoxide, methane, and other hydrocarbons, tar vapors, volatile sulfur compounds, and some noncombustible gases such as carbon dioxide and water vapor. The composition of the volatile matter varies greatly with different coals: the amount can vary with the rate of heating. The inert or noncombustible gas may range from 4 percent of the total volatile matter in low-volatile coals to 40 percent in subbituminous coals.

The standard method of determining the **fixed carbon** is to subtract from 100 the sum of the percentages of the moisture, volatile matter, and ash of the proximate analysis. It is the carbonaceous residue less ash remaining in the test crucible in the determination of the volatile matter. It does not represent the total carbon in the coal because a considerable part of the carbon is expelled as volatile matter in combination with hydrogen as hydrocarbons and with oxygen as carbon monoxide and carbon dioxide. It also is not pure carbon because it may contain several tenths percent of hydrogen and oxygen, 0.4 to 1.0 percent of nitrogen, and about half of the sulfur that was in the coal.

In the standard method, **ash** is the inorganic residue that remains after burning the coal in a muffle furnace to a final temperature of 700 to 750°C (1292 to 1382°F). It is composed largely of compounds of silicon, aluminum, iron, and calcium, with smaller quantities of compounds of magnesium, titanium, sodium, and potassium. Although the constituents are reported as oxides by the analyst, they occur in the ash largely as a mixture of silicates, oxides, and sulfates, with smaller quantities of other compounds. The silicates have their origin in the shale and clay minerals. The principal source of the iron oxide is pyrite, which burns to form ferric oxide and sulfur oxides. The calcium and magnesium oxides result from decomposition of the carbonate minerals. The sulfates are formed largely from interaction among carbonate minerals, pyrite, and oxygen. These ash-forming constituents consist of (1) "inherent" or "intrinsic" impurities that are present in an intimate mixture with the coal substance and are derived either from the original vegetable material or from external sources by sedimentation or precipitation during the process of accumulation of coal-forming vegetation; (2) impurities deposited either during the laying down of the coal bed or subsequently, that occur in the form of partings, veins, and nodules of clay, shale, pyrite, and calcite; and (3) impurities that become mechanically mixed with the coal in the process of mining, such as fragments of roof and floor.

The ash as determined is usually less than the inorganic mineral matter originally present in the coal. During incineration, various weight changes take place, such as loss of water of constitution of the silicate minerals, loss of carbon dioxide from carbonate minerals, oxidation of iron pyrites to iron oxide, and fixation of a part of the oxides of sulfur by bases such as calcium and magnesium.

The chemical composition of coal ash varies widely depending on the mineral constituents associated with the coal. Typical limits of ash composition of U.S. bituminous coals are as follows:[1]

| Constituent | Percent |
|---|---|
| Silica, $SiO_2$ | 20–60 |
| Alumina, $Al_2O_3$ | 10–35 |
| Ferric oxide, $Fe_2O_3$ | 5–35 |
| Calcium oxide, CaO | 1–20 |
| Magnesium oxide, MgO | 0.3–4 |
| Titanium dioxide, $TiO_2$ | 0.5–2.5 |
| Alkalies, $Na_2O$ and $K_2O$ | 1–4 |
| Sulfur trioxide, $SO_3$ | 0.1–12 |

The ash of subbituminous coals may have more CaO, MgO, and $SO_3$ than the ash of bituminous coals; the trend may be even more pronounced for lignite ash.

Ultimate analysis expresses the composition of coal as sampled in percentages of carbon, hydrogen, nitrogen, sulfur, oxygen, and ash. The carbon includes that present in the organic coal substance as well as a minor amount that may be present as mineral carbonates. In ASTM practice, the hydro-

[1]From Analyses of Ash from United States Coals, *BuMines Bull.* 567, 1956.

gen and oxygen values include those of the organic coal substance as well as those present in the form of moisture and the water of constitution of the silicate minerals. In certain other countries, the values for hydrogen and oxygen are corrected for the moisture in the coal and are reported separately. The ash is the same as reported in the proximate analysis; the sulfur, carbon, hydrogen, and nitrogen are determined chemically (see ASTM "Standards on Coal and Coke"). Because of the lack of a simple, direct method for determining oxygen in coal, it usually is estimated by subtracting the sum of carbon, hydrogen, nitrogen, sulfur, and ash from 100. If suitable corrections are made for the carbon, hydrogen, and sulfur derived from the inorganic material and for conversion of ash to mineral matter, the ultimate analysis represents the elemental composition of the organic coal substance in terms of carbon, hydrogen, nitrogen, sulfur, and oxygen, but this is time-consuming and generally not of practical value.

Sulfur occurs in three forms in coal: (1) **pyritic sulfur,** or sulfur combined with iron as pyrite or marcasite; (2) **organic sulfur,** or sulfur combined with coal substance as a heteroatom or as a bridge atom; (3) **sulfate sulfur,** or sulfur combined mainly with iron or calcium together with oxygen as iron sulfate or calcium sulfate. Pyrite and marcasite are recognized by their metallic luster and pale brass-yellowish color, although some marcasite is almost white. Organic sulfur may comprise from about 20 to 85 percent of the total sulfur in the coal. Most freshly mined coal contains only very small quantities of sulfate sulfur; it increases in weathered coal. The total sulfur content of coal mined in the United States varies from about 0.4 to 5.5 percent by weight on a dry coal basis.

The **gross calorific value** of a fuel expressed in Btu/lb of fuel is the heat produced by complete combustion of a unit quantity, at constant volume, in an oxygen-bomb calorimeter under standard conditions (see ASTM "Standards on Coal and Coke"). It includes the latent heat of the water vapor in the products of combustion. Since the latent heat is not available for making steam in actual operation of boilers, a net calorific value is sometimes determined, although not in usual U.S. practice, by the following formula:

**Net calorific value,** Btu/lb = gross calorific value, Btu/lb
$$- (92.70 \times \text{total hydrogen, percent in coal})$$

The gross calorific value may also be approximated by Dulong's formula, as follows:

$$\text{Btu/lb} = 14{,}544C + 62{,}028\left(H - \frac{O}{8}\right) + 4{,}050S$$

$$(\text{Btu/lb} \times 2{,}328 = \text{kJ/kg})$$

where C, H, O, and S are weight fractions from the ultimate analysis. For anthracites, semianthracites, and bituminous coals, the calculated values are usually within 1½ percent of those determined by the bomb calorimeter. For subbituminous and lignitic coals, the calculated values show deviations often reaching 4 and 5 percent.

Because coal ash is a mixture of various components, it does not have a definite melting point; the gradual softening and fusion of the ash is not merely the successive melting of the various ash constituents but is a more complicated process in which reactions involving the formation of new and more fusible compounds take place.

The **fusibility of coal ash** is determined by heating a triangular pyramid (cone), ¾ in high and ¼ in wide at each side of the base, made up of the ash together with a small amount of organic binder (ASTM "Standards on Coal and Coke"). As the cone is heated, three temperatures are noted: (1) the initial deformation temperature (IDT), or the temperature at which the first rounding of the apex or the edges of the cone occurs; (2) the softening temperature (ST), or the temperature at which the cone has fused down to a spherical lump; and (3) the fluid temperature (FT), or the temperature at which the cone has spread out in a nearly flat layer. The softening interval is the degrees of temperature difference between (2) and (1), the flowing interval the difference between (3) and (2), and the fluidity range the difference between (3) and (1). Of the three, the softening temperature is most widely used. Table 4 shows ash-fusion data typical of some important United States coals.

Data on ash-fusion characteristics are useful to the combustion engineer concerned with evaluation of the clinkering tendencies of coals used in combustion furnaces and with corrosion of metal surfaces in boilers due to slag deposits. The kinds of mineral matter occurring in different coals are not well related to rank or geographic location, although there is a tendency for midcontinent coals (Indiana to Oklahoma) to have low ash-fusion temperatures. Significance is attached to all of the previously indicated fusion temperatures and the intervals between them. The IDT is sometimes identified with surface stickiness, the ST with plastic distortion or sluggish flow, and the FT with liquid mobility. Long fusion intervals often produce tough, dense slags; short intervals favor porous, friable structures.

Most bituminous coals, when heated at uniformly increasing temperatures in the absence or partial absence of air, fuse and become plastic. These coals may be designated as either caking or coking coals; different coals possess this property of caking or coking in different degrees. **Caking** usually refers to the fusion process in a boiler furnace. **Coking coals** are those that make good coke, suitable for metallurgical purposes. Coals that are caking in a fuel bed do not necessarily make good coke in a coke oven. Subbituminous coal, lignite, and anthracite are noncaking.

The **free-swelling index** test measures the free-swelling properties of coal and gives an indication of the caking characteristics of the coal when burned on fuel beds. It is not intended to determine the expansion of coals in coke ovens. The test consists in heating 1 g of pulverized coal in a silica crucible over a gas flame under prescribed conditions to form a coke button, the size and shape of which are then compared with a series of standard profiles numbered 1 to 9 in increasing order of swelling. For details of the test and its significance, see ASTM "Standards on Coal and Coke" and An Investigation of a Laboratory Test for Determination of the Free-Swelling Index of Coal, *BuMines Rept. Inv.* 3989. See also Free-Swelling and Grindability Indexes of United States Coals, *BuMines Inf. Circ.* 8025.

The **specific gravity** of coal is the ratio of the weight of solid coal to the weight of an equal volume of water. It is useful in calculating the weight of solid coal as it occurs in the ground for estimating the tonnage of coal per acre of surface. An increase in ash-forming mineral matter increases the specific

**Table 4. Fusibility of Ash from Coals of the United States**

| State | Bed | Softening temp, °F | | | Fusibility range, °F | | |
|---|---|---|---|---|---|---|---|
| | | Lowest | Highest | Average† | Lowest | Highest | Average† |
| PA | Lower Kittanning | 1930 2910+ | 2910 | 2490 (605) 2910+ (34) | 70 | 700 | 361 (548) |
| | Pittsburgh | 1970 2910+ | 2910 | 2490 (538) 2910+ (16) | 80 | 620 | 313 (491) |
| | Upper Freeport | 1960 2910+ | 2910 | 2420 (521) 2910+ (8) | 90 | 760 | 332 (492) |
| | Lower Freeport | 2050 2910+ | 2910 | 2500 (242) 2910+ (11) | 90 | 580 | 317 (208) |
| WV | Pittsburgh | 1970 | 2630 | 2180 (523) | 100 | 620 | 379 (521) |
| | Sewell | 2040 2910+ | 2910 | 2560 (363) 2910+ (41) | 80 | 680 | 338 (328) |
| | Pocahontas No. 3 | 2140 2910+ | 2910 | 2540 (344) 2910+ (39) | 60 | 650 | 286 (289) |
| | Cedar Grove | 2210 2910+ | 2910 | 2650 (202) 2910+ (30) | 80 | 510 | 264 (148) |
| OH | No. 6 | 2010 2910+ | 2890 | 2310 (288) 2910+ (8) | 100 | 640 | 387 (281) |
| IL | No. 6 | 1850 | 2690 | 2170 (724) | 70 | 710 | 345 (721) |
| | No. 5 | 1840 | 2400 | 2030 (273) | 80 | 680 | 337 (272) |
| KY | No. 9 | 1940 | 2440 | 2070 (470) | 80 | 660 | 334 (469) |
| | No. 11 | 1970 | 2420 | 2110 (204) | 110 | 550 | 370 (204) |
| | Hazard No. 4 | 2000 2910+ | 2910 | 2620 (201) 2910+ (19) | 90 | 600 | 279 (155) |
| IN | No. 5 | 1940 | 2780 | 2150 (232) | 30 | 700 | 357 (232) |
| UT | Hiawatha | 2000 | 2770 | 2320 (81) | 70 | 680 | 282 (77) |
| CO | Brookside | 2190 | 2780 | 2440 (64) | 150 | 480 | 306 (62) |
| PA | Anthracite | | | 2900 (4) | | | |

*See Fusibility of Ash of United States Coals, *BuMines Inf. Circ.* 7923.
†Numbers in parentheses show number of samples averaged.

gravity; e.g., bituminous coals of Alabama, ranging from 2 to 15 percent ash and from 2 to 4.5 percent moisture, vary in specific gravity from 1.26 to 1.37.

**Bulk density** is the weight per cubic foot of broken coal. It varies according to the specific gravity of the coal, its size distribution, its moisture content, and the amount of orientation when piled. The range of weight from subbituminous coal to anthracite is from 44 to 59 lb/ft³ when loosely piled; when piled in layers and compacted, the weight per cubic foot may increase as much as 25 percent. The weight of fuel in a pile can usually be determined to within 10 to 15 percent by measuring its volume. Typical weights of coal, as determined by shoveling loosely into a box of 8 ft³ capacity, are as follows: anthracite, 50 to 58 lb/ft³; low- and medium-volatile bituminous coal, 49 to 57 lb/ft³; high-volatile bituminous and subbituminous coal, 42 to 57 lb/ft³.

The **grindability** of coal, or the ease with which it can be ground fine enough for use as a pulverized fuel, is a composite of several specific physical properties such as hardness, tensile strength, and fracture. A laboratory procedure adopted by ASTM for evaluating grindability, known as the **Hardgrove** machine method, uses a specially designed grinding apparatus to determine the relative grindability or ease of pulverizing coal in comparison with a standard coal, chosen as 100 grindability. Primarily, the ASTM Hardgrove grindability test is used for estimating how various coals affect the capacity of commercial pulverizers. A general relationship exists between grindability of coal and its rank. Coals that are easiest to grind (highest grindability index) are those of about 14 to 30 percent volatile matter on a dry, ash-free basis. Coals of either lower or higher volatile-matter content usually are more difficult to grind. The relationship of grindability and rank, however, is not sufficiently precise for grindability to be estimated from the chemical analysis, partly because of the variation in grindability of the various petrographic and mineral components. Grindability indexes of U.S. coals range from about 20 for an anthracite to 120 for a low-volatile bituminous coal.

### Mining

Coal is mined by either underground or surface methods. In underground mining the coal beds are made accessible through shaft, drift, or slope entries (vertical, horizontal, or inclined, respectively), depending on location of the bed relative to the terrain.

The most widely used methods of coal mining in the United States are termed continuous and conventional mining. The former makes use of **continuous miners** which break the coal from the face and load it onto conveyors, shuttle cars, or rail cars in one operation. Continuous miners are of ripping, boring, or milling types or hybrid combinations of these. In conventional mining the coal is usually broken from the face by means of blasting agents or by pressurized air or carbon dioxide devices. In preparation for breaking, the coal may be cut horizontally or vertically by cutting machines and holes drilled for charging explosives. The broken coal is then picked up by loaders and discharged to conveyors or cars. A newer mining method that is increasing in usage is termed longwall mining, which employs shearing or plowing machines to break coal from more extensive faces, as the name implies. Pillars are not used because the roof is caved under controlled conditions behind the working face. About half of the coal presently mined underground is being cut by machine and nearly all of the mined coal is loaded mechanically.

The broken coal is transported in the mine on shuttle cars or tractor-trailers, conveyor belts, or rail cars. The mobile shuttle cars or trailers may carry coal from the face to a main conveyor or rail-car line; auxiliary conveyors from face to main may be used. Electric-power-driven mining machines and hauling cars are operated by batteries, through trailing cables or trolley wires. Diesel power is used to a limited extent. Electric power is usually distributed through the mine at 4,000 to 7,000 V and the machine utilization voltage is commonly 440.

An important requirement in all mining systems is **roof support.** When the roof rock consists of strong sandstone or limestone, relatively uncommon, little or no support may be required over large areas. Most mine roofs consist of shales and must be reinforced. Permanent supports may consist of arches, crossbars and legs, or single posts made of steel or wood. Screw or hydraulic jacks, with or without crossbars, often serve as temporary supports. Long roof bolts, driven into the roof and anchored in sound strata above, are being used more widely for support, permitting much more freedom of movement for machines. Drilling and insertion of bolts is being done by continuous miners or separate drilling machines. Ventilation is another necessary factor in underground mining to provide a proper atmosphere for personnel and to dilute or remove dangerous concentrations of methane and coal dust. The ventilation system must be well designed so that adequate air is supplied across the working faces without stirring up more dust.

When coal occurs near the surface, **strip or open-pit mining** is often more economical than underground mining. This is especially true in states west of the Mississippi River where coal seams are many feet thick and relatively low in sulfur. The proportion of coal production from surface mining has been increasing rapidly and now amounts to about 50 percent. However, because of potential restrictions by environmental legislation, the rate of increase may be slower than in recent years

In preparation for surface mining, core drilling is conducted to survey the underlying coal seams, usually with dry-type rotary drills. The overburden must then be removed. It is first loosened by ripping or drilling and blasting. Ripping can be accomplished by bulldozers or scrapers. Drilling is done with horizontal or vertical rotary or auger drills from 5 to 15 in (0.127 to 0.381 m) in diameter. Overburden and coal are then removed by shovels, draglines, bulldozers, or wheel excavators. The first two may have bucket capacities of 200 yd³ (153 m³). Draglines are most useful for very thick cover or long dumping ranges. Hauling of stripped coal is usually done by trucks or tractor-trailers with capacities up to 240 short tons (218 metric tons). Reclamation of stripped coal land is becoming increasingly necessary. This involves returning the land to near its original contour, replanting with ground cover or trees, and sometimes providing water basins.

### Preparation

About half the coal presently mined in the United States is being cleaned mechanically to remove impurities and supply a marketable product. Mechanical mining has increased the proportion of noncoal minerals in the product. At the preparation plant run-of-mine coal is usually given a preliminary size reduction with roll crushers or rotary breakers. Large or heavy impurities are then removed by hand picking or screening. Tramp iron is usually removed by magnets. Before washing, the coal may be given a preliminary size fractionation by screening. Nearly all preparation practices are based on density differences between coal and its associated impurities. **Heavy-medium** separators using magnetite or sand suspensions in water come closest to ideal gravity separation conditions. **Mechanical devices** include jigs, classifiers, washing tables, cyclones, and centrifuges. Fine coal, less than $\frac{1}{4}$ in (6.3 mm) is usually treated separately, and may finally be cleaned by froth flotation. Dewatering of the washed and sized coal may be accomplished by screening, centrifuging, or filtering, and finally the fine coal may be heated to complete the drying. Before shipment the coal may be dustproofed and freeze-proofed with oil or salt.

Removal of sulfur from coal has become an important aspect of preparation because of recent emphasis on the role of sulfur dioxide in air pollution. Pyrite, the main inorganic sulfur mineral, is partly removed along with other minerals in conventional cleaning, but some experimental processes to improve pyrite removal are being developed. These include magnetic and electrostatic separation, chemical leaching, and specialized froth flotation.

Coal is transported to consumers mostly by **barges** to river and coastal ports or by **unit trains,** which are long trains reserved exclusively for coal hauling and operating on fast regular schedules. A small amount of pipeline transmission is done, mostly as water slurries or by pneumatic means. Pipeline costs are competitive with unit train tariffs.

### Storage

Coal may heat spontaneously, with the liability of self-heating greatest among coals of lowest rank. The heating begins as soon as the freshly broken coal is exposed to air. The process accelerates with increase in temperature, and active burning will result if the heat from oxidation is not dissipated as fast as it is produced. The finer sizes of coal, having more surface area per unit weight than the larger sizes, are more susceptible to spontaneous heating.

The **prevention of spontaneous heating** in storage poses a problem of minimizing oxidation and of dissipating any heat produced. Air may carry away heat, but it also brings oxygen to create more heat. Spontaneous heating can be prevented or lessened by (1) storing coal under water; (2) compressing the pile in layers, as with a road roller, to retard access of air; (3) storing large-size coal; (4) preventing any segregation of sizes in the pile (it is usually best to pile in layers); (5) storing in small piles; (6) keeping the storage pile as low as possible (6 ft is the limit for many coals); (7) keeping storage away from any external sources of heat; (8) avoiding any draft of air through the coal; (9) using older portions of the storage first and avoiding accumulations of old coal in corners. It is desirable to watch the temperature of the pile. A thermometer inserted in an iron tube that has been driven into the coal pile will reveal the temperature. When the coal reaches a temperature of 50°C (120°F), it should be moved. Using water to put out a fire, although effective for the moment, may only delay the necessity of moving the coal. Furthermore, this may be dangerous because steam and coal can react at high temperatures to form carbon monoxide and hydrogen. (See The Storage of Coal, *BuMines Inf. Circ.* 7235.)

### Sampling

Because coal is a heterogeneous material, collection and han-

dling of samples that adequately represent the bulk lot of coal are required if the analytical and test data are to be meaningful. Coal is best sampled when in motion, as it is being loaded or unloaded from belt conveyors or other coal-handling equipment, by collecting increments of uniform weight evenly distributed over the entire lot. Each increment should be sufficiently large and so taken as to represent properly the various sizes of the coal. Standard methods of the ASTM detail the sampling procedures for coals classed according to ash content as follows: under 8 percent, 8.0 to 9.9 percent, 10.0 to 14.9 percent, and 15 percent ash and over. For each ash classification, there are eight size groups; and for each size group the methods prescribe a minimum number of increments, each of a minimum weight that results in a specified minimum of gross sample.

Two procedures are recognized: (1) commercial sampling and (2) special-purpose sampling, such as classification by rank or performance. The **commercial-sampling** procedure is intended for an accuracy such that, if a large number of samples were taken from a large lot of coal, the test results in 95 out of 100 cases would fall within ±10 percent of the ash content of these samples. For commercial sampling of lots up to 1,000 tons, it is recommended that one gross sample represent the lot taken. For lots over 1,000 tons, the following alternatives may be used: (1) Separate gross samples may be taken for each 1,000 tons of coal or fraction thereof, and a weighted average of the analytical determinations of these prepared samples may be used to represent the lot. (2) Separate gross samples may be taken for each 1,000 tons or fraction thereof, and the −20 or −60 mesh samples taken from the gross samples may be mixed together in proportion to the tonnage represented by each sample and one analysis carried out on the composite sample. (3) One gross sample may be used to represent the lot, provided that at least four times the usual minimum number of increments are taken. In **special-purpose sampling,** the increment requirements used in the commercial sampling procedure are increased according to prescribed rules.

For details of sampling procedures and the preparation of coal samples for analysis, see the latest edition of ASTM "Standards on Coal and Coke."

### Specifications

Specifications for the purchase of coal vary widely. In general, a specification drawn up as simply as the individual case under consideration will permit is recommended. Many characteristics of coal cannot be adequately stated or even tested in the laboratory. The government, purchasing coal on a guaranteed-analysis basis, specifies (1) delivery point, (2) size, (3) analytical constituents (limits), and (4) tonnages. For example: (1) Capitol Power Plant, Washington, D.C., Pennsylvania R.R.; (2) nut-and-slack through not less than 1½-in screen; (3) moisture (as received) maximum, 4 percent; volatile maximum, 29 percent; ash maximum, 10 percent; gross heating-value minimum, 13,900 Btu/lb, all on dry basis; (4) 70,000 short tons. The bidder must state the name of the mine, its location, shipping point, the railroad at point of shipment, the coal seam, mine operator, and guaranteed analysis.

All bids are rejected which offer coal that does not meet the analytical specifications limits, if it is known from previous experience that the coal has characteristics that make it unsuitable, or the coal is not usable in a particular device by actual trial. Award is made to the bidder offering coal, the use of which results in the lowest yearly total operating cost at the plant. This determination includes consideration of the price of the coal, the efficiency of its use, the maintenance of the equipment, the labor involved in its use, and the handling of ash. This cannot be expressed by any universal formula.

Clauses included in government coal contracts cover wage clauses to compensate for changes in wage scales, the disposition of coal that does not meet specifications, and the penalties involved.

### Processing

**Coal production** in 1972 approximated 600 million short tons, with bituminous coal amounting to nearly 98 percent, and anthracite and lignite each about 1 percent. There were about 5,000 operating mines employing 140,000 miners. Output per man-day approximated 20 tons for bituminous coal, 7 tons for anthracite, and 70 tons for lignite. Productivity in surface mining may be two to five times greater than for underground mines. Recent estimates of coal reserves in the United States approximate 1,500 billion short tons, about half of which may be practically recovered. Assuming production increases of up to 1,500 million tons per year, a potential supply for several hundred years exists.

Coal in the United States is now used almost exclusively as a source of heat and power, the major exception being for the production of metallurgical coke. The major market is the electric-utility industry in which pulverized coal is burned to generate steam for turbines which drive electric generators. In the primary-metal industries coal is used to produce coke for blast furnaces and foundries and for power in steel and rolling mills. Significant amounts of coal are also used, largely for power, in the food, paper, chemical, and ceramic industries. Coal used for transportation and space heating is now insignificant and is not listed in official statistics.

Coal consumption in 1972 was distributed approximately as follows in percent: electric utilities, 59; metallurgical uses, including coke production, 18; other manufacturing, 11; retail sales, 2; and export, 10. Coal provided about 50 percent of the fuel used for power generation and about 17 percent of the total U.S. energy demand. Because of increasingly stringent limitations on sulfur emissions from fuel processing and utilization, the relative position of coal to other fuels for both power generation and energy demand will probably change in the near future. The rapidly increasing total needs should result in a considerable increase in coal production. Studies of projected coal needs by the National Petroleum Council and the U.S. Department of the Interior estimate a demand for at least 1,000 million tons by 1985.

## COKE
### by D. E. Wolfson

REFERENCES: Advance Data on Coke and Coal Chemicals in 1971, BuMines Mineral Industry Surveys, U.S. Department of the Interior. Davis, Selection of Coals for Coke Making, *BuMines Rept. Inv.* 3601, 1942. Wolfson, Birge, and Walters, Comparison of Coke Produced by BM-AGA and Industrial Methods, *BuMines Rept. Inv.* 6354, 1964.

**Coke** is the infusible, cellular, coherent, solid material obtained from coal, pitch, petroleum residues, and from some other carbonaceous materials such as the residue from destructive distillation. This residue has a characteristic structure resulting from the decomposition and polymerization of a fused or semiliquid mass. At present, specific varieties of coke other than those from coal are distinguished by prefixing a qualifying word to indicate their source, as "petroleum coke" and "pitch coke." A word may also be prefixed to indicate the process by which coke is manufactured, e.g., in the case of coke from coal "slot-oven coke," "beehive coke," "gashouse coke," and "formcoke." See Table 5.

**High-temperature coke** is the type most commonly used in the United States. In 1971, slot-type recovery ovens produced 98.7 percent of the total output of high-temperature coke, with the remainder accounted for by nonrecovery beehive and other types of ovens. Blast furnaces utilized 91.7 percent, foundries 5.5 percent, with the remainder by other industries and residential heating. **Low- and medium-temperature cokes** have limited production in the United States because of the absence of open fireplaces common to European homes and because of the limited market for low-temperature tar.

A 1959 survey of **blast-furnace coke** plants representing 30 percent of the total in the United States is indicative of the chemical and physical properties of blast-furnace coke presently used. The volatile matter of the coke ranged from 0.6 to 1.4 percent; ash, 7.5 to 10.7 percent; sulfur, 0.6 to 1.1 percent; 2-in shatter index, 59 to 82; 1½-in shatter index, 83 to 91; 1-in tumbler (stability factor), 35 to 57; and ¼-in tumbler (hardness factor), 61 to 68. The quality of blast-furnace coke has improved during the last decade. Comparison of this survey with a previous survey made in 1949 shows that ash and sulfur contents of the coke have been reduced and that the average tumbler stability, which is the principal strength index for evaluating the physical properties of blast-furnace coke, has increased from 39 to 52.

During the formation of coke several products of commercial value are produced. If the plant is large enough to recover these products, their value is about 35 percent of the coal cost. The more valuable products are fuel gas with a heating value

**Table 5. Analyses of Cokes**

| Kind of process | "As-received" basis | | | | | | | | | High heat value, Btu/lb* |
|---|---|---|---|---|---|---|---|---|---|---|
| | Proximate, % | | | | Ultimate, % | | | | | |
| | Moisture | Volatile matter | Fixed carbon | Ash† | Hydrogen | Carbon | Nitrogen | Oxygen | Sulfur | |
| By-product coke | 0.4 | 1.0 | 89.6 | 9.0 | 0.7 | 87.7 | 1.5 | 0.1 | 1.0 | 13,200 |
| Beehive coke | 0.5 | 1.2 | 88.8 | 9.5 | 0.7 | 87.5 | 1.1 | 0.2 | 1.0 | 13,100 |
| Low-temperature coke | 0.9 | 9.6 | 80.3 | 9.2 | 3.1 | 81.0 | 1.9 | 2.8 | 1.0 | 12,890 |
| Pitch coke | 0.3 | 1.1 | 97.6 | 1.0 | 0.6 | 96.6 | 0.7 | 0.6 | 0.5 | 14,100 |
| Petroleum coke | 1.1 | 7.0 | 90.7 | 1.2 | 3.3 | 90.8 | 0.8 | 3.1 | 0.8 | 15,050 |

*Btu/lb × 2.328 = kJ/kg.
†Ash is part of both the proximate and ultimate analyses.

of 550 Btu/ft³ (20,500 kJ/m³), tar, and light oils which contain benzene, toluene, xylene, and naphthalene, ammonia, phenols, etc.

The requirements for **foundry coke** are somewhat different from those for blast-furnace coke. Chemically, in the cupola the only function of the coke is to furnish heat to melt the iron, whereas in the blast furnace the function is twofold—to supply carbon monoxide for reducing the ore and to supply heat to melt the iron. Usually, it should be of large size (more than 75 mm or 3 in) and strong enough to prevent excessive degradation by impact of the massive iron charged into the cupola shaft. The following chemical characteristics are desired in foundry coke: volatile matter, not over 2.0 percent; fixed carbon, not under 86.0 percent; ash, not over 12.0 percent; and sulfur, not over 1.0 percent. In the 1959 survey, coke from two plants showed the following properties: volatile matter, 0.6 and 1.4 percent; fixed carbon, 89.6 and 91.4 percent; ash, 8.7 and 7.5 percent; and sulfur, 0.6 percent. The 1½- and 2-in shatter indexes, which measure the ability of coke to withstand breakage by impact, were 98 and 97, respectively, for both cokes.

There appear to be no rigid specifications for the physical and chemical properties of **domestic coke** (coke for residential heating). The main chemical requirements are that the ash content be as low as possible, preferably below 12 percent, and that the softening temperature of the ash be as high as possible, preferably above 2300°F (1260°C). Low- and medium-temperature cokes are preferred for some domestic heating purposes. The main chemical characteristics distinguishing them from high-temperature cokes are a higher volatile-matter content (3 to 18 percent) and higher reactivity or combustibility, with consequent greater ease of ignition. Coke with 3 to 18 percent volatile matter can be burned without smoke formation.

The most serious operating problems encountered with water-gas manufacture are due to the presence of clinker and its removal. The ash content of **water-gas coke** should be low (8 to 9 percent), and best results are obtained with a softening temperature of about 2500°F (1370°C). Very high- or low-fusion ash can be troublesome in this application. A low sulfur content is desirable because if the sulfur is high, more of it will pass into the gas and thus increase the cost of purification. The coke should be larger than 2 in.

**Pitch coke** is made from coal-tar pitch, whereas petroleum coke is made from residues from refining petroleum. Both are characterized by high carbon and low ash content and are used mainly for the production of electrode carbon.

## PEAT, WOOD, AND MISCELLANEOUS SOLID FUELS
### by M. D. Schlesinger

REFERENCES: Sheridan: Peat, Bureau of Mines Minerals Yearbook 1971. DeCarlo: Peat, Mineral Facts and Problems, *BuMines Bull.* 660, 1972. Fryling: *Combustion Engineering*, Combustion Engineering, Inc., New York, 1967. Standard Classification of Peats, Mosses, Humus and Related Products, ASTM D2607-69. Lowry, "Chemistry of Coal Utilization," Wiley, 1945.

**Peat,** an early stage in the metamorphism of vegetable matter into coal, is the product of partial decomposition and disintegration of plant remains in water-bogs, swamps, or marshlands—and in the absence of air. There are four major types of peat, classified according to their generic origin and fiber content: sphagnum moss peat (peat moss), hypnum moss peat, reed-sedge peat, and peat humus. Other forms of peat are not specifically classified by ASTM. Like all material of vegetable origin, peat is a complex mixture of carbon, hydrogen, and oxygen in a ratio similar to that of a mixture of cellulose and lignin. Generally, peat is low in sulfur, nitrogen, and ash from the parent vegetation. These contaminants enter the bog by intrusion or leaching from adjacent strata. The Federal Trade Commission specifies that to be so labeled, the peat must contain at least 75 percent peat with the rest composed of normally associated soil materials.

Peat is used as a fuel in many countries, but in the United States the abundance of higher-rank fuels has kept peat out of the energy market. About half of the 13.8 billion short ton ($12.5 \times 10^9$ tons) reserve can be found in Minnesota, and although they account for 80 percent, commercial production occurs in 23 other states. After air drying, the moisture content is sometimes only 55 percent and its heating value is about 4,000 Btu/lb (9,300 kJ/kg). Before 1950, the annual output of peat was less than 100,000 tons, but since 1963 production has increased appreciably, 605,383 tons in 1971 which sold for an average $11.69 per ton. The main use for peat in the United States is soil improvement, a mulch, a filler for fertilizers, or litter for domestic animals. A Finnish company is now marketing a compressed peat in 170-litre plastic bags for environmental rescue missions such as cleaning oil spills on land and sea.

The **moisture content** of undrained bog peat is usually between 92 and 95 percent, but it is reduced to between 10 and 55 percent when used as a commercial fuel. Normally, peat is harvested by large earthmoving equipment from a drained bog that was dried by exposure to the wind and sun. The most popular method used in Ireland and the USSR today involves harrowing of drum-cut peat and allowing it to field-dry before being picked up mechanically or pneumatically. In Ireland, where about 99 percent of the peat output is burned, it accounts for about 50 percent of power production. Smaller percentages (but larger tonnage) go to power production in the Soviet Union, and large tonnages are used for agricultural purposes and as a litter material. World production of peat in 1971 was about 215 million short tons; the USSR produced 96 percent of the total.

The **chemical and physical properties** of peat vary considerably depending on its source and method of processing. Below are typical ranges:

| Processed peat | Air-dried | Mulled | Briquettes |
|---|---|---|---|
| Moisture, wt % | 25–50 | 50–55 | 10–12 |
| Bulk density, lb/ft³ | 15–25 | | 30–60 |
| Calorific value, Btu/lb | 6,200 | 3,700–5,300 | 8,000 |

Proximate analyses of samples, calculated back to a dry basis, follow a similar broad pattern: 55 to 70 percent volatile matter, 30 to 40 percent fixed carbon, and 2 to 10 percent ash. The dry, ash-free, ultimate analysis is generally: 53 to 63 percent carbon, 5.5 to 7.0 percent hydrogen, 30 to 40 percent oxygen, 0.3 to 0.5 percent sulfur, and 1.2 to 1.5 percent nitrogen.

**Wood,** when used as a fuel, is usually a by-product of the sawmill or papermaking industry. The conversion of logs to lumber results in 50 percent waste in the form of bark, shavings, and sawdust. Fresh timber contains 30 to 50 percent moisture, mostly in the cell structure of the wood, and after air drying for a year, the moisture content reduces to 18 to 25 percent. Kiln-dried wood contains about 8 percent moisture. Most of the waste from the conversion of wood to useful products is used to provide energy for the driven machinery. A typical analysis range is given in Table 6. When additional fuel is required, supplemental firing of coal, oil, or gas can be used.

**Table 6. Typical Analysis of Dry Wood Fuels**

|  | Most woods range | Typical analysis hemlock |
|---|---|---|
| Proximate analysis, %: |  |  |
| Volatile matter | 74–82 |  |
| Fixed carbon | 17–23 |  |
| Ash | 0.5–2.2 |  |
| Ultimate analysis, %:* |  |  |
| Carbon | 49.6–53.1 | 51.7 |
| Hydrogen | 5.8–6.7 | 5.9 |
| Oxygen | 39.8–43.8 | 42.4 |
| Heating value |  |  |
| Btu/lb | 8,560–9,130 | 8,650 |
| kJ/kg | 19,900–21,250 | 20,100 |
| Moisture, as received, % | 36–58 |  |
| Ash-fusion |  |  |
| temperature, °F: |  |  |
| Initial | 2650–2760 |  |
| Fluid | 2730–2830 |  |

*Typically, wood contains no sulfur and about 0.1 percent nitrogen. Cellulose = 44.5 percent C, 6.2 percent H, 49.3 percent O.

Combustion systems for wood are generally designed specially for the material or mixture of fuels to be burned. When the moisture content is high, 70 to 80 percent, the wood must be mixed with low-moisture fuel so that enough energy enters the boiler to support combustion. Dry wood may have a heating value of 8,750 Btu/lb but at 80 percent moisture a pound of wet wood has a heating value of only 1,750 Btu/lb. The heat required just to heat the fuel and evaporate the water is over 900 Btu, and combustion may not occur. Table 7 shows the moisture-energy relationship. The usual practice when burning wood is to propel the wood particles into the furnace through injectors along with preheated air with the purpose of inducing high turbulence in the boiler. Furthermore, the wood is injected high enough in the combustion chamber so that it is dried, and all but the largest particles are burned before they reach the grate at the bottom of the furnace. Spreader stokers and cyclone burners work well.

Wood for processing or burning is usually sold by the **cord,** an ordered pile 8 ft long, 4 ft high, and 4 ft wide or 128 ft³ (3.625 m³). Its actual solid content is only about 70 percent, or 90 ft³. Other measures for wood are the **cord run,** which is measured only by the 8-ft length and 4-ft height; the width may vary. Sixteen-inch-long wood is called **stovewood** or **blockwood.**

**Wood charcoal** is made by heating wood to a high temperature in the absence of air. Wood loses up to 75 percent of its weight and 50 percent of its volume owing to the elimination of moisture and volatile matter. As a result, charcoal has a higher heating value per cubic foot than the original wood, especially if the final product is compacted in the form of briquettes. Charcoal is marketed in the form of lumps, powder, or briquettes and finds some use as a fuel for curing, restaurant cooking, and a picnic fuel. Its nonfuel uses, particularly in the chemical industry, are as an adsorption medium for purifying gas and liquid streams and as a decolorizing agent.

**Table 7. Available Energy in Wood**

| Moisture, % | Heating value, Btu/lb | Wt water/wt wood |
|---|---|---|
| 0 | 8,750 | 0 |
| 20 | 7,000 | 0.25 |
| 50 | 4,375 | 1.00 |
| 80 | 1,750 | 4.00 |

In addition to peat and wood, several **lesser-known fuels** are in common use for the generation of industrial steam and power. Aside from their value as a fuel, the burning of wastes minimizes a troublesome disposal problem that could have serious environmental impact. Nearly all these waste fuels are cellulosic in character, and the heating value is a function of the carbon content. Ash content is generally low, but a lot of moisture could be present from processing, handling, and storage. On a moisture- and ash-free basis the heating values can be estimated at 8,000 Btu/lb; more resinous materials about 9,000 Btu/lb. Table 8 is a list of some typical by-product solid fuels.

**Bagasse** is the fibrous material left after pressing the juice from sugarcane. The ground-up waste usually contains about 50 percent moisture and is burned in much the same manner as wood waste, spreader stokers, or cyclone burners are used; supplemental fuel is added sometimes to maintain steady

**Table 8. By-product Fuels**

| | Heating value, Btu/lb (dry) | Moisture, % as received | Ash, % moisture-free |
|---|---|---|---|
| Black liquor (sulfate) | 6,500 | 35 | 40–45 |
| Cattle manure | 7,400 | 50–75 | 17 |
| Coffee grounds | 10,000 | 65 | 1.5 |
| Corncobs | 9,300 | 10 | 1.5 |
| Cottonseed cake | 9,500 | 10 | 8 |
| Municipal refuse | 9,500 | 43 | 8 |
| Pine bark | 9,500 | 40–50 | 5–10 |
| Rice straw or hulls | 6,000 | 7 | 15 |
| Scrap tires | 16,400 | 0.5 | 6 |
| Wheat straw | 8,500 | 10 | 4 |

combustion and to provide energy for the elimination of moisture. Bagasse can usually supply all the fuel requirements of raw-sugar mills. A typical analysis of dry bagasse from Puerto Rico is 44.47 percent C, 6.3 percent H, 49.7 percent O, and 1.4 percent ash. Its heating value is 8,390 Btu/lb.

Furnaces have been developed to burn particular wastes, and some preferences emerge by virtue of particular operating characteristics. Spreader stokers are preferred for wood waste and bagasse. Tangential firing seems to be used for coffee grounds, rice hulls, some wood waste, and chars from coal or lignite. Traveling-grate stokers are used for industrial wastes and coke breeze.

Recent research has demonstrated that **cellulosic wastes** can be converted to fuels that are more readily burned than those containing high moisture and volatile matter. Some of the materials listed in Table 8, particularly the large-volume wastes such as cattle manure and municipal refuse, have been converted. Reaction of cellulose with **carbon monoxide and steam** at 380°C and 100 to 250 atm yields about 2 barrels of oil per ton of dry feed. The product oil is low in sulfur, about half the present environmental limit, and it has a heating value of 15,000 Btu/lb. Heavy fuel oil has a heating value of about 18,000 Btu/lb.

**Pyrolysis,** the method used to make coke and charcoal, has also been applied to wastes. Depending on the temperature, the products can be only gas and residue (900°C), or at lower temperatures some liquids are produced. The gas is useful primarily for industrial purposes and has a heating value of 400 to 500 Btu/ft³ (15,000 to 18,600 kJ/m³). The oil has about 10,500 Btu/lb (24,500 kJ/kg). Specific gravity of the oil is about 1.0. Demonstration plants are being constructed to process wastes in several areas.

# PETROLEUM AND OTHER LIQUID FUELS
## by C. C. Ward

REFERENCES: Dunstan et al., "The Science of Petroleum," 4 vols., Oxford University Press. ASTM STP 7-B, "Significance of ASTM Tests for Petroleum Products," 1955. Gruse and Stevens, "The Chemical Technology of Petroleum," McGraw-Hill. Gruse, "Motor Fuels," Reinhold. Bell, "American Petroleum Refining," Van Nostrand. Bland and Davidson, "Petroleum Processing Handbook," McGraw-Hill.

Detailed information on sampling of petroleum and petroleum products is given in Method 8001, Federal Test Method Standard VV-L-791, and in ASTM D270. Complete descriptions of methods of analysis and testing are given in ASTM Standards, vols. 17 and 18, issued annually.

### Crude Oils and Petroleum Products

Deposits of petroleum occur throughout the world, and commercial fields have been located on every continent. Most of these deposits are several thousand feet deep and the oil is produced through wells that are drilled to penetrate the oil-bearing formations. Production in 1971 was estimated at 17.7 billion barrels, or 48 million barrels per day.

Crude oils are extremely complex mixtures, consisting predominantly of hydrocarbons and compounds containing sulfur, nitrogen, oxygen, and trace metals as minor constituents. The physical and chemical characteristics of crude oils vary widely, depending on the percentages of the various compounds that are present. The specific gravities cover a wide range, but most crude oils are between 0.80 and 0.97 g/ml or gravity between 45 and 15 degrees API. There also is a wide variation in viscosities, but most crudes are in the range from 2.3 to 23 centistokes. The ultimate composition shows 84 to 86 percent carbon, 10 to 14 percent hydrogen, and small percentages of sulfur, nitrogen, and oxygen. The sulfur content is usually below 1.0 percent, but it may be as high as 5.0 percent. Table 9 gives data on several crude oils and distillates.

**Refining Crude Oil** Crude oils are seldom used as fuel because they are more valuable when refined to petroleum products. Distillation separates the crude oil into fractions equivalent in boiling range to gasoline, kerosine, gas oil, lubricating oil, and a residual. Thermal or catalytic cracking is used to convert kerosine, gas oil, or residual to gasoline, lower-boiling fractions, and a residual coke. Catalytic reforming, isomerization, alkylation, polymerization, hydrogenation, and combinations of these catalytic processes are used to upgrade the various refinery intermediates into improved gasoline stocks or distillates. The major finished products are usually blends of a number of stocks, plus additives. Distillation curves for these products are shown in Fig. 2.

**Physical Properties of Petroleum Products** Petroleum products are sold in the United States by barrels of 42 gal corrected to 60°F, Table 10. Their *specific gravity* is expressed on an arbitrary scale termed degrees API (see Sec. 1).

The **high heat value** (hhv) of petroleum products is determined by combustion in a bomb with oxygen under pressure (ASTM D240). It may also be calculated, in products free from impurities, by the formula

$$Q_v = 22,320 - 3,780d^2$$

in which $Q_v$ is the hhv at constant volume (see Sec. 4) in Btu/lb

**Table 9. Analyses and High Heat Values of Crude Petroleum, Typical Distillates, and Fuel Oils**

| Product | Gravity, deg API | Specific gravity at 60 F | Wt per gal, lb | High heat value, Btu per lb* | Ultimate analysis, percent | | | | |
|---|---|---|---|---|---|---|---|---|---|
| | | | | | C | H | S | N | O |
| California crude............... | 22.8 | 0.917 | 7.636 | 18,910 | 84.00 | 12.70 | 0.75 | 1.70 | 1.20 |
| Kansas crude................... | 22.1 | 0.921 | 7.670 | 19,130 | 84.15 | 13.00 | 1.90 | 0.45 | |
| Oklahoma crude.... ..... | 31.3 | 0.869 | 7.236 | 19,502 | 85.70 | 13.11 | 0.40 | 0.30 | |
| Oklahoma crude............... | 31.0 | 0.871 | 7.253 | 19,486 | 85.00 | 12.90 | 0.76 | | |
| Pennsylvania crude........... | 42.6 | 0.813 | 6.769 | 19,505 | 86.06 | 13.88 | 0.06 | 0.00 | 0.00 |
| Texas crude................... | 30.2 | 0.875 | 7.286 | 19,460 | 85.05 | 12.30 | 1.75 | 0.70 | 0.00 |
| Wyoming crude............... | 31.5 | 0.868 | 7.228 | 19,510 | | | | | |
| Mexican crude............... | 13.6 | 0.975 | 8.120 | 18,755 | 83.70 | 10.20 | 4.15 | | |
| Gasoline...................... | 67.0 | 0.713 | 5.935 | ........ | 84.3 | 15.7 | | | |
| Gasoline...................... | 60.0 | 0.739 | 6.152 | 20,750 | 84.90 | 14.76 | 0.08 | | |
| Gasoline-benzene blend........ | 46.3 | 0.796 | 6.627 | ........ | 88.3 | 11.7 | | | |
| Kerosene...................... | 41.3 | 0.819 | 6.819 | 19,810 | | | | | |
| Gas oil....................... | 32.5 | 0.863 | 7.186 | 19,200 | | | | | |
| Fuel oil (Mex.)............... | 11.9 | 0.987 | 8.220 | 18,510 | 84.02 | 10.06 | 4.93 | | |
| Fuel oil (mid-continent)....... | 27.1 | 0.892 | 7.428 | 19,376 | 85.62 | 11.98 | 0.35 | 0.50 | 0.60 |
| Fuel oil (Calif.)............... | 16.7 | 0.9554 | 7.956 | 18,835 | 84.67 | 12.36 | 1.16 | | |

*Btu/lb × 2,328 = kJ/kg.

and $d$ is the specific gravity at 60/60°F (*NBS Misc. Pub.* 97, 1929).

The low heat value at constant pressure $Q_p$ may be calculated by the relation

$$Q_p = Q_v - 90.8H$$

where $H$ is the weight percentage of hydrogen and can be obtained from the relation

$$H = 26 - 15d$$

For the slight difference between heat values at constant volume and at constant pressure see Sec. 4.

Typical heats of combustion of petroleum oils free from water, ash, and sulfur are given in Table 11 with an estimated accuracy of 1 percent for normal products.

The heat value should be corrected for appreciable percent-

ages of noncombustible material present. When the oil contains much sulfur, the results may be corrected by using a hhv of 4,050 Btu/lb for sulfur. Calculation of net heat of combustion for aviation gasolines and aircraft turbine fuels is detailed in ASTM D1405. The calculation is based on the correlation of determined values with the aniline-gravity products for these types of fuels.

The **specific heat** $c$ of petroleum products of specific gravity $d$ and at temperature $t$ (°F) is given by the equation

$$c = (0.388 + 0.00045t)/\sqrt{d}$$

Values from this equation are in good agreement for oils from intermediate-base crude oils, about 2 percent high for oils from paraffin-base crudes, and about 2 percent low for oils from naphthene-base crudes. Values are not more than 4 percent high for highly cracked oils.

The **heat of vaporization** $L$ (Btu/lb) may be calculated from the equation

$$L = (110.9 - 0.09t)/d$$

The heat of vaporization per gallon (measured at 60°F) is

$$8.34Ld = 925 - 0.75t$$

indicating that the heat of vaporization per gallon is dependent only on the temperature of vaporization, $t$. Typical data for petroleum products are given in Table 12 (*NBS Misc. Pub.* 97, 1929, p. 34).

The estimated accuracy of the preceding data is 10 percent, when the vaporization is at sensibly constant temperature and at pressures below 50 lb/in², without chemical change. The values are too high for vaporization at high pressures and too low by more than 10 percent for products containing large quantities of lower members of the aromatic series (benzol).

**Gasoline** is a complex mixture of hydrocarbons that distills within the range 100 to 400°F. Commercial gasolines are blends of straight-run, cracked, reformed, and natural gasolines.

**Straight-run gasoline** is recovered from crude petroleum by simple distillation and contains a large proportion of normal hydrocarbons of the paraffin series. Its octane number (see Sec. 9) usually is too low for use in modern engines, and it is reformed and blended with other products to improve its combustion properties.

**Cracked gasoline** is manufactured by heating crude-petroleum distillation fractions or residues under pressure, or by heating with or without pressure in the presence of a catalyst. Heavier hydrocarbons are broken into smaller molecules, some of which distill in the gasoline range. The octane number of cracked gasoline is usually above that of straight-run gasoline.

**Reformed gasoline** is made by passing gasoline fractions over catalysts in such a manner that low-octane-number hydrocarbons are molecularly rearranged to high-octane-number components. Many of the catalysts use platinum and other metals deposited on a silica and/or alumina support.

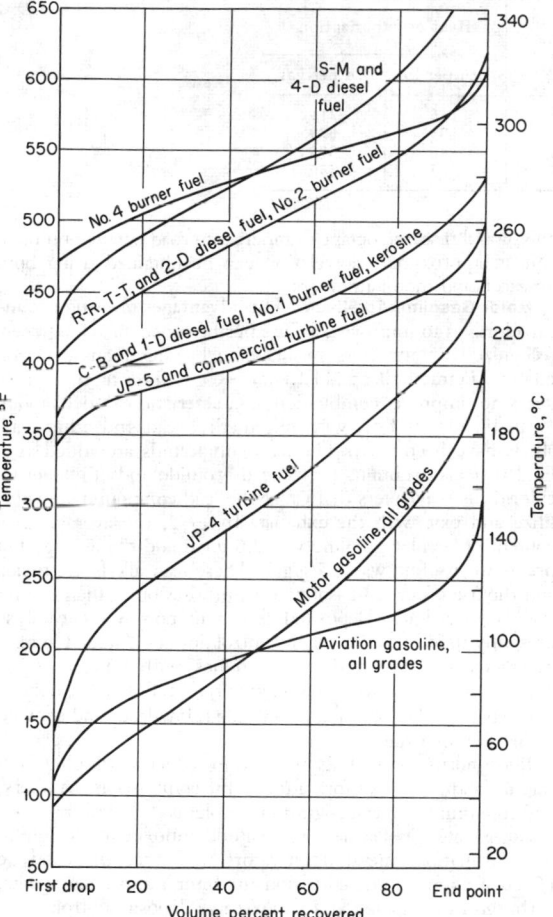

**Fig. 2** Typical distillation curves.

**Table 10. Coefficients of Expansion of Petroleum Products at 60°F**

| Deg API | <15 | 15–34.9 | 35–50.9 | 51–63.9 | 64–78.9 | 79–88.9 | 89–93.9 | 94–100 |
|---|---|---|---|---|---|---|---|---|
| Coef of expansion, F | 0.00035 | 0.0004 | 0.0005 | 0.0006 | 0.0007 | 0.0008 | 0.00085 | 0.0009 |

**Table 11. Heat Values of Petroleum Oils**

| Deg API at 60 F | Density, lb per gal[a] | High heat value at constant volume $Q_v$, Btu | | Low heat value at constant pressure $Q_p$, Btu | |
|---|---|---|---|---|---|
| | | Per lb | Per gal | Per lb | Per gal |
| 10 | 8.337 | 18,540 | 154,600 | 17,540 | 146,200 |
| 20 | 7.787 | 19,020 | 148,100 | 17,930 | 139,600 |
| 30 | 7.305 | 19,420 | 141,800 | 18,250 | 133,300 |
| 40 | 6.879 | 19,750 | 135,800 | 18,510 | 127,300 |
| 50 | 6.500 | 20,020 | 130,100 | 18,720 | 121,700 |
| 60 | 6.160 | 20,260 | 124,800 | 18,900 | 116,400 |
| 70 | 5.855 | 20,460 | 119,800 | 19,020 | 112,500 |
| 80 | 5.578 | 20,630 | 115,100 | 19,180 | 107,000 |

*Btu/lb × 2.328 = kJ/kg; Btu/gal × 279 = kJ/m³.

**Table 12. Latent Heat of Vaporization of Petroleum Products**

| Product | Gravity, deg API | Average boiling temp, deg F | Heat of vaporization | |
|---|---|---|---|---|
| | | | Btu per lb | Btu per gal |
| Gasoline........ | 60 | 280 | 116 | 715 |
| Naphtha........ | 50 | 340 | 103 | 670 |
| Kerosine........ | 40 | 440 | 86 | 595 |
| Fuel oil........ | 30 | 580 | 67 | 490 |

**Natural gasoline** is obtained from natural gas by liquefying those constituents which boil in the gasoline range either by compression and cooling or by absorption in oil. Natural gasoline is too volatile for general use, but proper characteristics can be secured by distillation or by blending. It is often blended with gasolines to adjust their volatility characteristics to meet climatic conditions.

**Catalytic hydrogenation** is used extensively to upgrade gasoline and cracking stocks for blending or further refining. Hydrogenation improves octane number, removes sulfur and nitrogen, and increases storage stability.

**Properties and Specifications for Motor Gasoline** The specifications for gasoline, ASTM D439, provide for five volatility classes of gasoline and six levels of antiknock performance. The detailed requirements for gasoline are shown in Table 13, and the antiknock performance-index requirements are shown in Table 14. Class A gasolines have a low volatility to minimize vapor lock resulting from excessive vapor formation during use in hot weather. Class E is a high-volatility gasoline that will permit easy starting during cold weather. The other classes are of intermediate volatility. D439 includes a schedule indicating the appropriate volatility class that should be used in each state during each month. Altitude is also taken into consideration.

**Antiknock characteristics** of gasolines are of utmost importance because engine power output and fuel economy are limited by the antiknock characteristics of the fuel. ASTM D439 includes six number designations of gasoline based on the minimum antiknock index, which currently is defined as the average of the research (RON) and motor (MON) octane numbers. The RON is a measure of antiknock performance under mild operating conditions at low to medium engine speeds. The MON is indicative of antiknock performance under more severe conditions, such as those encountered during power acceleration at relatively high engine speeds. There is no completely satisfactory method for translating motor and research octane numbers into road octane numbers, but an approximate correlation can be obtained using both research and motor numbers.

**Motor-Gasoline Additives** The advantages of using chemical additives to improve gasoline performance have long been recognized as attractive means of achieving high levels of quality. Tetraethyllead (TEL) has been used for more than 40 years to improve combustion characteristics. Since about 1960, blends of TEL with tetramethyl lead and other lead alkyls have been used. The lead compounds are added as a fluid that also contains ethylene dibromide and ethylene dichloride as scavengers that cause the lead compounds to volatilize and exit with the exhaust. In 1972, the average lead content of regular gasoline was 2.0 g/gal and the average for premium gasoline was 2.5 g/gal. These lead alkyls are toxic, and the use of leaded gasoline for purposes other than engine fuel is hazardous. Deposited lead will poison the catalyst presently used in the emission control devices installed in new cars. Accordingly, lead-free gasolines with an acceptable octane number are marketed nationally. ASTM D 439 defines the maximum lead content of unleaded, low-lead, and conventionally leaded fuels.

Phosphorus compounds are used to reduce spark-plug fouling, to modify deposit formation in the combustion chambers, and to eliminate surface-ignition problems.

**Other motor-gasoline additives** include antioxidants to inhibit gum formation, metal deactivators to reduce the catalytic effect of copper ions, corrosion inhibitors, anti-icing agents, carburetor detergents, and intake-valve-deposit controls.

The test methods for motor gasolines are described in detail in the annual ASTM "Book of Standards." The significance of these tests is described in the appendix to ASTM D439, "Significance of ASTM Specifications for Motor Gasolines." The characteristics of gasolines marketed in the United States are shown in semiannual surveys (Bureau of Mines Petroleum Products Survey 78).

**Table 13. Gasoline Specifications (ASTM D439)**

| Test | Volatility class | | | | | ASTM method |
|---|---|---|---|---|---|---|
| | A | B | C | D | E | |
| Distillation temp, °F, at percent evaporated: | | | | | | D86 |
| 10, max | 158 | 149 | 140 | 131 | 122 | |
| 50, min | 170 | 170 | 170 | 170 | 170 | |
| 50, max | 250 | 245 | 240 | 235 | 230 | |
| 90, max | 374 | 374 | 365 | 365 | 365 | |
| End point, max | 437 | 437 | 437 | 437 | 437 | |
| Residue, max | 2 | 2 | 2 | 2 | 2 | |
| Test temp for vapor/liquid 20, °F, max | 140 | 133 | 124 | 116 | 105 | D2533 |
| Vapor pressure, max lb/in² | 9.0 | 10.0 | 11.5 | 13.5 | 14.5 | D323 |
| Lead level, g/gal, max: | | | | | | |
| Unleaded | 0.07 | 0.07 | 0.07 | 0.07 | 0.07 | D2547 |
| Low lead | 0.50 | 0.50 | 0.50 | 0.50 | 0.50 | D2547 |
| Conventional | 4.2 | 4.2 | 4.2 | 4.2 | 4.2 | D526 |
| Corrosion, max | No. 1 | No. 1 | No. 1 | No. 1 | No. 1 | D130 |
| Gum, mg/100 ml, max | 5 | 5 | 5 | 5 | 5 | D381 |
| Sulfur, wt %, max | 0.10 | 0.10 | 0.10 | 0.10 | 0.10 | D1266 |

**Aviation Gasolines** Gasolines for aircraft piston engines have a narrower boiling range than motor gasolines; i.e., they have fewer low-boiling and fewer high-boiling components. Low-boiling components are eliminated to avoid vapor lock, prevent carburetor icing, and reduce evaporation losses on the ground and in the air. High-boiling components are removed to provide better distribution to the cylinders and reduce crankcase dilution. The military specifications (MIL-G-5572E, August 1972) for three grades of aviation gasolines differ in some details from those of the three grades of commercial aviation gasolines specified in ASTM D910. The three grades are indicated in Table 15. Specifications applicable to all three grades of military gasoline are given in Table 16.

The first number in the grade designation specified applies to lean-mixture cruise conditions to attain maximum economy, and the second number applies to rich-mixture takeoff

conditions to provide maximum power. Values below 100 are expressed as octane numbers, and values over 100 are given as performance numbers. Test method ASTM D2700, Knock Characteristics of Motor and Aviation-Type Fuels by the Motor Method, was developed to measure antiknock characteristics under lean-mixture conditions, and ASTM D909, Knock Characteristics of Aviation Fuels by the Supercharge Method, measures antiknock characteristics under rich-mixture conditions.

The production of aviation gasoline reached a peak of 192 million barrels in 1944, but production has dropped since the advent of jet-propelled aircraft, and only about 18 million barrels were produced in 1971.

**Jet or aviation turbine fuels** are not limited by antiknock requirements, and they have wider boiling ranges to provide greater availability. Military Specification MIL-T-5624H,

**Table 14. Antiknock Requirements for Gasoline**

| Gasoline antiknock designation | Antiknock (octane) index, min, one-half the sum of research octane number (RON) and motor octane number (MON): (RON + MON)/2 | Application |
|---|---|---|
| 1 | Less than 87 | For cars with low antiknock needs |
| 2 | 87* | Meets antiknock needs of most 1971 and later-model cars |
| 3 | 89 | For most 1970 and prior-model cars designed to operate on "regular" gasoline, and for 1971 and later-model cars that require higher antiknock performance than provided for in designation 2 |
| 4 | 91.5 | A "mid-premium" or "intermediate" designation which meets the lower antiknock needs of some cars designed to run on "premium" gasolines and the higher antiknock needs of some cars designed to run on "regular" |
| 5 | 95 | For most 1970 and prior-model cars with high-compression-ratio engines designed to run on "premium" gasolines, and for later-model cars with high-compression-ratio engines |
| 6 | 97.5 | For cars with high-compression-ratio engines designed to run on "premium" gasoline, but requiring higher antiknock performance than designation 5 |

*In addition the minimum motor octane number must be 82.

**Table 15. Aviation-Gasoline Grades (MIL-G-5572E)**

| Grade | Color | Tetraethyl lead content, ml/gal, max | Knock rating, min* | |
|-------|-------|------|------|------|
| | | | Lean mixture | Rich mixture |
| 80/87 | Red | 0.50 | 80 | 87 |
| 100/130 | Green | 4.60 | 100 | 130 |
| 115/145 | Purple | 4.60 | 115 | 145 |

*Knock ratings above 100.0 reported as performance numbers.

July 1971, covers two grades of aviation turbine fuels (Table 17). Fuel JP-4 is a relatively wide-boiling-range distillate that encompasses the boiling range of gasoline and kerosine. The average initial boiling point is about 140°F and the end point about 455°F. Fuel JP-5 is a high-flash-point kerosine-type fuel, initial boiling point of 360°F and end point 490°F. Most military procurement is for Grade JP-4. ASTM D1655 covers three grades of aviation turbine fuel: Type A is similar to JP-5, Type B is similar to JP-4, and Type A-1 is similar to Type A, except for special low-temperature operational characteristics. Commercial fuels are primarily Types A and A-1.

The conformance of aviation turbine fuels to current specifications is shown by data in annual surveys of these fuels (Bureau of Mines Petroleum Products Survey 79). The specifications and characteristics of aviation gasoline are stabilized, so survey reports are published at 5-year intervals (Bureau of Mines Petroleum Products Survey 64).

**Kerosine**    Kerosine fuel is less volatile than gasoline and has a higher flash point to provide greater safety in handling. Other quality tests are specific gravity, color, odor, distillation range, sulfur content, and burning quality. Most kerosine is used for heating, ranges, and illumination; so it is treated with sulfuric acid to reduce the content of aromatics, which burn with a smoky flame. Specification tests for quality control include flash point (min 115°F), distillation end point (max 572°F), sulfur (max 0.13 percent), and color (min + 16). The burning quality in a wick-type burner is described in ASTM D187.

**Gas oil** is a general term applied to distillates boiling between kerosine and lubricating oils, or in some cases between gasoline and lubricating oils. The name was derived from its initial use for making illuminating gas, but it is now used as burner fuel, diesel fuel, and catalytic-cracker charge stock.

**Diesel Fuel**    Diesel engines range from small, high-speed engines used in trucks and buses, to large, low-speed stationary engines for power plants; so several grades of diesel fuel are needed. ASTM D975 defines three grades as shown in Table 18. Also shown in Table 18 are the requirements for a high-quality marine diesel fuel as described in MIL-F-16884G, dated March 1973. The ASTM grades of fuels suitable for different classes of service:

Grade 1-D. A volatile distillate fuel for engines in service requiring frequent speed and load changes

Grade 2-D. A distillate fuel of lower volatility for engines in industrial and heavy mobile service

Grade 4-D. A fuel for low- and medium-speed engines

The annual fuel surveys (Bureau of Mines Petroleum Products Survey 77) provide an additional guide to fuel selection by grouping the fuels according to the following types of service:

Type C-B. Diesel fuel oils for city-bus and similar operations

Type T-T. Fuels for diesel engines in trucks, tractors, and similar service

Type R-R. Fuels for railroad diesel engines

**Table 16. Aviation-Gasoline Specifications—All Grades (MIL-G-5572E)**

| Test | Test limit | Test method |
|------|-----------|-------------|
| Distillation: | | |
|     Fuel evaporated, 10% min at | 167°F (75°C) | D86 |
|     Fuel evaporated, 40% max at | 167°F (75°C) | |
|     Fuel evaporated, 50% min at | 221°F (105°C) | |
|     Fuel evaporated, 90% min at: | 275°F (135°C) | |
|         End point, max | 338°F (170°C) | |
|         Sum of 10% and 50% evaporated temp, min | 307°F | |
|     Residue, vol, max % | 1.5 | |
|     Distillation loss, vol, max % | 1.5 | |
| Existent gum, max, mg/100 ml | 3.0 | D381 |
| Potential gum, 16 h aging, max, mg/100 ml | 6.0 | D873 |
| Precipitate, max, mg/100 ml | 2.0 | D873 |
| Sulfur, max, wt % | 0.05 | D1266 or D2622 |
| Reid vapor pressure at 100°F, min, lb/in² | 5.5 | D323 or D2551 |
| Reid vapor pressure at 100°F, max, lb/in² | 7.0 | D323 or D2551 |
| Freezing point, max | −76°F (−60°C) | D2386 |
| Copper corrosion, max | No. 1 | D130 |
| Water reaction: | | |
|     Interface rating, max | 2 | D1094 |
|     Vol. change, max, ml | 2 | D1094 |
| Heating value: | | |
|     Net heat of combustion, min, Btu/lb | 18,700* | D240 |
|     Aniline-gravity product, min | 7,500 | D611 and D287 |

*For grade 115/145, the minimums are 18,900 and 9,800. Refer to MIL-G-5572E and ASTM D910 for full details.

Table 17. Jet or Aviation Turbine Fuels (MIL-T-5624H)*

| Test | Test limits JP-4 | Test limits JP-5 | Test method |
|------|------|------|------|
| Distillation: | | | D86 |
|   Fuel recovered, 10% min at | | 400°F (204.4°C) | |
|   Fuel recovered, 20% min at | 290°F (143.3°C) | | |
|   Fuel recovered, 50% min at | 370°F (187.8°C) | | |
|   Fuel recovered, 90% min at | 470°F (243.3°C) | | |
|   End point, max | | 500°F (287.8°C) | |
|   Residue, vol % max | 1.5 | 1.5 | |
|   Distillation loss, vol % max | 1.5 | 1.5 | |
| Gravity °API, min (sp. gr. max) | 45.0 (0.802) | 36.0 (0.845) | D287 |
| Gravity °API, max (sp. gr. min) | 57.0 (0.751) | 48.0 (0.788) | D287 |
| Existent gum, mg/100 ml max | 7 | 7 | D381 |
| Sulfur, total, % weight max | 0.4 | 0.4 | D1266 |
| Mercaptan sulfur, % weight max | 0.001 | 0.001 | D1219 or D1323 |
| Vapor pressure, 100°F, lb/in² min (g/cm², min) | 2.0 (140.6) | | D323 or D2551 |
| Vapor pressure, 100°F, lb/in², max (g/cm², max) | 3.0 (210.9) | | D323 or D2551 |
| Freezing point, max | −72°F (−58°C) | −51°F (−46°C) | D2386 |
| Heating value: | | | |
|   Net heat of combustion, BTU/lb, min | 18,400 | 18,300 | D240 or D2382 |
|   Aniline-gravity product, min | 2,250 | 4,500 | D1405 |
| Viscosity, centistokes at −30°F (−34.4°C), max | | 16.5 | D445 |
| Aromatics, vol %, max | 25.0 | 25.0 | D1319 |
| Olefins, vol %, max | 5.0 | 5.0 | D1319 |

*Refer to MIL-T-5624H and ASTM 1655 specifications for full details.

Type S-M. Heavy-distillate and residual fuels for large stationary and marine diesel engines

The combustion characteristics of diesel fuels are expressed in terms of **cetane number,** a measure of ignition delay. A short ignition delay, i.e., the time period between injection and ignition, is desirable for a smooth-running engine. Some diesel fuels contain cetane improvers, which usually are alkyl nitrates. The cetane number is determined using an engine test (ASTM D613), or an approximate value, termed "cetane index" (ASTM D976), can be calculated for fuels that do not contain cetane improvers. In 1972, the average cetane number of Type C-B fuels was 50; for Type T-T, 49; for Type R-R, 46; and for Type S-M, 47.

The list of **additives** used in diesel fuels has grown in recent years because of the increased use of catalytically cracked fuels, rather than exclusively straight-run distillates. In addition to cetane improvers, the list includes antioxidants, corrosion inhibitors, and dispersants. The dispersants are added to prevent agglomeration of gum or sludge deposits so these deposits can pass through filters, injectors, and engine parts without plugging them.

**Gas-Turbine Fuels**  Four grades of gas-turbine fuels are specified by ASTM D2880 to indicate fuels for different classes of engines and types of service. They are described as follows:

Grade 1-GT. A volatile distillate for gas turbines that require a fuel that burns cleaner than Grade 2-GT.

Grade 2-GT. A distillate fuel of low ash and medium

Table 18. Specifications for Diesel Fuels

| Test | ASTM method | ASTM grade of diesel fuel 1-D | ASTM grade of diesel fuel 2-D | ASTM grade of diesel fuel 4-D | U.S. Military spec. MIL-F-16884G |
|------|------|------|------|------|------|
| | | Limit | Limit | Limit | |
| Flash point, min, °F | D93 | 100 or legal | 125 or legal | 130 or legal | 140 |
| Water and sediment, vol %, max | D1796 | Trace | 0.10 | 0.50 | |
| Viscosity, kinematic, centistokes, 100°F | D445 | | | | |
|   Min | | | 1.4 | 2.0 | 5.8 | 1.8 |
|   Max | | 2.5 | 4.3 | 26.4 | 4.5 |
| Carbon residue on 10% residuum, % max | D524 | 0.15 | 0.35 | | 0.20 |
| Ash, wt %, max | D482 | 0.01 | 0.01 | 0.10 | 0.005 |
| Sulfur, wt %, max | D129 | 0.50 | 0.50 | 2.0 | 1.00 |
| Ignition quality, cetane number, min | D613 | 40 | 40 | 30 | 45 |
| Distillation temp, °F, 90% evaporated: | D86 | | | | |
|   Min | | | 540 | | |
|   Max | | 550 | 640 | | 675 |

See ASTM D975 and MIL-F-16884G specifications for full details.

volatility, suitable for turbines not requiring Grade 1-GT.

Grade 3-GT. A low-volatility, low-ash fuel that may contain residual components.

Grade 4-GT. A low-volatility fuel containing residual components and having higher vanadium content than Grade 3-GT

Grade 1-GT corresponds in physical properties to No. 1 burner fuel and Grade 1-D diesel fuel. Grade 2-GT corresponds in physical properties to No. 2 burner fuel and Grade 2-D diesel fuel. The viscosity ranges of Grades 3-GT and 4-GT bracket the viscosity ranges of No. 4, No. 5 (light), No. 5 (heavy), and No. 6 burner fuels. In addition, ASTM D2880 limits the content of certain trace metals in gas-turbine fuels that experience has shown will form deposits of low-melting compounds that cause corrosion of the turbine blades.

**Fuel Oils**  The physical characteristics of the first three grades of fuel oil specified in ASTM D396 are similar to the same numbered grades of diesel fuels and gas-turbine fuels. Many refiners produce only one product of each grade to meet the requirements of a multipurpose fuel; i.e., Grade 1 burner fuel can also be used as Grade 1-D diesel fuel if it meets the cetane-number requirements and as a Grade 1-GT if the trace-metals content is low enough. The Navy is considering adoption of a high-boiling distillate (400 to 850°F), MIL-F-24397, for use as a multipurpose fuel that can be used as a boiler, gas turbine, or diesel fuel.

The ASTM grades are as follows:

No. 1. A distillate oil intended for vaporizing pot-type burners and other burners requiring this grade of fuel.

No. 2. A distillate oil for general-purpose domestic heating for use in burners not requiring No. 1 fuel oil.

No. 4. Preheating not usually required for handling or burning.

No. 5. (Light). Preheating may be required depending upon climate and equipment.

No. 5 (Heavy). Preheating may be required for burning and, in cold climates, may be required for handling.

No. 6. Preheating required for burning and handling.

ASTM D396 limits the sulfur content of No. 1 and No. 2 fuel oils to 0.5 percent. The sulfur content of fuels heavier than No. 2 must meet the legal requirements of the locality in which they are to be used. The additional refinery processing needed by some residual fuels to meet low-sulfur-content regulations may lower the viscosity enough to cause them to fall into a different grade classification.

The properties of fuel oils marketed in the United States in 1972 are detailed in the annual fuel survey (Bureau of Mines Petroleum Products Survey 76).

**Miscellaneous Liquid Fuels**

**Coal tar** is produced as a by-product of the destructive distillation of bituminous coal to produce coke. Its characteristics are determined by the type of coal, the type of equipment used, and the temperature and duration of heating. Most coal tars in the United States are produced under high-temperature conditions (1800 to 2200°F), which yield about 8 gal/ton of coal. At high coking temperatures, the tars are highly aromatic, have a relatively high naphthalene content, low tar-acid content, and a high specific gravity.

Coal tars find limited use in the United States as heavy fuel oils in equipment designed for preheating the fuel to reduce its viscosity and aid in atomization. In Europe, tar distillates, particularly the neutral-oil fraction, are used as fuel for low-speed diesel engines. It has been reported that engines using this kind of fuel are difficult to start because of the long ignition delay.

**Benzene** formerly was fractionated from coal tar, but now most benzene is derived from petroleum. Benzene has a high octane number, and at one time it was added to gasoline to improve combustion characteristics. Most modern gasolines contain significant quantities of both benzene and toluene produced in catalytic reformers.

**Ethyl alcohol** has been used for many years as a motor fuel when blended with gasoline, particularly in Europe. From 5 to 20 percent anhydrous alcohol is used in commercial blends. Alcohol has an octane number of 99 and a higher heat of vaporization than gasoline, both of which are technologically desirable. Despite alcohol having only two-thirds the calorific value of gasoline, blends of up to about 15 percent alcohol provide slightly more power and lower fuel consumption than straight gasoline. This anomaly is explained on the basis of a higher fuel/air ratio for the alcohol blend. In countries that imposed compulsory mixing, gasoline/alcohol blends were not readily acceptable by the public. However, gasoline/alcohol/benzene blends were acceptable, as indicated by experience in France with a blend consisting of 75 percent gasoline, 15 percent alcohol, and 10 percent benzene.

**Alcohol/benzene** blends will separate if the water tolerance of the blend is exceeded; so it is essential that care be exercised to exclude water. Despite the efforts of some agriculturally rich, petroleum-poor countries to promote alcohol/gasoline blends, the use of blends has never been popular because the higher cost of alcohol requires a premium price for the blends or a government subsidy.

**Methyl alcohol** has only 48 percent of the calorific value of gasoline, but it has been used as a motor fuel in blends with gasoline in countries that have tried to reduce dependence on imported gasoline. Methyl alcohol, because of its high heat of evaporation, sometimes is used in highly supercharged racing-car engines.

**Coal-in-oil** suspensions, sometimes referred to as colloidal fuels, are fluid mixtures of pulverized coal in fuel oil or coal-tar oil. Tests indicate a practical fuel can be prepared by dispersing 40 percent by weight of pulverized coal (98 to 99 percent through 230-mesh sieve) in heavy fuel oil (Barkley et al., *Trans. ASME*, **66**, 1944, p. 185). The best opportunity for successful use of coal-in-oil suspensions is in large installations.

**Synthetic Liquid Fuels**  Production of liquid fuels from coal can be achieved by four major processes: (1) Fischer-Tropsch synthesis, (2) pressure hydrogenation (Synthoil and Berguis process), (3) solvent refining, or (4) staged pyrolysis. A commercial plant using Fischer-Tropsch synthesis is in operation in South Africa, but the selling price is controlled. Research is continuing on the other processes to produce liquid fuels to supplement supplies of petroleum, particularly with a low sulfur content for environmental protection.

**Shale oil** is produced by retorting oil shale. **Oil shale** is a sedimentary rock containing solid organic material called kerogen. High temperature decomposes the kerogen, yielding an oil suitable for use as a refinery charge stock. The oil is relatively low in sulfur, but high in nitrogen and olefins; so it must be hydrogenated to produce satisfactory products. The tremendous deposits of oil shale in Colorado, Utah, and Wyoming, equivalent to about one trillion barrels of shale oil, present a challenge for research to develop methods for mining and retorting oil shale that will be competitive with petroleum. Shale oil has been produced in Europe, Africa, and Asia, and experimental production has been demonstrated in Colorado. In addition to conventional aboveground retorting, research is in progress on in situ production, or retorting without mining.

**Tar sands** are consolidated or unconsolidated rocks that contain bitumen that is too viscous to be recovered by conventional petroleum-production techniques. The largest deposit of tar sands is along the Athabasca River, Alberta, Canada. A plant producing 45,000 barrels/day of synthetic crude in this region operates on the principle of excavation, extraction with hot water, dehydration, coking, and hydrogenation. Methods for in situ recovery of the synthetic crude also have been investigated.

## GASEOUS FUELS
### by C. C. Ward

REFERENCES: Lewis and von Elbe, "Combustion, Flames and Explosions of Gases," Academic Press. Katz et al., "Handbook of Natural Gas Engineering," McGraw-Hill. AGA, "Natural Gas Handbook." Publications of AGA, API, ASTM, BuMines, and IGT.

**Natural gas,** the primary gaseous fuel in the United States, accounts for 98 percent of all gas deliveries to ultimate consumers by gas utilities and pipelines. Natural gas occurs in underground reservoirs separately or in association with crude petroleum. Annual marketed production is below 20 trillion ft³, including the liquefied methane being imported by tanker.

**Manufactured gas** is a fuel gas made from other solid, liquid, or gaseous materials, such as coal, coke, oil, or natural gas. The principal types of manufactured gas are retort coal gas, coke-oven gas, water gas, carbureted water gas, producer gas, oil gas, reformed natural gas, and reformed propane or liquefied petroleum gas. Several processes for making substitute natural gas (SNG) from coal are being developed.

**Mixed gas** is a gas prepared by adding natural gas or liquefied petroleum gas to a manufactured gas, giving a product of better utility and higher heat content or Btu value.

**Liquefied petroleum gas** (LP gas) is a hydrocarbon mixture usually containing propane, butane, isobutane, and to a lesser extent propylene, or butylene. The most common commercial products are propane, butane, or some mixture of the two. The propane and butanes generally are extracted from natural gas or crude petroleum. Propylene and butylenes result from cracking other hydrocarbons in a petroleum refinery and are two important chemical feedstocks. Liquefied petroleum gas used in some special industrial applications is supplied directly to consumers by a few gas utilities and is sold widely as bottled gas for use in homes, trailers, and other establishments where natural gas is not available. It is the major supplement now used during periods of peak demand.

The use of other fuels and equipment to supplement the regular supply of gas during periods of peak demand or in emergencies is known as **peak shaving.** Most gas utilities, particularly natural-gas utilities at the end of long-distance transmission lines, maintain peak-shaving or standby equipment. Also, gas utilities in many cases have established natural-gas storage facilities close to their distribution systems. This allows gas to be stored underground near the point of consumption during periods of low demand, as in summer, and then withdrawn to meet peak or emergency demands, as may occur in winter. Propane-air mixtures are the major supplements to natural gas for peak-shaving use. Gas manufactured by cracking various petroleum distillates supplies most of the remaining peak requirements.

**Composition of Gaseous Fuels** The principal constituent of natural gas is the simple hydrocarbon methane ($CH_4$). Other constituents are the heavier paraffinic hydrocarbons such as ethane, propane, and the butanes. Many natural gases contain from 5 to 20 percent of nitrogen, and some contain appreciable quantities of carbon dioxide and hydrogen sulfide. Trace quantities of argon, hydrogen, and helium usually are present. A portion of the heavier hydrocarbons, carbon dioxide, and hydrogen sulfide are removed from natural gas prior to its use as a fuel. Typical natural-gas analyses are given in Table 19 (for additional analyses, see *BuMines* IC8443 and *Bull.* 617). Manufactured gases contain methane, ethane, ethylene, propylene, hydrogen, carbon monoxide, carbon dioxide, and nitrogen, with low concentrations of water vapor, oxygen, and other gases. For descriptions of individual gas-manufacturing processes and analyses of the produced gases, see Sec. 7. When natural gas or LP gases are mixed with air to modify the performance characteristics or the heating value, the resulting mixtures contain nitrogen and oxygen.

**Specifications** Since the compositions of natural, manufactured, and mixed gases can vary so widely, no single set of specifications could cover all situations. The requirements are usually based on performance in burners and equipment, on minimum heat content, and on maximum sulfur content. Gas utilities in most states come under the supervision of state commissions or regulatory bodies. Some of these commissions set standards for the gas to be delivered to the consumer. Additionally, local authorities have set requirements, and in the case of natural gas, pipelines set certain requirements for gas before it is delivered to the pipeline. The utilities must provide a gas that is acceptable to all types of consumers and that will give satisfactory performance in all kinds of consuming equipment. The Federal Power Commission has not promulgated quality requirements for natural gas in interstate commerce.

**Odorization** Since natural gas as delivered to pipelines has practically no odor, the addition of an odorant is required by most regulations in order that the presence of the gas can be detected readily in case of accidents and leaks. This odorization is provided by the addition of trace amounts of some organic sulfur compounds to the gas before it reaches the consumer. The standard requirement is that a normal person will be able to detect the presence of the gas by odor when the concentration reaches 1 percent of gas in air. Since the lower limit of flammability of natural gas is approximately 5 percent, this 1 percent requirement is essentially equivalent to one-fifth the lower limit of flammability. The combustion of these trace amounts of odorant does not create any serious problems of sulfur content or toxicity.

Specifications for liquefied petroleum gases are set forth in ASTM D1835, as shown in Table 20. Commercial propane is used where high volatility is required, commercial butane is used where lower volatility is needed, and commercial PB mixtures cover a broad range of related products. Special-duty

**Table 19. Composition of Typical Natural Gases***

| Sample No. | Natural gas from oil or gas wells | | | | | | Natural gas from pipelines | | | | |
|---|---|---|---|---|---|---|---|---|---|---|---|
| | 299 | 318 | 393 | 522 | 732 | 1177 | 1214 | 1225 | 1249 | 1276 | 1358 |
| Composition, mole percent: | | | | | | | | | | | |
| Methane................ | 92.1 | 96.3 | 67.7 | 63.2 | 43.6 | 96.9 | 94.3 | 72.3 | 88.9 | 75.4 | 85.6 |
| Ethane................. | 3.8 | 0.1 | 5.6 | 3.1 | 18.3 | 1.7 | 2.1 | 5.9 | 6.3 | 6.4 | 7.8 |
| Propane................ | 1.0 | 0.0 | 3.1 | 1.7 | 14.2 | 0.3 | 0.4 | 2.7 | 1.8 | 3.6 | 1.4 |
| Normal butane.......... | 0.3 | 0.0 | 1.5 | 0.5 | 8.6 | 0.1 | 0.2 | 0.3 | 0.2 | 1.0 | 0.1 |
| Isobutane............. | 0.3 | 0.0 | 1.2 | 0.4 | 2.3 | 0.0 | 0.0 | 0.2 | 0.1 | 0.6 | 0.1 |
| Normal pentane........ | 0.1 | 0.0 | 0.6 | 0.1 | 2.7 | 0.3 | Tr | Tr | 0.0 | 0.1 | 0.0 |
| Isopentane............ | Tr | 0.0 | 0.4 | 0.2 | 3.3 | 0.0 | Tr | 0.2 | Tr | 0.2 | 0.1 |
| Cyclopentane.......... | Tr | 0.0 | 0.2 | Tr | 0.9 | Tr | Tr | 0.0 | Tr | Tr | 0.0 |
| Hexanes plus.......... | 0.2 | 0.0 | 0.7 | 0.1 | 2.0 | 0.1 | Tr | Tr | Tr | 0.1 | Tr |
| Nitrogen.............. | 0.9 | 1.0 | 17.4 | 27.9 | 3.0 | 0.6 | 0.0 | 17.8 | 2.2 | 12.0 | 4.7 |
| Oxygen................ | 0.2 | 0.0 | Tr | 0.1 | 0.5 | Tr | Tr | Tr | Tr | Tr | Tr |
| Argon................. | Tr | Tr | 0.1 | 0.1 | Tr | 0.0 | 0.0 | Tr | 0.0 | Tr | Tr |
| Hydrogen.............. | 0.0 | 0.2 | 0.0 | 0.0 | 0.1 | 0.0 | Tr | 0.1 | 0.1 | 0.0 | 0.0 |
| Carbon dioxide........ | 1.1 | 2.3 | 0.1 | 0.4 | 0.5 | 0.0 | 2.8 | 0.1 | 0.1 | 0.1 | 0.2 |
| Helium................ | Tr | Tr | 1.4 | 2.1 | Tr | Tr | Tr | 0.4 | 0.1 | 0.4 | 0.1 |
| Heating value† | 1062 | 978 | 1044 | 788 | 1899 | 1041 | 1010 | 934 | 1071 | 1044 | 1051 |
| Origin of sample...... | La. | Miss. | N. Mex. | Okla. | Tex. | W. Va. | Colo. | Kan. | Kan. | Okla. | Tex. |

*Analyses from *BuMines Bull.* 617 (Tr = trace).
†Calculated total (gross) Btu/ft³, dry, at 60°F and 30 in Hg.

propane for use as engine fuel and for various other special purposes is described in ASTM D2154.

**Analysis** The different methods for gas analysis include absorption, distillation, combustion, mass spectroscopy, infrared spectroscopy, and gas chromatography. Absorption methods, such as the Orsat and Hempel, involve absorbing individual constituents one at a time in suitable solvents and recording of contraction in volume measured. Distillation methods, such as use of a Podbielniak column, depend on the separation of constituents by fractional distillation and measurement of the volumes distilled. In combustion methods, certain combustible elements are caused to burn to $CO_2$ and $H_2O$, and the volume changes are used to calculate composition. Infrared spectroscopy is useful in particular applications. For the most accurate analyses, mass spectroscopy and gas chromatography are the preferred methods.

ASTM has adopted a number of methods for gas analysis, including ASTM D1302 "Analysis of Carbureted Water Gas by the Mass Spectrometer." The mass spectrometer is the instrument used in D1137 for the analysis of natural gases and related mixtures. Gas chromatography is used in ASTM D1945 and D1946 for analyzing natural gas and reformed gas. ASTM D1136 covers a volumetric-chemical method for natural gas that is limited in accuracy. Gas chromatography is specified in ASTM D2163 for analysis of LP gases, and D1717 covers commercial butane-butylene mixtures. Recent developments are typified by the complete gas-chromatographic analysis of fixed gases using argon as the carrier gas (Manka, *Anal. Chem.*, **36**, 1964, pp. 480–482).

**Physical Constants** The specific gravity of gases, including LP gases, may be determined conveniently by a number of methods and a variety of instruments. An accurate (±2 percent) direct-weighing procedure and the use of two commercial instruments are detailed in ASTM D1070.

The heat value of gases is generally determined at constant pressure in a flow calorimeter in which the heat released by

**Table 20. Liquefied-Petroleum-Gas Specifications***

| | Product designation | | | |
|---|---|---|---|---|
| | Propane | Butane | PB mixtures | Test method |
| Vapor pressure at 100°F, max, psig | 210 | 70 | | D1267 or D2598 |
| Volatile residue: | | | | |
| Butane and heavier, %, max | 2.5 | | | D2163 |
| Pentane and heavier, %, max | | 2.0 | 2.0 | D2163 |
| Residual matter: | | | | |
| Residue on evaporation, +100 ml, max ml | 0.05 | 0.05 | 0.05 | D2158 |
| Oil-stain observation | Pass | Pass | Pass | D2158 |
| Specific gravity at 60°/60°F | | To be reported | | D1657 or D2598 |
| Corrosion, copper strip, max | No. 1 | No. 1 | No. 1 | D1838 |
| Sulfur, grains/100 ft³, max | 15 | 15 | 15 | D1266 |
| Moisture content | | To be reported | | |
| Free-water content | | None | None | D1657 |

*Refer to ASTM D1835 for full details.

the combustion of a definite quantity of gas is absorbed by a measured quantity of water or air. The heat value is determined from the rise in temperature of the water or air (ASTM D900). A continuous-recording calorimeter is available for measuring heat values of natural gases (ASTM D1826). See Sec. 4 for a discussion of the combustion of fuels and for physical constants of gases and gas mixtures. More extensive data for gases have been published in many reference works: "Handbook of Chemistry and Physics," The Chemical Rubber Publishing Co., Cleveland; Selected Values of Properties of Hydrocarbons and Related Compounds, *API Research Project* 44, Texas A & M University, College Station, Texas; and *Bull.* 522, Phillips Petroleum Co., Bartlesville, Okla.

**Flammability** The lower and upper limits of flammability indicate the percentage of combustible gas in air below which and above which flame will not propagate. When flame is initiated in mixtures having compositions within these limits, it will propagate and therefore the mixtures are flammable. A knowledge of flammable limits and their use in establishing safe practices in handling gaseous fuels is important, e.g., when purging equipment used in gas service, in controlling factory or mine atmospheres, or in handling liquefied gases.

Many factors enter into the experimental determination of flammable limits of gas mixtures, including the diameter and length of the tube or vessel used for the test, the temperature and pressure of the gases, and the direction of flame propagation—upward or downward. For these and other reasons, great care must be used in the application of the data. In monitoring closed spaces where small amounts of gases enter the atmosphere, often the maximum concentration of the combustible gas is limited to one-fifth of the concentration of the gas at the lower limit of flammability of the gas-air mixture.

Table 21 lists the limits of flammability in air of a number of individual fuel gases and gas mixtures. Because the composition of natural and manufactured gases varies, the limits shown in Table 21 for these gases are illustrative. For further details and tabulations of other flammable limits, refer to *BuMines Bulls.* 503 and 627.

The calculation of flammable limits is accomplished by Le Chatelier's modification of the mixture law, which is expressed in its simplest form as

$$L = \frac{100}{(p_1/N_1 + p_2/N_2) + \cdots + p_n/N_n}$$

where $L$ is the volume percentage of fuel gas in a limited mixture of air and gas; $p_1, p_2, \ldots, p_n$ are the volume percentages of each combustible gas present in the fuel gas, calculated on an air- and inert-free basis so that $p_1 + p_2 + \cdots + p_n = 100$; and $N_1, N_2, \ldots, N_n$ are the volume percentages of each combustible gas in a limit mixture of the individual gas and air. The foregoing relation may be applied to gases with inert content of 10 percent or less without introducing an absolute error of more than 1 or 2 percent in the calculated limits. However, when the inert content exceeds 10 percent, the calculation must be modified as shown in the *BuMines Tech. Paper* 450 and *Bull.* 503.

The **rate of flame propagation or burning velocity** in gas-air mixtures is of importance in utilization problems, including those dealing with burner design and rate of energy release. There are several methods that have been used for measuring such burning velocities, in both laminar and turbulent flames. Results by the various methods do not agree, but any one

**Table 21. Limits of Flammability of Gases in Air**

| Gas | Flammable limits in air, vol % | |
|---|---|---|
| | Lower | Upper |
| Methane | 5.0 | 15.0 |
| Ethane | 3.0 | 12.4 |
| Propane | 2.1 | 9.5 |
| Butane | 1.8 | 8.4 |
| Isobutane | 1.8 | 8.4 |
| Pentane | 1.4 | 7.8 |
| Isopentane | 1.4 | 7.6 |
| Hexane | 1.2 | 7.4 |
| Ethylene | 2.7 | 36.0 |
| Propylene | 2.4 | 11.0 |
| Butylene | 1.7 | 9.7 |
| Acetylene | 2.5 | 100.0 |
| Hydrogen | 4.0 | 75.0 |
| Carbon monoxide | 12.5 | 74.2 |
| Ammonia | 15.0 | 28.0 |
| Hydrogen sulfide | 4.0 | 44.0 |
| Natural | 4.8 | 13.5 |
| Producer | 20.2 | 71.8 |
| Blast-furnace | 35.0 | 73.5 |
| Water | 6.9 | 70.5 |
| Carbureted-water | 5.3 | 40.7 |
| Coal | 4.8 | 33.5 |
| Coke-oven | 4.4 | 34.0 |
| High-Btu oil | 3.9 | 20.1 |

method does give relative values of utility. Maximum burning velocities for turbulent flames are greater than those for laminar flames. The bunsen flame method gives results that have significance in gas-utilization problems. In this method, the burning velocity is determined by dividing the volume rate of gas flow from the bunsen burner by the area of the inner cone of the flame. Changes in gas composition, in the ratio of fuel to air, and in temperature of the gas affect burning velocities. The burning velocity of some individual gases and gaseous fuels is shown in Fig. 3; all measurements are made at atmo-

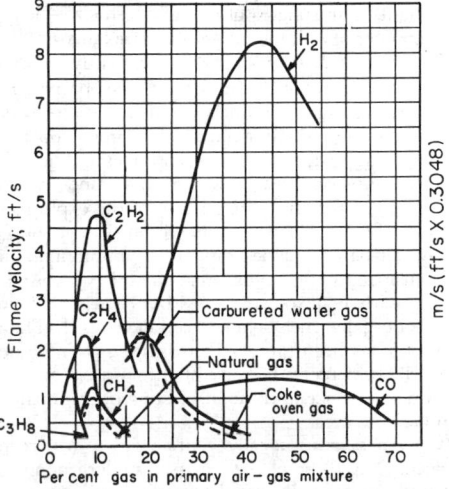

**Fig. 3** Variation of flame velocities in air-gas mixtures with variation of gas content, at atmospheric pressure. ft/s = 0.3048 m/s.

spheric temperature and pressure. A study of laminar bunsen flames is reported in Burning Velocities of Hydrocarbon Flames, by S. A. Weil, *IGT Research Bull.* 30, Chicago, 1961; and a study of turbulent bunsen flames is included in *BuMines Bull.* 604.

**Underground Gasification**  The gasification of coal in place underground was pioneered in the Soviet Union in 1931 and was used on a commercial scale in the USSR for some years after 1938. Experiments were conducted in other countries, including Belgium, Great Britain, France, and the United States. The feasibility of underground gasification was shown, but the economics were not favorable. The heating value of the produced gas seldom exceeded 100 Btu/ft³. A bibliography with 800 abstracts covers the literature on underground gasification from 1945 to 1960 (*BuMines Inf. Circ.* 8193). A shortage of energy has revived interest in underground gasification processes.

# EXPLOSIVES
## by Robert W. Van Dolah

REFERENCES: Cook, "The Science of High Explosives," Reinhold. Johansson and Persson, "Detonics of High Explosives," Academic Press. Davis, "The Chemistry of Powder and Explosives," Wiley. "Blasters Handbook," E. I. Du Pont de Nemours. "Manual on Rock Blasting," Aktiebolaget Atlas Diesel, Stockholm. Safety Recommendations for Sensitized Ammonium Nitrate Blasting Agents, *BuMines Inf. Circ.* 8179, 1963.

The technology of commercial blasting has undergone revolutionary changes in recent years. Until the mid-1950s, most blasting was done using dynamites and blasting agents, the latter known in the trade as **nitrocarbonitrates (NCN)**. Blasting agents are distinguished from explosives on the basis of their relative sensitivity; blasting agents cannot be detonated by means of a No. 8-test blasting cap when unconfined. **Liquid-oxygen explosives** accounted for a few percent of the total explosives consumed, and **black powder** was used for only a very small amount of blasting. The availability of low-cost, fertilizer-grade **ammonium nitrate (AN)** in convenient form—usually prilled roughly into spheres of 6- to 8-mesh size—led first to the development of a simple **AN–carbon black** mixture (Lee and Akre, U.S. patent 2,703,528) and subsequently to **AN–fuel oil (FO) mixtures.** The AN-FO could be readily prepared at the blasting site, leading to reduced transportation costs, and because it gave surprisingly good results when properly used, it was widely adopted, first in open-pit mines and quarries and subsequently in a variety of underground mines. Currently, AN-FO accounts for about 75 percent of the explosives used in the United States. More recently still, new types of blasting agents—water-based slurries or gels of ammonium nitrate, that are sensitized with explosives such as TNT or smokeless powder or with nonexplosive sensitizers such as aluminum powder and organic fuels—have been developed. The use of these slurries has been rapidly increasing, currently amounting to nearly 10 percent, because of their water compatibility and good performance in difficult blasting operations, such as those in taconite ores.

**AN-FO blasting agents** usually contain 5.5 to 6 percent fuel oil, typically No. 2 diesel fuel. If used underground, the oil content must be carefully regulated to minimize the produc-

tion of toxic fumes. Some AN-FO compositions contain aluminum or densifying agents. Premixed AN-FO is usually shipped in 50- to 100-lb bags, although bulk shipment and storage are practiced in certain operations. The mixed material is poured, blown, or augered into vertical boreholes or blown with air into other boreholes. Some AN-FO is packaged in 2- to 6-in polyethylene bags for use in vertical boreholes. AN-FO has a density of 0.85 to 1.0 g/cm³ and a detonation velocity in the range of 10,000 to 14,000 ft/s (3,000 to 4,300 m/s). Under most conditions of blasting, AN-FO cannot be initiated simply by a blasting cap; primers are required. Commonly used **primers** are cartridges of ordinary dynamite or specially cast charges of ¼ to ¾ lb of military-type explosives such as **Composition B or pentolite.** The efficiency of AN-FO lies in the method of loading, which fills the borehole completely and provides good coupling with the burden.

**Nitrocarbonitrate blasting agents** consist of non-cap-sensitive mixtures of fuel and oxidizer, containing no ingredient in itself classed as an explosive. Besides AN-FO, other NCN compositions have been widely used for many years because of their lower cost and decreased hazards in storage and handling, particularly when packaged in metal cans. Density may vary from 0.9 to 1.5 g/cm³; detonation velocity may range from 10,000 to 16,000 ft/s (3,000 to 4,800 m/s). Again, powerful primers must be used because of the low sensitivity of these materials.

The **water-based slurries** fall into two classes. The most common type uses an explosive sensitizer such as **TNT or nitrostarch.** When available, surplus Composition B and smokeless powders have been used in many compositions. The second type of slurry relies on the very energetic **aluminum-ammonium nitrate** reaction. No explosive sensitizer is used, although in some cases other organic fuels may be incorporated. Both types of slurries contain a thickening agent to prevent segregation of suspended solids. For large operations, mobile mixing trucks capable of high-speed mixing and loading of the composition directly into the large-diameter vertical boreholes have become popular. Another advantage of this on-site mixing and loading of slurries is the ability to change composition and strength between bottom and top loads. The density of the explosive-sensitized slurries is usually about 1.4 g/cm³ but may be as high as 1.7 g/cm³ for the aluminum-sensitized slurry; detonation velocities vary from 10,000 to 17,000 ft/s (3,000 to 5,200 m/s).

**Dynamite** is a generic term covering a multitude of nitroglycerine-sensitized mixtures of carbonaceous materials (wood, flour, starch) and oxygen-supplying salts such as ammonium nitrate and sodium nitrate. The nitroglycerin contains ethylene glycol dinitrate or other nitrated compounds to lower its freezing point, and antacids, such as chalk or zinc oxide, are added to dynamites for improved storage life. Dynamites are divided into nongelatinous or granular and gelatinous types, the latter containing nitrocellulose. All dynamites are cap-sensitive.

**Straight dynamites** are graded by the percentage of explosive oil they contain; this may be as low as 15 percent and as high as 60 percent. A typical percentage formulation for a 40 percent straight dynamite is: nitroglycerin, 40; sodium nitrate, 44; antacid, 2; carbonaceous material, 14. The rate of detonation increases with grade from 9,000 to 19,000 ft/s (2,700 to 5,800 m/s). Straight dynamites now find common use only in ditching where propagation by influence is practiced.

**Ammonia dynamites** differ from straight dynamites in that some of the sodium nitrate and much of the explosive oil have been replaced by ammonium nitrate. Strength of ammonia dynamites ranges from 15 to 60 percent, each grade having the same weight strength as the corresponding straight dynamite when compared in the ballistic mortar. A typical percentage formula for a 40 percent ammonia dynamite is: explosive oil, 14; ammonium nitrate, 36; sodium nitrate, 33; antacid, 1; carbonaceous material, 16. The rate of detonation, 4,000 to 17,000 ft/s, (1,200 to 5,200 m/s), again increases with grade. Low-density, high-weight-strength compositions are popular in many applications but AN-FO has displaced them in numerous operations.

**Blasting gelatin** is the strongest and highest-velocity explosive used in industrial operations. It consists essentially of explosive oil (nitroglycerin plus ethylene glycol dinitrate) colloided with about 7 percent nitrocellulose. It is completely water-resistant but has a poor fume rating and consequently finds only limited use.

**Gelatin dynamites** correspond to straight dynamites except that the explosive oil has been gelatinized by nitrocellulose; this results in a cohesive mixture having improved water resistance. Under confinement, the gelatins develop high velocity, ranging from 8,500 to 22,000 ft/s (2,600 to 6,700 m/s) and increasing between the grades of 20 and 90 percent. An approximate percentage composition for a 40 percent grade is: explosive oil, 32; nitrocellulose, 0.7; sulfur, 2; sodium nitrate, 52; antacid, 1.5; and carbonaceous material, 11. In the common grades of 40 and 60 percent, fume characteristics are good, making these types useful for underground hard-rock blasting.

**Ammonia gelatin dynamites** are similar to the ammonia dynamites except for their nitrocellulose content. These used to be popular in quarrying and hard-rock mining. Their excellent fume characteristics make them suitable for use underground, but again AN-FO is widely used. The rates of detonation of 7,000 to 20,000 ft/s (2,000 to 6,000 m/s) are somewhat less than the straight gelatins. A typical percentage composition for the 40 percent grade is: gelatinized explosive oil, 21; ammonium nitrate, 14; sodium nitrate, 49; with antacid and combustible making up the remainder.

The **semigels** are important variants of the ammonia gels; these contain less explosive oil, sodium nitrate, and nitrocellulose and more ammonium nitrate than the corresponding grade of ammonia gel. Rates of detonation fall in the limited range of 10,000 to 13,000 ft/s (3,000 to 4,000 m/s). These powders are cohesive and have good water resistance and good fume characteristics.

**Permissible explosives** are powders especially designed for use in underground coal mines, which have passed a series of tests established by the Bureau of Mines. The most important of these tests concern incendivity of the explosives—their tendency to ignite methane-air or methane-coal dust-air mixtures. Permissible explosives are either granular or gelatinous; the granular type makes up the bulk of the powders used today. Typically, a granular permissible contains the following, in percent: explosive oil, 9; ammonium nitrate, 65; sodium nitrate, 5; sodium chloride, 10; carbonaceous material, 10; and antacid, 1. Gels contain nitrocellulose for improved water resistance and more explosive oil. Detonation velocities for the granular grades vary from 4,500 to 11,000 ft/s (1,400 to 3,400 m/s), and for the gels, from 10,500 to 18,500 ft/s (3,200 to 5,600 m/s). Recently, several water-based permissible formulations have appeared on the commercial scene. Like the conventional permissibles, they are cap-sensitive but contain no nitroglycerin.

**Liquid-oxygen explosives (LOX)** once saw considerable use in coal strip mines but have been almost completely displaced by AN-FO or water-based slurries. LOX consists of bags of pressed carbon black or specially processed char that are saturated with liquid oxygen just before loading into the borehole. The rate of detonation ranges from 12,000 to 18,000 ft/s (3,700 to 5,500 m/s).

**Military explosives,** originally developed for such uses as bomb, shell, and mine loads and demolition work, have been adapted to many industrial explosive applications. The more common military explosives are listed in Table 22, with their compositions, ballistic mortar strengths, and detonation velocities.

**Amatol** was used early in World War II, largely because of the short supply of TNT. Modifications of amatol have been used as industrial blasting agents. **Explosive D,** or ammonium picrate, by virtue of its extreme insensitivity, was used in explosive-filled armor-piercing shells and bombs.

**RDX,** a very powerful explosive compound, was widely used during World War II in many compositions, of which Compositions A, B, and C were typical. **Composition A** was used as a shell loading; **B** was used as a bomb and shaped-charge filling; **C**, being plastic enough to allow molding to desired shapes, was developed for demolition work. Compositions that also contained aluminum powder were developed for improved underwater performance **(Torpex, HBX).** RDX has found limited commercial application as the base charge in some detonators, the filling for special-purpose detonating fuses or cordeau detonants, and the explosive in small shaped charges used as oil-well perforators and tappers for open-hearth steel furnaces.

**PETN** has never found wide military application because of its sensitivity and relative instability. It is used extensively, however, as the core of detonating fuses and in caps and boosters.

**Tetryl,** once widely used by the military as a booster loading and commercially as a base charge in detonators, has been displaced by other compositions. **Tetrytol** found limited application as a demolition charge.

**TNT** is a very widely used military explosive. Its stability, insensitivity, convenient melting point (81°C), and relatively low cost have made it the explosive of choice either alone or in admixture with other materials for loadings which are to be cast. A free-flowing pelletized form has found application in certain types of blasting requiring high loading density, where it is used to fill the cavity formed in sprung holes or the free space around the column of other explosives in the borehole.

## DUST EXPLOSIONS
### by Harry C. Verakis and John Nagy

REFERENCES: "Combustible Solids, Dusts, and Explosives," National Fire Codes, vol. 3, National Fire Protection Assoc. "Dust Explosions in Factories," New Series No. 22, H.M. Stationery Office, Great Britain. *BuMines Rept. Inv.* 4725, 5624, 5753, 5971, 7132, 7208, 7279, 7507.

A dust-explosion hazard exists where combustible dusts are

**Table 22. Physical Characteristics of Military Explosives**

| Explosive | Composition, percent | Ballistic mortar strength (TNT = 100) | Density, g per cc | Rate of detonation, m per sec |
|---|---|---|---|---|
| 80/20 amatol............ | Ammonium nitrate, 80; TNT, 20 | 117 | Cast | 4,500 |
| 50/50 amatol............ | Ammonium nitrate, 50; TNT, 50 | 122 | Cast | 5,600 |
| Composition A (pressed).... | RDX, 91; wax, 9 | 134 | 0.80 | 4,560 |
| | | | 1.20 | 6,340 |
| | | | 1.50 | 7,680 |
| | | | 1.60 | 8,130 |
| Composition B (cast)....... | RDX, 59.5; TNT, 39.5; wax, 1 | 130 | 1.65 | 7,660 |
| Composition C-3 (plastic)... | RDX, 77; tetryl, 3; mononitrotoluene, 5; dinitrotoluene, 10; TNT, 4; nitrocellulose, 1 | 145 | 1.55 | 8,460 |
| Composition C-4 (plastic)... | RDX, 91; dioctyl sebacate, 5.3; polyisobutylene, 2.1; oil, 1.6 | ... | 1.59 | 8,000 |
| Explosive D.............. | Ammonium picrate | 97 | 0.80 | 4,000 |
| | | | 1.20 | 5,520 |
| | | | 1.50 | 6,660 |
| | | | 1.60 | 7,040 |
| HBX-1.................. | RDX, 40; TNT, 38; aluminum, 17; desensitizer, 5 | 130 | 1.70 | 7,310 |
| Lead azide*............. | Lead azide | ... | 2.0 | 4,070 |
| | | | 3.0 | 4,630 |
| | | | 4.0 | 5,180 |
| PETN.................. | Pentaerythritol tetranitrate | 145 | 0.80 | 4,760 |
| | | | 1.20 | 6,340 |
| | | | 1.50 | 7,520 |
| | | | 1.60 | 7,920 |
| 50/50 Pentolite........... | PETN, 50; TNT, 50 | 120 | 1.20 | 5,410 |
| | | | 1.50 | 7,020 |
| | | | 1.60 | 7,360 |
| | | | Cast | 7,510 |
| Picric acid............... | Trinitrophenol | 108 | 1.20 | 5,840 |
| | | | 1.50 | 6,800 |
| | | | Cast | 7,350 |
| RDX (cyclonite).......... | Cyclotrimethylene trinitramine | 150 | 0.80 | 5,110 |
| | | | 1.20 | 6,550 |
| | | | 1.50 | 7,650 |
| | | | 1.60 | 8,000 |
| | | | 1.65 | 8,180 |
| Tetryl.................. | Trinitrophenylmethylnitramine | 121 | 0.80 | 4,730 |
| | | | 1.20 | 6,110 |
| | | | 1.50 | 7,160 |
| | | | 1.60 | 7,510 |
| 75/25 tetrytol............ | Tetryl, 75; TNT, 25 | 113 | 1.60 | 7,400 |
| TNT................... | Trinitrotoluene | 100 | 0.80 | 4,170 |
| | | | 1.20 | 5,560 |
| | | | 1.50 | 6,620 |
| | | | 1.60 | 6,970 |
| | | | Cast | 6,790 |

*Primary compound for blasting caps.

processed, handled, accumulate, or are stored. Expanding industrialization, combined with development of new products, has increased the variety and quantity of materials used in dust form. Dust explosions have caused serious industrial disasters. The number of severe dust explosions, exclusive of those occurring in mines, in the United States averages about 10 per year, with about four fatalities and property loss of approximately $2,000,000. In recent years, most dust explosions have involved wood, resins and plastics, starch, and aluminum. Most of the incidents occurred during crushing or pulverizing, buffing or grinding, conveying, and at dust collectors.

Despite the well-recognized hazards inherent with explosible dusts, the vast amount of technical data accumulated, and standards for prevention and protection of dust explosions, severe property damage and loss of life occur every year.

A **dust explosion** is the rapid combustion of a cloud of particulate matter where heat is generated at a higher rate than it is dissipated. In a confined space, the explosion is characterized by relatively rapid development of pressure with the evolution of large quantities of heat and reaction products. The condition necessary for a dust explosion is the simultaneous presence of a dust cloud of proper concentration in air or gas that will support combustion and an ignition source.

The Bureau of Mines defines **dust** as particles of materials smaller than 0.016 in in diameter, or those particles passing a No. 40 U.S. Standard Sieve, 420 $\mu$m (this definition relates to the limiting size, not to the average particle size of the material); and "explosible" dust means a dust which, when dispersed, is ignited by spark, flame, heated coil, or in the Godbert-Greenwald furnace at or below 730°C, when tested in accordance with the equipment and procedures described in *BuMines Rept. Inv.* 5624.

### Explosibility Factors

Empirical and experimental data are the chief guides in evaluating relative dust-explosion hazards. A mathematical model correlating some of the numerous interrelated factors affecting dust-explosion development in closed vessels has been developed. Details of the model are presented in *BuMines Repts. Inv.* 7279 and 7507.

**Dust Composition** (see Sec. 7)  Many industrial dusts are not pure compounds. The severity of a dust explosion varies with the chemical constitution and certain physical properties of the dust. High percentages of incombustibles, such as mineral matter or moisture, reduce the hazards of ignition and explosion. Heat of combustion, specific heat, ease of oxidation, and oxygen requirements influence the explosibility of dusts. Volatile combustible components in such materials as coals, asphalts, and pitches increase explosibility.

Dust composition also affects the amount and type of products produced in an explosion. Organic materials evolve new gaseous products, whereas most metals form solid oxides during combustion in an air atmosphere.

**Particle Size and Surface Area**  Explosibility of dusts increases with a decrease in particle size. Fine dust particles have greater surface area, more readily disperse into a cloud, mix better with air, remain longer in suspension, and oxidize more rapidly and completely than coarse particles. Decrease of particle size generally results in lower ignition temperature, lower igniting energy, lower minimum explosive concentration, and higher pressure and rates of pressure rise. Some metals, such as chromium, become explosive only at very fine particle sizes (average particle diameter of 3 $\mu$m), and almost all metals become pyrophoric if reduced to very fine powder.

**Range of Explosibility**  Most combustible dusts have a well-defined lower limit, but the upper limit is usually indefinite. The upper limit has been determined for only a few dusts, but these data have only limited importance in practice. The range of dust explosibility is normally 0.015 to greater than 10 oz/ft³

(10 kg/m³). The optimum concentration producing the strongest dust explosions is about 0.5 to 1.0 oz/ft³. Table 23 gives explosion characteristics for a number of dusts at a concentration of 0.5 oz/ft³. A typical example of the effect of dust concentration on maximum pressure and maximum rate of pressure rise for closed vessels of various size and shape is shown in Figs. 4 and 5.

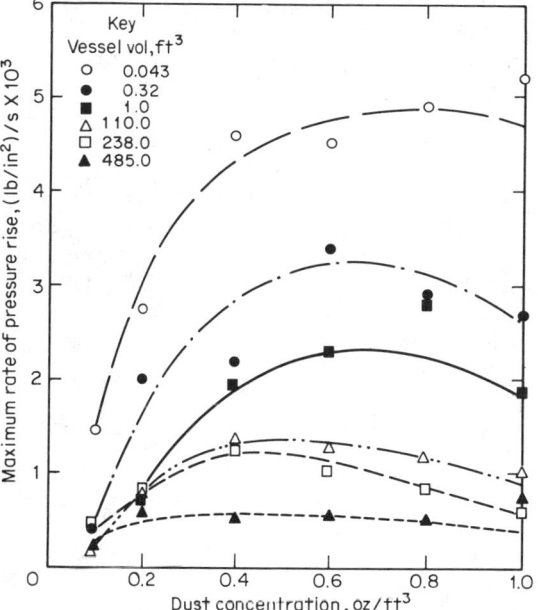

**Fig. 5**  Effect of dust concentration on maximum rate of pressure rise produced by explosion of cellulose acetate dust in different-sized vessels.

**Ignition Source**  Ignition sources known to have initiated dust explosions in industry include electric sparks and arcs in fuses, faulty wiring, motors and other appliances, static electrical discharges, open flames, frictional or metallic sparks, glowing particles, overheated bearings and other machine parts; hot electric bulbs, overheated driers, and other hot surfaces; dust layers may also ignite by these sources as well as by spontaneous ignition. Ignition temperatures of many dust clouds are given in Table 23. Normally the ignition temperature of a dust layer is considerably less than for a dust cloud. The position and intensity of the ignition source affect dust-explosion development; detailed information on these factors is presented in *BuMines Rept. Inv.* 7507.

**Turbulence**  Turbulence has a slight effect on maximum pressure, but a marked effect on the rates of pressure rise for dust explosions. Experiments show the maximum rate of pressure rise in highly turbulent dust-air mixture can be as much as eight times higher than in a nonturbulent mixture (*BuMines Repts. Inv.* 5815 and 7507).

**Moisture and Other Inerts**  Moisture in a dust absorbs heat and tends to reduce the explosibility of a dust. A high concentration of moisture in the dust also tends to reduce the dispersibility of a dust. An increase in moisture content causes an increase in ignition temperature and a reduction in maximum

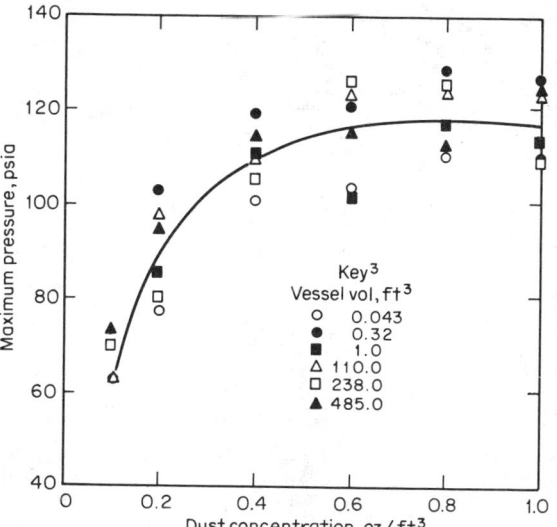

**Fi. 4**  Effect of dust concentration on maximum pressure produced by explosion of cellulose acetate dust in different-sized vessels.

## Table 23. Explosive Characteristics of Various Dusts*

| Type of dust | Ignition temperature of dust cloud, °C | Minimum igniting energy, J | Minimum explosive concentration, oz/ft³ | Maximum explosion pressure, psig | Maximum rate of pressure rise, lb/in²/s | Terminal oxygen concentration, %† | Relative explosion hazard |
|---|---|---|---|---|---|---|---|
| Agricultural: | | | | | | | |
| Alfalfa | 530 | 0.320 | 0.105 | 92 | 2,200 | | Moderate |
| Cereal grass | 550 | 0.800 | 0.250 | 52 | 500 | | Weak |
| Cinnamon | 440 | 0.030 | 0.060 | 114 | 3,900 | | Strong |
| Citrus peel | 730 | 0.045 | 0.065 | 71 | 2,000 | | Moderate |
| Cocoa | 500 | 0.120 | 0.065 | 55 | 900 | | Moderate |
| Coffee | 720 | 0.160 | 0.085 | 53 | 300 | | Weak |
| Corn | 400 | 0.040 | 0.055 | 95 | 6,000 | | Strong |
| Corn cob | 480 | 0.080 | 0.040 | 110 | 3,100 | | Strong |
| Corn dextrine | 410 | 0.040 | 0.040 | 105 | 7,000 | | Severe |
| Cornstarch | 390 | 0.030 | 0.040 | 115 | 9,000 | | Severe |
| Cotton linters | 520 | 1.920 | 0.500 | 48 | 150 | | Moderate |
| Cottonseed | 530 | 0.120 | 0.055 | 96 | 3,000 | | Moderate |
| Egg white | 610 | 0.640 | 0.140 | 58 | 500 | | Weak |
| Flax shive | 430 | 0.080 | 0.080 | 81 | 800 | | Moderate |
| Garlic | 360 | 0.240 | 0.100 | 80 | 2,600 | | Moderate |
| Grain, mixed | 430 | 0.030 | 0.055 | 115 | 5,500 | | Strong |
| Grass seed | 490 | 0.260 | 0.290 | 34 | 400 | | Weak |
| Guar seed | 500 | 0.060 | 0.040 | 98 | 2,400 | | Strong |
| Gum, Manila (copal) | 360 | 0.030 | 0.030 | 88 | 5,600 | | Severe |
| Hemp hurd | 440 | 0.035 | 0.040 | 103 | 10,000 | | Severe |
| Malt, brewers | 400 | 0.035 | 0.055 | 92 | 4,400 | | Strong |
| Milk, skim | 490 | 0.050 | 0.050 | 83 | 2,100 | | Strong |
| Pea flour | 560 | 0.040 | 0.050 | 95 | 3,800 | | Strong |
| Peanut hull | 460 | 0.050 | 0.045 | 82 | 4,700 | | Strong |
| Peat, sphagnum | 460 | 0.050 | 0.045 | 84 | 2,200 | | Strong |
| Pecan nutshell | 440 | 0.050 | 0.030 | 106 | 4,400 | | Strong |
| Pectin | 410 | 0.035 | 0.075 | 112 | 8,000 | | Severe |
| Potato starch | 440 | 0.025 | 0.045 | 97 | 8,000 | | Severe |
| Pyrethrum | 460 | 0.080 | 0.100 | 82 | 1,500 | | Moderate |
| Rauwolfia vomitoria root | 420 | 0.045 | 0.055 | 106 | 7,500 | | Strong |
| Rice | 440 | 0.050 | 0.050 | 93 | 2,600 | | Strong |
| Safflower | 460 | 0.025 | 0.055 | 84 | 2,900 | | Strong |
| Soy flour | 550 | 0.100 | 0.060 | 111 | 1,600 | 15 | Moderate |
| Sugar, powdered | 370 | 0.030 | 0.045 | 91 | 1,700 | | Strong |
| Walnut shell, black | 450 | 0.050 | 0.030 | 97 | 3,300 | | Strong |
| Wheat flour | 440 | 0.060 | 0.050 | 104 | 4,400 | | Strong |
| Wheat, untreated | 500 | 0.060 | 0.065 | 98 | 4,400 | | Strong |
| Wheat starch | 430 | 0.025 | 0.045 | 100 | 6,500 | | Severe |
| Wheat straw | 470 | 0.050 | 0.055 | 99 | 6,000 | | Strong |
| Yeast, torula | 520 | 0.050 | 0.050 | 105 | 2,500 | | Strong |
| Carbonaceous: | | | | | | | |
| Asphalt, resin, volatile content 57.5% | 510 | 0.025 | 0.025 | 94 | 4,600 | | Severe |
| Charcoal, hardwood mix, volatile content 27.1% | 530 | 0.020 | 0.140 | 100 | 1,800 | 18 | Strong |
| Coal, Colo., Brookside, volatile content, 38.7% | 530 | 0.060 | 0.045 | 88 | 3,200 | | Strong |
| Coal, Ill., No. 7, volatile content 48.6% | 600 | 0.050 | 0.040 | 84 | 1,800 | 15 | Strong |
| Coal, Ky., Breek, volatile content 40.6% | 610 | 0.030 | 0.050 | 88 | 4,000 | | Strong |
| Coal, Pa., Pittsburgh, volatile content 37.0% | 610 | 0.060 | 0.055 | 83 | 2,300 | 17 | Strong |
| Coal, Pa., Thick Freeport, volatile content 35.6% | 595 | 0.060 | 0.060 | 77 | 2,200 | | Moderate |
| Coal, W. Va., No. 2 Gas, volatile content 37.1% | 600 | 0.060 | 0.060 | 82 | 1,600 | | Moderate |
| Coal, Wyo., Laramie No. 3, volatile content 43.3% | 575 | 0.050 | 0.050 | 92 | 2,000 | | Strong |
| Gilsonite, Utah, volatile content 86.5% | 580 | 0.025 | 0.020 | 78 | 3,700 | | Severe |
| Lignite, Calif., volatile content 60.4% | 450 | 0.030 | 0.030 | 90 | 8,000 | | Severe |

| Type of dust | Ignition temperature of dust cloud, °C | Minimum igniting energy, J | Minimum explosive concentration, oz/ft³ | Maximum explosion pressure, psig | Maximum rate of pressure rise, lb/in²/s | Terminal oxygen concentration, %† | Relative explosion hazard |
|---|---|---|---|---|---|---|---|
| Pitch, coal tar, volatile content 58.1% | 710 | 0.020 | 0.035 | 88 | 6,000 | | Severe |
| Chemical compounds: | | | | | | | |
| Benzoic acid, $C_6H_5COOH$ | 620 | 0.020 | 0.030 | 74 | 5,500 | | Severe |
| Phosphorus pentasulfide, $P_2S_5$, slowly cooled to give single crystals | 280 | 0.015 | 0.050 | 54 | 10,000+ | | Severe |
| Phosphorus pentasulfide, $P_2S_5$, cooled quickly | 290 | 0.015 | 0.050 | 58 | 7,500 | 13 | Severe |
| Phthalimide, $C_6H_4(CO)_2NH$ | 630 | 0.050 | 0.030 | 79 | 4,500 | | Strong |
| Potassium bitartrate, $KHC_4H_4O_6$ | 520 | | | | | | Fire |
| Salicylanilide, $o$-$HOC_6H_4CONHC_6H_5$ | 610 | 0.020 | 0.040 | 61 | 4,400 | | Strong |
| Sodium thiosulfate, anhydrous, $Na_2S_2O_3$ | 510 | | | | | | Fire |
| Sorbic acid, $CH_3(CH:CH)_2 COOH$ | 470 | 0.015 | 0.020 | 88 | 10,000+ | | Severe |
| Sucrose, $C_{12}H_{22}O_{11}$ | 420 | 0.040 | 0.045 | 82 | 4,200 | 14 | Strong |
| Sulfur, $S_8$ 100% finer than 44 $\mu$m | 210 | 0.020 | 0.045 | 56 | 3,100 | | Strong |
| Sulfur, $S_8$, avg particle size 4 $\mu$m | 190 | 0.015 | 0.035 | 78 | 4,700 | 12 | Severe |
| Drugs: | | | | | | | |
| Aspirin (acetylsalicylic acid), $o$-$CH_3COOC_6H_4COOH$, fine | 660 | 0.025 | 0.050 | 83 | 10,000+ | | Severe |
| Mannitol (hexahydric alcohol), $CH_2OH(CHOH)_4CH_2OH$ | 460 | 0.040 | 0.065 | 82 | 2,800 | | Strong |
| Secobarbital sodium, $C_{12}H_{17}N_2O_3Na$ | 520 | 0.960 | 0.100 | 54 | 500 | | Weak |
| Vitamin C, ascorbic acid, $C_6H_8O_6$ | 460 | 0.060 | 0.070 | 88 | 4,800 | 15 | Strong |
| Explosives and related compounds: | | | | | | | |
| Dinitrobenzamide | 500 | 0.045 | 0.040 | 94 | 6,500 | | Strong |
| Dinitrobenzoic acid | 460 | 0.045 | 0.040 | 92 | 4,300 | | Strong |
| Dinitro-sym-diphenyl-urea (dinitrocarbanilide) | 550 | 0.060 | 0.095 | 87 | 2,500 | | Moderate |
| Dinitrotoluamide (3,5-dinitro-ortho-toluamide) | 500 | 0.015 | 0.050 | 106 | 10,000− | 13 | Severe |
| Metals: | | | | | | | |
| Aluminum | 650 | 0.015 | 0.045 | 100 | 10,000+ | 2 | Severe |
| Antimony | 420 | 1.920 | 0.420 | 8 | 100 | 16 | Weak |
| Boron | 470 | 0.060 | 0.100 | 90 | 2,400 | | Moderate |
| Cadmium | 570 | 4.000 | | | | | |
| Chromium | 580 | 0.140 | 0.230 | 56 | 5,000 | 14 | Moderate |
| Cobalt | 760 | | | | | | Fire |
| Copper | 900 | | | | | | Fire |
| Iron | 420 | 0.020 | 0.100 | 46 | 6,000 | 10 | Strong |
| Lead | 710 | | | | | | |
| Magnesium | 520 | 0.020 | 0.020 | 94 | 10,000+ | 0 | Severe |
| Molybdenum | 720 | | | | | | Fire |
| Nickel | 950+ | | | | | | |
| Selenium | 950+ | | | | | | |
| Silicon | 780 | 0.080 | 0.100 | 106 | 10,000+ | 12 | Severe |
| Tantalum | 630 | 0.120 | 0.200 | 51 | 3,700 | | Moderate |
| Tellurium | 550 | | | | | | |
| Thorium | 270 | 0.005 | 0.075 | 48 | 3,300 | 0 | Severe |
| Tin | 630 | 0.080 | 0.190 | 37 | 1,300 | 15 | Moderate |
| Titanium | 460 | 0.010 | 0.045 | 80 | 10,000+ | 0 | Severe |

Table 23. Explosive Characteristics of Various Dusts* (Continued)

| Type of dust | Ignition temperature of dust cloud, °C | Minimum igniting energy, J | Minimum explosive concentration, oz/ft³ | Maximum explosion pressure, psig | Maximum rate of pressure rise, lb/in²/s | Terminal oxygen concentration, %† | Relative explosion hazard |
|---|---|---|---|---|---|---|---|
| Tungsten | 950+ | | | | | | |
| Uranium | 20 | 0.045 | 0.060 | 53 | 3,400 | 0 | Severe |
| Vanadium, 86% | 500 | 0.060 | 0.220 | 48 | 600 | 13 | Weak |
| Zinc | 600 | 0.640 | 0.480 | 48 | 1,800 | 9 | Weak |
| Zirconium | 20 | 0.005 | 0.045 | 65 | 9,000 | 0 | Severe |
| Alloys and compounds: | | | | | | | |
| Aluminum-cobalt | 950 | 0.100 | 0.180 | 78 | 8,500 | | Moderate |
| Aluminum-copper | 930 | 1.920 | 0.280 | 27 | 500 | | Weak |
| Aluminum-iron | 550 | 0.720 | 0.500 | 21 | 100 | | Weak |
| Aluminum-magnesium | 430 | 0.020 | 0.020 | 90 | 10,000 | 0 | Severe |
| Aluminum-nickel | 940 | 0.080 | 0.190 | 79 | 10,000 | 14 | Moderate |
| Aluminum-silicon, 12% Si | 670 | 0.060 | 0.040 | 74 | 7,500 | | Strong |
| Calcium silicide | 540 | 0.130 | 0.060 | 73 | 10,000+ | 8 | Strong |
| Ferrochromium, high-carbon | 790 | | 2.000 | | | 19 | |
| Ferromanganese, medium-carbon | 450 | 0.080 | 0.130 | 47 | 4,200 | | Moderate |
| Ferrosilicon, 75% Si | 860 | 0.400 | 0.420 | 87 | 3,600 | 16 | Weak |
| Ferrotitanium, low-carbon | 370 | 0.080 | 0.140 | 53 | 9,500 | 13 | Strong |
| Ferrovanadium | 440 | 0.400 | 1.300 | | | 17 | |
| Thorium hydride | 260 | 0.003 | 0.080 | 60 | 6,500 | 6 | Severe |
| Titanium hydride | 440 | 0.060 | 0.070 | 96 | 10,000+ | 13 | Strong |
| Uranium hydride | 20 | 0.005 | 0.060 | 43 | 6,500 | 0 | Severe |
| Zirconium hydride | 350 | 0.060 | 0.085 | 69 | 9,000 | 8 | Strong |
| Plastics: | | | | | | | |
| Acetal resin (polyformaldehyde) | 440 | 0.020 | 0.035 | 89 | 4,100 | 11 | Severe |
| Acrylic polymer resin Methyl methacrylate-ethyl acrylate | 480 | 0.010 | 0.030 | 85 | 6,000 | 11 | Severe |
| Alkyd resin Alkyd molding compound | 500 | 0.120 | 0.155 | 15 | 150 | 15 | Weak |
| Allyl resin, allyl alcohol derivative, CR-39 | 500 | 0.020 | 0.035 | 106 | 10,000+ | 13 | Severe |
| Amino resin, urea-formaldehyde molding compound | 450 | 0.080 | 0.075 | 89 | 3,600 | 17 | Strong |
| Cellulosic fillers, wood flour | 430 | 0.020 | 0.035 | 110 | 5,500 | 17 | Severe |
| Cellulosic resin, ethyl cellulose molding compound | 320 | 0.010 | 0.025 | 102 | 6,000 | 11 | Severe |
| Chlorinated polyether resin, chlorinated polyether alcohol | 460 | 0.160 | 0.045 | 66 | 1,000 | | Moderate |
| Cold-molded resin, petroleum resin | 510 | 0.030 | 0.025 | 94 | 4,600 | | Severe |
| Coumarone-indene resin | 520 | 0.010 | 0.015 | 93 | 10,000+ | 14 | Severe |
| Epoxy resin | 530 | 0.020 | 0.020 | 86 | 6,000 | 12 | Severe |
| Fluorocarbon resin, fluorethylene polymer | 600 | | | | | | Fire |
| Furane resin, phenol furfural | 520 | 0.010 | 0.025 | 90 | 8,500 | 14 | Severe |
| Ingredients, hexamethylenetetramine | 410 | 0.010 | 0.015 | 98 | 10,000+ | 14 | Severe |
| Miscellaneous resins, petrin acrylate monomer | 220 | 0.020 | 0.045 | 104 | 10,000+ | | Severe |
| Natural resin, rosin, DK | 390 | 0.010 | 0.015 | 87 | 10,000+ | 14 | Severe |
| Nylon polymer resin | 500 | 0.020 | 0.030 | 89 | 7,000 | 13 | Severe |
| Phenolic resin, phenol-formaldehyde molding compound | 500 | 0.020 | 0.030 | 92 | 10,000+ | 14 | Severe |
| Polycarbonate resin | 710 | 0.020 | 0.025 | 78 | 4,700 | 15 | Severe |

| Type of dust | Ignition temperature of dust cloud, °C | Minimum igniting energy, J | Minimum explosive concentration, oz/ft³ | Maximum explosion pressure, psig | Maximum rate of pressure rise, lb/in²/s | Terminal oxygen concentration, %† | Relative explosion hazard |
|---|---|---|---|---|---|---|---|
| Polyester resin, polyethylene terephthalate | 500 | 0.040 | 0.040 | 91 | 5,500 | 13 | Strong |
| Polyethylene resin | 410 | 0.010 | 0.020 | 83 | 5,000 | 12 | Severe |
| Polymethylene resin, carboxypolymethylene | 520 | 0.640 | 0.115 | 70 | 5,500 | | Moderate |
| Polypropylene resin | 420 | 0.030 | 0.020 | 76 | 5,000 | | Severe |
| Polyurethane resin, polyurethane foam | 510 | 0.020 | 0.025 | 88 | 3,700 | | Severe |
| Rayon (viscose) flock | 520 | 0.240 | 0.055 | 88 | 1,700 | | Moderate |
| Rubber, synthetic | 320 | 0.030 | 0.030 | 93 | 3,100 | 15 | Severe |
| Styrene polymer resin, polystyrene latex | 500 | 0.020 | 0.020 | 91 | 7,000 | 13 | Severe |
| Vinyl polymer resin, polyvinyl butyral | 390 | 0.010 | 0.020 | 84 | 2,000 | 14 | Severe |

*Data taken from the following Bureau of Mines Reports of Investigations: RI 5753, "Explosibility of Agricultural Dusts"; RI 5971, "Explosibility of Dusts Used in the Plastics Industry"; RI 6516, "Explosibility of Metal Powders"; RI 7132, "Dust Explosibility of Chemicals, Drugs, Dyes and Pesticides"; RI 7208, "Explosibility of Miscellaneous Dusts." The data were obtained using the equipment described in RI 5624, "Laboratory Equipment and Test Procedures for Evaluating Explosibility of Dusts."

†The terminal oxygen concentration is the limiting oxygen concentration in air-$CO_2$ atmosphere required to prevent ignition of dust clouds by electric spark.

pressure and rates of pressure rise. However, the amount of moisture required to produce a marked lowering of the explosibility parameters is higher than can ordinarily be tolerated in industrial processes. Most mineral inert dusts admixed with a combustible absorb heat during the combustion reaction and reduce explosibility similar to the action of water. Some chemical compounds, such as sodium and potassium carbonates, act as inhibitors and are more effective than mineral inerts; the limiting inert dust concentration required to prevent ignition and explosion depends on the strength of the igniting source.

**Atmospheric Oxygen Concentration** The pressure and rate of pressure development decrease as the oxygen concentration in the atmosphere decreases. The ignition sensitivity of dusts decreases with decrease in oxygen concentration and for most dusts, ignition and explosion can be prevented by reducing the oxygen concentration to a safe value. Carbon dioxide, nitrogen, argon, helium, and water vapor are effective diluents. For highly reactive metal powders, only argon and helium are chemically inert. Limiting oxygen concentrations using carbon dioxide as a diluent are given in Table 23 for many dusts. With carbon dioxide as a diluent, a reduction of oxygen in the atmosphere to 11 percent is sufficient to prevent ignition by sparks for all dusts tested except the metallic powders. With nitrogen as the diluent, ignition of nonmetallic dusts is prevented by diluting the atmosphere to 8 percent oxygen. Some metal dusts, such as magnesium, titanium, and zirconium, ignite by spark in a pure carbon dioxide atmosphere. Freon and halons are sometimes used as diluent gases, but if metal dusts are involved, they can intensify rather than suppress ignition. The limiting oxygen concentration decreases as the dust becomes finer in particle size; limiting oxygen concentration varies slightly with dust concentration and is lowest at concentrations two to five times the stoichiometric mixture.

### Relative Dust Explosion Hazards

Table 23 gives test results of selected dusts whose explosive characteristics have been evaluated in the laboratory by the Bureau of Mines. The data were obtained for dusts passing a No. 200 sieve and represent the most hazardous of the specific materials tested. The values are relative rather than absolute since the test apparatus and experimental procedures affect the results to some degree. The samples were dried before testing only if the moisture content exceeded 5 percent. Detailed description of the equipment and procedures for the small-scale testing are given in *BuMines Rept. Inv. 5624*.

**Ignition Temperature** The ignition temperature of a dust cloud was determined by dispersing dust through a heated cylindrical furnace. The ignition temperature is the minimum furnace temperature at which flame appears at the bottom of the furnace in one or more trials in a group of four.

**Minimum Energy** The minimum electrical spark energy required to ignite a dust cloud was determined by dispersing the dust in a vertically mounted, 2¾-in-diameter, 12-in-long tube. The dust is dispersed by an air blast and then a condenser of known capacitance and voltage is discharged through a spark gap located within the dust dispersion. The top of the tube is enclosed with a paper diaphragm. The minimum energy for ignition of the dust cloud is the least required to produce flame propagation 4 in or longer in the tube.

**Minimum Concentration** The minimum explosive concentration or the lower flammable limit of a dust cloud was determined in a vertically mounted, 2¾-in diameter, 12-in-long tube using a continuous spark-igniting source. A known weight of dust was dispersed within the tube by an air blast. The lowest weight of dust at which sufficient pressure develops to burst a paper diaphragm enclosing the tube is used to calculate the minimum explosive concentration; this calculation is made utilizing the tube volume.

**Maximum Pressure and Rates of Pressure Rise** The maximum pressure and rates of pressure rise developed by a dust explosion were determined by dispersing dust in a closed steel tube. A continuous spark is used for ignition. A time record of pressure development is obtained. The maximum rate is the steepest slope of the pressure-time curve. Normally, explosion tests are made at dust concentrations of 0.10 to 2.0 oz/ft³.

Maximum pressure is primarily dependent on dust composition and independent of vessel size and shape. The maximum rate of pressure rise increases as vessel size decreases.

**Explosibility Index**  The ignition sensitivity of a dust cloud depends on the ignition temperature, minimum energy, and minimum concentration. The explosion severity of a dust depends on the maximum pressure and maximum rate of pressure rise. The overall explosion hazard of a dust is related to the ignition sensitivity and to explosion severity and is characterized by empirical indexes. The indexes are not derived from theoretical considerations, but provide a numerical rating consistent with research observations and practical experience. Results obtained for a sample dust are compared with values for a standard Pittsburgh-seam coal dust. The indexes are defined as follows:

$$\text{Ignition sensitivity} = \frac{(\text{ign temp} \times \text{min energy} \times \text{min conc}) \text{ Pittsburgh coal dust}}{(\text{ign temp} \times \text{min energy} \times \text{min conc}) \text{ sample dust}}$$

$$\text{Explosion severity} = \frac{\begin{array}{c}(\text{max explosive pressure} \\ \times \text{ max rate of pressure rise}) \text{ sample dust}\end{array}}{\begin{array}{c}(\text{max explosive pressure} \\ \times \text{ max rate of pressure rise}) \text{ Pittsburgh coal dust}\end{array}}$$

$$\text{Explosibility index} = \\ \text{ignition sensitivity} \times \text{explosion severity}$$

A dust having ignition and explosion characteristics equivalent to the standard Pittsburgh-seam coal has an explosibility index of unity. The relative hazard of dusts is further classified by the following adjective ratings: fire, weak, moderate, strong, or severe. The notation 0.1 designates a combustible dust presenting primarily a fire hazard as ignition of the dust cloud is not obtained by a spark or flame source, but only by an intense, heated surface source. These ratings, also shown in Table 23, are correlated with the empirical indexes as follows:

| Type of explosion | Ignition sensitivity | Explosion severity | Index of explosibility |
|---|---|---|---|
| Fire | | | <<0.1 |
| Weak | <0.2 | <0.5 | <0.1 |
| Moderate | 0.2–1.0 | 0.5–1.0 | 0.1–1.0 |
| Strong | 1.0–5.0 | 1.0–2.0 | 1.0–10 |
| Severe | >5.0 | >2.0 | >10 |

**Prevention of Dust Explosions**

(See also Sec. 12)

The following codes published by the National Fire Protection Association contain recommendations for a number of dust-producing industries: National Fire Code, vol. 3, Combustible Solids, Dusts and Explosives; National Fire Code, vol. 5, Electrical; National Fire Code, vol. 7, Alarm and Special Extinguishing Systems.

Additional sources of information may be found in the NFPA "Handbook of Fire Protection," the Factory Mutual "Handbook of Industrial Loss Prevention," and *BuMines Rept. Inv.* 6543.

**Safeguards** against explosions include the following:

**Elimination of Ignition Sources**  All sources of ignition should be eliminated from equipment containing combustible dust and from adjacent areas. Open flames or lights and

smoking should be prohibited. The use of electric or gas cutting and welding equipment for repairs should be avoided unless dust-producing machinery is shut down and all dust has been removed from the machines and from their vicinity. Additional safety measures to follow are grounding and bonding of all equipment to prevent the accumulation of static electrical charges; strict adherence to the National Electrical Code when installing electrical equipment and wiring in hazardous locations; use of magnetic separators to prevent entrance of ferrous materials into dust-grinding mills; use of nonferrous blades in fans through which dust passes; and avoidance of spark-producing tools in certain industries and of high-speed shafting and belts. (See also NFPA No. 77, Static Electricity.)

**Building and Equipment Construction**  Buildings should be constructed to minimize the collection of dust on beams, ledges, and other surfaces, particularly overhead. Vacuum cleaning is preferable to other methods for dust removal, but soft push brooms may be used without serious hazard. Buildings, including inside partitions, where combustible dusts are handled or stored should be detached units of incombustible construction. Hazardous units within buildings should be separated by substantial fire walls.

Grinders, conveyors, elevators, collectors, and other equipment which may produce dust clouds should be as dusttight as possible; they should have the smallest practical interior volume and should be constructed to withstand dust-explosion pressures.

Dust collectors should preferably be located outside of buildings or detached rooms and near the dust source. The choice of a suitable dust collector depends on the particle size, dryness, explosibility, dust concentration, gas velocity and temperature, efficiency and space requirements, and economic considerations. (See also Sec. 7.)

**Inerted Atmosphere**  Equipment such as grinders, conveyors, pulverizers, mixers, dust collectors, and sacking machines can frequently be protected by using an inerted atmosphere or explosion-suppression systems. The inert gas for this purpose may be obtained by dilution of air with flue gases from boilers, internal-combustion engines, or other sources, or by dilution with carbon dioxide, nitrogen, helium from high-pressure cylinders, or gas from inert-gas generators. The amount and rate of application of inert gas required depend upon the permissible oxygen concentration, leakage loss, atmospheric and operating conditions, equipment to be protected, and application method. Addition of inert dusts to the combustible dust may also prevent explosive dust-air mixtures from forming in and around equipment (See NFPA No. 69, Explosion Prevention Systems.)

**Relief Venting**  To reduce structural damage and to protect personnel from dust explosions, dust collectors and other equipment and the rooms in which dust-producing machinery is located should be provided with relief vents. Relief vents properly designed and located will sufficiently relieve explosion pressures in most instances and direct explosion gases away from occupied areas. The vents may be unrestricted or free openings; hinged or pivoted sash that swing outward at a low internal pressure; fixed sash with light wall anchorages; scored glass panes; light wall panels; monitors or skylights; paper, metal foil, or other diaphragms that burst at low pressures; poppet-type vent closures; pullout diaphragms; or other similar arrangements. Vents should be located near potential sources of ignition to keep explosion pressure at a minimum and to prevent a dust explosion from developing into a detonation in long ducts.

Empirical formulas and methods for calculating relief vent size have been developed. Unfortunately, because of the multiplicity of factors involved in venting requirements, none of the methods can be considered entirely satisfactory to cover the real range of situations confronting the equipment or building designer. Recommendations and guides are given in NFPA No. 68, Guide for Explosion Venting; Hartman, Pressure Release for Dust Explosions, *Nat. Fire Protect. Assoc. Quar.*, **40**, July 1946, pp. 47–53; Nagy, Zeilinger, and Hartmann, Pressure-Relieving Capacities of Various Diaphragm Materials, *BuMines Rept. Inv.* 4636, 1950; Hartmann and Nagy, Venting Dust Explosions, *Ind. Eng. Chem.*, **49**, Oct. 1957, pp. 1734–1740; Palmer, Fire Research Note No. 830, Dust Explosion Venting—A Reassessment of the Data, Aug. 1970, Fire Research Station, Boreham Woods, Herts, United Kingdom; Maisey, Gaseous and Dust Explosion Venting, *Chem. Process Eng. (Great Britain)*, Part I, Oct. 1965, pp. 527–563, Part II, Dec. 1965, pp. 462–472.

Maximum pressures developed by explosions of various dusts in a vented laboratory chamber are shown in Fig. 6. The data indicate that maximum pressures, if below 3,000 lb/ft², decrease approximately exponentially with the vent area. The explosion pressures produced under controlled laboratory conditions are generally higher than those developed under operating conditions. Experience has shown that many dust explosions occur near the explosive-range limits in nonuniform dispersions or with dust partially distributed in the building volume. Such explosions are relatively weak compared with optimum. Relief venting recommended in NFPA No. 68, Guide for Explosion Venting, will prevent major damage in nearly all incidents.

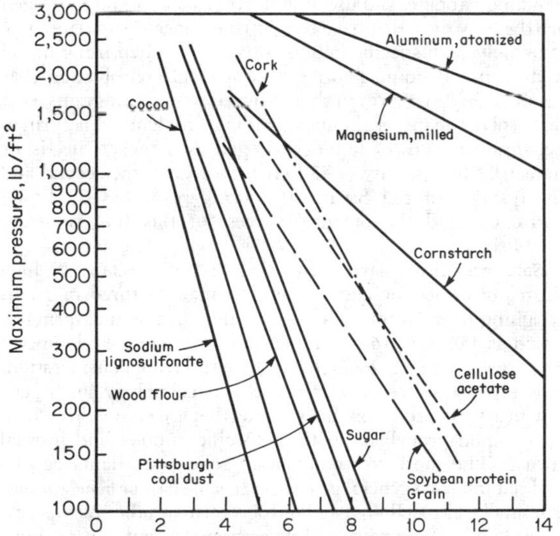

**Fig. 6** Maximum pressures produced by strong dust explosions in a 1-ft³ vented chamber.

### Combating Dust Fires

The following points should be observed in dealing with dust fires, in addition to the usual recommendations for fire prevention and fire fighting, including sprinkler protection (see also Sec. 12):

1. Attention should be directed to the potential hazard of spontaneous heating of dust products, particularly when grinding or pulverizing processes are used.

2. First-aid and fire-fighting equipment should be installed. Small hoses with spray nozzles or automatic sprinkler systems fitted with spray or fog nozzles are particularly satisfactory. The fine spray wets the dust and is not so likely to raise a dust cloud as with a solid stream. Portable extinguishers used to combat dust fires should be provided with similar devices for safe discharge.

3. Large hose of fire-department size giving solid water streams should be used with caution; a dust cloud may be formed with consequent risk of explosion. Plant employees and the fire department should be advised of this potential hazard in advance. Hose equipped with spray or fog nozzles should be provided and kept ready for an emergency.

4. Fires involving aluminum, magnesium, or some other metal powders are difficult to extinguish. Sand, talc, other dry inert materials, and special proprietary powders designed for this purpose should be used. These materials should be applied gently to smother the fire. Materials such as hard pitch can completely seal the dust from oxygen and may be used. (See NFPA Code Nos. 48, 65, 651, and 652.)

## ROCKET FUELS
### by Glenn H. Damon

REFERENCES: World Missile/Space Encyclopedia, *Missiles and Rockets*, July 29, 1963. De Rieux, Solid-Hybrid Propulsion, *Astronautics*, 1962, **7**, No. 3. Myers, Solid Propellant Rockets, *Astronautics*, 1962, **7**, No. 11. Penner, Propellants and Combustion, *Astronautics*, 1962, **7**, No. 11. Cohen, Evolving Solid Boosters for Space Missions, *Astronautics*, 1963, **8**, No. 1. Hendel, Chemical Rocket Propulsion Systems, *Chem. Eng.*, **68**, No. 5, 1961. Hendel, Advanced Rocket Propulsion, *Chem. Eng.*, **68**, No. 7, 1961. Warren, "Rocket Propellants," Reinhold, 1958. Hurden, Monopropellants, *Jour. Inst. of Fuels*, Feb. 1963. Robinson, Ethylene Oxide as a Monopropellant, *Jet Propulsion*, 1954. Brewer, Currie, and Knechthi, Ionic and Plasma Propulsion for Space Vehicles, *Proc. IRE*, 1961. Coar and King, Hydrogen for the Space Age, *Astronautics*, Mar. 1960. Slush Hydrogen, *Chem. Week*, Aug. 24, 1963. Gordon and Lee, Metals as Fuels in Multicomponent Propellants, *ARS Jour.*, April 1962. "Propellant Performance Data," Callery Chemical Co., 1961. Encyclopaedia Britannica, **19**, Rockets and Rocket Fuels, 1972. Von Braun and Ordway, "History of Rocketry and Space Travel," 1967. Schmidt, Handling and Use of Fluorine and Fluorine-Oxygen Mixtures in Rocket Systems, NASA, SP-3037, 1967. Smith, Literature of Rocket Propulsion, *Adv. in Chem. Series*, No. 78, 1968. Landel and Rembaum, "Chemistry in Space Research," 1972.

One of the **prime requirements** of a good rocket propellant is a high specific impulse. A successful rocket propellant should also exhibit most of the following characteristics: high heat of combustion; high density (high heat release per unit volume); low-molecular-weight combustion products; rapid conversion of chemical to thermal energy; high heat capacity of combustion products; combustion products that are mostly in gaseous form; chemical and physical stability to permit moderate-to-prolonged storage; relative insensitivity to shock; insignificant reactivity with common materials of construction below 150°F (66°C); low toxicity; and availability in large quantities at a reasonable cost. Admittedly, few if any rocket fuels exhibit all the desirable properties listed. Consequently, the selection of a propellant is a matter of compromise to achieve the maximum performance at a minimum hazard. Also, many propellants must be selected on the basis of their ability to perform a given mission even though they may have many undesirable characteristics. The ideal propellant has not, as yet, been discovered, but research continues toward more powerful, more exotic fuels for the rocket engines of tomorrow. (See also Sec. 11.)

The discussion in this section will be confined essentially to chemical rocket-propulsion systems. In general, propellant systems (fuel and oxidant) will be discussed rather than the fuel alone because most rocket fuels require an oxidizing agent and the overall performance of the rocket depends both upon the fuel and the oxidant. Experience has shown that fuel-rich systems generally give higher specific impulses because, for most fuels, the reaction products have lower molecular weight and the maximum temperatures are kept in a range where dissociation of the product molecules is of lesser importance. The marked influence of low-molecular-weight reaction products on performance effectively limits rocket fuels to those containing the light elements of the first two rows of the periodic table.

Instant readiness for action is another essential requirement for many military operations involving rockets. For this reason, cryogenic fuels and oxidizers are more suitable for space operations than for military operations. To meet military needs, attention has been focused on storable or prepackaged liquid propellants. Thus, the Titan II missile is fueled with a mixture of hydrazine and unsymmetrical dimethyl hydrazine (UDMH), and the oxidizer is nitrogen tetroxide; fuel and oxidizer are stored in separate tanks. Some prepackaged propellant components have been stored for two to three years with no deleterious effect on the propellant performance. Some of the most common storable liquid propellants are inhibited red fuming nitric acid (IRFNA), UDMH and its mixtures with hydrazine, monomethylhydrazine, UDMH + diethylenetriamine (DETA), chlorine trifluoride, and other halogen combinations.

Advances in propellant technology in the last few years make it difficult to give a meaningful comparison between the performance of **solid- and liquid-fueled** rocket engines. The main advantages of solid-fueled rockets are simplicity, reliability, ease of handling and storage, high density, high mass ratio, and relatively lower cost. Liquid propellants, on the other hand, have many advantages, such as: (1) combustion of large units can be more readily programmed; (2) thrust control and termination are relatively simple; (3) the fuels are relatively insensitive to temperature variations; (4) engines can be cut off and restarted on command; and (5) liquid-fuel systems are known that have higher specific impulse than that available in solid-fuel systems. A comparison of the advantages and disadvantages of each system indicates that the choice of propellant largely depends upon the mission to be accomplished. In the 1960s, much emphasis was placed on the development of higher-energy propellants. While such fuels are relatively expensive, they can be produced in moderate quantities. Unfortunately, the higher-energy products are generally unstable, and this lower stability results in materially increased hazard to personnel. More recent efforts are directed to improving the safety and performance of better-known propellant systems.

### Classification of Propellants

Rocket fuels can be classified according to their physical state and the method of furnishing the oxidant.

**Physical State**  The physical state of the fuel is an important factor in rocket-motor design. **Gaseous fuels** have never been considered suitable for rockets because of their low density. **Liquid propellants** are generally used for relatively long-range rockets only; **solid propellants** are used for most short-range rockets and increasingly for certain long-range missiles such as the Minuteman. Hybrid rockets, which use both liquid and solid fuels, are being developed.

There are many combinations of **liquid propellants** that function satisfactorily in a rocket engine. By definition, a liquid propellant is one in which the reacting material or materials are introduced into the combustion chamber in liquid form. These materials may serve as the fuel or oxidizer, and they may be in the form of a single compound or a mixture of two or more compounds. The number of potential liquid fuels is virtually limitless, but a really successful fuel should maintain relatively constant viscosity over a wide range of conditions, undergo handling without excessive loss or decomposition, be sufficiently stable to be used for regenerative cooling, and have a low freezing point. These relatively rigid specifications eliminate many otherwise promising liquid fuels. Some of the most promising fuel types at the present time are jet fuels, hydrogen, amines, hydrazine, monomethyl hydrazine, hydrides of light metals (such as boron hydrides), alcohols, unsymmetrical dimethyl hydrazine, acetylenic derivatives, ammonia, ethylene oxide, normal propyl nitrate, and hydrogen peroxide. Nitric acid, liquid oxygen, nitrogen tetroxide, fluorine, and flox (liquid fluorine plus liquid oxygen) are the most common oxidizers; hydrogen peroxide, chlorine trifluoride, ozone, and oxides of fluorine have potential importance.

Oxygen, fluorine, and hydrogen are the principal **cryogenic liquids being used in liquid-propulsion systems.** The extremely low boiling points of these liquids present many difficulties in handling, storage, and use; liquid oxygen and liquid hydrogen are the only ones that have been used extensively to date. The extremely high specific impulse obtainable when using liquid hydrogen and liquid fluorine in propellant combinations has resulted in broad research-and-development programs that have solved many of the most baffling problems. The principal disadvantages of liquid hydrogen as a rocket fuel is its unusually low density (4.43 lb/ft³). However, the second and third stages of the Saturn V rocket are fueled by liquid hydrogen, and the practical success of this fuel is firmly established.

**Solid propellants** have been developed to a relatively high degree of perfection and can now be manufactured in grains weighing over 50 tons. Solid-propellant grains measuring as much as 156 in (3.96 m) in diameter are under development, and sizes as large as 260 in (6.6 m) are receiving consideration. The rapid advance in solid-propellant technology in the last few years has produced fuel systems that are competitive with most liquid propellants both in specific impulse and in total thrust. The most commonly used solid propellants can be divided into two general groups: (1) **double-base or homogeneous,** essentially nitrocellulose and nitroglycerin or other nitroplasticizers; and (2) **composite or heterogeneous,** essentially an inorganic oxidizer in which the crystals are bonded together with a rubber or plasticlike material that serves as the fuel. Composite propellants can be subdivided into (1) rubberlike propellant compositions that bond to the case of the rocket motor, and (2) rigid propellant compositions that form a separate assembly inside the motor. The **binders** (fuels) most commonly used are asphalt, polysulfide, polyurethane, polyvinyl chloride + plasticizer, gums, and other resins. The most common **oxidizers** in use are ammonium perchlorate, lithium perchlorate, and ammonium nitrate. Higher specific impulses are obtained from solid propellants by the addition of light metals such as aluminum, magnesium, and beryllium in place of some of the fuel. Explosive materials such as HMX are also added in limited quantities.

The development of a practical hybrid (liquid-solid) rocket engine appears both possible and practical. Possibly this fuel system will combine the reliability and handling characteristics of the solid rocket with the controllability of the liquid-propellant engine.

**Method of Furnishing Oxidant**  Most liquid rocket-propellant combinations can be classified as monopropellants or bipropellants, depending upon how the combustible and the oxidant are mixed or combined. Although solid propellants have many of the characteristics of a monopropellant, they are sufficiently different to warrant restricting this discussion of monopropellants to liquid systems only.

**Monopropellants** are liquids containing both the fuel and the oxidizer in the same system. They decompose or react at a controllable rate, liberating heat and relatively simple gases. The propellant is normally a single compound, and only one storage space is required. Also, a monopropellant does not ignite at ambient temperatures but will ignite reliably with a short ignition delay when in contact with a catalyst or an ignition source.

Since monopropellants contain both the fuel and the oxidant, combustion or decomposition is initiated by a temperature rise or contact with a catalytic agent. Theoretically there are many fuel-oxidant systems that can function as monopropellants, but the number is limited because of two conflicting requirements, namely, they must be stable for storage and handling and yet readily combustible without the addition of an oxidant. Many otherwise good monopropellants cannot be used because of the hazard of sudden, uncontrolled decomposition (even detonation). Unfortunately, the specific impulse of most stable monopropellants is lower than many common bipropellant (fuel and oxidant separate) combinations. Although considerable improvement in the performance of this class of propellants has been achieved in the last few years, they still find their principal use in the auxiliary power systems of missiles and space vehicles. The most commonly employed monopropellants are hydrogen peroxide, $n$-propyl nitrate, ethylene oxide, ethyl nitrate, and hydrazine.

A large number of **bipropellant** systems have been proposed for rocket use, but factors such as safety in handling, shipping and storage, corrosive properties, freezing point, toxicity, availability, and physical characteristics materially limit the number of practical fuel combinations. However, continuing research will doubtless provide the rocket industry with new and better propellants, and it will surely improve the utilization of existing propellant systems. For the most part, bipropellant systems provide a degree of flexibility that cannot be obtained with solids or monopropellants. However, the complexity of the feed system required for the accurate transfer of huge volumes of fuel and oxidant from the storage tanks to the combustion chamber in a matter of seconds is a decided disadvantage because it increases cost and decreases reliability.

With **liquid bipropellants,** the combustible and the oxidant are mixed at the moment of use and the rate of energy release is relatively controllable. The combining materials are stored in separate tanks in the rocket and transferred to the combustion chamber by pressurizing the tanks or by separate metering pumps. Because the total combustion period for a rocket fuel is normally measured in seconds, the fuel-handling system must work with great precision.

Bipropellants are classed as **hypergolic** if they ignite spontaneously on mixing and **nonhypergolic** if an external ignition source is required. Hypergolic-propellant combinations are advantageous if the ignition delay is relatively short. Hypergolicity contributes greatly to simplicity and flexibility in spacecraft rockets by eliminating the need for a separate ignition system. Most of the hydrazine derivative fuels are hypergolic with most oxidizers, and fluorine is hypergolic with most fuels.

## Composition and Characteristics of Rocket Propellants

It is impossible to describe all the promising propellant systems now under investigation. Some of the newer propellants are security classified, but on many propellants, only limited information can be obtained by reference to the literature. Strictly comparable performance data are difficult to obtain. Specific impulse $I_{sp}$, a principal measure of the performance of a propellant system, varies over a considerable range depending on the conditions under which the data are obtained. Therefore, data on specific propellant systems may vary appreciably.

**Solid Propellants**  Double-base propellants, similar to smokeless gunpowder and made from nitrocellulose and nitroglycerin, were the principal fuels for military rockets in weapons such as the bazooka. A new process for manufacture, perfected during World War II, made it possible to produce relatively large rocket grains. While still suitable for some small, tactical military rockets, these propellants perform below many others developed for use in large rockets. A typical stabilized ballistite propellant may produce a specific impulse $I_{sp}$ of approximately 210 lb·s/lb(s).

**Most composite propellants** consist of an inorganic oxidizer with an organic fuel binder. The most commonly used oxidizers in composite propellants are ammonium, potassium, and lithium perchlorate. The original JATO (jet-assisted takeoff) units used ammonium or potassium perchlorate with a special grade of asphalt as the fuel. This propellant, although surprisingly successful for its intended use, produced large volumes of smoke, and it has been replaced by plastic-binder fuels such as polysulfides, nitropolymers, polyurethane, and hydrocarbon rubbers. The addition of light metals such as aluminum, magnesium, and beryllium to solid propellants improves the performance appreciably; such combinations are competitive with many liquid propellants. More energetic and faster-burning fuels using new nitroplasticizers show marked improvement in performance and will find increasing use in the future. The addition of limited quantities of explosives, such as HMX, has also improved performance. The actual performance of solid-fuel rockets is not generally revealed in the literature, but theoretical values of specific impulse $I_{sp}$ are reported up to about 300 lb·s/lb(s); some of the more energetic propellants now available should provide even higher specific impulses.

**Solid-propellant rockets always require an external source of ignition.** The principal problem is to obtain uniform, rapid ignition over the large surface areas where combustion is to occur without a violent pressure surge that may fracture the propellant grain. Areas where burning is not desired must be protected or inhibited. Igniters for solid-propellant rocket units are almost always pyrotechnics. The pyrotechnic igniters are initiated electrically with a hot wire surrounded by a small amount of a temperature-sensitive starter. The Minuteman and the Polaris are the best-known missiles employing solid propellants. Other missiles that have used solid propellants are the Nike-Zeus, Honest John, Scout, and Sargeant. Solid rockets are under development in sizes that should produce up to $3 \times 10^6$ lb of thrust.

**Liquid Propellants**   As stated previously, the number of liquid-propellant combinations is almost unlimited, but practical considerations tend to limit the field. Table 24 gives the characteristics of several well-known propellant combinations. The specific impulse $I_{sp}$ is given in lb·s/lb(s) and should be considered as an approximate value; the figures were taken from published reports. The $I_{sp}$ is determined normally at 1,000 psia expanded to 14.7 psia.

*Concentrated hydrogen peroxide.* With a solid catalyst such as calcium or sodium permanganate, this was used by the Germans in World War II as a monopropellant. Although the specific impulse is moderately low, high-strength hydrogen peroxide (90 to 100 percent) is useful for driving small auxiliary equipment used in the space program. When used as an oxidant with a suitable fuel, a relatively high $I_{sp}$ is attained. Hydrogen peroxide is more easily stored than cryogenic oxidants, but concentrated hydrogen peroxide is an extremely active oxidizing agent and must be handled with utmost care. Suitable container materials are aluminum, tin, borosilicate glass, and polyethylene. Highly purified hydrogen peroxide is moderately stable to both heat and shock; handling offers no serious problems provided the fuel is protected from all impurities.

*Normal propyl nitrate (NPN).* Normal propyl nitrate is a promising monopropellant for specific usage. Its performance is superior to that of hydrogen peroxide, but it is considered by some as hazardous to handle and use. NPN could probably

**Table 24. Characteristics of Some Liquid Rocket Propellants**

| Oxidizer | Fuel | Specific impulse $I_{sp}$ |
|---|---|---|
| Monopropellant | Ethylene oxide | 175 |
|  | Hydrogen peroxide | 166 |
|  | n-Propyl nitrate | 190 |
| Nitric acid | Turpentine | 221 |
| Red fuming nitric acid | Ethyl alcohol | 219 |
|  | Aniline | 221 |
|  | 50-50 UDMH-hydrazine | 283 |
|  | Jet fuel | 276 |
| Hydrogen peroxide | Ethyl alcohol | 230 |
|  | JP-4 | 233 |
|  | 50-50 UDMH-hydrazine | 280 |
| Oxygen | Ethyl alcohol | 242 |
|  | Ammonia | 294 |
|  | Kerosine | 300 |
|  | Hydrogen | 390 |
|  | UDMH | 295 |
|  | 50-50 UDMH-hydrazine | 312 |
| Fluorine | Hydrogen | 410 |
|  | Hydrazine | 298 |
|  | Methyl alcohol | 296 |
|  | JP-4 | 253 |
| Nitrogen tetroxide | 50-50 UDMH-hydrazine | 288 |
|  | Kerosine | 276 |
|  | Hydrazine | 292 |

be developed into a useful monopropellant if there were a demand for a propellant with its proved properties.

*Liquid oxygen with hydrocarbons.* Kerosine and the JP fuels with liquid oxygen form successful bipropellant systems for which the components are cheap and available in large quantities. The principal disadvantage from a military standpoint is the use of a cryogenic component. The system is used as the power source for the first stage of the Saturn V space vehicle.

*Liquid hydrogen and liquid oxygen.* The combination of liquid hydrogen and liquid oxygen gives one of the highest specific impulses (on a weight basis) of any propellant system now under development. Although these propellants are generally difficult to handle, most of the major problems have been solved. The two upper stages of the Saturn V space rocket employ this system. The relatively low-bulk density is a decided disadvantage for the lower stages of a rocket because of the large storage space required to hold a given weight of the propellant. Recent developments using a liquid hydrogen–liquid oxygen propellant under relatively high pressure have shown $I_{sp}$ as high as 450. The use of slush hydrogen instead of the lower-density liquid holds promise for future systems.

*Alcohol and liquid oxygen.* This was one of the first successful liquid propellants and was used to power the German V-2 rockets. Methyl, ethyl, and isopropyl alcohol can be used, but ethyl alcohol is preferred. Although the performance of this fuel is average, other propellants have largely replaced alcohol.

*Hydrazine, monomethyl hydrazine, and UDMH.* Hydrazine and its derivatives, monomethyl hydrazine and unsymmetrical dimethylhydrazine (UDMH) have great potential as rocket fuels. Liquid oxygen, hydrogen peroxide, fluorine, chlorine trifluoride, inhibited fuming nitric acid (IRFNA), and nitrogen tetroxide are all suitable oxidants, but the last two are the ones presently in use. The performance of anhydrous hydrazine is slightly superior to UDMH, but the relatively high freezing point [34.5°F (1.4°C)] of hydrazine is a decided disadvantage. UDMH has a satisfactory freezing point [−61.6°F (−52°C)], but its density is appreciably lower than anhydrous hydrazine. A 50-50 mixture of hydrazine and UDMH, known as **Aerozene 50,** has satisfactory physical properties and is used with nitrogen tetroxide as the propellant for the Titan II missile. This fuel system appears very satisfactory; both the fuel and the oxidant are storable, and the system is hypergolic. Other well-known rocket systems using UDMH and nitric acid are the Agena and Delta.

*Boron hydrides.* The hydrides of boron have a high heat release when used with oxidants. The most important compounds are diborane ($B_2H_6$) and pentaborane ($B_5H_9$). Both hydrides give high theoretical performance with oxidants such as oxygen and fluorine, but diborane, in particular, is very difficult to handle and use. The boron hydrides were once considered the most promising of the high-energy fuels, but their future usefulness is now uncertain. No missile system presently uses the hydrides of boron as a fuel.

*Ammonia.* The theoretically high specific impulse of ammonia when reacted with liquid oxygen or liquid fluorine gives it great potential as a rocket fuel. The performance of ammonia is only slightly less than that of hydrazine with the same oxidizers, and the cost and availability are most attractive. Mixtures of ammonia and hydrazine have been proposed for use because they have physical properties superior to either fuel alone. The X-15 is the only vehicle known to have employed ammonia as a propellant.

**Multicomponent Propellant Systems** The addition of finely divided, low-molecular-weight metals such as boron, beryllium, lithium, and aluminum to some high-energy bipropellant systems results in a significant increase in the theoretical impulse.

### Future Developments

The rapidly changing character of the chemical-propellant field makes it difficult to predict anything other than trends for the future. The performance of solid propellants should continue to be improved both by the development of new, higher-energy components and by the use of better combinations of presently known ingredients. A mixture of solid hydrogen and solid oxygen (or solid hydrogen and solid ozone) has been suggested as a superpropellant of the future. Unfortunately, more energetic propellant combinations generally tend to be less stable, and thereby the hazard in handling and use is increased.

It seems safe to predict that rocket systems will continue to employ high-energy fuels such as liquid hydrogen, hydrazine derivatives, boron derivatives, ammonia, and acetylene. Liquid fluorine, liquid ozone, oxygen difluoride, chlorine trifluo-ride, nitrogen trifluoride, perchloryl fluoride, and other similar oxidants will doubtless become increasingly more important. The use of a mixture of liquid and solid hydrogen at its triple point—the point of temperature and pressure at which liquid, vapor, and solid are in equilibrium ($-259.3°C$ and 1.04 lb/in² abs)—has been suggested as a means of improving the properties of liquid hydrogen. This slush hydrogen is reported to have a higher heat capacity and a higher density than liquid hydrogen, but it can still be handled by pumps.

Without a doubt, the specific impulse of purely chemical propellant systems is limited. Most authorities believe that a specific impulse of 500 is near the maximum limit. Several other systems are under consideration, e.g., the use of the energy from nuclear fission to heat a jet of a stable gas such as hydrogen. Other possible systems involve the energy of charged colloids, free radicals, ions, arc jets, and plasma jets. It is estimated that these systems could produce specific impulses of 1,000 to 10,000 or even higher. Although such performance is for the far-distant future, it will be a necessity if humanity is ultimately to achieve interplanetary travel (see also Sec. 4).

# CARBONIZATION OF COAL AND GAS MAKING

## by J. F. Farnsworth

REFERENCES: National Research Council, "Chemistry of Coal Utilization," Wiley. Porter, "Coal Carbonization," Reinhold. Morgan, "Manufactured Gas," J. J. Morgan, New York. BuMines Monograph 5 and other papers. Powell, "Future Possibilities in Methods of Gas Manufacture," and Russell, "The Selection of Coals for the Manufacture of Coke," papers presented to the AGA Production and Chemical Conference. Wilson and Wells, "Coal, Coke and Coal Chemicals," McGraw-Hill. Elliott, High-Btu Gas from Coal, *Coal Utilization*, Dec. 1961. Osthaus, Town Gas Production from Coal by the Koppers-Totzek Process, *Gas and Coke*, Aug. 1962. "Clean Fuels from Coal Symposium," IGT, Sept. 1973. Kirk-Othmar, Encyclopedia of Chemical Technology, 2d ed., Barker, Possible Alternate Methods for the Manufacture of Solid Fuel for the Blast Furnace, *Jour. Iron Steel Inst.*, Feb. 1971. Potter, Presidential Address, 1970, Formed Coke, *Jour. Inst. Fuel*, Dec. 1970, International Congress, Coke in Iron and Steel Industry. Charleroi, "1966 Gas Engineers Handbook," The Industrial Press.

**Carbonization of coal,** or the breaking down of its constituent substances by heat in the absence of air, is carried on for the production of coke for metallurgical, gas-making, and general fuel purposes; and gas of industrial and public-utility use. Coal chemicals recovered in this country include tar from which are produced crude chemicals and materials for creosoting, road paving, roofing, and waterproofing; light oils, mostly benzene and its homologues, used for motor fuels and chemical synthesis; ammonia, usually as ammonium sulfate, used mostly for fertilizer; to a lesser extent tar acids (phenol), tar bases (pyridine), and various other chemicals. Developments in the charging of preheated coal, pollution-control equipment, and the production of formed coke are discussed.

Gas making, as treated here, includes gas from coal carbonization, gasification of solid carbonaceous feedstocks via fixed and fluid-bed units, gasification in suspension or entrainment, and gasification of liquid hydrocarbons.

## CARBONIZATION OF COAL

**High-temperature carbonization** is carried on in ovens or retorts with flue-wall temperatures of $\pm1800°F$ (980°C) for the production of foundry coke and up to $\pm2550°F$ (1400°C) for the production of blast-furnace coke. Typical yields from carbonizing 2,204 lb (1.0 metric ton) of dry coal, containing 30 to 31 percent volatile matter, in a modern oven are: coke, 1,590 lb (720 kg); gas, 12,350 ft³ (330 m³); tar, 10 gal (37.85 l); water, 10.5 gal (39.8 l); light oil, 3.3 gal (12.5 l); ammonia, 4.9 lb (2.22 kg).

**Coal Characteristics** Most high-rank bituminous coals can be used to make coke and gas. Those well adapted to that purpose are not numerous. The coal must form a strong coherent coke and must not swell enough during coking to damage the oven walls. Coal for the best metallurgical coke should have a low ash and sulfur content (ash below 7.5 and sulfur below 1.0 percent is good). Most coking coals now exceed these figures. The coke produced should be strong and blocky.

The **coking quality** in bituminous coals is limited by the oxygen content. Coals of over 10.0 percent oxygen content (on the dry, ash-free basis) generally do not coke appreciably. A high oxygen content (7 to 10 percent) usually denotes a poor coke.

It is common practice in coal-chemical-recovery coking to mix two or more coals to make a better grade of coke or to avoid excessive expansion in the oven. One coal is usually of high (31 to 40 percent) volatile content and the other of low (15 to 22 percent). The low-volatile coal in the mix is generally in the range of 15 to 25 percent, with as much as 50 percent used in coal mixtures for producing foundry coke. High-volatile coals tend to shrink on coking while low-volatile coals tend to expand. Preliminary examination of the plastic properties, when heated, is of value in choosing the best types for blending, the high-volatile coals tending to be more fluid than the low volatiles. The rate of heating influences the coke quality obtained.

**Laboratory tests** forecast imperfectly the quality of coke producible in commercial practice. Gas and coal-chemical yields can be forecast fairly well. The small-scale **free-swelling index test** of ASTM, Committee D5, is sometimes used to estimate the coking properties. The **BM-AGA test,** developed jointly by the Bureau of Mines and a committee of the AGA (BuMines Monograph 5), gives yields close to those of commercial practice. The Geological Survey of Illinois (*Ind. Eng. Chem.*, **37**, 1945, 560–566) has an electrically heated oven for a 500-lb (226-kg) charge, the quality of whose resulting coke closely approximates that in commercial practice. **Box coking tests** (*Trans. AIME*, **74**, 1926, 500–639), made by placing a specially designed sheet-steel box containing about 60 lb (27 kg) of coal in a commercial oven, give the best small-scale information on coke quality. Accurate coke-quality determination is best had from a large-scale test in a commercial oven. Certain rapid small-scale tests for coal-chemical yields such as the **U.S. Steel Corp. tube test,** heating 20 g of coal in a glass tube under closely prescribed conditions, and the British **Gray-King assay** are useful for comparisons when checked against a standard coal of known industrial performance. The approximate estimation of coking properties of coals can be had from several plastometric methods, such as the **Giesler** (*Proc. ASTM*, **43**, 1943, 1176–1193), **Davis plastometer** (*Ind. Eng. Chem*, **3**, 1931, 43–45), and **Agde-Damm** (BuMines *Bull*. 445, 1942). Petrographic analysis can also be used to predict the coking properties of a coal or coal mixture. For the **agglomerating index** and the **agglutination test** see p. 7-2 et seq.

**Expansion tests** are made to indicate whether or not the expansion pressure of the coal or mixture of coals is sufficient to damage the oven walls during the coking process. The **Russell movable-wall oven** (*Proc. AGA*, 1928) cokes a 400-lb (180-kg) charge of coal in a horizontal, 12-in (30.48-mm)-wide oven, heated from both sides, but with one wall floating and balanced against scales. As pressure is developed in the chamber, it is counteracted and measured by continually rebalancing the scales, thus recording the expansion pressure on the whole wall. The results are correlated comparatively with the behavior of coals in commercial ovens. The Bethlehem tester (*Proc. ASTM*, **43**, 1943, 314–316), developed by W. T. Brown, cokes under a constant pressure of 40 lb (18 kg) in a flat, rectangular oven heated from the bottom, and measures vertical expansion. Test ovens developed by the Bureau of Mines at Tuscaloosa, AL, and U.S.S. Research at Monroeville, PA, merit study in connection with this subject.

**Coking Process** Coal produces coke because the particles of coal soften and fuse together when sufficiently heated. Initial softening of the coal, as determined by plastometer tests, occurs at 570 to 820°F (300 to 440°C). At or near the softening temperature gases of decomposition begin to appear in appreciable quantities, gas evolution increasing rapidly as the temperature is raised. This evolution of gases within the plastic mass causes the phenomenon that finally results in the cellular structure which is so characteristic of coke. Further temperature increase and decomposition cause hardening into coherent porous coke.

As a result of the low thermal conductivity of coal (less than

one-sixth of that of fire clay), and also of semicoke, heat penetrates slowly into the pieces and through the plastic layer; uniform plasticity or complete coalescence does not appear until the temperature (at the heated side of a softening layer) is considerably higher than the softening point of the coal. (See Fig. 1.)

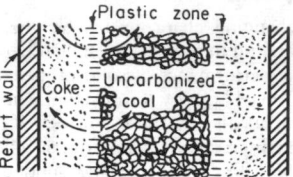

**Fig. 1** Diagrammatic illustration of the progress of carbonization and of the position of the plastic zone.

The **plastic zone** moves slowly from the hot wall of the oven toward the center, at a rate first decreasing with the distance from the wall and then increasing again at the middle of the oven. For several hours after charging a red-hot oven, the center of the charge remains cool. The plastic layer's temperature variation, from one border to the other, is from 700 to 875°F (370 to 468°C), and its thickness is ⅜ to ¾ in, depending on the coal, the charge density, and the oven temperature.

In the modern coal-chemical-recovery coke oven the average rate of travel of the plastic zone is about 0.70 in (17.8 mm) per h, and the average **coking rate,** to finished coke in the center, is 0.50 to 0.58 in (12.7 to 14.7 mm); i.e., a 17-in (432-mm) oven may be run on a net coking time of 16 to 17 h. (See Fig. 2.)

The gases and tar vapors travel chiefly outward toward the wall from the plastic layer and from the intermediate partly coked material, finding exit upward through coke and semicoke. Exit through the center core of uncoked coal, except for a very small fraction of the early formed gases, is barred by the relative impermeability of the plastic layer. The final chemical

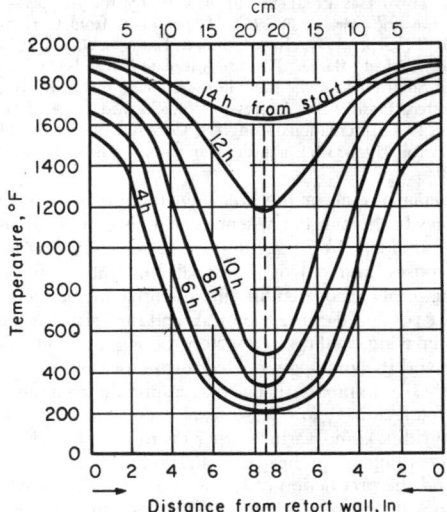

**Fig. 2** Temperature gradients in a cross section of a coal-chemical-recovery coke oven, 17 in wide, at about mid-depth on 17 h coking time.

**Table 1. Variation in Gas Yield during Carbonization**

| Period | Volume, m³/MT | kcal/m³ | Volume, ft³/ton | Btu/ft³ | Approx composition, % | | |
|---|---|---|---|---|---|---|---|
| | | | | | Hydrocarbons | Hydrogen | Oxides of carbon |
| First quarter | 1,130 | 5,800 | 3,630 | 651 | 41 | 46 | 7 |
| Second quarter | 1,000 | 3,400 | 3,190 | 610 | 37 | 53 | 7 |
| Third quarter | 1,010 | 5,050 | 3,250 | 567 | 32 | 59 | 6 |
| Fourth quarter | 585 | 3,230 | 1,875 | 363 | 8 | 82 | 5 |

products, including gas, are the resultant of secondary decompositions and interreactions in the course of this travel.

Average temperatures in various parts of the carbonizing system, for modern rapid coal-chemical-recovery oven operation, are about as follows: heating flues, at bottom, 2500 to 2600°F (1370 to 1425°C); heating flues, upper part, 2150 to 2450°F (1175 to 1345°C); oven wall, inner side (average final), 1850 to 2100°F (1010 to 1150°C).

**Temperature Effects during Carbonization**  In industrial carbonizing, higher temperatures at the oven wall and in the outer layers of the charge produce higher gas yield and less tar. The gases and tar vapors change in quality and quantity continuously during the carbonizing period. The percentage content of hydrocarbons and condensables in the oven gases decreases and that of hydrogen increases. Passage through the highly heated free space above the charge has the effect of increasing the yield of gas and light oil (reducing, however, the toluene and xylenes) and lowering tar yield, with increase of naphthalene, anthracene, and free carbon. Modern ovens tend to exercise control of the temperature in this free space. The rate of decomposition of ammonia increases above 1450°F (788°C).

The progressive change in gas yield and composition during the carbonizing period for a good gas-making coal is about as in Table 1. Some gas yields and composition are shown in Table 2.

The **overall thermal efficiency** of industrial coal carbonization (useful recovery of heat from total input of heat) is between 86 and 92 percent approx. External sensible-heat efficiencies are between 65 and 80 percent approx. Typical heat balances on coal-chemical-recovery ovens are shown in Table 3.

**Heat Used for Carbonization**  The total heating value of the gas burned in the flues to heat the ovens varies from 950 to 1250 Btu/lb (528 to 695 kcal/kg) of wet coal carbonized in efficient installations. Producer and blast-furnace gases are called "lean" gases. When underfiring with lean gas, both air and gas are regenerated in order to get sufficient flame temperature for the flues. Natural, refinery, and L.P. gases have been used to a limited extent for underfiring. These and coke-oven gas are rich gases and are not regenerated as their flame temperatures are sufficiently high and regeneration would crack their hydrocarbons. Air is regenerated in all cases.

## CARBONIZING APPARATUS

The current trend in design of **coal-chemical-recovery coke ovens** is to larger-capacity ovens and improved oven-wall liner brick and wall design to afford increased heat transfer from the heating flues to the oven chamber. Formerly, ovens were usually about 40 ft (12 m) long and from 12 to 16 ft (3.7 to 5.0 m) high. Modern ovens are about 50 ft (15 m) long and 20 to 23

**Table 2. Gas Yields and Composition from Various Coal Types with High-Temperature Carbonization**

| Coal | Temp on inner wall, °F (°C) | Gas yield, ft³/ton (m³/MT) | Gas composition, % | | | | | | |
|---|---|---|---|---|---|---|---|---|---|
| | | | Carbon dioxide | Carbon monoxide | Unsat'd hydrocarbons | Methane | Ethane, etc. | Hydrogen | Nitrogen |
| Pittsburgh bed, Fayette Co., Pa., V.M. 33.6 | 1950 (1065) | 11,700 (365) | 1.3 | 6.8 | 3.2 | 31.1 | 0 | 56.5 | 1.1 |
| Elkhorn bed, Letcher Co., Ky., V.M. 36.6 | 1950 (1065) | 11,500 (358) | 1.1 | 7.7 | 4.0 | 31.0 | 0.2 | 55.0 | 1.0 |
| Sewell bed, W. Va., V.M. 26.5 | 1950 (1065) | 12,000 (375) | 0.7 | 5.5 | 2.5 | 26.5 | 0 | 64.8 | 1.0 |
| Pocahontas No. 4, W. Va., V.M. 16.4 | 1950 (1065) | 11,900 (372) | 0.4 | 5.0 | 1.1 | 18.0 | 0 | 75.0 | 0.5 |
| Illinois, Franklin Co., V.M. 32.1 | 1950 (1065) | 12,000 (375) | 3.8 | 14.5 | 2.8 | 21.0 | 0 | 56.9 | 1.0 |
| Utah, Sunnyside, V.M. 38.8 | 1950 (1065) | 12,600 (394) | 3.0 | 14.5 | 3.7 | 26.0 | 0.5 | 51.3 | 1.0 |

V.M. = percentage of volatile matter.

Table 3. Heat Balance as a Function of the Volatile-Matter Content*

| % volatile matter (DAF) | 23.8 | 26.5 | 28.7 | 33.2 |
|---|---|---|---|---|
| Moisture, % | 10.6 | 10.3 | 9.7 | 10.1 |
| Consumption of heat, kcal/kg | 492 | 496 | 522 | 559 |
| Waste-gas loss, % | 10.9 | 10.7 | 10.8 | 9.6 |
| Surface loss, % | 10.1 | 9.9 | 9.5 | 8.8 |
| Total loss, % kcal/kg | 21.0 | 20.6 | 20.3 | 18.4 |
| | 103.3 | 102.2 | 106.0 | 102.9 |
| Effective heat (heat of coking), kcal/kg | 388.7 | 393.8 | 416.0 | 456.1 |
| Sensible heat in coke, kcal/kg | 284.1 | 278.2 | 281.0 | 260.2 |
| Sensible heat in gas, kcal/kg | 144.3 | 152.3 | 164.9 | 188.3 |
| Total sensible heat | 428.4 | 430.5 | 445.9 | 448.5 |
| Heat of reaction (exothermic) | +39.7 | +36.7 | +29.9 | −7.6 |

The heat of reaction is the difference between effective heat and loss by sensible heats.
*W. Weskamp, Influence of the Properties of Coking Coal as a Raw Material on High Temperature Coking in Horizontal Slot Ovens, *Glückauf*, **103** (5), 1967, 215–225.

ft (6 to 7 m) high and hold a charge of 35 tons (32 MT) or more. Average oven width is 16 to 19 in (400 to 475 mm), usually 18 in (450 mm), with a taper of 3.0 to 4.5 in (75 to 115 mm) from the pushing end to the coke-discharge end of the oven. Various oven designs are used, distinguished chiefly by their arrangements of the vertical heating flues and waste-heat ducts. Two basic designs are used for heating with rich gas (coke-oven gas): (1) the gun-flue design wherein the fuel gas is introduced via horizontal ducts atop the regenerators and thence through nozzles which meter the gas into the vertical flues, and (2) the underjet design which incorporates a basement, located underneath the regenerators, in the battery structure. Horizontal headers running parallel with the vertical heating walls convey the fuel gas via riser pipes through the regenerators to the vertical heating flues. One design recirculates waste combustion gas from the adjacent heating wall through a duct underneath the oven and regenerator by the jet action of the fuel gas through specially designed nozzles. This provides a leaner gas at the place combustion occurs and affords a more even vertical heat distribution for tall ovens. Designs for lean gas (blast-furnace gas, producer gas) employ sole flues beneath the regenerators for introducing the fuel gas. Ports control the quantity of gas fed from the sole flues into the regenerators. To prevent equalization of gas distribution after the gas leaves the sole flues, the regenerator chambers are divided into compartments. All modern ovens use regenerators for preheating the combustion air and lean gas. Average waste-gas (stack) temperatures range from 450 to 700°F (230 to 370°C).

Coke ovens are built in batteries of 15 to 106 and arranged so that each row of heating flues, or wall, heats half of two adjacent ovens. Modern practice is to build batteries of the maximum number of ovens in a single battery that can be operated by a single work crew to optimize productivity. This is in the range of 79 to 85 ovens per battery. Coal is charged from a larry car through openings in the top of the oven. After coking, doors are removed from both ends of the oven and coke is pushed out of the oven horizontally by a ram operated by a pusher machine. Gas is removed continuously at constant pressure (few millimeters of water column) via oven standpipes connected to a gas collecting main.

**Coal-Chemical Recovery** The gas is first cooled in either direct or indirect coolers which condense most of the tar and water from the gas. Some ammonia is absorbed in the water, forming a weak ammonia liquor. Exhausters (usually centrifugal) follow and operate from 6 to 12 in (150 to 300 mm) water column suction at the inlet to 50 to 80 in (1,250 to 2,000 mm) pressure discharge. Electrical precipitators remove the last traces of tar fog. The gas, combined with ammonia vapor stripped from the weak ammonia liquor, then passes through dilute sulfuric acid in saturators or scrubbers which recover the ammonia as ammonium sulfate. Phosphoric acid may be used as the absorbent to recover the ammonia as mono- or diammonium phosphate. Low-cost synthetic ammonia produced via reforming of natural gas and hydrocarbons has made recovery of coke-oven ammonia uneconomical. With the pending shortage of natural gas and oil, coke-oven ammonia will again become a valuable product.

After direct cooling with water, which removes much of the naphthalene from the gas, the gas is scrubbed of light oil (benzol, toluol, xylol, and solvents) with a petroleum oil. The enriched petroleum oil is stripped of the light oil by steam distillation, and the light oil is refined into its constituents by fractionation. Phenols and tar acids are recovered from the ammonia liquor and tar. Pyridine and tar bases are recovered from the ammonia saturator liquor and tar. Distillation of the tar produces cresols, naphthalene, and various grades of road tars and pitches. Acid gases (hydrogen sulfide, hydrogen cyanide) are removed from the gas. The hydrogen sulfide is converted to elemental sulfur. The cyanogen may be recovered as sodium cyanide.

**New Developments** Enactment of pollution-control laws and diminishing reserves of good coking coals have motivated developments in pollution-control devices, new operating techniques, and new coking processes. Of principal note are (1) smoke scrubbers being installed on larry cars to collect and scrub the smoke issuing from the charging holes during the coal-charging operation, (2) hooded arrangements over the hot-coke receiving car to collect and scrub the dust and smoke associated with the coke-pushing operation, and (3) enclosed coke-quenching systems to eliminate the plume evolved during quenching of the hot coke. **Systems for pipeline charging of preheated coal** are being installed/on commercial batteries. (Marting and Anvil, "Pipeline Charging Preheated Coal to Coke Ovens," U.N.E.C. Symposium, Rome, Mar. 1973.) Advantages include elimination of air pollution during charging, increased coking rates and hence increased coke production capacity for a given battery, and the use of greater amounts of weakly coking coals in the coking-coal blend. Preheating effects a moderate improvement in quality of cokes produced from strongly coking coal blends and a substantial improvement in quality of cokes produced from blends with large quantities of weakly coking coals. Figure 3 is a flow diagram of the **Coaltek system** for preheating and pipeline charging of coal.

**Formcoke,** the production of shaped coke pieces by extrusion or briquetting of coal fines followed by carbonization, has been practiced for many years in the United Kingdom and Europe to provide a "smokeless" fuel primarily for domestic heating. Formcoke for use in low-shaft blast furnaces is produced commercially from brown coal (lignite) at large plants in Lauchhammer and Schwarze-Pumpe, East Germany. As much as 60 to 70 percent of the formcoke is used in combination with conventional slot-oven coke. In the process, brown coal with a moisture content of about 40 percent is dried to 10 to 12 percent, crushed to minus 1 mm, and briquetted without the use of a binder. The briquettes are dried, preheated, and carbonized in combination predriers and vertical retorts in a semicontinuous operation. Figure 4 is a diagram of the Lauchhammer coking plant. Foundry coke briquettes are produced commercially in Poland by a process developed by the Institute of Chemical Processing of Coal. In this process, a char produced by low-temperature carbonization of lignitic-type coal in vertical shaft ovens is briquetted with tar as a binder. The briquettes are calcined in a tunnel kiln with hot oxidizing gases.

More recent developments are aimed at the production of metallurgical (blast-furnace) coke via continuous, completely enclosed processes. Advantages are the elimination of air pollution associated with conventional coke ovens and the use of greater amounts of weakly coking or noncoking coals. Major developments include the following: **FMC process** (Work, The FMC Coke Process, *Jour. Metals,* May 1966). Char and low-temperature tar are produced by pyrolysis of coal in a sequential series of fluid-bed carbonizers. Char and tar are recombined, blended, and briquetted in roll presses. Briquettes are thermally treated in two stages—a curing step followed by carbonization in a shaft kiln. The **Bergbau-Forschung process** [Peters, Status of Development of Barbau-Forschung Process for Continuous Production of Formed Coke, *Glückauf,* **103** (25), 1967]. This process involves devolatilization of low-rank coal to yield a hot char. The hot char is mixed with a fluid coking coal (about 70 percent char and 30 percent coking coal) which becomes plastic at the mixing temperature, and the mixture is briquetted hot in roll presses to produce "green" briquettes containing 7 to 8 percent volatile matter. If required, further devolatilization of the briquettes is accomplished in a vertical hot-sand carbonizer. **The Ancit Process** (Goosens and Hermann, "The Production of Blast Furnace Fuel by the Hot Briquetting Process of Eschweiler Bergwerks-Verein," ECEC, Rome, Mar. 1973). The process is similar to the B-F process in that it uses about 70 percent noncoking coal with about 30 percent coking coal as binder. The noncoking-coal component is conveyed pneumatically from bunkers and introduced at two locations into a horizontal, parallel-flight stream reactor heated by hot products of combustion. The coal is heated to 600°C in a fraction of a second and is thermally decomposed by the rapid evolution of water and volatile matter. Coal and gas are separated in a cyclone with the gas passing to a second reactor (installed in tandem arrangement) into which the coking coal is fed. Coal and gas pass to a second cyclone for separation. The two heated coals are fed by screw feeders into a vertical cylindrical mixer, and the mixture is fed to roll presses and briquetted. The **Consolidation Coal process** employs a heated rotary kiln to produce medium-temperature coke pellets from a mixture of char and coking coal. Pellets are then subjected to final high-temperature carbonization in a vertical-shaft unit. Other processes receiving attention are the Sumitomo process, Japan; Auscoke (BHP), Australia; the Sapozhnikov process, Russia; and INIEX, Belgium.

## GAS MAKING

Abundant supplies of natural gas spanning about three decades virtually eliminated the need for manufactured gas in the United States except for limited requirements for peak-load shaving and gas supply to isolated areas. These requirements have generally been supplied by gasification and reforming of oils and liquid petroleum gases or simply by propane-air mixture. Europe and the United Kingdom are presently enjoying new finds of natural gas on the continent and in the North Sea, with an accompanying decrease in manufactured gas. Rapid depletion of natural gas and oil reserves in the United States has reversed the cycle, and once again emphasis is on the production of gas from solid fuels.

**Gas Producers** A gas producer converts solid fuel into a combustible gas by continuous reaction with an oxygen-carrying gasifying agent. Producer gas may have a wide range of heating values [50 to 500 Btu/ft³ (445 to 4,450 kcal/m³)], depending on the fuel used and on the oxygen and steam concentrations and ratio in the gasifying agent. Gas producers are classified according to type of fuel bed employed:

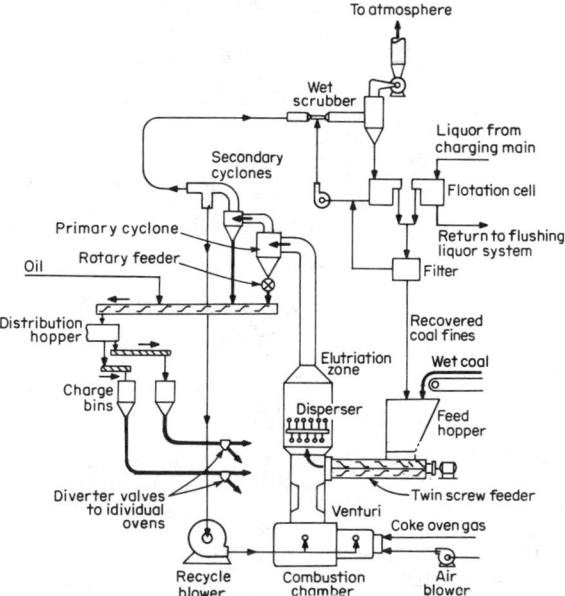

**Fig. 3** The Coaltek system, preheating and pipeline charging of coal to coke ovens.

1. Fixed or relatively static beds with fuel supported on a grate

2. Suspended or fluidized beds with fuel supported by the gaseous reactants

3. Entrained bed in which pulverized solid fuel is introduced concurrently with the gaseous reactants

Operation may be at atmospheric or elevated pressures. Oxygen-blown units are used to produce a gas of intermediate calorific value [300 to 500 Btu/ft³ (2,650 to 4,450 kcal/m³)] for use as a fuel gas or for conversion to substitute natural gas or chemicals.

Early **fixed-bed producers** used coke almost exclusively as the fuel, with the fines removed by screening. Anthracite was used to a limited extent. Units were air-blown. The upward-flowing blast of air and reactant gases meets the slowly descending fuel to produce a gas with a calorific value of about 125 Btu/ft³ (1,110 kcal/m³). Fuel is fed intermittently in batches. Fixed-bed units are characterized by a deep fuel bed and a fairly well defined sequence of zones. With respect to direction of gas flow these are: (1) the ash zone, (2) the oxidation zone, (3) the reduction zone, and (4) the devolatilization and preheating zone. Various designs have been used including Heurtey, Siemens, Lurgi, Wellman-Galusha, and Koppers-Kerpely. Units have been equipped with agitators designed to permit the use of bituminous coals.

**Water gas,** or **blue gas,** is produced by adapting fixed-bed producers to a blow (air blast) and run (steam-blowing) cycle. A typical cycle consists of attaining vigorous combustion of coke or coal by blowing air during the first part of the cycle (1 to 2 min). The gas produced, mostly $CO_2$ and $N_2$, is vented to atmosphere. Steam is then substituted for air (3 to 4 min—splitting the steam into uprun and downrun) and the resulting blue gas, mostly CO and $H_2$, is collected. Since the reaction is endothermic, the coke or coal bed becomes too cool to support the reaction after a time and it is necessary to reverse the flow from steam to air. Waste-heat boilers are incorporated to recover the sensible heat from both the blast gas and the blue gas. **Carbureted water gas** (blue gas) is made in water-gas sets equipped with a carburetor (for oil-gas making) and a superheater (for final cracking of oil vapors). The cycle employed is the same as that for the production of water gas, with oil being sprayed into the carburetor during the make part of the cycle to enrich the water gas. Typical percentage volumetric compositions are as follows:

anthracite, subbituminous coals, and lignite, have been successfully gasified in Lurgi units. While the gasifier has been primarily limited to the use of sized noncoking coals, tests are being made on a modified unit designed to utilize coking coals and a greater amount of coal fines.

Units may be air-blown to produce a low-calorific fuel gas [130 to 180 Btu/ft³ (1,155 to 1,600 kcal/m³)] or may be oxygen-blown to produce a gas of intermediate calorific value or a synthesis gas for converting to chemicals or upgrading to synthetic natural gas. Gasifier pressure is 350 to 450 lb/in² (23.8 to 30.6 atm). Coal is fed intermittently or batchwise via lock hoppers. Ash is removed by means of a revolving grate and lock hoppers. Oxygen (air) and steam are introduced through the revolving grate to effect a countercurrent gasification of the downward-moving coal bed. Reaction zones are fairly distinct. Coal is dried and preheated in the top section by the hot crude gas leaving the gasifier. Devolatilization and gasification occurs at temperatures ranging from 1150 to 1600°F (620 to 870°C) followed by gasification of the resulting char. The bottom of the bed is the combustion zone where oxygen is reacted with carbon to provide the heat for the reactions.

The crude gas leaves the gasifier at 700 to 1100°F (370 to 595°C), depending on the type of coal. It contains carbonization products such as tar, oil, naphtha, phenols, ammonia, and traces of coal and ash dust. The crude gas is scrubbed and cooled by circulating gas liquor and then further to 360°F (180°C) by a waste-heat boiler. Surplus gas liquor is routed to a tar-liquor separator. The liquor is further treated in a phenolsolvan unit to extract phenols, then treated to remove ammonia, carbon dioxide, and hydrogen sulfide, and finally routed to a biological treatment plant.

Typical composition of the gas with oxygen blowing is as follows:

|  | Vol %, Dry basis |
|---|---|
| $C_2H_4$ | 0.42 |
| $C_2H_6$ | 0.62 |
| $CH_4$ | 11.38 |
| CO | 20.24 |
| $H_2$ | 37.89 |
| $N_2$ + A | 0.33 |
| $CO_2$ | 28.63 |
| $H_2S$ + COS | 0.49 |

After removal of the acid gases, the gas may be used as a

| Gas | $CO_2$ | $O_2$ | Illuminants | CO | $H_2$ | $CH_4$ | $N_2$ | Sp. gr. | Btu/ft³ | kcal/m³ |
|---|---|---|---|---|---|---|---|---|---|---|
| Blue gas | 5.0 | 0.6 | | 38.0 | 48.0 | 1.2 | 7.2 | 0.56 | 290 | 2,570 |
| Carbureted water gas | 4.2 | 0.3 | 8.8 | 33.0 | 37.0 | 12.5 | 3.7 | 0.64 | 530 | 4,700 |

Present-day emphasis is on the complete gasification of coal to produce synthetic natural gas, low or intermediate calorific gases for industrial fuels, and synthesis gas for chemical production. Numerous coal-gasification processes have been developed to a limited extent; those presently developed to commercial status include Lurgi, Winkler, and Koppers-Totzek.

**The Lurgi process** (Fig. 5), employing a fixed-bed gasifier, has been demonstrated in 14 commercial plants (Rudolph, *Oil Gas Jour.*, Jan. 22, 1973). A variety of fuels, including coke,

fuel gas or may be upgraded to synthetic natural gas by applying the CO-shift conversion to a portion of the gas to obtain the required $H_2$ to CO molal ratio for methane synthesis.

**Fluidized-Bed Gasifiers** Supported by the gaseous reactants, solid particles are in random motion, exhibiting fluidlike characteristics of mobility, hydrostatic pressure, and a surface zone in which particle concentration decreases rapidly. In the fluidized bed, as compared with the fixed bed, there is an absence of temperature zones corresponding to oxidation and

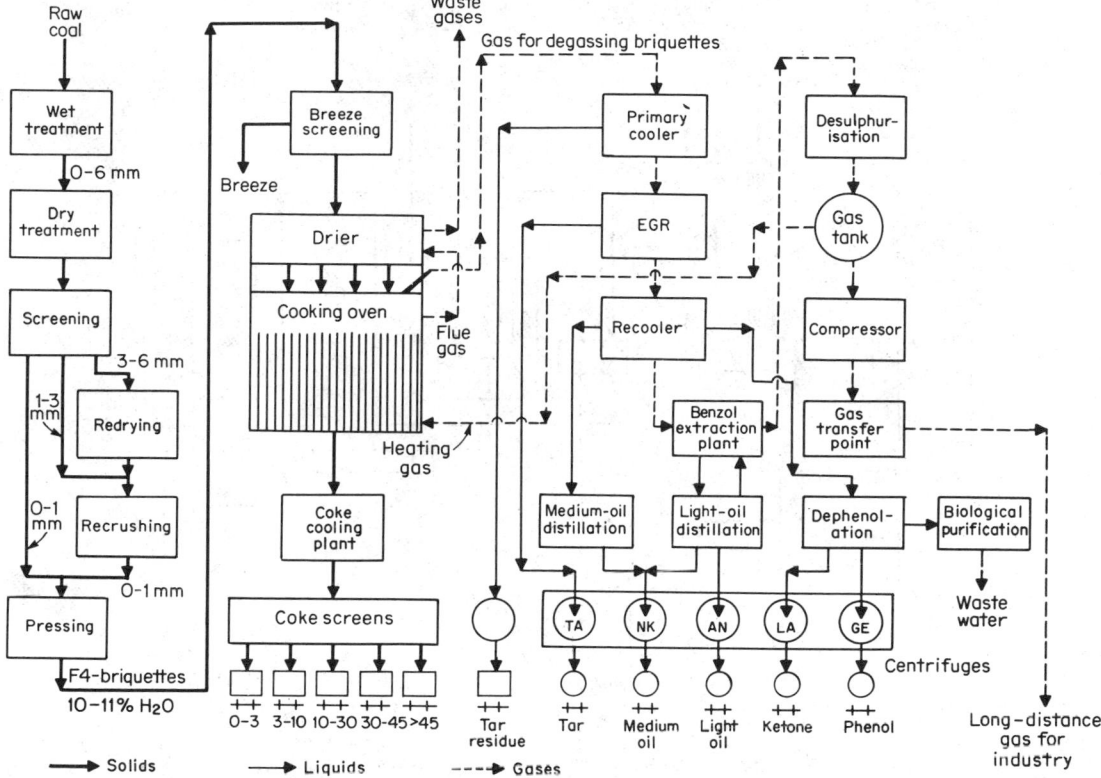

**Fig. 4** Lauchhammer coking plant, diagrammatic.

reduction reactions. The entire bed approaches a temperature determined by the relative rates of combustion and gasification reactions. Because of the comparatively low overall reaction temperature, 1800 to 2000°F (980 to 1095°C), the rate of carbon gasification is controlled by the chemical reaction rates. Because of the low temperature level, fluidized-bed units are successful primarily with highly reactive fuels such as lignites and noncoking subbituminous coals. Ash-softening temperature is a limiting factor. As soon as ash particles start to soften and agglomerate, fluidization is lost. **The Winkler process,** employing the fluid-bed technique, has also been commercialized (Banchik, "Clean Fuels from Coal," Symposium, IGT, Chicago, Sept. 1973). A total of 16 plants have been installed during the past 40 years; three plants are still operating. The process flow is shown in Fig. 6. Crushed coal (0 to ⅜ in/0 to 9.5 mm) is dried to about 8 percent moisture and fed to the gasifier by a variable-speed screw feeder. Coal reacts with oxygen and steam to produce a gas rich in carbon monoxide and hydrogen. The gasifier is operated at atmospheric pressure and a temperature range of 1500 to 1850°F (815 to 1010°C). At this temperature tars and hydrocarbons are reacted.

About 70 percent of the ash is carried out with the gas; the remainder is removed from the bottom by an ash screw. Secondary steam and oxygen is introduced above the bed to react the unconverted carbon carried over by the gas. The gas is cooled by a radiant boiler in the top section to prevent ash particles from fusing and depositing in the gas exit.

Raw gas leaving the gasifier is passed through a waste-heat boiler and then dedusted by cyclones, wet scrubber, and an electrostatic precipitator. The gas is then compressed and may be further treated via CO shift for ammonia production or for methanating to synthetic natural gas.

**Entrained gasifiers** are characterized by very short residence time in the reactor (less than 1.0 s). The fuel is gasified almost completely and instantaneously, and because of the concurrent flow of coal dust and oxygen, the gaseous hydrocarbons evolved from the coal at medium temperatures are passed through a zone of very high temperature, 3300 to 3500°F (1815 to 1927°C), in which they decompose so rapidly that coagulation of coal particles during the plastic stage does not occur. Thus, any carbonaceous fuel, including coking coals, which can be pulverized can be used in entrained gasifiers.

Entrained gasifiers are oxygen-blown. With air at ambient temperatures, the high temperatures required for the rapid reactions are not reached because of the heat absorbed by the nitrogen. The use of preheated air is technically feasible, but cost advantages afforded by the use of air instead of oxygen are offset by increased costs for preheating equipment, more gasifiers, larger lines, and larger equipment for a given heat requirement. The use of oxygen-enriched air for entrained gasifiers to produce a low-calorific-value gas shows promise.

The **Koppers-Totzek process** is the mostly highly commercialized of the entrained gasifiers (Farnsworth et al., "Production of Gas from Coal," Fifth Syn. Pipeline Gas Symposium, AGA/OCR, Chicago, Oct. 1973). A total of 16 plants have

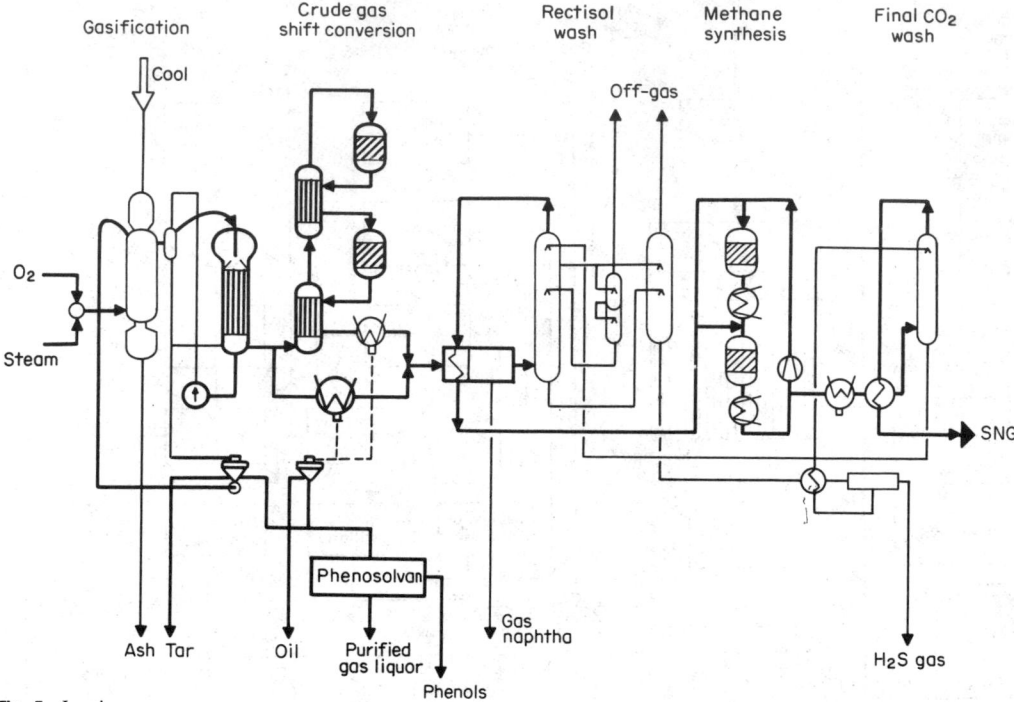

Gasification  Crude gas shift conversion  Rectisol wash  Methane synthesis  Final CO₂ wash

**Fig. 5** Lurgi process.

been built or are under construction. The process can gasify a variety of feedstocks, including all ranks of coals, char, petroleum coke, tars, heavy residuals, and light to heavy oils. The changeover from solids to liquids involves only a change of burner heads. Multiple-feed burners permit wide variations in

turndown ratio. The process is capable of instantaneous shutdown with full production resumed in 30 min.

At the high reaction temperatures the gasification products are a CO- and $H_2$-rich gas (free from tars, condensable hydrocarbons, ammonia, and phenol) and slag. Sulfur compounds

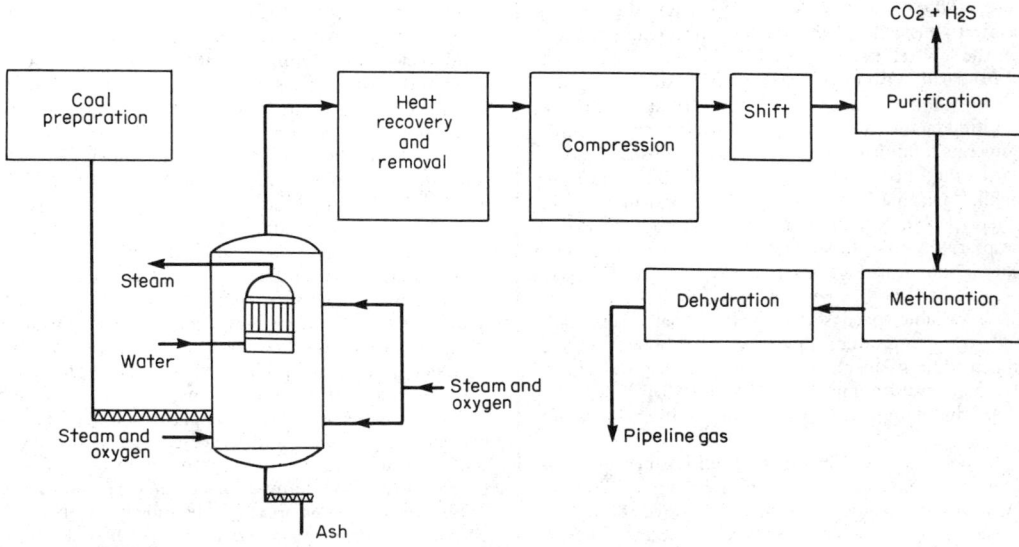

**Fig. 6** Winkler process.

in the raw gas are converted to elemental sulfur. The process is essentially pollution-free.

Figure 7 shows the process flow for the production of clean, desulfurized fuel gas or synthesis gas which can be converted to chemicals or synthetic natural gas.

Coal dried to 2 to 8 percent moisture, depending on rank, and pulverized to 70 to 80 percent minus 200 mesh is introduced with oxygen and steam into opposing burner heads. Gasifier pressure is slightly above atmospheric. Reaction temperatures are 3300 to 3500°F (1815 to 1927°C) at the burners. Gas exit temperature is about 2700 to 2800°F (1500°C).

About 50 to 70 percent of the coal ash drops from the bottom of the gasifier into a slag quench tank; the remaining ash is carried out with the gas as fine fly ash. Flux may be added to the feed to lower the ash fusion temperature or to adjust slag viscosity. Gas leaving the gasifier may be direct-water-quenched to solidify entrained slag droplets. The gas then passes through a waste-heat boiler to produce high-pressure steam up to 1,500 lb/in² gage (105 kg/cm²). After leaving the waste-heat boiler, the gas is cleaned and cooled in conventional wet-scrubbing equipment or by venturi scrubbers as shown in Fig. 7. If the gas is to be compressed to high pressures for chemical synthesis or synthetic-natural-gas production, electrostatic precipitators are used for further cleaning. The gas is then desulfurized to meet required specifications. Typical data are shown in Table 4.

**New Developments**   Current research is directed principally at developing high-pressure, 1,000 to 1,500 lb/in² (70 to 100 atm) coal-gasification processes for the production of synthetic natural gas. Many employ the multistage approach wherein the coal feed is devolatilized in the first stage to produce a gas rich in methane; the resulting char is gasified in the subsequent stages to produce a hot synthesis gas which is used to devolatilize the coal in the first stage. Apparent advantages in addition to the methane production are high coal throughputs per unit volume of reactor and having the product gas at high pressure for distribution. Major projects are briefly described.

**Hygas**   Institute of Gas Technology. Coal is crushed, dried, and pretreated (if necessary to destroy coking properties), and slurried with light oil. The slurry is fed to a three-stage gasifier operating at about 1,000 lb/in² (70 atm). Vaporization of the light oil, devolatilization of the coal, and hydrogasification take place in the first stage. Char is reacted with steam and $H_2$ in the second stage. Alternate processes are considered for the production of hydrogen (third stage). Heat for the reaction of steam with char to form $H_2$ and CO to be supplied by (1) electric current passing through a fluid bed or (2) combustion of a portion of the car with oxygen. Application of the **steam-iron process** is the third alternate to produce hydrogen.

Raw gas is quenched to knock out oil which is recovered, and the slurry oil is recycled. Acid gases are then removed, followed by the CO shift and final acid-gas clean-up before methanation to produce synthetic natural gas.

**$CO_2$ Acceptor Process**   Consolidation Coal Company. Heat for the reaction of carbon and steam is provided by reacting the $CO_2$ formed with dolomite. Removal of $CO_2$ enhances the exothermic CO shift and methanation reactions. The combined temperature effect affords operation of the process without oxygen or an external hydrogen supply.

Crushed and dried coal is fed to a fluid-bed devolatilizer, operating at about 1500°F (815°C) and 150 to 300 lb/in² (10 to 20 atm). Devolatilized char is conveyed by steam to the gasifier and gasified with steam at about 1600°F (870°C). Dolomite is used in both reactors to remove $CO_2$ and $H_2S$. Spent dolomite is conveyed by air and char by steam to the regenerator where the char is burned with air to provide heat for calcining the spent dolomite.

Raw gas passes through a heat-recovery section, water quench, and acid-gas removal. $H_2/CO$ ratio of the gas is about 3.2 so that no shift is required prior to methanation.

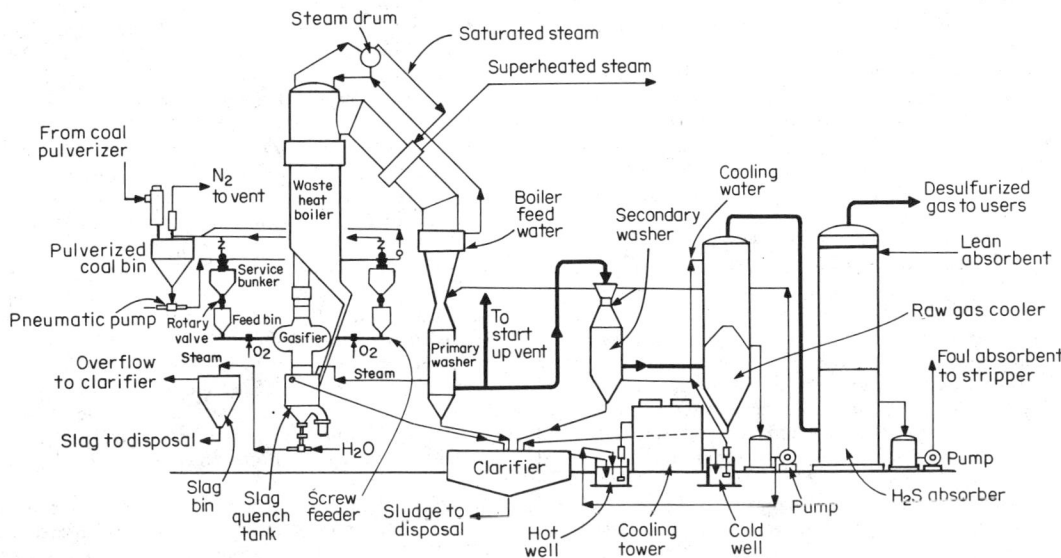

**Fig. 7**   Koppers-Totzek gasification process.

**Table 4. Koppers-Totzek Gasifier Data for U.S. Coals**

| Type of coal | Western coal | Illinois coal | Eastern coal |
|---|---|---|---|
| Gasifier feed: | | | |
| Dried coal to gasifier analysis, vol. % | | | |
| C | 56.76 | 61.94 | 69.88 |
| $H_2$ | 4.24 | 4.36 | 4.90 |
| $N_2$ | 1.01 | 0.97 | 1.37 |
| S | 0.67 | 4.88 | 1.08 |
| $O_2$ | 13.18 | 6.73 | 7.05 |
| Ash | 22.14 | 19.12 | 13.72 |
| Moisture | 2.00 | 2.00 | 2.00 |
| Gross heating value, Btu/lb | 9,888 | 11,388 | 12,696 |
| Oxygen NT/NT dried coal | 0.649 | 0.704 | 0.817 |
| Purity, % | 98.0 | 98.0 | 98.0 |
| Process steam, lb/NT dried coal | 272.9 | 541.3 | 587.4 |
| Gasifier Products: | | | |
| Jacket steam, lb/NT dried coal | 347.8 | 404.9 | 464.9 |
| High-pressure steam, lb/NT dried coal | 2,147.1 | 2,292.2 | 3,023.6 |
| Raw gas | | | |
| Analysis (dry basis), vol. % | | | |
| CO | 58.68 | 55.38 | 55.90 |
| $CO_2$ | 7.04 | 7.04 | 7.18 |
| $H_2$ | 32.86 | 34.62 | 35.39 |
| $N_2$ | 1.12 | 1.01 | 1.14 |
| $H_2S$ | 0.28 | 1.83 | 0.35 |
| COS | 0.02 | 0.12 | 0.04 |
| Total | 100.00 | 100.00 | 100.00 |
| Gross heating value, Btu/ft³ | 295.1 | 290.2 | 294.4 |
| Gas make, ft³/NT dried coal | 51,783 | 59,489 | 66,376 |
| Slag make, NT/NT dried coal | 0.222 | 0.190 | 0.138 |
| Process efficiency | 88.2 | 85.0 | 90.3 |
| Coal to gas efficiency | 77.3 | 75.8 | 77.0 |

**BI-Gas**  Bituminous Coal Research, Inc. Coal dried and pulverized is fed to the upper stage of the reactor where the coal is rapidly heated to 1700°F (925°C) and devolatilized by the upward-flowing hot gas from the lower stage. Entrained char, separated from the raw gas, is fed into the lower stage and gasified with oxygen and steam at a temperature of about 2700°F (1500°C), causing slagging of the ash which drains into a water-quench section. Raw gas from the upper stage is cooled, filtered, and then shifted. Acid gases are removed, followed by methanation to produce synthetic natural gas.

**Synthane**  U.S. Bureau of Mines. Crushed and dried coal is fed to a fluid-bed pretreater where about 12 percent of the steam and oxygen requirements are introduced. Coal is partially devolatilized and coking properties are destroyed in the pretreater operating at about 800°F (425°C). The partially devolatilized coal, together with volatile matter and excess steam, is then fed to the top of a fluid-bed gasifier operating at about 1800°F (980°C) and 500 to 1,000 lb/in² (34 to 68 atm). Oxygen and steam are introduced into the bottom of the gasifier. Char, containing about 30 percent of the carbon from the original coal, is removed, combined with char from the raw-gas cyclone separator, and used for boiler fuel. The raw product gas passes through cyclones to remove dust (char) and then to a water wash for cooling and tar removal. The clean gas goes to the CO shift converter, followed by acid-gas removal and methanation to synthetic natural gas.

Other processes under development include the **Atgas process** of Applied Technology Corporation which employs a bath of molten iron as the heat source. Coal, oxygen, and steam are introduced into the bath through lances. The carbon reacts with oxygen and steam to form carbon monoxide and hydrogen. The sulfur reports to the slag layer atop the bath. The **molten-salt process**—M. W. Kellogg Co.—employs preheated steam and oxygen to feed coal and recycled sodium carbonate into the molten-salt gasifier. The steam-coal reaction is catalyzed by the molten salt (sodium carbonate) to produce a gas of appreciable methane content and free of tars at comparatively low temperatures, 1700°F (927°C).

**Gasification of Liquid Hydrocarbons**  In the era of manufactured gas in the United States, both base-load and peak-shaving gases were produced by gasifying oils via thermal-cracking techniques. Most of the processes produced gases having calorific values compatible with coal gas (coke-oven gas). As natural gas became available, some processes were modified to produce a high-heating-value gas interchangeable with natural gas. These oil-gas units operated on a cyclic (heat-make) basis. To supply the heat for the endothermic thermal cracking of oil, a mass of checker brick was heated to 1300 to 1700°F (705 to 925°C) by burning oil and deposited carbon with air. During the "make" cycle steam and oil were introduced to produce a mixture of hydrogen, methane, saturated and unsaturated hydrocarbons, aromatic oils, tar, and carbon.

A variety of oil-gas sets in use employed the same basic operating principles—differing only in design details and principally in recuperator arrangements for recovering the sensible heat from the "heat" and "make" gases. Typical of the processes in use were the **Hall cyclic** reformer (with two or four shells) and the two-shell "inverted U" sets of the **Gas Machinery Co. (Petrogas).**

Further improvement in oil gasification resulted from the development of steam-reforming catalysts and the **catalytic-cracking processes.** Operation is cyclic and similar to that of a Hall or Petrogas with a catalyst bed replacing the checker-brick chamber. Typical commercial processes include the **Segas,** the **Onia-Gegi,** and the **UGI-CCR.**

Increased demand for synthesis gas for chemical production and later for synthetic natural gas led to extensive investigations and developments in methods using **partial oxidation, catalytic reforming,** and **hydrogasification.**

**Partial-oxidation** processes are used commercially to react liquid feedstocks (any weight) with oxygen to produce a synthesis gas consisting of carbon monoxide, carbon dioxide, and hydrogen. These processes are adaptable to the production of synthetic natural gas, but virtually all of the methane must be produced by catalytic methanation of carbon monoxide and hydrogen. Commercial processes of note are the **Texaco,** the **Shell,** and the **Koppers-Totzek (K-T).** The K-T process, designed primarily for solid feedstocks, has the flexibility of using liquid feedstocks and operates at atmospheric pressure, whereas the Texaco and Shell processes operate at elevated pressures. Recent developments by Texaco indicate that appreciable quantities of methane can be produced through partial oxidation of liquids by limiting reaction temperatures. Partial-oxidation processes are well adapted to the production of hydrogen for hydrogasification processes or other uses.

**Catalytic reforming** of naphtha, natural-gas liquids, and LPG is presently applied commercially for the production of synthetic natural gas. Over 30 plants with a total capacity of $6.5 \times 10^9$ standard cubic feet per day (184 million cubic metres per day) are planned, but the actual number of installations is limited by availability of feedstock. Commercial processes available are the **CR—G-catalytic rich gas**—British Gas Council; **MRG—methane rich gas**—Japan Gasoline Co.; and **Gasynthan**—BASF/Lurgi. Processes are similar in that each uses steam reforming of light hydrocarbons over a bed of nickel catalyst. The product gas is a mixture of methane, carbon monoxide, carbon dioxide, and hydrogen. Upgrading to synthetic natural gas requires methanation steps. The four basic steps of the process are desulfurization, gasification, methanation, and purification ($CO_2$ removal and drying).

**Hydrogasification** The British Gas Corporation has developed the **GRH—gas-recycle hydrogenator** for hydrogenating vaporizable oils to produce synthetic natural gas and the **FHB—fluid-bed hydrogenation** of gasifying crudes or heavy oils for synthetic-natural-gas production. In the **GRH process,** naphthas, middle distillates, and gas oils that need not be desulfurized are reacted directly with hydrogen-rich gas prepared by steam reforming a rich gas sidestream. Exothermic reactions decompose paraffins and naphthenes into methane and ethane. In the **FBH process,** crude or heavy oil is preheated and atomized in the presence of coke particles fluidized by a supply of preheated hydrogen-rich gas. Paraffins and naphthenes are hydrogenated to methane and ethane, and an aromatic condensate is recovered. Desulfurization, followed by secondary hydrogenation, allows reduction of hydrogen and ethane to produce synthetic natural gas.

# COMBUSTION FURNACES

## M. H. Mawhinney

REFERENCES: Trinks-Mawhinney, "Industrial Furnaces," vols. I and II, Wiley.

### FUELS

The selection of the best fuel should be based upon a study of the comparative prepared costs, cleanliness of operation, adaptability to temperature control, labor required, and the effect of each fuel upon the material to be heated and upon the furnace lining. Attention must be paid to the quantity to be burned in each burner, the atmosphere (fuel/air ratio) desired in the furnace, and the uniformity of temperature distribution required, which determines the number and the location of the burners. Common methods of burning furnace fuels are as follows.

### Solid Fuels (almost entirely bituminous coals)

Coal was once a common fuel for industrial furnaces, either hand-fired, stoker-fired, or with powdered coal burners. With the increasing necessity for accurate control of temperature and atmosphere in industrial heating, coal has been almost entirely replaced by liquid and gaseous fuels. With the reduced availability of natural gas for industrial purposes, it can be expected that methods will be developed for the production of a synthetic gas (natural-gas equivalent) from coal.

### Liquid Fuels (fuel oil and tar)

HIGH-PRESSURE BURNERS include all burners utilizing atomizing air or steam at pressures above 2 lb./in² gage (ca. 14 kN/m²) and are most common in high-capacity sizes. With 60 lb/in² gage (510 kN/m²) 90 percent of the air needed for combustion is induced from the atmosphere.

INTERMEDIATE-PRESSURE BURNERS utilize atomizing air at 1 to 2 lb/in² gage (7 to 14 kN/m²). Between 30 and 40 percent of the required combustion air is supplied mechanically, and the remainder induced.

FAN-BLAST BURNERS utilize atomizing air at 8 to 16 oz/in² (3.5 to 7 kN/m²), and induce about 50 percent of the combustion air.

### Gaseous Fuels

**Burners for refined gases** (natural gas, synthetic gas, coke-oven gas, clean producer gas, propane, butane):

TWO-PIPE SYSTEMS.  Include blast burners (open or closed setting), nozzle mixing, luminous flame, excess air (tempered flame), baffle, and radiant-tube burners, all for low-pressure gas and air.

PREMIX SYSTEMS.  Air and gas mixed in a blower and supplied through one pipe.

PORPORTIONING LOW-PRESSURE MIXERS.  Air and gas supplied under pressure and proportioned automatically (air aspirating gas or gas inspirating air). The resulting mixture is burned in tunnel burners, radiant-cup, baffle, radiant-tube, ribbon, and line burners.

Pilot flames should be supplied to ensure ignition for gas and oil burners, and insurance frequently requires additional safety provision in two main categories: an interconnected pressure system to prevent lighting if any burner in a zone is open, and burner monitors using heat or light to permit ignition.

**Burners for crude gases** (raw producer-gas, blast-furnace gas, or coke-oven gas):

SIMPLE MIXING SYSTEMS with large orifices and simple mechanisms which cannot become clogged by tar and dirt contained in these gases.

SEPARATE GAS AND AIR SUPPLIES to the furnace, with all mixture taking place within the furnace.

## TYPES OF INDUSTRIAL HEATING FURNACES

Heating furnaces are usually classified according to (1) the purpose for which the material is heated, (2) the nature of the transfer of heat to the material, (3) the method of firing the furnace, or (4) the method of handling material through the furnace.

**Purpose**  Primarily a metallurgical distinction, according as the furnace is intended for tempering, annealing, carburizing, cyaniding, case hardening, forging, heating for forming or rolling, enameling, or for some other purpose.

**Transfer of Heat**  The principal varieties are **oven furnaces,** in which the heat is transferred from the products of combustion of the fuel, in direct contact with the heated material, by convection and direct radiation from the hot gases or by reradiation from the hot walls of the furnace; **muffle furnaces,** in which the heat is conducted through a metal or refractory muffle which protects the heated material from contact with the gases, and is then transferred from the interior of the muffle by radiation to the heated material, which is sometimes surrounded by inert gases to exclude air; or **liquid-bath furnaces,** in which a metal pot is heated on the outside or by immersion.

This pot contains a liquid heating or processing medium which transfers heat to the material contained in it. This type includes low-temperature tempering furnaces with oil as the heating medium, hardening furnaces using a bath of lead, hardening and cyaniding furnaces with baths of special salts, and galvanizing or tinning furnaces for coating the heating material with zinc or tin. The generally accepted form of muffle is the **radiant-tube** fired furnace, in which the fuel is burned in metal or refractory tubes which radiate heat to the charge. An important form of furnace for temperatures below 1300°F (700°C) is the **recirculating** type, in which the atmosphere (products of combustion, air, or protective gases) is recirculated rapidly through the heating chamber. A recent development is **forced convection** heating by a large number of jets of hot gas at high velocity. In **high-speed** heating (or **patterned combustion**), premixed burners are arranged for close application of heat, and with a high-temperature head, very rapid heating is accomplished.

**Method of Firing**  This classification applies principally to the oven type of furnace, and it indicates whether the furnace is direct-fired, overfired, underfired, or heated by radiant tubes. Figure 1 shows the principles of each of these types. The **direct-fired** method finds increased utilization from constant improvement in the design and control of gas and oil burners, especially for temperatures above 1200°F (650°C). In **overfired** furnaces radiant burners fire through the roof and are arranged in patterns to obtain the best temperature distribution. The **underfired** furnace is excellent for temperatures between 800 and 1800°F (ca. 400 to 1000°C) because the heated product is protected from the burning fuel. The temperature and atmosphere can be closely controlled, but the temperature is limited by the life of the refractories to about 1800°F (1000°C). Many furnaces are now designed for the use of special protective atmospheres and involve the use of **radiant tubes** to avoid any contact with the combustion gases. These fuel-fired tubes of heat-resisting alloy may be horizontal across the furnace above and below the heated material or may be vertical on the sidewalls of the furnace.

**Method of Material Handling**  In the **batch** type, the heated material charged into the furnace remains in the same position until it is withdrawn after sufficient heating. In a **continuous furnace,** the material is moved through the furnace by mechanical means which include pushers, chain conveyors, reciprocating hearths, rotating circular hearths, cars, walking beams, and roller hearths. Continuous furnaces are principally labor-saving devices and may or may not save fuel.

## SIZE AND ECONOMY OF FURNACES

The size of furnace required depends upon the amount of material to be heated per hour, the heating time required, the

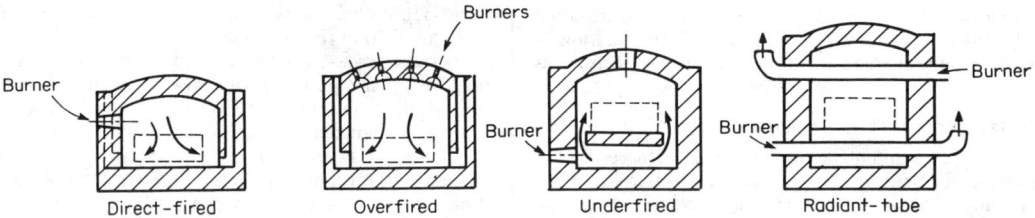

**Fig. 1**  Methods of firing oven furnaces.

**Table 1. Average Heat in Waste Products of Combustion at Various Temperatures, Percent of Low Heat Value of the Fuel**

| Temp of gases, °F | 1000 | 1200 | 1400 | 1600 | 1800 | 2000 | 2200 |
|---|---|---|---|---|---|---|---|
| Temp of gases, °C | 540 | 650 | 760 | 870 | 980 | 1090 | 1200 |
| % of low heat value in gases | 24 | 28 | 34 | 38 | 45 | 50 | 55 |

size of the pieces to be heated, and the amount of heat that can be liberated without excessive damage to the furnace. The efficiency and refractory life obtained depend upon the correctness of furnace size.

**Heating Time**  For the usual relation of refractory area to stock area, time to heat steel plate from one side for each ⅛ in (3.18 mm) of thickness varies from 3 min for high-speed heating and 6 to 12 min for heating for forming by usual methods to 20 min for heat treating. Steel cylinders will be heated in one-half these times per ⅛ in (3.18 mm) diam. Below 800°F (ca. 400°C), the time may be two to three times these values. Brass requires about one-half as long as steel to heat, copper 40 percent as long, and aluminum 85 percent as long. The preceding heating times are based on a furnace temperature 50 to 100°F (ca. 25 to 50°C) higher than the final temperature of the heated material. It is assumed that the material is fully exposed to the heat of the furnace. Piling of material in a furnace lengthens the heating time by an amount that must be determined by actual trial. In addition to simple heating, there is frequently additional **time required for soaking** (holding at furnace temperature) to cause metallurgical changes in the material or for some other reason.

The **weight of material in the furnace** at any time is the product of weight of material per hour multiplied by the heating time in hours. If the weight and sizes of pieces involved are known, the **area of the furnace** can then be fixed. The width and length of the furnace to produce this hearth are fixed by the method of firing to be used and by the method of handling material.

The life of a furnace at given temperature depends upon the rate of heating, which may be expressed in pounds per square foot of hearth area per hour. The maximum allowable **rate of heating steel** is about 35 lb/ft²/h for heat treating, 70 lb for in-and-out rolling-mill furnaces, 100 lb for single-zone continuous furnaces, and 150 lb for multiple-zone furnaces. These are upper limits which should not be used if long life of furnace refractories is expected. These rates are for heating mild steel; they may be about twice as great when heating brass, 2½ times as great for copper, 0.7 as great for alloy steel, and 1.1 times as great when heating aluminum. These maximum allowable rates should be used only for checking the calculation of size, because some shapes and sizes of pieces cannot be properly heated when piled in such a manner as to produce these rates. If the calculated size of the furnace corresponds to a rate of heating that is too great, it should be reduced by making the furnace larger. If the rate is too small, it can sometimes be increased by piling material differently in a smaller furnace.

EXAMPLE.  To determine furnace size. If a furnace is required to heat 20 pieces per h weighing 30 lb each and requiring a heating time of ½ h, the furnace must be large enough to hold ½ × 20 = 10 pieces. If each piece requires an area of 2 ft², the area of the hearth will be 2 × 10 = 20 ft² for a single layer of pieces in the furnace. If the furnace is of the batch type, a size of 4 ft wide × 5 ft deep would probably be about right for convenient handling. Upon checking, the rate of heating is 20 pieces per h × 30 lb/20 ft² = 30 lb/ft²/h. For this rate an underfired furnace would be satisfactory, although for other methods of firing, a

smaller furnace could be used if the pieces could be more densely piled without seriously interfering with the circulation in the furnace.

The heat released by the fuel in a furnace (heat input) is equal to the sum of the heat required in the heating process (useful heat) plus the heat losses from the furnace. **Heat input**

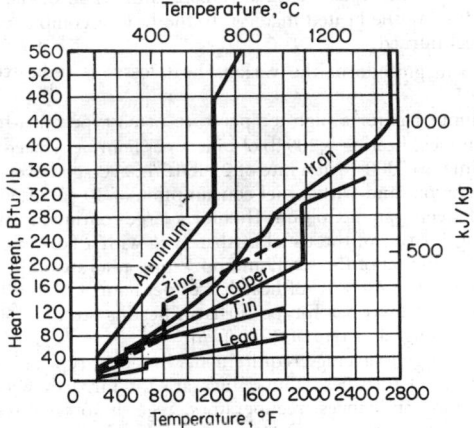

**Fig. 2**  Heat content of metals.

includes the heat of combustion of the fuel, sensible heat in preheated air or fuel, and heat in the material charged. Low-heat values of the fuel are used, and the sensible heat can be calculated from the specific heats of the preheated air, fuel, or material. **Useful heat** includes the heat absorbed by the material in the furnace. Figure 2 gives heat contents for different metals. In the simple heating of metals, the useful heat applied to the metal includes only the heat absorption, as given in Fig. 2; but there are many processes that include other requirements, such as drying, where moisture must be heated and evaporated, heating of chemical products where heat is utilized to cause chemical changes, and other special cases.

**Heat losses** in a heating furnace include heat lost in waste gases, radiation from and heat absorbed by refractories, heat carried out of the furnace by containers or conveyors, heat lost through openings, and heat in unburned fuel escaping with the products of combustion. The **heat contained in waste gases** depends upon the temperature of these gases as they leave the heating chamber. Table 1 gives the approximate percentage of heat contained in the flue gases from perfect combustion at different temperatures. These values are about the same for most fuels except producer gas and blast-furnace gas, the losses with which are higher than those given in the table.

**Radiation and heat absorption by refractories** depend also upon the rate of heating (which determines the interior temperature of the refractories) and upon the refractory area and thickness. Figure 3 shows the heat radiated through walls of different thickness at various furnace temperatures, for equilibrium conditions, when the wall has reached steady temperatures throughout (see also references at the beginning of this section and Keller, "Flow of Heat through Furnace Hearths," ASME, May 1928). The **heat carried out by containers and conveyors** is the sensible heat content of these items as they leave the heating chamber. Such losses include the heat in carburizing boxes, pans, chain conveyors, and furnace cars. **Radiation from furnace openings** depends upon the size and shape of the opening and the thickness of the walls in which they are located, as well as upon the temperature of the furnace. Some idea of the magnitude of these losses is given by the values in Table 2.

The **heat lost in unburned fuel** escaping with the flue gases is small in most furnaces because the fuel can be almost completely consumed.

The **efficiency of an industrial furnace** is the ratio of the heat absorbed by the heated material to the heat of combustion of the fuel burned.

The magnitude of the various heat losses is indicated in Table 3.

Column I is for a high-temperature batch-type billet-heating furnace, heating 4,200 lb of billets per hour, a furnace load at a time, to 2300°F, at a rate of 25 lb/ft²/h, averaged over 10 h of operation, and with a fuel consumption of 30 gal of oil per ton of steel heated. Column II is for a large continuous billet-heating furnace of the usual pusher type with a flow of gases opposite to that of the steel, and operating at a rate of 60 lb of steel heated to 2300°/ft² of hearth area per hour. Column III is for an underfired batch-type furnace, heating steel to 1600°F for annealing, at a rate of 30 lb/ft²/h.

Table 4 gives average requirements in fuel of typical industrial heating furnaces. The values are for furnaces without heat-saving appliances (recuperators, regenerators, or waste-heat boilers) except as noted and show the efficiency and the Btu required in the fuel per net pound of steel heated. To obtain the average amount of any fuel required, this latter figure is divided by the low heat value of the fuel. The values are for average rates of heating. Fuel economy is of small importance as compared with the quality of the product.

## FURNACE CONSTRUCTION

Furnace refractories are made up largely of standard bricks and shapes, and it is advisable to specify furnace dimensions that can be built with a minimum of cutting. Horizontal flues are made a multiple of 2½ or 3 in (63 or 76 mm) in height, and most other flue dimensions are multiples of 4½ in (114 mm) to correspond to the width and length of standard bricks. The area of furnace flues must be large enough to avoid excessive pressures at maximum fuel rates. Flues should be located so as to promote the circulation of gases in all parts of the furnace. Average allowable **velocities in flues** for furnaces without stacks are:

| Furnace temp, °F (°C) | 200 (93) | 1000 (538) | 1500 (816) | 2000 (1093) |
|---|---|---|---|---|
| Allowable velocity (hot gases), ft/s (m/s) | 9 (2.74) | 13 (3.97) | 15 (4.57) | 17 (5.2) |

The total **flue areas required** in in²/ft³ of fuel/h (or per gal/h for fuel oil) for furnaces without stacks at temperatures of the products of combustion of 1000 and 2000°F are as follows:

| Temp, °F (°C) | Fuel Oil | Natural Gas | Artificial Gas | Coke-Oven Gas | Raw Producer Gas |
|---|---|---|---|---|---|
| 1000 (538) | 14.0 | 0.11 | 0.06 | 0.05 | 0.02 |
| 2000 (1093) | 19.0 | 0.15 | 0.08 | 0.06 | 0.02 |

The **metal parts of a furnace** consist of the steel and cast-iron binding and of alloy parts exposed to the direct heat of the furnace. These alloy parts are of nickel or chromium alloys and must be made heavy enough to offset the loss of strength at high temperatures. They are resistant to oxidation at temperatures below 2000°F (1093°C).

## HEAT-SAVING METHODS

Methods of conserving heat include the use of recuperators or regenerators, waste-heat boilers (see Sec. 9), insulation of refractories, automatic control of temperature and atmo-

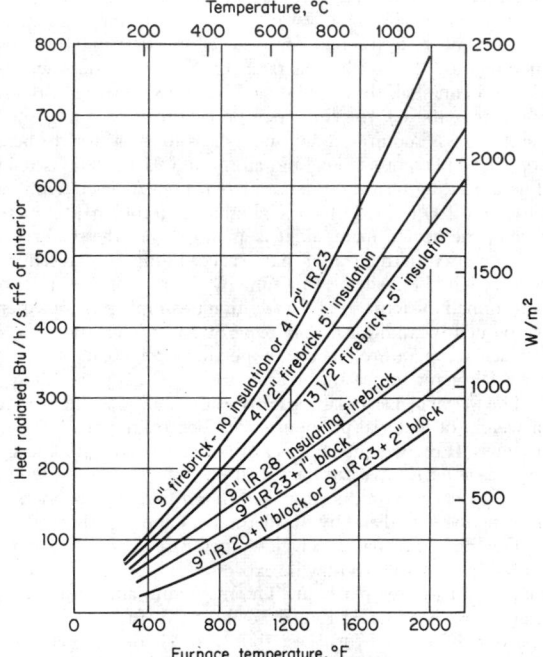

**Fig. 3**   Heat loss from thoroughly heated walls, based on interior area.

sphere, and special attention to the construction and operation of the furnace.

Recuperators and regenerators extract some heat from the escaping flue gases and return it to the furnace by preheating the combustion air or the entering fuel. In **recuperators,** continuous flow of hot gases and cold entering air or gas is maintained through metal or refractory ducts which keep the two gas streams apart but which conduct heat from the hotter stream to the colder. Recuperators are built in the form of self-

**Table 2. Radiation through Openings in Furnace Walls, Thousands of Btu/Hour**

| | Furnace temp, °F (°C) | | | | | |
| | 1400 (760) | | | 2200 (1200) | | |
| | Wall thickness, in* | | | Wall thickness, in* | | |
| Size of opening, in* | 4½ | 9 | 18 | 4½ | 9 | 18 |
|---|---|---|---|---|---|---|
| 4½ × 4½ | 1.4 | 1.1 | 0.8 | 5.1 | 4.1 | 2.8 |
| 9 × 9 | 7.8 | 6.1 | 4.5 | 28.5 | 22.7 | 16.8 |
| 18 × 18 | 37 | 30.5 | 24.3 | 137 | 114 | 90 |
| 24 × 24 | 71 | 60 | 48 | 264 | 225 | 180 |
| 36 × 36 | 173 | 150 | 124 | 650 | 560 | 465 |

*× 25.4 = mm.

enclosed units set above the ground or in pits below floor level, and are made of fire-clay tile, silicon carbide, or heat-resisting metal. Overall coefficients of heat transfer in metallic recuperators are between 2.5 and 6.0 Btu/(ft²)(h)(°F) (14 and 34 W/m²°C), and in silicon carbide recuperators about the same; the coefficient for fire-clay recuperators is considerably less than these values. Usual velocities of hot air in recuperators do not exceed 12 ft/s (3.6 m/s) in order to keep pressure drop to a reasonable value.

**Regenerators** are used where high temperature of air preheat is required to maintain high furnace temperature. They are usually constructed of firebrick and consist of two chambers completely filled with a checkerwork. The flow of flue gases and that of air or gas to be heated are periodically reversed, so that the hot gases and cold gases alternately flow through the two sets of chambers. The checkerwork retains the heat of the hot gases and gives it up to the cold gases with each reversal. Another regenerator design employs metal plates. Regenerators are more expensive than recuperators in most cases and are not frequently used with heating furnaces, but their use with open-hearth furnaces is almost universal. Overall coefficient of heat transfer in regenerators is from 1.5 to 2.5 Btu/ft² of checkerbrick surface per h per °F temperature difference (8.5 to 14 W/m²°C), and the usual mass velocity of hot gas through the openings of the checker is about 0.065 lb/ft²/s (0.32 kg/m²s).

The **saving effected by recuperators or regenerators** depends upon the temperature to which the incoming air or gas is preheated. With a flue-gas temperature of 1600°F, the theoretical saving in fuel with 200° preheat of combustion air is about 4 percent; with 400°, 11 percent; with 600°, 15 percent; and with 800°, 19 percent. A recuperator or regenerator installation, to be a good investment, must show a satisfactory net saving after all costs of repairs and shutdown time lost by such repairs are subtracted from the savings in fuel used.

**Automatic control** prevents the waste of heat by unnecessarily high temperatures, preventable cold periods, and excessive air or unburned fuel from poor combustion. Of even greater importance is the prevention of damage to the heated product from overheating, excessive oxidation, and objectionable chemical reaction between furnace atmosphere and the product (principally decarburization and recarburization). Automatic **temperature** controllers are actuated by thermocouples in the furnace. The thermocouple must not be located in the direct path of the flames, which are not only several hundred degrees hotter than the furnace temperature but are also of extremely variable temperature and not an indication of the average temperature. Automatic control of **atmosphere** for the consistent maintenance of good combustion is accomplished by properly proportioning the fuel and combustion air as they enter the furnace. This is accomplished by the utilization of some characteristic of the flow of one fluid to regulate the flow of the other fluid. Automatic **pressure** control operates the flue dampers of a furnace to maintain a constant predetermined pressure [usually about 0.01 to 0.05 in (0.25 to 1.25 mm) water] in the heating chamber, which excludes free oxygen from the surrounding atmosphere.

Care in furnace construction and operation is the simplest but most frequently neglected of all methods of heat saving. A large quantity of fuel can be saved by care in the construction of furnace refractories so that they will remain tight, by attention to the sealing of doors, and by taking care that the doors and other openings are kept closed when not in use.

## SPECIAL ATMOSPHERES

(See also Sec. 7)

In an increasing number of heat-treating operations the necessity for improved quality has created a demand for clean- or bright-heating furnaces, in which the heating material is surrounded by a suitable protective gas while it is heated by radiation from electric resistors, radiant tubes, or the walls of a muffle. Table 5 gives the chemical analysis of common protective gases used in the heat-treating industry.

**Type I** Purified hydrogen is used for annealing, brazing,

**Table 3. Heat Balances for Various Furnace Types**
(Percent of heat of combustion)

| Disposition of heat | I | II | III |
|---|---|---|---|
| Heat to material, or efficiency | 16 | 49 | 23 |
| Heat to refractories | 20 | 17 | 22 |
| Heat lost in flue gases | 44 | 19 | 40 |
| Heat to water cooling | | 5 | |
| Heat through openings | 20 | 10 | 15 |

**Table 4. Average Net Efficiencies and Fuel Requirements of Various Furnace Types with Good Operation**

| Type | Temp, °F | Temp, °C | Avg efficiency, % | Avg heat required from fuel, Btu/lb of steel* |
|---|---|---|---|---|
| Ingot heating, soaking pits, recuperative | 2000–2400 | 1100–1300 | 20 | 500 |
| Billet heating for forming: | | | | |
|    Batch, in-and-out | 2000–2400 | 1100–1300 | 20 | 1750 |
|    Continuous | 2000–2400 | 1100–1300 | 32 | 1100 |
| Wire annealing of coils, hood type | 1300–1500 | 700–800 | 16 | 1350 |
| Wire annealing of strands, in lead | 1300–1500 | 700–800 | 19 | 1100 |
| Wire patenting, strands | 1650 | 900 | 21 | 1250 |
| Wire baking, coils, continuous | 450 | 230 | 20 | 250 |
| Tube annealing, continuous, bright | 1300–1500 | 700–800 | 35 | 600 |
| Skelp heating, butt weld, continuous | 2900 | 1600 | 25 | 1500 |
| Slab heating, continuous, recuperative | 2400 | 1300 | 42 | 800 |
| Strip coil annealing, hood type | 1250–1400 | 680–760 | 30 | 600 |
| Hardening, continuous conveyor | 1650 | 900 | 21 | 1250 |
| Drawing, continuous conveyor | 900–1100 | 500–600 | 20 | 750 |
| Carburizing, gas, continuous | 1750 | 950 | 19 | 1500 |

\* × 2.326 = kJ/kg.

and other treatment of low-carbon steel; for the sintering of low-carbon ferrous powders; for the treatment of silicon iron (electrical sheets and strip); for the bright annealing of stainless steels, and the sintering of molybdenum, tungsten, and other metals.

**Type II**   Ammonia is dissociated by steam or electric heat, and is dried by chemical driers. By partial combustion the relative percentages of hydrogen and nitrogen may be varied as shown in Table 5. The resulting gases from this treatment are cheaper and are used for brazing and sintering copper alloys, and for annealing low-carbon steels. Dissociated ammonia without combustion is used for annealing stainless steels containing nickel, short-cycle heating of all carbon and alloy steels, treatment of silicon iron, and the treatment of cuprous products.

**Type III**   Rich hydrocarbon gas is produced by combustion with about 60 percent of theoretical air (6:1 air/gas ratio when using natural gas) in the presence of a nickel catalyst, followed by cooling to reduce the moisture content. It is used for the

annealing of low-carbon steels, for short-cycle hardening of low-carbon steels, for clean annealing of chrome-type stainless steels, for treatment of silicon iron, and for brazing of copper alloys.

**Type IV**   This gas is similar to type III except that about 90 percent of theoretical air is used for combustion. It is used for bright annealing of copper (straight $N_2$ and $CO_2$ can also be used for this purpose) and for clean heating of brass and bronze.

**Type V**   This gas is the same as type III but is conditioned by chemical removal of carbon dioxide by monoethanolamine and by drying in chemical driers. It is used for short-cycle treatment of all carbon, alloy, and high-speed steels; for sintering of all ferrous powders; and as a carrier gas for carburizing and carbon restoration with the addition of natural gas or propane.

**Type VI**   This gas is similar to type V except that about 90 percent of theoretical air is used in the combustion. The resulting gas is used for long-cycle treatment of all ferrous

**Table 5. Protective Gas Atmospheres**

| Type | Typical analysis | | | | | Dew point, °F (°C) |
|---|---|---|---|---|---|---|
| | $CO_2$ | CO | $CH_4$ | $H_2$ | $N_2$ | |
| I.  Hydrogen, purified | | | | 100.0 | | Minus 60 (−51) |
| II.  Dissociated ammonia | | | | 75.0–5.0 | 25.0–95.0 | |
| III.  Rich hydrocarbon gas, not conditioned | 5.5 | 9.0 | 0.8 | 15.0 | 69.7 | Plus 50 (10) |
| IV.  Lean hydrocarbon gas, not conditioned | 11.5 | 0.7 | | 0.7 | 87.1 | Plus 50 (10) |
| V.  Rich hydrocarbon gas, completely conditioned | 0.1 | 9.5 | 0.8 | 15.8 | 73.8 | Minus 60 (−51) |
| VI.  Lean hydrocarbon gas, completely conditioned | 0.1 | 2.8 | | 3.9 | 93.2 | Minus 60 (−51) |
| VII.  Endothermic generator gas | 0.5 | 20.0 | 1.0 | 38.0 | 40.5 | Plus 50 (10) |
| VIII.  Charcoal gas | 0.5 | 30.0 | | 2.0 | 67.5 | Plus 50 (10) |

materials except stainless steels containing nickel, and is effective in controlling decarburization in all carbon and alloy steels. It is also used for the annealing of brass and bronze.

**Type VII**  This endothermic gas is made in an externally heated generator with only 25 percent of theoretical air and is cooled to reduce moisture. It is used for short-cycle (under 2 h)

heat treating and brazing, usually with small furnace installations. It is also used for dry cyaniding and as a carrier gas for carburizing and carbon restoration.

**Type VIII**  By passing air over heated charcoal, a gas is produced which may also be used for short-cycle heat treating with small furnaces.

# INCINERATION

## by Charles R Velzy and Charles O. Velzy

REFERENCES: *Proc. ASME National Incinerator Conference*, 1964, 1966, 1968, 1970, 1972, 1974. APWA, "Municipal Refuse Disposal," Research Foundation Project 104. *Proc. APWA National Conference on Solid-Waste Research*, 1963. Design Considerations in Heat Recovery from Refuse, Intern. Symp. on Energy Recovery from Refuse, 1975. R. Hecklinger, "The Relative Value of Energy Derived from Municipal Refuse," *Proc. ASME Natl. Waste Processing Conference*, 1976.

Incineration is a method for processing of solid wastes by the burning of the combustible portions. It reduces the volume of solid wastes and eliminates the possibility of pollution of groundwater from putrescible organic waste, and the residue may serve as a source of mineral constituents and as a fill.

## NATURE OF THE FUEL

The refuse which is received at an incinerator today will contain a high proportion of paper; some wood; vegetable and animal waste; and varying amounts of cloth, leather, rubber, and plastics—together with metal cans, glass, and other noncombustible matter. Collections may also include metal appliances, furniture, tree limbs, waste building material, broken concrete, and other coarse waste matter, commonly classified as rubbish. With little or no regulation of the handling of refuse by the homeowner, there may be a wide variation in moisture content of refuse, depending on the weather. Thus, after a storm, the moisture content may be so high that it is difficult to sustain combustion.

## TYPES OF FURNACES

The type of furnace for incinerators is dictated largely by the type of grate around which the furnace is built. Except in small plants, the modern furnace is equipped with a mechanical grate.

**The Circular Furnace**  An early development in mechanical stoking was the monohearth circular furnace. The grate is stationary, with outer sections hinged for dumping of residue. At the center is a cone-shaped casting, and extending from the base of the cone is a pair of slowly revolving arms to stir the refuse gently during burning. The refuse is received through a gated opening in the roof, and the material is fed in a batching operation.

**The Rectangular Furnace**  The more recent mechanical-grate furnaces are rectangular in shape, with movement pro-

vided by travel of the grate or by a reciprocating or rocking action of the grate sections. Refuse is fed by gravity through a vertical chute, by a ram or similar arrangement.

## PLANT DESIGN

**Capacity**  The capacity to be provided is a function of (1) the area and population to be served; (2) the number of shifts (one, two, or three) the plant is to operate; and (3) the rate of refuse production for the population served. If records of collections have been kept, the capacity can be determined and forecasts made; lacking records, the quantity of refuse may be estimated as approximately 5 lb (2.3 kg) per capita per day. A small plant (100 tons per day) (90 metric tons per day) will probably operate one shift per day; for capacities above 400 tons (360 metric tons) per day, economic considerations usually dictate three-shift operation.

**Location**  An isolated site may be preferred to avoid the possible objections of neighbors to the proximity of a waste-disposal plant. However, well-designed and well-operated incinerators which do not present a nuisance are installed in many light industrial and commercial areas, thereby avoiding the economic burden of extended truck routes. Since considerable vertical distance is involved in passing refuse through an incinerator, there is an advantage in a sloping or hillside site. Collection trucks can then deliver refuse at the higher elevation while the residue trucks operate at the lower elevation with a minimum of site grading.

**Refuse-Handling Facilities**  Scales should be provided for recording the weight of material delivered by collection trucks. Trucks should then proceed to the tipping floor at the edge of the storage pit. This area, which may be open or enclosed, must be large enough to permit more than one truck at a time to maneuver to and from the dumping position.

Since collections may be limited to one 8-h daily shift (with partial weekend operation) while burning may be continuous over 24 h, ample **storage** must be provided. Seasonal and cyclic variations must also be considered in establishing the storage requirements.

A single pit extending along the front end of the furnace or two pits, one extending from each side of the front end of the furnace, prevail. Pits are relatively long, narrow, and deep. If much over 25 ft in width, it is generally necessary to rehandle refuse dumped from trucks.

**Feeding the Furnaces**  In a large incinerator (pit-and-crane type), burning continuously, refuse is transferred from storage pit to furnace hopper by a crane equipped with a grapple. (See Sec. 10.) The objectionable features of the batch-feed operation are the dumping of a large quantity of fuel into the furnace at one time and exposing the hot refractory to a blast of cold air.

In a more modern furnace using mechanized grates, a vertical charging chute, 12 to 14 ft (3.6 to 4.2 m) long, leads from the hopper to the front end of the furnace. This chute is kept full of refuse; feeding is accomplished by the operation of the mechanical grate, or by a ram; the front of the furnace is sealed from cold air; and the fuel is spread over the grate in a relatively thin bed.

In some newer plants, conveyors, live-bottom bins, and shredding and pneumatic handling of the combustible fraction of the refuse have been utilized. Continued development of these techniques will no doubt eliminate some of the problems that have been experienced to date.

**Flues and chambers**  beyond the furnace convey gases to the stack and house facilities for removal of fly ash. With limited furnace volume, the first chamber beyond the furnace is the combustion space in which the volatile gases escaping from the refuse are burned to completion. In the modern rectangular furnace, the necessary combustion volume may be provided in the furnace and the space formerly utilized as an expansion chamber may be used as a **spray chamber** where the floor may be covered with water to trap the larger particles of fly ash. Sprays and baffle walls remove some particles of fly ash; gas temperature is reduced as required by fans or other downstream equipment.

The draft for an incinerator furnace may be provided by a stack of adequate diameter and height or by an induced-draft fan. (See Secs. 4 and 14.) If the plant does not include heat-recovery equipment, if the flues and chambers are relatively short and simple, and with low-draft-loss air-pollution-control equipment, a stack of reasonable height will produce sufficient draft. With present-day air-pollution-control requirements, draft losses are usually higher and an induced-draft fan is usually essential.

**Air-Pollution Control**  The federal new-source emission standard for air-pollution control at new or enlarged municipal-sized incinerators is 0.08 grain per standard cubic foot (0.18 g/scm) corrected to 12 percent $CO_2$. (This is approximately 0.17 lb/1,000 lb of flue gas corrected to 50 percent excess air.) Some jurisdictions have promulgated emission requirements as low as 0.01 grain per standard cubic foot corrected to 12 percent $CO_2$. The designer should also be alert to the possible emission of noxious gases if certain wastes, such as some plastics, are burned in any quantity.

To meet current emission requirements, two basic techniques have been utilized on incinerators: electrostatic precipitation and wet scrubbing. Electrostatic precipitation has performed reliably at a number of installations and has given predictable emission-test results. Wet scrubbers produce a visible white plume at the stack, and corrosion may be a problem. It has been difficult to predict emission-test results with this equipment in incinerator service.

**Residue Discharge and Disposal**  The residue from refuse burning consists of relatively fine, light ash mixed with items such as burned tin cans, partly melted glass, and pieces of metal. Discharge from furnaces may be through manually operated dump grates or from mechanically operated grates to a hopper, where it is quenched and delivered to a truck through a bottom gate. The residue may also be discharged through a chute into a conveyor trough filled with water for quenching and then carried by flight conveyor to an elevated storage hopper for truck delivery. Usually there are two conveyor troughs, so arranged that the residue can be discharged to either, one trough being used at a time.

The lower end of the discharge chute leading to the trough is submerged in a water seal to prevent entrance of cold air to the furnace. In design of the conveyor mechanism, the proportions should be large because of the nature of the material handled, and the metal used should be selected to withstand severe abrasive service. Final disposal of the residue is by dumping at a suitable location; volume required for disposal is 5 to 15 percent of that required for dumping raw refuse.

**Miscellaneous Facilities**  Good working environment and reasonable comfort for the staff should be provided. Since handling refuse releases dust, a vacuum system with inlets well distributed about the plant will help keep the building clean.

## FURNACE DESIGN

**Capacity**  The basic design factors which determine furnace capacity are grate area and furnace volume. Both provision for and quantity of underfire air, and provision for quantity and method of applying overfire air influence capacity. The required grate area depends upon the selected burning rate, which varies between 60 and 90 lb of refuse/ft²/h in practice. Conservative design, with reasonable reserve capacity and reasonable refractory maintenance, calls for a burning rate between 60 and 70 lb/ft² of effective grate area.

Furnace volume is a function of the rate of heat release from the fuel. A commonly accepted minimum volume is that which results from a heat release of 20,000 Btu/ft²/h. Thus, at this rate, if the fuel has a heat content of 5,000 Btu/lb, the burning rate would be 4 lb/h/ft³ of furnace volume. A conservative design, allowing for some overload and possible quantities of refuse of high heat content, would be from 30 to 35 ft³/ton of rated capacity.

**Grates**  The primary objective of a mechanical grate is to convey the refuse automatically from the point of feed through the burning zone to the point of residue discharge with a proper depth of fuel and in a period of time to accomplish complete combustion. The rate of movement of the grate or its parts should be adjustable to meet varying conditions.

A secondary, but important, objective is to stir gently or tumble the refuse to aid in completeness of combustion. In the United States, there are three types of mechanical grates: (1) traveling, (2) rocking, and (3) reciprocating. With the traveling grate, stirring is accomplished by building the grate in two or more sections with a drop between sections to tumble the material. The reciprocating and rocking grates tumble the material by movement of the grate elements. In Europe, variations of the U.S. designs as well as other types have been developed. The Volund incinerator (Danish) uses a slowly rotating, refractory-lined cylinder or kiln through which the fuel passes as it is burned; the so-called Duesseldorf incinerator uses a series of rotating cylindrical grates in an inclined arrangement. (See Stabenow, ASME, 1964.)

**Configuration** Furnace configuration is largely dictated by the type of grate used. Thus the monohearth furnace uses a circular grate, and the furnace is a vertical cylinder. For the more recent mechanical grate, the furnace is rectangular in plan and the height is dependent upon the volume required by the limiting rate of heat release.

**Air Supply** The total air capacity provided in a refractory-walled incinerator must be more than the theoretical amount required for combustion in order to obtain complete combustion and to control temperatures—particularly with dry, high-heat-content refuse. The total combustion-air requirements may range to 10 lb of air/lb of refuse. For the modern mechanical-grate furnace chamber, two blower systems should be provided to supply combustion air to the furnace. Blower capacities can be divided, with approximately half from the underfire blower and half from the overfire blower and with dampers on fan inlets and air-distribution ducts for control. The pressure on the underfire system for most U.S. grate systems approximates 3 in of water. The pressure on the overfire air should be high enough so that the jet effect on passage through properly proportioned and distributed nozzles in the furnace roof and walls produces sufficient turbulence and retains the gases in the primary furnace chamber long enough to ensure complete combustion.

**Heat Calculation** Among the factors directly affecting design are moisture and combustible content of refuse as received, heat released by combustion, temperature control, and water requirements. The design of furnaces, chambers, flues, and other plant elements should be based on characteristics which result in large sizes. Controls should provide satisfactory operation for loads below the maximum. The computations which follow are for relatively high heat releases. (See also Kaiser, ASME, 1964.)

The prime factors in **heat calculations** are the moisture and combustible content of the refuse and heat released by burning the combustible portion of the refuse. The moisture content may vary from 20 to 50 percent by weight, and the combustible content may range from 25 to 70 percent. The combustible portion is composed largely of cellulose and similar materials, mixed with appreciable amounts of proteins, fats, oils, waxes, rubber, and plastics. The heat released by burning cellulose is approximately 8,000 Btu/lb, while that released by the fats, oils, etc., is approximately 17,000 Btu/lb. If cellulose, oil, and fat exist in the refuse in the ratio of 9:1, the heat content of the combustible matter will be 8,500 Btu/lb. The heat content per pound of refuse as received, for varying proportions of moisture and noncombustible, is given in Table 1 and Fig. 1.

Determination of the **air requirement** is illustrated by compu-

tation with refuse of 5,000 Btu/lb heat content where (from Fig. 1) the composition is: combustible, 58.6 percent; noncombustible, 19.0 percent.

Carbon and hydrogen are the essential fuel elements in combustion of refuse; sulfur and other elements which oxidize during combustion are present in trace amounts and do not contribute significantly to the heat of combustion. Carbon and hydrogen content can be determined from a complete analysis of the refuse, but such an analysis is of questionable value because of the variable character of refuse and the difficulty of obtaining representative samples. For the purpose of this computation, a typical analysis is used in which the total carbon is 28 lb and the hydrogen 0.6 lb/100 lb of refuse. It is probable that 1 to 3 lb of combustible material per 100 lb of

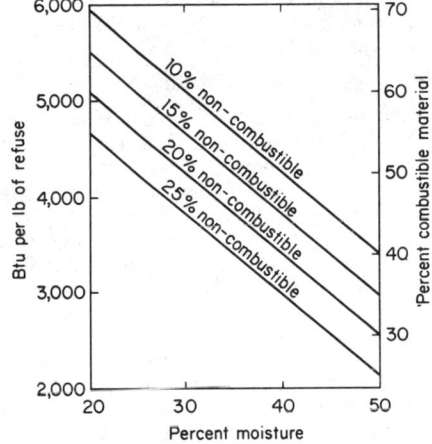

**Fig. 1** Moisture-heat-content relation with 8,500 Btu/lb combustible material.

refuse will escape unburned with the residue. For the sake of clarity in the illustrated computations, complete combustion is assumed.

Oxygen requirements and products of combustion can be determined from the reactions as follows:

**Cellulose**      $C_6H_{10}O_5 + 6O_2 \rightarrow 6CO_2 + 5H_2O$
Atomic wt:      $72 + 10 + 80 + 192 = 264 + 90$
$162 + 192 = 264 + 90$

Ratios:
Referred to carbon,   $1 + 0.14 + 1.11 + 2.667 = 3.667 + 1.25$
Referred to cellulose,       $1 + 1.185 = 1.63 + 0.555$

**Table 1**

| Noncombustible, % | 10 | | 15 | | 20 | | 25 | |
|---|---|---|---|---|---|---|---|---|
| Moisture, % | Comb., % | Heat content* | Comb., % | Heat content* | Comb., % | Heat content* | Comb., % | Heat content* |
| 50 | 40 | 3400 | 35 | 2975 | 30 | 2550 | 25 | 2125 |
| 40 | 50 | 4250 | 45 | 3825 | 40 | 3400 | 35 | 2975 |
| 30 | 60 | 5100 | 55 | 4675 | 50 | 4250 | 45 | 3825 |
| 20 | 70 | 5950 | 65 | 5525 | 60 | 5100 | 55 | 4675 |

*Btu/lb.

**Carbon**                      $C + O_2 \rightarrow CO_2$
  Atomic wt:                    $12 + 32 = 44$
  Ratio:                        $1 + 2.667 = 3.667$
**Hydrogen**                    $2H_2 + O_2 \rightarrow 2H_2O$
  Atomic wt:                    $4 + 32 = 36$
  Ratio:                        $1 + 8 = 9$

The theoretical air required per 100 lb of refuse follows from these figures where air is considered to contain 23.15 percent oxygen.

$$\text{Air required} = [(28 \times 2.667)/0.2315] + [(0.6 \times 8)/0.2315] = 343.3 \text{ lb/100 lb refuse}$$

For incineration, furnace temperature must be controlled to minimize refractory maintenance. With no other provision for heat absorption, it is necessary to introduce excess air well beyond the needs for complete combustion, e.g., 140 percent of theoretical so that, in the example cited:

$$\text{Total air} = 2.4 \times 343.3 = 824 \text{ lb/100 lb refuse}$$

To summarize the quantities for a computation of furnace temperature, a materials balance is given in Table 2, equating the input to the furnace and output for 100 lb of refuse. In this tabulation, allowance is made for moisture in the air at a commonly accepted rate of 0.0132 lb/lb of dry air. Some residue quench water will be evaporated, and the moisture added to the flue gases is estimated at 5 lb for each 100 lb of refuse burned. Since the assumed analyses are not precise, an exact balance is not obtained, but the indicated computations are sufficiently accurate for incinerator design.

In Table 2, total air is broken down into oxygen and nitrogen on the basis that 23.15 percent of the air is oxygen. To compute the air in the "output," or flue gas, the nitrogen is the same as the "input." Oxygen is diminished by the amount consumed in combustion. Since carbon and hydrogen unite

**Table 2. Materials Balance for Furnace**
(In lb/100 lb of refuse)

| Input: | | | |
|---|---|---|---|
| Refuse | | | |
| Combustible material | | | |
| Cellulose | 52.74 | | |
| Oils, fats, etc. | 5.86 | 58.6 | |
| Moisture | | 22.4 | |
| Noncombustible | | 19.0 | 100.0 |
| Total air, at 140% excess air | | | |
| Oxygen | 191.0 | | |
| Nitrogen | 633.0 | | 824.0 |
| Moisture in air | | | 11.0 |
| Residue quench water | | | 5.0 |
| Total | | | 940.0 |
| Output: | | | |
| $CO_2$ ($28 \times 3.667$) | | | 102.7 |
| Air—Oxygen ($191 - 80$) | 111.0 | | |
| Nitrogen | 633.0 | | 744.0 |
| Moisture | | | |
| In refuse | | 22.4 | |
| From burning cellulose | | 29.3 | |
| From burning hydrogen | | 5.4 | |
| In air | | 11.0 | |
| In residue quench water | | 5.0 | 73.1 |
| Noncombustible material | | | 19.0 |
| Unaccounted for | | | 1.2 |
| Total | | | 940.0 |

with oxygen during combustion, the oxygen consumed per 100 lb refuse is:

For carbon, $28 \times 2.667 = 74.68$ lb
For hydrogen, $0.6 \times 8 = 4.8$ lb
Total                         $79.48$, say 80 lb

The moisture from burning cellulose and hydrogen is: for cellulose, $0.555 \times 52.74 = 29.3$ lb; for hydrogen, $9 \times 0.6 = 5.4$ lb.

The **heat-balance computation** in Table 3 is predicated on base temperature = 80°F; vapor enthalpy = 1,048 Btu/lb; fly-ash temperature leaving furnace = 1680°F; fly ash = 2 percent of total refuse burned; residue is noncombustible introduced with the refuse; specific heat of fly ash and residue = 0.25. Total moisture in the flue gas is the sum of moisture in the refuse, bound water in fuel, moisture in the air at 0.0132 lb/lb of air, and water evaporated in quenching the residue. Heat loss through the furnace enclosure depends largely on furnace construction and is estimated as 3 percent of the total heat input.

**Flue-gas temperature** is calculated from the data of Tables 2 and 3 and the enthalpy data of Fig. 2 as follows:

| | | |
|---|---|---|
| Input at 1680°F (Table 3) | | 511,530 Btu |
| Losses: | | |
| Vaporization at 80°F, lb | | |
| Moisture in refuse | 22.4 | |
| Bound water (Table 2) | 34.7 | |
| Air moisture | 11.0 | |
| Residue quenching | 5.0 | |
| | 73.1 | |
| $73.1 \times 1,048$ | | 76,609 |
| Fly-ash carry-over (Table 3) | | 800 |
| Residue (Table 3) | | 475 |
| Through furnace walls (Table 3) | | 15,346 |
| Unaccounted for (Table 3) | | 400 |
| Total loss | | 93,630 |
| Balance to heat flue gas | | 417,900 Btu |

Enthalpy = $417,900/(846.7 + 73.1) = 455$ Btu/lb. Percent moisture, for use with Fig. 2 = $(73.1 \times 100)/(846.7 + 73.1) = 7.95$.

From Fig. 2, the gas temperature is 1680°F. This is satisfactory and indicates that the assumed excess air of 140 percent is reasonable.

## FLUE-GAS TEMPERING

In the design of an incinerator, consideration must be given to the condition of the gas discharged to atmosphere. Flue gas leaving the furnace contains appreciable fly ash, a substantial portion of which must be removed in order to meet the requirements of current air-pollution-control regulations. Equipment for fly-ash removal may be (1) wet type, or so-called wet scrubber; or (2) dry type, using electrostatic precipitators, requiring gas temperatures below approximately 600°F. With either type of equipment on refractory-walled furnaces, gas temperatures are reduced by evaporation of water and a computation of water quantity is required.

Table 5 gives the requisite heat-balance calculations for a *spray chamber* where flue gas is cooled to 600°F by water evaporating from sprays. There will be some air leakage into the chamber, estimated as 10 percent of the dry flue gas, or 85 lb of dry air. With this air, there will be 1.12 lb of moisture. Some minor losses will occur in sluicing the fly ash.

**Table 3. Heat Balance for Furnace**
(In Btu/100 lb of refuse)

| | | |
|---|---:|---:|
| Heat input: | | |
|   Refuse, 100 × 5,000 | | 500,000 |
|   Air moisture, 11 × 1,048 | | 11,530 |
|     Total | | 511,530 |
| Heat output: | | |
|   Heat of dry gas at 1680°F, 847 × 424 (Fig. 2) | 359,128 | |
|   Heat in water vapor, 73.1 × (1,900 − 48) (from steam tables) | 135,381 | |
|   Heat in fly ash, assuming 2 lb/100 lb of refuse and specific heat of 0.25, 2 × 0.25 × (1,680 − 80) | 800 | |
|   Heat in residue, 19 × 0.25 × (180 − 80) | 475 | |
|   Loss through furnace enclosure at 3% of total input | 15,346 | |
|   Unaccounted for | 400 | |
|     Total | | 511,530 |

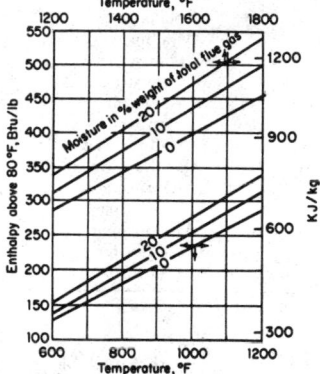

**Fig. 2** Enthalpy of flue gas above 80°F.

The computation in Table 4 shows that there will be total heat energy of 358,908 Btu to be absorbed by moisture in flue gas leaving the furnace and moisture evaporated from cooling sprays, or total moisture = 358,908/(1335 − 48) = 278.9 lb. The spray water added = 278.9 − 73.1 = 205.8 lb.

## RECOVERY

**Salvage**  A number of attempts have been made to salvage portions of the refuse received at disposal plants to help offset operating cost. Under favorable conditions, those efforts have met with some success. In general, the lack of a stable market and the cost of labor and equipment involved have made salvage unattractive to date.

With the current national interest in resource recovery, several demonstration-scale mechanized material-separation schemes in operation, and extensive continuing pilot-scale work being conducted by the U.S. Bureau of Mines, conditions may change in the future relative to salvage of materials from municipal solid wastes.

**Fly Ash and Residue**  Fly ash has been used to a limited extent as a concrete additive and as a base for fertilizer. Incinerator residue is widely used for land reclamation in low areas and, in some cases, as a road-base material. Additional potential uses are being developed by the U.S. Bureau of Mines and others.

**Heat Recovery**  Perhaps the most potentially attractive form of recovery, or extraction, of resources from the volume of municipal solid wastes is recovery of waste heat from the incineration process.

Several options exist when considering recovery of waste heat from incineration. These options include mass burning in a refractory-walled furnace with a waste-heat boiler inserted in the flue downstream; mass burning in a water-walled furnace with the convection surface immediately downstream; refuse preprocessing and separation of the combustible fraction with combustion taking place in a utility-type boiler partially in suspension and partially on a grate; and pyrolysis of preprocessed refuse. A list of plants in North America in start-up or operation as of the second half of 1975 extracting energy from combustion of municipal-type waste is shown below. This list excludes plants built for developmental or experimental purposes and plants that utilize specialized industrial wastes.

**Table 4**

| | | Units | | |
|---|---|---|---|---|
| Plant location | No. | Size, (tons/day) | Status | Type |
| Ames, IA | | 200 | Start-up 1975 | Refuse processing w/suspension burning |
| Baltimore, MD | 1 | 1,000 | Start-up 1975 | Pyrolysis |
| Braintree, MA | 2 | 120 each | Oper. 1970 | Mass burning |
| Chicago, IL | | | | |
|   Southwest | 4 | 300 each | Oper. 1962 | Waste-heat boiler |
|   Northwest | 4 | 400 each | Oper. 1970 | Mass burning with water-wall furnace |

**Table 4**   *(Continued)*

| Plant location | Units | | | |
| | No. | Size, (tons/day) | Status | Type |
| --- | --- | --- | --- | --- |
| Hamilton, Ontario | 2 | 600 each | Oper. 1972 | Refuse processing w/suspension burning |
| Harrisburg, PA | 2 | 360 each | Oper. 1973 | Mass burning with water-wall furnace |
| Hempstead, NY |  |  |  |  |
| Merrick | 4 | 150 each | Oper. 1952 | Waste-heat boilers |
| Oceanside | 2 | 350 each | Oper. 1974 | Mass burning with water-wall furnace |
| Miami, FL (20th St.) | 6 | 150 each | Oper. 1954 | Waste-heat boilers |
| Montreal, Quebec | 4 | 300 each | Oper. 1971 | Mass burning with water-wall furnace |
| Nashville, TN | 2 | 360 each | Oper. 1947 | Mass burning with water-wall furnace |
| Quebec City, Quebec | 4 | 250 each | Oper. 1974 | Mass burning with water-wall furnace |
| Saugus, MA | 2 | 600 each | Start-up 1975 | Mass burning with water-wall furnace |

Figure 3 illustrates a plant with heat recovery.

In considering the above options, one should take into account the overall energy balance in the various systems. These systems can be grouped under the following general categories: burning as-received refuse; burning mechanically processed refuse; burning thermochemically processed refuse; and burning biochemically processed refuse. In all the processing systems, less heat will be available for use than there was prior to processing.

The tabulation below from published data regarding the production of a fuel gas from refuse will partially illustrate the net energy loss in converting the available energy to another form:

| Composition | % by volume | % by weight | % of total carbon |
| --- | --- | --- | --- |
| $CO$ | 47 | 62.1 | 70 |
| $H_2$ | 33 | 3.1 |  |
| $CO_2$ | 14 | 29.1 | 21 |
| $CH_4$ | 4 | 3.0 | 6 |
| $C_2H_2$ | 1 | 1.4 | 3 |
| $N_2$ | 1 | 1.3 |  |

Note that 21 percent of the carbon is in $CO_2$ which will not

**Table 5. Heat Balance for Spray Chamber**
(In Btu)

| | | |
| --- | --- | --- |
| Input at 1680°F (from Table 3): | | |
| Heat of dry gas | 359,128 | |
| Heat in water vapor | 135,381 | |
| Heat in fly-ash carry-over | 800 | |
| Heat unaccounted for | 400 | |
| Total | | 495,709 |
| Output at 600°F: | | |
| Heat in dry gas $(847 + 85) \times 128$ | 119,296 | |
| Heat in air-leakage moisture, $1.12 \times (1335 - 48)$ | 1,441 | |
| Heat loss through walls at 3% input | 14,871 | |
| Minor losses from sluicing at 1% of 119,296 | 1,193 | |
| Heat in vapor from furnace and spray water | 358,908 | |
| Total | | 495,709 |

| Process | Energy loss, % | | | Total net available energy Btu/lb. |
|---|---|---|---|---|
| | Processing | Combustion | Total | |
| As received | 1 | 39 | 40 | 2,640 |
| Dry shredding | 18 | 30 | 48 | 2,288 |
| Wet shredding | 35 | 21 | 56 | 1,936 |
| Pyrolysis, oil | 62 | 9 | 71 | 1,276 |
| Pyrolysis, gas | 32 | 25 | 57 | 1,892 |
| Pyrolysis with oxygen | 37 | 15 | 52 | 2,112 |
| Anaerobic digestion | 72 | 6 | 78 | 968 |

burn, while 70 percent of the carbon is in CO where 30 percent of the elemental energy in carbon is no longer available. While this heat is not wasted, it is lost energy not available to do further work.

A tabulation of energy losses and total net available energy, based on information published in 1974–1975 for refuse with an initial heat content of 4,400 Btu/lb, is given above.

Of the 4,400 Btu/lb in the refuse as received, the tabulated data indicate the useful energy that may be made available through combustion. While the data are not absolute, the relative magnitudes are meaningful, provided similar degrees

of design efficiency and sophistication of control are used for each process.

Other factors to consider in selection and design of heat-recovery facilities include efficiency of boiler facilities, furnace-chamber design, and combustion air supply. While in older plants with waste-heat boilers installed in downstream flues, steam production averaged 1.5 to 1.8 lb/lb of refuse, in newer water-walled furnaces and suspension-fired units, steam production is of the order of 3.0 lb/lb of refuse. The lower efficiency in waste-heat boiler units is due to higher heat losses in the plant stack effluent, in turn caused by higher

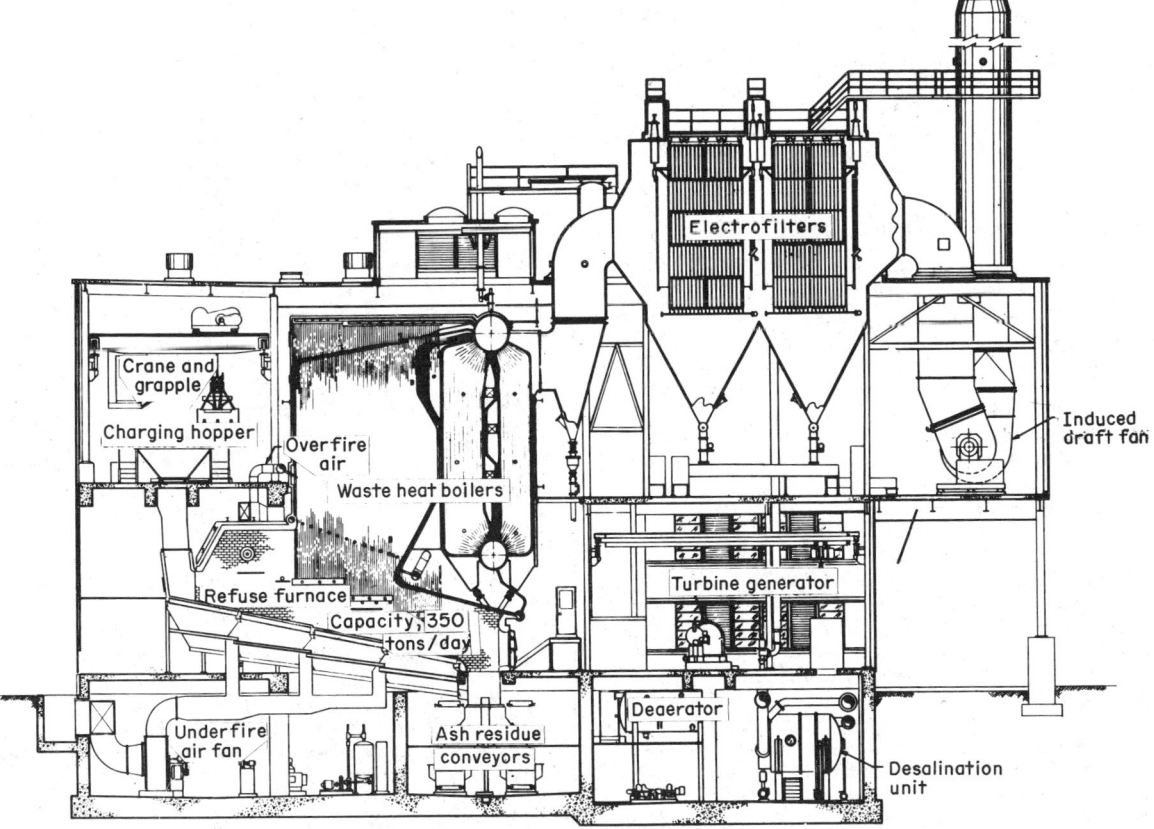

**Fig. 3** Section through one of the refuse furnaces, waste-heat boilers, and electrofilters at the Oceanside Incinerator, town of Hempstead, NY. *(Charles R Velzy Associates, Inc.)*

excess air levels required to control combustion temperatures properly in the refractory-lined primary furnace enclosure.

In most water-walled furnaces and furnaces in which shredded combustible refuse fractions are burned, the usual configuration is a tall primary chamber with the gases passing out the top and into the convection boiler surface after completion of combustion. It has been found desirable in mass-burning water-wall plants to coat a substantial height of the primary combustion chamber (where boiler-tube metal temperatures will exceed 500°F) with a refractory material and to limit average gas velocities to under 15 ft/s. Gas velocity entering the boiler convection bank should be less than 30 ft/s. Water-table studies have been found to be very useful in checking combinations of furnace configurations and introduction of combustion air.

In mass burning, combustion air is usually supplied from both under and over the grate. This is not necessarily the case when burning prepared refuse. In a water-wall mass-burning type of furnace, overfire air is utilized to enhance turbulence and mixing of combustion gases with the combustion air, and for completion of combustion. Accordingly, this air is best introduced through numerous relatively small (1½- to 3-in-diameter) nozzles, at pressures of 20 in of water and higher. Ideally, provision should be made for the introduction of the overfire air at several different elevations in the furnace.

As this nation's energy needs become more critical in the future, this readily available source of energy will be tapped more frequently. The technology is available now for successful application of these techniques if provision is made for adequate funding and properly trained operating staffs.

# ELECTRIC FURNACES AND OVENS

## by William E. Lewis

REFERENCES: Robiette, "Electric Melting and Smelting Practice," Griffin. Campbell, "High-Temperature Technology," Wiley. "Electric-Furnace Steel Proceedings," Annual, AIME. Paschkis, "Industrial Electric Furnaces and Appliances," Interscience. Stansel, "Induction Heating," McGraw-Hill. Ess, The Modern Arc Furnace, *Iron Steel Eng.*, Feb. 1944.

## CLASSIFICATION AND SERVICE

In **resistor furnaces and ovens** heat is developed by the passage of current through distributed resistors (heating units) mounted apart from the charge. Alternating current of a standard power frequency is used. The furnace service is for heat applications to solids such as heat treatment of metals, annealing glass, and firing of vitreous enamel. Oven service is limited to drying and baking processes usually below 500°F (260°C).

In **induction heaters** heat is developed by currents induced in the charge. The service is heating metals to temperatures below the melting points.

In **induction furnaces** heat is developed by currents induced in the charge. The service is melting metals and alloys.

In **arc furnaces** heat is developed by an arc, or arcs, drawn either to the charge or above the charge. Direct-arc furnaces are those in which the arcs are drawn to the charge itself. In indirect-arc furnaces the arc is drawn between the electrodes and above the charge. A standard power frequency is used in either case. The general service is melting and refining metals and alloys.

In **resistance furnaces** of the submerged-arc type, heat is developed by the passage of current from electrode to electrode through the charge. The manufacture of basic products, such as ferroalloys, graphite, calcium carbide, and silicon carbide, is the general service. Alternating current at a standard power frequency is used. An exception is the use of direct current where the product is obtained by electrolytic action in a molten bath, e.g., in the production of aluminum.

The **characteristics of electric heat** are:

1. Precision of the control of the development of heat and of its distribution.

2. The heat development is independent of the nature of the gases surrounding the charge. This atmosphere can be selected at will with reference to the nature of the charge and the chemistry of the heat process. This freedom is often a primary reason for the use of electric heat.

3. The maximum temperature is limited only by the nature of the material of the charge.

The first two characteristics underlie the design of all electric heating apparatus. The third is utilized in thermal processes for the production of certain materials not obtainable in any other way.

### Resistor Furnaces

Resistor furnaces may be either the batch or the continuous type. Batch furnaces include box furnaces, elevator furnaces, car-bottom furnaces, and bell furnaces. Continuous furnaces include belt-conveyor furnaces, chain-conveyor furnaces, rotary-hearth furnaces, and roller-hearth furnaces.

Standard resistor furnaces are designed to operate at temperatures within the range 1000 to 2000°F (550 to 1200°C). For higher heating chamber temperatures, see Resistors, Sec. 7.

The **heating chamber** of a standard furnace is an enclosure with a refractory lining, a surrounding layer of heat insulation, and an outer casing of steel plate, or for large furnaces an outer layer of brick or tile, as indicated by Figs. 1 and 2. The hearth of a batch furnace often is constructed of a heat-resisting alloy, made in sections to prevent warping. In some continuous furnaces the conveyor forms the hearth; in others a separate hearth is required.

Insulating firebrick—a semirefractory material—is commonly used for the inner lining of the heating chamber. This material has thermal and physical properties intermediate between those of fire-clay brick and heat-insulating materials. A lining of this kind has less heat-storage capacity than a fire-

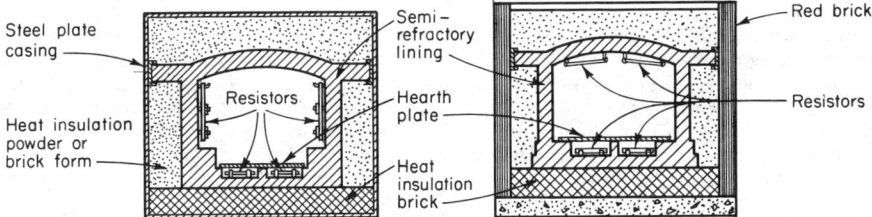

**Fig. 1** Heating chamber with sidewall and hearth resistors.    **Fig. 2** Heating chamber with roof and hearth resistors.

clay brick lining, and its use accordingly decreases the time periods of heating and cooling the chamber and also decreases the stored-heat loss for a given cycle of operation. Other advantages are its high heat-insulating value and light weight.

The maximum temperature of the inner face of the layer of heat insulation determines the character of material required for the insulation. Practically all resistor furnaces have insulation made of diatomite. Composite wall structures with a 4½-in (11-cm) semirefractory lining and a 9- to 13-in (23- to 33-cm) layer of heat insulation represent general practice for standard furnaces.

**Atmospheres** A mixture of air and the gases evolved from the charge constitutes a **natural atmosphere** in the heating chamber of a resistor furnace. The composition of such an atmosphere in a batch furnace is variable during a heating cycle. A natural atmosphere in the heating chamber of a continuous furnace is mainly air. Natural atmospheres are used where the extent of the action of oxygen on the charge during the heating cycle is not objectionable and for processes where that chemical action is desired.

The basis of an **artificial atmosphere** is the exclusion of oxygen (air) from the heating chamber by the substitution of some other gas or mixture of gases. This gas or mixture of gases is selected with reference to the chemical activity of that atmosphere on the charge at the temperature of the heat application. A definite chemical action may be desired, for example, the reduction of any metallic oxide present on the charge, or it may be required that the artificial atmosphere be chemically inactive. Thus artificial atmospheres are divided into (1) active or process atmospheres and (2) inactive or protective atmospheres. The term "controlled" atmosphere refers generally to a protective atmosphere, but it also includes artificial atmospheres of some degree of chemical activity. An example of a process atmosphere is the use of a hydrocarbon gas to carburize steel. Examples of controlled atmospheres are: the bright annealing of metals, the prevention of decarburization of steel during a heat application, the use of a reducing gas (hydrogen or carbon monoxide) in a copper brazing furnace, etc. In this last example the reducing gas serves to clean the faces of the joint to be made (by removal of any oxide present) and to maintain that cleanliness during the operation. The primary gases for controlled atmospheres are hydrogen and carbon monoxide and nitrogen. See also Sec. 7.

The main uses of controlled atmospheres are (1) the prevention of the formation of oxides on the material of the charge, or conversely the reduction of any oxides present, and (2) the prevention of a change in the carbon content of a steel undergoing a heat treatment. Each of these uses denotes a chemical system in which the reactions are reversible.

The chemical systems relating to metallic oxides are:

*A.* Oxide + hydrogen ⇌ metal + water vapor
*B.* Oxide + carbon monoxide ⇌ metal + carbon dioxide

The chemical systems relating to carbon in steel are:

*E.* Methane ⇌ hydrogen + carbon
*F.* Carbon monoxide ⇌ carbon dioxide + carbon

In artificial atmospheres the volume ratio of the two gases in the heating chamber should be so maintained as to correspond to the desired direction of the chemical activity of the system, or, if no chemical action is desired, to maintain that volume ratio at (or near) its equilibrium value for the temperature of the heat application. The equilibrium volume ratios for each of the four chemical systems *A*, *B*, *E*, and *F* for carbon steel over the usual range of temperature of heat-treatment processes and for atmospheric pressure are shown in Fig. 3. There is little tendency toward a change in the carbon content of a steel below the critical range. Oxidation is active down to about 1100°F (650°C).

Curves *E* and *F* of Fig. 3 show the volume ratios of systems *E* and *F* for equilibriums with graphite. The equilibrium volume ratios of these two chemical systems for carbon in solid solution in steel (austenite) depend in each case on the carbon content of the steel. For the methane-hydrogen-carbon system (*E*) the volume ratio of the two gases at equilibrium with carbon in an unsaturated steel at a given temperature is less than the value shown by curve *E*. For the carbon monoxide-carbon dioxide-carbon system (*F*) the volume ratio of the two gases at equilibrium with the carbon in low- and medium-carbon steel at a given temperature is somewhat greater than the value shown by curve *F*; for high-carbon steels the equilibrium volume-ratios approach the values of curve *F*.

In the case of the hydrogen-iron oxide reaction, curve *A*,

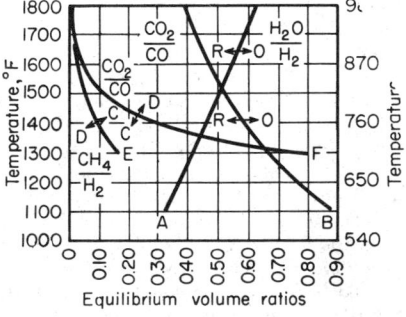

**Fig. 3** Equilibrium volume ratios of the chemical systems *A*, *B*, *E*, and *F* for steel. *C* = carburizing condition; *O* = oxidizing; *D* = decarburizing; *R* = reducing.

the water-vapor content of the mixture of gases at equilibrium decreases with decrease of temperature. Hence if a steel is to be cooled in a controlled atmosphere of this kind the permissible water-vapor content of the controlled atmosphere is dictated by the lowest temperature of the operation. The reverse is true of the carbon monoxide-iron oxide reaction, curve B. Thus if at a given temperature the carbon dioxide content of the mixture of carbon monoxide and carbon dioxide is less than the volume for equilibrium at that temperature it will be less than the volume for equilibrium at any lower temperatures and the steel can be cooled in that atmosphere without oxidation.

In the use of mixtures of the gases of the chemical systems noted to form controlled atmospheres for the heat treatment of steel, the interactions of the gases at elevated temperatures must be controlled by removal of all or nearly all the carbon dioxide and water vapor from the heating chamber.

The available data concerning controlled atmospheres for the protection of alloy steels during heat-treatment processes indicate that the technique for alloy steels is much the same as for carbon steels; i.e., a controlled atmosphere suitable for a carbon steel would, in general, be suitable for an alloy steel of the same carbon content.

In the heat treatment of nonferrous metals and alloys the use of either chemical system A or B requires for each oxide a knowledge of the equilibrium volume ratios of the chemical system used over the range of the operating temperature. Individual problems may arise. For example, copper can be bright-annealed in an atmosphere of dry steam—an inactive gas for this application—but the resultant staining of the copper during cooling may be objectionable. Copper usually contains a small percentage of oxide, and when annealing such copper in an atmosphere containing a reducing gas the temperature of the metal must be kept below about 750°F (420°C); otherwise the oxide will be reduced and the copper made brittle.

The foregoing discussion of atmosphere in heating chambers is intended to indicate the principles involved in the use of gases at elevated temperatures. The terms oxidation, reduction, carburization, and decarburization refer here to the chemical condition of a particular atmosphere and not to the extent of its effect on a charge. In all cases the concentration of the active gas or gases, time, temperature, in case of steel the carbon content and the gas pressure, and the catalytic action of hot surfaces within the chamber are important factors in the result obtained.

**Bath Heating**    Heating for local hardening of edge tools is the most general service. The lead-bath furnace has a working temperature range of 650 to 1700°F (360 to 950°C). The salt-bath furnace can be adapted to working-temperature ranges within a total range of 300 to 2350°F (170 to 1300°C) by the selection of suitable mixture of salts. The two salt baths most generally used are cyanide mixtures and chloride mixtures. The rate of heating by immersion is much faster than obtained by radiation. The rate of heat transfer in a salt bath is about one-half that in the lead bath. An additional use of the salt bath is for cyaniding, in effect a process atmosphere.

**Resistors**    The resistor of a standard furnace is a sinuous winding mounted on the inner surfaces of the heating chamber as shown in Figs. 1 and 2. The resistor winding covers practically the entire surface of the space chosen. Resistors are applied on the basis of 2 to 3 kW/ft² (20 to 30 kW/m²) of wall surface in general practice. The basis of resistor location is radiation to all surfaces of the charge. Hence, the height and width dimensions of the heating chamber indicate the choice between sidewall and roof resistors. In some cases both locations are used. Uniform distribution of heat flow to the charge is obtained by a designed distribution of the surfaces of the resistors supplemented by reradiation from the inner surfaces of the chamber.

Resistors for the majority of standard furnaces are made of 80 Ni, 20 Cr alloy. A nickel-chromium-iron alloy is used in some furnaces for operation only over the lower portion of the furnace temperature range. Both ribbon and cast shapes are in use. The effort in each case is to obtain the maximum surface area per unit length of resistor and at the same time retain sufficient mechanical strength in the resistor winding.

The 80 Ni, 20 Cr alloy is self-protecting against oxidation, but this protection decreases with rise of temperature. The operating temperature of a resistor should be no higher than is needed in each case and should always be at a safe margin below the softening point of the alloy, which is about 2500°F (1390°C). This corresponds to a maximum furnace temperature of about 2100°F (1170°C). The life of a resistor is also affected by the frequency of heating and cooling. Barring accidents, the resistor of a standard furnace under average conditions of operation has a long life, usually measured in years of service.

The nickel-chromium alloy resistor is used in artificial atmospheres as well as in natural atmospheres. This alloy is not resistant to compounds of sulfur and is affected to some extent by carbon monoxide.

The electric insulation of the resistor circuit is that of its refractory supports at elevated temperatures. This limits the voltage of the circuit to about 600 V. Small furnaces are usually designed for 110 V, medium sizes for 220 V, and the larger units for 440 V. Single phase up to 25 or 30 kW and three phase for higher ratings is general practice.

The resistivity-temperature coefficient of the nickel-chromium alloy permits the operation of resistors of this material on constant-voltage circuits. The rate of heat development in a resistor is proportional to the square of the applied voltage; hence maintenance of normal voltage is desirable. Voltage regulation is not as important as for other types of electrical apparatus because of the heat-storage capacity of the structure of the heating chamber. The power factor of the resistor circuit is practically unity.

**High-Temperature Furnaces**    **Silicon carbide** is the basis of a type of nonmetallic resistor for heating-chamber temperatures up to about 2800°F (1560°C). The material is formed into rods. Resistors of this material do not require protection against oxidation and are operated on constant-voltage circuits.

**Molybdenum** resistors are suitable for temperatures up to 3000°F (1670°C). Above that temperature the metal begins to vaporize. A molybdenum resistor cannot be operated in a natural atmosphere, and also it must be protected from reactions with silica and carbon. The metal is immune from reactions with sulfur compounds, nitrogen, and water vapor. Hydrogen is the most common artificial atmosphere used with molybdenum resistors. The difference between the cold and hot resistances of the circuit makes a starting device necessary.

Other materials used to some extent for resistors are iron, tungsten, and graphite. These require protection against oxidation.

**Temperature Regulation** The temperature of the heating chamber of a resistor furnace is in most cases regulated by a more or less intermittent application of current—the on-and-off method—which is made automatic by instrument control. This method utilizes the heat-storage capacity of the inner lining of the heating chamber as a temperature equalizer. The variation from the normal temperature of the chamber can be kept within less than 7°F (4°C), plus and minus, without undue wear of the temperature-control equipment. Temperature regulation by voltage control is equally applicable to resistor furnaces, and the trend is toward the use of this more accurate method particularly for the more important installations.

**Temperature protection** for resistor furnaces is obtained by means of a temperature fuse mounted in the heating chamber and connected in the control circuit of the power supply to the furnace.

**Multiple-Temperature Control** The resistors of the larger furnaces are divided into two or more circuits. Each circuit can be equipped with individual temperature control. That arrangement provides temperature regulation at more than one location in the heating chamber and is an aid toward maintaining uniform temperature distribution within the chamber.

The subdivision of resistor circuits is used also for zone heating—and zone cooling where needed—in continuous furnaces.

**Melting Pots** Resistor heating is applied to melting pots for the soft metals and alloys and for lead baths and for salt baths. The immersion heating unit is used for temperatures up to 950°F (530°C). For higher temperatures the metal pot is heated by resistors mounted outside and around the pot. The assembly in each case includes a heat-insulating wall similar to that of a resistor furnace. Another method of heating applicable only to salt baths is the passage of alternating current (of any frequency) between electrodes immersed in the bath.

**Tempering Furnaces** The temperature is comparatively low—below 1300°F (720°C). Electrically heated oil baths and salt baths are used for tempering many kinds of small parts. Another form of tempering furnace is a vertical resistor furnace with the addition of a removable metal cylinder (or basket) to contain the charge and to provide an annular passageway for the circulation of air (by a fan mounted on the furnace) over the resistors and thence through the charge—an application of forced convection heating.

**Sizes** The electrical rating is the general method of expressing the size of a resistor furnace. Sizes up to 100 kW

predominate, 100- to 500-kW furnaces are common, and others within the range 500 to 1,000 kW are in service. The data of Table 1 refer to common sizes of so-called box furnaces for general service.

The **losses** from a resistor furnace for a given heating chamber temperature are as follows: The open-door loss is a variable depending on the area of the door (or doors) and the percentage of the time that the door is open—from a continuous furnace this loss also varies with the type and speed of the conveyor; with artificial atmospheres, the loss of heat in the gases discharged for atmosphere control; the stored-heat loss, a variable that depends on the extent and frequency of the cooling of the furnace within a given period of operation; the heat dissipated from the outer surfaces of the furnace.

The **operating efficiency** is expressed as either pounds of material treated per kWh or kWh per ton. Representative values for average service for the heat treatment of steel range from 7 to 12 lb/kWh. Corresponding values for nonferrous metals and alloys are within the range 12 to 22 lb/kWh.

The general field of the batch furnace is defined by the following conditions: (1) intermittent and varied production; (2) long periods of heating (and in some cases slow cooling); (3) heating service beyond the range of the handling capacity of furnace conveyors; and (4) supplementary heating service. A continuous furnace is indicated where the flow of material to be heated is reasonably uniform and continuous, i.e., mass-production conditions. In some cases, batch furnaces with automatic charging and discharging equipment are essentially continuous furnaces.

### Resistor Ovens

The resistor oven is a modification of the resistor furnace to correspond to the low temperatures of drying and baking processes. The heating chamber is an insulated metal structure with a fresh-air inlet and an exhaust fan for ventilation (the removal of vapors and gases evolved from the charge). A refractory lining is not required. Ovens may be of the batch type with conventional methods of handling the charge or of the continuous type, usually with chain conveyors.

The most common type of electric oven is heated by resistors mounted in a separate compartment of the heating-chamber enclosure. The heat transfer is by forced convection which is accomplished by recirculation of the chamber atmosphere by a motor-driven fan.

A resistor oven with filament-type lamps as heating units provides what is generally known as infrared heating. The lamps, usually with self-contained reflectors, surround the

**Table 1. Box Resistor Furnaces, 1850°F (1030°C) Class**

| Connected load, kW | Power supply, 200 V | Lb of steel per h at 1500°F | Time in min to heat to 1500°F (830°C) when used previous day | Radiation in kWh per h at 1500°F | Inside Width | Inside Depth | Inside Height | Overall Width | Overall Depth | Overall Height, door closed | Overall Height, door open |
|---|---|---|---|---|---|---|---|---|---|---|---|
| 29 | 1-phase | 300 | 35 | 4.9 | 18 | 36 | 18 | 55 | 89 | 86 | 97 |
| 45 | 3-phase | 500 | 35 | 6.9 | 24 | 54 | 20 | 61 | 108 | 90 | 101 |
| 60 | 3-phase | 650 | 25 | 7.8 | 30 | 63 | 23 | 78 | 125 | 90 | 98 |
| 72 | 3-phase | 750 | 25 | 9.1 | 36 | 72 | 23 | 84 | 135 | 90 | 98 |

charge, and the heat transfer is by radiation, mainly in the infrared portion of the spectrum. This type of oven is best adapted to the continuous heating of charges which present a large surface area in proportion to the mass and which require only surface heating, e.g., baking finishes on sheet products.

**Ventilation**  The vapors and gases evolved from the charge during baking processes are often flammable, and the continuous discharge of these products from the oven chamber is essential for protection against explosions. For detailed recommendations, see *Pamphlet* 74 of the Assoc. Factory Mutual Fire Ins. Cos., Boston.

## Dielectric Heating

The term relates to the heat developed in dielectric materials, such as rubber, glue, textiles, paper, and plastics, when exposed to an alternating electric field. The material to be heated is placed between plate-form electrodes, as indicated in Fig. 4. It is not necessary that the electrodes be in contact with the charge; hence continuous heating is often practicable.

If the material of the charge is homogeneous and the electric field uniform, heat is developed uniformly and simultaneously throughout the mass of the charge. The thermal conductivity of the material is a negligible factor in the rate of heating. The temperatures and services are within the oven classification.

The frequency and voltage for this class of service depend in each case on the electrical properties of the material of the charge at the temperature specified for the heat application. The frequencies in use range from 2 to 40 MHz; the most common frequencies are from 10 to 30 MHz. It is advisable to select the frequency for heating by trial.

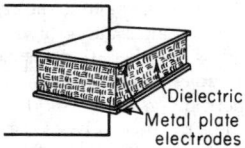

**Fig. 4**  Assembly for dielectric heating.

The upper limit of voltage across the electrodes is fixed by the spark-over value and by corona. The permissible voltage gradient depends on the nature of the material of the charge. Values within the range 2,000 to 6,000 V/in (790 to 2,400 V/cm) are found in practice; the voltage across the electrodes should not exceed 15,000 V.

Applications of dielectric heating include setting glue as in plywood manufacture, curing rubber, drying textiles, and the heat treatment of plastics.

## Induction Heating

In induction heating, the lateral surface of the charge is exposed to an alternating magnetic flux. The currents thus induced in the charge flow wholly within its mass. The term "eddy-current heating" is sometimes applied to the method.

A common assembly, if the charge is to be heated to a temperature below its melting point, is to place the charge within a coil as indicated in Fig. 5. An alternating current in the coil establishes the required alternating magnetic flux around the charge.

A peculiar feature of such assemblies, termed "induction heaters," is the absence of heat insulation; the coil is water-cooled. Thus, the charge is heated in the open air, or an artificial atmosphere can be used, if the assembly is enclosed. This requires rapid heating with heat cycles measured in minutes or seconds.

The frequency required is a function of the electric and magnetic properties of the charge at the temperature specified for the heat application and of the radius, or one-half the thickness, of the charge. This frequency for a given material increases with decrease of the dimension noted. The fre-

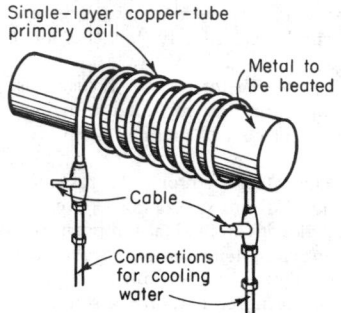

**Fig. 5**  Assembly for induction heating.

quency in any case is not critical. In practice 480, 960, 3,000, 9,600 Hz and around 450 kHz suffice for the entire range of induction heating. The highest frequencies needed are those for heating steel charges to temperatures above the Curie point. About ½ in (1¼ cm) diam in this case is the lower limit for 9,600 Hz. This limit dimension is decreased for steel heated to temperatures below the Curie point and for all charges of nonferrous materials.

The operation can be either batch heating or continuous heating as required. Applications include heating for forging, for annealing, for hardening steel, for brazing, soldering, and strain relief. As most of the heat is developed within the annular zone of the charge, the method is particularly well adapted to heating steel parts for surface hardening. A recent application of induction heating is the raising in temperature of billet-size ingots for rolling into merchant bars.

## Arc Furnaces

Two types of arc furnace are in common use: (1) the three-phase furnace and (2) the single-phase furnace. The general field of the three-phase furnace is the melting and refining of carbon and alloy steels; that of the single-phase furnace is the melting of nonferrous alloys. There is an increasing amount of arc-furnace capacity used for melting and refining various types of iron.

**Three-Phase Arc Furnaces**  The general design of this type of furnace is shown in Fig. 6. In operation, each heat is started by swinging the furnace roof aside and then loading the refractory-lined furnace body with scrap dropped from a crane-handled clamshell charging bucket. Arcs next are drawn between the lower ends of the graphite electrodes and the scrap; melting proceeds under automatic control until the hearth carries the molten metal. This fluidizing stage is effected at about 85 percent thermal efficiency. Several charges usually are needed to build up the bath—particularly in ingot practice. The furnace tilts forward for pouring; the back tilt serves in the removal of slag and permits the furnace hearth to be kept in proper condition. The working door is opposite to the pouring spout. Large furnaces frequently also have a side door.

**Refractories** Furnaces that produce foundry steels operate

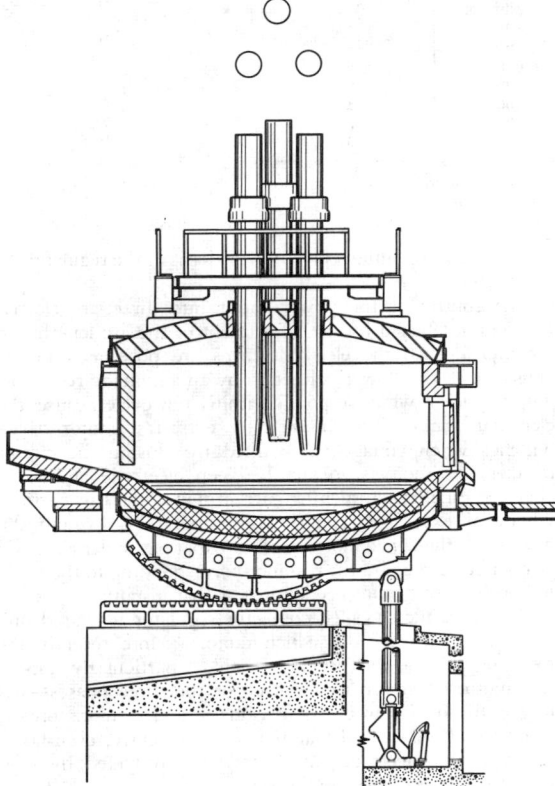

**Fig. 6** Three-phase arc furnace with basic lining.

with acid lining. This means silica brick form the walls; the hearth is of gannister or the equivalent. Silica-brick roofs are the more widely used although, for intermittent operation, clay brick may be preferred. The slags of acid-lining practice remove no phosphorus or sulfur. Essentially all ingot operations are carried on with basic linings. This means magnesite bottom and sidewalls, so that the limey slags employed will not erode them. Entry ports for the electrodes may be of extra-quality refractories to prolong roof life, particularly where the furnace is in continuous operation. In basic practice phosphorus joins the slag readily; sulfur can be removed next by a second slag, when this slag has been made highly reducing. Slag covering the molten bath serves in refining the metal and reduces the heating of wall and roof brick. Superrefractories find application in high-temperature, long-refining operations. Electric irons are made in acid-lined furnaces.

**Temperature** Arcs approximate 6300°F (3500°C); hence operation must be carried out so as to protect the refractories as much as possible. As the top-charge furnace now has supplanted the door-charge furnace in nearly all cold-melt work, the conditions for shielding the refractories during the melt-down stage of each heat are good. With the furnace filled

to the top with scrap, the electrodes bore down through that scrap, and the heat of the arcs is liberated right in the metallic charge itself. When the charge, and any back charges made, approach the fluid stage it is customary to reduce both the power input and the length of arcs employed. During the finishing stages, roof and sidewalls are protected both by the slag and by the "umbrella" effect of the electrodes themselves. Deserving mention is the expanding use of oxygen to gain speed in production, which makes for increasing furnace temperatures. The higher sidewalls of modern furnaces aid in obtaining good roof life.

**Charges** The three-phase arc furnace is primarily a unit for converting scrap charges into steel for pouring into ingots and castings. This type of equipment finds increasing use also in the cold melting and duplexing of gray and white irons. Hand and chute charging have practically disappeared, at least insofar as furnaces of a ton charge size upward are concerned. One of the main advantages of the top-charge furnace is that the scrap used does not need to be cut to door size, as was the case formerly.

Although first employed only for the more expensive grades of steel, the arc furnace now is used widely in making ingots for rolling into merchant bars and similar grades. The speed of production on this type of working—termed single-slag dephosphorizing basic practice—can be double that obtained with the same furnace used to make two slag dephosphorized and desulfurized basic alloy steels. Acid working on foundry steels generally approximates the same speed as single-slag dephosphorizing basic practice, and some alloy steels require about half again as much time. While most carbon steel for castings is made on an acid hearth, a basic bottom is regularly used for making manganese steels, for refining nickel and copper, and for the furnacing of many heat-resistant alloys. Sec. 6 discusses steel-foundry practice.

In general, approximately 320 kWh at 100 percent thermal efficiency will be needed to melt 1 ton of cold steel scrap. This means about 400 kWh will be needed to fluidize each ton. Additionally, about 100 kWh/ton will be needed to finish the heat and superheat the bath—this in the case of ordinary plain carbon steels made on single-slag acid or basic practice. Double-slag steel heats will require no more power than others for fluidizing the scrap charge, but the additional power needed for melting new slag, refining, melting added alloys, etc., may require as much as 250 kWh/ton of bath, or even more.

Three-phase arc furnaces are usually given an hourly productive rating in terms of acid foundry steels when these equipments are supplied in sizes up to and including the 11-ft (3.4-m) diam unit. However, with many furnaces extra-powered, quite a few shops exceed the normal hourly rating considerably—in some cases by essentially 100 percent. Representative sizes of furnaces are listed in Table 2.

**Arcs** The arc in each phase is maintained between the lower end of the electrode and the top of the charge (or bath, after the molten state is reached). Higher voltages can serve for melting as the size of the furnace increases; thus, where a 7-ft (2.1-m) diam furnace employs 215 V as its highest melting potential, a 15-ft furnace would use 290 V or higher as the top tap. The furnace transformer is universally of the motor-operated tap-changer type, and in the case of, say, a 10,000 kVA at 55°C rise substation, a secondary voltage variation of more than 150 V is customary. The range of lower voltages used for refining the molten metal is obtained by changing the

**Table 2. Sizes of Three-Phase Arc Furnaces**

| Diam of shell, ft | Normal charge, tons | Normal powering, kVA | Normal productive rate, single-slag steels, tons/h* |
|---|---|---|---|
| 5 | 1½ | 600 | ½ |
| 7 | 3½ | 1,500 | 1½ |
| 9 | 8 | 3,000 | 3 |
| 11 | 16 | 6,000 | 6 |
| 12½ | 27 | 9,000 | 9 |
| 15 | 50 | 12,500 | 13 |
| 20 | 115 | 25,000 | 27 |
| 24 | 225 | 36,000 | 40 |

*Many users exceed these outputs, particularly those using burners and oxygen to speed operations.

primary of the main transformer from delta to star connection; this reduces both voltage and capacity to 58 percent of their values with delta primary connection. If, say, 12 tons of steel scrap are to be melted down to fluid in 1 h, then the electrical energy needed will approximate 5,000 kWh. With 245 V used as the principal melt-down voltage, the current per phase will have to average close to 12,000 A. A 12-in (30-cm) diam graphite electrode would amply carry this current. Small furnaces operate with 600 kVA and even higher powering per ton of charge, whereas in the case of the larger equipments the electrical backing of the furnace normally does not exceed 300 kVA/ton of charge.

**Reactance** is required in the circuits of an arc furnace to give stability and to limit the current when an electrode makes contact with the metallic charge. The inherent reactance (impedance) in the instance of 10,000-kVA installations and above normally is sufficient. The total stabilizing reactance provided in the case of a 1,000-kVA load normally approximates 30 percent.

**Regulation** The characteristics of an arc furnace circuit for a given applied voltage are shown in Fig. 7. For each voltage there is a value of current that gives maximum power in the

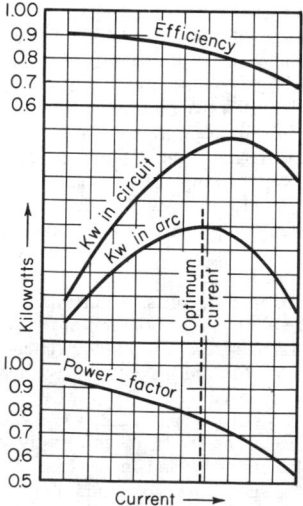

**Fig. 7** Characteristics of an arc-furnace circuit.

furnace. This optimum current is the basis of the regulation of the circuit.

The control of the power input into direct-arc electric furnaces is effected by the adjustment of the arc length. To accomplish this, the electrode arms are positioned in the "raise" or in the "lower" direction by an automatic regulator. This regulator, which responds within a few cycles, causes the electrode arms to be lowered by extra-fast motor-driven winches when voltage is obtained by closing the circuit breaker. As soon as contact between electrode and scrap charge is established, melting current flows, and this current, whenever excessive, functions immediately through the medium of the winch motor to elevate that particular electrode arm and electrode by the distance corresponding to the diminution in power input needed just at that instant.

Formerly, the so-called contactor regulator was used universally to energize the winch motors. More recently the rotary regulator—this, in effect, being a particularly responsive motor generator set for each of the three phases—has forged to the forefront by reason of giving more precise control with minimized maintenance. Currently, even faster response and electrode-travel speed are provided by low-inertia static-regulating equipment.

**Single-Phase Arc Furnaces** Single-phase arc furnaces usually are manufactured in the two-electrode type. When the electrodes operate vertically, the furnace melts much as a three-phase direct-arc furnace does. However, most vertical-electrode single-phase furnaces are of laboratory size—that is, up to 150 kVA in powering.

When two electrodes are mounted horizontally in a rocking furnace an indirect-arc unit is obtained. Many rocking furnaces serve well in the melting of brasses, bronzes, and in similar work. Volatiles are reincorporated in the metal since the bath washes over much of the interior of a rocking furnace. The oscillation approximates 200°.

Rocking furnaces usually do not exceed 500 kW in powering. A single operating voltage can suffice. In regulating a rocking furnace, only one electrode need be movable, on a carriage under automatic control, to maintain the requisite amperage by varying the length, and therefore the resistance, of the arc gap.

### Induction Furnaces

There are two basic types of metal-melting induction furnaces: (1) coreless and (2) core-type. Both types utilize the principle of a transformer. The high-voltage circuit is coupled with that of the low voltage without directly connecting the two circuits. The element responsible for this coupling effect is the magnetic field. Induction heating utilizes the property of the magnetic field, which enables heat to be transferred without direct contact. By correctly disposing the high-voltage winding, which in the case of the induction furnace would be an induction coil or inductor, the magnetic field is directed so that the metal to be heated or melted is made to absorb energy. The temperature attainable is limited solely by the resistance to heat of the surrounding lining material. Induction heating enables any temperature to be achieved while providing for excellent regulation of temperature and metallurgical properties. Any metal which will conduct electric current can be melted in an induction furnace.

**Coreless Induction Furnaces** (See Fig. 8.) This type of furnace consists of a crucible, copper coil, and framework on supports arranged for tilting and pouring. The specially designed induction coil acts as the primary of the transformer. The crucible conforms to conventional refractory practice. A rammed crucible is used for furnaces above 50 kW, and preformed crucibles are used on smaller furnaces such as laboratory units.

The principle of operation is essentially the same as that of the induction heater previously described. The initial charge in the furnace is cold scrap metal—pieces of assorted dimensions and shapes and a large percentage of voids. As the power is applied and the heat cycle progresses, the charge changes to a body of molten metal; additional cold metal is added until the molten-metal level is brought to the desired temperature and metallurgical chemistry. The furnace then is tapped.

When the metal in the furnace becomes fluid, depending on whether a line frequency or medium-frequency supply by means of convertors is used, a certain electromagnetic stirring action will occur. This stirring action is peculiar to the induction furnace and aids in the production of certain types of alloys. The stirring action increases as the frequency is reduced.

Line-frequency applications are generally reserved to fur-

naces having a metal-holding capacity of 800 lb (360 kg) and above. There is always an ideal relationship between the size of a coreless furnace and its operating frequency. As a general rule, a small furnace gives best results at high to medium frequencies and large furnaces work best at the lower frequencies. A frequency is suited to a given furnace when it yields good, fast melting with a gentle stirring action. Too high or too low frequencies are accompanied by undesirable side effects. The tabulation below gives the charge weights and frequencies generally to be used:

| Charge weight, lb | Frequency, Hz |
|---|---|
| 2-50 | 9,600 |
| 12-500 | 3,000 |
| 200-15,000 | 960 |
| 800-75,000 | 60 |

The coreless induction furnace is usually charged full and tapped empty, although at line frequencies, it may be necessary to retain a certain amount of metal in the furnace to continue the operation, since it is difficult to start the furnace with small metal particles, such as turnings and borings, in a cold crucible. As a result, it is general practice to retain a heel in the furnace of about one-third its molten-metal volume. This problem can be avoided in furnaces of higher frequencies, where start-up can be performed with small-size metal charges without carrying the heel.

Coreless induction furnaces are particularly attractive for melting charges and alloys of known analysis; in essence, the operation becomes one of metal melting with rapidly absorbed electric heat without disturbing the metallurgical properties of the initial charge.

These furnaces are supplied from a single-phase source. In order to obtain a balanced three-phase input, it is necessary specifically to design the electrical equipment for the inclusion of capacitors and suitable reactors, which are generally automatically switched (by inductance changes) during the operation in order to provide a reasonably high power factor. Power factors on such furnaces can be kept at or near unity. In high-frequency coreless induction furnaces, high power factors are necessary to prevent overburdening the motor-generator equipment.

**Core-Type Induction Furnaces** (See Fig. 9.) The transformer is actually wound to conform to a typical transformer design having an iron core and layers of wire acting as a primary circuit. The melting channel acts as a ring short circuit around this transformer in the melting chamber. According to the desired melting capacity, one, two, or three such transformers (or **inductors,** as they are called) may be added to the furnace shell. At all times, the channel must hold sufficient metal to maintain a short circuit around the transformer core. Air cooling is used as required to prevent undue heating of the inductor coils and magnetic cores.

The melting output is controlled by varying the voltage supplied to the inductors with the aid of a variable-voltage transformer connected to the primary circuit of the supply. Core-type furnaces always use line frequencies. Voltage or power-input regulation, therefore, can be performed by adjusting the tap setting of the transformer feeding the furnace transformer attached to the furnace shell. These transformers are single-phase units, and by using three such units, a balanced three-phase input can be obtained. The current flowing

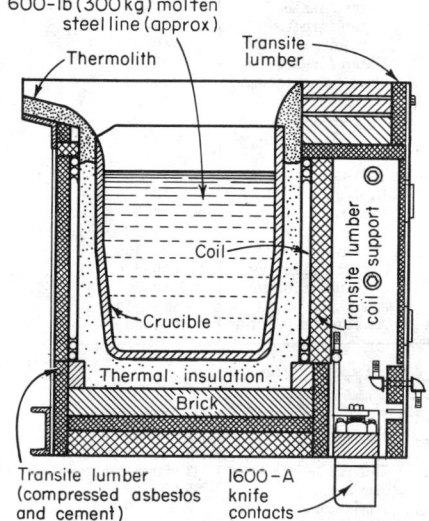

600-lb (300 kg) molten steel line (approx)

Thermolith

Transite lumber

Coil

Transite lumber coil support

Crucible

Thermal insulation

Brick

Transite lumber (compressed asbestos and cement)

1600-A knife contacts

**Fig. 8** Coreless induction furnace.

through the primary inductors by transformation causes a much larger current in the metal loop, whose resistance creates heat for melting.

The core-type furnace is the most efficient type of induction furnace because its iron core concentrates magnetic flux in the area of the magnetic loop, ensuring maximum power transfer from primary to secondary. Efficiency in the use of power can be as high as 95 to 98 percent.

The essential loop of metal must always be maintained in the core-type furnace. If this loop is allowed to freeze by cooling, extreme care is necessary in remelting because the

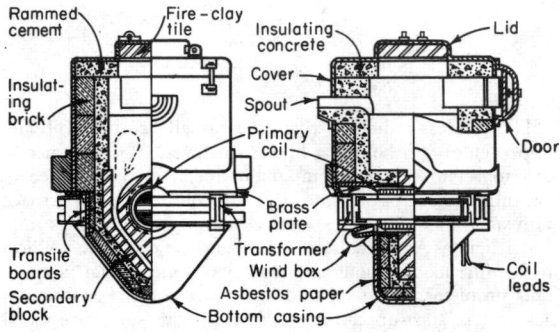

**Fig. 9**  Core-type induction furnace.

loop may rupture and disrupt the circuit. This could require extensive work in dismantling the coil and restoring the loop.

Consequently, core-type furnaces rarely are permitted to cool. This makes alloy changes difficult because a heel of molten metal always is required.

The relatively narrow melting channels must be kept as clean as possible since a high metal temperature exists in this loop. Nonmetallics or tramps in the charge metal tend to accumulate on the walls in the channel area, restricting the free flow of metal and ultimately closing the passage.

This furnace is particularly useful for melting of nonferrous metals such as aluminum, copper, copper alloys, and zinc.

**Power Requirements for Electric Furnaces**

The **energy required for melting metals** in electric furnaces varies for a given metal or alloy with the size of the furnace, the thickness of the refractory lining, the temperature of the molten metal, the rate of melting, and with the degree of the continuity of the operation of the furnace. An estimated efficiency of 50 to 60 percent is often used for preliminary purposes. As is well known, 3- to 6-ton direct-arc furnaces often are used to tap acid foundry steels with the consumption of less than 500 kWh to the ton, and large ingot furnaces of this same type, operating basic-lined on common steels for ingots, give even better results despite the call for several more charges of scrap per heat.

Average values in kWh/ton of molten metal are as follows: yellow brass, 200 to 350; red brass, 250 to 400; copper, 250 to 400; lead, 30 to 50; steel melting, when making high-quality double-slag basic heats, 650 to 800 (Table 3).

Electrode consumption varies considerably in arc furnaces because of their different constructions and operations. Aver-

**Table 3. Energy Consumption of Electric Furnaces**

| Process | Type of furnace | lb/kWh |
|---|---|---|
| Baking finishes on sheet metal | Batch oven | 10–18 |
| Baking finishes on sheet metal | Continuous oven | 25–30 |
| Baking bread | Continuous oven | 10–12 |
| Annealing brass and copper | Batch furnace | 10–25 |
| Annealing steel | Batch furnace | 5–15 |
| Hardening steel | Batch furnace | 7–11 |
| Tempering steel | Batch furnace | 15–25 |
| Annealing glass | Continuous furnace | 40–100 |
| Vitreous enameling, single coat | Batch furnace | 5–8 |
| Vitreous enameling, single coat | Continuous furnace | 10–15 |
| Galvanizing | Batch furnace | 12–20 |

| Melting metals | Type of furnace | kWh/ton (2,000 lb) |
|---|---|---|
| Lead | Resistor | 40–50 |
| Solder 50–50 | Resistor | 40–50 |
| Tin | Resistor | 35–50 |
| Zinc | Induction | 80–100 |
| Brass | Arc and induction | 250–400 |
| Steel, melting only | Arc and induction | 450–700 |
| Steel, melting and refining | Arc | 600–750 |
| Gray iron | Arc and induction | 450–600 |

| Furnace products | Type of furnace | kWh ton (2,000 lb) |
|---|---|---|
| Aluminum | Electrolytic | 22,000–27,000 |
| Calcium carbide | Resistance | 3,000–6,000 |
| Ferroalloys | Resistance | 4,000–8,000 |
| Graphite | Resistance | 3,000–8,000 |
| Phosphoric acid | Resistance | 5,000–6,000 |
| Silicon carbide | Resistance | 8,500–10,000 |
| Smelting iron ore | Resistance | 1,650–2,400 |

age values in pounds of electrode per ton of molten metal are: steel melting, with graphite electrodes, 5 to 10; brass melting, with graphite electrodes, 3 to 5. Graphite electrodes have largely superseded carbon electrodes.

### Submerged-Arc and Resistance Furnaces

The resistance furnace is essentially a refractory-lined chamber with electrodes—movable or fixed—buried in the charge. This simplicity permits a wide range of designs and much latitude in dimensions. The general service is heating charges of a refractory nature to bring about chemical reactions or changes in the physical structure of the material of the charge. The energy requirement of each of such processes is a large item in the cost of production. Large units and a favorable power location are the rule. Resistance furnaces also are termed submerged-arc furnaces and/or, in quite a few instances, smelting-type furnaces.

The only limit on the temperature to which a charge can be heated by this method is the temperature at which the materials of the charge are vaporized. For temperatures beyond the limit of refractory linings, the materials of the charge are used to form a protective layer between the core of the charge (through which the current passes) and the walls of the furnace.

**Resistance furnaces with movable electrodes** may be either single-phase or polyphase. The materials of the charge are fed more or less continuously, and the product is discharged intermittently or continuously as required. In some cases the product is in the molten state; in others the product is a vapor. The usual method of operation is the use of a single operating voltage and a constant power input. The power is regulated by adjustment of the depths of the electrodes in the charge. The load is fairly uniform and, if polyphase, is kept reasonably well balanced.

The **resistance furnace with fixed electrodes** is designed for heating materials in batches and is usually rectangular in shape with an electrode at each end for single-phase operation. The length and cross-sectional area of the path of the current are proportioned to suit the power characteristics of the charge. Refractory materials have negative temperature-resistance coefficients, and hence to maintain constant power in the furnace circuit the applied voltage must be reduced as the temperature of the charge rises in proportion to the square root of the ratio of the initial resistance of the furnace circuit to the resistance of the furnace circuit at the end of the heat cycle. If the materials of the charge are nonconductors at room temperature, a starting circuit is provided by means of a core of carbon—usually coke—placed in the charge. The heat cycles of furnaces of this class generally extend over a period of several days.

Some of the more common **uses of the resistance furnace** are:

**Calcium carbide furnaces** are charged continuously with lime and coke. These equipments can be either open or closed top. This type of furnace has been built up to 70,000 kVA in electrical powering—covered and sealed for gas collection.

**Ferroalloy furnaces** for the production of ferrochrome, ferrosilicon, ferromanganese, etc., are usually three-phase furnaces with movable electrodes and are similar in construction to the three-phase arc furnace. The charge is a mixture of the ore (oxide) of the selected metal, scrap iron, and a reducing agent, generally carbon except for very low carbon content alloys, for which some other reducing agent such as aluminum or silicon is required. Six-electrode furnaces often are used for power inputs of 15,000 kVA and more.

The **graphitizing furnace** is of the single-phase batch type. Artificial graphite is made by heating amorphous carbon (coal or coke) while shielded from air to a high temperature—around 4500°F (2500°C). The presence of some metallic impurity, such as iron oxide, in the charge appears to be necessary for the conversion of amorphous carbon to graphitic carbon. The raw material for making bulk graphite constitutes both the charge and the protective layer around the core of the charge. Graphite shapes are made from the corresponding shapes of amorphous carbon which are embedded—between the electrodes—in raw material as noted for the manufacture of bulk graphite.

The **silicon carbide furnace** is similar to the graphitizing furnace. The charge is a mixture of sand (silica), coke, sawdust, and a small amount of salt. This mixture is packed around a core of granulated coke to form the initial circuit between the electrodes. The sand and coke are the reacting materials. The sawdust serves to make the charge porous so that the gases formed during the heating of the charge can escape freely. The salt vaporizes and removes impurities, such as iron, in the form of chlorides. The temperature of the process is 2700 to 3400°F (1500 to 1880°C).

Section **8**

# Machine Elements

**BY**

**HEARD K. BAUMEISTER**  *Senior Engineer, International Business Machines Corporation.*
**EUGENE A. AVALLONE**  *Professor, Department of Mechanical Engineering, The City College, The City University of New York.*
**GEORGE W. MICHALEC**  *Assistant to Vice President, Operations Technology, American Can Co. Formerly Professor and Associate Dean of Engineering and Science, Stevens Institute of Technology.*
**DUDLEY D. FULLER**  *Professor of Mechanical Engineering, Columbia University.*
**CORT L. MILLER**  *Engineering Associate, Eastman Kodak Company, Kodak Park, Engineering Division.*
**MICHAEL J. WASHO**  *Engineer, Eastman Kodak Company, Kodak Park, Engineering Division.*
**EDWARD W. FISHER**  *Product Specialist, Retired, Garlock, Inc.*
**RENO C. KING**  *Associate Dean, Queensborough Community College, Bayside, N.Y.*
**C. H. BERRY**  *Late Gordon McKay Professor of Mechanical Engineering, Emeritus, Harvard University; and late Professor of Mechanical Engineering, Northeastern University.*

## MECHANISM
### by Heard K. Baumeister

| | |
|---|---|
| Linkages | 8-3 |
| Cams | 8-4 |
| Rolling Surfaces | 8-7 |
| Epicyclic Trains | 8-7 |
| Hoisting Mechanisms | 8-8 |

## MACHINE ELEMENTS
### by Eugene A. Avallone

| | |
|---|---|
| Screw Fastenings | 8-9 |
| Rivet Fastenings | 8-30 |
| Keys, Pins, and Cotters | 8-30 |
| Splines | 8-31 |
| Couplings | 8-35 |
| Clutches | 8-38 |
| Hydraulic Power Transmission | 8-39 |
| Brakes | 8-41 |
| Shrink, Press, Drive, and Running Fits | 8-45 |
| Shafts, Axles, and Cranks | 8-48 |
| Pulleys, Sheaves, and Flywheels | 8-49 |
| Belt Drives | 8-51 |
| Chain Drives | 8-59 |
| Rotary and Reciprocating Elements | 8-69 |
| Springs | 8-74 |
| Wire Rope | 8-85 |

| | |
|---|---|
| Fiber Lines | 8-89 |
| Nails and Spikes | 8-97 |
| Standard Cross-Section Symbols | 8-97 |

## GEARING
### by George W. Michalec

| | |
|---|---|
| Basic Gear Data | 8-98 |
| Fundamental Relationships of Spur and Helical Gears | 8-100 |
| Helical Gears | 8-104 |
| Nonspur Gear Types | 8-106 |
| Worm Gears and Worms | 8-110 |
| Design Standards | 8-111 |
| Strength and Durability | 8-111 |
| Gear Materials | 8-116 |
| Gear Lubricants | 8-116 |

## FLUID-FILM BEARINGS
### by Dudley D. Fuller

| | |
|---|---|
| Plain Cylindrical Journal Bearings | 8-120 |
| Elements of Journal Bearings | 8-126 |
| Thrust Bearings | 8-130 |
| Sliding Bearings | 8-133 |
| Gas-lubricated Bearings | 8-133 |

### BEARINGS WITH ROLLING CONTACT
#### by Cort L. Miller and Michael W. Washo

Components and Specifications ........................... 8-136
Principal Standard Bearing Types ...................... 8-136
Rolling-Contact Bearings Life, Load, and Speed Relationships 8-138
Procedure for Determining Size, Life, and Bearing Type .... 8-140
Bearing Closures ...................................... 8-141
Bearing Mounting ..................................... 8-141
Lubrication .......................................... 8-142

### PACKINGS AND SEALS
#### by Edward W. Fisher

Packing and Seals .................................... 8-143

### PIPE AND PIPE FITTINGS
#### by Reno C. King

Piping Standards ...................................... 8-147
Steel Pipe and Tubing ................................. 8-148
Cast-Iron Pipe ........................................ 8-155
Pipes and Tubes of Nonferrous Materials ................ 8-169
Vitrified, Wooden-Stave, and Concrete Pipe ............. 8-171
Fittings for Wrought-Iron and Steel Pipe ................ 8-174

### PREFERRED NUMBERS
#### by C. H. Berry

Preferred Numbers .................................... 8-201

# MECHANISM

## by Heard K. Baumeister

REFERENCES: Beggs, "Mechanism," McGraw-Hill. Hrones and Nelson, "Analysis of the Four Bar Linkage," Wiley. Jones, "Ingenious Mechanisms for Designers and Inventors," 4 vols., Industrial Press. Moliam, "The Design of Cam Mechanisms and Linkages," Elsevier. Chironis, "Gear Design and Application," McGraw-Hill.

**Definition** A mechanism is that part of a machine which contains two or more pieces so arranged that the motion of one compels the motion of the others according to a definite law depending upon the nature of the combination.

### LINKAGES

Links may be of any form so long as they do not interfere with the desired motion. The simplest form is that of four bars, A, B, C, D, fastened together at their ends by cylindrical pins, and which are all movable in parallel planes. If the links are of different lengths and each is fixed in turn, there will be four possible combinations; but as two of these are similar there will be produced three mechanisms having distinctly different motions. Thus, in Fig. 1, if D is fixed A can rotate and C oscillate, giving the **beam-and-crank** mechanism, as used on side-wheel steamers. If B is fixed, the same motion will result; if A is fixed (Fig. 2), links B and D can rotate, giving the **drag-link** mechanism used to feather the floats on paddle wheels. Fixing link C (Fig. 3), D and B can only oscillate, and a **rocker** mechanism sometimes used in straight-line motions is produced. It is customary to call a rotating link a **crank**; an

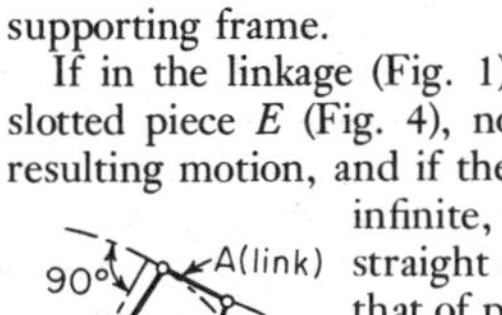

**Fig. 1** Beam-and-crank mechanism.

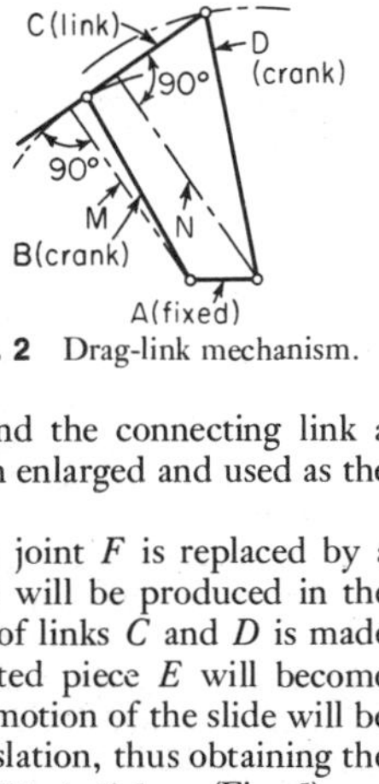

**Fig. 2** Drag-link mechanism.

oscillating link a **lever,** or beam; and the connecting link a **connecting rod.** The fixed link is often enlarged and used as the supporting frame.

If in the linkage (Fig. 1) the pin joint F is replaced by a slotted piece E (Fig. 4), no change will be produced in the resulting motion, and if the length of links C and D is made infinite, the slotted piece E will become straight and the motion of the slide will be that of pure translation, thus obtaining the engine, or **sliding-block, linkage** (Fig. 5).

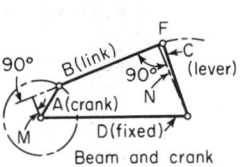

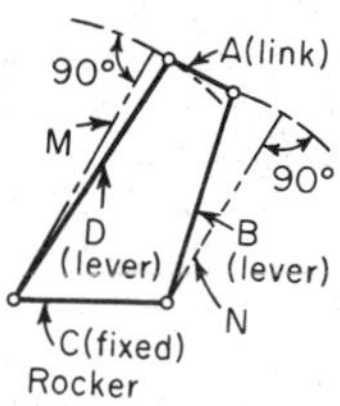

**Fig. 3** Rocker mechanism.

If in the sliding-block linkage (Fig. 5) the long link B is fixed (Fig. 6), A will rotate and E will oscillate and the infinite links C and D may be indicated as shown. This gives the **swinging-block linkage.** When used as a quick-return motion the slotted piece and slide are usually interchanged (Fig. 7)

which in no way changes the resulting motion. If the short link A is fixed (Fig. 8), B and E can both rotate, and the mechanism known as the **turning-block linkage** is obtained. This is better known under the name of the **Whitworth quick-return motion,** and is generally constructed as in Fig. 9. The **ratio of**

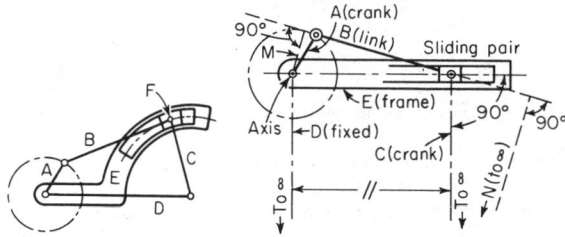

**Figs. 4 and 5** Sliding-block linkage.

**time of advance to time of return,** H/K, of the two quick-return motions (Figs. 7 and 9) may be found by locating, in the case of the swinging block (Fig. 7), the two tangent points (t) and measuring the angles H and K made by the two positions of the crank A. If H and K are known, the axis of E may be

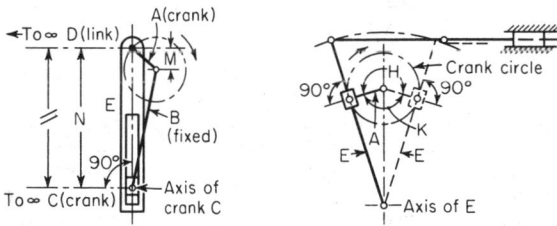

**Fig. 6** Swinging-block linkage.    **Fig. 7**

located by laying off the angles H and K on the crank circle and drawing the tangents E, their intersection giving the desired point. For the turning-block linkage (Fig. 9), determine the angles H and K made by the crank B when E is in the horizontal position; or, if the angles are known, the axis of

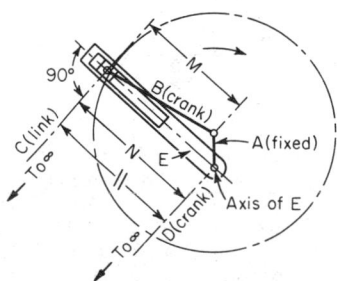

**Fig. 8** Turning-block linkage.

*E* may be determined by drawing a horizontal line through the two crankpin positions *(S)* for the given angle, and the point where a line through the axis of *B* cuts *E* perpendicularly will be the axis of *E*.

**Velocities** of any two or more points on a link must fulfill the following conditions (see Sec. 3). (1) Components along the

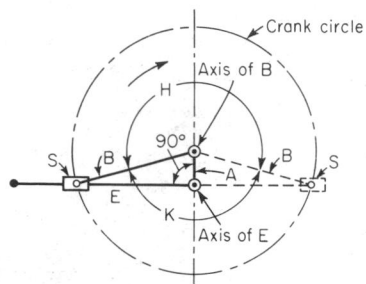

**Fig. 9** Whitworth quick-return motion.

link must be equal and in the same direction (Fig. 10): $V_a = V_b = V_c$. (2) Perpendiculars to $V_A$, $V_B$, $V_C$ from the points *A*, *B*, *C* must intersect at a common point *d*, the **instant center** (or instantaneous axis). (3) The velocities of points *A*, *B*, and *C* are directly proportional to their distances from this center (Fig. 11): $V_A/a = V_B/b = V_C/c$. For a straight link the tips of the vectors representing the velocities of any number of points on

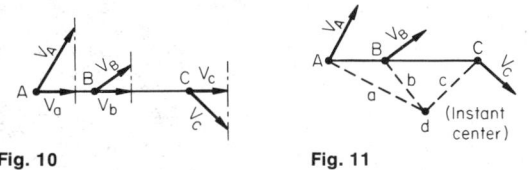

**Fig. 10**          **Fig. 11**

the link will be on a straight line (Fig. 12): *abc* = a straight line. To find the velocity of any point when the velocity and direction of any two other points are known, condition (2) may be used, or a combination of (1) and (3). The **linear velocity ratio** of any two points on a linkage may be found by determining the distances *e* and *f* to the instant center (Fig. 13); then $V_c/V_b$ = *e*/*f*. This may often be simplified by noting that a line drawn parallel to *e* and cutting *B* produced forms two similar triangles *efB* and *sAy*, which gives $V_c/V_b$ = *e*/*f* = *s*/*A*. The **angular velocity ratio** for any position of two oscillating or rotating links *A* and *C* (Fig. 1), connected by a movable link *B*, may be

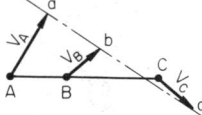

**Fig. 12**

determined by scaling the length of the perpendiculars *M* and *N* from the axes of rotation to the centerline of the movable link. The angular velocity ratio is inversely proportional to these perpendiculars, or $O_C/O_A$ = *M*/*N*. This method may be applied directly to a linkage having a sliding pair if the two infinite links are redrawn perpendicular to the sliding pair, as

indicated in Fig. 14. *M* and *N* are also shown in Figs. 1, 2, 3, 5, 6, 8. In Fig. 5 one of the axes is at infinity; therefore, *N* is infinite, or the slide has pure translation.

**Forces** A mechanism must deliver as much work as it receives, neglecting friction; therefore, the force at any point

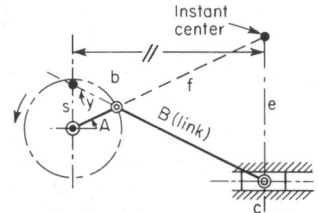

**Fig. 13**

*F* multiplied by the velocity $V_F$ in the direction of the force at that point must equal the force at some other point *P* multiplied by the velocity $V_P$ at that point; or the forces are inversely as their velocities and $F/P = V_P/V_F$. It is at times more convenient to equate the moments of the forces acting

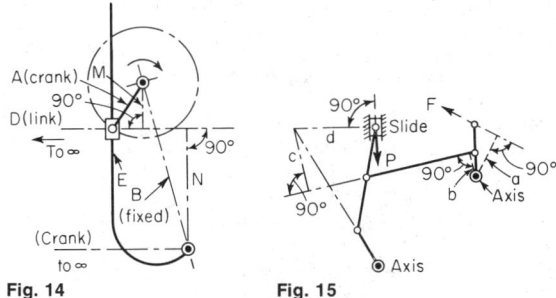

**Fig. 14**          **Fig. 15**

around each axis of rotation (sometimes using the instant center) to determine the force acting at some other point. In Fig. 15, $(F \times a \times c)/(b \times d) = P$.

## CAMS

**Cam Diagram** A cam is usually a plate or cylinder which communicates motion to a follower by means of its edge or a groove cut in its surface. In the practical design of cams the follower (1) must assume a definite series of positions while the driver occupies a corresponding series of positions or (2) must arrive at a definite location by the time the driver arrives at a particular position. The former design may be severely limited in speed because the interrelationship between the follower and cam positions may yield a follower displacement vs. time function that involves large values for the successive time derivatives, indicating large accelerations and forces and large impacts and noise. The second design centers about finding that particular interrelationship between the follower and cam positions that results in the minimum forces and impacts so that the speed may be made quite large. In either case, the desired interrelationship must be put into hardware as discussed below. In the case of high-speed machines, small irregularities in the cam surface or geometry may be severely detrimental.

A stepwise displacement in time for the follower running on

a cam driven at constant speed is, of course, impossible because the follower would require infinite velocities. A step in velocity for the follower would result in infinite accelerations; these in turn would bring into being forces that approach infinite magnitudes which would tend to destroy the machine. A step in acceleration causes a large jerk and large shock waves to be transmitted and reflected throughout the parts that generate noise and would tend to limit the life of the machine. A step in jerk, the third derivative of the follower displacement with respect to time, seems altogether acceptable. In those designs requiring or exhibiting clearance between the follower and cam (usually at the bottom of the stroke), as gentle and slow a ramp portion as can be tolerated must be inserted on either side of the clearance region to limit the magnitude of the acceleration and jerk to a minimum. The tolerance on the clearance adjustment must be small enough to assure that the follower will be left behind and picked up gradually by the gentle ramp portions of the cam.

Table 1 shows the comparable and relative magnitudes of velocity, acceleration, and jerk for several high-speed cams where the displacements are all taken as 1 at time 1 without any overshoot in any of the derivatives.

The three most common forms of motion used are uniform motion (Fig. 16), harmonic motion (Fig. 17), and uniformly accelerated and retarded motion (Fig. 18). In plotting the diagrams (Fig. 18) for this last motion, divide $ac$ into an even number of equal parts and $bc$ into the same number of parts with lengths increasing by a constant increment to a maximum

and then decreasing by the same decrement, as, for example, 1, 3, 5, 5, 3, 1, or 1, 3, 5, 7, 9, 9, 7, 5, 3, 1. In order to prevent shock when the direction of motion changes, as at $a$ and $b$ in the uniform motion, the harmonic motion may be used; if the cam is to be operated at high speed, the uniformly accelerated and retarded motion should preferably be employed; in either case there is a very gradual change of velocity.

**Pitch Line**    The actual pitch line of a cam varies with the type of motion and with the position of the follower relative to

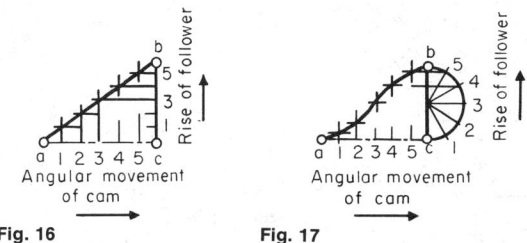

Fig. 16                Fig. 17

the cam's axis. Most cams as ordinarily constructed are covered by the following four cases.

FOLLOWER ON LINE OF AXIS (Fig. 19).    To draw the pitch line, subdivide the motion $bc$ of the follower in the manner indicated in Figs. 16, 17, 18. Draw a circle with a radius equal to the smallest radius of the cam $a0$ and subdivide it into angles $0a1'$, $0a2'$, $0a3'$, etc., corresponding with angular displace-

**Table 1. Displacement, Velocity, Acceleration, and Jerk for Some Cams***

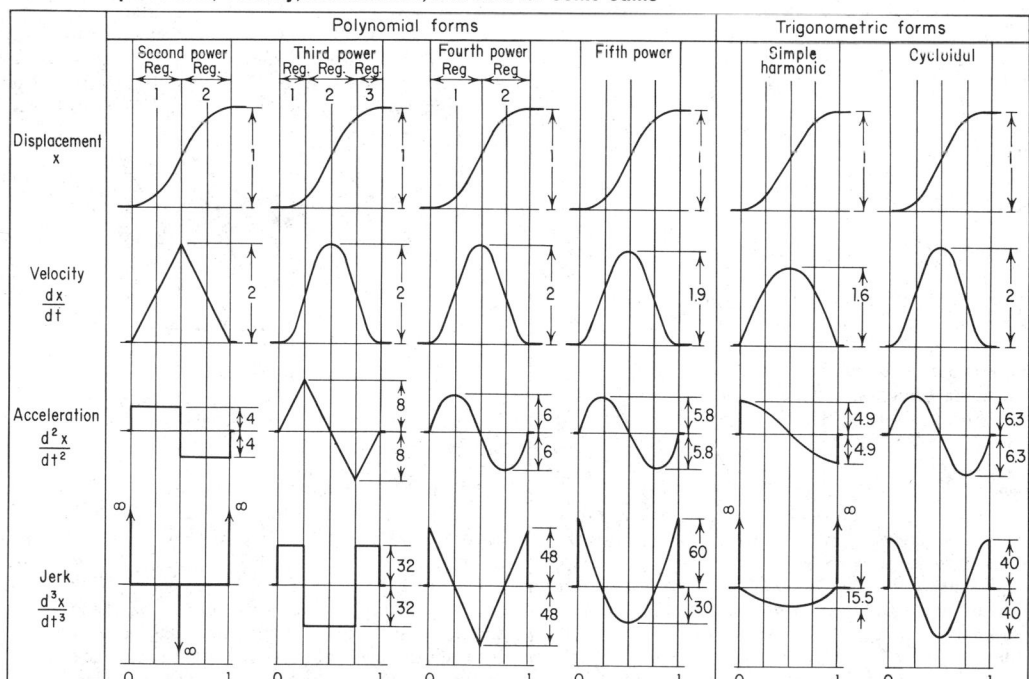

Time ⟶

*Adapted from Gutman, *Machine Design*, March, 1951.

ments of the cam for positions 1, 2, 3, etc., of the follower. With *a* as a center and radii *a*1, *a*2, *a*3, etc., strike arcs cutting radial lines at *d*, *e*, *f*, etc. Draw a smooth curve through points *d*, *e*, *f*, etc.

OFFSET FOLLOWER (Fig. 20).    Divide *bc* as indicated in Figs. 16, 17, and 18. Draw a circle of radius *ac* (highest point of rise

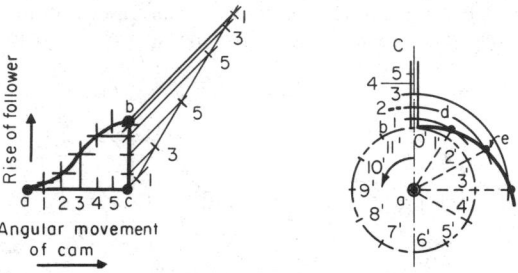

**Fig. 18**                      **Fig. 19**

of follower) and one tangent to *cb* produced. Divide the outer circle into parts 1′, 2′, 3′, etc., corresponding with the angular displacement of the cam for positions 1, 2, 3, etc., of the follower, and draw tangents from points 1′, 2′, 3′, etc., to the small circle. With *a* as a center and radii *a*1, *a*2, *a*3, etc., strike

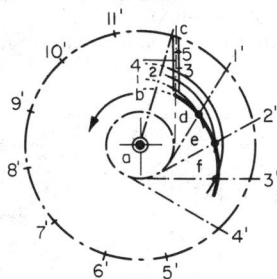

**Fig. 20**

arcs cutting tangents at *d*, *e*, *f*, etc. Draw a smooth curve through *d*, *e*, *f*, etc.

ROCKER FOLLOWER (Fig. 21).    Divide the stroke of the slide *S* in the manner indicated in Figs. 16, 17, and 18, and transfer these points to the arc *bc* as points 1, 2, 3, etc. Draw a circle of radius *ak* and divide it into parts 1′, 2′, 3′, etc., corresponding with angular displacements of the cam for positions 1, 2, 3, etc., of the follower. With *k*, 1′, 2′, 3′, etc., as centers and radius *bk*, strike arcs *kb*, 1′*d*, 2′*e*, 3′*f*, etc., cutting at *bdef* arcs struck with *a* as a center and radii *ab*, *a*1, *a*2, *a*3, etc. Draw a smooth curve through *b*, *d*, *e*, *f*, etc.

CYLINDRICAL CAM (Fig. 22).    In this type of cam more than one complete turn may be obtained, provided in all cases the follower returns to its starting point. Draw rectangle *wxyz* (Fig. 22) representing the development of cylindrical surface of the cam. Subdivide the desired motion of the follower *bc* horizontally in the manner indicated in Figs. 16, 17, and 18, and plot the corresponding angular displacement 1′, 2′, 3′, etc., of the cam vertically; then through the intersection of lines from these points draw a smooth curve. This may best be shown by an example, assuming the following data for the

diagram in Fig. 22: Total motion of follower = *bc*; circumference of cam = 2π*r*. Follower moves harmonically 4 units to right in 0.6 turn, then rests (or "dwells") 0.4 turn, and finishes with uniform motion 6 units to right and 10 units to left in 2 turns.

**Cam Design**    In the practical design of cams the following points must be noted. If only a small force is to be transmitted, sliding contact may be used, otherwise **rolling contact.** For the

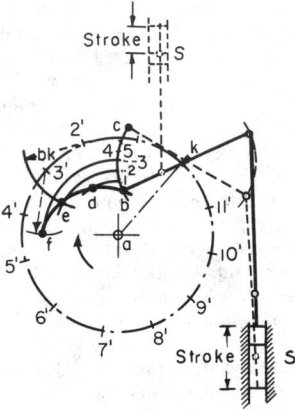

**Fig. 21**

latter the pitch line must be corrected in order to get the true slope of the cam. An approximate construction (Fig. 23) may be employed by using the pitch line as the center of a series of arcs the radii of which are equal to that of the follower roll to be used; then a smooth curve drawn tangent to the arcs will give the slope desired for a roll working on the periphery of the cam (Fig. 23*a*) or in a groove (Fig. 23*b*). For plate cams the

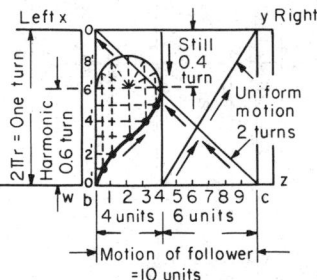

**Fig. 22**  Cylindrical cam.

roll should be a small cylinder, as in Fig. 24*a*. In cylindrical cams it is usually sufficiently accurate to make the roll conical, as in Fig. 24*b*, in which case the taper of the roll produced should intersect the axis of the cam. If the pitch line *abc* is made too sharp (Fig. 25) the follower will not rise the full amount. In order to prevent this **loss of rise,** the pitch line should have a radius of curvature at all parts of not less than the roll's diameter plus ⅛ in. For the same rise of follower, *a*, and angular motion of the cam, *O*, the slope of the cam changes considerably, as indicated by the heavy lines *A*, *B*, and *C* (Fig. 26). Care should be taken to keep a moderate slope and thereby keep down the side thrust on the follower, but this should not be carried too far, as the cam would become too large and the friction increase.

## ROLLING SURFACES

In order to connect two shafts so that they shall have a definite angular velocity ratio, rolling surfaces are often used; and in order to have no slipping between the surfaces they must

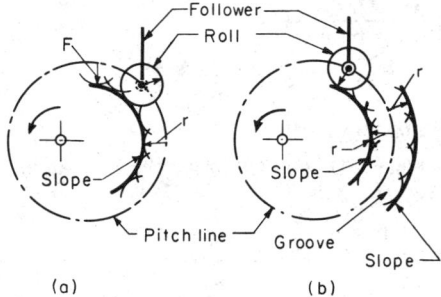

**Fig. 23**

fulfill the following two conditions: the line of centers must pass through the point of contact, and the arcs of contact must be of equal length. The angular velocities, expressed usually in r/min, will be inversely proportional to the radii: $N/n = r/R$. The two surfaces most commonly used in practice, and the

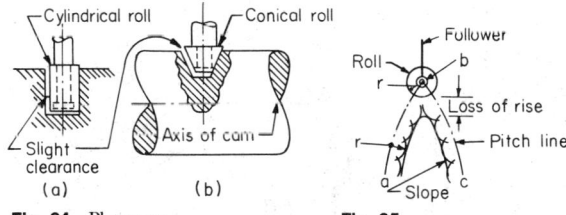

**Fig. 24**    Plate cam.                **Fig. 25**

only ones having a constant angular velocity ratio, are cylinders where the shafts are parallel, and cones where the shafts (produced) intersect at an angle. In either case there are two possible directions of rotation, depending upon whether the surfaces roll in opposite directions (external contact) or in the same direction (internal contact). In Fig. 27, $R = nc/(N + n)$ and $r = Nc/(N + n)$; in Fig. 28, $R = nc/(N - n)$ and $r = Nc/(N - n)$. In Fig. 29, $\tan B = \sin A/[(n/N) + \cos A]$, and $\tan C = \sin A/[(N/n) + \cos A]$; in Fig. 30, $\tan B = \sin A/[(N/n) - \cos A]$, and $\tan C = \sin A/[(n/N) - \cos A]$. With the above

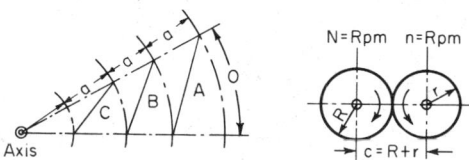

**Fig. 26**                **Fig. 27**

values for the angles $B$ and $C$, and the length $d$ or $e$ of one of the cones, $R$ and $r$ may be calculated.

## EPICYCLIC TRAINS

Epicyclic trains are combinations of gears in which some or all of the gears have a motion compounded of rotation about an axis and a translation or revolution of that axis. The gears are

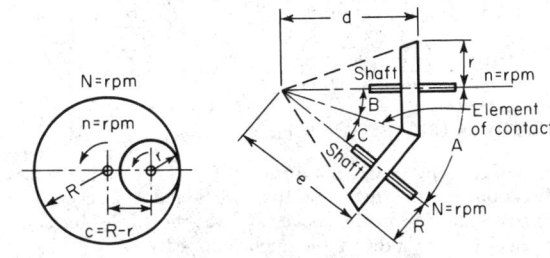

**Fig. 28**                **Fig. 29**

usually connected by a link called an arm, which often rotates about the axis of the first gear. Such trains may be calculated by first considering all gears locked and the arm turned; then the arm locked and the gears rotated. The algebraic sum of the

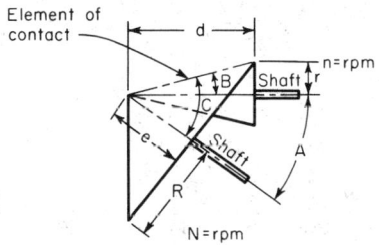

**Fig. 30**

separate motions will give the desired result. The following examples and method of tabulation will illustrate this. The figures on each gear refer to the number of teeth for that gear.

|  | A | B | C | D |
|---|---|---|---|---|
| Gear locked, Fig. 31 | +1 | +1 | +1 | +1 |
| Arm locked, Fig. 31 | 0 | −1 | $-1 \times {}^{50}\!/_{20}$ | $-1 \times {}^{50}\!/_{20} \times {}^{20}\!/_{40}$ |
| Addition, Fig. 31 | +1 | 0 | $+3\frac{1}{2}$ | $-\frac{1}{4}$ |
| Gears locked, Fig. 32 | +1 | +1 | +1 | +1 |
| Arm locked, Fig. 32 | 0 | −1 | $-1 \times {}^{30}\!/_{20}$ | $+1 \times {}^{30}\!/_{20} \times {}^{20}\!/_{70}$ |
| Addition, Fig. 32 | +1 | 0 | $+2\frac{1}{2}$ | $+1\frac{3}{7}$ |

In Figs. 31 and 32 lock the gears and turn the arm $A$ right-handed through 1 revolution ($+1$); then lock the arm and turn the gear $B$ back to where it started ($-1$); gears $C$ and $D$ will have rotated the amount indicated in the tabulation. Then the algebraic sum will give the relative turns of each gear. That is, in Fig. 31, for one turn of the arm, $B$ does not move and $C$ turns in the same direction 3½ revolutions, and $D$ in the opposite direction ¼ revolution; whereas in Fig. 32, for one

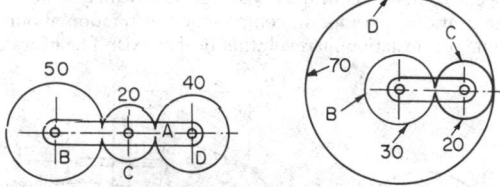

**Figs. 31 and 32.** Epicyclic trains.

turn of the arm, $B$ does not turn, but $C$ and $D$ turn in the same direction as the arm, respectively, 2½ and 1⅞ revolutions. (NOTE: The arm in the above case was turned +1 for convenience, but any other value might be used.)

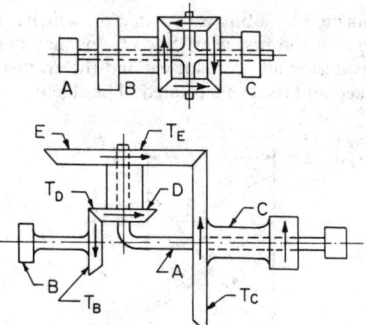

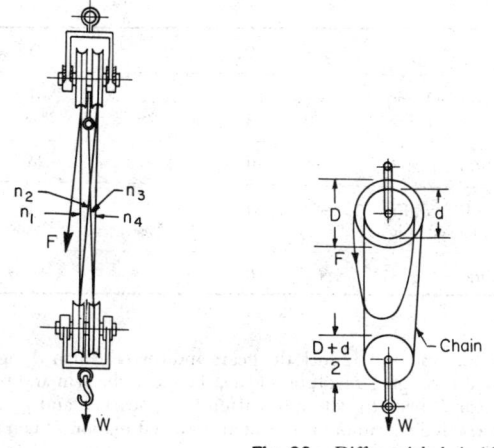

**Figs. 33 and 34.** Bevel epicyclic trains.

**Bevel epicyclic trains** are epicyclic trains containing bevel gears and may be calculated by the preceding method, but it is usually simpler to use the general formula which applies to all cases of epicyclic trains:

**Fig. 36.** Differential chain block.

$$\frac{\text{Turns of } C \text{ relative to arm}}{\text{Turns of } B \text{ relative to arm}}$$

$$= \frac{\text{absolute turns of } C - \text{turns of arm}}{\text{absolute turns of } B - \text{turns of arm}}$$

The left-hand term gives the value of the train and can always be expressed in terms of the number of teeth ($T$) on the gears. Care must be used, however, to express it as either plus (+) or minus (−), depending upon whether the gears turn in the same or opposite directions.

$$\frac{\text{Relative turns of } C}{\text{Relative turns of } B} = \frac{C - A}{B - A} = -1 \text{ (in Fig. 33)}$$

$$= +\frac{T_E}{T_C} \times \frac{T_B}{T_D} \text{ (in Fig. 34)}$$

### HOISTING MECHANISMS

**Pulley Block** (Fig. 35)  Given the weight $W$ to be raised, the force $F$ necessary is $F = V_W \times W/V_F = W/n = \text{load}/$

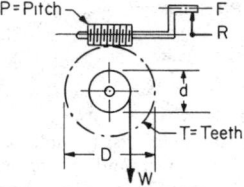

**Fig. 37.** Worm and wheel.

number of ropes, $V_W$ and $V_F$ being the respective velocities of $W$ and $F$.

**Differential Chain Block** (Fig. 36)  $F = V_W \times W/V_F = W(D - d)/2D$.

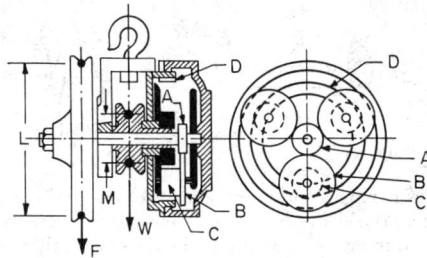

**Fig. 38.** Triplex chain block.

**Worm and Wheel** (Fig. 37)  $F = \pi d(n/T)W/2\pi R = WP(d/D)/2\pi R$, where $n$ = number of threads, single, double, triple, etc.

The **triplex chain block** (Fig. 38) is a geared hoist making use of the epicyclic train. $W = F \times L/M \{1 + [(T_D/T_C) \times (T_B/T_A)]\}$, where $T$ = number of teeth on gears.

**Toggle Joint** (Fig. 39)  $P = F \times s \cos A/t$.

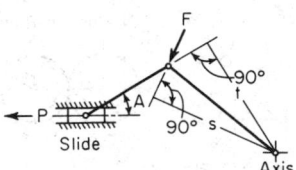

**Fig. 39.** Toggle joint.

# MACHINE ELEMENTS

## by Eugene A. Avallone

REFERENCES: ANSI Standards. Machinery's Handbook, Industrial Press. Power's Manual of Practical Engineering Data, McGraw-Hill. Manufacturers' bulletins, e.g., American Chain and Cable Co., American Steel and Wire Division of United States Steel Corp., Atlas Tack Corp., Bethlehem Steel Co., Republic Steel Corp., John A. Roebling's Sons Co., Townsend Co. Handbook of Mechanical Spring Design, Associated Spring Corp. Industrial Fasteners Institute Standards. Roller and Silent Chain Drives Assn. Design Manual. SAE Standards. French, "Engineering Drawing," McGraw-Hill. Maleev, "Machine Design," International Textbook. Kimball and Barr, "Elements of Machine Design," Wiley. Reuleaux, "The Constructor." Bach, "Die Maschinen Elemente," Kroener. Spotts, "Design of Machine Elements," Prentice-Hall. Peterson, "Stress Concentration Design Factors," Wiley.

NOTE: At this writing, conversion to metric hardware and machine elements is only beginning. SI units are introduced as appropriate, but the bulk of the material is still presented in the form in which the designer or reader will find it available in the field.

## SCREW FASTENINGS

By an accord signed in Washington, D.C., on Nov. 18, 1948, the screw-thread standardization committees of Canada, the United Kingdom, and the United States accepted a common standardization of screw threads for their respective countries and called it the **Unified Thread Standard.** This standard was additional to and was not a substitute for the earlier American standards. It was based on the earlier standards and incorporated changes suggested from experience in the manufacture, assembly, and use of screw threads.

Unified screw threads have largely replaced American standard threads. It is expected that after a transition period of some years, new standard metric threads will predominate. Signed into law, the Metric Conversion Act of 1975 will add impetus to the pace of metrication in the United States. At this writing some metric fasteners are already found in automotive vehicles manufactured for sale in the domestic market. Agreement has been reached with regard to metric screw threads and fastener sizes. Preparation of standards is being undertaken by the Industrial Fasteners Institute in this country; it is expected that ANSI standards for metric threaded fasteners will evolve shortly.

### Unified and American Screw Threads for Bolts, Nuts, and Machine Screws

The new **Unified screw-thread standards** are published by the ANSI as American Standard Unified and American Screw Thread Publication B1.1-1974. The thread series referred to in this standard are groups of diameter-pitch combinations distinguished from each other by the number of threads used with each diameter in the series. There are six Standard series and three Special series. The latter should not be employed until use of the Standard series is proved impossible. Threads in the Unified and American standard which are common to the British and Canadian standards are known as Unified and carry the prefix letter U in the thread series symbol. Those without this prefix are American standard only. The Standard series and their suggested uses are as follows:

**1. Coarse-thread Series (UNC and NC).** For general use, especially where rapid assembly is required and for gray iron, soft metals, and plastics.

**2. Fine-thread Series (UNF and NF).** For applications requiring greater strength or where the length of engagement is limited.

**3. Extra-fine-thread Series (UNEF and NEF).** For highly stressed parts and where internal threads are required in thin-walled fasteners.

**4. 8 Thread Series (8N).** A substitute for the coarse-thread series for diameters larger than 1 in.

**5. 12 Thread Series (12 UN and 12N).** A continuation of the fine-thread series for diameters larger than 1½ in.

**6. 16 Thread Series (16 UN and 16N).** A continuation of the extra-fine-thread series for diameters larger than 2 in.

There are three **Special series:** 8UN, UNS, and NS.

The Unified and American screw-thread standard recognizes eight **thread classes** distinguished from each other by the amount of allowance and/or tolerance. **Classes 1A, 2A, and 3A** apply to external threads only, **Classes 1B, 2B, and 3B** apply to internal threads only, and **Classes 2 and 3,** which are used with American standard threads only, apply to both external and internal threads.

1. **Classes 1A and 1B** provide liberal allowance for ease of assembly even when threads are dirty or slightly damaged.

2. **Classes 2A and 2B.** For production of bolts, screws, nuts, and other commercial fasteners. Permits external threads to be plated.

3. **Classes 3A and 3B.** For close-tolerance work where no allowance is required.

4. **Classes 2 and 3.** Retained as American standard, pending industry transition to the Unified classes.

The tolerances specified by each of these classes are based on a length of engagement equal to the nominal thread diameter.

The designation for a screw thread consists of a set of numbers and letter symbols which denote the diameter, pitch, thread series, class of fit, and hand as follows: **¼″-20 UNC-2A-LH**

¼″ = nominal size (fractional diameter or screw number)

20 = number of threads per inch

UNC = thread series symbol

2A = thread class symbol

LH = left-hand (no symbol required for right-hand threads)

Tables 1 through 6 give the screw diameters, allowances, tolerances, and important areas for the six Standard series. The largest diameter of a screw thread is called the **major diameter,** and the smallest diameter is called the **minor diameter.**

### Table 1. Coarse-thread Series—UNC and NC
(All dimensions in inches.  **Bold type indicates Unified threads.**)

| Size | Threads per in, $n$ | Major diam, $D^*$ max | Pitch diam, $E$ max | Basic minor diam, $K_s$ | Allowances†, classes 1A and 2A | Class 1A, major diam tolerances† | Class 2A and 3A, major diam tolerances‡ | Minor diam, $K_n$ min | Minor diam tolerances, classes 1B, 2B, and 3B§ | Basic min minor diam, sq in. | Stress area, sq in. |
|---|---|---|---|---|---|---|---|---|---|---|---|
| 1 (0.073) | 64 | 0.0730 | 0.0629 | 0.0538 | 0.0006 | ...... | 0.0038 | 0.0561 | 0.0062 | 0.0022 | 0.0026 |
| 2 (0.086) | 56 | 0.0860 | 0.0744 | 0.0641 | 0.0006 | ...... | 0.0041 | 0.0667 | 0.0070 | 0.0031 | 0.0036 |
| 3 (0.099) | 48 | 0.0990 | 0.0855 | 0.0734 | 0.0007 | ...... | 0.0045 | 0.0764 | 0.0081 | 0.0041 | 0.0048 |
| 4 (0.112) | 40 | 0.1120 | 0.0958 | 0.0813 | 0.0008 | ...... | 0.0051 | 0.0849 | 0.0090 | 0.0050 | 0.0060 |
| 5 (0.125) | 40 | 0.1250 | 0.1088 | 0.0943 | 0.0008 | ...... | 0.0051 | 0.0979 | 0.0083 | 0.0067 | 0.0079 |
| 6 (0.138) | 32 | 0.1380 | 0.1177 | 0.0997 | 0.0008 | ...... | 0.0060 | 0.1042 | 0.0098 | 0.0075 | 0.0090 |
| 8 (0.164) | 32 | 0.1640 | 0.1437 | 0.1257 | 0.0009 | ...... | 0.0060 | 0.1302 | 0.0087 | 0.0120 | 0.0139 |
| 10 (0.190) | 24 | 0.1900 | 0.1629 | 0.1389 | 0.0010 | ...... | 0.0072 | 0.1449 | 0.0106 | 0.0145 | 0.0174 |
| 12 (0.216) | 24 | 0.2160 | 0.1889 | 0.1649 | 0.0010 | ...... | 0.0072 | 0.1709 | 0.0098 | 0.0206 | 0.0240 |
| ¼ | 20 | 0.2500 | 0.2175 | 0.1887 | 0.0011 | 0.0122 | 0.0081 | 0.1959 | 0.0108 | 0.0269 | 0.0317 |
| 5⁄16 | 18 | 0.3125 | 0.2764 | 0.2443 | 0.0012 | 0.0131 | 0.0087 | 0.2524 | 0.0106 | 0.0454 | 0.0522 |
| 3⁄8 | 16 | 0.3750 | 0.3344 | 0.2983 | 0.0013 | 0.0142 | 0.0094 | 0.3073 | 0.0109 | 0.0678 | 0.0773 |
| 7⁄16 | 14 | 0.4375 | 0.3911 | 0.3499 | 0.0014 | 0.0155 | 0.0103 | 0.3602 | 0.0115 | 0.0933 | 0.1060 |
| ½ | 13 | 0.5000 | 0.4500 | 0.4056 | 0.0015 | 0.0163 | 0.0109 | 0.4167 | 0.0117 | 0.1257 | 0.1416 |
| ½ | 12 | 0.5000 | 0.4459 | 0.3978 | 0.0015 | 0.0172 | 0.0114 | 0.4098 | 0.0125 | 0.1205 | 0.1374 |
| 9⁄16 | 12 | 0.5625 | 0.5084 | 0.4603 | 0.0016 | 0.0172 | 0.0114 | 0.4723 | 0.0120 | 0.1620 | 0.1816 |
| 5⁄8 | 11 | 0.6250 | 0.5660 | 0.5135 | 0.0016 | 0.0182 | 0.0121 | 0.5266 | 0.0125 | 0.2018 | 0.2256 |
| ¾ | 10 | 0.7500 | 0.6850 | 0.6273 | 0.0018 | 0.0194 | 0.0129 | 0.6417 | 0.0128 | 0.3020 | 0.3340 |
| 7⁄8 | 9 | 0.8750 | 0.8028 | 0.7387 | 0.0019 | 0.0208 | 0.0139 | 0.7547 | 0.0134 | 0.4193 | 0.4612 |
| 1 | 8 | 1.0000 | 0.9188 | 0.8466 | 0.0020 | 0.0225 | 0.0150 | 0.8647 | 0.0150 | 0.5510 | 0.6051 |
| 1⅛ | 7 | 1.1250 | 1.0322 | 0.9497 | 0.0022 | 0.0246 | 0.0164 | 0.9704 | 0.0171 | 0.6931 | 0.7627 |
| 1¼ | 7 | 1.2500 | 1.1572 | 1.0747 | 0.0022 | 0.0246 | 0.0164 | 1.0954 | 0.0171 | 0.8898 | 0.9684 |
| 1⅜ | 6 | 1.3750 | 1.2667 | 1.1705 | 0.0024 | 0.0273 | 0.0182 | 1.1946 | 0.0200 | 1.0541 | 1.1538 |
| 1½ | 6 | 1.5000 | 1.3917 | 1.2955 | 0.0024 | 0.0273 | 0.0182 | 1.3196 | 0.0200 | 1.2938 | 1.4041 |
| 1¾ | 5 | 1.7500 | 1.6201 | 1.5046 | 0.0027 | 0.0308 | 0.0205 | 1.5335 | 0.0240 | 1.7441 | 1.8983 |
| 2 | 4½ | 2.0000 | 1.8557 | 1.7274 | 0.0029 | 0.0330 | 0.0220 | 1.7594 | 0.0267 | 2.3001 | 2.4971 |
| 2¼ | 4½ | 2.2500 | 2.1057 | 1.9774 | 0.0029 | 0.0330 | 0.0220 | 2.0094 | 0.0267 | 3.0212 | 3.2464 |
| 2½ | 4 | 2.5000 | 2.3376 | 2.1933 | 0.0031 | 0.0357 | 0.0238 | 2.2294 | 0.0300 | 3.7161 | 3.9976 |
| 2¾ | 4 | 2.7500 | 2.5876 | 2.4433 | 0.0032 | 0.0357 | 0.0238 | 2.4794 | 0.0300 | 4.6194 | 4.9326 |
| 3 | 4 | 3.0000 | 2.8376 | 2.6933 | 0.0032 | 0.0357 | 0.0238 | 2.7294 | 0.0300 | 5.6209 | 5.9659 |
| 3¼ | 4 | 3.2500 | 3.0876 | 2.9433 | 0.0033 | 0.0357 | 0.0238 | 2.9794 | 0.0300 | 6.7205 | 7.0992 |
| 3½ | 4 | 3.5000 | 3.3376 | 3.1933 | 0.0033 | 0.0357 | 0.0238 | 3.2294 | 0.0300 | 7.9183 | 8.3268 |
| 3¾ | 4 | 3.7500 | 3.5876 | 3.4433 | 0.0034 | 0.0357 | 0.0238 | 3.4794 | 0.0300 | 9.2143 | 9.6546 |
| 4 | 4 | 4.0000 | 3.8376 | 3.6933 | 0.0034 | 0.0357 | 0.0238 | 3.7294 | 0.0300 | 10.6084 | 11.0805 |

The above values are based on a length of engagement equal to the nominal diameter.

\* Major diameter at intersection of rounded root with flanks of threads.

† Allowances apply to external threads, Classes 1A and 2A only. The allowance for Class 3A threads is zero.

‡ Major diameter of internal threads may extend to a $P/24$ flat.

§ Minor diameter of external threads may extend to a $P/8$ flat.

The **pitch diameter** is the diameter to an imaginary line drawn through the thread profile such that the widths of the thread and groove are equal. These terms apply to both internal and external threads. The maximum diameters of an external thread are equal to the basic diameters minus the allowance. For Class 3A threads the allowance is zero. The minimum diameters of an internal thread are equal to the basic diameters.

The minimum minor diameters of an external thread and the maximum major diameter of an internal thread are not specified, being determined by the crest of an unworn tool. The Unified and American screw-thread proportions are shown in Fig. 1.

### American Standard Screw Threads for Bolts, Nuts, and Machine Screws

The **American standard thread** for machine screws, shown in Fig. 2, is proportioned by the formulas $p$ = pitch = 1/number of threads per inch, and $f$ = flat = $p/8$. The dimensions of American standard screw threads are published by the ANSI

**Table 2.  Fine-thread Series—UNF and NF**

(All dimensions in inches.   **Bold type indicates Unified threads.**)

| Size (Identification) | Threads per in., $n$ | Screw basic diameters | | External threads | | | | Internal threads | | Areas of sections | |
|---|---|---|---|---|---|---|---|---|---|---|---|
| | | Major diam, $D$* max | Pitch diam, $E$ max | Basic minor diam, $K$* | Allowances†, classes 1A and 2A | Class 1A, major diam tolerances‡ | Class 2A and 3A, major diam tolerances‡ | Minor diam, $K_n$ min | Minor diam tolerances, classes 1B, 2B, and 3B§ | Basic min minor diam, sq in. | Stress area, sq in. |
| 0 (0.060) | 80 | 0.0600 | 0.0519 | 0.0447 | 0.0005 | ...... | 0.0032 | 0.0465 | 0.0049 | 0.0015 | 0.0018 |
| 1 (0.073) | 72 | 0.0730 | 0.0640 | 0.0560 | 0.0006 | ...... | 0.0035 | 0.0580 | 0.0055 | 0.0024 | 0.0027 |
| 2 (0.086) | 64 | 0.0860 | 0.0759 | 0.0668 | 0.0006 | ...... | 0.0038 | 0.0691 | 0.0062 | 0.0034 | 0.0039 |
| 3 (0.099) | 56 | 0.0990 | 0.0874 | 0.0771 | 0.0007 | ...... | 0.0041 | 0.0797 | 0.0068 | 0.0045 | 0.0052 |
| 4 (0.112) | 48 | 0.1120 | 0.0985 | 0.0864 | 0.0007 | ...... | 0.0045 | 0.0894 | 0.0074 | 0.0057 | 0.0065 |
| 5 (0.125) | 44 | 0.1250 | 0.1102 | 0.0971 | 0.0007 | ...... | 0.0048 | 0.1004 | 0.0075 | 0.0072 | 0.0082 |
| 6 (0.138) | 40 | 0.1380 | 0.1218 | 0.1073 | 0.0008 | ...... | 0.0051 | 0.1109 | 0.0077 | 0.0087 | 0.0101 |
| 8 (0.164) | 36 | 0.1640 | 0.1460 | 0.1299 | 0.0008 | ...... | 0.0055 | 0.1339 | 0.0077 | 0.0128 | 0.0146 |
| 10 (0.190) | 32 | 0.1900 | 0.1697 | 0.1517 | 0.0009 | ...... | 0.0060 | 0.1562 | 0.0079 | 0.0175 | 0.0199 |
| 12 (0.216) | 28 | 0.2160 | 0.1928 | 0.1722 | 0.0010 | ...... | 0.0065 | 0.1773 | 0.0084 | 0.0026 | 0.0257 |
| ¼ | 28 | 0.2500 | 0.2268 | 0.2062 | 0.0010 | 0.0098 | 0.0065 | 0.2113 | 0.0077 | 0.0326 | 0.0362 |
| ⁵⁄₁₆ | 24 | 0.3125 | 0.2854 | 0.2614 | 0.0011 | 0.0108 | 0.0072 | 0.2674 | 0.0080 | 0.0524 | 0.0579 |
| ⅜ | 24 | 0.3750 | 0.3479 | 0.3239 | 0.0011 | 0.0108 | 0.0072 | 0.3299 | 0.0073 | 0.0809 | 0.0876 |
| ⁷⁄₁₆ | 20 | 0.4375 | 0.4050 | 0.3762 | 0.0013 | 0.0122 | 0.0081 | 0.3834 | 0.0082 | 0.1090 | 0.1185 |
| ½ | 20 | 0.5000 | 0.4675 | 0.4387 | 0.0013 | 0.0122 | 0.0081 | 0.4459 | 0.0078 | 0.1486 | 0.1597 |
| ⁹⁄₁₆ | 18 | 0.5625 | 0.5264 | 0.4943 | 0.0014 | 0.0131 | 0.0087 | 0.5024 | 0.0082 | 0.1888 | 0.2026 |
| ⅝ | 18 | 0.6250 | 0.5889 | 0.5568 | 0.0014 | 0.0131 | 0.0087 | 0.5649 | 0.0081 | 0.2400 | 0.2555 |
| ¾ | 16 | 0.7500 | 0.7094 | 0.6733 | 0.0015 | 0.0142 | 0.0094 | 0.6823 | 0.0085 | 0.3513 | 0.3724 |
| ⅞ | 14 | 0.8750 | 0.8286 | 0.7874 | 0.0016 | 0.0155 | 0.0103 | 0.7977 | 0.0091 | 0.4805 | 0.5088 |
| 1 | 12 | 1.0000 | 0.9459 | 0.8978 | 0.0018 | 0.0172 | 0.0114 | 0.9098 | 0.0100 | 0.6245 | 0.6624 |
| 1⅛ | 12 | 1.1250 | 1.0709 | 1.0228 | 0.0018 | 0.0172 | 0.0114 | 1.0348 | 0.0100 | 0.8118 | 0.8549 |
| 1¼ | 12 | 1.2500 | 1.1959 | 1.1478 | 0.0018 | 0.0172 | 0.0114 | 1.1598 | 0.0100 | 1.0237 | 1.0721 |
| 1⅜ | 12 | 1.3750 | 1.3209 | 1.2728 | 0.0019 | 0.0172 | 0.0114 | 1.2848 | 0.0100 | 1.2302 | 1.3137 |
| 1½ | 12 | 1.5000 | 1.4459 | 1.3978 | 0.0019 | 0.0172 | 0.0114 | 1.4098 | 0.0100 | 1.5212 | 1.5799 |

The above values are based on a length of engagement equal to the nominal diameter.

* Major diameter at intersection of rounded root with flanks of threads.

† Allowances apply to external threads, Classes 1A and 2A only. The allowance for Class 3A threads is zero.

‡ Major diameter of internal threads may extend to a $P/24$ flat.

§ Minor diameter of external threads may extend to a $P/8$ flat.

as B1.1-1974 and are based upon the earlier National screw-thread standard. The screw-thread profile is that previously known as the Sellers or U.S. standard. There are four thread series, as follows: coarse-thread, fine-thread, special-pitch, and 8, 12, and 16 pitch series. Four classes of fit are given, numbered 1 through 4.

**Foreign Standard Screw Threads for Bolts, Nuts, and Machine Screws; Forms and Proportions**

The **Whitworth Standard Thread** (Fig. 3) is still used in Great Britain and is based on the formulas $d = 0.6403p$, $r =$ radius = $0.1373p$ (see Table 7).

The **British Association screw-thread** (Fig. 4) dimensions are given in Table 8.

The **French (metric) screw thread** (Fig. 5) is based on the formulas $p =$ pitch, mm, $d = 0.6495p$, and $f = p/8$ (see Table 9).

**International (metric) standard screw thread,** as adopted by the "Congrés internationale pour l'unification des filetages," in Zurich, Oct. 24, 1898, is shown in Fig. 6, in which $d = 0.7036p$ and $t = 0.866p$ (see Table 10).

**Power-transmission Screw Threads: Forms and Proportions**

The **Acme thread** is obtained in four series (ANSI B1.5-1973 and B1.8-1973):

The **29° general-purpose thread** (Fig. 7) is used for all Acme thread applications except in special design cases. The general dimensions are given in Table 11.

The **29° stub thread** (Fig. 8) is used where heavy loads are encountered and where space limitations or other economic considerations make a shallow thread desirable (see Table 11).

The **60° stub thread** (Fig. 9) is used in special applications in the machine-tool industry (see Table 11).

The **10° modified square thread** (Fig. 10) is equivalent to a "square thread" for all practical purposes.

The range of threads per inch is as follows: 29° general-purpose threads, 16 to 1; 29° stub thread, 16 to 2; 60° stub thread, 16 to 4.

The number of threads per inch diameter is not standardized. The usual values are given in Table 12.

**American Standard Screw Threads for High-Strength Bolting** These threads are used with pressure vessels, steel pipe flanges, fittings, valves, and other services. They can be applied to either hot or cold surfaces where high tensile stresses are produced when the joints are made up. For sizes 1 in and smaller, the ANSI coarse-thread series is used. For larger sizes, ANSI 8-pitch thread series is used (see Table 13).

### Screw Threads for Pipes

**American standard taper pipe thread,** ANSI B2.1-1968 is shown in Fig. 11 and is made to the following specifications: The taper is 1 in in 16 or 0.75 in/ft. The basic length of the effective external taper thread is determined by $L_2 = p(0.8D + 6.8)$, where $D$ is the basic outside diameter of the pipe (see Table 14).

**American Standard Straight Pipe Thread** There are certain types of joints where straight pipe threads can be used to advantage. The ANSI has adopted five types of joints, the pertinent dimensions of which are shown in Table 15: (1) pressuretight joints with sealer, for pipe couplings, cols. 3 and 4; (2) pressuretight joints without sealer, for drain plugs, filler

**Table 3. Extra-fine-thread Series—UNEF and NEF**
(All dimensions in inches. **Bold type indicates Unified threads.**)

| Identification | | Screw basic diameters | | External threads | | | | Internal threads | | Areas of sections | |
|---|---|---|---|---|---|---|---|---|---|---|---|
| Size | Threads per in., $n$ | Major diam, $D$* max | Pitch diam, $E$ max | Basic minor diam, $K_s$ | Allowances†, classes 1A and 2A | Class 1A, major diam tolerances‡ | Class 2A and 3A, major diam tolerances‡ | Minor diam, $K_n$ min | Minor diam tolerances, classes 1B, 2B, and 3B§ | Basic min minor diam, sq in. | Stress area, sq in. |
| 12 (0.216) | 32 | 0.2160 | 0.1957 | 0.1777 | 0.0009 | ...... | 0.0060 | 0.1822 | 0.0073 | 0.0242 | 0.2269 |
| ¼ | 32 | 0.2500 | 0.2297 | 0.2117 | 0.0010 | ...... | 0.0060 | 0.2162 | 0.0067 | 0.0344 | 0.0377 |
| 5/16 | 32 | 0.3125 | 0.2922 | 0.2742 | 0.0010 | ...... | 0.0060 | 0.2787 | 0.0060 | 0.0581 | 0.0622 |
| ⅜ | 32 | 0.3750 | 0.3547 | 0.3367 | 0.0010 | ...... | 0.0060 | 0.3412 | 0.0057 | 0.0878 | 0.0929 |
| **7/16** | **28** | **0.4375** | **0.4143** | **0.3937** | **0.0011** | ...... | **0.0065** | **0.3988** | **0.0063** | **0.1201** | **0.1270** |
| **½** | **28** | **0.5000** | **0.4768** | **0.4562** | **0.0011** | ...... | **0.0065** | **0.4613** | **0.0063** | **0.1616** | **0.1695** |
| 9/16 | 24 | 0.5625 | 0.5354 | 0.5114 | 0.0012 | ...... | 0.0072 | 0.5174 | 0.0070 | 0.2030 | 0.2134 |
| ⅝ | 24 | 0.6250 | 0.5979 | 0.5739 | 0.0012 | ...... | 0.0072 | 0.5799 | 0.0070 | 0.2560 | 0.2676 |
| 11/16 | 24 | 0.6875 | 0.6604 | 0.6364 | 0.0012 | ...... | 0.0072 | 0.6424 | 0.0070 | 0.3151 | 0.3280 |
| **¾** | **20** | **0.7500** | **0.7175** | **0.6887** | **0.0013** | ...... | **0.0081** | **0.6959** | **0.0078** | **0.3685** | **0.3855** |
| **13/16** | **20** | **0.8125** | **0.7800** | **0.7512** | **0.0013** | ...... | **0.0081** | **0.7584** | **0.0078** | **0.4388** | **0.4573** |
| **⅞** | **20** | **0.8750** | **0.8425** | **0.8137** | **0.0013** | ...... | **0.0081** | **0.8209** | **0.0078** | **0.5153** | **0.5352** |
| **15/16** | **20** | **0.9375** | **0.9050** | **0.8762** | **0.0014** | ...... | **0.0081** | **0.8834** | **0.0078** | **0.5979** | **0.6194** |
| **1** | **20** | **1.0000** | **0.9675** | **0.9387** | **0.0014** | ...... | **0.0081** | **0.9459** | **0.0078** | **0.6866** | **0.7095** |
| **1 1/16** | **18** | **1.0625** | **1.0264** | **0.9943** | **0.0014** | ...... | **0.0087** | **1.0024** | **0.0081** | **0.7702** | **0.7973** |
| 1⅛ | 18 | 1.1250 | 1.0889 | 1.0568 | 0.0014 | ...... | 0.0087 | 1.0649 | 0.0081 | 0.8705 | 0.8993 |
| 1 3/16 | 18 | 1.1875 | 1.1514 | 1.1193 | 0.0015 | ...... | 0.0087 | 1.1274 | 0.0081 | 0.9770 | 1.0074 |
| 1¼ | 18 | 1.2500 | 1.2139 | 1.1818 | 0.0015 | ...... | 0.0087 | 1.1899 | 0.0081 | 1.0895 | 1.1216 |
| 1 5/16 | 18 | 1.3125 | 1.2764 | 1.2443 | 0.0015 | ...... | 0.0087 | 1.2524 | 0.0081 | 1.2082 | 1.2420 |
| 1⅜ | 18 | 1.3750 | 1.3389 | 1.3068 | 0.0015 | ...... | 0.0087 | 1.3149 | 0.0081 | 1.3330 | 1.3684 |
| 1 7/16 | 18 | 1.4375 | 1.4014 | 1.3693 | 0.0015 | ...... | 0.0087 | 1.3774 | 0.0081 | 1.4640 | 1.5010 |
| 1½ | 18 | 1.5000 | 1.4639 | 1.4318 | 0.0015 | ...... | 0.0087 | 1.4399 | 0.0081 | 1.6011 | 1.6397 |
| 1 9/16 | 18 | 1.5625 | 1.5264 | 1.4943 | 0.0015 | ...... | 0.0087 | 1.5024 | 0.0081 | 1.7444 | 1.7846 |
| 1⅝ | 18 | 1.6250 | 1.5889 | 1.5568 | 0.0015 | ...... | 0.0087 | 1.5649 | 0.0081 | 1.8937 | 1.9357 |
| 1 11/16 | 18 | 1.6875 | 1.6514 | 1.6193 | 0.0015 | ...... | 0.0087 | 1.6274 | 0.0081 | 2.0493 | 2.0929 |
| **1¾** | **16** | **1.7500** | **1.7094** | **1.6733** | **0.0016** | ...... | **0.0094** | **1.6823** | **0.0085** | **2.2873** | **2.2382** |
| **2** | **16** | **2.000** | **1.9594** | **1.9233** | **0.0016** | ...... | **0.0094** | **1.9323** | **0.0085** | **2.8917** | **2.9501** |

The above values are based on a length of engagement equal to the nominal diameter.

* Major diameter at intersection of rounded root with flanks of threads.

† Allowances apply to external threads, Classes 1A and 2A only. The allowance for Class 3A threads is zero.

‡ Major diameter of internal threads may extend to a $P/24$ flat.

§ Minor diameter of external threads may extend to a $P/8$ flat.

**Table 4. 8 Thread Series—8N**

(All dimensions given in inches.)

| Size | Threads per in., $n$ | Screw basic diameters | | External threads | | | | Internal threads | | Areas of sections | |
|---|---|---|---|---|---|---|---|---|---|---|---|
| | | Major diam, $D^*$ max | Pitch diam, $E$ max | Basic minor diam, $K_s$ | Allowances†, classes 1A and 2A | Class 1A, major diam tolerances‡ | Class 2A and 3A, major diam tolerances‡ | Minor diam, $K_n$ min | Minor diam tolerances, classes 1B, 2B, and 3B§ | Basic min minor diam, sq in. | Stress area, sq in. |
| 1⅛ | 8 | 1.1250 | 1.0438 | 0.9716 | 0.0021 | ...... | 0.0150 | 0.9897 | 0.0150 | 0.7277 | 0.7896 |
| 1¼ | 8 | 1.2500 | 1.1688 | 1.0966 | 0.0021 | ...... | 0.0150 | 1.1147 | 0.0150 | 0.9290 | 0.9985 |
| 1⅜ | 8 | 1.3750 | 1.2938 | 1.2216 | 0.0022 | ...... | 0.0150 | 1.2397 | 0.0150 | 1.1548 | 1.2319 |
| 1½ | 8 | 1.5000 | 1.4188 | 1.3466 | 0.0022 | ...... | 0.0150 | 1.3647 | 0.0150 | 1.4052 | 1.4899 |
| 1⅝ | 8 | 1.6250 | 1.5438 | 1.4716 | 0.0022 | ...... | 0.0150 | 1.4897 | 0.0150 | 1.6801 | 1.7723 |
| 1¾ | 8 | 1.7500 | 1.6688 | 1.5966 | 0.0023 | ...... | 0.0150 | 1.6147 | 0.0150 | 1.9796 | 2.0792 |
| 1⅞ | 8 | 1.8750 | 1.7938 | 1.7216 | 0.0023 | ...... | 0.0150 | 1.7397 | 0.0150 | 2.3036 | 2.4107 |
| 2 | 8 | 2.0000 | 1.9188 | 1.8466 | 0.0023 | ...... | 0.0150 | 1.8647 | 0.0150 | 2.6521 | 2.7665 |
| 2⅛ | 8 | 2.1250 | 2.0438 | 1.9716 | 0.0024 | ...... | 0.0150 | 1.9897 | 0.0150 | 3.0252 | 3.1469 |
| 2¼ | 8 | 2.2500 | 2.1688 | 2.0966 | 0.0024 | ...... | 0.0150 | 2.1147 | 0.0150 | 3.4228 | 3.5519 |
| 2½ | 8 | 2.5000 | 2.4188 | 2.3466 | 0.0024 | ...... | 0.0150 | 2.3647 | 0.0150 | 4.2917 | 4.4352 |
| 2¾ | 8 | 2.7500 | 2.6688 | 2.5966 | 0.0025 | ...... | 0.0150 | 2.6147 | 0.0150 | 5.2588 | 5.4164 |
| 3 | 8 | 3.0000 | 2.9188 | 2.8466 | 0.0026 | ...... | 0.0150 | 2.8647 | 0.0150 | 6.3240 | 6.4957 |
| 3¼ | 8 | 3.2500 | 3.1688 | 3.0966 | 0.0026 | ...... | 0.0150 | 3.1147 | 0.0150 | 7.4874 | 7.6738 |
| 3½ | 8 | 3.5000 | 3.4188 | 3.3466 | 0.0026 | ...... | 0.0150 | 3.3647 | 0.0150 | 8.7490 | 8.9504 |
| 3¾ | 8 | 3.7500 | 3.6688 | 3.5966 | 0.0027 | ...... | 0.0150 | 3.6147 | 0.0150 | 10.1088 | 10.3249 |
| 4 | 8 | 4.0000 | 3.9188 | 3.8466 | 0.0027 | ...... | 0.0150 | 3.8647 | 0.0150 | 11.5667 | 11.7995 |
| 4¼ | 8 | 4.2500 | 4.1688 | 4.0966 | 0.0028 | ...... | 0.0150 | 4.1147 | 0.0150 | 13.1228 | 13.3683 |
| 4½ | 8 | 4.5000 | 4.4188 | 4.3466 | 0.0028 | ...... | 0.0150 | 4.3647 | 0.0150 | 14.7771 | 15.0372 |
| 4¾ | 8 | 4.7500 | 4.6688 | 4.5966 | 0.0029 | ...... | 0.0150 | 4.6147 | 0.0150 | 16.5295 | 16.8042 |
| 5 | 8 | 5.0000 | 4.9188 | 4.8466 | 0.0029 | ...... | 0.0150 | 4.8647 | 0.0150 | 18.3802 | 18.6694 |
| 5¼ | 8 | 5.2500 | 5.1688 | 5.0966 | 0.0029 | ...... | 0.0150 | 5.1147 | 0.0150 | 20.3290 | 20.6330 |
| 5½ | 8 | 5.5000 | 5.4188 | 5.3466 | 0.0030 | ...... | 0.0150 | 5.3647 | 0.0150 | 22.3760 | 22.6945 |
| 5¾ | 8 | 5.7500 | 5.6688 | 5.5966 | 0.0030 | ...... | 0.0150 | 5.6147 | 0.0150 | 24.5211 | 24.8541 |
| 6 | 8 | 6.0000 | 5.9188 | 5.8466 | 0.0030 | ...... | 0.0150 | 5.8647 | 0.0150 | 26.7645 | 27.1118 |

The above values are based on a length of engagement equal to the nominal diameter.

\* Major diameter at intersection of rounded root with flanks of threads.

† Allowances apply to external threads, Classes 1A and 2A only.   The allowance for Class 3A threads is zero.

‡ Major diameter of internal threads may extend to a $P/24$ flat.

§ Minor diameter of external threads may extend to a $P/8$ flat.

plugs, etc., cols. 5 and 6; (3) free-fitting mechanical joints for fixtures, cols. 7 to 10; (4) loose-fitting mechanical joints with locknuts, cols. 11 to 14; (5) loose-fitting mechanical joints for hose couplings, cols. 11 to 14.

**American Standard Dryseal Pipe Threads** In some instances, especially in automotive work, pipe threads which can be made up without lubricant or sealer are desirable. This series fills that requirement and has the same general form as the Standard Taper threads, except that the form is truncated as shown in Fig. 12, and further that, in Fig. 11, $L_4 = L_2 + 1$. There is no clearance between external and internal threads when the joint is made up; thus the flanks and flats meet, forming a metal-to-metal joint. A lubricant may be used to prevent galling (see Table 16). (Refer to ANSI B2.2-1968.)

Tap drill sizes for taper and straight pipe threads are listed in Table 17.

**Wrench boltheads, nuts, and wrench openings** have been stan-dardized by the ANSI 18.2-1972 series. Wrench openings are given in Table 18.

Boltheads and nuts have been standardized to fit wrench sizes. These standards are given in ANSI B18.2-1972. The pertinent information is given in Table 19.

**Machine Screws**

Machine screws are defined according to head types as follows:

**Flat Head** This screw has a flat surface for the top of the head with a countersink angle of 82°. It is standard for machine screws, cap screws, and wood screws.

**Round Head** This screw has a semielliptical-shaped head and is standard for machine screws, cap screws, and wood screws except that for the cap screw it is called *button head*.

**Fillister Head** This screw has a rounded surface for the top

**Table 5. 12 Thread Series—12UN and 12N**
(All dimensions in inches. **Bold type indicates Unified threads.**)

| Identification | | Screw basic diameters | | External threads | | | | Internal threads | | Areas of sections | |
|---|---|---|---|---|---|---|---|---|---|---|---|
| Size | Threads per in., $n$ | Major diam, $D$* max | Pitch diam, $E$ max | Basic minor diam, $K_s$ | Allowances†, classes 1A and 2A | Class 1A, major diam tolerances‡ | Class 2A and 3A, major diam tolerances‡ | Minor diam, $K_n$, min | Minor diam tolerances, classes 1B, 2B, and 3B§ | Basic min minor diam, sq in. | Stress area, sq in. |
| ½ | 12 | 0.5000 | 0.4459 | 0.3978 | 0.0016 | ...... | 0.0114 | 0.4098 | 0.0127 | 0.1205 | 0.1374 |
| ⅝ | 12 | 0.6250 | 0.5709 | 0.5228 | 0.0016 | ...... | 0.0114 | 0.5348 | 0.0115 | 0.2097 | 0.2319 |
| 11/16 | 12 | 0.6875 | 0.6334 | 0.5853 | 0.0016 | ...... | 0.0114 | 0.5973 | 0.0112 | 0.2635 | 0.2883 |
| ¾ | 12 | 0.7500 | 0.6959 | 0.6478 | 0.0017 | ...... | 0.0114 | 0.6598 | 0.0109 | 0.3234 | 0.3508 |
| 13/16 | 12 | 0.8125 | 0.7584 | 0.7103 | 0.0017 | ...... | 0.0114 | 0.7223 | 0.0106 | 0.3895 | 0.4195 |
| ⅞ | 12 | 0.8750 | 0.8209 | 0.7728 | 0.0017 | ...... | 0.0114 | 0.7848 | 0.0108 | 0.4617 | 0.4943 |
| 15/16 | 12 | 0.9375 | 0.8834 | 0.8353 | 0.0017 | ...... | 0.0114 | 0.8473 | 0.0090 | 0.5000 | 0.5753 |
| 1 1/16 | 12 | 1.0625 | 1.0084 | 0.9603 | 0.0017 | ...... | 0.0114 | 0.9723 | 0.0100 | 0.7151 | 0.7556 |
| 1 3/16 | 12 | 1.1875 | 1.1334 | 1.0853 | 0.0017 | ...... | 0.0114 | 1.0973 | 0.0100 | 0.9147 | 0.9604 |
| 1 5/16 | 12 | 1.3125 | 1.2584 | 1.2103 | 0.0017 | ...... | 0.0114 | 1.2223 | 0.0100 | 1.1389 | 1.1898 |
| 1 7/16 | 12 | 1.4375 | 1.3834 | 1.3353 | 0.0018 | ...... | 0.0114 | 1.3473 | 0.0100 | 1.3876 | 1.4438 |
| 1⅝ | 12 | 1.6250 | 1.5709 | 1.5228 | 0.0018 | ...... | 0.0114 | 1.5348 | 0.0100 | 1.8067 | 1.8701 |
| 1¾ | 12 | 1.7500 | 1.6959 | 1.6418 | 0.0018 | ...... | 0.0114 | 1.6598 | 0.0100 | 2.1168 | 2.1853 |
| 1⅞ | 12 | 1.8750 | 1.8209 | 1.7728 | 0.0018 | ...... | 0.0114 | 1.7848 | 0.0100 | 2.4514 | 2.5250 |
| 2 | 12 | 2.0000 | 1.9459 | 1.8978 | 0.0018 | ...... | 0.0114 | 1.9098 | 0.0100 | 2.8106 | 2.8892 |
| 2⅛ | 12 | 2.1250 | 2.0709 | 2.0228 | 0.0018 | ...... | 0.0114 | 2.0348 | 0.0100 | 3.1943 | 3.2779 |
| 2¼ | 12 | 2.2500 | 2.1959 | 2.1478 | 0.0018 | ...... | 0.0114 | 2.1598 | 0.0100 | 3.6025 | 3.6914 |
| 2⅜ | 12 | 2.3750 | 2.3209 | 2.2728 | 0.0019 | ...... | 0.0114 | 2.2848 | 0.0100 | 4.0353 | 4.1291 |
| 2½ | 12 | 2.5000 | 2.4459 | 2.3978 | 0.0019 | ...... | 0.0114 | 2.4098 | 0.0100 | 4.4927 | 4.5916 |
| 2⅝ | 12 | 2.6250 | 2.5709 | 2.5228 | 0.0019 | ...... | 0.0114 | 2.5348 | 0.0100 | 4.9745 | 5.0784 |
| 2¾ | 12 | 2.7500 | 2.6959 | 2.6478 | 0.0019 | ...... | 0.0114 | 2.6598 | 0.0100 | 5.4810 | 5.5900 |
| 2⅞ | 12 | 2.8750 | 2.8209 | 2.7728 | 0.0019 | ...... | 0.0114 | 2.7848 | 0.0100 | 6.0119 | 6.1259 |
| 3 | 12 | 3.0000 | 2.9459 | 2.8978 | 0.0019 | ...... | 0.0114 | 2.9098 | 0.0100 | 6.5674 | 6.6865 |
| 3⅛ | 12 | 3.1250 | 3.0709 | 3.0228 | 0.0019 | ...... | 0.0114 | 3.0348 | 0.0100 | 7.1475 | 7.2714 |
| 3¼ | 12 | 3.2500 | 3.1959 | 3.1478 | 0.0019 | ...... | 0.0114 | 3.1598 | 0.0100 | 7.7521 | 7.8812 |
| 3⅜ | 12 | 3.3750 | 3.3209 | 3.2728 | 0.0019 | ...... | 0.0114 | 3.2848 | 0.0100 | 8.3812 | 8.5152 |
| 3½ | 12 | 3.5000 | 3.4459 | 3.3978 | 0.0019 | ...... | 0.0114 | 3.4098 | 0.0100 | 9.0349 | 9.1740 |
| 3⅝ | 12 | 3.6250 | 3.5709 | 5.5228 | 0.0019 | ...... | 0.0114 | 3.5348 | 0.0100 | 9.7132 | 9.8570 |
| 3¾ | 12 | 3.7500 | 3.6959 | 3.6478 | 0.0019 | ...... | 0.0114 | 3.6598 | 0.1000 | 10.4159 | 10.4649 |
| 3⅞ | 12 | 3.8750 | 3.8209 | 3.7728 | 0.0020 | ...... | 0.0114 | 3.7848 | 0.0100 | 11.1433 | 11.2970 |
| 4 | 12 | 4.0000 | 3.9459 | 3.8978 | 0.0020 | ...... | 0.0114 | 3.9098 | 0.0100 | 11.8951 | 12.0540 |
| 4¼ | 12 | 4.2500 | 4.1959 | 4.1478 | 0.0020 | ...... | 0.0114 | 4.1598 | 0.0100 | 13.4725 | 13.6411 |
| 4½ | 12 | 4.5000 | 4.4459 | 4.3918 | 0.0020 | ...... | 0.0114 | 4.4098 | 0.0100 | 15.1480 | 15.3265 |
| 4¾ | 12 | 4.7500 | 4.6959 | 4.6478 | 0.0020 | ...... | 0.0114 | 4.6598 | 0.0100 | 16.9217 | 17.1099 |
| 5 | 12 | 5.0000 | 4.9459 | 4.8978 | 0.0020 | ...... | 0.0114 | 4.9098 | 0.0100 | 18.7936 | 18.9916 |
| 5¼ | 12 | 5.2500 | 5.1959 | 5.1478 | 0.0020 | ...... | 0.0114 | 5.1598 | 0.0100 | 20.7636 | 20.9717 |
| 5½ | 12 | 5.5000 | 5.4459 | 5.3978 | 0.0020 | ...... | 0.0114 | 5.4098 | 0.0100 | 22.8319 | 23.0496 |
| 5¾ | 12 | 5.7500 | 5.6959 | 5.6478 | 0.0021 | ...... | 0.0114 | 5.6598 | 0.0100 | 24.9983 | 25.2257 |
| 6 | 12 | 6.0000 | 5.9459 | 5.8978 | 0.0021 | ...... | 0.0014 | 5.9098 | 0.0100 | 27.2628 | 27.4988 |

The above values are based on a length of engagement equal to the nominal diameter.

* Major diameter at intersection of rounded root with flanks of threads.

† Allowances apply to external threads, Classes 1A and 2A only. The allowance for Class 3A threads is zero.

‡ Major diameter of internal threads may extend to a $P/24$ flat.

§ Minor diameter of external threads may extend to a $P/8$ flat.

**Table 6. 16 Thread Series—16UN and 16N**
(All dimensions in inches.   **Bold type indicates Unified threads.**)

| Identification | | Screw basic diameters | | External threads | | | | Internal threads | | Areas of sections | |
| --- | --- | --- | --- | --- | --- | --- | --- | --- | --- | --- | --- |
| Size | Threads per in., $n$ | Major diam, $D$* max | Pitch diam, $E$ max | Basic minor diam, $K_s$ | Allowances†, classes 1A and 2A | Class 1A, major diam tolerances‡ | Class 2A and 3A, major diam tolerances‡ | Minor diam, $K_n$ min | Minor diam tolerances, classes 1B, 2B, and 3B§ | Basic min minor diam, sq in. | Stress area, sq in. |
| 1 3/16 | 16 | 0.8125 | 0.7719 | 0.7358 | 0.0015 | ..... | 0.0094 | 0.7448 | 0.0085 | 0.4200 | 0.4429 |
| 7/8 | 16 | 0.8750 | 0.8344 | 0.7983 | 0.0015 | ..... | 0.0094 | 0.8073 | 0.0085 | 0.4949 | 0.5197 |
| 1 5/16 | 16 | 0.9375 | 0.8969 | 0.8608 | 0.0015 | ..... | 0.0094 | 0.8698 | 0.0085 | 0.5759 | 0.6025 |
| **1** | 16 | 1.0000 | 0.9594 | 0.9233 | 0.0015 | ..... | 0.0094 | 0.9323 | 0.0085 | 0.6630 | 0.6916 |
| 1 1/16 | 16 | 1.0625 | 1.0219 | 0.9958 | 0.0015 | ..... | 0.0094 | 0.9948 | 0.0085 | 0.7563 | 0.7867 |
| **1 1/8** | 16 | 1.1250 | 1.0844 | 1.0483 | 0.0015 | ..... | 0.0094 | 1.0573 | 0.0085 | 0.8557 | 0.8880 |
| 1 3/16 | 16 | 1.1875 | 1.1469 | 1.1108 | 0.0015 | ..... | 0.0094 | 1.1198 | 0.0085 | 0.9612 | 0.9955 |
| **1 1/4** | 16 | 1.2500 | 1.2094 | 1.1733 | 0.0015 | ..... | 0.0094 | 1.1823 | 0.0085 | 1.0729 | 1.1090 |
| 1 5/16 | 16 | 1.3125 | 1.2719 | 1.2358 | 0.0015 | ..... | 0.0094 | 1.2448 | 0.0085 | 1.1907 | 1.2287 |
| **1 3/8** | 16 | 1.3750 | 1.3344 | 1.2983 | 0.0015 | ..... | 0.0094 | 1.3073 | 0.0085 | 1.3147 | 1.3545 |
| 1 7/16 | 16 | 1.4375 | 1.3969 | 1.3608 | 0.0016 | ..... | 0.0094 | 1.3698 | 0.0085 | 1.4448 | 1.4865 |
| **1 1/2** | 16 | 1.5000 | 1.4594 | 1.4233 | 0.0016 | ..... | 0.0094 | 1.4323 | 0.0085 | 1.5810 | 1.6246 |
| 1 9/16 | 16 | 1.5625 | 1.5219 | 1.4858 | 0.0016 | ..... | 0.0094 | 1.4948 | 0.0085 | 1.7234 | 1.7687 |
| **1 5/8** | 16 | 1.6250 | 1.5844 | 1.5483 | 0.0016 | ..... | 0.0094 | 1.5573 | 0.0085 | 1.8719 | 1.9191 |
| 1 11/16 | 16 | 1.6875 | 1.6469 | 1.6108 | 0.0016 | ..... | 0.0094 | 1.6198 | 0.0085 | 2.0265 | 2.0757 |
| 1 13/16 | 16 | 1.8125 | 1.7719 | 1.7358 | 0.0016 | ..... | 0.0094 | 1.7448 | 0.0085 | 2.3542 | 2.4070 |
| **1 7/8** | 16 | 1.8750 | 1.8344 | 1.7983 | 0.0016 | ..... | 0.0094 | 1.8073 | 0.0085 | 2.5272 | 2.5819 |
| 1 15/16 | 16 | 1.9375 | 1.8969 | 1.8608 | 0.0016 | ..... | 0.0094 | 1.8698 | 0.0085 | 2.7062 | 2.7269 |
| 2 1/16 | 16 | 2.0625 | 2.0219 | 1.9858 | 0.0016 | ..... | 0.0094 | 1.9948 | 0.0085 | 3.0831 | 3.1434 |
| 2 1/8 | 16 | 2.1250 | 2.0844 | 2.0483 | 0.0016 | ..... | 0.0094 | 2.0573 | 0.0085 | 3.2807 | 3.3427 |
| 2 3/16 | 16 | 2.1875 | 2.1469 | 2.1108 | 0.0016 | ..... | 0.0094 | 2.1198 | 0.0085 | 3.4844 | 3.5483 |
| **2 1/4** | 16 | 2.2500 | 2.2094 | 2.1733 | 0.0016 | ..... | 0.0094 | 2.1823 | 0.0085 | 3.6943 | 3.7601 |
| 2 5/16 | 16 | 2.3125 | 2.2719 | 2.2358 | 0.0017 | ..... | 0.0094 | 2.2448 | 0.0085 | 3.9103 | 3.9708 |
| 2 3/8 | 16 | 2.3750 | 2.3344 | 2.2983 | 0.0017 | ..... | 0.0094 | 2.3073 | 0.0085 | 4.1324 | 4.2018 |
| 2 7/16 | 16 | 2.4375 | 2.3969 | 2.3608 | 0.0017 | ..... | 0.0094 | 2.3696 | 0.0085 | 4.3606 | 4.4319 |
| **2 1/2** | 16 | 2.5000 | 2.4594 | 2.4233 | 0.0017 | ..... | 0.0094 | 2.4323 | 0.0085 | 4.4950 | 4.6682 |
| **2 5/8** | 16 | 2.6250 | 2.5844 | 2.5483 | 0.0017 | ..... | 0.0094 | 2.5573 | 0.0085 | 5.0822 | 5.1790 |
| **2 3/4** | 16 | 2.7500 | 2.7094 | 2.6733 | 0.0017 | ..... | 0.0094 | 2.6823 | 0.0085 | 5.5940 | 5.6745 |
| **2 7/8** | 16 | 2.8750 | 2.8344 | 2.7983 | 0.0017 | ..... | 0.0094 | 2.8073 | 0.0085 | 6.1303 | 6.2143 |
| **3** | 16 | 3.0000 | 2.9594 | 2.9233 | 0.0017 | ..... | 0.0094 | 2.9323 | 0.0085 | 6.6911 | 6.7789 |
| **3 1/8** | 16 | 3.1250 | 3.0844 | 3.0483 | 0.0017 | ..... | 0.0094 | 3.0573 | 0.0085 | 7.2765 | 7.3678 |
| **3 1/4** | 16 | 3.2500 | 3.2094 | 3.1733 | 0.0017 | ..... | 0.0094 | 3.1823 | 0.0085 | 7.8864 | 7.9814 |
| **3 3/8** | 16 | 3.3750 | 3.3344 | 3.2983 | 0.0017 | ..... | 0.0094 | 3.3073 | 0.0085 | 8.5209 | 8.6194 |
| **3 1/2** | 16 | 3.5000 | 3.4594 | 3.4233 | 0.0017 | ..... | 0.0094 | 3.4323 | 0.0085 | 9.1799 | 9.2821 |
| **3 5/8** | 16 | 3.6250 | 3.5844 | 3.5483 | 0.0017 | ..... | 0.0094 | 3.5573 | 0.0085 | 9.8634 | 9.9691 |
| **3 3/4** | 16 | 3.7500 | 3.7094 | 3.6733 | 0.0017 | ..... | 0.0094 | 3.6823 | 0.0085 | 10.5715 | 10.6809 |
| **3 7/8** | 16 | 3.8750 | 3.8344 | 3.7983 | 0.0018 | ..... | 0.0094 | 3.8073 | 0.0085 | 11.3042 | 11.4170 |
| **4** | 16 | 4.0000 | 3.9594 | 3.9233 | 0.0018 | ..... | 0.0094 | 3.9323 | 0.0085 | 12.0614 | 12.1779 |
| **4 1/4** | 16 | 4.2500 | 4.2094 | 4.1733 | 0.0018 | ..... | 0.0094 | 4.1823 | 0.0085 | 13.6494 | 13.7730 |
| **4 1/2** | 16 | 4.5000 | 4.4594 | 4.4233 | 0.0018 | ..... | 0.0094 | 4.4323 | 0.0085 | 15.3355 | 15.4662 |
| **4 3/4** | 16 | 4.7500 | 4.7094 | 4.6733 | 0.0018 | ..... | 0.0094 | 4.6823 | 0.0085 | 17.1199 | 12.2575 |
| **5** | 16 | 5.0000 | 4.9594 | 4.9233 | 0.0018 | ..... | 0.0094 | 4.9323 | 0.0085 | 19.0024 | 19.1470 |
| **5 1/4** | 16 | 5.2500 | 5.2094 | 5.1733 | 0.0018 | ..... | 0.0094 | 5.1823 | 0.0085 | 20.9831 | 31.1350 |
| **5 1/2** | 16 | 5.5000 | 5.4538 | 5.4233 | 0.0018 | ..... | 0.0094 | 5.4323 | 0.0085 | 23.0620 | 23.2208 |
| **5 3/4** | 16 | 5.7500 | 5.7094 | 5.6733 | 0.0019 | ..... | 0.0094 | 5.6823 | 0.0085 | 25.2390 | 25.4047 |
| **6** | 16 | 6.0000 | 5.9594 | 5.9233 | 0.0019 | ..... | 0.0094 | 5.9323 | 0.0085 | 27.5142 | 27.6868 |

The above values are based on a length of engagement equal to the nominal diameter.

* Major diameter at intersection of rounded root with flanks of threads.

† Allowances apply to external threads, Classes 1A and 2A only.   The allowance for Class 3A threads is zero.

‡ Major diameter of internal threads may extend to a $P/24$ flat.

§ Minor diameter of external threads may extend to a $P/8$ flat.

of the head, the remainder being cylindrical. The head is standard for machine screws and cap screws.

**Oval Head** This screw has a rounded surface for the top of the head and a countersink angle of 82°. It is standard for machine screws and wood screws.

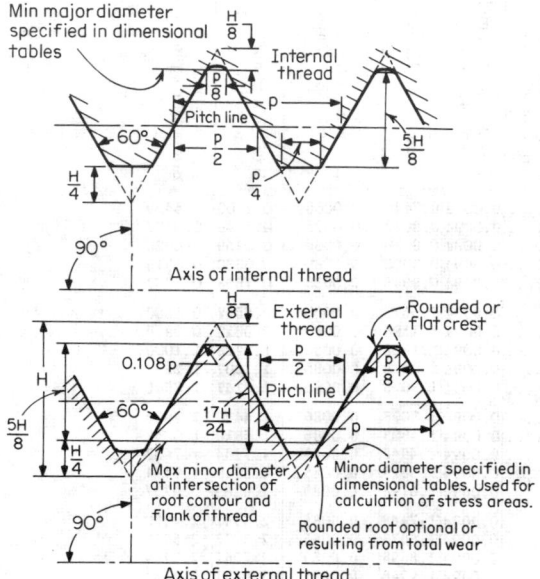

**Fig. 1** 60° Unified and American screw-thread forms.

**Hexagon Head** This screw has a hexagon-shaped head for use with external wrenches. It is standard for machine screws.

**Socket Head** This screw has an internal hexagon-shaped socket in the head for internal wrenching. It is standard for cap screws.

These screw heads are shown in Fig. 13; pertinent dimensions are in Table 20.

### Eyebolts

Eyebolts are classified as rivet, nut, or screw, and can be had on a swivel. See Fig. 14 and Tables 21 and 22. The safe working load may be obtained for each application by applying an appropriate factor of safety.

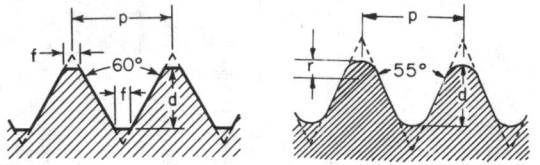

**Fig. 2** American standard screw threads.  **Fig. 3** Whitworth screw thread.

### Driving Recesses

Driving recesses for screws used in modern practice are shown in Fig. 15.

### Setscrews

Setscrews are used for fastening collars, sheaves, gears, etc., to shafts to prevent relative rotation or translation. They are

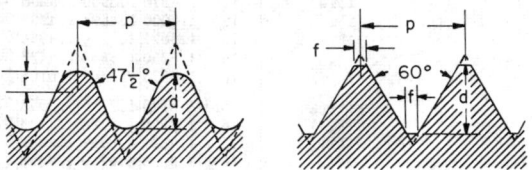

**Fig. 4** British Association screw thread.  **Fig. 5** Metric screw thread.

available in a variety of head and point styles, as shown in Fig. 16. A complete tabulation of dimensions is found in ANSI B18.3-1969 and B18.6.2-1972. Holding power for various sizes is given in Table 23.

### Locking Fasteners

Locking fasteners are used to prevent loosening of a threaded fastener in service and are available in a wide variety differing

**Table 7. Whitworth Standard Screw Threads**
(See Fig. 3)

| Diam of screw, in. | Threads per in. | Depth of thread, in. | Diam at root of thread, in. | Diam of screw, in. | Threads per in. | Depth of thread, in. | Diam at root of thread, in. |
|---|---|---|---|---|---|---|---|
| ¼ | 20 | 0.0320 | 0.1860 | 1⅝ | 5 | 0.1281 | 1.3689 |
| 5⁄16 | 18 | 0.0356 | 0.2414 | 1¾ | 5 | 0.1281 | 1.4939 |
| ⅜ | 16 | 0.0400 | 0.2950 | 2 | 4½ | 0.1423 | 1.7154 |
| 7⁄16 | 14 | 0.0457 | 0.3460 | 2¼ | 4 | 0.1601 | 1.9298 |
| ½ | 12 | 0.0534 | 0.3933 | 2½ | 4 | 0.1601 | 2.1798 |
| 9⁄16 | 12 | 0.0534 | 0.4558 | 2¾ | 3½ | 0.1830 | 2.3841 |
| ⅝ | 11 | 0.0582 | 0.5086 | 3 | 3½ | 0.1830 | 2.6341 |
| 11⁄16 | 11 | 0.0582 | 0.5711 | 3¼ | 3¼ | 0.1970 | 2.8560 |
| ¾ | 10 | 0.0640 | 0.6219 | 3½ | 3¼ | 0.1970 | 3.1060 |
| 13⁄16 | 10 | 0.0640 | 0.6844 | 3¾ | 3 | 0.2134 | 3.3231 |
| ⅞ | 9 | 0.0711 | 0.7327 | 4 | 3 | 0.2134 | 3.5731 |
| 1 | 8 | 0.0800 | 0.8399 | 4½ | 2⅞ | 0.2227 | 4.0546 |
| 1⅛ | 7 | 0.0915 | 0.9420 | 5 | 2¾ | 0.2328 | 4.5343 |
| 1¼ | 7 | 0.0915 | 1.0670 | 5½ | 2⅝ | 0.2439 | 5.0121 |
| 1⅜ | 6 | 0.1067 | 1.1616 | 6 | 2½ | 0.2561 | 5.4877 |
| 1½ | 6 | 0.1067 | 1.2866 | | | | |

**Table 8. British Association Screw Threads**
(See Fig. 4)

| Number | Diam of screw, mm | Approx diam, in. | Pitch, mm | Approx pitch, in. | Diam at root of thread, mm | Number | Diam of screw, mm | Approx diam, in. | Pitch, mm | Approx pitch, in. | Diam at root of thread, mm |
|---|---|---|---|---|---|---|---|---|---|---|---|
| 0 | 6.0 | 0.236 | 1.00 | 0.0394 | 4.8 | 13 | 1.20 | 0.047 | 0.25 | 0.0098 | 0.90 |
| 1 | 5.3 | 0.209 | 0.90 | 0.0354 | 4.22 | 14 | 1.00 | 0.039 | 0.23 | 0.0091 | 0.72 |
| 2 | 4.7 | 0.185 | 0.81 | 0.0319 | 3.73 | 15 | 0.90 | 0.035 | 0.21 | 0.0083 | 0.65 |
| 3 | 4.1 | 0.161 | 0.73 | 0.0287 | 3.22 | 16 | 0.79 | 0.031 | 0.19 | 0.0075 | 0.56 |
| 4 | 3.6 | 0.142 | 0.66 | 0.0260 | 2.81 | 17 | 0.70 | 0.028 | 0.17 | 0.0067 | 0.50 |
| 5 | 3.2 | 0.126 | 0.59 | 0.0232 | 2.49 | 18 | 0.62 | 0.024 | 0.15 | 0.0059 | 0.44 |
| 6 | 2.8 | 0.110 | 0.53 | 0.0209 | 2.16 | 19 | 0.54 | 0.021 | 0.14 | 0.0055 | 0.37 |
| 7 | 2.5 | 0.098 | 0.48 | 0.0189 | 1.92 | 20 | 0.48 | 0.019 | 0.12 | 0.0047 | 0.34 |
| 8 | 2.2 | 0.087 | 0.43 | 0.0169 | 1.68 | 21 | 0.42 | 0.017 | 0.11 | 0.0043 | 0.29 |
| 9 | 1.9 | 0.075 | 0.39 | 0.0154 | 1.43 | 22 | 0.37 | 0.015 | 0.10 | 0.0039 | 0.25 |
| 10 | 1.7 | 0.067 | 0.35 | 0.0138 | 1.28 | 23 | 0.33 | 0.013 | 0.09 | 0.0035 | 0.22 |
| 11 | 1.5 | 0.059 | 0.31 | 0.0122 | 1.13 | 24 | 0.29 | 0.011 | 0.08 | 0.0031 | 0.19 |
| 12 | 1.3 | 0.051 | 0.28 | 0.0110 | 0.96 | 25 | 0.25 | 0.010 | 0.07 | 0.0028 | 0.17 |

**Table 9. French Metric Standard Screw Threads**
(See Fig. 5)

| Nominal diam, mm | Pitch, mm French std | Pitch, mm German and Swiss std | Pitch diam, mm | Root diam, mm | Nominal diam, mm | Pitch, mm French std | Pitch, mm German and Swiss std | Pitch diam, mm | Root diam, mm |
|---|---|---|---|---|---|---|---|---|---|
| 2 | 0.40 | 0.40 | 1.740 | 1.48 | 24 | 3.00 | 3.00 | 22.051 | 20.10 |
| 3 | .... | 0.50 | 2.675 | 2.35 | 26 | 3.00 | .... | 24.051 | 22.10 |
| 3 | 0.60 | .... | 2.610 | 2.22 | 28 | 3.00 | .... | 26.051 | 24.10 |
| 4 | .... | 0.70 | 3.545 | 3.09 | 30 | 3.50 | 3.50 | 27.727 | 25.45 |
| 4 | 0.75 | .... | 3.513 | 3.03 | 32 | 3.50 | .... | 29.727 | 27.45 |
| 5 | .... | 0.80 | 4.480 | 3.96 | 34 | 3.50 | .... | 31.727 | 29.45 |
| 5 | 0.90 | .... | 4.415 | 3.83 | 36 | 4.00 | 4.00 | 33.402 | 30.80 |
| 10 | 1.50 | 1.50 | 9.026 | 8.05 | 38 | 4.00 | .... | 35.402 | 32.80 |
| 12 | 1.50 | .... | 11.026 | 10.05 | 40 | 4.00 | .... | 37.402 | 34.80 |
| 12 | .... | 1.75 | 10.863 | 9.73 | 42 | 4.50 | 4.50 | 39.077 | 36.15 |
| 14 | 2.00 | 2.00 | 12.701 | 11.40 | 44 | 4.50 | .... | 41.077 | 38.15 |
| 16 | 2.00 | 2.00 | 14.701 | 13.40 | 46 | 4.50 | .... | 43.077 | 40.15 |
| 18 | 2.50 | 2.50 | 16.376 | 14.75 | 48 | 5.00 | .... | 44.752 | 41.50 |
| 20 | 2.50 | 2.50 | 18.376 | 16.75 | 50 | 5.00 | .... | 46.752 | 43.50 |
| 22 | 2.50 | 2.50 | 20.376 | 18.75 | | | | | |

**Table 10. International Metric Standard Screw Threads**
(See Fig. 6)

| Diam of screw, mm | Pitch, mm | Diam of screw, mm | Pitch, mm | Diam of screw, mm | Pitch, mm | Diam of screw, mm | Pitch mm | Diam of screw, mm | Pitch, mm | Diam of screw, mm | Pitch, mm |
|---|---|---|---|---|---|---|---|---|---|---|---|
| 6 | 1.00 | 12 | 1.75 | 24 | 3.00 | 42 | 4.50 | 64 | 6.00 | 96 | 8.00 |
| 7 | 1.00 | 14 | 2.00 | 27 | 3.00 | 45 | 4.50 | 68 | 6.00 | 116 | 9.00 |
| 8 | 1.25 | 16 | 2.00 | 30 | 3.50 | 48 | 5.00 | 72 | 6.50 | 136 | 10.00 |
| 9 | 1.25 | 18 | 2.50 | 33 | 3.50 | 52 | 5.00 | 76 | 6.50 | | |
| 10 | 1.50 | 20 | 2.50 | 36 | 4.00 | 56 | 5.50 | 80 | 7.00 | | |
| 11 | 1.50 | 22 | 2.50 | 39 | 4.00 | 60 | 5.50 | 88 | 7.50 | | |

### Table 11. Acme Thread Series

($D$ = outside diam, $p$ = pitch. All dimensions in inches)
(See Figs. 7 to 10)

| Symbols | Thread dimensions | | | |
|---|---|---|---|---|
| | 29° general purpose | 29° stub | 60° stub | 10° modified |
| $t$ = thickness of thread.............. | $0.5p$ | $0.5p$ | $0.5p$ | $0.5p$ |
| $R$ = basic depth of thread........... | $0.5p$ | $0.3p$ | $0.433p$ | $0.5p*$ |
| $F$ = basic width of flat.............. | $0.3707p$ | $0.4224p$ | $0.250p$ | $0.4563p†$ |
| $G$ = (see Figs. 7, 8, 10).............. | $F - (0.52 \times$ clearance) | $F - (0.52 \times$ clearance) | $0.227p$ | $F - (0.17 \times$ clearance) |
| $E$ = basic pitch diam................. | $D - 0.5p$ | $D - 0.3p$ | $D - 0.433p$ | $D - 0.5p$ |
| $K$ = basic minor diam................. | $D - p$ | $D - 0.6p$ | $D - 0.866p$ | $D - p$ |

\* A clearance of at least 0.010 in. is added to $h$ on threads of 10-pitch and coarser, and 0.005 in. on finer pitches, to produce extra depth, thus avoiding interference with threads of mating parts at minor or major diameters.
† Measured at crest of screw thread.

### Table 12. Acme Thread Diameter-Pitch Combinations

(See Figs. 7 to 10)

| Size | Threads per in. | Size | Threads per in. | Size | Threads per in. | Size | Threads per in. | Size | Threads per in. |
|---|---|---|---|---|---|---|---|---|---|
| ¼ | 16 | ⅝ | 8 | 1¼ | 5 | 2¼ | 3 | 4 | 2 |
| 5⁄16 | 14 | ¾ | 6 | 1⅜ | 4 | 2½ | 3 | 4½ | 2 |
| ⅜ | 12 | ⅞ | 6 | 1½ | 4 | 2¾ | 3 | 5 | 2 |
| 7⁄16 | 12 | 1 | 5 | 1¾ | 4 | 3 | 2 | | |
| ½ | 10 | 1⅛ | 5 | 2 | 4 | 3½ | 2 | | |

### Table 13. Screw Threads for High-Strength Bolting

(All dimensions in inches)

| Size | Threads per in. | Allowance (minus) | Major diam | Major diam tolerance | Max pitch diam* | Max pitch diam tolerance | Minor diam max | Nut max minor diam | Nut max minor diam tolerance | Nut max pitch diam* | Nut max pitch diam tolerance |
|---|---|---|---|---|---|---|---|---|---|---|---|
| ¼ | 20 | 0.0010 | 0.2490 | 0.0072 | 0.2165 | 0.0026 | 0.1877 | 0.2060 | 0.0101 | 0.2211 | 0.0036 |
| 5⁄16 | 18 | 0.0011 | 0.3114 | 0.0082 | 0.2753 | 0.0030 | 0.2432 | 0.2630 | 0.0106 | 0.2805 | 0.0041 |
| ⅜ | 16 | 0.0013 | 0.3737 | 0.0090 | 0.3331 | 0.0032 | 0.2990 | 0.3184 | 0.0111 | 0.3389 | 0.0045 |
| 7⁄16 | 14 | 0.0013 | 0.4362 | 0.0098 | 0.3898 | 0.0036 | 0.3486 | 0.3721 | 0.0119 | 0.3960 | 0.0049 |
| ½ | 13 | 0.0015 | 0.4985 | 0.0104 | 0.4485 | 0.0037 | 0.4041 | 0.4290 | 0.0123 | 0.4552 | 0.0052 |
| 9⁄16 | 12 | 0.0016 | 0.5609 | 0.0112 | 0.5068 | 0:0040 | 0.4587 | 0.4850 | 0.0127 | 0.5140 | 0.0056 |
| ⅝ | 11 | 0.0017 | 0.6233 | 0.0118 | 0.5643 | 0.0042 | 0.5118 | 0.5397 | 0.0131 | 0.5719 | 0.0059 |
| ¾ | 10 | 0.0019 | 0.7481 | 0.0128 | 0.6831 | 0.0045 | 0.6254 | 0.6553 | 0.0136 | 0.6914 | 0.0064 |
| ⅞ | 9 | 0.0021 | 0.8729 | 0.0140 | 0.8007 | 0.0049 | 0.7366 | 0.7689 | 0.0142 | 0.8098 | 0.0070 |
| 1 | 8 | 0.0022 | 0.9978 | 0.0152 | 0.9166 | 0.0054 | 0.8444 | 0.8795 | 0.0148 | 0.9264 | 0.0076 |
| 1⅛ | 8 | 0.0024 | 1.1226 | 0.0152 | 1.0414 | 0.0055 | 0.9692 | 1.0045 | 0.0148 | 1.0517 | 0.0079 |
| 1¼ | 8 | 0.0025 | 1.2475 | 0.0152 | 1.1663 | 0.0058 | 1.0941 | 1.1295 | 0.0148 | 1.1771 | 0.0083 |
| 1⅜ | 8 | 0.0025 | 1.3725 | 0.0152 | 1.2913 | 0.0061 | 1.2191 | 1.2545 | 0.0148 | 1.3024 | 0.0086 |
| 1½ | 8 | 0.0027 | 1.4973 | 0.0152 | 1.4161 | 0.0063 | 1.3439 | 1.3795 | 0.0148 | 1.4278 | 0.0090 |
| 1⅝ | 8 | 0.0028 | 1.6222 | 0.0152 | 1.5410 | 0.0065 | 1.4688 | 1.5045 | 0.0148 | 1.5531 | 0.0093 |
| 1¾ | 8 | 0.0029 | 1.7471 | 0.0152 | 1.6659 | 0.0068 | 1.5937 | 1.6295 | 0.0148 | 1.6785 | 0.0097 |
| 1⅞ | 8 | 0.0030 | 1.8720 | 0.0152 | 1.7908 | 0.0070 | 1.7186 | 1.7545 | 0.0148 | 1.8038 | 0.0100 |
| 2 | 8 | 0.0031 | 1.9969 | 0.0152 | 1.9157 | 0.0073 | 1.8435 | 1.8795 | 0.0148 | 1.9294 | 0.0104 |
| 2⅛ | 8 | 0.0032 | 2.1218 | 0.0152 | 2.0406 | 0.0075 | 1.9682 | 2.0045 | 0.0148 | 2.0545 | 0.0107 |
| 2¼ | 8 | 0.0033 | 2.2467 | 0.0152 | 2.1655 | 0.0077 | 2.0933 | 2.1295 | 0.0148 | 2.1798 | 0.0110 |
| 2½ | 8 | 0.0035 | 2.4965 | 0.0152 | 2.4153 | 0.0082 | 2.3431 | 2.3795 | 0.0148 | 2.4305 | 0.0117 |
| 2¾ | 8 | 0.0037 | 2.7463 | 0.0152 | 2.6651 | 0.0087 | 2.5929 | 2.6295 | 0.0148 | 2.6812 | 0.0124 |
| 3 | 8 | 0.0038 | 2.9962 | 0.0152 | 2.9150 | 0.0092 | 2.8428 | 2.8795 | 0.0148 | 2.9318 | 0.0130 |
| 3¼ | 8 | 0.0039 | 3.2461 | 0.0152 | 3.1649 | 0.0093 | 3.0927 | 3.1295 | 0.0148 | 3.1820 | 0.0132 |
| 3½ | 8 | 0.0040 | 3.4960 | 0.0152 | 3.4148 | 0.0093 | 3.3426 | 3.3795 | 0.0148 | 3.4321 | 0.0133 |

The American standard form of thread shall be used. Pitch diameter tolerances include errors of lead and angle.

\* The maximum pitch diameters of screws are smaller than the minimum pitch diameters of nuts by these amounts.

### Table 14. ANSI Taper Pipe Thread
(All dimensions in inches)
(See Fig. 11)

| Nominal pipe size | O.D. of pipe | Threads per in. | Pitch of thread | Hand-tight engagement length, $L_1$ | Effective thread external length, $L_2$ | Wrench make-up length for internal thread length, $L_3$ | Over-all length external thread, $L_4$ |
|---|---|---|---|---|---|---|---|
| 1/16 | 0.3125 | 27 | 0.03704 | 0.160 | 0.2611 | 0.1111 | 0.3896 |
| 1/8 | 0.405 | 27 | 0.03704 | 0.180 | 0.2639 | 0.1111 | 0.3924 |
| 1/4 | 0.540 | 18 | 0.05556 | 0.200 | 0.4018 | 0.1667 | 0.5946 |
| 3/8 | 0.675 | 18 | 0.05556 | 0.240 | 0.4078 | 0.1667 | 0.6006 |
| 1/2 | 0.840 | 14 | 0.07143 | 0.320 | 0.5337 | 0.2143 | 0.7815 |
| 3/4 | 1.050 | 14 | 0.07143 | 0.339 | 0.5457 | 0.2143 | 0.7935 |
| 1 | 1.315 | 11½ | 0.08696 | 0.400 | 0.6828 | 0.2609 | 0.9845 |
| 1¼ | 1.660 | 11½ | 0.08696 | 0.420 | 0.7068 | 0.2609 | 1.0085 |
| 1½ | 1.900 | 11½ | 0.08696 | 0.420 | 0.7235 | 0.2609 | 1.0252 |
| 2 | 2.375 | 11½ | 0.08696 | 0.436 | 0.7565 | 0.2609 | 1.0582 |
| 2½ | 2.875 | 8 | 0.12500 | 0.682 | 1.1375 | 0.2500 | 1.5712 |
| 3 | 3.500 | 8 | 0.12500 | 0.766 | 1.2000 | 0.2500 | 1.6337 |
| 3½ | 4.000 | 8 | 0.12500 | 0.821 | 1.2500 | 0.2500 | 1.6837 |
| 4 | 4.500 | 8 | 0.12500 | 0.844 | 1.3000 | 0.2500 | 1.7337 |
| 5 | 5.563 | 8 | 0.12500 | 0.937 | 1.4063 | 0.2500 | 1.8400 |
| 6 | 6.625 | 8 | 0.12500 | 0.958 | 1.5125 | 0.2500 | 1.9462 |
| 8 | 8.625 | 8 | 0.12500 | 1.063 | 1.7125 | 0.2500 | 2.1462 |
| 10 | 10.750 | 8 | 0.12500 | 1.210 | 1.9250 | 0.2500 | 2.3587 |
| 12 | 12.750 | 8 | 0.12500 | 1.360 | 2.1250 | 0.2500 | 2.5587 |
| 14 O.D. | 14.000 | 8 | 0.12500 | 1.562 | 2.2500 | 0.2500 | 2.6837 |
| 16 O.D. | 16.000 | 8 | 0.12500 | 1.812 | 2.4500 | 0.2500 | 2.8837 |
| 18 O.D. | 18.000 | 8 | 0.12500 | 2.000 | 2.6500 | 0.2500 | 3.0837 |
| 20 O.D. | 20.000 | 8 | 0.12500 | 2.125 | 2.8500 | 0.2500 | 3.2837 |
| 24 O.D. | 24.000 | 8 | 0.12500 | 2.375 | 3.2500 | 0.2500 | 3.6837 |

### Table 15. American Standard Straight Pipe Threads
(All dimensions in inches)

| Nominal pipe size | Threads per in. | Pressure-tight with seals | | Pressure-tight without seals | | Free-fitting External | | Free-fitting Internal | | Loose-fitting External | | Loose-fitting Internal | |
|---|---|---|---|---|---|---|---|---|---|---|---|---|---|
| | | Pitch diam, max | Minor diam, min | Pitch diam, max | Minor diam, min | Pitch diam, max | Major diam, max | Pitch diam, max | Minor diam, min | Pitch diam, max | Major diam, max | Pitch diam, max | Minor diam, min |
| (1) | (2) | (3) | (4) | (5) | (6) | (7) | (8) | (9) | (10) | (11) | (12) | (13) | (14) |
| 1/16 | 27 | ...... | ...... | 0.2812 | 0.2491 | | | | | | | | |
| 1/8 | 27 | 0.3782 | 0.342 | 0.3736 | 0.3415 | 0.3748 | 0.399 | 0.3783 | 0.350 | 0.3840 | 0.409 | 0.3898 | 0.362 |
| 1/4 | 18 | 0.4951 | 0.440 | 0.4916 | 0.4435 | 0.4899 | 0.527 | 0.4951 | 0.453 | 0.5038 | 0.541 | 0.5125 | 0.470 |
| 3/8 | 18 | 0.6322 | 0.577 | 0.6270 | 0.5789 | 0.6270 | 0.664 | 0.6322 | 0.590 | 0.6409 | 0.678 | 0.6496 | 0.607 |
| 1/2 | 14 | 0.7851 | 0.715 | 0.7784 | 0.7150 | 0.7784 | 0.826 | 0.7851 | 0.731 | 0.7963 | 0.844 | 0.8075 | 0.753 |
| 3/4 | 14 | 0.9956 | 0.925 | 0.9889 | 0.9255 | 0.9889 | 1.036 | 0.9956 | 0.941 | 1.0067 | 1.054 | 1.0179 | 0.964 |
| 1 | 11½ | 1.2468 | 1.161 | 1.2386 | 1.1621 | 1.2386 | 1.296 | 1.2468 | 1.181 | 1.2604 | 1.318 | 1.2739 | 1.208 |
| 1¼ | 11½ | 1.5915 | 1.506 | ...... | ...... | 1.5834 | 1.641 | 1.5916 | 1.526 | 1.6051 | 1.663 | 1.6187 | 1.553 |
| 1½ | 11½ | 1.8305 | 1.745 | ...... | ...... | 1.8223 | 1.880 | 1.8305 | 1.764 | 1.8441 | 1.902 | 1.8576 | 1.692 |
| 2 | 11½ | 2.3044 | 2.219 | ...... | ...... | 2.2963 | 2.354 | 2.3044 | 2.238 | 2.3180 | 2.376 | 2.3315 | 2.265 |
| 2½ | 8 | 2.7739 | 2.650 | ...... | ...... | 2.7622 | 2.846 | 2.7739 | 2.679 | 2.7934 | 2.877 | 2.8129 | 2.718 |
| 3 | 8 | 3.4002 | 3.277 | ...... | ...... | 3.3885 | 3.472 | 3.4002 | 3.305 | 3.4198 | 3.503 | 3.4393 | 3.344 |
| 3½ | 8 | 3.9005 | 3.777 | ...... | ...... | 3.8888 | 3.972 | 3.9005 | 3.806 | 3.9201 | 4.003 | 3.9396 | 3.845 |
| 4 | 8 | 4.3988 | 4.275 | ...... | ...... | 4.3871 | 4.470 | 4.3988 | 4.304 | 4.4184 | 4.502 | 4.4379 | 4.343 |
| 5 | 8 | ...... | ...... | ...... | ...... | 5.4493 | 5.533 | 5.4610 | 5.366 | 5.4805 | 5.564 | 5.5001 | 5.405 |
| 6 | 8 | ...... | ...... | ...... | ...... | 6.5060 | 6.589 | 6.5177 | 6.423 | 6.5372 | 6.620 | 6.5567 | 6.462 |
| 8 | 8 | ...... | ...... | ...... | ...... | ...... | ...... | ...... | ...... | 8.5313 | 8.615 | 8.5508 | 8.456 |
| 10 | 8 | ...... | ...... | ...... | ...... | ...... | ...... | ...... | ...... | 10.6522 | 10.735 | 10.6717 | 10.577 |
| 12 | 8 | ...... | ...... | ...... | ...... | ...... | ...... | ...... | ...... | 12.6491 | 12.732 | 12.6686 | 12.574 |

**Table 16. American Standard Dryseal Pipe Threads**
(See Fig. 12)

| Threads per in., $n$ | Truncation, in. | | Width of flat | |
|---|---|---|---|---|
| | Min | Max | Min | Max |
| 27 Crest | 0.047p | 0.094p | 0.054p | 0.108p |
| Root | 0.094p | 0.140p | 0.108p | 0.162p |
| 18 C | 0.047p | 0.078p | 0.054p | 0.090p |
| R | 0.078p | 0.109p | 0.090p | 0.126p |
| 14 C | 0.036p | 0.060p | 0.042p | 0.070p |
| R | 0.060p | 0.085p | 0.070p | 0.098p |
| 11½ C | 0.040p | 0.060p | 0.046p | 0.069p |
| R | 0.060p | 0.090p | 0.069p | 0.103p |
| 8 C | 0.042p | 0.055p | 0.048p | 0.064p |
| R | 0.055p | 0.076p | 0.064p | 0.088p |

**Table 17. Pipe-Thread Tap-Drill Sizes, In**

| Nominal pipe size | Taper thread | Straight thread |
|---|---|---|
| 1/16 | 0.250 | 0.250 |
| 1/8 | 21/64 | 11/32 |
| 1/4 | 27/64 | 7/16 |
| 3/8 | 9/16 | 37/64 |
| 1/2 | 11/16 | 23/32 |
| 3/4 | 57/64 | 59/64 |
| 1 | 1⅛ | 1 5/32 |
| 1¼ | 1 15/32 | 1½ |
| 1½ | 1 23/32 | 1¾ |
| 2 | 2 3/16 | 2 7/32 |
| 2½ | 2 19/32 | 2 21/32 |

**Table 18. Wrench Boltheads, Nuts, and Wrench Openings**

(All dimensions in inches)

| Basic or max width across flats, boltheads and nuts | Wrench openings | | Basic or max width across flats, boltheads, and nuts | Wrench openings | | Basic or max width across flats, boltheads, and nuts | Wrench openings | | Basic or max width across flats, boltheads, and nuts | Wrench openings | |
|---|---|---|---|---|---|---|---|---|---|---|---|
| | Max | Min | | Max | Min | | Max | Min | | Max | Min |
| 5/32 | 0.163 | 0.158 | 13/16 | 0.826 | 0.818 | 1 13/16 | 1.835 | 1.822 | 3 | 3.035 | 3.016 |
| 3/16 | 0.195 | 0.190 | 7/8 | 0.888 | 0.880 | 1⅞ | 1.898 | 1.885 | 3⅛ | 3.162 | 3.142 |
| 1/4 | 0.257 | 0.252 | 15/16 | 0.953 | 0.944 | 2 | 2.025 | 2.011 | 3⅜ | 3.414 | 3.393 |
| 5/16 | 0.322 | 0.316 | 1 | 1.015 | 1.006 | 2 1/16 | 2.088 | 2.074 | 3½ | 3.540 | 3.518 |
| 11/32 | 0.353 | 0.347 | 1 1/16 | 1.077 | 1.068 | 2 3/16 | 2.225 | 2.200 | 3¾ | 3.793 | 3.770 |
| 3/8 | 0.384 | 0.378 | 1⅛ | 1.142 | 1.132 | 2¼ | 2.277 | 2.262 | 3⅞ | 3.918 | 3.895 |
| 7/16 | 0.446 | 0.440 | 1¼ | 1.267 | 1.257 | 2⅜ | 2.404 | 2.388 | 4⅛ | 4.172 | 4.147 |
| 1/2 | 0.510 | 0.504 | 1 5/16 | 1.331 | 1.320 | 2 7/16 | 2.466 | 2.450 | 4¼ | 4.297 | 4.272 |
| 9/16 | 0.573 | 0.566 | 1⅜ | 1.394 | 1.383 | 2 9/16 | 2.593 | 2.576 | 4½ | 4.550 | 4.524 |
| 19/32 | 0.605 | 0.598 | 1 7/16 | 1.457 | 1.446 | 2⅝ | 2.656 | 2.639 | 4⅝ | 4.676 | 4.649 |
| 5/8 | 0.636 | 0.629 | 1½ | 1.520 | 1.508 | 2¾ | 2.783 | 2.766 | 5 | 5.055 | 5.026 |
| 11/16 | 0.699 | 0.692 | 1⅝ | 1.646 | 1.634 | 2 13/16 | 2.845 | 2.827 | 5⅜ | 5.434 | 5.403 |
| 3/4 | 0.763 | 0.755 | 1 11/16 | 1.708 | 1.696 | 2 15/16 | 2.973 | 2.954 | 5¾ | 5.813 | 5.780 |
| 25/32 | 0.794 | 0.786 | | | | | | | 6⅛ | 6.192 | 6.157 |

Wrenches shall be marked with the "nominal size of wrench" which is equal to the basic or maximum width across flats of the corresponding bolthead or nut.

Allowance (min clearance) between maximum width across flats of nut or bolthead and jaws of wrench equals $(1.005W + 0.001)$. Tolerance on wrench opening equals plus $(0.005W + 0.004)$ from minimum ($W$ equals nominal size of wrench).

**Table 19. Width across Flats of Boltheads and Nuts**
(All dimensions in inches)

| Nominal size or basic major diam of thread | Dimensions of regular boltheads unfinished, square, and hexagon | | Dimensions of heavy boltheads unfinished, square, and hexagon | | Dimensions of cap-screw heads hexagon | | Dimensions of setscrew heads | | Dimensions of regular nuts and regular jam nuts, unfinished, square, and hexagon (jam nuts, hexagon only) | | Dimensions of machine-screw and stove-bolt nuts, square and hexagon | | Dimensions of heavy nuts and heavy jam nuts, unfinished, square, and hexagon (jam nuts, hexagon only) | |
|---|---|---|---|---|---|---|---|---|---|---|---|---|---|---|
| | Max | Min | Max | Min | Max | Min | Max | Min | Max | Min | Max | Min | Max | Min |
| No. 0 | | | | | | | | | | | 0.1562 | 0.150 | | |
| No. 1 | | | | | | | | | | | 0.1562 | 0.150 | | |
| No. 2 | | | | | | | | | | | 0.1875 | 0.180 | | |
| No. 3 | | | | | | | | | | | 0.1875 | 0.180 | | |
| No. 4 | | | | | | | | | | | 0.2500 | 0.241 | | |
| No. 5 | | | | | | | | | | | 0.3125 | 0.302 | | |
| No. 6 | | | | | | | | | | | 0.3125 | 0.302 | | |
| No. 8 | | | | | | | | | | | 0.3438 | 0.332 | | |
| No. 10 | | | | | | | | | | | 0.3750 | 0.362 | | |
| No. 12 | | | | | | | | | | | 0.4375 | 0.423 | | |
| ¼ | 0.3750 | 0.362 | | | 0.4375 | 0.428 | 0.2500 | 0.241 | 0.4375 | 0.425 | 0.4375 | 0.423 | 0.5000 | 0.488 |
| ⁵⁄₁₆ | 0.5000 | 0.484 | | | 0.5000 | 0.489 | 0.3125 | 0.302 | 0.5625 | 0.547 | 0.5625 | 0.545 | 0.5938 | 0.578 |
| ⅜ | 0.5625 | 0.544 | | | 0.5625 | 0.551 | 0.3750 | 0.362 | 0.6250 | 0.606 | 0.6250 | 0.607 | 0.6875 | 0.669 |
| ⁷⁄₁₆ | 0.6250 | 0.603 | | | 0.6250 | 0.612 | 0.4375 | 0.423 | 0.7500 | 0.728 | | | 0.7812 | 0.759 |
| ½ | 0.7500 | 0.725 | 0.8750 | 0.850 | 0.7500 | 0.736 | 0.5000 | 0.484 | 0.8125 | 0.788 | | | 0.8750 | 0.850 |
| ⁹⁄₁₆ | 0.8750 | 0.847 | 0.9375 | 0.909 | 0.8125 | 0.798 | 0.5625 | 0.545 | 0.8750 | 0.847 | | | 0.9375 | 0.909 |
| ⅝ | 0.9375 | 0.906 | 1.0625 | 1.031 | 0.8750 | 0.860 | 0.6250 | 0.606 | 1.0000 | 0.969 | | | 1.0625 | 1.031 |
| ¾ | 1.1250 | 1.088 | 1.2500 | 1.212 | 1.0000 | 0.983 | 0.7500 | 0.729 | 1.1250 | 1.088 | | | 1.2500 | 1.212 |
| ⅞ | 1.3125 | 1.269 | 1.4375 | 1.394 | 1.1250 | 1.106 | 0.8750 | 0.852 | 1.3125 | 1.269 | | | 1.4375 | 1.394 |
| 1 | 1.5000 | 1.450 | 1.6250 | 1.575 | 1.3125 | 1.292 | 1.0000 | 0.974 | 1.5000 | 1.450 | | | 1.6250 | 1.575 |
| 1⅛ | 1.6875 | 1.631 | 1.8125 | 1.756 | 1.5000 | 1.477 | 1.1250 | 1.096 | 1.6875 | 1.631 | | | 1.8125 | 1.756 |
| 1¼ | 1.8750 | 1.812 | 2.0000 | 1.938 | 1.6875 | 1.663 | 1.2500 | 1.219 | 1.8750 | 1.812 | | | 2.0000 | 1.938 |
| 1⅜ | 2.0625 | 1.994 | 2.1857 | 2.119 | | | 1.3750 | 1.342 | 2.0625 | 1.994 | | | 2.1875 | 2.119 |
| 1½ | 2.2500 | 2.175 | 2.3750 | 2.300 | | | 1.5000 | 1.464 | 2.2500 | 2.175 | | | 2.3750 | 2.300 |
| 1⅝ | 2.4375 | 2.356 | 2.5625 | 2.481 | | | | | 2.4375 | 2.356 | | | 2.5625 | 2.481 |
| 1¾ | 2.6250 | 2.538 | 2.7500 | 2.662 | | | | | 2.6250 | 2.538 | | | 2.7500 | 2.662 |
| 1⅞ | 2.8125 | 2.719 | 2.9375 | 2.844 | | | | | 2.8125 | 2.719 | | | 2.9375 | 2.844 |
| 2 | 3.0000 | 2.900 | 3.1250 | 3.025 | | | | | 3.0000 | 2.900 | | | 3.1250 | 3.025 |
| 2¼ | 3.3750 | 3.262 | 3.5000 | 3.388 | | | | | 3.3750 | 3.262 | | | 3.5000 | 3.388 |
| 2½ | 3.7500 | 3.625 | 3.8750 | 3.750 | | | | | 3.7500 | 3.625 | | | 3.8750 | 3.750 |
| 2¾ | 4.1250 | 3.988 | 4.2500 | 4.112 | | | | | 4.1250 | 3.988 | | | 4.2500 | 4.112 |
| 3 | 4.5000 | 4.350 | 4.6250 | 4.475 | | | | | 4.5000 | 4.350 | | | 4.6250 | 4.475 |
| 3¼ | | | | | | | | | | | | | 5.0000 | 4.838 |
| 3½ | | | | | | | | | | | | | 5.3750 | 5.200 |
| 3¾ | | | | | | | | | | | | | 5.7500 | 5.562 |
| 4 | | | | | | | | | | | | | 6.1250 | 5.925 |

Regular boltheads are for general use. Unfinished boltheads are not finished on any surface. Semifinished boltheads are finished under head.

Regular nuts are for general use. Semifinished nuts are finished on bearing surface and threaded. Unfinished nuts are not finished on any surface but are threaded.

vastly in design, performance, and function. Since each has special features which may make it of particular value in the solution of a given machine problem, it is important that great care be exercised in the selection of a particular design in order that its properties may be fully utilized. These fasteners may be divided into six groups, as follows: seating lock, spring stop nut, interference, wedge, blind, and quick-release. The **seating-lock type** locks only when firmly seated and is therefore free-running on the bolt. The **spring stop-nut type** of fastener functions by a spring action clamping down upon the bolt. The **prevailing torque type** locks by elastic or plastic flow of a portion of the fastener material. A recent development employs an adhesive coating applied to the threads. The **wedge type** locks by relative wedging of either elements or nut and bolt. The **blind type** usually utilizes spring action of the fastener, and the **quick-release type** utilizes a quarter-turn release device. An example of each is shown in Fig. 17.

There is a continuing effort to develop industry-wide standards for these fasteners. One such specification developed for prevailing torque fasteners by the Industrial Fasteners Institute is based on locking torque and may form a precedent for other types of fasteners as well.

**Wood screws** are made in lengths from ¼ to 5 in for steel and from ¼ to 3½ in for brass screws, increasing by ⅛ in up to 1 in, by ¼ in up to 3 in, and by ½ in up to 5 in. The **American standard** sizes are given in Table 25. Screws are made with flat, round, or oval heads. Figure 19 shows several screw heads.

**Washers** for bolts and lag screws, either round or square, are made to the dimensions given in Table 26.

**Self-tapping screws** are available in three types. **Thread-forming** tapping screws plastically displace material adjacent to the pilot hole. **Thread-cutting** tapping screws have cutting edges

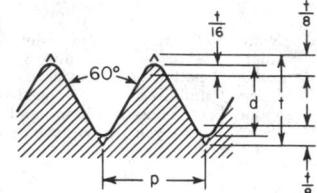

**Fig. 6**   International metric standard screw thread.

### Table 20.  Head Diameters (Maximum) in Inches

| | | | Machine screws | | | |
|---|---|---|---|---|---|---|
| Nominal size | Screw diam | Flat head | Round head | Fillister head | Oval head | Hexagon head across flats |
| 2 | 0.086 | 0.172 | 0.162 | 0.140 | 0.172 | 0.125 |
| 3 | 0.099 | 0.199 | 0.187 | 0.161 | 0.199 | 0.187 |
| 4 | 0.112 | 0.225 | 0.211 | 0.183 | 0.225 | 0.187 |
| 5 | 0.125 | 0.252 | 0.236 | 0.205 | 0.252 | 0.187 |
| 6 | 0.138 | 0.279 | 0.260 | 0.226 | 0.279 | 0.250 |
| 8 | 0.164 | 0.332 | 0.309 | 0.270 | 0.332 | 0.250 |
| 10 | 0.190 | 0.385 | 0.359 | 0.313 | 0.385 | 0.312 |
| 12 | 0.216 | 0.438 | 0.408 | 0.357 | 0.438 | 0.312 |
| ¼ | 0.250 | 0.507 | 0.472 | 0.414 | 0.507 | 0.375 |
| ⁵⁄₁₆ | 0.3125 | 0.636 | 0.591 | 0.519 | 0.636 | 0.500 |
| ⅜ | 0.375 | 0.762 | 0.708 | 0.622 | 0.762 | 0.562 |

| | | | Cap screws | | |
|---|---|---|---|---|---|
| Nominal size | Screw diam | Flat head | Button head | Fillister head | Socket head |
| ¼ | 0.250 | ½ | ⁷⁄₁₆ | ⅜ | ⅜ |
| ⁵⁄₁₆ | 0.3125 | ⅝ | ⁹⁄₁₆ | ⁷⁄₁₆ | ⁷⁄₁₆ |
| ⅜ | 0.375 | ¾ | ⅝ | ⁹⁄₁₆ | ⁹⁄₁₆ |
| ⁷⁄₁₆ | 0.4375 | ¹³⁄₁₆ | ¾ | ⅝ | ⅝ |
| ½ | 0.500 | ⅞ | ¹³⁄₁₆ | ¾ | ¾ |
| ⁹⁄₁₆ | 0.5625 | 1 | ¹⁵⁄₁₆ | ¹³⁄₁₆ | ¹³⁄₁₆ |
| ⅝ | 0.625 | 1⅛ | 1 | ⅞ | ⅞ |
| ¾ | 0.750 | 1⅜ | 1¼ | 1 | 1 |
| ⅞ | ..... | ..... | ..... | 1⅛ | 1⅛ |
| 1 | ..... | ..... | ..... | 1⁵⁄₁₆ | 1⁵⁄₁₆ |

### Table 21.  Regular Nut Eyebolts—Selected Sizes
(Thomas Laughlin Co., Portland, Me.)
(All dimensions in inches)
(See Fig. 14)

| Diam and shank length | Thread length | Eye dimension | | Approx breaking strength, lb | Diam and shank length | Thread length | Eye dimension | | Approx breaking strength, lb |
|---|---|---|---|---|---|---|---|---|---|
| | | I.D. | O.D. | | | | I.D. | O.D. | |
| ¼ × 2 | 1½ | ½ | 1 | 2,200 | ¾ × 6 | 3 | 1½ | 3 | 23,400 |
| ⁵⁄₁₆ × 2¼ | 1½ | ⅝ | 1¼ | 3,600 | ¾ × 10 | 3 | 1½ | 3 | 23,400 |
| ⅜ × 4½ | 2½ | ¾ | 1½ | 5,200 | ¾ × 15 | 5 | 1½ | 3 | 23,400 |
| ½ × 3¼ | 1½ | 1 | 2 | 9,800 | ⅞ × 8 | 4 | 1¾ | 3½ | 32,400 |
| ½ × 6 | 3 | 1 | 2 | 9,800 | 1 × 6 | 3 | 2 | 4 | 42,400 |
| ½ × 10 | 3 | 1 | 2 | 9,800 | 1 × 9 | 4 | 2 | 4 | 42,400 |
| ⅝ × 4 | 2 | 1¼ | 2½ | 15,800 | 1 × 18 | 7 | 2 | 4 | 42,400 |
| ⅝ × 6 | 3 | 1¼ | 2½ | 15,800 | 1¼ × 8 | 4 | 2½ | 5 | 67,800 |
| ⅝ × 10 | 3 | 1¼ | 2½ | 15,800 | 1¼ × 20 | 6 | 2½ | 5 | 67,800 |

**Table 22. Shoulder Nut Eyebolts—Selected Sizes**

(Thomas Laughlin Co., Portland, Me.)
(All dimensions in inches)
(See Fig. 14)

| Diam and shank length | Thread length | Eye dimension | | Approx breaking strength, lb | Diam and shank length | Thread length | Eye dimension | | Approx breaking strength, lb |
|---|---|---|---|---|---|---|---|---|---|
| | | I.D. | O.D. | | | | I.D. | O.D. | |
| $\frac{1}{4} \times 2$ | $1\frac{1}{2}$ | $\frac{1}{2}$ | $\frac{7}{8}$ | 2,200 | $\frac{3}{4} \times 6$ | 3 | $1\frac{1}{2}$ | $2\frac{3}{4}$ | 23,400 |
| $\frac{5}{16} \times 2\frac{1}{4}$ | $1\frac{1}{2}$ | $\frac{5}{8}$ | $1\frac{1}{8}$ | 3,600 | $\frac{7}{8} \times 5$ | $2\frac{1}{2}$ | $1\frac{3}{4}$ | $3\frac{1}{4}$ | 32,400 |
| $\frac{3}{8} \times 4\frac{1}{2}$ | $2\frac{1}{2}$ | $\frac{3}{4}$ | $1\frac{3}{8}$ | 5,200 | $1 \times 9$ | 4 | 2 | $3\frac{3}{4}$ | 42,400 |
| $\frac{1}{2} \times 6$ | 3 | 1 | $1\frac{3}{4}$ | 9,800 | $1\frac{1}{4} \times 8$ | 4 | $2\frac{1}{2}$ | $4\frac{1}{2}$ | 67,800 |
| $\frac{5}{8} \times 4$ | 2 | $1\frac{1}{4}$ | $2\frac{1}{4}$ | 15,800 | $1\frac{1}{2} \times 15$ | 6 | 3 | $5\frac{1}{2}$ | 98,400 |

**Table 23. Cup-point Setscrew Holding Power**

| Nominal screw size | Seating torque, lb·in | Axial holding power, lb |
|---|---|---|
| No. 0 | 0.5 | 50 |
| No. 1 | 1.5 | 65 |
| No. 2 | 1.5 | 85 |
| No. 3 | 5 | 120 |
| No. 4 | 5 | 160 |
| No. 5 | 9 | 200 |
| No. 6 | 9 | 250 |
| No. 8 | 20 | 385 |
| No. 10 | 33 | 540 |
| $\frac{1}{4}$ in. | 87 | 1,000 |
| $\frac{5}{16}$ in. | 165 | 1,500 |
| $\frac{3}{8}$ in. | 290 | 2,000 |
| $\frac{7}{16}$ in. | 430 | 2,500 |
| $\frac{1}{2}$ in. | 620 | 3,000 |
| $\frac{9}{16}$ in. | 620 | 3,500 |
| $\frac{5}{8}$ in. | 1,225 | 4,000 |
| $\frac{3}{4}$ in. | 2,125 | 5,000 |
| $\frac{7}{8}$ in. | 5,000 | 6,000 |
| 1 in. | 7,000 | 7,000 |

1. Torsional holding power in inch-pounds is equal to one-half of the axial holding power times the shaft diameter in inches.

2. Experimental data were obtained by seating an alloy-steel cup-point setscrew against a steel shaft with a hardness of Rockwell C 15. Screw threads were Class 3A, tapped holes were Class 2B. Holding power was defined as the minimum load necessary to produce 0.01 in. of relative movement between the shaft and the collar.

3. Cone points will develop a slightly greater holding power; flat, dog, and oval points, slightly less.

4. Shaft hardness should be at least 10 Rockwell C points less than the setscrew point.

5. Holding power is proportional to seating torque. Torsional holding power is increased about 6% by use of a flat on the shaft.

6. Data by F. R. Kull, Fasteners Book Issue, *Machine Design*, Mar. 11, 1965.

**Table 24. Coach and Lag Screws**

| Diam of screw, in | $\frac{1}{4}$ | $\frac{5}{16}$ | $\frac{3}{8}$ | $\frac{7}{16}$ | $\frac{1}{2}$ | $\frac{5}{8}$ | $\frac{3}{4}$ | $\frac{7}{8}$ | 1 |
|---|---|---|---|---|---|---|---|---|---|
| No. of threads per in | 10 | 9 | 7 | 7 | 6 | 5 | $4\frac{1}{2}$ | 4 | $3\frac{1}{2}$ |
| Across flats of hexagon and square heads, in | $\frac{3}{8}$ | $\frac{15}{32}$ | $\frac{9}{16}$ | $2\frac{1}{32}$ | $\frac{3}{4}$ | $\frac{15}{16}$ | $1\frac{1}{8}$ | $1\frac{5}{16}$ | $1\frac{1}{2}$ |
| Thickness of hexagon and square heads, in | $\frac{3}{16}$ | $\frac{1}{4}$ | $\frac{5}{16}$ | $\frac{3}{8}$ | $\frac{7}{16}$ | $\frac{17}{32}$ | $\frac{5}{8}$ | $\frac{3}{4}$ | $\frac{7}{8}$ |

Length of threads for screws of all diameters

| Length of screw, in | $1\frac{1}{2}$ | 2 | $2\frac{1}{2}$ | 3 | $3\frac{1}{2}$ | 4 | $4\frac{1}{2}$ |
|---|---|---|---|---|---|---|---|
| Length of thread, in | To head | $1\frac{1}{2}$ | 2 | $2\frac{1}{4}$ | $2\frac{1}{2}$ | 3 | $3\frac{1}{2}$ |

| Length of screw, in | 5 | $5\frac{1}{2}$ | 6 | 7 | 8 | 9 | 10–12 |
|---|---|---|---|---|---|---|---|
| Length of thread, in | 4 | 4 | $4\frac{1}{2}$ | 5 | 6 | 6 | 7 |

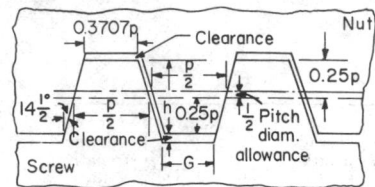

**Fig. 7**  29° general-purpose thread.

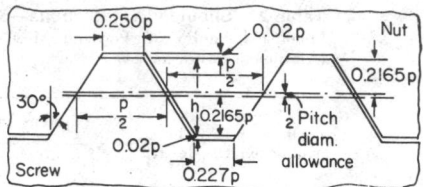

**Fig. 9**  60° stub thread.

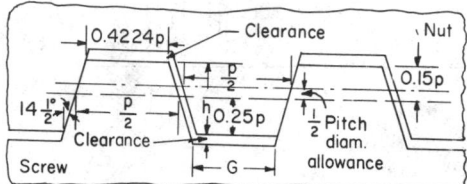

**Fig. 8**  29° stub thread.

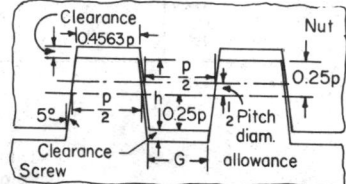

**Fig. 10**  10° modified square thread.

**Figs. 7–10**  Standard Acme threads.

and chip cavities (flutes) and form a mating thread by removing material adjacent to the pilot hole. Thread-cutting screws are generally used to join thicker and harder materials and require a lower driving torque than thread-forming screws. **Metallic drive** screws are forced into the material by pressure and are intended for making permanent fastenings. These three types are further classified on the basis of thread and point form as shown in Table 27. In addition to these body forms, a number of different head types are available. Basic dimensional data are given in Table 28.

**Carriage bolts** have been standardized in ANSI B18.5-1971. They come in styles shown in Fig. 20*a*. The range of bolt

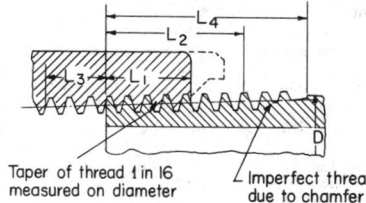

**Fig. 11**  American standard taper pipe thread.

diameters is No. 10 (=0.19 in) to 1 in, No. 10 to ¾ in, No. 10 to ½ in, and No. 10 to ¾ in, respectively.

**Stove bolts** (Fig. 20*b*) are made in the sizes given in Table 29.

**Materials, Strength, and Service Adaptability of Machine Bolts and Screws**

The **materials** to be used for bolts, screws, and nuts depend on service conditions and relative costs. The stresses to be

allowed in determining the proportions in any case depend on the nature of the loading and the material. Table 30 will serve as a guide to the selection of **bolts and nuts for fastenings.**

**Safe loads in tension for American standard bolts,** as determined by Harvey D. Williams, Bureau of Ships, U.S. Navy, are given in Table 31. Colvin and Stanley ("American Machinists' Handbook") computed the tensile and shearing strengths given in Table 32.

**General Notes on the Design of Bolted Joints**

**Bolts subjected to shock** and sudden change in load are found to be more serviceable when the unthreaded portion of the bolt is

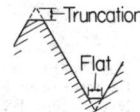

**Fig. 12**  American standard dryseal pipe threads.

turned down or drilled to the area of the root of the thread. The drilled bolt is stronger in torsion than the turned-down bolt.

When a **number of bolts** are **employed in fastening** together two parts of a machine, such as a cylinder and cylinder head, the load carried by each bolt depends on its relative tightness, the tighter bolts carrying the greater loads. When the conditions of assembly result in differences in tightness, lower working stresses must be used in designing the bolts than otherwise are necessary. On the other hand, it may be desirable to have the bolts the weakest part of the machine, since their breakage from overload in the machine may result in a minimum replacement cost. In such cases, the breaking load of the bolts

**Table 25. American Standard Wood Screws**

| Number | 0 | 1 | 2 | 3 | 4 | 5 | 6 | 7 | 8 |
|---|---|---|---|---|---|---|---|---|---|
| Threads per in | 32 | 28 | 26 | 24 | 22 | 20 | 18 | 16 | 15 |
| Diameter, in | 0.060 | 0.073 | 0.086 | 0.099 | 0.112 | 0.125 | 0.138 | 0.151 | 0.164 |

| Number | 9 | 10 | 11 | 12 | 14 | 16 | 18 | 20 | 24 |
|---|---|---|---|---|---|---|---|---|---|
| Threads per in | 14 | 13 | 12 | 11 | 10 | 9 | 8 | 8 | 7 |
| Diameter, in | 0.177 | 0.190 | 0.203 | 0.216 | 0.242 | 0.268 | 0.294 | 0.320 | 0.372 |

**Table 26. Dimensions of Steel Washers, In**

| Bolt size | Plain washer | | | Lock washer | | |
|---|---|---|---|---|---|---|
| | Hole diam | O.D. | Thickness | Hole diam | O.D. | Thickness |
| 3/16 | 1/4 | 9/16 | 3/64 | 0.194 | 0.337 | 0.047 |
| 1/4 | 5/16 | 3/4 | 1/16 | 0.255 | 0.493 | 0.062 |
| 5/16 | 3/8 | 7/8 | 1/16 | 0.319 | 0.591 | 0.078 |
| 3/8 | 7/16 | 1 | 5/64 | 0.382 | 0.688 | 0.094 |
| 7/16 | 1/2 | 1 1/4 | 5/64 | 0.446 | 0.784 | 0.109 |
| 1/2 | 9/16 | 1 3/8 | 3/32 | 0.509 | 0.879 | 0.125 |
| 9/16 | 5/8 | 1 1/2 | 3/32 | 0.573 | 0.979 | 0.141 |
| 5/8 | 11/16 | 1 3/4 | 1/8 | 0.636 | 1.086 | 0.156 |
| 3/4 | 13/16 | 2 | 1/8 | 0.763 | 1.279 | 0.188 |
| 7/8 | 15/16 | 2 1/4 | 5/32 | 0.890 | 1.474 | 0.219 |
| 1 | 1 1/16 | 2 1/2 | 5/32 | 1.017 | 1.672 | 0.250 |
| 1 1/8 | 1 1/4 | 2 3/4 | 5/32 | 1.144 | 1.865 | 0.281 |
| 1 1/4 | 1 3/8 | 3 | 5/32 | 1.271 | 2.058 | 0.312 |
| 1 3/8 | 1 1/2 | 3 1/4 | 11/64 | 1.398 | 2.253 | 0.344 |
| 1 1/2 | 1 5/8 | 3 1/2 | 11/64 | 1.525 | 2.446 | 0.375 |
| 1 5/8 | 1 3/4 | 3 3/4 | 11/64 | | | |
| 1 3/4 | 1 7/8 | 4 | 11/64 | | | |
| 1 7/8 | 2 | 4 1/4 | 11/64 | | | |
| 2 | 2 1/8 | 4 1/2 | 11/64 | | | |
| 2 1/4 | 2 3/8 | 4 3/4 | 3/16 | | | |
| 2 1/2 | 2 5/8 | 5 | 7/32 | | | |

may well be equal to the load which causes the weakest member of the machine connected to be stressed up to the elastic limit.

**Bolts screwed up tight** have an initial stress due to the tightening (preload) before any external load is applied to the machine member. The initial tensile load due to screwing up for a tight joint varies approximately as the diameter of the bolt, and may be estimated at 16,000 lb/in of diameter. The actual value depends upon the applied torque and the efficiency of the screw threads. Applying this rule to bolts of 1 in diam or less results in excessively high stresses, thus demonstrating why bolts of small diameter frequently fail during assembly. It is advisable to use as large diameter bolts as possible in pressure-tight joints requiring high tightening loads.

In pressuretight joints without a gasket the force on the bolt under load is essentially never greater than the initial tightening load. When a gasket is used, the total bolt force is approximately equal to the initial tightening load plus the external load. In the first case, deviations from the rule are a result of elastic behavior of the joint faces without a gasket, and inelastic behavior of the gasket in the latter case. The following generalization will serve as a guide. If the bolt is more yielding than the connecting members, it should be designed simply to resist the initial tension or the external load, whichever is greater. If the probable yielding of the bolt is 50 to 100 percent of that of the connected members, take the resultant bolt load as the initial tension plus one-half the external load. If the

yielding of the connected members is probably four to five times that of the bolt (as when certain packings are used), take the resultant bolt load as the initial tension plus three-fourths the external load.

In cases where bolts are subjected to cyclic loading, an increase in the initial tightening load decreases the operating stress range. In certain applications it is customary to fix the tightening load as a fraction of the yield-point load of the bolt.

In order to avoid the possibility of bolt failure in pressure-tight joints and to obtain uniformity in bolt loads, some means of determining initial bolt load (**preload**) is desirable. Calibrated torque wrenches are available for this purpose, reading directly in inch-pounds or inch-ounces. Inaccuracies in initial bolt load are possible even when using a torque wrench, owing to variations in the coefficient of friction between the nut and the bolt and, further, between the nut or bolthead and the abutting surface.

An exact method to determine the preload in a bolt requires that the bolt elongation be measured. For a through bolt in which both ends are accessible, the elongation is measured, and the preload force $P$ is obtained from the relationship

$$P = AEe \div l$$

where $E$ = modulus of elasticity, $l$ = original length, $A$ = cross-sectional area, $e$ = elongation. In cases where both ends of the bolt are not accessible, strain-gage techniques may be

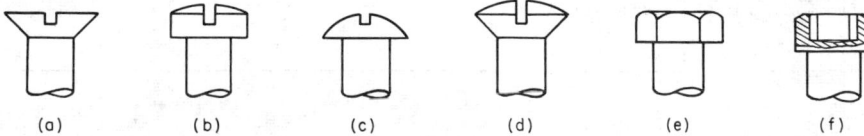

**Fig. 13**   Machine-screw heads. (*a*) Flat head; (*b*) fillister head; (*c*) round head; (*d*) oval head; (*e*) hexagon head; (*f*) socket head.

**Table 27. Tapping Screw Forms**

(Adapted, with permission, from *Machine Design*, Mar. 11, 1965.)

| ASA type and thread form | Description and recommendations |
|---|---|
| 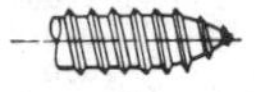 AB | Spaced thread, with gimlet point, designed for use in sheet metal, resin-impregnated plywood, wood, and asbestos compositions. Used in pierced or punched holes where a sharp point for starting is needed. |
| 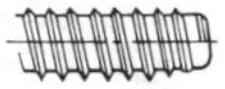 B | Type B is a blunt-point spaced-thread screw, used in heavy-gage sheet metal and non-ferrous castings. |
| 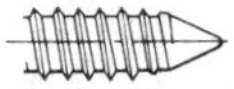 BP | Same as Type B, but has a 45 deg included angle unthreaded cone point. Used for locating and aligning holes or piercing soft materials. |
| 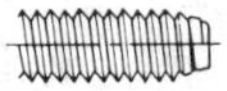 C | Blunt point with threads approximating machine-screw threads. For applications where a machine-screw thread is preferable to the spaced-thread form. Unlike thread-cutting screws, the Type C makes a chip-free assembly. |
|  U | Multiple-threaded drive screw with steep helix angle and a blunt, unthreaded starting pilot. Intended for making permanent fastenings in metals and plastics. Hammered or mechanically forced into work. Should not be used in materials less than one screw diameter thick. |
| 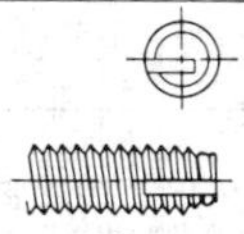 D | Blunt point with single narrow flute and threads approximating machine-screw threads. Flute is designed to produce a cutting edge which is radial to screw center. For low-strength metals and plastics; for high-strength brittle metals; and for rethreading clogged pretapped holes. |
| 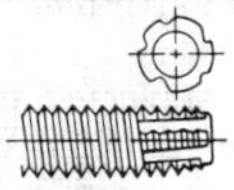 F | Approximate machine-screw thread and blunt point. |
| 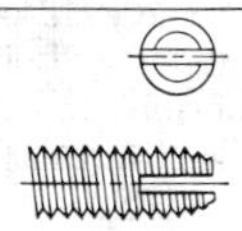 G | Approximate machine-screw thread with single through slot which forms two cutting edges. For low-strength metals and plastics. |
| 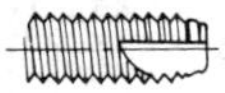 T | Same as Type D with single wide flute for more chip clearance. |
| 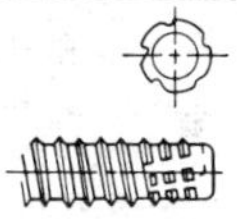 BF | Spaced thread with blunt point and five evenly spaced cutting grooves and chip cavities. Wall thickness should be $1\frac{1}{2}$ times major diameter of screw. Reduces stripping in brittle plastics and die castings. |
| 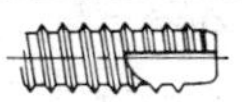 BT | Same as Type BF except for single wide flute which provides room for twisted, curly chips. |
| 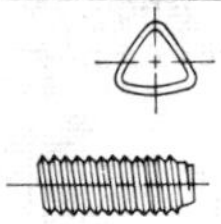 TT | Thread-rolling screws roll-form clean, screw threads. The plastic movement of the material it is driven into locks it in place. The "Taptite"® form is shown here. |

**Table 28. Self-tapping Screws**

| Screw size | Basic major diam, in. | Threads per in | | | | | Type U | |
|---|---|---|---|---|---|---|---|---|
| | | AB | B, BP | C | D, F, G, T | BF, BT | Max outside diam, in. | Number of thread starts |
| 00 | ...... | ... | ... | ........ | ........ | ... | 0.060 | 6 |
| 0 | 0.060 | 40 | 48 | ........ | ........ | 48 | 0.075 | 6 |
| 1 | 0.073 | 32 | 42 | ........ | ........ | 42 | | |
| 2 | 0.086 | 32 | 32 | 56 and 64 | 56 and 64 | 32 | 0.100 | 8 |
| 3 | 0.099 | 28 | 28 | 48 and 56 | 48 and 56 | 28 | | |
| 4 | 0.112 | 24 | 24 | 40 and 48 | 40 and 48 | 24 | 0.116 | 7 |
| 5 | 0.125 | 20 | 20 | 40 and 44 | 40 and 44 | 20 | | |
| 6 | 0.138 | 18 | 20 | 32 and 40 | 32 and 40 | 20 | 0.140 | 7 |
| 7 | 0.151 | 16 | 19 | ........ | ........ | 19 | 0.154 | 8 |
| 8 | 0.164 | 15 | 18 | 32 and 36 | 32 and 36 | 18 | 0.167 | 8 |
| 10 | 0.190 | 12 | 16 | 24 and 32 | 24 and 32 | 16 | 0.182 | 8 |
| 12 | 0.216 | 11 | 14 | 24 and 28 | 24 and 28 | 14 | 0.212 | 8 |
| 14 | 0.242 | 10 | ... | ........ | ........ | ... | 0.242 | 9 |
| ¼ | 0.250 | ... | 14 | 20 and 28 | 20 and 28 | 14 | | |
| 16 | 0.268 | 10 | | | | | | |
| 18 | 0.294 | 9 | | | | | | |
| ⁵⁄₁₆ | 0.3125 | ... | 12 | 18 and 24 | 18 and 24 | 12 | 0.315 | 11 |
| 20 | 0.320 | 9 | | | | | | |
| 24 | 0.372 | 9 | | | | | | |
| ³⁄₈ | 0.375 | ... | 12 | 16 and 24 | 16 and 24 | 12 | 0.378 | 12 |
| ⁷⁄₁₆ | 0.4375 | ... | 10 | ........ | ........ | 10 | | |
| ½ | 0.500 | ... | 10 | ........ | ........ | 10 | | |

employed to determine the strain in the bolt, and thence the preload.

The **load-indicator washer,** patented by Cooper & Turner, Inc., for use with high-strength bolts (ASTM A325 and A490), allows **bolt preload** to be applied rapidly and simply. The device is a hardened washer with embossed protrusions (Fig. 21a). Tightening the bolt causes the protrusions to flatten and results in a decrease in the gap between washer and bolt head. The prescribed degree of bolt-tightening load, or preload, is obtained when the gap is reduced to a predetermined amount (Fig. 21b.) A feeler gage of a given thickness is used to determine when the gap has been closed to the prescribed amount. With a paired bolt and load-indication washer, the degree of gap closure is proportional to bolt preload. The system is reported to provide bolt-preload force accuracy within +15 percent of that prescribed.

In **drilling and tapping cast iron for steel studs,** it is necessary to tap to a depth equal to 1½ times the stud diameter so that the strength of the cast-iron threads in shear may equal the tensile strength of the stud. Drill sizes and depths of hole and thread are given in Table 33.

It is not good practice to drill holes to be tapped through the metal into pressure spaces, for even though the bolt fits tightly, leakage will result that is difficult to eliminate.

**Table 29. Stove Bolts**

| Diam of bolt, in........ | ⅛ | ⁵⁄₃₂ | ³⁄₁₆ | ⁷⁄₃₂ | ¼ | ⁵⁄₁₆ | ³⁄₈ |
|---|---|---|---|---|---|---|---|
| No. of threads per in... | 32 | 28 | 24 | 22 | 18 | 18 | 16 |

**Screw-thread inserts** made of high-strength material (Fig. 22) are useful in many cases to provide increased thread strength and life. Soft or ductile materials tapped to receive thread inserts exhibit improved load-carrying capacity under static

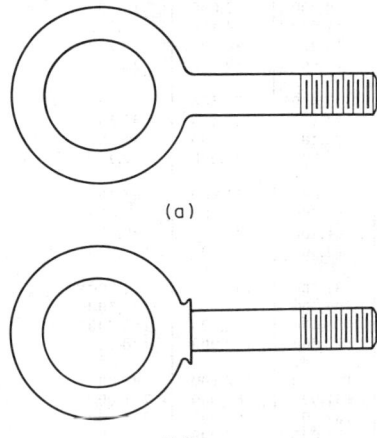

(a)

(b)

**Fig. 14**  (a) Regular nut and (b) shoulder nut eyebolts.

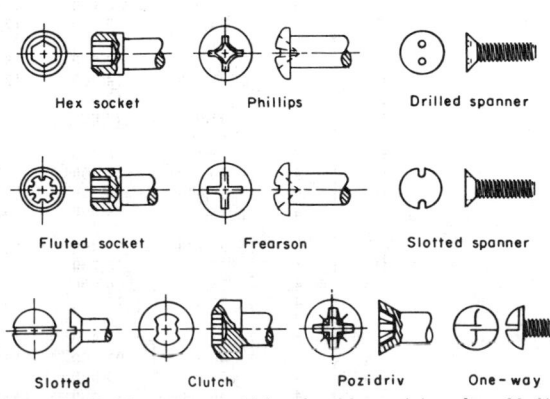

Hex socket       Phillips        Drilled spanner

Fluted socket    Frearson        Slotted spanner

Slotted    Clutch    Pozidriv    One-way

**Fig. 15**  Driving recesses. (*Adapted, with permission, from Machine Design.*)

**Table 30. Physical Requirements for Threaded Fasteners (Bolts and Capscrews)**
(Adapted with permission from *Machine Design*, Mar. 11, 1965.)

| SAE grade | Bolt size, diam, in. | Proof load, psi | Min tensile strength, psi | Hardness | |
|---|---|---|---|---|---|
| | | | | Brinell | Rockwell |
| 0 | All sizes | | | | |
| 1 | All sizes | ....... | 55,000 | 207 max | 95 B max |
| 2 | Up to ½ incl | 55,000 | 69,000 | 241 max | 100 B max |
| | Over ½ to ¾ incl | 52,000 | 64,000 | | |
| | Over ¾ to 1½ incl | 28,000 | 55,000 | 207 max | |
| 3 | Up to ½ incl | 85,000 | 110,000 | 207–269 | 95–104 B |
| | Over ½ to ⅝ incl | 80,000 | 100,000 | | |
| 5 | Up to ¾ incl | 85,000 | 120,000 | 241–302 | 23–32 C |
| | Over ¾ to 1 incl | 78,000 | 115,000 | 235–302 | 22–32 C |
| | Over 1 to 1½ incl | 74,000 | 105,000 | 223–285 | 19–30 C |
| 5.1 | Up to ⅜ incl | 85,000 | 120,000 | 241–375 | 23–40 C |
| 7 | Up to 1½ incl | 105,000 | 133,000 | 269–321 | 28–34 C |
| 8 | Up to 1½ incl | 120,000 | 150,000 | 302–352 | 32–38 C |

and dynamic loading conditions. Holes in which threads have been stripped or otherwise damaged can be restored through the use of thread inserts.

Holes for thread inserts are drilled oversize and specially tapped to receive the insert selected to mate with the threaded fastener used. The standard material for inserts is 18-8 stainless steel, but other materials are available, such as phosphor bronze and Inconel. Recommended insert lengths are given in Table 34.

**Drill Sizes for American Standard Thread Taps**  Table 35 gives tap-drill sizes for taps with American Standard and Unified threads.

**Table 31. Safe Loads for American Standard Bolts**

| Nominal diam, in. | No. of threads per in. | Ultimate strength, psi | | | | | | |
|---|---|---|---|---|---|---|---|---|
| | | 20,000 | 40,000 | 50,000 | 60,000 | 65,000 | 80,000 | 95,000 |
| ¼ | 20 | 57 | 115 | 143 | 172 | 186 | 229 | 272 |
| ⁵⁄₁₆ | 18 | 99 | 198 | 247 | 297 | 322 | 396 | 470 |
| ⅜ | 16 | 150 | 301 | 376 | 451 | 488 | 601 | 714 |
| ⁷⁄₁₆ | 14 | 207 | 415 | 519 | 623 | 675 | 830 | 986 |
| ½ | 13 | 282 | 564 | 704 | 845 | 915 | 1,125 | 1,340 |
| ⁹⁄₁₆ | 12 | 365 | 730 | 912 | 1,095 | 1,186 | 1,460 | 1,730 |
| ⅝ | 11 | 456 | 913 | 1,140 | 1,370 | 1,480 | 1,820 | 2,170 |
| ¾ | 10 | 690 | 1,380 | 1,725 | 2,070 | 2,240 | 2,760 | 3,280 |
| ⅞ | 9 | 964 | 1,930 | 2,410 | 2,900 | 3,140 | 3,860 | 4,580 |
| 1 | 8 | 1,265 | 2,530 | 3,170 | 3,800 | 4,120 | 5,060 | 6,010 |
| 1⅛ | 7 | 1,595 | 3,190 | 3,990 | 4,790 | 5,180 | 6,380 | 7,570 |
| 1¼ | 7 | 2,070 | 4,140 | 5,180 | 6,210 | 6,730 | 8,280 | 9,830 |
| 1⅜ | 6 | 2,440 | 4,890 | 6,110 | 7,330 | 7,940 | 9,780 | 11,600 |
| 1½ | 6 | 3,020 | 6,040 | 7,540 | 9,060 | 9,800 | 12,050 | 14,300 |
| 1⅝ | 5½ | 3,530 | 7,060 | 8,820 | 10,600 | 11,500 | 14,100 | 16,750 |
| 1¾ | 5 | 4,060 | 8,120 | 10,150 | 12,200 | 13,200 | 16,200 | 19,250 |
| 1⅞ | 5 | 4,800 | 9,600 | 12,000 | 14,400 | 15,600 | 19,200 | 22,800 |
| 2 | 4½ | 5,360 | 10,750 | 13,400 | 16,100 | 17,400 | 21,500 | 25,500 |
| 2¼ | 4½ | 7,120 | 14,200 | 17,800 | 21,400 | 23,100 | 28,500 | 33,800 |
| 2½ | 4 | 8,750 | 17,500 | 21,900 | 26,300 | 28,400 | 35,000 | 41,500 |
| 2¾ | 4 | 11,000 | 22,000 | 27,500 | 33,000 | 35,700 | 44,000 | 52,200 |
| 3 | 4 | 13,400 | 26,800 | 33,500 | 40,200 | 43,600 | 53,600 | 63,600 |
| 3¼ | 4 | 16,100 | 32,200 | 40,200 | 48,400 | 52,400 | 64,400 | 76,400 |
| 3½ | 4 | 19,000 | 38,100 | 47,600 | 57,200 | 61,900 | 76,200 | 90,400 |
| 3¾ | 4 | 22,200 | 44,500 | 55,600 | 66,700 | 72,300 | 89,000 | 105,500 |
| 4 | 4 | 25,700 | 51,400 | 64,200 | 77,000 | 83,400 | 102,800 | 122,000 |
| 4¼ | 4 | 29,350 | 58,700 | 73,400 | 88,100 | 95,400 | 117,400 | 139,300 |
| 4½ | 4 | 33,300 | 66,600 | 83,200 | 100,000 | 108,000 | 133,000 | 158,000 |
| 4¾ | 4 | 37,400 | 75,000 | 93,700 | 112,000 | 122,000 | 150,000 | 178,000 |
| 5 | 4 | 41,900 | 83,800 | 105,000 | 126,000 | 136,000 | 167,500 | 199,000 |
| 5¼ | 4 | 46,600 | 93,200 | 116,500 | 140,000 | 151,000 | 186,000 | 221,000 |
| 5½ | 4 | 51,500 | 103,000 | 129,000 | 154,500 | 167,900 | 206,000 | 244,500 |
| 5¾ | 4 | 56,700 | 113,500 | 142,000 | 170,000 | 184,000 | 227,000 | 269,000 |
| 6 | 4 | 62,000 | 124,000 | 155,000 | 186,000 | 202,000 | 248,000 | 295,000 |

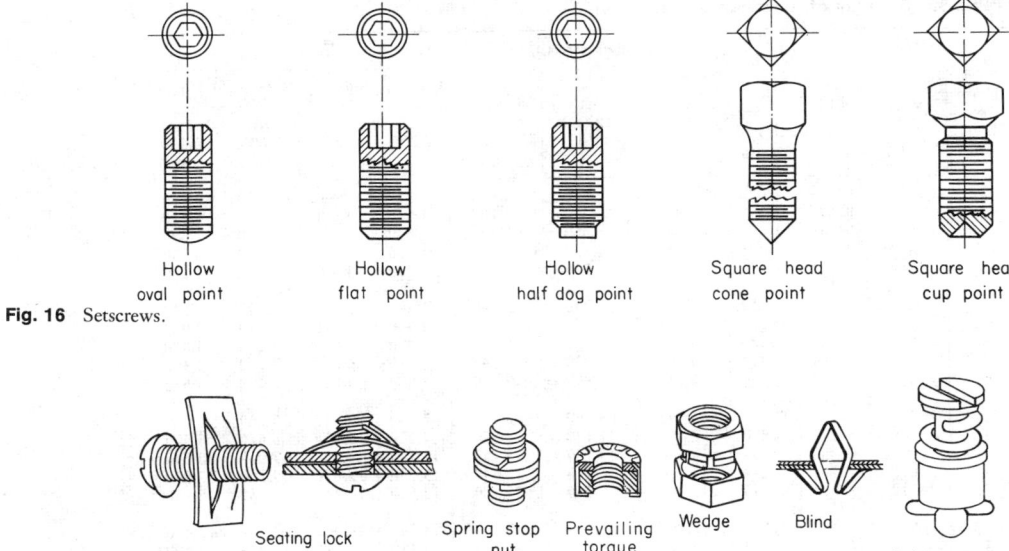

**Fig. 16**   Setscrews.

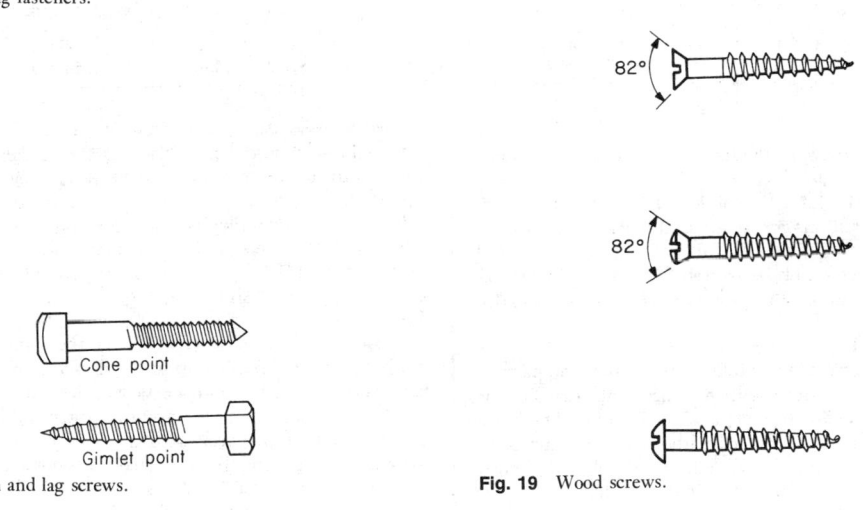

**Fig. 17**   Locking fasteners.

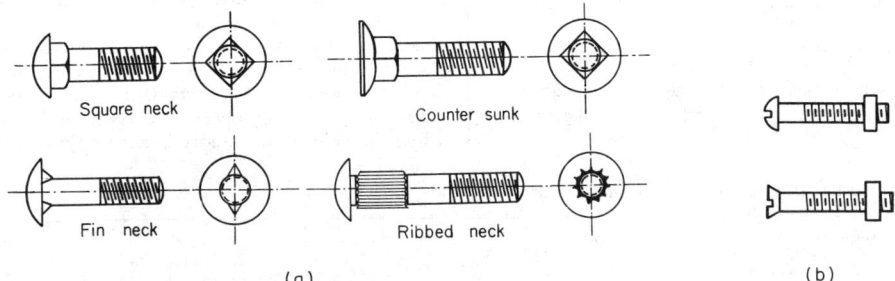

**Fig. 18**   Coach and lag screws.

**Fig. 19**   Wood screws.

(a)

(b)

**Fig. 20**   (*a*) Carriage and (*b*) stove bolts.

**Table 32. Strength of American Standard Bolts**

| Bolt | | | Areas | | Tensile strength, lb | | | Shearing strength, lb | | | |
|---|---|---|---|---|---|---|---|---|---|---|---|
| | | | | | | | | Full bolt | | Bottom of thread | |
| Diam of bolt, in. | No. of threads per in. | Full bolt sq in. | Bottom of thread, sq in. | At 10,000 psi | At 12,500 psi | At 17,500 psi | At 7,500 psi | At 10,000 psi | At 7,500 psi | At 10,000 psi |
| $\frac{1}{4}$ | 20 | 0.049 | 0.027 | 270 | 340 | 470 | 380 | 490 | 200 | 270 |
| $\frac{5}{16}$ | 18 | 0.077 | 0.045 | 450 | 570 | 790 | 580 | 770 | 340 | 450 |
| $\frac{3}{8}$ | 16 | 0.110 | 0.068 | 680 | 850 | 1,190 | 830 | 1,100 | 510 | 680 |
| $\frac{7}{16}$ | 14 | 0.150 | 0.093 | 930 | 1,170 | 1,630 | 1,130 | 1,500 | 700 | 930 |
| $\frac{1}{2}$ | 13 | 0.196 | 0.126 | 1,260 | 1,570 | 2,200 | 1,470 | 1,960 | 940 | 1,260 |
| $\frac{9}{16}$ | 12 | 0.248 | 0.162 | 1,620 | 2,030 | 2,840 | 1,860 | 2,480 | 1,220 | 1,620 |
| $\frac{5}{8}$ | 11 | 0.307 | 0.202 | 2,020 | 2,520 | 3,530 | 2,300 | 3,070 | 1,510 | 2,020 |
| $\frac{3}{4}$ | 10 | 0.442 | 0.302 | 3,020 | 3,770 | 5,290 | 3,310 | 4,420 | 2,270 | 3,020 |
| $\frac{7}{8}$ | 9 | 0.601 | 0.419 | 4,190 | 5,240 | 7,340 | 4,510 | 6,010 | 3,150 | 4,190 |
| 1 | 8 | 0.785 | 0.551 | 5,510 | 6,890 | 9,640 | 5,890 | 7,850 | 4,130 | 5,510 |
| $1\frac{1}{8}$ | 7 | 0.994 | 0.693 | 6,990 | 8,660 | 12,130 | 7,450 | 9,940 | 5,200 | 6,930 |
| $1\frac{1}{4}$ | 7 | 1.227 | 0.890 | 8,890 | 11,120 | 15,570 | 9,200 | 12,270 | 6,670 | 8,900 |
| $1\frac{3}{8}$ | 6 | 1.485 | 1.054 | 10,540 | 13,180 | 18,450 | 11,140 | 14,850 | 7,910 | 10,540 |
| $1\frac{1}{2}$ | 6 | 1.767 | 1.294 | 12,940 | 16,170 | 22,640 | 13,250 | 17,670 | 9,700 | 12,940 |
| $1\frac{5}{8}$ | $5\frac{1}{2}$ | 2.074 | 1.515 | 15,150 | 18,940 | 26,510 | 15,550 | 20,740 | 11,360 | 15,150 |
| $1\frac{3}{4}$ | 5 | 2.405 | 1.745 | 17,450 | 21,800 | 30,520 | 18,040 | 24,050 | 13,080 | 17,440 |
| $1\frac{7}{8}$ | 5 | 2.761 | 2.049 | 20,490 | 25,610 | 35,860 | 20,710 | 27,610 | 15,370 | 20,490 |
| 2 | $4\frac{1}{2}$ | 3.142 | 2.300 | 23,000 | 28,750 | 40,250 | 23,560 | 31,420 | 17,250 | 23,000 |
| $2\frac{1}{4}$ | $4\frac{1}{2}$ | 3.976 | 3.021 | 30,210 | 37,770 | 52,870 | 29,820 | 39,760 | 22,660 | 30,210 |
| $2\frac{1}{2}$ | 4 | 4.909 | 3.716 | 37,160 | 46,450 | 65,040 | 36,820 | 49,090 | 27,870 | 37,160 |
| $2\frac{3}{4}$ | 4 | 5.940 | 4.620 | 46,200 | 57,750 | 80,840 | 44,580 | 59,400 | 34,650 | 46,200 |
| 3 | $3\frac{1}{2}$ | 7.069 | 5.428 | 54,280 | 67,850 | 94,990 | 53,020 | 70,690 | 40,710 | 54,280 |

## RIVET FASTENINGS

**Forms and Proportion of Rivets** The forms and proportions of small and large rivets have been standardized and conform to ANSI B18.1.1-1972 and B18.1.2-1972 (Fig. 23).

**Materials Specifications for Rivets and Plates** See Sec. 6.

**Conventional signs** to indicate the form of the head to be used and whether the rivet is to be driven in the shop or the field at the time of erection are given in Fig. 24. **Rivet lengths and grips** are shown in Fig. 25.

For **structural riveting,** see Sec. 12.

**Punched vs. Drilled Plates** Holes in plates forming parts of riveted structures are punched, punched and reamed, or drilled. Punching, while cheaper, is objectionable. The holes in different plates cannot be spaced with sufficient accuracy to register perfectly on being assembled. If the hole is punched out, say $\frac{1}{16}$ in smaller than is required and then reamed to size, the metal injury by cold flow during punching will be removed. Drilling, while more expensive, is more accurate and does not injure the metal.

## KEYS, PINS, AND COTTERS

Keys and keyseats have been standardized and are listed in ANSI B17.1-1967. Descriptions of the principal key types follow.

**Woodruff keys** are made to facilitate removal of pulleys from shafts. They should not be used as sliding keys. Cutters for milling out the keyseats, as well as special machines for using the cutters, are to be had from the manufacturer. Where the hub of the gear or pulley is relatively long, two keys should be used. Slightly rounding the corners or ends of these keys will obviate any difficulty met with in removing pulleys from shafts. The key is shown in Fig. 26 and the dimensions in Table 36.

**Square and flat plain taper keys** have the same dimensions as gib-head keys (Table 37) up to the dotted line of Fig. 27. **Gib-head keys** (Fig. 27) are necessary when the smaller end is inaccessible for drifting out and the larger end is accessible. It can be used, with care, with all sizes of shafts. Its use is forbidden in certain jobs and places for safety reasons. Proportions are given in Table 37.

The minimum stock length of keys is four times the key width, and maximum stock length of keys is sixteen times the key width. The increments of increase of length are two times the width.

**Sunk keys** are made to the form and dimensions given in Fig. 28 and Table 38. These keys are adapted particularly to the case of fitting adjacent parts with neither end of the key accessible. **Feather keys** prevent parts from turning on a shaft while allowing them to move in a lengthwise direction. They

**Table 33. Depths to Drill and Tap Cast Iron for Studs**

| Diam of stud, in. | $\frac{1}{4}$ | $\frac{5}{16}$ | $\frac{3}{8}$ | $\frac{7}{16}$ | $\frac{1}{2}$ | $\frac{9}{16}$ | $\frac{5}{8}$ | $\frac{3}{4}$ | $\frac{7}{8}$ | 1 |
|---|---|---|---|---|---|---|---|---|---|---|
| Diam of drill, in. | $\frac{13}{64}$ | $\frac{17}{64}$ | $\frac{5}{16}$ | $\frac{3}{8}$ | $\frac{27}{64}$ | $\frac{31}{64}$ | $\frac{17}{32}$ | $\frac{41}{64}$ | $\frac{3}{4}$ | $\frac{55}{64}$ |
| Depth of thread, in. | $\frac{3}{8}$ | $\frac{15}{32}$ | $\frac{9}{16}$ | $\frac{21}{32}$ | $\frac{3}{4}$ | $\frac{27}{32}$ | $\frac{15}{16}$ | $1\frac{1}{8}$ | $1\frac{5}{16}$ | $1\frac{1}{2}$ |
| Depth to drill, in. | $\frac{7}{16}$ | $\frac{17}{32}$ | $\frac{5}{8}$ | $\frac{23}{32}$ | $\frac{27}{32}$ | $\frac{15}{16}$ | $1\frac{3}{32}$ | $1\frac{1}{4}$ | $1\frac{7}{16}$ | $1\frac{5}{8}$ |

are of the forms shown in Fig. 29 with dimensions as given in Table 38.

In **transmitting large torques,** it is customary to use two or more keys, as shown in Figs. 30 and 31. The arrangement shown in Fig. 30 permits more ready cutting of the keyway. If but one key is used with the arrangement shown in Fig. 31, torque can be taken in only one direction.

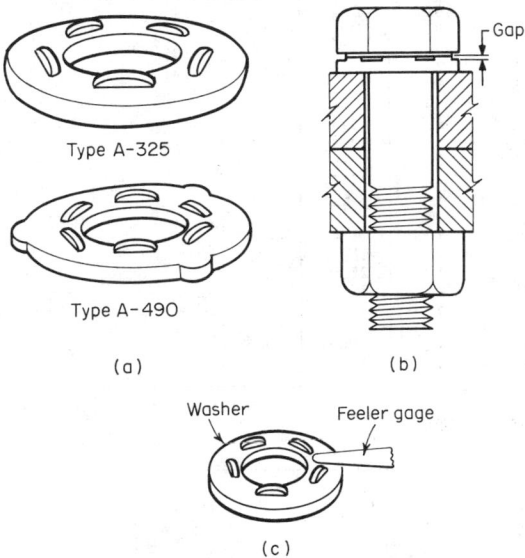

Fig. 21 Load-indicator washers.

Another means for fastening gears, pulleys flanges, etc., to shafts is through the use of mating pairs of tapered sleeves known as **Grip-springs.** A set of sleeves is shown in Fig. 32. For further references see data issued by the Ringfeder Corp., Westwood, N. J.

**Tapered pins** (Fig. 33) can be used to transmit very small torques or for positioning. They should be fitted so that the parts are drawn together to prevent their working loose when the pin is driven home. Table 39 gives dimensions of Morse tapered pins.

The Groov-Pin Corp., New Jersey, has developed a special **grooved pin** (Fig. 34) which may be used instead of smooth taper pins in certain cases.

**Straight·pins,** likewise, are used for transmission of light torques or for positioning. **Spring pins** have come into wide use recently. Two types shown in Figs. 35 and 36 deform elasti-

**Table 34. Screw-Thread Insert Lengths**
(Heli-Coil Corp.)

| Shear strength of parent material, psi | Bolt material ultimate tensile strength, psi | | | | |
|---|---|---|---|---|---|
| | 60,000 | 90,000 | 125,000 | 170,000 | 220,000 |
| | Length in terms of nominal insert diameter | | | | |
| 15,000 | 1½ | 2 | 2½ | 3 | |
| 20,000 | 1 | 1½ | 2 | 2½ | 3 |
| 25,000 | 1 | 1½ | 2 | 2 | 2½ |
| 30,000 | 1 | 1 | 1½ | 2 | 2 |
| 40,000 | 1 | 1 | 1½ | 1½ | 2 |
| 50,000 | 1 | 1 | 1 | 1½ | 1½ |

cally in the radial direction when driven; the resiliency of the pin material locks the pin in place. They can replace straight and taper pins and combine the advantages of both, i.e., simple tooling, ease of removal, reusability, ability to be driven from either side.

**Cotter pins** (Fig. 37) are used to secure or lock nuts, clevises, etc. Driven into holes in the shaft, the eye prevents complete passage, and the split ends, deformed after insertion, prevent withdrawal.

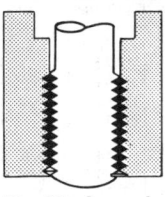

**Cottered joints** may be employed for fastening rods to other rods, rods to pistons and crossheads, yokes to rods (as in the case of connecting rods), and for services of similar kinds. Some

Fig. 22 Screw-thread insert.

forms of such joints and proportions recommended are shown in Figs. 38 to 41.

When two rods are to be joined so as to permit movement at the joint, a round pin is used in place of a cotter. In such cases, the proportions may be as shown in Fig. 42 (knuckle joint).

## SPLINES

**Involute spline** proportions, dimensions, fits, and tolerances are given in detail in ANSI B92.1-1970. External and internal involute splines (Fig. 43) have the same general form as involute gear teeth, except that the teeth are one-half the depth of

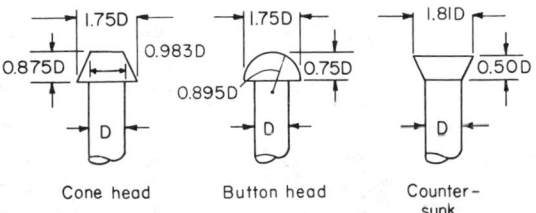

Fig. 23 Rivet heads.

standard gear teeth and the pressure angle is 30°. The spline is designated by a fraction in which the numerator is the diametral pitch and the denominator is always twice the numerator.

There are 15 series, as follows: 1/2, 2.5/5, 3/6, 4/8, 5/10, 6/12, 8/16, 10/20, 12/24, 16/32, 20/40, 24/48, 32/64, 40/80, 48/96. The number of teeth within each series varies from 6 to 50. Both a flat-root and a fillet-root type are provided. There are three **types of fits:** (1) **major diameter**—fit controlled by varying the major diameter of the external spline; (2) **sides of teeth**—fit controlled by varying tooth thickness and customarily used for fillet-root splines; (3) **minor diameter**—fit controlled by varying the minor diameter of the internal spline. Each type of fit is further divided into three classes: (*a*) **sliding**—clearance at all points; (*b*) **close**—close on either major diameter, sides of teeth, or minor diameter; (*c*) **press**—interference on either the major diameter, sides of teeth, or minor diameter. Important basic formulas for tooth proportions are:

$$D = \text{pitch diam}$$
$$N = \text{number of teeth}$$
$$P = \text{diametral pitch}$$
$$p = \text{circular pitch}$$
$$t = \text{circular tooth thickness}$$
$$a = \text{addendum}$$

## Table 35. Tap-drill Sizes for American Standard Screw Threads

(The sizes listed are the commercial tap drills to produce approx 75 percent full thread)

| Size | Coarse-thread series Threads per in. | Tap drill size | Fine-thread series Threads per in. | Tap drill size | Size | Coarse-thread series Threads per in. | Tap drill size | Fine-thread series Threads per in. | Tap drill size |
|---|---|---|---|---|---|---|---|---|---|
| No. 0 | .. | ...... | 80 | $\frac{3}{64}$ | $\frac{3}{4}$ | 10 | $2\frac{1}{32}$ | 16 | $1\frac{1}{16}$ |
| No. 1 | 64 | No. 53 | 72 | No. 53 | $\frac{7}{8}$ | 9 | $\frac{49}{64}$ | 14 | $1\frac{3}{16}$ |
| No. 2 | 56 | No. 50 | 64 | No. 50 | 1 | 8 | $\frac{7}{8}$ | 14 | $1\frac{5}{16}$ |
| No. 3 | 48 | No. 47 | 56 | No. 45 | $1\frac{1}{8}$ | 7 | $\frac{63}{64}$ | 12 | $1\frac{3}{64}$ |
| No. 4 | 40 | No. 43 | 48 | No. 42 | $1\frac{1}{4}$ | 7 | $1\frac{7}{64}$ | 12 | $1\frac{11}{64}$ |
| No. 5 | 40 | No. 38 | 44 | No. 37 | $1\frac{3}{8}$ | 6 | $1\frac{7}{32}$ | 12 | $1\frac{19}{64}$ |
| No. 6 | 32 | No. 36 | 40 | No. 33 | $1\frac{1}{2}$ | 6 | $1\frac{21}{64}$ | 12 | $1\frac{27}{64}$ |
| No. 8 | 32 | No. 29 | 36 | No. 29 | $1\frac{3}{4}$ | 5 | $1\frac{35}{64}$ | | |
| No. 10 | 24 | No. 25 | 32 | No. 21 | 2 | $4\frac{1}{2}$ | $1\frac{25}{32}$ | | |
| No. 12 | 24 | No. 16 | 28 | No. 14 | $2\frac{1}{4}$ | $4\frac{1}{2}$ | $2\frac{1}{32}$ | | |
| $\frac{1}{4}$ | 20 | No. 7 | 28 | No. 3 | $2\frac{1}{2}$ | 4 | $2\frac{1}{4}$ | | |
| $\frac{5}{16}$ | 18 | F | 24 | I | $2\frac{3}{4}$ | 4 | $2\frac{1}{2}$ | | |
| $\frac{3}{8}$ | 16 | $\frac{5}{16}$ | 24 | Q | 3 | 4 | $2\frac{3}{4}$ | | |
| $\frac{7}{16}$ | 14 | U | 20 | $\frac{25}{64}$ | $3\frac{1}{4}$ | 4 | 3 | | |
| $\frac{1}{2}$ | 13 | $\frac{27}{64}$ | 20 | $\frac{29}{64}$ | $3\frac{1}{2}$ | 4 | $3\frac{1}{4}$ | | |
| $\frac{9}{16}$ | 12 | $\frac{31}{64}$ | 18 | $\frac{33}{64}$ | $3\frac{3}{4}$ | 4 | $3\frac{1}{2}$ | | |
| $\frac{5}{8}$ | 11 | $\frac{17}{32}$ | 18 | $\frac{37}{64}$ | 4 | 4 | $3\frac{3}{4}$ | | |

## Table 36. Woodruff Key Dimensions [ANSI B17.2-1967(R1972)]

(All dimensions in inches)

| Key No. | Nominal key size, $A \times B$ | Width of key, A Max | Min | Diam of key, B Max | Min | Height of key C Max | C Min | D Max | D Min | Distance below E |
|---|---|---|---|---|---|---|---|---|---|---|
| 204 | $\frac{1}{16} \times \frac{1}{2}$ | 0.0635 | 0.0625 | 0.500 | 0.490 | 0.203 | 0.198 | 0.194 | 0.188 | $\frac{3}{64}$ |
| 304 | $\frac{3}{32} \times \frac{1}{2}$ | 0.0948 | 0.0928 | 0.500 | 0.490 | 0.203 | 0.198 | 0.194 | 0.188 | $\frac{3}{64}$ |
| 305 | $\frac{3}{32} \times \frac{5}{8}$ | 0.0948 | 0.0938 | 0.625 | 0.615 | 0.250 | 0.245 | 0.240 | 0.234 | $\frac{1}{16}$ |
| 404 | $\frac{1}{8} \times \frac{1}{2}$ | 0.1260 | 0.1250 | 0.500 | 0.490 | 0.203 | 0.198 | 0.194 | 0.188 | $\frac{3}{64}$ |
| 405 | $\frac{1}{8} \times \frac{5}{8}$ | 0.1260 | 0.1250 | 0.625 | 0.615 | 0.250 | 0.245 | 0.240 | 0.234 | $\frac{1}{16}$ |
| 406 | $\frac{1}{8} \times \frac{3}{4}$ | 0.1260 | 0.1250 | 0.750 | 0.740 | 0.313 | 0.308 | 0.303 | 0.297 | $\frac{1}{16}$ |
| 505 | $\frac{5}{32} \times \frac{5}{8}$ | 0.1573 | 0.1563 | 0.625 | 0.615 | 0.250 | 0.245 | 0.240 | 0.234 | $\frac{1}{16}$ |
| 506 | $\frac{5}{32} \times \frac{3}{4}$ | 0.1573 | 0.1563 | 0.750 | 0.740 | 0.313 | 0.308 | 0.303 | 0.297 | $\frac{1}{16}$ |
| 507 | $\frac{5}{32} \times \frac{7}{8}$ | 0.1573 | 0.1563 | 0.875 | 0.865 | 0.375 | 0.370 | 0.365 | 0.359 | $\frac{1}{16}$ |
| 606 | $\frac{3}{16} \times \frac{3}{4}$ | 0.1885 | 0.1875 | 0.750 | 0.740 | 0.313 | 0.308 | 0.303 | 0.297 | $\frac{1}{16}$ |
| 607 | $\frac{3}{16} \times \frac{7}{8}$ | 0.1885 | 0.1875 | 0.875 | 0.865 | 0.375 | 0.370 | 0.365 | 0.359 | $\frac{1}{16}$ |
| 608 | $\frac{3}{16} \times 1$ | 0.1885 | 0.1875 | 1.000 | 0.990 | 0.438 | 0.433 | 0.428 | 0.422 | $\frac{1}{16}$ |
| 609 | $\frac{3}{16} \times 1\frac{1}{8}$ | 0.1885 | 0.1875 | 1.125 | 1.115 | 0.484 | 0.479 | 0.475 | 0.469 | $\frac{5}{64}$ |
| 807 | $\frac{1}{4} \times \frac{7}{8}$ | 0.2510 | 0.2500 | 0.875 | 0.865 | 0.375 | 0.370 | 0.365 | 0.359 | $\frac{1}{16}$ |
| 808 | $\frac{1}{4} \times 1$ | 0.2510 | 0.2500 | 1.000 | 0.990 | 0.438 | 0.433 | 0.428 | 0.422 | $\frac{1}{16}$ |
| 809 | $\frac{1}{4} \times 1\frac{1}{8}$ | 0.2510 | 0.2500 | 1.125 | 1.115 | 0.484 | 0.479 | 0.475 | 0.469 | $\frac{5}{64}$ |
| 810 | $\frac{1}{4} \times 1\frac{1}{4}$ | 0.2510 | 0.2500 | 1.250 | 1.240 | 0.547 | 0.542 | 0.537 | 0.531 | $\frac{5}{64}$ |
| 811 | $\frac{1}{4} \times 1\frac{3}{8}$ | 0.2510 | 0.2500 | 1.375 | 1.365 | 0.594 | 0.589 | 0.584 | 0.578 | $\frac{3}{32}$ |
| 812 | $\frac{1}{4} \times 1\frac{1}{2}$ | 0.2510 | 0.2500 | 1.500 | 1.490 | 0.641 | 0.636 | 0.631 | 0.625 | $\frac{7}{64}$ |
| 1008 | $\frac{5}{16} \times 1$ | 0.3135 | 0.3125 | 1.000 | 0.990 | 0.438 | 0.433 | 0.428 | 0.422 | $\frac{1}{16}$ |
| 1009 | $\frac{5}{16} \times 1\frac{1}{8}$ | 0.3135 | 0.3125 | 1.125 | 1.115 | 0.484 | 0.479 | 0.475 | 0.469 | $\frac{5}{64}$ |
| 1010 | $\frac{5}{16} \times 1\frac{1}{4}$ | 0.3135 | 0.3125 | 1.250 | 1.240 | 0.547 | 0.542 | 0.537 | 0.531 | $\frac{5}{64}$ |
| 1011 | $\frac{5}{16} \times 1\frac{3}{8}$ | 0.3135 | 0.3125 | 1.375 | 1.365 | 0.594 | 0.589 | 0.584 | 0.578 | $\frac{3}{32}$ |
| 1012 | $\frac{5}{16} \times 1\frac{1}{2}$ | 0.3135 | 0.3125 | 1.500 | 1.490 | 0.641 | 0.636 | 0.631 | 0.625 | $\frac{7}{64}$ |
| 1210 | $\frac{3}{8} \times 1\frac{1}{4}$ | 0.3760 | 0.3750 | 1.250 | 1.240 | 0.547 | 0.542 | 0.537 | 0.531 | $\frac{5}{64}$ |
| 1211 | $\frac{3}{8} \times 1\frac{3}{8}$ | 0.3760 | 0.3750 | 1.375 | 1.365 | 0.594 | 0.589 | 0.584 | 0.578 | $\frac{3}{32}$ |
| 1212 | $\frac{3}{8} \times 1\frac{1}{2}$ | 0.3760 | 0.3750 | 1.500 | 1.490 | 0.641 | 0.636 | 0.631 | 0.625 | $\frac{7}{64}$ |

Numbers indicate the nominal key dimensions. The last two digits give the nominal diameter (B) in eighths of an inch and the digits preceding the last two give the nominal width (A) in thirty-seconds of an inch. Thus, 204 indicates a key $\frac{2}{32} \times \frac{4}{8}$ or $\frac{1}{16} \times \frac{1}{2}$ in.; 1210 indicates a key $1\frac{2}{32} \times 1\frac{0}{8}$ or $\frac{3}{8} \times 1\frac{1}{4}$ in.

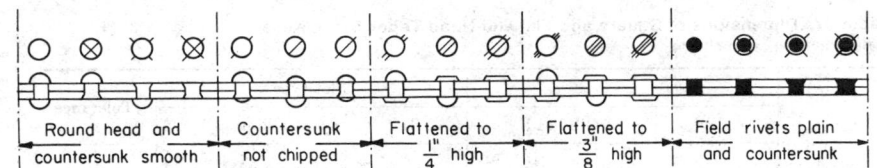

Fig. 24    Conventional signs for rivets.

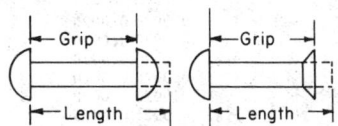

Fig. 25    Rivet length and grip.

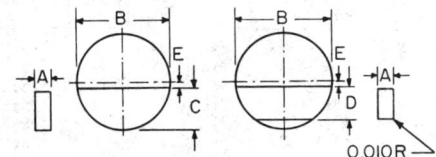

Fig. 26    Woodruff key.

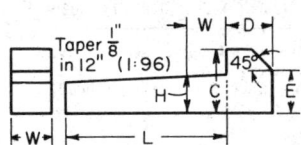

Fig. 27    Gib-head taper stock key.

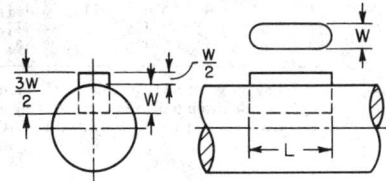

Fig. 28    Sunk key.

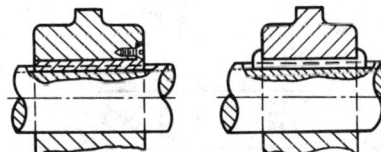

Fig. 29    Feather key.

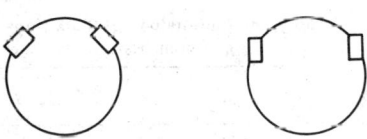

Figs. 30 and 31    Double keying of shafts.

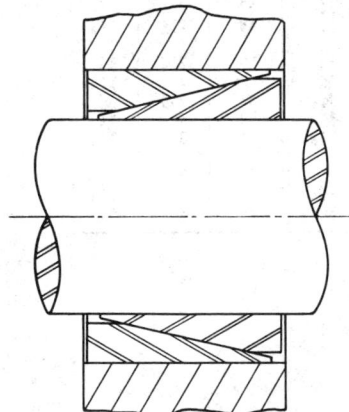

Fig. 32    Grip-springs.

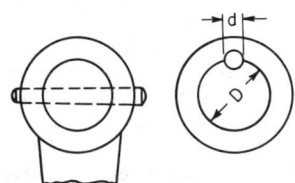

Fig. 33    Taper pins.

Fig. 34    Grooved pins.

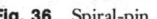

Fig. 35    Roll-pin.

Fig. 36    Spiral-pin.

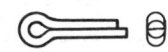

Fig. 37    Cotter pin.

**Table 37. Dimensions of Square and Flat Gib-Head Taper Stock Keys***
(All dimensions in inches)

| Shaft diam | Square type | | | | | Flat type | | | | | Tolerance | |
|---|---|---|---|---|---|---|---|---|---|---|---|---|
| | Key | | Gib-head | | | Key | | Gib-head | | | | |
| | Max width, *W* | Height at large end,† *H* | Height, *C* | Length, *D* | Height edge of chamfer, *E* | Max width, *W* | Height at large end,† *H* | Height, *C* | Length, *D* | Height edge of chamfer, *E* | On width (−) | On height (+) |
| 1/2 – 9/16 | 1/8 | 1/8 | 1/4 | 7/32 | 5/32 | 1/8 | 3/32 | 3/16 | 1/8 | 1/8 | 0.0020 | 0.0020 |
| 5/8 – 7/8 | 3/16 | 3/16 | 5/16 | 9/32 | 7/32 | 3/16 | 1/8 | 1/4 | 3/16 | 5/32 | 0.0020 | 0.0020 |
| 15/16–1 1/4 | 1/4 | 1/4 | 7/16 | 11/32 | 11/32 | 1/4 | 3/16 | 5/16 | 1/4 | 3/16 | 0.0020 | 0.0020 |
| 1 5/16 –1 3/8 | 5/16 | 5/16 | 9/16 | 13/32 | 13/32 | 5/16 | 1/4 | 3/8 | 5/16 | 1/4 | 0.0020 | 0.0020 |
| 1 7/16 –1 3/4 | 3/8 | 3/8 | 11/16 | 15/32 | 15/32 | 3/8 | 1/4 | 7/16 | 3/8 | 5/16 | 0.0020 | 0.0020 |
| 1 13/16–2 1/4 | 1/2 | 1/2 | 7/8 | 19/32 | 5/8 | 1/2 | 3/8 | 5/8 | 1/2 | 7/16 | 0.0025 | 0.0025 |
| 2 5/16 –2 3/4 | 5/8 | 5/8 | 1 1/16 | 23/32 | 3/4 | 5/8 | 7/16 | 3/4 | 5/8 | 1/2 | 0.0025 | 0.0025 |
| 2 7/8 –3 1/4 | 3/4 | 3/4 | 1 1/4 | 7/8 | 7/8 | 3/4 | 1/2 | 7/8 | 3/4 | 5/8 | 0.0025 | 0.0025 |
| 3 3/8 –3 3/4 | 7/8 | 7/8 | 1 1/2 | 1 | 1 | 7/8 | 5/8 | 1 1/16 | 7/8 | 3/4 | 0.0030 | 0.0030 |
| 3 7/8 –4 1/2 | 1 | 1 | 1 3/4 | 1 3/16 | 1 3/16 | 1 | 3/4 | 1 1/4 | 1 | 13/16 | 0.0030 | 0.0030 |
| 4 3/4 –5 1/2 | 1 1/4 | 1 1/4 | 2 | 1 7/16 | 1 7/16 | 1 1/4 | 7/8 | 1 1/2 | 1 1/4 | 1 | 0.0030 | 0.0030 |
| 5 3/4 –6 | 1 1/2 | 1 1/2 | 2 1/2 | 1 3/4 | 1 3/4 | 1 1/2 | 1 | 1 3/4 | 1 1/2 | 1 1/4 | 0.0030 | 0.0030 |

* Stock keys are applicable to the general run of work and the tolerances have been set accordingly. They are not intended to cover the finer applications where a closer fit may be required.

† This height of the key is measured at the distance *W* equal to the width of the key, from the gib head.

**Table 38. Dimensions of Sunk Keys**
(All dimensions in inches. Letters refer to Fig. 28)

| Key No. | *L* | *W* | Key No. | *L* | *W* | Key No. | *L* | *W* | Key No. | *L* | *W* |
|---|---|---|---|---|---|---|---|---|---|---|---|
| 1 | 1/2 | 1/16 | 13 | 1 | 3/16 | 22 | 1 3/8 | 1/4 | 54 | 2 1/4 | 1/4 |
| 2 | 1/2 | 3/32 | 14 | 1 | 7/32 | 23 | 1 3/8 | 5/16 | 55 | 2 1/4 | 5/16 |
| 3 | 1/2 | 1/8 | 15 | 1 | 1/4 | F | 1 3/8 | 3/8 | 56 | 2 1/4 | 3/8 |
| 4 | 5/8 | 3/32 | B | 1 | 5/16 | 24 | 1 1/2 | 1/4 | 57 | 2 1/4 | 7/16 |
| 5 | 5/8 | 1/8 | 16 | 1 1/8 | 3/16 | 25 | 1 1/2 | 5/16 | 58 | 2 1/2 | 5/16 |
| 6 | 5/8 | 5/32 | 17 | 1 1/8 | 7/32 | G | 1 1/2 | 3/8 | 59 | 2 1/2 | 3/8 |
| 7 | 3/4 | 1/8 | 18 | 1 1/8 | 1/4 | 51 | 1 3/4 | 1/4 | 60 | 2 1/2 | 7/16 |
| 8 | 3/4 | 5/32 | C | 1 1/8 | 5/16 | 52 | 1 3/4 | 5/16 | 61 | 2 1/2 | 1/2 |
| 9 | 3/4 | 3/16 | 19 | 1 1/4 | 3/16 | 53 | 1 3/4 | 3/8 | 30 | 3 | 3/8 |
| 10 | 7/8 | 5/32 | 20 | 1 1/4 | 7/32 | 26 | 2 | 3/16 | 31 | 3 | 7/16 |
| 11 | 7/8 | 3/16 | 21 | 1 1/4 | 1/4 | 27 | 2 | 1/4 | 32 | 3 | 1/2 |
| 12 | 7/8 | 7/32 | D | 1 1/4 | 5/16 | 28 | 2 | 5/16 | 33 | 3 | 9/16 |
| A | 7/8 | 1/4 | E | 1 1/4 | 3/8 | 29 | 2 | 3/8 | 34 | 3 | 5/8 |

**Table 39. Morse Standard Taper Pins**
(Taper, 1/8 in ft. Lengths increase by 1/4 in. Dimensions in inches)

| Size No. | 0 | 1 | 2 | 3 | 4 | 5 | 6 | 7 | 8 | 9 | 10 |
|---|---|---|---|---|---|---|---|---|---|---|---|
| Diam at large end | 0.156 | 0.172 | 0.193 | 0.219 | 0.250 | 0.289 | 0.341 | 0.409 | 0.492 | 0.591 | 0.706 |
| Length | 0.5–3 | 0.5–3 | 0.75–3.5 | 0.75–3.5 | 0.75–4 | 0.75–4 | 0.75–5 | 1–5 | 1.25–5 | 1.5–6 | 1.5–6 |

$b$ = dedendum
$D_o$ = major diam
$TIF$ = true involute form diam
$D_R$ = minor diam

**Flat and fillet roots**

$$D = N/P$$
$$p = \pi/P$$
$$t = p/2$$
$$a = 0.5000/P$$
$$D_o \text{ (external)} = \frac{N + 1}{P}$$
$$TIF \text{ (internal)} = \frac{N + 1}{P}$$
$$D_R = \frac{N + 1}{P} \text{ (minor-diameter fits only)}$$
$$TIF \text{ (external)} = \frac{N - 1}{P}$$

$b = 0.600/P + 0.002$ (For major-diameter fits, the internal spline dedendum is the same as the addendum; for minor-diameter fits, the dedendum of the external spline is the same as the addendum.)

**Fillet root only**

½ through $^{12}/_{24}$

$$D_o \text{ (internal)} = \frac{N + 1.8}{P}$$
$$D_R \text{ (external)} = \frac{N - 1.8}{P}$$
$$b \text{ (internal)} = 0.900/P$$
$$b \text{ (external)} = 0.900/P$$

$^{16}/_{32}$ through $^{48}/_{96}$

$$D_o \text{ (internal)} = \frac{N + 1.8}{P}$$
$$D_R \text{ (external)} = \frac{N - 2}{P}$$
$$b \text{ (internal)} = 0.900/P$$
$$b \text{ (external)} = 1.000/P$$

Internal spline dimensions are basic while external spline dimensions are varied to control fit.

The advantages of involute splines are. (1) maximum strength at the minor diameter, (2) self-centering equalizes bearing and stresses among all teeth, and (3) ease of manufacture through the use of standard gear-cutting tools and methods.

The design of involute splines is critical in shear. The torque capacity may be determined by the formula $T = LD^2S_s/1.2732$, where $L$ = spline length, $D$ = pitch diam, $S_s$ = allowable shear stress.

**Parallel-side splines** have been standardized by the SAE for 4, 6, 10, and 16 spline fittings. They are shown in Fig. 44; pertinent data are in Tables 40 and 41.

## COUPLINGS

A **coupling** makes a semipermanent connection between two shafts. They are of three main types: **rigid, flexible,** and **fluid.**

### Rigid Couplings

Rigid couplings are used only on shafts which are perfectly aligned. The **flanged-face coupling** (Fig. 45) is the simplest of these. The flanges must be keyed to the shafts. The **keyless compression coupling** (Fig. 46) affords a simple means for connecting abutting shafts without the necessity of key seats on the shafts. When drawn over the slotted tapered sleeve the two flanges automatically center the shafts and provide sufficient contact pressure to transmit medium or light loads. **Ribbed-clamp couplings** (Fig. 47) are split longitudinally and are bored to the shaft diameter with a shim separating the two halves. It is necessary to key the shafts to the coupling.

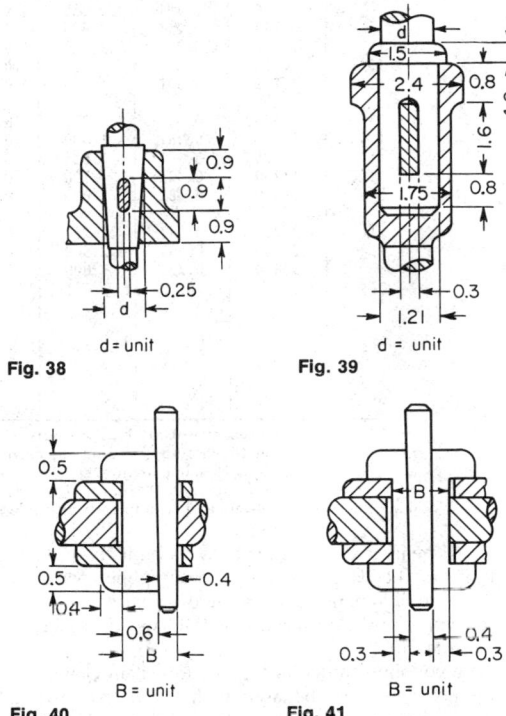

**Fig. 38**

**Fig. 39**

**Fig. 40**

**Fig. 41**

**Figs. 38–41** Cottered joints.

### Flexible Couplings

Flexible couplings are designed to connect shafts which are misaligned either laterally or angularly. A secondary benefit is the absorption of impacts due to fluctuations in shaft torque or angular speed. The **Oldham,** or **double-slider, coupling** (Fig. 48) may be used to connect shafts which have only lateral misalignment. The **"Fast" flexible coupling** (Fig. 49) consists of two hubs each keyed to its respective shaft. Each hub has generated splines cut at the maximum possible distance from the shaft end. Surrounding the hubs is a casing or sleeve which is split transversely and bolted by means of flanges. Each half of

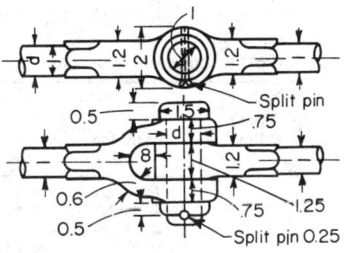

**Fig. 42** Knuckle joint.

**Table 40. Dimensions of Spline Fittings**
(SAE Standard.   All dimensions in inches)

| Nominal diam | 4-spline for all fits | | 6-spline for all fits | | 10-spline for all fits | | 16-spline for all fits | |
|---|---|---|---|---|---|---|---|---|
| | $D$ max* | $W$ max† | $D$ max* | $W$ max† | $D$ max* | $W$ max† | $D$ max* | $W$ max† |
| ¾ | 0.750 | 0.181 | 0.750 | 0.188 | 0.750 | 0.117 | | |
| ⅞ | 0.875 | 0.211 | 0.875 | 0.219 | 0.875 | 0.137 | | |
| 1 | 1.000 | 0.241 | 1.000 | 0.250 | 1.000 | 0.156 | | |
| 1⅛ | 1.125 | 0.271 | 1.125 | 0.281 | 1.125 | 0.176 | | |
| 1¼ | 1.250 | 0.301 | 1.250 | 0.313 | 1.250 | 0.195 | | |
| 1⅜ | 1.375 | 0.331 | 1.375 | 0.344 | 1.375 | 0.215 | | |
| 1½ | 1.500 | 0.361 | 1.500 | 0.375 | 1.500 | 0.234 | | |
| 1⅝ | 1.625 | 0.391 | 1.625 | 0.406 | 1.625 | 0.254 | | |
| 1¾ | 1.750 | 0.422 | 1.750 | 0.438 | 1.750 | 0.273 | | |
| 2 | 2.000 | 0.482 | 2.000 | 0.500 | 2.000 | 0.312 | 2.000 | 0.196 |
| 2¼ | 2.250 | 0.542 | 2.250 | 0.563 | 2.250 | 0.351 | | |
| 2½ | 2.500 | 0.602 | 2.500 | 0.625 | 2.500 | 0.390 | 2.500 | 0.245 |
| 3 | 3.000 | 0.723 | 3.000 | 0.750 | 3.000 | 0.468 | 3.000 | 0.294 |
| 3½ | ..... | ..... | ..... | ..... | 3.500 | 0.546 | 3.500 | 0.343 |
| 4 | ..... | ..... | ..... | ..... | 4.000 | 0.624 | 4.000 | 0.392 |
| 4½ | ..... | ..... | ..... | ..... | 4.500 | 0.702 | 4.500 | 0.441 |
| 5 | ..... | ..... | ..... | ..... | 5.000 | 0.780 | 5.000 | 0.490 |
| 5½ | ..... | ..... | ..... | ..... | 5.500 | 0.858 | 5.500 | 0.539 |
| 6 | ..... | ..... | ..... | ..... | 6.000 | 0.936 | 6.000 | 0.588 |

\* Tolerance allowed of −0.001 in. for shafts ¾ to 1¾ in., inclusive; of −0.002 for shafts 2 to 3 in., inclusive; −0.003 in. for shafts 3½ to 6-in., inclusive, for 4-, 6-, and 10-spline fittings; tolerance of −0.003 in. allowed for all sizes of 16-spline fittings.

† Tolerance allowed of −0.002 in. for shafts ¾ in., to 1¾ in., inclusive; of −0.003 for shafts 2 to 6 in., inclusive, for 4-, 6-, and 10-spline fittings; tolerance of −0.003 allowed for all sizes of 16-spline fittings.

this sleeve has generated internal splines cut on its bore at the end opposite to the flange. These internal splines permit a definite error of alignment between the two shafts.

Another type, the Waldron coupling (Midland-Ross Corp.), is shown in Fig. 50.

The chain coupling shown in Fig. 51 uses silent chain, but standard roller chain can be used with the proper mating sprockets. Nylon links enveloping the sprockets are another variation of the chain coupling.

**Steelflex couplings** (Fig. 52) are made with two grooved steel hubs keyed to their respective shafts. Connection between the two halves is secured by a specially tempered alloy-steel member called the "grid."

In the rubber flexible coupling shown in Fig. 53, the torque is transmitted through a comparatively soft rubber section acting in shear. The type in Fig. 54 loads the intermediate rubber member in compression. Both types permit reasonable shaft misalignment and are recommended for light loads only.

**Universal joints** are used to connect shafts with much larger values of misalignment than can be tolerated by the other types of flexible couplings. Shaft angles up to 30° may be used. The Hooke's-type joint (Fig. 55) suffers a loss in efficiency with increasing angle which may be approximated for

angles up to 15° by the following relation: efficiency = $100(1 − 0.003\theta)$, where $\theta$ is the angle between the shafts. The velocity ratio between input and output shafts with a single universal joint is equal to

$$\omega_2/\omega_1 = \cos \theta/1 − \sin^2 \theta \sin^2 (\alpha + 90°)$$

where $\omega_2$ and $\omega_1$ are the angular velocities of the driven and driving shafts respectively, $\theta$ is the angle between the shafts,

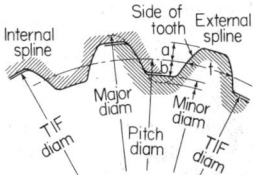

**Fig. 43**  Involute spline.

and $\alpha$ is the angular displacement of the driving shaft from the position where the pins on the drive shaft yoke lie in the plane of the two shafts. A velocity ratio of 1 may be obtained at any angle using two Hooke's-type joints and an intermediate shaft. The intermediate shaft must make equal angles with the main

**Table 41. Spline Proportions**

| No. of splines | $W$ for all fits | Permanent fit | | To slide when not under load | | To slide under load | |
|---|---|---|---|---|---|---|---|
| | | $h$ | $d$ | $h$ | $d$ | $h$ | $d$ |
| 4 | 0.241$D$ | 0.075$D$ | 0.850$D$ | 0.125$D$ | 0.750$D$ | | |
| 6 | 0.250$D$ | 0.050$D$ | 0.900$D$ | 0.075$D$ | 0.850$D$ | 0.100$D$ | 0.800$D$ |
| 10 | 0.156$D$ | 0.045$D$ | 0.910$D$ | 0.070$D$ | 0.860$D$ | 0.095$D$ | 0.810$D$ |
| 16 | 0.098$D$ | 0.045$D$ | 0.910$D$ | 0.070$D$ | 0.860$D$ | 0.095$D$ | 0.810$D$ |

shafts, and the driving pins on the yokes attached to the intermediate shaft must be set parallel to each other.

The Bendix-Weiss "rolling-ball" universal joint provides constant angular velocity. Torque is transmitted between two yokes through a set of four balls such that the centers of all four balls lie in a plane which bisects the angle between the shafts. Other variations of constant velocity universal joints are found in the Rzeppa, Tracta, and double Cardan types.

speed, hundreds of r/min, and $D$ is the outside diameter, ft. The output torque is equal to the input torque over the entire range of input-output speed ratios. Thus the prime mover can be operated at its most effective speed regardless of the speed of the output shaft. Other advantages are that the prime mover cannot be stalled by application of load and that there is no transmission of shock loads or torsional vibration between the connected shafts.

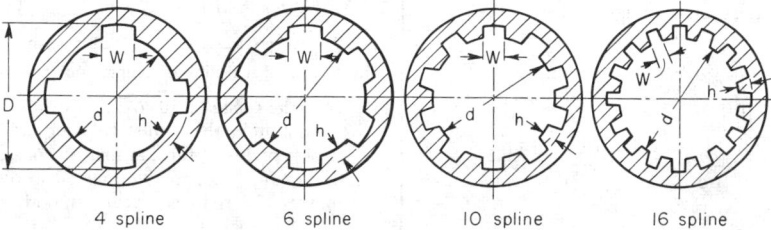

4 spline    6 spline    10 spline    16 spline

**Fig. 44**  Parallel-side splines.

### Fluid Couplings

(See also Sec. 11)

Fluid couplings (Fig. 56) have two basic parts—the input member, or impeller, and the output member, or runner. There is no mechanical connection between the two shafts, power being transmitted by kinetic energy in the operating fluid. The impeller B is fastened to the flywheel A and turns at

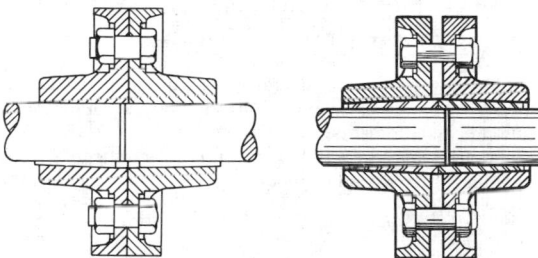

**Fig. 45**  Flanged face coupling.    **Fig. 46**  Keyless compression coupling.

engine speed. As this speed increases, fluid within the impeller moves toward the outer periphery because of centrifugal force. The circular shape of the impeller directs the fluid toward the runner C, where its kinetic energy is absorbed as torque delivered by shaft D. The positive pressure behind the fluid causes flow to continue toward the hub and back through the impeller. The toroidal space in both the impeller and runner is divided into compartments by a series of flat radial vanes.

The torque capacity of a fluid coupling with a full-load slip of about 2.5 percent is $T = 0.09n^2D^5$, where $n$ is the impeller

A **hydraulic torque converter** (Fig. 57) is similar in form to the hydraulic coupling, with the addition of a set of stationary guide vans, the reactor, interposed between the runner and the impeller. All blades in a converter have compound curvature. This curvature is designed to control the direction of fluid flow. Kinetic energy is therefore transferred as a result of both a scalar and vectorial change in fluid velocity. The blades are designed such that the fluid will be moving in a direction parallel to the blade surface at the entrance (Fig. 58) to each

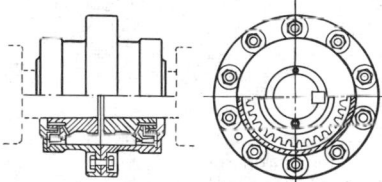

**Fig. 49**  "Fast" flexible coupling.

section. With a design having fixed blading, this can be true at only one value of runner and impeller velocity, called the design point. Several design modifications are possible to overcome this difficulty. The angle of the blades can be made adjustable, and the elements can be divided into sections operating independently of each other according to the load requirements. Other refinements include the addition of mul-

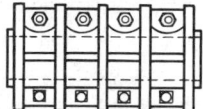

**Fig. 47**  Ribbed-clamp coupling.

**Fig. 48**  Double-slider coupling.

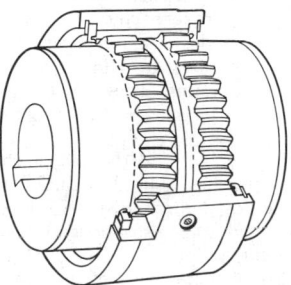

**Fig. 50**  Waldron coupling.

tiple stages in the runner and reactor stages as in steam reaction turbines (see Sec. 9). The advantages of a torque converter are the ability to multiply starting torque five to six times and to serve as a stepless transmission. As in the coupling, torque varies as the square of speed and the fifth power of diameter.

Optimum efficiency (Fig. 59) over the range of input-output speed ratios is obtained by a combination converter coupling. When the output speed rises to the point where the torque multiplication factor is 1.0, the clutch point, the torque reac-

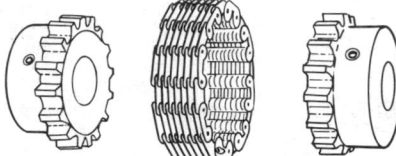

**Fig. 51**  Chain coupling.

tion on the reactor element reverses direction. If the reactor is mounted to freewheel in this opposite direction, the unit will act as a coupling over the higher speed ranges. An automatic friction clutch (see Clutches, below) set to engage at or near the clutch point will also eliminate the poor efficiency of the converter at high output speeds.

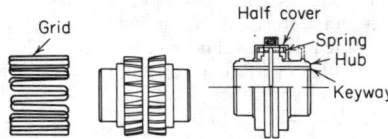

**Fig. 52**  Falk Steelflex coupling.

## CLUTCHES

Clutches are couplings which permit the disengagement of the coupled shafts during rotation.

**Positive clutches** are designed to transmit torque without slip. The **jaw clutch** is the most common type of positive clutch. These are made with **square jaws** (Fig. 60) for driving in both directions or **spiral jaws** (Fig. 61) for unidirectional drive.

Engagement speed should be limited to 10 r/min for square jaws and 150 r/min for spiral jaws. If disengagement under load is required, the jaws should be finish-machined and lubricated.

**Fig. 53**  Rubber flexible coupling, shear type.

**Friction clutches** are designed to reduce coupling shock by slipping during the engagement period. They also serve as safety devices by slipping when the torque exceeds their maximum rating. They may be divided into two main groups, **axial** and **rim clutches,** according to the direction of contact pressure.

The **cone clutch** (Fig. 62) and the **disk clutch** (Fig. 63) are examples of axial clutches. The disk clutch may consist of either a single plate or multiple disks. Table 42 lists typical friction materials and important design data. The torque capacity of a disk clutch is given by $T = 0.5 i f F_a D_m$, where $T$ is the torque, $i$ the number of pairs of contact surfaces, $f$ the applicable coefficient of friction, $F_a$ the axial engaging force,

and $D_m$ the mean diameter of the clutch facing. The spring forces holding a disk clutch in engagement are usually of relatively high value, as given by the allowable contact pressures. In order to lower the force required at the operating lever, elaborate linkages are required, usually having lever

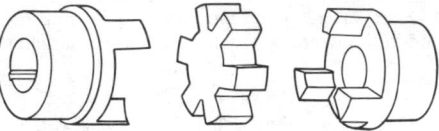

**Fig. 54**  Rubber flexible coupling, compression type.

ratios in the range of 10 to 12. As these linkages must rotate with the clutch, they must be adequately balanced and the effect of centrifugal forces must be considered. Disk clutches are often run wet, either immersed in oil or in a spray. The advantages are reduced wear, smoother action, and lower

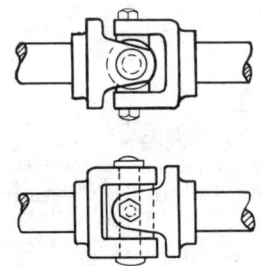

**Fig. 55**  Hooke's universal joint.

operating temperatures. Disk clutches are often operated automatically by either air or hydraulic cylinders as, for example, in automobile automatic transmissions.

**Rim clutches** may be subdivided into two groups: (1) those employing either a band or block (Fig. 64) in contact with the

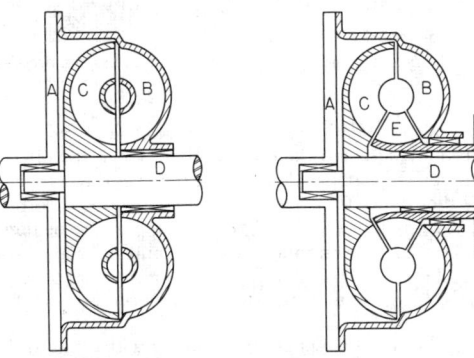

**Fig. 56**  Fluid coupling (A) Flywheel; (B) impeller; (C) runner; (D) output shaft.

**Fig. 57**  Hydraulic torque converter. (A) Flywheel; (B) impeller; (C) runner; (D) output shaft; (E) reactor.

rim and (2) overrunning clutches (Fig. 65) employing the wedging action of a roller or sprag. Clutches in the second category will automatically engage in one direction and freewheel in the other.

**Table 42. Friction Coefficients and Allowable Pressures**
(From Maleev, *Machine Design*, International Textbook, by permission)

| Materials in contact | Friction coefficient $f$ | | | Allowable pressure, psi |
| --- | --- | --- | --- | --- |
| | Dry | Greasy | Lubricated | |
| Cast iron on cast iron...................... | 0.2–0.15 | 0.10–0.06 | 0.10–0.05 | 150–250 |
| Bronze on cast iron....................... | | 0.10–0.05 | 0.10–0.05 | 80–120 |
| Steel on cast iron........................ | 0.30–0.20 | 0.12–0.07 | 0.10–0.06 | 120–200 |
| Wood on cast iron........................ | 0.25–0.20 | 0.12–0.08 | ......... | 60–90 |
| Fiber on metal............................ | ......... | 0.20–0.10 | ......... | 10–30 |
| Cork on metal............................ | 0.35 | 0.30–0.25 | 0.25–0.22 | 8–15 |
| Leather on metal......................... | 0.5–0.3 | 0.20–0.15 | 0.15–0.12 | 10–30 |
| Wire asbestos on metal.................... | 0.5–0.35 | 0.30–0.25 | 0.25–0.20 | 40–80 |
| Asbestos blocks on metal.................. | 0.48–0.40 | 0.30–0.25 | ......... | 40–160 |
| Asbestos on metal, short action............ | ......... | ......... | 0.25–0.20 | 200–300 |
| Metal on cast iron, short action........... | ......... | ......... | 0.10–0.05 | 200–300 |

## HYDRAULIC POWER TRANSMISSION

Hydraulic power transmission systems comprise machinery and auxiliary components which function to generate, transmit, control, and utilize hydraulic power. The **working fluid,** a pressurized incompressible liquid, is usually either a petroleum base or a fire-resistant type. The latter are water and oil

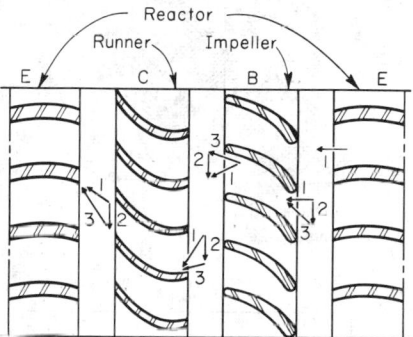

**Fig. 58** Schematic of converter blading. (1) Absolute fluid velocity; (2) velocity vector—converter elements; (3) fluid velocity relative to converter elements.

emulsions, glycol-water mixtures, or synthetic liquids such as silicones or phosphate esters.

Liquid is pressurized in a **pump** by virtue of its resistance to flow; the pressure difference between pump inlet and outlet results in flow. Most hydraulic applications employ positive-displacement pumps of the gear, vane, screw, or piston type; piston pumps are axial, radial, or reciprocating (see Sec. 14).

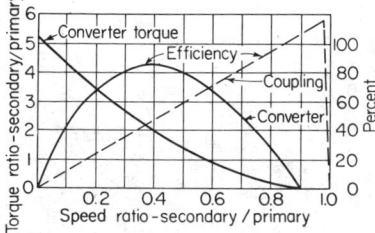

**Fig. 59** Hydraulic-coupling characteristic curves. (*Heldt, "Torque Converters and Transmissions," Chilton.*)

Power is transmitted from pump to controls and point of application through a combination of **conduit and fittings** appropriate to the particular application. Flow characteristics of hydraulic circuits take into account fluid properties, pressure drop, flow rate, and pressure-surging tendencies. Conduit

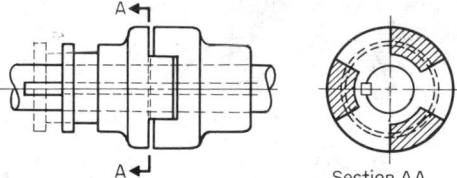

**Fig. 60** Square-jaw clutch.

systems must be designed to minimize changes in flow velocity, velocity distribution, and random fluid eddies, all of which dissipate energy and result in pressure drops in the circuit (see Sec. 3). Pipe, tubing, and flexible hose are used as hydraulic power conduits; suitable fittings are available for all types and for transition from one type to another.

**Controls** are generally interposed along the conduit between the pump and point of application (i.e., an actuator or motor), and act to control pressure, volume, or flow direction.

**Pressure control valves,** of which an ordinary safety valve is a common type (normally closed), include relief and reducing valves and pressure switches (Figs. 66, 67). Pressure valves,

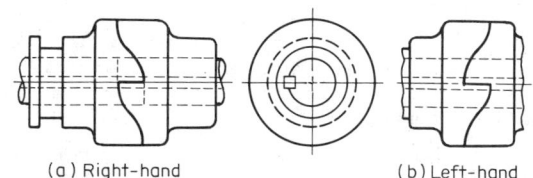

(a) Right-hand    (b) Left-hand

**Fig. 61** Spiral-jaw clutch.

normally closed, can be used to control sequential operations in a hydraulic circuit. **Flow control valves** throttle flow to or bypass flow around the unit being controlled, resulting in pressure drop and temperature increase as pressure energy is dissipated. Figure 68 shows a simple needle valve with variable orifice usable as a flow control valve. **Directional control valves** serve primarily to direct hydraulic fluid to the point of application. Directional control valves with rotary and sliding spools are shown in Figs. 69 and 70.

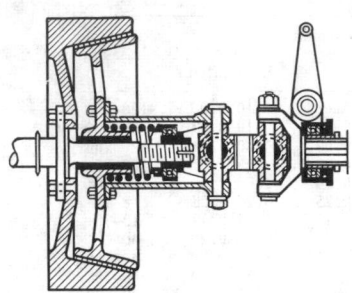

**Fig. 62**   Cone clutch.

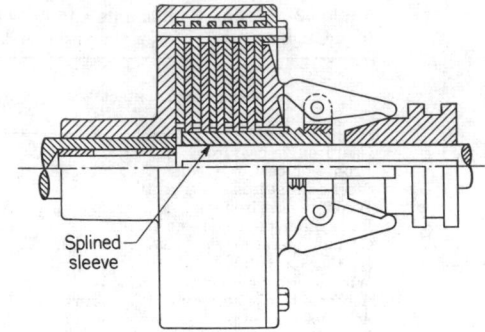

Splined
sleeve

**Fig. 63**   Multidisk clutch.

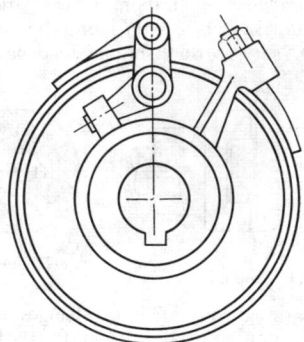

**Fig. 64**   Band clutch.

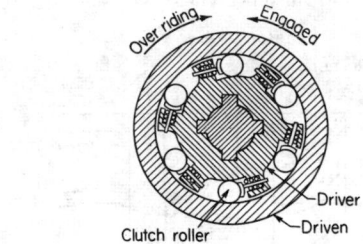

Over riding    Engaged

Driver

Clutch roller    Driven

**Fig. 65**   Overrunning clutch.

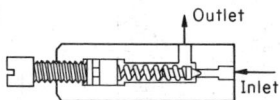

Outlet

Inlet

**Fig. 66**   Relief valve.

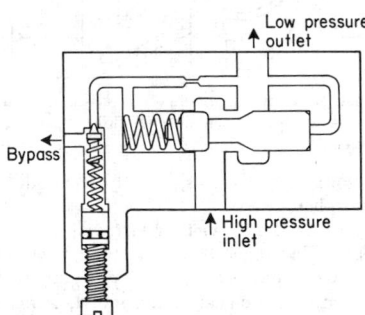

Low pressure
outlet

Bypass

High pressure
inlet

**Fig. 67**   Reducing valve.

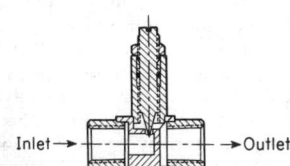

Inlet →    → Outlet

**Fig. 68**   Needle valve.

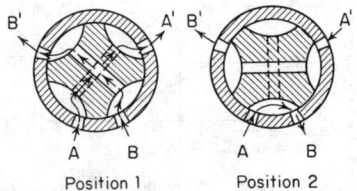

B'    A'    B'    A'

A    B    A    B

Position 1    Position 2

**Fig. 69**   Rotary-spool directional flow valve.

**Accumulators** are effectively "hydraulic flywheels" which store potential energy by accumulating a quantity of pressurized hydraulic fluid in a suitable enclosed vessel. The bag type shown in Fig. 71 uses pressurized gas inside the bag working against the hydraulic fluid outside the bag.

Pressurized hydraulic fluid acting against an **actuator** or motor converts fluid pressure energy into mechanical energy.

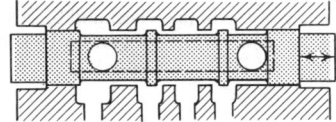

**Fig. 70**  Sliding-spool directional flow valve.

Motors providing continuous rotation have operating characteristics closely related to their pump counterparts. A linear actuator, or cylinder (Fig. 72), provides straight-line reciprocating motion; a rotary actuator (Fig. 73) provides arcuate oscillatory motion.

**Hydraulic fluids** (liquids and air) are conducted in **pipe, tubing,** or **flexible hose.** Hose is used when the lines must flex or in applications in which fixed, rigid conduit is unsuitable. Table 43 lists SAE standard hoses. Maximum recommended operating pressure for a broad range of industrial applications is approximately 25 percent of rated bursting pressure. Due consideration must be given to the operating-temperature range; most applications fall in the range from −40 to 200°F (−40 to 95°C). Higher operating temperatures can be accommodated with appropriate materials.

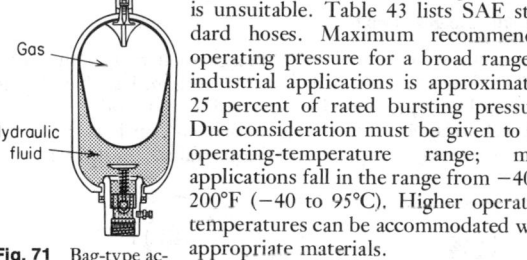

**Fig. 71**  Bag-type accumulator.

**Hose fittings** are of the screw-type or swaged, depending on the particular application and operating pressure and temperature. A broad variety of hose-end fittings is available from the industry.

**Pipe** has the advantage of being relatively cheap, is applied mainly in straight runs, and is usually of steel. Fittings for pipe are either standard pipe fittings for fairly low pressures or more elaborate ones suited to leakproof high-pressure operation.

**Tubing** is more easily bent into neat forms to fit between inlet and outlet connections. Steel and stainless-steel tubing is used for the highest-pressure applications; aluminum, plastic, and copper tubing is also used as appropriate for the operating conditions of pressure and temperature. Copper tubing hastens the oxidation of oil-base hydraulic fluids; accordingly, its

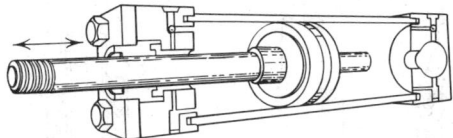

**Fig. 72**  Linear actuator or hydraulic cylinder.

use should be restricted either to air lines or with liquids which will not be affected by copper in the operating range.

**Tube fittings** for permanent connections allow for brazed or welded joints. For temporary or separable applications, **flared** or **flareless fittings** are employed (Figs. 74 and 75). ANSI B116.1-1974 and B116.2-1974 pertain to tube fittings. The variety of fittings available is vast; the designer is advised to refer to manufacturers' literature for specifics.

Parameters entering into the design of a hydraulic system are volume of flow per unit time, operating pressure and temperature, viscosity characteristics of the fluid within the operating range, and compatibility of the fluid/conduit material.

**Flow velocity** in suction lines is generally in the range of 1 to 5 ft/s (0.3 to 1.5 m/s); in discharge lines it ranges from 10 to 25 ft/s (3 to 8 m/s).

The pipe or tubing is under internal pressure. Selection of material and wall thickness follows from suitable equations (see Sec. 5). Safety factors range from 6 to 10 or higher, depending on the severity of the application (i.e., vibration, shock, pressure surges, possibility of physical abuse, etc.). JIC specifications provide a guide to the designer of hydraulic systems.

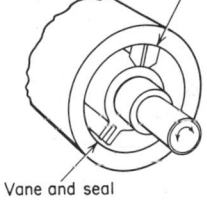

**Fig. 73**  Rotary actuator.

## BRAKES

(See also Secs. 10 and 11)

**Block brakes** are shown diagrammatically in Figs. 76 to 80. They consist of a block or shoe of wood or cast iron bearing

**Table 43. SAE Standard Hoses**

| | |
|---|---|
| 100R1A | One-wire-braid reinforcement, synthetic-rubber-covered |
| 100R1T | Same as R-1A except with a thin, nonskive cover |
| 100R2A | Two-wire-braid reinforcement, synthetic rubber cover |
| 100R2B | Two spiral wire plus one wire-braid reinforcement, synthetic rubber cover |
| 100R2AT | Same as R2A except with a thin, nonskive cover |
| 100R2BT | Same as R2B except with a thin, nonskive cover |
| 100R3 | Two rayon-braid reinforcement, synthetic rubber cover |
| 100R5 | One textile braid plus one wire-braid reinforcement, textile braid cover |
| 100R7 | Thermoplastic tube, synthetic-fiber reinforcement, thermoplastic cover (thermoplastic equivalent to SAE 100R1A) |
| 100R8 | Thermoplastic tube, synthetic-fiber reinforcement, thermoplastic cover (thermoplastic equivalent to SAE 100R2A) |
| 100R9 | Four-ply, light-spiral-wire reinforcement, synthetic-rubber cover |
| 100R9T | Same as R9 except with a thin, nonskive cover |
| 100R10 | Four-ply, heavy-spiral-wire reinforcement, synthetic-rubber cover |
| 100R11 | Six-ply, heavy-spiral-wire reinforcement, synthetic-rubber cover |

upon an iron or steel wheel. The force relations obtaining in the operation of these brakes may be formulated as follows:

In Fig. 76, let $F$ = load applied at end of lever arm; $A$ = distance from point of application of $F$ to block center; $B$ = distance from block center to center of fulcrum pin; $R$ =

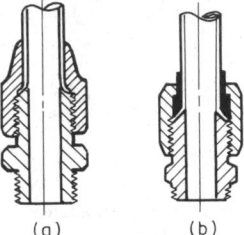

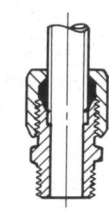

**Fig. 74**   Flared tube fittings. (*a*) 45° flared fitting; (*b*) Triple-lok, 37° flared fitting. (*Parker-Hannafin Co.*)

**Fig. 75**   Ferulok flareless tube fitting. (*Parker-Hannafin Co.*)

reaction between wheel and block; $f$ = coefficient of friction; $P$ = tangential frictional resistance. Then, for rotation in either direction,

$$F(A + B) = RB \qquad R = P/f \qquad \text{and} \qquad F = PB/f(A + B)$$

In Fig. 77, let $C$ = leverage distance from fulcrum pin to line of action of $P$. Then, for clockwise rotation, $F(A + B) =$

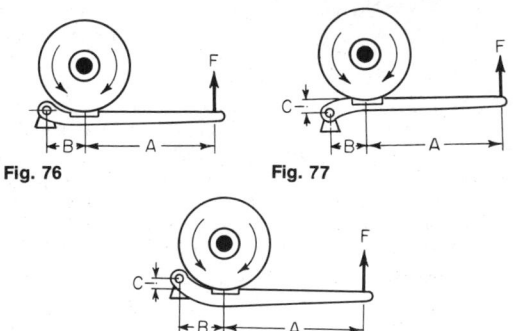

**Fig. 76**        **Fig. 77**

**Fig. 78**

**Figs. 76–78**   Block brakes.

$RB - fRC$. For counterclockwise rotation, $F(A + B) = RB + fRC$.

In the case of clockwise rotation, it will be noted that $C/B$ must be less than $1/f$, or the brake will be self-acting, i.e., will bind.

In the arrangement shown in Fig. 78, for clockwise rotation, $F(A + B) = RB + fRC$ and for counterclockwise rotation, $F(A + B) = RB - fRC$.

In the latter case (for counterclockwise rotation), $C/B$ must be less than $1/f$, or the brake will be self-acting, i.e, will bind.

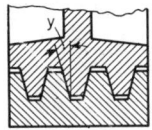

Double blocks are used to eliminate bending of the shaft. Figure 80 illustrates one way in which such brakes may be rigged. The force relations are as follows:

Let $F$ = load applied at end of lever, lb; $R$ = reaction between wheel and *each* block, lb; $P$ = tangential frictional resistance on each block surface, lb; $f$ = coefficient of

**Fig. 79**   Grooved block brake.

friction of materials in contact for a given condition of surface; $r$ = drum radius, in; $T$ = torque on drum shaft, in·lb; other notations as in the figure. Then,

$$R = FA(c + d)/\cos(x/2)2Bc \qquad P = Rf \qquad T = 2Rfr$$

The point $o$ should be a floating pivot to permit adjustment as the blocks wear.

Should the **face** of the brake wheel and blocks be **grooved**, as shown in Fig. 79, $f/(\sin y + f \cos y)$ must be substituted for $f$

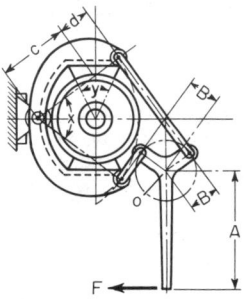

**Fig. 80**   Double block brake.

in the foregoing equations, $y$ being equal to half the angle included by the faces of the grooves and not less than 23°, to prevent binding; $y$ may have any value up to 30°.

**Band brakes** are shown diagrammatically in Figs. 81 to 83. The bands are usually of an asbestos fabric, sometimes rein-

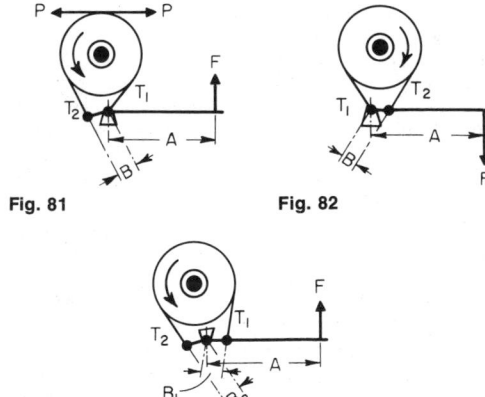

**Fig. 81**        **Fig. 82**

**Fig. 83**

**Figs. 81–83**   Band brakes.

forced with copper wire and impregnated with asphalt. The force relations obtaining in their operations are as follows:

In Fig. 81, let $F$ = force at end of brake handle; $P$ = tangential force at rim of wheel; $f$ = coefficient of friction of materials in contact; $a$ = angle of wrap of band, deg; $T_1$ = total tension in band on tight side; $T_2$ = total tension in band on slack side. Then $T_1 - T_2 = P$ and $T_1/T_2 = 10^{0.0076fa} = 10^b$ where $b = 0.0076fa$. Also, $T_2 = P/(10^b - 1)$ and $T_1 = P \times 10^b/(10^b - 1)$. The values of $10^{0.0076fa}$ are given in Fig. 105 for $a$ in radians. (See also Sec. 3.)

For the arrangement shown in Fig. 81,

$$FA = T_2 B = PB/(10^b - 1)$$

and

$$F = PB/A(10^b - 1)$$

For the construction illustrated in Fig. 82,

$$F = (PB/A)[10^b/(10^b - 1)]$$

For the differential brake shown in Fig. 83,

$$F = (P/A)[(B_2 - 10^b B_1)/(10^b - 1)]$$

In this arrangement, the quantity $10^b \times B_1$ must always be less than $B_2$, or the band will grip the wheel and the brake, or part of the mechanism to which it is attached, will be ruptured.

It is usual in practice to have the leverage ratio $A/B$ for block brakes about 5:1, and for band brakes about 10:1. The bands are faced with maple blocks.

If $f$ for wood on iron is taken at 0.3 and the angle of wrap for the band is 270°, i.e., subtends ¾ of the circumference, then $10^b = 4$ approx, and $C/B$ is taken equal to $1/0.5 = 2$, the loads required for a given torque will be as follows for the cases just considered and for the leverage ratios stated above:

Block brake, Fig. 76 . . . . . . . . . . . . . . . . . . . . . . . . . . . $F = 0.55P$
Block brake, Fig. 77 (clockwise rotation) . . . . . . . . . . . . $F = 0.22P$
Block brake, Fig. 77 (counterclockwise rotation) . . . . . . $F = 0.90P$
Block brake, Fig. 78 (clockwise rotation) . . . . . . . . . . . . $F = 0.90P$
Block brake, Fig. 78 (counterclockwise rotation) . . . . . . . $F = 0.22P$
Band brake, Fig. 81 . . . . . . . . . . . . . . . . . . . . . . . . . . . . $F = 0.033P$
Band brake, Fig. 82 . . . . . . . . . . . . . . . . . . . . . . . . . . . . $F = 0.133P$
Band brake, Fig. 83 . . . . . . . . . . . . . . . . . . . . . . . . . . . . $F = 0.016P$

In the case of Fig. 83, the dimension $B_2$ must be greater than $B_1 \times 10^b$. Accordingly, $B_1$ is taken as ¼, $A$ as 10, and, since $10^b = 4$, $B_2$ is taken as 1½.

The principal function of a brake is to absorb energy. This energy appears at the surface of the brake as heat, which must be carried away at a sufficiently rapid rate to prevent burning of the wooden blocks. Suitable proportions may be arrived at as follows:

Let $p$ = unit pressure on brake surface, lb/in² = $R$ (reaction against block)/area of block; $v$ = velocity of brake rim surface, ft/s = $2\pi rn/60$, where $n$ = r/min of brake wheel. Then $pv$ = work absorbed per in² of brake surface per second, and $pv \le 1,000$ for intermittent applications of load with comparatively long periods of rest and poor means for carrying away heat (wooden blocks); $pv \le 500$ for continuous application of load and poor means for carrying away heat (wooden blocks); $pv \le 1,400$ for continuous application of load with effective means for carrying away heat (oil bath).

**Cone brakes** may be made to the form shown in Fig. 84. The force relations are as follows:

Let $F$ = load applied at end of lever, lb; $Q$ = normal pressure on cone surface, lb; $P$ = tangential force on the rim of the brake, lb; $r$ = mean radius of cone surface, in; and $y$ = half angle of cone. Then $Q = F(b/a)/(\sin y + f \cos y)$ and $P = fF(b/a)/(\sin y + f \cos y)$; $F = P(a/b)(\sin y + f \cos y)/f$. For $a = 1$, $b = 10$, $y = 15°$, and $f = 0.2$ for cast iron on cast iron, $F = 0.23P$, approximately.

Such brakes are frequently used for lowering loads by the means shown in Fig. 85. The drum shaft is driven through the worm shaft by means of worm and wheel. In raising the load,

the ratchet runs free. As the load tends to lower, the ratchet engages and the worm thrusts the cone surfaces together. Lowering of the load must be accomplished by the application of torque to the worm shaft. In the case of worms of small pitch, no brake is required, for the wheel cannot turn the worm. This brake is therefore adapted to worms of large pitch, and the reason for employing a large pitch is that a more efficient drive is obtained. The force relations are approximately as follows (see Fig. 85):

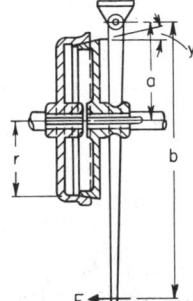

**Fig. 84** Cone brake.

Let $R$ = mean radius of cone surface, in; $Q$ = force normal to cone surface, lb; $P$ = tangential force at cone surface, acting with the leverage $R$; $F$ = axial force, lb; $f$ = coefficient of friction between cone surfaces; $a$ = angle between shaft axis and element of cone surface, deg; $b$ = angle of pitch of worm on worm wheel, deg; $f_1$ = coefficient of friction of worm and wheel teeth; $r$ = pitch radius of worm, in; $r_3$ = pitch radius of ratchet wheel, in; $L$ = load on ratchet teeth, lb. Let $x = f \cos a + \sin a$, and $y = (\tan b - f_1)$; then $Lr_3$ = torque on worm shaft (in·lb) due to $F$, where $F$ is the axial force due to load on drum = $Fry$ approximately.

The resisting moment of the clutch is $PR = QfR = RFf/x$. To hold the load and prevent its lowering, the actuating torque $Fry$ must be less than the reacting torque $RFf/x$. Accordingly, $Rf/x \ge ry$, or $R \ge ryx/f$. The angles $a$ and $b$ may be assumed 22°, $f$ at 0.08, and $f_1$ at 0.10, depending on the lubrication, whence $x = 0.45$ and $y = 0.30$. Then for the above assumptions $R \ge 1.7r$. If $a = 30°$, $b = 35°$, then, for $f$ and $f_1$ as before, $x = 0.57$, $y = 0.60$, and $R \ge 4.3r$.

**Disk brakes** of the same general type of construction but using a flat face instead of a cone are shown in Fig. 86. There are two faces to be moved against friction.

The force relations are $Lr_3 \le 2fFR = 2PR$, where $fF = P$. Also $R \ge ry/2f$.

Frequently disk brakes are made as shown in Fig. 87. The pinion $Q$ engages the gear in the drum (not shown). When the load is to be raised, power is applied through the gear and the connection between $B$ and $C$ is accomplished by the advancing of $B$ along $A$ and the clamping of the friction disks $D$ and $D$ and the ratchet wheel $E$. The reversal of the motor disconnects $B$ and $C$. In lowering the load, only as much reversal of rotation of the gear is given as is needed to reduce the force in the friction disks so that the load may be lowered under control.

The force relations are as follows (Fig. 87):

Let $R$ = mean radius of friction plates, in; $f$ = coefficient of friction between plates; $F$ = axial force along the screw =

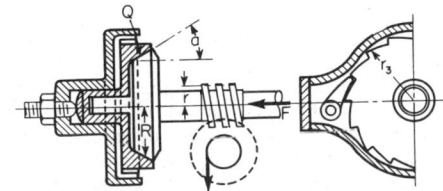

**Fig. 85** Cone brake for lowering loads.

force in friction plates, lb; $P$ = tangential force on friction plates at mean radius $R$, lb, $= fF$; $W$ = load on pinion teeth, lb; $r$ = radius of pinion pitch circle, in; $r_1$ = radius of pitch circle of screw, in; $a$ = angle of screw thread, deg; $f_1$ = coefficient of friction in thread; $z = (\tan a + f_1)$.

Then the load in lowering causes a moment $Wr = fFR + Fr_1z$, approx. To sustain the load, $Wr \leq 2fFR$ and $fR \geq r_1z$.

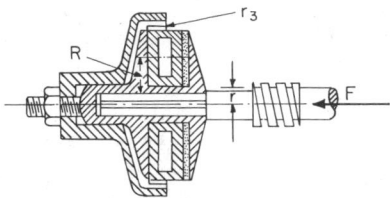

**Fig. 86**  Disk brake.

Acceptable values of the several factors are: $a = 10°$; $f = 0.08$; $f_1 = 0.10$; $R = 9$ in. Substituting these in the last equation, $r_1 \leq 2.5$ in $+$. Any radius of screw less than $2\frac{1}{2}$ in consistent with strength will be satisfactory for the above conditions.

A **multidisk brake** is shown in Fig. 88. This type of construction results in an increase in the number of friction faces. The

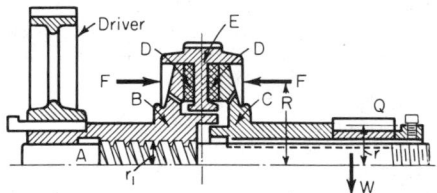

**Fig. 87**  Disk brake.

drum shaft is geared to the pinion $A$, while the motive power for driving comes through the gear $G$. In raising the load, direct connection is had between $G$, $B$, and $A$. In lowering, $B$ moves relatively to $G$ and forces the friction plates together, those plates fast to $E$ being held stationary by the pawl on $E$.

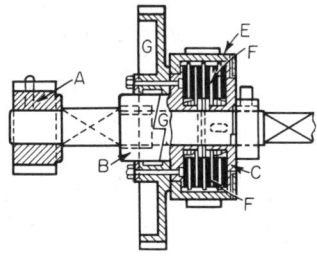

**Fig. 88**  Multidisk brake.

In the figure, there are three plates fast to $E$, one fast to $G$, and one fast to $C$.

In addition to the notation given in the previous case, let $n$ = number of faces in sliding contact when the part $C$ is moved relatively to part $E$ which carries the ratchet. This condition obtains upon the load beginning to lower, and $n_1$ = number of faces in sliding contact when the parts $G$ and $C$ both move relatively to $E$. This obtains when full gripping of the plates

takes place. The load in beginning to lower occasions the following force relation: $Wr = nfFR + Fr_1z$. To sustain the load, it is necessary that $Wr \leq fFR$. Hence, to prevent the load from dropping, $r_1z \leq fR(n_1 - n)$.

**Internal brakes** (Fig. 89) are used extensively on motor vehicles (see Sec. 11) where rotation occurs in both directions. The blocks are lined with a suitable friction material.

Brakes of the type illustrated are self-energizing; that is, friction makes the shoe tend to follow the rotating brake drum, wedging itself between the drum and the point at which it is

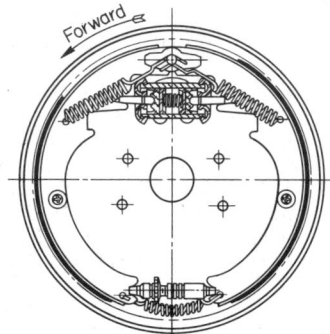

**Fig. 89**  Internal brake.

anchored. This action builds up a tremendous amount of friction, and for this reason gives a great deal of braking power without the use of excessive pedal pressure.

**Eddy-current brakes** (Fig. 90) are used with flywheels where quick braking is essential, and where large kinetic energy of the rotating masses precludes the use of block brakes due to excessive heating, as in reversible rolling mills. A number of

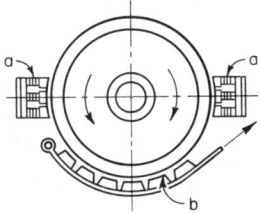

**Fig. 90**  Eddy-current brake.

poles $a$ are electrically excited (north and south in turn) and create a magnetic flux which permeates the gap and the iron of the rim, causing eddy currents. The flywheel energy is converted through these currents into heat. The hand brake $b$ may be used for quicker stopping when the speed of the wheel is considerably decreased; i.e., when the eddy-current brake is inefficient. Two brakes are provided to avoid bending forces on the shaft.

**Electric brakes** are often used in cranes, bridges, turntables, and machine tools, where an automatic application of the brake is important as soon as power is cut off. The brake force is supplied by an adjustable spring which is counteracted by the force of a solenoid or a centrifugal thruster. Interruption of current automatically applies the spring-activated brake shoes. Figures 91 and 92 show these types of electric brake.

## SHRINK, PRESS, DRIVE, AND RUNNING FITS

ANSI B4.1-1967 recommends preferred sizes, allowances, and tolerances for fits between plain cylindrical parts. Such fits include bearing, shrink and drive fits, etc. Terms used in describing fits are defined as follows: **Allowance:** minimum clearance (positive allowance) or maximum interference (negative allowance) between mating parts. **Tolerance:** total permis-

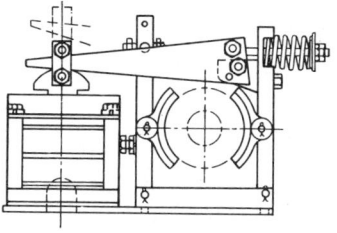

**Fig. 91** Solenoid-type electric brake.

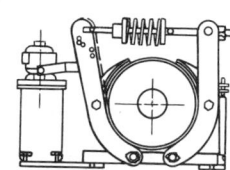

**Fig. 92** Thrustor-type electric brake.

sible variation of size. **Limits of size:** applicable maximum and minimum sizes. **Clearance fit:** one having limits of size so prescribed that a clearance always results when mating parts are assembled. **Interference fit:** in this case, limits are so prescribed that interference always results on assembly. **Transition fit:** this may have either a clearance or an interference on assembly. **Basic size:** one from which limits of size are derived by the application of allowances and tolerances. **Unilateral tolerance:** in this case a variation in size is permitted in only one direction from the basic size.

Fits are divided into the following general classifications: (1) running and sliding fits, (2) locational clearance fits, (3) transi-

tion fits, (4) locational interference fits, and (5) force or shrink fits.

**1. Running and Sliding Fits.** These are intended to provide similar running performance with suitable lubrication allowance throughout the range of sizes. These fits are further subdivided into the following classes:

**Class RC1. Close-sliding Fits.** Intended for accurate location of parts which must assemble without perceptible play.

**Class RC2. Sliding Fits.** Parts made with this fit move and turn easily but are not intended to run freely; also, in larger sizes they may seize under small temperature changes.

**Class RC3. Precision-running Fits.** These are intended for precision work at slow speeds and light journal pressures but are not suitable where appreciable temperature differences are encountered.

**Class RC4. Close-running Fits.** For running fits on accurate machinery with moderate surface speeds and journal pressures, where accurate location and minimum play is desired.

**Classes RC5 and RC6. Medium-running Fits.** For higher running speeds or heavy journal pressures, or both.

**Class RC7. Free-running Fits.** For use where accuracy is not essential, or where large temperature variations are likely to be present, or both.

**Classes RC8 and RC9. Loose-running Fits.** For use with materials such as cold-rolled shafting or tubing made to commercial tolerances.

**Limits of clearance** given in ANSI B4.1-1967 for each of these classes are given in Table 44. Hole and shaft tolerances are listed on a unilateral tolerance basis in this reference to give the clearance limits of Table 44, the hole size being the basic size.

**2. Locational Clearance Fits.** These are intended for normally stationary parts which can, however, be freely assem-

### Table 44. Limits of Clearance for Running and Sliding Fits (Basic Hole)

(Limits are in thousandths of an inch on diameter)

| Nominal size range, in. Over | To | Class | | | | | | | | |
|---|---|---|---|---|---|---|---|---|---|---|
| | | RC1 | RC2 | RC3 | RC4 | RC5 | RC6 | RC7 | RC8 | RC9 |
| 0.04 | 0.12 | 0.1 | 0.1 | 0.3 | 0.3 | 0.6 | 0.6 | 1.0 | 2.5 | 4.0 |
| | | 0.45 | 0.55 | 0.8 | 1.1 | 1.4 | 1.8 | 2.6 | 5.1 | 8.1 |
| 0.12 | 0.24 | 0.15 | 0.15 | 0.4 | 0.4 | 0.8 | 0.8 | 1.2 | 2.8 | 4.5 |
| | | 0.5 | 0.65 | 1.0 | 1.4 | 1.8 | 2.2 | 3.1 | 5.8 | 9.3 |
| 0.24 | 0.40 | 0.2 | 0.2 | 0.5 | 0.5 | 1.0 | 1.0 | 1.6 | 3.0 | 5.0 |
| | | 0.6 | 0.85 | 1.3 | 1.7 | 2.2 | 2.8 | 3.9 | 6.6 | 10.7 |
| 0.40 | 0.71 | 0.25 | 0.25 | 0.6 | 0.6 | 1.2 | 1.2 | 2.0 | 3.5 | 6.0 |
| | | 0.75 | 0.95 | 1.4 | 2.0 | 2.6 | 3.2 | 4.6 | 7.9 | 12.8 |
| 0.71 | 1.19 | 0.3 | 0.3 | 0.8 | 0.8 | 1.6 | 1.6 | 2.5 | 4.5 | 7.0 |
| | | 0.95 | 1.2 | 1.8 | 2.4 | 3.2 | 4.0 | 5.7 | 10.0 | 15.5 |
| 1.19 | 1.97 | 0.4 | 0.4 | 1.0 | 1.0 | 2.0 | 2.0 | 3.0 | 5.0 | 8.0 |
| | | 1.1 | 1.4 | 2.2 | 3.0 | 4.0 | 5.2 | 7.1 | 11.5 | 18.0 |
| 1.97 | 3.15 | 0.4 | 0.4 | 1.2 | 1.2 | 2.5 | 2.5 | 4.0 | 6.0 | 9.0 |
| | | 1.2 | 1.6 | 2.6 | 3.6 | 4.9 | 6.1 | 8.8 | 13.5 | 20.5 |
| 3.15 | 4.73 | 0.5 | 0.5 | 1.4 | 1.4 | 3.0 | 3.0 | 5.0 | 7.0 | 10.0 |
| | | 1.5 | 2.0 | 3.2 | 4.2 | 5.8 | 7.4 | 10.7 | 15.5 | 24.0 |
| 4.73 | 7.09 | 0.6 | 0.6 | 1.6 | 1.6 | 3.5 | 3.5 | 6.0 | 8.0 | 12.0 |
| | | 1.8 | 2.3 | 3.6 | 4.8 | 6.7 | 8.5 | 12.5 | 18.0 | 28.0 |
| 7.09 | 9.85 | 0.6 | 0.6 | 2.0 | 2.0 | 4.0 | 4.0 | 7.0 | 10.0 | 15.0 |
| | | 2.0 | 2.6 | 4.4 | 5.6 | 7.6 | 9.6 | 14.3 | 21.5 | 34.0 |
| 9.85 | 12.41 | 0.8 | 0.8 | 2.5 | 2.5 | 5.0 | 5.0 | 8.0 | 12.0 | 18.0 |
| | | 2.3 | 2.9 | 4.9 | 6.5 | 9.0 | 11.0 | 16.0 | 25.0 | 38.0 |
| 12.41 | 15.75 | 1.0 | 1.0 | 3.0 | 3.0 | 6.0 | 6.0 | 10.0 | 14.0 | 22.0 |
| | | 2.7 | 3.4 | 5.8 | 7.4 | 10.4 | 13.0 | 19.5 | 29.0 | 45.0 |
| 15.75 | 19.69 | 1.2 | 1.2 | 4.0 | 4.0 | 8.0 | 8.0 | 12.0 | 16.0 | 25.0 |
| | | 3.0 | 3.8 | 7.2 | 9.0 | 13.0 | 16.0 | 22.0 | 32.0 | 51.0 |

**Table 45. Limits of Interference for Force and Shrink Fits**
(Limits are in thousandths of an inch on diameter)

| Nominal size range, in. | | Class | | | | |
|---|---|---|---|---|---|---|
| Over | To | FN1 | FN2 | FN3 | FN4 | FN5 |
| 0.04 | 0.12 | 0.05 | 0.2 | .... | 0.3 | 0.5 |
| | | 0.5 | 0.85 | .... | 0.95 | 1.3 |
| 0.12 | 0.24 | 0.1 | 0.2 | .... | 0.4 | 0.7 |
| | | 0.6 | 1.0 | .... | 1.2 | 1.7 |
| 0.24 | 0.40 | 0.1 | 0.4 | .... | 0.6 | 0.8 |
| | | 0.75 | 1.4 | .... | 1.6 | 2.0 |
| 0.40 | 0.56 | 0.1 | 0.5 | .... | 0.7 | 0.9 |
| | | 0.8 | 1.6 | .... | 1.8 | 2.3 |
| 0.56 | 0.71 | 0.2 | 0.5 | .... | 0.7 | 1.1 |
| | | 0.9 | 1.6 | .... | 1.8 | 2.5 |
| 0.71 | 0.95 | 0.2 | 0.6 | .... | 0.8 | 1.4 |
| | | 1.1 | 1.9 | .... | 2.1 | 3.0 |
| 0.95 | 1.19 | 0.3 | 0.6 | 0.8 | 1.0 | 1.7 |
| | | 1.2 | 1.9 | 2.1 | 2.3 | 3.3 |
| 1.19 | 1.58 | 0.3 | 0.8 | 0.8 | 1.5 | 2.0 |
| | | 1.3 | 2.4 | 2.4 | 3.1 | 4.0 |
| 1.58 | 1.97 | 0.4 | 0.8 | 1.2 | 1.8 | 3.0 |
| | | 1.4 | 2.4 | 2.8 | 3.4 | 5.0 |
| 1.97 | 2.56 | 0.6 | 0.8 | 1.3 | 2.3 | 3.8 |
| | | 1.8 | 2.7 | 3.2 | 4.2 | 6.2 |
| 2.56 | 3.15 | 0.7 | 1.0 | 1.8 | 2.8 | 4.8 |
| | | 1.9 | 2.9 | 3.7 | 4.7 | 7.2 |
| 3.15 | 3.94 | 0.9 | 1.4 | 2.1 | 3.6 | 5.6 |
| | | 2.4 | 3.7 | 4.4 | 5.9 | 8.4 |
| 3.94 | 4.73 | 1.1 | 1.6 | 2.6 | 4.6 | 6.6 |
| | | 2.6 | 3.9 | 4.9 | 6.9 | 9.4 |
| 4.73 | 5.52 | 1.2 | 1.9 | 3.4 | 5.4 | 8.4 |
| | | 2.9 | 4.5 | 6.0 | 8.0 | 11.6 |
| 5.52 | 6.30 | 1.5 | 2.4 | 3.4 | 5.4 | 10.4 |
| | | 3.2 | 5.0 | 6.0 | 8.0 | 13.6 |
| 6.30 | 7.09 | 1.8 | 2.9 | 4.4 | 6.4 | 10.4 |
| | | 3.5 | 5.5 | 7.0 | 9.0 | 13.6 |
| 7.09 | 7.88 | 1.8 | 3.2 | 5.2 | 7.2 | 12.2 |
| | | 3.8 | 6.2 | 8.2 | 10.2 | 15.8 |
| 7.88 | 8.86 | 2.3 | 3.2 | 5.2 | 8.2 | 14.2 |
| | | 4.3 | 6.2 | 8.2 | 11.2 | 17.8 |
| 8.86 | 9.85 | 2.3 | 4.2 | 6.2 | 10.2 | 14.2 |
| | | 4.3 | 7.2 | 9.2 | 13.2 | 17.8 |
| 9.85 | 11.03 | 2.8 | 4.0 | 7.0 | 10.0 | 16.0 |
| | | 4.9 | 7.2 | 10.2 | 13.2 | 20.0 |
| 11.03 | 12.41 | 2.8 | 5.0 | 7.0 | 12.0 | 18.0 |
| | | 4.9 | 8.2 | 10.2 | 15.2 | 22.0 |
| 12.41 | 13.98 | 3.1 | 5.8 | 7.8 | 13.8 | 19.8 |
| | | 5.5 | 9.4 | 11.4 | 17.4 | 24.2 |
| 13.98 | 15.75 | 3.6 | 5.8 | 9.8 | 15.8 | 22.8 |
| | | 6.1 | 9.4 | 13.4 | 19.4 | 27.2 |
| 15.75 | 17.72 | 4.4 | 6.5 | 9.5 | 17.5 | 25.5 |
| | | 7.0 | 10.6 | 13.6 | 21.6 | 30.5 |
| 17.72 | 19.69 | 4.4 | 7.5 | 11.5 | 19.5 | 27.5 |
| | | 7.0 | 11.6 | 15.6 | 23.6 | 32.5 |

bled or disassembled. These are subdivided into various classes which run from snug fits for parts requiring accuracy of location, through medium clearance fits (spigots) to the looser fastener fits where freedom of assembly is of prime importance.

**3. Transition Fits.** These are for applications where accuracy of location is important, but a small amount of either clearance or interference is permissible.

**4. Locational Interference Fits.** Used where accuracy of location is of prime importance and for parts requiring rigidity and alignment with no special requirements for bore pressure.

Data on clearance limits, interference limits, and hole and shaft diameter tolerances for locational clearance fits, transition fits, and locational interference fits are given in ANSI B4.1-1967.

**5. Force or Shrink Fits.** These are characterized by approximately constant bore pressures throughout the range

of sizes; interference varies almost directly as the diameter, and the differences between maximum and minimum values of interference are small. These are divided into the following classes:

**Class FN1. Light-drive Fits.** For applications requiring light assembly pressures (thin sections, long fits, cast-iron external members).

**Class FN2. Medium-drive Fits.** Suitable for ordinary steel parts or for shrink fits on light sections. These are about the tightest fits that can be used on high-grade cast-iron external members.

**Class FN3. Heavy-drive Fits.** For heavier steel parts or shrink fits in medium sections.

**Classes FN4 and FN5. Force Fits.** These are suitable for parts which can be highly stressed. Shrink fits are used instead of press fits in cases where the heavy pressing forces required for mounting are impractical.

In Table 45 are listed the limits of interference (maximum and minimum values) for the above classes of force or shrink fits for various diameters, as given in ANSI B4.1-1967. Hole and shaft tolerances to give these interference limits are also listed in this reference.

**Stresses Produced by Shrink or Press Fit** STEEL HUB ON STEEL SHAFT. The maximum equivalent stress, pounds per square inch, set up by a given press-fit allowance (in inches per inch of shaft diameter) is equal to $3x \times 10^7$, where $x$ is the allowance per inch of shaft diameter (Baugher, *Trans. ASME*, 1931, p. 85). The press-fit pressures set up between a steel hub and shaft, for various ratios $d/D$ between shaft and hub outside diameters, are given in Fig. 93. These curves are accurate to 5

**Pressure Required in Making Press Fits** The force required to press a hub on the shaft is given by $\pi fpdl$, where $l$ is length of fit, $p$ the unit press-fit pressure between shaft and hub, $f$ the coefficient of friction, and $d$ the shaft diameter. Values of $f$ varying from 0.03 to 0.33 have been reported, the lower values being due to yielding of the hub as a consequence of too high a fit allowance; the average is around 0.10 to 0.15. (For additional data see Horger and Nelson, "Design Data and Methods," ASME, 1953, pp. 87–91.)

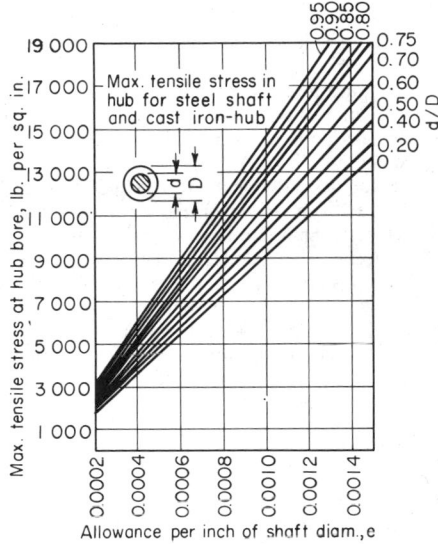

**Fig. 94** Variation in tensile stress in cast-iron hub in press-fit allowance.

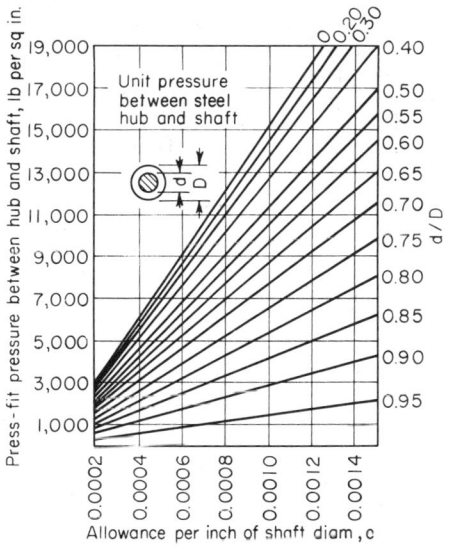

**Fig. 93** Press-fit pressures between steel hub and shaft.

percent even if the shaft is hollow, provided the inside shaft diameter is not over 25 percent of the outside. The equivalent stress given above is based on the maximum shear theory and is numerically equal to the radial-fit pressure added to the tangential tension in the hub. Where the shaft is hollow, with an inside diameter equal to more than about 25 percent of the outside diameter, the allowance in inches per inch to obtain an equivalent hub stress of 30,000 lb/in² may be determined by using Lame's thick cylinder formulas (*Jour. Applied Mechanics*, 1937, p. A-185). It should be noted that these curves hold only when the maximum equivalent stress is below the yield point; above the yield point, plastic flow occurs and the stresses are less than calculated.

CAST-IRON HUB ON STEEL SHAFT. Where the shaft is solid, or hollow with an inside diameter not over 25 percent of the outside diameter, Fig. 94 may be used to determine maximum tensile stresses in the cast-iron hub, resulting from the press-fit allowance; for various ratios $d/D$, Fig. 95 gives the press-fit pressures. These curves are based on a modulus of elasticity of $30 \times 10^6$ lb/in² for steel and $15 \times 10^6$ for cast iron. For a hollow shaft with an inside diameter more than about ¼ the outside, the Lamé formulas may be used.

**Torsional Holding Ability** The torque required to cause complete slippage of a press fit is given by $T = \frac{1}{2}\pi fpld^2$. Local slippage will usually occur near the end of the fit at much lower torques. If the torque is alternating, stress concentration and rubbing corrosion will occur at the hub face so that, eventually, fatigue failure may occur at considerably lower torques. Only in cases of static torque application is it justifiable to use ultimate torque as a basis for design.

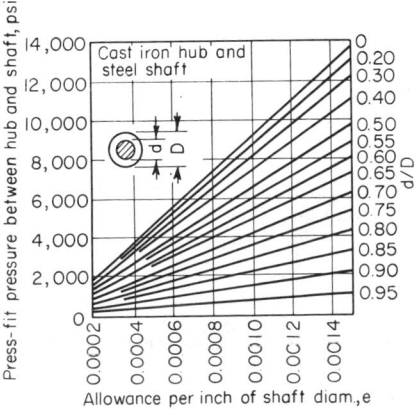

**Fig. 95** Press-fit pressures between cast-iron hub and shaft.

## SHAFTS, AXLES, AND CRANKS

(For critical speeds and torsion of shafts see Sec. 5.)

Most shafts are subject to combined bending and torsion stresses. These in turn may be steady, variable, or a combination of the two. For example, in operation, an electric-motor shaft may be subject to a steady torque on which are superimposed variations due to load fluctuations. An overhung pinion shaft may be subject to essentially constant torsion and completely reversed bending stress. In addition, shafts may be subject to repeated impact loading, as when electric motors are frequently started or plugged. In such cases the peak transient torques may reach several times the nominal values of the starting or plugging torques, depending on load inertia. An example of the latter is the roll-table drive used in steel-strip mills. (For calculating torques in such cases see Wahl, *Jour. Applied Mechanics*, March 1941, p. A-17.)

In the past, the design of shafts has frequently been based on the use of relatively low working stresses, to cover up unknown factors of ignorance. At present there is a trend toward more accurate evaluation of the fatigue strength of shafts under various operating conditions, but it is still necessary to use a factor of safety to cover various uncertainties in our knowledge of action of materials under various operating conditions and in knowledge of the applied loading. The formulas given under Cases 1 to 4 below for calculating the factors of safety of shafts under operating conditions are based on equations developed by R. E. Peterson, utilizing the Mises criterion of strength as basis (see Peterson, "Stress Concentration Design Factors," Wiley). This book also gives methods for calculating factors of safety for shafts under a limited number of cycles of alternating or shock loading.

Assuming a solid shaft of diameter $d$ inches, the nominal torsional stress $\tau$ and bending stress $\sigma$ are

$$\tau = \frac{16T}{\pi d^3} = 5.1 \frac{T}{d^3} \tag{1}$$

$$\sigma = \frac{32M}{\pi d^3} = 10.2 \frac{M}{d^3} \tag{2}$$

where $T$ and $M$ are the torque and bending moments, in·lb; and $\tau$ and $\sigma$ are stresses, lb/in². 

**Case 1** A shaft is subject to a **pure torque** consisting of a **steady torque** $T_0$, on which is superimposed an **alternating torque** $T_v$ (no bending being present). This loading imposes a steady torsional stress $\tau_0$, combined with an alternating torsional stress $\tau_v$. The stresses $\tau_0$ and $\tau_v$ are calculated from Eq. (1), using $T_0$ and $T_v$, respectively. In this case, the factor of safety $n$ is given by

$$n = \frac{1}{\sqrt{3}[\tau_0/(1.33\sigma_y) + K_{tfs}\tau_v/\sigma_e]} \tag{3}$$

where $\sigma_y$ is the yield stress of the shaft material in tension and $\sigma_e$ the endurance limit in reversed bending. (Typical $\sigma_y$ and $\sigma_e$ values for various materials are given in Sec. 5.) The factor

1.33 in the denominator of Eq. (3) is based on a limit design method (taking the redistribution of stress into account), while $K_{tfs}$ is the **fatigue-notch** factor in torsion. The factor $K_{tfs}$ where fillets or grooves are present is given by

$$K_{tfs} = q(K_{ts} - 1) + 1 \tag{4}$$

where $K_{ts}$ = stress-concentration factor for grooved or filleted shaft in torsion (see Sec. 5) and $q$ = notch-sensitivity factor, which depends on the notch or fillet radius and on the material. Average values of $q$ for quenched and tempered steels, and for annealed or normalized steels for various notch or fillet radii, as given by Peterson (*ibid.*), are listed in Table 46. The values in Table 46 are average and may vary considerably in individual cases, depending on the material and on the shape of the notch; in doubtful cases, $q = 1$ should be assumed; i.e., $K_{tfs}$ should be taken equal to $K_{ts}$.

The choice of the factor of safety $n$ for any given application is the prerogative of the designer and depends on such things as seriousness of failure, uncertainty in knowledge of applied loads or material properties, and other factors. Values of $n$ in the range 2 to 3 have been used frequently in the past.

If a value of $n$ is chosen and $T_0$ and $T_v$ are given, the shaft diameter $d$ can be calculated from

$$d = 2.17 \sqrt[3]{n \left( \frac{K_{tfs}T_v}{\sigma_e} + \frac{T_0}{1.33\sigma_y} \right)} \tag{5}$$

**Case 2** A round shaft is subject to a **steady bending moment** $M_0$ on which is superimposed an **alternating bending moment** $M_v$ (no torsion being present). These moments give a steady stress $\sigma_0$, superimposed on a variable stress $\sigma_v$ calculated from Eq. (2). In this case the equations for factor of safety $n$ and diameter $d$ are

$$n = \frac{1}{\sigma_0/(\sigma_u) + K_{tf}\sigma_v/\sigma_e} \tag{6}$$

$$d = 2.17 \sqrt[3]{n \left( \frac{M_0}{\sigma_u} + \frac{K_{tf}M_v}{\sigma_e} \right)} \tag{7}$$

where $\sigma_u$ = ultimate strength or $1.7\sigma_y$, whichever is smaller. The fatigue-notch factor $K_{tf}$ in Eqs. (6) and (7) is

$$K_{tf} = q(K - 1) + 1 \tag{8}$$

where the sensitivity index $q$ may be estimated from Table 46 or, in case of doubt, may be taken as unity. $K_t$ is the theoretical stress-concentration factor in bending; approximate values are given in Sec. 5 for fillets or grooves. (For more accurate values, see Peterson, *ibid.*) For press fits, the latter reference gives the values for fatigue-notch factor $K_f$ listed in Table 47. These may be used for estimating $K_{tf}$ in Eqs. (6) and (7). In general the $K_f$ values increase with shaft size. (For further details, see Peterson, *ibid.*)

**Case 3** A case frequently encountered in practice is that of a shaft subject to a **steady torque** $T_0$ and a completely **reversed**

### Table 46. Values* of Notch-Sensitivity Factor $q$

| Notch or fillet radius $r$, in | 0.01 | 0.03 | 0.05 | 0.10 | 0.20 |
|---|---|---|---|---|---|
| $q$ (for quenched and tempered steels) | 0.70 | 0.90 | 0.95 | 0.98 | 1.00 |
| $q$ (for annealed or normalized steels) | 0.40 | 0.62 | 0.73 | 0.86 | 0.93 |

* These values apply where the ratio $t/r < 3$ and $t$ = depth of notch or height of shoulder.

**Table 47. $K_f$ Values for Plain Press Fits**
(Obtained from fatigue tests in bending)

| Shaft material | Shaft diam, in | Collar or hub material | $K_f$ | Remarks |
|---|---|---|---|---|
| 0.42% carbon steel | 1⅝ | 0.42% carbon steel | 2.0 | No external reaction through collar |
| 0.45% carbon axle steel | 2 | Ni-Cr-Mo steel (casehardened) | 2.3 | No external reaction through collar |
| 0.45% carbon axle steel | 2 | Ni-Cr-Mo steel (casehardened) | 2.9 | External reaction taken through collar |
| Cr-Ni-Mo steel (310 Brinell) | 2 | Ni-Cr-Mo steel (casehardened) | 3.9 | External reaction taken through collar |
| 2.6% Ni steel (57,000 lb/in² fatigue limit) | 2 | Ni-Cr-Mo steel (casehardened) | 3.3–3.8 | External reaction taken through collar |
| Same, heat-treated to 253 Brinell | 2 | Ni-Cr-Mo steel (casehardened) | 3.0 | External reaction taken through collar |

bending moment $M_v$, giving a steady shear stress $\tau_0$ and an alternating bending stress $\sigma_v$. The corresponding expressions for $n$ and $d$ are

$$n = \frac{1}{\sqrt{3(\tau_0/1.33\sigma_y)^2 + (K_{tf}\sigma_v/\sigma_c)^2}} \tag{9}$$

$$d = 2.17 \sqrt[3]{n \sqrt{0.422\left(\frac{T_0}{\sigma_y}\right)^2 + \left(\frac{K_{tf}M_v}{\sigma_s}\right)^2}} \tag{10}$$

**Case 4** In the general case where a shaft is subject to a combination of a **steady** plus **alternating torsion stress** and a **steady** plus **alternating bending stress**, the factor of safety $n$ and shaft diameter $d$ are given by

$$n = \frac{1}{\sqrt{(\sigma_0/\sigma_u + K_{tf}\sigma_v/\sigma_e)^2 + 3(\tau_0/1.33\sigma_y + K_{tfs}\tau_v/\sigma_e)^2}} \tag{11}$$

$$d = 2.17\,(\sqrt[3]{n})$$
$$\times \sqrt[3]{\sqrt{\left(\frac{M_0}{\sigma_u} + \frac{K_{tf}M_v}{\sigma_e}\right)^2 + \frac{3}{4}\left(\frac{T_0}{1.33\sigma_y} + \frac{K_{tfs}T_v}{\sigma_e}\right)^2}}$$

where the symbols have the same meaning as for Cases 1 to 3 and $\sigma_u$ is taken as the ultimate strength or $1.7\sigma_y$, whichever is smaller. Case 4 corresponds, for example, to the case of a motor which is started and stopped frequently.

EXAMPLE. A shaft is subject to an alternating bending moment $M_v$ of 12,000 in·lb and a constant torque $T_0$ of 7,000 in·lb. Assume that at the point of maximum bending moment there is a press fit having a $K_{tf}$ value of 2.5. The material is a medium-carbon axle steel with a yield strength $\sigma_y$ of 70,000 lb/in² and an endurance limit $\sigma_e$ of 50,000 lb/in². The value of the factor of safety $n$ is to be taken as 2. These are the conditions of Case 3. Substituting these values in Eq. (**10**), the required diameter $d$ becomes

$$d = 2.17 \sqrt[3]{2\sqrt{\left(\frac{2.5(12,000)}{50,000}\right)^2 + 0.422\left(\frac{7,000}{70,000}\right)^2}} = 2.31 \text{ in}$$

For data on heat treatments, tensile strength, yield strength, and endurance limits of carbon and alloy steels commonly used in axles and shafts see Lessells, "Strength and Resistance of Metals," Wiley, pp. 393–395.

**Stiffness of shafting** may become important where critical speeds, vibration, etc., may occur. Also, the lack of sufficient stiffness in shafts may give rise to bearing troubles. Critical speeds of shafts in torsion or bending and shaft deflections may be calculated using the methods of Sec. 5. For shafts of variable diameter see Spotts, "Design of Machine Elements," Prentice-Hall. In order to avoid trouble where sleeve bearings are used, the angular deflections at the bearings in general must be kept within certain limits. One rule is to make the shaft deflection over the bearing width equal to a small fraction of the oil-film thickness. Note that since shaft stiffness is proportional to the modulus of elasticity, alloy-steel shafts are no stiffer than carbon-steel shafts of the same diameter.

**Crankshafts** For calculating the torsional stiffness of crankshafts, the formulas given in Sec. 5 may be used.

**Marine-engine shafts** and **diesel-engine crankshafts** (see Sec. 11) should be designed not only for strength but for avoidance of critical speed. (See *Trans. ASME*, Applied Mechanics, **50**, no. 8, for methods of calculating critical speeds of diesel engines.)

## PULLEYS, SHEAVES, AND FLYWHEELS

**Arms** of pulleys, sheaves, and flywheels are subjected to stresses due to condition of founding, to details of construction (such as split or solid), and to conditions of service, which are difficult to analyze. For these reasons, no accurate stress relations can be established, and the following formulas must be understood to be only approximately correct. It has been established experimentally by Benjamin (*Am. Mach.*, Sept. 22, 1898) that thin-rim pulleys do not distribute equal loads to the several pulley arms. For these, it will be safe to assume the tangential force on the pulley rim as acting on half the number of arms. Pulleys with comparatively thick rims, such as engine band wheels, have all the arms taking the load. Furthermore, while the stress action in the arms is similar to that in a beam fixed at both ends, the amount of restraint at the rim depending on the rim's elasticity, it may nevertheless be assumed for purposes of design that cantilever action is predominant. The bending moment at the hub in arms of thin-rim pulleys will be $M = PL/\tfrac{1}{2}N$, where $M$ = bending moment, in·lb; $P$ = tangential load on the rim, lb; $L$ = length of the arm, in; and $N$ = number of arms. For thick-rim pulleys and flywheels, $M = PL/N$.

For arms of elliptical section having a width of two times the thickness, where $E$ = **width of arm section** at the rim, in, and $s_t$ = intensity of tensile stress, lb/in²

$$E = \sqrt[3]{40PL/s_tN} \text{ (thin rim)} = \sqrt[3]{20PL/s_tN} \text{ (thick rim)}$$

For single-thickness belts, $P$ may be taken as $50B$ lb and for double-thickness belts $P = 75B$ lb, where $B$ is the width of pulley face, in. Then $E = k \times \sqrt[3]{BL/s_tN}$, where $k$ has the following values: for thin rim, single belt, 13; thin rim, double belt, 15; thick rim, single belt, 10; thick rim, double belt, 12. For cast iron of good quality, $s_t$ due to bending may be taken at 1,500 to 2,000. The arm section at the rim may be made from $\frac{2}{3}$ to $\frac{3}{4}$ the dimensions at the hub.

For high-speed pulleys and flywheels, it becomes necessary to check the arm for tension due to rim expansion. It will be safe to assume that each arm is in tension due to one-half the centrifugal force of that portion of the rim which it supports. That is, $T = As_t = Wv^2/2NgR$, lb, where $T$ = tension in arm, lb; $N$ = number of arms; $v$ = speed of rim, ft/s; $R$ = radius of pulley, ft; $A$ = area of arm section, in²; $W$ = weight of pulley rim, lb, and $s_t$ = intensity of tensile stress in arm section, lb/in², whence $s_t = WRn^2/5{,}800NA$, where $n$ = r/min of pulley. **Arms of flywheels** having heavy rims may be subjected to severe stress action due to the inertia of the rim at sudden load changes. There being no means of predicting the probable maximum to which the inertia may rise, it will be safe to make the arms equal in strength to $\frac{3}{4}$ of the shaft strength in torsion. Accordingly, for elliptical arm sections,

$$N \times 0.5E^3 s_t = \frac{3}{4} \times 2s_sd^3 \quad \text{or} \quad E = 1.4d\sqrt[3]{s_s/s_tN}$$

For steel shafts with $s_s$ = 8,000 and cast-iron arms with $s$ = 1,500,

$$E = 2.4d/\sqrt[3]{N} = 1.3d \text{ (for 6 arms)} = 1.2d \text{ (for 8 arms)}$$

where $2E$ = width of elliptical arm section at hub, in (thickness = $E$), and $d$ = shaft diameter, in.

**Rims** of belted pulleys cast whole may have the following proportions (see Fig. 96):

$$t_2 = \frac{3}{4}h + 0.005D \qquad t_1 = 2t_2 + C \qquad W = \frac{9}{8}B \text{ to } \frac{5}{4}B$$

where $h$ = belt thickness, $C$ = $\frac{1}{24}W$, and $B$ = belt width, all in inches.

Engine band wheels, flywheels, and pulleys run at **high speeds** are subjected to the following stress actions in the rim:

Considering the rim as a free ring, i.e., without arm restraint, and made of cast iron or steel, $s_t = v^2/10$ (approx),

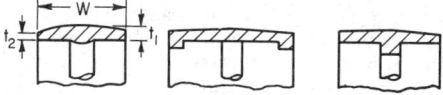

**Fig. 96** Rims for belted pulleys.

where $s_t$ = intensity of tensile stress, lb/in², and $v$ = rim speed, ft/s. For beam action between the arms of a solid rim, $M = Pl/12$ (approx), where $M$ = bending moment in rim, in·lb; $P$ = centrifugal force of that portion of the rim between arms, lb; and $l$ = length of rim between arms, in; from which $s_t = WR^2n^2/450N^2Z$, where $W$ = weight of entire rim, lb; $R$ = radius of wheel, ft; $n$ = r/min of wheel; and $Z$ = section

modulus of rim section, in³. In case the rim section is of the forms shown in Fig. 96, care must be taken that the flanges do not reduce the section modulus from that of the rectangular section. For **split rims** fastened with bolts as shown in Fig. 97, the stress analysis is as follows:

Let $w$ = weight of rim portion $L$ in in length, lb; $w_1$ = weight of lug, lb; $L_1$ = lever arm of lug, in; and $s_t$ = intensity

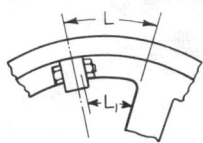

**Fig. 97** Split rim.

of tensile stress lb/in² in rim section joining arm. Then $s_t = 0.00034n^2R(w_1L_1 + wL/2)/Z$, where $n$ = r/min of wheel; $R$ = wheel radius, ft; and $Z$ = section modulus of rim section, in³. The above equation gives the value of $s_t$ for bending when the bolts are loose, which is the worst possible condition that may arise.

On this basis of analysis, $s_t$ should not be greater than 8,000 lb/in². The stress due to bending in addition to the stress due to rim expansion as analyzed previously will be the probable maximum intensity of stress for which the rim should be checked for strength. The flange bolts, because of their position, do not materially relieve the bending action. In case a tie rod leads from the flange to the hub, it will be *safe* to consider it as an additional factor of safety. When the tie rod is kept tight, it very materially strengthens the rim.

A more accurate method for calculating maximum stresses due to centrifugal force in flywheels with arms cast integral with the rim is given by Timoshenko, "Strength of Materials," Part II, 1941, p. 98. More exact equations for calculating stresses in the arms of flywheels and pulleys due to a combination of belt pull, centrifugal force, and changes in velocity are given by Heusinger, *Forschung*, 1938, p. 197. In both treatments, shrinkage stresses in the arms due to casting are neglected.

**The relative strengths of different types of wheel construction** are shown by the results of Benjamin's experiments on the bursting of flywheels (*Trans. ASME*, 1899, 1902). The types of wheels experimented with and the speeds at which they burst are shown in Fig. 98 and Table 48.

**Large flywheels for high rim speeds** and severe working conditions (as for rolling-mill service) have been made from flat-rolled steel plates with holes bored for the shaft. A group of such wheels may be welded together by circumferential welds to form a large flywheel. By this means, the welds do not carry direct centrifugal loads, but serve merely to hold the parts in position. Flywheels up to 15 ft diam for rolling-mill service have been constructed in this way.

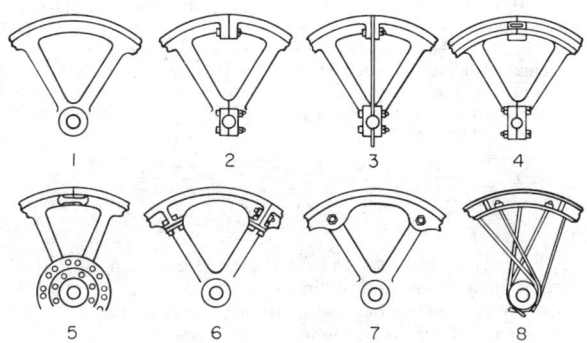

**Fig. 98** Types of flywheel construction (see Table 48).

**Table 48. Test Data on the Flywheels of Fig. 98**

| Number* | 1 | 2 | 3 | 4 | 5 | 6 | 7 | 8 |
|---|---|---|---|---|---|---|---|---|
| Number of arms | 6 | 6 | 6 | 6 | 8 | 6 | 6 | 24 |
| Rim speed at failure, ft/s = v | 395 | 194 | 225 | 305 | 256 | 223 | 393 | 424 |
| Comparative rim speeds at failure | 100 | 49 | 57 | 77 | 65 | 56.5 | 100 | 107 |
| Apparent rim tension at failure lb/in² = v²/10 | 15,625 | 3,764 | 5,062 | 9,302 | 6,502 | 4,973 | 15,445 | 17,978 |
| Efficiency† of construction | 0.85 | 0.19 | 0.265 | 0.49 | 0.34 | 0.26 | 0.84 | 0.94 |

*Construction of flywheels: No. 1 solid wheel; No. 2, in halves, with flange joints; No. 3, in halves, with reinforced joints; No. 4, in halves, with link joints; No. 5, segmental, with link joints; No. 6, in halves, with pad joints; No. 7, solid rim, separate spider; No. 8, solid rim, with tangent spokes.
†Efficiency assuming tensile strength at 19,000 lb/in².

## BELT DRIVES

(For the properties of belt materials, see Sec. 6)

**Leather belts** of the best quality are those made from the "butt" of the hide. The ultimate tensile **strength** of leather used in belts varies from 3,000 to 5,000 lb/in². Average values of the breaking strength of good oak-tanned belting, as determined by Benjamin, are as follows: For single (double) belts in solid leather 900 (1,400), at riveted joint 600 (1,200), at laced joint 350 lb/in of width.

The **weight** of leather belting is about 0.035 lb/in³. For thickness of single-, double-, and triple-leather belts as specified by National Industrial Leather Assoc., see Table 49. Rawhide, semirawhide (i.e., surface-tanned rawhide), and

**Table 49.   Theoretical Belt-Capacity Factors—$K$ (Leather Belts)**

| Belt speed, fpm | Single ply | | Double ply | | | Triple ply | |
|---|---|---|---|---|---|---|---|
| | $1\frac{1}{64}$ in. | $1\frac{3}{64}$ in. | $1\frac{8}{64}$ in. | $2\frac{0}{64}$ in. | $2\frac{3}{64}$ in. | $3\frac{0}{64}$ in. | $3\frac{3}{64}$ in. |
| | Med. | Heavy | Light | Med. | Heavy | Med. | Heavy |
| 600 | 1.1 | 1.2 | 1.5 | 1.8 | 2.2 | 2.5 | 2.8 |
| 800 | 1.4 | 1.7 | 2.0 | 2.4 | 2.9 | 3.3 | 3.6 |
| 1,000 | 1.8 | 2.1 | 2.6 | 3.1 | 3.6 | 4.1 | 4.5 |
| 1,200 | 2.1 | 2.5 | 3.1 | 3.7 | 4.3 | 4.9 | 5.4 |
| 1,400 | 2.5 | 2.9 | 3.5 | 4.3 | 4.9 | 5.7 | 6.3 |
| 1,600 | 2.8 | 3.3 | 4.0 | 4.9 | 5.6 | 6.5 | 7.1 |
| 1,800 | 3.2 | 3.7 | 4.5 | 5.4 | 6.2 | 7.3 | 8.0 |
| 2,000 | 3.5 | 4.1 | 4.9 | 6.0 | 6.9 | 8.1 | 8.9 |
| 2,200 | 3.9 | 4.5 | 5.4 | 6.6 | 7.6 | 8.8 | 9.7 |
| 2,400 | 4.2 | 4.9 | 5.9 | 7.1 | 8.2 | 9.5 | 10.5 |
| 2,600 | 4.5 | 5.3 | 6.3 | 7.7 | 8.9 | 10.3 | 11.4 |
| 2,800 | 4.9 | 5.6 | 6.8 | 8.2 | 9.5 | 11.0 | 12.1 |
| 3,000 | 5.2 | 5.9 | 7.2 | 8.7 | 10.0 | 11.6 | 12.8 |
| 3,200 | 5.4 | 6.3 | 7.6 | 9.2 | 10.6 | 12.3 | 13.5 |
| 3,400 | 5.7 | 6.6 | 7.9 | 9.7 | 11.2 | 12.9 | 14.2 |
| 3,600 | 5.9 | 6.9 | 8.3 | 10.1 | 11.7 | 13.4 | 14.8 |
| 3,800 | 6.2 | 7.1 | 8.7 | 10.5 | 12.2 | 14.0 | 15.4 |
| 4,000 | 6.4 | 7.4 | 9.0 | 10.9 | 12.6 | 14.5 | 16.0 |
| 4,200 | 6.7 | 7.7 | 9.3 | 11.3 | 13.0 | 15.0 | 16.5 |
| 4,400 | 6.9 | 7.9 | 9.6 | 11.7 | 13.4 | 15.4 | 16.9 |
| 4,600 | 7.1 | 8.1 | 9.8 | 12.0 | 13.8 | 15.8 | 17.4 |
| 4,800 | 7.2 | 8.3 | 10.1 | 12.3 | 14.1 | 16.2 | 17.8 |
| 5,000 | 7.4 | 8.4 | 10.3 | 12.5 | 14.3 | 16.5 | 18.2 |
| 5,200 | 7.5 | 8.6 | 10.5 | 12.8 | 14.6 | 16.8 | 18.5 |
| 5,400 | 7.6 | 8.7 | 10.6 | 12.9 | 14.8 | 17.1 | 18.8 |
| 5,600 | 7.7 | 8.8 | 10.8 | 13.1 | 15.0 | 17.3 | 19.0 |
| 5,800 | 7.7 | 8.9 | 10.9 | 13.2 | 15.1 | 17.5 | 19.2 |
| 6,000 | 7.8 | 8.9 | 10.9 | 13.2 | 15.2 | 17.6 | 19.3 |

Minimum pulley diameters for belt speeds, in.

| | | | | | | | |
|---|---|---|---|---|---|---|---|
| Up to 2,500 fpm | $2\frac{1}{2}$ | 3 | 4 | 5* | 8* | 16† | 20† |
| 2,500–4,000 fpm | 3 | $3\frac{1}{2}$ | $4\frac{1}{2}$ | 6* | 9* | 18† | 22† |
| 4,000–6,000 fpm | $3\frac{1}{2}$ | 4 | 5 | 7* | 10* | 20† | 24† |

* For belts 8 in. and over add 2 in. to minimum pulley diameter shown.
† For belts 8 in. and over add 4 in. to minimum pulley diameter shown.

chrome-tanned (green) belts are not as serviceable in dry places as oak-tanned belts. Rawhide and chrome-tanned belts give good service in damp places, such as dye houses.

Certain special **mineral-tanned leather belts** are now available, while for application where excessive moisture exists, special waterproofed leather belts can be had. Common leather belts are not suitable where exposed to steam, dampness, or dripping oil.

**Belt Joints**    **Cemented joints** when properly made have a strength equal to that of the solid leather. **Leather-laced** and **riveted joints** are about ⅓ and ⅔ as strong, respectively, as the solid leather. **Wire-laced joints** have 85 to 90 percent of the strength of solid leather. Heavy metal belt joints should not be used on belts run at high speeds.

**Rubber belting** is made from fabric or cord impregnated and bound together by vulcanized rubber compounds. The fabric or cord may be of cotton or rayon. Nylon cord and steel cord or cable are also available. Advantages are high tensile strength, strength to hold metal fasteners satisfactorily, and resistance to deterioration by moisture. The best-rubber fabric construction for most types of service is made from hard or tight-woven fabric with a "skim coat" or thin layer of rubber between plies. The cord type of construction allows the use of smaller pulley diameters than the fabric type, and also develops less stretch in service. It must be used in the endless form, except in cases where the oil-field type of clamp may be used.

Initial tensions in rubber belts run from 15 to 25 lb/ply/in width. A common rule is to cut belts 1 percent less than the minimum tape-line measurement around the pulleys. For heavy loads, a 1½ percent allowance is usually required, although, because of shrinkage less initial tension is required for wet or damp conditions. Initial tensions of 25 lb/ply/in may overload shafts or bearings. Maximum safe tight-side tensions for rubber belts are as follows:

| Duck weight, oz | 28 | 32 | 32.66 | 34.66 | 36 |
|---|---|---|---|---|---|
| Tension, lb/ply/in width | 25 | 28 | 30 | 32 | 35 |

Centrifugal forces at high speeds require higher tight-side tensions to carry rated horsepower.

Rubber belting may be bought in endless form or made endless in the field by means of a vulcanized splice produced by a portable electric vulcanizer. For endless belts the drive should provide take-up of 2 to 4 percent to allow for length variation as received and for stretch in service. The amount of take-up will vary with the type of belt used. For certain drives, it is possible to use endless belts with no provision for take-up, but this involves a heavier belt and a higher initial unit tension than would be the case otherwise. Ultimate tensile strength of rubber belting varies from 280 to 600 lb or more per inch width per ply. The weight varies from 0.02 to 0.03 or more lb/in width/ply. Belts with steel reinforcement are considerably heavier. For horsepower ratings of rubber belts, see Table 53.

**Arrangements for Belt Drives**    In belt drives, the centerline of the belt advancing on the pulley should lie in a plane passing through the midsection of the pulley at right angles to the shaft. Shafts inclined to each other require connections as shown in Fig. 99. In case guide pulleys are needed, their positions can be determined as shown in Figs. 100 to 102. In Fig. 102 the center circles of the two pulleys to be connected are set in correct relative position in two planes, *a* being the angle between the planes (= supplement of angle between shafts). If any two points as *E* and *F* are assumed on the line of

intersection *MN* of the planes, and tangents *EG*, *EH*, *FJ*, and *FK* are drawn from them to the circles, the center circles of the guide pulleys must be so arranged that these tangents are also tangents to them, as shown. In other words, the middle planes of the guide pulleys must lie in the planes *GEH* and *JFK*.

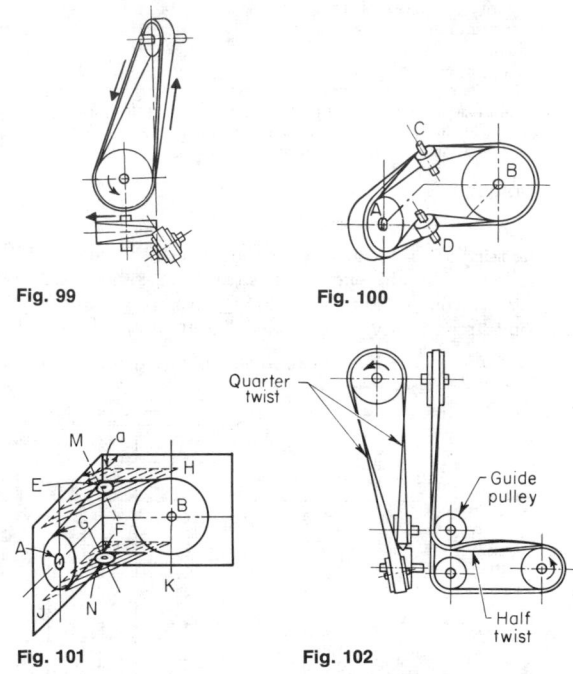

**Fig. 99**                              **Fig. 100**

**Fig. 101**                              **Fig. 102**

**Figs. 99–102**    Arrangements for belt drives.

When these conditions are met, the belts will run in either direction on the pulleys.

To avoid the necessity of taking up the **slack** in belts which have become stretched and permanently lengthened, a **belt tightener** such as shown in Fig. 103 may be employed. It should be placed on the slack side of the belt and nearer the driving pulley than the driven pulley. Pivoted motor drives may also be used to maintain belt tightness with minimum initial tension.

**Length of Belt for a Given Drive**    The length of an **open belt** for a given drive is equal to $L = 2C + 1.57(D + d) + (D - d)^2/4C$, where $L$ = length of belt, in; $D$ = diam of large pulley, in; $d$ = diam of small pulley, in; and $C$ = distance between pulley centers, in. Center distance $C$ is given by $C = 0.25b + 0.25\sqrt{b^2 - 2(D - d)^2}$, where $b = L - 1.57(D + d)$. When a **crossed belt** is used, the length in $L = 2C + 1.57(D + d) + (D + d)^2/4C$.

**Step or Cone Pulleys**    For belts operating on step pulleys, the pulley diameters must be such that the belt will fit over any pair with equal tightness. With **crossed belts,** it will be apparent from the equation for length of belt that the sum of the pulley diameters need only be constant in order that the belt may fit with equal tightness on each pair of pulleys. With open belts, the length is a function of both the sum and the difference of the pulley diameters; hence no direct solution of the problem is possible.

**A graphical method** devised by C. A. Smith (*Trans. ASME,* 10) is shown in Fig. 104. Let $A$ and $B$ be the centers of any pair of pulleys in the set, the diameters of which are known or assumed. Bisect $AB$ in $C$, and draw $CD$ at right angles to $AB$. Take $CD = 0.314$ times the center distance $L$, and draw a circle tangent to the belt line $EF$. The belt line of any other pair of pulleys in the set will then be tangent to this circle. If the angle $EF$ makes with $AB$ is greater than $18°$, draw a tangent to the circle $D$ making an angle of $18°$ with $AB$ and from a center on $CD$ distant $0.298L$ above $C$ draw an arc tangent to this $18°$ line. All belt lines with angles greater than $18°$ will be tangent to this last-drawn arc.

A very slight error in a graphical solution drawn to any scale much under full size will introduce an error seriously affecting the equality of belt tensions on the various pairs of pulleys in the set, and where much power is to be transmitted it is advisable to calculate the pulley diameters from the following **formulas** derived from Burmester's graphical method ("Lehrbuch der Mechanik").

Let $D_1$ and $D_2$ be, respectively, the diameters of the smaller and larger pulleys of a pair, $n = D_2/D_1$, and $l$ = distance between shaft centers, all in inches. Also let $m = 1.58114l -$

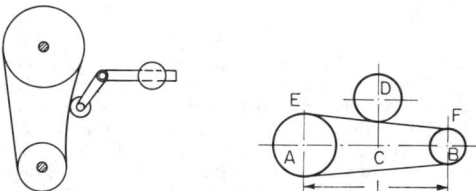

**Fig. 103**  Belt tightener.    **Fig. 104**

$D_0$, where $D_0$ = diam of both pulleys for a speed ratio $n = 1$. Then $(D_1 + m)^2 + (nD_1 + m)^2 = 5l^2$. First settle on values of $D_0$, $l$, and $n$, and then substitute in the equation and solve for $D_1$. The diameter $D_2$ of the other pulley of the pair will then be $nD_1$. The values are correct to the fourth decimal place.

The speeds given by cone pulleys should increase in a **geometrical ratio**; i.e., each speed should be multiplied by a constant $a$ in order to obtain the next higher speed. Let $n_1$ and $n_2$ be, respectively, the lowest and highest speeds (r/min) desired and $k$ the number of speed changes. Then $a = $

$\sqrt[k-1]{n_2/n_1}$. In practice, $a$ ranges from 1.25 up to 1.75 and even 2. The ideal value for $a$ in machine-tool practice, according to Carl G. Barth, would be 1.189. In the example below, this would mean the use of 18 speeds instead of 8.

EXAMPLE.    Let $n_1 = 16$, $n_2 = 400$, and $k = 8$, to be obtained with four pairs of pulleys and a back gear. From formula, $a = \sqrt[7]{25} = 1.584$, whence speeds will be 16, $(16 \times 1.584 =) 25.34$, $(25.34 \times 1.584 =) 40.14$, and similarly 63.57, 100.7, 159.5, 252.6, and 400. The first four speeds are with the back gear in; hence the back-gear ratio must be $100.7 \div 16 = 6.29$.

**Transmission of Power by Belts**  The turning force (tangential) on the rim of a pulley driven by a flat belt is equal to $T_1 - T_2$, where $T_1$ and $T_2$ are, respectively, the tensions in the driving (tight) side and following (slack) side of the belt. (For the relations of $T_1$ and $T_2$ at low peripheral speeds, see

Sec. 3.) Log $(T_1/T_2) = 0.0076fa$ when the effect of centrifugal force is neglected and $T_1/T_2 = 10^{0.0076fa}$. Figure 105 gives values of this function. When the speeds are high, however, the relations of $T_1$ to $T_2$ are modified by centrifugal stresses in the belt, in which case log $(T_1/T_2) = 0.0076f(1 - x)a$, where $f$ = coefficient of friction between the belt and pulley surface, $a$ = angle of wrap, and $x = 12wv^2/gt$, in which $w$ = weight of 1 $in^3$ of belt material, lb; $v$ = belt speed, ft/s; $g = 32.2$, and $t$ = allowable working tension, $lb/in^2$. Values of $x$ for leather belting (with $w = 0.035$ and $t = 300$) are as follows:

| $v$ | 30 | 40 | 50 | 60 | 70 | |
|---|---|---|---|---|---|---|
| $x$ | 0.039 | 0.070 | 0.118 | 0.157 | 0.214 | |

| $v$ | 80 | 90 | 100 | 110 | 120 | 130 |
|---|---|---|---|---|---|---|
| $x$ | 0.279 | 0.352 | 0.435 | 0.526 | 0.626 | 0.735 |

Researches by Barth (*Trans. ASME*, 1909) seem to show that $f$ is a function of the belt velocity, varying according to the formula $f = 0.54 - 140/(500 + V)$ for leather belts on iron pulleys, where $V$ = belt velocity in ft/min. For practical design, however, the following values of $f$ may be used: for leather belts on cast-iron pulleys, $f = 0.30$; on wooden pulleys, $f = 0.45$; on paper pulleys, $f = 0.55$. The treatment of belts with belt dressing, pulleys with cork inserts, and dampness are all factors which greatly modify these values, tending to make them higher.

The **arc of contact** on the smaller of two pulleys connected by an open belt, in degrees, is approximately equal to $180 - [60(D - d)/l]$, where $D$ and $d$ are the larger and smaller pulley diameters and $l$ the distance between their shaft centers, all in inches. This formula gives an error not exceeding 0.5 percent.

**Design of Leather Belts**  The National Industrial Leather Assoc. gives the following formulas for calculating the belt width or horsepower rating of oak-tanned flat leather belt drives.

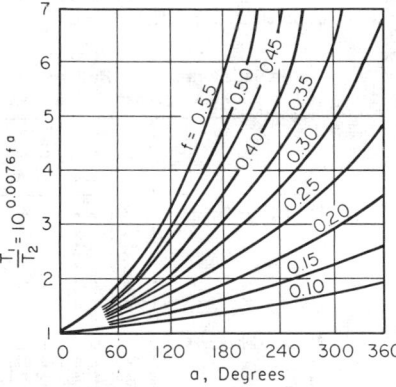

**Fig. 105**  Values of $10^{0.0076fa}$.

For **electric-motor drives**, $W = RM/KP$ and $H = WKP/M$, where $W$ = width of belt, in; $R$ = nameplate horsepower rating of electric motor; $H$ = horsepower rating of belt; $K$ = theoretical belt capacity factor, Table 49; $M$ = motor-load correction factor, Table 50; $P$ = pulley correction factor, Table 51. Values of minimum pulley diameters for various belt speeds are indicated in Table 49. Where special operating conditions are present, multiply belt width by a factor $F$ given in Table 52.

For **drives other than electric motors** the formulas are $W = HF/KP$ and $H = WKP/F$, where $H$ = horsepower rating of belt and $W$ = belt width. The factors $K$, $P$, and $F$ are given in Tables 49, 51, and 52, respectively.

$(\text{hp} \times S)/(W \times K)$, where hp = actual horsepower transmitted or, if not known, rated horsepower of driver unit as shown on nameplate; $W$ = belt width, in. Values of HP (horsepower per inch of width for 180° arc of contact for various belt speeds

**Table 50. Correction Factors $M$ for Type of Motor and Starting Method Used—Leather Belts**

| Motor type and starting method | Correction factor, $M$ |
|---|---|
| Squirrel cage, compensator starting | 1.5 |
| Squirrel cage, line starting | 2.0 |
| Slip ring and high starting torque | 2.5 |

**Table 51. Correction Factors $P$ for Diameter of Smaller Pulley—Leather Belts**

| Diameter of small pulley, in. | Correction factor, $P$ |
|---|---|
| 4 and under | 0.5 |
| 4½–8 | 0.6 |
| 9–12 | 0.7 |
| 13–16 | 0.8 |
| 17–30 | 0.9 |
| Over 30 | 1.0 |

**Table 52. Correction Factors $F$ for Special Operating Conditions—Leather Belts**

| Operating conditions | Correction factor, $F$ |
|---|---|
| Oily, wet, or dusty atmosphere | 1.35 |
| Vertical drives | 1.2 |
| Jerky loads | 1.2 |
| Shock and reversing loads | 1.4 |

For belt speeds over 6,000 ft/min or for pulley sizes less than those indicated, the supplier should be consulted.

EXAMPLE. A squirrel-cage induction-motor drive rated at 15 hp, 1,750 r/min, has the following operating conditions: across-the-line starting; motor pulley 8 in diam; driven pulley 16 in diam; belt speed = $1,750 \times \pi \times {}^8\!/_{12} = 3,665$ ft/min. Belt thickness is "medium double" for 8 in pulley diam, from bottom of Table 49. From this table, belt capacity factor $K$ is 10.2 for medium double belt at 3,665 ft/min. Factor $M$ from Table 50 is 2.0, for line-start squirrel-cage motor, and $P$ from Table 51 is 0.6 for 8 in diam pulley. Required belt width from equation above is $W = RM/KP = 15 \times 2.0/10.2 \times 0.6 = 4.9$ in. Use a 5-in-width medium double leather belt. If the motor is used on a vertical drive without special motor base, the above belt width should be multiplied by the factor $F = 1.2$ from Table 52, giving a 6-in-wide medium double belt.

**Horsepower Ratings of Rubber Transmission Belts**  For calculating horsepower rating of rubber transmission belts, the following formula is used by B. F. Goodrich Co.: HP =

and various materials) are given in Table 53. The ratings of Table 53 have been corrected for effects of centrifugal force and are based on a coefficient of friction of 0.3. $S$ = service factor (Table 54). $K$ = arc of contact factor (Table 55). Minimum pulley diameters are given in Table 56.

EXAMPLE. Assume a normal-torque squirrel-cage ac motor for a centrifugal-fan drive. Motor speed is 1,725 r/min and pulley diameter is 5 in. Horsepower transmitted at 1,725 r/min is 4.0. Arc of contact is 160°. From Table 54 factor $S$ = 1.2 and, from Table 55, factor $K$ = 0.93. Belt speed is $1,725 \times \pi \times {}^5\!/_{12} = 2,250$ ft/min. For a 5 in pulley diameter choose from Table 56 a 3-ply, 32-oz fabric belt, giving HP = 3.0 at 2,250 ft/min from Table 53. Required belt width from above equation is $W = (\text{hp} \times S)/(\text{HP} \times K) = (4 \times 1.2)/(3 \times 0.93) = 1.73$ in.

**V-Belt Drives**

V-belt drives are widely used in power transmission. Such drives consist essentially of endless belts of trapezoidal cross

**Table 53. Horsepower Ratings of Rubber Belts**
(HP = horsepower per inch of belt width for 180° wrap)

| | Ply | Belt speed, fpm | | | | | | | | | | |
|---|---|---|---|---|---|---|---|---|---|---|---|---|
| | | 500 | 1,000 | 1,500 | 2,000 | 2,500 | 3,000 | 4,000 | 5,000 | 6,000 | 7,000 | 8,000 |
| 32-oz fabric | 3 | 0.7 | 1.4 | 2.1 | 2.7 | 3.3 | 3.9 | 4.9 | 5.6 | 6.0 | | |
| | 4 | 0.9 | 1.9 | 2.8 | 3.6 | 4.4 | 5.2 | 6.5 | 7.4 | 7.9 | | |
| | 5 | 1.2 | 2.3 | 3.4 | 4.5 | 5.5 | 6.5 | 8.1 | 9.2 | 9.8 | | |
| | 6 | 1.4 | 2.8 | 4.1 | 5.4 | 6.6 | 7.8 | 9.6 | 11.0 | 11.7 | | |
| | 7 | 1.6 | 3.2 | 4.7 | 6.2 | 7.7 | 9.0 | 11.2 | 12.8 | 13.6 | | |
| | 8 | 1.8 | 3.6 | 5.3 | 7.0 | 8.7 | 10.2 | 12.7 | 14.6 | 15.5 | | |
| 32-oz hard fabric | 3 | 0.7 | 1.5 | 2.2 | 2.9 | 3.5 | 4.1 | 5.1 | 5.8 | 6.2 | 6.1 | 5.5 |
| | 4 | 1.0 | 2.0 | 3.0 | 3.9 | 4.7 | 5.5 | 6.8 | 7.8 | 8.3 | 8.1 | 7.3 |
| | 5 | 1.3 | 2.5 | 3.7 | 4.9 | 5.9 | 6.9 | 8.5 | 9.8 | 10.3 | 9.1 | 9.0 |
| | 6 | 1.5 | 3.0 | 4.5 | 5.9 | 7.1 | 8.3 | 10.2 | 11.7 | 12.3 | 12.1 | 10.7 |
| | 7 | 1.7 | 3.5 | 5.2 | 6.9 | 8.3 | 9.7 | 11.9 | 13.6 | 14.3 | 14.1 | 12.4 |
| | 8 | 1.9 | 4.0 | 5.9 | 7.9 | 9.5 | 11.1 | 13.6 | 15.5 | 16.3 | 16.0 | 14.1 |
| | 9 | 2.1 | 4.5 | 6.6 | 8.9 | 10.6 | 12.4 | 15.3 | 17.4 | 18.3 | 17.9 | 15.8 |
| | 10 | 2.3 | 5.0 | 7.3 | 9.8 | 11.7 | 13.7 | 17.0 | 19.3 | 20.3 | 19.8 | 17.5 |
| No. 70 rayon cord | 3 | 1.6 | 3.1 | 4.6 | 6.0 | 7.3 | 8.6 | 10.6 | 12.0 | 12.7 | 12.3 | 10.7 |
| | 4 | 2.1 | 4.1 | 6.1 | 8.0 | 9.8 | 11.5 | 14.5 | 16.6 | 17.8 | 17.8 | 16.4 |
| | 5 | 2.6 | 5.1 | 7.6 | 10.1 | 12.3 | 14.5 | 18.3 | 21.1 | 23.0 | 23.5 | 22.2 |
| | 6 | 3.1 | 6.2 | 9.2 | 12.1 | 14.8 | 17.5 | 22.1 | 25.7 | 28.1 | 28.9 | 27.9 |
| | 7 | 3.6 | 7.2 | 10.7 | 14.1 | 17.4 | 20.4 | 26.0 | 30.3 | 33.2 | 34.5 | 33.7 |
| | 8 | 4.1 | 8.2 | 12.2 | 16.2 | 19.9 | 23.4 | 29.8 | 34.8 | 38.4 | 40.0 | 39.4 |

**Table 54. Service Factors** $S$

| Application | Squirrel-cage ac motor | | Wound rotor a-c motor (slip ring) | Single-phase capacitor motor | d-c shunt-wound motor | Diesel engine, 4 or more cyl, above 700 rpm |
|---|---|---|---|---|---|---|
| | Normal torque, line start | High torque | | | | |
| Agitators................ | 1.0–1.2 | 1.2–1.4 | 1.2 | | | |
| Compressors.............. | 1.2–1.4 | | 1.4 | 1.2 | 1.2 | 1.2 |
| Belt conveyors (ore, coal, sand)................. | ... | 1.4 | ... | ... | 1.2 | |
| Screw conveyors.......... | ... | 1.8 | ... | ... | 1.6 | |
| Crushing machinery....... | ... | 1.6 | 1.4 | ... | ... | 1.4–1.6 |
| Fans, centrifugal.......... | 1.2 | ... | 1.4 | ... | 1.4 | 1.4 |
| Fans, propeller........... | 1.4 | 2.0 | 1.6 | ... | 1.6 | 1.6 |
| Generators and exciters.... | 1.2 | ... | ... | ... | 1.2 | 2.0 |
| Line shafts............... | 1.4 | ... | 1.4 | 1.4 | 1.4 | 1.6 |
| Machine tools............. | 1.0–1.2 | ... | 1.2–1.4 | 1.0 | 1.0–1.2 | |
| Pumps, centrifugal........ | 1.2 | 1.4 | 1.4 | 1.2 | 1.2 | |
| Pumps, reciprocating...... | 1.2–1.4 | ... | 1.4–1.6 | ... | ... | 1.8–2.0 |

section which ride in V-shaped pulley grooves. The belts are formed of cord and fabric, impregnated with rubber, the cord material being cotton, rayon, synthetic, or steel. V-belt drives are quiet, able to absorb shock and operate at low bearing

**Table 55. Arc of Contact Factor** $K$**—Rubber Belts**

| Arc of contact, deg......... | 140 | 160 | 180 | 200 | 220 |
|---|---|---|---|---|---|
| Factor $K$................... | 0.82 | 0.93 | 1.00 | 1.06 | 1.12 |

pressures. A V belt should ride with the top surface approximately flush with the top of the pulley groove; clearance should be present between the belt base and the base of the groove so that the belt rides on the groove walls.

**Light-duty or fractional-horsepower V-belt drives** are widely used on fractional-horsepower motors or engines. Also, they

are intended for use on small-diameter pulleys and on drives in which only one belt is required. Typical applications are washing machines, domestic ironers, small fans and blowers, centrifugal pumps, and stokers. The data given for light-duty V-belt drives in Tables 57 to 63 are based on information given in "Standards for Light-duty or Fractional-Horsepower V-Belts," Rubber Manufacturers Assoc. (RMA).

Nominal dimensions of light-duty V belts are given in Table 57. To calculate speed ratio and belt speed use the effective OD minus $2X$ (Table 58) as a basis.

For calculating belt lengths and center distances use the equations given above, taking $D$ and $d$ as the effective outside diameters of large and small pulleys, respectively (Fig. 106). In case of a V-flat drive, the OD values are to be taken as the OD of the pulley plus twice the nominal belt thickness. The RMA recommends that if the center distance $C$ is not dictated by other considerations, the following values be used: (1) for

**Table 56. Minimum Pulley Diameters—Rubber Belts (Inches)**

| | Ply | Belt speed, fpm | | | | | | | | | | |
|---|---|---|---|---|---|---|---|---|---|---|---|---|
| | | 500 | 1,000 | 1,500 | 2,000 | 2,500 | 3,000 | 4,000 | 5,000 | 6,000 | 7,000 | 8,000 |
| 32-oz fabric.... | 3 | 4 | 4 | 4 | 4 | 5 | 5 | 5 | 6 | 6 | | |
| | 4 | 4 | 5 | 6 | 6 | 7 | 7 | 8 | 9 | 10 | | |
| | 5 | 6 | 7 | 9 | 10 | 10 | 11 | 12 | 13 | 14 | | |
| | 6 | 9 | 10 | 11 | 13 | 14 | 16 | 16 | 18 | 19 | | |
| | 7 | 13 | 14 | 16 | 17 | 18 | 19 | 21 | 22 | 24 | | |
| | 8 | 18 | 19 | 21 | 22 | 23 | 24 | 25 | 27 | 29 | | |
| 32-oz hard fabric......... | 3 | 3 | 3 | 3 | 4 | 4 | 4 | 4 | 5 | 5 | 6 | 7 |
| | 4 | 4 | 4 | 5 | 5 | 6 | 6 | 7 | 7 | 8 | 9 | 12 |
| | 5 | 5 | 6 | 7 | 8 | 8 | 9 | 10 | 11 | 12 | 13 | 16 |
| | 6 | 6 | 8 | 10 | 11 | 11 | 12 | 13 | 15 | 16 | 18 | 21 |
| | 7 | 10 | 12 | 14 | 15 | 15 | 16 | 17 | 19 | 20 | 22 | 26 |
| | 8 | 14 | 16 | 17 | 18 | 19 | 20 | 21 | 23 | 24 | 27 | 31 |
| | 9 | 18 | 20 | 21 | 22 | 23 | 24 | 25 | 27 | 28 | 31 | 36 |
| | 10 | 22 | 24 | 25 | 26 | 27 | 28 | 29 | 31 | 33 | 35 | 41 |
| No. 70 rayon cord....... | 3 | 5 | 6 | 7 | 7 | 8 | 8 | 9 | 10 | 11 | 12 | 13 |
| | 4 | 7 | 8 | 9 | 9 | 10 | 11 | 12 | 12 | 14 | 15 | 17 |
| | 5 | 9 | 10 | 11 | 12 | 13 | 13 | 15 | 16 | 17 | 19 | 21 |
| | 6 | 13 | 14 | 15 | 16 | 16 | 17 | 18 | 19 | 21 | 23 | 25 |
| | 7 | 16 | 17 | 18 | 19 | 20 | 21 | 22 | 23 | 24 | 26 | 29 |
| | 8 | 19 | 20 | 22 | 23 | 23 | 24 | 25 | 26 | 28 | 30 | 33 |

speed ratios less than 3, $C = (D + d)/2 + d$; (2) for speed ratios of 3 or more, $C = D$.

The basic horsepower ratings of the 3L, 4L, and 5L cross-section V belts are given in Tables 59 to 61. These are nominal

**Table 57. Nominal Dimensions and Recommended Minimum Sheave Diameters—Light-Duty V Belts**

| Cross section | Nominal top width, in. | Nominal thickness, in. | Min sheave O.D., in. |
|---|---|---|---|
| 2L* | ¼ | 5⁄32 | 0.8 |
| 3L | ⅜ | 7⁄32 | 1.5 |
| 4L | ½ | 5⁄16 | 2.5 |
| 5L | 21⁄32 | ⅜ | 3.5 |

\* The 2L section is in limited usage and is not made by all manufacturers.

ratings based on 180° arc of contact on small pulley; for smaller arcs of contact these should be reduced by the correction factors of Table 62. In applying Tables 59 to 62, the nominal horsepower ratings of V-belt drives should be increased by the factors given in Table 63, depending on the type of service.

**Heavy service** consists of heavy starting, shock, or reciprocating loads; frequent starting and stopping; industrial production service; or a combination of these. In the case of **light service** none of these factors is present to any degree.

EXAMPLE.   A reciprocating pump is to be driven at approximately 550 r/min by a ½-hp motor having a speed of 1,725 r/min. From Table 63, the service factor taken as 1.6 gives a value of design horsepower equal to 0.8. Center distance $C = 12$ in; outside diameter of large and small sheaves are 12 and 4 in, respectively. Belt speed is $1,725 \times \pi \times 3.8/12 = 1,715$ ft/min. Arc of contact is $180 - (12 - 4)60/12 = 140°$. From Table 62, correction factor for V-to-V belt and 140° arc of contact is 0.89. From Table 60, rating for a 4L cross-section belt at 1,600 ft/min is 1.26 hp for 180° arc of contact. For 140° arc of contact, the horsepower rating is $0.89(1.26) = 1.12$ hp. Since the belt needs to transmit only 0.8 hp, this size should be ample. The length of belt is given by $L = 2C + 1.57(D + d) + (D - d)^2/4C = 24 + 1.57(12 + 4) + (12 - 4)^2/48 = 50.4$ in. A standard 4L belt with an outside length of 51 in may be used, with appropriate adjustment of center distance.

**Multiple V-belt drives** are customarily used for the transmission of power in heavy-duty applications. For such drives, the **Multiple V-belt Drive and Mechanical Power Transmission Assoc.**, in cooperation with the **Rubber Manufacturers Assoc.**, has published a booklet, "Engineering Standard—Multiple V-belt Drives," 1955. The data of Tables 64 to 72 and Figs. 107 to 109 are based on this booklet.

The standard V-belt cross sections and nominal dimensions

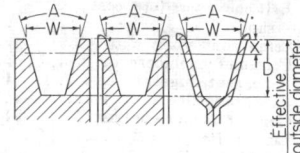

**Fig. 106**   Standard sheave groove dimensions—light-duty V belts.

for multiple V-belt drives are given in Table 64; standard groove dimensions and minimum sheave diameters, in Table 65. The face width of stock standard sheaves is equal to $S(N - 1) + 2E$, where $N$ = number of grooves; $S$ and $E$ are taken from Table 65. V-belt cross sections as recommended for various r/min values of the small sheave and various horse-

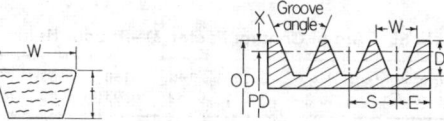

**Fig. 107**   V-belt cross sections.   **Fig. 108**   Standard groove dimensions for multiple V-belt sheaves.

power ratings of the drive are indicated in Fig. 109. Where a drive falls near the dividing line, it is desirable to investigate both cross sections.

Maximum horsepower ratings for multiple V belts of premium quality at various speeds are listed in Tables 68 to 72 (lower ratings for standard-quality belts are listed in the above-mentioned publication). These ratings are based on 180° arc of contact and assume belts of average length. For

**Table 58. Sheave Dimensions—Light-Duty V Belts**
(See Fig. 106)

| Belt cross section | Sheave effective O.D., in. | Groove angle, deg | $W$, in. | $D$, in. | $2X$, in. |
|---|---|---|---|---|---|
| 2L | Under 1.5 | 32 | 0.240 | 0.250 | 0.10 |
|    | 1.5–1.99 | 34 | 0.243 | | |
|    | 2.0–2.5 | 36 | 0.246 | | |
|    | Over 2.5 | 38 | 0.250 | | |
| 3L | Under 2.2 | 32 | 0.360 | 0.406 | 0.15 |
|    | 2.2–3.19 | 34 | 0.364 | | |
|    | 3.2–4.2 | 36 | 0.368 | | |
|    | Over 4.20 | 38 | 0.372 | | |
| 4L | Under 2.65 | 30 | 0.485 | 0.490 | 0.20 |
|    | 2.65–3.24 | 32 | 0.490 | | |
|    | 3.25–5.65 | 34 | 0.494 | | |
|    | Over 5.65 | 38 | 0.504 | | |
| 5L | Under 3.95 | 30 | 0.624 | 0.580 | 0.30 |
|    | 3.95–4.94 | 32 | 0.630 | | |
|    | 4.95–7.35 | 34 | 0.637 | | |
|    | Over 7.35 | 38 | 0.650 | | |

**Table 59. Horsepower Ratings of 3L Cross-Section V Belts**
(Based on 180° arc of contact on small sheave)

| Belt speed, fpm | Effective O.D. of small sheave, in. | | | |
|---|---|---|---|---|
| | 1½ | 2 | 2½ | 3 and larger |
| 200 | 0.05 | 0.08 | 0.10 | 0.11 |
| 400 | 0.08 | 0.14 | 0.18 | 0.20 |
| 600 | 0.11 | 0.20 | 0.25 | 0.29 |
| 800 | 0.12 | 0.24 | 0.31 | 0.36 |
| 1,000 | 0.13 | 0.28 | 0.37 | 0.43 |
| 1,200 | 0.14 | 0.32 | 0.43 | 0.50 |
| 1,400 | 0.15 | 0.35 | 0.48 | 0.56 |
| 1,600 | 0.15 | 0.38 | 0.52 | 0.62 |
| 1,800 | 0.14 | 0.41 | 0.57 | 0.67 |
| 2,000 | 0.13 | 0.43 | 0.60 | 0.72 |
| 2,200 | 0.12 | 0.44 | 0.64 | 0.77 |
| 2,400 | 0.10 | 0.45 | 0.66 | 0.81 |
| 2,600 | 0.07 | 0.46 | 0.69 | 0.84 |
| 2,800 | 0.04 | 0.46 | 0.70 | 0.87 |
| 3,000 | 0.01 | 0.45 | 0.72 | 0.89 |
| 3,200 | .... | 0.44 | 0.72 | 0.91 |
| 3,400 | .... | 0.42 | 0.72 | 0.92 |
| 3,600 | .... | 0.39 | 0.71 | 0.92 |
| 3,800 | .... | 0.36 | 0.69 | 0.92 |
| 4,000 | .... | 0.31 | 0.67 | 0.91 |
| 4,200 | .... | 0.26 | 0.64 | 0.89 |
| 4,400 | .... | 0.21 | 0.60 | 0.86 |
| 4,600 | .... | 0.14 | 0.55 | 0.82 |
| 4,800 | .... | 0.06 | 0.49 | 0.77 |
| 5,000 | .... | .... | 0.42 | 0.72 |
| 5,200 | .... | .... | 0.34 | 0.65 |
| 5,400 | .... | .... | 0.26 | 0.58 |
| 5,600 | .... | .... | 0.16 | 0.49 |
| 5,800 | .... | .... | 0.05 | 0.39 |
| 6,000 | .... | .... | .... | 0.29 |

other lengths and arcs of contact these ratings must be corrected by the factors given in Tables 62 and 66. The belt speeds are figured using the pitch diameter (Table 65). The equivalent diameters $d_e$ are obtained by multiplying the pitch diameter in inches of the small sheave by a "small-diameter factor" obtained from Table 67, depending on the speed ratio. Applications with belt speeds over 5,000 ft/min may require special material or construction as well as balancing; these should be referred to the manufacturer. In using Tables 68 to 72, the rated horsepower of the drive (nameplate rating) should be multiplied by a service factor (Table 54), depending

on the severity of service. Service factors for additional applications are given in the above-mentioned booklet.

EXAMPLE. The following example (due to B. F. Goodrich Co.) illustrates the use of the tables in designing a multiple V-belt drive. Assume a 25-hp squirrel-cage normal-torque line-start motor with a speed of 1,160 r/min to drive a centrifugal fan at 400 r/min. Center distance $C$ = 39.2 in. Sheave diameter = 24 in PD. From Table 54, service factor = 1.2; hence design horsepower = 1.2 × 25 = 30 and, from Fig. 109, a C-section V-belt is indicated. Speed ratio = 1,160/400 = 2.90. Size of small sheave = 24/2.9 = 8.3 PD. From above, belt pitch length is 130.7 in. From Table 67 the small-diameter factor is 1.13, giving $d_e$ = 8.3 × 1.13 = 9.4 in. Belt speed = 1,160 × 8.3 × $\pi$/12 = 2,520 ft/min. From Table 70 for a C section, premium-quality belt at 2,520 ft/min and $d_e$ = 9.4 in, by interpolation we obtain 9.3 hp per belt. From Table 62 the arc factor is 0.94 and from Table 66 the length factor is 0.98 (using interpolation). Hence the hp per belt is 9.3 × 0.94 × 0.98 = 8.6 hp, and four premium-quality belts are required for 30 hp. If standard-quality belts are used, five belts will be required.

Narrow V belts with cross sections designated 3V, 5V, and 8V overlap applications of the A, B, C, D, and E sizes.

**V-Flat Drives**    These consist of a grooved small pulley driving a large flat pulley and are practical when (1) the speed ratio is over 3:1 and (2) the center distance is equal to or slightly less than the diameter of the larger pulley.

**Quarter-Turn Drives**    V belts can also be used with quarter-turn drives, in which case special deep-groove sheaves are

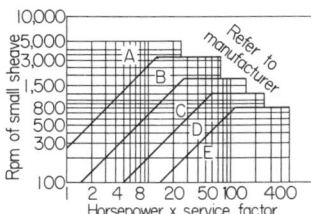

**Fig. 109**   V-belt cross section for required horsepower rating. Letters A, B, C, D, E refer to belt cross section (see Table 54 for service factor).

used. For further details, see the "Engineering Standards" booklet mentioned previously.

**Cogged V belts** have cogs molded integrally on the underside of the belt (Fig. 110). Sheaves can be up to 25 percent smaller in diameter with cogged belts because of the greater flexibility

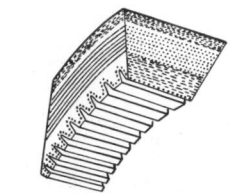

**Fig. 110** Cogged V belt.　　**Fig. 111** Ribbed V belt.

inherent in the cogged construction. An extension of the cogged belt mating with a sheave or pulley notched at the same pitch as the cogs leads to a drive particularly useful for timing purposes.

**Ribbed V belts** are really flat belts molded integrally with longitudinal ribbing on the underside (Fig. 111). Traction is provided principally by friction between the ribs and sheave grooves rather than by wedging action between the two, as in conventional V-belt operation. The flat upper portion trans-

mits the tensile belt loads. Ribbed belts serve well when substituted for multiple V-belt drives and for all practical purposes eliminate the necessity for belt-matching in multiple V-belt drives.

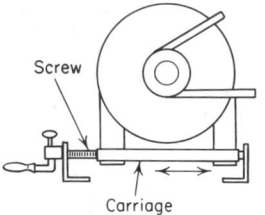

**Fig. 112** Adjustable motor base.

### Adjustable Motor Bases

To maintain proper belt tensions on short center distances, an adjustable motor base is often used. Figure 112 shows an embodiment of such a base made by the Automatic Motor Base Co., in which adjustment for proper belt tension is made by turning a screw which opens or closes the center distance between pulleys, as required. The carriage portion of the base is spring loaded so that after the initial adjustment for belt

**Table 60. Horsepower Ratings of 4L Cross-Section V Belts**
(Based on 180° arc of contact on small sheave)

| Belt speed, fpm | Effective O.D. of small sheave, in. | | | | |
|---|---|---|---|---|---|
| | 2 | 2½ | 3 | 3½ | 4 and larger |
| 200 | 0.07 | 0.13 | 0.16 | 0.18 | 0.21 |
| 400 | 0.12 | 0.23 | 0.31 | 0.36 | 0.40 |
| 600 | 0.15 | 0.32 | 0.43 | 0.51 | 0.57 |
| 800 | 0.17 | 0.40 | 0.54 | 0.64 | 0.73 |
| 1,000 | 0.18 | 0.46 | 0.65 | 0.78 | 0.88 |
| 1,200 | 0.17 | 0.51 | 0.74 | 0.89 | 1.01 |
| 1,400 | 0.16 | 0.56 | 0.82 | 1.01 | 1.14 |
| 1,600 | 0.14 | 0.60 | 0.90 | 1.11 | 1.26 |
| 1,800 | 0.11 | 0.62 | 0.96 | 1.19 | 1.37 |
| 2,000 | 0.08 | 0.64 | 1.02 | 1.28 | 1.47 |
| 2,200 | 0.04 | 0.67 | 1.08 | 1.37 | 1.58 |
| 2,400 | .... | 0.68 | 1.12 | 1.43 | 1.66 |
| 2,600 | .... | 0.66 | 1.16 | 1.50 | 1.75 |
| 2,800 | .... | 0.65 | 1.18 | 1.54 | 1.81 |
| 3,000 | .... | 0.63 | 1.19 | 1.58 | 1.87 |
| 3,200 | .... | 0.60 | 1.20 | 1.61 | 1.92 |
| 3,400 | .... | 0.55 | 1.19 | 1.63 | 1.96 |
| 3,600 | .... | 0.50 | 1.16 | 1.64 | 1.98 |
| 3,800 | .... | 0.43 | 1.13 | 1.63 | 2.00 |
| 4,000 | .... | 0.35 | 1.09 | 1.61 | 2.00 |
| 4,200 | .... | 0.24 | 1.03 | 1.58 | 1.98 |
| 4,400 | .... | 0.14 | 0.96 | 1.53 | 1.95 |
| 4,600 | .... | 0.01 | 0.87 | 1.46 | 1.91 |
| 4,800 | .... | .... | 0.76 | 1.39 | 1.85 |
| 5,000 | .... | .... | 0.65 | 1.30 | 1.78 |
| 5,200 | .... | .... | 0.51 | 1.19 | 1.70 |
| 5,400 | .... | .... | 0.36 | 1.07 | 1.59 |
| 5,600 | .... | .... | 0.18 | 0.91 | 1.46 |
| 5,800 | .... | .... | .... | 0.72 | 1.29 |
| 6,000 | .... | .... | .... | 0.54 | 1.11 |

**Table 61. Horsepower Ratings of 5L Cross-Section V Belts**
(Based on 180° arc of contact small sheave)

| Belt speed, fpm | Effective O.D. of small sheave, in. | | | | |
|---|---|---|---|---|---|
| | 3 | 3½ | 4 | 4½ | 5 and larger |
| 200 | 0.13 | 0.19 | 0.24 | 0.27 | 0.30 |
| 400 | 0.23 | 0.35 | 0.45 | 0.52 | 0.58 |
| 600 | 0.30 | 0.49 | 0.64 | 0.75 | 0.83 |
| 800 | 0.36 | 0.62 | 0.80 | 0.95 | 1.07 |
| 1,000 | 0.40 | 0.72 | 0.95 | 1.14 | 1.28 |
| 1,200 | 0.42 | 0.80 | 1.09 | 1.31 | 1.48 |
| 1,400 | 0.43 | 0.87 | 1.20 | 1.46 | 1.67 |
| 1,600 | 0.42 | 0.93 | 1.31 | 1.60 | 1.84 |
| 1,800 | 0.40 | 0.97 | 1.40 | 1.73 | 1.99 |
| 2,000 | 0.36 | 1.00 | 1.47 | 1.84 | 2.13 |
| 2,200 | 0.31 | 1.01 | 1.54 | 1.94 | 2.26 |
| 2,400 | 0.25 | 1.01 | 1.58 | 2.02 | 2.37 |
| 2,600 | 0.17 | 1.00 | 1.62 | 2.09 | 2.47 |
| 2,800 | 0.08 | 0.97 | 1.63 | 2.15 | 2.56 |
| 3,000 | .... | 0.93 | 1.64 | 2.19 | 2.63 |
| 3,200 | .... | 0.87 | 1.63 | 2.21 | 2.68 |
| 3,400 | .... | 0.79 | 1.60 | 2.22 | 2.72 |
| 3,600 | .... | 0.69 | 1.54 | 2.20 | 2.73 |
| 3,800 | .... | 0.57 | 1.47 | 2.17 | 2.72 |
| 4,000 | .... | 0.43 | 1.37 | 2.11 | 2.69 |
| 4,200 | .... | 0.26 | 1.26 | 2.03 | 2.64 |
| 4,400 | .... | 0.08 | 1.12 | 1.93 | 2.57 |
| 4,600 | .... | .... | 0.97 | 1.81 | 2.48 |
| 4,800 | .... | .... | 0.78 | 1.66 | 2.36 |
| 5,000 | .... | .... | 0.58 | 1.49 | 2.22 |
| 5,200 | .... | .... | 0.34 | 1.30 | 2.06 |
| 5,400 | .... | .... | 0.08 | 1.07 | 1.86 |
| 5,600 | .... | .... | .... | 0.82 | 1.64 |
| 5,800 | .... | .... | .... | 0.54 | 1.39 |
| 6,000 | .... | .... | .... | 0.23 | 1.11 |

tension has been made by the screw, the spring will compensate for a normal amount of stretch in the belts. When there is more stretch than can be accommodated by the spring, the screw is turned to provide the necessary belt tensions. The carriage can be moved while the unit is in operation, and the motor base is provided for vertical as well as horizontal mounting.

## CHAIN DRIVES

### Roller-Chain Drives

The advantages of finished steel roller chains are high efficiency (around 98 to 99 percent), no slippage, no initial tension required, and chains may travel in either direction. The basic construction of roller chain is shown in Fig. 113 and Table 73.

The shorter the pitch, the higher the permissible operating speed of roller chain. Horsepower capacity in excess of that provided by a single chain may be had by the use of multiple chains, which are essentially parallel single chains assembled on pins common to all strands. Because of its lightness in relation to tensile strength, the effect of centrifugal pull does not need to be considered; even at the unusual speed of 6,000 ft/min, this pull is only 3 percent of the ultimate tensile strength.

Sprocket wheels with fewer than 16 teeth may be used for relatively slow speeds, but 18 to 24 teeth are desirable for high-speed service. Sprockets with fewer than 25 teeth, running at speeds above 500 or 600 r/min, should be heat-treated to give a tough wear-resistant surface testing between 35 and 45 on the Rockwell C hardness scale.

If the speed ratio requires the larger sprocket to have as many as 128 teeth, or more than eight times the number on the smaller sprocket, it is usually better, with few exceptions, to make the desired reduction in two or more steps. The ANSI

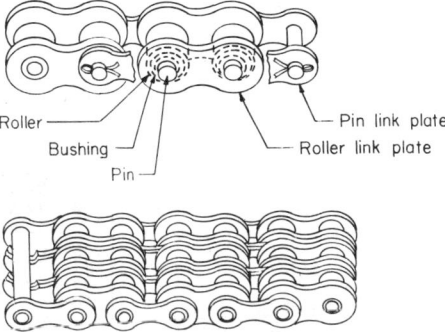

Roller — Pin link plate
Bushing — Roller link plate
Pin

**Fig. 113** Roller-chain construction.

**Table 62. Correction Factors for Arc of Contact—V-Belt Drives**

| Arc of contact, deg | Correction factor | | Arc of contact, deg | Correction factor | |
|---|---|---|---|---|---|
| | V to V | V to flat* | | V to V | V to flat* |
| 180 | 1.00 | 0.75 | 130 | 0.86 | 0.86 |
| 170 | 0.98 | 0.77 | 120 | 0.82 | 0.82 |
| 160 | 0.95 | 0.80 | 110 | 0.78 | 0.78 |
| 150 | 0.92 | 0.82 | 100 | 0.74 | 0.74 |
| 140 | 0.89 | 0.84 | 90 | 0.69 | 0.69 |

*A V-to-flat drive is one comprised of a small sheave and a larger diameter flat pulley.

**Table 63.   Service Factors for V-Belt Drives**

| Typical machines | Type of service | Service factors |
|---|---|---|
| Domestic washing machines<br>Domestic ironers<br>Advertising display fixtures<br>Small fans and blowers | Light | 1.0–1.2 |
| Fans and blowers (heavy rotors)<br>Centrifugal pumps<br>Oil burners<br>Home workshop machines | Medium | 1.2–1.4 |
| Stokers<br>Reciprocating pumps and compressors<br>Refrigerators<br>Drill presses, grinders<br>Lathes<br>Meat slicers<br>Machines for industrial use | Heavy | 1.4–1.6 |

**Table 64.  Nominal V-Belt Cross Sections**
(See Fig. 107)

| Belt | Width, $w$, in. | Thickness, $t$, in. |
|---|---|---|
| A | $\frac{1}{2}$ | $\frac{5}{16}$ |
| B | $2\frac{1}{32}$ | $13\frac{3}{32}$ |
| C | $\frac{7}{8}$ | $17\frac{7}{32}$ |
| D | $1\frac{1}{4}$ | $\frac{3}{4}$ |
| E | $1\frac{1}{2}$ | $29\frac{9}{32}$ |

**Table 65.   Standard Groove Dimensions for Multiple V-Belt Sheaves**
(See Fig. 108)

| Belt | Pitch diameter, in. | | Groove angle, deg | Standard groove dimensions, in. | | | | |
|---|---|---|---|---|---|---|---|---|
| | Min recommended | Range | | $W$ | $D$ | $X$ | $S$ | $E$ |
| A | 3.0 | 2.6–5.4<br>Over 5.4 | 34<br>38 | 0.494<br>0.504 | 0.490 | 0.125 | $\frac{5}{8}$ | $\frac{3}{8}$ |
| B | 5.4 | 4.6–7.0<br>Over 7.0 | 34<br>38 | 0.637<br>0.650 | 0.580 | 0.175 | $\frac{3}{4}$ | $\frac{1}{2}$ |
| C | 9.0 | 7.0–7.99<br>8.0–12.0<br>Over 12.0 | 34<br>36<br>38 | 0.879<br>0.887<br>0.895 | 0.780 | 0.200 | 1 | $1\frac{1}{16}$ |
| D | 13.0 | 12.0–12.99<br>13.0–17.0<br>Over 17.0 | 34<br>36<br>38 | 1.259<br>1.271<br>1.283 | 1.050 | 0.300 | $1\frac{7}{16}$ | $\frac{7}{8}$ |
| E | 21.0 | 18.0–24.0<br>Over 24.0 | 36<br>38 | 1.527<br>1.542 | 1.300 | 0.400 | $1\frac{3}{4}$ | $1\frac{1}{8}$ |

**Table 66. Length-Correction Factors—Multiple V Belts**

| Std length designation | Belt cross section | | | | |
|---|---|---|---|---|---|
| | *A* | *B* | *C* | *D* | *E* |
| 26 | 0.81 | | | | |
| 33 | 0.86 | | | | |
| 38 | 0.88 | 0.83 | | | |
| 46 | 0.92 | 0.87 | | | |
| 51 | 0.94 | 0.89 | 0.80 | | |
| 55 | 0.96 | 0.90 | | | |
| 62 | 0.99 | 0.93 | | | |
| 66 | 1.00 | 0.94 | | | |
| 71 | 1.01 | 0.95 | | | |
| 78 | 1.03 | 0.98 | | | |
| 81 | .... | 0.98 | 0.89 | | |
| 85 | 1.05 | 0.99 | 0.90 | | |
| 96 | 1.08 | .... | 0.92 | | |
| 105 | 1.10 | 1.04 | 0.94 | | |
| 120 | 1.13 | 1.07 | 0.97 | 0.86 | |
| 136 | .... | 1.09 | 0.99 | | |
| 158 | .... | 1.13 | 1.02 | 0.92 | |
| 173 | .... | 1.15 | 1.04 | 0.93 | |
| 195 | .... | 1.18 | 1.07 | 0.96 | 0.92 |
| 240 | .... | 1.22 | 1.11 | 1.00 | 0.96 |
| 300 | .... | 1.27 | 1.16 | 1.05 | 1.01 |
| 360 | .... | .... | 1.21 | 1.09 | 1.05 |
| 420 | .... | .... | 1.24 | 1.12 | 1.09 |
| 540 | .... | .... | .... | 1.18 | 1.14 |
| 660 | .... | .... | .... | 1.23 | 1.19 |

tooth form [ANSI B29.1-1963 (R1972)] allows roller chain to adjust itself to a larger pitch circle as the pitch of the chain elongates owing to natural wear in the pin-bushing joints. The greater the number of teeth, the sooner the chain will ride out too near the ends of the teeth.

Idler sprockets may be used on either side of the standard roller chain, to take up slack, to guide the chain around obstructions, to change the direction of rotation of a driven shaft, or to provide more wrap on another sprocket. Idlers should not run faster than the speeds recommended as maximum for other sprockets with the same number of teeth. It is desirable that idlers have at least two teeth in mesh with the chain, and it is advisable, though not necessary, to have an idler contact the idle span of chain.

**Table 67. Small-Diameter Factors—Multiple V Belts**

| Speed ratio range | Small-diameter factor |
|---|---|
| 1.000–1.019 | 1.00 |
| 1.020–1.032 | 1.01 |
| 1.033–1.055 | 1.02 |
| 1.056–1.081 | 1.03 |
| 1.082–1.109 | 1.04 |
| 1.110–1.142 | 1.05 |
| 1.143–1.178 | 1.06 |
| 1.179–1.222 | 1.07 |
| 1.223–1.274 | 1.08 |
| 1.275–1.340 | 1.09 |
| 1.341–1.429 | 1.10 |
| 1.430–1.562 | 1.11 |
| 1.563–1.814 | 1.12 |
| 1.815–2.948 | 1.13 |
| 2.949 and over | 1.14 |

**Horsepower ratings** are based upon the number of teeth and the rotative speed of the smaller sprocket, either driver or follower. The pin-bushing bearing area, as it affects allowable working load, is the important factor for medium and higher speeds. For extremely slow speeds, the chain selection may be based upon the ultimate tensile strength of the chain. For chain speeds of 25 ft/min and less, the chain pull should be not more than $\frac{1}{5}$ of the ultimate tensile strength; for 50 ft/min, $\frac{1}{6}$; for 100 ft/min, $\frac{1}{7}$; for 150 ft/min, $\frac{1}{8}$; for 200 ft/min, $\frac{1}{9}$; and for 250 ft/min, $\frac{1}{10}$ of the ultimate tensile strength.

Ratings for **multiple-strand chains** are proportional to the number of strands. The recommended numbers of strands for multiple chains are 2, 3, 4, 6, 8, 10, 12, 16, 20, and 24, with the maximum overall width in any case limited to 24 in.

The **horsepower ratings** in Table 74 are modified by the **service factors** in Table 75. Thus for a drive having a nominal rating of 3 hp, subject to heavy shock, abnormal conditions, 24-h-day operation, the chain rating obtained from Table 74 should be at least $3 \times 1.7 = 5.1$ hp.

**Chain-Length Calculations** Referring to Fig. 114, $L$ = length of chain, in; $P$ = pitch of chain, in; $R$ and $r$ = pitch radii of large and small sprockets, respectively, in; $D$ = center distance, in; $A$ = tangent length, in; $a$ = angle between tangent and centerline; $N$ and $n$ = number of teeth on larger and smaller sprockets, respectively; $(180 + 2a)$ and $(180 - 2a)$ = angles of contact on larger and smaller sprockets, respectively, deg.

$$a = \sin^{-1} [(R - r)/D] \qquad A = D \cos a$$
$$L = NP(180 + 2a)/360 + nP(180 - 2a)/360 + 2D \cos a$$

**Table 68. Horsepower Ratings for Premium-Quality A-Section V Belts**

| Belt speed, fpm | Equivalent diameter $d_e$ | | | | | | |
|---|---|---|---|---|---|---|---|
| | 2.6 | 3.0 | 3.4 | 3.8 | 4.2 | 4.6 | 5.0 and over |
| 1,200 | 0.69 | 1.02 | 1.27 | 1.46 | 1.62 | 1.76 | 1.87 |
| 1,400 | 0.74 | 1.12 | 1.42 | 1.65 | 1.83 | 1.99 | 2.12 |
| 1,600 | 0.78 | 1.22 | 1.56 | 1.82 | 2.03 | 2.21 | 2.36 |
| 1,800 | 0.82 | 1.31 | 1.68 | 1.98 | 2.22 | 2.42 | 2.59 |
| 2,000 | 0.84 | 1.38 | 1.80 | 2.13 | 2.40 | 2.62 | 2.80 |
| 2,200 | 0.85 | 1.45 | 1.91 | 2.27 | 2.57 | 2.81 | 3.01 |
| 2,400 | 0.85 | 1.51 | 2.01 | 2.40 | 2.72 | 2.99 | 3.21 |
| 2,600 | 0.84 | 1.55 | 2.09 | 2.52 | 2.87 | 3.15 | 3.40 |
| 2,800 | 0.82 | 1.58 | 2.17 | 2.63 | 3.00 | 3.31 | 3.57 |
| 3,000 | 0.78 | 1.60 | 2.23 | 2.72 | 3.12 | 3.45 | 3.73 |
| 3,200 | 0.74 | 1.61 | 2.28 | 2.81 | 3.23 | 3.59 | 3.88 |
| 3,400 | 0.68 | 1.60 | 2.31 | 2.88 | 3.33 | 3.70 | 4.02 |
| 3,600 | 0.60 | 1.59 | 2.34 | 2.93 | 3.41 | 3.81 | 4.14 |
| 3,800 | 0.52 | 1.55 | 2.35 | 2.97 | 3.48 | 3.90 | 4.25 |
| 4,000 | 0.41 | 1.51 | 2.34 | 3.00 | 3.54 | 3.98 | 4.35 |
| 4,200 | 0.30 | 1.44 | 2.32 | 3.01 | 3.57 | 4.04 | 4.43 |
| 4,400 | 0.17 | 1.37 | 2.29 | 3.01 | 3.60 | 4.09 | 4.49 |
| 4,600 | 0.02 | 1.27 | 2.23 | 2.99 | 3.61 | 4.11 | 4.54 |
| 4,800 | .... | 1.16 | 2.17 | 2.96 | 3.60 | 4.13 | 4.57 |
| 5,000 | .... | 1.04 | 2.08 | 2.90 | 3.57 | 4.12 | 4.59 |
| 5,200 | .... | 0.89 | 1.98 | 2.83 | 3.53 | 4.10 | 4.58 |
| 5,400 | .... | 0.73 | 1.85 | 2.74 | 3.47 | 4.06 | 4.56 |
| 5,600 | .... | 0.54 | 1.71 | 2.64 | 3.38 | 4.00 | 4.52 |
| 5,800 | .... | 0.34 | 1.55 | 2.51 | 3.28 | 3.92 | 4.46 |
| 6,000 | .... | 0.12 | 1.37 | 2.36 | 3.16 | 3.82 | 4.38 |

**Table 69. Horsepower Ratings for Premium-Quality B-Section V Belts**

| Belt speed, fpm | Equivalent diameter $d_e$ | | | | | | |
|---|---|---|---|---|---|---|---|
| | 4.6 | 5.0 | 5.4 | 5.8 | 6.2 | 6.6 | 7.0 and over |
| 1,200 | 1.91 | 2.20 | 2.45 | 2.66 | 2.85 | 3.01 | 3.16 |
| 1,400 | 2.12 | 2.46 | 2.75 | 3.00 | 3.22 | 3.41 | 3.58 |
| 1,600 | 2.31 | 2.70 | 3.03 | 3.32 | 3.57 | 3.78 | 3.98 |
| 1,800 | 2.49 | 2.93 | 3.30 | 3.62 | 3.90 | 4.14 | 4.36 |
| 2,000 | 2.64 | 3.13 | 3.54 | 3.90 | 4.21 | 4.48 | 4.72 |
| 2,200 | 2.78 | 3.32 | 3.77 | 4.17 | 4.51 | 4.81 | 5.07 |
| 2,400 | 2.90 | 3.48 | 3.98 | 4.41 | 4.78 | 5.11 | 5.40 |
| 2,600 | 3.00 | 3.63 | 4.17 | 4.63 | 5.04 | 5.39 | 5.70 |
| 2,800 | 3.08 | 3.76 | 4.34 | 4.84 | 5.27 | 5.65 | 5.99 |
| 3,000 | 3.14 | 3.86 | 4.49 | 5.02 | 5.49 | 5.90 | 6.26 |
| 3,200 | 3.17 | 3.95 | 4.61 | 5.18 | 5.68 | 6.12 | 6.50 |
| 3,400 | 3.19 | 4.01 | 4.72 | 5.32 | 5.85 | 6.31 | 6.73 |
| 3,600 | 3.18 | 4.05 | 4.80 | 5.44 | 6.00 | 6.49 | 6.92 |
| 3,800 | 3.15 | 4.07 | 4.85 | 5.53 | 6.12 | 6.64 | 7.10 |
| 4,000 | 3.09 | 4.06 | 4.89 | 5.60 | 6.22 | 6.77 | 7.25 |
| 4,200 | 3.00 | 4.02 | 4.89 | 5.64 | 6.29 | 6.87 | 7.37 |
| 4,400 | 2.90 | 3.97 | 4.88 | 5.66 | 6.34 | 6.94 | 7.48 |
| 4,600 | 2.75 | 3.87 | 4.82 | 5.64 | 6.36 | 6.99 | 7.54 |
| 4,800 | 2.59 | 3.75 | 4.75 | 5.60 | 6.35 | 7.00 | 7.58 |
| 5,000 | 2.39 | 3.60 | 4.64 | 5.53 | 6.31 | 6.99 | 7.59 |
| 5,200 | 2.16 | 3.43 | 4.50 | 5.43 | 6.24 | 6.95 | 7.57 |
| 5,400 | 1.90 | 3.21 | 4.33 | 5.29 | 6.13 | 6.87 | 7.52 |
| 5,600 | 1.61 | 2.97 | 4.13 | 5.13 | 6.00 | 6.76 | 7.44 |
| 5,800 | 1.28 | 2.69 | 3.89 | 4.93 | 5.83 | 6.62 | 7.32 |
| 6,000 | 0.92 | 2.38 | 3.62 | 4.69 | 5.62 | 6.44 | 7.17 |

**Table 70. Horsepower Ratings for Premium-Quality C-Section V Belts**

| Belt speed, fpm | Equivalent diameter $d_e$ | | | | | |
|---|---|---|---|---|---|---|
| | 7.0 | 8.0 | 9.0 | 10.0 | 11.0 | 12.0 and over |
| 1,200 | 3.65 | 4.48 | 5.13 | 5.65 | 6.07 | 6.43 |
| 1,400 | 4.06 | 5.03 | 5.79 | 6.39 | 6.89 | 7.30 |
| 1,600 | 4.44 | 5.55 | 6.41 | 7.10 | 7.67 | 8.14 |
| 1,800 | 4.79 | 6.03 | 7.00 | 7.78 | 8.42 | 8.95 |
| 2,000 | 5.10 | 6.48 | 7.56 | 8.42 | 9.13 | 9.72 |
| 2,200 | 5.38 | 6.90 | 8.09 | 9.04 | 9.82 | 10.5 |
| 2,400 | 5.62 | 7.28 | 8.58 | 9.61 | 10.5 | 11.2 |
| 2,600 | 5.83 | 7.63 | 9.93 | 10.2 | 11.1 | 11.8 |
| 2,800 | 6.00 | 7.94 | 9.45 | 10.7 | 11.6 | 12.5 |
| 3,000 | 6.13 | 8.21 | 9.83 | 11.1 | 12.2 | 13.1 |
| 3,200 | 6.23 | 8.45 | 10.2 | 11.5 | 12.7 | 13.6 |
| 3,400 | 6.29 | 8.64 | 10.5 | 11.9 | 13.1 | 14.1 |
| 3,600 | 6.30 | 8.80 | 10.7 | 12.3 | 13.6 | 14.6 |
| 3,800 | 6.27 | 8.91 | 11.0 | 12.6 | 13.9 | 15.1 |
| 4,000 | 6.20 | 8.97 | 11.1 | 12.9 | 14.3 | 15.4 |
| 4,200 | 6.08 | 8.99 | 11.3 | 13.1 | 14.5 | 15.8 |
| 4,400 | 5.92 | 8.97 | 11.3 | 13.2 | 14.8 | 16.1 |
| 4,600 | 5.70 | 8.88 | 11.4 | 13.3 | 15.0 | 16.3 |
| 4,800 | 5.43 | 8.75 | 11.3 | 13.4 | 15.1 | 16.5 |
| 5,000 | 5.10 | 8.57 | 11.3 | 13.4 | 15.2 | 16.7 |
| 5,200 | 4.73 | 8.33 | 11.1 | 13.4 | 15.2 | 16.7 |
| 5,400 | 4.30 | 8.04 | 11.0 | 13.3 | 15.2 | 16.8 |
| 5,600 | 3.80 | 7.69 | 10.7 | 13.1 | 15.1 | 16.7 |
| 5,800 | 3.25 | 7.27 | 10.4 | 12.9 | 14.9 | 16.7 |
| 6,000 | 2.64 | 6.80 | 10.0 | 12.6 | 14.7 | 16.5 |

**Table 71. Horsepower Ratings for Premium-Quality D-Section V Belts**

| Belt speed, fpm | Equivalent diameter $d_e$ | | | | | |
|---|---|---|---|---|---|---|
| | 12.0 | 13.0 | 14.0 | 15.0 | 16.0 | 17.0 and over |
| 1,200 | 8.26 | 9.32 | 10.2 | 11.0 | 11.7 | 12.3 |
| 1,400 | 9.22 | 10.5 | 11.5 | 12.4 | 13.2 | 13.9 |
| 1,600 | 10.1 | 11.5 | 12.7 | 13.8 | 14.7 | 15.5 |
| 1,800 | 10.9 | 12.5 | 13.9 | 15.1 | 16.1 | 17.0 |
| 2,000 | 11.7 | 13.4 | 15.0 | 16.3 | 17.4 | 18.4 |
| 2,200 | 12.4 | 14.3 | 16.0 | 17.4 | 18.7 | 19.8 |
| 2,400 | 13.0 | 15.1 | 16.9 | 18.5 | 19.8 | 21.1 |
| 2,600 | 13.5 | 15.8 | 17.8 | 19.5 | 21.0 | 22.3 |
| 2,800 | 14.0 | 16.4 | 18.6 | 20.4 | 22.0 | 23.4 |
| 3,000 | 14.3 | 17.0 | 19.3 | 21.2 | 23.0 | 24.5 |
| 3,200 | 14.6 | 17.5 | 19.9 | 22.0 | 23.8 | 25.4 |
| 3,400 | 14.9 | 17.9 | 20.4 | 22.7 | 24.6 | 26.3 |
| 3,600 | 15.0 | 18.2 | 20.9 | 23.3 | 25.3 | 27.2 |
| 3,800 | 15.1 | 18.4 | 21.3 | 23.8 | 26.0 | 27.9 |
| 4,000 | 15.0 | 18.5 | 21.6 | 24.2 | 26.5 | 28.5 |
| 4,200 | 14.9 | 18.6 | 21.8 | 24.5 | 26.9 | 29.0 |
| 4,400 | 14.6 | 18.5 | 21.9 | 24.7 | 27.3 | 29.5 |
| 4,600 | 14.3 | 18.4 | 21.8 | 24.9 | 27.5 | 29.8 |
| 4,800 | 13.9 | 18.1 | 21.7 | 24.9 | 27.6 | 30.1 |
| 5,000 | 13.3 | 17.7 | 21.5 | 24.8 | 27.6 | 30.2 |
| 5,200 | 12.6 | 17.2 | 21.2 | 24.6 | 27.6 | 30.2 |
| 5,400 | 11.9 | 16.6 | 20.7 | 24.2 | 27.3 | 30.1 |
| 5,600 | 10.9 | 15.9 | 20.1 | 23.8 | 27.0 | 29.8 |
| 5,800 | 9.93 | 15.0 | 19.4 | 23.2 | 26.6 | 29.5 |
| 6,000 | 8.77 | 14.1 | 18.6 | 22.5 | 26.0 | 29.0 |

**Table 72. Horsepower Ratings for Premium-Quality E-Section V Belts**

| Belt speed, fpm | Equivalent diameter $d_e$ | | | | | |
|---|---|---|---|---|---|---|
| | 18.0 | 20.0 | 22.0 | 24.0 | 26.0 | 28.0 and over |
| 1,200 | 14.7 | 16.4 | 17.9 | 19.1 | 20.1 | 21.0 |
| 1,400 | 16.5 | 18.6 | 20.2 | 21.6 | 22.8 | 23.8 |
| 1,600 | 18.3 | 20.6 | 22.5 | 24.1 | 25.5 | 26.6 |
| 1,800 | 19.9 | 22.5 | 24.7 | 26.5 | 28.0 | 29.3 |
| 2,000 | 21.4 | 24.4 | 26.7 | 28.7 | 30.4 | 31.9 |
| 2,200 | 22.9 | 26.1 | 28.7 | 30.9 | 32.8 | 34.4 |
| 2,400 | 24.2 | 27.7 | 30.6 | 33.0 | 35.0 | 36.7 |
| 2,600 | 25.4 | 29.2 | 32.3 | 34.9 | 37.1 | 39.0 |
| 2,800 | 26.5 | 30.6 | 34.0 | 36.8 | 39.1 | 41.1 |
| 3,000 | 27.5 | 31.9 | 35.5 | 38.5 | 41.0 | 43.2 |
| 3,200 | 28.4 | 33.1 | 36.9 | 40.1 | 42.8 | 45.1 |
| 3,400 | 29.2 | 34.2 | 38.2 | 41.6 | 44.5 | 46.9 |
| 3,600 | 29.8 | 35.1 | 39.4 | 43.0 | 46.0 | 48.6 |
| 3,800 | 30.4 | 35.9 | 40.5 | 44.2 | 47.4 | 50.2 |
| 4,000 | 30.7 | 36.6 | 41.4 | 45.4 | 48.7 | 51.6 |
| 4,200 | 31.0 | 37.1 | 42.2 | 46.3 | 49.9 | 52.9 |
| 4,400 | 31.1 | 37.6 | 42.8 | 47.2 | 50.9 | 54.1 |
| 4,600 | 31.1 | 37.8 | 43.3 | 47.9 | 51.8 | 55.1 |
| 4,800 | 30.9 | 37.9 | 43.6 | 48.4 | 52.5 | 55.9 |
| 5,000 | 30.5 | 37.8 | 43.8 | 48.8 | 53.0 | 56.6 |
| 5,200 | 30.0 | 37.6 | 43.8 | 49.0 | 53.4 | 57.2 |
| 5,400 | 29.3 | 37.2 | 43.7 | 49.1 | 53.6 | 57.5 |
| 5,600 | 28.5 | 36.7 | 43.4 | 48.9 | 53.7 | 57.7 |
| 5,800 | 27.5 | 35.9 | 42.9 | 48.6 | 53.5 | 57.7 |
| 6,000 | 26.2 | 35.0 | 42.2 | 48.2 | 53.2 | 57.6 |

**Table 73. Roller-Chain Data and Dimensions**

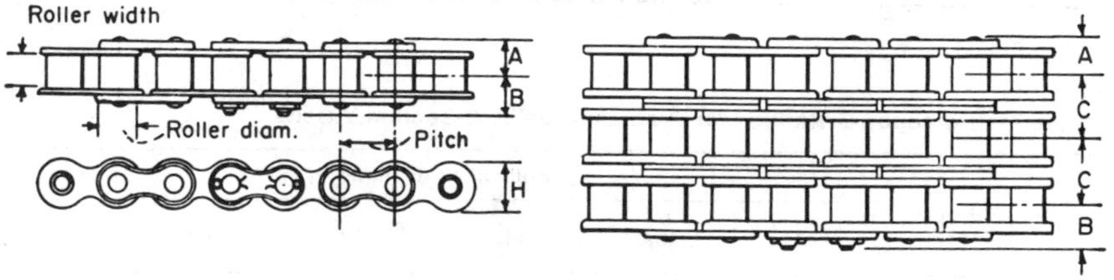


| ASA chain number | Pitch, in. | Roller | | Pin diam, in. | Roller link plate | | Dimensions, in. | | | Tensile strength per strand, lb | Recommended max rpm | | |
|---|---|---|---|---|---|---|---|---|---|---|---|---|---|
| | | Width, in. | Diam, in. | | Thickness, in. | Height $H$, in. | $A$ | $B$ | $C$ | | Teeth 12 | 18 | 24 |
| 25 | 1/4 | 1/8 | 0.130 | 0.091 | 0.030 | 0.230 | 0.150 | 0.190 | 0.260 | 875 | 5,000 | 7,000 | 7,000 |
| 35 | 3/8 | 3/16 | 0.200 | 0.141 | 0.050 | 0.344 | 0.224 | 0.290 | 0.400 | 2,100 | 2,380 | 3,780 | 4,200 |
| 41 | 1/2 | 1/4 | 0.306 | 0.141 | 0.050 | 0.383 | 0.256 | 0.315 | ...... | 2,000 | 1,750 | 2,725 | 2,850 |
| 40 | 1/2 | 5/16 | 0.312 | 0.156 | 0.060 | 0.452 | 0.313 | 0.358 | 0.563 | 3,700 | 1,800 | 2,830 | 3,000 |
| 50 | 5/8 | 3/8 | 0.400 | 0.200 | 0.080 | 0.594 | 0.384 | 0.462 | ...... | 6,100 | 1,300 | 2,030 | 2,200 |
| 50 | 5/8 | 3/8 | 0.400 | 0.200 | 0.080 | 0.545 | 0.384 | 0.462 | 0.707 | 6,600 | 1,300 | 2,030 | 2,200 |
| 60 | 3/4 | 1/2 | 0.469 | 0.234 | 0.094 | 0.679 | 0.493 | 0.567 | 0.892 | 8,500 | 1,025 | 1,615 | 1,700 |
| 80 | 1 | 5/8 | 0.625 | 0.312 | 0.125 | 0.903 | 0.643 | 0.762 | 1.160 | 14,500 | 650 | 1,015 | 1,100 |
| 100 | 1 1/4 | 3/4 | 0.750 | 0.375 | 0.156 | 1.128 | 0.780 | 0.910 | 1.411 | 24,000 | 450 | 730 | 850 |
| 120 | 1 1/2 | 1 | 0.875 | 0.437 | 0.187 | 1.354 | 0.977 | 1.123 | 1.796 | 34,000 | 350 | 565 | 650 |
| 140 | 1 3/4 | 1 | 1.000 | 0.500 | 0.220 | 1.647 | 1.054 | 1.219 | 1.929 | 46,000 | 260 | 415 | 500 |
| 160 | 2 | 1 1/4 | 1.125 | 0.562 | 0.250 | 1.900 | 1.250 | 1.433 | 2.301 | 58,000 | 225 | 360 | 420 |
| 180 | 2 1/4 | 1 13/32 | 1.406 | 0.687 | 0.281 | 2.140 | 1.421 | 1.770 | 2.530 | 76,000 | 180 | 290 | 330 |
| 200 | 2 1/2 | 1 1/2 | 1.562 | 0.781 | 0.312 | 2.275 | 1.533 | 1.850 | 2.800 | 95,000 | 170 | 260 | 300 |
| 240 | 3 | 1 7/8 | 1.875 | 0.937 | 0.375 | 2.850 | 1.722 | 2.200 | 3.375 | 135,000 | 120 | 190 | 210 |

**Table 74. Horsepower Ratings for Single-strand, Roller-Chain Drives**

**ANSI No. 25 — ¼-in. pitch**

| Teeth | \multicolumn{12}{Rpm of sprocket} |
|---|---|---|---|---|---|---|---|---|---|---|---|---|
| | 200 | 400 | 800 | 1,200 | 1,600 | 2,000 | 2,400 | 3,000 | 4,000 | 5,000 | 6,000 | 7,000 |
| 12 | 0.09 | 0.18 | 0.33 | 0.45 | 0.54 | 0.62 | 0.67 | 0.73 | 0.78 | 0.73 | | |
| 15 | 0.12 | 0.24 | 0.44 | 0.60 | 0.73 | 0.85 | 0.94 | 1.03 | 1.14 | 1.20 | 1.14 | |
| 18 | 0.15 | 0.29 | 0.54 | 0.74 | 0.92 | 1.05 | 1.16 | 1.31 | 1.48 | 1.56 | 1.55 | 1.48 |
| 21 | 0.18 | 0.34 | 0.63 | 0.88 | 1.09 | 1.26 | 1.39 | 1.57 | 1.77 | 1.86 | 1.88 | 1.81 |
| 24 | 0.21 | 0.39 | 0.72 | 0.99 | 1.21 | 1.40 | 1.57 | 1.77 | 2.01 | 2.11 | 2.14 | 2.07 |

**ANSI No. 35 — ⅜-in. pitch**

| Teeth | Rpm of sprocket | | | | | | | | | | | |
|---|---|---|---|---|---|---|---|---|---|---|---|---|
| | 200 | 400 | 800 | 1,200 | 1,600 | 2,000 | 2,400 | 2,800 | 3,200 | 3,600 | 4,000 | 4,500 |
| 12 | 0.34 | 0.60 | 1.01 | 1.31 | 1.53 | 1.66 | 1.72 | 1.73 | | | | |
| 15 | 0.43 | 0.78 | 1.35 | 1.78 | 2.12 | 2.37 | 2.54 | 2.65 | 2.70 | 2.69 | | |
| 18 | 0.52 | 0.96 | 1.65 | 2.21 | 2.65 | 2.98 | 3.24 | 3.43 | 3.52 | 3.57 | 3.55 | |
| 21 | 0.61 | 1.12 | 1.95 | 2.61 | 3.14 | 3.53 | 3.86 | 4.08 | 4.22 | 4.28 | 4.28 | |
| 24 | 0.70 | 1.28 | 2.22 | 2.98 | 3.57 | 4.04 | 4.38 | 4.65 | 4.81 | 4.86 | 4.87 | 4.75 |

**ANSI No. 40 — ½-in. pitch**

| Teeth | Rpm of sprocket | | | | | | | | | | | |
|---|---|---|---|---|---|---|---|---|---|---|---|---|
| | 200 | 400 | 600 | 800 | 1,000 | 1,200 | 1,600 | 1,800 | 2,000 | 2,400 | 2,800 | 3,200 |
| 12 | 0.77 | 1.34 | 1.81 | 2.16 | 2.46 | 2.71 | 2.99 | 3.07 | 3.10 | | | |
| 15 | 0.99 | 1.76 | 2.40 | 2.93 | 3.38 | 3.77 | 4.32 | 4.52 | 4.67 | 4.81 | | |
| 18 | 1.20 | 2.15 | 2.94 | 3.63 | 4.21 | 4.71 | 5.48 | 5.76 | 5.97 | 6.27 | 6.35 | |
| 21 | 1.41 | 2.52 | 3.47 | 4.27 | 4.97 | 5.57 | 6.50 | 6.86 | 7.13 | 7.50 | 7.63 | |
| 24 | 1.60 | 2.88 | 3.96 | 4.87 | 5.67 | 6.35 | 7.40 | 7.80 | 8.12 | 8.51 | 8.68 | 8.57 |

**ANSI No. 50 — ⅝-in. pitch**

| Teeth | Rpm of sprocket | | | | | | | | | | | |
|---|---|---|---|---|---|---|---|---|---|---|---|---|
| | 100 | 200 | 300 | 400 | 600 | 800 | 1,000 | 1,200 | 1,400 | 1,600 | 1,800 | 2,200 |
| 12 | 0.80 | 1.44 | 1.99 | 2.48 | 3.26 | 3.86 | 4.3 | 4.6 | 4.8 | | | |
| 15 | 1.02 | 1.87 | 2.61 | 3.27 | 4.39 | 5.31 | 6.0 | 6.8 | 7.0 | 7.3 | 7.5 | |
| 18 | 1.23 | 2.27 | 3.19 | 4.01 | 5.41 | 6.58 | 7.5 | 8.3 | 8.9 | 9.4 | 9.7 | |
| 21 | 1.45 | 2.66 | 3.75 | 4.70 | 6.38 | 7.77 | 8.9 | 9.8 | 10.6 | 11.1 | 11.6 | 11.9 |
| 24 | 1.65 | 3.05 | 4.27 | 5.37 | 7.28 | 8.85 | 10.2 | 11.2 | 12.1 | 12.6 | 12.1 | 13.6 |

**ANSI No. 60 — ¾-in. pitch**

| Teeth | Rpm of sprocket | | | | | | | | | | | |
|---|---|---|---|---|---|---|---|---|---|---|---|---|
| | 50 | 100 | 200 | 300 | 400 | 600 | 800 | 1,000 | 1,200 | 1,400 | 1,600 | 1,800 |
| 12 | 0.73 | 1.34 | 2.41 | 3.30 | 4.05 | 5.2 | 6.1 | 6.6 | 6.9 | | | |
| 15 | 0.92 | 1.72 | 3.14 | 4.34 | 5.39 | 7.1 | 8.5 | 9.5 | 10.2 | 10.6 | | |
| 18 | 1.12 | 2.10 | 3.82 | 5.31 | 6.63 | 8.9 | 10.6 | 12.0 | 13.0 | 13.7 | 14.1 | |
| 21 | 1.31 | 2.46 | 4.49 | 6.24 | 7.80 | 10.4 | 12.5 | 14.1 | 15.4 | 16.3 | 16.9 | |
| 24 | 1.50 | 2.80 | 5.11 | 7.12 | 8.90 | 11.9 | 14.3 | 16.1 | 17.6 | 18.6 | 19.2 | 19.5 |

**ANSI No. 80 — 1-in. pitch**

| Teeth | Rpm of sprocket | | | | | | | | | | | |
|---|---|---|---|---|---|---|---|---|---|---|---|---|
| | 50 | 100 | 150 | 200 | 300 | 400 | 500 | 600 | 700 | 800 | 1,000 | 1,160 |
| 12 | 1.68 | 3.07 | 4.28 | 5.3 | 7.2 | 8.7 | 9.8 | 10.7 | 11.4 | | | |
| 15 | 2.14 | 3.95 | 5.57 | 7.0 | 9.6 | 11.8 | 13.6 | 15.1 | 16.3 | 17.3 | | |
| 18 | 2.59 | 4.81 | 6.79 | 8.6 | 11.8 | 14.5 | 16.9 | 18.9 | 20.5 | 21.9 | 24.0 | |
| 21 | 3.03 | 5.62 | 7.96 | 10.1 | 13.9 | 17.1 | 19.9 | 22.3 | 24.3 | 26.0 | 28.5 | |
| 24 | 3.46 | 6.43 | 9.10 | 11.5 | 15.8 | 19.5 | 22.7 | 25.4 | 27.7 | 29.6 | 32.5 | 33.9 |

Table 74.　Horsepower Ratings for Single-strand, Roller-Chain Drives　(*Continued*)

| Teeth | ANSI No. 100 | | | | | | 1¼-in. pitch | | | | | |
|---|---|---|---|---|---|---|---|---|---|---|---|---|
| | Rpm of sprocket | | | | | | | | | | | |
| | 25 | 50 | 100 | 200 | 300 | 400 | 500 | 650 | 700 | 750 | 800 | 870 |
| 12 | 1.72 | 3.19 | 5.8 | 9.9 | 13.0 | 15.6 | 17.2 | | | | | |
| 15 | 2.19 | 4.10 | 7.5 | 13.1 | 17.5 | 21.3 | 24.0 | 27.2 | 28.1 | | | |
| 18 | 2.55 | 4.97 | 9.i | 16.0 | 21.6 | 26.6 | 30.2 | 34.5 | 35.7 | 36.8 | | |
| 21 | 3.08 | 5.80 | 10.7 | 18.9 | 25.5 | 31.4 | 35.7 | 40.9 | 42.3 | 43.5 | 44.6 | |
| 24 | 3.52 | 6.62 | 12.2 | 21.5 | 29.2 | 35.4 | 40.5 | 46.5 | 48.1 | 49.5 | 50.6 | 52.0 |

| Teeth | ANSI No. 120 | | | | | | 1½-in. pitch | | | | | |
|---|---|---|---|---|---|---|---|---|---|---|---|---|
| | Rpm of sprocket | | | | | | | | | | | |
| | 25 | 50 | 75 | 100 | 150 | 200 | 250 | 300 | 350 | 400 | 500 | 600 |
| 12 | 2.90 | 5.4 | 7.6 | 9.6 | 13.2 | 16.2 | 18.7 | 21.0 | 22.8 | 24.3 | | |
| 15 | 3.71 | 6.9 | 9.8 | 12.5 | 17.3 | 21.6 | 25.3 | 28.6 | 31.4 | 33.9 | 38.0 | |
| 18 | 4.74 | 8.4 | 12.0 | 15.3 | 21.3 | 26.6 | 31.3 | 35.4 | 39.2 | 42.4 | 47.9 | |
| 21 | 5.24 | 9.9 | 14.0 | 17.9 | 24.9 | 31.2 | 36.8 | 41.7 | 46.2 | 50.0 | 56.7 | 61.7 |
| 24 | 5.99 | 11.3 | 16.0 | 20.4 | 28.5 | 35.7 | 41.9 | 47.6 | 52.6 | 57.1 | 64.6 | 70.3 |

| Teeth | ANSI No. 140 | | | | | | 1¾-in. pitch | | | | | |
|---|---|---|---|---|---|---|---|---|---|---|---|---|
| | Rpm of sprocket | | | | | | | | | | | |
| | 20 | 30 | 50 | 100 | 150 | 200 | 250 | 300 | 350 | 400 | 450 | 475 |
| 12 | 3.72 | 5.4 | 8.4 | 14.8 | 20.1 | 24.5 | 28.1 | 31.0 | | | | |
| 15 | 4.73 | 6.9 | 10.8 | 19.3 | 26.6 | 32.8 | 38.2 | 42.8 | 46.7 | | | |
| 18 | 5.73 | 8.3 | 13.1 | 23.7 | 32.7 | 40.5 | 47.3 | 53.2 | 58.4 | 62.9 | | |
| 21 | 6.70 | 9.7 | 15.3 | 27.7 | 38.4 | 47.6 | 55.7 | 62.8 | 69.0 | 74.5 | 79.0 | |
| 24 | 7.65 | 11.1 | 17.5 | 31.7 | 43.7 | 54.3 | 63.6 | 71.6 | 78.7 | 84.8 | 89.9 | 92.4 |

| Teeth | ANSI No. 160 | | | | | | 2-in. pitch | | | | | |
|---|---|---|---|---|---|---|---|---|---|---|---|---|
| | Rpm of sprocket | | | | | | | | | | | |
| | 10 | 20 | 40 | 80 | 120 | 160 | 200 | 240 | 280 | 320 | 360 | 400 |
| 12 | 2.9 | 5.5 | 10.1 | 18.0 | 24.6 | 30.1 | 34.8 | 38.6 | | | | |
| 15 | 3.7 | 7.0 | 13.0 | 23.5 | 32.4 | 40.2 | 47.0 | 52.9 | 58.0 | 62.4 | | |
| 18 | 4.4 | 8.5 | 15.8 | 28.7 | 39.7 | 49.5 | 58.1 | 65.7 | 72.4 | 78.3 | | |
| 21 | 5.2 | 9.9 | 18.5 | 33.6 | 46.7 | 58.1 | 68.3 | 77.5 | 85.5 | 92.5 | 99 | |
| 24 | 5.9 | 11.3 | 21.1 | 38.4 | 53.5 | 66.5 | 78.0 | 88.3 | 97.4 | 105.4 | 112 | 118 |

| Teeth | ANSI No. 180 | | | | | | 2¼-in. pitch | | | | | |
|---|---|---|---|---|---|---|---|---|---|---|---|---|
| | Rpm of sprocket | | | | | | | | | | | |
| | 10 | 20 | 40 | 60 | 80 | 100 | 140 | 180 | 220 | 260 | 300 | 330 |
| 12 | 4.09 | 7.72 | 14.2 | 19.9 | 25.0 | 29.7 | 33.7 | 44.4 | | | | |
| 15 | 5.19 | 9.84 | 18.3 | 25.8 | 32.7 | 39.0 | 50.2 | 59.8 | 68.1 | | | |
| 18 | 6.27 | 11.9 | 22.1 | 31.4 | 39.9 | 47.7 | 61.8 | 73.9 | 84.5 | 93.7 | | |
| 21 | 7.33 | 14.0 | 26.0 | 36.9 | 47.0 | 56.3 | 73.0 | 87.6 | 100.3 | 111.4 | 121.1 | |
| 24 | 8.38 | 15.9 | 29.7 | 42.1 | 53.6 | 64.7 | 83.2 | 99.7 | 114.2 | 126.9 | 137.7 | 145.0 |

| Teeth | ANSI No. 220 | | | | | | 2½-in. pitch | | | | | |
|---|---|---|---|---|---|---|---|---|---|---|---|---|
| | Rpm of sprocket | | | | | | | | | | | |
| | 10 | 20 | 40 | 60 | 80 | 100 | 120 | 160 | 200 | 240 | 260 | 280 |
| 12 | 5.6 | 10.5 | 19.1 | 26.8 | 33.6 | 39.6 | 45.1 | 54.4 | | | | |
| 15 | 7.1 | 13.4 | 24.7 | 34.8 | 43.9 | 52.2 | 59.8 | 73.4 | 85 | | | |
| 18 | 8.6 | 16.2 | 30.0 | 42.4 | 53.7 | 64.1 | 73.7 | 90.7 | 105 | 118 | | |
| 21 | 10.0 | 18.9 | 35.1 | 49.7 | 63.1 | 75.3 | 86.6 | 106.9 | 124 | 139 | 146 | |
| 24 | 11.4 | 21.6 | 40.2 | 56.8 | 71.9 | 86.0 | 98.8 | 121.8 | 142 | 159 | 166 | 173 |

**Table 74. Horsepower Ratings for Single-strand, Roller-Chain Drives** (*Continued*)

| Teeth | ANSI No. 240 | | | | 3-in. pitch | | | | | | | |
|---|---|---|---|---|---|---|---|---|---|---|---|---|
| | Rpm of sprocket | | | | | | | | | | | |
| | 10 | 20 | 30 | 40 | 60 | 80 | 100 | 120 | 140 | 160 | 180 | 200 |
| 12 | 9.5 | 17.7 | 25.1 | 32.0 | 44.4 | 55.1 | 64.7 | 73.0 | | | | |
| 15 | 12.2 | 22.6 | 32.4 | 41.5 | 58.0 | 72.7 | 86.9 | 97.9 | 108.0 | 118.5 | | |
| 18 | 14.6 | 27.7 | 39.3 | 50.5 | 70.8 | 87.1 | 105.7 | 120.7 | 134.5 | 147.0 | 158.0 | |
| 21 | 17.0 | 32.1 | 46.0 | 59.1 | 83.1 | 104.7 | 124.2 | 142.2 | 158.7 | 173.5 | 187.2 | 199.4 |
| 24 | 19.4 | 36.7 | 52.5 | 67.5 | 94.8 | 119.4 | 141.8 | 162.3 | 180.8 | 197.6 | 212.9 | 226.9 |

If $L_p$ = length of chain in pitches, and $D_p$ = center distance in pitches,

$$L_p = (N + n)/2 + a(N - n)/180 + 2D_p \cos a$$

Avoiding the use of trigonometrical tables,

$$L_p = 2C + (N + n)/2 + K(N - n)^2/C$$

where $C$ is the center distance in pitches and $K$ is a variable depending upon the value of $(N - n)/C$. Values of $K$ are as follows:

| $(N - n)/C$ | 0.1 | 1.0 | 2.0 | 3.0 |
|---|---|---|---|---|
| $K$ | 0.02533 | 0.02538 | 0.02555 | 0.02584 |

| $(N - n)/C$ | 4.0 | 5.0 | 6.0 |
|---|---|---|---|
| $K$ | 0.02631 | 0.02704 | 0.02828 |

Formulas for chain length on multisprocket drives are cumbersome except when all sprockets are the same size and on the same side of the chain. For this condition, the chain length in pitches is equal to the sum of the consecutive center distances in pitches plus the number of teeth on one sprocket.

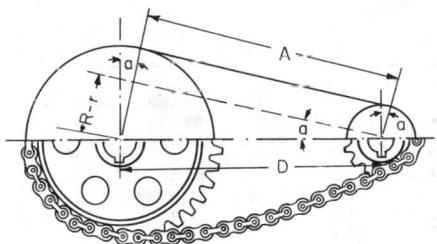

**Fig. 114**

Actual chain lengths should be in even numbers of pitches. When necessary, an odd number of pitches may be secured by the use of an offset link, but such links should be avoided if possible. An offset link is one pitch; half roller link at one end and half pin link at the other end. If center distances are to be nonadjustable, they should be selected to give an initial snug fit for an even number of pitches of chain. For the average application, a center distance equal to 40 ± 10 pitches of chain represents good practice.

There should be at least 120° of wrap in the arc of contact on a power sprocket. For ratios of 3:1 or less, the wrap will be 120° or more for any center distance or number of teeth. To secure a wrap of 120° or more, for ratios greater than 3:1, the center distance must be not less than the difference between the pitch diameters of the two sprockets.

**Sprocket Diameters**  $N$ = number of teeth; $P$ = pitch of chain, in; $D$ = diameter of roller, in. The pitch of a standard roller chain is measured from the center of a pin to the center of an adjacent pin.

$$\text{Pitch diam} = P/\sin\frac{180}{N}$$

$$\text{Bottom diam} = \text{pitch diam} - D$$

$$\text{Outside diam} = P\left(0.6 + \cot\frac{180}{N}\right)$$

$$\text{Caliper diam} = \left(\text{pitch diam} \times \cos\frac{90}{N}\right) - D$$

The exact bottom diameter cannot be measured for an odd number of teeth, but it can be checked by measuring the distance (caliper diameter) between bottoms of the two tooth spaces nearest opposite to each other. Bottom and caliper diameters must not be oversize—all tolerance must be negative. ANSI negative tolerance = $0.003 + (0.001 \times P\sqrt{N})$ in.

**Design of Sprocket Teeth for Roller Chains**  The section profile for the teeth of roller chain sprockets, recommended by ANSI, has the proportions shown in Fig. 115. Let $P$ = chain pitch; $W$ = chain width (length of roller); $n$ = number of strands of multiple chain; $M$ = overall width of tooth profile section; $H$ = nominal thickness of link plate, all in inches. Referring to Fig. 115, $T = 0.93W - 0.006$, for single-strand chain; $= 0.90W - 0.006$, for double- and triple-strand chains; $= 0.88W - 0.006$, for quadruple- or quintuple-strand chains; and $= 0.86W - 0.006$, for sextuple-strand chain and over. $C = 0.5P$. $E = \frac{1}{8}P$. $R(\text{min}) = 1.063P$. $Q = 0.5P$. $A = W + 4.15H + 0.003$. $M = A(n - 1) + T$. Tolerance on sprocket

**Table 75. Service Factors for Roller-Chain Drives**

| Type load | Service factor | |
|---|---|---|
| | 10-hr day | 24-hr day |
| Uniform.................. | 1.0 | 1.2 |
| Moderate shock.......... | 1.2 | 1.4 |
| Heavy shock............. | 1.4 | 1.7 |

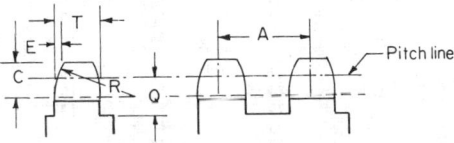

**Fig. 115**  Sprocket-tooth sections.

thickness = ±(0.02W + 0.002). For further data, see ANSI B29.1-1963 (R1972).

One of the most important requirements of a sprocket for roller chain is that the tooth space, or roller seat, should not be undersize. The size and shape of new straddle cutters, space cutters, or hobs should be checked carefully by cutting and testing a sample sprocket.

The method of laying out a standard sprocket tooth for roller chains is illustrated by Fig. 116. This form of tooth is recommended by the ANSI and is designed for maximum efficiency throughout the life of the drive. Because of the large pressure angle, the tendency of the teeth to wear hook-shaped is reduced, while the chain rides higher on the teeth as it elongates, thus accommodating itself to a larger pitch circle.

In Fig. 116, let $P$ = pitch; $D$ = nominal roller diameter, in;

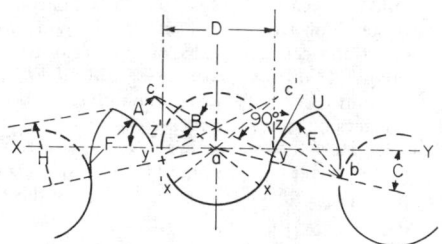

**Fig. 116**  Laying out a standard sprocket tooth.

$N$ = number of teeth; $D' = 1.005D + 0.003$; $A = 35 + 60/N$, deg; $B = 18 - 56/N$, deg; $C = 180/N$, deg.

In laying out the tooth, first draw line $XY$. Locate point $a$, and with that as a center and radius $ax$ equal to ½$D'$, draw a circular arc for the "seating curve" $xx'$.

Draw line $xac$ making angle $A$ with line $XY$, and locate point $c$ so that $ac = 0.8D$. Draw line $cy$ making angle $B$ with line $cx$. With center at $c$ and radius $cx$, draw arc $xy$ for the "working curve."

Draw line $yz$ perpendicular to line $cy$. Draw line $ab$ making angle $C$ with line $XY$, and locate point $b$ so that $ab = 1.24D$. Draw line $bz$ parallel to line $yc$. With $b$ as center and radius $bz$, draw the "topping curve," arc $zu$ tangent to line $zy$.

A similar construction for the other half will complete the tooth outline.

Outside diameter of sprocket when tooth is pointed = $P$ cot $(180/N) + 2H$.

The recommended value for $H$ is $0.3P$; and when this value is chosen, the outside diameter of the sprocket will be $P[0.6 + \cot(180/N)]$.

**Inverted-tooth (silent) chain drives** have a typical tooth form shown in Fig. 117. Such chains should be operated in an oil-retaining casing with provisions for lubrication. The use of offset links and chains with an uneven number of pitches should be avoided.

**Horsepower ratings** per inch of silent chain width, given in ANSI B29.2-1957 (R1971), for various chain pitches and speeds, are shown in Tables 76 and 77. These ratings are based on ideal drive conditions with relatively little shock or load variation, an average life of 20,000 h being assumed. In utilizing the horsepower ratings of the tables, the nominal horsepower of the drive should be multiplied by a service factor depending on the application. These service factors are listed in Table 78.

For further details on service factors for a large number of specific applications, see ANSI B29.2-1957 (R1971). This pamphlet also gives additional data on lubrication, sprocket dimensions, etc. In utilizing Tables 76 and 77 the required chain width is obtained by dividing the design horsepower by the horsepower ratings given. For calculating silent-chain lengths, the procedure for roller-chain drives may be used.

**Table 76.  Horsepower Ratings per Inch Width—Silent-Chain Drives**

| No. of teeth in smaller sprocket | Pitch, in. | Rpm of small sprocket | | | | | | |
|---|---|---|---|---|---|---|---|---|
| | | 500 | 1,000 | 1,500 | 2,000 | 3,000 | 4,000 | 5,000 |
| 21 | ⅜ | 2.2 | 4.1 | 5.8 | 7.2 | 9.1 | 9.9 | 9.5 |
| 25 | ⅜ | 2.6 | 4.9 | 7.0 | 8.8 | 11.0 | 12.0 | 12.0 |
| 29 | ⅜ | 3.0 | 5.8 | 8.2 | 10.3 | 13.0 | 15.0 | 15.0 |
| 33 | ⅜ | 3.5 | 6.6 | 9.4 | 12.0 | 15.0 | 17.0 | 17.0 |
| 37 | ⅜ | 3.9 | 7.3 | 11.0 | 13.0 | 17.0 | 19.0 | 19.0 |
| 21 | ½ | 4 | 7 | .... | 11 | 13 | | |
| 25 | ½ | 4 | 8 | .... | 14 | 17 | 16 | |
| 29 | ½ | 5 | 10 | .... | 17 | 20 | 20 | |
| 33 | ½ | 6 | 11 | .... | 19 | 23 | 23 | |
| 37 | ½ | 7 | 13 | .... | 21 | 26 | | |

| No. of teeth in smaller sprocket | Pitch, in. | Rpm of small sprocket | | | | | | |
|---|---|---|---|---|---|---|---|---|
| | | 500 | 1,000 | 1,200 | 1,800 | 2,000 | 2,500 | 3,000 | 3,500 |
| 21 | ⅝ | 6 | 10 | 12 | 15 | 16 | 16 | 16 | |
| 25 | ⅝ | 7 | 13 | 15 | 19 | 20 | 21 | 21 | 19 |
| 29 | ⅝ | 8 | 15 | 17 | 22 | 24 | 25 | 25 | 23 |
| 33 | ⅝ | 9 | 17 | 20 | 26 | 27 | 29 | 29 | 27 |
| 37 | ⅝ | 10 | 19 | 22 | 29 | 31 | 34 | 33 | |
| 21 | ¾ | 8 | 14 | 16 | 19 | 20 | 19 | | |
| 25 | ¾ | 10 | 17 | 20 | 25 | 25 | 24 | | |
| 29 | ¾ | 12 | 21 | 24 | 29 | 30 | 30 | | |
| 33 | ¾ | 13 | 24 | 27 | 34 | 35 | 35 | | |
| 37 | ¾ | 15 | 27 | 31 | 38 | 39 | 39 | | |

**Table 77. Horsepower Ratings per Inch Width—Silent-Chain Drives**

| No. of teeth in smaller sprocket | Pitch, in. | Rpm of small sprocket | | | | | | | |
|---|---|---|---|---|---|---|---|---|---|
| | | 300 | 500 | 700 | 1,000 | 1,200 | 1,500 | 1,800 | 2,000 |
| 21 | 1 | 9 | 14 | 18 | 23 | 25 | 26 | 26 | |
| 25 | 1 | 11 | 17 | 22 | 28 | 31 | 33 | 33 | 33 |
| 29 | 1 | 13 | 20 | 26 | 33 | 37 | 40 | 41 | 40 |
| 33 | 1 | 14 | 23 | 30 | 39 | 43 | 47 | 47 | 46 |
| 37 | 1 | 16 | 26 | 34 | 43 | 48 | 52 | 53 | |
| 21 | 1¼ | 14 | 21 | 26 | 32 | 33 | | | |
| 25 | 1¼ | 16 | 25 | 32 | 40 | 42 | 42 | | |
| 29 | 1¼ | 19 | 30 | 38 | 47 | 50 | 51 | | |
| 33 | 1¼ | 22 | 34 | 44 | 55 | 58 | 59 | | |
| 37 | 1¼ | 24 | 38 | 50 | 61 | 65 | | | |

| | | Rpm of small sprocket | | | | | | | |
|---|---|---|---|---|---|---|---|---|---|
| | | 200 | 300 | 400 | 500 | 600 | 700 | 800 | 900 | 1,000 |
| 21 | 1½ | 13 | 19 | 24 | 29 | 32 | 35 | 37 | 39 | 39 |
| 25 | 1½ | 16 | 23 | 30 | 35 | 40 | 44 | 47 | 49 | 52 |
| 29 | 1½ | 19 | 27 | 35 | 41 | 47 | 52 | 56 | 59 | 60 |
| 33 | 1½ | 22 | 31 | 40 | 47 | 54 | 60 | 64 | 68 | 70 |
| 37 | 1½ | 24 | 35 | 47 | 53 | 61 | 67 | 72 | 77 | 79 |
| 45 | 1½ | 30 | 43 | 54 | 65 | 74 | 81 | 86 | 90 | |
| 21 | 2 | 23 | 32 | 40 | 42 | 50 | 52 | | | |
| 25 | 2 | 28 | 39 | 49 | 56 | 62 | 66 | 68 | 68 | |
| 29 | 2 | 33 | 46 | 58 | 67 | 74 | 79 | 82 | 82 | |
| 33 | 2 | 37 | 53 | 66 | 77 | 85 | 91 | 94 | 94 | |
| 37 | 2 | 42 | 60 | 74 | 88 | 99 | 102 | 105 | | |
| 45 | 2 | 51 | 72 | 90 | 105 | 115 | 121 | | | |

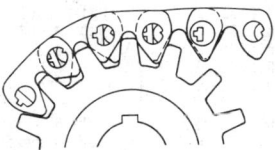

**Fig. 117**  Inverted-tooth (silent-chain) drive.

**Table 78. Service Factors for Silent-Chain Drives**

| Load type | Service factor | |
|---|---|---|
| | 10-h day | 24-h day |
| Uniform | 1.0 | 1.3 |
| Moderate shock | 1.4 | 1.7 |
| Heavy shock | 1.7 | 2.0 |

## ROTARY AND RECIPROCATING ELEMENTS

**Analysis of the Motion of the Reciprocating Parts**  Let $r =$ crank radius, in; $R =$ crank radius, ft; $s = 2r =$ stroke, in; $l =$ length of connecting rod, in; $x =$ displacement of piston from head end dead center, in; $a =$ crank angle corresponding to $x$; $b =$ connecting-rod angle corresponding to $x$; $v =$ crankpin velocity, ft/s; $c =$ piston velocity, ft/s; $N =$ r/min of the crank. Then, when the centerlines of the piston rod and crankshaft lie in the same plane,

$$x = r(1 - \cos a) + l(1 - \cos b)$$
$$= r\left(1 - \cos a + \frac{r}{2l}\sin^2 a\right) \quad \text{(approx)}$$

Table 79 gives correct piston positions corresponding to various crank angles. Table 80 gives crank angles corresponding to various piston positions.

When, however, the centerline of motion of the reciprocating masses is located a distance $h$ from the center of rotation of the crank, as shown in Fig. 118, the dead centers are unsymmetrical and the stroke is greater than $2r$, having the value

$$s = \sqrt{(l + r)^2 - h^2} - \sqrt{(l - r)^2 - h^2}$$

**Determination of the Velocity of the Reciprocating Parts**  The velocity of the reciprocating parts is

$$c = v\left(\sin a + r\sin 2a/2l\sqrt{1 - \frac{r^2}{l^2}\sin^2 a}\right)$$
$$= v(\sin a + r\sin 2a/2l) \quad \text{(approx)}$$

when the centerlines of the piston rod and crankshaft lie in the same plane. The ratio $c/v$ is the tangential factor; Table 81 gives the values of velocities corresponding to various crank angles. When, however, the crank rotates about a center off the centerline, as in Fig. 118 (**offset cylinder**), the above formula does not hold, and the graphical solution of Fig. 119 may be resorted to. $AH$ is the centerline of the piston rod produced, $AP$ is the connecting rod, $OP$ the crank with center at $O$.

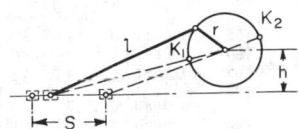

**Fig. 118**

Draw $OQ$ at right angles to $AH$. For any position $AP$ of the connecting rod cutting $OQ$ in $E$, the length $OE$ represents the velocity of the reciprocating parts to the same scale to which $OP$ represents the crankpin velocity. The graphical construction of Fig. 119 shows how to construct velocity ellipses for the whole revolution.

**Determination of the Acceleration of the Reciprocating Parts** The acceleration $p$ of the reciprocating parts is given by

$$p = v^2(\cos a + r \cos 2a/l)/R \quad \text{(approx)}$$

when the centerlines of the pistons and crankshaft lie in the same plane.

Table 82 gives accurate values of the acceleration for different crank angles.

If the cylinder is offset, as in Fig. 118, the acceleration of the reciprocating parts may be found by the approximate formula

$$p = (v^2/R)[\cos a + (r/l) \cos 2a + b/l \sin a]$$

The **inertia force** in pounds per in$^2$ of piston is $f = Mp/A = Wv^2(\cos a + r \cos 2a/l)/AgR = 0.00034WN^2R(\cos a + r \cos 2a/l)/A$, where $W$ = total weight of reciprocating masses, lb; $A$ =

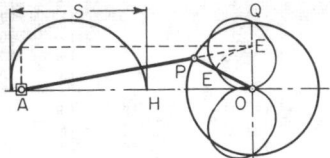

**Fig. 119**

piston area, in$^2$. The value of $W$ for any given engine is to be taken as the weight of the piston, piston rod, crosshead, and a portion of the connecting rod. It is usual to include $\frac{1}{2}$ to $\frac{2}{3}$ the connecting-rod weight as part of the reciprocating masses.

**Table 79. Piston Positions for Various Crank Angles**

(From beginning of stroke toward crankshaft. To find distance of piston from beginning of stroke, multiply tabular value by length of stroke)

$$\text{Calculated from } \frac{x}{2r} = \frac{1}{2}(1 - \cos a) + \frac{l}{2r}\left(1 - \sqrt{1 - \frac{r^2}{l^2}\sin^2 a}\right)$$

| Crank angles, deg $a$ | Ratio of length of connecting rod to length of crank | | | | | | | | | | |
|---|---|---|---|---|---|---|---|---|---|---|---|
| | $\frac{l}{r} = 2.5$ | 3 | 3.5 | 4 | 4.5 | 5 | 5.5 | 6 | 7 | 8 | ∞ |
| 5 | 0.0027 | 0.0025 | 0.0024 | 0.0023 | 0.0023 | 0.0022 | 0.0022 | 0.0022 | 0.0022 | 0.0021 | 0.0019 |
| 10 | 0.0106 | 0.0101 | 0.0098 | 0.0095 | 0.0093 | 0.0091 | 0.0090 | 0.0089 | 0.0087 | 0.0085 | 0.0076 |
| 15 | 0.0238 | 0.0226 | 0.0218 | 0.0212 | 0.0208 | 0.0204 | 0.0210 | 0.0198 | 0.0194 | 0.0191 | 0.0170 |
| 20 | 0.0419 | 0.0399 | 0.0385 | 0.0375 | 0.0367 | 0.0360 | 0.0355 | 0.0350 | 0.0343 | 0.0338 | 0.0302 |
| 25 | 0.0648 | 0.0618 | 0.0597 | 0.0580 | 0.0568 | 0.0558 | 0.0550 | 0.0543 | 0.0532 | 0.0524 | 0.0468 |
| 30 | 0.0922 | 0.0880 | 0.0849 | 0.0827 | 0.0809 | 0.0795 | 0.0784 | 0.0774 | 0.0759 | 0.0748 | 0.0670 |
| 35 | 0.1238 | 0.1181 | 0.1141 | 0.1111 | 0.1088 | 0.1069 | 0.1054 | 0.1042 | 0.1022 | 0.1007 | 0.0904 |
| 40 | 0.1590 | 0.1518 | 0.1467 | 0.1430 | 0.1401 | 0.1377 | 0.1358 | 0.1342 | 0.1318 | 0.1299 | 0.1170 |
| 45 | 0.1975 | 0.1887 | 0.1825 | 0.1779 | 0.1774 | 0.1716 | 0.1693 | 0.1674 | 0.1643 | 0.1621 | 0.1464 |
| 50 | 0.2387 | 0.2283 | 0.2210 | 0.2156 | 0.2114 | 0.2081 | 0.2054 | 0.2032 | 0.1996 | 0.1970 | 0.1786 |
| 55 | 0.2822 | 0.2702 | 0.2618 | 0.2556 | 0.2508 | 0.2470 | 0.2439 | 0.2413 | 0.2373 | 0.2342 | 0.2132 |
| 60 | 0.3274 | 0.3139 | 0.3044 | 0.2974 | 0.2921 | 0.2878 | 0.2843 | 0.2814 | 0.2769 | 0.2735 | 0.2500 |
| 65 | 0.3737 | 0.3588 | 0.3484 | 0.3407 | 0.3348 | 0.3301 | 0.3263 | 0.3231 | 0.3182 | 0.3144 | 0.2887 |
| 70 | 0.4207 | 0.4045 | 0.3932 | 0.3850 | 0.3786 | 0.3735 | 0.3694 | 0.3660 | 0.3607 | 0.3567 | 0.3290 |
| 75 | 0.4677 | 0.4505 | 0.4386 | 0.4298 | 0.4230 | 0.4177 | 0.4133 | 0.4097 | 0.4041 | 0.3999 | 0.3706 |
| 80 | 0.5142 | 0.4963 | 0.4839 | 0.4747 | 0.4777 | 0.4621 | 0.4576 | 0.4539 | 0.4480 | 0.4436 | 0.4132 |
| 85 | 0.5599 | 0.5415 | 0.5288 | 0.5194 | 0.5122 | 0.5065 | 0.5019 | 0.4981 | 0.4920 | 0.4876 | 0.4564 |
| 90 | 0.6044 | 0.5858 | 0.5729 | 0.5635 | 0.5563 | 0.5505 | 0.5458 | 0.5420 | 0.5359 | 0.5314 | 0.5000 |
| 95 | 0.6471 | 0.6287 | 0.6160 | 0.6066 | 0.5994 | 0.5937 | 0.5891 | 0.5852 | 0.5792 | 0.5747 | 0.5436 |
| 100 | 0.6879 | 0.6699 | 0.6575 | 0.6484 | 0.6414 | 0.6358 | 0.6313 | 0.6275 | 0.6216 | 0.6172 | 0.5868 |
| 105 | 0.7265 | 0.7093 | 0.6974 | 0.6886 | 0.6819 | 0.6765 | 0.6722 | 0.6685 | 0.6629 | 0.6587 | 0.6294 |
| 110 | 0.7627 | 0.7465 | 0.7353 | 0.7270 | 0.7206 | 0.7156 | 0.7114 | 0.7080 | 0.7027 | 0.6987 | 0.6710 |
| 115 | 0.7963 | 0.7814 | 0.7710 | 0.7633 | 0.7574 | 0.7527 | 0.7489 | 0.7457 | 0.7408 | 0.7371 | 0.7113 |
| 120 | 0.8274 | 0.8139 | 0.8044 | 0.7974 | 0.7921 | 0.7878 | 0.7843 | 0.7814 | 0.7776 | 0.7735 | 0.7500 |
| 125 | 0.8558 | 0.8438 | 0.8354 | 0.8292 | 0.8244 | 0.8206 | 0.8175 | 0.8149 | 0.8108 | 0.8078 | 0.7868 |
| 130 | 0.8815 | 0.8711 | 0.8638 | 0.8584 | 0.8542 | 0.8509 | 0.8482 | 0.8459 | 0.8424 | 0.8398 | 0.8214 |
| 135 | 0.9046 | 0.8958 | 0.8896 | 0.8851 | 0.8815 | 0.8787 | 0.8764 | 0.8745 | 0.8715 | 0.8692 | 0.8536 |
| 140 | 0.9250 | 0.9179 | 0.9128 | 0.9090 | 0.9061 | 0.9038 | 0.9019 | 0.9003 | 0.8978 | 0.8960 | 0.8830 |
| 145 | 0.9429 | 0.9372 | 0.9332 | 0.9302 | 0.9279 | 0.9261 | 0.9246 | 0.9233 | 0.9213 | 0.9199 | 0.9096 |
| 150 | 0.9583 | 0.9540 | 0.9510 | 0.9487 | 0.9469 | 0.9455 | 0.9444 | 0.9434 | 0.9420 | 0.9408 | 0.9330 |
| 155 | 0.9711 | 0.9681 | 0.9660 | 0.9643 | 0.9631 | 0.9621 | 0.9613 | 0.9606 | 0.9595 | 0.9587 | 0.9532 |
| 160 | 0.9816 | 0.9796 | 0.9782 | 0.9772 | 0.9764 | 0.9757 | 0.9752 | 0.9747 | 0.9740 | 0.9735 | 0.9698 |
| 165 | 0.9897 | 0.9886 | 0.9878 | 0.9872 | 0.9867 | 0.9863 | 0.9860 | 0.9858 | 0.9854 | 0.9851 | 0.9830 |
| 170 | 0.9954 | 0.9949 | 0.9946 | 0.9943 | 0.9941 | 0.9939 | 0.9938 | 0.9937 | 0.9935 | 0.9933 | 0.9924 |
| 175 | 0.9989 | 0.9987 | 0.9986 | 0.9986 | 0.9985 | 0.9984 | 0.9984 | 0.9984 | 0.9984 | 0.9983 | 0.9981 |
| 180 | 1.0000 | 1.0000 | 1.0000 | 1.0000 | 1.0000 | 1.0000 | 1.0000 | 1.0000 | 1.0000 | 1.0000 | 1.0000 |

**Table 80. Crank Angles and Piston Positions for Connecting Rods of Different Lengths**

Calculated from $\cos^{-1}\left(\dfrac{1-x_1}{1-2x_1\big/\left(1+\dfrac{l}{r}\right)}-x_1\right)$, where $x_1=\dfrac{x}{2r}$

| Fraction of stroke from commencement, $\dfrac{x}{2r}$ | Ratio of length of connecting rod to length of crank | | | | | | | | | | |
|---|---|---|---|---|---|---|---|---|---|---|---|
| | 2.5 | 3 | 3.5 | 4 | 4.5 | 5 | 5.5 | 6 | 7 | 8 | ∞ |
| | Angle through which crank has advanced from dead center, deg | | | | | | | | | | |
| 0.005 | 6.86 | 7.02 | 7.15 | 7.25 | 7.34 | 7.40 | 7.46 | 7.51 | 7.59 | 7.65 | 8.11 |
| 0.01 | 9.70 | 9.94 | 10.13 | 10.27 | 10.39 | 10.48 | 10.56 | 10.63 | 10.74 | 10.83 | 11.48 |
| 0.02 | 13.75 | 14.09 | 14.35 | 14.56 | 14.72 | 14.86 | 14.97 | 15.06 | 15.22 | 15.34 | 16.26 |
| 0.03 | 16.88 | 17.30 | 17.62 | 17.87 | 18.07 | 18.23 | 18.37 | 18.49 | 18.68 | 18.82 | 19.95 |
| 0.04 | 19.53 | 20.02 | 20.38 | 20.67 | 20.91 | 21.10 | 21.26 | 21.39 | 21.61 | 21.78 | 23.07 |
| 0.05 | 21.88 | 22.43 | 22.84 | 23.16 | 23.42 | 23.64 | 23.82 | 23.97 | 24.21 | 24.40 | 25.84 |
| 0.06 | 24.02 | 24.62 | 25.08 | 25.43 | 25.71 | 25.95 | 26.15 | 26.31 | 26.58 | 26.79 | 28.36 |
| 0.07 | 26.01 | 26.65 | 27.14 | 27.53 | 27.84 | 28.09 | 28.30 | 28.48 | 28.77 | 28.99 | 30.68 |
| 0.08 | 27.86 | 28.56 | 29.08 | 29.49 | 29.82 | 30.09 | 30.32 | 30.51 | 30.82 | 31.06 | 32.86 |
| 0.09 | 29.62 | 30.36 | 30.92 | 31.35 | 31.70 | 31.99 | 32.23 | 32.43 | 32.76 | 33.01 | 34.92 |
| 0.10 | 31.29 | 32.07 | 32.66 | 33.12 | 33.49 | 33.79 | 34.05 | 34.26 | 34.61 | 34.87 | 36.87 |
| 0.15 | 38.77 | 39.74 | 40.47 | 41.04 | 41.49 | 41.86 | 42.17 | 42.43 | 42.85 | 43.17 | 45.57 |
| 0.20 | 45.31 | 46.46 | 47.31 | 47.97 | 48.49 | 48.92 | 49.27 | 49.57 | 50.05 | 50.42 | 53.13 |
| 0.25 | 51.32 | 52.62 | 53.58 | 54.31 | 54.90 | 55.38 | 55.77 | 56.10 | 56.63 | 57.04 | 60.00 |
| 0.30 | 56.99 | 58.43 | 59.49 | 60.30 | 60.94 | 61.46 | 61.89 | 62.25 | 62.82 | 63.26 | 66.42 |
| 0.35 | 62.45 | 64.03 | 65.18 | 66.06 | 66.75 | 67.30 | 67.76 | 68.15 | 68.76 | 69.22 | 72.54 |
| 0.40 | 67.80 | 69.51 | 70.75 | 71.68 | 72.41 | 73.00 | 73.49 | 73.89 | 74.53 | 75.02 | 78.46 |
| 0.45 | 73.12 | 74.95 | 76.26 | 77.25 | 78.02 | 78.63 | 79.14 | 79.56 | 80.23 | 80.73 | 84.26 |
| 0.50 | 78.46 | 80.41 | 81.79 | 82.82 | 83.62 | 84.26 | 84.78 | 85.22 | 85.90 | 86.42 | 90.00 |
| 0.55 | 83.90 | 85.95 | 87.39 | 88.46 | 89.28 | 89.94 | 90.48 | 90.92 | 91.62 | 92.14 | 95.74 |
| 0.60 | 89.50 | 91.64 | 93.13 | 94.23 | 95.07 | 95.74 | 96.28 | 96.73 | 97.44 | 97.96 | 101.54 |
| 0.65 | 95.35 | 97.56 | 99.08 | 100.20 | 101.05 | 101.72 | 102.27 | 102.72 | 103.42 | 103.94 | 107.46 |
| 0.70 | 101.54 | 103.80 | 105.34 | 106.46 | 107.31 | 107.98 | 108.52 | 108.97 | 109.66 | 110.17 | 113.58 |
| 0.75 | 108.21 | 110.49 | 112.02 | 113.13 | 113.97 | 114.62 | 115.15 | 115.58 | 116.25 | 116.74 | 120.00 |
| 0.80 | 115.57 | 117.82 | 119.32 | 120.39 | 121.19 | 121.82 | 122.32 | 122.73 | 123.37 | 123.83 | 126.87 |
| 0.85 | 123.94 | 126.10 | 127.51 | 128.52 | 129.26 | 129.84 | 130.31 | 130.68 | 131.26 | 131.69 | 134.43 |
| 0.90 | 133.96 | 135.90 | 137.17 | 138.05 | 138.71 | 139.21 | 139.61 | 139.94 | 140.44 | 140.81 | 143.13 |
| 0.91 | 136.26 | 138.15 | 139.37 | 140.22 | 140.85 | 141.34 | 141.72 | 142.03 | 142.51 | 142.86 | 145.08 |
| 0.92 | 138.71 | 140.52 | 141.69 | 142.51 | 143.11 | 143.57 | 143.94 | 144.24 | 144.70 | 145.03 | 147.14 |
| 0.93 | 141.32 | 143.05 | 144.16 | 144.94 | 145.51 | 145.95 | 146.30 | 146.58 | 147.01 | 147.33 | 149.32 |
| 0.94 | 144.13 | 145.77 | 146.82 | 147.55 | 148.09 | 148.50 | 148.82 | 149.09 | 149.49 | 149.79 | 151.64 |
| 0.95 | 147.21 | 148.73 | 149.71 | 150.38 | 150.88 | 151.26 | 151.56 | 151.81 | 152.18 | 152.45 | 154.16 |
| 0.96 | 152.62 | 152.02 | 152.90 | 153.52 | 153.97 | 154.31 | 154.59 | 154.81 | 155.14 | 155.39 | 156.93 |
| 0.97 | 154.51 | 155.75 | 156.53 | 157.07 | 157.47 | 157.77 | 158.01 | 158.20 | 158.50 | 158.71 | 160.05 |
| 0.98 | 159.15 | 160.18 | 160.83 | 161.28 | 161.61 | 161.86 | 162.06 | 162.22 | 162.46 | 162.46 | 163.74 |
| 0.99 | 165.23 | 165.98 | 166.44 | 166.77 | 167.00 | 167.18 | 167.32 | 167.44 | 167.61 | 167.74 | 168.52 |
| 0.995 | 169.55 | 170.08 | 170.41 | 170.64 | 170.81 | 170.94 | 171.04 | 171.12 | 171.24 | 171.33 | 171.89 |
| 1.00 | 180.00 | 180.00 | 180.00 | 180.00 | 180.00 | 180.00 | 180.00 | 180.00 | 180.00 | 180.00 | 180.00 |

**Relation between the Force in the Line of Piston Travel and the Tangential Reaction at the Crankpin**  In Fig. 120, if $P$ is the force in the line of piston travel, the force acting along the connecting rod is $C = P/\cos b$, and that on the guides $N = P \tan b$. The force tangent to the crankpin circle at the crankpin is $T = C \sin (a + b) = P \sec b \sin (b + a) = P \sin a(l + r \cos a/\sqrt{l^2 - r^2 \sin^2 a})$. Table 81 gives values of $\sec b \sin (b + a)$ for various values of $l/r$. They are equal to the tangential factors of piston velocities for different crank angles.

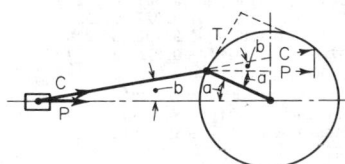

**Fig. 120**

**Determination of Flywheel Weight**

**Tangential Effort Curves**  The total force on the piston at each instant is obtained from an indicator card or other suitable instrumentation providing a pressure or force measurement during the stroke. In the case of a double-acting engine, this force is obtained by subtracting from the pressure on one face the corresponding pressure on the other face of the piston. In Fig. 121 the indicator card *ABCDEF* of a single-cylinder, single-acting, four-cycle diesel engine is shown. The **inertia force** of the reciprocating parts per square inch of piston area is obtained by multiplying the mass of the parts per square inch of piston area (piston, piston-rod, crosshead, and 0.35 to 0.45 of the connecting rod) by the acceleration of the piston (see Table 82). It is represented by the ordinates of the curves *xyz* and *x'yz'* for the forward and return strokes, respectively. The ordinates between the total gas pressure and the inertia curves ($p_1$, $p_2$) give the resultant pressures acting on the piston along the centerline of the engine.

The tangential reactions at the crankpin can be obtained by multiplying the resultant pressure on the piston by the tangential factors for the various crank angles (see Table 81). Figure 122 shows the tangential reaction for the diesel engine of Fig. 121 plotted against the travel of the crank.

In engine design, it is usual to construct the tangential effort curve from data obtained at normal or rated engine load and to assume that the engine is working against a constant and uniform torque. The uniform resisting torque is represented in Fig. 122 by $XX$, which is so located that the shaded areas above and below it are equal to one another. The areas above $XX$ represent in foot-pounds per square inch of piston area the amount of energy which is effective in increasing the kinetic energy of the moving masses; and the areas below $XX$ represent the energy per square inch of piston area liberated in the retardation of the moving masses for the production of useful work in the driven unit.

In the case of multicylinder engines, the tangential effort curves for each cylinder are superimposed, properly shifted with respect to each other, corresponding to the angular location of their respective cranks.

**Determination of Flywheel Weight**   Let $M$ = mass of wheel rim; $W$ = weight of wheel rim, lb; $v_1$ = max linear rim velocity, ft/s; $v_2$ = min rim velocity, ft/s; $v = \frac{1}{2}(v_1 + v_2)$ = average rim velocity, ft/s; $k = (v_1 - v_2)/v$ = coefficient of velocity fluctuation; $D$ = mean diam of rim, ft; $N$ = r/min; $A$ = piston area, in$^2$; $E_m$ = max energy in ft·lb per in$^2$ of piston area which must be absorbed or liberated by the wheel rim with velocity variation between $v_1$ and $v_2$. Then

$$E_m \times A = M \frac{v_1^2}{2} - M \frac{v_2^2}{2} = \frac{W}{g} kv^2 = \frac{kWD^2N^2}{11,744}$$

or

$$W = \frac{11,744 E_m A}{kD^2 N^2}$$

About $\frac{9}{10} W$ may be placed in the rim to account for the fly-

**Table 81. Tangential Factors and Piston Velocities**

| Crank angles forward, deg $a$ | \multicolumn{11}{c}{Ratio of connecting-rod length to crank length} |
| | $\frac{l}{r} = 2.5$ | 3 | 3.5 | 4 | 4.5 | 5 | 5.5 | 6 | 7 | 8 | ∞ |
|---|---|---|---|---|---|---|---|---|---|---|---|
| 5 | 0.1219 | 0.1161 | 0.1120 | 0.1089 | 0.1065 | 0.1045 | 0.1029 | 0.1016 | 0.0996 | 0.0980 | 0.0872 |
| 10 | 0.2442 | 0.2307 | 0.2226 | 0.2164 | 0.2117 | 0.2079 | 0.2048 | 0.2022 | 0.1981 | 0.1950 | 0.1736 |
| 15 | 0.3594 | 0.3425 | 0.3304 | 0.3215 | 0.3145 | 0.3089 | 0.3043 | 0.3005 | 0.2946 | 0.2901 | 0.2588 |
| 20 | 0.4718 | 0.4499 | 0.4343 | 0.4227 | 0.4136 | 0.4064 | 0.4006 | 0.3957 | 0.3880 | 0.3822 | 0.3420 |
| 25 | 0.5781 | 0.5516 | 0.5329 | 0.5189 | 0.5081 | 0.4995 | 0.4925 | 0.4866 | 0.4770 | 0.4706 | 0.4226 |
| 30 | 0.6768 | 0.6464 | 0.6250 | 0.6091 | 0.5968 | 0.5870 | 0.5791 | 0.5724 | 0.5620 | 0.5542 | 0.5000 |
| 35 | 0.7667 | 0.7331 | 0.7097 | 0.6923 | 0.6788 | 0.6682 | 0.6595 | 0.6522 | 0.6409 | 0.6325 | 0.5736 |
| 40 | 0.8466 | 0.8108 | 0.7859 | 0.7675 | 0.7533 | 0.7421 | 0.7329 | 0.7253 | 0.7134 | 0.7045 | 0.6428 |
| 45 | 0.9156 | 0.8786 | 0.8530 | 0.8341 | 0.8196 | 0.8081 | 0.7988 | 0.7910 | 0.7789 | 0.7699 | 0.7071 |
| 50 | 0.9730 | 0.9358 | 0.9102 | 0.8915 | 0.8771 | 0.8657 | 0.8565 | 0.8488 | 0.8368 | 0.8279 | 0.7660 |
| 55 | 1.0181 | 0.9820 | 0.9572 | 0.9392 | 0.9253 | 0.9144 | 0.9055 | 0.8982 | 0.8867 | 0.8782 | 0.8192 |
| 60 | 1.0507 | 1.0168 | 0.9937 | 0.9769 | 0.9641 | 0.9540 | 0.9457 | 0.9390 | 0.9284 | 0.9205 | 0.8660 |
| 65 | 1.0707 | 1.0402 | 1.0196 | 1.0046 | 0.9932 | 0.9842 | 0.9769 | 0.9709 | 0.9615 | 0.9545 | 0.9063 |
| 70 | 1.0784 | 1.0525 | 1.0350 | 1.0224 | 1.0127 | 1.0051 | 0.9990 | 0.9939 | 0.9860 | 0.9801 | 0.9397 |
| 75 | 1.0743 | 1.0539 | 1.0402 | 1.0303 | 1.0228 | 1.0169 | 1.0121 | 1.0081 | 1.0020 | 0.9974 | 0.9659 |
| 80 | 1.0592 | 1.0452 | 1.0357 | 1.0289 | 1.0238 | 1.0197 | 1.0164 | 1.0137 | 1.0095 | 1.0063 | 0.9848 |
| 85 | 1.0341 | 1.0269 | 1.0221 | 1.0186 | 1.0160 | 1.0139 | 1.0122 | 1.0109 | 1.0087 | 1.0071 | 0.9962 |
| 90 | 1.0000 | 1.0000 | 1.0000 | 1.0000 | 1.0000 | 1.0000 | 1.0000 | 1.0000 | 1.0000 | 1.0000 | 1.0000 |
| 95 | 0.9583 | 1.9655 | 0.9703 | 0.9738 | 0.9764 | 0.9785 | 0.9801 | 0.9815 | 0.9837 | 0.9853 | 0.9962 |
| 100 | 0.9104 | 0.9245 | 0.9339 | 0.9407 | 0.9459 | 0.9499 | 0.9532 | 0.9559 | 0.9601 | 0.9633 | 0.9848 |
| 105 | 0.8575 | 0.8779 | 0.8916 | 0.9015 | 0.9090 | 0.9150 | 0.9198 | 0.9237 | 0.9299 | 0.9344 | 0.9659 |
| 110 | 0.8010 | 0.8269 | 0.8444 | 0.8570 | 0.8667 | 0.8742 | 0.8804 | 0.8855 | 0.8934 | 0.8992 | 0.9397 |
| 115 | 0.7419 | 0.7724 | 0.7930 | 0.8080 | 0.8194 | 0.8284 | 0.8357 | 0.8417 | 0.8511 | 0.8581 | 0.9063 |
| 120 | 0.6814 | 0.7153 | 0.7383 | 0.7551 | 0.7680 | 0.7781 | 0.7863 | 0.7931 | 0.8037 | 0.8116 | 0.8660 |
| 125 | 0.6202 | 0.6564 | 0.6811 | 0.6991 | 0.7130 | 0.7239 | 0.7328 | 0.7401 | 0.7516 | 0.7601 | 0.8192 |
| 130 | 0.5591 | 0.5963 | 0.6219 | 0.6406 | 0.6550 | 0.6664 | 0.6756 | 0.6833 | 0.6953 | 0.7042 | 0.7660 |
| 135 | 0.4986 | 0.5356 | 0.5612 | 0.5801 | 0.5946 | 0.6061 | 0.6154 | 0.6232 | 0.6353 | 0.6444 | 0.7071 |
| 140 | 0.4390 | 0.4748 | 0.4997 | 0.5181 | 0.5322 | 0.5435 | 0.5526 | 0.5602 | 0.5721 | 0.5810 | 0.6428 |
| 145 | 0.3805 | 0.4140 | 0.4375 | 0.4549 | 0.4683 | 0.4790 | 0.4877 | 0.4949 | 0.5062 | 0.5147 | 0.5736 |
| 150 | 0.3232 | 0.3536 | 0.3750 | 0.3909 | 0.4032 | 0.4130 | 0.4209 | 0.4276 | 0.4380 | 0.4458 | 0.5000 |
| 155 | 0.2672 | 0.2937 | 0.3124 | 0.3263 | 0.3371 | 0.3457 | 0.3528 | 0.3586 | 0.3678 | 0.3747 | 0.4226 |
| 160 | 0.2122 | 0.2342 | 0.2498 | 0.2614 | 0.2704 | 0.2776 | 0.2835 | 0.2884 | 0.2961 | 0.3018 | 0.3420 |
| 165 | 0.1583 | 0.1752 | 0.1872 | 0.1962 | 0.2032 | 0.2088 | 0.2133 | 0.2171 | 0.2231 | 0.2276 | 0.2588 |
| 170 | 0.1051 | 0.1165 | 0.1247 | 0.1309 | 0.1356 | 0.1394 | 0.1425 | 0.1451 | 0.1492 | 0.1523 | 0.1736 |
| 175 | 0.0524 | 0.0582 | 0.0623 | 0.0654 | 0.0679 | 0.0698 | 0.0714 | 0.0727 | 0.0748 | 0.0763 | 0.0872 |
| 180 | 0.0000 | 0.0000 | 0.0000 | 0.0000 | 0.0000 | 0.0000 | 0.0000 | 0.0000 | 0.0000 | 0.0000 | 0.0000 |

Tangential pressure on crank = resultant horizontal pressure times tabular quantity.   Forward stroke is toward crankshaft.   Wrist-pin velocity = crankpin velocity times tabular quantity. Calculated from

$$\sin a + r \sin 2a/2l \sqrt{1 - \frac{r^2}{l^2}\sin^2 a}$$

### Table 82. Inertia Factors and Piston Accelerations

(Values of $\cos a + r \cos 2a/ly + r^3 \sin^2 2a/4l^3 y^3$, where $y = \sqrt{1 - \frac{r^2}{l^2}\sin^2 a}$. Algebraic signs relate to forward stroke; use opposite signs for return stroke)

| Crank angles forward deg, $a$ | Ratio of connecting-rod length to crank length | | | | | | | | | | |
|---|---|---|---|---|---|---|---|---|---|---|---|
| | $\frac{l}{r} = 2.5$ | 3 | 3.5 | 4 | 4.5 | 5 | 5.5 | 6 | 7 | 8 | ∞ |
| 0 | 1.4000 | 1.3333 | 1.2857 | 1.2500 | 1.2222 | 1.2000 | 1.1818 | 1.1667 | 1.1429 | 1.1250 | 1.0000 |
| 5 | 1.3908 | 1.3249 | 1.2778 | 1.2426 | 1.2152 | 1.1932 | 1.1753 | 1.1604 | 1.1369 | 1.1193 | 0.9962 |
| 10 | 1.3635 | 1.2997 | 1.2543 | 1.2204 | 1.1941 | 1.1731 | 1.1559 | 1.1416 | 1.1192 | 1.1024 | 0.9848 |
| 15 | 1.3183 | 1.2580 | 1.2155 | 1.1839 | 1.1594 | 1.1399 | 1.1239 | 1.1107 | 1.0899 | 1.0744 | 0.9659 |
| 20 | 1.2558 | 1.2006 | 1.1621 | 1.1335 | 1.1116 | 1.0941 | 1.0799 | 1.0682 | 1.0496 | 1.0357 | 0.9397 |
| 25 | 1.1770 | 1.1283 | 1.0948 | 1.0702 | 1.0514 | 1.0365 | 1.0244 | 1.0144 | 0.9987 | 0.9871 | 0.9063 |
| 30 | 1.0829 | 1.0423 | 1.0149 | 0.9950 | 0.9779 | 0.9681 | 0.9585 | 0.9505 | 0.9382 | 0.9290 | 0.8660 |
| 35 | 0.9750 | 0.9439 | 0.9236 | 0.9091 | 0.8983 | 0.8898 | 0.8830 | 0.8775 | 0.8688 | 0.8624 | 0.8192 |
| 40 | 0.8551 | 0.8349 | 0.8225 | 0.8140 | 0.8078 | 0.8031 | 0.7993 | 0.7963 | 0.7917 | 0.7883 | 0.7660 |
| 45 | 0.7252 | 0.7172 | 0.7133 | 0.7112 | 0.7100 | 0.7092 | 0.7086 | 0.7083 | 0.7078 | 0.7076 | 0.7071 |
| 50 | 0.5878 | 0.5929 | 0.5980 | 0.6026 | 0.6064 | 0.6097 | 0.6124 | 0.6148 | 0.6186 | 0.6215 | 0.6428 |
| 55 | 0.4455 | 0.4643 | 0.4787 | 0.4899 | 0.4988 | 0.5061 | 0.5121 | 0.5171 | 0.5250 | 0.5310 | 0.5736 |
| 60 | 0.3013 | 0.3338 | 0.3574 | 0.3751 | 0.3889 | 0.4000 | 0.4091 | 0.4167 | 0.4286 | 0.4375 | 0.5000 |
| 65 | 0.1583 | 0.2041 | 0.2363 | 0.2601 | 0.2785 | 0.2931 | 0.3050 | 0.3149 | 0.3305 | 0.3420 | 0.4226 |
| 70 | 0.0197 | 0.0776 | 0.1175 | 0.1468 | 0.1692 | 0.1869 | 0.2013 | 0.2132 | 0.2319 | 0.2458 | 0.3420 |
| 75 | −0.1117 | −0.0434 | 0.0030 | 0.0368 | 0.0625 | 0.0828 | 0.0993 | 0.1129 | 0.1341 | 0.1499 | 0.2588 |
| 80 | −0.2329 | −0.1567 | −0.1054 | −0.0682 | −0.0400 | −0.0178 | 0.0002 | 0.0150 | 0.0381 | 0.0553 | 0.1736 |
| 85 | −0.3417 | −0.2605 | −0.2062 | −0.1669 | −0.1372 | −0.1138 | −0.0949 | −0.0793 | −0.0550 | −0.0369 | 0.0872 |
| 90 | −0.4364 | −0.3536 | −0.2981 | −0.2582 | −0.2279 | −0.2041 | −0.1849 | −0.1690 | −0.1443 | −0.1260 | 0.0000 |
| 95 | −0.5160 | −0.4348 | −0.3805 | −0.3412 | −0.3115 | −0.2881 | −0.2692 | −0.2536 | −0.2293 | −0.1212 | −0.0872 |
| 100 | −0.5802 | −0.5040 | −0.4527 | −0.4155 | −0.3873 | −0.3651 | −0.3471 | −0.3323 | −0.3092 | −0.2920 | −0.1736 |
| 105 | −0.6292 | −0.5610 | −0.5146 | −0.4809 | −0.4551 | −0.4348 | −0.4184 | −0.4048 | −0.3835 | −0.3677 | −0.2588 |
| 110 | −0.6644 | −0.6064 | −0.5665 | −0.5373 | −0.5149 | −0.4971 | −0.4827 | −0.4708 | −0.4521 | −0.4382 | −0.3420 |
| 115 | −0.6869 | −0.6411 | −0.6090 | −0.5851 | −0.5667 | −0.5521 | −0.5402 | −0.5303 | −0.5148 | −0.5032 | −0.4226 |
| 120 | −0.6987 | −0.6662 | −0.6426 | −0.6249 | −0.6111 | −0.6000 | −0.5909 | −0.5833 | −0.5714 | −0.5625 | −0.5000 |
| 125 | −0.7016 | −0.6829 | −0.6685 | −0.6573 | −0.6483 | −0.6411 | −0.6351 | −0.6301 | −0.6221 | −0.6161 | −0.5736 |
| 130 | −0.6978 | −0.6927 | −0.6875 | −0.6830 | −0.6792 | −0.6759 | −0.6732 | −0.6708 | −0.6670 | −0.6641 | −0.6428 |
| 135 | −0.6890 | −0.6970 | −0.7009 | −0.7030 | −0.7043 | −0.7050 | −0.7056 | −0.7059 | −0.7064 | −0.7066 | −0.7071 |
| 140 | −0.6770 | −0.6971 | −0.7096 | −0.7181 | −0.7243 | −0.7290 | −0.7328 | −0.7358 | −0.7404 | −0.7438 | −0.7660 |
| 145 | −0.6633 | −0.6944 | −0.7147 | −0.7292 | −0.7400 | −0.7485 | −0.7553 | −0.7609 | −0.7695 | −0.7759 | −0.8192 |
| 150 | −0.6491 | −0.6898 | −0.7172 | −0.7370 | −0.7521 | −0.7640 | −0.7736 | −0.7815 | −0.7939 | −0.8030 | −0.8660 |
| 155 | −0.6356 | −0.6843 | −0.7178 | −0.7424 | −0.7612 | −0.7761 | −0.7882 | −0.7982 | −0.8139 | −0.8256 | −0.9063 |
| 160 | −0.6236 | −0.6788 | −0.7173 | −0.7458 | −0.7678 | −0.7853 | −0.7995 | −0.8113 | −0.8298 | −0.8436 | −0.9397 |
| 165 | −0.6136 | −0.6738 | −0.7163 | −0.7480 | −0.7725 | −0.7920 | −0.8079 | −0.8212 | −0.8419 | −0.8575 | −0.9659 |
| 170 | −0.6061 | −0.6700 | −0.7153 | −0.7492 | −0.7755 | −0.7965 | −0.8137 | −0.8280 | −0.8504 | −0.8673 | −0.9848 |
| 175 | −0.6015 | −0.6675 | −0.7146 | −0.7498 | −0.7772 | −0.7991 | −0.8171 | −0.8320 | −0.8555 | −0.8731 | −0.9962 |
| 180 | −0.6000 | −0.6667 | −0.7143 | −0.7500 | −0.7778 | −0.8000 | −0.8182 | −0.8333 | −0.8571 | −0.8750 | −1.0000 |

wheel effect of the arms and other rotating masses. Some acceptable values of $k$ are: pumps, 0.03–0.05; machine shops, 0.025–0.03; looms and paper mills, 0.025; spinning mills, 0.015; crushers, 0.2; electric generators, ac, 0.003; dc, 0.002.

**Wittenbauer's Analysis for Flywheel Performance** The method does not involve more computation work than the one described above, but it is more accurate where the reciprocating parts are comparatively heavy. Wittenbauer's method avoids the inaccuracy resulting from the evaluation of the inertia forces on the reciprocating parts on the basis of the uniform nominal speed of rotation for the engine.

Let the crankpin velocity be represented by $v_r$ and the velocity of any moving masses ($m_1$, $m_2$, $m_3$, etc.) at any instant or phase be represented, respectively, by $v_1$, $v_2$, $v_3$, etc. The kinetic energy of the entire engine system of moving masses may then be expressed as

$$E = \tfrac{1}{2}(m_1 v_1^2 + m_2 v_2^2 + m_3 v_3^2 + \cdots) = \tfrac{1}{2}M_r v_r^2$$

or, the single reduced mass $M_r$ at the crankpin which possesses the equivalent kinetic energy is

$$M_r = [m_1(v_1/v_r)^2 + m_2(v_2/v_r)^2 + m_3(v_3/v_r)^2 + \cdots]$$

In an engine mechanism, sufficiently accurate values of $M_r$ can be obtained if the weight of the connecting rod is divided between the crankpin and the wrist pin so as to retain the center of gravity of the rod in its true position; usually 0.55 to 0.65 of the weight of the connecting rod should be placed on the crankpin, and 0.45 to 0.35 of the weight on the wrist pin. $M_r$ is a variable in engine mechanisms on account of the reciprocating parts and should be found for a number of crank positions. It should include all moving masses except the flywheel.

The total energy $E$ used in accelerating reciprocating parts from the beginning of the forward stroke up to any crank position can be obtained by finding from the indicator cards the total work done in the cylinder (on both sides of the piston) up to that time and subtracting from it the work done in overcoming the resisting torque, which may usually be assumed constant. The mean energy of the moving masses is $E_0 = \frac{1}{2} M_r v_r^2$.

In Fig. 123, the reduced weights of the moving masses $G_F + G_{r5}$ are plotted on the $X$ axis corresponding to different crank

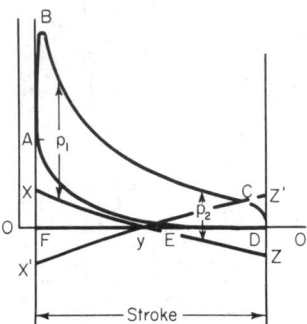

**Fig. 121**

positions. $G_F = gM_F$ is the reduced flywheel weight and $G_{r5} = gM_{r5}$ is the sum of the other reduced weights. Against each of these abscissas is plotted the energy $E$ available for acceleration measured from the beginning of the forward stroke. The curve 0123456 is the locus of these plotted points.

The diagram possesses the following property: Any straight line drawn from the origin $O$ to any point in the curve is a

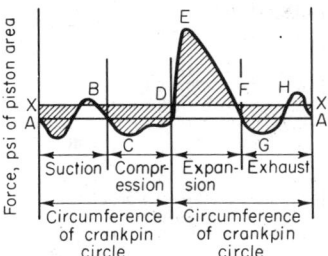

**Fig. 122**

measure of the velocity of the moving masses; tangents bounding the diagram measure the limits of velocity between which the crankpin will operate. The maximum linear velocity of the crankpin in feet per second is $v_2 = \sqrt{2g} \tan a_2$, and the minimum velocity is $v_1 = \sqrt{2g} \tan a_1$. Any desired change in $v_1$ and $v_2$ may be accomplished by changing the value of $G_F$, which means a change in the flywheel weight or a change in the flywheel weight reduced to the crankpin. As $G_F$ is very large compared with $G_r$ and the point 0 cannot be located on the diagram unless a very large drawing is made, the tangents are best formed by direct calculation:

$$\tan \alpha_2 = \frac{v_r^2}{2g}(1 + k) \qquad \tan \alpha_1 = \frac{v_r^2}{2g}(1 - k)$$

where $k$ is the coefficient of velocity fluctuation. The two tangents $ss$ and $tt$ to the curve 0123456, thus drawn, cut a distance $\Delta E$ and on the ordinate $E_0$. The reduced flywheel weight is then found to be

$$G_F(\Delta E)g/v_r^2 k$$

## SPRINGS

It is assumed in the following formulas that the springs are in no case stressed beyond the elastic limit (i.e., that they are perfectly elastic) and that they are subject to Hooke's law.

### Notation

$P$ = safe load, lb
$f$ = deflection for a given load $P$, in
$l$ = length of spring, in
$V$ = volume of spring, $in^3$
$S_s$ = safe stress (due to bending), $lb/in^2$
$S_v$ = safe shearing stress, $lb/in^2$
$U$ = resilence, in·lb

The **work** in inch-pounds performed in **deflecting a spring** from O to $f$ (spring duty) is $U = Pf/2 = s_s^2 V/CE$. This is based upon the assumption that the deflection is proportional to the load. $C$ is a constant dependent upon the shape of the springs.

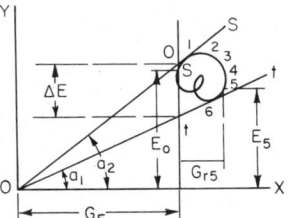

**Fig. 123**

The **time of vibration** $T$ (in seconds)) **of a spring** (weight not considered) is equal to that of a simple circular pendulum whose length $l_0$ equals the deflection $f$ (in ft) that is produced

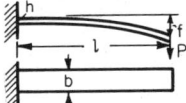

**Fig. 124**

in the spring by the load $P$. $T = \pi \sqrt{l_0/g}$, where $g$ = acceleration of gravity, $ft/s^2$.

### Springs Subjected to Bending

1. **Rectangular Plate Spring** (Fig. 124)

$$P = bh^2 S_s/6l \qquad I = bh^3/12 \qquad U = Pf/2 = VS_s^2/18E$$
$$f = Pl^3/3EI = 4Pl^3/bh^3E = 2l^2 S_s/3hE$$

2. **Triangular Plate Spring** (Fig. 125). The elastic curve is a circular arc.

$$P = bh^2 S_s/6l \qquad I = bh^3/12 \qquad U = Pf/2 = S_s^2 V/6E$$
$$f = Pl^3/2EI = 6Pl^3/bh^3E = l^2 S_s/hE$$

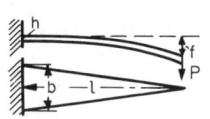

**Fig. 125**

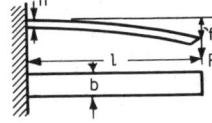

**Fig. 126**

**3. Rectangular plate spring with end tapered** in the form of a cubical parabola (Fig. 126). The elastic curve is a circular arc. $P$, $l$, and $f$ same as for triangular plate spring (Fig. 125). $U = S_s^2 V/9E$.

The strength and deflection of **single-leaf flat springs** of various forms are given (Bruce, *Am. Mach.*, July 19, 1900) by the formulas $h = al^2/f$ and $b = cPl/h^2$. The volume of the spring is given by $V = vlbh$. The values of the constants $a$ and $c$ and the resilience in inch-pounds per cubic inch are given in Table 83, in terms of the safe stress $S_s$. Values of $v$ are given also.

**4. Compound (Leaf or Laminated) Springs.** If several springs of rectangular section are combined, the resulting compound spring should (1) form a beam of uniform strength that (2) does not open between the joints while bending (i.e., elastic curve must be a circular arc). Only the type immediately following meets both requirements, the others meeting only the second requirement.

**5. Laminated Triangular Plate Spring** (Fig. 127). If the triangular plate spring shown at I is cut into an even number ($= 2n$) of strips of equal width (in this case eight strips of width $b/2$), and these strips are combined, a laminated spring will be formed whose carrying capacity will equal that of the original uncut spring; or $P = nbh^2 S_s/6l$; $n = 6Pl/bh^2S_s$.

**6. Laminated rectangular plate spring with leaf ends tapered** in the form of a cubical parabola (Fig. 128); see Case 3.

**7. Laminated Trapezoidal Plate Spring with Leaf Ends Tapered** (Fig. 129). The ends of the leaves are trapezoidal in form and are tapered according to the formula

$$z = \frac{h}{\sqrt[3]{1 + \frac{b_1}{b}\left(\frac{a}{x} - 1\right)}}$$

**8. Semielliptic Springs** (for locomotives, trucks, etc.). Referring to Fig. 130, the load $2P$ (lb) acting on the spring center band produces a tensional stress $P/\cos a$ in each of the inclined shackle links. This is resolved into the vertical force $P$ and the horizontal force $P \tan a$, which together produce a bending moment $M = P(l + p \tan a)$. The equations given in (1), (2), and (3) apply to curved as well as straight springs. The bearing force $= 2P = (2nbh^2/6)[S_s/(l + p \tan a)]$, and the deflection $= (6l^2/nbh^3)P(l + p \tan a)/E = l^2 S_s/hE$.

In addition to the bending moment, the leaves are subjected to the tension force $P \tan a$ and the transverse force $P$, which

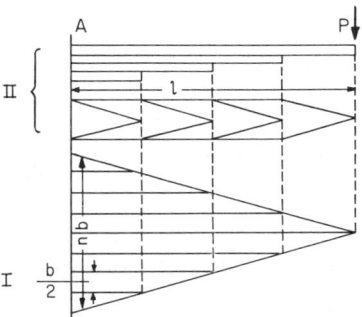

**Fig. 127**

produce in the upper leaf an additional stress $S = P \tan a/bh$, as well as a transverse shearing stress.

In determining the number of leaves $n$ in a given spring, allowance should be made for an excess load on the spring caused by the vibration. This is usually made by decreasing the allowable stress about 15 percent.

The foregoing does not take account of initial stresses caused by the band. For more detailed information, see Wahl, "Mechanical Springs," Penton.

**9. Elliptic Springs.** Safe load $P = nbh^2 S_s/6l$, where $l = \frac{1}{2}$ distance between bolt eyes (less $\frac{1}{2}$ length of center band, where used); deflection $f = 4l^2 S_s K/hE$, where

**Table 83. Strength and Deflection of Single-Leaf Flat Springs**

| Load applied at end of spring; $c = 6/S_s$ | | | | Load applied at center of spring; $c = 6/4S_s$ | | | |
|---|---|---|---|---|---|---|---|
| Plans and elevations of springs | a | U | v | Plans and elevations of springs | a | U | v |
| | $\dfrac{S_s}{E}$ | $\dfrac{S_s^2}{6E}$ | $\dfrac{1}{2}$ | | $\dfrac{S_s}{4E}$ | $\dfrac{S_s^2}{6E}$ | $\dfrac{1}{2}$ |
| Parabolic arc | $\dfrac{4S_s}{3E}$ | $\dfrac{S_s^2}{6E}$ | $\dfrac{2}{3}$ | | $\dfrac{0.87S_s}{4E}$ | $\dfrac{0.70S_s^2}{6E}$ | $\dfrac{5}{8}$ |
| | $\dfrac{2S_s}{3E}$ | $\dfrac{0.33S_s^2}{6E}$ | $1$ | Parabolic arcs | $\dfrac{S_s}{3E}$ | $\dfrac{S_s^2}{6E}$ | $\dfrac{2}{3}$ |
| | $\dfrac{0.87S_s}{E}$ | $\dfrac{0.70S_s^2}{6E}$ | $\dfrac{5}{8}$ | | $\dfrac{1.09S_s}{4E}$ | $\dfrac{0.725S_s^2}{6E}$ | $\dfrac{3}{4}$ |
| | $\dfrac{1.09S_s}{E}$ | $\dfrac{0.725S_s^2}{6E}$ | $\dfrac{3}{4}$ | | $\dfrac{3S_s}{6E}$ | $\dfrac{0.33S_s^2}{6E}$ | $1$ |

$$K = \frac{1}{(1-r)^3}\left[\frac{1-r^2}{2} - 2r(1-r) - r^2 \ln r\right]$$

$r$ being the number of full-length leaves $\div$ total number $(n)$ of leaves in the spring. All dimensions in inches. For semielliptic springs, the deflection is only half as great. Safe load $= nbh^2S_s/3l$. (Peddle, *Am, Mach.*, Apr. 17, 1913.)

**Coiled Springs** In these, the load is applied as a couple $Pr$ which turns the spring while winding or holds it in place when

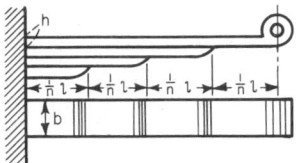

**Fig. 128**

wound up. If the spindle is not to be subjected to bending moment, $P$ must be replaced by two equal and opposite forces $(P/2)$ acting at the circumference of a circle of radius $r$. The formulas are the same in both cases. The springs are assumed to be fixed at one end and free at the other. The bending

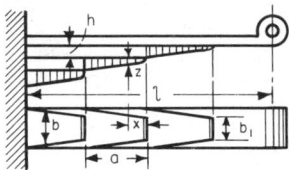

**Fig. 129**

moment acting on the section of least resistance is always $Pr$. The length of the straightened spring $= l$. See Benjamin and French, Experiments on Helical Springs, *Trans. ASME*, **23**, p. 298.

For **heavy closely coiled helical springs** the usual formulas are inaccurate and result in stresses greatly in excess of those

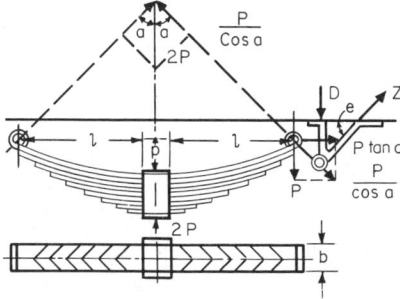

**Fig. 130**

assumed. See Wahl, Stresses in Heavy Closely-coiled Helical Springs, *Trans. ASME*, 1929. In springs 10 to 12 and 15 to 18, the quantity $k$ is unity for lighter springs and has the stated values (supplied by Wahl) for heavy closely coiled springs.

**10. Spiral Coiled Springs of Rectangular Cross Section** (Fig. 131)

$$P = bh^2S_s/6rk \qquad I = bh^3/12 \qquad U = Pf/2 = S_s^2V/6Ek^2$$
$$f = ra = Plr^2/EI = 12Plr^2/Ebh^3 = 2rlS_s/hEk$$

For heavy closely coiled springs $k = (3c - 1)/(3c - 3)$, where $c = 2R/h$ and $R$ is the minimum radius of curvature at the center of the spiral.

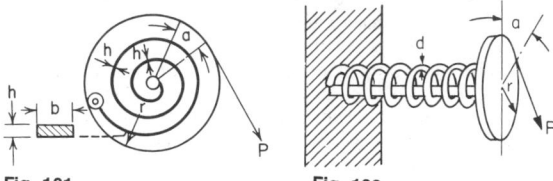

**Fig. 131**                    **Fig. 132**

**11. Cylindrical Helical Spring of Circular Cross Section** (Fig. 132)

$$P = \pi d^3S_s/32rk \qquad I = \pi d^4/64 \qquad U = Pf/2 = S_s^2V/8Ek^2$$
$$f = ra = Plr^2/EI = 64Plr^2/\pi Ed^4 = 2rlS_s/dEk$$

For heavy closely coiled springs, $k = (4c - 1)/(4c - 4)$ where $c = 2r/d$.

**12. Cylindrical Helical Spring of Rectangular Cross Section** (Fig. 133)

$$P = bh^2S_s/6rk \qquad I = bh^3/12 \qquad U = Pf/2 = S_s^2V/8Ek^2$$
$$f = ra = Plr^2/EI = 12Plr^2/Ebh^3 = 2rlS_s/hEk$$

For heavy closely coiled springs, $k = (3c - 1)/(3c - 3)$, where $c = 2r/h$.

**Springs Subjected to Torsion**

The statements made concerning coiled springs subjected to bending apply also to springs 13 and 14.

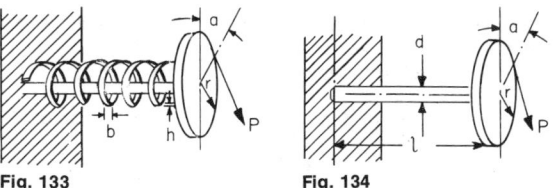

**Fig. 133**                    **Fig. 134**

**13. Straight Bar Spring of Circular Cross Section** (Fig. 134)

$$P = \pi d^3S_v/16r = 0.1963d^3S_v/r \qquad U = Pf/2 = S_v^2V/4G$$
$$f = ra = 32r^2lP/\pi d^4G = 2rlS_v/dG$$

**14. Straight Bar Spring of Rectangular Cross Section** (Fig. 135)

$$P = 2b^2hS_v/9r \qquad K = b/h$$
$$U = Pf/2 = 4S_v^2V(K^2 + 1)/45G \text{ (max when } K = 1)$$
$$f = ra = 3.6r^2lP(b^2 + h^2)/b^3h^3G = 0.8rlS_v(b^2 + h^2)/bh^2G$$

**Springs Loaded Axially Either in Tension or Compression**

NOTE. For springs 15 to 18, $r = $ mean radius of coil; $n = $ number of coils.

**15. Cylindrical Helical Spring of Circular Cross Section** (Fig. 136)

$$P = \pi d^3S_v/16rk = 0.1963d^3S_v/rk \qquad U = Pf/2 = S_v^2V/4Gk^2$$
$$f = 64nr^3P/d^4G = 4\pi nr^2S_v/dGk$$

For heavy closely coiled springs, $k = [(4c - 1)/(4c - 4)] + 0.615/c$, where $c = 2r/d$.

**16. Cylindrical Helical Spring of Rectangular Cross Section** (Fig. 137)

$$P = 2b^2hS_v/9rk; \quad K = b/h$$
$$U = Pf/2 = 4S_v^2V(K^2 + 1)/45Gk^2 \text{ (max when } K = 1)$$
$$f = 7.2\pi nr^3P(b^2 + h^2)/b^3h^3G = 1.6\pi nr^2S_v(b^2 + h^2)/bh^2Gk$$

For heavy closely coiled springs, $k = [(4c - 1)/(4c - 4)] + 0.615/c$, where $c = 2r/b$.

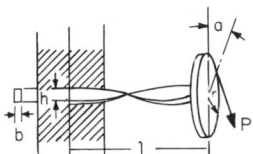

**Fig. 135**

**17. Conical Helical Spring of Circular Cross Section** (Fig. 138)

$l$ = length of developed spring, $d$ = diameter of wire, $r$ = maximum mean radius of coil

$$P = \pi d^3S_v/16rk = 0.1963d^3S_v/rk \qquad U = Pf/2 = S_v^2V/8Gk^2$$
$$f = 16r^2lP/\pi d^4G = 16nr^3P/d^4G = rlS_v/dGk = \pi nr^2S_v/dGk$$

For heavy closely coiled springs, $k = [(4c - 1)/(4c - 4)] + 0.615/c$, where $c = 2r/d$.

**18. Conical Helical Spring of Rectangular Cross Section** (Fig. 139)

$b$ = small dimension of section; $d$ = large dimension of section; $r$ = maximum mean radius of coil

$$K = b/h \ (\leq 1); \quad P = 2b^2hS_v/9rk$$
$$U = Pf/2 = 2S_v^2V(K^2 + 1)/45Gk^2 \text{ (max when } K = 1)$$
$$f = 1.8r^2lP(b^2 + h^2)/b^3h^3G = 1.8\pi nr^3P(b^2 + h^2)/b^3h^3G$$
$$= 0.4rlS_v(b^2 + h^2)/bh^2Gk = 0.4\pi nr^2S_v(b^2 + h^2)/bh^2Gk$$

For heavy closely coiled springs, $k = [(4c - 1)/(4c - 4)] + 0.615c$, where $c = 2r/r_o - r_i$.

**19. Truncated Conical Springs** (17 and 18). The formulas under 17 and 18 apply for truncated springs. In calculating

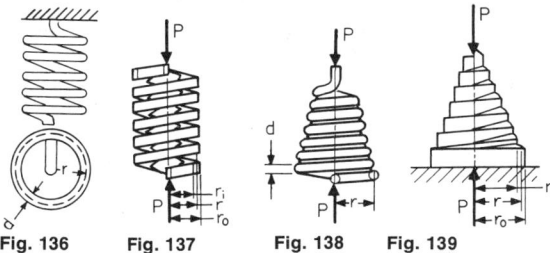

**Fig. 136       Fig. 137       Fig. 138       Fig. 139**

deflection $f$, however, it is necessary to substitute $(r_1^2 + r_2^2)$ for $r^2$, and $\pi n(r_1 + r_2)$ for $\pi nr$, $r_1$ and $r_2$ being, respectively, the greatest and least mean radii of the coils.

NOTE.    The preceding formulas for various forms of coiled springs are sufficiently accurate when the cross-section dimensions are small in comparison with the radius of the coil, and for small pitch. Springs 15

to 19 are for either tension or compression but formulas for springs 17 and 18 are good for compression only until the largest coil flattens out; then $r$ becomes a variable, depending on the load.

**Allowable Working Stresses for Springs**    Allowable working stresses depend upon the type of spring, the atmosphere, temperature, ratio of spring diameter to wire diameter, rapidity, and regularity of deflections, shock loads, wire size, and stress range. The values given in Table 84 are for **compression springs** not subjected to shock loads.

The fatigue life of a spring is considered "infinite" if it will withstand 10,000,000 cycles of deflections; "long" if between 1,000,000 and 10,000,000; "average" if between 100,000 and 1,000,000; "short" if between 10,000 and 100,000. **Severe service** comprises those applications in which springs are subjected to rapid deflections over long periods of time, such as in automobile valves, engine valves, or pneumatic hammers; **average service,** if subjected to normal operating deflections such as in brakes, motors, machine tools, or switches; **light service** if the springs are statically loaded or are subjected to less than 10,000 slow intermittent deflections during their lifetime, as in safety devices, emergency equipment, and springs used infrequently.

In Table 84, $E$ is used for springs subjected to tensile stresses due to bending, such as torsion, flat, cantilever, clock, and power springs. $G$ is used for springs subjected to torsional stresses, such as compression and extension helical springs.

**Safe working loads and deflections of cylindrical helical springs** of round steel wire in tension or compression are given in Table 85. The table is based on the formulas given for spring 15. $d$ = diameter of steel wire, in; $D$ = pitch diameter (center to center of wire), in; $P$ = safe working load for given unit stress, lb; $f$ = deflection of 1 coil for safe working load, in.

The table is based on the values of unit stress indicated, and $G = 12,500,000$. For any other value of unit stress, divide the tabular value by the unit stress used in the table and multiply by the unit stress to be used in the design. For any other value of $G$, multiply the value of $f$ in the table by 12,500,000 and divide by the value of $G$ chosen. For **square steel wire,** multiply values of $P$ by 1.06, and values of $f$ by 0.75. For **round brass wire,** take $S_s = 10,000$ to 20,000, and multiply values of $f$ by 2 (Howe).

EXAMPLES OF USE OF TABLE 85.    1. Required the safe load $(P)$ for a spring of $3/8$-in round steel with a pitch diameter $(D)$ of $3\frac{1}{2}$ in. In the line headed $D$, under $3\frac{1}{2}$, is given the value of $P$, or 678 lb. This is for a unit stress of 115,000 lb/in². The load $P$ for any other unit stress may be found by dividing the 678 by 115,000 and multiplying by the unit stress to be used in the design. To determine the number of coils this spring would need to compress (say) 6 in under a load of (say) 678 lb, take the value of $f$ under 678, or 0.938, which is the deflection of one coil under the given load. Therefore, 6/0.938 = 6.4, say 7, equals the number of coils required. The spring will therefore be $2\frac{5}{8}$ in long when closed ($7 \times 3/8$), counting the working coils only, and must be $8\frac{5}{8}$ in long when unloaded. Whether there is an extra coil at one end which does not deflect will depend upon the details of the particular design. The deflection in the above example is for a unit stress of 115,000 lb/in². The rule is, divide the deflection by 115,000 and multiply by the unit stress to be used in the design.

2. A $7/16$-in steel spring of $3\frac{1}{2}$ in OD has its coils in close contact. How much can it be extended without exceeding the limit of safety? The maximum safe load for this spring is found to be 1,074 lb and the deflection of one coil under this load is 0.810 in. This is for a unit stress

**Table 84. Physical Properties of Commonly Used Spring Materials**

(From *Handbook of Mechanical Spring Design*, Associated Spring Corp., by permission.)

| Material | Analysis | Tensile properties | | | | Torsional properties of wire | | | Process of manufacture, chief uses, special properties |
|---|---|---|---|---|---|---|---|---|---|
| | | Ultimate strength, psi | Elastic limit, psi | Modulus of elasticity ($E$) | Rockwell hardness | Ultimate strength, psi | Elastic limit, psi | Modulus in torsion ($G$) | |
| **Flat cold-rolled spring steel** | | | | | | | | | |
| Clock spring steel AS 100 SAE 1095 | C 0.90–1.05 %<br>Mn 0.30–0.50 % | 180,000–340,000 | 150,000–310,000 | 30,000,000 | C40–52 | Not used | Not used | Not used | Cold-rolled and heat-treated before forming. Clock and motor springs, miscellaneous flat springs |
| Flat spring steel AS 101 SAE 1074 | C 0.70–0.80 %<br>Mn 0.50–0.80 % | 160,000–320,000 | 125,000–280,000 | 30,000,000 | Annealed B70–85 temp'd C38–50 | Not used | Not used | Not used | Cold-rolled, annealed, or tempered. Miscellaneous flat springs. Most popular spring steel |
| Flat spring steel AS 102 SAE 1060 | C 0.50–0.65 %<br>Mn 0.60–0.90 %<br>P and S 0.04 % max | 160,000–280,000 | 120,000–180,000 | 30,000,000 | Annealed B70–85 temp'd C38–50 | Not used | Not used | 11,500,000 | Use cold-rolled and annealed. Miscellaneous flat springs, static loads |
| **Carbon-steel wires** | | | | | | | | | |
| High-carbon wire AS 8 | C 0.85–0.95 %<br>Mn 0.25–0.60 % | 200,000–250,000 | 160,000–210,000 | 30,000,000 | C44–48 | 160,000–200,000 | 110,000–150,000 | 11,500,000 | Cold-rolled or drawn. High-grade helical springs or wire forms |
| Oil-tempered wire AS 10 ASTM A229–41 | C 0.60–0.70 %<br>Mn 0.60–0.90 % | 155,000–300,000 | 120,000–250,000 | 30,000,000 | C42–46 | 115,000–200,000 | 80,000–130,000 | 11,500,000 | Cold-drawn and heat-treated before coiling. General spring use |
| Music wire AS 5 ASTM A228–47 | C 0.70–1.00 %<br>Mn 0.30–0.60 % | 250,000–500,000 | 150,000–350,000 | 30,000,000 | | 150,000–300,000 | 90,000–180,000 | 11,500,000–12,000,000, depending on size | Patented and cold-drawn. Miscellaneous small springs of various types—high quality |
| Hard-drawn spring wire AS 20 ASTM A227–47 | C 0.60–0.70 %<br>Mn 0.90–1.20 % | 150,000–300,000 | 100,000–200,000 | 30,000,000 | | 120,000–220,000 | 75,000–130,000 | 11,500,000 | Patented and cold-drawn. Same uses as music wire but lower-quality wire |
| **Hot-rolled alloy steel** | | | | | | | | | |
| Silico-manganese alloy steel AS 70 SAE 9260 | C 0.55–0.65 %<br>Mn 0.60–0.90 %<br>Si 1.80–2.20 % | 200,000–250,000 | 180,000–230,000 | 30,000,000 | C42–52 | 140,000–175,000 | 100,000–130,000 | 11,500,000 | Hot- or cold-rolled or drawn. Better heat resistance than Cr. Va |

Alloy and stainless spring materials

| Material | Composition | | | | | | | | Remarks |
|---|---|---|---|---|---|---|---|---|---|
| Chrome-vanadium alloy steel AS 32 SAE 6150 | C 0.48–0.53 %, Mn 0.70–0.90 %, P 0.04 max, S 0.04 max, Si 0.20–0.35 %, Cr 0.80–1.10 %, V 0.15 min, Subject to standard tolerances | 200,000–250,000 | 180,000–230,000 | 30,000,000 | C42–48 | 140,000–175,000 | 100,000–130,000 | 11,500,000 | Cold-rolled or drawn. Special applications |
| Chrome-silicon alloy steel AS 33 SAE 9254 | C 0.50–0.60 %, Mn 0.50–0.80 %, Si 1.20–1.60 %, Cr 0.50–0.80 % | 250,000–325,000 | 220,000–300,000 | 30,000,000 | C47–51 | 160,000–200,000 | 130,000–160,000 | 11,500,000 | Hot- or cold-rolled or drawn. Used at high stresses. Resists heat well to 450 F |
| 18-8 type stainless AS 35 SAE 30302 | C 17–20 %, Ni 6–10 %, C 0.08–0.15 %, Mn 2 % max, Si 0.30–0.75 % | 160,000–330,000 | 60,000–260,000 | 28,000,000 | C35–45 | 120,000–240,000 | 45,000–140,000 | 10,000,000 | Cold-rolled or drawn. Best corrosion resistance. Fair temperature resistance |
| Type 316 stainless SAE 30316 | C 16–18 %, Ni 10–14 %, Mo 2–3 %, Mn 2 % max, C 0.08 % max, Si 1 % max, P 0.04 % max, S 0.03 % max | 170,000–250,000 | 130,000–200,000 | 28,000,000 | C35–45 | 120,000–220,000 | 80,000–130,000 | 11,000,000 | Cold-rolled or drawn. Heat-treated after forming. Resists corrosion when polished. Good temperature resistance |

Nonferrous spring materials

| Material | Composition | | | | | | | | Remarks |
|---|---|---|---|---|---|---|---|---|---|
| Spring brass AS 55 AS 155 | Cu 64–72 %, Zn remainder | 100,000–130,000 | 40,000–60,000 | 15,000,000 | B90 | 45,000–90,000 | 30,000–60,000 | 5,500,000 | Cold-rolled or drawn. For electrical conductivity at low stresses. For corrosion resistance |
| Nickel silver | Cu 56 %, Zn 25 %, Ni 18 % | 135,000–150,000 | 80,000–110,000 | 16,000,000 | B95–100 | 85,000–100,000 | 60,000–70,000 | 5,500,000 | Cold-rolled or drawn. Better quality than brass. Also used for its color. Corrosion resistant |
| Phosphor-bronze AS 60 AS 160 | Cu 91–93 %, Sn 7–9 %  or  Cu 94–96 %, Sn 4–6 % | 100,000–150,000 | 60,000–110,000 | 15,000,000 | B90–100 | 80,000–105,000 | 50,000–85,000 | 6,250,000 | Cold-rolled or drawn. Used for corrosion resistance and electrical conductivity |
| Silicon bronze AS 46 AS 146 (Made under various trade names) | Si 2–3 %, small amounts of tin or mn, balance copper | 100,000–150,000 | 60,000–110,000 | 15,000,000 | B90–100 | 80,000–105,000 | 50,000–85,000 | 6,250,000 | Cold-rolled or drawn. Used as substitute for phosphor-bronze where lower cost is necessary |
| Monel AS 40 AS 140 | Ni (+Co) 63.0–70.0, Cu remainder, Fe 2.50 max, C 0.30 max, Mg 2.00 max, Si 0.50 max, S 0.024 max | 120,000–165,000 | 85,000–125,000 | 26,000,000 | C23–32 | 85,000–110,500 | 50,000–70,000 | 9,500,000 | Cold-rolled or drawn. Resists corrosion. Moderate stresses to 400 F |

**Table 84. Physical Properties of Commonly Used Spring Materials** (Continued)

| Material | Analysis | Tensile properties | | | Rockwell hardness | Torsional properties of wire | | | Process of manufacture, chief uses, special properties |
|---|---|---|---|---|---|---|---|---|---|
| | | Ultimate strength, psi | Elastic limit, psi | Modulus of elasticity (E) | | Ultimate strength, psi | Elastic limit, psi | Modulus in torsion (G) | |
| "K" monel AS 40 AS 140 | Ni (+Co) 63.0–70.0  Mg 1.50 max  Cu remainder  Si 1.00 max  Fe 2.00 max  S 0.01 max  Al 2.0–4.0  Ti 0.25–1.00  C 0.25 max | 120,000–180,000 | 85,000–140,000 | 26,000,000 | C23–35 | 85,000–130,000 | 50,000–75,000 | 9,500,000 | Same as monel except higher operational stresses can be employed to 450 F. Precipitation hardened by thermal treatment |
| Inconel AS 40 AS 140 | Ni (+Co) 72.00 min  Mg 1.00 max  Cu 0.50 max  Si 0.50 max  Fe 6.0–10.0  S 0.015 max  Cr 14.0–17.0  C 0.15 max | 140,000–185,000 | 110,000–140,000 | 31,000,000 | C25–37 | 95,000–130,000 | 55,000–80,000 | 11,000,000 | Cold-rolled or drawn. Resists corrosion. High stresses to 650 F. |
| Inconel "X" AS 40 AS 140 | Ni (+Co) 70.00 min  Mg 1.00 max  Cu 0.50 max  Si 0.50 max  Fe 5.0–9.0  S 0.01 max  Cr 14.0–17.0  Ti 2.00–2.50  Al 0.40–1.0  C 0.08 max  Cb (+Ti) 0.70–1.20 | 130,000–220,000 | 90,000–150,000 | 31,000,000 | C24–46 | 90,000–155,000 | 50,000–90,000 | 12,000,000 | Resists corrosion and oxidation. Can be used to 1000 F for prolonged periods of service; up to 1200 F for short periods of intermittent temperature exposure |
| Duranickel AS 40 AS 140 | Ni (+Co) 93.00 min  Mg 0.50 max  Cu 0.25 max  Si 1.00 max  Fe 0.60 max  S 0.01 max  Al 4.00–4.75  Ti 0.25–1.00  C 0.30 max | 125,000–205,000 | 80,000–140,000 | 30,000,000 | C25–43 | 85,000–145,000 | 50,000–85,000 | 11,000,000 | Cold-rolled or drawn. Precipitation hardened by heat-treatment. Resists corrosion. High stresses to 600 F. |
| Beryllium copper AS 45 AS 145 | Cu 98%  Be 2% | 160,000–200,000 | 100,000–150,000 | 16,000,000–18,500,000 subject to heat-treatment | C35–42 | 100,000–130,000 | 65,000–95,000 | 6,000,000–7,000,000 subject to heat-treatment | Cold-rolled or drawn Corrosion resistance like copper. High physicals for electrical work. Low hysteresis |

**Table 85. Safe Working Loads and Deflections of Cylindrical Helical Steel Springs of Circular Cross Section**

(For closely coiled springs divide given load and deflection values by the curvature factor *k*)

| Allowable unit stress, psi | Wire gage W. & M. | Diam, in. | D | 5/32 | 3/16 | 1/4 | 5/16 | 3/8 | 7/16 | 1/2 | 5/8 | 3/4 | 7/8 | 1 | 1 1/8 | 1 1/4 | 1 3/8 | 1 1/2 | 1 5/8 | 1 3/4 | 1 7/8 | 2 | 2 1/4 |
|---|---|---|---|---|---|---|---|---|---|---|---|---|---|---|---|---|---|---|---|---|---|---|---|
| 150,000 | 20 | .035 | P | 16.2 | 13.4 | 10.0 | 8.10 | 6.66 | 5.75 | 4.96 | 4.05 | 3.39 | | | | | | | | | | | |
| | | | f | .026 | .037 | .067 | .105 | .149 | .200 | .276 | .420 | .608 | | | | | | | | | | | |
| | 19 | .041 | P | 26.2 | 21.6 | 16.2 | 13.0 | 10.8 | 9.27 | 8.10 | 6.52 | 5.35 | 4.57 | | | | | | | | | | |
| | | | f | .023 | .032 | .057 | .089 | .128 | .175 | .229 | .362 | .512 | .697 | | | | | | | | | | |
| | 18 | .047 | P | 39.1 | 32.6 | 24.5 | 19.6 | 16.4 | 13.9 | 12.3 | 9.80 | 8.10 | 6.92 | 6.14 | | | | | | | | | |
| | | | f | .019 | .028 | .049 | .078 | .112 | .153 | .200 | .311 | .449 | .610 | .800 | | | | | | | | | |
| | 17 | .054 | P | 59.4 | 49.6 | 37.2 | 29.7 | 24.6 | 21.2 | 18.5 | 14.7 | 12.4 | 10.5 | 9.25 | 8.23 | | | | | | | | |
| | | | f | .016 | .024 | .043 | .067 | .098 | .133 | .174 | .273 | .390 | .532 | .695 | .878 | | | | | | | | |
| | 1/16″ | .062 | P | | 74.9 | 56.1 | 44.9 | 37.3 | 32.0 | 28.0 | 22.4 | 18.6 | 16.1 | 13.9 | 12.5 | 11.2 | | | | | | | |
| | | | f | | .021 | .037 | .058 | .084 | .115 | .151 | .235 | .340 | .460 | .605 | .760 | .947 | | | | | | | |
| | 16 | .063 | P | | 78.2 | 58.7 | 46.9 | 39.2 | 33.9 | 29.4 | 23.5 | 19.6 | 16.8 | 14.7 | 13.2 | 11.9 | 10.7 | | | | | | |
| | | | f | | .020 | .037 | .057 | .083 | .113 | .148 | .233 | .335 | .445 | .591 | .748 | .930 | 1.12 | | | | | | |
| | 15 | .072 | P | | | 80.7 | 70.0 | 58.7 | 50.2 | 43.6 | 35.2 | 29.0 | 25.0 | 21.9 | 19.5 | 17.5 | 16.0 | | | | | | |
| | | | f | | | .032 | .050 | .077 | .100 | .130 | .203 | .294 | .405 | .521 | .652 | .802 | .986 | | | | | | |
| | 14 | .080 | P | | | 121 | 96.6 | 80.5 | 69.1 | 60.4 | 48.2 | 40.1 | 34.6 | 30.1 | 26.7 | 24.2 | 22.1 | 20.2 | | | | | |
| | | | f | | | .029 | .045 | .065 | .090 | .117 | .183 | .262 | .359 | .470 | .593 | .735 | .886 | 1.105 | | | | | |
| 140,000 | 13 | .092 | P | | | 171 | 136 | 113 | 97.6 | 85.5 | 68.9 | 57.3 | 48.8 | 42.6 | 37.8 | 34.5 | 31.3 | 28.6 | 27.3 | | | | |
| | | | f | | | .023 | .037 | .053 | .072 | .098 | .148 | .214 | .291 | .388 | .481 | .593 | .720 | .854 | .986 | | | | |
| | 3/32″ | .093 | P | | | 178 | 142 | 118 | 99.5 | 89.1 | 71.2 | 59.1 | 50.9 | 44.3 | 39.6 | 35.7 | 32.3 | 29.6 | | | | | |
| | | | f | | | .023 | .036 | .052 | .071 | .093 | .146 | .211 | .286 | .376 | .473 | .585 | .707 | .841 | | | | | |
| | 12 | .105 | P | | | | 204 | 170 | 147 | 127 | 102 | 85.4 | 73.0 | 63.4 | 56.6 | 51.1 | 46.3 | 42.6 | 38.9 | | | | |
| | | | f | | | | .032 | .047 | .064 | .083 | .122 | .188 | .256 | .336 | .425 | .512 | .632 | .755 | .880 | | | | |
| | 11 | .120 | P | | | | 303 | 253 | 217 | 190 | 152 | 126 | 108 | 95.2 | 84.2 | 76.2 | 69.2 | 63.5 | 58.5 | 54.3 | | | |
| | | | f | | | | .028 | .041 | .055 | .073 | .114 | .174 | .223 | .296 | .368 | .449 | .551 | .657 | .768 | .893 | | | |
| | 1/8″ | .125 | P | | | | | 286 | 245 | 214 | 171 | 143 | 121 | 107 | 95.5 | 85.2 | 78.0 | 71.5 | 65.8 | 60.8 | 57.2 | | |
| | | | f | | | | | .039 | .053 | .069 | .109 | .169 | .213 | .278 | .333 | .437 | .528 | .626 | .731 | .855 | .981 | | |
| | 10 | .135 | P | | | | | 359 | 309 | 270 | 217 | 171 | 154 | 135 | 120 | 108 | 98.7 | 90.2 | 82.7 | 72.2 | 71.8 | 67.5 | |
| | | | f | | | | | .036 | .049 | .064 | .106 | .145 | .198 | .260 | .327 | .399 | .486 | .581 | .680 | .791 | .908 | 1.04 | |
| | 9 | .148 | P | | | | | | 408 | 356 | 285 | 237 | 207 | 178 | 158 | 142 | 130 | 118 | 109 | 102 | 95.0 | 89.0 | |
| | | | f | | | | | | .045 | .059 | .092 | .132 | .180 | .236 | .293 | .370 | .448 | .530 | .620 | .723 | .828 | .945 | |
| | 5/32″ | .156 | P | | | | | | 480 | 418 | 330 | 270 | 234 | 208 | 185 | 167 | 152 | 139 | 128 | 119 | 111 | 104 | 92.7 |
| | | | f | | | | | | .042 | .056 | .087 | .125 | .171 | .223 | .282 | .350 | .422 | .509 | .588 | .685 | .785 | .896 | 1.12 |
| | 8 | .162 | P | | | | | | | 468 | 376 | 311 | 276 | 234 | 207 | 187 | 170 | 156 | 143 | 134 | 125 | 117 | 103 |
| | | | f | | | | | | | .054 | .084 | .121 | .165 | .216 | .273 | .338 | .409 | .488 | .566 | .663 | .757 | .863 | 1.09 |

**Table 85.  Safe Working Loads and Deflections of Cylindrical Helical Steel Springs of Circular Cross Section**  (Continued)

(For closely coiled springs divide given load and deflection values by the curvature factor *k*)

| Allowable unit stress, psi | Wire gage W. & M. | Diam, in. | D | ½ | ⅝ | ¾ | ⅞ | 1 | 1⅛ | 1¼ | 1⅜ | 1½ | 1⅝ | 1¾ | 1⅞ | 2 | 2¼ | 2½ | 2¾ | 3 | 3½ | 4 | 4½ |
|---|---|---|---|---|---|---|---|---|---|---|---|---|---|---|---|---|---|---|---|---|---|---|---|
| 140,000 | 7 | .177 | P | 608 | 487 | 406 | 347 | 305 | 270 | 243 | 223 | 205 | 187 | 174 | 163 | 152 | 135 | 122 | | | | | |
| | | | f | .049 | .077 | .111 | .151 | .198 | .251 | .311 | .375 | .447 | .522 | .606 | .695 | .793 | 1.00 | 1.24 | | | | | |
| | 3⁄16″ | .187 | P | 642 | 522 | 426 | 367 | 320 | 284 | 256 | 233 | 213 | 197 | 183 | 170 | 160 | 142 | 128 | | | | | |
| | | | f | .041 | .065 | .093 | .127 | .166 | .210 | .260 | .314 | .373 | .447 | .510 | .584 | .665 | .832 | 1.04 | | | | | |
| | 6 | .192 | P | 696 | 556 | 465 | 396 | 348 | 309 | 278 | 254 | 233 | 214 | 199 | 186 | 174 | 154 | 139 | 126 | | | | |
| | | | f | .040 | .063 | .091 | .124 | .160 | .205 | .252 | .308 | .366 | .428 | .499 | .571 | .652 | .825 | 1.02 | 1.23 | | | | |
| | 5 | .207 | P | | 694 | 579 | 495 | 432 | 385 | 346 | 315 | 288 | 266 | 247 | 232 | 216 | 192 | 173 | 158 | 144 | | | |
| | | | f | | .059 | .085 | .115 | .151 | .191 | .236 | .286 | .342 | .396 | .462 | .533 | .607 | .757 | .943 | 1.11 | 1.36 | | | |
| 125,000 | 7⁄32″ | .218 | P | | 812 | 678 | 580 | 509 | 452 | 408 | 370 | 339 | 310 | 291 | 270 | 255 | 225 | 204 | 185 | 169 | | | |
| | | | f | | .055 | .080 | .109 | .142 | .180 | .223 | .269 | .321 | .374 | .437 | .488 | .570 | .710 | .891 | 1.08 | 1.28 | | | |
| | 4 | .225 | P | | 895 | 746 | 640 | 560 | 498 | 447 | 407 | 372 | 345 | 320 | 299 | 280 | 248 | 224 | 203 | 187 | | | |
| | | | f | | .055 | .078 | .106 | .138 | .175 | .213 | .262 | .312 | .372 | .425 | .486 | .565 | .691 | .866 | 1.05 | 1.24 | | | |
| | 3 | .244 | P | | 1120 | 950 | 811 | 711 | 632 | 570 | 527 | 475 | 438 | 406 | 381 | 356 | 316 | 284 | 259 | 237 | 222 | | |
| | | | f | | .049 | .071 | .098 | .131 | .161 | .200 | .240 | .287 | .336 | .391 | .449 | .537 | .646 | .800 | .965 | 1.14 | 1.53 | | |
| | ¼″ | .250 | P | | | 1027 | 880 | 760 | 685 | 617 | 560 | 513 | 476 | 440 | 410 | 385 | 342 | 308 | 281 | 266 | 256 | | |
| | | | f | | | .070 | .095 | .131 | .157 | .191 | .236 | .281 | .328 | .385 | .439 | .524 | .624 | .780 | .946 | 1.12 | 1.44 | | |
| | 2 | .263 | P | | | 1195 | 1125 | 895 | 795 | 717 | 652 | 598 | 551 | 501 | 478 | 448 | 400 | 359 | 326 | 298 | 256 | | |
| | | | f | | | .066 | .089 | .118 | .149 | .183 | .224 | .266 | .312 | .363 | .416 | .475 | .592 | .740 | .896 | 1.06 | 1.44 | | |
| | 9⁄32″ | .281 | P | | | 1450 | 1240 | 1087 | 969 | 863 | 794 | 724 | 665 | 620 | 580 | 543 | 482 | 437 | 395 | 362 | 310 | | |
| | | | f | | | .062 | .085 | .111 | .140 | .172 | .209 | .250 | .292 | .340 | .390 | .443 | .562 | .692 | .840 | 1.02 | 1.36 | | |
| | 1 | .283 | P | | | | 1264 | 1110 | 985 | 886 | 805 | 740 | 682 | 634 | 592 | 564 | 492 | 439 | 402 | 370 | 317 | | |
| | | | f | | | | .084 | .111 | .139 | .169 | .207 | .246 | .289 | .338 | .386 | .440 | .559 | .690 | .883 | .990 | 1.35 | | |
| 115,000 | 5⁄16″ | .312 | P | | | | 1575 | 1376 | 1220 | 1100 | 1000 | 915 | 845 | 775 | 733 | 687 | 610 | 550 | 500 | 460 | 392 | 343 | |
| | | | f | | | | .070 | .092 | .116 | .144 | .174 | .207 | .242 | .283 | .322 | .368 | .467 | .577 | .697 | .829 | 1.12 | 1.47 | |
| | 00 | .331 | P | | | | | 1636 | 1455 | 1316 | 1187 | 1090 | 1000 | 932 | 870 | 818 | 725 | 653 | 594 | 545 | 468 | 410 | |
| | | | f | | | | | .088 | .109 | .135 | .163 | .194 | .227 | .264 | .302 | .346 | .437 | .541 | .654 | .770 | 1.05 | 1.30 | |
| | 11⁄32 | .341 | P | | | | | 1820 | 1620 | 1452 | 1325 | 1214 | 1120 | 1040 | 970 | 910 | 808 | 728 | 661 | 608 | 520 | 454 | |
| | | | f | | | | | .082 | .105 | .127 | .156 | .186 | .218 | .256 | .293 | .330 | .413 | .522 | .625 | .745 | 1.02 | 1.32 | |
| | 000 | .362 | P | | | | | 2140 | 1910 | 1714 | 1560 | 1430 | 1318 | 1220 | 1147 | 1070 | 950 | 858 | 778 | 714 | 612 | 535 | |
| | | | f | | | | | .079 | .100 | .123 | .149 | .177 | .207 | .243 | .273 | .317 | .400 | .495 | .598 | .713 | .965 | 1.26 | |
| | ⅜″ | .375 | P | | | | | | 2110 | 1940 | 1780 | 1580 | 1458 | 1354 | 1265 | 1185 | 1058 | 950 | 860 | 790 | 678 | 592 | 528 |
| | | | f | | | | | | .097 | .117 | .144 | .172 | .201 | .234 | .268 | .308 | .382 | .478 | .579 | .688 | .938 | 1.22 | 1.54 |
| | 0000 | .393 | P | | | | | | 2430 | 2180 | 1984 | 1820 | 1680 | 1560 | 1458 | 1365 | 1212 | 1092 | 990 | 910 | 780 | 682 | 670 |
| | | | f | | | | | | .092 | .114 | .137 | .164 | .195 | .223 | .256 | .292 | .369 | .457 | .550 | .657 | .890 | 1.16 | 1.47 |

## Pitch diameter, D in.

| Allowable unit stress, psi | Wire gage W. & M. | Diam, in. | D | 1¼ | 1⅜ | 1½ | 1⅝ | 1¾ | 1⅞ | 2 | 2¼ | 2½ | 2¾ | 3 | 3½ | 4 | 4½ | 5 | 5½ | 6 |
|---|---|---|---|---|---|---|---|---|---|---|---|---|---|---|---|---|---|---|---|---|
| 115,000 | 13/32″ | .406 | P | 2400 | 2170 | 2000 | 1840 | 1710 | 1600 | 1500 | 1330 | 1200 | 1090 | 1000 | 855 | 750 | 666 | | | |
| | | | f | .108 | .134 | .159 | .185 | .217 | .248 | .284 | .353 | .444 | .525 | .640 | .867 | 1.13 | 1.43 | | | |
| | 00000 | .430 | P | 2875 | 2610 | 2400 | 2210 | 2050 | 1918 | 1798 | 1598 | 1440 | 1308 | 1200 | 1028 | 900 | 800 | | | |
| | | | f | .104 | .126 | .150 | .175 | .204 | .234 | .267 | .338 | .418 | .503 | .600 | .815 | 1.06 | 1.35 | | | |
| | 7/16″ | .437 | P | 3000 | 2730 | 2500 | 2310 | 2140 | 2000 | 1800 | 1665 | 1500 | 1365 | 1250 | 1074 | 940 | 835 | 750 | | |
| | | | f | .100 | .124 | .148 | .173 | .201 | .231 | .264 | .327 | .412 | .490 | .593 | .810 | 1.05 | 1.33 | 1.64 | | |
| 110,000 | 000000 | .460 | P | | 3065 | 2800 | 2580 | 2400 | 2230 | 2100 | 1865 | 1680 | 1530 | 1400 | 1200 | 1058 | 952 | 840 | | |
| | | | f | | .112 | .134 | .157 | .183 | .209 | .239 | .303 | .374 | .447 | .536 | .729 | .956 | 1.21 | 1.49 | | |
| | 15/32″ | .468 | P | | 3265 | 2940 | 2725 | 2530 | 2375 | 2210 | 1970 | 1770 | 1610 | 1472 | 1265 | 1110 | 935 | 885 | | |
| | | | f | | .111 | .132 | .154 | .182 | .206 | .235 | .295 | .368 | .444 | .530 | .720 | .943 | 1.19 | 1.47 | | |
| | 0000000 | .490 | P | | 3675 | 3270 | 3115 | 2890 | 2710 | 2535 | 2245 | 2025 | 1840 | 1690 | 1445 | 1268 | 1125 | 1015 | 920 | |
| | | | f | | .106 | .126 | .148 | .172 | .196 | .225 | .284 | .351 | .424 | .506 | .688 | .900 | 1.13 | 1.40 | 1.70 | |
| | ½″ | .500 | P | | | 3610 | 3320 | 3090 | 2890 | 2710 | 2410 | 2160 | 1970 | 1810 | 1550 | 1352 | 1205 | 1082 | 985 | |
| | | | f | | | .123 | .144 | .168 | .192 | .220 | .274 | .347 | .415 | .495 | .672 | .880 | 1.11 | 1.37 | 1.65 | |
| | 9/16″ | .562 | P | | | | 4700 | 4390 | 4090 | 3830 | 3420 | 3080 | 2790 | 2565 | 2190 | 1913 | 1710 | 1535 | 1395 | 1280 |
| | | | f | | | | .128 | .149 | .175 | .195 | .248 | .306 | .372 | .440 | .596 | .782 | .990 | 1.22 | 1.47 | 1.75 |
| | 5/8″ | .625 | P | | | | | 6100 | 5600 | 5260 | 4660 | 4210 | 3825 | 3505 | 3000 | 2630 | 2340 | 2110 | 1913 | 1750 |
| | | | f | | | | | .134 | .154 | .176 | .218 | .275 | .328 | .397 | .538 | .705 | .875 | 1.05 | 1.33 | 1.58 |
| 100,000 | 11/16″ | .687 | P | | | | | | | 6325 | 5660 | 5090 | 4630 | 4250 | 3625 | 3195 | 2825 | 2560 | 2330 | 2125 |
| | | | f | | | | | | | .145 | .183 | .228 | .274 | .327 | .443 | .580 | .733 | .908 | 1.00 | 1.30 |
| | ¾″ | .750 | P | | | | | | | | 7400 | 6640 | 6030 | 5540 | 4745 | 4150 | 3690 | 3325 | 3025 | 2770 |
| | | | f | | | | | | | | .178 | .218 | .252 | .299 | .402 | .532 | .671 | .832 | 1.00 | 1.19 |
| | 13/16″ | .812 | P | | | | | | | | | 8420 | 7660 | 7000 | 6000 | 5260 | 4675 | 4200 | 3825 | 3500 |
| | | | f | | | | | | | | | .192 | .232 | .276 | .376 | .490 | .620 | .766 | 1.10 | 1.10 |
| | ⅞″ | .875 | P | | | | | | | | | 10830 | 9550 | 8700 | 7500 | 6560 | 5740 | 5250 | 4770 | 4730 |
| | | | f | | | | | | | | | .179 | .218 | .257 | .348 | .456 | .577 | .712 | .860 | 1.02 |
| 90,000 | 15/16″ | .937 | P | | | | | | | | | | 10600 | 9700 | 8400 | 7160 | 6470 | 5810 | 5290 | 4850 |
| | | | f | | | | | | | | | | .179 | .217 | .290 | .383 | .480 | .591 | .715 | .855 |
| | 1″ | 1.000 | P | | | | | | | | | | | 11780 | 10100 | 8800 | 7850 | 7050 | 6330 | 5870 |
| | | | f | | | | | | | | | | | .206 | .276 | .360 | .454 | .561 | .680 | .803 |
| | 1⅛″ | 1.125 | P | | | | | | | | | | | | 14400 | 12600 | 11230 | 10100 | 9200 | 8400 |
| | | | f | | | | | | | | | | | | .244 | .320 | .405 | .496 | .600 | .718 |
| 80,000 | 1¼″ | 1.250 | P | | | | | | | | | | | | 24700 | 18200 | 15300 | 13250 | 12540 | 11500 |
| | | | f | | | | | | | | | | | | .260 | .287 | .364 | .442 | .545 | .648 |
| | 1⅜″ | 1.375 | P | | | | | | | | | | | | | 20400 | 18100 | 16150 | 14850 | 13600 |
| | | | f | | | | | | | | | | | | | .280 | .294 | .364 | .440 | .522 |

**Table 86.  Standard Hoisting Rope**

Composed of 6 strands and a fiber core, 19 wires to the strand.   (John A. Roebling's Sons Co.)

| Diam, in. | Approx weight per ft, lb | Breaking strength, tons | | |
|---|---|---|---|---|
| | | Blue center steel | Plow steel | Mild plow steel |
| 2¾ | 12.10 | 292.0 | 254.0 | |
| 2½ | 10.00 | 244.0 | 212.0 | |
| 2¼ | 8.10 | 200.0 | 174.0 | |
| 2⅛ | 7.23 | 179.0 | 156.0 | |
| 2 | 6.40 | 160.0 | 139.0 | 121.0 |
| 1⅞ | 5.63 | 141.0 | 123.0 | 107.0 |
| 1¾ | 4.90 | 124.0 | 108.0 | 93.6 |
| 1⅝ | 4.23 | 107.0 | 93.4 | 81.2 |
| 1½ | 3.60 | 92.0 | 80.0 | 69.6 |
| 1⅜ | 3.03 | 77.7 | 67.5 | 58.8 |
| 1¼ | 2.50 | 64.6 | 56.2 | 48.8 |
| 1⅛ | 2.03 | 52.6 | 45.7 | 39.8 |
| 1 | 1.60 | 41.8 | 36.4 | 31.6 |
| ⅞ | 1.23 | 32.2 | 28.0 | 24.3 |
| ¾ | 0.90 | 23.8 | 20.7 | 18.0 |
| ⅝ | 0.63 | 16.7 | 14.5 | 12.6 |
| 9⁄16 | 0.51 | 13.5 | 11.8 | 10.2 |
| ½ | 0.40 | 10.7 | 9.35 | 8.13 |
| 7⁄16 | 0.31 | 8.27 | 7.19 | 6.25 |
| ⅜ | 0.23 | 6.10 | 5.31 | 4.62 |
| 5⁄16 | 0.16 | 4.26 | 3.71 | 3.22 |
| ¼ | 0.10 | 2.74 | 2.39 | 2.07 |

**Table 87.  Extra-pliable Hoisting Rope**

(Six 37-wire strands and fiber core)

| Diam, in. | Approx weight per ft, lb | Breaking strength, tons | |
|---|---|---|---|
| | | Blue center steel | Plow steel |
| 3½ | 19.00 | 449.0 | 390.0 |
| 3¼ | 16.37 | 390.0 | 339.0 |
| 3 | 13.95 | 335.0 | 291.0 |
| 2¾ | 11.72 | 284.0 | 247.0 |
| 2½ | 9.69 | 236.0 | 205.0 |
| 2¼ | 7.85 | 193.0 | 168.0 |
| 2⅛ | 7.00 | 173.0 | 150.0 |
| 2 | 6.20 | 154.0 | 134.0 |
| 1⅞ | 5.45 | 136.0 | 118.0 |
| 1¾ | 4.75 | 119.0 | 103.0 |
| 1⅝ | 4.09 | 103.0 | 89.3 |
| 1½ | 3.49 | 87.9 | 76.4 |
| 1⅜ | 2.93 | 74.1 | 64.5 |
| 1¼ | 2.42 | 61.5 | 53.5 |
| 1⅛ | 1.96 | 50.1 | 43.5 |
| 1 | 1.55 | 39.8 | 34.6 |
| ⅞ | 1.19 | 30.6 | 26.6 |
| ¾ | 0.87 | 22.6 | 19.6 |
| ⅝ | 0.61 | 15.8 | 13.7 |
| 9⁄16 | 0.49 | 12.9 | 11.2 |
| ½ | 0.39 | 10.2 | 8.85 |
| 7⁄16 | 0.30 | 7.82 | 6.80 |
| ⅜ | 0.22 | 5.77 | 5.02 |
| 5⁄16 | 0.16 | 4.03 | 3.50 |
| ¼ | 0.10 | 2.59 | 2.25 |

**Table 88.  Extra-pliable Hoisting Rope**

(Eight 19-wire strands and fiber core)

| Diam, in. | Approx weight per ft, lb | Breaking strength, tons | |
|---|---|---|---|
| | | Blue center steel | Plow steel |
| 1½ | 3.26 | 79.4 | 69.1 |
| 1⅜ | 2.74 | 67.1 | 58.3 |
| 1¼ | 2.27 | 55.7 | 48.4 |
| 1⅛ | 1.84 | 45.3 | 39.4 |
| 1 | 1.45 | 36.0 | 31.3 |
| ⅞ | 1.11 | 27.7 | 24.1 |
| ¾ | 0.82 | 20.5 | 17.8 |
| ⅝ | 0.57 | 14.3 | 12.4 |
| 9⁄16 | 0.46 | 11.6 | 10.1 |
| ½ | 0.36 | 9.23 | 8.02 |
| 7⁄16 | 0.28 | 7.09 | 6.17 |
| ⅜ | 0.20 | 5.24 | 4.55 |
| 5⁄16 | 0.14 | 3.65 | 3.18 |
| ¼ | 0.09 | 2.35 | 2.04 |

of 115,000 lb/in². Therefore, 0.810 is the greatest admissible opening between any two coils. In this way, it is possible to ascertain whether or not a spring is overloaded, without knowledge of the load carried.

## WIRE ROPE

Wire ropes are built up of **strands** of wires laid together, the number of wires commonly used being 4, 7, 12, 19, and 37. Ordinarily the wires are laid into strands in the direction opposite to the twist of the strands into rope. When wires and

on larger hoisting installations with the 6 × 7 (6 × 19) rope, the diameters may be 96 (90) times the rope diameter. In certain cases, the tread diameters may be less but should not be below 42 (30) [18] {21} times the rope diameters. Larger tread diameters give increased rope life and more economical service.

The size and condition of the sheave grooves are most important in determining rope life. The clearances for new or remachined grooves and the minimum clearances before sheave replacement or remachining grooves should be as follows:

| Nominal rope diam, in | $\frac{1}{4}-\frac{5}{16}$ | $\frac{3}{8}-\frac{3}{4}$ | $\frac{13}{16}-1\frac{1}{8}$ | $\frac{13}{16}-1\frac{1}{2}$ | $1\frac{9}{16}-2\frac{1}{4}$ | $2\frac{5}{16}$ and larger |
|---|---|---|---|---|---|---|
| Recommended clearance, in | $\frac{1}{64}$ | $\frac{1}{32}$ | $\frac{3}{64}$ | $\frac{1}{16}$ | $\frac{3}{32}$ | $\frac{1}{8}$ |

strands are laid in the same direction, the rope is known as **lang-lay rope. Standard wire rope** is made of six wire strands and a sisal core. Wire strands are laid around the core, either to the right or to the left, and the resulting rope is designated as **right lay** or **left lay.** The lay may be long or short, the shorter lay forming the more flexible rope. The **core** of a wire rope is, as a rule, sisal saturated with a lubricant. It provides little additional strength but acts as a cushion to preserve the shape of the rope and helps to lubricate the wires. A wire-strand or wire-rope core adds 7 to 10 percent to the strength of the rope but will wear from the friction between it and the outer strands as rapidly as the outside of the rope. This does not apply to stationary ropes.

For great flexibility, the strands of a wire rope sometimes consist of wire ropes, which in turn are made of strands composed of wires, as in tiller rope. Running ropes and one construction of ship's hawsers are made with strands composed of 12 or 18 wires each, laid about a fiber core. Ropes so made are very pliable and present good resistance to outside friction. Individual strands of wires are employed as smoke-stack guys, span wires for trolley roads, and wherever only moderate flexibility is needed.

**Sizes of Ropes** The diameter of wire rope is the circle which will just contain the rope. In a rope classification the first number is the number of strands in the rope; the last number is the number of wires in a strand; and the middle number, if any, is the number of minor strands in a major strand. If there is a wire core, the rope is labeled IWRC (Independent Wire Rope Core), or if the core is the same as the main strands, it is counted as a strand.

**Strength and Working Loads** The test **strength of wire ropes** seldom exceeds 90 percent of the aggregate strength of all the wires, the average being about 82.5 percent.

The working load should never exceed $\frac{1}{5}$ of the breaking strength and for many conditions should not be greater than $\frac{1}{6}$ to $\frac{1}{8}$. The proper factor of safety for a wire rope demands consideration of all loads; acceleration; deceleration; rope speed; rope attachments; the number, size, and arrangements of sheaves and drums; conditions producing corrosion and abrasion; length of rope; etc. The desirable factor of safety for given conditions can best be obtained by consulting the manufacturers of the wire rope. See Tables 86 to 94.

**Sizes of Drums or Sheaves** Tread diameters for 6 × 7 (6 × 19) [6 × 37] {8 × 19} rope should be approximately 72 (45) [27] {31} times the rope diameters, for average conditions. For economical service, these should often be increased; e.g.,

If a wire rope is operated over grooves that are too small, the rope will be abraded rapidly; in addition, a tight or corrugated sheave groove may disturb the strand relationship of the rope and necessitate its premature removal. The pressure of wire rope against a sheave groove is calculated by dividing the rope tension in pounds by the product of the radius to the bottom of the sheave groove and the diameter of the rope, both in inches. This quantity should not have a value greater than 450 lb/in² for cast iron, or 850 lb/in² for cast steel. For greater pressures, a material of greater wear resistance, such as manganese steel, should be used.

**Handling** Wire rope must not be coiled or uncoiled like hemp rope. When it is received upon a reel, the latter should be mounted upon a spindle or turntable and the rope then run off. When shipped in a coil, it should be rolled along the ground like a wheel. All untwisting and kinking must be avoided. When a wire rope is to be cut, soft iron wire should be served on each side of the place where the division is to be made to keep the rope from untwisting. (See Seizing.)

**Materials** Rope made from iron wire is now used only for passenger elevators and similar service where the tendency to abrasion is comparatively slight, the speed is high, and the loads are moderate. The three grades of steel commonly used for the manufacture of wire rope are designated "blue center," or "improved plow steel," "plow steel," and "mild plow steel." These grades were established in the *NBS Bull.* R 198–43.

**Standard hoisting rope** (Fig. 140) is made of 6 strands, each of 19 wires, the strands being laid around a fiber core.

**Extra-pliable hoisting rope** is made of 6 strands of 37 wires each and a fiber core (Fig. 141). The wires in this rope are much finer than those used in the standard hoisting rope and consequently not as suitable to withstand abrasion. These ropes are used on electric cranes, dredges, and for similar service requiring a strong tough rope that will operate successfully over small sheaves.

**Extra-pliable hoisting rope** of 8 strands of 19 wires and a fiber core (Fig. 142) is much more pliable than the standard con-

 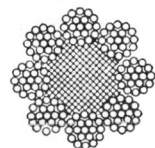

Fig. 140          Fig. 141          Fig. 142

**Table 89.  Standard Coarse-laid Rope for Haulages and Transmissions**
(Composed of six 7-wire strands and a fiber core)

| Diam, in. | Approx weight per ft, lb | Breaking strength, tons | | | Diam, in. | Approx weight per ft, lb | Breaking strength, tons | | |
|---|---|---|---|---|---|---|---|---|---|
| | | Blue center steel | Plow steel | Mild plow steel | | | Blue center steel | Plow steel | Mild plow steel |
| 1½ | 3.38 | 86.2 | 75.0 | 65.2 | ⅝ | 0.59 | 15.9 | 13.9 | 12.0 |
| 1⅜ | 2.84 | 73.1 | 63.6 | 55.3 | 9⁄16 | 0.48 | 13.0 | 11.3 | 9.82 |
| 1¼ | 2.34 | 61.0 | 53.0 | 46.1 | ½ | 0.38 | 10.3 | 8.96 | 7.79 |
| 1⅛ | 1.90 | 49.8 | 43.3 | 37.7 | 7⁄16 | 0.29 | 7.93 | 6.90 | 6.00 |
| 1 | 1.50 | 39.7 | 34.5 | 30.0 | ⅜ | 0.21 | 5.86 | 5.10 | 4.43 |
| ⅞ | 1.15 | 30.7 | 26.7 | 23.2 | 5⁄16 | 0.15 | 4.10 | 3.56 | 3.10 |
| ¾ | 0.84 | 22.7 | 19.8 | 17.2 | ¼ | 0.094 | 2.64 | 2.30 | 2.00 |

struction of 6 strands of 19 wires. The metallic area of an eight-strand rope is not as great as that of a six-strand rope, and the wires are smaller, but under severe bending stresses the decrease in strength is largely offset by the great pliability. It can be used over comparatively small sheaves and drums such as are frequently found on derricks. It is not good practice to use it except for comparatively light loads or where there is much overwinding, because it would flatten or lose shape more quickly than 6 × 19 rope. Moreover it stretches more than a 6 × 19 rope.

Galvanized extra-pliable cast-steel hoisting rope is much more flexible than the six-strand hoisting rope and is often used in preference to galvanized cast-steel running rope.

**Standard coarse-laid rope** (Fig. 143) is made of six strands and a fiber core with seven wires to the strand. It is much stiffer than standard hoisting rope and requires larger sheaves. On account of the smaller number of wires, this rope should also be used with a higher factor of safety, as the breaking of one or two wires materially reduces the strength of the rope. The

wires used are considerably larger in diameter than in hoisting rope, and consequently will stand greater wear. **Iron rope** of this construction is recommended for power transmissions equipped with large sheaves. **Cast-steel** and **extra-strong cast-steel rope** are recommended for mine haulages, tramways, sand lines, and similar service where conditions tend to severe abrasion. **Plow-steel** and **improved plow-steel ropes** are recommended in place of cast steel when it is desirable to reduce the dead weight of the rope itself, or where, by reason of increased loads, it is necessary to use a stronger rope without increasing its diameter. This rope is particularly adapted for very long mine haulages.

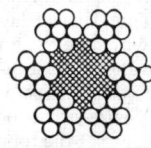

**Fig. 143**

**Flattened-strand wire rope** is designed to give increased contact or wearing surface. The wear is consequently lessened upon any one individual wire and the necessity of the use of heavier wire diminished, which results in greater flexibility.

**Tables 90.  Flat Rope**

| Width and thickness, in. | Approx weight per ft, lb | No. of ropes | Breaking strength, tons | | Width and thickness, in. | Approx weight per ft, lb | No. of ropes | Breaking strength, tons | |
|---|---|---|---|---|---|---|---|---|---|
| | | | Plow steel | Mild plow steel | | | | Plow steel | Mild plow steel |
| ⅞ × 8 | 10.69 | 10 | 271.0 | 236.0 | ½ × 3½ | 2.82 | 8 | 72.7 | 63.3 |
| ⅞ × 7 | 9.63 | 9 | 244.0 | 212.0 | ½ × 3 | 2.47 | 7 | 63.6 | 55.4 |
| ⅞ × 6 | 8.56 | 8 | 217.0 | 188.0 | ½ × 2½ | 2.13 | 6 | 54.5 | 47.4 |
| ⅞ × 5 | 7.50 | 7 | 190.0 | 165.0 | | | | | |
| | | | | | ⅜ × 6 | 3.63 | 18 | 94.1 | 81.9 |
| ¾ × 8 | 9.70 | 11 | 227.0 | 197.0 | ⅜ × 5½ | 3.42 | 17 | 88.9 | 77.3 |
| ¾ × 7 | 8.13 | 10 | 206.0 | 179.0 | ⅜ × 5 | 3.03 | 15 | 78.4 | 68.2 |
| ¾ × 6 | 7.31 | 9 | 185.0 | 161.0 | ⅜ × 4½ | 2.83 | 14 | 73.2 | 63.7 |
| ¾ × 5 | 6.50 | 8 | 165.0 | 143.0 | ⅜ × 4 | 2.44 | 12 | 62.7 | 54.6 |
| | | | | | ⅜ × 3½ | 2.23 | 11 | 57.5 | 50.0 |
| ⅝ × 8 | 8.32 | 15 | 214.0 | 186.0 | ⅜ × 3 | 1.84 | 9 | 47.1 | 40.9 |
| ⅝ × 7 | 7.23 | 13 | 186.0 | 162.0 | ⅜ × 2½ | 1.64 | 8 | 41.8 | 36.4 |
| ⅝ × 6 | 6.14 | 11 | 157.0 | 137.0 | ⅜ × 2 | 1.25 | 6 | 31.4 | 27.3 |
| ⅝ × 5½ | 5.59 | 10 | 143.0 | 124.0 | | | | | |
| ⅝ × 5 | 5.04 | 9 | 129.0 | 112.0 | 5⁄16 × 4 | 2.17 | 15 | 55.3 | 48.1 |
| ⅝ × 4½ | 4.50 | 8 | 114.0 | 99.5 | 5⁄16 × 3½ | 1.89 | 13 | 47.9 | 41.7 |
| ⅝ × 4 | 3.95 | 7 | 100.0 | 87.1 | 5⁄16 × 3 | 1.61 | 11 | 40.5 | 35.3 |
| ⅝ × 3½ | 3.40 | 6 | 85.8 | 74.6 | 5⁄16 × 2½ | 1.33 | 9 | 33.2 | 28.8 |
| | | | | | 5⁄16 × 2 | 1.05 | 7 | 25.8 | 22.4 |
| ½ × 7 | 5.85 | 16 | 145.0 | 126.0 | 5⁄16 × 1½ | 0.77 | 5 | 18.5 | 16.0 |
| ½ × 6 | 4.85 | 14 | 127.0 | 111.0 | | | | | |
| ½ × 5½ | 4.50 | 13 | 118.0 | 103.0 | ¼ × 3 | 1.34 | 13 | 31.3 | 27.2 |
| ½ × 5 | 4.16 | 12 | 109.0 | 94.9 | ¼ × 2½ | 1.15 | 11 | 26.5 | 23.0 |
| ½ × 4½ | 3.82 | 10 | 90.9 | 79.1 | ¼ × 2 | 0.88 | 9 | 21.7 | 18.8 |
| ½ × 4 | 3.16 | 9 | 81.8 | 71.2 | ¼ × 1½ | 0.69 | 7 | 16.8 | 14.6 |

**Table 91. Flattened-Strand Hoisting Rope, Type A**

(Five 28-wire strands and a fiber core)

| Diam, in. | Approx wt per ft, lb | Breaking strength, tons | | | |
|---|---|---|---|---|---|
| | | Improved plow steel | Plow steel | Extra-strong cast steel | Cast steel |
| ⅜ | 0.22 | 5.6 | 4.8 | 4.4 | 4. |
| ½ | 0.39 | 9.5 | 8.2 | 7.5 | 6.8 |
| 9⁄16 | 0.49 | 11.9 | 10.3 | 9.4 | 8.5 |
| ⅝ | 0.61 | 14.6 | 12.7 | 11.6 | 10.4 |
| ¾ | 0.87 | 20.9 | 18.2 | 16.5 | 14.8 |
| ⅞ | 1.19 | 28.5 | 24.8 | 22.4 | 20.1 |
| 1 | 1.55 | 37. | 32. | 29. | 26. |
| 1⅛ | 1.96 | 47. | 41. | 37. | 33. |
| 1¼ | 2.42 | 57. | 50. | 45. | 41. |

**Table 92. Flattened-Strand Hoisting Rope, Type B**

(Six 25-wire strands and a fiber core)

| Diam, in. | Approx wt per ft, lb | Breaking strength, tons | | | |
|---|---|---|---|---|---|
| | | Improved plow steel | Plow steel | Extra-strong cast steel | Cast steel |
| ⅜ | 0.25 | 6.9 | 6. | 5.5 | 4.9 |
| ½ | 0.45 | 11.8 | 10.3 | 9.3 | 8.4 |
| 9⁄16 | 0.57 | 14.8 | 12.8 | 11.6 | 10.5 |
| ⅝ | 0.70 | 18.2 | 15.8 | 14.4 | 12.9 |
| ¾ | 1.01 | 26. | 22.6 | 20.5 | 18.4 |
| ⅞ | 1.39 | 35.4 | 30. | 27.9 | 25. |
| 1 | 1.80 | 46. | 40. | 36. | 32. |
| 1⅛ | 2.28 | 58. | 50.5 | 45.5 | 40. |
| 1¼ | 2.81 | 71.5 | 62. | 56. | 50.5 |
| 1⅜ | 3.40 | 86. | 74. | 67. | 60. |
| 1½ | 4.05 | 101. | 88. | 79. | 71. |
| 1⅝ | 4.75 | 118. | 103. | 93. | 83. |
| 1¾ | 5.51 | 136. | 118. | 107. | 96. |
| 2 | 7.2 | 177. | 154. | 139. | 125. |
| 2¼ | 9.1 | 222. | 193. | 176. | 158. |
| 2½ | 11.2 | 270. | ..... | 214. | 193. |
| 2¾ | 13.6 | 323. | ..... | 257. | 233. |

**Table 93. Nonrotating Hoisting Rope**

(Composed of 18 strands and a fiber core, 7 wires to the strand)

| Diam, in. | Approx weight per ft, lb | Breaking strength, tons | | Diam, in. | Approx weight per ft, lb | Breaking strength, tons | |
|---|---|---|---|---|---|---|---|
| | | Blue center steel | Plow steel | | | Blue center steel | Plow steel |
| 1¾ | 5.30 | 114.0 | 98.8 | ⅞ | 1.32 | 29.5 | 25.7 |
| 1⅝ | 4.57 | 98.4 | 85.6 | ¾ | 0.97 | 21.8 | 19.0 |
| 1½ | 3.89 | 84.4 | 73.4 | ⅝ | 0.68 | 15.3 | 13.3 |
| 1⅜ | 3.27 | 71.3 | 62.0 | 9⁄16 | 0.55 | 12.4 | 10.8 |
| 1¼ | 2.70 | 59.2 | 51.5 | ½ | 0.43 | 9.85 | 8.57 |
| 1⅛ | 2.19 | 48.2 | 41.9 | 7⁄16 | 0.33 | 7.58 | 6.59 |
| 1 | 1.73 | 38.3 | 33.3 | ⅜ | 0.24 | 5.59 | 4.86 |

**Table 94. Galvanized Common Steel Wire Strand**

(Composed of 7 wires laid together)

| Diam, in. | Approx weight per 1,000 ft, lb | Approx strength, lb | Diam, in. | Approx weight per 1,000 ft, lb | Approx strength, lb | Diam, in. | Approx weight per 1,000 ft, lb | Approx strength, lb |
|---|---|---|---|---|---|---|---|---|
| ⅝ | 813 | 11,600 | ⅜ | 273 | 4,250 | 7⁄32 | 98.3 | 1,540 |
| 9⁄16 | 671 | 9,600 | 5⁄16 | 205 | 3,200 | 3⁄16 | 72.9 | 1,150 |
| ½ | 517 | 7,400 | 9⁄32 | 164 | 2,570 | 5⁄32 | 51.3 | 870 |
| 7⁄16 | 399 | 5,700 | ¼ | 121 | 1,900 | ⅛ | 31.8 | 540 |

**Table 95. Number of Seizings on Each Side of Cut**

(Bethlehem Steel Co.)

Steel wire ropes, form-set (preformed) . 1
Steel wire ropes, nonpreformed:
  Ropes ⅞ in diam and smaller ...... 2
  Ropes 15⁄16–1 1⁄16 in diam .......... 3
  Ropes 1⅛ in diam and larger ...... 4
  Lang-lay ropes 1½ in diam and larger  4
  Iron wire ropes, nonpreformed ...... 2

**Table 96. Recommended Lengths of Seizings**

(Bethlehem Steel Co.)

| Rope diameter, in. | Length of seizings, in. |
|---|---|
| ½ and smaller | ½ |
| 9⁄16–⅞ | 1 |
| 1–1¼ | 1½ |
| 1⅜–1⅝ | 2 |
| 1¾–2 | 3 |
| 2⅛ and larger | 4 |

The wearing surface is approximately 150 percent greater than that of a round-strand rope. Another feature of this type of rope is that, the interstices between the strands being lessened, a greater number of wires are used for the same diameter. It is always made lang-lay. Flattened-strand rope has little tendency to kink, and, owing to its smooth wearing surface, saves wear on pulleys, sheaves, and drums. It is not so flexible or so fatigue-resistant as round-strand rope of the same general classification. The strength is greater than that of round-strand ropes, but the weight is proportionately greater than the increase in strength. These ropes are made in "blue center" steel grade only and are available in three styles: D, B, and G (Fig. 144). Style D is used on haulages and consists of six

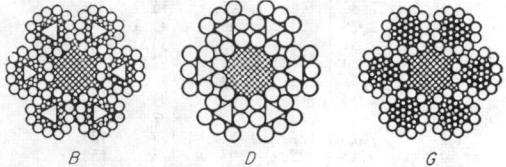

**Fig. 144** Flattened-strand wire rope.

strands, each containing seven outside wires laid around a triangular-shaped center wire. Styles B and G, used for hoisting purposes, consist of six strands, each strand of which is made up of two layers of 12 wires each.

**Nonspinning hoisting rope** (Fig. 145) is constructed of 6 strands of 7 wires each, lang-lay (wires in the strands and strands themselves laid to the left), laid around a fiber core, and covered with an outer layer composed of 12 strands, 7 wires, regular lay (wires in the strands laid to the left and strands themselves laid to the right). The object of this combination of lays is to prevent a free load suspended on the end of a single line from rotating. This type of rope is recommended for "back-haul" or single-line derricks; also for shaft sinking and mine hoisting where the bucket or cage swings free without guides. It works best where it does not overwind on the drum.

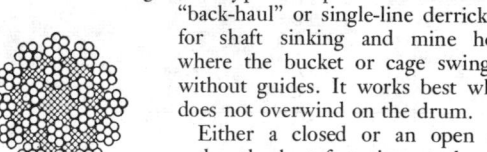

**Fig. 145** Nonspinning hoisting rope.

Either a closed or an open socket makes the best fastening on the end of nonspinning rope. These may be fastened in the same manner as any rope socket, but great care must be taken in attaching the socket to the rope to see that the strands do not untwist or allow any slack to work back into the rope. It is best to seize the end of the rope tightly for a distance of 4 or 5 in just outside the socket until the socketing is completed, when it may be taken off.

**Steel-clad ropes** (Fig. 146) are made in three constructions for the purpose of securing different degrees of flexibility: the 6 × 19, 6 × 37, and 6 × 61 types, respectively. Flat strips of steel

**Table 97. Seizing Wire Diameters**
(Bethlehem Steel Co.)

| Diam of rope, in. | Approx diam of seizing wire, in. |
|---|---|
| ¼ | 0.063 |
| ⅜–⅝ | 0.080 |
| ¾–1⅛ | 0.104 |
| 1¼–1½ | 0.124 |
| 1⅝ and larger | 0.138 |

wound spirally around each of the six strands composing the rope give additional wearing surface without sacrificing flexibility. When the outer flat-steel winding is worn through, a complete hoisting rope remains, with unimpaired strength. These ropes are designed to meet very severe conditions of service. The increased life obtained by the use of steel-clad

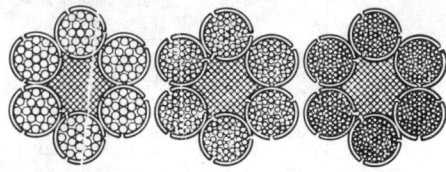

**Fig. 146** Steel-clad rope.

rope is, in places where conditions are suitable, from 50 to 100 percent. It is recommended particularly for such service as dredging. The breaking strength of these ropes is less than that of a round strand rope of the same diameter and grade.

**Galvanized wire rope** has almost entirely superseded manila rope for shrouds and stays aboard ship. It is cheaper in first cost, is relatively unaffected by the weather, does not stretch and contract with changes in atmospheric conditions, and thus saves a great deal of labor in setting up. There is great

**Fig. 147** Galvanized steel wire strand.

reduction in bulk and weight by its use, as it is only one-fifth or one-sixth as large as a manila rope of equal strength. Consequently, it offers only half as much surface to the wind. It is much less liable to accidents by being cut or chafed, and does not rot and give way suddenly without warning. Galvanized rope is better suited for guys for derricks than hemp rope or rods linked together.

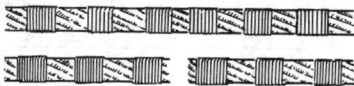

**Fig. 148** Wire rope with three seizings, before and after cutting.

**Galvanized steel wire strand** (Fig. 147) is used chiefly for guying poles and smokestacks, for supporting trolley wire, and for operating railroad signals. For overhead catenary construction of suspending trolley wire, the special grades of strand are preferable because they possess greater strength and toughness. The smallest sizes (sometimes called "galvanized

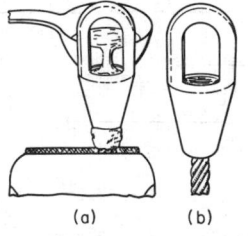

**Fig. 149** Wire-rope socket. (*a*) Pouring molten zinc into basket of socket; (*b*) finished socket.

**Fig. 150** Grommet sling.

seizing strand") are used for seizing or binding the ends of wire rope and thimble splices, and for tying rope into coils.

**Seizing** When wire rope is cut to make attachment to fittings or for splicing, it is important that it be properly seized. The wires and strands are laid under uniform tension, and the tension is maintained (Fig. 148). The seizing required

**Fig. 151** Single-leg sling.

will vary with the rope diameter; annealed iron, plain or galvanized, is used for seizing according to the recommendations in Tables 95 through 97.

**Wire-Rope Fittings** Attachment to a socket is made by separating and straightening the wires, cutting out the hemp center, cleansing with kerosene, dipping in one-half muriatic

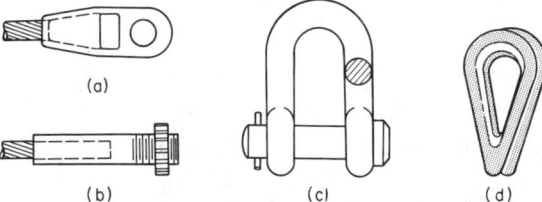

**Fig. 152** Wire-rope fittings. (a) Zinced-type socket; (b) swaged-type socket; (c) pin-type shackle; (d) standard thimble.

acid and one-half water (which must be cleaned from the wires after dipping), distributing the wires uniformly in the socket, and pouring in molten zinc after plugging up the lower end of the socket with fire clay. Such an attachment will develop the full strength of the rope (Fig. 149). Spliced eye connections when properly made will not pull out and will develop from 60 percent of the full strength of a 2½-in-diameter rope up to 95 percent for a ½-in rope. Clip and clamp connections are not

**Fig. 153** Galvanized steel hawsers.

**Fig. 154** Galvanized steel mooring lines and hawsers.

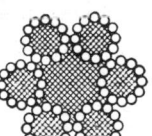

**Fig. 155** Standard 6x12 galvanized running rope and hawsers.

desirable as permanent fastenings on hoisting ropes and will develop 75 to 85 percent of the strength of the rope. Clips should be installed so that the U-bolt part is around the short, or dead, end of the rope (Table 98).

**Wire-rope slings** are made up in numerous styles, some of

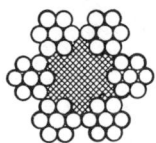

**Fig. 156** Galvanized iron riggings and guy rope.

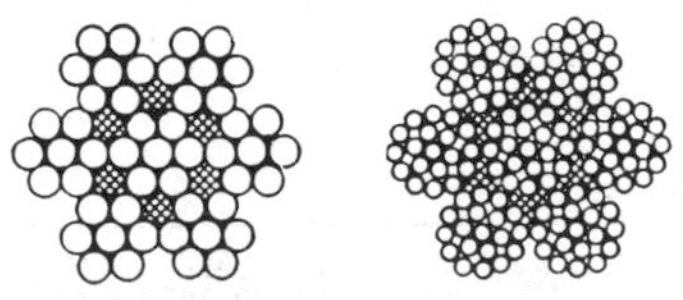

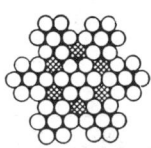

**Figs. 157 and 158** Galvanized steel bridge rope.

which are manufactured endless in grommet construction and develop the full strength of the rope (Figs. 150 and 151).

Examples of **end attachments** and miscellaneous wire-rope fittings are shown in Fig. 152.

The **efficiency of an end attachment** is measured by its ability to hold the wire rope. If the assembly reaches catalog strength of the rope before the end attachment loosens or gives way, the

**Table 98. Number of Clips and Spacing for Safe Application**
(Bethlehem Steel Co.)

| Rope diam, in. | Diam of U-bolt, in. | Approx weight, lb | Minimum no. clips for each rope end | Spacing of clips, in. |
|---|---|---|---|---|
| 3⁄16 | 11⁄32 | 0.09 | 2 | 2 |
| ¼ | 7⁄16 | 0.18 | 2 | 2 |
| 5⁄16 | ½ | 0.30 | 2 | 2 |
| 3⁄8 | 9⁄16 | 0.47 | 2 | 2¼ |
| 7⁄16 | 5⁄8 | 0.71 | 2 | 2⅝ |
| ½ | 11⁄16 | 0.73 | 3 | 3 |
| 5⁄8 | ¾ | 1.01 | 3 | 3¾ |
| ¾ | 7⁄8 | 1.57 | 4 | 4½ |
| 7⁄8 | 1 | 2.42 | 4 | 5¼ |
| 1 | 1⅛ | 2.64 | 4 | 6 |
| 1⅛ | 1¼ | 3.32 | 5 | 6¾ |
| 1¼ | 17⁄16 | 4.48 | 5 | 7½ |
| 1⅜ | 1½ | 4.88 | 6 | 8¼ |
| 1½ | 123⁄32 | 5.44 | 6 | 9 |
| 1⅝ | 1¾ | 7.02 | 6 | 9¾ |
| 1¾ | 115⁄16 | 9.28 | 7 | 10½ |
| 2 | 2⅛ | 12.04 | 8 | 12 |
| 2¼ | 2⅝ | 14.81 | 8 | 12 |
| 2½ | 2⅞ | 16.60 | 8 | 13 |

efficiency of the attachment is 100 percent. Table 99 shows approximate efficiencies developed by the various types of end attachments when properly applied.

**Galvanized mast-arm rope** is used for arc lights, mast arms, or other purposes where exposed to moisture. The rope is more durable than manila rope and does not shrink.

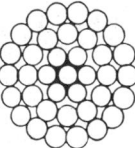

**Fig. 159** Galvanized steel bridge strand.

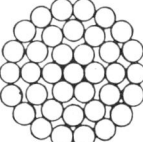

**Fig. 160** Smooth coil track strand.

## FIBER LINES

The breaking strength of various fiber lines is given in Table 112.

### Knots, Hitches, and Bends

No two parts of a knot which would move in the same direction if the rope were to slip should lie alongside of and touching each other. The knots shown in Fig. 161 are known by the following names:

*A*, bight of a rope; *B*, simple or overhand knot; *C*, figure 8

**Table 99. End-Attachment Efficiencies**

(Bethlehem Steel Co.)

| Fitting | Nominal efficiency, % |
|---|---|
| Wire rope sockets zinced attachments............. | Catalog rated strength of rope |
| Fittings (swaged or pressed)..................... | Catalog rated strength of rope |
| Open wedge sockets............................. | 80–90 |
| Clips (U-bolt type)............................. | 80 |
| Spliced-in thimbles: | |
| ¼ in. diam and smaller...................... | 90 |
| ⁵⁄₁₆ in. diam................................. | 89 |
| ⅜ in. diam................................. | 88 |
| ½ in. diam................................. | 86 |
| ⅝ in. diam................................. | 84 |
| ¾ in. diam................................. | 82 |
| ⅞ in., and up............................... | 80 |

**Table 100. Tensile Strength of Galvanized Steel Wire Strands, Lb**

(7-wire strand)

| Diam, in.................. | ⅝ | ⁹⁄₁₆ | ½ | ⁷⁄₁₆ | ⅜ | ⁵⁄₁₆ | ⁹⁄₃₂ | ¼ | ⁷⁄₃₂ | ³⁄₁₆ | ⁵⁄₃₂ | ⅛ |
|---|---|---|---|---|---|---|---|---|---|---|---|---|
| Wt of 1,000 ft, lb.......... | 813 | 671 | 517 | 399 | 273 | 205 | 164 | 121 | 98.3 | 72.9 | 51.3 | 31.8 |
| Siemens-Martin............ | 19,100 | 15,700 | 12,100 | 9,350 | 6,950 | 5,350 | 4,250 | 3,150 | 2,560 | 1,900 | 1,470 | 910 |
| High-strength............. | 29,600 | 24,500 | 18,800 | 14,500 | 10,800 | 8,000 | 6,400 | 4,750 | 3,850 | 2,850 | 2,140 | 1,330 |
| Extra-high strength........ | 42,400 | 35,000 | 26,900 | 20,800 | 15,400 | 11,200 | 8,950 | 6,650 | 5,400 | 3,990 | 2,940 | 1,830 |

Minimum elongation, percent, in 24 in.: S.M., 8; H.S., 5; E.H.S., 4.

**Table 101. Properties of Special Grades of Extra Galvanized Special Strands**

| Approx diam, in. | No. of wires in strand | Approx breaking strength, lb | | | | Approx wt per 1,000 ft, lb | Approx diam, in. | No. of wires in strand | Approx breaking strength, lb | | | | Approx wt per 1,000 ft, lb |
|---|---|---|---|---|---|---|---|---|---|---|---|---|---|
| | | Extra high strength | High strength | Siemens-Martin | Common soft strand | | | | Extra high strength | High strength | Siemens-Martin | Common soft strand | |
| 1 | 19 | 104,500 | 73,200 | 47,000 | 28,700 | 2073 | 1¼ | 37 | 162,200 | 113,600 | 73,000 | 44,600 | 3,248 |
| ⅞ | 19 | 79,700 | 55,800 | 35,900 | 21,900 | 1581 | 1⅛ | 37 | 130,800 | 91,600 | 58,900 | 36,000 | 2,691 |
| ¾ | 19 | 58,300 | 40,800 | 26,200 | 16,000 | 1155 | 1 | 37 | 102,700 | 71,900 | 46,200 | 28,300 | 2,057 |
| ⅝ | 19 | 40,200 | 28,100 | 18,100 | 11,000 | 796 | 1¾ | 61 | 315,000 | 220,600 | 141,800 | 86,600 | 6,346 |
| ⁹⁄₁₆ | 19 | 33,700 | 24,100 | 16,100 | 9,640 | 637 | 1⅝ | 61 | 271,300 | 189,900 | 122,100 | 74,600 | 5,497 |
| ½ | 19 | 26,700 | 19,100 | 12,700 | 7,620 | 504 | 1½ | 61 | 229,600 | 160,700 | 103,300 | 63,100 | 4,665 |
| | | | | | | | 1⅜ | 61 | 192,000 | 134,400 | 86,400 | 52,800 | 3,901 |

Minimum elongation, percent, in 24 in.: E.H.S., 4; H.S., 5; S.M., 8.

**Table 102. Galvanized Mast-arm Rope**

(Composed of nine 4-wire strands and a cotton center)

| Diam, in............................................... | ⅜ | ⁵⁄₁₆ | ¼ |
|---|---|---|---|
| Weight per ft, lb....................................... | 0.158 | 0.107 | 0.070 |
| Approx breaking stress, lb.............................. | 2200 | 1530 | 1100 |

## Table 103. Galvanized Steel Hawsers

(Fig. 153. Composed of 6 strands and a fiber core, 37 wires to the strand)

| Diam, in. | Approx weight per ft, lb | Breaking strength, tons | |
|---|---|---|---|
| | | Blue center | Plow steel |
| 2⅜ | 8.74 | 199.0 | 173.0 |
| 2⁵⁄₁₆ | 8.29 | 189.0 | 164.0 |
| 2¼ | 7.85 | 179.0 | 156.0 |
| 2⅛ | 7.00 | 160.0 | 139.0 |
| 2¹⁄₁₆ | 6.59 | 151.0 | 132.0 |
| 2 | 6.20 | 143.0 | 124.0 |
| 1¹⁵⁄₁₆ | 5.82 | 134.0 | 117.0 |
| 1¹³⁄₁₆ | 5.09 | 118.0 | 102.0 |
| 1¾ | 4.75 | 110.0 | 95.7 |
| 1¹¹⁄₁₆ | 4.41 | 103.0 | 89.2 |
| 1⅝ | 4.09 | 95.3 | 82.9 |
| 1½ | 3.49 | 81.5 | 70.9 |
| 1⁷⁄₁₆ | 3.20 | 75.0 | 65.3 |
| 1⅜ | 2.93 | 68.8 | 59.8 |
| 1¼ | 2.42 | 57.1 | 49.7 |
| 1³⁄₁₆ | 2.19 | 51.7 | 44.9 |
| 1⅛ | 1.96 | 46.5 | 40.4 |
| 1¹⁄₁₆ | 1.75 | 41.6 | 36.1 |
| 1 | 1.55 | 36.9 | 32.1 |
| ⅞ | 1.19 | 28.4 | 24.7 |
| ¹³⁄₁₆ | 1.02 | 24.5 | 21.3 |
| ¾ | 0.87 | 21.0 | 18.2 |

## Table 104. Galvanized Steel Mooring Lines and Hawsers

(Fig. 154. Composed of 6 strands and a fiber core, each strand composed of 24 wires and a fiber center)

| Diam, in. | Approx weight per ft, lb | Breaking strength, tons | |
|---|---|---|---|
| | | Blue center | Plow steel |
| 2¹⁄₁₆ | 5.87 | 134.0 | 116.0 |
| 2 | 5.52 | 126.0 | 110.0 |
| 1¹⁵⁄₁₆ | 5.18 | 119.0 | 103.0 |
| 1¹³⁄₁₆ | 4.53 | 104.0 | 90.8 |
| 1¾ | 4.23 | 97.5 | 84.8 |
| 1¹¹⁄₁₆ | 3.93 | 90.9 | 79.0 |
| 1⅝ | 3.64 | 84.5 | 73.4 |
| 1½ | 3.11 | 72.3 | 62.9 |
| 1⁷⁄₁₆ | 2.85 | 66.5 | 57.9 |
| 1⅜ | 2.61 | 61.0 | 53.1 |
| 1¼ | 2.16 | 50.7 | 44.1 |
| 1³⁄₁₆ | 1.95 | 45.9 | 39.9 |
| 1⅛ | 1.75 | 41.2 | 35.9 |
| 1¹⁄₁₆ | 1.56 | 36.9 | 32.1 |
| 1 | 1.38 | 32.8 | 28.5 |
| ⅞ | 1.06 | 25.2 | 21.9 |
| ¹³⁄₁₆ | 0.91 | 21.8 | 19.0 |
| ¾ | 0.78 | 18.6 | 16.2 |
| ⅝ | 0.54 | 13.0 | 11.3 |
| ½ | 0.35 | 8.40 | 7.30 |
| ⅜ | 0.194 | 4.77 | 4.14 |

## Table 105. Standard 6 × 12 Galvanized Running Rope and Hawsers

(Fig. 155. Composed of 6 strands and a fiber core. each strand consisting of 12 wires and a fiber center)

| Diam, in. | Approx weight per ft, lb | Breaking strength, tons | | |
|---|---|---|---|---|
| | | Blue center | Plow steel | Iron |
| 2¹⁄₁₆ | 4.47 | 93.6 | 81.4 | |
| 2 | 4.20 | 88.2 | 76.7 | |
| 1¹⁵⁄₁₆ | 3.94 | 83.0 | 72.2 | |
| 1¹³⁄₁₆ | 3.45 | 73.0 | 63.5 | |
| 1¾ | 3.22 | 68.3 | 59.4 | |
| 1¹¹⁄₁₆ | 2.99 | 63.6 | 55.3 | |
| 1⅝ | 2.77 | 59.2 | 51.4 | |
| 1½ | 2.36 | 50.7 | 44.1 | |
| 1⁷⁄₁₆ | 2.17 | 46.7 | 40.6 | |
| 1⅜ | 1.99 | 42.8 | 37.2 | |
| 1¼ | 1.64 | 35.6 | 30.9 | |
| 1³⁄₁₆ | 1.48 | 32.2 | 28.0 | |
| 1⅛ | 1.33 | 29.0 | 25.2 | |
| 1¹⁄₁₆ | 1.19 | 25.9 | 22.5 | 10.0 |
| 1 | 1.05 | 23.0 | 20.0 | 8.89 |
| ⅞ | 0.80 | 17.7 | 15.4 | 6.85 |
| ¹³⁄₁₆ | 0.69 | 15.3 | 13.3 | 5.92 |
| ¾ | 0.59 | 13.1 | 11.4 | 5.06 |
| ⅝ | 0.41 | 9.16 | 7.97 | 3.54 |
| ⁹⁄₁₆ | 0.33 | 7.45 | 6.48 | 2.88 |
| ½ | 0.26 | 5.91 | 5.14 | 2.28 |
| ⁷⁄₁₆ | 0.20 | 4.55 | 3.95 | 1.76 |
| ⅜ | 0.15 | 3.36 | 2.92 | 1.30 |
| ⁵⁄₁₆ | 0.10 | 2.34 | 2.04 | 0.905 |

## Table 106. Galvanized Iron Rigging and Guy Rope

(Fig. 156. Composed of 6 strands and a fibercore, 7 wires to the strand)

| Diam, in. | Approx weight per ft, lb | Breaking strength, tons | Circum of good-grade three-strand manila rope of nearest strength, in. |
|---|---|---|---|
| 1¼ | 2.34 | 21.2 | 7 |
| 1³⁄₁₆ | 2.12 | 19.2 | 6½ |
| 1⅛ | 1.90 | 17.3 | 6 |
| 1¹⁄₁₆ | 1.70 | 15.5 | 5½ |
| 1 | 1.50 | 13.8 | 5¼ |
| ⅞ | 1.15 | 10.7 | 4¾ |
| ¹³⁄₁₆ | 0.99 | 9.23 | 4¼ |
| ¾ | 0.84 | 7.90 | 3¾ |
| ⅝ | 0.59 | 5.54 | 3¼ |
| ⁹⁄₁₆ | 0.48 | 4.51 | 3 |
| ½ | 0.38 | 3.58 | 2½ |
| ⁷⁄₁₆ | 0.29 | 2.76 | 2¼ |
| ⅜ | 0.21 | 2.04 | 2 |
| ⁵⁄₁₆ | 0.15 | 1.42 | 1½ |
| ¼ | 0.094 | 0.918 | 1¼ |

When made with wire strand core add 10 percent to weights and 7½ percent to breaking strengths.

**Table 107. Galvanized Steel Bridge Rope**
(Figs. 157 and 158. All ropes contain 7 strands)

| Diam, in. | No. of wires per strand | Approx weight per ft, lb | Gross metallic area, sq in. | Min ultimate strength, tons | Diam, in. | No. of wires per strand | Approx weight per ft, lb | Gross metallic area, sq in. | Min ultimate strength, tons |
|---|---|---|---|---|---|---|---|---|---|
| 1 | 7 | 1.67 | 0.471 | 45.7 | 2⅛ | 19 | 7.73 | 2.17 | 210 |
| 1⅛ | 7 | 2.11 | 0.596 | 57.8 | 2¼ | 19 | 8.66 | 2.42 | 235 |
| 1¼ | 7 | 2.64 | 0.745 | 72.2 | 2⅜ | 19 | 9.61 | 2.69 | 261 |
| 1⅜ | 7 | 3.21 | 0.906 | 87.8 | 2½ | 19 | 10.60 | 2.97 | 288 |
| 1½ | 7 | 3.82 | 1.076 | 104 | 2⅝ | 19 | 11.62 | 3.27 | 317 |
| 1⅝ | 19 | 4.51 | 1.27 | 123 | 2¾ | 19 | 12.74 | 3.58 | 347 |
| 1¾ | 19 | 5.24 | 1.47 | 143 | 2⅞ | * | 13.90 | 3.91 | 379 |
| 1⅞ | 19 | 6.03 | 1.69 | 164 | 3 | 28 | 15.11 | 4.25 | 412 |
| 2 | 19 | 6.85 | 1.92 | 186 | | | | | |

\* Center strand, 28 wires; outside strand, 19 wires.

**Table 108. Galvanized Steel Bridge Strand**
(See Fig. 159)

| Diam, in. | Approx weight per ft, lb | Approx area, sq in. | Min ultimate strength, tons |
|---|---|---|---|
| 1 | 2.00 | 0.577 | 61 |
| 1¹⁄₁₆ | 2.30 | 0.663 | 69 |
| 1⅛ | 2.61 | 0.751 | 78 |
| 1³⁄₁₆ | 2.92 | 0.843 | 86 |
| 1¼ | 3.22 | 0.931 | 96 |
| 1⁵⁄₁₆ | 3.58 | 1.04 | 106 |
| 1⅜ | 3.89 | 1.12 | 116 |
| 1⁷⁄₁₆ | 4.29 | 1.24 | 126 |
| 1½ | 4.70 | 1.36 | 138 |
| 1⁹⁄₁₆ | 5.11 | 1.48 | 150 |
| 1⅝ | 5.52 | 1.60 | 162 |
| 1¹¹⁄₁₆ | 5.98 | 1.73 | 176 |
| 1¾ | 6.45 | 1.87 | 188 |

**Table 109. Smooth Coil Track Strand**
(See Fig. 160)

| Diam, in. | No. of wires in strand | Approx weight per 100 ft, lb | Breaking strength, tons | |
|---|---|---|---|---|
| | | | Extra-high-strength steel | High-strength steel |
| ½ | 19 | 55 | 15.3 | 12.6 |
| ⁹⁄₁₆ | 19 | 70 | 18.0 | 15.0 |
| ⅝ | 19 | 86 | 22.3 | 19.2 |
| ¾ | 19 | 124 | 32.5 | 27.6 |
| ⅞ | 19 | 169 | 44.4 | 37.6 |
| 1 | 19 | 220 | 58.0 | 49.2 |
| 1⅛ | 37 | 270 | 70.7 | 60.0 |
| 1¼ | 37 | 323 | 84.6 | 71.8 |
| 1⅜ | 37 | 401 | 105.0 | 88.8 |
| 1½ | 37 | 488 | 127.5 | 108.4 |
| 1⅝ | 61 | 563 | 146.0 | 124.0 |
| 1¾ | 61 | 659 | 171.0 | 145.8 |
| 1⅞ | 61 | 728 | 189.0 | 161.0 |
| 2 | 91 | 840 | 218.0 | 185.0 |
| 2⅛ | 91 | 935 | 240.0 | 204.0 |
| 2¼ | 91 | 1,036 | 266.0 | 233.0 |
| 2½ | 91 | 1,310 | 335.0 | 285.0 |

**Table 110. Copper, Iron, Tinned, and Galvanized Sash Cords**
(Six 7-wire strands and cotton core)

| Diam, in. | ¼ | ⁷⁄₃₂ | ³⁄₁₆ | ⁵⁄₃₂ | ⅛ | ³⁄₃₂ | ¹⁄₁₆ |
|---|---|---|---|---|---|---|---|
| Weight per ft, lb | | | | | | | |
| Copper | 0.108 | 0.083 | 0.061 | 0.044 | 0.026 | 0.015 | 0.007 |
| Iron | 0.094 | 0.072 | 0.053 | 0.038 | 0.023 | 0.013 | 0.006 |
| Breaking strength, lb | | | | | | | |
| Bright copper | 1,225 | 940 | 688 | 478 | 306 | 172 | 77 |
| Annealed copper | 760 | 580 | 425 | 295 | 190 | 105 | 48 |
| Galvanized iron | 1,836 | 1,413 | 1,035 | 756 | 504 | 283 | 126 |
| Bright iron | 2,040 | 1,570 | 1,150 | 840 | 560 | 315 | 140 |
| Annealed iron | 1,225 | 940 | 688 | 478 | 306 | 172 | 77 |

**Table 111. Tiller Rope or Hand Rope**
(6 strands of 42 wires each, 252 wires in all, 7 fiber cores)

| Diam, in. | ⅝ | ⁹⁄₁₆ | ½ | ⁷⁄₁₆ | ⅜ | ⁵⁄₁₆ | ¼ |
|---|---|---|---|---|---|---|---|
| Approx weight per ft, lb | 0.43 | 0.35 | 0.28 | 0.21 | 0.16 | 0.11 | 0.07 |
| Breaking strength, tons | | | | | | | |
| Plow steel | 8.04 | 6.53 | 5.18 | 3.98 | 2.93 | 2.05 | 1.31 |
| Iron | 3.57 | 2.90 | 2.30 | 1.77 | 1.30 | 0.908 | 0.584 |

**Table 112. Breaking Strength of Fiber Lines, Lb**
(Adapted, by permission of the U.S. Naval Institute, Annapolis Md., and Wall Rope Works, Inc., New York, N.Y.)

| Diam | Cir. | Manila | Composite | Sisal | Sisal mixed | Sisal hemp | Agave or jute | Nylon | Dacron | Poly-ethylene | Poly-propylene (mono-filament) | Esterlon (polyester) |
|---|---|---|---|---|---|---|---|---|---|---|---|---|
| 3/16 | 5/8 | 450 | ..... | 360 | 340 | 310 | 270 | 1,000 | 850 | 700 | 800 | 720 |
| 1/4 | 3/4 | 600 | ..... | 480 | 450 | 420 | 360 | 1,500 | 1,380 | 1,200 | 1,200 | 1,150 |
| 5/16 | 1 | 1,000 | ..... | 800 | 750 | 700 | 600 | 2,500 | 2,150 | 1,750 | 2,100 | 1,750 |
| 3/8 | 1 1/8 | 1,350 | ..... | 1,080 | 1,010 | 950 | 810 | 3,500 | 3,000 | 2,500 | 3,100 | 2,450 |
| 7/16 | 1 1/4 | 1,750 | ..... | 1,400 | 1,310 | 1,230 | 1,050 | 4,800 | 4,500 | 3,400 | 3,700 | 3,400 |
| 1/2 | 1 1/2 | 2,650 | ..... | 2,120 | 1,990 | 1,850 | 1,590 | 6,200 | 5,500 | 4,100 | 4,200 | 4,400 |
| 9/16 | 1 3/4 | 3,450 | ..... | 2,760 | 2,590 | 2,410 | 2,070 | 8,300 | 7,300 | 4,600 | 5,100 | 5,700 |
| 5/8 | 2 | 4,400 | ..... | 3,520 | 3,300 | 3,080 | 2,640 | 10,500 | 9,500 | 5,200 | 5,800 | 7,300 |
| 3/4 | 2 1/4 | 5,400 | ..... | 4,320 | 4,050 | 3,780 | 3,240 | 14,000 | 12,500 | 7,400 | 8,200 | 9,500 |
| 13/16 | 2 1/2 | 6,500 | ..... | 5,200 | 4,880 | 4,550 | 3,900 | 17,000 | 15,000 | 8,900 | 9,800 | 11,500 |
| 7/8 | 2 3/4 | 7,700 | ..... | ..... | ..... | ..... | ..... | 20,000 | 17,500 | 10,400 | 11,500 | 13,500 |
| 1 | 3 | 9,000 | ..... | 7,200 | 6,750 | 6,300 | 5,400 | 24,000 | 20,000 | 12,600 | 14,000 | 16,500 |
| 1 1/16 | 3 1/4 | 10,500 | ..... | 8,400 | 7,870 | 7,350 | 6,300 | 28,000 | 22,500 | 14,500 | 16,100 | 19,000 |
| 1 1/8 | 3 1/2 | 12,000 | ..... | 9,600 | 9,000 | 8,400 | 7,200 | 32,000 | 25,000 | 16,500 | 18,300 | 21,500 |
| 1 1/4 | 3 3/4 | 13,500 | ..... | 10,800 | 10,120 | 9,450 | 8,100 | 36,500 | 28,500 | 18,600 | 21,000 | 24,300 |
| 1 5/16 | 4 | 15,000 | ..... | 12,000 | 11,250 | 10,500 | 9,000 | 42,000 | 32,000 | 21,200 | 24,000 | 28,000 |
| 1 1/2 | 4 1/2 | 18,500 | 16,600 | 14,800 | 13,900 | 12,950 | 11,100 | 51,000 | 41,000 | 26,700 | 30,000 | 34,500 |
| 1 5/8 | 5 | 22,500 | 20,300 | 18,000 | 16,900 | 15,800 | 13,500 | 62,000 | 50,000 | 32,700 | 36,500 | 41,500 |
| 1 3/4 | 5 1/2 | 26,500 | 23,800 | 21,200 | 19,900 | 18,500 | 15,900 | 77,500 | 61,000 | 39,500 | 44,000 | 51,000 |
| 2 | 6 | 31,000 | 27,900 | 24,800 | 23,200 | 21,700 | 18,600 | 90,000 | 72,000 | 47,700 | 53,000 | 61,000 |
| 2 1/8 | 6 1/2 | 36,000 | ..... | ..... | ..... | ..... | ..... | 105,000 | 81,000 | 55,800 | 62,000 | 70,200 |
| 2 1/4 | 7 | 41,000 | 36,900 | 32,800 | 30,800 | 28,700 | ..... | 125,000 | 96,000 | 63,000 | 70,000 | 81,000 |
| 2 1/2 | 7 1/2 | 46,500 | ..... | ..... | ..... | ..... | ..... | 138,000 | 110,000 | 72,500 | 80,500 | 92,000 |
| 2 5/8 | 8 | 52,000 | 46,800 | 41,600 | 39,000 | 36,400 | ..... | 154,000 | 125,000 | 81,000 | 90,000 | 103,000 |
| 2 7/8 | 8 1/2 | 58,000 | ..... | ..... | ..... | ..... | ..... | 173,000 | 140,000 | 92,000 | 100,000 | 116,000 |
| 3 | 9 | 64,000 | 57,500 | 51,200 | 48,000 | 44,800 | ..... | 195,000 | 155,000 | 103,000 | 116,000 | 130,000 |
| 3 1/4 | 10 | 77,000 | 69,300 | 61,600 | 57,300 | 53,900 | ..... | 238,000 | 190,000 | 123,000 | 137,000 | 160,000 |
| 3 1/2 | 11 | 91,000 | ..... | ..... | ..... | ..... | ..... | 288,000 | 230,000 | 146,000 | 162,000 | 195,000 |
| 4 | 12 | 105,000 | 94,500 | 84,000 | 78,800 | 73,500 | ..... | 342,000 | 275,000 | 171,000 | 190,000 | 230,000 |

Breaking strength is the maximum load the line will hold at the time of breaking. The working load of a line is one-fourth to one-fifth the breaking strength.

**Table 113.  Wire Nails for Special Purposes**
(Steel wire gage)

| Length, in. | Barrel nails | | Barbed roofing nails | | Barbed dowel nails | |
|---|---|---|---|---|---|---|
| | Gage | No. per lb | Gage | No. per lb | Gage | No. per lb |
| 5/8 | 15½ | 1,570 | .. | ... | 8 | 394 |
| 3/4 | 15½ | 1,315 | 13 | 729 | 8 | 306 |
| 7/8 | 14½ | 854 | 12 | 478 | 8 | 250 |
| 1 | 14½ | 750 | 12 | 416 | 8 | 212 |
| 1⅛ | 14½ | 607 | 12 | 368 | 8 | 183 |
| 1¼ | 14 | 539 | 11 | 250 | 8 | 162 |
| 1⅜ | 13 | 386 | 11 | 228 | 8 | 145 |
| 1½ | 13 | 355 | 10 | 167 | 8 | 131 |
| 1¾ | .... | ..... | 10 | 143 | | |
| 2 | .... | ..... | 9 | 104 | | |

| Length, in. | Clout nails | | Slating nails | | Fine nails | | |
|---|---|---|---|---|---|---|---|
| | Gage | No. per lb | Gage | No. per lb | Length, in. | Gage | No. per lb |
| 3/4 | 15 | 999 | | | | | |
| 7/8 | 14 | 733 | | | | | |
| 1 | 14 | 648 | 12 | 425 | 1 | 16½ | 1,280 |
| 1⅛ | 14 | 580 | 10½ | 229 | | | |
| 1¼ | 13 | 398 | .... | ... | 1 | 17 | 1,492 |
| 1⅜ | 13 | 365 | | | | | |
| 1½ | 13 | 336 | 10½ | 190 | 1⅛ | 15 | 757 |
| 1¾ | .. | ... | 10 | 144 | 1⅛ | 16 | 984 |
| 2 | .. | ... | 9 | 104 | | | |

**Table 114.  Approximate Number of Boat Spikes to a Keg of 200 Lb**

| Size of spike, sq in. | Length of side, in. | | | | | | | | | |
|---|---|---|---|---|---|---|---|---|---|---|
| | 4 | 5 | 6 | 7 | 8 | 9 | 10 | 11 | 12 | 14 |
| 5/8 | ..... | ..... | ..... | ..... | 214 | 190 | 176 | ... | 144 | 122 |
| 1/2 | ..... | ..... | ..... | ..... | 324 | 286 | 258 | 244 | 220 | 192 |
| 7/16 | ..... | ..... | ..... | 480 | 438 | ... | 378 | | | |
| 3/8 | 1,114 | 930 | 816 | 690 | 622 | 532 | 492 | ... | 434 | |
| 5/16 | 1,776 | 1,342 | 1,124 | 978 | 858 | 776 | 706 | | | |
| 1/4 | 2,576 | 2,134 | 1,778 | 1,488 | 1,382 | | | | | |

**Table 115.  Wire Nails and Spikes**
(Steel wire gage)

| Size of nail | Length, in. | Casing nails | | Finishing nails | | Clinch nails | | Shingle nails | |
|---|---|---|---|---|---|---|---|---|---|
| | | Gage | No. per lb | Gage | No. per lb | Gage | No. per lb | Gage | No. per lb |
| 2d | 1 | 15½ | 940 | 16½ | 1,473 | 14 | 723 | 13 | 434 |
| 3d | 1¼ | 14½ | 588 | 15½ | 880 | 13 | 432 | 12 | 271 |
| 4d | 1½ | 14 | 453 | 15 | 634 | 12 | 273 | 12 | 233 |
| 5d | 1¾ | 14 | 389 | 15 | 535 | 12 | 234 | 12 | 203 |
| 6d | 2 | 12½ | 223 | 13 | 288 | 11 | 158 | | |
| 7d | 2¼ | 12½ | 200 | 13 | 254 | 11 | 140 | | |
| 8d | 2½ | 11½ | 136 | 12½ | 196 | 10 | 101 | | |
| 9d | 2¾ | 11½ | 124 | 12½ | 178 | 10 | 91.4 | | |
| 10d | 3 | 10½ | 90 | 11½ | 124 | 9 | 70 | | |
| 12d | 3¼ | 10½ | 83 | 11½ | 113 | 9 | 64.1 | | |
| 16d | 3½ | 10 | 69 | 11 | 93 | 8 | 50 | | |
| 20d | 4 | 9 | 51 | 10 | 65 | 7 | 36.4 | | |
| 30d | 4½ | 9 | 45 | | | | | | |
| 40d | 5 | 8 | 37 | | | | | | |

**Table 115. Wire Nails and Spikes** (*Continued*)

| Size of nail | Length, in. | Boat nails Heavy Diam. in. | Boat nails Heavy No. per lb | Boat nails Light Diam, in. | Boat nails Light No. per lb | Hinge nails Heavy Diam, in. | Hinge nails Heavy No. per lb | Hinge nails Light Diam, in. | Hinge nails Light No. per lb | Flooring nails Gage | Flooring nails No. per lb |
|---|---|---|---|---|---|---|---|---|---|---|---|
| 4d | 1½ | ¼ | 47 | 3/16 | 82 | ¼ | 53 | 3/16 | 90 | 11 | 168 |
| 6d | 2 | ¼ | 36 | 3/16 | 62 | ¼ | 39 | 3/16 | 66 | 10 | 105 |
| 8d | 2½ | ¼ | 29 | 3/16 | 50 | ¼ | 31 | 3/16 | 53 | 9 | 72 |
| 10d | 3 | ⅜ | 11 | ¼ | 24 | ⅜ | 12 | ¼ | 25 | 8 | 56 |
| 12d | 3¼ | ⅜ | 10.4 | ¼ | 22 | ⅜ | 11 | ¼ | 23 | 7 | 44 |
| 16d | 3½ | ⅜ | 9.6 | ¼ | 20 | ⅜ | 10 | ¼ | 22 | 6 | 32 |
| 20d | 4 | ⅜ | 8 | ¼ | 18 | ⅜ | 8 | ¼ | 19 |  |  |

| Size of nail | Length, in. | Common wire nails and brads Gage | Common wire nails and brads No. per lb | Barbed car nails Heavy Gage | Barbed car nails Heavy No. per lb | Barbed car nails Light Gage | Barbed car nails Light No. per lb | Spikes Size | Spikes Length, in. | Spikes Gage | Spikes Approx No. per lb |
|---|---|---|---|---|---|---|---|---|---|---|---|
| 2d | 1 | 15 | 847 | .. | ... | .. | ... | 10d | 3 | 6 | 43 |
| 3d | 1¼ | 14 | 548 | .. | ... | .. | ... | 12d | 3¼ | 6 | 39 |
| 4d | 1½ | 12½ | 294 | 10 | 179 | 12 | 284 | 16d | 3½ | 5 | 31 |
| 5d | 1¾ | 12½ | 254 | 9 | 124 | 10 | 152 | 20d | 4 | 4 | 23 |
| 6d | 2 | 11½ | 167 | 9 | 108 | 10 | 132 | 30d | 4½ | 3 | 18 |
| 7d | 2¼ | 11½ | 150 | 8 | 80 | 9 | 95 | 40d | 5 | 2 | 14 |
| 8d | 2½ | 10¼ | 101 | 8 | 72 | 9 | 88 | 50d | 5½ | 1 | 11 |
| 9d | 2¾ | 10¼ | 92 | 7 | 55 | 8 | 65 | 50d | 6 | 1 | 10 |
| 10d | 3 | 9 | 66 | 7 | 50 | 8 | 59 | ... | 7 | 5/16 in. | 7 |
| 12d | 3¼ | 9 | 61 | 6 | 39 | 7 | 46 | ... | 8 | ⅜ | 4.1 |
| 16d | 3½ | 8 | 47 | 6 | 36 | 7 | 43 | ... | 9 | ⅜ | 3.7 |
| 20d | 4 | 6 | 30 | 5 | 27 | 6 | 32 | ... | 10 | ⅜ | 3.3 |
| 30d | 4½ | 5 | 23 | 5 | 24 | 6 | 28 | ... | 12 | ⅜ | 2.7 |
| 40d | 5 | 4 | 18 | 4 | 18 | 5 | 22 |  |  |  |  |
| 50d | 5½ | 3 | 14 | 3 | 14 | 4 | 17 |  |  |  |  |
| 60d | 6 | 2 | 11 | 3 | 13 | 4 | 15 |  |  |  |  |

**Table 116. Cut Steel Nails and Spikes**
(Sizes, lengths, and approximate number per lb)

| Sizes | Length, inches | Common | Clinch | Finishing | Casing and box | Fencing | Spikes | Barrel | Slating | Tobacco | Brads | Shingle |
|---|---|---|---|---|---|---|---|---|---|---|---|---|
| 2d | 1 | 740 | 400 | 1,100 | ...... | ...... | ...... | 450 | 340 |  |  |  |
| 3d | 1¼ | 460 | 260 | 880 | ...... | ...... | ...... | 280 | 280 |  |  |  |
| 4d | 1½ | 280 | 180 | 530 | 420 | ...... | ...... | 190 | 220 |  |  |  |
| 5d | 1¾ | 210 | 125 | 350 | 300 | 100 | ...... | ...... | 180 | 130 |  |  |
| 6d | 2 | 160 | 100 | 300 | 210 | 80 | ...... | ...... | ...... | 97 | 120 |  |
| 7d | 2¼ | 120 | 80 | 210 | 180 | 60 | ...... | ...... | ...... | 85 | 94 |  |
| 8d | 2½ | 88 | 68 | 168 | 130 | 52 | ...... | ...... | ...... | 68 | 74 | 90 |
| 9d | 2¾ | 73 | 52 | 130 | 107 | 38 | ...... | ...... | ...... | 58 | 62 | 72 |
| 10d | 3 | 60 | 48 | 104 | 88 | 26 | ...... | ...... | ...... | 48 | 50 | 60 |
| 12d | 3¼ | 46 | 40 | 96 | 70 | 20 | ...... | ...... | ...... | ...... | 40 |  |
| 16d | 3½ | 33 | 34 | 86 | 52 | 18 | 17 | ...... | ...... | ...... | 27 |  |
| 20d | 4 | 23 | 24 | 76 | 38 | 16 | 14 |  |  |  |  |  |
| 25d | 4¼ | 20 | ...... | ...... | ...... | ...... |  |  |  |  |  |  |
| 30d | 4½ | 16½ | ...... | ...... | 30 | ...... | 11 |  |  |  |  |  |
| 40d | 5 | 12 | ...... | ...... | 26 | ...... | 9 |  |  |  |  |  |
| 50d | 5½ | 10 | ...... | ...... | 20 | ...... | 7½ |  |  |  |  |  |
| 60d | 6 | 8 | ...... | ...... | 16 | ...... | 6 |  |  |  |  |  |
| ...... | 6½ | ...... | ...... | ...... | ...... | ...... | 5½ |  |  |  |  |  |
| ...... | 7 | ...... | ...... | ...... | ...... | ...... | 5 |  |  |  |  |  |

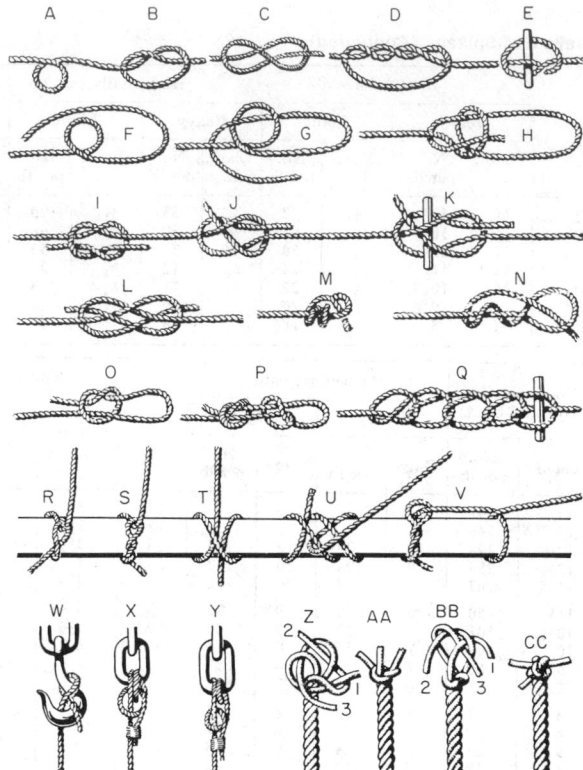

**Fig. 161**    Rope knots, hitches, and bends.

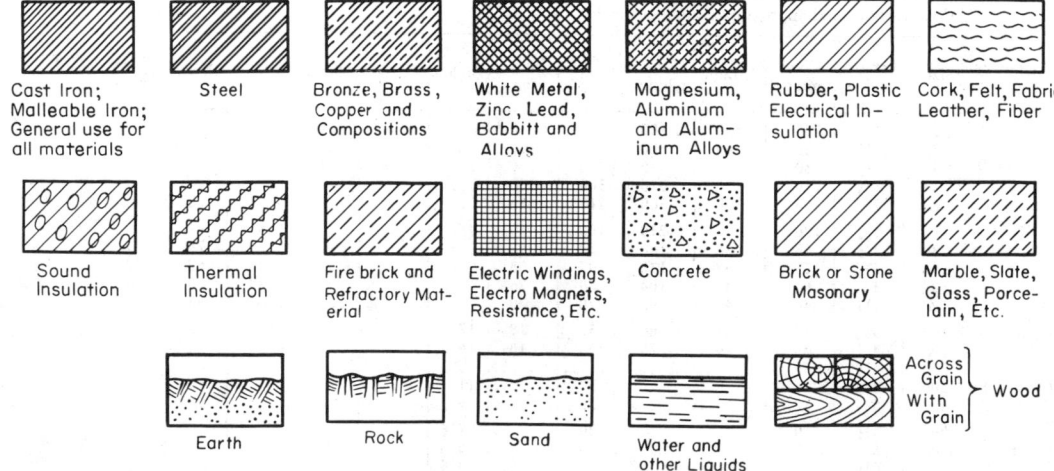

| | | |
|---|---|---|
| Cast Iron; Malleable Iron; General use for all materials | Steel | Bronze, Brass, Copper and Compositions |

White Metal, Zinc, Lead, Babbitt and Alloys

Magnesium, Aluminum and Aluminum Alloys

Rubber, Plastic Electrical Insulation

Cork, Felt, Fabric, Leather, Fiber

Sound Insulation

Thermal Insulation

Fire brick and Refractory Material

Electric Windings, Electro Magnets, Resistance, Etc.

Concrete

Brick or Stone Masonary

Marble, Slate, Glass, Porcelain, Etc.

Earth

Rock

Sand

Water and other Liquids

Across Grain / With Grain } Wood

**Fig. 162**    Standard cross sections.

**Table 117. Sizes of American Wire Tacks**

| Oz | Length, in. | Size of wire, steel wire gage | | | Oz | Length, in. | Size of wire, steel wire gage | | |
|---|---|---|---|---|---|---|---|---|---|
| | | Uphol-sterers | Carpet | Bill posters | | | Uphol-sterers | Carpet | Bill posters |
| 1 | $\frac{3}{16}$ | 18 | 18 | ....... | 10 | $\frac{5}{8}$ | 14½ | 15 | 12 |
| 1½ | $\frac{7}{32}$ | 18 | 18 | ....... | 12 | $\frac{11}{16}$ | 14½ | 15 | 12 |
| 2 | $\frac{1}{4}$ | 17 | 17 | 15 | 14 | $\frac{3}{4}$ | 14 | 14½ | 11½ |
| 2½ | $\frac{5}{16}$ | 17 | 17 | 15 | 16 | $\frac{13}{16}$ | 14 | 14½ | 11½ |
| 3 | $\frac{3}{8}$ | 16 | 16 | 14 | 18 | $\frac{7}{8}$ | 13½ | 13½ | 11 |
| 4 | $\frac{7}{16}$ | 16 | 16 | 13½ | 20 | $\frac{15}{16}$ | 13½ | 13½ | 11 |
| 6 | $\frac{1}{2}$ | 15 | 15 | 13 | 22 | 1 | 13½ | 13½ | 10½ |
| 8 | $\frac{9}{16}$ | 15 | 15 | 12½ | 24 | 1⅛ | 13 | 13 | 10½ |

knot; *D*, double knot; *E*, boat knot, *F*, bowline, first step; *G*, bowline, second step; *H*, bowline, completed; *I*, square or reef knot; *J*, sheet bend or weaver's knot; *K*, sheet bend with a toggle; *L*, carrick bend; *M*, "stevedore" knot completed; *N*, "stevedore" knot commenced; *O*, slip knot; *P*, Flemish loop; *Q*, chain knot with toggle; *R*, half hitch; *S*, timber hitch; *T*, clove hitch; *U*, rolling hitch; *V*, timber hitch and half hitch; *W*, blackwall hitch; *X*, fisherman's bend; *Y*, round turn and half hitch; *Z*, wall knot commenced; *AA*, wall knot completed; *BB*, wall-knot crown commenced; *CC*, wall-knot crown completed.

The bowline *H*, one of the most useful knots, will not slip, and after being strained is easily untied. Knots *H*, *K*, and *M* are easily untied after being under strain. The knot *M* is useful when the rope passes through an eye and is held by the knot, as it will not slip, and is easily untied after being strained. The wall knot is made as follows: Form a bight with strand 1 and pass the strand 2 around the end of it, and the strand 3 around the end of 2, and then through the bight of 1, as shown at *Z* in the figure. Haul the ends taut when the appearance is as shown at *AA*. The end of the strand 1 is now laid over the center of the knot, strand 2 laid over 1, and 3 over 2, when the end of 3 is passed through the bight of 1, as shown at *BB*. Haul all the strands taut, as shown at *CC*. The "stevedore" knot (*M*, *N*) is used to hold the end of a rope from passing through a hole. When the rope is strained, the knot draws up tight, but it can be easily untied when the strain is removed. If a knot or hitch of any kind is tied in a rope, its failure under stress is sure to occur at that place. The shorter the bend in the standing rope, the weaker is the knot. The approximate strength of knots compared with the full strength of (dry) rope ( = 100), based on Miller's experiments (*Mach.*, 1900, p. 198), is as follows: eye splice over iron thimble, 90; short splice in rope, 80; *S* and *Y*, 65; *H*, *O*, and *T*, 60; *I* and *J*, 50; *B* and *P*, 45.

## NAILS AND SPIKES

**Nails** are either **wire nails** of circular cross section and constant diameter or **cut nails** of rectangular cross section with taper from head to point. The larger sizes are called **spikes**. The length of the nail is expressed in the "**penny**" system, the equivalents in inches being given in Tables 113 to 117. The letter d is the accepted symbol for penny. A keg of nails

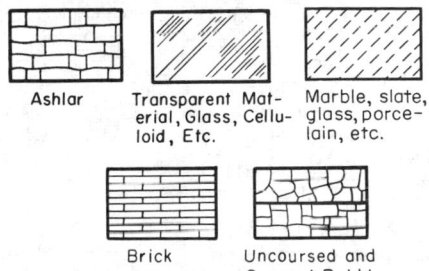

Ashlar          Transparent Material, Glass, Celluloid, Etc.          Marble, slate, glass, porcelain, etc.

Brick          Uncoursed and Coursed Rubble

**Fig. 163**  Standard outside views.

weighs 100 lb. **Heavy hinge nails** or **track nails** with countersunk heads have chisel points unless diamond points are specified. **Plasterboard nails** are smooth with circumferential grooves and have diamond points. **Spikes** are made either with flat heads and diamond points or with oval heads and chisel points.

## STANDARD CROSS-SECTION SYMBOLS

Figures 162 and 163 are abstracted from American Drafting Standards Manual (ANSI Y14.2-1973). Subdivisions of any of the materials may be made by taking one of these standard cross sections as a basis and making minor changes or by writing on the standard cross section the name of the material.

# GEARING

## by George W. Michalec

REFERENCES: Buckingham, "Manual of Gear Design," Industrial Press. Cunningham, F. W., Noncircular Gears, *Machine Design*, Feb. 19, 1957. Cunningham, F. E., and D. S. Cunningham, Rediscovering the Noncircular Gear, *Machine Design*, Nov. 1, 1973. Dudley, "Gear Handbook," McGraw-Hill. Dudley, "Practical Gear Design," McGraw-Hill. Michalec, "Precision Gearing: Theory and Practice," Wiley. Shigely, "Engineering Design," McGraw-Hill. AGMA Standards: 390.03, "Gear Handbook," etc. "Gleason Bevel and Hypoid Gear Design," Gleason Works, Rochester.

### Notation

$a$ = addendum
$b$ = dedendum
$B$ = backlash, linear measure along pitch circle
$c$ = clearance
$C$ = center distance
$d_b$ = base diam of pinion
$d_o$ = outside diam of pinion
$d_r$ = root diam of pinion
$D$ = pitch diameter
$D_P$ = pitch diam of pinion
$D_G$ = pitch diam of gear
$D_o$ = outside diam of gear
$D_b$ = base diam of gear
$D_t$ = throat diam of worm gear
$F$ = face width
$b_k$ = working depth
$b_t$ = whole depth
inv $\phi$ = involute function (tan $\phi - \phi$)
$l$ = lead (advance of worm or helical gear in 1 rev)
$l_P(l_G)$ = lead of pinion (gear) in helical gears
$L$ = lead of worm
$m$ = module
$m_G$ = gear ratio ($m_G = N_G/N_P$)
$m_p$ = contact ratio (of profiles)
$M$ = measurement of over pins
$n_p(n_G)$ = speed of pinion (gear), r/min
$N_P(N_G)$ = number of teeth in pinion (gear)
$n_w$ = number of threads in worm
$p$ = circular pitch
$p_b$ = base pitch (= normal pitch for helical gears)
$p_n$ = normal circular pitch of helical gear, worm mesh
$P_d$ = diametral pitch
$P_{dn}$ = normal diametral pitch
$R$ = pitch radius
$R_c$ = radial distance from center of gear to center of measuring pin
$R_P(R_G)$ = pitch radius of pinion (gear)
$R_T$ = testing radius when rolled on a variable center distance inspection fixture
$s$ = stress
$t$ = tooth thickness
$t_n$ = normal circular tooth thickness
$T_P(T_G)$ = formative number of teeth in pinion (gear) (in bevel gears)

$v$ = pitch line velocity
$X$ = correction factor for profile shift
$\alpha$ = addendum angle
$\gamma$ = pitch angle of bevel pinion
$\gamma R$ = face angle at root of pinion tooth
$\gamma o$ = face angle at tip of pinion tooth
$\Gamma$ = pitch angle of bevel gear
$\Gamma_R$ = face angle at root of gear tooth
$\Gamma g_o$ = face angle at tip of gear tooth
$\delta$ = dedendum angle
$\overline{\Delta C}$ = relatively small change in center distance $C$
$\phi$ = pressure angle
$\psi$ = helix or spiral angle
$\psi_P(\psi_G)$ = helix angle of teeth in pinion (gear)
$\Sigma$ = shaft angle

## BASIC GEAR DATA

**Gear Types** Gears are grouped in accordance with tooth forms, shaft arrangement, pitch, and quality. Tooth forms and shaft arrangements are:

| Tooth form | Shaft arrangement |
|---|---|
| Spur ......... | Parallel |
| Helical ........ | Parallel or skew |
| Worm ......... | Skew |
| Bevel ......... | Intersecting |
| Hypoid ........ | Skew |

Pitch divisions are: coarse (below $20P_d$) and fine ($20P_d$ and finer). Quality types are: commercial, precision, and ultraprecision.

**Pitch definitions** (see Fig. 1). **Diametral pitch** $P_d$ is the ratio of number of teeth in the gear to the diameter of the pitch circle $D$ measured in inches, $P_d = N/D$. **Circular pitch** $p$ is the linear measure in inches along the pitch circle between corresponding points of adjacent teeth. From these definitions, $P_d p = \pi$. The **base pitch** $p_b$ is the distance along the line of action between successive involute tooth surfaces. The base and

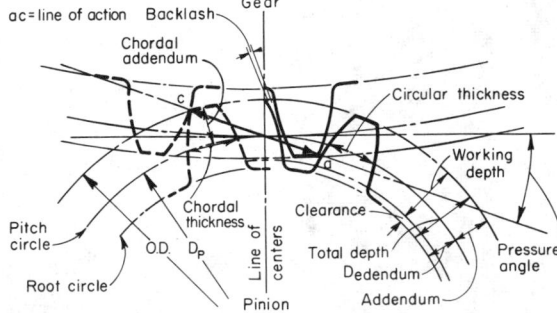

**Fig. 1** Gear-tooth nomenclature.

circular pitches are related as $p_b = p \cos \phi$, where $\phi$ = the pressure angle.

**Pitch circle** is the imaginary circle that rolls without slippage with a pitch circle of a mating gear. The pitch (circle) diameter equals $D = N/P_d = N p/\pi$.

**Pressure angle** $\phi$ for all gear types is the acute angle between the common normal to the profiles at the contact point and the common pitch plane. For spur gears it is simply the acute angle formed by the common tangent between base circles of mating gears and a normal to the line of centers. For standard gears pressure angles of $14\frac{1}{2}°$ and 20° have been adopted by the ANSI and the gear industry (see Fig. 1).

The **base circle** (or **base cylinder**) is the circle from which the involute tooth profiles are generated. The relationship between the base-circle and pitch-circle diameter is $D_b = D \cos \phi$.

**Tooth proportions** are established by the addendum, dedendum, working depth, clearance, tooth circular thickness, and pressure angle (see Fig. 1). In addition, gear face width $F$ establishes thickness of the gear measured parallel to the gear axis.

For involute teeth, proportions have been standardized by ANSI and AGMA into a limited number of systems using a basic rack for specification (see Fig. 2, Table 1). For nonspur and noninvolute types see Table 2.

**Gear ratio** (or **mesh ratio** ) $m_G$ is the ratio of number of teeth in a meshed pair, expressed as a number greater than 1; $m_G = N_G/N_P$, where the pinion is the member having the lesser number of teeth. For spur and parallel-shaft helical gears, the base-circle ratio must be identical to the gear ratio. The speed ratio of gears is inversely proportionate to their numbers of teeth. Only for standard spur and parallel-shaft helical gears is the

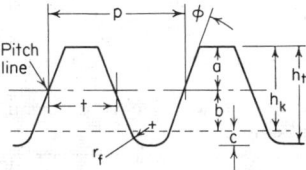

**Fig. 2** Basic rack for involute-gear systems. $a$ = addendum, $b$ = dedendum, $c$ = clearance, $b_k$ = working depth, $b_t$ = whole depth, $p$ = circular pitch, $r_f$ = fillet radius, $t$ = tooth thickness, $\phi$ = pressure angle.

### Table 1. Tooth Proportions of Basic Rack for Standard Involute Gear Systems

| Tooth parameter (of basic rack) | Symbol, Fig. 1 | Tooth proportions for various standard systems | | | | | |
|---|---|---|---|---|---|---|---|
| | | 1 | 2 | 3 | 4 | 5 | 6 |
| | | Full depth involute, $14\frac{1}{2}$ deg | Full depth involute, 20 deg | Stub involute, 20 deg | Coarse pitch involute spur gears, 20 deg | Coarse pitch involute spur gears, 25 deg | Fine pitch involute, 20 deg |
| 1. System sponsors | | ANSI and AGMA | ANSI | ANSI and AGMA | AGMA | AGMA | ANSI and AGMA |
| 2. Pressure angle | $\phi$ | $14\frac{1}{2}°$ | 20° | 20° | 20° | 25° | 20° |
| 3. Addendum | $a$ | $1/P_d$ | $1/Pd$ | $0.8/P_d$ | $1.000/P_d$ | $1.000/P_d$ | $1.000/P_d$ |
| 4. Min dedendum | $b$ | $1.157/P_d$ | $1.157/P_d$ | $1/P_d$ | $1.250/P_d$ | $1.250/P_d$ | $1.200/P_d + 0.002$ |
| 5. Min whole depth | $b_t$ | $2.157/P_d$ | $2.157/P_d$ | $1.8/P_d$ | $2.250/P_d$ | $2.250/P_d$ | $2.2002/P_d$ 0.002 in |
| 6. Working depth | $b_k$ | $2/Pd$ | $2/Pd$ | $1.6/P_d$ | $2.000/P_d$ | $2.000/P_d$ | $2.000/P_d$ |
| 7. Min clearance | $c$ | $0.157/P_d$ | $0.157/P_d$ | $0.200/P_d$ | $0.250/P_d$ | $0.250/P_d$ | $0.200/P_d + 0.002$ in |
| 8. Basic circular tooth thickness on pitch line | $t$ | $1.5708/P_d$ | $1.5708/P_d$ | $1.5708/P_d$ | $\pi/2P_d$ | $\pi/2P_d$ | $1.5708/P_d$ |
| 9. Fillet radius in basic rack | $r_f$ | $1\frac{1}{3} \times$ clearance | $1\frac{1}{2} \times$ clearance | Not standardized | $0.300/P_d$ | $0.300/P_d$ | Not standardized |
| 10. Diametral pitch range | | Not specified | Not specified | Not specified | 19.99 and coarser | 19.99 and coarser | 20 and finer |
| 11. Governing standard: ANSI | | B6.1 | B6.1 | B6.1 | | | B6.7 |
| AGMA | | 201.02A | | 201.02A | 201.02 | 201.02 | 207.05 |

### Table 2. Nonspur Gear System Standards

| Gear type | ANSI no. | AGMA no. | Title |
|---|---|---|---|
| Straight bevel....... | B6.13 | 208.02 | System for Straight Bevel Gears |
| Spiral bevel......... | ...... | 209.03 | System for Spiral Bevel Gears |
| Zerol bevel......... | ...... | 202.03 | Zerol Bevel Gear System |
| Face.............. | ...... | 203.03 | Fine-pitch On-center Face Gears for 20 Deg Involute Spur Pinions |
| Wormgearing........ | B6.9 | 374.04 | Design for Fine-pitch Wormgearing |

pitch diameter ratio equal to the gear ratio and inversely proportionate to the speed ratio.

### Metric Gears—Tooth Proportions and Standards

Metric gearing not only is based upon different units of length measure but also involves its own unique design standard. This means that metric gears and American-standard-inch diametral-pitch gears are not interchangeable.

In the metric system the *module m* is analogous to pitch and is defined as

$$m = \frac{D}{N} = \text{millimeters of pitch diameter per tooth}$$

Note that, for the module to have proper units, the pitch diameter must be in millimeters.

Tooth proportions for metric gears are specified by the International Standards Organization (ISO) metric module in terms of the basic rack (see Fig. 3). Specific size dimensions are obtained from multiplying by $m$ (the module). The ISO metric gear standard permits wide interchangeability of metric gears conforming to particular standards such as DIN and JIS.

The preferred standard gears of the metric system are not interchangeable with the preferred diametral-pitch sizes. Table 3 lists commonly used pitches and modules of both systems (preferred values are boldface).

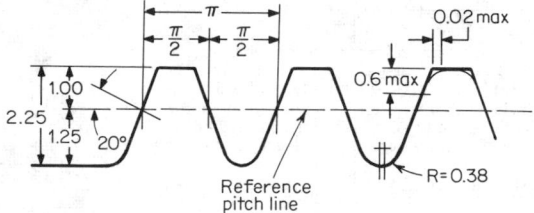

**Fig. 3**   The ISO basic rack for metric-module gears.

## FUNDAMENTAL RELATIONSHIPS OF SPUR AND HELICAL GEARS

**Center distance** is the distance between axes of mating gears and is determined from $C = (n_G + N_P)/2P_d$, or $C = (D_G + D_P)/2$. Deviation from ideal center distance of involute gears is not detrimental to proper (conjugate) gear action which is one of the prime superiority features of the involute tooth form.

**Contact Ratio**   Referring to the top part of Fig. 4 and assuming no tip relief, the pinion engages in the gear at $a$, where the outside circle of the gear tooth intersects the line of action $ac$. For the usual spur gear and pinion combination there will be two pairs of teeth theoretically in contact at engagement (a gear tooth and its mating pinion tooth considered as a pair). This will continue until the pair ahead (bottom part of Fig. 4) disengages at $c$, where the outside circle of the pinion intersects the line of action $ac$, the movement along the line of action being $ab$. After disengagement the pair behind will be the only pair in contact until another pair engages, the movement along the line of action for single-pair contact being $bd$. Two pairs are theoretically in contact during the remaining intervals, $ab + dc$.

**Contact ratio** expresses the average number of pairs of teeth theoretically in contact and is obtained numerically by divid-

ing the length of the line of action by the normal pitch. For full-depth teeth, without undercutting, the contact ratio is $m_p = (\sqrt{D_o^2 - D_b^2} + \sqrt{d_o^2 - d_b^2} - C \sin \phi)/2p \cos \phi$. The result will be a mixed number with the integer portion the number of pairs of teeth always in contact and carrying load, and the decimal portion the amount of time an additional pair of teeth are engaged and share load. As example, for $m_p$ between 1 and 2:

Load is carried by one pair, $(2 - m_p)/m_p$ of the time.

Load is carried by two pairs, $2(m_p - 1)/m_p$ of the time.

In Figs. 5 to 7 contact ratios are given for standard generated gears, the lower part of Figs. 5 and 6 representing the effect of undercutting.

These charts are applicable to both standard diametral-pitch gears made in accordance with American standards and also standard metric gears that have an addendum of one module.

**Tooth Thickness**   For standard gears, the tooth thickness $t$ of mating gears is equal, where $t = p/2 = \pi/2P_d$ measured linearly along the arc of the pitch circle. The tooth thickness $t_1$ at any radial point of the tooth (at diameter $D_1$) can be calculated from the known thickness $t$ at the pitch radius $(D/2)$ by the relationship $t_1 = t(D_1/D) - D_1 (\text{inv } \phi_1 - \text{inv } \phi)$, where inv $\phi = \tan \phi - \phi$ (see Buckingham and Dudley for detail values).

**Over-pins measurements** (spur gears) are another means of deriving tooth thickness. If cylindrical pins are inserted in tooth spaces diametrically opposite one another (or nearest space for odd number teeth), Fig. 8, the tooth thickness can be derived from the measurement $M$ by $t = D(\pi/N + \text{inv } \phi_1 - \text{inv } \phi - d_w/D_b)$, cos $\phi_1 = (D \cos \phi)/2R_c$, $R_c = (M - d_w)/2$ (for even number of teeth), $R_c = (M - d_w)/[2 \cos (90/N)]$ (for odd number of teeth), where $d_w$ = pin diameter, $R_c$ = distance from gear center to center of pin, and $M$ = measurement over pins.

For the reverse situation, the over-pins measurement $M$ can be found for a given tooth thickness $t$ at diameter $D$ and pressure angle $\phi$ by the following: inv $\phi_1 = t/D + \text{inv } \phi + d_w/(D \cos \phi) - \pi/N$, $M = (D \cos \phi)/\cos \phi_1 + d_w$ (for even number of teeth), $M = [(D \cos \phi)/\cos \phi] \cos (90°/N) + d_w$ (for odd number of teeth).

Table values of over-pins measures (see Dudley and Van

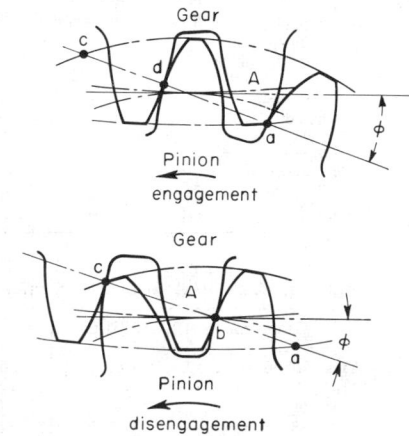

**Fig. 4**   Contact conditions at engagement and disengagement.

**Table 3. Metric and American Gear Equivalents**

| Diametral pitch $P_d$ | Module $m$ | Circular pitch | | Circular tooth thickness | | Addendum | |
|---|---|---|---|---|---|---|---|
| | | Inches | Millimeters | Inches | Millimeters | Inches | Millimeters |
| 1/2 | 50.8000 | 6.2832 | 159.593 | 3.1416 | 79.7965 | 2.0000 | 50.8000 |
| 0.5080 | 50 | 6.1842 | 157.080 | 3.0921 | 78.5398 | 1.9685 | 50 |
| 0.5644 | 45 | 5.5658 | 141.372 | 2.7850 | 70.6858 | 1.7730 | 45 |
| 0.6048 | 42 | 5.1948 | 131.947 | 2.5964 | 65.9734 | 1.6529 | 42 |
| 0.6513 | 39 | 4.8237 | 122.522 | 2.4129 | 61.2610 | 1.5361 | 39 |
| 0.7056 | 36 | 4.4527 | 113.097 | 2.2249 | 56.5487 | 1.4164 | 36 |
| 3/4 | 33.8667 | 4.1888 | 106.396 | 2.0943 | 53.1977 | 1.3333 | 33.8667 |
| 0.7697 | 33 | 4.0816 | 103.673 | 2.0400 | 51.8363 | 1.2987 | 33 |
| 0.8467 | 30 | 3.7105 | 94.248 | 1.8545 | 47.1239 | 1.1806 | 30 |
| 0.9407 | 27 | 3.3395 | 84.823 | 1.6693 | 42.4115 | 1.0627 | 27 |
| 1 | 25.4000 | 3.1416 | 79.800 | 1.5708 | 39.8984 | 1.0000 | 25.4001 |
| 1.0583 | 24 | 2.9685 | 75.398 | 1.4847 | 37.6991 | 0.9452 | 24 |
| 1.1546 | 22 | 2.7210 | 69.115 | 1.3600 | 34.5575 | 0.8658 | 22 |
| 1.2700 | 20 | 2.4737 | 62.832 | 1.2368 | 31.4159 | 0.7874 | 20 |
| 1.4111 | 18 | 2.2263 | 56.548 | 1.1132 | 28.2743 | 0.7087 | 18 |
| 1.5 | 16.9333 | 2.0944 | 53.198 | 1.0472 | 26.5988 | 0.6667 | 16.933 |
| 1.5875 | 16 | 1.9790 | 50.267 | 0.9894 | 25.1327 | 0.6299 | 16 |
| 1.8143 | 14 | 1.7316 | 43.983 | 0.8658 | 21.9911 | 0.5512 | 14 |
| 2 | 12.7000 | 1.5708 | 39.898 | 0.7854 | 19.949 | 0.5000 | 12.7000 |
| 2.1167 | 12 | 1.4842 | 37.699 | 0.7420 | 18.8496 | 0.4724 | 12 |
| 2.5 | 10.1600 | 1.2566 | 31.918 | 0.6283 | 15.9593 | 0.4000 | 10.1600 |
| 2.5400 | 10 | 1.2368 | 31.415 | 0.6184 | 15.7080 | 0.3937 | 10 |
| 2.8222 | 9 | 1.1132 | 28.275 | 0.5565 | 14.1372 | 0.3543 | 9 |
| 3 | 8.4667 | 1.0472 | 26.599 | 0.5235 | 13.2995 | 0.3333 | 8.4667 |
| 3.1416 | 8.0851 | 1.0000 | 25.400 | 0.5000 | 12.7000 | 0.3183 | 8.0851 |
| 3.1750 | 8 | 0.9895 | 25.133 | 0.4948 | 12.5664 | 0.3150 | 8.00 |
| 3.5 | 7.2571 | 0.8976 | 22.799 | 0.4488 | 11.3994 | 0.2857 | 7.2571 |
| 3.6286 | 7 | 0.8658 | 21.991 | 0.4329 | 10.9956 | 0.2756 | 7.000 |
| 3.9078 | 6.5 | 0.8039 | 20.420 | 0.4020 | 10.2101 | 0.2559 | 6.5 |
| 4 | 6.3500 | 0.7854 | 19.949 | 0.3927 | 9.9746 | 0.2500 | 6.3500 |
| 4.2333 | 6 | 0.7421 | 18.850 | 0.3710 | 9.4248 | 0.2362 | 6.0000 |
| 4.6182 | 5.5 | 0.6803 | 17.279 | 0.3401 | 8.6394 | 0.2165 | 5.5 |
| 5 | 5.0801 | 0.6283 | 15.959 | 0.3142 | 7.9794 | 0.2000 | 5.080 |
| 5.0802 | 5 | 0.6184 | 15.707 | 0.3092 | 7.8537 | 0.1968 | 5.000 |
| 5.3474 | 4.75 | 0.5875 | 14.923 | 0.2938 | 7.4612 | 0.1870 | 4.750 |
| 5.6444 | 4.5 | 0.5566 | 14.138 | 0.2783 | 7.0688 | 0.1772 | 4.500 |
| 6 | 4.2333 | 0.5236 | 13.299 | 0.2618 | 6.6497 | 0.1667 | 4.233 |
| 6.3500 | 4 | 0.4947 | 12.565 | 0.2473 | 6.2827 | 0.1575 | 4.000 |
| 6.7733 | 3.75 | 0.4638 | 11.781 | 0.2319 | 5.8903 | 0.1476 | 3.750 |
| 7 | 3.6286 | 0.4488 | 11.399 | 0.2244 | 5.6998 | 0.1429 | 3.629 |
| 7.2571 | 3.5 | 0.4329 | 10.996 | 0.2164 | 5.4979 | 0.1378 | 3.500 |
| 7.8154 | 3.25 | 0.4020 | 10.211 | 0.2010 | 5.1054 | 0.1279 | 3.250 |
| 8 | 3.1750 | 0.3927 | 9.974 | 0.1964 | 4.9886 | 0.1250 | 3.175 |
| 8.4667 | 3 | 0.3711 | 9.426 | 0.1855 | 4.7130 | 0.1181 | 3.000 |
| 9 | 2.8222 | 0.3491 | 8.867 | 0.1745 | 4.4323 | 0.1111 | 2.822 |
| 9.2364 | 2.75 | 0.3401 | 8.639 | 0.1700 | 4.3193 | 0.1082 | 2.750 |
| 10 | 2.5400 | 0.3142 | 7.981 | 0.1571 | 3.9903 | 0.1000 | 2.540 |
| 10.1600 | 2.50 | 0.3092 | 7.854 | 0.1546 | 3.9268 | 0.0984 | 2.500 |
| 11 | 2.3091 | 0.2856 | 7.254 | 0.1428 | 3.6271 | 0.0909 | 2.309 |
| 11.2889 | 2.25 | 0.2783 | 7.069 | 0.1391 | 3.5344 | 0.0886 | 2.250 |
| 12 | 2.1167 | 0.2618 | 6.646 | 0.1309 | 3.3325 | 0.0833 | 2.117 |
| 12.7000 | 2 | 0.2474 | 6.284 | 0.1236 | 3.1420 | 0.0787 | 2.000 |
| 13 | 1.9538 | 0.2417 | 6.139 | 0.1208 | 3.0696 | 0.0769 | 1.954 |
| 14 | 1.8143 | 0.2244 | 5.700 | 0.1122 | 2.8500 | 0.0714 | 1.814 |
| 14.5143 | 1.75 | 0.2164 | 5.497 | 0.1082 | 2.7489 | 0.0689 | 1.750 |
| 15 | 1.6933 | 0.2094 | 5.319 | 0.1047 | 2.6599 | 0.0667 | 1.693 |
| 16 | 1.5875 | 0.1964 | 4.986 | 0.0982 | 2.4936 | 0.0625 | 1.587 |
| 16.9333 | 1.5 | 0.1855 | 4.712 | 0.0927 | 2.3562 | 0.0591 | 1.500 |
| 18 | 1.4111 | 0.1745 | 4.432 | 0.0873 | 2.2166 | 0.0556 | 1.411 |
| 20 | 1.2700 | 0.1571 | 3.990 | 0.0785 | 1.9949 | 0.0500 | 1.270 |
| 20.3200 | 1.25 | 0.1546 | 3.927 | 0.0773 | 1.9635 | 0.0492 | 1.250 |

**Table 3. Metric and American Gear Equivalents** *(Continued)*

| Diametral pitch $P_d$ | Module $m$ | Circular pitch | | Circular tooth thickness | | Addendum | |
|---|---|---|---|---|---|---|---|
| | | Inches | Millimeters | Inches | Millimeters | Inches | Millimeters |
| 22 | 1.1545 | 0.1428 | 3.627 | 0.0714 | 1.8136 | 0.0455 | 1.155 |
| **24** | 1.0583 | 0.1309 | 3.325 | 0.0655 | 1.6624 | 0.0417 | 1.058 |
| 25.4000 | **1** | 0.1237 | 3.142 | 0.0618 | 1.5708 | 0.0394 | 1.000 |
| 28 | 0.90701 | 0.1122 | 2.850 | 0.0561 | 1.4249 | 0.0357 | 0.9071 |
| 28.2222 | **0.9** | 0.1113 | 2.827 | 0.0556 | 1.4137 | 0.0354 | 0.9000 |
| 30 | 0.84667 | 0.1047 | 2.659 | 0.0524 | 1.3329 | 0.0333 | 0.8467 |
| 31.7500 | **0.8** | 0.0989 | 2.513 | 0.04945 | 1.2566 | 0.0315 | 0.8000 |
| **32** | 0.79375 | 0.0982 | 2.494 | 0.04909 | 1.2468 | 0.0313 | 0.7937 |
| 33.8667 | **0.75** | 0.0928 | 2.357 | 0.04638 | 1.1781 | 0.0295 | 0.7500 |
| 36 | 0.70556 | 0.0873 | 2.217 | 0.04363 | 1.1083 | 0.0278 | 0.7056 |
| 36.2857 | **0.7** | 0.0865 | 2.199 | 0.04325 | 1.0996 | 0.0276 | 0.7000 |
| 40 | 0.63500 | 0.0785 | 1.994 | 0.03927 | 0.9975 | 0.0250 | 0.6350 |
| 42.3333 | **0.6** | 0.0742 | 1.885 | 0.03710 | 0.9423 | 0.0236 | 0.6000 |
| 44 | 0.57727 | 0.0714 | 1.814 | 0.03570 | 0.9068 | 0.0227 | 0.5773 |
| **48** | 0.52917 | 0.0655 | 1.661 | 0.03272 | 0.8311 | 0.0208 | 0.5292 |
| 50 | 0.50800 | 0.0628 | 1.595 | 0.03141 | 0.7976 | 0.0200 | 0.5080 |
| 50.8000 | **0.5** | 0.06184 | 1.5707 | 0.03092 | 0.7854 | 0.0197 | 0.5000 |
| 63.5000 | **0.4** | 0.04947 | 1.2565 | 0.02473 | 0.6283 | 0.0157 | 0.4000 |
| **64** | 0.39688 | 0.04909 | 1.2469 | 0.02454 | 0.6234 | 0.0156 | 0.3969 |
| 67.7333 | 0.375 | 0.04638 | 1.1781 | 0.02319 | 0.5890 | 0.0148 | 0.3750 |
| **72** | 0.35278 | 0.04363 | 1.1082 | 0.02182 | 0.5541 | 0.0139 | 0.3528 |
| 72.5714 | 0.35 | 0.04329 | 1.0996 | 0.02164 | 0.5498 | 0.0138 | 0.3500 |
| 78.1538 | 0.325 | 0.04020 | 1.0211 | 0.02010 | 0.5105 | 0.0128 | 0.3250 |
| **80** | 0.31750 | 0.03927 | 0.9975 | 0.01964 | 0.4987 | 0.0125 | 0.3175 |
| 84.6667 | **0.3** | 0.03711 | 0.9426 | 0.01856 | 0.4713 | 0.0118 | 0.3000 |
| 92.3636 | 0.275 | 0.03401 | 0.8639 | 0.01700 | 0.4319 | 0.0108 | 0.2750 |
| **96** | 0.26458 | 0.03272 | 0.8311 | 0.01636 | 0.4156 | 0.0104 | 0.2646 |
| 101.6000 | 0.25 | 0.03092 | 0.7854 | 0.01546 | 0.3927 | 0.00984 | 0.2500 |
| **120** | 0.21167 | 0.02618 | 0.6650 | 0.01309 | 0.3325 | 0.00833 | 0.2117 |
| 125 | 0.20320 | 0.02513 | 0.6383 | 0.01256 | 0.3192 | 0.00800 | 0.2032 |
| 127.0000 | **0.2** | 0.02474 | 0.6284 | 0.01237 | 0.3142 | 0.00787 | 0.2000 |
| 150 | 0.16933 | 0.02094 | 0.5319 | 0.01047 | 0.2659 | 0.00667 | 0.1693 |
| 169.3333 | 0.15 | 0.01855 | 0.4712 | 0.00928 | 0.2356 | 0.00591 | 0.1500 |
| 180 | 0.14111 | 0.01745 | 0.4432 | 0.00873 | 0.2216 | 0.00555 | 0.1411 |
| **200** | 0.12700 | 0.01571 | 0.3990 | 0.00786 | 0.1995 | 0.00500 | 0.1270 |
| 203.2000 | 0.125 | 0.01546 | 0.3927 | 0.00773 | 0.1963 | 0.00492 | 0.1250 |

Keuren) facilitate measurements for all standard gears including those with slight departures from standard. (For correlation with tooth thickness and testing radius, see Michalec, *Product Eng.*, May 1957, and "Precision Gearing: Theory and Practice," Wiley, 1966.)

**Testing radius** $R_T$ is another means of determining tooth thickness and refers to the effective pitch radius of the gear when rolled intimately with a master gear of known size calibration. (See Michalec, *Product Eng.*, November, 1956, and "Precision Gearing: Theory and Practice," *op. cit.*) For standard design gears the testing radius equals the pitch radius. The testing radius may be corrected for small departures $\Delta t$ from ideal tooth thickness by the relationship, $R_T = R + \Delta t/2 \tan \phi$ where $\Delta t = t_1 - t$ and is positive and negative respectively for thicker and thinner tooth thicknesses than standard value $t$.

**Backlash** $B$ is the amount by which the width of a tooth space exceeds the thickness of the engaging tooth measured on the pitch circle. Backlash does not adversely affect proper gear function except for lost motion upon reversal of gear rotation. Backlash inevitably occurs because of necessary fabrication

tolerances on tooth thickness and center distance plus need for clearance to accommodate lubricant and thermal expansion. Proper backlash can be introduced by a specified amount of tooth thinning or slight increase in center distance. The relationship between small change in center distance $\overline{\Delta C}$ and backlash is $B = 2\overline{\Delta C} \tan \phi$ (see Michalec, "Precision Gearing: Theory and Practice").

**Total composite error** (tolerance) is a measure of gear quality in terms of the net sum of irregularity of its testing radius $R_T$ due to pitch-circle runout and tooth-to-tooth variations (see Michalec, *op. cit.*).

**Tooth-to-tooth composite error** (tolerance) is the variation of testing radius $R_T$ between adjacent teeth caused by tooth spacing, thickness, and profile deviations (see Michalec, *op. cit.*).

**Profile shifted gears** have tooth thicknesses that are significantly different from nominal standard value; excluded are deviations caused by normal allowances and tolerances. They are also known as **modified gears, long and short addendum gears,** and **enlarged gears.** They are produced by cutting the teeth

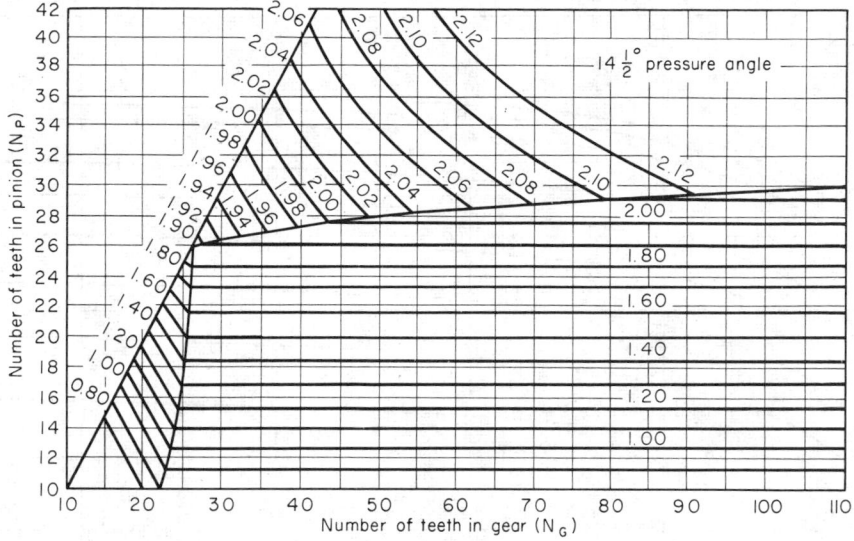

**Fig. 5**   Contact-ratio, spur-gear parts—full-depth standard generated teeth, 14½° pressure angle.

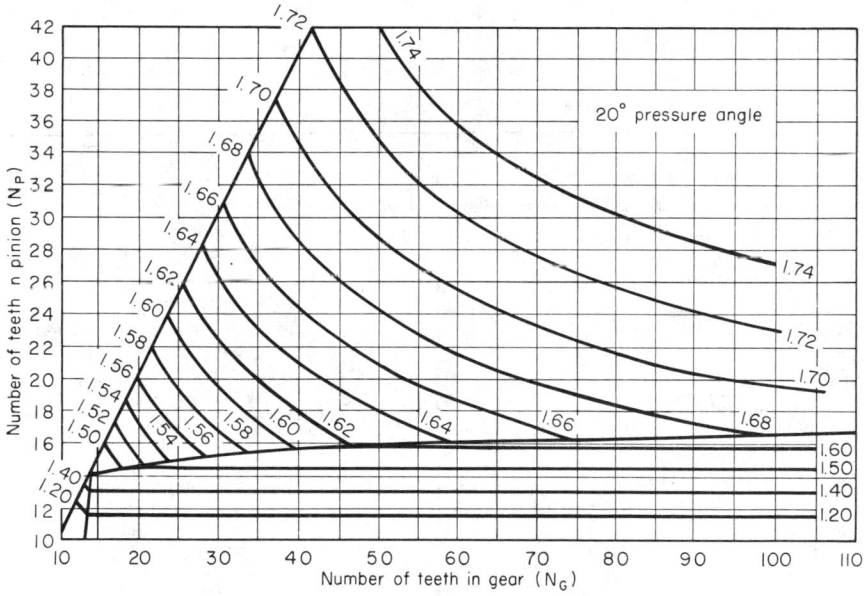

**Fig. 6**   Contact-ratio spur-gear pairs—full-depth standard generated teeth, 20° pressure angle.

with standard cutters at enlarged or reduced outside diameters. The result is a relative shift of the two families of involutes forming the tooth profiles, simultaneously with a shift of the tooth radially outward or inward (see Fig. 9). Calculation of operating conditions and tooth parameters are $C_1 = (C \cos \phi)/\cos \phi_1$, where inv $\phi_1 = $ inv $\phi + [N_P(t_G' + t_P') - \pi D_P]/[D_P (N_P + N_G)]$, $t_G' = t + 2X_G \tan \phi$, $t_P' = t + 2X_P \tan \phi$, $D_o' = (N_G + 2)/P_d + 2X_G$, $d_o' = (N_P + 2)/P_d + 2X_P$, $\phi = $

standard pressure angle, $\phi_1 = $ operating pressure angle for profile shifted gears, $C_1 = $ operating center distance for zero backlash mesh, $t_G' = $ tooth thickness of gear measured at standard pitch radius, $t_P' = $ tooth thickness of pinion measured at standard pitch radius, $X_G = $ correction for profile shift of gear, and $X_P = $ correction for profile shift of pinion. The quantity $X$ is positive for enlarged gears and negative for thinned gears.

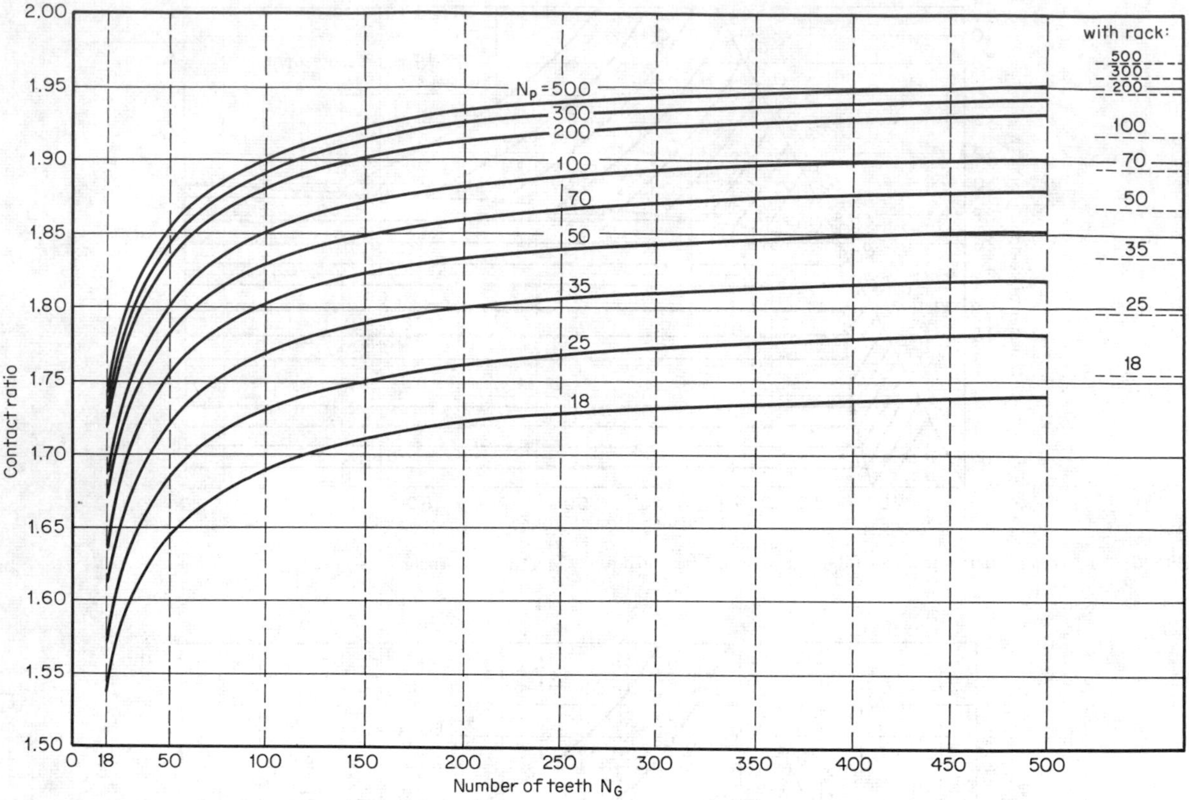

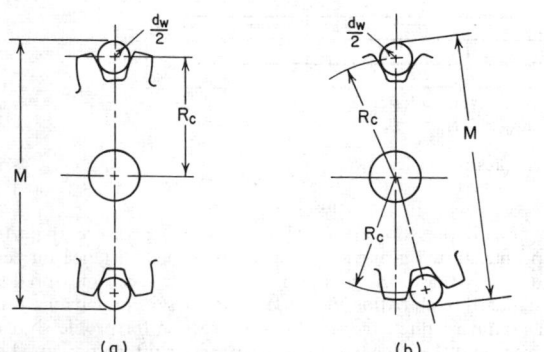

**Fig. 7**  Contact ratio for large numbers of teeth; spur-gear pairs, full-depth standard teeth; 20° pressure angle, *(Data by R. Feeney and T. Wall.)*

### Metric-Module Gear-Design Equations

Basic design equations for spur gearing utilizing the metric module are listed in Table 4.

### HELICAL GEARS

Helical gears divide into two general applications: (1) for

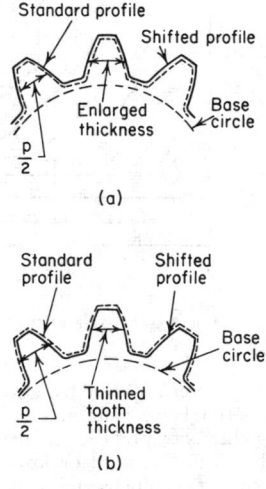

**Fig. 9**  Geometry of profile-shifted teeth. (*a*) Enlarged case; (*b*) thinned-tooth-thickness case.

**Fig. 8**  Geometry of over-pins measurements. (*a*) For even number of teeth; (*b*) for odd number of teeth.

**Table 4. Spur-Gear Design Formulas**

| To obtain | From known | Use this formula* |
|---|---|---|
| Pitch diameter $D$ | Module; diametral pitch | $D = mN$ |
| Circular pitch $p_c$ | Module; diametral pitch | $p_c = m\pi = \dfrac{D}{N}\pi = \dfrac{\pi}{P}$ |
| Module $m$ | Diametral pitch | $m = \dfrac{25.4}{P}$ |
| No. of teeth $N$ | Module and pitch diameter | $N = \dfrac{D}{m}$ |
| Addendum $a$ | Module | $a = m$ |
| Dedendum $b$ | Module | $b = 1.25m$ |
| Outside diameter $D_o$ | Module and pitch diameter or number of teeth | $D_o = D + 2m = m(N + 2)$ |
| Root diameter $D_r$ | Pitch diameter and module | $D_r = D - 2.5m$ |
| Base-circle diameter $D_b$ | Pitch diameter and pressure angle $\phi$ | $D_b = D \cos\phi$ |
| Base pitch $p_b$ | Module and pressure angle | $p_b = m\pi \cos\phi$ |
| Tooth thickness at standard pitch diameter $T_{std}$ | Module | $T_{std} = \dfrac{\pi}{2} m$ |
| Center distance $C$ | Module and number of teeth | $C = \dfrac{m(N_1 + N_2)}{2}$ |
| Contact ratio $m_p$ | Outside radii, base-circle radii, center distance, pressure angle | $m_p = \dfrac{\sqrt{{}_1R_o^2 - {}_1R_b^2} + \sqrt{{}_2R_o^2 - {}_2R_b^2} - C\sin\phi}{m\pi \cos\phi}$ |
| Backlash (linear) $B$ (along pitch circle) | Change in center distance | $B = 2(\Delta C)\tan\phi$ |
| Backlash (linear) $B$ (along pitch circle) | Change in tooth thickness, $T$ | $B = \Delta T$ |
| Backlash (linear) (along line-of-action) $B_{LA}$ | Linear backlash (along pitch circle) | $B_{LA} = B \cos\phi$ |
| Backlash (angular) $B_a$ | Linear backlash (along pitch circle) | $B_a = 6{,}880\,\dfrac{B}{D}$ (arc minutes) |
| Min. number teeth for no undercutting, $N_c$ | Pressure angle | $N_c = \dfrac{2}{\sin^2\phi}$ |

*All linear dimensions in millimeters.

driving parallel shafts and (2) for driving skew shafts (mostly at right angles), in which case they are referred to as *crossed-axis* helical gears. The helical tooth form may be imagined as consisting of an infinite number of staggered laminar spur gears, resulting in the curved cylindrical helix shape.

**Pitch of helical gears** is definable in two planes. The diametral and circular pitches measured in the plane of rotation (transverse) are defined as for spur gears. However, pitches measured normal to the tooth are related by the cosine of the helix angle; thus normal diametral pitch $= P_{dn} = P_d/\cos\psi$, normal circular pitch $= p_n = p \cos\psi$, and $P_{dn}p_n = \pi$. **Axial pitch** is the distance between corresponding sides of adjacent teeth measured parallel to the gear axis and is calculated as $p_a = p \cot\psi$.

**Pressure angle** of helical gears is definable in the normal and transverse planes by $\tan\phi_n = \tan\phi \cos\psi$. The transverse pressure angle, which is effectively the real pressure angle, is always greater than the normal pressure angle.

**Tooth thickness** $t$ of helical gears can be measured in the plane of rotation, as with spur gears, or normal to the tooth surface $t_n$. The relationship of the two thicknesses is $t_n = t \cos\psi$.

**Over-Pins Measurement of Helical Gears** Tooth thicknesses $t$ at diameter $d$ can be found from a known over-pins measurement $M$ at known pressure angle $\phi$, corresponding to diameter $D$, by $R_c = (M - d_w)/2$ (for even number of teeth), $R_c = (M - d_w)/[2 \cos(180/2N)]$ (for odd number of teeth), cos $\phi_1 = (D \cos\phi)/2R_c$, $\tan\phi_n = \tan\phi \cos\psi$, $\cos\psi_b = \sin\phi_n/\sin\phi$, and $t = D[\pi/N + \text{inv}\,\phi_1 - \text{inv}\,\phi - d_w/(D \cos\phi \cos\psi_b)]$.

**Parallel-shaft helical gears** must conform to the same conditions and requirements as spur gears with parameters (pressure angle and pitch) consistently defined in the transverse plane. Since standard spur-gear cutting tools are usually used, normal plane values are standard, resulting in nonstandard transverse pitches and nonstandard pitch diameters and center distances. For parallel shafts, helical gears must have identical helix angles but must be of opposite hand (left and right helix directions). The commonly used helix angles range from 15 to 35°. To make most advantage of the helical form, the advance of a tooth should be greater than the circular pitch; recommended ratio is 1.5 to 2 with 1.1 minimum. This overlap provides two or more teeth in continual contact with resulting greater smoothness and quietness than spur gears. Because of the helix, the normal component of the tangential pressure on the teeth produces end thrust of the shafts. To remove this objection, gears are made with opposite-handed helixes on each half of the face and are then known as herringbone gears (see Fig. 10).

**Crossed-axis helical gears,** also called *spiral* or *screw gears* (Fig. 11), are a simple type of involute gear used for connecting nonparallel, nonintersecting shafts. Contact is point and there is considerably more sliding than with parallel-axis helicals, which limits the load capacity. The individual gear of this mesh is identical in form and specification to a parallel-shaft helical gear. Crossed-axis helicals can connect any shaft angle $\Sigma$, although 90° is prevalent. Usually, the helix angles will be of the same hand, although for some extreme cases it is possible to have opposite hands, particularly if the shaft angle is small.

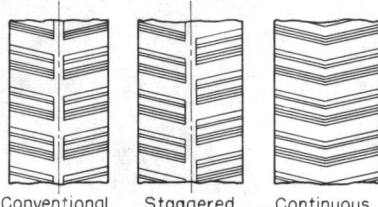

Conventional   Staggered   Continuous

**Fig. 10** Herringbone gears.

**Helical-Gear Calculations** For parallel shafts the center distance is a function of the helix angle as well as the number of teeth, that is, $C = (N_G + N_P)/2P_{dn} \cos \psi$. This offers a powerful method of gearing shafts at any specified center distance to a specified velocity ratio. For crossed-axis helicals the problem of connecting a pair of shafts for any velocity ratio admits of a number of solutions, since both the pitch radii and the helix angles contribute to establishing the velocity ratio. The formulas given in Table 5 are of assistance in calculations. The notation used in this table is as follows:

$N_P(N_G)$ = number of teeth in pinion (gear)
$D_P(D_G)$ = pitch diam of pinion (gear)
$p_P(p_G)$ = circular pitch of pinion (gear)
$p_n$ = normal circular pitch for both gears
$P_{a_n}$ = normal diametral pitch for both gears
$d_0(D_0)$ = outside diam of driver (follower)
$a$ = addendum of normal pitch
$\psi_G$ = tooth helix angle of gear
$\psi_P$ = tooth helix angle of pinion

$l_P(l_G)$ = lead of pinion (gear)
     = lead of tooth helix
$n_P(n_G)$ = r/min of pinion (gear)
$\Sigma$ = angle between shafts in plan
$C$ = center distance.

## NONSPUR GEAR TYPES

**Bevel gears** are used to connect two intersecting shafts in any given speed ratio. The tooth shapes may be designed in any of the shapes shown in Fig. 12. Bevels connecting nonintersecting shafts are called **skew bevel gears.** A special type developed by the Gleason Works and used widely in automotive products is known as **hypoid gearing** (see *Jour. SAE* **18**, no. 6), although this is not a true bevel gear. The "spherical involute" or, more accurately, **octoid** tooth form is almost universally used for bevel gears. This has a basic crown tooth with a straight-sided form.

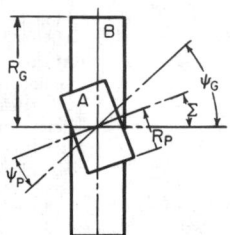

**Fig. 11** Crossed-axis helical gears.

Referring to Fig. 13, we see that the pitch surfaces of bevel gears are frustums of cones whose vertices are at the intersection of the axes; the essential elements and definitions are as follows.

    **Addendum Angle** $\alpha$ The angle between elements of the face cone and pitch cone

    **Back Angle** The angle between an element of the back cone and a plane of rotation. It is equal to the pitch angle.

    **Back Cone** The angle of a cone whose elements are tangent to a sphere containing a trace of the pitch circle.

    **Back-Cone Distance** The distance along an element of the back cone from the apex to the pitch circle.

    **Cone Distance** $A_o$ The distance from the end of the tooth (heel) to the pitch apex.

### Table 5. Formulas for Helical-Gear Calculations

| Pinion | | Gear | | |
|---|---|---|---|---|
| To find | Formula | To find | Formula | Remarks |
| $\psi_P$ | $\tan \psi_P = D_P n_P / D_G n_G$ | $\psi_G$ | 90° × | Axes at right angles only |
| $\psi_P$ | $\cos \psi_P = p_n / p_G$ | $\psi_G$ | $\Sigma - \psi_P$ | |
| $\psi_P$ | $\tan \psi_P = \pi D_P / l_P$ | $\psi_G$ | $\Sigma - \psi_P$ | |
| $p_n$ | $\pi D_P \cos \psi_P / N_P$ | $p_n$ | $\pi D_G \cos \psi_G / N_G$ | Same in both gears |
| $p_P$ | $D_P / N_P$ | $p_G$ | $\pi D_G / N_G$ | |
| $l_P$ | $\pi D_P / \tan \psi_P$ | $l_G$ | $\pi D_G / \tan \psi_G$ | Axes at right angles only |
| $N_P$ | $D_P P_d \cos \psi_P$ | $N_G$ | $D_G P_d \cos \psi_G$ | |
| $D_P$ | $2C \left/ \left( \dfrac{n_P}{n_G} \tan \psi_G + 1 \right) \right.$ | $D_G$ | $2C \left/ \left( \dfrac{n_P}{n_G} \tan \psi_P + 1 \right) \right.$ | Axes at right angles only |
| $D_P$ | $2C \left/ \left( \dfrac{n_P \cos \psi_P}{n_G \cos \psi_G} + 1 \right) \right.$ | $D_G$ | $2C - D_P$ | |
| $D_P$ | $N_P / P_{dn} \cos \psi_P$ | $D_G$ | $N_G / P_{dn} \cos \psi_G$ | |
| $d_0$ | $D_P + 2a$ | $D_0$ | $D_G + 2a$ | |
| $d_0$ | $D_P + (2/P_d)$ | $D_0$ | $D_G + (2/P_d)$ | |
| Cutter* | $N_P / \cos^3 \psi_P$ | | $N_G / \cos^3 \psi_G$ | For profile milling teeth |

Center distance $C = (N_P / 2P_d \cos \psi_P) + (N_G / 2P_d \cos \psi_G)$.
* Spur-gear cutter to be used which is correct for the number of teeth given by the formulas.

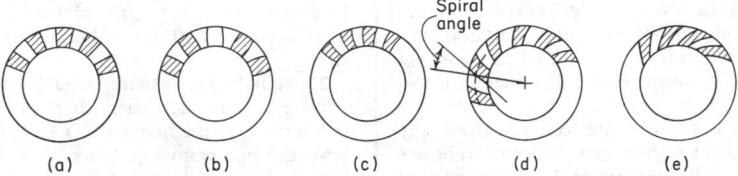

**Fig. 12** Bevel-gear types. (*a*) Old-type straight teeth; (*b*) modern coniflex straight teeth (exaggerated crowning); (*c*) Zerol teeth; (*d*) spiral teeth; (*e*) hypoid teeth.

**Crown**   The sharp corner forming the outside diameter.

**Crown-to-Back**   The distance from the outside diameter edge (crown) to the rear of the gear.

**Dedendum Angle** $\delta$   The angle between elements of the root cone and pitch cone.

**Face Angle** $\gamma_o$   The angle between an element of the face cone and its axis. It is equal to pitch angle.

**Face Width** $F$   The length of teeth along the cone distance.

**Front Angle**   The angle between an element of the front cone and a plane of rotation.

**Generating Mounting Surface,** GMS   The diameter and/or plane of rotation surface or shaft center which is used for locating the gear blank during fabrication of the gear teeth.

**Heel**   The portion of a bevel gear tooth near the outer end.

**Mounting Distance,** MD   For assembled bevel gears, the distance from the crossing point of the axes to the registering surface, measured along the gear axis. Ideally, it should be identical to the pitch apex to back.

**Mounting Surface,** MS   The diameter and/or plane of rotation surface which is used for locating the gear in the application assembly.

**Octoid**   The mathematical form of the bevel tooth profile. Closely resembles a spherical involute but is fundamentally different.

**Pitch Angle** $\Gamma$   The angle formed between an element of the pitch cone and the bevel-gear axis. It is the half angle of the pitch cone.

**Pitch Apex to Back**   The distance along the axis from apex of pitch cone to a locating registering surface on back.

**Registering Surface,** RS   The surface in the plane of rotation which locates the gear blank axially in the generating machine and the gear in application. These are usually identical surfaces, but not necessarily so.

**Root Angle,** $\gamma_R$   The angle formed between a tooth root element and the axis of the bevel gear.

**Shaft Angle,** $\Sigma$   The angle between mating bevel-gear axes; also, the sum of the two pitch angles.

**Spiral Angle,** $\psi$   The angle between the tooth trace and an element of the pitch cone, corresponding to helix angle in helical gears. The spiral angle is understood to be at the mean cone distance.

**Toe**   The portion of a bevel tooth near the inner end.

Bevel gears are described by the parameter dimensions at the large end (heel) of the teeth. Pitch, pitch diameter, and tooth dimensions, such as addendum are measurements at this point. At the large end of the gear, the tooth profiles will approximate those generated on a spur-gear pitch circle of

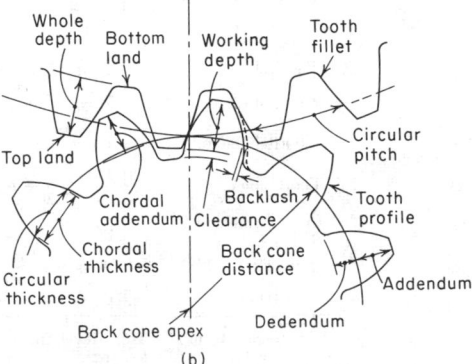

**Fig. 13**   Geometry of bevel-gear nomenclature. (*a*) Section through axes; (*b*) view along section Z-Z.

radius equal to the back-cone distance. The formative number of teeth is equal to that contained by a complete spur gear. For pinion and gear, respectively, this is $T_P = N_P/\cos\gamma$; $T_G = N_G/\cos\Gamma$, where $T_P$ and $T_G$ = formative number teeth, and $N_P$ and $N_G$ = actual number teeth.

Although bevel gears can connect intersecting shafts at any angle, most application is for right angles. When such bevels are in a 1:1 ratio, they are called **mitre gears.** Bevels connecting shafts other than 90° are called **angular bevel gears.** The speeds of the shafts of bevel gears are determined by $n_P/n_G = \sin\Gamma/\sin\gamma$, where $n_P(n_G)$ = r/min of pinion (gear), and $\gamma(\Gamma)$ = pitch angle of pinion (gear).

All standard bevel-gear designs in the United States are in accordance with the **Gleason bevel-gear system.** This employs a basic pressure angle of 20° with long and short addendums for ratios other than 1:1 to avoid undercut pinions and to increase strength.

**20° Straight Bevel Gears for 90° Shaft Angle.** Since straight bevel gears are the easiest to produce and offer maximum precision, they are frequently a first choice. Modern straight-bevel-gear generators produce a tooth with localized tooth bearing designated by the Gleason registered tradename Coniflex. These gears have a slightly crowned tooth form (see Fig. 12b). Because of the superiority of Coniflex bevel gears over the earlier "true-straight bevels" and because of their faster production, they are preferred. The design parameters of Fig. 13 are calculated by the formulas of Table 6. Backlash data are given in Table 7.

### Table 6. Straight-Bevel-Gear Dimensions*

(All linear dimensions in inches)

| | | | | | |
|---|---|---|---|---|---|
| 1. Number of pinion teeth† | $n$ | | 5 | Working depth | $h_k = \dfrac{2.000}{P_d}$ |
| 2. Number of gear teeth† | $N$ | | 6 | Whole depth | $h_t = \dfrac{2.188}{P_d} + 0.002$ |
| 3. Diametral pitch | $P_d$ | | 7 | Pressure angle | $\phi$ |
| 4. Face width | $F$ | | 8 | Shaft angle | $\Sigma$ |

| | Pinion | Gear |
|---|---|---|
| 9. Pitch diameter | $d = \dfrac{n}{P_d}$ | $D = \dfrac{N}{P_d}$ |
| 10. Pitch angle | $\gamma = \tan^{-1}\dfrac{n}{N}$ | $\Gamma = 90° - \gamma$ |
| 11. Outer cone distance | $A_O = \dfrac{D}{2\sin\Gamma}$ | |
| 12. Circular pitch | $p = \dfrac{3.1416}{P_d}$ | |
| 13. Addendum | $a_{OP} = h_k - a_{OG}$ | $a_{OG} = \dfrac{0.540}{P_d} + \dfrac{0.460}{P_d(N/n)^2}$ |
| 14. Dedendum‡ | $b_{OP} = \dfrac{2.188}{P_d} - a_{OP}$ | $b_{OG} = \dfrac{2.188}{P_d} - a_{OG}$ |
| 15. Clearance | $c = h_t - h_k$ | |
| 16. Dedendum angle | $\delta_P = \tan^{-1}\dfrac{b_{OP}}{A_O}$ | $\delta_G = \tan^{-1}\dfrac{b_{OG}}{A_O}$ |
| 17. Face angle of blank | $\gamma_O = \gamma + \delta_G$ | $\Gamma_O = \Gamma + \delta_P$ |
| 18. Root angle | $\gamma_R = \gamma - \delta_P$ | $\Gamma_R = \Gamma - \delta_G$ |
| 19. Outside diameter | $d_o = d + 2a_{OP}\cos\gamma$ | $D_o = D + 2a_{OG}\cos\Gamma$ |
| 20. Pitch apex to crown | $x_o = \dfrac{D}{2} - a_{OP}\sin\gamma$ | $X_O = \dfrac{d}{2} - a_{OG}\sin\Gamma$ |
| 21. Circular thickness | $t = p - T$ | $T = \dfrac{p}{2} - (a_{OP} - a_{OG})\tan\phi$ $-\dfrac{K \text{ (see Fig. 14)}}{P_d}$ |
| 22. Backlash | $B$ (see Table 7) | |
| 23. Chordal thickness | $t_c = t - \dfrac{t^3}{6d^2} - \dfrac{B}{2}$ | $T_C = T - \dfrac{T^3}{6D^2} - \dfrac{B}{2}$ |
| 24. Chordal addendum | $a_{CP} = a_{OP} + \dfrac{t^2\cos\gamma}{4d}$ | $a_{CG} = a_{OG} + \dfrac{T^2\cos\Gamma}{4D}$ |
| 25. Tooth angle | $\dfrac{3438}{A_O}\left(\dfrac{t}{2} + b_{OP}\tan\phi\right)$ minutes | $\dfrac{3438}{A_O}\left(\dfrac{T}{2} + b_{OG}\tan\phi\right)$ minutes |
| 26. Limit-point width (L.E.) | $W_{LOP} = (T - 2b_{OP}\tan\phi) - 0.0015$ | $W_{LOG} = (t - 2b_{OG}\tan\phi) - 0.0015$ |
| 27. Limit-point width (S.E.) | $W_{LiP} = \dfrac{A_O - F}{A_O}(T - 2b_{OP}\tan\phi)$ | $W_{LiG} = \dfrac{A_O - F}{A_O}(t - 2b_{OG}\tan\phi)$ |
| 28. Tool-point width | $W = W_{LiP} -$ stock allowance | $W = W_{LiG} -$ stock allowance |

*Abstracted from "Gleason Bevel and Hypoid Gear Design Handbook." Tables 6 to 10 and Figs. 14 and 15 from Gleason Works, Inc.

†Numbers of teeth: ratios with 16 or more teeth in pinion: 15/17 and higher; 14/20 and higher; 13/30 and higher.

‡The actual dedendum will be 0.002 in greater than calculated.

**Table 7. Recommended Backlash for Bevel-Gear Meshes***

| Diametral pitch | Backlash | |
|---|---|---|
| 1.00–1.25 | 0.020–0.030 | |
| 1.25–1.50 | 0.018–0.026 | For general-purpose bevel gears assembled ready to run. In case of choice, use the smaller backlash tolerances. |
| 1.50–1.75 | 0.016–0.022 | |
| 1.75–2.00 | 0.014–0.018 | |
| 2.00–2.50 | 0.012–0.016 | |
| 2.50–3.00 | 0.010–0.013 | In many instances these limits require modification to suit the special conditions of operation. For the finer pitches, such as precision instrument gears, it may be necessary to reduce the backlash values to a minimum. |
| 3.00–3.50 | 0.008–0.011 | |
| 3.50–4.00 | 0.007–0.009 | |
| 4–5 | 0.006–0.008 | |
| 5–6 | 0.005–0.007 | |
| 6–8 | 0.004–0.006 | |
| 8–10 | 0.003–0.005 | |
| 10–20 | 0.002–0.004 | |
| 20 and finer | 0.001–0.003 | |

*Abstracts from "Gleason Bevel and Hypoid Gear Design Handbook."

**Angular straight bevel gears** connect shaft angles other than 90° (larger or smaller), and the formulas of Table 6 are not entirely applicable, as shown in the following:

Item 8, Shaft angle, is the specified non-90° shaft angle.

Item 10, Pitch angles. Shaft angle $\Sigma$ less than 90°, $\tan \gamma = \sin \Sigma/(N/n + \cos \Sigma)$; shaft angle $\Sigma$ greater than 90°, $\tan \gamma = \sin (180 - \Sigma)/[N/n - \cos (180° - \Sigma)]$.

For all shaft angles, $\sin \gamma/\sin \Gamma = n/N$.

Item 13, Addendum, requires calculation of the equivalent 90° bevel gear ratio $m_{90}$, $m_{90} = [N \cos \gamma/n \cos \Gamma]^{1/2}$. The value $m_{90}$ is used as the ratio $N/n$ when applying the formula for addendum.

Item 20, Pitch apex to crown, $x_o = A_0 \cos \gamma - a_{op} \sin \gamma$, $X_o = A_0 \cos \Gamma - a_{oG} \sin \Gamma$.

Item 21, Circular thickness, except for high ratios, $K$ may be zero.

**Spiral Bevel Gears for 90° Shaft Angle**  The spiral curved teeth produce additional overlapping tooth action which results in smoother gear action, lower noise, and higher load capacity. The spiral angle has been standardized by Gleason at 35°. Design parameters are calculated by the formulas of Table 8.

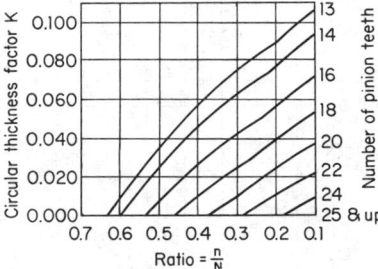

**Fig. 14**  Circular-thickness factor for straight and Zerol bevel gears.

**Angular Spiral Bevel Gears**  Several items deviate from the formulas of Table 8 in the same manner as angular straight bevel gears. Therefore, the same formulas apply for the deviating items with only the following exceptions:

Item 21, Circular thickness, the value of $K$ in Fig. 15 must be determined from the equivalent 90° bevel ratio ($m_{90}$) and the

**Table 8. Spiral-Bevel-Gear Dimensions***
(All dimensions in inches)

| | | | | | |
|---|---|---|---|---|---|
| 1 | Number of pinion teeth | $n$ | 5 | Working depth | $h_k = \dfrac{1.700}{P_d}$ |
| 2 | Number of gear teeth | $N$ | 6 | Whole depth | $h_t = \dfrac{1.888}{P_d}$ |
| 3 | Diametral pitch | $P_d$ | 7 | Pressure angle | $\phi$ |
| 4 | Face width | $F$ | 8 | Shaft angle | $\Sigma$ |

| | | Pinion | Gear |
|---|---|---|---|
| 9 | Pitch diameter | $d = \dfrac{n}{P_d}$ | $D = \dfrac{N}{P_d}$ |
| 10 | Pitch angle | $\gamma = \tan^{-1} \dfrac{n}{N}$ | $1 = 90° - \gamma$ |
| 11 | Outer cone distance | $A_0 = \dfrac{D}{2 \sin \Gamma}$ | |
| 12 | Circular pitch | $p = \dfrac{3.1416}{P_d}$ | |
| 13 | Addendum | $a_{OP} = h_k - a_{OG}$ | $a_{OG} = \dfrac{0.460}{P_d} + \dfrac{0.390}{P_d \left(\dfrac{N}{n}\right)^2}$ |
| 14 | Dedendum | $b_{OP} = h_t - a_{OP}$ | $b_{OG} = h_t - a_{OG}$ |
| 15 | Clearance | $c = h_t - h_k$ | |
| 16 | Dedendum angle | $\delta_P = \tan^{-1} \dfrac{b_{OP}}{A_o}$ | $\delta_G = \tan^{-1} \dfrac{b_{OG}}{A_o}$ |
| 17 | Face angle of blank | $\gamma_0 = \gamma + \delta_G$ | $\Gamma_0 = \Gamma + \delta_P$ |
| 18 | Root angle | $\gamma_R = \gamma - \delta_P$ | $\Gamma_R = \Gamma - \delta_G$ |
| 19 | Outside diameter | $d_O = d + 2a_{OP} \cos \gamma$ | $D_O = D + 2a_{OG} \cos \Gamma$ |
| 20 | Pitch apex to crown | $x_O = \dfrac{D}{2} - a_{OP} \sin \gamma$ | $X_O = \dfrac{d}{2} - a_{OG} \sin \Gamma$ |
| 21 | Circular thickness | $t = p - T$ | $T = \dfrac{p}{2}(a_{OP} - a_{OG}) \dfrac{\tan \phi}{\cos \phi} - \dfrac{K \text{ (Fig. 15)}}{P_d}$ |
| 22 | Backlash | See Table 7 | |

*Abstracted from "Gleason Bevel and Hypoid Gear Design Handbook."

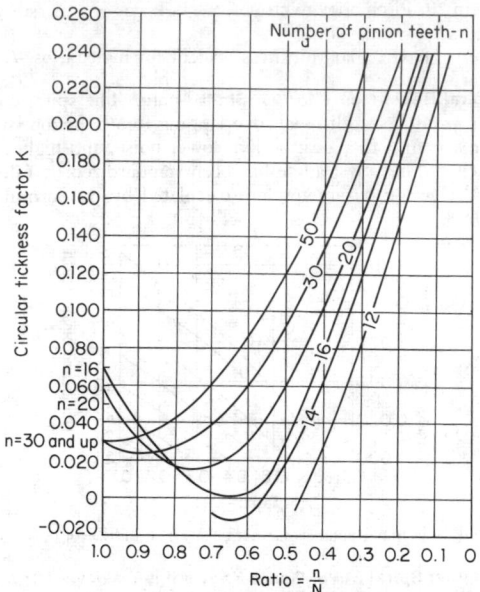

**Fig. 15** Circular-thickness factors for spiral bevel gears with 20° pressure angle and 35° spiral angle. Left-hand pinion driving clockwise or right-hand pinion driving counterclockwise.

equivalent 90° bevel pinion. The latter is computed as $n_{90} = (n \sin \Gamma_{90})/\cos \gamma$, where $\tan \Gamma_{90} = m_{90}$.

**Zerol Bevel Gears for 90° Shaft Angle** The Zerol bevel is essentially equivalent to the straight bevel in function. It is produced by the same equipment used for spiral bevels. Design parameters are calculated by the formulas of Table 9.

**Angular Zerol Bevel Gears** The formulas of Table 9 apply with the same deviations as for angular straight bevel gears.

## WORM GEARS AND WORMS

Worm gearing is used for obtaining large speed reductions between nonintersecting shafts making an angle of 90° with each other. If a gear such as shown in Fig. 16 engages a straightworm, as shown in Fig. 17, the combination is known as **single enveloping worm gearing.** If a gear of the kind shown in Fig. 16 engages a worm as shown in Fig. 18, the combination is known as **double enveloping worm gearing.**

With worm gearing, the **velocity ratio** is the ratio between the number of teeth on the gear and the number of threads on the worm. Thus, a 30-tooth worm gear meshing with a single-threaded worm will have a velocity ratio of 1 to 30; i.e., the worm must make 30 rev in order to revolve the worm gear once. For double-threaded worm, there will be 15 rev of the worm to one of the worm gear, etc. High velocity ratios are thus obtained with relatively small gears.

Tooth proportions of the worm in the central section (Fig. 17) follow standard rack designs, such as 14½, 20, and 25°.

**Table 9. Zerol Bevel Gear Dimensions***
(All linear dimensions in inches)

| | | | | | |
|---|---|---|---|---|---|
| 1 | Number of pinion teeth | $n$ | 5 | Working depth | $h_k = \dfrac{2.000}{P_d}$ |
| 2 | Number of gear teeth | $N$ | 6 | Whole depth | $h_t = \dfrac{2.188}{P_d} + 0.002$ |
| 3 | Diametral pitch | $P_d$ | 7 | Pressure angle | $\phi$ |
| 4 | Face width | $F$ | 8 | Shaft angle | $\Sigma$ |

| | | Pinion | Gear |
|---|---|---|---|
| 9 | Pitch diameter | $d = \dfrac{n}{P_d}$ | $D = \dfrac{N}{P_d}$ |
| 10 | Pitch angle.......... | $\gamma = \tan^{-1} \dfrac{n}{N}$ | $\Gamma = 90° - \gamma$ |
| 11 | Outer cone distance... | $A_o = \dfrac{D}{2 \sin \Gamma}$ | |
| 12 | Circular pitch......... | $p = \dfrac{3.1416}{P_d}$ | |
| 13 | Addendum.......... | $a_{OP} = h_k - a_{OG}$ | $a_{OG} = \dfrac{0.540}{P_d} + \dfrac{0.460}{P_d(N/n)^2}$ |
| 14 | Dedendum.......... | $b_{OP} = h_t - a_{OP}$ | $b_{OG} = h_t - a_{OG}$ |
| 15 | Clearance.......... | $c = h_t - h_k$ | |
| 16 | Dedendum angle...... | $\delta_P = \tan^{-1} \dfrac{b_{OP}}{A_o} + \Delta\delta$ (see Table 10) | $\delta_G = \tan^{-1} \dfrac{b_{OG}}{A_o} + \Delta\delta$ (see Table 10) |
| 17 | Face angle of blank... | $\gamma_O = \gamma + \delta_G$ | $\Gamma_o = \Gamma + \delta_P$ |
| 18 | Root angle.......... | $\gamma_R = \gamma - \delta_P$ | $\Gamma_R = \Gamma - \delta_G$ |
| 19 | Outside diameter...... | $d_o = d + 2a_{OP} \cos \gamma$ | $D_o = D + 2a_{OG} \cos \Gamma$ |
| 20 | Pitch apex to crown... | $x_o = \dfrac{D}{2} - a_{OP} \sin \gamma$ | $X_o = \dfrac{d}{2} - a_{OG} \sin \Gamma$ |
| 21 | Circular thickness..... | $t = p - T$ | $T = \dfrac{p}{2} - (a_{OP} - a_{OG}) \tan \phi$ $- \dfrac{K}{P_d}$ (see Fig. 14) |

*Abstracted from "Gleason Bevel and Hypoid Gear Design Handbook."

**Table 10. Dedendum Formula for Zerol Bevel Gears***

| Pressure angle, deg | Change in dedendum angle = $\Delta\delta$ (min) |
|---|---|
| 20 | $\Delta\delta = \dfrac{6,668}{N_c} - \dfrac{300\ \sqrt{d \sin \Gamma}}{N_c F} - \dfrac{14 P_d}{N_c}$ |
| $22\frac{1}{2}$ | $\Delta\delta = \dfrac{4,868}{N_c} - \dfrac{300\ \sqrt{d \sin \Gamma}}{N_c F} - \dfrac{14 P_d}{N_c}$ |
| 25 | $\Delta\delta = \dfrac{3,412}{N_c} - \dfrac{300\ \sqrt{d \sin \Gamma}}{N_c F} - \dfrac{14 P_d}{N_c}$ |

*Abstracted from Gleason, "Bevel and Hypoid Gear Design Handbook"

$N_c$ = number of teeth in crown gear − $2 P_d A_o$
$F$ = face width
$d$ = pinion pitch diameter
$A_o$ = outer cone distance
$P_d$ = diametral pitch
$\Gamma$ = gear pitch angle

The mating worm gear is cut conjugate for a unique worm size and center distance. The geometry and relating design equations for a straight-sided cylindrical worm are best seen from a development of the pitch plane (Fig. 19).

$$D_w = \text{pitch diameter of worm} = \frac{n_w p_n}{\pi \sin \lambda}$$

$$p_n = p \cos \lambda = \frac{\pi\, D_w}{n_w} \sin \lambda$$

$$L = \text{lead of worm} = n_w p$$

$$D_g = \text{pitch diameter of worm gear}$$
$$= \frac{N_g}{P_d} = \frac{p N_g}{\pi} = \frac{P_n N_g}{\pi \cos \lambda}$$

$$C = \text{center distance}$$
$$= \frac{D_w + D_g}{2} = \frac{p_n}{2\pi}\left(\frac{N_y}{\cos \lambda} + \frac{n_w}{\sin \lambda}\right)$$

where $n_w$ = number of threads in worm
$N_g$ = number of teeth in worm gear
$Z$ = velocity ratio = $\dfrac{N_g}{n_w}$

The pitch diameter of the worm gear is established by the number of teeth, which in turn comes from the desired gear ratio. The pitch diameter of the worm is somewhat arbitrary. The lead must match the worm gear's circular pitch, which can be satisfied by  an infinite number of worm diameters; but for a fixed lead value, each worm diameter has a unique lead angle. AGMA offers a design formula that provides near optimized geometry:

$$D_w = \frac{C^{.875}}{2.2}$$

where $C$ = center distance

Worm-gear width is also somewhat arbitrary. Generally it will be ⅗ to ⅔ of the worm's outside diameter.

**Worm mesh nonreversibility,** a unique feature of some designs, occurs because of the large amount of sliding in this type of gearing. For a given coefficient of friction there is a critical value of lead angle below which the mesh is nonreversible. This is generally 10° and lower but is related to the materials and lubricant. Most single-thread worm meshes are in this category. This locking feature can be a disadvantage or in some designs can be put to advantage.

**Other Gear Types**

Gears for special purposes include the following (details are to be found in the references).

**Spiroid** (Illinois Tool Works) gears, used to connect skew shafts, resemble a hypoid-type bevel gear but in performace are more like worm meshes. They offer very high ratios and a large contact ratio resulting in high strength. The **Helicon** (Illinois Tool Works) gear is a variation in which the pinion is not tapered, and ratios under 10:1 are feasible.

**Beveloid** (Vinco Corp.) gears are tapered involute gears which can couple intersecting shafts, skew shafts, and parallel shafts.

**Face gears** have teeth cut on the rotating face plane of the gear and mate with standard involute spur gears. They can connect intersecting or nonparallel, nonintersecting shafts.

**Noncircular gears** or **function gears** are used for special motions or as elements of analog computers. They can be made with elliptical, logarithmic, spiral, and other functions. See Cunningham references; also, Cunningham Industries, Inc., Stamford, Conn.

**DESIGN STANDARDS**

In addition to the ANSI and AGMA standards on basic tooth proportions, the AGMA sponsors a large number of national standards dealing with gear design, specification, and inspection. (Consult AGMA, 1901 N. Fort Myer Drive, Arlington, VA 22209, for details.) A helpful item is the AGMA "Gear Handbook," 390.03, which establishes a system of quality classes for all gear sizes and pitches, ranging from crude commercial gears to the highest orders of ultraprecision gears. There are 12 quality classes, numbered from 5 through 16 in ascending quality. Tolerances are given for key functional parameters: runout, pitch, profile, lead, total composite error, tooth-to-tooth composite error, and tooth thickness. Also, tooth-thickness tolerances and recommended mesh backlash are included. These are related to diametral pitch and pitch diameter in recognition of fabrication achievability. Data are available for spur, helical, herringbone, bevel, and worm gearing; and spur and helical racks. Special sections cover gear applications and suggested quality number; gear materials and treatment; and standard procedure for identifying quality, material, and other pertinent parameters. These data are too extensive for inclusion in this handbook, and the reader should refer to AGMA 390.03, "Gear Handbook," vol. 1.

**STRENGTH AND DURABILITY**

The capacity of a gear is measured in terms of tooth strength and surface durability. There have been many attempts to derive expressions for calculating safe beam strength and surface stress, starting with the original Lewis-Buckingham formulas and extending to the latest AGMA equations.

The **Lewis formula** is an old method that is still useful in the analysis of tooth beam strength. It is based upon a tooth layout (Fig. 20) where the load is assumed to be at the tip. By Lewis' formula, $W_b = FSY/P_d$, or, in circular pitch, $W_t = SFp_c y_c$, and $y_c = Y/\pi$. The factor $Y$ is derived from the layout as $Y = 2XP_d/3$.

For a given tooth design this value is constant with pitch and varies only with number of teeth (see Table 11). The transmitted load is calculated from horsepower by $W_t = 126,000P_t/[d(\text{r/min})_p]$, where $P_t$ = transmitted horsepower and $(\text{r/min})_p$ = pinion speed, r/min. For safe design, $W_b \geq W_t$.

The values of $S$ for various BHN are:

| BHN | Cast iron | 160 | 210 | 215 | 270 | 315 | 335 | 360 | 440 |
|---|---|---|---|---|---|---|---|---|---|
| Allowable stress, 1,000 lb/in² | 14 | 40 | 50 | 60 | 65 | 70 | 77 | 83 | 90 |

**Refinement of Lewis' formula** is necessary because the worse-load condition is not at the tip owing to multiple-tooth pairs in contact, and the stress concentration at the root fillet is neglected. One modified formula is $W_b = K_r m_p FSY/P_d$, where $m_p$ = contact ratio and $K_r$ = stress-concentration factor ($K_r = 1$ for precision gears, 0.75 for high commercial grade, and 0.5 for commercial grade).

**Dynamic tooth strength** takes into consideration dynamic forces in a gear train not accounted for by the Lewis formula. This requires the beam strength to be sufficiently large to accommodate an increased transmitted load, termed dynamic load $W_d$. For safe design, $W_b \geq W_d$; approximate values of dynamic load can be obtained by modifying the transmitted load with an appropriate factor: $W_d = W_t \cdot \overline{DF}$, where $\overline{DF}$ = $(600 + V)/600$ for general commercial-quality gears operating at a velocity less than 3,000 ft/min; $DF = (1,200 + V)/1,200$ for high commercial-quality gears operating at a velocity less than 6,000 ft/min; $DF = (250 + V)/150 + 0.25$ for nonmetallic gears, 600 to 5,000 ft/min, where $V$ = pitch-line velocity, ft/min.

**Buckingham's dynamic equation** is an attempt to be more accurate through combination of profile error magnitude, elastic properties, and empirical data, where $W_d = W_t + [0.05V(FC + W_t)]/[0.05V + (FC + W_t)^{1/2}]$ (for spur gears), and $W_d = W_t + [0.05V(FC \cos^2 \psi + W_t) \cos \psi]/[0.05V + (FC \cos^2 \psi + W_t)^{1/2}]$ (for helical gears). The constant $C$ is a deformation factor, values of which can be derived from Buckingham's special equations or taken from Table 12.

**Durability** Stresses generated in the surface layers of the teeth by the crushing action of the forces can exceed material limits and result in a failure in the form of pitting, scoring, scuffing, seizing, and plastic deformation. Based upon Hertz contact stresses, Buckingham developed the following durability equations, modified by a contact ratio factor, $W_w = D_p QKFm_p$ (for spur gears), and $W_w = D_p QKFm_p/\cos^2 \psi$ (for helical gears), where $Q$ = ratio factor = $2N_G/(N_G + N_p)$, and $K$ = Buckingham's durability factor as shown in Table 13. For safe design, $W_w \geq W_d$.

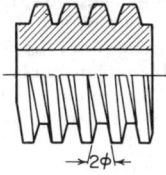

**Fig. 16**

**Fig. 17**

Other modifications of the strength and durability equations come from kinds of service. See Table 14 for useful service factors.

**AGMA Strength and Durability Ratings** The AGMA Gear Rating committee has developed new strength and durability rating equations suitable for modern gearing and arranged to

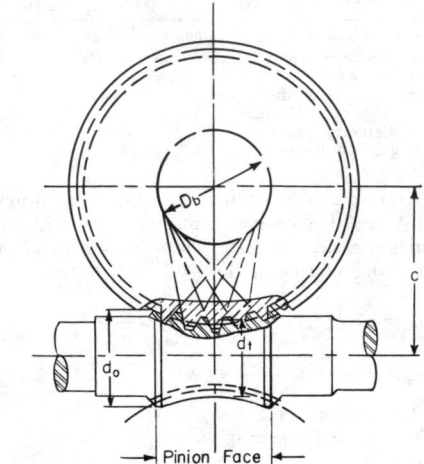

**Fig. 18** Double enveloping worm gear.

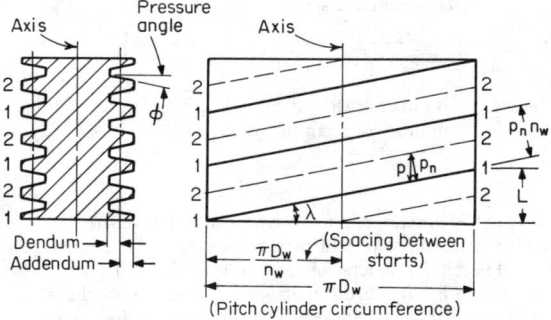

(a) Cylindrical worm (double-thread example)   (b) Development of worm's pitch cylinder

**Fig. 19** Cylindrical worm geometry and design parameters.

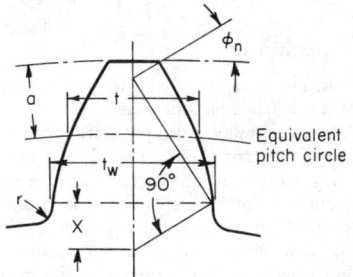

**Fig. 20** Layout for beam strength.

Table 11. Values of Form Factor Y in Lewis Formula for Spur Gears

| No. of teeth | *Full depth, composite 14½° | Full depth involute 20° | †Stub involute 20° | No. of teeth | *Full depth, composite 14½° | Full depth involute 20° | †Stub involute 20° |
|---|---|---|---|---|---|---|---|
| 12 | 0.210 | 0.245 | 0.311 | 30 | 0.320 | 0.358 | 0.437 |
| 14 | 0.226 | 0.276 | 0.339 | 43 | 0.346 | 0.396 | 0.464 |
| 16 | 0.242 | 0.295 | 0.361 | 60 | 0.358 | 0.421 | 0.484 |
| 18 | 0.261 | 0.308 | 0.377 | 100 | 0.371 | 0.446 | 0.506 |
| 20 | 0.283 | 0.320 | 0.393 | 150 | 0.377 | 0.459 | 0.518 |
| 25 | 0.305 | 0.339 | 0.418 | Rack | 0.390 | 0.484 | 0.550 |

* Use these values also for Brown and Sharpe full-depth teeth.
† Use these values also for Nuttall stub, AGMA stub, and Fellows stub, 4 or 8 pitch. For Fellows stub, 5, 6, and 7 pitch, values are slightly greater; for 9, 10, and 12 pitch, slightly less.

enable expansion or contraction of governing parameters and details in accordance with available data and needs of the application. The following AGMA formulas are intended for all sizes of spur, helical, and bevel gears (see Table 15). (Reference should be made to AGMA standards for details.)

In the equations below, $d$ is the pinion pitch diameter and subscript $p$ refers to the pinion.

TOOTH STRENGTH (bending stress). Stress = $S_t$ = $f$ (load, size, stress distribution) = $(W_t K_o/K_v)(P_d/F)(K_s K_m/J)$, and in terms of allowable bending stress, $S_t \leq S_{at} K_L/K_e K_T$, and the tooth strength power rating is $P_{at} = [(r/min)_p, dK_v/126,000 K_o \times [F/P_d]\ [J/K_m K_s]\ [S_{at}K_l/K_t K_r]$.

SURFACE DURABILITY. Stress = $S_c$ = $f$(load, size, stress distribution) = $C_p(W_t C_o/C_v)(C_s/dF)(C_f C_m/I)^{1/2}$, and in terms of allowable surface durability stress, $S_c = S_{ac}(C_L C_H/C_R C_T)$, while the durability power rating is $P_{ac} = [(r/min)_p F/126,000] [IC_v/C_s C_m C_f C_o] \times [(S_{ac}d/C_p)\ (C_l C_H)/(C_T C_R)]$.

Figures 21 through 23 and Tables 16 through 19 provide data for applying the AGMA rating formulas to spur gears.

This information is extracted from AGMA standards and compiled in the same manner as given in Shigley (see AGMA standards and Shigley, "Engineering Design," chaps. 11, 12, McGraw-Hill, for details). The allowable stress $S_{at}$ can be the yield strength, fatigue strength, or endurance limit, depending upon the gear application. The allowable durability stress $S_{ac}$ is surface-contact stress and is the endurance limit of this Hertz-type surface crushing stress. (See Tables 21 and 22 for recommended material endurance limits and yield strengths.)

The rating of helical gears requires $J$ and $I$ factors that depend upon specific helix angle. These data are too extensive to be included in this introductory treatment, and the reader is referred to the AGMA standards. Also, for the rating of all types of bevel gears consult the AGMA standards.

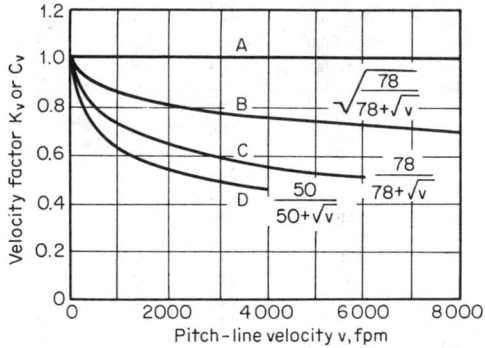

Fig. 21 Velocity-correction factor $K_v$ or $C_v$. Use curve A for high-precision shaved or ground gears when no significant dynamic load is developed. If the dynamic load is appreciable, use curve B when the gears are high-precision, with a shaved or ground finish. Curve C should be used for moderate dynamic loads and curve D for spur gears when the teeth are formed by hobbing or shaping. If the teeth are milled or are inaccurate, velocity factors even lower than those represented by curve D should be employed. (By permission, from "Gear Handbook," p. 13–22, McGraw-Hill, New York, 1962.)

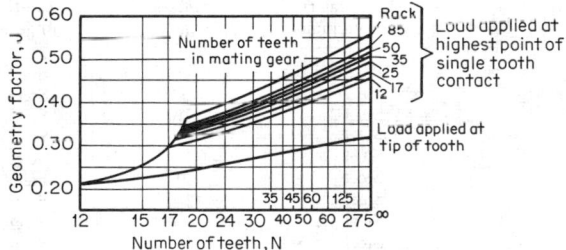

Fig. 22 Geometry factors $J$ for 20° spur gears, full depth, with a fillet radius of $r_f = 0.35/P$. The asbcissa is the number of teeth on the particular gear whose geometry factor is desired. (By permission, from "Gear Handbook," p. 13–32, McGraw-Hill, New York, 1962.)

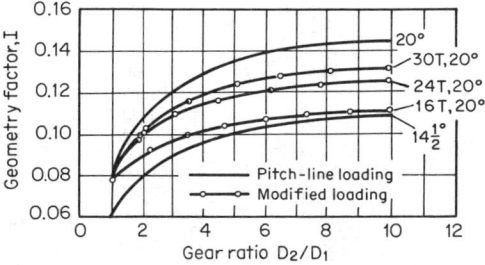

Fig. 23 Geometry factor $I$ for external spur gears with full-depth teeth; geometry factors are either given for loading at the pitch line or modified for the beginning of single-tooth contact. The tooth numbers refer to the pinion. (By permission, from "Gear Handbook," p. 13–24, McGraw-Hill, New York, 1962.)

### Table 12. Values of Deformation Factor $C$

| Materials, pinion and gear | Tooth form | Error in action, in | | | | | |
|---|---|---|---|---|---|---|---|
| | | 0.0005 | 0.001 | 0.002 | 0.003 | 0.004 | 0.005 |
| Cast iron and cast iron | $14\frac{1}{2}°$ | 400 | 800 | 1,600 | 2,400 | 3,200 | 4,000 |
| Steel and cast iron | | 550 | 1,100 | 2,200 | 3,300 | 4,400 | 5,500 |
| Steel and steel | | 800 | 1,600 | 3,200 | 4,800 | 6,400 | 8,000 |
| Cast iron and cast iron | 20° full depth | 415 | 830 | 1,660 | 2,490 | 3,320 | 4,150 |
| Steel and cast iron | | 570 | 1,140 | 2,280 | 3,420 | 4,560 | 5,700 |
| Steel and steel | | 830 | 1,660 | 3,320 | 4,980 | 6,640 | 8,300 |
| Cast iron and cast iron | 20° stub tooth | 430 | 860 | 1,720 | 2,580 | 3,440 | 4,300 |
| Steel and cast iron | | 590 | 1,180 | 2,360 | 3,540 | 4,720 | 5,900 |
| Steel and steel | | 860 | 1,720 | 3,440 | 5,160 | 6,880 | 8,600 |

### Table 13. Values of $K$ for Buckingham's Durability Equation*

| Material in pinion | Brinell hardness number | Material in gear | Brinell hardness number | $S_c$, surface endurance limit, lb/in² | $K$, $14\frac{1}{2}°$ | $K$, 20° |
|---|---|---|---|---|---|---|
| Steel | 150 | Steel | 150 | 50,000 | 30 | 41 |
| Steel | 200 | Steel | 150 | 60,000 | 43 | 38 |
| Steel | 250 | Steel | 150 | 70,000 | 58 | 79 |
| Steel | 200 | Steel | 200 | 70,000 | 58 | 79 |
| Steel | 250 | Steel | 200 | 80,000 | 76 | 103 |
| Steel | 300 | Steel | 200 | 90,000 | 96 | 131 |
| Steel | 250 | Steel | 250 | 90,000 | 96 | 131 |
| Steel | 300 | Steel | 250 | 100,000 | 119 | 162 |
| Steel | 350 | Steel | 250 | 110,000 | 144 | 196 |
| Steel | 300 | Steel | 300 | 110,000 | 144 | 196 |
| Steel | 350 | Steel | 300 | 120,000 | 171 | 233 |
| Steel | 400 | Steel | 300 | 125,000 | 186 | 254 |
| Steel | 350 | Steel | 350 | 130,000 | 201 | 275 |
| Steel | 400 | Steel | 350 | 140,000 | 233 | 318 |
| Steel | 500 | Steel | 350 | 145,000 | 250 | 342 |
| Steel | 400 | Steel | 400 | 150,000 | 268 | 366 |
| Steel | 500 | Steel | 400 | 175,000 | 364 | 497 |
| Steel | 600 | Steel | 400 | 180,000 | 385 | 526 |
| Steel | 500 | Steel | 500 | 190,000 | 430 | 588 |
| Steel | 600 | Steel | 600 | 230,000 | 630 | 861 |
| Steel | 150 | Cast iron | 180 | 50,000 | 44 | 60 |
| Steel | 200 | Cast iron | 180 | 70,000 | 87 | 119 |
| Steel | ⩾250 | Cast iron | 180 | 90,000 | 144 | 196 |
| Steel | 150 | Phosphor bronze | 100 | 50,000 | 46 | 62 |
| Steel | 200 | Phosphor bronze | 100 | 70,000 | 91 | 124 |
| Steel | ⩾250 | Phosphor bronze | 100 | 85,000 | 135 | 204 |
| Cast iron | 180 | Cast iron | 180 | 90,000 | 193 | 284 |
| Ga Meehanite | | GA Meehanite | | 80,000 | 105 | 144 |
| Steel | | GA Meehanite | | 80,000 | 90 | 123 |

*Abstracted from Buckingham, "Manual of Gear Design," Industrial Press; and Slaymaker, "Mechanical Design and Analysis," Wiley.

**Table 14. Service Factors for Enclosed Gear Drives***

| Prime mover | Duration of service | Driven machine load classifications | | |
|---|---|---|---|---|
| | | Uniform | Moderate shock | Heavy shock |
| Electric motor | Occasional ½ h/day | 0.50 | 0.80 | 1.25 |
| | Intermittent 3 h/day | 0.80 | 1.00 | 1.50 |
| | Up to 10 h/day | 1.00 | 1.25 | 1.75 |
| | 24 h/day | 1.25 | 1.50 | 2.00 |
| Multicylinder internal-combustion engine | Occasional ½ h/day | 0.80 | 1.00 | 1.50 |
| | Intermittent 3 h/day | 1.00 | 1.25 | 1.75 |
| | Up to 10 h/day | 1.25 | 1.50 | 2.00 |
| | 24 h/day | 1.50 | 1.75 | 2.25 |
| Single-cylinder internal-combustion engine | Occasional ½ h/day | 1.00 | 1.25 | 1.75 |
| | Intermittent 3 h/day | 1.25 | 1.50 | 2.00 |
| | Up to 10 h/day | 1.50 | 1.75 | 2.25 |
| | 24 h/day | 1.75 | 2.00 | 2.50 |

*Abstracted from AGMA 420.03, "Standard Practice for Helical and Herringbone Gear Speed Reducers and Increasers."

**Table 15.  Definitions of Terms**

| Term | Strength | Durability |
|---|---|---|
| Load: | | |
| Transmitted load.................... | $W_t$ | $W_t$ |
| Dynamic factor..................... | $K_v$ | $C_v$ |
| Overload factor..................... | $K_o$ | $C_o$ |
| Size: | | |
| Pinion pitch diameter............... | ... | $d$ |
| Net face width...................... | $F$ | $F$ |
| Transverse diametral pitch.......... | $P_d$ | |
| Size factor......................... | $K_s$ | $C_s$ |
| Stress distribution: | | |
| Load distribution factor............ | $K_m$ | $C_m$ |
| Geometry factor.................... | $J$ | $I$ |
| Surface condition factor............ | ... | $C_f$ |
| Stress: | | |
| Calculated stress................... | $s_t$ | $s_c$ |
| Allowable stress.................... | $s_{at}$ | $s_{ac}$ |
| Elastic coefficient.................. | ... | $C_P$ |
| Hardness-ratio factor............... | ... | $C_H$ |
| Life factor......................... | $K_L$ | $C_L$ |
| Temperature factor................. | $K_T$ | $C_T$ |
| Factor of safety.................... | $K_R$ | $C_R$ |

**Table 16. Overload Correction Factors $K_o$ and $C_o$***

| Source of power | Driven machinery | | |
|---|---|---|---|
| | Uniform | Moderate shock | Heavy shock |
| Uniform | 1.00 | 1.25 | 1.75 |
| Light shock | 1.25 | 1.50 | 2.00 |
| Medium shock | 1.50 | 1.75 | 2.25 |

*Darle W. Dudley, "Gear Handbook," p. 13-20, McGraw-Hill, New York, 1962.

**Table 17. Load-Distribution Factor $K_m$ and $C_m$ for Spur Gears***

| Characteristics of support | Face width, in | | | |
|---|---|---|---|---|
| | 0–2 | 6 | 9 | 16 up |
| Accurate mountings, small bearing clearances, minimum deflection, precision gears | 1.3 | 1.4 | 1.5 | 1.8 |
| Less rigid mountings, less accurate gears, contact across full face | 1.6 | 1.7 | 1.8 | 2.2 |
| Accuracy and mounting such that less than full face contact exists | | Over | 2.2 | |

*Darle W. Dudley, "Gear Handbook," p. 13-21, McGraw-Hill, New York, 1962.

**Table 18. Reliability Factors $K_R$ and $C_R$***

| Reliability requirements | $C_R$ | $K_R$ |
|---|---|---|
| High | 1.25 up | 1.50 up |
| More than 99% | 1.00 | 1.00 |
| 67 to 99% | 0.80† | 0.70 |

*Darle W. Dudley, "Gear Handbook," p. 13-15, McGraw-Hill, New York, 1962.
†Plastic deformation may occur at this load.

**Table 19. Values of the Elastic Coefficient $C_p$ for Spur and Helical Gears with Nonlocalized Contact*·†**

| Pinion material | Modulus of elasticity $E$ | Gear material and modulus of elasticity $E$ | | | |
|---|---|---|---|---|---|
| | | Steel $(30)(10)^6$ | Cast iron $(19)(10)^6$ | Aluminum bronze $(17.5)(10)^6$ | Tin bronze $(16)(10)^6$ |
| Steel | $(30)(10)^6$ | 2,300 | 2,000 | 1,950 | 1,900 |
| Cast iron | $(19)(10)^6$ | 2,000 | 1,800 | 1,800 | 1,750 |
| Aluminum bronze | $(17.5)(10)^6$ | 1,950 | 1,800 | 1,750 | 1,700 |
| Tin bronze | $(16)(10)^6$ | 1,900 | 1,750 | 1,700 | 1,650 |

*Darle W. Dudley, "Gear Handbook," p. 13-22, McGraw-Hill, New York, 1962.
†$\mu = 0.30$.

**GEAR MATERIALS** (See Table 20)

Plain carbon steels are most widely used as the most economical; similarly, cast iron is used for large units or intricate body shapes. Heat-treated carbon and alloy steels are used for the more severe load and wear resistant applications. Pinions are usually made harder to equalize wear. Strongest and most wear-resistant gears are a combination of heat-treated high-alloy steel cores with case-hardened teeth. (See Dudley "Gear Handbook," chap. 10.) Bronze is particularly recommended for worm gears and crossed-helical gears. Stainless steels are limited to special corrosion-resistant environment applications. Aluminum alloys are used for light-duty instrument gears and airborne lightweight requirements. Nonmetallic materials are for noise reduction, vibration damping, and economical large-quantity production (see also Sec. 6).

**GEAR LUBRICANTS** (See Table 23)

Proper lubrication is important to prevention of premature wear of tooth surfaces. Choice of lubricant is a function of gear load, speed, temperature, and type of lubricating system (see also Secs. 6 and 8).

**Table 20. Typical Gear Materials**

| Material and designation | Tensile strength, psi | Yield strength, psi | Hardness, BHN | Condition |
|---|---|---|---|---|
| **Cast irons:** | | | | |
| ASTM 20 | 22,000 | ....... | 156 | As cast |
| 30 | 31,000 | ....... | 201 | As cast |
| 60 | 62.500 | ....... | 262 | As cast |
| **Plain carbon steels:** | | | | |
| AISI 1020 | 55,000 | 30,000 | 110 | Hot-rolled |
| 1020 | 78,000 | 66,000 | 155 | Cold-worked |
| 1040 | 76,000 | 42,000 | 150 | Hot-rolled |
| 1040 | 123,000 | 93,000 | 350 | Cold-worked |
| 1080 | 112,000 | 61,000 | 230 | Hot-rolled |
| 1080 | 189,000 | 142,000 | 385 | Cold-worked |
| 1117 | 62,000 | 34,000 | 120 | Hot-rolled |
| 1117 | 80,000 | 68,000 | 163 | Cold-worked |
| **Alloy steels:** | | | | |
| AISI 3140 | 105,000 | 90,000 | 280 | Heat-treated |
| 3140 | 228,000 | 209,000 | 450 | Heat-treated |
| 4140 | 145,000 | 120,000 | 290 | Normalized |
| 4140 | 215,000 | 190,000 | 440 | Heat-treated |
| 4820 | 150,000 | 125,000 | 325 | Heat-treated |
| 4820 | 206,000 | 166,000 | 415 | Heat-treated |
| 6120 | 125,000 | 94,000 | ...... | Heat-treated |
| 8620 | 122,000 | 98,000 | 245 | Normalized |
| 8620 | 173,000 | 142,000 | 375 | Heat-treated |
| 9310 | 152,000 | 120,000 | 350 | Heat-treated |
| 9310 | 180,000 | 140,000 | 375 | Heat-treated |
| **Stainless steels:** | | | | |
| AISI 303 | 90.000 | 35.000 | 160 | Annealed |
| 303 | 110.000 | 75,000 | 240 | Cold-worked |
| 416 | 75.000 | 40,000 | 155 | Annealed |
| 416 | 160.000 | 140,000 | 350 | Heat-treated |
| **Bronzes:** | | | | |
| Aluminum bronze ASTMB139 | 105,000 | 60,000 | B100* | |
| Phosphor bronze ASTMB1397 | 60,000 | 45,000 | B70* | |
| Silicon bronze ASTMB99 | 58,000 | 25,000 | B100* | |
| **Aluminum alloys:** | | | | |
| 2024-T4 | 68,000 | 47,000 | 120 | Heat-treated ½ hard |
| 7075-T6 | 83,000 | 73,000 | 150 | Heat-treated ¾ hard |
| **Nonmetallics:** | | | | |
| Phenolic laminate | | | | |
| NEMA, Grade C | 11,000 | ....... | M-103* | |
| NEMA, Grade L | 14,000 | ....... | M-105* | |
| Nylon | | | | |
| ASTM6 | 8,700 | 6,000 | M-100* | 2.5% moisture |
| ASTM66 | 11,000 | 8,500 | M-108* | 2.5% moisture |

\* Rockwell.

**Table 21. Recommended Endurance Limits of Spur-, Helical-, and Bevel-Gear Materials\*·†**

| Material | Min hardness | Endurance limit,‡ lb/in² | | |
|---|---|---|---|---|
| | | Spur gears | Helical gears | Bevel gears |
| Steel | 140 | 20,000–22,000 | 20,000–22,000 | 11,000 |
| | 180 | 25,000–28,000 | 25,000–28,000 | 14,000 |
| | 300 | 35,000–40,000 | 35,000–45,000 | 19,000 |
| | 450 | 45,000–50,000 | 45,000–60,000 | 25,000 |
| | R55C§ | 55,000–65,000 | 55,000–65,000 | 30,000 |
| Cast iron: | | | | |
| AGMA grade 20 | | 5,000 | 5,000 | 2,700 |
| AGMA grade 30 | 175 | 8,500 | 8,500 | 4,600 |
| AGMA grade 40 | 200 | 13,000 | 13,000 | 7,000 |

\*Darle W. Dudley, "Gear Handbook," p. 13-31, McGraw-Hill, New York, 1962.

†It should be noted that the values listed here are *not the same* as the endurance limits of materials for completely reversed bending. The endurance limits given here are the limiting values of this stress when $K_L$, $K_T$, and $K_R$ are assumed to be unity.

‡Called "allowable fatigue stress" by the AGMA.

§Case-carburized; Rockwell hardness.

**Table 22. Surface-Endurance Limits $S_{fe}$ for Gear Materials*†**

| Material | Heat treatment | Min surface hardness | | Endurance limit $S_{fe}$, lb/in² |
|---|---|---|---|---|
| | | BHN | Rockwell C | |
| Steel | Case-carburized | 625 | 60 | 250,000 |
| Steel | Case-carburized | 575 | 55 | 200,000 |
| Steel | Flame-hardened | 500 | 50 | 190,000 |
| Steel | Hardened and tempered | 440 | 50 | 190,000 |
| Steel | Hardened and tempered | 360 | | 160,000 |
| Steel | Hardened and tempered | 300 | | 135,000 |
| Steel | Hardened and tempered | 240 | | 115,000 |
| Steel | Hardened and tempered | 180 | | 95,000 |
| AGMA grade 20 C.I. | As cast | | | 30,000 |
| AGMA grade 30 C.I. | As cast | 175 | | 50,000 |
| AGMA grade 40 C.I. | As cast | 300 | | 65,000 |

*Darle W. Dudley, "Gear Handbook," p. 13-18, McGraw-Hill, New York, 1962, and Wells Coleman, Bevel Gears, *Mach. Design*, vol. 33, p. 130, November 1961.

†The surface-endurance limit is called the allowable contact-stress number by Wellauer and the allowable contact stress by Coleman.

**Table 23. Typical Gear Lubricants**

| Lubricant type | Military specification | Useful temp range (deg F) | Commercial source and specification (a partial listing) | | Remarks and applications |
|---|---|---|---|---|---|
| | | | Source | Identification | |
| Oils: | | | | | |
| Petroleum | MIL-L-644B | −10 to 250 | Esso Standard Oil Co. / Franklin Oil and Gas Co. / Royal Lubricants Co. / Texaco | #4035 or Unvis P-48 / L-499B / Royco 380 / 1692 Low Temp. Oil | Good general-purpose lubricant for all quality gears having a narrow range of operating temperature |
| Diester | MIL-L-6085A | −67 to 350 | Anderson Oil Co. / Eclipse Pioneer Div., Bendix / Shell Oil Co. / E. F. Houghton and Co. | Windsor Lube I-245X / Pioneer P-10 / AeroShell Fluid 12 / Cosmolubric 270 | General-purpose, low-starting torque, and stable over a wide temperature range. Particularly suited for precision instrument gears and small machinery gears |
| Diester | MIL-L-7808C | −67 to 400 | Sinclair Refining Co. / Socony Mobil Oil Co. / Bray Oil Co. / Esso Standard Oil Co. | Aircraft Turbo S Oil / Avrex S Turbo 251 / Brayco 880 / Esso Turbine Oil 15 | Suitable for oil spay or mist system at high temperature. Particularly suitable for high-speed power gears |
| Silicone | | −75 to 350 | Dow-Corning Corp. | DC200 | Rated for low-starting torque and lightly loaded instrument gears |
| Silicone | | −100 to 600 | General Electric Co. | Versilube 81644 | Best load carrier of silicone oils with widest temperature range. Applicable to power gears requiring wide temperature ranges |
| Greases: | | | | | |
| Diester oil-lithium soap | MIL-G-7421A | −100 to 200 | Royco Lubricants Co. / Texaco | Royco 21 / Low Temp. No. 1888 | For moderately loaded gears requiring starting torques at low temperatures |
| Diester oil-lithium soap | MIL-G-3278A | −67 to 250 | Esso Standard Oil Co. / Shell Oil Co. / Sinclair Refining Co. / Bray Oil Co. | Beacon 325 / AeroShell Grease 11 / Sinclair 3278 Grease / Braycote 678 | General-purpose light grease for precision instrument gears, and generally lightly loaded gears |
| Petroleum oil-sodium soap | MIL-L-3545 | −20 to 300 | Esso Standard Oil Co. / Standard Oil Co. of Calif. | Andok 260 / RPM Aviation Grease #2 | A high-temperature lubricant for high speed and high loads |
| Mineral oil-sodium soap | .......... | −25 to 250 | Esso Standard Oil Co. | Andok C | Stiff grease that channels readily. Suitable for high speeds and highly loaded gears |

**Table 24. Recommended Yield Strengths for Steels***

| Heat treatment | BHN | $S_y$, lb/in² |
|---|---|---|
| Annealed or normalized | 150 | 30,000 |
| | 200 | 50,000 |
| | 250 | 75,000 |
| Quenched and tempered | 200 | 60,000 |
| | 250 | 85,000 |
| | 300 | 110,000 |
| | 350 | 135,000 |
| | 400 | 160,000 |

*Called "allowable yield stress" by the AGMA.

# FLUID-FILM BEARINGS

## by Dudley D. Fuller

REFERENCES: "General Conference on Lubrication and Lubricants," ASME, 1957. Shaw and Macks, "Analysis and Lubrication of Bearings," McGraw-Hill. Forbes, Pope, and Everitt, "Lubrication of Industrial and Marine Machinery," Wiley. Fuller, "Theory and Practice of Lubication for Engineers," Wiley. O'Connar and Boyd, "Standard Handbook of Lubrication Engineering," McGraw-Hill.

Plain bearings, according to their function, may be

**Journal bearings,** cylindrical in shape, carrying a rotating shaft.

**Thrust bearings,** the function of which is to prevent lengthwise motion of a rotating shaft.

**Guide bearings,** to guide a machine element in its lengthwise motion, usually without rotation of the element.

In exceptional cases of design, or with a complete **failure of lubrication,** a bearing may run dry. The coefficient of friction is then between 0.25 and 0.40, depending on the materials of the rubbing surfaces. With the **bearing barely greasy,** or when the bearing is well lubricated but the speed of rotation is very slow, boundary lubrication takes place. The coefficient of friction may vary from 0.08 to 0.14. This condition occurs also in any bearing when the shaft is starting from rest if the bearing is not equipped with an oil lift.

**Semifluid,** or **mixed,** lubrication exists between the journal and bearing when the conditions are not such as to form a load-carrying fluid film and thus separate the surfaces. Semifluid lubrication takes place at comparatively low speed, with intermittent or oscillating motion, heavy load, insufficient oil supply to the bearing (wick or waste lubrication, drop-feed lubrication). Semifluid lubrication may also exist in thrust bearings with fixed parallel-thrust collars, in guide bearings of machine tools, in bearings with copious lubrication where the shaft is bent or the bearing is misaligned, or where the bearing surface is interrupted by improperly arranged oil grooves. The coefficient of friction in such bearings may range from 0.02 to 0.08 (Fuller, Mixed Friction Conditions in Lubrication, *Lubrication Eng.*, 1954).

**Fluid** or **complete lubrication,** when the rubbing surfaces are completely separated by a fluid film, provides the lowest friction losses and prevents wear. A certain amount of oil must be fed to the oil film in order to compensate for end leakage and maintain its carrying capacity. Pressure lubrication, from a pump or gravity tank, is used, or automatic lubricating devices are provided in self-contained bearings (oil rings or oil disks), or the bearing is submerged in an oil bath (thrust bearings for vertical shafts).

## PLAIN CYLINDRICAL JOURNAL BEARINGS

### Notation

$R$ = radius of bearing, in (cm)

$r$ = radius of journal, in (cm)

$mr = R - r$ = radical clearance, in (mm)

$W$ = load on a bearing, lb (kg)

$\mu$ = viscosity, lb·s/in²

$Z$ = viscosity, centipoise (cP)

$\beta$ = angle between load and entering edge of oil film

$\eta$ = coefficient for side leakage of oil

$\nu$ = kinematic viscosity, in²/s (centistokes)

$\Lambda$ = compressibility number (gas bearings)

$R_e$ = Reynolds number, $\dfrac{umr}{v}$

$P_a$ = ambient pressure, lb/in² abs (kilonewtons/m²; kN/m²)

$P = W/ld$ = unit pressure, lb/in² (kilonewtons/m²; kN/m²)

$N$ = r/min of a shaft

$m$ = clearance ratio (diametral clearance/diameter)

$F$ = friction force, lb (kg)

$A$ = operating characteristic of a plain cylindrical bearing (see below)

$P'$ = alternate operating characteristic of a plain cylindrical bearing

$h_0$ = min film thickness, mm

$\epsilon$ = eccentricity ratio or ratio of eccentricity to radial clearance

$f$ = coefficient of friction

$f'$ = friction factor

$l$ = length of bearing, in (cm)
$d = 2r$ = diam of journal, in (cm)
$K_f$ = friction factor of a plain cylindrical bearing
$t_w$ = temp of bearing wall, °F (°C)
$t_0$ = temp of air, °F (°C)
$t_1$ = temp of oil film, °F (°C)
$u$ = surface speed, in/s (cm/s)
$\omega$ = angular velocity, rad/s
$\rho$ = mass density, lb·s²/in⁴ (g·s²/cm⁴)

Fluid lubrication in plain cylindrical bearings depends on the viscosity of the lubricant and on its adhesion to the surfaces of the journal and the bearing. The radial clearance provided in the bearing forms, automatically, a wedge-shaped film between the journal and the bearing. The oil is entrained by the journal into the film. A hydrodynamic pressure is created in the film, sufficient to float the journal and carry the load applied to it.

The **minimum film thickness** $h_0$ determines the closest approach of the journal·and bearing surfaces with complete lubrication (Fig. 1). The allowable closest approach depends on the degree of finish of these surfaces and on the rigidity of the journal and bearing structures. In practice, $h_0 = 0.00075$ in (0.019 mm) is common in electric motors and generators of medium speed, with steel shafts in babbitted bearings; $h_0 = 0.003$ (0.0762 mm) to 0.005 in (0.127 mm) for large steel shafts running at high speed in babbitted bearings (turbogenerators, fans), with pressure oil-supply for lubrication; $h_0 = 0.0001$ in (0.0025 mm) to 0.0002 in (0.005 mm) in automotive and aviation engines, with very fine finish of the surfaces.

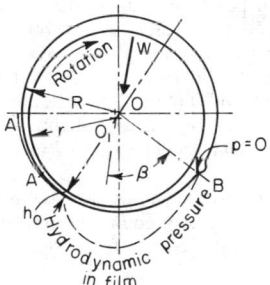

**Fig. 1** Journal bearing with perfect lubrication.

Figure 2 gives the relationship between $\epsilon$ and the load-carrying coefficient $A$ for a plain cylindrical journal. The operating characteristic of the bearing is

$$A = \frac{132}{\eta}(1,000m)^2 \frac{P}{ZN}$$

In Fig. 1, $\beta$ is the angle between the direction of the load $W$ and the entering edge of the load-carrying oil film, in degrees. The entering edge is at the place where the hydrodynamic pressure is equal or nearly equal to the atmospheric pressure and may be at the location of the oil-distributing groove $B$, or at the end of a machined recess pocket as at $AA$. For **complete bearings**, i.e., when the inner surface of the bearing is not interrupted by grooves, $\beta$ may be taken as 90°. For a 120° bearing with a central load, $\beta$ may be taken as 60°.

The coefficient $\eta$ corrects for side leakage. There is a loss of load-carrying capacity caused by the drop in the hydrodynamic gage pressure $p$ in the oil film from the midsection of the bearing toward its ends; $p = 0$ at the ends. The value of $\eta$

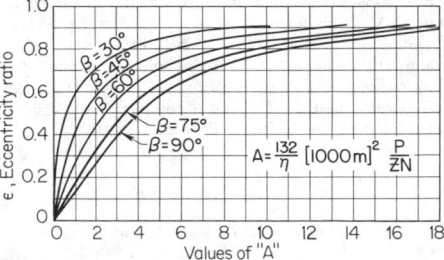

**Fig. 2** Eccentricity ratio for a plain cylindrical journal.

depends on the length-diameter ratio $l/d$ and $\epsilon$, the eccentricity ratio. Values of $\eta$ are given in Fig. 3.

EXAMPLE 1. A generator bearing, 6 in diam by 9 in long, carries a vertical downward load of 8,650 lb; $N$ = 720 r/min. The diametral clearance of the bearing is 0.012 in; the bearing is split on its horizontal diameter, and the lower half is relieved 40° down on each side, for oil distribution along journal; the bearing arc is therefore 100°; with the load vertical, $\beta = 50°$; bearing temperature 160°F. The absolute viscosity of the oil in the film is 12centipoises (medium turbine oil). $P = W/ld = 160$ lb/in²; $\mu = 12 \times 1.45 \times 10^{-7} = 17.4 \times 10^{-7}$ lb·s/in². The solution is one of trial and error. By using Fig. 3 in conjunction with Fig. 2, only a few trials are necessary to obtain the answer. As a first trial assume $\epsilon = 0.85$. For an $l/d$ ratio of 1.5 in Fig. 3, $\eta$, the end-leakage factor, will be 0.77. Compute $A$ using this value of $\eta$. $m = 0.012/6 = 0.002$.

$$A = \frac{132}{0.77}\left(2\right)^2 \frac{160}{12 \times 720} = 12.7$$

Enter Fig. 2 with this value of $A$ and at $\beta = 50°$, and find that $\epsilon - 0.9$. This value is larger than the initial assumption for $\epsilon$. As a second trial, $\epsilon = 0.88$. Then $\eta = 0.8$, $A = 12.2$, and $\epsilon = 0.89$. This is a sufficiently close check. The minimum film thickness is $h_0 = mr(1 - \epsilon) = 0.002 \times 3 \times 0.12 = 0.0007$ in (0.01778 mm).

For severe operating conditions the value of $A$ may exceed 18, the limit of Fig. 2. For complete journal bearings under extreme operating conditions, Fig. 4 should be used. The ordinate is $P'$. The curves are drawn for various values of $l/d$ instead of values of $\beta$ as in Fig. 2. Values of $\epsilon$ may thus be

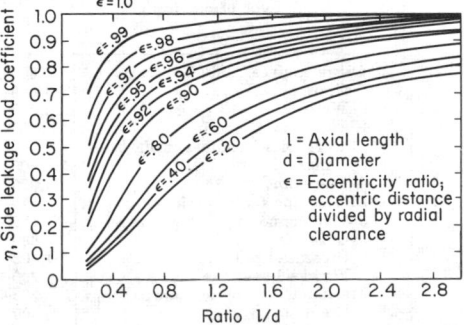

**Fig. 3**

obtained directly (Dennison, Film-lubrication Theory and Engine-bearing Design, *Trans. ASME*, **58**, 1936).

**EXAMPLE 2.**  A 360° journal bearing $2\frac{1}{2}$ in diam and $3\frac{7}{8}$ in long carries a steady load of 3,875 lb. Speed $N = 500$ r/min; diametral clearance, 0.0064 in; average viscosity of the oil in the film, 23.4 centipoises (SAE 20 light motor oil at 105°F). $P = 3,875/(2.5 \times 3.875) = 400$ lb/in². Value of $m = 0.0064/2.5 = 0.00256$. Value of $l/d = 1.55$. First, attempt to use Figs. 2 and 3 in this solution. Assume eccentricity ratio $\epsilon$ is 0.9. Then, in Fig. 3, with $l/d = 1.55$, value of $\eta$ is determined as 0.8. $A$ is calculated as 37. This is completely off scale in Fig. 2. Consider instead Fig. 4. Value of $P'$ is computed as

$$P' = 6.9\left(2.56\right)^2 \frac{400}{23.4 \times 500} = 1.54$$

In Fig. 4, enter the curves with $P' = 1.54$, and move left to intersect the curve for $l/d = 1.5$. Drop downward to read a value for $1/(1 - \epsilon)$ of 16. Then $\frac{1}{16} = 1 - \epsilon$, or the eccentricity ratio $\epsilon = \frac{15}{16}$, or 0.94. The minimum film thickness, as in Example 1 = $h_0 = mr(1 - \epsilon)$, or

$$h_0 = 0.00256 \times 1.25 \, (1 - 0.94) = 0.0002 \text{ in } (0.0051 \text{ mm})$$

**Allowable mean bearing pressures** in bearings with fluid-film lubrication are given in Table 1. If the load maintains the same magnitude and direction when the journal is at rest (heavily loaded shafts, heavy gears), the mean bearing pressure should be somewhat less than when bearings are loaded only when running.

For internal-combustion-engine bearing design, Etchells and Underwood (*Machine Design*, Sept. 1942) list the following maximum design pressures for bearing alloys, pounds per square inch of projected area. Lead-base babbitt (75 to 85 percent lead, 4 to 10 percent tin, 9 to 15 percent antimony) 600 to 800; tin-base babbitt (0.35 to 0.6 percent lead, 86 to 90 percent tin, 4 to 9 percent antimony, 4 to 6 percent copper) 800 to 1,000; cadmium-base alloy (0.4 to 0.75 percent copper, 97 percent cadmium, 1 to 1.5 percent nickel, 0.5 to 1.0 percent silver) 1,200 to 1,500; copper-lead alloy (45 percent lead, 55 percent copper) 2,000 to 3,000; copper-lead (25 percent lead, 3 percent tin, 72 percent copper) 3,000 to 4,000; silver (0.5 to 1.0 percent lead on surface, 99 percent silver) 5,000 up. The above pressures are based on fatigue life of 500 h at 300°F bearing temperature, and a bearing metal thickness 0.01 to 0.015 in for lead-, tin-, and cadmium-base metals and 0.25 in

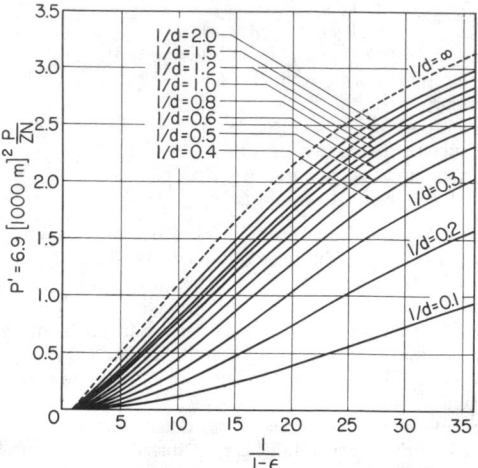

**Fig. 4**  Load-carrying parameter in terms of eccentricity.

for copper, lead, and silver. At lower temperatures the life will be greatly extended.

**Length-diameter ratios** are usually chosen between $l/d = 1$ and $l/d = 2$, although many engine bearings are designed with $l/d = 0.5$, or even less. In shorter bearings, the carrying capacity of the oil film is greatly impaired by the effect of side leakage. Longer bearings are used to restrain the shaft from vibration, as in line shafts, or to position the shaft accurately, as in machine tools. In power machines, the tendency is toward shorter bearings. Typical values are as follows: turbogenerators, 0.8 to 1.5; gasoline and diesel engines for main and crankpin bearings, 0.4 to 1.0, with most values between 0.5 and 0.8; generators and motors, 1.5 to 2.0; ordinary shafting, heavy, with fixed bearings, 2 to 3; light, with self-aligning bearings, 3 to 4; machine-tool bearings, 2 to 4; railroad journal bearings, 1.2 to 1.8.

For the **clearance between journal and bearing** see Fits in Section 8. Medium fits may be used for journals running at speeds under 600 r/min, and free fits for speeds over 600 r/min.

**Table 1.  Current Practice in Mean Bearing Pressures**

| Type of bearing | Permissible pressure, psi, of projected area | Type of bearing | Permissible pressure, psi, of projected area |
|---|---|---|---|
| Diesel engines, main bearings...... | 800–1,500 | Automotive gasoline engines, main bearings................... | 500–1,000 |
| Crankpin..................... | 1,000–2,000 | Crankpin.................... | 1,500–2,500 |
| Wrist pin.................... | 1,800–2,000 | Air compressors, main bearings... | 120–  240 |
| Electric motor bearings........... | 100–  200 | Crankpin.................... | 240–  400 |
| Marine Diesel engines, main bearings......................... | 400–  600 | Crosshead pin................ | 400–  800 |
| Crankpin.................... | 1,000–1,400 | Aircraft engine crankpin........ | 700–2,000 |
| Marine line-shaft bearings........ | 25–  35 | Centrifugal pumps............. | 80–  100 |
| Steam engines, main bearings...... | 150–  500 | Generators, low or medium speed.. | 90–  140 |
| Crankpin.................... | 800–1,500 | Roll-neck bearings............. | 1,500–2,500 |
| Crosshead pin................ | 1,000–1,800 | Locomotive crankpins.......... | 1,500–1,900 |
| Flywheel bearings............. | 200–  250 | Railway-car axle bearings........ | 300–  350 |
| Marine steam engine, main bearings | 275–  500 | Miscellaneous ordinary bearings... | 80–  150 |
| Crankpin.................... | 400–  600 | Light line shaft................ | 15–  25 |
| Steam turbines and reduction gears. | 100–  220 | Heavy line shaft............... | 100–  150 |

Kingsbury suggests for these journals a diametral clearance = $0.002 + 0.001d$ in. In journals running at high speed, diametral clearance = $0.002d$ should be used in order to lower the friction losses in the bearing. All units are in inches. The most satisfactory clearance should, of course, be based on a complete bearing analysis which includes both load-carrying capacity and heat generation due to friction. For example, a bearing designed to run at the extremely high speed of 50,000 r/min uses a diametral clearance of 0.0025 in for a journal 0.8 in in diameter, a clearance ratio, clearance/diameter of 0.00316.

For high-speed internal-combustion-engine bearings using forced-feed lubrication, medium fits are used. Federal-Mogul recommends the following diametral clearances in inches per inch of shaft diameter for insert-type bearings: tin-base and high-lead babbitts, 0.0005; cadmium-silver-copper, 0.0008; copper-lead, 0.001.

The dependence of the **coefficient of friction** for journal bearings on the bearing clearance, lubricant viscosity, rotational speed, and loading pressure, as reported by McKee and others, is shown in Section 3. A plot of the coefficient of friction against the parameter $ZN/P$ is a convenient method for showing this relationship. $ZN/P$ is a parameter based on mixed units. $Z$ is the viscosity in centipoises, $N$ is r/min, $P$ is the mean Pressure on the bearing due to the load, pounds per square inch of projected area, and $m$ is the clearance ratio. Values of $ZN/P$ greater than about 30 indicate fluid-film conditions in the bearings. If the viscosity of the lubricant becomes lower or if there is a reduction in rotational speed or an increase in load, the value of $ZN/P$ will become smaller until the coefficient of friction reaches a minimum value. Any further reduction in $ZN/P$ will produce breakdown of the oil film, marking the transition from fluid-film lubrication with complete separation of the moving surfaces to semifluid or mixed lubrication, where there is partial contact. As soon as semifluid conditions are initiated, there will be a sharp increase in the coefficient of friction. The critical value of $ZN/P$, where this transition takes place, will be lowest for a rigid bearing and shaft with finely finished surfaces.

Figure 5 shows a generalization of the relationship between the coefficient of friction for a journal bearing and the parameter $ZN/P$, indicating the various possible lubrication regimes that may be expected. For optimum design, a value of $ZN/P$ somewhere between 30 and 300 would be recommended, but, in any case, the determination of minimum film thickness $h_0$ should be the deciding parameter. For extremely large values of $ZN/P$, whirl instability may be developed or a condition of turbulence may be established in the fluid film.

The friction force in plain journal bearings may be estimated by the use of the expression $F = K_f \mu N r l / m$, where $\mu$ is in lb·s/in² units. The value of $K_f$ depends upon the magnitude of $\epsilon$ and the type of bearing. Figure 6 shows values of $K_f$ for a complete bearing, a 150° partial bearing, and a 120° partial

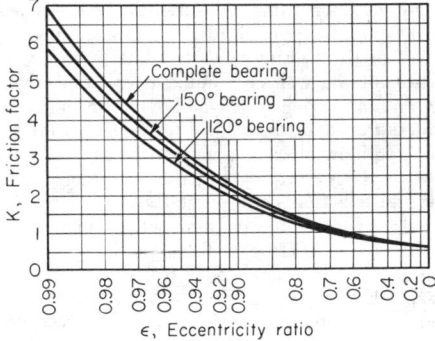

**Fig. 6** Variation of the friction factor of a bearing with eccentricity ratio.

bearing, assuming that the clearance space is at all times filled with lubricant.

EXAMPLE 3. As an illustration of the use of Fig. 6, determine the friction force in the bearing of Example 2. This is a complete journal bearing 2½ in diam by 3⅞ in. The value of $\epsilon$ was determined as 0.94. From Fig. 6, $K_f = 2.8$. Then

$$F = \frac{2.8 \times 23.4 \times 1.45 \times 10^{-7} \times 500 \times 1.25 \times 3.875}{0.00256}$$
$$= 8.97 \text{ lb } (4.08 \text{ kg})$$

The coefficient of friction $F/W = 8.97/3875 = 0.00231$. The mechanical loss in the bearing is $FV/33,000$ hp, where $V$ is the peripheral velocity of the journal, ft/min.

Friction hp = $(8.97 \times 500 \times \pi \times 2.5)/(33,000 \times 12)$
$$= 0.089 \text{ hp } (66.37 \text{ W})$$

Turbulence in the fluid film of a journal bearing will increase the friction loss. Figure 7 (Smith and Fuller, Journal Bearing Operation at Super-laminar Speeds, *Trans. ASME*, **78**, 1956) shows test results for such bearings, expressed in terms of a Reynolds number for the fluid film, $R_e = umr/v$. Laminar conditions hold up to a $R_e$ of about 1,000. Friction may be calculated for laminar flow by using Fig. 6 or the first branch of the curve in Fig. 7, where $f' = 2/R_e$. The values from Fig. 7 may be converted to friction torque $T$ by the use of the expression $T = f'\pi\rho U^2 r^2 l$, where $\rho$ is the mass density of the lubricant, lb·s²/in⁴. From $R_e$ of 1,000 to about 1,600, there is a transition zone, and at $R_e$ of 1,600, fully turbulent flow may be expected. Values of $f'$ are then obtained from $f' = 0.078/R_e^{0.43}$.

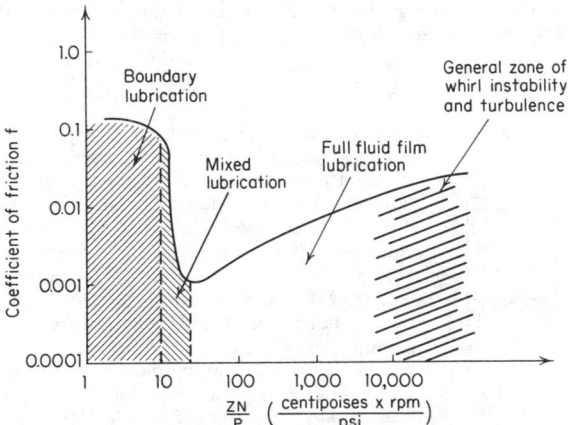

**Fig. 5** Various zones of possible lubrication for a journal bearing.

EXAMPLE 4. A journal bearing is 4.5 in diameter by 4.5 in long. Speed 22,000 r/min. $mr = 0.002$ in. Viscosity $\mu$, 1 centipoise (water) = $1.45 \times 10^{-7}$ lb·s/in²; mass density $\rho = 62.4/1,728 \times 386 = 9.35 \times 10^{-5}$ lb·s²/in⁴; $\nu = \mu/\rho = 1.45 \times 10^{-7}/9.35 \times 10^{-5} = 0.155 \times 10^{-2}$ in²/s; $U = 22,000 \times 2\pi \times 2.25/60 = 5,180$ in/s; $R_e = 5,180 \times 0.002/0.155 \times 10^{-2} = 6,680$. This would indicate turbulence in the film. Value of $f'$ is then $0.078/6,680^{0.43} = 0.078/44.2 = 1.765 \times 10^{-3}$. Friction torque $T = 1.765 \times 10^{-3} \times \pi \times 9.35 \times 10^{-5} \times 5,180^2 \times 2.25^2 \times 4.5$, $T = 317.5$ in·lb. Friction horsepower $= 2\pi TN/12 \times 33,000 = 2\pi \times 317.5 \times 22,000/12 \times 33,000$, FHP = 111 (82.77 kW).

In self-contained bearings (electric motor, line shaft, etc.) without external oil or water cooling, the **heat dissipation** is equal to the heat generated by friction in the bearing.

The heat dissipated from the outside bearing wall to the surrounding air, according to Karelitz (*Trans. ASME,* **52,** 1930), is $L = 2.2S(t_w - t_0)$ Btu/h for *quiet air,* where $S$ is the surface area of the bearing wall, ft², and $t_w$ and $t_0$ are the temperatures of the wall and ambient air, respectively, °F. With *moving air* having a velocity of 500 ft/min, the heat dissipation is $L = 6.5S(t_w - t_0)$. The surface area $S = (10$ to $15)$ $dl/144$ for pillow blocks carrying bushings, and $S = (18$ to $25)$ $dl/144$ for larger bearing pedestals carrying cast-iron or steel bearing shells (units are square feet).

The temperature of the oil film will be higher than the temperature of the bearing wall. Typical ranges of values according to Karelitz (*Trans. ASME,* **64,** 1942), Pearce (*Trans. ASME,* **62,** 1940), and Needs (*Trans. ASME,* **68,** 1948) for self-contained bearings with oil bath, oil ring, and waste-packed lubrication are shown in Fig. 8.

EXAMPLE 5. The frictional loss for the generator bearing of Example 1, computed by the method outlined in Example 3, is 0.925 hp with $\epsilon = 0.88$, $K_f = 1.6$, and $F = 27$ lb. Operating in moving air the heat dissipated by the bearing housing will be $L = 6.5S(t_w - t_0)$. Since this is a self-contained bearing, the heat dissipated is also equal to the heat generated by friction in the oil film, or $L = 0.925 \times 2,545 = 2,355$ Btu/h. With $S = 25 \times 6 \times 9/144 = 9.4$ ft², $t_w = t_0 = 2,355/6.5 \times 9.4 = 38.5$°F. This is the temperature rise of the bearing wall above the ambient room temperature. For an 80°F room, the wall temperature of the bearing would be about 118°F. In Fig. 8 an oil-ring bearing in moving air with a temperature rise of wall over ambient of 38°F should have a film temperature 50°F higher than that of the wall. The film temperature on the basis of Fig. 8 will then be $80 + 38 + 50$, or 168°F. This is close enough to the value of the film temperature of 160°F from Example 1, with which the friction loss in the bearing was computed, to indicate that this bearing can operate without the need for external cooling.

To predict the operating temperature of a self-contained

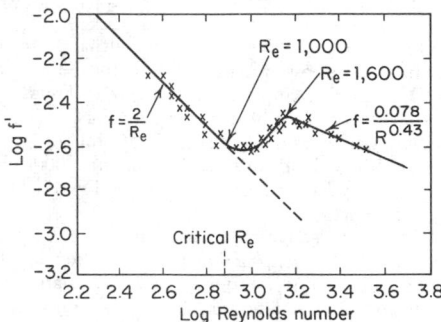

**Fig. 7** Friction $f'$ as a function of Reynolds number for an unloaded journal bearing with $l/d = 1$ (*Smith and Fuller.*)

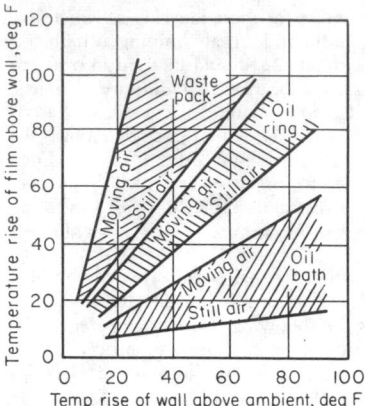

**Fig. 8** Temperature rise of the film.

bearing, the cut-and-try method shown above may be used. First, an oil-film temperature is assumed. Viscosity and friction losses are calculated. Then the temperature rise of the wall over ambient is computed so as to dissipate to the atmosphere an amount of heat equal to the friction loss. Lastly from Fig. 8 the corresponding oil-film temperature is estimated and compared to the value that was originally assumed. A few adjustments of the assumed film temperature will produce satisfactory agreement and indicate the leveling-off temperature of the bearing. Self-contained bearings have been built with diameters of 3, 8, and 24 in (7.62, 20.32, and 60.96 cm) to operate at shaft speeds of 3,600, 1,000, and 200 r/min, respectively. These designs indicate a rough limit for bearings with no external cooling. The highest bearing temperature permissible with normal lubricants is about 210°F (100°C).

The temperature of automotive-type bearings is held within safe limits by using a **pressure-feed oil supply.** Sufficient lubricant is forced through the bearing to act as a coolant and prevent overheating. One widely used practice is to place a circumferential groove at the center of the bearing to which the oil supply is fed. This is effective as far as cooling is concerned but has the disadvantage of interrupting the active length of the bearing and lowering its $l/d$ ratio (see Fig. 9). The axial flow through the bearing, in cubic inches per second (with English units), one side only, is given by

$$Q_1 = \frac{\Delta P m^3 r^4 \pi}{6\mu b} \left(1 + \frac{3}{2}\epsilon^2\right)$$

where $b$ is the effective axial length of the half bearing and $\Delta P$ is the difference between the oil pressure in the circumferential groove and the pressure at the ends of the bearing. The total flow will be twice this quantity. The value of the last term in this equation will vary from 1.0 for a concentric shaft and bearing indicated by $\epsilon = 0$ to a value of 2.5 for the extreme case of the shaft touching the bearing wall, indicated when $\epsilon = 1$. Most of the heat caused by friction in the bearing is carried away by the circulating oil. Permissible temperature rises for this type of bearing may range from 15 to 50°F (8 to 28°C). In extreme cases a rise of 100°F (55°C) can be tolerated for high-strength bearing materials. The lower val-

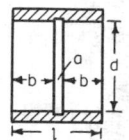

**Fig. 9** Bearing with central circumferential groove.

ues of temperature rise usually indicate needlessly large oil flow. Such a condition will result in an excessive friction loss in the bearing.

EXAMPLE 6.  The bearing of Examples 2 and 3 is lubricated by a circumferential groove with an oil-supply pressure of 30 lb/in² and, as before, $\epsilon = 0.94$, $m = 0.0026$, and $\mu = 23.4 \times 1.45 \times 10^{-7}$ lb·s/in². Length $b$ is about 1.93 in.

$$Q_1 \text{ flow out one side} = \frac{30 \times 0.0026^3 \times 1.25^4 \times \pi}{6 \times 23.4 \times 1.45 \times 10^{-7} \times 1.93}$$
$$\times [1 + 3/2(0.94)^2] = 0.240 \text{ in}^3/\text{s} \ (3.93 \text{ cm}^3/\text{s})$$

Total flow (two sides) = 0.48 in³/s = 53 lb/h for sp gr = 0.85. The friction loss from Example 3 = 0.089 hp = 226 Btu/h. With a specific heat of 0.5 Btu/(lb) (°F), and assuming that all the friction energy is given up to the oil in the form of heat, the temperature rise $\Delta t = 226/0.5 \times 53 = 8.5°F$ (4.72°C).

A definite **minimum rate of oil feed** is required to maintain a fluid film in journal bearings. This makes no allowance for the additional flow that may be needed to cool the bearings. However, many industrial bearings run at relatively low speeds with light loads and, as a consequence, additional oil flow to provide cooling is not necessary. But if a fluid film is desired, a definite minimum amount of lubricant is required. If the volume of lubricant fed to the bearing is less than this minimum requirement, there will not be a complete fluid film in the bearing. Friction will rise, wear will become greater, and the satisfactory service life of such a bearing will be reduced. This minimum lubricant supply can be evaluated by using the equation

$$Q_M = K_M u r m l$$

where $Q_M$ is in drops per min; $K_M$ is approximately 2.2; $u$ is the surface velocity of the shaft, in/min; $r$, $m$, and $l$ as before.

EXAMPLE 7.  The minimum feed rate for a journal bearing 2⅛ in diam by 2⅛ in long will be determined. Diametral clearance is 0.0045 in; speed, 1,230 r/min; load, 40 lb/in² based on projected area. $u = 1,230 \times \pi \times 2.125 = 10,220$ in/min, $r = 1.062$ in, $m = 0.0045/2.125 = 0.00212$, $l = 2.125$ in. Substituting,

$$Q_M = 2.2 \times 10,220 \times 1.062 \times 0.00212 \times 2.125$$
$$= 108 \text{ drops/min}$$

This would be equivalent to about 0.28 in³/min (4.59 cm³/min). (Fuller and Sternlicht, Preliminary Investigation of Minimum Lubricant Requirements of Journal Bearings, *Trans. ASME*, **78**, 1956.)

Many bearings are supplied with oil at low rates of feed by **felts, wicks,** and **drop-feed oilers.** Wicks can supply substantial rates of feed if they are properly designed. The two basic types of wick feed are siphon wicks, as shown in Fig. 10, and bottom wicks, as shown in Fig. 11. Data on oil delivery for these wicks are shown in Figs. 12 and 13. The data, from the American Felt Co., are for SAE Fl felts, based on a cross-

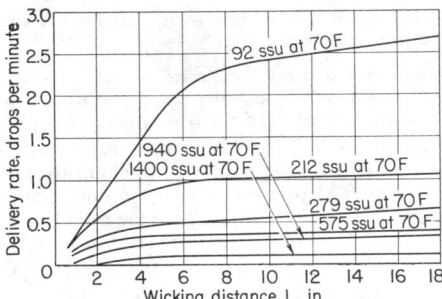

**Fig. 12**  Oil delivery with siphon wick (Fig. 10).

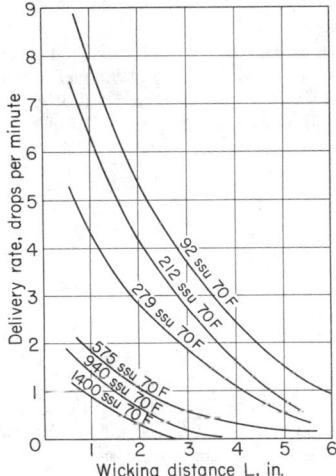

**Fig. 13**  Oil delivery with bottom wick (Fig. 11).

sectional area of 0.1 in². The flow rate is indicated in drops per minute.

EXAMPLE 8.  If it is desired to deliver 12.5 drops/min to a journal bearing, and if the viscosity of the oil is 212 s Saybolt Universal at 70°F, and if $L$, Fig. 10, is 5 in, what size of round wick would be required? From Fig. 12, for the stated conditions the delivery rate would be 0.9 drop/min for an area of 0.1 in². If 12.5 drops/min is needed, this would mean an area of 12.5 divided by 0.9 and multiplied by 0.1, or 1.4 in². For a round wick this would mean a diameter of 1⅜ in (3.49 cm).

If a **bottom wick** is considered with $L = 4$ in, Fig. 11, then in Fig. 13 the delivery rate using the same oil would be 1.6 drops/min; and if 12.5 drops/min is required, the area would be 12.5 divided by 1.6 and

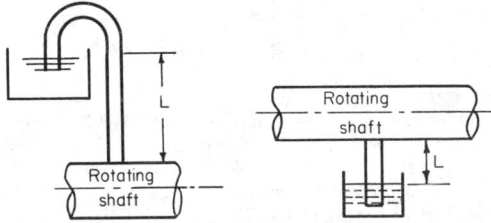

**Fig. 10**  Siphon wick.    **Fig. 11**  Bottom wick.

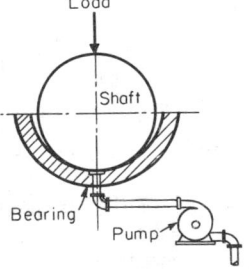

**Fig. 14**  Diagram of oil lift.

multiplied by 0.1, or 0.78 in². This would mean a bottom wick of 1 in diam if it is round (2.54 cm).

When journal **bearings** are **started, stopped,** or **reversed,** or whenever conditions are such that the operating value of $ZN/P$ falls below the critical value for that bearing, the oil film will be ruptured and metal-to-metal contact will increase friction and cause wear. This condition can be eliminated by using a **hydrostatic oil lift.** High-pressure oil is introduced to the area between the bottom of the journal and the bearing (Fig. 14). If the pressure and quantity of flow are great enough, the shaft, whether it is rotating or not, will be raised and supported by an oil film. Neglecting axial flow, which is small, the flow up one side is

$$Q_1 = \frac{Wr}{A\mu} (m)^3 \qquad \text{in}^3/\text{s}$$

(dimensions of $A$ and $B$ are keyed to English units) and the inlet pressure required, $P_o = \mu Q_1 B/br^2 (m)^3$, where $b$ is the axial length of the high-pressure recess. Values of $A$ and $B$ are given in the following table as a function of $\epsilon$.

| $\epsilon$ | 0 | 0.1 | 0.2 | 0.3 | 0.4 | 0.5 | 0.6 | 0.7 | 0.75 | 0.8 | 0.85 | 0.9 |
|---|---|---|---|---|---|---|---|---|---|---|---|---|
| $A$ | 24.0 | 28.1 | 33.8 | 41.6 | 53.3 | 72.0 | 105 | 173 | 237 | 360 | 613 | 1,320 |
| $B$ | 18.9 | 23.2 | 29.0 | 38.2 | 52.7 | 77.9 | 128 | 246 | 344 | 634 | 1,260 | 3,360 |

| $\epsilon$ | 0.91 | 0.92 | 0.93 | 0.94 | 0.95 | 0.96 | 0.97 | 0.98 | 0.99 |
|---|---|---|---|---|---|---|---|---|---|
| $A$ | 1,620 | 2,070 | 2,620 | 3,530 | 5,040 | 7,800 | 13,700 | 30,600 | 121,000 |
| $B$ | 4,340 | 5,810 | 8,040 | 11,800 | 18,400 | 32,100 | 65,300 | 179,000 | 348,000 |

Current practice is to make the total area of the high-pressure recess in a bearing 2½ to 5 percent of the projected area ($l \times d$) of the bearing. It is generally desirable to use a check valve in the supply line to the oil lift so that, when the journal builds up a hydrodynamic oil-film pressure, reverse flow of oil in the supply line will be prevented.

EXAMPLE 9. A 4.000-in-diam journal rests in a bearing of 4.012 in diam. SAE 30 oil at 100°F (105 centipoises) is supplied under pressure to a groove at the lowest point in the bearing. Length of bearing, 6 in, length of groove, 3 in, load on bearing, 3,600 lb. What inlet pressure and oil flow are needed to raise the journal 0.004 in?

$$b_o = mr(1 - \epsilon)$$
$$0.004 = 0.006(1 - \epsilon)$$
$$\epsilon = 0.333$$

From the table, $A = 44.5$, $B = 42$.

$$Q_1 = \frac{3,600 \times 2}{44.5 \times 105 \times 1.45 \times 10^{-7}} (0.003)^3$$
$$= 0.287 \text{ in}^3/\text{s, one side } (4.70 \text{ cm}^3/\text{s})$$

Flow from both sides = $(0.287 \times 2) \times {}^{60}/_{231} = 0.149$ gal/min (0.564 l/min).

Oil-supply pressure,

$$P_o = \frac{105 \times 1.45 \times 10^{-7} \times 0.287 \times 42}{3 \times 4} \times \frac{1}{0.003^3} = 566 \text{ lb/in}^2$$

An adjustable constant-volume pump or a spur-gear pump with a capacity of about 1,000 lb/in² (6,894 kN/m²) should be used to allow for pressure that may be built up in the line before the journal begins to rise.

Other configurations for hydrostatically lubricated journal bearings are shown in Fig. 15. These were obtained by means of electric analog solutions (Loeb, "Determination of Flow,

Film Thickness and Load-carrying Capacity of Hydrostatic Bearings through the Use of the Electric Analog Field Plotter," *Trans. ASLE,* vol. 1, 1958). The data from Fig. 15 are exact for a uniform film thickness corresponding to $\epsilon = 0$ but may be used with discretion for other values of $\epsilon$.

## ELEMENTS OF JOURNAL BEARINGS

Typical dimensions of solid and split **bronze bushings** are given in Table 2.

Bronze bushings made from hard-drawn sheets and rolled into cylindrical shape are made with a wall thickness of only $\frac{1}{32}$ in for bearings up to ½ in diam and with a wall thickness of $\frac{1}{16}$ in for bearings from 1 in diam up. The wall thickness of these bearings depends chiefly upon the strength of the material which supports them. Bushings of this type are pressed into place, and the bearing surface is finished by burnishing with a slightly tapered bar to a mirror finish. The allowable bearing pressures may exceed those of cast bronze shown in Table 1 by 10 to 20 percent.

**Babbitt linings** in larger bearings are generally employed in thickness of ⅛ in or over and must be provided with sufficient anchorage in the supporting shell. The anchors take the form of dovetailed grooves or holes drilled in the shell and counter-bored from the outside.

Improved conditions are obtained by sweating or bonding the babbitt to the shell by tinning the latter, using potassium chlorate as flux. Tin-base babbitts and other low-strength materials evidence some yielding when subjected to heavy pressures. This tendency may be alleviated by the use of a thinner layer of the bearing material, fused either to a bronze or to a steel shell. This improves the fatigue life of the bearing material. Standard bearing inserts of this type are available in tin-base babbitts, high-lead babbitts, cadmium alloys, and copper-lead mixtures in diameters up to about 6 in (15.24 cm) (Fig. 16). A few materials can be obtained in sizes up to 8 in (20.32 cm). Some types are available with flanges or with other special features. The bearing lining may vary from about 0.001 in (0.025 mm) to 0.1 in (2.5 mm) in thickness depending upon the size of the bearing.

Figure 17 shows the principal types of bonded babbitt linings. Figure 17a is for normal operating conditions. Figure 17b is for more severe operating conditions.

General practice for the **thickness of babbitt lining and shells** is as follows: Fig. 18, $b = \frac{1}{32}d + \frac{1}{8}$ in, $S = 0.18d$ for bronze or steel $= 0.2d$ for cast iron; Fig. 18a, $t = \frac{b}{2} + \frac{1}{16}$ in, $W = 1.8t$, $W_1 = 2.2t$.

Solid bronze or steel bushings, when pressed into the bearing housing, must be finished after pressing in. Light press fits and securing by setscrews or keys are preferable to heavy press fits and no keying, since heavy pressure, especially in thin-walled bushings, will set up stresses which will release

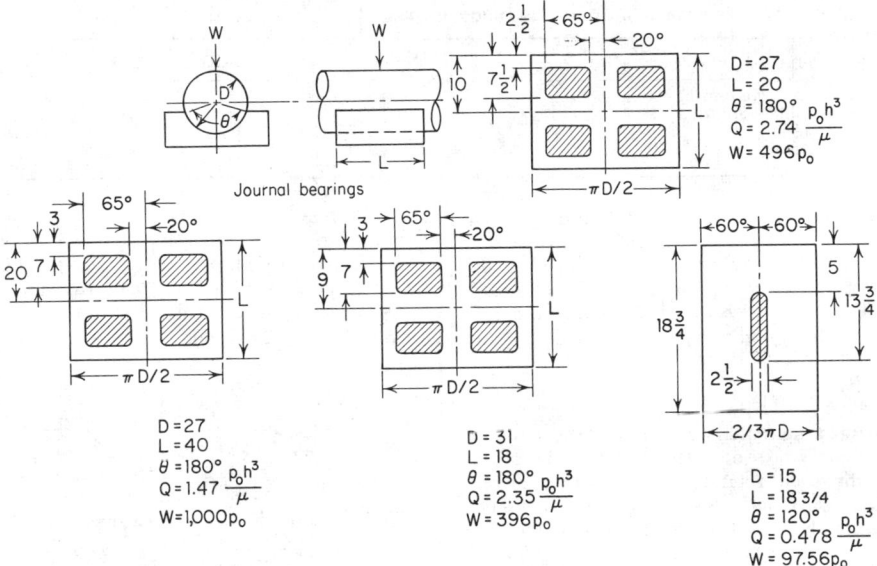

**Fig. 15**  Electric analog solutions for load-carrying capacity and flow for journal bearings (Loeb). English units.

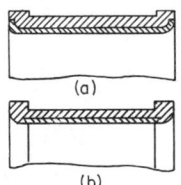

**Fig. 16**  Bearing insert.

themselves if bearings should run hot in service and will result in closing in on the journal and scoring when cooling.

**Uniform Load Distribution**  Misalignment between journal and bearing should never be so great as to cause metallic

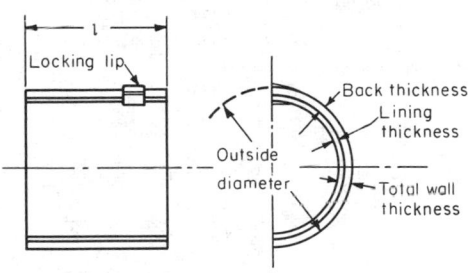

**Fig. 17**

contact. The maximum allowable inclination $\alpha$ of the shaft to the bearing is given by tan $\alpha = md/l$.

Whenever the deflection angle of the bearing installation is greater than $\alpha$, either the bearing length should be reduced or, if that is not feasible, the bearing should be mounted on a spherical seat to permit self-alignment.

**Oil grooves** are of two kinds, axial and circumferential; the former distribute the oil lengthwise in the bearing; the latter distribute it around the shaft at the oilhole, and also collect and return oil which would otherwise be forced out at the ends of the bearing. Grooves have often been put into bearings indiscriminatingly, with the result that they scrape off the oil and interrupt the film.

In Fig. 19, $W$ is the resultant force or load, pounds, on the

bearing or journal. The radial ordinates $P_1$, to the dotted curve, show the pressures, lb/in², of the journal on the oil film due to the load when there is no axial groove, while the ordinates $P_2$, to the solid curve, show the pressures with an incorrectly located groove. Since there is no oil pressure near the groove, the permissible load $W$ must be reduced or the film will be ruptured.

Groove dimensions (Fig. 20) are given by the following relations: $a = \frac{1}{3}$ wall thickness; $W_o = 2.5a$; $W_d = 3a$; $c = 0.5W_d$; $f = \frac{1}{16}$ in to $0.5W_d$.

In order to maintain the oil film, **the axial distributing groove should be placed in the unloaded sector** of the bearing. The location of grooves in a variety of cases is shown in Figs. 21 to 30.

### Horizontal Bearings, Rotational Motion

DIRECTION OF LOAD KNOWN AND CONSTANT

Load downward or inside the lower 60° segment as in the case of ring-oiling bearings (Fig. 21).

Load at an angle more than 45° to the vertical centerline (Fig. 22).

Force- or drop-feed oiling. The oil inlet may be anywhere within the no-load sector (Fig. 23).

Oil introduced through the center of the revolving shaft (Fig. 24).

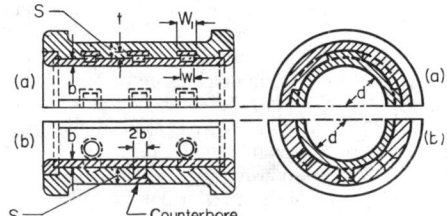

**Fig. 18**

Table 2.  Wall Thickness of Bronze Bushings, Inches

| Diam of journal, in | < ¼ | ¼–½ | ½–1 | 1–1½ | 1½–2½ | 2½–4 | 4–5½ |
|---|---|---|---|---|---|---|---|
| Solid bushing, normal | 1/16 | 3/32 | 1/8 | 3/16 | 1/4 | 3/8 | 1/2 |
| Split bushing, normal | 3/32 | 1/8 | 5/32 | 7/32 | 5/16 | 15/32 | 5/8 |
| Solid bushing, thin | 1/16 | 3/32 | 3/32 | 1/8 | 3/16 | 1/4 | 3/8 |
| Split bushing, thin | 1/16 | 3/32 | 1/8 | 3/16 | 1/4 | 3/8 | 1/2 |

Where oil-ring electric-motor bearings will be subjected by the purchaser to belt loads varying from vertical downward to horizontal, a continuous type of oil groove developed by the General Electric Co. has proved very successful (Fig. 25). There are no critical spots with this groove because only a small percentage of the babbitt surface is removed along any axial line.

### Rotating Load

For rotating shafts, a circumferential groove at the middle of the bearing and an axial groove on the no-load side (Fig. 26).

For stationary shafts and rotating bearings, a circumferential groove in the bearing and an axial groove on the no-load side. The oilhole is in the shaft at the midlength of the bearing (Fig. 27).

### Load Direction Uncertain

Oil-ring bearings (Figs. 21 and 22) may be used, although they have defects under certain load directions. With forced or drop feed, the oilhole enters a circumferential groove at the middle of the bearing and the axial groove is omitted (Fig. 28). Arrangements for introducing oil through the rotating shaft can be made.

### Bearings with Oscillatory Motion

#### Direction of Load Constant

No oil film can be built up owing to the small sliding velocity, and boundary lubrication will exist. Axial grooves in the loaded sector distribute the lubricant to all parts of the bearing and avoid dry spots (Fig. 29).

#### Load Direction Reversed during Oscillation

Fluid-film lubrication is possible, at least during part of the motion, owing to the vacuum caused by shaft moving back and forth. Figure 30 shows grooving which may be modified to suit local conditions. This arrangement is also advisable for bearings under a load which reverses in direction periodically without any rotation of the bearing. The lubrication may then provide an oil cushion to soften shocks.

**Bearing seals** are used to prevent oil leakage from the bearing housing and to protect the bearing from outside dust, water, vapors, etc. A drainage groove at the end of the bearing is effective to divert the oil passing through the bearing back into the oil well (Fig. 31a). The drain holes at the bottom of the groove must be ample for passage of the oil flow. An oil thrower mounted on the shaft is shown in Fig. 31b. The bearing housing may be provided with a single (Fig. 31c) or double collecting groove, or with brass or aluminum strip scrapers (Fig. 31d), to collect the oil creeping along the shaft.

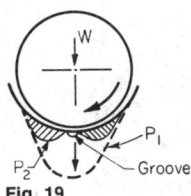

**Fig. 19**

For protection from dust, etc., felt packing rings are often used (Fig. 31e). The felt ring is soaked in oil to prevent charring by friction heat. In severe cases, additional protection by a labyrinth runner is very effective (Fig. 31f).

Standard seals are available for oil and grease retention as shown in Fig. 32a, b, and c. The seal material that is pressed

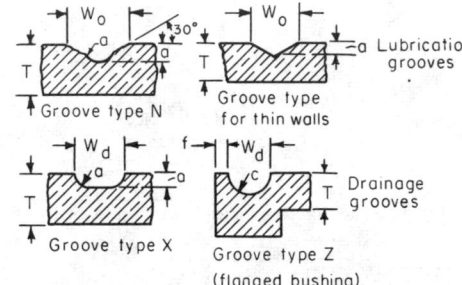

**Fig. 20**   Lubrication and drainage grooves.

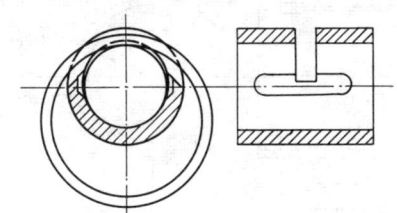

**Fig. 21**

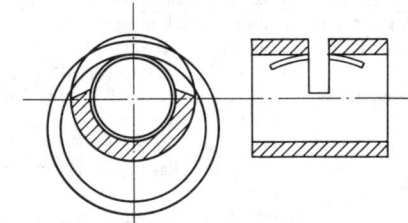

**Fig. 22**

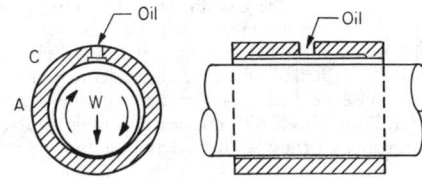

**Fig. 23**

against the rotating shaft may be leather or a synthetic rubber. Oak-tanned leather is suitable up to temperatures of 140°F (60°C) and rubbing speeds of 750 ft/min (229 m/min). Chrome-tanned leather may be used for temperatures up to approxi-

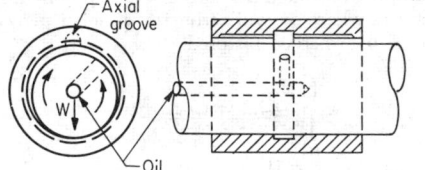

**Fig. 24**

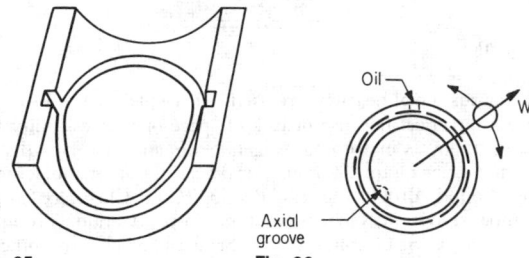

**Fig. 25**                    **Fig. 26**

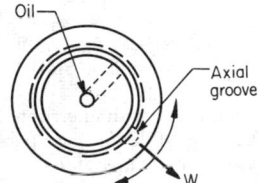

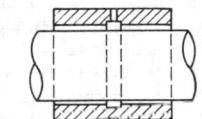

**Fig. 27**                    **Fig. 28**

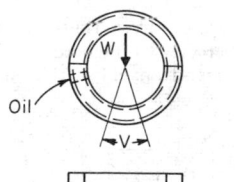

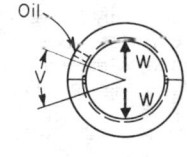

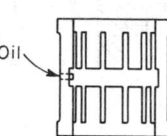

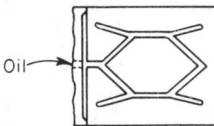

**Fig. 29**                    **Fig. 30**

mately 180°F (82.2°C) and speeds of 2,000 ft/min (610 m/min). Synthetic rubber is satisfactory for temperatures as high as about 250°F (121°C) and for faster rubbing speeds. Figure 32*a* shows the seal material pressed against the shaft by a series of flexible fingers or leaf springs. In Fig. 32*b* a helical garter spring provides the gripping force. In Fig. 32*c* a rubber spring is used for the same purpose.

**Line-shaft bearings** are purchasable in standard sizes by steps of ¼ in for the smaller diameters, and by steps of ½ in for the

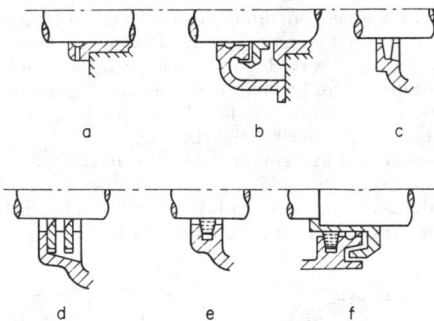

**Fig. 31**  Sealing end grooves.

larger sizes. **Hangers** for line-shaft bearings may be purchased in types permitting ceiling, floor, or wall suspension.

**Types of bearings** are shown in Figs. 33 to 38. They include the principal methods of lubrication and types of construction.

**Oilless bearings** is the accepted term for self-lubricating bearings containing lubricants in solid or liquid form in their material. Graphite, molybdenum disulfide, and Teflon are used as solid lubricants in one group, and another group consists of porous structures (wood, metal), containing oil, grease, or wax.

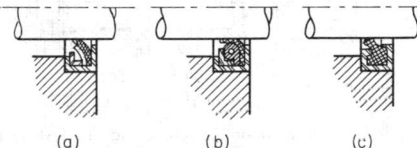

    (a)            (b)            (c)

**Fig. 32**  Seals for oil and grease retention.

**Graphite-lubricated bearings** (bridge bearings, sheaves, trolley wheels, high-temperature applications) consist generally of cast bearing bronze as a supporting structure containing various overlapping designs of grooves which are filled with graphite. The graphite is mixed with a binder, and the plastic mass is pressed into the cavities to the hardness of a lead pencil; 45 percent of the bearing area may be graphite. Drilled holes are also used for filling the bronze bearings with graphite in plastic form or with preformed graphite plugs. Strip bronze with indentations or grooves filled with graphite is used in sheet form or rolled to a butted cylindrical bushing. These bearings are manufactured by Bound Brook Oil-less Bearing Co., Johnson Bronze Co., Merriman Bros., Boston, and others.

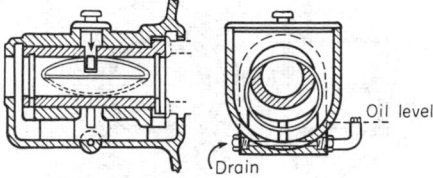

Oil grooves: Full lines for load in direction of arrow.
    Full and dotted lines for indeterminate
    load

**Fig. 33**  Ring-oiled bearing solid bushing.

**Hardwood bearings** containing lubricants are of lignum vitae which has a natural gum or hard maple which is impregnated with oil, grease, or wax. Lignum-vitae propeller bearings are used for ships and for bearings in the chemical industry, hard-maple bearings are used for loose pulleys, in textile machinery, wringer blocks, and the like.

**Porous-metal bearings,** compressed from metal powders and sintered, contain up to 35 percent of liquid lubricant. See ASTM B202-45T for sintered bronze and iron bearings, and also Army and Navy Specification AN-B-7G. The porous

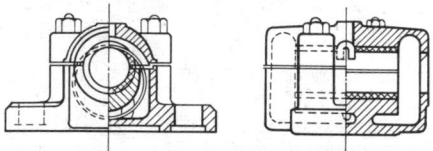

**Fig. 34**    Rigid ring-oiling pillow block. *(Link-Belt Co.)*

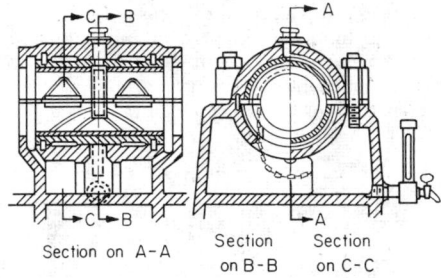

Section on A-A    Section on B-B    Section on C-C

**Fig. 35**    Split bearing with one chain. Main crank-shaft bearing. Vertical oil engine.

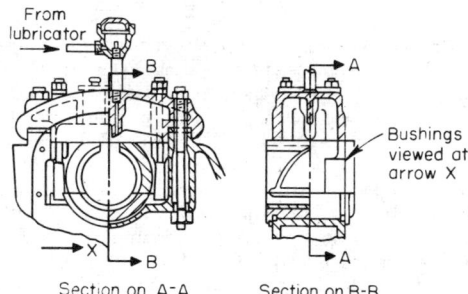

From lubricator

Bushings viewed at arrow X

Section on A-A    Section on B-B

**Fig. 36**    Crankshaft main bearing. Horizontal engine. Drop-feed lubrication.

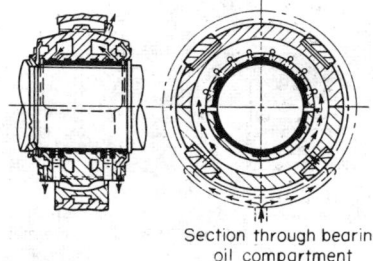

Section through bearing oil compartment

**Fig. 37**

metal generally consists of a 90-10 copper-tin bronze with 1½ percent graphite. These bearings do not require oil grooves since capillarity distributes the oil and maintains an oil film. If additional lubrication from an oil well should be provided, oil will be absorbed through the porous wall as required. For high temperatures where oil will carburize, a higher percentage of graphite (6 to 15 percent) is used.

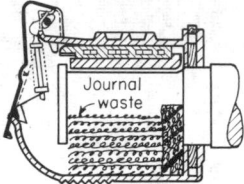

Journal waste

**Fig. 38**

Porous-metal bearings are used, where plain metal bearings are impractical because of lack of space or inaccessibility for lubrication, as in automotive generators and motors and vacuum-cleaner motors. Porous-metal bearings are manufactured by Bound Brook Oil-less Bearing Co.; Chrysler Corp., Ampex Division; General Motors Corp., Moraine Products Division; U.S. Graphite Corp., Saginaw, Mich., and others.

## THRUST BEARINGS

At low speeds, shaft shoulders or collars bear against flat bearing rings. The lubrication may be semifluid, and the friction is comparatively high.

For hardened-steel collars on bronze rings, with interrupted service, pressures up to 2,000 lb/in² (13,790 kN/m²) are permissible; for continuous low-speed operation, 1,500 lb/in² (10,341 kN/m²); for steel collars on babbitted rings, 200 lb/in² (1,378.8 kN/m²). In multicollar thrust bearings, the values are reduced considerably because of the difficulty in distributing the load evenly between the several collars.

The performance of the bearing thrust rings is much improved by the introduction of **grooves** with tapered lands as shown in Fig. 39. The lands extend on either side of the

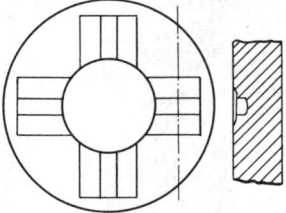

**Fig. 39**    Thrust collar with grooves fitted with tapered lands.

groove. The taper angle of the lands is very slight, so that a pressure oil film is formed between the bearing ring and the collar of the shaft. It is generally known that slightly tapered radial grooves will develop a hydrodynamic load-carrying film, when formed in the manner of Fig. 39. Leloup (Report on Investigations of Thrust Bearings, *Rev. universelle mines,* **5,** 1949) has evaluated, in a quantitative manner, the buildup of pressure associated with placing a radius on the edge of such a groove. Leloup gives the test results below for a flat steel thrust washer 2.36 in OD (5.99 cm) and 1.65 in ID (4.19 cm), with a unit load of about 100 lb/in² (689.4 kN/m²).

For high speeds or where low friction losses and a low wear rate are essential, **pivoted segmental thrust bearings** are used (Kingsbury thrust bearing, or Mitchell bearing in Europe). The bearing members in this type are tiltable shoes which rest on hard steel buttons mounted on the bearing housing. The shoes are free to form automatically a wedge-shaped oil film between the shoe surface and the collar of the shaft (Figs. 40 to 42).

| $N$, r/min | Coef of friction, flat washer | Coef of friction, grooved washer |
|---|---|---|
| 260 | 0.011 | 0.0065 |
| 520 | 0.0099 | 0.0067 |
| 900 | 0.015 | 0.0067 |

The **minimum oil-film thickness** $h_o$, in, between the shoe and the collar, at the trailing edge of the shoe, is approximately

$$h_o = 0.0341 \sqrt{\mu u l / P_{avg}}$$

where $\mu$ is the viscosity in lb·s/in²; $u$ is the velocity of the collar, on the mean diam, in/min; $l$ is the length of a shoe, at the mean diam of the collar, in the direction of sliding motion; $P_{avg}$ is the average load on the shoes, lb/in². As indicated in Fig. 40, $b = l$, approximately. The standard thrust bearings have six shoes. Load-carrying capacities of Kingsbury thrust bearings are given in Table 3.

The coefficient of friction in Kingsbury thrust bearings, referred to the mean diameter of the shoes, is approximately $f$

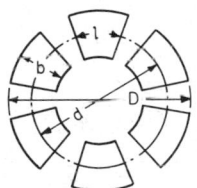

$= 11.7 h_o/l$, where $h_o$ is computed as shown above. Figures 41 and 42 show typical pivoted segmental thrust bearings. They usually embody a system of rocking levers which are used for alignment and equalization of load on the several shoes (Fig. 43).

**Fig. 40** Kingsbury thrust bearing with six shoes.

Thrust may be carried on a hydrostatic step bearing as shown schematically in Fig. 44, where high-pressure oil at $P_o$ lb/in² is supplied at the center of the bearing from an external pump. The lubricant flows radially outward through the annulus of depth $h_o$ and escapes at the periphery of the shaft at some pressure $P_1$ which is usually about atmospheric pressure. An oil film will be present whether the shaft rotates or not. Friction in these bearings can be made to approach zero, depending upon the rotational velocity and the viscosity of the lubricant film. Figure 45 shows the step bearing of a vertical turbogenerator. The load-carrying capacity

$$W = \frac{P_o \pi}{2} \left[ \frac{R^2 - R_o^2}{\ln (R/R_o)} \right]$$

This equation is valid even when the recess is eliminated, in which case $R_o$ becomes the radius of the inlet oil-supply pipe. The flow of lubricant in cubic inch per second, using English units, is $Q = P_o \pi b_o^3 / 6\mu \ln (R/R_o)$. The friction horsepower loss in the bearing is $H_f = N^2 \mu (R^4 - R_o^4)/383,000 h_o$, where $N$ is r/min.

The pumping horsepower loss in forcing the lubricant

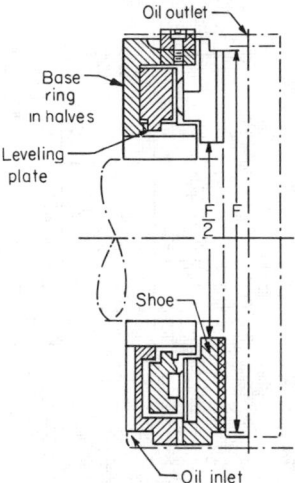

**Fig. 41** Left half of six-shoe self-aligning equalizing horizontal thrust bearing for load in either axial direction.

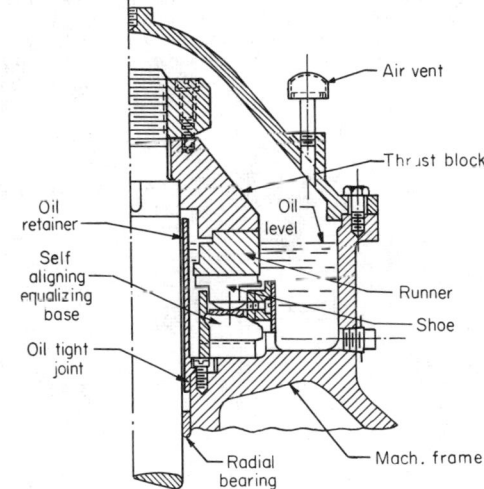

**Fig. 42** Half section of mounting for vertical thrust bearing.

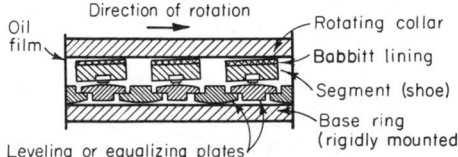

**Fig. 43** Kingsbury thrust bearings. (Developed cylindrical sections.)

through the bearing is $H_p = Q(P_o - P_1)/6,600\eta$, where $\eta$ is the efficiency of the pump and $Q$ the volume in in³/s.

EXAMPLE 10. A typical 5,000-kW vertical turbogenerator has a thrust load of about 101,000 lb; outside diameter of bearing, 16 in; diam

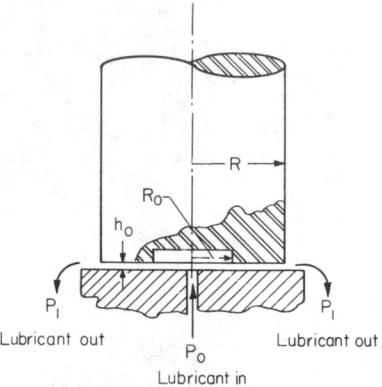

**Fig. 44**  Hydrostatic step bearing.

of recess, 10 in; pump efficiency, 0.5; speed, 750 r/min. Substituting these values, $101,000 = \dfrac{P_o\pi}{2}\left[\dfrac{8^2 - 5^2}{\ln(8/5)}\right]$, or $P_o = 774$ lb/in². In practice about 825 lb/in² is used on this step bearing to provide some margin of safety. Film thickness in the bearing should be from 0.001 to 0.01 in to protect the surfaces from metal-to-metal contact and allow passage of harmful grit that may find its way into the system. The film thickness determines the oil flow for a given viscosity and pressure. With $b_o = 0.006$ in (0.1524 mm) and SAE 20 oil at 130°F (29 centipoises), $Q = 825 \times \pi \times (0.006)^3/6 \times 29 \times 1.45 \times 10^{-7} \times 0.470 = 46.8$ in³/s (766.91 cm³/s). Flow $= 46.8 \times 60/231 = 12.15$ gal/min (45.99 l/min). The horsepower lost due to friction, $H_f = 750^2 \times 29 \times 1.45(8^4 - 5^4)/383,000 \times$

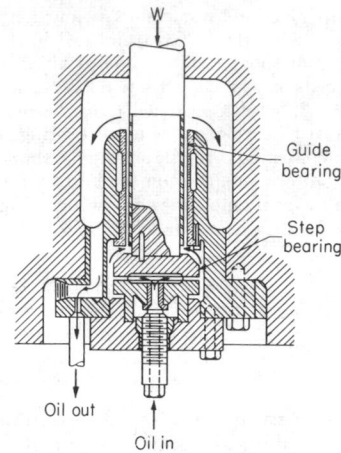

**Fig. 45**  Step bearing of a vertical turbo-generator.

$0.006 = 3.58$ hp (2.669 kW). The horsepower lost due to pumping with pump efficiency of 0.5, $H_p = 46.8 \times 825/6,600 \times 0.5 = 11.7$ hp (8.725 kW). The total energy lost $= 11.7 + 3.58 = 15.28$ hp (11.39 kW).

Evaluation of these equations for other film thicknesses will show that the minimum lost energy will occur between $b_o = 0.004$ and $b_o = 0.006$ in (0.1016 and 0.1524 mm). The coefficient of friction corresponding to an energy loss of 15.28 hp in the above example is 0.002.

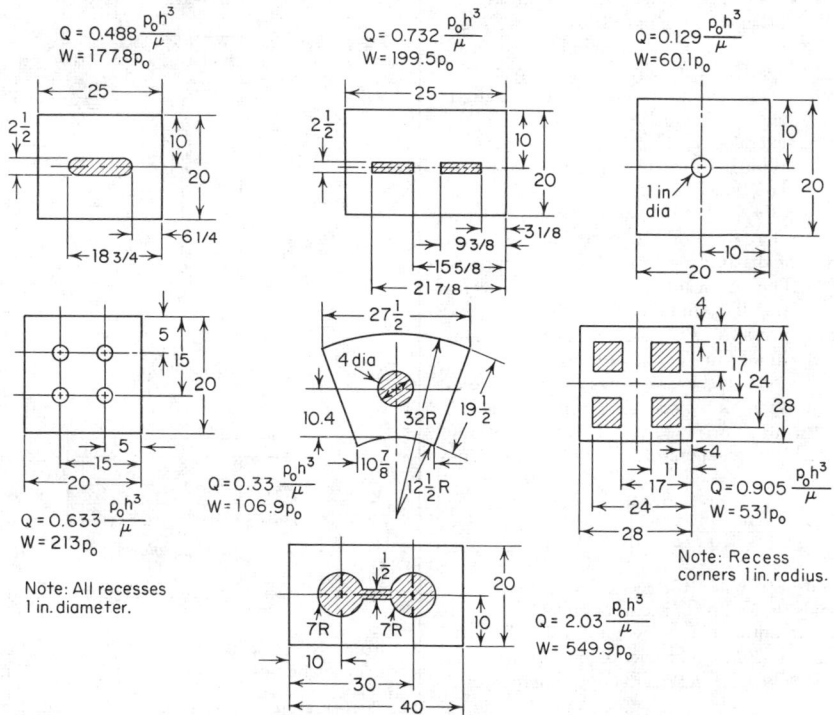

**Fig. 46**  Electric analog solutions for load-carrying capacity and flow for several flat thrust bearings (Loeb). English units.

Other configurations for hydrostatically lubricated thrust bearings are shown in Fig. 46 from Loeb. They may be used directly to obtain value of load-carrying capacity $W$ and flow $Q$ (English units).

## SLIDING BEARINGS

All sliding bearings (Fig. 47) to wear true must have the sliding parts of nearly equal lengths. Bearings made in this way will be found not to wear out of true. Oiling is accomplished in several ways, an acceptable method being that shown in Fig. 48. Short slides in many machine tools are lubricated by oil pads or direct oil application. The weight of the table and work and thrust of the tool cause wear on the bottom and sides of the guides. To compensate for the wear in both directions, bearings are sometimes made V-shaped, as shown in Fig. 49.

Simpler sliding bearings in machine tools are made with provision for adjustment (as shown in Fig. 50) of which there are many modifications. Recent applications involving hydrostatic lubrication on machine-tool ways have been very successful.

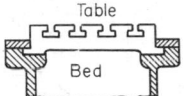

**Fig. 47**

## GAS-LUBRICATED BEARINGS

The fluid-film calculations included in Examples 1 through 10 have assumed that oil (or, in one case, water) was the lubricant. Actually, almost any **process fluid** may be used if proper recognition is given to the viscosity, corrosive action, change in state (where a liquid is close to its boiling point), toxicity, and in the case of a gas, its compressibility. Fluid-film journal and thrust bearings have run successfully, for example, on water, kerosene, gasoline, acid, liquid refrigerants, mercury, molten metals, and a wide variety of gases.

The previous equations for load-carrying capacity, film thickness, friction, and flow may be used for process liquids, but for gases, proper recognition must be made of the compressibility effects.

Because of the great value of gas-lubricated bearings for special applications, and to demonstrate the methods for handling the compressibility action, an introduction to the **design** of **gas-lubricated bearings** follows.

Naturally, if the change in pressure within the bearing

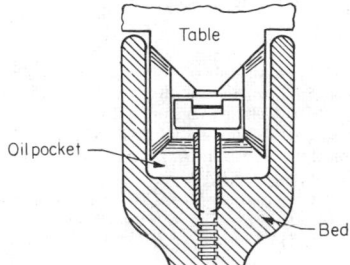

**Fig. 48**

clearance is small compared to ambient pressure, the compressibility effect will be likewise small, and lubrication equations based on liquids may be used. A **compressibility parameter** $\Lambda$ indicates the extent of this action. For hydrodynamic jour-

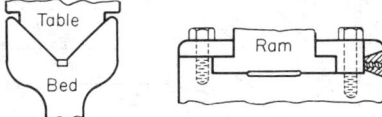

**Fig. 49**

nal bearings it has the form $\Lambda = 6\mu\omega/P_a m^2$. For values of $\Lambda$ less than one, the previous equations of this section for journal bearings may be used. For values of $\Lambda$ greater than one, compressibility effects are included through the use of Figs. 51 to 54. (Data from Elrod and Burgdorfer, Proceedings First International Symposium on Gas-lubricated Bearings, 1959, and Raimondi, *Trans. ASLE*, vol. IV, 1961.)

EXAMPLE 11. Determine the minimum film thickness for a journal bearing 0.5 in (1.27 cm) diameter by 0.5 in long. Ambient pressure 14.7 lb/in² abs (101.34 kN/m² abs). Speed 12,000 r/min. Load 0.4 lb (0.88 kg). Diametral clearance 0.0005 in (0.0127 mm). Lubricant, air at 100°F and 14.7 lb/in² abs (2.68 × 10⁻⁹ lb·s/in² from Fig. 55). $m = 0.0005/0.5 = 0.001$ in/in. $\omega = 12,000 \times 2\pi/60 = 1,256$ rad/s, $\Lambda = (6 \times 2.68 \times 10^{-9} \times 1,256)/14.7 \times 0.001^2 = 1.37$, and $W/dlP_a = 0.4/0.5 \times 0.5 \times 14.7 = 0.109$. Then, in Fig. 53 ($l/d = 1$), we find that $\epsilon = 0.22$, and the minimum film thickness $h_0 = 0.00025(1 - 0.22) = 0.000195$ in (0.00495 mm).

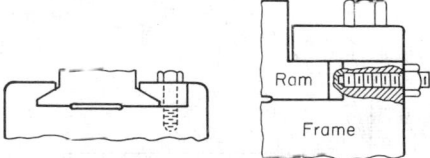

**Fig. 50**

Gas-lubricated journal bearings should be checked for **whirl stability**. Figure 56 is applicable with sufficient accuracy to bearings where $l/d$ is equal to or greater than one. It is used in conjunction with Fig. 51 for $l/d = \infty$. The stability parameter is $\omega_1^*$ which, for a bearing having only gravity loading, has the value $\omega_1^* = \omega\sqrt{mr/g}$.

EXAMPLE 12. To determine whether the bearing of Example 11 is stable at the running speed of 12,000 r/min, we compute $\omega_1^*$ as $1,256\sqrt{0.00025/386} = 1.015$. The value of eccentricity ratio $\epsilon_0$ for $l/d = \infty$ is computed from Fig. 51 on the basis of $W$ being the load per inch of bearing length. Thus $\overline{W}$(lb/in) = 0.4/0.5 = 0.8 lb/in. For Fig. 51, $W/dlP_a = 0.218$ and in Fig. 51, we determine $\epsilon_0 = 0.18$. Then (in Fig. 56), for $\omega_1^* = 1.015$ and $\Lambda = 1.37$, we find the intersection at about where a curve for $\epsilon_0 = 0.18$ would be found. The bearing should just be stable. An intersection point on the $\epsilon_0$ line or to the left should represent a stable condition. An intersection point to the right of the appropriate $\epsilon_0$ line would predict an unstable condition.

**Thrust bearings** of the tilting-pad variety are less susceptible to compressibility effects and may be considered as liquid-lubricated for values of $\Lambda$ (suitable for thrust bearings) less than about 30. $\Lambda = 6u\mu l/h_0^2 P_a$ where $l$ is the length of the shoe in the direction of sliding and $U$ is the linear velocity at the mean radius. However, the shoes should not be made flat for

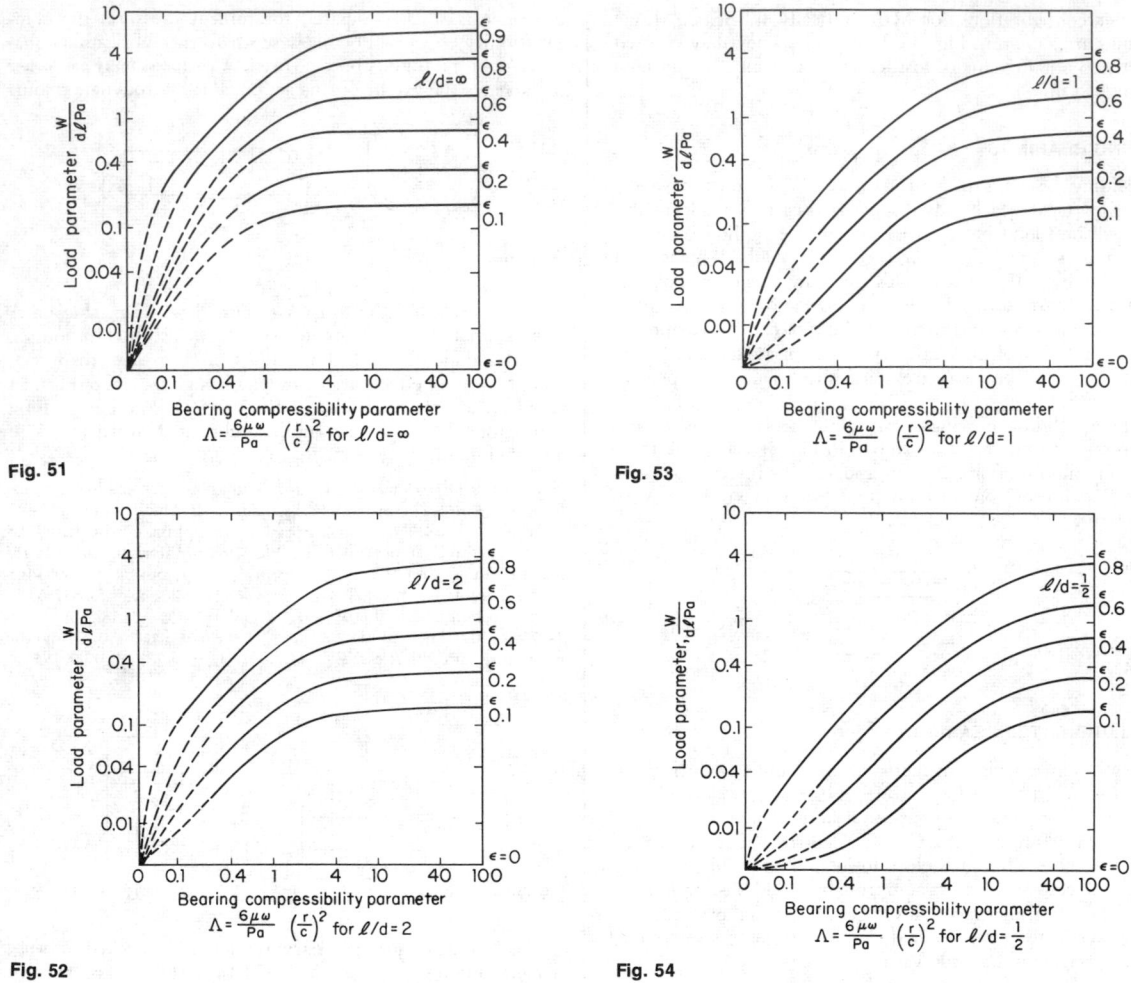

**Fig. 51**

**Fig. 52**

**Fig. 53**

**Fig. 54**

**Figs. 51–54** Theoretical load-carrying parameter vs. compressibility parameter for full journal bearing for $l/d = \infty$ (Fig. 51); $l/d = 2$ (Fig. 52); $l/d = 1$ (Fig. 53); $l/d = 0.5$ (Fig. 54). *(Elrod and Raimondi.)*

gas operation but should have a crowned contour (see Fig. 57). (Gross, "Gas Film Lubrication," Wiley, 1962.) An approximate value for the crown is to make $\delta = \frac{3}{4} h_0$.

Gas-lubricated hydrostatic bearings unlike liquid-lubricated bearings cannot be designed on the basis of constant flow volume. They are designed instead to have a pressure loss produced by an **orifice restrictor** in the supply line. Such throttling enables the bearing to have load-carrying capacity and stiffness. For maximum stiffness the pressure drop in the orifice may be about one-half of the manifold supply pressure. For a circular thrust bearing with a single circular orifice, the load-carrying capacity is given with sufficient accuracy by the equation previously used for liquids (see Fig. 44). $W = (P_R - P_a/2)[R^2 - R_0^2/\ln (R/R_0)]$, where $P_R$ is the recess pressure, lb/

in² abs. The flow volume, however, is given by $Q_0 = \pi h_0^3/[6\mu \ln (R/R_0)](P_0^2 - P_1^2)/2P_0$. $Q_0$ and $P_0$ refer to recess conditions, and $Q_1$ and $P_1$ refer to ambient conditions. Pressures are absolute.

EXAMPLE 13.  A circular thrust bearing 6 in (15.24 cm) diameter with a recess 2 in (5.08 cm) diameter has a film thickness of $h_0 = 0.0015$ in (0.0381 mm). $P_0 = 30$ lb/in² gage or 44.7 lb/in² abs (308.16 kN/m²). $P_1$ is room pressure, 14.7 lb/in² abs (101.34 kN/m² abs). Depth of recess is 0.02 in. Applied load is 375 lb. $Q_0 = (\pi \times 0.0015^3)/(6 \times 2.68 \times 10^{-9} \ln 3)(44.7^2 - 14.7^2)/(2 \times 44.7)$, $Q_0 = 12.3$ in³/s (201.6 cm³/s) at recess pressure. Converted to free air, $Q_1 = Q_0(P_0/P_1)$ with isothermal expansion, $Q_1 = 12.3(44.7/14.7) = 37.4$ in³/s (612.87 cm³/s), or $Q_1 = 37.4 \times 60 = 2,244$ in³/min (36.77 l/min). Actual measured flow = 2,440 in³/min (39.98 l/min).

**Table 3. Capacities of Six-Shoe Standard-Duty Horizontal and Vertical Thrust Bearings**

(Based on viscosity of 150 s Saybolt at operating temperatures.  Capacities given may be increased from 10 to 25% if viscosity is increased in same proportion)

| Bearing size, in | Area, sq in. | Rpm | | | | | | Bearing size, in | Area, sq in. | Rpm | | | | | |
|---|---|---|---|---|---|---|---|---|---|---|---|---|---|---|---|
| | | 100 | 200 | 400 | 800 | 1,800 | 3,600 | | | 100 | 150 | 200 | 300 | 500 | 700 |
| | | Safe load, thousands of pounds | | | | | | | | Safe load, thousands of pounds | | | | | |
| 5 | 12.5 | 1.44 | 1.7 | 2.0 | 2.4 | 2.9 | 3.5 | 19 | 180 | 40.00 | 44.0 | 48.0 | 53.0 | 60.0 | 65.0 |
| 6 | 18.0 | 2.30 | 2.7 | 3.2 | 3.8 | 4.6 | 5.5 | 21 | 220 | 51.00 | 57.0 | 61.0 | 68.0 | 77.0 | 84.0 |
| 7 | 24.5 | 3.30 | 3.9 | 4.7 | 5.6 | 6.8 | 8.0 | 23 | 264 | 65.00 | 72.0 | 77.0 | 85.0 | 97.0 | 105.0 |
| 8 | 32.0 | 4.60 | 5.5 | 6.6 | 7.8 | 9.6 | 11.4 | 25 | 312 | 80.00 | 88.0 | 95.0 | 105.0 | 119.0 | 123.0 |
| 9 | 40.5 | 6.20 | 7.4 | 8.8 | 10.4 | 13.0 | 15.0 | 27 | 364 | 97.00 | 107.0 | 115.0 | 127.0 | 144.0 | 146.0 |
| 10½ | 55.1 | 9.20 | 10.8 | 13.0 | 15.4 | 19.0 | 22.0 | 29 | 420 | 116.00 | 128.0 | 137.0 | 152.0 | 168.0 | 168.0 |
| 12 | 72.0 | 12.80 | 15.2 | 18.0 | 21.0 | 26.0 | 29.0 | 31 | 480 | 137.00 | 151.0 | 162.0 | 180.0 | 192.0 | 192.0 |
| 13½ | 91.1 | 17.20 | 20.0 | 24.0 | 29.0 | 35.0 | 36.0 | 33 | 544 | 160.00 | 177.0 | 189.0 | 210.0 | 220.0 | 220.0 |
| 15 | 112.5 | 22.00 | 26.0 | 32.0 | 37.0 | 45.0 | | 37 | 684 | 215.00 | 235.0 | 250.0 | 275.0 | 275.0 | |
| 17 | 144.5 | 30.00 | 36.0 | 43.0 | 51.0 | 58.0 | | 41 | 840 | 275.00 | 305.0 | 325.0 | 335.0 | 335.0 | |
| | | | | | | | | 45 | 1012 | 345.00 | 385.0 | 405.0 | 405.0 | | |

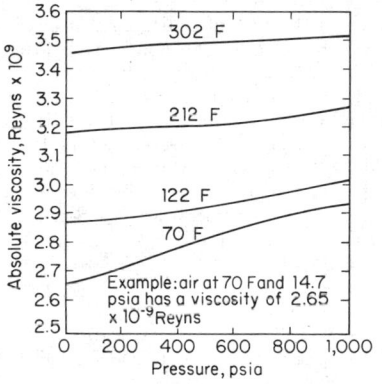

**Fig. 55** Absolute viscosity of air. *(Iwasaki, Sci. Rpts., Research Inst., Tohoku Univ., Ser. A; Kestin and Pilarczyk, Trans. ASME, vol. 56, 1954).* Reyns = $1.45 \times 10^{-7}$ centipoise.

The recess depth in Fig. 44 should be as shallow as possible, and the orifice restrictor in the supply line should be close to the entrance to the recess in order to avoid pneumatic instability. (Licht, Sternlicht and Fuller, Self-excited Vibrations of an Air-lubricated Thrust Bearing, *ASME Trans*, 1958, vol. 80.) It is good practice to limit the manifold pressure to less than 100 lb/in² gage (689.4 kN/m²) and the recess pressure to less than 50 lb/in² gage (344.7 kN/m²) in order to avoid the formation of shocks in the bearing film with extreme pressure losses.

A review of gas-bearing lubrication may also be found in the Proceedings of the First International Symposium on Gas-lubricated Bearings, as well as in the book by Gross.

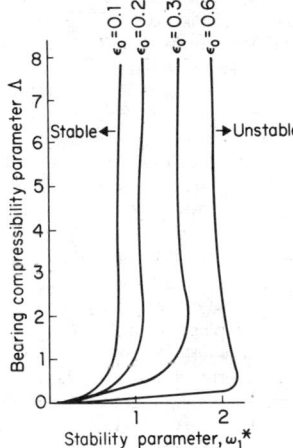

**Fig. 56** Half-frequency translatory whirl threshold for infinite length, 360° journal bearing. *(Castelli and Elrod.)*

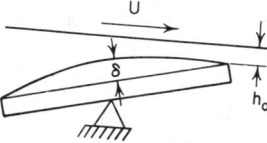

**Fig. 57** Schematic of tilting-pad shoe, showing crown height δ.

# BEARINGS WITH ROLLING CONTACT

### by Cort L. Miller and Michael W. Washo

REFERENCES: Anti-Friction Bearing Manufacturers Association, Inc. (AFBMA), Method of Evaluating Load Ratings. American National Standards Institute (ANSI), Load Ratings for Ball and Roller Bearings. AFBMA, "Mounting Ball and Roller Bearings." Tedric A. Harris, "Rolling Bearing Analysis."

## COMPONENTS AND SPECIFICATIONS

Rolling-contact bearings are designed to support and locate rotating shafts or parts in machines. They transfer loads between rotating and stationary members and permit relatively free rotation with a minimum of friction. They consist of **rolling elements (balls or rollers)** between an **outer** and **inner ring. Cages** are used to space the rolling elements from each other. Figure 1 illustrates the common terminology used in describing rolling-contact bearings.

**Rings** The inner and outer rings of a rolling-contact bearing are normally made of SAE 52100 steel, hardened to Rockwell C 60 to 67. The rolling-element raceways are accurately ground in the rings to a very fine finish (16 $\mu$in or less).

Rings are available for special purposes in such materials as stainless steel, ceramics, and plastic. These materials are used in applications where corrosion is a problem.

**Rolling Elements** Normally the rolling elements, balls or rollers, are made of the same material and finished like the rings. Other rolling-element materials, such as stainless steel, ceramics, Monel, and plastics, are used in conjunction with various ring materials where corrosion is a factor.

**Cages** Cages, sometimes called separators or retainers, are used to space the rolling elements from each other. Cages are furnished in a wide variety of materials and construction. Pressed-steel cages, riveted or clinched, are most common. Solid machined cages are used where greater strength or higher speeds are required. They are fabricated from bronze or synthetic-plastic-type materials. At high speeds, the synthetic type operates more quietly with a minimum amount of friction. Bearings without cages are referred to as full-complement.

A wide variety of rolling-contact bearings are normally manufactured to standard boundary dimensions (bore, OD, width) and tolerances which have been standardized by the AFBMA. All bearing manufacturers conform to these standards, thereby permitting interchangeability. ANSI has for the most part adopted these standards and published them as follows:

| | | | |
|---|---|---|---|
| Basic Boundary Plan | .... B3.14 | Instrument Bearings | .... B3.10 |
| Gaging Practice | ......... B3.4 | Ball Standards | ......... B3.12 |
| Terminology | ........... B3.7 | Ball Load Ratings | ...... B3.15 |
| Mounting Dimensions | .... B3.8 | Roller Load Ratings | ..... B3.16 |
| Mounting Accessories | ... B3.9 | | |

## PRINCIPAL STANDARD BEARING TYPES

The selection of the type of rolling-contact bearing depends upon many considerations, as evidenced by the numerous types available. Furthermore, each basic type of bearing is furnished in several **standard "series"** as illustrated in Fig. 2. Although the bore is the same, the outside diameter, width, and ball size are progressively larger. The result is that a wide range of load-carrying capacity is available for a given size shaft, thus giving designers considerable flexibility in selecting standard-size interchangeable bearings. Some of the more common bearings are illustrated below and their characteristics described briefly.

### Ball Bearings

**Single-Row Radial** (Fig. 3)   This bearing is often referred to as the **deep groove** or conrad bearing. Available in many variations—single or double shields or seals. Normally used for radial and moderate thrust loads.

**Maximum Capacity** (Fig. 4)   The geometry is similar to that of a deep-groove bearing except for a **filling slot**. This slot allows more balls in the complement and thus will carry heavier radial loads. However, because of the filling slot, the thrust capacity in both directions is reduced.

**Double-Row** (Fig. 5)   This bearing provides for heavy radial and light thrust loads without increasing the OD of the bearing. It is approximately 60 to 80 percent wider than a comparable single-row bearing. Because of the filling slot, thrust loads must be light.

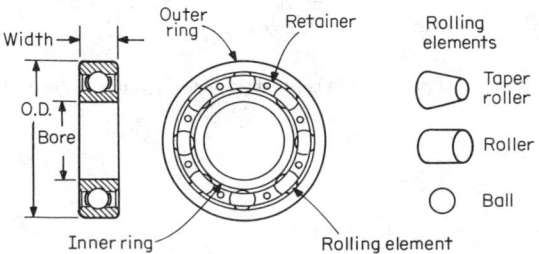

**Fig. 1**   Rolling-contact-bearing terminology.

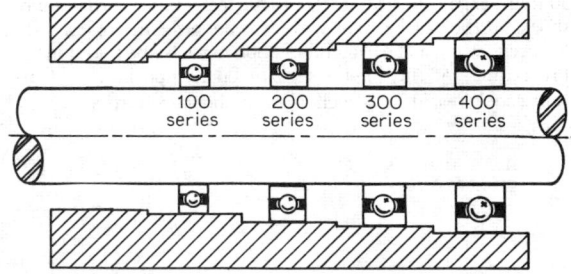

**Fig. 2**

**Internal Self-aligning Double-Row** (Fig. 6) This bearing may be used for primarily radial loads where **self-alignment** (0.003 to 0.005 in) is required. The self-aligning feature should not be abused, as excessive misalignment or thrust load (11 percent of radial) causes early failure.

**Angular-Contact Bearings** (Fig. 7) These bearings are designed to support **combined radial and thrust** loads or heavy thrust loads depending on the contact-angle magnitude. Bearings having large contact angles can support heavier thrust loads. They may be mounted in pairs which are referred to as **duplex bearings**: back-to-back, tandem, or face-to-face. These bearings may be preloaded to minimize axial movement and deflection of the shaft.

**Ball Bushings** (Fig. 8) This type of bearing is used for linear motions on hardened shafts (Rockwell C58 to 64) but cannot be used for rotary motion.

**Split-Type Ball Bearing** (Fig. 9) This type of ball or roller bearing has split inner ring, outer ring, and cage. They are assembled by screws. This feature is expensive but useful where it is difficult to install or remove a solid bearing.

### Roller Bearings

**Cylindrical Roller** (Fig. 10) These bearings utilize cylinders with approximate length-diameter ratio ranging from 1:1 to 1:3 as rolling elements. Normally used for heavy radial loads. Especially useful for free axial movement of the shaft. Highest speed limits for roller bearings.

**Needle Bearings** (Fig. 11) These bearings have rollers whose length is at least four times their diameter. They are most useful where space is a factor. Available with or without inner race. If shaft is used as inner race, it must be hardened and ground. Full-complement type is used for high loads, oscillating, or slow speeds. Cage type should be used for rotational motion. Cannot support thrust loads.

**Tapered-Roller** (Fig. 12) These bearings are used for heavy radial and thrust loads. The bearing is designed so that all elements in the rolling surface and the raceways intersect at a common point on the axis: thus **true rolling** is obtained. Where maximum system rigidity is required, the bearings can be adjusted for a preload.

**Spherical-Roller** (Fig. 13) These bearings are excellent for heavy radial loads and moderate thrust. Their internal self-aligning feature is useful in many applications but should not be abused.

### Thrust Bearings

**Ball Thrust Bearing** (Fig. 14) May be used for low-speed applications where other bearings carry the radial load. These bearings are made with shields, as well as the open type.

**Straight-Roller Thrust Bearing** (Fig. 15) These bearings are made of a series of short rollers to minimize the skidding, which causes twisting, of the rollers. They may be used for moderate speeds and loads.

**Tapered-Roller Thrust** (Fig. 16) Eliminates the skidding that takes place with straight rollers but causes a thrust load between the ends of the rollers and the shoulder on the race. Thus speeds are limited because the roller end and race flange are in sliding contact.

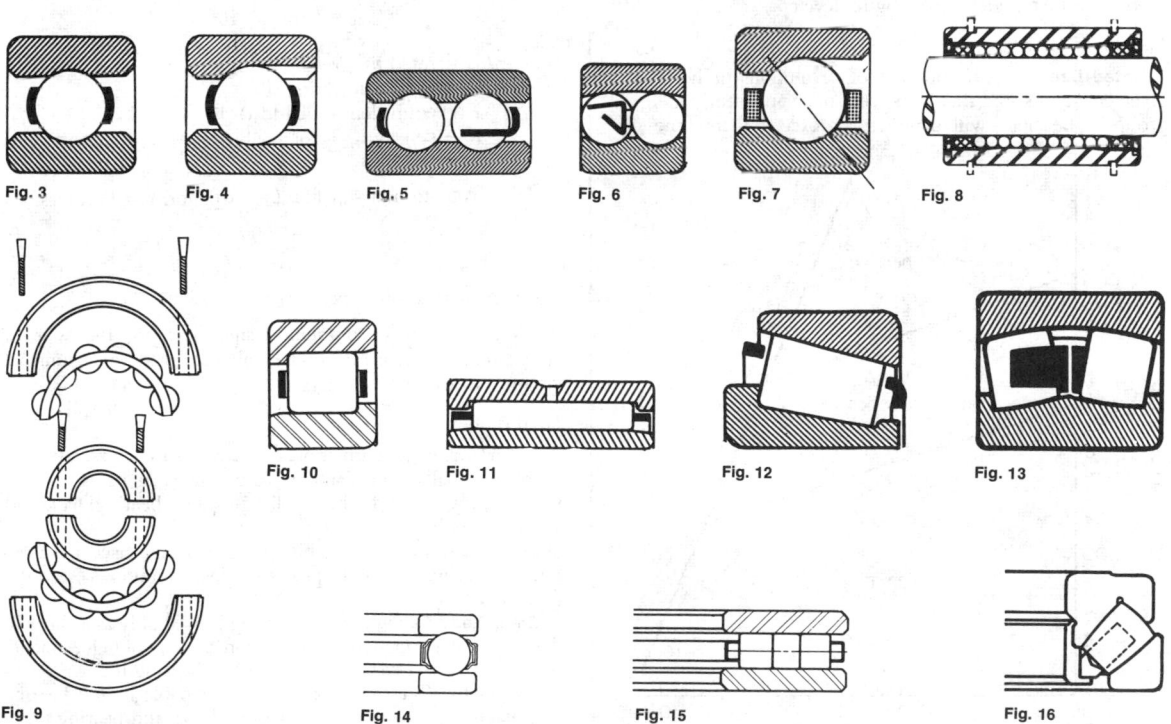

Fig. 3    Fig. 4    Fig. 5    Fig. 6    Fig. 7    Fig. 8

Fig. 9    Fig. 10    Fig. 11    Fig. 12    Fig. 13    Fig. 14    Fig. 15    Fig. 16

**Figs. 3 through 16**

### Selection of Ball or Roller Bearing

**Selection** of the type of rolling-element bearing is a function of many factors, such as load, speed, misalignment sensitivity, space limitations, and desire for precise shaft positioning. However, to determine if a ball or roller bearing should be selected, the following **general rules** apply:

1. Ball bearings function on theoretical point contact. Thus they are suited for higher speeds and lighter loads than roller bearings.

2. Roller bearings are generally more expensive except in larger sizes. Since they function theoretically on line contact, they will carry heavy loads, including shock, more satisfactorily, but are limited in speed.

Use Fig. 17 as a general guide to determine if a ball or roller bearing should be selected. This figure is based on a rated life of 30,000 h.

### ROLLING-CONTACT BEARINGS LIFE, LOAD, AND SPEED RELATIONSHIPS

An accurate knowledge of the load-carrying capacity and expected life is essential in the proper selection of ball and roller bearings. Bearings that are subject to millions of different stress applications fail owing to fatigue. In fact, **fatigue** is the only cause of **failure** if the bearing is properly lubricated, mounted, and sealed against the entrance of dust or dirt and is maintained in this condition. For this reason, the **life** of an individual bearing is defined as the total number of revolutions or hours at a given constant speed at which a bearing runs before the first evidence of fatigue develops.

### Definitions

**Rated Life** $L_{10}$   The number of revolutions or hours at a given constant speed that 90 percent of an apparently identical group of bearings will complete or exceed before the first

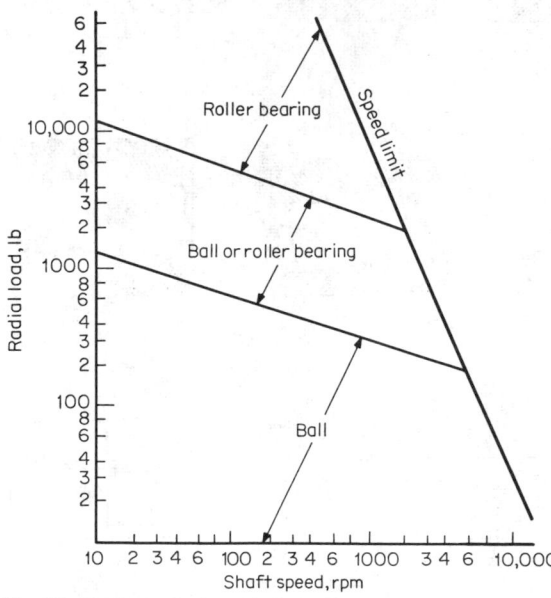

**Fig. 17**   Guide to selection of ball or roller bearing.

evidence of fatigue develops; i.e., 10 out of 100 bearings will fail before rated life. The names **Minimum life** and $L_{10}$ **life** are also used to mean rated life.

**Basic Load Rating** $C$   The radial load that a ball bearing can withstand for one million revolutions of the inner ring. Its value depends on bearing type, bearing geometry, accuracy of fabrication, and bearing material. The basic load rating is also called the **specific dynamic capacity**, the **basic dynamic capacity**, or the **dynamic load rating**.

**Equivalent Radial Load** $P$   Constant stationary radial load which, if applied to a bearing with rotating inner ring and stationary outer ring, would give the same life as that which the bearing will attain under the actual conditions of load and rotation.

**Static Load Rating** $C_0$   Static radial load which corresponds to a total permanent deformation of rolling element and race at the most heavily stressed contact of 0.0001 of the rolling-element diameter.

**Static Equivalent Load** $P_0$   Static radial load, if applied, would cause the same total permanent deformation at the most heavily stressed ball and race contact as that which occurs under actual condition of loading.

### Bearing Rated Life

Standard formulas have been developed to predict the statistical rated life of a bearing under any given set of conditions. These formulas are based on an exponential relationship of load to life which has been established from extensive research and testing.

$$L_{10} = \left(\frac{C}{P}\right)^K \times 10^6 \tag{1}$$

where $L_{10}$ = rated life, revolutions
$C$ = basic load rating, lb
$P$ = equivalent radial load, lb
$K$ = constant, 3 for ball bearings, 10/3 for roller bearings

To convert to hours of life ($L_{10}$), this formula becomes

$$L_{10} = \frac{16,700}{N} \left(\frac{C}{P}\right)^K \tag{2}$$

where $N$ = rotational speed, r/min

Table 1 lists some common design lives vs. the type of application. These may be altered to suit unusual circumstances.

### Load Rating

The **load rating** is a function of many parameters, such as number of balls, ball diameter, and contact angle. Two load ratings are associated with a rolling-contact bearing: **basic** and **static** load rating.

**Basic Load Rating** $C$   This rating is *always* used in determining bearing life for all speeds and load conditions [see Eqs. (**1**) and (**2**)].

**Static Load Rating** $C_0$   This rating is used only as a check to determine if permanent deformation of the rolling elements will occur. It is *never* used to calculate bearing life.

Values for $C$ and $C_0$ are readily attainable in any bearing manufacturer's catalog as a function of size and bearing type. Table 2 lists the basic and static load ratings for some common sizes and types of bearings.

**Table 1. Design-Life Guide**

| Application | Design life, h, $L_{10}$ | Application | Design life, h, $L_{10}$ |
|---|---|---|---|
| Agricultural equipment | 3,000–6,000 | Domestic appliances | 1,000–2,000 |
| Aircraft engines | 1,000–3,000 | Electric motors: | |
| Aircraft jet engines | 1,500–4,000 | Domestic | 1,000–2,000 |
| Automotive: | | Industrial | 20,000–30,000 |
| Bus, car | 2,000–5,000 | Elevator | 8,000–15,000 |
| Trucks | 1,500–2,500 | Fans: | |
| Blowers: | 20,000–30,000 | Industrial | 8,000–15,000 |
| Continuous 8-h service | 20,000–40,000 | Mine ventilation | 40,000–50,000 |
| Continuous 24-h service | 40,000–60,000 | Gearing units (multipurpose) | 8,000–15,000 |
| Continuous 24-h service (extreme reliability) | 100,000–200,000 | Intermittent service | 8,000–15,000 |
| Compressors | 40,000–60,000 | Paper machines | 50,000–60,000 |
| Conveyors | 20,000–40,000 | Pumps | 40,000–60,000 |

## Equivalent Load

There are two **equivalent-load** formulas. Bearings operating with some finite speed use the equivalent radial load $P$ in conjunction with $C$ [Eq. (1)] to calculate bearing life. The static equivalent load is used in comparison with $C_0$ in applications when a bearing is highly loaded in a static mode.

**Equivalent Radial Load $P$**   All bearing loads are converted to an equivalent radial load. Equation (3) is the general formula used for both ball and roller bearings.

$$P = XR + YT \qquad (3)$$

where $P$ = equivalent radial loads, lb
   $R$ = radial load, lb
   $T$ = thrust (axial) load, lb
$X$ and $Y$ = radial and thrust factors (Table 3)

The empirical $X$ and $Y$ factors in Eq. (3) depend upon the geometry, loads, and bearing type. Average $X$ and $Y$ factors can be obtained from Table 3. Two values of $X$ and $Y$ are listed. The set $X_1 Y_1$ or $X_2 Y_2$ giving the largest equivalent load should always be used.

**Static Equivalent Load $P_0$**   The static equivalent load may be compared directly to the static load rating $C_0$. If $P_0$ is greater than the $C_0$ rating, permanent deformation of the rolling element will occur. Calculate $P_0$ as follows:

$$P_0 = X_0 R + Y_0 T \qquad (4)$$

where $P_0$ = static equivalent load, lb
   $X_0$ = radial factor (see Table 4)
   $Y_0$ = thrust factor (see Table 4)
   $R$ = radial load, lb
   $T$ = thrust (axial) load, lb

If a load higher than the basic static load rating is imposed while rotating, the deformation is distributed evenly and no

**Table 2. Approximate Basic and Static Load Ratings vs. Types and Sizes**
(Ratings are in pounds) 1 lb = 0.45359 kg

| Bearing bore, mm | Ball single-row 200 series | | Ball single-row 300 series | | Ball double-row 200 series | | Roller cylindrical 300 series | | Roller spherical 22200 series | |
|---|---|---|---|---|---|---|---|---|---|---|
| | $C_0$ | $C$ | $C_0$ | $C$ | $C_0$ | $C$ | $C_0$ | $C$ | $C_0$ | $C$ |
| 10 | 600 | 1,040 | 850 | 1,430 | 800 | 1,210 | 1,020 | 1,960 | | |
| 12 | 680 | 1,180 | 1,040 | 1,650 | 1,250 | 1,820 | 1,350 | 2,540 | | |
| 15 | 780 | 1,330 | 1,220 | 1,970 | 1,430 | 2,030 | 1,520 | 2,820 | | |
| 17 | 1,000 | 1,660 | 1,470 | 2,340 | 1,840 | 2,510 | 2,070 | 3,700 | | |
| 20 | 1,390 | 2,220 | 1,760 | 2,730 | 2,540 | 3,480 | 2,560 | 4,490 | | |
| 25 | 1,560 | 2,420 | 2,350 | 3,550 | 2,858 | 3,780 | 3,720 | 6,360 | | |
| 30 | 2,250 | 3,360 | 3,120 | 4,600 | 4,110 | 5,140 | 5,070 | 8,460 | | |
| 35 | 3,070 | 4,430 | 4,020 | 5,770 | 5,600 | 6,700 | 6,400 | 10,400 | | |
| 40 | 3,520 | 5,040 | 5,020 | 7,060 | 6,430 | 7,680 | 7,930 | 12,500 | 11,800 | 15,200 |
| 45 | 4,000 | 5,660 | 6,130 | 8,430 | 7,320 | 8,620 | 9,310 | 14,700 | 12,600 | 15,900 |
| 50 | 4,450 | 6,070 | 8,010 | 10,750 | 8,130 | 9,220 | 11,600 | 17,900 | 13,600 | 16,800 |
| 55 | 5,630 | 7,500 | 9,400 | 12,410 | 10,300 | 11,400 | 12,600 | 19,100 | 16,500 | 20,300 |
| 60 | 6,950 | 9,070 | 10,902 | 14,179 | 12,700 | 13,800 | 15,200 | 22,800 | 20,800 | 25,200 |
| 65 | 7,660 | 9,900 | 12,516 | 16,051 | 14,000 | 15,000 | 19,900 | 29,000 | 25,500 | 30,200 |
| 70 | 8,410 | 10,714 | 14,240 | 18,030 | 15,400 | 16,300 | 21,400 | 30,800 | 27,500 | 31,900 |
| 75 | 9,190 | 11,610 | 16,080 | 19,600 | 16,900 | 17,300 | 23,200 | 32,900 | 29,100 | 33,100 |
| 80 | 10,010 | 12,550 | 18,020 | 21,230 | 18,300 | 19,100 | 27,000 | 38,100 | 32,100 | 36,800 |
| 85 | 11,750 | 14,490 | 20,080 | 22,880 | 19,500 | 19,700 | 30,900 | 43,300 | 38,200 | 43,200 |
| 90 | 13,630 | 16,540 | 22,250 | 24,580 | 22,100 | 22,600 | 35,200 | 48,800 | 44,500 | 49,800 |
| 95 | 15,650 | 18,740 | 24,530 | 26,300 | 28,600 | 28,600 | 39,500 | 54,200 | 48,800 | 54,700 |
| 100 | 17,800 | 21,130 | 29,430 | 29,940 | 32,500 | 32,100 | 44,700 | 60,800 | 55,700 | 61,900 |
| 110 | 20,100 | 23,000 | 32,040 | 31,800 | 30,500 | 30,700 | 53,200 | 70,500 | 72,000 | 78,400 |

**Table 3. Radial and Thrust Factors**

| Bearing type | $X_1$ | $Y_1$ | $X_2$ | $Y_2$ |
|---|---|---|---|---|
| Single-row ball | 1.0 | 0.0 | 0.56 | 1.40 |
| Double-row ball | 1.0 | 0.75 | 0.63 | 1.25 |
| Cylindrical roller | 1.0 | 0.0 | 1.0 | 0.0 |
| Spherical roller | 1.0 | 2.5 | 0.67 | 3.7 |

practical impairment occurs until the deformation becomes quite large. Some equipment operates with loads greatly exceeding the static capacity, such as bearings supporting artillery (twice static capacity), or aircraft control pulleys (four times static capacity). The load which will fracture a bearing is approximately eight times the static load rating.

Oscillating loads, where the motion is such that the rolling element rotates less than half a revolution, approach static load conditions. This type of load is conducive to rapid **false brinelling** and requires special lubrication techniques.

### Required Capacity

The basic load rating $C$ is very useful in the selection of the type and size of bearing. By calculating the required capacity needed for a bearing in a certain application and comparing this with known capacities, a bearing can be selected. To calculate the required capacity, the following formula can be used:

$$C_r = \frac{P \, (L_{10} N)^{1/K}}{Z} \tag{5}$$

where $C_r$ = required capacity, lb
  $L_{10}$ = rated life, h
  $P$ = equivalent radial load, lb
  $K$ = constant, 3 for ball bearings, 10/3 for roller bearings
  $Z$ = constant, 25.6 for ball bearings, 18.5 for roller bearings
  $N$ = rotation speed, r/min

### Speed Limits

Many factors combine to determine the limiting speeds of ball and roller bearings. It depends on several factors, like bearing size, inner- or outer-ring rotation, contacting seals, radial clearance and tolerances, operating loads, type of cage and cage material, temperature, and type of lubrication. A convenient check on speed limits can be made from a *dn* value. The *dn* value is a direct function of size and speed and is dependent on type of lubrication. It is calculated by multiplying the bore in millimeters (mm) by the speed in r/min.

$$dn = \text{bore (mm)} \times \text{speed (r/min)} \tag{6}$$

A guide for *dn* values is listed in Table 5. When these values

**Table 4. Radial and Thrust Factors**

| Type of bearing | $X_0$ | $Y_0$ |
|---|---|---|
| Single-row ball | 0.6 | 0.5 |
| Double-row ball | 0.6 | 0.5 |
| Spherical roller, 22200 series | 1.0 | 2.9 |
| Cylindrical roller | 1.0 | 0.0 |

**Table 5. *dn* Values vs. Bearing Types**

| Bearing type | Series | Max *dn* value | |
|---|---|---|---|
| | | Grease | Oil |
| Single-row ball | 100, 200, 300, 400, 30, in | 200,000 | 300,000 |
| Double-row ball | 200, 300 | 160,000 | 220,000 |
| Cylindrical roller | 200, 300 | 150,000 | 200,000 |
| Spherical roller | 22200 | 120,000 | 170,000 |

are exceeded, bearing life is shortened. The values are only a guide for approaching difficulties and can be exceeded by special bearings, lubrication, and application.

### Friction

One of the assets of rolling-contact bearings is their low friction. The **coefficient of friction** varies appreciably with the type of bearing, load, speed, lubrication, and sealing element. For rough calculations the following coefficients can be used for normal operating conditions and favorable lubrication:

  Single-row ball bearings  ....  0.0015
  Roller bearings  ...........  0.0018

Excess grease, contact seals, etc., will increase these values, and allowances should be made.

## PROCEDURE FOR DETERMINING SIZE, LIFE, AND BEARING TYPE

Basically, three common situations may be encountered in the analysis of a bearing system: bearing-size selection, bearing-type selection, and bearing-life determination. Each of these problems requires the following conditions to be known: radial load, thrust load, and speed. The static load capacity is not considered in the following procedures but should be analyzed if the bearing rotational speed is slow or if the bearing is idle for a period of time.

### Bearing-Size Selection

Known type and series:
  1. Select desired design life (Table 1).
  2. Calculate equivalent radial load $P$ [Eq. (3)].
  3. Calculate required capacity $C_r$ [Eq. (5)].
  4. Compare $C_r$ with capacities $C$ in Table 2. Select first bore size having a capacity $C$ greater than $C_r$.
  5. Check bearing speed limit [Eq. (6)].

### Bearing-Type Selection

Known bore size and life:
  1. Select ball or roller bearing (Fig. 18).
  2. Calculate equivalent load $P$ [Eq. (3)] for various bearing types (conrad, spherical, etc.).
  3. Calculate $C_r$ [Eq. (5)].
  4. Compare $C_r$ with capacities $C$ in Table 3, and select the type that has a capacity equal to or greater than $C_r$.
  5. Check bearing speed limit [Eq. (6)].

### Bearing-Life Determination

Known bearing size:
  1. Select ball or roller bearing (Fig. 18).

2. Calculate equivalent radial load $P$ [Eq. (**3**)].
3. Select basic load rating $C$ from Table 3.
4. Calculate rated life $L_{10}$ [Eq. (**1**) or (**2**)].
5. Check calculated life with design life.

## BEARING CLOSURES

Rolling-element bearings are made with a wide variety of **closures**. Basically, they are open, shielded, or sealed (Figs. 18 and 19). **Shielded bearings** have a small clearance between the

**Fig. 18**

**Fig. 19**

stationary shield and rotating ring. This provides reasonable exclusion of dirt without an increase in friction. **Sealed bearings** have a flexible lip (usually synthetic rubber) in contact with the inner ring. Friction is increased, but more effective retention of lubricant and exclusion of dirt is obtained. Seals should not be used to seal a fluid head or at high speeds.

## BEARING MOUNTING

Correct **mounting** of a rolling-contact bearing is essential to obtain its rated life. Many types of mounting methods are available. The selection of the proper method is a function of

the accuracy, speed, load, and cost of the application. The most common and best method of bearing retention is a press fit against a shaft shoulder secured with a locknut. End caps are used to secure the bearing against the housing shoulder (Fig. 20). Retaining rings are also used to fix a bearing on a shaft or in a housing (Fig. 21). Each shaft assembly normally must provide for expansion by allowing one end to float. This can be accomplished by allowing the bearing to expand linearly in the housing or by using a straight roller bearing on one end. Care must be exercised when designing a **floating installation** because it requires a slip fit. An excessively loose fit will cause the bearing to spin on the shaft or in the housing.

Table 6 lists shaft and housing tolerances for press fits with ABEC 1 precision applications (pumps, gear reducers, electric motors, etc.) and ABEC 7 precision applications (grinding spindles, etc.).

Mounted units such as **ball-bearing pillow blocks** (Fig. 22) are frequently used for fans and conveyors. Three common meth-

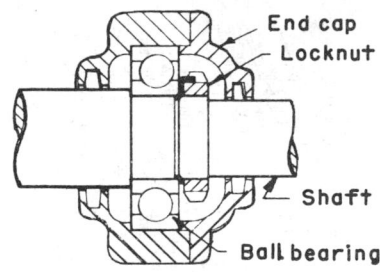
**Fig. 20**

### Table 6. Shaft and Housing Tolerances for Press Fit

| Bearing bore, mm | Shaft tolerances, in, ABEC 1 precision | Bearing bore, mm | Shaft tolerances, in, ABEC 7 precision |
|---|---|---|---|
| 4–6 | +0.0000 −0.0002 | 4–30 | +0.00000 −0.00015 |
| 7–17 | +0.0000 −0.0003 | 35–50 | +0.0000 −0.0002 |
| 20–50 | +0.0000 −0.0004 | 55–80 | +0.0000 −0.0003 |
| 55–80 | +0.0000 −0.0005 | 85–120 | +0.00000 −0.00035 |
| 85–120 | +0.0000 −0.0006 | | |

| Bearing OD, mm | Housing tolerances, in, ABEC 1 precision | Bearing OD, mm | Housing tolerances, in, ABEC 7 precision |
|---|---|---|---|
| 16–30 | +0.0008 −0.0000 | 16–80 | +0.0002 −0.0000 |
| 32–47 | +0.0010 −0.0000 | 85–120 | +0.0003 −0.0000 |
| 52–80 | +0.0012 −0.0000 | 125–225 | +0.0004 −0.0000 |
| 85–120 | +0.0014 −0.0000 | | |
| 125–180 | +0.0016 −0.0000 | | |

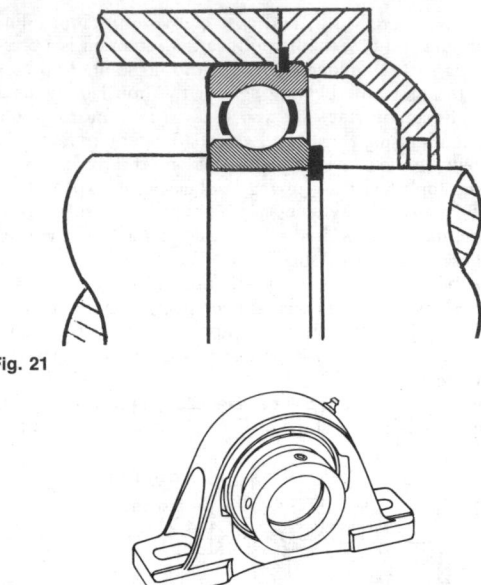

**Fig. 21**

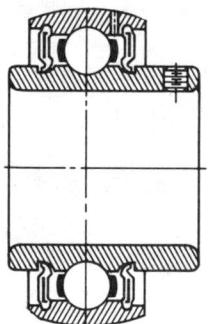

**Fig. 22**

ods are used to attach the bearing to the shaft: setscrew, eccentric locking collar, and taper-sleeve adapter.

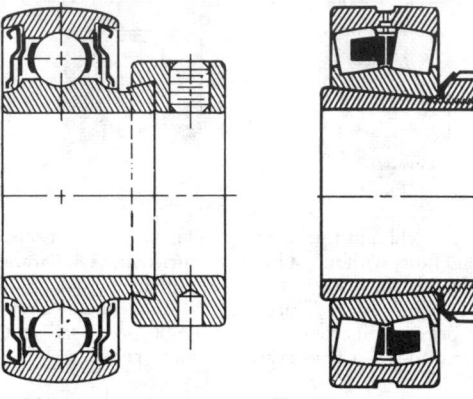

**Setscrew**  Figure 23 illustrates the use of an extended inner-ring bearing held to the shaft with a setscrew. This is a simple method and is suitable only for lightly loaded bearings.

**Eccentric Locking Collar**  Figure 24 illustrates the use of an extended inner-ring bearing held to the shaft with an eccentric collar. This method tends to keep the shaft centered in the bearing more concentrically than the setscrew method. It is suitable for light to moderate loads.

**Taper-Sleeve Adapter**  Figure 25 illustrates the use of a taper-sleeve adapter to mount the bearing on the shaft. It provides uniform concentric contact between the shaft and bearing

**Fig. 23**

bore. However, skill is required to tighten the locking nut enough to keep the sleeve from spinning on the shaft and yet not so tight that the inner race of the bearing is expanded to the point where the clearance is removed from the bearing. It is very difficult to obtain the correct setting with light-series bearings. They are excellent for heavy-duty spherical roller bearings.

## LUBRICATION

Rolling-contact bearings need a **fluid lubricant** to obtain or exceed their rated life. In the absence of high-temperature environment, only a small amount of lubricant is required for excellent performance. Excess lubricant will cause heating of the bearing and accelerate the deterioration of the lubricant. Optimum lubrication of rolling-contact bearings can be predicted by elastohydrodynamic theory (EHD). It has been shown that film thickness is sensitive to bearing speed of operation and lubricant viscosity properties and, moreover, that the film thickness is virtually insensitive to load.

**Fig. 24**          **Fig. 25**

**Grease** is commonly used for lubrication of rolling-contact bearings because of its convenience and minimum maintenance. A high-quality lithium-based NLGI 2 grease should be used for temperatures up to 180°F (82°C). In applications involving high speed, oil lubrication is often necessary. Table 7 can be used as a general guide in selecting oil of the proper viscosity for rolling-contact bearings.

In applications using grease, it is necessary to replenish the lubricant. Relubrication intervals in hours of operation are dependent on temperature, speed, and bearing size. Table 8 is a general guide which represents the time after which it is advisable to add a small amount of grease in order to safeguard the bearings. The intervals are valid up to 160°F (71°C) and should be divided by 2 for cylindrical roller bearings and by 10 for spherical roller bearings.

**Table 7. Oil-Lubrication Viscosity**
(Viscosity in ISO identification numbers*)

| Bearing bore, mm | Bearing speed, r/min | | | | |
|---|---|---|---|---|---|
| | 10,000 | 3,600 | 1,800 | 600 | 50 |
| 4–7 | 68 | 150 | 220 | | |
| 10–20 | 32 | 68 | 150 | 220 | 460 |
| 25–45 | 10 | 32 | 68 | 150 | 320 |
| 50–70 | 7 | 22 | 68 | 150 | 320 |
| 75–90 | 3 | 10 | 22 | 68 | 220 |
| 100 | 3 | 7 | 22 | 68 | 220 |

*ISO identification number = midpoint viscosity in centistokes at 40°C.

**Table 8. Ball-Bearing Grease Relubrication Intervals**
(Hours of operation)

| Bearing bore, mm | Bearing speed, r/min | | | | |
|---|---|---|---|---|---|
| | 5,000 | 3,600 | 1,750 | 1,000 | 200 |
| 10 | 8,700 | 12,000 | 25,000 | 44,000 | 220,000 |
| 20 | 5,500 | 8,000 | 17,000 | 30,000 | 150,000 |
| 30 | 4,000 | 6,000 | 13,000 | 24,000 | 127,000 |
| 40 | 2,800 | 4,500 | 11,000 | 20,000 | 111,000 |
| 50 | | 3,500 | 9,300 | 18,000 | 97,000 |
| 60 | | 2,600 | 8,000 | 16,000 | 88,000 |
| 70 | | | 6,700 | 14,000 | 81,000 |
| 80 | | | 5,700 | 12,000 | 75,000 |
| 90 | | | 4,800 | 11,000 | 70,000 |
| 100 | | | 4,000 | 10,000 | 66,000 |

# PACKINGS AND SEALS

## by Edward W. Fisher

REFERENCES: Staniar, "Plant Engineering Handbook," McGraw-Hill. Thorn, Rubber and Plastic Packings, *Rubber Age*, Jan. 1956. Roberts, Gaskets and Bolted Joints, *Jour. Applied Mechanics*, June 1950. Nonmetallic Gaskets, *Machine Design*, Nov. 1954. Elonka, "Basic Data on Seals," a *Power* reprint, McGraw-Hill.

**Packings** include those means used to prevent or minimize leakage of a fluid through mechanical clearances in either the static or dynamic state.

**Gaskets** are installed in static clearances which normally exist between parallel flanges or concentric cylinders. Sealing of flat flange gaskets is effected by compressive loading achieved through bolting or other mechanical means. The full face gasket (Fig. 1) is not recommended because the material outside the bolt holes is ineffective. The simple ring gasket (Fig. 2) is more efficient and economical. With irregularly contoured flanges, bolt holes may serve to locate the gasket, in which case they should be placed in lobes with full sealing flange width maintained between the inner edge of the holes and the inside of the gasket. **Metal-to-metal** fits require a recess whose volume is greater than that of the gasket to be used. The gasket, such as an O ring (Fig. 13), either rectangular or round cross-sectioned, extends above the groove sufficiently to provide a minimum cross-sectional compression of 15 percent for initial seating. In service, the fluid load automatically provides the sealing force. **Warped or wavy** flanges, often resulting from welding or as found in glass-lined equipment, require gaskets that are softer or thicker than normal in order to adjust to high and low areas. Excessive thickness, even though installed in a groove, must be avoided or "mushrooming" and inadequate loading will result. Tongue-and-groove joints (Fig. 4) confine the gasket and thus are well adapted to accommodating such extra thickness.

In addition to the types (Figs. 5, 6, and 7) shown, as defined in the table (Fig. 37), there are the machined metal-profile gasket (Fig. 8) and solid metal designs in flat, round, and either octagonal or oval API ring joint gaskets for extreme pressures and temperatures against steam, oil, and gases. Compressibilities are very low and depend on cross section. The envelope gasket (Fig. 3), usually polytetrafluoroethylene with a variety of cores, is particularly useful for extremely corrosive or noncontaminating service under average pressure conditions.

**Cylindrical** or **concentric** gasketing involves the use of a gland to provide retention and mechanical loading, as in the standard mechanical joint for cast-iron pipe (Fig. 10) or the condenser tube-sheet ferrule (Fig. 11). There are also gaskets shaped like cups and designed to be self-tightening under pressure (Fig. 12). The O ring (Fig. 13) located in an annular groove and precompressed as in the grooved flange, is a self-energized gasket. A cylindrical ring with internal single or double lips, also automatic in action, is quite common in pipe joints.

Beyond these types are many specialty gaskets designed for specific or proprietary use, for example, a seal for a removable drumhead.

The compressibility of various gasketing materials is shown in Fig. 37, and their common usage is listed in Table 1. Beyond rubber are many elastomeric materials generally similar in mechanical behavior but varying in temperature limits and fluid compatibility (see Sec. 6).

The **proper design** of a gasketed joint requires flange rigidity to avoid bowing, surface finish commensurate with gasket type and fluid pressure, and adequate bolt loading. The load must seat the gasket, i.e., cause the material to flow into and fill flange irregularities. It must also carry the fluid load with sufficient excess so that the residual pressure on the gasket exceeds the fluid pressure. These values, known respectively as the seating load $y$ in lb/in$^2$ and the gasket factor $m$, vary with gasket material and thickness. The *ASME* Code for Unfired Pressure Vessels, Section VIII, gives sufficient detail for typical joint design and tabulates values for $y$ and $m$ for various gasketing materials.

While high bolt loading is desirable for tight and enduring joints, it must not crush the gasket material. Crushing-strength values, which will vary with thickness and temperature, can be obtained from the gasket manufacturers. Consistent with condition of the flanges, the thinner the gasket the more efficient the joint.

Data on the design of **O-ring joints** are available from suppliers. The nominal pressure limit for O rings, based on typical mechanical clearances, is 1,500 lb/in² (10 MN/m²) without backup rings and 3,000 lb/in² (20 MN/m²) with backup rings. If clearances can be eliminated, as in a flanged joint with close metal-to-metal contact, no limit can be set. Other self-energizing joints, such as the boiler hand-hole plate (Fig. 9), need only sufficient load to effect an initial seal.

**Valve disks** are specialized gaskets designed for joints that are frequently broken and restored. Disks for globe valves (Fig. 14) are usually encased in a disk holder with a swivel mounting so that they do not rotate while closing. They are made of firm rubber for bib washers to hard rubber and phenolics for more severe service. Plastics such as nylon and polytetrafluoroethylene are also used. Pump valves (Fig. 15) are described in Sec. 14. **Valve seats** of rubber are used with metal valve disks on some pumps, e.g., the rotary drilling pump (Fig. 16). Plastics are also used for seats, notably in ball valves.

**Dynamic packings** include all packings that operate on moving surfaces. In functioning, to retain fluid under pressure, they carry the hydraulic load. When no pressure exists, as in many oil-seal applications, the packing is mechanically loaded as by a spring (Fig. 28) or by its own resiliency. Dynamic packings therefore operate as bearings, thereby indicating the need for lubrication to serve as both a separating film and a coolant. The presence of a film is vital for satisfactory service life, but it also means that leakage will occur. Low-viscosity fluids and high pressures add to leakage problems, as both require thin films to minimize leakage. This causes higher friction and results in heat, which is the one most detrimental factor in packing life. Deep packings reduce leakage but seriously increase frictional heat, particularly at high speeds. Normally the fluid being sealed serves as the lubricant. Thus, where oils are involved, maximum efficiency is obtained. Next in order are clean water, solvents, and fluids containing solids which progressively yield more unsatisfactory results unless supplemental lubrication is provided. Lubrication may be provided by using a lantern ring in the center of the set through which the more satisfactory lubricant is fed to the packings (Fig. 27). The preferred method of introducing the medium is to supply it at a slightly higher pressure, 5 to 10 lb/in² (3.5 to 7 kN/m²), than the material being sealed. The choice of fluid is governed by the media involved, since the two should be compatible. In cases of extreme contamination, the lantern is moved to the bottom of the box and is used as a means to introduce a flushing fluid. The lantern is also effective in excluding air from a device operating at negative heads. In centrifugal pumps equipped in this manner, it is called a **water seal.**

Dynamic packings are classified in three ways:

1. On the basis of shape of the surfaces: cylindrical, conical, spherical, or flat. Cylindrical packings are in turn classified according to whether they pack on the outside perimeter, as in piston packings (Figs. 17 to 20) or inside perimeter, as on rods or shafts (Figs. 21 to 28). Other examples are: conical, the

plug-cock lining (Fig. 29); spherical, the ball joint (Fig. 30); and flat, the mechanical seals (Figs. 31 and 32).

2. On the basis of the type of motion: rotary, oscillating, reciprocating, or helical (as in a rising-stem valve packing).

3. On the basis of nonautomatic: soft or jamb packings (tightened by external means, generally a gland); or automatic-preformed, molded shapes (self-tightening under pressure).

The **selection of a packing** is a matter of economics. In most cases several types are available, some of which, though expensive in the first place, yield exceptional service. A less costly choice might yield degraded service.

For **reciprocating elements,** the **O ring** (Fig. 20) is extremely simple; however, suppliers' data reveal it is a precision part requiring close tolerances. As an elastomeric material completely exposed to the operating fluid, it is subject to attack. Careful choice of elastomer is necessary to ensure compatibility or else either shrinkage or swelling may occur, causing early failure. It is best suited to medium pressure service from 1,500 to 3,000 lb/in² (10 to 20 MN/m²) with backup rings and intermittent movement, as in hydraulic cylinder or valve stem service. It is not a good choice for pump service. Backup rings are preferably of the heavy blocklike cross section in either leather or tetrafluoroethylene, avoiding the thin spiral type. The split **piston ring** (Fig. 17), generally of cast iron, is widely used in gas, oil, and steam engines. Large pistons frequently employ segmental rings similar to floating metal rod packings (Fig. 24) but facing outward. **Floating metal packing rings** are made of numerous radial or tangential segments, making it possible for them to contract on the shaft; they are assembled, generally, in sets of two to break the joints and are held together with garter springs. They are used for steam, gas, or air, in either engines or compressors under the most severe conditions of operation and pressure up to 35,000 lb/in². Normally oil lubrication is provided; however, for less severe service, filled polytetrafluoroethylene rings perform very well in dry gases without auxiliary lubrication. Step-, scarf-, or butt-cut rings of laminated cotton fabric, bonded with an elastomer or phenolic resin, are employed in water pumps, gasoline pumps, etc. They may float as cast-iron piston rings or be retained by a gland as in Fig. 18. **Cups** (Fig. 19) or their inverted form with the lip on the ID known as **flange packings,** are fully automatic and very tight. They are used principally for slow-speed applications such as air hoists. **Nested V and conical rings** (Figs. 22 and 23) are automatic, though often provided with a gland to effect initial fit. They are made of a wide range of materials from homogeneous elastomers through cotton or asbestos fabric reinforcement for severe duty. They range in hardness from soft and flexible to semirigid. Use of multiple rings allows them to be cut open for ease of installation and replacement. **Soft or jamb packings** are best suited for rod or plunger service, since an adjustable gland (Fig. 21) is required. They are normally formed in rectangular section with a butt joint staggered from ring to ring at installation. Many materials are employed, such as braided flax saturated with wax or viscous lubricants for water and aqueous solutions; braided asbestos similarly treated or often impregnated with polytetrafluoroethylene suspensoid for superior and more severe service; laminated rubberized cotton fabric for hot water, low-pressure steam and ammonia; rolled rubberized asbestos fabric for steam; and rolled or twisted metal foil for high-temperature and high-pressure conditions. Packings containing woven or braided asbestos fibers are also made

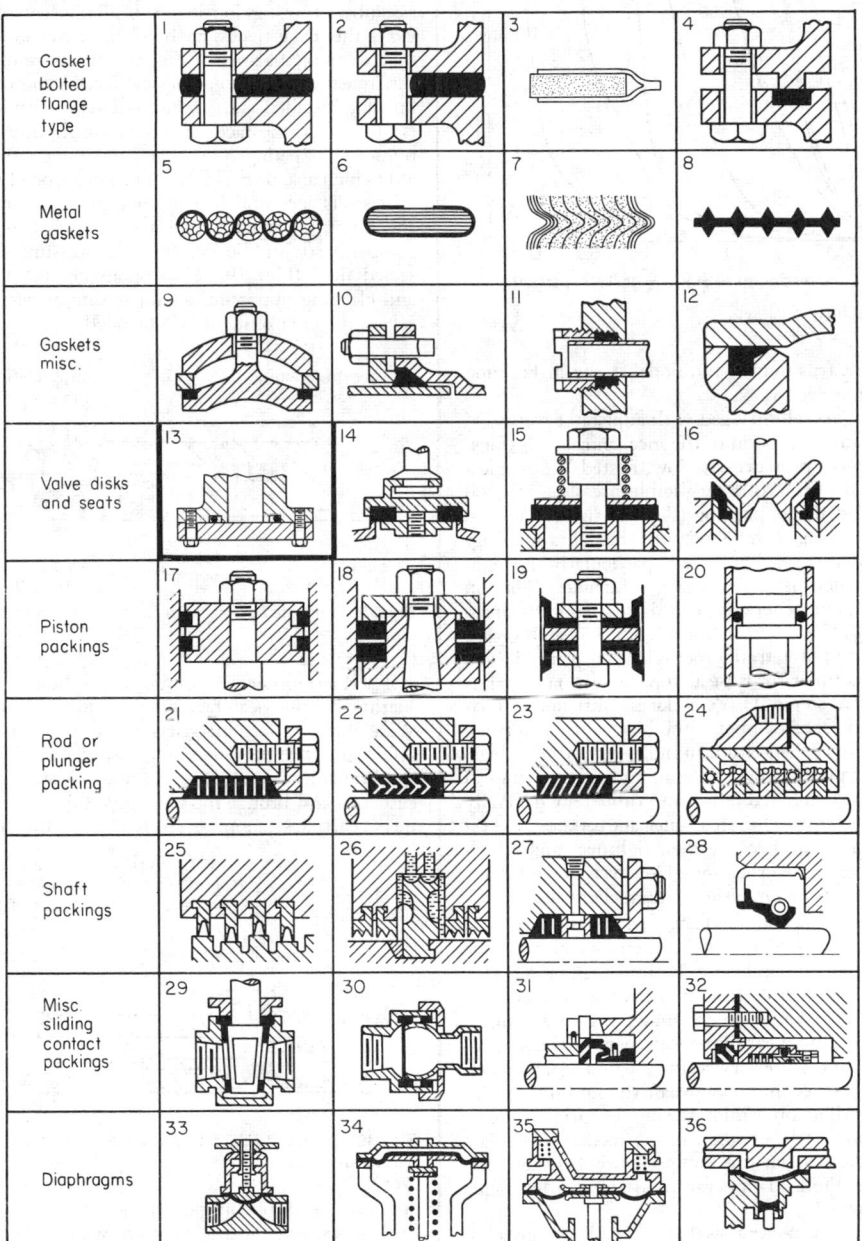

**Figs. 1 to 36**  Packings.

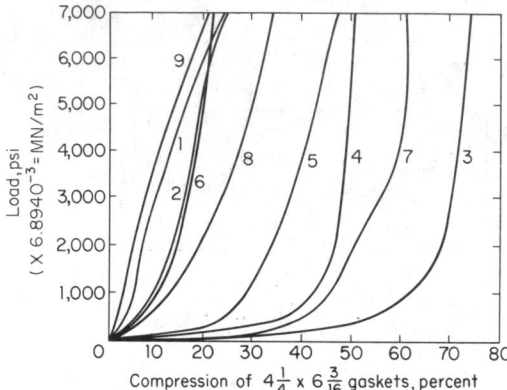

**Fig. 37**  Compressibility of gaskets.

from wire-inserted yarns to gain additional strength. For pipe expansion joints, see Sec. 8.

**Rotary shafts** are generally packed with adjustable soft packings, with the notable exception of the mechanical seals (Figs. 31 and 32), and where pressures are low, nested V or conical styles may be used. At zero or negligible pressures, the oil seal, a spring-loaded flange packing (Fig. 28), is very widely used. Where some leakage can be tolerated, the labyrinth (Fig. 25) and controlled gap seals are used, particularly on high-speed equipment such as steam and gas turbines. The **soft packings** are of the same general type as those used for reciprocating service, with the asbestos braid lubricated with grease and graphite or with polytetrafluoroethylene suspensoid. The latter is the most popular for typical applications on centrifugal pumps and valve stems. **Plastic packings** consisting of mixtures of unoriented asbestos fibers, graphite, soft metal particles, and a binder furnished either in bulk or in formed rings. To prevent loss of plastic from the stuffing box, end rings of metal foil, braid, or laminated fabric and rubber are generally used. For continuous rotary service, **automatic packings** are best restricted to low pressure because their tightness under high pressure tends to cause overheating. However, for intermittent service, as on valve stems, they are excellent.

**Oil seals** (Fig. 28) are unique flange packings having an elastomer lip generally bonded to a metal cup which is press-fitted into a smooth cylindrical bore. Basically, an oil seal is a flange packing with a flexible lip and a narrow contact area about ⅟₁₆ in (1.6 mm) wide which, under pressure, causes extreme local heating and wear. They can therefore be recommended only for nonpressure service and are best against good lubricating media. To accommodate shaft runout up to 0.020 in (0.5 mm) depending on r/min, the lip is spring-loaded. Since the lip is completely exposed to the media, particular care should be taken to ensure compatibility between the elastomer and the fluid. Temperature is another element important in this area.

**Mechanical, rotary, or end-face seals** (Figs. 31 and 32) consist of radial planar surfaces, normal to the shaft axis, cooperating as a thrust bearing. One face, the seat, is generally resiliently gasketed to the housing (Fig. 38). The mating face is driven by the shaft and sealed thereon by a secondary seal such as a bellows (Fig. 31) or automatic-type packing (Fig. 32) permitting axial movement to compensate for end play and wear.

Initial bearing contact is through spring loading augmented in service by fluid pressure (Fig. 38). For extreme speeds, often the seat is mounted on the shaft and the seal in the stationary housing eliminates critical dynamic balancing of the seal assembly, transferring it to the simple seat. Accurate centering of the seal is necessary but is more economical and positive than dynamic balance of the seal. Abrasive materials, such as silt, must be kept from the seal faces. One means is a throat bushing installed in the bottom of the seal cavity. A clean fluid is directed at the faces at a pressure slightly higher than the pumpage, creating a counterflow through the bushing (Fig. 38), which also provides cooling. Retention of hazardous fluids or gases is accomplished by mounting two seals back to back and filling the space between with a lubricious barrier liquid compatible with the system at a pressure slightly over the sealed fluid (Fig. 39). Continuous circulation offers cooling and cleaning opportunities for optimum performance. In essence, the environment is controlled.

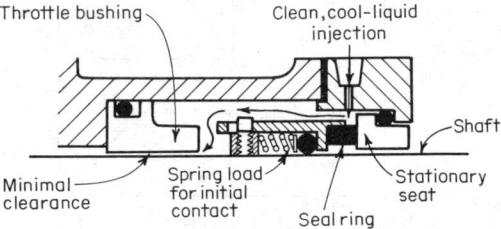

**Fig. 38**  Throttle bushing with restrictive clearance and counterflow.

There are other possibilities depending on the application. Mating bearing-seal faces may require chemical resistance. Wide choice of face materials, one of which is usually carbon, offers an extreme range of pressures, service, and chemical conditions at very high efficiency. Typical mating faces are cast iron and bronze for mild conditions, with ceramics, cermets, carbides, stainless steels, and related alloys for severe

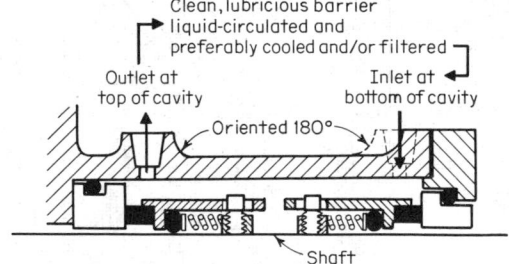

**Fig. 39**  Back-to-back seal installation with barrier liquid for hazardous fluids.

applications. Probably one of the simplest designs is the bellows-supported phenolic-cotton washer bearing against a machined face of the cast-iron pump body of an automotive water pump (Fig. 31). A more refined, flexible design is shown in Fig. 32, and detailed in part in Fig. 40, which illustrates unbalanced and balanced constructions and how they are respectively suitable for pressures up to 250 and 3,000 lb/in² (1.7 and 20 MN/m²). The purpose of balancing is to reduce the unit face load to approximately 60 to 70 percent of the unit fluid load. A complete balance is not practical.

For extremely high speeds, where it is desirable to eliminate all rubbing contact, the **labyrinth** seal (Fig. 25) is chosen. This seal is not fluid-tight but restricts a serious flow by means of a torturous path and induced turbulence. It is widely used on steam turbines (Sec. 9). Where no leakage is permissible, a liquid seal based on the U-tube principle (Fig. 26) may be used. The natural weight of the liquid is amplified by centrifugal force so that under high r/min a fair pressure differential can be sealed. Another noncontacting seal is the **controlled gap seal** which is being used on gas turbines where pressure differentials are not excessive and a small amount of leakage can be tolerated. The seal consists of a ring with a shaft clearance in the range of 0.0005 to 0.0015 in (0.013 to 0.038 mm) and is made of such materials that this clearance is maintained at all operating temperatures. Usually one end of the ring is faced to form an axial seal against the inside of its housing.

Diaphragms are a form of dynamic packing but include the requirements of a gasket where they are gripped or held in position. In service they are leakless, although generally limited in travel. By literally rolling one cylinder inside another, considerable increase in travel is possible. This type is often called a bellows, and a simple application is the mechanical seal suspension shown in Fig. 31. In the diaphragm valve (Fig. 33) the diaphragm replaces both the conventional stem packing and valve disk. **Diaphragms** of fabric such as cotton or nylon (except friable materials such as glass) covered with an elastomer suitable for the media and temperatures involved are used in pumps (fuel pump, Fig. 35) and in motors (Fig. 34) to operate valves, switches, and other controls. Correctly designed diaphragms are made with slack to permit a natural rolling action. Flat sheet stock should be used only where limited travel is desired. An unusual application is found in the Pulsa-feeder pump (Fig. 36) in that the diaphragm is under

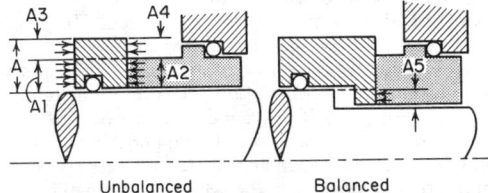

Unbalanced                Balanced

**Fig. 40**  Mechanical seal-face loading. $A$ = gross area under fluid load; $A_1$ = effective fluid-load area for closing the seal; $A_2$ = pressure-gradient area where average unit load equals that of the fluid on $A_1$ (shaft clearance ignored); $A_3$; $A_4$ = fluid-load areas which balance each other; $A_5$ = balancing area, normally such that $A_2 + A_5$ will reduce the average unit load on this face to 60 to 70 percent of the unit fluid load on $A_1$.

balanced fluid pressure on both sides and is therefore unstressed. Thin sheet metal, usually with concentric corrugations, is used where movement is limited and long life is desired. However, where considerable movement is involved, the possibility of fatigue must be considered.

### Table 1. Common Usage of Gasketing Materials

| No. | Type | Service principally for | Thickness tested, in ($\times 25.4$ = mm) |
|---|---|---|---|
| 1 | Sheet rubber | Water | $\frac{1}{16}$ |
| 2 | Cloth inserted sheet | Water | $\frac{1}{16}$ |
| 3 | Cork composition | Oil, low pressure | $\frac{1}{8}$ |
| 4 | Gasket paper | Oil, low pressure | $\frac{1}{16}$ |
| 5 | Rubberized asbestos cloth (Fig. 9) | Hot water (boiler manholes, etc.) | $\frac{1}{4}$ |
| 6 | Compressed asbestos sheet | All services up to 750°F (400°C) | $\frac{1}{16}$ |
| 7 | Corrugated sheet metal with filling (Fig. 5) | Steam, oil at high temperatures | $\frac{1}{4}$ |
| 8 | Metal jacket over asbestos center (Fig. 6) | Steam, oil at high temperatures | $\frac{1}{8}$ |
| 9 | Spirally wound steel strip with intervening asbestos layers (Fig. 7) | Steam, oil at high temperatures | $\frac{3}{16}$ |

# PIPE AND PIPE FITTINGS

## by Reno C. King

REFERENCES: King, "Piping Handbook," McGraw-Hill. ANSI Code for Power Piping. ASTM Specifications. Tube Turns Division, Natural Cylinder Gas Co., catalogs. Crane Co., catalogs and bulletins. Grinnell Co., Inc., "Piping Design and Engineering." M. W. Kellogg Co., "Design of Piping Systems," Wiley. United States Steel Co., catalogs and bulletins.

### PIPING STANDARDS

**Standardization** in the piping industry is the function of many groups, among whom are the American Society for Testing and Materials (ASTM), the American National Standards Institute (ANSI), the American Water Works Association

(AWWA), and the Pipe Fabrication Institute (PFI); in addition, specifications have been issued by several governmental agencies to cover work on federal installations.

The ASTM has as its aim the "promotion of knowledge of materials of engineering, and standardization of specifications and methods of testing." It is concerned with chemical and physical properties of piping (and other materials or products) as delivered from the fabricating mill.

The ANSI[1] deals with overall piping systems. It standardizes dimensions, sets permissible stress values as functions of temperature, establishes working formulas for determination of wall thickness as determined by pressure and material at a given temperature, specifies general character of construction of valves and fittings, deals with the support, anchoring, and flexibility of a piping system and, in general, sets up a code of minimum requirements for the safety and reliability of a system. The work done by this organization has culminated in hundreds of standards for individual materials, published in several issues of its Code for Power Piping and related publications.

The AWWA was one of the first groups to realize the importance and advantage of standardization of cast-iron pipe and fittings. Its standard, first published in 1908, has been superseded by ANSI publications.

PFI has been active in preparation of standards dealing with shop fabrication, particularly in standardizing the technique for preparation of pipe ends prior to welding.

The Manufacturers' Standardization Society of the Valve and Fittings Industry engages in preparation of standards for both valves and fittings, generally in the lower range of pressures and temperatures.

Piping is available in both ferrous and nonferrous materials. Cast iron and steel represent the ferrous class; brass, copper, aluminum, and lead are nonferrous metals which have wide applications as piping materials. Concrete, wood, tile, and plastic have found general acceptance for specific usages.

## STEEL PIPE AND TUBING

There is no definite rule for distinguishing among the general terms pipe, tube, and tubing: the proper designation has been determined by trade usage. The word **pipe** is generally used to apply to tubular products commonly used for pipelines and connections for conveying fluid from point to point. The word **tube** (or **tubing**) is generally used to apply to tubular products used in heat exchangers and boilers and in the machine and aircraft industries.

**Piping Fabrication Methods**   Piping is commercially available as manufactured by the following techniques: electric-fusion welding, electric-resistance welding, submerged-arc electric welding, seamless piercing and rolling, forging, turning, and boring, and hollow forging. In addition, some mills are prepared to extrude small-diameter pipe and tubing in a variety of geometric shapes or to cast large-diameter steel pipe.

**Electric-Fusion Welding**   Flat plate, known as **skelp,** is prepared in proper width and thickness for the desired pipe inside and outside diameters. It is then charged into an electric furnace and, when the proper welding temperature has been

[1]Throughout this section, tabular data have been reproduced or abstracted from ANSI Standards. This is by permission of the publisher, The American Society of Mechanical Engineers, United Engineering Center, 345 East 47th Street, New York, N.Y. 10017. Complete sets of any, or all, standards are available from that source.

reached, is drawn through a funnel-shaped die so shaped that the plate is gradually formed into the shape of a tube, with the edges of the plate being forced squarely together and fused. The formed pipe then passes through a series of rolls in which it is sized or drawn to final dimensions.

**Electric-Resistance Welding**   For pipes or tubes sized 4 in (10.2 cm) OD and under, strip is fed onto a set of forming rolls which consists of horizontal and vertical rollers so placed as to gradually form the flat strip into a tube. The tube form then passes to the welding electrodes. The electrodes are copper disks connected to the secondary of a revolving transformer assembly. The copper-disk electrodes make contact on each side of the seam of the tube form; a flow of current takes place across the seam, and temperature is raised to the welding point. Outside flash is removed by a cutting tool as the tube leaves the electrodes; inside flash is removed either by an air hammer or by passing a mandrel through the welded tube after the tube has been cooled.

**Submerged-Arc Electric Welding**   This process is used for pipes from 24 to 36 in (61.0 to 91.4 cm) OD. Flat plate is first pressed into a U and later into an O shape. The O shape is placed in an automatic welder and backed up on the inside by a water-cooled copper shoe. Two electrodes in close proximity are used. The electrodes are not in actual contact with the pipe. The current passes from one electrode through a granular flux and across the gap in the pipe to the second electrode. The high temperature of the arc heats the edges of the plate; a welding rod placed just over the seam is thereby melted and metal is deposited in the groove. After the outside weld has been made, the pipe is conveyed to an inside welder where a similar operation is carried on, except that no backup shoe is needed.

**Seamless Tubing and Pipe**   A heated billet is brought into contact with tapered revolving rolls in such a way that the billet is pulled into the space allowed between the rolls. A piercing mandrel is placed in this space; the soft center of the billet makes it possible for the rolls to draw the billet over the mandrel, producing a hollow shell. When the billet has entirely passed over the mandrel, it is in the form of a thick-walled seamless tube. The heavy-walled tube is then passed to a rolling mill which reduces the tube to piping of proper outside diameter and wall thickness.

The method of fabrication described above is limited as to diameter and thickness. For seamless alloy tubes and for heavy-walled carbon-steel tubes or piping, a process known as **cupping and drawing** is frequently used. A circular flat plate of proper diameter and thickness is heated, placed in a hydraulic press, and pressed by a ram through a die. The cup so formed is reheated and pressed through a smaller die, thus elongating the cup so that it becomes a short cylinder with one closed end. This short cylinder is then placed in a horizontal drawbench, and with reheating as necessary, is pushed by a ram through dies of successively smaller diameters until the desired outer diameter is reached.

**Forged, Turned, and Bored Tubing**   In this process, the ingot is heated and forged to a rough cylindrical shape, oversize in both diameter and length. The forging is then placed in a lathe and the outside turned down to the desired outer diameter. Rough ends are then removed so that the finally desired length is obtained. The cylinder is then placed in a boring mill, and the inside bored out until the desired wall thickness is secured.

**Hollow-Forged Pipe and Tubing**   In this process, ingots are cast and their ends cropped; then they are placed in a furnace and heated to a specified temperature. The heated ingot is placed in a press where it is pierced. This hollow cylinder, open at one end, is then descaled and drawn over a mandrel on a horizontal drawbench. The closed end is then burned off, and the hollow forging is chemically descaled. Following this, the forging is straightened, placed in a lathe, and the outer diameter machined to a true dimension. The inside is dressed to remove scale, but no machining is done on the inside.

**Code Designations**  Appropriate ASTM specifications list the physical and chemical properties of materials used in piping systems. The complete compilation of "Steel Piping, Tubing and Fittings" can be purchased from the ASTM, 1916 Race St., Philadelphia, Pa. 19103. The treatment in this section is a brief outline of frequently encountered materials.

**Carbon-steel piping** is most frequently used as manufactured in accordance with ASTM specifications A106 and A53. The chemical composition of these two materials is identical; both are subjected to physical tests, but those for A106 are more rigorous. For example, the Code for Pressure Piping permits the use of A53 for pressures of 600 lb/in² gage (22,137 N/m²) and less but excludes its use for higher pressures; A106 can be used for pressures not above 2,500 lb/in² gage (92,237 N/m²). A53 and A106 are made in Grades A and B; Grade B has higher strength properties but is less ductile and, for this reason, Grade A is permitted only for cold bending or close coiling. When carbon steel is intended for use in welded construction at temperatures in excess of 775°F (413°C), consideration should be given to the possibility of graphite formation.

**Carbon-molybdenum steel piping** may be obtained as A204 (electric-fusion-welded), A335 (seamless) or A369 (forged, turned, and bored). This material was developed in past years when steam temperatures were approaching, but not reaching, 1000°F (538°C) under which conditions carbon steel was both unsatisfactory and uneconomical. It has been found that there is a tendency for carbon-molybdenum to show graphitization at temperatures in excess of 800°F (427°C), and its use in welded construction above this value should be with caution.[1]

**Chromium-molybdenum steel** has been used for temperatures up to 1100°F (593°C). In the small diameters, the material is usually available in the seamless construction; because of the inability of the seamless mills to fabricate large-diameter and heavy-walled pipe, it may be necessary to resort to the more expensive hollow-forged or forged-and-bored piping for higher pressures and temperatures. The material for a high-temperature piping system should be selected after a careful review of technical and economic considerations; the following is intended only as being indicative of recent and current practice. For temperatures up to 950°F (510°C), ½ percent Cr–½ percent Mo (A335, Grade P2) is used; for temperatures 950 to 1000°F (510 to 538°C), 1 percent Cr–½ percent Mo (A335, Grade P12) is used; for temperatures 1000 to 1050°F (538 to 566°C), 1¼ percent Cr–½ percent Mo (A335, Grade P11) may be used; for temperatures 1050 to 1100°F (566 to 593°C), 2¼ percent Cr–1 percent Mo (A335, Grade P22) is frequently used. When there is a combination of high temperatures and erosive action, 5 percent Cr–½ percent Mo (A335, Grade 5) has been found desirable.

[1]Modern steel-making practices have reduced significantly the problem of graphitization. However, in pipe installed in the 1940s and early 1950s, there have been many failures.

**Stainless-steel piping** is available in a variety of compositions, most popular of which are ASTM A213, Grade TP321 (16 percent Cr–8 percent Ni, stabilized with titanium) and ASTM A213, Grade TP347 (18 percent Cr–8 percent Ni, stabilized with columbium). Either of these two materials may be used up to 1200°F (649°C); particular care must be given to choice of welding rod to avoid brittleness in the welds.

Refer to Tables 1 and 2, respectively, for permissible stress values for piping materials at low and elevated temperatures.

**Schedule Designations**  Many years ago piping was designated as standard, extra-strong, and double extra-strong. There was no provision for thin-walled pipe, and no intervening standard thicknesses between the three schedules, which covered too great a spread to be economical without intermediate weights. Table 3 lists piping as a function of the schedule number which is given, approximately, by the following relationship: Schedule no. = 1,000 × $P/SE$, where $P$ is operating pressure, lb/in² gage, and $SE$ is allowable stress range multiplied by joint efficiency, lb/in².

EXAMPLE.  Find the required schedule of ASTM A106 Grade B pipe operating at 1,150 lb/in² gage and 600°F.
Table 2 lists $SE$ value as 15,000 lb/in². Substituting, 1,000 (1,150/15,000) = 76.6. Use schedule no. 80, tentatively, but check with Eq. (1), below.

Commercial sizes of wrought-iron and steel pipe are known by their nominal inside diameter (ID) from ⅛ (0.3175 cm) to 12 in (30.5 cm). Above 12 in ID, pipe is usually known by its outside diameter (OD). All classes of pipe of a given nominal size have the same OD, the extra thickness for different weights being on the inside.

**Thickness of Pipe**  The following notes, covering power piping systems, have been abstracted from Part 2 of the Code for Power Piping (ANSI B31.1.0-1967).

For inspection purposes, the minimum thickness of pipe wall to be used for piping at different pressures and for temperatures not exceeding those for the various materials listed in Tables 1 and 2 shall be determined by the formula

$$t_m = \frac{PD}{2(SE + Py)} + A \qquad (1)$$

where $t_m$ = minimum pipe-wall thickness, in, allowable on inspection; $P$ = maximum internal service pressure, lb/in² gage (plus water-hammer allowance in case of cast-iron conveying liquids); $D$ = OD of pipe, in; $SE$ = maximum allowable stress in material due to internal pressure and joint efficiency, at the design temperature, lb/in²; values of $SE$ given in Tables 1 and 2 include allowance for joint efficiency; $y$ = a coefficient, values for which are listed in Table 4; $A$ = allowance for threading, mechanical strength, and corrosion, in, with values of $A$ listed in Table 5.

The thickness of cast-iron pipe conveying liquid may be taken from Table 14, using the pressure class next higher than the maximum internal service pressure in pounds per square inch. Where cast-iron pipe is used for steam service, the thickness should be calculated by Eq. (1), using $SE$ values listed in Table 1.

Plain-end pipe includes pipe joined by flared compression couplings, lapped joints, and by welding, i.e., by any method that does not reduce the wall thickness of the pipe at the joint.

**Physical and Chemical Properties of Pipes, Tubes, Etc.** The design of piping for operation above 750°F (399°C) presents many problems not encountered at lower temperatures.

**Table 1. Allowable Stress Values for Temperatures up to 650°F (343.4°C)**
(ANSI B36.1.0—1967)

| ASTM spec. No. | Grade | Nominal composition | Spec. min tensile | Longitudinal joint efficiency factor | P No.[a] | Max allowable stress value, lb/in², for metal temperatures not exceeding °F[b] | | | | | | | |
|---|---|---|---|---|---|---|---|---|---|---|---|---|---|
| | | | | | | −20 to 100 | 200 | 300 | 400[c] | 450 | 500 | 600 | 650 |
| **Pipe:** | | | | | | | | | | | | | |
| Welded carbon steel, | | | | | | | | | | | | | |
| butt-welded A120, | | | | | 1 | 6,500 | 6,350 | 6,100 | 5,850 | 5,700 | | | |
| Lap-welded A120, | | | | | 1 | 8,800 | 8,600 | 8,200 | 7,800 | 7,600 | | | |
| Automatically welded austenitic steel, A312 | TP304H[a] | 18Cr-8Ni | 75,000 | 0.85 | 8 | 15,950 | 14,050 | 13,200 | 12,700 | | 12,350 | 12,200 | 12,150 |
| | TP304H[e] | 18Cr-8Ni | 75,000 | 0.85 | 8 | 15,950 | 13,600 | 11,700 | 10,400 | | 9,700 | 9,200 | 9,050 |
| | TP316H[e,a] | 18Cr-12Ni-Mo | 75,000 | 0.85 | 8 | 15,950 | 14,850 | 14,350 | 13,850 | | 13,600 | 13,600 | 13,600 |
| | TP316H[e] | 18Cr-12Ni-Mo | 75,000 | 0.85 | 8 | 15,950 | 13,700 | 12,400 | 11,450 | | 10,700 | 10,100 | 9,850 |
| | TP321H[e,a] | 18Cr-10Ni-Ti | 75,000 | 0.85 | 8 | 15,950 | 14,300 | 13,450 | 13,100 | | 12,850 | 12,850 | 12,850 |
| | TP321H[e] | 18Cr-10Ni-Ti | 75,000 | 0.85 | 8 | 15,950 | 14,050 | 13,000 | 12,300 | | 11,450 | 10,900 | 10,600 |
| Electric-fusion-welded austenitic steel, | | | | | | | | | | | | | |
| A358 Class I[e,a] | TP304 | 18Cr-8Ni | 75,000 | 1.00 | 8 | 18,750 | 16,550 | 15,550 | 14,950 | | 14,550 | 14,350 | 14,300 |
| Class I[e] | TP304 | 18Cr-8Ni | 75,000 | 1.00 | 8 | 18,750 | 16,000 | 13,750 | 12,250 | | 11,400 | 10,800 | 10,650 |
| Class II[e,a] | TP304 | 18Cr-8Ni | 75,000 | 0.90 | 8 | 16,900 | 14,900 | 14,000 | 13,450 | | 13,100 | 12,900 | 12,900 |
| Class II[e] | TP304 | 18Cr-8Ni | 75,000 | 0.90 | 8 | 16,900 | 14,400 | 12,400 | 11,000 | | 10,250 | 9,700 | 9,600 |
| Class I[e] | TP316 | 18Cr-12Ni-Mo | 75,000 | 1.00 | 8 | 18,750 | 17,500 | 16,900 | 16,300 | | 16,000 | 16,000 | 16,000 |
| Class I[e,a] | TP316 | 18Cr-12Ni-Mo | 75,000 | 1.00 | 8 | 18,750 | 16,100 | 14,600 | 13,500 | | 12,600 | 11,900 | 11,600 |
| Class II[e] | TP316 | 18Cr-12Ni-Mo | 75,000 | 0.90 | 8 | 16,900 | 16,550 | 15,200 | 14,700 | | 14,400 | 14,400 | 14,400 |
| Class II[e] | TP316 | 18Cr-12Ni-Mo | 75,000 | 0.90 | 8 | 16,900 | 14,500 | 13,150 | 12,150 | | 11,350 | 10,700 | 10,450 |
| Class I[e,a] | TP321 | 18Cr-10Ni-Ti | 75,000 | 1.00 | 8 | 18,750 | 16,800 | 15,850 | 15,400 | | 15,100 | 15,100 | 15,100 |
| Class I[e] | TP321 | 18Cr-10Ni-Ti | 75,000 | 1.00 | 8 | 18,750 | 16,550 | 15,300 | 14,450 | | 13,600 | 12,800 | 12,500 |
| Class II[e,a] | TP321 | 18Cr-10Ni-Ti | 75,000 | 0.90 | 8 | 16,900 | 15,100 | 14,250 | 13,900 | | 13,600 | 13,600 | 13,600 |
| Class II[e] | TP321 | 18Cr-10Ni-Ti | 75,000 | 0.90 | 8 | 16,900 | 14,900 | 13,750 | 13,000 | | 12,150 | 11,500 | 11,250 |
| **Seamless:** | | | | | | | | | | | | | |
| A120 carbon steel | | | | | 1 | 10,800 | 10,600 | 10,200 | 9,800 | 9,600 | | | |
| A335 ferritic alloy[e] | P5 | 5Cr-½Mo | 60,000 | 1.00 | 5 | 15,000 | 15,000 | 15,000 | 15,000 | | 14,500 | 14,000 | 13,700 |
| A335 ferritic alloy[e] | P5b | 5Cr-½Mo-Si | 60,000 | 1.00 | 5 | 15,000 | 15,000 | 15,000 | 15,000 | | 14,500 | 14,000 | 13,700 |
| A369 ferritic alloy[e] | FP5 | 5Cr-½Mo | 60,000 | 1.00 | 5 | 15,000 | 15,000 | 15,000 | 15,000 | | 14,500 | 14,000 | 13,700 |
| A312 Austenitic[e,a] | TP304H | 18Cr-8Ni | 75,000 | 1.00 | 8 | 18,750 | 16,000 | 15,550 | 14,950 | | 14,550 | 14,350 | 14,350 |
| Austenitic[e,a] | TP304H | 18Cr-8Ni | 75,000 | 1.00 | 8 | 18,750 | 16,000 | 13,750 | 12,250 | | 11,400 | 10,800 | 10,650 |
| Austenitic[e,a] | TP316H | 18Cr-12Ni-Mo | 75,000 | 1.00 | 8 | 18,750 | 17,500 | 16,900 | 16,300 | | 16,000 | 16,000 | 16,000 |
| Austenitic[e,a] | TP316H | 18Cr-12Ni-Mo | 75,000 | 1.00 | 8 | 18,750 | 16,000 | 14,600 | 13,500 | | 12,600 | 11,900 | 11,600 |
| Austenitic[e,a] | TP321H | 18Cr-10Ni-Ti | 75,000 | 1.00 | 8 | 18,750 | 16,550 | 15,850 | 15,400 | | 15,100 | 15,100 | 15,100 |
| Austenitic[e,a] | TP321H | 18Cr-10Ni-Ti | 75,000 | 1.00 | 8 | 18,750 | 16,550 | 15,300 | 14,950 | | 13,500 | 12,800 | 12,500 |
| A376 Austenitic[e,a] | TP304H | 18Cr-8Ni | 75,000 | 1.00 | 8 | 18,750 | 16,800 | 15,550 | 14,950 | | 14,550 | 14,350 | 14,300 |
| Austenitic[e] | TP304H | 18Cr-8Ni | 75,000 | 1.00 | 8 | 18,750 | 16,550 | 13,750 | 12,250 | | 11,400 | 10,800 | 10,650 |
| Austenitic[e,a] | TP316H | 18Cr-12Ni-Mo | 75,000 | 1.00 | 8 | 18,750 | 16,000 | 16,900 | 16,300 | | 16,000 | 16,000 | 16,000 |
| Austenitic[e,a] | TP316H | 18Cr-12Ni-Mo | 75,000 | 1.00 | 8 | 18,750 | 16,100 | 14,600 | 13,500 | | 12,600 | 11,900 | 11,600 |
| Austenitic[e] | TP321H | 18Cr-10Ni-Ti | 75,000 | 1.00 | 8 | 18,750 | 16,800 | 15,850 | 15,400 | | 15,100 | 15,100 | 15,100 |
| Austenitic[e] | TP321H | 18Cr-10Ni-Ti | 75,000 | 1.00 | 8 | 18,750 | 16,550 | 15,300 | 14,950 | | 13,500 | 12,800 | 12,500 |
| A430 Austenitic[e,a] | FP304H | 18Cr-8Ni | 70,000 | 1.00 | 8 | 17,500 | 15,450 | 14,500 | 13,950 | | 13,550 | 13,350 | 13,300 |
| Austenitic[e] | FP304H | 18Cr-8Ni | 70,000 | 1.00 | 8 | 17,500 | 15,450 | 13,750 | 12,250 | | 11,400 | 10,800 | 10,650 |
| Austenitic[e] | FP316H | 18Cr-12Ni-Mo | 70,000 | 1.00 | 8 | 17,500 | 16,350 | 15,750 | 15,200 | | 14,950 | 14,950 | 14,950 |
| Austenitic[e,a] | FP316H | 18Cr-12Ni-Mo | 70,000 | 1.00 | 8 | 17,500 | 16,100 | 14,600 | 13,500 | | 12,600 | 11,900 | 11,600 |
| Austenitic[e] | FP321H | 18Cr-10Ni-Ti | 70,000 | 1.00 | 8 | 17,500 | 15,700 | 14,800 | 14,350 | | 14,100 | 14,100 | 14,100 |
| Austenitic[e] | FP321H | 18Cr-10Ni-Ti | 70,000 | 1.00 | 8 | 17,500 | 15,700 | 14,800 | 14,350 | | 13,500 | 12,800 | 12,500 |

| Material | | | | | |
|---|---|---|---|---|---|
| Nonferrous: | | | | | |
| B43 seamless red-brass pipe[e] | | 8,000 | 8,000 | 8,000 | 7,000 | 3,000 |
| B42 seamless copper pipe[e] | | 6,000 | 5,500 | 5,500 | 4,750 | 3,000 |
| B75 seamless copper tube[e] | | 6,000 | 5,500 | 5,500 | 4,750 | 3,000 |
| B68 bright annealed seamless copper tube[e] | 30,000 | 6,000 | 5,500 | 5,500 | 4,750 | 3,000 |
| B88 annealed seamless copper water tube[e] | 30,000 | 6,000 | 5,500 | 5,500 | 4,750 | 3,000 |
| A254 copper brazed steel tube class I[c] | 42,000 | 6,000 | 5,500 | 5,500 | 4,750 | 3,000 |
| A254 copper brazed steel tube class II[c] | 42,000 | 3,600 | 3,300 | 3,300 | 2,850 | 1,800 |
| Cast iron: | | | | | |
| FSB and WW-P-421, centrifugally cast types I and II[d] | | 6,000 | 6,000 | 6,000 | 6,000 | 6,000 |
| ANSI A21.6 metal molds[d] | | 6,000 | 6,000 | 6,000 | 6,000 | 6,000 |
| ANSI B21.8 sand-lined molds[d] | | 6,000 | 6,000 | 6,000 | 6,000 | 6,000 |
| ANSI A21.2 pit-cast[d] | | 4,000 | 4,000 | 4,000 | 4,000 | 4,000 |

The stress values tabulated include a longitudinal joint efficiency factor where applicable.

The stress values in this table may be interpolated to determine values for intermediate temperatures.

Materials listed in Table 126.1 of the ANSI Standard for which allowable stresses are not tabulated may be used. Allowable stresses for such materials should be taken from Sections I and VIII of the ASME Boiler and Pressure Vessel Code or, for aluminum-alloy materials, from ANSI B31.3, Petroleum Refinery Piping.

[a] The grouping of materials as to P-number classification is made on the basis of hardenability characteristics. The P numbers indicated in this table are identical to those adopted by the ASME Boiler and Pressure Vessel Code. Qualification of welding procedures, welders, and welding operators is required and should comply with the ASME Boiler and Pressure Vessel Code (Section IX) except as modified by Par. 127.5.

[b] The several types and grades of pipe and tube tabulated should not be used at temperature in excess of the maximum temperatures for which the allowable stress values are indicated.

[c] For saturated steam at 250 lb/in² (406°F) the values given for 400°F may be used.

[d] Cast-iron pipe may not be used for oil lines.

[e] For allowable stress values above 650°F, see Table 2.

[f] In sizes 8 in and larger and schedule 140 or heavier, the minimum tensile strength is 70,000 lb/in². In these sizes and thicknesses, the indicated allowable stresses should be reduced by the ratio of 70 divided by 75.

[g] Because of the relatively low yield strength of these materials, these higher stress values were established at temperatures where the short-time tensile properties govern to permit the use of these alloys where slightly greater deformation is acceptable. The stress values in this range exceed 62½ percent but do not exceed 90 percent of the yield strength at temperature. Use of these stresses may result in dimensional changes due to permanent strain. These stress values are not recommended for the flanges of gasketed joints or other applications where slight amounts of distortion can cause leakage or malfunction.

**Table 2. Allowable Stress Values for Temperatures 650 to 1200°F (343.4 to 649°C)**
(ANSI B36.1.0-1967)

| ASTM spec. No. | Grade | Nominal composition | Spec. min tensile | Longitudinal joint efficiency factor | P No.[a] | 650[c] | 700 |
|---|---|---|---|---|---|---|---|
| Pipe: | | | | | | | |
| Butt-welded: | | | | | | | |
| A53 carbon steel | | | 45,000 | 0.60 | 1 | 6,750 | |
| A72 wrought iron | | | 42,000 | 0.60 | 2 | 6,300 | |
| Lap-welded A72 wrought iron | | | 42,000 | 0.80 | 2 | 8,400 | |
| Electric-fusion-welded | | | | | | | |
| A134 Carbon steel | A245A | | 48,000 | | 1 | 8,000 | |
| | A245B | | 52,000 | | 1 | 9,600 | |
| | A245C | | 55,000 | | 1 | 10,100 | |
| | A283A | | 45,000 | | 1 | 8,300 | |
| | A283B | | 50,000 | | 1 | 9,200 | |
| | A283C | | 55,000 | | 1 | 10,100 | |
| | A283D | | 60,000 | | 1 | 10,100 | |
| A139 | A | | 48,000 | 0.80 | 1 | 9,600 | 9,250 |
| | B | | 60,000 | 0.80 | 1 | 12,000 | 11,350 |
| A155 Class I[a,j] | C45 | | 45,000 | 1.00 | 1 | 11,250 | 10,900 |
| Class II[a,j] | C45 | | 45,000 | 0.90 | 1 | 10,100 | 9,800 |
| Class I[a,j] | C50 | | 50,000 | 1.00 | 1 | 12,500 | 12,100 |
| Class II[a,j] | C50 | | 50,000 | 0.90 | 1 | 11,250 | 10,900 |
| Class I[a,j] | C55 | | 55,000 | 1.00 | 1 | 13,750 | 13,250 |
| Class II[a,j] | C55 | | 55,000 | 0.90 | 1 | 12,400 | 11,900 |
| Class I[a,j] | KC55 | C-Si | 55,000 | 1.00 | 1 | 13,750 | 13,250 |
| Class II[a,j] | KC55 | C-Si | 55,000 | 0.90 | 1 | 12,400 | 11,900 |
| Class I[a,j] | KC60 | C-Si | 60,000 | 1.00 | 1 | 15,000 | 14,350 |
| Class II[a,j] | KC60 | C-Si | 60,000 | 0.90 | 1 | 13,500 | 12,900 |
| Class I[a,j] | KC65 | C-Si | 65,000 | 1.00 | 1 | 16,250 | 15,500 |
| Class II[a,j] | KC65 | C-Si | 65,000 | 0.90 | 1 | 14,600 | 13,950 |
| Class I[a,j] | KC70 | C-Si | 70,000 | 1.00 | 1 | 17,500 | 16,600 |
| Class II[a,j] | KC70 | C-Si | 70,000 | 0.90 | 1 | 15,750 | 14,950 |
| Class I[e,j] | CM65 | C-½Mo | 65,000 | 1.00 | 3 | 16,250 | 16,250 |
| Class II[e,j] | CM65 | C-½Mo | 65,000 | 0.90 | 3 | 14,600 | 14,600 |
| Class I[e,j] | CM70 | C-½Mo | 70,000 | 1.00 | 3 | 17,500 | 17,500 |
| Class II[e,j] | CM70 | C-½Mo | 70,000 | 0.90 | 3 | 15,750 | 15,750 |
| Class I[e,j] | CM75 | C-½Mo | 75,000 | 1.00 | 3 | 18,750 | 18,750 |
| Class II[e,j] | CM75 | C-½Mo | 75,000 | 0.90 | 3 | 16,850 | 16,850 |
| Class I[d,j] | CMS75 | C-Mn-Si | 75,000 | 1.00 | 1 | 18,750 | 17,700 |
| Class I[f,j] | ½CR | ½Cr-½Mo | 65,000 | 1.00 | 3 | 16,250 | 16,250 |
| Class II[f,j] | ½CR | ½Cr-½Mo | 65,000 | 0.90 | 3 | 14,600 | 14,600 |
| Class I[j] | 1CR | 1Cr-½Mo | 60,000 | 1.00 | 4 | 15,000 | 15,000 |
| Class II[j] | 1CR | 1Cr-½Mo | 60,000 | 0.90 | 4 | 13,500 | 13,500 |
| Class I[j] | 1¼CR | 1¼Cr-½Mo-Si | 60,000 | 1.00 | 4 | 15,000 | 15,000 |
| Class II[j] | 1¼CR | 1¼Cr-½Mo-Si | 60,000 | 0.90 | 4 | 13,500 | 13,500 |
| Class I[j] | 2¼CR | 2¼Cr-1Mo | 60,000 | 1.00 | 5 | 15,000 | 15,000 |
| Class II | 2¼CR | 2¼Cr-1Mo | 60,000 | 0.90 | 5 | 13,500 | 13,500 |
| Class I | 5Cr | 5Cr-½Mo | 60,000 | 1.00 | 5 | 13,700 | 13,400 |
| Class II | 5CR | 5Cr-½Mo | 60,000 | 0.90 | 5 | 12,300 | 12,050 |
| A358 Class I[g,h,i,k] | 304 | 18Cr-8Ni | 75,000 | 1.00 | 8 | | 14,200 |
| Class I[g,h,i] | 304 | 18Cr-8Ni | 75,000 | 1.00 | 8 | | 10,500 |
| Class II[g,h,i,k] | 304 | 18Cr-8Ni | 75,000 | 0.90 | 8 | | 12,800 |
| Class II[g,h,i] | 304 | 18Cr-8Ni | 75,000 | 0.90 | 8 | | 9,450 |
| Class I[g,i,k] | 316 | 18Cr-12Ni-Mo | 75,000 | 1.00 | 8 | | 16,000 |
| Class I[g,i] | 316 | 18Cr-12Ni-Mo | 75,000 | 1.00 | 8 | | 11,300 |
| Class II[g,i,k] | 316 | 18Cr-12Ni-Mo | 75,000 | 0.90 | 8 | | 14,400 |
| Class II[g,i] | 316 | 18Cr-12Ni-Mo | 75,000 | 0.90 | 8 | | 10,150 |
| A135 carbon steel | A | | 48,000 | 0.85 | 1 | 10,200 | 9,900 |
| | B | | 60,000 | 0.85 | 1 | 12,750 | 12,200 |

| Max allowable stress value, lb/in², for metal temperatures not exceeding °F[b] | | | | | | | | | |
|---|---|---|---|---|---|---|---|---|---|
| 750 | 800 | 850 | 900 | 950 | 1,000 | 1,050 | 1,100 | 1,150 | 1,200 |
| 8,300 | | | | | | | | | |
| 9,950 | | | | | | | | | |
| | | | | | | | | | |
| 9.700 | 8,300 | | | | | | | | |
| 8,700 | 7,500 | | | | | | | | |
| 11,000 | 9,400 | | | | | | | | |
| 9,900 | 8,450 | | | | | | | | |
| 12,050 | 10,200 | | | | | | | | |
| 10,850 | 9,200 | | | | | | | | |
| 12,050 | 10,200 | | | | | | | | |
| 10,850 | 9,200 | | | | | | | | |
| 12,950 | 10,800 | | | | | | | | |
| 11,650 | 9,700 | | | | | | | | |
| 13,850 | 11,400 | | | | | | | | |
| 12,450 | 10,250 | | | | | | | | |
| 14,750 | 12,000 | | | | | | | | |
| 13,750 | 10,800 | | | | | | | | |
| 16,250 | 15,650 | 14,400 | 12,500 | | | | | | |
| 14,600 | 14,100 | 12,950 | 11,250 | | | | | | |
| 17,500 | 16,900 | 15,000 | 12,750 | | | | | | |
| 15,750 | 15,200 | 13,500 | 11,500 | | | | | | |
| 18,750 | 18,000 | 15,900 | 13,000 | | | | | | |
| 16,850 | 16,200 | 14,300 | 11,700 | | | | | | |
| 15,650 | 12,000 | 7,800 | | | | | | | |
| 16,250 | 15,650 | 14,400 | 12,500 | 10,000 | 6,200 | | | | |
| 14,600 | 14,100 | 12,950 | 11,250 | 9,000 | 5,600 | | | | |
| 15,000 | 14,750 | 14,200 | 13,100 | 11,000 | 6,550 | 4,050 | | | |
| 13,500 | 13,300 | 12,800 | 11,800 | 9,900 | 5,900 | 3,650 | | | |
| | | | | | | | | | |
| 15,000 | 15,000 | 14,400 | 13,100 | 11,000 | 6,550 | 4,050 | 3,000 | | |
| | | | | | | | | | |
| 13,500 | 13,500 | 12,950 | 11,800 | 9,900 | 5,900 | 3,650 | 2,700 | | |
| 15,000 | 15,000 | 14,400 | 13,100 | 11,000 | 7,800 | 5,800 | 4,200 | | |
| 13,500 | 13,500 | 12,950 | 11,800 | 9,900 | 7,000 | 5,200 | 3,800 | | |
| 13,100 | 12,800 | 12,400 | 11,500 | 10,000 | 7,300 | 5,200 | 3,300 | 2,200 | |
| 11,800 | 11,500 | 11,150 | 10,350 | 9,000 | 6,550 | 4,650 | 2,950 | 1,950 | |
| 14,100 | 14,000 | 13,850 | 13,600 | 13,400 | 12,500 | 11,100 | 8,700 | 6,900 | 5,500 |
| 10,400 | 10,250 | 10,150 | 10,000 | 9,750 | 9,450 | 9,000 | 8,250 | 6,900 | 5,500 |
| 12,700. | 12,600 | 12,450 | 11,250 | 12,050 | 11,250 | 10,000 | 7,850 | 6,200 | 4,950 |
| 9,350 | 9,250 | 9,150 | 9,000 | 8,750 | 8,500 | 8,100 | 7,400 | 6,200 | 4,950 |
| 15,850 | 15,550 | 15,250 | 14,850 | 14,550 | 14,000 | 13,300 | 12,000 | 9,500 | 6,950 |
| 11,000 | 10,800 | 10,600 | 10,300 | 10,100 | 9,800 | 9,650 | 9,450 | 8,650 | 6,950 |
| 14,250 | 14,000 | 13,700 | 13,350 | 13,100 | 12,600 | 11,950 | 10,800 | 8,550 | 6,250 |
| 9,900 | 9,700 | 9,550 | 9,250 | 9,100 | 8,800 | 8,700 | 8,500 | 7,800 | 6,250 |
| 9,100 | | | | | | | | | |
| 11,000 | | | | | | | | | |

| ASTM spec. No. | Grade | Nominal composition | Spec. min tensile | Longitudinal joint efficiency factor | P No.[a] | 650° | 700 |
|---|---|---|---|---|---|---|---|
| Automatically welded austenitic steel, A312 | TP304H | 18Cr-8Ni[p,i,k] | 75,000 | 0.85 | 8 | | 12,050 |
| | TP304H | 18Cr-8Ni[p,i] | 75,000 | 0.85 | 8 | | 8,900 |
| | TP316H | 18Cr-12Ni-Mo[p,i,k] | 75,000 | 0.85 | 8 | | 13,600 |
| | TP316H | 18Cr-12Ni-Mo[p,i] | 75,000 | 0.85 | 8 | | 9,600 |
| | TP321H | 18Cr-10Ni-Ti[p,i,k] | 75,000 | 0.85 | 8 | | 12,850 |
| | TP321H | 18Cr-10Ni-Ti[p,i] | 75,000 | 0.85 | 8 | | 10,350 |
| Seamless A53 carbon steel | A[d] | | 48,000 | 1.00 | 1 | 12,000 | 11,650 |
| | B[d] | | 60,000 | 1.00 | 1 | 15,000 | 14,350 |
| A106 carbon steel | A[d] | | 48,000 | 1.00 | 1 | 12,000 | 11,650 |
| | B[d] | | 60,000 | 1.00 | 1 | 15,000 | 14,350 |
| | C[d] | | 70,000 | 1.00 | 1 | 17,500 | 16,600 |

The stress values tabulated include a longitudinal joint efficiency factor where applicable.

The stress values in this table may be interpolated to determine values for intermediate temperatures.

Materials listed in Table 126.1 of the ANSI Standard for which allowable stresses are not tabulated may be used. Allowable stresses for such materials shall be taken from Sections I and VIII of the ASME Boiler and Pressure Vessel Code.

[a] The grouping of materials as to P-number classification is made on the basis of hardenability characteristics. The P numbers indicated in this table are identical to those adopted by the ASME Boiler and Pressure Vessel Code. Qualification of welding procedures, welders, and welding operators is required and should comply with the ASME Boiler and Pressure Vessel Code (Section IX) except as modified by Par. 127.5.

[b] The several types and grades of material tabulated should not be used at temperatures in excess of the maximum temperatures for which the allowable stress values are indicated.

[c] For stress values below 650°F which are not tabulated in Table 1, refer to Section I, Table PG-23.1 of the ASME Boiler and Pressure Vessel Code.

[d] Upon prolonged exposure to temperatures above about 775°F, the carbide phase of carbon steel may be converted to graphite.

[e] Upon prolonged exposure to temperatures above about 875°F, the carbide phase of carbon-molybdenum steel may be converted to graphite.

[f] Upon prolonged exposure to temperatures above about 975°F, the carbide phase of chrome-molybdenum steel (with chromium under 0.60) may be converted to graphite.

[g] At temperatures over 1000°F these stress values apply only when the carbon is 0.04 percent or higher.

[h] In size 8 in and larger and schedule 140 or heavier, the minimum tensile strength may be 70,000 lb/in². In these sizes and thicknesses, the indicated allowable stresses should be reduced by the ratio of 70 divided by 75.

[i] For allowable stress values below 700°F, see Table 1.

[j] The values tabulated apply to firebox quality material.

[k] Because of the relatively low yield strength of these materials, these higher stress values were established at temperatures where the short-time tensile properties govern to permit the use of these alloys where slightly greater deformation is acceptable. The stress values in this range exceed 62½ percent but do not exceed 90 percent of the yield strength at temperature. Use of these stresses may result in dimensional changes due to permanent strain. These stress values are not recommended for the flanges of gasketed joints or other applications where slight amounts of distortion can cause leakage or malfunction.

For the properties of steel applicable to high-temperature service (as well as to ordinary service) for pipes, tubes, fittings, bolting material, etc., see Sec. 6. For a discussion of creep properties, see Sec. 5.

Piping of thickness designed in accordance with Eq. (1) may be used for any combination of pressure and temperature for which SE values are listed in Tables 1 and 2. The following summarizes piping-industry practice.

**Steam Pressures above 250 (9,224 N/m²) and not above 2,500 lb/in² (92,237 N/m²) Temperatures not above 1100°F (593°C)** For pressures in excess of 100 lb/in² (3,690 N/m²), the pipe may be seamless steel (A106), (A312), (A335), or (A376); or electric-fusion-welded steel (A155); or forged-and-bored steel (A369); or automatic-welded steel (A312). For pressures between 250 and 600 lb/in² (9,224 and 22,137 N/m²) the pipe may be seamless steel (A106) or (A53); electric-fusion-welded steel (A155); electric-resistance-welded steel (A135) or (A53). For pressures of 250 lb/in² (9,224 N/m²) and lower and for service up to 750°F (399°C), any of the following may be used: electric-fusion-welded steel (A134) or (A139); electric-resistance-welded steel (A135); seamless or welded steel (A53); or wrought-iron pipe (A72). Grade A seamless pipe (A106) or (A53); or Grade A electric-welded pipe (A53), (A135), or (A139) is used for close coiling or cold bending. Pipe permissible for services specified may be used for temperatures higher than 750°F (399°C), unless otherwise prohibited, if the SE values of Tables 1 and 2 are used when calculating the required wall thickness.

Valves and fittings must have flange openings or welded ends, and valves must have external stem threads. Valves must be of cast or forged steel or of forged or cast nonferrous material. Forged and cast-steel screwed valves and fittings may be used up to 300 lb/in² and 500°F for 3 (2) [1½] in pipe, and pressure from 250 to 400 (400 to 600) [600 to 2,500] lb/in². Malleable-iron screwed fittings (300 lb MSS SP-31) may be used for pressures not greater than 300 lb/in² and temperatures not over 500°F. Valves 8 in and larger should have the bypass of at least ¾ in, commercial size.[1] Welded fittings may be used of the same material and thickness as the pipe with which they are to be used.

**Steam Pressures from 125 to 250 lb/in² (4,612 to 9,224 N/m²),**

[1] See Manufacturere Standardization Society SP-31 for recommended size of bypass valves.

| Max allowable stress value, lb/in², for metal temperatures not exceeding °F[b] | | | | | | | | | |
|---|---|---|---|---|---|---|---|---|---|
| 750 | 800 | 850 | 900 | 950 | 1,000 | 1,050 | 1,100 | 1,150 | 1,200 |
| 12,000 | 11,900 | 11,750 | 11,550 | 11,400 | 10,600 | 9,450 | 7,400 | 5,850 | 4,650 |
| 8,850 | 8,700 | 8,650 | 8,500 | 8,300 | 8,050 | 7,650 | 7,000 | 5,850 | 4,650 |
| 13,450 | 13,200 | 12,950 | 12,600 | 12,350 | 11,900 | 11,300 | 10,200 | 8,050 | 5,900 |
| 9,350 | 9,200 | 9,000 | 8,750 | 8,600 | 8,350 | 8,200 | 8,050 | 7,350 | 5,900 |
| 12,850 | 12,850 | 12,750 | 12,650 | 12,600 | 12,300 | 11,650 | 8,000 | 6,550 | 4,650 |
| 10,200 | 10,100 | 9,850 | 9,600 | 9,500 | 9,350 | 9,200 | 8,500 | 6,550 | 4,650 |
| 10,700 | 9,000 | | | | | | | | |
| 12,950 | 10,800 | | | | | | | | |
| 10,700 | 9,000 | | | | | | | | |
| 12,950 | 10,800 | | | | | | | | |
| 14,750 | 12,000 | | | | | | | | |

**Temperature not above 450°F (232°C)** Pipe may be electric-fusion-welded steel (A134 or A139), electric-resistance-welded steel (A135), seamless or welded steel (A53), or wrought iron (A72). Copper and brass may be used if the temperature does not exceed 406°F. Cast iron may also be used. For close coiling or cold bending, Grade A seamless steel (A53); or Grade A electric-welded steel (A53), (A135), or (A139) is suitable. Pipe permissible for this service may be used for temperatures above 450°F (232°C) if the proper *SE* is used in calculating the pipe-wall thickness.

Valves below 3 in may have inside stem screws. Stop valves 8 in and over must be bypassed. Bodies, bonnets, and yokes are of cast iron, malleable iron, steel, bronze, brass, or Monel. Flanged-steel fittings must conform to the 300 lb ANSI Standard B16.5, if of cast iron, to the 250 lb ANSI Standard B16.2; or, for screwed fittings, to the ANSI Standard B16.4. Malleable-iron screwed fittings must conform to the 300 lb MSS SP-31 specifications, except that the 150 lb ANSI Standard B16.3 may be used for pressures not greater than 150 lb. Welded fittings may be used.

**Steam Pressures from 25 to 125 lb/in², Temperatures not above 450°F** Pipe may be of steel, wrought iron, cast iron, copper, or brass; valve bodies of cast iron, malleable iron, steel, or brass. Fittings are of 125 lb or 150 lb American Standard cast iron with screwed or flanged ends, or of malleable iron with screwed ends.

**Steam Pressures 25 lb/in² and less, Temperature up to 450°F** Pipe may be of steel, wrought iron, spiral-riveted steel, brass, copper, or cast iron. Flanged fittings conform to the 25 lb ANSI Standard B16.2. Screwed fittings are of the 125 ANSI Standard B16.4 or of the 150 lb ANSI Standard B16.3 for cast iron or malleable iron, respectively, or the B16.15 for bronze. Welded fittings may be used.

**Pipe coils** are made from any of the commercial sizes of iron, steel, brass, and copper pipe and tubing. Limiting center-to-center dimensions, to which pipe coils can be fabricated in sizes ¾ to 2 in, are given in Table 10. Steel tubing cannot be bent to the absolute limits of brass or copper.

**Seamless mechanical tubing** is obtainable in outside diameters ranging from ¼ to 10¾ in in wall thickness from 20 gage to 2 in (0.091 to 5.08 cm), and in standard pipe weights and dimensions up to 24 in (60.96 cm). Oval, square, rectangular, and other special shapes can be obtained in various sizes and wall thicknesses. Mechanical tubing is available either hot-finished or cold-drawn, but is furnished principally cold-drawn. It is readily adaptable to varied treatment by expansion, cupping, tapering, swaging, flanging, coiling, welding, and similar manipulations. Typical of the many uses are aircraft tubing, automobile axle housings, drive shafts, drive-shaft housings, tie rods, steering columns, piston rods and pins, gear rings, roller-bearing cases and cones, cylinders for various purposes, machine parts, sleeves, bushings, spacers, surgical instruments, and hypodermic needles. Table 11 gives weights and dimensions of round seamless-steel tubing for sizes that have by common usage become standard. Detailed information on mechanical tubing for any particular applications can be obtained from manufacturers.

Dimensions and weights of **condenser** and **heat-exchanger tubes** are given in Table 12 and of **boiler** tubes in Table 13.

**Spiral Pipe** Spiral pipe is strong lightweight steel pipe with a single continuous welded helical seam from end to end stiffening it throughout. It is listed in sizes 6 to 42 in ID (15.24 to 106.7 cm), in various thicknesses, and in lengths up to 40 ft (12.19 m). It is used for high- and low-pressure water lines, vacuum lines, exhaust-steam lines, low-pressure air lines, sand and gravel conveying and similar services. It is also used extensively by the petroleum industry, for oil and gas lines, for low-pressure steam lines, etc.

Spiral pipe may be asphalt-coated or galvanized. The pipe is designed for special joints, flanges, and lightweight fittings, but the ANSI flanges and fittings can be furnished, if desired. For further details, refer to catalog of Taylor Forge and Pipe Works, Chicago.

The **sleeve-type coupling** illustrated in Fig. 1 is particularly suitable for plain-end pipe and is widely used. A gasket is used to make a tight joint. Advantages of this coupling are low cost, the use of unskilled labor in making the connections, and the fact that small changes in alignment and grade can be made with regular straight lengths of pipe by a movement in the coupling. This type of coupling is used extensively in long oil lines.

## CAST-IRON PIPE

**Cast-iron pipe** is extensively used for water, gas, sewage, culverts, drains, etc., in a wide range of sizes and for varying

**Table 3. Physical Properties of Pipe***

(Grinnell Co., Inc.)

| Nominal pipe size, O.D., in. | Schedule number† a | b | c | Wall thickness, in. | I.D., in. | Inside area, sq in. | Metal area, sq in. | Sq ft outside surface, per ft | Sq ft inside surface, per ft | Weight per ft, lb | Weight of water per ft, lb | Moment of inertia, in.⁴ | Section modulus, in.³ | Radius gyration, in. |
|---|---|---|---|---|---|---|---|---|---|---|---|---|---|---|
| 1/8 0.405 | ... | ... | 10S | 0.049 | 0.307 | 0.0740 | 0.0548 | 0.106 | 0.0804 | 0.186 | 0.0321 | 0.00088 | 0.00437 | 0.1271 |
|  | 40 | Std | 40S | 0.068 | 0.269 | 0.0568 | 0.0720 | 0.106 | 0.0705 | 0.245 | 0.0246 | 0.00106 | 0.00525 | 0.1215 |
|  | 80 | XS | 80S | 0.095 | 0.215 | 0.0364 | 0.0925 | 0.106 | 0.0563 | 0.315 | 0.0157 | 0.00122 | 0.00600 | 0.1146 |
| 1/4 0.540 | ... | ... | 10S | 0.065 | 0.410 | 0.1320 | 0.0970 | 0.141 | 0.1073 | 0.330 | 0.0572 | 0.00279 | 0.01032 | 0.1694 |
|  | 40 | Std | 40S | 0.088 | 0.364 | 0.1041 | 0.1250 | 0.141 | 0.0955 | 0.425 | 0.0451 | 0.00331 | 0.01230 | 0.1628 |
|  | 80 | XS | 80S | 0.119 | 0.302 | 0.0716 | 0.1574 | 0.141 | 0.0794 | 0.535 | 0.0310 | 0.00378 | 0.01395 | 0.1547 |
| 3/8 0.675 | ... | ... | 10S | 0.065 | 0.545 | 0.2333 | 0.1246 | 0.177 | 0.1427 | 0.423 | 0.1011 | 0.00586 | 0.01737 | 0.2169 |
|  | 40 | Std | 40S | 0.091 | 0.493 | 0.1910 | 0.1670 | 0.177 | 0.1295 | 0.568 | 0.0827 | 0.00730 | 0.02160 | 0.2090 |
|  | 80 | XS | 80S | 0.126 | 0.423 | 0.1405 | 0.2173 | 0.177 | 0.1106 | 0.739 | 0.0609 | 0.00862 | 0.02554 | 0.1991 |
| 1/2 0.840 | ... | ... | 10S | 0.083 | 0.674 | 0.357 | 0.1974 | 0.220 | 0.1765 | 0.671 | 0.1547 | 0.01431 | 0.0341 | 0.2692 |
|  | 40 | Std | 40S | 0.109 | 0.622 | 0.304 | 0.2503 | 0.220 | 0.1628 | 0.851 | 0.1316 | 0.01710 | 0.0407 | 0.2613 |
|  | 80 | XS | 80S | 0.147 | 0.546 | 0.2340 | 0.320 | 0.220 | 0.1433 | 1.088 | 0.1013 | 0.02010 | 0.0478 | 0.2505 |
|  | 160 | ... | ... | 0.187 | 0.466 | 0.1706 | 0.383 | 0.220 | 0.1220 | 1.304 | 0.0740 | 0.02213 | 0.0527 | 0.2402 |
|  | ... | XXS | ... | 0.294 | 0.252 | 0.0499 | 0.504 | 0.220 | 0.0660 | 1.714 | 0.0216 | 0.02425 | 0.0577 | 0.2192 |
| 3/4 1.050 | ... | ... | 5S | 0.065 | 0.920 | 0.665 | 0.2011 | 0.275 | 0.2409 | 0.684 | 0.2882 | 0.02451 | 0.0467 | 0.349 |
|  | ... | ... | 10S | 0.083 | 0.884 | 0.614 | 0.2521 | 0.275 | 0.2314 | 0.857 | 0.2661 | 0.02970 | 0.0566 | 0.343 |
|  | 40 | Std | 40S | 0.113 | 0.824 | 0.533 | 0.333 | 0.275 | 0.2157 | 1.131 | 0.2301 | 0.0370 | 0.0706 | 0.334 |
|  | 80 | XS | 80S | 0.154 | 0.742 | 0.432 | 0.435 | 0.275 | 0.1943 | 1.474 | 0.1875 | 0.0448 | 0.0853 | 0.321 |
|  | 160 | ... | ... | 0.218 | 0.614 | 0.2961 | 0.570 | 0.275 | 0.1607 | 1.937 | 0.1284 | 0.0527 | 0.1004 | 0.304 |
|  | ... | XXS | ... | 0.308 | 0.434 | 0.1479 | 0.718 | 0.275 | 0.1137 | 2.441 | 0.0641 | 0.0579 | 0.1104 | 0.284 |
| 1 1.315 | ... | ... | 5S | 0.065 | 1.185 | 1.103 | 0.2553 | 0.344 | 0.310 | 0.868 | 0.478 | 0.0500 | 0.0760 | 0.443 |
|  | ... | ... | 10S | 0.109 | 1.097 | 0.945 | 0.413 | 0.344 | 0.2872 | 1.404 | 0.409 | 0.0757 | 0.1151 | 0.428 |
|  | 40 | Std | 40S | 0.133 | 1.049 | 0.864 | 0.494 | 0.344 | 0.2746 | 1.679 | 0.374 | 0.0874 | 0.1329 | 0.421 |
|  | 80 | XS | 80S | 0.179 | 0.957 | 0.719 | 0.639 | 0.344 | 0.2520 | 2.172 | 0.311 | 0.1056 | 0.1606 | 0.407 |
|  | 160 | ... | ... | 0.250 | 0.815 | 0.522 | 0.836 | 0.344 | 0.2134 | 2.844 | 0.2261 | 0.1252 | 0.1903 | 0.387 |
|  | ... | XXS | ... | 0.358 | 0.599 | 0.2818 | 1.076 | 0.344 | 0.1570 | 3.659 | 0.1221 | 0.1405 | 0.2137 | 0.361 |
| 1 1/4 1.660 | ... | ... | 5S | 0.065 | 1.530 | 1.839 | 0.326 | 0.434 | 0.401 | 1.107 | 0.797 | 0.1038 | 0.1250 | 0.564 |
|  | ... | ... | 10S | 0.109 | 1.442 | 1.633 | 0.531 | 0.434 | 0.378 | 1.805 | 0.707 | 0.1605 | 0.1934 | 0.550 |
|  | 40 | Std | 40S | 0.140 | 1.380 | 1.496 | 0.669 | 0.434 | 0.361 | 2.273 | 0.648 | 0.1948 | 0.2346 | 0.540 |
|  | 80 | XS | 80S | 0.191 | 1.278 | 1.283 | 0.881 | 0.434 | 0.335 | 2.997 | 0.555 | 0.2418 | 0.2913 | 0.524 |

| Nominal size & OD | Sched. No. | Desig. | Sched. | (1) | (2) | (3) | (4) | (5) | (6) | (7) | (8) | (9) | (10) | (11) |
|---|---|---|---|---|---|---|---|---|---|---|---|---|---|---|
| 1¼ 1.066 | 160 | … | … | 0.250 | 1.160 | 1.057 | 1.107 | 0.434 | 0.304 | 3.765 | 0.458 | 0.2839 | 0.342 | 0.506 |
|  | … | XXS | … | 0.382 | 0.896 | 0.631 | 1.534 | 0.434 | 0.2346 | 5.214 | 0.2732 | 0.341 | 0.411 | 0.472 |
| 1½ 1.900 | … | … | 5S | 0.065 | 1.770 | 2.461 | 0.375 | 0.497 | 0.463 | 1.274 | 1.067 | 0.1580 | 0.1663 | 0.649 |
|  | … | … | 10S | 0.109 | 1.682 | 2.222 | 0.613 | 0.497 | 0.440 | 2.085 | 0.962 | 0.2469 | 0.2599 | 0.634 |
|  | 40 | Std | 40S | 0.145 | 1.610 | 2.036 | 0.799 | 0.497 | 0.421 | 2.718 | 0.882 | 0.310 | 0.326 | 0.623 |
|  | 80 | XS | 80S | 0.200 | 1.500 | 1.767 | 1.068 | 0.497 | 0.393 | 3.631 | 0.765 | 0.391 | 0.412 | 0.605 |
|  | 160 | … | … | 0.281 | 1.338 | 1.406 | 1.429 | 0.497 | 0.350 | 4.859 | 0.608 | 0.483 | 0.508 | 0.581 |
|  | … | XXS | … | 0.400 | 1.100 | 0.950 | 1.885 | 0.497 | 0.288 | 6.408 | 0.412 | 0.568 | 0.598 | 0.549 |
| 2 2.375 | … | … | 5S | 0.065 | 2.245 | 3.96 | 0.472 | 0.622 | 0.588 | 1.604 | 1.716 | 0.315 | 0.2652 | 0.817 |
|  | … | … | 10S | 0.109 | 2.157 | 3.65 | 0.776 | 0.622 | 0.565 | 2.638 | 1.582 | 0.499 | 0.420 | 0.802 |
|  | 40 | Std | 40S | 0.154 | 2.067 | 3.36 | 1.075 | 0.622 | 0.541 | 3.653 | 1.455 | 0.666 | 0.561 | 0.787 |
|  | 80 | XS | 80S | 0.218 | 1.939 | 2.953 | 1.477 | 0.622 | 0.508 | 5.022 | 1.280 | 0.868 | 0.731 | 0.766 |
|  | 160 | … | … | 0.343 | 1.689 | 2.240 | 2.190 | 0.622 | 0.442 | 7.444 | 0.971 | 1.163 | 0.979 | 0.729 |
|  | … | XXS | … | 0.436 | 1.503 | 1.774 | 2.656 | 0.622 | 0.393 | 9.029 | 0.769 | 1.312 | 1.104 | 0.703 |
| 2½ 2.875 | … | … | 5S | 0.083 | 2.709 | 5.76 | 0.728 | 0.753 | 0.709 | 2.475 | 2.499 | 0.710 | 0.494 | 0.988 |
|  | … | … | 10S | 0.120 | 2.635 | 5.45 | 1.039 | 0.753 | 0.690 | 3.531 | 2.361 | 0.988 | 0.687 | 0.975 |
|  | 40 | Std | 40S | 0.203 | 2.469 | 4.79 | 1.704 | 0.753 | 0.646 | 5.793 | 2.076 | 1.530 | 1.064 | 0.947 |
|  | 80 | XS | 80S | 0.276 | 2.323 | 4.24 | 2.254 | 0.753 | 0.608 | 7.661 | 1.837 | 1.925 | 1.339 | 0.924 |
|  | 160 | … | … | 0.375 | 2.125 | 3.55 | 2.945 | 0.753 | 0.556 | 10.01 | 1.535 | 2.353 | 1.637 | 0.894 |
|  | … | XXS | … | 0.552 | 1.771 | 2.464 | 4.03 | 0.753 | 0.464 | 13.70 | 1.067 | 2.872 | 1.998 | 0.844 |
| 3 3.500 | … | … | 5S | 0.083 | 3.334 | 8.73 | 0.891 | 0.916 | 0.873 | 3.03 | 3.78 | 1.301 | 0.744 | 1.208 |
|  | … | … | 10S | 0.120 | 3.260 | 8.35 | 1.274 | 0.916 | 0.853 | 4.33 | 3.61 | 1.822 | 1.041 | 1.196 |
|  | 40 | Std | 40S | 0.216 | 3.068 | 7.39 | 2.228 | 0.916 | 0.803 | 7.58 | 3.20 | 3.02 | 1.724 | 1.164 |
|  | 80 | XS | 80S | 0.300 | 2.900 | 6.61 | 3.02 | 0.916 | 0.759 | 10.25 | 2.864 | 3.90 | 2.226 | 1.136 |
|  | 160 | … | … | 0.437 | 2.626 | 5.42 | 4.21 | 0.916 | 0.687 | 14.32 | 2.348 | 5.03 | 2.876 | 1.094 |
|  | … | XXS | … | 0.600 | 2.300 | 4.15 | 5.47 | 0.916 | 0.602 | 18.58 | 1.801 | 5.99 | 3.43 | 1.047 |
| 3½ 4.000 | … | … | 5S | 0.083 | 3.834 | 11.55 | 1.021 | 1.047 | 1.004 | 3.47 | 5.01 | 1.960 | 0.980 | 1.385 |
|  | … | … | 10S | 0.120 | 3.760 | 11.10 | 1.463 | 1.047 | 0.984 | 4.97 | 4.81 | 2.756 | 1.378 | 1.372 |
|  | 40 | Std | 40S | 0.226 | 3.548 | 9.89 | 2.680 | 1.047 | 0.929 | 9.11 | 4.28 | 4.79 | 2.394 | 1.337 |
|  | 80 | XS | 80S | 0.318 | 3.364 | 8.89 | 3.68 | 1.047 | 0.881 | 12.51 | 3.85 | 6.28 | 3.14 | 1.307 |
| 4 4.500 | … | … | 5S | 0.083 | 4.334 | 14.75 | 1.152 | 1.178 | 1.135 | 3.92 | 6.40 | 2.811 | 1.249 | 1.562 |
|  | … | … | 10S | 0.120 | 4.260 | 14.25 | 1.651 | 1.178 | 1.115 | 5.61 | 6.17 | 3.96 | 1.762 | 1.549 |
|  | 40 | Std | 40S | 0.237 | 4.026 | 12.73 | 3.17 | 1.178 | 1.054 | 10.79 | 5.51 | 7.23 | 3.21 | 1.510 |
|  | 80 | XS | 80S | 0.337 | 3.826 | 11.50 | 4.41 | 1.178 | 1.002 | 14.98 | 4.98 | 9.61 | 4.27 | 1.477 |
|  | 120 | … | … | 0.437 | 3.626 | 10.33 | 5.58 | 1.178 | 0.949 | 18.96 | 4.48 | 11.65 | 5.18 | 1.445 |
|  | 160 | … | … | 0.531 | 3.438 | 9.28 | 6.62 | 1.178 | 0.900 | 22.51 | 4.02 | 13.27 | 5.90 | 1.416 |
|  | … | XXS | … | 0.674 | 3.152 | 7.80 | 8.10 | 1.178 | 0.825 | 27.54 | 3.38 | 15.29 | 6.79 | 1.374 |

\* See footnote at end of table.
† See footnote at end of table.

**Table 3. Physical Properties of Pipe\*** *(Continued)*

| Nominal pipe size, O.D., in. | Schedule number a | b | c | Wall thickness, in. | I.D., in. | Inside area, sq in. | Metal area, sq in. | Sq ft outside surface, per ft | Sq ft inside surface, per ft | Weight per ft, lb | Weight of water per ft, lb | Moment of inertia, in.⁴ | Section modulus, in.³ | Radius gyration, in. |
|---|---|---|---|---|---|---|---|---|---|---|---|---|---|---|
| 5<br>*5.563* | | | 5S | 0.109 | 5.345 | 22.44 | 1.868 | 1.456 | 1.399 | 6.35 | 9.73 | 6.95 | 2.498 | 1.929 |
| | | | 10S | 0.134 | 5.295 | 22.02 | 2.285 | 1.456 | 1.386 | 7.77 | 9.53 | 8.43 | 3.03 | 1.920 |
| | 40 | Std | 40S | 0.258 | 5.047 | 20.01 | 4.30 | 1.456 | 1.321 | 14.62 | 8.66 | 15.17 | 5.45 | 1.878 |
| | 80 | XS | 80S | 0.375 | 4.813 | 18.19 | 6.11 | 1.456 | 1.260 | 20.78 | 7.89 | 20.68 | 7.43 | 1.839 |
| | 120 | | | 0.500 | 4.563 | 16.35 | 7.95 | 1.456 | 1.195 | 27.04 | 7.09 | 25.74 | 9.25 | 1.799 |
| | 160 | | | 0.625 | 4.313 | 14.61 | 9.70 | 1.456 | 1.129 | 32.96 | 6.33 | 30.0 | 10.80 | 1.760 |
| | | XXS | | 0.750 | 4.063 | 12.97 | 11.34 | 1.456 | 1.064 | 38.55 | 5.62 | 33.6 | 12.10 | 1.722 |
| 6<br>*6.625* | | | 5S | 0.109 | 6.407 | 32.2 | 2.231 | 1.734 | 1.677 | 5.37 | 13.98 | 11.85 | 3.58 | 2.304 |
| | | | 10S | 0.134 | 6.357 | 31.7 | 2.733 | 1.734 | 1.664 | 9.29 | 13.74 | 14.40 | 4.35 | 2.295 |
| | 40 | Std | 40S | 0.280 | 6.065 | 28.89 | 5.58 | 1.734 | 1.588 | 18.97 | 12.51 | 28.14 | 8.50 | 2.245 |
| | 80 | XS | 80S | 0.432 | 5.761 | 26.07 | 8.40 | 1.734 | 1.508 | 28.57 | 11.29 | 40.5 | 12.23 | 2.195 |
| | 120 | | | 0.562 | 5.501 | 23.77 | 10.70 | 1.734 | 1.440 | 36.39 | 10.30 | 49.6 | 14.98 | 2.153 |
| | 160 | | | 0.718 | 5.189 | 21.15 | 13.33 | 1.734 | 1.358 | 45.30 | 9.16 | 59.0 | 17.81 | 2.104 |
| | | XXS | | 0.864 | 4.897 | 18.83 | 15.64 | 1.734 | 1.282 | 53.16 | 8.17 | 66.3 | 20.03 | 2.060 |
| 8<br>*8.625* | | | 5S | 0.109 | 8.407 | 55.5 | 2.916 | 2.258 | 2.201 | 9.91 | 24.07 | 26.45 | 6.13 | 3.01 |
| | | | 10S | 0.148 | 8.329 | 54.5 | 3.94 | 2.258 | 2.180 | 13.40 | 23.59 | 35.4 | 8.21 | 3.00 |
| | 20 | | | 0.250 | 8.125 | 51.8 | 6.58 | 2.258 | 2.127 | 22.36 | 22.48 | 57.7 | 13.39 | 2.962 |
| | 30 | | | 0.277 | 8.071 | 51.2 | 7.26 | 2.258 | 2.113 | 24.70 | 22.18 | 63.4 | 14.69 | 2.953 |
| | 40 | Std | 40S | 0.322 | 7.981 | 50.0 | 8.40 | 2.258 | 2.089 | 28.55 | 21.69 | 72.5 | 16.81 | 2.938 |
| | 60 | | | 0.406 | 7.813 | 47.9 | 10.48 | 2.258 | 2.045 | 35.64 | 20.79 | 88.8 | 20.58 | 2.909 |
| | 80 | XS | 80S | 0.500 | 7.625 | 45.7 | 12.76 | 2.258 | 1.996 | 43.39 | 19.80 | 105.7 | 24.52 | 2.878 |
| | 100 | | | 0.593 | 7.439 | 43.5 | 14.96 | 2.258 | 1.948 | 50.87 | 18.84 | 121.4 | 28.14 | 2.847 |
| | 120 | | | 0.718 | 7.189 | 40.6 | 17.84 | 2.258 | 1.882 | 60.63 | 17.60 | 140.6 | 32.6 | 2.807 |
| | 140 | | | 0.812 | 7.001 | 38.5 | 19.93 | 2.258 | 1.833 | 67.76 | 16.69 | 153.8 | 35.7 | 2.777 |
| | | XXS | | 0.875 | 6.875 | 37.1 | 21.30 | 2.258 | 1.800 | 72.42 | 16.09 | 162.0 | 37.6 | 2.757 |
| | 160 | | | 0.906 | 6.813 | 36.5 | 21.97 | 2.258 | 1.784 | 74.69 | 15.80 | 165.9 | 38.5 | 2.748 |
| 10<br>*10.750* | | | 5S | 0.134 | 10.482 | 86.3 | 4.52 | 2.815 | 2.744 | 15.15 | 37.4 | 63.7 | 11.85 | 3.75 |
| | | | 10S | 0.165 | 10.420 | 85.3 | 5.49 | 2.815 | 2.728 | 18.70 | 36.9 | 76.9 | 14.30 | 3.74 |
| | 20 | | | 0.250 | 10.250 | 82.5 | 8.26 | 2.815 | 2.683 | 28.04 | 35.8 | 113.7 | 21.16 | 3.71 |
| | | | | 0.279 | 10.192 | 81.6 | 9.18 | 2.815 | 2.668 | 31.20 | 35.3 | 125.9 | 23.42 | 3.70 |
| | 30 | | | 0.307 | 10.136 | 80.7 | 10.07 | 2.815 | 2.654 | 34.24 | 35.0 | 137.5 | 25.57 | 3.69 |
| | 40 | Std | 40S | 0.365 | 10.020 | 78.9 | 11.91 | 2.815 | 2.623 | 40.48 | 34.1 | 160.8 | 29.90 | 3.67 |
| | 60 | XS | 80S | 0.500 | 9.750 | 74.7 | 16.10 | 2.815 | 2.553 | 54.74 | 32.3 | 212.0 | 39.4 | 3.63 |
| | 80 | | | 0.593 | 9.564 | 71.8 | 18.92 | 2.815 | 2.504 | 64.33 | 31.1 | 244.9 | 45.6 | 3.60 |
| | 100 | | | 0.718 | 9.314 | 68.1 | 22.63 | 2.815 | 2.438 | 76.93 | 29.5 | 286.2 | 53.2 | 3.56 |

| Nom. size / OD | Sched. No. | Std/XS | 5S–80S | Wall | ID | | | | | | | | | |
|---|---|---|---|---|---|---|---|---|---|---|---|---|---|---|
| **10** *10.750* | 120 | | | 0.843 | 9.064 | 64.5 | 26.24 | 2.815 | 2.373 | 89.20 | 28.0 | 324 | 60.3 | 3.52 |
| | 140 | | | 1.000 | 8.750 | 60.1 | 30.6 | 2.815 | 2.291 | 104.13 | 26.1 | 368 | 68.4 | 3.47 |
| | 160 | | | 1.125 | 8.500 | 56.7 | 34.0 | 2.815 | 2.225 | 115.65 | 24.6 | 399 | 74.3 | 3.43 |
| **12** *12.750* | | | 5S | 0.165 | 12.420 | 121.2 | 6.52 | 3.34 | 3.25 | 19.56 | 52.5 | 129.2 | 20.27 | 4.45 |
| | | | 10S | 0.180 | 12.390 | 120.6 | 7.11 | 3.34 | 3.24 | 24.20 | 52.2 | 140.5 | 22.03 | 4.44 |
| | 20 | | | 0.250 | 12.250 | 117.9 | 9.84 | 3.34 | 3.21 | 33.38 | 51.1 | 191.9 | 30.1 | 4.42 |
| | 30 | | | 0.330 | 12.090 | 114.8 | 12.88 | 3.34 | 3.17 | 43.77 | 49.7 | 248.5 | 39.0 | 4.39 |
| | | Std | 40S | 0.375 | 12.000 | 113.1 | 14.58 | 3.34 | 3.14 | 49.56 | 49.0 | 279.3 | 43.8 | 4.38 |
| | 40 | | | 0.406 | 11.938 | 111.9 | 15.74 | 3.34 | 3.13 | 53.53 | 48.5 | 300 | 47.1 | 4.37 |
| | | XS | 80S | 0.500 | 11.750 | 108.4 | 19.24 | 3.34 | 3.08 | 65.42 | 47.0 | 362 | 56.7 | 4.33 |
| | 60 | | | 0.562 | 11.626 | 106.2 | 21.52 | 3.34 | 3.04 | 73.16 | 46.0 | 401 | 62.8 | 4.31 |
| | 80 | | | 0.687 | 11.376 | 101.6 | 26.04 | 3.34 | 2.978 | 88.51 | 44.0 | 475 | 74.5 | 4.27 |
| | 100 | | | 0.843 | 11.064 | 96.1 | 31.5 | 3.34 | 2.897 | 107.20 | 41.6 | 562 | 88.1 | 4.22 |
| | 120 | | | 1.000 | 10.750 | 90.8 | 36.9 | 3.34 | 2.814 | 125.49 | 39.3 | 642 | 100.7 | 4.17 |
| | 140 | | | 1.125 | 10.500 | 86.6 | 41.1 | 3.34 | 2.749 | 139.68 | 37.5 | 701 | 109.9 | 4.13 |
| | 160 | | | 1.312 | 10.126 | 80.5 | 47.1 | 3.34 | 2.651 | 160.27 | 34.9 | 781 | 122.6 | 4.07 |
| **14** *14.000* | 10 | | | 0.250 | 13.500 | 143.1 | 10.80 | 3.67 | 3.53 | 36.71 | 62.1 | 255.4 | 36.5 | 4.86 |
| | 20 | | | 0.312 | 13.376 | 140.5 | 13.42 | 3.67 | 3.50 | 45.68 | 60.9 | 314 | 44.9 | 4.84 |
| | 30 | Std | | 0.375 | 13.250 | 137.9 | 16.05 | 3.67 | 3.47 | 54.57 | 59.7 | 373 | 53.3 | 4.82 |
| | 40 | | | 0.437 | 13.126 | 135.3 | 18.62 | 3.67 | 3.44 | 63.37 | 58.7 | 429 | 61.2 | 4.80 |
| | | XS | | 0.500 | 13.000 | 132.7 | 21.21 | 3.67 | 3.40 | 72.09 | 57.5 | 484 | 69.1 | 4.78 |
| | | | | 0.562 | 12.876 | 130.2 | 23.73 | 3.67 | 3.37 | 80.66 | 56.5 | 537 | 76.7 | 4.76 |
| | 60 | | | 0.593 | 12.814 | 129.0 | 24.98 | 3.67 | 3.35 | 84.91 | 55.9 | 562 | 80.3 | 4.74 |
| | | | | 0.625 | 12.750 | 127.7 | 26.26 | 3.67 | 3.34 | 89.28 | 55.3 | 589 | 84.1 | 4.73 |
| | | | | 0.687 | 12.626 | 125.2 | 28.73 | 3.67 | 3.31 | 97.68 | 54.3 | 638 | 91.2 | 4.71 |
| | 80 | | | 0.750 | 12.500 | 122.7 | 31.2 | 3.67 | 3.27 | 106.13 | 53.2 | 687 | 98.2 | 4.69 |
| | | | | 0.875 | 12.250 | 117.9 | 36.1 | 3.67 | 3.21 | 122.66 | 51.1 | 781 | 111.5 | 4.65 |
| | 100 | | | 0.937 | 12.126 | 115.5 | 38.5 | 3.67 | 3.17 | 130.73 | 50.0 | 825 | 117.8 | 4.63 |
| | 120 | | | 1.093 | 11.814 | 109.6 | 44.3 | 3.67 | 3.09 | 150.67 | 47.5 | 930 | 132.8 | 4.58 |
| | 140 | | | 1.250 | 11.500 | 103.9 | 50.1 | 3.67 | 3.01 | 170.22 | 45.0 | 1,017 | 146.8 | 4.53 |
| | 160 | | | 1.406 | 11.188 | 98.3 | 55.6 | 3.67 | 2.929 | 189.12 | 42.6 | 1,127 | 159.6 | 4.48 |
| **16** *16.000* | 10 | | | 0.250 | 15.500 | 188.7 | 12.37 | 4.19 | 4.06 | 42.05 | 81.8 | 384 | 48.0 | 5.57 |
| | 20 | | | 0.312 | 15.376 | 185.7 | 15.38 | 4.19 | 4.03 | 52.36 | 80.5 | 473 | 59.2 | 5.55 |
| | 30 | Std | | 0.375 | 15.250 | 182.6 | 18.41 | 4.19 | 3.99 | 62.58 | 79.1 | 562 | 70.3 | 5.53 |
| | | | | 0.437 | 15.126 | 179.7 | 21.37 | 4.19 | 3.96 | 72.64 | 77.9 | 648 | 80.9 | 5.50 |
| | 40 | XS | | 0.500 | 15.000 | 176.7 | 24.35 | 4.19 | 3.93 | 82.77 | 76.5 | 732 | 91.5 | 5.48 |
| | | | | 0.562 | 14.876 | 173.8 | 27.26 | 4.19 | 3.89 | 92.66 | 75.4 | 813 | 106.6 | 5.46 |
| | | | | 0.625 | 14.750 | 170.9 | 30.2 | 4.19 | 3.86 | 102.63 | 74.1 | 894 | 112.2 | 5.44 |
| | 60 | | | 0.656 | 14.688 | 169.4 | 31.6 | 4.19 | 3.85 | 107.50 | 73.4 | 933 | 116.6 | 5.43 |
| | | | | 0.687 | 14.626 | 168.0 | 33.0 | 4.19 | 3.83 | 112.36 | 72.7 | 971 | 121.4 | 5.42 |

* See footnote at end of table.
† See footnote at end of table.

**Table 3. Physical Properties of Pipe\*** *(Continued)*

| Nominal pipe size, O.D., in. | Schedule number† a | b | c | Wall thickness, in. | I.D., in. | Inside area, sq in. | Metal area, sq in. | Sq ft outside surface, per ft | Sq ft inside surface, per ft | Weight per ft, lb | Weight of water per ft, lb | Moment of inertia, in.⁴ | Section modulus, in.³ | Radius gyration, in. |
|---|---|---|---|---|---|---|---|---|---|---|---|---|---|---|
| 16 16.000 | .... | .... | .... | 0.750 | 14.500 | 165.1 | 35.9 | 4.19 | 3.80 | 122.15 | 71.5 | 1,047 | 130.9 | 5.40 |
|  | 80 | .... | .... | 0.842 | 14.314 | 160.9 | 40.1 | 4.19 | 3.75 | 136.46 | 69.7 | 1,157 | 144.6 | 5.37 |
|  | .... | .... | .... | 0.875 | 14.250 | 159.5 | 41.6 | 4.19 | 3.73 | 141.35 | 69.1 | 1,193 | 154.1 | 5.36 |
|  | 100 | .... | .... | 1.031 | 13.938 | 152.6 | 48.5 | 4.19 | 3.65 | 164.83 | 66.1 | 1,365 | 170.6 | 5.30 |
|  | 120 | .... | .... | 1.218 | 13.564 | 144.5 | 56.6 | 4.19 | 3.55 | 192.29 | 62.6 | 1,556 | 194.5 | 5.24 |
|  | 140 | .... | .... | 1.437 | 13.126 | 135.3 | 65.7 | 4.19 | 3.44 | 223.50 | 58.6 | 1,760 | 220.0 | 5.17 |
|  | 160 | .... | .... | 1.593 | 12.814 | 129.0 | 72.1 | 4.19 | 3.35 | 245.11 | 55.9 | 1,894 | 236.7 | 5.12 |
| 18 18.000 | 10 | .... | .... | 0.250 | 17.500 | 240.5 | 13.94 | 4.71 | 4.58 | 47.39 | 104.3 | 549 | 61.0 | 6.28 |
|  | 20 | .... | .... | 0.312 | 17.376 | 237.1 | 17.34 | 4.71 | 4.55 | 59.03 | 102.8 | 678 | 75.5 | 6.25 |
|  | .... | Std | .... | 0.375 | 17.250 | 233.7 | 20.76 | 4.71 | 4.52 | 70.59 | 101.2 | 807 | 89.6 | 6.23 |
|  | 30 | .... | .... | 0.437 | 17.126 | 230.4 | 24.11 | 4.71 | 4.48 | 82.06 | 99.9 | 931 | 103.4 | 6.21 |
|  | .... | XS | .... | 0.500 | 17.000 | 227.0 | 27.49 | 4.71 | 4.45 | 93.45 | 98.4 | 1,053 | 117.0 | 6.19 |
|  | 40 | .... | .... | 0.562 | 16.876 | 223.7 | 30.8 | 4.71 | 4.42 | 104.75 | 97.0 | 1,172 | 130.2 | 6.17 |
|  | .... | .... | .... | 0.625 | 16.750 | 220.5 | 34.1 | 4.71 | 4.39 | 115.98 | 95.5 | 1,289 | 143.3 | 6.15 |
|  | .... | .... | .... | 0.687 | 16.626 | 217.1 | 37.4 | 4.71 | 4.35 | 127.03 | 94.1 | 1,403 | 156.3 | 6.13 |
|  | 60 | .... | .... | 0.750 | 16.500 | 213.8 | 40.6 | 4.71 | 4.32 | 138.17 | 92.7 | 1,515 | 168.3 | 6.10 |
|  | .... | .... | .... | 0.875 | 16.250 | 207.4 | 47.1 | 4.71 | 4.25 | 160.04 | 89.9 | 1,731 | 192.8 | 6.06 |
|  | 80 | .... | .... | 0.937 | 16.126 | 204.2 | 50.2 | 4.71 | 4.22 | 170.75 | 88.5 | 1,834 | 203.8 | 6.04 |
|  | 100 | .... | .... | 1.156 | 15.688 | 193.3 | 61.2 | 4.71 | 4.11 | 207.96 | 83.7 | 2,180 | 242.2 | 5.97 |
|  | 120 | .... | .... | 1.375 | 15.250 | 182.6 | 71.8 | 4.71 | 3.99 | 244.14 | 79.2 | 2,499 | 277.6 | 5.90 |
|  | 140 | .... | .... | 1.562 | 14.876 | 173.8 | 80.7 | 4.71 | 3.89 | 274.23 | 75.3 | 2,750 | 306 | 5.84 |
|  | 160 | .... | .... | 1.781 | 14.438 | 163.7 | 90.7 | 4.71 | 3.78 | 308.51 | 71.0 | 3,020 | 336 | 5.77 |
| 20 20.000 | 10 | .... | .... | 0.250 | 19.500 | 298.6 | 15.51 | 5.24 | 5.11 | 52.73 | 129.5 | 757 | 75.7 | 6.98 |
|  | 20 | .... | .... | 0.312 | 19.376 | 294.9 | 19.30 | 5.24 | 5.07 | 65.40 | 128.1 | 935 | 93.5 | 6.96 |
|  | .... | Std | .... | 0.375 | 19.250 | 291.0 | 23.12 | 5.24 | 5.04 | 78.60 | 126.0 | 1,114 | 111.4 | 6.94 |
|  | 30 | .... | .... | 0.437 | 19.126 | 287.3 | 26.86 | 5.24 | 5.01 | 91.31 | 124.6 | 1,286 | 128.6 | 6.92 |
|  | .... | XS | .... | 0.500 | 19.000 | 283.5 | 30.6 | 5.24 | 4.97 | 104.13 | 122.8 | 1,457 | 145.7 | 6.90 |
|  | 40 | .... | .... | 0.562 | 18.876 | 279.8 | 34.3 | 5.24 | 4.94 | 116.67 | 121.3 | 1,624 | 162.4 | 6.88 |
|  | .... | .... | .... | 0.593 | 18.814 | 278.0 | 36.2 | 5.24 | 4.93 | 122.91 | 120.4 | 1,704 | 170.4 | 6.86 |
|  | .... | .... | .... | 0.625 | 18.750 | 276.1 | 38.0 | 5.24 | 4.91 | 129.33 | 119.7 | 1,787 | 178.7 | 6.85 |
|  | .... | .... | .... | 0.687 | 18.626 | 272.5 | 41.7 | 5.24 | 4.88 | 141.71 | 118.1 | 1,946 | 194.6 | 6.83 |
|  | 60 | .... | .... | 0.750 | 18.500 | 268.8 | 45.4 | 5.24 | 4.84 | 154.20 | 116.5 | 2,105 | 210.5 | 6.81 |
|  | .... | .... | .... | 0.812 | 18.376 | 265.2 | 48.9 | 5.24 | 4.81 | 166.40 | 115.0 | 2,257 | 225.7 | 6.79 |
|  | 80 | .... | .... | 0.875 | 18.250 | 261.6 | 52.6 | 5.24 | 4.78 | 178.73 | 113.4 | 2,409 | 240.9 | 6.77 |
|  | .... | .... | .... | 1.031 | 17.938 | 252.7 | 61.4 | 5.24 | 4.70 | 208.87 | 109.4 | 2,772 | 277.2 | 6.72 |
|  | 100 | .... | .... | 1.281 | 17.438 | 238.8 | 75.3 | 5.24 | 4.57 | 256.10 | 103.4 | 3,320 | 332 | 6.63 |

| Nominal size | O.D. | Schedule No.[a] | Wall designation[b] | Wall thickness, $t$, in. | I.D., $d$, in. | Inside area, sq in. | Area of metal, sq in. | Sq ft outside surface, per ft | Sq ft inside surface, per ft | Weight of pipe per ft, lb | Weight of water per ft, lb | Moment of inertia, in.⁴ | Section modulus, in.³ | Radius of gyration, in. |
|---|---|---|---|---|---|---|---|---|---|---|---|---|---|---|
| 20 | 20.000 | 120 | ... | 1.500 | 17.000 | 227.0 | 87.2 | 5.24 | 4.45 | 296.37 | 98.3 | 3,760 | 376 | 6.56 |
|  |  | 140 | ... | 1.750 | 16.500 | 213.8 | 100.3 | 5.24 | 4.32 | 341.10 | 92.6 | 4,220 | 422 | 6.48 |
|  |  | 160 | ... | 1.968 | 16.064 | 202.7 | 111.5 | 5.24 | 4.21 | 379.01 | 87.9 | 4,590 | 459 | 6.41 |
| 24 | 24.000 | 10 | ... | 0.250 | 23.500 | 434 | 18.65 | 6.28 | 6.15 | 63.41 | 188.0 | 1,316 | 109.6 | 8.40 |
|  |  | ... | ... | 0.312 | 23.376 | 430 | 23.20 | 6.28 | 6.12 | 78.93 | 186.1 | 1,629 | 135.8 | 8.38 |
|  |  | 20 | Std | 0.375 | 23.250 | 425 | 27.83 | 6.28 | 6.09 | 94.62 | 183.8 | 1,943 | 161.9 | 8.35 |
|  |  | ... | ... | 0.437 | 23.126 | 420 | 32.4 | 6.28 | 6.05 | 109.97 | 182.1 | 2,246 | 187.4 | 8.33 |
|  |  | ... | XS | 0.500 | 23.000 | 415 | 36.9 | 6.28 | 6.02 | 125.49 | 180.1 | 2,550 | 212.5 | 8.31 |
|  |  | 30 | ... | 0.562 | 22.876 | 411 | 41.4 | 6.28 | 5.99 | 140.80 | 178.1 | 2,840 | 237.0 | 8.29 |
|  |  | ... | ... | 0.625 | 22.750 | 406 | 45.9 | 6.28 | 5.96 | 156.03 | 176.2 | 3,140 | 261.4 | 8.27 |
|  |  | 40 | ... | 0.687 | 22.626 | 402 | 50.3 | 6.28 | 5.92 | 171.17 | 174.3 | 3,420 | 285.2 | 8.25 |
|  |  | ... | ... | 0.750 | 22.500 | 398 | 54.8 | 6.28 | 5.89 | 186.24 | 172.4 | 3,710 | 309 | 8.22 |
|  |  | 60 | ... | 0.968 | 22.064 | 382 | 70.0 | 6.28 | 5.78 | 238.11 | 165.8 | 4,650 | 388 | 8.15 |
|  |  | 80 | ... | 1.218 | 21.564 | 365 | 87.2 | 6.28 | 5.65 | 296.36 | 158.3 | 5,670 | 473 | 8.07 |
|  |  | 100 | ... | 1.531 | 20.933 | 344 | 108.1 | 6.28 | 5.48 | 367.40 | 149.3 | 6,850 | 571 | 7.96 |
|  |  | 120 | ... | 1.812 | 20.376 | 326 | 126.3 | 6.28 | 5.33 | 429.39 | 141.4 | 7,830 | 652 | 7.87 |
|  |  | 140 | ... | 2.062 | 19.876 | 310 | 142.1 | 6.28 | 5.20 | 483.13 | 134.5 | 8,630 | 719 | 7.79 |
|  |  | 160 | ... | 2.343 | 19.314 | 293 | 159.4 | 6.28 | 5.06 | 541.94 | 127.0 | 9,460 | 788 | 7.70 |
| 30 | 30.000 | 10 | ... | 0.312 | 29.376 | 678 | 29.1 | 7.85 | 7.69 | 98.93 | 293.8 | 3,210 | 214 | 10.50 |
|  |  | 20 | ... | 0.500 | 29.000 | 661 | 46.3 | 7.85 | 7.59 | 157.53 | 286.3 | 5,040 | 336 | 10.43 |
|  |  | 30 | ... | 0.625 | 28.750 | 649 | 57.6 | 7.85 | 7.53 | 196.08 | 281.5 | 6,220 | 415 | 10.39 |

\* The ferritic stainless steels may be about 5 percent less, and the austenitic stainless steels about 2 percent greater than the values shown in this table, which are based on weights for carbon steel. The following formulas were used in the computation of the values shown in the table:

Weight of pipe per ft, lb $= 10.6802t(D - t)$

Weight of water per ft, lb $= 0.3405d^2$

Sq ft outside surface, per ft $= 0.2618D$

Sq ft inside surface, per ft $= 0.2618d$

Inside area, sq in. $= 0.785d^2$

Area of metal, sq in. $= 0.785(D^2 - d^2)$

Moment of inertia, in.⁴ $= 0.0491(D^4 - d^4)$
$= A_M R_g^2$

Section modulus, in.³ $= \dfrac{0.0982(D^4 - d^4)}{D}$

Radius of gyration, in. $= 0.25\sqrt{D^2 + d^2}$

$A_M$ = Area of metal, sq in.
$d$ = I.D., in.
$D$ = O.D., in.
$R_g$ = Radius of gyration, in.
$t$ = Pipe wall thickness, in.

†a. ASA B36.10 steel-pipe schedule numbers.
b. ASA B36.10 steel-pipe nominal wall-thickness designations.
c. ASA B36.19 stainless-steel-pipe schedule numbers (5S is not an approved standard).

**Table 4. Values of** $y$
(Interpolate for intermediate values) (ANSI B31.1.0-1967)

|  | Temp, °F (°C) | | | | | |
| --- | --- | --- | --- | --- | --- | --- |
|  | 900 (482) and below | 950 (510) | 1000 (538) | 1050 (566) | 1100 (593) | 1150 (621) and above |
| Ferritic steels | 0.4 | 0.5 | 0.7 | 0.7 | 0.7 | 0.7 |
| Austenitic steels | 0.4 | 0.4 | 0.4 | 0.4 | 0.5 | 0.7 |

**Table 5. Values of** $A$
(ANSI B31.1.0-1967)

| Type of pipe | Value of $A$, in |
| --- | --- |
| Cast-iron pipe, centrifugally cast ........................... | 0.14 |
| Cast-iron, pit-cast ........................................ | 0.18 |
| Threaded-steel, wrought-iron, or nonferrous pipe: |  |
| ⅜ in and smaller ...................................... | 0.05 |
| ½ in and larger ....................................... | Depth of thread |
| Grooved-steel, wrought-iron or nonferrous pipe ............. | Depth of groove |
| Plain-end steel, wrought-iron or tube: ...................... |  |
| 1 in and smaller ..................................... | 0.05 |
| 1¼ in and larger .................................... | 0.065 |
| Plain-end nonferrous pipe or tube ........................ | 0.000 |

**Table 6. Specifications for Tensile Strength of Pipe**[a]

| ANSI designation | ASTM designation | Style of pipe | Tensile strength, min, lb/in² | Scope |
| --- | --- | --- | --- | --- |
| B36.1 | A53 | Welded and seamless steel | Furnace welded: 45,000 Electric-resistance-welded or seamless: Grade A, 48,000; Grade B, 60,000 | [b] |
| B36.2 | A72 | Welded wrought iron | 40,000 |  |
| B36.3 | A106 | Lap-welded and seamless steel for high temperatures | Seamless: Grade A, 48,000; Grade B, 60,000 | [c] [d] |
| B36.4 | A134 | Electric-fusion-welded steel; sizes 16 in and over | See ASTM Standard A245, A283, or A285 | [e] |
| B36.5 | A135 | Electric-resistance-welded steel | Grade A, 48,000. Grade B, 60,000 |  |
| B36.9 | A139 | Electric-fusion-welded steel, sizes 4 in and over | Grade A, 48,000. Grade B, 60,000 | [f] [g] |
| B36.11 | A155 | Electric-fusion-welded steel pipe for high-temperature high-pressure service | Varies with material. See ASTM Standard A155 | [h] |
| G8.7 | A120 | Black and hot-dipped galvanized welded and seamless steel | Same as ANSI B36.1 and ASTM A53 | [i] |

Multiply lb/in² values listed by 36.895 to obtain values in N/m².

[a]Asbstracted from indicated standards of the American Society for testing and Materials, 1973.

[b]Commercial steel pipe for general uses, also for coiling, bending, flanging, and similar forming operations when so specified.

[c]Commercial wrought-iron pipe for general uses, also for coiling, bending, flanging, and other special purposes.

[d]Lap-welded and seamless steel pipe for high-temperature service. Suitable for bending, flanging, and similar forming operations.

[e]Cover pipe 16 in (40.64 cm) diam and over in wall thicknesses up to ¾ in (1.905 cm), fabricated from steel plates by electric-fusion welding.

[f]Pipe up to 30 in (76.2 cm) intended for conveying liquids, gas, or vapor at temperatures below 450°F (232°C). Adapted for flanging, bending, and similar forming operations in Grade A class.

[g]Covers sizes 4 to <16 in (10.16 to <40.64 cm) in wall thicknesses not over ⅝ in (1.59 cm), fabricated from steel plates by electric-fusion welding. Intended for conveying liquids, gas, or vapor at temperatures below 450°F (232°C).

[h]Electric-fusion-welded steel pipe having an outside diameter of 18 in (45.72 cm) and over for high-temperature and high-pressure service. Suitable for bending, flanging, corrugating, and similar forming operations. Welding in accordance with Par. U-68 of the ASME code for unfired pressure vessels.

[i]Commercial steel pipe for ordinary uses such as low-pressure steam, liquid, or gas lines. Not intended for coiling or close bending, or for high-temperature service.

**Table 7. Dimensions of Welded and Seamless Steel Pipe***

| Nominal pipe size, in | Outside diam, in | Nominal wall thickness, in, for stated schedule numbers | | | | | | | | | |
|---|---|---|---|---|---|---|---|---|---|---|---|
| | | 10 | 20 | 30 | 40 | 60 | 80 | 100 | 120 | 140 | 160 |
| ⅛ | 0.405 | | | | **0.068** | | **0.095** | | | | |
| ¼ | 0.540 | | | | **0.088** | | **0.119** | | | | |
| ⅜ | 0.675 | | | | **0.091** | | **0.126** | | | | |
| ½ | 0.840 | | | | **0.109** | | **0.147** | | | | |
| ¾ | 1.050 | | | | **0.113** | | **0.154** | | | | |
| 1 | 1.315 | | | | **0.133** | | **0.179** | | | | |
| 1¼ | 1.660 | | | | **0.140** | | **0.191** | | | | |
| 1½ | 1.900 | | | | **0.145** | | **0.200** | | | | |
| 2 | 2.375 | | | | **0.154** | | **0.218** | | | | 0.344 |
| 2½ | 2.875 | | | | **0.203** | | **0.276** | | | | 0.375 |
| 3 | 3.5 | | | | **0.216** | | **0.300** | | | | 0.438 |
| 3½ | 4.0 | | | | **0.226** | | **0.318** | | | | |
| 4 | 4.5 | | | | **0.237** | | **0.337** | | 0.438 | | 0.531 |
| 5 | 5.563 | | | | **0.258** | | **0.375** | | 0.500 | | 0.625 |
| 6 | 6.625 | | | | **0.280** | | **0.432** | | 0.562 | | 0.718 |
| 8 | 8.625 | | 0.250 | **0.277** | **0.322** | 0.406 | **0.500** | 0.593 | 0.718 | 0.812 | 0.906 |
| 10 | 10.75 | | 0.250 | **0.307** | 0.365 | 0.500 | 0.593 | 0.718 | 0.843 | 1.000 | 1.125 |
| 12 | 12.75 | | 0.250 | **0.330** | 0.406 | 0.562 | 0.687 | 0.843 | 1.000 | 1.125 | 1.312 |
| 14 OD | 14.0 | 0.250 | 0.312 | 0.375 | 0.437 | 0.593 | 0.750 | 0.937 | 1.062 | 1.250 | 1.406 |
| 16 OD | 16.0 | 0.250 | 0.312 | 0.375 | 0.500 | 0.656 | 0.843 | 1.031 | 1.218 | 1.437 | 1.594 |
| 18 OD | 18.0 | 0.250 | 0.312 | 0.437 | 0.562 | 0.718 | 0.937 | 1.156 | 1.343 | 1.562 | 1.781 |
| 20 OD | 20.0 | 0.250 | 0.375 | 0.500 | 0.593 | 0.812 | 1.031 | 1.250 | 1.500 | 1.750 | 1.969 |
| 24 OD | 24.0 | 0.250 | 0.375 | 0.562 | 0.687 | 0.937 | 1.218 | 1.500 | 1.750 | 2.062 | 2.344 |

The schedule numbers are approximate values of $1,000P/SE$ (see text for the symbols). Thicknesses include a mill tolerance of 12.5 percent. Thicknesses in bold type in schedules 30 and 40 agree with those of standard-weight pipe, those in schedules 60 and 80 with extra-strong pipe. Multiply values shown in inches by 2.54 to obtain corresponding values in centimeters.

*Abstracted from Table A2 of ASTM Standard A53-72a.

**Table 8. Dimensions of Welded Wrought-Iron Pipe***

| Nominal pipe size, in | Outside diam, in | Nominal wall thickness, in | |
|---|---|---|---|
| | | Standard weight pipe | Extra-strong pipe |
| ¼ | 0.540 | **0.090** | **0.122** |
| ⅜ | 0.675 | **0.093** | **0.129** |
| ½ | 0.840 | **0.111** | **0.151** |
| ¾ | 1.050 | **0.115** | **0.157** |
| 1 | 1.315 | **0.136** | **0.183** |
| 1¼ | 1.660 | **0.143** | **0.195** |
| 1½ | 1.900 | **0.148** | **0.204** |
| 2 | 2.375 | **0.158** | **0.223** |
| 2½ | 2.875 | **0.208** | **0.282** |
| 3 | 3.5 | **0.221** | **0.306** |
| 3½ | 4.0 | **0.231** | **0.325** |
| 4 | 4.5 | **0.242** | **0.344** |
| 5 | 5.563 | **0.263** | **0.383** |
| 6 | 6.625 | **0.286** | **0.441** |
| 8 | 8.625 | **0.329** | **0.510** |
| 10 | 10.75 | **0.372** | 0.510 |
| 12 | 12.75 | 0.383 | 0.510 |

Wrought-iron pipe contains about 3 percent of slag, 0.5 percent C, and other impurities. It is more resistant to corrosion than is steel pipe.

*Abstracted from earlier editions of ASTM Standard A72. The welded wrought-iron pipe covered by that standard is no longer manufactured, and ASTM specifications were withdrawn in 1971.

pressures, and is particularly adapted to underground and submerged service because of its comparatively high corrosion-resistance qualities. It is more durable than bare wrought-iron and steel pipe; however, steel pipe, when properly coated and wrapped, has been found to be resistant to corrosion when placed in certain soils. For any particular installation, relative cost figures for cast iron vs. coated-and-wrapped steel must be obtained to determine the most economical material. The tensile strength of commercial cast-iron pipe is uncertain, and, because of its low elasticity, it is not suitable for lines subject to the strains of expansion, contraction, and vibration. This pipe may be obtained in various thicknesses and weights with (1) flanges cast on, (2) ends threaded for screwed-on flanges, (3) ends prepared for mechanical joint, (4) ends grooved or shouldered for patented coupling, (5) one end bell, other end spigot, and (6) one end

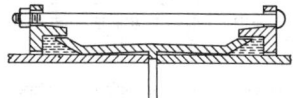

**Fig. 1** Sleeve-type, plain-end coupling.

hub, other end spigot. **Bell-and-spigot** ends are most popular for underground work; hub-and-spigot ends are most frequently used for sewage systems in enclosed spaces. Spigot-end joints are prepared by tightly tamping in hemp or jute at the bottom of the recess with a yarning iron and then pouring in molten lead; the lead, when cooled, is calked in tightly with a calking iron and makes a gastight joint. For exposed piping, flanged ends are used, the joints being made up with gaskets. Flanged pipe has superior strength and tightness of the joint and is used where pipelines can be well supported. The bell-and-spigot joint possesses greater flexibility and provides for expansion and contraction. It is therefore suitable for water

**Table 9. API Standard Line Pipe, Threaded**
(All weights and dimensions are nominal. United States Steel Corporation)

| Size, in. | Weight per ft, lb: Threads and coupling | Weight per ft, lb: Plain end | Thickness, in. | Diameter, in.: O.D. | Diameter, in.: I.D. | Threads per in. | Couplings: Length, in. | Couplings: O.D., in. | Couplings: Weight, lb | Test pressure,* psi: Lap-welded or Grade A seamless | Test pressure,* psi: Grade B seamless | Test pressure,* psi: Grade C seamless |
|---|---|---|---|---|---|---|---|---|---|---|---|---|
| 1/8 | 0.25 | 0.24 | 0.068 | 0.405 | 0.269 | 27 | 1 1/16 | 0.563 | 0.04 | 700 | 700 | 700 |
| 1/4 | 0.43 | 0.42 | 0.088 | 0.540 | 0.364 | 18 | 1 5/8 | 0.719 | 0.09 | 700 | 700 | 700 |
| 3/8 | 0.57 | 0.57 | 0.091 | 0.675 | 0.493 | 18 | 1 5/8 | 0.875 | 0.13 | 700 | 700 | 700 |
| 1/2 | 0.86 | 0.85 | 0.109 | 0.840 | 0.622 | 14 | 2 1/8 | 1.063 | 0.24 | 700 | 700 | 700 |
| 3/4 | 1.14 | 1.13 | 0.113 | 1.050 | 0.824 | 14 | 2 1/8 | 1.313 | 0.34 | 700 | 700 | 700 |
| 1 | 1.70 | 1.68 | 0.133 | 1.315 | 1.049 | 11 1/2 | 2 5/8 | 1.576 | 0.54 | 700 | 700 | 700 |
| 1 1/4 | 2.30 | 2.27 | 0.140 | 1.660 | 1.380 | 11 1/2 | 2 3/4 | 2.054 | 1.03 | 1,000 | 1,100 | 1,300 |
| 1 1/2 | 2.75 | 2.72 | 0.145 | 1.900 | 1.610 | 11 1/2 | 2 3/4 | 2.200 | 1.17 | 1,000 | 1,100 | 1,300 |
| 2 | 3.75 | 3.65 | 0.154 | 2.375 | 2.067 | 11 1/2 | 3 1/4 | 2.875 | 2.13 | 1,000 | 1,100 | 1,300 |
| 2 1/2 | 5.90 | 5.79 | 0.203 | 2.875 | 2.469 | 8 | 4 1/8 | 3.375 | 3.27 | 1,000 | 1,100 | 1,300 |
| 3 | 7.70 | 7.58 | 0.216 | 3.500 | 3.068 | 8 | 4 1/4 | 4.000 | 4.09 | 1,000 | 1,100 | 1,300 |
| 3 1/2 | 9.25 | 9.11 | 0.226 | 4.000 | 3.548 | 8 | 4 3/8 | 4.625 | 5.92 | 1,200 | 1,300 | 1,600 |
| 4 | 11.00 | 10.79 | 0.237 | 4.500 | 4.026 | 8 | 4 1/2 | 5.200 | 7.59 | 1,200 | 1,300 | 1,600 |
| 5 | 15.00 | 14.62 | 0.258 | 5.563 | 5.047 | 8 | 4 5/8 | 6.296 | 9.98 | 1,200 | 1,300 | 1,600 |
| 6 | 19.45 | 18.97 | 0.280 | 6.625 | 6.065 | 8 | 4 7/8 | 7.390 | 12.92 | 1,200 | 1,300 | 1,600 |
| 8 | 25.55 | 24.70 | 0.277 | 8.625 | 8.071 | 8 | 5 1/4 | 9.625 | 23.18 | 1,200 | 1,300 | 1,600 |
| 8 | 29.35 | 28.55 | 0.322 | 8.625 | 7.981 | 8 | 5 1/4 | 9.625 | 23.18 | 1,200 | 1,300 | 1,600 |
| 10 | 32.75 | 31.20 | 0.279 | 10.750 | 10.192 | 8 | 5 3/4 | 11.750 | 31.55 | 1,000 | 1,200 | 1,400 |
| 10 | 35.75 | 34.24 | 0.307 | 10.750 | 10.136 | 8 | 5 3/4 | 11.750 | 31.55 | 1,000 | 1,200 | 1,400 |
| 10 | 41.85 | 40.48 | 0.365 | 10.750 | 10.020 | 8 | 5 3/4 | 11.750 | 31.55 | 1,000 | 1,200 | 1,400 |
| 12 | 45.45 | 43.77 | 0.330 | 12.750 | 12.090 | 8 | 6 1/8 | 14.000 | 49.27 | 1,000 | 1,200 | 1,400 |
| 12 | 51.15 | 49.56 | 0.375 | 12.750 | 12.000 | 8 | 6 1/8 | 14.000 | 49.27 | 1,000 | 1,200 | 1,400 |
| 14 O.D. | 57.00 | 54.57 | 0.375 | 14.000 | 13.250 | 8 | 6 3/8 | 15.000 | 45.83 | 950 | 1,100 | 1,400 |
| 15 O.D. | 61.15 | 58.57 | 0.375 | 15.000 | 14.250 | 8 | 6 5/8 | 16.000 | 51.26 | 900 | 1,000 | 1,400 |
| 16 O.D. | 65.30 | 62.58 | 0.375 | 16.000 | 15.250 | 8 | 6 3/4 | 17.000 | 55.83 | 850 | 1,000 | 1,300 |
| 17 O.D. | 73.20 | 69.70 | 0.393 | 17.000 | 16.214 | 8 | 7 | 18.000 | 61.67 | 850 | 950 | 1,200 |
| 18 O.D. | 81.20 | 76.84 | 0.409 | 18.000 | 17.182 | 8 | 7 1/8 | 19.000 | 66.53 | 800 | 950 | 1,200 |
| 20 O.D. | 90.00 | 85.58 | 0.409 | 20.000 | 19.182 | 8 | 7 5/8 | 21.000 | 79.37 | 750 | 850 | 1,100 |

The permissible variation in weight for any length of pipe is 10 percent above and 3½ percent below; but the carload weight should not be more than 1¾ percent under the calculated weight.
Furnished with threads and couplings and in random lengths, unless otherwise ordered.
The weight per foot of pipe with threads and couplings is based on a length of 20 ft, including the coupling.
*Test pressure butt-welded pipe ⅛ to 1 in = 700 lb/in$^2$; 1¼ to 3 in = 800 lb/in$^2$.

**Table 10. Center-to-Center Dimensions of Pipe Coils**

| Size of pipe, in | Recommended and advisable minimum, in: Schedule 40 | Recommended and advisable minimum, in: Schedule 80 |
|---|---|---|
| 3/4 | 3 1/2 | 2 1/2 |
| 1 | 4 | 3 |
| 1 1/4 | 5 | 4 |
| 1 1/2 | 6 | 5 |
| 2 | 8 | 6 |

pipe and is largely used for that purpose. Figure 2 shows a typical form of this joint for ordinary pressures. Figure 3 shows one form of **mechanical joint** suitable for water, gas, or oil. Other forms of joint, plain-end pipe with couplings, and threaded pipe also are manufactured. Cast-iron pipe, fittings, and valves have been found unsuitable for superheated steam service. The Code for Pressure Piping, ANSI B31.1.0-1967, states that cast iron may be used for steam service not over 250 lb/in$^2$ or 406°F (9,224 N/m$^2$ or 208°C) provided that it meets the requirements as dictated by Eq. (1).

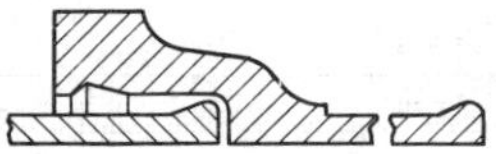

**Fig. 2** Standard bell-and-spigot joint.

Wall thicknesses for the various conditions which cast-iron pipe is designed to meet are determined in accordance with the requirements of ANSI A21.1-1967, Thickness Design of Cast-Iron..

Cast-iron pipe is made by **centrifugal casting** in which molten iron is admitted to the interior of a sand-lined or metal-lined cast-iron flask, the mold being rotated at high speeds so that

**Table 11. Approx Weight of Round Seamless Cold-finished Carbon-Steel Mechanical Tubing, Lb/Ft***

(Carbon, 0.25% max. Standard sizes for warehouse stocks random lengths. United States Steel Corporation)

| Wall thickness | | O.D., in. | | | | | | | | | | | | | | | | | |
| In. | G or in. | ⅜ | ½ | ⅝ | ¾ | ⅞ | 1 | 1⅛ | 1¼ | 1⅜ | 1½ | 1⅝ | 1¾ | 1⅞ | 2 | 2⅛ | 2¼ | 2⅜ | 2½ |
|---|---|---|---|---|---|---|---|---|---|---|---|---|---|---|---|---|---|---|---|
| 0.035 | 20 G | 0.127 | 0.174 | 0.221 | 0.267 | 0.314 | 0.361 | 0.407 | 0.454 | ..... | 0.548 | ..... | ..... | ..... | ..... | | | | |
| 0.049 | 18 G | 0.171 | 0.236 | 0.301 | 0.367 | 0.432 | 0.498 | 0.563 | 0.629 | 0.694 | 0.759 | 0.825 | 0.890 | ..... | 1.02 | | | | |
| 0.058 | 17 G | ..... | 0.274 | 0.351 | 0.429 | ..... | 0.584 | | | | | | | | | | | | |
| 0.065 | 16 G | 0.215 | 0.302 | 0.389 | 0.476 | 0.562 | 0.649 | 0.736 | 0.823 | 0.909 | 0.996 | 1.08 | 1.17 | 1.26 | 1.34 | 1.43 | 1.52 | 1.60 | 1.69 |
| 0.083 | 14 G | | 0.370 | 0.480 | 0.591 | 0.702 | 0.813 | 0.924 | 1.03 | 1.15 | 1.26 | ..... | ..... | ..... | 1.70 | | | | |
| 0.095 | 13 G | | 0.411 | 0.538 | 0.665 | 0.791 | 0.918 | 1.05 | 1.17 | 1.30 | 1.43 | 1.55 | 1.68 | 1.81 | 1.93 | 2.06 | 2.19 | 2.31 | 2.44 |
| 0.109 | 12 G | | 0.455 | 0.601 | 0.746 | 0.892 | 1.04 | 1.18 | 1.33 | ..... | 1.62 | ..... | ..... | ..... | ..... | | | | |
| 0.120 | 11 G | | 0.487 | 0.647 | 0.807 | 0.968 | 1.13 | 1.29 | 1.45 | 1.61 | 1.77 | 1.93 | 2.09 | 2.25 | 2.41 | 2.57 | 2.73 | 2.89 | 3.05 |
| 0.134 | 10 G | | | 0.703 | 0.882 | ..... | 1.24 | 1.42 | 1.60 | 1.78 | 1.96 | 2.13 | 2.31 | ..... | 2.67 | ..... | ..... | ..... | ..... |
| 0.156 | 5/32 | | | 0.781 | 0.990 | 1.20 | 1.41 | 1.61 | 1.82 | 2.03 | 2.24 | 2.45 | 2.66 | 2.86 | 3.07 | 3.28 | 3.49 | 3.70 | 3.91 |
| 0.188 | 3/16 | | | ..... | 1.13 | 1.38 | 1.63 | 1.88 | 2.13 | 2.38 | 2.63 | 2.89 | 3.14 | 3.39 | 3.64 | 3.89 | 4.14 | 4.39 | 4.64 |
| 0.219 | 7/32 | | | ..... | ..... | 1.53 | 1.83 | 2.12 | 2.41 | 2.70 | 3.00 | 3.29 | 3.58 | 3.87 | 4.17 | 4.46 | 4.75 | 5.04 | 5.34 |
| 0.250 | ¼ | | | | 1.34 | 1.67 | 2.00 | 2.34 | 2.67 | 3.00 | 3.34 | 3.67 | 4.01 | 4.34 | 4.67 | 5.01 | 5.34 | 5.67 | 6.01 |
| 0.281 | 9/32 | | | | ..... | ..... | ..... | ..... | 2.91 | ..... | 3.66 | 4.03 | 4.41 | 4.78 | 5.16 | ..... | 5.91 | 6.28 | 6.66 |
| 0.313 | 5/16 | | | | ..... | ..... | ..... | ..... | 3.13 | 3.55 | 3.97 | 4.39 | 4.80 | 5.22 | 5.64 | 6.06 | 6.48 | 6.89 | 7.31 |
| 0.375 | ⅜ | | | | ..... | ..... | ..... | ..... | 3.50 | 4.01 | 4.51 | 5.01 | 5.51 | 6.01 | 6.51 | 7.01 | 7.51 | 8.01 | 8.51 |
| 0.438 | 7/16 | | | | ..... | ..... | ..... | ..... | ..... | ..... | 4.97 | 5.53 | 6.14 | 6.72 | 7.31 | 7.89 | 8.48 | 9.06 | 9.65 |
| 0.500 | ½ | | | | ..... | ..... | ..... | ..... | ..... | ..... | 5.34 | 6.01 | 6.68 | 7.34 | 8.01 | 8.68 | 9.35 | 10.0 | 10.7 |
| 0.625 | ⅝ | | | | ..... | ..... | ..... | ..... | ..... | ..... | ..... | ..... | ..... | ..... | 9.18 | ..... | 10.8 | 11.7 | 12.5 |

*Other standard sizes, in certain standard wall thicknesses, vary by ⅛-in increments for 2½ to 3½ in; by ¼-in increments from 3½ to 7½ in; by ½-in increments from 7½ to 10½ in OD. There are also standard sizes for every 1/16-in from ⅜ to 1⅝ in OD.
To obtain weights in kg/m, multiply tabular values shown by 1.42.

**Table 12. Steel Condenser and Heat-Exchanger Tubes**
(Dimensions and weights. United States Steel Corporation)

| OD, in | Thickness, in | Avg wall | | | Min wall | | |
|---|---|---|---|---|---|---|---|
| | | ID, in | Area of metal, in$^2$* | Weight per ft, lb† | ID, in | Area of metal, in$^2$* | Weight per ft, lb† |
| ½ | 0.035 | 0.430 | 0.0511 | 0.1738 | 0.423 | 0.0558 | 0.1898 |
| | 0.050 | 0.400 | 0.0707 | 0.2403 | 0.390 | 0.0769 | 0.2614 |
| | 0.065 | 0.370 | 0.0888 | 0.3020 | 0.357 | 0.0963 | 0.3272 |
| ⅝ | 0.035 | 0.555 | 0.0649 | 0.2205 | 0.548 | 0.0709 | 0.2412 |
| | 0.050 | 0.525 | 0.0903 | 0.3071 | 0.515 | 0.0985 | 0.3348 |
| | 0.065 | 0.495 | 0.1144 | 0.3888 | 0.482 | 0.1243 | 0.4227 |
| | 0.085 | 0.455 | 0.1442 | 0.4902 | 0.438 | 0.1561 | 0.5308 |
| ¾ | 0.050 | 0.650 | 0.1100 | 0.3738 | 0.640 | 0.1201 | 0.4082 |
| | 0.065 | 0.620 | 0.1399 | 0.4755 | 0.607 | 0.1524 | 0.5181 |
| | 0.085 | 0.580 | 0.1776 | 0.6037 | 0.563 | 0.1928 | 0.6556 |
| | 0.095 | 0.560 | 0.1955 | 0.6646 | 0.541 | 0.2119 | 0.7204 |
| ⅞ | 0.050 | 0.775 | 0.1296 | 0.4406 | 0.765 | 0.1417 | 0.4817 |
| | 0.065 | 0.745 | 0.1654 | 0.5623 | 0.732 | 0.1805 | 0.6136 |
| | 0.085 | 0.705 | 0.2110 | 0.7172 | 0.688 | 0.2296 | 0.7804 |
| | 0.095 | 0.685 | 0.2328 | 0.7914 | 0.666 | 0.2530 | 0.8599 |
| 1 | 0.050 | 0.900 | 0.1492 | 0.5073 | 0.890 | 0.1633 | 0.5551 |
| | 0.065 | 0.870 | 0.1909 | 0.6491 | 0.857 | 0.2086 | 0.7090 |
| | 0.085 | 0.830 | 0.2443 | 0.8306 | 0.813 | 0.2663 | 0.9052 |
| | 0.095 | 0.810 | 0.2701 | 0.9182 | 0.791 | 0.2940 | 0.9994 |
| 1¼ | 0.050 | 1.150 | 0.1885 | 0.6408 | 1.140 | 0.2065 | 0.7020 |
| | 0.065 | 1.120 | 0.2420 | 0.8226 | 1.107 | 0.2647 | 0.8999 |
| | 0.085 | 1.080 | 0.3111 | 1.058 | 1.163 | 0.3397 | 1.155 |
| | 0.095 | 1.060 | 0.3447 | 1.172 | 1.041 | 0.3761 | 1.278 |
| | 0.105 | 1.040 | 0.3777 | 1.284 | 1.019 | 0.4117 | 1.399 |
| 1½ | 0.050 | 1.400 | 0.2278 | 0.7743 | 1.390 | 0.2497 | 0.8488 |
| | 0.065 | 1.370 | 0.2930 | 0.9962 | 1.357 | 0.3209 | 1.091 |
| | 0.085 | 1.330 | 0.3779 | 1.285 | 1.313 | 0.4053 | 1.378 |
| | 0.095 | 1.310 | 0.4193 | 1.426 | 1.291 | 0.4581 | 1.557 |
| | 0.105 | 1.290 | 0.4602 | 1.564 | 1.269 | 0.5024 | 1.708 |
| 1¾ | 0.065 | 1.620 | 0.3441 | 1.170 | 1.606 | 0.3803 | 1.293 |
| | 0.085 | 1.580 | 0.4446 | 1.512 | 1.561 | 0.4907 | 1.668 |
| | 0.095 | 1.560 | 0.4939 | 1.679 | 1.539 | 0.5448 | 1.852 |
| | 0.105 | 1.540 | 0.5426 | 1.845 | 1.517 | 0.5902 | 2.007 |
| | 0.120 | 1.510 | 0.6145 | 2.089 | 1.484 | 0.6766 | 2.300 |
| 2 | 0.065 | 1.870 | 0.3951 | 1.343 | 1.856 | 0.4370 | 1.486 |
| | 0.085 | 1.830 | 0.5114 | 1.738 | 1.811 | 0.5649 | 1.920 |
| | 0.095 | 1.810 | 0.5685 | 1.933 | 1.789 | 0.6275 | 2.133 |
| | 0.105 | 1.790 | 0.6251 | 2.125 | 1.767 | 0.6896 | 2.344 |
| | 0.120 | 1.760 | 0.7087 | 2.409 | 1.734 | 0.7812 | 2.656 |

*Multiply values shown by 0.0645 to obtain areas in cm$^2$.
†Multiply values shown by 1.42 to obtain weights in kg/m.

the molten metal is thrown by centrifugal force against the lining. ANSI specifications have been prepared for the various combinations of fabrication procedure and intended end use.

Table 14 lists thicknesses and weight data for pipe centrifugally cast in sand-lined molds and intended for use with water or other liquids. For pit-cast pipe, or for centrifugally cast pipe with metal molds or intended for gas service, consult the

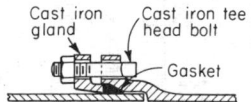

**Fig. 3**  Mechanical joint.

appropriate ANSI specification as listed above.

The employment of cast-iron pipe for gas supply and distribution is second in importance only to its use for carrying water. Bell-and-spigot gas pipe is similar in design to bell-and-spigot water pipe (Fig. 2). For flanged gas pipe, the 25 lb ANSI standard flanges are approved for maximum gas pressures of 25 lb/in$^2$ (922 N/m$^2$). The 125 lb ANSI standard flanges are approved for gas pressures of 125 lb/in$^2$ (4,612 N/m$^2$), up to 4 in nominal pipe size; 100 lb/in$^2$ (3,690 N/m$^2$), 6 to 12 in; and 80 lb/in$^2$ (2,952 N/m$^2$), 16 to 48 in. The type of joint shown in Fig. 2 is also widely used for gas.

**Flexible-Joint Pipe**  The necessity for crossing streams and other waterways and of laying pipelines into them has developed various forms of flexible-joint pipe adapted to laying

**Table 13. Seamless-Steel Boiler Tubes**
(United States Steel Corporation)

| Outside diam, in | Thickness BWG | Thickness In | Mfg.* wt, lb/ ft | Outside diam, in | Thickness BWG | Thickness In | Mfg.* wt, lb/ ft | Outside diam, in | Thickness BWG | Thickness In | Mfg.* wt, lb/ ft |
|---|---|---|---|---|---|---|---|---|---|---|---|
| 1 | 13 | 0.095 | 1.037 | 2½ | 12 | 0.109 | 3.171 | 4½ | 10 | 0.134 | 7.103 |
|  | 12 | 0.109 | 1.168 |  | 11 | 0.120 | 3.457 |  | 9 | 0.148 | 7.817 |
|  | 11 | 0.120 | 1.263 |  | 10 | 0.134 | 3.835 |  | 8 | 0.165 | 8.702 |
|  | 10 | 0.134 | 1.384 |  | 9 | 0.148 | 4.207 |  | 7 | 0.180 | 9.447 |
| 1¼ | 13 | 0.095 | 1.323 | 2¾ | 12 | 0.109 | 3.504 | 5 | 9 | 0.148 | 8.720 |
|  | 12 | 0.109 | 1.502 |  | 11 | 0.120 | 3.823 |  | 8 | 0.165 | 9.711 |
|  | 11 | 0.120 | 1.628 |  | 10 | 0.134 | 4.244 |  | 7 | 0.180 | 10.550 |
|  | 10 | 0.134 | 1.793 |  | 9 | 0.148 | 4.658 |  | 6 | 0.203 | 11.810 |
| 1½ | 13 | 0.095 | 1.619 | 3 | 12 | 0.109 | 3.838 | 5½ | 9 | 0.148 | 9.622 |
|  | 12 | 0.109 | 1.836 |  | 11 | 0.120 | 4.189 |  | 8 | 0.165 | 10.720 |
|  | 11 | 0.120 | 1.994 |  | 10 | 0.134 | 4.652 |  | 7 | 0.180 | 11.650 |
|  | 10 | 0.134 | 2.201 |  | 9 | 0.148 | 5.110 |  | 6 | 0.203 | 13.050 |
| 1¾ | 13 | 0.095 | 1.910 | 3¼ | 11 | 0.120 | 4.555 | 6 | 7 | 0.180 | 12.750 |
|  | 12 | 0.109 | 2.169 |  | 10 | 0.134 | 5.061 |  | 6 | 0.203 | 14.290 |
|  | 11 | 0.120 | 2.360 |  | 9 | 0.148 | 5.061 |  | 5 | 0.220 | 15.410 |
|  | 10 | 0.134 | 2.610 |  | 8 | 0.165 | 6.179 |  | 4 | 0.238 | 16.640 |
| 2 | 13 | 0.095 | 2.201 | 3½ | 11 | 0.120 | 4.921 |  |  |  |  |
|  | 12 | 0.109 | 2.503 |  | 10 | 0.134 | 5.469 |  |  |  |  |
|  | 11 | 0.120 | 2.726 |  | 9 | 0.148 | 6.012 |  |  |  |  |
|  | 10 | 0.034 | 3.018 |  | 8 | 0.065 | 6.683 |  |  |  |  |
| 2¼ | 13 | 0.095 | 2.492 | 4 | 10 | 0.134 | 6.286 |  |  |  |  |
|  | 12 | 0.109 | 2.837 |  | 9 | 0.148 | 6.915 |  |  |  |  |
|  | 11 | 0.120 | 3.092 |  | 8 | 0.165 | 7.693 |  |  |  |  |
|  | 10 | 0.134 | 3.427 |  | 7 | 0.180 | 8.347 |  |  |  |  |

*Multiply values shown by 1.42 to obtain weights in kg/m.

under water, which are capable of motion through several degrees without leakage. Figure 4 shows one style of such joint which has an adjustment of about 15° in standard sizes.

In selecting the thickness of a pipe for a submerged line, the internal-service pressure is seldom the determining factor, as ample allowance should be made to minimize the risk of breakage in laying and to withstand external shocks from floating ice or other objects. The dimensions and weights given in Table 15 are typical of those listed by several manufacturers.

"**Universal**" **pipe** (Fig. 5) is cast-iron pipe with hub-and-spigot ends, the contact surfaces of which are machined on a taper, giving an iron-to-iron joint. By making the tapers of slightly different pitch, the joint provides for flexibility while remaining tight. Two bolts to the joint are sufficient, except for pressures above 175 lb/in² (6,457 N/m²). Universal cast-iron pipe is used largely for carrying gas and water and is suitable for all pressures and services. The pipe is tested with hydrostatic pressure of 300 to 500 lb/in² (11,068 to 18,447 N/m²). All universal pipe and special castings of a given diameter and of any class are interchangeable with those of a different class. Standard laying lengths, 6 ft (1.83 m). Thicknesses and

weights of standard types up to 16 in are given in Table 17. Information on other types and sizes and on fittings may be obtained from Central Foundry Co.

Pressure ratings of 150 and 250 lb/in² correspond, respectively, to 5,534 and 9,224 N/m². Multiply listed values of weights in pounds by 0.4536 to obtain corresponding weights in kilograms.

**Fittings for Cast-Iron Water Pipe** Flanged fittings of the dimensions of the ANSI standard for steam are not often used with cast-iron water pipe. The longer fittings of the AWWA are generally preferred because of low friction loss. The dimensions of the flanged fittings of this class conform very closely to the dimensions of the bell-and-spigot fittings of the AWWA. The flange thicknesses and drillings conform to those of the ANSI Standards. These fittings, both flange, and bell-and-spigot type, are made in a great variety of forms known as "standard special fittings." For dimensions and weights, see manufacturers' catalogs or standard specifications of the AWWA.

**Cast-iron soil pipe** and fittings are of the hub-and-spigot form, similar in design to the cast-iron water pipe shown in Fig. 2.

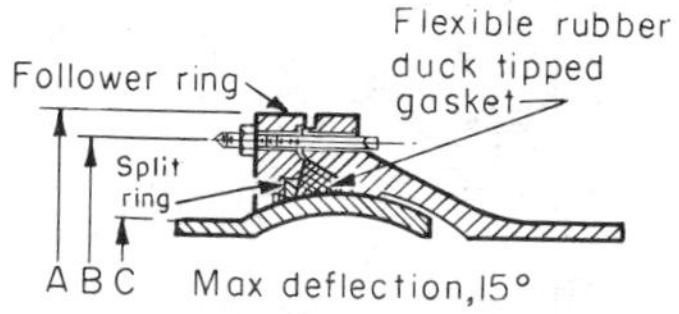

Follower ring
Split ring
Flexible rubber duck tipped gasket
A B C   Max deflection, 15°

**Fig. 4** Flexible joint (see Table 15 for dimensions).

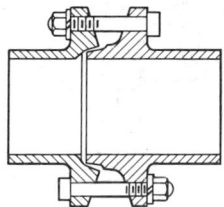

**Fig. 5** "Universal" cast-iron pipe.

**Table 14. Standard Weights and Thicknesses of Cast-Iron Bell-and-Spigot Pipe for Water***

| Nominal size, in | Class 50, 50 lb/in² (1,845 N/m²), 115 ft (35.05 m) head | | | Class 100, 100 lb/in² (3,690 N/m²), 231 ft (70.41 m) head | | | Class 150, 150 lb/in² (5,535 N/m²), 346 ft (105.5 m) head | | | Approx lb lead per joint 2 in thick | Approx lb hemp or jute per joint |
|---|---|---|---|---|---|---|---|---|---|---|---|
| | Thickness, in | OD, in | Wt, lb per avg ft | Thickness, in | OD, in | Wt, lb per avg ft | Thickness, in | OD, in | Wt, lb per avg ft | | |
| 3 | 0.32 | 3.96 | 12.4 | 0.32 | 3.96 | 12.4 | 0.32 | 3.96 | 12.4 | 6.2 | 0.17 |
| 4 | 0.35 | 4.80 | 16.5 | 0.35 | 4.80 | 16.5 | 0.35 | 4.80 | 16.5 | 7.5 | 0.21 |
| 6 | 0.38 | 6.90 | 25.9 | 0.38 | 6.90 | 25.9 | 0.38 | 6.90 | 25.9 | 10.3 | 0.31 |
| 8 | 0.41 | 9.05 | 37.0 | 0.41 | 9.05 | 37.0 | 0.41 | 9.05 | 37.0 | 13.3 | 0.44 |
| 10 | 0.44 | 11.10 | 49.1 | 0.44 | 11.10 | 49.1 | 0.44 | 11.10 | 49.1 | 16.0 | 0.53 |
| 12 | 0.48 | 13.20 | 63.7 | 0.48 | 13.20 | 63.7 | 0.48 | 13.20 | 63.7 | 19.0 | 0.61 |
| 14 | 0.48 | 15.30 | 74.6 | 0.51 | 15.30 | 78.8 | 0.51 | 15.65 | 80.7 | 22.0 | 0.81 |
| 16 | 0.54 | 17.40 | 95.2 | 0.54 | 17.40 | 95.2 | 0.54 | 17.80 | 97.5 | 30.0 | 0.94 |
| 18 | 0.54 | 19.50 | 107.6 | 0.58 | 19.50 | 114.8 | 0.58 | 19.92 | 117.2 | 33.8 | 1.00 |
| 20 | 0.57 | 21.60 | 125.9 | 0.62 | 21.60 | 135.9 | 0.62 | 22.06 | 138.9 | 37.0 | 1.25 |
| 24 | 0.63 | 25.80 | 166.0 | 0.68 | 25.80 | 178.1 | 0.73 | 26.32 | 194.0 | 44.0 | 1.50 |
| 30 | 0.79 | 32.00 | 257.6 | 0.79 | 32.00 | 257.6 | 0.85 | 32.00 | 275.4 | 54.3 | 2.06 |
| 36 | 0.87 | 38.30 | 340.9 | 0.87 | 38.30 | 340.9 | 0.94 | 38.30 | 365.9 | 64.8 | 3.00 |
| 42 | 0.97 | 44.50 | 442.0 | 0.97 | 44.50 | 442.0 | 1.05 | 44.50 | 475.3 | 75.3 | 3.62 |
| 48 | 1.06 | 50.80 | 551.6 | 1.06 | 50.80 | 551.6 | 1.14 | 50.80 | 589.6 | 85.5 | 4.37 |

Pipe weights indicated are approximate and include allowance for bell based on a 16 ft laying length. Calculations are for pipe laid without blocks, on flat-bottom trench, with tamped backfill under 5 ft of cover. For other conditions, see ANSI A21.1. Thicknesses given above include allowance for water hammer and factory tolerance.

To obtain weights in kg/m, multiply value. shown in lb per avg ft by 1.42.

*Condensed from Table 8.2 of Specification AWWA C 108-70.

**Table 15. Dimensions and Weights of Flexible-Joint Pipe***
(Dimensions refer to Fig. 4)

| Nominal diam, in | Class | Dimensions, in | | | Bolts | | | Average metal thickness, in | Weight† of pipe, incl. bell, lb per 12-ft length |
|---|---|---|---|---|---|---|---|---|---|
| | | A | B | C | No. required | Size, in | Length, in | | |
| 4 | B | 12.13 | 9.75 | 5.00 | 8 | 0.75 | 4.50 | 0.45 | 290 |
| 4 | C | 12.13 | 9.75 | 5.00 | 8 | 0.75 | 4.50 | 0.48 | 305 |
| 4 | D | 12.13 | 9.75 | 5.00 | 8 | 0.75 | 4.50 | 0.52 | 325 |
| 6 | B | 14.25 | 11.75 | 7.10 | 12 | 0.75 | 4.50 | 0.48 | 440 |
| 6 | C | 14.25 | 11.75 | 7.10 | 12 | 0.75 | 4.50 | 0.51 | 460 |
| 6 | D | 14.25 | 11.75 | 7.10 | 12 | 0.75 | 4.50 | 0.55 | 490 |
| 8 | B | 17.25 | 14.75 | 9.30 | 12 | 0.75 | 5.25 | 0.51 | 635 |
| 8 | C | 17.25 | 14.75 | 9.30 | 12 | 0.75 | 5.25 | 0.56 | 680 |
| 8 | D | 17.25 | 14.75 | 9.30 | 12 | 0.75 | 5.25 | 0.60 | 720 |
| 10 | B | 20.56 | 18.00 | 11.40 | 16 | 0.75 | 5.25 | 0.57 | 905 |
| 10 | C | 20.56 | 18.00 | 11.40 | 16 | 0.75 | 5.25 | 0.62 | 965 |
| 10 | D | 20.56 | 18.00 | 11.40 | 16 | 0.75 | 5.25 | 0.68 | 1035 |
| 12 | B | 23.75 | 21.00 | 13.50 | 16 | 0.75 | 6.25 | 0.62 | 1200 |
| 12 | C | 23.75 | 21.00 | 13.50 | 16 | 0.75 | 6.25 | 0.68 | 1280 |
| 12 | D | 23.75 | 21.00 | 13.50 | 16 | 0.75 | 6.25 | 0.75 | 1375 |

*United States Pipe and Foundry Co.
†Weights do not include follower rings, bolts, or gaskets. For sizes above 12 in, see manufacturers' catalogs. Multiply weights shown by 38.7 to obtain weight in kg per m or by 0.4536 to obtain weight in kg per 12-ft length.

Tapped openings and pipe plugs are threaded in accordance with the taper pipe thread requirements of ANSI B2.1-1968.

The ANSI standard, Threaded Cast-iron Pipe for Drainage, Vent, and Waste Services," ANSI A40.5-1943, covers two types of pipe having threaded joints in nominal pipe sizes 1¼ to 12 in and in lengths 5 to 27 ft. One type has external threads on both ends; the other type has external threads on one end and an internal threaded drainage hub on the other end.

## PIPES AND TUBES OF NONFERROUS MATERIALS

**Brass tubing** is commercially available in the form known as yellow brass, an alloy consisting of approximately 65 percent copper and 35 percent zinc, and is used principally for ornamental work and handrailings. It has a density of approximately 0.3 lb/in³ (0.0085 kg/cm³) the exact density being dependent upon the specific chemical composition. **Brass piping** is most frequently encountered as red brass, an alloy consisting of approximately 85 percent copper and 15 percent zinc. This alloy, having a density of about 0.32 lb/in³ (0.009 kg/cm³), has been found to be structurally superior to the yellow brasses and is used where the fluid being conveyed has corrosive properties.

**Copper** is available either as pipe or as tubing. In the form of piping, it has the same outer diameter as that of standard steel pipe. As tubing, it is used for a variety of purposes, such as for compressed-air instrumentation lines, hydraulic control lines around machinery, domestic oil-burner and heating systems, and for general plumbing purposes. **Copper tubing** is furnished in 12-ft (3.7-m) and 20-ft (6.1-m) straight lengths or in coils of 100-ft (30.5-m) length. **Type K** tubing, in coils, is used for underground work where the minimum number of joints, combined with greater thickness of type K tubing, is of distinct advantage. **Type L** tubing, usually in straight lengths,

is used as a principal piping material for plumbing systems in homes and buildings; this is largely due to the economy of installation made possible by the use of soldered fittings. Copper deteriorates rapidly under high temperatures and repeated stresses. At a temperature of 360°F (182°C) its strength is reduced 15 percent, and on this account it should never be used for high steam pressures and temperatures.

Commercial sizes of **aluminum tubing** are listed by the manufacturers in even outside diameters and in wall thicknesses conforming to stubs gage. Aluminum pipe is available as listed in Table 21. To obtain the approximate weight per foot of aluminum pipe or tubing, a weight of 0.098 lb/in³ (2.71 g/cm³) may be used.

**Lead pipe** is supplied in straight lengths, in coils, or in reels. Sizes and weights are given in Tables 22 and 23. The data in Table 23 conform to the standards advocated by the Lead Industries Assoc. for water service and plumbing.

**Block tin** is a term used in the metal trade to refer to products made wholly from strictly pure high-grade tin. While tin pipe has many and varied uses, its most important applications are in types of equipment handling liquids intended for human consumption. Tin pipe does not rust, or corrode, and therefore does not contaminate most of the liquids passing through it.

**Plastic pipes** and **tubes** are available in a wide range of diameters and thicknesses, with Table 25 generally representative. The plastic used is resistant to attack by many chemicals, light in weight, flexible, and available in coiled form so that installation time is low. It is used for a variety of purposes including drainage, irrigation, sewage, and for conveying chemical solutions or waters that would attack metal piping. Caution should be used in selection of plastic piping insofar as service temperature is concerned: e.g., polyethylene is suitable for a maximum temperature of 120°F (49°C). Table 26 lists corrosion-resistance data for polyethylene plastic piping.

**Pipes with Special Linings** For use in lines through which

**Table 16. Cast-Iron Flanged Pipe**

(ANSI 125- and 250-lb Standards from ANSI B 16.1-1967)

| Size | Actual OD, in | Diam of flange, in | Diam of raised face, in | Diam of bolt circle, in | No. of bolts | Diam of bolts, in | Thickness of flange, in | Max working pressure, lb/in² | Wall thickness, in | Wt per ft plain end, lb | Wt of one flange, lb | Wt 16-ft length with 2 flanges, lb |
|---|---|---|---|---|---|---|---|---|---|---|---|---|
| 3 | 3.96 | 7.50 | | 6.00 | 4 | ⅝ | 0.75 | 150 | 0.38 | 13.3 | 7 | 225 |
| 3 | 3.96 | 8.25 | 5.69 | 6.62 | 8 | ¾ | 1.12 | 250 | 0.38 | 13.3 | 12 | 235 |
| 4 | 4f.80 | 9.00 | | 7.50 | 8 | ⅝ | 0.94 | 150 | 0.38 | 16.5 | 13 | 290 |
| 4 | 4.80 | 10.00 | 6.94 | 7.88 | 8 | ¾ | 1.25 | 250 | 0.38 | 16.5 | 20 | 305 |
| 6 | 6.90 | 11.00 | | 9.50 | 8 | ¾ | 1.00 | 150 | 0.38 | 24.3 | 17 | 425 |
| 6 | 6.90 | 12.50 | 9.69 | 10.62 | 12 | ¾ | 1.44 | 250 | 0.38 | 24.3 | 34 | 455 |
| 8 | 9.05 | 13.50 | | 11.75 | 8 | ¾ | 1.12 | 150 | 0.41 | 34.7 | 27 | 610 |
| 8 | 9.05 | 15.00 | 11.94 | 13.00 | 12 | ⅞ | 1.62 | 250 | 0.41 | 34.7 | 50 | 655 |
| 10 | 11.10 | 16.00 | | 14.25 | 12 | ⅞ | 1.19 | 150 | 0.44 | 46.0 | 38 | 810 |
| 10 | 11.10 | 17.50 | 14.06 | 15.25 | 16 | 1 | 1.88 | 250 | 0.44 | 46.0 | 70 | 875 |
| 12 | 13.20 | 19.00 | | 17.00 | 12 | ⅞ | 1.25 | 150 | 0.48 | 59.8 | 58 | 1,075 |
| 12 | 13.20 | 20.50 | 16.44 | 17.75 | 16 | 1⅛ | 2.00 | 250 | 0.52 | 64.6 | 102 | 1,240 |
| 14 | 15.30 | 21.00 | | 18.75 | 12 | 1 | 1.38 | 150 | 0.51 | 73.9 | 72 | 1,325 |
| 14 | 15.30 | 23.00 | 18.94 | 20.25 | 20 | 1⅛ | 2.12 | 250 | 0.59 | 85.1 | 130 | 1,620 |
| 16 | 17.40 | 23.50 | | 21.25 | 16 | 1 | 1.44 | 150 | 0.54 | 89.2 | 90 | 1,605 |
| 16 | 17.40 | 25.50 | 21.06 | 22.50 | 20 | 1¼ | 2.25 | 250 | 0.63 | 103.6 | 162 | 1,980 |
| 18 | 19.50 | 25.00 | | 22.75 | 16 | 1⅛ | 1.56 | 150 | 0.58 | 107.6 | 90 | 1,900 |
| 18 | 19.50 | 28.00 | 23.31 | 24.75 | 24 | 1¼ | 2.38 | 250 | 0.68 | 125.4 | 200 | 2,405 |
| 20 | 21.60 | 27.50 | | 25.00 | 20 | 1⅛ | 1.69 | 150 | 0.62 | 127.5 | 115 | 2,270 |
| 20 | 21.60 | 30.50 | 25.56 | 27.00 | 24 | 1¼ | 2.50 | 250 | 0.72 | 147.4 | 245 | 2,850 |
| 24 | 25.80 | 32.00 | | 29.50 | 20 | 1¼ | 1.88 | 150 | 0.73 | 179.4 | 160 | 3,190 |
| 24 | 25.80 | 36.00 | 30.25 | 32.00 | 24 | 1½ | 2.75 | 250 | 0.79 | 193.7 | 370 | 3,840 |
| 30 | 32.00 | 38.75 | | 36.0 | 28 | 1¼ | 2.12 | 150 | 0.85 | 259.5 | 240 | 4,630 |
| 30 | 32.00 | 43.0 | 37.19 | 39.25 | 28 | 1¾ | 3.00 | 250 | 0.99 | 300.9 | 530 | 5,875 |
| 36 | 38.30 | 46.00 | | 42.75 | 32 | 1½ | 2.38 | 150 | 0.94 | 344.2 | 350 | 6,205 |
| 36 | 38.30 | 50.0 | 43.69 | 46.0 | 32 | 2 | 3.38 | 250 | 1.10 | 401.1 | 710 | 7,840 |
| 42 | 44.50 | 53.00 | | 49.5 | 36 | 1½ | 2.62 | 150 | 1.05 | 447.2 | 500 | 8,155 |
| 42 | 44.50 | 57.0 | 50.44 | 52.75 | 36 | 2 | 3.69 | 250 | 1.22 | 517.6 | 900 | 10,080 |
| 48 | 50.80 | 59.50 | | 56.0 | 44 | 1½ | 2.75 | 150 | 1.14 | 554.9 | 625 | 10,130 |
| 48 | 50.80 | 65.0 | 58.44 | 60.75 | 40 | 2 | 4.00 | 250 | 1.33 | 644.9 | 1350 | 13,020 |

Pressure ratings of 150 and 250 lb/in² correspond, respectively, to 5,534 and 9,224 Nm². Multiply listed values of weights in pounds by 0.4536 to obtain the corresponding weights in kilograms.

are passed solutions containing more or less free acid or other corrosive agents, standard pipe, valves, and fittings may be **lead-lined, tin-lined,** or **rubber-lined,** to resist corrosive action. This lining prolongs the life of the pipe and also gives it additional strength. For mine service in coal districts where the drainage water is more or less impregnated with sulfur or free sulfuric acid, **wood-lined pipe** and fittings are sometimes used. For special service, **seamless-copper-lined pipe** is also used. The **cement lining** of cast-iron and steel pipe for water and other services is regarded with increasing favor because of its protec-

**Table 17. Standard Weights and Thicknesses of Universal Cast-Iron Pipe**

(Central Foundry Co.)

| Nominal ID, in | Class 150, 150 lb/in² (5,534 N/m²) | | | Class 250, 250 lb/in² (9,224 N/m²) | | |
|---|---|---|---|---|---|---|
| | Approx thickness, in | Estimated wt, lb per | | Approx thickness, in | Estimated wt, lb per | |
| | | Ft | 6-ft length | | Ft | 6-ft length |
| 2 | 0.25 | 7 | 42 | 0.31 | 8 | 48 |
| 3 | 0.30 | 11¼ | 67½ | 0.34 | 12½ | 75 |
| 4 | 0.32 | 16½ | 99 | 0.37 | 18 | 108 |
| 6 | 0.36 | 26.6 | 160 | 0.43 | 30 | 180 |
| 8 | 0.39 | 38½ | 231 | 0.47 | 44¼ | 265½ |
| 10 | 0.43 | 53½ | 321 | 0.50 | 60½ | 363 |
| 12 | 0.47 | 69 | 414 | 0.53 | 77½ | 465 |
| 14 | 0.50 | 87 | 522 | 0.565 | 98½ | 591 |
| 16 | 0.53 | 106 | 636 | 0.60 | 121 | 726 |

| ID, in | 2 | 3 | 4 | 6 | 8 | 10 | 12 | 14 | 16 |
|---|---|---|---|---|---|---|---|---|---|
| Bolt sizes, in | ½ × 4 | ½ × 4¼ | ⅝ × 5 | ¾ × 6 | ⅞ × 6¾ | 1 × 7½ | 1 × 8 | 1⅛ × 9 | 1¼ × 9½ |

**Table 18. Cast-Iron Soil Pipe and Fittings**
(Approximate weights, in pounds*)

| Fittings | Size of fitting, in | | | | | | | | | | | | | | |
|---|---|---|---|---|---|---|---|---|---|---|---|---|---|---|---|
| | 2 | 3 | 4 | 5 | 6 | 3 × 2 | 4 × 2 | 4 × 3 | 5 × 2 | 5 × 3 | 5 × 4 | 6 × 2 | 6 × 3 | 6 × 4 | 6 × 5 |
| Pipe, per ft | 5 | 9 | 12 | 15 | 19 | | | | | | | | | | |
| ¼ bends, regular | 5 | 10 | 15 | 19 | 24 | | | | | | | | | | |
| ¼ bends, short sweep | 6 | 13 | 18 | 23 | 28 | | | | | | | | | | |
| ¼ bends, long sweep | 8 | 16 | 22 | 28 | 34 | | | | | | | | | | |
| ⅕ bends | 5 | 10 | 14 | 18 | 22 | | | | | | | | | | |
| ⅙ bends | 5 | 9 | 13 | 17 | 20 | | | | | | | | | | |
| ⅛ bends | 4 | 8 | 12 | 15 | 18 | | | | | | | | | | |
| 1/16 bends | 4 | 8 | 11 | 13 | 16 | | | | | | | | | | |
| Return bends | 7 | 14 | 21 | 27 | 34 | | | | | | | | | | |
| T branches | 8 | 15 | 21 | 26 | 32 | 13 | 16 | 19 | 19 | 22 | 24 | 22 | 25 | 27 | 30 |
| Tapped T branches† | 7 | | | | | 12 | 15 | | 18 | | | 20 | | | |
| Sanitary T branch | 8 | 16 | 22 | 28 | 34 | 14 | 17 | 20 | 20 | 23 | 26 | 23 | 26 | 29 | 32 |
| Tapped sanitary T branch† | 8 | | | | | 12 | 15 | | 18 | | | 21 | | | |
| Y branch | 8 | 17 | 24 | 32 | 40 | 14 | 17 | 20 | 20 | 24 | 27 | 23 | 27 | 31 | 35 |
| Inverted Y branch | 9 | 18 | 25 | 33 | 41 | 15 | 18 | 22 | 22 | 25 | 29 | 25 | 29 | 33 | 37 |
| Combination Y and ⅛ bend | 10 | 20 | 29 | 38 | 50 | 15 | 18 | 24 | 21 | 27 | 33 | 24 | 30 | 36 | 42 |
| Upright Y branch | 10 | 20 | 28 | 37 | 47 | 16 | 19 | 23 | 22 | 27 | 32 | 25 | 30 | 35 | 40 |
| Vent branch | 9 | 18 | 25 | 32 | 41 | 14 | 18 | 21 | 21 | 24 | 28 | 22 | 27 | 31 | 36 |
| Double hubs | 5 | 8 | 11 | 13 | 15 | | | | | | | | | | |
| Reducers | | | | | | 6 | 7 | 9 | 8 | 10 | 11 | 9 | 11 | 12 | 13 |
| Increasers | | | | | | 9 | 10 | 12 | 12 | 14 | 15 | 13 | 15 | 16 | 18 |
| Tapped increasers | | | | | | 9 | 11 | | 12 | | | 14 | | | |

*Weights of pipe include the hub. Laying lengths of pipe are 5 ft. From the data given for staple fittings, weights of other fittings may be estimated. For data on pipe sizes 8, 10, and 15 in and other data, see ANSI A40.1.

†Tapped up to 2 in.

tion against unusual destructive agencies and its ability to prevent tuberculation. Standard **hard-rubber-pipe** and fittings have been developed for working pressures of 50 lb/in² (1,845 N/m²) at normal temperatures. Standard sizes run from ¼ to 4 in diam (0.635 to 10.2 cm), in 10-ft (3.05-m) lengths. For temperature above 120°F (49°C), the use of hard-rubber-lined steel pipe is recommended. This pipe is suitable for conveying strong acids and chemicals.

## VITRIFIED, WOODEN-STAVE, AND CONCRETE PIPE

**Vitrified pipe** is used extensively for drains and sewerage systems. Burnt-clay tile, being rendered impervious to water by glazing, is by far the best material for sewage purposes as it is not attacked by acids. Dimensions are given in Table 27. For sizes larger than 36 in (91.4 cm) and other data, refer to the

**Table 19. Sizes and Weights of S.P.S. Copper and 85 Red Brass Pipe***
(American Brass Co.)

| Standard pipe size (S. P. S.), in | OD, in | ID, in | Wall thickness, in | Weight, lb/ft | |
|---|---|---|---|---|---|
| | | | | 85 red brass | Copper |
| ⅛ | 0.405 | 0.281 | 0.0620 | 0.2533 | 0.2590 |
| ¼ | 0.540 | 0.375 | 0.0825 | 0.4496 | 0.4596 |
| ⅜ | 0.675 | 0.494 | 0.0905 | 0.6302 | 0.6441 |
| ½ | 0.840 | 0.625 | 0.1075 | 0.9381 | 0.9588 |
| ¾ | 1.050 | 0.822 | 0.1140 | 1.271 | 1.299 |
| 1 | 1.315 | 1.062 | 0.1265 | 1.791 | 1.831 |
| 1¼ | 1.660 | 1.368 | 0.1460 | 2.633 | 2.692 |
| 1½ | 1.900 | 1.600 | 0.1500 | 3.127 | 3.196 |
| 2 | 2.375 | 2.062 | 0.1565 | 4.136 | 4.228 |
| 2½ | 2.875 | 2.500 | 0.1875 | 6.003 | 6.136 |
| 3 | 3.500 | 3.062 | 0.2190 | 8.560 | 8.750 |
| 3½ | 4.000 | 3.500 | 0.2500 | 11.17 | 11.42 |
| 4 | 4.500 | 4.000 | 0.2500 | 12.66 | 12.94 |
| 5 | 5.563 | 5.062 | 0.2505 | 15.85 | 16.20 |
| 6 | 6.625 | 6.125 | 0.2500 | 18.99 | 19.41 |
| 8 | 8.625 | 8.000 | 0.3125 | 30.95 | 31.63 |
| 10 | 10.750 | 10.019 | 0.3655 | 45.22 | 46.22 |
| 11 | 11.750 | 11.000 | 0.3750 | 50.82 | 51.94 |
| 12 | 12.750 | 12.000 | 0.3750 | 55.28 | 56.51 |

***85 % copper, 15 % zinc.**

**Table 20. Sizes and Weights of Copper Tubes**
(American Brass Co.)

| Nominal size, in | OD, in, types* K, L | ID, in | | Wall thickness, in | | Permissible variation of mean OD, in Types K, L | | Weight,† lb/ft | |
|---|---|---|---|---|---|---|---|---|---|
| | | Type K | Type L | Type K | Type L | Annealed | Hard-drawn | Type K | Type L |
| ⅜ | 0.500 | 0.402 | 0.430 | 0.049 | 0.035 | 0.0025 | 0.001 | 0.269 | 0.198 |
| ½ | 0.625 | 0.527 | 0.545 | 0.049 | 0.040 | 0.0025 | 0.001 | 0.344 | 0.285 |
| ⅝ | 0.750 | 0.652 | 0.666 | 0.049 | 0.042 | 0.0025 | 0.001 | 0.418 | 0.362 |
| ¾ | 0.875 | 0.745 | 0.785 | 0.065 | 0.045 | 0.003 | 0.001 | 0.641 | 0.455 |
| 1 | 1.125 | 0.995 | 1.025 | 0.065 | 0.050 | 0.0035 | 0.0015 | 0.839 | 0.655 |
| 1¼ | 1.375 | 1.245 | 1.265 | 0.065 | 0.055 | 0.004 | 0.0015 | 1.04 | 0.884 |
| 1½ | 1.625 | 1.481 | 1.505 | 0.072 | 0.060 | 0.0045 | 0.002 | 1.36 | 1.14 |
| 2 | 2.125 | 1.959 | 1.985 | 0.083 | 0.070 | 0.005 | 0.002 | 2.06 | 1.75 |
| 2½ | 2.625 | 2.435 | 2.465 | 0.095 | 0.080 | 0.005 | 0.002 | 2.93 | 2.48 |
| 3 | 3.125 | 2.907 | 2.945 | 0.109 | 0.090 | 0.005 | 0.002 | 4.00 | 3.33 |
| 3½ | 3.625 | 3.385 | 3.425 | 0.120 | 0.100 | 0.005 | 0.002 | 5.12 | 4.29 |
| 4 | 4.125 | 3.857 | 3.905 | 0.134 | 0.110 | 0.005 | 0.002 | 6.51 | 5.38 |
| 5 | 5.125 | 4.805 | 4.875 | 0.160 | 0.125 | 0.005 | 0.002 | 9.67 | 7.61 |
| 6 | 6.125 | 5.741 | 5.845 | 0.192 | 0.140 | 0.005 | 0.002 | 13.9 | 10.2 |
| 8 | 8.125 | 7.583 | 7.725 | 0.271 | 0.200 | 0.006 | 0.003 | 25.9 | 19.3 |
| 10 | 10.125 | 9.449 | 9.625 | 0.338 | 0.250 | 0.008 | 0.004 | 40.3 | 30.1 |
| 12 | 12.125 | 11.315 | 11.565 | 0.405 | 0.280 | 0.008 | 0.004 | 57.8 | 40.4 |

*Type K recommended for underground service and general plumbing. Type L suitable for interior plumbing and other services.
†Multiply these values by 1.48 to obtain weight in kg/m.

publications of the Clay Products Assoc., Chicago (see ASTM Standard C13).

**Wood-stave pipe** (Fig. 7) is used to a large extent for municipal water supply, outfall sewers, mining, irrigation, and various other uses providing for the transportation of water. The water carried may be hot, cold, or acid. It is made either untreated or creosoted by a vacuum and pressure process. This process uses 8 lb of creosote per cubic foot of wood treated. The untreated pipe is most used where the pipe is constantly full of water, and the wood therefore completely saturated, although in many such instances the creosoted wood is used to give assurance of permanence. (See also Sec. 6.)

Wood-stave pipe is made in two types: machine-banded pipe and continuous-stave pipe. **Machine-banded pipe** is banded with wire and is made with wood or metal collars, or with inserted joints. **Continuous-stave pipe** is manufactured in units consisting of staves, bands, and shoes, shipped in knocked-down form, and constructed in the trench. In building this type of pipe, the staves are laid so as to break joints and the completed pipe is without joints. Continuous-stave pipe is banded with individual bands, ranging in size from ⅜ to 1 in (0.95 to 2.54 cm), depending upon the size of the pipe. A factor of safety of 4 is maintained in the band, based on an ultimate strength of 60,000 lb/in² (2,214,000 N/m²) of cross section. The maximum pressure to which a continuous-stave pipe may be subjected depends upon the size of the pipe. The head for small pipes may run as high as 400 ft (121.9 m) while in the largest sizes the head would be less than 200 ft (15.2 to 121.9 m).

Machine-banded pipe is made for pressures of 50 to 400 ft (61 m). The staves are made from redwood or Douglas fir lumber, dried and carefully selected. The inside and outside of the staves are dressed to conform to the circumferential lines; and the edges of the staves to conform to the radial lines.

Wooden pipe is most largely built in western sections of the United States, close to the natural lumber market. The sizes of machine-banded pipe range from 2 to 24 in (5.08 to 61 cm), and of the continuous-stave pipe from 6 in to 20 ft (15.2 cm to 6.1 m) inside diameter.

Pipe made from **plywood** is molded in lengths up to 11 ft (3.35 m) in diameters 3 in (7.62 cm) and up, and in wall thicknesses to specifications. Tubes made from fiber, by a molding process, are obtainable in a variety of sizes and lengths.

**Concrete pipe** is an important factor in sewer, conduit, railroad, culvert, and water-pipe construction. The pipe, as usually made, is constructed of concrete reinforced longitudinally with bars and transversely with wire mesh or steel bands. It is made in sections of definite length, with the longitudinal reinforcement so disposed as to provide for the interlocking of one section with another, and so formed that when these are locked together and cemented they form a continuous line of pipe free from leakage or seepage. Various forms of joints are used, all capable of taking care of expansion. Figure 8 shows one type of construction. Concrete pipe is manufactured in a great variety of diameters, thicknesses, and lengths to suit almost any requirement arising in practice. (See also Sec. 6.)

**Asbestos-cement pipe,** known by the trade name **Transite** pipe in this country, was developed initially in Europe. It has had a rapidly increasing use in this country for a wide variety of services. Made of a mixture of cement and asbestos fiber, it has great resistance to corrosion and is used in mine drainage systems, waterworks systems, gas lines, sewerage systems, etc. It is manufactured in diameters 3 to 36 in (7.62 to 91.4

**Table 21. Aluminum Piping***

| Nominal pipe size, in | Schedule number | OD, in | ID, in | Wall thickness, in | Weight per linear foot, lb, plain ends† | Cross-sectional wall area, in² | Inside cross-sectional area, in² | Moment of inertia, in⁴ | Section modulus, in³ | Radius of gyration, in |
|---|---|---|---|---|---|---|---|---|---|---|
| ⅛ | 40‡ | 0.405 | 0.269 | 0.068 | 0.085 | 0.0720 | 0.0568 | 0.0011 | 0.0053 | 0.1215 |
| | 80§ | 0.405 | 0.215 | 0.095 | 0.109 | 0.0925 | 0.0363 | 0.0012 | 0.0060 | 0.1146 |
| ¼ | 40‡ | 0.540 | 0.364 | 0.088 | 0.147 | 0.1250 | 0.1041 | 0.0033 | 0.0123 | 0.1628 |
| | 80§ | 0.540 | 0.302 | 0.119 | 0.185 | 0.1574 | 0.0716 | 0.0038 | 0.0139 | 0.1547 |
| ⅜ | 40‡ | 0.675 | 0.493 | 0.091 | 0.196 | 0.1670 | 0.1909 | 0.0073 | 0.0216 | 0.2090 |
| | 80§ | 0.675 | 0.423 | 0.126 | 0.256 | 0.2173 | 0.1405 | 0.0086 | 0.0255 | 0.1991 |
| ½ | 40‡ | 0.840 | 0.622 | 0.109 | 0.294 | 0.2503 | 0.3039 | 0.0171 | 0.0407 | 0.2613 |
| | 80§ | 0.840 | 0.546 | 0.147 | 0.376 | 0.3200 | 0.2341 | 0.0201 | 0.0478 | 0.2505 |
| ¾ | 10 | 1.050 | 0.884 | 0.083 | 0.297 | 0.2521 | 0.6138 | 0.0297 | 0.0566 | 0.3432 |
| | 40‡ | 1.050 | 0.824 | 0.113 | 0.391 | 0.3326 | 0.5333 | 0.0370 | 0.0705 | 0.3337 |
| | 80§ | 1.050 | 0.742 | 0.154 | 0.510 | 0.4335 | 0.4324 | 0.0448 | 0.0853 | 0.3214 |
| 1 | 5 | 1.315 | 1.185 | 0.065 | 0.300 | 0.2553 | 1.103 | 0.0500 | 0.0760 | 0.4425 |
| | 10 | 1.315 | 1.097 | 0.109 | 0.486 | 0.4130 | 0.9452 | 0.0757 | 0.1151 | 0.4382 |
| | 40‡ | 1.315 | 1.049 | 0.133 | 0.581 | 0.4939 | 0.8643 | 0.0873 | 0.1328 | 0.4205 |
| | 80§ | 1.315 | 0.957 | 0.179 | 0.751 | 0.6388 | 0.7193 | 0.1056 | 0.1606 | 0.4066 |
| 1¼ | 5 | 1.660 | 1.530 | 0.065 | 0.383 | 0.3257 | 1.839 | 0.1037 | 0.1250 | 0.5644 |
| | 10 | 1.660 | 1.442 | 0.109 | 0.625 | 0.5311 | 1.633 | 0.1605 | 0.1934 | 0.5497 |
| | 40‡ | 1.660 | 1.380 | 0.140 | 0.786 | 0.6685 | 1.496 | 0.1947 | 0.2346 | 0.5397 |
| | 80§ | 1.660 | 1.278 | 0.191 | 1.037 | 0.8815 | 1.283 | 0.2418 | 0.2913 | 0.5238 |
| 1½ | 5 | 1.900 | 1.770 | 0.065 | 0.441 | 0.3747 | 2.461 | 0.1579 | 0.1662 | 0.6492 |
| | 10 | 1.900 | 1.682 | 0.109 | 0.721 | 0.6133 | 2.222 | 0.2468 | 0.2598 | 0.6344 |
| | 40‡ | 1.900 | 1.610 | 0.145 | 0.940 | 0.7995 | 2.036 | 0.3099 | 0.3262 | 0.6226 |
| | 80§ | 1.900 | 1.500 | 0.200 | 1.256 | 1.068 | 1.767 | 0.3912 | 0.4118 | 0.6052 |
| 2 | 5 | 2.375 | 2.245 | 0.065 | 0.555 | 0.4717 | 3.958 | 0.3149 | 0.2652 | 0.8170 |
| | 10 | 2.375 | 2.157 | 0.109 | 0.913 | 0.7760 | 3.654 | 0.4992 | 0.4204 | 0.8021 |
| | 40‡ | 2.375 | 2.067 | 0.154 | 1.264 | 1.074 | 3.356 | 0.6657 | 0.5606 | 0.7871 |
| | 80§ | 2.375 | 1.939 | 0.218 | 1.737 | 1.477 | 2.953 | 0.8679 | 0.7309 | 0.7665 |
| 2½ | 5 | 2.875 | 2.709 | 0.083 | 0.856 | 0.7280 | 5.764 | 0.7100 | 0.4939 | 0.9876 |
| | 10 | 2.875 | 2.635 | 0.120 | 1.221 | 1.039 | 5.453 | 0.9873 | 0.6868 | 0.9750 |
| | 40‡ | 2.875 | 2.469 | 0.203 | 2.004 | 1.704 | 4.788 | 1.530 | 1.064 | 0.9474 |
| | 80§ | 2.875 | 2.323 | 0.276 | 2.650 | 2.254 | 4.238 | 1.924 | 1.339 | 0.9241 |
| 3 | 5 | 3.500 | 3.334 | 0.083 | 1.048 | 0.8910 | 8.730 | 1.301 | 0.7435 | 1.208 |
| | 10 | 3.500 | 3.260 | 0.120 | 1.498 | 1.274 | 8.346 | 1.822 | 1.041 | 1.196 |
| | 40‡ | 3.500 | 3.068 | 0.216 | 2.621 | 2.228 | 7.393 | 3.017 | 1.724 | 1.164 |
| | 80§ | 3.500 | 2.900 | 0.300 | 3.547 | 3.016 | 6.605 | 3.894 | 2.225 | 1.136 |
| 3½ | 5 | 4.000 | 3.834 | 0.083 | 1.201 | 1.021 | 11.55 | 1.960 | 0.9799 | 1.385 |
| | 10 | 4.000 | 3.760 | 0.120 | 1.720 | 1.463 | 11.10 | 2.755 | 1.378 | 1.372 |
| | 40‡ | 4.000 | 3.548 | 0.226 | 3.151 | 2.680 | 9.887 | 4.788 | 2.394 | 1.337 |
| | 80§ | 4.000 | 3.364 | 0.318 | 4.326 | 3.678 | 8.888 | 6.281 | 3.140 | 1.307 |
| 4 | 5 | 4.500 | 4.334 | 0.083 | 1.354 | 1.152 | 14.75 | 2.810 | 1.249 | 1.562 |
| | 10 | 4.500 | 4.260 | 0.120 | 1.942 | 1.651 | 14.25 | 3.963 | 1.761 | 1.549 |
| | 40‡ | 4.500 | 4.026 | 0.237 | 3.733 | 3.174 | 12.73 | 7.232 | 3.214 | 1.510 |
| | 80§ | 4.500 | 3.826 | 0.337 | 5.183 | 4.407 | 11.50 | 9.611 | 4.272 | 1.477 |
| 5 | 40‡ | 5.563 | 5.047 | 0.258 | 5.057 | 4.300 | 20.01 | 15.16 | 5.451 | 1.878 |
| | 80§ | 5.563 | 4.813 | 0.375 | 7.188 | 6.112 | 18.19 | 20.67 | 7.432 | 1.839 |
| 6 | 40‡ | 6.625 | 6.065 | 0.280 | 6.564 | 5.581 | 28.89 | 28.14 | 8.496 | 2.246 |
| | 80§ | 6.625 | 5.761 | 0.432 | 9.884 | 8.405 | 26.07 | 40.49 | 12.22 | 2.195 |
| 8 | 30 | 8.625 | 8.071 | 0.277 | 8.543 | 7.265 | 51.16 | 63.35 | 14.69 | 2.953 |
| | 40‡ | 8.625 | 7.981 | 0.322 | 9.878 | 8.399 | 50.03 | 72.49 | 16.81 | 2.938 |
| | 80§ | 8.625 | 7.625 | 0.500 | 15.01 | 12.76 | 45.66 | 105.7 | 24.51 | 2.878 |
| 10 | | 10.750 | 10.192 | 0.279 | 10.79 | 9.178 | 81.59 | 125.9 | 23.42 | 3.704 |
| | 30 | 10.750 | 10.136 | 0.307 | 11.84 | 10.07 | 80.69 | 137.4 | 25.57 | 3.694 |
| | 40‡ | 10.750 | 10.020 | 0.365 | 14.00 | 11.91 | 78.85 | 160.7 | 29.90 | 3.674 |
| | 80§ | 10.750 | 9.750 | 0.500 | 18.93 | 16.10 | 74.66 | 211.9 | 39.43 | 3.628 |
| 12 | 30 | 12.750 | 12.090 | 0.330 | 15.14 | 12.88 | 114.8 | 248.5 | 38.97 | 4.393 |
| | ‡ | 12.750 | 12.000 | 0.375 | 17.14 | 14.58 | 113.1 | 279.3 | 43.81 | 4.377 |
| | § | 12.750 | 11.750 | 0.500 | 22.63 | 19.24 | 108.4 | 361.5 | 56.71 | 4.335 |

**Table 21. Aluminum Piping\*** *(Continued)*

| Nominal pipe size, in | Schedule number | OD, in | ID, in | Wall thickness, in | Weight per linear foot, lb, plain ends† | Cross-sectional wall area, in² | Inside cross-sectional area, in² | Moment of inertia, in⁴ | Section modulus, in³ | Radius of gyration, in |
|---|---|---|---|---|---|---|---|---|---|---|
| | | | | | Construction pipe | | | | | |
| | | 2.00 | 1.900 | 0.050 | 0.3602 | 0.3063 | 2.835 | 0.1457 | 0.1457 | 0.6897 |
| | | 3.00 | 2.900 | 0.050 | 0.5449 | 0.4634 | 6.605 | 0.5042 | 0.3361 | 1.043 |
| | | 4.00 | 3.900 | 0.050 | 0.7297 | 0.6205 | 11.95 | 1.210 | 0.6051 | 1.397 |
| | | 5.00 | 4.896 | 0.052 | 0.9506 | 0.8083 | 18.83 | 2.474 | 0.9896 | 1.749 |
| | | 6.00 | 5.876 | 0.062 | 1.360 | 1.157 | 2.712 | 5.098 | 1.699 | 2.100 |
| | | 7.00 | 6.856 | 0.072 | 1.843 | 1.567 | 36.92 | 9.403 | 2.687 | 2.450 |
| | | 8.00 | 7.812 | 0.094 | 2.745 | 2.335 | 47.93 | 18.24 | 4.561 | 2.795 |

See end of Table 3 for conversion to metric units.
\*Aluminum Co. of America.
†Weights calculated for 6061 and 6063. For 3003 multiply by 1.010.
‡Also designated as standard pipe.
§Also designated as extra-heavy or extra-strong pipe. All calculations based on nominal dimensions.

cm), in 13 ft (3.96 m) lengths, and in pressure classes of 50, 100, 150, and 200 lb/in² (1,845, 3,690, 5,535, and 7,379 N/m²). Complete information as to specifications and recommended uses can be obtained from the Johns-Manville Co.

### FITTINGS FOR WROUGHT-IRON AND STEEL PIPE

**American National Standard Cast-Iron Pipe Flanges and Flanged Fittings for Max Working Saturated Steam Pressure of 25, 125, and 250 lb/in² (922, 4,612, and 9,224 N/m²)**

INTRODUCTORY NOTES

**Sizes** The sizes of the fittings in the following tables are nominal pipe sizes. In the 25-lb standard, the nominal pipe size is the same as the port diameter of the fittings for *all* sizes. In the 125- and 250-lb standards the nominal pipe size is the same as the port diameter of fittings for pipe having inside diameters of 12 in (30.5 cm) and smaller. For pipe 14 in (35.6 cm) and larger, the corresponding outside diameter of the pipe is given, and consequently the fittings will have a smaller port diameter.

**Pressure Rating** In the 25-lb standard the sizes 36 in (91.4 cm) and smaller may also be used for maximum nonshock working hydraulic pressures of 43 lb/in² gage (1,587 N/m²) or a maximum gas pressure of 25 lb/in² gage, at or near the ordinary range of air temperatures. In the 125-lb standard, the sizes 12 in (30.5 cm) and smaller may also be used for maxi-

mum nonshock working hydraulic pressure of 175 lb/in² gage (6,457 N/m²) at or near the ordinary range of air temperatures. In the 250-lb standard, the sizes 12 in and smaller may be used for maximum nonshock working hydraulic pressures of 400 lb/in² gage (14,758 N/m²) at or near the ordinary range of air temperatures.

**Facing** All 25- and 125-lb cast-iron flanges and flanged fittings are plain faced, i.e., without projection or raised face. All 250-lb cast-iron flanges and flanged fittings have a raised face 1/16 in (0.16 cm) high, of the diameters given in Table 28. The raised face is included in the minimum flange thickness and center-to-face dimensions.

An **inspection limit** of ±1/32 in (0.08 cm) is allowed on all center-to-contact-surface dimensions for sizes up to and including 10 in (25.4 cm) and ±1/16 in on sizes larger than 10 in. An inspection limit of ±1/16 in is allowed on all contact-surface to contact-surface dimensions for sizes up to and including 10 in and ±1/8 in (0.32 cm) on sizes larger than 10 in.

**Dimensions** In the 25-lb standard, the flange diameters, bolt circles, and number of bolts are the same as in the 125-lb ANSI Standard B16.1-1967, with a reduction in the thickness of flanges and bolt diameters, thereby maintaining interchangeability between the two standards.

The center-to-face and face-to-face dimensions of 25-lb standard fittings are the same as for the 125-lb standard.

**Bolting** Drilling templates are in multiples of four, so that fittings may be made to face in any quarter. **Boltholes** straddle the centerline. For bolts smaller than 1¾ in (4.45 cm) the

**Table 22. Weights and Dimensions of Lead Tubing\***
(National Lead Co.)

| ID, in | OD, in | Weight† per ft, oz | Id, in | OD, in | Weight† per ft, oz | ID, in | OD, in | Weight† per ft, oz |
|---|---|---|---|---|---|---|---|---|
| 1/16 | 1/8 | 0.75 | 3/16 | 5/16 | 3.00 | 1/4 | 1/2 | 12.00 |
| 1/8 | 13/64 | 1.00 | 1/4 | 11/32 | 4.00 | 5/16 | 3/8 | 2.00 |
| 1/8 | 3/16 | 1.50 | 1/4 | 3/8 | 5.00 | 3/8 | 7/16 | 2.00 |
| 1/8 | 1/4 | 2.00 | 1/4 | 13/32 | 6.00 | 3/8 | 1/2 | 7.00 |
| 3/16 | 1/4 | 2.00 | 1/4 | 7/16 | 8.00 | 7/16 | 1/2 | 4.00 |

\*Furnished in coils of approx 25 lb, or on reels carrying approx 50 or 100 lb.
†Multiply weight in oz/ft by 0.93 to obtain weight in g/cm.

**Table 23. Weights and Dimensions of Lead Pipe**
(National Lead Co.)

| ID, in | Classification | | OD, in | Weight* per ft, lb | ID, in | Classification | | OD, in | Weight* per ft, lb |
| --- | --- | --- | --- | --- | --- | --- | --- | --- | --- |
| | East | West | | | | East | West | | |
| ⅜ | E | AQ | 0.520 | 0.50 | 1 | E | AQ | 1.192 | 1.63 |
| | D | XL | 0.549 | 0.63 | | D | XL | 1.232 | 2.00 |
| | C | L | 0.577 | 0.75 | | C | L | 1.284 | 2.50 |
| | B | M | 0.631 | 1.00 | | B | M | 1.356 | 3.25 |
| | A | S | 0.725 | 1.50 | | A | S | 1.428 | 4.00 |
| | AA | XS | 0.811 | 2.00 | | AA | XS | 1.492 | 4.75 |
| | AAA | XXS | 0.888 | 2.50 | | AAA | XXS | 1.596 | 6.00 |
| ½ | E | AQ | 0.628 | 0.56 | 1¼ | E | AQ | 1.442 | 2.00 |
| | D | XL | 0.666 | 0.75 | | D | XL | 1.486 | 2.50 |
| | C | L | 0.712 | 1.00 | | C | L | 1.528 | 3.00 |
| | B | M | 0.756 | 1.25 | | B | M | 1.592 | 3.75 |
| | A | S | 0.798 | 1.50 | | A | S | 1.670 | 4.75 |
| | AA | XS | 0.876 | 2.00 | | AA | XS | 1.765 | 6.00 |
| | AAA | XXS | 1.012 | 3.00 | | AAA | XXS | 1.889 | 7.75 |
| ⅝ | E | AQ | 0.765 | 0.75 | 1½ | E | AQ | 1.740 | 3.00 |
| | D | XL | 0.803 | 1.00 | | D | XL | 1.776 | 3.50 |
| | C | L | 0.881 | 1.50 | | C | L | 1.830 | 4.25 |
| | B | M | 0.953 | 2.00 | | B | M | 1.882 | 5.00 |
| | A | S | 1.019 | 2.50 | | A | S | 1.984 | 6.50 |
| | AA | XS | 1.082 | 3.00 | | AA | XS | 2.076 | 8.00 |
| | AAA | XXS | 1.137 | 3.50 | | AAA | XXS | 2.272 | 11.25 |
| ¾ | E | AQ | 0.906 | 1.00 | 1¾ | D | XL | 2.024 | 4.00 |
| | D | XL | 0.940 | 1.25 | | C | L | 2.086 | 5.00 |
| | C | L | 1.006 | 1.75 | | B | M | 2.146 | 6.00 |
| | B | M | 1.068 | 2.25 | | A | S | 2.193 | 6.75 |
| | A | S | 1.156 | 3.00 | | AA | XS | 2.404 | 10.50 |
| | AA | XS | 1.212 | 3.50 | | AAA | XXS | 2.624 | 14.75 |
| | AAA | XXS | 1.336 | 4.75 | | | | | |

Additional standard sizes are 2, 2½, 3, 4, 5, and 6 in.
*Multiply weight in lb/ft by 14.88 to obtain weight in gm/cm.

boltholes are drilled ⅛ in larger in diameter than the nominal size of the bolt. Holes for bolts 1¾ in and larger are drilled ¼ in (0.64 cm) larger than nominal diameter of bolts. **Bolts** of steel are with standard "rough square heads" and the nuts are of steel with standard "rough hexagonal" dimensions; all as given in the American Standard on Wrench Head Bolts and Nuts and Wrench Openings of the National Screw Thread Commission. For bolts, 1¾ in (4.45 cm) diam and larger, bolt studs with a nut on each end are recommended.

**Hexagonal nuts** for pipe sizes 1 to 48 in (2.54 to 122 cm) in the 125-lb standard and 1 to 16 in (40.6 cm) in the 250-lb standard can be conveniently pulled up with open wrenches of mini-

**Table 24. Sizes and Weights of Block-Tin Pipe**

| ID, in | OD, in | Weight* per ft, oz | ID, in | OD, in | Weight* per ft, oz | ID, in | OD, in | Weight* per ft, oz |
| --- | --- | --- | --- | --- | --- | --- | --- | --- |
| 3/16 | ¼ | 1½ | ⅜ | 9/16 | 7½ | ⅝ | ¾ | 7 |
| 3/16 | 5/16 | 2½ | ⅜ | 19/32 | 8½ | ⅝ | 25/32 | 9 |
| ¼ | ⅜ | 3 | ⅜ | ⅝ | 9½ | ⅝ | 13/16 | 10½ |
| ¼ | 13/32 | 4 | ⅜ | ⅝ | 10½ | ⅝ | ⅞ | 15 |
| ¼ | 7/16 | 5 | ⅜ | 21/32 | 12 | ¾ | ⅞ | 8 |
| ¼ | 15/32 | 6 | 7/16 | 9/16 | 4½ | ¾ | 29/32 | 10 |
| ¼ | ½ | 7 | 7/16 | ⅝ | 8 | ¾ | 29/32 | 11 |
| ¼ | ½ | 8 | ½ | 19/32 | 4 | ¾ | 15/16 | 12½ |
| 5/16 | 7/16 | 4 | ½ | ⅝ | 5 | ¾ | 1 | 17 |
| 5/16 | ½ | 5½ | ½ | ⅝ | 5½ | ¾ | 1 1/32 | 20 |
| 5/16 | 17/32 | 7½ | ½ | 21/32 | 7 | ¾ | 1 1/16 | 22½ |
| ⅜ | ½ | 4½ | ½ | 21/32 | 7½ | 1 | 1 3/16 | 16 |
| ⅜ | ½ | 5 | ½ | 11/16 | 9 | 1 | 1 3/16 | 17 |
| ⅜ | 17/32 | 5½ | ½ | 23/32 | 10½ | 1 | 1 7/32 | 19½ |
| ⅜ | 9/16 | 6½ | ½ | ¾ | 12½ | | | |

*Multiply weight in oz/ft by 0.93 to obtain weight in g/cm.

**Table 25. Polyethylene Plastic Pipe Dimensions**

| Nominal size, in | OD, in | ID, in | Wall thickness, in | Max operating press at 75°F, lb/in² | Wt* per ft, lb | Shipping length, ft |
|---|---|---|---|---|---|---|
| 1/2 | 0.782 | 0.622 | 0.080 | 75 | 0.0704 | 100 and 400 (coiled) |
| | 0.842 | 0.622 | 0.110 | 100 | 0.101 | 100 and 400 (coiled) |
| 3/4 | 1.024 | 0.824 | 0.100 | 75 | 0.1157 | 100 and 400 (coiled) |
| | 1.114 | 0.824 | 0.145 | 100 | 0.176 | 100 and 400 (coiled) |
| 1 | 1.300 | 1.050 | 0.125 | 75 | 0.1838 | 100 and 400 (coiled) |
| | 1.410 | 1.050 | 0.180 | 100 | 0.277 | 100 and 400 (coiled) |
| 1 1/4 | 1.660 | 1.380 | 0.140 | 70 | 0.266 | 100 and 400 (coiled) |
| | 1.710 | 1.380 | 0.165 | 75 | 0.3190 | 100 and 400 (coiled) |
| | 1.860 | 1.380 | 0.240 | 100 | 0.487 | 100 and 400 (coiled) |
| 1 1/2 | 1.900 | 1.610 | 0.145 | 60 | 0.318 | 100 and 400 (coiled) |
| | 2.000 | 1.610 | 0.195 | 75 | 0.441 | 100 and 400 (coiled) |
| | 2.170 | 1.610 | 0.280 | 100 | 0.662 | 100 and 400 (coiled) |
| 2 | 2.375 | 2.067 | 0.154 | 50 | 0.428 | 100 and 400 (coiled) |
| | 2.567 | 2.067 | 0.250 | 75 | 0.725 | 100 and 400 (coiled) |
| | 2.777 | 2.067 | 0.355 | 100 | 1.077 | 100 and 400 (coiled) |
| 3 | 3.500 | 3.068 | 0.216 | 50 | 0.888 | 100 (coiled) |
| | 3.670 | 3.068 | 0.301 | 75 | 1.280 | 100 (coiled) |
| | 4.068 | 3.068 | 0.500 | 100 | 2.230 | 100 (coiled) |
| 4 | 4.500 | 4.026 | 0.237 | 40 | 1.265 | 25 (straight) |
| | 4.820 | 4.026 | 0.397 | 75 | 2.200 | 25 (straight) |
| | 5.386 | 4.026 | 0.680 | 100 | 4.000 | 25 (straight) |
| 6 | 6.625 | 6.065 | 0.280 | 35 | 2.200 | 25 (straight) |

*Multiply weight in lb/ft by 14.88 to obtain g/cm.

mum design of heads. Hexagonal nuts for pipe sizes 48 to 96 in (244 cm) in the 125-lb standard and 18 to 48 in (45.7 to 122 cm) in the 250-lb standard can be conveniently pulled up with box wrenches.

**Spot Facing** The boltholes of 25-, 125-, and 250-lb cast-iron flanges and flanged fittings are not spot-faced for ordinary service. When required, the flanges and fitting in sizes 30 in (76.2 cm) and larger may be spot-faced or back-faced to the minimum thickness of flange with a plus tolerance of 1/8 in (0.32 cm).

**Reducing Fittings** Reducing elbows and side-outlet elbows carry same dimensions center to face as straight-size elbows corresponding to the size of the larger opening.

Tees, side-outlet tees, crosses, and laterals sizes 16 in (40.6 cm) and smaller, reducing on the outlet or branch, have the same dimension center to face and face to face as straight-size fittings corresponding to the size of the larger opening. Sizes 18 in (45.7 cm) and larger, reducing on the outlet or branch, are made in two lengths depending on the size of the outlet as given in the tables of dimensions.

Tees, crosses, and laterals, reducing on the run only, have the same dimensions center to face and face to face as straight-size fittings corresponding to the size of the larger opening.

Reducers and eccentric reducers for all reductions have the same face-to-face dimensions for the larger opening.

Special double-branch elbows whether straight or reducing have the same dimension center to face as straight-size elbows corresponding to the size of the larger opening.

Side-outlet elbows and side-outlet tees have all openings on intersecting centerlines.

**Elbows** Special degree elbows ranging from 1 to 45° have the same center-to-face dimension given for 45° elbows, and those over 45 and up to 90° have the same center-to-face dimensions given for 90° elbows. The angle designation of an elbow is its deflection from straight-line flow and is the angle between its flange faces.

**Screwed Companion Flanges** Screwed companion flanges in the 25-lb standard should not be thinner than those in the 125-lb standard on sizes 24 in (61 cm) and smaller. Other types of flanges may have thicknesses as given in Table 28.

**Laterals** Laterals (Y branches) both straight and reducing sizes 8 in and larger are reinforced to compensate for the inherent weakness in the casting design.

The American National Standard covers also dimensions (not included in the tables) of base elbows and base tees and anchorage bases for straight tees and reducing tees.

American National Standard cast-iron pipe flanges and flanged fittings are available for maximum nonshock working hydraulic pressure of 800 lb/in² gage (29,516 N/m²) at ordinary air temperatures.

**Assembly of Flanged Joints** The optimun degree of tightening occurs when a stress of 30,000 lb/in² (1,100,000 N/m²) is uniformly reached in each flange stud or bolt. For a modulus of elasticity of 30,000,000 lb/in² (1.1 × 10⁹ N/m²) a stress of 30,000,000 lb/in² occurs when the elongation, determined with a dial indicator or micrometer, is 0.001 in/in of stud length measured between centers of nuts. Uniform tension in flange bolts may also be obtained by use of a torque wrench; bearing surfaces of nuts must have a good machine finish, and threads must be properly lubricated for reliable results with a torque wrench. Torque values below have been found to give 30,000 lb/in² stress in studs:

**Table 26. Corrosion-Resistance Data, Polyethylene Pipe**

| Reagent | Performance at | | Reagent | Performance at | |
|---|---|---|---|---|---|
| | 75°F (24°C) | 120°F (49°C) | | 75°F (24°C) | 120°F (49°C) |
| Acetic acid, glacial* | F | NG | Lactic acid, 90 percent | E | E |
| Acetic acid, 10 percent* | E | E | Linseed oil | E | E |
| Acetone | NG | NG | Lubricating oil | NG | NG |
| Ammonia, dry gas | E | E | Magnesium chloride | E | E |
| Ammonium hydroxide, 10 percent | E | E | Magnesium sulfate | E | E |
| Ammonium hydroxide, 28 percent | E | E | Methyl bromide | NG | |
| Amyl acetate | NG | NG | Methyl isobutyl ketone | F | NG |
| Aniline | E | F | Nitric acid, 10 percent | E | E |
| Benzene | NG | NG | Nitric acid, 30–50 percent | E | E |
| Bromine | NG | NG | Nitric acid, 70 percent | E | E |
| Butyraldehyde | E | G | Oleic acid | F | NG |
| Calcium chloride, saturated | E | E | Phosphoric acid, 30 percent | E | E |
| Calcium hydroxide | E | E | Phosphoric acid, 90 percent | E | NG |
| Calcium hypochlorite | E | E | Photographic developer | E | E |
| Carbon disulfide | NG | NG | Potassium borate | E | E |
| Carbon tetrachloride | NG | NG | Potassium carbonate | E | E |
| Carbonic acid | E | E | Potassium chloride, saturated | E | E |
| Chlorine, dry gas | F | NG | Potassium dichromate | E | E |
| Chlorine, liquid | NG | NG | Potassium hydroxide | E | E |
| Chlorosulfonic acid | NG | NG | Potassium nitrate | E | E |
| Citric acid, saturated | E | E | Potassium permanganate | E | E |
| Copper sulfate | E | E | Silicic acid | E | E |
| Cyclohexanone | NG | NG | Silver nitrate | E | E |
| Diethylene glycol | E | E | Sodium benzoate | E | E |
| Dioxane | E | G | Sodium bisulfite | E | E |
| Ethyl acetate | F | NG | Sodium carbonate, concentrated | E | E |
| Ethyl alcohol, 35 percent | NG | NG | Sodium chloride, saturated solution | E | E |
| Ethyl butyrate | F | NG | Sodium hydroxide, 10 percent | E | E |
| Ethylene dichloride | NG | NG | Sodium hydroxide, 50 percent | E | E |
| Ferric chloride | E | E | Sodium sulfate | E | E |
| Ferrous sulfate, 15 percent aq. | E | E | Stannic chloride, saturated | E | E |
| Fluorine | E | NG | Stearic acid, 100 percent | E | E |
| Fluosilicic acid, concentrated | E | F | Sulfuric acid, 10 percent | E | E |
| Formaldehyde, 40 percent | E | E | Sulfuric acid, 30 percent | E | E |
| Formic acid, 50 percent | E | E | Sulfuric acid, 60 percent | E | F |
| Furfuryl alcohol | NG | NG | Sulfuric acid, 98 percent | F | NG |
| Gasoline | NG | NG | Tannic acid | E | E |
| Hydrobromic Acid | E | E | Toluene | NG | NG |
| Hydrochloric acid, 10 percent | E | E | Trichlorobenzene | F | NF |
| Hydrochloric acid, 37 percent | E | E | Trichloroethylene | NG | NG |
| Hydrofluoric acid, 48 percent | E | E | Vinegar | E | E |
| Hydrofluoric acid, 75 percent | E | F | Xylene | NG | NG |
| Hydrogen peroxide, 30 percent | E | G | Zinc chloride | E | E |
| Hydrogen peroxide, 90 percent | G | NG | Zinc sulfate | E | E |

Corrosion-resistance data given in this table based on laboratory tests conducted by the manufacturers of the materials covered, and are indicative only of the conditions under which the tests were made. This information may be considered as a basis for recommendation, but not as a guarantee. Materials should be tested under actual service to determine suitability for a particular purpose.

E = excellent, G = good, F = fair, NG = not good.

*Polyethylene is permeable to acetic acid.

| Stud Diam, in | Threads per in | Torque, lb·ft* |
|---|---|---|
| ⅝ | 11 | 89 |
| ¾ | 10 | 107 |
| ⅞ | 9 | 162 |
| 1 | 8 | 244 |
| 1⅛ | 8 | 322 |
| 1¼ | 8 | 410 |
| 1⅜ | 8 | 510 |
| 1½ | 8 | 615 |

*Multiply by 0.149 to obtain torque in kilogram-metres.

## American National Standard Steel Pipe Flanges and Flanged Fittings

INTRODUCTORY NOTES

**Pressure Ratings and Tests**   These standards are known as "American 150, 300, 400, 600, 900, 1,500, and 2,500 lb Steel Flange Standards" (ANSI B16.5-1968), said pressure designation being the recommended rating at the temperatures given in Table 29. This table shows recommended ratings for various temperatures, together with hydrostatic shell test pres-

**Table 27. Standard-Strength Vitrified Clay Pipe**
(Dimensions refer to Fig. 6)

| Size, $D$, in | Laying length $L$: Nominal, ft | Laying length $L$: Limit of minus variation,* in/ft length | Max difference in length of two opposite sides, in | OD of barrel, in: Min | OD of barrel, in: Max | ID of socket ½ in above base, $D_s$, in: Min | ID of socket ½ in above base, $D_s$, in: Max | Depth of socket $L_s$, in: Nominal | Depth of socket $L_s$, in: Min | Thickness of barrel, $T$, in: Nominal | Thickness of barrel, $T$, in: Min | Thickness of socket at ½ in from outer end, $T_s$, in: Nominal | Thickness of socket at ½ in from outer end, $T_s$, in: Min |
|---|---|---|---|---|---|---|---|---|---|---|---|---|---|
| 4 | 2, 2½, 3 | ¼ | 5/16 | 4⅞ | 5⅛ | 5¾ | 6⅛ | 1¾ | 1½ | ½ | 7/16 | 7/16 | ⅜ |
| 6 | 2, 2½, 3 | ¼ | ⅜ | 7 1/16 | 7 7/16 | 8 3/16 | 8⅝ | 2½ | 2 | ⅝ | 9/16 | ½ | 7/16 |
| 8 | 2, 2½, 3 | ¼ | 7/16 | 9¼ | 9¾ | 10½ | 11 | 2½ | 2¼ | ¾ | 11/16 | 9/16 | ½ |
| 10 | 2, 2½, 3 | ¼ | 7/16 | 11½ | 12 | 12¾ | 13¼ | 2⅝ | 2⅜ | ⅞ | 13/16 | ⅝ | 9/16 |
| 12 | 2, 2½, 3 | ¼ | 7/16 | 13¾ | 14 5/16 | 15⅛ | 15¾ | 2¾ | 2½ | 1 | 15/16 | ¾ | 11/16 |
| 15 | 3, 4 | ¼ | ½ | 17 3/16 | 17 13/16 | 18⅝ | 19¼ | 2⅞ | 2⅝ | 1¼ | 1⅛ | 15/16 | ⅞ |
| 18 | 3, 4 | ¼ | ½ | 20⅝ | 21 7/16 | 22¼ | 23 | 3 | 2¾ | 1½ | 1⅜ | 1⅛ | 1 1/16 |
| 21 | 3, 4 | ¼ | 9/16 | 24⅛ | 25 | 25⅞ | 26¾ | 3¼ | 3 | 1¾ | 1⅝ | 1 5/16 | 1 3/16 |
| 24 | 3, 4 | ⅜ | 9/16 | 27½ | 28½ | 29⅜ | 30⅜ | 3⅜ | 3⅛ | 2 | 1⅞ | 1½ | 1⅜ |
| 27 | 3, 4 | ⅜ | 9/16 | 31 | 32⅛ | 33 | 34⅛ | 3½ | 3¼ | 2¼ | 2⅛ | 1 11/16 | 1 9/16 |
| 30 | 3, 4 | ⅜ | ⅝ | 34⅜ | 35⅝ | 36½ | 37¾ | 3⅝ | 3⅜ | 2½ | 2⅜ | 1⅞ | 1¾ |
| 33 | 3, 4 | ⅜ | ⅝ | 37⅝ | 38 15/16 | 39⅞ | 41¼ | 3¾ | 3½ | 2⅝ | 2½ | 2 | 1 13/16 |
| 36 | 3, 4 | ⅜ | 11/16 | 40¾ | 42¼ | 43¼ | 44¾ | 4 | 3¾ | 2¾ | 2⅝ | 2 1/16 | 1⅞ |

When ordering standard-strength vitrified-clay pipe, give the size of pipe (ID) and the laying strength wanted, and refer to ASTM Specification C 13. Standard lengths of pipe shown meet normal practice in various sections of the country. Manufacturers' stocks include those lengths conforming to local practice.

*There is no limit for plus variation.

sures for one set of conditions. For similar tables for other conditions, refer to ANSI B16.5-1968.

**Sizes** The size of the fittings and companion flanges in the tables is identified by the corresponding nominal pipe size.

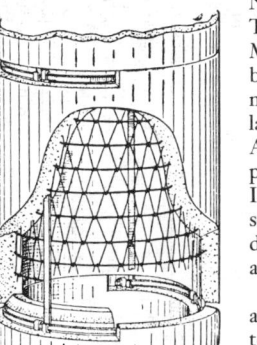

For pipe 14 in (35.6 cm) and larger, the corresponding outside diameter of the pipe is given.

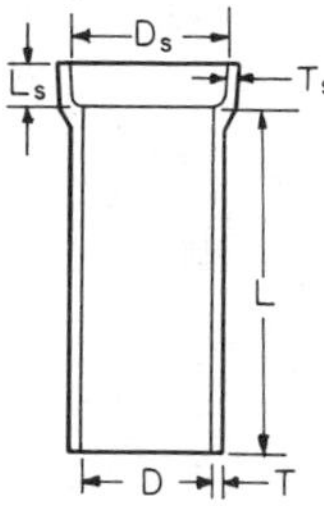

**Materials** The flanged fittings and flanges should be either steel castings or steel forgings of the grade complying with the ASTM specifications recommended under these standards for the various pressure-temperature ratings for which these standards are designed. A few of these characteristics from ANSI B16.5-1968 are given in Table 30.

**Fig. 6** Joint for vitrified pipe.

**Bolting material** including nuts and washers are based on a high-grade product equal to that given in ASTM Standard Specifications for Alloy-Steel Bolting Material for High Temperature Service and with physical and chemical requirements in accordance with the tables given under ANSI B16.5-1968. **Commercial steel bolts should not be used at steam pressures over 250 lb/in² (9,224 N/m²) and temperatures over 450°F (232°C).** Nuts should be of carbon or alloy steel. Washers when used under nuts should be of forged or rolled carbon steel.

**Bolting** Drilling templates are in multiples of four, so that fittings may be made to face in any quarter. Boltholes straddle the centerlines. Boltholes are drilled ⅛ in (0.32 cm) larger in

diameter than the nominal size of bolt. Bolts or bolt studs threaded at both ends may be used and should be equipped with cold-punched or cold-pressed semifinished nuts of American National Standard rough dimensions, chamfered and trimmed.

All bolts and bolt studs having diameters 1 in and smaller, and the corresponding nuts are threaded with the American National Standard Screw Thread, Coarse Thread Series, Medium Fit, Class 3, while those bolts and bolt studs whose diameters are 1⅛ in (2.86 cm) and larger have special threads of the American National form whose pitch is ⅛ in (8 threads per in). It is recommended that these special threads be allowed a pitch-diameter tolerance of −0.006 in and a lead tolerance of ±0.002 in.

Bolt studs with a nut at each end are recommended for high-temperature service.

The allowable working fiber stress, considering internal allowable working pressure only, in bolting material for value bonnet flanges, cleanout flanges, etc., is not to exceed 9,000 lb/in² assuming the pressure to act upon an area circumscribed by the periphery of the outside of the contact surface.

**Fig. 8** Reinforced-concrete pipe.

**Metal Thickness** Minimum metal thicknesses specified in the tables are based on an allowable fiber stress of 7,000 lb/in², using the modified Barlow formula of the ASME Boiler Construction Code for cylindrical sections and adding 50 percent to the thickness thus determined to compensate for the shape of the fittings. The minimum commercial casting thickness is considered to be ¼ in; therefore, the standards do not show thicknesses less than this. The minimum thickness in these standards means the minimum thickness in any part of the finished casting.

The modified Barlow formula is as follows: For pipes having nominal diameters of ¼ to 5 in, $P + 125 = 2S(t - 0.065)/D$. For pipes of nominal diameters over 5 in, $P = 2S(t - 0.1)/D$, where $P$ is the working pressure, lb/in², $t$ is the thickness of wall of pipe, in; $D$ is the actual outside diameter of pipe, in; and $S$ is 7,000 lb/in².

**Ring Joints** The dimensions used for ring and groove joint facings were developed by a committee of the API. The corresponding dimensions and ring numbers incorporated in the ANSI Standard are identical with those given in API Standard 5G. The dimension for the depth of groove is added to the basic flange thickness, which makes it necessary to include separate tables of dimensions for fittings having the ring joint facing.

**Fitting Dimensions** An inspection limit of ±1/32 in is allowed on all center-to-contact surface dimensions for sizes up to and including 10 in, and ±1/16 in on sizes larger than 10 in. An inspection limit of ±1/16 in is allowed on all contact-surface to contact-surface dimensions for sizes up to and including 10 in, and ±⅛ in, on sizes larger than 10 in.

When elbows having longer radii than specified in the standards are required, the use of pipe bends is recommended.

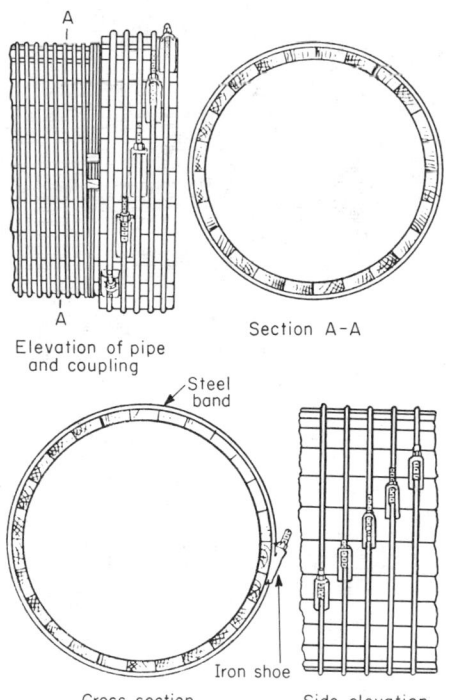

Elevation of pipe and coupling

Section A-A

Cross section

Side elevation

**Fig. 7** Wood-stave pipes.

**Table 28. Templates for Drilling Cast-Iron Pipe Flanges, Flanged Valves, and Fittings***
(ANSI B16.1-1967)

| Nominal pipe size, in | Diameter of flange, in | Thickness of flange (minimum), in | Diameter of raised face, in | Diameter of bolt circle, in | Number of bolts | Diameter of bolts, in | Diameter of drilled boltholes, in | Length of bolts, in | Length of bolt stud with two nuts, in | Total effective area bolt metal, in² | Stress in bolt metal, lb/in²† | Size of ring gasket, in |
|---|---|---|---|---|---|---|---|---|---|---|---|---|
| 25-lb standard (922 N/m²) | | | | | | | | | | | | |
| 4 | 9 | 3/4 | | 7½ | 8 | 5/8 | 3/4 | 2¼ | | 1.616 | 570 | 4 × 6⅞ |
| 5 | 10 | 3/4 | | 8½ | 8 | 5/8 | 3/4 | 2¼ | | 1.616 | 750 | 5 × 7⅞ |
| 6 | 11 | 3/4 | | 9½ | 8 | 5/8 | 3/4 | 2¼ | | 1.616 | 930 | 6 × 8¾ |
| 8 | 13½ | 3/4 | | 11¾ | 8 | 5/8 | 3/4 | 2¼ | | 1.616 | 1470 | 8 × 11 |
| 10 | 16 | 7/8 | | 14¼ | 12 | 5/8 | 3/4 | 2½ | | 2.424 | 1440 | 10 × 13⅜ |
| 12 | 19 | 1 | | 17 | 12 | 5/8 | 3/4 | 2¾ | | 2.424 | 2195 | 12 × 16⅛ |
| 14 | 21 | 1⅛ | | 18¾ | 12 | 3/4 | 7/8 | 3¼ | | 3.620 | 1750 | 14 × 18 |
| 16 | 23½ | 1⅛ | | 21¼ | 16 | 3/4 | 7/8 | 3¼ | | 4.830 | 1710 | 16 × 20½ |
| 18 | 25 | 1¼ | | 22¾ | 16 | 3/4 | 7/8 | 3½ | | 4.830 | 1965 | 18 × 22 |
| 20 | 27½ | 1¼ | | 25 | 20 | 3/4 | 7/8 | 3½ | | 6.040 | 1920 | 20 × 24½ |
| 24 | 32 | 1⅜ | | 29½ | 20 | 3/4 | 7/8 | 3¾ | | 6.040 | 2690 | 24 × 28¾ |
| 30 | 38¾ | 1½ | | 36 | 28 | 7/8 | 1 | 4¼ | | 11.760 | 2030 | 30 × 35⅛ |
| 36 | 46 | 1⅝ | | 42¾ | 32 | 7/8 | 1 | 5 | | 13.440 | 2610 | 36 × 41⅞ |
| 42 | 53 | 1¾ | | 49½ | 36 | 1 | 1⅛ | 5¼ | | 19.800 | 2315 | 42 × 48½ |
| 48 | 59½ | 2 | | 56 | 44 | 1 | 1⅛ | 5½ | | 24.200 | 2475 | 48 × 55 |
| 54 | 66¼ | 2¼ | | 62¾ | 44 | 1 | 1⅛ | 5¾ | | 24.200 | 3195 | 54 × 61¾ |
| 60 | 73 | 2¼ | | 69¼ | 52 | 1⅛ | 1¼ | 6 | | 36.020 | 2515 | 60 × 68⅛ |
| 72 | 86½ | 2½ | | 82½ | 60 | 1⅛ | 1¼ | 6¼ | | 41.570 | 3120 | 72 × 81⅜ |
| 84 | 99¾ | 2¾ | | 95½ | 64 | 1¼ | 1⅜ | 7¼ | | 57.140 | 3005 | 84 × 94¼ |
| 96 | 113¼ | 3 | | 108½ | 68 | 1¼ | 1⅜ | 7¾ | | 60.570 | 3705 | 96 × 107¼ |
| 125-lb standard (4,612 N/m²) | | | | | | | | | | | | |
| 1 | 4¼ | 7/16 | | 3⅛ | 4 | 1/2 | 5/8 | 1¾ | | | | 1 × 2⅝ |
| 1¼ | 4⅝ | 1/2 | | 3½ | 4 | 1/2 | 5/8 | 2 | | | | 1¼ × 3 |
| 1½ | 5 | 9/16 | | 3⅞ | 4 | 1/2 | 5/8 | 2 | | | | 1½ × 3⅜ |
| 2 | 6 | 5/8 | | 4¾ | 4 | 5/8 | 3/4 | 2¼ | | | | 2 × 4⅛ |
| 2½ | 7 | 11/16 | | 5½ | 4 | 5/8 | 3/4 | 2½ | | | | 2½ × 4⅞ |
| 3 | 7½ | 3/4 | | 6 | 4 | 5/8 | 3/4 | 2½ | | | | 3 × 5⅜ |
| 3½ | 8½ | 13/16 | | 7 | 8 | 5/8 | 3/4 | 2¾ | | | | 3½ × 6⅜ |
| 4 | 9 | 15/16 | | 7½ | 8 | 5/8 | 3/4 | 3 | | | | 4 × 6⅞ |
| 5 | 10 | 15/16 | | 8½ | 8 | 3/4 | 7/8 | 3 | | | | 5 × 7¾ |
| 6 | 11 | 1 | | 9½ | 8 | 3/4 | 7/8 | 3¼ | | | | 6 × 8¾ |
| 8 | 13½ | 1⅛ | | 11¾ | 8 | 3/4 | 7/8 | 3½ | | | | 8 × 11 |
| 10 | 16 | 1 3/16 | | 14¼ | 12 | 7/8 | 1 | 3¼ | | | | 10 × 13⅜ |

| | | | | | | | | | | |
|---|---|---|---|---|---|---|---|---|---|---|
| 12 | 19 | 1 1/4 | | 17 | 12 | 7/8 | 1 | 3 3/4 | | 12 × 16 1/8 |
| 14 OD | 21 | 1 3/8 | | 18 3/4 | 12 | 1 | 1 1/8 | 4 1/4 | | 14 × 17 3/4 |
| 16 OD | 23 1/2 | 1 7/16 | | 21 1/4 | 16 | 1 | 1 1/8 | 4 1/2 | | 16 × 20 1/4 |
| 18 OD | 25 | 1 9/16 | | 22 3/4 | 16 | 1 1/8 | 1 1/4 | 4 3/4 | | 18 × 21 5/8 |
| 20 OD | 27 1/2 | 1 11/16 | | 25 | 20 | 1 1/8 | 1 1/4 | 5 | | 20 × 23 7/8 |
| 24 OD | 32 | 1 7/8 | | 29 1/2 | 20 | 1 1/4 | 1 3/8 | 5 1/2 | | 24 × 28 1/4 |
| 30 OD | 38 3/4 | 2 1/8 | | 36 | 28 | 1 1/4 | 1 3/8 | 6 1/4 | | 30 × 34 5/8 |
| 36 OD | 46 | 2 3/8 | | 42 3/4 | 32 | 1 1/2 | 1 5/8 | 7 | | 36 × 41 1/4 |
| 42 OD | 53 | 2 5/8 | | 49 1/2 | 36 | 1 1/2 | 1 5/8 | 7 1/2 | | 42 × 48 |
| 48 OD | 59 1/2 | 2 3/4 | | 56 | 44 | 1 1/2 | 1 5/8 | 7 3/4 | | 48 × 54 1/2 |
| 54 OD | 66 1/4 | 3 | | 62 3/4 | 44 | 1 3/4 | 2 | 8 1/2 | 10 1/2 | 54 × 61 |
| 60 OD | 73 | 3 1/8 | | 69 1/4 | 52 | 1 3/4 | 2 | 8 3/4 | 11 | 60 × 67 1/2 |
| 72 OD | 86 1/2 | 3 1/2 | | 82 1/2 | 60 | 1 3/4 | 2 | 9 1/2 | 12 | 72 × 80 3/4 |
| 84 OD | 99 3/4 | 3 7/8 | | 95 1/2 | 64 | 2 | 2 1/4 | 10 1/2 | 13 | 84 × 93 1/2 |
| 96 OD | 113 1/4 | 4 1/4 | | 108 1/2 | 68 | 2 1/4 | 2 1/2 | 11 1/2 | 14 1/2 | 96 × 106 1/4 |
| 250-lb standard (9,224 N/m²) | | | | | | | | | | |
| 1 | 4 7/8 | 11/16 | 2 11/16 | 3 1/2 | 4 | 5/8 | 3/4 | 2 1/4 | | 1 × 2 7/8 |
| 1 1/4 | 5 1/4 | 3/4 | 3 1/16 | 3 7/8 | 4 | 5/8 | 3/4 | 2 1/2 | | 1 1/4 × 3 1/4 |
| 1 1/2 | 6 1/8 | 13/16 | 3 9/16 | 4 1/2 | 4 | 3/4 | 7/8 | 2 1/2 | | 1 1/2 × 3 3/4 |
| 2 | 6 1/2 | 7/8 | 4 3/16 | 5 | 8 | 5/8 | 3/4 | 2 1/2 | | 2 × 4 3/8 |
| 2 1/2 | 7 1/2 | 1 | 4 15/16 | 5 7/8 | 8 | 3/4 | 7/8 | 3 | | 2 1/2 × 5 1/8 |
| 3 | 8 1/4 | 1 1/8 | 5 11/16 | 6 5/8 | 8 | 3/4 | 7/8 | 3 1/4 | | 3 × 5 7/8 |
| 3 1/2 | 9 | 1 3/16 | 6 5/16 | 7 1/4 | 8 | 3/4 | 7/8 | 3 1/4 | | 3 1/2 × 6 1/2 |
| 4 | 10 | 1 1/4 | 6 15/16 | 7 7/8 | 8 | 3/4 | 7/8 | 3 1/2 | | 4 × 7 1/8 |
| 5 | 11 | 1 3/8 | 8 5/16 | 9 1/4 | 8 | 3/4 | 7/8 | 3 3/4 | | 5 × 8 1/2 |
| 6 | 12 1/2 | 1 7/16 | 9 11/16 | 10 5/8 | 12 | 3/4 | 7/8 | 3 3/4 | | 6 × 9 7/8 |
| 8 | 15 | 1 5/8 | 11 15/16 | 13 | 12 | 7/8 | 1 | 4 1/4 | | 8 × 12 1/8 |
| 10 | 17 1/2 | 1 7/8 | 14 1/16 | 15 1/4 | 16 | 1 | 1 1/8 | 5 | | 10 × 14 1/4 |
| 12 | 20 1/2 | 2 | 16 7/16 | 17 3/4 | 16 | 1 1/8 | 1 1/4 | 5 1/2 | | 12 × 16 5/8 |
| 14 OD | 23 | 2 1/8 | 18 15/16 | 20 1/4 | 20 | 1 1/8 | 1 1/4 | 5 3/4 | | 13 1/4 × 19 1/8 |
| 16 OD | 25 1/2 | 2 1/4 | 21 1/16 | 22 1/2 | 20 | 1 1/4 | 1 3/8 | 6 | | 15 1/4 × 21 1/4 |
| 18 OD | 28 | 2 3/8 | 23 5/16 | 24 3/4 | 24 | 1 1/4 | 1 3/8 | 6 1/4 | | 17 × 23 1/2 |
| 20 OD | 30 1/2 | 2 1/2 | 25 9/16 | 27 | 24 | 1 1/4 | 1 3/8 | 6 1/2 | | 19 × 25 3/4 |
| 24 OD | 36 | 2 3/4 | 30 5/16 | 32 | 24 | 1 1/2 | 1 5/8 | 7 1/2 | 9 1/2 | 23 × 30 1/2 |
| 30 OD | 43 | 3 | 37 3/16 | 39 1/4 | 28 | 1 3/4 | 2 | 8 1/4 | 10 1/2 | 29 × 37 1/2 |
| 36 OD | 50 | 3 3/8 | 43 11/16 | 46 | 32 | 2 | 2 1/4 | 9 1/4 | 11 1/2 | 34 1/2 × 44 |
| 42 OD | 57 | 3 11/16 | 50 7/16 | 52 3/4 | 36 | 2 | 2 1/4 | 9 3/4 | 12 | 40 1/4 × 50 3/4 |
| 48 OD | 65 | 4 | 58 7/16 | 60 3/4 | 40 | 2 | 2 1/4 | 10 1/2 | 13 | 46 × 58 3/4 |

*To obtain linear dimensions in cm, multiply tabular values in in by 2.54. To obtain areas in cm², multiply tabular values in in² by 6.45. To obtain stress in N/m², multiply values in lb/in² by 36.895.

†The stress shown is that of internal pressure assumed to act only on a circular area equal in diameter to the outside diameter of the ring gasket covering the flange to the inside of the bolts for the 25-lb standard.

**Table 29. Nonshock Pressure and Temperature Ratings for Carbon-Steel Flanges and Flanged Fittings**
(ANSI B16.5-1968)

| Primary pressure rating, lb/in² gage | 150 | 300 | 400 | 600 | 900 | 1,500 |
|---|---|---|---|---|---|---|
| Hydrostatic shell test pressure, lb/in² gage | 425 | 1,100 | 1,450 | 2,175 | 3,250 | 5,400 |
| Service temp, °F | | | Max nonshock pressures, lb/in² gage | | | |
| 100 | 275 | 720 | 960 | 1,440 | 2,160 | 3,600 |
| 150 | 255 | 710 | 945 | 1,420 | 2,130 | 3,550 |
| 200 | 240 | 700 | 930 | 1,400 | 2,100 | 3,500 |
| 250 | 225 | 690 | 920 | 1,380 | 2,070 | 3,450 |
| 300 | 210 | 680 | 910 | 1,365 | 2,050 | 3,415 |
| 350 | 195 | 675 | 900 | 1,350 | 2,025 | 3.375 |
| 400 | 180 | 665 | 890 | 1,330 | 2,000 | 3,330 |
| 450 | 165 | 650 | 870 | 1,305 | 1,955 | 3,255 |
| 500 | 150* | 625 | 835 | 1,250 | 1,875 | 3,125 |
| 550 | 140 | 590 | 790 | 1,180 | 1,775 | 2,955 |
| 600 | 130 | 555 | 740 | 1,110 | 1,660 | 2,770 |
| 650 | 120 | 515 | 690 | 1,030 | 1,550 | 2,580 |
| 700 | 110 | 470 | 635 | 940 | 1,410 | 2,350 |
| 750 | 100 | 425 | 575 | 850 | 1,275 | 2,125 |
| 800 | 92 | 365 | 365 | 730 | 1,100 | 1,830 |
| 850 | 82 | 300* | 400* | 600* | 900* | 1,500* |
| 900 | 70 | 225 | 280 | 445 | 670 | 1,115 |
| 950 | 55 | 155 | 220 | 310 | 465 | 770 |
| 1000 | 40 | 85 | 160 | 170 | 255 | 430 |

*Primary service-pressure rating.

**Laterals**   The 45° laterals of the larger sizes may require additional reinforcement to compensate for the inherent weakness in this shape of casting.

**Valve Dimensions**   The center to contact-surface and contact-surface to contact-surface dimensions of valves for the various pressures are in accordance with the requirements of ANSI B16.10-1957 for ferrous flanged valves.

**Reducing Fittings**   Reducing fittings have the same center-to-flange edge dimensions as those of straight-size fittings of the largest opening.

**Side-Outlet Fittings**   All side-outlet fittings have all openings on the intersecting centerlines.

**Welding Neck Flanges**   The materials, facings, spot facings, etc., conform to the requirements given for other flanges,

with the additional provision that the carbon content of the steel shall not exceed 0.35 percent.

Templates for drilling and center to contact-surface dimensions of the American Standard 150-lb steel flanges and flanged fittings are the same as for the American Standard 125-lb cast-iron flanged fitting standard.

Templates for drilling and center to contact-surface dimensions of the ANSI 300-lb steel flanges and flanged fittings are the same as for the ANSI 600-lb steel flanged fitting standard for sizes ½ to 1½ in (Table 33); and the same as for the ANSI 250-lb cast-iron flanged fitting standard for sizes 2 to 24 in.

**Flanged Pipe Joints**

The usual form of pipe joint is that made up by bolting

**Table 30. Physical Characteristics of Steels for Flanges and Fittings**
(ASTM Spec. A193-63T)

| | Steel castings* | Alloy-steel bolt material† | | | Forged-steel flanges‡ | |
|---|---|---|---|---|---|---|
| | | Class A | Class B | Class C | Class I | Class II |
| Tensile strength, lb/in² | 70,000 | 95,000 | 105,000 | 125,000 | 60,000 | 70,000 |
| Yield point, min, lb/in² | 36,000 | 70,000 | 80,000 | 105,000 | 30,000 | 36,000 |
| Elongation in 2 in, min, percent | 22 | 20 | 20 | 16 | 22 | 18 |
| Reduction of area, min, percent | 30 | 50 | 50 | 50 | 35 | 24 |

Phosphorus max, 0.05 percent; sulfur max, 0.05 percent.
*Carbon steel (ASTM A216-60T).
†ASTM A193-62T.
‡ASTM A181-61T.

**Table 31. Facing Dimensions for the American 150-, 300-, 400-, 600-, 900-, 1,500- and 2,500-lb Steel Flanges (ANSI B16.5-1968)**

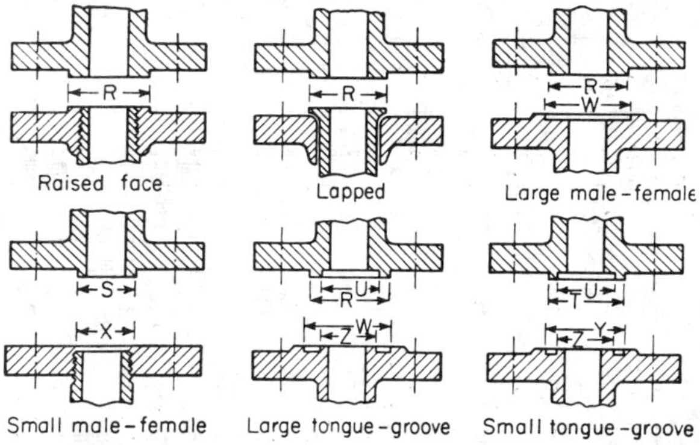


| Nominal pipe size, in. | Outside diameter, in: Raised face, lapped, large male, and large tongue, *R* | Small male, *S* | Small tongue, *T* | ID of large and small tongue, *U* | Outside diameter, in: Large female and large groove, *W* | Small female, *X* | Small groove, *Y* | ID of large and small groove, *Z* | Height, in: Raised face, 150 and 300 lb std* | Raised face, large and small male and tongue 400-, 600-, 900-, 1,500- and 2,500-lb studs | Depth of groove or female companion flanges |
|---|---|---|---|---|---|---|---|---|---|---|---|
| 1/2 | 1 3/8 | 23/32 | 1 3/8 | 1 | 1 7/16 | 25/32 | 1 7/16 | 15/16 | 1/16 | 1/4 | 3/16 |
| 3/4 | 1 11/16 | 15/16 | 1 11/16 | 1 5/16 | 1 3/4 | 1 | 1 3/4 | 1 1/4 | 1/16 | 1/4 | 3/16 |
| 1 | 2 | 1 3/16 | 1 7/8 | 1 1/2 | 2 1/16 | 1 1/4 | 1 13/16 | 1 7/16 | 1/16 | 1/4 | 3/16 |
| 1 1/4 | 2 1/2 | 1 1/2 | 2 1/4 | 1 7/8 | 2 9/16 | 1 9/16 | 2 5/16 | 1 13/16 | 1/16 | 1/4 | 3/16 |
| 1 1/2 | 2 7/8 | 1 3/4 | 2 1/2 | 2 1/8 | 2 15/16 | 1 13/16 | 2 9/16 | 2 1/16 | 1/16 | 1/4 | 3/16 |
| 2 | 3 5/8 | 2 1/4 | 3 1/4 | 2 7/8 | 3 11/16 | 2 5/16 | 3 5/16 | 2 13/16 | 1/16 | 1/4 | 3/16 |
| 2 1/2 | 4 1/8 | 2 11/16 | 3 3/4 | 3 3/8 | 4 3/16 | 2 3/4 | 3 13/16 | 3 5/16 | 1/16 | 1/4 | 3/16 |
| 3 | 5 | 3 5/16 | 4 5/8 | 4 1/4 | 5 1/16 | 3 3/8 | 4 11/16 | 4 3/16 | 1/16 | 1/4 | 3/16 |
| 3 1/2 | 5 1/2 | 3 13/16 | 5 1/8 | 4 3/4 | 5 9/16 | 3 7/8 | 5 3/16 | 4 11/16 | 1/16 | 1/4 | 3/16 |
| 4 | 6 3/16 | 4 5/16 | 5 11/16 | 5 3/16 | 6 1/4 | 4 3/8 | 5 3/4 | 5 1/8 | 1/16 | 1/4 | 3/16 |
| 5 | 7 5/16 | 5 3/8 | 6 13/16 | 6 5/16 | 7 3/8 | 5 7/16 | 6 7/8 | 6 1/4 | 1/16 | 1/4 | 3/16 |
| 6 | 8 1/2 | 6 3/8 | 8 | 7 1/2 | 8 9/16 | 6 7/16 | 8 1/16 | 7 7/16 | 1/16 | 1/4 | 3/16 |
| 8 | 10 5/8 | 8 3/8 | 10 | 9 3/8 | 10 11/16 | 8 7/16 | 10 1/16 | 9 5/16 | 1/16 | 1/4 | 3/16 |
| 10 | 12 3/4 | 10 1/2 | 12 | 11 1/4 | 12 13/16 | 10 9/16 | 12 1/16 | 11 3/16 | 1/16 | 1/4 | 3/16 |
| 12 | 15 | 12 1/2 | 14 1/4 | 13 1/2 | 15 1/16 | 12 9/16 | 14 5/16 | 13 7/16 | 1/16 | 1/4 | 3/16 |
| 14 O.D. | 16 1/4 | 13 3/4 | 15 1/2 | 14 3/4 | 16 5/16 | 13 13/16 | 15 9/16 | 14 11/16 | 1/16 | 1/4 | 3/16 |
| 16 O.D. | 18 1/2 | 15 3/4 | 17 5/8 | 16 3/4 | 18 9/16 | 15 13/16 | 17 11/16 | 16 11/16 | 1/16 | 1/4 | 3/16 |
| 18 O.D. | 21 | 17 3/4 | 20 1/8 | 19 1/4 | 21 1/16 | 17 13/16 | 20 3/16 | 19 3/16 | 1/16 | 1/4 | 3/16 |
| 20 O.D. | 23 | 19 3/4 | 22 | 21 | 23 1/16 | 19 13/16 | 22 1/16 | 20 15/16 | 1/16 | 1/4 | 3/16 |
| 24 O.D. | 27 1/4 | 23 3/4 | 26 1/4 | 25 1/4 | 27 5/16 | 23 13/16 | 26 5/16 | 25 3/16 | 1/16 | 1/4 | 3/16 |

*Included in the minimum flange thickness dimensions. A 1/16-in raised face is also permitted on the 400-, 600-, 900-, 1,500-, and 2,500-lb flange standards, but it must be added to the minimum flange thicknesses. Regular facing for 400-, 600-, 900-, 1,500-, and 2,500-lb flange standards is a 1/4-in raised face not included in minimum flange thicknesses dimensions.

A tolerance of 1/64 in is allowed on the inside and outside diameters of all facings.

Gaskets for male-female and tongue-groove joints should cover the bottom of the recess with minimum clearance taking into account the tolerances stated above.

together flanges cast or forged integral with the pipe or fitting, threaded flanges, loose flanges on pipes with lapped ends, and flanges arranged for welding. These forms are illustrated above in Table 31 and in Fig. 9. The threaded joint is satisfactory for low and medium steam pressures. The lapped joint is permitted in the same sizes and service ratings as for joints with integral flanges. It is extensively used in high-class work. With the ring joint a higher pressure can be maintained with the same total bolt stress than is possible with the flat gasket type of joint. The welded joint eliminates possibility of leak-

**Table 32. Templates for Drilling, American National Standard Steel Pipe Flanges and Flanged Fittings (ANSI B16-5-1968)***

(All dimensions in inches)

| Nominal pipe size | 400 lb standard | | | | | 600 lb standard | | | | | 900 lb standard | | | | | 1,500 lb standard | | | | |
|---|---|---|---|---|---|---|---|---|---|---|---|---|---|---|---|---|---|---|---|---|
| | Outside diam of flange | Thickness of flange, minimum | Diam of bolt circle | Number of bolts | Size of bolts | Outside diam of flange | Thickness of flange, minimum | Diam of bolt circle | Number of bolts | Size of bolts | Outside diam of flange | Thickness of flange, minimum | Diam of bolt circle | Number of bolts | Size of bolts | Outside diam of flange | Thickness of flange, minimum | Diam of bolt circle | Number of bolts | Size of bolts |
| ½ | For sizes below 4 in. use dimensions of 600 lb fittings | | | | | 3¾ | 9/16 | 2⅝ | 4 | ½ | For sizes below 3 in. use dimensions of 1,500 lb fittings | | | | | 4¾ | ⅞ | 3¼ | 4 | ¾ |
| ¾ | | | | | | 4⅝ | ⅝ | 3¼ | 4 | ⅝ | | | | | | 5⅛ | 1 | 3½ | 4 | ¾ |
| 1 | | | | | | 4⅞ | 11/16 | 3½ | 4 | ⅝ | | | | | | 5⅞ | 1⅛ | 4 | 4 | ⅞ |
| 1¼ | ...... | ...... | ...... | ...... | ...... | 5¼ | 13/16 | 3⅞ | 4 | ⅝ | ...... | ...... | ...... | ...... | ...... | 6¼ | 1⅛ | 4⅜ | 4 | ⅞ |
| 1½ | ...... | ...... | ...... | ...... | ...... | 6⅛ | ⅞ | 4½ | 4 | ¾ | ...... | ...... | ...... | ...... | ...... | 7 | 1¼ | 4⅞ | 4 | 1 |
| 2 | ...... | ...... | ...... | ...... | ...... | 6½ | 1 | 5 | 8 | ⅝ | ...... | ...... | ...... | ...... | ...... | 8½ | 1½ | 6½ | 8 | ⅞ |
| 2½ | ...... | ...... | ...... | ...... | ...... | 7½ | 1⅛ | 5⅞ | 8 | ¾ | ...... | ...... | ...... | ...... | ...... | 9⅝ | 1⅝ | 7½ | 8 | 1 |
| 3 | ...... | ...... | ...... | ...... | ...... | 8¼ | 1¼ | 6⅝ | 8 | ¾ | 9½ | 1½ | 7½ | 8 | ⅞ | 10½ | 1⅞ | 8 | 8 | 1⅛ |
| 3½ | ...... | ...... | ...... | ...... | ...... | 9 | 1⅜ | 7¼ | 8 | ⅞ | | | | | | | | | | |
| 4 | 10 | 1⅜ | 7⅞ | 8 | ⅞ | 10¾ | 1½ | 8½ | 8 | ⅞ | 11½ | 1¾ | 9¼ | 8 | 1⅛ | 12¼ | 2⅛ | 9½ | 8 | 1¼ |
| 5 | 11 | 1½ | 9¼ | 8 | ⅞ | 13 | 1¾ | 10½ | 8 | 1 | 13¾ | 2 | 11 | 8 | 1¼ | 14¾ | 2⅞ | 11½ | 8 | 1½ |
| 6 | 12½ | 1⅝ | 10⅝ | 12 | ⅞ | 14 | 1⅞ | 11½ | 12 | 1 | 15 | 2 3/16 | 12½ | 12 | 1⅛ | 15½ | 3¼ | 12½ | 12 | 1⅜ |
| 8 | 15 | 1⅞ | 13 | 12 | 1 | 16½ | 2 3/16 | 13¾ | 12 | 1⅛ | 18½ | 2½ | 15½ | 12 | 1⅜ | 19 | 3⅝ | 15½ | 12 | 1⅝ |
| 10 | 17½ | 2⅛ | 15¼ | 16 | 1⅛ | 20 | 2½ | 17 | 16 | 1¼ | 21½ | 2¾ | 18½ | 16 | 1⅜ | 23 | 4¼ | 19 | 12 | 1⅞ |
| 12 | 20½ | 2¼ | 17¾ | 16 | 1¼ | 22 | 2⅝ | 19¼ | 20 | 1¼ | 24 | 3⅛ | 21 | 20 | 1⅜ | 26½ | 4⅞ | 22½ | 16 | 2 |
| 14 O.D. | 23 | 2⅜ | 20¼ | 20 | 1¼ | 23¾ | 2¾ | 20¾ | 20 | 1⅜ | 25¼ | 3⅜ | 22 | 20 | 1½ | 29½ | 5¼ | 25 | 16 | 2¼ |
| 16 O.D. | 25½ | 2½ | 22½ | 20 | 1⅜ | 27 | 3 | 23¾ | 20 | 1½ | 27¾ | 3½ | 24¼ | 20 | 1⅝ | 32½ | 5¾ | 27¾ | 16 | 2½ |
| 18 O.D. | 28 | 2⅝ | 24¾ | 24 | 1⅜ | 29¼ | 3¼ | 25¾ | 20 | 1⅝ | 31 | 4 | 27 | 20 | 1⅞ | 36 | 6⅜ | 30½ | 16 | 2¾ |
| 20 O.D. | 30½ | 2¾ | 27 | 24 | 1½ | 32 | 3½ | 28½ | 24 | 1⅝ | 33¾ | 4¼ | 29½ | 20 | 2 | 38¾ | 7 | 32¾ | 16 | 3 |
| 24 O.D. | 36 | 3 | 32 | 24 | 1¾ | 37 | 4 | 33 | 24 | 1⅞ | 41 | 5½ | 35½ | 20 | 2½ | 46 | 8 | 39 | 16 | 3½ |

*See Introductory Notes, p. 8–?.

**Table 33. Dimensions of American National Standard Companion Flanges (ANSI B16.5-1968)***
(All dimensions in inches)

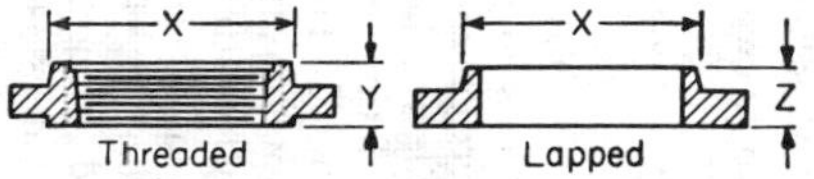


| Nom pipe size | 150 lb | | | 300 lb | | | 400 lb | | | 600 lb | | | 900 lb | | | 1,500 lb | | |
|---|---|---|---|---|---|---|---|---|---|---|---|---|---|---|---|---|---|---|
| | X | Y | Z | X | Y | Z | X | Y | Z | X | Y | Z | X | Y | Z | X | Y | Z |
| 1/2 | 1 3/16 | 5/8 | 5/8 | 1 1/2 | 7/8 | 7/8 | For sizes below 4 in., | | | 1 1/2 | 7/8 | 7/8 | For sizes below 3 | | | 1 1/2 | 1 1/4 | 1 1/4 |
| 3/4 | 1 1/2 | 5/8 | 5/8 | 1 7/8 | 1 | 1 | use dimensions of | | | 1 7/8 | 1 | 1 | in., use dimen- | | | 1 3/4 | 1 3/8 | 1 3/8 |
| 1 | 1 15/16 | 11/16 | 11/16 | 2 1/8 | 1 1/16 | 1 1/16 | the 600 lb flanges | | | 2 1/8 | 1 1/16 | 1 1/16 | sions of 1,500 lb | | | 2 1/16 | 1 5/8 | 1 5/8 |
| 1 1/4 | 2 5/16 | 13/16 | 13/16 | 2 1/2 | 1 1/16 | 1 1/16 | .... | ..... | ..... | 2 1/2 | 1 1/8 | 1 1/8 | flanges | | | 2 1/2 | 1 5/8 | 1 5/8 |
| 1 1/2 | 2 9/16 | 7/8 | 7/8 | 2 3/4 | 1 3/16 | 1 3/16 | .... | ..... | ..... | 2 3/4 | 1 1/4 | 1 1/4 | .... | ... | .... | 2 3/4 | 1 3/4 | 1 3/4 |
| 2 | 3 1/16 | 1 | 1 | 3 5/16 | 1 5/16 | 1 5/16 | .... | ..... | ..... | 3 5/16 | 1 7/16 | 1 7/16 | .... | ... | .... | 4 1/8 | 2 1/4 | 2 1/4 |
| 2 1/2 | 3 9/16 | 1 1/8 | 1 1/8 | 3 15/16 | 1 1/2 | 1 1/2 | .... | ..... | ..... | 3 15/16 | 1 5/8 | 1 5/8 | .... | ... | .... | 4 7/8 | 2 1/2 | 2 1/2 |
| 3 | 4 1/4 | 1 3/16 | 1 3/16 | 4 5/8 | 1 11/16 | 1 11/16 | .... | ..... | ..... | 4 5/8 | 1 13/16 | 1 13/16 | 5 | 2 1/8 | 2 1/8 | 5 1/4 | 2 7/8 | 2 7/8 |
| 3 1/2 | 4 13/16 | 1 1/4 | 1 1/4 | 5 1/4 | 1 3/4 | 1 3/4 | .... | ..... | ..... | 5 1/4 | 1 15/16 | 1 15/16 | | | | | | |
| 4 | 5 5/16 | 1 5/16 | 1 5/16 | 5 3/4 | 1 7/8 | 1 7/8 | 5 3/4 | 2 | 2 | 6 | 2 1/8 | 2 1/8 | 6 1/4 | 2 3/4 | 2 3/4 | 6 3/8 | 3 9/16 | 3 9/16 |
| 5 | 6 7/16 | 1 7/16 | 1 7/16 | 7 | 2 | 2 | 7 | 2 1/8 | 2 1/8 | 7 7/16 | 2 3/8 | 2 3/8 | 7 1/2 | 3 1/8 | 3 1/8 | 7 3/4 | 4 1/8 | 4 1/8 |
| 6 | 7 9/16 | 1 9/16 | 1 9/16 | 8 1/8 | 2 1/16 | 2 1/16 | 8 1/8 | 2 1/4 | 2 1/4 | 8 3/4 | 2 5/8 | 2 5/8 | 9 1/4 | 3 3/8 | 3 3/8 | 9 | 4 11/16 | 4 11/16 |
| 8 | 9 11/16 | 1 3/4 | 1 3/4 | 10 1/4 | 2 7/16 | 2 7/16 | 10 1/4 | 2 11/16 | 2 11/16 | 10 3/4 | 3 | 3 | 11 3/4 | 4 | 4 1/2 | 11 1/2 | 5 5/8 | 5 5/8 |
| 10 | 12 | 1 15/16 | 1 15/16 | 12 5/8 | 2 5/8 | 3 3/4 | 12 5/8 | 2 7/8 | 4 | 13 1/2 | 3 3/8 | 4 3/8 | 14 1/2 | 4 1/4 | 5 | 14 1/2 | 6 1/4 | 7 |
| 12 | 14 3/8 | 2 3/16 | 2 3/16 | 14 3/4 | 2 7/8 | 4 | 14 3/4 | 3 1/8 | 4 1/4 | 15 3/4 | 3 5/8 | 4 5/8 | 16 1/2 | 4 5/8 | 5 5/8 | 17 3/4 | 7 1/8 | 8 5/8 |
| 14 O.D. | 15 3/4 | 2 1/4 | 3 1/8 | 16 3/4 | 3 | 4 3/8 | 16 3/4 | 3 5/16 | 4 5/8 | 17 | 3 11/16 | 5 | 17 3/4 | 5 1/8 | 6 1/8 | 19 1/2 | ..... | 9 1/2 |
| 16 O.D. | 18 | 2 1/2 | 3 7/16 | 19 | 3 1/4 | 4 3/4 | 19 | 3 11/16 | 5 | 19 1/2 | 4 3/16 | 5 1/2 | 20 | 5 1/4 | 6 1/2 | 21 3/4 | ..... | 10 1/4 |
| 18 O.D. | 19 7/8 | 2 11/16 | 3 13/16 | 21 | 3 1/2 | 5 1/8 | 21 | 3 7/8 | 5 3/8 | 21 1/2 | 4 5/8 | 6 | 22 1/4 | 6 | 7 1/2 | 23 1/2 | ..... | 10 7/8 |
| 20 O.D. | 22 | 2 7/8 | 4 1/16 | 23 1/8 | 3 3/4 | 5 1/2 | 23 1/8 | 4 | 5 3/4 | 24 | 5 | 6 1/4 | 24 1/2 | 6 1/2 | 8 1/4 | 25 1/4 | ..... | 11 1/2 |
| 24 O.D. | 26 1/8 | 3 1/4 | 4 3/8 | 27 5/8 | 4 3/16 | 6 | 27 5/8 | 4 1/2 | 6 1/4 | 28 1/4 | 5 1/2 | 7 1/4 | 29 1/2 | 8 | 10 1/2 | 30 | ..... | 13 |

* Other dimensions are given in Tables 31 and 32. Finished bore on lapped flange to be such as method of attachment of pipe requires.

age between flange and pipe. It is very successful in lines subject to high temperatures and pressures and heavy expansion strains. The welding-neck flange is available in the various pipe sizes. Specific requirements covering the application of all the types of joints in common use are outlined in the Code for Pressure Piping (ANSI B31.1.0-1967).

**Facing of Flanges**   Various styles of finish are used on the faces of flanges, having for their purpose the retention of the gasket used to make a tight joint. Those in general use are as follows (see Table 31): plain straight face, plain face corrugated or scored, male and female, tongue and groove, and raised face.

The **plain straight face** has the entire face of the flange faced straight across and may be used with either a full face or ring gasket. The **plain face, serrated or V-grooved,** is a plain face upon which concentric grooves have been cut with either a round-nose or V-shaped tool. This finish is sometimes of advantage when the service demands an exceptionally thick, loosely woven fibrous or soft metallic gasket, because the roughening of the faces of the flanges tends to keep the gasket from blowing out. The **male-and-female** facing consists of a recess in one flange and a corresponding raised face or projection on the other, extending from the inside of the pipe nearly to the inside of the boltholes. In the **tongue-and-groove** facing, the tongue or raised face and the groove or recess are narrow rings located between the boltholes and the port. The male-and-female and the tongue-and-groove facing have been extensively used, particularly on hydraulic lines, and to a more limited extent on high-pressure steam lines. Both these types, however, have in common several objectionable features from the standpoint of manufacture, erection, and maintenance. These objections are removed by the use of the **raised-face** facing, which consists of a high narrow raised ring on each of

the mating flanges, whose inside diameter is the same as that of the pipe or port. It is particularly recommended for high-pressure steam and hydraulic lines. **Gaskets** used in this type of joint are either soft fibrous material or soft metal and extend from the inside of the pipe to the boltholes, and only the small portion in contact with the narrow raised face is subjected to the compressive effect of the bolts. The following advantages are claimed for the raised-face type of facing: all mating of flanges has been eliminated; any valve or fitting may be removed from the line without springing the line apart; the gasket is automatically centered by its outer edge coming in contact with the bolts; the outside edges of the flanges are far enough apart to make it possible to determine whether the joint has been properly made.

**Unions** may be classified as **screw** and **flange.** Typical designs are shown in Fig. 10, where at the top left is represented a female screw union of the gasket type, at the top right a female screw union having a brass to iron seat that is noncorrosive and a ground joint that eliminates the need for a gasket, and at the bottom a flange union of the gasket type. As in the case of other pipe fittings, unions and union fittings are available in the various pipe sizes and in materials and designs suitable for any service conditions. Very large flange unions can be made by bolting together two screwed companion flanges.

### Screwed Fittings

Screwed fittings are made of cast iron, malleable iron, cast steel, forged steel, or brass. Plain standard fittings are generally used for low-pressure gas and water, as in house plumbing and railing work, while the beaded fitting is the standard steam, air, gas, or oil fitting. Screwed fittings are supplied with a large factor of safety. The questions of strength involve much more than the pressure from within the pipe which induces a comparatively low stress in the material. The greater strains come from expansion, contraction, weight of piping, settling, water hammer, etc. Dimensions of cast-iron and malleable-iron screwed fittings of the American National Standard are given in Tables 34 and 35.

The dimensions of ferrous plugs, bushings, locknuts, and caps with pipe threads are covered by ANSI B16.14-1971. The dimensions of pipe plugs from this standard are given in Table 42.

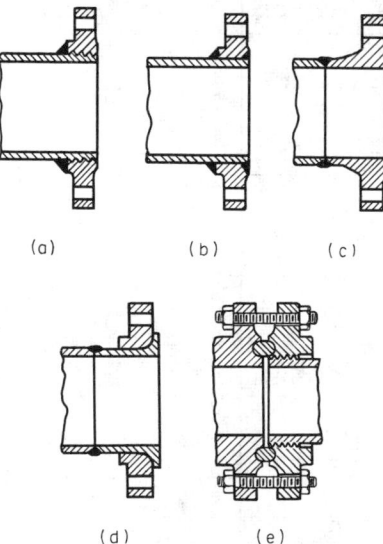

**Fig. 9**   Welded flange joints and ring joint. (*a*) Forged steel, screwed flange, back-welded, and refaced; (*b*) forged steel, slip-on welding flange, welded front and back and refaced; (*c*) forged steel, welding neck flange, butt-welded to pipe; (*d*) lap welding nipple, butt-welded to pipe; (*e*) ring joint.

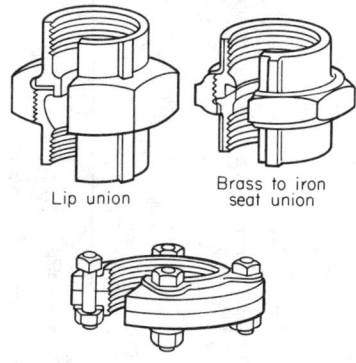

Lip union

Brass to iron
seat union

Flange union

**Fig. 10**   Types of pipe unions.

The normal **amount of thread engagement** necessary to make a tight joint for ANSI Standard Pipe Thread joints as recommended by Crane Co. is as follows:

| | | | | | | | | | |
|---|---|---|---|---|---|---|---|---|---|
| Size of pipe, in. | 1/8 | 1/4 | 3/8 | 1/2 | 3/4 | 1 | 1 1/4 | 1 1/2 | 2 |
| Length of thread, in. | 1/4 | 3/8 | 3/8 | 1/2 | 9/16 | 11/16 | 11/16 | 11/16 | 3/4 |
| Size of pipe, in. | 2 1/2 | 3 | 3 1/2 | 4 | 5 | 6 | 8 | 10 | 12 |
| Length of thread, in. | 15/16 | 1 | 1 1/16 | 1 1/8 | 1 1/4 | 1 5/16 | 1 7/16 | 1 5/8 | 1 3/4 |

The Manufacturers' Standardization Society of Valve and Fitting Industry (MSS) has standardized malleable-iron and bronze screw fittings for several pressures.

**Brass screwed fittings** are made in both the 125- and 250-lb standards. They are used for any water pipe where bad water makes steel pipe undesirable. Brass fittings may be had in iron pipe sizes. Forged-steel screwed fittings are made for cold water or oil-working pressures up to 6,000 lb/in² hydrostatic. The ANSI has approved standard B16.26-1967 for brass fittings for flared copper tubes for maximum cold-water service pressure of 175 lb/in².

**Railing Fittings** Fittings of special construction and of lighter weight than standard steam, gas, and water pipe fittings are widely used for hand railings around areaways, on stairs, for office enclosures with gates, and for permanent ladders. Railing fittings are made in various styles, generally globe-shaped in body, with ends reduced to take thread and recessed to cover all threads. They are furnished in malleable iron, black and galvanized, and in brass.

Special railing-fitting joints are available, such as the slip-and-screwed joint, where the post connection is screwed and the rim of the fitting is so made that the rail will slip into the fitting and allow for an angular variation of several degrees, being fastened by pins which are riveted over and filed smooth. The flush-joint stair-rail fitting is another special style of fitting which provides a hand rail with even surfaces at the joints.

**Drainage fittings,** as shown in the figures accompanying Table 36, have no pockets for the lodgment of solids, and the

**Table 34. Dimensions of ANSI 150-lb Standard Malleable-Iron Screwed Fittings***
(ANSI B16.3-1971)
(All dimensions in inches)

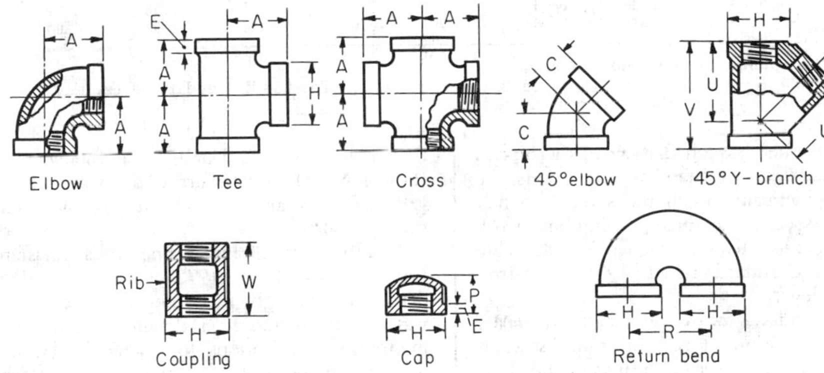


| Size | A | H | E | C | V | U | W | P | R Close | R Medium | R Open |
|---|---|---|---|---|---|---|---|---|---|---|---|
| 1/8 | 0.69 | 0.693 | 0.200 | | | | 0.96 | | | | |
| 1/4 | 0.81 | 0.844 | 0.215 | 0.73 | | | 1.06 | | | | |
| 3/8 | 0.95 | 1.015 | 0.230 | 0.80 | 1.93 | 1.43 | 1.16 | | | | |
| 1/2 | 1.12 | 1.197 | 0.249 | 0.88 | 2.32 | 1.71 | 1.34 | 0.87 | 1.000 | 1.25 | 1.50 |
| 3/4 | 1.31 | 1.458 | 0.273 | 0.98 | 2.77 | 2.05 | 1.52 | 0.97 | 1.250 | 1.50 | 2.00 |
| 1 | 1.50 | 1.771 | 0.302 | 1.12 | 3.28 | 2.43 | 1.67 | 1.16 | 1.500 | 1.875 | 2.50 |
| 1 1/4 | 1.75 | 2.153 | 0.341 | 1.29 | 3.94 | 2.92 | 1.93 | 1.28 | 1.750 | 2.25 | 3.00 |
| 1 1/2 | 1.94 | 2.427 | 0.368 | 1.43 | 4.38 | 3.28 | 2.15 | 1.33 | 2.188 | 2.50 | 3.50 |
| 2 | 2.25 | 2.963 | 0.422 | 1.68 | 5.17 | 3.93 | 2.53 | 1.45 | 2.625 | 3.00 | 4.00 |
| 2 1/2 | 2.70 | 3.589 | 0.478 | 1.95 | 6.25 | 4.73 | 2.88 | 1.70 | | | 4.50 |
| 3 | 3.08 | 4.285 | 0.548 | 2.17 | 7.26 | 5.55 | 3.18 | 1.80 | | | 5.00 |
| 3 1/2 | 3.42 | 4.843 | 0.604 | 2.39 | | | 3.43 | 1.90 | | | |
| 4 | 3.79 | 5.401 | 0.661 | 2.61 | 8.98 | 6.97 | 3.69 | 2.08 | | | |
| 5 | 4.50 | 6.583 | 0.780 | 3.05 | | | | 2.32 | | | |
| 6 | 5.13 | 7.767 | 0.900 | 3.46 | | | | 2.55 | | | |

*The complete standard (ANSI B16.3-1971) covers also reducing couplings, elbows, tees, crosses, and service or street elbows and tees.

**Table 35. Dimensions of 125- and 250-lb Standard Cast-Iron Screwed Fittings ***
(ANSI B16.4-1971)
(All dimensions in inches)

Elbow          Tee          Cross          45° elbow

| Size | 125 lb | | | | 250 lb | | | |
| | A | H | E | C | A | H | E | C |
|------|------|------|------|------|------|------|------|------|
| ¼ | 0.81 | 0.93 | 0.38 | 0.73 | 0.94 | 1.17 | 0.49 | 0.81 |
| ⅜ | 0.95 | 1.12 | 0.44 | 0.80 | 1.06 | 1.36 | 0.55 | 0.88 |
| ½ | 1.12 | 1.34 | 0.50 | 0.88 | 1.25 | 1.59 | 0.60 | 1.00 |
| ¾ | 1.13 | 1.63 | 0.56 | 0.98 | 1.44 | 1.88 | 0.68 | 1.13 |
| 1 | 1.50 | 1.95 | 0.62 | 1.12 | 1.63 | 2.24 | 0.76 | 1.31 |
| 1¼ | 1.75 | 2.39 | 0.69 | 1.29 | 1.94 | 2.73 | 0.88 | 1.50 |
| 1½ | 1.94 | 2.68 | 0.75 | 1.43 | 2.13 | 3.07 | 0.97 | 1.69 |
| 2 | 2.25 | 3.28 | 0.84 | 1.68 | 2.50 | 3.74 | 1.12 | 2.00 |
| 2½ | 2.70 | 3.86 | 0.94 | 1.95 | 2.94 | 4.60 | 1.30 | 2.25 |
| 3 | 3.08 | 4.62 | 1.00 | 2.17 | 3.38 | 5.36 | 1.40 | 2.50 |
| 3½ | 3.42 | 5.20 | 1.06 | 2.39 | 3.75 | 5.98 | 1.49 | 2.63 |
| 4 | 3.79 | 5.79 | 1.12 | 2.61 | 4.13 | 6.61 | 1.57 | 2.81 |
| 5 | 4.50 | 7.05 | 1.18 | 3.05 | 4.88 | 7.92 | 1.74 | 3.19 |
| 6 | 5.13 | 8.28 | 1.28 | 3.46 | 5.63 | 9.24 | 1.91 | 3.50 |
| 8 | 6.56 | 10.63 | 1.47 | 4.28 | 7.00 | 11.73 | 2.24 | 4.31 |
| 10 | 8.08 | 13.12 | 1.68 | 5.16 | 8.63 | 14.37 | 2.58 | 5.19 |
| 12 | 9.50 | 15.47 | 1.88 | 5.97 | 10.00 | 16.84 | 2.91 | 6.00 |

* This applies to elbows and tees only.
The 125 lb standard covers also reducing elbows and tees.  The 250 lb standard covers only the straight sizes.

length of the thread chamber is such that when the pipe is threaded to the American National Standard dimensions, the end of the pipe will practically touch the shoulder when screwed in. They are especially adapted to plumbing work and vacuum-cleaning pipe installations. Dimensions in Table 36 conform to ANSI Standard B16.12-1971, Cast-Iron Threaded Drainage Fittings.

The development of standards for **cast-iron long-turn sprinkler fittings** was begun by the National Fire Protection Assoc. in 1914 with a study of the peculiar needs of fittings intended for fire-protection purposes. These fittings (screwed and flanged) are rated at 175 and 250 lb.

American National Standard Air Gaps and Backflow Preventers in Plumbing Systems, ANSI A40.4-1942 and A40.6-1943, was set up to establish minimum requirements for plumbing, including water-supply distributing systems, drainage and venting systems, fixtures, apparatus, and devices, and the standardization of plumbing equipment in general.

**Ammonia valves and fittings** must provide a high margin of safety against accidents. Flanged valves and fittings have tongue-and-groove faces to assure tightness at the joints and against blowing out gaskets. Gaskets are of lead or compressed asbestos sheet. Screwed valves and fittings have long threads and are recessed so that the joints may be soldered. These valves and fittings are made of malleable iron, ferrosteel, or forged steel, depending on the size and style. Valves are all iron, with steel stems, and have special lead disk faces or steel disks. No copper or brass must be used in their construction. Flanged valves are generally interchangeable with flanged fit-

tings. All valves and fittings for ammonia are tested to 300 lb (11,068 N/m²) air pressure under water. For dimensions of valves, fittings, and specialties for ammonia, refer to manufacturers' catalogs.

**Soldered-Joint Fittings**  The American standard for these fittings (ANSI B16.18-1972) covers certain dimensions of soldered-joint wrought metal and cast brass fittings for copper water tubing including (1) detailed dimensions of the bore, (2) minimum specifications for materials, (3) minimum inside diameter of the fitting, (4) metal thickness for both wrought metal and cast brass fittings, and (5) general dimensions for cast brass fittings including center-to-shoulder dimensions for both straight and reducing cast fittings. Pressure and temperature ratings also are given. Sizes of the fittings are identified by the nominal tubing size as covered by the American National Standard Specifications for Copper Water Tube, ANSI H23.1-1972 (ASTM B85-1969). Dimensions of some of the fittings from this standard are given in Table 37.

## Valves

The face-to-face dimensions of ferrous flanged and welding end valves are given in ANSI B16.10-1957. The types covered are

**Wedge-Gate Valves**  Cast iron, for 125-, 175-, and 250-lb steam service pressure and 800 lb hydraulic pressure, and steel, for 150-, 300-, 400-, 600-, 900-, and 1,500-lb steam service pressures (see Fig. 11).

**Double-Disk Gate Valves**  Cast iron, for 125- and 250-lb steam service pressure and 800-lb hydraulic pressure.

**Globe and Angle Valves**  Cast iron, for 125- and 250-lb

steam service pressure, and steel, for 150-, 300-, 400-, 600-, 900-, 1,500- and 2,500-lb steam service pressures (see Fig. 12).

**Swing-Check Valves** Cast iron, for 125- and 250-lb steam service pressure and 800-lb hydraulic pressure, and steel, for 150-, 300-, 400-, and 600-lb steam service pressures.

Except for ring-joint facings the face-to-face dimension for flanged valves is the distance between the faces of the connecting end flanges upon which the gaskets are actually compressed, i.e., the "contact surfaces."

All flanges for 125-lb cast-iron valves are plain-faced. The facings of the 250-lb cast-iron, and the 150- and 300-lb steel valves have a 1/16-in raised face which is included in the contact-surface to contact-surface dimensions. The contact-surface to contact-surface dimensions of steel valves for 400-lb and higher pressures and for cast-iron valves for 800-lb hydraulic pressure include a 1/4-in raised face.

The end-to-end dimensions for **welding-end** valves for sizes 1 to 8 in are the same as the contact-surface to contact-surface dimensions given in the tables for steel valves. For details of welding bevel see ANSI B16.10-1957 and Fig. 15.

A plus or minus **tolerance** of 1/16 in is allowed on all face-to-face dimensions of valves 10 in and smaller, and a tolerance of 1/8 in on sizes 12 in and larger. This standard agrees with MSS SP32, 1937, API 5G2-36 for pipelines, and API 600A-38 for steel-flanged wedge-gate valves wherever valves of similar type, size, pressure, and material appear.

**Cocks** The ordinary plug cock operated by a handle or wrench is a form of valve in comparatively small sizes suitable for ordinary service only. The Code for Pressure Piping requires that where cocks are used for high-temperature service they shall be so designed as to prevent galling, either by making the plugs of different material from the body of the cock or by treating the plugs to ensure different physical properties. By means of special design features that eliminate the tendency to leak and stick, the plug-cock type of valve has become available in large sizes and for severe service conditions. Sizes are listed as high as 30 in and are gear-operated in the larger sizes. For further details, refer to manufacturers' catalogs.

### Expansion and Flexibility

Piping systems must be designed so that they (1) will not fail because of excessive stresses, (2) will not produce excessive thrusts or moments at connected equipment, or (3) will not

**Table 36. Dimensions of American National Standard Cast-Iron Screwed Drainage Fittings**
(ANSI B16.12-1971)
(All dimensions in inches)

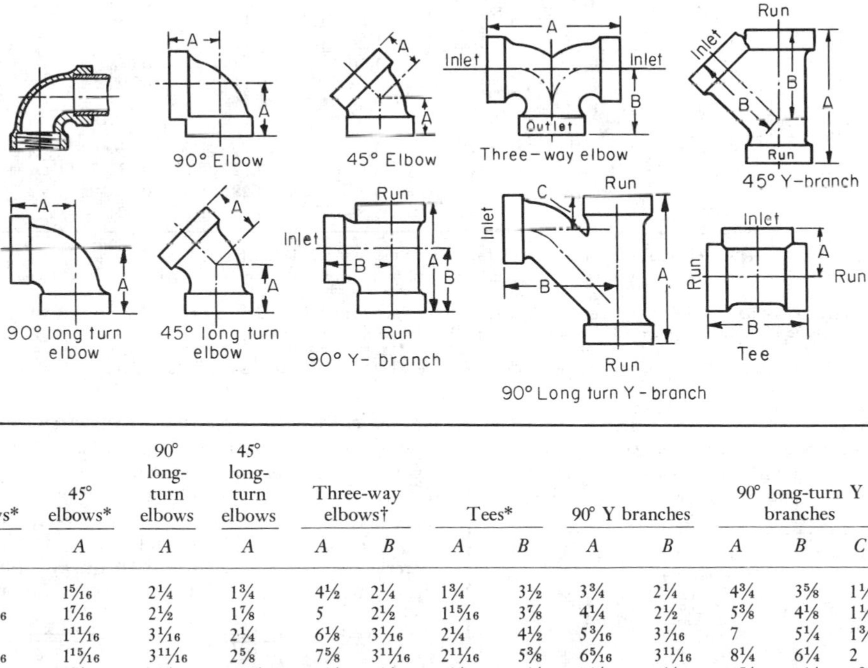


| Size, in | 90° elbows* | 45° elbows* | 90° long-turn elbows | 45° long-turn elbows | Three-way elbows† | | Tees* | | 90° Y branches | | 90° long-turn Y branches | | | 45° Y branches* | |
|---|---|---|---|---|---|---|---|---|---|---|---|---|---|---|---|
| | A | A | A | A | A | B | A | B | A | B | A | B | C | A | B |
| 1¼ | 1¾ | 15/16 | 2¼ | 1¾ | 4½ | 2¼ | 1¾ | 3½ | 3¾ | 2¼ | 4¾ | 3⅝ | 1⅛ | 5 | 3¼ |
| 1½ | 1 15/16 | 1 7/16 | 2½ | 1⅞ | 5 | 2½ | 1 15/16 | 3⅞ | 4¼ | 2½ | 5⅜ | 4⅛ | 1¼ | 5½ | 3⅝ |
| 2 | 2¼ | 1 11/16 | 3 1/16 | 2¼ | 6⅛ | 3 1/16 | 2¼ | 4½ | 5 3/16 | 3 1/16 | 7 | 5¼ | 1¾ | 6½ | 4⅜ |
| 2½ | 2 11/16 | 1 15/16 | 3 11/16 | 2⅝ | 7⅝ | 3 11/16 | 2 11/16 | 5⅜ | 6 5/16 | 3 11/16 | 8¼ | 6¼ | 2 | 7⅞ | 5⅜ |
| 3 | 3 1/16 | 2 3/16 | 4¼ | 2 15/16 | 8½ | 4¼ | 3 1/16 | 6⅛ | 7¼ | 4¼ | 9⅞ | 7½ | 2⅜ | 9 | 6 3/16 |
| 4 | 3 13/16 | 2⅝ | 5 3/16 | 3½ | 10⅜ | 5 3/16 | 3 13/16 | 7⅝ | 8¾ | 5 3/16 | 13 | 9⅞ | 3⅛ | 10⅞ | 7 11/16 |
| 5 | 4½ | 3 1/16 | 6⅛ | 4⅛ | 12¼ | 6⅛ | 4½ | 9 | 10 5/16 | 6⅛ | 15¾ | 12¼ | 3½ | 12 15/16 | 9¾ |
| 6 | 5⅛ | 3 7/16 | 7⅛ | 4⅞ | 14¼ | 7⅛ | 5⅛ | 10¼ | 11 15/16 | 7⅛ | 18¾ | 14⅝ | 4⅛ | 14⅞ | 10¾ |

*Same as adopted for 125-lb Cast-iron Threaded Fittings, ANSI B16.4-1971.
†Three-way elbows have same dimensions as 90° long-radius elbows.
Double Y branches have the same dimensions as single Y branches.
Other fittings which are available are as follows: 5⅝, 11¼, and 60° elbows; basin tees and crosses; double 90° Y branches; double 90° long-turn Y branches; 45° double Y branches; S traps; half S traps; offsets, couplings, increasers, and reducing sizes.

**Table 37. Soldered-Joint Fittings—Dimensions of Elbows, Tees, and Crosses**
(ANSI B16.18-1972)
(All dimensions in inches)

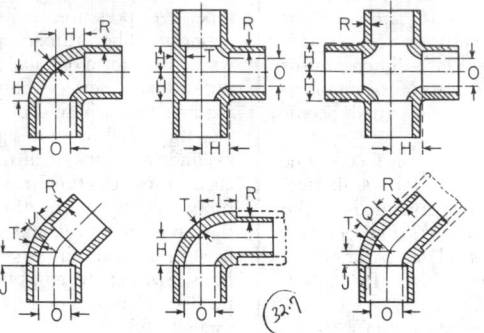

| Nominal size | Cast brass† | | | | | | | Wrought metal |
|---|---|---|---|---|---|---|---|---|
| | H* | I | J | Q | O‡ | T | R | (T and R)§,¶ |
| ¼ | ¼ | ⅜ | 3/16 | ¼ | 0.31 | 0.08 | 0.048 | 0.030 |
| ⅜ | 5/16 | 7/16 | 3/16 | 5/16 | 0.43 | 0.08 | 0.048 | 0.035 |
| ½ | 7/16 | 9/16 | 3/16 | 5/16 | 0.54 | 0.09 | 0.054 | 0.040 |
| ¾ | 9/16 | 11/16 | ¼ | ⅜ | 0.78 | 0.10 | 0.060 | 0.045 |
| 1 | ¾ | ⅞ | 5/16 | 7/16 | 1.02 | 0.11 | 0.066 | 0.050 |
| 1¼ | ⅞ | 1 | 7/16 | 9/16 | 1.26 | 0.12 | 0.072 | 0.055 |
| 1½ | 1 | 1⅛ | ½ | ⅝ | 1.50 | 0.13 | 0.078 | 0.060 |
| 2 | 1¼ | 1⅜ | 9/16 | ¾ | 1.98 | 0.15 | 0.090 | 0.070 |
| 2½ | 1½ | 1⅝ | ⅝ | ⅞ | 2.46 | 0.17 | 0.102 | 0.080 |
| 3 | 1¾ | 1⅞ | ¾ | 1 | 2.94 | 0.19 | 0.114 | 0.090 |
| 3½ | 2 | 2⅛ | ⅞ | 1⅛ | 3.42 | 0.20 | 0.120 | 0.100 |
| 4 | 2¼ | 2⅜ | 15/16 | 1¼ | 3.90 | 0.22 | 0.132 | 0.110 |
| 5 | 3⅛ | | 1 7/16 | | 4.87 | 0.28 | 0.168 | 0.125 |
| 6 | 3⅝ | | 1⅝ | | 5.84 | 0.34 | 0.204 | 0.140 |

Wrought fittings as well as cast fittings, must be provided with a shoulder or stop at the bottom end of socket.
*Dimensions for reducing elbows, reducing crosses, reducing tees, couplings, caps, bushings, adapters, and fittings with pipe thread on one end are also included in this standard.
†These dimensions may be used for wrought metal fittings as well as for cast brass fittings at manufacturer's option.
‡This dimension is the same as the inside diameter Class L tubing, ANSI H23.1-192 (ASTM B88-1969).
§This dimension has the same thickness as Class L tubing.
¶These dimensions are minimum, but in every case the thickness of wrought fittings should be at least as heavy as the tubing with which it is to be used.

leak at joints because of expansion of the pipe. Flexibility is provided by changes of direction in the piping through the use of bends or loops, or provision may be made to absorb thermal strains by use of expansion joints. All, or portions, of the pipe may be corrugated to improve flexibility; in many systems, however, sufficient change is provided by the geometry of the layout to make unnecessary the use of either expansion joints or corrugated sections of piping. Judicious cold springing is beneficial in assisting the piping system to attain its most favorable condition. Because of plastic flow of the piping material, hot stresses tend to decrease with time while cold stresses tend to increase with time; their sum, called the stress

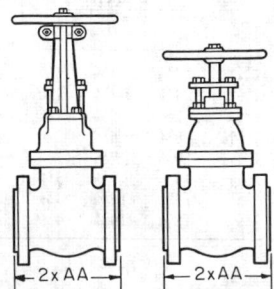

**Fig. 11**   Wedge-gate valves.

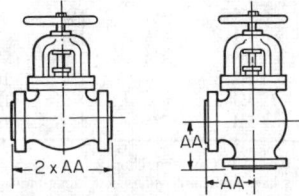

**Fig. 12**   Globe valve and angle valve.

### Table 38. Dimensions of Long-Radius 90° Butt-welding Elbows
(Standard weight—ANSI B16.9-1971, ASTM A234)
(All dimensions are in inches)

| Nominal pipe size | OD | ID | Wall thickness | Center to face | Pipe schedule numbers | Approx wt, lb |
|---|---|---|---|---|---|---|
| 2½ | 2.875 | 2.469 | 0.203 | 3¾ | 40 | 2.92 |
| 3 | 3.500 | 3.068 | 0.216 | 4½ | 40 | 4.58 |
| 3½ | 4.000 | 3.548 | 0.226 | 5¼ | 40 | 6.43 |
| 4 | 4.500 | 4.026 | 0.237 | 6 | 40 | 8.70 |
| 5 | 5.563 | 5.047 | 0.258 | 7½ | 40 | 14.7 |
| 6 | 6.625 | 6.065 | 0.280 | 9 | 40 | 22.9 |
| 8 | 8.625 | 7.981 | 0.322 | 12 | 40 | 46.0 |
| 10 | 10.750 | 10.020 | 0.365 | 15 | 40 | 81.5 |
| 12 | 12.750 | 12.000 | 0.375 | 18 | ST* | 119 |
| 14 | 14.000 | 13.250 | 0.375 | 21 | 30 | 154 |
| 16 | 16.000 | 15.250 | 0.375 | 24 | 30 | 201 |
| 18 | 18.000 | 17.250 | 0.375 | 27 | ST* | 256 |
| 20 | 20.000 | 19.250 | 0.375 | 30 | 20 | 317 |
| 22 | 22.000 | 21.250 | 0.375 | 33 | ST* | 385 |
| 24 | 24.000 | 23.250 | 0.375 | 36 | 20 | 458 |
| 26 | 26.000 | 25.250 | 0.375 | 39 | ST* | 539 |
| 30 | 30.000 | 29.250 | 0.375 | 45 | ST* | 720 |
| 34 | 34.000 | 33.250 | 0.375 | 51 | ST* | 926 |
| 36 | 36.000 | 35.250 | 0.375 | 54 | ST* | 1,040 |
| 42 | 42.000 | 41.250 | 0.375 | 63 | ST* | 1,420 |

*Standard weight.

range, remains substantially constant. For this reason no credit is warranted with regard to stresses; for calculation of forces and moments, the effect of cold spring is recognized by use of a cold-spring factor varying from 0 to 1 for cold spring varying from 0 to 100 percent.

The allowable stress range $S_A$ is calculated by

$$S_A = f\,(1.25\,S_c + 0.25\,S_h)$$

where $S_c$ and $S_h$ are the $S$ values for the mimimum cold and maximum hot conditions, respectively, as given in Tables 1

### Table 39. Dimensions of Straight Butt-welding Tees
(Standard weight—ANSI B16.9-1971, ASTM A234)
(Dimensions are in inches)

| Nominal pipe size | OD | ID | Wall thickness | Center to end | Center to end | Pipe schedule numbers | Approx wt, lb |
|---|---|---|---|---|---|---|---|
| 2½ | 2.875 | 2.469 | 0.203 | 3 | 3 | 40 | 5.21 |
| 3 | 3.500 | 3.068 | 0.216 | 3⅛ | 3⅜ | 40 | 7.44 |
| 3½ | 4.000 | 3.548 | 0.226 | 3¾ | 3¾ | 40 | 9.85 |
| 4 | 4.500 | 4.026 | 0.237 | 4⅛ | 4⅛ | 40 | 12.6 |
| 5 | 5.563 | 5.047 | 0.258 | 4⅞ | 4⅞ | 40 | 19.8 |
| 6 | 6.625 | 6.065 | 0.280 | 5⅝ | 5⅝ | 40 | 29.3 |
| 8 | 8.625 | 7.981 | 0.322 | 7 | 7 | 40 | 53.7 |
| 10 | 10.750 | 10.020 | 0.365 | 8½ | 8½ | 40 | 91.2 |
| 12 | 12.750 | 12.000 | 0.375 | 10 | 10 | ST* | 132 |
| 14 | 14.000 | 13.250 | 0.375 | 11 | 11 | 30 | 172 |
| 16 | 16.000 | 15.250 | 0.375 | 12 | 12 | 30 | 219 |
| 18 | 18.000 | 17.250 | 0.375 | 13½ | 13½ | ST* | 282 |
| 20 | 20.000 | 19.250 | 0.375 | 15 | 15 | 20 | 354 |
| 22 | 22.000 | 21.250 | 0.375 | 16½ | 16½ | ST* | 437 |
| 24 | 24.000 | 23.250 | 0.375 | 17 | 17 | 20 | 493 |
| 26 | 26.000 | 25.250 | 0.375 | 19½ | 19½ | ST* | 634 |
| 30 | 30.000 | 29.250 | 0.375 | 22 | 22 | ST* | 855 |
| 34 | 34.000 | 33.250 | 0.375 | 25 | 25 | ST* | 1,136 |
| 36 | 36.000 | 35.250 | 0.375 | 26½ | 26½ | ST* | 1,294 |

*Standard weight.

**Table 40. Dimensions of Long-Radius 45° Butt-welding Elbows**
(Standard weight—ANSI B16.9-1971, ASTM A234)
(Dimensions are in inches)

| Nominal pipe size | O.D. | I.D. | Wall thickness | Center to face | Radius | Pipe schedule numbers | Approx wt, lb |
|---|---|---|---|---|---|---|---|
| 2½ | 2.875 | 2.469 | 0.203 | 1¾ | 3¾ | 40 | 1.64 |
| 3 | 3.500 | 3.068 | 0.216 | 2 | 4½ | 40 | 2.43 |
| 3½ | 4.000 | 3.548 | 0.226 | 2¼ | 5¼ | 40 | 3.29 |
| 4 | 4.500 | 4.026 | 0.237 | 2½ | 6 | 40 | 4.31 |
| 5 | 5.563 | 5.047 | 0.258 | 3⅛ | 7½ | 40 | 7.30 |
| 6 | 6.625 | 6.065 | 0.280 | 3¾ | 9 | 40 | 11.3 |
| 8 | 8.625 | 7.981 | 0.322 | 5 | 12 | 40 | 22.8 |
| 10 | 10.750 | 10.020 | 0.365 | 6¼ | 15 | 40 | 40.4 |
| 12 | 12.750 | 12.000 | 0.375 | 7½ | 18 | ST* | 59.5 |
| 14 | 14.000 | 13.250 | 0.375 | 8¾ | 21 | 30 | 76.5 |
| 16 | 16.000 | 15.250 | 0.375 | 10 | 24 | 30 | 100 |
| 18 | 18.000 | 17.250 | 0.375 | 11¼ | 27 | ST* | 128 |
| 20 | 20.000 | 19.250 | 0.375 | 12½ | 30 | 20 | 158 |
| 22 | 22.000 | 21.250 | 0.375 | 13½ | 33 | ST* | 192 |
| 24 | 24.000 | 23.250 | 0.375 | 15 | 36 | 20 | 229 |
| 26 | 26.000 | 25.250 | 0.375 | 16 | 39 | ST* | 269 |
| 30 | 30.000 | 29.250 | 0.375 | 18½ | 45 | ST* | 358 |
| 34 | 34.000 | 33.250 | 0.375 | 21 | 51 | ST* | 463 |
| 36 | 36.000 | 35.250 | 0.375 | 22¼ | 54 | ST* | 518 |
| 42 | 42.000 | 41.250 | 0.375 | 26 | 63 | ST* | 707 |

*Standard weight.

**Table 41. Dimensions of Concentric and Eccentric Butt-welding Reducers**
(Standard weight—ANSI B16.9-1971, ASTM A234)
(Dimensions are in inches)

| Nominal pipe size | Length | Approx wt, lb | Nominal pipe size | Length | Approx wt, lb | Nominal pipe size | Length | Approx wt, lb Conc. | Approx wt, lb Ecc. |
|---|---|---|---|---|---|---|---|---|---|
| 2½ × 1 | 3½ | 1.30 | 8 × 6 | 6 | 13.4 | 22 × 20 | 20 | 157 | |
| 2½ × 1¼ | 3½ | 1.47 | 8 × 6 | 6 | 13.9 | 24 × 16 | 20 | 160 | |
| 2½ × 1½ | 3½ | 1.51 | 10 × 4 | 7 | 21.1 | 24 × 18 | 20 | 163 | |
| 2½ × 2 | 3½ | 1.60 | 10 × 5 | 7 | 21.8 | 24 × 20 | 20 | 167 | |
| 3 × 1¼ | 3½ | 1.70 | 10 × 6 | 7 | 22.3 | 26 × 18 | 24 | 200 | |
| 3 × 1½ | 3½ | 1.89 | 10 × 8 | 7 | 23.2 | 26 × 20 | 24 | 200 | |
| 3 × 2 | 3½ | 2.00 | 12 × 5 | 8 | 30.5 | 26 × 22 | 24 | 200 | |
| 3 × 2½ | 3½ | 2.16 | 12 × 6 | 8 | 31.1 | 26 × 24 | 24 | 200 | |
| 3½ × 1¼ | 4 | 2.35 | 12 × 8 | 8 | 32.1 | 30 × 20 | 24 | 220 | |
| 3½ × 1½ | 4 | 2.52 | 12 × 10 | 8 | 33.4 | 30 × 24 | 24 | 220 | |
| 3½ × 2 | 4 | 2.71 | 14 × 6 | 13 | 55.8 | 30 × 26 | 24 | 220 | |
| 3½ × 2½ | 4 | 2.96 | 14 × 8 | 13 | 57.2 | 30 × 28 | 24 | 220 | |
| 3½ × 3 | 4 | 3.05 | 14 × 10 | 13 | 60.4 | | | | |
| 4 × 1½ | 4 | 2.73 | 14 × 12 | 13 | 63.4 | | | Conc. | Ecc. |
| 4 × 2 | 4 | 3.17 | 16 × 8 | 14 | 70.2 | 34 × 24 | 24 | 270 | 229 |
| 4 × 2½ | 4 | 3.34 | 16 × 10 | 14 | 72.9 | 34 × 26 | 24 | 270 | 237 |
| 4 × 3 | 4 | 3.50 | 16 × 12 | 14 | 75.6 | 34 × 30 | 24 | 270 | 253 |
| 4 × 3½ | 4 | 3.61 | 16 × 14 | 14 | 77.5 | 34 × 32 | 24 | 270 | 261 |
| 5 × 2 | 5 | 5.05 | 18 × 10 | 15 | 86.9 | 36 × 24 | 24 | 340 | 237 |
| 5 × 2½ | 5 | 5.52 | 18 × 12 | 15 | 89.2 | 36 × 26 | 24 | 340 | 245 |
| 5 × 3 | 5 | 5.73 | 18 × 14 | 15 | 90.9 | 36 × 30 | 24 | 340 | 261 |
| 5 × 3½ | 5 | 5.86 | 18 × 16 | 15 | 94.0 | 36 × 32 | 24 | 340 | 269 |
| 5 × 4 | 5 | 5.99 | 20 × 12 | 20 | 134 | 36 × 34 | 24 | 340 | 277 |
| 6 × 2½ | 5½ | 7.61 | 20 × 14 | 20 | 135 | 42 × 24 | 24 | 260 | |
| 6 × 3 | 5½ | 8.00 | 20 × 16 | 20 | 138 | 42 × 26 | 24 | 270 | |
| 6 × 3½ | 5½ | 8.14 | 20 × 18 | 20 | 142 | 42 × 30 | 24 | 285 | |
| 6 × 4 | 5½ | 8.19 | 22 × 14 | 20 | 148 | 42 × 32 | 24 | 295 | |
| 6 × 5 | 5½ | 8.65 | 22 × 16 | 20 | 151 | 42 × 34 | 24 | 300 | |
| 8 × 3½ | 6 | 12.8 | 22 × 18 | 20 | 154 | 42 × 36 | 24 | 310 | |
| 8 × 4 | 6 | 13.1 | | | | | | | |

**Table 42. Dimensions of 125-, 150-, and 250-lb Pipe Plugs***
(ANSI B16.14-1971)
(All dimensions in inches)

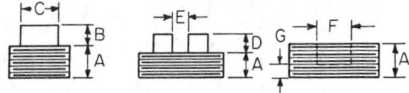

| Nominal pipe size | Square-head pattern | | | Slotted pattern | | | Countersunk pattern† (square sockets) | | |
|---|---|---|---|---|---|---|---|---|---|
| | A | B | C | A | D | E | A | F | G |
| ⅛ | 0.37 | 0.24 | ⁹⁄₃₂ | | | | | | |
| ¼ | 0.44 | 0.28 | ⅜ | | | | | | |
| ⅜ | 0.48 | 0.31 | ⁷⁄₁₆ | | | | | | |
| ½ | 0.56 | 0.38 | ⁹⁄₁₆ | | | | 0.56 | ⅜ | 0.16 |
| ¾ | 0.63 | 0.44 | ⅝ | | | | 0.63 | ½ | 0.18 |
| 1 | 0.75 | 0.50 | ¹³⁄₁₆ | | | | 0.75 | ½ | 0.20 |
| 1¼ | 0.80 | 0.56 | ¹⁵⁄₁₆ | | | | 0.80 | ¾ | 0.22 |
| 1½ | 0.83 | 0.62 | 1⅛ | | | | 0.83 | ¾ | 0.24 |
| 2 | 0.88 | 0.68 | 1⁵⁄₁₆ | | | | 0.88 | ⅞ | 0.26 |
| 2½ | 1.07 | 0.74 | 1½ | | | | 1.07 | 1⅛ | 0.29 |
| 3 | 1.13 | 0.80 | 1¹¹⁄₁₆ | | | | 1.13 | 1⅜ | 0.31 |
| 3½ | 1.18 | 0.86 | 1⅞ | | | | 1.18 | 1½ | 0.34 |
| 4 | | | | 1.22 | 1.00 | 0.88 | 1.22 | 2 | 0.37 |
| 5 | | | | 1.31 | 1.00 | 0.88 | 1.31 | 2¼ | 0.46 |
| 6 | | | | 1.40 | 1.25 | 1.25 | 1.40 | 2½ | 0.52 |
| 8 | | | | 1.57 | 1.38 | 1.50 | | | |

*The material of (ANSI B16.14-1971) is to be cast iron, malleable iron, or steel, for use in connection with fittings covered by the American National Standard 125-lb cast-iron screwed fittings (ANSI B16.4) and the American National Standard 150-lb malleable-iron screwed fittings (ANSI B16.3).
†Hexagon sockets (sizes ⅛ to 1 in) have dimensions to fit regular wrenches used with hexagon socket setscrews.

and 2. The stress-reduction factor $f$ is a function of the number of hot-to-cold-to-hot (full) temperature cycles anticipated over the life of the plant, as follows:

| *Total No. of Full Temp Cycles over Expected Life* | *Stress-Reduction Factor* |
|---|---|
| 7,000 and less | 1.0 |
| 14,000 and less | 0.9 |
| 22,000 and less | 0.8 |
| 45,000 and less | 0.7 |
| 100,000 and less | 0.6 |
| Over 100,000 | 0.5 |

The bending and torsional stresses calculated (see paragraph 119.6.4 of ANSI B31.1.0-1967) are used to determine the maximum computed expansion stress $S_E = \sqrt{S_b^2 + 4 S_t^2}$, where $S_b$ and $S_t$ are bending and torsional stresses, respectively. $S_E$ must not exceed the allowable stress range $S_A$.

In recent years, many principal high-temperature steam lines have either been analyzed, tested in a model-testing machine, or both. No rigid rule is stipulated for the requirement of analysis or model test; however, the Code for Pressure Piping suggests that when the following criterion is not satisfied, need for an analysis is indicated: $DY/(L - U)^2 \leqslant 0.03$ where $D$ is the nominal pipe size, in, $Y$ is the resultant of movements to be absorbed by pipeline, in, $U$ is the length of straight line joining the anchor points, ft, and $L$ is the length of the developed line axis, ft.

**Expansion Joints for Steam Pipelines** In many instances it may be economical to care for thermal expansion by use of expansion joints. For low-pressure steam lines, the use of packed expansion joints may be feasible; experience has indicated that packed joints are difficult to maintain when used on high-pressure lines. Figure 13 shows a type of joint that has been successfully used for high-pressure, high-temperature service. The bellows is designed to take either axial, lateral, or combined axial and lateral deflections. The internal sleeve guides movement of the joint and also protects the flexible bellows from direct contact with the fluid being handled. Face-to-face dimensions, as well as permissible axial and lateral deflections, are indicated in Table 43.

Where large lateral deflections are to be absorbed, two expansion joints separated by a length of pipe as shown in Fig. 14 may be used. With such an arrangement, the lateral deflection permissible with one joint only may be increased many times. Tie rods, as shown, should always be installed to protect the joint against overtravel and externally to guide movement of the joint.

Table 44, extracted from the Code for Pressure Piping, lists thermal-expansion data for both ferrous and nonferrous pip-

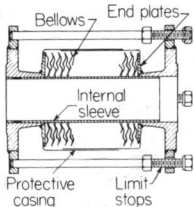

**Fig. 13** Expansion joint for steam line. (*Croll-Reynolds, Inc.*)

**Table 43. Dimensions of Expansion Joints***

| Pipe size | Pressure series, lb/in² gage | Face-to-face dimensions,† axial movements of | | | Lateral movements‡ equivalent to axial movements of | | |
|---|---|---|---|---|---|---|---|
| | | 1 in | 2 in | 3 in | 1 in | 2 in | 3 in |
| 4 | 150 | 8½ | 11 | 15½ | 1/16 | ¼ | 9/16 |
| | 300 | 12 | 17 | 24 | ⅛ | ½ | 1 |
| | 600 | 17½ | 26½ | | ¼ | 1 | |
| | 900 | 31½ | | | 11/16 | | |
| 6 | 150 | 9½ | 12 | 16½ | 1/16 | 3/16 | ⅜ |
| | 300 | 13 | 18 | 25 | 3/32 | 13/32 | 15/16 |
| | 600 | 18½ | 27½ | | 7/32 | 27/32 | |
| | 900 | 33½ | | | ½ | | |
| 8 | 150 | 10½ | 13 | 17½ | 1/32 | 3/16 | ⅜ |
| | 300 | 14 | 19 | 26 | 3/32 | ⅜ | 27/32 |
| | 600 | 20 | 29 | | 3/16 | ¾ | |
| | 900 | 35½ | | | 15/32 | | |
| 10 | 150 | 10½ | 13 | 17½ | 1/32 | ⅛ | 5/16 |
| | 300 | 14 | 19 | 26 | 1/16 | 5/16 | 23/32 |
| | 600 | 21½ | 30½ | | 5/32 | ⅝ | |
| | 900 | 37 | | | ⅜ | | |
| 12 | 150 | 22½ | 14 | 18½ | 1/32 | ⅛ | ¼ |
| | 300 | 15 | 20 | 27 | 1/16 | ¼ | 19/32 |
| | 600 | 21½ | 30½ | | ⅛ | ½ | |
| | 900 | 38½ | | | 5/16 | | |
| 14 | 100 | 12½ | 15½ | | 1/32 | ⅛ | |
| | 150 | 15 | 20 | | 1/16 | 7/32 | |
| | 300 | 20½ | | | 3/32 | | |
| 16 | 100 | 12½ | 15½ | | 1/32 | 3/32 | |
| | 150 | 15 | 20 | | 1/32 | 3/16 | |
| | 300 | 20½ | | | 3/32 | | |
| 18 | 100 | 13½ | 16½ | | 1/32 | 3/32 | |
| | 150 | 16 | 21 | | 1/32 | 3/16 | |
| | 300 | 21½ | | | 3/32 | | |
| 20 | 100 | 14½ | 16½ | | | 3/32 | |
| | 150 | 16 | 21 | | 1/32 | 3/16 | |
| | 300 | 21½ | | | 3/32 | | |
| 24 | 100 | 14½ | 17½ | | | 1/16 | |
| | 150 | 17 | 22 | | 1/32 | 5/32 | |
| | 300 | 22½ | | | 1/16 | | |
| 30 | 100 | 9½ | 12½ | | | 1/16 | |
| | 150 | 12 | 17 | | 1/32 | ⅛ | |
| | 300 | 19½ | | | 1/16 | | |

*Croll-Reynolds, Inc.
†For welding ends, add 4 in to face-to-face dimension shown.
‡Consult manufacturer for permissible combined axial and lateral deflection.

ing. For expansion at temperatures intermediate between those shown, straight-line interpolation is permitted.

The **rubber expansion joint** has become an established part of pipeline equipment. Its special field of application is on low-pressure and vacuum lines in condenser applications, etc., and it is recommended for pressures up to 25 lb/in² gage where the maximum temperature does not exceed 250°F. Standard joints for pressure installations are reinforced to withstand working pressures up to 125 lb/in² gage and temperatures up to 200°F. Joints are available in all standard pipe sizes.

### Welding in Power-Plant Piping

(For dimensions of welding fittings see Tables 38-41; for welding techniques see also Sec. 13)

The majority of main-cycle and service steel piping in modern steam power plants is of welded construction. Steel pipe of 2 in size and smaller is generally socket-welded; larger-size pip-ing is usually butt-welded. Frequently, depending on location and scheduling, piping larger than 2 in size is prefabricated; smaller piping is shipped to the construction site in random lengths and is fabricated concurrently with installation. Small-sized chromium-molybdenum piping requiring bending is frequently also shop-fabricated so as to avoid high field preheat, welding, and stress-relieving costs. It is desirable to schedule shipment of hangers so that they will be available at the jobsite upon arrival of the prefabricated piping; this avoids the expense of providing, installing, and later removing temporary hangers and supports. Aside from the economy of welded construction, it is a virtual necessity in high-pressure, high-temperature work because of danger of leakage if joints are flanged.

Shop welds are frequently made by automatic or semiautomatic submerged-arc or inert-gas shielded-arc processes; field welds are generally of the manual type and may be done by the shielded metal-arc, inert-gas metal-arc, or gas-welding processes. Welding in power piping systems, whether in the shop or at the jobsite, must be done by welders who have qualified under provisions of the Code for Pressure Piping or the ASME Boiler and Pressure Vessel Code.

**End Preparation for Butt Welds**  Figure 15 shows the end preparation recommended (not required) for piping whose wall thickness is ¾ in or less, and Fig. 16 shows that required for piping with wall thickness above ¾ in. During the welding process, to avoid entrance of welding material into the pipe, backing rings may be used as shown in Fig. 17a,b, and c.[1] Note that thick-walled pipes (over ¾ in) are taper-bored on the inside in order that they may receive a tapered, machined backing ring.

**Preheating**  Prior to start of welding, many materials require preheat to a specified temperature; preheat may be done by electrical-resistance or induction heating or by ring-type gas burners placed concentrically with the pipe. The preheat temperature is measured by indicating crayons or by thermocouple pyrometers and must be maintained during the welding operation. Table 1, Appendix D, of the Code for Pressure Piping lists materials used in piping systems and the appropriate temperatures for preheat. In general, the following is indicative of the intent only; for specific instances, the code must be consulted.

**Carbon steel** and **wrought iron** should be preheated to a "hand-hot" condition if the ambient temperature at time of field installation is 32°F (0°C) or less: carbon steels which have minimum tensile properties of 70,000 lb/in² (2,580,000N/m²) or higher should be preheated to 250°F (121°C); under other conditions, preheat is not mandatory, but some purchasers insist that the contractor preheat heavy-walled piping such as boiler feed.

**Low-alloy steels** with a chromium content not exceeding ¾ percent and low-alloy steels with a total alloy content not exceeding 2 percent are required to be preheated to a minimum temperature of 300°F (149°C).

**Alloy steels** with a chromium content between ¾ and 2 percent and low-alloy steels with a total alloy content not exceeding 2¾ percent require preheating to 375°F (191°C) minimum. Those with a total alloy content greater than 2¾ percent but not exceeding 10 percent require preheating to a temperature of 450°F (232°C) minimum.

**High-alloy** steels containing the martensitic phase require preheating to 450°F (232°C) minimum; preheating is a matter of agreement between the purchaser and contractor in the case of welding high-alloy ferritic steels (ASTM A240 and A268). The possible advantages of preheat have not been established in the case of welding high-alloy austenitic steels, and for this reason the Code for Pressure Piping states that preheat is optional for these materials.

**Welding procedure** varies with material and welding process. In general, the pipe ends must be cleaned of oil or grease, and excessive amounts of scale or rust should be removed. The size and type of welding rod must be stated; the number of layers or passes is determined by the thickness of the pieces being joined. All slag or flux remaining on any bead of welding must be removed before laying down the next successive bead; any cracks or blowholes that appear on the surface of any bead

[1]Consumable inserts are also available. They are recommended for installation in piping systems which require a smooth, unobstructed interior surface.

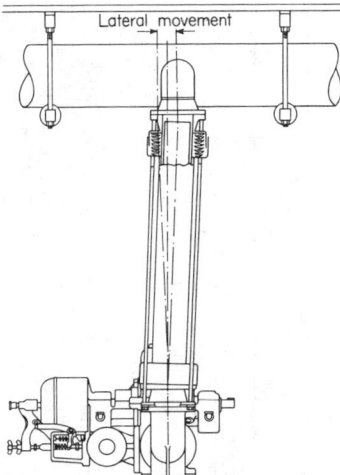

**Fig. 14**  Arrangement of expansion joints for large lateral deflection. (*Croll-Reynolds, Inc.*)

must be chipped or ground away before the next bead of weld material is deposited. Throughout the welding process, it is essential that the minimum specified preheat temperature be maintained.

**Stress Relieving**  Welded joints in all carbon-steel material whose thickness is ¾ in (1.91 cm) or greater must be stress-relieved at a temperature of 1100°F (593°C) or over for a period of time proportioned on the basis of at least 1 h/in of pipe-wall thickness (but in no case less than ½ h) and then allowed to cool slowly (generally under a blanket) and uniformly. No stress relief is required for joints in carbon-steel piping whose wall thickness is less than ¾ in.

Welded joints in alloy steels with a wall thickness of ½ in (1.27 cm) or greater, having a chromium content not exceeding ¾ percent, and low-alloy steels with a total alloy content not exceeding 2 percent require stress-relieving at a temperature of 1200°F (649°C) or over for a period of time proportioned on the basis of at least 1 h/in (0.4 h/cm) of wall thickness, but in no case less than ½ h.

Welded joints in alloy steels having a chromium content exceeding ¾ percent, or a total alloy content exceeding 2

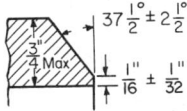

**Fig. 15**  Recommended end preparation for pipe wall thickness, ¾ in or less.

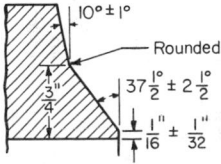

**Fig. 16**  Recommended end preparation for pipe wall thickness greater than ¾ in.

**Table 44. Thermal-Expansion Data**
(ANSI B31.1.0-1967)

| Material | Coefficient | Temp range: 70°F (21°C) to | | | | | | | | | | | | | |
| --- | --- | --- | --- | --- | --- | --- | --- | --- | --- | --- | --- | --- | --- | --- | --- |
| | | 70 (21) | 200 (93) | 300 (149) | 400 (205) | 500 (260) | 600 (316) | 700 (371) | 800 (427) | 900 (482) | 1000 (538) | 1100 (593) | 1200 (649) | 1300 (705) | 1400 (760) |
| Carbon steel: carbon-moly steel low-chrome steels (through 3% Cr) | A | | 6.38 | 6.60 | 6.82 | 7.02 | 7.23 | 7.44 | 7.65 | 7.84 | 7.97 | 8.12 | 8.19 | 8.28 | 8.36 |
| | B | 0 | 0.99 | 1.82 | 2.70 | 3.62 | 4.60 | 5.63 | 6.70 | 7.81 | 8.89 | 10.04 | 11.10 | 12.22 | 13.34 |
| Intermediate alloy steels: 5 Cr Mo-9 Cr Mo | A | | 6.04 | 6.19 | 6.34 | 6.50 | 6.66 | 6.80 | 6.96 | 7.10 | 7.22 | 7.32 | 7.41 | 7.49 | 7.55 |
| | B | 0 | 0.94 | 1.71 | 2.50 | 3.35 | 4.24 | 5.14 | 6.10 | 7.07 | 8.06 | 9.05 | 10.00 | 11.06 | 12.05 |
| Austenitic stainless steels | A | | 9.34 | 9.47 | 9.59 | 9.70 | 9.82 | 9.92 | 10.05 | 10.16 | 10.29 | 10.39 | 10.48 | 10.54 | 10.60 |
| | B | 0 | 1.46 | 2.61 | 3.80 | 5.01 | 6.24 | 7.50 | 8.80 | 10.12 | 11.48 | 12.84 | 14.20 | 15.56 | 16.92 |
| Straight chromium stainless steels: 12 Cr, 17Cr, and 27 Cr | A | | 5.50 | 5.66 | 5.81 | 5.96 | 6.13 | 6.26 | 6.39 | 6.52 | 6.63 | 6.72 | 6.78 | 6.85 | 6.90 |
| | B | 0 | 0.86 | 1.56 | 2.30 | 3.08 | 3.90 | 4.73 | 5.60 | 6.49 | 7.40 | 8.31 | 9.20 | 10.11 | 11.01 |
| 25 Cr-20 Ni | A | | 7.76 | 7.92 | 8.08 | 8.22 | 8.38 | 8.52 | 8.68 | 8.81 | 8.02 | 9.00 | 9.08 | 9.12 | 9.18 |
| | B | 0 | 1.21 | 2.18 | 3.20 | 4.24 | 5.33 | 6.44 | 7.60 | 8.78 | 9.95 | 11.12 | 12.31 | 13.46 | 14.65 |
| Monel 67: Ni-30 Cu | A | | 7.84 | 8.02 | 8.20 | 8.40 | 8.58 | 8.78 | 8.96 | 9.16 | 9.34 | 9.52 | 9.70 | 9.88 | 10.04 |
| | B | 0 | 1.22 | 2.21 | 3.25 | 4.33 | 5.46 | 6.64 | 7.85 | 9.12 | 10.42 | 11.77 | 13.15 | 14.58 | 16.02 |
| Monel 66: Ni-29 CuAl | A | | 7.48 | 7.68 | 7.90 | 8.09 | 8.30 | 8.50 | 8.70 | 8.90 | 9.10 | 9.30 | 9.50 | 9.70 | 9.89 |
| | B | 0 | 1.17 | 2.12 | 3.13 | 4.17 | 5.28 | 6.43 | 7.62 | 8.86 | 10.16 | 11.50 | 13.00 | 14.32 | 15.78 |
| Aluminum | A | | 12.95 | 13.28 | 13.60 | 13.90 | 14.20 | | | | | | | | |
| | B | 0 | 2.00 | 3.66 | 5.39 | 7.17 | 9.03 | | | | | | | | |
| Gray cast iron | A | | 5.75 | 5.93 | 6.10 | 6.28 | 6.47 | 6.65 | 6.83 | 7.00 | 7.19 | | | | |
| | B | 0 | 0.90 | 1.64 | 2.42 | 3.24 | 4.11 | 5.03 | 5.98 | 6.97 | 8.02 | | | | |
| Bronze | A | | 10.03 | 10.12 | 10.23 | 10.32 | 10.44 | 10.52 | 10.62 | 10.72 | 10.80 | 10.90 | 11.00 | | |
| | B | 0 | 1.56 | 2.79 | 4.05 | 5.33 | 6.64 | 7.95 | 9.30 | 10.68 | 12.05 | 13.47 | 14.92 | | |
| Brass | A | | 9.76 | 10.00 | 10.23 | 10.47 | 10.69 | 10.92 | 11.16 | 11.40 | 11.63 | 11.85 | 12.09 | | |
| | B | 0 | 1.52 | 2.76 | 4.05 | 5.40 | 6.80 | 8.26 | 9.78 | 11.35 | 12.98 | 14.65 | 16.39 | | |
| Wrought iron | A | | 7.32 | 7.48 | 7.61 | 7.73 | 7.88 | 8.01 | 8.13 | 8.29 | 8.39 | | | | |
| | B | 0 | 1.14 | 2.06 | 3.01 | 3.99 | 5.01 | 6.06 | 7.12 | 8.26 | 9.36 | | | | |
| Copper-nickel (70/30) | A | | 8.54 | 8.71 | 8.90 | | | | | | | | | | |
| | B | 0 | 1.33 | 2.40 | 3.52 | | | | | | | | | | |

$A$ = mean coefficient of thermal expansion $\times 10^6$, in/in/°F in going from 70°F (21°C) to indicated temperature
$B$ = linear thermal expansion, in/100 ft in going from 70°F (21°C) to indicated temperature
Multiply values of $A$ shown by 1.8 to obtain coefficient of expansions in cm/cm/°C.
Multiply values of $B$ shown by 8.33 to obtain linear expansion in cm per 100 m.

percent, except high-alloy ferritic (ASTM A240, A268) and austenitic steels, regardless of wall thickness, require stress relief at a temperature of 1200°F or over for a period of time proportioned on the basis of at least 1 h/in of wall thickness, but in no case less than ½ h. Stress relief of high-alloy ferritic steel (A240, A268) and austenitic steels is not required but may be performed as agreed upon by purchaser and contractor. In welds between austenitic and ferritic materials, stress relieving is optional and, if used, shall be a matter of agreement between the purchaser and contractor. Because of the difference between the coefficients of thermal expansion of the two dissimilar materials, careful consideration should be given to the selection of a heat treatment, if any, that will be beneficial to the welded joint.

**Graphitization** is precipitation of carbon at the grain boundaries in the heat-affected zone during the welding process. Such a phenomenon occurs when some metals operate at high temperatures for extended periods of time. It has been observed particularly in carbon-molybdenum steels that operate at 900°F (482°C) or higher. Recent practice indicates that chromium in both the parent metal and welding rod reduces, or perhaps eliminates entirely, the possibility of graphitization. Graphitization can be corrected by (1) replacing the carbon or carbon-molybdenum installation with the properly selected chromium-molybdenum alloy, or (2) removing the various graphitized welded joints and replacing with chrome-bearing alloy spool pieces, or (3) chipping and grinding away that material at the welded joints which has become graphitized and subsequently rewelding with the proper chrome-bearing electrode, or (4) heat-treating the graphitized joints at a specified temperature for a specified time (a temporary cure); this will redissolve the graphite in the parent metal and in the weld metal. Since graphitization has been most usually observed at the joints, method 1 is perhaps more drastic than necessary. Method 3 does not give assurance that all graphitized material has been removed because it is necessary to avoid cutting through the backing ring. Method 4 will give short relief (measured in months or years) but is not a permanent remedy.

### Pipe Supports

The Code for Pressure Piping includes many types of supports and gives directions for their application. A proper pipe support must have a strong rigid base properly supported, and an adjustable roll construction which will maintain the alignment in any direction. It is important to avoid friction caused by the movement of the pipe in the support and to have all parts of sufficient strength to maintain alignment at all times. Wire hangers, band iron hangers, wooden hangers, hangers made from small pipe, and hangers having one vertical pipe support do not maintain alignment.

The direction of expansion in a pipe run can be predetermined by anchoring one end, both ends, or the middle. **Anchors** must be firmly fastened to a rigid and heavy part of the power-plant structure, and must also be securely fastened to the pipe; otherwise the equipment for absorbing expansion is useless, and severe stresses may be thrown on parts of the piping system. Some methods of support are shown in Figs. 18 and 19. Welded steel brackets (Fig. 18a) are available in light, medium, and heavy weights. Many types of supports can be mounted on these brackets, such as the anchor chair shown on the bracket at (a), pipe roller supports of the type at

(c), pipe roll stands of various types such as shown in Fig. 19, pipe seats, etc. Figure 18b illustrates one of the many types of adjustable ring hangers in use. The split ring hanger can be applied after the pipeline is in place. At (c) in Fig. 18 is shown a spring cushion pipe roll hanger recommended for service where constant support[1] is required and compensation must be made for movement of the piping. The springs provide an efficient means of absorbing the vibration. Figure 18d shows one of the many types of pipe saddle supports available. Figure 19 shows a cast-iron pipe roll stand designed for cases where vertical adjustment is not necessary but where provision must be made for expansion and contraction of the pipeline. Several designs of such stands with provision for vertical adjustment and of the same general dimensions are also available. One type of cast-iron roll and plate, illustrated in Fig. 19, provides for expansion and contraction where vertical adjustment is not necessary. If necessary, the base-plate can be raised or lowered by use of shims. Detailed information and dimensions of a great variety of pipe supports can be found in manufacturers' catalogs.

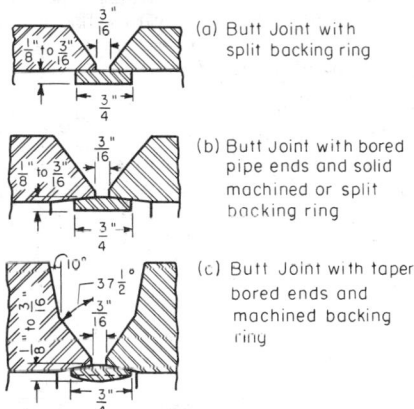

**Fig. 17**  Recommended backing-ring types. (a) Butt joint with split backing ring; (b) butt joint with bored pipe ends and solid machined or split backing ring; (c) butt joint with taper-bored ends and machined backing ring.

In supporting a high-temperature piping system, it is necessary to provide for expansion and contraction due to cyclic changes. It is often possible to find a point of zero movement along the run of a long line and to support a considerable portion of the total load by a rigid hanger or support of the type shown in Figs. 18 and 19. However, for other portions of the run, some form of spring support is often indicated. For relatively light lines, which are not subjected to excessive movements from hot to cold positions, a variable spring hanger will frequently suffice; for heavy lines, or those in which expansion movements are great, it is advisable to use constant support or counterweighted hangers so that transfer of weight to other hangers or equipment connections is prevented. Parts (a) and (b) of Fig. 20 indicate, respectively, a horizontal and vertical run of piping supported by a **constant-support hanger.** Figures 20c and 21a indicate horizontal runs supported by

[1] The support afforded by the hanger of Fig. 18c is constant only in the sense that some degree of support is always present. It might be more appropriately termed a variable-support device.

**Table 45. Maximum Spacing of Pipe Supports at 750°F (399°C)\***

| Nominal pipe size, in | 1 | 1½ | 2 | 3 | 4 | 6 | 8 | 10 | 12 | 14 | 16 | 18 | 20 | 24 |
|---|---|---|---|---|---|---|---|---|---|---|---|---|---|---|
| Maximum span, ft | 7 | 9 | 10 | 12 | 14 | 17 | 19 | 22 | 23 | 25 | 27 | 28 | 30 | 32 |
| Maximum span, m | 2.13 | 2.74 | 3.05 | 3.66 | 4.27 | 5.18 | 5.79 | 6.71 | 7.01 | 7.62 | 8.23 | 8.53 | 9.14 | 9.75 |

\*This tabulation assumes that concentrated loads, such as valves and flanges, are separately supported. Spacing is based on a combined bending and shear stress of 1,500 lb/in² when pipe is filled with water; under this condition, sag in pipeline between supports will be approximately 0.1 in.

**variable-spring hangers.** Figure 21*b* shows a riser supported by a variable spring beneath a base elbow. Figure 21*c* indicates a **sway brace** that is used to control vibration and undesirable movement in a piping system.

The **distance between supports** will vary with the kind of piping and the number of valves and fittings. Supports should be provided near changes in direction, branch lines, and particularly near valves. The weight of piping must not be carried through valve bodies. In establishing the location of pipe supports, the designer should be guided by two requirements: (1) the horizontal span must not be so long that sag in the pipe will impose an excessive stress in the pipe wall and (2) the pipeline must be pitched downward so that the outlet of each span is lower than maximum sag in the span. Table 45 lists spacing for standard-weight pipe supports.

### Pipe Insulation

(See Secs. 4 and 6 for heat-transmission data)

The value of a steam-pipe covering is measured by its ability to reduce heat losses. This might range from 50 percent for small, low-temperature lines to 90 percent for large, high-temperature lines. Many **pipe-insulating materials** are available: 85 percent magnesia, foam glass, calcium silicate, and various forms of diatomaceous earths. Some of these materials are suited for relatively low temperatures only; others are best suited for high temperatures, and still others are suitable over a considerable temperature range.

**Pipe insulation is applied** in molded sections 3 ft long. For high-temperature work, the insulation is applied in at least two layers with the joints staggered so as to prevent a direct channel for heat loss. Because of its maximum-temperature limitation of about 600°F (316°C), 85 percent magnesia is used as the second layer with a high-temperature-resistant material placed in direct contact with the pipe. The molded insulation is fastened securely in place with copper or galvanized wire and is then given a surface finish; indoor pipes are first sheathed with resin paper and covered with canvas, either pasted or sewed; outdoor pipes may be weather-protected by a coating of asphaltic-type waterproofing compound, they may be sheathed and canvased and then given a weather-proof surface, or they may be encased in metallic (steel or aluminum) jackets.

The heat loss from an insulated pipe appears in three phases: heat passes by *conduction* through the metallic pipe walls and through the insulating material; it then is dissipated from the outside surface of the insulation by *convection* and by *radiation*. Extremely accurate calculations must also take into account the temperature drop by convection through the film on the inside surface of the pipe. The task of accurately calculating heat losses is somewhat tedious, since the convec-

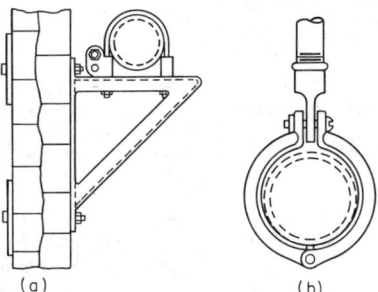 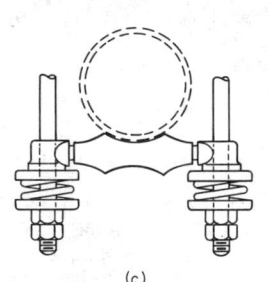 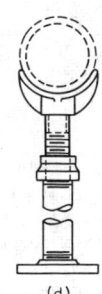

(a)          (b)          (c)          (d)

**Fig. 18**   Methods of supporting pipes.

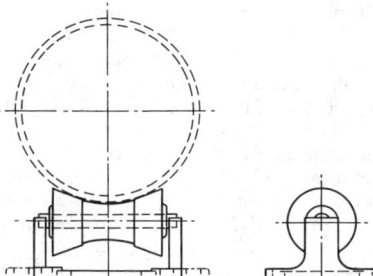 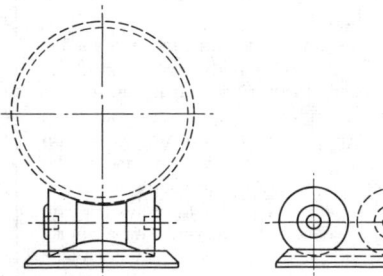 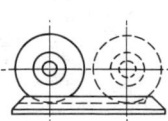

**Fig. 19**   Pipe supports on cast-iron rolls.

tion and radiation losses are related to the surface temperature (outside of insulation), which is unknown until conduction losses are balanced against surface losses.

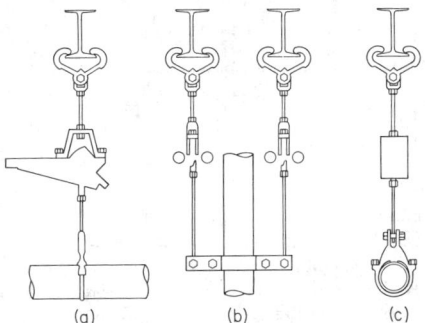

**Fig. 20**   Spring hangers.

For combined convection and radiation coefficients for bare pipes, and all necessary formulas to permit trial-and-error calculations, see Sec. 4. Insulation manufacturers publish data which give heat losses for wide ranges of pipe size and temperature.

### Identification of Piping

The American National Standards Institute has approved a **Scheme for the Identification of Piping Systems** (ANSI A13.1-1956). This scheme is limited to the identification of piping systems in industrial plants, not including pipes buried in the ground, and electric conduits. Fittings, valves, and pipe coverings are included, but not supports, brackets, or other accessories.

**Classification by Color**   All piping systems are classified by the nature of the material carried. Each piping system is placed, by the nature of its contents, in the following classifications:

| *Class* | *Color* |
| --- | --- |
| F—Fire-protection equipment .... | Red |
| D—Dangerous materials ........ | Yellow (or orange) |
| S—Safe materials ............. | Green (or the achromatic colors, white, black, gray, or aluminum) |
| P—Protective materials ........ | Bright blue |
| V—Extra valuable materials ..... | Deep purple |

**Method of Identification**   At conspicuous places throughout a piping system, color bands should be painted on the pipes to designate to which one of the five main classes it belongs. If desired, the entire length of the piping system may be painted the main classification color.

Further, the actual contents of a piping system may be indicated by, preferably, a stenciled legend of standard size giving the name of the contents in full or abbreviated form. These legends should be placed on the color bands. The identification scheme may be extended by the use of colored stripes placed at the edges of the colored bands.

The bands, legends, and stripes should be placed at intervals throughout the piping system, preferably adjacent to valves and fittings to ensure ready recognition during operation, repairs, and at times of emergency.

A recommended classification, under this color scheme of materials carried in pipes, includes, as dangerous, combustible gases and oils, hot water and steam above atmospheric pressure; as safe, compressed air, cold water, and steam under vacuum.

### Pressure Hose

Hose with durable rubber lining may be obtained to withstand any needed pressure. If the rubber compound is properly made, the life of a hose will be 7 to 10 years, while a cheaper hose, lined with inferior material, will probably not last more than 3 or 4 years. See also Secs. 3 and 12.

**American National Fire-Hose Coupling Screw Thread** (ANSI B26-1953)   This standard is intended to cover the threaded part of fire-hose couplings, hydrant outlets, standpipe connections, and all other special fittings on fire lines, where fittings of the nominal diameters given in Table 46 are used. It also includes the limiting dimensions of the field inspection gages. The American National Standard form of thread must be used.

**American National Standard Hose-Coupling Screw Threads** (ANSI B2.4 1966)   These standards apply to the threaded parts of hose couplings, valves, nozzles, and all other fittings used in direct connection with hose intended for fire protection or for domestic, industrial, or general service in nominal sizes given in Table 47. The American National Standard thread form is used. This coupling is similar in design to the fire-hose couplings illustrated in Fig. 22.

**Flexible metal hose and tubing** are available for a wide range of conditions of temperature, pressure, vibration, and corrosion, and are made in two basic constructions, corrugated or interlocked, and in either bronze or steel.

The corrugated type (Fig. 23) may have either annular or helical corrugated formations, usually covered with metal braid, and is adapted to high-pressure high-temperature leakproof service. Some typical applications include diesel-engine

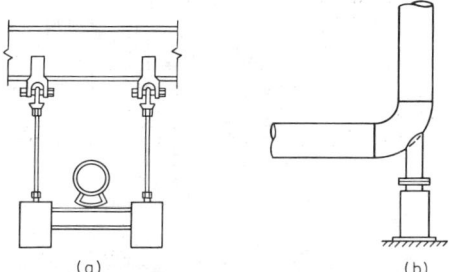

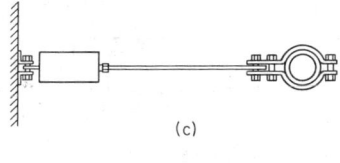

**Fig. 21**   Spring hangers and sway brace.

**Table 46. Dimensions of Standard Fire-Hose Couplings**
(ANSI B2.4-1966)
(All dimensions in inches. Letters refer to Fig. 22)

| Inside diam, *C* | Diam of thread, *D* | No. of threads per in. | *L* | *I* | *H* | *J* | *T* |
|---|---|---|---|---|---|---|---|
| 2½ | 3¹⁄₁₆ | 7½ | 1 | ¼ | 1⁵⁄₁₆ | ³⁄₁₆ | 1¹⁄₁₆ |
| 3 | 3⅝ | 6 | 1⅛ | ⁵⁄₁₆ | 1¹⁄₁₆ | ¼ | 1³⁄₁₆ |
| 3½ | 4¼ | 6 | 1⅛ | ⁵⁄₁₆ | 1¹⁄₁₆ | ¼ | 1³⁄₁₆ |
| 4½ | 5¾ | 4 | 1¼ | ⁷⁄₁₆ | 1³⁄₁₆ | ⅜ | 1⁵⁄₁₆ |

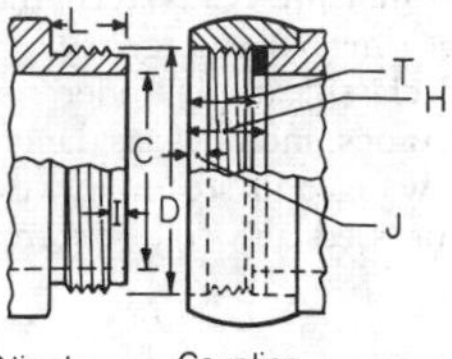


**Fig. 22** Typical form of standard coupling.

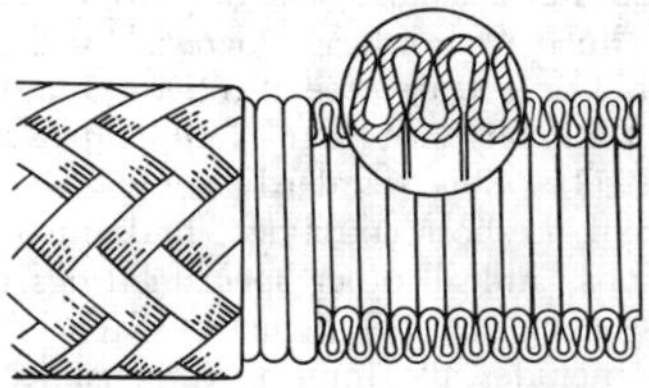

**Fig. 23** Flexible metal hose.

exhaust hose, reciprocating flexible connections, loading and unloading hose, saturated and superheated steam lines, lubricating lines, gas and oil lines, vibration connections, etc.

The interlocked type is made in several ways; the fully interlocked type is illustrated in Fig. 24. Typical applications include wiring conduit, cable armor, decorative wiring covering, dust-collective tubing, grease and oil connections, flexible spouts, and moderate-pressure oil lines.

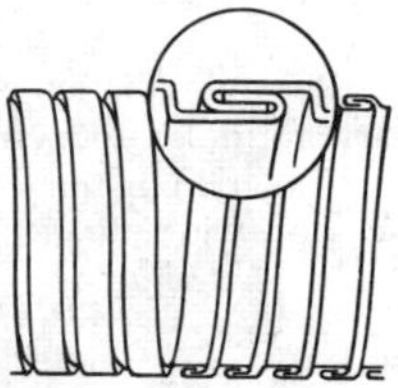

**Fig. 24** Interlocked flexible metal hose.

Standard couplings and fittings can be attached to flexible metal hose or tubing by various methods such as brazing or welding. Each type of hose construction has limits of service use and proved application usages. Information and recommendations as to the type and size to use under any given conditions should be obtained from the manufacturers.

**Table 47. Dimensions of Standard Hose Couplings**
(ANSI B2.4-1966)
(All dimensions in inches. Letters refer to Fig. 22)

| Service and nominal size | Inside diam, *C* | Diam of thread, *D* | No. of threads per in. | *L* | *I* | *H* | *T* |
|---|---|---|---|---|---|---|---|
| Garden: | | | | | | | |
| ½, ⅝, ¾ | ²⁵⁄₃₂ | 1¹⁄₁₆ | 11½ | ⁹⁄₁₆ | ⅛ | ¹⁷⁄₃₂ | ⅜ |
| Chemical: | | | | | | | |
| ¾, 1 | 1¹⁄₃₂ | 1⅜ | 8 | ⅝ | ⁵⁄₃₂ | ¹⁹⁄₃₂ | ¹⁵⁄₃₂ |
| Fire: | | | | | | | |
| 1½ | 1¹⁷⁄₃₂ | 2 | 9 | ⅝ | ⁵⁄₃₂ | ¹⁹⁄₃₂ | ¹⁵⁄₃₂ |
| Other connections: | | | | | | | |
| ½ | ¹⁷⁄₃₂ | ¹³⁄₁₆ | 14 | ½ | ⅛ | ¹⁵⁄₃₂ | ⁵⁄₁₆ |
| ¾ | ²⁵⁄₃₂ | 1¹⁄₃₂ | 14 | ⁹⁄₁₆ | ⅛ | ¹⁷⁄₃₂ | ⅜ |
| 1 | 1¹⁄₃₂ | 1⁹⁄₃₂ | 11½ | ⁹⁄₁₆ | ⁵⁄₃₂ | ¹⁷⁄₃₂ | ⅜ |
| 1¼ | 1⁹⁄₃₂ | 1⅝ | 11½ | ⅝ | ⁵⁄₃₂ | ¹⁹⁄₃₂ | ¹⁵⁄₃₂ |
| 1½ | 1¹⁷⁄₃₂ | 1⅞ | 11½ | ⅝ | ⁵⁄₃₂ | ¹⁹⁄₃₂ | ¹⁵⁄₃₂ |
| 2 | 2¹⁄₃₂ | 2¹¹⁄₃₂ | 11½ | ¾ | ³⁄₁₆ | ²³⁄₃₂ | ¹⁹⁄₃₂ |

# PREFERRED NUMBERS

## by C. H. Berry

REFERENCES: Hirshfeld and Berry, Size Standardization by Preferred Numbers, *Mech. Eng.*, Dec. 1922. Schlink, A New Tool for Standardizers, *Am. Mach.*, July 12, 1923. Tornebohm, The Development and Importance of Preferred Number Series, *Mech. Eng.*, Oct. 1923. Schlink, Use of Preferred Numbers, *Jour. SAE*, Feb. 1925. Table of Preferred Numbers, ANSI Standard Z17.1, 1973. Steczynski, Preferred Numbers for American Practice, *Mech. Eng.*, Nov. 1928. Von Dobbeler, Preferred Numbers, *Mech. Eng.*, March 1929.

Many manufactured articles are made in several sizes which may be designated by some dimension, speed, capacity, or other feature. Each such series of products may be paralleled by a series of numbers.

It is generally agreed that such number series should be **geometric progressions;** i.e., each term should be a fixed percentage larger than the preceding. A geometric series provides small steps for small numbers, large steps for large numbers, and this best meets most requirements. The small steps in the diameter of the numbered twist drills would be absurd in drills of 1 in diameter and larger.

In the case of sized objects that are used principally as raw material, e.g., steel rod, an arithmetic progression may be preferable because it tends to reduce the cost of machining. It is desirable to be able to buy raw material a fixed amount (rather than a fixed percentage) larger than the finished article.

**Preferred numbers** is the name given to various series proposed for general use. These are either geometric progressions or approximations thereto. A geometric series is defined by one term and the ratio of each term to the preceding. On the choice of these elements for a preferred number series, there is as yet no general agreement. The same value would hardly be satisfactory for all cases. The idea of preferred numbers is to provide a master series from which terms can be chosen to suit any needs. This would ultimately lead to a comprehensive plan in all fields of manufacture, so that, for example, the sizes of shafting would be in accord with the sizes of bearings, and indeed with all manner of cylindrical machine elements.

An advantage of a geometric series is that if linear dimensions are chosen in the series, areas, volumes, and other functions of powers of dimensions are also members of the same series.

In one of the most carefully considered systems of preferred numbers the base term is 1, and the ratio is $\sqrt[80]{10}$. In this series, the 81st term is 10, and accordingly the series from 10 to 100 or from 0.01 to 0.1, or, in general, from $10^n$ to $10^{n+1}$ is identical with the series from 1 to 10 with the decimal point shifted. This series will rarely be used in full; some will choose alternate terms, some every fourth, fifth, tenth, or twentieth term. The index of the root, 80, has as factors $2^4$ and 5, so that the series readily yields subseries having as ratios the roots of 10 with indices 2, 4, 8, 16, 5, 10, 20, 40, thus giving a wide range of choice.

The strict logic of this series has been somewhat impaired by the adoption of rounded values that are slightly different in the 1 to 10 and 10 to 100 intervals. For the United States, the ANSI has adopted a Table of Preferred Numbers (ANSI Z17.1-1973) which differs slightly from the system described in the preceding paragraph.

Another type of series is the **semigeometric series** (Steczynski, *loc. cit.*) consisting of a basic geometric series with 1 as the base term, and a ratio of 2, giving a series . . . ⅛, ¼, ½, 1, 2, 4,. . . . Between consecutive terms are inserted arithmetic series of 2, 4, 8, or 16 terms, in general using different numbers of terms in different intervals.

**Table 1. Basic Series of Preferred Numbers: R 80 Series***

| | | | |
|---|---|---|---|
| 1.00 | 1.80 | 3.15 | 5.60 |
| 1.03 | 1.85 | 3.25 | 5.80 |
| 1.06 | 1.90 | 3.35 | 6.00 |
| 1.09 | 1.95 | 3.45 | 6.15 |
| 1.12 | 2.00 | 3.55 | 6.30 |
| 1.15 | 2.06 | 3.65 | 6.50 |
| 1.18 | 2.12 | 3.75 | 6.70 |
| 1.22 | 2.18 | 3.87 | 6.90 |
| 1.25 | 2.24 | 4.00 | 7.10 |
| 1.28 | 2.30 | 4.12 | 7.30 |
| 1.32 | 2.36 | 4.25 | 7.50 |
| 1.36 | 2.43 | 4.37 | 7.75 |
| 1.40 | 2.50 | 4.50 | 8.00 |
| 1.45 | 2.58 | 4.62 | 8.25 |
| 1.50 | 2.65 | 4.75 | 8.50 |
| 1.55 | 2.72 | 4.87 | 8.75 |
| 1.60 | 2.80 | 5.00 | 9.00 |
| 1.65 | 2.90 | 5.15 | 9.25 |
| 1.70 | 3.00 | 5.30 | 9.50 |
| 1.75 | 3.07 | 5.45 | 9.75 |

*Reproduced from American National Standard Preferred Numbers Z17.1, with permission of ANSI.

# Power Generation

**BY**

**FRANK A. RITCHINGS**  *Vice President, Ebasco Services Incorporated, New York.*

**R. R. BENNETT**  *Chief Consulting Mechanical Engineer, Ebasco Services, Inc., New York.*

**GEORGE W. KESSLER**  *Retired Vice President, The Babcock & Wilcox Company.*

**FREDERICK G. BAILY**  *Manager, Turbine Application Engineering, Large Steam Turbine-Generator Department, General Electric Company.*

**JOSEPH R. SPENCER**  *Senior Design Engineer, Heat Transfer Engineering, Westinghouse Electric Co.*

**RICHARD M. STEPHANI**  *Senior Design Engineer, Heat Transfer Engineering, Westinghouse Electric Co.*

**J. A. BOLT**  *Professor of Mechanical Engineering, University of Michigan.*

**D. E. COLE**  *Associate Professor of Mechanical Engineering, University of Michigan.*

**W. MIRSKY**  *Professor of Mechanical Engineering, University of Michigan.*

**D. J. PATTERSON**  *Associate Professor of Mechanical Engineering, University of Michigan.*

**REEVES MORRISSON**  *Corporate Technical Staff, United Technologies Corporation.*

**LOUIS H. RODDIS, JR.**  *Consulting Engineer, Charleston, SC.*

**JOHN E. GRAY**  *President, International Energy Associates, Ltd.*

**CARL L. NEWMAN**  *Vice President, Engineering, Consolidated Edison Co. of New York, Inc.*

**N. J. PALLADINO**  *Dean of Engineering, Pennsylvania State University.*

**W. G. WHIPPEN**  *Manager, Product Development, Allis Chalmers, York Division.*

**WILBUR D. MARSH**  *Senior Application Engineer, Electric Utility Systems Engineering Department, General Electric Company.*

**JOHN I. YELLOTT**  *College of Architecture, Arizona State University, Tempe, AZ.*

**V. F. ESTCOURT**  *Consulting Engineer, Bechtel Power Corporation.*

**E. S. KRENDEL**  *Professor of Operations Research and Statistics, University of Pennsylvania.*

**ERICH A. FARBER**  *Professor and Research Professor of Mechanical Engineering; Director, Solar Energy and Energy Conversion Laboratory, University of Florida.*

**D. K. McLAUGHLIN**  *Associate Professor of Mechanical Engineering, Oklahoma State University.*

**W. L. HUGHES**  *Professor of Electrical Engineering, Oklahoma State University.*

**SHERWOOD B. MENKES**  *Professor of Mechanical Engineering, The City College, The City University of New York.*

**SOURCES OF ENERGY**
**by Frank A. Ritchings**
**Revised by R. R. Bennett**

Fossil Fuels............................................ 9-3
Nuclear Fuels.......................................... 9-4
Other Sources of Energy ............................. 9-5
Energy Requirements ................................. 9-6

**STEAM BOILERS**
**by George W. Kessler**

Fuels Available for Boiler Firing........................ 9-7
Ash and Slag .......................................... 9-7
Stokers ............................................... 9-10
Pulverizers ........................................... 9-11
Burners ............................................... 9-12

Cyclone Furnaces .................................... 9-13
Unburned Combustible Loss .......................... 9-13
Boiler Types ....................................... 9-14
Furnaces ........................................... 9-16
Superheaters and Reheaters ......................... 9-20
Economizers ........................................ 9-21
Air Heaters ........................................ 9-22
Steam Temperature, Adjustment and Control .......... 9-23
Operating Controls ................................. 9-24
Boiler Circulation ................................. 9-24
Flow of Gas through Boiler Unit .................... 9-26
Performance ........................................ 9-26
Steam Purification ................................. 9-28
Water Treatment .................................... 9-29
Care of Boilers .................................... 9-33
Codes .............................................. 9-34
Nuclear Boilers .................................... 9-34

### STEAM ENGINES
Work and Dimensions of the Steam Engine ............ 9-35

### STEAM TURBINES
#### by Frederick G. Baily
Steam Flow through Nozzles and Buckets in Impulse Turbines   9-40
Low-Pressure Elements of Turbines .................. 9-42
Turbine Buckets, Blading, and Parts ................ 9-45
Industrial and Auxiliary Turbines .................. 9-47
Large Central-Station Turbines ..................... 9-49
Steam-Turbine Performance .......................... 9-52
Installation, Operation, and Maintenance Considerations ..... 9-57

### POWER-PLANT HEAT EXCHANGERS
#### by Joseph R. Spencer and Richard M. Stephani
Surface Condensers ................................. 9-59
Air-Cooled Condensers .............................. 9-64
Direct-Contact Condensers .......................... 9-65
Air Ejectors ....................................... 9-66
Vacuum Pumps ....................................... 9-71
Cooling Towers ..................................... 9-71
Dry Cooling Towers, with Direct-Contact Condensers ...... 9-74
Spray Ponds ........................................ 9-74
Closed Feedwater Heaters ........................... 9-74
Open, Deaerating, and Direct-Contact Heaters ....... 9-77
Evaporators ........................................ 9-78

### INTERNAL-COMBUSTION ENGINES
#### by J. A. Bolt, D. E. Cole, W. Mirsky, and D. J. Patterson
General Features .................................... 9-78
Wankel (Rotary) Engines ............................ 9-80
Aircraft Engines ................................... 9-80
United States Automobile Engines ................... 9-82
Foreign Automobile Engines ......................... 9-83
Truck and Bus Engines .............................. 9-83
Tractor Engines .................................... 9-85
Stationary Engines ................................. 9-86
Locomotive Engines ................................. 9-86
Marine Engines ..................................... 9-86
Small Industrial, Utility, and Recreation-Vehicle Gaso-
line Engines ....................................... 9-87
Combustion Chambers ................................ 9-88
Fuels .............................................. 9-92
Combustion Knock ................................... 9-92
Carburetion ........................................ 9-97
Fuel Injection ..................................... 9-98
Ignition and Spark Advance ......................... 9-100

Air and Fuel Lines, Manifolds, Mixture Distribution,
and Mufflers ....................................... 9-101
Supercharging ...................................... 9-102
Scavenging Two-Stroke-Cycle Engines ................ 9-103
Regulation ......................................... 9-104
Cooling Systems .................................... 9-104
Lubrication ........................................ 9-105
Analysis of the Engine Process ..................... 9-106
Internal-Combustion Engines and Air Pollution ...... 9-109
Special Developments ............................... 9-111

### GAS TURBINES
#### by Reeves Morrisson
Gas-Turbine Cycles ................................. 9-114
Gas-Turbine Types .................................. 9-115
Gas-Turbine Components ............................. 9-116
Gas-Turbine Arrangements ........................... 9-117
General Characteristics ............................ 9-117
Fuels .............................................. 9-118
Applications ....................................... 9-118

### NUCLEAR (ATOMIC) POWER
#### by Louis H. Roddis, Jr., John E. Gray, Carl L. Newman, and N. J. Palladino
Fission and Fusion Energy .......................... 9-119
Nuclear Physics .................................... 9-119
Utilization of Fission Energy ...................... 9-121
Properties of Materials ............................ 9-125
Fission-Reactor Design ............................. 9-127
Nuclear-Power-Plant Economics ...................... 9-130
Nuclear-Power-Plant Safety ......................... 9-133
Nuclear-Power-Plant Licensing ...................... 9-133
Other Power Applications ........................... 9-134
Fusion Nuclear Energy .............................. 9-135

### HYDRAULIC TURBINES
#### by W. G. Whippen
General ............................................ 9-136
Reaction Turbines .................................. 9-138
Impulse Turbines ................................... 9-143
Reversible Pump/Turbines ........................... 9-145
Model Tests ........................................ 9-146
Cavitation ......................................... 9-147
Speed Regulation ................................... 9-147
Auxiliaries ........................................ 9-149
Turbine Tests ...................................... 9-149

### DIRECT ENERGY CONVERSION
#### by Wilbur D. Marsh
Direct Energy Conversion ........................... 9-149

### POWER MISCELLANY
Introduction ....................................... 9-153
Solar Energy (By JOHN I. YELLOTT) .................. 9-153
Geothermal Power (By V. F. ESTCOURT) ............... 9-159
Man- and Animal-Generated Power (By E. S. KRENDEL) ... 9-162
Hot-Air Engines (By ERICH A. FARBER) ............... 9-163
Power from Vegetation .............................. 9-163
Wind Power (By D. K. McLAUGHLIN AND W. L. HUGHES) ... 9-164
Power from the Tides ............................... 9-168
Utilization of the Energy of the Waves ............. 9-168
Utilization of Heat Energy of the Sea .............. 9-169
Power from Hydrogen ................................ 9-169
Flywheel Energy Storage (By SHERWOOD B. MENKES) .... 9-170

# SOURCES OF ENERGY

## by Frank A. Ritchings
## Revised by R. R. Bennett

REFERENCES: Reserves of Crude Oil, Natural Gas Liquids, and Natural Gas in the U.S. and Canada as of December 31, 1976, American Petroleum Institute. Annual Statistical Review—Petroleum Industry Statistics 65-74, American Petroleum Institute. Worldwide Issue, *Oil & Gas Jour.* **74,** no. 52, Dec. 27, 1976. Potential Supply of Natural Gas in the U.S., Mineral Resources Institute, Colorado School of Mines, 1973. Coal resources in the United States, *U.S. Geol. Surv. Bull.* 1412. Geological Estimates of Undiscovered Recoverable Oil & Gas Resources in the U.S., *U.S. Geol. Surv. Circ.* 725. United Nations Statistical Office, 1973, "Statistical Yearbook," 1974: New York, U.N. Department of Economic and Social Affairs. Bureau of Mines, 1976, Metals, Minerals, and Fuels, vol. I of "Minerals Yearbook 1974," published annually. Coal Data 1975, National Coal Association, 1976. International Coal 1976, National Coal Association, 1976. Foreign Uranium Sources—Status and Developments, John A. Patterson, Div. of Uranium Resources and Enrichment, U.S. Energy Research and Development Administration, Jan. 1977.

Energy requirements are supplied primarily by fossil fuels, nuclear fuels, hydroelectric, and other. About 93 percent of United States needs in 1974 were provided by fossil fuels, less than 2 percent by nuclear fuels, and about 5 percent by hydroelectric with negligible amounts from other sources. About 97 percent of world requirements came from fossil fuels, with the other 3 percent largely from hydroelectric and nuclear sources.

## FOSSIL FUELS

**Petroleum** Proved reserves of crude oil and natural-gas liquids in the United States, based upon estimated discovered quantities which geological and engineering data demonstrate with reasonable certainty to be recoverable in future years from presently known reservoirs under existing economic and operating conditions, are published annually by the American Petroleum Institute. Estimates of additional remaining producible reserves which will be discovered, proved, and produced in the future from the total original oil in place are derived by the *U.S. Geol. Surv. Circ.* 725 from present and projected conditions in the industry.

### U.S. Reserves
Billions of barrels (Billions of cubic metres)

| | Proved as of Jan. 1, 1977 | Additional producible remaining to be discovered (as of 1/1/75) |
|---|---|---|
| Crude oil | 30.9 (4.91) | 82 (13.04) |
| Natural-gas liquids | 6.4 (1.02) | 16 (2.54) |
| Total liquid hydrocarbons | 37.3 (5.93) | 98 (15.58) |

Estimates of proved crude-oil reserves in all countries of the world as of Jan. 1, 1977, were published by *Oil and Gas Journal.* New discoveries are continually adding to and changing proved reserves in many parts of the world, and this latest estimate is considered indicative of current producible quantities. No valid estimates are available for additional reserves remaining to be discovered for countries other than the United States.

### World Crude-Oil Reserves
Billions of barrels (Billions of cubic metres)

| | Proved as of Jan. 1, 1977 | | % of world |
|---|---|---|---|
| Asia Pacific | 19 | (2.5) | 3.2 |
| Europe | 25 | (2.5) | 4.2 |
| Middle East | 326 | (55.6) | 54.4 |
| Africa | 61 | (10.6) | 10.2 |
| North America, inc. U.S. | 46 | (7.9) | 7.7 |
| Central and South America | 21 | (4.5) | 3.5 |
| U.S.S.R. | 78 | (12.7) | 13.0 |
| Republic of China | 20 | (3.2) | 3.3 |
| Other | 3 | (0.5) | 0.5 |
| Total world | 599 | (100) | 100.0 |

**Natural Gas** Proved reserves of natural gas in the United States, based upon the same definition as for crude oil and natural-gas liquids, are estimated annually by the American Gas Association. The estimated total additional potential supply remaining to be discovered is prepared by the Potential Gas Committee, sponsored by the Potential Gas Agency, Colorado School of Mines Foundation, Inc.

### U.S. Natural-Gas Reserves
Trillions of cubic feet (Trillions of cubic metres)

| | Proved as of Jan. 1, 1977 | Additional potential supply remaining to be discovered (as of 1/1/75) |
|---|---|---|
| Natural gas | 216 (6.1) | 1,146 (32.5) |

Estimates of proved reserves of natural gas in all countries of the world as of Jan. 1, 1977, were published by *Oil and Gas Journal.* As with crude oil, large additional natural-gas reserves are currently being discovered and developed in Alaska, the arctic regions, offshore areas, northern Africa, and other locations remote from consuming markets. Valid estimates of additional probable remaining reserves in the world are not available.

**Coal** Authoritative information about reserves of coal is presented in *Geological Survey Bulletin* 1412, Coal Resources of the United States, Jan. 1, 1974. Remaining United States reserves of bituminous, subbituminous, lignite, and anthracite, which have been estimated by mapping and exploration of areas with 0 to 3,000 ft overburden, total 1,731 billion short tons (1,570 billion kg). The U.S. Geological Survey estimates probable additional resources in unmapped and unexplored areas with 0 to 3,000 ft overburden amount to 1,850 billion tons (1,678 trillion kg) and in areas with 3,000 to 6,000 ft overburden an additional 389 billion tons (353 trillion kg).

The estimates of coal reserves in the various states indicate the location of the total 1,731 billion tons.

### World Natural-Gas Reserves
Trillions of cubic feet (Trillions of cubic metres)

| | Proved as of Jan. 1, 1977 | % of world |
|---|---|---|
| Asia Pacific | 120 ( 3.40) | 5.2 |
| Europe | 142 (14.02) | 6.2 |
| Middle East | 513 (14.53) | 22.3 |
| Africa | 209 ( 5.92) | 9.1 |
| North America, inc. U.S. | 288 ( 8.16) | 12.5 |
| South America | 79 ( 2.24) | 3.4 |
| U.S.S.R. | 918 (25.99) | 39.8 |
| Republic of China | 25 ( 0.71) | 1.1 |
| Other | 10 ( 0.28) | 0.4 |
| Total world | 2,304 (65.24) | 100.0 |

### U.S. Coal Reserves Remaining Jan. 1, 1974
Billions of short tons (trillions of kilograms)

| **Bituminous Coal** | | **Subbituminous Coal** | |
|---|---|---|---|
| Illinois | 146 (132) | Montana | 177 (161) |
| West Virginia | 100 ( 91) | Alaska | 111 (101) |
| Kentudky | 54 ( 49) | Wyoming | 123 (111) |
| Colorado | 109 ( 99) | New Mexico | 51 ( 46) |
| Pennsylvania | 64 ( 58) | Other states | 24 ( 22) |
| Ohio | 41 ( 37) | Total | 486 (441) |
| Indiana | 32 ( 29) | | |
| Missouri | 31 ( 29) | **Lignite** | |
| Other states | 170 (154) | North Dakota | 351 (318) |
| Total | 747 (678) | Montana | 113 (102) |
| | | Other states | 14 ( 13) |
| **Total All Ranks** | | Total | 478 (433) |
| Bituminous | 747 ( 678) | | |
| Subbituminous | 486 ( 441) | **Anthracite and Semi anthracite** | |
| Lignite | 478 ( 433) | Pennsylvania | 19 ( 17) |
| Anthracite | 20 ( 18) | Other states | 1 ( 1) |
| Total | 1,731 (1,570) | Total | 20 ( 18) |

About half of the 1,731 billion tons are considered producible because of favorable depth of overburden and thickness of coal beds.

Only about 520 billion tons (362 trillion kg), or 30 percent of all ranks of coal, are commercially available in beds generally less than 1,000 ft deep, including bituminous and anthracite in beds 42 in or more thick, and subbituminous and lignite in beds 10 ft or more thick. The U.S. Geological Survey estimates that of the 1,731 billion tons of reserves, 65 percent contains 1.0 percent or less sulfur, 15 percent contains 1.1 to 3.0 percent, and 20 percent contains 3.0 or more percent sulfur. East of the Mississippi River, 20 percent is low sulfur, 37 percent is medium, and 43 percent is high.

### Coal Resources of the World
Billions of short tons (trillions of kilograms)

| | By mapping and exploration | Probable additional |
|---|---|---|
| Asia (incl. U.S.S.R.) | 4,000 (3,632) | 7,000 (6,356) |
| North America | 1,900 (1,725) | 2,500 (2,270) |
| Europe | 300 ( 272) | 500 ( 454) |
| Africa | 90 ( 82) | 160 ( 145) |
| Oceania | 70 ( 64) | 60 ( 54) |
| South and Central America | 30 ( 27) | 10 ( 9) |
| Total | 6,390 (5,802) | 10,230 (9,288) |

The estimated total remaining original coal resources of the world by continents were reported in *U.S. Geological Survey Bulletin* 1412 based on estimates from about 50 countries, which at best are gross approximations. The tonnages indicate the general magnitude of reserves in each area.

**Shale Oil** The portion of total U.S. reserves of oil from oil shale, measured or proved, which are considered minable and amenable to processing under present (1977) economic conditions are estimated to be 125 billion barrels (25.4 billion m³), based upon grades averaging 30 gal/ton in beds at least 100 ft thick (*U.S. Geological Survey Bulletin* 1412). Most of the oil shale occurs in Colorado. No extensive commercial production is expected for many years. World reserves occur largely in the United States and Brazil, with small quantities in other countries.

**Tar Sands** Large deposits are in the Athabasca area of northern Alberta, Canada, estimated capable of producing 100 to 300 billion barrels (15.9 to 47.7 billion m³) of oil. About 6.3 billion barrels (1.0 billion m³) have been proved as economically recoverable within the radius of the present large mining and recovery plant in Athabasca. About 15 million barrels (2.4 million m³) were produced in this plant in 1971.

### NUCLEAR FUELS

**Uranium** Reserves of uranium in the United States and certain countries in the world have been presented in a report by the U.S. ERDA. Div. of Uranium Resources and Enrichment, dated Jan. 1977. Reasonably assured resources refer to uranium which occurs in ore deposits of such grade, quantity, and configuration that it can, within the given price range, be profitably recovered with concurrently proved mining and processing technology. Estimated additional resources refers to uranium surmised to occur in unexplored extensions of known deposits or in undiscovered deposits in known uranium districts, and which are expected to be discoverable and economically exploitable in the given price range. United States uranium resources are largely located in New Mexico, Wyoming, and Colorado.

**Estimated Uranium, $U_3O_8$, Resources, Jan. 1, 1976 (excluding Eastern Block Countries)**
Thousands of tons (trillions of kilograms)

| | Less than $30/lb* | |
|---|---|---|
| | Reasonably assured | Estimated additional |
| *North America* | | |
| U.S. | 640 (581) | 1,060 (962) |
| Canada | 225 (204) | 787 (715) |
| Mexico | 8 ( 7) | |
| Denmark (Greenland) | 8 ( 7) | 13 ( 12) |
| *Africa* | | |
| South & SW Africa | 359 (326) | 96 ( 87) |
| Niger | 65 ( 59) | 39 ( 35) |
| Algeria | 36 ( 33) | |
| Gabon | 26 ( 24) | 13 ( 12) |
| C.A.R. | 10 ( 9) | 10 ( 9) |
| Zaire | 2 ( 2) | 2 ( 2) |
| *Europe* | | |
| Sweden | 390 (354) | |
| France | 72 ( 65) | 52 (47) |
| Spain | 30 ( 27) | 56 ( 51) |
| Yugoslavia | 9 ( 8) | 20 ( 18) |
| Portugal | 9 ( 8) | |
| Finland | 3 ( 3) | |
| Germany | 1 ( 1) | 5 ( 5) |
| Italy | 2 ( 2) | 1 ( 1) |
| U.K. | 2 ( 2) | 5 ( 5) |
| *Australia* | 430 (390) | 100 |
| *Asia* | | |
| India | 38 ( 35) | 30 ( 27) |
| Japan | 10 ( 9) | |
| Korea | 3 ( 3) | |
| Turkey | 4 ( 4) | 1 ( 1) |
| *South America* | | |
| Argentina | 27 ( 25) | 51 ( 46) |
| Brazil | 14 ( 13) | 11 ( 10) |
| *Total* | 2,423 (2,201) | 2,351 (2,045) |

*1976 dollars.

**Thorium** Total known resources of thorium, the availability of which is considered reasonably assured, are estimated at more than half a million short tons of thorium oxide (450 trillion kg). This figure includes only tonnages which may be exploited under present technology at prices less than $10 per pound of $ThO_2$. Most of the known resources are in placer deposits in India. There seems to be little prospect of significant requirements for thorium in the near future.

**OTHER SOURCES OF ENERGY**

**Hydroelectric and Pumped Storage for Electric Generation** About 55,000 MW of hydroelectric-plant capacity and

9,000 MW of pumped-storage capacity were available in the United States in 1975, representing 12.7 percent and 1.8 percent of total electric generating capacity of all types. Although most available sites for economical production of hydroelectric energy have been developed, some additional hydroelectric capacity will be provided at new sites or by additions at existing plants. Pumped-storage-increased capacity will be limited by the availability of suitable sites and a dependable supply of economical pumping energy. The flexibility of operation of a pumped-storage plant in meeting sudden load changes and its ability to provide high-inertia spinning reserve at low operating cost are additional benefits that can weigh heavily in favor of this type of installation, particularly in the future as the proportion of nuclear capacity in service increases. Hydro and pumped storage generated 12.8 percent of electricity generated by all sources of energy in 1976 in the United States.

World installed hydropower capacity at present is located about 40 percent in North America and 40 percent in Europe.

**Wood** The amounts of wood fuels consumed in the United States for energy sources are negligible when considering the overall energy requirements from all sources, and are not included in present energy statistical reports. Wood-waste and refuse materials from lumber mills and paper mills are used for fuels, particularly in the Pacific Northwest. In some parts of the world, wood in the form of peat or charcoal is of some minor local importance. Wood fuels are not likely to be of significance in the future.

**Tidal Power** Although the potential for tidal hydroelectric power is of interest, little has been developed, principally because of the great cost of constructing dams in deep water where velocities resulting from tides are high. It has been estimated that the maximum amount of tidal power that would be possible to develop in the few locations in the world where the tidal range is as much as 25 ft—the minimum required to generate power is only 45 billion kW/h/year, a small fraction of world electricity requirements.

**Wind Power** The maximum energy practically obtainable from the wind is less than 3 percent of the world's 1976 requirements of energy, and the amount actually obtained in 1976 was probably not more than a small fraction of percent of the potential.

**Solar Energy** About one two-billionth of the sun's radiation impinges on the earth, but about half of this is radiated into interstellar space by our atmosphere. Energy received at the earth's surface is about $7^{18}$ kWh/year. It has been estimated that 330,000 kWh/year/acre might reasonably be generated in Arizona from solar energy. Small solar water-heating plants and solar distillation plants have been functioning satisfactorily for years. The most successful application of solar energy to date has taken place in the space programs. Increased emphasis is being placed on solar energy for water and space heating. Solar energy for the production of electricity will probably not be used in significant amounts in this century.

**Geothermal Energy** The only presently exploited geothermal field in the United States is at The Geysers, near San Francisco. Geothermal wells are currently being developed in Arizona and planned for other areas. By 1974, capacity of geothermal projects was approximately 2 percent of capacity of hydroelectric plants. The source of most geothermal energy is the molten rock, or magma, in the earth's interior. When underground water comes into contact with the magma, hot water and steam are produced. When this occurs in large quantities and within a few miles of the surface, the steam and hot water can be tapped and used to turn the turbines that generate electricity. The most promising areas for exploration lie near earthquake faults, volcanic regions, and hot springs and geysers. Italy, Japan, Iceland, New Zealand, and Mexico have geothermal power plants.

**Hydrogen**  The use of hydrogen gas to replace declining resources of fossil fuels recently has been suggested for consideration. Its primary disadvantages, other than economic, are its low density, more difficult handling characteristics, and probably greater hazards. Present methods for producing hydrogen make it more expensive for heating than fossil fuels.

## ENERGY REQUIREMENTS

**Total Consumption**  The U.S. Bureau of Mines tabulates annual gross consumption of energy resources by major sources and consuming sectors, based on the estimated total consumption of coal, gas and oil, and hydro and nuclear electric generation. The 1975 consumption, in quadrillions of Btu (quintillions of joules, J), for each consuming sector amounted to the following: For household and commercial, coal supplied 0.3 (0.3), gas supplied 7.6 (8.1), and oil supplied 5.6 (5.9), for a total of 13.5 (14.3). In the industrial sector, coal supplied 3.6 (3.8), gas supplied 8.6 (9.1), and oil supplied 6.7 (7.1) for a total of 18.9 (20.0). For the transportation sector, gas supplied 0.5 (0.5), and oil supplied 18.0 (19.0), for a total of 18.5 (19.5). Raw input-energy requirement for the electric-generation sector was supplied by coal, 9.6 (10.1); gas, 3.1 (3.3); oil, 2.9 (3.1); hydro 3.0 (3.2); and nuclear 1.5 (1.6), for a total of 20.1 (21.3). Miscellaneous accounted for oil, 0.1 (0.1). Total consumption for all sectors was 71.1 quadrillion Btu (75.0 quintillions of joules).

Electric generation in 1975 required 404 million tons (366 trillions of kg) of coal, 3.1 trillion ft³ (0.08 trillion m³) of natural gas, and 525 million barrels (83 million m³) of oil.

The United Nations "Statistical Yearbook" tabulates annual consumption of energy sources in the world for all countries.

### 1972 World Energy Consumption as Coal Equivalent Total

(Million metric tons or billion kg per capita, kg)

| Region | Total energy | Energy per capita |
|---|---|---|
| Africa | 134 | 363 |
| North America | 2,660 | 11,526 |
| Central America | 158 | 1,228 |
| South America | 126 | 759 |
| Asian Middle East | 95 | 857 |
| Asia, except Middle East | 565 | 481 |
| Europe, except Eastern Europe | 1,435 | 4,000 |
| Oceania | 85 | 4,275 |
| U.S.S.R. | 1,179 | 4,767 |
| Republic of China | 445 | 567 |
| Eastern Europe | 527 | 4,659 |
| Total world | 7,409 | 1,984 |

**Petroleum**  Crude-oil production in the United States was about 3.1 billion barrels (0.49 billion m³) in 1975, which is less than the 1972–73 peak rate, since it is not expected that a greater production level will be supported by the level of reserves which will be maintained by new discoveries in the

future. Production will continue near present rates for many years but eventually will decline as economically producible reserves are used up. Imports supplied about 31 percent of U.S. petroleum supply in 1975, and by 1985–1990, probably 50 percent will be imported, largely from the Middle East and Africa, to meet increasing needs.

Crude-oil production in the world reached 19.4 billion barrels (3.08 billion m³) in 1975. Reserves in the Middle East, Africa, and the U.S.S.R. will be available to supply increasing requirements in the future, probably through 1990.

**Natural Gas**  Production in the United States in 1975 amounted to about 20.1 trillion ft³ (0.57 trillion m³), which is less than the peak rate that occurred in 1972–73, owing to declining reserves. Annual new discoveries are now considerably less than requirements, which will result in depletion of this irreplaceable fuel and decline of production rate in the near future. By 1990, perhaps 35 to 40 percent of U.S. gas consumption will be obtained from imports (pipelines from Canada and LNG tankers from the Middle East, Africa, and other reserve surplus areas) and from synthetic gas (SNG) manufactured in many large plants from oil, coal, and possibly oil-shale feedstocks.

Consumption of natural gas in Europe and other parts of the world has been increasing in recent years because of large new discoveries of reserves in areas where natural gas was not previously available.

**Coal**  There is an abundant supply of coal reserves in the United States to meet expected increased tonnages that will be required principally for electric generation. The rate of increase in future production will be influenced by problems of sulfur removal and other environmental regulations, difficulties with establishing additional large-scale mining operations, and on the plus side, requirements for conversion to synthetic gas and oil. Petroleum and natural gas are currently replacing and reducing coal requirements in the world.

**Nuclear**  Nuclear fuels produced 8.2 percent of the electricity generated in the United States in 1975; however, by 1990 it is generally expected that nuclear sources will be used to generate 40 to 50 percent of the electricity produced, or 16 to 20 percent of total energy consumption from all types of sources for all purposes.

U.S. known reserves of uranium suitable for use in light-water reactors amount to about 640 thousand short tons of $U_3O_8$ (581 trillions of kilograms) at present cost under \$30/lb. Additional probable resources when identified and fully delineated would raise these reserves to about 1,700 thousand tons $U_3O_8$ (1,544 trillions of kilograms) at present cost under \$30/lb and more speculative resources might further raise the total to about 3700 thousand tons (3360 trillions of kilograms) at present cost under \$30/lb.

If there is a permanent ban on reprocessing spent reactor fuel, 1,700 thousand tons of $U_3O_8$ (1,544 trillions of kilograms) would be sufficient for the lifetime (30 years) requirements of about 300 million kilowatt light-water reactor plants. If fuel is reprocessed and the recovered uranium and plutonium recycled in light water reactors the same amount of uranium could support somewhat over 400 million kilowatts of nuclear power. Inasmuch as foreseen demand for electricity are likely to require about 400 million kilowatts of nuclear capacity by the year 2000, it is apparent that either the more speculative resources will have to be confirmed or advanced reactor types must be developed and deployed. Breeder reactors have the potential of extending the nuclear capability almost indefinitely.

# STEAM BOILERS

## by George W. Kessler

REFERENCES: The Babcock & Wilcox Co., "Steam—Its Generation and Use." De Lorenzi, "Combustion Engineering," Combustion Engineering, Inc. Staniar, "Plant Engineering Handbook," McGraw-Hill. Powell, "Water Conditioning for Industry," McGraw-Hill. "Boiler and Pressure Vessel Code," "Power Test Code for Steam Generating Units," ASME. "1958 Manual, Steam Generating Equipment—Industry Standards and Engineering Information," ABMA.

### FUELS AVAILABLE FOR BOILER FIRING

(See also Sec. 7)

Boilers in the United States are usually fired by the most economical fuel available. Natural gas and residual oil are burned in the Southwest and on the West Coast. Residual oil is used in the Southeast and, with favorable prices, on the East

Black liquor from the cooking of wood pulp is burned in many paper mills. Blast-furnace and coke-oven gases are used to fire steel-mill boilers. Lean CO gas, a waste product from oil-refinery operation, is burned in combination with a richer gas.

### ASH AND SLAG

Ash and slag from solid and liquid fuels cause many problems in boiler operation. The ash, when sintered or fused, forms clinkers in fuel beds and deposits on the furnace walls, superheaters, and boiler surfaces which reduce heat absorption and increase draft loss. Fly-ash collectors are required to minimize atmospheric pollution, and the disposal of slag and fly ash is costly.

**Table 1. Burning Equipment for Solid Fuels***

| Fuel | Source | Stokers | | | | PC | | Crushed Cyclone (where fusion temp is suitable) | Cell furnaces |
| | | Underfeed | Traveling grate | Chain grate, jet ignition | Spreader | U flame, pulverized fuel | Horizontal burners, pulverized fuel | | |
|---|---|---|---|---|---|---|---|---|---|
| Coke breeze | | | x | | | | | | |
| Anthracite | E. Pa. | | x | | | x | | | |
| Bituminous coal: | | | | | | | | | |
|   17–27% volatile | W. Va., Cent. Pa. | x | | | x | x | x | x | |
|   27–35% volatile: | | | | | | | | | |
|     Strongly coking | W. Pa., W. Va., Ky., Ohio, Utah | x | x | x | x | x | x | x | |
|     Weakly coking | Ind., Iowa, Ill., Colo., W. Ky. | | x | x | x | x | x | x | |
| Subbituminous coal | Mont., Wyo. | | | | | x | x | x | |
| Pipeline slurry | | | | | | x | x | x | |
| Lignite | N. Dak., S. Dak., Mont., Wyo., Tex. | | x | x | x | x | x | x | |
| Low-temp fluid-coal char | | | | | | x† aux | x† aux | x† aux | |
| Petroleum coke, 9–14% volatile | | | | | | x† aux | x† aux | x† aux | |
| Fluid-petroleum coke, 4–5% volatile | | | | | | x | x† aux | x | |
| Wood and bark‡ | | | | | x | | | x† aux | x |
| Bagasse | | | | | x | | | | x |

*Equipment indicated will usually result in a good application, but there are many factors affecting the burning of fuel, and variations of fuel properties that guide the individual selection of burning equipment.

†aux = auxiliary fuel-coal, oil, or gas.

‡Bark and wood are also burned on inclined grates and in Dutch-oven pile furnaces.

Coast. Solid fuels serve most of the boiler needs on the East Coast and in the central section of the country. However, the development of pipelines and unitrains, the availability of fuels, and the Environmental Protection Agency's (EPA) restrictions pertaining to the gaseous emissions of $SO_2$ and $NO_x$ will greatly affect and, probably, change regional use patterns. Table 1 lists the equipment used to burn different types of solid fuels.

**Pulverized-coal dry-ash-type furnaces** are usually provided with water-cooled wall surfaces to keep the ash relatively dry, but some ash may adhere to the furnace walls or may deposit on the superheater and other boiler-component surfaces. Pulverized-coal **slag-tap-type furnaces** have the burners located so as to maintain a high-gas-temperature zone near the furnace floor and thus keep the ash molten for tapping. The suitability of a coal ash for slag-tap operation is dependent upon its fluid, or flow, temperature. The cone-fusion method of determining the fluid temperature of the ash is widely used, but the viscosity of the ash is a better guide.

The **viscosity** of the ash can be determined by a viscometer. The test requires considerable time and special equipment. The approximate viscosity can be calculated from the chemical, or spectrographic, analysis of the ash after establishing the ratio of the silica to the sum of the silicon, iron, calcium, and magnesium oxides in the ash (Reid and Cohen, *Trans. ASME,* 1944). Figure 1 shows, in full lines, the viscosity, the temperature, and the silica ratio of slags from two coals as determined by viscometer cooling tests. The dashed lines show the corresponding values calculated from spectrographic analyses. At the **critical viscosity** (where some of the constituents start to freeze), the viscosity curve rises steeply and the upper portions of the dashed lines are drawn from a temperature point on the ratio curve 200°F (93°C) higher than the cone-softening temperature and at a slope of 10° from the vertical. Slag, for satisfactory tapping, should have a viscosity of 250 poises (P) at a temperature less than 2600°F (1426°C). Thus, in Fig. 1, coal A is satisfactory for tapping but coal B is questionable. The viscosity and the fusing temperature are greatly affected by silica and, to a lesser degree, ferric oxide (Fig. 2). A **low ash fluid temperature** facilitates ash tapping as slag but may lead to deposits on the furnace walls and superheaters.

Some of the sodium and potassium in coal ash vaporizes in the furnace and later condenses on the superheater tubes, forming a cementlike accumulation of slag and fly ash. The outer layer of this deposit is similar to fly ash, but the inner layer contains much greater amounts of $Na_2O$, $K_2O$, and $SO_3$. The $SO_3$ content is usually so large that a portion of the sulfur exists as complex sulfates of sodium, potassium, iron, and alumina, all of which have lower melting temperatures than the normal sulfates.

The temperature at which **sintering of fly ash** starts may be much lower than the initial deformation temperature of the ash determined from coal samples using conventional laboratory methods. However, by using a special laboratory furnace, fly ash may be produced from coal samples with a sintered strength comparable with that of fly ash from an operating boiler. The **sintered strength** of slag deposits can be established by crushing cylinders of fly ash after heating for 10 h at temperatures ranging from 1400 to 2000°F (760 to 1093°C) (Barnhart and Williams, *Trans. ASME,* 1956). Figure 3 shows the wide difference in the sintered strength of fly ash from two coals whose initial deformation temperatures are identical and whose fusing temperatures are almost the same. Figure 4 illustrates the effect of the sodium oxide content of the coal upon sintered strength, and experience indicates that fouling of the convection heat-absorbing surfaces is reduced when firing coals having a sodium oxide content of less than 0.3 percent of the dry coal weight.

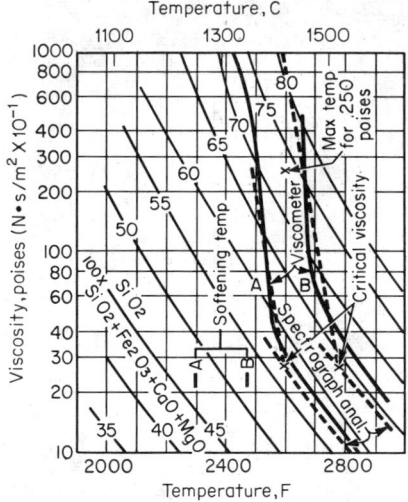

**Fig. 1**  Viscosity of slags by viscometer and as calculated from analysis.

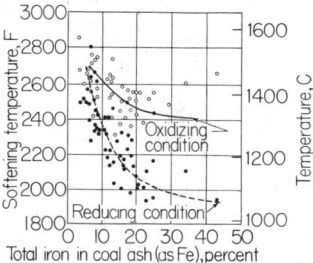

**Fig. 2**  Influence of iron on the ash-softening temperature.

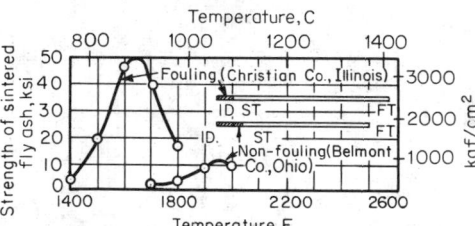

**Fig. 3**  Comparative sintered strengths and temperatures of initial deformation (ID), start of sintering (ST), and ash fusion (FT) for a fouling and a nonfouling coal.

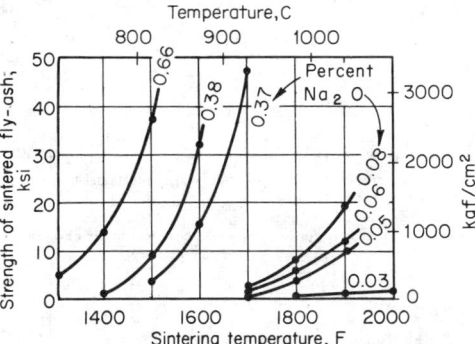

**Fig. 4**  Effect of sodium oxide content of coal on the sintered strength of fly ash.

**Additives,** such as dolomite, lime, and magnesia, are effective in reducing the sintered strength of fly ash, as shown in Fig. 5. Dolomite also is effective in neutralizing the acid in flue gases and in eliminating condensation and subsequent plugging at the cold end of air heaters.

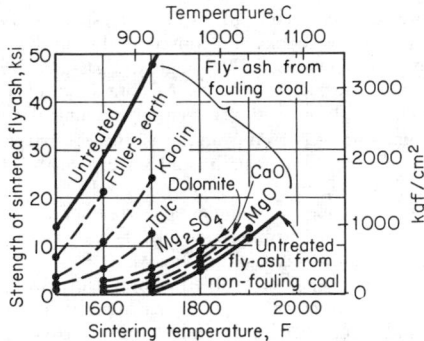

**Fig. 5**  Effect of additives on the sintered strength of fly ash (1 part additive to 4 parts fly ash).

Although fuel oils have a low ash content (0.05 to 0.20 percent), slagging and high-temperature fuel-ash corrosion are experienced in oil-fired units. The fuel-oil analyses in Table 2 indicate that high-alkali fuel-oil ash, with or without vanadium, can cause trouble. Thus dolomite, added in quantities equal to the fuel-oil ash weight, is used in many oil-fired boilers to produce a softer slag which can be removed easily by soot blowing (McIlroy, Holler, and Lee, *Trans. ASME*, 1954). In some installations, air-heater corrosion and pluggage and acid stack discharge also are minimized by the use of dolomite additives (Huge and Piotter, *Trans. ASME*, 1955).

**Table 2. Analyses of Ash from Heavy Fuel Oil**

|  | Analysis, % | | |
|---|---|---|---|
|  | Trouble-free fuel oil | Troublesome fuel oils | |
| Ferric oxide, $Fe_2O_3$ | 56 | 8 | 6 |
| Silica, $SiO_2$ | 25 | 9 | 5 |
| Alumina, $Al_2O_3$ | 6 | 4 | 1 |
| Lime, $CaO$ | 3 | 10 | 1 |
| Magnesia, $MgO$ | 9 | 3 | 2 |
| Vandium pentoxide, $V_2O_5$ |  | 1 | 39 |
| Alkali sulfates | 1 | 65 | 46 |

| Constituent | Melting point in air | |
|---|---|---|
|  | °F | °C |
| $V_2O_5$ | 1274 | 690 |
| $NaSO_4$ | 1625 | 885 |
| $MgSO_4$ | 2165 | 1185 |
| $CaSO_4$ | 2640 | 1449 |
| $Fe_2O_3$ | 2850 | 1566 |

**Soot-blower systems** are used to maintain boiler efficiency and capacity by the periodic removal of ash and slag from the heat-absorbing surfaces. Steam or air jets from the soot-blower nozzles dislodge the dry or sintered ash and slag, which then fall into hopper or travel along with the gaseous products of combustion to the removal equipment.

Types of soot blowers vary with their location in the boiler unit, the severity of ash or slag conditions, and the arrangement of the heat-absorbing surfaces.

Furnace walls are generally cleaned with **wall blowers** (Fig. 6a) which project a nozzle assembly into the furnace for blowing and then retract it behind the wall tubes for protection after operation.

Tube banks in high-gas-temperature zones, such as slag screens, superheaters, and reheaters, where slag or sintered ash may accumulate, are generally cleaned by **long-lance retracting-type blowers** (Fig. 6b). The lance, which rotates or oscillates as it advances into the boiler, is fitted with large nozzles to supply a powerful cleaning action and is retracted from the boiler for protection when it is not operating.

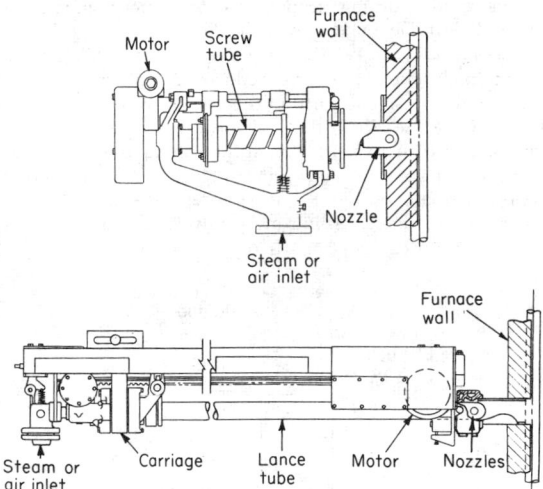

**Fig. 6**  Retracting soot blowers. (a) Furnace-wall blower; (b) long-lance blower.

Tube banks located in low-gas-temperature zones, including the economizer and boiler sections, where uncooled metals have satisfactory life and ash removal is easier, usually can be cleaned by **multiple-nozzle rotating-type soot blowers** (Fig. 7). However, long-lance retracting-type blowers may be necessary for very wide boilers, for extended cleaning ranges, or where the ash tends to pack or cake.

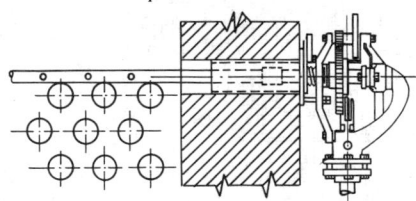

**Fig. 7**  Rotating soot blower with multiple nozzles.

**Soot blowers for air heaters** generally are arranged to blow through the plate or tube assemblies with single- or multiple-nozzle elements moving in an arc or straight-line motion. These blowers may also supply water for washing the air-heater surface.

**Shot cleaning** is effective in removing dry ash from the heat-absorbing surfaces by raining steel or iron shot over a horizontal area where the shot can then fall by gravity through or over the tubes. The shot is collected in hoppers and recycled for further use.

Additives to soften oil slag for easier removal by soot blowers, to protect against high-temperature fuel-ash corrosion, to reduce the bonding strength of the slag to tubes, or to overcome low-end-temperature corrosion may be introduced as a slurry spray through long-lance retractable blowers or through separate spraying equipment.

**Automatic controls** for soot-blower systems often are used and can be arranged to operate the blowers in a prescribed sequence at time intervals adjusted to boiler-cleaning requirements or to receive signals from the boiler unit's instruments and controls so as to operate the soot blowers selectively in the various heat-absorbing sections in order to maintain the required cleanliness and heat absorption.

**Dust collectors** (see also Sec. 18) are required for all large coal-burning boiler units in order to reduce atmospheric pollution. The amount of ash entrained in the flue gas varies from about 80 percent of the ash in the coal for dry-ash pulverized-coal firing to approximately 50 percent for slag-tap pulverized-coal firing, and from 15 to 30 percent for cyclone-furnace firing. **Mechanical separators** (Fig. 8) and **electrostatic dust collectors** (Fig. 9) may be used in series, but most pulverized-coal-fired units

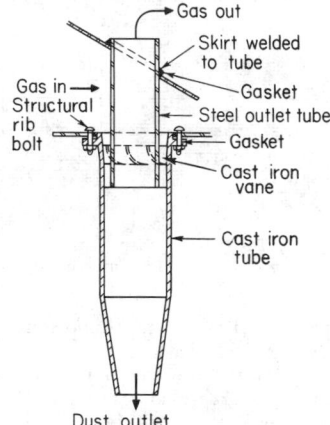

**Fig. 8** Mechanical dust-collector element.

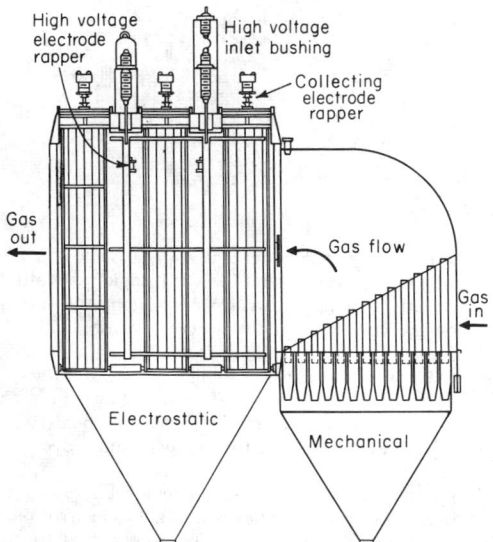

**Fig. 9** Mechanical and electrostatic dust collectors in series.

use only electrostatic collectors. The fly ash from spreader-stoker-fired units is coarse, and consequently, mechanical separators generally are used. Although the gas-dust loading is low in cyclone-furnace boilers, electrostatic dust collectors are usually required to meet the restrictions on air pollution.

The **bottom ash** recovered from the ashpits of chain-grate and spreader-type stokers is usually sold for cement-block aggregate. **Slag** from slag-tap furnaces can be used as a black granular coating for asbestos shingles and roofing, as a mixture containing slag, fly ash, lime, and water for Poz-o-pac roads, or as an antiskid material for icy roads.

**Fly ash** presents disposal problems because of its low density and the large volume which must consequently be handled. It is not suitable for fill material unless quickly covered. However, it can be utilized as an admixture, replacing 20 to 30 percent of portland cement, and as a lightweight aggregate after sintering.

## STOKERS

Very few stokers are used for new installations, primarily because of ash-disposal and air-pollution problems. However, many stokers of various types are still operating in older plants.

**Chain-** and **traveling-grate stokers** have been extensively used to burn noncoking coals, but only a few installations have been made in recent years because of their slow response to load changes, possible loss of ignition on swinging loads, high ashpit heat losses, high excess-air requirements, and limitations on size.

**Spreader stokers** with continuous-ash-discharge traveling grates (Fig. 10), intermittent-cleaning dump grates, or reciprocating continuous-cleaning grates (Fig. 24) are capable of burning all types of bituminous and lignitic coals. The fines are burned in suspension, and the larger fuel particles are burned on the grate. The use of a thin, fast-burning fuel bed provides rapid response to variations in load. Rotating mechanical feeding and distributing devices are generally used with spreader stokers. These stokers operate with low excess air and high efficiencies when the carbon in the fly ash is

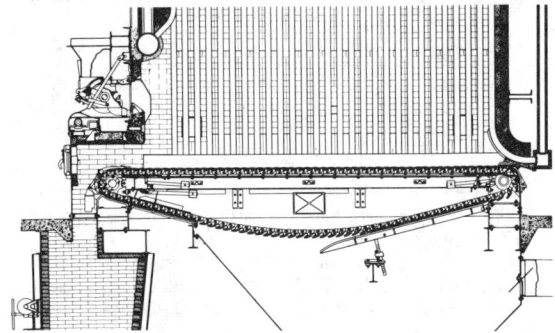

**Fig. 10** Spreader stoker with traveling grate.

reinjected above the grate. However, relatively low gas velocities through the boiler are necessary to prevent fly-ash erosion, and fly-ash collectors should be used to reduce air pollution.

Single- or double-retort **underfeed stokers** with side ash dump and multiple-retort underfeed stokers with rear ash discharge are well suited for the burning of coking coals. These stokers operate best at steady loads. Both the ashpit heat loss and the maintenance are high.

## PULVERIZERS

Pulverizers supply coal, for the firing of boilers, in a stream of air which enters the pulverizer at a temperature ranging from 300 to 750°F (150 to 400°C), depending upon the amount of moisture in the coal. The pulverizer provides the active mixing necessary for drying, and the temperature of the coal and air mixture in and leaving the pulverizer ranges from 130 to 180°F (55 to 80°C). In **bin-storage systems,** the coal and air (or flue gas) from the pulverizers are separated in cyclones, and the coal is then stored in bins and fed to the burners as needed. In **direct-fired systems,** the coal and the air pass directly from the pulverizer to the burners and the desired firing rate is regulated by the rate of pulverizing. Most installations utilize the direct-fired system.

Three general types of pulverizers are used with coal. The **slow-speed type** (Fig. 11) consists of a rotating drum with a tumbling charge of steel balls. It may be used for all types of coal but is particularly adaptable to highly abrasive fuels having a high silica content. **Medium-speed pulverizers** are used for all grades of bituminous coal; and their low power requirements, consistently high fineness, and quick response to load change are well suited to large industrial and utility boiler applications. Figure 12 shows the **ball-race type** and Fig. 13 the **bowl and roller type** of medium-speed pulverizers. **High-speed pulverizers** also are used for all grades of bituminous coal. This type of pulverizer (Fig. 14) utilizes a crushing section, a grinding section, and finally, an exhauster.

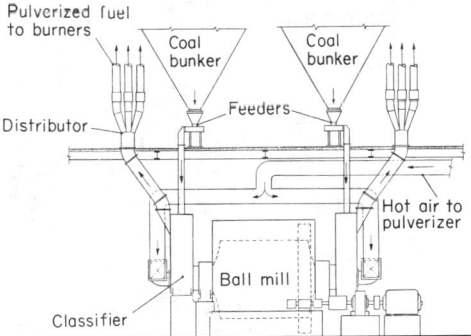

**Fig. 11** Slow-speed pulverizer, ball-mill type.

The required **fineness** of pulverization varies with the type of coal and with the size and kind of furnace, and usually ranges from 65 to 80 percent through a 200-mesh screen. (The U.S. Standard sieve 200-mesh screen has 200 openings per linear inch, resulting in a nominal aperture of 0.0029 in, 0.074 mm.) The ASTM equivalent is 74 $\mu$m. See Secs. 6 and 18. The **capacity** of a pulverizer is affected by the **grindability** of the coal and the fineness produced (Fig. 15), and capacities are established by testing with coals of different grindability.

Pulverized-coal firing is rarely used for boilers of less than 100,000 lb (45 tonnes) of steam per h capacity, since the use of stokers is more economical. Direct-fired pulverizers for large boilers have been built for capacities up to 70 tons (63 tonnes) per h.

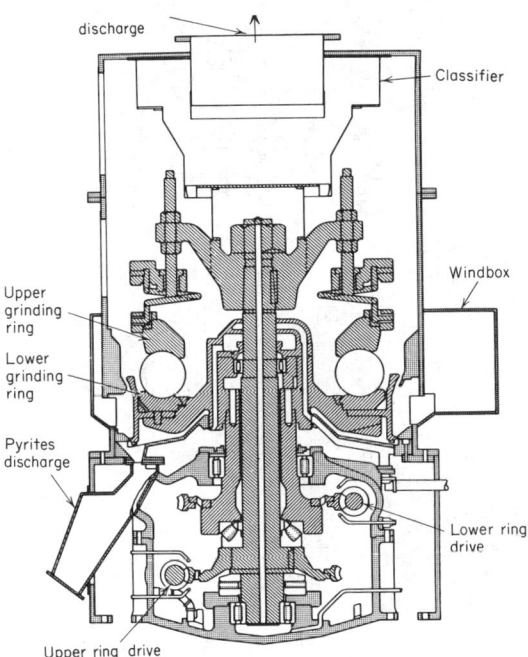

**Fig. 12** Medium-speed pulverizer, contrarotation ball-race type.

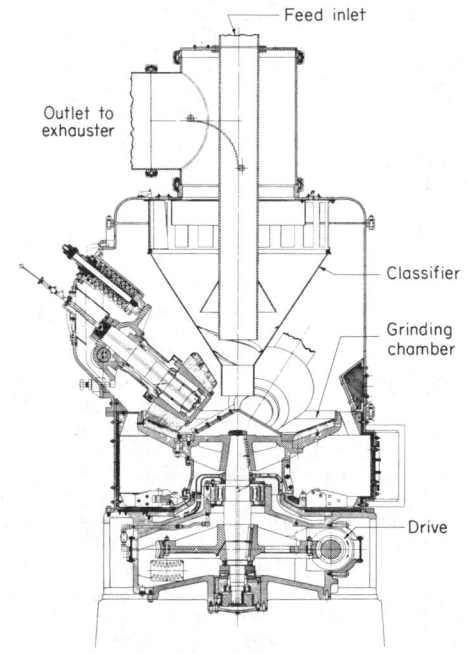

**Fig. 13** Medium-speed pulverizer, roller type.

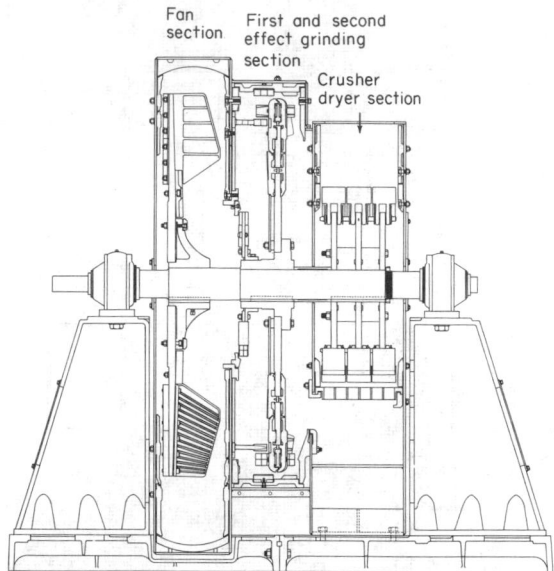

**Fig. 14**  High-speed pulverizer, attrition type.

## BURNERS

The primary purpose of a fuel burner is to mix and direct the flow of fuel and air so as to ensure rapid ignition and complete combustion. In pulverized-coal burners, a part (15 to 25 percent) of the air, called **primary air,** is initially mixed with the

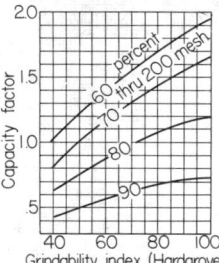

**Fig. 15.**  Pulverizer-capacity factors for varying fineness and grindability; medium-speed, ball-race-type pulverizers.

fuel to obtain rapid ignition and to act as a conveyor for the fuel. The remaining portion, or **secondary air,** is introduced to the burners outside of the primary-air ports.

Circular-type burners (Fig. 16) may be used for firing coal, oil, or gas. They are built in capacities as high as 165 million Btu/h (41,600 kcal/h) for coal, and higher for oil or gas.

Oil, when fired, can be atomized by the fuel pressure or by a compressed gas, usually steam or air. Atomizers utilizing fuel pressure generally are of the **uniflow or return-flow mechanical types.** The uniflow type uses an oil pressure of 300 to 600lb/in$^2$ (21 to 42 kgf/cm$^2$) at the maximum flow rate and is limited to an operating range of about 2 to 1. If a load range greater than 2 to 1 is required, the return-flow type of atomizer is used. This type of atomizer uses oil pressures up to 1,000 lb/in$^2$ (70 kgf/cm$^2$) and provides an operating range of as much as 10 to 1 under favorable conditions. Steam- and air-type atomizers also provide an operating range of approximately 10 to 1, but with a relatively low oil pressure (300 lb/in$^2$; 21 kgf/cm$^2$). The steam consumption required for good atomization usually is less than 1 percent of the boiler's steam output.

Natural gas and some process gases [provided they are sufficiently clean and have a calorific heating value of more than 500 Btu/ft$^3$ (4.45 kcal/m$^3$)] can be burned by admission

through a perforated ring, through radial spuds (Fig. 16), or through a centrally located center-fire type of fuel element. The center-fire fuel element can be removed for cleaning, and consequently, restrictions on gas cleanliness are less severe for this type of burner.

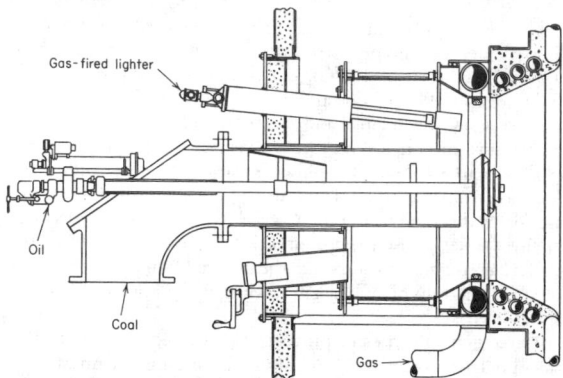

**Fig. 16**  Circular-type burner for pulverized coal, oil, or gas.

The cell-type pulverized-coal burner (Fig. 17) uses two or three fuel nozzles and provides the excellent ignition characteristics of the circular burner. Gas, when fired in these burners, is introduced through fixed-spud-type elements located in the burner throat, and return-flow or steam- or compressed-air-type atomizers are used for firing oil.

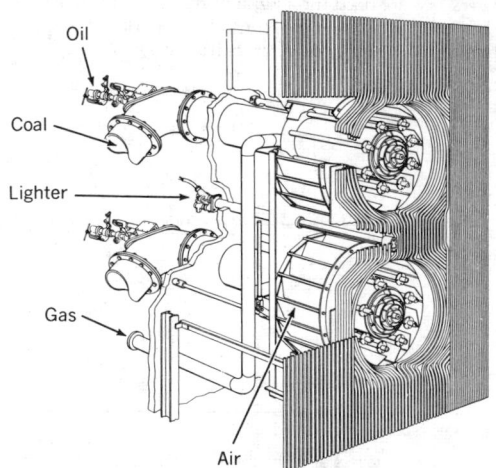

**Fig. 17**  Cell-type burner for pulverized coal, oil, or gas.

When corner-fired burners (Fig. 18) are used, the mixing of the fuel and the air for combustion takes place in the furnace (Fig. 18a). Oil and gas also can be fired in these burners by inserting fuel elements in the corner ports adjacent to the pulverized-coal nozzles. The burner tips can be tilted, as shown in Fig. 18b and c, to control the steam temperature.

A properly designed pulverized-coal installation should operate satisfactorily over a range of 2 or 3 to 1 without the need of auxiliary fuel to maintain ignition and without increasing or decreasing the number of burners in service. If some of the pulverizers and burners on large units are taken out of

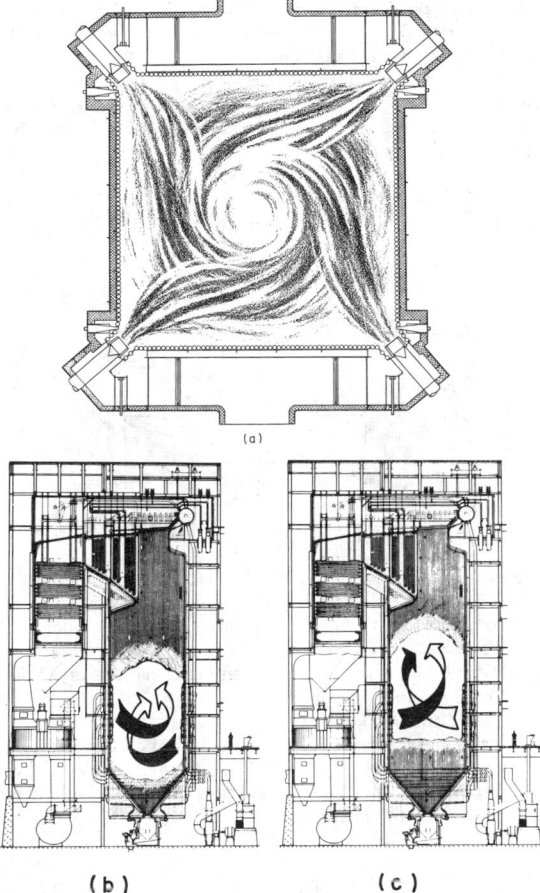

**Fig. 18**  Corner-fired tangential tilting burners for pulverized coal, oil, or gas. *(a)* Plan section; *(b)* burners tilted down; *(c)* burners tilted up.

service as the load decreases, it is possible to operate at ratings down to one-sixth of full-load steam flow without the use of auxiliary fuel.

Blast-furnace and coke-oven gas burners are usually of the circular type, in which the gas is introduced either through a centrally positioned nozzle or through an annular port surrounding the coal nozzle, or of the intertube type, where the fuel and air ports are alternated across the width of the burner.

## CYCLONE FURNACES

The cyclone furnace is designed to burn low-ash-fusion coals and to retain most of the coal ash in the slag, which is then tapped from the furnace, thus preventing the passage of the ash through the heat-absorbing surfaces. The coal, crushed to 4-mesh size, is admitted with primary air in a tangential manner to the primary burner (Fig. 19). The finer particles burn in suspension while the coarser particles are thrown by centrifugal force to the outer wall of the cyclone furnace. The wall surface with its sticky coating of molten slag retains most of the particles of coal until they burn and leave their molten

ash on the wall. The molten ash drains into the boiler furnace and then, through an opening in the boiler furnace floor, into the slag-collecting tank.

The secondary air, which is admitted tangentially at the top of the cyclone, vigorously scrubs the coal particles on the wall, and combustion is completed at a firing rate of about ½ million Btu/(ft³)(h) [4,450 kcal/(m³)(h)].

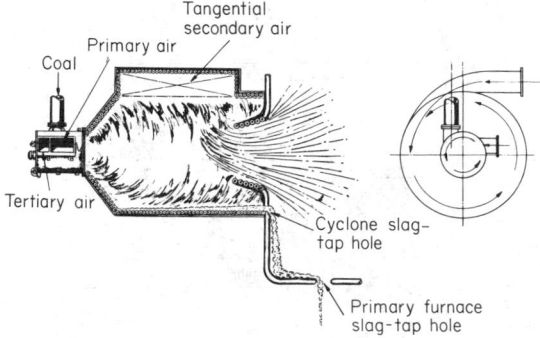

**Fig. 19**  Cyclone furnace.

Figure 27 shows a boiler fired with twenty-three 10-ft-diam cyclone furnaces. The primary-furnace walls (consisting of fully studded tubes) also are wetted with molten ash and help to catch ash particles that are not retained in the cyclone furnaces. Gas, when available, can be burned by injection through openings at the bottom of the secondary-air ports. Oil is burned by spraying it axially into the cyclone through the primary burner or by firing it tangentially through an oil element located in the secondary-air port.

## UNBURNED COMBUSTIBLE LOSS

The unburned combustible loss in the fly ash from pulverized-coal firing varies with the furnace-heat liberation, the type of furnace cooling, the use of slag-tap or dry-ash removal, the volatility and fineness of the coal, the excess air, and the type of burner (see Fig. 20). There is practically no combustible in the fluid slag from slag-tap furnaces. Although the hopper refuse from dry-ash furnaces usually is low in combustible, the combustible may be appreciable in some cases. The fly ash from cyclone-furnace boilers has a very low combustible content, varying from an equivalent 0.03 percent efficiency loss when burning Illinois coal to 0.15 percent for Ohio coal.

Fly-ash combustible from spreader stokers varies widely with the rating, the size, and the type of coal burned. The combustible carry-over at high rates of firing is relatively large, but reinjection of the fly ash is common and the loss in efficiency can be reduced as shown in Fig. 21.

The combustible loss in the fly ash from solid fuels is determined by withdrawing a representative sample of fly ash and flue gas from the boiler outlet flue, or stack, at the same velocity in the sampling tip as the gas velocity in the flue (see ASME Test Code). The rates of flue-gas flow and fly-ash collection are measured, and the dust loading in pounds per 1,000 lb of flue gas is then calculated. The combustible in the fly ash also can be measured. The data in Fig. 22 can be used for rapid determination of the efficiency loss when the dust loading and the amount of combustible are known.

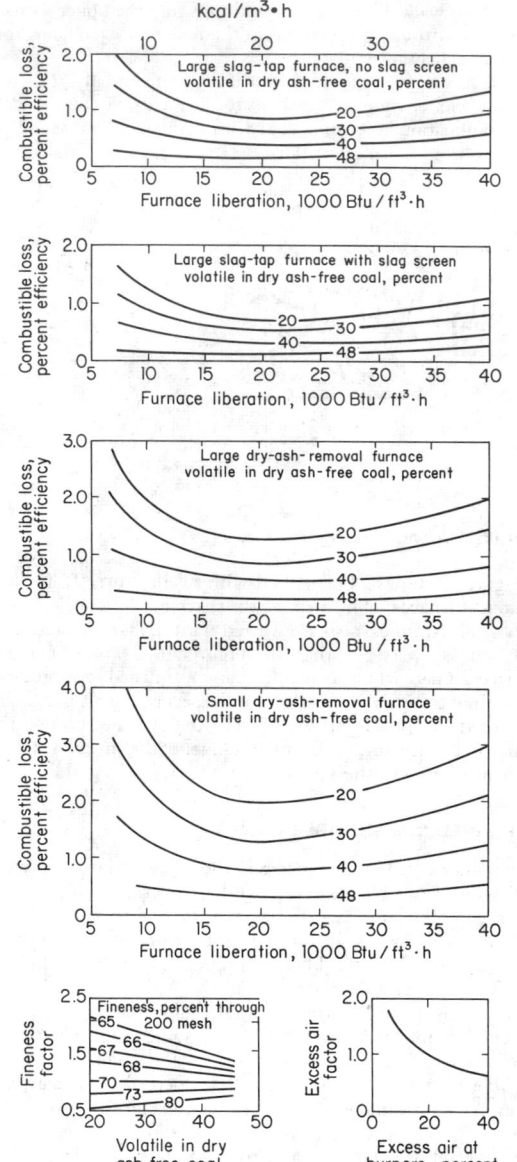

**Fig. 20** Principal factors affecting combustible loss for pulverized-coal firing in various types of furnaces.

## BOILER TYPES

The greater safety of water-tube-type boilers was recognized more than 100 years ago, and water-tube-type boilers have generally superseded the fire-tube type except in special cases such as small package-boiler designs and waste-heat boiler designs for medium- and low-pressure applications.

In the package-boiler field [5,000 to approximately 300,000 lb (2.3 to 140 tonnes) of steam per h], water-tube boilers are almost invariably used, particularly in the larger sizes. These boilers are oil- or gas-fired and generally utilize two drums with a water-cooled furnace adjacent to the side of a vertical single-gas-pass tube bank (Fig. 23).

The field-assembled two-drum, three-gas-pass, integral-furnace-type boiler is generally used (Fig. 24) for coal-fired boilers in the size range of 30,000 to 100,000 lb (14 to 45 tonnes) of steam per h.

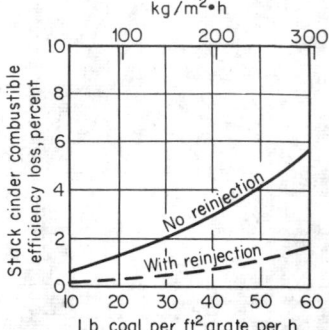

**Fig. 21** Combustible-efficiency loss, spreader stoker.

Many types of boilers are used for capacities ranging from 100,000 to 1,000,000 lb (45 to 450 tonnes) of steam per h, but most of them are of the single-gas-pass, two-drum, bent-tube construction. However, for some applications, integral-furnace-type boilers are used for steaming capacities up to 500,000 lb (230 tonnes) per h.

Boilers utilizing banks of tubes directly connected to the steam and water drums are, in general, limited to a maximum steam pressure of 1,650 lb/in² (116 kgf/cm²), since the wide tube spacing required to maintain high drum-ligament efficiency reduces the effectiveness of heat absorption.

Many designs of high-pressure, high-temperature, utility-type boilers, ranging from capacities of 500,000 to 9,000,000 lb (230 to 4,100 tonnes) of steam per h, are used, but they can generally be classed as radiant-type boilers. In radiant boilers, little or no steam is generated by convection heat-absorbing surface since virtually all the steam is generated in the tubes forming the furnace-enclosure walls by the heat radiated to these tubes from the hot combustion gases. Figure 25 illustrates a large oil- and/or gas-fired natural-circulation-type radiant-boiler unit, and Fig. 26 shows a tangentially pulverized coal-fired, supercritical-pressure, combined-circulation type of boiler. The cyclone furnace forced-flow, once-through unit shown in Fig. 27 is known as a universal pressure boiler, since it can be designed for operation at pressures either above or below the critical pressure (3,206 lb/in²; 225 kgf/cm²).

Drum-type natural or assisted-circulation boilers are restricted to a maximum pressure of about 2,600 lb/in² (183 kgf/cm²) at the superheater outlet because of circulation and steam-separation characteristics. However, once-through forced-flow-type boilers are not restricted to any pressure plateau by circulation limits.

In once-through forced-flow boilers, the water generally flows from the economizer to the furnace-wall tubes, then to the gas-convection-pass enclosure tubes and the primary superheater. Usually, the transition to the vapor phase (if operation is below the critical pressure) begins in the furnace circuits and, depending upon the operating conditions and the

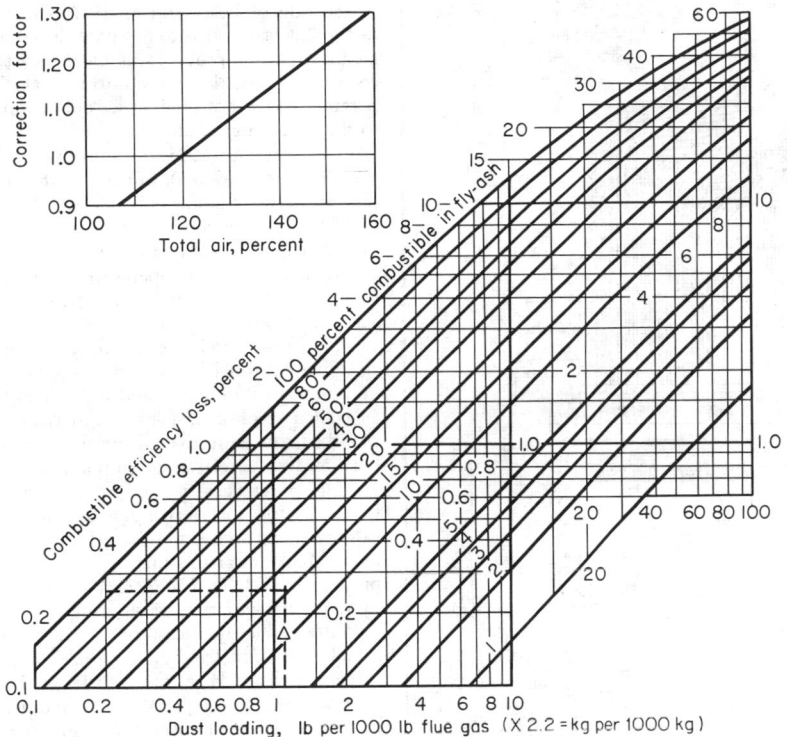

**Fig. 22** Chart for determining combustible-efficiency loss in fly ash. Heat loss in percent from combustible in fly ash =

$$\left[ \frac{DL \times (C/100) \times 14{,}600}{\dfrac{10{,}000{,}000}{7.6 \times (TA/100) + 0.9} + [DL \times (C/100) \times 14{,}600]} \right] \times 100$$

where $DL$ = dust loading, lb/1,000 lb flue gas ($\times$ 2.2 = kg/1,000 kg); $C$ = combustible in fly ash, percent; $TA$ = total air, percent.

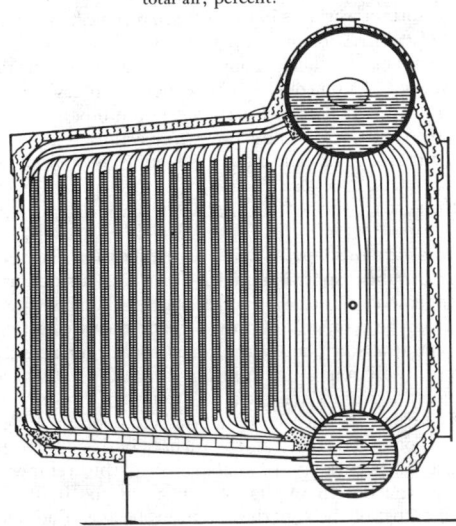

**Fig. 23** Water-tube, package boiler; single gas pass.

design, is completed either in the gas-convection-pass enclosure or in the primary superheater. The steam from the primary superheater passes to the secondary (and possibly to a tertiary) superheater. One or more reheaters are provided to reheat the low-pressure steam.

In addition to boilers for the conversion of energy in conventional fuels (coal, oil and gas) to steam for power, heating or process use, many boilers have been developed for special requirements.

Waste-heat and exhaust-gas boilers utilize the sensible heat in the gas to generate steam. In the recovery of heat from the gas, water-tube boilers, often in conjunction with superheaters and economizers, are generally used; but fire-tube boilers may be used for cooling process or other gases when the containment of pressurized gas is a factor and the steam requirements are small.

High-temperature-water boilers provide hot water, under pressure, for space heating of large areas. Water is circulated at pressures up to 450 lb/in² (31.6 kgf/cm²) through the generator and the heating system. The water leaves the generator at subsaturated temperatures ranging up to 400°F (200°C). The boilers usually incorporate a water-cooled furnace and convec-

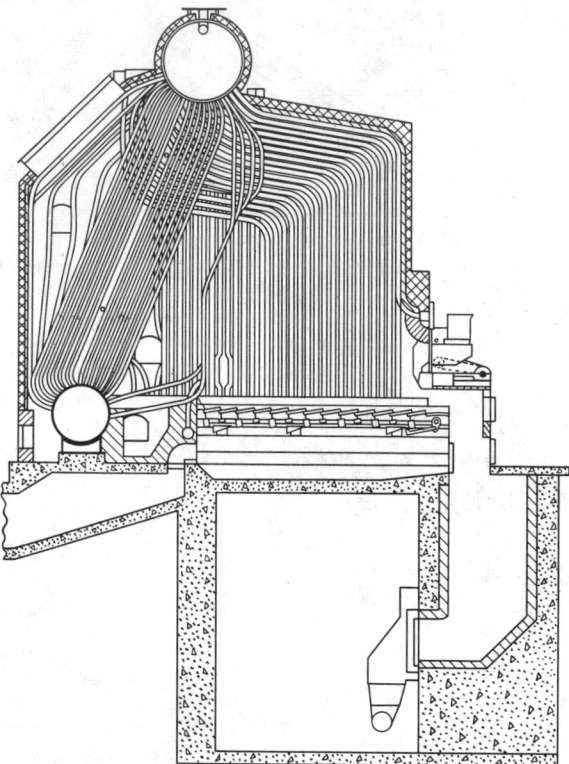

**Fig. 24** Integral-furnace boiler; three-gas-pass; reciprocating-grate, spreader-stoker-fired.

tion gas-pass enclosure, with the convection-heat-absorbing surface arranged in sections similar to those of an economizer. Sizes generally range up to 60 million Btu/h (15,120 kcal/h) for package units, and field-erected units can be designed for much higher capacities.

The exhaust CO gas from oil- refinery fluid catalytic-cracking units is used as the fuel for carbon monoxide boilers. Generally, a cylindrical furnace is used to contain the pressurized gas and the CO burners are arranged tangentially to increase the residence time of the gas in the furnace. The furnace walls are water-cooled and the tubes are refractory-covered to promote ignition. Conventional-type gas and/or oil burners are provided for start-up, for continuous pilots, and for the generation of steam when the cracker is out of service.

Recovery-type boilers are designed specifically for the recovery of chemicals in the spent cooking liquors from kraft, sulfite, soda, and other papermaking processes. The liquor is fired in a water-cooled furnace, either in suspension or in a smelt bed on the furnace floor. The chemicals, depending upon the process, are recovered from the smelt or the flue gas in a form which permits economical conversion for reuse.

## FURNACES

A furnace is an enclosure provided for the combustion of fuel. The enclosure confines the products of combustion and is capable of withstanding the high temperatures developed and the pressures used. Its dimensions and geometry are adapted

to the rate of heat release, to the type of fuel, and to the method of firing so as to promote the complete burning of the combustible and provide suitable disposal of the ash. In water-cooled furnaces, the heat-absorbed materially affects the temperature of gases at the furnace outlet and contributes directly to the generation of steam.

Prior to about 1925, most furnaces were constructed of firebrick (see Section 6). But as the steaming capacities and the physical sizes of boiler units increased, and as the suspension burning of pulverized coal was developed, limits were reached in the heights of refractory walls that could be made self-supporting at high temperatures. Limits also were reached because of the inability of refractories to resist the fluxing action of molten fuel ash.

These limits can be extended by cooling the brickwork with air flowing through channels in the structure or by sectionalizing the walls into panels and transferring the load to external air-cooled steel or cast-iron supporting members. The heat absorbed can be recovered by using the cooling air for combustion, thus accelerating ignition and burning of the fuel.

The low tensile strength of refractories makes it difficult to provide shapes for overhanging contours or roof closures. As a result, sprung arches and shaped tiles suspended from steel are used. Many refractory mixtures have air or hydraulic-setting properties, and they can be used to form monolithic structures by ramming, guniting, or pouring into forms.

Water-cooled furnaces are used with most boiler units and for all types of fuel and methods of firing. Water cooling of the furnace walls reduces the transfer of heat to the structural members and, consequently, their temperature can be limited to that which will meet the requirements of strength and resistance to oxidation. Water-cooled tube constructions facilitate large furnace dimensions and optimum arrangements of roofs, hoppers, arches, and mountings for burners; and the use of tubular screens, platens, or division walls to increase the amount of heat-absorbing surface in the combustion zone. The use of water-cooled furnaces reduces the external heat losses, and these losses for conventional-type furnaces are shown in Fig. 40, p. 9-28.

Heat-absorbing surfaces in the furnace receive heat from the products of combustion and, consequently, lower the furnace exit-gas temperature. The principal mechanisms of heat transfer take place simultaneously. These include intersolid radiation from the fuel bed or fuel particles, nonluminous radiation from the products of combustion, convection heat transfer from the furnace gases, and heat conduction through deposits and tube metals. (See also Sec. 4.) The absorption effectiveness of the furnace surfaces is influenced by the deposits of ash or slag.

Furnaces vary in shape and size, in the location and spacing of burners, in the disposition of heat-absorbing surface, and in the arrangement of arches and hoppers. Flame shape and length affect the geometry of radiation and the rate and distribution of heat absorption by the water-cooled surfaces.

Analytical solutions of the transfer of heat in the furnaces of steam-generating units are extremely complex, and it is most difficult to calculate furnace outlet-gas temperature by theoretical methods. Nevertheless, the furnace outlet-gas temperature must be accurately predicted, since this temperature determines the design of the remainder of the boiler unit, particularly that of the superheater and reheater. The calculations must therefore be based upon test results supplemented

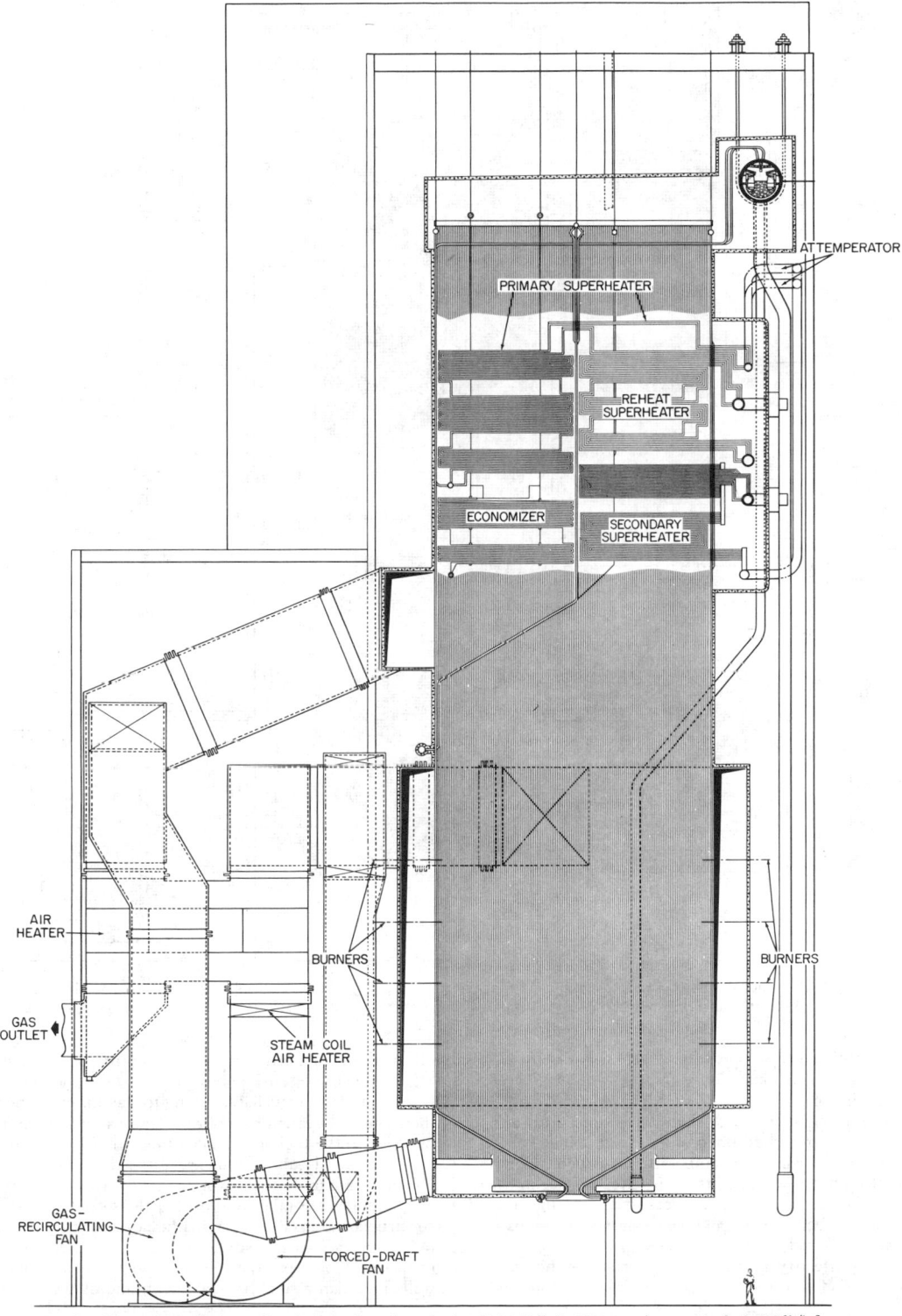

**Fig. 25** Natural-circulation radiant boiler; oil- and gas-fired; 4,200,000 lb (1,900 tonnes) steam per h; 2,600 lb/in² (183 kgf/cm²); 1005°F (540°C) steam temperature; 1005°F (540°C) reheat steam temperature.

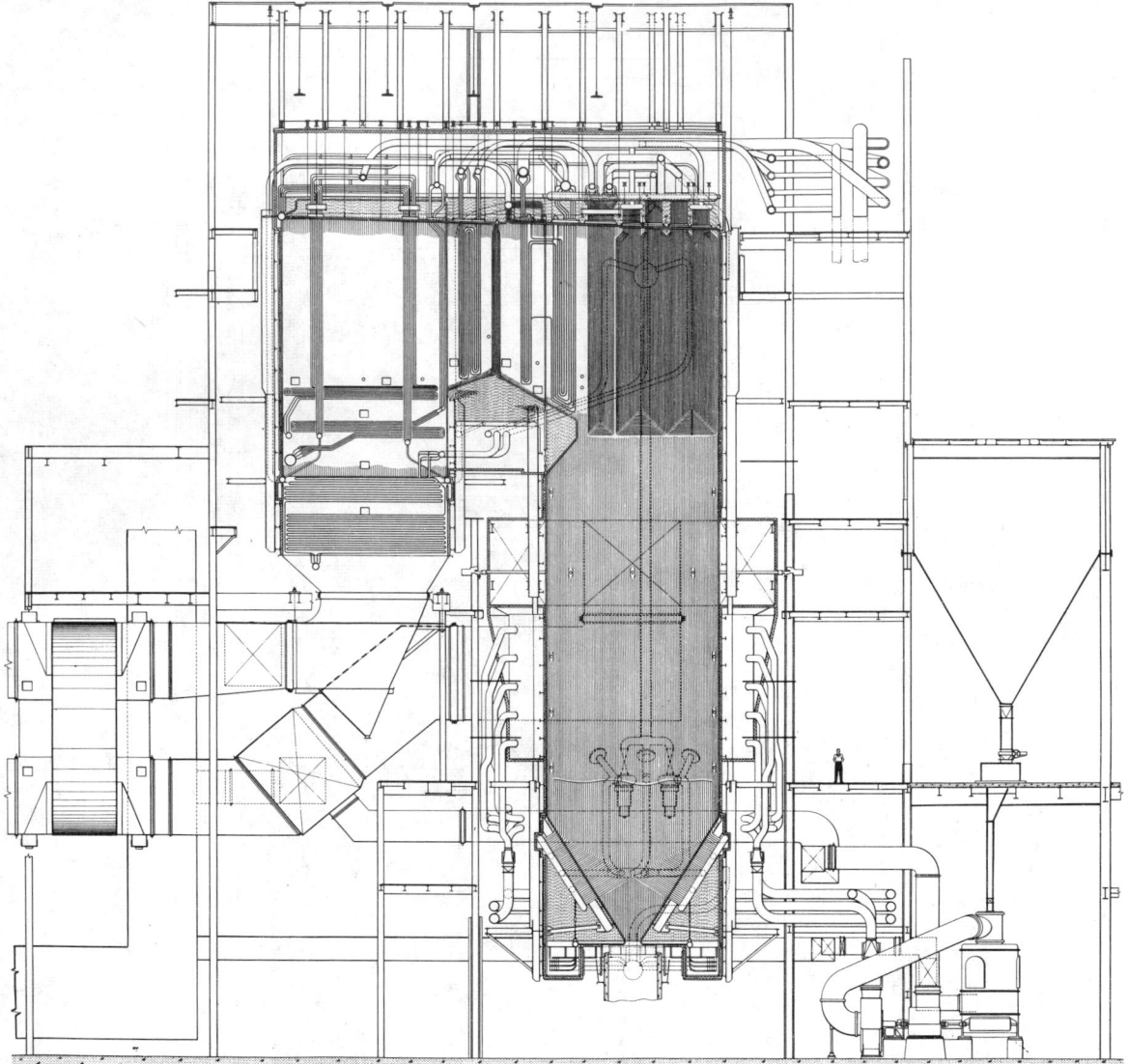

**Fig. 26**  Once-through boiler with combined circulation; twin pressurized furnaces; pulverized coal tangentially fired; 6,400,000 lb (2,900 tonnes) steam per h; 3,650 lb/in² (257 kgf/cm²); 1003°F (539°C) steam temperature; 1003°F (539°C) reheat steam temperature.

by data accumulated from operating experience and by judgments predicated upon knowledge of the principles of heat transfer and the characteristics of fuels and slags.

The curves in Figs. 28 and 29 show the gas temperatures at the furnace outlet of typical boiler units when firing coal, oil, and gas. The furnace exit-gas temperatures vary considerably with coal firing because of the insulating effect of ash and slag deposits on the heat-absorbing surfaces. The amount of surface is the major factor in overall furnace heat absorption, and the heat released and available for absorption per hour per square foot of effective heat-absorbing surface is therefore a satisfactory basis for correlation. The heat released and availa-

ble for absorption is the sum of the calorific heat content of the fuel fired and the sensible heat of the combustion air, less the sum of the heat unavailable owing to the unburned portion of the fuel and the latent heat of the water vapor formed from the moisture in the fuel and the combustion of hydrogen.

Furnace-wall tubes often are pitched on close centers to obtain maximum heat absorption and to facilitate ash removal. The arrangement takes the form of the so-called tangent-tube construction (Fig. 30) wherein the adjacent tubes are almost touching with only a small clearance provided for erection purposes. However, most boilers now use membrane tube walls in which a steel bar, or membrane, is welded between

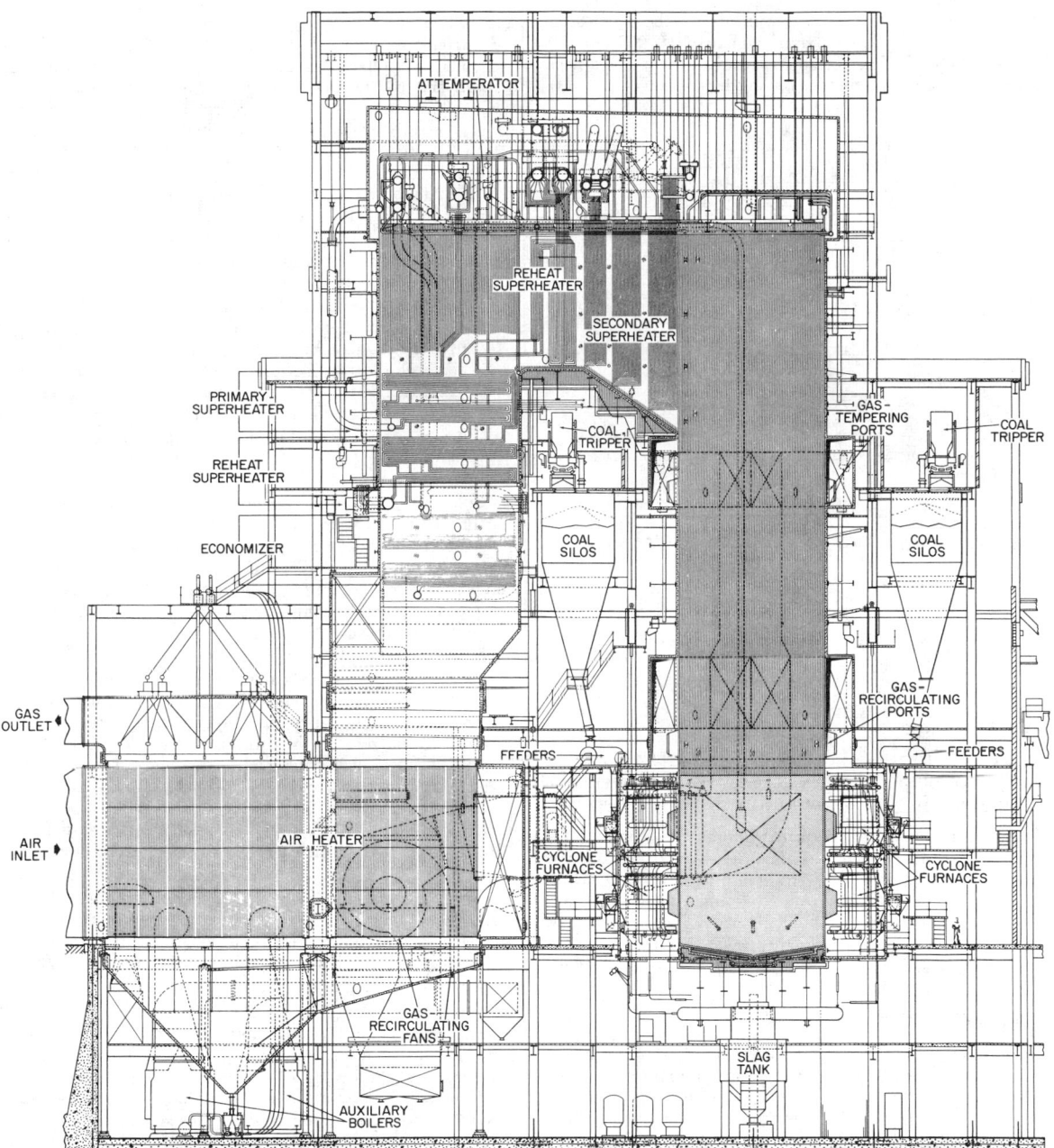

**Fig. 27**   Universal Pressure® boiler; pressurized cyclone furnace coal-fired; 8,000,000 lb (3,640 tonnes) steam per h; 3,650 lb/in² (257 kgf/cm²); 1003°F (539°C) steam temperature; 1003°F (539°C) reheat steam temperature.

adjacent tubes. This construction facilitates the fabrication of water-cooled walls in large shop-assembled tube panels. Less effective cooling is obtained, at a lower cost, by placing the tubes on wider spacing and using extended metal surface in the form of flat studs welded to the tubes. If even less cooling is desired, the tube spacing can be increased and refractory installed between or behind the tubes to form the wall enclosure.

Additional furnace cooling in the form of tubular platens, division walls, or wide-spaced tubular screens often is used; and, in high-heat input zones, the tubes may be protected by refractory coverings anchored to the tubes by studs. Peak heat-absorption rates of furnace-wall tubes in the combustion zone may, in some designs, approximate 200,000 Btu/(h)(ft²) [542 kcal/(h)(m²)] of projected surface, but the average heat-absorption rate for the furnace is considerably lower (Fig. 31).

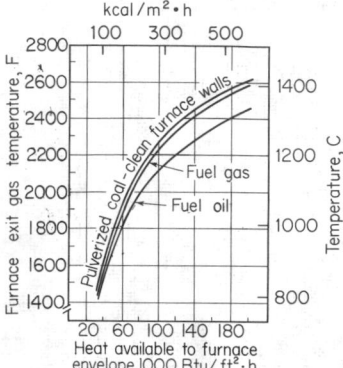

**Fig. 28**  Approximate gas temperatures at water-cooled furnace outlet with different fuels.

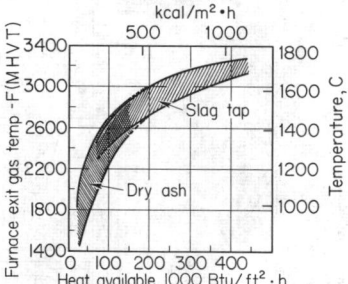

**Fig. 29**  General range of furnace exit-gas temperatures, pulverized-coal firing (MHVT = multiple-shield high-velocity thermocouple).

Furnace walls must be adequately supported with provision for thermal expansion and with reinforcing buckstays to withstand the lateral forces caused by the difference between the furnace pressure and the surrounding atmosphere. The furnace enclosure must prevent air infiltration when the furnace is operated under suction and gas leakage when the furnace is operated at pressures above atmospheric.

## SUPERHEATERS AND REHEATERS

The addition of heat to steam after evaporation, or change of state, is accompanied by an increase in the temperature and the enthalpy of the fluid. The heat is added to the steam in boiler components called superheaters and reheaters, which are comprised of tubular elements exposed to the high-temperature gaseous products of combustion.

The advantages of superheat and reheat in power generation result from thermodynamic gain in the Rankine cycle (see Sec. 4) and from the reduction of heat losses due to moisture in the low-pressure stages of the turbine. With high steam pressures and temperatures, more useful energy is available, but the advances to high steam temperature often are restricted by the strength and the oxidation resistance of the steel and the ferrous alloys currently available and economically practical

for use in boiler pressure-part and turbine-blade constructions.

The term superheating is applied to the higher-pressure steam and the term reheating to the lower-pressure steam which has given up some of its energy during expansion in the high-pressure turbine. With high initial steam pressure, one or more stages of reheating may be employed to improve the thermal efficiency.

Separately fired superheaters may be used, but superheaters usually are installed as an integral part of the steam-generating unit and broadly classified as radiant convection types, depending upon the predominant method of heat transfer to the heat-absorbing surfaces.

The quantity of heat absorbed and the amount of superheat attained are dependent upon the size, location, and arrangement of the heat-absorbing surfaces; the temperature differentials between the gas, the tube metal, and the steam; and the heat-transfer coefficients. Steam-temperature characteristics of radiant- and convection-type superheaters are shown in Fig. 32, as well as the effect of using a combination of these types to produce a more uniform steam temperature over a wide operating range.

Superheaters of the predominantly radiant type usually are arranged for direct exposure to the furnace gases and, in some designs, form a part of the furnace enclosure. In other designs, the surface is arranged in the form of tubular loops or platens, on wide lateral spacing, extending into the furnace. Such surface is exposed to high-temperature furnace gases traveling at relatively low velocities, and the transfer of heat is principally by radiation.

Convection-type superheaters are installed beyond the furnace exit where the gas temperatures are lower than those in the zones where radiant-type superheaters are used. The tubes are usually arranged in the form of parallel elements on close lateral spacing and in tube banks extending partially or completely across the width of the gas stream, with the gas flowing through the relatively narrow spaces between the tubes. High rates of gas flow and, thus, high convection heat-transfer rates are obtained at the expense of gas-pressure drop through the tube bank.

Superheaters, shielded from the furnace combustion zone by arches or wide-spaced screens of steam-generating tubes, which receive heat by radiation from the high-temperature gases in cavities or intertube spaces and also by convection due to the relatively high rate of gas flow through the tube banks, have both radiant and convection characteristics.

Superheaters may utilize tubes arranged in the form of hairpin loops connected in parallel to inlet and outlet headers; or they may be of the continuous-tube type, where each element consists of a number of tube loops in series between the inlet and outlet headers. The latter arrangement permits the use of large tube banks, thus increasing the amount of heat-absorbing surface that can be installed and providing economy of space and reduction of cost. Either type may be designed for the drainage of the condensate which forms within the tubes during outages of the unit, or they may be used in pendent arrangements which are not drainable but, usually, have simpler and better supports. Nondrainable superheaters require additional care during start-up to remove the condensate by evaporation; thus the gas temperature entering the superheater must be kept below 1000°F (538°C) to prevent overheating of the tube metal before steam flow is established in all elements of the tube banks.

The heat transferred from high-temperature gases by radiation and convection is conducted through the metal tube wall and imparted by convection to the high-velocity steam in the tubes. The removal of heat by the steam is necessary to keep the tube metals within a safe temperature range consistent with the temperature limits of oxidation and the creep or rupture strength of the materials (see Sec. 5). Allowable design stresses for various steels and alloys, as established by the ASME Code, are listed in Table 3 (see also Sec. 6). The practical use limit for each material also is indicated. For economic reasons, it is customary to use low-carbon steel in the steam inlet sections of the superheater and, progressively, more costly alloys as the metal temperatures increase.

The rate of steam flow through superheater tubes must be sufficiently high to keep the metal temperatures within a safe operating range and to ensure good distribution of flow through all the elements connected in parallel circuits. This can be accomplished by arrangements which provide for multiple passes of the steam flow through the superheater tube banks. Excessive steam-flow rates, while providing lower tube-metal temperatures, should be avoided, since they result in high pressure drop with consequent loss of thermodynamic efficiency. As a general guide, the range of the steam-flow rates required for various steam and gas temperature conditions is shown in Table 4.

The spacing of the tabular elements in the tube bank and, consequently, the rate of gas flow and convection heat transfer are governed primarily by the types of fuel fired, draft-loss considerations, and the fouling and erosive characteristics of fuel ash carried in the gas stream. With clean gases, or in the low-gas-temperature zones of coal-fired units, a gas-flow rate of about 6,000 lb/(h)(ft²) (8.2 kg/m²s) of free-flow area is generally within economic limits. In the higher-gas-temperature zones, 1600 to 2300°F (871 to 1260°C), the adherence and the accumulation of ash deposits can reduce the gas-flow area and, in some cases, may completely bridge the space between tubes. Thus, as gas temperatures increase, it is customary to increase the tube spacings in the tube banks to avoid excessive draft loss and to facilitate ash removal.

## ECONOMIZERS

Economizers remove heat from the moderately low-temperature combustion gases after the gases leave the steam-generating and superheating/reheating sections of the boiler unit.

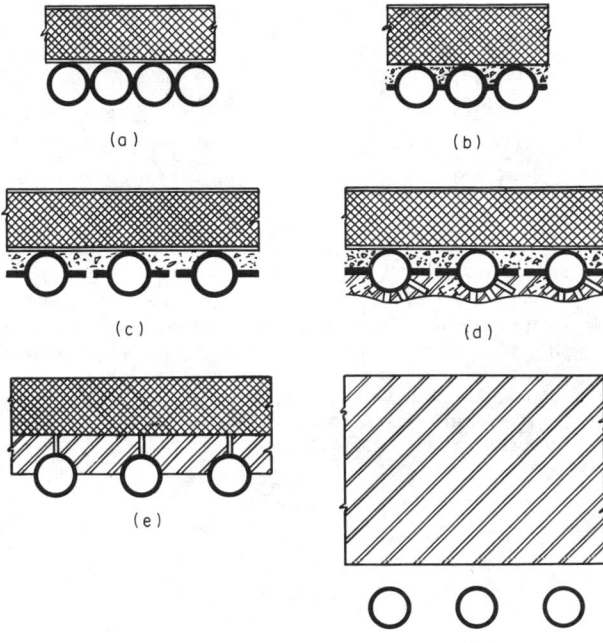

**Fig. 30**  Water-cooled furnace-wall construction types. (*a*) Tangent tube wall; (*b*) welded-membrane tube wall; (*c*) flat studs welded to sides of tubes; (*d*) full-stud tube wall, refractory-covered; (*e*) tube and tile wall; (*f*) tubes spaced from refractory wall.

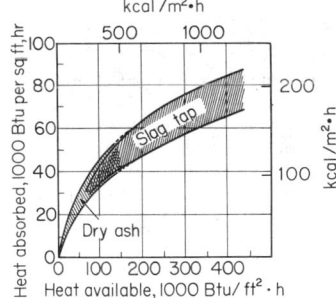

**Fig. 31**  General range of average heat-absorption rate in water-cooled pulverized-coal-fired furnaces.

**Table 3. Superheater and Reheater Tubes—Maximum Allowable Design Stress, lb/in² (× 0.070307 = kgf/cm²)**

| Material | ASME spec. No. and type | Temp, °F (°C) | | | | | | | |
|---|---|---|---|---|---|---|---|---|---|
| | | 900 (482) | 950 (510) | 1000 (538) | 1050 (566) | 1100 (593) | 1150 (621) | 1200 (649) | 1300 (704) |
| Carbon steel | SA210 | 5,000 | 3,000 | | | | | | |
| Carbon moly | SA209, T1a | 12,500 | 8,500 | | | | | | |
| Croloy ½ | SA213, T2 | 12,500 | 10,000 | 6,250 | | | | | |
| Croloy 1¼ | SA213, T11 | 13,100 | 11,000 | 6,550 | 4,050 | | | | |
| Croloy 2¼ | SA213, T22 | | 11,000 | 7,800 | 5,800 | 4,200 | | | |
| Croloy 5 | SA213, T5 | | | | 4,150 | 3,050 | 2,000 | | |
| Croloy 9 | SA213, T9 | | | | 5,500 | 3,300 | 2,200 | 1,500 | |
| Croloy 304H | SA213, TP 304H | | | | 9,500 | 8,850 | 7,700 | 6,050 | 3,700 |
| Croloy 321H | SA213, TP 321H | | | | 10,100 | 8,800 | 6,900 | 5,350 | 3,150 |

**Table 4. Range of Steam-Mass-Flow Values for Convection Superheaters**

| Temp, °F (°C) | | Steam mass flow | |
|---|---|---|---|
| Steam | Gas | lb/(h)(ft² flow area) | kg/m²s |
| Less than 750 (399) | 1200 (649) | 75,000–150,000 | 102–204 |
| 700–800 (371–426) | 1600 (871) | 250,000–350,000 | 340–475 |
| 800–900 (426–482) | 2400 (1316) | 400,000–500,000 | 545–680 |
| 900–1000 (482–538) | 2400 (1316) | 500,000–600,000 | 680–816 |
| 1000–1100 (538–593) | 2400 (1316) | 700,000 and higher | 950 |

Economizers are, in effect, feedwater heaters which receive water from the boiler feed pumps and deliver it at a higher temperature to the steam generator. Economizers are used instead of additional steam-generating surface, since the feedwater and, consequently, the heat-receiving surface are at temperatures below the saturated-steam temperature and thus the gases can be cooled to lower temperature levels for greater heat recovery and economy.

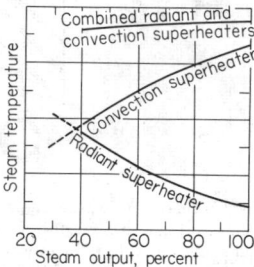

**Fig. 32** Comparative radiant and convection superheater characteristics.

Economizers are forced-flow, once-through, convection heat-transfer devices, usually consisting of steel tubes, to which feedwater is supplied at a pressure above that in the steam-generating section and at a rate corresponding to the steam output of the boiler unit. They are classed as horizontal- or vertical-tube types, according to geometrical arrangement; as longitudinal or crossflow, depending upon the direction of gas flow with respect to the tubes; as parallel or counterflow, with respect to the relative direction of gas and water flow; as steaming or nonsteaming, depending on the thermal performance; as return-bend or continuous-tube, depending upon the details of design; and as bare-tube or extended-surface, according to the type of heat-absorbing surface. Staggered or in-line tube arrangements may be used. The arrangement of tubes affects the gas flow through the tube bank, the draft loss, the heat-transfer characteristics, and the ease of cleaning.

The size of an economizer is governed by economic considerations involving the cost of fuel, the comparative cost and thermal performance of alternate steam-generating or air-heater surface, the feedwater temperature, and the desired exit-gas temperature. In many cases, it is more economical to use both an economizer and an air heater.

The temperatures of the economizer tube metals generally approximate those of the water flowing within the tubes, and thus with low feedwater temperatures condensation and external corrosion are encountered in those locations where the tube-metal temperature is below that of the acid or water dew point of the gas (see Fig. 33). Internal corrosion and pitting also may be experienced if the feedwater contains more than

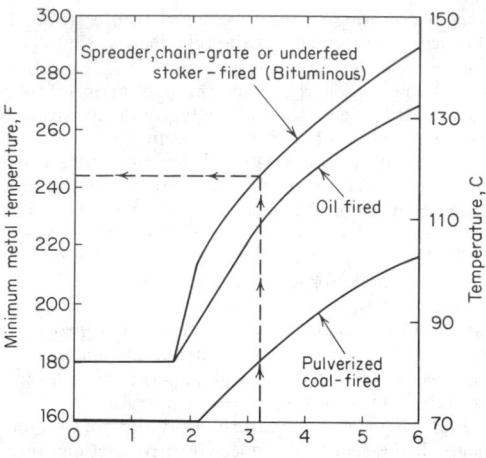

**Fig. 33** Limiting metal temperatures to avoid external corrosion in economizers or air heaters when burning coal, fuel oil (grades 1 to 6) and natural gas containing sulfur. For anthracite coal or natural gas without sulfur, minimum metal temperature is 160°F (71°C). To avoid internal corrosion of economizer tubes, the temperature of the feedwater entering should not be less than 212°F (100°C).

0.007 ppm of dissolved oxygen. Therefore, it is imperative to maintain feedwater temperatures above the dew-point temperature of the gas and to provide suitable deaeration of the feedwater for the removal of oxygen.

## AIR HEATERS

Air heaters, like economizers, remove heat from the relatively low-temperature combustion gases. The temperature of the inlet air is less than that of the water to the economizer, and thus it is possible to reduce the temperature of the gaseous products of combustion further before they are discharged to the stack.

The heat recovered from the combustion gases is recycled to the furnace with the combustion air and, when added to the thermal energy released from the fuel, is available for absorption by the steam-generating unit, with a resultant gain in overall thermal efficiency. The use of preheated combustion air accelerates ignition and promotes rapid and complete burning of the fuel.

Air heaters are usually classed as recuperative or regenerative. Both types utilize the convection transfer of heat from the gas stream to a metal or other solid surface and the convection transfer of heat from this surface to the air. In recuperative air heaters, exemplified by the tubular or plate types (Fig. 34), the stationary metal parts form a separating boundary between the heating and cooling fluids, and the heat passes by conduction through the metal wall. There are two commonly used types of regenerative air heaters (Fig. 35). In one, the heat-transferring members are moved alternately through the gas and air streams undergoing successive heating and cooling cycles and transferring heat by the thermal-storage capacity of the members. The other type of regenerative air heater has stationary elements, and the alternate flow of gas and air is controlled by rotating the inlet and outlet connections.

Recuperative and regenerative air heaters may be arranged

either vertically or horizontally and for either parallel or counterflow of the gas and air. The gases are usually passed through the tubes of tubular air heaters to facilitate cleaning, although in some designs, particularly for marine installations, the air flows through the tubes.

Improved heat transfer and better utilization of the heat-absorbing surfaces are obtained with a counterflow of the gases and the use of small flow channels. Regenerative-type air heaters readily lend themselves to these two principles and thus offer high capability in minimum space. However, regenerative air heaters have the disadvantages of air leakage into the gas stream and the transport of fly ash into the combustion air system. Tubular-type recuperative air heaters do not encounter these problems.

The products of combustion from most fuels contain a high percentage of water vapor, and thus condensation will be experienced in air heaters if the exposed metal surfaces are cooled below the dew point of the gas. Minute concentrations of sulfur trioxide in the gases, originating from the combustion of sulfur and varying with the sulfur content of the fuel and the method of firing, combine with the water vapor in the combustion gases to form sulfuric acid, which may condense on the metal surfaces at acid-dew-point temperatures as high as 250 to 300°F (121 to 149°C), well above the water dew point (Huge and Piotter, *Trans. ASME*, 1955). Such condensation leads to corrosion and/or the fouling of the gas-flow area. It is most likely to occur during the winter when the entering-air temperature is low, and at low operating loads or in localized sections at the cold-air inlet if there is poor distribution of the air or the gas flowing through the air heater. Corrosion and fouling can be prevented by the use of auxiliary steam-heated air heaters located ahead of the air inlet, by recirculating heated air from the outlet duct, or by bypassing a portion of the cold air to reduce the airflow through the air heater. Both recuperative and regenerative air heaters often are designed with separate corrosion sections arranged to facilitate the replacement of the vulnerable cold-end portions.

## STEAM TEMPERATURE, ADJUSTMENT AND CONTROL

The control of steam temperatures is vital to the life of high-temperature equipment and to the economy of power generation. Actual, or operating, steam temperatures below the design temperature reduce the thermodynamic efficiency and increase fuel cost, and temperatures above the design temperature reduce the margins of reserve in the strength of tubes, headers, piping, valves, and turbine elements. Further, sudden or extreme temperature variations may cause destructive stresses, particularly in rotating equipment.

It is sometimes necessary, because of the complexities involved in the design evaluation of heat-transfer rates and fuel characteristics, to modify installed equipment so as to obtain the required steam temperatures. Such changes might involve the installation of baffles for the distribution of gas through the superheater and the removal, or addition, of tubular elements in the superheater or in those components preceding the superheater which affect the temperature of the gas to the superheater. Therefore, it is desirable, and usually essential, to provide some means of controlling steam temperature so as to compensate for the variations in fuel, heat transfer, and surface-cleanliness conditions encountered during operation.

These may include (1) damper control of the gases to the superheater and/or reheater; (2) recirculation of the low-temperature gaseous products of combustion to the furnace so as to change the relative amounts of heat absorbed in the furnace, in the superheater, and/or in the reheater; (3) selective use of burners at different elevations in the furnace or the use of tilting burners to change the location of the combustion zone with respect to the furnace heat-absorbing surface; (4) attemperation or controlled cooling of the steam at superheater inlet, at superheater outlet, or between the primary and secondary stages of the superheater; (5) control of the firing rate in divided furnaces; and (6) control of the firing rate relative to the pumping rate of water in forced-flow once-through boilers.

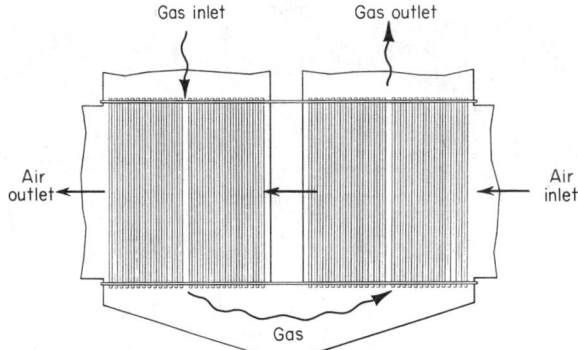

**Fig. 34**  Tubular air heater, two-gas-pass, single-air-pass.

The speed of response differs for the various methods, and the control of steam temperature by gas bypass or flame position is slower than that by spray-water attemperation. The operating controls for these methods can be arranged for manual, automatic, or combination adjustment, and the use of more than one method often facilitates the maintaining of constant steam temperature over a wider range of boiler load (Fig. 36).

The attemperation of superheated steam by direct-contact water spray (Fig. 37) results in an equivalent increase in high-pressure steam generation without thermal loss. Spray attemperation requires the use of essentially pure water, such as condensate, to avoid impurities in the steam. Submerged-type attemperators, when used, are generally restricted to relatively low-pressure boilers operating with steam temperatures of 850°F (454°C) or less. Usually, spray attemperators are not used for the control of reheat-steam temperature since their use reduces the overall thermal-cycle efficiency. They are, however, often installed for the emergency control of reheat-steam temperatures.

Ash and slag deposits on superheater and reheater surfaces reduce heat transfer and lower steam temperatures. Similar deposits on the furnace walls and steam-generating surface ahead of the superheater and/or reheater also reduce heat transfer to those surfaces, resulting in higher-temperature gas to the superheater and reheater and, consequently, increased steam temperatures. Thus the control of surface cleanliness is an important factor in the control of steam temperature.

Increased excess air results in higher steam temperatures because of reduction in furnace radiant-heat absorption, the greater amount of gas, and the increased convection heat transfer in the superheater and/or reheater. Operation with feedwater temperatures below that anticipated also results in increased steam temperatures because of the greater firing rate required to maintain steam generation.

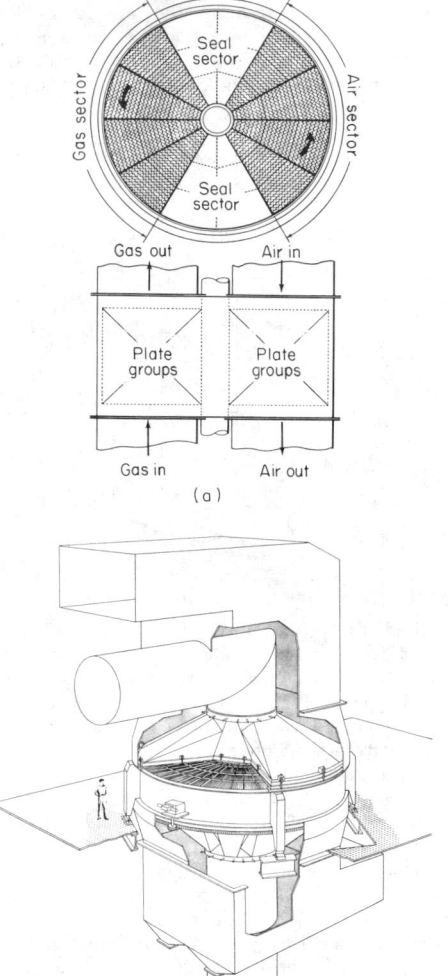

(a)

(b)

**Fig. 35**  Two designs of rotary regenerative air heaters.

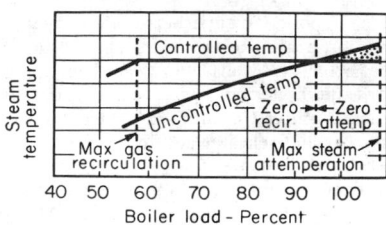

**Fig. 36**  Steam-temperature control by flue-gas recirculation and attemperation.

## OPERATING CONTROLS

(See also Sec. 16)

The need for operating instruments and manual or automatic controls varies with the size and type of equipment, the method of firing, and the proficiency of the operating personnel.

Safe operation and efficient performance require information relative to the (1) water level in the boiler drum; (2) burner performance; (3) steam and feedwater pressures; (4) superheated and reheated steam temperatures; (5) pressures of the gas and air entering and leaving principal components; (6)

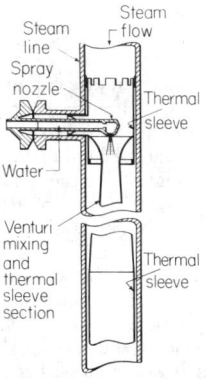

**Fig. 37**  Spray-type direct-contact attemperator.

feedwater and boiler-water chemical conditions and particle carry-over; (7) operation of feed pumps, fans, and fuel-burning and fuel-preparation equipment; (8) relationship of the actual combustion air passing through the furnace to that theoretically required for the fuel fired; (9) temperatures of the fuel, water, gas, and air entering and leaving the principal components of the boiler unit; and (10) fuel, feedwater, steam, and air flows so as to monitor operating conditions continuously and to make such adjustments as might be necessary.

Control of the various functions to maintain the desired operating conditions may be accomplished on small-capacity boilers by the manual adjustment of valves, dampers, and motor speeds. Most oil- and gas-fired package-type boilers are equipped with automatic controls to purge the furnace, to start and stop the burners, and to maintain the required steam pressure and water level. The operating requirements of utility and large industrial boilers dictate the use of automatic controls for the major variables, such as feedwater, flow, firing rate, and steam temperature. The type of boiler and its components generally establishes the basic mode of control, and analog controls of either the pneumatic or the electric type are available.

Sequence controls often are applied in the start-up of utility boilers to program the furnace purge, burner light-off, and burner control. Interlocks are essential to ensure the proper starting and firing sequence and to alarm or automatically shut down the unit in the event of the failure of essential auxiliaries.

## BOILER CIRCULATION

Adequate circulation in the steam-generating section of a boiler is required to prevent overheating of the heat-absorbing surfaces, and it may be provided naturally by gravitational forces, mechanically by pumps, or by a combination of both methods.

Natural circulation is produced by the difference in the densities of the water in the unheated downcomers and the steam-water mixture in the heated steam-generating tubes. This density differential provides a large circulating force (curve A, Fig. 38). The downcomers and the heated circuits are so designed that the friction, or resistance to flow, through the system balances the circulating force at the desired total circulating flow.

The forced-recirculation or assisted-circulation type of boiler uses a steam drum similar to that used with natural-circulation boilers. The supply of water to the furnace walls and the boiler surfaces flows from this drum to a circulating pump, which supplies the pressure necessary to force the water through the water-steam-mixture circuits and then back

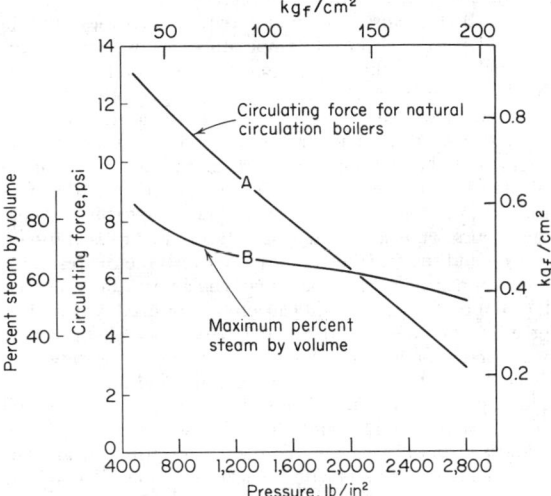

**Fig. 38**  Maximum percent steam by volume and circulating force for natural-circulation boilers.

to the drum, where the steam and water are separated. The total quantity of water pumped usually is four to six times the amount of steam evaporated, as shown in Fig. 39. The recirculating pump produces a differential pressure of 30 to 40 lb/in² (2.1 to 2.8 kgf/cm²), and the power required is equivalent to about 0.5 percent of the heat input to the boiler. Resistor orifices are required at the entrance to each tube, or circuit, to control flow distribution.

In assisted-circulation designs, the velocities or flows are independent of boiler rating, thus facilitating the use of smaller connecting piping and, sometimes, smaller-diameter furnace-wall tubing than that used with natural-circulation-type units. Both drum-type natural- and assisted-circulation boilers operate with essentially saturated steam temperatures in all parts of the steam-evaporating sections, and they can be used for drum operating pressures ranging up to approximately 2,800 lb/in² (197 kgf/cm²).

In natural- and assisted-circulation boilers, it is essential to wet the inside surfaces continually by the water of the two-phase water-steam mixture so as to prevent the overheating of these heat-absorbing surfaces.

Satisfactory cooling of the heat-absorbing steam-generating surfaces is dependent upon the pressure, heat flux, water-steam mass velocity, percent steam by weight (quality), tube diameter, and the tube's internal geometry. The heat flux is one of the most predominant of these parameters, and the rate of furnace heat absorption at the maximum firing rate therefore, generally dictates design considerations.

In forced-flow once-through boilers, the water from the feed supply is pumped to the inlet of the heat-absorbing circuits. Evaporation, or change of state, takes place along the length of the circuit, and when evaporation is completed the

steam is superheated. These units do not require steam or water drums and, in most cases, use relatively small-diameter tubes. The boilers can be started rapidly owing to the elimination of the drums and the reduced amount of metal. The water flow to the unit is the same as the steam output (Fig. 39), and fluid velocities greater than those needed for natural- or assisted-circulation units must be used at full load so as to maintain adequate velocities at the low loads and, thus, satisfactory tube-metal temperatures at all loads.

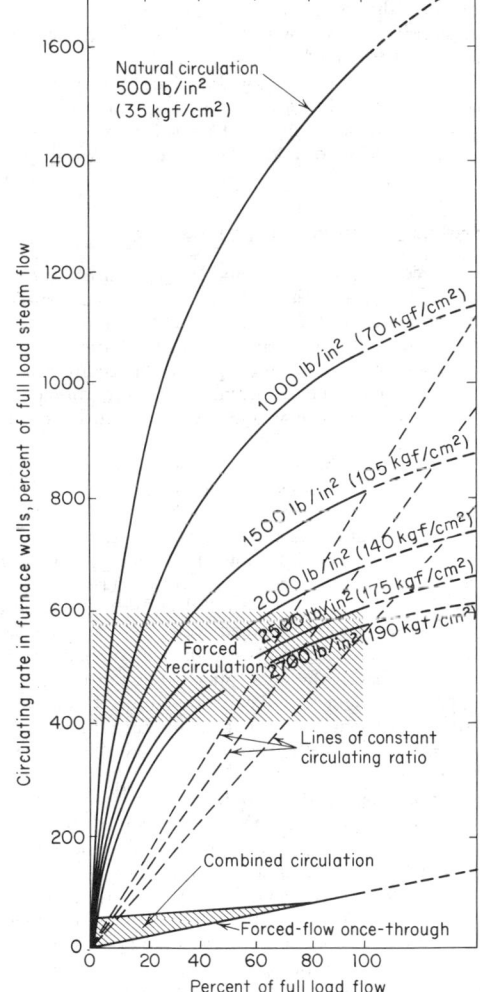

**Fig. 39**  Circulating flow for natural- and forced-circulation systems.

The transition from a liquid to a vapor at, or above, the critical steam pressure of 3,206 lb/in² (225 kgf/cm²) is dependent upon temperature and takes place without a change in density. Thus separation of steam and water is impossible and forced-flow once-through boilers must be used.

Forced-flow once-through boilers must be operated above a specified minimum flow—usually one-quarter to one-third of full-load flow—in order to maintain adequate water velocities in the furnace-wall tubes. However, the turbogenerator can be operated at any load by the use of a bypass system that diverts the excess flow to a flash tank for heat recovery. The bypass system also can be used as a pressure-relieving system, as the source of low-pressure steam to the turbine during start-up, and as a means of controlling steam temperature to the turbine during hot restarts.

Combined-circulation units utilize forced once-through flow with flow recirculation in the furnace walls to provide satisfactory water velocities during start-up and low-load operations. In this design, some of the water at the exit of the furnace circuits is mixed with the incoming feedwater, flows to and through a circulating pump, and then passes to the furnace-wall inlet headers. The use of combined circulation increases the water velocities in the furnace tubes at low loads, and since recirculation is not used at the higher loads, there is no increase in velocity or in the resistance to flow at the higher loads.

### FLOW OF GAS THROUGH BOILER UNIT

During the combustion of fuel and the transfer of heat to the heat-absorbing surfaces, it is necessary to maintain sufficient pressure to overcome the resistance to flow imposed by the burning equipment, tube banks, directional turns, and the flues and dampers in the system. The resistance to air and gas flows depends upon the arrangement of the equipment and varies with the rates of flow and the temperatures of the air and gas.

The term draft denotes the difference between atmospheric pressure and some lower pressure existing in the furnace or gas passages of a steam-generating unit. Draft loss is defined as the difference in static pressure of a gas between two points in a system, both of which are below atmospheric pressure, and is the result of the resistance to flow. These terms originated with the use of the so-called natural-draft units in which the pressure differentials are obtained from a chimney or stack which produces static pressures throughout the boiler setting that are below that of the atmosphere. The terms are rather loosely applied to modern boilers using induced draft and/or forced draft, mechanically produced by fans, in which the pressures throughout the boiler unit may be well above atmospheric pressure.

Forced-draft fans, handling cold, clean air, provide the most economical source of energy to produce flow through high-capacity units (see Sec. 14). Induced-draft fans, handling hot flue gases, require more power and are subject to fly-ash erosion. However, they facilitate operation by providing a draft in the boiler setting and thus prevent the outward leakage of gas through joints or crevices in the boiler-unit enclosure. As a result of advances in furnace and boiler-setting designs to eliminate gas leakage, modern units often are built for positive-pressure operation, thus eliminating the need for induced-draft fans. Such units are generally referred to as pressure-fired units, while those using induced-draft fans are classed as suction units. When both forced- and induced-draft fans are used, the boilers are designated as balanced-draft units.

The pressure drop throughout the unit is caused by the fluid friction in the gas stream and the shock losses at the turns or the contractions and enlargements of sections. It can be calculated as a function of the fluid-mass flow and fluid properties in accordance with the principles of fluid flow (see Sec.

3). It is essential in the design of a boiler unit to determine the sum of all component resistances in the flow system at the maximum load in order to establish the fan requirements. It is customary to specify test-block static head, temperature, and capacity requirements of the fan in excess of those calculated so as to allow for departure from ideal flow conditions and to provide a satisfactory margin of reserve.

**Stack effect** is caused by the difference in densities resulting from the difference in the temperatures of two vertical columns of gas. In a chimney, or stack, the stack effect is due to the difference between the confined hot gas and the cooler surrounding air and the equal static pressure at the top or free outlet of the stack. The stack effect, which varies with the height and the mean temperature of the columns, can be calculated from the data in Table 5. The effect is the static draft produced by a stack, at sea level, with no gas flow. When flow occurs, a portion of the stack effect is used to establish gas velocity and the remainder is used to overcome the resistance of the connected system, including the dampers and the stack. The limit of natural-draft capacity is reached when these forces are in balance with the dampers in a wide-open position. Stack performance may be favorably or adversely affected by external factors such as the wind and the atmospheric conditions. The available draft varies directly with the barometric pressure for altitudes above sea level.

Stack effects also exist within the boiler setting and are most pronounced in tall units with vertical gas passes. The individual gas columns within the setting may aid the head produced by the fan or chimney if the flow is upward or may reduce it if the flow is downward. The net stack effect, and its overall influence on the performance of the fan, may be calculated from the data in Table 5, taking into account the positive or negative effects. The relationship between local static pressures and the atmospheric pressure is most important, since gas may blow into the room through an open inspection door at the top of a furnace, even though a strong draft, or negative pressure, exists at some lower elevation.

### PERFORMANCE

Steam-generating units are designed for specific operating conditions and are generally sold with a guarantee of performance. The boiler rating is usually specified and guaranteed in terms of steam output (lb/h) at a given pressure and temperature at full load or maximum continuous operation. When the steam is reheated, the rating includes this requirement in terms of the quantity of reheat steam at stated inlet and outlet steam pressures and temperatures.

Generally, either the efficiency or the gas temperature leaving the unit is guaranteed at a specified rate of operation, and the draft loss and the quality or purity of the steam also may be guaranteed at this rate. When component equipment such as stokers, pulverizers, burners, and air heaters are supplied by different manufacturers, the performance of the individual components is usually guaranteed by the various manufacturers and then, in turn, guaranteed by the prime contractor.

Anticipated-performance data for several rates of operation may be given to the purchaser in addition to the guaranteed-performance data. Guarantees may be demonstrated by acceptance tests, conducted in accordance with established codes, as agreed upon by the parties to the contract. However, acceptance tests are more difficult to perform as unit size and capacity increase, and overall performance usually is deter-

**Table 5. Stack Effect or Pressure Difference, in of Water for Each Foot of Vertical Height (mm of Water for Each Metre of Vertical Height)**
Barometer = 29.92 in Hg (759.97 mm Hg)

| Avg temp in flue | | Air temp outside flue | | | | | | | |
|---|---|---|---|---|---|---|---|---|---|
| | | 40°F | 5°C | 60°F | 15°C | 80°F | 25°C | 100°F | 35°C |
| 250°F | 125°C | 0.0041 in H₂O/ft | 0.346 mm H₂O/m | 0.0035 in H₂O/ft | 0.303 mm H₂O/m | 0.0030 in H₂O/ft | 0.262 mm H₂O/m | 0.0025 in H₂O/ft | 0.224 mm H₂O/m |
| 500°F | 250°C | 0.0070 | 0.563 | 0.0064 | 0.520 | 0.0058 | 0.480 | 0.0053 | 0.442 |
| 1000°F | 500°C | 0.0098 | 0.788 | 0.0092 | 0.744 | 0.0086 | 0.708 | 0.0081 | 0.665 |
| 1500°F | 750°C | 0.0111 | 0.902 | 0.0106 | 0.858 | 0.0100 | 0.818 | 0.0095 | 0.780 |
| 2000°F | 1000°C | 0.0120 | 0.972 | 0.0114 | 0.925 | 0.0108 | 0.887 | 0.0103 | 0.850 |
| 2500°F | 1250°C | 0.0125 | 1.018 | 0.0119 | 0.975 | 0.0114 | 0.934 | 0.0109 | 0.896 |

mined from the operating data. Guarantees of materials and the quality of manufacture and erection are usually considered separately from those pertaining to operating performance.

Heat balances account for the thermal energy entering the system in terms of its ultimate useful heat absorption or thermal loss. Methods of measuring and calculating the quantities involved in heat balances are presented in the ASME Power Test Code for Stationary Steam Generating Units.

The heat input is predicated upon the hourly firing rate, the calorific heating value of the fuel, and any additional heat supplied from an outside source. Heat in the preheated combustion air obtained from an air heater integral with the boiler unit is not considered in the determination of heat input, since this heat is recycled within the system.

The heat absorption in a boiler is calculated from the rate of steam output and the increase in fluid enthalpy from feedwater conditions to that at the superheater outlet. The amount of heat absorbed by the steam passing through the reheater, if used, is added to the heat absorbed in the boiler, economizer, and superheater. The total heat absorption also must take into account any steam generated which bypasses the superheater. Usually, the heat absorption is determined on an hourly basis.

In its simplest form,

Efficiency (percent)
   = heat absorbed, Btu (cal)/h/heat input, Btu (cal)/h × 100

Both the heat input and the heat absorption may be very large quantities. Therefore, unless elaborate precautions are taken in the sampling and the measurement of fuel and steam quantities, it is difficult to obtain test data having the degree of accuracy required to determine the actual efficiency of the boiler unit. For this reason, boiler efficiency usually is established from the heat losses, since each of the thermal losses is a relatively small percentage of the heat entering the system and reasonable errors in measurement will not appreciably affect the final result.

The principal thermal losses are those due to the sensible heat in the gases leaving the unit, the latent-heat losses associated with the evaporation of fuel moisture and with the formation of water vapor resulting from the burning of the hydrogen in the fuel, the unburned-combustible loss, the loss from the boiler setting or enclosure due to external convection and radiation, and the ashpit loss. The first two losses can be derived from the fuel analysis, the exit-gas temperature, and the analysis of the flue gas (see Sec. 4). The unburned-combustible loss can be established by a qualitative and quantitative sampling of the refuse and the fly ash. The radiation from the boiler setting can be estimated in detail, but it also can be approximated from Fig. 40. The heat loss to the ashpit of large units can be determined by measuring the quantity of quenching water evaporated from the furnace ash hopper or the slag tank; and for small units the ashpit loss is included in the heat loss from the boiler setting. The sum of these heat losses, expressed as a percentage of the total heat input, is the total measurable loss. The anticipated or guaranteed performance data include a tolerance for the so-called manufacturer's margin (unaccountable losses) in the order of 0.5 to 0.75 percent, depending upon the type of fuel fired. The thermal efficiency of the unit is established by subtracting the sum of all these losses from 100 percent.

Tests of the individual components of the boiler unit, such as the furnace, superheater, economizer, or air heater, may be conducted for the determination of heat-transfer and gas-flow characteristics, for comparisons with other units, or to facilitate changes in operating procedures or equipment. Available ASME codes delineate such tests.

## STEAM PURIFICATION

In drum-type boilers, using either natural or assisted circulation, a mixture of steam and water is delivered to the upper or steam drum where separation of the steam and water takes place and a water level is maintained. The water is recycled through downcomers to the heat-absorbing circuits, and the steam is discharged from the top of the drum for use as saturated steam or the supply to the superheater.

Separation of the steam and water by buoyant or gravitational force requires a relatively large cross-sectional area and, consequently, a low fluid velocity within the drum as well as an effective difference in the fluid densities, which decreases as pressure is increased. Steam entrainment in the recycled water impedes circulation; and water entrainment in the outlet steam transports dissolved or suspended particulate matter into the superheater, steam piping, and turbine, where particle deposition can cause overheating of the tubes or flow obstructions in the turbine blading with subsequent loss of capacity, efficiency, and dynamic balance.

Gravity separation of steam and water may be satisfactory in low-pressure, low-duty boilers. This type of separation can be augmented by the use of baffles which utilize a change in direction to throw out the water droplets, or by dry pipes which impose a pressure drop that promotes evaporation of the moisture and reduces the tendency of solids to deposit in the superheater.

In high-pressure boiler units, particularly those employing high evaporative ratings, a part of the circulating head can be utilized to provide a separating force, many times greater than that of gravity, in centrifugal separating devices such as the cyclone steam separators shown in Fig. 41. These separators deliver steam-free water to the drum and downcomers, and discharge steam with a minimum of water entrainment. Secondary steam-and-water separation is accomplished by passing the steam at a low velocity through sinuously shaped passages between closely spaced corrugated plates, which provide a large surface area for intercepting entrained boiler-water particles. In modern high-capacity boilers, the steam leaving the steam drum contains less than 0.1 ppm of total solids.

Mechanical steam separators do not prevent the transport of silica in a vapor solution. The amount of silica dissolved in the steam is dependent upon its concentration in the boiler water, and for a given concentration, the ratio of silica in the steam to the silica in the water increases rapidly with an increase in the operating pressure (see Fig. 42). Silica can be removed by steam washers which provide a large surface area for contact with the relatively pure feedwater and which reabsorb the silica and return it to the boiler-water system. Turbine deposits can be practically eliminated and the requirements of boiler blowdown materially reduced by the use of steam washers for steam purification. Steam washers are used to best advantage in medium-pressure boilers operating with large amounts of makeup water, particularly if the makeup water contains silica in an insoluble form.

Impurities in the water entering a forced-flow once-through boiler leave with the steam unless they are deposited in the boiler unit. Therefore, the equivalent of steam purification must be accomplished by treatment of the feedwater before the water enters the boiler so as to prevent the accumulation of deposits in the boiler unit and the turbine. In the treatment, most of the feedwater, including the steam condensate, is passed through a mixed-bed demineralizer that removes suspended as well as dissolved impurities.

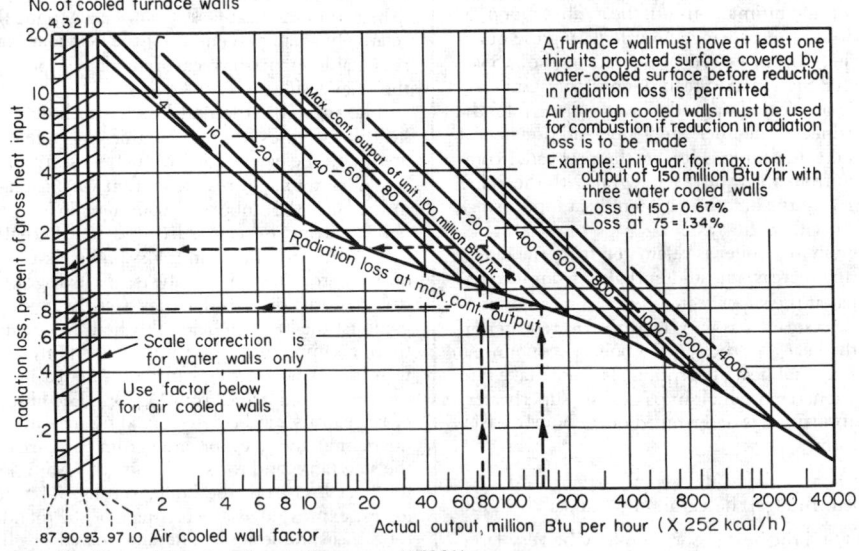

**Fig. 40** External heat loss from boiler setting (ABMA).

## WATER TREATMENT

(See also Sec. 6).

All natural waters contain impurities, many of which may be harmful in boiler operation. These impurities originate from the earth and the atmosphere (or from municipal and industrial wastes) and are broadly classed as suspended or dissolved organic and inorganic matter and as dissolved gases.

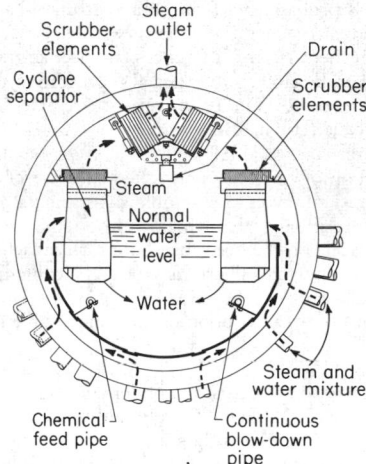

**Fig. 41** Drum internal arrangement with cyclone steam separators and scrubber elements.

The concentration of the impurities is customarily expressed in terms of the parts by weight of the constituent per million parts of water (ppm). This is equivalent to the percentage of concentration multiplied by $10^4$ and has the advantage of using positive whole numbers for small concentrations. However, for exceedingly small concentrations, especially those involving gases, the quantities are sometimes expressed as parts per billion (ppb). Concentrations also may be expressed in terms of the number of grains per gallon, but in boiler practice this has generally been superseded by the gravimetric relation, ppm. One grain per gallon is equal to 17.1 ppm.

In boilers, water is converted into steam and the steam leaves the boiler drum in a relatively pure state. Impurities, other than the gases which enter with the feedwater, are thus retained and concentrated in the boiler water. High concentrations of foam-producing solids in the boiler water contribute to particle and water carry-over and contamination of the steam. Chemical and solubility changes also take place in the boiler, particularly as temperature is increased.

With few exceptions, the waters found in nature are not suitable for use as boiler feedwater; but they can be used after proper treatment. In essence, this entails the removal from the raw water of those constituents which are known to be harmful; supplementary treatment, within the boiler or connected system, of residual impurities to convert them into harmless forms; and systematic removal, by blowdown of boiler-water concentrates, to prevent excessive accumulation of solids within the boiler unit.

The ultimate purpose of feedwater and boiler-water treatment is to prevent deposits of scale or sludge on and corrosion of the internal boiler surfaces. Hard-scale formations, formed by certain constituents in the zones of high-heat input, retard the flow of heat and raise the metal temperatures. This can lead to overheating and the failure of pressure parts. Sludge or solid particles normally carried in suspension may settle locally and restrict the flow of cooling water or, in some cases, deposit in the form of insulating layers with a resultant effect similar to that of hard scale. Oil and grease prevent adequate wetting of internal boiler surfaces and, in areas of high heat input, may cause overheating. Further, oil and grease may carbonize and form a tightly adherent insulating coating. Corrosion (see also Sec. 6), due to acidic conditions or to dissolved gases, can weaken a boiler because of the loss of metal. Corrosion usually occurs as cavities and pits in localized areas which, as they deepen, may penetrate the metal. Frequently, corrosion occurs under internal deposits because of elevated temperatures and solids concentrations. Therefore, corrosion and solids deposits are closely related. Certain chemical reactions produce an intergranular attack of the metal that may lead to embrittlement and ultimate fracture.

When condensate is used as the feedwater to a boiler, water treatment is minimized, since it is required only for the small amounts of raw water that may leak into the system and the makeup water needed (usually $\frac{1}{2}$ to 3 percent) to replace the loss of steam and condensate from the system. However, in industrial plants using a large portion of the steam generated for process work, the makeup-water requirements may be 90 to 100 percent of the total feedwater flow. Such plants require a considerable amount of water treatment.

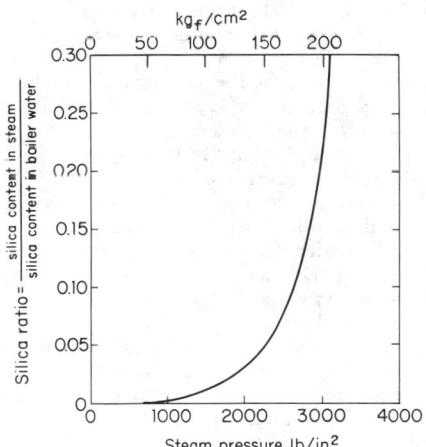

**Fig. 42** Equilibrium relationship of silica ratio and operating pressure for a given concentration of silica in boiler water.

The treatment best suited, or economically justified, for any given plant depends upon the characteristics of the water supply, the amount of makeup water, and the design of the steam-generating and related equipment. Usually, the feedwater and boiler-water treatment is supervised by a chemist, and often it is desirable to engage a reputable feedwater specialist to prescribe specific procedures. However, the results obtained depend upon the diligence and integrity of routine sampling and the control carried out by plant personnel.

### Raw Water

The treatment of raw water for makeup and boiler feedwater involves one or more of the following:

1. The removal of suspended solids. Large particles in the water are removed by settling and decantation or by filtering through screens, fabrics, or beds of granular material. Small particles which settle slowly or colloidal particles which do not settle can be removed by coagulation using floc-forming chemicals, such as alum or ferrous sulfate, to trap the particles in the floc. The floc is then removed by settling or filtration. The solids can be removed intermittently or on a continuous basis.

2. Chemical treatment for removal of hardness. Calcium, magnesium, and silica are the principal scale-forming impurities in water and, if present in the boiler water, may form compounds whose solubility decreases with an increase in temperature.

In the lime-soda process for softening water, lime (calcium hydroxide) reacts with the soluble calcium and magnesium bicarbonates to form precipitates of calcium carbonate and magnesium hydroxide which can be removed as sludge. The soda ash (sodium carbonate) reacts with the scale-forming calcium sulfate and magnesium sulfate, and precipitates the calcium and magnesium as insoluble carbonates. Both reactions produce sodium sulfate, a soluble and non-scale-forming compound. When the hot-lime-soda process is carried out at temperatures of 200 to 250°F (93 to 121°C), the reactions are accelerated and some of the silica may be removed.

The reactions, as in all chemical processes, tend to approach equilibrium but they are affected by time, the completeness of mixing, and the removal of the products. Therefore, in either the intermittent batch or the continuous process some unreacted hardness is left in the treated water.

3. Cation exchange for removal of hardness. Certain naturally occurring minerals, such as sodium aluminum silicate, or synthetic resins, such as the polystyrenes or phenolic-type materials, have the ability to exchange sodium ions for calcium and magnesium ions if present in a water solution. Thus softening can be accomplished by passing raw or filtered water through beds of granulated zeolite particles. The calcium and magnesium ions are retained by the zeolite material, while their equivalents of non-scale-forming sodium ions are released to the water solution. Before complete exhaustion of the sodium is reached, the softening equipment must be isolated from the system and regenerated by the passage of a strong brine of sodium chloride through the softener. Sodium ions are thus restored to the zeolite, and calcium, or magnesium, is removed as a soluble chloride and drained to waste. After the regenerating cycle, the equipment is purged of the brine by flushing with filtered water and then returned to softening service.

The most popular system today combines chemical treatment and cation exchange, and utilizes hot lime (with or without magnesium for silicate removal) followed by hot sodium-cation exchange.

4. Demineralization for complete removal of dissolved solids. Several types of synthetic organic resins are capable of selectively removing undesirable cations or anions from water solutions by their exchange for hydrogen or hydroxyl ions. When used in combination, as separate or mixed beds of small-sized beads or particles through which the water flows, they can produce an effluent that is virtually free of mineral solutes and satisfactory for boiler feedwater. The cation exchanger is regenerated by acid which restores hydrogen ions to the resin in exchange for the calcium, sodium, or other metallic cations removed from the water. The anion exchanger is regenerated by the use of caustic soda, or another appropriate base, which restores the hydroxyl ions in exchange for the chloride, sulfate, or other negative chemical radicals previously removed from the water. The hydrogen and hydroxyl ions released from the resins during the heating process combine to form pure water. The greatest effectiveness is attained by a mixed-bed arrangement of resins, since the interchange of cation and anion components proceeds in minute increments and with less probability of the escape of unexchanged ions. With individual regeneration, the mixed resins are separated hydraulically because of differences in specific gravity. The resins can then be sluiced to external regeneration facilities or regenerated in place. The resins must be remixed before the demineralizer is returned to service.

5. Evaporation. Essentially pure water can be obtained by the evaporation of raw water and the collection of the distillate. Evaporation leaves the soluble constituents as concentrates in the residual water which can be removed by blowdown, or as scale on the heat-absorbing surfaces which can be mechanically removed. There may be some contamination in the distillate because of the carry-over of water particles with the vapor and the reabsorption of noncondensable gases.

### Feedwater

Boiler feedwater may consist of condensate, treated water, or a mixture of both. Usually, there is only a small amount of dissolved and suspended solids as a result of the treatment and, generally, the removal of additional solids is not required. However, any dissolved gases present must be removed to prevent corrosion in the boiler and the preboiler system.

Dissolved oxygen is, perhaps, the greatest factor in the corrosion of steel surfaces in contact with water. It may be present in the makeup water or the feedwater because of previous contacts with atmospheric air, or it may be added to the water by the leakage of air into the system through low-pressure-pump seals, storage tanks, etc.

Oxygen may be partially removed (to a residual of 0.2 to 0.3 ppm) by heating the water to boiling temperature in open-type feedwater heaters. Tray or spray-type deaerating heaters are more effective in removing oxygen (to residuals of 0.02 to 0.04 ppm), and the amount of oxygen in the water can be reduced to 0.007 ppm or less by the use of multistage deaerators arranged for the countercurrent scavenging of noncondensable gases. (See Sec. 9.)

It is customary to supplement feedwater deaeration by adding a scavenging agent, such as sodium sulfite or hydrazine, to effect the complete removal of residual oxygen. Sodium sulfite combines with oxygen to form sodium sulfate, but it should not be used at operating pressures in excess of 1,800 lb/in² (127 kgf/cm²), since it decomposes to corrosive products at high temperatures. Thus hydrazine should be used at high pressures.

In boiler plants using high-purity feedwater, corrosion may be experienced in the condensate piping and the preboiler system because of dissolved gases such as carbon dioxide, sulfur dioxide, or hydrogen sulfide in the water. These gases may originate from the atmosphere or from constituents in the boiler water. They are released in the steam generators, intimately mixed with the outgoing steam, and with the exception of those partially removed by the vacuum pumps, returned to solution in the condenser. These gases in the condensate produce an acidic reaction leading to corrosion, even in the absence of dissolved oxygen. The corrosion products in the preboiler cycle often are carried into the boiler and may deposit on the heat-absorbing surfaces, with resultant overheating of these surfaces.

A small amount of alkaline boiler water is sometimes recirculated to the feedwater heaters to raise the pH of the feedwater and thus prevent corrosion in the preboiler system. However, this procedure may precipitate sludge in the feedwater piping if appreciable hardness is present in the boiler water.

The pH of the feedwater can be increased by the addition of ammonia or volatile amines, such as morpholine or cyclohexylamine. Generally, these compounds are added as early as possible in the preboiler system. This procedure prevents corrosion in the early stages of moisture formation in the turbine and the condenser, as well as in the entire condensate-return system.

Filming amines also can be used and are generally introduced to the system through chemical pumps in the feedwater or steam lines. These materials do not change the pH of the fluid but protect against corrosion by forming a monomolecular coating on the metal surfaces. However, caution must be exercised since excessive use of filming amines has been known to agglomerate boiler sludge and produce strongly adherent internal deposits.

### Boiler Water

Boiler water is treated internally to prevent corrosion, the fouling of heat-absorbing surfaces, and the contamination of steam. Internal treatment also aids in maintaining water conditions within satisfactory limits. The internal treatment requires the introduction of chemicals in suitable amounts to react with the residual impurities in the feedwater.

Corrosion in boilers is prevented or minimized by maintaining alkaline boiler water. This condition may be expressed in terms of pH or as total alkalinity.

Acid or alkaline reactions of aqueous solutions are due to the presence of free or excess hydrogen ($H^+$) or hydroxyl ($OH^-$) ions, and the strength of the reaction varies with the concentration or activity of the excess ions. Some compounds enter into solution without dissociation while others dissociate partially or completely into ions carrying positive or negative electrical charges. If such ionizable compounds contribute hydrogen ($H^+$) ions to the solution (e.g., HCl), they add to the strength of its acid reaction; if they contribute hydroxyl ($OH^-$) ions (e.g., NaOH), they add to its alkaline or base reaction. When the ions of many different compounds are present, as is the usual case with boiler waters, their interaction or buffering affects the resulting concentration of the specific ions, and the solution tends to approach a balance or equilibrium in accordance with the principles of chemical mass action. It is therefore possible by the addition of some compounds which in themselves contain neither hydrogen nor hydroxyl components to suppress or release these ions from other constituent solutes and thereby change the acidity or alkalinity of the solution.

The pH value of a solution, which designates its acidity or alkalinity, refers to a logarithmic scale proposed by Sorenson in 1909. The symbol p is derived from the German word Potenz, meaning power or exponent; and the symbol H represents the hydrogen-ion concentration. Thus, by definition, the pH value is equal to the logarithm of the reciprocal of the hydrogen-ion concentration measured in gram mols per litre.

Pure water, which may be considered as composed primarily of molecular $H_2O$, exhibits a slight degree of dissociation to hydrogen (+) and hydroxyl (−) ions in the equilibrium amounts, at room temperature, of 0.0000001 gmol each per litre of water. It thus has the somewhat unusual capability of reacting, under proper conditions, as a weak acid or as a weak base and is said to be amphoteric. The $H^+$ and $OH^-$ ions are in exact balance, and the water is electrically neutral.

The equation which expresses the equilibrium dissociation of water and also applies to water solutions is

$$H^+OH^- = K^{H_2O} = 10^{-14} \quad \text{at } 25°C$$

$H^+$ and $OH^-$ represent the respective concentrations of the ionized hydrogen and hydroxyl groups, and the dissociation product $K^{H_2O}$ is found by experimental methods to be $1/10^{14}$ at 25°C.

Since the product of the two concentrations is a constant, some $OH^-$ ions are present even in a highly acidic solution and some $H^+$ ions are present in a basic solution, and the relationship of these factors can be determined from the measurement of either term. Thus, in the case of neutral water, the value of each is $10^{-7}$, or 0.0000001, gmol/l.

In a solution having a hydrogen-ion concentration of $10^{-3}$ gmol/l, the corresponding hydroxyl-ion concentration is $10^{-11}$, as a result of the dominating influence of the solvent, which is present in great excess and maintains the product equilibrium. Although either factor in the equation can be used, the conventional scale is based upon the measurement of the hydrogen ion.

Numerical values of this relationship extend over an extremely wide range and can best be expressed in terms of logarithms or exponents; thus:

$$\log H^+ + \log OH^- = -14$$
$$-\log H^+ - \log OH^- = 14$$
$$\log 1/H^+ = 14 - \log 1/OH^- = pH$$

The term pH is used to represent $\log 1/H^+$ and is therefore the logarithm of the reciprocal of the hydrogen-ion concentration (more properly, the hydrogen-ion activity which is equal to the concentration multiplied by an activity factor that approaches unity in dilute solutions).

For neutral water, the $pH = \log 1/H^+ = \log 1/0.0000001 = \log 1/10^{-7} = 7.0$.

For an acid solution in which $H^+$ exceeds $OH^-$, say $H^+ = 10^{-3}$, the $pH = \log 1/10^{-3} = 3.0$.

For an alkaline solution in which $OH^-$ exceeds $H^+$, say $OH^- = 10^{-3}$, the $pH = \log 1/10^{-11} = 11.0$.

In practical terms, the pH scale extends from 0 to 14, as shown in Table 6. The value 7.0, corresponding to pure water, is considered the neutral point; values below 7.0 are increasingly acidic, and values above 7.0 are increasingly alkaline. Since the pH scale is logarithmic, a change from one number to the next in series is equivalent to a change of ten times the activity. Beyond the range of the scale, the strength of acid or alkaline solutions is expressed in terms of normality or percentage of concentration.

The pH of a water sample can be determined accurately by the measurement of its electrical potential. It also can be approximated by indicators that change color in certain pH ranges as the result of their reaction with the solution. The pH of boiler water usually is maintained within the range of 10.2

**Table 6. Relationship of pH and Hydrogen-Ion Concentration**

| | pH | Hydrogen-ion concentration, gmol/l | |
|---|---|---|---|
| Acid range | 0 | 1.0 | $10^0$ |
| | 1 | 0.1 | $10^{-1}$ |
| | 2 | 0.01 | $10^{-2}$ |
| | 3 | 0.001 | $10^{-3}$ |
| | 4 | 0.000,1 | $10^{-4}$ |
| | 5 | 0.000,01 | $10^{-5}$ |
| | 6 | 0.000,001 | $10^{-6}$ |
| Neutral | 7 | 0.000,000,1 | $10^{-7}$ |
| Alkaline range | 8 | 0.000,000,01 | $10^{-8}$ |
| | 9 | 0.000,000,001 | $10^{-9}$ |
| | 10 | 0.000,000,000,1 | $10^{-10}$ |
| | 11 | 0.000,000,000,01 | $10^{-11}$ |
| | 12 | 0.000,000,000,001 | $10^{-12}$ |
| | 13 | 0.000,000,000,000,1 | $10^{-13}$ |
| | 14 | 0.000,000,000,000,01 | $10^{-14}$ |

to 11.5 for boilers operating at pressures compatible with an 1,800 lb/in² (127 kgf/cm²) turbine throttle pressure. Above these pressures, mixed-bed demineralizers are generally used to treat the makeup water, the boiler-water treatment is low in solids, and the pH ranges from 9.0 to 10.0.

In forced-flow once-through units the recommended pH range is 8.8 to 9.2 for preboiler systems using copper alloys. If the preboiler system does not incorporate the use of cooper alloys, the recommended pH is 9.2 to 9.5.

Total alkalinity (expressed in ppm) is a measure of all reactives that have the ability to neutralize acids. It is determined by titrating a water sample with a standard acid, and it is frequently expressed as equivalent calcium carbonate, which has a molecular weight of 100. Total alkalinity, as determined in this manner, is not exactly the same as the pH measurement of alkalinity because of the buffering action which occurs in complex solutions. However, it is widely used as a reference and in the case of low-pressure boilers where higher concentrations of greater diversity of solids can be tolerated, it often is more satisfactory than the measurement of pH as an index of boiler-water conditions.

The elimination of hardness in boiler water is necessary to prevent scale. Hardness can be removed by introducing one of the forms of sodium or potassium phosphate and thoroughly mixing it with the boiler water. The residual calcium ions entering with the feedwater are precipitated as an insoluble phosphate sludge and the magnesium is precipitated as a nonadherent magnesium hydroxide, if the alkalinity is maintained at a pH of 10 or higher. A lower pH may result in the formation of magnesium phosphate, an adherent type of sludge. Routine control facilitates the adjustment of the pH by the addition of sodium hydroxide, or its equivalent, and the maintenance of a moderate excess of phosphate ions in the boiler water.

Early methods of internal treatment employed the use of soda ash for hardness removal. However, the hydrolysis of soda ash at the temperatures encountered with high operating pressures releases carbon dioxide into the steam, making it difficult to maintain an excess of carbonate and promoting corrosion in the condensate system. In some services, sodium carbonate in combination with the hydroxides and phosphates of sodium is used for hardness removal. A phosphate sludge is preferred since it is less adherent and more easily kept in suspension.

Silica may enter the system in the form of soluble compounds or as finely divided particles which are not removed by filtration. It dissolves in alkaline boiler waters and will, with unreacted calcium or magnesium hardness in the water, form an adherent scale. Under some conditions, it may produce complex scale-forming silicates with soluble or colloidal iron oxide (acmite) or alumina (analcite). The crystalline matrix of these deposits tends to trap sludge particles and contributes to the accumulation of scale on heated surfaces.

Silica also is soluble in steam, and its solubility increases rapidly at temperatures above 500°F (260°C). Thus it can be transported in a vapor phase into the turbine and deposited on the turbine blading. This characteristic necessitates the limiting of the silica concentration in the boiler water in order to avoid turbine deposits, and the limits, varying with operating pressure, range from about 10 ppm at 1,000 lb/in² (70 kgf/cm²) to 0.3 ppm at 2,500 lb/in² (176 kgf/cm²).

Silica is partially removed from raw water by the hot lime-soda softening process and can be completely removed by the evaporation of the makeup water. Soluble silica can be removed by demineralization, but in colloidal forms it may pass through the treating beds. The silica concentration in the boiler water can be controlled by blowdown.

Many operators of industrial boilers use the Chelant methods of water treatment. Chelants react with the residual divalent metal ions (calcium, magnesium, and iron) in the boiler water to form soluble complexes. The resultant soluble complexes are removed by use of continuous blowdown. One of the most popular methods uses the sodium salt of ethylenediamine-tetraacetic acid ($Na_4$ EDTA). Chelant methods of treatment have been used in boilers operating at pressures as high as 1,500 lb/in² (105 kgf/cm²).

The recommended limits of boiler-water concentration, as defined in the ABMA manual, are listed in Table 7. These data are not applicable to forced-flow once-through boilers. The total solids content can be determined by weighing the residue of a water sample which has been evaporated to dryness. The dissolved-solids content can be determined in a similar manner from a filtered sample, but for immediate determinations and control purposes, it can be quickly approximated by an electrical-conductivity measurement and the use of conversion factors previously established by comparisons with gravimetric determinations.

Solids concentration also can be controlled by intermittent

**Table 7. Recommended Limits of Boiler-water Concentration (ABMA)**

| Pressure at outlet of steam-generating unit, psig ($\times$ 0.07037 = kgf/cm²) | Total solids, ppm | Total alkalinity, ppm | Suspended solids, ppm |
|---|---|---|---|
| 0–300 | 3,500 | 700 | 300 |
| 301–450 | 3,000 | 600 | 250 |
| 451–600 | 2,500 | 500 | 150 |
| 601–750 | 2,000 | 400 | 100 |
| 751–900 | 1,500 | 300 | 60 |
| 901–1,000 | 1,250 | 250 | 40 |
| 1,001–1,500 | 1,000 | 200 | 20 |
| 1,501–2,000 | 750 | 150 | 10 |
| 2,001 and higher | 500 | 100 | 5 |

or continuous blowdown. The amount of blowdown and the time interval between blows should be coordinated with operation and should consider or anticipate load changes, water conditioning, and chemical treatment.

In forced-flow once-through boilers, the impurities entering with the feedwater must leave with the steam or be deposited within the unit. Thus, such units require high-purity feedwater and the control of corrosion by volatile bases, such as ammonia, which will prevent or minimize deposits in the boiler unit or the turbine. Raw-water leakage into the system must be prevented, and the makeup must be evaporated or demineralized water.

When sampling water from high-pressure, high-temperature sources, cooling is required to prevent the flashing or selective loss of water vapor at atmospheric pressure. The approved methods for water sampling and analysis are discussed in the Annual Book of ASTM Standards, Part 23.

## CARE OF BOILERS

The care of boilers is delineated in Section VII of the ASME Boiler and Pressure Vessel Code. Principal considerations include the initial preparation of new equipment for service; normal operation, including routine start-up and shutdown, emergency operations; inspection and maintenance; and idle storage. In all these phases, the handling of equipment is the responsibility of the operator, but recommendations and operating instructions supplied by the manufacturer should be thoroughly understood and followed.

The initial preparation of new boiler units for service, or of old equipment after completing major alterations or repairs, involves the removal of construction or foreign material from the setting and from the interior of pressure parts, hydrostatic testing and inspection for leaks, and the boiling out of the unit with a caustic solution for the removal of grease and other deposits in the steam-generating pressure parts. The boiler unit is fired at a low rate during boil-out. This procedure facilitates the desired slow drying of any refractories used in the setting. Boil-out pressure should be approximately 50 percent of normal operating pressure but should not exceed 600 lb/in² (42 kgf/cm²). During the boil-out period, ranging from 12 to 36 h, the unit is blown down periodically through all the blowdown connections so as to eject any sediment removed from the surfaces. The boil-out often is supplemented, particularly in high-pressure boilers, by inhibited-acid cleaning for the removal of mill scale.

It is general practice, following the boil-out, to reduce the concentration of boil-out chemicals to a satisfactory level for operation by blowing down and replenishing the amount of

water blown down with fresh water. The pressure is then raised to test and set the safety valves to code requirements; the superheater and the steam piping are blown out to remove any foreign material; and the boiler is placed on the line for a period of low-load operation, during which the auxiliary equipment, controls, and interlocks are test-operated. After these operations, it is advisable to shut down, cool, and drain the boiler unit prior to a thorough internal and external inspection and any adjustments or modifications required to the equipment.

Normal operation involves the orderly start-up and shutdown of equipment and operation, under controlled conditions, to meet plant requirements. Statistics show that about 80 percent of the recorded furnace explosions occur during start-up and low-load operation, and particular care must be taken during such operations to prevent such explosions. The National Fire Protection Association's Committee on Boiler Furnace Explosions has prepared standards for the prevention of furnace explosions. These standards delineate the preferred sequence of starting fuel-burning equipment, the recommended minimum flame-monitoring equipment and safety interlocks, the recommended fuel-transport piping systems, the recommended purging procedures, and the procedures to be taken in the event of burner or furnace flame-out.

Normal operation also entails the maintenance of specified feedwater and boiler-water conditioning, designed steam and metal temperatures, clean gas passages and heat-absorbing surfaces, and fuel-ash removal.

The rate of firing during start-up is limited by the allowable metal temperatures in the superheater and reheater until steam flow through the turbine is established. Then, after the steam flow is established, the temperatures and the temperature differentials in the various parts of the boiler and turbine control the permissible rate of firing.

Emergency operations are usually the direct result of abnormal conditions such as the failure of the feedwater supply, the rupture of a pressure part, the interruption of the fuel supply, the loss of air, or a burner flame-out. Automatic safety interlocks usually are installed which trip the fuel supply and shut down the unit if these or other hazardous conditions are experienced. Abnormal operating conditions, which might become hazardous if allowed to persist, such as low (fuel-rich) or high (air-rich) air-fuel ratios or the failure of essential auxiliaries, require appropriate action to correct operating conditions. An operator who cannot correct an abnormal condition must determine whether operation can continue and, if not, must shut down the unit in the proper manner or activate the emergency trip through the automatic interlock system.

Inspection and maintenance should be performed during regularly scheduled outages. A list of the known items requiring repair or maintenance should be prepared before the outage and should be supplemented by any additional items noted in thorough inspections of the boiler and auxiliary equipment during the outage. A major item on the work list should be the maintenance of internal and external cleanliness. External cleaning is usually accomplished by water washing or air lancing. The internal surfaces of small boiler units are usually mechanically cleaned, but the large-capacity units are generally chemically cleaned. Under competent supervision, which can be obtained from several firms specializing in chemical cleaning, this method can be used with complete confidence for boilers of all sizes. The chemical-cleaning solution normally is composed of a 3 to 5 percent solution of hydrochloric acid, wetting and complexing agents for the removal of silica or other hard-to-remove deposits (such as iron and copper oxides), and a suitable inhibitor to prevent excessive chemical attack on the pressure parts of the boiler unit. Hydrochloric acid, however, should not be used for cleaning stainless-steel surfaces, since it can cause stress-corrosion cracking. Thus other organic or inorganic acids are used depending upon the type of material to be cleaned and the composition of the deposit to be removed.

The chemical cleaning of drum-type boilers involves filling the boiler [which has previously been uniformly heated to a temperature of about 175°F (79°C)] with the cleaning solution at 150 to 160°F (65 to 71°C) and allowing it to soak for 6 to 8 h or until samples of the solution show no appreciable further reduction in acid strength. The boiler should never be fired while it contains an acid solution, and open lights or other ignition sources must be prohibited in the area to avoid the ignition of the explosive gases, usually hydrogen, evolved during the cleaning operation. The unit is then drained and flushed several times, preferably under a nitrogen blanket, with neutral or slightly acidic water to remove the loosened deposits and to displace the acid solution and any corrosive gases. The flushing is followed by boil-out with an alkaline solution to neutralize any residual acid and to passivate the surfaces. The unit is then flushed with clean water to remove the remaining loose deposits, and it is inspected before being returned to service.

When chemically cleaning forced-flow once-through units, the solvent is continuously circulated through the unit for 4 to 6 h. Generally, the solvent is an inhibited solution of hydroxyacetic-formic acids at a temperature of 200°F (93°C).

Boiler units removed from service for long periods of time may be stored wet or dry. It is practically impossible to drain and dry modern high-pressure utility boilers completely with their complex furnace and superheater circuitry. Thus wet lay-up of the unit with water treated with 10 ppm ammonia and 200 ppm hydrazine is the best means of protection for both short- and long-term lay-up. If, however, dry storage is utilized, the system must be kept dry; and low humidity within the pressure parts and setting can be maintained by the use of trays of moisture-absorbing materials, such as silica gel or lime, which must be replenished at intervals to retain their effectiveness. When using either wet or dry storage, the system should be pressurized to a few pounds above atmospheric pressure with nitrogen gas.

## CODES

The ASME Boiler and Pressure Vessel Code, initiated in 1914 and supplemented by continuing revisions, contains the basic rules for the safe design, the construction, and the materials for steam-generating units. Its legal status depends upon its adoption by state or municipal authority. The Code is administered by the National Board of Boiler and Pressure Vessel Inspectors. This organization also has established the "Recommended Rules for Repairs by Fusion Welding to Power Boilers and Unfired Pressure Vessels." The adoption of both codes has been widespread, and they form the basis for the pertinent legal requirements in all but a few localities throughout the country. The National Bureau of Casualty and Surety Underwriters' book titled "Synopsis of Boiler Rules and Regulations" lists the states and the communities having laws which govern the installation and the operation of steam boilers.

## NUCLEAR BOILERS

(See also p. 9-119 et seq.)

Nuclear-power boilers are usually identified by the primary fluid used as the reactor coolant. A number of coolants and system-design concepts have been studied, but only three basic coolants have been used in U.S. power plants—liquid metals, gases, and water.

Most of the nuclear systems for commercial power production use water in some form as the primary coolant, principally the pressurized-water-reactor (PWR) and the boiling-water-reactor (BWR) systems. The preference for water as the reactor coolant is due to the fact that its physical, chemical, and thermodynamic characteristics are well known and materials and equipment are available for its handling and containment.

The steam-generating unit shown in Fig. 43 is typical of those used in PWR installations. In this design, hot primary

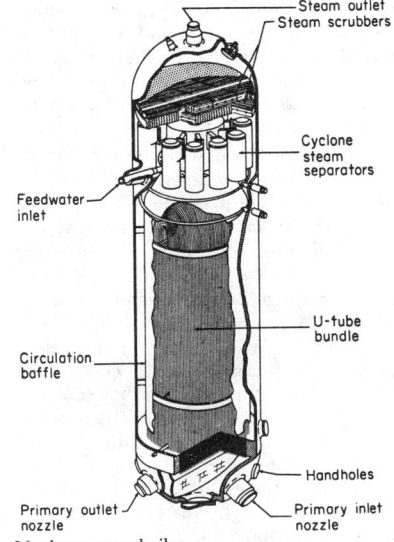

**Fig. 43**  Nuclear-power boiler.

fluid enters one side of the divided primary head, passes through the U-type tubes, and leaves through the primary outlet nozzle. Boiling takes place on the outside, or secondary side, of the tubes, and the steam-water mixture passes upward through the riser section and then the steam-and-water separators. The steam is discharged from the scrubbers into the outlet connection. The separated water flows downward, mixes with the incoming feedwater, and is then circulated downward through the annulus around the tube bundle to reenter the tube bundle at the bottom.

Generally, in units of this type, stainless-steel or Inconel tubes are used to minimize corrosion, the structural components are of carbon or low-alloy steel, and the primary head and primary tube-sheet faces are completely clad with stainless steel or Inconel.

Design considerations are similar to those for fossil-fuel boiler units, but particular emphasis is placed upon possible hazards, codes and specifications, and system economics.

1. **Hazards.** In the design of nuclear systems, great consideration must be given to the damage and the loss of life that could result from an accident, and also the psychological effect of such an accident. Potential dangers that must be considered by the boiler designer include the following.

**Radioactivity.** The primary fluid and any materials transported in the primary passages may become radioactive from neutron bombardment. This radioactivity can hold at a high level for varying lengths of time, and since the radioactive products may be deposited anywhere in the primary system, the boiler designer must design the equipment for minimum exposure of personnel to radiation.

**Chemical Poisons.** Some fission products are lethal, not only because they are radioactive but because they are chemically poisonous by both ingestion and inhalation. Personnel must be protected against exposure to these poisons.

**Chemical Reactions.** In some nuclear systems, possible contact between the primary and secondary fluids could cause strong chemical reactions. Sodium and water, for example, react violently, yet sodium has been selected as the primary fluid in some systems because of its excellent nuclear properties and good heat-transfer and heat-transport characteristics. Therefore, steam generators in which sodium is used for the primary fluid must be designed so that direct contact between the sodium and the water is unlikely and so that the effects of such a contact, if one should occur, are minimized.

2. **Codes and Specifications.** The requirements for safe design and fabrication are delineated in the ASME Boiler and Pressure Vessel Code, augmented by Special Code Cases. The special cases applying specifically to nuclear components establish design, inspection, and fabrication rules which are more stringent than those usually required for other types of equipment. The construction of commercial nuclear components is governed by the ASME Code, Section III, "Nuclear Power Plant Components" while that of equipment for military use is governed by special military specifications.

3. **Economics.** Basic design considerations of PWR and the BWR systems differ greatly from those of fossil-fuel systems. The temperature difference that produces the flow or transfer of heat between the primary and secondary water is dependent upon the difference in the pressures of the two fluids, and the hotter primary fluid is maintained at the higher pressure. Therefore, an optimized steam-generator design using a minimum amount of heat-absorbing surface should have the boiling secondary fluid on the outside of the tubes since a tube, or cylinder, can withstand greater internal pressures. This contrasts with fossil-fuel-fired water-tube boiler designs in which the boiling fluid is contained within the tubes.

Both internal and external heat-transfer coefficients are high in nuclear steam generators, and consequently, in most cases the tube metal and the fouling film coefficients control the overall rate of heat transfer. Thus tube diameters should be small and tube walls as thin as practical. Further, the water must be conditioned to minimize scale and sludge formations and, thus, fouling coefficients.

Primary-water systems are designed for relatively high pressures, 1,200 to 2,500 lb/in² (84 to 176 kgf/cm²), and extremely compact systems are economically justified in the efforts to minimize the size and the weight of the steam generator, reactor, and other pressure vessels. Fluid temperature differences are necessarily small when operating with the highest practical secondary-steam pressure; and this tends to increase surface requirements and, consequently, the size of the steam generator.

# STEAM ENGINES

## (Staff Contribution)

REFERENCES: Ewing, "The Steam Engine and Other Engines," Cambridge. Ripper, "Steam Engine Theory and Practice," Longmans. Heck, "The Steam Engine and Turbine," Van Nostrand. Allen, "Uniflow, Back Pressure, and Steam Extraction Engines," Pitman. Peabody, "Valve Gears for Steam Engines," Wiley. Zeuner, "Treatise on Valve Gears," Spon. Spangler, "Valve Gears," Wiley. Dalby, "Valves and Valve Gear Mechanisms," Longmans. Tubman, An Appreciation of the Triple Expansion Engine, CME, I, *Mech. Eng.*, Mar. 1974.

The reciprocating steam engine was the heart of the early industrial era. It dominated power generation for stationary and transportation service for more than a century until the development of the steam turbine and the internal-combustion engine. The mechanisms were numerous (see Patent Office listings), but practicality essentially standardized the positive-displacement, double-acting, piston and cylinder design in a vertical or horizontal configuration. These engines were heavy, cast-iron structures, e.g., 50 to 100 lb/hp; they had low piston speed (600 to 1,200 ft/min); long stroke (up to 6 ft); low turning speeds (50 to 500 r/min); steam conditions less than 300 lb/in² (dry saturated, or 100 to 200°F superheat); noncondensing or condensing (25 ± in Hg vacuum); sized from children's toys to 25,000 hp. Diversity of valve gear was an inventor's paradise with many suitable for reversible operation, as in locomotive, rolling-mill, and ship applications. Most of the machine elements known today, such as cylinders, piston rods, crossheads, connecting rods, crankshafts, flywheels, and governors, were developed in steam engines.

These engines utilize the expansive power of steam. Theoretically, the more the steam can be expanded in the engine cylinder, the better will be the economy. Practical losses, which occur in every steam engine, limit the expansion ratio and result in a minimum steam rate for a definite degree of expansion. Cylinder dimensions preclude the practical utilization of high-vacuum conditions because of the high specific volume (e.g., 339 ft³/lb dry and saturated at 2 inHg abs). The steam turbine can effectively utilize maximum vacuum.

## WORK AND DIMENSIONS OF THE STEAM ENGINE

Thermodynamic principles define the limits of the conversion efficiency of heat into work (see Rankine and Carnot cycles, Sec. 4). In fact, the historical development of thermodynamic principles was primarily aimed in the nineteenth century at defining the thermal performance of steam engines and steam power plants. It can be said that the steam engine did more for thermodynamics than thermodynamics did for the steam engine. Steam tables and steam charts (e.g., temperature-entropy; enthalpy-entropy; and pressure-volume) are essential to evaluate theoretical and actual equipment performance.

The pressure-volume diagram, or indicator card, is of primary significance in both the design and operation of the reciprocating piston and cylinder mechanism—not only steam engines, but also internal-combustion engines, air compressors, and pumps. The utility of the p-v diagram is often lost in attempts to improve fluid-dynamic reciprocating mechanisms. The work and power are the consequence of the difference in pressure on the two sides of the piston, expressed as mean effective pressure (mep or $p_m$). This is applied to the cylinder dimensions and rotating speed in the **"plan" equation**

$$\text{hp} = \frac{p_m L a n}{33,000} \qquad (1)$$

where $p_m$ = mep, lb/in²
$L$ = stroke, ft
$a$ = effective piston area, in²
$n$ = number of cycles completed per min
33,000 = mechanical equivalent of horsepower, ft·lb/min, by definition

A typical steam-engine indicator card is shown in Fig. 1 identifying the established nomenclature for the events of the cycle. Superimposed is a theoretical diagram, with expansion, but without clearance or compression. The terminal pressure controls the steam, or water, rate. The term "back pressure" is used for pressures above the atmosphere, while "condenser pressure" is used when engines operate with negative exhaust pressure. The volume ratio $v_j/v_h = R$ is called the **ratio of expansion.**

The average ordinate, or **mean effective pressure** $p_m$ [applicable in Eq. (1)] for the ideal cycle diagram $ghjkl$ is

$$p_m = P[(1 + \log_e R)/R] - p \qquad (2)$$

The expansion phase $h$ to $j$ of the cycle is *logarithmic* where $pv$

= *constant*. It is not isothermal but reasonably approximates expansion (and compression) in actual engines where condensation and evaporation on cylinder walls, heads, pistons, and valves are cyclically recurrent.

The value of $R$ is commonly around 4 in simple engines. It should increase as $P$ increases and as $p$ decreases, usually between 3 and 5. It should be higher in jacketed than in unjacketed engines. The efficiency of the engine depends largely on the value chosen for $R$. This is dictated by service requirements, e.g., load variation, load fluctuation, engine governing type (cutoff vs. throttle), overload capacity, steam cost, and economics. Efficiency, however, is not the sole requirement for all installations. The high starting torque of a steam locomotive dictates a valve gear that allows full-stroke admission of steam with zero expansion, a rectangular indicator card with consequent maximum $p_m$.

Figure 1 shows that the theoretical card has a larger area than the actual card. The value of $p_m$ obtained from Eq. (2) must be multiplied by the **diagram factor** to obtain the actual $p_m$ under the assumed conditions. This factor may have a value between 0.7 and 0.95. The actual $p_m$ is obtained from the card drawn on the indicator drum under running conditions of the

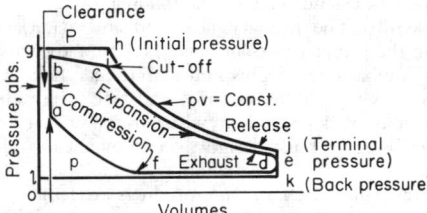

**Fig. 1** Typical steam-engine indicator diagram.

engine. The card area, graphically measured by a planimeter (see Sec. 16), is divided by the card length to get the average equivalent height of the card. The spring scale of the instrument, applied to this average height, gives the mean effective pressure actually obtaining within the engine cylinder. This value, when introduced in Eq. (1), gives the indicated horsepower (ihp) of the engine.

**Losses in steam-engine cylinders** are (1) incomplete expansion; (2) initial condensation; (3) throttling, affected by valve and port area; and (4) radiation, which can be considered constant. The point of best steam economy occurs with the $p_m$ at which the total of all losses is a minimum.

### Mechanical Efficiency and Shaft Output

$$\text{Brake horsepower (bhp)} = \text{ihp} \times \text{mechanical efficiency} \qquad (3)$$
$$\text{Friction horsepower} = \text{ihp} - \text{bhp} \qquad (4)$$

Friction horsepower is substantially constant over the load range. Mechanical efficiency may be calculated for the whole load range if the friction horsepower is known for one load point. Representative values for mechanical efficiency (at full load) vary from 0.80 to 0.94:

| bhp | 25 | 50 | 75 | 100 |
|---|---|---|---|---|
| Friction hp | 6 | 6 | 6 | 6 |
| ihp | 31 | 56 | 81 | 106 |
| Mechanical efficiency, % | 81 | 89.5 | 92.5 | 94 |

Figure 2 shows the effect of mechanical efficiency and

generator efficiency on the steam rate of an engine-generator set, both as to relative magnitude and as to location of the minimum values.

**Engine economy** may be improved by a number of methods: (1) **separation of inlet and outlet ports;** (2) **steam jackets,** applied to cylinders and heads to keep surfaces hot and dry; (3) **multiple expansion.** Condensation losses are related to the temperature

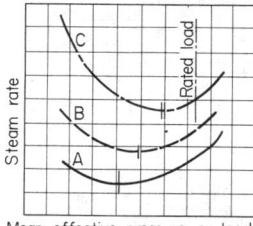

**Fig. 2**   Engine steam-rate curves. *(A)* At the steam cylinder, lb/ihp·h; *(B)* at the engine shaft, lb/bhp·h; *(C)* at the generator terminals, lb/kWh.

difference existing in the cylinder. Cylinders in series reduce this temperature difference and allow more complete overall expansion of the steam. Two, three, and four cylinders in series, as in **compound-, triple-,** and **quadruple-expansion** engines were common constructions, but the successive improvement is smaller for each additional stage. (4) **Superheating** gives the vapor the properties of a gas, reduces cylinder condensation, and necessitates decisive changes in engine design. The overall improvement in performance and water rate is so substantial that superheat is prevalent in practice. (5) The **uniflow** arrangement was the last great improvement in design (Figs. 3 and 4).

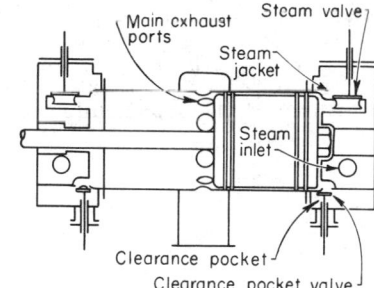

**Fig. 3**   Condensing uniflow-engine cylinder, with clearance pockets.

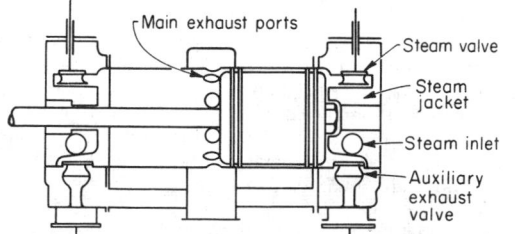

**Fig. 4**   Noncondensing uniflow-engine cylinder, with auxiliary exhaust valves.

Its high economy results from the high temperature of the residual steam at the end of compression. This temperature, aided by jackets, is higher than the live steam temperature, so

that initial condensation is reduced to a negligible amount. For **condensing operation** the engines are built with steam valves only (two-valve type), Fig. 3. Clearance pockets are provided, with either hand-operated or automatic valves to permit operation with atmospheric exhaust or against back pressure. **Noncondensing uniflow** engines have, in addition, auxiliary exhaust valves to reduce the otherwise large clearance required (four-valve type). If back pressure is variable, exhaust-valve *gears* are designed to change the length of compression while the engine is in operation (Fig. (4).

**Engine Steam (Water) Rates**   The efficiency of steam engines has generally been expressed in terms of pounds of steam per horsepower-hour. The term water rate was frequently used because of the convenient accuracy in weighing liquid water in the condensing plant. Figure 5 shows, on a percentage basis, two types of performance curves, one in lb/hph and the other in lb/h. The latter is identified as the "total

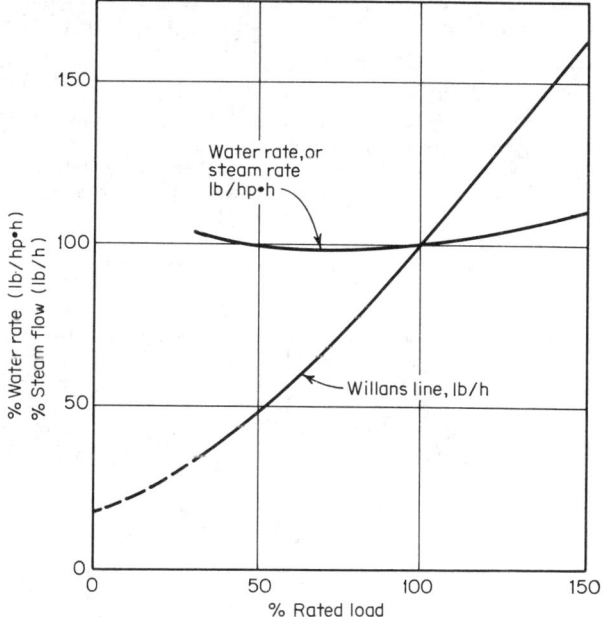

**Fig. 5**   Actual steam (water) rate (lb/hph) and Willans line (lb/h) for a noncondensing uniflow engine, percentage basis.

steam" or **Willans line** and is straight for an engine with fixed cutoff and variable initial pressure. Figure 6 reflects the impact of steam pressure and superheat on the steam consumption of a condensing uniflow engine.

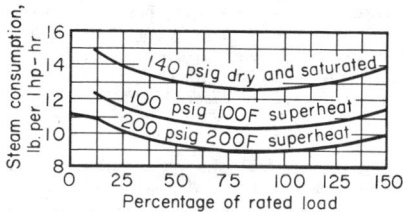

**Fig. 6**   Steam consumption of uniflow engine with 27.5 in vacuum.

The **Rankine cycle** (see Sec. 4) is the accepted thermodynamic standard for comparing the performance of steam prime movers (engines and turbines). It is predicated on complete isentropic (reversible adiabatic) expansion of the steam from initial to back pressure. It is shown on the $p$-$v$ basis in Fig. 7.

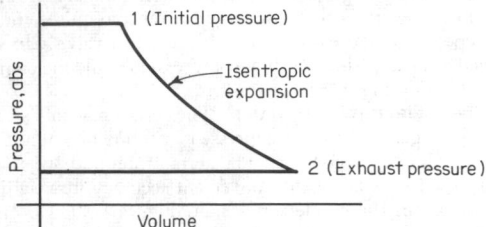

**Fig. 7** Rankine cycle, $p$-$v$ diagram.

There is no compression or clearance. The water rate of this cycle is most conveniently calculated by use of the Mollier chart (Sec. 4) where

$$\text{Rankine steam rate} = 2{,}545/(h_1 - h_2) \quad \text{lb/hph} \quad (5)$$

$h_1$ and $h_2$ are the initial and exhaust enthalpies, Btu/lb (constant entropy). The lowest point of the actual steam-rate curve (Fig. 6) is usually referred to as the Rankine rate, and the ratio of Rankine rate to actual rate is the **Rankine efficiency ratio (RER)**. This ratio may vary from 0.5 to 0.9, depending on the type of engine.

When clearance, compression, and incomplete expansion are introduced, the methods of Sec. 4 should be used to evaluate steam rate. This involves steam tables and charts to find the net work of the indicator card from the positive and negative areas of the several component phases. Figure 8 shows a theoretical steam-rate curve, computed by such methods, together with the actual test curve for the engine.

**Engine Details** Valves and valve-gear types range from the simplest D slide to gridiron, double slide (Meyer or Rider), rocking, piston, releasing and nonreleasing Corliss, and poppet.

**Volumetric clearance** should be made as small as possible. Slide and piston valves bring clearance volumes to 12 to 15 percent, Corliss valves 6 to 10 percent, and poppet valves 4 to 8 percent.

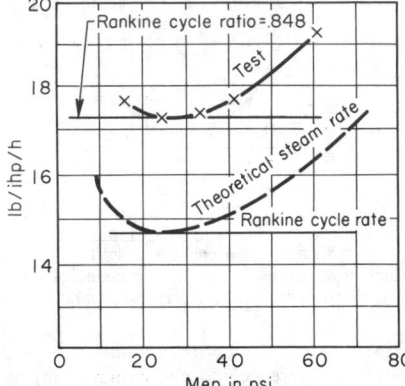

**Fig. 8** Theoretical and test steam-rate curves. Uniflow engine 20 × 24 at 200 r/min; throttle conditions 150 lb/in² saturated; exhaust to atmosphere.

**Valve and Port Sizes** Flow area of valves and cross section of ports are usually determined by port area = $AS/C$ in², where $A$ is net piston area, in², $S$ is mean piston speed, ft/min, and $C$ is a constant, ft/min. Values of $C$ are approximately 9,000 to 15,000 ft/min for inlet and 6,000 to 7,000 ft/min for outlet. Small valves and ports represent lost work.

**Superheated Steam** With high-temperature steam, the cylinder and parts design must allow for free expansion. Poppet and piston-valve cylinders can easily meet these requirements. The orthodox type of Corliss cylinder, however, is not suitable for highly superheated steam.

# STEAM TURBINES

## by Frederick G. Baily

REFERENCES: Stodola, "Steam and Gas Turbines," trans. by L. C. Loewenstein, McGraw-Hill. Bartlett, "Steam Turbine Performance and Economics," McGraw-Hill, 1958. Warren, Development of Steam Turbines for Main Propulsion of High-Powered Combatant Ships, *Trans. Soc. Naval Architects Marine Engrs.*, 1946. Newman, Modern Extraction Turbines, *Power Plant Eng.*, Jan.–Apr., 1945. Salisbury, "Steam Turbines and Their Cycles," Wiley. Campbell and Heckman, Tangential Vibration of Steam Turbine Buckets, *Trans. ASME*, **47**, 1925, pp. 643–671. Campbell, Protection of Steam Turbine Disk Wheels from Axial Vibration, *Trans. ASME*, **46**, 1924, pp. 31–139. Deák and Baird, A Procedure for Calculating the Packet Frequencies of Steam Turbine Exhaust Blades, *Trans. ASME*, **85**, series A, Oct. 1963.

Steam turbines have established a wide usefulness as prime movers, and are manufactured in many different forms and arrangements. They are used to drive many different types of apparatus, e.g., electric generators, pumps, compressors, and for driving ship propellers, through suitable gears. When designed for variable-speed operation, a turbine may be operated over a considerable speed range, which may be of advantage in many applications. Steam turbines range in output capacity from a few horsepower to well over 1,000 MW. The largest ones are used for generator drive in central power stations.

Turbines are classified descriptively in various ways:

**1. By steam supply and exhaust conditions**, e.g., condensing, noncondensing, automatic extraction, mixed pressure (in which steam is supplied from more than one source at more than one pressure), regenerative extraction, reheat.

**2. By casing or shaft arrangement**, e.g., single casing, tandem compound (two or more casings with the shaft coupled together in line), cross compound (two or more shafts not in line, often at different r/min).

**3. By number of exhaust stages in parallel** as regards steam flow, e.g., double flow, triple flow.

**4. By details of stage design,** e.g., **impulse** or **reaction.**

**5. By direction of steam flow** in the turbine, e.g., axial flow, radial flow, tangential flow. In this country, radial-flow steam turbines have not been used; there are quite a few such machines abroad. Axial-flow units predominate; some small turbines in this country operate on the tangential-flow principle.

**6. Whether single-stage or multistage.** Small turbines, or those designed for small energy drop, may have only one stage; larger units are always multistage.

**7. By type of driven apparatus,** e.g., generator, mechanical, or ship drive.

**8. By nature of steam supply,** e.g., fossil-fuel-fired boiler, or light-water nuclear reactor.

Any particular turbine unit may be described under one or more of these classifications, e.g., a single-casing condensing regenerative-extraction fossil unit, or a tandem-compound three-casing four-flow steam-reheat nuclear unit.

**Turbine-Stage Design** A turbine stage consists of a **stationary set of blades,** often called **nozzles,** and a moving set adjacent thereto, called **buckets,** or **rotor blades.** These stationary and rotating blades act together to allow the steam flow to do work on the rotor, which can be transmitted to the **load** through the **shaft** on which the rotor assembly is carried. Classical turbine-stage design recognized two distinct designs of turbine stage, "impulse" and "reaction" (see classification 4 above). In the **impulse** stage, the total pressure drop for the stage is taken across the nozzles or stationary element, the flow through the buckets or rotor blades then being substantially at constant static pressure. This may be extended to include flow through an additional set of stationary "intermediate" blades and another row of buckets, or rotor blades (Curtis or two-row stages). See velocity diagrams (Figs. 1 and 2).

In the **reaction** stage, the total pressure drop assigned to the stage is divided equally between the stationary blades and the rotor blades, giving rise to a velocity diagram, as shown in Fig. 3. As can be seen, there arises a marked difference in the shapes of the rotor blades in the two classical designs; the impulse buckets do much more turning of the stream; the reaction-bucket shape is more nearly the same as the nozzle-blade shape.

Modern fluid-flow theory recognizes that only in rare cases can an axial-flow turbine stage be either pure impulse or pure reaction. The annulus following the nozzle exit is filled with steam flowing with a high tangential velocity, i.e., a vortex, confined between inner and outer boundaries, and for equilibrium to exist, there must be a gradient in static pressure from a lower-than-average value at the inner boundary to a higher-than-average value at the outer boundary. The amount of this depends upon the boundary **radius ratio,** $R_{outer}/R_{inner}$. Thus it can be seen that only for radius ratio near 1.0 (small-height blades) can it be said that any one pressure condition exists for the stage. All axial-flow turbine stages of larger radius ratios tend to be more nearly impulse at the inner diameter and more nearly reaction at the outer diameter.

Differences in basic mechanical construction of axial-flow turbine stages exist. Generally, the reaction type of turbine has continued as in the past with a "drum rotor" and stationary blades fixed in the casing, while the impulse type continues as a "diaphragm-and-wheel" construction. However, the mechanical construction bears no fixed relation to the degree of impulse or reaction adopted in the blading design. Designers employ the mechanical construction which they deem suitable for best reliability and efficiency.

The impulse element, when employed for the first expansion, permits nozzle-group control, i.e., steam admission to each group, opening and closing, successively, in response to load changes. This improves the efficiency at low loads through the reduction of the loss due to throttling. With reaction elements, a somewhat similar effect may be obtained by bypassing certain stages as the load increases. The same practice may also be employed in impulse elements, for the purpose of sustaining overload.

A greater enthalpy drop can be employed per stage with the impulse element, particularly in the case of multivelocity elements, thus reducing the number of stages in a turbine. This is of special importance in the first expansion if the nozzle chamber is not integral with the turbine casing, so that the

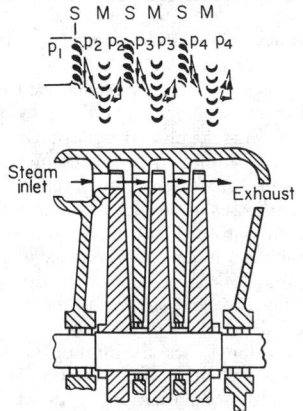

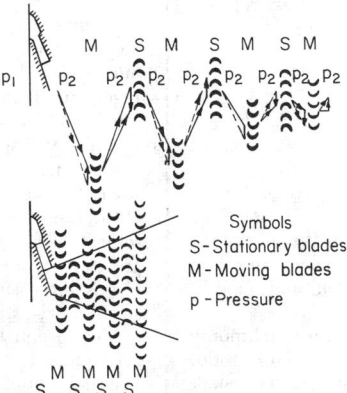

Symbols
S - Stationary blades
M - Moving blades
p - Pressure

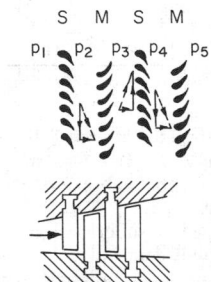

**Fig. 1** Impulse turbine with single-velocity stages.

**Fig. 2** Impulse turbine with multivelocity stages.

**Fig. 3** Reaction turbine.

**Figs. 1 to 3** Diagrammatic illustrations of turbine elements and corresponding bucket-velocity diagrams.

casing is not subjected to the initial steam conditions. If turbine elements of equal blade speed could have equal efficiency, one 2 row impulse element would equal four 1 row impulse elements or 16 rows (8 pairs) of reaction elements.

**General Advantages of Steam Turbines**  Compared with reciprocating engines steam turbines require less floor space, lighter foundations, and less attendance; have a lower lubricating-oil consumption, with no internal lubrication, the exhaust steam being free from oil; have no reciprocating masses with their resulting vibrations; have uniform torque; have no rubbing parts excepting the bearings; have great overload capacity, great reliability, low maintenance cost, and excellent regulation; are capable of operating with higher steam temperature and of expanding to lower exhaust pressure than the reciprocating steam engine. Their efficiencies may be as good as steam engines for small powers, and much better at large capacities. Single units can be built of greater capacity than can any other type of prime mover. Small turbines cost about the same as reciprocating engines; larger turbines cost much less than corresponding sizes of reciprocating engines, and they can be built in capacities never reached by reciprocating engines. Combustion-gas turbines possess many of the advantages of steam turbines but are not available in ratings much exceeding 100 MW.

## STEAM FLOW THROUGH NOZZLES AND BUCKETS IN IMPULSE TURBINES

**Nozzles**  For the general treatment of the flow of steam and for maximum weight of flow of saturated steam, see Sec. 4.

The **theoretical work** obtainable from the expansion of 1 lb of steam is equal to the enthalpy drop in isentropic expansion $h_1 - h_{s2}$, in Btu/lb, and the spouting velocity is 223.8 $\sqrt{h_1 - h_{s2}}$ ft/s (m/s = 44.7 $\sqrt{h'_1 - h'_{s2}}$, where $h'$ is in kJ/kg). The actual expansion is not isentropic but follows a path such as $h_1h_2$ on the enthalpy-entropy diagram (Fig. 4), and the available work becomes $h_1 - h_2$.

The **nozzle efficiency** is $(h_1 - h_2)/(h_1 - h_{s2})$.

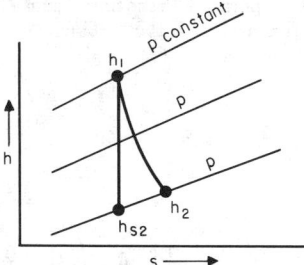

**Fig. 4**

The required **throat area** of the nozzle is $A_t = W v_t/V_t$ and the **mouth area** is $A_m = W v_m/V_m$, where $v$ is specific volume, $V$ is velocity, and the subscripts relate to throat and mouth, respectively.

In the case of a subcritical pressure ratio, throat and mouth conditions will coincide; with a supercritical pressure ratio, the mouth area will be larger than the throat area. For **nozzle velocity coefficients** based on tests, see Keenan and Kraft, *Trans. ASME*, **71**, 1949, pp. 773–787. The **bucket-velocity coefficient** is the ratio of the average exit velocity from the bucket divided

by the velocity equivalent of the total energy available to the bucket, i.e., the sum of inlet-velocity energy and pressure-drop energy. Typical values of tests on impulse buckets are shown in Fig. 5, when the bucket inlet angle is 3 to 5° larger than the exit angle. These are for subsonic flow. Reaction buckets can have the same coefficients as nozzles. For intermediate cases between pure impulse and pure reaction, interpolate between the values for these limiting cases.

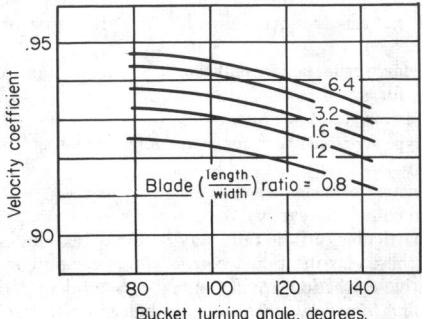

**Fig. 5**  Impulse bucket-velocity coefficients.

If, in the velocity diagram for a usual turbine stage, $V_1$ = actual nozzle exit velocity, $V_2$ = bucket relative entrance velocity, $V_3$ = bucket relative exit velocity, $V_4$ = absolute leaving velocity, $V_0$ = velocity corresponding to total stage available energy, then the **diagram efficiency** is $(V_1^2 - V_2^2 + V_3^2 - V_4^2)/V_0^2$.

Diagram efficiencies calculated with nozzle and bucket coefficients given above will be higher than efficiencies derived by turbine tests, because of losses not existing in stationary tests of nozzles and buckets.

For supersonic bucket velocities, the impulse bucket-velocity coefficient is lower by the factors in Table 1.

**Table 1. Impulse Bucket-Velocity Coefficient Factors**

| Mach No. | < | 1.0 | 1.2 | 1.3 | 1.5 | 1.75 | 2.0 |
|---|---|---|---|---|---|---|---|
| Factor | | 1.0 | 0.997 | 0.995 | 0.978 | 0.928 | 0.816 |

**Turbine-Stage Efficiency**  Single-row stages of short blade length have relatively lower efficiency, owing to inner and outer sidewall losses; stages with longer blades are therefore higher in efficiency. Figure 6a shows typical values of stage efficiency for single-row stages, plotted against the wheel-speed–steam-speed ratio, with pitch diam, inches/nozzle area, square inches, as a parameter. These curves reflect the net total of losses: (1) friction losses in nozzles (stationary blades); (2) friction losses in buckets (rotor blades); (3) rotation loss of rotor; (4) leakage loss between inner circumference of stationary element and rotor; (5) leakage loss between tip of rotor blades and casing; (6) moisture and supersaturation losses, if steam is wet (not included in curves of Fig. 6a).

Nozzle and bucket friction losses are minimized by good aerodynamic design and by increasing **aspect ratio** (blade length/steam passage width). Rotation losses depend upon disk or rotor dimensions and surrounding stationary parts. Exact values for rotation loss depend on several factors; the following formulas may be relied upon for all usual purposes:

$$L_d = 0.042D^2w \times (U/100)^{2.9} = 1.4 \times 10^{-12}(D')^2w'(U')^{2.9}$$
$$L_b = 0.187Dwh^{1.25} \times (U/100)^{2.9} = 3.34$$
$$\times 10^{-11}D'w'(h')^{1.25}(U')^{2.9}$$

where $L_d$ = rotation loss of disk carrying buckets, kW
$L_b$ = rotation loss of one row of buckets, kW
$U$ = wheel speed at pitch diameter, ft/s; $U'$, m/s
$D$ = pitch diameter (at center line of nozzle), ft; $D'$, mm
$w$ = density of steam, lb/ft³; $w'$, kg/m³
$h$ = mean bucket height, in; $h'$, mm

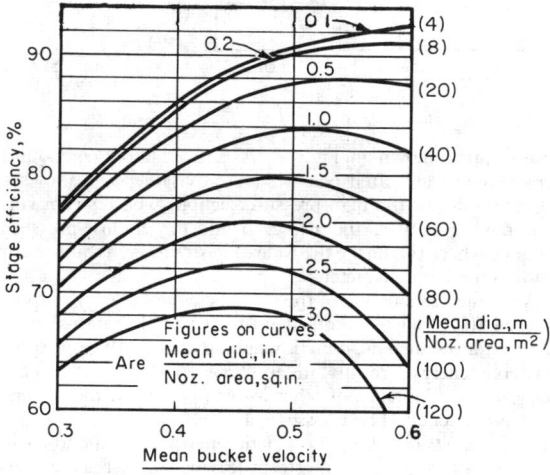

**Fig. 6a** Turbine single-row stage efficiency.

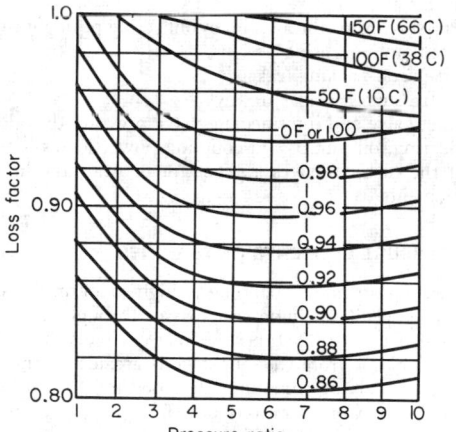

**Fig. 6b** Supersaturation and moisture loss.

$L_b$ must be figured for each row of buckets. $L_d$ plus the sum of the values of $L_b$ gives the total rotation loss in kilowatts for dry saturated steam. The formula for $L_b$ is approximate.

**Leakage loss** of steam between inner circumference of stationary element and rotor is minimized by maintaining minimum practical clearance and by use of labyrinth packings (see Figs. 12 and 13 and accompanying text). Leakage loss of steam between tip of rotor blades and casing is similar to that

through labyrinths between shaft and stationary parts; the magnitude depends upon the clearance area and the amount of reaction; other things being equal, the larger the percentage of reaction, the larger the leakage (see Fig. 7). Thus designers often employ considerable amounts of reaction in stages with long blades where such losses are small; this improves the net efficiency. In stages with short blades, the best net efficiency obtains with near impulse design. The curves in Fig. 6a reflect the effects of these practices.

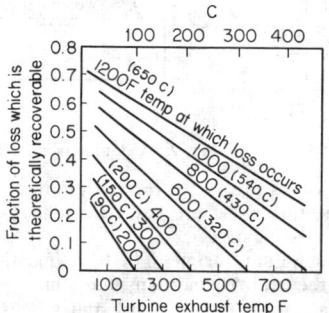

**Fig. 6c** Energy-regain chart.

The presence of moisture in the steam causes extra losses. These are probably mainly due to three factors:

1. Effect of supersaturation; that is, the steam in expanding rapidly does not remain in equilibrium but tends to be more or less supercooled; thus less than theoretical equilibrium energy is available.

2. The presence of water drops increases friction losses in the steam itself.

3. Water drops tend to move more slowly than the vapor; they strike the rotor blades at unfavorable velocities and exert a braking effect.

Figure 6b gives correction factors which may be applied to the values from Fig. 6a to arrive at stage efficiencies in the "wet" region. Curves are identified by initial superheat, °F, or by initial quality, percent.

**Two- or three-row stages,** with one set of nozzles, have lower basic efficiency than single-row stages; they are useful in smaller turbines, and for the first, or governing, stage in medium units. They are generally no longer employed in large central-station designs. Approximate relative efficiency level of these stages is shown in Fig. 7.

The losses occurring in a turbine stage are partially recoverable in succeeding stages in a multistage turbine because the energy available to succeeding stages is increased above that resulting from isentropic expansion. The amount of this factor depends upon the temperature at which the loss takes place, and the turbine exhaust temperature. Values for the reheat or energy-regain factor are shown in Fig. 6c.

**Turbine Steam-Flow Requirements** The steam flow required by a turbine is related to its power output and steam conditions by the following expressions:

Flow, lb/h = (TSR/efficiency) × power output, hp or kW

where TSR = theoretical steam rate, lb/hph = 2,545/available energy, Btu/lb, or, TSR, lb/kWh = 3,412.14/available energy, Btu/lb.

Values of TSR are given in tables or may be calculated. In

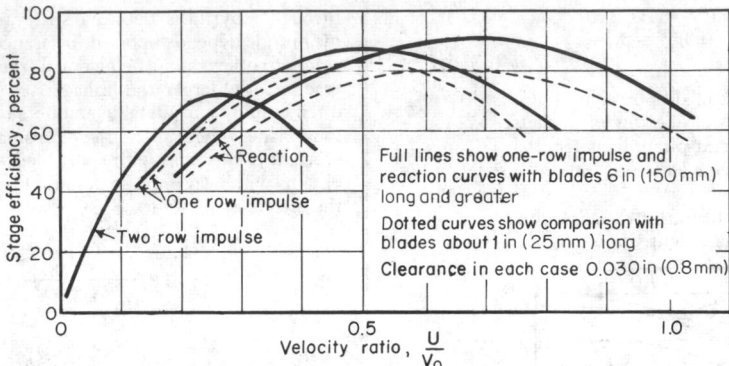

**Fig. 7** Variation of the efficiency of turbine elements with velocity ratio.

case of mixed-pressure or extraction turbines, the various sections of the turbine where flows are added or subtracted must be treated separately.

**Turbine Steam-Path Design** The basic quantities required for this are the steam conditions (i.e., inlet pressure and temperature and exhaust pressure), the required flow (see above), and the turbine r/min. The latter is often fixed by the requirements of the driven machine; if not, the choice of r/min by the designer is based upon experience, and factors such as space or weight limitations, efficiency requirements, stress limits, or required exhaust area. Often several preliminary layouts are needed to arrive at the best design. If it appears that a single stage will suffice, the problem is simple, since the entire available energy is allotted to the one stage. If a multistage machine is required, the total available energy must be divided properly between the various stages.

For a given wheel-speed–steam-speed ratio, the energy to be allotted to the stage is directly proportional to the square of the product (r/min $\times$ D), D being the rotor mean diameter. If available energy AE is in Btu/lb, D, in, and velocity ratio, 0.50, then

$$AE = (\text{r/min} \times D)^2 / 6.57 \times 10^8$$

The sum of the AE's per stage for all the stages must equal the total energy on the turbine, which is greater than the isentropic available energy by the amount of the reheat factor.

Having determined the energy to be assigned to each stage, the steam pressure in each stage is fixed, and this, together with the enthalpy determined from the efficiency of the machine from inlet, determines the steam specific volume. The velocity ratio and the degree of reaction decided upon fix the velocity diagram, from which the velocities through the stationary and rotating blades are determined. With the flow Q, lb/h, the velocity V, ft/s, and the volume v, ft³/lb, determined, theoretical areas required are given by

$$A = Qv/25V \quad \text{in}^2$$

The actual area required will be larger than theoretical because of friction losses; a value of 0.98 for nozzle flow coefficient is reasonable.

There is no fixed criterion for the number of stages to be used in a steam-turbine design, and experienced designers differ in the number they will choose for any particular design. The stage velocity ratio, the mean diameter, and often the r/min are subject to judgment, bearing in mind the general

relationships shown in Fig. 6a, b, and c. Cross-compound arrangements are often possible and advisable; this allows a higher r/min for the high-pressure section, where steam volume flow is smaller, and a lower r/min for the low-pressure section, where the final exhaust area desired may require long blades on a large diameter.

A detailed calculation of the efficiencies and energy outputs of each stage can be summed up to the total "internal used energy" of the turbine, which, when divided by the isentropic available energy, results in an **"internal efficiency."** Then, account must be taken of other losses to arrive at the turbine overall efficiency. These losses are

1. Exhaust loss, i.e., the kinetic energy corresponding to the absolute velocity of the steam leaving the last stage, plus the pressure drop through the exhaust connection to the turbine outlet flange, where exhaust pressure is by custom measured as a static pressure
2. Pressure drops through interconnecting piping if turbine has more than one casing
3. Shaft end-packing leakages
4. Valve-stem leakages, if any
5. Inlet-valve and intermediate-valve pressure-drop losses
6. Bearing, oil-pump, and coupling power losses
7. If the turbine drives a generator or gear, the losses of these elements

## LOW-PRESSURE ELEMENTS OF TURBINES

In condensing turbines expanding to high vacuum, the steam increases in specific volume as it passes through the stages, exhausting at about 1,000 times the inlet volume. The increase in specific volume from stage to stage is greatest in the latter few stages, which are commonly designed for pressure ratios of about 2 : 1. Considerations of efficiency and economics dictate providing reasonably low interstage steam velocity in these stages, and reasonable leaving loss from the last-stage blades. These requirements are satisfied by providing a steam path whose cross-sectional area increases in proportion to the specific volume. The area increase may be achieved by increasing blading length, by increasing the mean diameter of the steam path, by arranging the last expansions in multiple flow, or by some combination of two or more. The relatively small single casing unit of Fig. 8 is provided with increasing area by the first two techniques.

The ratio of the last-stage blade height to the mean diameter

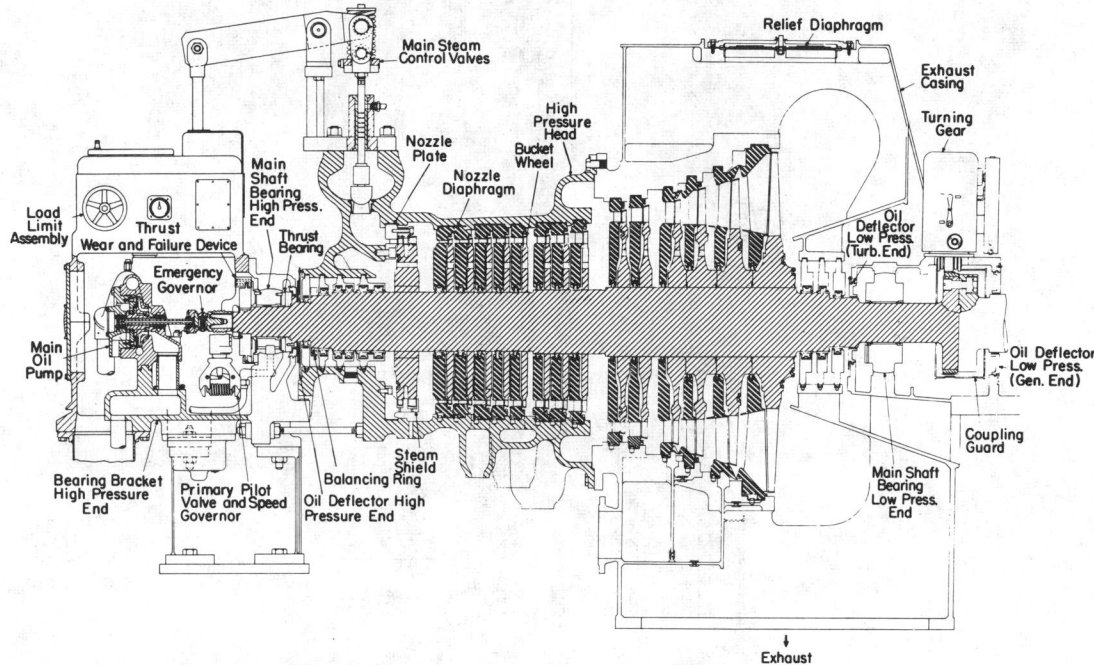

**Fig. 8**  Cross section of a multistage impulse condensing turbine rated 30,000 kW. (*General Electric Co.*)

of the steam path may be as high as 0.35 in the last stage of large central-station units. With such a ratio, there is a variation in blade velocity between the inner and the outer radii of the steam path that cannot be properly satisfied by any one steam velocity. Losses incidental to this may be minimized or eliminated by providing blades with warped surfaces, the sections at the inner radius partaking of impulse form with relatively small inlet angles, and those at the outer radius of reaction form with large inlet angles. The longest last-stage blades in service today have tips whose speed exceeds sonic velocity. Many are of transonic design employing subsonic profiles toward the inner radius, blending into supersonic profiles at the tip.

Among the methods of obtaining increased low-pressure blade areas through multiple flowing are:

1. A single-casing turbine with the last elements arranged for double flow.

2. The steam expansion divided between two casings, coupled in tandem, and driving a single main generator, the low-pressure casing being arranged for double flow. This arrangement is commonly extended to provide tandem-compound turbines with four and six exhaust ends. Figure 9 illustrates the application of a double-flow low-pressure casing to a unit of medium rating.

3. Cross-compound turbines, in which the steam expansion is divided between two or more separate casings driving separate generators, electrically synchronized. The low-pressure casings are usually double-flow and can be arranged so as to provide two, four, or six exhaust ends. This system permits the turbine elements to be operated at different speeds, selected as appropriate to the respective steam volumes. It lends itself to geared applications, such as marine-propelling

machinery, when two or more pinions of different diameters and speeds drive a single-output gear wheel.

4. Divided-flow turbines in which steam expands in a series of elements in a single flow to a point where the flow is divided, with, perhaps, one-third continuing expansion within the same casing to condenser pressure, the remaining two-thirds expanding to condenser pressure in a separate double-flow casing. This construction has been used to provide triple-flow exhaust ends.

The **leaving loss** at the exit from the last row of blades is $V_{c2}^2/50,100$ Btu/lb of steam, where $V_{c2}$ is the absolute terminal velocity in feet per second [or $(V'_{c2})^2/2,000$ kJ/kg, where $V'_{c2}$ is in m/s]. The presence of moisture in the steam causes **moisture loss.** The acceleration of moisture particles is less as the density of the steam decreases; hence the difference between the velocities of the particles and the steam increases. As indicated in Fig. 10, with steam velocity $V_{s1}$ and moisture velocity $V_m$ leaving the stationary blades and with bucket velocity $V_b$, the velocities relative to the moving blades of the steam are $V_{c1}$ and of the moisture $V_{m1}$. The component $V_{m2}$ of the moisture relative to the moving blades is opposite to the direction of their motion and is proportional to the force acting on the back of the blades, needed to accelerate the water to blade speed, and results in negative work. This negative work can be calculated when the weight of moisture per pound of steam and $V_m$ are known. The results of many tests indicate that the *efficiency of a stage is reduced about 1 percent for each 1 percent moisture present in the steam.*

The presence of moisture particles will result in erosion of the blades along their inlet edges if $V_{m1}$ is too large—unless the moisture content is very small. The rate of erosion is reduced by using protective shielding of hardened materials

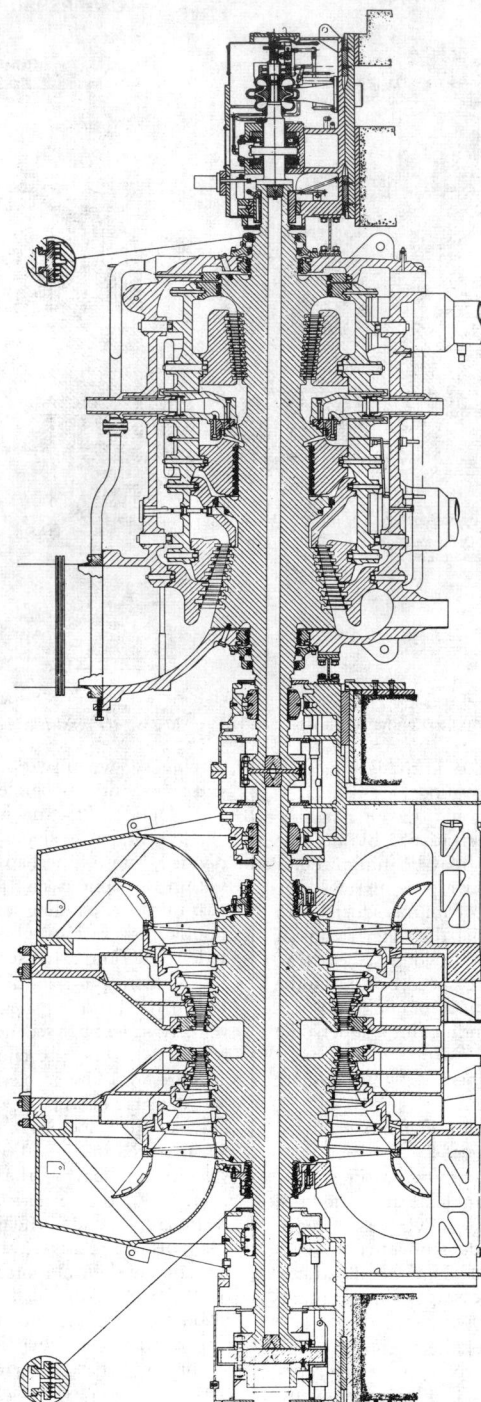

**Fig. 9** Cross section of a 160,000-kW tandem-compound double-flow reheat turbine. (*Westinghouse Electric Corporation.*)

along the inlet edge. Attached Stellite shields and thermally hardened edges are commonly employed. Protective materials and processes are carefully chosen so as to minimize the possibility of corrosion. Experience with high-alloy bucket steels indicates that a threshold velocity exists at about $V_b = 900$ ft/s (270 m/s), below which impact erosion does not normally occur. With Stellite shields, satisfactory service has been obtained with values of $V_b$ in excess of 1,900 ft/s (580 m/s). The severity of erosion penetration is dependent upon the thermodynamic properties of the stage and the effectiveness of reducing moisture content by means of interstage collection and drainage as well.

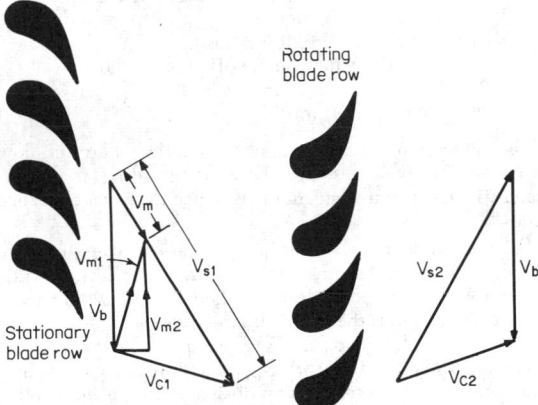

**Fig. 10**  Velocity of steam and of the moisture in steam.

**Rotative Speed**  The r/min selected greatly influences weight and cost. With two geometrically similar turbines, one having twice the linear dimensions of the other, the steam-path areas of the larger, and hence its capacity, would be four times and its weight eight times as great as those of the smaller. The weight per unit of capacity with similar machines increases inversely as the speed with strict geometrical similarity. For this reason, the highest possible low-pressure blade speeds and r/min are selected. Machines of different speeds are not usually made strictly geometrically similar, and the reduction of specific weight is not so rapid as the above rule would indicate. With large-capacity turbines, blade speeds, at the outer radius, have reached over 1,900 ft/s (580 m/s). With high speeds and small dimensions, the turbine can operate with higher steam temperature and greater temperature fluctuations because of lighter casing walls and less mass of rotor; the amount of distortion is less with more uniform heating; the turbine can be heated and put in service more quickly; space requirements are less; and dynamic loadings on foundations are less.

**Balancing**  (see Secs. 3 and 5)  With a rotor, or a component of a rotor, of relatively short axial length (such as a disk), static balancing may suffice. Single bodies of more than half the diameter in axial length are usually dynamically balanced by the use of balancing machines. The balancing may be done at less than the running speed, since a bladed turbine rotor cannot be rotated in air at a speed approaching the running speed, or at high speed in an evacuated spin facility, or with a combination of both. Balance at full speed may not be satisfactory unless the balance weights are applied at points diametri-

cally opposite the errors in balance, so that balance corrections must frequently be provided along the length of the rotor. Five balance planes are commonly used for the rotors of large central-station turbines, of which three are in the rotor body between bearings and two are in the overhung couplings.

Unsatisfactory operation of turbine rotors in service, resembling a simple unbalance, may be caused by nonuniform material or nonuniform heating of the rotor. The latter may be caused by permitting a rotor to remain stationary in a hot casing, or by packing rubbing which may apply frictional heating to the "high" side of the rotor, leading to further bowing into the rub. Care must be taken to see that nonuniformity of material which could cause rotor distortion with heat is avoided, and to avoid rubbing of the packings on the rotor. Turning gears are generally used to keep the rotor turning at low speed to maintain uniform temperature when the turbine is shut down and cooling, and for a prolonged period before starting.

## TURBINE BUCKETS, BLADING, AND PARTS

Figure 11 shows various blade fastenings. Blades are subject to vibration and possible fatigue fracture if their natural frequency is resonant with some applied vibration force. Forced vibrations may arise from the following causes (see also Sec. 5):

1. Variations in steam forces. The blade frequencies should not be even multiples of the running speed, nor should they be resonant with the frequency of passing nozzle partitions or exhaust-hood struts.

2. Shock, the result of blades being subjected to discontinuous steam flow, such as may be caused by incomplete peripheral steam admission or extraction.

3. Torsional vibrations of the rotor.

**Fig. 11**  Turbine-blade fastenings.

High-speed, low-pressure blades of condensing turbines are usually of tapering section and have a warped surface in order to provide appropriate blade angles throughout their length. Long blades of this type frequently have their natural frequency between three and four times the running speed or even lower. Such blades should always be specially tuned to have a margin in frequency away from running-speed stimuli.

Margins from running speed to assure freedom from fatigue due to resonant vibration within and transverse to the plane of the wheel are as follows:

| Frequency, cycles per revolution | 2 | 3 | 4 | 5 |
|---|---|---|---|---|
| Margin between critical and running speeds, %: | | | | |
| Within wheel plane (tangential) | 15 | 10 | 5 | 5 |
| Transverse to wheel plane (axial) | 20 | 10 | 10 | 5 |

Higher-frequency buckets whose frequencies cannot assuredly be made nonresonant should be designed with adequate strength to resist such stimuli as may occur under service conditions.

**Blade Materials**   The material in most general use is a low-carbon stainless steel of the following composition: Cr, 12 to 14 percent; C, 0.10 to 0.12; Mn, 0.08 max; P, 0.03 max; S, 0.05 max; Si, 0.25 max. Its physical characteristics in the heat-treated condition at room temperature may be tensile strength, 100,000 lb/in² (690 MPa); yield point, 80,000 lb/in² (550 MPa); elongation, 21 percent; reduction of area, 60 percent. For the higher-temperature blades, particularly on large machines, it is practice to use alloyed chrome steel to achieve the required strength and oxidation-erosion resistance (see also Sec. 6).

**Rotor Materials**   Since steam turbines operate at high speeds, rotor materials must be of very high integrity and of basically high strength. In addition, the material should be "tough" at the temperatures at which it is to be highly stressed. A measure of this toughness may be obtained by running Charpy notch impact tests at various temperatures (see Sec. 5). In large modern machines the rotor forgings are almost exclusively made of steel melted in basic electric furnaces and vacuum poured to achieve freedom from internal defects. Turbine-rotor forgings are usually made with small amounts of alloying elements such as Ni, Cr, Va, or Mo.

**Casing and Bolting Materials**   Casings are almost always made of castings in order to achieve the complicated shapes required by these components. The alloy compositions used are selected so as to provide good weldability and castability as well as good physical properties. Bolts are made of forged or rolled materials.

The practical use of higher steam pressures and temperatures is limited by the strength and cost of available materials.

**Leakage**   Metallic labyrinth packings are employed to (1) reduce internal steam leakage from stage to stage, (2) prevent steam from escaping the turbine from elevated-pressure ends, and (3) prevent air from leaking into the turbine at subatmospheric-pressure shaft ends. Interstage packings usually employ single rings with multiple teeth. End packings are arranged in multiple rings. At the high-pressure end, the leakage of steam past the first few rings may be carried to a lower-pressure stage of the turbine so that the outer rings need only prevent the leakage of low-pressure steam to atmosphere. The annulus above the last ring, or group of rings, is connected to a packing exhauster which maintains a pressure slightly below atmospheric. In consequence, no steam leaks out along the shaft, while a small amount of air is drawn past the final packing to the exhauster. Vacuum packings are provided with an inner annulus which is supplied with steam above atmospheric pressure. Steam flows inward from that annulus to supply the leakage toward the vacuum end, and outward toward a packing-exhauster connection. Thus steam is prevented from escaping along the shaft, while air is prevented from being drawn into the turbine.

Some turbines use carbon end packings or water seals. Carbon packings consist of one or more rings of pure carbon made in sections of 90 or 120° and held toward the shaft with small clearances by means of springs. The springs should have an axial component of force to hold the rings against the side of the box.

**Labyrinths** dependent upon radial clearances are shown in Fig. 12. In the design with "high-and-low" teeth, heavy teeth are cut on the turbine shaft, and thin teeth are part of a renewable packing ring which is made in segments backed and held inward by flat springs. The key indicated in the figure prevents turning of the segments. These types require that the rotor remain sensibly concentric with the stator but do not require a close axial adjustment.

Labyrinths dependent upon axial clearances are shown in Fig. 13. These require the maintenance of a close axial adjustment of the rotor.

The flow through a labyrinth may be approximately determined by the formula

$$W = 25KA \sqrt{\frac{(P_1/V_1)[1 - (P_2/P_1)^2]}{N - \ln(P_2/P_1)}}$$

where $W$ = mass flow of steam, lb/h
$\quad K$ = experimentally determined coefficient
$\quad A$ = area through packing clearance space, in²
$\quad P_1$ = initial pressure, lb/in² abs
$\quad V_1$ = initial specific volume of steam, ft³/lb
$\quad P_2$ = final pressure, lb/in² abs
$\quad N$ = number of throttlings

The value of $K$ for interlocking labyrinths where the flow velocity is effectively destroyed between throttlings is approximately 50 and is independent of clearance for usual clearance values.

For noninterlocking labyrinths, i.e., stationary teeth against a straight cylindrical shaft, the value of $K$ varies with the ratio of tooth spacing to radial clearance, being about 120 for tooth spacing of five times the radial clearance, reducing to approximately 50 for a tooth spacing fifty times the clearance.

**Turning Gears**   Large turbines are equipped with turning gears to rotate the rotors slowly during warming up, cooling off, and particularly during shutdown periods of several days when it may be necessary to start the turbine again at short notice. The object is to maintain the shaft or rotor at an approximately uniform temperature circumferentially, so as to maintain straightness and preserve the balance. Turning gears permit an appreciable reduction in starting time, particularly following a relatively short shutdown.

It is seldom necessary to use high-pressure oil to lift the journals off their bearings when using a turning gear. A low-pressure motor-driven oil pump is used which floods the bearings with about half their usual flow of oil. The turning gear is made powerful enough to start the rotor and rotate it at low speed.

**High-Temperature Bolting**   The bolting of high-pressure high-temperature joints, particularly turbine-shell or valve-bonnet joints, is very exacting. It is worthwhile to taper the threads of either the male or the female element so that the engagement of the threads throughout the length of the engaged threaded portion will give approximately uniform bearing. The reliability of taper-threaded bolts is superior to that of parallel-threaded bolts.

**Thrust bearings** must usually be designed to carry axial rotor thrust in either direction, with sufficient margin to take care of unusual operating conditions. Thrust runners may be machined solid on the shaft or can be a separate piece shrunk on and secured from endwise motion. The stationary bearing surfaces may be of the pivoted-shoe type, or made in solid plates with babbitt or other bearing-material facing, with grooves for oil supply and "lands" to carry the thrust load.

Axial thrust on the turbine rotor is caused by pressure and velocity differences across the rotor blades, pressure differences from one side to the other on wheels or rotor bodies, and

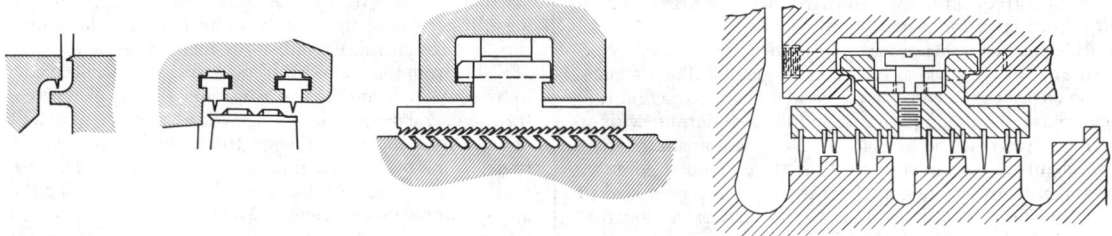

**Fig. 12**   Labyrinths with radial clearance.

pressure differences across shaft labyrinths which have steps in diameter. The net thrust is the sum of all these effects, some of which may be in one direction, and some in the opposite direction. Rotor-blade and wheel or body thrust are usually in the direction of steam flow. It is usual to balance this thrust either partially or completely by proper choice of shaft packing diameters and pressure differences so that the net thrust is not too large. The thrust bearing must be made large enough so that it is not overloaded by the net thrust. In this respect it is necessary to foresee all operating conditions which may influence the net thrust and allow for these; in addition some margin must be allowed for abnormal or unforeseen circumstances which may occur in service. (See also Sec. 8.)

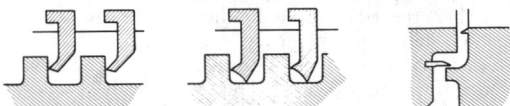

**Fig. 13**   Labyrinths with axial clearance.

**Governors**   Steam turbines are nearly always equipped with speed-control governors and with separate overspeed governors. The only exceptions are special cases where it is judged that the possibility of overspeed due to loss of load is exceedingly remote. The speed-control governor may be arranged for a wide range of speed setting, in the case of a variable-speed turbine. The steam flow-controlling valve or valves are operated by this governor, usually through a hydraulic relay mechanism. The overspeed governor is usually of an overisochronous type, arranged to trip at 10 percent over normal full speed (on some small turbines, 15 percent), actuating quick-closing stop valves to shut off the steam supply to the turbine. Speed-control governing systems are usually designed so that the overspeed-governing system is not brought into action on sudden loss of full load.

On automatic extraction machines, the speed governor must be correlated with the extraction-pressure controlling-valve system.

On fossil-reheat turbines, because of the large stored steam volume in the reheater and piping, and on nuclear turbines, because of the moisture separator/steam reheater and piping volumes, it is necessary to protect the turbine from overspeed on sudden loss of load by shutting off this stored steam ahead of the lower-pressure stages. It is done by intermediate intercept valves, actuated by a governor set slightly higher than the speed-controlling governor. An intermediate stop valve actuated by the overspeed trip is usually added in series for additional protection.

Speed-governing systems can be supplied of various sensitivities and speed ranges, to suit the requirements of the driven apparatus. Both mechanical-hydraulic and electrohydraulic systems are employed, with the latter representing the more recent practice for large units (see also Sec. 16).

## INDUSTRIAL AND AUXILIARY TURBINES

**Low-capacity turbines** are employed for services such as engine-room auxiliaries and small generating sets. Usually they comprise a single turbine element. Their efficiency may be less than that of a corresponding reciprocating engine, but they are employed because of their compactness and because they require no internal lubrication. The exhaust steam is free from oil and grease and is available for heating purposes. They are frequently coupled to the driven machine by means of speed-reducing gears. Turbines of this type are usually of the axial-flow type, but the **tangential helical-flow** turbine, in which the steam is directed tangentially and radially inward by nozzles against buckets milled in the wheel rim and made to flow in a helical path reentering the buckets one or more times, is also used. Such machines have been limited to small single-stage designs, and are very simple and rugged.

In **back-pressure turbines,** the exhaust steam is employed for some heating process, and the turbine work may be a by-product. If all the exhaust steam is condensed in heat-absorbing apparatus and returned to the system, the thermal efficiency of the system may be over 90 percent. One application is the **superposition** of a high-pressure system on lower-pressure power units, with the exhaust from the high-pressure turbine going to the low-pressure steam mains. By this device, an old power station can be rehabilitated and its capacity increased. Two methods of operation are in use: (1) with constant intermediate pressure as when the lower-pressure power units operate also with steam from existing lower-pressure boilers and (2) with variable intermediate pressure as when the low-pressure units receive steam only from the back-pressure turbine.

**Boiler-feed–pump-drive turbines** have been used extensively as part of the power-plant system, especially for large, high-pressure plants where the required feed-pump power may amount to 4 percent of the gross plant output, and for large nuclear units. The turbine and pump can be matched as to rotative speed. These turbines are variously integrated into the main cycle. The most common present practice is to use condensing, nonextracting turbines supplied with steam in the range of 150 to 200 lb/in² abs (1,000 to 1,400 kPa) taken from the exhaust of the intermediate sections of fossil turbines, or from the inlet of the low-pressure sections of nuclear units. These turbines normally have a connection to the main steam supply for starting and low-load operation. Similar auxiliary turbines are frequently used to drive the forced or induced-draft fans of large fossil-fuel-fired boilers.

With **extraction turbines,** partly expanded steam is extracted for external process use at one or more points. The turbines may be either condensing or noncondensing. Extraction turbines are usually designed to sustain full rated output, with or without extraction, and are provided with automatic regulating mechanisms to deliver steam from the extraction points at constant pressure, as long as there is sufficient power load to permit the necessary flow. The use of such extraction turbines, particularly with high initial pressures in connection with many industrial processes requiring moderate- or low-pressure steam, results frequently in a high efficiency of power production, i.e., the only heat required in such a plant over and above that to provide the required process steam is the heat equivalent of the power generated by the steam before extraction. This means that such power can be produced at nearly 100 percent thermal efficiency.

Figure 14 illustrates a typical double-automatic, condensing extraction turbine, providing two controlled extraction pressures. In this case, the unit is equipped with internal spool valves at both extraction points. Grid and poppet-type valves have been used for this purpose. The extraction-stage valves are under the control of an extraction-pressure-actuated governor; they determine the flow to the subsequent stages of the turbine and maintain the pressure in the extraction stage. The operation of the valves is by means of a diaphragm or piston controlling the admission of high-pressure steam or oil to an actuating cylinder, which, in turn, opens or closes the valves to the nozzle ports to the succeeding stage. Extractions of this kind are called pressure-controlled extractions, and the pressure is maintained practically constant over a wide range if the load is sufficient to permit the required steam flow.

**Mechanical-drive turbines** are commonly applied where moderate to high power and/or precise speed control of the driven machine are needed. Typical applications include the powering of papermaking machines and the driving of fluid compressors in petrochemical plants. So many sizes and types are available from the manufacturers, and they have been adapted to so many applications, that is is impossible to give here more than a general description. These turbines are commonly built in sizes from a few to several thousand horsepower. If the speed of the driven machine is low, a reduction gear may be used in order to reduce the size and cost of the driving turbine and to improve its efficiency. Mechanical-drive turbines have a wide range of application, being adaptable for any steam conditions and a wide variety of speeds. They can be equipped with speed governors suited to the requirements, i.e., of very good stability and accuracy, if this is desirable, and arranged for various constant-speed settings over a speed range as wide as 10:1.

**Main propulsion marine turbines** (see also Sec. 11) are basically the same as central-station or industrial turbines except that usually the turbine is divided into a high-pressure and a low-pressure element, each geared through a common low-speed gear to the propeller shaft. The advantages of this compound arrangement are that two high-speed pinions divide the load on a common low-speed gear, thus reducing gear weight as compared with a single turbine. The high-pressure turbine can be made higher-speed than the low-pressure turbine and so be better adapted to the low volume flow. Each turbine can have a short rugged shaft, and either turbine can be used to propel the ship in an emergency.

Geared marine turbines require a reversing element for operating the vessel astern. This is typically a two-stage impulse turbine with two 2-row, or one 2-row and one 1-row, velocity stages arranged in the exhaust space of the low-pressure ahead turbine, so as to operate under vacuum under normal ahead conditions. The rotation loss of such an astern turbine is about ½ percent under normal ahead conditions.

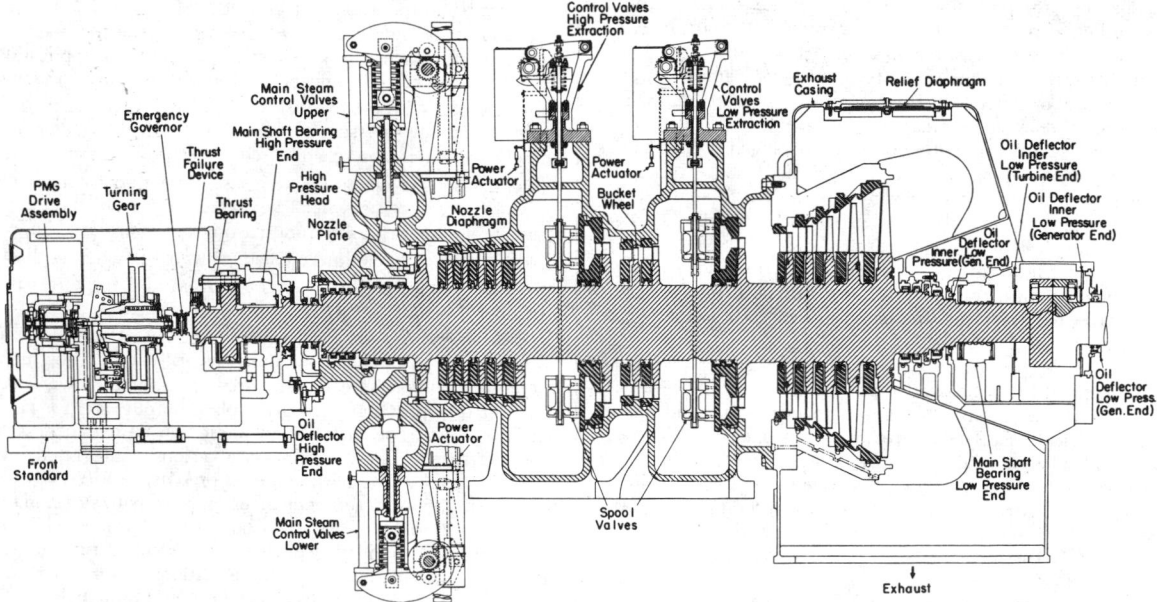

**Fig. 14**   Double-automatic condensing extracting turbine, 25,000 kW. *(General Electric Co.)*

Being directly geared to the propeller shaft, marine turbines must work at variable speeds. Overspeed governors are not required but are sometimes applied as a precautionary measure. Control is effected in most cases by sequentially operated nozzle valves. A typical marine turbine rated 16,000 hp is shown in Fig. 15. Ratings of 70,000 hp have been built, and larger sizes are realistic.

## LARGE CENTRAL-STATION TURBINES

Figures 16 and 17 show examples of large central-station turbines as built by two manufacturers.

Figure 16 illustrates a 3,600-r/min tandem-compound four-flow unit rated 500 MW. It is a single-reheat fossil unit for nominal inlet-steam conditions of 2,400 lb/in² gage (16,650 kPa) 1000°F (538°C), reheat to 1000°F (538°C). The right-hand casing is a combined high-pressure and reheat section. Steam flows from the left center to the right through the impulse-type governing stage, then reverses, flowing to the left through the nine reaction-type high-pressure stages, and exhausts from the casing to the reheat section of the boiler. Reheated steam reenters that casing at its right center, flowing

to the right through five reheat stages, turning once more to flow to the left between the inner and outer casings, finally exhausting up and to the left to the two double-flow low-pressure casings on the left end of the unit. The inlet stop and control valves, the reheat stop and intercepter valves, and the generator are not shown.

Four-flow units of this general type employ last-stage rotor blades from 25 to 35 in (600 to 900 mm) in length and in ratings up to 600 MW. Significantly higher ratings require dividing the functions of the combined casing into a separate single-flow high-pressure casing and a separate two-flow reheat casing, for a total of four. Ratings much in excess of 800 MW exceed the flow capacity of a four-flow low-pressure turbine, necessitating the addition of a third double-flow element, for a total of five casings. Tandem-compound six-flow 3,600 r/min units are commonly offered for ratings up to about 1,100 MW. Somewhat larger ratings can be accommodated by 3,600/3,600- or 3,600/1,800-r/min cross-compound units, but economic considerations makes their present application rather infrequent.

Figure 17 illustrates an 1,800-r/min tandem-compound six-flow turbine designed for steam from light-water nuclear reac-

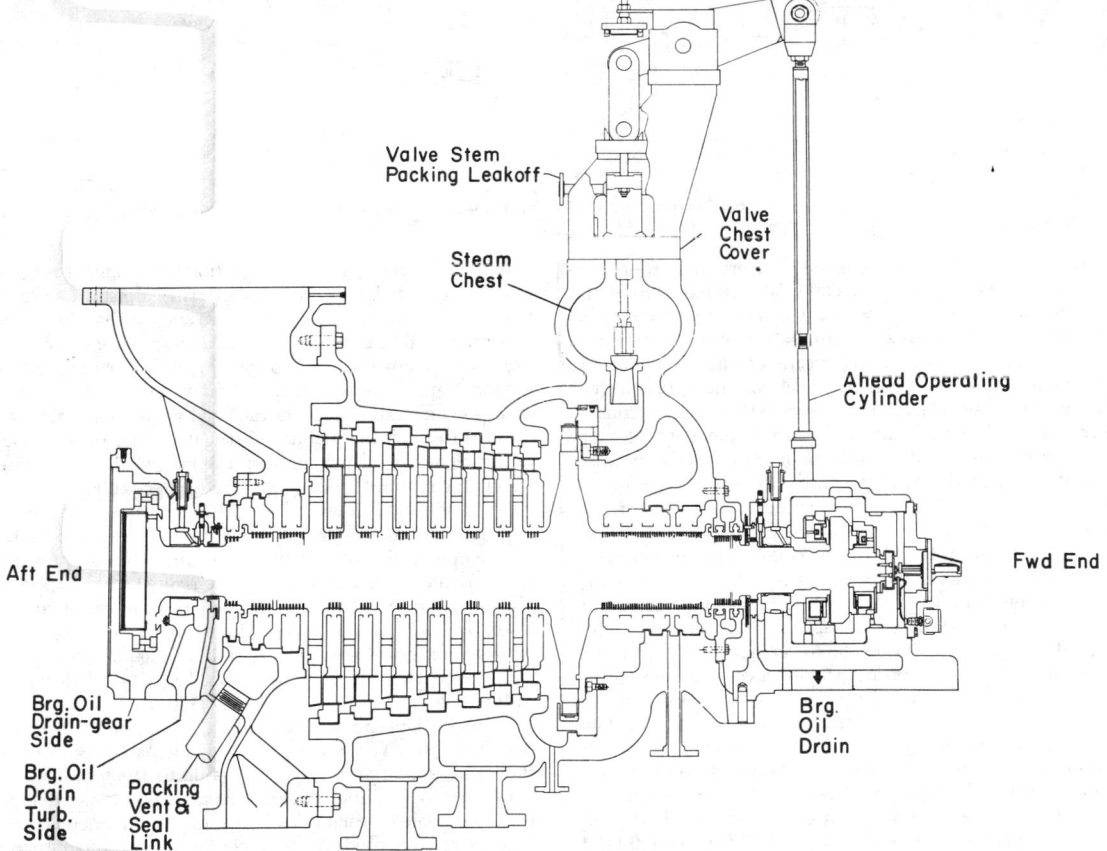

**Fig. 15a**  16,000-hp cross-compound marine turbine designed for steam conditions of 600 lb/in² gage, 850°F, 1½ inHg abs. High-pressure section for 6,550 r/min normal speed. Low pressure section shown in Fig. 15b. *(General Electric Co.)*

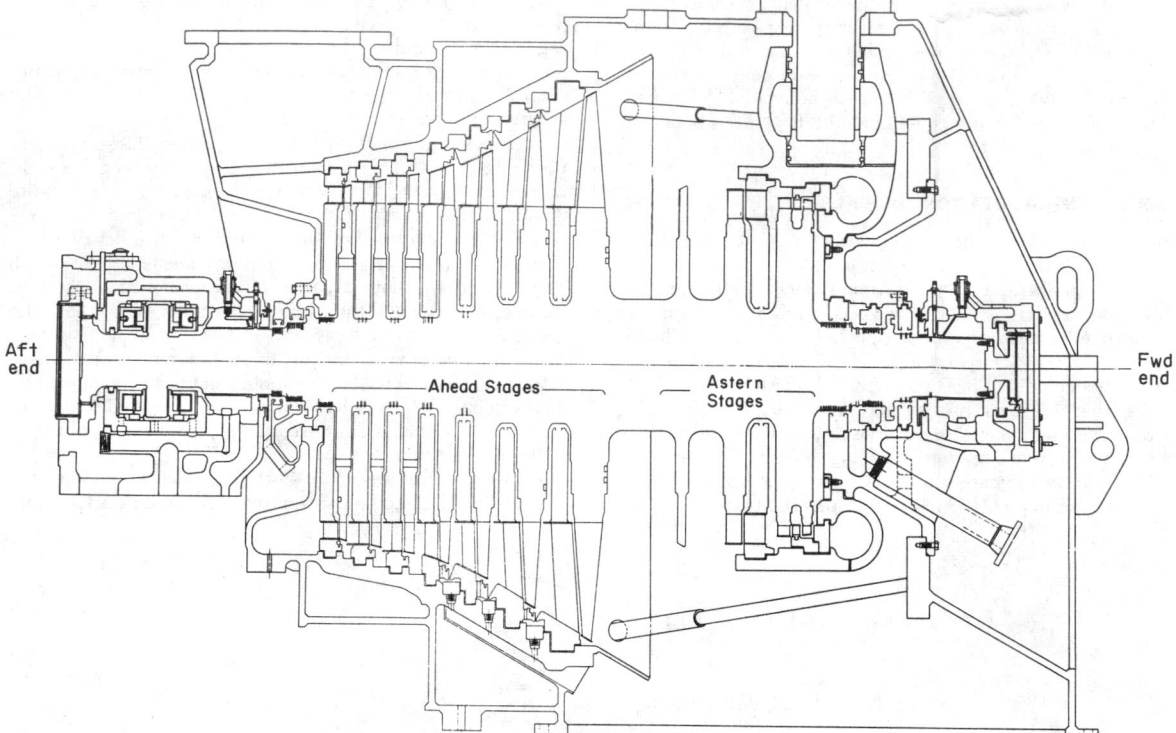

**Fig. 15b**  Same marine turbine as in Fig. 15a, but low-pressure section for 3,750 r/min normal speed. Low-pressure section contains two-stage reversing element. (General Electric Co.)

tors. Reactors of both the boiling and pressurized-water types raise steam at 1,000 lb/in² abs (7,000 kPa) approximately with little or no initial superheat, so that the initial temperature is about 545°F (285°C), with a fraction of 1 percent of moisture frequently present. The poorer steam conditions result in higher steam rates than seen by fossil turbines. The lower initial pressure causes larger initial specific volume. In consequence, a typical nuclear turbine must accommodate 2.5 to 4 times the initial volume flow, and about 1½ times the exhaust volume flow of a fossil unit of the same rating. These considerations and the fact that the low temperature of the steam results in high moisture content in the expansion lead to the choice of 1,800 r/min. In halving the speed, diameters are less than doubled, balancing the advantages of larger steam-path area to accommodate large flow, while reducing velocities to minimize the occurrence of impact moisture erosion. The shortened energy range due to the lower initial conditions requires only two kinds of casings, high-pressure and low-pressure, compared with the three needed by fossil-reheat units.

Referring to Fig. 17, the steam enters the double-flow nozzle boxes of the high-pressure section, to the left, through stop and control valves which are not shown. It flows in both directions through the impulse-type stages, exhausting through four connections on each end of the shell. At the exhaust, the pressure is reduced to 200 lb/in² abs (1,400 kPa) approximately, and the moisture content is increased to 12 percent. That moisture poses an erosion risk and a perfor-

mance loss to the low-pressure section following. It is current practice to dry the steam in an external moisture separator, frequently combined with one or two stages of steam-to-steam reheating, before admission to the low-pressure casings. Figure 18 is a cross section through a combination moisture separator and two-stage steam reheater. The exhaust from the high-pressure turbine enters the shell at the bottom, flowing upward through the inclined corrugated-plate moisture-separating elements, which remove essentially all the entrained water. It continues upward, flowing over the tubes of the first-stage bundle, which are filled with steam extracted from the high-pressure turbine, approaching to within 25°F (14°C)± of that temperature. Next, it flows over the tubes of the second-stage bundle, which are supplied with initial steam, approaching to within 25°F (14°C)± of that temperature, or to 520°F (270°C)±. The steam leaves the vessel at the top and is admitted to the low-pressure sections of Fig. 17, through stop and intercept valves, not shown. The last-stage blade length is in the range of 38 to 52 in (960 to 1,320 mm).

Units of the type described are built to ratings of approximately 1,300 MW with larger sizes anticipated. Similar four-flow units are employed for ratings up to 1,000 ±MW.

Other types of nuclear reactors, such as the high-temperature gas-cooled reactor and the liquid-metal-cooled fast-breeder reactor, produce steam conditions at temperature and pressure levels comparable with fossil-fuel-fired boilers, leading to the use of 3,600-r/min units similar to fossil practice.

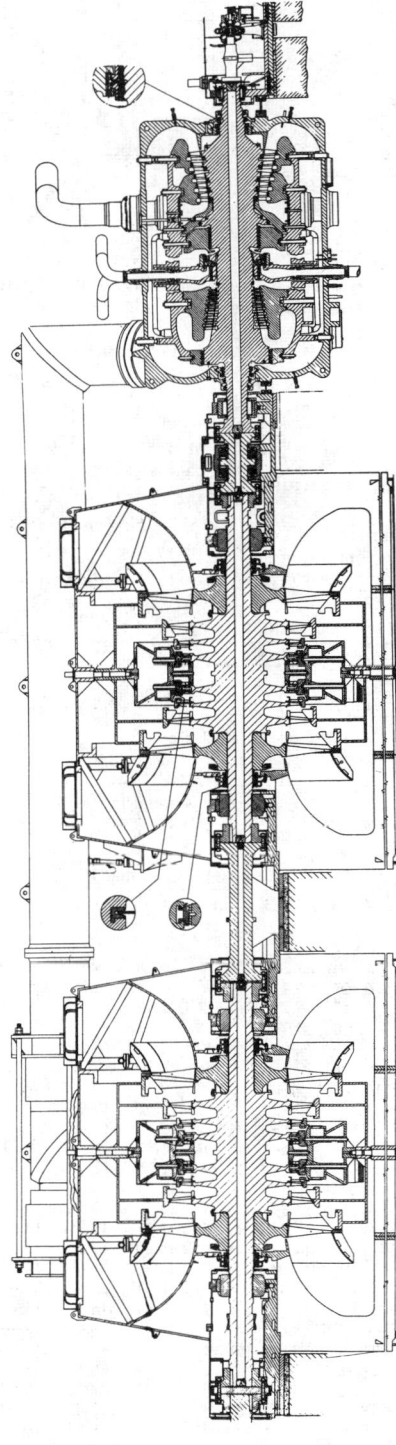

**Fig. 16** Cross section of a 500-MW tandem-compound quadruple-flow 3,600-r/min reheat turbine. (*Westinghouse Electric Corp.*)

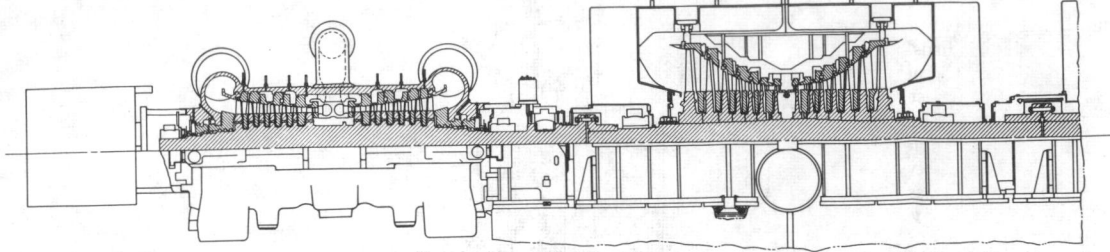

**Fig. 17** Cross section of a 1,300-MW class tandem-compound six-flow 1,800-r/min turbine for steam from light-water nuclear reactors, showing one of three low-pressure sections. (*General Electric Co.*)

## STEAM-TURBINE PERFORMANCE

The ideal **steam rate** (steam consumption, lb/kWh) of a simple turbine cycle is $3,412.14/(h_1 - h_{s2})$, where $h$ is in Btu/lb [or kg/kWs $= 1/(h_1' - h_{s2}')$, $h'$ in kJ/kg]. (See Fig. 4.)

The actual steam rate is $3,412.14/\eta_t (h_1 - h_{s2})$, where $\eta_t$ is the **engine efficiency** of the turbine only, inclusive of mechanical losses [or kg/kWs $= 1/\eta_t(h_1' - h_{s2}')$].

The actual steam rate of turbine and generator is $3,412.14/\eta_e (h_1 - h_{s2})$, where $\eta_e$ is the engine efficiency including all mechanical and electrical losses [or kg/kWs $= 1/\eta_e (h_1' - h_{s2}')$].

The enthalpy of the steam leaving the turbine elements $h_2$ is

$$h_2 = h_1 - \eta_s (h_1 - h_{s2}) = (Wh_1 - 3,412.14P_g)/W$$

where $\eta_s$ is the engine efficiency of the steam path, inclusive of leakages and losses but exclusive of mechanical and electrical losses; $W$ is the steam flow, lb/h; and the gross output $P_g$ is the net output plus mechanical and electrical losses, kW [or $h_2' = h_1' - \eta_s (h_1' - h_{s2}') = (W'h_1' - P_g)/W'$, where $W'$ is flow in kg/s].

Table 2 gives steam rates for the ideal simple turbine cycle

## Table 2. Theoretical Steam Rates for Typical Steam Conditions, lb/kWh*

| | | | | | | | | Initial pressure, lb/in² gage | | | | | | | |
|---|---|---|---|---|---|---|---|---|---|---|---|---|---|---|---|
| 150 | 250 | 400 | 600 | 600 | 850 | 850 | 900 | 900 | 1,200 | 1,250 | 1,250 | 1,450 | 1,450 | 1,800 | 2,400 |
| | | | | | | | | Initial temp, °F | | | | | | | |
| 365.9 | 500 | 650 | 750 | 825 | 825 | 900 | 825 | 900 | 825 | 900 | 950 | 825 | 950 | 1000 | 1000 |
| | | | | | | | | Initial superheat, °l | | | | | | | |
| 0 | 94.0 | 201.9 | 261.2 | 336.2 | 297.8 | 372.8 | 291.1 | 366.1 | 256.3 | 326.1 | 376.1 | 232.0 | 357.0 | 377.9 | 337.0 |
| | | | | | | | | Initial enthalpy, Btu/lb | | | | | | | |
| 1,195.5 | 1,261.8 | 1,334.9 | 1,379.6 | 1,421.4 | 1,410.6 | 1,453.5 | 1,408.4 | 1,451.6 | 1,394.7 | 1,438.4 | 1,468.1 | 1,382.7 | 1,461.2 | 1,480.1 | 1,460.4 |

| Exhaust pressure | 150 | 250 | 400 | 600 | 600 | 850 | 850 | 900 | 900 | 1,200 | 1,250 | 1,250 | 1,450 | 1,450 | 1,800 | 2,400 |
|---|---|---|---|---|---|---|---|---|---|---|---|---|---|---|---|---|
| **inHg abs** | | | | | | | | | | | | | | | | |
| 2.0 | 10.52 | 9.070 | 7.831 | 7.083 | 6.761 | 6.580 | 6.282 | 6.555 | 6.256 | 6.451 | 6.133 | 5.944 | 6.408 | 5.900 | 5.668 | 5.633 |
| 2.5 | 10.88 | 9.343 | 8.037 | 7.251 | 6.916 | 6.723 | 6.415 | 6.696 | 6.388 | 6.584 | 6.256 | 6.061 | 6.536 | 6.014 | 5.773 | 5.733 |
| 3.0 | 11.20 | 9.582 | 8.217 | 7.396 | 7.052 | 6.847 | 6.530 | 6.819 | 6.502 | 6.699 | 6.362 | 6.162 | 6.648 | 6.112 | 5.862 | 5.819 |
| 4.0 | 11.76 | 9.996 | 8.524 | 7.644 | 7.282 | 7.058 | 6.726 | 7.026 | 6.694 | 6.894 | 6.541 | 6.332 | 6.835 | 6.277 | 6.013 | 5.963 |
| **lb/in² gage** | | | | | | | | | | | | | | | | |
| 5 | 21.69 | 16.57 | 13.01 | 11.05 | 10.42 | 9.838 | 9.288 | 9.755 | 9.209 | 9.397 | 8.820 | 8.491 | 9.218 | 8.351 | 7.874 | 7.713 |
| 10 | 23.97 | 17.90 | 13.83 | 11.64 | 10.95 | 10.30 | 9.705 | 10.202 | 9.617 | 9.797 | 9.180 | 8.830 | 9.593 | 8.673 | 8.158 | 7.975 |
| 20 | 28.63 | 20.44 | 15.33 | 12.68 | 11.90 | 11.10 | 10.43 | 10.982 | 10.327 | 10.490 | 9.801 | 9.415 | 10.240 | 9.227 | 8.642 | 8.421 |
| 30 | 33.69 | 22.95 | 16.73 | 13.63 | 12.75 | 11.80 | 11.08 | 11.67 | 10.952 | 11.095 | 10.341 | 9.922 | 10.801 | 9.704 | 9.057 | 8.799 |
| 40 | 39.39 | 25.52 | 18.08 | 14.51 | 13.54 | 12.46 | 11.66 | 12.304 | 11.52 | 11.646 | 10.831 | 10.380 | 11.309 | 10.134 | 9.427 | 9.136 |
| 50 | 46.00 | 28.21 | 19.42 | 15.36 | 14.30 | 13.07 | 12.22 | 12.90 | 12.06 | 12.16 | 11.284 | 10.804 | 11.779 | 10.531 | 9.767 | 9.442 |
| 60 | 53.90 | 31.07 | 20.76 | 16.18 | 15.05 | 13.66 | 12.74 | 13.47 | 12.57 | 12.64 | 11.71 | 11.20 | 12.22 | 10.90 | 10.08 | 9.727 |
| 75 | 69.4 | 35.77 | 22.81 | 17.40 | 16.16 | 14.50 | 13.51 | 14.28 | 13.30 | 13.34 | 12.32 | 11.77 | 12.85 | 11.43 | 10.53 | 10.12 |
| 80 | 75.9 | 37.47 | 23.51 | 17.80 | 16.54 | 14.78 | 13.77 | 14.55 | 13.55 | 13.56 | 12.52 | 11.95 | 13.05 | 11.60 | 10.67 | 10.25 |
| 100 | | 45.21 | 26.46 | 19.43 | 18.05 | 15.86 | 14.77 | 15.59 | 14.50 | 14.42 | 13.27 | 12.65 | 13.83 | 12.24 | 11.21 | 10.73 |
| 125 | | 57.88 | 30.59 | 21.56 | 20.03 | 17.22 | 16.04 | 16.87 | 15.70 | 15.46 | 14.17 | 13.51 | 14.76 | 13.01 | 11.84 | 11.28 |
| 150 | | 76.5 | 35.40 | 23.83 | 22.14 | 18.61 | 17.33 | 18.18 | 16.91 | 16.47 | 15.06 | 14.35 | 15.65 | 13.75 | 12.44 | 11.80 |
| 160 | | 86.8 | 37.57 | 24.79 | 23.03 | 19.17 | 17.85 | 18.71 | 17.41 | 16.88 | 15.41 | 14.69 | 16.00 | 14.05 | 12.68 | 12.00 |
| 175 | | | 41.16 | 26.29 | 24.43 | 20.04 | 18.66 | 19.52 | 18.16 | 17.48 | 15.94 | 15.20 | 16.52 | 14.49 | 13.03 | 12.29 |
| 200 | | | 48.24 | 29.00 | 26.95 | 21.53 | 20.05 | 20.91 | 19.45 | 18.48 | 16.84 | 16.05 | 17.39 | 15.23 | 13.62 | 12.77 |
| 250 | | | 69.1 | 35.40 | 32.89 | 24.78 | 23.08 | 23.90 | 22.24 | 20.57 | 18.68 | 17.81 | 19.11 | 16.73 | 14.78 | 13.69 |
| 300 | | | | 43.72 | 40.62 | 28.50 | 26.53 | 27.27 | 25.37 | 22.79 | 20.62 | 19.66 | 20.89 | 18.28 | 15.95 | 14.59 |
| 400 | | | | 72.2 | 67.0 | 38.05 | 35.43 | 35.71 | 33.22 | 27.82 | 24.99 | 23.82 | 24.74 | 21.64 | 18.39 | 16.41 |
| 425 | | | | 84.2 | 78.3 | 41.08 | 38.26 | 38.33 | 35.65 | 29.24 | 26.21 | 24.98 | 25.78 | 22.55 | 19.03 | 16.87 |
| 600 | | | | | | 78.5 | 73.1 | 68.11 | 63.4 | 42.10 | 37.03 | 35.30 | 34.50 | 30.16 | 24.06 | 20.29 |

* From Theoretical Steam Rate Tables—Compatible with the 1967 ASME Steam Tables, ASME, 1969.

through a wide range of operating conditions. The performance of a turbine is usually expressed as a steam rate in the case of machines having no extraction or admission of steam between inlet and exhaust, which is generally true of small units and most noncondensing turbines.

Turbines having automatic pressure-controlled extractions or admissions of steam between inlet and exhaust usually have their performance expressed by a chart showing required throttle flow vs. load for varying amounts of steam extracted or admitted at specified conditions.

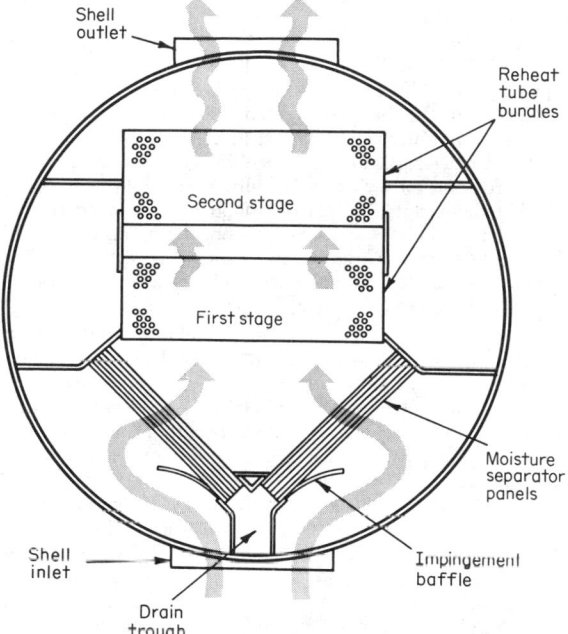

**Fig. 18**  Cross section of a combination moisture separator and two-stage steam reheater for use with nuclear turbines.

Turbines working on regenerative and/or reheat cycles, condensing, usually have performance expressed as a heat rate, based upon a carefully specified heat-cycle arrangement. This is usually illustrated by a diagram that defines all the surrounding conditions. See, for example, Fig. 23, actual cycle.

The above methods for expressing turbine performance are more satisfactory for most applications than the use of turbine "engine efficiencies." However, it is useful to know the general range of turbine efficiency realized in practice. The engine efficiency of a turbine depends mainly upon the flow areas and diameter of stages, the average velocity ratio, as can be deduced from Fig. 6, the number of turbine stages, and the steam conditions. With so many variables, it is not possible to do more than show a general picture of efficiency as a function of rating, as in Fig. 19, for multistage condensing turbines. Noncondensing turbines will usually have similar efficiency levels; automatic extraction turbines will generally be slightly lower because of extra losses in the control-stage sections.

Approximate steam rates for turbines operating without auxiliary admissions or extractions of steam between inlet and

exhaust may be estimated for any turbine rating by dividing theoretical steam rate, corresponding to inlet steam pressure and temperature and exhaust pressure, by the appropriate turbine efficiency from Fig. 19.

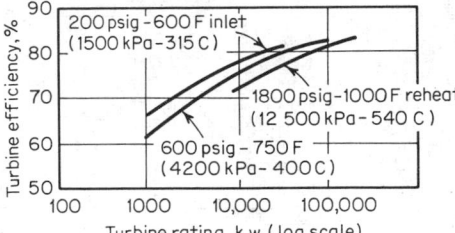

**Fig. 19**  Turbine efficiencies vs. capacity.

**A short method for calculating extraction-turbine performance** is illustrated by the following example:

Assume a 12,500-kW automatic-extraction-condensing unit operating at 10,000 kW, with 175,000 lb/h extraction for process at 150 psig with no extraction for feedwater heating, and throttle steam conditions of 850 lb/in², 825°F, exhaust at 2 inHg abs.

Procedure.   Find theoretical steam rates (TSR) from Table 2 or steam charts; TSR$_1$ for 850 lb/in², 825°F, 2 in Hg is 6.58 lb/kWh; TSR$_2$ for 850 lb/in², 825°F to 150 lb/in² is 18.61 lb/kWh.

Turbine-generator efficiency from Table 3, single autoextraction at 80 percent rating (10,000 kW on a 12,500-kW unit), is 78 percent. Efficiency correction for autoextraction (see Table 3) is 0.92. Then actual steam rate (ASR) is TSR/(efficiency × correction); ASR$_1$ = 6.58/78% × 0.92 = 9.17 lb/kWh; ASR$_2$ = 18.61/78% × 0.92 = 25.9 lb/kWh; kW generation from extraction flow = extraction flow/ASR$_2$ = 175,000/25.9 = 6,760 kW.

kW to be generated by condenser flow = 10,000 − 6,760 = 3,240.
Condenser flow required is 3,240 × 9.17 = 29,700 lb/h, or (say) 30,000 lb/h. Total steam flow to throttle then is 175,000 + 30,000 = 205,000 lb/h.

Figure 20 gives correction factors for speed which differ from 3,600 r/min and is representative of units designed for about 4 inHg abs exhaust pressure.

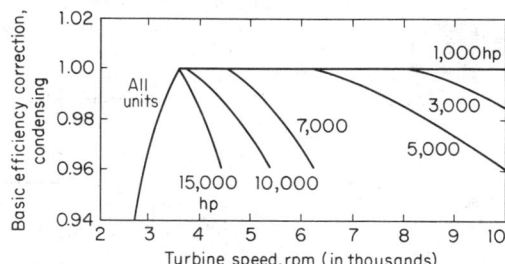

**Fig. 20**  Correction factor for condensing mechanical-drive turbines.

**Mechanical-Drive Turbines**  Table 3 and Figs. 19 and 20 provide efficiency-estimating data for typical condensing turbines, primarily for 3,600-r/min generator drive. Figure 21 shows approximate values for turbine efficiency to be expected for noncondensing mechanical-drive units designed for a broad range of horsepower rating, speed, and inlet-steam conditions.

**Large Central-Station Turbine-Generators**

References:   Spencer, Cotton, and Cannon, A Method for Predict-

**Table 3. Basic Efficiency for Steam Turbines, Straight Condensing at Rated Load**

| kW capacity | Equivalent mechanical drive, hp | Initial steam conditions (gage pressure and temp.) | | | | |
|---|---|---|---|---|---|---|
| | | 250 lb/in² 500°F | 400 lb/in² 650°F | 600 lb/in² 750°F | 800 lb/in² 825°F | 1,250 lb/in² 900°F |
| 875 | 1,200 | 63 | 63 | 62 | | |
| 1,875 | 2,600 | 76 | 67 | 66 | | |
| 2,500 | 3,500 | 69 | 69 | 68 | | |
| 5,000 | 6,900 | | 74 | 73 | 73 | |
| 7,500 | 10,300 | | 76 | 75 | 75 | |
| 12,500 | 17,200 | | 78 | 78 | 78 | 77 |
| 15,625 | 21,500 | | 79 | 79 | 79 | 77 |
| 20,000 | 27,100 | | 79 | 80 | 79 | 79 |

Efficiency correction factors, mechanical drive and autoextraction-condensing turbines—multiply basic efficiencies by:

| | At 80% rating | At 100% rating |
|---|---|---|
| Single autoextraction-condensing | 0.92 | 0.96 |
| Double autoextraction-condensing | 0.88 | 0.92 |

ing the Performance of Steam Turbine-Generators . . . 16,500 kW and Larger, *Trans. ASME*, ser. A, Oct. 1963. Baily, Cotton, and Spencer, Predicting the Performance of Large Steam Turbine-Generators Operating with Saturated and Low Superheat Steam Conditions, *Proc. American Power Conference*, 1967; discussion of foregoing, *Combustion*, Sept. 1967. Spencer and Booth, Heat Rate Performance of Nuclear Steam Turbine-Generators, *Proc. American Power Conference*, 1968. Baily, Booth, Cotton, and Miller, "Predicting the Performance of 1800-rpm Large Steam Turbine-Generators Operating with Light Water-Cooled Reactors," General Electric publication GET-6020, 1973. "Heat Rates for Fossil Reheat Cycles Using General Electric Steam Turbine-Generators 150,000 kW and Larger," General Electric publication GET-2050C, 1974.

The performance of central-station turbine-generators is generally expressed as **heat rate,** Btu/kWh, the ratio of the heat added to the cycle in Btu/h, to generation, in kW. Heat rate may be converted to **thermal efficiency** using the relationship, Efficiency = 3,412.14/heat rate (or 1/heat rate expressed in kJ/kWs). Heat rates are calculated by the preparation of a **heat balance,** which considers steam conditions, steam flow, tur-

bine-expansion efficiency, packing leaking losses, exhaust loss at the end of the low-pressure expansion (perhaps other casings as well), mechanical losses, electrical losses associated with the generator, moisture separation and reheat if present, and extraction for feedwater heating. **Gross heat rate** is calculated without consideration of the power consumed by the boiler feedpump. **Net heat rate** does consider pump power and is higher (poorer) than gross by a factor related to the initial steam pressure. If the pump is driven by an auxiliary turbine, as is the present usual practice, net heat rate is the natural result of the heat-balance calculation, and gross heat rate has little meaning. **Net station heat rate** considers the auxiliary power required by the rest of the power-plant equipment, and the boiler efficiency of fossil plants. It is generally about 3 percent higher than net heat rate in nuclear plants (3 percent auxiliary power, 100 percent "boiler" efficiency), and about 16 percent higher than net heat rate in the case of a coal-fired plant (4 percent auxiliary power, 90 percent boiler efficiency).

A typical current value of net station heat rate for the fossil steam conditions is 9,000 Btu/kWh (2.64 kJ/kWs), equivalent to a thermal efficiency of 38 percent. A typical net station heat rate for a large light-water nuclear-reactor plant is about 10,-100 Btu/kWh (2.96 kJ/kWs), or about 34 percent thermal efficiency.

Table 4 lists representative net heat rates for large fossil turbines of today's types and steam conditions. Steam pressures in excess of 3,500 lb/in² gage (24,200 kPa), and initial and reheat temperatures in excess of 1000°F (538°C) were frequently employed in the past. However, a number of operat-

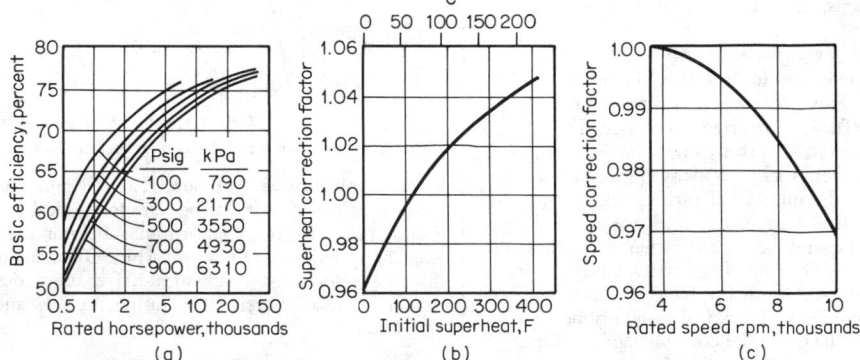

**Fig. 21** Mechanical-drive turbine efficiencies. *(a)* Basic efficiency, 3,600 r/min. Figures on curve are inlet-steam pressure in lb/in² gage. *(b)* Superheat correction factor. *(c)* Rated-load speed-correction factor.

**Table 4. Representative Net Heat Rates for Large Fossil Central-Station Turbine-Generators**

| Nominal rating, MW, at 1.5 inHg abs | Steam conditions | | | Tandem compound, 3,600 r/min, last-stage buckets | | | | Boiler feed-pump drive | Net heat rate, Btu/kWh, at rated load and steam conditions, and at exhaust pressure, inHg abs | | | | |
|---|---|---|---|---|---|---|---|---|---|---|---|---|---|
| | Throttle pressure, lb/in² gage | Temp, °F | Reheat temp, °F | No. of rows | Length, in | Exhaust area, ft² | Approx kW/ft² | | 1.5 | 2 | 3 | 4 | 5 |
| 150 | 1,800 | 1000 | 1000 | 2 | 26 | 82 | 1,820 | Motor | 8,010 | 8,060 | 8,230 | 8,440 | 8,630 |
| 235 | 1,800 | 1000 | 1000 | 2 | 26 | 82 | 2,860 | Motor | 8,240 | 8,240 | 8,290 | 8,380 | 8,500 |
| 250 | 1,800 | 1000 | 1000 | 2 | 30 | 111 | 2,250 | Motor | 8,080 | 8,100 | 8,220 | 8,400 | 8,620 |
| 250 | 1,800 | 1000 | 1000 | 2 | 30 | 111 | 2,250 | Turbine | 8,030 | 8,060 | 8,200 | 8,390 | 8,610 |
| 250 | 2,400 | 1000 | 1000 | 4 | 30 | 222 | 2,250 | Turbine | 7,850 | 7,890 | 8,030 | 8,240 | 8,450 |
| 500 | 2,400 | 1000 | 1000 | 4 | 33.5 | 264 | 2,650 | Turbine | 7,790 | 7,830 | 7,970 | 8,170 | 8,370 |
| 700 | 2,400 | 1000 | 1000 | 6 | 30 | 334 | 3,000 | Turbine | 7,860 | 7,870 | 7,970 | 8,130 | 8,320 |
| 1,000 | 2,400 | 1000 | 1000 | 4 | 30 | 222 | 2,250 | Turbine | 7,920 | 7,930 | 8,000 | 8,100 | 8,250 |
| 500 | 3,500 | 1000 | 1000 | 4 | 33.5 | 264 | 2,650 | Turbine | 7,620 | 7,660 | 7,820 | 8,030 | 8,220 |
| 700 | 3,500 | 1000 | 1000 | 6 | 30 | 334 | 3,000 | Turbine | 7,670 | 7,690 | 7,810 | 7,980 | 8,170 |
| 1,000 | 3,500 | 1000 | 1000 | 6 | 33.5 | 397 | 2,770 | Turbine | 7,710 | 7,730 | 7,810 | 7,940 | 8,090 |
| 1,100 | 3,500 | 1000 | 1000 | | | | | Turbine | 7,680 | 7,700 | 7,810 | 7,960 | 8,140 |

ing and economic considerations have led to near standardization on the single reheat cycle with initial pressure of 2,400 or 3,500 psig (16,650 or 24,240 kPa), with initial and reheat temperature of 1000°F (538°C).

Table 5 lists some representative net heat rates for large nuclear turbines for service with steam from boiling-water reactors (BWR), at 950 psig (6,650 kPa), ½ percent initial moisture. Values for other light-water reactors may be approximated by reducing heat rate by 1 percent for each 100 lb/in² (690 kPa) pressure increase, reducing heat rate by 0.15 percent for reducing initial moisture to 0 percent, reducing heat rate by 0.3 percent for each 10°F (6°C) of initial superheat provided.

### Reheating with Regenerative Cycle

REFERENCES: Reynolds, Reheating in Steam Turbines, *Trans. ASME*, **71**, 1949, p. 701. Harris and White, Developments in Resuperheating in Steam Power Plants, *Trans. ASME, 71, 1949, p. 685.*

Reheating is currently used on all new large fossil central-station turbines. It is accomplished by taking the steam from the turbine after partial expansion, reheating it in a separate section of the boiler, and returning it to the next lower-pressure section of the turbine. Reheating results in lowering of the turbine heat rate by approximately 5 percent; the exact improvement is dependent on several factors. Roughly speaking, 40 percent of the improvement comes from having added heat to the cycle at a higher-than-average temperature (thermodynamic gain), and the remaining 60 percent comes from improvement in turbine efficiency due to reduced moisture loss and increased reheat factor.

Reheating can theoretically be done any number of times, but because of extra cost of apparatus and piping, and the steam pressure drops required in practice (8 to 10 percent of the reheat pressure), the economic gains diminish rapidly with more than one reheating (see Fig. 22). In a few cases, two reheatings are employed. (See also Sec. 4).

The throttle and condenser steam-flow rates for a given turbine output are reduced approximately 17 and 13 percent, respectively, by reheating once to the initial temperature, as compared with no reheat with the same initial steam conditions.

The maximum gain in heat rate from one reheating with a fixed-percentage pressure drop through the reheating system occurs when the reheat pressure is about 0.15 of the initial pressure. In practice, however, the reheat pressure is higher, 0.20 to 0.30 times initial pressure, because of the extra cost of larger piping, valves, etc., required for lower reheater pressures owing to the larger steam volume.

### Regenerative Feedwater Heating
(See also Sec. 4)

The heat consumption of a turbine may be reduced by heating the condensate (feedwater) in stages by the condensation of

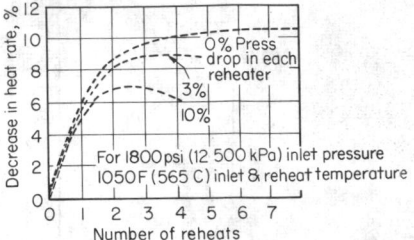

**Fig. 22**  Approximate gains due to reheating.

steam extracted at various points from the turbine. This is shown diagrammatically for an ideal cycle and a more practical cycle in Fig. 23. The difference between the two is in the use of mixing heaters in the ideal cycle with each discharge pumped back while the practical cycle has closed heaters with cascaded drains in the upper and pumped drains in the lowest heater, together with some pressure drop between turbine and heaters and a terminal temperature difference between saturated-steam temperature in the heater and feedwater temperature coming out. Usually the difference between such an ideal and a practical cycle is about 1½ percent. A deaerating type of contact heater with no terminal difference may be substituted for one of the closed heaters as is shown in Fig. 23. Other variations are the use of (1) all open-contact heaters, or (2) drain coolers to reduce the loss due to cascading the drips, or (3) a desuperheating section on the top heater to get a higher final feed temperature, thereby approaching most closely to the ideal cycle.

Figures 24 and 25, and Table 6 supply data on the results of regenerative heating based on the ideal cycle of Fig. 23. Figure 24 shows the reduction in heat rate for various initial pressures and temperatures at 1 in Hg abs (3.4 kPa) exhaust pressure, for various feedwater temperatures and number of heaters. The increase in throttle flow necessary to maintain the same power

**Table 5. Representative Net Heat Rates for Large Nuclear Central-Station Turbine-Generators**

| Warranted reactor thermal power, MWt | Nominal turbine rating, MWe at 2 inHg abs | Tandem Compound, 1,800 r/min last-stage buckets | | | | Net heat rate, Btu/kWh, at warranted reactor thermal power, at rated steam conditions, and at exhaust pressure, inHg abs | | | | |
|---|---|---|---|---|---|---|---|---|---|---|
| | | No. of rows | Length, in | Exhaust area, ft² | Approx kW/ft² | 1.5 | 2 | 3 | 4 | 5 |
| 2440 | 840 | 4 | 38 | 423 | 1,980 | 9,950 | 9,950 | 10,030 | 10,190 | 10,410 |
| 2440 | 850 | 4 | 43 | 495 | 1,720 | 9,810 | 9,820 | 9,950 | 10,170 | 10,440 |
| 2890 | 1,010 | 6 | 38 | 634 | 1,590 | 9,750 | 9,780 | 9,950 | 10,200 | 10,480 |
| 2890 | 990 | 4 | 43 | 495 | 2,000 | 9,980 | 9,980 | 10,050 | 10,200 | 10,410 |
| 3580 | 1,230 | 6 | 38 | 634 | 1,940 | 9,910 | 9,920 | 10,000 | 10,170 | 10,380 |
| 3580 | 1,250 | 6 | 43 | 743 | 1,680 | 9,780 | 9,790 | 9,930 | 10,160 | 10,430 |
| 3830 | 1,310 | 6 | 38 | 634 | 2,070 | 9,990 | 9,990 | 10,050 | 10,190 | 10,390 |
| 3830 | 1,330 | 6 | 43 | 743 | 1,790 | 9,840 | 9,850 | 9,960 | 10,170 | 10,420 |

All units assume boiling-water reactor steam conditions of 965 lb/in² abs, 1,190.8 Btu/lb, and two-stage steam reheat with 25°F approach to reheating steam temperature.

**Table 6. Total Steam Bled, Percent of Throttle Flow**

| Final feed temp | | Steam pressure and temperature | | | | | | | |
| --- | --- | --- | --- | --- | --- | --- | --- | --- | --- |
| | | 400 psig, (2,900 kPa) | | 600 psig 825°F (4,200 kPa, 440°C) | | 1.250 psig, 950°F (8,700 kPa, 510°C) | | 1,500 psig, 1050°F (10,400 kPa, 565(°C) | |
| | | Stages of feedwater heating | | | | | | | |
| °F | °C | 2 | 10 | 2 | 10 | 2 | 10 | 2 | 10 |
| 150 | 65 | 7.0 | 7.1 | 6.9 | 7.0 | | | | |
| 200 | 93 | 11.4 | 11.8 | 11.3 | 11.6 | 11.2 | 11.5 | 10.7 | 11.0 |
| 250 | 121 | 15.6 | 16.2 | 15.5 | 16.0 | 15.5 | 15.9 | 12.6 | 13.1 |
| 300 | 149 | 19.6 | 20.6 | 19.5 | 20.4 | 19.5 | 20.3 | 18.8 | 19.5 |
| 350 | 177 | 23.6 | 24.8 | 23.5 | 24.6 | 23.6 | 24.6 | 20.7 | 21.6 |
| 400 | 204 | 27.1 | 29.0 | 27.1 | 28.8 | 27.4 | 28.9 | 26.4 | 27.8 |
| 450 | 232 | | | 30.2 | 32.5 | 31.1 | 33.2 | 30.0 | 32.0 |
| 500 | 260 | | | | | 35.0 | 37.4 | 33.9 | 36.1 |

output when extracting steam for feedwater heating is shown in Fig. 25.

## INSTALLATION, OPERATION, AND MAINTENANCE CONSIDERATIONS

Steam turbines are capable of long life and high reliability with relatively little maintenance, if proper attention is paid to their installation, operation, and preventive maintenance. This section considers four areas proved to be of particular importance by recent operating experience.

### Steam Temperature—Starting and Loading

REFERENCES: Spencer and Timo, Starting and Loading of Large Steam Turbines, *Proc. American Power Conference*, 1974. Ipsen and Timo, The Design of Turbines for Frequent Starting, *Proc. American Power Conference*, 1969. Timo and Sarney, "The Operation of Large Steam Turbines to Limit Cyclic Shell Cracking," ASME Paper 67-WA/PWR-4, 1967.

Changing steam temperature at constant load or changing load at constant temperature subjects rotors and shells to thermal transients. Whereas temperature and load may be changed in seconds, it may take heavy metal sections hours to reach equilibrium with the new temperatures imposed on their surfaces. Parts are subjected to transient thermal stresses which may deplete their low-cycle thermal fatigue life. Repeated thermal cycles may lead to full expenditure of the available fatigue life of the part, followed by surface cracking. Further cycles tend to drive the cracks deeper into the affected part, leading to steam leakage through shells or vibration problems with rotors. Cracks tend to be driven deeper by downward steam-temperature changes, since the surface chills faster than the underlying material and is stressed in tension.

Steam-turbine manufacturers publish specific starting-and-loading instructions for their units. Data are provided such that the operator may select loading rates that stay within an acceptable expenditure of total low-cycle fatigue life per starting or loading cycle. For example, if a unit is expected to be started and loaded daily for a 30-year life, total cycles will be about 10,000, and it would be desirable to avoid exceeding 0.01 percent life expenditure per daily cycle.

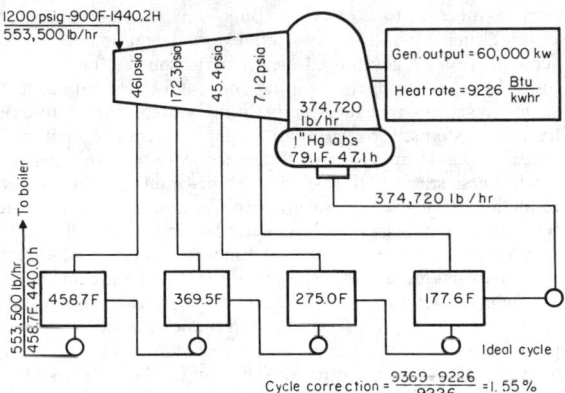

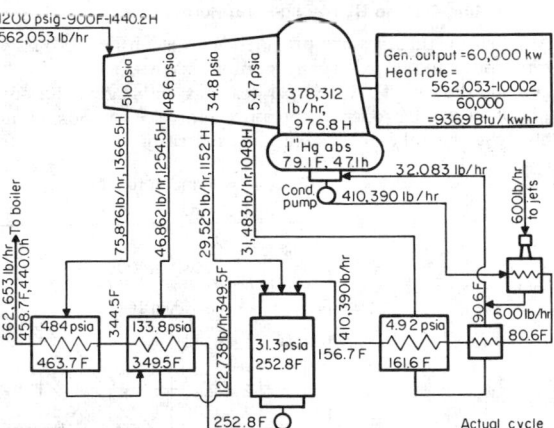

**Fig. 23**  Comparison of ideal and actual cycles for regenerative feedwater heating.

### Water-Induction Damage

REFERENCES: "Recommended Practices for the Prevention of Water Damage to Steam Turbines Used for Electric Power Generation, Part 1—Fossil Fueled Plants, Part 2—Nuclear Fueled Plants," ASME Standard TWDPS-1, 1972 and 1973.

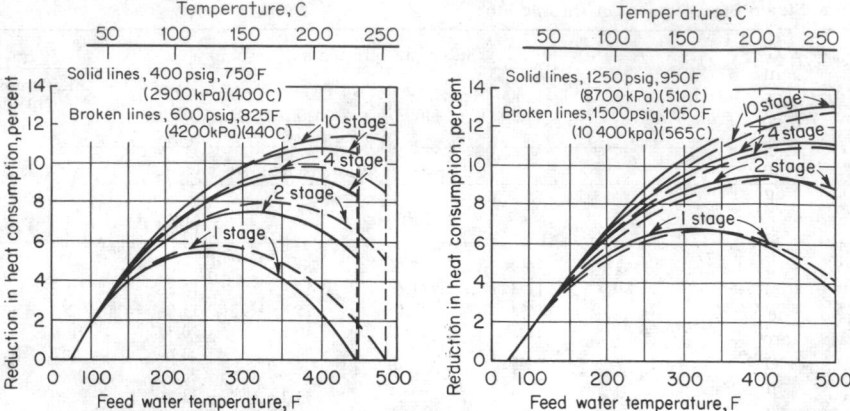

**Fig. 24**  Reduction in heat rate by use of ideal regenerative cycle, with 1 inHg back pressure.

Any connection to a steam turbine is a potential source of water either by induction from external equipment or by accumulation of condensed steam. The sources include the following along with their piping and drains: main and reheat steam systems; reheat attemperating system (fossil units); bypass systems, crossaround piping, moisture separator/reheater system (nuclear units); extraction system and feedwater heaters; steam-seal system; turbine-drain system. Water induction may lead to steam-path damage such as broken buckets, thrust-bearing failure, rotor bowing, and shell distortion, which may be indicated by abnormal vibration or differential expansion, or inability to turn the rotor on turning gear.

Water induction may be prevented by proper system installation, provision of protective and indicating devices, and periodic testing, inspection, and maintenance. Detailed recommendations are given in ASME Standard TWDPS-1, and manufacturers' instructions.

### Lubricating-Oil and Hydraulic-Fluid Purity

Most steam turbines are provided with a lubricating-oil system consisting of reservoir, pumps, coolers, and piping to provide the thrust and journal bearings with a generous supply of oil at the proper temperature and viscosity. Some units also use the lube-oil system as a source of fluid power for control devices and steam-valve actuation. It is most important to assure the cleanliness and purity of the lube oil at all times to avoid bearing and journal damage, or control-system malfunction. Bearings have failed because of oil starvation caused by clogged lines. At least one overspeed failure has resulted from the silting of control devices with rust caused by the entry of water into the oil system.

Units employing an electrohydraulic control system frequently use a synthetic fire-resistant fluid for the high-pressure control hydraulics, separate from the petroleum oil used in the bearing lubrication system. High-pressure hydraulic systems employing synthetic fluid offer advantages in size reduction, speed of response, and fire safety. However, because of the need for very close clearances between small parts at high pressures, and because of the poorer rust-preventive properties of the fluid, cleanliness is of even greater importance than with oil-based systems.

Turbine manufacturers provide instructions for cleaning oil systems by oil flushing between installation and first operation, and for maintaining the required oil and hydraulic-fluid purity. It is important that these be followed carefully.

### Steam Purity

Boiler-feedwater treatments have traditionally been designed

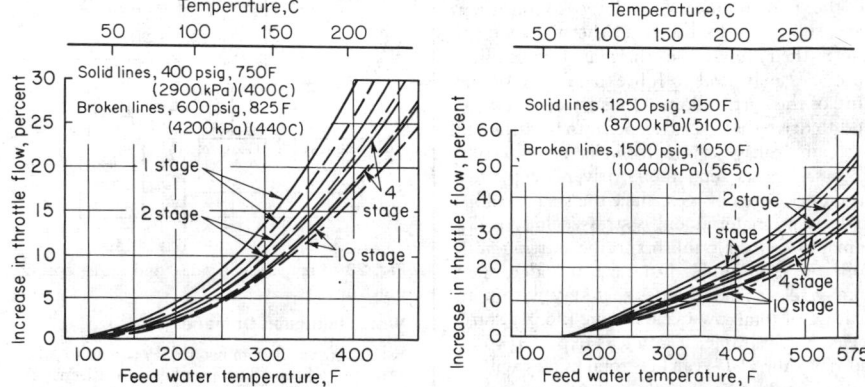

**Fig. 25**  Increase in steam flow necessary to maintain the same power output when using the ideal regenerative cycle, with 1 inHg back pressure.

to remove solids that might clog steam passages in the boiler, remove salts that could cause scaling of tube surfaces and interfere with heat transfer, prevent corrosion of tube surfaces by reducing oxygen content and by maintaining pH at a high level, and provide "clean" steam to the turbine.

In the past the main concern with steam quality in the turbine was the level of silica present, since that chemical tends to deposit in the steam passages and causes reduction in capacity and efficiency. In general, the monitoring and control of silica has been well developed, and as steam turbines have increased in rating, the passages have increased in area, so that the net effect has been a reduction in the extent of losses in efficiency and capacity from deposits.

On the other hand, the growth in unit ratings has been accomplished without a proportional growth in physical size, and has resulted in greater power densities per casing, per stage, and per pound of material. Such increased duty has required the use of higher-strength alloys operating at higher stresses. As a result, modern turbine components are more susceptible to stress-corrosion cracking than those in older, smaller units, and therefore require better control of contaminants in the steam.

The **feedwater treatment** in most stations is designed to provide a sufficiently low level of contaminants so that stress-corrosion cracking of turbine components should not be a problem. The "zero solids" treatment is particularly successful, but even the "coordinated phosphate" treatment can be controlled so as to provide steam of acceptable chemistry. Unfortunately, there are a number of situations in which undesirable chemicals can be introduced in the steam in spite of well-intentioned "normal" water-control practices. For example, the composition of coatings put on turbine components for corrosion protection during shipment, storage, and installation must be controlled. The chemistry of solutions used for the removal of coatings during installation, and the methods used, must be carefully regulated. The turbine must be protected during the chemical cleaning of related components such as the boiler, condenser, and feedwater heaters. Critical components have been damaged by fumes from cleaning. The feedwater system must be designed so that only water of high purity is used for boiler desuperheater sprays, and for the turbine exhaust-hood sprays used to limit temperature during light-load operation. Condensate demineralizers must be operated and regenerated so as to ensure that they do not introduce the harmful chemicals they are intended to remove. In the event of a leak into the condenser of impure cooling water, it is important to avoid changing the feedwater treatment in such a way that the turbine is subjected to harmful contaminants introduced to protect other station components.

In each of these undesirable instances, the average concentration of chemicals in the steam can be quite low, but high local concentrations can be developed through several mechanisms. For example, dilute solutions can enter crevices not washed by flowing steam; as water evaporates on heating, the concentration of the solution wetting the surfaces tends to increase. In the case of expansion-joint bellows, chemicals contained in steam condensing on shutdown or cold start-up tend to be trapped and concentrated in the bottom of the bellows convolutions. Succeeding cycles can lead to dry residue or to concentrated solutions during operating conditions which provide moisture. In the case of the steam path, as the expansion crosses into the moisture region, the first droplets of water condensed from the steam will tend to contain most of the contaminants. Concentration-enhancement factors of 100 to 1,000 can be achieved. In modern reheat turbines the early-moisture region occurs in one of the later few stages of the low-pressure section, and at light loads can occur on the most highly stressed last-stage buckets.

Extreme care must be exercised in protecting turbines from chemical contamination during installation, operation, and maintenance. Any deviation from sound feedwater-treatment practice during condenser leaks should be done with the full realization that damage to the turbine may result.

# POWER-PLANT HEAT EXCHANGERS

## by Joseph R. Spencer and Richard M. Stephani

NOTE: Standards for this industry retain English units except as indicated in the text.

## SURFACE CONDENSERS

REFERENCES: Heat Exchange Institute Standards for Surface Condensers.

The power-plant surface condenser is attached to the low-pressure exhaust of a steam turbine (see Figs. 1 and 2). Its purposes are (1) to produce a vacuum or desired back pressure at the turbine exhaust for the improvement of plant heat rate, (2) to condense turbine exhaust steam for reuse in the closed cycle, and (3) to deaerate the condensate.

An economical turbine back pressure is from 1.0 to 3.5 in Hg abs. The factors involved in establishing this pressure are involved and will not be discussed here.

An equipment diagram of a closed power-plant cycle is shown in Fig. 1.

For a condenser to deaerate the condensate, it must remove oxygen and other noncondensable gases to an acceptable level compatible with material selection and/or chemical treatment of the feedwater (condensate). Depending on materials and treatment, the dissolved $O_2$ level must normally be kept below $0.005$ cm³/l for turbine units operating with high-pressure and -temperature steam.

**Deaeration** in a condenser is accomplished by applying

Henry's law, which states that the concentration of the dissolved gas in a solution is directly proportional to the partial pressure of that gas in the free space above the liquid, with the exception of those gases (e.g., $CO_2 + NH_3$) which unite chemically with the solvent. In a condenser droplets of condensate are continually scrubbed with steam, liberating the $O_2$ and permitting it to flow to the low-pressure air-removal section, where it is discharged to the atmosphere by the air-removal equipment.

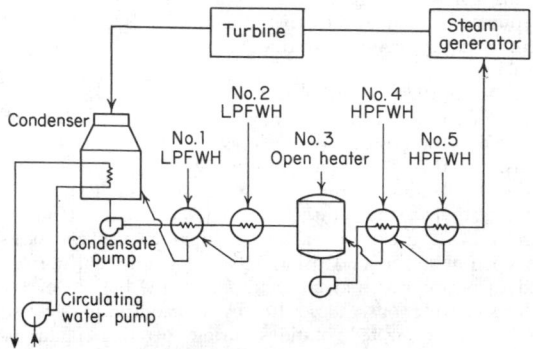

**Fig. 1**  Equipment arrangement, schematic.

To remove the last traces of $O_2$ from the condensate, an ammonia compound such as hydrazine is normally added. Free ammonia is liberated in the cycle and is either removed with the noncondensables as a gas or is condensed and retained in the condensate, depending on the detailed design of the condenser air-removal section. If the ammonia is concentrated as a liquid, it can be very corrosive to certain copper-base materials.

Most condenser manufacturers have **tube-bundle configurations** unique to their design philosophy. Basically, pressure losses from turbine exhaust to the air offtake are kept to a minimum and tubes are arranged to promote good heat-transfer rates. Small condensers are usually cylindrical, whereas large ones are rectangular for better utilization of space. Most turbines exhaust downward into the condenser, but condensers are also built to accommodate side as well as axial exhaust turbines.

Because of the inherent strength of cylindrical shapes as opposed to flat plates, condenser **water boxes** are generally made with curved surfaces. This has come about because of the increased pressure resulting from cooling towers, which in turn, are the result of environmental influences. With a cooling tower, pressures are in the 60 to 80 lb/in² range, whereas with water from lakes, rivers, etc., where a siphon system can be employed, water-box design pressures are in the 20 to 30 lb/in² range.

As a general rule, **tube selection** is based on economics; 18 BWG admiralty metal has been satisfactory for fresh-water service and 20 BWG 90 to 10 copper-nickel material likewise for seawater. Material prices fluctuate greatly, and selection can be influenced by first cost. At this writing, considerable emphasis is placed on lost revenue due to downtime caused by tube leaks as well as other causes. To eliminate or greatly reduce the possibility of tube leaks, titanium tube material is being used.

Low-pressure feedwater heaters are frequently located in the steam-inlet neck of a condenser. This is done to minimize pressure drop in the extraction steam piping and to utilize floor space surrounding the condenser better.

A sufficient number of tube supports must be provided so that tubes do not vibrate excessively. **Flow-induced vibration** is caused by high-velocity steam entering the tube bundle.

Where once-through boiler or nuclear steam generators are used, it is imperative to dispose of large quantities of steam during starting and stopping of a turbine unit. The condenser, because of its large volume, has been used as a convenient dumping place for this steam. Means must be provided within the condenser to accommodate the high-energy steam without damage to condenser tubing, structural members, or the low-pressure end of the turbine.

A big factor in condenser performance is **tube cleanliness.** Tubes can be plugged with leaves, marine life, and other debris. By valving, various arrangements for backwashing or flushing are achieved. For more thorough cleaning of inside tube surfaces, rubber plugs or wire brushes are manually inserted and propelled through each tube, or abrasive balls or brushes are automatically recirculated through a condenser if proper auxiliary equipment is provided.

Movement due to **temperature differences** between turbine and condenser is usually accommodated by a stainless-steel bellow or rubber-belt-type **expansion joint.** For small units the condenser may be supported on springs and rigidly connected to the turbine. To accommodate differential expansion between condenser shell and tubes, a flexible diaphragm or other expansion element may be installed.

The **tube-to-tube sheet joint** is usually rolled but has been welded in certain installations. Proper material selection must be made regardless of whether the joint is rolled or welded. Currently, considerable emphasis is placed on inward leakage of circulating water into the condenser steam space. Double tube sheets are being made for some units to ensure against such leakage. Again, as with tube-material selection, the decision is an economic one, and other factors such as feedwater treatment and downtime must be considered.

Condensers for some large turbine units having two or three low-pressure ends are designed for **multipressure application.** Usually there will be a gain, in the form of either heat-rate improvement or a reduction in capital investment, with a multipressure condenser. A rather complex analysis must be made for each application before a decision can be made.

The current "Standards for Steam Surface Condensers," 6th ed., 1970, is published by the Heat Exchange Institute. An addendum entitled "Condenser Construction Standard" was issued in 1975 to provide guidance in solving most structural problems encountered.

### Performance Calculations, Sizing

#### Notation

$A$ = condenser-surface area, ft²
$C_c$ = cleanliness factor
$C_1$ = heat-transfer-rate constant
$C_m$ = material and gage factor
$C_t$ = temperature correction factor
$c_p$ = specific heat, Btu/lb, °F
$G$ = circulating-water quantity, gal/min
$h$ = enthalpy, Btu/lb
$h_r$ = heat rejected by steam, Btu/lb

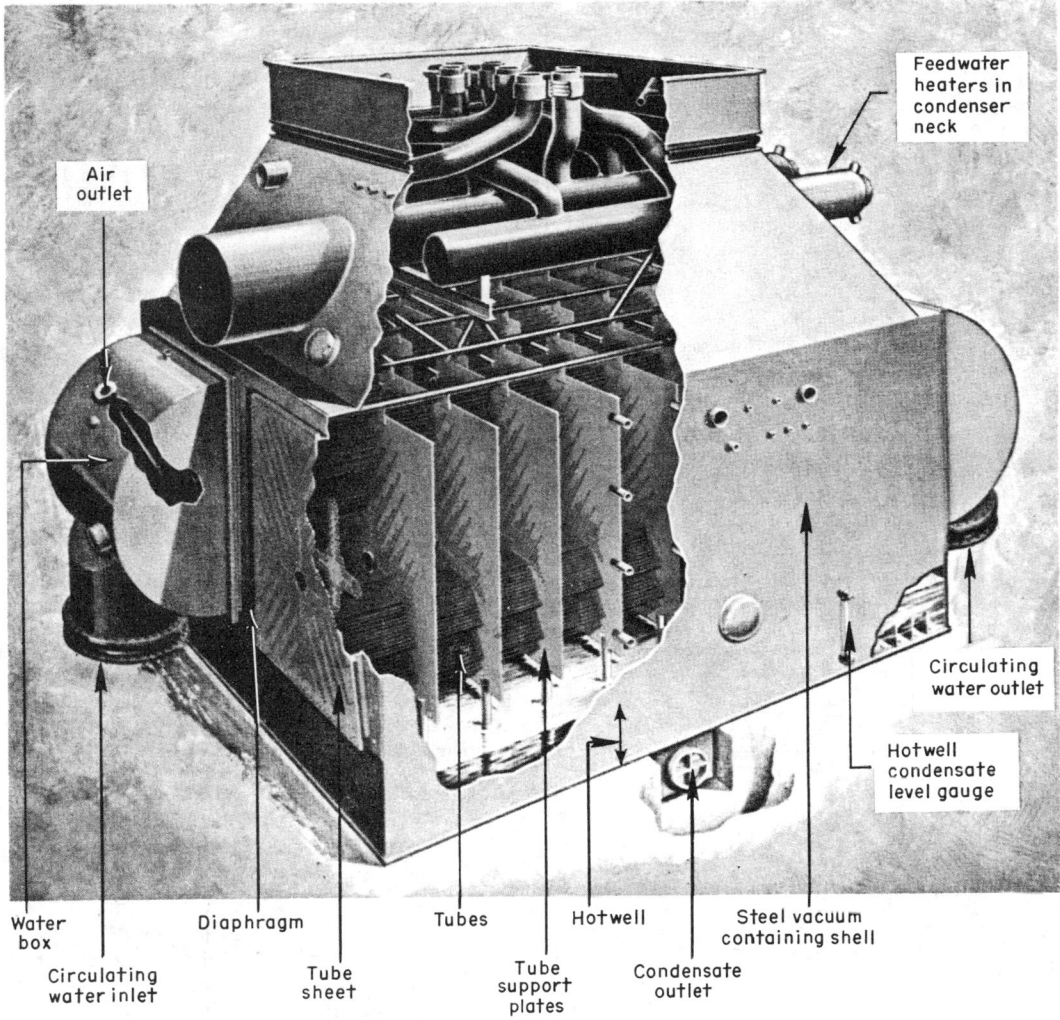

Feedwater
heaters in
condenser
neck

Air
outlet

Circulating
water outlet

Hotwell
condensate
level gauge

Water
box

Diaphragm

Tubes

Hotwell

Steel vacuum
containing shell

Circulating
water inlet

Tube
sheet

Tube
support
plates

Condensate
outlet

**Fig. 2**  One-pass rectangular surface condenser.

$k$ = tube diameter and gage factor, Table 12
$L$ = length of water travel, ft
NPSH = net positive suction head, ft
OD = tube outside diameter, in
$Q$ = heat transferred, Btu/h
$R$ = temperature rise $(t_o - t_i)$, °F
TDH = total dynamic head, ft
$t$ = tube thickness, in; temperature, °F
$ttd$ = terminal temperature difference = $t_v - t_o$
$t_i$ = inlet-water temperature, °F
$t_o$ = outlet-water temperature, °F
$t_v$ = saturation temperature in condenser, °F
$U_o$ = overall heat-transfer rate, Btu/(ft²)(h)(°F)
$V$ = water velocity, ft/s
$W_s$ = steam to be condensed, lb/h
$\Delta t_m$ = logarithmic mean temperature difference, °F
$\epsilon$ = effectiveness

In **sizing** a condenser, the steam flow and heat rejected to the condenser are obtained from the turbine heat balance. Table 1

**Table 1. Steam Flow to Condensers***

| Turbine throttle conditions | Weight flow, lb/kWh |
|---|---|
| 600 lb/in², 825°F | 7.08 |
| 850 lb/in², 900°F | 6.45 |
| 1,250 lb/in², 950°F | 5.80 |
| 1,450 lb/in², 1000°F | 5.44 |
| 1,450 lb/in², 1000°F, 1000°F | 4.70 |

*Approximate values for unit sizes to 100,000 kW and exhaust at 1.0 in Hg abs.

NOTE: For exhaust pressures other than 1.0 in Hg abs, multiply tabular values by 1.02 (1.04)[1.06] for 1.5 (2.0)[2.5] in.

gives representative steam flows; heat rejected to the condenser is approximately 950 Btu/lb of steam for nonreheat turbines and 975 Btu/lb for reheat machines. Figure 3 shows approximate average water temperatures for the United States; local water temperature should be used, when known. The number of passes is usually dictated by the plant arrangement, with total length of water travel and tube diameter dictated by economic considerations. Normally, small-diameter tubes, single-pass condensers are used where water is tiful, and larger-diameter tubes, two-pass condensers when water is scarce. The vacuum, or back pressure, is determined by economic evaluation, but Table 2 gives normal recommended values for average water temperatures. Table 3 is a pressure-temperature conversion table.

A **cleanliness factor** is applied to the heat-transfer rate of new, clean tubes to allow for gradual decrease by fouling. A standarized cleanliness factor of 85 percent is frequently used, but this can often be misleading and even erroneous. The fouling is attributable to (1) sedimentation, (2) scaling, (3) reaction, (4) corrosion, and (5) biological growth. The fouling is correctly determined by use; in some cases the cleanliness factor will be 90 percent, whereas in other cases it may never rise above 75 percent.

**Velocities** normally used are: for clean water, 7.0 ft/s; for very clean water with cooling towers, 8.0 ft/s; and for seawater, with entrained sand, as low as 6.0 ft/s to minimize erosion. Prevalent velocities are 6.5 ft/s with aluminum-brass tubes, 7.0 ft/s with admiralty metal, and 8+ ft/s with stainless steel. **Water-temperature rise** is about 10°F for single-pass condensers and 15°F for two-pass condensers, with a minimum 5° terminal temperature difference (TTD).

**Approximate general rules for condensers serving turbines rated up to 100 MW.** The **surface area**, ft², is equal to the steam flow, lb/h, divided by 10 for single-pass condensers and by 7.5 for two-pass condensers. **Circulating-water quantity,** gal/min, is equal to the area, ft², for a two-pass condenser and is twice the area for a single-pass condenser. Condenser **proportions** are given in Table 4. Empty **weight** of an installed condenser is 5 to 6 lb/ft² of surface. (See Fig. 9 for relationships.)

### Condenser Calculations

It is normally a tedious process to calculate the size and performance by the use of the logarithmic mean temperature difference. With the use of several aids, the calculations can be simplified. (See Heat Exchange Institute Standards for details.)

The basic diagram is shown in Fig. 4, and the applicable equations are

**Table 2. Normal Condenser Pressures and Circulating-Water Temperatures**

| Water inlet temp $t_b$, °F | Normal back pressure, inHg abs |
|---|---|
| 55 | 1.0 |
| 70 | 1.5 |
| 80 | 2.0 |
| 85 | 2.5 |
| 90 | 3.0 |
| 95 | 3.5 |

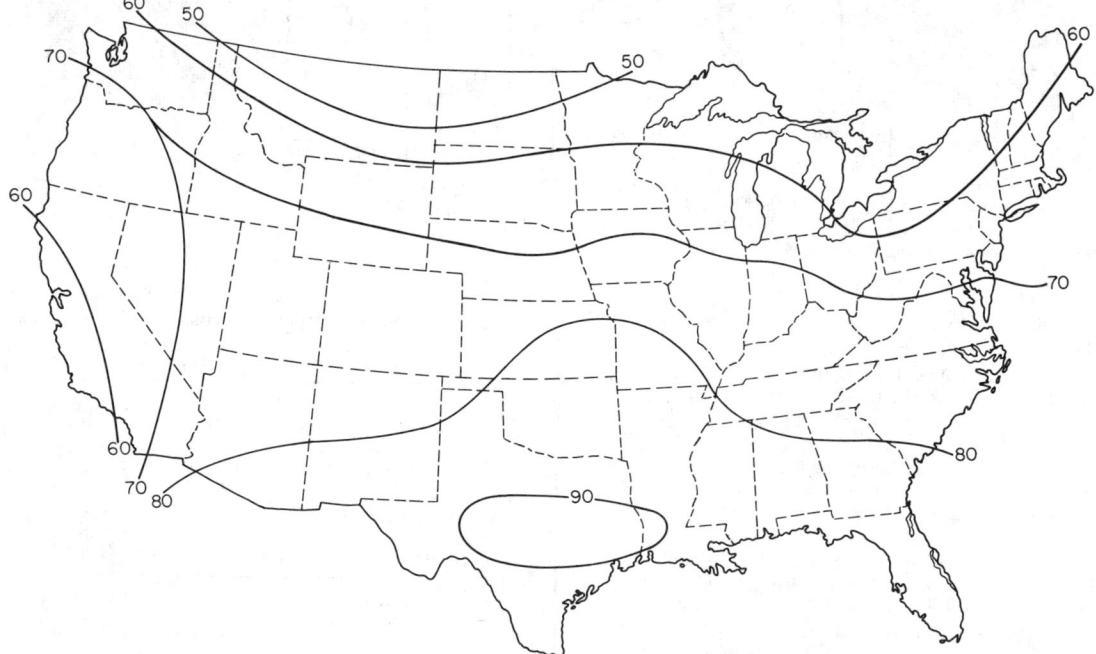

**Fig. 3**  Average inlet temperatures of circulating water, United States.

**Table 3. Pressure-Temperature Conversion Table***

| Abs press, inHg | Sat temp, °F | Abs press, inHg | Sat temp, °F | Abs press, inHg | Sat temp, °F | Abs press, inHg | Sat temp, °F |
|---|---|---|---|---|---|---|---|
| 0.20 | 34.57 | 1.35 | 88.36 | 2.50 | 108.71 | 8.00 | 152.24 |
| 0.25 | 40.23 | 1.40 | 89.51 | 2.55 | 109.38 | 9.00 | 157.09 |
| 0.30 | 44.96 | 1.45 | 90.64 | 2.60 | 110.06 | 10.00 | 161.49 |
| 0.35 | 49.06 | 1.50 | 91.72 | 2.65 | 110.72 | 11.00 | 165.54 |
| 0.40 | 52.64 | 1.55 | 92.77 | 2.70 | 111.37 | 12.00 | 169.28 |
| 0.45 | 55.89 | 1.60 | 93.81 | 2.75 | 112.01 | 13.00 | 172.78 |
| 0.50 | 58.80 | 1.65 | 94.80 | 2.80 | 112.63 | 14.00 | 176.05 |
| 0.55 | 61.48 | 1.70 | 95.78 | 2.85 | 113.25 | 15.00 | 179.14 |
| 0.60 | 63.96 | 1.75 | 96.73 | 2.90 | 113.86 | 16.00 | 182.05 |
| 0.65 | 66.26 | 1.80 | 97.65 | 2.95 | 114.45 | 17.00 | 184.82 |
| 0.70 | 68.41 | 1.85 | 98.56 | 3.00 | 115.06 | 18.00 | 187.45 |
| 0.75 | 70.43 | 1.90 | 99.43 | 3.20 | 117.35 | 19.00 | 189.96 |
| 0.80 | 72.32 | 1.95 | 100.30 | 3.40 | 119.51 | 20.00 | 192.37 |
| 0.85 | 74.13 | 2.00 | 101.14 | 3.60 | 121.57 | 21.00 | 194.68 |
| 0.90 | 75.84 | 2.05 | 101.96 | 3.80 | 123.53 | 22.00 | 196.90 |
| 0.95 | 77.48 | 2.10 | 102.77 | 4.00 | 125.43 | 23.00 | 199.03 |
| 1.00 | 79.03 | 2.15 | 103.56 | 4.20 | 127.21 | 24.00 | 201.09 |
| 1.05 | 80.53 | 2.20 | 104.33 | 4.40 | 128.94 | 25.00 | 203.08 |
| 1.10 | 81.96 | 2.25 | 105.09 | 4.60 | 130.61 | 26.00 | 205.00 |
| 1.15 | 83.33 | 2.30 | 105.85 | 4.80 | 132.20 | 27.00 | 206.87 |
| 1.20 | 84.64 | 2.35 | 106.58 | 5.00 | 133.76 | 28.00 | 208.67 |
| 1.25 | 85.93 | 2.40 | 107.30 | 6.00 | 140.78 | 29.00 | 210.43 |
| 1.30 | 87.17 | 2.45 | 108.01 | 7.00 | 146.86 | 29.921 | 212.00 |

*See also Sec. 4 for further data.

$$Q = UA\,\Delta t_m \tag{1}$$
$$Q = 500Gc_p(t_o - t_i) \tag{2}$$
$$R = W_s b_r / 500G \tag{3}$$
$$A = GkL/V \tag{4}$$
$$k = 0.107\,OD/(OD - 2t)^2 \tag{5}$$
$$\epsilon = (t_o - t_i)/(t_v - t_i) \tag{6}$$
$$c = 1 - e^{-R/\Delta t_m} \tag{7}$$
$$R/\Delta t_m = [kL(U_o C_t C_c C_m)]/500V \tag{8}$$

EXAMPLE.  Calculate the surface and circulating-water requirements to condense 445,000 lb of steam per h to an absolute pressure of 2.00 in Hg abs. The heat rejected is 980 Btu/lb of steam and is absorbed by circulating seawater at an average inlet temperature of 75°F. The condenser is to be two-pass with $\frac{7}{8}$-in 18-gage 26-ft-active-length aluminum-brass tubes; cleanliness factor = 85 percent.

SOLUTION.  Obtain the material and gage factor $C_m$ from Table 5. Calculate the tube diameter and gage factor $k$ by Eq. (5). Use a velocity of 7 ft/s; $(t_v - t_i) = 101.1 - 75.0 = 26.1°$.

To calculate the temperature rise $R$ in Eq. (3), the quantities $R/\Delta t_m$ and $\epsilon$ must first be computed from Eqs. (8) and (7).

The heat-transfer rate $U_o$ is a function of velocity and tube diameter; $U_o = C_1 \sqrt{V}$, with velocity limits of 3.0 and 10.0 ft/s. In Eq. (8), reduce the quantity to $U_o/V$ to $C_1/\sqrt{V}$. For $\frac{5}{8}$- and $\frac{3}{4}$-in tubes, $C_1 = 267$; for $\frac{7}{8}$- and 1-in tubes, $C_1 = 263$; and for $1\frac{1}{8}$- and $1\frac{1}{4}$-in tubes, $C_1 = 259$. Inlet-water-temperature correction factor $C_t$ is obtained from Table 6. By Eqs. (8) and (7),

$$\frac{R}{\Delta t_m} = \frac{0.1550 \times 52(263 \times 1.025 \times 0.85 \times 0.97)}{500 \times \sqrt{7.0}} = 1.35$$

and $\epsilon = 0.740$. In Eq. (6), multiply $(t_v - t_i)$ by $\epsilon$ to get a rise of 19.3°. By Eq. (3), $G = 1.96W_s/R = 45,200$ gal/min; and by Eq. (4), surface $= 52,100$ ft².

**Performance curves** (Fig. 5) are drawn for a condenser to show the back pressure for various condenser steam loads and inlet-water temperatures maintaining a minimum TTD of 5°. The zero-load back pressure and the cutoff pressure are shown in Fig. 6, where the cutoff limitation is set by the air-removal equipment. A combined turbine-condenser performance curve is sometimes drawn in which heat rejected vs. back pressure for a given turbine load is superimposed on the condenser performance.

**Circulating-water pumps** are usually half capacity, with 3 or 4 percent additional allowance on each pump for miscellaneous heat exchangers. The total dynamic head (TDH) is the sum of

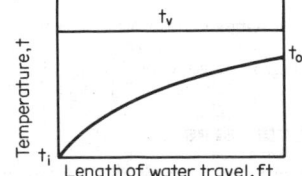

**Fig. 4**  Temperature vs. water travel, surface condenser.

**Table 4. Condenser Proportions**

| Surface, area, sq ft | Effective tube lengths, ft | | |
|---|---|---|---|
| | ¾ O.D. | ⅞ in O.D. | 1 in O.D. |
| 1,000–1,750 | 8, 10, 12, 14 | | |
| 2,000–2,500 | 10, 12, 14, 16 | | |
| 2,750–4,750 | 12, 14, 16, 18 | 12, 14, 16, 18 | |
| 5,000–7,000 | 14, 16, 18, 20 | 14, 16, 18, 20 | 14, 16, 18, 20 |
| 7,250–14,000 | 16, 18, 20, 22 | 16, 18, 20, 22 | 16, 18, 20, 22 |
| 15,000–19,000 | | 18, 20, 22, 24 | 18, 20, 22, 24 |
| 20,000–27,500 | | 20, 22, 24, 26 | 20, 22, 24, 26 |
| 30,000–47,500 | | 22, 24, 26, 28 | 22, 24, 26, 28 |
| 50,000 and over | | 24, 26, 28, 30 | 24, 26, 28, 30 |

**Table 5. Material and Gage Factor**

| Tube material | $C_m$ | | | | | | |
|---|---|---|---|---|---|---|---|
| | BWG No. | | | | | | |
| | 24 | 22 | 20 | 18 | 16 | 14 | 12 |
| Admiralty metal | 1.06 | 1.04 | 1.02 | 1.00 | 0.96 | 0.92 | 0.87 |
| Arsenical copper | 1.06 | 1.04 | 1.02 | 1.00 | 0.96 | 0.92 | 0.87 |
| Aluminum | 1.06 | 1.04 | 1.02 | 1.00 | 0.96 | 0.92 | 0.87 |
| Aluminum brass | 1.03 | 1.02 | 1.00 | 0.97 | 0.94 | 0.90 | 0.84 |
| Aluminum bronze | 1.03 | 1.02 | 1.00 | 0.97 | 0.94 | 0.90 | 0.84 |
| Muntz metal | 1.03 | 1.02 | 1.00 | 0.97 | 0.94 | 0.90 | 0.84 |
| 90-10 Cu-Ni | 0.99 | 0.97 | 0.94 | 0.90 | 0.85 | 0.80 | 0.74 |
| 70-30 Cu-Ni | 0.93 | 0.90 | 0.87 | 0.82 | 0.77 | 0.71 | 0.64 |
| Cold-rolled, low-carbon steel | 1.00 | 0.98 | 0.95 | 0.91 | 0.86 | 0.80 | 0.74 |
| Stainless steels: | | | | | | | |
|     Type 410/430 | 0.88 | 0.85 | 0.82 | 0.76 | 0.70 | 0.65 | 0.59 |
|     Type 304/316 | 0.83 | 0.79 | 0.75 | 0.69 | 0.63 | 0.56 | 0.49 |
|     Type 329 | 0.78 | 0.76 | 0.74 | 0.69 | 0.65 | 0.60 | 0.54 |
| Titanium (tentative) | 0.85 | 0.81 | 0.77 | 0.71 | | | |

the condenser and water-box friction (Figs. 7 to 11), the pipe loss, and any unrecovered static head. Normal heads are about 20 ft; for cooling-tower installations, 60 ft. (See also Sec. 14 and Fig. 12.)

**Condensate pumps** are usually full capacity, determined by condenser flow plus any heater drains dumped into the condenser and 5 percent additional margin. Vertical-pit-type

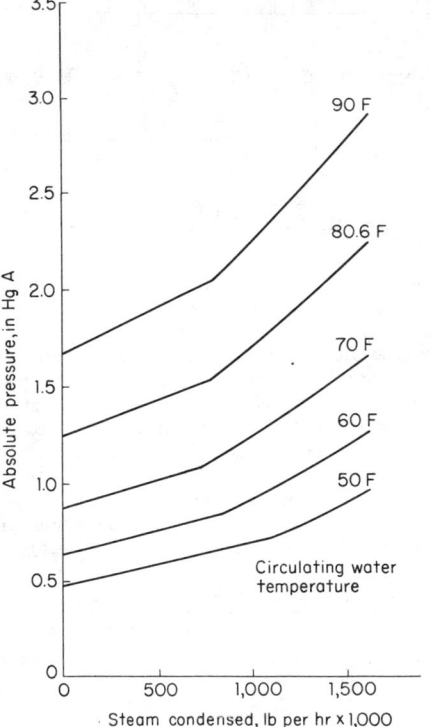

**Fig. 5** Representative performance curves for a surface condenser. 157,500 ft² one-pass surface condenser; 1-in 18-gage, 37.8-ft-active-length aluminum-brass tubes. Performance is based on 85 percent clean tubes, 221,400 gal/min at a velocity of 7 ft/s, and 968 Btu/lb rejected to the circulating water.

**Table 6. Inlet-Water-Temperature Correction Factors**

| Inlet temp $t_i$, deg F | Correction factor, $C_t$ |
|---|---|
| 40 | 0.685 |
| 50 | 0.810 |
| 60 | 0.915 |
| 70 | 1.000 |
| 80 | 1.045 |
| 90 | 1.075 |
| 100 | 1.100 |

designs are commonly used because of low net-positive-suction-head (NPSH) requirements with water at the boiling point. TDH ranges from a normal value of 100 ft on small plants to 800 ft on larger plants. (See Sec. 14.)

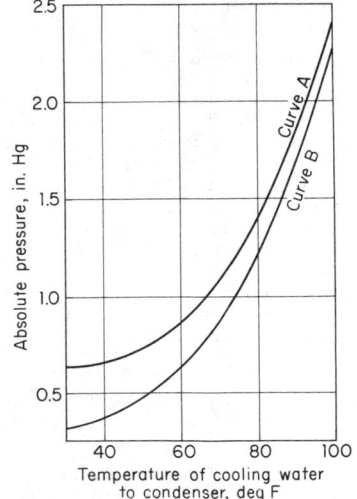

**Fig. 6** Cutoff and zero-load vacuums. Curve *A*, cutoff (except where the absolute pressure is limited to 5°F TTD); curve *B*, zero load.

### AIR-COOLED CONDENSERS

The air-cooled condenser is used where adequate water is not available and to minimize the number of pieces of equipment.

It is adaptable to power plants mounted on trains or made up in modules and transported to isolated locations. Installations have been limited to plant sizes below 1,000 kW. The steam condenses inside the tubes, cooling air flows over the external finned surface, and louvers control the airflow, generally origi-

coefficient; steam-side pressure drop, in both horizontal and vertical tubes, with uniformity of steam distribution to all tubes; air-side pressure drop; fin-tube geometry; and bundle geometry.

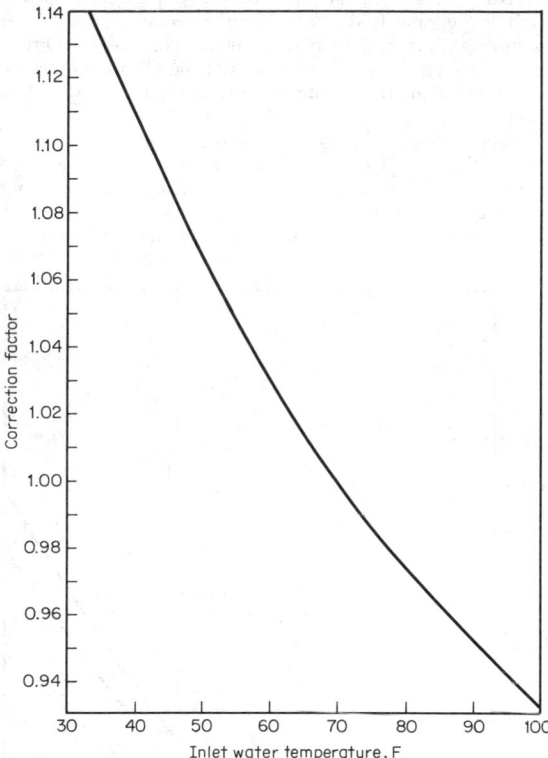

**Fig. 9** Temperature correction for friction losses in tubes (see Figs. 7 and 8). *(Heat Exchange Institute, 1970.)*

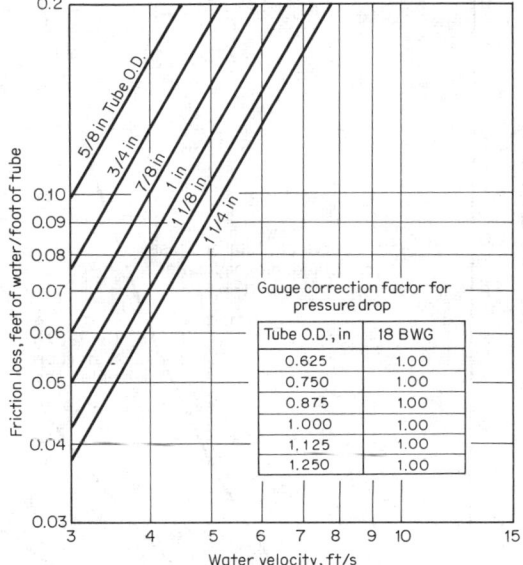

**Fig. 7** Friction loss for water flowing in 18 BWG tubes, low velocity. *(Heat Exchange Institute, 1970.)*

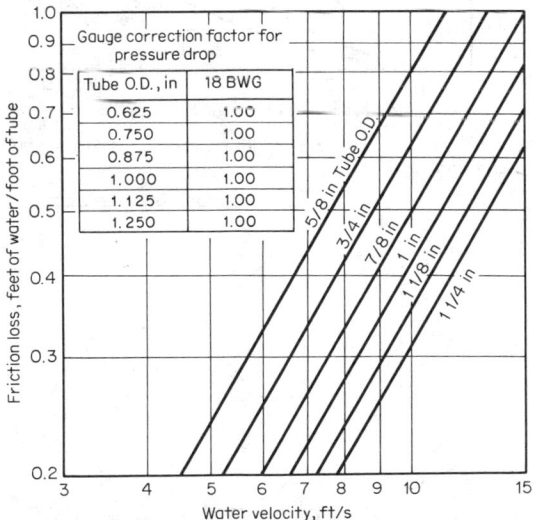

**Fig. 8** Friction loss for water flowing in 18 BWG tubes, high velocity. *(Heat Exchange Institute, 1970.)*

nating with propeller fans (see Sec. 14). Freezing is a problem which needs rigorous attention. Air-recirculation potentials are similar to those on cooling towers. Aluminum is generally used for tubes, fins, frames, and louvers to reduce weight. Items to be considered in design are overall heat-transfer

### DIRECT-CONTACT CONDENSERS

REFERENCES: Heat Exchange Institute Standards for Direct Contact and Low Level Condensers. How to Design Barometric Condensers, *Chem. Eng.*, Feb. 1956.

When (1) low investment is desired and (2) condensate recovery is not a factor, direct-contact condensers are effective. They are relatively simple to build and operate, are limited to sizes less than 250,000 lb of steam per h, and are built in three types: (1) barometric, (2) low-level, and (3) jet.

Figure 13 shows a self-supported counterflow **barometric condenser** with tail pipe; hot well, and air ejector. Steam and cooling water flow in opposite directions, with the coldest water for final condensation and cooling of noncondensables. The air pump must handle that part of the air disengaged from the cooling water as well as air leakage. The head required to pump water is pipe friction plus static head, minus 75 percent approx of the design vacuum. The barometric condenser is usually placed outdoors, and the water leg must be more than 34 ft high.

The **low-level condenser** substitutes a pump for the water leg of the barometric condenser to remove the liquid from the vacuum space. The **jet condenser** utilizes the aspirating effect of a jet for the entrainment of noncondensables and the consequent elimination of a separate air pump.

In the usual direct-contact condenser, where steam and raw circulating water are mixed, the recovery of pure condensate is precluded; greater feedwater makeup is necessary, and poorer vacuums are attained than with surface condensers. Direct-contact-condenser installations are not found in large plants, but there is some recent interest in their use with dry cooling towers.

Table 7 gives values of expected air content of cooling waters; Fig. 14 gives typical performance curves; Fig. 15 defines flows $W$, temperatures $t$, and enthalpies $h$ needed for the basic heat-balance equation $W_s(h_s - h_2) = W_w(t_2 - t_1)$. The latent heat $(h_s - h_2)$ is frequently taken as 950 Btu/lb of steam.

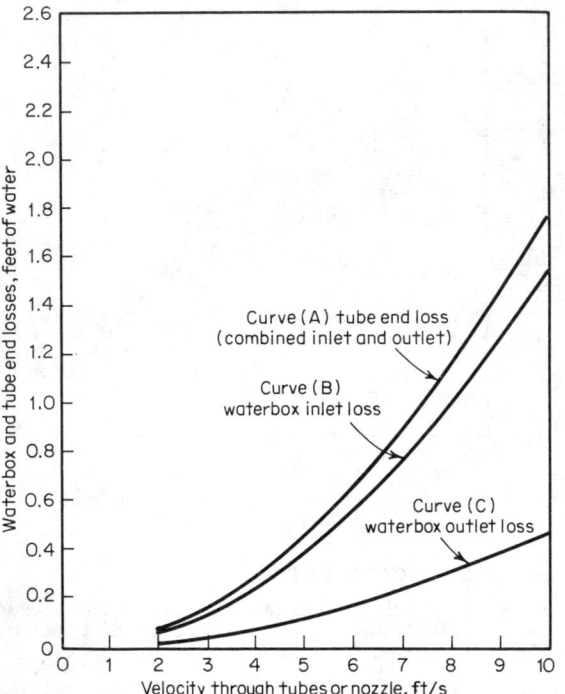

**Fig. 10** Water-box and tube-end losses, single-pass condensers. (*Heat Exchange Institute, 1970.*)

Shell materials are usually steel plate; with dirty or corrosive water, bronze, stainless steel, or linings of ceramic, plastic, or rubber may be used.

## AIR EJECTORS

REFERENCE: Heat Exchange Institute, Standards for Steam Jet Ejectors.

The steam-jet air ejector is used to remove noncondensable air and gases from condensers. It consists of a suction chamber, diffuser, and steam nozzle. The high-velocity jet of steam issuing from the nozzle entrains the gas, and the kinetic energy

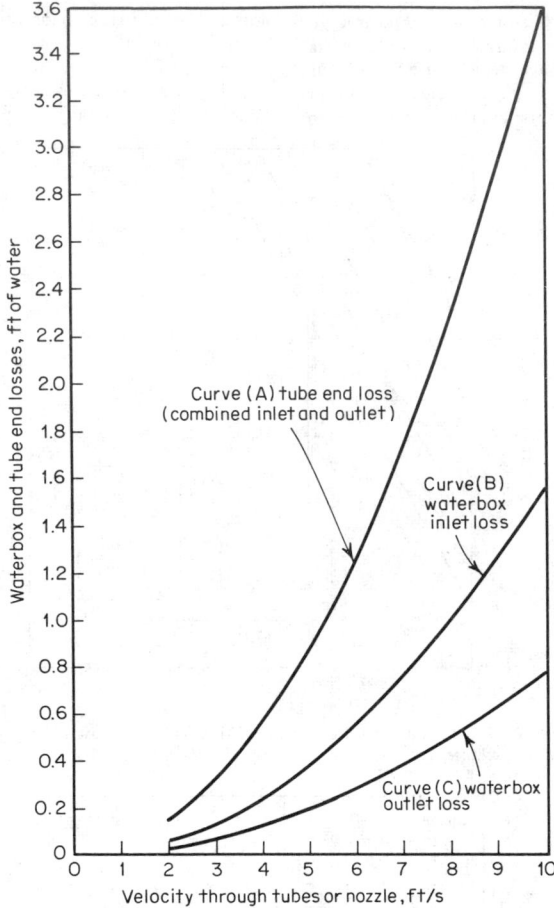

**Fig. 11** Water-box and tube-end losses, two-pass condensers. (*Heat Exchange Institute, 1970.*)

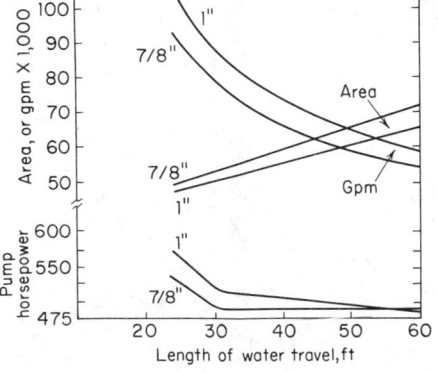

**Fig. 12** Condenser-configuration selection for a given heat load. $W_s$ = 477,355 lb/h; $h$ = 975 Btu/lb; $t_i$ = 70.0°F; 1.50 inHg abs; 18-gage admiralty-metal tubes; 7.0 ft/s water velocity; 85 percent cleanliness factor.

of the mixture is converted to pressure energy in the diffuser. Ratings are in lb/h at a given suction pressure. Ejectors are operated in series or are staged for absolute suction pressures of 1± inHg abs. They will handle wet or dry mixtures, they are easy to operate, installation costs are low, and there are no moving parts—with consequent long life, high sustained efficiency, and low maintenance.

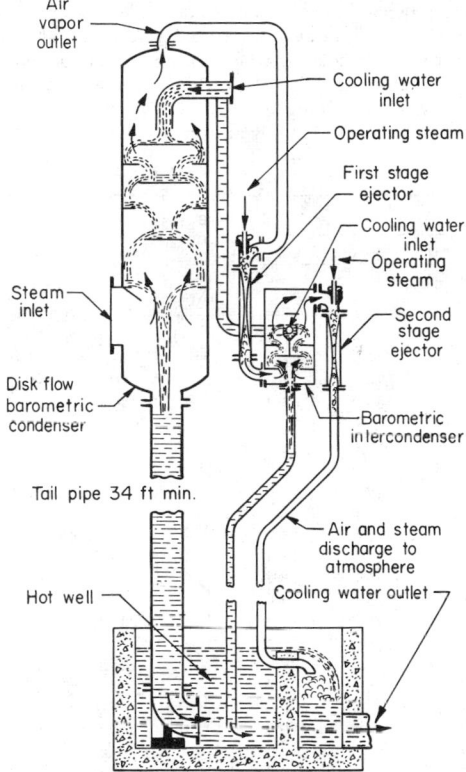

**Fig. 13** Counterflow barometric condenser with two-stage condensing ejector. (*Ingersoll-Rand Co.*)

**Two-stage ejectors,** with surface-type inter- and aftercondensers (Fig. 16), are common in steam-surface-condenser application. The main condensate serves as a coolant, returning heat regeneratively to the boiler feed system. Two-stage ejectors have a shutoff pressure of 0.5 inHg abs approx. Two elements are usually provided, one serving as a spare.

Computation of **ejector performance** requires application of Dalton's law of mixtures (see Sec. 4), where basically the total pressure is the sum of the condensable vapor (steam) pressure and the noncondensable gas (air) pressure. In power-plant condensers, the mixture is saturated with steam so that the temperature of the mixture fixes the partial pressure of the vapor.

Air ejectors are usually rated with suction conditions of 1 inHg abs and 7.5°F subcooling. Since the saturation temperature of steam at 1 inHg abs is 79°F, the mixture entering the ejector is at 71.5°F. The partial pressure of the vapor at 71.5°F is 0.78 inHg abs, and the partial pressure of the air is 1 − 0.78

**Table 7. Air in Cooling Water**

| Water temp, °F | Air, ft³/min per 1,000 gal/min |
|---|---|
| 35 | 4.0 |
| 40 | 3.78 |
| 50 | 3.35 |
| 60 | 2.97 |
| 70 | 2.68 |
| 80 | 2.41 |
| 90 | 2.21 |
| 100 | 2.0 |

= 0.22 inHg abs. Therefore, an ejector with a standard rating (see Table 8) of 12.5 ft³/min of free dry air at 30 inHg abs handles, by the gas laws, 12.5(30/0.22) = 1,704 ft³/min at the ejector suction.

Motive steam to the ejector is usually supplied from turbine throttle conditions, e.g., 600 lb/in², 750°F; 1,500 lb/in², 1000°F; 2,500 lb/in², 1050°F, through a reducing valve to a pressure not higher than 600 lb/in².

The **inter- and aftercondenser,** one- or two-pass, is a small heat exchanger which condenses the motive steam and allows the noncondensable gases, such as $O_2$ and $CO_2$, to be expelled to

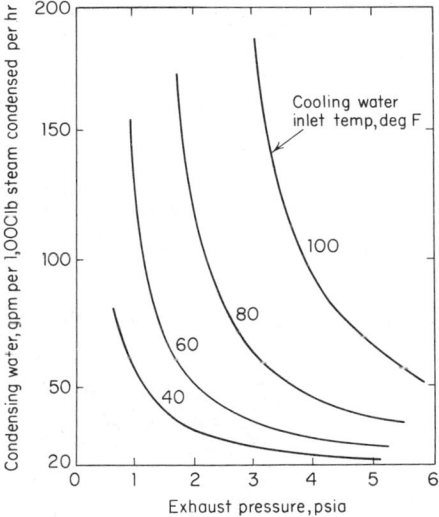

**Fig. 14** Representative performance curves of a direct-contact condenser.

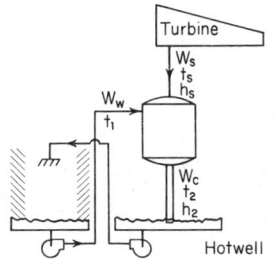

**Fig. 15** Direct-contact-condenser performance terms.

the atmosphere. However, condensable vapors such as $NH_3$ may be returned to the feedwater system. The water-box pressure of tube-side pressure must be designed for the condensate-pump shutoff head. The pressure drop through the inter- and aftercondensers varies from 1 to 10 ft of water, with 3 ft as a reasonable average.

Normal **materials** used are: nozzle, stainless steel type 303; steam chest, steel; inlet-air chamber, cast steel; diffuser, aluminum bronze; inter- and aftercondenser shell, steel; inter- and aftercondenser tubes, stainless steel type 304; inter- and aftercondenser tube sheet, carbon steel.

### Hogging and Priming Ejectors

Single-stage steam-jet ejectors (Fig. 17) are used for evacuat-

ing, or **hogging**, the air from the steam side of a surface condenser, in 15 or 20 min time, so that steam flow can be established and the condenser brought on the line. These ejectors are usually noncondensing and exhaust to the atmosphere. Ratings (Table 9) are typically in lb/h of free dry air at 70°F, suction conditions of 15 inHg abs, shutoff pressure about 2 inHg abs, motive steam pressures and temperatures as with multistage ejectors.

**Priming ejectors** are also single-stage noncondensing units and are used to withdraw air from the water side and to fill the condenser with circulating water during start-up periods. The hogging ejector may also serve this priming function by a suitable system of piping and valves.

**Materials** normally employed are: for the nozzle, stainless

**Table 8. Venting-Equipment Capacities***

| Effective steam flow each main exhaust opening, lb/h | | One condenser shell | | | | | | | | |
|---|---|---|---|---|---|---|---|---|---|---|
| | | Total number of exhaust openings | | | | | | | | |
| | | 1 | 2 | 3 | 4 | 5 | 6 | 7 | 8 | 9 |
| Up to 25,000 | scfm† | 3.0 | 4.0 | 5.0 | 5.0 | 7.5 | 7.5 | 7.5 | 10.0 | 10.0 |
| | Dry air, lb/h | 13.5 | 18.0 | 22.5 | 22.5 | 33.8 | 33.8 | 33.8 | 45.0 | 45.0 |
| | Water vapor, lb/h | 29.7 | 39.6 | 49.5 | 49.5 | 74.4 | 74.4 | 74.4 | 99.0 | 99.0 |
| | Total mixture, lb/h | 43.2 | 57.6 | 72.0 | 72.0 | 108.2 | 108.2 | 108.2 | 144.0 | 144.0 |
| 25,001–50,000 | scfm† | 4.0 | 5.0 | 7.5 | 7.5 | 10.0 | 10.0 | 10.0 | 12.5 | 12.5 |
| | Dry air, lb/h | 18.0 | 22.5 | 33.8 | 33.8 | 45.0 | 45.0 | 45.0 | 56.2 | 56.2 |
| | Water vapor, lb/h | 39.6 | 49.5 | 74.4 | 74.4 | 99.0 | 99.0 | 99.0 | 123.6 | 123.6 |
| | Total mixture, lb/h | 57.6 | 72.0 | 108.2 | 108.2 | 144.0 | 144.0 | 144.0 | 179.8 | 179.8 |
| 50,001–100,000 | scfm† | 5.0 | 7.5 | 10.0 | 10.0 | 12.5 | 12.5 | 15.0 | 15.0 | 15.0 |
| | Dry air, lb/h | 22.5 | 33.8 | 45.0 | 45.0 | 56.2 | 56.2 | 67.5 | 67.5 | 67.5 |
| | Water vapor, lb/h | 49.5 | 74.4 | 99.0 | 99.0 | 123.6 | 123.6 | 148.5 | 148.5 | 148.5 |
| | Total mixture, lb/h | 72.0 | 108.2 | 144.0 | 144.0 | 179.8 | 179.8 | 216.0 | 216.0 | 216.0 |
| 100,001–250,000 | scfm† | 7.5 | 12.5 | 12.5 | 15.0 | 17.5 | 20.0 | 20.0 | 25.0 | 25.0 |
| | Dry air, lb/h | 33.8 | 56.2 | 56.2 | 67.5 | 78.7 | 90.0 | 90.0 | 112.5 | 112.5 |
| | Water vapor, lb/h | 74.4 | 123.6 | 123.6 | 148.5 | 175.1 | 198.0 | 198.0 | 247.5 | 247.5 |
| | Total mixture, lb/h | 108.2 | 179.8 | 179.8 | 216.0 | 251.8 | 288.0 | 288.0 | 360.0 | 360.0 |
| 250,001–500,000 | scfm† | 10.0 | 15.0 | 17.5 | 20.0 | 25.0 | 25.0 | 30.0 | 30.0 | 35.0 |
| | Dry air, lb/h | 45.0 | 67.5 | 78.7 | 90.0 | 112.5 | 112.5 | 135.0 | 135.0 | 157.5 |
| | Water vapor, lb/h | 99.0 | 148.5 | 173.1 | 198.0 | 247.5 | 247.5 | 297.0 | 297.0 | 346.5 |
| | Total mixture, lb/h | 144.0 | 216.0 | 251.8 | 288.0 | 360.0 | 360.0 | 432.0 | 432.0 | 504.0 |
| 500,001–1,000,000 | scfm† | 12.5 | 20.0 | 20.0 | 25.0 | 30.0 | 30.0 | 35.0 | 40.0 | 40.0 |
| | Dry air, lb/h | 56.2 | 90.0 | 90.0 | 112.5 | 135.0 | 135.0 | 157.5 | 180.0 | 180.0 |
| | Water vapor, lb/h | 123.6 | 198.0 | 198.0 | 247.5 | 297.0 | 297.0 | 346.5 | 396.0 | 396.0 |
| | Total mixture, lb/h | 179.8 | 288.0 | 288.0 | 360.0 | 432.0 | 432.0 | 504.0 | 576.0 | 576.0 |
| 1,000,001–2,000,000 | scfm† | 15.0 | 25.0 | 25.0 | 30.0 | 35.0 | 40.0 | 40.0 | 45.0 | 50.0 |
| | Dry air, lb/h | 67.5 | 112.5 | 112.5 | 135.0 | 157.5 | 180.0 | 180.0 | 202.5 | 225.0 |
| | Water vapor, lb/h | 148.5 | 247.5 | 247.5 | 297.0 | 346.5 | 396.0 | 396.0 | 445.5 | 495.0 |
| | Total mixture, lb/h | 216.0 | 360.0 | 360.0 | 432.0 | 504.0 | 576.0 | 576.0 | 648.0 | 720.0 |
| 2,000,001–3,000,000 | scfm† | 17.5 | 25.0 | 30.0 | 35.0 | 40.0 | 45.0 | 50.0 | 55.0 | 60.0 |
| | Dry air, lb/h | 78.7 | 112.5 | 135.0 | 157.5 | 180.0 | 202.5 | 225.0 | 247.5 | 270.0 |
| | Water vapor, lb/h | 173.1 | 247.5 | 297.0 | 346.5 | 396.0 | 445.5 | 495.0 | 544.5 | 594.0 |
| | Total mixture, lb/h | 251.8 | 360.0 | 432.0 | 504.0 | 576.0 | 648.0 | 720.0 | 792.0 | 864.0 |
| 3,000,001–4,000,000 | scfm† | 20.0 | 30.0 | 35.0 | 40.0 | 45.0 | 50.0 | 55.0 | 60.0 | 65.0 |
| | Dry air, lb/h | 90.0 | 135.0 | 157.5 | 180.0 | 202.5 | 225.0 | 247.5 | 270.0 | 292.5 |
| | Water vapor, lb/h | 198.0 | 297.0 | 346.5 | 396.0 | 445.5 | 495.0 | 544.5 | 594.0 | 643.5 |
| | Total mixture, lb/h | 288.0 | 432.0 | 504.0 | 576.0 | 648.0 | 720.0 | 792.0 | 864.0 | 936.0 |

**Table 8. Venting-Equipment Capacities** *(Continued)*

| | | Two condenser shells | | | | | | | | | | | | |
| | | Total number of exhaust openings | | | | | | | | | | | | |
| Effective steam flow each main exhaust opening, lb/h | | 2 | 3 | 4 | 5 | 6 | 7 | 8 | 9 | 10 | 11 | 12 | 13 | 14 |
|---|---|---|---|---|---|---|---|---|---|---|---|---|---|---|
| 100,000–250,000 | scfm† | 15.0 | 20.0 | 20.0 | 20.0 | 25.0 | 25.0 | 30.0 | 30.0 | 35.0 | 35.0 | 40.0 | 40.0 | 40.0 |
| | Dry air, lb/h | 67.5 | 90.0 | 90.0 | 90.0 | 112.5 | 112.5 | 135.0 | 135.0 | 157.5 | 157.5 | 180.0 | 180.0 | 180.0 |
| | Water vapor, lb/h | 148.5 | 198.0 | 198.0 | 198.0 | 247.5 | 247.5 | 297.0 | 297.0 | 346.5 | 346.5 | 396.0 | 396.0 | 396.0 |
| | Total mixture, lb/h | 216.0 | 288.0 | 288.0 | 288.0 | 360.0 | 360.0 | 432.0 | 432.0 | 504.0 | 504.0 | 576.0 | 576.0 | 576.0 |
| 250,001–500,000 | scfm† | 20.0 | 20.0 | 25.0 | 30.0 | 30.0 | 35.0 | 40.0 | 40.0 | 50.0 | 50.0 | 50.0 | 60.0 | 60.0 |
| | Dry air, lb/h | 90.0 | 90.0 | 112.5 | 135.0 | 135.0 | 157.5 | 180.0 | 180.0 | 225.0 | 225.0 | 225.0 | 270.0 | 270.0 |
| | Water vapor, lb/h | 198.0 | 198.0 | 247.5 | 297.0 | 297.0 | 346.5 | 396.0 | 396.0 | 495.0 | 495.0 | 495.0 | 594.0 | 594.0 |
| | Total mixture, lb/h | 288.0 | 288.0 | 360.0 | 432.0 | 432.0 | 504.0 | 576.0 | 576.0 | 720.0 | 720.0 | 720.0 | 864.0 | 864.0 |
| 500,001–1,000,000 | scfm† | 25.0 | 25.0 | 30.0 | 35.0 | 40.0 | 50.0 | 50.0 | 50.0 | 60.0 | 60.0 | 70.0 | 70.0 | 70.0 |
| | Dry air, lb/h | 112.5 | 112.5 | 135.0 | 157.5 | 180.0 | 225.0 | 225.0 | 225.0 | 270.0 | 270.0 | 315.0 | 315.0 | 315.0 |
| | Water vapor, lb/h | 247.5 | 247.5 | 297.0 | 346.5 | 396.0 | 495.0 | 495.0 | 495.0 | 594.0 | 594.0 | 693.0 | 693.0 | 693.0 |
| | Total mixture, lb/h | 360.0 | 360.0 | 432.0 | 504.0 | 576.0 | 720.0 | 720.0 | 720.0 | 864.0 | 864.0 | 1,008.0 | 1,008.0 | 1,008.0 |
| 1,000,001–2,000,000 | scfm† | 30.0 | 35.0 | 40.0 | 40.0 | 50.0 | 50.0 | 60.0 | 60.0 | 70.0 | 70.0 | 80.0 | 80.0 | 90.0 |
| | Dry air, lb/h | 135.0 | 157.5 | 180.0 | 180.0 | 225.0 | 225.0 | 270.0 | 270.0 | 315.0 | 315.0 | 360.0 | 360.0 | 405.0 |
| | Water vapor, lb/h | 297.0 | 345.5 | 396.0 | 396.0 | 495.0 | 495.0 | 594.0 | 594.0 | 693.0 | 693.0 | 792.0 | 792.0 | 891.0 |
| | Total mixture, lb/h | 432.0 | 504.0 | 576.0 | 576.0 | 720.0 | 720.0 | 864.0 | 864.0 | 1,008.0 | 1,008.0 | 1,152.0 | 1,152.0 | 1,296.0 |
| 2,000,001–3,000,000 | scfm† | 35.0 | 40.0 | 40.0 | 50.0 | 60.0 | 60.0 | 70.0 | 70.0 | 80.0 | 80.0 | 90.0 | 100.0 | 100.0 |
| | Dry air, lb/h | 157.5 | 180.0 | 180.0 | 225.0 | 270.0 | 270.0 | 315.0 | 315.0 | 360.0 | 360.0 | 405.0 | 450.0 | 450.0 |
| | Water vapor, lb/h | 346.5 | 396.0 | 396.0 | 495.0 | 594.0 | 594.0 | 693.0 | 693.0 | 792.0 | 792.0 | 891.0 | 990.0 | 990.0 |
| | Total mixture, lb/h | 504.0 | 576.0 | 576.0 | 720.0 | 864.0 | 864.0 | 1,008.0 | 1,008.0 | 1,152.0 | 1,152.0 | 1,296.0 | 1,440.0 | 1,440.0 |
| 3,000,001–4,000,000 | scfm† | 40.0 | 50.0 | 50.0 | 60.0 | 70.0 | 70.0 | 80.0 | 80.0 | 90.0 | 100.0 | 100.0 | 110.0 | 120.0 |
| | Dry air, lb/h | 180.0 | 225.0 | 225.0 | 270.0 | 315.0 | 315.0 | 360.0 | 360.0 | 405.0 | 450.0 | 450.0 | 495.0 | 540.0 |
| | Water vapor, lb/h | 396.0 | 495.0 | 495.0 | 594.0 | 693.0 | 693.0 | 792.0 | 792.0 | 891.0 | 990.0 | 990.0 | 1,089.0 | 1,188.0 |
| | Total mixture, lb/h | 576.0 | 720.0 | 720.0 | 864.0 | 1,008.0 | 1,008.0 | 1,152.0 | 1,152.0 | 1,296.0 | 1,440.0 | 1,440.0 | 1,584.0 | 1,728.0 |

**Table 8. Venting Equipment Capacities** *(Continued)*

| | | | | | | Three condenser shells | | | | | | |
| | | | | | | Total number of exhaust openings | | | | | | |
| Effective steam flow each main exhaust opening, lb/h | | 3 | 4 | 5 | 6 | 7 | 8 | 9 | 10 | 11 | 12 | 13 | 14 |
|---|---|---|---|---|---|---|---|---|---|---|---|---|---|
| 250,000–500,000 | scfm[†] | 30.0 | 30.0 | 37.5 | 37.5 | 37.5 | 45.0 | 52.5 | 52.5 | 60.0 | 60.0 | 75.0 | 75.0 |
| | Dry air, lb/h | 135.0 | 135.0 | 168.8 | 168.8 | 168.8 | 202.5 | 236.3 | 236.3 | 270.0 | 270.0 | 337.5 | 337.5 |
| | Water vapor, lb/h | 297.0 | 297.0 | 371.4 | 371.4 | 371.4 | 445.5 | 519.9 | 519.9 | 594.0 | 594.0 | 742.5 | 742.5 |
| | Total mixture, lb/h | 432.0 | 432.0 | 540.2 | 540.2 | 540.2 | 648.0 | 756.2 | 756.2 | 864.0 | 864.0 | 1,080.0 | 1,080.0 |
| 500,001–1,000,000 | scfm[†] | 30.0 | 37.5 | 45.0 | 45.0 | 52.5 | 52.5 | 60.0 | 75.0 | 75.0 | 75.0 | 90.0 | 90.0 |
| | Dry air, lb/h | 135.0 | 168.8 | 202.5 | 202.5 | 236.3 | 236.3 | 270.0 | 337.5 | 337.5 | 337.5 | 405.0 | 405.0 |
| | Water vapor, lb/h | 297.0 | 371.4 | 445.5 | 445.5 | 519.9 | 519.9 | 594.0 | 742.5 | 742.5 | 742.5 | 891.0 | 891.0 |
| | Total mixture, lb/h | 432.0 | 540.2 | 648.0 | 648.0 | 756.2 | 756.2 | 864.0 | 1,080.0 | 1,080.0 | 1,080.0 | 1,296.0 | 1,296.0 |
| 1,000,001–2,000,000 | scfm[†] | 37.5 | 45.0 | 52.5 | 52.5 | 60.0 | 75.0 | 75.0 | 75.0 | 90.0 | 90.0 | 105.0 | 105.0 |
| | Dry air, lb/h | 168.8 | 202.5 | 236.3 | 236.3 | 270.0 | 337.5 | 337.5 | 337.5 | 405.0 | 405.0 | 472.5 | 472.5 |
| | Water vapor, lb/h | 371.4 | 445.5 | 519.9 | 519.9 | 594.0 | 742.5 | 742.5 | 742.5 | 891.0 | 891.0 | 1,039.5 | 1,039.5 |
| | Total mixture, lb/h | 540.2 | 648.0 | 756.2 | 756.2 | 864.0 | 1,080.0 | 1,080.0 | 1,080.0 | 1,296.0 | 1,296.0 | 1,512.0 | 1,512.0 |
| 2,000,001–3,000,000 | scfm[†] | 45.0 | 52.5 | 60.0 | 75.0 | 75.0 | 75.0 | 90.0 | 90.0 | 105.0 | 105.0 | 120.0 | 120.0 |
| | Dry air, lb/h | 202.5 | 236.3 | 270.0 | 337.5 | 337.5 | 337.5 | 405.0 | 405.0 | 472.5 | 472.5 | 540.0 | 540.0 |
| | Water vapor, lb/h | 445.5 | 519.9 | 594.0 | 742.5 | 742.5 | 742.5 | 891.0 | 891.0 | 1,039.5 | 1,039.5 | 1,188.0 | 1,188.0 |
| | Total mixture, lb/h | 648.0 | 756.2 | 864.0 | 1,080.0 | 1,080.0 | 1,080.0 | 1,296.0 | 1,296.0 | 1,512.0 | 1,512.0 | 1,728.0 | 1,728.0 |
| 3,000,001–4,000,000 | scfm[†] | 52.5 | 60.0 | 75.0 | 75.0 | 90.0 | 90.0 | 105.0 | 105.0 | 120.0 | 120.0 | 135.0 | 135.0 |
| | Dry air, lb/h | 236.3 | 270.0 | 337.5 | 337.5 | 405.0 | 405.0 | 472.5 | 472.5 | 540.0 | 540.0 | 607.5 | 607.5 |
| | Water vapor, lb/h | 519.9 | 594.0 | 742.5 | 742.5 | 891.0 | 891.0 | 1,039.5 | 1,039.5 | 1,188.0 | 1,188.0 | 1,336.5 | 1,336.5 |
| | Total mixture, lb/h | 756.2 | 864.0 | 1,080.0 | 1,080.0 | 1,296.0 | 1,296.0 | 1,512.0 | 1,512.0 | 1,728.0 | 1,728.0 | 1,944.0 | 1,944.0 |

*Heat Exchange Institute, 1970.

†14.7 psia at 70°F.

NOTE: These tables are based on air leakage only and the air-vapor mixture at 1 inHg abs and 71.5°F.

steel, 18-8 chrome nickel; steam chest, steel; air-inlet chamber and diffuser, cast steel.

## VACUUM PUMPS

REFERENCE: Woodman, Rotary or Steam-Jet Air Pumps, *Power*, Aug. 1948.

**Mechanical vacuum pumps,** used for hogging, holding, and priming, are of several types for power-plant service: (1) reciprocating—piston, diaphragm; (2) rotary—sliding-vane, oval-water-seal, eccentric-rotor. (See Sec. 14 for details.) Multistage

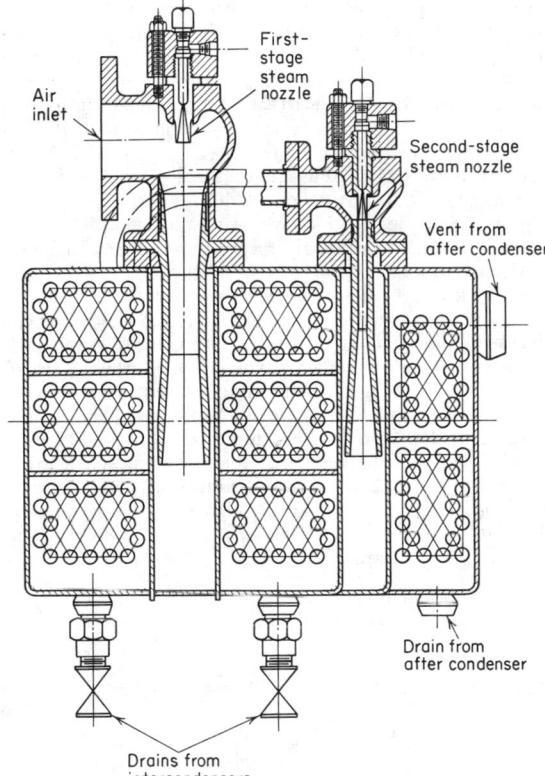

**Fig. 16**   Two-stage air ejector with surface-type inter- and aftercondensers. *(Westinghouse Corp.)*

pumps usually need intercooling. A silencer is generally provided on the air discharge. Normally, two half-capacity pumps are installed, both operating during hogging and one adequate for holding service. The relative capabilities of a vacuum pump and hogging and holding ejectors are illustrated in Fig. 18 (note the considerable hogging capacity of the vacuum pump on start-up).

Advantages of vacuum pumps compared to steam-jet air ejectors include (1) system independent of a steam supply; (2) start-up when steam is not available, as on a once-through boiler system; (3) system capable of completely automatic operation; (4) high-pressure steam piping eliminated; (5) recycling of noncondensables and ammonia eliminated; (6) quieter operation. Disadvantages include (1) damage by water enter-

ing the intake (except when pumps having a liquid rotating seal); (2) higher initial cost; (3) higher maintenance cost. Operating costs are approximately equal. Various combination arrangements are found in practice, e.g., an air-operated air-ejector first stage followed by a vacuum-pump second stage, with consequent savings in motor-horsepower and space requirements.

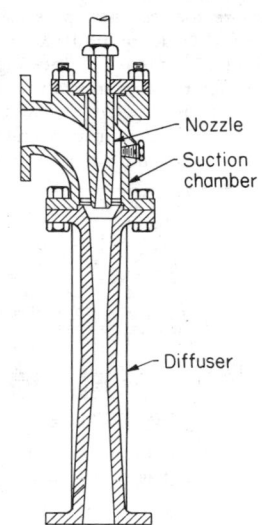

**Fig. 17**   Single-stage air ejector. *(Ingersoll-Rand Co.)*

## COOLING TOWERS

REFERENCES: "Evaluated Weather Data for Cooling Equipment Design," Fluor Products Co. Cooling Towers, *Power*, Mar. 1963. Baker and Shryock, A Comprehensive Approach to the Analysis of Cooling Tower Performance, *Trans. ASME*, 1961.

The prevalent type of cooling tower (see Figs. 19 and 22)

**Table 9. Hogger Capacities***

| Total steam condensed, lb/h | scfm†—dry air at 10 inHg abs design suction pressure |
|---|---|
| Up to 100,000 | 50 |
| 100,001–250,000 | 100 |
| 250,001–500,000 | 200 |
| 500,001–1,000,000 | 350 |
| 1,000,001–2,000,000 | 700 |
| 2,000,001–3,000,000 | 1,050 |
| 3,000,001–4,000,000 | 1,400 |
| 4,000,001–5,000,000 | 1,750 |
| 5,000,001–6,000,000 | 2,100 |
| 6,000,001–7,000,000 | 2,450 |
| 7,000,001–8,000,000 | 2,800 |
| 8,000,001–9,000,000 | 3,150 |
| 9,000,001–10,000,000 | 3,500 |

NOTE: In the range of 500,000 lb/h steam condensed and above, the table provides evacuation of the air in the condenser and low-pressure turbine from atmospheric pressure to 10 inHg abs in about 30 min if the volume of condenser and low-pressure turbine is assumed to be 26 ft³/1,000 lb/h of steam condensed.
*Heat Exchange Institute, 1970.
†14.7 psia at 70°F.

dissipates heat by the evaporation of some of the water sprayed into the air circulated through the tower. It is used where water is in limited supply, where temperature pollution of natural water bodies is to be avoided, where water conservation is to be effected, or where otherwise polluted sources must be avoided. Figure 20 illustrates the functional cycle and the significant basic terms (see also Sec. 4).

**Wet-bulb temperature,** for design purposes, should not exceed the maximum expected value more than 5 percent of the time during summer (June to September). (See Fig. 21).

**Approach** is the difference in temperature between the cold water leaving the tower and the ambient wet bulb.

**Cooling range** is the difference in temperature between the hot water entering and the cold water leaving the tower.

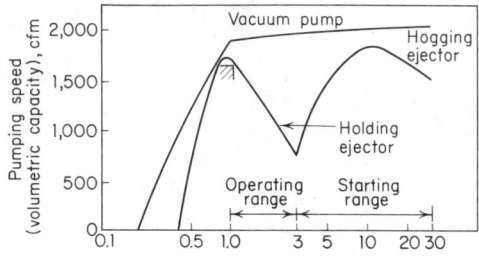

**Fig. 18** Relative capabilities of air-removal pumps.

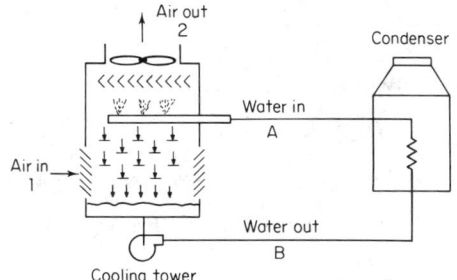

**Fig. 19** Schematic for induced-draft, counterflow, cooling-tower installation.

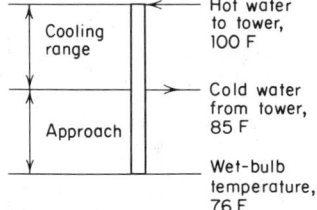

**Fig. 20** Cooling-tower terms.

**Drift** is the water lost as mist or droplets entrained by the circulating air and discharged to the atmosphere. It is in addition to the evaporative loss and is minimized by good design.

**Makeup** is the water required to replace total losses by evaporation, drift, blowdown, and small leaks.

The early, simple atmospheric towers have, because of reliance on natural air circulation, high pumping heads, excessive spray losses, and makeup, been largely superseded by

three important types: (1) forced-draft, (2) induced-draft, and (3) hyperbolic (Fig. 22).

The **mechanical forced-draft tower** (Fig. 22*a*) gives (1) a controllable air supply from fans conveniently located for inspection and maintenance at ground level, and (2) reduced water-pumping head. Nonuniform distribution of air over the ground area of the tower cell, recirculation of vapor from the tower discharge to the tower inlet, with its deleterious effects and fan icing in cold weather, and limitations on the physical diameter of fans are all problems with forced-draft towers.

The **induced-draft tower** (Fig. 22*b*) is prevalent in U.S. practice. The fan is mounted on the top (discharge) of the cell, with consequent improved air distribution within the cell; drift eliminators reduce makeup requirements; spray nozzles, downspouts, splash plates, and splash bars ensure ample evaporative surface for the water, with maximum volumetric heat-transfer rates. In the **counterflow** design, air is introduced beneath the cell fill, but in the **crossflow** design (Fig. 22*b*), air is introduced at the sides of the fill.

The **hyperbolic tower** (Fig. 22*c*) utilizes the chimney effect (height, 300± ft) for natural circulation. It has been favored in large European installations where the prevalent lower dry-bulb ambient temperature gives a greater difference in density between entrance to and exit from the tower. The wide approach keeps the unit size within practical bounds, and the savings in fan power support the higher investment. Recent installations in the United States attest to its attractiveness.

**Precautions** are necessary to avoid freezing troubles in cold weather, fire hazards with intermittent operations, corrosion, scale, and microbiological growth problems. **Location** must avoid recirculation: for tower lengths to 250 ft, the long-axis orientation should be parallel to the prevailing wind; towers longer than 250 ft should be arranged broadside to the prevailing wind. The tower should be isolated as much as possible; adjacent heat sources compel the specification of higher design wet-bulb temperatures.

### Performance Calculations

(See Fig. 19)

Applicable equations are

$$W_{a1}h_{a1} + W_{v1}h_{v1} + W_{wA}h_{fA}$$
$$= W_{a2}h_{a2} + W_{v2}h_{v2} + W_{wB}h_{fB} \quad \textbf{(9)}$$

$$W_{wB} = W_{wA} - (W_{v2} - W_{v1}); \ W_{a1} = W_{a2} \quad \textbf{(10)}$$

$$W_{wA}(h_{fA} - h_{fB}) = (W_{a2}h_{a2} + W_{v2}h_{v2})$$
$$- (W_{a1}h_{a1} + W_{v1}h_{v1}) - (W_{v2} - W_{v1})h_{fB} \quad \textbf{(11)}$$

$$h_{fA} - h_{fB} = t_{wA} - t_{wB} \quad \textbf{(12)}$$

and

$$W_a h_a + W_v h_v \quad \textbf{(13)}$$

= total heat, from psychrometric chart    (See Sec. 4.)

$$W_{wA}(t_{wA} - t_{wB}) = \text{total heat at 2}$$
$$- \text{total heat at 1} - (W_{v2} - W_{v1})h_{fB} \quad \textbf{(14)}$$

where $W$ = flow, lb/h; $a$ = air; $w$ = water; $v$ = vapor; $h$ = enthalpy, Btu/lb; $f$ = fluid; $t$ = temperature, °F; 1, 2, $A$, $B$ = locations.

EXAMPLE.  Given: flow = 80,000; wet bulb = 76°F; dry bulb = 86°F; range (also condenser rise) = 100 − 85 = 15°F; approach = 9°F. Find: economical tower size, circulating-pump power, fan capacity and power, and makeup water.

SOLUTION.  (1) *Economical tower size:* A study must be made to tie in with condenser. Heat to be dissipated is constant. Vary rise (range),

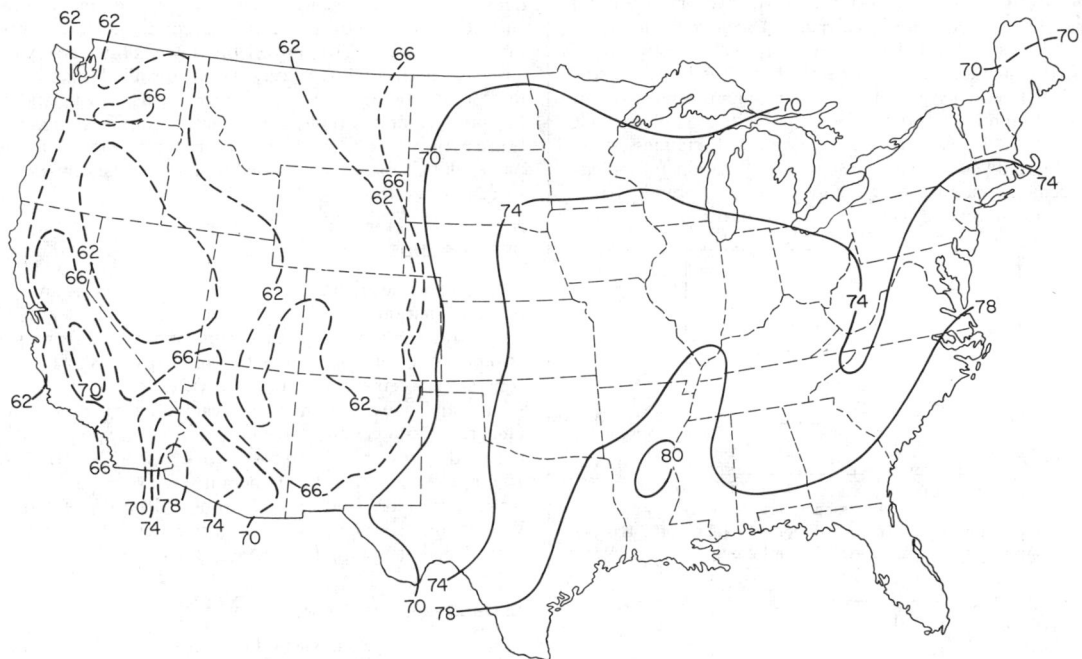

**Fig. 21**  Wet-bulb temperature isolines at the 5 percent level, United States. *(Adapted from Fluor Products Co., by permission.)*

approach, and then compare capital costs of cooling tower, circulating pump, and condenser with operating costs for fan and pump power. (2) *Circulating-water-pump power;* bhp = gal/min × TDH/3,960 × eff. TDH = condenser friction + pipe friction + static head. (3) *Fan capacity and power:* If air leaves tower saturated at 95°F, then from psychrometric charts (Sec. 4),

$$\text{Total heat} = 1 \text{ lb dry air } C_p(t_a - 0°) + W_v h_g \text{ at } t_a \qquad (15)$$

For air and vapor at 1, vapor = 120 grains/lb dry air. By Eqs. **(13)** and **(15)**, total heat = 1 × 0.24(86 − 0) + (120/7000)1,099.2 = 39.5 Btu/lb dry air. For air and vapor at 2, vapor = 256 grains/lb dry air. Total heat = 63.1 per lb dry air. Sp vol (air and vapor) = 14.80 ft³/lb dry air. By Eq. **(14)**, $W_{wA}$(100 − 85) = 63.1 − 39.5 − (256 − 120)$^{63}/_{7,000}$ = 1.50 lb water/lb dry air. 80,000 gal/min/1.50 × 500 = 26.6 × 10⁶ lb dry air/h, or 26.6 × 10⁶ × 14.80/60 = 6.58 × 10⁶ ft³/min. With test data on static-pressure requirements for the tower and with the fan characteristics, the horsepower to drive the fan can be calculated. (4) *Makeup:* $W_{v2} - W_{v1}$ = (256 − 120) 26.6 × 10⁶/7,000 = 515,000 lb/h.

Normally, a cooling tower is purchased for only one guarantee point. It is well, however, to have **performance** curves (Fig. 23) showing operation for various wet-bulb temperatures and cooling ranges. The **investment** for a cooling tower is essentially a matter of water flow and is influenced by approach, range, and wet bulb (Fig. 24). The **cost evaluation** should include consideration of tower frame and fill, fans, motors, basin and pump pit, pump head, fan horsepower, freight, labor, and erection. In the choice of a tower for a power plant, there should be coordinated study and evaluation of the turbine and condenser for best overall economy.

The **height** of a field-erected induced-draft tower, from basin curb to fan deck, ranges from 8 to 50 ft; widths vary from 6 to 60 ft; lengths from 8 to 500 ft; fan-stack height between 2 and 15 ft.

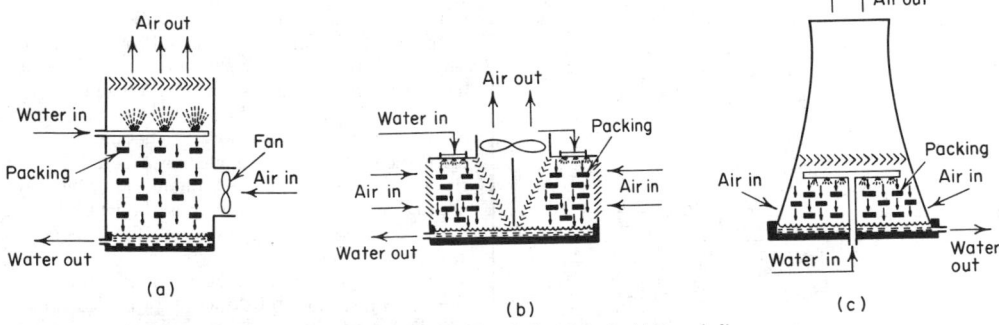

**Fig. 22**  Cooling-tower types. *(a)* Forced-draft; *(b)* crossflow induced-draft; *(c)* hyperbolic.

**Materials** used are: frame—redwood (treated or untreated), Douglas fir (treated), steel (galvanized), concrete; fill—redwood, plastics (polyethylene, polypropylene), cement asbestos board (extruded); casing—cement asbestos board (corrugated, slat), redwood, fiber glass, aluminum, concrete; fan blades—aluminum, glass-reinforced polyester, stainless steel, monel; fan hubs—cast iron, galvanized steel, stainless steel; fan stack—redwood, steel, masonite, drift eliminators—redwood, cement, asbestos board; spray nozzles—ceramic, bakelite; louvers—redwood, cement asbestos board.

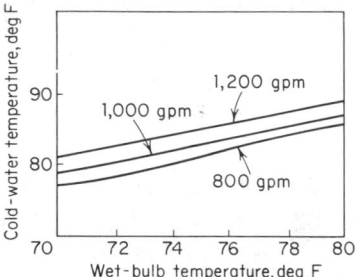

**Fig. 23**  Cooling-tower performance: design 85°F cold water, 95°F hot water; 78°F wet bulb; 10° range; 1,000 gal/min cell.

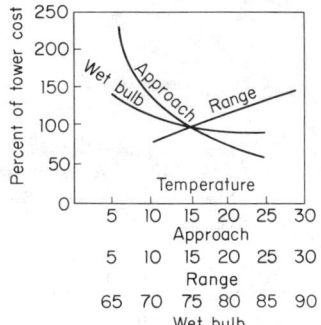

**Fig. 24**  Comparative cooling-tower costs. *(Foster Wheeler Corp.)*

### DRY COOLING TOWERS, WITH DIRECT-CONTACT CONDENSERS

REFERENCE: Ritchings and Lotz, Economics of Closed vs. Open Cooling Water Cycles, *Power Eng.*, May and June 1963.

The thermodynamic requirement for a heat sink can generally best be met with a natural body of water such as a river, lake, or ocean. When water supplies are limited, a **"wet" cooling tower** (water losses less than 2 or 3 percent) may be used for reclamation. A **"dry" cooling tower** will further reduce this loss. Figure 25 shows an application with a direct-contact con-

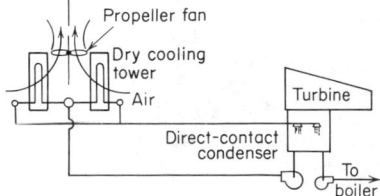

**Fig. 25**  Dry cooling-tower system.

denser. The finned surfaces of the dry cooling tower are substituted for the tubes of a surface condenser. Heat-transfer rates (see Sec. 4) are appreciably lower, but the extended surface, relatively low in cost, may be installed in an induced-draft tower or in a hyperbolic tower, as in England for a 125,000-kW unit. Problems of recirculation, freezing, and air leakage must be solved. If a minor evaporative loss is permissible, peaking capacity is obtained by wetting the heat-exchange surface in the tower.

### SPRAY PONDS

If land area is available, a spray pond may serve as an alternative reclamation system. Pipelines are typically arranged on 50-ft centers, with five nozzles (50± gal/min each) every 12 ft of pipeline. Cooling is limited by the relatively short period of contact between spray and air. Drift, especially with adverse wind conditions, can make losses higher than on towers. Generally, if cooling efficiencies above 50 percent are required, a tower is favored, where efficiency = $(t_h - t_c) \times 100/(t_h - t_{wb})$, where the temperature $t$ subscripts are $h$ = hot water, $c$ = cold water, $wb$ = wet bulb of the ambient air. (See Wright and Kirsopp, Pond Surface Cooling of a Chemical Plant Cooling Water, *Proc. Am. Power Conference*, 1960.)

### CLOSED FEEDWATER HEATERS

REFERENCES: Heat Exchange Institute Standards for Closed Feedwater Heaters. ASME Boiler and Pressure Vessel Code, Sec. VIII.

Feedwater heaters are used (1) to effect the thermodynamic gains of the regenerative steam cycle (see Sec. 4 and 9), and (2) to raise water temperatures to a sufficient value for the avoidance of thermal shock to the boiler metal. The **number and types of heaters** usually employed are: (1) plant sizes of up to 70 mW, two closed low-pressure heaters, one open heater, one closed high-pressure heater; (2) plant sizes of 75 to 300 mW, two or three closed low-pressure heaters, one open heater, two closed high-pressure heaters; (3) fossil-fueled plant sizes above 300 mW, three or four closed low-pressure heaters, one open heater, two or three high-pressure heaters. Nuclear units require very large feedwater flows, necessitating double or triple parallel strings of heaters. There are generally five or six low-pressure and one high-pressure heaters (no open heaters) in each string.

Minimum heater cost prevails with minimum restrictive specifications, e.g., horizontal, two-pass, high water velocity (10 ft/s at 60°F), no length limits. Overall heater length is limited by maximum available tube lengths of 100 ft for copper alloys, admiralty metal, and copper-nickel and 85 ft for Monel. With U-tube construction, this results in heater lengths of about 48 and 40 ft, respectively. A general rule, to ensure good steam distribution, is that the length in feet shall not exceed the shell diameter in inches plus 2; i.e., with a 30-in diam shell, the length should not exceed 32 ft. **Pressure drop** through the tubes must be economically evaluated as it varies approximately with the square of the water velocity.

If a **length restriction** is imposed, the designer may have to substitute a four-pass arrangement for the two-pass design, with consequent large-diameter shell and water chamber, heavier walls, more tube holes to be drilled, more tubes to be installed, and a cost increase. If a **pressure-drop restriction** is imposed, a lower water velocity results, with more tubes, larger shell and chamber diameter, and more surface because of the lower heat-transfer rate. Vertical heaters, with appropriate construction details, are also higher in cost.

The **condensing heater** (Fig. 26), like a surface condenser, performs at constant saturation temperature outside the tubes with no condensate subcooling. Feedwater is generally heated to within 5°F of the saturation temperature. The addition of a **drain-cooling section** (Fig. 27), with a cover or enveloping baffle around the tubes of the inlet pass, can reduce the drip temperature to within 10 or 15°F of the incoming water. When the steam is at sufficient pressure (above 125 lb/in² abs) and contains enough superheat (more than 250°), a **desuperheating section** (Fig. 27), using an enveloping baffle around the tubes at feedwater outlet, can raise the water temperature above the saturation temperature of the entering steam. Drain coolers

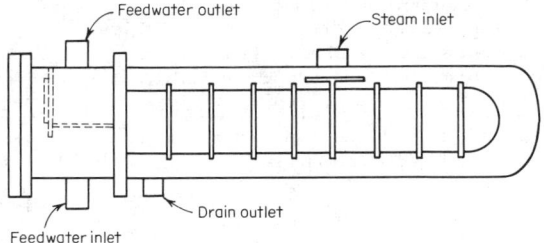

**Fig. 26** Basic heater section for condensing heating.

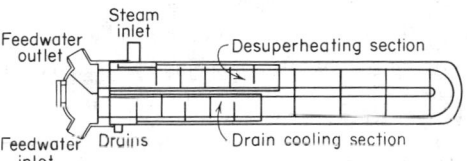

**Fig. 27** Basic heater section with drain cooling and desuperheater sections added.

and desuperheating sections improve overall plant heat rate by reduction of energy degradation. Subcooling of drains reduces flashing, erosion, vibration, and noise when drains are dropped to a lower-pressure heater. Special attention should be given to drains flashed to the main condenser, as this is a direct thermodynamic loss. As an alternate, these drains can be pumped forward into the feedwater line.

**Heat-transfer rates** for a condensing feedwater heater range, overall, between 500 and 900 Btu/(h)(ft²)(°F). (See Figs. 28 and 29.) Heat-transfer rates for drain-cooling and desuperheating sections must be separately evaluated, with overall ranges

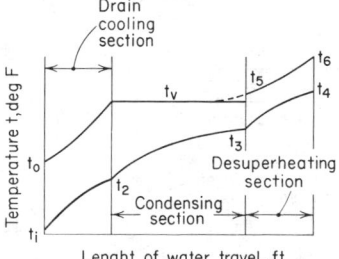

**Fig. 28** Temperature vs. water travel, closed-feed heater.

from 300 to 500 on the former and 80 to 140 on the latter. (See Figs. 30 and 31.) Normally, the three sections are included in the same shell (Fig. 27); when the cooler becomes too large, as with the lowest-pressure heater, a separate shell-and-tube exchanger may be justified.

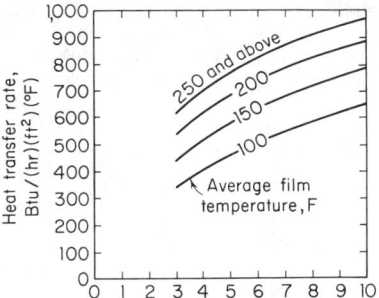

**Fig. 29** Approximate condensing, heat-transfer rates with 18 BWG admiralty-metal tubes. Average film temp = $t_v - 0.8\delta t_m$.

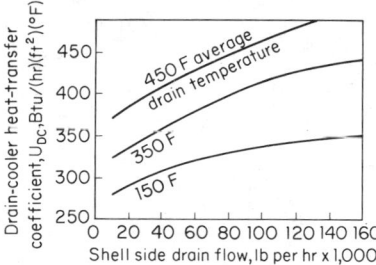

**Fig. 30** Approximate drain-cooler heat-transfer rates.

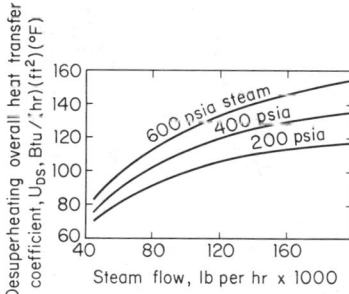

**Fig. 31** Approximate desuperheating section heat transfer rates.

**Low-pressure heaters** (upstream of the boiler feed pump) are usually designed for tube-side pressure of less than 900 psig. This is based on the shutoff head of the condensate pump plus 10 percent. For pressures up to 300 to 400 psig, a bolted cover with flange and gasket (Fig. 32a) is used, while higher pressures usually justify a welded joint to eliminate the gasket (Fig. 32b and c). A manway is utilized for access to the tube sheet. Tubes are generally rolled into the tube sheet.

**High-pressure heaters** (downstream of the boiler feed pump) are designed for shutoff head of the pump plus 10 percent, resulting in about 1,500 psig for a nuclear plank and up to 5,000 psig for fossil-fuel plants. Hemispherical head construction (Fig. 32c) and welded tube-to-tube sheet joints are almost exclusively used. Since the cost of the tubes represents the largest portion of the heater cost, material selection is important (see chart below). Depending on the desired water-chemistry requirements, low-pressure heaters usually use admiralty metal, 90-10 Cu-Ni, stainless or carbon steel. High-pressure heaters require the use of the higher-strength alloys of 70-30 Cu-Ni, monel, stainless, or carbon steel. The thinnest wall possible [$t = Pd/(2s + 0.8p)$] or the industry minimum standard of 18 BWG (20 BWG for stainless steel) is utilized.

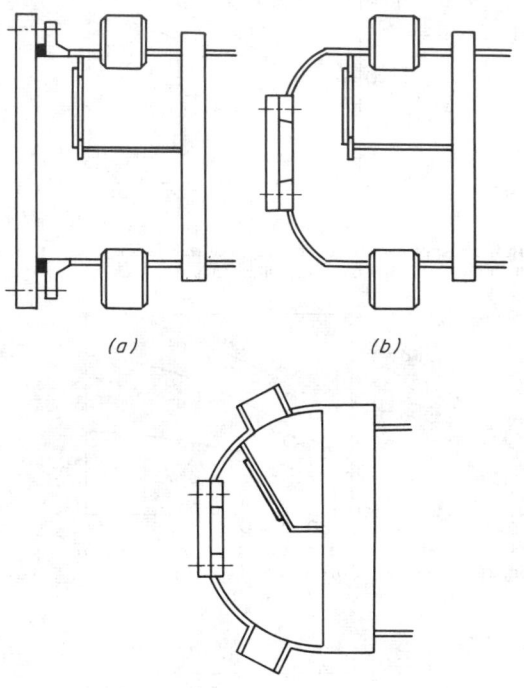

**Fig. 32** Feedwater heater channels: (a) Bolted; (b) Dished; (c) hemispherical.

### Performance Calculations

#### Notation

$A$ = heat-transfer surface, ft²
$C_m$ = material and gage correction factor, Table 11
$D$ = tube ID, in, Table 13
$d$ = OD of tube, in (⅝ and ¾ in are most common)
$F_1$ = friction factor, Table 14

$F_2$ = water-temperature correction factor, Table 15
$h$ = enthalpy, Btu/lb
$k$ = tube diameter and gage factor, Table 12
$L$ = active length of tubes per pass, ft
$N$ = number of passes
$n$ = number of tubes
$P$ = design pressure, lb/in²
$p$ = pressure, lb/in²
$Q$ = total heat transferred, Btu/h
$S$ = allowable design stress at tube design temperature, lb/in², from ASME Code
$t$ = wall thickness, in (see Table 10); temperature, °F
$U$ = overall heat-transfer rate, Btu(h)(ft²)(°F) (the material correction factor, Table 11, must be applied to the overall heat-transfer rate)
$V$ = water velocity, ft/s
$W$ = steam flow, lb/h
$W_w$ = feedwater flow, lb/h
$\Delta p$ = pressure drop, lb/in²
$\Delta t_m$ = logarithmic mean temperature difference, °F (Fig. 28)

Heat-balance data will provide the major parameters of steam pressure, feedwater flows, and temperatures. The choice of tube diameter and material must be made first. Tube thickness is calculated by

$$t = Pd/(2S + 0.8P)$$

Test pressure is 1.5 times design pressure.

**Condensing-heater-size** equations are

$$Q = UA\ \Delta t_m = W_w(h_3 - h_2) = W_s(\Delta h_s)$$

(See Figs. 28 and 29.) The minimum terminal temperature difference is 2°F.

The active length of tubes is calculated from $L = 500AV/NW_wk$ where maximum velocity is 10 ft/s at 60°F; the number of tubes, from $n = 3.82A/Ld$; the pressure drop, from $\Delta p = (L + 5.5D)F_1F_2N/D^{1.24}$.

The shell design pressure is 20 percent greater than the

**Table 10. Tube-Wall Thickness**

| Gage, BWG | Thickness, in |
|---|---|
| 15 | 0.072 |
| 16 | 0.065 |
| 17 | 0.058 |
| 18 | 0.049 |
| 19 | 0.042 |
| 20 | 0.035 |
| 22 | 0.028 |

**Table 11. Material and Gage Correction Factor** $C_m$
(For ⅝- to 1-in tubes)

| | Tube material | | | | | | | |
|---|---|---|---|---|---|---|---|---|
| BWG | Ars-Cu | Admiralty | 90-10 Cu-Ni | 80-20 Cu-Ni | 70-30 Cu-Ni | Monel | 18-8 SS | Carbon steel |
| 18 | 1.00 | 1.00 | 0.97 | 0.95 | 0.92 | 0.89 | 0.85 | 0.96 |
| 17 | 1.00 | 1.00 | 0.94 | 0.91 | 0.87 | 0.85 | 0.80 | 0.92 |
| 16 | 1.00 | 1.00 | 0.91 | 0.88 | 0.84 | 0.82 | 0.77 | 0.89 |
| 15 | 1.00 | 0.99 | 0.89 | 0.86 | 0.82 | 0.79 | 0.74 | 0.87 |
| 14 | 1.00 | 0.96 | 0.85 | 0.82 | 0.77 | 0.75 | 0.70 | 0.83 |
| 13 | 0.98 | 0.93 | 0.81 | 0.78 | 0.73 | 0.70 | 0.65 | 0.79 |

**Table 12. Tube Diameter and Gage Factor** $k$

| Diam, in | BWG | | | | |
|---|---|---|---|---|---|
| | 20 | 18 | 17 | 16 | 15 |
| $\frac{5}{8}$ | 0.217 | 0.2407 | 0.2580 | 0.2728 | 0.289 |
| $\frac{3}{4}$ | 0.1732 | 0.1887 | 0.1995 | 0.2089 | 0.218 |
| $\frac{7}{8}$ | 0.1445 | 0.1550 | 0.1624 | 0.1686 | 0.1748 |
| 1 | 0.1238 | 0.1314 | 0.1369 | 0.1413 | 0.1458 |

**Table 13. Feedwater-Heater Tube Constants**

| Tube BWG | $\frac{5}{8}$ in OD | | $\frac{3}{4}$ in OD | | $\frac{7}{8}$ in OD | |
|---|---|---|---|---|---|---|
| | $D$ | $D^{1.24}$ | $D$ | $D^{1.24}$ | $D$ | $D^{1.24}$ |
| 18 | 0.527 | 0.452 | 0.652 | 0.589 | 0.777 | 0.731 |
| 17 | 0.509 | 0.433 | 0.634 | 0.567 | 0.759 | 0.710 |
| 16 | 0.495 | 0.418 | 0.620 | 0.554 | 0.745 | 0.695 |
| 15 | 0.481 | 0.404 | 0.606 | 0.538 | 0.731 | 0.678 |
| 14 | 0.458 | 0.380 | 0.584 | 0.514 | 0.709 | 0.652 |
| 13 | ..... | ..... | 0.560 | 0.488 | 0.685 | 0.625 |

**Table 14. Friction Loss, $lb/in^2$ per ft of Straight Travel**

| Water velocity (at 60°F), ft/s | $F_1$ |
|---|---|
| 3.0 | 0.018 |
| 4.0 | 0.031 |
| 5.0 | 0.047 |
| 6.0 | 0.065 |
| 7.0 | 0.085 |
| 8.0 | 0.108 |
| 9.0 | 0.134 |
| 10.0 | 0.162 |

**Table 15. Water-Temperature Correction Factor\***

| Average water temp, deg F | $F_2$ |
|---|---|
| 100 | 0.905 |
| 150 | 0.840 |
| 200 | 0.795 |
| 250 | 0.770 |
| 300 | 0.755 |
| 350 | 0.750 |
| 400 | 0.755 |
| 450 | 0.757 |
| 500 | 0.795 |
| 550 | 0.830 |
| 600 | 0.885 |

\*Average water temperature = $t_s - \Delta t_m$.

maximum operating pressure, rounded to the higher 25 $lb/in^2$ as a minimum.

A **drain-cooling section,** when added, is treated as a separate heat exchanger where the quantities $t_v$, $t_o$, $t_i$, $W_s$, and $W_w$ are known; $t_2$ is the only unknown and is to be calculated from a heat balance on the section or $Q = W_s(h_v - h_o) = W_w(h_2 - h_1)$. Compute $\Delta t_m$; find the area of the drain-cooler section $A_{dc}$ from $Q = UA_{dc}\,\Delta t_m$, with overall $U$ from Fig. 30.

A **desuperheating section** is similarly calculated with $t_v$, $t_6$, $t_4$, $p_6$, $h_6$, $h_4$, $W_s$ known. Allow approximately 1 percent pressure drop through the section to get $p_v$, and 80 to 90 percent desuperheating (30°F superheat entering condensing section) to obtain $t_5$ and $h_5$. By a heat balance on section, calculate $h_3$ from $Q = W_s(h_6 - h_5) = W_w(h_4 - h_3)$. With equivalent $t_3$,

calculate $\Delta t_m$ and substitute in $Q = UA\,\Delta t_m$ for area of desuperheating section (overall $U$ from Fig. 31).

**Materials**

Heaters are designed in accordance with the Heat Exchange Institute Standards for Closed Feedwater Heaters and the ASME Boiler and Pressure Vessel Code, Sec. VIII, using the following materials:
Tubes:

| Material | | Max temp, °F | |
|---|---|---|---|
| | | Rolled joint | Welded joint |
| Admiralty metal | | 350 | |
| 90-10 Cu-Ni | ASME SB-395 | 400 | 450 |
| 80-20 Cu-Ni | | 400 | 475 |
| 70-30 Cu-Ni | | 400 | 525 |
| 70-30 Cu-Ni, stress-relieved | | 400 | 525 |
| Monel, ASME SB-163 | | 400 | 600 |
| Carbon steel, ASME SA-556, Grades A, B, and C | | 350 | 650 |
| Stainless steel, ASME SA-249; TP 304 (welded) | | 350 | 650 |

Shell, nozzles: carbon-steel pipe and plant, ASME SA-285 Grade C; SA-515 Grade 70; SA-106 Grade B.
Channels, or heads, channel covers: carbon-steel plate, forgings, and pipe, ASME SA-515 Grade 70; SA-105 Grade II; SA-106 Grade B.
Baffles and tube supports: steel plate, ASME SA-283 Grade C.
Tube sheets: carbon-steel plate and forgings, ASME SA-515 Grade 70; SA-105 Grade II.

Carbon steel is used for temperatures up to 800°F. Chromium-molybdenum steel (ASME SA-387 Grade C) is used for higher temperatures.

## OPEN, DEAERATING, AND DIRECT-CONTACT HEATERS

REFERENCE: Heat Exchange Institute Standards for Deaerators and Deaerating Heaters.

Deaerators or deaerating heaters serve (1) to degasify feedwater and thus reduce equipment corrosion (see Sec. 6), (2) to heat feedwater regeneratively and improve thermodynamic efficiency (see Sec. 9), and (3) to provide storage, positive submergence, and surge protection on the boiler feed-pump suction (see Sec. 14).

Removal of oxygen and carbon dioxide from boiler feedwater and process water at elevated temperature is essential for adequate conditioning. In some power-plant applications, a well-designed surface condenser gives adequate deaeration and the accompanying exclusive use of closed feedwater heaters.

A modern **deaerator** will, by mechanical action, reduce $O_2$ content of effluent to less than 0.005 $cm^3/l$ and $CO_2$ content to a negligible amount. Water must (1) be heated to and kept at saturation temperature, as the gas solubility is zero at the boiling point of the liquid, and (2) be mechanically agitated by spraying or cascading over trays for effective scrubbing, release, and removal of gases. Gases must be swept away by an adequate supply of steam. Since the water is heated to saturation conditions, the terminal temperature difference is zero with maximum improvement in associated turbine heat rate. Extremely low partial gas pressures, dictated by Henry's law, call for large volumes of scrubbing steam. **Vent condensers** of the shell-and-tube or direct-contact types, cooled by incoming feed, serve to recover heat and water before release of gases to the atmosphere.

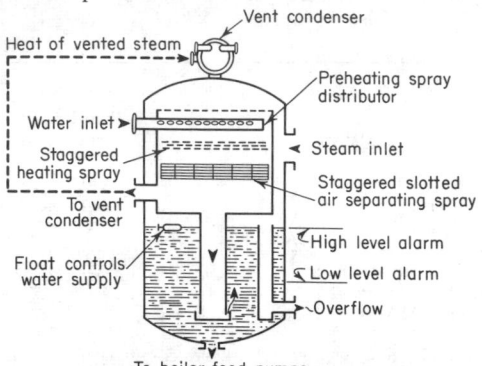

**Fig. 33**   Tray-type deaerating heater section.

The **tray-type deaerator** (Fig. 33) is prevalent. While it has some tendency to scale, it will operate at wide load conditions and is practically independent of water-inlet temperature. Trays can be loaded to some 10,000 lb/(ft²)(h), and the deaerator seldom exceeds 8 ft in height.

The **spray type** uses a high-velocity steam jet to atomize and scrub the preheated water. Applications are (1) marine service, where it is unaffected by ship roll and pitch, and (2) industrial plants where operating pressures are stable. It requires a temperature gradient, e.g., 50°F minimum, to produce the fine sprays and vacuums with the cold water required.

### Materials

Open-feed heaters are designed to the ASME Boiler and Pressure Vessel Code. Materials used are: shell—steel, ASME SA-285 Grade C and SA-515 Grade 70; trays—stainless steel, type 304; baffles in vent condenser—stainless steel, type 304; spray valves—stainless steel, type 304.

### EVAPORATORS

For general treatment of evaporators used in the preparation of makeup water, see Sec. 6.

# INTERNAL-COMBUSTION ENGINES

## by J. A. Bolt, D. E. Cole, W. Mirsky, and D. J. Patterson

REFERENCES: Heldt, "High-Speed Diesel Engines." Lichty, "Internal-Combustion Engines," McGraw-Hill. Obert, "Internal-Combustion Engines," International Textbook. Blackie. Schweitzer, "Scavenging of Two-Stroke-Cycle Engines," Macmillan. Taylor and Taylor, "The Internal-Combustion Engine," International Textbook. Vincent, "Supercharging the Internal-Combustion Engine," McGraw-Hill. Crouse, "Automotive Engine Design," McGraw-Hill. Patterson and Henein, "Emissions from Combustion Engines," Ann Arbor Science. Norbye, "The Wankel Engine," Chilton. Lichty, "Combustion Engine Processes," McGraw-Hill. Schmidt, "The Internal Combustion Engine," Chapman & Hall. Starkman, "Combustion Generated Air Pollution," Plenum. *Trans. ASME. Trans. SAE. Proc. I Mech E*, Automobile Division, *Automotive Inds*.

### GENERAL FEATURES

#### Cycles

**(See also Sec. 4)**   In piston-type **internal-combustion engines** the combustion process is assumed to occur at constant volume (Fig. 1), at constant pressure (Fig. 2), or by some combination of these (Fig. 3). The constant-volume process is characteristic of the **spark-ignition** or **Otto cycle**; the constant-pressure is found only in the slow-speed **compression-ignition** or **diesel cycle**; with both processes, the cycle is sometimes called a **mixed, combination,** or **limited-pressure cycle.** Actually, the indicator card obtained from a high-speed, mixed-cycle, compression-ignition engine may be similar to that obtained from a spark-ignition engine. The fundamental differences are the

methods of mixing the air and fuel (before compression in the Otto cycle, and usually near the end of compression in the diesel cycle) and the methods of ignition.

The nominal **compression ratio** (usually specified) is the displacement plus clearance volume divided by the clearance volume. The actual compression ratio is appreciably less than the nominal value because of late intake valve or port closing.

The **compression pressure** may be estimated from the relation $p = r_a^{1.33} p_m$, where $p_m$ is the intake-manifold pressure and $r_a$ is the actual compression ratio.

The actual compression pressure can be determined with a compression-pressure gage that traps the gases by means of a check valve, thus indicating the maximum pressure under motoring conditions.

**Spark-ignition engines** use volatile liquids or gases as fuel, have compression ratios between 6:1 and 12:1 (limited by combustion knock of the fuel-air mixture) and compression pressures from below 150 to above 300 lb/in² (1034–2068 kPa), use carburetors, gas-mixing valves, or fuel-injection systems, and operate on the Otto cycle. Gasoline is the fuel commonly used in airplane, automobile, small marine, small stationary, and tractor engines. Commercial gases such as blast-furnace gas, coal gas. coke-oven gas, carbureted water gas, producer gas, and natural gas are used in large stationary engines. The mixtures used generally are near chemically correct. Combustion pressures are usually 3.5 to 5 times the compression pressures. Load and speed are usually controlled by throttling the charge; piston speeds above 3,000 ft/min (15.24 m/s) are permissible.

**Advantages** are low first cost, low specific weight, low cranking effort required, wide variation obtainable in speed and load, high mechanical efficiency, and fairly low specific fuel consumption at high compression ratios and wide-open throttle.

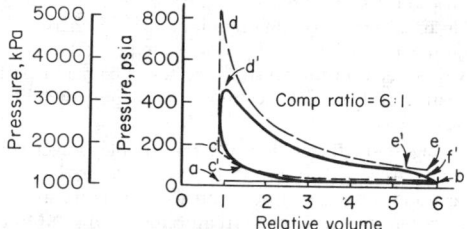

**Fig. 1**  Otto cycle (constant-volume combustion, no excess air).
Processes of the Ideal Cycle
*a-b*  Suction stroke; admission of the charge
*b-c*  Compression stroke
at *c*  Ignition of the compressed charge
*d-e*  Expansion
at *e*  Exhaust valve opens
*b-a*  Exhaust stroke

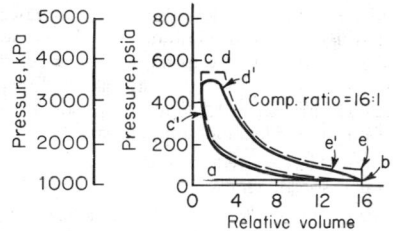

**Fig. 2**  Diesel cycle (constant-pressure combustion, 50 percent excess air).
Processes of the Ideal Cycle
*a-b*  Suction stroke; admission of air
*b-c*  Compression of air
*c-d*  Injection and combustion of the fuel
*d-e*  Expansion
at *e*  Exhaust valve opens
*b-a*  Exhaust stroke

**Compression-ignition engines** use liquid fuels of low volatility varying from fuel oil and distillates to crude oil (see Sec. 7), have compression ratios between 11.5:1 and 22:1 and compression pressures 400 to 700 lb/in² (2,760 to 4,830 kPa), and operate on the diesel or mixed cycle. Generally no ignition devices are used although some lower compression ratio and multiple chamber engines may require starting aids. Load and speed are controlled by varying the fuel quantity injected.

The **dual-fuel engine** is a diesel engine with a compression ratio which is too low to result in ignition, at the desired time, of the gas-air mixture inducted into the cylinder. A pilot (small) injection of liquid fuel with good ignition quality is used to initiate the combustion process.

**Advantages** are low specific fuel consumption, high thermal efficiency at part loads, possibly lower fuel cost, no preignition, low CO and hydrocarbon emission at low and moderate loads, suitability for two-stroke operation, and excellent durability. The lower-compression engines are of simpler con-

struction (usually valveless two-stroke cycle), lighter weight, lower first cost, lower operating expense, and higher mechanical efficiency than the higher-compression engines.

The **four-stroke cycle** (four-cycle) requires four piston strokes or two crankshaft revolutions per cycle (Figs. 1 and 2). This engine cycle is used almost exclusively in automobile, tractor, and aircraft engines in all types and sizes, and also in engines of other classifications with the exception of most outboard engines. Small four-cycle engines are always single-acting (combustion on only one side of the piston). The pumping loss, indicated by the area between the exhaust and the induction curves (Fig. 61) is more than the negative loop of the indicator card and at wide-open throttle depends on valve openings and speed and amounts to 3 to 7 percent of the engine indicated power. Throttling increases the pumping loss and reduces the positive indicator card area.

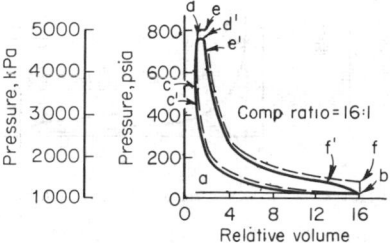

**Fig. 3**  Mixed cycle (constant-volume and constant-pressure combustion, 50 percent excess air).
Processes of the Ideal Cycle
*a-b*  Suction stroke; admission of air
*b-c*  Compression of air
*c-d*  Ignition and constant-volume combustion
*e-f*  Expansion
at *f*  Exhaust begins
*b-a*  Exhaust stroke
Broken lines represent the ideal cycle; solid lines show characteristics of the actual cycle. Primes attached to letters show actual locations of events, although some lower-compression-ratio and multiple-chamber engines may require starting aids.

**Advantages** compared with two-cycle crankcase-compression spark-ignition engines: wider variation in speed and load, cooler pistons, common crankcase in multicylinder engines, good lubrication secured more easily, no fuel loss during exhaust, lower specific fuel consumption, lower pumping losses, less exhaust dilution, lower hydrocarbon emissions, positive-displacement inlet and exhaust processes, and easier power regulation.

The **two-stroke cycle** (two-cycle) engine requires two piston strokes or only one revolution for each cycle. Exhaust ports in the cylinder wall are uncovered by the piston, or exhaust valves in the cylinder head are opened near the end (at 60 to 88 percent) of the expansion stroke, permitting the escape of exhaust gases and reducing the pressure in the cylinder (Fig. 4). The charge of air or combustible mixture flows into and is compressed in a separate crankcase compartment for each cylinder, by a compressor or a blower to slightly above atmospheric pressure. Intake ports are uncovered by the piston or intake valves, are opened soon after the opening of the exhaust, and the compressed charge flows into the cylinder, expelling most of the exhaust products, some charge escaping with the exhaust.

Blowers or compressors, driven either mechanically or by an exhaust turbine, are often used to supply sufficient air to scavenge the cylinder and clearance space and, in some cases, to supercharge the engine. Although small engines use only ports and crankcase compression (Fig. 5), large engines usually have separate compressors or blowers and either ports, or both valves and ports. With crankcase compression, low scavenging of the cylinder occurs, and the volumetric efficiency is low (30 to 70 percent). The crankcase compression work amounts to 7 to 12 percent of the total work, but scavenging blowers with displacements 20 to 80 percent greater than piston displacement may require as high as 30 percent of the indicated work in high-speed engines.

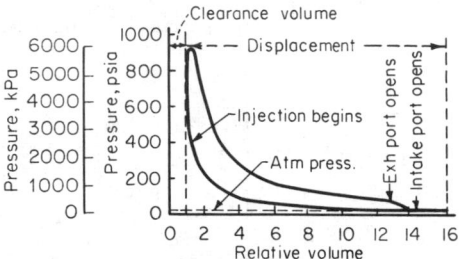

**Fig. 4**  Indicator card for two-cycle diesel engine (16:1 compression ratio).

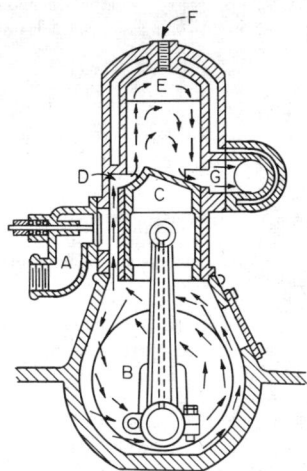

**Fig. 5**  Small two-cycle gasoline engine with crankcase compression.

**Advantages** compared with four-cycle engines: 50 to 80 percent greater power output per unit piston displacement and same speed, depending on scavenging, twice as many power impulses per cylinder per revolution, low cost for valveless designs employing crankcase compression, low $NO_x$ emissions (spark ignited), and light weight.

### WANKEL (ROTARY) ENGINES

The Wankel (rotary) engine is a relatively new engine invented by Felix Wankel. Present applications include several foreign passenger cars and small engines. Operation can be as either a spark-ignited or diesel engine. Currently all are spark-ignited. A triangular rotor rotates on an eccentric shaft inside an epitrochoidal housing. The rotor tips are in constant contact with the housing and form three working chambers. The operating cycle (Fig. 6) follows the four-stroke cycle principle, although the output shaft rotates at three times rotor speed, providing one power stroke per crankshaft revolution (as in the single-cylinder two-cycle). A fixed gear meshes with an internal gear in the rotor to maintain the proper relationship between the rotor and housing. Housing cooling is by either water or air and rotor cooling by oil or fuel/air charge. Auxiliary equipment (carburetor, ignition system, etc.) is similar to that used on reciprocating piston engines. The major problem is the sealing grid (90° intersection of seals and single-line seal at apexes of rotor).

Advantages: small size, low weight, simple, fewer components (compared with piston-type four-cycle engines), superior breathing, no valves (rotor and its seals serve as valves), lower friction, low fuel-octane requirement, low $NO_x$ emissions, no reciprocating unbalance.

### AIRCRAFT ENGINES

Small aircraft piston engines use the Otto cycle, are air-cooled and invariably four-stroke. All types are naturally aspirated or supercharged. These engines have the lowest specific weight of piston engines (Table 1). Aircraft engines always have valve-in-head construction; they also have low specific fuel consumption; the guaranteed minimum may be as low as 0.40 lb (0.18 kg)/bhph. Maximum cruising bmep ranges from 100 to 150 lb/in² (689 to 1,030 kPa), while takeoff bmep ranges from 120 to 200 lb/in² (827 to 1,380 kPa) depending principally on compression ratio, fuel, and supercharge.

Aircraft engines are valved to obtain high mean effective pressure at rated speed. High-output engines usually operate with compression ratios and supercharging levels that prohibit full-throttle operation at or near sea level with the specified fuel. Such engines can be operated at full throttle at or above a critical altitude beyond which the power decreases with an increase in altitude. Below the critical altitude, the power decreases with a decrease in altitude, at constant intake-manifold pressure, because of the increase in exhaust back pressure and ambient temperature. Turbocharged engines, however, maintain essentially constant power with constant manifold pressure from sea level to a critical altitude, which can be as high as 20,000 ft (6,100 m).

The performance of an aircraft engine varies with speed, manifold pressure, and altitude. At any given speed and with a fixed supercharger or blower ratio (supercharger r/min to engine r/min and wide-open throttle the bhp of an engine varies almost linearly with the density of the atmosphere. This characteristic makes possible the estimating of the power curves at altitude if the sea-level output is known. The power at 20,000 ft (6,100 m) altitude is about 50 percent of the sea-level wide-open throttle output for any given speed.

Low specific fuel consumption is obtained by operating with reduced r/min for the desired power with a lean-mixture setting, the optimum output at this point being about one-half to two-thirds of the normal rated output of the engine.

Aircraft-engine cylinders are limited to a maximum diameter of about 6 in (0.5 m) by piston cooling and knocking difficulties that arise with large cylinders, high compression ratios, and high intake-manifold pressures. The stroke-bore

**Table 1. Selected Piston-Type Aircraft Engines**

| Make and model | Cylinders | | | | Displacement, in³/cm³ | Compression ratio | Ratings (METO)† | | | Weights§ | | Fuel octane No. required |
|---|---|---|---|---|---|---|---|---|---|---|---|---|
| | Number | Arrangements* | Bore, in/cm | Stroke, in/cm | | | bhp‡ | r/min | At sea level or altitude, ft | bmep at max bhp, lb/in²/kPa | Engine, dry, lb/kg | |
| Avco Lycoming | | | | | | | | | | | | |
| 0-320-E | 4 | H | 5.13/13.03 | 3.88/9.86 | 319.8/5,240.6 | 7.0 | 150 | 2,700 | SL | 138/952 | 244/110.4 | 80/87 |
| I0-360-A | 4 | H | 5.13/13.03 | 4.38/9.86 | 351.0/5,915.7 | 8.7 | 200 | 2,700 | SL | 163/1,120 | 293/132.6 | 100/130 |
| TIGO-541-E | 6 | H | 5.13/13.03 | 4.38/9.86 | 541.5/8,873.6 | 7.3 | 425 | 3,200 | 15,000 | 195/1,340 | 700/316.8 | 100/130 |
| Teledyne Continental | | | | | | | | | | | | |
| 0-200 | 4 | H | 4.06/10.31 | 3.88/9.86 | 201/3,293.8 | 7.1 | 100 | 2,750 | SL | 143/986 | 190/86.0 | 80/87 |
| I0-470 | 6 | H | 5.0/12.7 | 4.0/10.16 | 471/7,718.3 | 8.6 | 260 | 2,600 | SL | 166/1,140 | 465/210.5 | 100/130 |
| GTSIO-520-F | 6 | H | 5.25/13.34 | 4.0/10.16 | 520/8,521.3 | 7.5 | 435 | 3,400 | 19,000 | 195/1,340 | 647/292.8 | 100/130 |

*H = horizontal (production of radial and vertical cylinder arrangements has been discontinued).

†METO = maximum except during takeoff.

‡Takeoff power is usually higher than the recommended maximum continuous power.

§Weights are for dry engines without hub or starter.

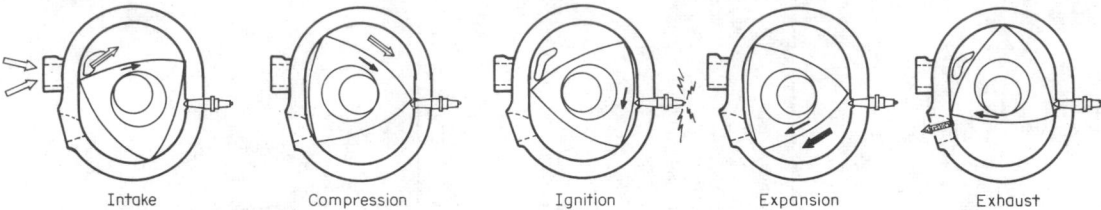

Intake          Compression          Ignition          Expansion          Exhaust

**Fig. 6**  Operating cycle—Wankel rotating combustion engine.

ratio is commonly below 1.0. The larger piston-engine sizes are being replaced by turbine-type engines.

Piston aircraft engines are mostly of the horizontally opposed design (desirable profile) with an even number of cylinders, usually four, six, or eight. They are equipped with a dual-ignition system, resulting in more reliable combustion.

## UNITED STATES AUTOMOBILE ENGINES

The size (displacement) of U.S. automobile engines varies from 97 in³ (1,590 cm³) to 500 in³ (8,194 cm³). Rated horsepower per unit of engine displacement has decreased considerably because of emission control and more realistic rating procedures. Average displacement has leveled off, stroke has increased, and bore has decreased since 1970 (Fig. 7).

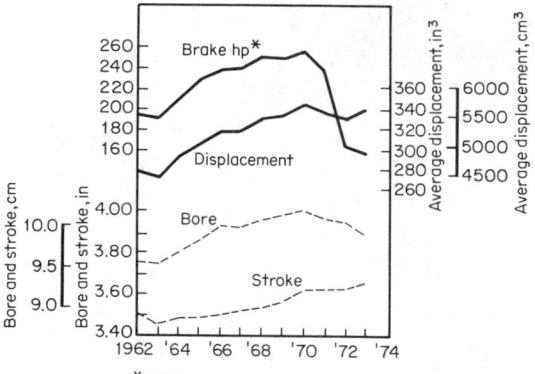

*1972 and 1973 based on net hp

**Fig. 7**  Trend of average American automobile-engine characteristics.

Emission control has brought about (1) reduced compression ratios, (2) improved valves, (3) retarded ignition timing, (4) more precise and accurate fuel metering (carburetion), and (5) controlled inlet-air temperatures. Current advertised horsepower is based on net rather than gross horsepower. Net horsepower is measured with a fully equipped engine, including fuel pump, oil pump, intake-air system, exhaust its intended function unaided.

Valve timing, particularly the closure of intake valves (varying from 35 to 105° crankshaft rotation after bottom dead center for 1973 cars, with 67° average), is selected to give a rising volumetric efficiency and torque as the r/min is increased. At high speeds, restrictions in the intake system, developed primarily by the carburetor venturi and the intake valves begin to throttle the engine. The 1973 base engines developed maximum torque at approximately 2,300 r/min. Despite reduced torque at higher speeds, the power peaked at

3,200 to 4,600 r/min. Higher powers were obtained with four-barrel (compound) carburetors and other modifications.

Engines have been supplied with various numbers of cylinders over the years (from 2 to 16), but the conventional engine has become the V-8 (Fig. 8) for larger cars (80 percent of all 1972 car models), the remainder being 6-cylinder (10 percent) and 4-cylinder (10 percent). For a given displacement, fewer cylinders increases economy and lowers HC emissions but with an increase in unbalance and torque fluctuation. Advantages of the V-8 arrangement include short length and low height with low-hood designs; rigid, compact construction, with carburetors located centrally in the V, giving intake-manifold designs with short branches; five-bearing crankshafts; cylinders and upper crankcase of integral construction; detachable cylinder heads; and overhead valves. The crankshaft is arranged with the crankpins in two planes providing excellent dynamic balance. The connecting rods for the corresponding cylinders on each side of the V are side by side on the same crankpin, with the right and left cylinder blocks suitably staggered. The most recently designed engines have four cylinders and are for subcompact cars. Several feature a belt-driven, overhead camshaft and one, Chevrolet Vega, employs a high-silicon aluminum crankcase in which the pistons run directly on the aluminum bore. The in-line "fours" are not dynamically balanced.

While the average cylinder bore increased from 3.27 in (8.31 cm) in 1946 to 3.88 in (9.85 cm) in 1973, the stroke decreased from about 4.11 in (10.44 cm) to 3.65 in (9.27 cm) in the same period, with reduction in the width and weight of V-8 engines and the added rigidity necessary for satisfactory bearing life in high-speed engines.

Quiet operation is essential in automobile engines and requires close manufacturing control of clearances. Most engines have hydraulic valve lifters and means of rotating the exhaust valves, at least, for prolonged life. Exhaust-valve seats are hardened to prevent recession with unleaded fuel.

It is standard practice to provide pistons with three rings above the wrist pin—two narrow compression rings above one oil-scraper ring, with several drain holes from the back of the ring groove to return oil to the crankcase. All compression rings are narrow, of cast iron from 0.063 to 0.0787 in (0.160 to 0.200 cm) wide, and have a wear-resistant coating, such as chromium (particularly for the top ring) and tin, or compounds (iron oxide, phosphates), which minimizes the running-in period without danger of scuffing or scoring the cylinders. Oil-scraper rings are of steel or cast iron, generally a little less than 3/16 in (0.476 cm) wide, and are provided with internal expander springs to increase pressure against the cylinder walls. Many oil-scraper rings, especially those of steel, also use wear-resistant coatings.

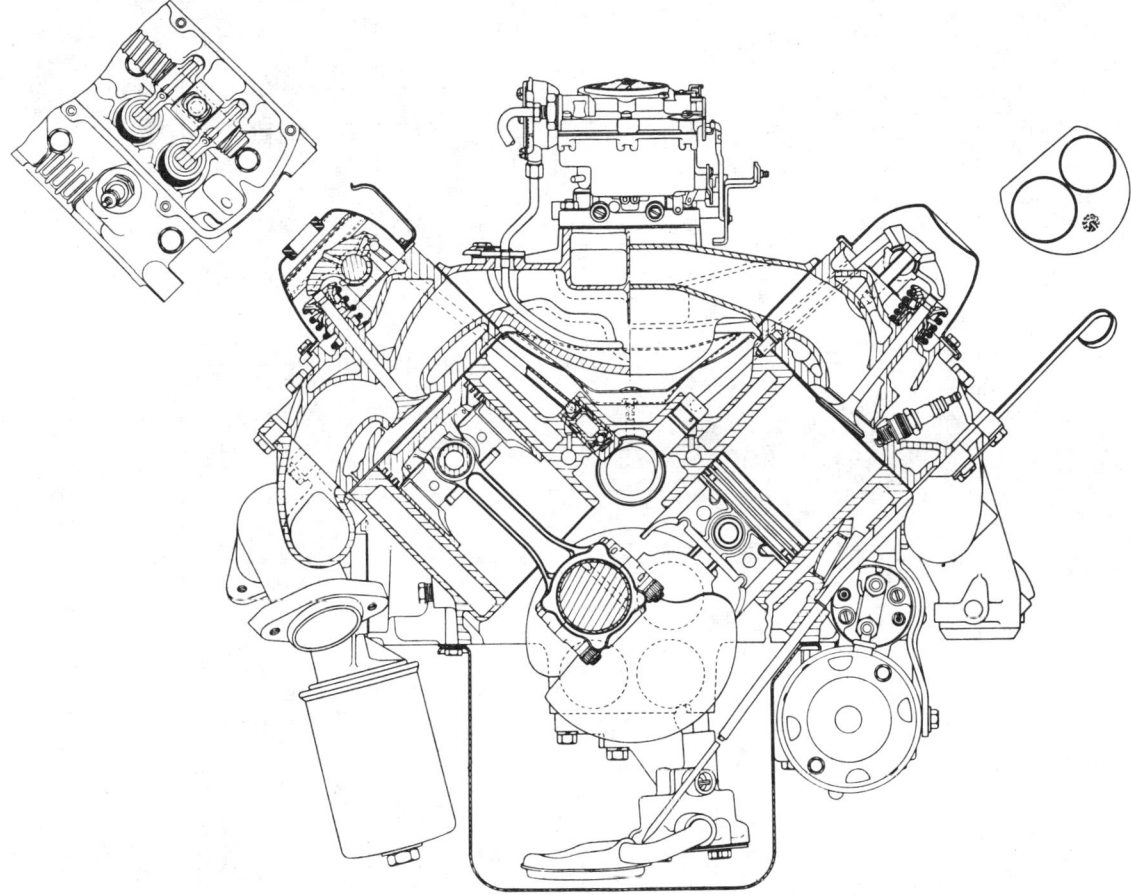

**Fig. 8** Oldsmobile V-8 engine.

**Water cooling** under a pressure of 15 lb/in² (103 kPa) is standard for U.S. cars (several smaller European cars are air-cooled). Water-jacket temperatures are controlled by thermostats which open at 195°F (90.6°C), permitting water from the cylinder heads to flow to the radiator. All engines are designed either with water all around each cylinder or "siamesed" and with jackets the full length of the piston travel.

### FOREIGN AUTOMOBILE ENGINES

Foreign automobile engines (Table 3) are principally of the four-stroke Otto-cycle type, with some engines of the diesel type. The engines are principally water-cooled, using both thermosiphon and pump systems, and combinations of both. Most horizontal-opposed type of engines are air-cooled. The bhp values range from 32 to 352, compared to 54 to 250 in the United States. Most engines are normally aspirated. The displacements range from 2 cylinders, 36.5 in³ (598 cm³); 4 cylinders, 48 to 135 in³ (787 to 2,212 cm³); 6 cylinders, 122 to 288 in³ (1,999 to 4,719 cm³); 8 cylinders, 182 to 400 in³ (2,982 to 6,555 cm³); 12 cylinders, 240 to 325 in³ (3,933 to 5,326 cm³); most of the engines being of the 4-cylinder in-line type.

The bmep at maximum power ranges from 94 to 168 lb/in² (648 to 1,158 kPa) compared with 82 to 134 lb/in² (565 to 924 kPa) in the United States, for four-stroke Otto-cycle carbureted engines. These are attained at 1,813 ft/min (9.21 m/s) (Volkswagen) and 3,364 ft/min (17.09 m/s) (Alfa Romeo) mean piston speeds, compared with 2,320 ft/min (11.79 m/s) Checker and 2,286 ft/min (11.71 m/s) Pontiac in the United States.

### TRUCK AND BUS ENGINES

Truck and bus engines (Figs. 9 and 10) are similar to automobile engines but, in general, are larger, having 300 to 1,000 in³ (4,916 to 16,387 cm³) displacement, are more rugged, and run at lower speeds. Both Otto and diesel cycles are used, with the diesel predominant in the larger sizes. The two-cycle blower-scavenged diesel engine with or without a turbocharger is in common use (Fig. 10). Water cooling is universally used. The gasoline engines are naturally aspirated and have compression ratios slightly lower than automobile engines but are valved similarly, have similar performance curves, and attain maximum bhp (55 to above 300) at a mean piston speed averaging

**Table 2. Selected United States Automobile Engines***

| | Cylinders | | | | Com-pres-sion ratio | Published maximums | | | | |
| Make and model | Type and No. of cylinders† | Bore, in/cm | Stroke, in/cm | Displacement, in³/cm³ | | Net bhp | at r/min | Net torque (lb·ft)/ N·m | at r/min | Car weight, lb/kg |
|---|---|---|---|---|---|---|---|---|---|---|
| American Motors Corp.: | | | | | | | | | | |
| Gremlin | IL-6 | 3.75/9.53 | 3.50/8.89 | 232/3,802 | 8.0 | 100 | 3,600 | 185/251 | 1,800 | 2,642/1,198 |
| Javelin | V-8 | 3.75/9.53 | 3.44/8.74 | 304/4,983 | 8.4 | 150 | 4,200 | 245/332 | 2,500 | 3,104/1,408 |
| Chrysler Corp.: | | | | | | | | | | |
| Plymouth Valiant | IL-6 | 3.40/8.64 | 3.64/9.25 | 198/3,245 | 8.4 | 95 | 4,000 | 150/203 | 1,600 | 2,865/1,299 |
| Plymouth Fury | V-8 | 3.91/9.93 | 3.31/8.41 | 318/5,212 | 8.6 | 150 | 3,600 | 265/350 | 2,000 | 3,860/1,751 |
| Dodge Charger | IL-6 | 3.40/8.64 | 4.12/10.47 | 225/3,688 | 8.4 | 105 | 3,600 | 185/251 | 1,600 | 3,395/1,540 |
| Chrysler New Yorker | V-8 | 4.32/10.97 | 3.75/9.53 | 440/7,212 | 8.2 | 215 | 3,600 | 345/468 | 2,000 | 4,355/1,975 |
| Ford Motor Comp.: | | | | | | | | | | |
| Ford Pinto | IL-4 | 3.19/8.10 | 3.06/7.77 | 97.6/1,600 | 8.0 | 54 | 4,600 | 80/108 | 2,400 | 2,115/959 |
| Ford Torino | IL-6 | 3.68/9.35 | 3.91/9.93 | 250/4,098 | 8.0 | 92 | 3,200 | 197/267 | 1,600 | 3,577/1,622 |
| Ford LTD | V-8 | 4.00/10.16 | 3.50/8.89 | 351/5,753 | 8.0 | 158 | 3,800 | 264/358 | 2,400 | 4,107/1,863 |
| Thunderbird | V-8 | 4.36/11.07 | 3.59/9.12 | 429/7,031 | 8.0 | 208 | 4,400 | 327/443 | 2,600 | 4,505/2,043 |
| Mercury Comet | IL-6 | 3.68/9.35 | 3.13/7.95 | 200/3,288 | 8.3 | 84 | 3,800 | 151/205 | 1,800 | 2,904/1,317 |
| Lincoln | V-8 | 4.36/11.07 | 3.85/9.78 | 460/7,539 | 8.0 | 219 | 4,400 | 360/488 | 2,800 | 5,049/2,290 |
| General Motors: | | | | | | | | | | |
| Chevrolet Vega | IL-4 | 3.50/8.89 | 3.63/9.22 | 140/2,295 | 8.0 | 72 | 4,400 | 100/136 | 2,400 | 2,313/1,049 |
| Chevrolet Nova | IL-6 | 3.88/9.86 | 3.53/8.97 | 250/4,098 | 8.5 | 100 | 3,600 | 175/237 | 1,600 | 3,065/1,390 |
| Chevrolet Monte Carlo | V-8 | 4.00/10.16 | 3.48/8.84 | 350/5,737 | 8.5 | 145 | 4,000 | 255/346 | 2,400 | 3,713/1,684 |
| Pontiac Firebird | V-8 | 3.88/9.86 | 3.75/9.53 | 350/5,737 | 7.6 | 150 | 4,000 | 270/366 | 2,000 | 3,380/1,533 |
| Pontiac | | | | | | | | | | |
| Grandville | V-8 | 4.12/10.46 | 3.75/9.53 | 400/6,556 | 8.0 | 200 | 4,000 | 310/420 | 2,400 | 4,333/1,965 |
| Buick Century | V-8 | 3.80/9.65 | 3.85/9.78 | 350/5,737 | 8.5 | 150 | 3,800 | 265/359 | 2,400 | 3,780/1,714 |
| Buick Electra 225 | V-8 | 4.31/10.95 | 3.90/9.91 | 455/7,457 | 8.5 | 225 | 4,000 | 360/488 | 2,400 | 4,581/2,078 |
| Oldsmobile | | | | | | | | | | |
| Cutlass | V-8 | 4.06/10.31 | 3.39/8.61 | 350/5,737 | 8.5 | 180 | 3,800 | 275/373 | 2,800 | 3,786/1,717 |
| Cadillac Eldorado | V-8 | 4.30/10.92 | 4.30/10.92 | 500/8,196 | 8.5 | 235 | 3,800 | 385/522 | 2,400 | 4,880/2,213 |

*Auto. Ind., **48**, no. 7, Apr. 1, 1973, p. 63.
†IL = in line; V = V type.

**Table 3. Selected Foreign Automobile Engines***

| Make and model | Type and No. of cylinders | Bore, in/cm | Stroke, in/cm | Displacement, in³/cm³ | Compression ratio | Maximum bhp at r/min | | Car shipping weight, lb/kg |
|---|---|---|---|---|---|---|---|---|
| | | | | Great Britain | | | | |
| Austin Marina 2 door | IL-4 | 3.16/8.03 | 3.50/8.89 | 109.7/1,275 | 8.0 | 60 | 5,500 | 1,650/748 |
| Jaguar XJ12 | V-12 | 3.50/8.89 | 2.70/6.86 | 326/5,343 | 9.0 | 258 | 6,000 | 3,900/1,769 |
| Jensen Interceptor III | V-8 | 4.32/10.97 | 3.75/9.53 | 440/7,212 | 10.3 | 330 | 6,000 | |
| MG MGB | IL-4 | 3.16/8.03 | 3.50/8.89 | 109.6/1,798 | 8.0 | 95 | 5,400 | 1,920/871 |
| Rolls Royce Silver Shadow | V-8 | 4.10/10.41 | 3.90/9.91 | 412/6,750 | 9.0 | | | 4,600/2,087 |
| Triumph TR6 | IL-6 | 2.94/7.47 | 3.74/9.50 | 152/2,498 | 7.75 | 106 | 4,900 | 2,280/1,034 |
| | | | | France | | | | |
| Citroen D Special | IL-4 | 3.39/8.61 | 3.37/8.56 | 121.1/1,985 | 8.75 | 98 | 5,750 | 2,975/1,349 |
| Peugeot 504 | IL-4 | 3.46/8.79 | 3.20/8.13 | 120.3/1,974 | 7.6 | 92 | 5,500 | 2,525/1,145 |
| Renault 12 USA | IL-4 | 3.03/7.70 | 3.31/8.41 | 95.5/1,565 | 8.6 | 73 | 5,000 | 2,050/930 |
| | | | | Germany | | | | |
| Audi 100 | IL-4 | 3.31/8.41 | 3.32/8.43 | 114.4/1,871 | 8.2 | 91 | 5,200 | 2,405/1,091 |
| BMW Bavaria | IL-6 | 3.50/8.89 | 3.15/8.00 | 182/2,985 | 8.3 | 170 | 5,800 | |
| Ford Capri 2600 | V-6 | 3.54/8.99 | 2.63/6.68 | 155/2,600 | 8.2 | 107 | 5,000 | 2,389/1,084 |
| Mercedes Benz 220 | IL-4 | 3.43/8.71 | 3.64/9.25 | 134/2,296 | 8.0 | 85 | 4,500 | 2,960/1,343 |
| Mercedes 220D (diesel) | IL-4 | 3.43/8.71 | 3.64/9.25 | 134/2,296 | 21.0 | 57 | 4,200 | 3,041/1,379 |
| Opel Manta 2 door | IL-4 | 3.66/9.30 | 2.75/6.99 | 115.8/1,898 | 7.6 | 75 | 4,800 | |
| Porsche 9115 | OP-6 | 3.31/8.41 | 2.77/7.04 | 142.8/2,341 | 8.5 | 181 | 6,500 | 2,425/1,100 |
| Volkswagen Beetle | OP-4 | 3.37/8.56 | 2.72/6.91 | 96.6/1,584 | 7.3 | 46 | 4,000 | 1,742/790 |
| | | | | Italy | | | | |
| Alfa Romeo 2000 | IL-4 | 3.30/8.40 | 3.48/8.85 | 119.7/1,962 | 9.0 | 148 | 5,800 | 2,442/1,108 |
| Ferrari 365 6TC-4 | V-12 | 3.19/8.10 | 2.80/7.11 | 267.8/4,390 | 8.8 | 340 | 6,800 | 3,190/1,447 |
| Fiat 128 Sedan | IL-4 | 3.15/8.00 | 2.18/5.54 | 68.1/1,116 | 8.8 | 55 | | 1,697/770 |
| Maserati Bora 4900 | V-8 | 3.70/9.40 | 3.51/8.92 | 300.5/4,930 | 8.5 | 310 | 6,000 | 3,200/1,451 |
| | | | | Japan | | | | |
| Datsun PL510 | IL-4 | 3.27/8.31 | 2.90/7.37 | 87.3/1,595 | 8.5 | 100 | 6,000 | 2,017/915 |
| Honda S 800 | IL-4 | 2.36/5.99 | 2.76/7.01 | 48.3/791 | 9.2 | 70 | 8,000 | 1,687/765 |
| Mazda RX-2 | R-2 | | | 140/1,146 | 9.4 | 120 | 6,500 | 2,116/960 |
| Subaru 1400 | OP-4 | 3.35/8.51 | 2.36/5.99 | 83.6/1,361 | 9.0 | 61 | 5,600 | 1,860/844 |
| Toyota Corolla | IL-4 | 2.95/7.50 | 2.60/6.60 | 71.1/1,166 | 9.0 | 73 | 6,000 | 1,653/750 |
| | | | | Sweden | | | | |
| Saab 96 | V-4 | 3.54/8.99 | 2.63/6.68 | 104/1,698 | 8.0 | 65 | 4,700 | 2,030/921 |
| Volvo 142 E | IL-4 | 3.50/8.89 | 3.15/8.00 | 121.1/1,986 | 8.7 | 125 | 6,000 | 2,463/1,117 |

*Auto. Ind., **48**, no. 7, Apr. 1, 1973, pp. 66, 68.
†IL = in line; V = vee type; OP = opposed; R = rotary combustion.

about 2,600 ft/min (13.21 m/s). Maximum torque is obtained at about 1,500 r/min, appreciably lower than for automobile engines. Diesel engines, which are often supercharged and/or turbocharged, have piston speeds ranging from 1,500 to 2,100 ft/min (7.62 to 10.67 m/s) operating at somewhat lower speeds than the gasoline engines. Truck and bus gasoline engines can operate at speeds slightly higher than those at which maximum power is attained. Some diesel engines can be operated near the maximum power speed. Most are operated below (Fig. 11).

Truck and bus engines are usually built with four or six cylinders in line, or with eight cylinders in V-type construction. The stroke-bore ratio ranges from about 0.75 to 1.50, the later engines having the smaller values (see Tables 4 and 5).

**TRACTOR ENGINES**

Both spark-ignition (gasoline or liquefied petroleum gas) and compression-ignition (diesel) engines are used in tractors. Almost all engines above 100 hp are diesels. Most engines are four-cycle, water-cooled, with valve-in-head arrangement. Integral-bore, dry-sleeve, or wet-sleeve cylinder-liner arrangements are utilized. One-, two-, three-, four-, five-, six-, and eight-cylinder engines are used. All except the eight-cylinder are usually in-line arrangement in order to minimize the width of the hood in the interest of driver visibility. Tractor applications have a high load factor and relatively low speed, and high low-speed torque (Fig. 12).

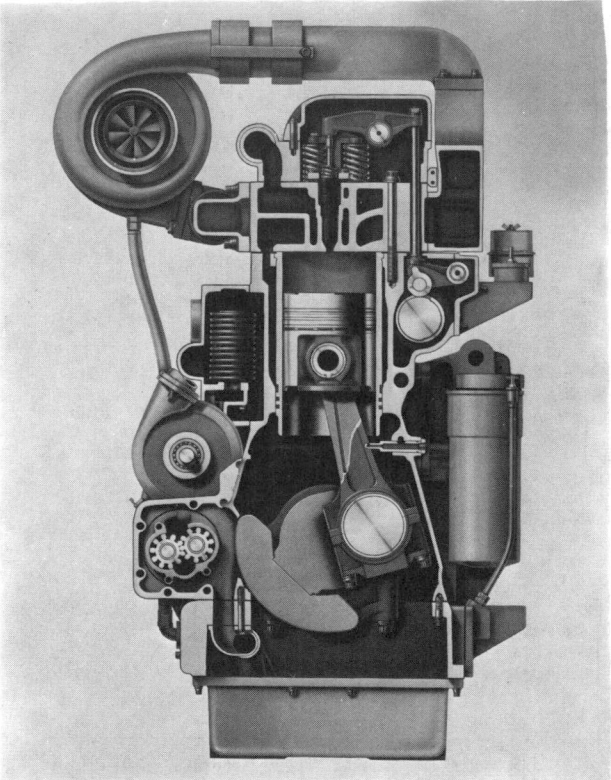

**Fig. 9** Cummins Model KT 450-hp turbocharged, four-cycle diesel engine.

## STATIONARY ENGINES

Stationary engines may be either Otto or diesel cycle, use either liquid or gaseous fuel, and use the two- or four-stroke cycle. The power output ranges up to about 48,000 bhp per engine. Piston rods, which permit the crankcase to be isolated from blowby gases, and also permit the bottom face of the piston to function as an air pump, are commonly used. Nearly all engines are now single-acting. In the larger sizes, cylinders are built up of several parts, cylinder liners are used, and long pistons for single-acting diesel engines usually have about five piston rings. The valves are usually massive and in the largest sizes may be cooled. The exhaust pipe is sometimes water-cooled, and the pistons for large-bore engines are always liquid-cooled with oil or water. In general, there is much less integral construction (Fig. 13) in the large stationary engine than in the automotive type.

Stationary engines usually operate at constant speed and are governed by throttling the charge of the Otto engine and by varying the amount of fuel injected into the diesel engine.

Compression ratios as low as 6:1 are used with spark-ignition engines and as low as 12:1 with compression-ignition engines.

Stationary engines are usually built with 1 to 12 cylinders, with cylinders in a vertical line, or with a V-type arrangement having up to 16 or more cylinders.

## LOCOMOTIVE ENGINES

(See also Sec. 11)

Diesel engines are used for locomotive service because of high thermal efficiency (overall efficiency of 25 percent for a diesel locomotive compared with 6 percent for a steam locomotive), high availability for service, elimination of destructive pounding of rails, and elimination of shutdown fuel losses. The locomotive diesels are usually lighter, have lower displacement per cylinder, but higher outputs per cubic inch of displacement than the corresponding stationary diesel engine. These engines are built with four to sixteen cylinders. One or two engines are used per locomotive, and one or more locomotives are used per train. (See Table 7.)

## MARINE ENGINES

(See also Sec. 11)

Marine engines may be either Otto or diesel cycle and may be normally aspirated or supercharged. Outboard engines range in power from less than 1 to more than 150 bhp and are generally of the two-cycle gasoline type. Modified automotive gasoline engines are also used in larger boats. Diesel engines range in power from below 50 to 4,800 bhp and may be either two- or four-cycle. Both automotive and stationary types are used in marine work and, consequently, wide variations in specific weights are found. (See Tables 4, 5, and 8.)

The outboard two-cycle engines (see Table 8) have crankcase compression (Fig. 14) and may have either two ports, three ports, or two ports with a rotary or reed crankcase inlet valve for the highest outputs (Fig. 15). The Otto four-cycle engines have L-head or valve-in-head construction, and the diesel four-cycle engines are invariably of valve-in-head construction.

Regular gasoline is required for the Otto-cycle engines, the fuel and lubricating oil being mixed for the outboard two-cycle engines. A comparison of a number of fuel-consumption curves for various types of engines shows the large low-speed marine diesel engine as having the lowest specific fuel consumption (Fig. 16).

Low-speed engines are direct-connected to the propellers, and high-speed engines are connected through reduction gearing. Outboard engines have maximum speeds of 4,500 to 6,000 r/min, the piston speeds ranging from 1,000 to about 2,500 ft/min (5.08 to 12.7 m/s).

Maximum engine speed is obtained at the intersection of the propeller-horsepower and the maximum-horsepower curves (Fig. 17). This may occur well below the speed for maximum horsepower or beyond this speed. The effect of throttling to reduce speed is to increase the specific fuel consumption considerably.

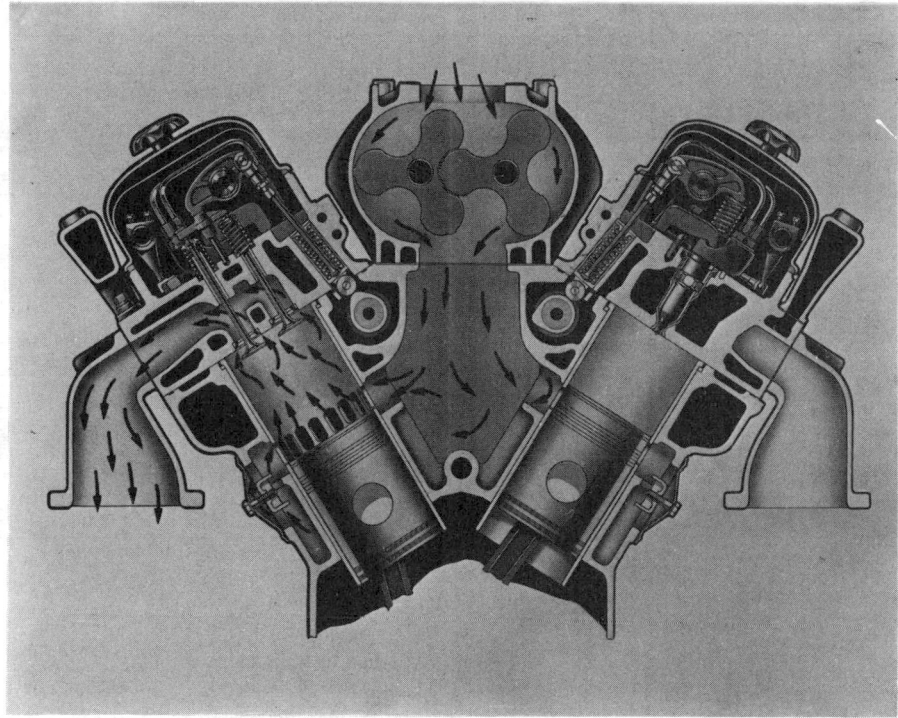

**Fig. 10**  Blower-scavenged General Motors Model 8V-7IN two-cycle diesel engine.

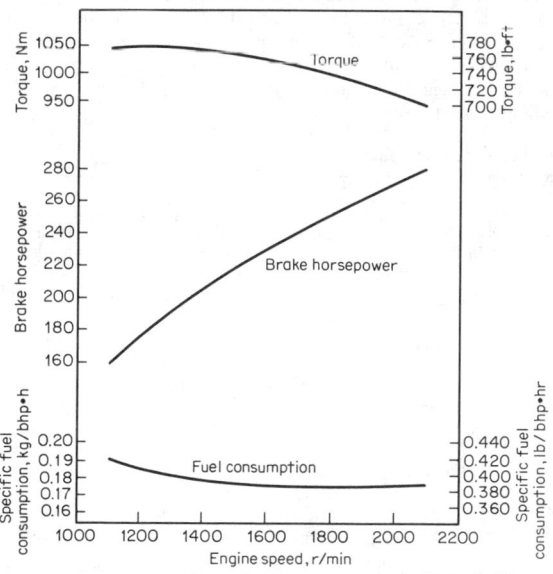

**Fig. 11** Performance data of blower-scavenged General Motors Model 8V-7IN two-cycle diesel engine.

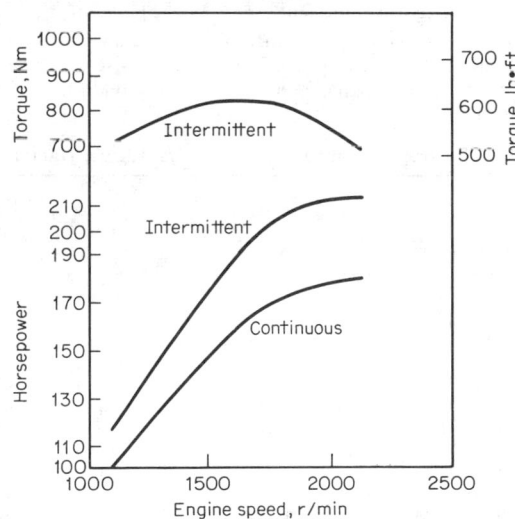

**Fig. 12**  Performance of John Deere Model 6531 A diesel engine.

## SMALL INDUSTRIAL, UTILITY, AND RECREATION-VEHICLE GASOLINE ENGINES

The Otto-cycle engine with two or fewer cylinders is predominant in this class (Table 6). Fuel economy is generally not of major concern, whereas simplicity and performance are important. For reciprocating engines the L-head design is predominant. Both high- and low-speed engines are used, depending on the application. Engines of this type generally

**Table 4. Selected Diesel Engines for Truck, Bus, Tractor, Marine, and Industrial Use***

| Make and model | Designed for† | No. | Cylinders | | Displacement, in³/cm³ | Compression ratio | Max continuous bhp‡ at r/min | Max torque, lb·ft/N·m at r/min |
| | | | Bore, in/cm | Stroke, in/cm | | | | |
|---|---|---|---|---|---|---|---|---|
| Allis-Chalmers 2800 | Tr,I | 6 | 3.88/9.85 | 4.25/10.80 | 301/4,932 | 16.25 | 68 2,200 | 225/305 1,400 |
| Case 336 BD | Tr,I | 4 | 4.63/11.76 | 5.00/12.70 | 336/5,506 | 16.50 | 72 2,200 | 281/381 1,400 |
| Caterpillar D 348 | T,I | 12 | 5.41/13.74 | 6.50/16.51 | 1,786/29,267 | 16.50 | 725 1,800 | 2,840/3,850 1,500 |
| Chrysler Nissan IN675 | I | 6 | 4.33/10.99 | 5.13/13.03 | 452/7,407 | 16.00 | 168 1,800 | 664/900 1,400 |
| Cummins V-785-C | I | 8 | 5.50/13.97 | 4.13/10.49 | 785/12,864 | 15.90 | 220 2,300 | 560/759 1,600 |
| Deere 3164 D | Tr,I | 3 | 4.02/10.21 | 4.33/11.0 | 164/2,687 | 16.3§ | 40 2,200 | 122/165 1,500 |
| Detroit Diesel 8V-71T | T.Tr,I | 8 | 4.25/10.80 | 5.00/12.70 | 568/9,308 | 17.00§ | 335 2,100¶ | 965/1,308 1,600 |
| Deutz F6L-714 | T,B,Tr,I | 6 | 4.75/12.07 | 5.50/13.97 | 579/9,488 | 17.80 | 117 2,000 | 376/510 1,300 |
| Ford 401-D | Tr, | 6 | 4.41/11.20 | 4.41/11.20 | 401/6,571 | 16.50 | 108 2,300¶ | 284/385 1,400 |
| GMC DH478 | T,B | 6 | 5.13/13.03 | 3.86/9.80 | 478/7,833 | 17.50 | 155 2,800¶ | 337/457 2,000 |
| Hercules D-3000 | Tr,I | 6 | 3.75/9.53 | 4.50/11.43 | 298/4,883 | 17.50 | 85 2,400 | 244/331 1,400 |
| International DT-407 | Tr,I | 6 | 4.33/10.99 | 4.88/12.40 | 407/6,670 | 17.00§ | 130 2,400 | 393/533 1,800 |
| Mack END 673E | T | 6 | 4.88/12.40 | 6.00/15.24 | 672/11,012 | 16.11 | 168 2,100¶ | 540/732 1,400 |
| Murphy MP-321 | I | 6 | 5.75/14.61 | 6.55/16.64 | 1,013/16,600 | 16.00 | 207 1,400 | 946/1,283 800 |
| Oliver 1855 | Tr | 6 | 3.88/9.85 | 4.38/11.13 | 31.0/5,080 | 16.00 | 98 2,400 | 298/404 1,600 |
| Perkins 6-354 | Tr,I | 6 | 3.88/9.85 | 5.00/12.70 | 354/5,801 | 16.00 | 88 2,250 | 288/390 1,400 |

*Auto. Ind., **48,** no. 7, Apr. 1, 1973.
†B = bus; T = truck; I = industrial; Tr = tractor.
‡With standard accessories.
§Turbocharged.
¶Maximum intermittent bhp.

employ air cooling of the cylinder and charge cooling of the interior. Higher mep air-cooled engines use forced fan cooling. Two-cycle engines are often used in applications where minimum weight is important.

Carburetion, ignition systems, and other auxiliary equipment are usually less sophisticated than those on larger engines. A given engine is often used in a wide range of applications.

## COMBUSTION CHAMBERS
### Spark-Ignition Engines

Most combustion chambers for four-cycle engines, especially automotive, are of the open-chamber design and employ overhead valves actuated by either pushrods or overhead cams. This design provides fast burning of the charge because of the concentrated combustion volume and minimizes surface area and thus heat losses and wall quenching. Figures 18 to 21 show typical designs.

**Table 5. Selected Gasoline Engines for Truck, Bus, Tractor, Marine, and Industrial Use**

| Make and model | Designed for* | No. | Cylinders | | Displacement in³/cm³ | Compression ratio | Rating,† bhp at r/min | Max torque lb·ft/N·m r/min |
| | | | Bore, in/cm | Stroke, in/cm | | | | |
|---|---|---|---|---|---|---|---|---|
| Allis-Chalmers G-153 | Ind | 4 | 3.44/8.74 | 4.13/10.49 | 153/2,507 | 8.0 | 52 2,400 | 132/179 1,460 |
| American Motors Jeep 232-16 | T,Tr,Gp,Ind | 6 | 3.75/9.53 | 3.50/8.89 | 232/3,801 | 8.0 | 128 4,000 | 215/292 1,400 |
| Case 201 G | Tr,Gp,Ind | 4 | 4.00/10.16 | 5.00/12.70 | 251/4,113 | 7.5 | 87 2,400 | 221/300 1,400 |
| Chevrolet 454 | T | 8 | 4.25/10.79 | 4.00/10.16 | 454/7,440 | 8.25 | 250 4,000 | 365/495 2,800 |
| Chrysler H-225 | M,Ind | 6 | 3.41/8.66 | 4.13/10.49 | 225/3,687 | 8.4 | 132 4,000 | 203/275 2,000 |
| Continental R 821-46 | Tr,Ind | 4 | 3.03/7.70 | 3.31/8.41 | 95.5/1,565 | 8.6 | 67.2 5,000 | 84/114 2,400 |
| Diamond Reo 8-250 | T,B | 8 | 4.24/10.79 | 4.13/10.49 | 468/7,669 | 7.5 | 250 3,400 | 420/569 2,400 |
| Dodge CT 900 | T | 8 | 4.19/10.64 | 3.75/9.53 | 413/6,768 | 7.5 | 238 3,600 | 355/481 2,000 |
| Ford 158-G | Tr | 3 | 4.20/10.67 | 3.81/9.68 | 158/2,591 | 7.75 | 45.7 2,100 | 120/163 1,350 |
| Hercules G-3400 | Tr,Ind | 6 | 4.00/10.16 | 4.50/11.43 | 339/5,555 | 7.5 | 143 2,800 | 311/422 1,400 |
| International V5-401 | T | 8 | 4.13/10.49 | 3.75/9.53 | 400/6,555 | 7.69 | 226 3,600 | 355/481 2,000 |
| Minneapolis-Moline HD 425 | Tr | 6 | 4.25/10.79 | 5.00/12.70 | 425/6,965 | 8.1 | 133 2,000 | 378/512 800 |
| Oliver 1655 | Tr | 6 | 3.75/9.53 | 4.00/10.16 | 265/4,343 | 8.5 | 84 2,200 | 220/298 1,300 |
| Volkswagen 124A | Ind | 4 | 3.38/8.59 | 2.72/6.91 | 96.7/1,585 | 7.7 | 53 3,600‡ | 75.5/102 2,500 |

*B = bus; T = trucks; Tr = tractor; M = marine; Gp = general purpose; Ind = industrial.
†hp without accessories.
‡hp with accessories.

**Table 6. Selected Small Gasoline Engines***

| Make and model | Cycle | Type | Cylinders | | | | Compression ratio | Valve location† | Continuous rating, bhp at r/min | Max torque lb·ft/ N·m at r/min | Weight, lb/kg |
| | | | No. | Bore, in/ cm | Stroke, in/ cm | Displacement, in³/cm³ | | | | | |
|---|---|---|---|---|---|---|---|---|---|---|---|
| Briggs & Stratton 100900 | 4 | Ver | 1 | 2.5/6.50 | 2.13/5.41 | 10.43/170.9 | 6.20 | L | 3.40 3,600 | 5.9/8.6 3,100 | 30.5/13.9 |
| Briggs & Stratton 80400 | 4 | Hor | 1 | 2.38/6.04 | 1.75/4.45 | 7.75/126.9 | 6.20 | L | 2.55 3,600 | 4.6/6.2 3,100 | 26.25/11.9 |
| Chrysler 500 | 2 | Hor and Ver | 1 | 2.00/6.45 | 1.59/4.04 | 5.00/81.9 | 6.50 | | 3.33 7,000 | 3.0/4.1 5,000 | 8.75/3.9 |
| Homelite XL-123 | 2 | Hor | 1 | 1.81/4.50 | 1.38/3.50 | 3.60/58.9 | 8.30 | | | | 13.0/5.9 |
| Kohler K-482 | 4 | Hor and Ver | 2 | 3.25/8.26 | 2.88/7.32 | 48.0/786.6 | 6.00 | L | 15.30 3,600 | 31.7/43.0 2,400 | 180/82 |
| O & R 13A | 2 | Ver | 1 | 1.25/3.18 | 1.09/2.77 | 1.34/21.9 | 9.00 | | 1.00 6,300 | .88/1.2 5,200 | 3.75/1.70 |
| Onan AI | 4 | Ver | 1 | 2.75/6.99 | 2.50/6.35 | 14.90/244.2 | 6.25 | L | 3.86 3,600 | 8.0/10.8 2,100 | 150/68.2 |
| Rockwell Zf-400 | 2 | Ver | 2 | 2.56/6.50 | 2.56/6.50 | 24.29/398 | 7.5 | | 34.0 6,250 | 28.0/38.0 6,250 | 62/28.2 |
| Tecumseh LAV 35-1A | 4 | Ver | 1 | 2.50/6.35 | 1.84/4.67 | 9.06/148.5 | | L | 2.27 2,400 | 5.25/7.12 3,100 | 21.5/9.77 |
| Tecumseh HH160 | 4 | Hor | 1 | 3.5/8.89 | 2.88/7.32 | 27.66/453.3 | 8.7 | I | 16.00 3,600 | 25.0/33.9 2,400 | 84/38.2 |
| Wisconsin TRA-10 D | 4 | Ver | 1 | 3.13/7.96 | 2.88/7.32 | 22.05/361.3 | 6.47 | L | 9.90 3,400 | 15.58/21.1 2,000 | 80/36.4 |
| Fichtel & Sachs KM 914 | 4 | Hor | 1 | | | 18.5/303 | 8.0 | RC | 20 5,000 | | 56/25.5 |

*Auto. Ind., **48**, no 7, Apr. 1, 1973, pp. 104–106.
†I = valve in head; L = L head; RC = rotary combustion.

**Table 7. Selected Large Diesel Engines***

| Make and model | Cylinder arrangement† | Cycle | Cylinders | | | | Compression ratio | Max rating, bhp at r/min | Turbo-charger | Weight, lb/kg |
| | | | No. | Bore, in/ cm | Stroke, in/ cm | Displacement, in³/l | | | | |
|---|---|---|---|---|---|---|---|---|---|---|
| Alco 251 | V | 4 | 12 | 9.0/22.86 | 10.5/26.67 | 8,016/131.36 | 11.5 | 3,150 1,000 | Yes | 35,581/16,152 |
| De Laval GVB-16 | V | 4 | 16 | 13.0/33.02 | 15./38.1 | 31,856/522.03 | | 2,900 | Yes | 141,000/64,005 |
| Electro-Motive 12-645 | V | 2 | 12 | 9.06/22.01 | 10./25.4 | 7,740/126.84 | 20.0 | 2,475 900 | Yes | 25,000/11,350 |
| Fairbanks-Morse 12-38 D 8 1/8 | OP | 2 | 12 | 8.13/20.65 | 10./25.4 | 12,443/203.90 | 16.1 | 3,840 900 | Yes | 45,000/20,430 |
| White-Superior 40-VX-12 | V | 4 | 12 | 10.0/25.4 | 10.5/26.67 | 9.896/162.17 | | 1,800 1,000 | Yes | |

*Diesel and Gas Turbine Catalog, 1971.
†V = V type; OP = opposed.

**Table 8. Selected Outboard Engines***

| Make and model | Cylinders | | | | Engine type† | Rating, bhp at r/min | Weight, lb/kg |
| | No. | Bore, in/cm | Stroke, in/cm | Displacement, in³/cm³ | | | |
| --- | --- | --- | --- | --- | --- | --- | --- |
| Chrysler 3.6 | 1 | 2.06/5.23 | 1.56/3.96 | 5.18/84.88 | IL | 3.6 4,500 | 33/15.0 |
| Chrysler 35 | 2 | 3.0/7.62 | 2.54/6.45 | 35.9/588.3 | AF | 35 4,750 | 127/57.6 |
| Evinrude Mate 2 | 1 | 1.56/3.96 | 1.37/3.48 | 2.64/43.26 | IL | 2 4,500 | 24/10.9 |
| Evinrude Sportster 25 | 2 | 2.50/6.35 | 2.25/5.72 | 22.0/360.5 | AF | 25 5,500 | 85/38.6 |
| Evinrude Triumph 70 | 3 | 3.00/7.62 | 2.34/5.94 | 49.7/814.4 | IL | 70 5,000 | 202/91.6 |
| Johnson 4R74 | 2 | 1.56/3.96 | 1.37/3.48 | 5.28/86.52 | AF | 4 4,500 | 32/14.5 |
| Johnson 15 R74 | 2 | 2.19/5.56 | 1.76/4.47 | 13.2/216.3 | AF | 15 6,000 | 65/29.5 |
| Johnson 135ESL74 | 4 | 3.50/8.89 | 2.59/6.58 | 99.6/1,632.6 | V | 135 5,000 | 271/122.9 |
| Mercury 40 | 1 | 2.0/5.08 | 1.75/4.44 | 5.5/90.13 | IL | 4 | 56/25.4 |
| Mercury 402 | 2 | 2.88/7.32 | 2.56/6.50 | 33.3/545.7 | AF | 40 | 141/64.0 |
| Mercury 1500 | 6 | 2.88/7.32 | 2.56/6.50 | 99.8/1,635.4 | IL | 150 | 273/123.8 |

*Auto. Ind., revised periodically.
†IL = in line; AF = alternate firing; V = V 4.

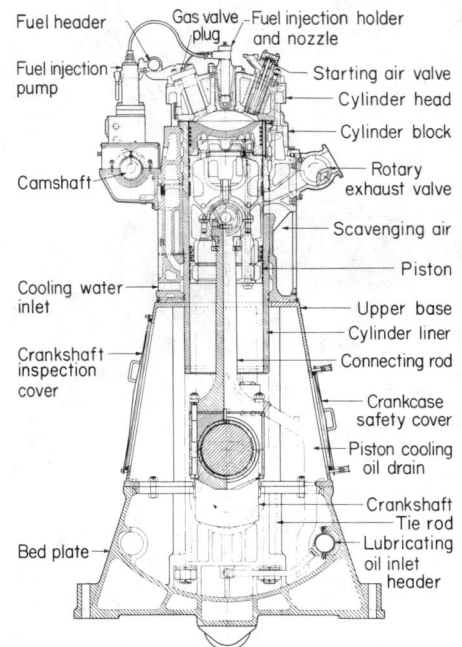

**Fig. 13** Nordberg 12-cylinder, 29 × 40 in (73.7 × 101.6 cm), 12,000-bhp diesel engine.

In each case, the spark-plug location and chamber shape yield a relation between flame-front area and flame travel similar to that in Fig. 22. For a given engine, the maximum rate of pressure development is proportional to the peak flame-front area viewed at TDC. For the chambers of Figs. 18 and 19 virtually all the volume is in the head. This design allows large valves and minimizes piston inertia. The hemispherical or domed head of Fig. 20 allows very large valves and with a domed piston provides a combustion chamber of normal compression ratio. Aircraft engines usually have this chamber shape. The chamber of Fig. 21 is entirely in the bore and piston. This permits considerable design latitude within a

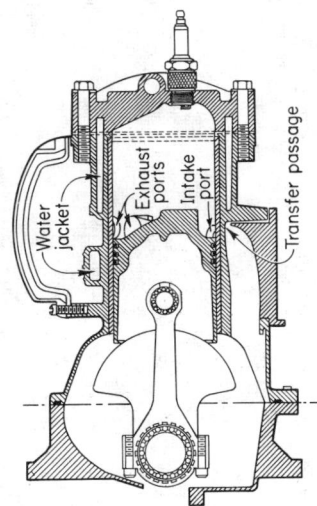

**Fig. 14** Johnson two-cycle, 30-hp, outboard motor.

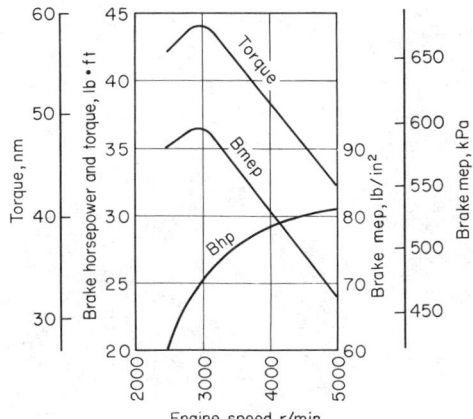

**Fig. 15** Performance of Johnson outboard motor.

given engine. To suppress knock, each chamber has a quench region in the end gas far from the spark plug. In the open-chamber design compression-induced turbulence (squish) is low and mixture motion depends on intake-induced swirl, which is generally quite considerable.

Where cost or engine height are of principal concern, a design such as the L head (Fig. 23), in which the valves are actuated from below directly by the cam, is commonly used.

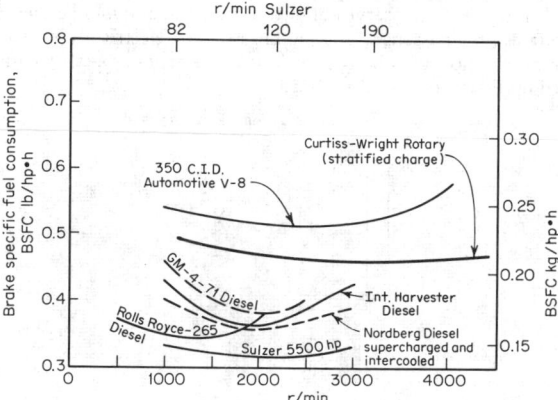

Fig. 16  Specific fuel-consumption curves.

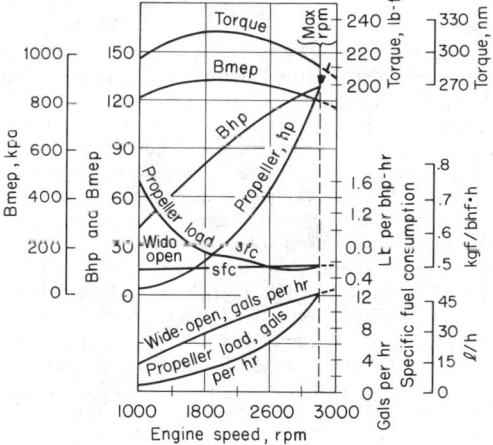

Fig. 17  Performance of six-cylinder 3⁷⁄₁₆ × 4³⁄₄ in (8.73 × 12.07 cm) Chrysler marine engine with 7:1 compression ratio.

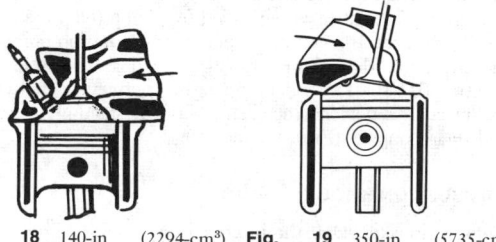

**Fig.    18**  140-in    (2294-cm³) Chevrolet four-cylinder engine.   **Fig.    19**  350-in    (5735-cm³) Buick V-8 engine.

Spark-plug location can compensate to some extent for the slow burn rate of the elongated chamber. The L head has a relatively high surface area, and thus greater heat loss and wall quenching, and a small space for valves.

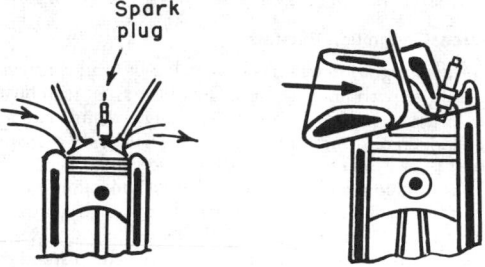

**Fig.  20**  Hemispherical  com-bustion chamber, Dodge truck engine.    **Fig. 21**  Ford truck engine.

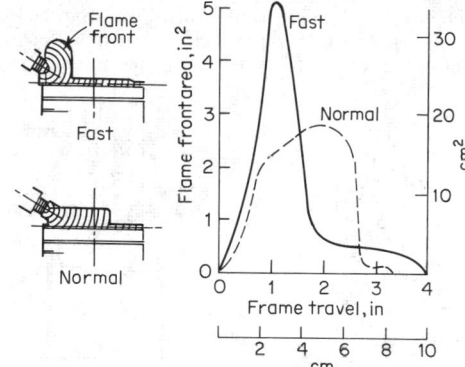

**Fig. 22**  Flame-front-area curves for fast and normal-burn combustion chambers.

**Combustion Roughness**  The high rate of pressure rise during combustion at relatively low engine speeds (1,000 r/min) and the maximum combustion pressure near peak indicated torque engine speeds subject the engine mechanism to severe stresses which may result in combustion-roughness noise unless the pressure rate is controlled. Andon and Marks (*SAE Trans.*, **72**, 1964, p. 636) found that spark-plug location, combustion-chamber shape, and the degree of turbulence are the principal factors that can be varied to control combustion rate and the resulting noise. In a given engine, the maximum combustion rate of pressure rise can be reduced by retarding the spark timing or decreasing the volumetric efficiency.

**Fig. 23**  L-head engine.

A rate of pressure rise of 30 lb/in² (207 kPa) per degree of crank travel results in near maximum efficiency, and higher rates of pressure rise usually result in combustion-roughness noise. Increasing the rigidity of the crankshaft and crankcase increases the natural frequency of vibration and reduces the deflection of the structure caused by combustion and inertial forces and thereby reduces roughness noise.

Die-cast aluminum transmits roughness noise more readily than cast iron. Structurally stiff compact engines such as the V-block are better than the in-line engines in this regard.

### Compression-Ignition Engines

Compression-ignition engines have valve-in-head construction if operating with four cycles. Two types of combustion-chamber arrangements are used: (1) a single chamber, referred to as the open or direct-injection type, and (2) a two-chamber, referred to as a prechamber or divided-chamber type.

**Open Combustion Chambers (Direct Injection)** Fuel is injected under high pressure, usually through a multiple-orifice nozzle, directly into the clearance space or chamber between the piston and the cylinder head. The piston head is usually conformed (Figs. 24 and 25) to fit the fuel spray, and swirl moves the air into the fuel spray. Air swirl is accomplished by intake-port design, by shrouding the intake valve, or in the two-stroke-cycle engine, by using tangential intake ports. High turbulence is accomplished by having the piston closely approach part of the cylinder head. This forces the gases out of the small clearances and agitates the mixture.

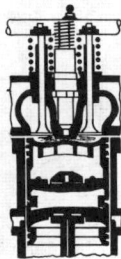

**Fig. 24**  General Motors combustion chamber (swirl produced by intake ports).

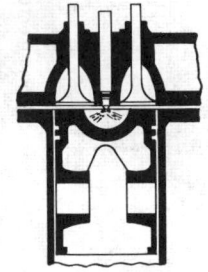

**Fig. 25**  Open combustion chamber with high turbulence and some swirl.

**Precombustion chambers** are divided into two parts, the major volume being between the piston and the cylinder head and connected by a small passageway to the minor volume located in the cylinder head (Fig. 26). Fuel is injected only into the smaller chamber and, except under light loads, partial combustion occurs and discharges the burning mixture into the larger chamber in which combustion is completed. This type of combustion chamber produces smooth combustion but has fairly high fluid-friction and heat-transfer losses.

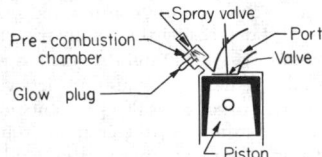

**Fig. 26**  Precombustion chamber.

**Turbulence** or **divided combustion chambers** are modifications of the precombustion chamber, having the major chamber in the cylinder head and usually only a small clearance space between the piston and the cylinder head (Fig. 27). The passage between the two chambers is considerably larger than the passage in the precombustion chamber. Close piston clearance produces high turbulence in the prechamber and promotes rapid combustion. Part of the prechamber containing the transfer passage is commonly thermally insulated. The chamber (Fig. 26) is typical of those used in automotive diesel engines for passenger automobile use. An electrically heated hot-wire glow plug is used to assist cold starting, together with compression ratios up to 24:1.

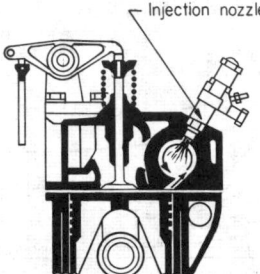

**Fig. 27**  Ricardo divided combustion chamber.

### FUELS
(See also Sec. 7)

Fuels for internal-combustion engines are predominantly petroleum products. Natural and manufactured gases are also used in limited quantities.

Gasoline consists of various amounts of many hydrocarbons, each having its vapor-pressure-temperature characteristic. Also included are small amounts of additives such as knock suppressors, deposit modifiers, antioxidants, metal deactivators, antirust agents, anti-icing agents, detergents, upper-cylinder lubricants, and dyes. The overall volatility of a gasoline sample is specified in U.S. industry by the ASTM Distillation Test (D86) and the ASTM Vapor Pressure Test (D323, Reid Method). The distillation test records the increasing temperature of the vapor in the neck of a flask vs. percent evaporated as 100 ml of the fuel is gradually heated and distilled. The vapor-pressure test gives the pressure in a split-chamber bomb which contains the fuel sample and room air in thermal equilibrium with a 100°F (311°K) temperature bath. The pressure, corrected to standard initial bomb-air conditions, gives the Reid vapor pressure (RVP) of the gasoline. Typical results for summer- and winter-grade gasolines are given in Table 9.

In general, volatility is specified by giving the initial boiling point (IBP), the 10, 50, and 90 percent evaporation points, and the end-point (EP) temperatures as well as the Reid vapor pressure (RVP). These data indicate the effects of fuel volatility on engine performance, e.g., ease of starting, warm-up, and fuel economy (Figs. 28 and 29).

### COMBUSTION KNOCK

Combustion knock in the internal-combustion engine is produced by the spontaneous combustion or autoignition of an appreciable portion of the charge (Rassweiler and Withrow,

**Table 9. Average Volatility Characteristics of U.S. Motor Gasolines**

| | Summer | | Winter | |
| --- | --- | --- | --- | --- |
| | Premium | Regular | Premium | Regular |
| IBP temp | 93°F (34°C) | 94°F (34°C) | 82°F (28°C) | 82°F (28°C) |
| 10% temp | 126°F (52°C) | 125°F (52°C) | 107°F (42°C) | 107°F (42°C) |
| 50% temp | 216°F (102°C) | 207°F (97°C) | 207°F (97°C) | 198°F (92°C) |
| 90% temp | 321°F (160°C) | 338°F (170°C) | 316°F (157°C) | 333°F (167°C) |
| End-point temp | 402°F (205°C) | 414°F (212°C) | 397°F (203°C) | 407°F (209°C) |
| RVP: | | | | |
| lb/in² | 8.9 | 8.8 | 12.3 | 12.1 |
| kPa | 61.3 | 60.6 | 84.8 | 83.4 |

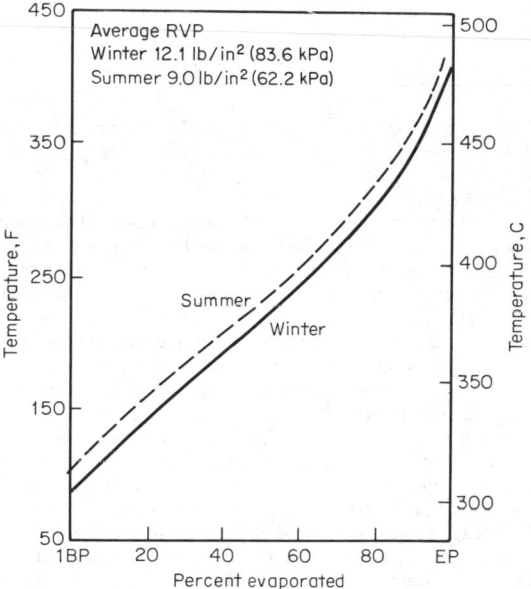

**Fig. 28** Distillation curves for typical summer and winter gasolines. (*Ethyl Corp.*)

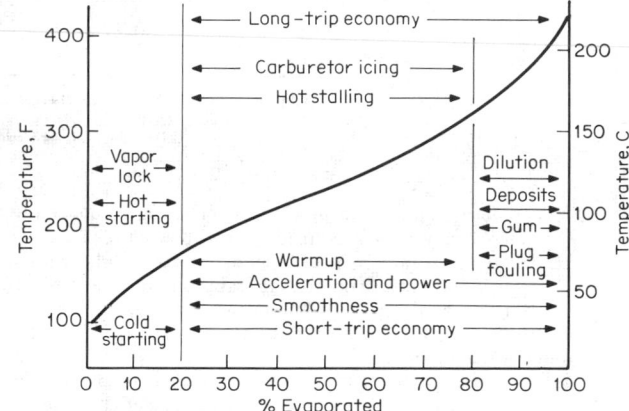

**Fig. 29** Effects of volatility characteristics on engine performance.

*Trans. SAE*, **31**, 1936, p. 297), which results in an extremely rapid local pressure rise and sharp metallic knock. In the spark-ignition engine this occurs when the last fraction of the charge to burn is compressed appreciably above its autoignition temperature by the fraction that has burned. However, if the lag in time (ignition lag) between the attainment of this temperature and the spontaneous local appearance of the flame is increased by the use of a knock-suppressor additive, the flame from the spark plug will be permitted to travel through the unburned fraction before it can knock. Thus knocking tendency depends on the autoignition temperature, ignition lag, and flame speed of the air-fuel mixture.

In the compression-ignition engine, combustion knock occurs when the ignition lag is comparatively long, thus permitting the evaporation of a considerable portion of the injected fuel which suddenly inflames and produces the characteristic knock. Thus knock reduction requires long ignition lags in spark-ignition engines and short ignition lags in compression-ignition engines.

**Autoignition Temperatures and Ignition Lag**  The autoignition temperature of an air-fuel mixture is the lowest tempera-

ture at which chemical reaction proceeds at a rate sufficient to result eventually (long time lag) in inflammation. This temperature depends principally on air-fuel mixture, fuel properties, mixture pressure, and the test apparatus. Subjecting a mixture to a temperature higher than the autoignition temperature results in inflammation after a shorter time lag, there being a temperature for each mixture which results in practically instantaneous ignition (Table 10).

An increase in pressure decreases the ignition temperature of an air-fuel mixture (Table 10). See CRC Report on Project CFD-37-53 (CRC, 30 Rockefeller Plaza, New York 10020, NY).

**Flame Speed**  Flame speeds in spark-ignition engines are low immediately after ignition, attain maximum values when about half the combustion chamber has been inflamed, and decrease toward the end of the process. Mean flame speeds are a maximum for mixtures usually 10 to 20 percent richer than the chemically correct air-fuel ratio, and vary with fuels, engine speed, and turbulence.

At 900 ft/min (4.57 m/s) mean piston speed, mean flame speeds with maximum-power gasoline-air mixtures and optimum spark advance range from about 50 to 100 ft/s (15.24 to 30.48 m/s). Mean flame speeds under the same mixture, spark, and speed conditions in the CFR Waukesha engine are estimated at 55 (70) [80] ft/s (16.76, 21.34, 24.38 m/s) for a 4.6 (6.5) [8]:1 compression ratio, while with ethyl alcohol-air mixtures the corresponding mean flame speeds are 50 (65) [75] ft/s (15.24, 19.81, 22.86 m/s).

**Table 10. Ignition Temperatures of Air-Gas Mixtures at Atmospheric Pressure**

| Gas | Chemical symbol | With ignition lag[a] | | Instantaneous[b] Ignition | |
|---|---|---|---|---|---|
| | | % gas in mixture | Avg temp, °F/K | % gas in mixture | °F/K |
| Hydrogen | $H_2$ | 8–24 | 1130/893 | 10 | 1377/1020 |
| Carbon monoxide | CO | 13–47 | 1204/924 | 50 | 1708/1204 |
| Methane | $CH_4$ | 5–38 | 1202[c]/923 | 25 | >1832/1273 |
| Ethane | $C_2H_6$ | [d] | 968[c]/793 | | |
| Propane | $C_3H_8$ | [d] | 914[c]/763 | | |
| Acetylene | $C_2H_2$ | 4–22 | 804/702 | | |
| Ethylene | $C_2H_4$ | 6–19 | 1004/813 | 10 | 1832/1273 |
| Benzene | $C_6H_6$ | | | 50 | 1943/1335 |
| Ether | $C_4H_{10}O$ | | | 50 | 1892/1306 |
| Gasoline: | | | | | |
| 92 octane | | | | | 734/663[e] |
| 100 octane | | | | | 804/702[e] |

[a]Dixon and Coward, *Trans. Chem. Soc.*, **95**, 1909, p. 514.
[b]David, *Trans. Chem. Soc.*, **111**, 1917, p. 1003.
[c]Minimum value, since spread in values was appreciable.
[d]Composition not reported.
[e]Jones, *U.S. Bur. Mines Bull.*, 1946.

The mean flame speed in an automotive engine under the foregoing conditions is estimated from the optimum spark advance (approximately ⁵⁄₉ of combustion time) to be 90 ft/s (27.43 m/s) for a 6.5:1 compression ratio. In the same engine, mean flame speeds with optimum conditions vary from about 60 ft/s (18.29 m/s) at 500 ft/min (2.54 m/s) to about 170 ft/s (0.86 m/s) at 3,000 ft/min (15.24 m/s) mean piston speed. The higher flame speeds in the automotive engine at comparable piston speeds are due to greater turbulence than that obtained with the usual flat valve-in-head (CFR) design.

Because of the time required for combustion, it is necessary, in order to develop maximum torque, to ignite the mixture in the cylinder before the piston reaches the end of its compression stroke. Spark advance is measured by the number of degrees the crankshaft rotates between the time of the spark and the end of the compression stroke. **Optimum spark advance** is the timing which develops maximum torque. The torque lost by a spark timing earlier or later than optimum is at first only slight but increases rapidly as the timing is further removed from optimum. The tendency of an engine to knock increases almost directly with spark advance (Fig. 30). Before the introduction of exhaust-emission-control requirements, it was general practice in automobile-engine design to use a spark advance a few degrees less than optimum at wide-open throttle. This materially reduces the tendency to knock with only a negligible loss of torque. At part-throttle operation, where combustion knock is usually not a problem, the spark is advanced closer to optimum by manifold vacuum acting through a diaphragm on the spark-timing mechanism. However, in many cases a compromise must be made between spark advance for good torque and spark retard for control of exhaust oxides of nitrogen.

**Knock Suppression** The antiknock characteristics of a fuel can be improved by adding another fuel of better antiknock characteristics or a knock-suppressor fuel additive (Fig. 31). Tetraethyllead (TEL) is the most effective knock suppressor (Table 12) and has been used extensively in small amounts (up to 3 ml/gal of automotive fuels) in many "regular" gasolines and most "premium" gasolines. The addition of TEL to automotive gasolines is being eliminated because of its adverse effects on exhaust emissions. TEL contributes to engine deposits, spark-plug fouling, and degradation of exhaust catalytic reactors. New fuels will be assigned designation numbers according to the applications shown in Table 13, and will be identified by symbols (Fig. 32) showing both the designation number and lead content.

**Table 11. Ignition Temperatures at Various Air Pressure, °F (K)**

| Gas | Air pressure, atm | | | | | | | | |
|---|---|---|---|---|---|---|---|---|---|
| | 1 | 3 | 5 | 7 | 10 | 15 | 20 | 25 | 30 |
| Hydrogen* | 1134 (885) | 1132 (884) | 1112 (873) | 1100 (866) | | | | | |
| Methane* | 1343 (1001) | 1269 (960) | 1215 (930) | 1175 (908) | | | | | |
| Gasoline† | | | 590 (583) | | 480 (522) | 420 (489) | | | |
| Kerosene† | | | 670 (628) | | 490 (528) | 430 (494) | 400 (478) | 385 (469) | |
| Gas oil† | | | 580 (578) | | 500 (533) | 450 (505) | 435 (497) | 415 (486) | |
| Machine oil† | | | 710 (650) | | 610 (594) | 550 (561) | 520 (544) | | |
| Benzene† | | | 685 (636) | | 585 (580) | 545 (558) | 530 (550) | 515(541) | 500 (533) |
| Cylinder oil† | | | 825 (714) | | 710 (650) | 645 (614) | 620 (600) | 595 (586) | 572 (573) |
| Benzol† | | | | | 1095 (864) | 970 (794) | 930 (772) | 905 (758) | 880 (744) |

*Bone and Townsend. "Flame and Combustion in Gases," p. 69.
†Tausz and Schulte, *Zeit. V.D.I.*, 1924, p. 574.

**Table 12. Relative Effect of Antiknock Compounds***

| Compound | Chemical symbol | Weight for a given effect, gm |
|---|---|---|
| Tetraethyllead (TEL) | $Pb(C_2H_5)_4$ | 0.0295 |
| Aniline | $C_6H_5NH_2$ | 1.0000 |
| Ethyl iodide | $C_2H_5I$ | 1.55 |
| Ethyl alcohol | $C_2H_5OH$ | 4.75 |
| Xylene | $C_6H_4(CH_3)_2$ | 8.0 |
| Toluene | $C_6H_5CH_3$ | 8.8 |
| Benzene | $C_6H_6$ | 9.8 |

*"International Critical Tables," vol. 2, pp. 162–163.

Definition of Lead Content of Gasoline:

*Unleaded:* No intentional addition of lead compounds. May not contain more than 0.07 g of lead per gallon (0.018 g/l). (This value is subject to change as more information is obtained about the effect of lead on the potential catalytic-converter system under development.)

*Low Lead:* May contain lead compounds up to 0.5 g of lead per gallon (0.13 g/l).

*Leaded:* May contain lead compounds up to 4.2 g of lead per gallon (1.1 g/l).

**Permissible Compression Ratios**  Data obtained on single-cylinder test engines and multicylinder automotive engines show wide variations in the compression ratios producing incipient knocking of fuels of various octane numbers. High compression ratios in engines using fuels of given octane number do not necessarily indicate excellent design and high performance, since valve timing, manifolding, and porting

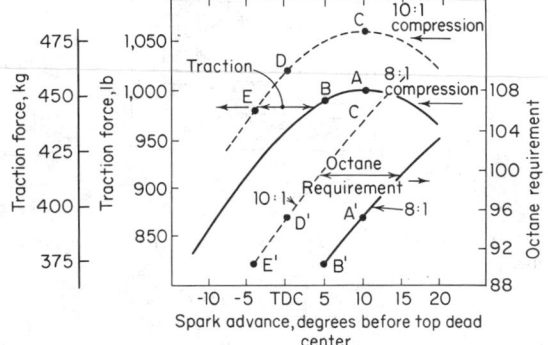

**Fig. 30**  Effect of spark advance on the octane requirement and traction force.

**Table 13. Antiknock Requirements for Gasolines**

| No. designation | Application | Antiknock index* min (RON + MON)/2 |
|---|---|---|
| 1 | For cars with low antiknock needs | < 87 |
| 2 | Meets antiknock needs of most 1971 and later model cars | 87† |
| 3 | For most 1970 and prior model cars designed to operate on "regular" gasoline, and for 1971 and later model cars that require higher antiknock performance than provided for in "designation 2" | 89‡ |
| 4 | A "midpremium" or "intermediate" designation which meets the lower antiknock needs of some cars designed to run on "premium" gasolines and the higher antiknock needs of some cars designed to run on "regular" | 91.5 |
| 5 | For most 1970 and prior model cars with high-compression-ratio engines designed to run on "premium" gasolines, and for later model cars with high-compression-ratio engines | 95‡ |
| 6 | For cars with high-compression-ratio engines designed to run on "premium" gasoline, but requiring higher antiknock performance than "designation 5" | 97.5 |

*One-half the sum of the research octane number (RON) and motor octane number (MON). The antiknock index of gasoline for use in areas where altitude is greater than 2,000 ft (600 m) may be reduced one-half (0.5) number for each succeeding 500 ft (150 m) but not to exceed a total of three numbers.

†In addition, the minimum motor octane number must be 82.

‡In the following states, this minimum may be reduced by one-half (0.5) number: Arkansas, Iowa, Kansas, Minnesota, Missouri, Montana, Nebraska, North Dakota, Oklahoma, South Dakota, Wisconsin, and Texas west of 99° longitude.

may act as throttling devices, or spark timing may be retarded. Average compression ratios vary considerably depending on design and fuel used. An increase in cylinder size usually necessitates a lower compression ratio (Fig. 33), which may be due in some cases to the use of lower grades of fuels with the larger spark-ignition engines. The larger diesel engines require

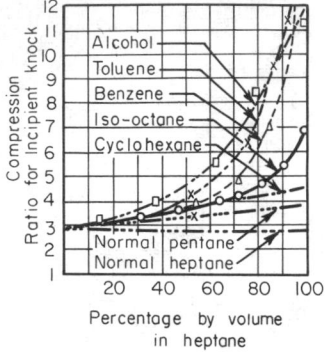

**Fig. 31** Effect of fuel composition on knock-limited compression ratio.

**Fig. 32** Symbols for indicating lead content and antiknock quality of gasoline.

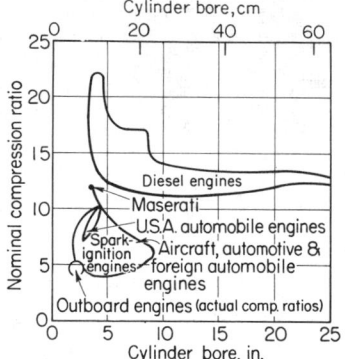

**Fig. 33** Range of compression ratios as a function of cylinder bore (1956).

less compression to obtain the desired temperatures since the heat-transfer effect is less than in the smaller engine. High-compression-ratio spark-ignition engines operate at higher combustion temperatures and thereby create higher concentrations of nitrogen oxides ($NO_x$). Because of the need to reduce $NO_x$, nominal compression ratios have dropped from a high of 10.5:1 to 8:1.

### Knock Rating of Fuels

For detailed information regarding methods, apparatus, reference fuels, and suppliers of apparatus and reference fuels, see ASTM Manual of Engine Test Methods for Rating Fuels, 1952.

**Octane Number**  The octane number (O.N.) of a fuel is the percentage by volume of iso-octane (2,2,4 trimethylpentane) in a mixture of iso-octane and normal heptane which matches the unknown fuel in knock tendency when compared by a specified procedure in the ASTM-CFR knock-testing engine. A fuel that knocks less than iso-octane is rated by the amount of tetraethyllead in iso-octane required to match the knock of the unknown fuel. The relationship between octane number and the tetraethyllead content of gasoline which indicates antiknock quality of gasoline above 100 O.N. is given in Fig. 34.

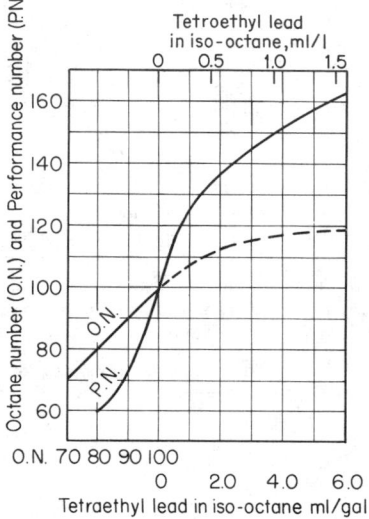

**Fig. 34** Octane number and performance-number scales.

The **performance number** scale for aviation gasolines (Fig. 34) is designed to relate fuel rating to average knock-limited performance (imep). A gasoline of the 100/130 grade indicates that in the aviation test (lean mixture, normally aspirated) the fuel is equivalent to iso-octane in imep while in the super-charge test (rich mixture) it permits an imep of 130 percent of that of iso-octane.

There are four methods, **research, motor, aviation** (lean mixture), and **supercharge** (rich mixture) (ASTM D908, 357, 614, and 909, respectively) in general use in the rating of gasolines according to knocking tendency. These methods vary principally in the engine operating conditions and the means for indicating knock intensity (see ASTM Manual).

The engine conditions, as indicated by the air or mixture temperature and by the coolant temperature, are less severe in the **research method** than in the other methods. Thus, some fuels rate lower by the motor method than by the research method. The **sensitivity of a fuel** is defined quantitatively as the difference between the research and motor octane numbers. Although most gasolines rate lower by the motor method than by the research method, straight-run gasolines have about the same rating by both methods. With present fuels and engines, the average of the research and motor method numbers provides the best correlation with the road ratings.

The use of the aviation and supercharge methods for aviation gasolines results in two performance numbers (such as 100/130). This grading indicates that the fuel is less sensitive than iso-octane to the difference in engine severity in the two test methods.

**Road ratings** (CRC, 30 Rockefeller Plaza, New York) of gasoline are determined by means of borderline knock curves for the test fuel and various blends of the reference fuels at various spark advances. The car is accelerated with wide-open throttle, and the speed at which steady knock ceases is recorded as the knock-die-out speed (CRC procedure, E-2). A plot of these data on a spark-advance vs. knock-die-out speed chart (Fig. 35) makes possible the determination of the octane number rating of the test fuel over the desired speed range.

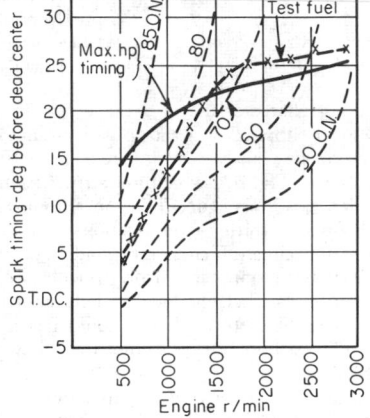

**Fig. 35** Octane number and performance-number scales.

**Mechanical octane number** is the resulting decrease in fuel octane requirements brought about by mechanical changes in engine design or control. Effective design variables include ignition control, combustion-chamber design, valve timing, carburetion, and engine-transmission combinations.

**Lead Susceptibility** Most gasolines have their antiknock characteristics appreciably improved by the addition of tetraethyllead. The susceptibility of the fuel to improved antiknock quality with lead additions varies with different fuels and decreases with the amount of lead added.

**Cetane Number** The cetane number (C.N.) of a diesel fuel is determined by matching its ignition quality with that of a blend of two reference fuels, normal cetane (C.N. = 100) and heptamethylnonane (C.N. = 15). The comparison is made in an ASTM-CFR diesel engine (Waukesha Motor Co.) using a specified procedure (ASTM D613 or CRC-F-5). Cetane number for the reference blend is calculated by

C.N. = (vol. percent *n*-cetane)
        + 0.15 (vol. percent heptamethylnonane)

The **ignition quality** of a fuel for a compression-ignition engine is indicated by the time lag between beginning of injection and start of rapid pressure rise caused by combustion. The compression ratio in the test engine is adjusted to result in a 13° crank-angle lag for the unknown fuel when injection begins at 13° before top center. The mixture of reference fuels which has the same lag under the same conditions indicates the cetane number of the fuel.

The factors that tend to aggravate knocking in a spark-ignition gasoline engine tend to suppress it in a diesel engine. Thus, octane has a low while heptane has a high cetane number. High C.N. fuels burn more smoothly and start cold engines more readily than low C.N. fuels.

Cetane number has been correlated with the mid-boiling point and the API gravity of the fuel, the relation for computing the approximate cetane number being (*Jour. Inst. Petroleum*, **30**, 193, 1944),

C.N. = 175.4 (log mid-boiling pt, °F) + 1.98 (API gr) − 496

The deviation from engine test C.N. for 579 fuels was less than ±5 C.N. for 94 percent of the fuels (see also Sec. 7).

Ignition accelerators increase the cetane number of diesel fuels, but the susceptibility of the fuel to the accelerator decreases with an increase in the amount of the accelerator (Fig. 36). Commercial amyl nitrate is slightly less effective than the pure isoamyl nitrate (Bogen and Wilson, *Petroleum Refiner*, July 1944).

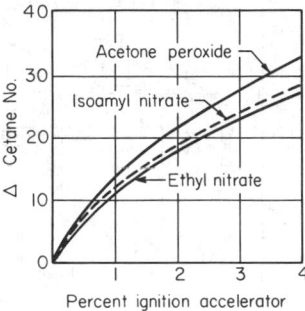

**Fig. 36** Improvement in C.N. with ignition accelerators. (*Bogen and Wilson.*)

## CARBURETION

The carburetor meters, atomizes (if liquid), and mixes the fuel with the air flowing to the engine. A rich mixture is required for idling and small throttle openings because of dilution of the small incoming charge with the exhaust gases in the clearance space. A maximum-economy mixture is desired at intermediate loads, and a maximum-power mixture is usually desired (automotive practice) at wide-open throttle (Fig. 37).

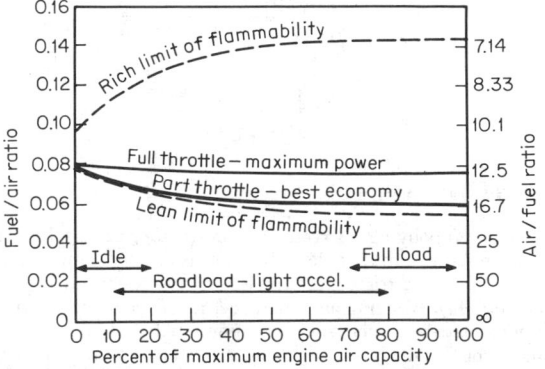

**Fig. 37** Mixture ratios for spark-ignition engines.

Maximum economy is obtained with lean mixtures. Maximum power is obtained with rich mixtures which permit optimum utilization of the oxygen with maximum energy liberation per unit of volume of the mixture. Mixture ratios for minimum specific fuel consumption and for maximum power are also indicated on Fig. 37. The rich mixture which provides maximum power also gives large amounts of carbon monoxide and unburned fuel in the exhaust.

In the carburetor, a float maintains a constant fuel level in a float chamber (Fig. 38) which is vented to the atmosphere or air entrance of the carburetor. The float-chamber fuel level is slightly below the outlet of the fuel jet. The engine air flows through a venturi tube which has a throat diameter of 0.75 to 0.85 of the diameter of the carburetor bore. The reduction in pressure at the venturi throat causes fuel to flow from the float chamber through the main fuel jet into the airstream. Fuel atomization is accomplished by the velocity difference

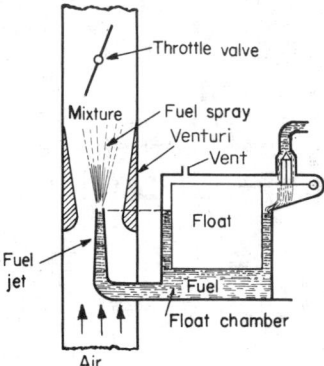

**Fig. 38**  Air-fuel ratios for gasoline.

between the air and fuel. Vaporization of the fuel continues while flowing to the engine cylinder but is usually not completed in the intake manifold at wide-open throttle.

The idling jet (Fig. 39) has maximum suction with the engine idling and the throttle closed. The main jet begins to supply fuel at a small throttle opening. The increasing rich-

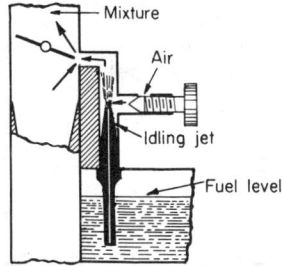

**Fig. 39**  Elementary updraft carburetor.

ness accompanying increase in throttle opening, which is characteristic of the simple carburetor, is usually compensated (Fig. 40) by a restricted air-bled jet or an unrestricted air-bled jet (Fig. 41). Bleeding air into a fuel jet reduces the effective head that causes fuel flow. The unrestricted air bleed eliminates the suction head due to venturi action, and the compensating fuel flow occurs only because of fuel head ($h$, Fig. 41).

Restricted air bleeding only partly reduces the head on the fuel jet. At wide-open throttle, the main fuel orifice area may be increased or another fuel orifice opened, by either throttle action or manifold pressure, to provide the maximum-power mixture.

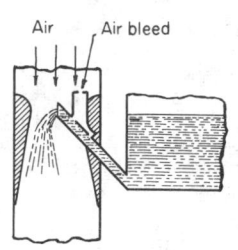

**Fig. 40**  Single-jet down-draft carburetor with restricted air bleeding.

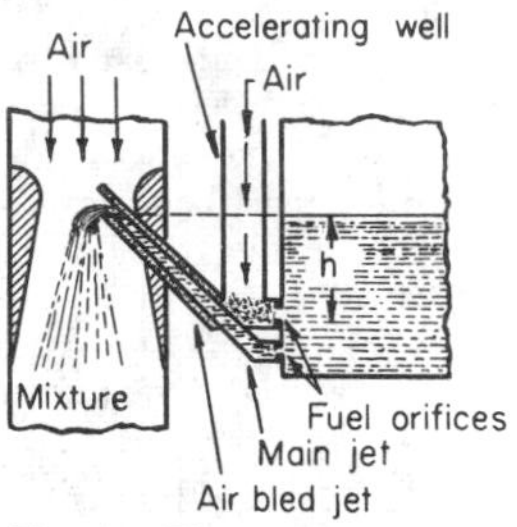

**Fig. 41**  Unrestricted air-bled jet and single-jet downdraft

The metering characteristic of a carburetor is the weight ratio of fuel to air supplied over the operating range of airflow. These carburetor flow characteristics are commonly shown on a graph, similar to Fig. 37, which gives the fuel-air ratios at idle, along the "road-load curve" for an automobile engine, and at wide-open throttle. Computations can be made to establish the approximate fuel-air ratio under specified conditions. These computations are made difficult because air is usually bled into the fuel channels to help control the fuel flow. Also, the airflow through the venturi is pulsating in nature and makes both the air and fuel flow somewhat unpredictable.

Actual carburetor characteristics approach the desirable characteristics indicated in Fig. 37. When the throttle is opened quickly, there is a momentary lag of the fuel flow in relation to the airflow because of the formation of a liquid film on the intake-manifold walls. This necessitates the use of some device (accelerating pump or well) for enriching the mixture and avoiding engine backfiring. At light loads the accelerating well of Fig. 41 contains fuel which is available on rapid throttle opening.

Carburetors for piston-type air-cooled aircraft engines provide fuel-air ratios that vary from 0.11 at idle to 0.085 for cruising rich (cruising lean being about 0.072). As the full-power condition is approached, the mixture richness is increased to aid cylinder cooling, becoming about 0.108 at maximum output. Aircraft carburetors must compensate for the effect of altitude on the air density and the resulting change in mixture ratio. Compensation for the enrichment that occurs with increasing altitude is usually provided by reducing the area of the main fuel-metering orifice. This is accomplished by moving a tapered needle in the orifice, which may be done manually or automatically by means of an air-pressure-sensitive bellows. A carburetor providing an 0.067 fuel-air ratio at sea level would provide a fuel-air ratio of about 0.091 at an altitude of 20,000 ft.

## FUEL INJECTION

**Diesel Engines**  The fuel-injection system for diesel engines usually consists of a pump, fuel line, and nozzle. This system provides control of the beginning, rate and duration of injection, and the desired fuel quantity per injection, and atomizes and partially distributes the fuel in the combustion chamber.

**Methods of Fuel Injection** In the pump-injection system (Fig. 42) a pump forces a desired quantity of fuel through a high-pressure fuel line and an atomizing nozzle into the combustion chamber. Unit injectors combine the plunger pump and spray nozzle, thus eliminating the high-pressure fuel line. Injection pressures usually range from 1,500 to 7,000 lb/in² (10,300 to 48,000 kPa) but may run above 25,000 lb/in² (172,-000 kPa) at high injection rates. Fuel-quantity control with this type of injection is usually effected by controlling a bypass valve in the pump or the pump discharge line or by varying the time of closure or opening of the fuel-pump inlet port. A separate fuel pump is commonly required for each engine cylinder, although rotating plunger pumps serving from two to eight cylinders per plunger are in common use, especially on small engines. In the common-rail system (Fig. 43), a constant pressure is maintained in the fuel-discharge line, and the discharge of fuel to the engine cylinder is controlled by the lift, timing, and duration of opening of the fuel valve. Considerable capacity of the discharge line is necessary in this case, a triplex pump being considered desirable to supply fuel at a constant rate.

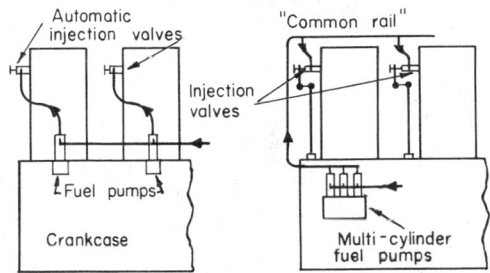

**Fig. 42**  Pump injection.         **Fig. 43**  Common rail.

**Pumps** The volume of liquid fuel to be metered per cycle is about $7 \times 10^{-5}$ in³/in³ (cm³/cm³) of displacement at full load for a normally aspirated engine and only a fraction of this for idling conditions.

Fuel pumps for pump-injection systems require rugged construction, are usually actuated by a cam, and are of constant mechanical stroke. The pump plunger and barrel (Fig. 44) are a very close lapped fit to minimize leakage. The inlet port is uncovered near the bottom of the plunger stroke, the low vapor pressure in the pump cylinder causing fuel flow into the barrel. On the upward plunger stroke, fuel flows past the discharge valve and out the spring-loaded nozzle. Fuel spray from the nozzle continues until the pump bypass port is uncovered. The pressure then subsides, the nozzle closes, and the pump delivery valve closes to maintain a desired residual line pressure until the next injection. Fuel-quantity control is obtained by turning the pump plunger in the barrel, which varies the time of uncovering the bypass port by the plunger helix. This in turn varies the duration of the effective plunger stroke and the amount of fuel injected. The high pressures which occur in these injection systems, together with the compressibility of the fuel oil, give rise to transient pressure-wave phenomena in the connecting lines which change injection characteristics.

**Fuel Lines** The tubing connecting the injection pump with the spray nozzle must be thick-walled, of smooth and uniform bore, and be made of ductile material to permit bending. Tubes range in size from ¼ in (0.64 cm) OD and 1/16 in (0.16 cm) ID to ½ in (1.3 cm) OD and 3/16 in (0.48 cm) ID.

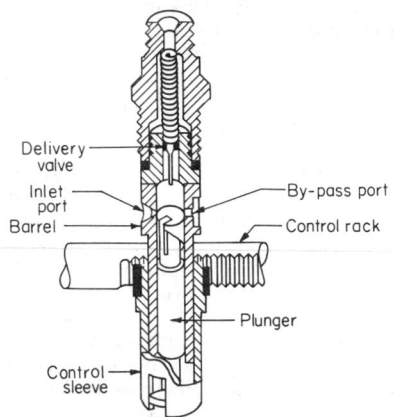

**Fig. 44**  Bosch plunger, barrel, and delivery-valve assembly.

Fuel lines should be of the same length for each cylinder to obtain the same transient hydraulic characteristics. Pressure waves due to plunger motion and nozzle efflux characteristics are developed, travel back and forth, and create disturbances during the injection process (De Juhasz, *Trans. ASME, Oil and Gas Power*, **60**, p.2; Bolt et al., *SAE Trans.*, 1971, paper 710569), making short lines desirable.

**Fuel-Injection Nozzles** Injection nozzles are usually hydraulically operated, having a differential-diameter valve (Fig. 45), held on its seat by a spring, which opens when the fuel-line pressure is increased by the injection pump. The valve closing pressure is always lower than the opening pressure, depending on the relative effective valve areas exposed to the fuel pressure. The valves may be either of pintle or multiple orifice type (Fig. 45). The former has a single orifice through which the end of the valve protrudes, and is used mostly with precombustion chambers. A pintle may be designed to restrict the fuel flow as it opens; this results in a throttling nozzle. The pintle nozzle provides a hollow conical spray with an included angle which varies from 4 to 60° with various pintle sizes. The multiple-orifice type nozzle (c) is used principally with the open combustion chamber.

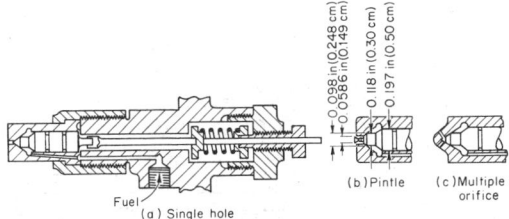

**Fig. 45**  Solid-injection closed nozzle.

**Fuel Sprays** The fuel spray may be a hollow cone or consist of a core and surrounding envelope of atomized fuel.

**Fuel-Injection-System Characteristics** At any given setting most fuel-injection systems deliver more fuel per cycle with decrease in speed, except under the idling condition of very low speed and small fuel delivery. The rising fuel-delivery curve with decrease in speed at full load contributes to the good torque rise and lugging ability of the high-speed diesel. At the idle condition, decrease in speed reduces fuel delivery and causes the engine to run still more slowly, etc. Thus the diesel engine is usually unstable and requires a governor for idling as well as at other loads.

**Spark-Ignition Engines** Fuel is injected into either the supercharger, the intake manifold, the intake valve ports, or the combustion chambers of spark-ignition engines. The advantages of fuel injection over carburetion are more uniform distribution of fuel between the engine cylinders, the reduction of combustion knock because of nonuniform fuel distribution (see also Combustion Knock, above), the elimination of heat to the intake manifold to obtain the desirable fuel vaporization, thus obtaining higher power outputs because of higher charge density, and the use of less volatile fuel. Carburetor (automobiles) and induction-system (aircraft) icing are eliminated. With an injection system it is easier to cut off the fuel flow during deceleration (coasting) than is the case with the carburetor with its fuel-wetted intake manifold.

Injection systems for spark-ignition engines may have mechanical or electronic control systems, with the electronic type being predominant. For the gasoline engine, in contrast to the diesel engine, the airflow must be sensed to establish the correct fuel-air ratio for various engine conditions. This can be done by sensing the airflow with a venturi or equivalent, as used in a carburetor, or by sensing the principal engine variables which control engine air consumption. The latter is referred to as a "speed-density" control system. A system of this type is shown schematically on Fig. 46, and the principal components are indicated. This system is similar to that used on some Cadillac and Mercedes automobile engines. Mechanical and/or hydraulic systems have also been used on automobiles (Chevrolet Corvette) and aircraft engines (Wright and Continental).

The principal disadvantages compared with carburetor systems are the added complications and increased cost, which result in part from sensing difficulties and the problem of metering very small fuel quantities per cycle.

Injection of fuel into the intake port or into the engine cylinder usually occurs early in the intake stroke, to provide the maximum time for fuel evaporation and mixing to occur. Injection into the intake port provides better mixing as the fuel and air are drawn at high velocity through the intake valve passage, and isolates the injector from combustion gases at high pressure and temperature.

## IGNITION AND SPARK ADVANCE

(For details of Ignition Systems, see Sec. 15)

The ignition system consists usually of a 12-V storage battery, induction coil, and high-tension distributor, or a high-tension magneto with a distributor. Recent practice employs electronic means to interrupt primary current in place of a contact breaker. For aircraft the low-tension distribution system to transformer spark plugs or other apparatus for obtaining a high-tension spark at the plug appears to eliminate some of the troubles of the conventional ignition system in aircraft engines at high altitude.

The usual firing order for in-line engines is 1, 3, 4, 2 or 1, 2, 4, 3 for four cylinders; 1, 5, 3, 6, 2, 4 (U.S. automobile engines) or 1, 4, 2, 6, 3, 5 for six cylinders in line: 1, 6, 5, 4, 3, 2 for V-6 engines. A common procedure for V-8 engines is to number the cylinders from front to rear, with the odd numbers in the left bank, as viewed from the driver's seat. For this arrangement a typical firing order is 1, 8, 4, 3, 6, 5, 7, 2. Radial engines have firing orders that skip every other cylinder. For a nine-cylinder single-row engine the order would be 1, 3, 5, 7, 9, 2, 4, 6, 8.

Spark-plug gaps vary from 0.020 to 0.080 in (0.5 to 2.0 mm). The smaller gaps require 4,000 to 8,000 V while the larger gaps require 10,000 to 34,000 V.

Spark plugs are usually located near the exhaust valve (hottest spot) so that the flame progresses toward the cooler part of the combustion chamber. This permits higher compression ratios without combustion knock. Part-throttle operation may require a slight change in location to provide a combustible mixture at the spark plug. Engines with lower than optimum compression ratios may have central spark-plug location which usually results in minimum combustion time. Dual ignition (two spark plugs located opposite each other) also reduces combustion time and results in a slight gain in efficiency (usually less than 1 percent) compared with single ignition with optimum spark advance. Rotary (Wankel) engines employ one or two spark plugs per rotor. Single plugs are centrally located either above or below the trochoid waist.

Since combustion requires appreciable time, the spark advance should distribute the combustion process before and after top center in order to obtain maximum power. **Optimum spark advance** depends principally on air-fuel mixture, amount of residual gas, emission-control requirements, combustion-chamber design, turbulence, engine speed, number of spark plugs, and spark-plug location. Maximum power air-fuel ratios require the minimum spark advance. Low-speed engines require 10 to 20° (crank travel) spark advance, and

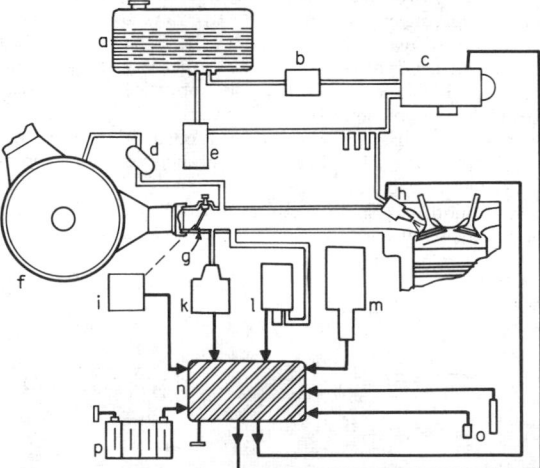

**Fig. 46** Schematic of electronic gasoline-injection system for an automobile. (*a*) Gasoline tank; (*b*) fuel filter; (*c*) fuel-supply pump; (*d*) auxiliary idling air valve; (*e*) fuel-pressure regulator; (*f*) air filter; (*g*) throttle plate; (*h*) fuel-injection valve; (*i*) throttle-position switch; (*k*) full-load pressure switch; (*l*) manifold pressure sensor; (*m*) ignition distributor with trigger contacts; (*n*) electronic control unit; (*o*) temperature sensor; (*p*) battery. (*From Robert Bosch diagram.*)

high-speed automotive engines require 30 to 40° spark advance. Racing engines require still more. Full load requires less spark advance than part throttle. Spark advance in most automotive engines is controlled automatically by engine speed and manifold vacuum; increases in both of these independently increase the spark advance.

## AIR AND FUEL LINES, MANIFOLDS, MIXTURE DISTRIBUTION, AND MUFFLERS

**Air lines,** extending from the engine to the outside air, should be designed for air velocities of 50 to 100 ft/s (15–30 m/s). Air cleaners for the removal of dirt particles reduce piston, ring, and cylinder wear. Air silencers, used at the entrance to the air lines, are combined with cleaners for automotive purposes.

**Gas lines** should be designed for gas velocities of 30 to 60 ft/s (9 to 18 m/s). A pressure drop of 0.07 to 0.12 in (1.8 to 3 mm) of water should be allowed per 100 ft (30 m) of pipe. Close calculation of the **size of gas pipe** for industrial gas installations is undesirable, because of the reduction of the normal cross section of the pipe by tar and dust deposits. A storage chamber of large diameter should be located in the pipe near the engine. The gas line should be free from traps and as free as possible from changes in direction of flow, and should be pitched slightly from both ends toward a low point in the center, where a sealed drain should be located. All water seals in the gas line should be as cool as possible to minimize evaporation and the breaking of the seal.

**Gasoline** and **fuel-oil lines** should be designed for maximum velocities under 1 ft/s (0.3 m/s) for gravity feed. Gasoline lines should be in a cool location, as heating above 100°F (38°C) may vaporize the lighter fractions and cause **vapor lock.** Sediment and water traps as well as strainers or filters should be located in the fuel line ahead of the carburetor, fuel pump, or injection pump.

**Intake Manifolds**  The intake manifold distributes the air (diesel) or the air and fuel (Otto) to the various cylinders of a multicylinder engine. The updraft manifold (carburetor below manifold) should have a **riser velocity** of not less than 40 ft/s (12 m/s) to entrain or lift the liquid fuel particles from the carburetor to the **distributor** section of the manifold. This determines the minimum wide-open throttle speed for satisfactory operation. The **downdraft** manifold (Fig. 47) permits larger areas and lower velocities between the carburetor and manifold, the average for automotive practice being about 50 ft/s (15 m/s) for a mean piston speed of 700 ft/min (3.5 m/s) (500 to 900 r/min). The corresponding mean gas velocity of about 250 ft/s (76 m/s) at a top speed of 3,500 ft/min (18 m/s) mean piston speed indicates a maximum loss of volumetric efficiency of about 3 percent due to pressure drop required to attain this velocity. Higher velocities are undesirable because of greater losses in volumetric efficiency, a drop in intake pressure of 1 lb/in² (6.9 kPa) below atmospheric reducing the volumetric efficiency about 7 percent.

**Intake manifolds** are either **streamlined** (always with diesel), of the **rake type,** or a combination of both. The sections may be either circular or rectangular. In some cases, the circular sections are combined with a flat **floor** or bottom to provide more surface on which the liquid fuel may spread and be vaporized. **Dams** are also used to prevent liquid fuel from flowing along the manifold floor.

The overlapping of the intake-valve periods produces com-

plicated disturbances in the intake manifold and has resulted in the use of dual manifolds in some engines with six or more cylinders. V-8 automobile engines have complicated dual manifolds, integrally cast, each manifold usually feeding two

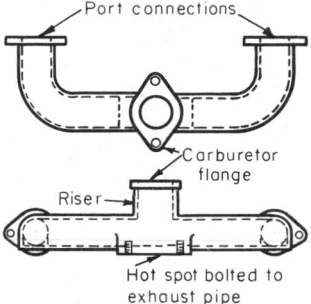

**Fig. 47**  Downdraft two-port intake manifold with flat floor and hot spot.

cylinders of each bank. Rotary engines have a simple intake-system requirement. A separate carburetor (or venturi) may be used to feed each chamber.

Preheating the carburetor air and controlling its temperature is desirable in that it reduces air-density variation and thereby allows better control of mixture ratio for economy and emission control. Carburetor icing is minimized and the need for an exhaust-manifold crossover valve reduced. Some heat must be supplied to vaporize the fuel. Preheating is undesirable in that it reduces volumetric efficiency, heats the carburetor and evaporates fuel in the float chamber, and may increase intake-system coking. Preheating the fuel is undesirable, since the light fuel fractions will be lost by vaporization. Consequently, most intake manifolds for liquid fuels have a **hot spot** (Fig. 47), which receives heat from the exhaust (usually thermostatically controlled) and on which the fuel particles impinge upon leaving the manifold riser. The dew points or condensation temperatures for fuel in an air-fuel mixture depend on the fuel, air-fuel ratio, and total pressure (Table 14), and may be determined from equilibrium air-distillation data.

**Mixture Distribution**  Perfect distribution consists of equal distribution of both air and fuel to the various cylinders of a multicylinder engine for each cycle. The compression pressure of each cylinder is an indication of the air distribution if all cylinders have identical valve timing and compression ratio and no leakage. The effect of unequal distribution of fuel or both air and fuel on the performance of each cylinder is suggested from the performance of a single-cylinder engine with variations in air-fuel ratio (Fig. 48). Cylinders receiving lean mixtures develop less power than those receiving correspondingly rich mixtures. Poor distribution can be masked by enriching the mixture and high power output attained at the cost of high fuel-consumption rates and HC and CO emission increases.

The CO, $CO_2$, and $O_2$ analysis of the exhaust gases from the individual cylinders and spark-plug temperatures (thermocouple in the center electrode) may be used for determination of mixture distribution. Typical mixture distribution with liquid fuel shows a variation of 10 to 25 percent in fuel

**Table 14. Condensation Temperatures of Liquid Fuels**

| Fuel | Deg API | ASTM Distillation, deg F (deg C) | | | | Condensation temp, deg F (deg C) | | | |
|---|---|---|---|---|---|---|---|---|---|
| | | | | | | Atmospheric pressure | | Pressure, ⅔ atmospheric | |
| | | | | | | Air-fuel ratio | | Air-fuel ratio | |
| | | Initial point | 50 percent | 90 percent | End point | 12:1 | 15:1 | 12:1 | 15:1 |
| Gasoline | 60 | 100(38) | 249(121) | 354(179) | 400(204) | 118(48) | 114(46) | 111(44) | 107(42) |
| Kerosene | 43 | 350(177) | 450(232) | 510(266) | 550(288) | 204(96) | 200(93) | 201(94) | 197(92) |

supplied the various cylinders. The use of a gaseous fuel in the same engine resulted in almost perfect distribution.

Distribution of load (and fuel) and variation in injection timing between cylinders of a diesel engine are similarly indicated by the exhaust-gas composition and temperature of the individual cylinders, the mixtures in the various cylinders usually being lean under all conditions.

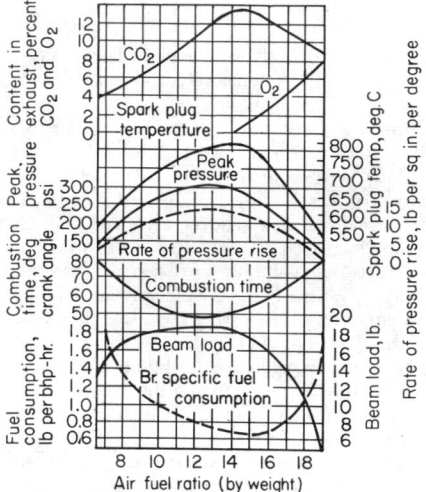

**Fig. 48** Influence of air-fuel ratio in an overhead-valve engine with cylindrical combustion chamber. (*Rabezzana and Kalmar, Auto. Ind., 66, 1932, p. 13.*)

**Exhaust Manifold** In addition to adequate flow area, the most important consideration in exhaust-manifold design is expansion. Small multicylinder engines employ a one-piece exhaust manifold. This is anchored at the center of the engine, and elongated holes are provided at other points for expansion. Large engines employ multipiece manifolds with expansion joints. They are usually water-cooled in marine applications to prevent strain and for safety reasons. In turbocharged engines exhaust manifolds are usually insulated to conserve energy.

Manifold flow area should be large enough to avoid local back-pressure buildup during exhausting and in connection with the total exhaust system should not raise back pressure excessively at maximum flow. A flow area of 0.7 times intake manifold area may be used for four-cycle engines. For two-cycle engines exhaust flow area commonly exceeds intake flow area. The overlapping of the exhaust-valve periods in multicylinder engines makes multiple manifolding desirable.

**Mufflers** (See Sec. 12) Exhaust mufflers, to be efficient as sound silencers, must decrease the exhaust-gas velocity and also absorb the sound waves or cancel them by interference with other waves from the same source. Mufflers should have volumes six to eight times the piston displacement and may contain baffles with or without holes. Mufflers that cancel sound waves by interference usually break the waves into two parts which follow different paths and meet again out of phase before leaving the muffler (Davis et al., *NACA Report* 1192).

Exhaust **muffle pits** are used with large engines, are usually made of concrete, have a volume about twenty times the piston displacement, and may be open or provided with baffling or partly filled with loose stone through which the gases must pass. A stack is usually provided for the escape of the gases and a drain for the discharge of the condensed water from the exhaust products.

Exhaust **back pressure** should be kept to a minimum since an increase of 1 lb/in² (6.9 kPa) in back pressure decreases the maximum power output about 2 percent, about 1 percent being due to more exhaust work and the balance to the effect of increased clearance gas pressure on volumetric efficiency.

### SUPERCHARGING

Supercharging increases the amount of charge per cycle above that of the normally aspirated engine. It therefore permits more fuel to be burned and is a practical means to greater engine power. It also increases mechanical and thermal stresses in the engine. For aircraft piston engines it is a means to high power output for takeoff and to offset the rare atmosphere at high altitude. For diesel engines supercharging is a means to smaller and lighter power plants. It is especially practical for diesel engines because supercharging does not add to the fuel quality required, and because diesel engines do not use combustion air as effectively as spark-ignited engines.

Two types of superchargers are in general use: the positive-displacement type (Roots blower) and the centrifugal type (see Sec. 14). The positive type is desirable for variable-speed engines (diesel rail car) where high torque is required at various speeds, since the pressure and capacity characteristics of this type do not decrease with speed. The centrifugal type is particularly adaptable to aircraft engines because of high capacity, small size, and low weight. It is also used with truck diesel engines to improve the volumetric efficiency and power at high engine speeds. Centrifugal blowers may have one or two stages with cooling between stages or after the single or last stage. The rotors run at maximum speeds of 16,000 to 50,000 r/min.

Superchargers may be geared directly to the engine, or driven by an exhaust turbine (Fig. 49).

Supercharging a carbureted engine which has the optimum

compression ratio for the available fuel causes combustion knock and necessitates lowering of the compression ratio, enriching the mixture, or increasing the octane number of the fuel. Supercharging a diesel engine increases the maximum pressure and necessitates lowering the compression ratio or changing the fuel-injection timing, if the limiting pressure is already attained without supercharging. The use of variable engine compression ratio together with supercharging is very desirable but difficult and costly.

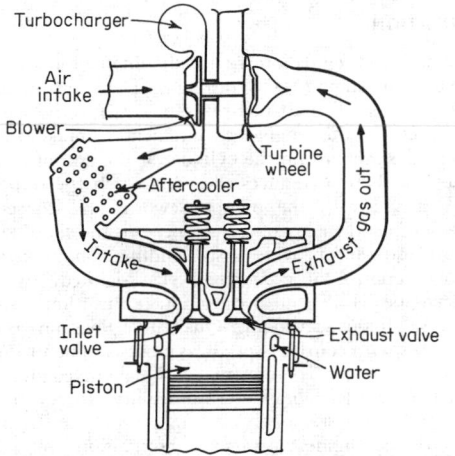

**Fig. 49**   Schematic of exhaust turbosupercharger installation on Worthington diesel-engine cylinder.

Turbocharging of diesel engines is usual for most applications, including truck, locomotive, and marine applications. The exhaust-gas turbine provides more complete expansion of the combustion gases than is practical in the engine cylinder and contributes to improved thermal efficiency. Higher output also increases the engine mechanical efficiency. (Buchi, *ASME Trans.*, **59**, pp. 85–96.) Supercharging assists with $NO_x$ control without a fuel-economy penalty. The fuel economy and other performance characteristics of a supercharged Sulzer marine diesel engine of 30 in (760 mm) cylinder diameter are shown in Fig. 50.

## SCAVENGING TWO-STROKE-CYCLE ENGINES

Crankcase compression for each cylinder or an engine-driven blower may be used for scavenging and charging the cylinders of a two-stroke-cycle engine (Fig. 51). With crankcase compression, used on small gasoline outboard marine engines, a third port uncovered by the piston near the top of the stroke, a rotary valve, or an automatic poppet valve may be used. The top of the piston is shaped to deflect the gases entering the cylinder and prevent short circuiting to the exhaust port. A "dead" spot of gases may remain in the lower center of the cylinder or in the upper corners.

The intake may be inclined and tangential, giving the entering charge an upward helical motion (Fig. 52). This motion keeps the entering charge near the cylinder walls, forcing the exhaust gases to the center of the cylinder and down to the exhaust port. Some engines employ loop scavenging (Fig. 53).

The General Motors (Fig. 10), and some large diesel engines

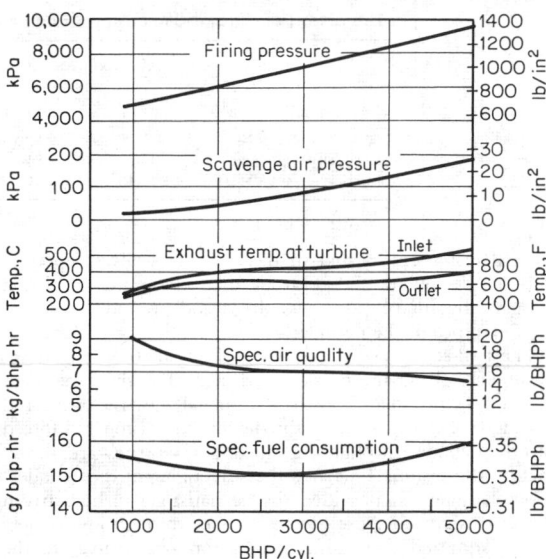

**Fig. 50**   Performance values of a turbocharged diesel engine (Sulzer).

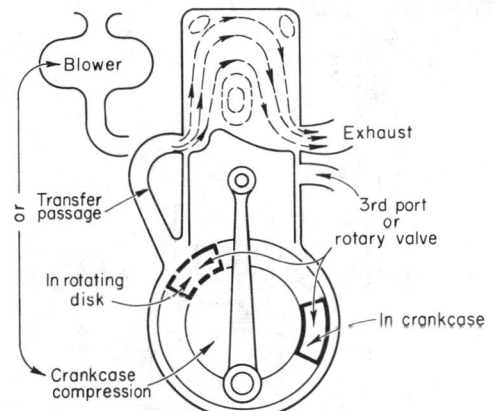

**Fig. 51**   Common scavenging systems.

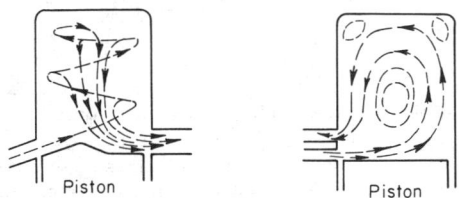

**Fig. 52**   Hesselman helical loop   **Fig. 53**   M-A-N loop scavenging. system.

have poppet exhaust valves in the cylinder head, eliminating the hot exhaust ports from the cylinder walls. Intake ports can then completely surround the cylinder. Typical blower pressures and related conditions are as shown in Table 15 for the General Motors engines.

The Fairbank-Morse opposed-piston engines have intake

**Table 15. Two-Stroke-Cycle Diesel Engine**

| Engine type | Air-box pressure 500 to 2,300 r/min | | Max bmep* | | Max imep* | |
|---|---|---|---|---|---|---|
| | in Hg | cm Hg | lb/in² | kPa | lb/in² | kPa |
| Naturally aspirated | 1–9 | 2.54–22.86 | 101 | 697 | 129 | 890 |
| Turbocharged | 1.5–40 | 3.81–101.62 | 136 | 938 | 163 | 1,125 |

*At 2,300 r/min.

and exhaust ports located in the cylinder walls at opposite ends of the cylinder. A blower forces the entering charge through the intake ports which extend around the entire circumference of the cylinder.

The two-stroke-cycle Sulzer diesel engine (Fig. 54) makes use of a crosshead and piston-rod design. The air in the space between the piston and the diaphragm carrying the piston-rod stuffing box is displaced by the descending piston and forced into the cylinder through the intake ports as they are uncovered by the piston. This aids the scavenging of the cylinder. Supercharging is provided by exhaust-gas-turbine-driven blowers. Air from these blowers passes through intercoolers and is admitted through automatic one-way valves to the scavenging chambers. Electric-motor-driven air blowers are also sometimes used on large engines to assist scavenging at idle and low power (Fig. 54).

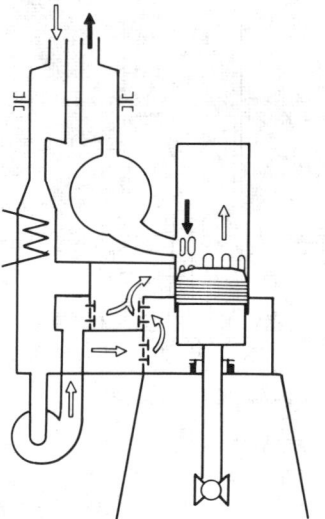

**Fig. 54**  Sulzer RND two-cycle diesel engine (schematic).

The crosshead design, with its stuffing box, also prevents combustion products from contaminating the crankcase oil system, and is commonly used on large engines. This RND design is used for cylinder diameters of 680 760, 900, and 1,050 mm bore (27, 30, 35, and 42 in). The largest cylinder, with a stroke of 1,800 mm (71 in), delivers a normal output of 4,000 bhp at 108 r/min corresponding to a bmep of 1,030 kPa (150 lb/in²) and a piston speed of 6,480 m/s (1,300 ft/min). Figure 50 shows the performance curves of this type up to an overload of 5,000 bhp per cylinder. The smaller cylinders produce correspondingly lower outputs with roughly the same mep and piston speed.

## REGULATION

Speed and load regulation are usually obtained in four-stroke-cycle gasoline engines by throttling the charge or by quantitative governing. The air-fuel ratio may vary some with load, depending on the engine requirements. Intake throttling reduces the pressure in the cylinder at the end of the intake stroke and thereby reduces the mass of charge trapped per cycle. It increases the pumping work, and consequently decreases the efficiency of the engine. It is the simplest means for regulation and provides a stable idling condition.

Power control of carbureted, spark-ignited, two-stroke-cycle engines is difficult because intake throttling reduces or prevents cylinder scavenging. Qualitative governing is used in some cases with constant-speed gas engines. The fuel is throttled while the air is unrestricted. Qualitative governing is limited by the lean limit of flammability. In addition, lean mixtures burn more slowly and may lower the thermal efficiency of the engine. Advancing the ignition timing, as the mixture is made leaner, reduces this loss of efficiency.

Governing by variation of ignition timing, as used on some small two-stroke-cycle gasoline engines, is undesirable since this is regulation by variation of thermal efficiency and may result in overheating the engine. Variation of ignition timing is required for each change in speed and load to obtain optimum results (see Ignition and Spark Advance).

Power control or regulation of diesel engines is obtained by varying the amount of fuel injected per cycle, which provides increasing richness of the fuel-air mixture as the engine load is increased. Minimum specific fuel consumption usually occurs at about 85 percent of the maximum load.

Diesel engines require a speed governor to provide a stable idle. Mechanical, hydraulic, and electronic speed-control governors are used on diesel engines. These control the fuel-pump injection quantity and provide speed regulations at all operating conditions.

## COOLING SYSTEMS

Engine cylinders must be cooled to maintain a lubricant film on the cylinder walls and other sliding surfaces. The cylinder heads, pistons, and exhaust valves are cooled to prevent combustion knock or destruction of these parts due to overheating. The lubricant must be cooled to maintain the desirable viscosity under operating conditions. Water- or air-cooling systems are used. Pistons, exhaust valves, and lubricants in comparatively small or low-duty engines are sufficiently cooled by contact with other engine parts or the lubricant between them and do not require separate cooling systems.

**Water Cooling**  The heat removed by the cooling water from the cylinders and cylinder heads ranges from 15 to 20 percent of the energy input for large diesel engines, 20 to 35 percent for automotive engines, and may run as high as 40 percent for automotive engines at one-third load. An increase in speed from 1,000 to 3,000 r/min decreases the heat loss to the cooling water about 20 to 30 percent. These values indicate a heat loss ranging from 40 to 50 percent of the brake output for large diesel engines and 100 to 150 percent of the brake output for automotive engines.

Pistons in small engines are cooled by heat transfer to the cylinder walls and lubricant. In some cases, an appreciable quantity of oil is directed against the piston head to maintain desirable head temperatures.

Water is fed into and out of water-cooled pistons usually by means of swing joints or telescopic water connections. To keep the piston jackets of large engines filled, the cooling water must be under a pressure of 60 to 80 lb/in² (415 to 552 kPa). For the other jackets, 10 to 15 lb/in² (69 to 104 kPa) is sufficient, and therefore two cooling systems at the different pressures may be used or the water may be throttled at the entrance to the other jackets. Valves on the discharge sides of the jackets should not be used. Even with the high pressure of the jacket water, there will be a water hammer in the piston jackets unless the area of the discharge is restricted to about 20 percent of the area of the inlet opening. The average percentage distribution of the water to the various jackets is approximately as follows: 60 to the cylinder jacket and 40 to the cylinder-head jacket with uncooled piston; 50 to the cylinder jacket, 25 to the cylinder-head jacket, and 25 to the piston and piston rod with water-cooled piston.

The cooling system may be noncirculating, in which case cold water is supplied, flows through the water jacket, and is wasted. With a temperature rise of 90°F (32°C) from 2 (7.6) (in large engines) to 4 gal (15.2 l) of water per bhp-h must be supplied. In some engines, cooling-water outlet temperatures as low as 120°F (49°C) must be maintained, which with high inlet temperatures may increase the required quantity of water to as much as 10 gal (38 l) per bhp-h. Water is circulated at a rate of 25 to 50 gal (95 to 190 l) per bhp-h with a temperature rise of 10 to 20°F (5.6 to 11.2°C) while flowing through the water jacket.

Natural circulation (**thermosiphon**), with low velocities, requiring large connections, may be used if the water forms a complete circuit. In this case, hot water rises in the engine jacket, flows to the radiator, is cooled, descends, and flows to the engine jacket.

Small stationary engines are often **hopper-cooled**. This is an **evaporative cooling system** requiring about 4 to 6 lb (1.8 to 2.7 kg) of makeup water per bhp-h. The ASTM-CFR knock-testing engines are cooled by evaporation, the vapor being condensed by a **reflex condenser** (copper coil with cold water flowing through) and returned to the cylinder jacket.

In high-output engines, the coolant is directed at the hottest spot, usually the exhaust-valve seat, either with an external or internal inlet manifold for a common cylinder-block jacket; otherwise vapor bubbles will form, cling to the surface, and cause overheating. In vertical engines, the water usually flows upward, around the cylinder barrel into the cylinder-head jacket, and to the outlet. Higher output and reduced knocking tendency have resulted from directing the entering coolant at the hot spots in the cylinder head and then flowing the coolant downward around the cylinders.

The **size of piping** required for the inlet to the jacket may be calculated for 3 ft/s (0.9 m/s) water velocity, and for the discharge 2 ft/s (0.6 m/s) for large engines. In engines with recirculation, these values are increased up to 10 or 15 ft/s (3 or 4.5 m/s) and the size of the inlet line to the circulating pump is usually larger than the outlet from the water jacket. It is desirable to have an open or visible outlet from each separately cooled part, showing the flow of cooling water. In order to ensure totally filled systems which are as air-free as possible, overflow bottles may be used. Materials such as ethylene glycol are used as the cooling medium when higher jacket temperatures are desired (see Sec. 6) without pressure increase and for freeze protection.

**Air Cooling** Small industrial engines, motorcyle engines, and many aircraft engines are air-cooled. Air cooling eliminates the necessity of the water or other liquid cooling medium, water jackets, pumps, radiators, and water connections, but necessitates single-cylinder construction, finning, baffles, and, in some cases, blowers. The permissible compression ratio and output of air-cooled aircraft engines depend on the efficiency of the cooling of the cylinder head, exhaust valves, and seats. Long fins [1 to 2 in (25 to 50 mm) depending on cylinder size], closely spaced [0.10 to 0.20 in (2.5 to 5 mm)], with baffles directing the air at high velocity at the hot spots, have made high outputs of aircraft engines possible at the expense of considerable air resistance or drag. Properly designed cooling ducts may make use of the heat added to the cooling air to obtain a jet effect, thereby reducing the drag.

**Oil Cooling** The shearing of the various oil films by the moving parts and the contact with hot parts of an engine cause a rise in oil temperature until energy input to the oil equals the energy transferred by it to the cooler parts it contacts. Oil coolers are required for high-duty engines to maintain oil temperatures of 200°F (94°C) or under. Temperatures of 250°F (121°C) are not uncommon in automobile engines on hot days; 300°F (150°C) is considered too high, particularly for those oils which decompose rapidly under conditions causing the high oil temperatures. The desired oil temperature is maintained by circulating the oil through a radiator or cooler to an oil sump and then through the engine. Transmission-oil cooling is also accomplished by placing plate coolers in the end tanks of radiators or with separate radiators.

## LUBRICATION
(See also Sec. 6)

**General** A suitable liquid lubricant is vital to the satisfactory operation of an internal-combustion engine. It prevents excessive wear and deposit accumulation, and removes heat from the areas of relatively high temperature within the engine.

Most engine oils are composed of base oils and additives. The base oils are usually mineral oils, but some are synthetic oils. Chemical additives are incorporated in engine-oil formulations to impart desirable performance characteristics which are not provided by base oils alone.

**Physical Properties** Both the physical and chemical properties of an engine oil affect its performance. The principal physical property involved is viscosity. It must be low enough at low temperatures to permit cranking and starting, and high enough at high temperatures to provide an adequate oil film between rubbing surfaces. Sufficiently high viscosity is also required to help prevent excessive oil consumption at high temperatures. Another physical property affecting consumption is volatility. Consumption can be undesirably high with an oil of excessive volatility. Both viscosity and volatility are influenced by base-oil selection. Viscosity can also be modified by means of additives.

**Chemical additives** also perform a number of other functions in an engine oil. For example, oxidation inhibitors reduce oxidative and thermal degradation, which can lead to varnish and sludge deposition, and to excessive thickening. Detergent-dispersant additives suspend insoluble products formed during engine operation, so that they can be removed from the engine when the used oil is drained. Antiwear agents help protect rubbing surfaces, especially those subject to boundary lubrication. Antifoam agents prevent foam and air entrainment, and their adverse effects on oil pressure and heat transfer. Rust inhibitors eliminate the formation of corrosion products on ferrous surfaces, and reduce corrosive wear.

**SAE Classifications**    Engine oils are classified by the Society of Automotive Engineers in two general groupings, viscosity and performance. Viscosity is measured at both 0°F (−18°C) and 210°F (98.9°C). SAE 5W, 10W, and 20W viscosity numbers are related to 0°F (−18°C) measurements; and SAE 20, 30, 40, and 50 to 210°F (98.9°C) measurements. Multigrade oils, such as SAE 5W-20 and SAE 10W-30, are those which have viscosity characteristics meeting requirements at both temperatures.

The SAE performance classification includes engine oils ranging in performance quality from that of a straight mineral oil plus a small amount of pour-point and/or foam depressants, to those required in severe passenger-car service and heavy-duty truck and off-highway operation. Letter designations such as SD and SE apply to oils for gasoline-engine service; designations CC and CD apply to oils for diesel-engine and heavy-duty gasoline-engine service. Performance of these oils is evaluated in both single- and multicylinder engine tests. Factors measured include rust, wear, deposit formation, oil consumption, and oil thickening.

REFERENCES: Most of the foregoing information on engine lubrication is covered in more detail, and references are provided, in several pages on engine oil in the "SAE Handbook." Additional information may be found in two books by Schilling, "Motor Oils and Engine Lubrication" (1968), and "Automobile Engine Lubrication" (1972), Scientific Publications (G.B.) Ltd.

### ANALYSIS OF THE ENGINE PROCESS

**Cycle Analysis**    The theoretical indicated thermal efficiency $\eta_a$ of the air-standard Otto cycle depends only on the volumetric compression ratio $r_c$ (Sec. 4).

$$\eta_a = 1 - (1/r_c)^{0.4} \qquad (1)$$

For the air-standard diesel cycle (constant-pressure combustion),

$$\eta_a = 1 - (r_d^{1.4} - 1)/[1.4r_c^{0.4}(r_d - 1)] \qquad (2)$$

where $r_d$ is the volumetric expansion ratio during the constant-pressure combustion.

However, the diesel engine is more closely approximated by the limited-pressure combustion (dual) cycle for which the thermal efficiency is given by

$$\eta_a = 1 - (r_p r_d^{1.4} - 1)/\{r_c^{0.4}[(r_p - 1) + 1.4r_p(r_d - 1)]\} \qquad (3)$$

where $r_p$ is the pressure ratio during the constant-volume heat addition.

The efficiencies indicated by the air-standard analysis are much higher than can be attained, principally because of variable specific heats, dissociation, and heat and time losses. The fuel-air cycle analysis takes into consideration variable fuel-air mixture properties and dissociation of combustion products and results in lower, more realistic values for efficiencies (Figs. 55 and 56). Real cycle analyses include, in addition, combustion losses and heat losses and leakages, resulting in thermal efficiencies approximately 80 percent of those predicted by the fuel-air cycle analyses.

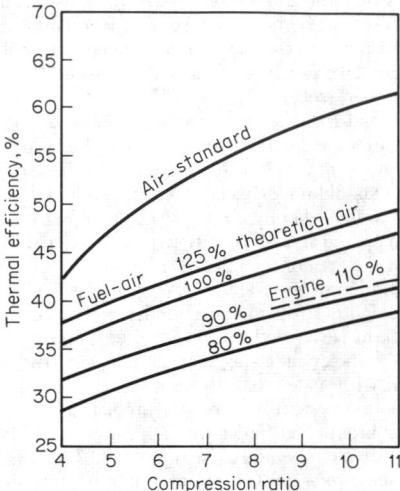

**Fig. 55**    Indicated thermal efficiencies of various Otto-cycle analyses and engine test results.

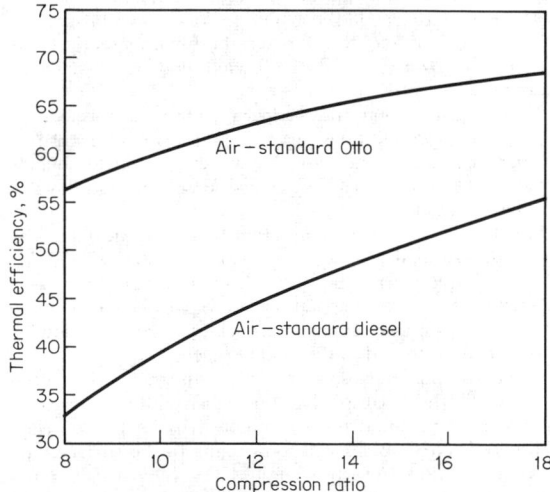

**Fig. 56**    Indicated thermal efficiency of the air-standard diesel cycle with constant-pressure combustion.

The mean effective pressure is equal to net work divided by volumetric displacement. Horsepower follows from

$$hp = (mep) \, Lan/K \qquad (4)$$

where mep = mean effective pressure, lb/in² (kPa)

$L$ = stroke, ft (m)
$a$ = total piston area, in² (m²)
$n$ = number of cycles completed per min
$K$ = 33,000 (0.4566 for SI units)

Increasing the compression ratio and the air-fuel ratio increases the thermal efficiency of the Otto cycle (Fig. 55). Increasing the compression ratio and the air-fuel ratio also increases the thermal efficiency of the diesel cycle with constant-pressure combustion (Fig. 56). Mean effective pressures depend on charge input and thermal efficiency, both of which depend on compression ratio and air and fuel supply (Figs. 57 and 58).

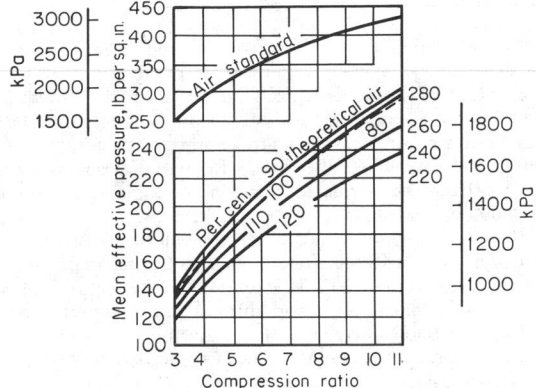

**Fig. 57** mep for Otto-cycle engine [air and liquid iso-octane supplied at 60°F (15.6°C)].

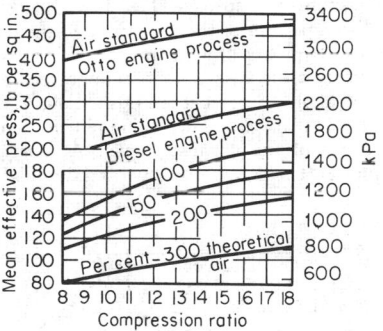

**Fig. 58** mep for diesel-cycle engine [air and liquid dodecane supplied at 60°F (15.6°C)].

### Deviations from Ideal Processes

Changes in atmospheric pressure and temperature change power output although not appreciably affecting thermal efficiency. The standard correction formulas adopted by the SAE for brake power output are, for spark-ignition engines,

$$\text{bhp}_s = \text{bhp}_o \, (P_s/P_o) \sqrt{T_o/T_s}$$

and for diesel engines,

$$\text{bhp}_s = \text{bhp}_o \, (P_s/P_o) \, (T_o/T_s)^{0.7}$$

where subscripts $o$ and $s$ indicate observed and standard conditions, respectively, $P_o$ being the dry barometric pressure.

**Pumping work** (area $EFGAE$, Fig. 59) is required to overcome the resistance to flow of fresh charge into and exhaust products out of the cylinder. It varies from almost 0 for an engine running at slow speed with wide-open throttle to about 3 lb/in² (20.7 kPa) mep at high speed, and to 10 lb/in² with idle throttle position. The negative loop (area $AXFGA$, Fig. 59) in the normally aspirated engine represents about 60 percent of the pumping work.

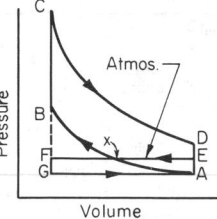

**Fig. 59** Throttled Otto cycle.

**Volumetric efficiency** is the ratio of the volume of air (for a liquid-fuel engine) and charge (for a gas engine) actually admitted, measured at atmospheric pressure and temperature, to the displacement volume. High manifold and cylinder temperatures and resistance to flow reduce volumetric efficiency. High-output aircraft engines have maximum volumetric efficiencies of 85 to 90 percent at rated speeds. Supercharged engines can have volumetric efficiencies well above 100 percent or can be designed to maintain high volumetric efficiencies with an increase in speed.

Volumetric efficiencies of automobile engines (Fig. 60) diminish appreciably as piston speeds exceed 1,200 to 1,600 ft/min (6.1–8.1 m/s), decreasing to 60 percent approximately at top speeds of 2,500 to 3,200 ft/min (12.7–16.3 m/s). These vary appreciably with valve timing.

**Fig. 60** Volumetric efficiencies of normally aspirated engines with poppet valves.

Two-cycle crankcase compression engines have volumetric efficiencies of about 60 percent at low speeds and 50 percent or below at high speeds.

Fuels with high latent heats and low boiling temperatures reduce the charge temperature and result in high volumetric efficiencies (Table 16).

Tests by Ziurys and Phelps (*Ind. Eng. Chem.*, **28**, 1936, p. 1094; **29**, 1937, p. 495) using ethyl alcohol and alcohol-gasoline blends resulted in volumetric efficiencies 5 to 7 percent higher for ethyl alcohol (see also Table 16), and about 1 percent higher for a 10 percent alcohol-gasoline blend than for gasoline.

Air cleaners, governors, and carburetors increase the resistance to flow and decrease the volumetric efficiency.

The **compression stroke** begins with heat transfer from the cylinder walls to the charge but ends with heat transfer from the charge to the walls. The net result is a heat loss of about ½

**Table 16. Relative Ideal Volumetric Efficiencies with Liquid Fuels**
(Air and liquid fuel at 60°F)

| Fuel | Chemically correct air-fuel ratio | Heat required per lb of fuel for complete evaporation Btu/lb (J/gm) | Percent evap with 200 Btu supplied per lb of fuel | Suction temperature, deg F (deg C) | | Relative volumetric efficiency | |
|---|---|---|---|---|---|---|---|
| | | | | Complete evaporation | 200 Btu per lb of fuel | Complete evaporation | 200 Btu per lb of fuel |
| Gasoline............ | 15.11 | 330 (768) | 81 | 106 (41) | 81 (27) | **100** | **100** |
| Benzene............ | 13.26 | 149 (347) | 100+ | 50 (10) | 68 (20) | **111** | **103** |
| Ethyl alcohol....... | 8.99 | 425 (989) | 55 | 71 (22) | 55 (13) | **107** | **105** |
| Methyl alcohol..... | 6.46 | 519 (1207) | 49 | 68 (20) | 45 (7) | **108** | **107** |

to 1½ percent of the heat value of the charge. Heat loss during compression lowers the compression curve and work, but lowers the expansion work a greater amount, thus slightly reducing the net work output.

The mean value of the exponent in the approximate equation $pv^n$ = constant for the polytropic compression of a correct mixture of octane and air is about 1.33. For the diesel cycle in which the gas under compression is mainly air, the exponent is about 1.38 for a 15:1 compression ratio. Heat transfer into the charge increases the exponent, but leakage lowers its value.

The **combustion process** begins before the end of the compression stroke and ends after the beginning of the expansion stroke. Rassweiler, Withrow, and Cornelius (*Trans. SAE*, **34**, 1939) found a maximum variation of 30 percent in combustion time for a single-cylinder engine while variations of both maximum cylinder pressure and rate of pressure rise of about 30 percent were found by Patterson (SAE paper 660129) in a V-8 engine. These effects are thought to be due primarily to cyclic variations in swirl and turbulence at the spark plug but can arise also from variations in air-fuel ratio and homogeneity of the mixture. Poor distribution in multicylinder engines also causes variations between cylinders. Marvin (*NACA Rept.* 276, 1927) has shown that the loss due to the actual combustion time is 3 to 6 percent of the ideal efficiency with constant-volume combustion. Any energy liberated before or after top center piston position has an availability corresponding to the expansion ratio for the piston position at which the energy is liberated. Heat loss during the combustion process amounts to 5 to 7 percent of the heat supply of the charge at rated outputs and speeds for automotive engines.

**Flame Travel**   Flame begins at the spark plug in the spark-ignition engine or at various points in the combustion chamber of the compression-ignition engine and travels in all directions through the mixture. At the end of combustion, the first part of the charge to burn has reached a higher temperature than the last portion to burn, except in the constant-pressure process. Rassweiler and Withrow (*Trans. SAE*, **30**, 1935, p. 125) observed temperature differences of over 400°F (222°C) (nonknocking) and over 600°F (333°C) (knocking) in a nonturbulent combustion chamber.

The combustion of part of the charge compresses the remainder. Thus with 30 percent of the mass burned, the volume of the unburned portion will be about 35 percent of the total volume. This increases the temperature of the last fraction to burn in a constant-volume process by about 500°F (278°C) above its temperature at the beginning of combustion.

The mean adiabatic exponent for the **expansion process** is about 1.22, varying from about 1.20 at the beginning to 1.25

at the end of the process. Heat transfer and leakage increase the exponent, an average value of 1.33 being found by Rassweiler and Withrow (*Trans. SAE*, **33**, 1938, p. 185) in an L-head 2⅞ by 4¾ in (7.3 cm × 12.1 cm) cylinder. The heat transfer during the expansion stroke amounts to 8 to 12 percent of the heat value of the charge. Because of heat losses and real gas effects the thermal efficiency of the engine is decreased by about 40 percent.

Blowdown begins at 80 to 90 percent of the expansion stroke. This reduces the work about 1 to 2 percent. High gas velocities (1,200 to 1,500 ft/s, 366 to 457 m/s) are attained and heat transfer, including that through exhaust-port walls, amounts to 10 to 20 percent of the heat value of the charge but represents practically no loss of work or availability.

The **exhaust process** is the discharging of some of the gases from the cylinder by the piston. The exhaust pressure is usually above atmospheric (Fig. 61) but may run below atmospheric for part of the stroke. Heat loss during exhaust amounts to 3 to 5 percent of the heat value of the charge but does not represent any loss of work or availability. At full load about 80 percent of the gases in the cylinder escape during blowdown, and about 15 percent are pushed out of the cylinder during exhaust, the remainder being left in the clearance space at the end of the exhaust stroke. At part load the blowdown pulse is smaller and residual fraction higher.

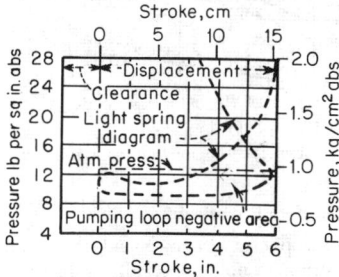

**Fig. 61**   Pumping loop of Otto-cycle engine.

**Exhaust-gas temperatures** vary with speed and load (Figs. 62 and 63), high loads and speeds resulting in the highest temperatures. The first gases escaping during release are at the highest temperature.

**Exhaust-Gas Analysis**   The composition of exhaust varies depending upon the equivalence ratio and the hydrogen-carbon ratio of the fuel. Figure 64 shows calculated composition including water (wet basis) for complete combustion. Charts of theoretical exhaust composition for various fuels have been prepared by D'Alleva (*General Motors Research Rept.* 372, 1960) and Eltinge (*SAE Trans.*, 1968). In practice nonequilibrium products, namely, unburned hydrocarbons and nitrogen oxide, appear in trace quantities. Figure 65 shows measured exhaust composition on a dry basis for a fuel whose hydrogen-carbon ratio was 1.85.

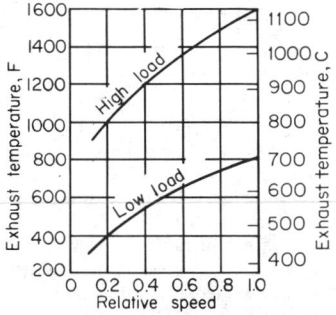

**Fig. 62** Exhaust temperatures for Otto-cycle engines.

**Fig. 63** Exhaust temperatures for diesel-cycle engines.

Equations for calculating air-fuel ratio from exhaust CO, $CO_2$, HC, and $O_2$ have been developed by Spindt (*SAE Trans.*, 1966) and others. Such determinations and those for calculation of remaining but unmeasured exhaust species depend upon carbon, hydrogen, oxygen, and nitrogen balances employing measured values for some exhaust constituents (Obert, "Internal Combustion Engines").

Table 17 gives a summary of the principal exhaust-gas

constituents and combustion efficiency. HC and NO were not measured.

## INTERNAL-COMBUSTION ENGINES AND AIR POLLUTION

(See also Sec. 18)

### General Relationship

Air pollution from automobile engines (smog) was first detected about 1942 in Los Angeles, CA (see Haagen-Smit, *Sci. Am.*, **210**, no. 1, 1964, p. 25). Smog arises from sunlight-induced photochemical reactions between nitrogen dioxide and the several hundred hydrocarbons in the atmosphere (see Caplan, SAE PT-12, p. 20). Undesirable products of the reactions include ozone, aldehydes, and peroxyacylnitrates (PAN). These are highly oxidizing in nature and cause eye

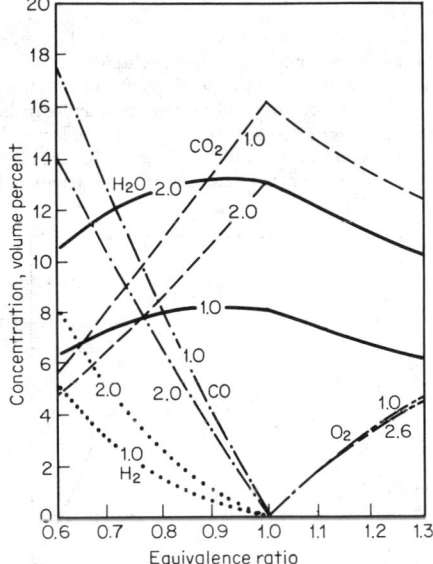

**Fig. 64** Theoretical mole percent of the principal combustion products of hydrocarbon fuels for fuel hydrogen-carbon ratios.

**Table 17. Summary of Exhaust-Gas Constituents and Combustion Efficiency***

| Air/Fuel | Percent by volume | | | | | | H2O/CO2 | H2O/Fuel | Combustion† eff., percent |
|---|---|---|---|---|---|---|---|---|---|
| | CO2 | O2 | CO | H2 | N2 | H2O | | | |
| 11 | 8.76 | 0.15 | 9.14 | 4.66 | 77.08 | 13.78 | 1.57 | 0.972 | 66.7 |
| 12 | 10.18 | 0.44 | 6.65 | 3.39 | 79.13 | 13.93 | 1.37 | 1.043 | 73.8 |
| 13 | 11.60 | 0.59 | 4.31 | 2.20 | 81.09 | 14.16 | 1.22 | 1.122 | 81.5 |
| 14 | 13.02 | 0.63 | 2.09 | 1.07 | 82.99 | 14.46 | 1.11 | 1.205 | 89.6 |
| 15 | 13.23 | 1.35 | 0.99 | 0.50 | 83.72 | 14.09 | 1.06 | 1.247 | 93.8 |
| 16 | 12.62 | 2.49 | 0.68 | 0.35 | 83.65 | 13.30 | 1.05 | 1.256 | 94.8 |
| 17 | 12.00 | 3.55 | 0.48 | 0.25 | 83.51 | 12.54 | 1.05 | 1.261 | 95.5 |
| 18 | 11.45 | 4.49 | 0.30 | 0.16 | 83.39 | 11.88 | 1.04 | 1.267 | 96.2 |
| 19 | 10.90 | 5.36 | 0.20 | 0.10 | 83.23 | 11.25 | 1.03 | 1.269 | 96.5 |
| 20 | 10.40 | 6.15 | 0.11 | 0.06 | 83.07 | 10.68 | 1.03 | 1.272 | 96.9 |
| 21 | 9.92 | 6.86 | 0.08 | 0.04 | 82.90 | 10.16 | 1.03 | 1.271 | 96.9 |
| 22 | 9.44 | 7.55 | 0.06 | 0.03 | 82.71 | 9.65 | 1.02 | 1.268 | 96.8 |
| 23 | 9.00 | 8.18 | 0.05 | 0.03 | 82.53 | 9.19 | 1.02 | 1.266 | 96.7 |
| 24 | 8.60 | 8.74 | 0.06 | 0.03 | 82.37 | 8.78 | 1.02 | 1.264 | 96.6 |

*Gerrish and Voss, *NACA Rept.* 616, 1937.
†Low-heat-value basis.

**Table 18. Estimated Total Annual U.S. Emissions from Man-Made Transportation-Engine Sources***

| | Carbon monoxide | Hydrocarbons | Sulfur oxides | Nitrogen oxides | Particulates |
|---|---|---|---|---|---|
| Million tons (kg)/year | 100.1 (91,000) | 32.0 (29,000) | 26 (23,600) | 20.7 (18,800) | 12 (11,000) |
| Gasoline, % | 59.0 | 47.5 | } 3.9† | 31.8 | } 8.5† |
| Diesel, % | 0.2 | 0.7 | | 2.9 | |

*U.S. Department of Health, Education and Welfare.
†Estimated average, for gasoline and Diesel, combined.

and throat irritation. Visibility-decreasing nitrogen dioxide and aerosols are also formed.

Five categories of air pollutants and the percent contribution from gasoline and diesel engines involved in transportation are listed in Table 18. Slight additional emissions arise from stationary-engine sources.

Transportation-engine emissions comprise 42 percent by weight of all U.S. man-made air pollutants. Most of the emitted mass is carbon monoxide. Locally, man-made pollutants can exceed those from natural sources and can become dominant. However, emitted mass does not reflect pollution severity. For example, one unit mass of oxides of sulfur may equal 30 units of carbon monoxide or more in terms of health effects.

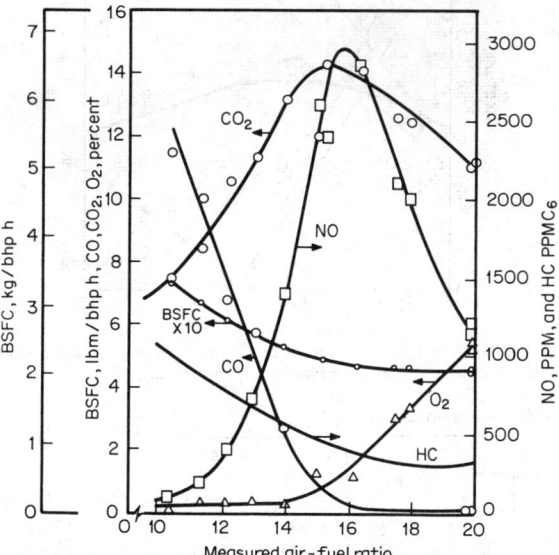

**Fig. 65**  Gasoline-engine basic specific fuel consumption and emissions vs. air-fuel ratio.

Emissions from internal-combustion engines include those from blowby, evaporation, and exhaust. These can vary considerably in amount and composition depending on engine type, design, and condition, fuel-system type, fuel volatility, and engine operating point. For an automobile without emission control it is estimated that of the hydrocarbon emissions, 20 to 25 percent arise from blowby, 60 percent from the exhaust, and the balance from evaporative losses primarily from the fuel tank and to a lesser extent from the carburetor.

All other nonhydrocarbon emissions emanate from the exhaust.

At least 200 hydrocarbon (HC) compounds have been identified in exhaust. Some such as the olefin compounds react rapidly and to a great extent in the atmosphere to form smog products. These are termed reactive hydrocarbons. Others such as the paraffins are virtually unreactive. Of several reactivity scales, the HEW scale (Table 19) has become most widely used. It ranks hydrocarbons on a scale from 0 to 8 reactivity units per mole of hydrocarbon. The reactivity of a mixture of hydrocarbon gases is computed as the sum of the reactivity and mole-fraction products.

Atmospheric oxides of nitrogen (commonly termed $NO_x$) are a mixture of nitric oxide (NO) and nitrogen dioxide ($NO_2$). Oxides of sulfur ($SO_x$) arising from fuel sulfur are a mixture of $SO_2$, $SO_3$, and $SO_4$. Sulfuric acid may be a product in some oxidizing catalytic exhaust converters.

**Table 19. HEW Scale of Reactivity**

| Substance or subclass | Averaged responses on 0 to 10 scale |
|---|---|
| $C_1$–$C_5$ paraffins | 0 |
| Acetylene | 0 |
| Benzene | 0 |
| $C_6$ paraffins | 1 |
| Toluene (and other monoalkylbenzenes) | 3 |
| Ethylene | 4 |
| 1-Alkenes | 7 |
| Diolefins | 6 |
| Dialkyl and trialkylbenzenes | 6 |
| Internally double-bonded olefins | 8 |

Emission quantities are most often reported as mole percent for the principal exhaust constituents (CO, $H_2O$, $CO_2$, $H_2$, $O_2$, $N_2$) and as moles per million moles (ppm) for trace constituents (HC, NO). Ppm may be assumed to be equal to volume percent and ppm by volume. Hydrocarbons are reported commonly as either ppm equivalent hexane ($C_6H_{14}$) or as ppm carbon ($C_1$). One ppm $C_6$ is equivalent to 6ppm C. Other reporting parameters are:

Specific emission rate, g pollutant/bhp-h or g pollutant/vehicle mile (km)
Emission index, g pollutant/1,000 g fuel

### Emission Sources

#### Homogeneous Combustion and Gasoline Engines

HYDROCARBON EMISSIONS

Hydrocarbon emissions arise from five sources when the

fuel-air mixtures are premixed. These are wall quenching, incomplete combustion, wall deposits, leakage, and scavenging loss.

Daniel described the result of wall quenching to be (6th Symposium on Combustion, Reinhold, 1957, p. 886) a visibly dark layer of unburned and partially burned fuel-air mixture left adjacent to all engine combustion-chamber wall surfaces. In normally aspirated four-cycle engines this unburned layer, which is typically 0.002 to 0.015 in (0.051 to 0.38 mm), can account for virtually the entire exhaust hydrocarbon content, typically 300 to 500 ppm $C_6$.

Increasing mixture pressure or temperature or increasing wall temperature can reduce the thickness of the layer. No practical technique is known to eliminate the quenched layer.

Under engine operating conditions where cylinder-exhaust residual is high, incomplete flame propagation may leave a portion of the combustion-chamber contents unburned. This can produce hydrocarbon emissions of several thousand ppm $C_6$. When residual exceeds 15 percent by weight, some incomplete combustion may occur. Commonly, incomplete combustion is the major hydrocarbon-emission source under vehicle coast-down conditions. Deposits increase hydrocarbon emissions.

Wankel-engine hydrocarbon emissions arise from the above sources, but in addition appreciable hydrocarbons may arise from mixture leakage through the leading rotor-apex seal directly into the exhaust.

In carbureted engines where intake pressure exceeds exhaust (as in two-cycle engines) raw-mixture loss to the exhaust during the valve-overlap period creates very high hydrocarbon emissions. Emissions from two-cycle carbureted engines may be 10 times higher than four-cycle engine emissions.

### Carbon Monoxide Emissions

The principal source of exhaust carbon monoxide (CO) is rich-mixture combustion. Figure 65 shows exhaust composition as mixture strength is varied. Cycle-to-cycle and cylinder-to-cylinder maldistribution increase CO. Additional CO may arise from incomplete combustion of HC in an exhaust-treatment device.

### Oxides of Nitrogen Emissions

The principal oxide of nitrogen emitted from engines is nitric oxide, NO. NO forms relatively slowly in postflame-combustion gases. The formation depends primarily on peak combustion-gas temperature and to a lesser extent on oxygen content. Once formed, NO tends to be frozen and persists in high nonequilibrium concentrations during expansion and exhaust. When mixtures are more than 30 percent lean or where air is injected into the exhaust, about 5 percent of NO may be oxidized to $NO_2$ prior to leaving the exhaust pipe. Nitric oxide variation with mixture strength is shown in Fig. 65.

### Aldehyde Emissions

Aldehydes are incomplete-combustion products of hydrocarbon oxidation. They arise from the same wall-quenching and incomplete-combustion mechanisms as the unburned HC themselves but in addition may be formed in low-temperature exhaust-system oxidation reactions. Typically gasoline-engine exhaust contains 100 ppm C aldehyde, less for rich and more for lean mixtures.

### Particulate Emissions

Solid-particle emissions from gasoline engines consist primarily of carbon, metals, metal oxides, and metal halides when TEL scavengers are added to fuel. Metals arise from fuel and lubricant additives such as lead, phosphorus, zinc, and barium as well as engine wear and rusting. Liquid aerosols of heavy hydrocarbons are emitted primarily during starting but also during running. Of these, polynuclear aromatics (PNA) are of concern to health.

**Heterogeneous Combustion and Diesel Engines**  The hydrocarbon, carbon monoxide, and nitric oxide emission formation in heterogeneous combustion processes is from the same basic mechanisms as in the gasoline engine. However, the relative importance of each is different. In the diesel engine, the emission formation in combustion can be considered to arise from a broad distribution of individual fuel-air elements, some of which burn rich and some lean. Patterson and Henein (Emissions from Combustion Engines and Their Control, *Ann Arbor Science*, 1972, p. 260) have suggested the various emission sources in the combustion of a fuel spray injected into swirling air as shown in Fig. 66. In addition, fuel impingement on chamber surfaces contributes to unburned hydrocarbons and carbon smoke particles. Hydrocarbon emissions from diesels contain many heavy fuel types. Normally, heated sampling lines are required to avoid HC condensation during sample analysis.

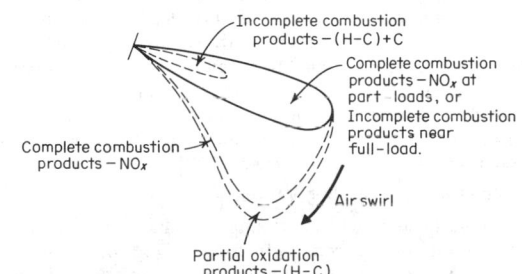

**Fig. 66**  Emission formation in a fuel spray injected in swirling air.

The amount of carbon, carbon monoxide, hydrocarbon, and nitric oxide emission from diesel engines varies considerably with engine and fuel-system design as well as operating point. Moreover because of the high air-mass flow rate at part load, emissions from diesel and gasoline engines cannot be compared directly except on a specific emission basis. Typically diesel emissions as mole percent or ppm are corrected for dilution to an equivalence ratio of unity for comparison with gasoline-engine data. On a corrected basis diesel engines emit hydrocarbons in the range of 50 to 500 ppm $C_6$ and $NO_x$ up to 3,000 ppm. Carbon monoxide emission from the diesel is very low, typically under 1,000 ppm (0.1 percent).

Diesel engines emit considerable amounts of carbon smoke and odor constituents. Most carbon arises from pyrolysis in rich combustion regions. Spindt, Barnes, and Somers (SAE paper 710605, 1971) have identified low concentrations of many potential oxidation products and certain fuel fractions as being the odor-producing compounds. Human-panel techniques are commonly used to measure odor.

### Effects of Gasoline-Engine Design Variables

**Combustion-Chamber Shape**  The shape of the combustion chamber affects hydrocarbon emissions through a change in surface area and chamber volume. When viewed at top dead center, the ratio of volume of the unburned quench layer relative to the entire clearance volume is an indicator of the volume fraction hydrocarbon emission. This may be stated as

$$HC, ppm \propto A_s/V_c$$

where $A_s$ is chamber surface area and $V_c$ clearance volume at TDC. According to Sheffler (SAE, PT 12, p. 60), the following change surface-volume ratio, thereby changing hydrocarbon emissions: chamber shape, stroke-bore ratio, compression ratio, and cylinder displacement. In automotive reciprocating engines surface-volume ratios range from 5 to 8 in²/in³ (2 to 3 cm²/cm³). Wankel rotary engines are about 50 percent higher. In well-designed engines, combustion-chamber shape has little if any effect on CO. A fast-burning chamber increases $NO_x$ because of increase in peak temperature.

**Ignition System**  Under most running conditions improvement in spark energy, rise time, and spark duration produce little improvement in emissions. However, a high-energy, long-duration spark in conjunction with extended electrodes and wider spark gap has been found to minimize the increase in HC emissions due to misfire, as mixtures are leaned well beyond chemically correct.

Exhaust back pressure and valve overlap affect HC and $NO_x$ through changes in exhaust residual. Increasing overlap or back pressure lowers $NO_x$ and may lower HC emissions. At very light loads excessive residual leads to incomplete combustion, thereby increasing hydrocarbon emissions.

**Induction Systems and Carburetion**  Optimum mixture preparation and distribution are a key to minimum emissions. High-velocity manifolds, increased manifold heating, and staged carburetion with small primary venturi have shown emission gains (see Bartholomew, SAE PT-12, 1971).

### Gasoline-Engine Operating Variables

Mixture ratio and ignition timing are the most important operating variables affecting emission. Figure 65 shows the effect of the mixture ratio on major emission amounts. Retarding ignition timing from the best efficiency setting reduces HC and $NO_x$ emissions significantly. Excessive retard increases HC and CO emissions. Increasing engine speed reduces HC emissions as ppm. $NO_x$ emissions increase as load increases. Increasing coolant temperature normally reduces HC emissions and increases $NO_x$ emissions.

Choke calibration is critical to low emissions, especially with exhaust-treatment devices. A lean choke setting with fast opening is often best when exhaust treatment is not used. With exhaust treatment, emissions may be reduced to virtually zero once the system is warmed up. Thus the only significant emissions from such systems are those generated during starting and warm-up. With exhaust treatment, a fast-opening choke initially set rich may accelerate exhaust-reactor light-off.

### Gasoline-Engine Exhaust-Treatment Devices

Catalytic or noncatalytic exhaust-treatment devices may be used to clean up remaining exhaust emissions. Thermal reactors are noncatalytic devices which rely on homogeneous bulk gas reactions to oxidize CO and HC. Commonly they appear to be enlarged exhaust manifolds, but they may have internal baffling. $NO_x$ is unaffected in thermal reactors. Reactions are enhanced by increased exhaust temperature (reduced compression ratio, retarded timing) or by increasing exhaust combustibles (rich mixtures). Typically temperatures of 1500°F (800°C) or more are required for peak efficiency. Commonly the engine is run rich to give 1 percent CO, and air is injected into the exhaust. A typical thermal-reactor residence time is 100 ms. Required reactor volume is commonly 1.5 times engine displacement. Maximum allowable reactor temperature is governed by materials and is typically 1750°F (950°C).

Catalytic systems are capable of reducing $NO_x$ as well as oxidizing CO and HC. However, a reducing environment for $NO_x$ treatment is required which necessitates a richer than chemically correct engine mixture ratio. A two-bed converter may be used in which air is injected into the second stage to oxidize CO and HC. Experimental single-stage three-component catalysts have been demonstrated but require extremely precise carburetion to be effective.

Catalyst support beds may be of the pellet type or honeycomb (monolith). Materials for reducing catalysts include rhodium, Monel, and ruthenium. Materials for oxidizing catalysts include platinum and palladium. The amount of active catalyst material required may be as low as 0.05 troy oz (1.4 g) per vehicle system. A typical catalyst volume is about one-fourth to one-half engine displacement. The weight of the catalyst bed typically is about 2.2 to 5 lb (1 to 2.3 kg) excluding metallic container. Many aged catalysts can oxidize quite efficiently at temperatures as low as 800 to 1000°F (427 to 538°C). Above 1500°F (816°C), they deteriorate rapidly. Compared with thermal reactors catalytic converters warm up slowly because of the larger mass of material. Catalytic converters may be located farther from the engine than thermal reactors, since they operate efficiently at lower temperatures. Monolithic converters, which can normally be operated hotter, can be smaller and placed closer to the engine.

One effective exhaust-treatment system employs a thermal reactor as a first-stage element. At start-up air is injected into the thermal reactor. This provides rapid light-off. During steady operation the thermal reactor acts as a mixing volume before the exhaust is passed to a dual-bed reducing-oxidizing catalyst (Fig. 67).

### Emission Testing

**Automotive—Light-Duty Vehicles**  Gaseous emissions from automobiles are measured on a light-duty driving cycle termed the Federal Test Procedure (FTP). This involves various accelerations, decelerations, and cruise modes on a chassis dynamometer. The car is started after a 12-h soak in a 60 to 86°F (16 to 30°C) ambient. Top cycle speed is 56.7 mi/h (91.3 km/h). Test procedures, fuel, and instrumentation are detailed in the Federal Register (**37**, no. 221, part II, 1972).

In the FTP procedure the exhaust is introduced into a constant-volume sampler (CVS) in which ambient air is mixed with raw exhaust. Emissions are calculated in grams per mile from average concentrations and total sampler volume flow. In theory the constant-volume sampler yields true mass-emission results. Prior to 1975, a single collection bag was employed (CVS-1). Since then three bags have been used (CVS-3) to determine emissions during key cycle portions. The emission

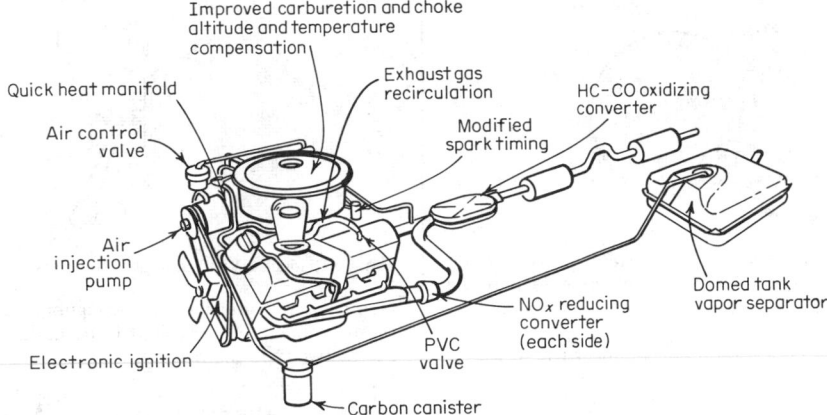

**Fig. 67** Dual catalytic-converter system.

concentrations from each bag are weighted and the sums used for certification purposes.

**Heavy-Duty Engines** Emissions from heavy-duty engines, both gasoline and diesel, are determined in steady-engine dynamometer tests. For diesels 13 operating modes are assessed including idle and various loads at the peak-torque and peak-power engine speeds. For gasoline engines a 9-mode test consisting of various loads at 2,000 r/min is specified. Gaseous-emission results are calculated in grams per horse-power-hour by applying weighing factors to the modal results. Smoke particulates from diesels are assessed using a light-beam-extinction procedure to determine opacity.

## SPECIAL DEVELOPMENTS

The **stratified-charge spark-ignited engine** has potential for burning lean mixtures for improved fuel economy and reduced emissions. In theory, an overall lean mixture is formed in the combustion chamber but is controlled to be fuel-rich in the vicinity of the spark plug at the time of ignition. The rich portion is thus easily ignited, and this in turn ignites the remaining lean mixture. Some of the benefits claimed are (1) lower flame temperature, (2) improved cycle efficiency, (3) less heat loss, (4) less dissociation, and (5) less pumping loss. In addition, less $NO_x$ would be expected because of the lower combustion temperatures and less unburned hydrocarbon, and CO would result because of the overall excess of oxygen.

Two rather distinct means for accomplishing the stratified-charge condition are under consideration.

1. A single combustion chamber with a well-controlled rotating air motion. This arrangement is illustrated (Fig. 68) by the Texaco Combustion Process (TCP), patented in 1949. The Ford Motor Company programmed combustion (PROCO) is a similar single-chamber concept.

**Table 21. Emission Standards for Heavy Trucks (over 6,000 lb GVW), Gasoline and Diesel Engines**

| Test procedure (engine dynamometer) | 1975–1976 Calif. | 1975–1977 federal | 1977 Calif., 1978 Fed.† |
|---|---|---|---|
| | Mass (g/bhp h)* | | |
| Hydrocarbons | 10 | 16 | 5 |
| Nitrogen oxides | | | |
| Carbon monoxide | 30 | 40 | 25 |
| Evaporation, g | ‡ | ‡ | ‡ |

*Standards combined for HC + $NO_x$.
†Standards assumed.
‡Evaporative controls required on all heavy-duty trucks, effective 1973 California (January 1974 on units with fuel tanks over 50 gal), assumed nationwide in 1978. New intermediate-duty truck (6,000 to 10,000 GVW) requirements expected starting 1977. Diesel engines subject to federal smoke regulation. Smoke opacity, 1974: acceleration 20 percent, lug 15 percent, maximum any mode 50 percent.

**Table 20. Emission Standards for Passenger Cars and Light Trucks (under 6,000 lb GVW)**

| Emission, g/mi | Pre-1966 (uncontrolled) | | 1975–1977 Calif. Fed. | | 1978 Nation |
|---|---|---|---|---|---|
| Test procedure | CVS-1 | CVS-3 | CVS-3 | | |
| Hydrocarbons | 17 | 15 | 0.9 | 1.5 | 0.41 |
| Carbon monoxide | 125 | 90 | 9.0 | 15 | 3.4 |
| Nitrogen oxides | 4.6 | 4.6 | 2.0 | 3.1 | 2.0 |
| Evaporation, g/test | 40 | | 2 | | 2 |

CVS-1: cold-start test. CVS-3: CVS cold/hot weighted test.
Light-truck requirements (g/mi) 1975–1976 federal: HC 2.0/CO 20/$NO_x$ 3.1, 1975 California: HC 2.0/CO 20/$NO_x$ 2.0; 1976 California: HC 0.9/CO 17/$NO_x$ 2.0.

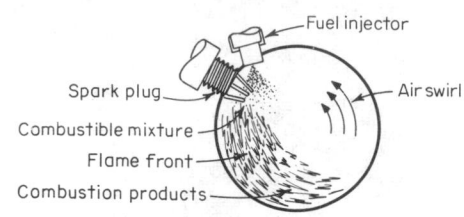

**Fig. 68**  Texaco combustion process (TCP).

2. A prechamber or two-chamber system. This is illustrated by Fig. 69, which shows the general arrangement of the Honda compound-vortex controlled-combustion (CVCC) system.

For both systems, very careful development has proved to

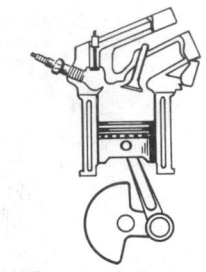

**Fig. 69**  Honda CVCC combustion process.

be necessary to obtain complete combustion of the fuel under the wide range of speed and load conditions required of an automotive-type engine.

# GAS TURBINES

## by Reeves Morrisson

REFERENCES: Vincent, "The Theory and Design of Gas Turbines and Jet Engines," McGraw-Hill. Zucrow, "Principles of Jet Propulsion and Gas Turbines," Wiley. "Gas Turbine Engineering Handbook," *The Gas Turbine Catalog* (annual), *The Gas Turbine International Magazine* (bimonthly), Gas Turbine Publications, Inc. *Gas Turbine World* (bimonthly), Pequot Publishing, Inc. *ASME Trans.*, Gas Turbine Power Division, and *Journal for Power*.

## GAS-TURBINE CYCLES

The basic gas-turbine cycle (Fig. 1) is the Brayton, or Joule, cycle (see Sec. 4), consisting of adiabatic compression, constant-pressure heating, and adiabatic expansion. Because the expanding gases are hotter, the work obtainable in the expansion process is greater than the work of compression. The net work of the cycle is the difference between the two. By adding a **regenerator** (Fig. 2) to recover heat from the turbine exhaust, the efficiency is improved. The additions of **intercooling** (Fig. 3) in the compressor and **reheat** (Fig. 4) of the working fluid during expansion increase the output of a given size of gas

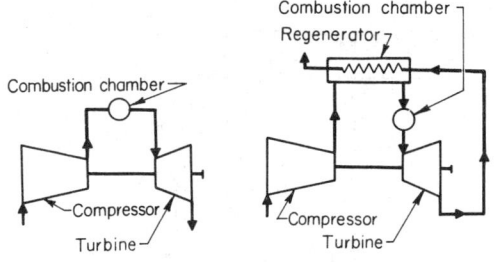

**Fig. 1**  Simple gas-turbine cycle.    **Fig. 2**  Gas turbine with regeneration.

turbine, and with the addition of a regenerator a further improvement is made in efficiency. The explosion or constant-

volume combustion cycle was used in early gas turbines (Holzwarth) but is not suited to a continuous-flow turbine process and results in very complicated machinery.

The **thermal efficiency** of the Brayton cycle depends mainly on the pressure ratio, the turbine inlet temperature, and the parasitic losses—especially the efficiency of the compressor and turbine. In the theoretical case of an ideal air standard with no internal losses, the thermal efficiency can be shown to depend only on compressor pressure ratio ($p_2/p_1$).

$$\text{Efficiency} = 1 - 1/(p_2/p_1)^{(k-1)/k}$$

For a real cycle with losses, thermal efficiency depends also on turbine inlet temperature, since higher temperature means increased useful work output and a proportionate reduction in the effect of internal losses.

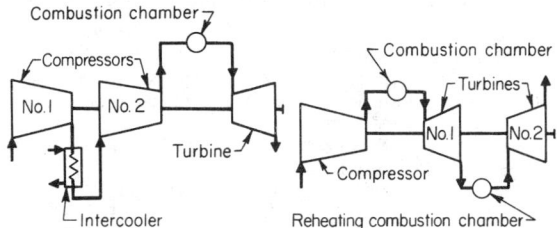

**Fig. 3**  Gas turbine with intercooling.    **Fig. 4**  Gas turbine with reheating.

Figure 5 shows the influence of the **pressure ratio** on the performance of the simple gas-turbine cycle over a range of turbine inlet temperature from 1500 to 3000°F (1089 to 1922 K). Ambient conditions and component efficiencies represent the best that can be done with today's level of technology: e.g., compressor and turbine efficiency vary with pressure ratio as determined by constant polytropic efficiency of 90 percent; air is bled from the compressor to cool the turbine blades and fixed nozzle vanes in sufficient amount to maintain safe metal temperature, 6 percent of overall pressure ratio is used in generating turbulence in the combustion chamber, and an additional 6 percent is lost in flow losses in ducting and unused exhaust kinetic energy.

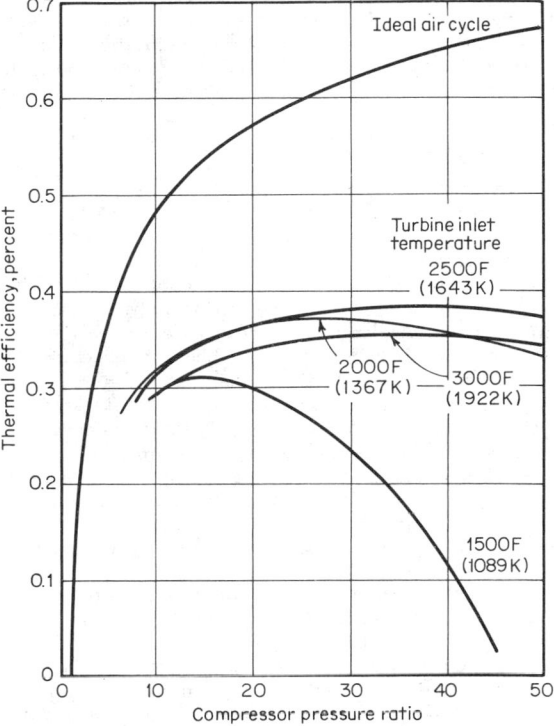

**Fig. 5**  Effect of pressure ratio on thermal efficiency of a simple gas-turbine cycle at various turbine inlet temperatures.

The **real cycle** falls short of the ideal air cycle partly because real air properties ($k$, $c_p$) are not constant over this range of temperature, and mainly because of internal losses. Losses to the cycle of cooling air begin to be significant above 2000°F (1367 K) and are so serious at 3000°F (1922 K) as to make this temperature impractical at present levels of cooling technology.

At a pressure ratio significantly less than the optimum value as indicated in Fig. 5, the temperature of the turbine exhaust is higher than that of the compressor exhaust and it becomes possible to transfer energy to the compressed airstream by means of a heat exchanger (see Fig. 2), thus saving fuel and improving efficiency. The amount of improvement depends on the amount of heat-exchange surface, and frictional and leakage losses. Commonly, the **regenerative engine** is designed to do about as well as the simple-cycle engine of higher pressure ratio. The choice is between the difficulty of achieving high compressor efficiency at high pressure ratio and the complexity, bulk, and cost of the heat exchanger.

In the **combined cycle** (Fig. 6) the hot turbine exhaust gases are used to generate steam to drive a steam turbine. Combined-cycle power plants may be **unfired** or **supplementary fired.** In the latter type additional fuel is supplied to the steam boiler

and the gas turbine may be a rather small part of the total plant. In the unfired system, the steam section is auxiliary to the gas turbine and usually controlled automatically by the gas turbine. As much as 40 percent additional power can be obtained in the unfired system without burning additional fuel. The overall thermal efficiency is thus very high—around 40 percent with today's turbines (Fig. 7).

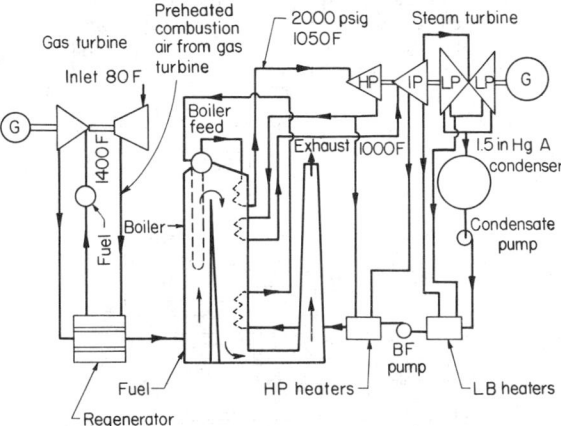

**Fig. 6**  Combined steam and gas turbine using gas turbine to supply preheated combustion air to boiler.

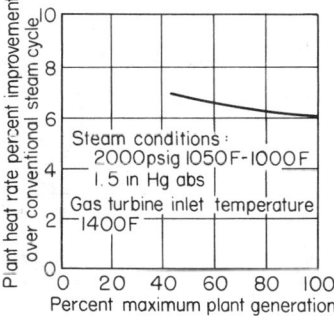

**Fig. 7**  Overall plant-efficiency improvement using combined steam and gas-turbine cycle (Fig. 6) as compared with conventional steam plant.

## GAS-TURBINE TYPES

Gas-turbine plants may be **open** or **closed.** In the **open** type the working fluid (air and combustion products) passes through the engine only once (Figs. 1 to 4). This design offers the advantage of simple control and sealing systems. It also can be designed for high power-weight ratios (aircraft units) and for operation without cooling water. Most of the gas-turbine plants in operation are of this type.

In the **closed** type (Fig. 8) the working fluid is continuously recycled. The heat, from an external source, is transferred through the walls of a closed heater. Several plants of this type have been in operation in Europe burning fossil fuels. The closed cycle is also being studied for use with high-temperature gas-cooled nuclear reactors. The advantages of the closed cycle are clean working fluid and the ability to control its density. High working-fluid density reduces the size of the compressor and turbine, and control of gas density makes it possible to vary power output without changing pressure ratio or turbine temperature so that a wide range of load can be carried at practically constant speed and efficiency. The major disadvantage of the fossil-fuel-fired closed cycle is the size and cost of the heat exchangers—especially the high-temperature heater. The closed cycle can use working fluids other than air, e.g., helium in the case of high-temperature nuclear reactors. The different thermodynamic properties affect the pressure ratio for maximum efficiency but have no big effect on the overall thermal efficiency (Fig. 9).

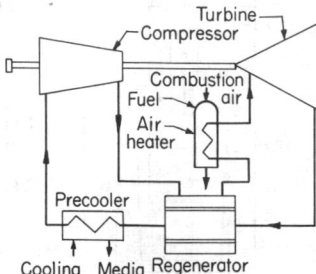

**Fig. 8**   Closed-cycle gas turbine.

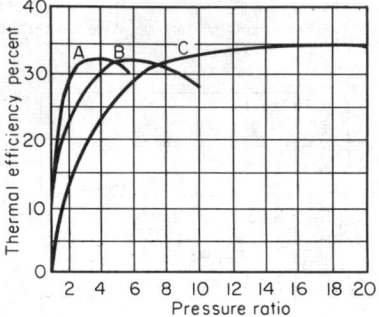

**Fig. 9**   Effect of pressure ratio on thermal efficiency for various working fluids. *(A)* Monatomic gas, $k = 1.67$; *(B)* diatomic gas, $k = 1.40$; *(C)* triatomic gas, $k = 1.28$.

## GAS-TURBINE COMPONENTS

The **net output** of the gas turbine is the difference between the power produced in the turbine and that absorbed by the compressor. Typically about two-thirds of the turbine power is used to drive the compressor. *Performance is thus very sensitive to the efficiency of these components.* For example, a loss of 5 percentage points in compressor efficiency will result in over 10 percent reduction in net output. **Inlet-air filters** are used to protect industrial and marine gas turbines from dust, fumes, and salt which could erode, corrode, or form deposits on compressor or turbine and reduce their efficiency. Water washing and cleaning by blowing mild abrasives (e.g., ground walnut shells) through the engine are used to restore performance of dirty compressors.

**Axial-flow compressors** are used on all large gas turbines because of their high efficiency and capacity. They have been constructed in sizes handling airflows up to 1,000 lb/s (800,000 ft³/min; 380 m³/s). A problem in the design of high-pressure compressors is to avoid operation in the **instability** or **surge**

**region.** A number of high-pressure-ratio gas turbines have compressors which are in this region during starting and low-load and low-speed operation. Methods of avoiding compressor instability under these conditions include interstage bleed, discharge bleed, and controllable-angle stator vanes. Where a two-stage compressor is used, the relative speed of the two compressors is varied to control surge. In small sizes [1 to 4 lb/s (0.45 to 1.8 Kg/s) airflow] it becomes difficult to control tolerances and clearances in the axial-flow compressor, and their performance suffers. Notable advances in efficiency and pressure ratio per stage have been demonstrated with centrifugal compressors in recent years, and they tend to be preferred for small gas turbines (see also Sec. 14).

**Turbines** are almost all axial-flow except for a few smaller sizes which are radial-inward-flow. The incentive to operate at ever higher gas temperatures has led to intricate designs of cooling passages inside the turbine blades and nozzle vanes. Air bled from the compressor is commonly used as coolant; liquid cooling has been tested experimentally. Work is actively proceeding on the development of ceramic materials for turbine blades and other hot parts. The goal is to increase efficiency by reducing the use of cooling air and to reduce the basic cost of the hot parts. Much is still to be learned about the techniques of design for brittle materials. High-temperature alloys—mainly nickel and cobalt base—are still prevalent practice. Coatings to protect the materials from corrosion are very important and continually being researched. Fuel additives are being developed which react in the combustion process with corrosive elements, e.g., sulfur, salt, vanadium, to inhibit their attack of hot turbine parts.

The **combustion chamber** or **combustor** must bring the gases to a controlled uniform temperature with a minimum loss of pressure. Extremely high rates of heat release are common [5 million Btu/(h)(ft³)(atm), $2 \times 10^{11}$ J/(h)(m³)(atm)]. The major problems in addition to efficiency and proper mixing of the gases are flame stabilization, elimination of pulsation and noise, and control of polluting emissions—especially nitrogen oxides. Combustors are usually made of metal, cooled by the incoming air (see Fig. 10), but ceramic materials have been used.

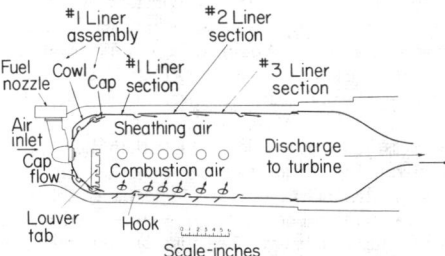

**Fig. 10**   Cross section of metal-cooled combustor arranged for residual oil firing.

**Regenerators** for transferring heat from the turbine exhaust to the air leaving the compressors are useful in increasing efficiency but increase the weight, volume, and cost of a gas-turbine plant. They must withstand rapid large temperature changes and must have low pressure drop. Their large size makes them unsuitable for aircraft. Compressor intercoolers are subject to the same limitations as regenerators and in addition require a supply of cooling fluid. In an effort to reduce the size of heat exchangers, extended surface designs have been used. Rotary regenerators, offering high performance and low weight, are under development for automotive applications (see also Sec. 4).

## GAS-TURBINE ARRANGEMENTS

The various gas-turbine cycles and types can be constructed in single- or multiple-shaft arrangements. The **single-shaft** machines are the simplest with all rotating elements operating as a single assembly. Figures 1 to 4 show various cycle arrangements of single-shaft machines. This type is best suited for constant speed and load operation. Figure 11 shows a typical load-speed range for a single-shaft machine. The

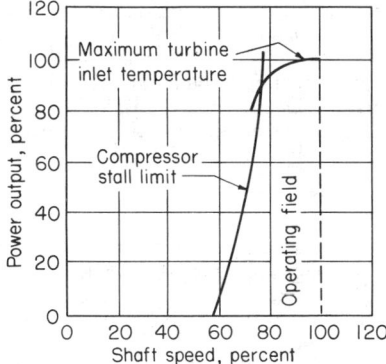

**Fig. 11**  Single-shaft gas-turbine operating range for various speeds and loads.

minimum operating speed is set by the stall (pumping-limit) characteristic of the compressor. The maximum power is set by the maximum allowable gas-turbine inlet temperature. The **multiple-shaft** machines may have a separate output turbine in series or parallel, with one or more turbines driving the compressor. The latter may operate at any speed desired, independent of the output turbine. Figure 12 illustrates a two-shaft arrangement with the power turbine in series. The major

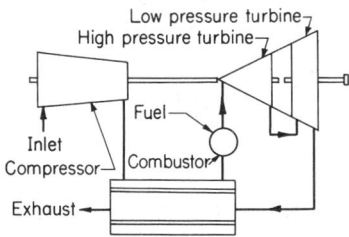

**Fig. 12**  Regenerative-cycle gas turbine, two-shaft arrangement with separate power turbine in series.

advantage of the multiple-shaft arrangement is the wide load-speed range possible, as shown in Fig. 13. This makes it most suitable for mechanical-drive application where variable speed and load operation are required. Figure 14 shows a regenerative, intercooled gas turbine with a two-shaft arrangement. The high-pressure turbine drives both the load and the high-pressure compressor while the low-pressure turbine drives the

low-pressure compressor. Multiple-shaft units can be arranged for improved part-load thermal efficiency.

The **aircraft jet engine** (Fig. 15) is a compressor-combustor-turbine system whose output is in the form of a high-energy exhaust stream. With the addition of a separate power turbine in series, it provides an extremely light and compact source of power. The availability of well-proved jet engines in a wide range of power has led to extensive use of this arrangement for industrial purposes. Nearly one-third of industrial gas turbines are of this type.

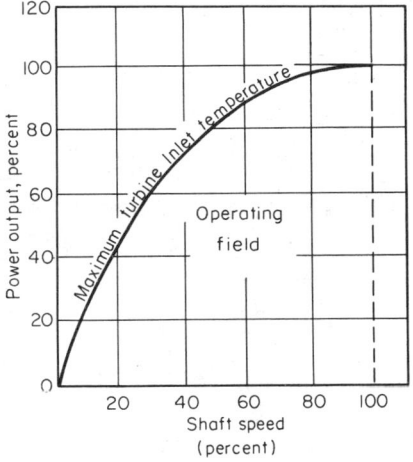

**Fig. 13**  Two-shaft gas-turbine operating range for various speeds and loads.

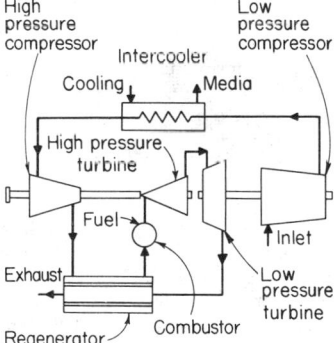

**Fig. 14**  Regenerative, intercooled gas turbine with two-shaft arrangement.

The **fan engine** (Fig. 16) is a development of the simple jet engine in which a large-diameter compressor stage is added to pump additional air for propulsion. The additional air bypasses the main engine. The turbine must be enlarged to provide power to drive the fan, and the thrust of the main jet is thereby reduced, but the net effect is increased thrust and improved efficiency.

## GENERAL CHARACTERISTICS

In its simplest form (Fig. 1), the gas turbine is small, light, requires only a modest foundation and building, does not require cooling water, runs unattended, and can be remotely or automatically controlled. It is capable of rapid start-up and loading, low standby losses, low maintenance, and long life. The simple-cycle gas turbine has good efficiency at full load when operating temperature and pressure ratio are high (Fig. 5); but as load is reduced, by reducing turbine temperature or r/min (i.e., pressure ratio), the efficiency falls. Efficiency can be maintained in a multiple-engine installation by cutting out engines as the load is reduced. A relatively flat efficiency curve can be obtained at the cost of some complexity using regeneration and regulation of airflow by variable-area turbine nozzles (Fig. 17).

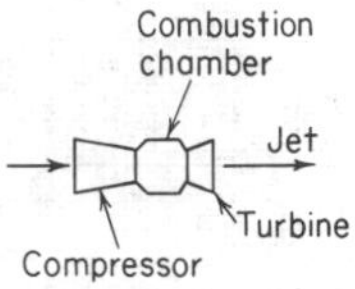

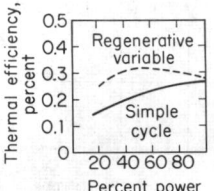

**Fig. 15**  Jet engine. 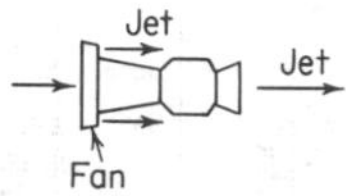          **Fig. 16**  Fan engine.

**Fig. 17**  Efficiency at part load.

Gas turbines can be arranged to supply power, high-pressure air, or hot exhaust gases either singly or in combination. They have been used in the Houdry process to charge the catalytic-cracking reactor and in Velox boilers to pressurize the furnace. The exhaust gases are used in many installations to make steam for process use. This flexibility of application is a characteristic feature of the gas turbine.

### FUELS

(See also Sec. 7)

**Natural gas** (essentially methane) has proved to be an ideal fuel. Light **distillate oils** are also satisfactory. Fuels containing sulfur, salt, or as in the case of residual oil, vanadium and other metals will cause corrosion unless water-washed to remove salt and treated with additives which inhibit the corrosion mechanism. A special problem of marine engines or those located near the sea is airborne salt which enters the engine through the air intake unless filters are installed.

Extensive work has been done on firing gas turbines directly with powdered **coal,** but no successful results yet obtained. Conversion of coal to liquid or gaseous fuels is actively under development, and preliminary tests indicate that if these synthetic fuels are free of such contaminants as sulfur, they can be used in gas turbines.

### APPLICATIONS

About 85 manufacturers, concentrated in the principal industrial countries, but mainly in the United States and Western Europe, produce over 500 models of gas-turbine engines, mainly in the size range from 100 to 100,000 hp (75 to 75,000

kW). Gas turbines have been used most successfully in rather large sizes and especially in aviation, which accounts for two-thirds of gas-turbine sales. In recent years industrial gas-turbine sales have been growing at 30 to 50 percent per year. Whether this trend can continue depends on the availability of clean fuel, which is currently in short supply.

### Military Aviation

(See also Sec. 11)

New military aircraft are almost exclusively gas-turbine-powered. Simple-cycle engines producing shaft power are used for helicopters and to drive propellers in small or moderate-speed aircraft. Most combat types now in use have jet engines, often using afterburning (reheat ahead of the jet nozzle) to get up to 100 percent additional thrust for short periods. The fan engine is increasingly favored for both long-range and fighter aircraft, and most new jet engines are of this type.

Future trends in military aircraft are toward higher speed and vertical takeoff. Both require the use of higher turbine-inlet temperature. Development of materials and design techniques for higher temperature will permit higher output and will improve thermal efficiency (Fig. 5).

### Commercial Aviation

(See also Sec. 11)

The jet airliner has demonstrated exceptionally high utilization and profitability, owing in large part to the good reliability record and low operating cost of the turbojet and turbofan engines. This, in turn, has contributed to very rapid growth of commercial aviation.

Propeller, fan, and jet-type engines are in use, with the trend more and more to the fan engine and to higher power. Engines rated at 40,000 lb (180,000 N) of thrust are now in general use. A rapidly growing segment of commercial aviation is air cargo, carried to a large extent in passenger aircraft. Specialized, very large cargo aircraft powered by very large gas turbines are coming into use.

### Electric Generation

Over two-thirds of the large industrial gas turbines are in electric-generating use. Electric utility companies in the United States use gas turbines for peak-load duty primarily. Installed and operating costs even using refined fuels are favorable for such intermittent duty, and gas turbines are well suited to automation and remote control. Aircraft engines adapted for this service offer very rapid start-up—as short as 2 min from start to full load. Peak-load power plants up to 150 MW have been installed, with a single generator driven by a battery of engines.

Gas turbines are used for base-load electric generation where additional capacity is needed quickly, where refined fuel, such as natural gas, is available at low cost, or where turbine exhaust energy can be utilized. The combined cycle (Figs. 6 and 7) makes possible improved efficiency and opens up the intermediate-load market to this power-plant type. Again, however, current worry over clean fuel supply is discouraging the electric utilities from heavy commitment to this system.

### Gas-Pipeline Transmission

(See also Sec. 11)

This industry is a major user of gas turbines—about one-sixth of all industrial units of large size. Gas turbines have been installed as compressor drives in sizes up to 30,000 hp (22,500 kW). This is an excellent application for the gas turbine, since natural gas is an ideal fuel and large powers are required. Exploitation of the arctic gas fields is expected to create a market for big gas-turbine compressor drives.

### Transportation

(See also Sec. 11)

In ships the high specific power of the gas turbine permits designing for higher speed. For merchant ships this can mean more revenue-producing voyages per year and better utilization of the investment. This is particularly true of container ships and other rapid-loading vessels which spend relatively more time at sea. The first ships of this type are now operating and will demonstrate the reality of the anticipated advantages.

The British Navy pioneered the use of gas turbines in fast **motor-torpedo boats** and in **larger warships** used together with steam or diesel power plants. New designs with all gas-turbine power are coming into service in the U.S., Canadian, British, and Russian navies.

Gas turbines have been used in **railroad** freight locomotives but seem not to offer significant advantage over diesel engines. Lightweight passenger trains, on the other hand, benefit from the light weight of gas turbines, and several trains are operating successfully in Europe and America.

The U.S. **automobile** companies have invested much effort on gas turbines for trucks, buses, and automobiles but have not yet developed a fully satisfactory competitor to the piston engine.

### Other

The gas turbine is a key element in what have come to be known as **total energy** units. In such a system, waste-heat recovery supplies heat, and/or refrigeration which together with electrical power from the engine itself can supply all the energy needs for a shopping center, institution, or factory. The present worldwide interest in conserving energy will encourage use of "total energy." In addition to pumping natural gas in pipelines, gas turbines are used extensively in the petroleum industry for compressor and pump drive in refineries, for oil-well pressurization, and for many other uses.

# NUCLEAR (ATOMIC) POWER

## by Louis H. Roddis, Jr., John E. Gray, Carl L. Newman, and N. J. Palladino

REFERENCES: Etherington, "Nuclear Engineering Handbook," McGraw-Hill. Hogerton, "The Atomic Energy Deskbook," Reinhold. "The New Force of Atomic Energy" and "Reactor Handbook," AEC. Kramer, "Nuclear Energy—What It Is and How It Acts," Technical Publishing Co. Blatz, "Radiation Hygiene Handbook," McGraw-Hill. "Civilian Nuclear Power—A Report to the President," AEC. Ellis, "Nuclear Technology for Engineers," McGraw-Hill. Glasstone, "Source Book on Atomic Engineering," Van Nostrand. Murray, "Nuclear Reactor Physics," Prentice-Hall. "Nuclear Terms—A Brief Glossary," AEC.

NOTE: The Atomic Energy Commission (AEC), by act of Congress, has been abolished and superseded by the Energy Research and Development Administration (ERDA), assuming the research and development duties, and the Nuclear Regulatory Commission (NRC), assuming the regulatory duties. Further legislation is pending to make ERDA a part of a new Department of Energy (DOE), leaving NRC unchanged as an independent regulatory agency.

### FISSION AND FUSION ENERGY

**Nuclear power** is the energy derived from the fission (or splitting) of the nuclei of heavy elements, such as uranium or thorium, or from the fusion (or combining) of the nuclei of light elements, such as deuterium or tritium. Particles set in motion by these processes yield heat energy virtually instantaneously. The amount of energy released per atom exceeds by a factor of several million the amount of energy obtainable per atom in a chemical reaction, such as the burning of fossil fuels. While control of the fusion process is still surrounded by tremendous technical problems, considerable success has been achieved in utilizing heat energy produced by fission for power generation, propulsion, industrial production, and scientific experiments.

### NUCLEAR PHYSICS

A **nuclear reaction** occurs when the particles making up the nucleus of an atom are rearranged—as distinct from a chemical reaction, which is caused by changes in the electron structure surrounding the nucleus.

**Fundamental (elementary) particles** are, theoretically, the irreducible constituents of the material world. More than 20 particles are regarded (1972) as elementary, the exact number being dependent on how they are classified. New experimental techniques lead to discovery of more particles; advances in theory indicate that some are compounds of particles. For example, atoms and atomic nuclei were at one time believed to be noncomposite. Except for the electron and proton, all the fundamental particles are unstable.

An **electron** carries one unit of negatively charged electricity [defined as $1.6 \times 10^{-19}$ coulomb (C)] and has a mass at rest equal to 1/1,836 times the mass of the hydrogen atom. Electrons emitted by radioactive atoms are called **beta ($\beta$) rays** and have energies up to several MeV (million electron volts).

A **proton** carries one unit of positively charged electricity equal in magnitude to that of the electron and has a mass, at rest, of 1.0072764 atomic mass units [u when 1 u ($1.66057 \times 10^{-27}$ kg) is defined as 1/12 the mass of a neutral atom of the most abundant isotope of carbon]. It is identical to the nucleus of the hydrogen atom. Protons are not emitted spontaneously from atomic nuclei.

A **neutron** has a mass approximately equal to that of the proton but lacks an electric charge. It cannot be detected by subjecting it to electric or magnetic fields, nor can its presence be shown by its passage through cloud chambers. The presence of neutrons can be shown only by their interaction with other particles.

An **atom,** which is the basic unit of any chemical element, consists of a central nucleus surrounded by planetary electrons. The number of its electrons determines its chemical characteristics and in electrically neutral atoms equals the positive charge on its nucleus. The nuclei of all atoms except the hydrogen atom are composed of protons and neutrons. The nucleus of the hydrogen atom consists of a single proton, which gives it an atomic number of 1. This number is designated by the letter $Z$. The number of neutrons in the nucleus is represented by the letter $N$. From this it follows that the mass number $A$, which represents the atomic weight, is equal to $N + Z$. Currently (1977) 105 elements are known. All with atomic numbers above 92 (uranium) must be produced artificially and have a short half-life.

**Isotopes** are elements which have identical chemical characteristics but different atomic weights; *i.e.*, certain atoms of the same element have different numbers of neutrons in their nuclei. There are considerably more than a thousand isotopes of the known elements. Only 320 of these exist in nature, of which approximately 40 are unstable and radioactive.

For example, hydrogen has three isotopes. One is ordinary hydrogen with a proton nucleus. Another is deuterium, whose nucleus consists of a proton and a neutron. The third is tritium, which consists of one proton and two neutrons. At the heavy end of the natural periodic table is uranium, with 92 protons. When it has 146 neutrons, it is $_{92}U^{238}$, which comprises 99.28 percent of natural uranium. The remainder of natural uranium consists of 0.006 percent $U^{234}$ and 0.71 percent $U^{235}$. The latter isotope, with 92 protons and 143 neutrons, is the only one of the three which is readily fissionable and is the one utilized in most of the nuclear-power reactors operating in the United States. It is separated from the natural element in the enrichment desired by a gaseous-diffusion process in government facilities.

**Energy and mass** are used interchangeably in nuclear physics, and the total energy of a nucleus can be determined by measurement of its exact mass $m$, which is related to its energy $E$ by the Einstein equation $E = mc^2$, where $c$ is the velocity of light, $3 \times 10^{10}$ cm/s. This law of equivalence of mass and energy unifies two long-accepted laws: the conservation of energy and the conservation of matter. Singly, the two older laws must be regarded as high-order approximations, adequate in all engineering fields except atomic energy. The exact mass $m$ of a nucleus differs by only a few hundredths of a percent from an integral number $A$ of proton masses, but this small deviation is of significance in nuclear theory, since it measures the difference in energy between the nucleus and its separate components, i.e., its binding energy. The energies of nuclides are recorded in the tables in terms of their masses.

Physical atomic weights do not quite equal chemical atomic weights because the oxygen used in the chemical determinations is a mixture of isotopes, whereas in mass spectrographic measurements the single isotope $C^{12}$ is used for reference. The relationship is: Physical atomic weight = $1.000280 \times$ chemical atomic weight.

**Planck's Constant** Since radiation has the properties of both waves and particles, it is theorized that in the emission or absorption of energy by atoms or molecules, the process takes place by steps, each step being the emission or absorption of an **indivisible quantity of electromagnetic energy,** which is consid-

ered the elementary quantum of action. It is known as **Planck's constant** and has the value $h = 6.624 \times 10^{-27}$ erg·s. It is used in virtually all quantum relationships, including Schrödinger's equation and Heisenberg's uncertainty principle, and in the formula for the energy levels of atomic hydrogen. The standard notation is $\hbar = h/2\pi$.

**Relativistic Mass** When moving particles approach the speed of light, the appropriation for kinetic energy, $\Sigma \frac{1}{2}m_0v^2$, fails; it then becomes necessary to use the equation $E = mc^2$ and to use the relativistic mass, a function that increases with velocity according to the law $m - m_0/\sqrt{1 - (v/c)^2}$, where $m_0$ is the rest mass (i.e., the mass of newtonian mechanics) and $v$ is the coordinate velocity of the particle.

**Binding energy** is the work required to disintegrate an atom completely into $Z$ protons and $A$-$Z$ neutrons. It is a measure of the total kinetic and potential energy of the nucleons in the nucleus and may be calculated from the equations

$$B = C^2 \Delta \quad \text{and} \quad \Delta = Zm_H + (A - Z)m_n - m$$

where $m$ is the mass of the nuclide, $m_H$ the mass of the hydrogen atom, and $m_n$ the mass of the neutron, all on the physical atomic-mass scale (u). The quantity of $\Delta$ is known as the **mass defect.** The energy involved is usually stated in MeV.

**Fusion** Figure 1 shows that if two deuterium ($H^2$) nuclei react to produce one $He^3$ nucleus plus one neutron, the total binding energy of the four particles (two neutrons plus two protons) involved in the reaction is increased from approximately 4.4 to 7.6 MeV, releasing 3.2 MeV for the reaction. Such a process is called **fusion.** The deuterium-deuterium fusion reaction may also produce an $H^3$ plus an $H^1$ nucleus, with a release of 4.0 MeV.

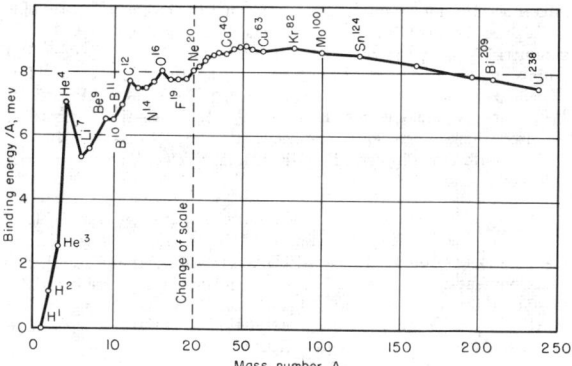

**Fig. 1** Binding energy per nuclear particle for stable nuclei.

For fusion to occur, the two nuclei must approach each other with exceptionally high kinetic energy in order to overcome their electrostatic repulsion; and since kinetic energy is proportional to absolute temperature, effective fusion can occur only at extremely high temperatures. For fusion to continue as a chain reaction, it is further necessary that nuclear collisions be sufficiently frequent to maintain the high temperature in spite of heat radiation. The condition necessary for frequent collisions is high pressure.

A light-element chain reaction is the principal source of heat in the sun, hydrogen being converted to helium as the net result of a multistage process. The temperature and pressure at which this reaction is sustained at the center of the sun are 35 million °F and $1.5 \times 10^{11}$ lb/in² respectively.

A deuterium-deuterium reaction can be sustained at temperatures and pressures which are considerably lower than those in the solar reactions but still fantastically high by terrestrial standards. Achievement of a controlled light-element chain reaction for power generation by fusion depends on producing high particle velocity and high density locally by electromagnetic means, without development of correspondingly high temperatures and pressures at the container walls.

**Fission** is the division of an atomic nucleus into parts of comparable mass, either naturally (spontaneously) or under bombardment (induced) with neutrons, $\alpha$ particles, $\gamma$ rays, deutrons, or protons. In elements with $Z > 90$, fission can be produced by neutrons of low or moderate energy.

Fission is important because in the process neutrons are emitted that may, under certain conditions, be utilized to produce further fissions, thus leading to a self-sustaining, or "chain," reaction. It is thus possible to "burn" uranium to obtain an energy release per atom approximately $10^8$ greater than from a chemical fuel. The energy release in the fission of $U^{235}$ is:

|  | MeV |
| --- | --- |
| Kinetic energy of fission products | 168 |
| Kinetic energy of fission neutrons | 5 |
| Energy of $\gamma$ rays | 10 |
| Energy of $\beta$ rays | 5 |
| Energy of neutrinos | 11 |
| Total | 199 |

The total of 199 MeV is equivalent to $3.2 \times 10^{-11}$ W·s, which indicates that it requires the fissioning of $1.12 \times 10^{17}$ atoms to release a kilowatthour of energy. This number of atoms forms only a speck of matter six-thousands of an inch in diameter.

**Radioactivity** occurs when an unstable nucleus undergoes atomic disintegration by emitting $\alpha$, $\beta$, or $\beta+$ particles, $\gamma$ or X-ray electromagnetic radiation.

The **alpha** ($\alpha$) **particle** is identical with the nucleus of a helium atom. The emission of an $\alpha$ particle creates a new nucleus, with $Z$ reduced by 2 and $A$ decreased by 4 mass units.

**Beta** ($\beta$) **decay** involves emission of an electron from the nucleus of an atom, thereby increasing $Z$ by unity.

The **gamma** ($\gamma$) **ray** is electromagnetic radiation originating in the nucleus—as differentiated from X-rays, which usually are less energetic and arise from energy adjustments as electrons move between orbital shells outside the nucleus. Gamma rays are emitted instantaneously when a neutron is captured in a nucleus and are frequently emitted following ejection of a particle from a radioactive nucleus.

The **positron** ($\beta+$) is a particle whose mass is the same as that of an electron and whose charge is equal in magnitude but opposite in sign. Positron emission decreases $Z$ by unity.

**Electron capture (EC)** results when an unstable atom decays by capturing an orbital electron in the nucleus, resulting also in a decrease of $Z$ by unity. This capture produces a vacancy in the orbital shell which is filled by an electron moving from an outer shell, giving rise to X-ray characteristics.

**Half-life** is the time required for half the original nuclei in a sample of an isotope to decay; it is given as $0.693/\lambda$ s. Half-lives of the radioactive nuclides vary from $10^{-7}$ s to $10^{10}$ years.

## UTILIZATION OF FISSION ENERGY

### Nonpower Types

Nuclear reactors having a purpose other than the generation of usable power may be classified into three general types: teaching reactors, research reactors, and test reactors.

**Teaching reactors** have the characteristics of very little excess reactivity, very low power capability (a few to a few thousand watts of heat energy), and relatively easy access to the core. Teaching reactors are used mainly for the purpose of instruction in the behavior of a nuclear reactor. With a teaching reactor novice reactor operators may experience the response of reactor instrumentation as reactivity is inserted and criticality is approached, "just critical" behavior, and power increases or decreases.

Teaching reactors are generally owned and operated by universities. In addition to being used for operator training, they are used in a variety of nuclear-engineering instructional experiments. Typical experiments include measurements of neutron-flux distribution; measurement of neutronic characteristics such as generation time, leakage fraction, thermal-utilization factor, resonance escape probability, and other characteristics affecting criticality and kinetics; and measurement of parameters affecting control such as control-rod worth and reactivity feedback coefficients. Although the neutron flux available from a teaching reactor is relatively small, it is nevertheless significant and can be used for limited production of radioactive isotopes.

**Research reactors** generally have a power capability of 10 thermal megawatts approximately. They are for the purpose of providing an intense radiation source, both neutrons and gamma rays. A research reactor is usually designed to have a high leakage flux in order to facilitate its use as a radiation source.

The neutrons from a research reactor have a variety of uses including neutron-activation analysis, neutron radiography, neutron diffractometry, and production of radioactive isotopes. The gamma rays are applied in materials testing and in studies using the interaction of gamma rays with the nuclei of various elements.

In **neutron-activation analysis** an unknown sample is irradiated in a calibrated neutron flux and the constituents of the sample become radioactive through neutron absorption. The radiation from the sample is then analyzed. Since the radiation from a specific radioactive isotope is unique to that isotope, a qualitative and quantitative determination of the constituents of the sample is possible. Extremely small quantities of radioactivity can be detected; therefore, neutron-activation analysis is a very sensitive analytical tool.

**Neutron radiography** is a technique similar to X-radiography. The perturbation of a collimated beam of neutrons that has passed through the radiographed item is determined by film exposure or other means. Unlike X-rays that serve to outline dense, highly absorptive materials, neutron radiography serves to outline light materials that preferentially scatter neutrons from the incident beam.

**Neutron diffractometry** is a technique used to determine the molecular and crystalline structure of materials. The analysis is accomplished through a precise determination of the diffraction pattern from a collimated monoenergetic beam of neutrons that has interacted with the sample.

Certain radioactive isotopes are useful and effective in the **treatment of some diseases.** Isotopes that do not occur naturally can be produced in large quantities by exposure of the precursor element to the neutron fluxes available from research or test reactors.

**Test reactors** generally have a power capability ranging from 25 to 100 thermal megawatts. They are used largely for the purpose of providing an experimental radiation environment similar to that which exists in the core of a power reactor. They are also used for large-scale radioactive-isotope production.

Some test reactors that have been operated by the U.S. AEC (and ERDA) are the materials test reactor (MTR), the engineering test reactor (ETR), and the advanced test reactor (ATR). Another is the General Electric test reactor (GETR) operated by the General Electric Company. These reactors have experimental facilities in which prototype power-reactor fuel assemblies may be operated at the power density, pressure, temperature, and coolant flow that would be experienced in actual application. Other power-reactor components that may be exposed to a core-type environment can be similarly tested to verify design characteristics.

### Power Cycles

Although the energy of fission appears as kinetic energy (85 percent) and photon energy of $\gamma$ rays, there is no practical method of converting this energy directly into useful work on a large scale. Therefore, a nuclear reactor must be treated as a heat source which differs from a chemical heat source in that no oxygen is required and the heat does not have to be removed from gaseous combustion products which possess poor heat-transfer properties.

In the large power capacities associated with nuclear power plants, the **Rankine vapor cycle** is the one in use. The Rankine cycle appears to be best suited for nuclear power because (1) the maximum practicable operating temperature for a reactor corresponds to or is lower than the temperature used in the conventional Rankine vapor cycle; (2) the problem of containment of radioactivity is better solved by this cycle than by other cycles; (3) reactors are most economical in large sizes, as are Rankine engines.

Three variations of the Rankine cycle (Fig. 2) are being used. The **direct cycle** is the least complicated and thermodynamically the most desirable. Its disadvantages are the facts that the boiling process does not permit the high power densities which are attainable with liquid cooling and that radioactive steam is carried into the turbine and condenser. The **indirect cycle** gives greater power density and eliminates radioactivity in the turbine, although it too produces steam at lower pressures and temperatures than those considered most efficient for modern turbines. Both concepts are highly developed and commercially available. Plants utilizing sodium or other liquid metal as the heat-exchange medium use an intermediate-link system to isolate the water system from radioactive sodium, but the ability to operate at high temperature with liquid-metal coolants permits the generation of high-temperature steam to utilize the present state of steam-turbine technology.

The only other cycle which has been given consideration is the **Brayton cycle.** This cycle is being examined with renewed interest because of the rapid advances in gas-cooled-reactor technology and the availability of large gas turbines suitable for reliable utility-industry service. The use of an open cycle using air has been discarded because of the spread of activation products from the turbine exhaust and the corrosion problems in the reactor core and fuel elements. The use of a closed Brayton cycle with the latest high-temperature gas-cooled reactors (HTGR) has potential advantages in that the gas temperatures can match the latest gas-turbine designs and a very compact plant can be built.

The ability of a nuclear reactor to produce high temperatures not limited by chemical-reaction equilibrium has generated interest in advanced concepts such as magnetohydrodynamics and thermionic generators. These systems are limited by the availability of suitable materials of construction and appear destined to remain in the research phase until suitable materials are developed.

### Power-Reactor Types

During the early years in the development of power reactors, a large number of possible reactor designs were investigated. In the present state of the industry, the designs which have gained the greatest acceptance all are solid fuel in the form of uranium dioxide. The moderator may be either graphite or water and the coolant may be water, gas, or liquid sodium. Experimental power reactors using a solution of uranium compounds in water or molten salt have been built but have not assumed industrial importance because of many unsolved corrosion problems. The **water-cooled reactors** presently represent the greatest fraction of installed and planned power capacity, with the **gas-cooled** systems in second place. The liquid-metal-cooled reactor systems appear destined to be confined to fast breeder reactors, which must be considered as being in an early stage of development. For the efficient utilization of the energy represented by the world's uranium reserves, a successful breeder reactor is required since the present generation of reactors are all consumers of only the naturally occurring fissionable isotope of uranium, $U^{235}$.

For utilization of natural uranium as a fuel, $D_2O$ is substituted for $H_2O$ as the coolant-moderator. The reactor must be refueled more frequently than enriched reactors to maintain an adequate operating reactivity margin. However, $D_2O$ reactors such as the Canadian designs utilized on-line refueling to minimize shutdown time. The success of this type of reactor depends on the availability and cost of $D_2O$.

**Boiling-water reactors** have a simpler design and can utilize relatively thin-walled vessels and pipes because they operate at moderate pressures compared with pressurized-water reactors. Fuel cladding temperatures are only slightly higher than steam temperatures, and there is an inherent safety factor because the steam-void volume increases on a transient power increase. Some of the problems caused by radioactivity carry-over, such as maintenance on the turbine and condenser and the prevention of radioactive leakage, are offset by the cost of the boiler in pressurized-water reactors. However, boiling-water reactor vessels despite their lower design pressure are larger and heavier than pressurized-water vessels of similar rating and extensive water-purification equipment is required to remove corrosion products and other impurities from the feedwater before introduction into the reactor.

Limitations on the **power density** imposed by exit voids (i.e., the vapor volume of the exiting steam-water mixture from the reactor core) are a disadvantage in that low power density contributes to high fuel-inventory charges. Load changes are accomplished by steam bypass control or rods since adjustment of the turbine throttle and consequent reduction in steam flow will cause an increase in pressure in the reactor, which in turn will collapse the steam bubbles and increase reactivity. The control system functions to maintain constant reactor pressure, and reactivity control is achieved in part by varying the recirculation rate in the reactor. The net decomposition of the water into O and H is much greater than in a pressurized-water reactor.

**Pressurized-Water Reactors**    Properties of water and steam which led to their predominance as general-purpose heat-transfer mediums have also caused their widespread application as a reactor coolant. A major disadvantage of water results from its relatively high vapor pressure. However, this can be partially overcome by allowing boiling in the reactor. Thermal efficiencies up to 36 percent are possible.

The use of $H_2O$ as a coolant and moderator is based on well-developed technology which indicates that the ultimate size (or capacity) will be dictated primarily by heat-transfer requirements. The average heat flux $q/A$ is around 200,000 Btu/(h) (ft²) (630,000 W/m²), with a maximum of 600,000 Btu/(h) (ft²) approximately (1,890,000 W/m²).

The **light-water** ($H_2O$) **reactors** require slightly enriched fuel. However, the cost of enrichment is economically justified because the increased power density reduces inventory charges. To provide sufficient neutron moderation, the $H_2O$-U volume ratio is kept slightly above 2:1.

The use of oxide fuel minimizes corrosion. Design problems are reasonably well understood, although costly structural provisions must be made to load and unload fuel because of the high pressures. Fuel enrichment usually runs between 1.5 and 4.5 percent, depending on the alloy used for corrosion resistance.

Control rods, burnable neutron absorbers, and soluble boron in the coolant are used to control excess reactivity, achieve high fuel burnup, and maintain power distribution within the design limits.

**Gas-cooled reactors** offer low fuel and operating costs because of their ability to utilize natural uranium as fuel. This advantage is offset by higher capital costs caused by the large system size, in turn resulting from the lower power density usually used. More advanced designs such as the high-temperature gas-cooled reactor (**HTGR**) and its variants can use low enriched uranium and are able to use higher temperatures and the required structural material for such temperatures. These newer gas-cooled designs are much more compact than the natural-uranium reactors which they are beginning to supplant.

Carbon dioxide is the coolant usually used in natural-uranium reactors, as it is nontoxic, only mildly radioactive, nonflammable, noncontaminating in case of leakage, and relatively inexpensive. Helium gives indication of being an ideal coolant, but its limited availability and high cost make it attractive in a power reactor only where the inventory can be kept small and its superior performance at high temperatures can be utilized. Helium, being inert, does not chemically react with graphite and the reactor structure at high temperatures as does carbon dioxide.

The latest gas-cooled reactor designs are prestressed-concrete vessels which contain the core, the gas circulators, and the steam generators. The inclusion of as much equipment and piping as possible inside the vessel structure greatly simplifies the control of gas leakage.

**Fast breeder reactors** operate at extremely high power densities (as they are unmoderated) and are liquid-metal-cooled. Their attractiveness stems from an ability to "breed," i.e., produce more fuel than they consume. Aside from savings in fuel cost, this characteristic is believed necessary to conserve the world's supply of energy.

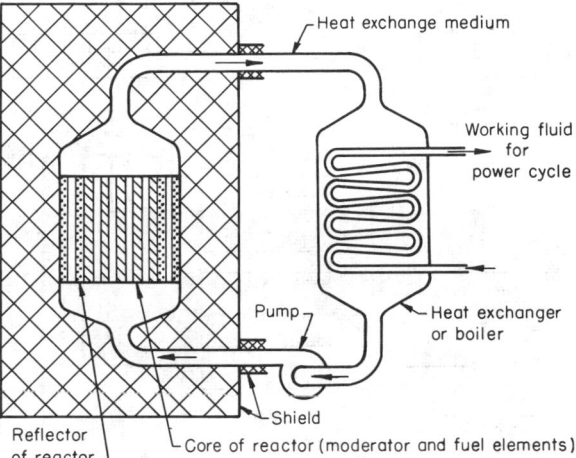

**Fig. 2**    Plant cycles—coupling with working fluid. (*A. Amorosi, Selection of Reactors, "Nuclear Engineering Handbook," McGraw-Hill, 1957.*)

Two conflicting definitions of a breeder reactor are in common use: (1) a nuclear reactor that produces more fissionable material than it consumes, regardless of type of fuel used; (2) a nuclear reactor that produces the same species of material as it consumes, regardless of the net gain or loss. The first definition has wider acceptance in the United States.

The selection of a coolant for a breeder reactor is restricted by the requirement that it not act as a moderator. This requirement rules out the use of water, although the attainment of some breeding in a water-cooled reactor has been under investigation. The choice of coolant is limited to helium or liquid metal.

Choice of a liquid-metal coolant is based mainly on nuclear properties and on the engineering difficulties associated with melting temperatures and corrosion effects. Those of prime interest are sodium, sodium-potassium alloy, bismuth, lead, and lead-bismuth alloy. Separated $Li^7$ would be very attractive were it not for its high cost. Sodium has been selected as the preferred coolant because of its availability, good nuclear properties, and the short half-life of its induced radioactivity. However, in the presence of oxygen, it burns and can react violently with water.

Advantages of operating in the fast-neutron spectrum include (1) structural material for the reactor core can be selected without consideration of neutron absorption; (2) very little excess reactivity is needed to compensate for fission-product buildup; and (3) reprocessed fuel in which fission-product removal may not be complete can be used.

The disadvantages are the increased damage to structural

materials by the high fast flux and the control problems resulting from the short neutron lifetime.

### Fuel and Waste Cycle

The steps and processes in a typical fuel cycle for light-water reactors are described below. Figure 3 illustrates the relationship of these steps and processes.

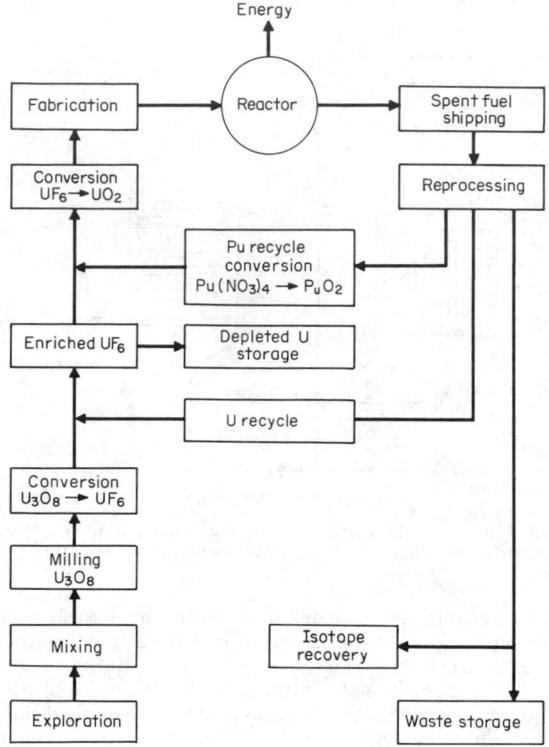

**Fig. 3** Nuclear fuel cycle.

**Natural Uranium**  Uranium occurs naturally in ores in very low concentrations, less than ½ percent by weight. In certain very unusual cases, the uranium content in the ore, called ore grade, may be as high as 10 percent, but these occurrences are extremely rare.

Uranium ores are mined by conventional techniques and then crushed and ground in a mill, which has special equipment for the recovery and purification of the ore into uranium concentrates. The product is in the form of a powder called yellow cake, which has the chemical form of either ammonium or sodium diuranate, both of which are yellow, hence the name.

**Conversion of $U_3O_8$ to $UF_6$**  The process of enriching the $U^{235}$ isotope (from 0.711 percent to the 2 or 3 percent required by water reactors) requires that the uranium be in gaseous form. The gas $UF_6$ is used; so the natural $U_3O_8$ must be converted to natural $UF_6$.

**Enrichment** alters the relative amounts of $U^{235}$ and $U^{238}$ in uranium, raising the $U^{235}$ concentration from 0.711 percent to the percentage required for use in a reactor. By so doing, the input (feed) stream of natural $UF_6$ is broken into two output (product plus tails) streams, one enriched and one depleted.

The process of enrichment makes use of the mass difference between $U^{235}$ and $U^{238}$ isotopes. In a gaseous form, the average energies of all $UF_6$ molecules are equal whether the U atom is $U^{235}$ or $U^{238}$. Since the energies are equal, the average velocity of the lighter $U^{235}F_6$ molecule is greater than that of the $U^{238}F_6$. Because of this greater velocity and, hence, a greater momentum, the $U^{235}F_6$ molecules can diffuse through a barrier more easily. The gaseous-diffusion plants work on this principle.

**Fabrication**  The gaseous enriched $UF_6$ that comes out of the diffusion plant is delivered in standard cylindrical gas bottles. The $UF_6$ is then converted to $UO_2$, the fuel form in which it is used. The $UO_2$ is in the form of a powder which is then milled to provide suitable sintering properties. The powder is formed into pellets by cold pressing, sintered and ground to final dimensions. The pellets are loaded into a zirconium-alloy tube (cladding) which is backfilled with helium and sealed at each end with a welded-end plug. The fuel rods are assembled into complete fuel assemblies. The number of fuel rods in an assembly varies from 50 to 200 or more, depending on the reactor type and specific design.

**Spent-Fuel Shipping**  When fuel is discharged from the reactor, it is cooled for about 3 to 6 months and then shipped to the reprocessing plant for recovery of residual fuel values. A shipping cask loaded with fuel will make several trips between the reactor and the reprocessing plant.

**Reprocessng and Reconversion**  Spent nuclear fuel has substantial residual value in its uranium and plutonium, and possibly in other fission products or transuranic elements. The spent fuel is mechanically chopped and dissolved in acid, and the spent-fuel solution proceeds through several chemical steps to purify the uranium and plutonium. Provision can also be made to recover desired spent-fuel by-products. The plutonium product of the plant is in the form of a nitrate, and the uranium might be converted to $UF_6$ by the reprocessor. The reprocessing of spent fuel results in relatively large quantities of high-level liquid wastes. At present and for the foreseeable future, these wastes will be stored in above-ground tanks. However, AEC policy (November 1970) indicates that liquid high-level wastes must be solidified within 5 years from the generation of these wastes and sent to an ERDA repository within 10 years of their generation. At the time of the deposition of these wastes with the AEC, a one-time fee will be paid for their permanent disposal and care on government-owned and controlled property. While it is likely that perpetual storage as liquid waste is slightly more economic than solidification, there is general agreement that solidification is preferable from the point of view of overall safety and public benefit. Several processes for solidification of wastes to inert refractory form have been developed to the point of demonstrating technical feasibility. However, additional development work is needed in this area to demonstrate economically preferable methods of treatment. Reprocessing plants also generate significant quantities of solid wastes which are disposed of by burial at licensed facilities. Smaller amounts of gaseous wastes are also generated. The primary wastes are iodine and the noble gas krypton 85. Iodine is removed by filtration through activated-charcoal filters. Very small quantities of krypton are released. However, technical feasibility of the krypton-removal process has been demonstrated, and it could be applied if necessary.

**Uranium and Plutonium Recycle**  As described above, the components of value in the spent fuel are the remaining uranium (now of lower enrichment) and plutonium generated through neutron capture in $U^{238}$. The reclaimed uranium may be converted from nitrate directly to $UO_2$ and blended with new feed material in the fabrication process, or it may be converted to $UF_6$ and used as feed to the enrichment plant. The latter course is usually more economical.

## PROPERTIES OF MATERIALS

Materials used in nuclear reactors can be divided into six basic categories: fuel, radiation, control, coolant, structural, and shielding. While the nature of the materials may be different, they must be physically and chemically stable and compatible with their environment for a sufficient period to enable the reactor to operate predictably, stably, and economically under the particular set of parameters imposed by the design. (See also Sec. 6.)

In addition to the environmental conditions to which materials may ordinarily be exposed, reactor materials must also sustain the extraordinary bombardment of X-rays (photons) and other atomic particles which are products of nuclear fission. In their passage through matter, these photons and atomic particles may inflict severe damage, since the energy of these particles is spent mainly in ejecting or pulling electrons from their orbits, i.e., producing **ionization**. Gamma photons and neutrons are highly penetrating, whereas fission fragments have a path under 0.001 in in solids and liquids (but cause intense ionization over that path).

The **behavior of materials** following breakage of a bond by ionization may differ markedly. The fragments of simple, highly stable molecules such as water ($H_2O$) or carbon dioxide ($CO_2$) will most probably recombine. However, it is highly probable that the parts of complex **organic compounds** such as lubricants and electrical insulation will suffer permanent fragmentation and/or recombination in a different manner to produce new molecules.

**Metals,** on the other hand, are essentially unaffected by ionization since the probability of displacement of excited atoms in a metal lattice is small and the atoms will merely revert to their original status on ridding themselves of their excess energy. The atoms of **nonconducting crystals** and glasses are similarly immobile, and most of their properties are insensitive to the ionization. However, since these crystals are nonconducting, some of the electrons displaced by ionization become trapped in lattice vacancies, producing changes of optical properties (such as darkening of glass by gamma irradiation) and change of electrical resistivity.

On direct collision, **fast neutrons** may displace atoms from a chemical compound without affording opportunity for immediate recombination. Similarly, **energetic neutrons** may displace atoms from their normal lattice positions in a metal or other crystalline material and push them into abnormal interstitial positions in the lattice, generally producing an effect on metals similar to work hardening. The hardness and tensile strength of the metal will increase, but the metal becomes more brittle. **Thermal neutrons** affect the properties of metals only in those cases where the probability of absorption is high enough to cause significant transmutation to other elements.

**Most severe damage** occurs when the atoms of a fuel material fission and produce two or more new atoms from each of the original fissioned atoms. Volumetric changes occur by the substitution of these additional atoms in the lattice for the original atom, and the lattice is further damaged by the intrusion of the high-velocity fission fragments.

### Fuel Materials

Fuel materials are generally divided into two categories: fissionable and fertile. The **fissionable materials** are those isotopes of uranium and plutonium which fission upon interaction with thermal neutrons. These are the isotopes $U^{233}$, $U^{235}$, $Pu^{239}$, and $Pu^{241}$, which have odd atomic weights. The **fertile materials** are those isotopes of uranium and plutonium which have even atomic weights. These absorb neutrons under proper conditions and undergo a series of nuclear reactions, with the eventual formation of fissionable isotopes. Thus $U^{238}$ forms $Pu^{239}$, and $Pu^{240}$ forms $Pu^{241}$. In addition, thorium$^{232}$ forms the fissionable isotope $U^{233}$.

Both fissionable and fertile materials fall short of possessing the **properties considered ideal** for a nuclear fuel, namely, corrosion resistance, good thermal conductivity, strength, and ductility. To overcome these problems as well as to prevent fission products from entering the coolant, various techniques are used, including plating, cladding, and alloying.

Materials for this use must be corrosion-resistant, be compatible with the fuel materials, be physically stable under irradiation, have good heat-transfer characteristics, have good nuclear properties, and be reasonably fabricable. From a consideration of cross section alone, only aluminum, beryllium, magnesium, and zirconium are attractive for use in thermal reactors. Magnesium usually is ruled out on the basis of strength and corrosion resistance but has been widely used in some gas-cooled systems. In fast reactors, a number of other materials become attractive, the most important being stainless steel, which also is used under special conditions in thermal reactors. Consideration also is being given to ceramic materials, BeO, $Be_2C$, SiC, and $MoSi^2$, which are very attractive in all phases except resistance to thermal shock.

**Moderating material** must be capable of reducing neutron energy very rapidly and should have good strength at high temperatures, corrosion resistance, thermal and radiation stability, and reasonable cost. Such properties are not available in a single material. Graphite is the most widely used solid moderator. Attention also is being given to beryllium and its compounds, deuterium, oxygen, and hydrogen.

The nuclear and physical properties considered desirable for **fuel diluents and cladding** are, in general, desirable for structural materials. For the latter, however, more emphasis is placed on strength and corrosion resistance than on nuclear properties. Stainless steel, aluminum, zirconium, molybdenum, titanium, niobium, and their alloys are used. As operating temperatures continue to rise, ceramics and cermets will have to be developed for this purpose.

Materials used for **control purposes** must have a high cross section for absorption of neutrons, adequate strength, low mass to permit movement, good corrosion resistance, chemical and dimensional stability under heat and irradiation, and low cost. Boron, cadmium, and hafnium are given the most consideration in various metallic and ceramic forms and as compounds dissolved in the coolant.

### Coolant Properties

The choice of a coolant also is a compromise, since no known material possesses all the **desirable characteristics.** These are good heat-transfer coeffient and good heat capacity on a volume basis, low absorption cross section, low vapor pressure at

**Table 1. General Properties of Reactor Materials before Irradiation***

| Materials | Density, g per cc | Melting point, deg C | Specific heat, cal per (mol)(deg C) | Thermal conductivity, cal per (cm)(sec)(deg C) | Coefficient of thermal expansion, $10^{-6}$ per deg C | Ultimate tensile strength, $10^3$ psi | Modulus of elasticity, $10^6$ psi | Capture cross sections, 0.025 eV ($\sigma_a$ barns) |
|---|---|---|---|---|---|---|---|---|
| **FUELS AND FERTILE MATERIALS** | | | | | | | | |
| Uranium, normal | 19.3 | 1133 ± 1 | 6.649; 100 C, 6.59 | 100 C, 0.063 | 25–125 C, 45.8 | 56 | 24 | 7.68 |
| Thorium, normal | 11.71 | 1690 ± 10 | | 100 C, 0.090 | 30–100 C, 11.5 | 30.6 | 10.6 | 7.56 |
| Plutonium$^{239}$ | 19.60 | 623 ± 7 | | 100 C, 0.061 | | 125 | 13–16 | 1,026 |
| Uranium oxide ($UO_2$) | 10.96 | 2500–2600 | 15.38 | 100 C, 0.022 | 22–363 C, 9.3–10.8 | | 21 | 24.3 |
| Al-16 weight % U alloy | | ~650 | | 200 C, 0.42 | 20–100 C, 20 | 17.7 | 10.6 | |
| Zr-4 weight % U alloy | 6.72 | 1840 ± 25 | | 100 C, 0.033 | 190–300 C, 6.66 | 63 | 14.4 | |
| **DILUENTS AND CLADDING MATERIALS** | | | | | | | | |
| Zirconium | 6.50 | 1845 ± 25 | 6.31 | 50 C, 0.05 | 5.82 | 30–38 | 13.8 | 0.18 |
| Zircaloy 2 | 6.55 | 1820.2 | | <0.04 | 20–100 C, 5.2 ± 0.6 | 65 | 14 | |
| Aluminum (2S) | 2.694 | 660.2 | 100 C, 6.07 | 0.53 | 20–100 C, 23.8 | 13 | 10.3 | 0.26 |
| Magnesium | 1.74 | 650 | 25 C, 6.08 | 18 C, 0.376 | 49 C, 26 | Annealed, 27 | Annealed, 6.5 | 0.069 |
| $MoSi_2$ | 6.24 | 1870 | | 370 C, 0.088 | 0–1500 C, 5.1 | 27 C, 22 | | |
| **SOLID MODERATORS** | | | | | | | | |
| Graphite | 2.27 | Sublimes at 3650 ± 25 C | 2.066 | 100 C, 0.37–0.48 | 20–250 C, 1.9–4.0 | 0.500–2.4 | 0.5–1.2 | 0.0032 |
| $Be_2C$ | 6.24 | 1870 | 100 C, 13.69 | 30 C, 0.02 | 25–200 C, 7.7 | | 400 C, 23 | |
| Beryllium | 1.847 | 1315 | 100 C, 7.7 | 100 C, 0.34 | 20–200 C, 11.54 | 29 | 44 | 0.01 |
| Sintered BeO† | 2.2–2.8 | 2550 ± 25 | | 200 C, 0.19 | 25–100 C, 5.5 | 400 C, 15 | 400 C, 39 | 0.00072 |
| SiC | 3.21 | Decomposes at 2500 C | 327 C, 10.06 | 400 C, 0.060 | 0–1700 C, 4.4 | 1700 C, 50 | | |
| **STRUCTURAL MATERIALS** | | | | | | | | |
| Stainless steel (347) | 8.027 | 1427 | 100 C, 0.12‡ | 100 C, 0.037 | 20–100 C, 16.5 | 85 | 29 | 3 |
| Inconel X | 8.51 | 1395–1425 | 25–100 C, 0.109‡ | 0–100 C, 0.036 | 20–100 C, 11.5 | Annealed, 100–120 | | 4.1 |
| Molybdenum | 10.2 | 2622 ± 10 | 100 C, 6.24 | 0 C, 0.32 | 5.1 | 67 | 48 | 2.7 |
| Titanium (commercial purity) | 4.507 | 1690 | 0–500 C, 6.61 | | 25 C, 8.5 | 85–100 | 15.5 | 5.8 |
| Niobium | 8.57 | 2415 | 0 C, 6.01 | 25 C, 0.41 | | 1200 C, 14.7 | | 1.1 |
| **CONTROL MATERIALS** | | | | | | | | |
| Cadmium | 8.694 | 321 | 25–321 C, 6.19 | 0.22 | 20–100 C, 31.8 | 10.3 | 7.1–10 | 2,450 |
| Hafnium | 13.36 | 2130 ± 15 | 25–2227 C, 6.16 | | 0–100 C, 5.9 | 67.5 | 14 | 105 |

Sources: H. A. Saller, Properties of Reactor Materials, "Nuclear Engineering Handbook," McGraw-Hill, 1958; "Reactor Physics Constants," ANL 5800, 1963.

*All properties change with irradiation at room temperature except when specified.

†Thermal conductivity, strength, and the modulus of elasticity vary directly with the density; however, the coefficient of thermal expansion appears unaffected.

‡Cal/(g) (°C).

**Table 2. Properties of Nuclear Coolants***

| AQUEOUS AND ORGANIC LIQUIDS | | | | | |
| --- | --- | --- | --- | --- | --- |
| | Water | Heavy water | Diphenyl | Dowtherm-A | Santowax R |
| Density, lb/ft³ | 42.4 (600 F) | 46.2 (600 F) | 46.4 (675 F) | 46.6 (675 F) | 52.29 (675 F) |
| Viscosity, lb/hr/ft | 0.20 (600 F) | 0.220 (600 F) | 0.353 (675 F) | 0.7986 (675 F) | 0.630 (675 F) |
| Melting point, deg F | 32 | 39 | 156 | 53.2 | 293 |
| Boiling point, deg F | 212 | 214 | 49 | 496 | 687–784 |
| Heat capacity, Btu/lb/deg F | 1.54 (600 F) | 1.68 (600 F) | 0.637 (675 F) | 0.675 (675 F) | 0.539 (675 F) |
| Thermal conductivity, Btu/hr/ft/deg F | 0.294 (600 F) | 0.294 (600 F) | 0.0763 (675 F) | 0.098 (675 F) | 0.064 (675 F) |
| Capture cross section, $\sigma_a$ 0.025 eV (barns) | 0.66 | 0.0011 | 3.3 | 3.3 | 4.6 |

| GASES | | | | | | | |
| --- | --- | --- | --- | --- | --- | --- | --- |
| | Hydrogen | Helium | CO₂ | Air | Nitrogen | Steam | Neon |
| Density, lb/ft³ | 0.00185 | 0.00375 | 0.041 | 0.0275 | 0.026 | 0.0167 | 0.0190 |
| Viscosity, lb/hr/ft | 0.042 | 0.094 | 0.081 | 0.089 | 0.086 | 0.069 | 0.1445 |
| Heat capacity, Btu/lb/deg F | 2.625 | 1.245 | 0.283 | 0.263 | 0.268 | 0.518 | 0.246 |
| Thermal conductivity, Btu/hr/ft deg F | 0.224 | 0.159 | 0.0328 | 0.0337 | 0.0327 | 0.0341 | 0.0536 |
| Prandtl No | 0.660 | 0.735 | 0.6988 | 0.6898 | 0.7097 | 1.048 | 0.6632 |

| LIQUID METALS | | | | | | | | |
| --- | --- | --- | --- | --- | --- | --- | --- | --- |
| | Gallium | Lithium | Potassium | Rubidium | Sodium | Tin | Na(56%) K(44%) | Pb(44.5%) Bi(55.5%) |
| Density, lb/ft³ | 359 | 29.9 | 44.6 | 84.4 | 51.2 | 421 | 48.8 | 625.5 |
| Viscosity, lb/hr/ft | 1.836 | 1.188 | 0.3816 | 0.415 | 0.5040 | 2.736 | 0.435 | 2.88 |
| Melting point, deg F | 85.86 | 354 | 147 | 102 | 208 | 449 | 66.2 | 257 |
| Boiling point, deg F | 3,601 | 2,403 | 1,400 | 1,270 | 1,621 | 4,118 | 1,518 | 3,038 |
| Heat capacity, Btu/lb/deg F | 0.082 | 1.0 | 0.180 | 0.0877 | 0.3005 | 0.0639 | 0.2484 | 0.035 |
| Thermal conductivity, Btu/hr/ft/deg F | 18.0 | 18.0 | 21.15 | 13.2 | 37.7 | 19.0 | 16.35 | 8.05 |
| Capture cross section, $\sigma_a$ 0.025 eV (barns) | 2.77 | 71.0† | 1.97 | 0.70 | 0.505 | 0.60 | | 0.094 |

*At 1000°F unless otherwise specified. Adapted from L. Green, Reactor Coolant Properties, *Nucleonics*, **19**, No. 11, which contains a detailed bibliography.

the operating temperature, minimum of chemical reaction with the material contacted by the coolant, good resistance to forming γ emitters with long half-lives, stability under irradiation and high temperatures, and workable melting points.

Coolants are divided into three broad **classes:** aqueous and organic liquids, liquid metals, and gases. **Water** has found the widest application, largely because it is readily available and economical, has fairly good heat-transfer properties, and offers no corrosion problems which cannot be handled. Its most serious drawback is its high vapor pressure. **Heavy water** has superior qualifications except for cost, which is high. Diphenyl and other **organic liquids** have demonstrated an ability to stand up under irradiation but are handicapped by a tendency to deposit decomposition products on the fuel-element plates.

**Liquid metals** are attractive because they give high heat-transfer rates and do not require pressurizing to operate at high temperatures. Sodium is most attractive because of its high boiling point and high heat conductivity, but it reacts violently with water and has poor lubricating qualities. Lithium 7 better meets qualifications but is harder to obtain and is costly.

**Gas coolants** offset their advantage of being inert at elevated temperatures with poor heat-transfer qualities. They must be operated at high pressures to be useful in power reactors.

Sodium hydroxide and yellow phosphorus are among other materials considered as coolants, but their acceptability has yet to be demonstrated.

## FISSION-REACTOR DESIGN

**Neutron Balance** The main objective of reactor design is to achieve neutron balance during reactor operation. At steady state, the number of neutrons produced during fission is equal to the neutrons lost by absorption and leakage processes. The net rate of change of neutron density $n$ (neutrons per cubic centimeter) is given by the neutron-conservation equation:

$$\frac{\partial n}{\partial t} = \text{production} - \text{absorption} - \text{leakage}$$

If $\partial n/\partial t$ is positive, the reactor is said to be supercritical; if it is zero, the reactor is at steady state and is said to be critical; if it is negative, the reactor is subcritical.

**Cross Section** The microscopic cross section $\sigma$ for a particular nuclear process is the effective target area of the nucleus with which a neutron must interact to produce the given reaction. The unit of $\sigma$ is barn (1 barn = $10^{-24}$ cm²). Neutron cross sections (absorption, scattering, and fission) constitute the basic nuclear data for reactor design. Absorption cross sections for thermal neutrons range from 0.00046 barn for deuterium to $3.3 \times 10^6$ barns for xenon[135]. Moderating atoms for the slowing down of fast neutrons have low atomic mass, small absorption, and large scattering cross sections. Examples are hydrogen, deuterium beryllium, and carbon. Table 3 gives thermal-neutron cross sections in barns for selected elements (based upon the naturally occurring combinations of isotopes).

**Multiplication Factor**  The number of neutrons produced in any one generation in a reactor for each neutron produced in a previous generation is called the multiplication factor $k$. If $k = 1$, the reactor is critical; if $k < 1$, it is subcritical; and if $k > 1$, it is supercritical. For an infinite thermal reactor, multiplication factor $k_\infty = \eta f p \epsilon$. $\eta$ is the number of fission neutrons produced per neutron absorption in fuel, $f$ is the thermal-utilization factor (neutron-absorption rate in fuel/total absorption rate), $\epsilon$ is the fast-fission factor, and $p$ is the resonance-capture escape probability. The leakage of neutrons from the reactor core is accounted for by multiplying $k_\infty$ by the factor $1/(1 + M^2B^2)$, where $M^2$ is the migration area (sum of the squares of the diffusion length $L$ and the slowing-down length $\tau$) and $B^2$ is the geometrical buckling ($B^2 = 0$ for an infinite reactor).

**Breeding Ratio**  If the ratio of a number of fissile nuclides produced by the capture of neutrons by the fertile nuclides to the number of fissile nuclides consumed is greater than unity, it is called the breeding ratio; and if it is less than unity, it is called the conversion ratio. For breeding, the value of $\eta$ (defined above) must be greater than 2, as one neutron is required to maintain the chain reaction by fission, another neutron maintains breeding by its capture by the fertile nuclide, and the remaining to compensate for the loss of neutrons due to absorption and leakage. Neutron economy favors the breeding of Pu[239] fissile nuclides from the fast-neutron capture by U[238] fertile nuclides. This breeding process is the basis of LMFBR (liquid-metal fast breeder reactor). On the other hand, thermal breeders are based upon the breeding of uranium 233 fissile nuclides from thorium 232 fertile nuclides by the capture of a thermal neutron.

**Diffusion Theory**  Multienergy-group diffusion theory is presently the basis of reactor design. For a critical reactor the neutron-balance equation reduces to

$$D_g \nabla^2 \phi_g - \Sigma R_g{}^{\phi_g} + S_g = 0$$

where $\phi_g$ and $S_g$ are the $g$th energy-group neutron flux and the slowing-down neutron source; $D_g$ and $R_g$ are the diffusion coefficient and removal cross section. The latter are called group constants, which are obtained by the analysis of basic fuel lattice. The solution of the above equations provides the reactor multiplication factor, reactor critical size, neutron fluxes, power distributions, isotopics of elements, and fuel-cycle length.

**Control**  Extra fuel is required for depletion to compensate fuel lost in producing energy, to account for the loss of reactivity due to the buildup of fission products and the change in reactivity due to changes in fuel and moderator temperatures. The control of excess reactivity is undertaken by the control rods of strongly absorbing materials having large absorption cross sections (such as boron, hafnium, or silver-indium-cadmium) and the burnable poison rods containing boron or gadolinium. In PWR, soluble boron is also employed. Control rods are inserted to shut down the reactor and are withdrawn to make the reactor critical.

A fraction of prompt neutrons (0.0065) are delayed neutrons. The presence of the delayed neutrons makes the control of reactor easier.

### Thermal

**Sources of Heat**  The energy produced by fissioning appears in several forms, each of which eventually degrades to heat. To provide for removal of this heat, the core, or active portion of the reactor, usually consists of uranium-bearing fuel elements with passages around these elements for flow of coolant. The rate of coolant flow and the temperature of the coolant must be such as to permit removal of all the heat generated in the elements without exceeding allowable materials properties.

**Hot-Spot and Hot-Channel Factors**  The problem of heat removal is complicated by the fact that heat is not generated uniformly throughout the core. The geometrical configuration of most power reactors is approximately represented by a finite cylinder of height $H$ and radius $R$. Diffusion theory predicts the following thermal neutron flux pattern for an unreflected, unrodded, and homogenized cylindrical reactor as a function of axial $Z$ and radial $r$ variables.

$$\phi(z,r) = \phi_0 \cos\frac{\pi Z}{H} J_0 \frac{2.405r}{R}$$

where $\phi_0$ is peak flux and $J_0$ is the Bessel function.

In addition, the power generation is influenced by mechanical factors such as variations in fuel-element dimensions, fuel concentrations, and flow. These nuclear and mechanical factors are generally called hot-spot factors. Thus the peak power density or heat flux in the core is greater than the average by an amount indicated by the hot-spot factor. In addition, hot-

**Table 3. Absorption ($\sigma_\alpha$) and Scattering ($\sigma_s$) Cross Sections for Thermal Neutrons**

| Element | $\sigma_\alpha$ | $\sigma_s$ | Element | $\sigma_\alpha$ | $\sigma_s$ | Element | $\sigma_\alpha$ | $\sigma_s$ |
|---|---|---|---|---|---|---|---|---|
| Aluminum | 0.23 | 1.4 | Helium | 0.007 | 0.8 | Potassium | 1.97 | 1.5 |
| Beryllium | 0.01 | 7 | Hydrogen | 0.33 | 38 | Plutonium[239] | 1.026 | 9.6 |
| Bismuth | 0.032 | 9 | Iron | 2.53 | 11 | Sodium | 0.505 | 4.0 |
| Boron | 755 | 4 | Lead | 0.17 | 11 | Thorium | 7.56 | 12.6 |
| Cadmium | 2,450 | 7 | Magnesium | 0.063 | 3.6 | Tin | 0.6 | 4 |
| Carbon | 0.0032 | 4.8 | Molybdenum | 2.7 | 7 | Titanium | 5.6 | 4 |
| Chromium | 2.9 | 3.0 | Nickel | 4.6 | 17.5 | Uranium | 7.68 | 8.3 |
| Copper | 3.7 | 7.2 | Niobium | 1.1 | 5 | Zirconium | 0.18 | 8 |
| Hafnium | 105 | 8 | Oxygen | <0.0002 | 4.2 | | | |

NOTE: Individual isotopes have markedly different cross section. For example, U[235] has total $\sigma_a = 682$ barns and fission cross section $\sigma_f = 580$ barns.

channel factors are identified which indicate how much greater the temperature rise of coolant is in the hot channel than in the average channel. As an example of how such factors are used, consider a reactor in which the axial heat-generation pattern is a cosine and the total flow passes uniformly once through all channels of the core; the maximum fuel-element surface temperature $T_{sm}$ depends on the hot-channel and hot-spot factors $F_{\Delta T}$ and $F_{\theta}$ and on the three temperatures—the coolant temperature at the core inlet $T_1$, the coolant-temperature rise in the core $\Delta T$, and the average film-temperature drop in the core $\theta_a$. ($F_{\theta}$ includes the axial peak-to-average of $\pi/2$.)

$$T_{sm} = T_1 + \tfrac{1}{2}F_{\Delta T}\,\Delta T + \tfrac{1}{2}\sqrt{F_{\Delta T}^2\,\nabla T^2 + 4F_{\theta}^2\theta_a^2}$$

**Design Criteria**  The maximum fuel-element surface temperature $T_{sm}$ is of interest because it may be limited by corrosion effects or by a desire to avoid film boiling, a condition which lowers the heat transfer from the fuel to the coolant. However, other limiting conditions also exist. In addition the internal fuel-element temperature during all anticipated reactor transients is generally limited to a maximum value (generally below the melting point of the fuel) or to a maximum average value.

Coolant-temperature rise generally is limited indirectly by fuel-temperature limits. Occasionally, however, thermal stresses whthin the fuel element may impose a direct limit. In pressurized-water reactors, the coolant temperature is limited not to exceed the saturation temperature of the fluid; thus it is more significant to speak of the maximum enthalpy rise, which is related directly to the steam quality of the fluid leaving the hot channel.

Density changes within the coolant may be limited to provide control stability in reactors that use boiling heat transfer.

Coolant velocity usually is limited to maximum value by erosion or by vibration resonances and sometimes is required to exceed a minimum value because of crud deposition. The burnout condition is determined in a given fuel-element geometry for a particular combination of fuel-element heat flux, primary coolant flow, and primary coolant enthalpy. Steady-state, local, or bulk boiling may have to be limited, even when burnout is not a danger, because of pitting of fuel-element surfaces or crud deposition on the fuel surfaces. The burnout problem is most acute during reactor transients; hence steady-state burnout generally is not limiting.

## Mechanical

The mechanical design of nuclear reactors is influenced by three factors not encountered with other apparatus: the need to control criticality, the effects of irradiation, and the high power density produced in the core.

**Control of Criticality**  The following requirements are introduced by the need to control criticality:

1. Not only must components perform reliably, but also if they fail to function as designed, the reactor will shut down. Small parts located elsewhere in the core should be so designed that if they fail they do not interfere with the action of control rods in shutting down the reactor.

2. Emergency shutdown mechanisms must be fast-acting devices to reduce the time delay in initiating shutdown.

3. During normal operation, components of the core must be rigid enough so that deflections due to mechanical or thermal stresses do not introduce unpredicted criticality variations.

**Effects of Irradiation**  Neutron and gamma irradiations cause the following effects, which must be considered in the design of reactor systems and reactor components:

1. Radiation effects on properties of materials—particularly on the brittle-fracture characteristics of the reactor vessel, on the relaxation characteristics of bolts, and on crevice corrosion

2. Heat generation brought about by attenuation of radiation within components materials—particularly insofar as such heat influences heat-removal requirements, thermal stresses, and thermal distortions of parts

3. Induced radioactivity in corrosion and wear products that could be carried by the coolant and deposited in such regions as pumps and valves

**High Power Density**  A nuclear reactor requires a large amount of heat-transfer surface for removal of the high power which can be produced per unit volume. Thus the reactor consists of many small, closely packed fuel elements. Mechanical tolerances are often appreciable fractions of the dimensions and spacing of these elements and therefore influence thermal performance. The influence of such tolerances on performance must be evaluated. Parts also must be rigid and vibration-free in the face of the high fluid-flow rates required for power removal. The high power density may also lead to large thermal stresses in fuel elements.

Furthermore, moving parts may have to operate without benefit of organic lubricants because such lubricants tend to sludge under irradiation, and if deposited on heat surfaces could interfere with removal of heat from the core.

Residual radioactivity also influences programs and facilities for refueling of the core and for periodic maintenance of components in the reactor system. In addition, personnel exposure by the above radiation requires adequate shielding.

**Shielding**  To protect reactor operating personnel against damaging biological effects of neutrons and gamma rays, shielding is required around a nuclear reactor. Neutron and gamma ray fluxes in the range of $10^{13}$ to $10^{14}$ must be attenuated to $10^3$ particles/cm$^2$/s to meet the tolerance radiation level.

To attenuate gamma rays, which interact primarily with the orbital electrons of atoms, a material with a high atomic number containing a high density of these electrons is required. Examples are lead, tungsten, depleted uranium, or concrete containing high-Z elements in the form of scrap or heavy ore.

To attenuate neutrons, they must be slowed down and then absorbed. Hydrogenous materials such as water, concrete, or polyethylene are excellent moderators. The slowed neutrons must be absorbed without producing high-energy-capture gamma rays by using boron 10. Therefore, some gamma shielding outside the neutron shield is generally required.

## Electrical

**Neutron-level instruments,** located adjacent to but outside the core or core vessel, are used for the basic control of nuclear reactors. Thermocouples, flowmeters, pressure taps, and fission chambers are used inside the core for power calibration of nuclear instruments and for determining the distribution of power and/or temperature within the core; in turn, the power distribution is used to compute heat fluxes and fuel temperatures for comparison with limiting values.

Neutron-level instruments utilize the fact that neutrons can produce ionizing particles to indicate their presence. The two most common processes used for this purpose are the $(n,\alpha)$ reaction with boron and the fission reaction. The resulting ion pair is used to produce ionization of a gas in a tube containing a central anode and outer electrode. The electric discharge produced by the ionization is used for measurement purposes. The detailed design features vary widely with the range of the instrument and the type of reactor in which it is to be used.

Different neutron-level instruments are employed during the reactor operation at various power levels, because of the wide range of sensitivity required to monitor neutron fluxes from start-up to full-power condition. In power reactors at least three instrument ranges are used with the overlap between the adjacent ranges. These are (1) source range, (2) intermediate range, and (3) power range. During the reactor start-up (source range) the $BF_3$ proportional counters are used to measure neutron flux in the range of $10^{-1}$ to $5\times10^4$ neutrons/cm²/s. The compensated ionization chambers are employed to monitor neutron flux in the intermediate power range of $2.5\times10^2$ to $2.5\times10^{10}$ neutrons/cm²/s. In the power range, where neutron flux ranges from $2.5\times10^7$ to $2.5\times10^{10}$ neutrons/cm²/s, uncompensated ionization chambers are employed.

Compensated ionization chambers are designed to count neutrons against a background of gamma radiation. Uncompensated chambers are useful for providing axial power distribution.

All the above instruments are located outside the core and monitor neutron flux, hence reactor power. The latter is also obtained from the measure of $\Delta T$ (change in temperature) across the reactor core from the temperature-measuring instruments located in the primary loop of the reactor. However, these instruments do not provide local flux peaking and local anomalous coolant-temperature situations. In-core instrumentation must be used for this purpose.

**In-core instrumentation** is employed to provide information on neutron-flux and fuel-assembly-coolant outlet temperature at a few selected locations in the reactor core. Thermocouples measure the coolant outlet temperature and thus provide the rise in local coolant temperature, a measure of the increase in local reactor power. Miniature, fixed, and movable fission chambers containing $U_3O_8$ which is 90 to 95 percent enriched $U^{235}$ constitute neutron-flux detectors. The experimental data obtained during the reactor operation from the in-core instrumentation provide the check against the reactor design parameters such as the assembly power distributions and hot-channel factors.

In addition to thermocouples and fission chambers, other in-core instruments such as pressure, displacement, and strain transducers are also desirable. None of these instruments except thermocouples have operated satisfactorily over a long period. The installation of in-core instrumentation is vital to the safe and economic operation of the reactor.

## NUCLEAR POWER-PLANT ECONOMICS

Decisions by electric utilities as to what type of generating plants to order to meet future base-load requirements are based largely, though by no means wholly, on a comparison of the costs for which the alternative types can produce electricity over their lifetimes. Other factors, which have become relatively more important in recent years, include adequacy and reliability of fuel supply, including dependence on foreign sources; environmental and safety considerations; availability of suitable sites; public acceptance as influenced particularly by environmental opposition; the costs, uncertainties, and delays involved in governmental regulation; lead times required from initial decision to get plants on-line; and plant reliability and availability.

As a practical matter in the United States, utilities seeking to add to their base-load capacity have the choice only of plants utilizing nuclear fuel or one of the fossil fuels coal, gas, or oil. Most usable hydrosites have already been developed.

The discussion of nuclear generating costs here is directed to light-water reactors of either the pressurized-water (PWR) or boiling-water (BWR) types. Although other types are in use (e.g., high temperature gas-cooled) or under development (e.g., fast breeder), the overwhelming preponderance of reactors in commercial use are, and will for some time continue to be, light-water types.

It is customary to compare generating costs for nuclear and fossil-fueled plants in three broad categories: (1) fixed costs on capital investment, (2) fuel costs, and (3) operating, maintenance, and insurance costs.

### Fixed Costs on Capital Investment

In standard accounting practice, the capital investment in a nuclear power plant is considered to be the total cost of building the plant and placing it in commercial operation. Accounting classification systems employed by both NRC and FPC provide separately for "direct" and "indirect" costs. Direct costs are associated on an item-by-item basis with land and land rights, the equipment and structures which comprise the complete power plant, and coolant and moderator materials. Equipment and structures are customarily subdivided into reactor plant and equipment, turbine plant and equipment, other electrical equipment, equipment and facilities of a general nature, and transmission plant. Indirect costs consist mainly of expenses which apply to all portions of the physical plant, including engineering and design costs, construction facilities, taxes, interest during construction, and the owner's administrative and overhead costs.

The total of these items tends to be higher for nuclear than for fossil-fueled plants largely because of (1) the need for containment shells, shielding, instrumentation, and other measures to contain radioactivity and assure safety; (2) the greater complexity, hence greater costs, or reactor equipment as compared with conventional boiler equipment; and (3) the lower turbine steam temperatures and pressures, and the lower thermal efficiencies of nuclear plants as compared with conventional plants, requiring the use of larger, hence costlier, turbines. The discrepancy is wider in warm climates, which permit fossil-fuel plants to utilize outdoor (unenclosed) construction.

An analysis performed by AEC early in 1972 indicated that, for modern central-station power plants assumed to begin operation around 1980, nuclear plants will cost 10 to 15 percent more than coal-fired plants; which, in turn, will cost roughly 10 percent more than oil-fired plants and one-third more than gas-fired plants.

Mitigating factors which prevent the construction-cost discrepancy between nuclear and coal plants, in particular, from being still larger are (1) the lower cost of fuel-handling equipment in nuclear plants and (2) the need in coal plants, but not in nuclear plants, for smoke-control and ash-removal equipment.

Unit construction costs (expressed in dollars per kilowatt) for both nuclear and fossil-fueled plants are lower at any given time as plants increase in size because of various economies of scale. The decreases are sharper in the case of nuclear plants, largely because of the prominence of certain minimum and relatively fixed costs of assuring safety. The existence of these inescapable costs places small nuclear plants at a particularly great economic disadvantage.

Geographical location can have a significant effect on capital costs. An analysis made in 1969 for the AEC by Battelle Northwest identified the following as potentially significant site-dependent cost variables: local labor rates, local costs of construction materials, shipping charges for major equipment items, site-preparation costs, state sales tax on materials and equipment, local and state property taxes during construction, land and land-right costs, transmission equipment, and special facilities not required at all sites, such as long circulating water lines at some coastal sites, or cooling towers or ponds at others. The incremental capital cost of cooling towers is currently (1973) estimated at from $10 to $20 per kilowatt.

Adding a nuclear unit to a generating station which already has one or more nuclear units is less costly than constructing a first nuclear unit of otherwise identical characteristics. The principal savings in multiple-unit plants result from use of a developed site; reduced engineering and design efforts; and use of joint facilities and equipment such as cooling-water intake and discharge facilities, control rooms, warehouses, shops, offices, roads, fuel-handling and storage facilities, and temporary construction facilities.

The unit capital costs of nuclear plants often can be reduced substantially over their lifetimes by using improved fuel cores to achieve outputs higher than the original rating, which is usually conservative. To take advantage of this opportunity, allowance for higher output must be made at the outset in the capacity of turbine generators. Conversely, should it be necessary (as seemed possible early in 1973) to derate plants to take account of safety uncertainties such as the effectiveness of emergency core-cooling systems, the effect would be to increase unit capital costs.

Total construction costs of nuclear plants have risen rapidly since about 1963, when it was believed that a large light-water plant could be built for $125 per kilowatt or less. Currently (1977) plants are being estimated to cost $800 per kilowatt or more. The principal factors making for such increases in cost have been inflation, stretchouts of plant schedules (adding to interest during construction), lower labor productivity, increases in the scope of projects to add safety and environmental features, and more stringent quality control.

Some of these factors may continue to push unit construction costs upward in the future. Counterbalancing factors tending toward a moderation in unit costs include continued increases in plant size, the likelihood of greater standardization of systems and components, a larger sales volume over which to spread manufacturing overhead costs, and improved field assembly and construction methods. One should take note of the fact that the historical tendency in the industry has been to underestimate construction costs by a wide margin.

The fixed-cost component of generating costs (mills per kilowatthour) can be obtained from construction costs (dollars per kilowatt) by applying an annual fixed-charge rate and an annual plant-capacity factor.

Annual fixed charges cover all costs that are in direct proportion to the initial capital investment and are independent of the extent to which the facilities represented by this investment are used. A simple procedure for evaluation of a plant, which approximates the levelized annual charges obtained through utility accounting procedures, is to use constant annual fixed charges, i.e., to apply a constant fixed-charge rate each year to the initial investment.

The fixed-charge rate for a utility will vary because of a number of factors including amount of state and local taxes, plant life and accounting method for depreciation purposes, and debt-equity composition of the utility's financial structure. A range of fixed-charge rates for depreciating capital of an investor-owned utility illustrates the components and relative importance of these items:

| Component | Fixed-charge range, % |
|---|---|
| Return on investment | 7.0–9.0 |
| Depreciation | 1.0–1.3 |
| Interim replacements of short-lived assets | 0.3–0.5 |
| Property insurance | 0.3–0.4 |
| Federal income taxes | 2.0–2.8 |
| State and local taxes | 2.9–3.5 |
| Total | 13.5–17.5 |

A lower fixed charge rate would be applied to the nondepreciating capital (land and land rights and working capital) because depreciation, interim replacements, and property insurance are eliminated and associated adjustments are made in taxes. Not included in annual fixed charges are relatively constant fixed costs, such as nuclear liability insurance or plant staffing expense, which are only indirectly related to plant investment.

The annual capacity factor is calculated by dividing the kilowatthours generated by the plant in a year by the kilowatthours that would have been generated had the plant operated at full capacity throughout all the hours in the year (8,760 except in leap years).

The conversion from capital to generating costs may be illustrated as follows: Assume a plant of 1,000,000 kW capacity costing $400 million, with annual fixed changes of 14 percent and operating at 80 percent capacity factor. The construction-cost contribution to generating costs would then be 8 mills/kWh, calculated as follows:

$$\frac{\$400,000,000 \times 14\%}{1,000,000 \text{ kW} \times 8,760 \text{ h} \times 80\%} = \frac{\$56,000,000}{7,008,000,000 \text{ kWh}} = \frac{8 \text{ mills}}{\text{kWh}}$$

## Fuel Costs

The economic advantage of nuclear power plants lies essentially in fuel costs which are substantially lower than those of fossil-fueled plants. An AEC estimate as of mid-1972 indicated that nuclear fuel-cycle costs for a 1,000-MW power plant were in the range of 17 to 18 cents per million Btu. Rapid increases in recent years have placed average fossil-fuel costs at two to three times this level.

The fuel-cost advantage is implicit in the compactness of

nuclear fuel: 685 lb of uranium, if fully consumed, contains the energy equivalent of 1.7 million tons of coal, 7.2 million barrels of oil, or 32 billion ft³ of natural gas. It has not yet proved possible in any water-cooled reactor to consume more than a small fraction of the uranium used in nuclear fuel; yet this fraction is high enough to yield the cost advantage noted above.

Determining nuclear-fuel costs is a far more complicated matter than determining fossil-fuel costs because of the need to take account of a complex sequence of events which begins long before the uranium fuel is inserted in the reactor and ends long after the spent fuel has been removed from the reactor. The entire sequence, which is called the nuclear fuel cycle, includes mining and milling of uranium, refining of uranium and conversion to uranium hexafluoride, enrichment of the uranium in the isotope uranium 235, conversion of enriched uranium to fuel material, fabrication of reactor-fuel elements, use of fuel elements in nuclear power plants, recovery and marketing of by-product plutonium, chemical reprocessing of spent fuel to obtain reusable fuel material, and disposal of radioactive wastes.

It is difficult to predict accurately the future course of nuclear-fuel costs because of uncertainties related to government policies, ore reserves, changes in technology, etc. There are indications, nevertheless, that the discrepancy between nuclear and fossil-fuel costs may continue to widen. (This statement applies to light-water reactors. The fuel costs for fast breeder reactors, when developed, are expected to be substantially less than those of light-water reactors, mainly because of the credit for the substantial quantities of bred plutonium.)

Some principal items of nuclear-fuel cost and possibilities of change include the following:

**Fabrication**  The cost of fuel-element fabrication is taken to include all the steps necessary to change a starting material (UF⁶ in the case of enriched uranium elements) into a usable element. These steps involve chemical conversion to a powder or metal, metallurgical and mechanical processing to form and clad the elements, inspection, testing, and scrap recovery. Fabrication costs are expected to continue declining as greater quantities are handled, processing becomes more efficient, and fuel design is improved.

**Recycling of Bred Plutonium**  Uranium-fueled light-water reactors in normal operation produce a certain amount of fissionable plutonium 239 through capture of neutrons in the nonfissionable isotope uranium 238. The value of this bred plutonium is deducted as a credit from the total of net fuel costs. Development now underway to recycle the plutonium in light-water reactors and to use it as an initial fuel charge in fast breeder reactors should increase its future value and thereby reduce overall fuel costs.

**Shipping Charge**  The costs of transporting irradiated-fuel elements from a reactor site to a chemical-processing plant include charges for freight, shipping casks, handling, and insurance. The shape of the fuel element (having an effect on the size of cask required), the irradiation level, the cooling time required, the shipping distance, the route and type of transport, and the regulatory requirements are all factors which affect the cost.

**Enrichment**  Enrichment in the fissionable isotope uranium 235 is accomplished at present (1973) only in three ERDA-owned plants. This is the only part of the nuclear power industry not yet in private hands. ERDA is making technology available and otherwise encouraging private entry into this field. ERDA enrichment charges in 1972 were $32 per kilogram of separative work. There is a likelihood that these costs may increase in step with the price of electricity, which is the major ingredient in enrichment costs.

**Burnup**  The amount of heat that can be obtained from a given weight of nuclear fuel before it is discharged from the reactor is known as the fuel's burnup. Higher burnups are expected as a result of technological improvements. This would have the consequences of reducing the amount of fuel that must be fabricated, reprocessed, and shipped per unit of electrical output. Cost reductions should be achieved as a result, but they may be limited by the fact that fuel elements capable of achieving the higher burnups may be more expensive to fabricate and process and may require higher enrichment.

**Chemical Reprocessing**  Until 1966 all chemical reprocessing (dissolving irradiated fuel, followed by separation and storing of fission products and production of purified solutions of uranium and plutonium nitrates) was performed at AEC sites. Since that time private facilities able to sustain the current requirements for the nuclear-power industry have been built and placed in operation, and the ERDA facilities are being utilized as backup and for research and development. Lower reprocessing charges may result from higher volume and improvements in technology.

### Operating and Maintenance Costs

The main cost items included in this category are station labor, training of replacement personnel, expendable materials, and supplies other than fuel, maintenance operations, and insurance.

The novelty and complexity of nuclear plants, a conservative approach to health and safety problems, and the need for specialized knowledge are all factors requiring a more skilled staff in nuclear plants and a longer period of training. Problems of coal and ash handling and maintenance of tubes and related boiler components require a larger working force in coal plants. While insurance costs have been higher for nuclear plants, it is expected that a continuance of safe-operating experience will eventually equalize this item.

Numbers of operating personnel per generating unit are less for a plant having two or more nuclear units than for a single-unit plant. Increasing the size of a nuclear unit does not materially affect the number of people required to operate it. For these two reasons, one can expect declining personnel requirements per kilowatt of capacity with future increases in nuclear power. As a result of the high inflation rates experienced in the mid-1970s, all these costs are now nearly doubled for 1980–82 operational dates.

### Total Generating Costs

In mid-1972, AEC estimated that a 1,000-MWe light-water reactor ordered for operation in 1979 would produce energy for a cost of 11.9 mills/kWh, including provision for cooling towers and escalation during construction. Of this total, 8.9 mills was ascribed to capital costs ($390 per kW), 2.0 mills to fuel-cycle costs, and 1.0 mills to operating and maintenance costs.

Although many uncertainties becloud cost estimates for both nuclear and fossil-fueled plants, comparisons at this time

(1977) tend to favor nuclear power in many parts of the United States on either economic or noneconomic bases, or both. The possibility of using gas has been reduced by supply shortages. Use of oil presents a problem, since domestic supplies of low-sulfur oil are diminishing and dependence on foreign sources makes for an uncertain supply and the possibility of sharp price increases. While coal is plentiful, most of what is available, particularly in the Eastern United States, has a sulfur content greater than is allowed by state air-polution regulations. Coal-treatment technology is not yet adequate to cope economically with this problem. Thus many utilities feel obliged to turn in part to nuclear power notwithstanding its own considerable technical environmental and regulatory uncertainties.

Nuclear power's share of the nation's central-station electric-generating capacity is therefore increasing rapidly. Whereas in 1970 nuclear-power plants comprised less than 2 percent of this total capacity, Federal Power Commission estimates for the year 1990 are for an increase of the nuclear share to about 37 percent. Orders for nuclear plants placed during 1973 confirmed this trend, amounting to approximately 42,000 MW.

### Effect on Fuel Resources

The emergence of nuclear power as a viable energy source comes at a time when additional environmentally acceptable sources of energy are sorely needed, both in the United States and in other countries, to meet continued rapid increases in demand. As indicated above, domestic production of both natural gas and oil appears to be approaching its peak and may decline in the future. Dependence on foreign sources entails risks with regard to dependability of supply and the U.S. balance of payments. New technology is needed in order to be able to mine and use this country's extensive coal reserves in an environmentally and economically acceptable manner.

Uranium supplies are limited and, if utilized only in light-water reactors, would rapidly diminish. Breeder reactors, now under active development by the AEC, can utilize most (60 percent or more) of our uranium (and/or thorium) resources rather than just a minute percentage. M. K. Hubbert (U.S. Geological Survey) has estimated that the energy potentially available from the fissioning of uranium and thorium by means of breeders is "at least a few orders of magnitude greater than that from all the fossil fuels combined." Commercial operation of breeder reactors is forecast for the 1980s.

A practically unlimited source of energy is available from the deuterium in the world's oceans should development work now underway in several countries lead to production of useful power from controlled fusion reactions. The technical problems are formidable, however, and the harnessing of fusion as an energy source is not expected until perhaps the end of the century.

### NUCLEAR-POWER-PLANT SAFETY

A reactor cannot be made to explode like a bomb. The fuel used is not of sufficient purity, nor is it arranged in such a way that generation of a large explosive force is possible. On the contrary, reactors are so designed that if a runaway reaction should occur, the fuel would tend to disperse and the nuclear reaction would come to a halt.

Reactors do present a large potential hazard because of the generation within the fuel of highly radioactive fission products. The essential safety challenge is to assure that these fission products remain confined at all times.

Through 1972, U.S. civilian nuclear-power plants had accumulated more than 120 reactor years of operation without an accident affecting the public. This record was achieved primarily with reactors of less power and complexity than those now being ordered by utilities.

Confidence that newer reactors can continue the impressive safety record is based on several mutually reinforcing aspects of reactor design and operation. Maximum advantage is taken of natural laws which provide inherent safety features. Reactors designed for central-station service have a self-correcting tendency (the so-called "Doppler effect") to slow down their rate of fission when a nuclear excursion begins. Further, as fuel and moderator water become hotter, their density decreases, also lowering reactivity.

Reactors are equipped with neutron-absorbing control rods which, by movement in and out of the reactor core, change the rate at which neutrons are fed to the fuel and thus slow down or speed up the chain reaction. If an abnormal condition develops, sensitive instrument systems can cause the rods to be plunged instantaneously into the core, capturing so many neutrons that the chain reaction stops.

Electronic systems which can activate the control rods are tied into instruments that monitor various aspects of reactor operation for abnormalities such as leaks, excessive temperatures, and unusual pressures. These safety systems are installed to prevent minor mishaps from escalating into major accidents.

The various natural and engineered protective features, along with the high quality standards which regulatory authorities require in the design, construction, and operation of reactors make a major accident extremely unlikely.

Nevertheless, such accidents are postulated to occur, and various additional safety features are installed to mitigate their consequences. These include thick containment shells to prevent the escape of radioactivity and emergency core-cooling systems to prevail over the most serious accident which can reasonably be hypothesized. This is an accident in which a main coolant pipe breaks, causing all the coolant fluid to be lost.

Further assurance of nuclear-power-plant safety is provided by the regulatory requirements summarized below.

### NUCLEAR-POWER-PLANT LICENSING

It is federal law, stemming from the Atomic Energy Act of 1954, as amended, that no nuclear power plant may be constructed without a construction permit or operated without an operating license from the NRC. (The only exception is NRC-owned plants, for which parallel authorization procedures are in effect.)

The applicable regulations appear in Title 10, Code of Federal Regulations. Steps involved in the review of license applications are summarized below:

**Site Review** Informal discussions with NRC staff about the proposed site(s) generally precede filing of license applications. Factors considered include the population density, seismology, meteorology, geology, and hydrology of the area. Location remote from densely populated areas was required for the first generation of nuclear power plants. Subsequent practice has permitted location closer to population centers if compensated for by proper engineered safety features.

**Preliminary Safety-Analysis Report** An application for a construction permit must be accompanied by a Preliminary Safety Analysis Report, a formidable document which may exceed 1,000 pages in length. It describes the plant's design and safety features together with measures which would be taken to cope with credible accidents.

**NRC and ACRS Review** Following a review of the Safety Analysis Report by the NRC staff, it is forwarded, with the staff's evaluation, to the Advisory Committee on Reactor Safeguards (ACRS), an independent group of experts established by Congress. When it has reached a conclusion on whether the proposed plant can be built and operated safely, the ACRS reports its views in a letter to the NRC which is made public.

**Environmental Review** Also required, pursuant to the National Environmental Policy Act of 1969, is an Environmental Report in which the applicant discusses the environmental impact of the proposed plant. The report includes a cost-benefit analysis which balances the environmental costs of the proposed plant (including effects on air quality, water quality, land use, historical values, and aesthetics), and of various less harmful alternatives against the benefits of the proposed and alternative plants.

The NRC staff prepares a detailed environmental statement based on its evaluation of the applicant's report and forwards both documents to appropriate federal, state, and local agencies for comment. A final NRC environmental statement, incorporating NRC's appraisal of environmental impact, along with comments received and responses thereto, is then made public.

**Public Hearing** Safety and environmental aspects are next brought together for consideration in a public hearing conducted by a three-member Atomic Safety and Licensing Board (ASLB). The hearing is held near the proposed reactor site, and public participation by persons whose interests may be affected by the project is provided for. If warranted by the ASLB's decision, the NRC may then issue a construction permit.

**Final Safety-Analysis Report** When final design is completed, the applicant submits a Final Safety Analysis Report in support of an application for an operating license. Once again the NRC staff and ACRS make independent reviews and the ACRS reports its opinion to the NRC. A public hearing is not mandatory at this stage, but the NRC may order one if there is substantial public interest or an important and difficult safety problem.

**Operating License** The first operating license for any plant may be provisional for an initial period of operation, usually for 5 years, at the end of which time a further review is made to determine whether to issue a full-term license for up to 40 years of operation. Each operating license contains technical specifications which must be met to minimize danger to public safety and the environment.

**Generic Review** In an effort to achieve the benefits of standardization (including among others, greater assurance of safety, reduced costs through quantity production, and more rapid licensing review) the NRC adopted in 1973 a plan for generic licensing. Under this plan designs for components, systems, and complete plants can be given licensing approval, independent of any specific site but for a specified range of site

conditions, following review and approval by the NRC Regulatory staff, by the ACRS, and an Atomic Safety and Licensing Board. On the basis of this approval, items can then be incorporated in plants at specific sites meeting the reference conditions without further detailed licensing review. The NRC will also issue licenses to manufacture the items without reference to a specific installation. In taking this action, the NRC announced that it would not, until further notice, accept plants above 3,800 MWt (about 1,300 MWe) for licensing review. The upward spiraling of plant size with each succeeding generation of plants had been one of the factors impeding standardization.

**Continuing Review** A licensed reactor is subject to NRC surveillance throughout its operating life to assure that AEC regulations and license conditions are being observed. Significant changes in design are evaluated and authorized by the NRC. Periodic inspections are made. Significant incidents are investigated. Should the NRC find a safety or environmental hazard, it has ample statutory authority to issue corrective orders, including shutting down or even dismantling the reactor.

**Antitrust Review** Before issuing an operating license, the NRC must make a finding that operation of the plant will not be in violation of antitrust laws.

**Operator Licenses** Persons who operate the controls of a power reactor must be individually licensed by the NRC following written examinations of their knowledge of the facility and its operating procedures.

Licensing procedures have been a source of considerable cost and delay, especially since the Calvert Cliffs decision of the U.S. Court of Appeals for the District of Columbia (Aug. 23, 1971), which widened NRC's environmental responsibilities.

NRC has attempted to ameliorate the situation by augmenting its regulatory staff and streamlining its licensing procedures. Steps taken have included prescribing standard contents and format for license applications, eliminating delay tactics in hearings through modified rules, and encouraging the development of engineering standards governing reactor components and systems. NRC's current (1973) goals are to complete action on a construction-permit application within 12 months, and on an operating-license application prior to the fuel-loading date.

Nuclear plants are subject also to regulation by state and local authorities, primarily with respect to land-use, water-quality, and air-quality matters.

## OTHER POWER APPLICATIONS

**Nuclear Ships** The earliest and most successful of the efforts to develop nuclear power for ship propulsion were those by the U.S. AEC and the U.S. Navy. These initial efforts toward this end predated and contributed to the development of large electric-power generators. The U.S. Navy launched the *USN Nautilus* in 1954 and followed it with about 100 other operational nuclear submarines and four operational surface vessels. Additional submarines and surface vessels have been committed by the U.S. Navy. Nuclear submarines are also being operated by the British, Russian, and French navies.

Although a ship can be propelled by diesel engines, steam engines, gas turbines, and other devices, the steam turbine has emerged as the principal modern marine propulsion unit. When a nuclear reactor of a type developed for generating stations is used to produce the steam, the combustion process is eliminated. Such a propulsion plant is especially advantageous for a submarine, since it is then completely free from dependence on the earth's atmosphere. The U.S. Navy's nuclear ships and submarines all employ the pressurized-water type of nuclear reactor.

On commercial ships, the advantages of nuclear power are related to higher-speed operation over long-run routes. Experience indicates that an increase in speed usually produces substantial increases in cargoes, thereby increasing a ship's load factor. In conventional ships, the power required for higher speed is increased by the third power of the speed ratio. This means larger engines and greater fuel-storage capacity. On the other hand, a nuclear ship can operate continuously at the maximum level of its hull form without penalty in pay load. Such advantages are especially desirable for icebreakers, large bulk-cargo vessels, and high-speed long-distance passenger service. By 1972, the earlier U.S. effort, which resulted in the *N. S. Savannah* and the West German effort, which resulted in the *N.S. Otto Hahn*, had not yet produced an acceptable basis for further commitments of commercial nuclear ships. The U.S.S.R. icebreaker *Lenin* continues in service.

**Other Vehicles** The use of nuclear power in airplanes, automobiles, and railroad trains has been deterred by the requirement for heavy shielding to protect against radioactivity and collisions.

**Space Travel** In space and other difficult-to-reach regions, where a compact, dependable source of energy is required, nuclear power holds much promise. Under the SNAP (Systems for Nuclear Auxiliary Power) and subsequent primary power-system programs, much research and development has been done to develop radioisotope heat-energy capsules, thermoelectric generators, and compact reactors. By the end of 1972, sixteen nuclear-power systems had been launched.

**Isotopic power generators** take advantage of the fact that radiations of an appreciable amount of radioactive material sealed in a container are absorbed by the container. This action creates high temperatures, which are utilized to create an electric current by thermocouples or by direct-energy conversion—thermionic emission.

Plutonium[238] and strontium[90] were the radioisotopes first used. Eight radioisotopes are now being utilized in this work.

Technology programs exist for development of advanced hydride (thermoelectric and Brayton cycle), advanced liquid-metal-cooled (Brayton and potassium Rankine), and thermionic reactor systems.

## FUSION NUCLEAR ENERGY

In order to develop practical nuclear-fusion reactors for the production of electrical and thermal energy in large central-station installations, significant developments from the present state of the art must be obtained.

There are a number of possible fusion fuel cycles, the most promising of which are shown in Table 6.

Because of its high energy gain and its low threshold temperature, the DT reaction is considered most attractive for first-generation fusion reactors. The inherent features of the reaction and resulting ionized gas (plasma) will determine many of the basic characteristics of DT fusion reactors.

The basic problem of containment of plasma at practical temperatures and densities obviously must involve confinement systems which depend on magnetic forces or other forms of nonmaterial containment (inertial, for example). Four general approaches were receiving attention worldwide in 1973:

1. Steady-state toroidal systems of the tokamak principle. In these systems, the plasma is retained by a magnetic field in toroidal form. A low-plasma-density system with containment times in the millisecond region.

2. Magnetic-mirror systems. Here the plasma is retained in a double-mirror configuration cyclically. A low-plasma-density system also.

3. The pinch system wherein a cyclically increasing magnetic field forces the plasma together. If in the radial direction, this is called a theta pinch; if axially, a Z pinch. A high-plasma-density system with containment times in the microsecond region.

4. Laser-fusion systems wherein small pellets of fusionable material are triggered by laser-beam compression successively. An extremely high plasma density system with containment times in the nano- to picosecond region.

The DT plasma temperature should be over 10 keV. The combination of density and containment time should achieve a product of approximately $10^{14}$. The simultaneous achievement of the three required parameters of plasma temperature, plasma density, and containment time is required for the establishment of scientific feasibility. The scientific, technical, and size limitations of present-day fusion experiments preclude any of them from simultaneously achieving all three of the plasma parameters (temperature, density, and confinement time) required for successful fusion reactor. This

**Table 4. Radioisotopic Materials**

| Radioisotope | Half-life, years | Typical power density, watts per cc* | Radiation present† | Shielding required |
|---|---|---|---|---|
| Strontium[90] | 28 | 1.4 | Beta and bremsstrahlung | Heavy |
| Cesium[137] | 30 | 0.21 | Beta and gamma | Heavy |
| Cerium[144] | 0.78 | 24.5 | Beta and gamma | Heavy |
| Promethium[147] | 2.7 | 1.8 | Beta | Minor |
| Polonium[210] | 0.38 | 1,210 | Alpha | Minor |
| Plutonium[238] | 89 | 3.5 | Alpha | Minor |
| Curium[242] | 0.45 | 1,150 | Alpha | Minor |
| Curium[244] | 18 | 26.4 | Alpha and neutron | Moderate |

Source: AEC TID-20079.
*Reduced below theoretical maxima in order to account for expected isotopic impurities.
†Of dominant importance in heat generation and for consideration of radiation shielding.

**Table 5. Summary of Space Nuclear-Power Systems Launched by the United States (1960–1972)**

| System | Mission | Launch date | Fate |
|---|---|---|---|
| SNAP-3A | TRANSIT-4A | 6/29/61 | Successfully achieved > 1,000-year earth orbit |
| SNAP-3A | TRANSIT-4B | 11/15/61 | Successfully achieved > 1,000-year earth orbit |
| SNAP-9A | TRANSIT-5BN-1 | 9/28/63 | Successfully achieved > 1,000-year earth orbit |
| SNAP-9A | TRANSIT-5BN-2 | 12/5/63 | Successfully achieved > 1,000-year earth orbit |
| SNAP-9A | TRANSIT-5BN-3 | 4/21/64 | Failed to achieve orbit |
| SNAP-10A | SNAPSHOT | 4/3/65 | Successfully achieved ≈ 2,800-year earth orbit |
| SNAP-19B2 | NIMBUS-B-1 | 5/18/68 | Failed to achieve orbit, retrieved from ocean floor |
| SNAP-19B3 | NIMBUS-III | 4/14/69 | Successfully achieved ≈ 3,000-year earth orbit |
| SNAP-27 | APOLLO-12 | 11/14/69 | Successfully placed on lunar surface |
| SNAP-27 | APOLLO-13 | 4/11/70 | Failed to reach moon, returned to Pacific Ocean |
| SNAP-27 | APOLLO-14 | 1/31/71 | Successfully placed on lunar surface |
| SNAP-27 | APOLLO-15 | 7/26/71 | Successfully placed on lunar surface |
| SNAP-27 | APOLLO-16 | 4/16/72 | Successfully placed on lunar surface |
| SNAP-19 | PIONEER | 3/2/72 | Spacecraft successfully operated on fly by Jupiter |
| TRANSIT-RTG | TRANSIT | 9/2/72 | Spacecraft successfully operating in earth orbit |
| SNAP-27 | APOLLO-17 | 12/7/72 | Successfully placed on lunar surface |

SOURCE: President's Annual Report to Congress on Space and Aeronautics (1972).

**Table 6**

| Reactor equation | Approx threshold plasma temp* | Approx avg energy gain per fusion |
|---|---|---|
| $D + T \rightarrow {}^4He\ (3.5\,MeV) + n(14.1\ MeV)$ | 10 keV | 1,800 |
| $D + D \begin{cases} {}^3He\ (0.82\ MeV) + n(2.45\ MeV) \\ T\ (1.01\ MeV) + p(3.02\ MeV) \end{cases}$ | 50 keV | 70 |
| $D + {}^3He \rightarrow {}^4He\ (3.6\ MeV) + p(14.7\ MeV)$ | 100 keV | 180 |

*1 keV = 11,000,000 K approximately.

achievement, which would demonstrate scientific feasibility, will require larger, more complex facilities than presently (1973) available unless some breakthrough occurs.

Once scientific feasibility is established, the practical removal of energy for the plasma into a usable form in a cyclical or steady-state output is a formidable obstacle in itself. The commercial feasibility would then require economic solution.

From the present state of research most estimates indicate that the first commercial fusion power plant would not be available before the year A.D. 2000, with its principal impact on energy production occurring thereafter.

# HYDRAULIC TURBINES

## by W. G. Whippen

REFERENCES: Daugherty, "Hydraulic Turbines," McGraw-Hill. Daugherty and Ingersoll, "Fluid Mechanics," McGraw-Hill. Barrows, "Water Power Engineering," McGraw-Hill. Creager and Justin, "Hydroelectric Handbook," Wiley.

## GENERAL

### Notation

$B$ = width of distributor or height of wicket gates, in (mm)
$D$ = diameter of runner, in (mm)
$D_1$ = diameter of runner, at centerline of guide case, in (mm)
$D_{th}$ = throat diameter of runner, in (mm)
$D_r$ = relative diameter of runner, in (mm) (= entrance diameter for low-speed Francis turbines; throat diameter for high-speed Francis turbines; bore of throat ring at centerline of blade for propeller turbines)
$D_d$ = diameter top of draft tube, in (mm)
$D_p$ = pitch diameter of impulse-turbine runner, in (mm)
$d$ = jet diameter of impulse turbine, in (mm)
$e$ = overall efficiency of turbine = $e_h \times e_m$
$e_h$ = hydraulic efficiency (including draft tube)
$e_m$ = mechanical efficiency
$g$ = acceleration of gravity, ft/s² (m/s²)
$H$ = net effective head, ft (m)
$h$ = head change due to load change, ft (m)
$n$ = r/min
$n_1$ = r/min at 1 ft (m) head = $n/\sqrt{H}$
$n_s$ = specific speed = $n\sqrt{P}/H^{5/4}$

$P$ = horsepower = 0.746 kW

$P_1$ = horsepower (kW) at 1 ft (m) head = $P/H^{3/2}$

$Q$ = discharge, ft$^3$/s (m$^3$/s)

$Q_1$ = discharge at 1 ft (m) head = $Q/\sqrt{H}$, ft$^3$/s (m$^3$/s)

$t$ = thickness, in (mm), or time, s

$u$ = circumferential velocity of a point on runner, ft/s (m/s)

$V$ = absolute velocity of water, ft/s (m/s)

$V_u$ = tangential components of absolute velocity = $V \cos \alpha$

$V_r$ = component of $V$ in radial plane

$v$ = velocity of water relative to runner, ft/s (m/s)

$w$ = weight of water lb/ft$^3$(kg/m$^3$)

$z$ = number of runner buckets (blades for propeller turbines)

$\alpha$ = angle between $V$ and $u$ (measured between positive directions)

$\beta$ = angle between $v$ and $u$ (measured between positive directions)

$\theta$ = angle of water deflection relative to bucket

$\phi$ = peripheral coefficient = $\pi Dn/720\sqrt{2gH}$ $(\pi Dn/60\sqrt{2gH})$

$\sigma$ = cavitation coefficient

**Fundamental Formulas**  The theoretical power $P_T$ of a hydraulic turbine can be expressed as

$$P_T = HQw/550 \; (HQw/100)$$

The actual power of a hydraulic turbine is the theoretical power multiplied by the turbine efficiency $e$,

$$P = eP_T = eHQw/550 \; (eHQw/100)$$

The laws of proportionality for homologous turbines are:

| For constant runner diam | For constant head | For variable diam and head |
|---|---|---|
| $P \propto H^{3/2}$ | $P \propto D^2$ | $P \propto D^2 H^{3/2}$ |
| $n \propto H^{1/2}$ | $n \propto 1/D$ | $n \propto H^{1/2}/D$ |
| $Q \propto H^{1/2}$ | $Q \propto D^2$ | $Q \propto D^2 H^{1/2}$ |

**Nomenclature**  The nomenclature used throughout this section is based on NEMA's standards publication HT1-1957, "Hydraulic Turbines, Governors and Accessory Equipment," which also contains definitions.

**Types of Turbines**  Three characteristic types of hydraulic turbines are now in general use: the **impulse** type (Fig. 12); the **Francis reaction** type (Fig. 1); and the **propeller reaction** type (Fig. 3). The propeller type may be further divided into fixed and adjustable blade types. All three types have in common a stationary guide case (or nozzle in the case of the impulse type) in which the static head is transformed partly, or wholly, into velocity, and a revolving part, the runner.

In the guide case of the impulse turbine, the static head is completely transformed into velocity, so that air surrounds both the jet issuing from the nozzle and the runner.

In the guide case of the reaction types, the static head is only partly transformed into velocity, leaving an overpressure between the guide case and the runner. This overpressure causes an acceleration of the relative velocity of the water passing through the runner, the discharge area of which is smaller than the entrance area. Except when operating vented at low loads, the water passages are completely filled with water from the intake to the end of the draft tube.

The impulse type is of relatively low (specific) speed, suitable for the higher heads; the Francis type is of relatively medium speed, suitable for medium heads; while the propeller type is of relatively high speed, suitable for low heads.

**Selection of Turbine Type and Casing**  Impulse turbines receive their water supply directly from a pipeline. Francis and propeller reaction types are set in a case of concrete or metal. While the limits of head to which the impulse and reaction types are adopted may be fairly well defined in practice, as roughly outlined in Table 1, there is no definite line where the application of one type ends and the other begins. The limits given in the table should not be taken as representing absolute points beyond which the respective types are not suitable, but rather as an indication of general practice. The choice between impulse and reaction types depends upon the size of the unit as well as upon the head and other considerations.

**Specific Speed**  The common basis of comparison between turbine runners of different types and between runners of the same type but different design and characteristics is termed *specific speed* $n_s$. This is the constant relationship between the speed of a runner at the point of highest efficiency and the maximum power output at this speed regardless of size. However, since both power and speed vary with head, specific speed is defined as the relationship between the speed $n_1$ and power $P_1$ at 1 ft (m) head. Subscript 1 denotes that the value is

**Table 1. General Arrangements of Turbine Installations and Usual Head Limits Employed**

| Type | Setting | Construction | No. of runners | Usual head limits for direct-connected units | |
|---|---|---|---|---|---|
| | | | | ft | m |
| Reaction turbines, 6 to 1,000 ft (2 to 300 m) head | Axial flow | Vertical or horizontal or slanted | 1 | 6–60 | 2–20 |
| | Encased | Concrete vertical | 1 | 15–130 | 5–40 |
| | | Cast or welded plate steel. Vertical or horizontal | 1 or 2 | 30–1,600 | 10–500 |
| Impulse wheels 500 to 6,000 ft (150 to 1,800 m) head | Encased | Horizontal or vertical | 1–6 | 500–6,000 | 150–1,800 |

reduced by the similarity law to a 1 ft (m) head basis. Since $n \propto 1/D$ and $P \propto D^2$, the product $n_1\sqrt{P_1}$ remains a constant regardless of the size of the runner and is designated the specific speed of the runner. The term *specific speed* for this relationship stems from the fact that $n_1\sqrt{P_1}$ also is the value of the speed in r/min at best efficiency which the runner would have if operated under 1 ft (m) head, the runner being of such size as to develop 1 hp (kW) ($P_1 = 1$).

Since $n \propto \sqrt{H}$ and $P \propto H^{3/2}$, the specific speed of any runner operating under head $H$ will be $n_s = n\sqrt{P}/H^{5/4}$, where $n$ is the best efficiency speed and $P$ the maximum power output at this speed, all at head $H$.

**Selecting the Speed** Hydraulic turbines are usually connected to ac generators. The turbine speed must agree with one of the synchronous speeds required for the system frequency. The prevailing frequency for most systems in America is 60 Hz. Synchronous speeds are determined by the formula $n = 120 \times$ frequency/number of poles in generator. The number of poles must be even.

The speed should be as high as practicable, as the higher the speed the less expensive will be the turbine and generator and the more efficient will be the generator.

A convenient way of determining the highest practicable speed is by the relation of the specific speed to the head. For Francis turbines this may be taken as $n_s = 800/\sqrt{H}(1,900/\sqrt{H})$; for propeller-type turbines as $n_s = 1,000/\sqrt{H}(2,100/\sqrt{H})$. The adoption of these values of specific speed is predicated on the use of a reasonable setting of the unit with reference to tailwater level, i.e., selection of proper cavitation coefficient.

**Number of Units** From the standpoint of reducing the number of auxiliaries and the amount of associated equipment and also reducing initial and maintenance costs for the entire plant, the number of units should be kept to a minimum. Also the larger the unit, the higher the efficiency and generally at a lower cost. However, other considerations, such as flexibility of operation, higher efficiency operation during low-load demands, and minimum loss of capacity during shutdown for repair or maintenance, might dictate the use of multiple units where one unit would be feasible in terms of physical size. For some projects, the physical size of the unit has been limited to the maximum size runner that could be shipped in one piece; this is largely due to the extra manufacturing costs involved in furnishing split runners. However, since split runners present no serious mechanical difficulties, the tendency in recent years has been to disregard this limitation. Another limitation on size is the availability of machine tools for machining large turbine parts.

## REACTION TURBINES

**Francis** Figure 1 shows a Francis type inward flow reaction turbine for medium head. The runner consists of a relatively large number of shrouded buckets. Movable wicket gates with axes parallel to the turbine shaft control the flow. This type of turbine is normally used for heads ranging from 75 to 1,600 ft (25 to 500 m). Specific speeds vary from 15 to 100 (50 to 400). Low specific speed runners have narrow inlet passages. For high specific speed runners, the width of the runner inlet is increased (Fig. 2).

**Propeller Turbines, Fixed and Adjustable Blade** The propeller turbine has a runner which is normally provided with from 3 to 10 unshrouded blades, either fixed or adjustable. This type of turbine usually is used for heads from 10 ft (3 m) up to 120 ft (40 m), although in a few cases they have been used for heads up to 200 ft (60 m). The higher the head, the

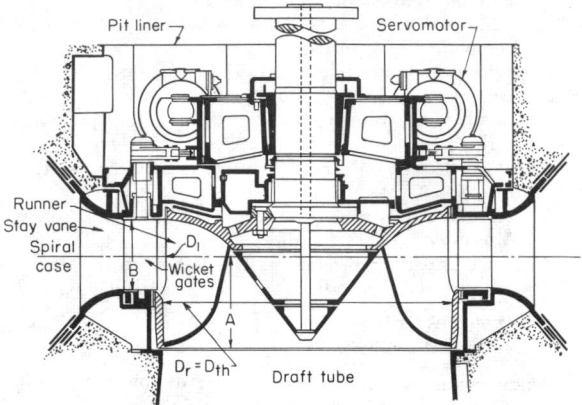

**Fig. 1** Medium-head Francis turbine.

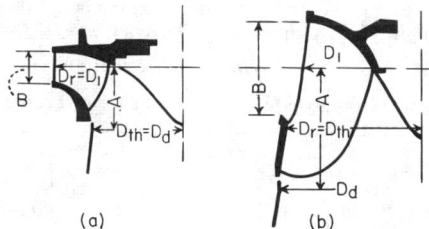

**Fig. 2** Typical profiles of Francis-type runners. (*a*) Low specific speed; (*b*) high specific speed.

greater the number of blades used. Specific speeds vary from 80 to 250 (300 to 1,000). Propellers have very steep efficiency vs.-power curves (4 and 5, Fig. 5). Adjustable-blade propeller runners are used to produce a flat efficiency curve over a wide range of power (6, Fig. 5) and to produce considerably more power beyond the maximum efficiency point than can be obtained with a fixed-blade runner of equal diameter. For fixed blade runners the blade angle is usually set between 16 and 28°, where maximum efficiency occurs. For adjustable blade runners, the blade angle may vary from 10° min to 40° max. The blades may be adjusted by hand or motor. However, these methods have been largely abandoned in favor of the automatically oil pressure operated blades commonly called a **Kaplan turbine** (Fig. 3). The blades are adjusted by means of an oil operated piston located either within the main shaft or in the runner hub above or below the runner blades. The oil is admitted to and discharged from the piston by means of a distributor head either on top of the generator shaft or surrounding the main shaft below the generator. The oil pressure is supplied from the governor oil pressure system. The controls are so arranged that the blade tilt varies automatically with the wicket gate opening so as to produce a maximum efficiency envelope curve (Fig. 9). A large operating-cylinder capacity is required. The $S$, ft·lb (J) may be approximated from the formula $S = 20Pn_8^{1/4}/\sqrt{H}(14Pn_8^{1/4}/\sqrt{H})$. If antifriction bearings are used in the hub instead of bronze bearings for the blade trunnions, this formula becomes $S = 10Pn_8^{1/4}\sqrt{H}(7Pn_8^{1/4}/\sqrt{H})$. However, since the hydraulic unbalance on the blades varies with speed, gate opening, and the location of the blade pivot axis, the cylinder size should be calculated from model test data.

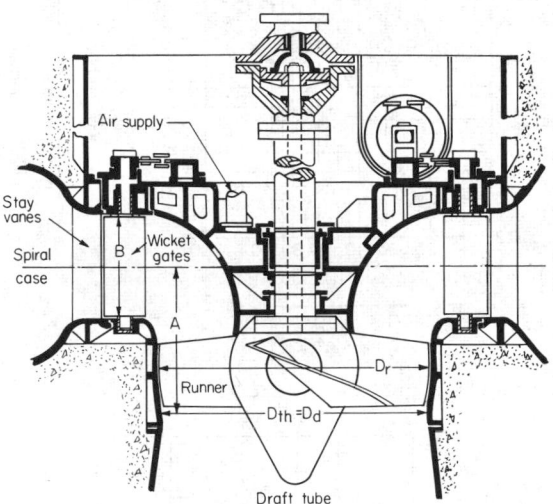

**Fig. 3**  Adjustable-blade Kaplan-type turbine.

**Axial-Flow Turbines**  Axial-flow turbines use the propeller type runner with either fixed or adjustable blades. Their characteristic feature is the straight-through, or nearly straight-through, water passageway from intake to discharge. The shaft is, therefore, either horizontal or inclined (Fig. 4). The spiral or semispiral case and elbow draft tube, which require substantial widths and depth of excavation, are eliminated. Therefore, with a reduction in height and area of the powerhouse and the turbine's suitability for location directly within the dam, an overall construction-cost savings may be obtained for the power plant part of the project compared with conventional vertical units. This may make it possible to build or redevelop power plants for low heads or in small sizes that have previously been considered uneconomical.

In recent years, axial-flow turbines have been installed for use with tidal power since they can be arranged to operate with water flowing in either direction and, when required, to pump with water flowing in either direction in addition to operating as turbines.

There are four general types of axial-flow turbines. The first type, having the generator rotor mounted around the periphery of the turbine runner, has a problem in sealing the large diameter gap between the rotor and the turbine water passageway. Such installations have had high maintenance costs and excessive outages. The **pit** and **bulb** types locate the generator in series with the turbine runner at a submerged elevation. In the pit type, a watertight pit is used to house the generator and speed increaser, when used. The bulb type has the generator enclosed in a streamlined, watertight housing located in the water passageway, on either the upstream or the downstream side of the runner (Fig. 4).

The fourth type of axial-flow turbine is the tube-type turbine, which has the generator located outside the water passages. With this type, a slight bend in the water passageway permits extending the turbine shaft externally. While the unit can be arranged so the generator is either upstream or downstream, the latter arrangement is more practical for large, low head units. To reduce excavation, the shaft may be inclined, thereby raising the generator higher above tailwater elevation. In some cases, gear type speed increasers are used to reduce the combined equipment cost and the generator size and weight.

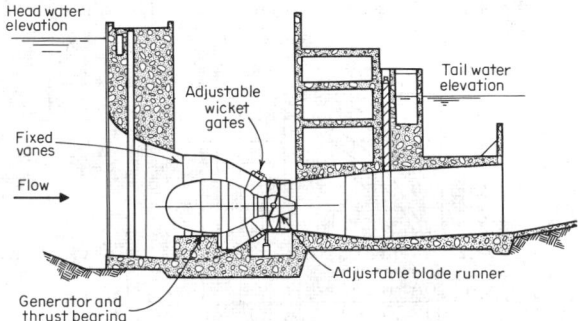

**Fig. 4**  Axial-flow bulb-type turbine.

Axial flow turbines are suitable for heads up to at least 60 ft (20 m) with basically the same limitations as apply to the conventional Kaplan or other propeller type turbines. Maximum unit capacity is limited, however, by maximum practical speed increaser torques and maximum practical horizontal generator capacities. This appears to be approx 100 MW at present. Either adjustable blade or fixed blade propeller runners and either fixed position or movable radial wicket gates can be used. Power and efficiency performance is comparable to conventional vertical shaft propeller turbines.

**Design**  The water entering the runner is given a whirl or tangential component by the guide vanes. This whirl is taken out by the runner so that, at the design point (point of best efficiency), the water leaves the runner without appreciable whirl. According to Euler's theorem, the power of a turbine in steady motion equals the angular velocity multiplied by the change in angular momentum experienced by the mass of water flowing in a unit time in its passage through the turbine. This is expressed in the following equations:

$$P = [62.4Q(u_1 V_{u_1} - u_2 V_{u_2})]/550g = (QHe_h)/8.81 \quad \text{(1)}$$

and
$$u_1 V_{u_1} - u_2 V_{u_2} = e_h gH \quad \text{(2)}$$

$$P = [1,000Q(u_1 v_{u_1} - u_2 v_{u_2})]/100g$$
$$= QHe_h w/100$$

(See notation at beginning of chapter.) Subscript 1 refers to the inlet edge of the bucket and 2 to the discharge edge.

In practice $e_h$ may be taken as 0.92 to 0.94, and the term $u_2 V_{u_2}$ may be omitted as it becomes zero for discharge in a radial plane, $\alpha_2 = 90°$.

The right hand member of Eq. **(2)** may be regarded as a fixed value for a given head. Consequently, in order to have a high speed type runner ($u_1$, large) the whirl $V_{u_1}$ placed in the water by the guide case must be small. This means that for the best load, low speed low capacity (small $n_s$) runners will have relatively small gate openings and short wicket gates, whereas high speed high capacity (large $n_s$) runners will have relatively large gate openings and long wicket gates.

Equations (**1**) and (**2**) are used in designing the runner and guide case. The $P$ and $Q$ used in Eq. (**1**) are for the point of best efficiency and not for the rated values. For specific speeds up to about 60 (250) the power at best efficiency may be taken as 85 percent of the rated power. Above that specific speed the point of best efficiency occurs at higher loads as indicated in Fig. 5. Adjustable blade propeller turbine-runner blades are laid out for an intermediate load of 75 to 80 percent of the rated load. The actual design of a runner should be worked out by an experienced designer on the basis of coordinated test data.

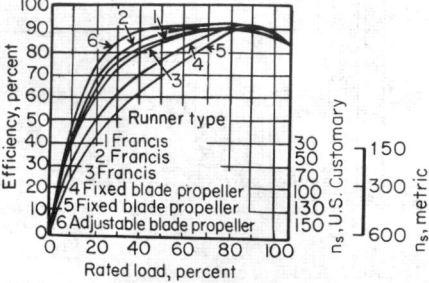

**Fig. 5** Efficiency-load relations for reaction turbines.

The general proportions of a normal line of turbine runners may be determined as follows for any combination of net head $H$ and rated power $P$. (Rated power is usually considered to be 95 percent of full gate power.)

First select the specific speed appropriate to the head (see page 9-137). Next calculate $n$ from the formula $n_s = n\sqrt{P}/H^{5/4}$. If $n$ is not a synchronous speed, select the nearest synchronous speed and calculate the corresponding $n_s$. This value of $n_s$ may be used in Fig. 6 to determine the peripheral coefficient $\phi$ for $D_r$ and $D_1$. Then the diameter may be determined from the formula

$$\phi = \pi D_n/720\sqrt{2gH}(\pi D_n/60\sqrt{2gH}).$$

Other proportions of the runner may be determined from Fig. 7, which shows the ratio to $D_r$ of width of distributor $B$; diameter at top of draft tube $D_d$; and centerline of distributor to top of draft tube $A$. See Fig. 1 for $D_r$, $B$, and $A$.

The values of unit power $P_1$, and unit speed $n_1$, for $D_r = 100$ (3m), which correspond to the specific speed of the runner may be determined from Fig. 6. The following formulas may then be used: $P = P_1H^{3/2}(D_r/100)^2$ [$= P_1H^{3/2}(D_r/3)^2$] and $n = n_1\sqrt{H}(100/D_r)$ [$n_1\sqrt{H}(3/D_r)$]. These formulas may be used to verify the diameter calculated from the peripheral coefficient, and will prove useful in selecting runner sizes for units which must meet power requirements other than rated power at design head. In such cases the values of $n_1$ and $P_1$ from Fig. 6 should be used in conjunction with Fig. 8.

For preliminary powerhouse layouts these ratios of diameter to $D_r$ may be used for all specific speeds: gate circle, 1.20; stay ring, 1.65; pit liner, 1.50; unit spacing, 3.5 to 4.5.

The clearance space between the discharge tips of the wicket gates and the entrance edge of the runner buckets is treated as a free vortex space in which the tangential compo-

nent of the absolute velocity of discharge from the gates ($V_u$) increases in inverse proportion to the radius while the component in a radial plane ($V_r$) increases in inverse proportion to the area.

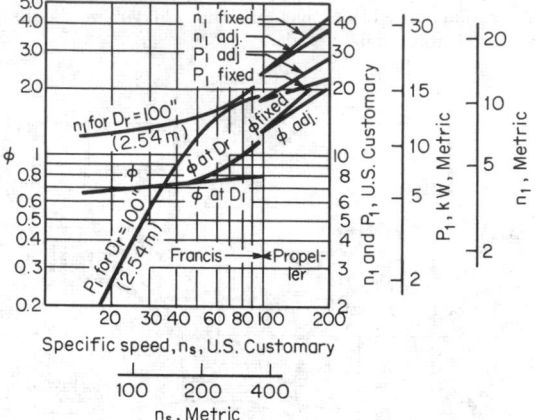

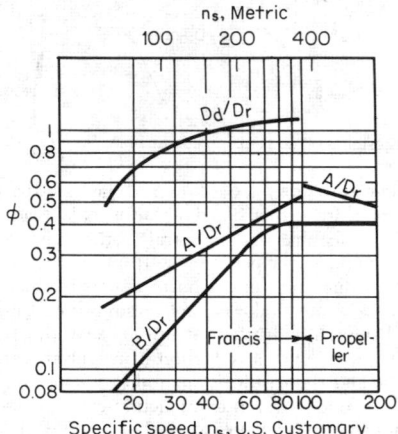

**Figs. 6 and 7** Turbine characteristics and specific speed.

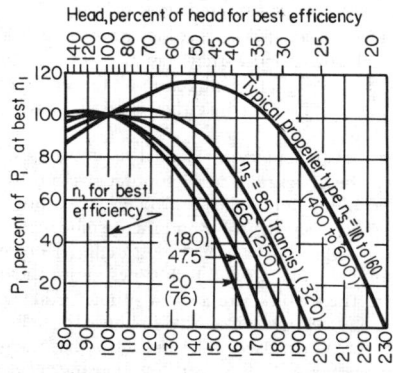

**Fig. 8** Variation of unit power with unit speed and specific speed.

To obtain the best efficiency, the water must enter the runner without shock and leave it with as little velocity as possible; i.e., the entrance angles of the buckets must be approximately in line with the relative direction of the entering flow and the direction of the absolute discharge velocity should be approximately in a radial plane (axial in the case of propeller runners). This condition prevails only at the most efficient load. Above that load the water enters the draft tube with reverse whirl and below that load with a forward whirl.

The shape of the buckets should be kept as simple as possible, consistent with meeting the proper angles. The curvature of the bucket along a flow line is made greatest near the entrance and reduces as the discharge edge is approached, somewhat the shape of a portion of a parabola.

**Turbine Characteristics** Speed, power, and efficiency characteristics vary widely with specific speed as shown in Figs. 5 and 6. The variation of unit power $P_1$ with the unit speed (as a percent of normal basis) is shown in Fig. 8 for typical specific speeds ranging from 20 to 160 (70 to 600). For low speed Francis type turbines $P_1$ decreases as $n_1$ increases, as when operating at heads below normal. This is due to the centrifugal effect, which tends to decrease the flow as $n_1$ increases. This characteristic of low specific speed Francis runners makes them less suitable for operation under extreme variations in head. With the higher specific speeds and particularly with the propeller type, $P_1$ increases as $n_1$ increases above normal. This makes the higher specific speeds particularly suitable for operation at widely varying heads.

Figure 9 shows the advantage of blade adjustment for propeller type turbines. By angular adjustment of the blades as the load changes, the peaked power efficiency curve of the fixed blade propeller is transformed into a flat power efficiency curve, and the maximum power output is considerably increased.

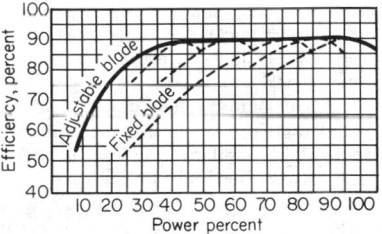

**Fig. 9** Efficiency-power relations for fixed- and adjustable-blade propeller turbines.

**Runaway Speed** If a turbine runner is allowed to revolve freely without load and with the wicket gates wide open, it will overspeed to a value called the *runaway speed*. The runaway speed, at normal head, varies with the specific speed and for Francis turbines ranges from 170 percent (normal speed = 100 percent) at low specific speed [$n_s = 20$ (76)] to some 195 percent at high specific speed [$n_s = 100$ (383)]. For propeller turbines, the runaway speed varies with blade angle—the steeper the blade angle, the lower the runaway speed. For fixed blade propellers with the blades set from 16 to 28°, where maximum efficiency is usually obtained, the runaway speed ranges approx from 255 to 180 percent, respectively. For adjustable blade turbines, where the minimum blade angle is sometimes as low as 6 to 16° in order to obtain high efficiency at part load, the maximum possible runaway speed

will be about 290 to 270 percent, respectively. However, with adjustable-blade propeller turbines, there is from the standpoint of efficiency an optimum relationship between runner-blade angle and wicket gate opening, usually controlled by a cam in the operating mechanism; the higher the gate opening, the steeper the blade angle. Thus the combination of wide-open gate and minimum design blade angle can only occur in the so called "off cam" position, which is an extremely rare possibility. In most units, this maximum possible off cam runaway speed is reduced by limiting the minimum design blade angle to 16°. Another method is the use of a runaway speed limiter. There are several designs in use. One of the most reliable is a valve designed to open by centrifugal force at speeds of 130 percent approx. The open valve bypasses the runner blade servomotor piston, thus equalizing the oil pressure on both sides of this piston. If the runner blades are designed to open because of hydraulic unbalance at overspeeds, they will then go to a higher blade angle under off cam overspeed conditions. The off-cam runaway speed thus can be limited to some 185 percent if the runner is designed for a maximum blade angle of 28 to 40°.

For all turbines, if the maximum head is higher than the normal head, the runaway speed will be increased in proportion to the square root of the head. Therefore, runaway speeds should be based on the maximum operating head rather than on the normal head. Any runaway speed above 180 percent can increase appreciably the cost of the generator.

**Turbine Thrust and** $WR^2$ The turbine thrust is usually carried by a thrust bearing furnished with the generator. For vertical units the thrust $T$ consists of the hydraulic thrust $T_h$ plus the weight of the turbine runner and shaft $T_w$, lb (kg).

The hydraulic thrust is obtained by using a thrust coefficient $K_t$ multiplied by the weight, lb (kg), of a circular column of water whose diameter is the diameter of the runner $D_r$, in (m), and whose height is equal to the maximum head $H_m$, ft (m), under which the turbine must operate, or $T_h = K_t D_r^2 H_m / 2.94$.

The thrust coefficient for Francis turbines varies with the specific speed and may be taken as approximately $K_t = n_s/250$ (= $n_s/65$). In using this value of $K_t$ it is assumed that the runner seals are properly arranged and that the crown of the runner inside the seals is properly drained; otherwise the thrust may be much higher.

For propeller turbines, with either fixed or adjustable blades, the thrust coefficient $K_t$ may be taken as 0.90 for all specific speeds.

**Reaction-Turbine Elements**

**Runner and Wearing Rings** The number of buckets $z$ for Francis runners ranges from about 21 for low specific speed to 12 for high specific speed. The approximate number may be taken as $z = 55/n_s^{1/3} (250/n_s^{1/3})$.

The number of blades for propeller type turbines ranges from 8 for low specific speeds to 3 for high specific speeds.

Most runners are made of cast steel, which can readily be repaired by welding with either mild or stainless steel electrodes. Built-up runners with pressed steel or cast steel buckets welded to the crown and band are sometimes used for Francis turbines. The welded type is suitable for very large runners. Bronze may be used for smaller runners of all types and for medium high heads. The use of cast stainless steel runners is increasing for high heads and for conditions where pitting may be troublesome.

Propeller runners are practically always made of cast steel, and the surfaces over which pitting may be expected are overlaid with stainless-steel welding, before finishing to the final contour. In place of overlay welding, solid stainless steel inserts are sometimes welded into the Francis buckets or propeller blades.

The functions of the runner seals for Francis turbines are to prevent excessive leakage loss and thus improve the efficiency and control the hydraulic thrust.

Rolled, forged, and cast steels make excellent wearing rings for low and medium head units. To prevent seizure in case of contact the rotating ring material should differ from that of the stationary rings. Stainless, steel, bronze, or steel with bronze inserts make excellent wearing rings. For extremely high heads, stainless steel should be used to prevent undue wear and erosion. Wearing rings should be made renewable.

Seal clearances are made as small as practicable to reduce leakage, particularly on high head units. Larger clearances are required with water lubricated main bearings, subject to considerable wear before being readjusted.

**Main-Shaft Bearing**   The main shaft must be rigid and is made of a medium grade of forged steel, with torsional stress ranging from 4,000 to 6,000 lb/in$_2$ (27,600 to 41,400 N/m$^2$).

A stainless steel sleeve is installed on the bearing surface of the shaft. Such a sleeve is also put on the shaft at the packing box.

Turbines are usually provided with one main bearing located in the head cover as near the runner as practicable. This is usually babbitted and in halves, with an independent low pressure oiling system. Self-lubricated babbitted bearings, containing an oil reservoir and pumping grooves in the babbitt, are sometimes used.

Water lubricated bearings of lignum vitae, rubber, or special composition materials are sometimes preferred, particularly for small and medium sized propeller type turbines where the bearing is located at the bottom of the head cover cone and where the packing box of an oil lubricated bearing would be inaccessible and where it would be difficult to avoid water contamination. With the water lubricated bearing the packing box is placed above the bearing.

**Spiral Case**   The spiral case must be proportioned so as to cause relatively low friction losses, as well as to prevent eddying which would travel into the runner and affect its efficiency. The cases generally used are (1) metal case; cast steel, cast iron, or steel plate, and (2) concrete case. Metal cases are made as complete spirals, customarily with uniform or slightly increasing velocity from the throat to the small end. When a valve is used at the case entrance, the net area at the centerline of the valve should be at least equal to the case entrance area.

Concrete cases may be complete spirals or semispirals, rectangular or oval in cross section. There should be no piers close to the turbine.

Rated load velocities for general practice for cases of various types are shown in Fig. 10. Higher velocities are sometimes used with the larger units to reduce size and cost.

**Stay Ring**   That part of the guide apparatus between the spiral case and the wicket gates, containing stationary stay vanes, is called the stay ring. The water is accelerated within this space as it approaches the gates. The number of stay vanes

employed is usually made equal to or one-half the number of gates. They are placed at that angle which will cause the least obstruction to the flow.

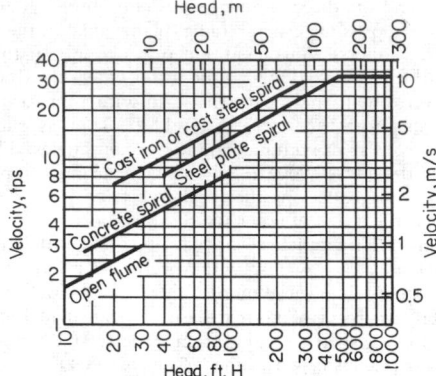

**Fig. 10**   Case velocities.

The stay ring is cast integral with cast cases and is made separately of welded or cast steel for concrete or steel plate spiral cases. It should be a continuous ring to facilitate erection, and very rigid, because it serves as a foundation for the rest of the turbine and generator.

**Wicket Gates and Operating Mechanism**   Wicket gates control the power and speed of the turbine. The number of gates $m$ ranges from 12 for small units to 28 for large units. The overall dimensions of the turbine decrease as the number of gates increase.

To prevent interference between the gates and the runner buckets, which may cause noise and vibration, the discharge tips of the fully open gates should be kept well away from the inlet edges of the runner buckets of Francis turbines, the radial clearance being large enough to prevent the gate tips from overhanging the curved part of the discharge ring. Gate tip diameter = 1.1 × (maximum inlet diameter).

The height and angular movement of the gates increase with the specific speed. The angular movement varies from 15° for low specific speed to 80° for high specific speed.

Most wicket gates are made of cast steel; a few, for higher heads, are made of stainless steel; some, for lower heads, are weldments built from rolled materials and castings.

Each gate connection to the operating ring should be provided with a breaking element to protect the gate and other mechanism in case of an obstruction. Because of the inherent instability of a free pump/turbine gate and its tendency to flutter, it is advisable to install gate-restraining devices to hold the gate if a breaking element fails. This device also avoids potential resonant flutter with the natural frequencies in the system. Each gate should also be provided with stops to prevent it from striking the runner or reversing after the breaking element fails.

The capacity $G$, ft·lb (J), of the oil pressure cylinder or cylinders (sometimes known as governor capacity) may be computed from the formula

$$G = 18Pn_s^{1/4}/\sqrt{H} \qquad (13Pn_s^{1/4}/\sqrt{H})$$

Some units now provide individual servomotors on each

wicket gate. This synchronized system allows the hydraulic control system to dampen a tendency to gate flutter.

One or more vacuum breakers or air valves are installed in the head cover to admit air to the runner or draft tube, to improve efficiency at low gate openings or to alleviate draft tube vortex cavitation. An air valve is also necessary on propeller-type units to break the vacuum under the head cover and to help prevent the runner from "screwing up" in the water when the gates are suddenly closed. The air valve is piped to the outside of the powerhouse above the floodwater level.

**Draft Tubes**  The draft tube may serve the double purpose of (1) allowing the turbine to be set above tailwater level, without loss of head, to facilitate inspection and maintenance, and (2) regaining, by diffuser action, the major portion of the kinetic energy delivered to it from the runner.

At rated load the velocity at the upstream end of the tube for modern units ranges from 24 to 30 ft/s (7 to 10 m/s), representing from 9 to 16 ft (3 to 5 m) head. As the specific speed is increased and the head reduced, it becomes increasingly more important to have an efficient draft tube. Good practice limits the velocity at the discharge end of the tube to 5 to 7 ft/s (2 to 3 m/s), representing less than 1 ft (0.3 m) velocity head loss.

The elbow type of tube is now used with most turbine installations. With this type the vertical portion begins with a conical section which gradually flattens in the elbow section and then discharges horizontally through substantially rectangular sections to the tailrace. Most of the regain of energy takes place in the vertical portion, very little in the elbow section which is shaped to deliver the water to the horizontal portion so that the regain may be efficiently completed. Figure 11 shows proportions of a good elbow tube. One or two vertical piers are placed in the horizontal portion of the tube for structural reasons.

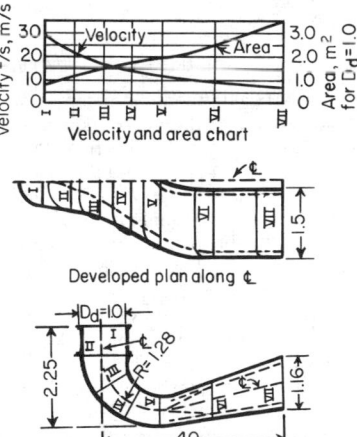

**Fig. 11**  Elbow draft-tube layout.

Most tubes are made of concrete with a steel plate lining extending from the upper end to a point where the velocity has been sufficiently reduced [say 20 ft/s (7 m/s)] to prevent erosion of the concrete. Sometimes the liner is carried around the elbow. Pier noses are also lined where necessary to prevent erosion and for structural reasons.

## IMPULSE TURBINES

Impulse turbines are utilized when the head is too high for the practical use of Francis turbines, which is normally a head exceeding 1,600 ft (500 m). Impulse turbines are also sometimes used for heads below 1,600 ft (500 m) where excessive erosion due to foreign materials in the water presents a problem. The main disadvantage of impulse turbines, especially for low heads, is their low specific speed. In the past, this was overcome on conventional horizontal shaft units by the use of two runners or two jets per runner. In recent years, the vertical shaft multijet impulse turbine (Fig. 12) has become popular.

The efficiency obtainable from a horizontal shaft impulse turbine is about 90 percent. Field tests on multijet vertical units have shown efficiencies as high as 91.5 percent. The use of multiple jets on the vertical units reduces the percentage loss due to windage of the runner. The unit can be operated with a reduced number of jets at part load, thus increasing part load efficiency. Six jets are about the maximum number that can be used on one runner without jet interference.

**Selection of Speed**  The first consideration is the selection of suitable $n_s$. As with reaction turbines, there is a relation which determines approximately the limiting value of specific speed $n_s$ (per jet) for any head acting. For heads around 1,000 ft (300 m), $n_s = 5.0$ to $5.5$ (19 to 21) gives high efficiency. For heads around 2,000 ft (600 m), maximum efficiency is attained near $n_s = 4.0$ to $5.0$ (15 to 19). Care has to be exercised in selecting an $n_s$ (per jet), especially for the higher heads, so that the proper number of buckets can be used on the wheel disk. The higher the $n_s$, the fewer the number of buckets required (Fig. 13) but the smaller the pitch circle. The latter can create stress problems in the attachment of the buckets to the disk. If the resulting speed $n$ is quite low for the power to be developed, the $n_s$ of the unit and consequently the speed $n$ can be increased by increasing the number of runners or the number of jets per runner. The $n_s$ of the unit will then be $n_s$ (per jet) times the square root of the number of jets.

**Basic Dimensions**  The pitch diameter $D_p$ of the runner is twice the distance from the center of the runner to the axial centerline of the jet and is determined by the value of the coefficient $\phi$ selected. $D_p = (1,840\phi \sqrt{H})/n$  $(60\sqrt{2gH}/\pi n)$. The variation of $\phi$ with $n_s$ is established by experience and model tests (Fig. 13).

The quantity of water $Q$ discharged per jet is $550P/wHe$ $(100P/wHe)$, where $P$ is the horsepower developed by each jet.

The diameter of the jet is $4.95\sqrt{Q}/H^{1/4}$ in $(0.55\sqrt{Q}/H^{1/4}\text{m})$, using a velocity coefficient of 0.97 for the jet.

The **velocity** in the inlet should not exceed $0.09\sqrt{2gH}$ approx, and it is considered good practice not to exceed an absolute value of 30 ft/s (10 m/s).

The **valve** employed in the inlet pipe should be of a type which, when open, permits a smooth, uninterrupted flow, free of eddies or obstructions. Modern preference is for rotary sphere valves for this purpose.

The **housing** serves primarily to carry off the discharged water to the tail pit below and to support the nozzles. On horizontal shaft units, at the place where the runner receives the discharge from the jet, the housing should be about 10 to 12 times the jet diameter. At the place where the runner has been cleared of the discharge, the housing should be as narrow

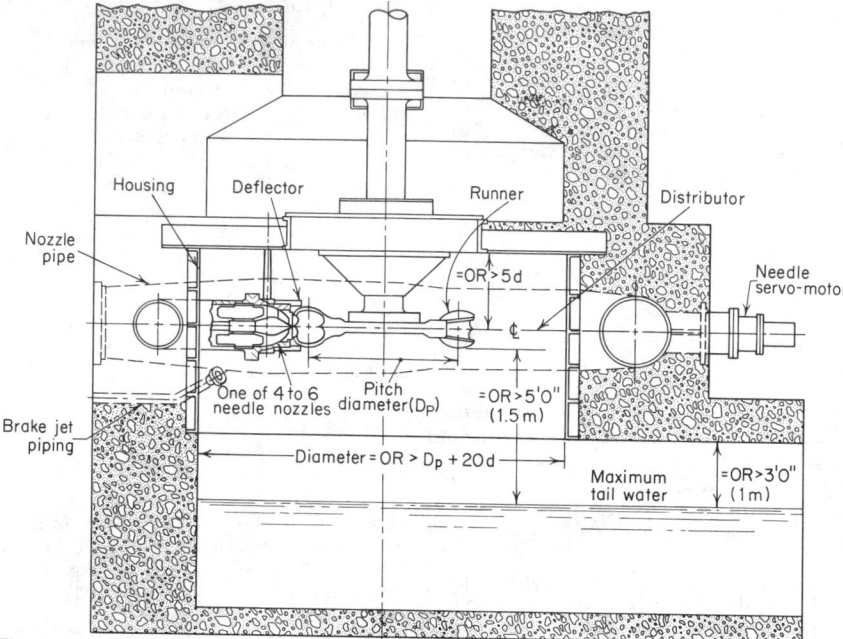

**Fig. 12**  Vertical-shaft multijet impulse-turbine installation.

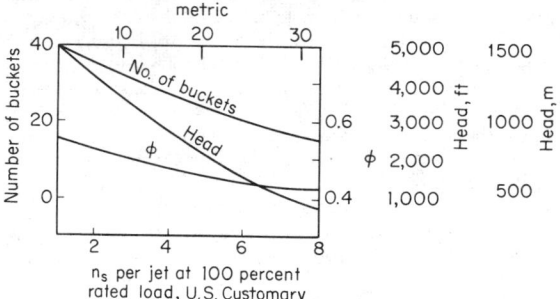

**Fig. 13**  Impulse-turbine characteristics and proportions.

as possible to decrease windage. Suitable baffling should be used to carry the discharge away from the buckets. The housing should be vented adequately near the center of the runner.

For vertical shaft units, the housing should be of ample size to prevent discharge water from interfering with the buckets. The housing should also be vented adequately near the center of the runner.

The **setting** of the horizontal shaft unit should be such that the lower edges of the buckets are at least 3 ft (1 m) above maximum tailwater elevation. For vertical shaft units, this distance should be at least 5 ft (1.5 m). Impulse turbine discharge entrains large amounts of air, and unless this is completely replaced, a vacuum will be produced in the housing which will draw the tailwater up and drown out the runner. Thus the roof of the discharge tunnel or passageway, for both horizontal- and vertical shaft units, should be at least 3 ft (1 m) above the maximum operating tailwater elevation to permit the free circulation of air to the runner (Fig. 12).

In cases where extremely high tailwater occurs for a short period, compressed air can be used to depress the tailwater.

Experience has shown that the ft³/s (m³/s) of air (at the discharge pressure of the compressor) required to maintain the depressed tail water is about 15 percent of the ft³/s (m³/s) of water ($Q$) discharged by the turbine.

**Runner**  The force $F$ exerted by a stream upon a stationary bucket (Fig. 14) is

$$F = MV(1 - \cos \theta) = wQV(1 - \cos \theta)/g$$

where $M$ = mass per s and $\theta$ = angle in deg through which the water is turned relative to the bucket (see section $Y$-$Y$, Fig. 14). If the bucket is moving in the direction of the stream with a velocity $u$, $F = wQ(V - u)(1 - \cos \theta)/g$. The work done by the stream = $Fu = wQn(V - u)(1 - \cos \theta)/g$. When $\theta$ is greater than 90°, the cosine becomes negative. Thus the closer the angle $\theta$ is to 180°, the greater the amount of work done with the same $Q$ and $V$. However, in order to prevent the water from striking the succeeding bucket, $\theta$ should never be 180° (see section $Y$-$Y$, Fig. 14). Also, since the water discharged from the bucket must have some velocity, the value of $u$ is made somewhat less than $V/2$.

The angle $\beta_1$, at the centerline of the bucket should be calculated from the velocity triangle (Fig. 14). The value of $\beta_1$, is greatest as the entrance lip $E$ enters the jet. The angle of the underside of the bucket should be greater than $\beta_1$ to avoid pitting.

In terms of the jet diameter which will give maximum guaranteed or rated capacity, the approximate proportions of the buckets are $B = 3d$, $L = 2.6d$, $D = 0.85d$ (Fig. 14). These proportions depend on the bucket contour and may vary greatly. The contour of the entrance lip $E$ is important, since the manner in which this lip comes in contact with the jet

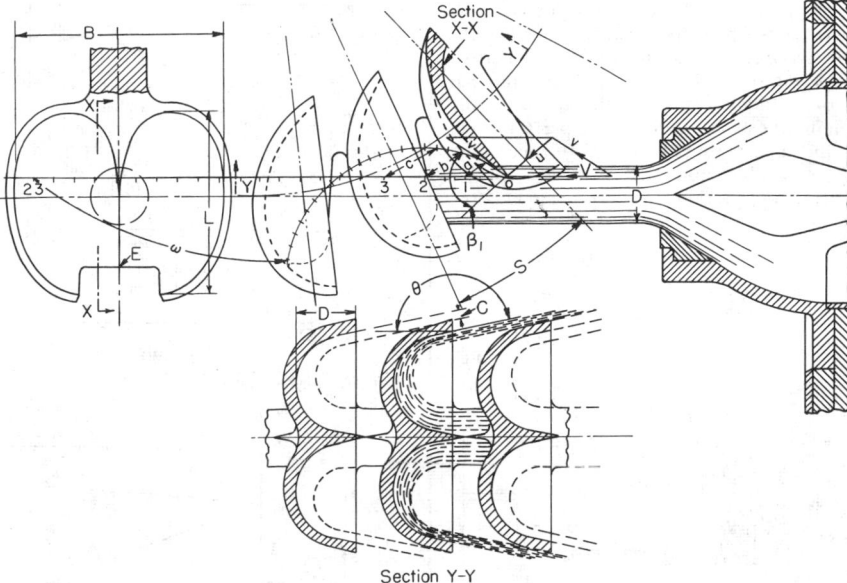

Section Y-Y

**Fig. 14** Impulse-turbine bucket diagram.

affects the path of the water passing to the preceding bucket. The size of the bucket with relation to the jet diameter also determines the percentage of full load at which maximum efficiency occurs. The larger the bucket, the higher this percentage of load.

The number of buckets on a runner should be such that no water can pass through the buckets without being deflected by the buckets. Figure 13 indicated approximately the most efficient number.

The **needle nozzle** should be placed as close to the buckets as possible, as the jet tends to lose its compactness of form shortly after emerging from the nozzle. The needle and the seat of the nozzle should be designed for easy replacement and should be of a material highly resistant to erosion. The needle should terminate in a cone of 30 to 45°. The nozzle tip should have a subtended angle of 45 to 60° and a diameter at discharge about 20 percent greater than the calculated diameter of the jet. The maximum diameter of the needle should be about 15 percent greater than the discharge diameter of the nozzle tip. The diameter of the upstream portion of the nozzle should be such that the velocity does not exceed $0.10 \sqrt{2gH}$. (See Lowy, Efficiency Analysis of Pelton Wheels, *Trans. ASME*, Aug. 1944, for additional information on the design of impulse turbines.)

**Regulation** The inertia of the water flowing through the long penstocks usually employed with impulse turbines prohibits rapid reduction in velocity because of the pressure rise which would occur. Therefore, to minimize the speed rise following a sudden load rejection, it is necessary to reduce the hydraulic power delivered to the runner without changing the flow in the penstock too rapidly. This is usually accomplished by placing a governor controlled jet deflector between the needle nozzle and the runner. The governor moves this deflector rapidly into the jet, cutting off the load. It is not unusual for the deflector to cut off the entire jet in 1½ s. Since the

deflector acts on the jet after it leaves the nozzle, there is no change of flow in the penstock; hence there is no pressure rise. The governor then moves the needle at a permissible rate (in terms of pressure rise), with simultaneous automatic withdrawal of the deflector. The jet is finally reduced the necessary amount to correspond to the reduced load. The needle must also move slowly in the opening direction for oncoming loads to avoid penstock collapse because of large pressure drops.

**Runaway Speed** The runaway speed for impulse turbines ranges from 160 to 190 percent of normal speed, depending upon the specific speed of the runner; the higher the specific speed, the higher the runaway speed.

## REVERSIBLE PUMP/TURBINES

There has been a trend in recent years toward the increased use of **pumped storage** hydro facilities for seasonal storage and peaking capacity. In this type of project, surplus low value energy from either hydro or thermal plants is used to pump water during off-peak periods to an elevated reservoir, where it becomes available for generating high-value peaking energy. While separate pumps and hydraulic turbines of conventional design can and have been used for this purpose, the development of single reversible pump/turbines has made many pumped storage projects economically feasible.

Figure 15 is a cross section of a reversible pump/turbine, showing that, with the exception of the runner, it is essentially a conventional turbine. It has a spiral case, stay ring, movable wicket gates, head cover, discharge ring, and draft tube. The wicket gates must be designed for flow in both directions. A few units have been built without movable wicket gates. The runner is essentially a pump runner modified for optimum performance while generating power. (See Sec. 14 for the design of centrifugal pumps and their characteristics.) Conventional turbine runners, because of their short blades, are not well suited for the pumping cavitation requirements.

The reversible pump/turbines have certain fundamental performance characteristics which are inherent in the design. The relationship between pumping and generating performance for a given specific speed is more or less fixed and can be modified only to a minor degree by alterations in the design. Figure 16 shows the expected generating performance, and Fig. 17 shows the expected pumping performance based on model tests of a reversible pump/turbine with a specific speed $n_s = 47.3$ (180) generating and $n_s = 2,600$ (600) pumping.

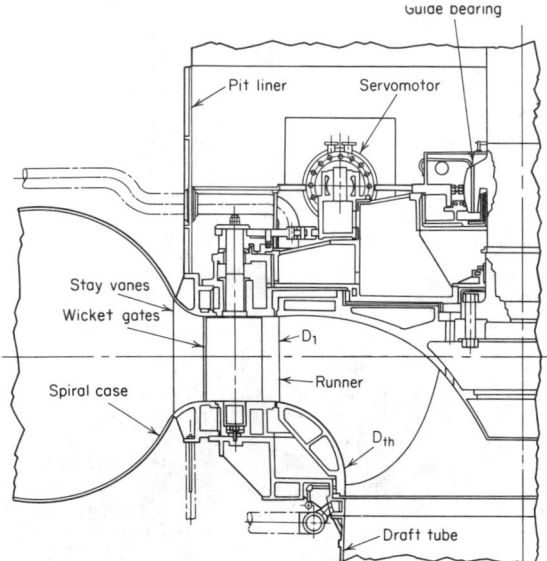

**Fig. 15** Reversible pump turbine.

The best efficiency, with reversible pump/turbines, occurs at a lower speed when generating than when pumping. This can be compensated for by using a generator/motor capable of operating at two speeds with a constant frequency. Several units of this type have been built, but since two speed generator/motors cost considerably more than those built for single speed, a careful study should be made of the advantages to be obtained in operating at two speeds.

Single reversible pump/turbines can be built for any heads up to 2,000 ft (600 m). Beyond this, either multiple stage reversible units or separate pumps and turbines should be

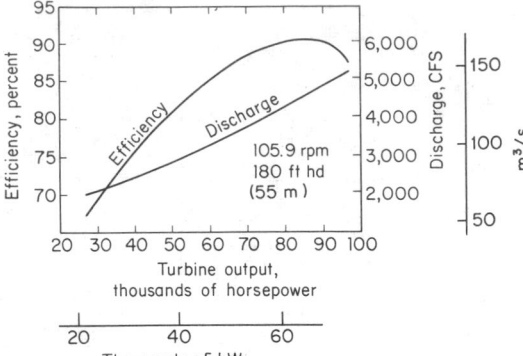

**Fig. 16** Generating.

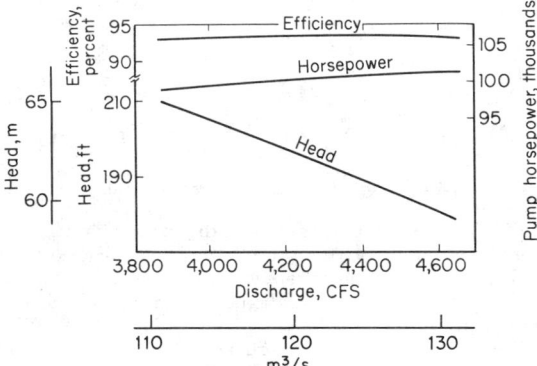

**Fig. 17** Pumping.
**Figs. 16 and 17** Typical performance curves for a pump turbine.

used. Some of the more notable reversible pump/turbine installations are shown in the table below.

**Runaway Speed** The runaway speed for pump/turbines is considerably lower than that for conventional turbines. It ranges from 150 percent of normal for low specific speed runners to 175 percent for high specific speed runners.

## MODEL TESTS

Model tests serve several purposes. They are primarily used to check turbine runner, wicket gate, draft tube, casing, and

| Plant | Generating | | | | Pumping | | | | Year in operation |
| | Power | | Head | | Capacity | | Head | | |
| | 1,000 hp | MW | ft | (m) | ft³/s | (m³/s) | ft | (m) | |
|---|---|---|---|---|---|---|---|---|---|
| Pedreira "A" | 5.2 | 3.9 | 49 | (15) | 690 | (19.5) | 49 | (15) | 1939 |
| Hiwassee | 62 | 46 | 190 | (58) | 3,900 | (111) | 205 | (62.5) | 1956 |
| Lewiston | 21 | 15.5 | 75 | (22.9) | 3,400 | (96.4) | 85 | (25.9) | 1961 |
| Taum Sauk | 220 | 165 | 790 | (241) | 2,650 | (75.0) | 765 | (233) | 1963 |
| Cruachan | 100 | 74.6 | 1,130 | (343) | 1,010 | (28.6) | 1,175 | (358) | 1966 |
| Cabin Creek | 166 | 125 | 1,190 | (363) | 1,375 | (39.0) | 1,060 | (323) | 1967 |
| Robiei | 41 | 30.5 | 1,280 | (390) | 344 | (9.75) | 1,115 | (340) | 1967 |
| Villarino | 135 | 100 | 1,250 | (382) | 1,020 | (29) | 1,325 | (404) | 1969 |
| Kisenyami | 240 | 180 | 755 | (220) | 3,880 | (110) | 645 | (197) | 1969 |
| Raccoon Mt. | 465 | 350 | 940 | (287) | 3,850 | (110) | 1,000 | (305) | 1974 |

(sometimes) inlet works designs for optimum performance. Correctly interpreted, they may also be used as a reliable indication of the performance of the units in the field. In many cases, purchasers specify the performance of a homologous model test, which is used as an acceptance test of the unit in lieu of field tests. In such cases, the field conditions, particularly the casing and draft tube, must be reproduced faithfully. Model tests should be run in accordance with the International Code for Model Acceptance Tests of Hydraulic Turbines, Publication 193, and the International Code for Model Acceptance Tests of Storage Pumps, Publication 193 of the International Electrotechnical Commission (IEC).

Figure 18 shows typical model test results of a reaction type turbine, in which the power $P$ at 1 ft (0.3 m) head and the efficiency are each plotted against the speed $n$ at 1 ft (0.3 m) head, for each of several gate openings.

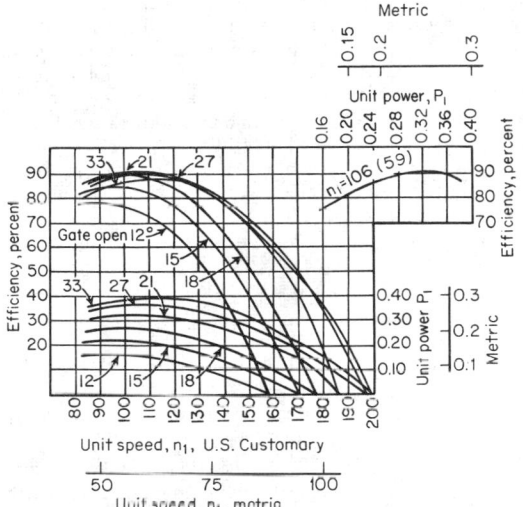

**Fig. 18**   Model-test curves for Francis runner [$D_r = D_{th} = 16.2$ in (0.410 m), $n_s$ about 65 (250)].

**Laws of Proportionality for Homologous Turbines**   The laws of proportionality shown on p. 9-137 are used to calculate from model tests the power, speed, and discharges of a homologous turbine of different diameter under a different head. The field unit will have a somewhat higher efficiency and power owing to proportionally smaller frictional and bearing losses. Expected field efficiency is customarily computed from the model efficiency by the Moody formula $e' = 1 - (1 - e)(D/D')^{1/5}$, in which the prime letters refer to the field installation. The stepup efficiency is usually computed for the point of best efficiency only, and the corresponding differential is applied as a constant value from, say, half load to full load.

## CAVITATION

Cavitation occurs when the pressure at any point in the flowing water drops below the vapor pressure of the water, which varies with temperature.

The relationship which produces cavitation is between vapor pressure, barometric pressure, setting of the runner with respect to tailwater, and net effective head on the turbine

and is expressed by the **Thoma cavitation coefficient** $\sigma = (H_b - H_v - H_s)/H$, where $H_b$ = barometric head, ft (m) of water; $H_v$ = vapor pressure of water, abs; $H_s$ = elevation, ft (m) of the runner above tailwater, measured at the centerline of the distributor of a Francis runner and at the centerline of the blades of a propeller runner (if the runner is submerged, $H_s$ becomes negative); and $H$ = total or net effective head, ft (m), on the turbine. From the above formula, $H_s = H_b - H_v - \sigma H$. Thus the setting of the runner depends upon the value of $\sigma$, which varies with specific speed $n_s$ of the runner and the individual characteristics of a particular runner design. In practice, the model of the proposed runner is first tested with relatively high back pressure ($H_s$ small or negative). Then the back pressure is reduced in increments until the breaking point, as indicated by a drop in power, efficiency, and discharge, is reached. This breaking point is designated as the critical $\sigma$ and will vary with gate opening and speed and, on propeller turbines, with blade angle. Consequently, $\sigma$ must be determined for a range of limiting conditions.

In the absence of cavitation tests, the value of $\sigma$ should not be lower than $\sigma = n_s^{3/2}/2,000$ ($n_s^{3/2}/15,000$) for Francis and propeller runners and $\sigma = n_s^2/25,000$ ($n_s^2/350,000$) for adjustable blade propeller runners.

The value of $\sigma$ at which a plant operates, depending largely upon the setting of the runner with respect to tailwater, is called the **plant** $\sigma$. To avoid excessive cavitation, the plant $\sigma$ should exceed the critical $\sigma$. The greater this margin, the less possibility of cavitation during operation. For a general discussion of cavitation phenomena, see Knapp, Recent Investigations of the Mechanics of Cavitation and Cavitation Damage, *Trans. ASME*, Oct. 1955. Laboratory tests and experience have shown that materials having a high resistance to cavitation erosion (pitting) and suitable for use in hydraulic turbines are the stainless steels and aluminum bronzes, especially when used as welding overlays. (See Rheingans, "Resistance of Various Materials to Cavitation Damage," ASME Report of 1956 Cavitation Symposium.)

## SPEED REGULATION

(See also Sec. 16)

Regulation is accomplished by changing the flow of water to the turbine. The flow is controlled by the wicket gates of reaction turbines and by the needle valve or jet deflector of impulse turbines. The governor, usually supplied with the turbine, moves the gates or needle in response to speed changes resulting from load or head changes.

A schematic diagram of a governor is shown in Fig. 19. The parts consist of a speed responsive device, a power element that changes the gates or needle position, and a follow-up or compensating device that prevents hunting.

The **speed responsive device** is usually a pair of spring-loaded flyballs mounted directly on the turbine shaft, or driven from the shaft by belt or gears, or driven by an electric motor that receives its power either from the bus line or from an independent generator driven from the main turbine generator shaft.

The **power element** consists of one or two oil operated power cylinders or servomotors which operate the turbine gates or needle. Oil pumps and a pressure tank or accumulator maintain a supply of oil under pressure. A valve operated by the flyballs controls the flow of oil to the servomotors or acts as a pilot valve controlling a larger relay valve, which in turn controls the oil to the servomotors. With a plant consisting of several units, a unit system may be used for each turbine, or there may be a central pumping unit for the turbines. The pump capacity is usually $3\frac{1}{3}$ servomotor volumes per minute. The capacity of the pressure tank is generally made 20 times the servomotor volume, allowing for 8 volumes of oil and 12 volumes of air. The velocity of oil in the pipelines is kept below 15 ft/s (5 m/s). When using individual servomotors on each gate, these values are usually increased 50 percent to balance the usual damping effect of the common massive two-servomotor system.

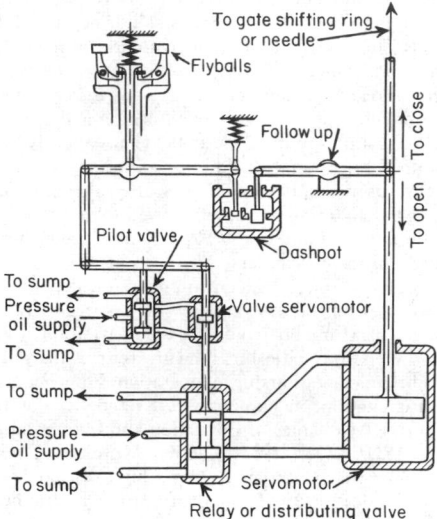

**Fig. 19**  Schematic diagram of governor.

The **follow-up** or **compensating device** connects the power piston of the servomotor to the control valve, usually through a dashpot, and causes the motion of gates or needle to stop when they have moved sufficiently to compensate for the load change.

The time for a full stroke or traverse of the governor is controlled by the rate of flow of oil to the servomotors; most governors have provisions for varying this time. The gate opening changes at a uniform rate over the major portion of the stroke and at a somewhat slower rate at the ends of the stroke. The governor *dead time*, or the elapsed time from the initial speed change to the first movement of the gates, is usually less than 0.2 s.

**Electric Governor**   In recent years, the functional requirements placed upon the hydraulic turbine governor have increased to the point where electrical control of the hydraulic turbine is attractive in view of the simplicity with which electrical signals can be manipulated. The basic elements of an electric-hydraulic governing system are (1) permanent magnet generator (PMG) or equivalent for measurement of turbine speed and the means for transmittal of such speed signals to the electrical portion of the governor; (2) an electric circuit sensitive to speed variations about some adjustable reference point; (3) amplifying circuits to convert speed reference changes, speed error signals, and auxiliary signals into a useful electric current; (4) an electrohydraulic transducer to transform the electric current into a hydraulic output signal; (5)

hydraulic amplifying equipment to deliver suitable power and the desired signal to the gate servomotors as a function of the output of the electrohydraulic transducer; (6) power supplies for the electric and hydraulic portions of the control. For further particulars of the electric governor, see Leum, Electric Governors for Hydro Turbines, *ASME Paper* 62-WA165.

**Speed-Regulation Requirements**   Usually, a sufficient measure of the regulation provided is the maximum speed rise resulting from sudden rejection of full load, as from the breaker tripping. A maximum speed rise of 30 percent of normal speed for this condition is a common limitation.

**Speed Rise Following Load Reduction**   For sudden load reductions, the approximate speed rise is

$$n_x/n = [1 + 1{,}620{,}000 T_x P_x (1 + h/H)^{3/2}/WR^2 n^2]^{1/2}$$
$$\{(1 + 22{,}850 T_x P_x (1 + h/H)^{3/2}/GD^2 N^2]^{1/2}\}$$

where $n_x$ is the r/min at the end of time $T_x$; $n$ is the speed before the load decrease; $T_x$ is the time interval, s, for the governor to adjust the flow to the new load; $P_x$ is the reduction in load; $h$ is the head rise caused by the retardation of the flow; $H$ is the net effective head before the load change; $WR^2$ is the product of the revolving parts, lb (kg), and the square of their radius of gyration, ft (m). For values of $h$, see below. Very rapid gate closure produces a reduction of pressure in the draft tube and the possibility of breaking the water column, with subsequent violent resurge which may damage the turbine.

**Speed Drop Following Load Increase**   For sudden load increases, the approximate speed drop is

$$n_x/n = [1 - 1{,}620{,}000 T_x P_x/WR^2 n^2 (1 - h/H)^{3/2}]^{1/2}$$
$$\{[1 - 22{,}850 T_x P_x/GD^2 N^2 (1 - h/H)^{3/2}]^{1/2}\}$$

where $P_x$ is the actual load increase and $h$ is the head drop caused by the increase of the flow. If the speed drop is to be determined for a given increase in gate opening, the governor time $T_x$ for making this increase and the normal change in load for the change in gate opening, under constant head $H$, can be used in the following formula:

$$n_x/n = [1 - 1{,}620{,}000 T_x P_x (1 - h/H)^{3/2}/WR^2 n^2]^{1/2}$$
$$\{[1 - 22{,}850 T_x P_x (1 - h/H)^{3/2}/GR^2 N^2]^{1/2}\}$$

The actual change in load, however, will be $P_x(1 - h/H)^{3/2}$.

For derivation of the above speed variation formulas and for a more accurate determination, see Strowger and Kerr, Speed Changes of Hydraulic Turbines for Sudden Changes of Load, *Trans. ASME*, 1926; and Rich, "Hydraulic Transients," McGraw-Hill.

**Water Hammer in Penstocks**   If a gate movement is considered as a series of instantaneous movements with a very small interval between each movement, the pressure variation in the penstock following the gate movement will be the effect of a series of pressure waves, each caused by one of the instantaneous small gate movements. For a steel penstock, the velocity of the pressure wave $a = 4{,}660/$ $\sqrt{1 + (d/100t)} \times (142/\sqrt{1 + d/100t})$, where $d$ is the penstock diameter, in (m), and $t$ is the penstock wall thickness, in (m). The pressure change at any point along the penstock at any time after the start of the gate movement may be calculated by summing up the effect of the individual pressure waves. See "Symposium on Water Hammer," ASME, 1933; and Parmakian, "Water Hammer Analysis," Dover.

Approximate formulas (De Sparre) for the increase in pressure $h$, ft (m), following gate closure, are given below. They are quite accurate for pressure rises not exceeding 50 percent of the initial pressure, which includes most practical cases.

| | |
|---|---|
| $h = aV/g$ | (for $K < 1$ and $N < 1$) |
| $h = aV/g[N + K(N - 1)]$ | (for $K < 1$ and $N > 1$) |
| $h = aV/g(2N - K)$ | (for $K > 1$ and $N > 1$) |

where $K = aV/2gH$; $N = aT/2L$; $V$ and $H$ are the penstock velocity, ft/s (m/s), and head, ft (m), prior to closure; $L$ is the penstock length, ft (m), and $T$ is the time of gate closure. For full load rejection, $T$ may be taken as 85 percent of the total gate traversing time to allow for nonuniform gate motion.

For pressure drop following a complete gate opening, the following formula (S. Logan Kerr) may be used with $T$ not less than $2L/a$:

$$h = \frac{aV}{g}\left(\frac{-K + \sqrt{K^2 + N^2}}{N^2}\right) = \text{pressure drop, ft (m)}$$

Pressure variations exceeding 40 percent rise and about 25 percent drop should be avoided. When the control, directly by the governor, causes undesirable pressure variations, a surge tank, a pressure regulator, or a jet deflector may be used. A **surge tank** is a standpipe with an atmospheric tank, attached to the penstock as close as possible to the casing inlet. The tank provides a reservoir and expansion chamber for the water demand or the water rejection following sudden gate movements, so that sudden accelerations or decelerations of the flow in the penstock are avoided.

**Pressure regulators** may be of either the water-wasting or water saving type. The **water wasting type** is a synchronous bypass, generally attached to the turbine casing. It is operated directly from the governor, or the gate mechanism of the turbine, and wastes such an amount as to keep the total water discharge equal at all times to the full-load discharge of the turbine. The bypass is a needle nozzle or a mushroom-shaped disk valve or a cone valve which opens and is partly balanced hydraulically by a piston under pipeline pressure. The **water saving type** permits the regulator to open upon rapid closure of the turbine gates, and then close slowly, so that the total water discharge is gradually reduced and finally limited to that through the turbine, adjusted for the new load.

## AUXILIARIES

Valves and head gates are provided for shutting off the water from each turbine for safety, for ease of maintenance, and to reduce water leakage losses. Motor operated steel head gates are generally used for low and medium head plants with concrete scroll cases, although a few still use stop logs. The butterfly type of valve placed close to the turbine casing is suitable for medium and high head units with metal casings of circular inlet diameter. Butterfly valves of 8 ft (m) diameter for 1,000 ft (m) head and 27 ft (m) diameter for 100 ft (m) head have been built. The new flow through valve or biplane valve is becoming more popular for this application. In recent years, rotary sphere valves have replaced gate valves for high heads and where the loss through the butterfly valve is excessive because of the obstruction to flow by the valve wicket.

## TURBINE TESTS

Field testing of hydraulic turbines to determine the absolute efficiency and output involves careful and accurate measurement of the power available in the water supplied to the turbine [water hp (kW)] and the turbine output [developed hp (kW)]; $e$ = developed hp (kW)/water hp (kW) = $550P/wQH$ ($75P/QHw$). The tests should be conducted in accordance with the International Code for Field Acceptance Tests of Hydraulic Turbines and/or Storage Pumps, IEC Publications 41 and 198.

Because of the difficulties and costs involved in making accurate measurements of horsepower, net head, and discharge in the field, there has been a trend in recent years to dispense with the field test, especially where a laboratory test on an homologous model turbine is available. Instead, an index test is made on the unit in the field, which measures the turbine output and relative discharge under various conditions. Index tests should be conducted in accordance with the International Code for Field Acceptance Tests of Hydraulic Turbines, IEC Publication 41.

# DIRECT ENERGY CONVERSION

## by Wilbur D. Marsh

REFERENCES: Kaye and Welsh, "Direct Conversion of Heat to Electricity," Wiley. Chang, "Energy Conversion," Prentice-Hall. Shive, "Properties, Physics and Design of Semiconductor Devices," Van Nostrand. Bredt, Thermoelectric Power Generation, *Power Eng.*, Feb.–Apr. 1963. Wilson, Conversion of Heat to Electricity by Thermionic Conversion, *Jour. Applied Physics*, Apr. 1959. Angrist, "Direct Energy Conversion," Allyn and Bacon. Harris and Moore, Combustion—MHD Power Generation for Central Stations, *IEEE Trans. Power Apparatus and Systems*, vol. 90, 1971. Roberts, Energy Sources and Conversion Techniques, *Am. Scientist*, Jan.–Feb. 1973. Poule, Fuel Cells: Today and Tomorrow, *Heating, Piping, and Air Conditioning*, Sept. 1970.

In contrast to the conventional thermal cycle for the conversion of heat into electricity are several more direct methods of converting thermal and chemical energy into electrical power. The methods which seem to have the greatest potential possibilities are thermoelectric, thermionic, magnetohydrodynamic (MHD), fuel-cell, and photovoltaic. The principles of operation of these processes have long been known, but technological and economic obstacles have thwarted their use. New applications, materials, and technology are now providing increased impetus to the development of these processes.

**Thermoelectric generation** is based on the phenomenon, discovered by **Seebeck** in 1821, that current is produced in a closed circuit of two dissimilar metals if the two junctions are maintained at different temperatures, as in thermocouples for measuring temperature. Typical thermocouples produce potentials in the order of 50 to 70 $\mu$V/°C and power at efficiencies in the order of 1 percent.

Certain semiconductors have thermoelectric properties superior to conductor materials, with resultant improved efficiency. The criterion for evaluating material characteristics for thermoelectric generation is the **figure of merit**, $Z$, measured in (°C)$^{-1}$ and defined as $Z = S^2/PK$, where $S$ = Seebeck coefficient, V/°C; $P$ = electrical resistivity, $\Omega$/cm$^3$; $K$ = thermal conductivity, W/(°C)(cm$^3$).

An ideal thermoelectric material would have a high Seebeck coefficient, low electrical resistivity, and low thermal conductivity. Unfortunately, materials having a low electrical resistivity have a high thermal conductivity since both properties are dependent, to some extent, on the number of free electrons in the material. The maximum conversion efficiency of a thermoelectric generator is a function of the figure of merit, the hot-junction temperature, and the temperature difference between the hot and cold junctions.

In some types of thermoelectric materials, the voltage difference between the hot and cold junctions results from the flow of negatively charged electrons ($n$ type, hot-junction positive), whereas in other types, the voltage difference between the cold and hot junctions results from the flow of positively charged voids vacated by electrons ($p$ type, cold-junction positive). Since the voltage output of a typical semiconductor thermoelectric couple is low (about 100 to 300 $\mu$V/°C, temperature difference between the hot and cold junctions), it is advantageous to use both $p$- and $n$-type materials in constructing a thermoelectric generator. The two types of materials make it possible to connect the thermojunctions in series electrically and in parallel thermally (Fig. 1).

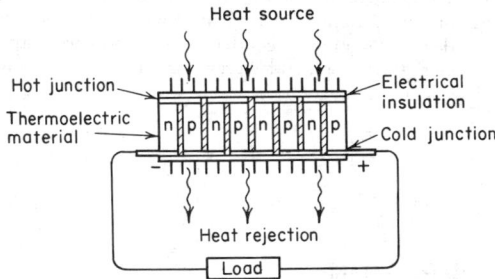

**Fig. 1**

Typical semiconductor thermoelectric materials are compounds and alloys of lead, zinc, tellurium, antimony, bismuth, germanium, arsenic, manganese, cobalt, and silicon. To these materials, minute quantities of "dopants" such as boron and iodine are sometimes added to improve properties. Optimized designs of thermoelectric junctions using semiconductor materials have resulted in experimental conversion efficiencies as high as 13 percent (Fig. 2); however, the efficiency of practical thermoelectric generators is lower, e.g., 4 to 9 percent. Materials which have higher figures of merit (2 or 3 × 10$^{-3}$) and which are capable of operating at higher temperatures (800 to 1000°C) are required for an appreciable

improvement in efficiency. High-temperature operation increases the tendency of materials to sublimate, oxidize, or become poisoned. It also increases the problem of hot-junction fabrication and the differential expansion of components. Thermoelectric-generation technology has matured considerably through its application to auxiliary power systems for space vehicles where modules as large as 150 W have been used. Although development work has continued, it is anticipated that applications will be limited to those special cases where light weight and long life are more important than efficiency and first cost.

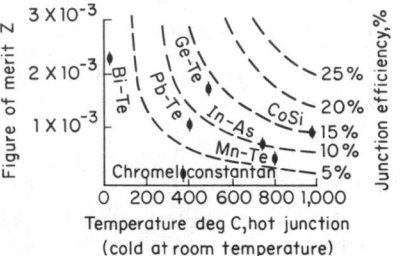

**Fig. 2**

The **Peltier effect,** discovered in 1834, is the inverse of the Seebeck effect. It involves the heating or cooling of the junction of two thermoelectric materials by passing current through the junction. The effectiveness of thermojunction as a cooling device has been greatly increased by the application of semiconductor thermoelectric materials.

**Thermionic generation,** proposed by **Schlicter** in 1915, uses a thermionic converter (Fig. 3), which is a vacuum or gas-filled device with a hot electron "emitter" (cathode) and a cold electron "collector" (anode) in or as part of a suitable gastight enclosure, with electrical connections to the anode and cathode, and with means for heating the cathode and cooling the anode.

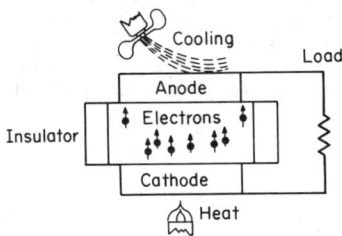

**Fig. 3**

Figure 4 is a plot of the electron energy at various places in the converter. The abscissa is cathode-anode spacing, and the ordinate is electron energy. The base line corresponds to the energy of the electrons in the cathode before heating. Heating the cathode imparts sufficient energy to some of the electrons to lift them over the **work-function barrier** (retaining force) at the surface of the cathode into the interelectrode space. (The lower the work function, the easier it is for an electron to escape from the surface of the cathode.) If it is assumed that the electrons can follow path $a$ to the anode with only a small loss of energy, they will "drop down" the work-function barrier as they join the electrons in the anode still retaining some of their potential energy (**Fermi level),** which is available to cause an electric current to flow in the external circuit. The work function of the anode should be as small as possible. The anode should be maintained at a lower temperature to prevent anode emission or back current. This pattern presumes that the electrons could follow path *a* from the cathode to the anode with little interference. Since, however, electrons are charged particles, those in the space between the cathode and anode form a space-charge barrier, as shown by *b*. This space-charge barrier limits the electrons emitted from the cathode. Space-charge formation can be reduced by close spacing of the cathode and anode surfaces or by the introduction of a suitable gas atmosphere that can be ionized by heating and thus neutralize the space charge. In vacuum-type thermionic converters, the spacing between cathode and anode must be less than 0.02 mm to get as many as 10 percent of the electrons over to

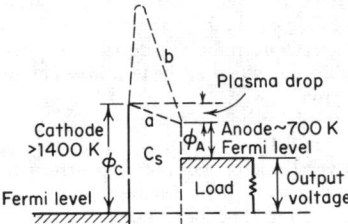

**Fig. 4**

the cathode and to achieve an efficiency of 4 to 5 percent. In gas-filled converters, the negative-electron space charge is neutralized by positive ions. Cesium vapor is used for this purpose. At low pressure, it will also lower the work function of the anode, and at high pressure, it can, in addition, be used to adjust the work function of the cathode. An anode work function of 1.0 V has been achieved at 400 K for a surface cesium on silver oxide. Efficiencies as high as 17 percent have been obtained with gas-filled converters operating at a cathode temperature of 1900°C (2173 K). The output voltage is 1 to 2 V, so the units must be connected in series for reasonable utilization voltages.

Thermionic converters have been used in special military and space applications where the light weight (1.8 kg/kW for gas-filled converters) is important and the efficiency is acceptable. Studies of thermionic converters incorporated into nuclear reactors and fossil-fired steam generators for topping the Rankine cycle indicate potential improvement in overall efficiency and cost, but only if reliable converters become available with higher power density and efficiency.

The **fuel cell** is an electrochemical device in which electrical energy is generated by chemical reaction without altering the basic components (electrodes and electrolyte) of the cell itself. The fact that electrode and electrolyte are invariant distinguishes the fuel cell from the primary cell and storage battery. The fuel cell dates back to 1839, when **Grove** demonstrated that the electrolysis of water could be reversed using platinum electrodes. The fuel cell is unique in that it converts chemical energy to electrical energy without an intermediate conversion to heat energy; its efficiency is therefore independent of the thermodynamic limitation of the Carnot cycle. In practical units, however, its efficiency is comparable with the efficiency of Carnot-limited engines.

Figure 5 is a simplified version of a hydrogen or hydrocarbon fuel cell with air or oxygen as the other reactant. The fuel

is supplied to the anode, where it is oxidized, freeing electrons, which flow in the external circuit, and hydrogen ions, which pass through the electrolyte to the cathode, where they combine with oxygen and electrons to form water. Electrodes for this type of cell are usually porous and impregnated with a catalyst. In a simple cell of this type, chemical and catalytic action take place only at the line (**notable surface of action**) where the electrolyte, gas, and electrode meet. One of the objectives in designing a practical fuel cell is to increase the notable surface of the action. This has been accomplished in a number of ways, but usually by the creation of porous electrodes within which, in the case of gas-diffusion electrodes, the fuel and oxidant in gaseous state can come in contact with the electrolyte at many sites. If the electrolyte is a liquid, a delicate balance must be achieved in which surface tension and density of the liquid must be considered and gas pressure and electrode pore size must be chosen to hold their interface inside the electrode. If the gas pressure is too high, the electrolyte is excluded from the electrode, gas leaks into the electrolyte, and ion flow stops; if the gas pressure is too low, drowning of the electrode occurs and electron flow stops.

Fuel cells may be broadly classified by operating temperature level, type of electrolyte, and type of fuel. Low-temperature (less than 150°C) fuel cells are characterized by the need for good and expensive catalysts, such as platinum and relatively simple fuel, such as hydrogen. High temperatures (500 to 1000°C) offer the potential for use of hydrocarbon fuels and lower-cost catalysts. Electrolytes may be either acidic or alkaline in liquid, solid, or solid-liquid composite form. In one type of fuel cell, the electrolyte is a solid polymer.

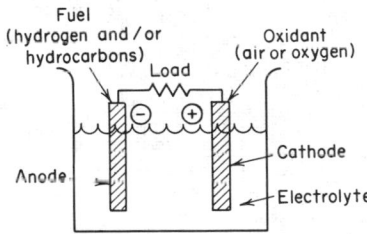

**Fig. 5**

Low-temperature fuel cells of the hydrogen-oxygen type, one a solid-polymer electrolyte type, and the other using free KOH as an electrolyte, have been successfully applied in generating systems for U.S. space vehicles. High-temperature fuel-cell development has been primarily in molten carbonite cells (500 to 700°C) and solid-electrolyte (zirconia) cells (1000°C), but no significant practical applications have resulted.

Considerable study and development work has been done toward the application of fuel-cell generating systems to bulk utility power systems. Low-temperature cells of the phosphoric acid matrix and solid-polymer electrolyte types using petroleum fuels and air have been considered. Cell efficiencies of about 50 percent have been achieved; but with losses in the fuel reformers and electrical inverters, the overall system efficiency becomes of the order of 37 percent. In this application fuel cells have environmental advantages, such as low noise, low atmospheric emissions, and low heat-rejection requirements. Additional development work is necessary to overcome the disadvantage of high catalyst costs and requirements for expensive fuels.

**Magnetohydrodynamic (MHD) generation** utilizes the movement of electrically conducting gas through a magnetic field. In the simple open-cycle MHD generator (Fig. 6), hot, partially ionized, compressed gas, which is the product of combustion, is expanded in a duct and forced through a strong magnetic field. Electrodes in the sides of the duct pick up the potential generated in the gas, so that current flows through the gas, electrodes, and external load. Temperature in excess of 3000 K is necessary for the required ionization of gas, but this can be reduced by the addition of a seeding material such as potassium or cesium. With seeding, the gas temperature may be reduced to the order of 2750 K. The temperature of the gas leaving the generator is about 2250 K. Although the efficiency of the basic MHD channel is of the order of 70 percent, only a portion of the available thermal energy can be removed in the channel. The remainder of the energy contained in the hot exhaust gas must be removed by a more conventional steam cycle. In this combined-cycle plant, the exhaust gas from the MHD generator is passed in turn through an air preheater, the steam superheater and boiler, and an economizer and stack-gas cooler. The air preheater is necessary to raise the temperature of incoming combustion air to some 1900 K in order to obtain the initial gas temperature of 2750 K.

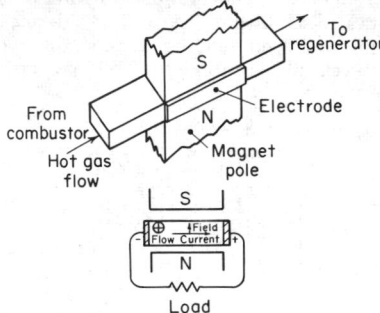

**Fig. 6**

The potential improvement in efficiency from the use of MHD in a combined-cycle plant is in the order of 15 to 30 percent. An overall steam-plant efficiency of 38 percent could be raised to some 45 to 54 percent. Contrasted to other methods for direct conversion, MHD appears best suited to the generation of large blocks of power. For example, an MHD generator 75 m long with an average magnetic field of 5 T (attained by means of a superconductive magnet) would have a net output of about 1,000 MW dc at 5 to 10 kV. Typically, this would provide topping energy for a steam plant of about 500 MW.

Closed-cycle MHD generators are also under study for bulk power generation. They are of two types: first, one in which the working fluid is an inert gas such as argon seeded with

cesium; and secondly, the liquid-metal type in which the working fluid is a helium-sodium mixture. Closed-cycle MHD offers the potential for high efficiency with considerably lower peak cycle temperatures, lower pressure ratios, and lower average magnetic-flux density.

**Photovoltaic generation** utilizes the direct conversion of light energy into electrical energy and stems from the discovery by **Becquerel** in 1839 that a voltage is generated when light is directed on one of the electrodes in an electrolyte solution. Subsequent work using selenium led to the development of the photoelectric cell and the exposure meter.

Photovoltaic effect is the generation of electric potential by the ionization by light energy (photons) of the area at or near the *p-n* junction of a semiconductor. The *p-n* junction constitutes a one-way potential barrier which permits the passage of photon-generated (−) electrons from the *p* to the *n* material and (+) "holes" from the *n* to the *p* material. The resulting excess of (−) electrons in the *n* material and (+) holes in the *p* material produces a voltage at the terminals comparable to the junction potential.

A commercially available solar cell consists of a silicon wafer of *n* material 1 cm by 2 cm by ½ mm thick and having a thin layer (several microns) of boron (*p* material) diffused on the side to be exposed to light. Connections are made by nickel plating and soldering. The nickel plating on the light-exposure side is restricted to a narrow strip so as not to interfere with the transmission of light.

The efficiency of a photovoltaic cell varies with the spectrum of the light. The maximum theoretical efficiency of a single-junction, single-transition cell with solar illumination is about 22 percent. The efficiency increases substantially with monochromatic light. The practical commercial efficiency of a silicon cell with sun illumination is about 10 percent; with tungsten-lamp illumination, it is about 12 percent.

The open-circuit voltage of a cell varies with the material used and the operating temperature. Open-circuit voltage of a silicon cell (10 percent efficiency) is about 0.6 V at room temperature. This increases or decreases inversely with temperature at a rate of $2 \times 10^{-3}$ V/°C approx. The power output of a 1- by 2-cm, 10 percent efficiency, silicon cell is in the order of $2 \times 10^{-2}$ W at room temperature. The power output varies inversely as temperature at a rate of $2 \times 10^{-5}$ W/°C approx.

Silicon cells have been widely used in space power systems where their long life and light weight are important considerations. Their cost of $500 to $600 per watt, however, has prohibited their use in terrestrial power systems. An added difficulty for any kind of solar-energy conversion process is the intermittent nature of the solar source, which poses the added cost of some form of energy-storage facility.

Other types of cells which have been proposed are cadmium sulfide and gallium arsenide. The former has the potential for lower cost, but at the sacrifice of efficiency and reliability. Gallium arsenide cells offer promise of higher efficiencies, but with costs probably much higher than silicon.

# POWER MISCELLANY

REFERENCES: AAAS, *Science*, Energy Issue, Apr. 19, 1974. Hottel and Howard, "New Energy Technology—Some Facts and Assessments," MIT Press. Fisher, "Energy Crises in Perspective," Wiley-Interscience. Hammond, Metz, and Maugh, "Energy and the Future," AAAS. Energy Self-Sufficiency: An Economic Evaluation, *Technol. Rev.*, May 1974, pp. 23–58. Szego and Kemp, *Chem. Tech.*, May 1973, pp. 275–284. Energy, the Worldwide Crisis, *Consulting Engineer*, Oct. 1974, pp. 53–76. A/A Report on Solar Energy (including photovoltaic, solar electric, satellite, wind, ocean thermal, biomass), *Astronautics & Aeronautics*, Nov. 1975, pp. 16–64.

## INTRODUCTION
### (Staff Contribution)

Many sources of raw energy have been variously proposed or used for the generation of power. Only a few sources—fossil fuels, nuclear fission, and elevated water—are dominant in practical applications today.

A more complete list of sources would include fossil fuels (coal, petroleum, natural gas); nuclear (fission and fusion); wood and vegetation; elevated-water supply; solar; winds; tides; waves; geothermal; muscles (man, animal); industrial, agricultural, and domestic wastes; atmospheric electricity; oceanic thermal gradients; oceanic currents.

Historically, wood, muscles, elevated water, and wind were prominent. These sources were superseded in the industrial era by fossil fuels, with nuclear energy the most recent addition. This dominance rests in the suitability of the thermal sources for practical stationary and transportation power plants. Features of acceptability include reliability, flexibility, portability, maneuverability, size, bulk, weight, efficiency, economy, maintenance, and costs. The plant for transportation service must be self-contained. For stationary service there is wider latitude for choice.

The dominant end product, especially for stationary applications, is electricity because of its favorable distribution and control features. However, there is no practical way of storing electrical energy. Electricity must be generated at the instant of its use. Reliability and continuity of service consequently dictate the need for reserve, alternate, and interconnection supports. Pumped storage, coal piles, and tanks of liquid and gaseous fuels variously offer the necessary continuity, flexibility, and reliability.

Raw-energy sources, other than fuels (fossil and nuclear) and elevated water, are particularly deficient in this storage aspect. For example, wind power is best for jobs that can wait for the wind, e.g., pumping water or grinding grain. Solar power, to avoid foul weather and the darkness of night, could call for desert locations or extraterrestrial satellites.

Despite such limitations an energy-intensive society can expect to see increasing efforts to harness many of the raw-energy sources cited. Several of these topics are treated in the following pages to show the factual and technical progress that has been made to adapt sources to practicality.

## SOLAR ENERGY
### by John I. Yellott

REFERENCES: Daniels, "Direct Use of the Sun's Energy," Yale. Zarem and Erway, "Introduction to the Utilization of Solar Energy," McGraw-Hill. Thekaekara, Solar Energy outside the Earth's Atmosphere, *Solar Energy*, **14**, no. 2, 1973. Moon, Standard Solar Radiation Curves, *Jour. Franklin Inst.*, Nov. 1940. ASHRAE, "Handbook of Fundamentals," 1972. Whillier, Plastic Covers for Solar Collectors, *Solar Energy*, **7**, no. 3, 1964. Yellott, Selective Reflectance, *Trans. ASHRAE*, **69**, 1963. ASHRAE, "Solar Energy Use for Heating and Cooling of Buildings," 1977. Hollands, Honeycomb Devices in Flat-Plate Solar Collectors, *Solar Energy*, **9**, no. 3, 1965. Hay, Natural Air Conditioning with Roof Ponds and Movable Insulation, *Trans. ASHRAE*, **75**, part I, p. 165, 1969. "Solar Cells," National Academy of Science, Washington, D.C., 1972

### Notation

$A, R, T$ = subscripts denoting absorbed, reflected, and transmitted solar radiation

$Btu$ = Btu/(h) (ft²)

$C$ = concentration ratio

$c$ = subscript denoting collector cover

$c_p$ = specific heat of fluid, Btu/(lb) (°F)

$I_{DN}$ = direct normal solar intensity, $Btu$

$I_d$ = diffuse radiation, $Btu$

$I_o$ = radiation intensity beyond earth's atmosphere, $Btu$

$I_r$ = reflected solar radiation, $Btu$

$I_{sc}$ = solar constant; normal incidence intensity at average earth-sun distance, $Btu$

$I_t$ = total solar radiation, $Btu$

$L$ = latitude, deg

$m$ = air mass

$o, i$ = subscripts denoting outgoing and incoming fluid conditions

$q$ = rate of heat flux, $Btu$

$q_I$ = heat flow through insulation, $Btu$

$T_p$ = temperature of absorbing surface, °R

$U$ = overall coefficient of heat transfer, $Btu/°F$

$w_f$ = flow rate of collecting fluid, lb/(h) (ft²)

$\phi$ = solar azimuth, deg from south

$\alpha, \rho, \tau$ = absorptance, reflectance, and transmittance for solar radiation

$\beta$ = solar altitude, deg

$\delta$ = solar declination, deg

$\epsilon$ = emittance for long-wave radiation

$\gamma$ = wall-solar azimuth, deg

$\lambda$ = unit of wavelength, $\mu m$

$\Sigma$ = angle of tilt from horizontal, deg

$\theta$ = incident angle, deg

**Solar-Energy Utilization** Solar energy reaches the earth's surface as short-wave electromagnetic radiation in the wavelength band between 0.3 and 3.0 $\mu$m; its peak spectral sensitivity occurs at 0.48 $\mu$m (Fig. 1). Total solar-radiation intensity on a horizontal surface at sea level varies from zero at sunrise and sunset to a noon maximum which can reach 340 *Btu* (1,070 W/m²) on clear summer days. This endless source of energy, despite its variability in magnitude and direction, can be used in three major processes (Daniels, "Direct Use of the Sun's Energy," Yale. Zarem and Erway, "Introduction to the Utilization of Solar Energy, McGraw Hill.): (1) **Heliothermal,** in which the sun's radiation is absorbed and converted into heat which can then be used for many purposes, such as evaporating seawater to produce salt or distilling it into potable water; heating domestic hot-water supplies; house heating by warm air or hot water; cooling by absorption refrigeration; cooking; generating electricity by vapor cycles and thermoelectric processes; attaining temperatures as high as 6500°F (3600°C) in solar furnaces. (2) **Heliochemical,** in which the shorter wavelengths can cause chemical reactions, sustain growth of plants and animals, convert carbon dioxide to oxygen by photosynthesis, and cause degradation and fading of fabrics, plastics, and paint. (3) **Helioelectrical,** in which part of the energy between 0.33 and 1.3 $\mu$m can be converted directly into electricity by photovoltaic cells. Silicon solar batteries have become the standard power sources for communication satellites, orbiting laboratories, and space probes. Their use for terrestrial power generation is currently under intensive study, with primary emphasis upon cost reduction.

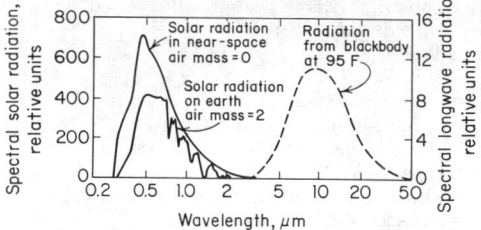

**Fig. 1**  Spectral distribution of solar radiation and radiation emitted by blackbody at 95°F (35°C).

**Solar-Radiation Intensity**  In space at the average earth-sun distance, 92.957 million miles (150 million km), solar-radiation intensity on a surface normal to the sun's rays is 429.2 ± 7 Btu (1,353 ± 21 W/m²). This quantity, called the solar constant $I_{sc}$, undergoes small (±2 percent) periodic variations which affect primarily the short-wave portion of the spectrum (Abbott, in Moon, Standard Solar Radiation Curves, *Jour. Franklin Inst.*, Nov. 1940). Since the earth-sun distance varies throughout the year, the intensity beyond the earth's atmosphere $I_o$ also varies by ±3.3 percent (Table 1). The great seasonal variations in terrestrial solar-radiation intensity are due to the earth's tilted axis, which causes the solar declination

$\delta$ (the angle between the earth's equatorial plane and the earth-sun line) to change from 0° on Mar. 21 and Sept. 21 to −23.5° on Dec. 21 and +23.5° on June 21.

In passing through the earth's atmosphere, the sun's radiation is partially and selectively absorbed, scattered, and reflected by water vapor and ozone, air molecules, natural dust, clouds, and man-made pollutants. Some of the scattered and reflected energy reaches the earth as diffuse or sky radiation $I_d$.

The intensity of the direct normal radiation $I_{DN}$ depends upon the clarity and the amount of precipitable moisture in the atmosphere and the length of the atmospheric path, which is determined by the solar altitude $\beta$ and expressed in terms of the air mass $m$, which is the ratio of the existing path length to the path length when the sun is at the zenith. Except at very low solar altitudes, $m = 1.0/\sin \beta$.

Figure 1 shows relative values of the spectral intensity of solar radiation in space for $m = 0$ (Thekaekara, Solar Energy outside the Earth's Atmosphere, *Solar Energy*, **14**, no. 2, 1973.) and at sea level (Moon, Standard Solar Radiation Curves, *Jour. Franklin Inst.*, Nov. 1940) for a solar altitude of 30° ($m = 2.0$). Table 2 shows the variation at 40° north latitude throughout typical clear summer (June 21) and winter (Dec. 21) days of: solar altitude and azimuth (measured from the south), direct normal radiation, total solar irradiation of horizontal and vertical south-facing surfaces.

The total solar irradiation reaching a terrestrial surface is the sum of the direct, diffuse, and reflected components: $I_t = I_{DN} \cos \theta + I_d + I_r$, where $\theta$ is the incident angle between the sun's rays and a line perpendicular to the receiving surface. $I_r$ is the short-wave radiation reflected from adjacent surfaces.

Direct-beam solar-radiation intensity is measured by **pyrheliometers** with collimating tubes to exclude all but the direct rays from their sensors, which may use calorimetric, thermoelectric, or photovoltaic means to produce a response proportional to the irradiation rate. Similar but uncollimated instruments called **pyranometers** are used to measure the total radiation from sun and sky; when their sensors are shaded from the sun's direct rays, they can also measure the diffuse component.

**Incident-Angle Determination**  The incident angle $\theta$ affects both the direct solar intensity and the solar-optical properties of the irradiated surface. For a flat surface tilted at an angle $\Sigma$ from the horizontal, $\cos \theta = \cos \beta \cos \gamma \cos \phi + \sin \beta \cos \Sigma$. For vertical surfaces, $\Sigma = 90°$; so $\cos \theta = \cos \beta \cos \gamma$; for horizontal surfaces, $\Sigma = 0°$ and $\theta = 90° - \beta$. (See ASHRAE, "Handbook of Fundamentals," chap. 22, for values of solar altitude, azimuth, and direct normal radiation throughout the year for 24 to 56° north latitude.)

**Solar-Optical Properties of Transparent Materials**  When solar radiation with total intensity $I_t$ falls on a transparent material, part of the energy is reflected, part is absorbed, and

**Table 1. Annual Variation in Solar Declination and Solar-Radiation Intensity beyond the Earth's Atmosphere**

| Date | Jan. 1 | Feb. 1 Nov. 10 | Mar. 1 Oct. 13 | Apr. 1 Sept. 12 | May 1 Aug. 12 | June 1 July 12 | July 1 |
|---|---|---|---|---|---|---|---|
| Declination, deg | −23.0 | −17.1 | −7.7 | + 4.4 | + 15.0 | + 22.0 | + 23.1 |
| Ratio, $I_o/I_{sc}$ | 1.033 | 1.029 | 1.017 | 1.000 | 0.983 | 0.971 | 0.967 |
| Intensity, $I_o$: | | | | | | | |
| $Btu$(W/m²) | 443.4 | 441.6 | 436.5 | 429.2 | 421.9 | 416.7 | 415.0 |
| | 1,398 | 1,392 | 1,376 | 1,353 | 1,330 | 1,314 | 1,308 |

**Table 2. Solar Altitude and Azimuth, Direct Normal Radiation, and Total Solar Radiation on Horizontal and Vertical South-Facing Surfaces, June 21 and Dec. 21, for 40° North Latitude**

| June 21, declination = +23.45° | | | | | | | |
|---|---|---|---|---|---|---|---|
| Time: A.M.; P.M. | 6; 6 | 7; 5 | 8; 4 | 9; 3 | 10; 2 | 11; 1 | 12; 12 |
| Solar altitude, deg | 14.8 | 26.0 | 37.4 | 48.8 | 59.8 | 69.2 | 73.5 |
| Solar azimuth, deg | 108.4 | 99.7 | 90.7 | 80.2 | 65.8 | 41.9 | 0.0 |
| Direct normal irradiation, *Btu* | 154 | 215 | 246 | 262 | 272 | 276 | 278 |
| Total irradiation, *Btu:* | | | | | | | |
| On horizontal surface | 60 | 123 | 182 | 233 | 273 | 296 | 304 |
| On vertical south surface | 10 | 14 | 16 | 47 | 74 | 92 | 98 |

| Dec. 21, declination = −23.45° | | | | | | | |
|---|---|---|---|---|---|---|---|
| Time: A.M.; P.M. | 6; 6 | 7; 5 | 8; 4 | 9; 3 | 10; 2 | 11; 1 | 12; 12 |
| Solar altitude, deg | | | 5.5 | 14.0 | 20.7 | 25.0 | 26.6 |
| Solar azimuth, deg | | | 53.0 | 41.9 | 29.4 | 15.2 | 0.0 |
| Direct normal irradiation, *Btu* | | | 88 | 217 | 261 | 279 | 284 |
| Total irradiation, *Btu:* | | | | | | | |
| On horizontal surface | | | 14 | 65 | 107 | 119 | 143 |
| On vertical south surface | | | 56 | 163 | 221 | 252 | 263 |

Values adapted from ASHRAE, "Handbook of Applications," 1974.

the remainder is transmitted. At any instant,

$$I_t = q_T + q_A + q_R = I_t(\tau + \alpha + \rho)$$

The sum of the solar-optical properties $\tau$, $\alpha$, and $\rho$ must equal unity, but the individual values depend upon the incident angle and wavelength of the radiation, the composition of the material, and the nature of any coatings which may be applied to the surfaces.

For uncoated single-strength (³/₃₂-in or 2.4-mm) clear window glass (Fig. 2), solar transmittance at normal incidence ($\theta = 0°$) is approximately 0.90, but the transmittance for long-wave thermal radiation ( 5 $\mu$m) is virtually zero. Thus glass acts as a "heat trap" by admitting solar radiation readily but retaining most of the heat produced by the absorbed sunshine. This "greenhouse effect," which is also exhibited but to a lesser degree by some plastic films (see Whillier, Plastic Covers for Solar Collectors, *Solar Energy*, **7**, no. 3, 1964), is the basis for most heliothermal processes. Heat-absorbing glass (¼ in (6.3 mm) thick (Fig. 2), which absorbs more than 50 percent of the incident solar radiation, is widely used by architects to reduce the heat and glare admitted through unshaded windows. Reflective coatings (Yellott, Selective Reflectance, *Trans. ASHRAE*, **69**, 1963) have been developed to serve similar purposes.

For all types of glass, transmittance falls and reflectance rises as $\theta$ exceeds about 30°. Absorptance increases somewhat owing to the increased path length and then drops off sharply toward zero as $\theta$ exceeds 60°.

**Absorptance and Emittance of Opaque Surfaces**  Opaque materials absorb or reflect all the incident sunshine. The absorptance $\alpha$ for solar radiation and the emittance $\epsilon$ for long-wave radiation at the temperature of the receiving surface are particularly important in heliotechnology. For a true blackbody, the absorptance and emittance are equal and do not change with wavelength. Most real surfaces have reflectances and absorptances which vary with wavelength (Fig. 3). Aluminum foil has a consistently low absorptance and high reflectance over the entire spectrum from 0.25 to 25 $\mu$m, while black paint has a high absorptance and low reflectance. White

paint, however, has low short-wave (solar) absorptance, but beyond 3 $\mu$m its absorptance and reflectance are virtually the same as for black paint.

Solar collectors require a high $\alpha/\epsilon$ ratio, while surfaces which should remain cool, such as rooftops or space vehicles, should have low ratios since their objective usually is to absorb as little solar-radiation and emit as much long-wave radiation

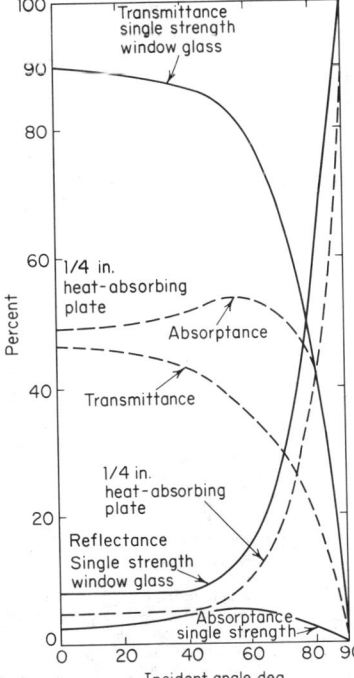

**Fig. 2**  Variation with incident angle of solar-optical properties of clear and heat-absorbing glass (¼ in = 6.3 mm).

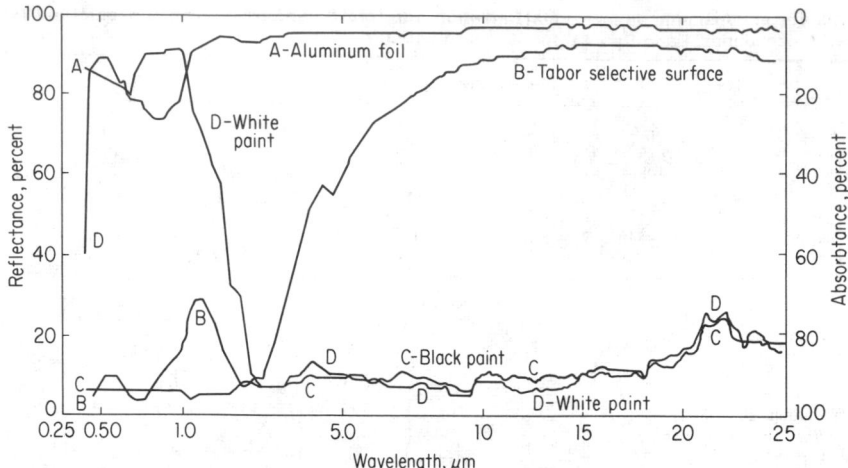

**Fig. 3**   Variation with wavelength of reflectance and absorptance for opaque surfaces.

as possible. Special surface treatments have been developed (see ASHRAE, "Solar Energy Use for Heating and Cooling of Buildings," 1977) for which the ratio $\alpha/\epsilon$ is above 7.0, making them suitable for solar collectors; others with ratios as low as 0.15 are useful as heat rejectors for space applications (see Table 3).

**Equilibrium Temperatures for Concentrating Collectors**   When a surface is irradiated, its temperature rises until the rate of solar-radiation absorption equals the rate at which heat is removed from the surface. If no heat is intentionally removed, the maximum temperature which can be attained by a blackbody ($\alpha = \epsilon$) is found from $I_{DN} C \alpha = 0.1713 \epsilon (T_p/100)^4$, where $C$ is the concentration ratio. Figure 4 shows the variation of blackbody equilibrium temperatures for earth and near-space where $I_{DN} = 320$ and $I_o = 430$ Btu (1,000 and 1,353 W/m²).

For flat-plate collectors, $C = 1.0$; so their maximum attainable temperatures are below 212°F (100°C) unless a selective surface is used with $\alpha/\epsilon > 1.0$, or both radiation and convection loss are suppressed by the use of multiple glass cover plates. Only the direct component of the total solar radiation can be concentrated, and concentrating collectors must follow the sun's apparent motion across the sky or use heliostats which serve the same function. Diffuse radiation cannot be concentrated effectively.

**Flat-Plate Collectors**   Direct, diffuse, and reflected solar radiation can be collected and converted into heat by flat-plate collectors (Fig. 5). These generally use blackened metal plates which are finned, tubed, or otherwise provided with passages through which water, air, or other fluids may flow and be heated to temperatures as much as 100 to 150°F (55 to 86°C) above the ambient air. The actual temperature rise may be estimated from the heat balance for a unit area of collector surface:

$$q_A = I_t \tau_c \alpha_p = w_f c_p (t_o - t_i) + q_I + U (t_p - t_a)$$

**Table 3. Solar Absorptance, Long-wave Emittance, and Radiation Ratio for Typical Surfaces**

| Surface or material | Shortwave (solar) absorptance, $\alpha$ | Long-wave emittance, $\epsilon$ | Radiation ratio, $\alpha/\epsilon$ |
|---|---|---|---|
| Flat, oil-based paints: | | | |
| Black | 0.90 | 0.90 | 1.00 |
| Red | 0.74 | 0.90 | 0.82 |
| Green | 0.50 | 0.90 | 0.55 |
| Aluminum | 0.45 | 0.90 | 0.50 |
| White | 0.25 | 0.90 | 0.28 |
| Whitewash on galvanized iron | 0.22 | 0.90 | 0.25 |
| Building materials: | | | |
| Asbestos slate | 0.81 | 0.96 | 0.84 |
| Tar paper, black | 0.93 | 0.93 | 1.00 |
| Brick, red | 0.55 | 0.92 | 0.59 |
| Concrete | 0.60 | 0.88 | 0.68 |
| Sand, dry | 0.82 | 0.90 | 0.92 |
| Glass | 0.04–0.70 | 0.84 | |
| Metals: | | | |
| Copper, polished | 0.18 | 0.04 | 4.50 |
| Copper, oxidized | 0.64 | 0.60–0.90 | 1.03–0.71 |
| Aluminum, polished | 0.30 | 0.05 | 6.00 |
| Selective surfaces: | | | |
| Tabor, electrolytic | 0.90 | 0.12 | 7.50 |
| Silicon cell, uncoated | 0.94 | 0.30 | 3.13 |
| Black cupric oxide on copper | 0.91 | 0.16 | 5.67 |

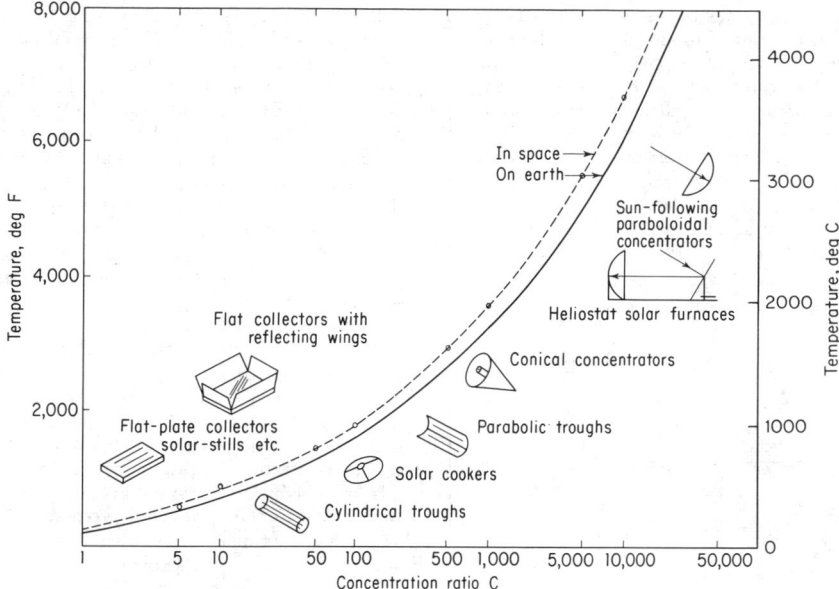

**Fig. 4**  Variation with concentration ratio of equilibrium temperatures for earth and space.

The loss from the back of the collector plate $q_l$ can be minimized by the use of adequate insulation. The radiation component of the loss from the upper surface can be reduced (Zaren and Erway, "Introduction to the Utilization of Solar Energy," McGraw-Hill; ASHRAE, "Solar Energy Case for Heating and Cooling of Buildings") by using selective-reflectance coatings with high $\alpha/\epsilon$ ratios and by using covers which are transparent to solar radiation but opaque to long-wave emissions (see Table 4). Both convection and radiation can be reduced by the use of honeycomb structures in the air space between the cover and the collector plate (see Hollands, "Honeycomb Devices in Flat-Plate Solar Collectors," *Solar Energy*, **9**, no. 3, 1965).

### Table 4. Transmittance and Overall Heat-Transfer Coefficients for Collectors with Glass and Plastic Covers

| Type and number of covers | None | One glass | One plastic | Two glass | Two plastic |
|---|---|---|---|---|---|
| Solar transmittance | 1.00 | .90 | 0.92 | 0.81 | 0.85 |
| Overall coefficient $U$ | 3.90 | 1.12 | 1.30 | 0.71 | 0.87 |

### Applications of Heliotechnology

**Solar Stills**  The glass-covered roof-type solar still (Fig. 6*a*) is in wide use in arid areas for the production of drinking water from salty or brackish sources. The sun's rays enter through the cover glasses, warm the water, and thus produce vapor which condenses on the inner surface of the cover. The water droplets coalesce and flow downward into the discharge troughs, while the remaining brine is periodically replaced with a new supply of nonpotable water. Daily yield ranges from 0.4 lb/ft² (2 kg/m²) of water surface in winter to 1.0 lb/ft² (5 kg/m²) in summer (Hay, Natural Air Conditioning with Roof Ponds and Movable Insulation, *Trans. ASHRAE*, **75**, part I, p. 165, 1969).

Inflated plastic films (Fig. 6*b*) have also been used to cover solar stills, but their greatest success has been achieved in controlled-environment greenhouses where the vapor which transpires from plant leaves is condensed and reused at the plant roots.

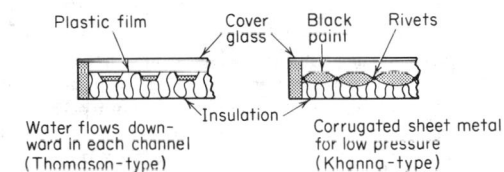

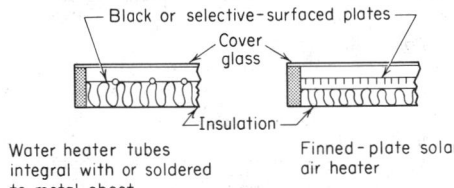

**Fig. 5**  Typical flat-plate solar-radiation collectors.

**Solar Water Heaters**  The simple thermosiphon solar water heater (Fig. 7) with a glass-covered flat-plate collector is used in thousands of homes in Australia, Japan, and North Africa. Under favorable climatic conditions (abundant sunshine and moderate winter temperatures) they can produce 30 to 50 gal (110 to 190 l) of water at temperatures up to 160°F (70°C) in summer and 120°F (50°C) in winter. Auxiliary electric heaters are often used to produce higher temperatures during unfavorable winter weather.

**Solar House Heating, Heat Storage**  House heating can be

accomplished in temperate climates by collecting solar radiation with flat-plate devices (Fig. 5) mounted on south-facing roofs or walls (in the northern hemisphere). Water or air, warmed by solar radiation, can be used in conventional heating systems, with small auxiliary fuel-burning apparatus available for use during protracted cloudy periods. Excess heat collected during the day can be stored for use at night in insulated tanks of hot water or beds of heated gravel. Heat-of-fusion storage systems, which use salts that melt and freeze at moderate temperatures, may also be used to improve heat-storage capacity per unit volume.

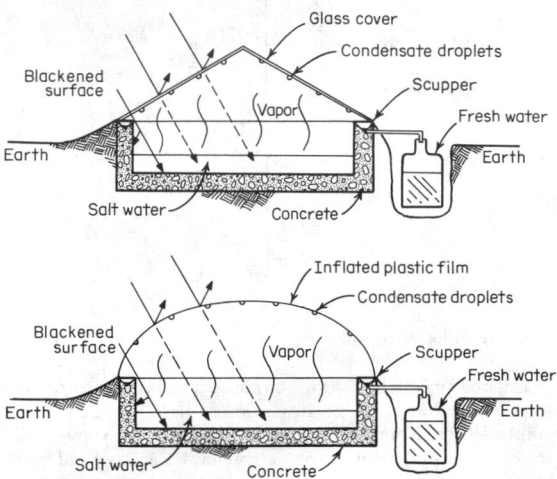

**Fig. 6**  Shallow basin horizontal-surface solar stills. *(a)* Glass-covered roof type; *(b)* inflated-plastic type.

**Solar airconditioning and refrigeration** can be done with absorption systems supplied with moderately high-temperature (250°F or 120°C) working fluids from high-performance flat-plate collectors. The economics of solar-energy utilization for domestic purposes become much more favorable when the same collection and storage apparatus can be used for both summer cooling and winter heating. One such system (see Hay, Natural Air Conditioning with Roof Ponds and Movable Insulation, *Trans. ASHRAE*, **75**, part I, p. 165, 1969) uses a combination of shallow ponds of water on horizontal rooftops with panels of insulation which may be moved readily to cover or uncover the water surfaces. During the winter, the ponds are uncovered during the day to absorb solar radiation and covered at night to retain the absorbed heat. The house is warmed by radiation from metallic ceiling panels which are in thermal contact with the roof ponds. During the summer, the ponds are covered at sunrise to shield them from the daytime sun, and uncovered at sunset to enable them to dissipate heat by radiation, convection, and evaporation to the sky.

**Solar cooking** utilizes (1) a sun-following broiler-type device with a metallized parabolic reflector and a grid in the focal area where cooking pots can be placed; (2) an oven-type cooker comprising an insulated box with glass covers over an open end which is pointed toward the sun. When reflecting wings are used to increase the solar input, midday temperatures as high as 400°F are reached.

**Solar Furnaces** Precise paraboloidal concentrators can focus the sun's rays upon small areas, and if suitable receivers are used, temperatures up to 6500°F can be attained. The concentrator must be able to follow the sun, either through movement of the paraboloidal reflector itself (Fig. 4) or by the use of a heliostat which tracks the sun and reflects the rays along a horizontal or vertical axis into the concentrator.

**Power from Solar Energy**  During the past century (Zarem and Erway, "Introduction to the Utilization of Solar Energy," McGraw-Hill) many attempts have been made to generate power from solar radiation through the use of both flat-plate and concentrating collectors. Hot-air and steam engines have operated briefly, primarily for pumping irrigation water, but none of these attempts has succeeded commercially because of high cost, intermittent operation, and lack of a suitable means for storing energy in large quantities. With the rapid rise in the cost of conventional fuels and the increasing interest in finding pollution-free sources of power, attention has again turned to parabolic-trough concentrators and selective surfaces (high $\alpha/\epsilon$ ratios) for producing high-temperature working fluids for Rankine and Brayton cycles.

**Direct Conversion of Solar Radiation into Electricity** Photovoltaic cells made from silicon, cadmium sulfide, gallium arsenide, and other semiconductors (see "Solar Cells," National Academy of Science, Washington, D.C., 1972) can convert solar radiation directly into electricity without the intervention of thermal cycles. Of primary importance today are the silicon solar batteries which are used in large numbers to provide power for space probes, orbiting laboratories, and communication satellites. Their extremely high cost and relatively low efficiency have thus far made them noncompetitive with conventional power sources for terrestrial applications, but intensive research is currently underway to reduce their production cost and to improve their efficiency.   Generation

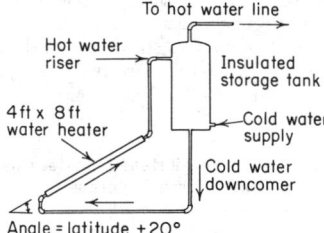

**Fig. 7**  Thermosiphon type of solar water heater.

of power from solar radiation on the earth's surface encounters the inherent problems of intermittent availability and relatively low intensity. At the maximum noon intensity of 340 *Btu* (1,080 mW/m²) and 100 percent energy conversion, 10 ft² (1.1 m²) of collection area would produce 1 thermal kilowatt, but with a conversion efficiency of 10 percent, the area required for an electrical kilowatt approaches 100 ft² (9.3 m²). Thus very large collection areas are essential, regardless of what method of conversion may be employed. However, the total amount of solar radiation falling on the arid southwestern section of the United States is great enough to supply all the nation's electrical needs, provided that the necessary advances are made in collection, conversion, and storage of the unending supply of energy from the sun.

## GEOTHERMAL POWER
### by V. F. Estcourt

REFERENCES: Austin, "Prospects for Advances in Energy Conversion Technologies for Geothermal Energy Development," *Proceedings of Second United Nations Symposium in Development and Use of Geothermal Resources*, 1975. Beaulaurier, "Binary Thermodynamic Power Cycles Applied to Geothermal Fluids," Bechtel Corp., 1974. Kruger and Otte, "Geothermal Energy," Stanford University Press, 1973. Brewer, Eriksen, and Prindle, "Discovery and Development of Geothermal Resources," Joint Power Conference, 1971. Wood, "Geothermal Power," UNESCO *Earth Sciences*, **12**, 1973. Armstead, *ibid*.

Although the generation of electric power from certain categories of geothermal heat sources has been well established for a good many years as a result of installations especially in Italy, New Zealand, and the United States, progress in the utilization of other known sources has been relatively slow. Steady growth in interest in the development of geothermal power has occurred since the United Nations Conference on New Sources of Energy in Rome in 1961. Further stimulation resulted from the General Conference of UNESCO in 1966 by a resolution designed to promote, among other things, international cooperation in research and exploration of geothermal-energy sources; and recently (1975) from the Second United Nations Symposium on the Development and Use of Geothermal Resources, in San Francisco.

As a result of the present critical energy situation, there has been an intensified interest and public pressure for more vigorous development of this resource for both power and direct use in heating cities, industries, and agriculture.

Traditionally the concept of geothermal power has emphasized dry or wet steam sources (partly flashed steam from hot water at or near saturation temperature). These categories are commonly referred to, respectively, as vapor-dominated and liquid-dominated. It is only in recent years that more serious thought has been given to the optimal utilization of geothermal hot water, geopressured deep reservoirs of hot water, hot dry rock, and magma formations.

**Geothermal Sources** Existing sources which emit dry and slightly superheated steam are exemplified by the installations at Lardarello in Italy and the Geysers in California. One liquid-dominated well in the Salton Sea area reportedly had a bottom-hole temperature of 800°F (427°C) at a depth of 8,100 ft (2,469 m). This is believed to be the hottest geothermal steam discovered to date. Examples of liquid-dominated wells which tend to produce dry steam (because of low fluid content) are Valles Caldera Region, New Mexico; the Monte Amiata Region, Italy; and Matsukawa and Onikobe, Japan. Naturally occurring two-phase mixtures of water or brine and steam (liquid-dominated) at elevated temperatures are exemplified in fields located at Wairakei and Broadlands in New Zealand, Otake in Japan, Pauzhetsk and Paratunka in U.S.S.R., and Cerro Prieto in Mexico and Imperial Valley in California (both in the same geologic province known as the Salton Trough).

Table 1 provides a list of *existing and planned electrical capacity*. These figures must be considered as approximate only, and frequently changing, because of modifications in plans and significant discrepancies in the various published data. Another factor is the changing scene with respect to available steam from the wells caused by loss of pressure and new wells added. For example, it is of interest to note that the present capacity of 160 MW for New Zealand is made up of 150 MW

for Wairakei and 10 MW for Kawarau. This is in contrast to the earlier capability of approximately 192 MW for Wairakei only. The reduction at this time is principally the result of a judgment decision based upon declining well pressure. It has been reported (*Electrical World*, Jan. 1, 1973) that the plant is "undergoing major modifications" in order to "keep the plant at 152 MW in 1982, instead of letting it drop to 100 MW as has been forecast." A more recent report, however, indicates that this program has been dropped, but not for technical reasons.

**Table 1. Geothermal Power, Approximate Capacity, MWe, 1975 and Planned to 1980\***

| Country (1) | 1975 (2) | Planned additional† prior to 1980 (3) | Total cols. 2 + 3 (4) |
|---|---|---|---|
| U.S.A, | 502 | 641 | 1,143 |
| Italy | 405.6 | 75 | 480.6 |
| New Zealand | 160‡ | 370 | 530 |
| Mexico | 75.5 | 75 | 150.5 |
| Japan | 43 | 100 | 143 |
| U.S.S.R. | 5.4 | 20 | 25.4 |
| Iceland | 2.5 | 32 | 34.5 |
| El Salvador | 30 | 60 | 90 |
| Guadalupe | | 30 | 30 |
| Turkey | | 20 | 20 |
| Totals | 1,224 | 1423 | 2,647 |

\*The figures in col. 2 are intended to represent the existing net capability of the plant, taking into consideration any limitations in available steam supply from the wells.

†Planned additional capacity (col. 3) represents an estimate of those planned expansions most likely to occur by 1980. Numerous other additions are in the planning stage, and progressive revisions (upward or downward) may be expected.

‡Presently restricted by reduced well pressure.

Undeveloped-resource categories include hot dry rock, geopressured zones, and magma. Of these three, and with the possible exception of liquid-dominated reservoirs, hot dry rock was considered by Congress as having the best potential of relatively short-range development. Subsequent congressional actions established a Geothermal Energy Coordination and Management Project under a proposed six-member task force to explore the feasibility and practicability of generating electricity from hot dry rock, geopressured zones, and hot-water systems. However, it is now reported that a broader study will be carried out under the more recent Geothermal Energy Research, Development and Demonstration Act with responsibilities presently falling under the Energy Research and Development Administration (ERDA). This exemplifies the political uncertainties involved in the development of commercial applications of geothermal power in the short term.

The *hot dry rock* category is defined as a geological formation having a high heat content but not containing meteoric or magmatic waters to provide a heat-transport medium. Thus the injection of water is required in order to carry heat to the surface, but the thermal conductivity of rock may limit power output. Vast amounts of energy stored in this form have been discovered in such localities as the Craters of the Moon, Idaho; and Marysville, Montana. In any event, because of the claimed large amount of this heat source, there is an immediate need to determine whether it is technically exploitable.

*Geopressured zones* occurring at levels between 5,000 and 20,000 ft (1,524 and 6,096 m) contain water varying widely in salinity. The fluid temperatures in such reservoirs vary from about 140°F (60°C) to 360°F (182°C) at fluid pressures from about 3,000 to 14,000 psig ($21 \times 10^6$ to $96 \times 10^6$ N/m²). The largest zones in this category in the United States exist in the northern Gulf of Mexico basin, in the area of the Texas, Louisiana, and Mississippi coasts. These are beneath the Gulf Continental Shelf. There are other zones of lesser extent in scattered locations throughout the United States.

*Magma* formations of molten rock have temperatures as high as 1100°F (593°C). In most regions in the continental United States such resources occur at depths of 100,000 ft (30,500 m) or more. However, in Hawaii and Alaska, magma systems exist at considerably shallower depths, although proximity to volcanic regions might be a deterrent.

**Exploration Technology** Geothermal sites historically have been identified from obvious surface manifestations such as hot springs, fumaroles, and geysers. Some discoveries also have been made accidentally while exploring or drilling for other natural resources. This approach essentially has been replaced by more scientific prospecting methods with techniques for appraising the extent, as well as the physical and thermodynamic properties, of the reservoir. Modern methods include geological studies involving aerial, surface, and subsurface investigations (including remote infrared sensing), and geochemical analyses which provide a guide for selecting specific drilling sites. Geophysical methods include drilling, measuring the temperature gradient in the drill hole, and measuring the thermal conductivity of rock samples taken at various depths.

Comprehensive discussions of these exploration methods are given by Brewer and others. The use of such techniques can be very costly, and the findings do not necessarily support a conclusion that the development of a particular geothermal field would be economically feasible. Thus, even though successfully developed fields have usually demonstrated the low cost of geothermal power, the financial outlay and risk of undertaking the initial exploration are sometimes very high. This offers a powerful deterrent to uninhibited prospecting activity—at least comparable with similar risks in oil and gas exploration.

**Environmental Considerations** Geothermal wells in different areas contain steam and water which vary over a wide range as to pressure, temperature, relative amounts of water and steam, mineral content, and noncondensable gases. Effluents carry a variety of noncondensable gases which may constitute a potentially unacceptable *environmental impact*. The most common and abundant are $CO_2$, $H_2S$, $NH_3$, and $CH_4$. Release of $H_2S$ (hydrogen sulfide) to the atmosphere produces an objectionable odor and, in concentrated amounts, can be dangerous to human and animal life. $CH_4$ (methane) is in solution in substantial amounts in the liquids of some geopressured zones and boils out of solution when the pressure is released. If present in greater than nuisance amounts, it might be recovered as an economic asset. The extent to which contaminants exist varies significantly as to both type and quantity in the known geothermal sources around the world.

It appears that very little work has been done on the release of radioactive gases from geothermal sources. The state of California has placed a limit of 3.0 picocuries/litre, and it is reported that measurements at geothermal discharges are well below this value. Studies in the Wairakei geothermal region were initiated early in 1975. An interesting exchange of letters on this subject has been reported in *Science* **189**, pp. 328–330, Aug. 1975).

The commercial development of these resources, particularly for electric power, also involves environmental considerations relative to land use, waste disposal, groundwater contamination, heat rejection, noise emissions, land subsidence, and induced earthquakes. In general, liquid wastes from geothermal plants are charged with dissolved solids such as sodium, calcium and potassium chloride, boron, arsenic, and other chemicals in a wide range of combinations and concentrations which in many instances preclude disposal in natural watercourses.

The potential environmental degradation varies greatly with both the process cycle and the type of energy-conversion process. Thus the foregoing discussion represents only a broad summary of the scope of the problems which may be encountered. The nature of the specifics applicable to a particular site may vary widely.

**Electrical Energy Conversion** The most appropriate conversion process in each instance is related to the chemical and thermodynamic properties of the effluents. Broadly speaking, there are two basic processes—direct and binary. Specifically, there are various concepts of each (Fig. 1), particularly with respect to the *direct process*. One example involves a system wherein a liquid-dominated reservoir delivers steam mixed with hot water or brine. Additional steam is produced in a separator by flashing the hot liquid at a reduced pressure, thereby separating the two phases. The steam fraction is then fed to a turbine. If pressure is sufficiently high in the first flash, the remaining liquid may be economically flashed a second time (and possibly a third time) at a lower pressure and can be fed to a lower-pressure turbine or a lower-pressure stage of the same turbine. The separated brine fraction is discarded.

Some investigators have been of the opinion that the generation of power from hot water by flashing it into steam at one or more temperature levels results in a relatively low cycle efficiency, partly because of the heat lost in the discarded brine fraction. However, because of the present lack of more definitive test information, the economics remain to be demonstrated.

Another form of the direct process is described as the *total-flow concept* (Fig. 1). The total discharge of steam and water is delivered directly to a mixed-phase expander. In one scheme, the enthalpy of the mixture is converted into kinetic energy in the form of low-temperature, high-velocity jets to drive a single-stage axial-or tangential-flow impulse turbine. Since all the available energy in the field is utilized in this manner, this method has the potential for achieving higher thermal efficiency than either the flashed-steam or binary process.

It is concluded by Austin that "the total flow concept offers the potential for more efficient utilization of water-dominated resources. . . . Over the entire temperature range, the total flow process has 60 to 30 percent greater specific power output than the single flash and double flash systems, respectively." The three basic systems are illustrated in Fig. 1, *a*–*b*, *c*, and *d* and the comparative dynamics are shown in Fig. 2, which compares an optimized single- and double-flash system, an optimized binary-cycle system, and total-flow method.

Some of the relative disadvantages of the flashed-steam systems can be avoided by application of the binary process (Fig. 1*c*), which employs a secondary working fluid such as

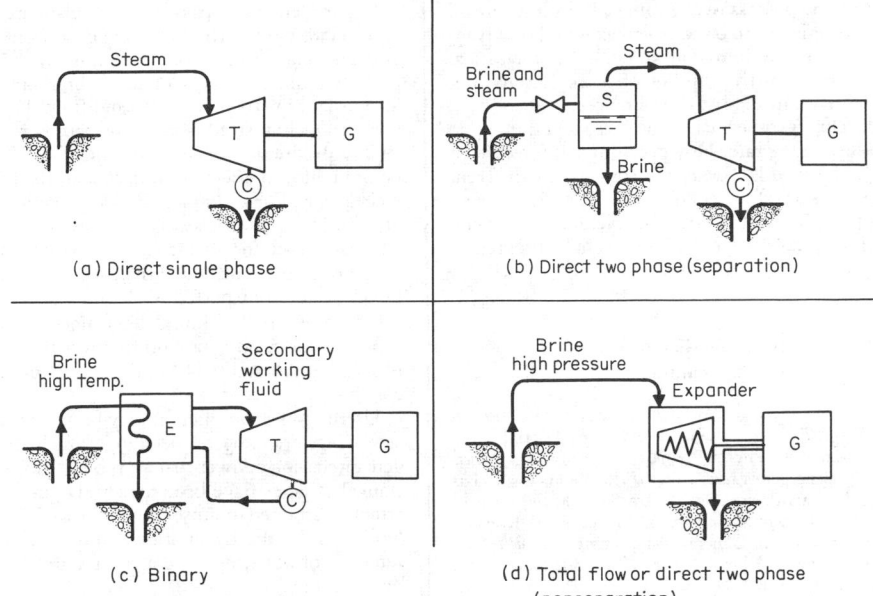

**Fig. 1** Geothermal conversion processes. T = turbine. G = generator. C = condenser. S = separator. E = heat exchanger. (*Bechtel.*)

isobutane or Freon. The fluid passes through a heat exchanger in which it absorbs heat from the well effluent (steam and/or water) without becoming contaminated by the effluent itself. It then expands through the turbine to convert its energy into useful power. The disadvantages relate to the presence of

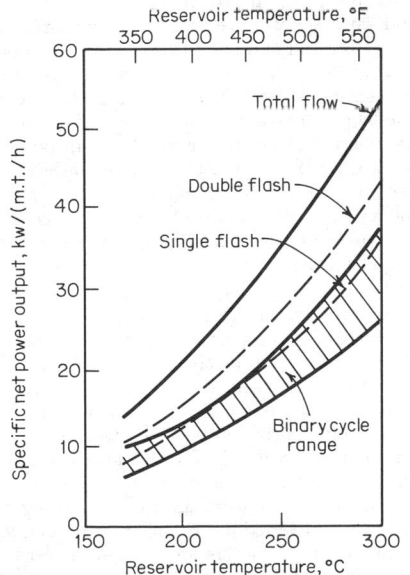

**Fig. 2** Relative performance characteristics of geothermal energy-conversion concepts. Binary system assumes down-hole pumping. Others work from self-pumped well (flashing flow). All systems work to 49°C (120.2°F) sink conditions. m.t./h = metric tons per hour = 2,200 lb/h. (*Arthur L. Austin.*)

significant quantities of noncondensables which provide a potential problem of scaling and erosion of moving parts.

The binary-cycle concept has the unique advantage of providing the means to accommodate large quantities of noncondensable gases. On the other hand, corrosion and plugging of the heat exchanger can be a formidable problem with geothermal effluents containing large quantities of total dissolved or entrained solids. The use of a down-well pump may be required to inhibit precipitation. A fairly recent (1967) demonstration of the binary process was the Kamchatka plant (U.S.S.R.), having a net capacity of 0.44 MWe.

The practical problems of corrosion and fouling in the total-flow process would be focused in the turbine as compared with the heat exchanger in the binary cycle. The relative economics of all systems could be significantly affected by the ultimate requirements for materials tolerant to the various brines, and other special provisions to combat these problems. These advanced concepts are currently being investigated by both privately funded and government-funded development programs. They are directed toward small-scale (approximately 10 MWe) pilot plants prior to full-scale demonstration systems. The problems are severe and depend on the physical and chemical characteristics of the effluent at a particular site. In general, it is easier to establish theoretical validity and pilot-scale performance than to solve the full-scale problems of a particular installation. It is likely that no one universal system will emerge. Instead, a variety of systems may be needed, each tailored for a specific resource.

Economic considerations probably will be the determining factor and should include, in addition to capital investment, an analysis of expected operating availability, maintenance costs, environmental impact, and other factors relating to cost effectiveness.

The generalized evaluations given above are not intended as

an endorsement of one process over another, but rather are an effort to point out some of the basic advantages and disadvantages that must be taken into consideration in their final development. More exhaustive studies of these various processes can be identified in the list of references.

Efforts to develop geothermal power are continuing to intensify at an accelerating rate. However, current forecasts of potential energy from this source, according to different authorities, vary by several orders of magnitude. More accurate predictions probably will not be available until the several ongoing research and development programs have progressed to a more advanced stage.

## MAN- AND ANIMAL-GENERATED POWER
### by E. S. Krendel

REFERENCES: Bink, The Physical Working Capacity in Relation to Working Time and Age, *Ergonomics*, June 1962. Bonjer, Actual Energy Expenditure in Relation to the Physical Working Capacity, *Ergonomics*, June 1962. Brody, "Bioenergetics and Growth," Reinhold, 1945. Harrison, Maximizing Human Power Output by Suitable Selection of Motion Cycle and Load, *Human Factors*, **12**, 1970, p. 3. Krendel, Design Requirements for Man-Generated Power, *Ergonomics*, Oct. 1960. Krendel, Man-Generated Power, *Mech. Eng.*, **82**, 1960, p. 36. McMahon, Size and Shape in Biology, *Science*, **179**, 1973. Suggs, The Effect of Load on Muscle Output, *Human Factors*, **11**, 1969, p. 3. Wilkie, Man as a Source of Mechanical Power, *Ergonomics*, Jan. 1960.

NOTE: $kW = hp \times 0.746$. $kg = 16 m \times 0.454$.

The use of human muscles to generate work will be examined from two points of view. The first is that of measuring the energy expended in gross, long-duration physical activities such as marching, forestry work, freight handling, and factory work. The second is that of determining the useful mechanical work which can be performed by specified muscle groups for brief or extended periods of time in well-defined work situations, such as pedaling or cranking.

Over an 8-h day for a 48-h week, a useful norm for a thirty-five-year-old European laborer for total power expenditure, including basal-metabolism energy, is 0.49 hp. Of this total expenditure, approximately 0.1 hp is available for useful work. A twenty-year-old man can generate about 15 percent more power than this norm, and a sixty-year-old man about 20 percent less. The total energy or power expenditure is needed for determining nutritional requirements for classes of labor. A rule of thumb for power developed by European males can be expressed as a function of age and duration of effort in minutes for work lasting from 4 to about 480 min, assuming that 20 percent of the total output is useful power.

| Age of man | Useful horsepower ($t$ in min) |
|---|---|
| 20 | $hp = 0.39 - 0.104 \log t$ |
| 35 | $hp = 0.35 - 0.092 \log t$ |
| 60 | $hp = 0.29 - 0.077 \log t$ |

Work scheduling, either as rhythmic work activity or with rest stops for recuperation, the temperature and humidity of the environment, and the detailed nature of the laborer's diet are factors which influence ability to generate and maintain the above nominal power values. These considerations should be factored in for specific work situations.

When man and a passive mechanism are working together to generate power, the following conditions obtain: Energy is available both from stores residing in the muscles (a total usefully available energy of about 0.6 hp-min, usually applied in transient bursts of activity) and from the oxidation of foods (for producing steady-state power). For transient activity, energy production depends on the mass of muscle which can be brought into effective contact with the power-transmission mechanism. For example, bicycle pedaling is an effective use of a large muscle mass. For steady-state activity, assuming adequate food for fuel energy, production depends on the oxygen supply and the efficiency with which oxygenated blood can be transported to the muscles.

The physiological limit, determined by oxygen-respiration capacity, for steady-state useful mechanical-power generation is between 0.4 and 0.54 hp, depending on the man's physical condition.

Useful power production may be achieved by such methods as rowing, cranking, or pedaling. The highest values of man-generated horsepower using robust subjects but not well-trained athletes have been achieved using a rowing assembly which restrained nonuseful motions of the torso and major limbs. Under these conditions up to 2 hp was generated over intervals of 0.6 s, and averages of about 1 hp were generated over 2 min.

The data from diverse sources of data for human power production over intervals from 6 to 120 s may be summarized to provide a rule of thumb as follows ($t$ in seconds):

$$hp = 2.5 t^{-0.33}$$

For pedaling efforts of from 1 to about 100 min, the useful power generated may be expressed as $hp = 0.53 - 0.13 \log (t$ in min).

In order to approach an optimal conversion (mechanical work/food energy) of 25 percent, a mechanism would be required to store and to transmit energy from the body muscle masses when they were operating at optimal efficiency. This condition occurs when the force exerted by the muscle is about one-half its maximum and the speed of muscle movement one-quarter of its maximum. Data on both force and speed for a given set of muscles are best measured in situ. Optimal conversion efficiency and maximum output power do not occur together. Maximum power output occurs at a load impedance of five to ten times the size of the human being's source impedance.

Brody has developed detailed nomograms for determining the energetic cost of muscular work by farm animals; these nomograms are useful for precise cost-effectiveness comparisons between animal and mechanical power-generation methods. A 1,500- to 1,900-lb (680- to 860-kg) horse can work continuously for up to 10 h a day at a rate of 1 hp, or equivalently pull 10 percent of its body weight for a total of 20 miles/day, and retain its vigor to an advanced age. Brody's work allows the following approximations for estimating the useful power output of work animals of varying sizes: The ratio of the power exerted in maximal energy production for a few seconds to the maximum steady-state power maintained for 5 to 30 min to the power produced in sustained heavy work over a 6- to 10-h day is approximately 25:4:1. For any one of these conditions, it has been found that, for healthy mature specimens,

$$hp_{animal} = hp_{man} \, (mass\ of\ animal/mass\ of\ man)^{0.73}$$

which is consistent with Kleiber's law. Thus, from the previously given horsepower magnitudes for men, one can compute the power generated by work animals.

## HOT-AIR ENGINES
### by Erich A. Farber

Hot-air engines are heat engines in which air or other gases such as $H_2$, He, and $N_2$ are used as the working fluid operating on the Stirling or Ericsson cycles (see Sec. 4) or modifications of them. While the earlier engines of this type were bulky, slow in speed, and low in efficiency, a number of new developments have rectified these deficiencies. Hot-air engines are multifuel engines and have been driven by solid, liquid, and gaseous fuels and with concentrated solar energy. They are quiet-running, relatively simple in construction, and if used with solar energy, without waste products.

**The Philips Hot-Air Engines** (*Philips Tech. Rev.*, **8**, 1936, pp. 129–136; **9**, 1947, pp. 97–104, 125–134; **20**, 1959, pp. 245–262; **31**, 1970, pp. 168–185; ASME 72-WA/Ener-9) The Philips Laboratory (in Holland) seems to have developed the first efficient, compact hot-air engine. It operates at 3,000 r/min, with hot chamber temperature of 1200°F (650°C), maximum pressure of 50 atm, and mep of 14 atm (14.1 bar). The regenerator consists of a porous coil of thin wires having 95 percent efficiency, saving about three-fourths of the heat required by the working fluid. The exhaust gases preheat the air, saving about 70 percent of this loss.

Single-cylinder engines, up to 90 hp (67 kw), and multi-cylinder engines of several hundred horsepower have been constructed with mechanical efficiencies of 90 percent and thermal efficiencies of 40 percent. Heat pipes incorporated in the design improve the heat-transfer characteristics. Philips hot-air engines have been installed in clean-air buses on an experimental basis. Exhaust estimates for an 1,800-kg car are $C_xH_y$ 0.02 g/mile (0.012 g/km), CO 1.00 g/mile (0.62 g/km), NO (25 percent recirculated) 0.16 g/mile (0.099 g/km).

**GMR Stirling Thermal Engine** (*Trans. SAE*, **68**, 1960, pp. 665–684; GM GPU-3, 69; GM STIR-LEC I, 1969). A cooperative program between the Philips Research Laboratory and the General Motors Corp. resulted in the development of two engines. One, weighing 450 lb (200 kg) and operating at a mean pressure of 1,500 lb/in² (103.4 bar), produces 30 hp (22 kN) at 1,500 r/min with 39 percent efficiency and 40 hp (30 kW) at 2,500 r/min with 33.3 percent efficiency. The other weighing 127 lb (57 kg) and operating at a mean pressure of 1,000 lb/in² (6.9 MN/m²), produces 6 hp (4.5 kW) at 2,400 r/min with 29.6 percent efficiency and 8.63 hp (6.4 kW) at 3,600 r/min with 26.4 pecent efficiency.

One such engine was used for a portable Stirling engine–electric-generator set; another was installed in a Stirling engine–electric hybrid car.

**Internally Focusing Regenerative Gas Engine** (ASME 61-WA-297.) This type of engine, conceived at the Solar Energy Laboratory of the University of Wisconsin, uses solar energy concentrated by a parabolic reflector and directed through a quartz dome upon an internally located absorber. This reduces the heat losses, since the engine has no external high-temperature heat-transfer surfaces. A small working model of this type of engine has been built at Battelle Memorial Institute and was demonstrated driving a small fan.

**Fractional-Horsepower Solar Hot-Air Engines** (ASME 64-WA/SOL-5; ASME 69-WA/SOL-5. BSD, June 1972, pp. 25–33, *ISES J.*, Feb. 1973, pp. 241–252.) The Solar Energy Laboratory of the University of Florida has developed small (¼ to ⅓ hp) (0.186 to 0.25 kW) solar hot-air engines (converted lawnmower engines). Some of these engines were self-supercharging to increase power. Water injection, self-acting, increased power by 19 ± percent. The average speed of closed-cycle engines is about 500 r/min and the average conversion efficiency 9 percent. Open-cycle engines separate the heating process from the working cycle so that they can be operated at much higher speeds. Any heat source can be used with these engines. They are simple, rugged, and designed for possible use in developing countries.

**The Stirling Engine for Space Power** (SAE 594C, 1962.) The General Motors Corp., under contract to the USAF Aeronautical Systems Command, adapted the GMR Stirling engine to space applications. A 3-kW engine has been built utilizing NaK heated to 1250°F (677°C) as heat source and water at 150°F (66°C) as the cooling medium. The engine is pressurized to a mean pressure of 1,500 lb/in² (103.4 bar), giving an efficiency of 27 percent at 2,500 r/min or 24.4 percent at 3,000 r/min. The weight of this solar-energy conversion system is given as 550 lb (249 kg). Chemical, nuclear, or other sources could also be used.

## POWER FROM VEGETATION
### (Staff Contribution)

Vegetation offers, by photosynthesis, a natural process for the storage of solar energy. The efficiency of the photosynthetic process for the conversion of sun's rays into a usable fuel form is low (less than 2 percent is probably realistic). Wood, wood waste, sawdust, hogged fuel, bagasse, straw, and tanbark have heating values ranging to 10,000 ± Btu/lb (see Sec. 7). They may be incinerated for disposal as waste material or burned directly for the subsequent production of steam. Tree farming, with controlled growth and cutting, proposes to balance harvesting plans to load demands; e.g., Szego and Kemp (*Chem. Tech.*, May 1973) project a 400 sq mile "energy plantation" to serve a 400 MW steam-electric plant. Such proposals would utilize proved steam power-plant cycles and equipment for novel breeding, growing, harvesting, preparation, and combustion of vegetation.

The **photosynthesis process** is basic to all agricultural practice. The human animal has long known how to convert grain to alcohol. So it can be said that as long as we can grow green stuff we should be able to harness some of the sun's energy. The prohibition era in the United States saw many efforts to use the alcohol production capacity of the nation to offer alcohol as a substitute or supplementary fuel for internal combustion engines. Ethanol ($C_2H_5OH$) and methanol ($CH_3OH$) have properties that are basically attractive for internal combustion engines, to wit, smokeless combustion, high volatility, high octane ratings, high compression ratios ($R_v > 10$). Heating values are 9,600 Btu/lb for methanol and 12,800 Btu/lb for ethanol. On a volume basis these translate, respectively, to 63,000 and 85,000 Btu/gal for methanol and ethanol. Gasoline, for comparison, runs 126,000 Btu/gal (20,-

700 Btu/lb). (See Section 7 for values.) The blending of ethanol and methanol with gasolines (9± gasoline to 1± alcohol) has been used particularly in Europe since the 1930s as a suitable internal-combustion-engine fuel. The miscibility of the lighter alcohols with water and gasoline introduces corrosion problems for engine parts, and lowers the octane number. Higher carbon alcohols (e.g., butyl) which are immiscible with water are possible blending substitutes, but their availability and cost are not presently attractive. Such properties as flash point would introduce further problems. While these constitute some of the unsolved technical problems, the basic principle of harnessing the sun's energy through vegetation will continue as a provocative challenge not only in the field of power generation but also as a solution for the perennial farm problem.

## WIND POWER
### by D. K. McLaughlin and W. L. Hughes

REFERENCES: Betz, A., "Introduction to the Theory of Flow Machines," Pergamon, New York, 1966. Eldridge, F. R. (ed.), "Proceedings of the Second Workshop on Wind Energy Conversion Systems," Washington, D.C., 1975. Glauert, H., [W. F. Durand (editor in chief)], "Aerodynamic Theory" vol. 6, div. L, p. 324, Springer, Berlin, 1935. Golding, E. W., "The Generation of Electricity by Wind Power," Philosophical Library, New York, 1956. Hutter, V., "Optimum Design Concept for Windelectric Converters," presented at the Workshop for Advanced Windenergy Systems, Stockholm, 1975. Putnam, P. C., "Power from the Wind," Van Nostrand, New York, 1948. Rankin, W. J., "Transactions, Institute of Naval Architects," vol. 6, p. 13, 1865. Wilson R. E., and P. B. S. Lissaman, "Applied Aerodynamics of Wind Power Machines," Oregon State University Report, 1974.

Wind is one of the oldest widely used sources of energy. Although its use is many centuries old, it has not been a dominant factor in the energy picture of developed countries for the past 50 years because of the abundance of fossil fuels. Recently, the realization that fossil fuels are in limited supply has awakened the need to develop wind power with modern technology on a large scale. Consequently, there has been a tremendous resurgence of effort in wind power in just the past few years. The state of knowledge will be rapidly increasing, and the reader must soon look in the current literature for information on the latest technology. There are, however, fundamental principles in wind-power technology which will not change. These fundamentals are discussed in the subsequent paragraphs.

**Wind Turbines**  The essential ingredient in a wind-energy conversion system (WECS) is the wind turbine, traditionally called the windmill. Today, wind-axis and cross-wind-axis turbines are the predominant configurations in use and under study throughout the world. In the performance analysis of wind turbines, the wind-axis devices were studied first, and their analysis set the present-day conventions for the evaluation of all turbines.

**General Momentum Theory for Wind-Axis Turbines**  Conventional analysis of wind-axis turbines begins with an axial-momentum balance originated by Rankine using the control volume depicted in Fig. 1. In this nomenclature, $V$ is wind speed decelerated to $V(1-a)$ at the turbine disk and to $V(1-2a)$ in the wake of the turbine. ($a$ is called the interference

factor.) Momentum analysis predicts the axial thrust on the turbine of radius $R$ to be

$$T = 2\pi R^2 \rho \, V^2 a(1-a) \qquad (1)$$

where $\rho$, air density, equals 0.00237 lbf· s²/ft⁴ (or 1.221 kg/m³) at sea-level standard-atmosphere conditions.

Application of the mechanical-energy equation to the con-

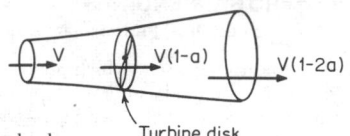

**Fig. 1**  Control volume.

trol volume depicted in Fig. 1 yields the prediction of power to the turbine of

$$P = 2\pi R^2 \rho \, V^3 a(1-a)^2 \qquad (2)$$

This power can be nondimensionalized with the energy flux $E$ in the upstream wind covering an area equal to the rotor disk, i.e.,

$$E = \tfrac{1}{2} \rho \, V^3 \, \pi \, R^2 \qquad (3)$$

The resulting power coefficient is

$$Cp = \frac{P}{E} = 4a \, (1-a)^2 \qquad (4)$$

This power coefficient has a theoretical maximum at $a = \frac{1}{3}$ of $Cp = 0.593$. This result was first predicted by Betz.

This derivation includes some important assumptions which limit its accuracy and applicability. First, the turbine must be a wind-axis configuration such that an average stream tube (Fig. 1) can be identified. Second, the portion of kinetic energy in the swirl component of velocity in the wake is neglected. Third, the effect of the radial pressure gradient is excluded. Partial accounting for the rotation in the wake has been included in the analysis of Glauert with the resulting prediction of power coefficient as a function of turbine-tip speed ratio $X = \Omega R/V$ (where $\Omega$ is the angular velocity of the turbine) shown in Fig. 2.

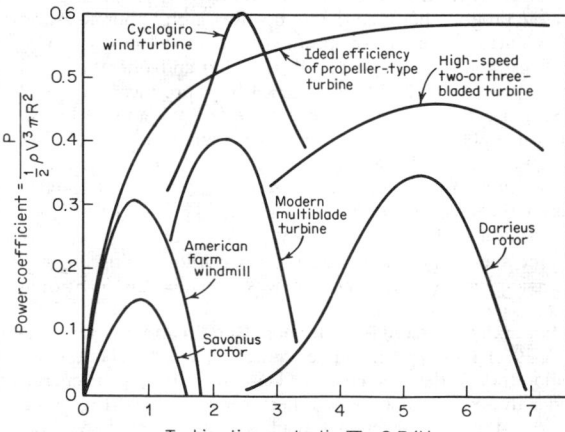

**Fig. 2**  Performance curves for wind turbines.

**Blade-Element Theory for Wind-Axis Turbines**  Blade-element theory provides the mechanism for analyzing the relationship between the individual airfoil properties and the interference factor $a$, the power produced $P$, and the axial thrust $T$ of the turbine. Rather than the stream tube of Fig. 1, the control volume consists of the annular ring bounded by streamlines depicted in Fig. 3. It is assumed that the flow in each annular ring is independent of the flow in all other rings.

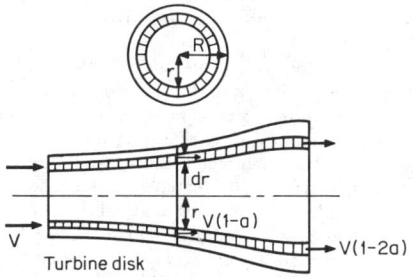

**Fig. 3**  Annular-ring control volume.

A schematic of the velocity and force-vector diagrams is given in Fig. 4. The elemental torque which acts on all blade elements in an annular ring is

$$dQ = \frac{B}{2} c\, r\rho\, W^2 (C_L \sin \varphi + C_D \cos \varphi)\, dr \tag{5}$$

The turbine is defined by the number $B$ of its blades, by the variation of the chord $c$, by the variation in blade angle $\theta$, and by the shape of the blade sections. $a' = \omega/2\Omega$, where $\omega$ is the angular velocity of the air just behind the turbine and $\Omega$ is the turbine angular velocity. $W$ is the velocity of the wind relative

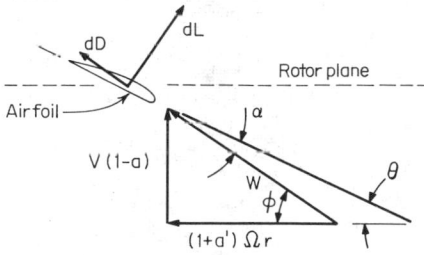

**Fig. 4**  Velocity and force-vector diagrams.

to the airfoil. Equation (5) is derived for the case with no turbine coning, which can be accounted for if appropriate. The sectional lift and drag coefficients $C_L$ and $C_D$ are obtained from empirical airfoil data and are unique functions of the local flow angle of attack $\alpha = \theta - \varphi$, and the local Reynolds number of the flow $Re = Wc/\nu$. Lift and drag coefficients are defined from

$$dL = C_L\ (\tfrac{1}{2}\rho\ W^2)c\ dr \tag{6}$$
$$dD = C_D\ (\tfrac{1}{2}\rho\ W^2)\ c\ dr \tag{7}$$

where $dL$ is the lift force on the element of blade and $dD$ is the drag force on the element of blade. In the Reynolds number $\nu$ is the kinematic viscosity of air, $160 \times 10^{-6}$ ft²/s ($14.9 \times 10^{-6}$ m²/s).

Power is computed by integrating Eq. (5) after multiplying it by the turbine angular velocity $\Omega$. The result is

$$P = \rho\, \frac{B\Omega}{2} \int_0^R crW^2\ (C_L \sin \varphi - C_D \cos \varphi)dr \tag{8}$$

Similarly the total thrust force on the turbine is

$$T = \rho\, \frac{B}{2} \int_0^R cW^2\ (C_L \cos \varphi + C_D \sin \varphi)dr \tag{9}$$

The relative velocity of the wind with respect to the airfoil section $W$ and the local angle of attack $\alpha$ are computed from the vector diagram of Fig. 3. To do this, the axial interference factor $a$ and the angular-velocity fraction $a' = \omega/2\Omega$ must be calculated by relating the blade-element forces to the momentum and energy equations applied to the annular control volume. The solution cannot be obtained in closed form, and a trial-and-error technique must therefore be used. The idea is that the blade forces are responsible for blocking the wind and for instilling swirl in the wake. However, to compute the blade forces, the amount of blockage $a$ and swirl $a'$ must be known. Hence the trial-and-error requirement.

A typical solution for steady-state operation of a two- or three-bladed wind-axis turbine is shown in Fig. 2. When optimized, these turbines run at high tip-speed ratios and are thus referred to in this manner. The curve shown in Fig. 2 for the two- or three-bladed wind turbine is for constant blade pitch angle. These turbines typically have pitch-change mechanisms which are used to feather the blades in extreme wind conditions. In some instances the blade pitch is continuously controlled to assist the turbine in maintaining constant speed. Turbines with continuous pitch control typically have flatter and hence more desirable operating curves than the one depicted in Fig. 2.

The traditional American farm windmill has a large number of blades with a high solidity ratio $\sigma$. ($\sigma$ is the ratio of area of the blades to swept area of the turbine $\pi R^2$.) It typically operates at slower speed with a lower power coefficient than high-speed turbines. However, the lower power coefficients result from poor airfoil lift properties on these turbines, not necessarily from the low-speed operation. Consequently, the power production of a modern multiblade turbine is also shown in Fig. 2 to be comparable with high-speed turbines and much greater than the American farm windmill. The particular turbine whose performance curve is shown is configured much like a bicycle wheel, with an outer rim held in compression with numerous spokes, about half of which are covered with airfoils. The blades of this turbine have a fixed pitch, as do most multiblade turbines.

The curves depicted in Fig. 2 representing the performance of high- and low-speed wind-axis turbines are theoretically predicted performance curves, with accompanying experimental confirmation. Hence they can be regarded as true performance curves.

**Cross-Wind-Axis Turbines**  There are three types of cross-wind-axis turbines of major importance as WECSs. The generally accepted advantage of cross-wind-axis turbines is the elimination of the requirement to drive the axis of the turbine into the wind. The poorest performer of the three is the Savonius rotor composed of two semicylindrical offset cups rotating about a vertical axis. It is a slow-speed turbine with characteristically high starting torque accompanied by its low overall aerodynamic efficiency. Perhaps its most important future use will be in conjunction with a Darrieus rotor to supply the necessary starting torque.

The Darrieus rotor looks somewhat like an "eggbeater" (Fig. 5). The blades are high-performance symmetric airfoils formed into a gentle curve called a "troposkien." This shape is selected to minimize the bending stresses in the blades. There are usually two or three blades in a turbine, and as shown in Fig. 2, the turbines operate efficiently at high speed.

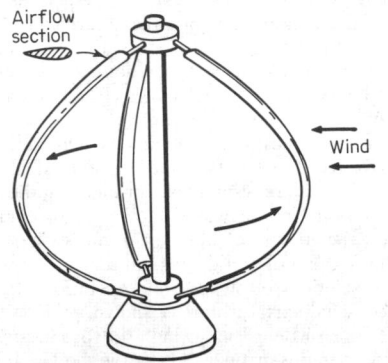

**Fig. 5** Darrieus rotor.

The third important cross-wind-axis turbine, the cyclogiro wind turbine (Fig. 6), is similar to the Darrieus rotor but with two important differences. First, the airfoils are straight, and second, the orientation (pitch) of the blades is continuously changed during rotation to maximize wind force. The peak power coefficients predicted for these turbines are greater than for any turbine (Fig. 2). Notice that the cyclogiro-turbine performance curve slightly exceeds the theoretical maximum for wind-axis turbines because the cyclogiro turbine decelerates a larger portion of the wind than does a wind-axis turbine of the same projected area of rotor disk.

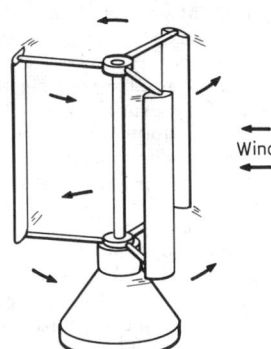

**Fig. 6** Cyclogiro rotor.

A major advantage of the Darrieus rotor—insensitivity to wind direction—is not a property of the cyclogiro turbine. For maximum efficiency the cam mechanism which controls local blade pitch must be continuously oriented into the wind.

Care must be taken not to overemphasize the aerodynamic

efficiency of wind-turbine configurations. The most important criterion in evaluating WECSs is the power produced on a per-unit-cost basis. The relationships between the turbine configurations discussed here and their potential production costs are not well known at this time. However, this information will be forthcoming.

Wind-augmentor systems are also being studied in conjunction with wind turbines. Two of the most important are the diffuser-augmentation and the vortex-augmentation systems. The cost effectiveness of these systems is not known at this time.

**Use of Wind-Energy Conversion System** Historically, wind-energy conversion systems were first used for milling grain and for pumping water. These tasks were ideally suited for wind-power sources, since the intermittent nature of the wind did not adversely affect the operation. Because water pumping is such an ideal load match for a wind-power system, there will no doubt be a resurgence and an expansion of irrigation and stock-watering usage powered by WECS.

The largest impact of wind power on the energy picture in the developed countries of the world is expected to be in the generation of electric power. This will probably involve pumping power directly into the power grid. Some applications may tie in with energy-storage systems involving compressed air, storage batteries, heated or pumped water, production of hydrogen, etc.

The generation of electricity will probably be accomplished by one of two systems: (1) synchronous ac generators, which require the wind turbine to operate at constant r/min; (2) variable r/min constant-frequency systems, which may be alternators rectified to dc (and then perhaps inverted to ac) or field-modulated generator systems.

**Power in the Wind** There are three major considerations

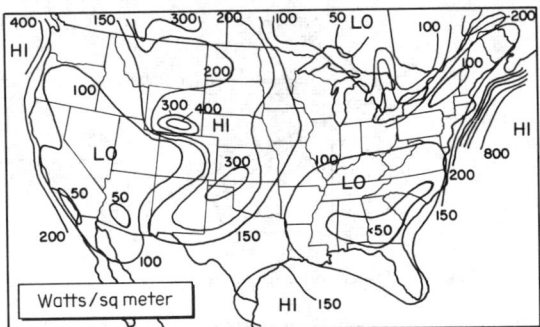

**Fig. 7** Annual average available wind power in the United States.

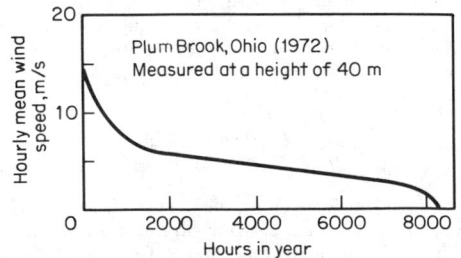

**Fig. 8** Wind variability at Plum Brook, OH (1972).

**Table 1. Wind Velocities in the United States**

| Station | Avg veloc-ity, mph | Pre-vailing direc-tion | Fast-est mile | Station | Avg veloc-ity, mph | Pre-vailing direc-tion | Fast-est mile |
|---|---|---|---|---|---|---|---|
| Albany, N.Y. | 9.0 | S | 71 | Louisville, Ky. | 8.7 | S | 68 |
| Albuquerque, N.M. | 8.8 | SE | 90 | Memphis, Tenn. | 9.9 | S | 57 |
| Atlanta, Ga. | 9.8 | NW | 70 | Miami, Fla. | 12.6 | ...... | 132 |
| Boise, Idaho | 9.6 | SE | 61 | Minneapolis, Minn. | 11.2 | SE | 92 |
| Boston, Mass. | 11.8 | SW | 87 | Mt. Washington, N.H. | 36.9 | W | 150 |
| Bismarck, N.Dak. | 10.8 | NW | 72 | New Orleans, La. | 7.7 | ...... | 98 |
| Buffalo, N.Y. | 14.6 | SW | 91 | New York, N.Y. | 14.6 | NW | 113 |
| Burlington, Vt. | 10.1 | S | 72 | Oklahoma City, Okla. | 14.6 | SSE | 87 |
| Chattanooga, Tenn. | 6.7 | .... | 82 | Omaha, Neb. | 9.5 | SSE | 109 |
| Cheyenne, Wyo. | 11.5 | W | 75 | Pensacola, Fla. | 10.1 | NE | 114 |
| Chicago, Ill. | 10.7 | SSW | 87 | Philadelphia, Pa. | 10.1 | NW | 88 |
| Cincinnati, Ohio | 7.5 | SW | 49 | Pittsburgh, Pa. | 10.4 | WSW | 73 |
| Cleveland, Ohio | 12.7 | S | 78 | Portland, Maine | 8.4 | N | 76 |
| Denver, Colo. | 7.5 | S | 65 | Portland, Ore. | 6.8 | NW | 57 |
| Des Moines, Iowa | 10.1 | NW | 76 | Rochester, N.Y. | 9.1 | SW | 73 |
| Detroit, Mich. | 10.6 | NW | 95 | St. Louis, Mo. | 11.0 | S | 91 |
| Duluth, Minn. | 12.4 | NW | 75 | Salt Lake City, Utah | 8.8 | SE | 71 |
| El Paso, Tex. | 9.3 | N | 70 | San Diego, Calif. | 6.4 | WNW | 53 |
| Galveston, Tex. | 10.8 | .... | 91 | San Francisco, Calif. | 10.5 | WNW | 62 |
| Helena, Mont. | 7.9 | W | 73 | Savannah, Ga. | 9.0 | NNE | 90 |
| Kansas City, Mo. | 10.0 | SSW | 72 | Spokane, Wash. | 6.7 | SSW | 56 |
| Knoxville, Tenn. | 6.7 | NE | 71 | Washington, D.C. | 7.1 | NW | 62 |

U.S. Weather Bureau records of the average wind velocity, and fastest mile, at selected stations. The period of record ranges from 6 to 84 years, ending 1954. No correction for height of station above ground.

used in evaluating a potential site for a WECS. The most important is the average wind speed at that location. However, since WECSs produce power proportional to the third power of wind speed, it is more appropriate to characterize the potential of a site in terms of the wind power available per unit area of cross section. Figure 7 is a map of the United States showing regions of annual average available wind power.

Variability of the wind is perhaps its second most important property. This type of information is used for determining appropriate peak generator capacity and assessing the variability to be expected in power production. An example of the wind variability at Plum Brook, OH, for 1972 is shown in Fig. 8.

A third characteristic of the wind is its established and

**Table 2. Beaufort Scale of Wind Force**
(Compiled by U.S. Weather Bureau, 1955)

| Beau-fort number | Miles per hour | Knots | Wind effects observed on land | Terms used in USWB forecasts |
|---|---|---|---|---|
| 0 | Less than 1 | Less than 1 | Calm; smoke rises vertically | |
| 1 | 1–3 | 1–3 | Direction of wind shown by smoke drift; but not by wind vanes | Light |
| 2 | 4–7 | 4–6 | Wind felt on face; leaves rustle; ordinary vane moved by wind | |
| 3 | 8–12 | 7–10 | Leaves and small twigs in constant motion; wind extends light flag | Gentle |
| 4 | 13–18 | 11–16 | Raises dust, loose paper; small branches are moved | Moderate |
| 5 | 19–24 | 17–21 | Small trees in leaf begin to sway; crested wavelets form on inland waters | Fresh |
| 6 | 25–31 | 22–27 | Large branches in motion; whistling heard in telegraph wires; umbrellas used with difficulty | Strong |
| 7 | 32–38 | 28–33 | Whole trees in motion; inconvenience felt walking against wind | |
| 8 | 39–46 | 34–40 | Breaks twigs off trees; generally impedes progress | Gale |
| 9 | 47–54 | 41–47 | Slight structural damage occurs; (chimney pots, slates removed) | |
| 10 | 55–63 | 48–55 | Seldom experienced inland; trees uprooted; considerable structural damage occurs | Whole gale |
| 11 | 64–72 | 56–63 | Very rarely experienced; accompanied by widespread damage | |
| 12 or more | 73 or more | 64 or more | Very rarely experienced; accompanied by widespread damage | Hurricane |

predictable variation of velocity with height. For the variation (typical) shown in Fig. 9, it is advantageous to construct a high support tower for a WECS. Variation of wind speed with height is dependent on the exact features of the landscape and

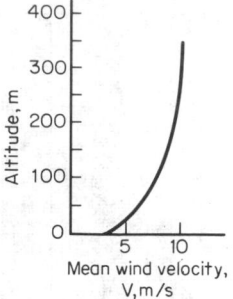

**Fig. 9**  Typical variation of mean wind velocity with height.

surrounding buildings. Consequently, siting is an important aspect of any WECS.

Table 1 shows average and peak wind velocities at locations within the continental United States. Table 2 lists the Beaufort scale of wind force commonly applied in weather maps.

## POWER FROM THE TIDES
### (Staff Contribution)

REFERENCES: Mosonyi, "Water Power Development," Hungarian Academy of Sciences, 1963; chap. 4, pp. 56–77, bibliography, pp. 1020–1025. Rose, "Energy Policy in the U.S.," *Sci. Am.*, Jan. 1974. David, "Energy; A Strategy of Diversity," *Technol. Rev.*, June 1973. The Rance Estuary Tidal Power Project, *Pub. Util. Ftly.*, Dec. 3, 1964. Norman Davey, "Studies in Tidal Power," London, 1923.

The tides are a renewable source of energy originating in the gravitational pull of the moon and sun, coupled with the rotation of the earth. The consequent portion of the earth's rotation is a mean ocean tide of $2\pm$ ft (0.6 m). The seashore periodic variation of the tides averages 12 h 25 min.

Tidal power is derivable from the large periodic variations in tidal flows and water levels in certain oceanic coastal basins. Suitable configurations of the continental shelves and of the coastal profiles result in reflection and resonance that amplify normally small bulges to ranges as high as $50\pm$ ft (15± m).

**Principal tidal-power sites** include the North Sea [12 ft (3.6 m)] average tidal range]; the Irish Sea [22 ft (6.7 m)]; the west coast of India [23 ft (7 m)]; the Kimberly coast of Western Australia [40 ft (12 m)]; San Jose Bay on the east coast of the Argentine [23 ft (7 m)]; the Kislaya Guba (Kisgalobskaia Bay) near the White Sea (no data); St. Michel (including the Rance estuary) on the Brittany coast of France [26 ft (8 m)]; the Bristol Channel (Severn) in England [32 ft (9.8 m)]; the Bay of Fundy (including the Chignecto Bay between New Brunswick and Nova Scotia and the Minas basin in Nova Scotia [40 ft (12 m)]; Passamaquoddy Bay between Maine and New Brunswick [18 ft (5.5 m)].

The harnessing of the tides reaches back into ancient history. Tidal mills, typically with undershot water wheels, were used in New England raceway estuaries, with reversible features for ebb and flood conditions. These power applications

were suitable for purposes such as grinding grain, but their number and size were small. In recent times the unique tidal ranges to $50\pm$ ft (15± m) have prompted many studies, proposals, and projects for most of the regions cited above. Despite these efforts for the generation of electricity, there are only two tidal-power developments in actual service (1974)—the Rance estuary in France (240,000 kW) and the Kislaya Guba in Russia (no data).

Developments take one of two general forms: **single-basin** or **multiple-basin.** A single-basin project, such as the Rance, has a dam, sluices, locks, and generating units in a structure separating a tidal basin from the sea. Water is trapped in the basin after a high tide. As the water level outside the basin falls with the tide, flow from the basin through turbines generates power. Power also may be generated when a basin emptied during a low tide is refilled on a rising tide. Numerous variations in operation are possible, depending on tide conditions and the relationship between the tide cycles and the load cycles. Pumping into or out of the basin increases the availability of the installed capacity for peak-load service. A multiple-basin development, such as projected for Chignecto or Passamaquoddy, generally has the power house between two basins. Sluices between the sea and the basins are so arranged that one basin is filled twice a day on high tide and the other emptied twice a day on low tide. Power output can be made continuous.

The amount of **energy available** from a tidal development is proportional to the basin area and to the square of the tidal range. Head variations are large in tidal projects during generating cycles and on a daily, monthly, and annual basis owing to various cosmic factors. Intermittent power, as from all single-basin plans and from two-basin plans with low-capacity factor, implies that the output can best be utilized as peaking capacity. Because of low heads, particularly toward the end of any generating cycle when pools have been drawn down, the cost of adding generating units only to tidal projects is well over the total installation cost of alternative peaking capacity. To be economically competitive with alternative capacity, the tidal projects must produce enough energy to pay the power-plant costs and also to pay for the dams and other costs, such as general site development, transmission, operation, and replacements.

The **risks and uncertainties** involved in designing, pricing, building, and operating capital-intensive tidal works, and the technological developments in alternative types of generating capacity, have tended to defeat tidal developments. Civil works are too extensive; transmission distances to load centers are too great; the required scale of development is too large for existing loads; the coordination of system demands and tidal generation requires interconnections for economic loading; the ultimate capacity of all the world's tidal potential is practically insignificant to meet the world's demands for electricity.

## UTILIZATION OF THE ENERGY OF THE WAVES
### (Staff Contribution)

According to Albert W. Stahl, USN (*Trans. ASME,* **13,** p. 438), the total energy of a series of **trochoidal deep-sea waves** may be expressed as follows: hp per ft of breadth of wave = 0.0329 $\times H^2 \sqrt{L}[1 - 4.935 (H^2/L^2)]$, where $H$ = height of wave, ft, and $L$ = length of wave between successive crests, ft. For example, with $L$ = 25 ft and $L/H$ = 50, hp = 0.04; with $L$ = 100 ft and $L/H$ = 10, hp = 31.3. Not much more than a quarter of the total energy of such waves would probably be available after reaching shallow water, and apparatus rugged enough for this purpose would doubtless be unable to utilize more than a third of this amount. **Wave motors** brought out from time to time have depended for their operation largely on the lifting power of the waves. One installed at Atlantic City, NJ (*Power*, Jan. 17, 1911), consisted of six 4-ft cylindrical floats 4 ft high. These, each weighing about 3,100 lb, were lifted 2 ft by the waves about 11 times per minute, and drove a horizontal shaft by means of chains and ratchets, developing but 12 hp, steadiness being obtained by the use of heavy flywheels. The fixed charges on the excess cost of wave motors thus far proposed, over conventional fuel-burning power plants of equal capacity, have been more than sufficient to care for the fuel and other additional costs necessitated by the use of the latter.

A **wave motor employing a hydraulic ram** for raising a portion of the water to a high level has been proposed by Smith (*Mech. Eng.*, Sept. 1927, p. 995). The waves enter a scoop which is connected to the ram by a long drive pipe. The apparatus is automatically adjusted for vertical level as the tide changes.

**Gravity waves** may be only a few feet high yet develop as much as 50 kW/ft of wave front. Most historical wave motors utilize (1) the kinetic energy of the waves by a device such as a paddle wheel, or (2) the potential energy by such a device as a series of floats. Few devices proposed utilize both forms of energy. Jacobs (*Power Eng.*, Sept. 1956) has analyzed the periodic fluctuation or **"seiching"** of the water level of harbors or basins where, with a resonant port, a 1,000-ft wave front might be used to achieve a liquid-piston effect for the compression of air, the air to be subsequently used in an air turbine.

## UTILIZATION OF HEAT ENERGY OF THE SEA
### (Staff Contribution)

REFERENCES: Claude and Boucherot, *Compt. rend.*, **183**, 1926, pp. 929–933. *The Engineer*, 1926, p. 584. Anderson and Anderson, *Mech Eng.*, Apr. 1966. Othmer and Roels, Power, Fresh Water, and Food from Cold, Deep Sea Water, *Science*, Oct. 12, 1973. Roe and Othmer, *Mech. Eng.*, May 1971.

Deep seawater, e.g., at 1 mile (1.6 m) depth in some tropical regions, may be as much as 50°F (28°C) colder than the surface water. This difference in temperature is a fundamental challenge to the power engineer, as it offers a potential for the conversion of heat into work. The Carnot cycle (see Sec. 4) specifies the limits of conversion efficiency. Typically, with a heat-source surface temperature $T_1$ = 85°F (545°R, 29.4°C, 303 K) and a heat-sink temperature $T_2$ 50° lower, or $T_2$ = 35°F (495°R, 1.7°C, 275 K) the ideal Carnot-cycle thermal efficiency = $(T_1 - T_2)/T_1$ = [(545 − 495)/545] 100 = 9.2 percent.

The actual overall efficiency of a real plant would be only a fraction of this ideal value, or about 2 or 3 percent. These values, both ideal and real, are far lower than the thermal efficiency of fuel-burning power plants. However, the unique heat capacity of water, coupled with the vast extent of the

oceans, presents a challenge for the development of machinery practically to harness an essentially limitless source of raw energy.

A representative analysis of some of the problems was exemplified in *The Engineer*, (1926, p. 584) using a surface temperature of 87°F (30.6°C), cold water at 30°F (1.1°C), and a fall of temperature to 78°F (25.6°C) in the evaporator (boiler). The steam produced would be at 0.55 psia (3,800 Pa) and condenser pressure 0.117 psia (800 Pa). With turbine efficiency of 75 percent, the steam consumption would be 53 lb/kWh (24 kg/kWh). It would require 116 lb (52 kg) of warm water to produce 1 lb (0.45 kg) of steam. For a 10,000-kW steam turbine, the steam-pipe diameter would be 23 ft (7 m) for a steam velocity of 200 ft/s (61 m/s), and other dimensions would be equally enormous.

The magnitude of these physical dimensions, coupled with the problems of high vacuum in both boiler and condenser, have prompted proposals to substitute a thermodynamic fluid with positive pressures and high densities at the prevailing low temperatures, e.g., propane, Freon, or ammonia. New problems are thus introduced tending to offset the advantages despite recent progress in underwater technology, e.g., (1) the sacrifice of availability by thermodynamic irreversibility in the transfer of heat in both the boiler and the condenser, (2) materials selection, (3) corrosion, and (4) maintenance. If multipurpose plants are projected, as for desalinization of seawater and mariculture as well as power generation, the economic pattern may favor the use of steam as the thermodynamic fluid with a surface-type condenser for the recovery of sweet water.

## POWER FROM HYDROGEN
### (Staff Contribution)

REFERENCE: Stewart and Edeskuty, Alternate Fuels for Transportation, *Mech. Eng.*, June 1974.

Hydrogen offers many attractive properties for use as fuel in a power plant. Fundamentally it is a "clean" fuel, smokeless in combustion with no particulate products, and if burned with oxygen, water vapor is the sole end product. If, however, it is burned with air, some of the nitrogen may combine at elevated temperature to form $NO_x$, a troublesome contaminant. If carbon is present as a fuel constituent, or if it can be picked up from a source such as a lubricant, the carbon introduces further contaminant potentials, e.g., carbon monoxide and cyanogens.

Basically the potential cleanliness of combustion is supported by other properties that make hydrogen a significant fuel, to wit, prevalence as a chemical element, calorific value, ignition temperature, explosibility limits, diffusivity, flame emissivity, flame velocity, ignition energy, and quenching distance.

Hydrogen offers a unique calorific value of 61,000 Btu/lb (140,000 kJ/kg). With a specific volume of 190 ft³/lb (12 m³/kg) this translates to 319 Btu/ft³ (12,000 kJ/m³) at normal pressure and temperature, 14.7 psia and 32°F (1 bar at 0°C). These figures, particularly on the volume basis, introduce many practical problems because hydrogen, with a critical point of −400°F (33 K) at 12.8 atm, is a gas at all normal, reasonable temperatures. When compared with alternative fuels, results are as shown in the table on p. 9-169.

**Bulk and Calorific Power of Selected Fuels (Approximate and Comparative)**

| Fuel | State | Sp. wt., lb/ft³ | Sp. gr. | Btu/lb | Btu/ft³ | Btu/gal |
|------|-------|-----------------|---------|--------|---------|---------|
| Hydrogen | Gas(NTP) | 0.0052 | 0.07 | 61,000 | 320 | (40) |
| Natural gas | Gas(NTP) | 0.042 | 0.67 | 24,000 | 1,000 | (130) |
| Gasoline (reg., 90 oct.) | Liquid | 46 | 0.72 | 20,500 | 950,000 | 125,000 |
| Ethanol (99 oct.) | Liquid | 49 | 0.79 | 12,800 | 620,000 | 82,000 |
| Methanol (98 oct.) | Liquid | 49 | 0.79 | 9,600 | 480,000 | 64,000 |
| Hydrogen | Liquid (36°R, 14.7 psia) | 4.4 | 0.07 | 56,000 | 240,000 | 32,000 |
| Coal | Piled | 50 | 0.8 | 12,000 | 600,000 | 80,000 |

These figures demonstrate the volumetric deficiency of gaseous hydrogen. High-pressure storage (50 to 100 atm) is a dubious substitute for the gasoline tank of an automobile. Liquefaction calls for cryogenic elements (Secs. 11 and 19). Chemical compounds, metallic hydrides, hydrazene, and alcohols are potential alternates, but practicality and cost are presently disadvantageous.

## FLYWHEEL ENERGY STORAGE
### by Sherwood B. Menkes

REFERENCES: The Oerlikon Electrogyro: Its Development and Application for Omnibus Service, *Auto. Eng.*, Dec. 1955. Beams, J. W., Magnetic Bearings, *SAE Automotive Congress Proc.*, Jan. 1964. Beachley, N. H., and A. A. Frank, Electric and Electric-Hybrid Cars, SAE Paper 730619, Mar. 1973. Clerk, R. C., The Utilization of Flywheel Energy, SAE Paper 711A, June 1963. Lawson, L. J., and K. W. Hellman, Design and Testing of High Energy Density Flywheels for Application to Flywheel/Heat Engine Hybrid Drives, SAE Paper 719150, Aug. 1971. Post, R. F., and S. F. Post, Flywheels, *Sci. Am.*, Dec. 1973. Rabenhorst, D. W., Primary Energy Storage and the Super Flywheel, Johns Hopkins University, TG 1081, Oct. 1970. Weber, R., and S. B. Menkes, Flywheel Energy Propulsion and the Electric Vehicle, Paper 7458, Electric Vehicle Symposium, Feb. 1974.

For many years a **flywheel** has been defined as a heavy wheel which is used to oppose and moderate by its inertia any fluctuation of speed in the machinery with which it revolves. Shafts in many different kinds of machinery are subjected to torque loading that is not uniform throughout a work cycle. By utilizing a flywheel, the designer can incorporate a smaller driving motor, and achieve a smoother operation.

Until recently, design of flywheels has not posed any serious difficulties, for work cycles have been relatively short, and the flywheel functioned solely to regulate speed. The kinetic energy has been relatively small. Concentration of material in a massive rim provides the maximum moment of inertia for a given amount of material.

Within the last few years, as a result of concern about fuel shortages and environmental pollution suggestions have been made to utilize unconventional energy sources. Accordingly there is much interest in the use of flywheels to store large amounts of kinetic energy. *Thus the flywheel is proposed as a major storage device, rather than as a means to effect speed regulation.*

As Post and Post observed, old concepts often reappear in technology as our needs change. Flywheels were probably first used as energy-storage devices in the potter's wheel, perhaps 5,000 years ago. The spindle was vertical; there was a head, on which clay was placed, and a separate flywheel below. The flywheel was used to store enough energy to turn rapidly.

A power plant is designed to operate most efficiently under a set of stated conditions. When it is necessary to operate the plant at off-design conditions, efficiency decreases, often quite severely. If the plant is operated only at high efficiency, and the excess energy is stored until it is needed, fuel is conserved. In addition, certain sources of energy (solar radiation, wind, etc.) become attractive provided that we can deal effectively with the question of storing energy thus *freely available* until such time as it can be used.

The flywheel is an attractive energy-storage concept for several reasons: (1) it is simple; (2) it is possible to store and abstract energy readily, either by mechanical means or by using electric motors and generators; (3) high power rates are practicable; (4) there is no stringent limitation on the number of charge and discharge cycles that can be used; (5) reliability promises to be high; and (6) maintenance costs should be low.

**Modern flywheel technology** is in its **infancy.** The **first** symposium on the state of the art was held in **November 1975.** Any specific application will require consideration of technical alternatives and a cost analysis. The following must be evaluated in each case: (1) how much energy can be stored per unit weight or volume of flywheel material, which in turn controls (2) the size flywheel required, (3) relative importance of friction losses and associated inefficiency, (4) system safety, and (5) nature of controls and systems needed to provide the proper interface between source of energy and the demand for it.

A uniform flat disk with a central hole was suggested to replace the massive-rim flywheel, but the resultant dynamic stress distribution limits its use.

Improved stress distribution (for an **isotropic material**) can be effected by thickening the flywheel toward the center and making it possible to achieve a constant tangential stress distribution. The energy-density capability of a flywheel in which constraints other than those due to stress considerations are removed can be calculated from

$$T = K_s \, \sigma/\rho$$

in which $T$ is the specific energy, $K_s$ a flywheel shape factor, $\sigma$ the material working stress, and $\rho$ the material density. For a solid metal wheel, the ideal shape is one in which $K_s$ is unity.

Lawson reports that Lockheed has achieved a shape factor of 0.832; such a wheel constructed of *maraging steel* results in a *T* value of 52 Wh/kg.

The parameter *T* is useful to compare candidate energy-storage concepts. Table 1, prepared by Weber and Menkes as part of a feasibility study of a flywheel-powered local-duty automobile, indicates the range of possible values. Note the inclusion of the *Oerlikon gyrobus*, the first vehicular application of flywheel energy storage. Advanced anisotropic materials offer great promise as flywheel materials, and many organizations are now engaged in the design and development of fiber-composite flywheels. These high-strength fibers, which include E glass, PRD −49 (Kevlar), S glass, fused silica, and others, dictate radical changes in design concepts.

The size range for suggested flywheels is considerable, as are also the recommended speed and energy capacity. Some applications are discussed below.

**Table 1. Energy Density for Various Storage Elements**

| Storage element | Wh/kg |
| --- | --- |
| Internal-combustion-engine system | 550* |
| Electrochemical storage: | |
|   Lead-acid | 18–33 |
|   Nickel-cadmium | 26–40 |
|   Silver-zinc | 66–132 |
|   Zinc-air (experimental) | 110–176 |
|   Sodium-sulfur (experimental) | 154–220 |
|   Lithium-halide (experimental) | 220 |
| Flywheels: | |
|   Gyreacta transmission | 0.7† |
|   Oerlikon Gyrobus | 7.0† |
|   4340 steel | 26 |
|   Maraging steel | 55 |
|   Advanced anisotropic materials | 190–870 |
| Hydraulic accumulator | 7–15 |
| Natural rubber band | 9 |

*Based on specific fuel consumption of 0.5 lb/bhp·h and engine weight equal to that of gasoline carried.
†Systems actually operated.

**Central Stations**  Long-range energy storage in central stations is accomplished by storing fuel (coal, oil, or gas), using a hydro reservoir and, more recently, cryogenic tanks. The basic problem is brought about by **highly fluctuating power demand.** A typical electric-utility load cycle has a peak on weekdays nearly double the demand at night, while there are no comparable peak demands on weekends.

Considerations of economy and efficiency make it attractive to increase the base capacity of the central station, to generate and store excess energy when it is available, and to draw on the stored energy when it is needed. One technique, in limited use, employs a pumped hydro-storage installation. The principal advantage there is that while the potential energy of fluid is stored at a higher elevation, there is no continuous loss of energy; this cannot be said for flywheel energy storage. Furthermore, pumped hydro-storage systems are completely safe and make use of existing technology both to store and to abstract the energy.

Unfortunately, *severe geographical constraints* limit the use of pumped hydro storage as a *universal solution.*

*Flywheels* offer a good alternative to *pumped hydro storage*, on the grounds of (1) compactness, (2) high power density, (3) reliability and low maintenance, (4) unlimited cycle life, and (5) good thermal compatibility with the environment. Several technological advances must be achieved, however, before flywheel energy storage becomes cost effective. These *necessary* improvements are: (1) development of low-cost, high-energy-density composite rotors, (2) development of very low friction bearing systems, and (3) development of improved motor-generator systems and controls.

The first two factors are self-evident; the last is not. A generator must extract energy from a constantly decelerating flywheel, and then feed it into a power network at constant voltage and frequency. The generator must either invert the variable-frequency input to the desired frequency or use some other scheme to accomplish the same result. Several systems are being developed which will do this.

**Transportation Applications**  Ground transport vehicles are powered, by and large, exclusively by internal-combustion engines. In passenger vehicles in particular, the thermal efficiency of the cycle is of the order of 10 to 15 percent. The waste of fossil-fuel distillates and the concomitant problem of air pollution are well documented. Accordingly, it is attractive to consider the possibility of generating electricity at a remote site, and *providing on-board energy storage*. Under certain circumstances, an auxiliary supply can be maintained external to the vehicle (as in a third rail), but for reasonable route flexibility, a self-contained store of energy is required.

A number of suggestions have been made which are in various stages of development. At one extreme is an *all-electric local-duty vehicle;* at the other is a *hybrid heat engine and flywheel energy storage* without electric energy utilization at all. An intermediate arrangement would use a *heat engine, a flywheel, and an electric traction motor* drive system.

In the **all-electric vehicle,** major design problems include (1) development of a passive bearing system with an ultra-low-energy drain, (2) an increase in energy-density capability of flywheels to provide reasonable *range* and *speed*, (3) a design safe enough to withstand collisions, and (4) development of a compact and efficient motor-generator unit.

In the **heat-engine flywheel hybrid** with entirely mechanical means of using flywheel energy, no new technology is needed. Such a vehicle can make fairly impressive gains in fuel economy, especially by means of *regenerative dynamic braking*. The major difference between this and conventional vehicles lies in the need for a *continuously variable transmission unit* coupled to the flywheel.

A **modification of the all-electric** vehicle would require the addition of a small heat engine, perhaps 25 percent the size of those now in use. This heat engine can be operated at maximum efficiency, with the storage element being used to supply energy for acceleration. The driver could switch to all-electric mode for urban driving or short trips.

**Regenerative dynamic braking** is in use in the New York subway system; a **flywheel-trolley coach** is under development for the city of San Francisco.

The output from an **exotic energy source** —sun or wind—is cyclical in nature. Exploitation of this type of energy source, especially for generating electric power, must be accompanied by suitable "flywheel" energy-storage devices. Toward that end, rotating flywheels hold high promise for small units adaptable to residential use, especially in remote areas.

# Materials Handling

BY

**HAROLD V. HAWKINS**  *Manager, Product Standards and Services, Columbus McKinnon Corporation, Tonawanda, N.Y.*

### METHODS OF MOVING MATERIALS

Materials-handling Systems .......................... 10-2
Selection of Type of Method ......................... 10-2
Influence of Codes and Specifications ................... 10-2

### LIFTING

Chains (BY E. R. BEHNKE) ........................... 10-4
Wire Rope ........................................ 10-9
Holding Mechanisms ................................ 10-11
Hoists (BY C. J. MANNEY) ........................... 10-14
Mine Hoists, Skips ................................. 10-16
Elevators, Dumbwaiters, Escalators ................... 10-19

### DRAGGING, PULLING, AND PUSHING

Hoists, Pullers, and Winches ......................... 10-21
Locomotive Haulage, Coal Mines ...................... 10-21
Industrial Cars ..................................... 10-22

Dozers, Draglines .................................. 10-23
Moving Sidewalks .................................. 10-24
Car-unloading Machinery ............................ 10-24

### CARRYING AND LIFTING

Containerization ................................... 10-25
Surface Handling .................................. 10-25
  Lift Trucks and Palletized Loads .................... 10-25
  Off-highway Vehicles and Earthmoving Equipment (BY P. G.
  KUCHURIS, JR.) ................................. 10-26
Above Surface Handling ............................. 10-28
Below-Surface Handling (Excavation) .................. 10-37

### CONVEYING AND CONTINUOUS FLOW

Overhead Conveyors (BY IVAN L. ROSS) ................. 10-40
Noncarrying Conveyors .............................. 10-44
Carrying Conveyors ................................. 10-50
Automatic Metering ................................. 10-63

# METHODS OF MOVING MATERIALS

## by Harold V. Hawkins

## MATERIALS-HANDLING SYSTEMS

**Materials-handling systems** often comprise many individual mechanisms integrated into a network which becomes a dominant factor in the design of the plant as well as in the manufacturing process involved. Individual motions may be required such as **lifting, translating,** or a complicated pattern of the two. The sequence of motions may be back and forth over the same path or may be unidirectional over a circulating or **continuous conveying system.** The type of commodity carried may be classified as liquid, solid, or a combination of the two; the solids may be granular, bulk material, or packaged. Liquids or granular solids may also be in containers, thus becoming packaged types. The motion needed, kind and consistency of item to be moved, quantity to be moved, and type of process to be accommodated all make for a complex of requirements that demands careful study to assure economic justification as well as technical sufficiency. The basic materials-handling elements available are outlined here to provide (1) the **lifting,** (2) the **dragging, towing,** and **pushing,** and (3) the **carrying and lifting** functions most often used in reciprocating motions, or (4) the **conveying systems** required for continuous movement of bulk or packaged materials.

## SELECTION OF TYPE OF METHOD

When making a choice of the type of handling system to be used, ask:

1. Is the job to be done an isolated motion (lifting, dragging, towing, or pushing), or should it be integrated into a sequence (conveying)?

2. What is the rate of weight and volume to be moved? (See Table 1.) Is the movement periodic (reciprocating), or can it be averaged out (continuous)?

3. What type of commodity is to be handled—bulk, granular, packaged, liquid? Its handling characteristics, such as **angle of repose** or **angle of slide,** will be important.

The **angle of repose** of any material is the angle with the horizontal at which material will stand when piled. For anthracite coal, it is about 27°; for coke of the same size, about 40°. Moisture content is often the controlling factor. The percentage of fine material in the mass has a decided influence on the angle, as the fines carry the bulk of the moisture. Screened material has an angle of repose of 35 to 40°, depending on the shape, smoothness, and method of storing. The average angle for crushed and screened limestone, iron and copper ore, and similar materials has been found to be 37°. Mine-run soft coal will stand at 35 to 37°

The **angle of slide** is the angle at which material will flow on an inclined surface. Anthracite coal will flow on steel plate inclined at about 20°, coke at about 25 to 30°. Ore, stone, etc., will slide at about 30° where the fine material is removed and at 35 to 40° for mine-run material. It is customary to build chutes on an angle of 45° for such material as coal, stone, and ore, this angle being increased when there is a large percentage of fine, damp material. Where gentle handling is required to prevent breakage, special tests should be made to determine the minimum angle of slide. A chute found too steep may be provided with cross angles to retard the flow.

The answers to these questions, together with the experience in conveying materials included in Table 2, should provide a basis for choice of one or more mechanical devices to accomplish alternate methods of handling. A careful study of the economics of each to minimize capital investment, maintenance, and operating costs will determine the best method.

Another important influence on industrial-equipment design stems from regulations of the Occupational Safety and Health Administration (OSHA) of the United States Department of Labor. While these regulations are directed at the employer, they restrict his usage of equipment so as to dictate its design. Because they apply differentially to some industries classed as especially hazardous, and to some equipment types which are also especially hazardous, no generalizations may be given here other than to warn of their existence and the influence they will have on materials-handling equipment.

## INFLUENCE OF CODES AND SPECIFICATIONS

A dominant factor in the choice of handling method or in the design of its parts may be the **local** or **national specifications** or **codes.** A review should be made of such general codes as

### Table 1. Weights of Materials*

| Material | Weight/volume | | Material | Weight/volume | | Material | Weight/volume | |
| | lb/ft³ | kg/m³ | | lb/ft³ | kg/m³ | | lb/ft³ | kg/m³ |
|---|---|---|---|---|---|---|---|---|
| Barley | 37–40 | 600–640 | Earth | 75–115 | 1,200–1,840 | Ore | 105–215 | 1,680–3,440 |
| Charcoal | 17–27 | 270–430 | Lignite | 31–47 | 500–750 | Rye | 44–50 | 700–800 |
| Clay | 95–169 | 1,520–2,700 | Limestone | 90–110 | 1,440–1,760 | Slag, blast-furnace | 37–63 | 600–1,000 |
| Coke | 26–30 | 420–480 | Oats | 28–31 | 450–500 | Stone, broken | 90–120 | 1,440–1,920 |
| | | | | | | Wheat | 44–50 | 700–800 |

*See also Table 2.

"Safety Code for Cranes, Derricks, and Hoists," ANSI B30.2; "Elevators, Dumbwaiters, and Escalators," ANSI A17.1; "Conveyors, Cableways, and Related Equipment," ANSI B20.1; "Manlifts," ANSI A90.1. These all affect design parameters, kinds of acceptable mechanisms, acceptable safety factors, materials that can be transported, etc. Codes are especially restrictive if people are to be transported. In many instances, local or state codes will be even more restrictive than national codes. If safety is not jeopardized, deviations from codes can be requested in advance of usage.

### Table 2. Preferred Types of Conveyors and Elevators for Bulk and Packaged Materials

| Material | Physical condition | Av wt/volume lb/ft³ | Av wt/volume kg/m³ | Reaction on conveyor | Preferred conveyors* | Preferred elevators* | Comment |
|---|---|---|---|---|---|---|---|
| Acid phosphate | Damp | 90 | 1,440 | Adheres | a, e | b | Sticky |
| Alum | Granular | 60–65 | 960–1,040 | Abrasive | a, b, c, e | g, b | |
| Aluminum oxide | Pulv. | 60 | 960 | Abrasive | a, e | g | |
| Ammonium nitrate | Pulv. | 62 | 990 | Hygroscopic | b, c, e | g, b | Explosive |
| Ammonium nitrate | Damp | 65+ | 1,040+ | Adheres | c, e | g, b | Sticky |
| Arsenic salts | Pulv. | 100 | 1,600 | Heavy | c, e | g, b | Poisonous |
| Ashes: dry | Granular | 35–40 | 560–640 | Abrasive | d, f | b | Dusty |
| wet | Sticky | 45–50 | 720–800 | Abrasive | f | b | Corrosive |
| Bone meal | Pulv. | 55–60 | 880–960 | | a, b, c, d, e | g, b, c | |
| Borax | Pulv. | 50–70 | 800–1,120 | Abrasive | a, b, c, d, e | g, b | |
| Bran | Granular | 16–20 | 260–320 | | a, b, c, d, e | g, b | Sometimes sticky |
| Brewers grains, hot | Granular | 55 | 880 | Corrosive | c, e | g, b | |
| Carbon black (pellets) | Granular | 40 | 640 | | a, e | g, b | Fragile |
| Cement, dry | Pulv. | 90–118 | 1,440–1,890 | | a, c, d, e | g, b | Packs |
| Clays | Pulv. | 35–60 | 560–960 | Adheres | a, b, c, e | g, b | Sluggish |
| Coal: anthracite | Lumpy | 50–54 | 800–860 | | a, b, c, e | g, b | |
| steam sizes | Granular | 50–60 | 800–960 | | a, b, c, d, e | g, b, c | |
| bituminous, lump | Lumpy | 50–60 | 800–960 | | a, b, e | b | |
| bituminous, slack | Granular | 50–60 | 800–960 | | a, b, c, d, e | g, b, c | |
| Chalk | Pulv. | 70–75 | 1,120–1,200 | | a, b, c, d, e | g, b, c | Sluggish |
| Coffee beans | Granular | 40–45 | 640–720 | | a, c, e | g, b | Fragile |
| Copra, ground | Pulv. | 40 | 640 | May be abrasive | a, b, c, e | g, b | Sticky |
| Cork, ground | Pulv. | 5–15 | 80–240 | | a, b, c, d, e | g, b | Sluggish |
| Corn, shelled | Granular | 45 | 720 | Abrasive shell | a, c, e | g, b, c | |
| Cottonseed | Granular | 35–40 | 560–640 | Sometimes sticky | a, b, c, d, e | g, b | |
| Cullet | Granular | 80–100 | 1,280–1,600 | Abrasive | a, b, e | g, b | Corrosive |
| Flaxseed | Granular | 45 | 720 | Shell abrasive | a, b, c, d, e | g, b, c | Free-flowing |
| Flue dirt | Pulv. | 100 | 1,600 | Abrasive | b, d, e, f | g, b | |
| Fly ash, clean | Pulv. | 35–45 | 560–720 | Mild abrasive | a, b, c, d, e | g, b, c | Free-flowing |
| Glass batch | Granular | 80+ | 1,280+ | Abrasive | a, b, e | g, b | |
| Glue | Granular | 45 | 720 | | a, c, e | g, b, c | Keep cool |
| Graphite (flour) | Pulv. | 40 | 640 | Lubricant | a, b, c, d, e | g, b, c | |
| Gravel | Granular | 95–135 | 1,520–2,160 | Abrasive | a, e, f | g, b | |
| Gypsum | Pulv. | 60 | 960 | | a, b, c, e | g, b | |
| Heavy ores | Lumpy | 100+ | 1600+ | | a, b, f | g, b | May be tough |
| Hog fuel | Stringy | 15–30 | 240–480 | May jam | a, b, d, e | g | |
| Lead salts | Pulv. | 60–150 | 960–2,400 | Sluggish | a, b, c, e | g, b | Poisonous |
| Lime, pebble | Granular | 55–80 | 880–1,280 | | a, b, c, e | g, b | |
| Limestone dust | Pulv. | 85–95 | 1,360–1,520 | Abrasive | a, b, e | g | |
| Malt | Dry | 45 | 720 | May be sticky | a, b, c, d, e | g, b | |
| Manufactured products | Boxed | 1–200 | 16–3,200 | | a, i, j | | |
| Merchandise: | | | | | | | |
| Packaged | Boxed | 15 | 240 | | a, b, i, j | | |
| Garments | Hanging | 5 | 80 | | i, j | | |
| Metallic dusts | Pulv. | 50–100 | 800–1,600 | Abrasive | a, b, c, d, e | g, b | Sometimes difficult |
| Mica, pulverized | Pulv. | 20–30 | 320–480 | Free-flowing | a, b, c, d, e | g, b, c | Dusty |
| Molybdenum conc'ts | Pulv. | 110 | 1,760 | Abrasive | a, b, d | b | Sticky |
| Petroleum coke | Lumpy | 42 | 670 | Mild abrasive | a, b, c, e | g, b | |
| Pumice | Pulv. | 45 | 720 | Mild abrasive | a, b, c, d, e | g, b, c | Polisher |
| Quartz (ground) | Pulv. | 110 | 1,760 | Very abrasive | a, b, c, d | g | |
| Rubber scrap | Stringy | 50 | 800 | Sluggish | a, b, e | g, b | Difficult |
| Salt: coarse | Granular | 50 | 800 | Hygroscopic | a, b, c, e | g, b | Corrosive if wet |
| cake | Pulv. | 75–95 | 1,200–1,520 | Flows freely | a, b, c, d, e | g, b | |
| Sand: dry | Granular | 90–110 | 1,440–1,760 | Abrasive | a, e, f | g, b | |
| damp | Granular | 90–110 | 1,440–1,760 | Sticky | a, e, f | g, b | |
| Sawdust | Granular | 15–20 | 240–320 | | a, b, c, d, e | g, b, c | |
| Sewage sludge | Pulv. | 60 | 960 | Sticky if wet | a, b, e, f | g | Abrasive |

**Table 2. Preferred Types of Conveyors and Elevators for Bulk and Packaged Materials** *(Continued)*

| Material | Physical condition | Av wt/volume | | Reaction on conveyor | Preferred conveyors* | Preferred elevators* | Comment |
|---|---|---|---|---|---|---|---|
| | | lb/ft³ | kg/m³ | | | | |
| Silica flour | Pulv. | 80 | 1,280 | Sluggish | a, d, e | g | Abrasive |
| Soap flakes | Granular | 10–20 | 160–320 | Fragile | a, c, e | g | Sticky if hot |
| Soda ash: light | Pulv. | 25–35 | 400–560 | Flows freely | a, b, c, d, e | g, c | Caustic |
| heavy | Pulv. | 55–65 | 880–1,040 | Flows freely | a, b, c, d, e | g, c | Caustic |
| Soybean flour | Pulv. | 30 | 480 | Sticky | a, b, c, e | g, c | Explosive dust |
| Starch | Pulv. | 30–40 | 480–640 | | a, b, c, e | g, c | Explosive dust |
| Sugar: raw | Granular | 55–65 | 880–1,040 | Sticky | a, b, c, e | g | |
| refined | Granular | 50–55 | 800–880 | | a, b, c, e | g | Handle gently |
| Sulfur | Pulv. | 55 | 880 | Corrosive if wet | a, b, c, e | g, h | Explosion risk |
| Talc | Pulv. | 50–60 | 800–960 | Mild abrasive | a, b, c, d, e | g, h | Adheres to metal |
| Tobacco stems | Stringy | 25 | 400 | Sluggish | a, b, d, e | g | |
| Wheat | Granular | 48 | 770 | Free-flowing | a, c, d, e | g, c | Keep clean |
| Wood chips | Granular | 18–20 | 290–320 | May arch | a, c, d, e | g, c | Corrosive if wet |
| Zinc oxide | Pulv. | 20–35 | 320–560 | May pack | a, b, c, d, e | g | Avoid discoloration |
| Zinc sulfate | Pulv. | 70 | 1,120 | May pack | a, b, c, d, e | g | |

*Explanation of letter symbols:
   *a*—belt. *b*—flight. *c*—continuous flow. *d*—pneumatic. *e*—screw. *f*—drag chain. *g*—belt and bucket. *h*—chain and bucket. *i*—overhead straight power. *j*—overhead power and free.

# LIFTING

## by Harold V. Hawkins and Associates

### CHAINS
### by E. R. Behnke

*Columbus McKinnon Corporation*

**Sling Chains**

Until 1933, slings for use where failure could be dangerous to personnel or equipment were usually made from wrought iron. It had good shock-absorbing capability and was easily welded using forge-welding techniques.

However, wrought iron had serious disadvantages. Most important were its tendency to work-harden in service and its inability to be welded electrically. Work hardening had to be corrected by frequent annealing to remove brittleness. The requirement for forge welding prevented the use of higher-quality and more consistent methods such as electric-resistance welding.

Research led to the introduction, in 1933, of the first alloy-steel, electrically welded link chain. Since then, alloy chain, which should never be annealed, has come into universal acceptance. Although iron is still specified for a handful of applications, it is no longer of commercial importance, and data have accordingly been omitted from this edition.

Most **alloy chains** are made from water-hardening grades of one of the constructional alloy steels such as 4615 or 8620 (see Sec. 6). Higher-carbon grades containing up to 0.30 percent carbon are becoming commercially available.

Current commercial alloy sling chain varies among producers from 250 to 450 Brinell in hardness. This is equivalent to a material tensile-strength range of 125,000 to 230,000 lb/in². In terms of $d^2$ short tons of breaking strength, where $d$ is the chain wire diameter in inches, these material tensile strengths provide a breaking-strength range of $62d^2$ to $120d^2$ short tons.

In metric nomenclature the breaking-strength range would be 544 to 1,054 newtons per square millimeter (N/mm²), based upon the actual breaking load and the sum of the cross-sectional areas of both link barrels. The approximate conversion relationship is $1.0d^2$ ton = 8.78 N/mm². For comparison, ASTM specifications for minimum breaking strengths average $60d^2$ short tons based upon nominal wire sizes or $56d^2$ short tons based upon actual wire diameters. Table 1 gives the ASTM specifications plus working-load limits for nominal and actual sizes. Link inside dimensions have not been standardized among manufacturers. However, there is a trend toward standardization under the influence of specification studies being conducted by various industry and standards committees in the United States and by the ISO (International Standards Organization). Working-load limits recommended by the NACM (National Association of Chain Manufacturers) for standard alloy chains in simple, straight tension average about $20d^2$ short tons (based upon the nominal size). Under a proposed new NACM specification the working-load ratings of a new alloy chain (to be designated Grade 80) would be increased 30 percent. Table 2 gives proposed specifications for the new alloy grade. Endurance limits are about 18 percent of breaking strengths. Thus, for chains with actual strengths near the specified minimums, endurance limits may be 45 percent less than working-load ratings. This fact should be taken into account where an application calls for many thousands of cycles of loading.

Present product specifications (NACM and ASTM) call for a minimum elongation of 15 percent at failure. This figure was established in 1924 as a supposed guarantee against brittle failure and to give visible warning of overloading and impending failure. Actually, it was an updating of the elongation requirements first specified in 1915 for both iron and steel

**Table 1.\* ASTM Specifications for Alloy-Steel Chain**
(NACM Old Standard 1001)

| Nominal size of chain,† in | Working-load limit, lb | Min proof-test load, lb | Min break-test load, lb | Max length, 100 links, in | Max weight, 100 ft, lb |
|---|---|---|---|---|---|
| ¼ | 3,250 | 6,500 | 10,000 | 98 | 84 |
| ⅜ | 6,600 | 13,200 | 19,000 | 134 | 175 |
| ½ | 11,250 | 22,500 | 32,500 | 156 | 288 |
| ⅝ | 16,500 | 33,000 | 50,000 | 182 | 453 |
| ¾ | 23,000 | 46,000 | 69,500 | 208 | 655 |
| ⅞ | 28,750 | 57,500 | 93,500 | 234 | 910 |
| 1 | 38,750 | 77,500 | 122,000 | 277 | 1,170 |
| 1⅛ | 44,500 | 89,000 | 143,000 | 332 | 1,425 |
| 1¼ | 57,500 | 115,000 | 180,000 | 371 | 1,765 |
| 1⅜ | 67,000 | 134,000 | 207,000 | 396 | 2,010 |
| 1½ | 80,000 | 160,000 | 244,000 | 432 | 2,185 |
| 1¾ | 100,000 | 200,000 | 325,000 | 503 | 3,020 |

\*Note applicable to all specification tables: To obtain metric values multiply inches by 25.4 for mm; lb by 0.45 for kg; feet by 0.3 for m.

†Most manufacturers make their chains in sizes up to ⅞ in, from bars that are ¹⁄₃₂ in larger than the nominal-size rating of the chain. CM Chain Division produces most sizes of its Here-Alloy sling chain from wire which is the same as the nominal-size rating.

classes of chain. They, in turn, were selected to describe, not what was required since this would have been the same for all types, but rather what each type was capable of achieving. Thus Class A iron was required to elongate 15 percent at failure, Class B iron 10 percent, and Class B steel 12 percent.

For soft chains (under 200 Brinell) in use before the advent of hardened alloy steel or even for some of the softer alloy chains sold today, the requirement may conceivably have provided some usefulness as an overload warning signal. However, it no longer has merit for this purpose when considering modern, high-quality chains of 300 Brinell or over. More than 50 percent of their total elongation at failure occurs during the final 10 percent of loading. It is unlikely, therefore, that the untrained eye will notice stretch from overloading until it reaches a hazardous amount. Promulgation of the philosophy of 15 percent elongation for overload detection can therefore instill a dangerously false sense of security.

Total deformation (plastic plus elastic) at failure is impor-

tant as one of the determinants of energy-absorption capability and therefore of impact resistance. However, its companion determinant—breaking strength—is equally important. The only practical way to measure their composite effect is by actual impact tests on actual chain samples—not by tests on Charpy or other prepared specimens of material. Special impact testers equipped with high-speed measuring and integrating devices have been developed for this purpose.

**High-test** chain is made from heat-treatable plain-carbon steel, usually with a carbon content of 0.15 to 0.22 percent. It provides good reliability and is widely used for load binding, tie-downs, and similar applications where failures would be costly but the maximum security provided by alloy chain is not required. It may have a typical breaking strength as high as $60d^2$ short tons, although NACM specifications require an average minimum of only $46d^2$ short tons. Hardness is about 170 to 250 Brinell. Table 3 gives ASTM specifications plus working-load limits for different sizes.

**Table 2. Alloy-Steel Chain**
(NACM Grade 80)

| Nominal size of chain, in | Working-load limit, lb | Min proof-test load, lb | Min break-test load, lb | Max length, 100 links, in | Max weight, 100 ft, lb |
|---|---|---|---|---|---|
| ⁷⁄₃₂ | 2,500 | 5,000 | 8,700 | 76 | 50 |
| ⁹⁄₃₂ | 4,100 | 8,200 | 14,400 | 98 | 84 |
| ⁵⁄₁₆ | 5,100 | 10,200 | 17,800 | 110 | 120 |
| ⅜ | 7,300 | 14,600 | 25,600 | 134 | 176 |
| ½ | 13,000 | 26,000 | 45,600 | 160 | 300 |
| ⅝ | 20,300 | 46,600 | 71,200 | 200 | 453 |
| ¾ | 29,300 | 58,600 | 102,500 | 235 | 655 |
| ⅞ | 39,900 | 79,800 | 139,500 | 270 | 910 |
| 1* | 52,100 | 104,200 | 182,300 | 277 | 1,170 |
| 1 ¼* | 81,400 | 162,800 | 284,800 | 371 | 1,765 |

\*Additional sizes manufactured by CM Chain Division but not included in NACM specification.

**Table 3. ASTM Specifications for High-Test Steel Chain**
(NACM Grade 43)*

| Trade size, in | Material size, in | | Nominal link inside dimension, in | | Max length, 100 links, in | Max weight, 100 ft, lb | Working-load limit, lb | Min proof-test load, lb | Min break-test load, lb |
|---|---|---|---|---|---|---|---|---|---|
| | Fraction | Decimal | Length | Width | | | | | |
| ¼ | ⁹⁄₃₂ | 0.281 | 0.82 | 0.39 | 86 | 80 | 2,500 | 4,100 | 7,750 |
| ⁵⁄₁₆ | ¹¹⁄₃₂ | 0.343 | 1.01 | 0.48 | 105 | 123 | 4,000 | 6,700 | 11,500 |
| ⅜ | ¹³⁄₃₂ | 0.406 | 1.15 | 0.56 | 121 | 175 | 5,100 | 8,500 | 16,200 |
| ⁷⁄₁₆ | ¹⁵⁄₃₂ | 0.468 | 1.29 | 0.65 | 135 | 235 | 6,600 | 11,200 | 20,700 |
| ½ | ¹⁷⁄₃₂ | 0.531 | 1.43 | 0.75 | 150 | 300 | 8,200 | 13,700 | 26,000 |
| ⅝ | ²¹⁄₃₂ | 0.656 | 1.79 | 0.90 | 186 | 450 | 11,500 | 19,500 | 36,900 |
| ¾ | ²⁵⁄₃₂ | 0.781 | 1.96 | 1.06 | 205 | 655 | 16,200 | 27,000 | 50,400 |
| ⅞† | ²⁹⁄₃₂ | 0.906 | 2.25 | 1.09 | 234 | 814 | 22,500 | 45,000 | 90,000 |
| 1† | 1¹³⁄₃₂ | 1.031 | 2.63 | 1.25 | 273 | 1,064 | 26,500 | 53,000 | 106,000 |

*Note that under proposed NACM specifications for Grades 43 and 28, the dimensions and strengths of certain chain sizes have been modified to improve uniformity.
†Specifications for these sizes have not yet been adopted by ASTM.

### General-Purpose Chains

**BBB chain,** which was similar in strength to proof coil chain, was dropped from NACM specifications as a standard in 1970. However, certain sizes are still produced for a few special applications where the changeover to proof coil link dimensions would create problems.

Proof coil chain is made from non-heat-treatable plain-carbon steel containing about 0.08 percent carbon. Its strength is about $32d^2$ short tons at 125 Brinell. Applications include logging, agricultural-equipment hitching, tow chains, animal leashes, and boat mooring. It is not recommended for critical security or overhead lifting. Some manufacturers weld electrically through ⅝ in, others through ¾ in, and some through 1 in. Larger sizes are fire-welded. Table 4 gives pertinent ASTM data plus working-load limits.

### High-Strength Binding Chain

Many specialty chains are used in specific industries for which standard specifications have not been established. One which is important enough to be mentioned here is made from a high-carbon (about 0.25 to 0.30 percent carbon), plain-carbon, or lean-alloy steel, fortified with elements such as boron or manganese in quantities sufficient to provide a uniformly hardened cross section. This chain is sold under various trade names as a very high strength chain for critical load-securing applications such as tie-down and binding chains on log and steel-transport trucks, and heavy-equipment towing chains. Capacity ratings may approach those of alloy sling chain with values up to about $20d^2$ tons. However, manufacturing control standards and total performance characteristics are not designed to qualify these chains to meet the critical demands of overhead lifting. Mechanical properties of one such proprietary product are given in Table 5. Under the proposed new NACM grading system, this chain will be designated as Grade 70.

### Chain Strength

Welded chain links are complex, statically indeterminate structures subjected to combinations of bending, shear, and tension under a normal axial load (see Seely and Smith, "Advanced Mechanics of Materials," Wiley). The maximum tensile stress occurs in the outside fiber at the intersection with the long axis (see Fig. 1). The maximum shear stress occurs approximately 45° away from this axis and on a radial line through the center of curvature of the link end.

For chains with low to medium hardness (under 400 Brinell), failure is typically due to shear. As chain hardness rises, there is a tendency for the typical mode of failure to shift from

**Table 4. ASTM Specifications for Proof Coil Steel Chain**
(NACM Grade 28)

| Trade size in | Material size | | Nominal inside link dimensions in | | Max length, 100 links, in | Max weight, 100 feet, lbs | Working-load limit, lbs | Min proof test load, lb | Min break-test load, lb |
|---|---|---|---|---|---|---|---|---|---|
| | Inches | Decimal | Length | Width | | | | | |
| ³⁄₁₆ | ⁷⁄₃₂ | 0.218 | 0.95 | 0.40 | 99 | 42 | 750 | 1,500 | 3,000 |
| ¼ | ⁹⁄₃₂ | 0.281 | 1.00 | 0.50 | 104 | 76 | 1,250 | 2,500 | 5,000 |
| ⁵⁄₁₆ | ¹¹⁄₃₂ | 0.343 | 1.10 | 0.50 | 114 | 115 | 1,875 | 3,750 | 7,500 |
| ⅜ | ¹³⁄₃₂ | 0.406 | 1.23 | 0.62 | 128 | 166 | 2,625 | 5,250 | 10,500 |
| ⁷⁄₁₆ | ¹⁵⁄₃₂ | 0.468 | 1.37 | 0.75 | 142 | 225 | 3,450 | 6,900 | 13,800 |
| ½ | ¹⁷⁄₃₂ | 0.531 | 1.50 | 0.81 | 156 | 286 | 4,500 | 9,000 | 18,000 |
| ⅝ | ²¹⁄₃₂ | 0.656 | 1.87 | 1.00 | 194 | 425 | 6,800 | 13,600 | 27,200 |
| ¾ | ²⁵⁄₃₂ | 0.781 | 2.12 | 1.12 | 220 | 605 | 9,500 | 19,000 | 38,000 |
| ⅞ | ²⁹⁄₃₂ | 0.906 | 2.50 | 1.37 | 260 | 811 | 11,375 | 22,750 | 45,500 |
| 1 | 1¹⁄₃₂ | 1.031 | 2.75 | 1.50 | 286 | 1,045 | 13,950 | 27,900 | 55,800 |

**Table 5. High-Strength Binding Chain**
(NACM Grade 70)*

| Trade size, in | Max length, 100 links, in | Max weight, 100 ft, lb | Working-load limit, lb | Min proof load, lb | Min break test, lb |
|---|---|---|---|---|---|
| ¼ | 98 | 84 | 3,150 | 6,300 | 12,600 |
| ⁵⁄₁₆ | 110 | 123 | 4,700 | 9,400 | 18,800 |
| ⅜ | 134 | 176 | 6,600 | 13,200 | 26,400 |
| ⁷⁄₁₆ | 135 | 235 | 8,750 | 17,500 | 35,000 |
| ½ | 160 | 300 | 11,300 | 22,600 | 45,200 |

*CM Chain Division's version is designated Dimension-70.

shear into tension due to bending. The failure location will simultaneously shift from the maximum shear plane to a plane in the long axis of the link.

The effect of combined stresses is to reduce the breaking strength from that computed by considering the applied load uniformly distributed in simple tension across the two circular

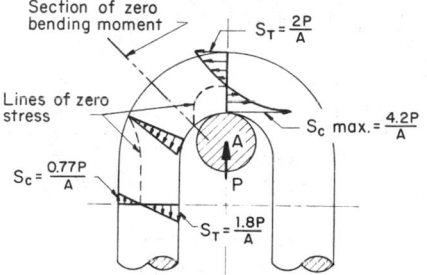

**Fig. 1** Magnitude of tensile and compressive stresses in a chain link under load. *(Reprint from Product Engineering, Nov. 25, 1963, copyright 1963 by McGraw-Hill.)*

areas of the straight side of a link. The actual breaking strength is about two-thirds of the value that would be obtained from such computation.

### Chain End Fittings

Most industrial chains must be equipped with some type of end fitting. These usually consist of oblong links or rings (called **masters**) on one end to fit over a crane hook and some variety of hook or enlarged link at the other end to engage the load. The hooks are normally drop-forged from heat-treated carbon or alloy steel. They are designed (using curved-beam theories) to be compatible in strength with the chain for which they are recommended. Master links or rings must fit over the rather large section thicknesses of crane hooks, and they therefore must have large inside dimensions. For this reason, they must be designed on the basis of bending strength and their section diameters will consequently be much larger than would be required on a straight tensile-strength basis.

CM Chain Division recommends that oblong master links be designed with an inside width of $3.5d$ and an inside length of $7.0d$, where $d$ is the section diameter in inches. For this condition and assuming a material yield strength of 100,000 lb/in², $d$ can be calculated from the relationship $d = \sqrt{WLL/20,000}$, where $WLL$ is the working-load limit in pounds.

Rings require a somewhat greater section diameter than oblong links to withstand the same load without deforming. **Master rings** with an inside diameter of $4d$ can be sized from the relationship $d = \sqrt{WLL/15,000}$ inches. Rings require a 15 percent larger section diameter and a 33 percent greater inside width than oblong links for the same load-carrying capacity, and the use of the less bulky oblong links is therefore preferred.

**Pear-shaped master links** were once the most commonly used design. However, they are less versatile than oblong links and can be inadvertently reversed, leading to bending of the narrow end due to jamming around the thick saddle of the crane hook. In recent years, their use has declined greatly.

**Hooks and end links** are attached to sling chains by means of coupling links, either welded or mechanical. Welded couplers require special equipment and skill to produce quality and reliability compatible with the other components of a sling and, consequently, must be assembled by manufacturers in plants. This could result in delays in obtaining new slings or repairs to existing ones. The advent of reliable mechanical couplers such as **Hammerloks**® (CM Chain Division), made from high-strength alloy forgings, relieved this problem greatly. With such units, customized slings can be assembled by users from component parts carried in local-distributor stocks

### Welded-Link Wheel Chains

These differ from sling chains and general-purpose chains in two principal respects: First, they are precisely calibrated to function in pocket wheels. Second, they are usually provided with considerably higher surface hardness to provide adequate wear life.

The most widely used variety is **Alloy Load wheel chain,** a short-link style used as the lifting chain in hand, electric, and air-powered hoists. Some roller chain is still used for this purpose, but its use is steadily declining because of the higher strength/weight ratio and three-dimensional flexibility of welded-link chain. Wire rope is also sometimes used in hoist applications. However, it is much less flexible then either welded-link or roller chain and consequently requires a drum of twenty to fifty times the rope diameter to maintain bending stresses within safe limits. Welded-link chain can operate over a three-pocket wheel with a pitch diameter of only six times the chain wire diameter. This permits the use of much smaller gear reductions and thus reduces hoist weight, bulk, and cost.

Welded Alloy Load wheel chain also has three times as much impact-absorption capability as wire rope of equal static tensile strength. As a result, wire rope of equal impact

**Table 6. Pocket-Wheel Specifications for Flite Chain**
(Columbus McKinnon Chain Division)

| Chain size, in | No. of pockets | Pitch diam. in | Outside diam. in | Thick ness, in | Max bore diam | | Max key size, in |
|---|---|---|---|---|---|---|---|
| | | | | | With square key, in | No key. in | |
| ⅜ | 4 | 4.777 | 5⅝ | 1⅞ | 2¼ | 2¾ | ½ |
| ⅜ | 5 | 5.971 | 6¹³⁄₁₆ | 1⅞ | 2¹⁵⁄₁₆ | 3¹¹⁄₁₆ | ¾ |
| ⅜ | 6 | 7.165 | 7⅝ | 1⅞ | 3¾ | 4⅝ | ⅞ |
| ½ | 4 | 5.095 | 5⅝ | 2¼ | 2¼ | 2¾ | ½ |
| ½ | 5 | 6.369 | 7 | 2¼ | 3⅛ | 3⅞ | ¾ |
| ½ | 6 | 7.643 | 8¼ | 2¼ | 3¹⁵⁄₁₆ | 4¹⁵⁄₁₆ | 1 |
| ⅝ | 4 | 6.369 | 7 | 2¾ | 2¾ | 3⅜ | ⅝ |
| ⅝ | 5 | 7.961 | 8¾ | 2¾ | 3¾ | 4⅝ | ⅞ |
| ⅝ | 6 | 9.554 | 10¼ | 2¾ | 4¹⁵⁄₁₆ | 6 | 1¼ |
| ¾ | 4 | 8.117 | 9 | 3⅛ | 3⅝ | 4½ | ⅞ |
| ¾ | 5 | 10.149 | 11⅛ | 3⅛ | 5 | 6 | 1¼ |
| ¾ | 6 | 12.170 | 13¹⁄₁₆ | 3⅛ | 6 | 6 | 1½ |

strength costs much more than Alloy Load wheel chain. Since average wire-rope life in a hoist is only about 5 percent of the life of chain, overall economics are greatly in favor of chain for hoisting purposes.

To achieve maximum flexibility, Alloy Load wheel chain is made with link inside dimensions of pitch = $3d$ and width = $1.25d$. Breaking strength is $90d^2$ tons, and endurance limit is about $18d^2$ tons. In hand-operated hoists, it can be used safely at working loads providing a design factor of 4 on ultimate strength. In powered hoists, the higher operating speeds and expectancy of more lifts during the life of the hoist require a somewhat higher design factor of safety. CM Chain Division recommends a factor of 7 for such units where starting up, stopping, and resonance effects do not cause dynamic load to exceed the static chain load by more than 25 percent. For this condition, chain fatigue life will exceed 500,000 lifts, the normal maximum requirement for a powered hoist. Standard Alloy Load chains, now produced in sizes from 0.125 through 0.562 in diameter, meet requirements for working-load limits up to 5 or 6 tons for hand-operated hoists and 3 short tons for power-operated hoists. More complete technical information on this special chain product and on chain-hoist design considerations is available from CM Chain Division.

Conveyor chains for use in high-load and/or high-speed applications are made to the same quality level as Alloy Load chain. However, they are usually designed with longer links, both for economy and to accommodate attachments such as flights. CM Chain Division reports that its brand, **Flite chain**, is produced to inside-width/section-diameter and inside-length/section-diameter ratios of 1.25 and 4, respectively. This chain has been designed to operate on pocket wheels having pitch diameters of $13d$ through $19d$. Figure 2 gives horsepower ratings for periodic manual and drip lubrication.

### Power Transmission

Power transmission, in the sense used to describe the transfer of power in machines from one shaft to another one close by, is a relatively new field of application for welded-link chain. Roller chain and other pin-link chains have been widely used for this purpose for many years. However, recent studies have shown that welded-link chain can be operated at speeds to 3,000 ft/min. Its three-dimensional flexibility, which can

sometimes eliminate the need for direction-change components, and its high strength-weight ratio suggest the possibility for significant cost savings in some power-transmission applications.

### Miscellaneous Special Chains

Special requirements calling for corrosion or heat resistance, nonmagnetic properties, noncontamination of dyes and foodstuffs, and spark resistance have resulted in the development of many special chains. They have been produced from beryl-

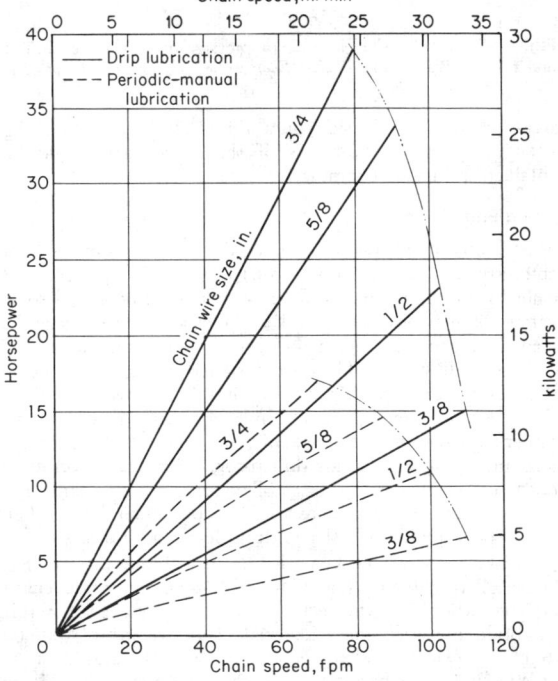

**Fig. 2**   Flite chain horsepower ratings. (*Reprint from Product Engineering, Nov. 25, 1963, copyright 1963 by McGraw-Hill.*)

lium copper, bronze, monel, Inconel, Hadfield's manganese, and aluminum and from a wide variety of AISI analyses of stainless and nonstainless alloys. However, the need for such special chains, other than those of stainless steel, is so infrequent that they are seldom carried as stock items.

# WIRE ROPE
### by Harold V. Hawkins

### Load Suspension and Haulage

**Wire rope** (see Sec. 8) used for suspending loads is usually required to be as flexible as possible to minimize the diameters of the drums or sheaves involved. Thus, a rope having six strands of 19 wires each on a hemp core is used (Section 8). Extrapliable ropes made with six strands of 37 wires each or eight strands of 19 wires each on a hemp core are also available but are much less durable because of the finer individual wires used. Hoisting ropes are constructed with the relative twist of the wires in the strands the reverse of the twist of the strands about the core (Fig. 3).

Based upon the service to be expected, special attention should be given to the ratio of drum and sheave diameters (Sec. 8) to cable diameter. For example, hoists having moderate-duty cycle may have a ratio of diameters as low as 20:1, but for extensive duty or where there is need for great safety, such as in elevators, this should be at least 45:1 or larger. Often the need for storing enough cable to obtain sufficient lift length may require a larger drum diameter than would otherwise be needed.

**Fig. 3**  Haulage rope.

**Haulage ropes** are of the same contruction as **suspension ropes** or are of the **lang-lay** type shown in Fig. 4, with the twist of the wire and strand in the same direction. This lang-lay construction increases the wear resistance of the rope, but it tends to untwist and should not be used where the load is in free suspension. Lang-lay rope is difficult to splice. By preforming individual wires and strands before laying up, secondary stresses due to bending are reduced and longer life is obtained. Preformed ropes have less tendency to kink and are easier to handle.

**Fig. 4**  Lang-lay rope.

**Hoisting** and **Haulage** ropes should be frequently **greased** to minimize wear and to prevent corrosion; either a special commercial lubricant or boiled linseed oil may be used on ropes subjected to atmospheric action, and a tacky petroleum and graphite on hoisting ropes in wet places. Crude oil or other lubricants having an acid or basic characteristic should not be used because of corrosive action on both wire and sisal core. To ensure penetration, lubricant can be applied hot, or a volatile solvent can be used. Ropes should be inspected frequently for broken wires and excessive wear.

For **strength** and **working loads** of wire rope, see Sec. 8.

### Track Cables

Cables used as **tracks to support loads suspended on trolleys** are either the locked-coil type for longest life (Fig. 5) or the round-

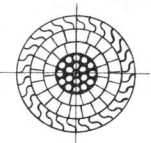

**Fig. 5**  Locked-coil track strand wire rope. (*United States Steel*)

wire track strand shown in Section 8. The strength of the locked-coil type is given in Table 7, while that for the round-wire or smooth-coil type is given in Section 8 also. This type of wire minimizes the impact loads on the outer wires which result from the rolling of the trolley.

### Fittings

**Fittings** for ropes are attached at the ends by (1) passing the rope around a minimum-radius thimble and then (*a*) attaching the rope to itself with rope clips (approximately 80 percent efficient), (*b*) splicing the rope to itself (80 to 95 percent efficient), (*c*) attaching the rope to itself by a metal ring which is swaged or crimped on (90 to 95 percent efficient), (2) using a fitting part of which is a steel tube which is pressed or swaged over the rope (90 to 100 percent efficient), (3) using zinc to embed the end of the rope in a fitting having a socket to receive it (100 percent efficient).

### Drums

**Drums** are made with smooth surfaces on hand-powered hoists and on power hoists subject to light-duty operation. Medium- and heavy-duty drums are normally grooved. Drums can be welded or cast, depending upon the quantity to be manufactured, since cast drums are economical when mass-produced. Large drums frequently have separate shells welded to the spider or end plate. Drum shells may be made from steel plates which are bent to a cylindrical shape and welded to the end plates with welded hubs before they are grooved for the rope. Steel-plate shells are stronger than cast shells, better balanced, and free from hidden initial defects. The thickness can be less, thus reducing the inertia of the rotating drum and the resulting acceleration-peak loads. Conical and cylindroconical drums are frequently used on large mine hoists. Faces of drums for medium and heavy duty are made wide enough to hold the rope in one layer plus two to four holding turns. The hole for attachment of the rope should be as shown in Fig. 6 to prevent excessive bending; this method of anchoring is normally done on cast drums which have a limited face width. Figure 7 shows an alternate, preferred method of anchoring the rope on welded and cast drums when space is not a problem. The pitch diameter of the drum should be at least twenty times the rope diameter in order to obtain reasonable

**Fig. 6**

**Table 7. Locked-Coil Track Strand Wire Rope**
(United States Steel Corp.)

| Diameter | | Special grade | | Standard grade | | Weight | |
|---|---|---|---|---|---|---|---|
| in | mm | Short ton | tonne | Short ton | tonne | lb/ft | kg/m |
| ¾ | 19.1 | 31.5 | 28.6 | 25 | 22.7 | 1.41 | 2.10 |
| ⅞ | 22.2 | 41.5 | 37.6 | 32 | 29.0 | 1.92 | 2.86 |
| 1 | 25.4 | 52.5 | 47.6 | 42 | 38.1 | 2.50 | 3.72 |
| 1⅛ | 28.6 | 66.0 | 59.9 | 54 | 49.0 | 3.16 | 4.70 |
| 1¼ | 31.8 | 81.0 | 73.5 | 65 | 59.0 | 3.91 | 5.82 |
| 1⅜ | 34.9 | 100.0 | 90.7 | 78 | 70.8 | 4.73 | 7.04 |
| 1½ | 38.1 | 120.5 | 109.3 | 93 | 84.4 | 5.63 | 8.38 |
| 1⅝ | 41.3 | 140.0 | 127.0 | 108 | 98.0 | 6.60 | 9.82 |
| 1¾ | 44.5 | 165.0 | 150 | 125 | 113.4 | 7.66 | 11.4 |
| 1⅞ | 47.6 | 187.5 | 170 | 138 | 125.2 | 8.79 | 13.1 |
| 2 | 50.8 | 215 | 195 | 158 | 143 | 10.00 | 14.9 |
| 2¼ | 57.2 | 280 | 254 | | | 12.50 | 18.6 |
| 2½ | 63.5 | 345 | 313 | | | 15.2 | 22.6 |
| 2¾ | 69.9 | 420 | 381 | | | 18.3 | 27.2 |
| 3 | 76.2 | 500 | 454 | | | 22.2 | 33.0 |
| 3¼ | 82.6 | 580 | 526 | | | 25.6 | 38.1 |
| 3½ | 88.9 | 690 | 626 | | | 29.9 | 44.5 |
| 3¾ | 95.3 | 785 | 712 | | | 33.9 | 50.4 |
| 4 | 101.6 | 880 | 798 | | | 38.4 | 57.1 |

life for both drum and rope. Long life requires forty-five to sixty times the rope diameter.

Where there is side draft on the rope, movable **idlers** are provided to align the rope and groove. The idlers may be moved parallel to the face of the drum by the side pressure of

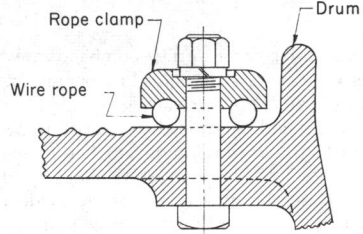

**Fig. 7**   Rope-anchoring attachment. *(McDowell Wellman.)*

the rope or may be driven positively sideways, thus eliminating friction and increasing the life of rope. In Fig. 8, idler sheave *c* revolves between fixed collars on shaft *a*, which is connected to the drum shaft by sprocket and chain. The shaft

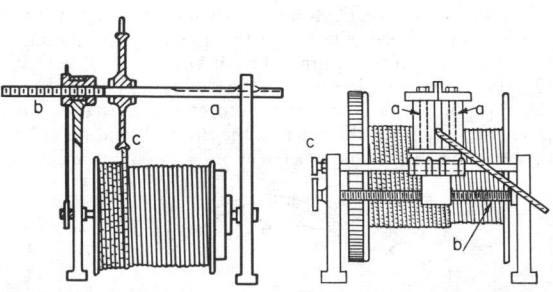

**Fig. 8**   Hoisting drum.      **Fig. 9**   Hoisting drum.

is prevented from rotating by a feather key. On sprocket *b*, a nut is held from moving sideways by flanges; the shaft *a* with sheave *c* moves in the direction of its axis. In an alternative construction, the idler shaft is threaded but held stationary, and the sheave hub is a nut. The sheave is turned by the friction of the rope, which causes it to travel back and forth. Figure 9 shows a construction used when the side draft is excessive. The upright rollers *a* are moved sideways by screw *b*, which is driven by sprocket and chain *c* from the drum shaft. For winding the rope on the drum in several layers, a clutch is provided which reverses the direction at the end of travel.

**Sheaves**

**Sheaves** should be grooved to fit the rope as closely as possible in order to prevent the rope from assuming an oval or elliptical shape under heavy load. They should be balanced and properly aligned to prevent swaying of the rope and abrasion against the sheave flanges. Sheaves and drums should be as large as possible to obtain maximum rope life, but factors such as weight of machinery for easy transport, minimizing headroom, and high-speed operation call for small sheaves. Hence rope life is sometimes sacrificed for overall economy. Undue wear on sheaves is avoided by flame-hardening them and by properly aligning them with the drum. The fleet angle of the rope should not exceed 1½°, but sometimes with grooved drums up to 2° is acceptable. Sheaves of any diameter can be welded or cast; most manufacturers make cast sheaves since they are economical. To avoid rope damage, worn sheaves should be replaced or the grooves turned before the sheaves are used with a new rope. In some cases, especially in mines, the grooves are lined with renewable, well-seasoned hardwood blocks.

Tackle blocks (Sec. 8) consist of one or two blocks, each carrying one or more sheaves. A single-sheave block (generally used to change the direction of a lead line and frequently arranged for easy removal of the loop of rope) is called a **snatch**

**block.** Blocks are made for both manila and wire rope. Those for manila rope are usually made with wooden cheeks to prevent chafing of the rope and have sheaves of smaller diameter than wire-rope blocks for the same size of rope. Heavy hoisting is almost universally done with wire-rope blocks. Tests by the American Bridge Co. found the following approximate efficiencies for well-designed and properly maintained ¾-in (19.1-mm) wire-rope tackle.

| Number of ropes supporting load | 1 | 2 | 3 | 4 | 5 | 6 | 7 | 8 | 9 | 10 | 11 | 12 |
|---|---|---|---|---|---|---|---|---|---|---|---|---|
| Efficiency, % | 86 | 96 | 91 | 87 | 82 | 78 | 74 | 71 | 68 | 65 | 62 | 59 |

Each snatch block between the hoisting blocks and the hoist or winch will have an efficiency of about 86 percent.

### Brakes

Small hoists are provided with hand-operated band brakes or electrically operated disk brakes; larger hoists have mechanically operated post brakes (see Sec. 8). On electric hoists, it is usual to apply the brake with a weight or spring and to remove it by a solenoid. Where controlled rate of application of a brake is desirable, a **thrustor** is frequently used. This is a self-contained unit consisting of a vertical motor, a centrifugal pump in a piston, and a cylinder filled with oil. Starting the motor raises the piston and connected counterweight. Stopping the motor permits the weight to fall. An adjustable range of several seconds in falling sets the brake slowly. Thrustors are built with considerably greater load capacities and stroke lengths than solenoids. Should the current fail, the brakes are automatically applied. On steam hoists, the brake is taken off by a steam piston and cylinder instead of by a solenoid. On overhead cranes, load brakes are used to sustain the load automatically at any point and occasionally to regulate the speed when lowering.

In one type of **load brake** (Fig. 10), the motor *A* drives drum *B* through the load brake on intermediate shaft *D*. A spider *C* is keyed to shaft *D*, its inner end supporting one end of a coiled bronze spring of square section *E*. The opposite end of the spring is fixed in flange *F*, which is loosely fitted to shaft *D* and directly attached to pinion *G*. Any relative angular motion of flange *F* and spider *C* alters the closeness of the coiling of spring *E*, consequently altering its outside diameter (considered as a drum). This outer surface is one of the friction surfaces of the brake, the other being provided by the internal face of drum *H*, which revolves loosely on shaft *D* at one end and on flange *F* at the other. Drum *H* is restrained from moving in one direction by ratchet *I* and pawls *J*. The exterior of drum *H* is grooved for heat dissipation.

The action of the **load brake** in Fig. 10 is as follows: When hoisting, the brake revolves as shown by the arrow. Pawls *J* permit drum *H* to revolve; consequently, the whole mechanism is locked and revolves as one piece. When stopping the load, the downward pull of the load reacts to drive drum *H* against pawl *J*. Flange *F* therefore moves slightly in an angular direction relative to spider *C*, and spring *E* consequently untwists until it grips the interior of drum *H*, thus locking the load. The action is such that the grip is slightly more than necessary to hold the load. Reversing the motor for lowering the load drives the interior of the brake surface against drum *H* so that the power consumed is the amount necessary to

overcome the excess holding power of the brake over the load reaction.

Load brakes of the **disk** and **cone** types (see Sec. 8) are also used and embody the same principle of pawl locks and differential action. The choice of brake type should be based on considerations of smoothness of working and lack of chatter, as well as on the power requirements for lowering at different values of load within the range of the crane.

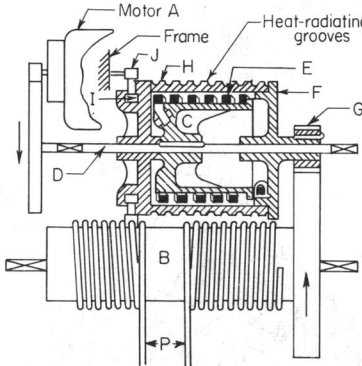

**Fig. 10**   Load brake.

In **regenerative braking**, the motor, when overhauled, acts as a generator to pump current back into the line. (See also Sec. 11.)

## HOLDING MECHANISMS
### by Harold V. Hawkins

### Lifting Tongs

Figure 11 shows the type of tongs used for lifting plates. The cams *a* grip the plate when the chains tighten, preventing

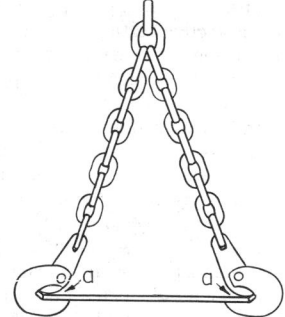

**Fig. 11**   Lifting tongs.

slipping. They are safer than plain hooks. Self-closing tongs (Fig. 12) are used for handling logs, manure, straw, etc. The rope *a* is attached to and makes several turns around drum *b*; chains *c* are attached to the bucket head *e* and to drum *b*. When power is applied to rope *a*, drum *b* is revolved, winding itself upon chains *c* and closing the tongs. To open, slack off on rope *a*, holding tongs on rope *d*, attached to head *e*.

## Lifting Magnets

**Lifting magnets** (see also Sec. 15) are materials-handling devices used for handling pig iron, scrap iron, castings, billets, tubes, rails, plates, skull-cracker balls, and other magnetic material.

At temperatures above dull-red heat, ordinary magnetic materials lose their magnetic properties, while certain stainless and high-manganese steels are nonmagnetic even at normal temperatures. Such materials cannot be handled by lifting magnets.

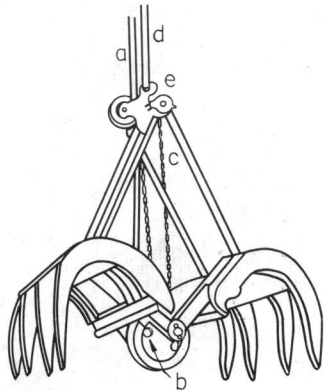

**Fig. 12**   Lifting tongs.

Most lifting magnets are not designed for continuous operation but for operation with the normal off times generally associated with materials-handling applications. In addition, attention should be given to operation of magnets on high-temperature loads so as to keep the magnet-coil temperature within the design limits of the insulation. Lifting magnets can be used for underwater operation when they are supplied with watertight cases and specially designed lead connections.

To obtain optimum magnet performance, a suitable **magnetic controller** should be used. It is necessary in most cases to provide means to reverse the current in the magnet in order to release the materials efficiently from the magnet. The controller should also have protective features to absorb the stored energy of the magnet during its discharge, especially when the dc power requirements for a magnet are supplied from a **rectifier-type** power supply.

Lifting magnets can be classifed as follows:

1. **Circular Magnets.** This configuration makes the most efficient use of materials, is extremely rugged, and is best suited for general lifting applications. Recently this category of magnets has been subdivided into two distinct types: a relatively large diameter magnet, which is especially efficient on low-permeability material such as scrap; and a magnet which, for equal weight, has a smaller diameter but a much deeper magnetic field, making it more suitable for high-permeability loads typically found in steel mills. Dings Elektrolift series of 8, 13, and 18 in diameters (203, 330, and 457 mm) is designed for machine-shop usage from 2,500 to 13,500 lb (1,130 to 6,120 kg).

2. **Rectangular Magnets.** Many types of material such as rails, beams, and plates can be handled more efficiently with a rectangular magnet. The tendency of these materials to pull away from the face of the magnet because of deflection rather than total weight is a limiting factor when these magnets are applied. Two or more small rectangular magnets mounted on a spreader beam usually give a better lifting performance than a single large magnet of equal weight. This is especially true in the case of thin plates.

3. **Specialty Magnets.** Loads such as coiled steel present unique problems when lifted by a magnet. Special magnets are available for loads of this type, but the scope of this text does not permit detailed discussion.

Construction features of a typical **circular magnet** are shown in Fig. 13. A winding of either strap aluminum or copper is located inside the cast-steel case. Strap material with insulating tape between turns, rather than insulated wire, is used for windings since it permits a greater number of turns in the same space. Recent advances in the processing of anodized aluminum have permitted the elimination of the turn-to-turn insulation, thereby allowing an increase in the number of turns, which results in increased lifting capacity. A nonmagnetic manganese-steel bottom plate is welded to the case to make the coil cavity watertight. The center and outer pole shoes are made so that they can be replaced, since they will wear in severe service.

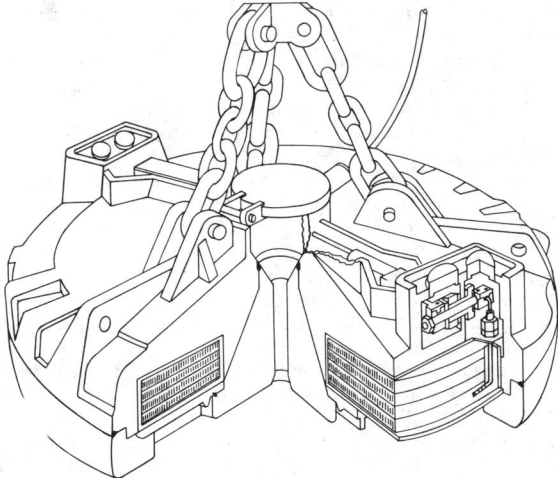

**Fig. 13**   Circular lifting magnet. (*Dings.*)

**Rectangular-magnet** construction features are shown in Fig. 14, which illustrates a typical plate-handling magnet. Coils for this type of magnet are generally wound with wire and formed to fit over the center pole. A manganese bottom plate is used to hold the coils in place and seal the coil cavity. The external flux path is from the center pole to each of the outer poles.

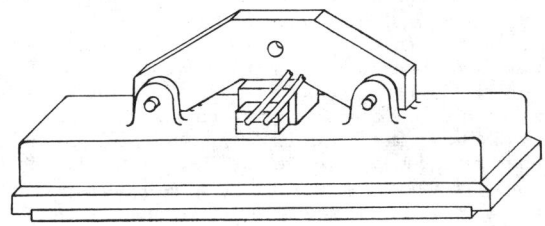

**Fig. 14**   Rectangular lifting magnet. (*Dings.*)

## Buckets and Scrapers

One type of **Williams self-filling dragline bucket** (Fig. 15) consists of a bowl with the top and digging end open. The shackles at *c* are so attached that the cutting edge will penetrate the material. When filled, the bucket is raised by keeping the line *a* taut and lifting on the fall line *d*. In this position, the bucket is carried to the dumping position, where by slacking off on line *a*, it is dumped. Line *b* holds up line *a* and the bridle chains while dumping. Such buckets are built in sizes from ¾ yd³ to as large as 85 yd³ (0.6 to 65 m³), weighing 188,000 lb (85,000 kg).

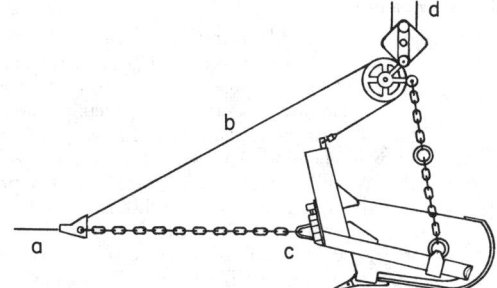

**Fig. 15** Self-filling dragline bucket.

Figure 16 shows an **open-bottom-type scraper.** Inhauling on cable *a* causes the sloping bottom plate to dig and load until upward pressure of the material against the top prevents further loading. The scraper continues its forward haul to the dumping point. Pulling on cable *b* deposits the load and returns the scraper to the excavation point. Scrapers are made in sizes from ⅓ to 20 yd³ capacity (0.25 to 15 m³).

The approximate **weights, capacities,** and **dimensions** of the more important types of **grab buckets** described below are given in Table 8. The **Hayward clamshell** type (Fig. 17) is used for handling coal, sand, gravel, etc., and for other flowable materials. The holding rope *a* is made fast to the head of the

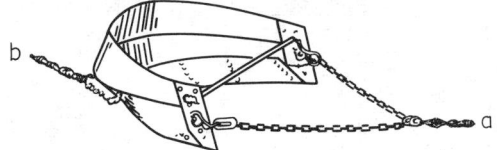

**Fig. 16** Scraper bucket. *(Sauerman.)*

bucket. The closing rope *b* makes several wraps around and is made fast to drum *d*, mounted on the shaft to which the scoops are pivoted. The chains *c* are made fast to the head of the bucket and to the small diameter of drum *d*. When power is applied to rope *b*, it causes the drum to wind itself up on chain *c*, raising the drum and closing the bucket. To dump, hold rope *a* and slack off rope *b*. The digging power of the bucket is determined by its weight and by the ratio of the diameters of the large and small parts of the drum. Hayward also has an **orange-peel** bucket that operates like the clamshell type (Fig. 17) but has four blades pivoted to close.

The **Hayward electrohydraulic** single-rope type (Fig. 18) listed in Table 8 is used not only for handling ore and sand but also where it is desirable to hook the bucket on a derrick or crane hook. The bucket requires only to be hung from the crane or derrick hook by the eye and an electric line plugged in. No ropes are fed into the bucket and no time-consuming shifting of lines is required—thus facilitating changing from magnet to bucket. The bowls are suspended from the head, which contains a complete electrohydraulic power unit consisting of an ac or dc motor, hydraulic pump, directional valve, sump, filters, and cylinders. When current is turned on, the motor drives the pump which provides system pressure. Energizing

**Table 8. Weights, Dimensions, and Capacities of Self-filling Grab Buckets\***
(Hayward Company)

| Make | Capacity, yd³† | Weight, lb | Dimensions, in | | | | | Rope to close, ft—in |
|---|---|---|---|---|---|---|---|---|
| | | | A | B | C | D | E | |
| General-purpose | ½ | 1,750 | 71 | 69 | 49 | 63 | 36 | 11—3 |
| bucket with ore | ¾ | 2,600 | 83 | 84 | 61 | 73 | 40 | 15—0 |
| bowl (Fig. 17) | 1 | 3,200 | 97 | 94 | 68 | 83 | 40 | 16—0 |
| | 1¼ | 3,400 | 97 | 94 | 68 | 83 | 47 | 16—0 |
| | 1½ | 4,600 | 106 | 104 | 74 | 91 | 50 | 18—0 |
| | 1¾ | 5,100 | 106 | 104 | 74 | 92 | 54 | 18—0 |
| | 2 | 5,750 | 106 | 107 | 76 | 94 | 62 | 20—0 |
| | 2½ | 7,500 | 117 | 120 | 84 | 105 | 63 | 21—0 |
| | 3 | 8,000 | 118 | 120 | 84 | 106 | 74 | 21—0 |
| | 4‡ | 13,500 | 136 | 140 | 103 | 119 | 74 | 23—6 |
| Electro-hydraulic | 1 | 7,200 | 78 | 56 | 67 | 64 | 39 | |
| single-rope | 2 | 7,800 | 87 | 56 | 67 | 65 | 52 | |
| clamshell | 3 | 10,200 | 120 | 67 | 90 | 83 | 68 | |
| bucket (Fig. 18) | 4 | 14,500 | 134 | 67 | 93 | 89 | 75 | |
| | 5 | 16,400 | 136 | 74 | 98 | 94 | 78 | |
| | 6 | 18,000 | 140 | 76 | 103 | 96 | 82 | |

\*Multiply yd³ by 0.76 to get m³, lb by 0.45 for kg, inches by 25.4 for mm, and feet by 0.30 for m.

†These ratings are for 1½-in crushed stone; increase 25 percent for 1½-in gravel and 50 percent for moist building sand. Those marked "special" are for coke or other light work.

‡Larger buckets are more or less special.

the solenoids in the directional valve directs the fluid flow to either end of the cylinders, which are attached to the blade

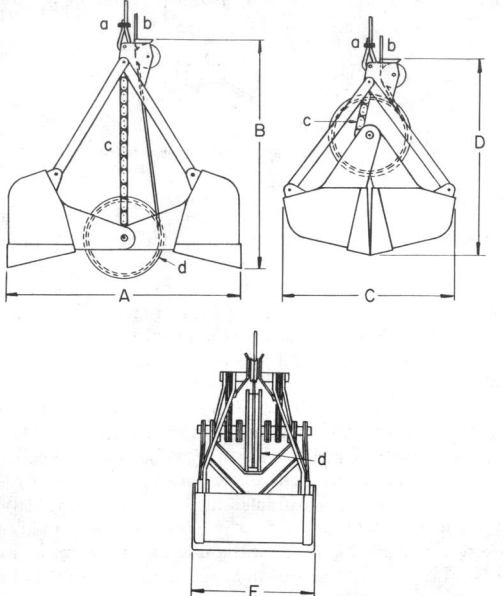

**Fig. 17**  Hayward grab bucket.

arms, thus opening or closing the bucket. The system uses a pressure-compensated variable-displacement pump which enters a no-delivery mode when system pressure is reached during the closing or opening action, thus avoiding any heat generation and reducing the load on the motor while maintain-

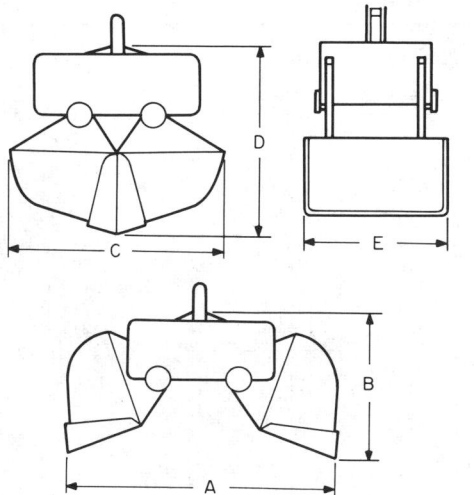

**Fig. 18**  Hayward electro-hydraulic single-rope-type grab bucket.

ing system pressure. The 1 yd³ (0.76 m³) bucket is provided with a 15-hp (11.2-kW) motor and 15 gal/min (0.95 l/s) pump. The 2 and 3 yd³ (1.5 to 3.5 m³) buckets have motors of 20 to 25 hp (15 to 19 kW), according to duty.

## HOISTS
### by C. J. Manney
*Consultant, Columbus McKinnon Corporation*

### Hand-Chain Hoists

Hand-chain hoists are portable lifting devices suspended from a hook and operated by a hand chain. They are used in various lifting and holding operations, especially for erection and maintenance work, and are available to 50 tons (45 tonnes) capacity and with almost unlimited lift. The principal type in current use is the *high-speed hoist* (Fig. 19); earlier types (such as the differential hoist) are obsolete. Because of its relatively high mechanical efficiency (65 to 80 percent) a high-speed hoist requires means to suspend the load at rest and during lowering; usually this is a brake of the *Weston self-energizing type* (see cone and disk brakes, Sec. 8), which provides a holding force proportional to the load. The separate hand and load chains operate over pocket wheels connected by a gear train. The brake is disengaged during hoisting by a one-way ratchet mechanism. To lower, the hand chain must be pulled continuously in the reverse direction to overcome the brake torque. Table 9 gives typical data for hand-chain hoists.

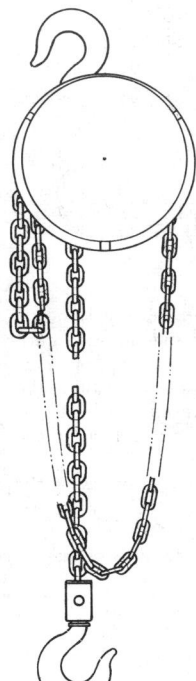

**Fig. 19**  High-speed hand hoist. *(C. M. Hoist Div., Columbus McKinnon.)*

### Pullers or Come-Alongs

Pullers or come-alongs are lever-operated chain or wire-rope hoists (see Fig. 20) for lifting or pulling at any angle. A reversible ratchet mechanism in the lever permits short-stroke operation for both tensioning and relaxing. The load is held by a Weston-type friction brake or a releasable ratchet. Much smaller and lighter than chain hoists of equal capacities, pullers are used for short-travel distances where the lever is within reach of the operator (e.g., tensioning wires, skidding machinery). Table 10 gives data for lever-operated hoists.

### Electric Hoists

Electric hoists are used for repetitive or high-speed lifting. Two types are available: chain (see Fig. 21) (both link and roller), in capacities from ⅛ to 5 tons; wire rope (see Fig. 22), in ratings from ⅛ to 20 tons. The typical hoist has a drum or sprocket centered in the frame, with the motor and gearing at opposite ends, the motor shaft passing through or alongside the drum or sprocket.

Electric hoists are equipped with at least two independent braking means. An electrically released brake causes spring-loaded disk brake plates to engage when current is off. When the hoist motor is activated, a solenoid overcomes the springs to release the brake. In the lowering direction, the motor acts as a generator, putting current back into the line and controlling the lowering speed. Some electric hoists use the same type of Weston brake as is used in hand hoists, but with this type, the motor must drive the load downward so as to try to release the brake. This type of brake generates considerable heat that must be dissipated—usually through an oil bath. The heat generated may also lower the useful-duty cycle of the hoist. If the Weston brake is used, an additional auxiliary hand-released or electrically released friction brake must be provided since the Weston brake will not act in the raising direction. All electric hoists have upper-limit switches; lower-limit switches are standard on chain hoists, optional on wire-rope hoists. Control is usually by push button: pendant ropes from the controller are obsolescent. Control is "dead-man" type, the hoist stopping instantly upon release. Modern multiple-speed ac controls for larger units have made dc hoists obsolete. Single-phase ac hoists are available to 1 hp, polyphase in all sizes.

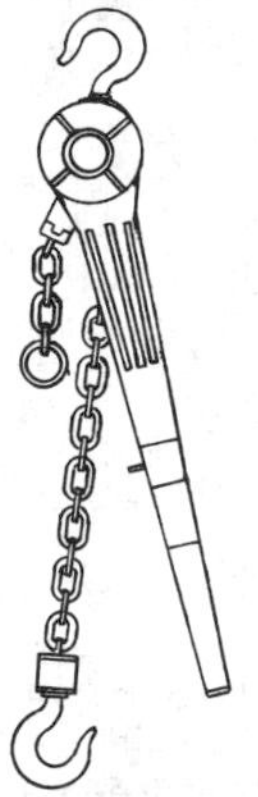

The hoist may be suspended by an integral hook or bolt-type lug or may be attached to a trolley rolling on an I beam or monorail. The trolley may be plain (push type), geared (operated by a hand chain), or motor-driven; the latter types are essential for heavier loads. Table 11 gives data for electric **chain hoists,** and Table 12 for **wire-rope hoists.**

### Air Hoists

Air hoists are similar to electric hoists except that air motors are used. Hoists with roller chain are available to 1 ton capacity, with link chain to 3 short tons (2.7 tonnes) and with wire rope to 15 short tons (13.6 tonnes) capacity. The motor may be of the rotary-vane or piston type. The piston motor is more costly but provides the best starting and low-speed performance and is preferred for larger-capacity hoists. A brake, interlocked with the controls, automatically holds the load in neutral; control movement releases the brake, either mechanically or by air pressure. Some air hoists also include a Weston-type load brake. The hoist may be suspended by a hook, lug, or trolley; the latter may be plain, geared, or air-motor-driven. Horizontal movement is limited to about 25 ft (7.6 m) because of the bulky air hose, although a runway system is available with a series of normally closed ports that are opened by a special trolley to supply air to the hoist.

Air hoists provide **infinitely variable speed,** according to the movement of the control valve. Very high speeds are possible with light loads. When severely overloaded, the air motor stalls without damage. Air hoists are smaller and lighter than electric hoists of equal capacity and can be operated in explosive atmospheres. They are more expensive than electric hoists, require mufflers for reasonably quiet operation and normally are fitted with automatic lubricators in the air supply.

### Jacks

Jacks are portable, hand-operated devices for moving heavy loads through short distances. There are three types in common use: screw jacks, rack-and-lever jacks, and hydraulic jacks. Bell-bottom **screw jacks** (Fig. 23) are available in capacities to 24 tons and lifting ranges to 14 in. The screw is rotated by a bar inserted in holes in the screw head or by a ratchet lever fitted to the head. Geared bridge jacks will lift up to 50 short tons (45 tonnes). A lever ratchet mechanism turns a bevel pinion; an internal thread in the gear raises the nonrotating screw. **Rack-and-lever jacks** (Fig. 24) consist of a cast-steel or malleable-iron housing in which the lever pivots. The rack toothed bar passes through the hollow housing; the load may be lifted either on the top or on a toe extending from the

**Fig. 20** Puller or come-along hoist. (*C. M. Hoist Div., Columbus McKinnon.*)

### Table 9. Data for Hand-Chain Hoists*

| Capacity | | Retracted distance between hooks | | Net weight | | Efficiency, % |
|---|---|---|---|---|---|---|
| Short tons | tonnes | in | mm | lb | kg | |
| ¼ | 0.23 | 14 | 335 | 44 | 20 | 80 |
| ½ | 0.45 | 15 | 380 | 47 | 21 | 80 |
| 1 | 0.91 | 18 | 455 | 50 | 23 | 80 |
| 1½ | 1.36 | 21 | 535 | 85 | 39 | 80 |
| 2 | 1.81 | 23 | 585 | 90 | 41 | 80 |
| 3 | 2.7 | 27 | 685 | 144 | 65 | 75 |
| 4 | 3.6 | 37 | 940 | 148 | 67 | 75 |
| 5 | 4.5 | 45 | 1,145 | 215 | 98 | 75 |
| 6 | 5.4 | 45 | 1,145 | 195 | 88 | 75 |
| | | 49 | 1,245 | 305 | 140 | 75 |
| 8 | 7.3 | 54 | 1,370 | 322 | 145 | 70 |

*From specification HMI 200, Hoist Manufacturers Institute.

The table covers Type C, Class II (lightweight) hook suspension units. Hoists are also available in 12, 16, 20, 25, 30, 40, and 50 tons capacity (11, 14.5, 18, 23, 27, 36, 45 tonnes).

Standard hook travel is 8 ft (2.4 m) for all capacities.

Efficiency = output × 100/input

**Table 10. Data for Lever-operated Chain Hoists***

| Capacity | | Retracted distance between hooks | | Net weight | |
|---|---|---|---|---|---|
| Short tons | tonnes | in | mm | lb | kg |
| ¾ | 0.68 | 14 | 355 | 16 | 7 |
| 1½ | 1.4 | 16 | 405 | 27 | 12 |
| 3 | 2.7 | 19 | 485 | 36 | 16 |
| 6 | 5.4 | 24 | 610 | 66 | 30 |

*From specification HMI 300, Hoist Manufacturers Institute.
The table lists capacities widely available in both link- and roller-chain styles and in load brake or ratchet and pawl types. Some types and styles are available from ¼ to 15 tons (13.6 tonnes) capacity.
Standard hook travel is 52 in (1,320 mm) minimum for all capacities.

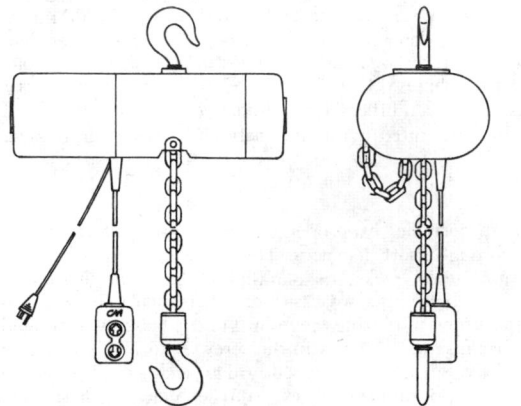

**Fig. 21**   Electric chain hoist. *(C. M. Hoist Div., Columbus McKinnon.)*

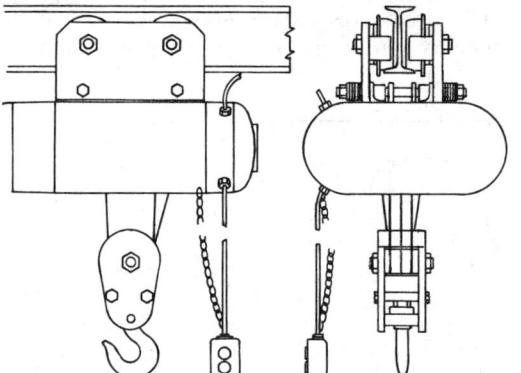

**Fig. 22**   Electric wire-rope hoist. *(C. M. Hoist Div., Columbus McKinnon.)*

bottom of the bar. The lever pawl may be biased either to raise or to lower the bar, the housing pawl holding the load on the return lever stroke. Rack-and-lever jacks to 20 short tons (18 tonnes) are direct-acting. Lever-operated geared jacks range up to 35 short tons (32 tonnes). Lifting heights to 18 in (0.46 m) are provided. **Track jacks** are rack-and-lever jacks which may be tripped to release the load. They are used for railroad-track work but not for industrial service where the tripping

features might be hazardous. **Hydraulic jacks** (Fig. 25) consist of a cylinder, a piston, and a lever-operated pump. Capacities to 100 short tons (91 tonnes) and lifting heights to 22 in (0.56 m) are available. Jacks 25 short tons (22.7 tonnes) and larger may be provided with two pumps, the second pump being a high-speed unit for rapid travel at partial load.

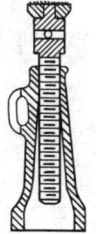   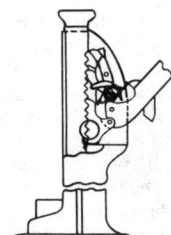

**Fig. 23**   Bell-bottom screw jack.     **Fig. 24**   Rack-and-lever jack.

### MINE HOISTS, SKIPS

**Large hoists** for such uses as mines and blast-furnace skips were formerly steam-engine-driven but are now almost entirely electrically driven. Such hoists are of either the single- or the double-drum type, with drums keyed or bushed on the shaft and driven through friction clutches. Drums are generally driven through single- or double-reduction gears, almost exclusively of the herringbone type and enclosed in oiltight casings. Each drum is equipped with a **post-type brake** (Fig. 26), usually operated by hand for small installations but on large machines applied by weights suspended from the operating lever and released through the action of an air- or oil-pressure cylinder. The clutches on large hoists are also equipped with air or oil cylinders. Large hoists operating at high rope speeds [drums more than 10 ft (3 m) in diameter and rope speeds 2,500 to 3,000 ft/min (12.7 to 15.2 m/s)] are frequently connected directly to the driving-motor shaft; they are designated as **first-motion** hoists.

Mine hoists are generally divided into (1) **Metal-mine hoists** (e.g., iron, copper, zinc, salt, gypsum, silver, gold, ores) and (2) **coal-mine hoists**. These classes subdivide into main hoists (for handling ores or coal) and hoists for men, timbers, and supplies. They are designed for operating (1) mine shafts, vertical and inclined, balanced and unbalanced; and (2) slopes, balanced and unbalanced. When an empty cage or platform

**Table 11. Data for Electric Chain Hoists***

| Capacity, | | Lifting speed | | Retracted distance between hooks | | Net weight | | Motor | |
|---|---|---|---|---|---|---|---|---|---|
| Short tons | tonnes | ft/min | m/min | in | mm | lb | kg | hp | kW |
| ⅛ | 0.11 | 32 | 10 | 15 | 380 | 61 | 28 | ¼ | 0.19 |
| ¼ | 0.23 | 16 | 5 | 15 | 380 | 61 | 28 | ¼ | 0.19 |
| | | 32 | 10 | 15 | 380 | 68 | 31 | ½ | 0.37 |
| ½ | 0.45 | 8 | 2.5 | 19 | 485 | 72 | 33 | ¼ | 0.19 |
| | | 16 | 5 | 15 | 380 | 68 | 31 | ½ | 0.37 |
| | | 32 | 10 | 16 | 405 | 105 | 48 | 1 | 0.75 |
| 1 | 0.91 | 8 | 2.5 | 19 | 485 | 78 | 35 | ½ | 0.37 |
| | | 16 | 5 | 16 | 405 | 106 | 48 | 1 | 0.75 |
| | | 32 | 10 | 17 | 430 | 113 | 51 | 2 | 1.49 |
| 2 | 1.81 | 8 | 2.5 | 24 | 610 | 126 | 57 | 1 | 0.75 |
| | | 16 | 5 | 25 | 635 | 128 | 58 | 2 | 1.49 |

*CM Hoist Div., Columbus McKinnon.
Hook suspension hoists, single-speed, 230/460 volts, 3φ, 60-Hz current.

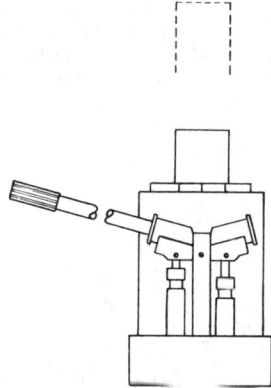

**Fig. 25**  Hydraulic jack,

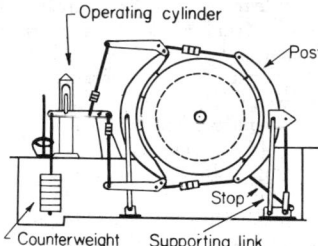

**Fig. 26**  Post-type brake.

descends while the loaded cage or platform ascends, as when both cables are wound on a single drum, the machine is referred to as a **balanced hoist.** Most medium- and large-sized hoists normally operate in balance, as the tonnage obtained for a given load and rope speed is about double that for an unbalanced hoist and the power consumption per ton hoisted is lower. Balanced operation can also be obtained by a counterweight. The counterweight (approximately equal to all the

**Table 12. Data for Electric Wire-Rope Hoists***

| Capacity | | Lifting speed | | Beam to high hook | | Net weight | | Motor | |
|---|---|---|---|---|---|---|---|---|---|
| Short tons | tonnes | ft/min | m/min | in | mm | lb | kg | hp | kW |
| ½ | 0.45 | 60 | 18 | 26 | 660 | 430 | 195 | 2 | 1.5 |
| ¾ | 0.68 | 37 | 11 | 26 | 660 | 430 | 195 | 2 | 1.5 |
| 1 | 0.91 | 30 | 9 | 25 | 635 | 465 | 210 | 2 | 1.5 |
| | | 37 | 11 | 26 | 660 | 440 | 200 | 3 | 2.2 |
| | | 60 | 18 | 26 | 660 | 445 | 200 | 4 | 3.0 |
| 1½ | 1.36 | 18 | 5 | 25 | 635 | 465 | 210 | 2 | 1.5 |
| | | 30 | 9 | 25 | 635 | 465 | 210 | 3 | 2.2 |
| | | 37 | 11 | 26 | 660 | 450 | 205 | 4 | 3.0 |
| 2 | 1.81 | 18 | 5 | 25 | 635 | 465 | 210 | 3 | 2.2 |
| | | 30 | 9 | 25 | 635 | 480 | 220 | 4 | 3.0 |
| 3 | 2.72 | 18 | 5 | 25 | 635 | 560 | 255 | 4 | 3.0 |
| 5 | 4.54 | 13 | 4 | 28 | 710 | 710 | 320 | 4½ | 3.4 |

*CM Hoist Div., Columbus McKinnon.
Hoists with plain trolley, single-speed, 230/460 volts, 3φ, 60-Hz current.

dead loads plus one-half the live load) is usually installed in guides within a single shaft compartment. The average depth of ore mines is about 2,000 ft (610 m), and that of coal mines is close to 500 ft (152 m). Most ore-mine hoists are of the double-drum type, normally hoisting in balance, each drum being provided with a friction clutch for changing the relative positions of the two skips when operating from various levels.

Hoists for coal mines are principally of the keyed-drum type, for operating in balance from one level. For high rope speeds in shallow shafts, it is generally advantageous to use **combined cylindrical and conical drums.** The cylindroconical drum places the maximum rope pull (weight of rope and loaded skip) on the small diameter, so that during the acceleration period of the cycle, the weight of the opposing skip is offering the greatest counterbalance torque, reducing motor peak loads and slightly reducing the power consumption. The peak reduction obtained becomes greater when shafts are shallower and hoisting speeds are higher, provided that the proportion between the smaller and the large diameters is increased as these conditions increase. By varying the ratio of diameters and the distribution of rope on the drum profiles with respect to the periods of acceleration, retardation, and constant speed, static and dynamic torques can be modified to produce the most economical power consumption and the minimum size of motor. The conical drum is not applicable for multiple-level operation, and except in very special cases, only a single layer of rope can be used.

**Skip hoists** for industrial purposes such as power-plant fuel handling and blast-furnace charging are similar to shallow-lift slow-speed coal-mine hoists in that they operate from a single level. Speeds of 100 to 400 ft/min (0.5 to 2 m/s) are usual. For blast-furnace charging with combined bucket and load weights up to 31,000 lb (14,000 kg) and a speed of 500 ft/min (2.5 m/s), modern plants consist of straight-drum geared engines, frequently with Ward Leonard control (Sec. 15).

Industrial skip hoists may be specified where the lift is too high for a bucket elevator, where the lumps are too large for elevator buckets, or where the material is pulverized and extremely abrasive or actively corrosive. For high lifts having a vertical or nearly vertical path, the skip with supporting structure usually costs less than a bucket elevator or an inclined-belt conveyor with bridge. Typical paths are shown in Fig. 27, paths *C* and *D* being suitable when the load is received through a track hopper.

The skip may be **manually loaded** direct from a wheelbarrow or dump car or **automatically loaded** by a pivoted chute, which is actuated by the bucket and which, when upturned, serves as a cutoff gate (Fig. 28).

For small capacities, the skip can be manually loaded with semiautomatic control. When the bucket has been filled, the operator pushes the start button and the bucket ascends, dumps, and returns to loading position. With automatic loading and larger capacity, the skip may have full automatic control. For economy, the bucket is counterbalanced by a weight, usually equaling the weight of the empty bucket plus half the load. For large capacity, a balanced skip in which one bucket rises as the other descends may be used. High-speed skips usually have automatic slowdown (two-speed motor) as the bucket nears the loading and discharge points.

**Ordinary-lay wire rope** (see Sec. 8) is employed for mine-hoisting duty, except that flat ropes are used for reel hoists.

Plow steel and improved plow steel are the most commonly used grades; the latter is used where the service is severe. Some state mining regulations require higher factors of safety than the usual hoisting requirements. The working capacity of new ropes is usually computed by using the minimum break-

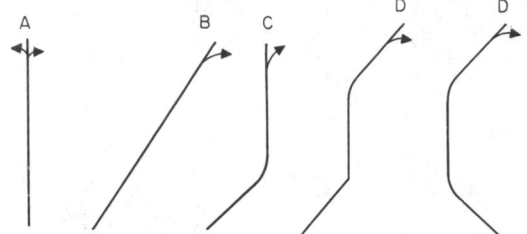

**Fig. 27** Typical paths for skip hoists: (*a*) vertical, with discharge to either side; (*b*) straight inclined run; (*c*) incline and vertical; (*d*) incline, vertical, incline.

ing strength given in the manufacturer's tables and the following factors of safety: rope lengths of 500 ft (152 m) or less, minimum factor 8; 500 to 1,000 ft (152 to 305 m), 7; 1,000 to 2,000 ft (305 to 610 m), 6; 2,000 to 3,000 ft (610 to 914 m), 5; 3,000 ft (914 m) or over, 4½. These are gross factors between

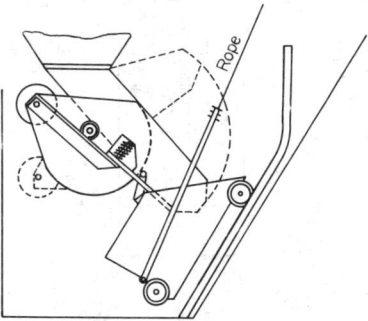

**Fig. 28** Automatic loader (in loading position).

the rated minimum breaking strength of the rope and the maximum static pull due to suspended load plus rope weight. The net factor, which should be used in dealing with large capacities and great depth, must take into consideration stresses due to acceleration and bending around the drum, together with suitable allowances for shock. With 6-by-19 wire construction, the pitch diameter of the drums is generally not less than sixty times the diameter of the rope. Drums are made of either cast iron or cast steel machine-grooved to suit the size of rope (see above). In large hoists, a lifting device is installed at the free rope end of the drum to assist the rope in doubling back over the first layer.

**Brakes** The band-type brake is restricted to small hoists for light duty. Two types of post brakes are used: the parallel-acting type (Fig. 26) and the anchored type. The bottom ends of the beams on the latter type are anchored to hinge-pin castings secured to the foundation. Asbestos composition blocks are rapidly replacing wood. Their coefficient of friction is greater and more uniform, and considerably higher unit pressures can be used. They eliminate screeching and arrest motion quickly without jerking. Blocks vary from 2½ to 5 in (64 to 127 mm) in thickness. The brake lining for medium-sized hoists consists of asbestos and cotton-fabric bands, varying in thickness from ½ to 1 in (13 to 25 mm) and secured to the beam by copper rivets.

**Hoist Motors**   Determining the proper size of motor for driving a hoist calls for setting up a definite cycle of duty based upon the required daily or hourly tonnage.

The **permissible hoisting speed** for mine hoists largely depends upon the depth of the shaft; the greater the depth, the higher the allowable speed. Conservative maximum hoisting speeds, as recommended by *Bu Mines Bull.* 75, are as follows:

| Depth of shaft | | Hoisting speed | |
|---|---|---|---|
| ft | m | ft/min | m/s |
| 500 or less | 150 or less | 1,200 | 6 |
| 500–1,000 | 150–300 | 1,600 | 8 |
| 1,000–2,000 | 300–600 | 2,000 | 10 |
| 2,000–3,000 | 600–900 | 3,000 | 15 |

High hoisting speeds call for rapid **acceleration** and **retardation.** For small hoists, the rate of acceleration may be made as low as 0.5 ft/s² (0.15 m/s²). An average value of 3 ft/s² is adopted for large hoists with fairly high speeds. Exceptional cases may require up to 6 ft/s² (1.9 m/s²). The speed should also be considered with regard to the weight of the material to be hoisted per trip. The question of whether the load should be increased and the speed reduced or vice versa is controlled by local conditions, mining laws, and practical experience. The rest period assigned to the duty cycle, i.e., the requisite time for loading at the bottom and unloading at the top, is dependent upon the equipment employed. With skips loaded from underground ore-storage hoppers, 5 to 6 s is the minimum that can be assumed. Unless special or automatic provision is made, the loading time should be taken as 8 to 10 s minimum. When the (1) hoisting speed, (2) weight of skip or cage, (3) weight of load, (4) periods of acceleration, (5) retardation, and (6) time for loading have been decided upon, the next step is to ascertain the "root-mean-square" equivalent continuous load-heating effect on the motor, taking into account rope and load weights, acceleration and deceleration of all hanging and rotating masses, and friction of sheaves, machines, etc. The friction load is usually taken as constant throughout the running period of the cycle. The overall efficiency of the mechanical parts of a single-reduction-geared hoist averages 80 percent; that of a first-motion hoist is closer to 85 percent. The motor selected must have sufficient starting torque to meet the temporary peaks of any cycle, including, in the case of balanced hoists, the requirements of trips out of balance.

**Electrical equipment for driving mine hoists** is of four classes:

1. *Direct-current motors* with resistance control for small hoists, usually series-wound but occasionally compound-wound in conjunction with dynamic braking control.

2. *Alternating-current slip-ring-type motor* with secondary resistance.

3. *Ward Leonard system of control* (Sec. 15) for higher efficiency, particularly on short lifts at high rope speeds, where the rheostatic losses during acceleration and retardation represent a large proportion of the net work done during the cycle;

for accuracy of speeds, with high-speed hoists; and for equalization of power demands. Complete control of the speed from standstill to maximum is obtained for all values of load from maximum positive to maximum negative. The lowering of unbalanced loads without the use of brakes is as readily accomplished as hoisting.

4. The *Ilgner Ward Leonard system* consists of a flywheel directly connected to a Ward Leonard motor-generator set and a device for automatically varying the speed through the secondary rheostatic control of the slip-ring induction motor driving the set. This form of equipment is used under conditions that prohibit the carrying of heavy loads or where power is purchased under heavy reservation charges for peak loads. It limits the power taken from the supply circuit to a certain predetermined value; whatever is required in excess of this value is produced by the energy given up by the flywheel as its speed is reduced.

## ELEVATORS, DUMBWAITERS, ESCALATORS

All **hydraulic elevators** except the short-rise slow-speed plunger type are obsolete. Drum-type electric elevators have been superseded by **traction-type** elevators, which are inherently safest because the reduction of tractive effort when the car or counterweight bottoms minimizes the possibility of these members being drawn into overhead work. The design of this type of equipment must follow closely the requirements of specification ANSI 17.1 or applicable local codes, which impose specific safety requirements, including use of limit switches, type of braking means, and number of supporting cables.

**Geared traction elevators** are used for car speeds up to 250 to 300 ft/min (1.3 to 1.5 m/s); above 300 ft/min (1.5 m/s), gearless machines have decided advantages. **Gearless machines** (Fig. 29) have either 2:1 or 1:1 roping. With 2:1 machines, the car speed is one-half the rope speed, with 1:1 machines, the car speed is the same as the rope speed. The 2:1 machines are adapted to

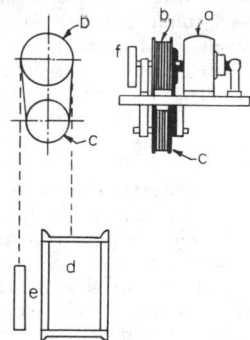

**Fig. 29**   Electric traction elevator.

car speeds from 300 to 500 ft/min (1.5 to 2.5 m/s); 1:1 machines are adapted to car speeds of 600 ft/min (3.0 m/s) and higher. Gearless machines are operating satisfactorily for car speeds of 1,000 ft/min (5.1 m/s).

**Motors**   Direct-current motors have the advantages of good starting torque and ease of speed control. Elevator motors are obliged to develop double rated torque at 125 percent rated current and have frequent starting, stopping, reversing, and running at constant speed. With sparkless commutation under all conditions, commercial motors cannot as a rule meet these requirements. For constant voltage, cumulative-compound motors with heavy series fields are used. The series fields are gradually short-circuited as the motor comes up to speed, after which the motor operates shunt. The shunt field is excited permanently, the current being reduced to a low value with increased resistance rather than the circuit opening when the motor is not in operation.

Squirrel-cage induction motors with high-resistance end rings to reduce starting current and increase starting torque are preferable to slip-ring motors because of their greater simplicity and ruggedness. For speeds greater than 150 ft/min (0.8 m/s), two-speed squirrel-cage induction motors are preferable. Two separate stator windings having different numbers of poles and giving speed ratios of 3:1 and greater are in most common use.

**Control**  There are two types of control in use: resistance control and Ward Leonard, or voltage, control. With resistance control, reduced voltage for starting is obtained by resistance in series with the armature. With voltage control, an individual dc generator is used for each elevator. The generator may be driven by either a dc or an ac motor (see Ward Leonard and Ilgner systems, above). The generator voltage is controlled through its field current, which gives the highest rates of acceleration and retardation. This system is equally efficient with either dc or ac supply. When resistance control is used, sections of the resistance are short-circuited successively, usually by contactors, until the armature is directly across the line. The objection to this method is the large energy loss in the starting resistance. With several elevators in operation, starting energy loss is reduced by using a multivoltage supply obtained from balancer sets across the line. With several elevators in operation, the unbalanced current carried by the balancer set is small.

Resistance control is used with geared machines for car speeds up to about 400 ft/min (2.0 m/s). For speeds up to 150 ft/min (0.8 m/s), single-speed motors are usually used; for speeds above 150 ft/min (0.8 m/s), two-speed motors are used. For speeds of 300 to 400 ft/min (1.5 to 2.0 m/s), voltage control is frequently used instead of resistance control, particularly with ac supply. For gearless machines, voltage control is used.

**Counterweights** are used with elevators, skip hoists, and mine hoists to equalize the engine load. The engine or motor can be smaller, since it must only drive against the unbalanced load and overcome friction. The brakes of a counterbalanced system retard greater masses and must be more powerful than for the same hoist without counterweights. Skip and mine hoists are counterbalanced by means of duplicate cars, one traveling up while the other goes down; the effective load is the difference between the weights of a full and an empty car, corrected for the weight of the rope. Elevators are usually overbalanced; i.e., the counterweight exceeds the light weight of the car by the average expectation of load. With high lifts, the weight of rope is neutralized by compensating chains hung from the top of the shaft and looped to the bottom of car, producing no pull when the car is down and maximum pull when the car is at the top.

**Loads**  The rated load of passenger elevators is as follows:

| Effective area of platform | | | | | |
| --- | --- | --- | --- | --- | --- |
| ft² | 40 | 60 | 80 | 100 | 120 |
| m² | 3.7 | 5.6 | 7.4 | 9.3 | 11.2 |
| Total capacity: | | | | | |
| lb | 3,400 | 5,600 | 7,700 | 9,800 | 12,000 |
| kg | 1,540 | 2,540 | 3,490 | 4,450 | 5,440 |

Codes in various cities specify that provision made for loads shall not be less than 75 lb/ft² (365 kg/m²), based on an average person occupying 2 ft² (0.19 m²) and weighing 150 lb (68 kg). Practice shows that for department stores, roof gardens, and subway stations, large cars may be loaded as high as 120 lb/ft² (585 kg/m²) effective area.

**Efficiencies and power consumption per car mile** of load of 2,500 lb (1,130 kg) for geared elevators are as follows:

Direct-current supply:

150 ft/min (0.76 m/s), efficiency 45 percent, power consumption 3.6 kWh (13.0 MJ).
300 ft/min (1.52 m/s), efficiency 55 percent, power consumption 5.0 kWh (18.0 MJ).

Alternating-current supply:

150 ft/min (0.76 m/s), efficiency 40 percent, power consumption 5.0 kWh (18.0 MJ).
300 ft/min (1.52 m/s) (two-speed), efficiency 50 percent, power consumption 7.5 kWh (27.0 MJ).

Gearless elevators, resistance-control dc supply:

600 ft/min (3.05 m/s), efficiency 60 to 65 percent, power consumption 5.5 kWh (19.8 MJ).

Efficiencies are based upon net load, i.e., full load minus overbalance.

**Car Mileage and Stops** (per elevator—8-h day)  **Office buildings—local elevators,** intensive service 12 to 20 car miles (19 to 32 car km), making about 150 regular stops per mile; **express elevators,** 20 to 40 car miles (32 to 64 car km), making about 75 to 100 regular stops per mile. **Department store elevators,** 4 to 8 car miles (6 to 13 car km), make about 350 regular stops per mile.

**Automatic push-button elevators** without microdrive are limited to speeds of 100 to 150 ft/min (0.5 to 0.8 m/s) with moderate loads owing to the difficulty of making accurate stops at higher speeds. With microdrive, speeds of 250 to 300 ft/min (1.27 to 1.52 m/s) can be used.

**Collective automatic control** is adapted to higher car speeds than the ordinary push-button elevator and is capable of handling more passengers, as the car responds to all calls made in the direction in which it is traveling. Signal control, in combination with microdrive and voltage control, is adapted to intensive service and high car speeds and is as fully automatic as a high-speed elevator can be. With signal control, a car automatically stops at a floor and levels with the landing in response to the pressing of a hall button. Signal-control elevators have been installed for car speeds of 1,000 ft/min (5.1 km/s). Manual control is limited to car speeds of 700 ft/min (3.6 km/s).

**Dumbwaiters** follow the general design requirements of elevators except that code specifications are somewhat more relaxed. For example, roller-link chain can be used for support instead of wire rope. Moreover, a single strand can be used instead of the mandatory minimums of three for traction-type elevators and two for drum-type elevators.

**Escalators** have the advantages of continuity of motion, great capacity, and small amounts of space occupied and current consumed for each passenger carried. Escalators are built 2, 3, and 4 ft (0.6, 0.9, and 1.2 m) wide between balustrading geared to move at the same speed as the element. The angle of incline is 30° from the horizontal, and the speed is 90 ft/min (0.46 m/s). The normal carrying capacity of 2- (3-) [4-] ft (0.6-, 0.9-, 1.2-m) escalators is 4,000 (6,000) [8,000] passengers per h.

# DRAGGING, PULLING, AND PUSHING

## by Harold V. Hawkins

### HOISTS, PULLERS, AND WINCHES

Many of the fundamental portable lifting mechanisms such as hoists or pullers (see above) can also be used forcefully to drag or pull materials. In addition, nonmobile versions, called **winches,** utilizing hoisting drums like that shown in Fig. 9 can also be used.

### LOCOMOTIVE HAULAGE, COAL MINES

**Rubber-tired haulage** at the coal face was introduced in 1935 and received further impetus with the introduction of the rubber-tired shuttle car in 1938. Crawler-mounted loaders and rubber-tired universal coal cutters completed the equipment needed for complete off-track mining. This off-track mining caused a revolution in face haulage, since it eliminated the expense of laying track in the rooms and advancing the track as the face of the coal advanced. It also made gathering locomotives of the cable-reel, crab-reel, and battery types obsolete. Practically all the gathering duty is now performed by rubber-tired shuttle cars, chain conveyors, extensible belt conveyors, and other methods involving off-track equipment. Some mines eliminated track completely by having belt conveyors carry the coal to the outside.

Most coal operators today employ a combination system, using conveyors on panel development and locomotives for the main-line haulage. A **track haulage system,** properly installed with 85- or 100-lb (39- or 45-kg) rail, well bonded, and carefully graded, ballasted, and drained, is still the safest and most dependable type of mine haulage. Grades in favor of or against the loads should not exceed 3½ percent and grades for long distances against the load should not exceed 2 percent. The trend in **main-line haulage** is toward heavier and faster locomotives, since the hauls are becoming longer and longer. The sleek, streamlined, fast, and easy-riding **Jeffrey** eight-wheel four-motor locomotives, available in 27-, 37-, and 50-short ton (24-, 34-, and 45-tonne) sizes, are ideally suited to these conditions. This type of locomotive has two four-wheel trucks, each having two motors. The trucks, having Pullman or longitudinal-type equalizers with snubbers, will go around a 50 ft (15 m) radius curve, have short overhang, and provide a very easy ride. This construction is easy on the track, with consequent low track-maintenance cost. Speed at full load ranges from 10 to 12.5 mi/h (4.5 to 5.6 m/s), and the maximum safe speed is approximately 30 mi/h (13.4 m/s).

The **eight-wheel locomotive** usually has a box frame; series-wound motors with single-reduction spur gearing; 10 steps of straight parallel, full electropneumatic contactor control; dynamic braking; 32-V battery-operated control and headlights, with the battery charged automatically from the trolley; straight air brakes; air-operated sanders; air horn; automatic couplers; one trolley pole; two headlights at each end; and blowers to ventilate the traction motors. The equipment is located so that it is easily accessible for repair and maintenance.

The eight-wheel type of locomotive has, to a great extent, superseded the tandem type, consisting of two four-wheel two-motor locomotives coupled together and controlled from the cab of one of the units of the tandem.

The older Jeffrey **four-wheel-type locomotive** is available in 11-, 15-, 20-, and 27-short ton (10-, 13.6-, 18-, 24.5-tonne) nominal weights. The 20- and 27-short ton (18- and 24.5-tonne) sizes have electrical equipment very much like that of the eight-wheel-type locomotive. Speeds are also comparable. The 15-short ton (13.6-tonne) locomotive usually has semi-electropneumatic contactor control, rather than full electropneumatic control, and dynamic braking but usually does not have the 32-V battery-operated control. The 11-short ton (10-tonne) locomotive has manual control, with manual brakes and sanders, although contactor control, air brakes, blowers, etc., are optional.

**Locomotive motors** have a horsepower rating on the basis of 1 h at 75°C above an ambient temperature of 40°C. Sizes range from a total of 100 hp (750 W) for the 11-ton (10-tonne) to a total of 720 hp for the 50-ton.

The following formulas are recommended by the Jeffrey Mining Machinery Co. to **determine the weight of a locomotive** required to haul a load. Table 1 gives haulage capacities of various weights of locomotives on grades up to 5 percent. The tabulation of haulage capacities shows how drastically the tons of trailing load decrease as the grade increases. For example, a 50-ton locomotive can haul 1,250 tons trailing load on the level but only 167 tons up a 5 percent grade.

The following formulas are based on the use of steel-tired or rolled-steel wheels on clean, dry rail.

**Weight of locomotive required on level track:**

$$W = L(R + A)/(0.3 \times 2,000 - A)$$

**Weight of locomotive required to haul train up the grade:**

$$W = L(R + G)/(0.25 \times 2,000 - G)$$

**Weight of locomotive necessary to start train on the grade:**

$$W = L(R + G + A)/(0.30 \times 2,000 - G - A)$$

where $W$ is the weight in tons of locomotive required; $R$ is the frictional resistance of the cars in pounds per ton and is taken as 20 lb for cars with antifriction bearings and 30 lb for plain-bearing cars; $L$ is the weight of the load in tons; $A$ is the acceleration resistance [this is 100 for 1 mi/(h)(s) and is usually taken at 20 for less than 10 mi/h or at 30 from 10 to 12 mi/h, corresponding to an acceleration of 0.2 or 0.3 mi/(h)(s)]; $G$ is the grade resistance in pounds per ton or 20 lb/ton for each percent of grade (25 percent is the running adhesion of the locomotive, 30 percent is the starting adhesion using sand); 2,000 is the factor to give adhesion in pounds per ton.

**Where the grade is in favor of the load:**

$$W = L(G - R)/(0.20 \times 2,000 - G)$$

**To brake the train to a stop on grade:**

$$W = L(G + B - R)/(0.20 \times 2,000 - G - B)$$

where $B$ is the braking (or decelerating) effort in pounds per ton and equals 100 lb/ton for a braking rate of 1 mi/(h)(s) or 20 lb/ton for a braking rate of 0.2 mi/(h)(s) or 30 lb for a braking rate of 0.3 mi/(h)(s). The adhesion is taken from a safety standpoint as 20 percent. It is not advisable to rely on using sand to increase the adhesion, since the sandboxes may be empty when sand is needed.

**Time in seconds to brake the train to stop:**

$$s = \frac{\text{mi/h (start)} - \text{mi/h (finish)}}{\text{deceleration in mi/(h)(s)}}$$

**Distance in feet to brake the train to a stop:**

$$\text{ft} = [\text{mi/h (start)} - \text{mi/h (finish)}] \times s \times 1.46/2$$

**Storage-battery locomotives** are used for hauling muck cars in tunnel construction where it is inconvenient to install trolley wires and bond the track as the tunnel advances. They are also used to some extent in metal mines and in mines of foreign countries where trolley locomotives are not permitted. They are seldom used anymore in coal mines of the United States. Their first cost is frequently less than that for a trolley installation. They also possess many of the advantages of the trolley locomotive and eliminate the danger and obstruction of the trolley wire. Storage-battery locomotives are limited by the energy that is stored in the battery and should not be used on steep grades or where large, continuous overloads are required. Best results are obtained where light and medium loads are to be handled intermittently over short distances with a grade of not over 3 percent against the load.

The general construction and mechanical features are similar to those of the four-wheel trolley type, with battery boxes located either on top of the locomotive or between the side frames, according to the height available. The motors are rugged, with high efficiency. Storage-battery locomotives for coal mines are generally of the explosion-tested type approved by the Bureau of Mines for use in gaseous mines.

The **battery** usually has sufficient capacity to last a single shift. For two- or three-shift operation, an extra battery box with battery is required so that one battery can be charging while the other is working on the locomotive. Motor-generator sets or rectifiers are used for charging the batteries. The overall efficiency of the battery, motor, and gearing is approximately 63 percent. The speed varies from 3 to 7 mi/h, the average being 3½ to 4½ mi/h. Battery locomotives are available in sizes from 2 to 50 tons. They are usually manufactured to suit individual requirements, since the sizes of motors and battery are determined by the amount of work that the locomotive has to do in an 8-h shift.

## INDUSTRIAL CARS

Various types of narrow-gage industrial cars are used for handling bulk and package materials inside and outside of

**Table 1. Haulage Capacities of Locomotives with Steel-Tired or Rolled-Steel Wheels***

| Grade | | Weight of locomotive, tons† | | | | | |
|-------|---|------|------|------|------|------|------|
| | | 11 | 15 | 20 | 27 | 37 | 50 |
| Level | Drawbar pull, lb | 5,500 | 7,500 | 10,000 | 13,500 | 18,500 | 25,000 |
| | Haulage capacity, gross tons | 275 | 375 | 500 | 675 | 925 | 1,250 |
| 1% | Drawbar pull, lb | 5,280 | 7,200 | 9,600 | 12,960 | 17,760 | 24,000 |
| | Haulage capacity, gross tons | 132 | 180 | 240 | 324 | 444 | 600 |
| 2% | Drawbar pull, lb | 5,260 | 6,900 | 9,200 | 12,420 | 17,020 | 23,000 |
| | Haulage capacity, gross tons | 88 | 115 | 153 | 207 | 284 | 384 |
| 3% | Drawbar pull, lb | 4,840 | 6,600 | 8,800 | 11,880 | 16,280 | 22,000 |
| | Haulage capacity, gross tons | 67 | 82 | 110 | 149 | 204 | 275 |
| 4% | Drawbar pull, lb | 4,620 | 6,300 | 8,400 | 11,340 | 15,540 | 21,000 |
| | Haulage capacity, gross tons | 46 | 63 | 84 | 113 | 155 | 210 |
| 5% | Drawbar pull, lb | 4,400 | 6,000 | 8,000 | 10,800 | 14,800 | 20,000 |
| | Haulage capacity, gross tons | 31 | 50 | 67 | 90 | 123 | 167 |

*Jeffrey Mining Machinery Co.
†Haulage capacities are based on 20 lb/ton rolling friction, which is conservative for roller-bearing cars. Multiply lb by 0.45 to get kg and tons by 0.91 to get tonnes.

buildings. Those used for bulk material are usually of the dumping type, the form of the car being determined by the duty. They are either pushed by men or drawn by mules, locomotives, or cable. The **rocker side-dump car** (Fig 1) consists of a truck on which is mounted a V-shaped steel body supported on rockers so that it can be tipped to either side, discharging material. This type is mainly used on construction work. Capacities vary from ⅔ to 5 tons for track gages of 18, 20, 24, 30, 36, and 56½ in. In the **gable-bottom car** (Fig. 2), the side doors $a$ are hinged at the top and controlled by levers $b$

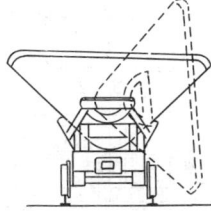 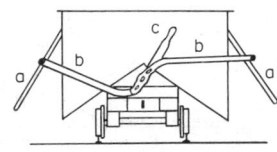

**Fig. 1** Rocker side-dump car.   **Fig. 2** Gable-bottom car.

and $c$, which lock the doors when closed. Since this type of car discharges material close to the ground on both sides of the track simultaneously, it is used mainly on trestles. Capacities vary from 29 to 270 ft³ for track gages of 24, 36, 40, and 50 in. The **scoop dumping car** (Fig. 3) consists of a scoop-shaped steel body pivoted at $a$ on turntable $b$, which is carried by the truck. The latch $c$ holds the body in a horizontal position, being released by chain $d$ attached to handle $e$. Since the body is mounted on a turntable, the car is used for service where it is desirable to discharge material at any point in the circle. This car is made with capacities from 12 to 27 ft³ to suit local requirements. The **hopper-bottom car** (Fig. 4) consists of a hopper on wheels, the bottom opening being controlled by

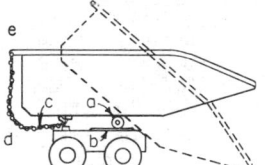 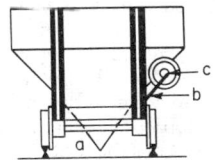

**Fig. 3** Scoop dumping car.   **Fig. 4** Hopper-bottom car.

door $a$, which is operated by chain $b$ winding on shaft $c$. The shaft is provided with handwheel and ratchet and pawl. The type of door or gate controlling the bottom opening varies with different materials.

The **box-body dump car** (Fig. 5) consists of a rectangular body pivoted on the trucks at $a$ and held in horizontal position by chains $b$. The side doors of the car are attached to levers so that the door is automatically raised when the body of the car is tilted to its dumping position. The cars can be dumped to either side. On the large sizes, where rapid dumping is required, dumping is accomplished by compressed air. This type of car is primarily used in excavation and quarry work, being loaded by power shovels. The greater load is placed on the side on which the car will dump, so that dumping is automatic when the operator releases the chain or latch. The car bodies may be steel or steel-lined wood. **Mine cars** are usually of the four-wheel type, with low bodies, the doors being at one end, and pivoted at the top with latch at the bottom. Industrial **tracks** are made with rails from 12 to 45 lb/yd (6.0 to 22 kg/m) and gages from 24 in to 4 ft 8½ in (0.6 to 1.44 m). Either steel or wooden ties are used. Owing to its lighter weight, the steel tie is preferred where tracks are

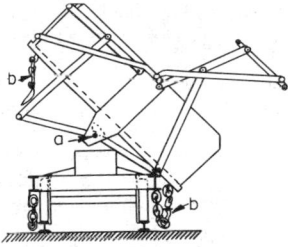

**Fig. 5** Box-body dump car.

frequently moved, the track being made up in sections. Industrial cars are frequently built with one wheel attached to the axle and the other wheel loose to enable the car to turn on short-radius tracks. Capacities vary from 4 to 50 yd³ (0.11 to 5.7 m³) for track gages of 36 to 56½ in (0.9 to 1.44 m), with cars having weights from 6,900 to 80,300 lb (3,100 to 36,000 kg). The frictional resistance per ton (2,000 lb) (8,900 N) for different types of mine-car bearings is as follows (see also Sec. 11):

| Types of bearings | Drawbar pull | | | | | |
|---|---|---|---|---|---|---|
| | Level track | | 2% grade | | 4% grade | |
| | lb/short ton | N/tonne | lb/short ton | N/tonne | lb/short ton | N/tonne |
| Spiral roller | 13 | 58 | 15 | 67 | 46 | 205 |
| Solid roller | 14 | 62 | 18 | 80 | 53 | 236 |
| Self-oiling | 22 | 98 | 31 | 138 | | |
| Babbitted, old style | 24 | 107 | 40 | 178 | | |

## DOZERS, DRAGLINES

The dual capability of some equipment, such as **dozers** and **draglines,** suggests that it should be mentioned as prime machinery in the area of materials handling by dragging, pulling, or pushing. Dozers are described in the discussion on earthmoving equipment since their basic frames are also used for power shovels and backhoes. In addition, dozers form the auxiliary function of pushing carryall earthmovers to assist them in scraping up their load. Dragline equipment is discussed with below-surface handling or excavation. The same type of equipment that would drag or scrape may also have a lifting function.

## MOVING SIDEWALKS

Moving horizontal belts with synchronized balustrading have been introduced to expedite the movement of passengers to or from railroad trains in depots or planes at airports (see belt conveyors). A necessary feature is the need to prevent the clothing of anyone (e.g., a child) sitting on the moving walk from being caught in the mechanism at the end of the walk. Use of a comblike stationary end fence protruding down into longitudinal slots in the belt is an effective preventive.

## CAR-UNLOADING MACHINERY

Four types of devices are in common use for unloading material from all types of open-top cars: crossover and horn dumps, used to unload mine cars with swinging end doors; rotary car dumps, for mine cars without doors; and tipping car dumps, for unloading standard-gage cars where large unloading capacity is required.

**Crossover Dump**   Figure 6 shows a car in the act of dumping. Figure 7 shows a loaded car pushing an empty car off the dump. A section of track is carried by a platform supported on rockers *a*. An extension bar *b* carries the weight *c* and the brake friction bar *d*. A hand lever controls the brake, acting on the friction bar and placing the dumping under the control of the operator. A section of track *e* in front of the dump is pivoted on a parallel motion and counterbalanced so that it is normally raised. The loaded car depresses the rails *e* and, through levers, pivots the horns *f* around the shafts *g*, releasing the empty car. The load car strikes the empty car, starting it down the inclined track. After the loaded car has passed the rails *e*, the springs return the horns *f* so that they stop the

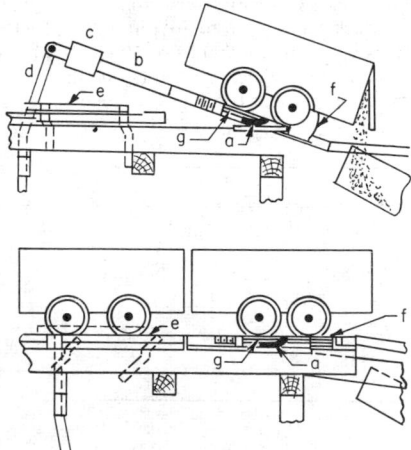

**Figs. 6 and 7**   Crossover dump.

loaded car in the position to dump. Buffer springs on the shaft *g* absorb the shock of stopping the car. Since the center of gravity of the loaded car is forward of the rockers, the car will dump automatically under control of the brake. No power is required for this dump, and one operator can dump three or four cars per minute.

**Rotary Gravity Dump** (Fig. 8)   This consists of a steel cylinder supported by a shaft *a*, its three compartments carrying three tracks. The loaded car *b* is to one side of the center and causes the cylinder to rotate, the material rolling to the chute

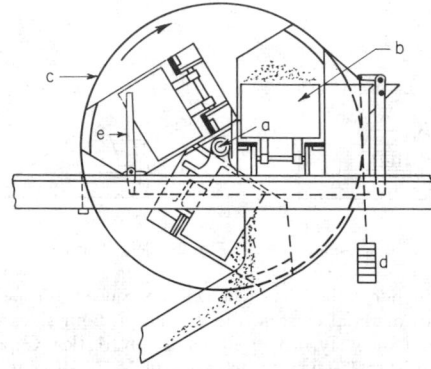

**Fig. 8**   Rotary gravity dump.

beneath. The band brake *c*, with counterweight *d*, is operated by lever *e*, putting the dumping under control of the operator. No power is required; one operator can dump two or three cars per minute.

**Rotary-power dumpers** are also built to take any size of open-top railroad car and are frequently used in power plants, coke plants, ports, and ore mines to dump coal, coke, ore, bauxite, and other bulk material. They are mainly of two types: (1) single barrel, and (2) tandem.

The McDowell-Wellman Engineering Co. dumper consists of a revolving cradle supporting a platen (with rails in line with the approach and runoff car tracks in the upright position), which carries the car to be dumped. A blocking on the dumping side supports the side of the car as the cradle starts rotating. Normally, the platen is movable and the blocking is fixed, but in some cases the platen is fixed and the blocking movable. Where there is no variation in the car size, the platen and blocking are both fixed. The cradle is supported on two end rings, which are bound with a rail and a gear rack. The rail makes contact with rollers mounted in sills resting on the foundation. Power through the motor rotates the cradle by means of two pinions meshing with the gear racks. The angle of rotation for a complete dump is 155° for a normal operation, but occasionally, a dumper is designed for 180° rotation. The clamps, supported by the cradle, start moving down as the dumper starts to rotate. These clamps are lowered, locked, released, and raised either by a gravity-powered mechanism or by hydraulic cylinders.

With the advent of the **unit-train system,** the investment and operating costs for a dumper have been reduced considerably.

The design of dumpers for unit train has improved and results in fewer maintenance problems. The use of rotary couplers on unit train eliminates uncoupling of cars while dumping because the center of the rotary coupler is in line with the center of rotation of the dumper.

The McDowell-Wellman Engineering Co. will, when it is specified, provide an automatic electronic weighing system with a car dumper. The platen, car, and load are raised from the platen supports, and the weight is automatically recorded at the beginning and end of each dump cycle.

**Car Shakers**  As alternatives to rotating or tilting the car, several types of car shakers are used to hasten the discharge of the load. Usually the shaker is a heavy yoke equipped with an unbalanced pulley rotated at 2,000 r/min by a 20-hp (15-kW) motor. The yoke rests upon the car top sides, and the load is actively vibrated and rapidly discharged. While a car shaker provides a discharge rate about half that of a rotary dumper, the smaller investment is advantageous.

**Car Positioner**  As the popularity of unit-train systems consisting of rail cars connected by rotary couplers has increased, more rotary dumping stations have been equipped with an automatic train positioner developed by McDowell-Wellman Engineering Company. This device consists of a carriage moving parallel to the railroad track actuated by either hydraulic cylinders or wire rope driven by a winch, which carries an arm that rotates in a vertical plane to engage the coupling between the cars. The machines are available in many sizes, the largest of which are capable of indexing 200-car trains in one- or two-car increments through the rotary dumper. These machines or similar are also available from FMC/Materials Handling System Division, Heyl and Patterson, Inc., and Whiting Corp.

# CARRYING AND LIFTING

## by Harold V. Hawkins and Associates

### CONTAINERIZATION

The proper packaging of material to assist in handling can significantly minimize the handling cost and can also have a marked influence on the type of handling equipment employed. For example, partial carload lots of liquid or granular material may be shipped in rigid or nonrigid containers equipped with proper lugs to facilitate in-transit handling. Heavy-duty rubberized containers that are inert to most cargo are available for repeated use in shipping partial carloads. The nonrigid container reduces return shipping costs, since it can be collapsed to reduce space. Disposable lightweight corrugated-cardboard shipping containers for small and medium-sized packages both protect the cargo and permit stacking to economize on space requirements. The type of container to be used should be planned or considered when the handling mechanism is selected.

### SURFACE HANDLING

#### Lift Trucks and Palletized Loads

The use of lift trucks is advantageous for movement of materials in industrial processing and warehousing where batches or unit quantities of materials of varying amounts are moved on schedules that are not well ordered to destinations that are quite varied. The most common type of lift truck is equipped with a fork on the front, which is projected under the load to be moved. The load is first lifted and then transported; the lifting action is either a direct lift or a direct lift coupled with a tilt to secure against quick stops. **Hand-powered vehicles** (Fig. 1) utilize a towing handle that operates both as a hydraulic pump for the lifting process and as a means of towing during the movement. Similar **electric-powered,** battery-energized units are available (Fig. 2). Loads to 6,000 lb (2,700 kg) can be moved in this fashion on pallets that have an initial clearance under them of about 3½ in (89 mm), with the added lifting

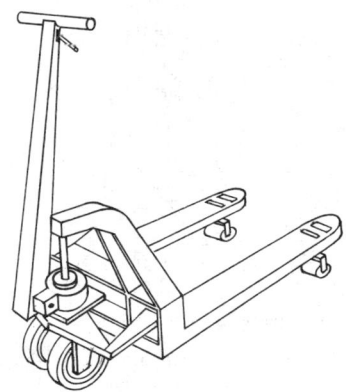

**Fig. 1**  Hand-powered pallet truck.

height 4 in (10 mm). The lifting forks vary upward from 24 in (0.61 m) in length and from 21 in (0.53 m) in width.

**Powered lift trucks** provide for lifting loads from the floor to as high as 17 ft (5.2 m) or more for stacking. Usual capacities are from 1,000 to 6,000 lb (450 to 2,700 kg). With one type of lift truck, the operator walks beside the truck, commanding it through controls on a handle; another type provides a platform for a standing operator; a third type has a conventional driver's seat (Fig. 3). All types have extremely short turning radii for mobility in narrow aisles. Motive power is by electric motor and battery or by internal-combustion engine generally operating from liquid propane gas. In some units, the gas engine drives a dc generator to provide current for an electric motor

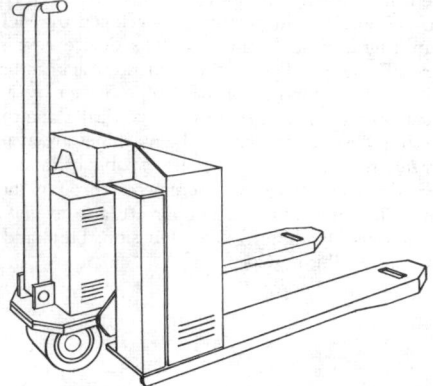

**Fig. 2**   Electric-powered, battery-energized hand truck.

in order to retain the convenience of electric controls. Large forklift trucks are also available for yard work to lift loads as heavy as 40 tons (36 tonne) to heights of 11 ft (3.4 m).

**Fig. 3**   Lift truck powered by an internal-combustion engine.

To facilitate the use of forklift trucks, the **load is generally piled on a pallet** [usually 36 to 42 in (0.91 to 1.07 m)] consisting of a low platform having sufficient clearance to the floor to permit the truck's lifting forks to protrude under the load. Some trucks are equipped with straddle forks, which also go under the load and press against the floor to take the overturning reaction as the load is lifted. This minimizes the need for a large counterbalance in the back of the lift truck, but it also makes movement with the load elevated somewhat more awkward. The pallet moves with the load and remains under it when the load is left at its destination. Pallets are so inexpensive as to be considered expendable.

**Customized load-handling mechanisms** are available in place of standard forks for lift trucks. One type provides a surface either curved (for rolls of newsprint) or flat (for large cardboard containers), depending upon the shape of the usual load,

and is equipped with a vacuum pump, which sucks the load to the surface sufficiently to permit the load to be lifted. Special hooks are available for lifting 30- or 55-gal (114- to 208-l) steel drums. The hydraulic lifting mechanism provides a source of power for manipulating many special clamps, pivots, etc. Hydraulically actuated clamps provide for squeezing rolls to permit lifting. Spindles or rams are available for inserting in steel coils to permit lifting.

### Off-Highway Vehicles and Earthmoving Equipment
*by P. G. Kuchuris, Jr., International Harvester Company*

The movement of large quantities of bulk materials, earth, gravel, and broken rock in road building, mining, construction, quarrying, and land clearing may be handled by **off-highway vehicles.** Such vehicles are mounted on large pneumatic tires or on crawler tracks if heavy pulling and pushing are required on poor or steep terrain. Width and weight of the rubber-tired equipment often exceed highway legal limits, and use of grouser tracks on highways is prohibited.

Proper selection of size and type of equipment depends on the amount, kind, and density of the material to be moved in a specified time and on the distances, direction, and steepness of grades, footing for traction, and altitude above sea level. Time cycles and pay loads for production per hour can then be estimated from manufacturers' performance data and job experience. This production per hour, together with the corresponding owning, operating, and labor costs per hour, enables selection by favorable cost per cubic yard, ton, or other pay unit (see Fig. 4a).

Current rapid progress in the development of off-highway equipment will soon make any description of size, power, and productivity obsolete. However, the following brief description of major off-highway vehicles will serve as a guide to their applications.

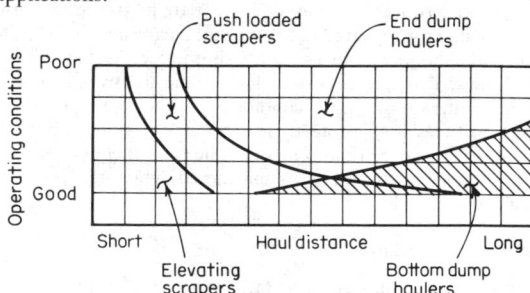

**Fig. 4a.**   Economical application zones: truck vs. scrapers.

**Crawler Tractors**   These are track-type prime movers for use with mounted bulldozers, rippers, winches, cranes, cable layers, and side booms rated by net engine horsepower in sizes from 40 to over 500 hp; maximum traveling speeds, 5 to 7 mi/h (8 to 11 km/h). Diesel-powered tractors with torque-converter drive and power-shift transmission develop drawbar pulls up to 90 percent or more of their weight with mounted equipment.

**Wheel Tractors** (Fig. 5)   Sizes range from rubber-tired industrial tractors for small scoops, loaders, and backhoes to large, diesel-powered, two-wheel or four-wheel pneumatic-tired prime movers for propelling scrapers and wagons. Large, four-wheel-drive, articulated-steering types also power bulldozers.

**Bulldozer—Crawler Type** (Fig. 4b)   This is a crawler tractor with a front-mounted blade, which is lifted by hydraulic or

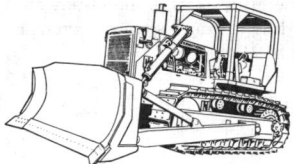

**Fig. 4b**  Bulldozer on crawler tractor. (*International Harvester.*)

cable power control. There are four basic types of moldboards: straight, semi-U and U (named by top-view shape), and angling. The angling type, often called **bullgrader** or **angledozer,** can be set for side casting 25° to the right or left of

**Fig. 5**  Wheeltractor, four-wheel-drive loader backhoe. (*International Harvester.*)

perpendicular to the tractor centerline, while the other blades can be tipped forward or back through about 10° for different digging conditions. All blades can be tilted for ditching, with hydraulic-power tilt available for all blades.

APPLICATION.   This is the best machine for pioneering access roads, for boulder and tree removal, and for short-haul earthmoving in rough terrain. It push-loads self-propelled scrapers and is often used with a rear-mounted ripper to loosen firm or hard materials, including rock, for scraper loading. U blades drift 15 to 20 percent more loose material than straight blades but have poor digging ability. Angling blades expedite sidehill benching and backfilling of trenches. Loose-material capacity of straight blades varies approximately as the blade height squared, multiplied by length. Average capacity of digging blades is about 1 yd³ loose measure per 30 *net* hp rating of the crawler tractor. Payload is 60 to 90 percent of loose measure, depending on material swell variations.

**Bulldozer—Wheel Type**   This is a four-wheel-drive, rubber-tired tractor, generally of the hydraulic articulated-steering type, with front-mounted blade that can be hydraulically raised, lowered, tipped, and tilted. Its operating weights range to 150,000 lb, with up to 700 hp, and its traveling speeds range from stall to about 20 mi/h for pushing and mobility.

APPLICATION.   It is excellent for push-loading self-propelled scrapers, for grading the cut, spreading and compacting the fill, and for drifting loose materials on firm or sandy ground for distances up to 500 ft. Useful tractive effort on firm earth surfaces is limited to about 60 percent of weight, as compared with 90 percent for crawler dozers.

**Loader—Crawler Type** (Fig. 6)   This is a track-type prime mover with front-mounted bucket that can be raised, dumped, lowered, and tipped by power control. Figure 6 illustrates an optional multipurpose bucket that can be used as a shovel,

bulldozer, clamshell, or scraper. Capacities range from 0.7 to 5.0 yd³, SAE rated. It is also available with grapples for pulpwood, logs, and lumber.

APPLICATION.   It is used for digging basements, pools, ponds, and ditches; for loading trucks and hoppers; for placing, spreading, and compacting earth over garbage in sanitary fills; for stripping sod; for removing steel-mill slag; and for carrying and loading pulpwood and logs.

**Loader—Wheel Type** (Fig. 7)   This is a four-wheel, rubber-tired prime mover with front-mounted hydraulic-operated shovel, often called a **Payloader.** It is available in SAE-rated capacities from ½ to 20 yd³ (0.4 to 15 m³) for materials

**Fig. 6**  Crawler-type loader with multipurpose bucket. (*International Harvester.*)

weighing 3,000 lb/yd³ (1,800 kg/m³), with larger buckets available for lighter materials.

APPLICATION.   This highly mobile equipment is used for

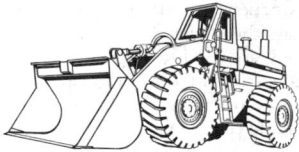

**Fig. 7**  Four-wheel-drive loader. (*International Harvester.*)

handling and loading materials of all kinds on firm surfaces.

**Scrapers—Tractor-drawn** (Fig. 8)   These are four-wheel,

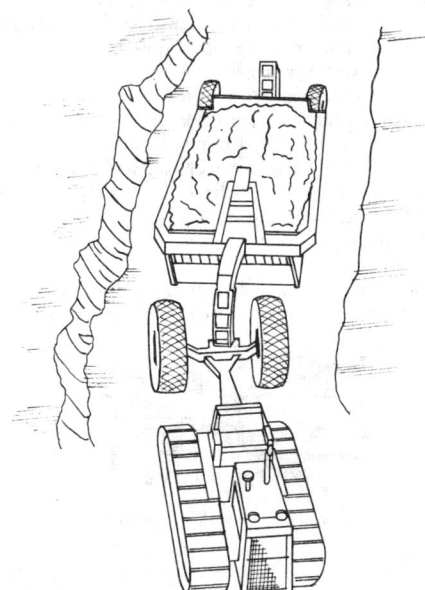

**Fig. 8**  Crawler tractor and drawn scraper. (*International Harvester.*)

rubber-tired trailers for loading, hauling, dumping, and spreading earth and for bank-sloping and finishing. They are self-loading when matched to crawler tractors having at least 12 *net* hp/yd³ struck capacity and operating weights 10 to 20 percent more than scraper pay-load weight. Popular sizes range from 7 to 27 yd³ struck capacity (5.4 to 20.6 m³).

APPLICATION. They are used for earthmoving for one-way distances up to 1,000 ft (300 m) or under traction and terrain conditions unsuitable for faster, self-propelled scrapers.

**Scrapers—Self-propelled** (Fig. 9) These are similar to the tractor-drawn type, but the front wheels are replaced by diesel-powered, two-wheel or four-wheel tractors with 148 to 650 + hp and with numerous load-matching speeds to 30 mi/h

**Fig. 9** Two-axle self-propelled twin-engine scraper with Pay-mate attachment. *(International Harvester.)*

and faster. Scraper rear wheels may also be driven by a separate rear-mounted engine. Ratings are given in cubic-yard struck/heaped capacities, such as 7/9, 14/18, 24/32, and 40/54. Pay-load capacities depend on loadability and swell of materials but approximate the struck capacity.

APPLICATION. They are used for high-speed earthmoving in road building for construction work. However, some sizes are available with an elevator or with all-wheel drive for self-loading under favorable conditions. Two all-wheel-drive units can also load each other through a Pay-mate (push-pull) type of attachment. Two-axle, four-wheel types have best maneuverability, but the three-axle type is sometimes preferred for easier riding on long hauls at higher speed.

**Bottom-dump Wagons** These are available in sizes to 150 short tons and are used in place of scrapers on large-wheel tractors or trucks for hauling earth, sand, gravel, and coal over longer distances. They require top loading by conveyor, dragline, shovel, or overhead grizzly.

**Off-Highway End-dump Truck** (Fig. 10) This is a heavy-duty, deisel-powered, load-on-back hauler with rear-dump body. It is regularly produced in rated capacities of 12 to 350 tons. The body cubic-yard struck capacity is about two-thirds tonnage rating. The all-wheel-drive, 50-ton-capacity, 537-hp Payhauler illustrated in Fig. 10 has power-shift transmission, with ten speeds to over 36 mi/h, and gross weight of 172,800 lb.

APPLICATION. It is used for hauling and dumping blasted rock, ore, and other hard and abrasive shovel-loaded materials

**Fig. 10** Off-highway end-dump truck. *(International Harvester.)*

in road and dam construction and in quarries and mines.

**Owning and Operating Costs** These include depreciation;

interest, insurance, taxes; parts, labor, repairs, and tires; fuel, lubricant, filters, hydraulic-system oil, and other operating supplies. This is reduced to cost per hour over a service life of 4 to 6 years of 2,000 h each—average 5 years, 10,000 h. Owning and operating costs of diesel-powered bulldozers and scrapers, excluding operator's wages, average 3 to 4 times the delivered price in 10,000 h.

## ABOVE-SURFACE HANDLING

### Monorails

Materials can be carried on light, rigid trackage, as described for overhead conveyors (see below). Trolleys are supported by structural I beams, H beams, or I-beam-like rails with special flat flanges to improve rolling characteristics of the wheels. Size of wheels and smoothness of tread are important in reducing rolling resistance. Figure 11 shows a typical rigid trolley for traversing short-radius track curves. Typical

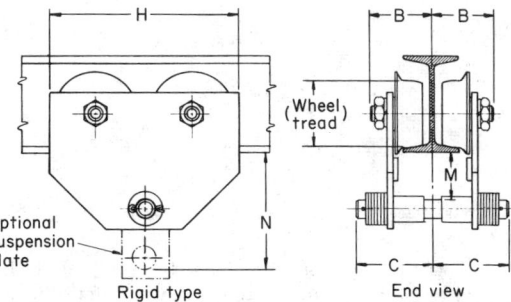

**Fig. 11** Monorail trolleys. *(C.M. Hoist Div., Columbus McKinnon.)*

dimensions for both types are given in Table 1. These trolleys may be plain, with geared handwheel and hand chain, or motor-driven. For very low headroom, the trolley can be built into the hoist; this is known as a **trolley hoist**.

### Overhead Traveling Cranes

An **overhead traveling crane** is a vehicle for lifting, transporting, and lowering loads. It consists of a bridge supporting a hoisting unit and is equipped with wheels for operating on an elevated runway or track. The hoisting unit may be fixed relative to the bridge but is usually supported on wheels, permitting it to traverse the length of the bridge.

The motions of the crane—hoisting, trolley traversing, and bridging—may be powered by hand, electricity, air, hydraulics, or a combination of these. Hand-powered cranes are generally built in capacities under 50 tons (45 tonnes) and are used for infrequent service where slow speeds are acceptable. Pneumatic cranes are used where electricity would be hazardous or where advantage can be taken of existing air supply. Electric cranes are the most common overhead type and can be built to capacities of 500 tons (454 tonnes) or more and to spans of 150 ft (46 m) and over.

**Single-Girder Cranes** (Fig. 12) In its simplest form, this consists of an I beam *a* supported by four wheels *b*. The trolley *c* traveling on the lower flanges carries the chain hoist, forming the lifting unit. The crane is moved by hand chain *d* turning sprocket wheel *e*, which is keyed to shaft *f*. The pinions on shaft *f* mesh with gears *g*, keyed to the axles of two

**Table 1. Typical Monorail Trolley Dimensions***

| Capacity, short tons | I-beam range (depth), in | Wheel-tread diam, in | Net weight, lb | B,† in | C, in | H, in | M, in | N, in | Min beam radius,† in |
|---|---|---|---|---|---|---|---|---|---|
| ½ | 5–10 | 3½ | 32 | 3¼ | 4⅛ | 9⅞ | 2⅜ | 6¼ | 21 |
| 1 | 5–10 | 3½ | 32 | 3¼ | 4⅛ | 9⅞ | 2⅜ | 6¼ | 21 |
| 1½ | 6–10 | 4 | 52 | 3⅞ | 4⅝ | 11⅜ | 2¹⁵⁄₁₆ | 7⁷⁄₁₆ | 30 |
| 2 | 6–10 | 4 | 52 | 3⅞ | 4⅝ | 11⅜ | 2¹⁵⁄₁₆ | 7⁷⁄₁₆ | 30 |
| 3 | 8–15 | 5 | 88 | 4⁷⁄₁₆ | 5⅜ | 13½ | 2¹³⁄₁₆ | 7¹⁵⁄₁₆ | 42 |
| 4 | 8–15 | 5 | 88 | 4⁷⁄₁₆ | 5⅜ | 13½ | 2¹³⁄₁₆ | 7¹⁵⁄₁₆ | 42 |
| 5 | 10–18 | 6 | 137 | 5³⁄₁₆ | 6³⁄₁₆ | 15⅜ | 3⁵⁄₁₆ | 10⅛ | 48 |
| 6 | 10–18 | 6 | 137 | 5³⁄₁₆ | 6³⁄₁₆ | 15⅜ | 3⁵⁄₁₆ | 10⅛ | 48 |
| 8 | 12–24 | 8 | 279 | 5½ | 7⁷⁄₁₆ | 21⅜ | 4³⁄₁₆ | 13¾ | 60 |
| 10 | 12–24 | 8 | 279 | 5½ | 7⁷⁄₁₆ | 21⅜ | 4³⁄₁₆ | 13¾ | 60 |

Metric values, multiply tons by 907 for kg, inches by 25.4 for mm, and lb by 0.45 for kg.
*CM Hoist Div., Columbus McKinnon.
†These dimensions are given for minimum beam.

wheels. An underslung construction may also be used, with pairs of wheels at each corner which ride on the lower flange of I-beam rails. Single-girder cranes may be hand-powered by pendant hand chains or electric-powered as controlled by a pendant push-button station.

**Electric Traveling Crane, Double-Girder Type** (Fig. 13) This consists of two bridge girders *a*, on the top of which are rails on which travels the self-contained hoisting unit *b*, called the **trolley**. The girders are supported at the ends by trucks with two or four wheels, according to the size of the crane.

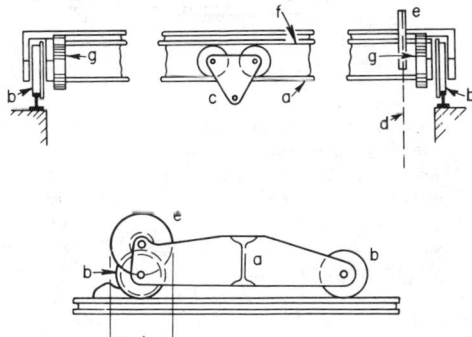

**Fig. 12** Hand-powered crane.

The crane is moved along the track by motor *c*, through shaft *d* and gearing to the truck wheels. Suspended from the girders on one side is the operator's cab *e*, containing the controller, or master switches, master hydraulic brake cylinder, warning device, etc. The bridge girders for small cranes are of the I-beam type, but on the longer spans, box girders are used to give torsional and lateral stiffness. The girders are rigidly attached to the truck end framing, which carries the double-flanged wheels for supporting the bridge. The end frames project over the rails so that in case of a broken wheel or axle, the frame will rest on the rail, preventing the crane from dropping. One wheel axle on each truck is fitted with gears for driving the crane or is coupled directly to the shaft which transmits power from the gear reducer. On a cab-operated bridge, a brake, usually hydraulic, is applied to the motor

shaft to stop the crane. Floor-operated cranes generally utilize spring-engaged, electrically released brakes.

The trolley consists of a frame which carries the hoisting machinery and is supported on wheels for movement along the bridge rails. The wheels are coupled to the trolley traverse motor through suitable gear reduction. Trolleys are frequently equipped with a second set of hoisting machinery to provide dual lifting means or an auxiliary of smaller capacity. The hoisting machinery consists of motor, motor brake, load brake, gear reduction, and rope drum. Wire rope winding in helical grooves on the drum is reeved over sheaves in the upper block and lower hook block for additional mechanical advantage. Limit switches are provided to stop the motors when limits of travel are reached. Current is brought to the crane by sliding or rolling collectors in contact with conductors attached to or parallel with the runway and preferably located at the cab end of the crane. Current from the runway conductors and cab is carried to the trolley in a like manner from conductors mounted parallel to the bridge girder. Festooned multiconductor cables are also used to supply current to crane or trolley.

Electric cranes are built for either alternating or direct current, with the former predominating. The motors for both

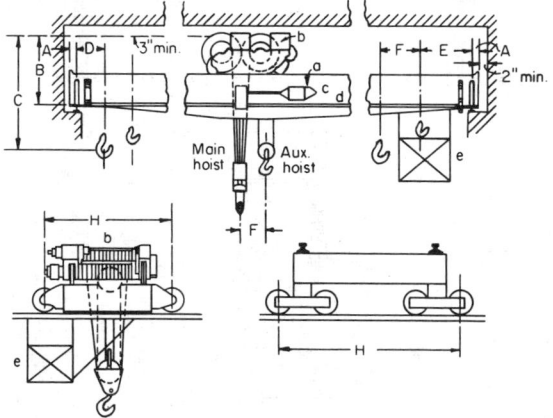

**Fig. 13** Electric traveling crane.

kinds of current are designed particularly for crane service. Direct-current motors are usually series-wound, and ac motors are generally of the wound-rotor or two-speed squirrel-cage type. The usual ac voltages are 230 and 460, the most common being 460. The **capacities** and other dimensions for standard electric cranes are given in Table 2.

### Gantry Cranes

**Gantry cranes** are modifications of traveling cranes and are generally used outdoors where it is not convenient to erect on overhead runway. The bridge (Fig. 14) is carried at the ends by the legs *a*, supported by trucks with wheels so that the crane can travel. As with the traveling crane, the bridge carries a hoisting unit; a cover to protect the machinery from the weather is often used. The crane is driven by motor *b* through a gear reduction to shaft *c*, which drives the vertical shafts *d* through bevel gears. Bevel- and spur-gear reductions connect the axles of the wheels with shafts *d*. Many gantry cranes are built without the cross shaft, employing separate motors, brakes, and gear reducers at each end of the crane. Gantry cranes are made in the same sizes as standard traveling cranes.

### Special-Purpose Overhead Traveling Cranes

A wide variety can be built to meet special conditions or handling requirements; examples are stacker cranes to move material into and out of racks, wall cranes using a runway on only one side of a building, circular running or pivoting cranes, and semi-gantries. Load-weighing arrangements can be incorporated, as well as special load-handling devices such as lifting beams, grapples, buckets, forks, and vacuum grips.

### Rotary Cranes and Derricks

**Rotary cranes** are used for lifting material and moving it to

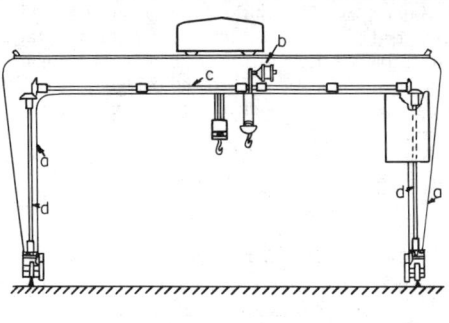

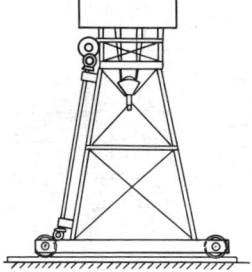

**Fig. 14**  Gantry crane.

points covered by a boom pivoted to a fixed or movable structure. **Derricks** are used outdoors (e.g., in quarries and for construction work), being built so that they can be easily moved. Pillar cranes are always fixed and are used for light, infrequent service. Jib cranes are used in manufacturing plants. Locomotive cranes mounted on car wheels are used to handle loads by hook or bulk material by means of tubs, grab buckets, or magnets. Wrecking cranes are of the same general type as locomotive cranes and are used for handling heavy loads on railroads.

**Derricks** are made with either wood or steel masts and booms, are of the guyed or stiff-leg type, and are either hand-slewed or power-swung with a bull wheel. Figure 15 shows a guyed wooden derrick of the bull-wheel type. The mast *a* is carried at the foot by pivot *k* and at the top by pivot *m*, held by rope guys *n*. The boom *b* is pivoted at the lower end of the

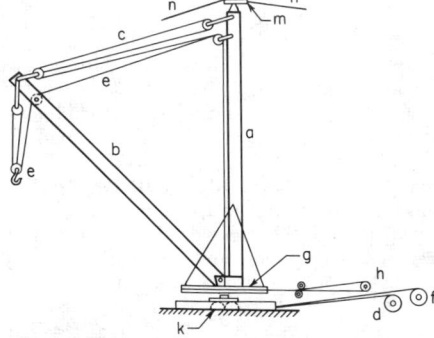

**Fig. 15**  Guyed wooden derrick.

mast. The rope *c*, passing over sheaves at the top of the mast and at the end of the boom and through the pivot *k*, is made fast to drum *d* and varies the angle of the boom. The hoisting rope *e*, from which the load is suspended, is made fast to drum *f*. The bull wheel *g* is attached to the mast and swings the derrick by a rope made fast to the bull wheel and passing around the reversible drum *h*. In derricks of the self-slewing type, the engine is mounted on a platform attached to the mast and the derrick is swung by a pinion meshing with a gear attached to the foundation. Either the bull-wheel or the self-slewing type may be made of steel or wood construction and may be of the guyed or stiff-leg type. Figure 16 shows a **column jib crane,** consisting of pivoted post *a* and carrying boom *b*, on which travels either an electric or a hand hoist *c*. The post *a* is attached to building column *d* so that it can swing through approximately 270°. Cranes of this type are rapidly being replaced by such other methods of handling materials as the mobile lift truck or the automotive-type crane. Column jib cranes are built with radii up to 20 ft (6 m) and for loads up to 5 tons (4.5 tonnes). Yard jib cranes are generally designed to meet special conditions.

### Locomotive Cranes

The locomotive crane (Fig. 17) is self-propelled and provided with trucks, brakes, automatic couplers, fittings, and clearances which will permit it to be used or hauled in a train; it can function as a complete unit on any railroad. Locomotive cranes are of the rotating-deck type, consisting of a hinged boom attached to the machinery deck, which is turntable-

**Table 2. Dimensions, Loads, and Speeds, Industrial Type Cranes**[a,e,f,g]

| Capacity main hoist, tons, 2,000 lb | Span, ft | Std. lift, main hoist,[b] ft | Std. hoist speed, ft/min | Dimensions, refer to Fig. 13 | | | | | | | Max load per wheel, lb[d] | Runway rail, lb/yd | X, in | No. of bridge wheels |
|---|---|---|---|---|---|---|---|---|---|---|---|---|---|---|
| | | | | A | Min B | C | D | E[c] | F | H | | | | |
| 5 | 40 | 36 | 46 | 7 | 2'9" | 3'1" | 3'0" | 2'6" | | 8'0" | 14,100 | 30 | 6 | 4 |
| | 60 | 53 | 46 | 7 | 3'9" | 3'1" | 3'0" | 2'6" | | 9'6" | 17,000 | 30 | 6 | 4 |
| | 80 | 86 | 46 | 7 | 3'9" | 3'1" | 3'0" | 2'6" | | 12'0" | 20,400 | 30 | 6 | 4 |
| | 100 | 118 | 46 | 7 | 3'9" | 3'1" | 3'0" | 2'6" | | 14'6" | 24,100 | 60 | 6 | 4 |
| 10 | 40 | 53 | 40 | 7 | 4'4" | 4'5" | 2'10" | 3'5" | | 9'6" | 19,700 | 40 | 6 | 4 |
| | 60 | 53 | 40 | 7 | 4'4" | 4'5" | 2'10" | 3'5" | | 10'0" | 23,000 | 60 | 6 | 4 |
| | 80 | 78 | 40 | 7 | 4'4" | 4'5" | 2'10" | 3'5" | | 12'0" | 26,600 | 60 | 6 | 4 |
| | 100 | 120 | 40 | 7 | 4'4" | 4'5" | 2'10" | 3'5" | | 14'6" | 30,800 | 80 | 6 | 4 |
| 15 | 40 | 46 | 26 | 7 | 4'4" | 5'0" | 2'11" | 3'4" | | 10'6" | 24,800 | 60 | 12 | 4 |
| | 60 | 46 | 26 | 7 | 4'4" | 5'0" | 2'11" | 3'4" | | 11'0" | 28,500 | 60 | 12 | 4 |
| | 80 | 46 | 26 | 7 | 4'4" | 5'0" | 2'11" | 3'4" | | 11'6" | 32,200 | 100 | 12 | 4 |
| | 100 | 80 | 26 | 8 | 4'6" | 5'0" | 2'11" | 3'4" | | 14'6" | 37,400 | 80 | 12 | 4 |
| 20, 5-ton auxiliary | 40 | 34 | 28 | 7 | 4'5" | 5'0" | 3'0" | 3'4" | 4'1" | 11'0" | 39,500 | 80 | 18 | 4 |
| | 60 | 34 | 28 | 8 | 4'6" | 5'0" | 3'0" | 3'4" | 4'1" | 11'0" | 33,600 | 60 | 18 | 4 |
| | 80 | 34 | 28 | 8 | 4'6" | 5'0" | 3'0" | 3'4" | 4'1" | 11'6" | 37,800 | 80 | 18 | 4 |
| | 100 | 60 | 28 | 8 | 4'6" | 5'0" | 3'0" | 3'4" | 4'1" | 14'6" | 43,700 | 100 | 18 | 8 |
| 30, 5-ton auxiliary | 40 | 34 | 17 | 9 | 5'1" | 5'10" | 3'11" | 3'5" | 4'10" | 11'6" | 51,600 | 100 | 18 | 4 |
| | 60 | 34 | 17 | 9 | 5'3" | 5'10" | 3'11" | 3'5" | 4'10" | 11'6" | 55,600 | 100 | 18 | 4 |
| | 80 | 34 | 17 | 9 | 5'3" | 5'10" | 3'11" | 3'5" | 4'10" | 12'0" | 59,500 | 100 | 18 | 4 |
| | 100 | 54 | 17 | 8 | 5'0" | 5'10" | 4'0" | 3'5" | 4'10" | 14'5" | 33,600 | 100 | 18 | 8 |
| 40, 5-ton auxiliary | 40 | 31 | 13 | 9 | 5'1" | 5'11" | 3'11" | 3'5" | 4'10" | 11'6" | 52,600 | 100 | 18 | 4 |
| | 60 | 31 | 13 | 9 | 5'3" | 5'11" | 3'11" | 3'5" | 4'10" | 11'6" | 57,500 | 100 | 18 | 4 |
| | 80 | 31 | 13 | 9 | 5'3" | 5'11" | 3'11" | 3'5" | 4'10" | 12'0" | 62,000 | 100 | 18 | 4 |
| | 100 | 49 | 13 | 8 | 5'0" | 5'11" | 4'1" | 3'5" | 4'10" | 14'5" | 34,892 | 60 | 18 | 8 |
| 50, 10-ton auxiliary | 40 | 27 | 11 | 9 | 5'6" | 6'9" | 4'5" | 3'8" | 5'1" | 12'0" | 61,000 | 100 | 18 | 4 |
| | 60 | 27 | 11 | 8 | 5'3" | 6'9" | 4'6" | 3'8" | 5'1" | 11'10" | 33,700 | 60 | 18 | 8 |
| | 80 | 27 | 11 | 8 | 5'3" | 6'9" | 4'6" | 3'8" | 5'1" | 12'1" | 36,800 | 80 | 18 | 8 |
| | 100 | 40 | 11 | 8 | 5'3" | 6'9" | 4'6" | 3'8" | 5'1" | 14'5" | 40,400 | 100 | 18 | 8 |
| 60, 10-ton auxiliary | 40 | 27 | 9 | 8 | 5'4" | 6'9" | 4'6" | 3'8" | 5'1" | 11'10" | 35,100 | 60 | 18 | 8 |
| | 60 | 27 | 9 | 8 | 5'4" | 5'9" | 4'6" | 3'8" | 5'1" | 12'1" | 38,800 | 80 | 18 | 8 |
| | 80 | 27 | 9 | 8 | 5'4" | 6'9" | 4'6" | 3'8" | 5'1" | 12'1" | 42,300 | 100 | 18 | 8 |
| | 100 | 40 | 9 | 10 | 5'5" | 6'9" | 4'5" | 3'8" | 5'1" | 14'9" | 46,500 | 80 | 18 | 8 |

[a] Crane & Hoist Division, Dresser Industries, Inc.
[b] For each 10-ft extra lift, increase H by X.
[c] Main hook, where no auxiliary hoist.
[d] Direct loads, no impact.
[e] The above figures should be used for preliminary work only as the data vary with different manufacturers.
[f] For 30-ton and smaller cranes, standard trolley speed is 100 to 150 ft/min, bridge speed is 75 to 300 ft/min, 40 to 60 ton, 100 ft/min trolley speed and 250 ft/min bridge speed.
[g] Multiply ft by 0.30 for m, ft/min by 0.0051 for m/s, inch by 25.4 for mm, lb by 0.45 for kg, lb/yd by 0.50 for kg/m.

mounted and operated either by mechanical rotating clutches or by a separate electric or hydraulic swing motor. The boom is operated by powered topping line, with a direct-geared hoisting mechanism to raise and lower it. Power to operate the machinery is deck-mounted, and the machinery deck is completely housed. The crane may be powered by internal-combustion engine or electric motor. The combination of internal-combustion engine, generator, and electric motor makes up the power arrangement for the diesel-electric locomotive crane. Another power arrangement is made up of internal-combustion engines driving hydraulic pumps for hydraulically powered swing and travel mechanisms. The car body and machinery deck are ballasted, thereby adding stability to the crane when it is rotated under load. The basic boom is generally 50 ft (15 m) in length (see Table 3); however, booms range to 130 ft (40 m) in length. Locomotive cranes are so designed that power-shovel, pile-driver, hook, bucket, or magnet attachments can be installed and the crane used in such service. Locomotive cranes are used most extensively in railroad work, steel mills, and scrap yards. The cranes usually have sufficient propelling power not only for the crane itself, but also for switching service and hauling cars.

### Truck Cranes

The advent of the truck crane has changed significantly the methods of lifting and placing heavy items such as concrete buckets, logs, pipe, and bridge or building members. Truck cranes can, without assistance, be rapidly equipped with accessory booms to reach to 260 ft (79 m) vertically—or 180 ft (55 m) vertically with 170 ft of horizontal reach.

#### Mechanical Models

TOWER CRANE (Fig. 18a and b).   Has vertical and horizontal members together with a boom and jib. Permits location close

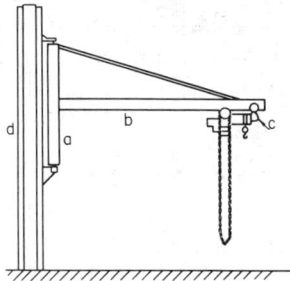

**Fig. 16**   Column jib crane.

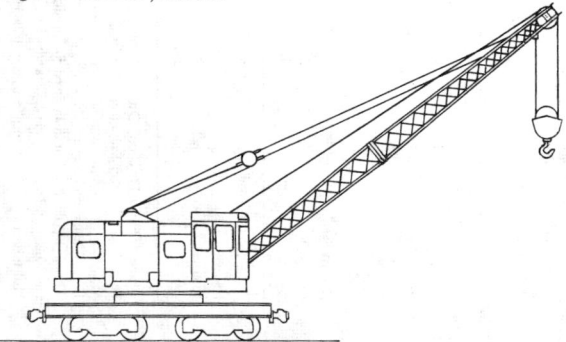

**Fig. 17**   Locomotive crane. *(American Hoist and Derrick.)*

to building with horizontal reach. Without jib, capacity to 27 tons (24.5 tonnes). With jib, reach to 180 ft (55 m). With jib, vertically to 190 ft (58 m). Lifting capacity based upon using outriggers.

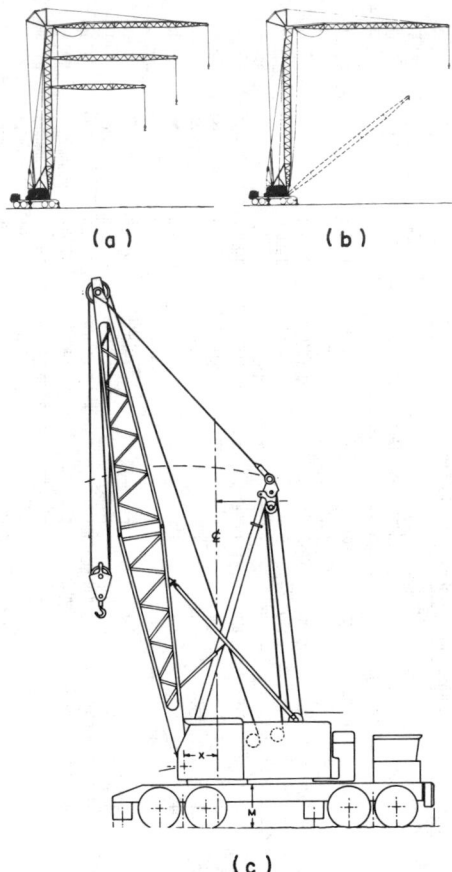

**Fig. 18**   Mechanical tower crane with vertical and horizontal extensions. (a) Tower working heights; (b) normal crane position; (c) crane with conventional boom, *(FMC Corp.)*

CONVENTIONAL CRANE (Fig. 18c).   With boom and with or without jib. Boom plus jib to 260 ft (79 m) and 125 tons (113 tonnes) capacity. Maximum working weight 230,000 lb (104,-000 kg).

GENERAL CHARACTERISTICS (Table 4).   8 × 4 drive wheels with air brakes on all eight wheels. Power hydraulic steering. Removable-pin-connected counterweights front or rear removable for roadability.

#### Hydraulic Models

SELF-PROPELLED (Fig. 19a).   Short wheelbase, two axle, single cab; 18½ tons (16.8 tonnes) capacity with two telescoping sections to 64 ft. Addition of a jib to 104 ft (32 m) reach.

HYDRAULIC TRUCK (Fig. 19b).   Three or four axle, two cabs, crane functions from upper cab; 45 tons (41 tonnes) capacity with three telescoping sections to 96 ft (29 m). Addition of a jib and boom extension to 142 ft (43 m).

GENERAL CHARACTERISTICS.   Hydraulic extensions save setup

**Table 3. Data on Locomotive Cranes\*†**

| Size, short tons | No. of wheels | Min working radius, 10-ft boom | | | Max working radius, 50-ft boom | | | Approx working wt, lb | Wheel base | Max on any wheel, lb |
|---|---|---|---|---|---|---|---|---|---|---|
| | | Radius A, ft | Free capacity, lb | Outrigger capacity, lb | Radius A, ft | Free capacity, lb | Outrigger capacity, lb | | | |
| 25 | 8 | 12 | 50,000 | 50,000 | 50 | 8,200 | 9,600 | 132,000 | 18'10" | 44,250 |
| 30 | 8 | 12 | 60,000 | 60,000 | 50 | 9,800 | 11,500 | 156,000 | 18'10" | 53,500 |
| 40 | 8 | 12 | 80,000 | 100,000 | 50 | 13,600 | 20,000 | 216,000 | 19'8" | 70,000 |
| 50–60 | 8 | 12 | 100,000 | 120,000 | 50 | 16,100 | 24,000 | 261,000 | 21'8" | 87,100 |
| 50–80 | 8 | 12 | 100,000 | 160,000 | 50 | 16,100 | 32,000 | 261,000 | 21'8" | 87,100 |
| 60–110 | 8 | 20 | 65,000 | 136,000 | 50 | 20,700 | 43,000 | 312,500 | 21'8" | 88,500 |

\*American Hoist & Derrick Co.
†Multiply short ton by 0.9 for tonne, ft by 0.30 for m, lb by 0.45 for kg.

**Table 4. Conventional Crane (Fig. 3)\* Capacity and Limits of Operation**

| Boom | | | | On outriggers | |
|---|---|---|---|---|---|
| Length ft | Radius ft | Angle deg | Point height, ft | Rear lb | Side lb |
| 30 | 11 | 81.0 | 33.5 | 250,000 | 250,000 |
| 30 | 25 | 46.3 | 24.5 | 123,300 | 123,300 |
| 60 | 16 | 80.7 | 63.1 | 145,100 | 145,100 |
| 60 | 50 | 39.7 | 41.0 | 52,200 | 50,100 |
| 90 | 20 | 81.3 | 92.9 | 131,600 | 131,600 |
| 90 | 80 | 31.5 | 49.8 | 30,500 | 26,000 |
| 180 | 40 | 79.2 | 180.6 | 46,700 | 46,700 |
| 180 | 170 | 21.6 | 68.9 | 8,300 | 6,200 |
| 230 | 50 | 79.0 | 229.6 | 21,000 | 21,000 |
| 230 | 220 | 18.9 | 77.5 | 2,900 | 1,800 |

SOURCE: FMC Corporation.

\*Multiply ft by 0.30 for m, lb by 0.45 for kg.

time and provide job-to-job mobility. Equipment such as this comes under Commercial Standard specification CS90-58, "Power Cranes and Shovels." Similar equipment, called **utility cranes,** without the highway-truck-type cabs, is also available.

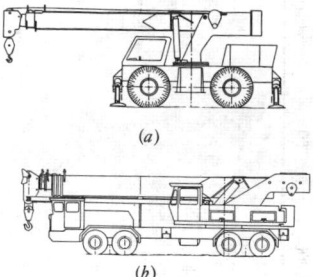

(a)

(b)

**Fig. 19**  Hydraulic crane with (a) self-propelled and (b) truck-type bases. (*FMC Corp.*)

### Cableways

Cableways are **aerial hoisting and conveying devices** using suspended steel cable for their tracks, the loads being suspended from carriages and moved by gravity or power. The most common uses are transporting material from open pits and quarries to the surface; handling construction material in the building of dams, docks, and other structures where the construction of tracks across rivers or valleys would be uneconomical; and loading logs on cars. The maximum clear *span* is 2,000 to 3,000 ft (610 to 914 m); the usual spans, 300 to 1,500 ft (91 to 457 m). The gravity type is limited to conditions where a grade of at least 20 percent is obtainable on the track cable. Transporting cableways move the load from one point to another. Hoisting transporting cableways hoist the load as well as transport it.

A **transporting cableway** may have one or two fixed track cables, inclined or horizontal, on which the carriage operates by gravity or power. The gravity transporting type (Fig. 20-I) will either raise or lower material. It consists of one track cable *a* on which travels the wheeled carriage *b* carrying the bucket. The traction rope *c* attached to the carriage is made fast to power drum *d*. The inclination must be sufficient for the carriage to coast down and pull the traction rope after it. The carriage is hauled up by traction rope *c*. Drum *d* is provided with a brake to control the lowering speed, and material may be either raised or lowered. When it is not possible to obtain sufficient fall to operate the load by gravity, traction rope *c* (Fig. 20-II) is made endless so that carriage *b* is drawn in either direction by power drum *d*. Another type of inclined cableway, shown in Fig. 20-III, consists of two track cables *aa*, with an endless traction rope *c*, driven and controlled by drum *d*. When material is being lowered, the loaded bucket *b* raises the empty carriage *bb*, the speed being controlled by the brake on the drum. When material is being raised, the drum is driven by power, the descending empty carriage assisting the engine in raising the loaded carriage. This type has twice the capacity of that shown in Fig. 20-I.

A **hoisting and conveying cableway** (Fig. 20-IV) hoists the material at any point under the track cable and transports it to any other point. It consists of a track cable *a* and carriage *b*, moved by the endless traction rope *c* and by power drum *d*. The hoisting of the load is accomplished by power drum *e* through fall rope *f*, which raises the fall block *g* suspended from the carriage. The fall-rope carriers *h* support the fall rope; otherwise, the weight of this sagging rope would prevent

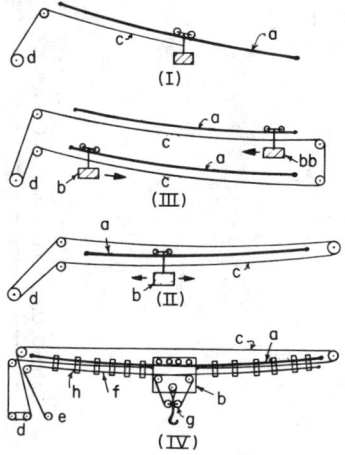

**Fig. 20**  Cableways.

fall block $g$ from lowering when without load. Where it is possible to obtain a minimum inclination of 20° on the track cable, the traction-rope drum $d$ is provided with a brake and is not power-driven. The carriage then descends by gravity, pulling the fall and traction ropes to the desired point. Brakes are applied to drum $d$, stopping the carrier. The fall block is lowered, loaded, and raised. If the load is to be carried up the incline, the carriage is hauled up by the fall rope. With this type, the friction of the carriage must be greater than that of the fall block or the load will run down. A novel development is the use of self-filling grab buckets operated from the carriages of cableways, which are lowered, automatically filled, hoisted, carried to dumping position, and discharged.

The **carriage speed** is 300 to 1,400 ft/min (1.5 to 7.1 m/s) [in special cases, up to 1,800 ft/min (9.1 m/s)]; average hoisting speed is 100 to 700 ft/min. The **average loads** for coal and earth are 1 to 5 tons (0.9 to 4.5 tonnes); for rock from quarries, 5 to 20 tons; for concrete, to 12 yd³ (9.1 m³) at 50 tons.

Two parallel traveling cableways (built by Lidgerwood Manufacturing Co.) at Glen Canyon Dam have 2,050- and 1,800-ft (625- and 549-m) spans with 910- and 810-ft (277- and 247-m) tracks, one above the other to permit separate operation with 50-ton (45-tonne) loads (using 12-yd³ concrete buckets) or operation together with 100-ton loads. Hoisting by 1⅛-in four-part-reeved wire rope to 700 ft/min and transporting by 1½-in (38-mm) wire rope to 1,400 ft/min (7.1 m/s) can be done simultaneously. The main track cable is a single locked-coil cable 4 in (102 mm) in diameter, having a breaking strength in excess of 880 tons (798 tonnes). The track cables are supported on traveling towers up to 189 ft high. The hoisting drum, powered by dual 500-hp motors, is 9 ft in diameter and 8 ft long and stores 2,200 ft of rope, with hoist action four-part-reeved (see Fig. 20-IV).

The **deflection of track cables** with their maximum gross loads at midspan is usually taken as 5½ to 6 percent of the span. Let $S$ = span between supports, ft; $L$ = one-half the span, ft; $w$ = weight of rope, lb/ft; $P$ = total concentrated load on rope, lb; $b$ = deflection, ft; $H$ = horizontal tension in rope, lb. Then $b = (wL + P)L/2H$; $P = (2b - wL^2)/L = (8bH - wS^2)/2S$

For track cables, a **factor of safety** of at least 4 is advised, though this may be as low as 3 for locked smooth-coil strands that use outer wires of high ultimate strength. For traction and fall ropes, the sum of the load and bending stress should be well within the elastic limit of the rope or, for general hoisting, about two-thirds the elastic limit (which is taken at 65 percent of the breaking strength). Let $P$ = load on the rope, lb; $A$ = area of metal in rope section, in²; $E$ = 29,500,000; $R$ = radius of curvature of hoisting drum or sheave, whichever is smaller, in; $d$ = diameter of individual wires in rope, in (for six-strand 19-wire rope, $d$ = 1/15 rope diam; for six-strand 7-wire rope, $d$ = 1/9 rope diam). Then load stress per in² = $T_1 = P/A$, and bending stress per in² = $T_b = Ed/2R$. The radius of curvature of saddles, sheaves, and driving drums is thus important to fatigue life of the cable (Sec. 8). In determining the horsepower required, the load on the traction ropes or on the fall ropes will govern, depending upon the degree of inclination.

### Cable Tramways

Cable tramways are aerial conveying devices using suspended cables, carriages, and buckets for transporting material over level or mountainous country or across rivers, valleys, or hills (they transport but do not hoist). They are used for handling small quantities over long distances, and their construction cost is insignificant compared with the construction costs of railroads and bridges. Five types are in use:

**Monocable, or Single-Rope, Saddle-Clip Tramway** Operates on grades to 50 percent gravity grip or on higher grades with spring grip and has capacity of 250 tons/h (63 kg/s) in each direction and speeds to 500 ft/min. Single section lengths to 16 miles without intermediate stations or tension points. Can operate in multiple sections without transshipment to any desired length [monocables to 170 miles (274 km) over jungle terrain are practical]. Construction costs from $15,000 per mile. Loads automatically leave the carrying moving rope and travel by overhead rail at angle stations and transfer points between sections with no detaching or attaching device required. Main rope constantly passes through stations for inspection and oiling. Cars are light and safe for passenger transportation.

**Single-Rope Fixed-Clip Tramway** Endless rope traveling at low speed, having buckets or carriers fixed to the rope at intervals. Rope passes around horizontal sheaves at each terminal and is provided with a driving gear and constant-tension device.

**Bicable, or Double-Rope, Tramway** Standing track cable and a moving endless hauling or traction rope traveling up to 500 ft/min (2.5 m/s). Used on excessively steep grades. A detacher and attacher is required to open and close the car grip on the traction rope at stations. Track cable is usually in sections of 6,000 to 7,000 ft (1,830 to 2,130 m) and counterweighted because of friction of stiff cable over tower saddles.

**Jigback, or Two-Bucket, Reversing Tramway** Usually applied to hillside operations for mine workings so that on steep slopes loaded bucket will pull unloaded one up as loaded one descends under control of a brake. Loads to 10 tons (9 tonnes) are carried using a pair of track cables and an endless traction rope fixed to the buckets.

**To-and-Fro, or Single-Bucket, Reversing Tramway** A single track rope and a single traction rope operated on a winding or hoist drum. Suitable for light loads to 3 tons (2.7 tonnes) for intermittent working on a hillside, similar to a hoisting and conveying cableway without the hoisting feature.

The **monocable tramway** (Fig. 21) consists of an endless cable $a$ passing over horizontal sheaves $d$ and $e$ at the ends and supported at intervals by towers. This cable is moved continu-

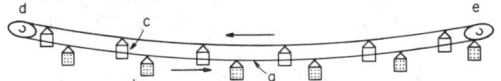

**Fig. 21**  Single-rope cable tramway.

ously, and it both supports and propels carriages $b$ and $c$. The carriages either are attached permanently to the cable (as in the **single-rope fixed-clip tramway**), in which case they must be loaded and dumped while in motion, or are attached by friction grips so that they may be connected automatically or by hand at the loading and dumping points. When the tramway is lowering material from a higher to a lower level, the grade is frequently sufficient for the loaded buckets $b$ to raise the empty buckets $c$, operating the tramway by gravity, the speed being controlled by a brake on grip wheel $d$.

The **bicable tramway** (Fig. 22) consists of two stationary track cables $a$, on which the wheeled carriages $c$ and $d$ travel. The endless traction rope $b$ propels the carriages, being attached by friction grips. Figure 23 shows the arrangement of the **overhead type.** The track cable *a* is supported at intervals by towers *b*, which carry the saddles *c* in which the track cable

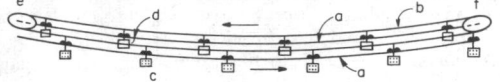

**Fig. 22**   Double-rope cable tramway.

rests. Each tower also carries the sheave *d* for supporting traction rope *e*. The self-dumping bucket *f* is suspended from carriage *g*. The grip *h*, which attaches the carriage to traction rope *e*, is controlled by lever *k*. In the **underhung type,** shown in Fig. 24, track cable *a* is carried above traction rope *e*. Saddle *c* on top of the tower supports the track cable, and sheave *d*

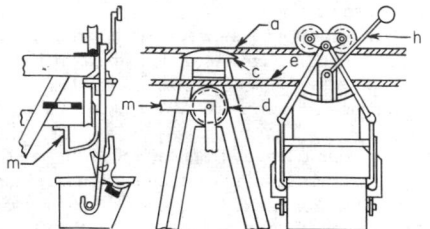

**Fig. 23**   Overhead-type double-rope cable tramway.

supports the traction cable. The sheave is provided with a rope guard *m*. The lever *h*, with a roller on the end, automatically attaches and detaches the grip by coming in contact with guides at the loading and dumping points. The carriages move in only one direction on each track. On steep downgrades, special hydraulic speed controllers are used to fix the speed of the carriages.

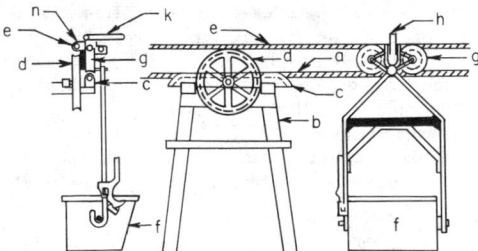

**Fig. 24**   Underhung-type double-rope cable tramway.

The **track cables** are of the special locked-joint smooth-coil, or tramway, type. Nearly all wire rope is made of plow steel, with the old cast-steel type no longer being in use. The track cable is usually provided with a smooth outer surface of Z-shaped wires for full lock type or with a surface with half the wires *H*-shaped and the rest round. Special tramway couplers are attached in the shops with zinc or are attached in the field by driving little wedges into the strand end after inserting the end into the coupler. The second type of coupling is known as **a dry socket** and, though convenient for field installation, is not held in as high regard for developing full cable strength. The usual spans for level ground are 200 to 300 ft (61 to 91 m). One end of the track cable is anchored; the other end is counterweighted to one-quarter the breaking strength of the rope so

that the horizontal tension is a known quantity. The **traction ropes** are made six-strand 7-wire or six-strand 19-wire, of cast or plow steel on hemp core. The maximum diameter is 1 in, which limits the length of the sections. The traction rope is endless and is driven by a drum at one end, passing over a counterweighted sheave at the other end.

Figure 25 shows a **loading terminal.** The track cables *a* are anchored at *b*. The carriage runs off the cable to the fixed track *c*, which makes a 180° bend at *d*. The empty buckets are loaded by chute *e* from the loading bin, continue around track

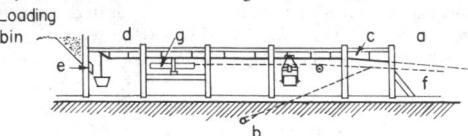

**Fig. 25**   Cable-tramway loading terminal.

*c*, are automatically gripped to traction cable *f*, and pass on the track cable *a*. Traction cable *f* passes around and is driven by drum *g*. When the carriages are permanently attached to the traction cable, they are loaded by a moving hopper, which is automatically picked up by the carriage and carried with it a short distance while the bucket is being filled. Figure 26 shows a **discharge terminal.** The carriage rolls off from the track cable

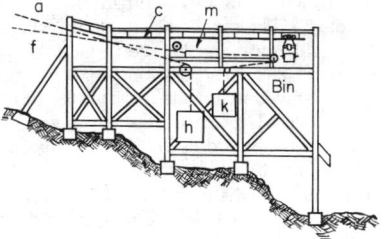

**Fig. 26**   Cable-tramway discharge terminal.

*a* to the fixed track *c*, being automatically ungripped. It is pushed around the 180° bend of track *c*, discharging into the bin underneath and continuing on track *c* until it is automatically gripped to traction cable *f*. The counterweights *h* are attached to track cables *a*, and the counterweight *k* is attached to the carriage of the traction-rope sheave *m*. The **supporting towers** are A frames of steel or wood. At abrupt vertical angles, the supports are placed close together and steel tracks installed in place of the cable. Spacing of towers will depend upon the capacity of the track cables and sheaves and upon the terrain as well as the bucket spacing.

**Stress in Ropes** (Roebling)   The deflection for track cables of tramways is taken as one-fortieth to one-fiftieth of the span to reduce the grade at the towers. Let $S$ = span between supports, ft; $h$ = deflection, ft; $P$ = gross weight of buckets and carriages, lb; $Z$ = distance between buckets, ft; $W_1$ = total load per ft of rope, lb; $H$ = horizontal tension of rope, lb. The formulas given for cableways then apply. When several buckets come in the span at the same time, special treatment is required for each span. For large capacities, the buckets are spaced close together, the load may be assumed to be uniformly distributed, and the live load per linear foot of span = $P/Z$. Then $H = W_1 S^2 / 8h$, where $W_1$ = (weight of rope per ft) + $(P/Z)$. When the buckets are not spaced closely, the equilibrium curve can be plotted with known horizontal tension and vertical reactions at points of support.

For figuring the traction rope, $t_0$ = tension on counterweight rope, lb; $t_1$, $t_2$, $t_3$, $t_4$ = tensions, lb, at points shown in Fig. 27; $n$ = number of carriers in motion; $a$ = angle subtended between the line connecting the tower supports and

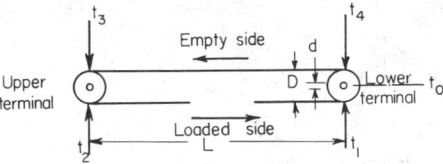

**Fig. 27**

the horizontal; $W_1$ = weight of each loaded carrier, lb; $W_2$ = weight of each empty carrier, lb; $w$ = weight of traction rope, lb/ft; $L$ = length of tramway of each grade $a$, ft; $D$ = diameter of end sheave, ft; $d$ = diameter of shaft of sheave, ft; $f_1$ = 0.015 = coefficient of friction of shaft; $f_2$ = 0.025 = rolling friction of carriage wheels. Then, if the loads descend, the maximum stress on the loaded side of the traction rope is

$$t_2 = t_1 + \Sigma(Lw \sin a + \tfrac{1}{2}nW_1 \sin a) \\ - f_2\Sigma(Lw \cos a + \tfrac{1}{2}nW_1 \cos a)$$

where $t_1 = \tfrac{1}{2}t_0[1 - f_1(d/D)]$. If the load ascends, there are two cases: (1) driving power located at the lower terminal, (2) driving power at the upper terminal. If the line has no reverse grades, it will operate by gravity at a 10 percent incline to 10 tons/h capacity and at a 4 percent grade for 80 tons/h. The preceding formula will determine whether it will operate by gravity.

The **power required** or developed by tramways is as follows: Let $V$ = velocity of traction rope, ft/min; $P$ = gross weight of loaded carriage, lb; $p$ = weight of empty carriage, lb; $N$ = number of carriages on one track cable; $P/50$ = friction of loaded carriage; $p/50$ = friction of empty carriage; $W$ = weight of moving parts, lb; $E$ = length of tramway divided by difference in levels between terminals, ft. Then, power required is

$$\text{hp} = \frac{NV}{33,000}\left(\frac{P-p}{E} \pm \frac{P+p}{50}\right) \pm 0.0000001WV$$

Where power is developed by tramways, use 80 instead of 50 under $P + p$.

## BELOW-SURFACE HANDLING (EXCAVATION)
### Power Shovels

**Power shovels** stand upon the bottom of the pit being dug and dig above this level. Small machines are used for road grading, basement excavation, clay mining, and trench digging; larger sizes are used in quarries, mines, and heavy construction; and the largest are used for removing overburden in opencut mining of coal and ore. The uses for these machines may be divided into two groups: (1) **loading,** where sturdy machines with comparatively short working ranges are used to excavate material and load it for transportation; (2) **stripping,** where a machine of very great dumping and digging reaches is used to both excavate the material and transport it to the dump or

wastepile. The **full-revolving shovel,** which is the only type built at the present (having entirely displaced the old railroad shovel), is usually composed of a crawler-mounted truck frame with a center pintle and roller track upon which the revolving frame can rotate. The revolving frame carries the swing and hoisting machinery and supports, by means of a socket at the lower end and cable guys at the upper end, a boom carrying guides for the dipper handles and machinery to thrust the dipper into the material being dug.

Figure 28 shows a full-revolving shovel. The dipper $a$, of cast or plate steel, is provided with special wear-resisting teeth. It is pulled through the material by a steel cable $b$ wrapped on a main drum $c$. Gasoline engines are used almost exclusively in the small sizes, and diesel, diesel-electric, or electric power units, with Ward Leonard control, in the large machines. The commonly used sizes are from ½ to 5 yd³ (0.4

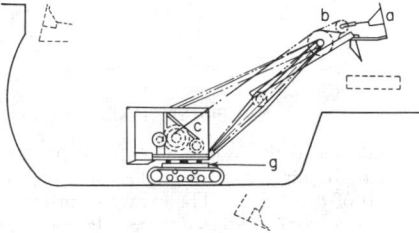

**Fig. 28** Revolving power shovel.

to 3.8 m³) capacity, but special machines for coal-mine stripping are built with buckets holding up to 33 yd³ (25 m³) or even more. The very large machines are not suited for quarry or heavy rock work. Sizes up to 5 yd³ (3.8 m³) are known as **quarry machines.** Stripping shovels are crawler-mounted, with double-tread crawlers under each of the four corners and with power means for keeping the turntable level when traveling over uneven ground. The crowd motion consists of a chain which, through the rack-and-pinion mechanism, forces the dipper into the material as the dipper is hoisted and withdraws it on its downward swing. On the larger sizes, a separate engine or motor is mounted on the boom for crowding. A separate engine working through a pinion and horizontal gear $g$ swings the entire frame and machinery to bring the dipper into position for dumping and to return it to a new digging position. Dumping is accomplished by releasing the hinged dipper bottom, which drops upon the pulling of a latch. With gasoline-engine or diesel-engine drives, there is only one prime mover, the power for all operations being taken off by means of clutches.

Practically all **power shovels** are readily **converted** for operation as **dragline excavators,** or **cranes.** The changes necessary are very simple in the case of the small machines; in the case of the larger machines, the installation of extra drums, shafts, and gears is required, in addition to the boom and bucket change.

The **telescoping-boom,** hydraulically operated excavator shown in Fig. 29 is a versatile machine that can be quickly converted from the rotating-boom power shovel shown in Fig. 29*a* to one with a crane boom (Fig. 29*b*) or backhoe shovel boom (Fig. 29*c*). It can dig ditches reaching to 22 ft (6.7 m) horizontally and 9 ft 6 in (2.9 m) below grade; it can cut slopes, rip, scrape, dig to a depth of 12 ft 6 in (3.8 m), and load to a height of 11 ft 2 in (3.4 m). It is completely hydraulic in all powerized functions.

### Dredges

**Placer dredges** are used for the mining of gold, platinum, and tin from placer deposits. The usual maximum digging depth of most existing dredges is 65 to 70 ft (20 to 21 m), but one dredge is digging to 125 ft (38 m). The dredge usually works are used principally for the excavation of sand and gravel beds from rivers, lakes, or ocean deposits. Since this type of dredge is not as a rule required to cut its own flotation, the bow corners of the hull may be made square and the digging ladder need not extend beyond the bow. The bucket chain may be of

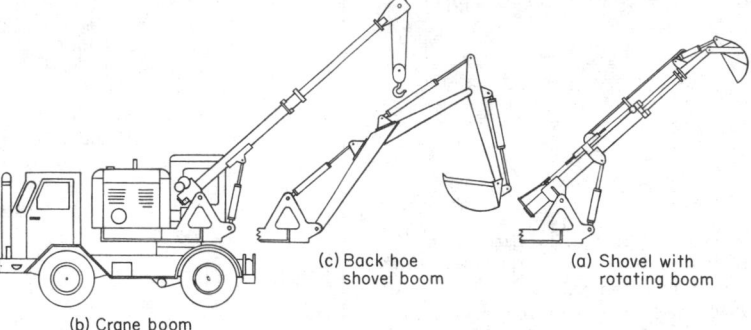

(b) Crane boom

(c) Back hoe shovel boom

(a) Shovel with rotating boom

**Fig. 29** Hydraulically operated excavator. *(Link-Belt.)*

with a bank above the water of 8 to 20 ft (2.4 to 6 m). Sometimes hydraulic giants are employed to break down these banks ahead of the dredge. The excavated material is deposited astern, and as the dredge advances, the pond in which the dredge floats is carried along with it.

The digging element consists of a chain of closely connected buckets passing over an idler tumbler and an upper or driving tumbler. The chain is mounted on a structural-steel ladder which carries a series of rollers to provide a bearing track for the chain of buckets. The upper tumbler is placed 10 to 40 ft (3 to 13 m) above the deck, depending upon the size of the dredge. Its fore-and-aft location is about 65 percent of the length of the ladder from the bow of the dredge. The ladder operates through a well in the hull, which extends from the bow practically to the upper-tumbler center. The material excavated by the buckets is dumped by the inversion of the buckets at the upper tumbler into a hopper, which feeds it to a revolving screen.

Placer dredges are made with buckets ranging in capacity from 2 to 20 ft³ (0.06 to 0.6 m³). The usual speed of operation is 15 to 30 buckets per min, in the inverse order of size.

The digging reaction is taken by stern spuds, which act as pivots upon which the dredge, while digging, is swung from side to side of the cut by swinging lines which lead off the dredge near the bow and are anchored ashore or pass over shore sheaves and are dead-ended near the lower tumbler on the digging ladder. By using each spud alternately as a pivot, the dredge is fed forward into the bank. Table 5 gives average capacities for placer dredges of various sizes working under normal conditions 24 h/day.

**Elevator dredges,** of which dredges are a special classification,

the close-connected placer-dredge type or of the open-connected type with one or more links between the buckets. The dredge is more of an elevator than a digging type, and for this reason the buckets may be flatter across the front and much lighter than the placer-dredge bucket.

The excavated material is usually fed to one or more revolving screens for classification and grading to the various commercial sizes of sand and gravel. Sometimes it is delivered to sumps or settling tanks in the hull, where the silt or mud is washed off by an overflow. Secondary elevators raise the material to a sufficient height to spout it by gravity or to load it by belt conveyors to the scows.

**Hydraulic dredges** are used most extensively in river and harbor work, where extremely heavy digging is not encountered and spoil areas are available within a reasonable radius of the dredge. This radius may vary from a few hundred feet to a mile or more, and with the aid of booster pumps in the pipeline, hydraulic dredges have pumped material through distances in excess of 2 mi (3.2 km), at the same time elevating it more than 100 ft (30 m). This type of dredge is also used for sand-and-gravel-plant operations and for land-reclamation work. Levees and dams can be built with hydraulic dredges. The usual maximum digging depth is about 50 ft (16 m). Hydraulic dredges are reclaiming copper stamp-mill tailings from a depth of 115 ft (35 m) below the water, and a depth of 165 ft (50 m) has been reached in a land-reclamation job.

The usual type of hydraulic dredge has a **digging ladder** suspended from the bow at an angle of 45° for the maximum digging depth. This ladder carries the suction pipe and cutter, with its driving machinery, and the swinging-line sheaves. The cutter head may have applied to it 25 to 1,000 hp (3.7 to

**Table 5. Capacities of Placer Dredges**

| Size of buckets, ft³ (m³) | 2(0.06) | 4(0.11) | 6(0.17) | 8(0.23) |
|---|---|---|---|---|
| Capacity, thousands of yd³/month (m³/h) | 34(36) | 70(74) | 106(113) | 148(157) |
| Size of buckets, ft³ (m³) | 10(0.28) | 13(0.37) | 16(0.45) | 20(0.57) |
| Capacity, thousands of yd³/month (m³/h) | 186(198) | 248(263) | 312(331) | 395(419) |

746 kW). The 20-in (0.5-m) dredge, which is the standard, general-purpose machine, has a cutter drive of about 300 hp. The usual operating speed of the cutter is 5 to 20 r/min.

The material excavated by the cutter enters the mouth of the suction pipe, which is located within and at the lower side of the cutter head. The material is sucked up by a centrifugal pump, which discharges it to the dump through a pipeline. The shore discharge pipe is usually of the telescopic type, made of No. 10 to ³⁄₁₀-in (3- to 7.5-mm) plates in lengths of 16 ft (5 m) so that it can be readily handled by the shore crew. Floating pipelines are usually made of plates from ¼ to ½ in (6 to 13 mm) thick and in lengths of 40 to 100 ft (12 to 30 m), which are floated on pontoons and connected together through rubber sleeves or, preferably, ball joints. The floating discharge line is flexibly connected to the hull in order to permit the dredge to swing back and forth across the cut while working without disturbing the pipeline.

Pump efficiency is usually sacrificed to make an economical unit for the handling of material, which may run from 2 to 25 percent of the total volume of the mixture pumped. Most designs have generous clearances and will permit the passage of stone which is 70 percent of the pipeline diameter. The pump efficiencies vary widely but in general may run from 50 to 70 percent.

**Commercial dredges** vary in size and discharge-pipe diameters from 12 to 30 in (0.3 to 0.8 m). Smaller or larger dredges are usually special-purpose machines. A number of 36-in (0.9-m) dredges are used to maintain the channel of the Mississippi River. The power applied to pumps varies from 100 to 3,000 hp (75 to 2,200 kW). The modern 20-in (0.5-m) commercial dredge has about 1,350 bhp (1,007 kW) applied to the pump.

**Diesel dredges** are built for direct-connected or electric drives, and modern steam dredges have direct-turbine or turboelectric drives. The steam turbine and the dc electric motor have the advantage that they are capable of developing full rating at reduced speeds.

The **hydraulic-dredge** installation (1938) used in the construction of the Fort Peck dam had four duplicate units. At maximum service, each unit (dredge, floating booster, and land booster), with five pumps in series, delivered 750,000 yd³ (570,000 m³) of fill per month against a lift of 240 ft (73 m) and through 17,000 ft of 28-in (5,200 m of 0.7-m) discharge line. The suction line between the cutter head and the first pump was 33 in (0.84 m) in diameter. The cutter head, 7 ft (2.1 m) in diameter by 6 ft (1.8 m) long, was carried by a 75-ft (23-m) ladder. (*Eng. News-Rec.*, Dec. 12, 1935.)

Within its scope, the hydraulic dredge can work more economically than any other excavating machine or combination of machines.

### Dragline Excavators

**Dragline excavators** are typically used for digging open cuts, drainage ditches, canals, sand, and gravel pits, where the material is to be moved 20 to 1,000 ft (6 to 305 m) before dumping. They cannot handle rock unless the rock is blasted. Since they are provided with long booms and mounted on turntables, permitting them to swing through a full circle, these excavators can deposit material directly on the spoil bank farther from the point of excavation than any other type of machine. Whereas a shovel stands below the level of the material it is digging, a dragline excavator stands above and can be used to excavate material under water.

Figure 30 shows a self-contained dragline mounted on crawler treads. The drive is almost exclusively gasoline in the small sizes and diesel, diesel-electric, or electric, frequently with Ward Leonard control, in the large sizes. The boom *a* is

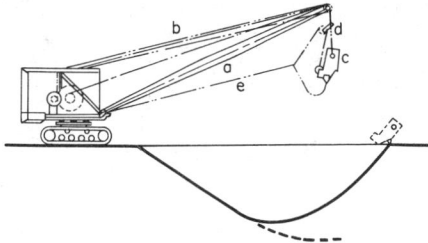

**Fig. 30**  Dragline excavator.

pivoted at its lower end to the turntable, the outer end being supported by cables *b* so that it can be raised or lowered to the desired angle. The scraper bucket *c* is supported by cable *d*, which is attached to a bail on the bucket, passes over a sheave at the head of the boom, and is made fast to the engine. A second cable *e* is attached to the front of the bucket and made fast to the second drum of the engine. The bucket is dropped and dragged along the surface of the material by cable *e* until filled. It is then hoisted by cable *d*, drawn back to its dumping position, *e* being kept tight until the dumping point is reached, when *e* is slacked, allowing the bucket to dump by gravity. After the bucket is filled, the boom is swung to the dumping position while the bucket is being hauled out. A good operator can throw the bucket 10 to 40 ft (3 to 12 m) beyond the end of the boom, depending on the size of machine and the working conditions. The depth of the cut varies from 12 to 75 ft (3.7 to 23 m), again depending on the size of machine and the working conditions. With the smaller machines and under favorable conditions, two or even three trips per minute are possible; but with the largest machines, even one trip per minute may not be attained. The more common sizes are for handling ³⁄₄- to 4-yd³ (0.6 to 3 m³) buckets with boom lengths up to 100 or 125 ft (30 to 38 m), but machines have been built to handle an 8-yd³ (6 m³) bucket with a boom length of 200 ft (60 m). The same machine can handle a 12-yd³ (9 m³) bucket with the boom shortened to 165 ft (50 m).

### Slackline Cableways

Used widely in sand-and-gravel plants, the **slackline cableway** employs an open-ended dragline bucket suspended from a carrier (Fig. 31) which runs upon a track cable. It will dig, elevate, and convey materials in one continuous operation.

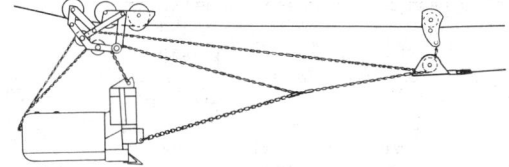

**Fig. 31**  Slackline-cable bucket and trolley.

**Table 6. Slackline Cableway***
(Sauerman)

| Kind of power | Approx. hp | Size of bucket, yd³ | Length of span, ft | Handling capacity in yd³/h when digging 30 ft below mast on an average haul of | | | | | |
|---|---|---|---|---|---|---|---|---|---|
| | | | | 150 ft | 200 ft | 250 ft | 300 ft | 400 ft | 500 ft |
| Electric | 20 | ½ | 400 | 25 | 22 | 20 | 17 | | |
| Diesel | 30 | ½ | 400 | 25 | 22 | 20 | 17 | | |
| Electric | 40 | ¾ | 500 | 39 | 37 | 36 | 33 | | |
| Diesel | 92 | ¾ | 500 | 57 | 54 | 51 | 46 | | |
| Electric | 60 | 1 | 600 | 51 | 50 | 46 | 43 | 37 | |
| Diesel | 140 | 1 | 600 | 83 | 80 | 70 | 65 | 55 | |
| Electric | 100 | 1½ | 700 | | 79 | 76 | 73 | 66 | 57 |
| Diesel | 140 | 1½ | 700 | | 97 | 91 | 84 | 72 | 61 |
| Electric | 150 | 2 | 800 | | | 94 | 90 | 80 | 70 |
| Diesel | 207 | 2 | 800 | | | 114 | 110 | 98 | 84 |
| Electric | 200 | 2½ | 900 | | | 113 | 112 | 105 | 95 |
| Diesel | 320 | 2½ | 900 | | | 150 | 147 | 125 | 107 |
| Electric | 250 | 3½ | 1,000 | | | | 156 | 140 | 129 |
| Diesel | 410 | 3½ | 1,000 | | | | 182 | 161 | 143 |

*Multiply hp by 0.75 for kW, yd⁴ by 0.8 for m³, ft by 0.30 for m.

Figure 32 shows a typical slackline-cableway operation. The bucket and carrier is *a; b* is the track cable, inclined to return the bucket and carrier by gravity; *c* is a tension cable for raising or lowering the track cable; *d* is the load cable; and *e* is a power unit with two friction drums having variable speeds. A mast or tower *f* is used to support guide and tension blocks at the high end of the track cable; a movable tail tower *g* supports the lower end of the track cable. The bucket is raised and lowered by tensioning or slacking off the track cable. The bucket is loaded, after lowering, by pulling on the load cable. The loaded bucket, after raising, is conveyed at high speed to the dumping point and is returned at a still higher speed by gravity to the digging point. The cableway can be operated in radial lines from a mast or in parallel lines between two moving towers. It will not dig rock unless the rock is blasted.

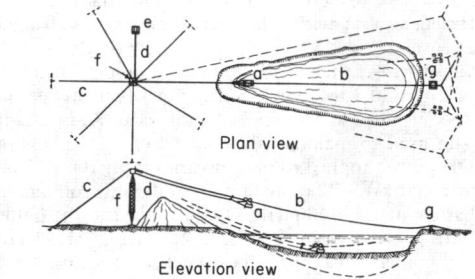

**Fig. 32** Slackline-cable plant *(Sauerman.)*

The depth of digging may vary from 5 to 100 ft. Capacities are shown in Table 6.

# CONVEYING AND CONTINUOUS FLOW

## by Harold V. Hawkins and Associates

### OVERHEAD CONVEYORS
#### by Ivan L. Ross
*ACCO Chain Conveyor Division*

Overhead conveyor systems are defined in two general classifications: the basic trolley conveyor and the power-and-free conveyor, each of which serves a definite purpose.

Trolley conveyors, often referred to as overhead power conveyors, consist of a series of trolleys or wheels supported from or within an overhead track and connected by an endless propelling means, such as chain, cable, or other linkages. Individual loads are usually suspended from the trolleys or wheels (Fig. 1). Trolley conveyors are utilized for transportation or storage of loads suspended from one conveyor which follows a single fixed path. They are normally used in applications where a balanced, continuous production is required. Track sections range from lightweight "tee" members or tubular sections, to medium- and heavy-duty I-beam sections. The combinations and sizes of trolley propelling means and track sections are numerous. Normally this type of conveyor is continually in motion at a selected speed to suit its function.

Power-and-free conveyor systems consist of at least one power conveyor, but usually more, where the individual loads are suspended from one or more free trolleys (not permanently connected to the propelling means) which are conveyor-propelled through all or part of the system. Additional portions of the system may have manual or gravity means of propelling the trolleys.

Worldwide industrial, institutional, and warehousing requirements of in-process and finished products have affected the considerable growth and development of power-and-free conveyors. Endless varieties of size, style, color, and all imagi-

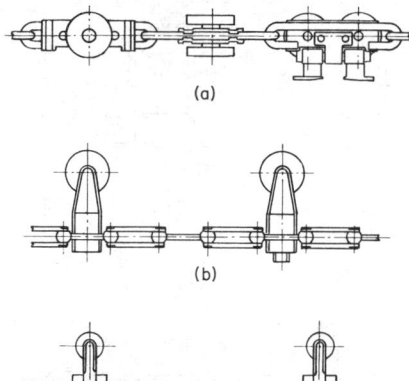

**Fig. 1** Typical conveyor chains or cable.

nable product combinations have extended the use of power-and-free conveyors. The power-and-free system combines the advantages of continuously driven chains with the versatile traffic system exemplified by traditional monorail unpowered systems. Thus, high-density load-transportation capabilities are coupled with complex traffic patterns and in-process or work-station requirements to enable production requirements to be met with a minimum of manual handling or transferring. Automatic dispatch systems for coding and programming of the load routing are generally used.

### Trolley Conveyors

The load-carrying member of a trolley conveyor is the trolley or series of wheels. The wheels are sized and spaced as a function of the imposed load, the propelling means, and the track capability. The load hanger (carrier) is attached to the conveyor and generally remains attached unless manually removed. However, in a few installations, the load hanger is transferred to and from the conveyor automatically.

The trolley conveyor can employ any chain length consistent with allowable propelling means and drive(s) capability. The track layout always involves horizontal turns and commonly has vertical inclines and declines.

When a dimensional layout, load spacing, weights of moving loads, function, and load and unload points are determined, the chain or cable pull can be calculated. Manufacturer's data should be used for frictional values. A classical point-to-point analysis should be made, using the most unfavorable loading condition.

In the absence of precise data, the following formulas can be used to find the approximate drive effort:

$$\text{Max drive effort, lb} = A + B + C$$
$$\text{Net drive effort, lb} = A + B + C - D$$
$$\text{Max drive effort, N} = (A + B + C)\,9.81$$
$$\text{Net drive effort, N} = (A + B + C - D)\,9.81$$

where $A = fw$ [where $w$ = total weight of chain, carriers, and live load, lb (kg) and $f$ = coefficient of friction]

$B = wS$ [where $w$ = average carrier load per ft (m), lb/ft (kg/m) and $S$ = total vertical rise, ft (m)]

$C = 0.017f\,(A + B)N$ (where $N$ = sum of all horizontal and vertical curves, deg)

$D = w'S'$ [where $w'$ = average carrier load per ft (m), lb/ft (kg/m) and $s'$ = total vertical drop, ft (m)]

For conveyors with antifriction wheels, with clean operating condition, the coefficient of friction $f$ may be 0.13.

Where drive calculations indicate that the allowable chain or cable tension may be exceeded, multiple-drive units are used. When multiple drives are used for constant speed, high-slip motors or fluid couplings are commonly used. If variable speed is required with multiple drives, it is common to use direct-current motors with direct-current supply and controls. For both constant and variable speed, drives are balanced to share drive effort. Other than those mentioned, many various methods of balancing are available for use.

Complexities of overhead conveyors, particularly with varying loads on inclines and declines, as well as other influencing factors (e.g., conveyor length, environment, lubrication, and each manufacturer's design recommendations), usually require detail engineering analysis to select the proper number of drives and their location. Particular care is also required to locate the take-up properly.

The following components or devices are used on trolley-conveyor applications:

**Trolley Assembly**  Wheels and their attachment portion to the propelling means (chain or cable) are adapted to particular applications, depending upon loading, duty cycle, environment, and manufacturer's design.

**Carrier Attachments**  These are made in three main styles: (1) enclosed tubular type, where the wheels and propelling means are carried inside; (2) semienclosed tubular type, where the wheels are enclosed and the propelling means is external; and (3) open-tee or I-beam type, where the wheels and propelling means are carried externally.

**Sprocket or Traction Wheel Turns**  Any arc of horizontal turn is available. Standards usually vary in increments of 15 to 180°.

**Roller Turns**  Any arc of horizontal turn is available. Standards usually vary in increments of 15° from 15 to 180°.

**Track Turns**  These are horizontal track bends without sprockets, traction wheels, or rollers; they are normally used on enclosed-track conveyors where the propelling means is fitted with horizontal guide wheels.

**Track Hangers, Brackets, and Bracing**  These conform to track size and shape, spaced at intervals consistent to allowable track stress and deflection applied by loading and chain or cable tensions.

**Track Expansion Joints**  For use in variable ambient conditions, such as ovens, these are also applied in many instances where conveyor track crosses building expansion joints.

**Chain Take-Up Unit**  Required to compensate for chain wear and/or variable ambient conditions, this unit may be traction-wheel, sprocket, roller, or track-turn type. Adjustment is maintained by screw, screw spring, counterweight, or air cylinder.

**Incline and Decline Safety Devices**  An "anti-back-up"

device will ratchet into a trolley or the propelling means in case of unexpected reversal of a conveyor on an incline. An "anti-runaway" device will sense abnormal conveyor velocity on a decline and engage a ratchet into a trolley or the propelling means. Either device will arrest the uncontrolled movement of the conveyor. The anti-runaway is commonly connected electrically to cut the power to the drive unit.

*Drive Unit* Usually sprocket or caterpillar type, these units are available for constant-speed or manual variable-speed control. Common speed variation is 1:3; e.g., 5 to 15 ft/min (1.5 to 4.5 m/min). Drive motors commonly range from fractional to 15 hp (11 kW).

*Drive-Unit Overload* Overload is detected by electrical, torque, or pull detection by any one of many available means. Usually overload will disconnect power to the drive, stopping the conveyor. When the reason for overload is determined and corrected, the conveyor may be restarted.

*Equipment Guards* Often it is desirable or necessary to guard the conveyor from hostile environment and contaminants. Also employees must be protected from accidental engagement with the conveyor components.

**Transfer Devices** Usually unique to each application, automatic part or carrier loading, unloading, and transfer devices are available. With growth in the use of power-and-free, carrier transfer devices have become rare.

### Power-and-Free Conveyor

The power-and-free conveyor has the highest potential application wherever there is a requirement for other than a single fixed-path flow (trolley conveyor). Power-and-free conveyors may have any number of automatic or manual switch points. A system will permit scheduled transit and delivery of work to the next assigned station automatically. Accumulation (storage) areas are designed to accommodate in-process inventory between operations.

The components and chain-pull calculation discussed for powered overhead conveyors are basically applicable to the power-and-free conveyor. Addition of a secondary free track surface is provided for the work carrier to traverse. This free track is usually disposed directly below the power rail but is sometimes found alongside the power rail. (This arrangement is often referred to as a "side pusher" or "drop finger.") The power-and-free rails are joined by brackets for rail (free-track) continuity. The power chain is fitted with pushers to engage the work-carrier trolley. Track sections are available in numerous configurations for both the power portion and the free portion. Sections will be enclosed, semienclosed, or open in any combinations. Two of the most common types are shown in Fig. 2.

As an example of one configuration of power-and-free, Fig. 3 shows the Acco Chain Conveyor Division enclosed-track power-and-free rail. In the cutaway portion, the pushers are shown engaged with the work-carrier trolley.

The pushers are pivoted on an axis parallel to the chain path and swing aside to engage the pusher trolley. The pusher trolley remains engaged on level and sloped sections. At automatic or manual switching points, the leading dispatch trolley head which is not engaged with the chain is propelled through the switch to the branch line. As the chain passes the switching point, the pusher trolley departs to the right or left from pusher engagement and arrives on a free line, where it is subject to manual or controlled gravity flow.

The distance between pushers on the chain for power-and-free use is established in accordance with conventional practice, except that the minimum allowable pusher spacing must take into account the wheelbase of the trolley, the bumper

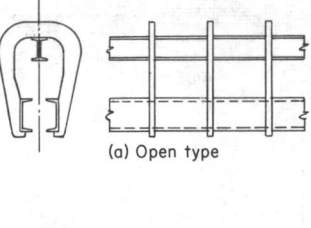

(a) Open type

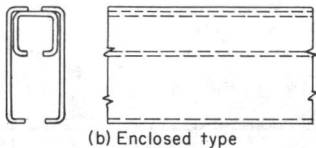

(b) Enclosed type

**Fig. 2** Conveyor tracks.

length, the load size, the chain velocity, and the action of the carrier at automatic switching and reentry points. A switching headway must be allowed between work carriers. An approximation of the minimum allowable pusher spacing is that the pusher spacing will equal twice the work-carrier bumper length. Therefore, a 4-ft (1.2-m) work carrier would indicate a minimum pusher spacing of 8 ft (2.4 m).

The load-transmission capabilities are a function of velocity

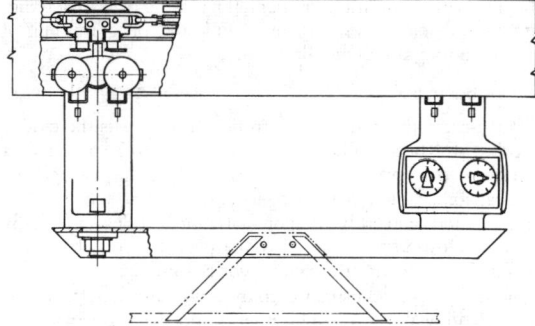

**Fig. 3** ACCO Chain Conveyor Division power-and-free conveyor rail and trolley heads.

and pusher spacing. At the 8-ft (2.4-m) pusher spacing and a velocity of 40 ft/min (0.22 m/s), five pushers per minute are made available. The load-transmission capability is five loads per minute, or 300 loads per hour.

**Method of Automatic Switching from a Powered Line to a Free Line** Power-and-free work carriers are usually switched automatically. To do this, it is necessary to have a code device on the work carrier and a decoding (reading) device along the track in advance of the track switch. Figure 4 shows the equipment relationship. On each carrier, the free trolley carries the code selection, manually or automatically introduced, which identifies it for a particular destination or routing. As the free trolley passes the reading station, the trolley intelligence is decoded and compared with a preset station code and its current knowledge of the switch position and branch-line condition; a decision is then reached which results in correct positioning of the rail switch.

In Fig. 4, the equipment illustrated includes a transistorized readout station *a*, which supplies 12 V direct current at stainless-steel code brushes *b*. The code brushes are

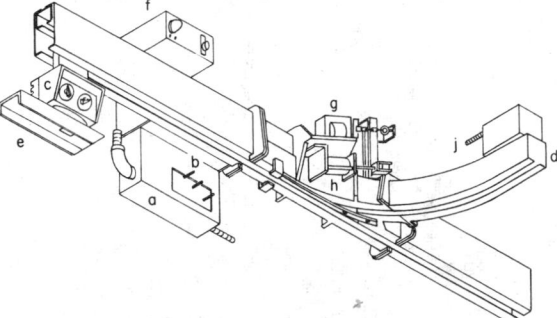

**Fig. 4** Switch mechanism exiting trolley from power rail to free rail. (*ACCO Chain Conveyor Division.*)

"matched" by contacts on the encoded trolley head *c*. When a trolley is in register and matches the code-brush positioning, it will be allowed to enter branch line *d* if the line is not full. If carrier *e* is to be entered, the input signal is rectified and amplified so as to drive a power relay at junction box *f*, which in turn actuates solenoid *g* to operate track switch *h* to the branch-line position. A memory circuit is established in the station, indicating that a full-line condition exists. This condition is maintained until the pusher pin of the switched carrier clears reset switch *j*.

The situations which can be handled by the decoding stations are as follows: (1) A trolley with a matching code is in register; there is space in the branch line. The carrier is allowed to enter the branch line. (2) A trolley with a matching code is in register; there is no space in the branch line. The carrier is automatically recirculated on the powered system and will continue to test its assigned destination until it can be accommodated. (3) A trolley with a matching code is in register; there is no space in the branch line. The powered conveyor is automatically stopped, and visual and/or audible signals are started. The conveyor can be automatically restarted when the full-line condition is cleared, and the waiting carrier will be allowed to enter. (4) A trolley with a nonmatching code is in register; in this case, the decoding station always returns the track switch to the main-line position, if necessary, and bypasses the carrier.

**Use and Control of Carriers on Free and Gravity Lines** Free and gravity lines are used as follows: (1) To connect multiple power-and-free conveyors, thus making systems easily extensible and permitting different conveyor designs for particular use. (2) To connect two auxiliary devices such as vertical conveyors and drop-lift stations. (3) As manned or automatic work stations. In this case, the size of the station depends on the number of carriers processed at one time, with additional space provision for arriving and departing carriers. (Considerable knowledge of work-rate standards and production-sched-

ule requirements is needed for accurate sizing.) (4) As manned or automatic storage lines, especially for handling production imbalance for later consumption.

Nonmanned automatic free lines require that the carrier be controlled in conformance with desired conveyor function and with regard to the commodity being handled. The two principal ways to control carriers in nonmanned free rails are (1) to slope the free rail so that all carriers will start from rest and use incremental spot retarders to check velocity, and (2) to install horizontal or sloped rails and use auxiliary power conveyor(s) to accumulate carriers arriving in the line.

In manned free lines, it is usual to have slope at the automatic arrival and automatic departure sections only. These sections are designed for automatic accumulation of a finite number of carriers, and retarding or feeding devices may be used. Throughout the remainder of the manned station, the carrier is propelled by hand.

**Method of Automatic Switching from a Free Line to a Powered Line** Power-and-free work carriers can be reentered into the powered lines either manually or automatically. The carrier must be integrated with traffic already on the powered line and must be entered so that it will engage with a pusher on the powered chain. Figure 5 illustrates a typical method of automatic reentry. The carrier *a* is held at a rest on a slope in the demand position by the electric trolley stop *b*. The demand to enter enables the sensing switches *c* and *d* mounted on the powered rail to test for the availability of a pusher. When a pusher that is not propelling a load is sensed, all conditions are met and the carrier is released by the electric trolley stop such that it arrives in the pickup position in advance of the pusher. A retarding or choke device can be used to keep the entering carrier from overrunning the next switch position or another carrier in transit. The chain pusher engages the pusher trolley and departs the carrier. The track switch *f* can remain in the branch-line position until a carrier on the main line *g* would cause the track switch to reset. Automatic reentry control ensures that no opportunity to use a pusher is overlooked and does not require the time of an operator.

**Power-and-Free Conveyor Components** All components used on trolley conveyors are applied to power-and-free conveyors. Listed below are a few of the various components unique to power-and-free systems.

TRACK SWITCH. This is used for diverting work carriers either automatically or manually from one line or path to another. Any one system may have both. Switching may be either to the right or to the left. Automatically, stops are usually operated pneumatically or electromechanically. Track switches are also used to merge two lines into one.

TROLLEY STOPS. Used to stop work carriers, these operate either automatically or manually on a free track section or on a powered section. Automatically, stops are usually operated pneumatically or electromechanically.

STORAGE. Portions or spurs of power-and-free conveyors are usually dedicated to the storage (accumulation) of work carriers. Unique to the design, type, or application, storage may be accomplished on (1) level hand-pushed lines, (2) gravity sloped lines (usually with overspeed-control retarders), (3) power lines with spring-loaded pusher dogs, or (4) powered lines with automatic accumulating free trolleys.

INCLINES AND DECLINES. As in the case of trolley conveyors, vertical inclines and declines are common to power-and-free.

In addition to safety devices used on trolley conveyors, similar devices may be applied to free trolleys.

LOAD BARS AND CARRIERS.    The design of load bars, bumpers, swivel devices, index devices, hooks, or carriers is developed at the time of the initial power-and-free investigation. The

seldom greater than about 40°. Their principal application is in handling coal. The flight conveyor of usual construction should not be specified for a material that is actively abrasive, such as damp sand and ashes. The **drag-chain conveyor** (Fig. 6) has an open-link chain, which serves, instead of flights, to

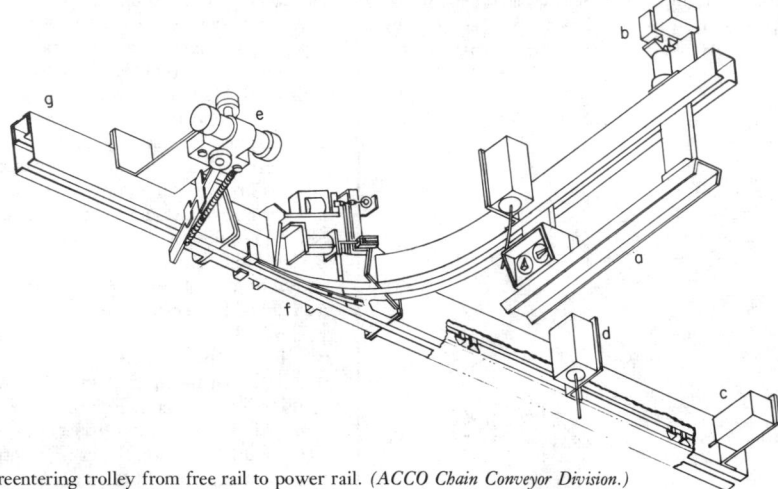

**Fig. 5**  Switch mechanism reentering trolley from free rail to power rail. *(ACCO Chain Conveyor Division.)*

system will see only the carrier, and all details of system design are a function of its design. How the commodity is being handled on or in the carrier is carefully considered to facilitate its use throughout the system, manually, automatically, or both.

VERTICAL CONVEYOR SECTIONS.    Vertical conveyor sections are often used as an accessory to power-and-free. For practical purposes, the vertical conveyor can be divided into two classes of devices:

DROP (LIFT) SECTION.    This device is used to drop (or lift) the work carrier vertically to a predetermined level in lieu of vertical inclines or declines. One common reason for its use is to conserve space. The unit may be powered by a cylinder or hoist, depending on the travel distance, cycle time, and load. One example of the use of a drop (lift) section is to receive a carrier on a high level and lower it to an operations level. The lower level may be a load-unload station or processing station. Automatic safety stops are used to close open rail ends.

INTERFLOOR VERTICAL CONVEYOR.    When used for interfloor service and long lifts, the vertical conveyor may be powered by high-speed hoists or elevating machines. In any case, the carriers are automatically transferred to and from the lift, and the dispatch control on the carrier can instruct the machine as to the destination of the carrier. Machines can be equipped with a variety of speeds and operating characteristics. Multiple carriers may be handled, and priority-call control systems can be fitted to suit individual requirements.

## NONCARRYING CONVEYORS
### by Harold V. Hawkins

**Flight Conveyors**

**Flight conveyors** are used for moving granular, lumpy, or pulverized materials along a horizontal path or on an incline

push the material along. With a hard-faced concrete or cast-iron trough, it serves well for handling ashes. The return run is, if possible, above the carrying run, so that the dribble will be back into the loaded run. A feeder must be provided unless the feed is otherwise controlled, as from a tandem elevator or conveyor. As the feeder is interlocked, either mechanically or electrically, the feed stops if the conveyor stops. Flight conveyors may be classified as **scraper type** (Fig. 7), in which the element (chain and flights) rests on the trough; **suspended-flight type** (Fig. 8), in which the flights are carried clear of the trough

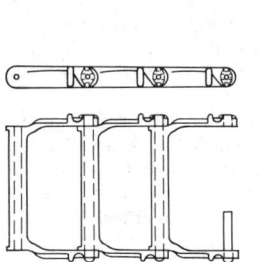

**Fig. 6**  Drag chain.

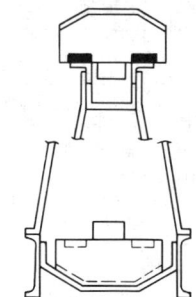

**Fig. 7**  Single-strand scraper-flight conveyor.

by shoes resting on guides; and **suspended-chain type** (Fig. 9), in which the chain rests on guides, again carrying the flights clear of the trough. These types are further differentiated as **single-strand** (Figs. 7 and 8) and **double-strand** (Fig. 9). For lumpy material, the latter has the advantage since the lumps will enter the trough without interference. For heavy duty also, the double strand has the advantage, in that the pull is divided between two chains. A special type for simultaneous handling of several materials may have the trough divided by longitudinal partitions. The material having the greatest coefficient of friction is then carried, if possible, in the central zone to equalize chain wear and stretch.

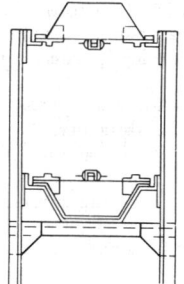

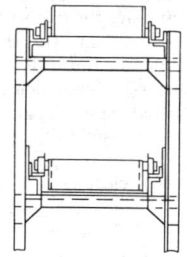

**Fig. 8**  Single-strand suspended-flight conveyor.

**Fig. 9**  Double-strand roller-chain flight conveyor.

Improvements in the welding and carburizing of welded-link chain have made possible its use in flight conveyors, offering several significant advantages, including economy and flexibility in all directions. Figure 10 shows a typical scraper-type flight cast from malleable iron incorporated onto a slotted conveyor bed. The small amount of fines that fall through the slot are returned to the top of the bed by the returning flights. Figure 11 shows a double-chain scraper conveyor in which the ends of the flights ride in a restrictive channel. These types of flight conveyors are driven by pocket wheels.

Flight conveyors of small capacity operate usually at 100 to 150 ft/min (0.51 to 0.76 m/s). Large-capacity conveyors operate at 100 ft/min (0.51 m/s) or slower; their long-pitch chains hammer heavily against the drive-sprocket teeth or pocket wheels at higher speeds. A conveyor steeply inclined should have closely spaced flights so that the material will not avalanche over the tops of the flights. The capacity of a given conveyor diminishes as the angle of slope increases. For the heaviest duty, hardened-face rollers at the articulations are essential.

The capacity of an inclined conveyor is found by multiplying the capacities in Table 1 by the following factors:

| Inclination, deg | 20 | 25 | 30 | 35 |
|---|---|---|---|---|
| Factor | 0.9 | 0.8 | 0.7 | 0.6 |

For flight shoes or flights against guides (not lubricated) or

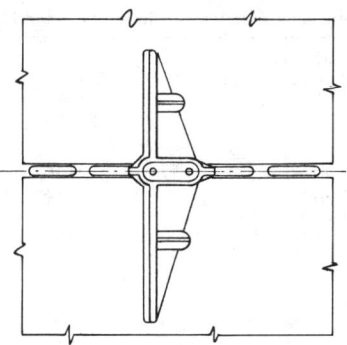

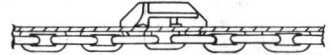

**Fig. 10**  Scraper flight with welded chain.

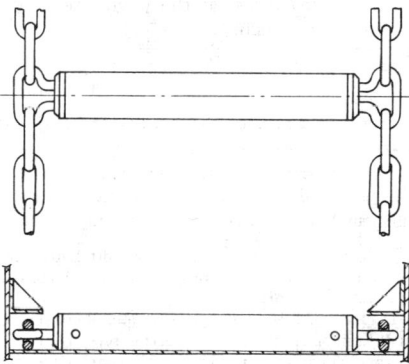

**Fig. 11**  Scraper flight using parallel welded chains.

trough, $F = 0.33$; if rollers are used instead of shoes, $F = 0.1$; for roller with antifriction bearings, 0.05.

The chain pull for a horizontal conveyor $= 2WLF +$

## Table 1. Capacity and Size of Lumps of Flight Conveyors

| Flight width and depth | | Quantity of material | | Approx capacity* | | Lump size, not to exceed 10% of total volume | | | |
|---|---|---|---|---|---|---|---|---|---|
| | | | | At 100 ft/min with 50 lb/ft³ material | At 0.5 m/s with 800 kg/m³ material | Single strand | | Double strand | |
| in | mm | ft³/ft | m³/m | Short ton/h | tonne/h | in | mm | in | mm |
| 12 × 6 | 300 × 150 | 0.40 | 0.037 | 60 | 54 | 31½ | 89 | 4 | 102 |
| 15 × 6 | 380 × 150 | 0.49 | 0.046 | 73 | 66 | 41½ | 114 | 5 | 127 |
| 18 × 6 | 460 × 150 | 0.56 | 0.052 | 84 | 76 | 5 | 127 | 6 | 152 |
| 24 × 8 | 610 × 200 | 1.16 | 0.108 | 174 | 158 | | | 10 | 254 |
| 30 × 10 | 760 × 250 | 1.60 | 0.149 | 240 | 218 | | | 14 | 356 |
| 36 × 12 | 910 × 300 | 2.40 | 0.223 | 360 | 327 | | | 16 | 406 |

*For material weighing other than 50 lb/ft³ (800 kg/m³) the capacity is proportional to weight.

**Table 2. Coefficients of Friction on Steel Plate**

| | |
|---|---|
| Anthracite | 0.33 |
| Bituminous coal | 0.59 |
| Cement | 0.93 |
| Clay | 0.60–0.70 |
| Coke | 0.36 |
| Copra | 0.40 |
| Grains | 0.30–0.40 |
| Hog fuel, dry | 0.65 |
| Hydrated lime | 0.65 |
| Limestone, pulv | 0.53 |
| Sawdust, wet | 0.60 |
| Soda ash | 0.65 |

$W_1 L F_1$, where $W$ = weight per ft of element, lb; $L$ = length of conveyor, ft; $F$ = coefficient of friction of element; $W_1$ = weight of material per ft of conveyor, lb; $F_1$ = material coefficient from Table 2.

In an **inclined conveyor,** the power requirement is the sum of that required for the horizontal run plus that required to lift the load. The turning effort at the pitch line of the head sprocket for a conveyor inclined

$$\theta° = WL(F \cos \theta + \sin \theta) + W_1 L(F_1 \cos \theta + \sin \theta) + WL(F \cos \theta - \sin \theta) \quad \text{lb}$$

The third term may be a minus quantity, indicating that the return run then is assisting the turning effort.

For the motor horsepower, add 10 percent for friction in the terminal bearings, plus 10 percent for loss in the speed-reduction gearing, plus 10 percent for starting or surge loads.

EXAMPLE. Horizontal conveyor, 150-ft (46-m) centers; capacity, 175 tons/h (1.3 tonnes/s) of bituminous coal. Determine conveyor size, chain pull, speed, and motor.

From Table 1, use a 24- by 8-in (93- by 31-mm) flight size at a speed of 100 ft/min (0.51 m/s). Assume an engineering-type chain with sleeve-bearing rollers and weighing, with flights, 60 lb/running ft (60 kg/m).

$$W \times L = 60 \times 150 = 9,000$$
$$W_1 \times L = 1.16 \times 50 \times 150 = 8,700$$
$$F = 0.01 \quad F_1 = 0.59$$
Chain pull = $(2 \times 9,000 \times 0.10) + (8,700 \times 0.59) = 6,933$ lb
$$hp = (6,933 \times 100)/33,000 = 21 \ (15.7 \text{ kW})$$
Use 27.9-hp motor.

**Cautions in Flight-Conveyor Selections** With abrasive material, the trough design should provide for renewal of the bottom plate without disturbing the side plates. If the conveyor is inclined and will reverse when halted under load, a solenoid brake or other automatic backstop should be provided. Chains may not wear or stretch equally. In a double-strand conveyor, it may be necessary to shift sections of chain from one side to the other to even up the lengths.

Intermediate slide gates should be set to open in the opposite direction to the movement of material in the conveyor.

The **continuous-flow conveyor** serves as a conveyor, as an elevator, or as a combination of the two. It is a slow-speed machine in which the material moves as a continuous core within a duct. Except with the **Redler** conveyor, the element is formed by a single strand of chain with closely spaced impellers, somewhat resembling the flights of a flight conveyor.

The **Bulk flo** (Fig. 12) has peaked flights designed to facilitate the outflow of the load at the point of discharge. The load, moved by a positive push of the flights, tends to provide self-

clearing action at the end of a run, leaving only a slight residue.

The **Redler** (Fig. 13) has skeletonized or U-shaped impellers which move the material in which they are submerged because the resistance to slip through the element is greater than the drag against the walls of the duct.

Materials for which the continuous-flow conveyor is suited are listed below in groups of increasing difficulty. The constant $C$ is used in the power equations below, and in Fig. 15.

$C = 1$: clean coal, flaxseed, graphite, soybeans, copra, soap flakes
$C = 1.2$: beans, slack coal, sawdust, wheat, wood chips (dry), flour
$C = 1.5$: salt, wood chips (wet), starch
$C = 2$: clays, fly ash, lime (pebble), sugar (granular), soda ash, zinc oxide
$C = 2.5$: alum, borax, cork (ground), limestone (pulverized)

Among the materials for which special construction is advised are bauxite, brown sugar, hog fuel, wet coal, shelled

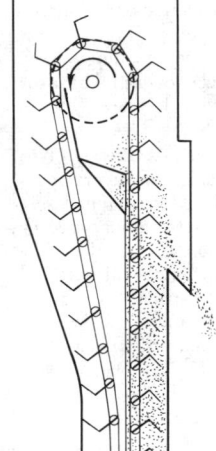

**Fig. 12** Bulk-flow continuous-flow elevator.

corn, foundry dust, cement, bug dust, hot brewers' grains. The machine should not be specified for ashes, bagasse, carbon-black pellets, sand and gravel, sewage sludge, and crushed stone. Fabrication from corrosion-resistant materials such as brass, monel, or stainless steel may be necessary for use with some corrosive materials.

Where a single **runaround conveyor** is required with multiple feed points and some recirculation of excess load, the Redler serves. The U-frame flights do not squeeze the loads, as they resume parallelism after separating when rounding the terminal wheels. As an elevator, this machine will also handle sluggish materials that do not flow out readily. A pusher plate opposite the discharge chute can be employed to enter between the legs of the U flights to push out such material. When horizontal or inclined, continuous-flow elevators such as the Redler are normally self-cleaning. When vertical, the Redler type can be made self-cleaning (except with sticky materials) by use of special flights.

Continuous-flow conveyors and elevators do not require a feeder (Fig. 14). They are self-loading to capacity and will not overload, even though there are several open- or uncontrolled-feed openings, since the duct fills at the first opening and automatically prevents the entrance of additional material at subsequent openings. Some special care may be required with free-flowing material.

The duct is easily insulated by sheets of asbestos cement or similar material to reduce cooling in transit. As the duct is completely sealed, there is no updraft where the lift is high. The material is protected from exposure and contamination or contact with lubricants. The handling capacity for horizontal or inclined lengths (nearly to the angle of repose of the material) approximates 100 percent of the volume swept through by the movement of the element. For steeper inclines or elevators, it is between 50 and 90 percent.

If the material is somewhat abrasive, as with wet bituminous coal, the duct should be of corrosion-resistant steel, of extra thickness, and the chain pins should be both extremely hard and of corrosion-resistant material.

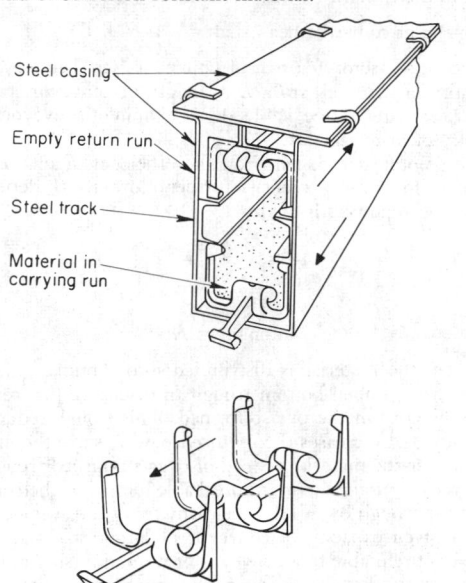

**Fig. 13**  Redler U-type continuous-flow conveyor.

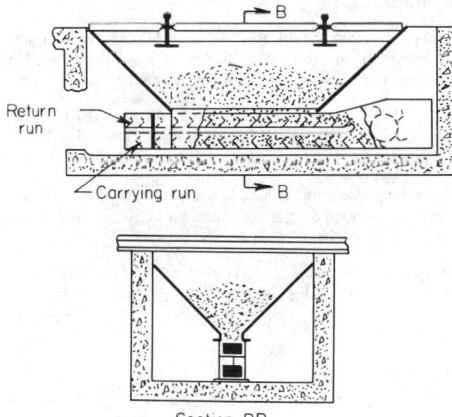

**Fig. 14**  Shallow-track hopper for continuous-flow conveyor with feed to return run.

A long horizontal run followed by an upturn is inadvisable because of radial thrust. Lumpy material is difficult to feed from a track hopper. An automatic brake is unnecessary, as an elevator will reverse only a few inches when released.

The motor horsepower $P$ required by continuous-flow conveyors for the five arrangements shown in Fig. 15 is given in the accompanying formulas in terms of the capacity $T$, in tons per h; the horizontal run $H$, ft; the vertical lift $V$, ft; and the constant $C$, values for which are given above. If loading from a track hopper, add 10 percent.

## Screw Conveyors

The screw, or spiral, conveyor is used quite widely for pulverized or granular, noncorrosive, nonabrasive materials when the required capacity is moderate, when the distance is not

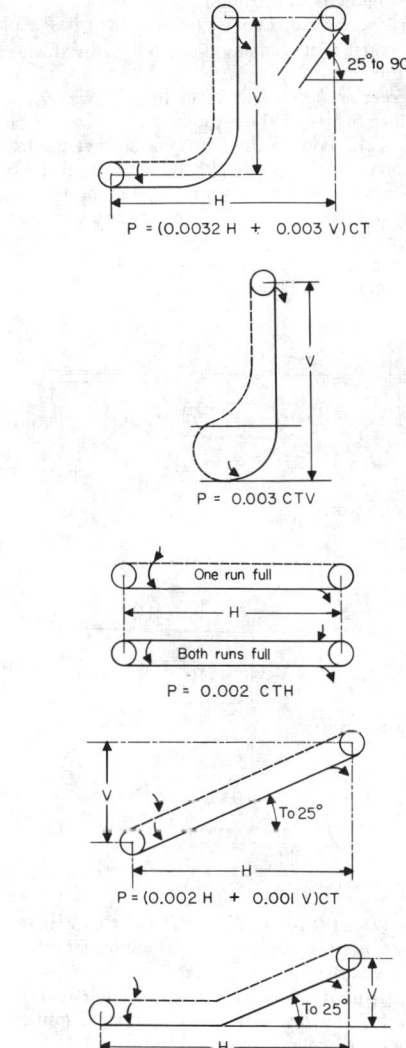

**Fig. 15**  Continuous-flow-conveyor arrangements,

more than about 200 ft (61 m), and when the path is not too steep. It usually costs substantially less than any other type of conveyor and is readily made dust-tight by a simple cover plate.

The conveyor will handle lumpy material if the lumps are not large in proportion to the diameter of the helix. If the length exceeds that advisable for a single conveyor, separate or tandem units are readily arranged. Screw conveyors may be inclined. A standard-pitch helix will handle material on inclines up to 35°. The reduction in capacity as compared with the capacity when horizontal is indicated in the following table:

| Inclination, deg | 10 | 15 | 20 | 25 | 30 | 35 |
|---|---|---|---|---|---|---|
| Reduction in capacity, percent | 10 | 26 | 45 | 58 | 70 | 78 |

Abrasive or corrosive materials can be handled with suitable construction of the helix and trough.

The standard screw-conveyor helix (Fig. 16) has a pitch approximately equal to its outside diameter. Other forms are used for special cases.

**Short-pitch screws** are advisable for inclines above 29°.

**Variable-pitch screws,** with short pitch at the feed end, automatically control the flow to the conveyor so that the load is correctly proportioned for the length beyond the feed point. With a short section either of shorter pitch or of smaller diameter, the conveyor is self-loading to capacity and does not require a feeder.

**Cut flights** (Fig. 17) are used for conveying and mixing cereals, grains, and other light materials.

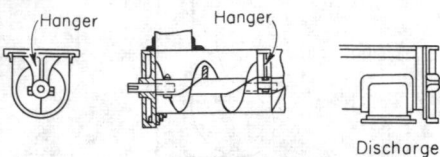

**Fig. 16**   Spiral conveyor.

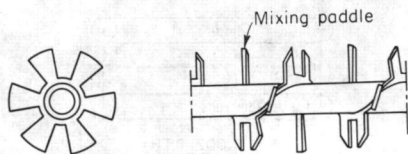

**Fig. 17**   Cut-flight conveyor.

**Fig. 18**   Ribbon conveyor.

**Ribbon screws** (Fig. 18) are used for wet and sticky materials, such as molasses, hot tar, and asphalt, which might otherwise build up on the spindle.

**Paddle screws** are used primarily for mixing such materials as mortar and bitulithic paving mixtures. One typical application is to churn ashes and water to eliminate dust.

Standard constructions have a plain or galvanized-steel helix and trough. For abrasives and corrosives such as wet ashes, both helix and trough may be of hard-faced cast iron. For simple abrasives, the outer edge of the helix may be faced with a renewable strip of Stellite or similar extremely hard material. For food products, aluminum, bronze, monel metal, or stainless steel is suitable but expensive.

Table 3 gives the capacities, allowable speeds, percentages of helix loading for five groups of materials, and the factor $F$ used in estimating the power requirement.

Table 4 gives the handling capacities for standard-pitch screw conveyors in each of the five groups of materials when the conveyors are operating at the maximum advised speeds and in the horizontal position. The capacity at any lower speed is in the ratio of the speeds.

**Power Requirements**   The power requirements for horizontal screw conveyors of standard design and pitch are determined by the Link-Belt Co. by the formula that follows. Additional allowances should be made for inclined conveyors, for starting under load, and for materials that tend to stick or pack in the trough, as with cement.

$$H = \text{hp at conveyor head shaft} = (ALN + CWLF) \times 10^{-6}$$

where $A$ = factor for size of conveyor (see Table 5); $C$ = quantity of material, ft³/h; $L$ = length of conveyor, ft; $F$ = factor for material (see Table 3); $N$ = r/min of conveyor; $W$ = density of material, lb/ft³.

The motor size depends on the efficiency $E$ of the drive (usually close to 90 percent); a further allowance $G$, depending on the horsepower, is made:

| $H$ | 1 | 1–2 | 2–4 | 4–5 | 5 |
|---|---|---|---|---|---|
| $G$ | 2 | 1.5 | 1.25 | 1.1 | 1.0 |

$$\text{Motor hp} = HG/E$$

When the material is distributed into a bunker, the conveyor has an open-bottom trough to discharge progressively over the crest of the pile so formed. This trough reduces the capacity and increases the required power, since the material drags over the material instead of over a polished trough.

If the material contains unbreakable lumps, the helix should clear the trough by at least the diameter of the average lump. For a given capacity, a conveyor of larger size and slower speed is preferable to a conveyor of minimum size and maximum speed. For large capacities and lengths, the alternatives—a flight conveyor or a belt conveyor—should receive consideration.

EXAMPLES.   1. Slack coal 50 lb/ft³ (800 kg/m³); desired capacity 50 tons/h (2,000 ft³/h) (45 tonnes/h); conveyor length, 60 ft (18 m); 14-in (0.36-m) conveyor at 80 r/min. $F$ for slack coal = 0.9 (group 2).

$$H = (255 \times 60 \times 80 + 2{,}000 \times 60 \times 60 \times 0.9)/1{,}000{,}000 = 6.6$$
$$\text{Motor hp} = (6.6 \times 1.0)/0.90 = 7.3 \quad (5.4 \text{ kW})$$

Use 7½-hp motor.

2. Limestone screenings, 90 lb/ft³; desired capacity, 10 tons/h (222 ft³/h); conveyor length, 50 ft; 9-in conveyor at 50 r/min. $F$ for limestone screenings = 2.0 (group 4).

$$H = (96 \times 50 \times 50 + 222 \times 90 \times 50 \times 2.0)/1{,}000{,}000 = 2.24$$
$$\text{Motor hp} = (2.24 \times 1.25)/0.90 = 2.8$$

Use 3-hp motor.

### Chutes

**Bulk Material**   If the material is fragile and cannot be set through a simple **vertical chute,** a **retarding chute** may be specified. Figure 19 shows a ladder chute in which the material trickles over shelves instead of falling freely. If it is necessary to minimize breakage when material is fed from a bin, a vertical box chute with flap doors opening inward, as shown in Fig. 20, permits the material to flow downward only from the top surface and eliminates the degradation that results from a converging flow from the bottom of the mass.

Straight inclined chutes for coal should have a slope of 40 to 45°. If it is found that the coal accelerates objectionably, the chute may be provided with cross angles over which the material cascades at reduced speed (Fig. 21).

**Lumpy material** such as coke and large coal, difficult to control when flowing from a bin, can be handled by a chain-

### Table 3. Capacities and Speeds of Spiral Conveyors

| Group | Max percent of cross section occupied by the material | Max density of material, lb/ft³ (kg/m³) | Max r/min for diameters | |
|---|---|---|---|---|
| | | | 6 in (152 mm) | 20 in (508 mm) |
| 1 | 45 | 50 ( 800) | 170 | 110 |
| 2 | 38 | 50 ( 800) | 120 | 75 |
| 3 | 31 | 75 (1,200) | 90 | 60 |
| 4 | 25 | 100 (1,600) | 70 | 50 |
| 5 | 12½ | | 30 | 25 |

Group 1 includes light materials such as barley, beans, brewers grains (dry), coal (pulv.), corn meal, cottonseed meal, flaxseed, flour, malt, oats, rice, wheat. The value of the factor $F$ is 0.5.

Group 2 includes fines and granular material. The values of $F$ are alum (pulv.), 0.6; coal (slack or fines), 0.9; coffee beans, 0.4; sawdust, 0.7; soda ash (light), 0.7; soybeans, 0.5; fly ash, 0.4.

Group 3 includes materials with small lumps mixed with fines. Values of $F$ are alum, 1.4; ashes (dry), 4.0; borax, 0.7; brewers grains (wet), 0.6; cottonseed, 0.9; salt, coarse or fine, 1.2; soda ash (heavy), 0.7.

Group 4 includes semiabrasive materials, fines, granular and small lumps. Values of $F$ are acid phosphate (dry), 1.4; bauxite (dry), 1.8; cement (dry), 1.4; clay, 2.0; fuller's earth, 2.0; lead salts, 1.0; limestone screenings, 2.0; sugar (raw), 1.0; white lead, 1.0; sulfur (lumpy), 0.8; zinc oxide, 1.0.

Group 5 includes abrasive lumpy materials which must be kept from contact with hanger bearings. Values of $F$ are wet ashes, 5.0; flue dirt, 4.0; quartz (pulv.), 2.5; silica sand, 2.0; sewage sludge (wet and sandy), 6.0.

controlled feeder chute with a screen of heavy endless chains hung on a sprocket shaft (Fig. 22). The weight of the chain curtain holds the material in the chute. When a feed is desired,

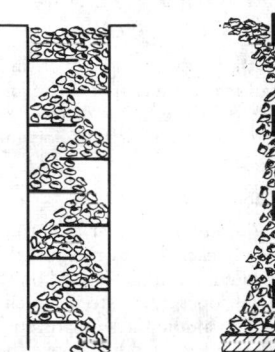

**Fig. 19** Ladder chute.

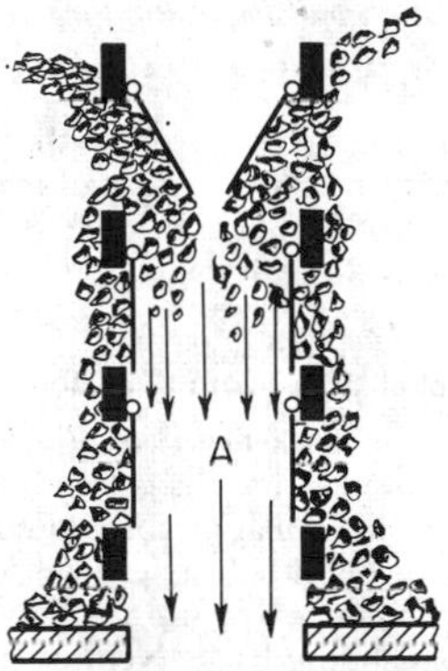

**Fig. 20** Box chute with flap doors. Chute is always full up to discharging point.

the sprocket shaft is revolved slowly, either manually or by a motorized reducer.

**Unit Loads    Mechanical handling of unit loads,** such as boxes, barrels, packages, castings, crates, and palletized loads, calls for methods and mechanisms entirely different from those adapted to the movement of bulk materials.

**Spiral chutes** are adapted for the direct lowering of unit loads of various shapes, sizes, and weights, so long as their slide characteristics do not vary widely. If they do vary, care must be exercised to see that items accelerating on the selected helix pitch do not crush or damage those ahead.

A spiral chute may extend through several floors, e.g., for lowering parcels in department stores to a basement shipping department. The opening at each floor must be provided with

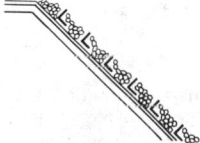
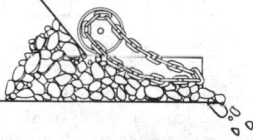

**Fig. 21** Inclined chute with cross angles.

**Fig. 22** Chain-controlled feeder chute.

automatic closure doors, and the design must be approved by the Board of Fire Underwriters.

At the discharge end, it is usual to extend the chute plate horizontally to a length in which the loads can come to rest. A tandem gravity roll conveyor may be advisable for distribution of the loads.

The **sheet-metal spiral** (Fig. 23) has a fixed blade and can be furnished in varying diameters and pitches, both of which determine the maximum size of package that can be handled.

### Table 4. Screw-Conveyor Capacities
(ft³/h)

| Group | Conveyor size, in* | | | | | | | |
|---|---|---|---|---|---|---|---|---|
| | 6 | 9 | 10 | 12 | 14 | 16 | 18 | 20 |
| 1 | 350 | 1,100 | 1,600 | 2,500 | 4,000 | 5,500 | 7,600 | 10,000 |
| 2 | 220 | 700 | 950 | 1,600 | 2,400 | 3,400 | 4,500 | 6,000 |
| 3 | 150 | 460 | 620 | 1,100 | 1,600 | 2,200 | 3,200 | 4,000 |
| 4 | 90 | 300 | 400 | 650 | 1,000 | 1,500 | 2,000 | 2,600 |
| 5 | 20 | 68 | 90 | 160 | 240 | 350 | 500 | 650 |

*Multiply by 25.4 to obtain mm.

**Table 5. Factor** *A*
(Self-lubricating bronze bearings assumed)

| Diam of conveyor, in | 6 | 9 | 10 | 12 | 14 | 16 | 18 | 20 | 24 |
|---|---|---|---|---|---|---|---|---|---|
| mm | 152 | 229 | 254 | 305 | 356 | 406 | 457 | 508 | 610 |
| Factor *A* | 54 | 96 | 114 | 171 | 255 | 336 | 414 | 510 | 690 |

These chutes may have receive and discharge points at any desired floors. There are certain kinds of commodities, such as those made of metal or bound with wire or metal bands, that cannot be handled satisfactorily unless the spiral chute is

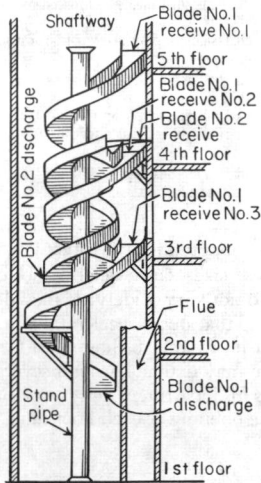

**Fig. 23**   Metal spiral chute.

designed to handle only that particular commodity. Sheet-metal spirals can be built with double or triple blades, all mounted on the same standpipe. Another form of sheet-metal spiral is the open-core type, which is especially adaptable for handling long and narrow articles or bulky classes of merchandise or for use where the spiral must wind around an existing column or pass through floors in locations limited by beams or girders that cannot be conveniently cut or moved. For handling bread or other food products, it is customary to have the spiral tread made from monel metal or aluminum.

## CARRYING CONVEYORS
### by Harold V. Hawkins

### Apron Conveyors

**Apron Conveyors** are specified for granular or lumpy materials. Since the load is carried and not dragged, less power is required than for screw or scraper conveyors. Apron conveyors may have stationary skirt or side plates to permit increased depth of material on the apron, e.g., when used as a feeder for taking material from a track hopper (Fig. 24) with controlled rate of feed. They are not often specified if the length is great, since other types of conveyor are substantially lower in cost. Sizes of lumps are limited by the width of the pans and the ability of the conveyor to withstand the impact of loading. Only end discharge is possible. The apron conveyor (Fig. 25) consists of two strands of roller chain separated by overlap-

ping apron plates, which form the carrying surface, with sides 2 to 6 in (51 to 152 mm) high. The chains are driven by sprockets at one end, take-ups being provided at the other end. The conveyors always pull the material toward the driving end. For light duty, flangeless rollers on flat rails are used; for heavy duty, single-flanged rollers and T rails are used. Apron conveyors may be run without feeders, provided that the opening of the feeding hopper is made sufficiently narrow to prevent material from spilling over the sides of the conveyor after passing from the opening. When used as a conveyor, the **speed** should not exceed 60 ft/min (0.30 m/s); when used as a feeder, 30 ft/min (0.15 m/s). Table 6 gives the capacities of apron feeders with material weighing 50 lb/ft³ (800 kg/m³) at a speed of 10 ft/min (0.05 m/s).

*Chain pull for horizontal-apron conveyor:*

$$2LF(W + W_1)$$

*Chain pull for inclined-apron conveyor:*

$$L(W + W_1)(F \cos \theta + \sin \theta) + WL(F \cos \theta - \sin \theta)$$

where $L$ = conveyor length, ft; $W$ = weight of chain and pans per ft, lb; $W_1$ = weight of material per ft of conveyor, lb; $\theta$ = angle of inclination, deg; $F$ = coefficient of rolling friction, usually 0.1 for plain roller bearings or 0.05 for antifriction bearings.

### Bucket Conveyors and Elevators

**Open-top bucket carriers** (Fig. 26) are similar to apron conveyors, except that dished or bucket-shaped receptacles take the place of the flat or corrugated apron plates used on the apron conveyor. The carriers will operate on steeper inclines than apron conveyors (up to 70°), as the buckets prevent material from sliding back. Neither sides extending above the tops of buckets nor skirtboards are necessary. **Speed,** when loaded by a feeder, = 60 ft/min (max) (0.30 m/s) and when dragging the load from a hopper or bin, ≦ 30 ft/min. The **capacity** should be calculated on the basis of the buckets being three-fourths full, the angle of inclination of the conveyor determining the loading condition of the bucket. For **power required,** see Fig. 32.

**V-bucket carriers** are used for elevating and conveying nonabrasive materials, principally coal when it must be elevated and conveyed with one piece of apparatus. The length and height lifted are limited by the strength of the chains and seldom exceed 75 ft (22.9 m). These carriers can operate on any incline and can discharge at any point on the horizontal run. The size of lumps carried is limited by the size and spacing of the buckets. The carrier consists of two strands of roller chain separated by V-shaped steel buckets. Figure 27 shows the most common form, where material is received on the lower horizontal run, elevated, and discharged through openings in the bottom of the trough of the upper horizontal run. The material is scraped along the horizontal trough of the conveyor, as in a flight conveyor. The steel guard plates *a* at the right prevent spillage at the bends. Figure 28 shows a different form, where material is dug by the elevator from a

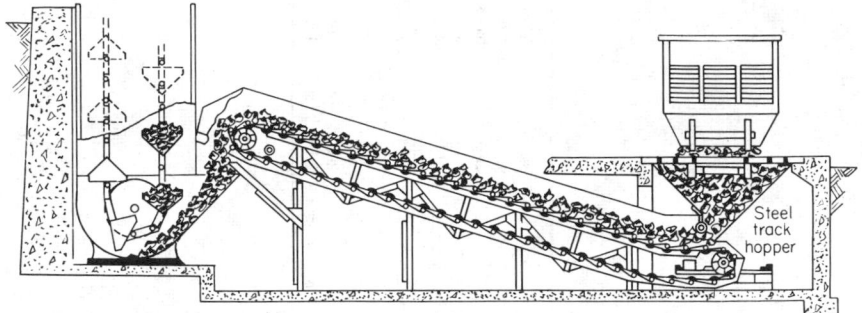

**Fig. 24**   Track hopper and apron feeder supplying a gravity-discharge bucket-elevator boot.

boot, elevated vertically, scraped along the horizontal run, and discharged through gates in the bottom of the trough. Figure 29 shows a variation of the type shown in Fig. 28, requiring one less bend in the conveyor. The troughs are of steel or steel-lined wood. When feeding material to the horizontal run,

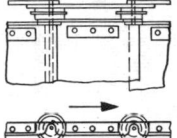

**Fig. 25**   Apron conveyor.

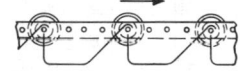

**Fig. 26**   Open-top carrier.

it is advisable to use an automatic feeder driven by power from one of the bend shafts to prevent overloading. Should the buckets of this type of conveyor be overloaded, they will spill on the vertical section. The drive is located at *b*, with take-up at *c*. The speed should not exceed 100 ft/min (0.51 m/s) when large material is being handled, but when material is small, speed may be increased to 125 ft/min (0.64 m/s). The best results are obtained when speeds are kept low. Table 7 gives the capacities and weights based on an even and continuous feed.

**Pivoted-bucket carriers** are used primarily where the path is a runaround in a vertical plane. Their chief application has been for the dual duty of handling coal and ashes in boiler plants. They require less power than V-bucket carriers, as the material is carried and not dragged on the horizontal run. The length and height lifted are limited by the strength of the chains. The length seldom exceeds 500 ft (152 m) and the height lifted 100 ft (30 m). They can be operated on any incline and can discharge at any point on the horizontal run. The size of lumps is limited by the size of buckets. The maintenance cost is extremely low. Many carrier installations are still in operation after 40 years of service. Other applica-

tions are for hot clinker, granulated and pulverized chemicals, cement, and stone.

The carrier consists of two strands of roller chain, with flanged rollers, between which are pivoted buckets, usually of

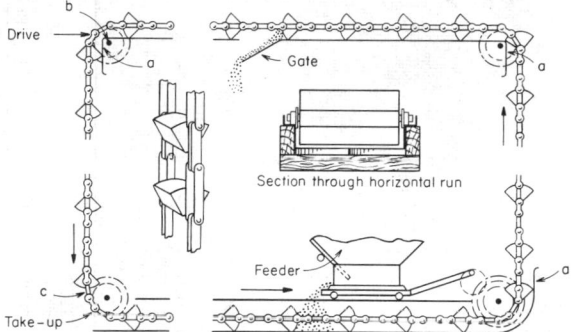

**Fig. 27**   V-bucket carrier.

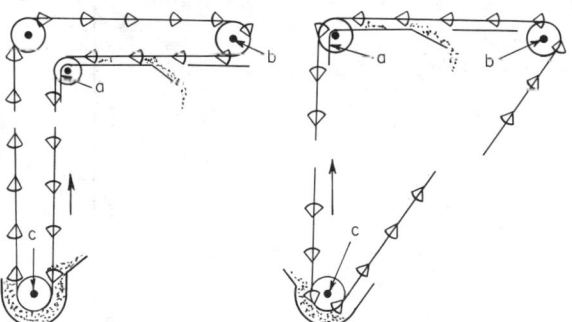

**Figs. 28 and 29**   V-bucket carriers.

**Table 6. Capacities of Apron Conveyors**

| Width between skirt plates | | Capacity, 50 lb/ft³ (800 kg/m³) material at 10 ft/min (0.05 m/s) speed | | | |
| | | Depth of load, in (mm) | | | |
| in | mm | 12(305) | 16(406) | 20(508) | 24(610) |
|---|---|---|---|---|---|
| 24 | 610 | 22(559) | 30(762) | | |
| 30 | 762 | 26(660) | 37(940) | 47(1,194) | 56(1,422) |
| 36 | 914 | 34(864) | 45(1,143) | 56(1,422) | 67(1,702) |
| 42 | 1,067 | 39(991) | 52(1,321) | 65(1,651) | 79(2,007) |

**Table 7. Capacities and Weights of V-Bucket Carriers***

| Buckets | | | | Capacity, tons of coal per hour at 100 ft/ min | Weight per ft of chains and buckets, lb | Buckets | | | | Capacity, tons of coal per hour at 100 ft/ min | Weight per ft of chains and buckets, lb |
| Length, in | Width, in | Depth, in | Spacing, in | | | Length, in | Width, in | Depth, in | Spacing, in | | |
|---|---|---|---|---|---|---|---|---|---|---|---|
| 12 | 12 | 6 | 18 | 29 | 36 | 30 | 20 | 10 | 24 | 126 | 70 |
| 16 | 12 | 6 | 18 | 32 | 40 | 36 | 24 | 12 | 30 | 172 | 94 |
| 20 | 15 | 8 | 24 | 43 | 55 | 42 | 24 | 12 | 30 | 200 | 105 |
| 24 | 20 | 10 | 24 | 100 | 65 | 48 | 24 | 12 | 36 | 192 | 150 |

*Multiply in by 25.4 for mm, ton/h by 0.25 for kg/s or by 0.91 for tonnes/h.

malleable iron. The drive (Fig. 30) is located at *a* or *a'*, the take-up at *b*. The material is fed to the buckets by a feeder at any point along the lower horizontal run, is elevated, and is discharged on the upper horizontal run. The tripper *c*, mounted on wheels so that it can be moved to the desired

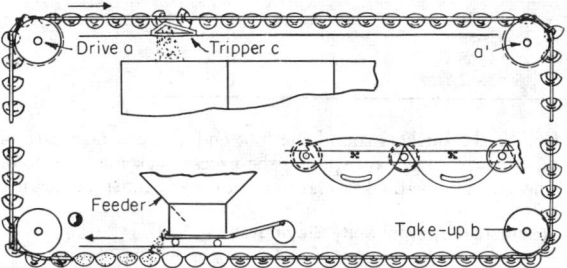

**Fig. 30**   Pivoted-bucket carrier.

dumping position, engages the cams on the buckets and tips them until the material runs out. The buckets always remain vertical except when tripped. The chain rollers run on T rails on the horizontal sections and between guides on the vertical runs. Speeds range from 30 to 60 ft/min (0.15 to 0.30 m/s).

After dumping, the overlapping bucket lips are in the wrong position to round the far corner; after rounding the take-up wheels, the lap is wrong for making the upturn. The Link Belt **Peck** carrier eliminates this by suspending the buckets from trunnions attached to rearward cantilever extensions of the inner links (Fig. 31). As the chain rounds the turns, the buckets swing in a larger-radius curve, automatically unlatch, and then lap correctly as they enter the straight run.

The pivoted-bucket carrier requires little attention beyond periodic lubrication and adjustment of take-ups. For the dual service of coal and ash handling, its only competitor is the skip hoist.

Table 8 shows the capacities of pivoted-bucket carriers with materials weighing 50 lb/ft³ (1.42 m³), with carriers operating at 40 to 50 ft/min (0.20 to 0.25 m/s), and with buckets loaded to 80 percent capacity. Figure 32 gives the motor size for specified horizontal and vertical centers with material weighing 50 lb/ft³ (1.42 m³) and with speed 40 ft/min (0.20 m/s).

EXAMPLE.   For a carrier with 240-ft (73-m) horizontal and 50-ft (15-m) vertical centers to handle 65 tons (59 tonnes) of coal per h, select from Table 8 a 24- by 30-in (610- by 762-mm) carrier at 40 ft/min (0.20 m/s). By Fig. 32, the lines for the horizontal and vertical centers intersect between 15 and 20 hp (11 and 15 kW). Specify a 20-hp (15-kW) motor.

**Bucket elevators** are of two types: (1) chain-and-bucket, where the buckets are attached to one or two chains; and (2) belt-and-bucket, where the buckets are attached to canvas or rubber belts. Either type may be vertical or inclined and may have continuous or noncontinuous buckets. Bucket elevators are used to elevate any bulk material that will not adhere to the bucket. Belt-and-bucket elevators are particularly well adapted to handling abrasive materials which would produce excessive wear on chains. Chain-and-bucket elevators are frequently used with perforated buckets when handling wet

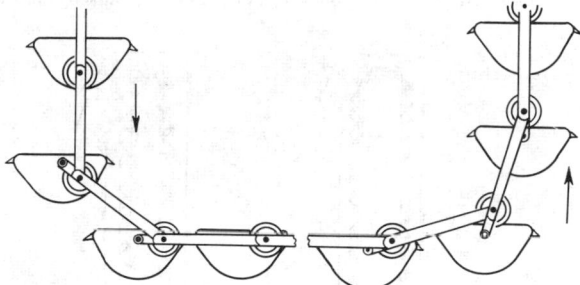

**Fig. 31**   Link-Belt Peck carrier buckets.

material, to drain off surplus water. The length of elevators is limited by the strength of the chains or belts. They may be built up to 100 ft (30 m) long, but they average 25 to 75 ft (7.6 to 23 m). Inclined-belt elevators operate best on an angle of about 30° to the vertical. At greater angles, the sag of the return belt is excessive, as it cannot be supported by rollers between the head and foot pulleys. This applies also to single-strand chain elevators. Double-strand chain elevators, however, if provided with roller chain, can run on an angle, as both the upper and return chains are supported by rails. The

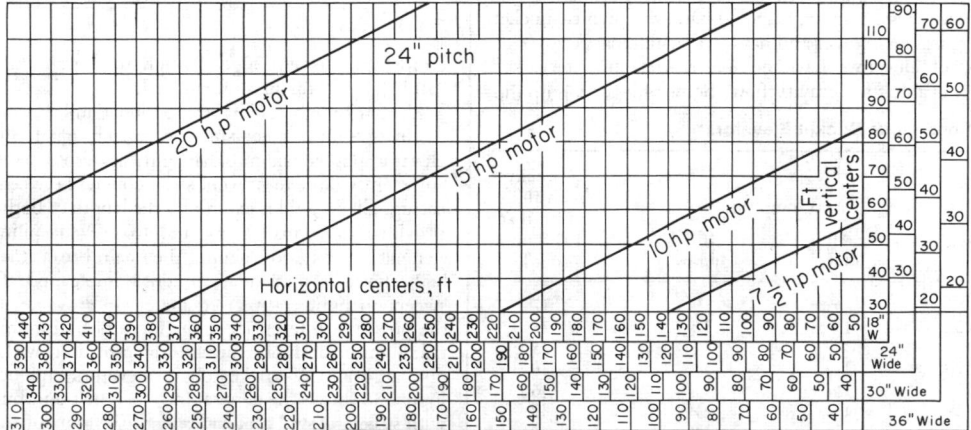

**Fig. 32**   Motor determinations for pivoted-bucket carriers. For metric values multiply feet by 0.30 for m, inches by 25.4 for mm, hp by 0.75 for kW. *(FMC Corp./MHS Div.)*

**Table 8. Capacities of Pivoted-Bucket Carriers with Coal or Similar Materials Weighing 50 lb/ft³ (800 kg/m³) at Speeds Noted**

| Bucket pitch × width | | Capacity of coal | | Speed | |
|---|---|---|---|---|---|
| in | mm | Short ton/h | tonne/h | ft/min | m/s |
| 24 × 18 | 610 × 457 | 35–45 | 32–41 | 40–50 | 0.20–0.25 |
| 24 × 24 | 610 × 610 | 50–60 | 45–54 | 40–50 | 0.20–0.25 |
| 24 × 30 | 610 × 762 | 60–75 | 54–68 | 40–50 | 0.20–0.25 |
| 24 × 36 | 610 × 914 | 70–90 | 63–82 | 40–50 | 0.20–0.25 |

size of lumps is limited by the size and spacing of the buckets and by the speed of the elevator.

**Continuous-bucket elevators** (Fig. 33 and Table 9) usually operate at 100 ft/min (0.51 m/s) or less and are single- or double-strand. The contents of each bucket discharge over the back of the preceding bucket. For maximum capacity and a

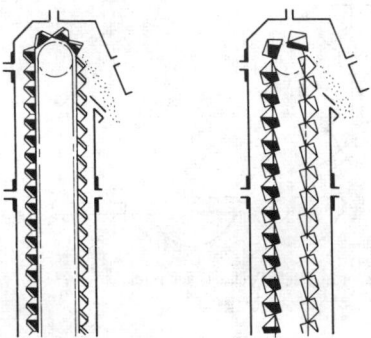

**Fig. 33** Continuous bucket.    **Fig. 34** Supercapacity.

large proportion of lumps, the buckets extend rearward behind the chain runs. The elevator is then called a supercapacity elevator (Fig. 34 and Table 10).

**Gravity-discharge elevators** operate at 100 ft/min (0.51 m/s) or less and are double-strand, with spaced V buckets. The path may be an L, an inverted L, or a runaround in a vertical plane (Fig. 27). Along the horizontal run, the buckets function as pushers within a trough. An elevator with a tandem flight conveyor costs less. For a runaround path, the pivoted-bucket carrier requires less power and has lower maintenance costs.

As bucket elevators have no **feed control,** an interlocked feeder is desirable for a gravity flow. Some types scoop up the

**Table 9. Continuous Bucket Elevators***

| Bucket size in | Max lump size, in | | Capacity with 50 lb material at 100 ft/min tons/h |
|---|---|---|---|
| | All lumps | 10% lumps | |
| 10 × 5 × 8 | ¾ | 2½ | 17 |
| 10 × 7 × 12 | 1 | 3 | 21 |
| 12 × 7 × 12 | 1 | 3 | 25 |
| 14 × 7 × 12 | 1 | 3 | 30 |
| 14 × 8 × 12 | 1¼ | 4 | 36 |
| 16 × 8 × 12 | 1½ | 4½ | 42 |
| 18 × 8 × 12 | 1½ | 4½ | 46 |

*Multiply in by 25.4 for mm, lb by 0.45 for kg, tons by 0.91 for tonne.

load as the buckets round the foot end and can take care of momentary surges by spilling the excess back into the boot. The continuous-bucket elevator, however, must be loaded

**Table 10. Supercapacity Elevators***
(Link Belt Co.)

| Bucket, in length × width × depth | Max lump size, large lumps not more than 20%, in | Capacity with 50 lb material tons/h |
|---|---|---|
| 16 × 12 × 18 | 8 | 115 |
| 20 × 12 × 18 | 8 | 145 |
| 24 × 12 × 18 | 8 | 175 |
| 30 × 12 × 18 | 8 | 215 |
| 24 × 17 × 24 | 10 | 230 |
| 36 × 17 × 24 | 10 | 345 |

*Multiply in by 25.4 for mm, lb by 0.45 for kg, tons by 0.91 for tonne.

after the buckets line up for the lift, i.e., when the gaps between buckets have closed.

**Belt-and-bucket elevators** are advantageous for grain, cereals, glass batch, clay, coke breeze, sand, and other abrasives if the

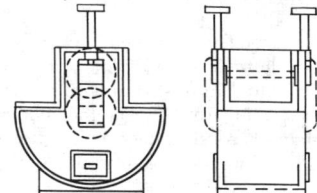

**Fig. 35** Cast-iron boot.

temperature is not high enough to scorch the belt [below 250°F (121°C) for natural rubber].

**Elevator casings** usually are sectional and dusttight, either of ³⁄₁₆-in (4.8-mm) sheet steel or, better, of aluminum. If the elevator has considerable height, its cross section must be sufficiently large to prevent sway contact between buckets and casing. Chain guides extending the length of both runs may be provided to control sway and to prevent piling up of the element, at the boot, should the chain break. **Caution:** Indoor high elevators may develop considerable updraft tending to sweep up light, pulverized material. Provision to neutralize the pressure differential at the top and bottom may be essential. Figure 35 shows the **cast-iron boot** used with centrifugal-discharge and V-bucket chain elevators and belt elevators. Figure 36 shows the general form of a belt-and-bucket elevator with **structural-steel boot and casing.** Elevators of this type must be run at sufficient speed to throw the discharging material clear of the bucket ahead.

**Capacity** Elevators are rated for capacity with the buckets 75 percent loaded. The buckets must be large enough to accommodate the lumps, even though the capacity is small. The bucket spacing and sizes given in Table 11 are standard with most manufacturers.

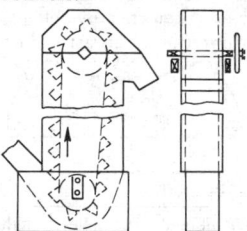

**Fig. 36** Structural-steel boot and casing.

**Power Requirements** The motor horsepower for the continuous-bucket and supercapacity elevators can be approximated as

$$\text{Motor hp} = (2 \times \text{tons/h} \times \text{lift, ft})/1{,}000$$

The motor horsepower of gravity-discharge elevators can be approximated by using the same formula for the lift and adding for the horsepower of the horizontal run the power as

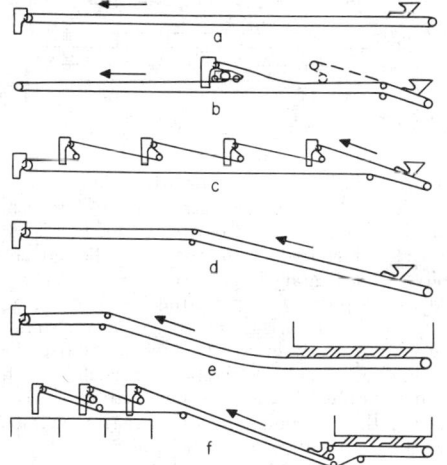

**Fig. 37** Typical arrangements of belt conveyors.

estimated for a flight conveyor. For a vertical runaround path, add a similar allowance for the lower horizontal run.

**Belt Conveyors**

The belt conveyor is a heavy-duty conveyor available for transporting large tonnages over paths beyond the range of any other type of mechanical conveyor. The capacity may be several thousand tons per hour, and the distance several miles. It may be horizontal or inclined upward or downward, or it may be a combination of these, as outlined in Fig. 37. The limit of incline is reached when the material tends to slip on the belt surface. There are special belts with molded designs to assist in keeping material from slipping on inclines. They will handle pulverized, granular, or lumpy material. Special compounds are available if material is hot or oily.

In its simplest form, the conveyor consists of a head or drive pulley, a take-up pulley, an endless belt, and carrying and return idlers. The spacing of the carrying idlers varies with the width and loading of the belt and usually is 5 ft (1.5 m) or less. Return idlers are spaced on 10-ft (3.0-m) centers or slightly less with wide belts. Sealed antifriction idler bearings are used almost exclusively, with pressure-lubrication fittings requiring attention about once a year.

**Belts** Belt width is governed by the desired conveyor capacity and maximum lump size. The standard rubber belt construction (Fig. 38) has several plies of square woven cotton duck or synthetic fabric such as rayon, nylon, or polyester

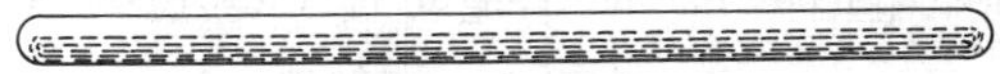

**Fig. 38** Rubber-covered conveyor belt.

cemented together with a rubber compound called the friction and covered both top and bottom with rubber to resist abrasion and keep out moisture. Top cover thickness is determined by the severity of the job and varies from $\frac{1}{16}$ to $\frac{3}{4}$ in (1.6 to 19 mm). The bottom cover is usually $\frac{1}{16}$ in (1.6 mm). By placing a layer of loosely woven fabric, called the breaker strip, between the cover and outside fabric ply, it is often possible to double the adhesion of the cover to the carcass. The belt is rated according to the tension to which it may safely be subjected, and this is a function of the length and lift of the conveyor. The standard Rubber Manufacturers Association (RMA) multiple ply ratings in lb/in (kg/mm) of width per ply are as follows (see also Secs. 6 and 8):

| RMA multiple ply No. | MP35 | MP43 | MP50 | MP60 | MP70 |
|---|---|---|---|---|---|
| Permissible pull, lb/in (kg/mm) of belt width per ply, vulcanized splice | 35 (0.63) | 43 (0.77) | 50 (0.89) | 60 (1.07) | 70 (1.25) |
| Permissible pull, lb/in (kg/mm) of belt width per ply, mechanical splice | 27 (0.48) | 33 (0.60) | 40 (0.71) | 45 (0.80) | 55 (0.98) |

Thus, for a pull of 4,200 lb (1,905 kg) and 24-in (0.61-m) belt

**Table 11. Gravity-Discharge Elevators**
[Capacity with bituminous coal, at 100 ft/min (0.51 m/s)]

| Bucket spacing, in (mm) | Bucket length × width, in (mm) | | | | |
|---|---|---|---|---|---|
| | 20 × 15 (508 × 381) | 24 × 15 (610 × 381) | 20 × 20 (508 × 508) | 24 × 20 (610 × 508) | 30 × 20 (762 × 508) |
| 18(457) | 58(1,473) | 70(1,778) | 104(2,642) | 125(3,175) | 159(4,039) |
| 24(610) | 44(1,118) | 52(1,321) | | | |
| 36(914) | 29(737) | 35(889) | 52(1,321) | 63(1,600) | 79(2,007) |

width, a five-ply MP35 could be used with a vulcanized splice or a five-ply MP50 could be used with a mechanical splice.

**High-Strength Belts** For belt conveyors of extremely great length, a greater strength per inch of belt width is available now through the use of improved wearing techniques that provide straight-warp synthetic fabric to support the tensile forces (Uniflex by Uniroyal, Inc., or Flexseal by B. F. Goodrich). Strengths go to 1,500 lb/in width tension rating. They are available in most cover and bottom combinations and have good bonding to carcass. The number of plies is reduced to two instead of as many as eight so as to give excellent flexibility. Widths to 60 in are available. Other conventional high-strength fabric belts are available to somewhat lower tensile ratings of 90(1.61), 120(2.14), 155(2.77), 195(3.48), and 240(4.29) lb/in (kg/mm) per ply ratings.

The B. F. Goodrich Company has developed a steel-cable-reinforced belt rated 700 to 4,400 lb/in (12.5 to 78.6 kg/mm) of belt width. The belt has parallel brass-plated 7 by 19 steel airplane cables ranging from $\frac{5}{32}$ to $\frac{3}{8}$ in (4.0 to 9.5 mm) diameter placed on $\frac{1}{2}$- to $\frac{3}{4}$-in (12.7- to 19.0-mm) centers.

These belts are used for long single-length conveyors and for high-lift, extremely heavy duty service, e.g., for taking ore from deep open pits, thus providing an alternative to a spiraling railway or truck route.

**Synthetic rubber** is in use for belts. Combinations of synthetic and natural rubbers have been found satisfactory. Synthetics are superior under special circumstances, e.g., neoprene for flame resistance and resistance to petroleum-based oils, Buna N for resistance to vegetable, animal, and petroleum oils, and butyl for resistance to heat (per RMA).

**Belt Life** With lumpy material, the impact at the loading point may be destructive. Heavy lumps, such as ore and rock, cut through the protective cover and expose the carcass. The impact shock is reduced by making the belt supports flexible. This can be done by the use of idlers with cushion or pneumatic tires (Fig. 39) or by supporting the idlers on rubber mountings. Chuting the load vertically against the belt should

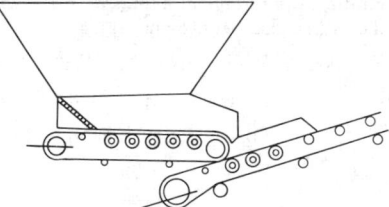

**Fig. 39** Pneumatic-tired idlers applied to belt feeder at loading point of belt conveyor.

be avoided. Where possible, the load should be given a movement in the direction of belt travel. When the material is a mixture of lumps and fines, the fines should be screened through to form a cushion for the lumps. Other destructive factors are belt overstressing, belts running out of line and rubbing against supports, broken idlers, and failure to clean the belt surface thoroughly before it comes in contact with the snub and tripper pulleys. Introduction of a 180° twist in the return belt (B. F. Goodrich Co.) at both head and tail ends can be used to keep the clean side of the belt against the return idlers and to prevent buildup. Using one 180° twist causes both sides of the belt to wear evenly. For each twist, 1 ft of length/in of belt width is required.

**Idler Pulleys** Troughing idlers are usually of the three-

pulley type (Fig. 40), with the troughing pulleys at 20°. There is a growing tendency toward the use of 35 and 45° idlers to increase the volume capacity of a belt; 35° idlers will increase the volume capacity of a given belt 25 to 35 percent over 20° idlers, and 45° idlers, 35 to 40 percent. The bearings, either roller or ball type, are protected by felt or labyrinth grease seals against the infiltration of abrasive dust. A belt running

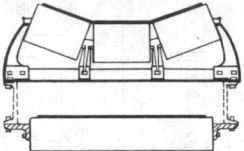

**Fig. 40** Standard assembly of three-pulley troughing idler and return idler.

out of line may be brought into alignment by shifting slightly forward one end or the other with a few idler sets. **Self-aligning idlers** (Fig. 41) should be spaced not more than 75 ft (23 m) apart. These call attention to the necessity of lining up the belt and should not serve as continuing correctives.

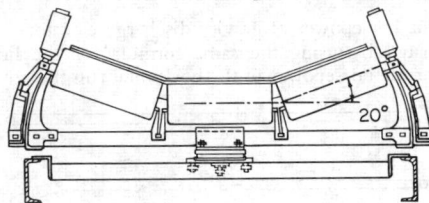

**Fig. 41** Self-aligning idler.

**Drive** Belt slip on the conveyor-drive pulley is destructive. There is little difference in tendency to slip between a bare pulley and a rubber-lagged pulley when the belt is clean and dry. A wet belt will adhere to a lagged pulley much better, especially if the lagging is grooved. Heavy-duty conveyors exposed to the possibility of wetting the belt are generally driven by a head pulley lagged with a ½-in (12.7-mm) rubber belt and with ¼- by ¼-in (6.4- by 6.4-mm) grooves spaced ½ in (12.7 mm) apart and, preferably, diagonally as a herringbone gear. A snub pulley can be employed to increase the arc of contact on the head pulley, and since the pulley is in contact with the dirty side of the belt, a belt cleaner is essential. The belt cleaner may be a high-speed bristle brush, a spiral rubber wiper (resembling an elongated worm pinion), circular disks mounted slantwise on a shaft to give a wiping effect when rotated, or a scraper. Damp deposits such as clay or semi-frozen coal dirt are best removed by multiple diagonal scrapers of stainless steel.

A belt conveyor should be emptied after each run to avoid a heavy starting strain. The motor should have high starting torque, moderate starting-current inrush, and good characteristics when operating under full load. The double-squirrel-cage ac motor fulfills these requirements.

**Heady-Duty Belt-Conveyor Drives** For extremely heavy duty, it is essential that the drive torque be built up slowly, or serious belt damage will occur. The hydraulic clutch, derived from the hydraulic automobile clutch, serves nicely. The best drive developed to date is the **dynamatic clutch** (Fig. 42). This has a magnetized rotor on the extended motor shaft, revolving within an iron ring keyed to the reduction gearing of the conveyor. The energizing current is automatically built up

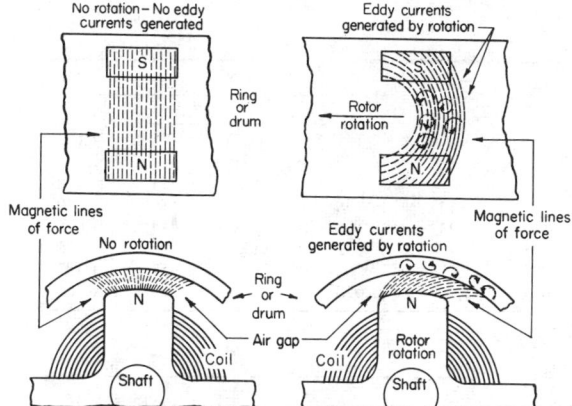

**Fig. 42** Operating principle of the dynamatic clutch.

over a period that may extend to 2 min, and the increasing magnetic pull on the ring builds up the belt speed.

**Take-ups** For short conveyors, a screw take-up is satisfactory. For long conveyors, the expansion and contraction of the belt with temperature changes and the necessity of occasional cutting and resplicing make a weighted gravity take-up preferable, especially if a vulcanized splice is used. The take-up should, if possible, be located where the slack first occurs, usually just back of the drive except in a conveyor inclined downward (retarding conveyor), when the take-up is located at the downhill end.

**Trippers** The load may be removed from the belt by a diagonal or V plow, but a tripper that snubs the belt backward is standard equipment. Trippers may be (1) stationary, (2) manually propelled by crank, or (3) propelled by power from one of the snubbing pulleys (Fig. 43) or by an independent motor. The discharge may be chuted to either side or back to the belt by a deflector plate. When the tripper must move back to the load-receiving end of the conveyor, it is usual to incline the belt for about 15 ft (4.6 m) to match the slope up to the tripper top pulley. As the lower tripper snub pulleys are in contact with the dirty side of the belt, a cleaner must be provided between the pulleys. A scraper in light contact with the face of the pulley may be advisable.

**Belt Slope** The slopes (in degrees) given in Table 12 are the maximum permissible angles for various materials.

**Determination of Motor Horsepower** The power required to drive a belt conveyor is the sum of the powers required (1) to move the empty belt, (2) to move the load horizontally, and (3) to lift the load if the conveyor is inclined upward. If (3) is

**Table 12. Maximum Belt Slopes for Various Materials, Degrees\***

| | |
|---|---:|
| Coal: anthracite, sized; mined, 50 mesh and under; or mined and sized | 16 |
| Coal, bituminous, mined, run of mine | 18 |
| Coal: bituminous, stripping, not cleaned; or lignite | 22 |
| Earth, as excavated, dry | 20 |
| Earth, wet, containing clay | 23 |
| Gravel, bank run | 20 |
| Gravel, dry, sharp | 15–17 |
| Gravel, pebbles | 12 |
| Grain, distillery, spent, dry | 15 |
| Sand, bank, damp | 20–22 |
| Sand, bank, dry | 16–18 |
| Wood Chips | 27 |

\*Uniroyal, Inc.

larger than the other two, an automatic brake must be provided to hold the conveyor if the current fails. A solenoid brake is usual. The power required to move the empty belt is given by Fig. 44a. The power to move 100 tons/h horizontally is given by the formula hp $= 0.4 + 0.00325L$, where $L$ is the

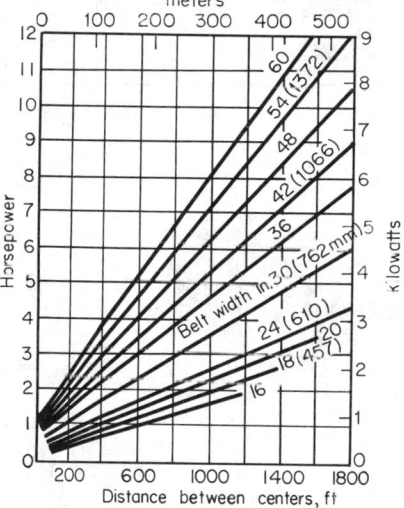

**Fig. 44a** Horsepower required to move belt conveyor empty at 100 ft/min (0.51 m/s).

distance between centers, ft. For other capacities the horsepower is proportional.

The capacity in tons per hour for materials of various

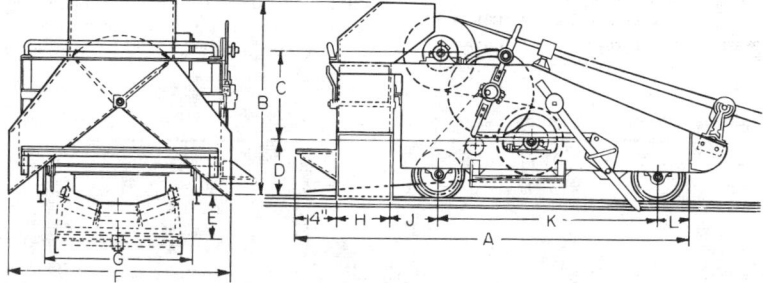

**Fig. 43** Self-propelled tripper with automatic reverse and auxiliary hand control.

**Table 13. Troughed Conveyor-Belt Capacities\***

Tons per hour (TPH) with belt speed of 100 ft/min (0.51 m/s)

| Belt width, in | Belt shape | Equal length, 3 roll | | | | | | | | | |
|---|---|---|---|---|---|---|---|---|---|---|---|
| | Idler angle | 20° | | | 35° | | | 45° | | | |
| | SCA† | 0° | 10° | 30° | 0° | 10° | 30° | 0° | 10° | 30° | CED‡ |
| 12 | | 10 | 14 | 24 | 16 | 20 | 28 | 19 | 29 | 35 | .770 |
| 24 | | 52 | 74 | 120 | 83 | 102 | 143 | 98 | 115 | 150 | 1.050 |
| 30 | | 86 | 121 | 195 | 137 | 167 | 232 | 161 | 188 | 244 | 1.095 |
| 42 | | 177 | 249 | 400 | 282 | 345 | 476 | 332 | 386 | 500 | 1.130 |
| 60 | | 375 | 526 | 843 | 598 | 729 | 1,003 | 702 | 815 | 1,053 | 1.187 |
| 72 | | 548 | 768 | 1,232 | 874 | 1,064 | 1,464 | 1,026 | 1,190 | 1,535 | 1.205 |

| Belt width, in | Belt shape | Long center, 3 roll | | | | | | |
|---|---|---|---|---|---|---|---|---|
| | Idler angle | 35° | | | 45° | | | |
| | SCA† | 0° | 10° | 30° | 0° | 10° | 30° | CED‡ |
| 12 | | | | | | | | |
| 24 | | 82 | 101 | 141 | 96 | 113 | 149 | 1.05 |
| 30 | | 111 | 144 | 212 | 133 | 163 | 225 | 1.12 |
| 42 | | 179 | 248 | 394 | 216 | 281 | 417 | 1.22 |
| 60 | | 266 | 417 | 734 | 324 | 468 | 772 | 1.35 |
| 72 | | 291 | 516 | 987 | 356 | 573 | 1,030 | 1.44 |

\*Tons per hour (TPH) = value from table $\times \dfrac{\text{(actual material wt., lbs/ft}^3\text{)}}{100} \times \dfrac{\text{(actual belt speed, ft/min)}}{100}$

†Surcharge angle (see Fig. 44b).

‡Capacity calculated for standard distance of load from belt edge: $(0.55\ b + 0.9)$, where $b$ = belt width, inches. For constant 2-in edge distance (CED) multiply by CED constant as given in this table.

For slumping materials (very free flowing), use capacities based upon 2-in CED. This includes dry silica sand, dry aerated portland cement, wet concrete, etc., with surcharge angle 5° or less. For metric units multiply in by 25.4 for mm; tons per hour by 0.91 for tonne per hour; ft/min by 0.0051 for m/s.

weights per cubic foot is given by Table 13. Table 14 gives minimum belt widths for lumps of various sizes. Table 15 gives advisable maximum belt speeds for various belt widths

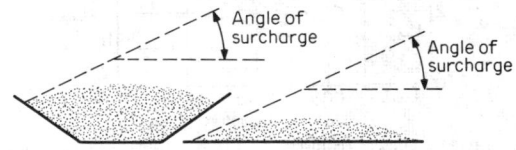

**Fig. 44b**   (*a*) Troughed belt; (*b*) flat belt. (*Uniroyal, Inc.*)

**Drive Calculations**   From the standpoint of the application of power to the belt, a conveyor is identical with a power belt (see Sec. 3). The determining factors are the coefficient of friction between the drive pulley and the belt, the tension in the belt, and the arc of contact between the pulley and the belt. The arc of contact is increased up to about 240° by using a snub pulley and up to 410° by using two pulleys geared together or driven by separate motors and having the belt wrapped around them in the form of a letter S. The resistance to be overcome is the sum of all the frictional resistances throughout the length of the conveyor plus, in the case of a rising conveyor, the resistance due to lifting the load. The sum of the conveyor and load resistances determines the working pull that has to be transmitted to the belt at the drive pulley. The total pull is increased by the slack-side tension necessary to keep the belt from slipping on the pulley. (See Section 3, for the ratio of the total pull to the slack-side tension.) Other factors adding to the belt pull are the component of the weight of the belt if the conveyor is inclined and a take-up pull to keep the belt from sagging between the idlers at the loading point. These, however, do not add to the working pull. The maximum belt pull determines the length of conveyor that can be used. If part of the conveyor runs downgrade, the load on it will reduce the working pull. In moderate-length conveyors,

**Table 14. Minimum Belt Width for Lumps\***

| Belt width, | in | 12 | 18 | 24 | 30 | 42 | 60 | 72 |
|---|---|---|---|---|---|---|---|---|
| | mm | 305 | 457 | 610 | 762 | 1,067 | 1,524 | 1,829 |
| Sized material, | in | 2 | 4 | 5 | 6 | 8 | 12 | 14 |
| | mm | 51 | 102 | 127 | 152 | 203 | 305 | 356 |
| Unsized material, | in | 4 | 6 | 8 | 10 | 14 | 20 | 24 |
| | mm | 102 | 152 | 203 | 254 | 356 | 508 | 610 |

\*Uniroyal, Inc.

Table 15. Maximum Belt Speeds, ft/min, for Various Materials*

| Width of belt, in | Light or free-flowing materials, grains, dry sand, etc. | Moderately free-flowing sand, gravel, fine stone, etc. | Lump coal, coarse stone, crushed ore | Heavy sharp lumpy materials, heavy ores, lump coke |
|---|---|---|---|---|
| 12–14 | 400 | 250 | | |
| 16–18 | 500 | 300 | 250 | |
| 20–24 | 600 | 400 | 350 | 250 |
| 30–36 | 750 | 500 | 400 | 300 |
| 42–60 | 850 | 550 | 450 | 350 |

*Multiply in by 25.4 for mm, ft/min by 0.005 for m/s.

stresses due to acceleration or deceleration are safely carried by the factor of safety used for belt-life calculations.

The total or maximum tension $T_{max}$ must be known to specify a suitable belt. The effective tension $T_e$ is the difference between tight-side tension and slack-side tension, or $T_e = T_1 - T_2$. The coefficient of friction between rubber and steel is 0.25; with a lagged pulley, between rubber and rubber, it is 0.55 for ideal conditions but should be taken as 0.35 to allow for loss due to dirty conditions.

Except for extremely heavy belt pulls, the tandem drive is seldom used since it is costly; the lagged-and-grooved drive pulley is used for most industrial installations.

leaves the pulley. Light-duty permanent-magnet types [for pulleys 12 to 24 in (305 to 610 mm) in diameter] will separate material through a 2-in (51-mm) layer on the belt. Heavy-duty permanent-magnet units (12 to 24 in in diameter) will separate material if the belt carries over 2 in of material or if the magnetic content is very fine. Even larger units are available for special applications. So effective and powerful are the permanent-magnet types that **electromagnetic pulleys** are available only in the larger sizes, from 18 to 48 in in diameter. The permanent-magnet type requires no slip rings, external power, or upkeep. Table 17 gives approximate performance data.

Table 16. Ratio of $T_1$ to $T_e$ for Various Arcs of Contact with Bare Pulleys and Lagged Pulleys

(Coefficients of friction 0.25 and 0.35, respectively)

| Belt wrap, deg | 180 | 200 | 210 | 215 | 220 | 240 |
|---|---|---|---|---|---|---|
| Bare pulley | 1.85 | 1.72 | 1.67 | 1.64 | 1.62 | 1.54 |
| Lagged pulley | 1.50 | 1.42 | 1.38 | 1.36 | 1.35 | 1.30 |

For a belt with 6,000-lb (26,700-N) max tension running on a bare pulley drive with 180° wrap (Table 16), $T = 1.85T_e = 6,000$ lb; $T_e = 3,200$ lb (14,200 N). Such a belt, 30 in wide, might be a 5-ply MP50, a reduced-ply belt rated at 200 lb/in, or a steel-cable belt with 5/32-in (4-mm) cables spaced on 0.650-in (16.5-mm) centers. The last is the most costly.

In an inclined belt with single pulley drive, the $T_{max}$ is lowest if the drive is at the head end and increases as the drive shifts toward the foot end.

**Allowance for Tripper**   The belt lifts about 5 ft to pass over the top snub pulley of the tripper (Fig. 43). Allowance should be made for this lift in determining the power requirement of the conveyor. If the tripper is propelled by the belt, an allowance of 1 hp (0.75 kW) for a 16-in (406-mm) belt, 3 hp (2.2 kW) for a 36-in (914-mm) belt, or 7 hp (5.2 kW) for a 60-in (1,524-mm) belt is ample. If a rotary cleaning brush is driven from one of the snub shafts, an allowance should be made which is approximately the same as that for the propulsion of the tripper.

**Magnetic pulleys** are frequently used as head pulleys on belt conveyors to remove tramp iron, such as stray nuts or bolts, before crushing; to concentrate magnetic ores, such as magnetite or nickeliferous pyrrhotite, from nonmagnetic material; and to reclaim iron from foundry refuse. A chute or hopper automatically receives the extracted material as it is drawn down through the other nonmagnetic material, drawn around the end pulley on the belt, and finally released as the belt

**Overhead magnetic separators (Dings),** both electromagnetic and Ceramox permanent-magnet types, for suspension above a belt conveyor are also available for all commercial belt widths to pull magnetic material from burden as thick as 40 in and at belt speeds to 750 ft/min. These may or may not be equipped with a separately encompassing belt to be self-cleaning. Wattages vary from 1,600 to 17,000. The permanent-magnet type require no electric power and have nonvarying magnet strength. An alternate type of belt protection is to use a Ferro Guard Detector (Dings) to stop belt motion if iron is detected.

**Trippers** of the fixed or movable type are used for discharging material between the ends of a belt conveyor. Figure 43 shows a **self-propelling tripper** which consists of two pulleys, over which the belt passes, the material being discharged into the chute as the belt bends around the upper pulley. The pulleys are mounted on a frame carried by four wheels and power-driven. A lever on the frame and stops alongside the rails enable the tripper, taking power from the conveyor belt, to move automatically between the stops, thus distributing the material. Rail clamps are provided to hold the tripper in a fixed position when discharging. Motor-driven trippers are used when it is desirable to move the tripper independently of the conveyor belt. Fixed trippers have their pulleys mounted on the conveyor framework instead of on a movable carriage.

**Shuttle conveyors** are frequently used in place of trippers for distributing materials. They consist of a reversible belt con-

Table 17. Data on Magnetic Pulleys (Dings)*

| Diameter, in | Approx shipping weight per ft width, lb | | | Light-duty permanent magnet—2-in burden depth | | Heavy-duty permanent or electromagnet, over 2-in burden depth | | Electro-magnet, approx watts/ft belt width |
|---|---|---|---|---|---|---|---|---|
| | Permanent magnet | | Electro-magnet | Belt speed, ft/min | Capacity, tons/h /ft belt width† | Belt speed, ft/min | Capacity, tons/hr /ft belt width† | |
| | Light | Heavy | | | | | | |
| 12 | 220 | 240 | | 135 | 23 | 175 | 28 | |
| 15 | 200 | 270 | | 165 | 28 | 200 | 33 | |
| 18 | 260 | 350 | 535 | 190 | 35 | 225 | 40 | 600 |
| 24 | 550 | 750 | 820 | 250 | 52 | 285 | 75 | 770 |
| 30 | | | 1,185 | | | 315 | 93 | 1,020 |
| 36 | | | 1,620 | | | 375 | 95 | 1,200 |
| 42 | | | 2,320 | | | 385 | 100 | 1,360 |
| 48 | | | 3,200 | | | 440 | 110 | 1,620 |

*Multiply in by 25.4 for mm, lb/ft by 1.5 for kg/m, ft/min by 0.0051 for m/s, ton/h/ft by 0.83 for kg/s/m, watts/ft by 3.3 for watts/m.

†Figures given are for level conveyors. Improvement factors vary from 4.5 percent at 5° to 18 percent at 20° of elevation of belt. Capacities shown are for coal at 50 lb/ft³ when computing volume conveyed. Volume of any other material will be reduced in proportion of its unit weight to that of coal.

veyor mounted upon a movable frame and discharging over either end.

**Belt-Conveyor Arrangements** Figure 37 shows **typical arrangements of belt conveyors.** *a* is a level conveyor receiving material at one end and discharging at the other. *b* shows a level conveyor with traveling tripper. The receiving end of the conveyor is depressed so that the belt will not be lifted against the chute when the tripper is at its extreme loading end. *c* is a level conveyor with fixed trippers. *d* shows an inclined end combined with a level section. *e* is a combination of level conveyor, vertical curve, and horizontal section. The radius of the vertical curve depends upon the weight of the belt and the tension in the belt at the point of tangency. This must be figured in each case and is found by the formula: min radius, ft = belt tension at lowest point of curve divided by weight per ft of belt. The belt weight should be for the worn belt with not over $\frac{1}{16}$-in (1.6-mm) top cover. *f* is a combination of level conveyor receiving material from a bin, a fixed dump, and inclined section, and a series of fixed trippers.

**Portable conveyors** are widely used around retail coal yards, small power plants, and at coal mines for storing coal and reclaiming it for loading into trucks or cars. They are also used for handling other bulk materials. They consist of a short section of chain or belt conveyors mounted on large wheels or crawler treads and powered with a gasoline engine or electric motor. They vary in length from 20 to 90 ft and can handle up to 250 tons/h (63 kg/s) of coal. For capacities greater than what two men can shovel onto a belt, some form of power loader is necessary.

**Sectional-belt conveyors** have come into wide use in coal mines for bringing the coal from the face and loading it into cars in the entry. They consist of short sections (6 ft or more) of light frame of special low-type construction. The sections are designed for ease of connecting and disconnecting for transfer from one part of the mine to another. They are built in lengths up to 1,000 ft (305 m) or more under favorable conditions and can handle 125 tons/h (32 kg/s) of coal.

**Sliding-belt conveyors** use belts sliding on decks instead of troughed belts carried on rollers. Sliding belts are used in the

shipping rooms of department stores for handling miscellaneous parcels, in post offices for handling mail bags, in chemical plants for miscellaneous light waste, etc. The decking preferably is of maple strips. If of steel, the deck should be perforated at intervals to relieve the vacuum effect between the bottom of the belt and the deck. Cotton or balata belts are best. The speed should be low. The return run may be carried on 4-in straight-face idlers. The power requirement is greater than with idler rollers.

The **oscillating conveyor** is a horizontal trough varying in width from 12 to 48 in (305 to 1,219 mm), mounted on rearward-inclined cantilever supports, and driven from an eccentric at 400 to 500 r/min. The effect is to "bounce" the material along at about 50 ft/min (0.25 m/s) with minimum wear on the trough. The conveyor is adapted to abrasive or hot fragmentary materials, such as scrap metals, castings, or metal chips. The trough bottom may be a screen plate to cull fine material, as when cleaning sand from castings, or the trough may have louvers and a ventilating hood to cool the moving material. These oscillating conveyors may have unit lengths up to 100 ft (30 m). Capacities range from a few tons to 100 tons/h (25 kg/s) with high efficiency and low maintenance.

### Feeders

When material is drawn from a hopper or bin to a conveyor, an **automatic feeder** should be used (unless the material is dry and free-running e.g., grain). The satisfactory operation of any conveyor depends on the material being fed to it in an even and continuous stream. The automatic feeder not only ensures a constant and controlled feed, irrespective of the size of material, but saves the expense of a man who would otherwise be required at the feeding point. Figure 45 shows a **reciprocating-plate feeder,** consisting of a plate mounted on four wheels and forming the bottom of the hopper. When the plate is moved forward, it carries the material with it; when moved back, the plate is withdrawn from under the material, allowing it to fall into the chute. The plate is moved by connecting rods from cranks or eccentrics. The capacity of this feeder is determined by the length and number of strokes, width of plate, and location of the adjustable gate. The number of strokes should not exceed 60 to 70 per min. When used under a track hopper, the material remaining on the plate may freeze in winter, as this type of feeder is not self-clearing.

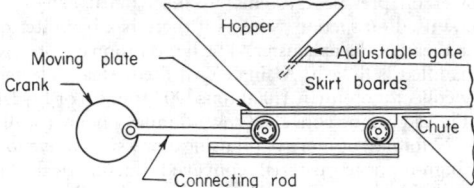

**Fig. 45**  Reciprocating-plate feeder.

**Vibrating Feeders**  The vibrating feeder consists of a plate inclined downward slightly and vibrated (1) by a high-speed unbalanced pulley, (2) by electromagnetic vibrations from one or more solenoids, as in the Jeffrey Manufacturing Co. feeder, or (3) by the slower pulsations secured by mounting the plate on rearward-inclined leaf springs.

The **electric vibrating feeder** (Fig. 46) operates magnetically with a large number of short strokes (7,200 per min from an alternating current in the small sizes and 3,600 from a pulsating direct current in the larger sizes). It is built to feed from a few pounds per minute to 1,250 tons/h (315 kg/s) and will handle any material that does not adhere to the pan. It is self-

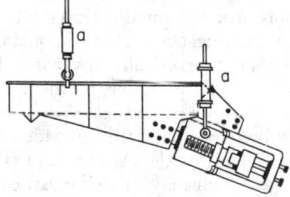

**Fig. 46**  Electric vibrating feeder.

cleaning, instantaneously adjustable for capacity, and controllable from any point near or remote. It is usually supported from above with spring shock absorbers *a* in each hanger, but it be may supported from below with similar springs in the supports. A modified form can be set to feed a weighed **constant amount** hourly for process control.

### Roller Conveyors

The principle involved in **gravity roller conveyors** is the control of motion due to gravity by interposing an antifriction trackage set at a definite grade. Roller conveyors are used in the movement of all sorts of package goods with smooth surfaces which are sufficiently rigid to prevent sagging between rollers—in warehouses, brickyards, building-supply yards, department stores, post offices, and the manufacturing and shipping departments of industrial manufacturers.

The **rollers vary in diameter** and strength from 1 in, with a capacity of 5 lb (2.3 kg) per roller, up to 4 in (102 mm), with a capacity of 1,800 lb (816 kg) per roller. The heavier rollers are generally used in foundries and steel mills for moving large molds, castings, or stacks of sheet steel. The small roller is used for handling small, light objects. The spacing of the rollers in the frames varies with the size and weight of the

objects to be moved. Three rollers should be in contact with the package to prevent hobbling. The grade of fall required to move the object varies from 1½ to 7 percent, depending on the weight and character of the material in contact with the rollers.

Figure 47 shows a typical cross section of a roller conveyor. Curved sections are similar in construction to straight sections, except that in the majority of cases multiple rollers (Fig. 48) are used to keep the package properly lined up and in the center of the conveyor.

**Fig. 47**  Gravity roller conveyor.    **Fig. 48**  Multiple-roller conveyor.

Figure 49 illustrates a **wheel conveyor,** used for handling bundles of shingles, fruit boxes, bundles of fiber cartons, and large, light cases. The wheels are of ball-bearing type, bolted to flat-bar or angle-frame rails.

When an installation involves a **trunk line with several tributary runs,** a simple two-arm deflector at each junction point holds

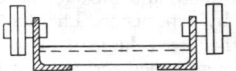

**Fig. 49**  Wheel-type conveyor.

back the item on one run until the item on the other has cleared. **Power-operated** roller conveyors permit handling up an incline. Usually the rolls are driven by sprockets on the spindle ends. An alternative of a smooth deck and pusher flights should be considered as costing less and permitting steeper inclines.

**Platform conveyors** are single- or double-strand conveyors (Fig. 50) with plates of steel or hardwood forming a continu-

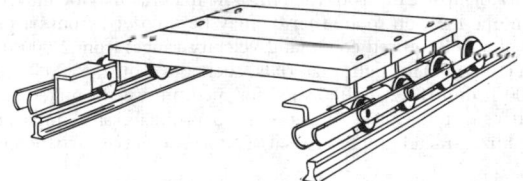

**Fig. 50**  Double-strand platform conveyor.

ous platform on which the loads are placed. They are adapted to handling heavy drums or barrels and miscellaneous freight.

### Pneumatic Conveyors

The pneumatic conveyor transports dry, free-flowing, granular material in suspension within a pipe or duct by means of a high-velocity airstream or by the energy of expanding compressed air within a comparatively dense column of fluidized or aerated material. Principal uses are (1) dust collection; (2) conveying soft materials, such as grain, dry foodstuff (flour and feeds), chemicals (soda ash, lime, salt cake), wood chips, carbon black, and sawdust; (3) conveying hard materials, such as fly ash, cement, silica metallic ores, and phosphate. The need in processing of bulk-transporting plastic pellets, powders, and flour under contamination-free conditions has increased the use of pneumatic conveying.

**Dust Collection**    All pipes should be as straight and short as possible, and bends, if necessary, should have a radius of at least three diameters of the pipe. Pipes should be proportioned to keep down friction losses and yet maintain the air velocities that will prevent settling of the material. Frequent cleanout openings must be provided. Branch connections should go into the sides of the main and deliver the incoming stream as nearly as possible in the direction of flow of the main stream. Sudden changes in diameter should be avoided to prevent eddy losses. (See also Sec. 7).

When **vertical runs** are short in proportion to the horizontal runs, the size of the riser is locally restricted, thereby increasing the air velocity and producing sufficient lifting power to elevate the material. If the vertical pipes are comparatively long, they are not restricted, but the necessary lifting power is secured by increased velocity and suction throughout the entire system.

The area of the main at any point should be 20 to 25 percent in excess of the sums of the branches entering it between the point in question and the dead end of the main. Floor sweepers, if equipped with efficient blast gates, need not be included in computing the main area. The diameter of the connecting pipe from machine to main and the suction required at each hood are determined by experience. The sum of the volumes of each branch gives the total volume to be handled by the fan.

**Fan Suction**    The maintained resistance at the fan is composed of (1) suctions of the various hoods, which must be chosen from experience (see Sec. 12), (2) collector loss, and (3) loss due to pipe friction.

The **pipe loss** for any machine is the sum of the losses in the corresponding branch and in the main from that branch to the fan. For each elbow, add a length equal to 10 diameters of straight pipe. The total loss in the system, or static pressure required at the fan, is equal to the sum of (1), (2), and (3).

**For conveying soft materials,** a fan is used to create a suction. The suspended material is collected at the terminal point by a separator upstream from the fan. The material may be moved from one location to another or may be unloaded from barge or rail car. Required conveying velocity ranges from 2,000 ft/min (10.2 m/s) for light materials, such as sawdust, to 3,000 to 4,000 ft/min (15.2 to 20.3 m/s) for medium-weight materials, such as grain. Since abrasion is no problem, steel pipe or galvanized-metal ducts are satisfactory. Unnecessary bends and fittings should be avoided to minimize power consumption.

**For conveying hard materials,** a water-jet exhauster or steam exhauster is used on suction systems, and a positive-displacement blower on pressure systems. A mechanical exhauster may also be used on suction systems if there is a bag filter or air washer ahead of the exhauster. The largest tonnage of hard material handled is fly ash. A single coal-fired, steam-electric plant may collect more than 1,000 tons (907 tonnes) of fly ash per day. Fly ash can be conveyed several miles pneumatically at 30 tons (27 tonnes) or more per h using a pressure conveyor. Another high-tonnage material conveyed pneumatically is cement. Individual transfer conveyors may handle several hundred tons per hour. Hard materials are usually also heavy and abrasive. Required conveying velocities vary from 4,000 to 5,000 ft/min (20.3 to 25.4 m/s). Heavy cast-iron or alloy pipe and fittings are required to prevent excessive wear.

**Vacuum pneumatic ash-handling systems** have the airflow induced by steam- or water-jet exhausters, or by mechanical blowers. Cyclone-type Nuveyor receivers collect the ash for storage in a dry silo. A typical Nuveyor system is shown in Fig. 51. The conveying velocity is dependent upon material handled in the system. Fly ash is handled at approximately 3,800 ft/min (19.3 m/s) and capacity up to 60 tons/h (15.1 kg/s). Positive-pressure pneumatic ash systems are becoming more common because of higher capacities required. These systems can convey fly ash up to 1½ mi (2.4 km) and capacities of 100 tons/h (25.2 kg/s) for shorter systems.

The **power requirement** for pneumatic conveyors is much greater than for a mechanical conveyor of equal capacity, but the duct can be led along practically any path. There are no moving parts and no risk of injury to the attendant. The vacuum-cleaner action provides dustless operation, sometimes important when pulverized material is unloaded from boxcars through flexible hose and nozzle. A few materials build up a static-electric charge which may introduce an explosion risk. Sulfur is an outstanding example. Sticky materials tend to pack at the elbows and are unsuitable for pneumatic handling.

The performance of a pneumatic conveyor cannot be predicted with the accuracy usual with the various types of mechanical conveyors and elevators. It is necessary to rely on the advice of experienced engineers.

The **Fuller-Kinyon system** for transporting dry pulverized

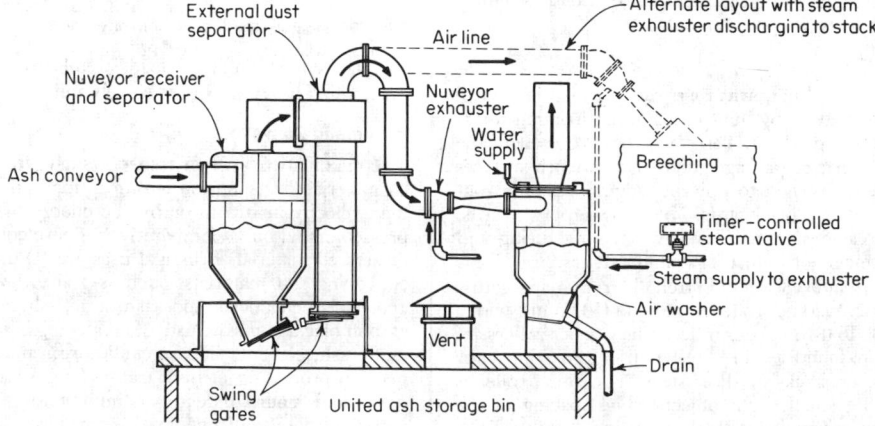

**Fig. 51**  Nuveyor pneumatic ash-handling system. (*United Conveyor Corp.*)

material consists of a motor- or engine-driven pump, a source of compressed air for fluidizing the material, a conduit or pipeline, distributing valves (operated manually, electropneumatically, or by motor), and electric bin-level indicators (high-level, low-level or both). The impeller is a specially designed differential-pitch screw normally turning at 1,200 r/min. The material enters the feed hopper and is compressed in the decreasing pitch of the screw flights. At the discharge end of the screw, the mass is introduced through a check valve to a mixing chamber, where it is aerated by the introduction of compressed air. The fluidized material is conveyed in the transport line by the continuing action of the impeller screw and the energy of expanding air. Practical distance of transportation by the system depends upon the material to be handled. Cement has been handled in this manner for distances up to a mile. The most important field of application is the handling of portland cement. For this material, the Fuller-Kinyon pump is used for such operations as moving both raw material and finished cement within the cement-manufacturing process; loading out; unloading ships, barges, and railway cars; and transferring from storage to mixer plant on large construction jobs.

The **Airslide** (registered trademark of the Fuller Company) **conveyor system** is an air-activated gravity-type conveyor using low-pressure air to aerate or fluidize pulverized material to a degree which will permit it to flow on a slight incline by the force of gravity. The conveyor comprises an enclosed trough, the bottom of which has an inclined air-permeable surface. Beneath this surface is a plenum chamber through which air is introduced at low pressure, depending upon the application. Various control devices for controlling and diverting material flow and for controlling air supply may be provided as part of complete systems. For normal conveying applications, the air is supplied by an appropriate fan; for operation under a head of material (as in a storage bin), the air is supplied by a rotary positive blower. The Airslide conveyor is widely used for horizontal conveying, discharge of storage bins, and special railway-car and truck-trailer transport, as well as in stationary blow-tank-type conveying systems. An important feature of this conveyor is low power requirement.

### Hydraulic Conveyors

Hydraulic conveyors are used for handling boiler-plant ash or slag from an ash hopper or slag tank located under the furnace. The material is flushed from the hopper to a grinder, which discharges to a jet pump or a mechanical pump for conveying to a disposal area or a dewatering bin (Fig. 52). Water requirements average 1 gal/lb of ash.

**Pipeline Transportation of Coal**   (See Sec. 11.)

**Pipeline Transportation of Concrete**   The **Pumpcrete** method of handling concrete by Rex Chain Belt Equipment provides for the pumping of mixed concrete by positive-displacement hydraulic rams in either 6- or 8-in (152- or 203-mm) steel pipes at heights up to 120 ft (37 m) or horizontal distances to 1,000 ft (or equivalent combinations at 1:8 ratio). Delivery rates are 15 to 65 yd³/h (11.5 to 50 m³/h) using 25 to 60 hp (18.7 to 45 kW) at the pump and up to 17 hp (12.7 kW) for the remixer at pump point. To ensure that no concrete is wasted and to simplify cleanout of the pipe, a "go-devil" plug is forced with water pressure (using the Pumpcrete ram) behind the concrete, cleaning the pipe and ejecting the last yard simultaneously. Sometimes air pressure is used if the disposal of the water is especially difficult.

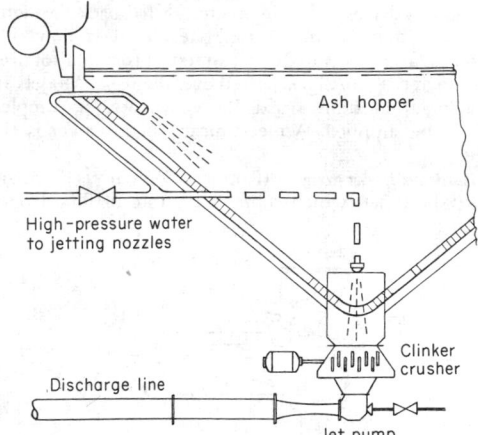

**Fig. 52**   Ash-sluicing system with jet pump. *(United Conveyor.)*

## AUTOMATIC METERING
### by Harold V. Hawkins

**Automatic scales** are often used for **weighing** and **recording** the weight of material being carried by belt conveyors, apron conveyors, open-top carriers, or pivoted-bucket carriers. They may also be adapted to recording the gross weights of loaded cars, such as are used on industrial cable railways or overhead transporters, without stopping the cars on the weighing track. By returning the empty cars, the mechanism can be made to deduct automatically the tare weight from the gross weight, leaving the net weight of the material transported. The output from the scale can be used to control the processing of material carried. The scales may be either continuous or batch weighers. Some of the continuous weighers function through constant volume control, and others through constant weight control, and thus they are automatic feeders as well as weighers.

Figure 53 shows the **Merrick Weightometer** attached to a belt conveyor. One or more of the conveyor idlers are carried on a

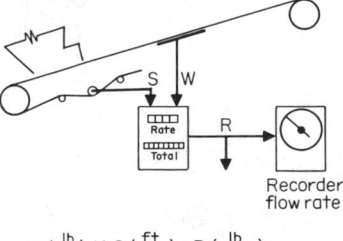

$$W \left( \frac{lb}{ft} \right) \times S \left( \frac{ft}{min} \right) = R \left( \frac{lb}{min} \right)$$

**Fig. 53**   Merrick Weightometer for belt conveyors. *(Merrick Scale Mfg. Co.)*

suspension to sense the weight $W$ of material on the belt. A digital tachometer driven by a bend pulley or the idling terminal provides a signal $S$ proportional to the conveyor-belt speed. The weight $W$ and speed $S$ signals are multiplied electronically to generate a signal proportional to the rate of flow of material on the belt. The rate signal may be monitored on a chart recorder or indicating meter, or fed back through a control system to maintain a desired rate of material flow. The rate signal is also integrated with time to provide a totalizer weight of material which has passed over the scale. Models are available for application to existing conveyors, or complete feeders can be supplied. A mechanical Weightometer is also available.

The **Hardinge feeder-weigher** (Koppers Company) (Fig. 54) is a pivoted short belt which controls the rate of flow from a

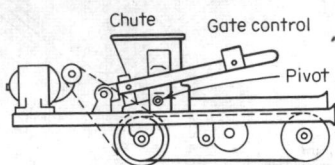

**Fig. 54**   Hardinge feeder-weigher.

hopper by weight per cubic foot. Any desired rate of feed is attained by adjusting the counterweight or by varying the speed. A constant-weight feed is maintained, and a revolution counter records the distance the belt travels and, when calibrated, records the weight. Its error should not exceed 0.5 percent.

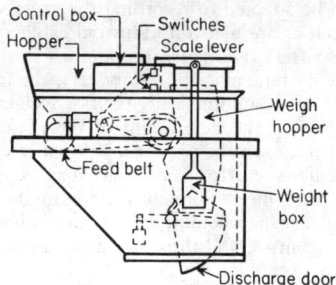

**Fig. 55**   Richardson automatic scale.

The **Richardson automatic scale** (Fig. 55) weighs batches of from 200 to 1,000 lb (91 to 454 kg) and records the weight total on a counter. A counterweighted hopper is served by a belt feeder. When the beam is poised, the feed is stopped and the

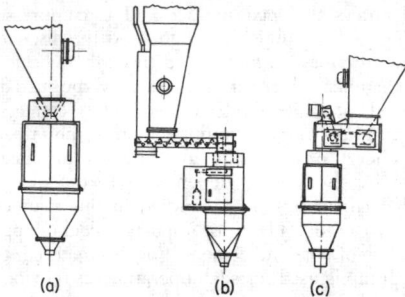

**Fig. 56**   Richardson steady feeders.

hopper empties. Then the cycle repeats. An accuracy within 0.5 percent is usual. Any flowable material without lumps larger than 3 in (76 mm) can be handled.

**Feeders for Automatic Batch Weighers**   To ensure constant accuracy in automatic batch weighing, the scale must be fed in a steady stream. This becomes obvious when automatic batch weighing is viewed as "cutting up" the stream into equal batches by weight. Figure 56 indicates how this is accomplished by Richardson for various kinds of bulk materials: (*a*) Granular free-flowing materials can be fed directly from an overhead bin. For many ground, sluggish materials, an agitator assists the flow. (*b*) Powdered materials which tend to flush from supply bins should be fed by means of a screw feeder long enough to ensure uniform flow. A gate switch on the scale starts and stops the screw feeder. (*c*) Many materials such as large pellets and briquettes should be fed to the scale by a vibrator or belt feeder. The scale inlet cutoff gate will then act as a "catch gate." The scale control system starts and stops the feeding mechanism and controls the "catch-gate" operation automatically.

Section **11**

# Transportation

BY

**HAROLD M. NELSON,** *Managing Editor, Automotive Industries.*
**JOHN F. PARTRIDGE,** *Manager, Industrial Engineering, Consolidated Rail Corp.*
**ANDERS T. ANDERSON,** *Commander, U.S. Navy (CDR), USN), SSBN Project Officer, Submarine Logistics Division, Naval Sea Systems Command.*
**WILLIAM BOLLAY,** *President, Aerophysics Development Corp.*
**J. J. CORNISH, III,** *Director of Engineering, Lockheed-Georgia Company.*
**MAURICE J. ZUCROW,** *Professor Emeritus, Purdue University.*
**BRUCE A. REESE,** *Professor and Head, School of Aeronautics and Astronautics, Purdue University.*
**NEIL A. ARMSTRONG,** *University Professor of Aerospace Engineering, University of Cincinnati.*
**GEORGE H. EWING,** *Vice President—Supplemental Fuels Development, Texas Eastern Transmission Corporation.*

## AUTOMOBILES
### by Harold M. Nelson

| | |
|---|---|
| General | 11-3 |
| Vehicle Dimensions | 11-3 |
| Traction Required | 11-3 |
| Transmission Mechanisms | 11-5 |
| Automatic Transmissions | 11-7 |
| Suspensions | 11-9 |
| Wheel Alignment | 11-10 |
| Steering | 11-11 |
| Brakes | 11-13 |
| Air Conditioning | 11-17 |
| Automobile Engines | 11-18 |

## RAILWAY ENGINEERING
### by John F. Partridge

| | |
|---|---|
| Electric Traction | 11-19 |
| Diesel-Electric Locomotives | 11-21 |
| Diesel-Hydraulic Locomotives | 11-26 |
| Gas-Turbine Locomotives | 11-27 |
| Freight and Passenger Cars | 11-28 |
| Train Resistance | 11-29 |
| Braking | 11-30 |
| Track | 11-31 |
| Car Retarders | 11-33 |
| Monorail Systems | 11-33 |
| Subway Trains | 11-34 |

## MARINE ENGINEERING
### by Anders T. Anderson

| | |
|---|---|
| The Marine Environment | 11-35 |
| Marine Vehicles | 11-35 |
| Seaworthiness | 11-36 |
| Engineering Constraints | 11-42 |
| Propulsion Systems | 11-44 |
| Main Propulsion Plants | 11-44 |
| Propulsors | 11-49 |
| Propulsion Transmission | 11-52 |
| High-performance Ship Systems | 11-55 |

## AIR RESISTANCE OF TRAINS, AUTOMOBILES, AND SHIPS
### by William Bollay

| | |
|---|---|
| Summary | 11-57 |

## AERONAUTICS
### by J. J. Cornish, III

| | |
|---|---|
| Definitions | 11-58 |
| Standard Atmosphere | 11-58 |
| Upper Atmosphere | 11-58 |
| Subsonic Aerodynamic Forces | 11-60 |
| Airfoils | 11-61 |
| Stability | 11-71 |
| Ground-Effect Machines (GEM) | 11-72 |
| Helicopters | 11-73 |
| Supersonic and Hypersonic Aerodynamics | 11-73 |
| Acoustic Theory | 11-79 |

### JET PROPULSION AND AIRCRAFT PROPELLERS
#### by Maurice J. Zucrow
#### in collaboration with Bruce A. Reese

Essential Features of Airbreathing or Thermal-Jet Engines . . . 11-85
Essential Features of Rocket Engines . . . . . . . . . . . . . . . . . . . . . . 11-87
Notation . . . . . . . . . . . . . . . . . . . . . . . . . . . . . . . . . . . . . . . . . . 11-90
Thrust Equations for Jet-propulsion Engines . . . . . . . . . . . . . . 11-92
Power and Efficiency Relationships . . . . . . . . . . . . . . . . . . . . . . 11-92
Performance Characteristics of Airbreathing Jet Engines . . . . . 11-93
Criteria of Rocket-motor Performance . . . . . . . . . . . . . . . . . . . . 11-96
Aircraft Propellers . . . . . . . . . . . . . . . . . . . . . . . . . . . . . . . . . . . 11-98

### ASTRONAUTICS
#### by The Department of Aerospace Engineering and Applied Mechanics, University of Cincinnati

Space Flight (by N. A. Armstrong) . . . . . . . . . . . . . . . . . . . . . . . 11-104
Astronomical Constants of the Solar System (by Helmut G. L.
    Krause) . . . . . . . . . . . . . . . . . . . . . . . . . . . . . . . . . . . . . . . . . 11-104
Space-vehicle Environments (by J. H. Farrow and R. E. Jewell) 11-107
Space-vehicle Trajectories, Flight Mechanics, and Performance

(by O. Elnan, W. R. Perry, J. W. Russell, A. G. Kromis,
    and D. W. Fellenz) . . . . . . . . . . . . . . . . . . . . . . . . . . . . . . . 11-110
Orbital Mechanics (by O. Elnan and W. R. Perry) . . . . . . . . . 11-112
Lunar- and Interplanetary-flight Mechanics (by J. W. Russell) 11-113
Atmospheric Entry (by D. W. Fellenz) . . . . . . . . . . . . . . . . . . 11-114
Attitude Dynamics, Stabilization, and Control of Spacecraft (by
    M. R. M. Crespo da Silva) . . . . . . . . . . . . . . . . . . . . . . . . . . 11-117
Space-vehicle Structures (by Robert J. Kroll) . . . . . . . . . . . . . . 11-119
Materials for Aerospace Applications (by Stephen D. Anto-
    lovich) . . . . . . . . . . . . . . . . . . . . . . . . . . . . . . . . . . . . . . . . . . 11-119
Space Environment (by W. R. Lucas) . . . . . . . . . . . . . . . . . . . . 11-122
Finite Element Analysis of Structures (by Lawrence H. Sobel) 11-122
Stresses and Stability (by Robert J. Kroll) . . . . . . . . . . . . . . . . 11-124
Vibration of Structures (by Lawrence H. Sobel) . . . . . . . . . . . 11-125
Space-vehicle Propulsion (by W. Tabakoff) . . . . . . . . . . . . . . . 11-127

### PIPELINE TRANSMISSION
#### by George H. Ewing

Natural Gas . . . . . . . . . . . . . . . . . . . . . . . . . . . . . . . . . . . . . . . . 11-130
Crude Oil and Oil Products . . . . . . . . . . . . . . . . . . . . . . . . . . . . 11-133
Solids . . . . . . . . . . . . . . . . . . . . . . . . . . . . . . . . . . . . . . . . . . . . . 11-134

# AUTOMOBILES
## by Harold M. Nelson

REFERENCES: Lichty, "Internal Combustion Engines," McGraw-Hill. Heldt, "The Automotive Chassis," Heldt. Heldt, "The Gasoline Automobile," Heldt. Annual (April) specification number of *Automotive Inds.* Annual (November) specification and show number of *Motor.* Molloy, "Automobile Engineer's Reference Book," George Newnes, Ltd., London. Annual "Automobile Facts and Figures," Motor Vehicle Manufacturers Assn.

## GENERAL

(See also Sec. 9, Internal Combustion Engines)

The use of automotive vehicles has increased to such an extent that by 1971 over 89 percent of all intercity traffic in passenger miles was carried by motor vehicles. Railroads carried less than 1 percent and airways 9.7 percent. Intercity motor carriers of freight handled half as many ton-miles as the railroads. In 1972 some 97 million automobiles and 22 million trucks and buses were **registered** in the United States, including almost 9 million automobiles manufactured that year. The average age in use was slightly less than 6 years, which has changed but little during the preceding 10 years. If it can be assumed that the average of these vehicles, in good condition, can maintain 60 mi/h (96.6 km/h) on modern highways, this represents a total of at least 3.3 billion horsepower available for use from the engines.

**Characteristics of the cars** purchased in 1972 are indicated by their optional and special equipment: Engines—four cylinders, 9.5 percent; six cylinders, 10.7 percent; eight cylinders V, 79.8 percent. Automatic transmissions, 92 percent. Power brakes, 67 percent. Power steering, 84 percent. Air conditioners, 69 percent. A representative distribution of car use is given in Fig. 1. Almost 55 percent of all car trips are less than 5 mi (8.0 km).

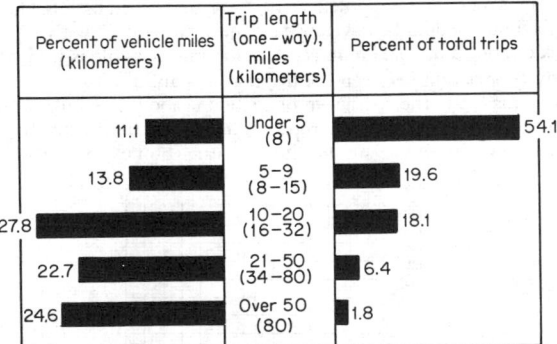

| Percent of vehicle miles (kilometers) | Trip length (one–way), miles (kilometers) | Percent of total trips |
|---|---|---|
| 11.1 | Under 5 (8) | 54.1 |
| 13.8 | 5–9 (8–15) | 19.6 |
| 27.8 | 10–20 (16–32) | 18.1 |
| 22.7 | 21–50 (34–80) | 6.4 |
| 24.6 | Over 50 (80) | 1.8 |

**Fig. 1** Distribution of car use. (*"Automobile Facts and Figures."*)

## VEHICLE DIMENSIONS

The average dimensions of the three 1972 cars with the highest production were: length, bumper to bumper, 221 in (5.6 m); wheelbase, 121 in (3.1 m); width, 80 in (2.0 m), increasing to 168 in (4.3 m) with opposite doors open; height, 55 in (1.4 m); ground clearance, 5.7 in (0.14 m). For the largest cars, these dimensions may be as great as 235 in (6 m) in length and 127 in (3.2 m) wheelbase, the width and height being little different than for the large-production cars. Turning diameter, as measured to the outside of the outside front tire, is about 42 ft (12.8 m), although the largest cars may require 55 ft (16.8 m). The overhang of the front fenders requires about 3 ft (1 m) more if the turn is made between walls.

## TRACTION REQUIRED

The total resistance, or traction force, required for steady motion of a vehicle on a level road, is the sum of (1) air resistance and (2) friction resistance. The latter is dominated to such an extent by the rolling resistance of tires that the friction of bearings may be disregarded in a first approximation. **Tire resistance,** as reported by Billingsley, Evans et al., *Trans. SAE,* 1942, was about 1 percent of the load carried at low speeds, increasing to about 1.5 percent at 60 mi/h (96.6 km/h). For modern passenger-car tires, with various percentages of synthetic rubbers, these values are about 1.2 to 1.4 percent at 30 mi/h (48.3 km/h), increasing to 1.6 to 2.0 percent at 70 mi/h (112.7 km/h). This is about nine times the rolling resistance reported for rail cars on steel rails (see *Univ. Illinois Exp. Sta. Bull.* 167). Greater **tire deflection,** by deviation from manufacturers' recommendations (see Table 1) on load and air pressure, increases tire resistance. Low temperatures do likewise, especially with high synthetic-rubber content. Lundstrom (*Trans. SAE,* 1957, p. 725) reports that a run of approximately 20 mi (32.2 km) at 30 mi/h (48.3 km/h), in 70°F (21.5°C) air, was necessary to reach temperature equilibrium in the tires. The **total air-and-rolling resistance** for a 4,000-lb (1,815 kg) car fell approximately from 117 lb (53 kg), after a 7-mi (11.3 km) run, to 92 lb (42 kg), after 20 mi (32.2 km).

**Air resistance** varies closely with the square of the car speed and has a value between 120 and 170 lb (55 and 77 kg), at 60 mi/h (96.6 km/h), depending on the body size and design. The total car drag at 60 mi/h (96.6 km/h) varies from 150 to 230 lb (68 to 104 kg), with an average of about 200 lb (90.7 kg) (equivalent to 32 hp) for the majority of modern American cars. Rolling drag and air drag are about equal at 40 mi/h (64.4 km/h). (See also Sec. 11, Air Resistance of Trains, automobiles and ships.)

Average values for **traction requirements** of several high-production 1956 cars with an average weight of 4,000 lb (1,815 kg) (including two passengers) are shown in Fig. 2, as reported by Kosier and McConnell in *Trans. SAE,* 1957, p. 730. Curves A, R, and T represent the air, rolling, and total drag, respectively, on a level road. Curves T', parallel to curve T, represent the displacement of the latter for gravity effects on the grades indicated, the additional traction being equal to the car weight [4,000 lb (1,815 kg)] × percent grade.

Curve EM-56 shows the traction available in high gear from the engines in this average car, which was advertised to

**Table 1. Passenger-Car Tires (Four-Ply, Tubeless)**

| Tire size | Recommended inflation pressure | | Max load | | Overall diam | | Static load radius | | r/mi (km)* |
|---|---|---|---|---|---|---|---|---|---|
| | lb/in² | kPa | lb | kg | in | cm | in | cm | |
| 6.00–13 | 24 | 165 | 725 | 329 | 23.9 | 60.7 | 11.1 | 28.2 | 889   1,430 |
| 6.50–13 | 24 | 165 | 835 | 379 | 24.9 | 63.2 | 11.6 | 29.5 | 852   1,371 |
| 7.00–13 | 24 | 165 | 920 | 417 | 25.4 | 64.5 | 11.8 | 30.0 | 837   1,347 |
| 6.50–14 | 24 | 165 | 880 | 399 | 25.6 | 65.0 | 11.9 | 30.2 | 830   1,335 |
| 7.00–14 | 24 | 165 | 975 | 442 | 26.3 | 66.8 | 12.2 | 31.0 | 809   1,302 |
| 7.50–14 | 24 | 165 | 1,085 | 492 | 27.1 | 68.8 | 12.5 | 31.8 | 790   1,271 |
| 8.00–14 | 24 | 165 | 1,175 | 533 | 27.6 | 70.1 | 12.7 | 32.3 | 778   1,252 |
| 8.50–14 | 24 | 165 | 1,265 | 574 | 28.2 | 71.6 | 13.0 | 33.0 | 760   1,223 |
| 9.00–14 | 24 | 165 | 1,355 | 615 | 29.0 | 73.7 | 13.2 | 33.5 | 748   1,204 |
| 6.00–15 | 26 | 179 | 900 | 408 | 26.3 | 66.8 | 12.2 | 31.0 | 809   1,302 |
| 6.50–15 | 26 | 179 | 1,000 | 454 | 26.8 | 68.1 | 12.4 | 31.5 | 796   1,281 |
| 6.70–15 | 26 | 179 | 1,115 | 506 | 27.5 | 69.9 | 12.8 | 32.5 | 772   1,242 |
| 7.10–15 | 26 | 179 | 1,195 | 542 | 28.1 | 71.4 | 13.0 | 33.0 | 760   1,223 |
| 7.60–15 | 26 | 179 | 1,310 | 594 | 28.7 | 72.9 | 13.3 | 33.8 | 742   1,194 |
| 8.00–15 | 26 | 179 | 1,395 | 633 | 29.2 | 74.2 | 13.4 | 34.0 | 737   1,186 |
| 8.20–15 | 24 | 165 | 1,415 | 642 | 29.8 | 75.7 | 13.6 | 34.5 | 726   1,168 |

*r/mi (km) (approx) at 35 mi/h (56.3 km/h) = 9,882/static load radius, in (cm).

develop about 200 hp under laboratory test conditions. Curve E-48 gives corresponding data for 1948 engines. The intersection of such curves with any of the constant-gradient curves indicates the top speed which may be attained on the selected grade. For example, the 1956 car should negotiate a 12.5 percent grade in high gear at 60 mi/h (97 km/h), or a maximum grade of 14.5 percent at 40 mi/h (64 km/h), whereas the 1948

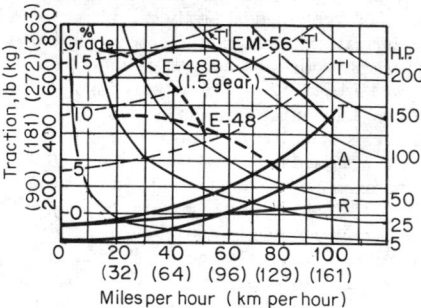

**Fig. 2** Traction required for a typical car and traction available from different engines.

car could negotiate only an 8 percent grade at 40 mi/h (64 km/h) and had a top speed of about 77 mi/h (124 km/h), on a level road.

**Effects of Transmission Gear Ratios** Constant-horsepower parabolas, which apply to any vehicle, are shown as light curves in Fig. 2. Except for small effects of friction losses, changes in gear ratio move points of a curve for traction available from an engine, such as EM-56, or E-48, along these constant-power parabolas to traction values multiplied by the change in gear reduction. In this way the curve E-48B was developed from the curve E-48 for an additional gear reduction of 1.5. Such shifting of the traction values available from engines follows gear changes in the rear axle as well as in the gear box.

**Acceleration** The difference between the traction available from an engine and that required for a steady speed on a given grade may be used for acceleration [or acceleration, mi/(h)(s) = 21.9 × (surplus traction force/total effective car weight)]. Car weight should include a factor for the rotating parts of the engine, which may be of considerable magnitude when a high gear reduction is used. The effective mass of engine rotating parts increases as the square of the engine revolutions per mile and, for a typical example, may equal the car weight at a gear ratio giving about 15,000 engine r/mi (Kosier and McDonnell, *Trans. SAE*, 1957, p. 730). This would result from a 5.33 gear ratio with a 3.6 axle ratio. Since the effective mass of engine rotating parts increases with the square of the gear reduction, whereas the traction force at the wheels increases directly, there is an optimum gear reduction for maximum acceleration (Fig. 3). For the hypothetical 1956 car under discussion, this maximum would be developed at a gear ratio of 4.2 with a 3.6 axle ratio. The maximum acceleration rate possible is limited by the friction between the driving tires and the road. For a dry roadway, the coefficient of friction is about 1.0; but when the roadway is wet, this drops to 0.6 or even 0.4, depending on the polish of the surface due to wear. (Stonex, *SAE Paper*

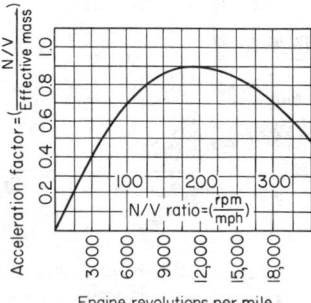

**Fig. 3** Equivalent mass added to car by rotating engine parts at various gear ratios. (*Trans. SAE*, 1957, *p.* 734.)

539A, 1962.) Thus for a car with 2,000 lb (907 kg) on the driving wheels, the limiting traction force would be about 2,000 lb (907 kg) on a dry pavement and less than half this value on a fairly wet pavement. The maximum accelerations possible under these two conditions for a car of 4,000 lb (1,814 kg) would be equivalent to a speed change of from 0 to 60 mi/h (97 km/h) in about 5.5 and 11 s, respectively.

**Fuel Consumption** It is convenient to plot a modified "Willans line" with the fuel consumption of a car, gallons per mile, versus rear-wheel traction, pounds (Fig. 4). A single straight line will generally represent the fuel consumption at a given fuel-air ratio over a wide range of loads and speeds. The points shown in Fig. 4 are for different speeds on two different cars. On such a plot the group of radial lines from the origin represent constant brake specific fuel-consumption (bsfc) values based on power available at the wheels. A fuel consumption of 0.10 gal/mi (0.35 l/km) at a traction of 225 lb (102 kg) represents a bsfc of 1.0, with a fuel weight of 6 lb/gal (0.7 kg/l). A study of the effects of gear changes may be approximated by displacing the fuel consumption line so that corresponding

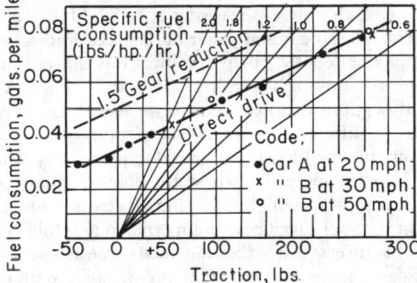

**Fig. 4** Effect of traction load on fuel consumption.

points, as their traction is changed by gear changes, will remain on the same bsfc lines. This is shown by the dotted line for an additional gear reduction of 1.5 and emphasizes the importance of axle ratio on fuel consumption. Such plots of fuel consumption may be combined with Fig. 2 by means of the traction scales which are common to both.

The fuel consumption obtainable from modern cars is indicated by the results of the annual **Mobilgas Economy Run** for stock cars, discontinued in 1970. In Fig. 5 the results are shown for the winners in the various classes plotted against overall car weight. This race, over a distance of about 1,300 mi (2,092 km) under a wide range of driving conditions, demonstrates that excellent fuel economy is possible when modern

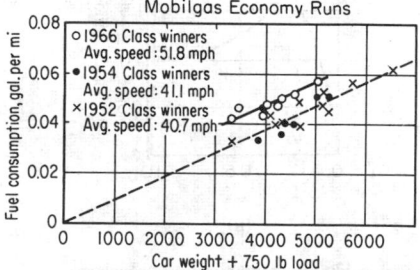

**Fig. 5** Effect of car weight plus 750-lb load on fuel consumption of American stock cars in Mobilgas Economy Run.

cars are driven at moderate speeds. The fuel-consumption rate rises at higher speeds and may be estimated from Figs. 2 and 4.

## TRANSMISSION MECHANISMS

**Friction clutches** are either (1) the single-disk type (Fig. 6),

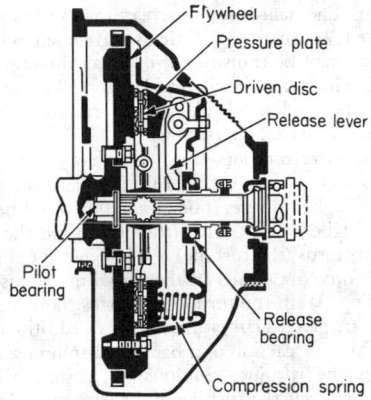

**Fig. 6** Single-plate dry-disk friction clutch.

connecting the engine to a manual transmission, or (2) the hydraulically operated multiple-disk type (Fig. 7, schematic), for control of the various planetary-gear changes in automatic transmissions. In (1), the area of the friction facing is usually based on a pressure of 30 lb/in² (207.8 kPa), and the torque

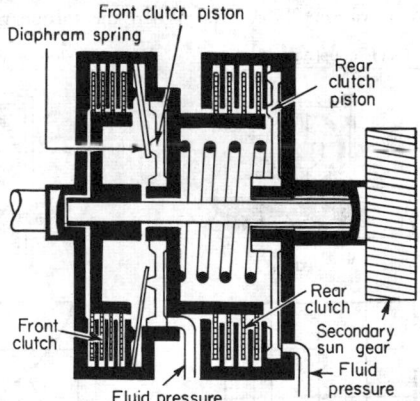

**Fig. 7** Schematic of two hydraulically operated multiple-disk clutches in an automatic transmission. *(Ford Motor Co.)*

rating on a friction coefficient of 0.25. The clutch is held in engagement by several coiled springs or a diaphragm spring and is disengaged by means of a pedal with such leverage that 30 to 40 lb (13 to 18 kg) will overcome the clutch springs.

**Fluid couplings** between the engine and transmission provide a smooth drive by the flow of oil between the flat radial blades in two adjacent toroidal casings (Fig. 8a, schematic). The difference in centrifugal force between the mass of oil contained in each toroid, when either is running at a speed higher than the other, causes a flow of oil from the periphery of the faster one to the slower one. Since this mass of oil is also rotating around the shaft at the speed of the driving torus, its impact on the blades of the slower torus develops a torque on the latter. The developed torque is equal to, and cannot exceed, the torque of the driving torus. In this respect it is similar to a slipping friction clutch. The driven member must always run at a lower speed, though at high rotative speeds, and when the torque demand is small, the slip may be only 2 or 3 percent. The **stalled torque** increases with the square of the engine speed, so that very little is developed when idling. Since torque may be transmitted in either direction, depending only on which member is rotating at the higher speed, the engine may be used as a brake as with friction clutches, and the car may be started by pushing.

**Torque-converter couplings** (Fig. 10a) have largely replaced fluid couplings because the torque transmitted can be increased at high slippage. The circulation of oil between the driving, or higher-speed, torus (the pump) and the driven, or lower-speed, torus (the turbine) results from the difference in centrifugal force developed in these two units, just as in the fluid coupling. With the torque converter, however, the turbine blades are given a curvature so that an additional torque is developed by the reaction of a backward-spinning mass of oil as it leaves the turbine. Stationary, or stator, blades are interposed between the turbine and the pump to change the direction of the oil spin. The entrance angle of the stator blades required for tangential flow varies widely with the slip ratio. For a given blade angle there is a hydraulic shock loss at any slip ratio greater or less than that which provides tangential flow. This is reflected in the rapid fall of the efficiency curve in Fig. 9b on each side of the maximum. The essential parts of a torque converter with its stationary stator are shown in Fig. 9a. A **stalled-torque** multiplication of 2.0 to 2.7 was developed in various 1964 cars. When the torque ratio is almost unity, the slip is such that oil from the turbine starts to impinge on the back of the stator blades. By mounting the stator assembly on a **sprag**, or one-way clutch (Fig. 10a), it remains stationary while subject to the reversing action of the backward-spinning oil mass as it leaves the turbine. When the slip reaches the point where the oil flow from the turbine begins to spin forward, the stator is free to turn with it. When the slip is further reduced, the unit acts as a fluid coupling with improved efficiency (Fig. 10b). Such a unit is a **fluid torque-converter coupling.** Some designs eliminate all slippage by the inclusion of a friction clutch which carries the load when a predetermined car speed is reached. These clutches are hydraulically operated, the engagement being controlled automatically by accelerator position and car speed.

Figure 11 compares a torque-converter coupling to a friction clutch on **car performance** in direct drive. The increase in traction available for acceleration from a standing start substantiates its public acceptance. An axle gear is generally used which gives a propeller shaft speed about 90 percent as great as with a manual transmission at the same car speed. The gain in engine efficiency compensates for the losses of the automatic transmissions under steady cruising speeds. However, there may be a considerable **loss of power during acceleration** (Fig. 12) unless supplemented by such modifications as one or more auxiliary gear ratios or variable-angle stator vanes. Various design modifications of torque-converter couplings have been introduced by different manufacturers, such as two or more stators, each independently mounted on one-way clutches, and variable pitch angles for the stator blades. These provide compromises in blade angles for the development of rapid acceleration without sacrifice of high efficiency while cruising. Most manufacturers warn that **automatic transmissions may be damaged** if cars are towed a considerable distance with the rear wheels turning. Starting an engine by pushing a car with an

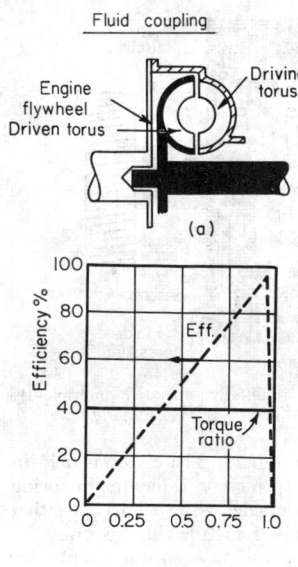

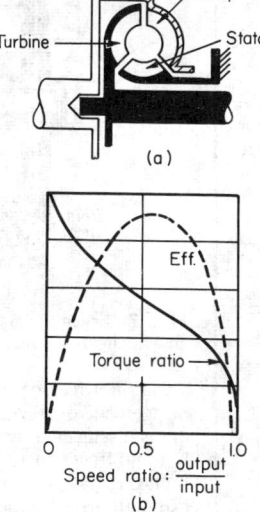

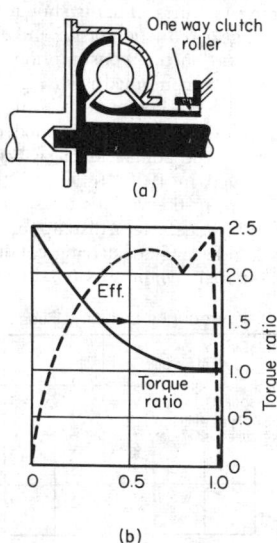

**Fig. 8** Fluid coupling: (a) section; (b) characteristics.

**Fig. 9** Torque converter: (a) section; (b) characteristics.

**Fig. 10** Torque-converter coupling: (a) section; (b) characteristics.

automatic transmission should follow the manufacturer's directions.

**Most manual transmissions** installed as standard equipment on American cars have three forward speeds, including direct drive, and one reverse. These speeds are obtained by sliding either one of two gears along a splined shaft to bring it into mesh with a corresponding gear on a countershaft which is, in turn, driven by a pair of gears in constant mesh. Helical gears are used to minimize noise. A **"synchromesh"** device (Fig. 13), acting as a friction clutch, brings the gears to be meshed approximately to the correct speed just before meshing and minimizes **"clashing,"** even with inexperienced drivers. Gear

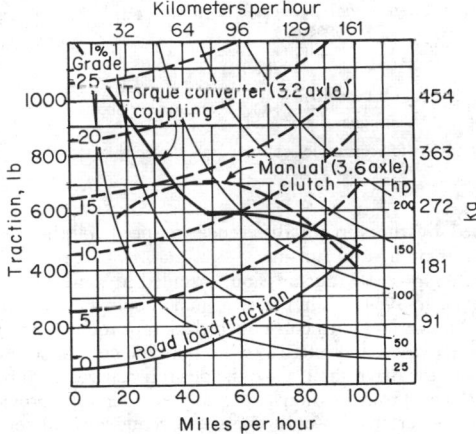

**Fig. 11** Comparative traction available in the performance of a fluid torque-converter coupling and a friction clutch. (*Trans. SAE,* 1957, *p.* 724.)

changes are generally in geometric ratios. Transmission ratios average about 2.76 in first gear, 1.64 in second, 1.0 in third or direct drive, and 3.24 in reverse. The shift lever is generally located on the steering column. Four-speed transmissions are offered as optional equipment at extra cost on most cars, with the shift lever on the floor. Average gear ratios are about 2.67 in first, 1.93 in second, 1.45 in third, and 1.0 in fourth or direct drive.

**Overdrives** have been available for some cars equipped with manual transmissions. These are supplemental planetary gear units with three planetary pinions driven around a stationary sun gear. The surrounding internal gear is coupled to the propeller shaft, which thus turns faster than the engine. The

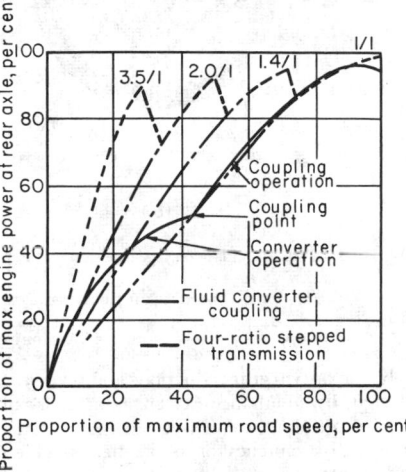

**Fig. 12** Comparison of the power transmission of a fluid torque-converter coupling with that obtained by use of a four-speed gearbox. (*Giles, Proc. IME,* 1957.)

gear ratio is selected to permit the engine to slow down to about 70 percent of the propeller-shaft speed and operate with less noise and friction. These units automatically come into action when the driver momentarily releases the accelerator pedal at a car speed above 25 to 28 mi/h (40 to 45 km/h). It may be thrown out of action through a solenoid by pressing the accelerator pedal to the floor.

## AUTOMATIC TRANSMISSIONS

All automatic transmissions, except the Hydra-Matic, use torque-converter couplings with planetary-gear units which

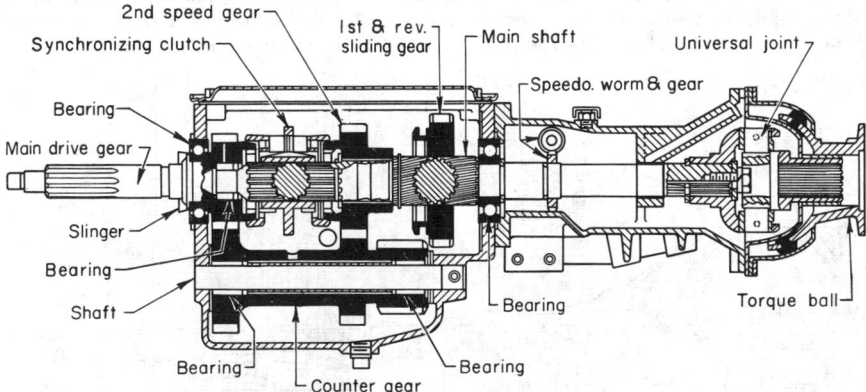

**Fig. 13** Three-speed synchromesh transmission (Buick).

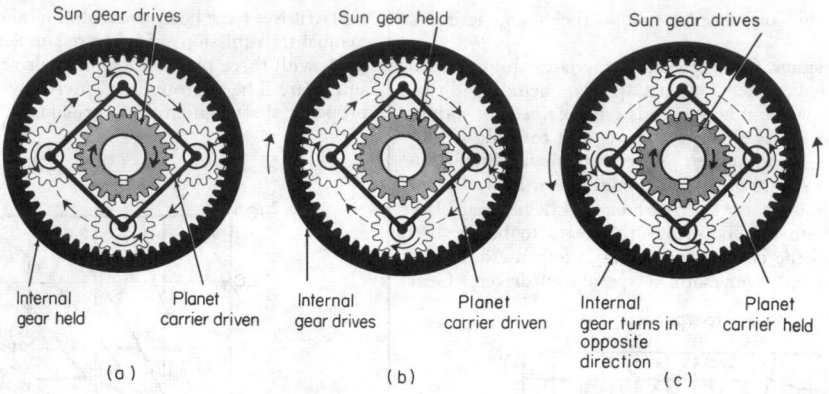

**Fig. 14**   Planetary gear action. (*a*) Large speed reduction: ratio $= 1 + \dfrac{\text{int gear diam}}{\text{sun gear diam}} = 3.33$ for example shown.

(*b*) Small speed reduction: ratio $= 1 + \dfrac{\text{sun gear diam}}{\text{int gear diam}} = 1.428$ for example shown. (*c*) Reverse:

ratio $= \dfrac{\text{int gear diam}}{\text{sun gear diam}} = -2.33$ for example shown.

can supply one or two gear reductions and reverse, depending on the design, by simultaneously engaging or locking various elements of planetary systems (Fig. 14). Automatic control is provided by disk clutches or brake bands which lock the various elements, operated by oil pressure as regulated by governors at car speeds where shifts are made from one speed to another. Mechanical losses in these transmissions are sufficient to require water cooling of the oil on all the larger units. Oil from the pump for the control system, after passing through a pressure regulator and the converter, passes through a cooling tank submerged in the engine-jacket water radiator.

A schematic of a representative automatic transmission, combining a three-element torque converter and a compound planetary gear (as used on several of General Motors' larger cars in 1964), is given in Fig. 15. The speed reductions and reverse are provided by a compound planetary system consisting of two simple systems in series. The two sun gears are an integral unit with the same number of teeth, and the forward internal gear carries the planets of the rear unit. The internal gears of both systems are all of the same size, and consequently all planets are of equal size. This arrangement, together with three clutches, two brake bands, and suitable one-way sprags, makes possible three forward gear or torque ratios besides direct drive and reverse, all of which may be doubled by the slip, or difference in speed, of the torque-converter pump and turbine.

The **Hydra-Matic** transmission, installed in some 15 million cars prior to 1964, made use of a fluid coupling rather than a torque converter, with consequent lack of torque boost. Two separate planetary systems, with both planet carriers connected to the drive shaft, gave the desired gear reductions.

Automatic transmissions for many smaller cars provide a torque converter with only one speed reduction and reverse; this is accomplished with the simplicity of a single planetary system.

Manufacturing precision and the use of helical gears contribute greatly to the comparative quietness of planetary systems.

The **differential** is a unit attached to the ring gear (Figs. 16 and 18) which equalizes the traction of both wheels and permits one wheel to turn faster than the other, as needed on curves. Each axle is driven by a bevel gear meshing with pinions on a cross-shaft pinion pin secured to the differential case. The case also carries the ring gear. An undesirable feature of the conventional differential is that no more traction may be developed on one wheel than on the other. If one wheel slips on ice, there is no traction to move the car. **Limited-slip differentials** are offered as optional equipment on most cars.

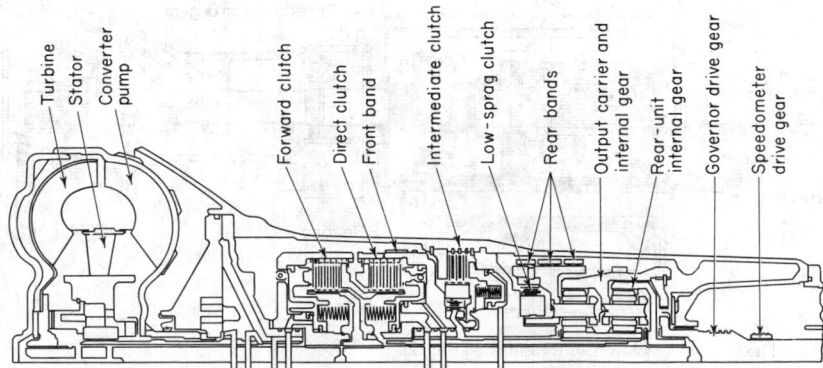

**Fig. 15**   Three-element torque converter and planetary gear (General Motors).

Figure 17 shows one design, in which four pinions are carried on two separate cross shafts at right angles to each other, each being driven by V-shaped notches in the carrier. As torque is developed to drive either axle, one pinion cross shaft or the

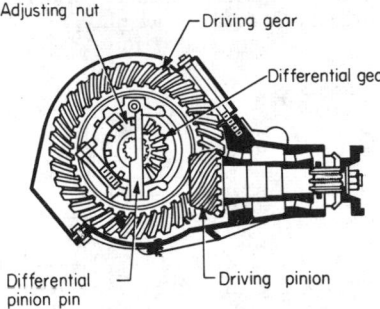

**Fig. 16** Rear-axle hypoid gearing.

other moves axially and locks the corresponding disk-clutch plates between that axle drive gear and the differential housing. In another design, similar disk clutches are locked by spring pressure, which prevents differential action until a differential torque greater than the limit established by the springs is developed.

The **semifloating rear axle** (Fig. 18) used on almost all cars has a bearing for each drive axle at the outer end of the housing as well as near the differential carrier, with the full load on each wheel taken by the drive axles in combined bending and shear. The **full-floating axle,** generally used on commercial vehicles, supports each wheel on two bearings carried by the axle housing or an extension to it. Each wheel is bolted to a flange on one of the axle shafts. The axle shafts carry none of the vehicle weight and may be withdrawn without jacking up the wheel.

## SUSPENSIONS

**Rear Suspensions** Torque reactions may be taken through longitudinal leaf springs, as in the **Hotchkiss drive,** or through radius rods when coil springs are used. Some designs in the past used a torque tube around the propeller shaft, bolted to the axle housing, with universal joints for both at the forward ends. **Leaf springs,** with one to nine leaves, were installed on about half the 1972 models, and coil springs on the other half. Spring stiffness (at the rear wheels) ranged from about 82 lb/in (14.6 kg/cm) deflection to some 160 lb (28.6 kg/cm). Shock absorbers, to dampen road vibrations, were standard equipment on all cars.

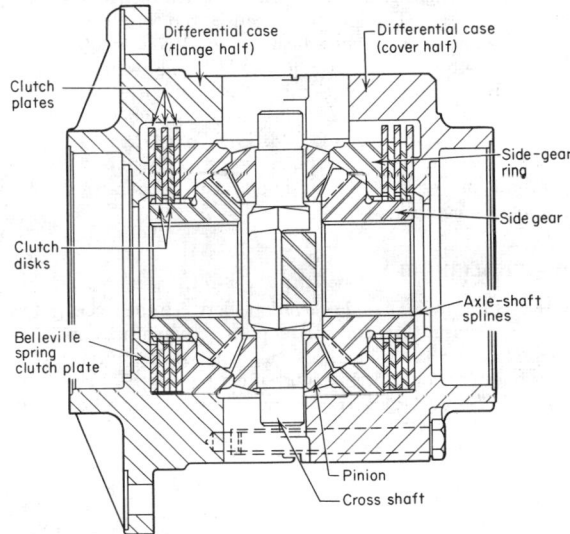

**Fig. 17** Limited-slip differential (Buick).

**Front-wheel Suspensions** Independent front-wheel suspensions are used on all cars, with the steering knuckle held directly between the wishbones by spherical joints (Fig. 19). The upper wishbone is shorter than the lower, to allow the springs to deflect without lateral movement of the tire at the

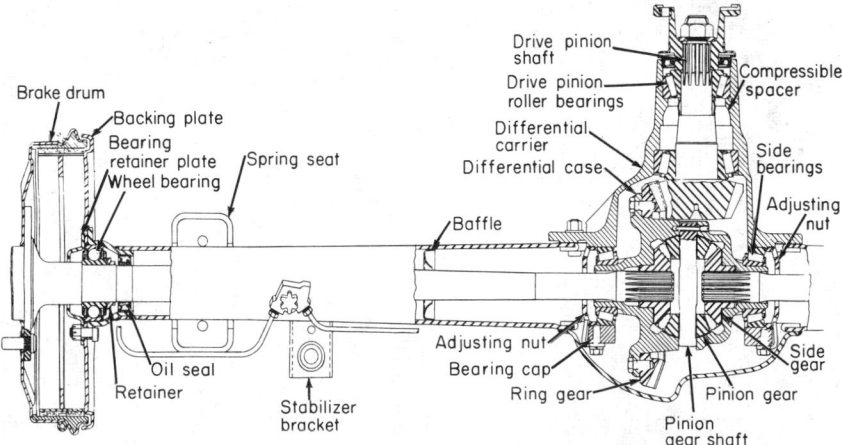

**Fig. 18** Rear axle (Oldsmobile).

point of ground contact. If the steering head in Fig. 20 is lifted by an amount $d$, the upper end is shifted sideways by an amount $d^2/2A$ and the lower by an amount $d^2/2B$. For a point on this steering head, extended to the point of contact between the tire and the road, to have no lateral movement,

$$hB = (h + s)\,A \qquad \text{or} \qquad h/(h + s) = A/B$$

A modification of the conventional suspension consists of sloping the upper wishbones down toward the rear, so that the steering spindle is given more "caster" when the front springs are compressed. This geometry causes the torque produced from braking at the front wheels to develop a couple on the inclined wishbones, which tends to raise the front of the car frame. By suitable proportioning of the parts it is possible by this means to reduce "nose diving" of the car when the brakes are applied (*Engineering*, Oct. 28, 1955, p. 611).

The load on these wishbones is generally taken by coil springs acting on the lower wishbone or by torsion-bar springs mounted longitudinally.

## WHEEL ALIGNMENT

**Caster** is the angle, in side elevation, between the steering axis and the vertical. It is considered positive when the upper end of the steering axis is inclined rearward. Manufacturers' specifications vary considerably, with the range for 1972 cars from $1\frac{1}{2}$ to $-2\frac{1}{4}$ deg. **Camber,** the inclination of the wheel plane from the vertical, is positive when the wheel leans outward, and varies from 2 to $-\frac{1}{2}$ deg, with many preferring 0 deg. **Toe-in** of a pair of wheels is the difference in transverse distance between the wheel planes taken at the extreme rear and front points of the tire treads. It is limited to $\frac{1}{2}$ in (6.4 mm), with $\frac{1}{8}$ in (3.2 mm) or less generally preferred.

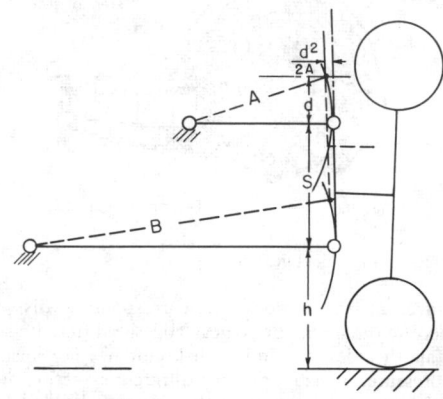

**Fig. 20**   Geometry of linkage for independent front-wheel suspension.

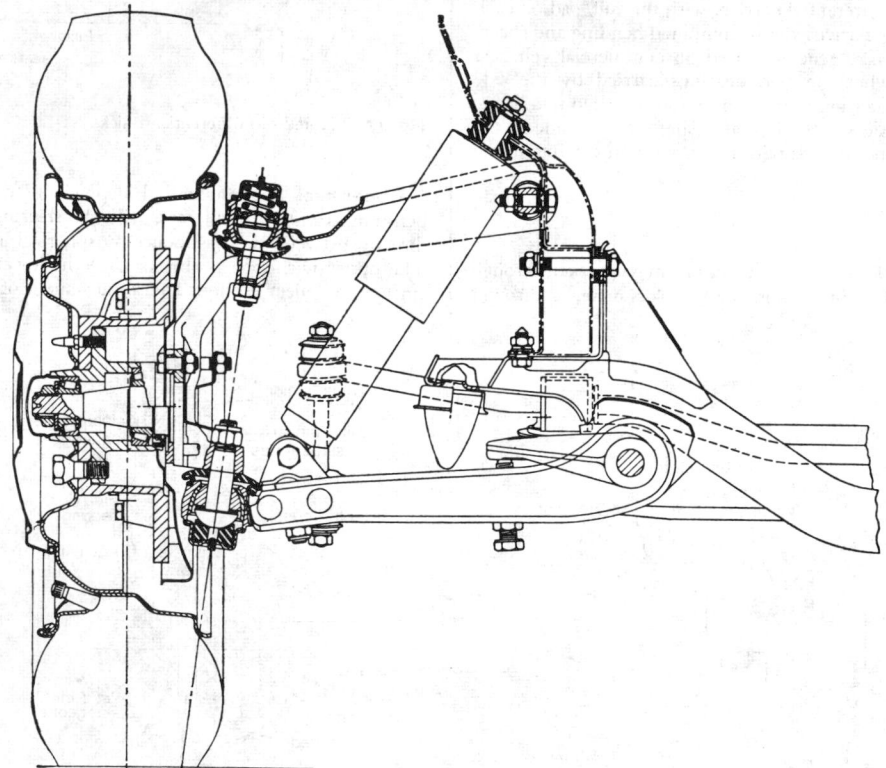

**Fig. 19**   Front-wheel suspension with spherical steering swivels. Design includes disk brakes, torsion-bar springs, and shock absorber.

### STEERING

The force applied to the steering wheel is generally multiplied through a **worm-and-roller** (Fig. 21) or a **recirculating-ball** (Fig. 22) type of steering gear. The overall ratios are such that 20° to 33° rotation of the steering wheel results in 1° turn of the front wheels for manual steering and 17.6° to 25.0° for power steer-

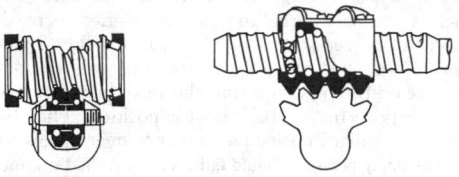

**Fig. 21** Roller-cam steering gear.    **Fig. 22** Recirculating-ball steering gear.

ing. Figure 23 illustrates the geometry of the prevalent **Acker-mann steering gear**. To avoid slippage of the wheels when turning a curve of radius $r$, the point of intersection $M$ for the

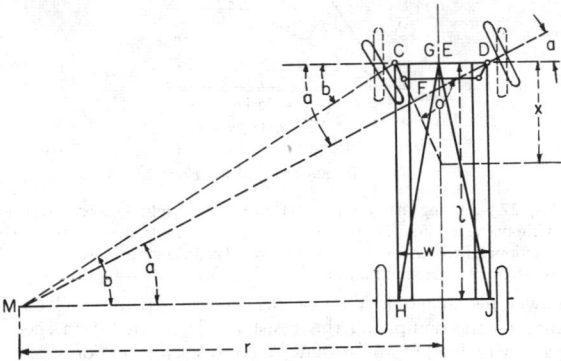

**Fig. 23** Geometry of the Ackermann steering gear.

projected front-wheel axes must fall in a vertical plane through the center of the rear axle. The torque, in foot-pounds, required to turn the wheels of a vehicle standing on smooth concrete varies with the angle of turn, from about 6 percent of the weight on the front axle, in pounds, to start a turn, to 17 percent for a 30° turn (Davis, *Trans. SAE*, 1945, p. 241).

### Power-Assisted Steering

The physical effort required to steer an automobile, especially when parking, is appreciably lessened by the power-assisted steering device. This permits reduction in the gear ratio between the steering wheel and the car wheels from some 29 to 20, with consequent reduction in the number of turns of the steering wheel for the complete movement of the front wheels from extreme right to left from 5.5 to 3.5. Power-assisted steering has been offered for many years on all U.S. cars as standard or optional equipment; public acceptance is such that 84 percent of the cars sold in 1972 were so equipped.

All systems provide (1) steering control in case of failure of the hydraulic-power assistance, and (2) a "feel of the road," by which the driver's effort on the steering wheel is proportional to the force needed to turn the front wheels and by which the

tendency of a car to straighten out from a turn or the drag of a soft front tire may be felt at the steering wheel.

Power assistance is effected by hydraulic pressure from an engine-driven pump, acting on a piston in the steering linkage. The piston and its cylinder may be incorporated in the steering-gear housing (see Fig. 28a) or installed as a supplemental link between the car frame and the steering cross-link. Oil pressure on the piston is controlled by a valve, such as the balanced spool valve of Fig. 24.

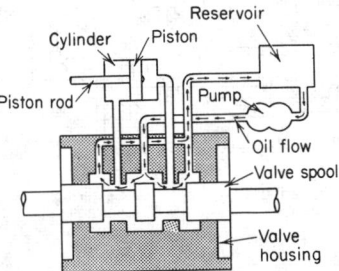

**Fig. 24** Schematic of the spool-type control valve in neutral position.

When the spool is moved slightly to the right (Fig 25), the lands on the spool restrict the return of oil from the pump through both return circuits, thus building up delivery pressure. Since the pump delivery is still open to the left end of the power cylinder while the right end is open to the pump suction, a force is developed to move the piston to the right. The greater the restriction imposed on the return of oil to the pump, the greater will be the pressure developed and the resulting force on the piston. Figure 25 shows the spool in a position where all flow of oil from the pump is cut off from its return and where the oil pressure would be raised to the maximum delivery pressure of the pump.

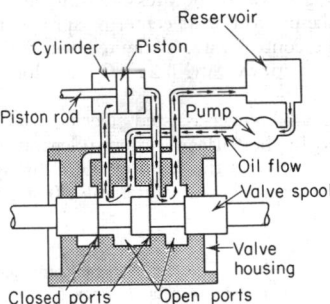

**Fig. 25** Control valve positioned for full-turn power assistance.

The spool (Fig. 26) is centered to the neutral position by suitable **centering springs**. These provide an increasing effort on the steering wheel for increasing steering angle. Although they aid in straightening out from a turn, they do not give the driver a feel of the force required to provide the steering direction. **Hydraulic reaction** against the spool, which is felt at the steering wheel and is proportional to the force developed by the steering gear, is developed by subjecting the ends of the spool to the oil pressure on either side of the power piston.

The valve is held in its neutral position by **preloading** the centering springs. Steering effort at the wheel overcomes this preload. During normal, straight-highway driving, the steering effort is less than the preload and there is no hydraulic assistance; the steering gear is freely reversible, and the driver can "feel the road" and correct for elements such as road camber and crosswinds. The caster action of the front wheels

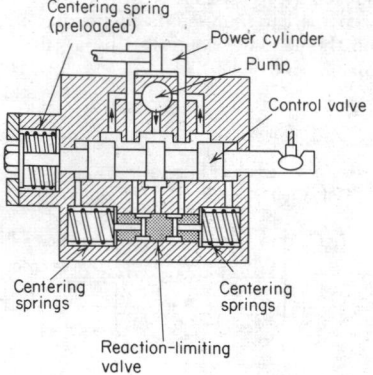

**Fig. 26** Hydraulic reaction-limiting valve (schematic).

straightens the path of the car when it is coming out of a turn. Any steering effort greater than the preload of the centering springs allows the spool movement to develop a steering assistance proportional to the steering effort and to correspondingly reduce, but not eliminate, the road reactions and shocks felt by the driver.

When the hydaulic reaction is sufficient to provide satisfactory driver acceptance during normal driving, it may be too great for parking at a curb. This is avoided by the Bendix **reaction-limiting valve** (Fig. 26). When steering effort develops a hydraulic pressure difference between the two sides of the power piston, a corresponding pressure difference moves the reaction-limiting valve. If the pressure difference is sufficient to overcome the preload of its centering springs, the pressure reaction on the control valve is limited without altering the pressures in the power cylinder. Comparable results are shown in Fig. 27.

Other designs of power-assisted steering gears accomplish similar results. In the **Saginaw** system (General Motors), the steering column operates the steering worm through a torsion bar. A rotary three-way open-center valve (Fig. 28b) is incorporated, with its outer element or body attached to the lower end of the torsion bar and its inner element or spool attached to the upper, or steering-wheel, end. When driving with negligible steering effort on a straight, smooth road, the steering is accomplished manually through the torsion bar with no power assistance. When steering resistance develops, e.g., requiring a force of about 1 lb (0.45 kg) at the rim of the steering wheel, the torsion bar deflects to start power assistance. Increasing force and deflection give increasing assistance; full power is developed at a rim pull of 3 lb (1.36 kg) approx, with a 4° twist. As the steering rack responds to the pressure on the power piston, the valve body rotates to bring the valve ports back to their neutral position. This "follow-up" motion is similar in principle to that of steering gears on ships.

**If oil-supply pressure should fail,** as by a stalled engine, manual steering is effected by stops on the valve elements, which engage when the torsion bar is deflected a few degrees. The

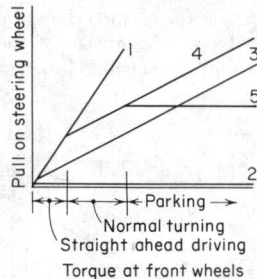

**Fig. 27** Braking characteristics: (1) manual steering; (2) power steering without reaction; (3) power-assisted with hydraulic and spring reaction; (4) power-assisted with preloaded centering springs; (5) power-assisted with reaction-limiting valve.

power piston would then be driven so as to force oil back through the pump, but this resistance is avoided by a check valve which allows oil to flow freely from the suction side of the pump to the delivery side but not in the reverse direction. To minimize deflections of the valve elements and binding from high oil pressures, the valve is designed with four sets of lands at 90° around the circumference (Fig. 29).

**Oil pumps** for power-assisted steering gears are generally driven from the engine by V belts, though in some instances they have been driven at higher speeds directly from the electric generator. A typical unit delivers 1.75 gal/min (6.62 l/

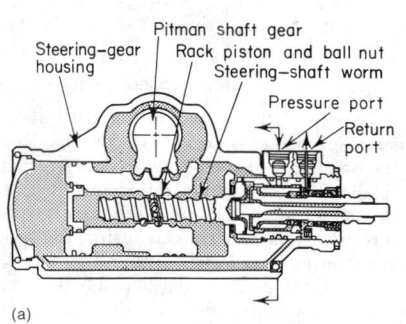

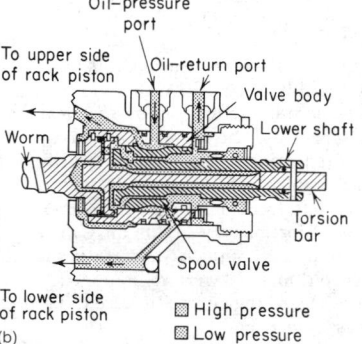

**Fig. 28** (*a*) In-line power-steering gear with rotary control valve. (*b*) Rotary-valve assembly with torsion-bar drive (Saginaw).

min) at engine idling speed, at any pressure up to 1,200 lb/in² (8,274 kPa) as may be required while parking. To avoid excessive power consumption and heating of the oil at high engine speeds where the volume and pressure requirements are small, a **flow-control valve** (Fig. 30) bypasses the pump when large volumes are pumped. The opening of this valve is

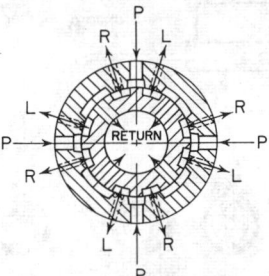

**Fig. 29**   Section of a rotary control valve with multiple ports to balance side thrust; $P$ = pressure in, R = right cylinder, L = left cylinder.

controlled by the pressure drop across flow-control orifice no. 1 while oil flows freely through the steering-control valve in its neutral position. This valve changes the oil-delivery character-

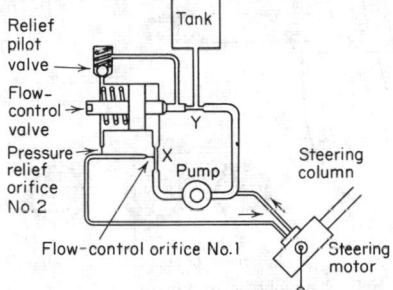

**Fig. 30**   Constant-flow valve for a power-steering pump.

istic from curve A to curve B in Fig. 31. When oil flow through the steering-control valve is interrupted by the steering-gear valve, the pressure drop across orifice no. 1 diminishes and allows the spring to close the flow-control valve, thus raising the discharge pressure. If the pressure rises beyond the

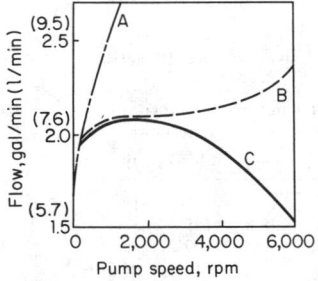

**Fig. 31**   Power-steering pump characteristics: (A) output with no flow-control valve; (B) typical output for a 1956-model flow-controlled pump; (C) dropping flow characteristic of the 1959 and 1960 pumps. Drop-off at high speeds results in reduced back pressure through the steering system and reduction in operating temperature.

setting of the relief pilot, the flow through it develops a pressure drop through pressure-relief orifice no. 2, also opening the flow-control valve. The drooping curve C has been developed by connecting the flow-control orifice no. 1 to a restriction X, in the pump bypass circuit, which acts as a venturi.

**Cavitation** at the pump inlet when running at high speeds is minimized by connecting the oil reservoir to another restricted passage, Y, also equivalent to a venturi in the pump circuit, thus permitting the circuit to be "supercharged," or operated at a pressure correspondingly above atmospheric.

Three types of **rotary pumps** for the high pressures required are shown in Fig. 32 (see also Sec. 14). Centrifugal force holds the sliding elements against a cam-shaped or eccentric case at high speeds. At low speeds, the sliding elements are held

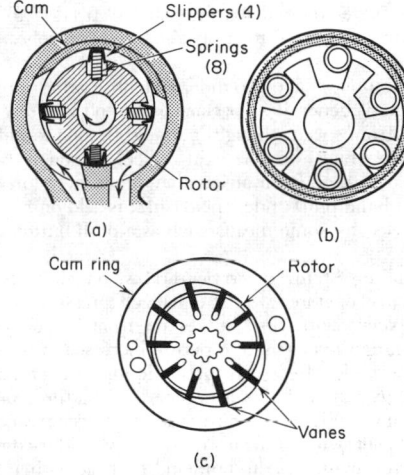

**Fig. 32**   Rotary-pump types: (a) Chrysler, (b) Ford (Eaton), (c) General Motors (Saginaw).

against the case—in design a by springs and in design c by oil pressure admitted to the base of the vanes. The double cam of design c, in addition to doubling the normal volumetric displacement, provides for balancing the oil pressure on each side of the rotor and on the bearings. The cam is contoured for uniform acceleration.

## BRAKES

The maximum retarding force which can be applied to a vehicle though its wheels is limited by the friction between the tires and the road, equal to the coefficient of friction times the vehicle weight. With a coefficient of 1.0, which is about the maximum for *dry* pavements, this force can equal the car weight and can develop a retardation of $1.0g$ (32.2 ft/s²), with a **stopping distance** $S = V^2/29.9$, where $V$ is in miles per hour. Tests conducted by the U.S. Bureau of Public Roads on **dry pavements** with vehicles picked at random on the U.S. highways (*SAE Trans.*, 1957, pp. 403–426) indicated that some vehicles and trucks of all classes developed such a retarding rate for at least part of the stopping distance but that most vehicles did not. Fifteen percent of the automobiles failed to develop a retardation rate greater than $0.84g$, and a corresponding percentage of heavy trucks failed to exceed $0.53g$.

Table 2 presents maximum stopping distances permissible under the 1962 Revision of the Uniform Vehicle Code which has been adopted by several states as the basis of their vehicle braking-performance requirements. A column is included for distances traversed during the ¾ s an average alert driver may require to start applying his brakes after perceiving the need to stop. This time interval varies widely for different drivers and under different conditions; many drivers' reactions may be considerably slower than this at least some of the time. Also, the braking distances will of course be greater on downhill slopes. This table is based on tests run by the Bureau of Public Roads (U.S. Department of Commerce) on more than 1,300 vehicles stopped from 20 mi/h (32 km/h). Suitable expansion factors were developed for the higher speeds shown. In the 1963 tests, 85 percent of the automobiles were stopped within 22.5 ft (6.86 m) and three-axle trucks weighing over 10,000 lb (4,536 kg) were stopped within 41 to 49 ft (12.5 to 14.9 m) after the drivers started to apply the brake controls at 20 mi/h (32 km/h).

All automobiles have two independent systems of brakes for safety. One is generally a **parking brake** and is rarely used to stop a car from speed, though it should be able to. The brake manually operates on the rear wheels through cables or mechanical linkage from an auxiliary foot lever (or, less frequently, a hand pull) under the dash; it is held on by a ratchet until released by some means such as a push button or a foot pedal.

The main system, or **service brakes,** on all U.S. cars is hydraulically operated, with equalized pressure to all four wheels, except with disk brakes on front wheels, where a proportioning valve is used to permit increased pressure to the disk calipers. Rubber seals preclude the use of petroleum products; hydraulic fluids are generally mixtures of glycols with inhibitors. Figure 33 shows the usual system, and Fig. 34 shows the **split system,** for improved safety, with two independent master cylinders in tandem, each actuating half the brakes, either front or rear or one front and the opposite rear. Failure of either hydraulic section allows stopping of the car by brakes on two wheels.

Figure 35 shows the customary design of a **brake master cylinder** and a **wheel-brake cylinder,** by which the brake shoes are applied in the conventional internal-expanding brakes (Fig.

36). When the brakes are released, a spring in the master cylinder returns the piston, with its flexible primary cup, to a stop plate. This uncovers the compensating port, permitting brake fluid to enter from the reservoir or to escape from the

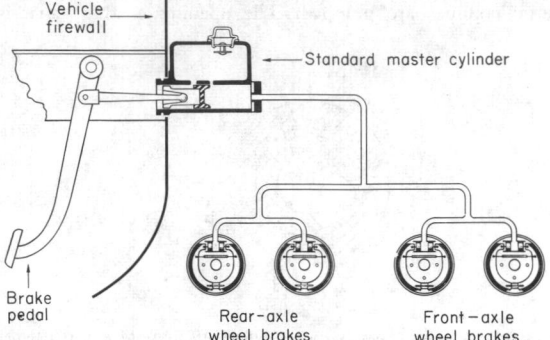

**Fig. 33**   Conventional hydraulic brake system.

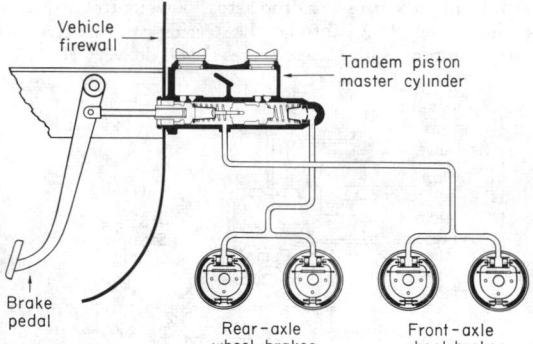

**Fig. 34**   "Split" hydraulic brake system.

wheel cylinders after brake application. The check valve facilitates the maintenance of 8- to 16-lb/in² (55- to 110-kPa) line pressure to prevent the entrance of air into the system.

Three **types of internal-expanding brakes** (Fig. 36) have been

**Table 2. Maximum Stopping Distances Permissible under Uniform Vehicle Code of 1962, on Level, Dry, and Smooth Highway Surfaces, ft (m)**

| Speed | | Distance for driver to react in ¾ s | | Automobiles | | | | Buses and trucks over 10,000 lb (4,536 kg) | | | |
|---|---|---|---|---|---|---|---|---|---|---|---|
| | | | | Braking distance | | Total distance | | Braking distance | | Total distance | |
| mi/h | km/h | ft | m | ft | m | ft | m | ft | m | ft | m |
| 20 | 32 | 22 | 6.7 | 25 | 7.6 | 47 | 14.3 | 40 | 12.2 | 62 | 18.9 |
| 30 | 48 | 33 | 10.1 | 55 | 16.8 | 88 | 26.8 | 92 | 28.0 | 125 | 38.1 |
| 40 | 64 | 44 | 13.3 | 105 | 32.0 | 149 | 45.4 | 165 | 50.3 | 209 | 63.7 |
| 50 | 80 | 55 | 16.8 | 188 | 57.1 | 243 | 74.1 | 255 | 77.7 | 310 | 94.5 |
| 60 | 97 | 66 | 20.1 | 300 | 91.4 | 366 | 111.6 | 370 | 112.8 | 436 | 132.9 |
| 70 | 113 | 77 | 23.5 | 453 | 138.1 | 530 | 161.5 | | | | |
| 80 | 129 | 88 | 26.8 | 650 | 198.1 | 738 | 224.9 | | | | |

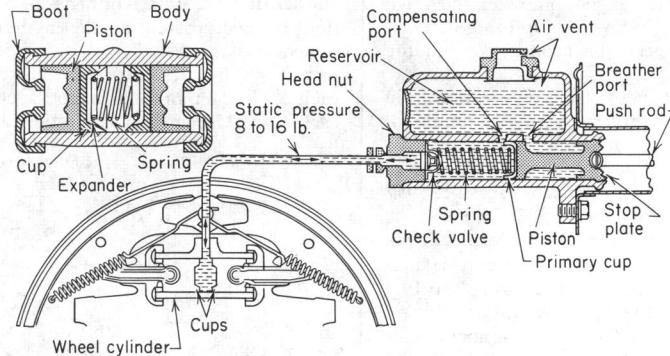

**Fig. 35**  Conventional master cylinder and wheel cylinder, brakes released.

accepted in service. All are **self-energizing,** where the drum rotation increases the applying force supplied by the wheel cylinder. Huck (*Trans. SAE*, Pt. I, 1926, p. 455) showed that self-energization, expressed as the percentage of change of effort required for a given braking effect because of the turning of the drum, is a product of the coefficient of friction and a design constant which depends on the location of the anchor pin, the arc subtended, and the location of the lining. Such a brake will self-lock if the friction coefficient exceeds a value also determined by this design constant. Burkhardt (*Trans. SAE*, Pt. II, 1925, p. 282) showed that the brake lining should be limited to a 120° arc because the self-energization of increments near the ends (especially near the toe) is so great that squealing may develop.

With the **trailing shoe** (Fig. 36*a*), friction is opposed to the actuating force. The resulting deenergization of this shoe causes it to do about one-third the work of the leading shoe. Its tendency to lock or squeal is much less, and the length and position of the lining are not so critical. The type of brake shown in Fig. 36*a*, with one leading and one trailing shoe, has been used by one U.S. manufacturer (Chrysler) for the rear wheels. The braking work and wear of the two shoes can be equalized by use of a larger bore for that half of the wheel cylinder which operates the trailing shoe.

The design shown in Fig. 36*b* has two leading shoes, each actuated by a single-piston wheel cylinder and each self-energizing. This design has been used for the front wheels where the Fig. 36*a* design was used for the rear wheels. A variation of this design has been used abroad, with two trailing shoes, and is equivalent to running Fig. 36*b* in reverse. It is less sensitive to changes in friction coefficient than other designs, and the required greater braking force is generally provided by a power booster.

Figure 36*c* shows the **Bendix Duo-Servo design,** now used on all U.S. cars, in which the self-energizing action of **two leading shoes** is much increased by operating them **"in series";** the braking force developed by the primary shoe becomes the actuating force for the secondary shoe. The action reverses with rotation. This design is even more sensitive than that of Fig. 36*b* to variations in the coefficient of friction. The lowered coefficient resulting from the heating effects of operation on long grades or repeated severe stops tends to develop brake **"fade."** Conversely, a large increase in friction coefficient, e.g., following standing with wet brakes, may develop a tendency to **"grab."**

**Adjustment for lining wear** is effected automatically on most cars. If sufficient wear has developed, a linkage may turn the notched wheel on the adjusting screw (Fig. 36*c*) by movement

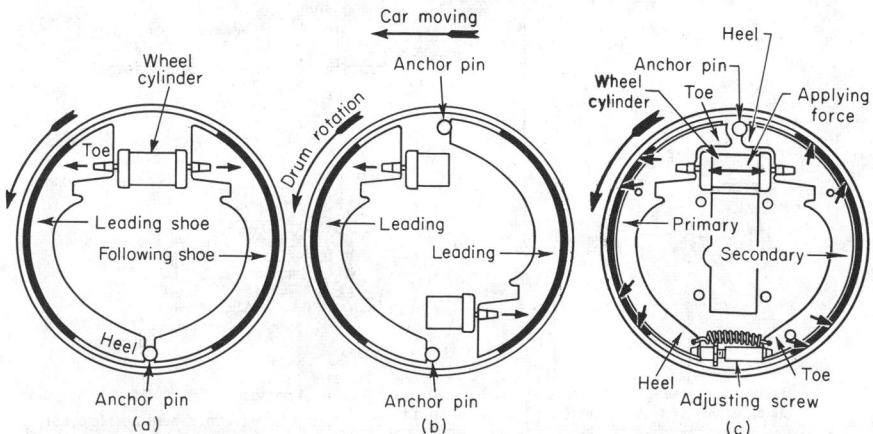

**Fig. 36**  Three types of internal-expanding brakes.

of the primary shoe relative to the anchor pin when the brake is applied with the car moving in reverse. On other designs, adjustment is by linkage between the hand brake and the adjusting wheel.

**Brake drums** are designed to be as large as practicable in order to develop the necessary torque with the minimum application effort and to limit the temperature developed in dissipating the heat of friction. The 14- and 15-in (36 and 38 cm) wheel-rim diameters prevalent on standard cars since 1957 limit the drum diameters to 10 to 12 in (25 to 30 cm), and the 13-in (33 cm) rims of "compact" cars limit the diameters to 9 to 9½ in (23 to 24 cm). Drum widths limit unit pressures between the linings and the drums to 16 to 23 lb (7.2 to 10.4 kg) of car weight per in². Drum **friction surfaces** are usually cast iron or iron alloy; **brake shoes** are lined with compounds or mixtures of asbestos, resin, and filler materials, formed relatively hard blocks and riveted or thermally bonded to the shoes; friction coefficients range from 0.3 to 0.4. Where identical brakes are used on front and rear wheels, the rear-wheel cylinders are smaller, so that about 40 to 45 percent of the total braking force is developed at the rear wheels. With the split system (Fig. 34), a smaller master cylinder for the rear brakes gives a similar division. Master cylinders are about 1 in (25.4 mm) in diameter, and other parts of the brake system are so proportioned that a 100-lb (45.4-kg) brake-pedal pressure develops 600- to 1,200-lb/in² (4.1- to 8.3-mPa) fluid pressure. Air in a hydraulic system makes the brakes feel spongy, and it must be bled wherever it accumulates, as at each wheel cylinder. Air is introduced when the system is opened for servicing or when the reservoir-fluid level is allowed to fall too low.

**Caliper disk brakes** (Fig. 37), favored particularly on racing cars, are found also on passenger cars, especially in Europe. They offer better heat dissipation by direct contact with moving air; they are not self-energizing, so that there is less drop in the friction coefficient with temperature rise of the brake shoes. Contrarily, the absence of self-energization requires higher hydraulic-system pressures and consequent power boosters on heavier cars. **Wear** of the friction pads is normally greater because of the smaller area of contact and the greater exposure to road dirt. The pads are consequently made

thicker than the linings of drum brakes, and automatic retraction is incorporated in the hydraulic cylinders. Typical dimensions are: disk diameter, 11½ in (29.2 cm); thickness, ⅜ to ½ in (9.5 to 12.7 mm); hydraulic cylinders, two per brake, each 2⅛ in (54 mm) in diameter; friction pads, two per brake, each 4 in² (25.8 cm²) in area [about 1 in² (6.5 cm²) per 152 lb

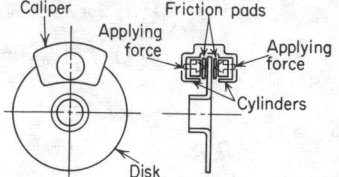

**Fig. 37**   Caliper disk brake (schematic).

(69 kg) vehicle weight]; usable thickness, 0.40 in (10 mm).

**Power-assisted brakes** relieve the driver of much physical effort in retarding or stopping a car. They are either standard or optional equipment on all major car models and some

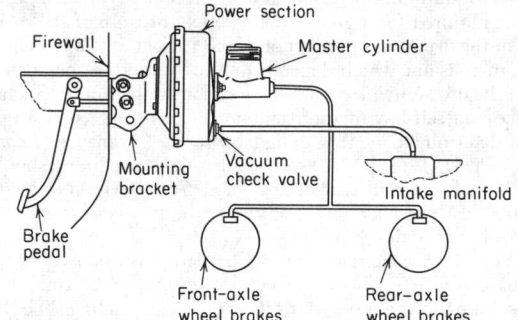

**Fig. 38**   Power-assisted brake installation.

compact cars. The supplemental force is developed on a diaphragm by vacuum from the engine intake manifold, either

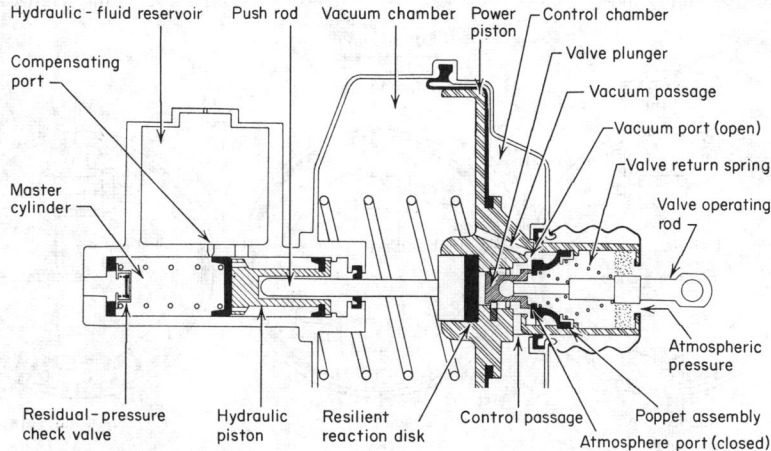

**Fig. 39**   Bendix Master Vac power brake, at rest position (schematic).

mechanically to the master cylinder or hydraulically, to boost (1) the force between the pedal and the master cylinder, or (2) the hydraulic pressure between the master cylinder and the brakes. Common characteristics are (1) a braking force which is related to pedal pressure so that the driver can feel a pedal reaction proportional to the force applied, and (2) ability to apply the brakes in the absence of the supplemental power.

Figures 38 and 39 illustrate a passenger-car **vacuum-suspended type** of power brake, where vacuum exists on both sides of the main power element when the brakes are released. In the released position, as shown, there is contact between the valve plunger and the poppet; thus the port is closed between the power cylinder and the atmosphere but there is clearance between the poppet and the piston hub, which provides a passage to equalize the pressure between both sides of the piston.

**Physical effort** applied to the brake pedal moves the valve operating rod toward the master-cylinder section. Initial movement of this rod, which carries with it both the valve plunger and the poppet (held in contact with each other by a spring), closes the port between the poppet and the power piston. This closes the vacuum passage and brings the valve plunger into contact with the resilient reaction disk. Additional movement of the valve rod then separates the valve plunger from the poppet, thus opening the atmospheric port and admitting air to the control chamber at the right of the piston. The air pressure in this chamber depends upon the amount of physical effort applied to the pedal. The pressure differences between the two sides of the power piston causes it to move toward the master cylinder, closing the vacuum port and transferring its force through the resilient reaction disk to the hydraulic piston of the master cylinder. This force tends to extrude the reaction disk against the valve plunger and react against the valve operating rod, thus reducing the pedal effort required. The hydraulic pressure is directly proportional to the effort applied by the driver, and so is the reaction force transmitted by the reaction disk, which reduces the pressure required on the brake pedal. The hydraulic pressure created by the master cylinder is the sum of the physical effort exerted on the brake pedal and the force developed by the power piston. When vacuum is not available, the unassisted effort on the pedal can still apply the brakes, though less effectively. An inherent feature of the vacuum-suspended type of power brake is the existence of vacuum, without an additional reservoir, for at least one brake stop after the engine is stopped. Figure 40 exemplifies the relationship between pedal effort and hydraulic line pressure.

The **booster-brake system** (Fig. 41), where the line pressure between the master cylinder and the brakes is boosted by a vacuum-operated hydraulic piston, has been largely superseded except for some large commercial vehicles and some foreign cars.

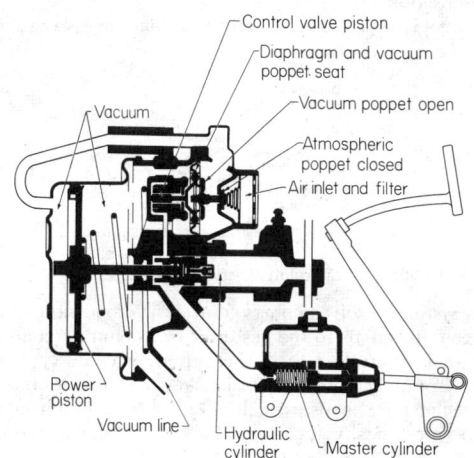

**Fig. 41** Bendix Hydra-Vac in released position.

## AIR CONDITIONING
(See also Sec. 12)

Automobiles are generally **ventilated** through an opening near the windshield with a plenum chamber to separate rain from air. Airflow developed by car motion is augmented, especially at low speed, by a variable-speed, electrically driven blower. When **heat** is required, this air is passed through a finned core served by the engine-jacket water. Core design typically calls for delivery of 20,000 Btu/h [125 ft³/min (3.5 m³/min) approx at 130°F (55°C) with 0°F (−18°C) ambient]. Car **temperature** is controlled by (1) mixing ambient with heated air, (2) mixing heated with recirculated air, or (3) variation of blower speed. Provision is always made to direct heated air against the interior of the windshield to prevent formation of ice or fog. Figure 42 illustrates schematically a three-speed blower which

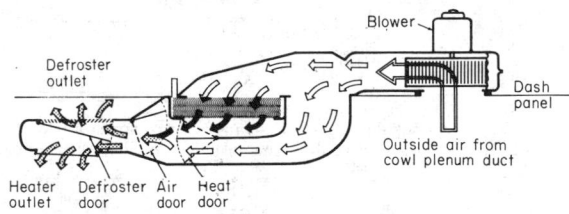

**Fig. 42** Heater airflow.

drives fresh air through (1) a radiator core, or (2) a bypass. The degree and direction of air heating are further regulated by the doors.

Almost 70 percent of the cars built in the United States today are equipped with air-conditioning systems. The **refrigeration capacity** of a typical installation is 20,000 Btu/h, or 1.5 tons, to lower the car temperature 25 to 30°F (14 to 17°C) at a

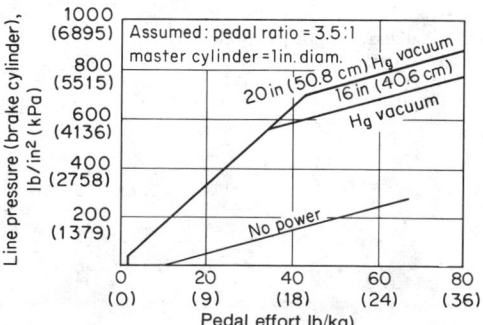

**Fig. 40** Performance chart of a typical power-assisted brake.

car speed of 30 mi/h (48 km/h). Two- or three-speed motor-driven blowers give about 1.5 air changes per minute and circulate 200 to 300 ft³/min (5.6 to 8.5 m³/min), of which 25 to 100 percent is filtered outside air. Circulation is directed along the underside of the roof or by outlets, adjustable to minimize draft sensation.

Figure 43 shows schematically a **combined air-heating and air-**

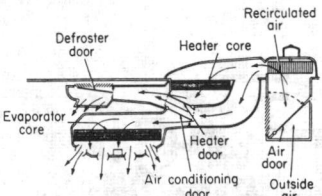

**Fig. 43**   Combined heater and air conditioner (Chevrolet).

**cooling system;** various dampers control the proportions of fresh and recirculated air to the heater or evaporator core; air temperature is controlled by a thermostat, which switches the compressor on and off through a magnetic clutch. More elaborate systems are in use which eliminate manual changeover and thermostatically actuate the heating and cooling cores.

A widely used Freon **compressor** (Fig. 44) has three horizontal cylinders arranged around the compressor shaft at 120°; the pistons are double-acting, giving the equivalent of a 6-cylinder arrangement with the merit of small torque variations. The pistons are driven by a swash plate of such thickness that its mass balances the reciprocating masses of the pistons. Pistons are 1½ in (38 mm) in diameter with a 1⁹⁄₁₆-in (30 mm) stroke, giving a displacement of 12.6 in³ (206.5 cm³) per revolution. At 3,000 r/min, this unit can develop about 3 tons of refrigeration. For smaller cars, this capacity is reduced by counterboring the piston heads for sufficient reexpansion to reduce the equivalent displacement to 10.8 in³ (177.0 cm³) without altering other dimensions. The unit is driven through a magnetic clutch with no moving electrical contacts. Refrigeration capacity is controlled by intermittent operation or by continuous operation with suction throttling. To reduce the drain on the battery, the compressor is locked out while the car engine is started.

## AUTOMOBILE ENGINES

For details on engines, see Sec. 9. For antifreeze-protection data, see Sec. 6.

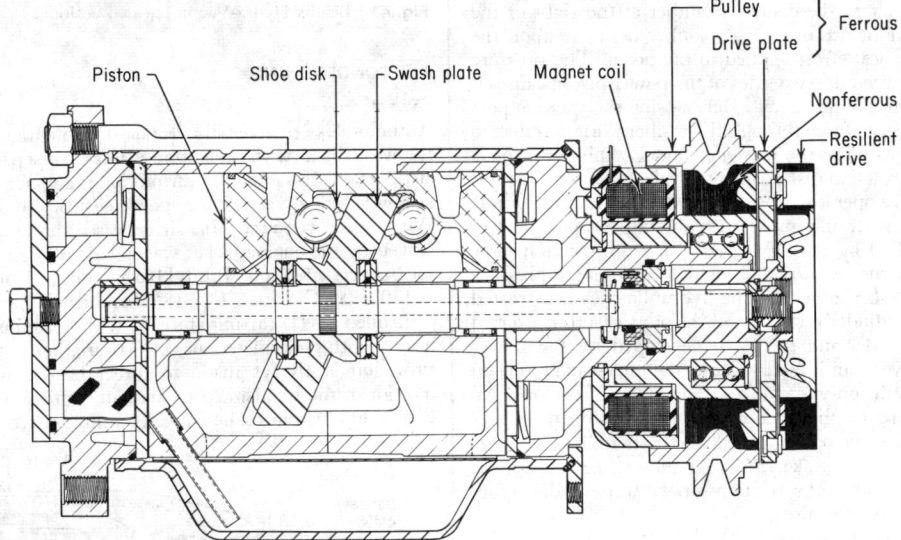

**Fig. 44**   Air-conditioning Freon compressor with three parallel cylinders; double-acting pistons driven by a swash plate; drive through a magnetic clutch with a stationary coil (Frigidaire).

# RAILWAY ENGINEERING
## by John F. Partridge

REFERENCES: *Proc. AAR*, Div. U, Mechanical. *Proc. AREA*. Bureau of Railway Economics. "Car Builder Cyclopedia," Simmons-Boardman. Sillcox, "Mastering Momentum," Simmons-Boardman. Schmidt, "High Horsepower Diesel-Hydraulic Locomotives in Heavy Duty Freight Service," *ASME Paper* 62-WA-245. Botzow, "Monorails," Simmons-Boardman. Pamphlet 1025A, Vol. I, Sec. K, General Railway Signal Co.

## ELECTRIC TRACTION

(See also Sec. 15)

Electric and diesel-electric locomotives are classified by wheel arrangement; letters represent the number of adjacent driving axles in a rigid truck (A for one axle, B for two axles, C for 3 axles, etc). Idler axles between drivers are designated by numerals. A plus sign indicates articulated trucks or motive power units. A minus sign indicates separated swivel trucks not articulated. For example, A1A − A1A + A1A − A1A represents a locomotive consisting of two articulated units mounted on six-wheel swivel trucks with motors on the end axles of each truck and one idler in each truck. This nomenclature has been adopted as standard by the AAR. (See Tables 1 and 2.)

**Drives** The principal types of drives for electric locomotives are as follows: Fig. 1a, "nose suspension," one side of the motor being carried on the driving axle and one side on the truck frame, driving through a pinion on the armature shaft and a gear on the axle; Fig. 1b "quill," the pinion on the

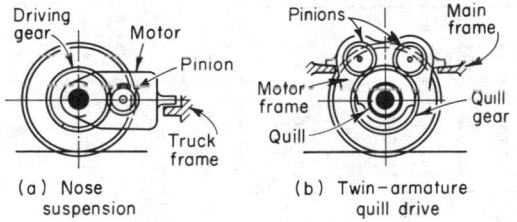

**Fig. 1** Electric-locomotive drives.

armature shaft (usually two armatures per motor are used) engaging a gear mounted on a sleeve, the latter normally concentric with the axle but capable of permitting the axle to move vertically to accommodate itself to the track surface and with the locomotive frame carrying the total weight of the motor and quill. The nose-suspension drive is used almost exclusively in new locomotives. Other forms of drives have been used, including a linkage drive, jack shaft and side rods, and a yoke drive, but have not been generally adopted by American designers. The characteristics of a satisfactory drive are: maximum proportion of weight carried on the frame rather than on the axle; motor weight well above the axle center; freedom for the driving axle to move vertically to accommodate variation in the track surface; a connection with

some flexibility between the armature shaft and the driving axle to absorb shocks.

**Traction Motors** Fundamental relations for motors used in railway service take the forms shown in Fig. 2.

The **series motor,** either ac or dc, has the inherent advantage of the combination of low speed, large torque, and high

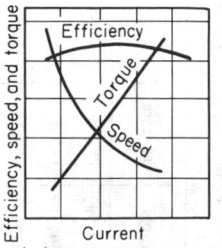 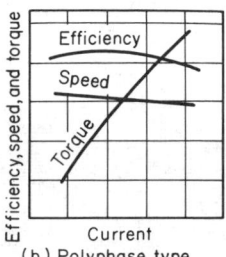

(a) Series type a-c or d-c      (b) Polyphase type

**Fig. 2** Characteristics of electric-locomotive motors.

starting current. At speeds below that at which maximum current can be imposed on the armature, the current is limited by controller resistance, which is cut out, step by step, as the motor speeds up. The normal motor connections do not permit a wide enough range of speeds. With the dc motor, variable connections are provided through the controller to change from series or series parallel (at low speeds) to full parallel at high speed. Field shunting is also used in some cases. Characteristics of the series motor are more fully described in the section on diesel locomotives. Curves shown in Fig. 3 afford a comparison of the types of motors used and show relations between speed, tractive force, and current. For the **ac motor,** the motors are placed permanently in parallel, and speed control is obtained by "tapping" the transformer to secure desirable fractional voltages (Fig. 3b). The **polyphase motor** has an important advantage in its lack of commutator and brushes, but being a constant-speed motor it has its principal application in slow freight service with infrequent stops. Speed variation, at considerable expense in efficiency, can be obtained for starting by the insertion of resistance in the short-circuited rotor circuit as shown in Fig. 3c.

Two (sometimes three) running speeds can be obtained by changing the number of poles, the speed varying inversely with the number operative, or by reconnection in concatenation, giving speeds half as great as the normal characteristic.

**Ignitron Rectifier** As a result of improvements in the design of the mercury-pool ignitron rectifier, it has been adopted for use in electric locomotives. Its use permits combining the economic advantages of the high-voltage ac overhead catenary and distribution systems with the efficiency of dc traction motors and their desirable operating characteristics. The ac voltage at the pantograph is stepped down by a liquid-cooled transformer. Transformer output is fed through the rectifiers before going to the traction motors. Control of

**Table 1. Electric Locomotives**

| Road | Date built | Service | Arrangement | Starting tractive force, limited by motors or 25% adhesion | Weight 1,000 lb[a] | | Wheelbase | | Volts | | Hp rating | | Transformer or other equipment | Number of traction motors | Drive | Max speed, mph |
|---|---|---|---|---|---|---|---|---|---|---|---|---|---|---|---|---|
| | | | | | Total | On drivers | Rigid | Total | Supply | Motors | 1 hr | Continuous | | | | |
| Various.... | ..... | Sw | (1)[b] (2)[d] | 25,000 | 100 | 100 | 6'6" | 22'–25' | 600–1,500 d-c | 600–750 d-c | 500 | 400 | None | 4 | Nose susp | 40–45 |
| N.Y.C.... | 1930[c] | Pass. | | 72,125 | 388.3 | 288.5 | 15'0" | 69'0" | 600 d-c | 600 d-c | ..... | 4,750 | [c] | 6 | Nose susp | 70 |
| P.R.R.... | 1951 | Frt | | 187,750[a] | 751[a] | 751 | 9'6" | 43'2" | 11,000 a-c | 600 d-c | ..... | 5,984 | Ign rectif | 12 | Nose susp | 63 |
| N.Y.N.H. &H. | 1954 | Pass. | | 87,000 | 348 | 348 | 15'0" | 52'6" | 11,000 a-c | 600 d-c | ..... | 4,000 | Ign rectif | 6 | Nose susp | 90 |
| N.Y.N.H. &H. | 1938 | Pass. | | 68,000 | 432 | 272 | 13'8" | 66'0" | 11,000 a-c 600 d-c | 600 d-c[f] or a-c S.P. | ..... | 3,600 | Transf | 6 twin | Quill | 90 |
| P.R.R.... | 1935 | Pass. | | 72,300 | 460 | 300 | 13'8" | 69'0" | 11,000 a-c | 600 a-c S.P. | 8,500[g] | 4,620 | Transf | 6 twin | Quill | 100 |
| Great Northern | 1947 | P&F | | 180,000 | 735 | 735 | 16'9" | 85'9" | 11,000 a-c | 600 d-c | ..... | 5,000 | Motor-gen set | 12 | Nose susp | 65 |
| N.&W.... | 1948 | Frt | | 260,000[a] | 1,033.8[a] | 1,033.8 | 9'0" | 133'10"[a] | 11,000 a-c | 600 d-c | ..... | 6,800[a] | Motor-gen set | 16 | Nose susp | 50 |
| N.Y.N.H. &H. | 1956 | Frt | | 118,000 | 394 | 394 | 13'0" | 52'9¾" | 11,000 a-c | 1,000 d-c approx | ..... | 3,300 | Ign rectif | 6 | Nose susp | 65 |
| P.R.R.... | 1960–1963 | Frt | | 89,000 | 386 | 386 | 13'0" | 52'9¾" | 11,000 a-c | 1,000 d-c approx | ..... | 4,400 | Ign rectif or sel rectif | 6 | Nose susp | 70 |
| Butte Anacon. & Pacific | 1957 | Frt | | 75,000 | 250 | 250 | 10'0" | 30'3½" | 2,400 d-c | 1,200 d-c | ..... | 2,480 | None | 4 | Nose susp | 45 |

[a] For entire locomotive.
[b] ● = driving wheel.
[c] Rebuilt 1955.
[d] ○ = non-driving wheel.
[e] Three series—two parallel and full parallel.
[f] Motors operate both d-c and a-c.
[g] "Short-time rating."

tractive effort is accomplished by cutting in accelerating resistors between rectifier and motors and by changing transformer taps. Motors are also reconnected at higher locomotive speeds (transition) to maintain proper level of armature current. (See also Sec. 15.)

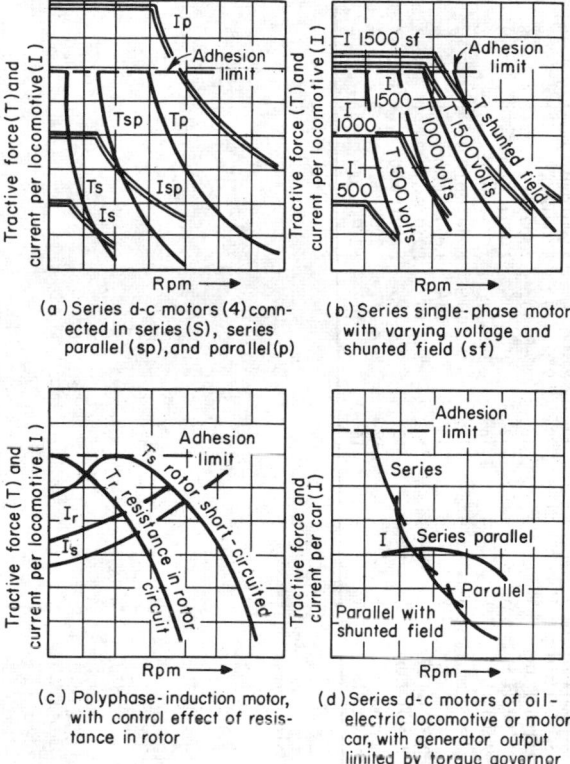

(a) Series d-c motors (4) connected in series (S), series parallel (sp), and parallel (p)

(b) Series single-phase motor with varying voltage and shunted field (sf)

(c) Polyphase-induction motor, with control effect of resistance in rotor

(d) Series d-c motors of oil-electric locomotive or motor car, with generator output limited by torque governor

**Fig. 3**  Speed–tractive-force relations of electric locomotives.

Recently, ignitron locomotives, gas-turbine locomotives, and some diesel-electric locomotives have been built with three-axle swing bolster trucks and three nose-suspension-type dc traction motors. To provide room for the motor on the center axle the center plate is located between two adjacent axles. Two horizontal spring-loaded sliding plates on each side of the truck frame also carry underframe load and provide proper distribution of weight on the axles.

## DIESEL-ELECTRIC LOCOMOTIVES

**General**  The high efficiency of the diesel engine is an important factor in its selection as a prime mover for traction. It is a constant-torque machine that cannot be started under load and hence requires a variably coupled transmission arrangement. Being essentially a constant-horsepower machine it is ideally suited to switching service since it can make use of its full rated-power output at low track speeds. Examples of the various types and service designs are shown in Table 2.

Diesel-electric locomotives have a dc generator coupled directly to the diesel-engine crankshaft. The generator is electrically connected to series traction motors having a nose-suspension mounting. The tooth ratio of the axle-mounted gears to the motor-pinion gears which they engage determines the speed range and is related to the type of service. A high ratio is used where high tractive efforts and low speeds are required, whereas high-speed passenger locomotives have a low ratio.

Most locomotives have four traction motors per unit and either four or six axles. Where high tractive efforts are required, six motors may be used to advantage, especially where track conditions require low axle loading.

### Diesel Engines
(See also Sec. 9)

**Types**  Diesel engines as developed for locomotive applications include the V type, two-cycle; V type, four-cycle; 6 or 8 cylinder in-line, four-cycle; and 8, 10, or 12 cylinder opposed-piston, two-cycle. Two-cycle engines are equipped with an air-intake blower or supercharger, gear-driven from the crankshaft. Four-cycle engines may be normally aspirated or supercharged. The supercharger turbine is driven by expansion of engine exhaust gases and usually accounts for 50 percent increase in the horsepower rating. In the V-type, two-cycle engine, the fuel injector mounted in each cylinder head is used to meter the fuel as well as atomize it, whereas in four-cycle engines a separate injector pump for each cylinder delivers a metered amount of fuel per stroke under high pressure to the injector.

The engine-control **governor** is an electrohydraulic device used to regulate the speed and horsepower output of the diesel engine. It is a self-contained unit mounted on the engine and driven off one of the engine camshafts. It is equipped with its own oil supply and oil pressure pump.

The governor contains four solenoids which are actuated individually or in combination from the 74-V auxiliary generator supply by a series of throttle switches actuated by the engineer's throttle. There are eight power positions of the throttle, each corresponding to a specific value of engine rpm. The governor maintains the predetermined value of engine rpm through a mechanical linkage to the engine fuel racks which controls the amount of fuel metered to the cylinders.

Since the engine–main-generator power plant functions as a constant-horsepower source of energy, the main-generator voltage must be controlled to provide a constant value of power output for a specific throttle position under varying conditions of locomotive speed, track gradient, atmospheric quality, and quality of fuel. The **load regulator,** which is an integral part of the governor, accomplishes this within the maximum safe values of main-generator voltage and current. For example, when the locomotive experiences an increase in track gradient with a consequent reduction in speed, traction-motor counter emf decreases, causing a change in traction-motor and main-generator current. Because this alters the load demand on the engine, speed of the engine tends to change to compensate. As the speed changes, the governor begins to reposition the fuel racks, but at the same time a pilot valve in the governor directs hydraulic pressure into a load-regulator vane motor which changes the resistance value of the load-regulator rheostat in series with the main-generator excitation circuit. This alters the main-generator excitation current and, consequently, main-generator voltage and returns the value of

**Table 2. Diesel-Electric Locomotives**

| Builder | Service | Arrangement* | Weight, lb | Tractive force, lb | | Engines | | | | | | | Wheelbase, ft-in | |
|---|---|---|---|---|---|---|---|---|---|---|---|---|---|---|
| | | | | Starting | Continuous, at speed (mi/h) | Type | No. of cycles | Cylndr. per eng. | Cylinder size, in | hp rating at (r/min) | Engs. per unit | No. of motors | Truck | Center plate |
| EMD | Sw | | 248,000 | 62,000 | 31,200 (12.0) | V | 2 | 12 | 8½ × 10 | 1,200 (800) | 1 | 4 | 8-0 | 22-0 |
| Alco | Sw | | 230,000 | 57,500 | 34,000 (8.0) | Inline | 4 | 6 | 12½ × 13 | 1,000 (740) | 1 | 4 | 8-0 | 22-6 |
| EMD | Rd sw | | 247,500 | 74,250 | 55,400 (10.8)† | V | 2 | 16 | 9 9/16 × 10 | 2,000 (900) | 1 | 4 | 9-0 | 34-0 |
| EMD | Rd sw | | 254,500 | 76,350 | 55,400 (11.3)† | V | 2 | 16 | 9 9/16 × 10 | 3,000 (900) | 1 | 4 | 9-0 | 34-0 |
| EMD | Rd sw | | 364,250 | 109,000 | 83,100 (11.1)† | V | 2 | 16 | 9 9/16 × 10 | 3,000 (900) | 1 | 6 | 13-7⅞ | 43-6 |
| EMD | Rd sw | | 364,250 | 109,000 | 83,100 (11.3)† | V | 2 | 20 | 9 9/16 × 10 | 3,600 (900) | 1 | 6 | 13-7⅞ | 43-6 |
| EMD | Rd sw‡ | | 488,000 | 122,000 | 102,800 (12.0) | V | 2 | 16 | 8½ × 10 | 2,500 (900) | 2 | 8 | 17-0¼ | 55-0 |
| Alco | Rd sw | | 240,000 | 60,000 | 53,000 (11.5) | V | 4 | 12 | 9 × 10½ | 2,000 (1,025) | 1 | 4 | 9-4 | 34-5 |
| Alco | Rd sw | | 340,000 | 85,000 | 79,500 (11.5) | V | 4 | 16 | 9 × 10½ | 2,750 (1,050) | 1 | 6 | 12-6 | 41-6 |
| GE | Rd sw | | 246,000 | 80,000 | .... | V | 4 | 12 | 9 × 10½ | 2,250 (1,025) | 1 | 4 | 9-4 | 36-2 |
| GE | Rd sw | | 363,000 | 120,000 | .... | V | 4 | 16 | 9 × 10½ | 3,000 (1,025) | 1 | 6 | 13-9 | 40-11 |
| EMD | Frt | | 246,000 | 61,500 | 40,000 (11.5) | V | 2 | 16 | 8½ × 10 | 1,750 (835) | 1 | 4 | 9-0 | 30-0 |
| Alco | Frt | | 240,000 | 60,000 | 38,000 (12.0) | V | 4 | 12 | 9 × 10½ | 1,600 (1,000) | 1 | 4 | 9-4 | 29-2 |
| EMD | Pass | | 331,800 | 55,300 | § | V | 2 | 12 | 8½ × 10 | 1,200 (800) | 2 | 4 | 14-1 | 43-0 |
| Alco | Pass | | 303,000 | 50,500 | 33,600 (21.0) | V | 4 | 16 | 9 × 10½ | 2,250 (1,000) | 1 | 4 | 15-6 | 34-2 |

* ● = driving wheel; ○ = nondriving wheel.
†With gear ratio of 62:15, maximum speed 65 mi/h.
‡Has no operating cab; must be operated in road service in multiple with other units.
§In most cases, rating will be determined by adhesion.

main-generator power output to normal. Engine fuel racks return to normal consistent with constant values of engine rpm.

The load regulator is effective within maximum and minimum limit values of the rheostat. Beyond these limits the power output of the engine is reduced. However, protective devices in the main-generator excitation circuit limit values of voltage output to ensure that values of current and voltage in the traction-motor circuits are within safe limits.

### Electric Traction Equipment

The **main generator** has several field windings incorporated in the stator; i.e., series, shunt, differential, battery (or separately excited from a smaller generator, or "exciter"), and starting. The last is employed when feeding power from the storage batteries into the main generator to start the engine.

Generator power output is controlled by (1) varying engine speed through movement of the engineer's controller, and (2) controlling the flow of current in the main-generator battery field or in the field of a separate exciter generator. The shunt and differential fields (if used) are designed to contribute toward a constant generator power output for a given engine speed as the armature current and voltage vary. They do not completely accomplish this. The battery or separately excited field is controlled by the load regulator to provide the final adjustment in excitation to load the engine properly. It is also automatically deenergized to reduce or remove the load, when certain undesirable conditions occur, to prevent damage to the power plant or other traction equipment.

**Auxiliary Generating Apparatus** DC power for battery charging, lighting, and control is furnished by a separate generator, geared or belt-driven from the main generator. Voltage output is regulated within 1 percent over the full range of engine speeds at 74 V. A separate generator may be provided to supply power for motor-driven blowers to ventilate the traction motors and for the radiator-fan motors. It may be a dc machine or a three-phase alternator. The alternator is mounted on the end of the main generator and directly coupled to it.

**Traction-motor blowers** are mounted above the locomotive underframe. Air from the centrifugal-blower housings is carried through the underframe and into the motor housings through flexible ducts.

**Traction motors** employ series- and commutating-field poles. The current in the series fields is reversed to change locomotive direction and may be partially shunted through resistors to reduce counter emf as locomotive speed is increased. Armature shafts are equipped with grease-lubricated roller bearings. Axle bearings are of the friction type and have lubricant wells and wool waste or spring-loaded felt wicks which maintain constant contact with the axle surface.

### Electrical Controls

Electropneumatic **contactors** are used to make and break the proper circuits between traction motors and the main generator. They are equipped with interlocks for various control-circuit functions. Magnetically operated contactors are used for other power and excitation circuits of lower amperage. An electropneumatic **cam switch,** consisting of a two-position drum with copper segments moving between spring-loaded fingers, is used to reverse traction-motor field current ("reverser") or to

set up the circuits for dynamic braking. This switch is not designed to operate under load.

Electrically actuated **relays** are used frequently in control and alarm circuits. Their contacts usually handle 75 V and 1 A or less. Relay-operating coils may be energized by the 75-V control system, or they may be preset to respond to a particular voltage level. The latter is sometimes used to activate transition and motor-shunting equipment in response to the level of main-generator voltage. Wheel-slip relays operate by sensing the difference in magnitude of the current in two traction-motor armatures paralleling each other or by sensing the difference in voltage drop across two motor armatures that are connected in series.

**Propulsion control circuits** transmit the engineer's movements of the throttle lever, reverse lever, and transition or dynamic brake control lever in the controlling unit to the power-producing equipment of each unit operating in multiple in the group. Before power is applied, all reversers must move to provide proper motor connections for the movement desired, taking into account the direction in which each individual unit is heading. Power contactors complete the proper circuits between generators and motors. Excitation circuits are then allowed to function to provide the proper main-generator field current while the engine speed increases to correspond to the engineer's controller position.

**Wheel slip** is detected by some form of speed-sensing equipment connected electrically to the motor circuits or connected mechanically to the axles. When slipping occurs, relays automatically reduce main-generator excitation or engine speed until slipping ceases, whereupon power is gradually reapplied. A warning light and buzzer in the operating cab inform the engineer, who, if the condition persists, must then notch back on the throttle.

### Batteries

Lead-acid storage batteries of 280 or 420 A·h capacity are usually used for starting the diesel engine. Thirty-two cells on each locomotive unit are used to provide the standard 64-V system. (See also Sec. 15.)

### Braking

**Air-brake System** The "independent" brake-valve handle at the engineer's position controls the air pressure supplied from the locomotive reservoirs to the brake cylinders on the locomotive itself. The "automatic" handle controls the air pressure in the brake pipe to the train.

Air for braking and for various pneumatic controls on the locomotive is supplied by a two-stage three-cylinder compressor, usually connected directly to the main-generator shaft. Since it is constantly moving with the rotation of the engine, a pressure-sensitive unloader is provided to maintain a pressure of approximately 130 to 140 lb/in² (9.1 to 9.8 kPa) in the main reservoirs. When charging an empty train line with the locomotive at standstill (maximum compressor demand), the engineer may increase engine (and compressor) rpm without feeding power to the traction motors.

**Dynamic Braking** On some locomotives dynamic brakes supplement the air-brake system. The traction motors are used as generators to convert the kinetic energy of the locomotive and cars into electric energy, which is dissipated through resistance grids located near the roof. Motor-driven blowers, designed to utilize some of this energy, force outside air over the grids and out the top of the roof. By directing a generous and evenly distributed air stream over the grids their physical size is reduced in keeping with the relatively small space available in the locomotive.

By means of a cam switch, the traction motors are connected to the resistance grids; their fields are usually connected in series across the main generator to supply the necessary high current excitation. By controlling the engine speed and/or main-generator excitation, the magnitude of the braking force is determined. Dynamic braking is not usually effective below 10 mi/h (16 km/h) on most locomotives, but it is very useful at 20 to 30 mi/h (32 to 48 km/h), particularly on long grades, where brake-shoe wear would otherwise be considerable. Other advantages are relieving crews from setting up air-brake retainers on freight cars, smoother control of train speed, and reduction in the number of wheel-rim thermal cracks.

An automatic protective device may be incorporated in the excitation circuit which limits motor currents to a maximum safe value. A light at the engineer's position provides a warning that further braking effort cannot be obtained and the air-brake system must be used for additional retardation.

## Performance

**Engine Indicated Horsepower** The horsepower delivered at the diesel-locomotive drawbar is the end result of a series of subtractions from the original indicated horsepower of the engine, in order to take into account efficiency of transmission equipment and losses due to the power requirements of various auxiliaries. The formula for the engine's indicated horsepower is ihp = $PLAN/33,000$, where $P$ = mean effective pressure in the cylinder, lb/in²; $L$ = length of piston stroke, ft; $A$ = piston area, in²; $N$ = total number of cycles completed per min. The factor $P$ is governed by the overall condition of the engine, quality of fuel, rate of fuel injection, completeness of combustion, compression ratio, etc. Factors $L$ and $A$ are fixed with the design of the engine. Factor $N$ is a function of engine speed, number of working chambers, and strokes needed to complete a cycle. (See also Secs. 4 and 9.)

**Engine Brake Horsepower** In order to obtain horsepower delivered to the crankshaft coupling (main-generator connection), frictional losses in bearings and gears must be subtracted from ihp. Also some power is used to drive lubricating-oil pumps, governor, water pump, and scavenging blower. The resultant horsepower at the coupling is brake horsepower (bhp).

**Thermal efficiency** of the diesel engine at the crankshaft, or the ratio of bhp output to the rate at which energy of the fuel is delivered to the engine, is about 30 percent. Thermal efficiency at the rail is about 23 percent.

**Rail Horsepower** A portion of the engine bhp is transmitted mechanically through the generator armature shaft via couplings or belts and pulleys to traction-motor blowers, air-brake compressor, auxiliary generator, and radiator cooling-fan generator or alternator. In some cases part of the generator electric output is used to run some of the auxiliaries. The remainder of the engine bhp transmitted to the main generator for traction purposes must be multiplied by generator efficiency (usually about 91 percent), and the result again multiplied by the efficiency of the traction motors (including power circuits) and gears to arrive at rail horsepower. Power output of the main generator for traction may be expressed as

$Watts_{traction} = E_g \times I_m$, where $E_g$ = main-generator voltage, $I_m$ = traction-motor current, A, multiplied by the number of parallel paths. Rail horsepower may be expressed as $hp_{rail}$ = mi/h × $TE/375$; where $TE$ = tractive effort at the rail, lb.

**Drawbar horsepower** represents power available at the rear of the locomotive to move the cars and may be expressed as

$$hp_{drawbar} = hp_{rail}$$
$$- \text{locomotive running resistance} \times mph/375$$

Running-resistance calculations are discussed below under Train Resistance. Theoretically, drawbar horsepower available, therefore, is power output of the diesel engine less parasitic losses described above.

**Speed–Tractive Effort** At full throttle the losses vary somewhat at different values of speed and tractive effort, but a curve of tractive effort plotted against speed is nearly hyperbolic. Figure 4 is a typical speed–tractive-effort curve for a 1,500-hp freight locomotive. The diesel-electric has full horsepower available over the entire speed range (within limits described below).

**Adhesion** In Fig. 4 the maximum value of tractive effort represents the point of wheel slippage with average rail condi-

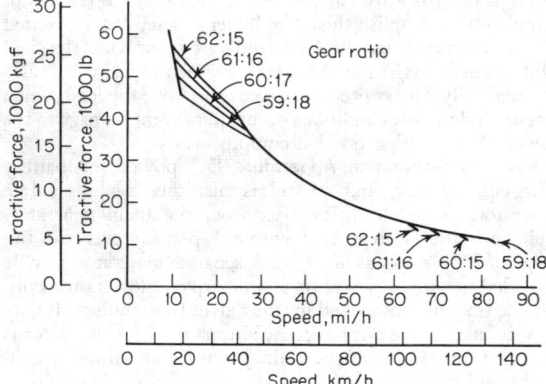

**Fig. 4** Speed–tractive-force relations of a 1,500-hp diesel-electric locomotive.

tions. It is usually taken at 25 percent adhesion or 25 percent of total locomotive weight on drivers. Actually the point of wheel slip will vary considerably with rail conditions, being as low as 10 percent or as high as 35 percent. Adhesion is greatly reduced by lubricants which spread as thin films in the presence of moisture on running surfaces.

**Traction-motor Characteristics** Motor torque is a function of armature current and field flux (which itself is a function of field current). Since traction motors are series-connected, armature and field current are the same (except when field-shunting circuits are introduced), and therefore tractive effort is a function solely of motor current. Figure 5 is a typical group of curves of tractive effort plotted against motor amperes. The indicated values of percent represent degrees of field shunting (see Saturation Curves). Wheel diameter and gear ratio must be specified when plotting torque in terms of tractive effort. (See also Sec. 15.)

Traction motors are usually rated in terms of maximum continuous current. This represents the current whose **heating effect** due to $I^2R$ losses in the armature and field windings is sufficient to raise the temperature of the motor to its maximum safe limit when cooling air at maximum expected ambient temperature is forced through it by the blowers at the prescribed rate. Continuous operation at this current level results in the motor operating at its maximum safe level, heat gener-

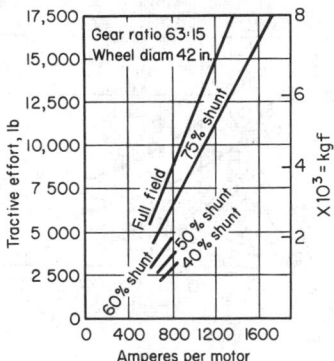

**Fig. 5**  Motor transmission.

ated being just equal to heat dissipated. The tractive effort corresponding to this current is usually somewhat lower than the adhesion limit above. Operation must be restricted by the engineer when hauling heavy trains up long grades. Somewhat higher current values may be permitted for short periods of time. These ratings are specified in terms of load-meter amperes and corresponding time intervals and are posted on or near the meter dial in the cab.

**Maximum Speed**  Traction motors are also rated in terms of maximum safe rpm, which in turn limits locomotive speed. The gear ratio and wheel diameter also alter speed and tractive effort at which full horsepower is developed through the motors at their continuous rating. Maximum locomotive speed may be expressed as follows:

$$\text{mph}_{max} = \frac{\text{wheel diam (in)} \times \text{motor rpm}_{max}}{\text{gear ratio} \times 336}$$

where gear ratio is the number of teeth on the motor gear mounted on the axle divided by the number of teeth on the pinion mounted on the armature shaft.

**Transition**  The primary function of the load-regulating equipment is to control main-generator excitation to utilize rated engine horsepower for traction over the full range of locomotive speeds. Since current is high at the high-tractive-effort end of the *TE*-speed curve and low at the opposite end, main-generator voltage must be regulated accordingly to hold the product of volts and amperes nearly constant. The design of generator and motors imposes practical limits in both directions. Figure 6 is a series of typical main-generator voltage-current curves for the eight throttle positions. To live within these limits and still accomplish the desired objective, two principal types of power circuits are employed: (1) transition and (2) field shunting. Sometimes they are used in combination. In Fig. 7, motors are connected in series when starting, by closing only contactor S-1. As current demand falls off, the S-1 contactor opens and P-1 and P-2 contactors close, providing a series-parallel connection.

**Field Shunting**  In Fig. 8, motor connectons are permanently in series parallel. As current demand decreases, shunt-

ing contactors M-1 and M-2 close at specific points on the generator voltage-current curve, reducing motor field current. For example, if M-1 and M-2 close at about 25 mi/h, motor-current demand suddenly increases to maintain the same tractive effort. The shunting effect also reduces required generator voltage at that speed and therefore permits a repositioning higher on the curve. As locomotive speed further increases, current demand is gradually reduced until a second set of contactors close (M-3 and M-4).

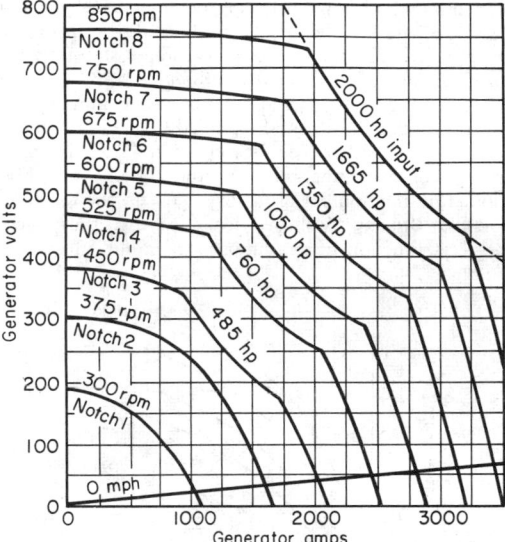

**Fig. 6**  Characteristic curves of a generator for a 2,000-hp diesel-electric locomotive.

**Saturation Curves**  Figure 9 shows a series of curves for a traction motor with internal voltage per rpm plotted against armature current. The uppermost curve represents equal field and armature currents (full field strength). Remaining curves are designated in terms of percent expressed as field current/armature current $\times$ 100%. The ratio may be expressed in terms of resistance values: $R_s/(R_f + R_s) \times 100\%$, where $R_f$ = motor series field resistance, $\Omega$, and $R_s$ = shunting resistance in parallel with the field, $\Omega$. The voltage-current relationship in a typical diesel-locomotive power circuit may be expressed as

$$E_g = [(\text{emf/rpm}) \times \text{mi/h} \times C + I_m R_m]N + I_m R_c$$

or    $\text{mi/h} = [E_g - I_m(R_c + R_m N)]/(\text{emf/rpm}) \times N \times C$

where $E_g$ = main-generator voltage; emf/rpm = motor inter-

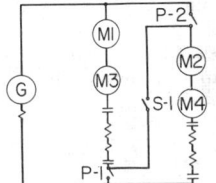

**Fig. 7**  Transition circuit.

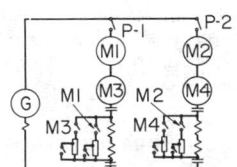

**Fig. 8**  Field shunting circuit.

nal volts per rev of the armature; mi/h = locomotive speed; $I_m$ = motor-armature current, A; $R_m$ = motor-armature and field resistance, $\Omega$; $N$ = number of motors in series; $R_c$ = other resistance in one path of the circuit through the motors, including cables, motor brushes, contactors, etc., $\Omega$; $C$ = constant represented by

$$\frac{336 \times \text{gear ratio (gear)/(pinion)}}{\text{wheel diam (in)}}$$

Speed–tractive-effort curves may be determined by using this formula, together with the appropriate curves illustrated in Figs. 5, 6, and 9. When fields are shunted, the appropriate emf/rpm curve must be used, and the proper shunting resistance taken into account in determining $R_m$. By assuming values of motor current, main-generator voltage may be determined from Fig. 6. Emf/rpm is selected from Fig. 9, and the formula used to find speed. The curve in Fig. 5 is used to determine corresponding tractive effort for one motor and multiplied by the number of motors to obtain the total.

**Dynamic Braking** Voltage generated by the traction motors is a function of revolutions per minute (locomotive

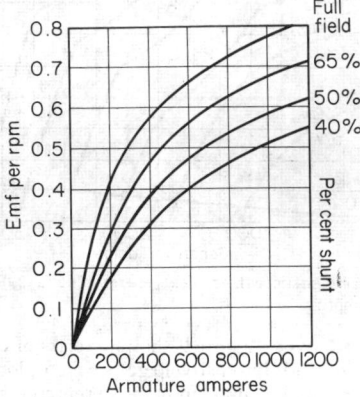

**Fig. 9**  Motor load saturation.

speed) and field current. Since the resistance of the grids is fixed, the current output and torque, or braking force, are also functions of these variables. Therefore, at low speeds, braking effort is limited by the current that can be passed through the motor field coils. In Fig. 10 this limit is represented by the portion of the curve from A to B. At high locomotive speeds, the limit is set by the current-carrying capacity of the resistance grids, or the portion of the curve from B to C. At speeds above point B, the value of field current must be lowered as speed increases, to limit the voltage and to stay within the grid capacity. Since the braking effect is a function of armature current and field current, the result is a gradual decrease in braking force. Since the kinetic energy of a train increases as the square of the speed, it may be seen that as speeds increase above 35 or 40 mi/h (56 or 65 km/h) the engineer depends more on the air-brake equipment.

## DIESEL-HYDRAULIC LOCOMOTIVES

In 1961, some of the Western railroads in the United States began using 4,000-hp (engine bhp) locomotives of European

manufacture and powered by two diesel engines driving six axles through two hydraulic transmissions. Three of these units are commonly used at the head end of a train of 4,500 gross tons to negotiate a long 2 percent grade. For the same tonnage and grade, five diesel-electric locomotives would be

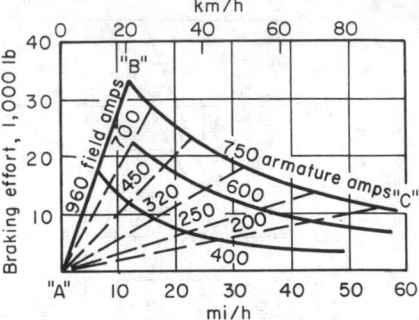

**Fig. 10**  Dynamic brake limits.

required, each rated at 1,750 hp (1,300 kW) (into main generator for traction). Each unit weighs 172 tons and is 67 ft long, with a maximum speed of 70 mi/h. Figure 11 is a graph of drawbar pull versus track speed. It also shows the braking effort of the hydrodynamic brake plotted against track speed.

The two **diesel engines** are 16-cylinder V types, with four-stroke cycle; each engine is rated at 2,000 bhp (1,500 kW), has 16.2:1 compression ratio, 7.3-in bore, and 7.9-in stroke, and is equipped with a water-cooled exhaust-gas supercharger. Unit injectors combine the functions of fuel pump and injection valve.

Each engine is connected by a Cardan shaft to a **hydraulic transmission** containing three torque converters and a built-in reversing gear. Mechanical energy at the transmission input

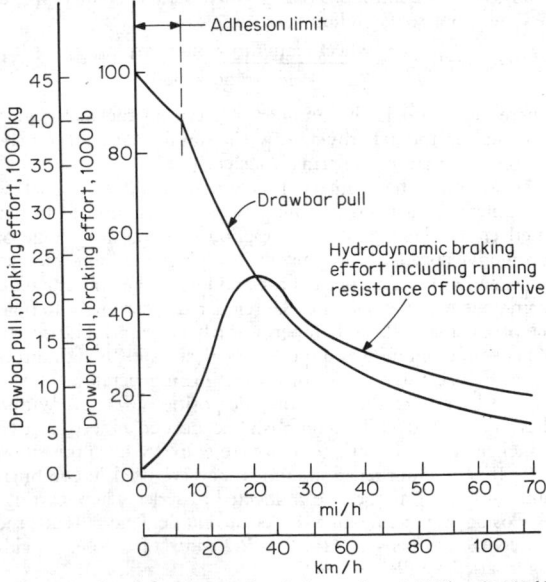

**Fig. 11**  Speed-drawbar pull relations of a 4,000-hp (3,000-kW) diesel-hydraulic locomotive. (*Krauss-Maffei A. G. München-Allach.*)

shaft is converted to kinetic energy in the "pump impellers" and changed into mechanical energy in the turbine runners connected to the output shaft. Each of the three torque converters, which contain their own impellers and runners, is assigned a well-defined speed range of the locomotive. The filling and emptying of the oil circuit for each converter are accomplished automatically, being controlled by track speed and the position of the locomotive throttle. Figure 12 is a graph of transmission efficiency plotted against track speed and shows the locomotive speed ranges of the three converters. Transition from one to the other is gradual because the oil from one is emptying while the other is filling as the locomotive speed changes. The efficiency of the transmission and gearing between the diesel engine and the axles is comparable to the efficiency of the generator–traction-motor transmission system in a diesel-electric locomotive.

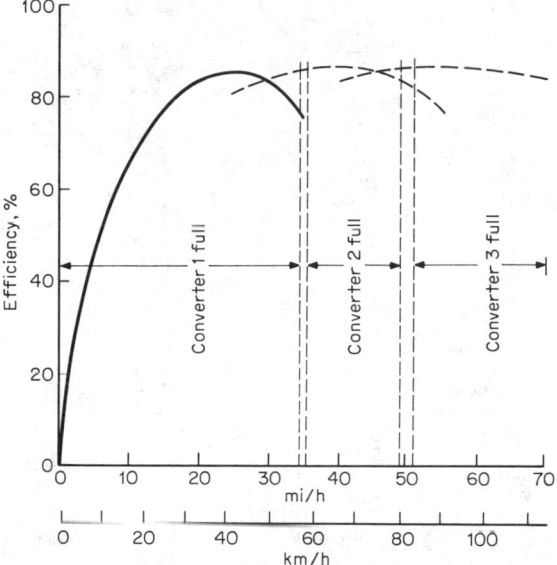

**Fig. 12** Curve of transmission efficiency and track speed for a diesel-hydraulic locomotive.

The output shaft of each transmission drives an **intermediate-gear assembly** mounted on the three-axle truck through a Cardan shaft. This assembly contains an encased spur gear which drives three double-reduction spur-bevel gears, one mounted on each axle.

The **hydrodynamic brake** operates on the principle of the Froude brake. One brake is mechanically connected to the output shaft of each transmission unit. With the engines idling, activation of the brake is accomplished by filling the hydraulic coupling circuit. Heat is dissipated in eddy formations, causing the temperature of the fluid in the circuit to increase. A heat exchanger is located in the circuit, which dissipates the heat through the locomotives water-cooling system. This system is also used to cool the engine-jacket, engine, and gear oil and the transmission oil.

Hydrostatically driven cooling fans, powered by multi-pistoned oil pumps and belt-connected to the engines, pull outside air through the radiators to maintain water temperature approximately constant. When the engines are idling down long mountain grades, the hydrodynamic braking heat helps keep the engines at the desired operating temperature.

## GAS-TURBINE LOCOMOTIVES

(See also Sec. 9)

**Design** Use of gas-turbine locomotives in the United States has been confined chiefly to one portion of one of the Western railroads. Locomotives constructed in 1956 are built in three sections, each with its own independent underframe. The first unit contains the operating cab and controls, auxiliary diesel engine (also used for hostling and switching), auxiliary generators, air compressor (air brake), diesel-fuel tank, miscellaneous auxiliaries, and one-half the total number of traction motors. The second portion contains dc traction generators, the gas turbine, fuel-heating equipment, and the remaining traction motors. The third unit is used to carry bunker C fuel oil. The gas turbine is rated at 10,700 hp (8,000 kW) at 1,000 ft (300 m) elevation and 80°F (21°C), and the locomotive rail horsepower rating is 8,500 (6,300 kW).

The diesel engine supplies power through an auxiliary generator to one of the traction generators for cranking the turbine. When sufficient rpm is attained, diesel fuel is fed into the combustion chambers. When all burners are firing properly and up to temperature, preheated, treated bunker C oil is fed to the burners, and diesel fuel is cut off.

The **turbine** is the single-shaft type having the air compressor at one end, combustion chambers in the middle, and turbine on the opposite end. Filtered-air intake and exhaust are on the roof. The turbine is connected to the traction generators through gear-reduction units. The traction generators are housed in a positively ventilated, sealed compartment and deliver power to the traction motors on all axles of the first and second units. The fuel oil is preheated at the fueling station before being fed into the third section, which is insulated with glass wool.

The air-brake compressor is motor-driven, power being supplied from one of the diesel-engine auxiliary generators. Lubricating oil for the gas turbine is filtered and pumped through cooling radiators. Dynamic-brake apparatus is housed in the first section. The first and second portions are each equipped with two 3-axle swing bolster trucks.

**Comparison with Diesel-Electric** Full-load tractive-effort–speed characteristics are similar to those of diesel-electric locomotives. The turbine is much more efficient at high values of power output than at the lower values and hence is best used where high-horsepower output is required over most of the run, as when hauling a nonstop freight train up a long sustained grade. Gas-turbine locomotives have fewer moving parts in the main power equipment than diesel-electric locomotives. They also have more horsepower per foot of length. The above-described three-section turbine locomotive is 165 ft (50 m) in overall length, whereas diesel units of the same total horsepower would measure about 260 ft (79 m). Gas-turbine locomotive weight per axle is 68,000 lb (31,000 kg) versus a diesel-electric's fully loaded weight of about 60,000 lb (27,000 kg). This turbine's weight on drivers does not decrease appreciably as the fuel supply is depleted, since bunker C oil is carried in the third section. The diesel-electric's weight with an empty fuel tank is reduced to about 59,000 lb (26,700 kg) per axle. When fully loaded with supplies, the diesel-electric

### Table 3. Dimensions of Freight Cars

| Kind and description | Weight empty, kg lb | Capacity kg lb | Capacity m³ ft³ | Length over striking castings, m ft-in | Max width inside, m ft-in | Rail to top of floor, m ft-in | Rail to top of roof or sides, m ft-in |
|---|---|---|---|---|---|---|---|
| Box, all steel* | 24,262 | 45,350 | 138.2 | 15.80 | 2.794 | 1.111 | 4.562 |
| | 53,500 | 100,000 | 4882 | 51-10 | 9-2 | 3-7¾ | 14-11¹⁹⁄₃₂ |
| Box, all steel | 21,617 | 45,350 | 118.0 | 12.70 | 2.896 | 1.092 | 4.597 |
| (welded)† | 43,300 | 100,000 | 4136 | 41-7⅞ | 9-6 | 3-7 | 15-1 |
| Box, HiCube | 51,926 | 47,844 | 283.2 | 27.74 | 2.794 | 1.070 | 5.182 |
| | 114,500 | 105,500 | 10,000 | 91-0 | 9-2 | 3-6⅛ | 17-0 |
| Box, auto parts | 39,137 | 79,816 | 170.2 | 20.09 | 2.794 | 1.124 | 4.674 |
| | 86,300 | 176,000 | 6013 | 65-11 | 9-2 | 3-8¼ | 15-4 |
| Stock, steel frame‡ | 24,806 | 36,280 | 75.6 | 12.75 | 2.591 | 1.111 | 4.400 |
| | 54,700 | 80,000 | 2670 | 41-10 | 8-6 | 3-7¾ | 14-5⁷⁄₃₂ |
| Refrigerator | 28,525 | 45,350 | 64.2 | 12.79 | 2.591 | 1.324 | . . . |
| | 62,900 | 100,000 | 2270 | 41-11¾ | 8-6 | 4-4⅛ | . . . |
| Hopper, all steel | 19,909 | 45,350 | 58.1 | 9.85 | 2.972 | . . . | 3.480 |
| | 43,900 | 100,000 | 2054 | 32-4 | 9-9 | . . . | 11-5 |
| Hopper, all steel | 21,779 | 63,490 | 78.5 | 12.70 | 3.150 | . . . | 3.251 |
| | 48,400 | 140,000 | 2773 | 41-8 | 10-4 | . . . | 10-8 |
| Ore, all steel | 24,761 | 86,165 | 36.3 | . . . | 2.927 | 1.119 | 2.643 |
| | 54,600 | 190,000 | 1283 | . . . | 9-7¼ | 3-8¹⁄₁₆ | 8-8¹⁄₁₆ |
| Gondola, all steel | 19,183 | 45,350 | 56.0 | 12.74 | 2.896 | 1.056 | 2.623 |
| | 42,300 | 100,000 | 1980 | 41-9¾ | 9-6 | 3-5⁹⁄₁₆ | 8-7¼ |
| Gondola, all steel | 27,936 | 65,032 | 56.5 | 16.61 | 2.896 | 1.081 | 2.299 |
| | 61,600 | 143,400 | 1995 | 54-6 | 9-6 | 3-6⁹⁄₁₆ | 7-6⁹⁄₁₆ |
| Covered hopper, 100T | 30,974 | 88,296 | 101.1 | 14.34 | 3.229 | 0.279 bot. of | 4.470 |
| | 68,300 | 194,700 | 3572 | 47-0¾ | 10-7⅛ | 0-11 hopper | 14-8 |
| Tank | 20,725 | 37.85 m³ | 37.8 | 11.43 | 2.235 diam | . . . | 4.308 |
| | 45,700 | 10,000 gal | 1337 | 37-6 | 7-4 | . . . | 14-1⅝ |
| Tank | 79,363 | 189.3 m³ | . . . | 27.13 | 2.896 outside | | 4.483 |
| | 175,000 | 50,000 gal | . . . | 89-0 | 9-6 @ center | | 14-8½ |
| Flat | 23,809 | 63,490 | . . . | 15.24 | 2.819 | 1.045 | . . . |
| | 52,500 | 140,000 | . . . | 50-0 | 9-3 | 3-5¹⁄8 | . . . |
| Piggyback | 28,344 | 52,833 | . . . | 27.41 | 2.800 outside | 0.800 | 1.003 |
| | 62,500 | 116,500 | | 89-11 | 9-2¼ | 2-7½ | 3-3½ |
| Auto, trilevel | 45,259 | 12 std. autos or 15 compacts | | 26.12 | 2.591 | 0.991 first deck | 5.128 5.588 top of load 16-9⅞ |
| | 99,800 | | | 85-8¼ | 8-6 | 3-3 | 18-4   top of load |

*T&G lining.
†Plywood lined.
‡Adjustable deck.

locomotive's ratio of weight on drivers to horsepower would be 141 lb/hp (86 kg/kW) against 96 lb/hp (58 kg/kW) for the gas turbine.

## FREIGHT AND PASSENGER CARS

Lightweight passenger cars have been built for service in diesel-powered trains with weights as low as 60,000 lb (27,000 kg) by making use of special designs and articulating units to reduce the number of trucks. High-strength steels, including cold-drawn stainless, are extensively used in welded construction. Until recently the use of aluminum alloys has been limited to riveted construction.

In freight-car construction the trend is to follow AAR specifications, although there are regional variances on account of special service requirements. Low-alloy high-strength steels are extensively used to reduce weight and corrosion losses and because of their weldability. Aluminum alloys are less extensively used but are being developed and adapted to freight-car structural requirements. Table 3 shows the dimensions of typical freight cars.

**Wheels and Axles**  Cast-steel and wrought-steel wheels are used on about 90 percent of the freight cars in the United States; data are given in Table 4. All passenger cars are equipped with wrought-steel wheels, except in Canada, where some steel-tired wheels are used.

Wrought-steel wheels are available in diameters from 28 to 42 in (71 to 122 cm) for passenger cars and diesel locomotives. Standard and temporary standards are shown in Sec. G of the AAR Manual of Standard and Recommended Practice (Div. V).

Axles have been standardized for freight-car interchange on

**Table 4. One-Wear Cast-Steel and Wrought-Steel Wheels for Freight Cars**

| | | | | | | |
|---|---|---|---|---|---|---|
| Load limit, eight-wheeled car | lb | 103,000 | 142,000 | 177,000 | 220,000 | 263,000 |
| | kg | 46,720 | 64,410 | 80,285 | 99,792 | 119,294 |
| Size of journal,   in | | $4\frac{1}{4} \times 8$ | $5 \times 9$ | $5\frac{1}{2} \times 10$ | $6 \times 11$ | $6\frac{1}{2} \times 12$ |
| | cm | 10.8 × 20.3 | 12.7 × 22.9 | 14.0 × 25.4 | 15.2 × 27.9 | 16.5 × 30.5 |
| Nominal 33-in-wheel weight, | lb | 660 | 660 | 650 | 670 | |
| cast steel | kg | 299.4 | 299.4 | 294.8 | 303.9 | |
| Nominal 36-in-wheel weight, | lb | | | | | 760 |
| cast steel | kg | | | | | 344.7 |
| Nominal 33-in-wheel weight, | lb | 612 | 577 | 575 | 640 | |
| wrought steel | kg | 277.6 | 261.7 | 260.8 | 290.3 | |

the basis of load limit at the rail (Table 5), and a series of passenger-car axles with enlarged wheel fits has been designed for high-speed passenger service for use with conventional or roller bearings. A recent development is a forged tubular axle.

Dimensions, load limits, limits of wear, and inspection and shop practice as recommended by the AAR will be found in the AAR Wheel and Axle Manual (Div. V). The dimensions of conventional bearings are fixed by the same interchange requirements as the standard axles. Roller bearings are generally applied to all new passenger-train cars. Some applications of roller bearings have been made on special-service freight cars.

**Design Fundamentals**  The Manual of the Mechanical Division of the AAR (Div. V) defines a large number of standards for the design of car details to which freight cars must conform on account of the universal interchange of cars on American railways. The manual is corrected annually by the AAR, and many of the standards are reproduced in the "Car Builders Cyclopedia," which is revised about every 4 years. Safety appliances, end ladders, steps, uncoupling levers, running boards, and other similar items must comply with the Safety Appliance Act as administered by the Interstate Commerce Commission.

## TRAIN RESISTANCE

(See Schmidt, *Bull.* 43, and Tuthill, *Bull.* 376, *Univ. Ill. Eng. Exp. Sta.*)

The resistance offered by a train to motion along the track could be expressed as a coefficient of friction, but it is more convenient to discuss it in terms of "pounds per ton" of weight. **Gross train resistance** is the force exerted by the locomotive at the rear drawbar. It overcomes two classes of resistances in producing and maintaining motion: "inherent,"

"true," "rolling," or "net" resistance, or merely "train" resistance, which is always present, and certain "incidental" or "accidental" resistances. The **inherent resistance** includes the resistance due to journal and flange friction, the action of the tread on the rail, and the resistance due to motion through still air. It may be defined as "the force necessary to maintain motion at constant speed on straight level track in still air." The **incidental resistances** include the resistances due to grade, curvature, acceleration, and wind resistance. Additional resistance due to extreme cold may well be considered an incidental effect.

**Inherent Resistance**  Of the elements of inherent resistance, at low speeds, journal friction is probably more than half the total, but at high speeds, air resistance is the predominant factor. Attempts to differentiate and evaluate the various elements through the speed range have not been very satisfactory, and it is more convenient to treat the total as a unit. At very high speeds, the effect of air resistance can be approximately determined; this is an aid to studies in its reduction by means of cowling and fairing. The resistance of a car moving in still air on straight level track at constant speed increases with the speed, but the resistance in pounds per ton decreases as the weight of the car increases. The total resistance of a 50-ton car is much less than twice as great as that of a 25-ton car under similar conditions. With known conditions of speed and car weight, inherent resistance can be predicted with reasonable accuracy; knowledge of track conditions will permit further refining of the estimate, but for very rough track or extreme cold, generous allowances must be made. Under such conditions, normal resistance may be doubled. A formula proposed by W. J. Davis, Jr. (*Gen. Elec. Rev.*, Oct., 1926) has been used extensively for inherent freight-train resistances at speeds up to 40 mi/h:

**Table 5. Dimensions of Standard Freight-car Axles**

| Size of journal, in | Carrying capacity of axle | | Diameter | | | | Length | | | | Weight of rough-turned axle | |
|---|---|---|---|---|---|---|---|---|---|---|---|---|
| | | | At wheel seat | | Minimum (at midlength) | | Overall | | Center to center of journals | | | |
| | lb | kg | in | cm | in | cm | in | cm | in | cm | lb | kg |
| $4\frac{1}{4} \times 8$ | 25,750 | 11,680 | $5\frac{3}{4}$ | 14.6 | $4\frac{3}{4}$ | 12.1 | $84\frac{1}{4}$ | 214.0 | 75 | 190.5 | 520 | 236 |
| $5 \times 9$ | 35,500 | 16,102 | $6\frac{1}{2}$ | 16.5 | $5\frac{3}{8}$ | 13.7 | $86\frac{1}{2}$ | 219.7 | 76 | 193.0 | 695 | 315 |
| $5\frac{1}{2} \times 10$ | 44,250 | 20,071 | 7 | 17.8 | $5\frac{7}{8}$ | 14.9 | $88\frac{1}{2}$ | 224.8 | 77 | 195.6 | 825 | 374 |
| $6 \times 11$ | 55,000 | 24,948 | $7\frac{5}{8}$ | 19.4 | $6\frac{7}{16}$ | 16.3 | $90\frac{3}{4}$ | 230.5 | 78 | 198.1 | 1,005 | 456 |
| $6\frac{1}{2} \times 12$ | 65,750 | 29,824 | $8\frac{1}{8}$ | 20.6 | $7\frac{3}{8}$ | 18.7 | $92\frac{3}{4}$ | 235.6 | 79 | 200.7 | 1,185 | 538 |

$$R = 1.3 + (29/w) + 0.045V + 0.0005 \, AV^2/wn$$

where $R$ = resistance, lb/ton; $w$ = average weight per axle, tons; $V$ = speed, mi/h; $n$ = total number of axles; and $A$ = cross-sectional area, ft$^2$. With the recent trend toward freight-train speeds of 50 to 70 mi/h, it has been found that actual resistance values fall considerably below calculations based on the above formula at these high speeds.

For passenger cars the Davis formula is satisfactory, giving good results at high speed with vestibuled passenger equipment (not streamlined). This formula is for coaches pulled by a locomotive and does not include head-end air resistance.

The introduction of the ultralightweight Talgo design, with the reduction in the number of axles, equipment cross section, and car unit length, has made it impossible to use the standard Davis formula. Actual road tests have indicated that the following modified Davis formula is quite reliable:

Locomotive: $R_L = 1.3 + 29/w + 0.03V + 0.002AV^2/wn$

Cars: $R_C = 1.3 + 29/w + 0.03V + 0.00034AV^2/wn^1$

where $n^1$ = number of axles per length of train equivalent to 80-ft passenger cars; $A$ = height of unit, ft, divided by 14.9 and multiplied by 120.

For conventional streamlined diesel and electric locomotives the Davis formula for vestibuled passenger-car equipment gives reasonable results, using a coefficient of 0.0016 for the last term in place of 0.00034 to include head-end air resistance. The resistance thus calculated may be subtracted from locomotive tractive effort available at the rail to obtain drawbar pull on tangent level track in still air at constant speed.

For **wind resistance** see Sec. 11 and Lipetz, *Trans. ASME*, Oct., 1937.

**Curve Resistance**  Resistance due to track curvature varies with speed and curvature; it is a minimum at the speed corresponding to the superelevation of the outer rail, for at this speed there would be the least flange contact. Curve resistance is also decreased by the practice of relieving (widening) the gage on curves from 1/8 to 1 in (3.2 to 25.4 mm), in proportion to their sharpness. For general estimates of car resistance and locomotive hauling capacity, speed and gage relief may be ignored and a figure of 0.8 lb per ton per degree used.

**Grade resistance** depends only on the angle of ascent or descent. It is 20 lb/ton for each foot-in-one-hundred rise, or "percent of grade," or 0.379 lb for each foot-per-mile rise. A car cannot be left standing with brakes off on a grade of more than about 0.3 of 1 percent [15 ft/mi (2.8 m/km)].

**Acceleration Resistance**  The force required to produce acceleration is the sum of the forces required to produce accelerated translation and that required to produce accelerated rotation of the wheels about their axle centers. A translatory acceleration of 1 mi/(h)(s) is produced by a force of 91.1 lb/ton. The rotary acceleration requirement adds 6 to 12 percent to this force, so that the total is nearly 100 lb/ton (the figure commonly used) for each mile per hour per second. If greater accuracy is required,

$$R_a = A(91.05W + 36.36n)$$

where $R_a$ is the total accelerating force, lb; $A$ the acceleration in mi/(h)(s); $W$ the weight of train, tons; and $n$ the number of axles.

**Acceleration and Distance**  If in a distance of $S$ ft the speed of a car or train changes from $V_1$ to $V_2$ mi/h, the force required to produce the acceleration (or the braking force if the speed is reduced) is $R_a = 74(V_2^2 - V_1^2)/S$, the coefficient 74 corresponding to the use of 100 lb/ton above. This formula is useful in the calculation of climbing a grade with the assistance of stored energy. In any train-resistance calculation or analysis, assumptions with regard to acceleration will generally submerge all other variables; e.g., an acceleration of 0.1 mi/(h)(s) requires a greater tractive force than that required to overcome inherent resistance for any car at moderate speeds.

**Starting Resistance**  The force required to produce an oil film on a journal is considerable, and if a train is fully stretched, so that all the wheels must be started rolling simultaneously (an unusual condition for freight trains, but approached in passenger trains), 35 to 45 lb/ton of weight of train will be required for starting. In practice, slack in couplers of draft gears results in more gradual starting which permits a freight train to move with a force of 15 to 20 lb; a passenger train requires 25 to 30 lb.

**Roller Bearings**  Little information is available on the resistance of roller-bearing freight cars. Only the journal friction is affected; hence the variation is important at starting, reducing the starting force to 5 or 6 lb/ton. The reduction at high speeds becomes relatively small. (See also Sec. 8.)

## BRAKING

(See also Sec. 8)

The forces effective in retarding the motion of a railway vehicle are (1) inherent resistance, (2) any incidental resistances effective at the time such as wind, curvature, or ascending grade, (3) friction of the brake shoes on the wheel rims. The brake-shoe pressure is the major factor in stopping and, aside from shoe friction and grade, other forces may be neglected. Brakes on American trains are operated by air pressure. A continuous pipe through the train (the train-line brake pipe) carries pressure from the compressors on the locomotive to a reservoir on each car through a "control valve." Reduction in the pressure in this pipe is effected by manipulation of the brake valve and results in the movement of a balanced piston in the control valve of each car. The piston movement carries with it a slide valve, through the ports of which compressed air flows from the "auxiliary reservoir" on each car into the brake cylinder, and the brakes are applied. The action thus initiated travels rapidly through the train. When the pressure in the train line is rebuilt, the balanced piston again moves into a position where the air in the brake cylinder is exhausted and the pressure in the auxiliary reservoir is returned to its normal value.

The foregoing describes briefly the functions of the fundamental automatic air brake based on the functions of the control valve. Brake equipment designed to specifications per ICC Docket 13528 is now required on all freight cars used in interchange service. The functions of the control valve have been refined to permit the handling of longer trains by more uniform brake performance. Important improvements in this design are (1) the time required to apply the brakes on the last car of a train has been reduced; (2) a more uniform and faster release of the brakes is possible; (3) emergency application is always available, irrespective of the state of brake application or release; this feature eliminates the possibility of trains being out of control on long grades.

In high-speed passenger service, a greater brake-shoe pressure may be applied at high speeds than at low speeds without the danger of sliding the wheels. This reduces stopping distances. It may be accomplished pneumatically, using a higher brake-pipe pressure, or by electropneumatic systems, using speed-governor control to proportion brake-shoe pressure according to train speed.

The total retarding force in pounds per ton may be taken as $F = (PLef/W) + 20G$, where $P$ = total brake-cylinder piston pressure, lb, $L$ = multiplying ratio of the leverage between cylinder pistons and wheel rims; $ef$ = product of the coefficient of brake-shoe friction and foundation-gear efficiency; $W$ = loaded weight of vehicle, lb; $G$ = grade, percent. Stopping distance in feet may be found from $b = 0.035 WV^2 p_n/W_1 R p_0 \cdot ef$, where $W$ = loaded weight of vehicle, lb; $V$ = speed, mi/h; $p_n$ = basic brake-cylinder pressure on which $R$ is selected, lb/in²; $W_1$ = light weight of vehicle, lb; $R$ = nominal braking ratio (expressed as a decimal), representing the ratio of total brake-shoe pressure at a stated brake-cylinder pressure (and with 100 percent brake foundation-gear efficiency) to vehicle light weight; $p_a$ = actual maintained brake-cylinder pressure from point of equivalent instantaneous maximum pressure development, lb/in², $e$ = brake foundation-gear efficiency (a decimal); $f$ = coefficient of brake-shoe friction. To this value of $b$ must be added the distance traveled at initial speed from the time the brake-valve handle is moved to application position until the point of maximum pressure development is reached.

The friction coefficient $f$ varies with the speed (usually lower at high speed), unit pressure (lower at high pressure), and with the material of the wheel and shoe. For stops below 60 mi/h (97 km/h) a conservative figure for an ordinary shoe on steel wheels is $ef = 0.15$. $p_n$ is based on 50 lb/in² (345 kPa) air pressure in the cylinder for freight-car and locomotive brakes, and 60 lb/in² (414 kPa) for passenger cars. The braking ratios $R$ commonly used are: freight cars, 75 percent of empty weight with 50 lb/in² (345 kPa) cylinder pressure; passenger cars, 90 percent of the empty weight based on 60 lb/in² (414 kPa) cylinder pressure and full service application (150 percent based on 100 lb/in² (690 kPa) cylinder pressure and emergency application); diesel locomotives, 60 to 80 percent. Seventy pounds per square inch is a typical value for the brake-pipe pressure of a fully charged freight train. This will give 50 lb/in² brake-cylinder pressure during a full service application on AB equipment, and 60 lb/in² brake-cylinder pressure with an emergency application. Total brake-shoe pressure at 60 lb/in² cylinder pressure may be determined from the braking ratio $R$ and $p_a$. Assuming $R$ = 70 percent, and car light weight is 50,000 lb, brake-shoe pressure = $RW_1 p_a/p_n$, or 0.70 × 50,000 × 60/50, or 42,000 lb.

To prevent wheel sliding, $F \leq \phi w$, where $F$ = force at the wheel rims resisting rotation of any pair of connected wheels, lb; $\phi$ = coefficient of wheel-rail adhesion or friction (a decimal); and $w$ = static weight upon a pair of wheels, lb. Actual or adhesive weight on wheels when the vehicle is in motion is affected by weight transfer (moment transmitted to the trucks by the inertia of the car body through the truck center plates) and vertical oscillation of body weight upon truck springs. The value of $\phi$ remains constant at its static value at all speeds, being about 0.25 for a dry or sanded moist rail. A value of 0.20 would provide reasonable assurance against sliding when used to include car-body oscillation and weight transfer.

The relationship between required coefficient $\phi_1$ of wheel-rail adhesion to prevent wheel sliding and rate of retardation in miles per hour per second $A$ may be expressed by the equation $\phi_1 = 0.04555A$.

## TRACK

**Gage**  The gage of railway track is the distance between the inner sides of the rail heads. Practically all American railways are laid with a **gage** of 4 ft 8½ in (1.435 m), which is known as **standard gage**. Rail wear causes an increase in gage; and on sharp curves, because of rail wear, it is usual practice to begin widening the gage on curves over 8°, the limiting value being ¾ in. In America the gage of **narrow-gage** track is usually 3 ft, the variations being 2½ and 3½ ft. In Europe, 1 m is the prevailing narrow gage.

**Track Spacing**  The distance between centers of main-line track varies between 12 and 14 ft. Twelve-foot spacing has, however, become unusual. The prevailing spacing is 14 ft 0 in (4.267 m) and has the implied endorsement of the AREA.

**Clearances**  The AREA standard clearance diagrams provide for a clear height of 22 ft (6.7 m) above the tops of the rail heads and for a width of 16 ft (4.9 m) on bridges and 15 ft (4.6 m) elsewhere. If obstructions are kept outside these limits, they will clear a person riding on the top or sides of the largest cars. Where it is certain that personnel will be kept off the cars, or where conditions require the minimum clearance, as in tunnels, these dimensions are much reduced. For tracks entering buildings an opening 12 ft (3.7 m) wide and 17 ft (5.2 m) high will ordinarily suffice to pass the largest locomotives and cars. A standard clearance diagram for railway bridges is Fig. 13. The diagram represents the AAR recommendation for new construction. For specific clearance limitations on a particular railroad, refer to the official publication, "Railway Line Clearances," Railway Equipment and Publishing Co.

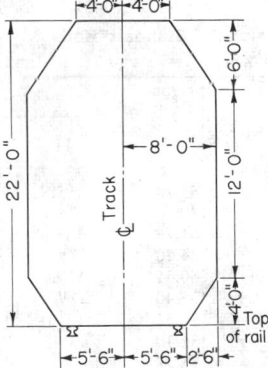

**Fig. 13**  Standard clearance diagram for railway bridges on tangent track.

**Curvature**  The curvature of track is designated in terms of "degree of curve." **Degree of curve** is the number of degrees of central angle subtended by a chord of 100-ft length (measured on the track centerline). Table 6 gives the radii for ordinary railway curves.

On important main lines where trains are operated at high speed, the curves are ordinarily not sharper than 6 or 8°. In

**Table 6. Radii of Steam Railway Curves**

| Degree of curve | Radius to track centerline | | Degree of curve | Radius to track centerline | | Degree of curve | Radius to track centerline | |
|---|---|---|---|---|---|---|---|---|
| | m | ft | | m | ft | | m | ft |
| 1 | 1746.504 | 5730 | 9 | 194.249 | 637.3 | 17 | 103.114 | 338.3 |
| 2 | 873.252 | 2865 | 10 | 174.864 | 573.7 | 18 | 97.414 | 319.6 |
| 3 | 582.168 | 1910 | 11 | 159.014 | 521.7 | 19 | 92.324 | 302.9 |
| 4 | 436.778 | 1433 | 12 | 145.786 | 478.3 | 20 | 87.752 | 287.9 |
| 5 | 349.301 | 1146 | 13 | 134.630 | 441.7 | 21 | 83.637 | 274.4 |
| 6 | 260.726 | 955.4 | 14 | 125.059 | 410.3 | 22 | 79.858 | 262.0 |
| 7 | 249.631 | 819.0 | 15 | 116.769 | 383.1 | | | |
| 8 | 218.481 | 716.8 | 16 | 109.515 | 359.3 | | | |

mountainous territory, in rare instances, main-line curves as sharp as 18° occur. Most diesel and electric locomotives are designed to traverse curves of 21°. Most cars alone will pass considerably sharper curves, the limiting factor in this case being the flexibility of the brake connections to the trucks.

**Rails** The AREA has defined standards for the cross section of rails of various weights per yard. The AREA standards provide for rail weights varying from 90 to 140 lb. The dimensions of these standard rail sections are shown in Table 7. A few railroads use special sections of their own design for heavy rails. (See Fig. 14 for dimensions.)

The **length of rails** has been largely determined by the length of the cars available for their transportation. The standard length in use is 39 ft (11.89 m).

The prevailing weight of rails used in main-line track is 90 to 133 lb/yd; on lines with very heavy traffic, weights up to 155 lb are in use. Secondary and branch lines are generally laid with 75- to 100-lb rail which has been previously used and partly worn in main-line service. A considerable mileage of welded rail has recently been laid and is in successful opera-

tion. In general, rails are first welded into lengths up to 20 rails, then laid, and later welded into lengths up to 2 mi (3.2 km). Electric-flash-, gas-fusion-, and thermite-welding processes are used.

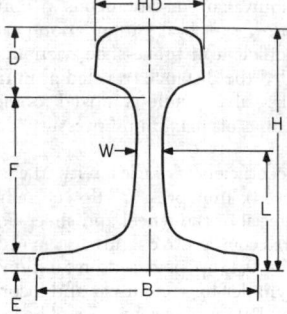

**Fig. 14** Standard rail section.

**Table 7. Dimensions of Standard Rails**
(Letters refer to Fig. 14)

| Standard and nominal weight lb | Weight per yd kg lb | Dimensions mm in | | | | | | | |
|---|---|---|---|---|---|---|---|---|---|
| | | H | B | HD | W | D | F | E | L |
| AREA 140 | 63.78 | 185.8 | 152.4 | 76.2 | 19.1 | 52.4 | 103.2 | 30.2 | 101.6 |
| | 140.60 | 7⁵/₁₆ | 6 | 3 | ³/₄ | 2¹/₁₆ | 4¹/₁₆ | 1³/₁₆ | 4 |
| AREA 136 | 61.78 | 185.8 | 152.4 | 74.6 | 17.5 | 49.2 | 106.4 | 30.2 | 98.6 |
| | 136.20 | 7⁵/₁₆ | 6 | 2¹⁵/₁₆ | ¹¹/₁₆ | 1¹⁵/₁₆ | 4⁹/₁₆ | 1³/₁₆ | 3⁷/₈ |
| AREA 133 | 60.51 | 179.4 | 152.4 | 76.2 | 17.5 | 49.2 | 100.0 | 30.2 | 95.3 |
| | 133.40 | 7¹/₁₆ | 6 | 3 | ¹¹/₁₆ | 1¹⁵/₁₆ | 3¹⁵/₁₆ | 1³/₁₆ | 3³/₄ |
| AREA 132 | 59.52 | 181.0 | 152.4 | 76.2 | 16.7 | 44.5 | 106.4 | 30.2 | 98.6 |
| | 132.10 | 7¹/₁₆ | 6 | 3 | ²¹/₃₂ | 1³/₄ | 4⁹/₁₆ | 1³/₁₆ | 3⁷/₈ |
| CB 122 | 55.57 | 172.3 | 152.4 | 74.6 | 16.7 | 49.2 | 91.7 | 31.3 | 84.6 |
| | 122.5 | 6²⁵/₃₂ | 6 | 2¹⁵/₁₆ | ²¹/₃₂ | 1¹⁵/₁₆ | 3³⁹/₆₄ | 1¹⁵/₆₄ | 3²¹/₆₄ |
| AREA 119 | 53.89 | 173.1 | 139.7 | 67.5 | 15.9 | 22.2 | 96.9 | 28.6 | 82.6 |
| | 118.80 | 6¹³/₁₆ | 5¹/₂ | 2²¹/₃₂ | ⁵/₈ | 1⁷/₈ | 3¹³/₁₆ | 1¹/₈ | 3¹/₄ |
| AREA 115 | 52.07 | 168.3 | 139.7 | 69.1 | 15.9 | 42.9 | 96.9 | 28.6 | 82.6 |
| | 114.75 | 6⁵/₈ | 5¹/₂ | 2²³/₃₂ | ⁵/₈ | 1¹¹/₁₆ | 3¹³/₁₆ | 1¹/₈ | 3¹/₄ |
| AREA 100 | 46.04 | 152.4 | 136.6 | 68.3 | 14.3 | 42.1 | 85.8 | 27.0 | 75.4 |
| | 101.50 | 6 | 5⁵/₈ | 2¹¹/₁₆ | ⁹/₁₆ | 1²¹/₃₂ | 3⁹/₃₂ | 1¹/₁₆ | 2⁸¹/₃₂ |
| ARA-A 100 | 45.54 | 152.4 | 139.7 | 69.9 | 14.3 | 39.7 | 85.8 | 27.0 | 74.6 |
| | 100.40 | 6 | 5¹/₂ | 2³/₄ | ⁹/₁₆ | 1⁹/₁₆ | 3³/₈ | 1¹/₁₆ | 2¹⁵/₁₆ |
| ARA-A 90 | 40.82 | 142.9 | 130.2 | 65.1 | 14.3 | 37.3 | 80.2 | 25.4 | 73.5 |
| | 90.0 | 5⁵/₈ | 5¹/₈ | 2⁹/₁₆ | ⁹/₁₆ | 1¹⁵/₃₂ | 3⁹/₃₂ | 1 | 2²⁹/₃₂ |

## CAR RETARDERS

In classification yards where cars roll from a "hump" into their assigned tracks by gravity, their speed is controlled by retarders. Each retarder is a rail brake electrically or pneumatically operated and equipped with heat-treated alloy-steel shoes mounted on each side of the rail to provide a controlled pressure on the inner and outer surfaces of the car wheels.

Modern yards are equipped with **automatic retarder control,** which predicts the rolling resistance of each separate car or "cut" and determines the speed at which it must leave the retarders in order to couple to the next car in its assigned track at a safe speed (2 to 4 mi/h) (3.2 to 6.4 km/h).

Figure 15 shows an elevation view of a typical installation with a hump region (single track) and a group region, which represents one of several tracks branching off at the junction switch. From the group retarders, each track branches into several classification tracks.

After leaving the crest of the hump, the cut passes over a weight-detecting device, which classifies it as "light," "medium," or "heavy." Next it passes through the **test section** of specified track length $S_{1-2}$, where its velocity is measured as it enters ($V_1$) and as it leaves ($V_2$). $Y$ is the difference in elevation between the entrance and the exit points. The rolling resistance $RR_{1-2}$ of the cut passing through the test section is calculated by an analog or digital computer on the basis of the formula

$$-RR_{1-2} = \frac{1}{S_{1-2}} \left[ Y - \frac{1}{2g_o} (V_2^2 - V_1^2) \right]$$

where $g_0$ is the acceleration of the cut due to gravity, modified by its weight classification and adjusted for the rotational energy of the car wheels.

The calculated value of $RR_{1-2}$ is transmitted to a second part of the computer, which determines the desired speed $V_h$ of the cut leaving the **hump retarders.** The positions of the track switches (junction-track-switch circuit) are also taken into account by the computer to allow for variations in curve and frog resistance, depending on the route selected between the hump and group retarders. As the cut enters the hump retarders, shoe pressure is at a maximum for the weight classification involved. A radar unit measures the speed of the cut while it moves through retarders and transmits the information

continuously to the retarder-control discriminators. The latter compare the actual speed with the calculated value of the desired leaving speed and reduce the shoe pressure accordingly until the value of $V_1$ is attained, at which time shoe pressure is released completely.

Control of the **group retarders** is accomplished in the same way. The desired speed $V_g$ for leaving the group retarders is calculated by a third part of the computer, using the formula

$$V_g = \sqrt{V_c^2 + 2g_o[C_{o-c} - (Y_{gc} - Y_c) + RR_{o-c}S_{o-c}]}$$

where $V_c$ is the desired velocity of the cut at coupling. It is preset and is the same for each cut. $C_{o-c}$ is the equivalent head loss due to curved-track rolling resistance, frog loss, and elevation difference between zero distance to coupling and the point-of-grade change. It depends on the route selected and is derived on the basis of measured past performance of a large number of cars through each specific route. $Y_{gc} - Y_c$ is the elevation difference between point-of-grade change and point of coupling. $Y_{gc}$ is fixed, but $Y_c$ is governed by the distance to coupling $S_{o-c}$ and the route selected. $S_{o-c}$ is measured by a track circuit whose electrical resistance is varied by the location of the next car ahead in the designated classification track. $RR_{o-c}$ is the predicted rolling resistance of the cut after it leaves the retarder. It is derived by computer from the rolling resistance $RR_{1-2}$ and is modified by the distance to coupling for light cars. It may also be modified slightly if desired by the retarder operator for unusual conditions.

## MONORAIL SYSTEMS

The monorail has been developed for high-speed passenger service in heavily populated urban localities, where the land area for right-of-way is at a premium. Operating 16 ft (4.9 m) or more above the ground level, it provides minimum interference with surface traffic. There are two general types: the "suspended" and the "supported."

The **suspended system** has been used successfully in Japan and Germany. The car body is suspended, pendulum fashion, from the running gear and rides below the beamway, which is supported 35 ft (10.7 m) above the ground. The running gear, equipped with pneumatic tires (or flanged wheels), travels over the top surface of the beamway. Auxiliary guide wheels attached to the truck frame bear against the sides of the

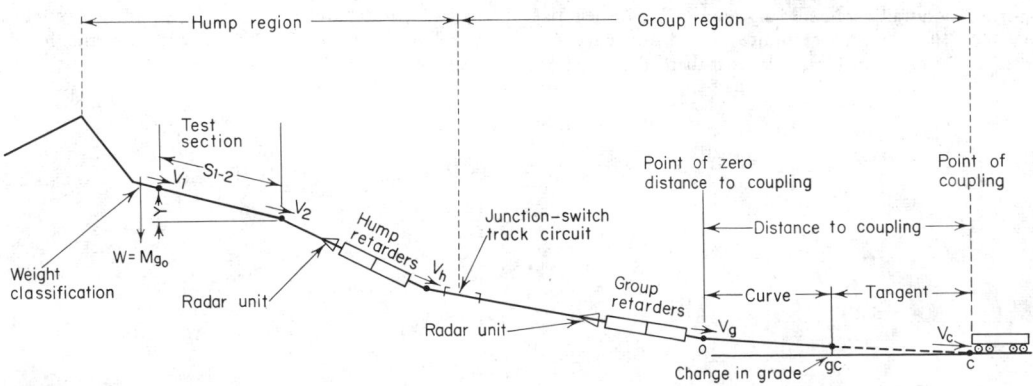

**Fig. 15**  Automatic car-retarder installation, elevation. *(General Railway Signal Co., Rochester, N.Y.)*

beamway to control sway. This type allows full use of the space within the car for pay load, but the car body must be designed to transmit the weight of car and passengers to the running gear above.

The **supported system,** as used at Disneyland and at the Seattle Fair, employs a prefabricated, prestressed, hollow concrete girder supported from below by reinforced-concrete T columns. A cross section of the beamway and car is shown in Fig. 16. Since the center of gravity of the car is above the

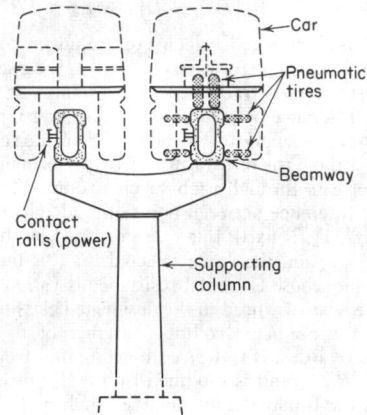

**Fig. 16** Supported monorail system, sectional view.

beamway, the guide wheels, which rotate in a horizontal plane, contact the beamway's vertical side to keep the car upright.

Trains of the design used at the Seattle Fair are composed of two 60-ft (18.3 m) cars, each weighing 50,000 lb (22,000 kg) (empty), with a 62-passenger seating capacity. Each car is mounted on four axles equipped with dual pneumatic tires, and each axle is driven by a 600 -V 100-hp dc motor through a double-reduction gearbox. Power is picked up by current-collector shoes from two contact rails mounted on one side of the beamway.

Maximum acceleration is automatically controlled at a rate of 2.5 mi/(h)(s). Deceleration rates are 3.3 mi/(h)(s) at the service rate and 5.7 in emergency. A dynamic brake is effective above speeds of 10 mi/h and is used in conjunction with a pneumatic system.

Supported monorail cars generally are capable of negotiating curves of 100-ft radius at limited speed and curves of 1,000-ft radius at 60 mi/h. Grades steeper than 10 percent can

be negotiated when pneumatic tires are used. Speeds of 90 mi/h have been attained with equipment under test. However, the Seattle installation was operated at 53 mi/h maximum over the 1.2-mi distance.

In **comparison with subway equipment,** the monorail car weighs 50 percent less per passenger (seating capacity). With pneumatic tires, it has almost double the rate of deceleration. Motor capacity and acceleration rates are similar.

### SUBWAY TRAINS

Subway cars such as those used by the New York City Transit Authority are self-propelled. They are equipped with contact shoes which take 600-V nominal dc electric power from a third rail. The cars are designed to operate in multiple, with as many as 12 cars controlled from one position. Each car is 51 ft 4 in long over drawbar faces and seats 44 passengers. It is designed to negotiate curves of 90-ft minimum radius and operate at speeds up to 55 mi/h on standard-gage track.

**Acceleration** is automatically controlled at 2.5 mi/(h)(s) between the limits of 37.5 tons empty and 51.5 tons fully loaded weight. Deceleration is accomplished by a blended system of dynamic and electropneumatic brake. Automatic dynamic braking provides selective rates of deceleration between 1.5 and 3.0 mi/(h)(s) for empty and fully loaded cars. Full dynamic braking is effective from 45 to 10 mi/h, being supplemented by air brakes for the final stop. The emergency-brake application, employing both systems, is designed to provide an average retardation rate of 3 mi/(h)(s) from 30 mi/h in order to stop a fully loaded car within 220 to 250 ft.

Acceleration and braking are both controlled by the operator's five-position master controller, which has the following positions: (1) "minimum power"—motors in full series; (2) "series"; (3) "series-parallel"; (4) "coast"—no power, dynamic-brake loop circuit established; and (5) "braking." With the controller in positions 2 and 3, nine steps of resistance commutation are progressed automatically from maximum resistance to full field. In position 5, dynamic braking is effected, uniformly and at a selected rate, in 17 steps down to 10 mi/h before the air-brake system automatically assumes full control. Traction motors are the four-pole dc commutating-pole type, rated at 100 shaft hp at 300 V, 280 A, and 1,175 r/min. Motor and axle are connected through a gear-drive unit equipped with a 123-tooth gear (mounted on the axle) and a 17-tooth pinion. Motor and drive unit are joined through a flexible coupling. Each car is equipped with four traction-motor and gear assemblies.

# MARINE ENGINEERING
## by Anders T. Anderson

REFERENCES: Harrington, "Marine Engineering," SNAME, 1971. Comstock, "Principles of Naval Architecture," SNAME, 1967. Myers, "Handbook of Ocean and Underwater Engineering," McGraw-Hill, 1969. "Rules for Building and Classing Steel Vessels," American Bureau of Shipping. Rawson and Tupper, "Basic Ship Theory," American Elsevier, 1968. Gillmer, "Modern Ship Design," Naval Institute, 1970. Barnaby, "Basic Naval Architecture," Hutchinson, London, 1967. *Jour. Inst. Environ. Sci.* D'Arcangelo, "Ship Design and Construction," SNAME, 1969. *Trans. Soc. Naval Architects and Marine Engrs.*, SNAME. *Naval Engrs. Jour.*, Am. Soc. of Naval Engrs., ASNE. *Trans. Royal Inst. of Naval Architects*, RINA. Figures and examples herein credited to SNAME have been included by permission of The Society of Naval Architects and Marine Engineers.

**Marine engineering** is an integration of many engineering disciplines directed to the development and design of systems of transport, warfare, exploration, and natural-resource retrieval which have one thing in common: operation in or on a body of water. Marine engineers are responsible for the engineering systems required to propel, work, or fight ships. They are responsible for the main propulsion plant; the powering and mechanization aspects of ship functions such as steering, anchoring, cargo handling, heating, ventilation, air conditioning, etc.; and other related requirements. They usually have joint responsibility with naval architects in areas of propulsor design; hull vibration excited by the propeller or main propulsion plant; noise reduction and shock hardening, in fact, dynamic response of structures or machinery in general; cargo-handling pumping systems; and environmental control and habitability. Marine engineering is a distinct multidiscipline and characteristically a dynamic, continuously advancing technology.

## THE MARINE ENVIRONMENT

Marine engineers must be familiar with their environment so that they may understand fuel and power requirements, vibration effects, and propulsion-plant strength considerations. The outstanding characteristic of the open ocean is its irregularity in storm winds as well as under relatively calm conditions. The **irregular sea** can be described by statistical mathematics based on the superposition of a large number of regular waves having different lengths, directions, and amplitudes. The characteristics of idealized regular waves are fundamental for the description and understanding of realistic, irregular seas. Actual sea states consist of a combination of many sizes of waves often running in different directions, and sometimes momentarily superimposing into an exceptionally large wave.

The effects of the marine environment also vary with water depth. As a ship passes from deep to shallow water, there is an appreciable change in the potential flow around the hull and a change in the wave pattern produced. Additionally, silt, sea life, and bottom growth may affect seawater systems or foul heat exchangers.

## MARINE VEHICLES

The platform is additionally a part of marine engineers' environment. Ships are supported by a buoyant force equal to the weight of the volume of water displaced. For surface ships, this weight is equal to the total weight of the structure, machinery, outfit, fuel, stores, crew, and useful load. The principal sources of resistance to propulsion are skin friction and the energy lost to surface waves generated by moving in the interface between air and water. Minimization of one or both of these sources of resistance has generally been a primary objective in the design of marine vehicles.

### Displacement Hull Forms

Displacement hull forms are the familiar monohull, the catamaran, and the submarine. The moderate-to-full-displacement, **monohull** form provides the best possible combinations of high-payload-carrying ability, economical powering characteristics, and good seakeeping qualities. A more slender hull form achieves a significant reduction in wave-making resistance, hence increased speed; however, it is limited in its ability to carry topside weight because of the low transverse stability of its narrow beam. The **catamaran** provides a solution to the problem of low transverse stability. It is increasingly popular in sailing yachts and is under limited development for research craft and small support ships. Sailing catamarans, with their superior transverse stability permitting large sail-plane area, have gained a speed advantage over monohull craft of comparable size. The advantages of a powered catamaran over a single monohull ship are increased deck space and relatively low roll angles. The **submarine,** operating at sufficient depths to avoid the formation of surface waves, experiences significant reductions in resistance as compared with a well-designed surface ship of equal displacement.

### Planing Hull Forms

The planing hull form, although most commonly used for yachts and racing craft, is used increasingly in small, fast commercial craft and in escort naval ships. The weight of the planing-hull ship is partially borne by the dynamic lift of the water streaming against a relatively flat or V-shaped bottom. The effective displacement is hence reduced below that of the ship supported statically, with a significant reduction in wave-making resistance at higher speeds.

### High-performance Ships

In a search for high performance and higher speeds in rougher seas, several advanced concepts to minimize wave-making resistance have been investigated. These concepts have been or are being developed in hydrofoil craft, surface-effect vehicles, and small water-plane-area twin hull (SWATH) forms (Fig. 1). The **hydrofoil craft** has a displacement planing hull that is raised clear of the water through dynamic lift generated by an underwater foil system. The **surface-effect vehicles** ride on a cushion of compressed air generated and maintained in the

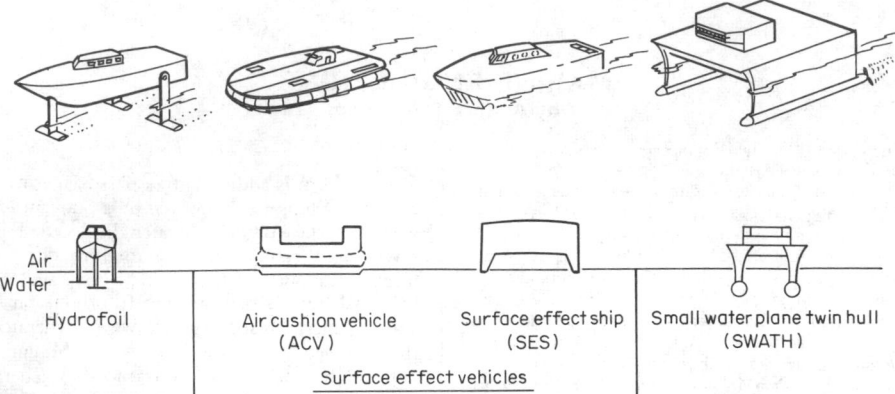

**Fig. 1**  Family of high-performance ships.

space between the vehicle and the surface over which it hovers or moves. The most practical vehicles employ a peripheral-jet principle, with flexible skirts for obstacle or wave clearance. A rigid sidewall craft, achieving some lift from water displacement, is more adaptable to marine construction techniques. The **SWATH** gains the advantages of the catamaran, twin displacement hulls, with the further advantage of minimized wave-making resistance achieved by submarine-shaped hulls beneath the surface and the small water-plane area of the supporting struts at the interface.

## SEAWORTHINESS

Seaworthiness is the quality of a marine vehicle being fit to accomplish its intended mission. In meeting their responsibilities to produce seaworthy vehicles, marine engineers must have a basic understanding of the effects of the marine environment with regard to the vehicle's (1) **structure,** (2) **stability** and **motions,** and (3) **resistance** and **powering** requirements.

### Units and Definitions

The introduction of SI units to the marine engineering field presents somewhat of a revolutionary change. Displacement, for instance, is a force and therefore is expressed in newtons (N) or meganewtons (MN); what for many years has been known as a 10,000-ton ship therefore becomes a 99.64-MN ship. To assist in the change, examples and data have been included in both USCS and SI units.

The **displacement** $\Delta$ is the weight of the water displaced by the immersed part of the vehicle. It is equal (1) to the buoyant force exerted on the vehicle and (2) to the weight of the vehicle (in equilibrium) and everything on board. Displacement is expressed in long tons equal to 2,240 lb (MN). The specific weight of seawater averages about 64 lb/ft³ (1 ton = 35 ft³); hence, the displacement in seawater is measured by the displaced volume $\nabla$ divided by 35. In fresh water, 1 ton = 35.9 ft³. (Specific weight of seawater = 10,053 N/m³; 1 MN = 99.47 m³; $\Delta = \nabla/99.47$ MN; in fresh water 1 MN = 102 m³.)

Two measurements of a merchant ship's earning capacity that are of significant importance to its design and operation are deadweight and tonnage. The **deadweight** of a ship is the weight of cargo, stores, fuel, water, personnel, and effects that

the ship can carry when loaded to a specific load draft. Deadweight is the difference between the load displacement, up to a minimum permitted freeboard, and the light displacement, which comprises hull weight and machinery. Deadweight is expressed in long tons (2,240 lb each) or MN. The volume of a ship is expressed in tons of 100 ft³ (2.83 m³) each and is referred to as its **tonnage.** Charges for berthing, docking, passage through canals and locks, and for many other facilities are based on tonnage. **Gross tonnage** is based on cubic capacity below the tonnage deck, plus allowances for certain compartments above, which are used for cargo, passengers, crew, and navigating equipment. Deduction of spaces for propulsion machinery, crew quarters, and other prescribed volumes from the gross tonnage leaves the **net tonnage.**

The dimensions of a ship may refer to the molded body (or form defined by the outside of the frames), to general outside or overall dimensions, and to dimensions, on which the determination of tonnage or of classification is based. There are thus (1) molded dimensions, (2) overall dimensions, (3) tonnage dimensions, and (4) classification dimensions. The published rules and regulations of the classification societies and the U.S. Coast Guard should be consulted for detailed information.

The **designed load waterline** (DWL) is the waterline at which a ship would float freely, at rest in still water, in its normally loaded or designed condition. The keel line of most ships is parallel to DWL. Some keel lines are designed to slope downward toward the stern, termed **designed drag.**

A vertical line through the intersection of DWL and the foreside of the stem is called the **forward perpendicular, FP.** The vertical line through the intersection of DWL with the afterside of the straight portion of the rudder post, with the afterside of the stern contour, or with the centerline of the rudder stock (depending upon stern configuration), is called the **after perpendicular, AP.**

The **length** on the designed load waterline, $L_{WL}$, is the length measured at the DWL, which, because of the stern configuration, may be equal to the **length between perpendiculars,** $L_{pp}$. The **classification society length** is commonly noted as $L_{pp}$. The extreme length is the **length overall,** $L_{OA}$.

The **molded beam** $B$ is the extreme breadth of the molded form. The extreme or overall breadth is occasionally used, referring to the extreme transverse dimension taken to the outside of the plating.

The **draft** $T$ (molded) is the distance from the top of the keel plate or bar keel to the load waterline. It may refer to draft amidships, forward, or aft.

**Trim** is the longitudinal inclination of the ship usually expressed as the difference between the draft forward, $T_F$, and the draft aft, $T_A$.

**Coefficients of Form** Assume the following notation: $L$ = length on waterline; $B$ = beam; $T$ = draft; $\nabla$ = volume of displacement; $A_{WP}$ = area of water plane; $A_M$ = area of midship section at draft $T$; $v$ = speed in ft/s (m/s); and $V$ = speed in knots.

**Block coefficient,** $C_B = \nabla/LBT$, may vary from about 0.38, for high-powered yachts and destroyers, to about 0.80 for slow-speed sea-going cargo ships.

**Midship section coefficient,** $C_M = A_M/BT$, varies from about 0.75 for tugs or trawlers to about 0.98 for cargo ships and is a measure of fullness of the maximum section.

**Prismatic coefficient,** $C_P = \nabla/LA_M$, ranges from 0.55 to 0.80, is a measure of fullness of the ends of the hull, and is an important parameter in powering estimates.

**Water-plane coefficient,** $C_{WP} = A_{WP}/LB$, ranges from about 0.67 to 0.87, is a measure of the fullness of the water plane, and may be estimated by the relation $C_{WP} \approx \frac{2}{3}C_B + \frac{1}{3}$.

**Displacement/length ratio,** $\Delta = \Delta/(L/100)^3$, ranges from about 50 for the slender destroyer hull to 250 for a slow cargo ship of very full form, is a measure of the slenderness of the hull, and is used in calculating the power of ships and in recording the resistance data of models.

Table 1 presents typical values of the coefficients with representative values for Froude number = $v/\sqrt{gL}$.

## Structure

The structure of a ship is a complex assembly of small pieces of material. Common hull structural materials for small boats are wood, aluminum, and fiber glass-reinforced plastic. Large ships are nearly always constructed of steel, although current practice is to use an increasing amount of aluminum, especially in deckhouses and other superstructures.

The **analysis** of a ship structure is accomplished through the following simplified steps: (1) Assume that the ship behaves like a box-shaped girder supported on a simple wave system; (2) estimate the loads acting on the ship, using simplified assumptions regarding weight and buoyancy distribution; (3) calculate the static shear forces and bending moments; (4) analyze the resulting stresses; and (5) iterate the design until the stresses are acceptable. The maximum longitudinal bend-

ing stresses which result from such simplified loading assumptions are used as an indicator of the maximum stress that will be developed, and an approximate factor of safety is introduced to allow for stresses induced from other types of external loading, from local loadings, from stress concentration, and from material fatigue over the life of the ship.

Weight, buoyancy, and load curves (Figs. 2 and 3) are developed for the ship for the determination of shear force and

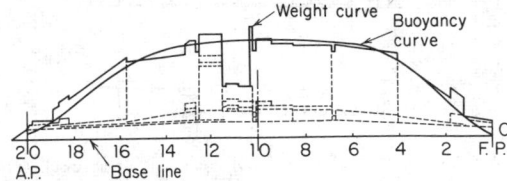

**Fig. 2** Representative buoyancy and weight curves for a ship in still water.

bending moment. Several extreme conditions of loading may be analyzed. The weight curves include the weights of the hull, superstructure, rudder, and castings, forgings, masts, booms, all machinery and accessories, solid ballast, anchors, chains, cargo, fuel, supplies, passengers, and baggage. Each individual weight is distributed over a length equal to the distance between frames at the location of that particular weight.

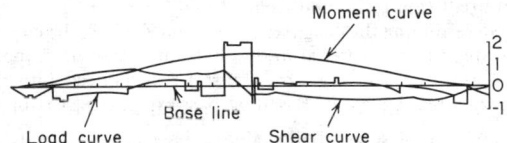

**Fig. 3** Representative moment and net-load curves for a ship in still water.

The traditional method of calculating the **maximum design bending moment** involved the determination of weight and buoyancy distributions with the ship poised on a wave of trochoidal form whose length $L_w$ is equal to the length of the ship, $L_{pp}$, and whose height $H_w = L_w/20$. More recently, other wave height standards have appeared. $H_w = 1.1\sqrt{L}$ is widely used (including standard use by the U.S. Navy), and others have been suggested and applied. With the wave crest amidships, the ship is in a **hogging** condition (Fig. 4a); the deck is in tension and the bottom shell in compression. With the trough amidships, the ship is in a **sagging** condition (Fig. 4b);

**Table 1. Coefficients of Form**

| Type of vessel | $\dfrac{v}{\sqrt{gL}}$ | $C_B$ | $C_M$ | $C_P$ | $C_{WP}$ | $\dfrac{\Delta}{(L/100)^3}$ |
|---|---|---|---|---|---|---|
| Great Lakes ore ships | 0.116–0.128 | 0.85–0.87 | 0.99–0.995 | 0.86–0.88 | 0.89–0.92 | 70–95 |
| Slow ocean freighters | 0.134–0.149 | 0.77–0.82 | 0.99–0.995 | 0.78–0.83 | 0.85–0.88 | 180–200 |
| Moderate-speed freighters | 0.164–0.223 | 0.67–0.76 | 0.98–0.99 | 0.68–0.78 | 0.78–0.84 | 165–195 |
| Fast passenger liners | 0.208–0.313 | 0.56–0.65 | 0.94–0.985 | 0.59–0.67 | 0.71–0.76 | 75–105 |
| Fast cruisers | 0.387–0.506 | 0.45–0.53 | | | | |
| Destroyers | 0.536–0.744 | 0.44–0.53 | 0.72–0.83 | 0.62–0.71 | 0.67–0.73 | 40–65 |
| Tugs | 0.268–0.357 | 0.45–0.53 | 0.71–0.83 | 0.61–0.66 | 0.71–0.77 | 200–420 |

the deck is in compression and the bottom in tension. For normal cargo ships with the machinery amidships, the hogging condition produces the highest bending moment, whereas for normal tankers and ore carriers with the machinery aft, the sagging condition gives the highest bending moment.

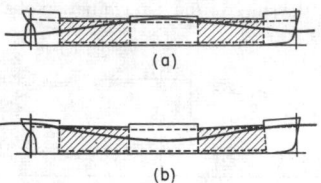

**Fig. 4** Hogging (*a*) and sagging (*b*) conditions of a vessel. (*From "Principles of Naval Architecture." SNAME,* 1967.)

The artificiality of the above assumption should be apparent; nevertheless, it has been an extremely useful one for many decades, and a great deal of ship data have been accumulated upon which to base refinements. Many advances have been made in the theoretical prediction of the actual loads a ship is likely to experience in a realistic, confused seaway over its life span. Today it is possible to predict reliably the bending moments and shear forces a ship will experience over a short term in irregular head seas. In the future, the reliability of predictions on other headings should be improved and long-term prediction techniques refined.

In determining the **design section modulus** *Z* which the continuous longitudinal material in the midship section must meet, an **allowable bending stress** $\sigma_{all}$ must be introduced into the bending stress equation. Based on past experience, an appropriate choice of such an allowable stress is $1.19\sqrt[3]{L}$, tons/in ($27.31\sqrt[3]{L}$, MN/m ), with *L* in ft (m).

For ships under 200 ft (61 m) in length, strength requirements are dictated more on the basis of locally induced stresses than longitudinal bending stresses. This may be accounted for by reducing the allowable stress based on the above equation or by reducing the *K* values indicated by the trend in Table 2.

**Table 2. Constants for Bending Moment Approximations**

| Type | Δ tons (MN) | L ft (m) | K* |
|---|---|---|---|
| Tankers | 35,000–150,000 (349–1495) | 600–900 (183–274) | 35–41 |
| High-speed cargo passenger | 20,000–40,000 (199–399) | 500–700 (152–213) | 29–36 |
| Great Lakes ore carrier | 28,000–32,000 (279–319) | 500–600 (152–183) | 54–67 |
| Destroyers | 3,500–7,000 (35–70) | 400–500 (122–152) | 23–30 |
| Destroyer escorts | 1,600–4,000 (16–40) | 300–450 (91–137) | 22–26 |
| Trawlers | 180–1,600 (1.79–16) | 100–200 (30–61) | 12–18 |
| Crew boats | 65–275 (0.65–2.74) | 80–130 (24–40) | 10–16 |

*Based on allowable stress of $\sigma_{all.} = 1.19\sqrt[3]{L}$, tons/in² ($27.31\sqrt[3]{L}$, MN/m²) with *L* in ft (m).

For aluminum construction, *Z* must be twice that obtained for steel construction. Minimum statutory values of section modulus are published by the U.S. Coast Guard in "Load Line Regulations." Shipbuilding classification societies such as the American Bureau of Shipping have section modulus standards somewhat greater.

The **maximum bending stress** at each section can be computed by the equation

$$\sigma = M/Z$$

where $\sigma$ = maximum bending stress, lb/in² (N/m²) or tons/in² (MN/m²)

$M$ = maximum bending moment, ft·lb (N·m) or ft·tons

$Z$ = section modulus, in²·ft (m³); $Z = I/y$, where $I$ = minimum vertical moment of inertia of section, in²·ft² (m⁴), and $y$ = maximum distances from neutral axis to bottom and strength deck, ft (m)

The **maximum shear stress** can be computed by the equation

$$\tau = Vac/It$$

where $\tau$ = horizontal shear stress, lb/in² (N/m²)

$V$ = shear force, lb (N)

$ac$ = moment of area above shear plane under consideration taken about neutral axis, ft³ (m³)

$I$ = vertical moment of inertia of section, in²·ft² (m⁴)

$t$ = thickness of material at shear plane, ft (m)

For an **approximation,** the bending moment of a ship may be computed by the equation

$$M = \Delta L/K, \text{ ft·tons (MN·m)}$$

where $\Delta$ = displacement, tons (MN); $L$ = length, ft (m); and $K$ is as listed in Table 2.

For ships, the maximum longitudinal bending stresses occur in the vicinity of the midship section at the deck edge and in the bottom plating. Maximum shear stresses occur in the shell plating in the vicinity of the quarter points at the neutral axis. For long, slender girders, such as ships, the maximum shear stress is small compared with the maximum bending stress.

The structure of a ship consists of a grillage of stiffened plating supported by longitudinals, longitudinal girders, transverse beams, transverse frames, and web frames. Since the primary stress system in the hull arises from longitudinal bending, it follows that the longitudinally continuous structural elements are the most effective in carrying and distributing this stress system. The strength deck, particularly at the side, and the keel and turn of bilge are highly stressed regions. The shell plating, particularly deck and bottom plating, carry the major part of the stress. Other **key longitudinal elements,** as shown in Fig. 5, are longitudinal deck girders, main deck stringer plate, gunwale bar, sheer strake, bilge strake, inner bottom margin strake, garboard strake, flat plate keel, center vertical keel, and rider plate.

**Transverse elements** include deck beams and transverse frames and web frames which serve to support and transmit vertical and transverse loads and resist hydrostatic pressure.

Good structural design minimizes the structural weight while providing adequate strength, minimizes interference with ship function, provides for effective continuity of the structure, facilitates stress flow around deck openings and other stress obstacles, and avoids square corner discontinuities in the plating and other stress concentration "hot spots."

A longitudinally framed ship is one which has closely

spaced longitudinal structural elements and widely spaced transverse elements. A transversely framed ship has closely spaced transverse elements and widely spaced longitudinal elements. Longitudinal framing systems are generally more efficient structurally, but because of the deep web frames supporting the longitudinals, it is less efficient in the use of

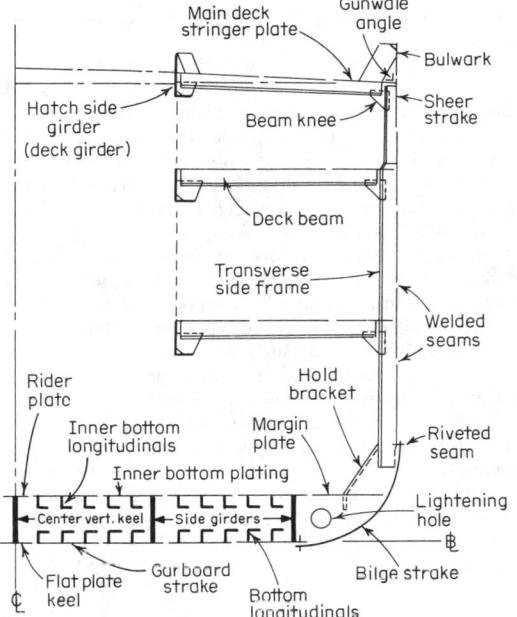

**Fig. 5**  Key structural elements.

internal space than the transverse framing system. Where interruptions of open internal spaces are unimportant, as in tankers and bulk carriers, longitudinal framing is universally used. However, modern practice tends increasingly toward longitudinal framing in other types of ships also.

### Stability

A ship afloat is in vertical equilibrium when the force of **gravity,** acting at the ship's **center of gravity** $G$, is equal, opposite, and in the same vertical plane as the force of **buoyancy,** acting upward at the **center of buoyancy** $B$. Figure 6 shows that an upsetting, transverse couple acting on the ship causes it to rotate about a longitudinal axis, taking a list $\phi$. $G$ does not change; however, $B$ moves to $B_1$, the centroid of the new

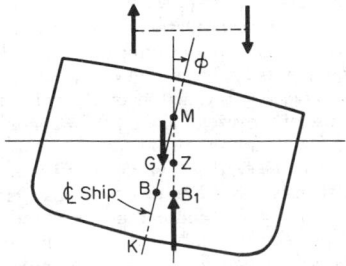

**Fig. 6**  Ship stability.

underwater volume. The resulting couple, created by the transverse separation of the two forces ($\overline{GZ}$), opposes the upsetting couple, thereby righting the ship. A static stability curve (Fig. 7), consisting of values for the righting arm $\overline{GZ}$ plotted against angles of inclination ($\phi$), gives a graphic representation of the static stability of the ship.

For small angles of inclination ($\phi < 10°$), centers of buoyancy follow a locus whose instantaneous center of curvature $M$ is known as the transverse **metacenter.** When $M$ lies above $G$ ($\overline{GM} > 0$), the resulting gravity-buoyancy couple will right the ship; the ship has positive stability. When $G$ and $M$ are coincident, the ship has neutral stability. When $G$ is above $M$ ($\overline{GM} < 0$), negative stability results. Hence $\overline{GM}$, known as the **transverse metacentric height,** is an indication of **initial stability** of a ship.

The transverse **metacentric radius** $\overline{BM}$ and the vertical location of the center of buoyancy are determined by the design of the ship and can be calculated. Once the vertical location of the ship's center of gravity is known, then $\overline{GM}$ can be found. The vertical center of gravity of practically all ships varies with the condition of loading and must be determined either by a careful calculation or by an inclining experiment.

Minimum values of $\overline{GM}$ ranging from 1.5 to 3.5 ft (0.46 to 1.07 m), corresponding to small and large seagoing ships, respectively, have been accepted in the past. The $\overline{GM}$ of passenger ships should be under 6 percent of the beam to ensure a reasonable (comfortable) rolling period.

The question of **longitudinal stability** affects the trim of a ship. As in the transvere case, a longitudinal metacenter exists, and a longitudinal metacentric height $\overline{GM_L}$ can be determined.

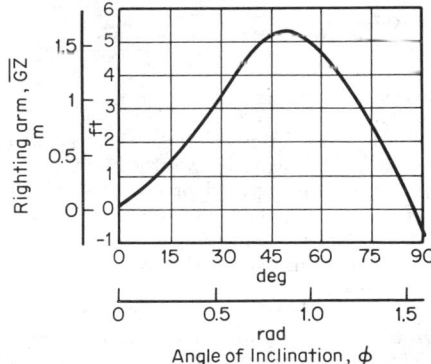

**Fig. 7**  Static stability curve.

The moment to alter trim 1 in, $\overline{MT1}$, is computed by $\overline{MT1} = \Delta \overline{GM}_L/12L$, ft·tons/in and moment to alter trim 1 m is $\overline{MT1} = 10^6 \Delta \overline{GM}_L/L$, N·m/m. Displacement $\Delta$ is in tons and MN, respectively.

The **location of a ship's center of gravity** changes as small weights are shifted within the system. The vertical, transverse, or longitudinal component of movement of the center of gravity is computed by $\overline{GG_1} = w \times d/\Delta$, where $w$ is the small weight, $d$ is the distance the weight is shifted in a component direction, and $\Delta$ is the displacement of the ship, which includes $w$.

Vertical changes in the location of $G$ caused by **weight addition** or **removal** can be calculated by

$$KG_1 = \frac{\Delta \overline{KG} \pm w \check{K}g}{\Delta_1}$$

where $\Delta_1 = \Delta \pm w$ and $\check{K}g$ is the height of the center of gravity of $w$ above the keel.

The free surface of the liquid in fuel oil, lubricating oil, and water storage and service tanks is deleterious to ship stability. The weight shift of the liquid as the ship rolls can be represented as a virtual rise in $G$, hence a reduction in $\overline{GM}$ and the ship's initial stability. This virtual rise, called the **free surface effect**, is calculated by the expression

$$GG_v = \gamma_t i / \gamma_w \nabla$$

where $\gamma_t,\ \gamma_w$ = specific gravities of liquid in tank and sea, respectively
$i$ = moment of inertia of free surface area about its longitudinal centroidal axis
$\nabla$ = volume of displacement of ship

### Ship Motions

The seakeeping qualities of ships are based on their motions in waves. These motions in turn directly affect the practices of marine engineers. The motion of any floating object has six degrees of freedom. Figure 8 shows the conventional coordinate axis for a ship. Oscillatory movement along axis $xx$ is called **surge**; along axis $yy$, **side sway**; and along $zz$, **heave**. Rotation about $xx$ is called **roll**; about $yy$, **pitch**; and about $zz$, **yaw**.

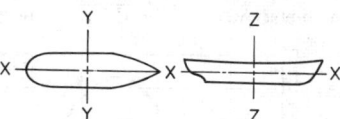

**Fig. 8**   Conventional ship coordinate axes.

**Rolling** has a major effect on crew comfort and on the structural and bearing requirements for machinery and its foundations. The natural period of roll of a ship is $T = 2\pi K / \sqrt{g\,\overline{GM}}$, where $K$ is the radius of gyration of ship mass about a longitudinal axis through $G$. $K$ varies from 0.4 to 0.5 of the beam, depending on the ship depth and transverse distribution of weights.

**Angular acceleration** of roll, if large, has a very distressing effect on crew, passengers, machinery, and structure. This can be minimized by increasing the period or by decreasing the roll amplitude. Maximum angular acceleration of roll is

$$d^2\phi/dt^2 = -4\pi^2(\phi_{max}/T)$$

where $\phi_{max}$ is the maximum roll amplitude. The period of the roll can be increased effectively by decreasing $\overline{GM}$; hence the lowest value of $\overline{GM}$ compatible with all stability criteria should be sought.

**Pitching** is in many respects analogous to rolling. The natural period of pitching, bow up to bow down, can be found by using the same expression used for the rolling period, with the longitudinal radius of gyration $K_L$ substituted for $K$. A good approximation is $K_L = L/4$, where $L$ is the length of the ship. The natural period of pitching is usually between one-third and one-half the natural period of roll. A by-product of pitching is **slamming**, the reentry of bow sections into the sea with heavy force.

**Heaving** and **yawing** are the other two principal rigid-body motions caused by the sea. The amplitude of heave associated with head seas, which generate pitching, may be as much as 15 ft (4.57 m); that arising from beam seas, which induce rolling, can be even greater. Yawing is started by unequal forces acting on a ship as it quarters into a sea. Once a ship begins to yaw, it behaves as it would at the start of a rudder-actuated turn. As the ship travels in a direction oblique to its plane of symmetry, forces are generated which force it into heel angles independent of sea-induced rolling.

**Analysis of ship movements** in a seaway is performed by using probabilistic and statistical techniques. The seaway is defined by a mathematically modeled wave energy density spectrum based on data gathered by oceanographers. This wave spectrum is statistically calculated for various sea routes and weather conditions. A ship response amplitude operator or transfer function is determined by linearly superimposing a ship's responses to varying regular waves, both experimentally and theoretically. The wave spectrum multiplied by the ship response transfer function yields the ship response spectrum. The ship response spectrum is an energy density spectrum from which the statistical character of ship motions can be predicted. (See Edward V. Lewis, The Motion of Ships in Waves, "Principles of Naval Architecture," SNAME, 1967.)

### Resistance and Powering

Resistance to ship motion through the water is the aggregate of (1) wave-making, (2) frictional, (3) pressure or form, and (4) air resistances.

**Wave-making resistance** is primarily a function of Froude's number, $\text{Fr} = v/\sqrt{gL}$, where $v$ = ship speed, ft/s (m/s); $g$ = acceleration of gravity, ft/s² (m/s²); and $L$ = ship length, ft (m). In many instances, the dimensional speed-length ratio $V/\sqrt{L}$ is used for convenience, where $V$ is given in knots. A ship makes at least two distinct wave patterns, one from the bow and the other at the stern. There also may be other patterns caused by abrupt changes in section. These patterns combine to form the total wave system for the ship. At various speeds there is mutual cancellation and reinforcement of these patterns. Thus, a plot of total resistance of the ship versus Fr or $V/\sqrt{L}$ is not smooth but shows humps and hollows corresponding to the wave cancellation or reinforcement. Normal procedure is to design the operating speed of a ship to fall at one of the low points in the resistance curve.

**Frictional resistance** is a function of Reynolds number (see Sec. 3). Because of the size of a ship, the Reynolds number is large and the flow is always turbulent.

**Pressure** or **form resistance** is a viscosity effect but is different from frictional resistance. The principal observed effects are boundary-layer separation and eddying near the stern.

It is usual practice to combine the wave-making and pressure resistances into one term, called the **residual resistance,** assumed to be a function of Froude's number. The combination, although not strictly legitimate, is practical because the pressure resistance is usually only 2 to 3 percent of the total resistance. The frictional resistance is then the only term which is considered to be a function of Reynolds number and can be calculated. Based on an analysis of the water resistance of flat, smooth planes, Schoenherr gives the formula

$$R_f = 0.5\rho S v^2 C_f$$

where $R_f$ = frictional resistance, lb (N); $\rho$ = mass density, lb/$s^2 \cdot ft^4$ (kg/m³); $S$ = wetted surface area, ft² (m²); $v$ = velocity, ft/s (m/s); and $C_f$ is the frictional coefficient computed from the ITTC formula

$$C_f = \frac{0.075}{(\log_{10} \text{Re} - 2)^2}$$

and Re = Reynolds number = $vL/\nu$.

Through towing tests of ship models at a series of speeds for which Froude numbers are equal between the model and the ship, **total model resistance** ($R_{tm}$) is determined. Residual resistance ($R_{rm}$) for the model is obtained by subtracting the frictional resistance ($R_{fm}$). By **Froude's law of comparison**, the residual resistance of the ship ($R_{rs}$) is equal to $R_{rm}$ multiplied by the ratio of ship displacement to model displacement. Total ship resistance ($R_{ts}$) then is equal to the sum of $R_{rs}$, the calculated ship frictional resistance ($R_{fs}$), and a correlation allowance that allows for the roughness of ship's hull opposed to the smooth hull of a model.

$$R_{tm} - \text{measured}$$
$$R_{rm} = R_{tm} - R_{fm}$$
$$R_{rs} = R_{rm}(\Delta s/\Delta m)$$
$$R_{ts} = R_{rs} + R_{fs} + R_a$$

For $R_a$, a nominal value of 0.0004 is used as the correlation allowance coefficient $C_a$. $R_a = 0.5\rho S v^2 C_a$.

From $R_{ts}$, the **total effective power** $P_E$ required to propel the ship can be determined:

$$P_E = \frac{R_{ts}v}{550} \quad \text{ehp} \quad \left(P_E = \frac{R_{ts}v}{1,000} \quad \text{kW}_E\right)$$

where $R_{ts}$ = total ship resistance, lb (N)
$v$ = velocity, ft/s (m/s)
$P_E = P_S \times P.C.$

where $P_S$ is the shaft power (see Propulsion Systems) and $P.C.$ is the propulsive coefficient, a factor which takes into consideration mechanical losses, propeller efficiency, and the flow interaction with the hull. $P.C. = 0.45$ to 0.53 for high-speed craft; 0.50 to 0.60, for tugs and trawlers; 0.55 to 0.65, for destroyers; and 0.63 to 0.72, for merchant ships.

Figure 9 illustrates the specific effective power for various displacement hull forms and planing craft over their appropriate speed regimes. Figure 10 shows the general trend of specific resistance versus Froude number and may be used for coarse powering estimates.

In early design stages, the hull form is not yet defined, and model testing is not feasible. Reasonably accurate power calculations are made by using preplotted model data from Taylor's Standard Series and the Series 60. The Standard Series data were originally compiled by Admiral David W. Taylor and were based on model tests of a series of uniformly varied

models of similar geometry. Revised Taylor's Series data are available in "A Reanalysis of the Original Test Data for the Taylor Standard Series," Taylor Model Basin, Rept. no. 806.

**Air Resistance** The wind resistance parallel to the ship's axis is roughly 30 percent greater when the wind direction is about 30° ($\pi/6$ rad) with the keel than when it is dead ahead, since the projected, above-water area is greater. The wind resistance can be approximated by $R_A = 0.002B^2V_R^2$ lb, where

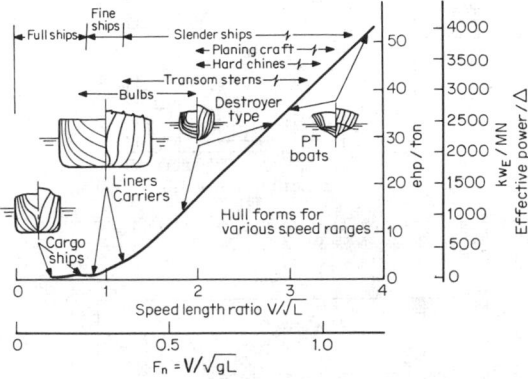

**Fig. 9** Specific effective power for various speed-length regimes. (*From "Handbook of Ocean and Underwater Engineering," McGraw-Hill, 1969.*)

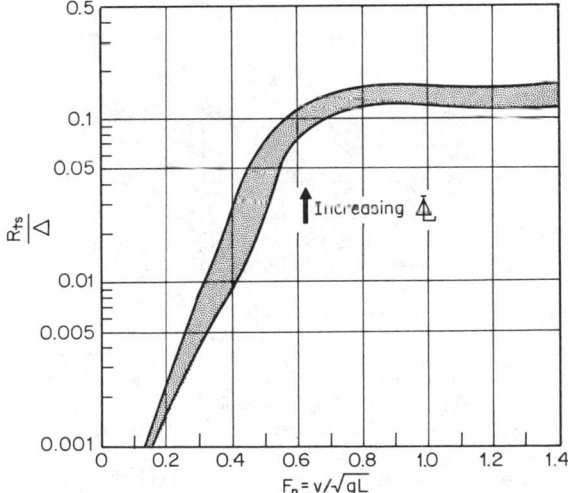

**Fig. 10** General trends of specific resistance versus Froude number.

$B$ = ship's beam, ft, and $V_R$ = ship speed relative to the air, in knots. ($R_A = 0.36B^2V_R^2$, where $B$ is in metres and $V_R$ in m/s).

**Powering of Small Craft** The American Boat and Yacht Council, Inc. (ABYC) provides the following guide for determining the maximum safe brake power for pleasure craft.

Compute a length-width factor by multiplying the overall boat length in feet by the overall stern width in feet (widest part of stern excluding fins and sheer). For SI units, use length and width in metres times 10.76. Locate the factor in Fig. 11 or Fig. 12 and read the boat brake power capacity.

For powering canoes, the maximum should be 3.0 bhp (2.2 kW$_B$) for lengths under 15 ft (4.6 m); 5 bhp (3.7 kW$_B$) for 15 to 18 ft (4.6 to 5.5 m); and 7.5 bhp (5.6 kW$_B$) for over 18 ft (5.5 m).

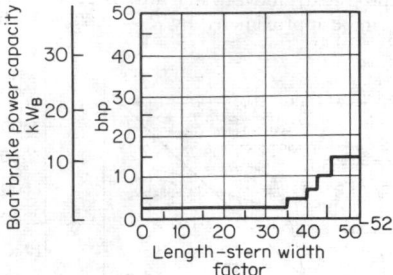

**Fig. 11**  Boat brake-power capacity length-width factor under 52. (*From "Safety Standards for Small Craft," ABYC, 1974.*)

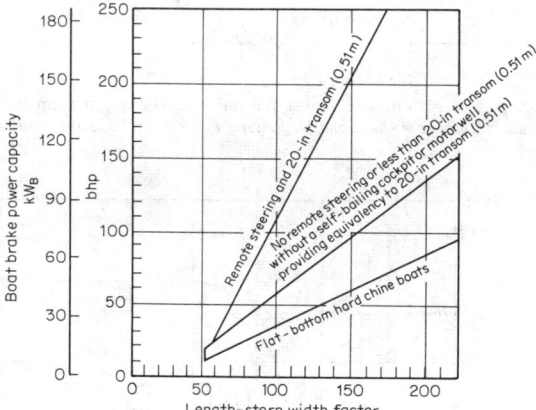

**Fig. 12**  Boat brake-power capacity for a length-width factor over 52. (*From "Safety Standards for Small Craft," ABYC, 1974.*)

## ENGINEERING CONSTRAINTS

The constraints affecting marine engineering design are too numerous, and some too obvious, to include in this section. Four significant categories, however, are discussed. The geometry of the hull forms immediately suggests **physical constraints.** The interaction of the vehicle with the marine environment suggests **dynamic constraints,** particularly vibration and shock. Machinery **noise** is not only a concern in submarine design but also a concern in its relation to human comfort. And the very broad topic of **environmental protection** is one of the foremost engineering constraints of today, having a very pronounced effect on the operating systems of a marine vehicle.

### Physical Constraints

Until recently, tonnage laws in effect made it economically desirable for the propulsion machinery spaces of a merchant ship to exceed 13 percent of the gross tonnage of the ship so that 32 percent of the gross tonnage could be deducted in computing net tonnage. In most design configurations, however, a great effort is made to minimize the space required for the propulsion plant in order to maximize that available to the mission or the money-making aspects of the ship.

Specifically, **space** is of extreme importance as each component of support equipment is selected. The dimensions of each component must fit into the master compact arrangement scheme to provide the most efficient operation and maintenance by engineering personnel.

**Weight** constraints for a main propulsion plant vary with the application. In a tanker where cargo capacity is limited by draft restrictions, the weight of machinery represents lost cargo. Cargo ships, on the other hand, rarely operate at full load draft and additionally the low weight of propulsion machinery somewhat improves inherent cargo ship stability deficiencies. The weight of each component of equipment is constrained by structural support and shock resistance considerations. Naval shipboard equipment, in general, is carefully analyzed to effect weight reduction.

### Dynamic Constraints

Dynamic effects, principally mechanical vibration, noise, shock, and ship motions, are considered in determining the dynamic characteristics of a ship and the dynamic requirements for equipment.

**Vibration** (See Sec. 9) Vibration analyses are especially important in the design of the propulsion shafting system and its relation to the excitation forces resulting from the propeller. The main propulsion **shafting** can vibrate in longitudinal, torsional, and lateral modes. Modes of **hull vibration** may be vertical, horizontal, torsional, or longitudinal; may occur separately or coupled; may be excited by synchronization with periodic harmonics of the propeller forces acting either through the shafting, by the propeller force field interacting with the hull afterbody, or both; and may also be set up by unbalanced harmonic forces from the main machinery, or, in some cases, by impact excitation from slamming or periodic-wave encounter.

It is most important to reduce the excitation forces at the source. Very objectionable and serious vibrations may occur when the frequency of the exciting force coincides with one of the hull or shafting-system natural frequencies.

**Vibratory forces generated by the propeller** are (1) alternating pressure forces on the hull due to the alternating hydrodynamic pressure fields of the propeller blades; (2) alternating propeller-shaft bearing forces, primarily caused by wake irregularities; and (3) alternating forces transmitted throughout the shafting system, also primarily caused by wake irregularities.

The most effective means to ensure a satisfactory level of vibration is to provide adequate clearance between the propeller and the hull surface and to select the propeller revolutions or number of blades to avoid synchronism. Replacement of a four-blade propeller, for instance, by a three- or five-blade, or vice versa, may bring about a reduction in vibration. Singing propellers are due to the vibration of the propeller blade edge about the blade body, producing a disagreeable noise or hum. Thinning the edge is sometimes helpful.

Vibration due to variations in **engine torque reaction** is hard to overcome. In ships with large diesel engines, the torque reaction tends to produce an athwartship motion of the upper ends of the engines. The motion is increased if the engine foundation strength is inadequate.

**Foundations** must be designed to take both the thrust and torque of the shaft and be sufficiently rigid to maintain alignment when the ship's hull is working in heavy seas. Engines designed for foundations with three or four points of support are relatively insensitive to minor working of foundations. Considerations should be given to using flexible couplings in cases when alignment cannot be assured.

**Torsional vibration** frequency of the prime mover–shafting-propeller system should be carefully computed, and necessary steps taken to ensure that its natural frequency is clear of the frequency of the main-unit or propeller heavy-torque variations; serious failures have occurred. Problems due to resonance of engine torsional vibration (unbalanced forces) and foundations or hull structure are usually found after ship trials, and correction consists of local stiffening of the hull structure. If possible, engines should be located at the nodes of hull vibrations.

Although more rare, a **longitudinal vibration** of the propulsion shafting has occurred when the natural frequency of the shafting agreed with that of a pulsating axial force, such as thrust variation.

**Other forces** inducing vibrations may be vertical inertia forces due to the acceleration of reciprocating parts of an unbalanced reciprocating engine or pump; longitudinal inertia forces due to reciprocating parts or an unbalanced crankshaft creating unbalanced rocking moments; and horizontal and vertical components of centrifugal forces developed by unbalanced rotating parts. Rotating parts can be balanced.

**Noise** The noise characteristics of shipboard systems are increasingly important, particularly in naval combatant submarines where remaining undetected is essential, and also from a human-factors point of view on all ships. Desired noise levels must be analyzed to ascertain that they can be practicably met or are justified. Each operating system and each piece of rotating or reciprocating machinery installed aboard a submarine are subjected to intensive airborne (noise) and structureborne (mechanical) vibration analyses and tests. Depending on the noise attenuation required, similar tests and analyses are also conducted for all surface ships, military and merchant.

**Shock** In naval combatant ships, shock loading due to noncontact underwater explosions is a major design parameter. Methods of qualifying equipment as "shock-resistant" might include "static" shock analysis, "dynamic" shock analysis, physical shock tests, or a combination.

**Motions** Marine lubricating systems are specifically distinguished by the necessity of including list, trim, roll, and pitch as design criteria. The American Bureau of Shipping requires satisfactory functioning of lubricating systems when the vessel is permanently inclined to an angle of 15° (0.26 rad) athwartship and 5° (0.09 rad) fore and aft. In addition, electric-generator bearings must not spill oil under a momentary roll of 22½° (0.39 rad). Military specifications cite the same permanent trim and list as for surface ships but add 45° (0.79 rad) roll and 10° (0.17 rad) pitch requirements. For submarines, a requirement of 30° (0.52 rad) trim, 15° (0.6 rad) list, 60° (1.05 rad) roll, and 10° (0.17 rad) pitch is imposed.

### Environmental Constraints

The **Refuse Act of 1899** (33 U.S.C. 407) prohibits discharge of any refuse material from marine vehicles into navigable waters of the United States or into their tributaries or onto their banks if the refuse is likely to wash into navigable waters. The term **refuse** includes nearly any substance. The Environmental Protection Agency (EPA) may grant permits for the discharge of refuse into navigable waters.

The **Oil Pollution Act of 1961** (33 U.S.C.1001-1015) prohibits the discharge of oil or oily mixtures (over 100 mg/l) from vehicles generally within 50 mi of land; some exclusions, however, are granted. (The Oil Pollution Act of 1924 was repealed in 1970 because of supersession by subsequent legislation.)

The **Port and Waterways Safety Act of 1972** (PL92-340) grants the U.S. Coast Guard authority to establish and operate mandatory vehicle traffic control systems. The control system must consist of a VHF radio for ship-to-shore communications, as a minimum. The Act, in effect, also extends the provisions of the Tank Vessel Act, which protects against hazards to life and property, to include **protection of the marine environment.** Regulations stemming from this Act govern standards of tanker design, construction, alteration, repair, maintenance, and operation.

The most substantive marine environmental protection legislation is the **Federal Water Pollution Control Act (FWPCA) of 1948** (PL 80-854), as amended in 1956 (PL 84-660), 1961 (PL 87-883), 1965 (PL 89-234), and 1970 (PL 91-224), and by the FWPCA Amendments of 1972 (PL 92-500). The 1972 Amendments contain provisions which greatly expand federal authority to deal with **marine pollution,** providing authority for control of pollution by oil and hazardous substances other than oil, and for the assessment of penalties. **Navigable waters** are now defined as " . . . the waters of the United States, including territorial seas."

"Hazardous substances other than oil" and harmful quantities of these substances will be defined and categorized as removable or nonremovable by EPA in regulations. The Coast Guard (EPA in inland areas) must ensure that any discharge of oil or removable hazardous substance into U.S. waters is removed. If not properly removed by the owner or operator of the responsible vehicle, they are liable for removal costs. Liability is not incurred if the discharge is due solely to an act of God, an act of war, or negligence of the U.S. government.

**Penalties** now may be assessed by the Coast Guard for any discharge. The person in charge of a vehicle is to notify the appropriate U.S. government agency of any discharge upon knowledge of it.

The Coast Guard regulations will establish **removal procedures** in coastal areas, contain regulations to **prevent discharges** from vehicles and transfer facilities, and regulations governing the **inspection of cargo vessels** to prevent hazardous discharges. The EPA regulations govern inland areas and non-transportation-related facilities.

Section 312 of the FWPCA as amended in 1972 deals directly with **discharge of sewage.** The EPA must issue standards for marine sanitation devices, and the Coast Guard must issue regulations for implementing those standards. In June, 1972, EPA published standards that will eventually prohibit discharge of any sewage from vessels, whether it is treated or not. Interim standards allow existing vehicles to use certain approved devices that discharge treated sewage for as long as the device is operative, if installed within 3 years of the issuance of the Coast Guard regulations. If the device is installed between the third and fifth year after the regulations are issued, it may be used until 8 years after the regulations are issued. If no device is installed before the end of the fifth year, the no-discharge standards must be met. New vehicles must meet the no-discharge standard after the second year following issuance of the regulations. The standards for sewage discharge which have been established by individual states are preempted if a vessel meets federal standards during the interim period, even though the state standards may be more restrictive. There may be exceptions to this where a state can show that no-discharge standard is necessary for protection and enhancement of specified waters.

## PROPULSION SYSTEMS

(See Sec. 9 for component details)

The basic operating requirement for the main propulsion system is to propel the vehicle at the required sustained speed for the range or endurance required and to provide suitable maneuvering capabilities. In meeting this basic requirement, the marine propulsion system integrates the power generator/prime mover, the transmission system, the propulsor, and other shipboard systems with the ship's hull or vehicle platform. Figure 13 shows propulsion system alternatives with the most popular drives for fixed-pitch and controllable-pitch propellers.

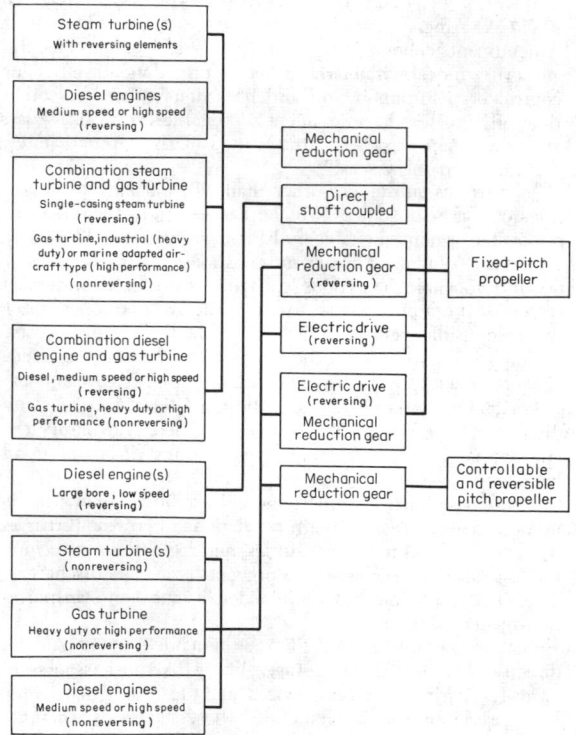

**Fig. 13**  Alternatives in the selection of a main propulsion plant. (*From "Marine Engineering," SNAME,* 1971.)

### Definitions for Propulsion Systems

**Brake power** $P_B$  bhp (kW$_B$), is the power delivered by the output coupling of a prime mover before passing through speed-reducing and transmission devices and with all continuously operating engine-driven auxiliaries in use.

**Shaft power** $P_S$, shp (kW$_S$), is the net power supplied to the propeller shafting after passing through all reduction gears or other transmission devices and after power for all attached auxiliaries has been taken off. Shaft power is measured in the shafting within the ship by a torsionmeter as close to the propeller or stern tube as possible.

**Delivered power** $P_D$, dhp (kW$_D$), is the power actually delivered to the propeller, somewhat less than $P_S$ because of the power losses in the stern tube and other bearings between the point of measurement and the propeller. Sometimes called **propeller power,** it is also equal to the effective power $P_E$, ehp (kW$_E$), plus the power losses in the propeller and the losses in the interaction between the propeller and the ship.

**Normal shaft power** or **normal power** is the power at which a marine vehicle is designed to run most of its service life.

**Maximum shaft power** is the highest power for which propulsion machinery is designed to operate continuously.

**Service speed** is the actual speed maintained by a vehicle at load draft on its normal route and under average weather and sea conditions typical of that route and with average fouling of bottom.

**Designed service speed** is the speed expected on trials in fair weather at load draft, with clean bottom, when machinery is developing a specified fraction of maximum shaft power. This fraction may vary but is of order of 0.9.

## MAIN PROPULSION PLANTS

### Steam Plants

The basic steam propulsion plant contains main boilers, steam turbines, a condensate system, a feedwater system, and numerous auxiliary components necessary for the plant to function. A heat balance calculation, the basic tool for determining the effect of various configurations on plant thermal efficiency, is demonstrated for the basic steam cycle. Both fossil-fuel and nuclear energy sources are successfully employed for marine applications.

**Main Boilers** (See Sec. 9)   The pressures and temperatures achieved in steam generating equipment have increased steadily over recent years, permitting either a higher-power installation for a given space or a reduction in the size and weight of a given propulsion plant.

The **trend in steam pressures** and **temperatures** has been an increase from 600 psig (4.14 MN/m²) and 850°F (454°C) during World War II to 1,200 psig (8.27 MN/m²) and 950°F (510°C) for naval combatants in and since the postwar era. For merchant ships, the progression has been from 400 psig (2.76 MN/m²) and 750°F (399°C) gradually up to 850 psig (5.86 MN/m²) and 850°F (454°C) in the 1960s, with some boilers at 1,500 psig (10.34 MN/m²) and 1000°F (538°C) in the 1970s.

The **quantity of steam** produced by a marine boiler ranges from approximately 1,500 lb/h (680 kg/h) in small auxiliary boilers to over 400,000 lb/h (181,500 kg/h) in large main propulsion boilers. Outputs of 750,000 lb/h (340,200 kg/h) or more per boiler are practical for high-power installations.

Most marine boilers are **oil-fired,** with wood, gas, and coal-fired boilers less common. Compared with other fuels, oil is easily loaded aboard ship, stored, and introduced into the furnace, and does not require the ash-handling facilities required for coal firing.

**Gas-fired boilers** are used primarily on power or drill barges which are fixed in location and can be supplied from shore (normally classed in the Ocean Engineering category). At sea, tankers designed to carry liquefied natural gas (LNG) may use the natural boil-off from their cargo gas tanks as a supplemental fuel (**dual fuel system**). The cargo gas boil-off is collected and pumped to the boilers where it is burned in conjunction with oil. The quantity of boil-off available is a function of the ambient sea and air temperatures, the ship's motion, and the cargo loading; thus, it may vary from day to day.

For economy of space, weight, and cost and for ease of operation, the trend in **boiler installations** is for fewer boilers of high capacity rather than a large number of boilers of lower capacity. The minimum installation is usually two boilers, to ensure propulsion if one boiler is lost; one boiler per shaft for twin-screw ships. Some large ocean-going ships operate on single boilers, requiring exceptional reliability in boiler design and operation.

Boilers which had either no air casing or only partial air casings were placed in separate firerooms, frequently kept under positive pressure to provide draft. Modern marine boilers are completely air-encased, with the outer casing cooled by the incoming combustion air; hence they are often located in the same machinery room as the main turbines. Space must be allowed for the maintenance of boiler tubes, drum internals, and components.

**Combustion systems** include forced-draft fans or blowers, the fuel oil service system, burners, and combustion controls. Operation and maintenance of the combustion system are extremely important to the efficiency and reliability of the plant. The best combustion with the least possible excess air should be attained.

**Main Turbines** (See Sec. 9) Single-expansion (i.e., single casing) marine steam turbines are fairly common at lower powers, usually not exceeding 4,000 to 6,000 shp (2,983 to 4,474 kW$_S$). Above that power range, the turbines are usually double-expansion (cross-compound) with high- and low-pressure turbines each driving the main reduction gear through its own pinion. The low-pressure turbine normally contains the reversing turbine in its exhaust end. The main condenser is either underslung and supported from the turbine, or the condenser is carried on the foundations, with the low-pressure turbine supported on the condenser.

The inherent **advantages** of the steam turbine have favored its use over the reciprocating steam engine for all large, modern marine steam propulsion plants. Turbines are not size-limited, and their high temperatures and pressures are accommodated safely. Rotary motion is simpler than reciprocating motion; hence unbalanced forces can be eliminated in the turbine. The turbine can efficiently utilize a low exhaust pressure; it is light-weight and requires minimum space and low maintenance.

The **reheat cycle** is the best and most economical means available to improve turbine efficiencies and fuel rates in marine steam propulsion plants. In the reheat cycle, steam is withdrawn from the turbine after partial expansion and is passed through a heat exchanger where its temperature is raised; it is then readmitted to the turbine for expansion to

condenser pressure. Marine reheat plants have more modest steam conditions than land-based applications because of lower power ratings and a greater reliability requirement for ship safety.

The reheat cycle is applied mostly in high-powered units above 25,000 shp (18,642 kW$_S$) and offers the maximum economical thermal efficiency that can be provided by a steam plant. Reheat cycles are not used in naval vessels because the improvement in efficiency does not warrant the additional complexity; hence a trade-off is made.

**Turbine foundations** must have adequate rigidity to avoid vibration conditions. This is particularly important with respect to periodic variations in propeller thrust which may excite longitudinal vibrations in the propulsion system.

**Condensate System** (See Sec. 9) The condensate system provides the means by which feedwater for the boilers is recovered and returned to the feedwater system. The major components of the condensate system of a marine propulsion plant are the **main condenser,** the **condensate pumps,** and the **deaerating feed tank** or **heater.** Both single-pass and two-pass condenser designs are used; the single-pass design, however, allows somewhat simpler construction and lower water velocities. The single-pass condenser is also adaptable to scoop circulation, opposed to pump circulation, which is practical for higher-speed ships. The deaerating feedwater heater is supplied from the condensate pumps, which take suction from the condenser hot well, together with condensate drains from steam piping and various heaters.

The deaerating feed heater is large and usually difficult to locate. It is normally maintained at about 35 psig (0.241 MN/m²) and 280°F (138°C) by auxiliary exhaust and turbine extraction steam. The condensate is sprayed into the steam atmosphere at the top of the heater, and the heated feedwater is pumped from the bottom by the feed booster or main feed pumps. In addition to removing oxygen or air, the heater also acts as a surge tank to meet all demands during maneuvering conditions. Since the feed pumps take suction where the water is almost saturated, the heater must be located 30 to 50 ft (9.14 to 15.24 m) above the pumps in order to provide enough positive suction head to prevent flashing from pressure fluctuations during sudden plant load change.

**Feedwater System** The feedwater system comprises the pumps, piping, and controls necessary to transport feedwater to the boiler or steam generator, to raise water pressure above boiler pressure, and to control flow of feedwater to the boiler. **Main feed pumps** are so vital that they are usually installed in duplicate, providing a standby pump capable of feeding the boilers at full load. Auxiliary steam-turbine-driven centrifugal pumps are usually selected. A typical naval installation consists of three main feed booster pumps and three main feed pumps for each propulsion plant. Two of the booster pumps are turbine-driven and one is electric. Additionally, an electric-motor-driven emergency booster pump is usually provided. The total capacity of the main feed pumps must be 150 percent of the boiler requirement at full power plus the required recirculation capacity. Reliable feedwater regulators are important.

**Steam Plant—Nuclear** (See Sec. 9) The compact nature of the energy source is the most significant characteristic of nuclear power for marine application. The fission of one gram of uranium per day produces about one megawatt of power.

(One pound produces 608,579 horsepower.) In other terms, the fission of 1 lb of uranium is equivalent to the combustion of about 86 tons (87,380 kg) of 18,500 Btu/lb (43.03 MJ/kg) fuel oil. This characteristic permits utilization of large power plants on board ship without the necessity for frequent refueling or large bunker storage. Economic studies, however, show that nuclear power, as presently developed, is best suited for military purposes, where the advantages of high power and endurance override the pure economic considerations. As the physical size of nuclear propulsion plants is reduced, their economic attractiveness for commercial marine application will increase.

**Major differences** between the nuclear power and fossil-fuel plants are: (1) The **safety aspect** of the nuclear reactor system—operating personnel must be shielded from fission product radiation, hence the size and weight of the reactor are increased and maintenance and reliability become more complicated; (2) the steam produced by a pressurized-water reactor plant is saturated and, because of the **high moisture content** in the turbine steam path, the turbine design requires careful attention; (3) the steam pressure provided by a pressurized-water reactor plant varies with output, the maximum pressure occurring at no load and decreasing approximately linearly with load to a minimum at full power; hence, **blade stresses** in a nuclear turbine increase more rapidly with a decrease in power than in a conventional turbine. Special attention must be given to the design of the control stage of a nuclear turbine. The *N.S. Savannah* was designed for 700 psig (4.83 MN/m²) at no load and 400 psig (2.76 MN/m²) at full power.

**Steam Plant—Heat Balance** The heat-balance calculation is the basic analysis tool for determining the effect of various steam cycles on the thermal efficiency of the plant and for determining the quantities of steam and feedwater flow.

The thermal arrangement of a **simple steam cycle,** illustrated in Fig. 14, and the following simplified analysis are taken from "Marine Engineering," SNAME, 1971:

The unit is assumed to develop 30,000 shp (22,371 kWₛ). The steam rate of the main propulsion turbines is 5.46 lb/(shp)(h) [3.32 kg/(kWₛ)(h)] with throttle conditions of 850 psig (5.86 MN/m²) and 950°F (510°C) and with the turbine exhausting to the condenser at 1.5 inHg abs (5,065 N/m²). To develop 30,000 shp (22,371 kWₛ), the throttle flow must be 163,800 lb/h (74,298 kg/h).

The generator load is estimated to be 1,200 kW, and the turbogenerator is thus rated at 1,500 kW with a steam flow of 10,920 lb/h (4,953 kg/h) for steam conditions of 850 psig (5.86 MN/m²) and 950°F (510°C) and a 1.5 inHg abs (5,065 N/m²) back pressure. The total steam flow is therefore 174,720 lb/h (79,251 kg/h).

Now trace the steam and water flow through the cycle. The flow exhausting from the main turbine is 163,800 − 250 = 163,550 lb/h (74,298 − 113 = 74,185 kg/h), 250 lb/h (113 kg/h) to the gland condenser, and from the auxiliary turbine is 10,870 lb/h (4,930 kg/h), 50 lb/h (23 kg/h) to the gland condenser. The two gland leak-off flows return from the gland leak-off condenser to the main condenser. The condensate flow leaving the main condenser totals 174,720 lb/h (79,251 kg/h).

It is customary to allow a 1°F (0.556°C) hot-well depression in the condensate temperature. Thus, at 1.5 inHg abs (5,065 N/m²) the condenser saturation temperature is 91.7°F (33.2°C) and the condensate is at 90.7°F (32.6°C). Entering the gland

condenser there is a total energy flow of 174,720 × 58.7 = 10,256,064 Btu/h (10,820,557,760 J/h). The gland condenser receives gland steam at 1281 Btu/lb (2,979,561 J/kg) and drains at a 10°F (5.6°C) terminal difference or 100.7°F (38.2°C). This adds a total of 300 × (1,281 − 68.7) or 363,690 Btu/h (383,-707,497 J/h) to the condensate, making a total of 10,619,754 Btu/h (11,204,265,257 J/h) entering the surge tank. The feed leaves the surge tank at the same enthalpy with which it enters.

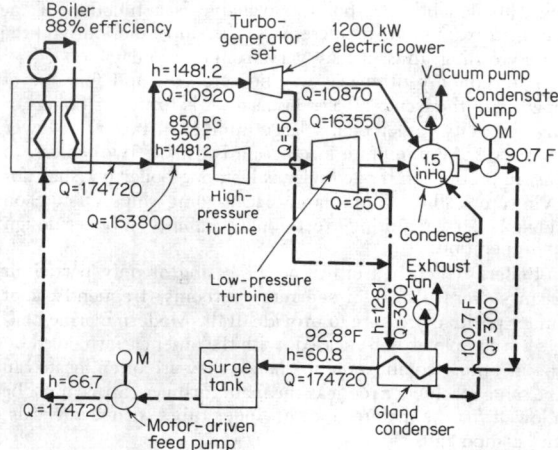

**Fig. 14** Simple steam cycle. $Q$ = flow, lb/h; $h$ = enthalpy, Btu/lb (see text for SI values).

The feed pump puts an amount of heat into the feedwater equal to the total power of the pump, less any friction in the drive system. This friction work can be neglected but the heat from the power input is a significant quantity. The power input is the total pump head in feet of feedwater, times the quantity pumped in pounds per hour, divided by the mechanical equivalent of heat and the efficiency. Thus,

Heat equivalent of feed pump work

$$= \frac{144 \Delta P v_f Q}{778 E} \quad \text{Btu/h} \quad \left( \frac{\Delta P v_f Q}{E} \quad \text{J/h} \right)$$

where $\Delta P$ = pressure change, lb/in² (N/m²)
 $v_f$ = specific volume of fluid, ft³/lb (m³/kg)
 $Q$ = mass rate of flow, lb/h (kg/h)
 $E$ = efficiency of pump

Assuming the feed pump raises the pressure from 15 to 1,015 psia (103,421 to 6,998,178 N/m²), the specific volume of the water is 0.0161 ft³/lb (0.001005 m³/kg), and a 50 percent pump efficiency; the heat equivalent of the feed pump work is 1,041,313 Btu/h (1,098,626,868 J/h). This addition of heat gives a total of 11,661,067 Btu/h (12,302,892,125 J/h) entering the boiler. Assuming no leakage of steam, the steam leaves the boiler at 1481.2 Btu/lb (3,445,219 J/kg), with a total thermal energy flow of 1481.2 × 174,720 = 258,795,264 Btu/h (273,-039,355,300 J/h). The difference between this total heat and that entering (258,795,264 − 11,661,067 = 247,134,197 Btu/h or 260,736,463,175 J/h) is the net heat added in the boiler by the fuel. With a boiler efficiency of 88 percent and a fuel having a higher heating value of 18,500 Btu/lb (43,030,353 J/kg), the quantity of fuel burned is

$$\text{Fuel flow rate} = \frac{247,134,197}{(18,500)(0.88)} = 15,180 \text{ lb/h}$$

or

$$\frac{260,736,463,175}{(43,030,353)(0.88)} = 6,886 \text{ kg/h}$$

The specific fuel rate is the fuel flow rate divided by the net shaft power [15,180/30,000 = 0.506 lb/(shp)(h) or 6886/22,371 = 0.308 kg/(kW$_\text{S}$)(h)]. The heat rate is the quantity of heat expended to produce one horsepower per hour (one kilowatt per hour) and is calculated by dividing the net heat added to the plant, per hour, by the power produced.

$$\text{Heat rate} = \frac{247,134,197}{30,000} = 8238 \text{ Btu/(shp)(h)}$$

or

$$\frac{260,736,463,175}{22,371} = 11,655,110 \text{ J/(kW}_\text{S})(h)$$

This simple cycle omits many details that must necessarily be included in an actual steam plant. A continuation of the example, developing the details of a complete analysis, is given in "Marine Engineering," SNAME, 1971. A heat balance is usually carried through several iterations until a particular level of accuracy is achieved. The first heat balance may be done from approximate data given in the SNAME Technical and Research Publication No. 3-11, "Recommended Practices for Preparing Marine Steam Power Plant Heat Balances," and then updated as equipment data are known.

### Diesel Engines

(See Sec. 9)

Diesel engines are referred to as being high-, medium-, or low-speed and, in general, are categorized as follows:

### Engine Speed Classifications

|  | Piston speed, ft/min (m/min) | Shaft speed, r/min |
|---|---|---|
| Low speed | 1,000–1,500 (305–457) | 100–514 |
| Medium speed | 1,200–1,800 (366–549) | 700–1,200 |
| High speed | 1,800–3,000 (549–914) | 1,800–4,000 |

A well-designed high-speed engine which is not overloaded can give equally good service as a slow-speed engine. Slow-speed engines are larger than high-speed engines, but wear rates are comparable; hence, slow-speed engine parts take longer to wear to the same percentage of their original dimension. The high-speed engine is smaller, lighter, and generally less expensive than the slow-speed engine and, although it costs less initially, it has higher fuel, operating, and maintenance costs.

**Medium- and High-speed Diesels** The number of medium- and high-speed diesel engines used in marine applications is relatively small compared with the total number of such engines produced. The medium- and high-speed marine engine of today is almost universally an adaptation of engines built in quantities for service in automotive and stationary applications. The truck and bus field contributes the high-speed engines in the range of 300 to 400 hp (224 to 298 kW). For intermittent use, maximum speed of these engines approx-

imates 3,000 r/min; for continuous service, speeds of 1,800 r/min are common. The off-highway equipment engines are in the power range from 500 to upward of 1,200 hp (373 to 895 kW) and speeds in the 1,200 to 1,800 r/min range. Diesel locomotive engines are available in units from 6 to 20 cylinders and ratings up to 4,000 hp (2,983 kW) at speeds from 850 to 1,100 r/min. Another group of engines coming from the stationary field includes units in the speed range from 300 to 514 r/min and powers up to 7,500 hp (5,593 kW).

Some of the engines were designed with **marine applications** in mind, and others require some modification in external engine hardware. The changes are those needed to suit the engine to the marine environment, meaning salt-laden air, high humidity, use of corrosive seawater for cooling, and operating from a pitching and rolling platform. It also may mean an installation made in confined spaces. In order to adapt to this environment, the prime requisite of the marine diesel engine is the ability to resist corrosion. Because marine engines may be installed with their crankshafts at an angle to the horizontal and because they are subjected to more motion than in many other applications, changes are also necessary in the lubricating oil system. The air intake to a marine engine may not be dust-free and dirt-free when operating in harbors, inland waters, or close offshore; therefore, it is as important to provide a good air cleaner as in any automotive or stationary installation.

Diesel engines are utilized in all types of marine vehicles, both in the merchant marine and in the navies of the world. The power range in which diesel engines have been used has increased directly with the availability of higher-power engines. The line of demarkation in horsepower between what is normally assigned to diesel and to steam has continually moved upward, as has the power installed in ships in general.

Small high-speed engines are commonly used in pleasure boats where the owner is safety-conscious and wants to avoid the use of gasoline. For the same reason, small boats in use in the Navy are usually powered by diesel engines, although the gas turbine is being used in special boats where high speed for short periods of time is the prime requirement. Going a little higher in power, diesels are used in many kinds of workboats such as fishing boats, tugs, ferries, dredges, river towboats and pushers, and smaller types of cargo ships and tankers. They are used in the naval counterparts of these ships and, in addition, for military craft such as minesweepers, landing ships, patrol and escort ships, amphibious vehicles, tenders, submarines, and special ships such as salvage and rescue ships and icebreakers. In the nuclear-powered submarine, the diesel is relegated to emergency generator-set use; however, it is still the best way to power a nonnuclear submarine when not operating on the batteries.

Diesel engines are used either singly or in multiple to drive propeller shafts. For all but high-speed boats, the rpm of the modern diesel is too fast to drive the propeller directly with efficiency, and some means of speed reduction, either mechanical or electrical, is necessary. If a single engine of the power required for a given application is available, then a decision must be made whether it or several smaller engines should be used. The decision may be dictated by the available space. The diesel power plant is flexible in adapting to specific space requirements. When more than one engine is geared to the propeller shaft, the gear serves as both a speed reducer and combining gear. The same series of engines could be used in an electric-drive propulsion system, with even greater flexibility. Each engine drives its own generator and may be located independently of other engines and the propeller shaft.

**Low-speed, Direct-coupled Diesels** Of the more usual prime mover selections, only low-speed diesel engines are directly coupled to the propeller shaft. This is due to the low rpm required for efficient propeller operation and the high rpm inherent with other types of prime movers.

A rigid hull foundation, with a high resistance to vertical, athwartship, and fore-and-aft deflections, is required for the low-speed diesel. The engine room must be designed with sufficient overhaul space and with access for large and heavy replacement parts to be lifted by cranes.

Because of the low-frequency noise generated by low-speed engines, the operating platform often can be located at the engine itself. Special control rooms are often preferred as the noise level in the control room can be significantly less than that in the engine room.

Electric power may be produced by a generator mounted directly on the line shafting. Operation of the entire plant may be automatic and remotely controlled from the bridge. The engine room is often completely unattended for 16 h a day.

### Gas Turbines

(See Sec. 9)

The gas turbine has developed since World War II to join the steam turbine and the diesel engine as alternative prime movers for various shipboard applications.

In gas turbines, the efficiency of the components is extremely important since the compressor power is very high compared with its counterpart in competitive thermodynamic cycles. For example, a typical marine propulsion gas turbine rated at 20,000 bhp (14,914 $kW_B$) might require a 30,000 hp (22,371 kW) compressor and, therefore, 50,000 hp (37,285 kW) in turbine power to balance the cycle.

The basic **advantages** of the gas turbine for marine applications are its simplicity and light weight. As an internal-combustion engine, it is a self-contained power plant in one package with a minimum number of large supporting auxiliaries. It has the ability to start and go on line very quickly. Having no large masses that require slow heating, the time required for a gas turbine to reach full speed and accept the load is limited almost entirely by the rate at which energy can be supplied to accelerate the rotating components to speed. A further feature of the gas turbine is its low personnel requirement and ready adaptability to automation.

The relative simplicity of the gas turbine has enabled it to attain outstanding records of reliability and maintainability when used for aircraft propulsion and in industrial service. The same level of reliability and maintainability can be expected in marine service if the unit is properly applied and installed.

**Marine units derived from aircraft engines** usually have the gas generator section, comprising the compressor and its turbine, arranged to be removed and replaced as a unit. Maintenance on the power turbine, which usually has the smallest part of the total maintenance requirements, is performed aboard ship. Because of their light weight, small gas turbines used for auxiliary power or the propulsion of small boats, can also be readily removed for maintenance.

**Units designed specifically for marine use** and those derived from industrial gas turbines are usually maintained and overhauled in place. Since they are somewhat larger and heavier than the aviation-type units, removal and replacement are not as readily accomplished. For this reason, they usually have split casings and other provisions for easy access and maintenance. The work can be performed by the usual ship repair forces.

Both **single-shaft** and **two-shaft** gas turbines can be used in marine service. Single-shaft units are most commonly used for generator drives. When used for main propulsion, where the propeller must operate over a very wide speed range, the single-shaft unit must have a controllable and reversible-pitch propeller or some equivalent variable-speed transmission, such as an electric drive, because of its limited speed range and poor acceleration characteristics. A multishaft unit is normally used for main propulsion, with the usual arrangement being a two-shaft unit with an independent variable-speed power turbine; the power turbine and propeller can be stopped, if necessary, and the gas producer kept in operation for rapid load pickup. The use of variable-area nozzles on the power turbine increases flexibility by enabling the compressor to be maintained at or near full speed and the air flow at low-power turbine speeds. Nearly full power is available by adding fuel, without waiting for the compressor to accelerate and increase the air flow. Where low-load economy is of importance, the controls can be arranged to reduce the compressor speed at low loads and maintain the maximum turbine inlet and/or exhaust temperature for best efficiency. Since a gas turbine inherently has a poor part-load fuel rate performance, this variable-area nozzle feature can be very advantageous.

The **physical arrangement** of the various components, i.e., compressors, combustion systems, and turbines, that make up the gas turbine is influenced by the thermodynamic factors (i.e., the turbine connected to a compressor must develop enough power to drive it), by mechanical considerations (i.e., shafts must have adequate bearings, seals, etc.), and also by the necessity to conduct the very high air and gas flows to and from the various components with minimum pressure losses.

In marine applications, the gas turbines usually **cannot be mounted rigidly** to the ship's structure. Normal movement and distortions of the hull when under way would cause distortions and misalignment in the turbine and cause internal rubs or bearing and/or structural failure. The turbine components can be mounted on a subbase built up of structural sections of sufficient rigidity to maintain the gas-turbine alignment when properly supported by the ship's hull. In cases where aircraft gas turbines have been adapted to marine use, some form of tubular structure may be used.

Since the gas turbine is a high-speed machine with output shaft speeds ranging from about 3,600 r/min for large machines up to 100,000 r/min for very small machines (approximately 25,000 r/min is an upper limit for units suitable for the propulsion of small boats), a **reduction gear is necessary** to reduce the speed to the range suitable for a propeller. Smaller units suitable for boats or driving auxiliary units, such as generators in larger vessels, frequently have a reduction gear integral with the unit. Larger units normally require a separate reduction gear, usually of the double- or triple-reduction type.

A gas turbine, in common with all turbine machinery, is **not inherently reversible** and must be reversed by external means. Electric drives offer ready reversing but are usually ruled out on the basis of weight, cost, and to some extent efficiency. From a practical standpoint there are two alternatives, a reversing gear or a controllable and reversible pitch (CRP) propeller. Both have been used successfully in gas-turbine-driven ships, with the CRP favored.

### Combined Propulsion Plants

In some shipboard applications, diesel engines, gas turbines, and steam turbines can be employed effectively in various combinations. The prime movers may be combined either mechanically, thermodynamically, or both.

The gas turbine is a very flexible power plant and consequently figures in most possible combinations which include combined diesel and gas-turbine plants (**CODAG**); combined steam- and gas-turbine plants (**COSAG** or **COGAS**); and combined gas-turbine and gas-turbine plants (**COGAG**). In these cycles, gas turbines and other engines or gas turbines of two different sizes or types are combined in one plant to give optimum performance over a very wide range of power and speed. In addition, combinations of diesel or gas-turbine plants (**CODOG**), or gas-turbine or gas-turbine plants (**COGOG**), where one plant is a diesel or a small gas turbine for use at low or cruising powers and the other a large gas turbine which operates alone at high powers, are also possibilities.

**COSAG**   Steam and gas turbines are connected to a common reduction gear but are thermodynamically independent. The chief application has been in high-speed naval vessels of the destroyer type which normally require operation at speeds above half power for only a small portion of their operating life. The boost power is furnished by a light-weight, simple-cycle, aircraft-type gas turbine. This combination produces a significant reduction in machinery weight.

**COGAS**   This type is sometimes designated **STAG** for steam turbine and gas turbine. Both mechanical and thermodynamic interconnections exist between the gas-turbine and steam cycles. The principal advantage gained by the thermodynamic interconnection is the potential for improved overall efficiency and for space and weight savings. In this arrangement, the gas turbine discharges to a heat-recovery boiler where the large quantity of heat in the exhaust gases is used to generate steam. The boiler supplies the steam turbine that is geared to the propeller. The steam turbine may be coupled to provide part of the power for the gas-turbine compressor. The gas turbine may be used to provide additional propulsion power, or the gas-turbine exhaust may be recovered to supply heat for various ship's services.

**CODOG**   Combined diesel or gas-turbine plants are in use in the Navy's PG-84 class of patrol gunboats and in the Coast Guard's *Hamilton* class cutters.

**COGOG**   A combined gas turbine or gas-turbine plant is installed in the Royal Canadian Navy's DDH-280 class of helicopter destroyers.

### PROPULSORS

The force to propel a marine vehicle arises from the rate of change of momentum induced in either the water or air. Since the force produced is directly proportional to the mass density of the fluid used, the reasonable choice is to induce the momentum change in water. If air were used, either the cross-sectional area of the jet must be large or the velocity must be high. A variety of propulsors are used to generate this stream of water aft relative to the vehicle: screw propellers, controllable and reversible-pitch propellers, water jets, vertical-axis

propellers, and other thrust devices. Figure 15 indicates the type of propulsor which provides the best efficiency for a given vehicle type.

### Screw Propellers

The screw propeller may be regarded as part of a helicoidal surface which, as it rotates, appears to "screw" its way through the water, driving water aft and the vehicle forward. A propeller is termed **"right-handed"** if it turns clockwise (viewed from aft) when producing ahead thrust; if counter-

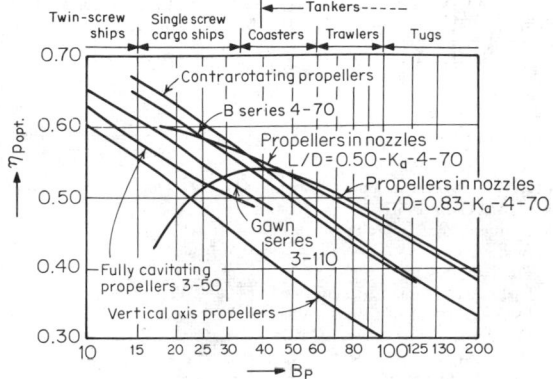

**Fig. 15**   Comparison of optimum efficiency values for different types of propulsors. (*From "Marine Engineering," SNAME*, 1971.)

clockwise, **"left-handed."** In a twin-screw installation, the starboard propeller is normally right-handed and the port propeller left-handed. The surface of the propeller blade facing aft, which experiences the increase in pressure, producing thrust, is the **face** of the blade; the forward surface is the **back**. The face is commonly constructed as a true helical surface of constant pitch; the back is not a helical surface. A **true helical surface** is generated by a line rotated about an axis normal to itself and advancing in the direction of this axis at constant speed. The distance the line advances in one revolution is the **pitch**. For simple propellers, the pitch is constant on the face; but in practice, it is common for large propellers to have a reduced pitch toward the hub and less usually toward the tip. The pitch at 0.7 times the maximum radius is usually a representative mean pitch; maximum lift is generated at that approximate point. Pitch may be expressed as a dimensionless ratio, $P/D$.

The **shapes** of blade outlines and sections vary greatly according to the type of ship for which the propeller is intended and to the designer's ideas. Figure 16 shows a typical design and defines many of the terms in common use. The **projected area** is the area of the projection of the propeller contour on a plane normal to the shaft, and the **developed area** is the total face area of all the blades. If the variation of helical cord length is known, then the true blade area, called **expanded area**, can be obtained graphically or analytically by integration.

Consider a section of the propeller blade at a radius $r$ with a pitch angle $\phi$ and pitch $P$ working in an unyielding medium; in one revolution of the propeller it will advance a distance $P$. Turning $n$ revolutions in unit time it will advance a distance $P \times n$ in that time. In a real fluid, there will be a certain amount of yielding when the propeller is developing thrust and the screw will not advance $P \times n$, but some smaller distance. The difference between $P \times n$ and that smaller distance is called the **slip velocity. Real slip ratio** is defined in Fig. 17.

A wake or a frictional belt of water accompanies every hull

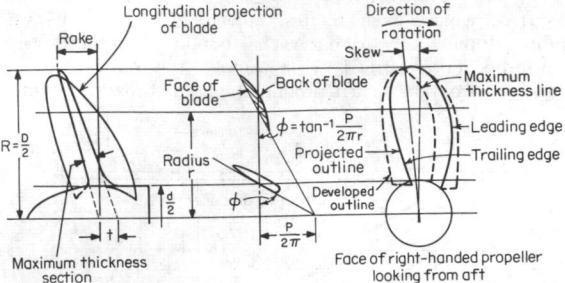

**Fig. 16** Typical propeller drawing. (*From "Principles of Naval Architecture," SNAME*, 1967.)

Diameter $D$      Pitch ratio = $P/D$      Pitch $P$
Blade thickness ratio = $t/D$      No. of blades 4
Pitch angle = $\phi$
Disk area = area of tip circle = $\pi D^2/4 = A_O$
Developed area of blades, outside hub = $A_D$
Developed area ratio = DAR = $A_D/A_O$
Projected area of blades (on transverse plane)
  outside hub = $A_P$
Projected area ratio = PAR = $A_P/A_O$
Blade width ratio = BWR = Max blade width/$D$
Mean width ratio = MWR = [$A_D$/length of blades (outside hub)]/$D$

in motion; its velocity varies as the ship's speed, the form of the ship, and the distance from the ship's side and from the bow, together with the condition of the hull surface. For ordinary propeller design the **wake velocity** is a fraction $w$ of the ship's speed. Wake velocity = $wV$. The **wake fraction** $w$ for a ship may be obtained from Fig. 18. The velocity of the ship relative to the ship's wake at the stern is $V_A = (1 - w)V$.

The **apparent slip ratio** $S_A$ is given by

$$S_A = (Pn - V)/Pn = 1 - V/Pn$$

Although real slip ratio, which requires knowledge of the wake fraction, is a real guide to ship performance, the apparent slip ratio requires only ship speed, revolutions, and pitch to calculate and is therefore often recorded in ships' logs.

For $P$ in ft (m), $n$ in r/min, and $V$ in knots, $S_A = (Pn - 101.3V)/Pn$; [$S_A = (Pn - 30.9V)/Pn$].

**Propeller Design** The design of a marine propeller is usually carried out by one of two methods. In the first, the design is based upon charts giving the results of open-water tests on a series of model propellers. These cover variations in a number of the design parameters such as pitch ratio, blade area, number of blades, and section shapes. A propeller which conforms with the characteristics of any particular series can be rapidly designed and drawn to suit the required ship conditions.

The second method is used in cases where a propeller is heavily loaded and liable to cavitation or has to work in a very uneven wake pattern; it is based on purely theoretical calculations. Basically this involves finding the chord width, section shape, pitch, and efficiency at a number of radii to suit the average circumferential wake values and give optimum efficiency and protection from cavitation. By integration of the

resulting thrust and torque-loading curves over the blades, the thrust, torque, and efficiency for the whole propeller can be found.

Using the first method and **Taylor's propeller** and **advance coefficients** $B_p$ and $\delta$, a convenient practical design and an initial estimate of propeller size can be obtained

$$B_p = \frac{n(P_D)^{0.5}}{(V_A)^{2.5}} \quad \left[\frac{1.158n(P_D)^{0.5}}{(V_A)^{2.5}}\right]$$

$$\delta = \frac{nD}{V_A} \quad \left(\frac{3.281nD}{V_A}\right)$$

$$\eta_o = \frac{TV_A}{325.7P_D\eta_R} \quad \left(\frac{TV_A}{1,942.5P_D\eta_R}\right)$$

where $B_p$ = Taylor's propeller coefficient
$\delta$ = Taylor's advance coefficient
$n$ = r/min
$P_D$ = delivered power, dhp (kW$_D$)
$V_A$ = speed of advance, knots
$D$ = propeller diameter, ft (m)
$T$ = thrust, lb (N)
$\eta_o$ = open-water propeller efficiency
$\eta_R$ = relative rotative efficiency ($0.95 < \eta_R < 1.0$, twin-screw; $1.0 < \eta_R < 1.05$, single-screw), a factor which corrects $\eta_o$ to the efficiency in the actual flow conditions behind the ship

Assume a reasonable value for $n$ and, using known values for $P_D$ and $V_A$ (for a useful approximation, $P_D = 0.98P_S$), calculate $B_P$. Then enter Fig. 19, or an appropriate series of Taylor or Troost propeller charts, to determine $\delta$, $\eta_o$, and $P/D$ for optimum efficiency. The charts and parameters should be varied, and the results plotted to recognize the most suitable propeller.

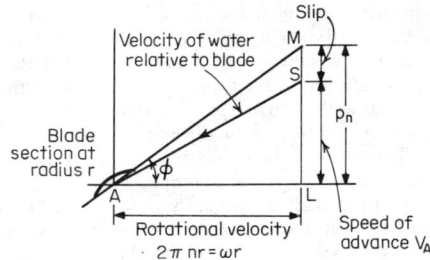

**Fig. 17** Definition of slip. (*From "Principles of Naval Architecture," SNAME*, 1967.)

$$\tan \phi = Pn/2\pi nr = P/2\pi r$$
Real slip ratio $S_R = MS/ML = (Pn - V_A)/Pn = 1 - V_A/Pn$

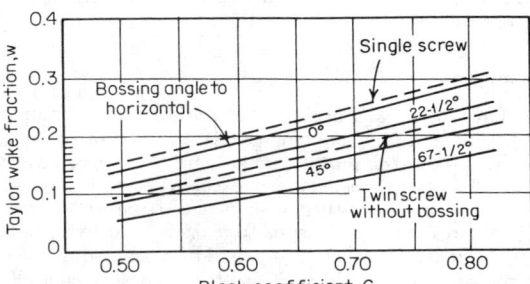

**Fig. 18** Wake fractions for single- and twin-screw models. (*From "Principles of Naval Architecture," SNAME*, 1967.)

**Propeller cavitation,** when severe, may result in marked increase in rpm, slip, and shaft power with little increase in ship speed or effective power. As cavitation develops, noise, vibration, and erosion of the propeller blades, struts, and

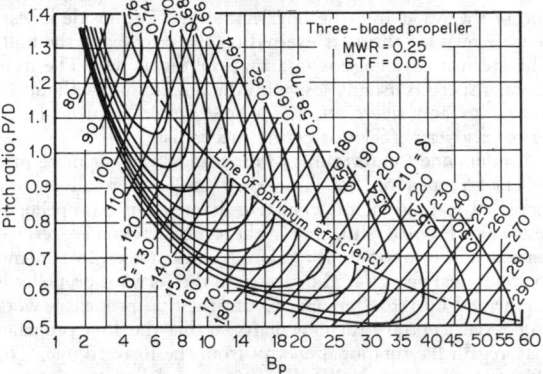

**Fig. 19** Typical Taylor propeller characteristic curves. (*From "Handbook of Ocean and Underwater Engineering," McGraw-Hill, 1969.*)

rudders are experienced. It may occur either on the face or on the back of the propeller. Although cavitation of the face has little effect on thrust and torque, extensive cavitation of the back can materially affect thrust and, in general, requires either an increase in blade area or a decrease in propeller rpm to combat. The erosion of the backs is caused by the collapse of cavitation bubbles as they move into higher pressure regions toward the trailing edge.

Avoidance of cavitation is an important requirement in propeller design and selection. The Netherlands Ship Model Basin suggests the following **criterion** for the minimum blade area required **to avoid cavitation:**

$$A_p^2 = \frac{T^2}{1,360(p_0 - p_v)^{1.5} V_A} \quad \text{or} \quad \left[ \frac{T^2}{5.44(p_0 - p_v)^{1.5} V_A} \right]$$

where $A_p$ = projected area of blades, ft² (m²)
$T$ = thrust, lb (N)
$p_0$ = pressure at screw centerline, due to water head plus atmosphere, lb/in² (N/m²)
$p_v$ = water-vapor pressure, lb/in² (N/m²)
$V_A$ = speed of advance, knots (1 knot = 0.515 m/s)

or
$$p_0 - p_v = 14.45 + 0.45b, \text{ lb/in}^2$$
$$(p_0 - p_v = 99,629 + 10,179b, \text{ N/m}^2)$$

where $b$ = head of water at screw centerline, ft (m).

A four-blade propeller of 0.97 times the diameter of the three-blade, the same pitch ratio, and 4/3 the area will absorb the **same power** at the same rpm as the three-blade propeller. Similarly, a two-blade propeller of 5 percent greater diameter is approximately equivalent to a three-blade unit. Figure 19 shows that propeller efficiency increases with decrease in value of the propeller coefficient $B_p$. A low value of $B_p$ in a slow-speed ship calls for a large-diameter (optimum) propeller. It is frequently necessary to limit the diameter of the propeller and accept the accompanying loss in efficiency.

The number of propeller blades is usually three or four. Four blades have been universal with single-screw merchant ships. Recently five, and even six, blades have been used to reduce vibration. Highly loaded propellers of fast ships and naval vessels call for large blade area, preferably three blades, to reduce blade interference.

Propeller fore-and-aft clearance from the stern frame of large single-screw ships at 70 percent of propeller radius should be greater than 18 in. The stern frame should be streamlined. The clearance from the propeller tips to the shell plating of twin-screw vessels ranges from about 20 in in low-powered ships to 4 or 5 ft in high-powered ships. U.S. Navy practice generally calls for a propeller tip clearance of $0.25D$. Many high-speed motorboats have a tip clearance of only several inches. The immersion of the propeller tips should be sufficient to prevent the drawing in of air.

EXAMPLE. Consider a three-blade propeller for a single-screw installation with $P_s$ = 2,950 shp (2,200 kW$_S$), $V$ = 17 knots, and $C_B$ = 0.52. Find the propeller diameter $D$ and efficiency $\eta_o$ for $n$ = 160 r/min.

From Fig. 18, the wake fraction $w$ = 0.165. $V_A$ = $V(1 - 0.165)$ = 14.19 knots. $P_D$ = 0.98$P_s$ = 2,891 dhp (2,156 kW$_D$). At 160 r/min, $B_P$ = 160(2,891)$^{0.5}$/(14.19)$^{2.5}$ = 11.34. [1.158 × 160(2,156)$^{0.5}$/(14.19)$^{2.5}$ = 11.34.] From Fig. 19, the following is obtained or calculated:

| r/min | $B_p$ | $P/d$ | $\eta_o$ | $\delta$ | $D$, ft (m) | $P$, ft (m) |
|---|---|---|---|---|---|---|
| 160 | 11.34 | 0.95 | 0.69 | 139 | 12.33 (3.76) | 11.71 (3.57) |

For a screw centerline submergence of 9 ft (2.74 m) and a selected projected area ratio PAR = 0.3, investigate the cavitation criterion. Assume $\eta_R$ = 1.0.
Minimum projected area:

$T = 325.7\eta_o\eta_R P_D/V_A = 325.7 \times 0.69 \times 1.0 \times 2,891/14.19$
$= 45,786$ lb
$(T = 1942.5 \times 0.69 \times 2,156/14.19 = 203,646$ N$)$
$p_0 - p_v = 14.45 + 0.45 (9) = 18.5$ lb/in²
$(p_0 - p_v = 99,629 + 10,179 (2.74) = 127,519$ N/m²$)$
$A_p^2 = (45,786)^2/1,360 (18.5)^{1.5}(14.19) = 1,365$, $A_p$ = 36.95 ft² minimum
$A_p^2 = (203,646)^2/5.44 (127,519)^{1.5}(14.19) = 11.79$, $A_p$ = 3.43 m² minimum

Actual projected area: PAR = $A_p/A_0$ = 0.3, $A_p$ = 0.3$A_0$ = $0.3\pi$ (12.33)²/4 = 35.82 ft² (3.33 m²)

Since the selected PAR does not meet the minimum cavitation criterion, either the blade width can be increased or a four-blade propeller can be adopted. To absorb the same power at the same rpm, a four-blade propeller of about 0.97 (12.33) = 11.96 ft (3.65 m) diameter with the same blade shape would have an $A_p$ = 35.82 × ⁴/₃ × (11.96/12.33)² = 44.94 ft² (4.17 m²), or about 25 percent increase. The pitch = 0.95 (11.96) = 11.36 ft (3.46 m). The real and apparent slip ratios for the four-blade propeller are:
$S_R = 1 - (101.3)(14.19)/(11.36)(160) = 0.209$
$S_A = 1 - (30.9)(17.0)/(3.46)(160) = 0.051$
that is, 20.9 and 5.1 percent ($S_A$ calculated using SI values).

Since the projected area in the three-blade propeller of this example is only slightly under the minimum, a three-blade propeller with increased blade would be the more appropriate choice.

**Controllable-reversible-pitch Propellers**

Controllable-reversible-pitch (CRP) propellers are screw propellers in which the blades are separately mounted on the hub, each on an axis, and in which the pitch of the blades can be changed, and even reversed, while the propeller is turning. The pitch is changed by means of an internal mechanism consisting essentially of hydraulic pistons in the hub acting on crossheads. CRP's are most suitable for vehicles which must meet different operating conditions, such as tugs, trawlers, ferries, minesweepers, and landing craft. As the propeller pitch is changed, the engine can still run at its most efficient speed. Maneuvering is more rapid since the pitch can be changed more rapidly than could the shaft revolutions. By use of CRP's, neither reversing mechanisms are necessary in reciprocating engines, nor astern turbines in turbine-powered vehicles, especially important in gas-turbine installations. Except for the larger hub needed to house the pitch-changing mechanism, the CRP can be made almost as efficient as the solid, fixed-blade propeller. The newest application is in the U.S. Navy's DD963 Class destroyer, developing approximately 40,000 shp (19,828 kW$_S$) per shaft.

### Water Jet

This method usually consists of an impeller or pump inside the hull, which draws water from outside, accelerates it, and discharges it astern as a jet at a higher velocity. It is a reaction device like the propeller but in which the moving parts are contained inside the hull, desirable for shallow-water operating conditions and maneuverability. The overall efficiency is lower than that of the screw propeller of diameter equal to the jet orifice diameter, principally because of inlet and ducting losses. Other disadvantages include the loss of volume to the ducting and impeller and the danger of fouling of the impeller. Water jets are currently used in several of the U.S. Navy's hydrofoil and surface-effect vehicles.

### Vertical-axis Propellers

There are two types of vertical-axis propulsor systems consisting of one or two vertical-axis rotors located underwater at the stern. Rotor disks are flush with the shell plating and have five to eight streamline, spadelike, vertical impeller blades fitted near the periphery of the disks. The blades feather during rotation of the disk to produce a maximum thrust effect in any direction desired. In the **Kirsten-Boeing** system the blades are interlocked by gears so that each blade makes a half revolution about its axis for each revolution of the disk. The blades of the **Voith-Schneider** system make a complete revolution about their own axis for each revolution of the disk. A bevel gear must be used to transmit power from the conventional horizontal drive shaft to the horizontal disk; therefore, limitations exist on the maximum power that can be transmitted. Although the propulsor is 30 to 40 percent less efficient than the screw propeller, it has obvious maneuverability advantages. Propulsors of this type have also been used at the bow to assist in maneuvering.

### Other Thrust Devices

**Pump Jet** In a pump-jet arrangement, the rotating impeller is external to the hull with fixed guide vanes either ahead and/or astern of it; the whole unit is enclosed in a duct or long shroud ring. The duct diameter increases from the entrance to the impeller so that the velocity falls and the pressure increases. Thus the impeller diameter is larger, thrust loading less, and the efficiency higher; the incidence of cavitation and noise is delayed. A penalty is paid, however, for the resistance of the duct.

**Kort Nozzles** In the Kort nozzle system, the screw propeller operates in a ring or nozzle attached to the hull at the top.

The longitudinal sections are of airfoil shape, and the length of the nozzle is generally about one-half its diameter. Unlike the pump-jet shroud ring, the Kort nozzle entrance is much bigger than the propeller, drawing in more water than the open propeller and achieving greater thrust. Because of the acceleration of the water into the nozzle, the pressure inside is less; hence a forward thrust is exerted on the nozzle and the hull. The greatest advantage is in a tug, pulling at rest. The free-running speed is usually less with the nozzle than without. In some tugs and rivercraft, the whole nozzle is pivoted and becomes a very efficient steering mechanism.

**Tandem and Contrarotating Propellers** Two or more propellers arranged on the same shaft are used to divide the increased loading factor when the diameter of a propeller is restricted. Propellers turning in the same direction are termed **tandem,** and in opposite directions, **contrarotating.** In tandem, the rotational energy in the race from the forward propeller is augmented by the after one. Contrarotating propellers work on coaxial, contrary-turning shafts so that the after propeller may regain the rotational energy from the forward one. The after propeller is of smaller diameter to fit the contracting race and has a pitch designed for proper power absorption. Such propellers have been used for years on torpedoes to prevent the torpedo body from rotating. Hydrodynamically, the advantages of contrarotating propellers are increased propulsive efficiency, improved vibration characteristics, and higher blade frequency. Disadvantages are the complicated gearing, coaxial shafting, and sealing problems.

**Supercavitating Propellers** When cavitation covers the entire back of a propeller blade, an increase of rpm cannot further reduce the pressure at the back but the pressure on the face continues to increase as does the total thrust, though at a slower rate than before cavitation began. Advantage of such fully cavitating propellers are an absence of back erosion and less vibration. Although the characteristics of such propellers were determined by trial and error, they have long been used on high-speed racing motor boats. The blade section design must ensure clean separation of flow at the leading and trailing edges and provide good lift-drag ratios for high efficiency. Introducing air to the back of the blades (**ventilated propeller**) will ensure full cavitation and also enables use at lower speeds.

**Partially Submerged Propellers** The appendage drag presented by a propeller supported below high-speed craft, such as planing boats, hydrofoil craft, and surface-effect ships, led to the development of partially submerged propellers. Although vibration and strength problems, arising from the cyclic loading and unloading of the blades as they enter and emerge from the air-water interface, remain to be solved, it has been demonstrated that efficiencies in partially submerged operation, comparable to fully submerged noncavitating operation, can be achieved. The performance of these propellers must be considered over a wide range of submergences.

**Outboard Gasoline Engine** Outboard gasoline engines of 1 to 100 bhp (0.75 to 7.5 kW$_B$), combining steering and propulsion, are popular for small pleasure craft. Higher-powered units are available with 40 to 400 bhp (30 to 300 kW$_B$) diesel drive for commercial use (see Sec. 9).

### PROPULSION TRANSMISSION

In modern ships, only large-bore, slow-speed diesel engines are directly connected to the propeller shaft. Transmission devices such as mechanical speed-reducing gears or electric drives (generator/motor transmissions) are required to convert the relatively high rpm of a compact, economical prime mover to the relatively low propeller rpm necessary for a high propulsive efficiency. In the case of steam turbines, medium- and high-speed diesel engines, and gas turbines, speed reduction gears are used. Gear ratios vary from relatively low values for medium-speed diesels up to approximately 50:1 for a compact turbine design. Where the prime mover is unidirectional, the drive mechanism must also include a reversing mechanism.

### Reduction Gears

Speed reduction is usually obtained with reduction gears. The simplest arrangement of a marine reduction gear is the **single-reduction,** single-input type, i.e., one pinion meshing with a gear (Fig. 20). This arrangement is used for connecting a propeller to a diesel engine or to an electric motor but is not used for propelling equipment with a turbine drive.

The usual arrangement for turbine-driven ships is the **double-reduction,** double-input, articulated type of reduction gear (Fig. 21). The two input pinions are driven by the two ele-

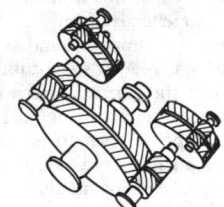

**Fig. 20** Single-reduction, single-input gear. (*From "Marine Engineering," SNAME,* 1971.)

**Fig. 21** Double-reduction, double-input, articulated gear. (*From "Marine Engineering," SNAME,* 1971.)

ments of a cross-compound turbine. The term *articulated* applies because a flexible coupling is generally provided between the first reduction or primary gear wheel and the second reduction or secondary pinion.

The **locked-train** type of double-reduction gear has become standard for high-powered naval ships and, because it minimizes the total weight and size of the assembly, is gaining increased popularity for higher-powered merchant ships (Fig. 22).

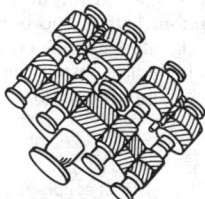

**Fig. 22** Double-reduction, double-input, locked-train gear. (*From "Marine Engineering," SNAME,* 1971.)

### Electric Drive

Electric propulsion drive is the principal alternative to direct- or geared-drive systems. The prime mover drives a generator or alternator which in turn drives a propulsion motor which is either coupled directly to the propeller or drives the propeller through a low-ratio reduction gear.

Among its **advantages** are the ease and convenience by which propeller speed and direction are controllable; the freedom of installation arrangement offered by the electrical connection between the generator and the propulsion motor; the flexibility of power use when not used for propulsion; the convenience of coupling several prime movers to the propeller without mechanical clutches or couplings; the relative simplicity of controls required to provide reverse propeller rotation when the prime mover is unidirectional; and the speed reduction that can be provided between the generator and the motor, hence between the prime mover and the propeller, without mechanical speed-reducing means.

The disadvantages are the inherently higher first cost, increased weight and space, and the higher transmission losses of the system. The advent of **superconductive electrical machinery** suitable for ship propulsion, however, has indicated order-of-magnitude savings in weight, greatly reduced volume, and the distinct possibility of lower costs. Superconductivity, a phenomenon occurring in some materials at temperatures near absolute zero, is characterized by the almost complete disappearance of electrical resistance. Research and development studies have shown that size and weight reductions by factors of 5 or more are possible. With successful development of superconductive electrical machinery and resolution of associated engineering problems, monohulled craft, such as destroyers, can benefit greatly from the location flexibility and maneuverability capabilities of superconductive electric propulsion, but the advantages of such a system will be realized to the greatest extent in the new high-performance ships where mechanical-drive arrangement is extremely complex.

In general, electric propulsion drives are employed in marine vehicles requiring a high degree of maneuverability such as ferries, icebreakers, and tugs; in those requiring large amounts of special-purpose power such as self-unloaders, fireboats, and large tankers; in those utilizing nonreversing, high-speed, and multiple prime movers; and in deep-submergence vehicles. Diesel-electric drive is ideally adapted to bridge control.

### Reversing

Reversing may be accomplished by stopping and reversing a reversible engine, as in the case of many reciprocating engines, or by adding reversing elements in the prime mover, as in the case of steam turbines. It is impracticable to provide reversing elements in gas turbines; hence a reversing capability must be provided in the transmission system or in the propulsor itself. Reversing reduction gears for such transmissions are available up to quite substantial powers, and CRP propellers are also used with diesel or gas-turbine drives. Electrical drives provide reversing by dynamic braking and energizing the electric motor in the reverse direction.

### Marine Shafting

A main propulsion shafting system consists of the equipment necessary to transmit the rotative power from the main propulsion engines to the propulsor; support the propulsor; transmit the thrust developed by the propulsor to the ship's hull; safely withstand transient operating loads (e.g., high-speed maneuvers, quick reversals); be free of deleterious modes of vibration; and provide reliable operation.

Figure 23 is a shafting arrangement typical of multishaft ships and single-shaft ships with transom sterns. The shafting

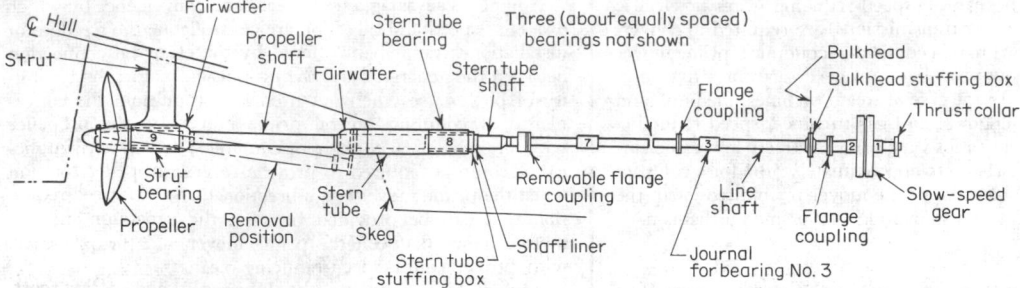

**Fig. 23** Shafting arrangement with strut bearings. (*From "Marine Engineering," SNAME, 1971.*)

must extend outboard a sufficient distance for adequate clearance between the propeller and the hull. Figure 24 is typical of single-screw merchant ships.

The shafting located inside the ship is **line shafting.** The outboard section to which the propeller is secured is the **propeller shaft** or **tail shaft.** The section passing through the stern tube is the **stern tube shaft** unless the propeller is supported by it. If there is a section of shafting between the propeller and stern tube shafts, it is an **intermediate shaft.**

The principal dimensions, kind of material, and material tests are specified by the classification societies for marine propulsion line shafting. For ships fitted with turbine machinery, the American Bureau of Shipping gives the following equation for the **minimum diameter of the line shaft:**

$d = C \sqrt[3]{K_1 P_s / n}$, inches (metres), where $P_s$ = shaft power at normal speed, shp (kW$_S$); $n$ = r/min, normal; $K_1$ = 64 (2.43) for ocean and coastwise service, 58 (2.20) for river and harbor service; and $C$ = constant for type of shaft, $C$ = 0.875 (adopted 1965).

Tail or propeller shafts are of larger diameter than the line or tunnel shafting because of bending moment from propeller weight; also inspection during service is possible only when the vessel is dry-docked. **Propeller shaft diameter** $d_e = d + D_p/c$, inches (metres), where $D_p$ = propeller diam, ft (m), $c$ = 12 (144) when a continuous bronze liner is fitted on the shaft, 8.3 (99.6) if liners are fitted only at bearings.

**Propellers are fitted** to shafts over 6 in (0.1524 m) diameter by a taper fit with a taper of 1:12. Aft of the propeller, the shaft is reduced to 0.58 to 0.68 of its specified diameter under the liner and is fitted with a relatively fine thread, nut, and keeper. Propeller torque is taken by a long flat key, the width and total depth of which are, respectively, about 0.21 and 0.11 times the shaft diameter under the liner. The forward end of the key

slot in the shaft should be tapered off in depth and terminate well aft of the large end of the taper to avoid a high concentrated stress at the end of the keyway. Every effort is made to keep saltwater out of contact with the steel tail shaft so as to prevent failure from corrosion fatigue.

A cast-iron, cast-steel, or wrought-steel-weldment **stern tube** is rigidly fastened to the hull. In single-screw ships it supports the propeller shaft and is normally provided with a packing gland at the forward end as seawater lubricates the stern-tube bearings. The aft stern-tube bearing is made four diameters in length; the forward stern-tube bearing is much shorter. A bronze liner, ¾ to 1 in (0.0191 to 0.0254 m) thick, is shrunk on the propeller shaft to protect it from corrosion and to give a good journal surface. Lignum-vitae inserts, arranged with the grain on end for medium and large shafts, are normally used for stern-tube bearings; the average bearing pressure is about 30 lb/in² (206,842 N/m²). Rubber and phenolic compound bearings, having longitudinal grooves, are also extensively used. White-metal oil-lubricated bearings and, in Germany, roller-type stern bearings have been fitted in a few installations.

The line or tunnel shafting is laid out, in long equal lengths, so that, in the case of a single-screw ship, withdrawal of the tail-shaft inboard for inspection every several years will require only the removal of the next inboard length of shafting. For large outboard or water-exposed shafts of twin-screw ships, a bearing spacing up to 30 shaft diameters has been used. A greater ratio prevails for steel shafts of power boats. Bearings inside the hull are spaced closer—normally under 15 diameters if the shaft is more than 6 in (0.1524 m) diameter. Usually, only the bottom half of the bearing is completely white-metaled; oil-wick lubrication is common, but oil-ring lubrication is used in high-class installations.

**Bearings** are used to support the shafting in essentially a

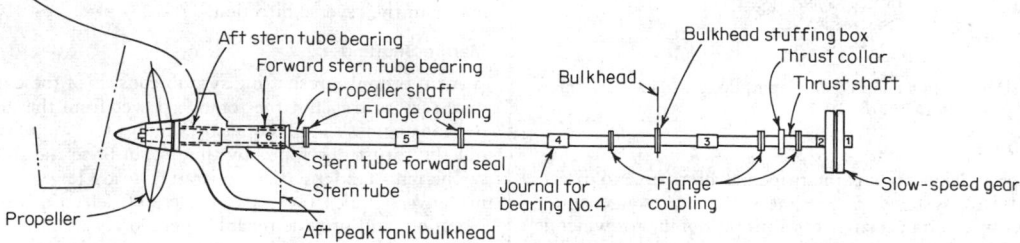

**Fig. 24** Shafting arrangement without strut bearings. (*From "Marine Engineering," SNAME, 1971.*)

straight line between the main propulsion engine and the desired location of the propeller. Bearings inside the ship are most popularly known as line-shaft bearings, steady bearings, or spring bearings. Bearings which support outboard sections of shafting are stern-tube bearings if they are located in the stern tube and strut bearings when located in struts.

The propeller thrust is transmitted to the hull by means of a main thrust bearing. The main thrust bearing may be located either forward or aft of the slow-speed gear.

## HIGH-PERFORMANCE SHIP SYSTEMS

High-performance ships have spawned new hull forms and propulsion systems which were relatively unknown just a generation ago. Because of their unique characteristics and requirements, conventional propulsors are rarely adequate. The prime thrust devices range from those using water, to water-air mixtures, to large air propellers.

The total propulsion power required by these vehicles is presented in general terms in the Gabrielli and Von Karman plot of Fig. 25. Basically, there are those vehicles obtaining lift by displacement and those obtaining lift from a wing or foil moving through the fluid. In order for marine vehicles to maintain a reasonable lift-to-drag ratio over a moderate range of speed, the propulsion system must provide increased thrust at lower speeds to overcome low-speed wave drag. The weight of any vehicle is converted into total drag and, therefore, total required propulsion and lift power.

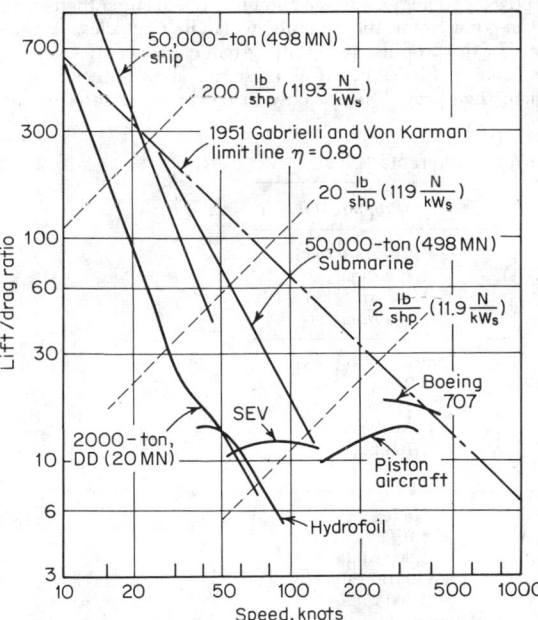

**Fig. 25** Lift-drag ratio versus calm-weather speed for various vehicles.

Light-weight, compact, and efficient propulsion systems to satisfy extreme weight sensitivity are achieved over a wide range of power by the marinized aircraft gas-turbine engine.

Marinization involves addition of a power turbine unit, incorporation of air inlet filtration to provide salt-free air, addition of silencing equipment on both inlet and exhaust ducts, and modification of some components for resistance to salt corrosion. The shaft power delivered is then distributed to the propulsion and lift devices by light-weight shaft and gear power transmission systems.

### Hydrofoil Craft

Under foilborne conditions, the support of the hydrofoil craft is derived from the dynamic lift generated by the foil system, and the craft hull is lifted clear of the water. Special problems relating to operation and directly related to the conditions presented by waves, high fluid density, surface piercing of foils and struts, stability, control, cavitation, and ventilation of foils have given rise to foil system configurations defined in Fig. 26.

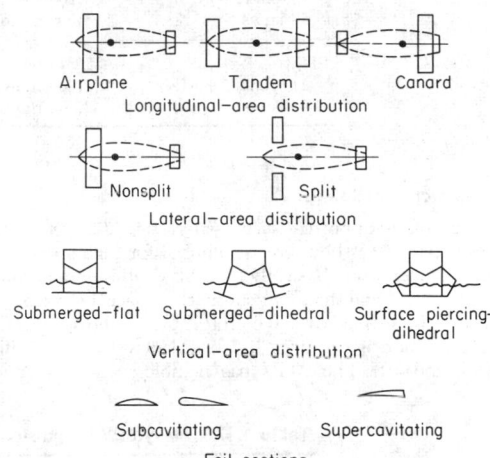

**Fig. 26** Some foil arrangements and sections.

The hydrofoil craft has three modes of operation: **hullborne, takeoff,** and **foilborne.** Design must be a compromise between the varied requirements of these operating conditions. This may be difficult, as with propeller design, where maximum thrust is often required at takeoff speed (normally about one-half the operating speed). A propeller designed for maximum efficiency at takeoff will have a smaller efficiency at design speeds, and vice versa. This is complicated by the need for supercavitating propellers at operating speeds over 50 knots. Two sets of propellers may be needed in such a case, one for hullborne operation and the other for foilborne operation. The selection of a surface-piercing or a fully submerged foil design is determined from specified operating conditions and the permissible degree of complexity. A **surface-piercing system** is inherently stable, as a deviation from the equilibrium position of the craft results in corrective lift forces. **Fully submerged foil** designs, on the other hand, are not stable by themselves and require a depth-control system for satisfactory operation. Such a control system, however, permits operation of the submerged-foil design in higher waves.

A summary of the U.S. Navy's hydrofoil propulsion system characteristics is given in Table 3.

**Table 3. U.S. Navy Hydrofoil Propulsion System Characteristics**

| Ship | Engine | Propulsor |
|------|--------|-----------|
| PCH-1<br>120 tons (1.2 MN)<br>45 knots | 2-Proteus<br>3,200 bhp (2,386 kW$_B$)<br>5,000 r/min | 4-Subcavitating propellers,<br>pod-mounted, fore and aft.<br>Twin aft retractable struts.<br>1,500 r/min. |
| AGEH<br>320 tons (3.2 MN) | 2-LM-1500<br>14,000 bhp (10,440<br>kW$_B$) 5,500 r/min | 2-Supercavitating propellers,<br>pod-mounted. Twin forward<br>retractable struts. 1,570 r/<br>min. |
| PGH-1<br>58 tons (0.6 MN)<br>50 knots | 1-TYNE<br>3,600 bhp (2,685 kW$_B$)<br>13,000 r/min | 1-Supercavitating CRP.<br>Single stern strut.<br>1,050 r/min. |
| PGH-2<br>58 tons (0.6 MN)<br>50 knots | 1-Proteus<br>3,200 bhp (2,386 kW$_B$)<br>5,000 r/min | 2-Centrifugal water jets.<br>Twin aft strut inlets.<br>1,500 r/min. |
| PHM<br>215 tons (2.1 MN)<br>50 knots | 1-LM-2500<br>22,000 bhp (15,405 kW$_B$)<br>3,600 r/min | 1-Two-stage axial flow water<br>jet. Twin aft strut inlets. |

### Surface-effect Vehicles (SEV)

The air-cushion-supported **surface-effect ship (SES)** and **air-cushion vehicle (ACV)** show great potential for high speed and amphibious operation. Basically, the SEV rides on a cushion of air generated and maintained in the space between the vehicle and the surface over which it moves or hovers. In the SES, the cushion is captured by fixed sidewalls and flexible seals fore and aft. The ACV has flexible seals all around.

Figure 27 illustrates different configurations of the surface-effect principle.

The U.S. Navy has developed four SEV's for demonstrating the technology to design and build much larger high-speed ocean-going ships and amphibious landing vehicles. A summary of the propulsion and lift system characteristics is given in Table 4. Since the ACV must rise above either land or water, their propulsors use only air for thrust, and the propor-

**Table 4. U.S. Navy SEV Propulsion and Lift System Characteristics**

| Ship/vehicle | Engines | Propulsor/lift |
|--------------|---------|----------------|
| SES-100A<br>100 tons (1.0 MN)<br>80 knots | 4-TF-35<br>2,800 bhp (2,088 kW$_B$)<br>14,500 r/min | 2-Axial flow water jets<br>4,200 r/min<br>3-Axial fans<br>2,880 r/min |
| SES-100B<br>100 tons (1.0 MN)<br>80 knots | 3-FT-12<br>4,500 bhp (3,356 kW$_B$)<br>9,000 r/min<br>3-ST-6<br>510 bhp (380 kW$_B$)<br>31,500 r/min | 2-Partially submerged<br>CRP's<br>2,613 r/min<br>8-Centrifugal fans<br>2,100 r/min |
| JEFF-A (ACV)<br>150 tons (1.49 MN)<br>50 knots | 6-TF-40<br>3,350 bhp (2,498 kW$_B$)<br>15,400 r/min | 4-Fan jets<br>8 ft (2.44 m) diam<br>1,800 r/min<br>8-Centrifugal fans<br>2,500 r/min |
| JEFF-B (ACV)<br>150 tons (1.49 MN)<br>50 knots | 6-TF-40<br>3,350 bhp (2,498 kW$_B$)<br>15,400 r/min | 2-Ducted-air propellers<br>12 ft (3.66 m) diam<br>1,250 r/min<br>4-Double-entry<br>centrifugal fans<br>1,900 r/min |

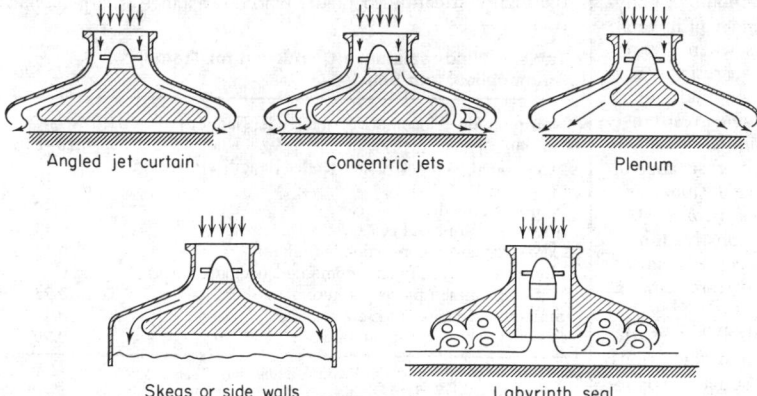

**Fig. 27**   Some configurations of the surface-effect principle.

tion of total power allocated to the lift fans is 30 percent rather than the 15 percent required by the SES designs.

### Small Water-plane Area Twin Hull

The **SWATH**, configured to be relatively insensitive to seaway motions, gains its buoyant lift from two parallel submerged, submarinelike hulls which support the main hull platform through relatively small vertical struts. In this configuration, a propulsion arrangement choice must be with regard to the location of the main engines: lower hull with the propulsor or upper hull with a suitable transmission system. The U.S. Navy has designed a 190-ton (1.89 MN), 25-knot version known as the SSP or semisubmerged platform. This development vehicle is propelled by two 6-ft (1.83 m) CRP propellers located at the aft end of each submerged hull. Power is provided by two independently operated T-64 gas turbines of 3,200 bhp (2,386 kW_B) each with a 1,000 r/min output speed from the attached gear box. Power is transmitted down each aft strut by a group of three wide-belt chain drives with speed reduction to deliver 450 r/min to the propller shaft.

# AIR RESISTANCE OF TRAINS, AUTOMOBILES, AND SHIPS
## by William Bollay

The fluid-dynamic resistance to motion of trains, automobiles, and ships is reflected by the data in Table 1 and Figs. 1 and 2. The wind-resistance coefficient of a body is $C_D = D/qS$, where $D$ is the resistance, lb (kg); $q = \frac{1}{2}\rho V^2$ the dynamic pressure; and $S$ the maximum cross-sectional area of the body. For a circular cylinder with wind blowing along the axis, $C_D \approx 0.90$. For a streamlined body such as an airship shape, $C_D \approx 0.040$. The values of $C_D$ for trains, automobiles, and ships lie in this range, as shown by Table 1.

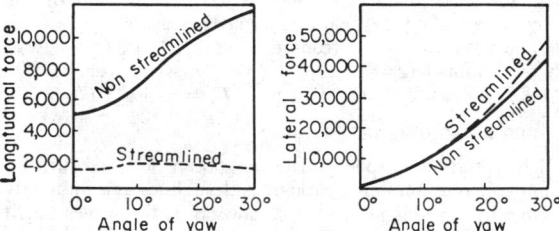

**Fig. 1**   Effect of yaw on a train with six coaches.

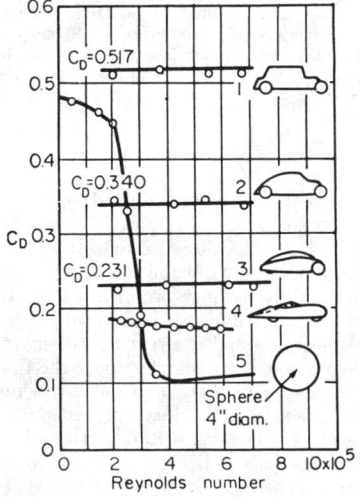

**Fig. 2**   Wind-resistance coefficients of various bodies.

Even though the air resistance due to the motion of a body such as a ship may be a negligibly small percentage of its total resistance, this may be magnified many times when a wind arises. For a ship with a speed of 10 knots, the air resistance is about 2½ percent of its total resistance. When the ship is proceeding against a 20-knot wind, the wind resistance becomes 22½ percent of the total resistance. Similarly, a small side wind may cause appreciable increase in the resistance of bodies such as ships or railroad trains. Figure 1 (Johansen, The Air-resistance of Passenger Trains, *Proc. I Mech. E,* **134,** 1936, p. 111) shows that the resistance of a nonstreamlined train may be doubled when the relative wind comes from an angle of yaw, although that of the streamlined train remains practically constant.

Figure 2 (Schmidt, *ZVdI,* **82,** 1938, p. 188) shows the variation of drag coefficient with Reynolds number for various automobile models compared with that of a sphere. Apparently there is no scale effect on the automobile tests except on the highly streamlined model whose resistance is largely skin friction.

**Table 1. Wind-resistance Coefficient for Trains, Automobiles, and Ships**

| | $C_D$ |
|---|---|
| Nonstreamlined locomotive* and tender [length 110–120 ft (36–40 m)] | 0.80–1.05 |
| Streamlined locomotive and tender [length 110–120 ft (36–40 m)] | 0.35–0.45 |
| Nonstreamlined railroad car | 0.40 |
| Streamlined railroad car | 0.15 |
| Conventional-type automobile (sedan)† | 0.52 |
| Moderately streamlined automobile† (rounded back) | 0.34 |
| Well-streamlined passenger automobile† | 0.23 |
| Streamlined racing-type automobile† | 0.17 |
| Passenger ship, nonstreamlined ‡ | 0.90 |

*Lipetz, Air Resistance of Railroad Equipment, *Trans. ASNE,* 1937.
†Schmidt, *ZVdI,* **82,** 1938, p. 188.
‡Taylor, "The Speed and Power of Ships," chap. 10.2, 1933.

# AERONAUTICS
## by J. J. Cornish, III

REFERENCES: National Advisory Committee for Aeronautics, *Technical Reports* (designated *NACA-TR* with number), *Technical Notes* (*NACA-TN* with number), and *Technical Memoranda* (*NACA-TM* with number). British Aeronautical Research Committee *Reports and Memoranda* (designated *Br. ARC-R & M* with number). *Ergebnisse der Aerodynamischen Versuchsanstalt zu Göttingen.* Diehl, "Engineering Aerodynamics," Ronald. Reid, "Applied Wing Theory," McGraw-Hill. Durand, "Aerodynamic Theory," Springer. Prandtl-Tietjens, "Fundamentals of Hydro- and Aeromechanics," and "Applied Hydro- and Aeromechanics," McGraw-Hill. Goldstein, "Modern Developments in Fluid Dynamics," Oxford. Millikan, "Aerodynamics of the Airplane," Wiley. Von Mises, "Theory of Flight," McGraw-Hill. Hoerner, "Aerodynamic Drag," Hoerner. Glauert, "The Elements of Airfoil and Airscrew Theory," Cambridge. Milne-Thompson, "Theoretical Hydrodynamics," Macmillan. Munk, "Fluid Dynamics for Aircraft Designers," and "Principles of Aerodynamics," Ronald. Abraham, "Structural Design of Missiles and Spacecraft," McGraw-Hill. "U.S. Standard Atmosphere, 1962," U.S. Government Printing Office.

## DEFINITIONS

**Aeronautics** is the science and art of flight and includes aviation, the operation of aircraft heavier than air (airplanes). An **aircraft** is any weight-carrying device designed to be supported by the air, either by buoyancy or by dynamic action. An **airplane** is a mechanically driven fixed-wing aircraft, heavier than air, which is supported by the dynamic reaction of the air against its wings. A **helicopter** is a kind of aircraft lifted and moved by a large propeller mounted horizontally above the fuselage. It differs from the **autogiro** in that this propeller is turned by motor power and there is no auxiliary propeller for forward motion. A **ground-effect machine (GEM)** is a heavier-than-air surface vehicle which operates in close proximity to the earth's surface (over land or water), never touching except at rest, being separated from the surface by a cushion or film of air,

however thin, and depending entirely upon aerodynamic forces for propulsion and control.

**Aerodynamics** is the branch of dynamics that treats of the motion of air and other gaseous fluids and of the forces acting on solids in motion relative to such fluids. Aerodynamics falls into velocity ranges, depending upon whether the velocity is below or above the local speed of sound in the fluid. The velocity range below the local speed of sound is called the **subsonic** regime. Where the velocity is above the local speed of sound, the flow is said to be **supersonic.** The term **transonic** refers to flows in which both subsonic and supersonic regions are present. The **hypersonic** regime is that speed range usually in excess of five times the speed of sound.

## STANDARD ATMOSPHERE

The standard atmosphere of Table 1 is a revised U.S. Standard Atmosphere, adapted by the United States Committee on Extension to the Standard Atmosphere (COESA) in 1962.

The values given up to about 65,000 ft are designated as **standard.** The region from 65,000 to 105,000 ft is designated **proposed standard.** U.S. Standard Atmosphere, 1962, gives data out to 2,320,000 ft; however, the region from 105,000 to 295,000 ft is designated **tentative,** and that portion of the atmosphere above 295,000 ft is termed **speculative.**

The assumed sea-level conditions are: pressure, $p_0 = 29.91$ in (760 mm) Hg $= 2,116.22$ lb/ft²; mass density, $\rho_0 = 0.002378$ slugs/ft³ (0.001225 g/cm³); $T_0 = 59°F$ (15°C).

## UPPER ATMOSPHERE

High-altitude atmospheric data have been obtained directly from balloons, sounding rockets, and satellites and indirectly from observations of meteors, aurora, radio waves, light absorption, and sound effects. At relatively low altitudes, the earth's atmosphere is, for aerodynamic purposes, a uniform

**Table 1. U.S. Standard Atmosphere, 1962**

| Altitude $h$, ft* | Temp $T$, °F† | Pressure ratio, $p/p_0$ | Density ratio, $\rho/\rho_0$ | $(\rho_0/\rho)^{0.5}$ | Speed of sound $V_s$, ft/s ‡ |
|---|---|---|---|---|---|
| 0 | 59.00 | 1.0000 | 1.0000 | 1.000 | 1,116 |
| 5,000 | 41.17 | 0.8320 | 0.8617 | 1.077 | 1,097 |
| 10,000 | 23.34 | 0.6877 | 0.7385 | 1.164 | 1,077 |
| 15,000 | 5.51 | 0.5643 | 0.6292 | 1.261 | 1,057 |
| 20,000 | −24.62 | 0.4595 | 0.5328 | 1.370 | 1,036 |
| 25,000 | −30.15 | 0.3711 | 0.4481 | 1.494 | 1,015 |
| 30,000 | −47.99 | 0.2970 | 0.3741 | 1.635 | 995 |
| 35,000 | −65.82 | 0.2353 | 0.3099 | 1.796 | 973 |
| 36,089 | −69.70 | 0.2234 | 0.2971 | 1.835 | 968 |
| 40,000 | −69.70 | 0.1851 | 0.2462 | 2.016 | 968 |
| 45,000 | −69.70 | 0.1455 | 0.1936 | 2.273 | 968 |
| 50,000 | −69.70 | 0.1145 | 0.1522 | 2.563 | 968 |
| 55,000 | −69.70 | 0.09001 | 0.1197 | 2.890 | 968 |
| 60,000 | −69.70 | 0.07078 | 0.09414 | 3.259 | 968 |
| 65,000 | −69.70 | 0.05566 | 0.07403 | 3.675 | 968 |
| 65,800 | −69.70 | 0.05356 | 0.07123 | 3.747 | 968 |
| 70,000 | −67.30 | 0.04380 | 0.05789 | 4.156 | 971 |
| 75,000 | −64.55 | 0.03452 | 0.04532 | 4.697 | 974 |
| 80,000 | −61.81 | 0.02725 | 0.03553 | 5.305 | 977 |
| 85,000 | −59.07 | 0.02155 | 0.02790 | 5.986 | 981 |
| 90,000 | −56.32 | 0.01707 | 0.02195 | 6.970 | 984 |
| 95,000 | −53.58 | 0.01354 | 0.01730 | 7.600 | 988 |
| 100,000 | −50.84 | 0.01076 | 0.01365 | 8.559 | 991 |

\* × 0.3048 = metres.
† × (°F −32)/1.8 = °C.
‡ × 0.3048 = m/s.

gas. Above 250,000 ft, day and night standards differ because of dissociation of oxygen by solar radiation. This difference in density is as high as 35 percent, but it is usually aerodynamically negligible above 250,000 ft since forces here will become less than 0.05 percent of their sea-level value for the same velocity.

**Temperature profile** of the COESA atmosphere is given in Fig. 1. From these data, other properties of the atmosphere can be calculated.

**Pressure and Density** For all practical purposes, both decrease exponentially with altitude. The variations in $T$ (Fig. 1) are accompanied by slight inflections in the curves of $p$ and $\rho$, but the deviations from a mean curve are far less than the scatter in test data from various sources. For altitudes above 100,000 ft, pressure and density may be approximated by

$$\log p_0/p = 0.00001910h + 0.0140$$
$$\log \rho_0/\rho = 0.00001890h - 0.1000$$

where $h$ is in ft. At $h$ = 320,000 ft, or about 60 mi (96 km), $p$ and $\rho$ are about one-millionth of the sea-level values.

**Speed** in aeronautics may be given in knots (now standard for USAF, USN, and FAA), miles per hour, feet per second, or metres per second. The international standard mile = 6,076.1155 ft (1,852 m). The basic relations are: 1 knot = 0.5144 m/s = 1.6877 ft/s = 1.1508 mi/h. For additional conversion factors see Sec. 1. Higher speeds are often given as **Mach number,** or the ratio of the particular speed to the speed of sound in the surrounding air. For additional data see Supersonic and Hypersonic Aerodynamics below.

**Axes** The forces and moments acting on an airplane (and the resultant velocity components) are referred to a set of three mutually perpendicular axes having the origin at the airplane center of gravity (cg) and moving with the airplane. The basic difference in these is in the direction taken for the longitudinal, or $x$, axis, as follows:

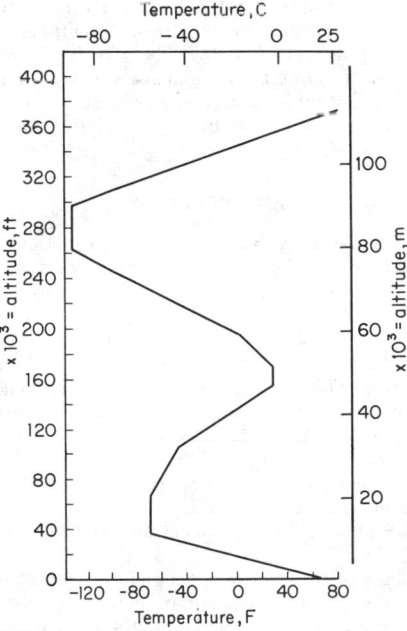

**Fig. 1** Temperature as a function of altitude. (*COESA.*)

**Wind axes:** The $x$ axis lies in the direction of the relative wind. This is the system most commonly used, and the one used in this section. It is shown on Fig. 2.

**Body axes:** The $x$ axis is fixed in the body, usually parallel to the thrust line.

**Stability axes:** The $x$ axis coincides with the principal inertial axis. This system eliminates the products of inertia in the equations of motion.

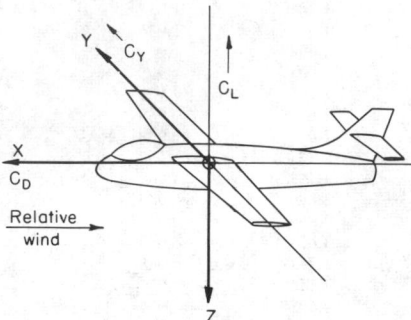

Positive moments: X→Y, Y→Z, Z→X

**Fig. 2** Wind axes.

**Absolute Coefficients** Aerodynamic force and moment data are usually presented in the form of absolute coefficients. Examples of force coefficients are: lift $C_L = L/qS$; drag $C_D = D/qS$; side force $C_C = C/qS$; where $q$ is the dynamic pressure $\frac{1}{2}\rho V^2$, $\rho$ = air mass-density, and $V$ = air speed. Examples of moment coefficients are $C_m = M/qSc$; roll $C_l = L/qSb$; and yaw $C_n = N/qSb$; where $c$ = mean wing chord and $b$ = wing span.

**Section Coefficients** NACA basic test data on wing sections are usually given in the form of section lift coefficient $C_l$ and section drag coefficient $C_d$. These apply directly to an infinite aspect ratio or to two-dimensional flow, but aspect-ratio and lift-distribution corrections are necessary in applying to a finite wing. For a wing having an elliptical lift distribution, $C_L = (\pi/4)C_l$.

## SUBSONIC AERODYNAMIC FORCES

When an airfoil is moved through the air, the motion produces a pressure at every point of the airfoil which acts normal to the surface. In addition, a frictional force tangential to the surface opposes the motion. The sum of these pressure and frictional forces gives the **resultant force** $R$ acting on the body. The point at which the resultant force acts is defined as the **center of pressure,** c.p. The resultant force $R$ will, in general, be inclined to the airfoil and the relative wind velocity $V$. It is resolved (Fig. 3) either along wind axes into

$L$ = **lift** = component normal to $V$

$D$ = **drag** or resistance = component along $V$

or along body axes into

$N$ = **normal force** = component perpendicular to airfoil chord

$T$ = **tangential force** = component along airfoil chord

Instead of specifying the center of pressure, it is convenient to specify the moment of the air forces about the so-called *aerodynamic center,* a.c. This point lies at a distance $a$ (about a quarter-chord length) back of the leading edge of the airfoil and is defined as the point about which the moment of the air forces remains constant when the angle of attack $\alpha$ is changed. Such a point exists for every airfoil. The force $R$ acting at the c.p. is equivalent to the same force acting at the a.c. plus a moment equal to that force times the distance between the c.p. and the a.c. (see Fig. 3). The location of the aerodynamic center, in terms of the chord $c$ and the section thickness $t$, is given approximately by

$$a/c = 0.25 - 0.40(t/c)^2$$

The distance $C_p$ from the leading edge to the center of pressure expressed as a fraction of the chord is, in terms of the moment $M_{c/4}$ about the quarter-chord point,

$$M_{c/4} = (\tfrac{1}{4} - C_p) \cdot c \cdot N$$

$$C_p = \frac{1}{4} - \frac{M_{c/4}}{c \cdot N}$$

From dimensional analysis it can be seen that the **air force on a body** of length dimension $l$ moving with velocity $V$ through air of density $\rho$ can be expressed as

$$F = \varphi \cdot \rho V^2 l^2$$

where $\varphi$ is a coefficient that depends upon all the dimensionless factors of the problem. In the case of a wing these are:

1. **Angle of attack** $\alpha$, the inclination between the chord line and the velocity $V$
2. **Aspect ratio** $A = b/c$, where $b$ is the span and $c$ the mean chord of the wing
3. **Reynolds number** Re $= \rho Vl/\mu$, where $\mu$ is the coefficient of viscosity of the air
4. **Mach number** $V/V_s$, where $V_s$ is the velocity of sound
5. Relative surface roughness
6. Relative turbulence

The dependence of the force coefficient $\varphi$ upon $\alpha$ and $A$ can be theoretically determined; the variation of $\varphi$ with the other parameters must be established experimentally, i.e., by model tests.

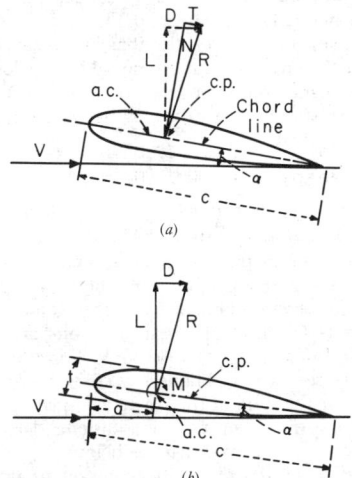

**Fig. 3** Forces acting on an airfoil; (a) actual forces; (b) equivalent forces through the aerodynamic center plus a moment.

## AIRFOILS

Applying Bernoulli's equation to the flow around a body, if $p$ represents the **static pressure,** i.e., the atmospheric pressure in the undisturbed air, and if $V$ is superposed as in Fig. 4 and if $p_1$, $V_1$ represent the pressure and velocity at any point 1 at the surface of the body,

$$p + \tfrac{1}{2}\rho V^2 = p_1 + \tfrac{1}{2}\rho V_1^2$$

The maximum pressure occurs at a point $s$ on the body at which the velocity is zero. Such a point is defined as the **stagnation point.** The maximum pressure increase occurs at this point and is

$$p_s - p = \tfrac{1}{2}\rho V^2$$

This is called the stagnation pressure, or **dynamic pressure,** and is denoted by $q$. It is customary to express all aerodynamic forces in terms of $\tfrac{1}{2}\rho V^2$, hence

$$F = \tfrac{1}{2}\varphi\rho V^2 l^2 = q\varphi l^2$$

For the case of a wing, the forces and moments are expressed as

$$\begin{aligned}
\textbf{Lift} &= L = C_L \tfrac{1}{2}\rho V^2 S = C_L q S \\
\textbf{Drag} &= D = C_D \tfrac{1}{2}\rho V^2 S = C_D q S \\
\textbf{Moment} &= M = C_M \tfrac{1}{2}\rho V^2 S c = C_M q S c
\end{aligned}$$

where $S$ = wing area and $c$ = wing chord.

The lift produced can be determined from the intensity of the circulatory flow or **circulation** $\Gamma$ by the relation $L' = \rho V \Gamma$,

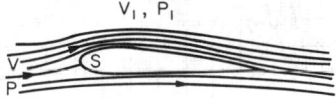

**Fig. 4**

where $L'$ is the lift per unit width of wing. In a **wing of infinite aspect ratio,** the flow is two-dimensional and the lift reaction is at right angles to the line of relative motion. The **lift coefficient** is a function of the angle of attack $\alpha$, and by mathematical analysis $C_L = 2\pi \sin \alpha$, which for small angles becomes $C_L = 2\pi\alpha$. Experiments show that $C_L = 2\pi\eta(\alpha + \alpha_0)$, where $\alpha_0$ is the angle of attack corresponding to zero lift. $\eta \approx 1 - 0.64(t/c)$, where $t$ is airfoil thickness.

In a **wing of finite aspect ratio,** the circulation flow around the wing creates a strong vortex trailing downstream from each wing tip (Fig. 5). The effect of this is that the direction of the

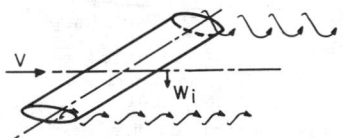

**Fig. 5** Vortex formation at wing tips.

resultant velocity at the wing is tilted downward by an **induced angle of attack** $\alpha_i = w_i/V$ (Fig. 6). If friction is neglected, the resultant force $R$ is now tilted to the rear by this same angle $\alpha_i$. The lift force $L$ is approximately the same as $R$. In addition, there is also a drag component $D_i$, called the **induced drag,** given by

$$D_i = L \tan \alpha_i = L w_i/V$$

For a given geometrical angle of attack $\alpha$ the **effective angle of attack** has been reduced by $\alpha_i$, and thus

$$C_L = 2\pi\eta(\alpha - \alpha_0 - \alpha_i)$$

According to the Lanchester-Prandtl theory, for **wings having an elliptical lift distribution**

$$\alpha_i = w_i/V = C_L/\pi A \qquad \text{rad}$$
$$D_i = L^2/\pi b^2 q \qquad \text{and} \qquad C_{Di} = D_i/qS = C_L^2/\pi A$$

where $b$ = span of wing and $A = b^2/S$ = aspect ratio of wing. These results also apply fairly well to wings not differing much from the elliptical shape. For **square tips,** correction factors are required:

$$\left.\begin{aligned}
\alpha_i &= \frac{C_L}{\pi A}(1 + \tau) \quad \text{where } \tau = 0.05 + 0.02A \\
C_{Di} &= \frac{C_L^2}{\pi A}(1 + \sigma) \quad \text{where } \sigma = 0.01A - 0.01
\end{aligned}\right\} \text{for } A < 12$$

These formulas are the basis for transforming the characteristics of rectangular wings from an **aspect ratio** $A_1$ to an aspect ratio $A_2$:

$$\begin{aligned}
\alpha_2 &= \alpha_1 + \frac{57.3 C_L}{\pi}\left(\frac{1 + \tau_2}{A_2} - \frac{1 + \tau_1}{A_1}\right) \qquad \text{deg} \\
C_{D2} &= C_{D1} + \frac{C_L^2}{\pi}\left(\frac{1 + \sigma_2}{A_2} - \frac{1 + \sigma_1}{A_1}\right)
\end{aligned}$$

For elliptical wings the values of $\tau$ and $\sigma$ are zero. Most wing-section data are given in terms of aspect ratios 6 and $\infty$. For other values, the preceding formulas must be used.

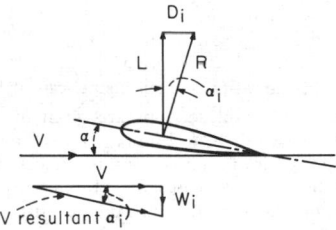

**Fig. 6** Induced angle of attack.

**Characteristics of Airfoils** Airfoil characteristics are expressed in terms of the dimensionless coefficients $C_L$, $C_D$, and $C_M$ and the angle of attack $\alpha$. The NACA presents the results for wings of aspect ratio 6 and also corrected to aspect ratio $\infty$ (see Fig. 10).

The lift coefficient $C_L$ is a linear function of the angle of attack up to a critical angle called the **stalling angle** (Fig. 7). The **maximum lift coefficient** $C_{Lmax}$ which can be reached is one of the important characteristics of a wing because it determines the landing speed of the airplane.

The **drag of a wing** is made up of two components: the **profile drag** $D_0$ and the **induced drag** $D_i$. The profile drag is due principally to surface friction. At aspect ratio $\infty$ or at zero lift the induced drag is zero, and thus the entire drag is profile drag. In Fig. 8 the coefficient of induced drag $C_{Di} = C_L^2/\pi A$ is also plotted. The difference $C_D - C_{Di} = C_{D0}$, the profile-drag coefficient. Among the desirable characteristics of an airfoil is a small value of the minimum profile-drag coefficient and a large value of $C_L/C_D$.

The moment characteristics of a wing are obtainable from the curve of the center of pressure as a function of $\alpha$ or by the **moment coefficient taken about the aerodynamic center** $C_{Ma.c.}$ as a function of $C_L$. A forward motion of the c.p. as $\alpha$ is increased corresponds to an unstable wing section. This instability is undesirable because it requires a large download on the tail to counteract it.

The characteristics of a wing section (Fig. 9) are determined principally by its **mean camber line,** i.e., the curvature of the median line of the profile, and secondly by the thickness distribution along the chord. In the NACA system of designation, when a four-digit number is used such as 2412, the significance is always first digit = maximum camber in percent of chord, second digit = location of position of maximum camber in tenths of chord measured from the leading edge (that is, 4 stands for 40 percent), and the last two figures indicate the maximum thickness in percent of chord. The NACA five-digit system is explained in *NACA-TR* 610.

### Selection of Wing Section

In selecting a wing section for a particular airplane the following factors are generally considered:
1. Max lift coefficient $C_{Lmax}$ (for landing speed)
2. Min drag coefficient $C_{Dmin}$
3. Ratio $C_{Lmax}/C_{Dmin}$
4. Moment coefficient at zero lift $C_{m0}$
5. Max value of the ratio $C_L/C_D$ (for range)

For certain special cases it is necessary to consider one or more factors from the following group:
6. Value of $C_L$ for max $C_L/C_D$
7. Value of $C_L$ for min profile drag
8. Ratio $C_L/C_D$ at $C_L = 0.70$ (for climb)
9. Max value of $C_L^3/C_D^2$ (for ceiling)
10. Type of lift-curve peak (stall characteristics)

Characteristics of airfoil sections are given in *NACA-TR* 586, 647, 669, 708, and 824. Figure 10 gives data on two typical sections: 0006 is often used for tail surfaces, and 4415 is especially suitable for the wings of subsonic airplanes. For Mach numbers greater than 0.6 thin wings must be used.

The dimensionless coefficients $C_L$ and $C_D$ are functions of the **Reynolds number** $\mathrm{Re} = \rho V l/\mu$. For wings, the characteristic length $l$ i, taken to be the chord. In standard air at sea level

$$\mathrm{Re} = 6{,}378 V_{ft/s} \cdot l_{ft} = 9{,}354 V_{mi/h} \cdot l_{ft}$$

For other heights, multiply by the coefficient $K = \dfrac{\rho/\mu}{\rho_0/\mu_0}$.

Values of $K$ are as follows:

The variations of $C_L$ and $C_D$ with the Reynolds number is known as **scale effect.** These are shown in Figs. 11 and 12 for some typical airfoils. Figure 12 shows in dotted lines also the theoretical variation of the drag coefficient for a smooth flat plate for laminar and turbulent flow, respectively, and also for the transition region.

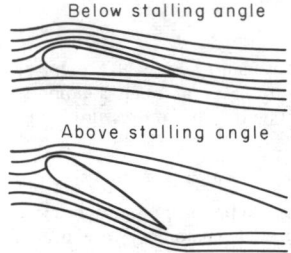

**Fig. 7**   Stalling angle of an airfoil.

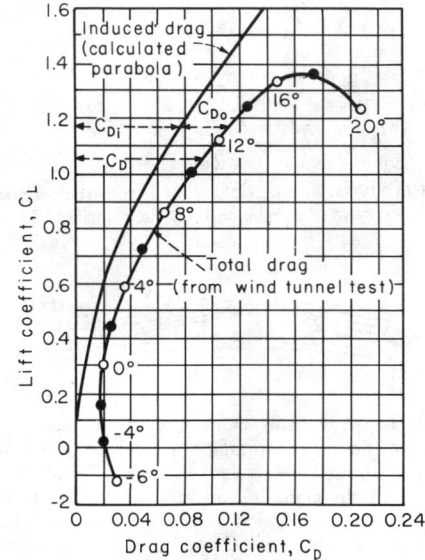

**Fig. 8**   Polar-diagram plot of airfoil data.

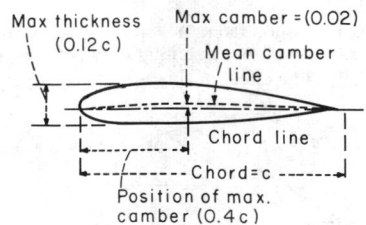

**Fig. 9**   Characteristics of a wing section.

| Altitude, ft* | 0 | 5,000 | 10,000 | 15,000 | 20,000 | 25,000 | 30,000 | 40,000 | 50,000 | 60,000 |
|---|---|---|---|---|---|---|---|---|---|---|
| $K$ | 1.000 | 0.884 | 0.779 | 0.683 | 0.595 | 0.515 | 0.443 | 0.300 | 0.186 | 0.116 |

* $\times 0.305$ = m.

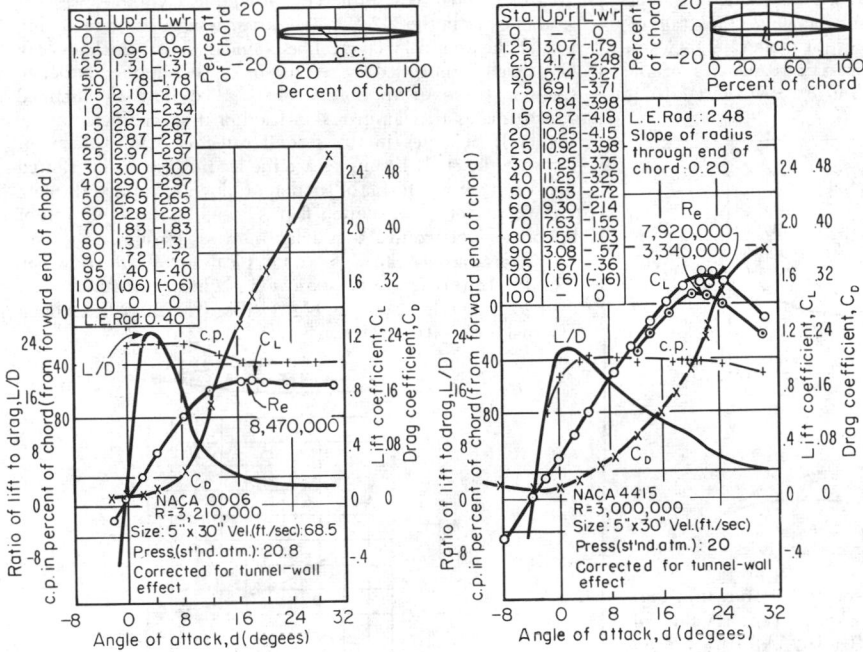

**Fig. 10**  Properties of an airfoil section.

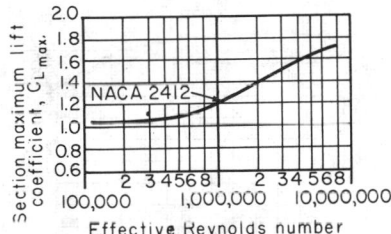

**Fig. 11**  Scale effect on section maximum lift coefficient.

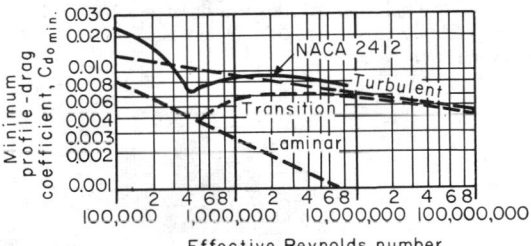

**Fig. 12**  Scale effect on minimum profile-drag coefficient.

In addition to the Reynolds number, airfoil characteristics also depend upon the **Mach number** (see Supersonic and Hypersonic Aerodynamics below).

**Laminar-flow Wings**  NACA has developed a series of **low-drag wings** in which the distribution of thickness along the chord is so selected as to maintain **laminar flow** over as much of the wing surface as possible. Typical low-drag wings are shown in Fig. 13.

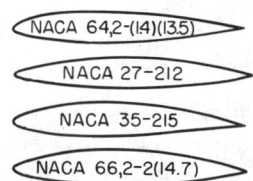

**Fig. 13**  Typical NACA low-drag airfoil sections.

The low-drag wing under controlled conditions of surface smoothness may have drag coefficients about 30 percent lower than those obtained on normal conventional wings. Low-drag airfoils are so sensitive to roughness in any form that the full advantage of laminar flow is unobtainable.

**Transonic Airfoils**  At airplane Mach numbers of about $M = 0.75$, normal airfoil sections begin to show a greater drag, which increases sharply as the speed of sound is approached. Airfoil sections known as **transonic airfoils,** shown in Fig. 14, have been developed which significantly delay this **drag rise** to Mach numbers of 0.9 or greater. Advantage may also be taken of these airfoil sections by increasing the thickness of the airfoil at a given Mach number without suffering an increase in drag.

### Flaps and Slots

The maximum lift of a wing can be increased by the use of **slots** on the leading edge or **flaps** on the trailing edge. **Fixed slots** are formed by rigidly attaching a curved sheet of metal or a small auxiliary airfoil to the leading edge of the wing. The trailing-edge flap is used to give increased lift at moderate angles of attack and to increase $C_{Lmax}$. Theoretical analyses of the effects are given in *NACA-TR* 938 and *Br. ARC-R&M* 1095.

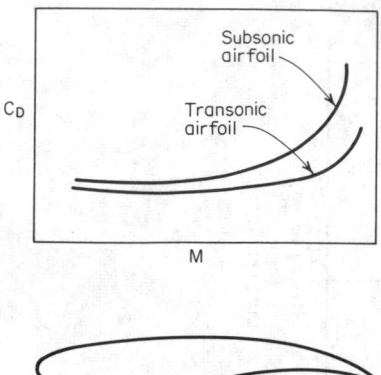

**Fig. 14**  Typical transonic airfoil-section characteristics.

Flight experience indicates that the four types shown in Fig. 15 have special advantages over other known types. Values of $C_{Lmax}$ for these types are shown in Table 2. All flaps cause a diving moment $(-C_m)$ which must be trimmed out by a download on the tail, and this download reduces $C_{Lmax}$. The correction is $\Delta C_{Lmax}$ (trim) = $\Delta C_{mo}(l/c)$, where $l/c$ is the tail length in mean chords.

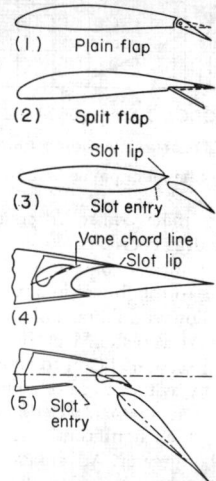

**Fig. 15**  Trailing-edge flaps: (1) plain flap; (2) split flap; (3) single-slotted flap; (4) double-slotted flap (retracted); (5) double-slotted flap (extended).

**Boundary-Layer Control (BLC)**  This includes numerous schemes for (1) maintaining laminar flow in the boundary layer in the flow over a wing or (2) preventing flow separation. Schemes in the first category try to obtain the lower frictional

drag of laminar flow either by providing favorable pressure gradients, as in the NACA "low-drag" wings, or by removing part of the boundary layer. The boundary-layer thickness can be partially controlled by the use of suction applied either to spanwise slots or to porous areas. The flow so obtained approximates the laminar skin-friction drag coefficients (see Fig. 12). Schemes in the second category try to delay or improve the stall. Examples are the leading-edge slot, slotted flaps, and various forms of suction or blowing applied through transverse slots. The slotted flap is a highly effective form of boundary-layer control that delays flow separation on the flap.

The **pressure distributions** on a typical airfoil are shown in Fig. 16. In this figure the ratio $p/q$ is given as a function of distance along the chord. In Fig. 17 the effect of addition of an external airfoil flap is shown.

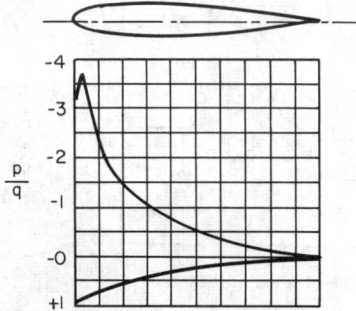

**Fig. 16**  Pressure distribution over a typical airfoil.

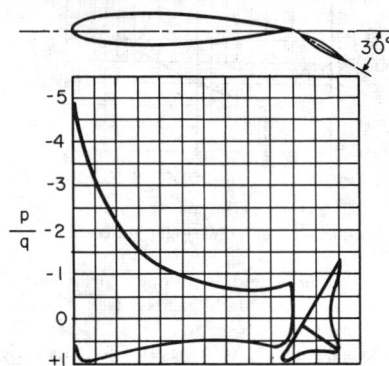

**Fig. 17**  Pressure distribution on an airfoil-flap combination.

### Airplane Performance

In **horizontal flight** the lift of the wings must be equal to the weight of the airplane, or $L = W$, where $W$ is the gross weight, lb (kg). This equation determines the minimum horizontal speed of an airplane since

$$W = C_{Lmax} \tfrac{1}{2}\rho V_{min}^2 S \qquad \text{or} \qquad V_{min} = \sqrt{2W/C_{Lmax}\rho S}$$

This corresponds to the **landing speed** without power, or the **stalling speed**.

In uniform horizontal flight the propeller thrust $T$ must be exactly equal to the drag of the airplane, or $T = D$. Then $TV$

**Table 2. Flaps***

| Type of flap | Section $C_{max}$ (approx) | Finite wing $C_{Lmax}$ (approx) | Remarks |
|---|---|---|---|
| Plain......................... | 2.4 | 1.9 | $C_f/c = 0.20$.  Sensitive to leakage between flap and wing |
| Split........................ | 2.6 | 2.0 | $C_f/c = 0.20$.  Simplest type of flap |
| Single slotted................. | 2.8 | 2.2 | $C_f/c = 0.25$.  Shape of slot and location of deflected flap are critical |
|  | 3.0 | 2.4 |  |
| Double slotted................. | 3.0 | 2.4 | $C_f/c = 0.25$.  Same comment as above |
|  | 3.4 | 2.7 |  |

*These data are for wing thickness $t/c = 0.12$ or $0.15$. Lift increment is dependent on the leading-edge radius of the wing.

= power available from the propeller and $DV$ = power required for overcoming the drag of the airplane; for horizontal flight $TV = DV$. If the airplane climbs at a rate $dh/dt$, where $h$ = height, then an additional power = $W (dh/dt)$ is required to increase the potential energy of the airplane. The equilibrium condition becomes

$$TV = DV + W(dh/dt)$$

If the **thrust power available** $P_{Ta}$ is measured in hp, $D$ and $W$ in lb, and $V$ and $dh/dt$ in ft/s,

$$P_{Ta} = \frac{1}{550} DV + \frac{1}{550} W(dh/dt) \qquad \text{(A)}$$

The thrust horsepower available $P_{Ta} = \eta P_u$, where $P_u$ = **available engine horsepower** and $\eta$ is the propeller efficiency. $P_{Ta}$ is determined as a function of $V$ from the engine and propeller characteristics. The **thrust horsepower required** to overcome the drag $D$ is $P_{Tr} = DV/550$. The **drag** is the sum of the wing profile drag $D_0$, the wing induced drag $D_i$, and the parasite drag of the other airplane parts $D_p$. $D_0$ can be obtained from wing-profile tests corrected to full-scale Reynolds number, $D_i$ from the induced drag formula $D_i = L^2/\pi qb^2$, and $D_p$ by summation of the parasite drag components due to fuselage, tail surfaces, landing gear, etc.

From Eq. (A) the performance of an airplane can be obtained either graphically or analytically. In the graphical determination the curves of power available and power required are plotted for a fixed altitude (Fig. 18). If $P_{Ta} > P_{Tr}$, the excess

horsepower can be used either for horizontal acceleration to a higher speed or for climbing. The maximum speed $V_{max}$ occurs when $P_{Ta} = P_{Tr}$. The rate of climb is determined by the equation $dh/dt = 550/W(P_{Ta} - P_{Tr})$. To calculate the **maximum velocity and rate of climb for another altitude,** another set of curves of $P_{Ta}$ and $P_{Tr}$ versus $V$ is constructed. At any altitude at which the air density is $\rho$ and for a given angle of attack (corresponding to an unchanged lift/drag ratio), $L/D = L_0/D_0$. As the lift is equal to the weight, $L = L_0 = W$, therefore $D = D_0$ and

$$P_{Tr}/P_{Tr0} = VD/V_0D_0 = V/V_0$$

But $C_L = C_{L0}$ and $C_L \cdot \frac{1}{2}\rho V^2 S = C_{L0}\frac{1}{2}\rho_0 V_0^2 S$. Therefore $V/V_0 = \sqrt{\rho_0/\rho}$. From this relation, new curves of $P_{Tr}$ versus $V$ may be constructed for various altitudes by multiplying both the ordinates and abscissas of the original curve by $\sqrt{\rho_0/\rho}$.

Figure 18 shows a performance chart for a 2,100-lb airplane. The maximum velocity at 10,000 ft altitude is 168.5 ft/s. At sea level $(P_{Ta} - P_{Tr})_{max} = 71$ and the rate of climb is $71 \times 33,000/2,100 = 1,120$ ft/min. At 10,000 ft altitude the rate is $30 \times 33,000/2,100 = 470$ ft/min.

When the curves of power required and power available become tangent to each other, there is only one speed at which the airplane can fly level and the rate of climb is zero. The corresponding altitude is the **absolute ceiling** $H$. It can be determined from the curve of maximum rate of climb as a function of altitude when this is approximated by a straight line (Fig. 19). The **service ceiling** $h_s$ is defined as the altitude at

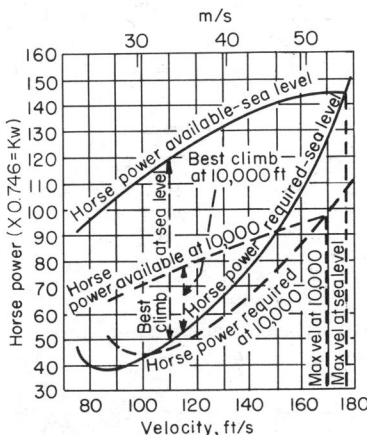

**Fig. 18**  Construction for determining the airplane ceiling.

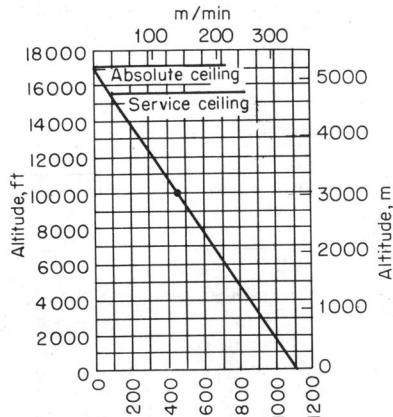

**Fig. 19**  Characteristic airplane curves.

which the rate of climb is 100 ft/min. For a linear decrease of rate of climb the following approximations hold:

Absolute ceiling: $\quad H = r_0 b/(r_0 - r)$

where $r_0$ is the rate of climb at sea level and $r$ is the rate of climb at altitude $b$.

Service ceiling: $\quad b_s = H(r_0 - 100)/r_0$

**Altitude climbed in $t$ min,**

$$b = H(1 - e^k)$$

where $k = -r_0 t/H$.

**Time to climb to altitude $b$,**

$$t = 2.303 \frac{H}{r_0} \log \frac{H}{H - b}$$

The maximum distance that an airplane can fly is called its **range,** and the length of time that it can remain flying, its **endurance.** If $W_0$ is the weight in pounds fully loaded and $W_1$ the weight after having consumed its fuel at the rate of $C$ lb/(bhp) (h), then

$$\text{Range} = 863.5 \frac{\eta}{C} \cdot \frac{L}{D} \cdot \log \frac{W_0}{W_1} \quad \text{mi}$$

$$\text{Endurance} = 750 \frac{\sqrt{W}}{V_c} \cdot \frac{\eta}{C} \cdot \frac{L}{D} \left( \frac{1}{\sqrt{W_1}} - \frac{1}{\sqrt{W_0}} \right) \quad \text{h}$$

where $\eta$ is the average propulsive efficiency and $L/D = C_L/C_D$ is the ratio of lift to drag. $V_c$ is the cruising speed, mi/h, at any gross weight $W$, lb. See Table 3 for dimension and performance of selected airplanes.

**Power Available**   The maximum efficiency $\eta_m$ and the diameter of a propeller to absorb a given power at a given speed and rpm are found from a propeller-performance curve. The thrust horsepower at maximum speed $P_{Tm}$ is found from $P_{Tm} = \eta_m P_m$. The **thrust horsepower at any speed** can be approximately determined from the ratios given in the following table:

| $V/V_{max}$ | | 20 | 30 | 40 | 50 | 60 | 70 | 80 | 90 | 1.00 | 1.10 |
|---|---|---|---|---|---|---|---|---|---|---|---|
| $P_T/P_{Tm}$ { | Fixed-pitch propeller | 0.29 | 0.44 | 0.57 | 0.68 | 0.77 | 0.84 | 0.90 | 0.96 | 1.00 | 1.03 |
| | constant-rpm propeller | 0.47 | 0.62 | 0.74 | 0.82 | 0.88 | 0.93 | 0.97 | 0.99 | 1.00 | 1.00 |

The **brake horsepower** of an engine decreases with increase of altitude. The variation of $P_T$ with altitude depends on engine and propeller characteristics with average values as follows:

| $b$, ft | 0 | 5,000 | 10,000 | 15,000 | 20,000 | 25,000 |
|---|---|---|---|---|---|---|
| $P_T/P_{T0}$ (fixed pitch) | 1.00 | 0.82 | 0.66 | 0.52 | 0.41 | 0.30 |
| $P_T/P_{T0}$ (controllable pitch) | 1.00 | 0.85 | 0.71 | 0.59 | 0.48 | 0.38 |

**Performance with Jet Thrust**   In all cases where jet thrust constitutes either a part or all of the power source, it is necessary to use graphical methods, plotting thrust and drag as functions of speed at each altitude to be investigated. The major thrust corrections as those due to losses in (1) the air intake, (2) the ducting system, and (3) the tailpipe. (See Sec. 6, Jet Propulsion and Aircraft Propellers.)

An excess thrust $T$, lb, at a speed $V$, mi/h, is equivalent to a thrust horsepower $P_T = TV/375$. The corresponding rate of climb is $dh/dt = 33,000P_T/W = 88TV/W$.

**Parasite Drag**

The drag of the nonlifting parts of an airplane is called the **parasite drag.** It consists of two components: the frictional and the eddy-making drag.

**Frictional drag,** or **skin friction,** is due to the viscosity of the fluid. It is the force produced by the viscous shear in the layers of fluid immediately adjacent to the body. It is always proportional to the wetted area, i.e., the total surface exposed to the air.

**Eddy-making drag,** sometimes called **form drag,** is due to the disturbance or wake created by the body. It is a function of the shape of the body.

The total drag of a body may be composed of the two components in any proportion, varying from almost pure skin friction for a plate edgewise or a good streamline form to 100 percent form drag for a flat plate normal to the wind. A **streamline form** is a shape having very low form drag. Such a form creates little disturbance in moving through a fluid.

When air flows past a surface, the layer immediately adjacent to the surface adheres to it, or the tangenitial velocity at the surface is zero. In the transition region near the surface, which is called the **boundary layer,** the velocity increases from zero to the velocity of the stream. When the flow in the boundary layer proceeds as if it were made up of laminae sliding smoothly over each other, it is called a **laminar boundary layer.** If there are also irregular motions in the layers normal to the surface, it is a **turbulent boundary layer.** Under normal conditions the flow is laminar at low Reynolds numbers and turbulent at high Re with a transition range of values of Re extending between $5 \times 10^5$ and $5 \times 10^7$. The profile-drag coefficients corresponding to these conditions are shown in Fig. 12.

For laminar flow the friction-drag coefficient is practically independent of surface roughness and for a flat plate is given by the Blasius equation,

$$C_{DF} = 2.656/\sqrt{\text{Re}}$$

The turbulent boundary layer is thicker and produces a greater frictional drag. For a smooth flat plate with a turbulent boundary layer

$$C_{DF} = 0.91/\log \text{Re}^{2.58}$$

For rough surfaces the drag coefficients are increased (see von Kármán, *Jour. Aeronaut. Sci.*, **1**, no. 1, 1934). The drag coefficients as given above are based on *projected area* of a double-surfaced plane. If *wetted* area is used, the coefficients must be divided by 2. The frictional drag is $D_F = C_{DF} qS$, where $S$ is the *projected area*, or $D_F = \frac{1}{2} C_{DF} qA$, where $A$ is the wetted area.

Values of $C_{DF}$ for double-surfaced planes may be estimated from the following tabulation:

Laminar flow (Blasius equation):

| Re | 10 | $10^2$ | $10^3$ | $10^4$ | $10^5$ | $10^6$ |
|---|---|---|---|---|---|---|
| $C_{DF}$ | 0.838 | 0.265 | 0.0838 | 0.0265 | 0.0084 | 0.00265 |

**Table 3. Principal Dimensions and Performance of Typical Airplanes***

| | Aero Commander | BAC | Beech | Bell | Boeing | Boeing | Boeing | Convair | Lockheed | Douglas | McDonnell Douglas | Piper | Sikorsky |
|---|---|---|---|---|---|---|---|---|---|---|---|---|---|
| Designation model no. | 680F-P | One-Eleven | Bonanza S35 | Ranger 47J2-A | 707-320 | 747-200B | 727 | 340 | 1011-1 | DC-8 Series 5D | DC-10-10 | Super Cub Pa 18-150 | S-58 |
| Type | Executive transport | Jet transport | Business plane | Executive helicopter | Jet transport | Passenger | Jet transport | Transport | Passenger | Jet Transport | Passenger | Utility plane | Helicopter |
| No. passengers | 5–7 | 63 | 4–6 | 4 | 133–186 | 374–500 | 70–114 | 44 | 256–400 | 116–176 | 255–380 | 2 | 12–18 |
| Cargo capacity, lb | 2,815 | N.A. | 270 | 1,120 | 17,750 | | 8,550 | N.A. | | 20,850 | | 50 | 50 |
| Span, ft | 49.5 | 83.5 | 33.4 | 37.0† | 145.9 | 195.6 | 108.6 | 105.4 | 155.3 | 142.3 | 155.3 | 35.3 | 56.0† |
| Overall length, ft | 35.1 | 92.1 | 26.4 | 43.3 | 145.5 | 231.3 | 134.3 | 79.2 | 178.6 | 150.5 | 181.4 | 22.5 | 65.8 |
| Overall height, ft | 14.5 | 23.8 | 6.5 | 9.3 | 42.6 | 63.4 | 34.0 | 28.2 | 55.3 | 42.3 | 58 | 6.7 | 14.3 |
| Wing area, ft² | 255 | 980 | 181 | Rotor | 2,892 | 5,500 | 1,650 | 817 | 3,456 | 2,868 | 3,861 | 178.5 | Rotor |
| Weight empty, lb‡ | 5,185 | 58,000 | 1,885 | 1,730 | 131,000 | 361,216 | 86,000 | 32,359 | 234,275 | 124,529 | 231,779 | 930 | 7,900 |
| Weight gross, lb | 8,000 | 73,800 | 3,300 | 2,850 | 328,000 | 775,000 | 142,000 | 47,000 | 430,000 | 310,000 | 430,000 | 1,750 | 13,000 |
| Power plant | (2) Lyc IGSO-540B1A | (2) RR Spey 2 | Con IO-520-B | Lyc VO540 | (-) PW JT3D-3 | JT9D-3, -7 | (3) PW JT8D-1 | (2) R2800 | RB-211-22B | (4) PW JT3D-3 | CF6-6D | Lyc O-320 | WR R1820 |
| High speed, knots | 252 | 469 | 184 | 9- | 524 | 528 | 535 | 273 | 530 | 521 | 530 | 113 | 106 |
| Cruise speed, knots | 220 | 441 | 178 | 8- | 500–525 | 478 | 513 | 247 | 473 | 472 | 473 | 100 | 84 |
| Landing speed, knots | 91 | N.A. | 54 | 0 | 128 | 140 | 121 | 75 | 125 | 133 | 138 | 37 | 0 |
| Range, nautical miles | 1,310 | 1,700 | 1,145 | 260 | 6,000 | 5,748 | 2,320 | 1,260 | 2,500 | 8,600 | 2,110 | 460 | 247 |

\* ft × 0.305 = m; ft² × 0.0929 = m²; lb m × 0.454 = kg.
† Rotor diameter.
‡ Operating weight empty (includes trapped oil).
N.A. Not available.

Turbulent flow:

| Re | $10^5$ | $10^6$ | $10^7$ | $10^8$ | $10^9$ | $10^{10}$ |
|---|---|---|---|---|---|---|
| $C_{DF}$ | 0.0148 | 0.0089 | 0.0060 | 0.0043 | 0.00315 | 0.0024 |

These are double-surface values which facilitate direct comparison with wing-drag coefficients based on projected area. For calculations involving wetted area, use one-half of double-surface coefficients. Interpolations in the foregoing tables must allow for the logarithmic functions; i.e., the variation in $C_{DF}$ is not linear with Re.

### Drag Coefficients of Various Bodies

For **bodies with sharp edges** the drag coefficients are almost independent of the Reynolds number, for most of the resistance is due to the difference in pressure on the front and rear surfaces. Table 4 gives $C_D = D/qS$, where $S$ is the maximum cross section perpendicular to the wind.

For **rounded bodies** such as **spheres, cylinders,** and **ellipsoids** the drag coefficient depends markedly upon the Reynolds number, the surface roughness, and the degree of turbulence in the air stream. A sphere and a cylinder, for instance, experience a sudden reduction in $C_D$ as the Reynolds number exceeds a certain critical value. The reason is that at low speeds (small

Re) the flow in the boundary layer adjacent to the body is laminar and the flow separates at about 83° from the front (Fig. 20). A wide wake thus gives a large drag. At higher speeds (large Re) the boundary layer becomes turbulent, gets additional energy from the outside flow, and does not separate on

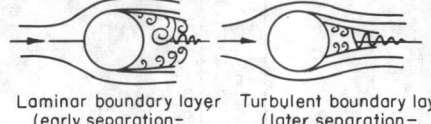

Laminar boundary layer         Turbulent boundary layer
(early separation—              (later separation—
wide wake)                      narrow wake)

**Fig. 20**   Boundary layer of a sphere.

the front side of the sphere. The drag coefficient is reduced from about 0.47 to about 0.08 at a critical Reynolds number of about 400,000 in free air. Turbulence in the air stream reduces the value of the critical Reynolds number (Fig. 21). The Reynolds number at which the sphere drag $C_D = 0.3$ is taken as a criterion of the amount of turbulence in the air stream of wind tunnels.

**Cylinders**   The drag coefficient of a cylinder with its axis normal to the wind is given as a function of Reynolds number in Fig. 22. Cylinder drag is sensitive to both Reynolds and

**Table 4.  Drag Coefficients**

| Object | Proportions | Attitude | $C_D$ |
|---|---|---|---|
| Rectangular plate, sides $a$ and $b$ | $\dfrac{a}{b} = \begin{matrix} 1 \\ 4 \\ 8 \\ 12.5 \\ 25 \\ 50 \\ \infty \end{matrix}$ | | 1.16<br>1.17<br>1.23<br>1.34<br>1.57<br>1.76<br>2.00 |
| Two disks, spaced a distance $l$ apart | $\dfrac{l}{d} = \begin{matrix} 1 \\ 1.5 \\ 2 \\ 3 \end{matrix}$ | | 0.93<br>0.78<br>1.04<br>1.52 |
| Cylinder | $\dfrac{l}{d} = \begin{matrix} 1 \\ 2 \\ 4 \\ 7 \end{matrix}$ | | 0.91<br>0.85<br>0.87<br>0.99 |
| Circular disk | | | 1.11 |
| Hemispherical cup, open back | | | 0.41 |
| Hemispherical cup, open front, parachute | | | 1.35 |
| Cone, closed base | | | $\alpha = 60°, 0.51$<br>$\alpha = 30°, 0.34$ |

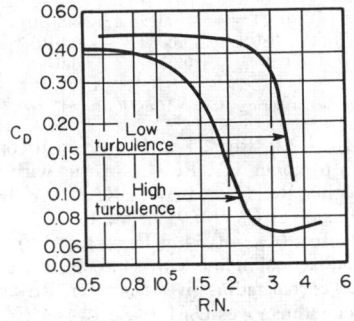

**Fig. 21** Drag coefficient of a sphere as a function of Reynolds number and of turbulence.

Mach number. Figure 22 gives the Reynolds number effect for $M = 0.35$. The increase in $C_D$ because of Mach number is approximately:

| M | 0.35 | 0.4 | 0.6 | 0.8 | 1.0 |
|---|---|---|---|---|---|
| $C_D$ increase, percent | 0 | 2 | 20 | 50 | 70 |

(see *NACA-TN* 2960)

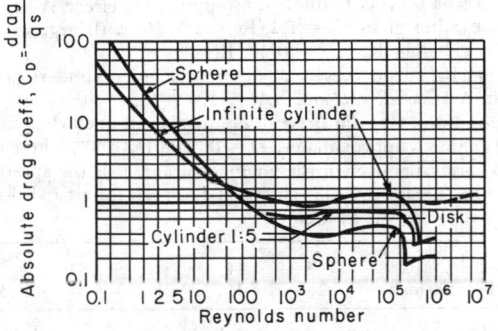

**Fig. 22** Drag coefficients of cylinders and spheres.

**Streamline Forms** The drag of a streamline body of revolution depends to a very marked extent on the Reynolds number. The difference between extreme types at a given Reynolds number is of the same order as the change in $C_D$ for a given form for values of Re from $10^6$ to $10^7$.

Tests reported in *NACA-TN* 614 indicate that the shape for minimum drag should have a fairly sharp nose and tail. At Re $= 6.6 \times 10^6$ the best forms for a fineness ratio (ratio of length to diameter) 5 have a drag

$$D = 0.040qA = 0.0175qV^{2/3}$$

where $A$ = max cross-sectional area and $V$ = volume. This value is equivalent to about 1.0 lb/ft² at 100 mi/h. For Re > 5 $\times 10^6$, the drag coefficients vary approximately as $Re^{0.15}$.

Minimum drag on the basis of cross-sectional or frontal area is obtained with a fineness ratio of the order of 2 to 3.

Minimum drag on the basis of contained volume is obtained with a fineness ratio of the order of 4 to 6 (see *NACA-TR* 291).

The following table gives the ordinates for good streamline shapes: the **Navy strut,** a two-dimensional shape; and the Class C **airships,** a three-dimensional shape. Streamline shapes for high Mach numbers have a fine entry and high fineness ratio.

| Percent length | 1.25 | 2.50 | 4.00 | 7.50 | 10.0 | 12.50 |
|---|---|---|---|---|---|---|
| Percent of max ordinate { Navy strut | 26 | 37.1 | 52.50 | 63.00 | 72.0 | 78.50 |
| Class C airship | 20 | 33.5 | 52.60 | 65.80 | 75.8 | 83.50 |

| Percent length | 20 | 40 | 60 | 80 | 90 | |
|---|---|---|---|---|---|---|
| Percent of max ordinate { Navy strut | 91.1 | 99.5 | 86.1 | 56.2 | 33.8 | |
| Class C airship | 94.7 | 99.0 | 88.5 | 60.5 | 49.3 | |

Test data on the *RM*-10 shape are given in *NACA-TR* 1160. This shape is a parabolic-arc type for which the coordinates are given by the equation $r_x = X/7.5(1.0 - X/L)$.

**Struts** Drag coefficients for streamline struts are given in the form

$$C_D = D/qS_f = D/qdl$$

where $S_f$ is the projected frontal area; $d$ is the thickness, ft; and $l$ is the length, ft. The variation of $C_D$ with Re for a Navy No. 1 strut of fineness ratio 3 is as follows:

| Re $\times 10^{-5}$ | 0.75 | 1.0 | 1.25 | 1.5 | 2.0 | 3.0 | 4.0 |
|---|---|---|---|---|---|---|---|
| $C_D$ | 0.114 | 0.102 | 0.093 | 0.088 | 0.085 | 0.077 | 0.073 |

**Wing Profile Drag** For accurate basic values of profile drag it is necessary to refer to test data on the wing section employed. In the absence of test data, the following approximate values may be used:

| Average $t/c$ | 0.10 | 0.12 | 0.14 | 0.16 |
|---|---|---|---|---|
| Basic { $C_{D0}$ | 0.0058 | 0.0060 | 0.0063 | 0.0067 |
| $D_0/S$ | 0.148 | 0.154 | 0.161 | 0.171 |
| Best wing { $C_{D0}$ | 0.0078 | 0.0080 | 0.0083 | 0.0087 |
| $D_0/S$ | 0.20 | 0.205 | 0.212 | 0.223 |
| Average roughness { $C_{D0}$ | 0.0098 | 0.0100 | 0.0103 | 0.0107 |
| $D_0/S$ | 0.25 | 0.256 | 0.264 | 0.274 |

The "basic" values are for a perfectly smooth wing of infinite aspect ratio, "best-wing" values are for smooth full flush-riveted construction, and "average-roughness" values are for flush-riveted leading-edge with brazier head rivets back of the 20 percent chord point. $C_{D0}$ is the profile-drag coefficient ($D_0 = C_{D0}qS$). The values of $D_0/S$ are drags in lb/ft² of projected wing area at 100 mi/h in standard air ($q_0 = 25.58$ lb/ft²). $t/c$ is the maximum wing thickness as a fraction of the chord.

Faired values of NACA data on **symmetrical sections** at Re = $8 \times 10^6$ give the variation of minimum $C_{D0}$ and $C_{DA}$ with thickness as follows:

| NACA section | 0006 | 0009 | 0012 | 0015 | 0018 | 0021 | 0025 | 0030 | 0035 |
|---|---|---|---|---|---|---|---|---|---|
| Min $C_{D0}$ | 0.0051 | 0.0056 | 0.0061 | 0.0067 | 0.0073 | 0.0080 | 0.0089 | 0.0103 | 0.0120 |
| $C_{DA}$ (see below) | 0.085 | 0.062 | 0.050 | 0.045 | 0.041 | 0.038 | 0.036 | 0.034 | 0.034 |

NOTE: $C_{DA}$ is the drag coefficient based on the frontal area. The 00 section is also suitable for use in struts or fairings. (See *NACA-TR* 628, 647, 669, 708.)

**Drag of Tail Surfaces** Owing to joints, control balances, and interference effects, the drag of a control surface is much higher than that of the basic section. The average value is about 0.40 lb/ft² at 100 mi/h in standard density.

**Streamline Wire** The drag coefficient of a standard streamline wire of lenticular or elliptical cross section is given as a function of Reynolds number in Fig. 23.

The drag coefficients of **wires and cables** are also shown in Fig. 23.

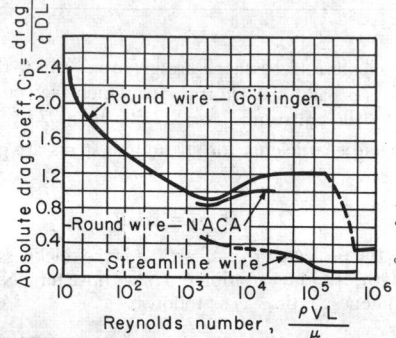

**Fig. 23** Drag coefficients of wires.

**Elliptic Cylinders**

| FR | $C_D$ when Re is | | | |
|---|---|---|---|---|
| | $3 \times 10^4$ | $6 \times 10^4$ | $1 \times 10^5$ | $2 \times 10^5$ |
| 1:1 | 1.20 | 1.22 | 1.22 | 1.23 |
| 2:1 | 0.62 | 0.57 | 0.46 | 0.35 |
| 4:1 | 0.32 | 0.32 | 0.30 | 0.24 |
| 8:1 | 0.27 | 0.23 | 0.22 | 0.21 |

The above data are for M < 0.4. For additional data see *NACA-TR* 619.

$C_L$ and $C_D$ for **inclined wires** are as follows:

| Angle of attack, deg | 0 | 15 | 30 | 45 | 60 | 75 | 90 |
|---|---|---|---|---|---|---|---|
| $C_L$ | 0 | 0.09 | 0.25 | 0.39 | 0.42 | 0.27 | 0 |
| $C_D$ | 0.01 | 0.05 | 0.17 | 0.46 | 0.77 | 1.01 | 1.12 |

$C_L$ and $C_D$ are based on the area $S = LD$, where $L$ is the length and $D$ the diam, ft.

Drag coefficients for **prismatic cylinders** of various cross sections:

Ellipse 1:2 (broadside) $C_D = 1.6$, Re < $6 \times 10^5$

Rectangular 2:1 (end-on) $C_D = 1.3$, Re < $1 \times 10^6$. A small corner radius reduces $C_D$ to 0.67 for Re < $8 \times 10^5$ with considerable scale effect at higher Re. A large corner radius gives $C_D = 0.30$ for Re < $2 \times 10^6$.

Rectangle 1:2 (broadside) $C_D = 2.2$, Re < $6 \times 10^5$. A small corner radius reduces $C_D$ to 1.8. A large corner radius reduces $C_D$ to 1.5 at Re < $2 \times 10^5$. At higher Re there is a large scale effect, $C_D$ falling to 0.5 at Re = $4 \times 10^5$.

Square 1:1 $C_D = 2.0$, Re < $1 \times 10^6$. A small corner radius reduces $C_D$ to about 1.1, Re < $6 \times 10^5$ with some scale effect at higher Re. $C_D = 0.55$ at Re = $1 \times 10^6$. A large corner radius gives $C_D = 0.95$ below Re = $2 \times 10^5$ with a large scale effect ($C_D = 0.54$ at Re = $1 \times 10^6$).

Square at 45° (diagonal in line with wind) $C_D = 1.6$, Re < $1 \times 10^6$. A small corner radius gives $C_D = 1.5$, Re < $1 \times 10^6$. A large corner radius gives $C_D$ 1.3, Re < $4 \times 10^5$ with large scale effect ($C_D = 0.40$, Re = $7 \times 10^5$).

Diamond 1:2 (broadside) $C_D = 1.6$, Re = $2 \times 10^4$. $C_D$ increases uniformly to 2.0 at Re = $7 \times 10^5$. A small corner radius has negligible effect on $C_D$. A large corner radius gives $C_D = 1.3$, Re = $2 \times 10$, increasing to $C_D = 1.7$, Re = 3 $\times 10^5$ with considerable scale effect at higher Re ($C_D = 0.44$, Re = $6 \times 10^5$).

Equilateral triangle (apex into wind) $C_D = 1.2 < 1 \times 10^6$. A small corner radius increased $C_D$ to 1.4, Re < $1 \times 10^6$. A large corner radius gives $C_D = 1.2$, Re = $4 \times 10^5$ with considerable scale effect at higher Re ($C_D = 0.35$, Re = 8 $\times 10^5$).

Equilateral triangle (apex downstream) $C_D = 2.0$, Re < $1 \times 10^6$. A small corner radius had no appreciable effect. A large corner radius gives $C_D = 1.2$, Re < $3 \times 10^5$ with large scale effect at higher Re ($C_D = 0.20$, Re = $6 \times 10^5$).

(Additional data on these and other prismatic cylinders are given in *NACA-TR* 619 and *NACA-TN* 3038.)

**Engine Drag, Nacelle Drag** The drag of an uncowled air-cooled engine is approximately $D = 0.050d^2(V/100)^2$, where $d$ is the overall diameter of the engine, in, and $V$ is the speed, mi/h. The average drag in pounds at 100 mi/h for a fixed-slot NACA type cowl is as follows:

| Engine diam, in | 40 | 44 | 48 | 52 | 56 |
|---|---|---|---|---|---|
| Drag, lb at 100 mi/h | 28 | 34 | 40 | 47 | 54 |

If adjustable **cowl flaps** or controlled airflow are used, the engine drag need not vary as the square of the speed; it may remain substantially constant. Theoretically, it is possible to cool an engine by forced cooling at an expenditure of about 2 percent of the engine power.

The average drag coefficient of a nacelle is $C_D = 0.20$, based on frontal area. This is equivalent to about 5.0 lb/ft² at 100 mi/h. A very clean form may be as low as 3 lb/ft² at 100 mi/h. The drag of a pure streamline form would be of the order of 1 lb/ft² at 100 mi/h (see *NACA-TR* 313, 314, and 415).

**Fuselage Drag** Owing to various projections and irregularities in the surface, the drag of an average airplane fuselage is considerably greater than the drag of a pure streamline form. Since the increase is due in effect to a substantially constant drag increment, the drag per unit of cross-sectional area tends to decrease with increase in the size of the fuselage. At 100 mi/h the drag in lb/ft² will be approximately as follows:

| A, ft² | 10 | 15 | 20 | 40 | 60 |
|---|---|---|---|---|---|
| D/A, average | 5.5 | 4.5 | 4.0 | 3.3 | 3.0 |
| D/A, lower limit | 4.0 | 3.3 | 3.0 | 2.4 | 2.3 |

Cockpit enclosures if properly designed and blended into fuselage lines, do not appreciably increase the fuselage drag coefficient. Sharp junctures must be avoided. The best fairing radius is approximately 25 percent of the enclosure height. Length of tail fairing should be 4 × height (see *NACA-TR* 730).

**Seaplane Floats, Flying-boat Hulls**  The drag of a seaplane float is between 3 and 6 lb/ft² at 100 mi/h, depending on the lines. The step accounts for 5 to 10 percent of the drag.

The drag of a flying-boat hull is comparable to, but slightly higher than, the drag of a fuselage of the same cross-sectional area. Average values at 100 mi/h are as follows:

| $A$, ft² | 40 | 60 | 80 | 100 |
|---|---|---|---|---|
| $D/A$, lb/ft² | 4.5 | 4.3 | 4.1 | 4.0 |

**Wire Mesh**  Measurements on square pieces of exposed wire mesh have given the following results:

| Percent area blocked | 100 | 80 | 60 | 50 | 40 | 30 | 20 |
|---|---|---|---|---|---|---|---|
| Percent flat plate, $C_D$ | 100 | 92 | 77 | 60 | 43 | 26 | 10 |

The pressure drop through a screen in a tube is given by

$$\Delta p = C_D q = \tfrac{1}{2} C_D \rho V^2$$

where $C_D$ is a function of the percent area blocked as follows:

| Percent area blocked | 20 | 30 | 40 | 50 | 60 | 70 |
|---|---|---|---|---|---|---|
| $C_D$ | 0.20 | 0.45 | 0.90 | 1.60 | 3.40 | 7.20 |

(see *Br. ARC-R&M* 1469.)

**Interference Drag**  The total drag of two objects in close proximity is generally greater than the sum of the individual free-flow drags. The increase (or decrease) is known as interference drag. It is especially important where the wing is one component and the fuselage or a nacelle the other component. In this case it may be reduced to a minimum by an appropriate fairing or "fillet" in the path of the expanding flow (see *NACA-TN* 460).

The interference drag of two parallel streamline struts is as follows:

| Spacing/thickness | 5 | 4 | 3 | 2 | 1.5 | 1.25 |
|---|---|---|---|---|---|---|
| $D/D_0$ | 1.00 | 1.06 | 1.12 | 1.25 | 2.25 | 2.87 |

The interference drag of a strut intersection with a flat surface is a function of the intersection angle $\theta$ (see Fig. 24). The drag increase, expressed as an equivalent length increase $\Delta L$ in diameters, is as follows:

| $\theta$, deg | 90 | 70 | 60 | 50 | 40 | 30 | 20 |
|---|---|---|---|---|---|---|---|
| $\Delta L$ | 0 | 1.5 | 2.5 | 4.0 | 6.0 | 9.5 | 14.0 |

## STABILITY

**Airplane Control**  An airplane is controlled in flight by imposing yawing, pitching, and rolling moments by use of rudders, elevators, and ailerons. This is known as the **three-control** system and is in almost universal use, although it is possible to dispense with either the ailerons or the rudders and so obtain a two-control system.

Controllability is separate and distinct from stability in physical significance, but not necessarily so in flight. An airplane may be made automatically stable by gyroscopic or other devices that actuate the controls mechanically; it is inherently stable if, on disturbance from any cause, the aerodynamic forces and moments induced always tend to return the airplane to its original attitude.

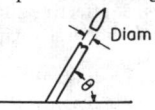

**Fig. 24**  Strut intersection with a flat surface.

**Stability** may be either static or dynamic. An airplane is **statically stable** if the moments tend to return it to the original attitude. It is **dynamically stable** if the oscillations produced by the static stability are rapidly damped out. If it is **statically unstable**, any departure tends to increase, there being an upsetting moment instead of a restoring moment. If it is **dynamically unstable,** the oscillations due to static stability tend to increase in amplitude with time. Static stability or a condition of stable equilibrium is necessary to obtain dynamic stability, but static stability does not ensure dynamic stability. Too much static stability may cause dynamic instability if damping is inadequate.

The common method of getting **static longitudinal stability** is by means of a tail surface. A wing alone has a lift force $L$ acting at the a.c. and a moment $M_0$ about the a.c. (Fig. 25). The moment is measured positively as a stalling moment. Most wings alone are "unstable" (positive lift-curve slope), though possessing a small negative moment (diving moment). If there were no other air forces acting, this could be balanced

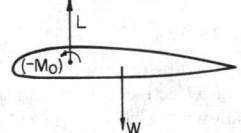

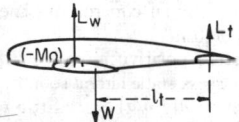

**Fig. 25**  Longitudinal stability of an airplane.

by putting the center of gravity of the airplane behind the a.c. a distance $\delta$ such that $W\delta = M_0$. This arrangement would be in equilibrium but would be unstable. Upon a small increase in the angle of attack such as by a gust, the lift force will be increased. This gives a moment tending to increase further the angle of attack. The curve of the moment of the air forces on a wing about the center of gravity thus starts from the value $M_0$ and becomes positive for larger angles of attack. The curve is shown in Fig. 26. The positive slope of $M$ versus $\alpha$ thus corresponds to an unstable moment. When a horizontal tail surface is added, a lift force also acts upon the tail. This gives a diving moment $M_t = -L_t \, l_t$. Upon an increase in the angle of attack $\alpha$, the diving tail moment is increased more rapidly than the stalling moment of the wing. Thus, the combination is stable. The resultant curve of pitching moment against angle of attack has a small negative slope. Too steep a slope would indicate longitudinal stiffness and difficulty in control.

**Rolling** (or **banking** ) does not produce any lateral shift in the center of the lift, so that there is no restoring moment as in pitch. However, when banked, the airplane sideslips toward the low wing. A fin placed above the center of gravity gives a lateral restoring moment that can correct the roll and stop the slip. The same effect can be obtained by **dihedral**, i.e., by raising the wing tips to give a transverse *V*. An effective dihedral of 1 to 3° on each side is generally required to obtain stability in roll. In low-wing monoplanes, 2° effective dihedral may require 8° or more of geometrical dihedral owing to interference between the wing and fuselage.

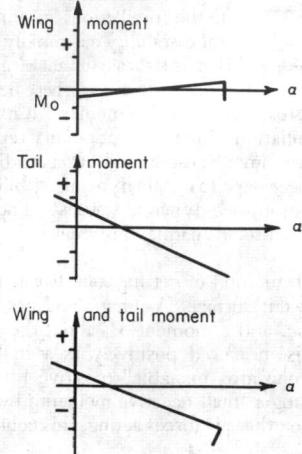

**Fig. 26**  Longitudinal stability. Wing and foil moments.

In a **yaw** or **slip** the line of action of the lateral force depends on the size of the effective vertical fin area. Insufficient fin surface aft will allow the skid or slip to increase. Too much fin surface aft will swing the nose of the plane around into a tight spiral. Sound design demands sufficient vertical tail surface for adequate directional control and then enough dihedral to provide lateral stability.

When moderate positive effective dihedral is present, the airplane will possess **static lateral stability,** and a low wing will come up automatically with very little yaw. If the dihedral is too great, the airplane may roll considerably in gusts, but there is little danger of the amplitude ever becoming excessive. **Dynamic lateral stability** is not assured by static stability in roll and yaw but requires that these be properly proportioned to the damping in roll and yaw.

**Spiral instability** is the result of too much fin surface and insufficient dihedral.

## HELICOPTERS

REFERENCES: *Br. ARC-R & M* 1111, 1127, 1132, 1157, 1730, and 1859. *NACA-TM* 827, 836, 858. *NACA-TN* 626, 835, 1192, 3323, 3236. *NACA-TR* 434, 515, 905, 1078. *NACA Wartime Reports* L-97, L-101, L-110. NACA, "Conference on Helicopters," May, 1954. Gessow and Myers, "Aerodynamics of the Helicopter," Macmillan.

Helicopters derive lift, propulsion force, and control effect from adjustments in the blade angles of the rotor system. At least two rotors are required, and these may be arranged in any form that permits control over the reaction torque. The common arrangements are main lift rotor, auxiliary torque-

control or tail rotor at 90°; two main rotors side by side; two main rotors fore and aft; and two main rotors, coaxial and oppositely rotating.

The helicopter rotor is an actuator disk or momentum device that follows the same general laws as a propeller. In calculating the rotor performance, the major variables concerned are diameter *D*, ft (radius *R*); tip speed, $V_t = \Omega R$, ft/s; angular velocity of rotor $\Omega = 2\pi n$, rad/s; and rotor solidity *q* (= ratio of blade area/disk area). The rotor performance is usually stated in terms of coefficients similar to propeller coefficients. The rotor coefficients are:

$$\text{Thrust coefficient } C_T = T/\rho(\Omega R)^2 \pi R^2$$
$$\text{Torque coefficient } C_Q = Q/\rho(\Omega R)^2 \pi R^3$$
$$\text{Torque } Q = 5,250 \text{ bhp/rpm}$$

**Hovering**  The **hovering flight** condition may be calculated from basic rotor data given in Fig. 27. These data are taken from full-scale rotor tests. Surface-contour accuracy can reduce the total torque coefficient 6 to 7 percent. The power required for hovering flight is greatly reduced near the ground. Observed flight-test data from various sources are plotted in Fig. 28. The ordinates are heights above the ground measured in rotor diameters.

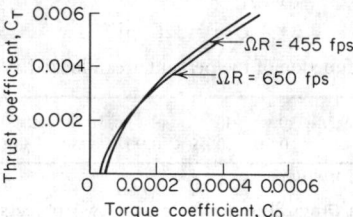

**Fig. 27**  Static thrust performance of NACA8-H-12 blades.

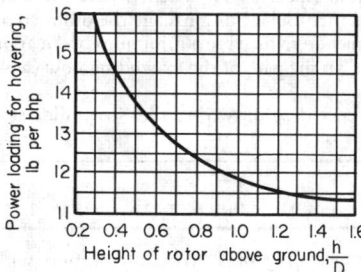

**Fig. 28**  Observed ground effect on Sikorsky-type helicopters.

**Effect of Gross Weight on Rate of Climb**  Figure 29 from Talkin (*NACA-TN* 1192) shows the rate of climb that can be obtained by reducing the load on a helicopter that will just hover. Conversely, given the rate of climb with a given load, the curves determine the increase in load that will reduce the rate of climb to zero; i.e., they determine the maximum load for which hovering is possible.

**Performance with Forward Speed**  The performance of a typical single-main-rotor-type helicopter may by shown by a curve of $C_{TR}/C_{QR}$ plotted against $\mu = v/\pi n D$, where *v* is the speed of the helicopter, as in Fig. 30. This curve includes the parasite-drag effects which are appreciable at the higher values of $\mu$. It is only a rough approximation to the experimentally determined values. $C_{TR}$ may be calculated to determine $C_{QR}$, from which $Q$ is obtained.

The performance of rotors at forward speeds involves a number of variables. For more complete treatment, see *NACA-TN* 1192 and *NACA Wartime Report* L-110.

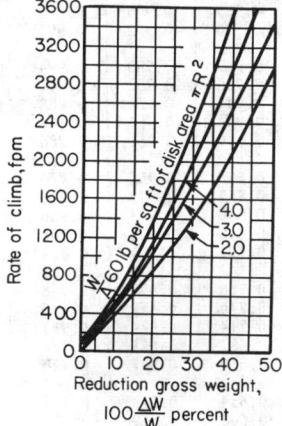

**Fig. 29**  Effect of gross weight on the rate of climb of a helicopter.

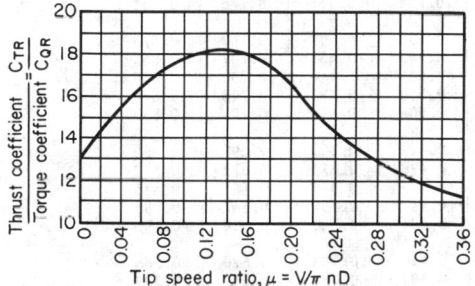

**Fig. 30**  Performance curve for a single-main-rotor-type helicopter.

### GROUND-EFFECT MACHINES (GEM)

For data on air-cushion vehicles and hydrofoil craft, see Sec. 6, Marine Engineering.

### SUPERSONIC AND HYPERSONIC AERODYNAMICS

REFERENCES: Liepmann and Roshko, "Elements of Gas Dynamics," Wiley. "High Speed Aerodynamics and Jet Propulsion" 12 vols., Princeton. Shapiro, "The Dynamics and Thermodynamics of Compressible Fluid Flow," Vols. I, II, Ronald. Howarth (ed.), "Modern Developments in Fluid Mechanics—High Speed Flow," Vols. I, II, Oxford. Kuethe and Schetzer, "Foundations of Aerodynamics," Wiley. Ferri, "Elements of Aerodynamics of Supersonic Flows," Macmillan. Bonney, "Engineering Supersonic Aerodynamics," McGraw-Hill.

The effect of the compressibility of a fluid upon its motion is determined primarily by the **Mach number** M.

$$M = V/V_s \tag{1}$$

where $V$ = speed of the fluid and $V_s$ = speed of sound in the fluid = $49.1 \sqrt{T}$, where $T$ is in °R and $V_s$ in ft/s. The Mach number varies with position in the fluid, and the compressibility effect likewise varies from point to point.

If a *body moves through the atmosphere*, the overall compressibility effects are a function of the Mach number $\overline{M}$ of the body, defined as

$\overline{M}$ = velocity of body/speed of sound in the atmosphere

Table 5 lists the useful **gas dynamics relations** between velocity, Mach number, and various fluid properties for isentropic flow. (See also Sec. 4, Thermal Properties and Thermodynamics.)

$$V^2 = 2(h_0 - h) = 2 C_p (T_0 - T)$$

$$\frac{V^2}{V_{s0}^2} = \frac{M^2}{1 + \dfrac{\gamma - 1}{2} M^2}$$

$$\left(\frac{p}{p_0}\right)^{(\gamma-1)/\gamma} = \frac{T}{T_0} = \left(\frac{\rho}{\rho_0}\right)^{\gamma-1} = \left(\frac{V_s}{V_{s0}}\right)^2 = \frac{1}{1 + \dfrac{\gamma - 1}{2} M^2}$$

$$\frac{\frac{1}{2}\rho V^2}{p_0} = \frac{(\gamma/2)M^2}{\left(1 + \dfrac{\gamma - 1}{2} M^2\right)^{\gamma/(\gamma-1)}} \tag{2}$$

$$\left(\frac{A}{A^*}\right)^2 = \frac{1}{M^2}\left[\frac{2}{\gamma + 1}\left(1 + \frac{\gamma - 1}{2} M^2\right)\right]^{(\gamma+1)/(\gamma-1)}$$

where $h$ = enthalpy of the fluid; $A$ = stream-tube cross section normal to the velocity; and the subscript $_0$ denotes the isentropic stagnation condition reached by the stream when stopped frictionlessly and adiabatically (hence isentropically). $V_{s0}$ denotes the speed of sound in that medium.

The superscript * denotes the conditions occurring when the speed of the fluid equals the speed of sound in the fluid. For $M = 1$, Eqs. (**2**) become

$$\frac{T^*}{T_0} = \left(\frac{V_s^*}{V_{s0}}\right)^2 = \frac{2}{\gamma + 1}$$

$$\frac{p^*}{p_0} = \left(\frac{2}{\gamma + 1}\right)^{\gamma/(\gamma-1)} \tag{3}$$

$$\frac{\rho^*}{\rho_0} = \left(\frac{2}{\gamma + 1}\right)^{1/(\gamma-1)}$$

For air $p^* = 0.52828 p_0$: $\rho^* = 0.63394\rho_0$; $T^* = 0.83333 T_0$.

For **subsonic regions** flow behaves similarly to the familiar hydraulics or incompressible aerodynamics: In particular, an increase of velocity is associated with a decrease of stream-tube area, and friction causes a pressure drop in a tube. There are no regions where the local flow velocity exceeds sonic speed. For **supersonic regions,** an increase of velocity is associated with an increase of stream-tube area, and friction causes a pressure rise in a tube. $M = 1$ is the dividing line between these two regions. For **transonic flows,** there are regions where the local flow exceeds sonic velocity, and this mixed-flow region requires special analysis. For **supersonic flows,** the entire flow field, with the exception of the regime near stagnation areas, has a velocity higher than the speed of sound.

The **hypersonic regime** is that range of very high supersonic speeds (usually taken as $M > 5$) where even a very streamlined body causes disturbance velocities comparable to the speed of

**Table 5.  Isentropic Gas Dynamics Relations***

| M | $\frac{p}{p_0}$ | $\frac{V}{V_{s0}}$ | $\frac{A}{A^*}$ | $\frac{\frac{1}{2}\rho V^2}{p_0}$ | $\frac{\rho V}{\rho_0 V_{s0}}$ | $\frac{\rho}{\rho_0}$ | $\frac{T}{T_0}$ | $\frac{V_s}{V_{s0}}$ |
|---|---|---|---|---|---|---|---|---|
| 0.00 | 1.000 | 0.000 | ∞ | 0.000 | 0.000 | 1.000 | 1.000 | 1.000 |
| 0.05 | 0.998 | 0.050 | 11.59 | 0.002 | 0.050 | 0.999 | 0.999 | 1.000 |
| 0.10 | 0.993 | 0.100 | 5.82 | 0.007 | 0.099 | 0.995 | 0.998 | 0.999 |
| 0.15 | 0.984 | 0.150 | 3.91 | 0.016 | 0.148 | 0.989 | 0.996 | 0.998 |
| 0.20 | 0.972 | 0.199 | 2.964 | 0.027 | 0.195 | 0.980 | 0.992 | 0.996 |
| 0.25 | 0.957 | 0.248 | 2.403 | 0.042 | 0.241 | 0.969 | 0.988 | 0.994 |
| 0.30 | 0.939 | 0.297 | 2.035 | 0.059 | 0.284 | 0.956 | 0.982 | 0.991 |
| 0.35 | 0.919 | 0.346 | 1.778 | 0.079 | 0.325 | 0.941 | 0.976 | 0.988 |
| 0.40 | 0.896 | 0.394 | 1.590 | 0.100 | 0.364 | 0.924 | 0.969 | 0.984 |
| 0.45 | 0.870 | 0.441 | 1.449 | 0.123 | 0.399 | 0.906 | 0.961 | 0.980 |
| 0.50 | 0.843 | 0.488 | 1.340 | 0.148 | 0.432 | 0.885 | 0.952 | 0.976 |
| 0.55 | 0.814 | 0.534 | 1.255 | 0.172 | 0.461 | 0.863 | 0.943 | 0.971 |
| 0.60 | 0.784 | 0.580 | 1.188 | 0.198 | 0.487 | 0.840 | 0.933 | 0.966 |
| 0.65 | 0.753 | 0.624 | 1.136 | 0.223 | 0.510 | 0.816 | 0.922 | 0.960 |
| 0.70 | 0.721 | 0.668 | 1.094 | 0.247 | 0.529 | 0.792 | 0.911 | 0.954 |
| 0.75 | 0.689 | 0.711 | 1.062 | 0.271 | 0.545 | 0.766 | 0.899 | 0.948 |
| 0.80 | 0.656 | 0.753 | 1.038 | 0.294 | 0.557 | 0.740 | 0.887 | 0.942 |
| 0.85 | 0.624 | 0.795 | 1.021 | 0.315 | 0.567 | 0.714 | 0.874 | 0.935 |
| 0.90 | 0.591 | 0.835 | 1.009 | 0.335 | 0.574 | 0.687 | 0.861 | 0.928 |
| 0.95 | 0.559 | 0.874 | 1.002 | 0.353 | 0.577 | 0.660 | 0.847 | 0.920 |
| 1.00 | 0.528 | 0.913 | 1.000 | 0.370 | 0.579 | 0.634 | 0.833 | 0.913 |
| 1.05 | 0.498 | 0.950 | 1.002 | 0.384 | 0.578 | 0.608 | 0.819 | 0.905 |
| 1.10 | 0.468 | 0.987 | 1.008 | 0.397 | 0.574 | 0.582 | 0.805 | 0.897 |
| 1.15 | 0.440 | 1.023 | 1.018 | 0.407 | 0.569 | 0.556 | 0.791 | 0.889 |
| 1.20 | 0.412 | 1.057 | 1.030 | 0.416 | 0.562 | 0.531 | 0.776 | 0.881 |
| 1.25 | 0.386 | 1.091 | 1.047 | 0.422 | 0.553 | 0.507 | 0.762 | 0.873 |
| 1.30 | 0.361 | 1.124 | 1.066 | 0.427 | 0.543 | •0.483 | 0.747 | 0.865 |
| 1.35 | 0.337 | 1.156 | 1.089 | 0.430 | 0.531 | 0.460 | 0.733 | 0.856 |
| 1.40 | 0.314 | 1.187 | 1.115 | 0.431 | 0.519 | 0.437 | 0.718 | 0.848 |
| 1.45 | 0.293 | 1.217 | 1.144 | 0.431 | 0.506 | 0.416 | 0.704 | 0.839 |
| 1.50 | 0.272 | 1.246 | 1.176 | 0.429 | 0.492 | 0.395 | 0.690 | 0.830 |
| 1.55 | 0.253 | 1.274 | 1.212 | 0.426 | 0.478 | 0.375 | 0.675 | 0.822 |
| 1.60 | 0.235 | 1.301 | 1.250 | 0.422 | 0.463 | 0.356 | 0.661 | 0.813 |
| 1.65 | 0.218 | 1.328 | 1.292 | 0.416 | 0.443 | 0.337 | 0.647 | 0.805 |
| 1.70 | 0.203 | 1.353 | 1.338 | 0.410 | 0.433 | 0.320 | 0.634 | 0.796 |
| 1.75 | 0.188 | 1.378 | 1.387 | 0.403 | 0.417 | 0.303 | 0.620 | 0.788 |
| 1.80 | 0.174 | 1.402 | 1.439 | 0.395 | 0.402 | 0.287 | 0.607 | 0.779 |
| 1.85 | 1.161 | 1.425 | 1.495 | 0.386 | 0.387 | 0.272 | 0.594 | 0.770 |
| 1.90 | 0.149 | 1.448 | 1.555 | 0.377 | 0.372 | 0.257 | 0.581 | 0.762 |
| 1.95 | 0.138 | 1.470 | 1.619 | 0.368 | 0.357 | 0.243 | 0.568 | 0.754 |
| 2.00 | 0.128 | 1.491 | 1.688 | 0.358 | 0.343 | 0.230 | 0.556 | 0.745 |
| 2.50 | 0.059 | 1.667 | 2.637 | 0.256 | 0.219 | 0.132 | 0.444 | 0.667 |
| 3.00 | 0.027 | 1.793 | 4.235 | 0.172 | 0.137 | 0.076 | 0.357 | 0.598 |
| 3.50 | 0.013 | 1.884 | 6.790 | 0.112 | 0.085 | 0.045 | 0.290 | 0.538 |
| 4.00 | 0.007 | 1.952 | 10.72 | 0.074 | 0.054 | 0.028 | 0.238 | 0.488 |
| 4.50 | 0.003 | 2.003 | 16.56 | 0.049 | 0.035 | 0.017 | 0.198 | 0.445 |
| 5.00 | 0.002 | 2.041 | 25.00 | 0.033 | 0.023 | 0.011 | 0.167 | 0.408 |
| 10.00 | 0.00002 | 2.182 | 536.00 | 0.002 | 0.001 | 0.005 | 0.048 | 0.218 |

*From Emmons, "Gas Dynamics Tables for Air," Dover Publications, Inc., 1947.

sound, and stagnation temperatures can become so high that the gas molecules dissociate and become ionized.

**Shock waves** may occur at locally supersonic speeds. At low velocities the fluctuations that occur in the motion of a body (at the start of the motion or during flight) propagate away from the body at essentially the speed of sound. At higher subsonic speeds, fluctuations still propagate at the speed of sound in the fluid away from the body in all directions, but now the waves cannot get so far ahead; therefore, the force coefficients increase with an increase in the Mach number.

When the Mach number of a body exceeds unity, the fluctuations instead of traveling away from the body in all directions are actually left behind by the body. These cases are illustrated in Fig. 31. As the supersonic speed is increased, the fluctuations are left farther behind and the force coefficients again decrease.

Where, for any reason, waves form an envelope, as at M > 1 in Fig. 31, a wave of finite pressure jump may result.

The speed of sound is higher in a higher-temperature region. For a compression wave, disturbances in the com-

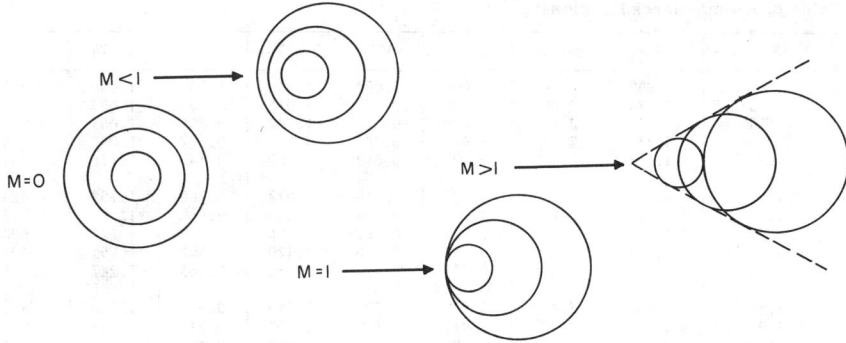

**Fig. 31**  Propagation of sound waves in moving streams.

pressed (hence high-temperature) fluid will propagate faster than and overtake disturbances in the lower-temperature region. In this way shock waves are formed.

For a **stationary normal shock wave**, the fluid velocities $V_1$ before and $V_2$ after the shock are related by

$$V_1 V_2 = (V_s^*)^2 \qquad (4)$$

Other properties are given by

$$p_2 - p_1 = \rho_1 V_1 (V_1 - V_2) \qquad (5)$$

$$\frac{T_2}{T_1} = \left(\frac{V_{s2}}{V_{s1}}\right)^2$$

$$= \left(\frac{2}{\gamma+1}\right)^2 \frac{1}{M_1^2}\left(1 + \frac{\gamma-1}{2}M_1^2\right)\left(\gamma M_1^2 - \frac{\gamma-1}{2}\right) \qquad (6)$$

$$M_2^2 = \left(1 + \frac{\gamma-1}{2}M_1^2\right)\left(\gamma M_1^2 - \frac{\gamma-1}{2}\right)^{-1} \qquad (7)$$

$$\frac{p_{20}}{p_1} = \left(\frac{\gamma+1}{2}\right)^{(\gamma+1)/(\gamma-1)} M^2 \left(\gamma - \frac{\gamma-1}{2M_1^2}\right)^{-1/(\gamma-1)} \qquad (8)$$

$$\frac{p_{20}}{p_{10}} = \left(\frac{\gamma+1}{2}\right)^{(\gamma+1)/(\gamma-1)} M^2$$

$$\times \left(1 + \frac{\gamma-1}{2}M^2\right)^{-\gamma/(\gamma-1)} \left(\gamma - \frac{\gamma-1}{2M^2}\right)^{-1/(\gamma-1)} \qquad (9)$$

The subscript 20 refers to the stagnation condition after the shock, and the subscript 10 refers to the stagnation condition ahead of the shock.

In a normal shock $M$, $V$, and $p_0$ decrease; $p$, $\rho$, $T$, and $s$ increase, whereas $T_0$ remains unchanged. These relations are given numerically in Table 6.

If the shock is moving, these same relations apply *relative to the shock*. In particular, if the shock advances into stationary fluid, it does so at the speed $V_1$ which is always greater than the speed of sound in the stationary fluid by an amount dependent upon the shock strength.

A **shock wave** may be **oblique** to a supersonic stream (see Fig. 32). If so, it behaves exactly like a normal shock to the normal component of the stream. The tangential component is left unchanged. Thus the resultant velocity not only drops abruptly in magnitude but also changes discontinuously in direction. Figure 33 gives the relations between $M_1$, $M_2$, $\theta_w$, and $\delta$.

If, for a given supersonic stream, the velocities following all possible oblique shocks are plotted, a **shock polar** is obtained, as in Fig. 32. For any given stream deflection there are two possible shock angles. The **supersonic flow past a wedge** can, in

principle, occur with either the strong shock B or the weak shock A attached to the leading edge; in practice only the weak shock occurs. The exact flow about the wedge may be computed by Eqs. (4) to (9) or more easily with the help of Fig. 33.

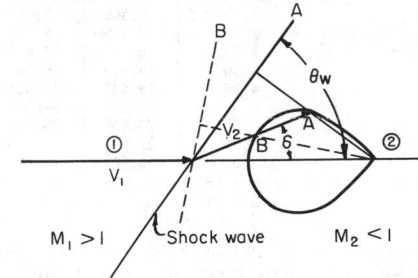

**Fig. 32**  Shock polar.

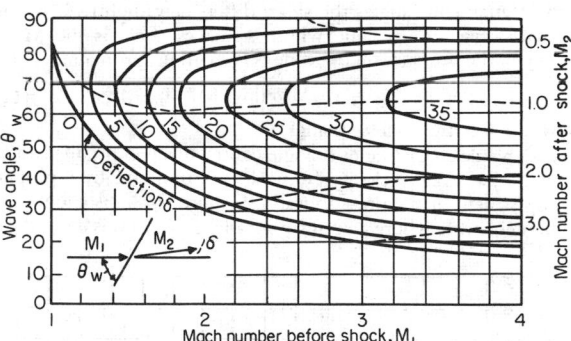

**Fig. 33**  Oblique-shock wave relations.

For small wedge angles, the shock angle differs only slightly from $\sin^{-1}(1/M)$, the **Mach angle**, and the velocity component normal to this **Mach wave** is the speed of sound; the pressure jump is small and is given approximately by

$$p - p_1 = (\gamma p_1 \overline{M}_1^2 / \sqrt{\overline{M}_1^2 - 1})\delta \qquad (10)$$

where $\delta$ is the wedge semiangle.

Exact solutions also exist for **supersonic flow past a cone**. Above a certain supersonic Mach number, a conical shock

#### Table 6. Normal Shock Relations*

| M | $p_2/p_1$ | $p_{20}/p_1$ | $p_{20}/p_{10}$ | $M_2$ | $V_{a2}/V_{a1}$ | $V_2/V_1$ | $T_2/T_1$ | $\rho_2/\rho_1$ |
|---|---|---|---|---|---|---|---|---|
| 1.00 | 1.000 | 1.893 | 1.000 | 1.000 | 1.000 | 1.000 | 1.000 | 1.000 |
| 1.05 | 1.120 | 2.008 | 1.000 | 0.953 | 1.016 | 0.923 | 1.033 | 1.084 |
| 1.10 | 1.245 | 2.133 | 0.999 | 0.912 | 1.032 | 0.855 | 1.065 | 1.169 |
| 1.15 | 1.376 | 2.266 | 0.997 | 0.875 | 1.047 | 0.797 | 1.097 | 1.255 |
| 1.20 | 1.513 | 2.408 | 0.993 | 0.842 | 1.062 | 0.745 | 1.128 | 1.342 |
| 1.25 | 1.656 | 2.557 | 0.987 | 0.813 | 1.077 | 0.700 | 1.159 | 1.429 |
| 1.30 | 1.805 | 2.714 | 0.979 | 0.786 | 1.091 | 0.660 | 1.191 | 1.516 |
| 1.35 | 1.960 | 2.878 | 0.970 | 0.762 | 1.106 | 0.624 | 1.223 | 1.603 |
| 1.40 | 2.120 | 3.049 | 0.958 | 0.740 | 1.120 | 0.592 | 1.255 | 1.690 |
| 1.45 | 2.286 | 3.228 | 0.945 | 0.720 | 1.135 | 0.563 | 1.287 | 1.776 |
| 1.50 | 2.458 | 3.413 | 0.930 | 0.701 | 1.149 | 0.537 | 1.320 | 1.862 |
| 1.55 | 2.636 | 3.607 | 0.913 | 0.684 | 1.164 | 0.514 | 1.354 | 1.947 |
| 1.60 | 2.820 | 3.805 | 0.895 | 0.668 | 1.178 | 0.492 | 1.388 | 2.032 |
| 1.65 | 3.010 | 4.011 | 0.876 | 0.654 | 1.193 | 0.473 | 1.423 | 2.115 |
| 1.70 | 3.205 | 4.224 | 0.856 | 0.641 | 1.208 | 0.455 | 1.458 | 2.198 |
| 1.75 | 3.406 | 4.443 | 0.835 | 0.628 | 1.223 | 0.439 | 1.495 | 2.279 |
| 1.80 | 3.613 | 4.670 | 0.813 | 0.617 | 1.238 | 0.424 | 1.532 | 2.359 |
| 1.85 | 3.826 | 4.902 | 0.790 | 0.606 | 1.253 | 0.410 | 1.573 | 2.438 |
| 1.90 | 4.045 | 5.142 | 0.767 | 0.596 | 1.268 | 0.398 | 1.608 | 2.516 |
| 1.95 | 4.270 | 5.389 | 0.744 | 0.586 | 1.284 | 0.386 | 1.647 | 2.592 |
| 2.00 | 4.500 | 5.640 | 0.721 | 0.577 | 1.299 | 0.375 | 1.688 | 2.667 |
| 2.50 | 7.125 | 8.526 | 0.499 | 0.513 | 1.462 | 0.300 | 2.138 | 3.333 |
| 3.00 | 10.333 | 12.061 | 0.328 | 0.475 | 1.637 | 0.259 | 2.679 | 3.857 |
| 3.50 | 14.125 | 16.242 | 0.213 | 0.451 | 1.821 | 0.235 | 3.315 | 4.261 |
| 4.00 | 18.500 | 21.068 | 0.139 | 0.435 | 2.012 | 0.219 | 4.047 | 4.571 |
| 4.50 | 23.458 | 26.539 | 0.092 | 0.424 | 2.208 | 0.208 | 4.875 | 4.812 |
| 5.00 | 29.000 | 32.653 | 0.062 | 0.415 | 2.408 | 0.200 | 5.800 | 5.000 |
| 10.00 | 116.500 | 129.220 | 0.003 | 0.388 | 4.515 | 0.175 | 20.388 | 5.714 |

*From Emmons, "Gas Dynamics Tables for Air," Dover Publications, Inc., 1947.

wave is attached to the apex of the cone. Figures 34 and 35 show these exact relations; they are so accurate that they are often used to determine the Mach number of a stream by measuring the shock-wave angle on a cone of known angle. For small cone angles the shock differs only slightly from the **Mach cone,** i.e., a cone whose semiapex angle is the Mach angle; the pressure on the cone is then given approximately by

$$p - p_1 = \gamma p_1 \overline{M_1^2} \delta^2 (\ln 2/\delta \sqrt{M_1^2 - 1}) \qquad \textbf{(11)}$$

where $\delta$ is the cone semiangle.

A **nozzle** consisting of a single contraction will produce at its exit a jet of any velocity from $M = 0$ to $M = 1$ by a proper adjustment of the pressure ratio. For use as a subsonic wind-tunnel nozzle where a uniform parallel gas stream is desired, it

is only necessary to connect the supply section to the parallel-walled or open-jet, test section by a smooth gently curving wall. If the radius of curvature of the wall is nowhere less than

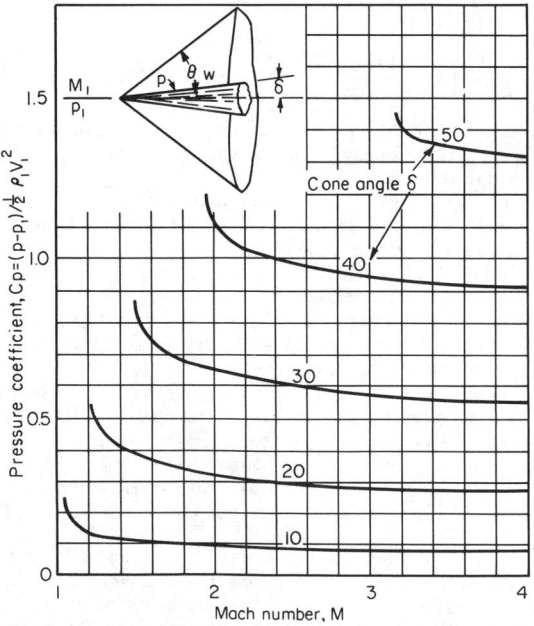

Fig. 35 Pressure coefficients for supersonic flow around cones.

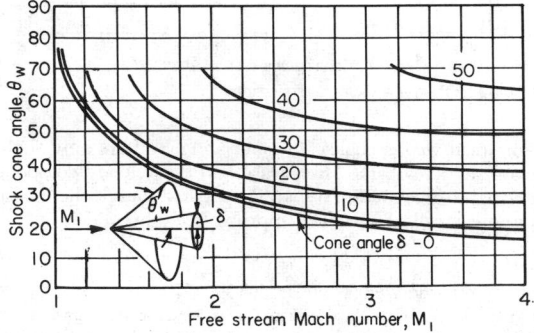

**Fig. 34** Wave angles for supersonic flow around cones.

the largest test-section cross-sectional dimension, no flow separation will occur and a good test gas stream will result.

When a **converging nozzle** connects two chambers with the pressure drop beyond the critical $[(p/p_0) < (p^*/p_0)]$, the Mach number at the exit of the nozzle will be 1; the pressure ratio from the supply section to the nozzle exit will be critical; and all additional expansion will take place outside the nozzle.

A nozzle designed to supply a supersonic jet at its exit must converge to a minimum section and diverge again. The area ratio from the minimum section to the exit is given in the column headed $A/A^*$ in Table 5. A converging-diverging nozzle with a pressure ratio $p/p_0$ = (gas pressure)/(stagnation pressure), Table 5, gradually falling from unity to zero will produce shock-free flow for all exit Mach numbers from zero to the subsonic Mach number corresponding to its area ratio. From this Mach number to the supersonic Mach number corresponding to the given area ratio, there will be shock waves in the nozzle. For all smaller pressure ratios, the Mach number at the nozzle exit will not change, but additional expansion to higher velocities will occur outside of the nozzle.

To obtain a uniform parallel shock-free supersonic stream, the converging section of the nozzle can be designed as for a simple converging nozzle. The diverging or supersonic portion must be designed to produce and then cancel the expansion waves. A series of designs are given in Table 7. These nozzles would perform as designed if it were not for the growth of the boundary layer. Experience indicates that these nozzles give a good first approximation to a uniform parallel supersonic stream but at a somewhat lower Mach number.

For **rocket nozzles** and other thrust devices the gain in thrust obtained by making the jet uniform and parallel at complete expansion must be balanced against the loss of thrust caused by the friction on the wall of the greater length of nozzle required. A simple conical diverging section, cut off experi-

mentally for maximum thrust, is generally used (see also Sec. 11, Jet Propulsion and Aircraft Propellers).

For transonic and supersonic flow, **diffusers** are used for the recovery of kinetic energy. They follow the test sections of supersonic wind tunnels and are used as inlets on high-speed planes and missiles for ram recovery. For the first use, the diffuser is fed a nonuniform stream from the test section (the nonuniformities depending on the particular body under test) and should yield the maximum possible pressure-rise ratio. In missile use the inlet diffuser is fed by a uniform (but perhaps slightly yawed) air stream. The maximum possible pressure-rise ratio is important but must provide a sufficiently uniform flow at the exit to assure good performance of the compressor or combustion chamber that follows.

In simplest form, a **subsonic diffuser** is a diverging channel, a nozzle in reverse. Since boundary layers grow rapidly with a pressure rise, subsonic diffusers must be diverged slowly, 6 to 8° equivalent cone angle, i.e., the apex angle of a cone with the same length and area ratio. Similarly, a **supersonic diffuser** in its simplest form is a supersonic nozzle in reverse. Both the convergent and divergent portions must change cross section gradually. In principle it is possible to design a shock-free diffuser. In practice shock-free flow is not attained, and the design is based upon minimizing the shock losses. Oblique shocks should be produced at the inlet and reflected a sufficient number of times to get compression nearly to M = 1. A short parallel section and a divergent section can now be added with the expectation that a weak normal shock will be formed near the throat of the diffuser.

An efficient diffuser for a supersonic inlet is illustrated in Fig. 36. The central body has stepped cones, each one of which produces an oblique conical shock wave. After two or three such weak shock compressions, the air flows at about M = 1 into an annular opening and is further compressed by an

**Table 7. Typical Nozzle Ordinates***

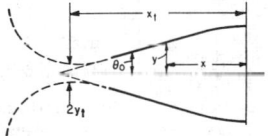

| M | 1.99 | | 2.42 | | 2.82 | | 3.24 | | 3.62 | | 4.04 | |
|---|------|------|------|------|------|------|------|------|------|------|------|------|
| $\theta_0$ | 7° | | 9° | | 12° | | 13° | | 14° | | 15° | |
| | $x$ | $y$ | $x$ | $y$ | $x$ | $y$ | $x$ | $y$ | $x$ | $y$ | $x$ | $y$ |
| | 0 | 7.50 | 0 | 7.50 | 0 | 7.50 | 0 | 7.50 | 0 | 7.50 | 0 | 7.50 |
| | 4.38 | 7.42 | 4.52 | 7.42 | 5.62 | 7.40 | 4.94 | 7.41 | 5.52 | 7.40 | 5.94 | 7.40 |
| | 8.36 | 7.28 | 8.69 | 7.28 | 9.70 | 7.26 | 9.43 | 7.26 | 10.39 | 7.23 | 11.20 | 7.21 |
| | 11.97 | 7.09 | 12.47 | 7.08 | 13.38 | 7.07 | 13.47 | 7.05 | 14.89 | 7.00 | 16.10 | 6.96 |
| | 15.19 | 6.88 | 15.91 | 6.84 | 16.64 | 6.84 | 17.15 | 6.79 | 18.75 | 6.73 | 20.22 | 6.67 |
| | 18.06 | 6.62 | 19.04 | 6.56 | 19.67 | 6.57 | 20 41 | 6.50 | 22.24 | 6.42 | 23.94 | 6.34 |
| | 20.65 | 6.35 | 21.80 | 6.27 | 22.35 | 6.29 | 23 41 | 6.19 | 25.39 | 6.09 | 27.17 | 6.00 |
| | 21.84 | 6.20 | 24.34 | 5.96 | 24.80 | 5.99 | 26.03 | 5.87 | 28.08 | 5.76 | 30.11 | 5.64 |
| | | | 26.57 | 5.65 | 26.95 | 5.69 | 28.42 | 5.53 | 30.56 | 5.41 | 32.69 | 5.28 |
| | $x_t$ | 4.48 | 27.48 | 5.49 | 28.90 | 5.37 | 30.46 | 5.21 | 32.73 | 5.07 | 34.95 | 4.92 |
| | | | | | 30.70 | 5.07 | 32.35 | 4.87 | 34.66 | 4.73 | 36.96 | 4.57 |
| | | | $x_t$ | 3.06 | 32.20 | 4.77 | 33.99 | 4.56 | 36.35 | 4.40 | 38.69 | 4.23 |
| | | | | | 32.90 | 4.62 | 35.44 | 4.25 | 37.87 | 4.08 | 40.26 | 3.90 |
| | | | | | | | 36.13 | 4.09 | 39.20 | 3.77 | 41.62 | 3.58 |
| | | | | | $x_t$ | 2.10 | | | 39.80 | 3.62 | 42.79 | 3.29 |
| | | | | | | | $x_t$ | 1.41 | | | 43.32 | 3.15 |
| | | | | | | | | | $x_t$ | 0.988 | | |
| | | | | | | | | | | | $x_t$ | 0.673 |

*From Puckett, Supersonic Nozzle Design, *Jour. Applied Mechanics*, 13, no. 4, 1948.

internal normal shock and by subsonic diffusion. *NACA-TM 1140* describes this diffuser. A ratio of pressure after diffusion to the total pressure in the atmosphere of as high as 0.6 is obtained with such diffusers at a Mach number of 3. The indications are that higher efficiencies are obtainable by careful design.

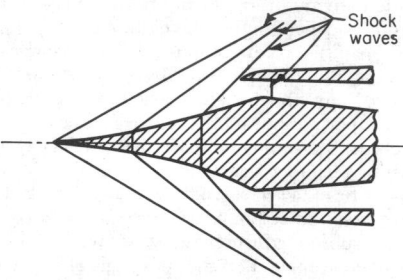

**Fig. 36** Oblique-shock diffuser for ram recovery on a supersonic-plane air intake (Oswatitsch diffuser).

For a **supersonic wind tunnel,** the best way to attain the maximum pressure recovery at a wide range of operating conditions is to make the diffuser throat variable. The ratio of diffuser-exit static pressure to the diffuser-inlet (test-section outlet) total pressure is given in Table 8. These pressure recoveries are attained by the proper adjustment of the throat section of a variable diffuser on a supersonic wind-tunnel nozzle.

A **supersonic wind tunnel** consists of a compressor or compressor system including precoolers or aftercoolers, a supply sec-

**Table 8**

| M | 1 | 1.5 | 2 | 3 | 4 |
|---|---|-----|---|---|---|
| $p_e/p_0$ | 0.83 | 0.69 | 0.50 | 0.23 | 0.10 |

tion, a supersonic nozzle, a test section with balance and other measuring equipment, a diffuser, and sufficient ducting to connect the parts. The **minimum pressure ratio** required from supply section to diffuser exit is given in Table 8. Any pressure ratio greater than this is satisfactory. The extra pressure ratio is automatically wasted by additional shock waves that appear in the diffuser. The compression ratio required of the compressor system must be greater than that of Table 8 by at least an amount sufficient to take care of the pressure drop in the ducting and valves. The latter losses are estimated by the usual hydraulic formulas.

After selecting a compressor system capable of supplying the required maximum pressure ratio, the test section area is computed from

$$A = 1.73 \frac{Q}{V_{s0}} \frac{A}{A^*} \frac{p_e}{p_0} \qquad \text{ft}^2 \qquad (12)$$

where $Q$ is the inlet volume capacity of the compressors, ft³/s; $V_{s0}$ is the speed of sound in the supply section ft/s; $A/A^*$ is the area ratio given in Table 5 as a function of M; $p_e/p_0$ is the pressure ratio given in Table 8 as a function of M.

The nozzle itself is designed for uniform parallel air flow in the test section. Such designs are given in Table 7, which covers only the part of the nozzle between the exit section and the maximum expansion angle. Since for each Mach number a different nozzle is required, the nozzle must be flexible or the

tunnel so arranged that fixed nozzles can be readily interchanged.

The Mach number of a test is set by the nozzle selection, and the Reynolds number is set by the inlet conditions and size of model. The Reynolds number is computed from

$$\text{Re} = \text{Re}_0 D \, (p_0/14.7) \, (540/T_0)^{1.268} \qquad (13)$$

where $\text{Re}_0$ is the Reynolds number per inch of model size for atmospheric temperature and pressure, as given by Fig. 37, $D$ is model diam, in; $p_0$ is the stagnation pressure, lb/in²; $T_0$ is the stagnation temperature, °R.

For a closed-circuit tunnel, the Reynolds number can be varied independently of the Mach number by adjusting the mass of air in the system, thus changing $p_0$.

**Intermittent-wind tunnels** for testing at high speeds do not require the large and expensive compressors associated with continuous-flow tunnels. They use either a large vacuum tank or a large pressure tank (often in the form of a sphere) to produce a pressure differential across the test section. Such tunnels may have steady flow for only a few seconds, but by careful instrumentation, sufficient data may be obtained in this time. A **shock tube** may be used as an intermittent-wind tunnel as well as to study shock waves and their interactions; it is essentially a long tube of constant or varying cross section

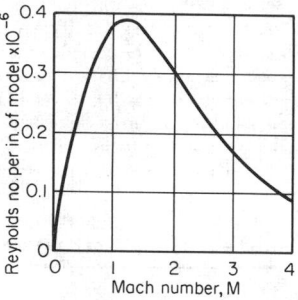

**Fig. 37** Reynolds number–Mach number relation (for a fixed model size and stagnation condition).

separated into two parts by a frangible diaphragm. High pressure exists on one side; by rupturing the diaphragm, a shock wave moves into the gas with the lower pressure. After the shock wave a region of steady flow exists for a few milliseconds. Very high stagnation temperatures can be created in a shock tube, which is not the case in a wind tunnel, so that it is useful for studying hypersonic flow phenomena.

Wind-tunnel force measurements are subject to errors caused by the model support strut. Wall interference is small at high Mach numbers for which the reflected model head wave returns well behind the model. Near $\bar{\text{M}} = 1$, the wall interference becomes very large. In fact, the tunnel chokes at Mach numbers given in Table 5, at

$$\frac{A}{A^*} = \cfrac{1}{1 - \cfrac{\text{area of model projected on test-section cross section}}{\text{area of test-section cross section}}}$$

There are two choking points: one subsonic and one supersonic. Between these two Mach numbers, it is impossible to test in the tunnel. As these Mach numbers are approached, the tunnel wall interference becomes very large.

For tests in this range of Mach number, specially constructed **transonic wind tunnels** with perforated or slotted walls have been built. The object here is to produce a mean flow velocity through the walls that comes close to that which would have existed there had the body been moving at that speed in the open. Often auxiliary blowers are needed to produce the necessary suction on the walls and to reinject this air into the tunnel circuit in or after the diffuser section.

Drag is difficult to predict precisely from wind-tunnel measurements, especially in the transonic regime. **Flight tests** of rocket-boosted models which coast through the range of Mach numbers of interest are used to obtain better drag estimates; from telemetered data and radar or optical sighting the deceleration can be determined, which in turn yields the drag.

## ACOUSTIC THEORY

As the speed of a body is increased from a low subsonic value (Fig. 38a), the local Mach number becomes unity somewhere in

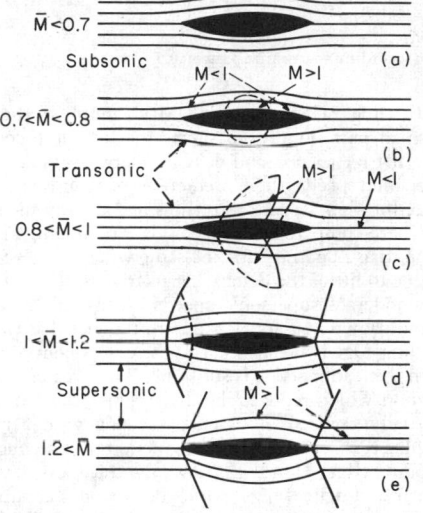

**Fig. 38** Regions of flow about an airfoil (Mach numbers are approximate).

the fluid along the surface of the body, i.e., the lower critical Mach number is reached. Above this there is a small range of transonic speed in which a supersonic region exists (Fig. 38b). Shock waves appear in this region attached to the sides of the body (Fig. 38c) and grow with increasing speed. At still higher transonic speeds a detached shock wave appears ahead of the

body, and the earlier side shocks either disappear or move to the rear (Fig. 38d). Finally, for a sharp-nosed body, the head wave moves back and becomes attached (Fig. 38e). The flow is now generally supersonic everywhere, and the transonic regime is replaced by the supersonic regime. With the appearance of shock waves there occurs a considerable alteration of the pressure distribution, and the center of pressure on airfoil sections moves from the one-fourth chord point back toward the one-half chord point. There is an associated increase of drag and, often, flow separation at the base of the shock.

The redistribution of pressures and the motion of shock waves over the wing surfaces through the transonic regime demands special consideration in the design of control surfaces so that they do not become ineffective by separation or inoperative by excessive loading.

If a body is slender (i.e., planes tangent to its surface at any point make small angles with the flight direction), the disturbance velocities caused by this body will be small compared with the flight speed and, excluding the hypersonic regime, small compared with the speed of sound. This permits the use of **acoustic theory** (also called linearized small-disturbance theory) to predict the approximate flow past the body. This theory relates the steady flow past the body at subsonic speeds to the incompressible flow past a distorted version of this body (the "generalized Prandtl-Glauert rule"). To find the velocity components in the **subsonic** flow about any slender body, first determine the velocity components, $u$, $v$, $w$ (in the $x$, $y$, $z$ directions, respectively), in the incompressible flow, at the same stream speed, about a stretched shape whose streamwise ($x$-axis direction) dimensions are $1/\beta$ times as great ($\beta = \sqrt{1 - \overline{M}^2}$). The desired velocity components are then $\beta^{-2}u$, $\beta^{-1}v$, $\beta^{-1}w$ at corresponding points of the stretched and unstretched bodies. For **thin airfoil sections** this theory predicts

$$C_L = 2\pi\alpha/\sqrt{1 - \overline{M}^2} \qquad \alpha \text{ in radians} \qquad (15)$$

where $L = C_L qS$ and $q = (\gamma/2)p_0\overline{M}^2$. For finite-span wings and for bodies, no such simple relations exist. However, an approximation to the overall **lift coefficient** for thin flat wings of rectangular platform is (see Fig. 40):

$$C_L = 2\pi A\alpha/[2 + \sqrt{4 + (\beta A)^2}] \qquad (16)$$

and the drag is made up of the induced drag due to lift and the skin-friction drag (see above, Airfoils).

$$C_D = (C_L/\pi A)(1 + 0.01\beta A) + C_{D_F} \qquad (17)$$

At supersonic speeds, the Prandtl-Glauert rule is also applicable if we replace $\beta$ by $\lambda = \sqrt{\overline{M}^2 - 1}$ and relate flow to flow past a stretched (or compressed) body at $\overline{M} = \sqrt{2}$, where $\lambda = 1$. For **thin airfoil sections** (see Fig. 39) this theory predicts

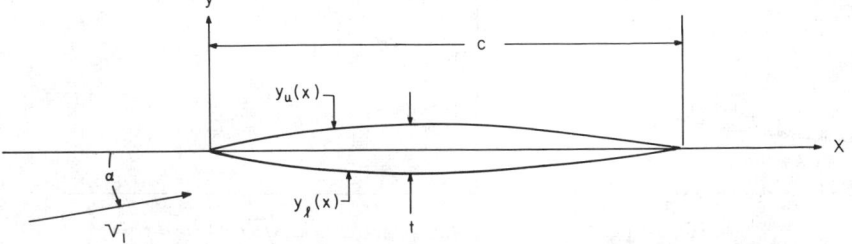

**Fig. 39** Supersonic airfoil section.

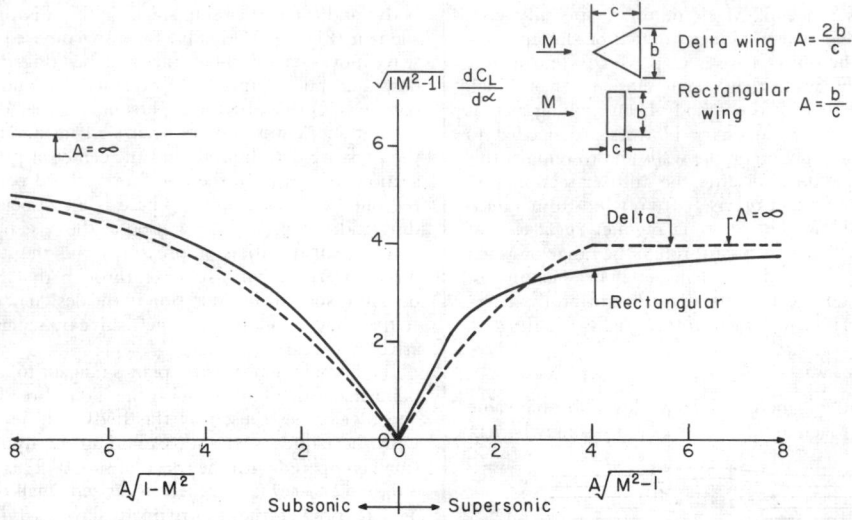

**Fig 40**   Lift-coefficient curve slope for rectangular and delta wings (according to linearized theory).

$$C_L = 4\alpha/\sqrt{M^2 - 1} \qquad \alpha \text{ in radians} \qquad (18)$$

$$C_D = \frac{2}{\lambda} \int_0^1 \left[ \left( \frac{dy_u}{dx} \right)^2 + \left( \frac{dy_l}{dx} \right)^2 \right] d\left( \frac{x}{c} \right) + \frac{4\alpha^2}{\lambda} + C_{D_F} \qquad (19)$$

where $dy_u/dx$ and $dy_l/dx$ are the slopes of the upper and lower surfaces of the airfoil, respectively, and $C_{D_F}$ is again the skin-friction drag coefficient. Note that there is a drag due to thickness and a drag due to lift at supersonic speeds for an airfoil section where there is none at subsonic speeds. For symmetric double-wedge airfoil sections, the thickness drag coefficient becomes $(4/\lambda)(t/c)^2$, and for symmetric biconvex airfoil sections it is $(16/3\lambda)(t/c)^2$, where $t$ is the maximum thickness.

Within the acoustic approximation, a disturbance at a point in supersonic flow can affect only the points in the downstream **Mach cone,** i.e., a conical region with apex at the point, axis parallel to the stream direction, and semicone angle equal to the Mach angle (see Fig. 31). Thus a rectangular wing with constant airfoil section has two-dimensional flow on all parts of the wing except the points within the tip Mach cones. At Mach numbers near unity these tip Mach cones cover nearly the entire wing, whereas at high Mach numbers they cover only a small part. Figure 40 shows the lift-curve coefficient slope for flat **rectangular** and **delta plan-form wings** at subsonic and supersonic speeds; these predictions of acoustic theory are inaccurate for high-aspect-ratio wings in the transonic regime, but elsewhere compare favorably with experiment. Figure 41 shows the drag due to lift for the same wings. The reduction in drag due to lift of the delta wing relative to the rectangular wing at moderate supersonic speeds, predicted by acoustic theory and shown in Fig. 41 is due to the fact that the delta-wing leading edge is **swept back,** and the component of velocity normal to the leading edge is subsonic. This creates a leading-edge suction which reduces the drag. This drag reduction is only partially realized in practice; a wing with a rounded leading edge realizes more of this reduction than a wing with a sharp leading edge. Figure 42 shows the thickness drag of the rectangular and delta wings, which occurs only at supersonic speeds. Note again that the leading-edge sweepback of the delta wing helps to reduce this drag at moderate supersonic speeds.

The lift of a slender **axially symmetric body** at small angle of

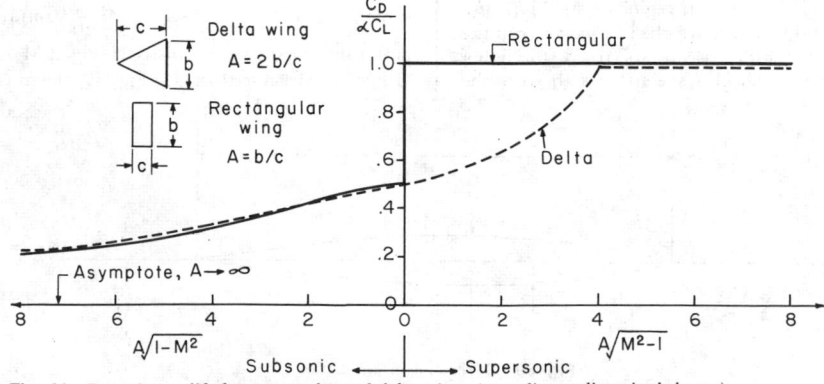

**Fig. 41**   Drag due to lift for rectangular and delta wings (according to linearized theory).

attack is nearly independent of Mach number and is given approximately by

$$C_L = 2\alpha \qquad (20)$$

where the lift coefficient is based on the cross-sectional area of the base. The drag for $\alpha = 0$ at subsonic speeds is made up of skin friction and **base drag**; a dead-air region exists just behind the blunt base, and the pressure here is below ambient, causing a rearward suction which is the base drag. At supersonic speeds a wave drag is added, which represents the energy dissipated in shock waves from the nose. Figure 43 shows a typical drag-coefficient curve for a body of revolution where the **fineness ratio** (ratio of length to diameter) is 12.2. The skin-friction drag coefficient based on wetted area is the same as that of a flat plate, within experimental error.

At transonic and moderate supersonic speeds, the wave drag of a **wing body combination** can be effectively reduced by making the cross-sectional area distribution (including wings) a smooth curve when plotted versus fuselage station. This is the **area rule**; its application results in a decided indentation in the fuselage contour at the wing juncture. **Lift interference** effects also occur, especially when the body diameter is not small compared with the wing span. If the wing is attached to a cylindrical body, and if the wing-alone lift coefficient would have been $C_L$, then the lift carried on the body is $K_B C_L$, and the lift carried on the wing is $K_W C_L$. Figure 44 shows the variation of $K_B$ and $K_W$ with body diameter to wing-span ratio.

The **effect of compressibility on skin-friction drag** is slight at subsonic speeds, but at supersonic speeds, a significant reduction in skin-friction coefficient occurs. Figure 45 shows the turbulent-boundary-layer mean skin-friction coefficient for a cone as a function of wall temperature and Mach number. An important effect at high speed is **aerodynamic heating.**

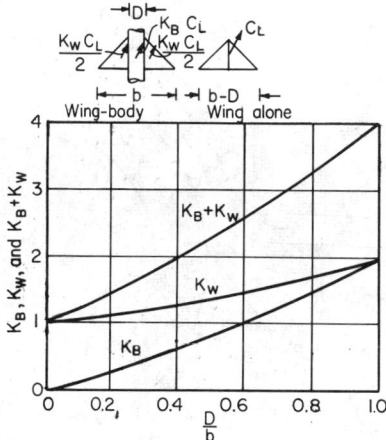

**Fig. 44**  Wing-body lift-interference factors.

Figure 46 illustrates the velocity profile behind the shock wave of a body traveling at hypersonic speed. The shock-wave front represents an area of high-temperature gas which radiates energy to the body, but boundary-layer convective heating is usually the major contributor. Behind this front is shown the velocity gradient in the boundary layer. The decrease in velocity in the boundary layer is brought about by the forces of interaction between fluid particles and the body (viscosity). This change in velocity is accompanied by a change in temperature and is dependent on the characteristics

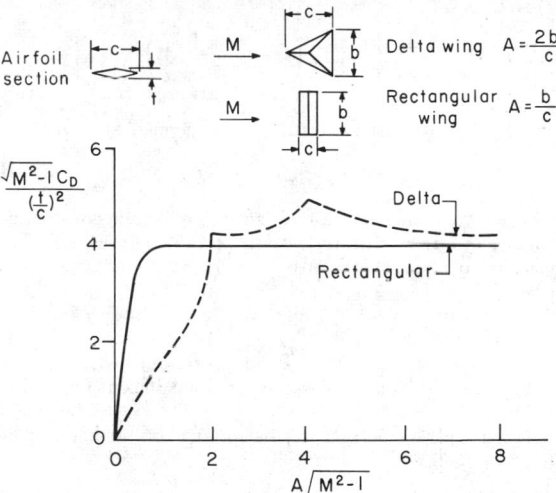

**Fig. 42**  Supersonic thickness drag coefficient for rectangular and delta wings with a symmetrical double-wedge airfoil section (according to linearized theory).

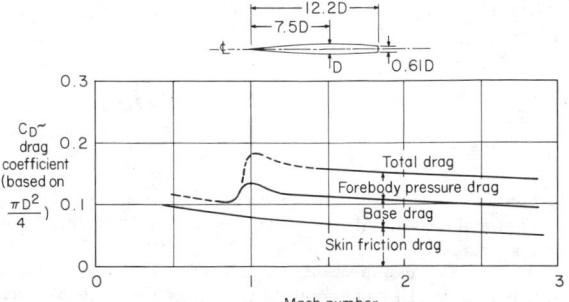

**Fig. 43**  Drag coefficient for parabolic-arc (*NACA RM*-10) body (calculated from experimental data for a 30,000-ft altitude, $D = 12$ in, from *NACR-TR* 1160 and 1161, 1954).

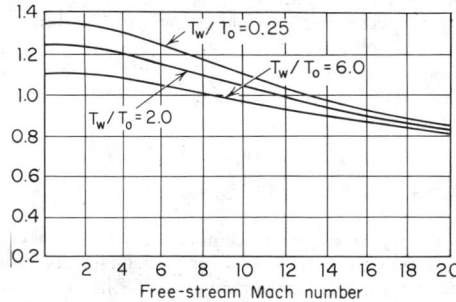

**Fig. 45**  Laminar-skin-friction coefficient for a cone.

of the boundary layer. For example, heat transfer from a turbulent boundary layer may be of an order of magnitude greater than for laminar flow.

If the gas is brought to rest instantaneously, the total energy will rise. The resulting temperature is known as the **stagnation temperature** (see stagnation point, Fig. 46).

$$T_S = T_\infty \left( 1 + \frac{\gamma - 1}{2} M^2 \right) \quad (21)$$

$T_\infty$ is the ambient temperature of the gas at infinity, and $\gamma$ is the specific-heat ratio of the gas. For undissociated air, $\gamma = 1.4$.

In general, this simple, one-dimensional relationship between velocity and temperature does not hold for tempera-

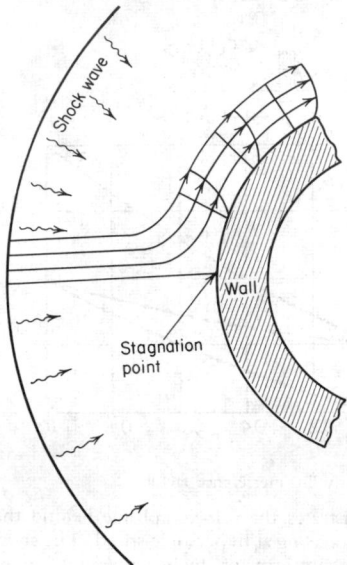

**Fig. 46** Velocity profile behind a shock wave.

ture in the boundary layer. The laminae of the boundary layer are not insulated from each other, and there is cross conduction. This is associated with the **Prandtl number** Pr, which is defined as

$$Pr = C_p \mu / k_r \quad (22)$$

where $k_r$ = thermal conductivity of fluid
$C_p$ = specific heat of fluid at constant pressure
$\mu$ = absolute viscosity of fluid

Defining the **recovery factor** $r$ as the ratio of the rise in the idealistic wall and stagnation temperature over the free-stream temperature,

$$r = (T_{aw} - T_\infty)/(T_s - T_\infty) = (Pr)^n \quad (23)$$

For laminar flow, $r = (Pr)^{1/2}$; and for turbulent flow, $r = (Pr)^{1/3}$.

Prandtl number Pr greatly complicates the thermal computations, but since it varies only over a small range of values, a recovery factor of 0.85 is generally used for laminar flow and 0.90 for turbulent flow.

For a thermally thin wall, the rate of change of the surface temperature is a function of the rate of total heat input and the surface's ability to absorb the heat.

$$dT_w / dt = \dot{q}_T / wcb \quad (24)$$

where $t$ = time
$q_T$ = forced convective heating + radiation heating − heat radiation from the skin
$w$ = density of skin material
$c$ = specific heat of skin material
$b$ = skin thickness
$T_w$ = skin temperature

The heat balance may then be written

$$wcb \left( \frac{dT_w}{dt} \right) = k_c \left[ T_0 \left( 1 + r \frac{\gamma_B - 1}{2} M_0^2 \right) - T_w \right] + \alpha G_s A_p - \epsilon \sigma T_w^4 \quad (25)$$

where $A_p$ = correction factor to account for area normal to radiation source; $\gamma_B$ = specific heat ratio of boundary layer; $\epsilon$ = radiative emissivity of surface; $\sigma$ = Stefan-Boltzmann constant [$17.3 \times 10^{-10}$ Btu/(h) (ft²) (°R⁴)]; $\alpha$ = surface absorptivity; $G_s$, = solar irradiation Btu/(h) (ft²).

For small time increments $\Delta t$,

$$\frac{dT_w}{dt} = \frac{T_{w2} - T_{w1}}{\Delta t}$$

Then

$$T_{w2} = T_{w1} + \frac{\Delta t}{wcb} \left\{ b_c \left[ T_0 \left( 1 + \frac{\gamma_B - 1}{2} M_0^2 \right) - T_w \right] + \alpha G_s A_p - T_w^4 \right\} \quad (26)$$

The local heat-transfer coefficient $b_c$ is defined as

$$b_c = \frac{k_r}{r^{4/3}} C_p C_f \rho_0 Vg \quad (27)$$

where $C_p$ = specific heat of air; $C_f$ = local skin-friction coefficient; $\rho_0$ = density outside of boundary layer; $V$ = velocity outside of boundary layer; $g$ = acceleration of gravity.

For a cone, $k_r = 1,800$ (laminar Re $< 2 \times 10^6$) and $k_r = 1,800 r^{4/3}$ (turbulent).

For a flat-plate transition, Reynolds number is $1 \times 10^6$.

Figure 45 gives laminar-skin-friction coefficient for a cone.

For a flat plate, multiply $C_f$ by $\sqrt{3/2}$. For turbulent skin-friction coefficient for a cone,

$$\frac{0.242}{\sqrt{A^2 C_f T_w / T_0}} (\sin^{-1} \psi + \sin^{-1} \theta)$$
$$= 0.41 \log \frac{Re}{2} C_f - 1.26 \log \frac{T_w}{T_0} \quad (28)$$

where

$$\psi = \frac{2A^2 - B}{\sqrt{B^2 + 4A^2}}, \qquad A^2 = \frac{(\gamma_0 - \frac{1}{2})M_0^2}{T_w/T_0}$$
$$\theta = \frac{B}{\sqrt{B^2 + 4A^2}} \qquad B = \frac{1 + (\gamma_0 - \frac{1}{2})M_0^2}{T_w/T_0} - 1$$

For a flat plate, use Re instead of Re/2 in Eq. (28).

Figure 47 shows data on stagnation and adiabatic wall temperatures.

The primary **measurements in aerodynamics** are pressure measurements. A well-aligned pitot tube with an impact-pressure hole at its nose and static-pressure holes 10 diameters or more back from the nose will accurately measure the **impact pressure** and the **static pressure** of a uniform gas stream. Up to sonic speed the impact pressure is identical with stagnation pressure, but at supersonic speeds, a detached shock wave forms

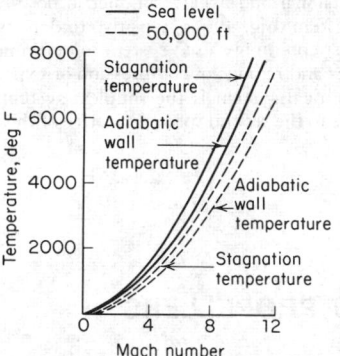

**Fig. 47** Variation of stagnation and adiabatic wall temperature with Mach number.

ahead of the probe, through which there is drop in stagnation pressure; the portion of the shock wave just ahead of the probe is normal, so that Eq. (9) gives the relation of the measured impact pressure to the isentropic stagnation pressure.

**Force measurements** of total lift, drag, side force, pitching, moment, yawing moment, and rolling moment on models are made on wind-tunnel balances just as at low speed. Small internal-strain-gage balances are often used to minimize strut interference.

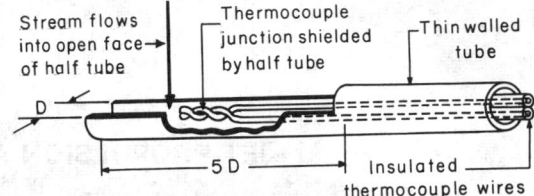

**Fig. 48**  Stagnation-temperature probe (recovery factor = 0.98).

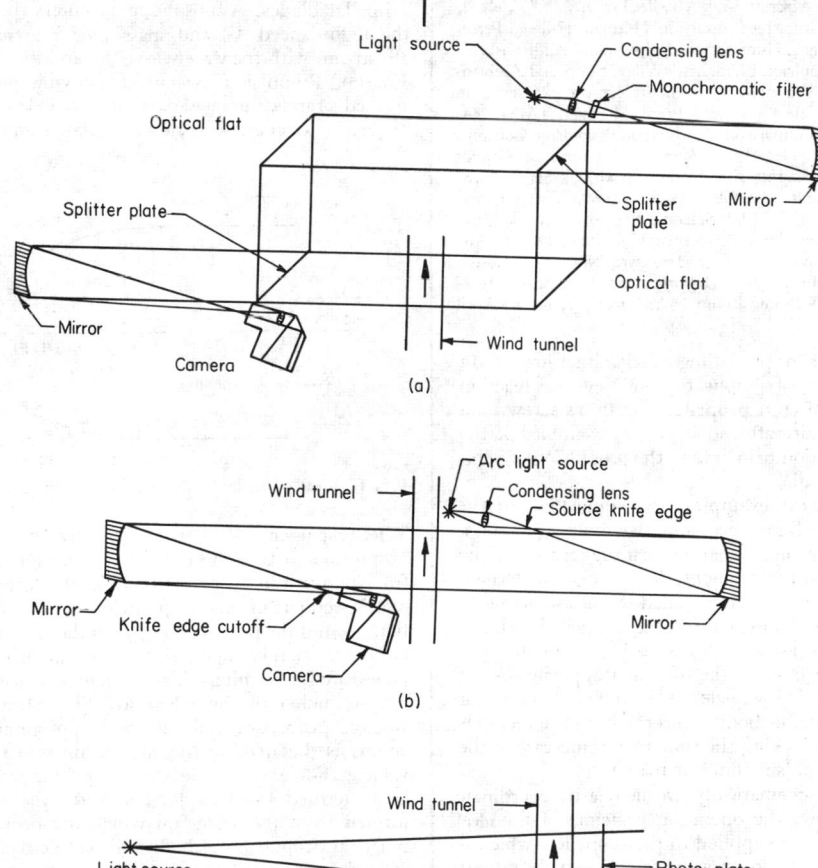

**Fig. 49**  Optical systems for observing high-speed flow phenomena: (*a*) interferometer; (*b*) Schlieren (two-mirror) system; (*c*) shadowgraph.

An open thermocouple is unreliable for determining **stagnation temperature** if it is in a stream of high velocity. Figure 48 shows a simple temperature probe in which the fluid is decelerated adiabatically before its temperature is measured. The recovery factor used in Eq. (9c), for this probe, accurately aligned to the stream, is 0.98.

Optical measurements in high-speed flow depend on the variation of index of refraction with gas density. This variation is given by

$$n = 1 + k\rho \qquad (29)$$

where $k = 0.116$ ft³/slug for air. A Mach-Zehnder **interferometer** (Fig. 49a), is capable of giving accurate density information for two-dimensional and axially symmetric flows. The **Schlieren** optical system (Fig. 49b) is sensitive to density gradients and is the most commonly used system to determine location of shock waves and regions of compression or expansion. The **shadowgraph** optical system is the simplest system (Fig. 49c) and is sensitive to the second space derivative of the density.

# JET PROPULSION AND AIRCRAFT PROPELLERS
### by Maurice J. Zucrow
### in Collaboration with
### Bruce A. Reese

REFERENCES: Zucrow, "Aircraft and Missile Propulsion," Vol. 2, Wiley. Hesse and Mumford, "Jet Propulsion," Pitman. Hill and Peterson, "Mechanics and Thermodynamics of Propulsion," Addison-Wesley. Sutton, "Rocket Propulsion Elements," Wiley. Shorr and Zaehringer, "Solid Propellant Technology," Wiley. Zucrow, "Science in Progress," 14th Series, Yale University Press. Moeckel, TMX-1864, NASA, 1969. Moeckel, Comparison of Advanced Propulsion Concept, *Jour. Spacecraft*, **9**, no. 12, Dec. 1972. Zucrow, "Elements of Aircraft and Missile Propulsion," Engineering Design Handbook Series, U.S. Army Material Command, AMCP 706-285, 1969. Forestor and Kuskevics, "Ion Propulsion," AIAA Selected Reprints. Corliss, "Propulsion Systems for Space Flight," McGraw-Hill, 1960. Dugan, Airbreathing Propulsion Trends, *Astronaut. Aeronaut.*, Nov. 1971. Rosen, Trends in Aircraft Propulsion, *Jour. Aircraft*, May 1972. Sears, Aerodynamics, Noise, and the Sonic Boom, *AIAA Jour.*, **7**, no. 4, April 1969.

All the known methods for propelling a body either in (or on) a fluid medium or in space are applications of Newton's reaction principle. Thus, the aircraft propeller, the ship's screw, and the jet propulsion of aircraft and boats are examples of the application of the reaction principle to the propulsion of vehicles in (or on) fluid media.

In the aforementioned examples, the application of the reaction principle involves increasing the momentum of a flowing mass of fluid in such a manner that the reaction to the time rate of increase in the momentum of that fluid, termed the **propulsive fluid,** creates a force, called the **thrust,** acting in the direction of motion desired for the propelled vehicle. Accordingly, the thrust arises from increasing the momentum flux of the propulsive fluid in the direction opposite to that desired for the propelled vehicle. The known devices for achieving the propulsion of bodies differ only in the methods and mechanisms for achieving the time rate of increase in the momentum of the propulsive fluid or matter.

Figure 1 illustrates schematically, in the relative coordinate system for steady flow, the operating principle of the **ideal aircraft propeller.** Power is supplied to the propeller, which is assumed to be equivalent to an **actuator disk,** which imparts only an axial acceleration to the air flowing through it. The rotation of the actuator disk produces a **slipstream** composed of the entire mass of air flowing in the axial direction through the actuator-disk area, i.e., the area of the circle swept by the propeller blades. Atmospheric air enters the slipstream with the flight speed $V_0$ and mass flow rate $m_a$. It leaves the slipstream with the **wake velocity** $w$, and the thrust is given by Eq. (5). Propulsion systems employing propellers will be termed **propeller propulsion.** (For detailed discussions of the airplane propeller, see below, Aircraft Propellers.)

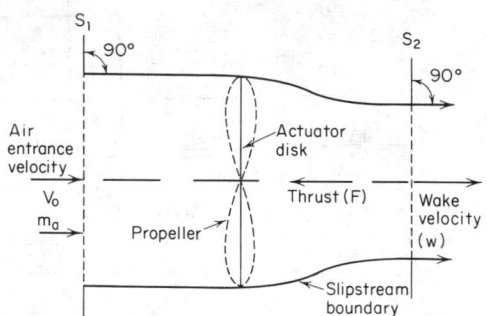

**Fig. 1** Ideal propeller in the relative coordinate system.

**Jet propulsion** differs from propeller and other methods of propulsion in that the propulsive fluid (or matter) is ejected from within the propelled body in the form of one or more high-speed jets of fluid or particles, instead of being caused to flow around the propelled body. In the abstract, at least, there are no restrictions upon either the type of matter, called the **propellant,** for forming a high-velocity exhaust jet or the means for producing the **propulsive jet.** The selection of the most suitable propellant and the most appropriate jet-propulsion engine is dictated by the specific mission for the propelled vehicle. For example, in the case of the jet propulsion of a boat, termed **hydraulic jet propulsion,** the propulsive jet is formed from the water on which the boat moves. For the practical propulsion of bodies through either the atmosphere of earth or in space, however, only two types of propulsive jets are suitable:

1. For propulsion within the atmosphere of earth, there is the jet formed by expanding a highly heated, compressed gas containing atmospheric air as either a major or a sole constitu-

ent. Such an engine is called either an **airbreathing** or a **thermal-jet engine.** If the heating is accomplished by burning a fuel in the air, the engine is a **chemical thermal-jet engine.** If the air is heated by direct or indirect heat exchange with a nuclear-energy source, the engine is termed a **nuclear thermal-jet engine.**

2. For propulsion both within and beyond the atmosphere of earth, there is the exhaust jet containing no atmospheric air. Such an exhaust jet is termed a **rocket jet,** and any matter used for creating the jet is called a **propellant.** A rocket may consist of a stream of gases, solids, liquids, ions, electrons, or a plasma. The assembly of all the equipment required for producing the rocket jet constitutes a **rocket engine.**

Modern airbreathing engines may be segregated into two principal types: (1) **ramjet engines,** and (2) **turbojet engines.** The turbojet engines are of two types: (1) the **simple turbojet engine,** and (2) the **turbofan,** or **bypass, engine.**

Rocket engines can be classified by the form of energy used for achieving the desired jet velocity. The three principal types of rocket engines are (1) **chemical rocket engines,** (2) **nuclear heat-transfer rocket engines,** and (3) **electric rocket engines.**

The propulsive element of a jet-propulsion engine, irrespective of type, is the **exhaust nozzle** or orifice. If the exhaust jet is gaseous, the assembly comprising all the other components of the jet-propulsion engine constitutes a gas generator for supplying highly heated, high-pressure gases to the exhaust nozzle.

These classifications of jet-propulsion engines apply to the basic types of engines. It is possible to have combinations of the different types of thermal-jet engines and also combinations of thermal-jet engines with rocket engines; only the principal types are discussed here.

## ESSENTIAL FEATURES OF AIRBREATHING OR THERMAL-JET ENGINES

In the subsequent discussions a *relative coordinate system* is employed wherein the atmospheric air flows toward the propulsion system with the flight speed $V_0$, and the gases leave the propulsion system with the velocity $w$, *relative to the walls of the propulsion system.* Furthermore, steady-state operating conditions are assumed.

**Ramjet Engine** Figure 2 illustrates schematically the essen-

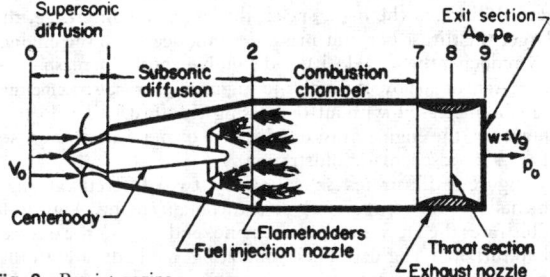

**Fig. 2** Ramjet engine.

tial features of a ramjet engine for propelling a vehicle at supersonic flight speeds. It comprises three major components: a diffusion system, consisting of a supersonic diffuser followed by a subsonic diffuser (0–2); a combustion chamber (2–7); and an exhaust nozzle (7–9). In the simplest ramjet engines, apart from the necessary control devices, the engine

has no moving parts. However, for accelerating the vehicle or for operating at different Mach numbers, a variable geometry diffuser and exhaust may be required.

The operating principle of the ramjet engine is as follows. The free-stream air flowing toward the engine with the supersonic Mach number $M_0$ is decelerated by the supersonic diffuser to approximately unity Mach number at the entrance to the subsonic diffuser; the deceleration is accompanied by the formation of shock waves and by an increase in the pressure of the air (diffusion). In the subsonic diffuser the air is further diffused so that it arrives at the entrance to the combustion chamber with a low Mach number ($M_2 = 0.2$, approximately); the value of $M_2$ is dictated by the combustion requirements. If $P_0$ is the total pressure of the free-stream air having the Mach number $M_0$, and $P_2$ that for the air entering the combustion system with the Mach number $M_2$, then it is desirable that $P_2/P_0$ be as large as possible.

In the combustion chamber a fuel is burned in the air, thereby raising the total temperature of the gases entering the exhaust nozzle (see Sec. 7) to approximately $T_7 = 4260°R$ (2366 K). Most generally a liquid hydrocarbon fuel is used, but experiments have been conducted with solid fuels, liquid hydrogen, liquid methane, and "slurries" of metallic fuels in a liquid fuel. The combustion process is not quite isobaric because of the pressure drops in the combustion chamber, because of the increase in the momentum of the working fluid due to heat addition, and because of friction. The hot gases are discharged to the atmosphere, after expanding in the exhaust nozzle, with the relative velocity $w = V_9$.

Since the ramjet engine can function only if there is a ram pressure rise at the entrance to the combustion chamber, it is not self-operating at zero flight speed. It must, therefore, be accelerated to a flight speed which permits the engine to develop sufficient thrust for accelerating the vehicle it propels to the design flight Mach number. Consequently, a ramjet-propelled missile, for example, must either be launched by dropping it from an airplane or be **boosted** to the required flight speed by means of **launching,** or **booster,** rockets. From our present knowledge it appears that the most appropriate flight regime for the ramjet engine is between $M_0 = 2$ and $M_0 = 5$, approximately; the upper limit is set by the problem of either cooling or protecting the outer skin of the engine body. If the fuel can be employed for cooling the metal parts exposed to high stagnation temperatures, then the Mach number range can be raised. In the past decade it has been demonstrated that it is possible to achieve combustion in a supersonic air stream so that it appears to be feasible to develop a supersonic combustion ramjet engine (**Scramjet**). Its successful development offers the potential of airplane flights in the hypersonic regime, $M_0 = 7$ to 12.

The advantage derived from supersonic combustion of the liquid fuel (usually liquid hydrogen) is that the diffuser of the Scramjet engine is required to decelerate the air entering the engine from $M_0$ to only approximately $M_2 = 0.35M_0$ instead of to $M_2 = 0.2$, which is essential with subsonic combustion. This elimination of the subsonic diffusion increases the diffuser efficiency, reduces the static pressure in the combustor (and, therefore, the engine weight and heat transfer rate), and increases the velocity in the combustor (thereby decreasing engine frontal area). It is the combination of the above effects that makes ramjet propulsion at high flight Mach numbers (> 7) feasible.

**Simple Turbojet Engine** Figure 3 illustrates schematically the principal features of a simple turbojet engine, which is basically a gas-turbine engine equipped with a propulsive nozzle and diffuser. Atmospheric air enters the engine and is partially compressed in the diffusion system, and further

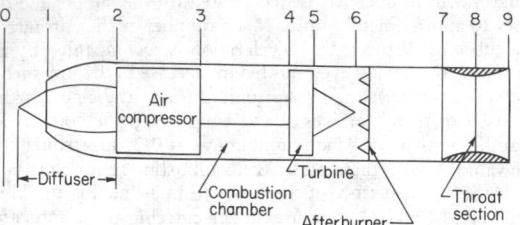

**Fig. 3** Simple turbojet engine.

compressed to a much higher pressure by the air compressor, which may be of either the axial-flow or centrifugal type. The highly compressed air then flows to a combustion chamber wherein sufficient fuel is burned to raise the total temperature of the gases entering the turbine to approximately $T_4 = 2160°R$ (1200 K) for an uncooled turbine. The maximum allowable value for $T_4$ is limited by metallurgical and stress considerations; it is desirable, however, that $T_4$ be as high as possible. The combustion process is approximately isobaric. The highly heated air, containing approximately 25 percent of combustion products, expands in the turbine, which is directly connected to the air compressor, and in so doing furnishes the power for driving the air compressor. From the turbine the gases pass through a tailpipe which may be equipped with an **afterburner.** The gases are expanded in a suitably shaped exhaust nozzle and ejected to the atmosphere in the form of a high-speed jet.

Like the ramjet engine, the turbojet engine is a continuous-flow engine. It has an advantage over the ramjet engine in that its functioning does not depend upon the ram pressure of the entering air, although the amount of ram pressure recovered does affect its overall economy and performance. The turbojet is the only airbreathing jet engine that has been applied in practice as the sole propulsion means for piloted aircraft. It appears to be eminently suited for propelling aircraft at speeds above 500 mi/h (805 km/h). As the design flight speed is increased, the ram pressure increases rapidly, and the characteristics of the turbojet engine tend to change over to those of the ramjet engine. Consequently, its top speed appears to be limited to that flight speed where it becomes more advantageous to employ the ramjet engine. No reliable figures can be given at this time, but estimates indicate that for speeds above approximately 2,000 mi/h (3,219 km/h) it will be more advantageous to use some form of ramjet engine.

The thrust of the simple turbojet engine increases rapidly with $T_4$, because increasing $T_4$ increases the jet velocity $V_j$. Actually, $V_j$ increases faster than the corresponding increase in $T_4$. It is also an inherent characteristic of the gas-turbine engine which produces shaft power, called a **turboshaft** engine, that its useful power increases proportionally faster than a corresponding increase in its turbine inlet temperature $T_4$. Because of the decrease in strength of turbine materials with increase in temperature, the turbine blades, stators, and disks require cooling at $T_4 > 2160°R$ (1200 K) approximately. Research engines are being operated with $T_4 = 3460°R$ (1922

K) and operational engines with $T_4 = 2860°R$ (1533 K). The cooling air is bled from the compressor at the appropriate stage (or stages) and used to cool the stator blades or rotor blades by convective, film, or transpiration heat transfer. Up to 10 percent of the compressor air may be bled for turbine cooling, and this air is "lost" for turbine work for that blade row where it is used for cooling. Consequently "trade" studies must be made to "weigh" the increased complexity of the engine and the turbine work loss due to air bleed against the increased engine performance associated with increased $T_4$.

As in the case of any gas-turbine power plant, the efficiencies of the components of the turbojet engine have an influence on its performance characteristics, but its performance is not nearly as sensitive to changes in the efficiency of its component machines as is a gas turbine which delivers shaft power. (See Sec. 9.)

It has been indicated that two types of compressors are currently employed, the axial-flow compressor and the centrifugal compressor. Irrespective of the type, the objectives are similar. The compressor must be reliable, compact, easy to manufacture, and have a small frontal area. Because of the limited air induction capacity of the centrifugal compressor, also called the radial compressor, engines for developing thrusts above 7,000 lb (31 kN) at static sea level, employ axial-flow compressors (see Sec. 14).

Because of its rather flat thrust versus speed curve, the turbojet engine introduces certain operational problems at takeoff because of the small ratio of takeoff thrust to thrust in flight. Since the exhaust gases from the turbine contain considerable excess air, the jet velocity, and consequently the thrust, can be increased by burning additional fuel in the tailpipe upstream from the exhaust nozzle. By employing **"tailpipe burning,"** or **"afterburning,"** as it is called, the thrust can be increased by 35 percent and at 500 mi/h, in a tactical emergency, by approximately 60 percent. With afterburning, the temperature of the gases entering the nozzle $T_7$ is of the order of 3800°R (2110 K).

**Turbofan, or Bypass, Engine** For a fixed turbine inlet temperature, the jet velocity from a simple turbojet engine propelling an airplane at subsonic speed is relatively constant. The propulsive efficiency depends on the ratio of the flight speed to the jet velocity and increases as the ratio increases. On the other hand, the thrust depends on the difference between the jet velocity and the flight speed; the larger the difference, the larger the thrust per unit mass of air induced into the engine. By reducing the jet velocity and simultaneously increasing the mass rate of airflow through the engine, the **propulsive efficiency** can be increased without decreasing the thrust. To do this, however, the engine must employ two or more airstreams (see Fig. 4 for a schematic illustration).

Figure 4 illustrates schematically two different arrangements for the components of a turbofan engine. Figure 4$a$ illustrates the aft-fan turbofan engine, and Fig. 4$b$ the ducted-fan turbofan engine. Their operating methods are similar basically. There are two turbines, a low-pressure turbine (LPT) and a high-pressure turbine (HPT); one drives the air compressor of the hot-gas generator and the other drives the fan. Air enters the fan at the rate of $\dot{m}_{aF}$ and is ejected through the nozzle of area $A_{7F}$ with the jet velocity $V_{jF} < V_j$, where $V_j$ is the jet velocity attained by the air flowing through the hot-gas generator with the mass flow rate $\dot{m}_{a1}$; the hot-gas generator is basically a turbojet engine. The fuel is added to $\dot{m}_{a1}$ at the rate $\dot{m}_f$. The hot-gas stream ($\dot{m}_{a1} + \dot{m}_f$) is discharged to the atmosphere through $A_7$ with the velocity $V_j$. Both types of turbofan engine produce an "overall" jet velocity $V_{jTF}$ which is smaller than that for a turbojet engine operating with the same $P_3/P_2$ and $T_4$. The arrangements shown in Fig. 4 are for

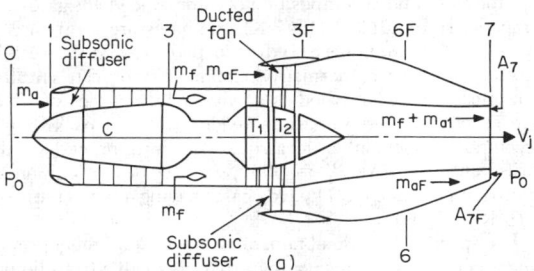

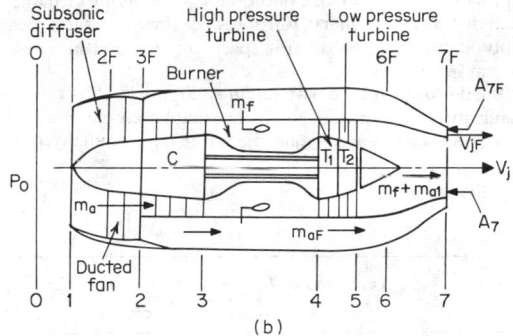

**Fig. 4** Two different schematic arrangements of components of turbofan engines (subsonic flight): (*a*) aft-fan turbofan engine; (*b*) ducted-fan turbofan engine.

subsonic propulsion. Fuel can, of course, be burned in the fan air $m_{aF}$ for increasing the thrusts of the engines. Practically all the newer commercial passenger aircraft are propelled by turbofan engines, and most of the older jet aircraft have been refurbished with new turbofan engines. The advantage of the lower effective jet velocity $V_{jTF}$ is twofold: (1) It increases the propulsive efficiency $\eta_p$ by reducing $\nu = V_{jTF}/V_0$ and, consequently, raises the value of $\eta_0$. (2) The reduced jet velocity reduces the jet noise; the latter increases with approximately the eighth power of the jet speed.

In the adaptation of the turbofan engine to supersonic propulsion, additional fuel is burned in the fan airstream, and the engine must be equipped with a supersonic diffuser and a variable-area supersonic exhaust nozzle.

## ESSENTIAL FEATURES OF ROCKET ENGINES

Figure 5 is a block-type diagram illustrating the essential features of a rocket engine, which comprises three main components: (A) a supply of propellant material contained in the rocket-propelled vehicle, (B) a propellant feed and metering system, and (C) a **thrust chamber**, also called a **rocket motor, thrustor,** or **accelerator.** In any rocket engine, energy must be added to the propellant as it flows through the thrustor. In Fig. 5, propellant material from the supply is metered and fed to the thrust chamber, where energy is added to it. As a

consequence of the energy addition, the propellant is discharged from the thrustor with the jet velocity $V_j$. The thrust $F$ acts in the direction opposite to that for the jet velocity $V_j$.

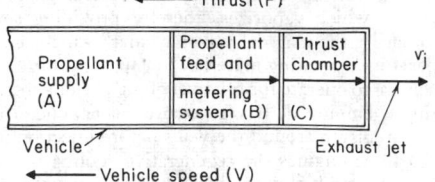

**Fig. 5** Essential features of a rocket engine.

### Chemical Rocket Engines

All chemical rocket engines have two common characteristics: (1) They utilize chemical reactions in a thrust chamber to produce a high-pressure, high-temperature gas at the entrance to a converging-diverging exhaust nozzle; (2) the hot propellant gas expands in flowing through the exhaust nozzle, and the expansion process converts a portion of the thermal energy, released by the chemical reaction, into the kinetic energy associated with a high-velocity gaseous-exhaust jet. Chemical rocket engines may be grouped into (1) liquid-bipropellant rocket engines, (2) liquid-monopropellant rocket engines, and (3) solid-propellant rocket engines.

**Liquid-bipropellant Rocket Engine** Figure 6 illustrates the essential features of a liquid-bipropellant rocket engine employing **turbopumps** for feeding two propellants, an **oxidizer** and a **fuel**, to a rocket motor. The motor comprises (1) an injector, (2) a combustion chamber, and (3) a converging-

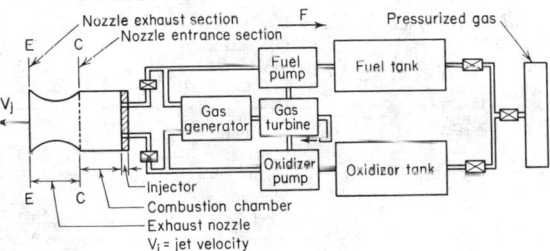

**Fig. 6** Essential features of a liquid-bipropellant rocket engine.

diverging exhaust nozzle. The liquid propellants are fed under pressure, through the injector, into the combustion chamber, where they react chemically to produce large volumes of high-temperature, high-pressure gases. For a given propellant combination, the **combustion temperature** $T_c$ depends primarily on the oxidizer-fuel ratio (by weight), termed the **mixture ratio,** and to a lesser extent upon the combustion static pressure $p_c$. When the mass rate of flow of the liquid propellants equals that of the exhaust gases, the combustion pressure remains constant—the mode of operation that is usually desired.

If the bipropellants react chemically when their liquid streams come in contact with each other, they are said to be **hypergolic.** Propellants which are not hypergolic are said to be **diergolic,** and some form of ignition system is required to initiate combustion.

Except in those cases where the operating duration of the rocket motor is very short or where the combustion temperature is low, means must be provided for protecting the interior walls of the motor. The two most common methods are ablative cooling and regenerative cooling. In **ablative cooling** the inner surfaces of the thrust chamber are covered with an ablative material which vaporizes, thereby providing some cooling. Ordinarily, the material leaves a "char" which acts as a high-temperature insulating material. For high-performance engines which are to operate for relatively long periods, regenerative cooling is employed. In **regenerative cooling,** one of the propellants is circulated around the walls before injection into the motor. In some engines the regenerative cooling is combined with local liquid film cooling at critical areas of the thrust chamber. See Table 1 for bipropellant combinations.

Refer to Fig. 6. By removing the oxidizer tank, the oxidizer pump, and the plumbing associated with the liquid oxidizer, one obtains the essential elements of a liquid-monopropellant rocket engine. Such an engine does not require a liquid oxidizer to cause the monopropellant to decompose and release its thermochemical energy.

There are basically three groups of monopropellants: (1) liquids which contain the fuel and oxidizer in the same molecule, e.g., hydrogen peroxide ($H_2O_2$) or nitromethane ($CH_3NO_2$); (2) liquids which contain either the oxidizer or the fuel constituent in an unstable molecular arrangement, e.g.,

hydrazine ($N_2H_4$); and (3) synthetic mixtures of liquid fuels and oxidizers. The most important liquid monopropellants are hydrazine and hydrogen peroxide (up to 98% $H_2O_2$). Hydrazine can be decomposed by a suitable metal catalyst, which is ordinarily packed in a portion of the thrust chamber at the injector end. The decomposition of hydrazine yields gases at a temperature of 2260°R (1265 K); the gases are a mixture of hydrogen and ammonia. Hydrogen peroxide can be readily decomposed either thermally, chemically, or catalytically. The most favored method is catalytically; a series of silver screens coated with samarium oxide is tightly packed in a decomposition chamber located at the injector end of the thrust chamber. At $P_c = 300$ psia, the decomposition of hydrogen peroxide (98%) yields gases having a temperature of 2260°R (1256 K).

The specific impulse obtainable from a liquid monopropellant is considerably smaller than that obtainable from liquid-bipropellant systems. Consequently, monopropellants are used for such auxiliary purposes as thrust vernier control, attitude reaction controls for space vehicles and missiles, and for gas generation.

**Solid-propellant Rocket Engine**  Figure 7 illustrates schematically a solid-propellant rocket engine employing an **internal-burning case-bonded grain;** the latter burns radially outward

**Table 1. Calculated Values of Specific Impulse for Liquid Bipropellant Systems**[a]

$[P_c = 1,000 \text{ lbf/in}^2 \ (6,895 \times 10^3 \text{ N/m}^2) \text{ to } P_e = 1 \text{ atm}]$

Shifting equilibrium; isentropic expansion; adiabatic combustion; one-dimensional flow

| Oxidizer | Fuel | $\dot{m}_o/\dot{m}_f$ | $\rho$ | Temperature $T_c$ | | $c^*$ | $I$ (s) |
|---|---|---|---|---|---|---|---|
| | | | | °R | K | | |
| Chlorine trifluoride | | | | | | | |
| (CTF) | Hydrazine | 2.80 | 1.50 | 6553 | 3640 | 5961 | 293.1 |
| (ClF$_3$) | MMH[b] | 2.70 | 1.41 | 5858 | 3254 | 5670 | 286.0 |
| | Pentaborane (B$_5$H$_9$) | 7.05 | 1.47 | 7466 | 4148 | 5724 | 289.0 |
| Fluorine (F$_2$) | Ammonia | 3.30 | 1.12 | 7797 | 4332 | 7183 | 359.5 |
| | Hydrazine | 2.30 | 1.31 | 8004 | 4447 | 7257 | 364.0 |
| | Hydrogen | 7.70 | 0.45 | 6902 | 3834 | 8380 | 411.1 |
| | Methane | 4.32 | 1.02 | 7000 | 3889 | 6652 | 343.8 |
| | RP-1 | 2.62 | 1.21 | 6839 | 3799 | 6153 | 318.0 |
| Hydrogen peroxide | Hydrazine | 2.00 | 1.26 | 4814 | 2674 | 5765 | 287.4 |
| (100% H$_2$O$_2$) | MMH | 3.44 | 1.26 | 4928 | 2738 | 5665 | 284.8 |
| IRFNA[c] | Hydine[d] | 3.17 | 1.26 | 5198 | 2888 | 5367 | 270.3 |
| | MMH | 2.57 | 1.24 | 5192 | 2885 | 5749 | 275.5 |
| Nitrogen tetroxide | Hydrazine | 1.30 | 1.22 | 5406 | 3002 | 5871 | 292.2 |
| (N$_2$O$_4$) | Aerozine-50[e] | 2.00 | 1.21 | 5610 | 3117 | 5740 | 289.2 |
| | MMH | 2.15 | 1.20 | 5653 | 3141 | 5730 | 288.7 |
| Oxygen (LOX) (O$_2$) | Hydrazine | 0.91 | 1.07 | 5667 | 3148 | 6208 | 312.8 |
| | Hydrogen | 4.00 | 0.28 | 4910 | 2728 | 7892 | 291.2 |
| | Methane | 3.35 | 0.82 | 6002 | 3333 | 6080 | 310.8 |
| | Pentaborane | 2.21 | 0.91 | 7136 | 3964 | 6194 | 318.1 |
| | RP-1 | 2.60 | 1.02 | 6164 | 3424 | 5895 | 300.0 |

[a]Based on "Theoretical Performance of Rocket Propellant Combinations," Rocketdyne Corporation, Canoga Park, CA.
[b]MMH: Monomethyl hydrazine (N$_2$H$_3$CH$_3$).
[c]IRFNA: Inhibited red fuming nitric acid; 84.4% (HNO$_3$), 14%(N$_2$O$_4$), 1%(H$_2$O), 0.6%(HF).
[d]Hydine: 60% UDMH[f], 40% Deta[g].
[e]Aerozine-50: 50% UDMH[f], 50% hydrazine (N$_2$H$_4$).
[f]UDMH: Unsymmetrical dimethylhydrazine (CH$_3$)$_2$N$_2$H$_4$.
[g]DETA: Diethylene triamine (C$_2$H$_{13}$N$_3$).

at a substantially constant rate. A solid propellant contains both its fuel and the requisite oxidizer. If the fuel and oxidizer are contained in the molecules forming the solid propellant, the propellant is termed a **double-base propellant.** Those solid propellants wherein a solid fuel and a solid oxidizer form an

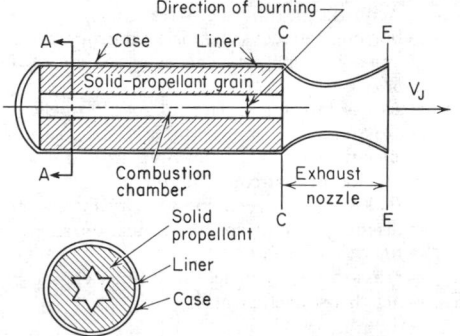

**Fig. 7** Essential features of an internal-burning case-bonded solid-propellant rocket engine.

intimate mechanical mixture are termed either **composite or heterogeneous** solid propellants. The chemical reaction of a solid propellant is initiated by an igniter. In general, the configuration of a propellant grain can be designed so that the area of the burning surface of the propellant varies to give a prescribed thrust-versus-time curve.

### Nuclear Heat-transfer Rocket Engine

Figure 8 illustrates schematically a nuclear heat-transfer rocket engine employing a **solid-core reactor.** The heat generated by

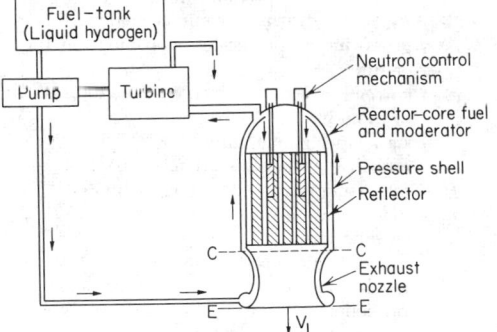

**Fig. 8** Schematic arrangement of a nuclear heat-transfer rocket engine.

the fissions of the uranium nucleus is utilized for heating a gaseous propellant, such as hydrogen, to a high temperature of 4000°R (2200 K) approx at the entrance section of the exhaust nozzle. The hot gas is ejected to the surroundings after expansion in a converging-diverging exhaust nozzle. The basic difference between the operating principles of a nuclear heat-transfer and a chemical rocket engine is the substitution of nuclear fission for chemical reaction as the source of heat for the propellant gas. Since the exhaust nozzle is the propulsive element, the remaining components of the nuclear heat-transfer rocket engine constitute the hot-gas generator.

The propellant in a nuclear heat-transfer rocket engine

functions basically as a fluid for cooling the solid-core nuclear reactor. Its selection is not based on energy considerations alone, but on such properties as specific heat, latent heat, molecular weight, and liquid density and on certain practical considerations. The feasibility of the aforementioned type of rocket engine was established under NASA's NERVA project, but because there was no mission which required the NERVA engine, the NERVA project was canceled in 1972.

### Electric Rocket Engines

Several different types of electric rocket engines have been conceived; the principal types are described below. All of them require some form of power plant for generating electricity. Thus the power plant may be nuclear, solar-cell batteries, thermoelectric, or other. The choice of the type of power plant depends upon the characteristics of both the electric rocket engine and the space-flight mission.

Figure 9 illustrates diagrammatically an electric rocket engine, comprising (1) a **nuclear power source,** (2) an energy-conversion unit for obtaining the desired form of electric energy, (3) a propellant feed and metering system, (4) an electrically operated thrustor, and (5) the requisite control devices. Electric rocket engines may be classified as (1) electrothermal, (2) electromagnetic, and (3) electrostatic.

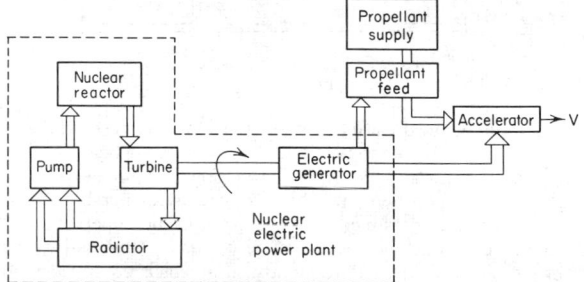

**Fig. 9** Essential features of an electric rocket engine.

**Electrothermal Rocket Engine** Figure 10a illustrates the type of engine which uses electric power for heating a gaseous propellant to a high temperature before ejecting it through a converging-diverging exhaust nozzle. If an electric arc is employed for heating the propellant, the engine is called a **thermal arc-jet rocket engine.**

**Electromagnetic Rocket Engine** Figure 10b illustrates an electromagnetic, or plasma, rocket engine. There is a wide variety of such engines, but all of them utilize the same operating principle. A plasma (a neutral ionized conducting gas) is accelerated by means of its interaction with either a stationary or a varying magnetic field. Basically, a plasma engine differs from a conventional electric motor by the substitution of a conducting plasma for a moving armature.

**Electrostatic Rocket Engine** Figure 10c illustrates schematically the electrostatic, or ion, rocket engine, comprising (1) a nuclear electric power plant; (2) a propellant supply; (3) ionization apparatus; (4) an ion accelerator; and (5) an electron emitter for **neutralizing** the ion beam ejected from the accelerator. Its operating principle is based on utilizing electrostatic fields for accelerating and ejecting electrically charged particles with extremely large velocities. The overall objective is to transform thermal (nuclear) energy into the kinetic energy associated with an extremely high-velocity stream of electrically neutral particles ejected from one or more thrustors.

All electric rocket engines are low-thrust devices and are in the research-and-development stage, with only limited applications. The recent interest in such engines stems from the

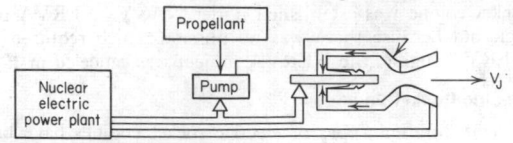

*(a)* Electrothermal rocket engine.

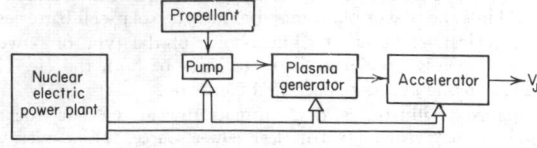

*(b)* Electromagnetic rocket engine.

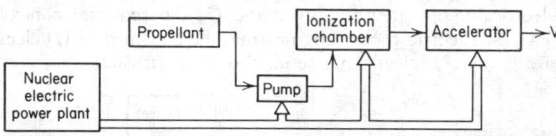

*(c)* Electrostatic (ion) engine.

**Fig. 10**   Electric rocket engines.

fact that chemical-rocket technology is now so advanced that it has become feasible to place heavy *pay loads,* such as a vehicle equipped with an electric rocket engine, into an earth orbit. An electric rocket engine which is virtually inoperable terrestrially can be positioned in space and can then operate effectively because of the absence of aerodynamic drag and strong gravitational fields. Consequently, if a vehicle equipped with an electric rocket engine is placed in an earth orbit, the low-thrust electric rocket engine can serve a useful purpose because, under the conditions in space, an exceedingly small thrust applied to a vehicle will accelerate it to a large vehicle velocity if the operating time for the engine is sufficiently long.

In a chemical rocket engine, the energy for propulsion, as well as the mass ejected through the exhaust nozzle, is provided by the propellants, but the energy which can be added to a unit mass of the propellant gas is a fixed quantity, limited by the nature of the chemical bonds of the reacting materials. For that reason, chemical rocket engines are said to be **energy-limited rocket engines.** Although a nuclear heat-transfer rocket engine is also energy-limited, the limitation is imposed by the amount of energy that can be added per unit mass of propellant without exceeding the maximum allowable temperature for the materials employed for the solid-core reactor.

In an electric rocket engine, on the other hand, the energy added to the propellant is furnished by a nuclear electric power plant. The power available for heating the propellant is limited by the maximum power output of the nuclear electric power plant accordingly, electric rocket engines are **power-limited.**

Only the chemical rocket engines have achieved operational realization, have a broad, well-developed technology, and,

potentially, can be developed for boosting any desired mass or pay load into an earth orbit or to the escape velocity.

## NOTATION

$a$ = acoustic speed
$a_0$ = Acoustic speed in free-stream air
$a_2$ = burning rate constant for solid propellant
$A_e$ = cross-sectional area of exit section of exhaust nozzle
$A_p$ = area of burning surface for solid propellant
$A_t$ = cross-sectional area of throat of exhaust nozzle
$c_p$ = specific heat at constant pressure
$c_v$ = specific heat at constant volume
$c^* = P_c A_t/\dot{m}$ = characteristic velocity for rocket motor
$C_d$ = discharge coefficient for exhaust nozzle
$C_F$ = thrust coefficient
$C_{Fg}$ = gross-thrust coefficient for ramjet engine
$C_{Fn}$ = net thrust coefficient
$D$ or $\mathfrak{D}$ = drag
$D_i = \dot{m}_a V_0$ = ram drag for airbreathing or thermal-jet engine
$d$ = diameter
$d_p$ = diameter of propeller
$E_{in}$ = rate at which energy is supplied propulsion system
$E_f$ = calorific value of fuel
$E_p$ = calorific value of rocket propellants
$f = \dot{m}_f/\dot{m}_a$ = fuel-air ratio
$f' = f/\eta_B$ = fuel-air ratio for ideal combustion chamber
$F$ = force or thrust
$F_i = m_e w - m_i V_0 + (p_e - p_0)A_e$ = thrust due to internal flow
$F_j = m_e V_j$ = jet thrust
$F_p = (p_e - p_0)A_e$ = pressure thrust
$F_g = F_i$ = gross thrust for ramjet engine
$F_g SFC$ = gross-thrust specific-fuel consumption for ramjet engine
$g_c$ = correction factor defined by Newton's second law of motion
$h$ = static specific enthalpy
$\Delta h_n$ = enthalpy change for exhaust nozzle
$H$ = total (stagnation) specific enthalpy
$\Delta H_c$ = lower heating value of fuel
$I$ = specific impulse
$I_a = F/g m_a$ = air specific impulse
$J$ = mechanical equivalent of heat; advance ratio, propeller
$k = c_p/c_v$ = specific heat ratio
$L$ or $\mathfrak{L}$ = lift
$\mathfrak{M}$ = mass
$m$ = mass rate of flow
$\dot{m}_a$ = mass rate of air consumption for thermal-jet engine
$\dot{m}_e$ = mass rate of flow of gas leaving propulsion system
$\dot{m}_f$ = mass rate of fuel consumption
$m_i$ = mass rate of flow of gas into propulsion system
$\dot{m}_o$ = mass rate of oxidizer consumption for rocket engine
$\dot{m}_p = m_o + m_f$ = mass rate of propellant consumption for rocket engine
$\bar{m}$ = molecular weight
$M_e$ = momentum of gases leaving propulsion system in unit time

$M_i$ = momentum of gases entering propulsion system in unit time

$M_0 = V_0/a_0$ = Mach number of free-stream air (flight speed)

$n$ = revolutions per unit time

$p$ = absolute static pressure; pitch of propeller blade

$p_c$ = absolute static pressure of gases in combustion chamber

$p_e$ = absolute static pressure in exit section of exhaust nozzle

$p_0$ = absolute static pressure of free-stream ambient air

$P$ = absolute total (stagnation) pressure

$P_c$ = absolute total pressure at entrance to exhaust nozzle

$P_0 = p_0\left(1 + \dfrac{k-1}{2}M_0^2\right)^{k/(k-1)}$ = absolute total pressure of free-stream air

$\mathcal{P}$ = propulsion power

$\mathcal{P}_L$ = leaving loss

$\mathcal{P}_T$ = thrust power

$q = \rho V^2/2$ = dynamic pressure

$q_0 = \rho_0 V_0^2/2 = k_0 p_0 M_0^2/2$ = dynamic pressure of free-stream air

$Q$ = torque

$Q_i$ = heat supplied to actual combustor

$Q_i'$ = heat supplied to ideal combustor (no losses)

$r_0$ = linear burning rate for solid propellant

$R = R_u/\bar{m}$ = gas constant

$R_u$ = universal gas constant = $\bar{m}R$

$t$ = absolute static temperature

$t_p$ = temperature of solid propellant prior to ignition

$T$ = absolute total (stagnation) temperature

$T_c$ = absolute total temperature of gas entering exhaust nozzle of rocket motor

$T_2$ = absolute total temperature at entrance to air compressor (see Fig. 3)

$T_3$ = absolute total temperature at exit section of air compressor (see Fig. 3)

$T_3'$ = absolute total temperature at exit section of ideal compressor (see Fig. 3) operating between some pressure limits as actual compressor

$T_4$ = absolute total temperature at entrance to turbine (see Fig. 3)

$T_5$ = absolute total temperature at exit from turbine (see Fig. 3)

$T_5'$ = absolute total temperature at exit from ideal turbine

$u = \pi n d_p$ = propeller tip speed

$V$ = velocity

$V_F$ = forward speed of propeller

$V_j$ = effective jet velocity

$V_0$ = velocity of free-stream air (flight speed)

$w$ = velocity of exit gases relative to walls of exhaust nozzle or velocity of air in ultimate wake of propeller

$\dot{W}$ = weight rate of flow

$\dot{W}_o$ = weight rate of flow of oxidizer

$\dot{W}_p$ = weight rate of flow of propellants

**Greek**

$\alpha = T_4/t_0 = \alpha_d \alpha_1$ = cycle temperature ratio; angle of attack

$\alpha_d = T_2/t_0$ = diffusion temperature ratio

$\alpha_1 = T_4/T_2$

$\beta$ = helix angle for propeller blade

$\Delta = gI_a/\sqrt{2gJc_p t_0}$ = thrust parameter for turbojet engine

$\delta = P/P_{std}$ corrected pressure

$\eta$ = efficiency

$\eta_B = f'/f$ = efficiency of combustion for thermal-jet engine

$\eta_c = (T_3' - T_2)/(T_3 - T_2)$ = isentropic efficiency of compressor

$\eta_d$ = isentropic efficiency of diffuser

$\eta_n = \varphi^2$ = isentropic efficiency of exhaust nozzle

$\eta_o = \mathcal{P}_T/E_{in}$ = overall efficiency of propulsion system

$\eta_P = \mathcal{P}_T/(\mathcal{P}_T + \mathcal{P}_L)$ = ideal propulsive efficiency

$\eta_t = (T_4 - T_5)/(T_4 - T_5')$ = isentropic efficiency of turbine

$\eta_{th} = \mathcal{P}/E_{in}$ = thermal efficiency of propulsion engine

$\lambda = \frac{1}{2} + \frac{1}{2}\cos\phi$ = divergence coefficient for exhaust nozzle

$\nu = V_0/w$ = speed ratio

$\rho$ = density

$\bar{\rho}$ = mean density

$\Omega = \sqrt{k}\left(\dfrac{2}{k+1}\right)^{(k-1)/(2k-1)}$

$\omega$ = angular velocity of propeller shaft

$\omega'$ = rate of rotation of slipstream at propeller

$\omega''$ = rate of rotation of slipstream in ultimate slipstream

$\Phi$ = fan velocity coefficient = $V_F/u$

$\phi$ = semiangle of exhaust-nozzle divergence; effective helix angle

$\varphi = \sqrt{\eta_n}$ = velocity coefficient for exhaust nozzle

$\rho$ = density

$\sigma$ = solidity of propeller

$\Theta = (P_3/p_0)^{(k-1)/k} = \Theta_d\Theta_c$ = cycle pressure ratio parameter

$\Theta_c = (P_3/P_2)^{(k-1)/k}$ = compressor pressure-ratio parameter

$\Theta_d = (P_2/p_0)^{(k-1)/k}$ = diffuser pressure-ratio parameter

$\Theta_n = (P_7/p_0)^{(k-1)/k} = (P_7/p_9)^{(k-1)/k}$ = nozzle pressure-ratio parameter

$\Theta_t = (P_4/P_5)^{(k-1)/k}$ = turbine pressure-ratio parameter

$\theta = T/T_{std}$ = corrected temperature

**Subscripts**

(a) numbered

0 = free stream

1 = entrance to subsonic diffuser

2 = exit from subsonic diffuser

3 = entrance to combustion chamber

4 = entrance to turbine of turbojet engine

5 = exit from turbine of turbojet engine

6 = tail-pipe entrance

7 = entrance to exhaust nozzle

8 = throat section of exhaust nozzle

9 = exit section of exhaust nozzle

(b) lettered

$a$ = air

$B$ = burner or combustion chamber
$b$ = blade
$c$ = compressor
$d$ = diffuser
$e$ = exit section; effective
$F$ = fan
$f$ = fuel
$h$ = hydraulic
$n$ = nozzle
$o$ = overall or oxidizer
$p$ = propellant
$P$ = propulsive
std = standard
$t$ = turbine or throat, as specified in text

**Statement on Units** The dynamic equations employed in the following sections are written for *consistent sets of units*, i.e., for sets in which 1 *unit of force* = 1 *unit of mass* × 1 *unit of acceleration*. Consequently, the gravitational correction factor $g_c$ in Newton's equation $F = (1/g_c)\, ma$ (see notation) has the numerical value unity and is omitted from the dynamic equations. When the equations are used for calculation purposes, $g_c$ should be included and its appropriate value employed.

## THRUST EQUATIONS FOR JET-PROPULSION ENGINES

Refer to Fig. 11, which illustrates schematically a rotationally symmetrical arbitrary propulsion engine immersed in a uni-

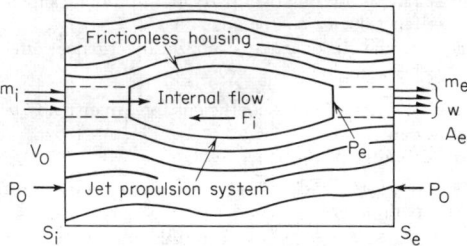

**Fig. 11**  Generalized jet-propulsion system.

form flow field. Because of the reactions between the fluid flowing through the engine, called the **internal flow,** and the interior surfaces wetted by the internal flow a resultant **axial force** is produced, that is, a force collinear with the longitudinal axis of the engine. If an axial force acts in the **forward direction,** employing a relative coordinate system, it is called a **thrust,** and if it acts in the **backward direction** it is called a **drag.** Similarly, the resultant axial force due to the **external flow,** the flow passing over the external surfaces of the propulsion system, is a thrust or drag depending upon whether it acts in the forward direction or the backward direction.

Application of the momentum equation of fluid mechanics to the generalized propulsion system illustrated in Fig. 11 gives the following equation for the thrust $F_i$ due to the **internal flow.** Thus, if it is assumed that the external flow is frictionless, there is no change in the rate of momentum for the external flow between $S_i$ and $S_e$. That is assumed here. Hence, the thrust $F = F_i$ is due entirely to the internal flow, and

$$F = F_i = \dot{m}_e w - \dot{m}_i V_0 + (p_e - p_0)A_e \qquad (1)$$

In Eq. (1), $\dot{m}_e w = F_j$ = **jet thrust,** $\dot{m}_i V_0 = D_i$ = **ram drag,** and $(p_e - p_0)A_e = F_p$ = **pressure thrust.** Hence,

$$F_i = F_j - D_i + F_p \qquad (2)$$

It is convenient to introduce a fictitious **effective jet velocity** $V_j$, which is defined by

$$F = \dot{m}_e V_j - \dot{m}_i V_0 = \dot{m}_e w - \dot{m}_i V_0 + (p_e - p_0)A_e \qquad (3)$$

If the gases are expanded completely, in the exhaust nozzle, then $w = V_j$ and $F_p = 0$. No appreciable error is introduced, in general, if it is assumed that $F_p = 0$. For **thermal jet engines,** $\dot{m}_i = \dot{m}_a$ (see notation), and $\dot{m}_e = \dot{m}_a + \dot{m}_f$. Let $f = \dot{m}_f/\dot{m}_a$, $\nu = V_0/V_j$, then

$$F_i = \left(\frac{1+f}{\nu} - 1\right)\dot{m}_a V_0 + (p_e - p_0)A_e \qquad (4)$$

In the case of uncooled turbojet engines, $\dot{m}_f$ is not significantly different from the fraction of $\dot{m}_a$ utilized for cooling the bearings and turbine disk. Consequently, no significant error is introduced by assuming that $\dot{m}_e = \dot{m}_a + \dot{m}_f \approx \dot{m}_a = \dot{m}$. Hence, for a simple **turbojet engine,** one may write

$$F_i = \dot{m}_a(V_j - V_0) = \dot{m}V_0\,(1/\nu - 1) \qquad (5)$$

Equation (5) is also the thrust equation for an **ideal propeller;** in that case $V_j$ is the **wake velocity** for the air leaving its slipstream. The thrust equation for a hydraulic jet propulsion system has the same form as Eq. (5). Let $V_0$ denote the speed of a boat, and $\dot{m}_w$ the mass rate of flow of the water entering the hydraulic pump and discharged by the exit nozzle with the velocity $w = V_j$ relative to the boat. Hence, for hydraulic jet propulsion

$$F = \dot{m}_s(V_j - V_0) = \dot{m}_w V_0(1/\nu - 1) \qquad (6)$$

where $\nu = V_0/V_j$.

In the case of **rocket engines,** since they do not consume atmospheric air, $\dot{m}_a = 0$, and the flow of gas out of the rocket motor, under steady-state conditions, is equal to $\dot{m}_p = \dot{m}_0 + \dot{m}_f$ (see notation).

The effective jet velocity $V_j$ is larger than $V_e$ = **exit velocity** if $p_e > p_0$; that is, if the gases are **underexpanded.** The effective jet velocity is a useful criterion because it can be determined accurately from the measured values of $F_i$ and $\dot{m}_p$ obtained from a static firing test of the rocket motor.

The ratio $F_i/\dot{m}_p$ is denoted by $I$ and called either the **specific thrust** or the **specific impulse.** Hence

$$I = F/\dot{m}_p = V_j/g_c \qquad (7)$$

Although the dimensions of specific impulse are force/(mass)(s), it is conventional to state its units as *seconds*.

Equation (5) for the thrust of a **turbojet engine,** when expressed in terms of the effective jet velocity $V_j$, becomes

$$F = \dot{m}_a(V_j - V_0) \qquad (8)$$

## POWER AND EFFICIENCY RELATIONSHIPS

In a jet-propulsion engine the **propulsion element** is the **exhaust nozzle,** and the rate at which energy is supplied to it is called the **propulsion power,** which is denoted by $\mathcal{P}$. The rate at which the propulsion system does useful work is termed the **thrust power** $\mathcal{P}_T$ and is given by

$$\mathcal{P}_T = FV_0 \qquad (9)$$

Assume that $p_e = p_0$, and that the only energy loss in a propulsion system is the **leaving loss** $\mathcal{P}_L = \dot{m}(V_j - V_0)^2/2$, that is, the kinetic energy associated with the jet gases discharged from the system; then the propulsive power is given by

$$\mathcal{P} = \mathcal{P}_L + \mathcal{P}_T \qquad (10)$$

The **ideal propulsive efficiency** is defined, in general, by

$$\eta_P = \mathcal{P}_T/(\mathcal{P}_T + \mathcal{P}_L) \qquad (11)$$
$$= \text{thrust power/propulsion power}$$

For a turbojet engine and hydraulic jet propulsion,

$$\eta_P = 2\nu/(1 + \nu) \qquad (12)$$

For a chemical rocket engine,

$$\eta_P = 2\nu/(1 + \nu^2) \qquad (13)$$

The propulsive efficiency $\eta_P$ is of more or less academic interest. Of more importance is the overall efficiency $\eta_o$, which for airbreathing and rocket engines is defined by

$$\eta_o = \eta_{th}\eta_P \qquad (14)$$

where $\eta_{th} = \mathcal{P}/E_{in} = $ thermal efficiency of system $\quad (15)$

and $E_{in}$ is the rate at which energy is supplied to the propulsion system.

The overall efficiency of a hydraulic jet-propulsion system is given by

$$\eta_o = \eta_{th}\eta_h\eta_p = \eta_h\eta_{th}[2\nu/(1 + \nu)] \qquad (16)$$

where $\eta_{th}$ is the thermal efficiency of the power plant which drives the water pump, and $\eta_h$ is the hydraulic efficiency of the water pump. To achieve a reasonable fuel consumption rate, $\eta_h$ must have a larger value.

For an airbreathing jet engine, $E_{in} = \dot{m}_f(E_f + V_0^2/2)$, and

$$\eta_o = \frac{2\nu}{1 + \nu} \frac{\mathcal{P}}{\dot{m}_f(E_f + V_0^2/2)} \qquad (17)$$

The ratio $\dot{m}_f/F$ is called the **thrust specific-fuel comsumption** (TSFC) and is measured in mass of fuel per hour per unit of thrust. Hence,

$$TSFC = \dot{m}_f/F = f\dot{m}_a/F \qquad (18)$$

For a rocket engine, $E_{in} = \dot{m}_p(E_p + V_0^2/2)$, so that

$$\eta_o = \frac{2\nu}{1 + \nu^2} \frac{\mathcal{P}}{\dot{m}_p(E_p + V_0^2/2)} \qquad (19)$$

## PERFORMANCE CHARACTERISTICS OF AIRBREATHING JET ENGINES

**The Ramjet Engine**   In ramjet technology the thrust due to the internal flow $F_i$ is called the **gross thrust** and denoted by $F_g$. Refer to Fig. 2 and assume $\dot{m}_0 \approx \dot{m}_9 \approx \dot{m}_a$. Then

$$F_g = F_i \approx \dot{m}_a(V_j - V_0) \qquad (20)$$

In level unaccelerated flight $F_g$ is equal to the external drag of the propelled vehicle and the ramjet body.

It is customary to express the thrust capabilities of the engine in terms of the **gross-thrust coefficient** $C_{Fg}$. If $A_m$ is the

maximum cross-sectional area of the ramjet engine, $q_0 = \rho_0 V_0^2/2 = k_0 p_0 M_0^2/2 = $ the **dynamic pressure** of the free-stream air, then

$$C_{Fg} = \frac{F_g}{q_0 A_m} = \frac{2F_g}{A_m k_0 p_0 M_0^2} \qquad (21)$$

In terms of the Mach numbers $M_0$ and $M_9$,

$$C_{Fg} = \frac{2A_9/A_m}{k_0 M_0^2}\left(\frac{P_9}{p_0}\frac{1 + k_9 M_9^2}{P_9/p_9} - 1\right) - 2\frac{A_0}{A_m} \qquad (22)$$

In a fixed-geometry engine, $M_9$ depends upon the total temperature $T_7$, the total pressure $P_7$, the fuel-air ratio $f$, the nozzle area ratio $A_e/A_t$, and the efficiency of the nozzle $\eta_n$. For estimating purposes, $k_0 = 1.4$ and $k_9 = 1.28$, when the engine burns a liquid-hydrocarbon fuel. The manner in which $C_{Fg}$ varies with altitude $M_0$ and fuel-air ratio $f$ is shown schematically in Fig. 12.

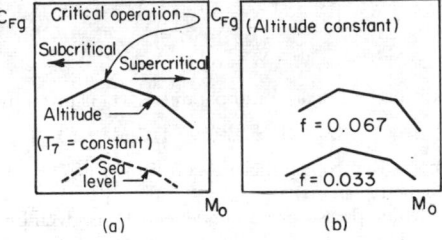

**Fig. 12**   Effect of altitude and fuel-air ratio on the gross-thrust coefficient of a fixed-geometry ramjet engine: (*a*) effect of altitude; (*b*) effect of fuel-air ratio.

If $\alpha = T_7/t_0 = $ cycle temperature ratio, and $k_B$ is the mean value of $k$ for the combustion gases, the rate at which heat is supplied to the engine is

$$Q_i = Q_i'/\eta_B = \dot{m}_a c_{pB} t_0(\alpha - T_2)/\eta_B \qquad (23)$$

It is readily shown that

$$Q_i = \frac{A_0 p_0 V_0}{\eta_B J} \frac{k_B}{k_B - 1}\left[\alpha - \left(1 + \frac{k_0 - 1}{2}M_0^2\right)\right] \qquad (24)$$

The **gross-thrust specific-fuel consumption** ($F_g SFC$) is, by definition,

$$F_g SFC = \dot{m}_f/F_g \qquad (25)$$

The principal sources of loss are aero-thermodynamic in nature and cause a decrease in total pressure between stations 0 and 7. For estimating purposes, assuming $M_0 = 2.0$ and $T_6 = 3800°R$ (2111 $K$), the total pressures across different sections of the engine may be assumed to be approximately those tabulated below.

| Part of Engine | Total pressure ratio |
|---|---|
| Supersonic diffuser (0–1) | $P_1/P_0 = 0.92$ |
| Subsonic diffuser (1–2) | $P_2/P_1 = 0.90$ |
| Flameholders (2–7) | $P_7/P_2 = 0.97$ |
| Combustion chamber (2–7) | $P_7/P_2 = 0.92$ |
| Exhaust nozzle (7–9) | $P_9/P_7 = 0.97$ |

The practical applications of the ramjet engine have been relatively few.

**The Simple Turbojet Engine**   A good insight into the design performance characteristics of the turbojet engine is obtained conveniently by making the following assumptions: (1) The mass rate of flow of working fluid is identical at all stations in the engine; (2) the thermodynamic properties of the working fluid are those for air; (3) the air is a perfect gas and its specific heats are constants; (4) there are no pressure losses due to friction or heat addition; (5) the exhaust nozzle expands the working fluid completely so that $p_9 = p_0$; and (6) the auxiliary power requirements can be neglected. Refer to Fig. 3 and let

$$\Theta = \left(\frac{P_3}{p_0}\right)^{(k-1)/k} \quad \Theta_d = \left(\frac{P_2}{p_0}\right)^{(k-1)/k} \quad \Theta_c = \left(\frac{P_3}{P_2}\right)^{(k-1)/k}$$

$$\Theta_t = \left(\frac{P_4}{P_5}\right)^{(k-1)/k} \quad \Theta_n = \left(\frac{P_7}{p_0}\right)^{(k-1)/k} = \left(\frac{P_7}{p_9}\right)^{(k-1)/k}$$

$$\alpha = T_4/t_0 \quad \alpha_d = T_2/t_0 \quad \alpha_1 = T_4/T_2$$

In view of the assumptions,

$$\Theta = \Theta_d\Theta_c = \Theta_t\Theta_n \qquad (26)$$

and

$$\alpha = \alpha_d\alpha_1 \qquad (27)$$

The diffuser pressure-ratio parameter $\Theta_d$ is given by

$$\Theta_d = 1 + \eta_d[(k-1)/2]M_0^2 \qquad (28)$$

where $\eta_d$ = isentropic efficiency of diffuser (0.75 to 0.90 for well-designed systems).

The turbine pressure-ratio parameter $\Theta_t$ is given by

$$1/\Theta_t = 1 - [(\Theta_c - 1)/\alpha_1\eta_c\eta_t] \qquad (29)$$

where $\eta_t$ = the isentropic efficiency of the turbine (0.90 to 0.95).

Heat supplied, per unit mass of air, is

$$Q_i = (c_p/\eta_B)(T_4 - T_3)$$

or

$$Q_i = \frac{c_p t_0}{\eta_B\eta_c}\,\alpha_d\left(\frac{\alpha}{\alpha_d}\,\eta_c - \eta_c - \Theta_c + 1\right) \qquad (30)$$

where $\eta_c$ is the isentropic efficiency of the air compressor (0.85 to 0.90), and $\eta_B$ is the efficiency of the burner (0.95 to 0.99.)

The enthalpy change for the exhaust nozzle $\Delta h_n$ is given by

$$\frac{\Delta h_n}{c_p t_0} = \frac{\eta_n}{\eta_c}[\alpha\eta_c - \alpha_d(\Theta_c - 1)] \times$$

$$\left\{1 - \frac{\alpha_1\eta_c\eta_t}{\Theta_c[1 + \eta_d(\alpha_d - 1)][\alpha_1\eta_c\eta_t - (\Theta_c - 1)]}\right\} \qquad (31)$$

The **specific thrust**, also called the **air specific impulse**, is

$$I_a = F_i/\dot{m}_a g = (\sqrt{2\Delta h_n} - V_0)/\dot{m}_a \qquad (32)$$

The overall efficiency $\eta_o$ is given by

$$\eta_o = \eta_{th}\eta_{p} = \frac{I_a V_0}{J Q_i}$$

$$= 2\eta_B\eta_c\frac{\lambda M_0\left(\frac{k-1}{2}\right)^{1/2}}{\alpha\eta_c - \left(1 + \frac{k.-1}{2}M_0^2\right)(\eta_c + \Theta_c - 1)} \qquad (33)$$

where $\eta_B$ ranges from 0.95 to 0.99.

The *TSFC* for a turbojet engine is given by

$$TSFC = \frac{\dot{m}_f}{F} \qquad (34)$$

It can be shown by dimensional analysis, if the effects of Reynolds numbers are neglected, that the variables entering into the performance of a given turbojet engine may be grouped as indicated in Table 2.

Figure 13 is a design-point chart presenting $\lambda$ and $\eta_0$ as functions of $\alpha$, $P_3/P_2$, and $\Theta_c$ for the subsonic performance of simple turbojet engines. The curves apply to propulsion at

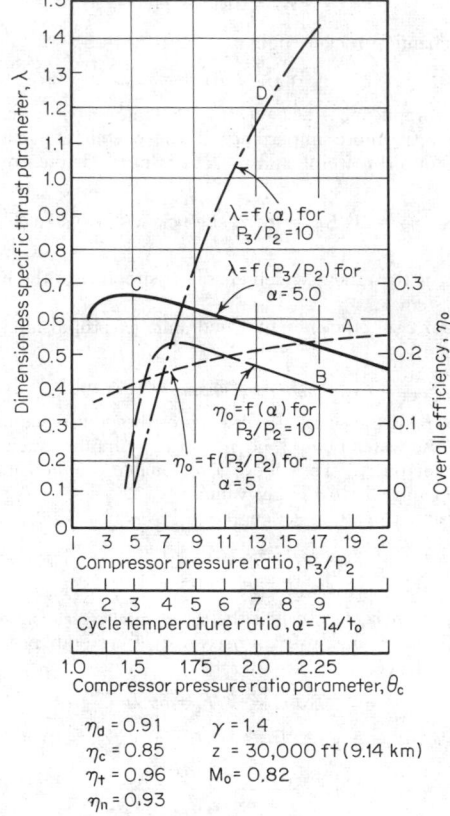

$$\begin{aligned}
\eta_d &= 0.91 & \gamma &= 1.4 \\
\eta_c &= 0.85 & z &= 30{,}000 \text{ ft (9.14 km)} \\
\eta_t &= 0.96 & M_0 &= 0.82 \\
\eta_n &= 0.93
\end{aligned}$$

**Fig. 13**  Dimensionless thrust parameter $\lambda$ and overall efficiency $\eta_o$ as functions of the compressor pressure ratio (simple turbojet engine).

30,000 ft (9.14 km) altitude for engines having characteristic data indicated in the figure. Curve A illustrates the effect of pressure ratio $P_3/P_2$ or $\Theta_c$ for engines operating with $\alpha = 5.0$. It is seen that increasing the compressor pressure ratio increases $\eta_o$ but it approaches a maximum value at $P_3/P_2$ approximately equal to 19. Curve B applies to the design point of engines having a fixed compressor pressure ratio of 10:1 but having different values of $\alpha$, that is, turbine inlet temperature. It shows that increasing $\alpha$ much above $\alpha = 5$ reduces the overall efficiency and consequently *TSFC*. Curve C presents

**Table 2**

| Non-dimensional group | Uncorrected form | Corrected form |
|---|---|---|
| Flight speed............................... | $V_0/\sqrt{t_0}$ | $V_0/\sqrt{\theta}$ |
| Rotational speed........................... | $N/\sqrt{T}$ | $N/\sqrt{\theta}$ |
| Air flow rate.............................. | $\dot{W}_a\sqrt{T}/D^2P$ | $\dot{W}_a\sqrt{\theta}/\delta$ |
| Thrust.................................... | $F/D^2P$ | $F/\delta$ |
| Fuel flow rate............................. | $\dot{W}_f J\Delta H_c/D^2P\sqrt{T}$ | $\dot{W}_f/\delta\sqrt{\theta}$ |

$\theta = T/T_{std} = T/519\ (T/288) =$ corrected temperature (exact value for $T_{std}$ is 518.699°R).
$\delta = P/p_{std} = P/14.7\ (P/1.013 \times 10^5) =$ corrected pressure.

the design-point characteristics for engines operating with $\alpha = 5$ but having different compressor pressure ratios. The maximum value of $\lambda$ is obtained at a relatively low pressure ratio: approximately $P_3/P_2 = 5$. Curve D presents the design-point characteristics of engines having a fixed compressor pressure ratio of 10:1 but different cycle temperature ratios. Increasing $\alpha$ gives significant increases in $\lambda$. Hence, large values of turbine inlet temperature $T_4$ offer the potential for large thrusts per unit of frontal area for such turbojet engines.

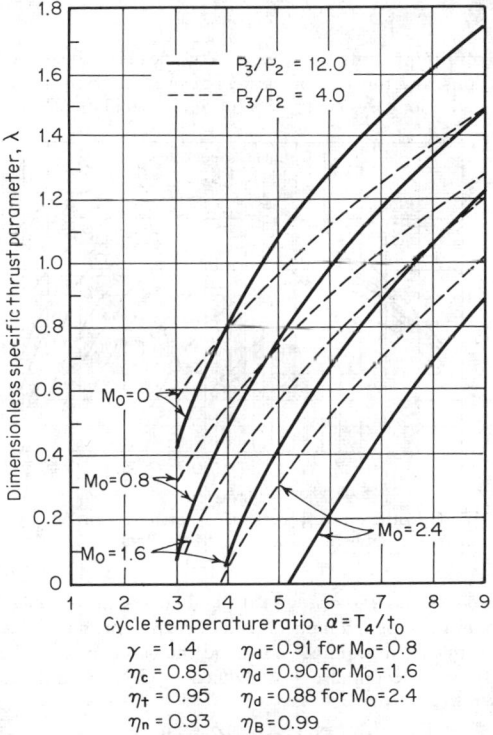

**Fig. 14** Dimensionless thrust parameter $\lambda$ for a simple turbojet engine and its overall efficiency $\eta_o$ as functions of the cycle temperature ratio $\alpha$, with the flight Mach number as a parameter.

Figure 14 presents the design-point characteristics for engines having $\eta_c = 0.85$, $\eta_t = 0.90$, $\eta_r = 0.93$, and $\eta_B = 0.99$. Different values of diffuser efficiency are presented for flight Mach numbers $M_0 = 0.8$ ($\eta_d = 0.91$), $M_0 = 1.60$ ($\eta_d = 0.90$) and $M_0 = 2.4$ ($\eta_d = 0.88$). Two different classes of

engines are considered, those with $P_3/P_2 = 4.0$ and those with $P_3/P_2 = 12.0$ The dimensionless thrust $\lambda$ is plotted as a function of $\alpha$ with $M_0$ as a parameter. It is evident that increasing $\lambda$, that is, the turbine inlet temperature $T_4$, offers the advantage of large increase in thrust per unit area of the engine under all the conditions considered in Fig. 14.

Figure 15 presents the effect of flight speed on the design-point performance, at 30,000 ft (9.14 km) altitude, for engines having $\eta_c = 0.85$, $\eta_t = 0.90$, $\eta_B = 0.95$, $T_4 = 2000°R$ (1111 K) and burning a fuel having a heating value $\Delta H_c = 18,700$ Btu/lbm (10,380 kcal/kgm). The magnitude of the compressor pressure ratio $P_3/P_2$ approaches unity when $V_0$ is approximately 1,500 mi/h, indicating that at higher speeds the compressor and turbine are superfluous; i.e., the propulsion requirements would be met more adequately by a ramjet engine. Raising $T_4$ tends to delay the aforementioned condition to a high value of $V_0$.

It is a characteristic of turbojet engines, since $\eta_o = \eta_{th}\eta_p$, that for a given $M_0$ increasing $\eta_{th}$ causes $\Delta h_n$ to increase, and hence the jet velocity $V_j$. As a consequence, $\eta_p$ is decreased. It is this characteristic which causes $\eta_o = f(\alpha)$, curve B of Fig. 13, to be rather flat for a wide range of values of $\alpha = T_4/t_0$.

The curves of Figs. 13 to 15 are design-point performance curves; each point on a curve is the design point for a different turbojet engine. The performance curves for a specific engine are either computed from experimental data pertaining to its components operating over a wide range of conditions or

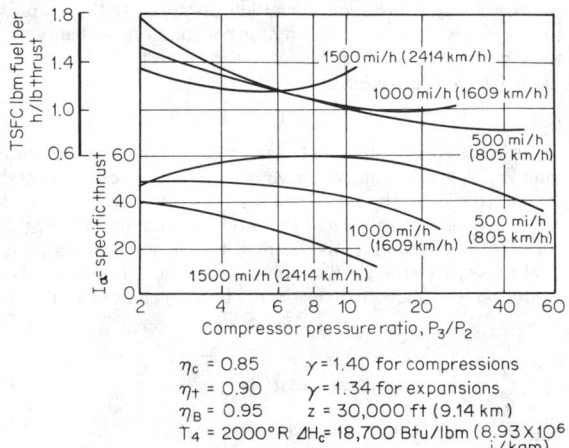

**Fig. 15** *TSFC* and specific thrust $I_a$ functions of the compressor pressure ratio, with flight speed as a parameter (simple turbojet engine).

obtained from testing the complete turbojet engine. It is customary to present the data in standardized form.

**The Turbofan or the Ducted-Fan Engine**   Refer to Fig. 4b. Assume that the hot-gas flow $(m_{a1} + m_f)$ is ejected through the converging nozzle, area $A_t$, with the effective jet velocity $V_j$, and that the airflow $m_{aF}$ is ejected through the converging annular nozzle, area $A_{tF}$, with the jet velocity $V_{jF}$. Let $\beta = \dot{m}_{aF}/\dot{m}_{a1}$ denote the bypass ratio, approximately 1.5 for current turbofan engines.

The thrust equation for the turbofan engine is

$$F = \dot{m}_{a1}(1 + f) V_j + \beta \dot{m}_{a1} V_{jF} - \dot{m}_{a1}(1 + \beta)V_0 \quad (35)$$

where $V_{jF} < V_j$. The thrust per unit mass of airflow, or air specific impulse, is

$$I_a = F/\dot{m}_{a1}(1 + \beta) \quad (36)$$

The *TSFC* is given by

$$TSFC = \dot{m}_f/F \quad (37)$$

For current turbofan engines the *TSFC* is approximately 0.75 of that for the turbojet counterparts. All studies indicate that long-range, long-endurance subsonic aircraft will require turbofan engines with larger bypass ratios. It should be noted that increasing the bypass ratio $\beta$ tends to increase the external drag of the engine. The optimum value of $\beta$ is that which makes the difference $(F - D_e)$ a maximum, where $D_e$ is the external drag of the engine.

To achieve the maximum potential for the turbofan engine, high values of turbine inlet temperature $T_4$ are essential; i.e., cooled turbine stator and rotor blades must be developed. There is an upper limit to $T_4$; above it the *TSFC* begins to increase. As $T_4$ is increased, however, the optimum value of $\beta$ must also be increased to avoid excessive leaving losses in the propulsive jet.

## CRITERIA OF ROCKET-MOTOR PERFORMANCE

In most applications of rocket motors the objective is to produce a large thrust for a limited period. The criteria of rocket-motor performance are, therefore, related to its thrust-producing capabilities and operating duration. If $\dot{W}_p$ denotes the rate at which the rocket motor consumes propellants, $P_c$ the total pressure at the entrance to the exhaust nozzle, and $C_w$ the **weight flow coefficient,** then

$$\dot{W}_p = C_w P_c A_t \quad (38)$$

Ordinarily, the values of $C_w$ as a function of $P_c$ and **mixture ratio** $\dot{W}_o/\dot{W}_f$ are determined experimentally. If no experimental data are available, the values of $C_w$ can be calculated with good accuracy from the calculated thermodynamic properties of the combustion gases. If $T_c$ is the total temperature, and $P_c$ the total pressure of the combustion gases at the entrance section of the nozzle, then, for steady operating conditions, $W_p = \dot{W}_g$,

$$\dot{W}_g = C_d A_t P_c \Omega \sqrt{\frac{g}{RT_c}} \quad (39)$$

where

$$\Omega = \left(\frac{2}{k + 1}\right)^{(k+1)/[2(k-1)]} \sqrt{k} \quad (40)$$

An equation similar to Eq. (38) can be written for the thrust developed by the rocket motor. Thus

$$F = C_F P_c A_t \quad (41)$$

The values of the **thrust coefficient** $C_F$ for specific propellant combinations are determined experimentally as functions of $P_c$ and the ratio $\dot{W}_o/\dot{W}_f$. When no experimental values are available, $C_F$ can be calculated from the following equation. Therefore

$$C_F = C_d \lambda \phi \Omega \sqrt{\frac{2k}{k - 1} Z_t} + \left(\frac{p_e}{P_c} - \frac{p_0}{P_c}\right)\frac{A_e}{A_t} \quad (42)$$

where

$$Z_t = 1 - \left(\frac{p_e}{P_c}\right)^{(k-1)/k}$$

Equation (42) shows that $C_F$ is independent of the combustion temperature $T_c$ and the molecular weight of the combustion gases. The nozzle area ratio $A_e/A_t$ is given by the formula

$$\frac{A_e}{A_t} = \frac{(P_c/p_e)^{1/k}}{\left(\dfrac{k + 1}{2}\right)^{1/(k-1)} \sqrt{[(k + 1)/(k - 1)]Z_t}} \quad (43)$$

Figure 16 presents $C_F$ calculated by means of the preceding equations, as a function of the area ratio $A_e/A_t$, with $P_c/p_0$ as a parameter, for gases having $k = 1.28$; the calculations assumed

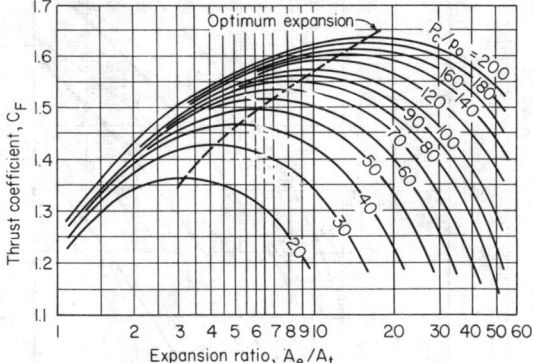

**Fig. 16**   Calculated thrust coefficient versus expansion ratio for different pressure ratios (based on $k = 1.28$, $\lambda = 0.9830$, $C_d = \varphi = 1.0$).

$\lambda = 0.983$ and $C_d = \varphi = 1.0$. The curves do not take into account the separation phenomena which occur when $A_e/A_t$ is larger than that required to expand the gases to $p_0$.

The specific impulse $I$ is defined by Eq. (8) and can be shown to be given by

$$I = F/\dot{m}_p = C_F/C_w = V_j/g_c \quad (44)$$

A criterion which is employed quite frequently for stating the performance of rocket motor is the characteristic velocity $c^*$. By definition

$$c^* = V_j/C_F = P_c A_t/\dot{m} = I/C_F \quad (45)$$

The values of $c^*$ for a given propellant combination are determined experimentally. It should be noted that the value of $c^*$ is independent of the thrust. Basically, $c^*$ measures the effectiveness with which the thermochemical energy of the propellants is converted into effective jet kinetic energy. When no experimental data are available the value of $c^*$ can be estimated quite closely from

$$c^* = \sqrt{RT_c}/\Omega \qquad (46)$$

Table 1 presents values of specific impulse for some possible rocket-propellant combinations.

Solid-propellant rocket motors may be segregated into (1) **double-base powders** and (2) **composite, or heterogeneous, propellants.** Double-base powders are gelatinized colloidal mixtures of nitroglycerin and cellulose to which certain stabilizers have been added. Heterogeneous, or composite, propellants are physical mixtures of a solid oxidizer in powder form and some form of solid fuel, such as a plastic or rubberlike material.

Solid-propellant rocket motor technology has made tremendous strides since the 1960s. It is possible to produce solid propellants with a wide range of linear burning rates and with good physical properties. It has been demonstrated experimentally that large thrust rocket motors with chamber diameters of approximately 22 ft (6.7 m) are feasible. As a result, the solid-propellant rocket motor has displaced the liquid rocket engine from a large number of applications, which heretofore were considered the province of the liquid rocket engine. In military applications, there is a strong tendency to supplant liquid rocket engines with solid rocket engines. In the 1970s, only in such areas as the large thrust launching rockets for boosting manned vehicles into space have the liquid rocket engines reigned supreme; the space shuttle will be equipped with liquid-bipropellant rocket engines, but it will be boosted with auxiliary solid-propellant rocket motors.

The following considerations are what make solid-propellant rocket motors attractive: (1) They are simpler in construction than liquid rocket engines; (2) they are easier to handle in the field; (3) they have instant readiness for use; (4) they have very good storage properties; (5) they have a larger average density $\bar{\rho}_P$ than most liquid-propellant combinations; (6) they have considerably fewer parts; and (7) their reliability in practice has been very good. Furthermore, once the performance characteristics and physical properties of a solid propellant have been established, it is a relatively straightforward engineering problem to design and develop solid-propellant rocket motors of widely different thrust levels and burning times $t_B$.

Rocket motors which burn double-base propellants find the greatest use in weaponry, e.g., artillery rockets, antitank weapons, etc.

Rocket motors burning heterogeneous, also called **composite**, solid propellants are used for propelling all kinds of military missiles and for sounding rockets. Modern composite propellants have three basic types of ingredients: (1) a fuel which is an organic polymer, called the **binder;** (2) a finely powdered oxidizer, ordinarily ammonium perchlorate ($NH_4NO_3$); and (3) various additives, for catalyzing the combustion process, for increasing the propellant density, for increasing the specific impulse (e.g., powdered aluminum), for improving the physical properties of the propellant, and the like. After the ingredients have been thoroughly mixed, the resulting viscous fluid is poured, usually under vacuum to prevent the formation of voids, directly into the chamber of the rocket motor which contains a suitable mandril for producing the desired configuration for the propellant grain. The propellant is cured

by polymerization to form an elastomeric material. Some of the more common binders are butadiene copolymers and polyurethane.

Modern composite solid propellants have densities ranging from 1.60 to 1.70, depending upon the formulation, and specific impulse values of approximately 245 s, based on a combustion pressure of 1000 psia and expansion to 14.7 psia.

Rocket motors using such propellants have been fabricated in chamber diameters as large as 260 in (660 cm).

In the case of solid-propellant rockets the rate of propellant consumption $\dot{m}_p$ is related to the **linear burning rate** for the propellant $r_0$, that is, the rate at which the burning surface of the propellant recedes normal to itself as it burns. For practical purposes it can be assumed that the linear burning rate $r_0$ is given by

$$r_0 = a_2 p_c^n \qquad (47)$$

where $a_2$ and $n$ are constants which are determined by experiment.

If $\rho_p$ denotes the density of the solid propellant and $A_p$ the area of the burning surface, then the propellant consumption rate $\dot{m}_p$ is given by

$$\dot{m}_p = A_p \rho_p r_0 = a_2 A_p \rho_p p_c^n \qquad (48)$$

Figure 17 presents the linear burning rate as a function of the combustion pressure for several propellants.

The linear burning rate $r_0$ is influenced by the temperature of the solid propellant $t_p$ prior to its ignition. Low values of $t_p$

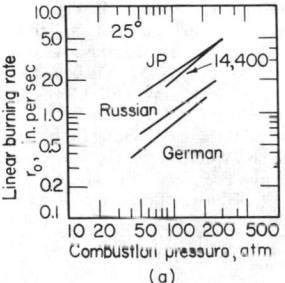

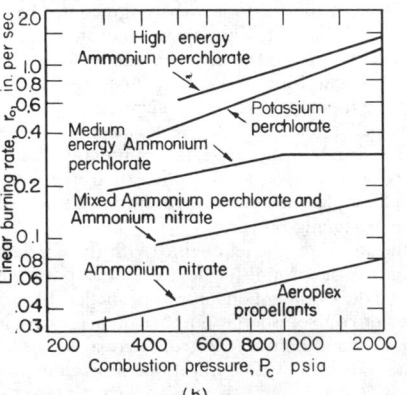

**Fig. 17** (a) Linear burning rate as a function of combustion pressure for double-base propellants. (b) Burning-rate characteristics of several heterogeneous propellants at 60°F. (*Aerojet-General Corp.*)

reduce the burning rate, and vice versa. Consequently, the temperature of the propellant must be given in presenting data on linear burning rates. Figure 18 presents $r_0$ as a function of $p_c$ for $t_p = -40$, 60, and 140°F for a composite propellant manufactured by the Aerojet-General Corp., Azusa, Calif.

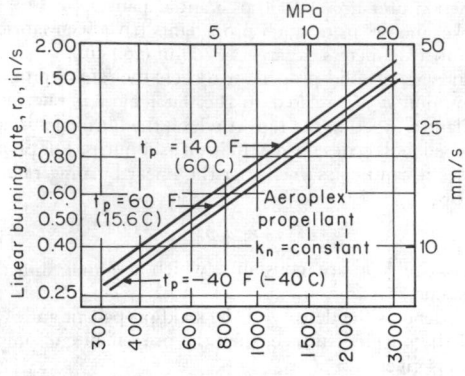

**Fig. 18**  Effect of propellant temperature upon the linear burning rate of a heterogeneous propellant. *(Aerojet-General Corp.)*

## AIRCRAFT PROPELLERS

REFERENCES: Theodorsen, "Theory of Propellers," McGraw-Hill. Smith, "Propellers for High Speed Flight," Princeton. Webb and Willer, "Propeller Performance at Zero Forward Speed," Wright Air Development Center Report 51–152. Kuchemann and Weber, "Aerodynamics of Propulsion," McGraw-Hill. Gray and Nicholas, "Representative Operating Charts of Propellers," NACA Wartime Report. Gutin, "On the Sound Field of a Rotating Propeller," *NACA-TM* 1195, Oct. 1948. Garrick and Watkins, "A Theoretical Study of the Effect of Forward Speed on the Free Space Pressure Field around Propellers," NACA, Oct. 1953. Ribner, "Propellers in Yaw," *NACA-TR* 820, 1945. Zucrow, "Principles of Jet Propulsion," Wiley. "Aircraft Propeller Handbook," Depts. of the Air Force, Navy, and Commerce, U.S. Government Printing Office, 1956.

The function of the aircraft propeller (or airscrew) is to convert the torque delivered to it by an engine into the thrust for propelling an airplane. If the airplane is in steady, level flight, the propeller thrust and the airplane drag equal each other. If the airplane is climbing, the thrust must overcome the drag plus the weight component of the airplane.

The propeller may be the sole thrust-producing element, but it can also be employed in conjunction with another thrust-producing device, such as the exhaust jet of a **turboprop** or **gas-turbine engine.** It may also be employed as a **windmilling** device for producing power.

Since the advent of the jet engine (with the nozzle replacing the propeller as the propulsive device), the importance and the research and development on aircraft propellers have declined. There are, of course, a large number of propellers in use on private aircraft and on the older commercial aircraft. Some of the new aircraft in STOL applications utilize the propeller as the propulsive device, and there is renewed interest in turboprop engines for energy-efficient cargo aircraft.

The conventional propeller consists of two or more equally spaced radial blades, which are rotated at a substantially uniform angular velocity. At any arbitrary radius, the section of a blade has the shape of an airfoil, but as the hub is approached, the sections become more nearly circular because of considerations of strength rather than aerodynamic performance. The portions of the blades in the vicinity of the hub contribute, at best, only a small portion of the thrust developed by the propeller blades.

Each section of the propeller blade experiences the aerodynamic reactions of an airfoil of like shape moving through the air in a similar manner. Furthermore, at each radius, the section of one blade forms with the corresponding sections on the other blades a series of similar airfoils, which follow one another as the propeller rotates.

The torque of the rotating propeller imparts a rotational motion to the air flowing through it and, furthermore, causes the pressure immediately behind the propeller to increase while that in front of it is reduced; i.e., the air is sucked toward the front of the propeller and pushed away behind it. A slipstream is created as illustrated in Fig. 19. The ratio of the cross-sectional area of the slipstream at any station to that of the actuator disk is termed the **slipstream contraction.**

The thrust produced by the propeller follows the relationships developed from the momentum theorem of fluid

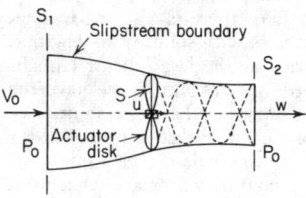

**Fig. 19**  Slipstream of a propeller.

mechanics for jet engines as presented in Fig. 11 and Eq. **(5)**. The thrust equation for the airbreathing engine (simplified) and the propeller is

$$F = \dot{m}(w - V_0) = \rho A V_0 (w - V_0) \qquad \textbf{(49)}$$

In the case of a propeller the mass flow $m$ is large and the velocity rise across the propeller is small, as compared with a jet engine of the same thrust. However, the thrust of propellers falls off rapidly as air compressibility effects on the propeller become large (flight speeds of 300 to 500 mi/h).

Propeller efficiency is defined as the useful work divided by the energy supplied to the propeller:

$$\eta_p = \frac{P_T}{P} = \frac{FV_0}{P} = \frac{m(v - V_0)V_0}{m/2\,(w^2 - V_0^2)} = \frac{2\nu}{1 + \nu} \qquad \textbf{(50)}$$

where $\nu = V_0/w$.

Since $w$ is only slightly larger than $V_0$, the ideal efficiency is of the order of 0.95, and actual efficiencies are in the range of 0.5 to 0.9.

### Propeller Theory

**Axial-Momentum Theory**  The axial-momentum theory of the propeller is due to Rankine and Froude, and the conception arose in the studies of propellers for ships. It is basically a special case of the generalized jet-propulsion system presented in Fig. 11.

The axial-momentum theory is only a first approximation of the action of the propeller and cannot be employed for design purposes. It neglects such factors as the drag of the blades, energy losses due to slipstream rotation, blade interference, and air compressibility effects. Because of those losses, an actual propeller requires power to rotate at zero thrust, and at zero thrust its propulsive efficiency is, of course, zero.

**Blade-Element Theory**  This theory, sometimes called the **strip theory**, takes into account the profile losses of the blade sections. It is the theory most commonly employed in designing a propeller blade and for assessing the off-design performance characteristics of the propeller.

Each section of the propeller is considered to be a rotating airfoil. It is assumed that the radial flow of air may be neglected, so that the flow over a blade section is two-dimensional. Figure 20 illustrates diagrammatically the velocity vec-

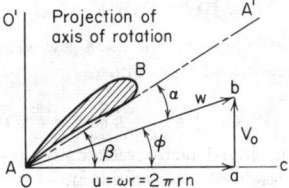

**Fig. 20**  Vector diagram for a blade element of a propeller.

tors pertinent to a blade section of length $dr$, located at an arbitrary radius $r$. The projection of the axis of rotation is $OO'$, the plane of rotation is $OC$, the blade angle is $\beta$, the angle of attack for the air flowing toward the blade (in the relative coordinate system) with the velocity $V_0$ is $\alpha$, the tangential velocity of the blade element is $u = 2\pi nr$, and the **pitch** or **advance angle** is $\phi$. From Fig. 20,

$$\alpha = \beta - \phi \qquad (51)$$

and

$$\phi = \tan^{-1}(V_0/\pi nd) \qquad (52)$$

Figure 21 illustrates the forces acting on the blade element. If $b$ denotes the width of the blade under consideration and $C_L$

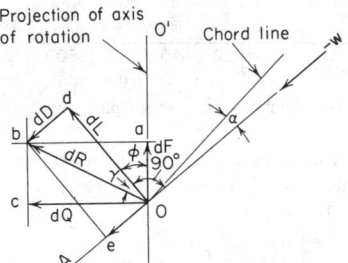

**Fig. 21**  Aerodynamic forces acting on a blade element.

and $C_D$ are the lift and drag coefficients of the blade (airfoil) section, then the thrust force $dF$ acting on the blade element is given by

$$dF = \tfrac{1}{2}\rho V_0^2 b \ dr C_L \frac{\cos(\phi + \gamma)}{\cos \gamma \sin^2 \phi} \qquad (53)$$

The corresponding torque force $dQ$ is given by

$$dQ = \tfrac{1}{2}\rho V_0^2 b \ dr C_L \frac{\sin(\phi + \gamma)}{\cos \gamma \sin^2 \phi} \qquad (54)$$

where

$$\tan \gamma = d\mathcal{D}/d\mathcal{L} = C_D/C_L = \mathcal{D}/\mathcal{L} \qquad (55)$$

where $\mathcal{D}$ and $\mathcal{L}$ denote drag and lift, respectively.

The propulsion efficiency of the blade element, termed the **blading efficiency**, is defined by

$$\eta_b = \frac{V_0 \ dF}{u \ dQ} = \frac{\tan \phi}{\tan(\phi + \gamma)} = \frac{\mathcal{L}/\mathcal{D} - \tan \phi}{\mathcal{L}/\mathcal{D} + \cot \phi} \qquad (56)$$

The value of $\phi$ which makes $\eta_b$ a maximum is termed the **optimum advance angle** $\phi_{opt}$. Thus

$$\phi_{opt} = \frac{\pi}{4} - \frac{\gamma}{2} = 45° - \frac{57.3}{2(\mathcal{L}/\mathcal{D})} \qquad (57)$$

The maximum blade efficiency is accordingly

$$(\eta_b)_{max} = \frac{2\gamma - 1}{2\gamma + 1} = \frac{2(\mathcal{L}/\mathcal{D}) - 1}{2(\mathcal{L}/\mathcal{D}) + 1} \qquad (58)$$

A force diagram similar to Fig. 21 can be constructed for each element of the propeller blade, taking into account the variation in $\phi$ with the radius. The resultant forces acting on the propeller are obtained by summing (integrating) those forces acting on each blade element.

The integrated lift coefficient is

$$C_{L_i} = \int_{r_L/R}^{r/R=1.0} C_{L_r} \left(\frac{r}{R}\right)^3 d\left(\frac{r}{R}\right) \qquad (59)$$

where $C_{Lr}$ = propeller-blade-section lift coefficient
$r/R$ = fraction of propeller-tip radius

The integrated lift coefficient is in the range 0.15 to 0.70. There is a dilemma in the design and selection of blade sections since high static thrust requires high $C_{L_i}$ which gives relatively low $\eta_p$ at cruise. Cruise conditions require low $C_{L_i}$ for good $\eta_p$. Variable camber propellers permit optimization of $C_{L_i}$ but at the expense of weight and complexity.

**Vortex Theory**  The simple two-dimensional blade-element theory can be further improved by taking account of the actual velocity distribution in the slipstream and thus determining the actual loading on the blading by the so-called *vortex theory*. The blades are represented by bound vortex filaments sweeping out vortex sheets to infinity. The vortex sheets are distorted both by the slipstream contraction and by the profile drag of the blades. If such distortions are neglected, the induced velocity in the wake can be calculated by assuming the vortex sheets to be tubular. Thus, in Fig. 20, the induced axial velocity becomes the arithmetic mean of the axial velocities at far distances upstream and downstream of the rotor disk; and the induced rotational velocity at the disk becomes $\omega - 0.5\omega'$, where $\omega'$ is the rotation of the slipstream in the plane of the propeller. Hence, the angle of attack $\alpha$ for a rotor element is given by

$$\alpha = \beta - \tan^{-1}[V_F/r(\omega - 0.5\omega')] \qquad (60)$$

where $V_F$ is the induced axial velocity in the plane of the propeller.

The addition of the vortex effects improves the representation of the propeller by the blade sections. There are, however, three-dimensional flow effects which are not included, and the following coefficients have been suggested to include this effect (which reduces the lift coefficient).

**Lift Curve Slope Correction Factors for Three-dimensional Airfoils***

| Radial station (ratio $r/R$) | Correction factor |
|---|---|
| (0.2)(0.3)(0.45)(0.6)(0.7) | 0.85 |
| 0.8 | 0.80 |
| 0.9 | 0.70 |
| 0.95 | 0.65 |

*Departments of the Air Force, Navy, and Commerce, "Propeller Handbook."

### Performance Characteristics

The pitch and the angle $\phi$ (Fig. 20) have different values at different radii along a propeller blade. It is customary, therefore, to refer all parameters determining the overall characteristics of a propeller to their values at either $0.7r$ or $0.75r$.

For a given blade angle $\beta$, the angle of attack $\alpha$ at any blade section is a function of the **velocity coefficient** $V_0/u$, where $u = \pi n d_p$; it is customary, however, to replace $V_0/u$ by its equivalent, the **advance ratio** $J = V_0/nd_p$.

For given values of $\beta$ and $n$, the angle of attack $\alpha$ attains its maximum value when $V_0 = 0$, with $\phi = 0$. In other words, the propeller thrust $F$ is maximum at takeoff. As $V_0$ increases, $\alpha$ decreases and so does the thrust. Finally, a forward speed is reached for which the thrust is zero.

Increasing $J = V_0/nd_p$, for a constant blade angle, causes the propeller to operate successively as a fan, propeller, brake, and windmill. Most of the operation, however, is conducted in the propeller state. At takeoff, the propeller is in the fan state. Windmilling must be avoided because it may overspeed the engine and damage it.

The lift coefficient $C_L$ is a linear function of $\alpha$ up to the stalling angle, and $C_D$ is a quadratic function of $\alpha$. Furthermore, the lift $\mathcal{L}$ is zero at a negative value for $\alpha$. Figure 22 presents the $\mathcal{L}/\mathcal{D}$ ratios of typical propeller sections. Figure 23 presents the blade (section) efficiency as a function of the pitch angle $\phi$ for different values of $\mathcal{L}/\mathcal{D}$. Propellers normally operate at $\mathcal{L}/\mathcal{D}$ ratios of 20 or more. Supersonic propellers operate with $\mathcal{L}/\mathcal{D}$ ratios of approximately 10, and from Fig. 22 it is apparent that they must operate close to the optimum pitch angle $\phi$ to obtain usable blade efficiencies.

**Propeller Coefficients**   It can be shown, neglecting the compressibility of the air, that

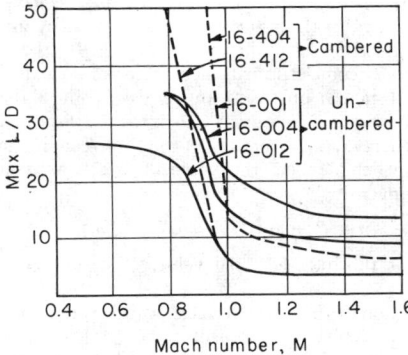

**Fig. 22**   Maximum lift-drag ratios for representative sections.

$$f(V_0, n, d_p, \rho, F) = 0 \tag{61}$$

By dimensional analysis, the following coefficients are obtained for expressing the performances of propellers having the same geometry:

$$F = \rho n^2 d_p^4 C_F \qquad Q = \rho n^2 d_p^5 C_Q \qquad \mathcal{P} = \rho n^3 d_p^5 C_P \tag{62}$$

$C_F$, $C_Q$, and $C_P$ are termed the **thrust, torque,** and **power coefficients,** respectively. They are independent of the size of the propeller. Consequently, tests of small-scale models can be employed for obtaining the values of $F$, $Q$, and $\mathcal{P}$ for geometrically similar full-scale propellers. Hence, the ideal propulsive efficiency of a propeller is given by

$$\eta_p = \frac{FV_0}{\mathcal{P}} = \frac{C_F}{C_P} \frac{V_0}{d_p n} = J \frac{C_F}{C_P} \tag{63}$$

Other useful coefficients are the **speed-power coefficient** $C_S$ and the **torque-speed coefficient** $C_{QS}$. Thus,

$$C_S = V_0 5 \sqrt{\rho/\mathcal{P}n^2} \qquad C_{QS} = V_0 \sqrt{\rho d_p^3/Q} \tag{64}$$

A commonly employed factor related to the geometry of a propeller is the **solidity** $\sigma$. By definition,

$$\sigma = c/p \tag{65}$$

The solidity varies from radius to radius; at a given radius, $\sigma$ is proportional to the power-absorption capacity of the annulus of blade elements.

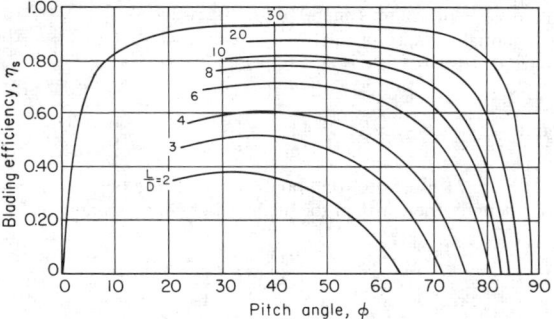

**Fig. 23**   Blade-element efficiency versus pitch angle.

The **activity factor** of a blade $AF$ is an arbitrary measure of the blade solidity and hence of its ability to absorb power. It takes the form

$$AF = \frac{100,000}{16} \int_{r_h/R}^{r/R\,=\,1.0} \frac{c}{d_p} \left(\frac{r}{R}\right)^3 d\left(\frac{r}{R}\right) \tag{66}$$

where $R = d_p/2$, $r$ = radius at any section, and $r_h$ = spinner radius.

### Performance

**Static Thrust**   From the simple axial-momentum theory, the static thrust $F_0$ of an actuator disk of diameter $d_p$ is given by

$$F_0 = (\pi\rho/2)^{1/3}(\mathcal{P}d_p)^{2/3} = 10.4\text{bhp} \\ \times d_p^{2/3} \quad \text{(at sea level)} \tag{67}$$

The **thrust horsepower** (thp) is accordingly

$$\text{thp} = \frac{F_0}{\text{bhp}} = (38,400/Nd_p)(1/C_p^{1/3}) \qquad (68)$$

Equations (67) and (68) assume uniform axial velocity, no rotation of the air, and no profile losses. Consequently, they may be in error by as much as 50 percent. The "Aircraft Propeller Handbook" utilizes a graphical method for obtaining the static thrust; Hesse suggests the use of the equations

$$\eta_{p_{\text{static}}} = \sqrt{\frac{2}{\pi}} \frac{C_F^{3/2}}{C_P} = 0.798 \frac{C_F^{3/2}}{C_P} \qquad (69)$$

and

$$F = 10.42[d_p\,(\eta_{p_{\text{static}}})\,\text{shp}]^{2/3}\,(\rho/\rho_0)^{1/3} \qquad (70)$$

together with data computed for a specific propeller, as illustrated in Fig. 24. The static thrust for a given power input

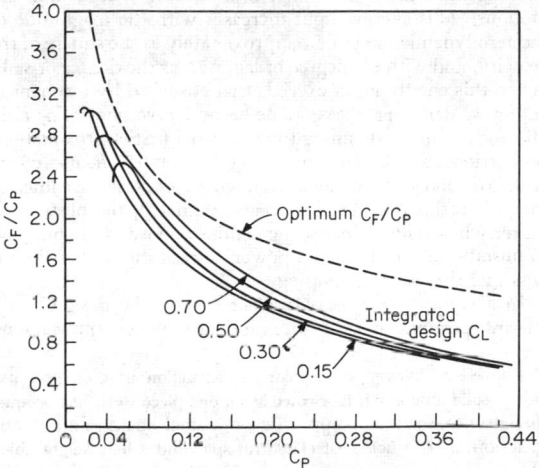

**Fig. 24** Static-thrust performance of a propeller; four-bladed, 100 activity propellers of various integrated design $C_L$'s. *(From Hamilton Standard Division, United Aircraft Corporation.)*

increases with the diameter and the number of blades. Controllable pitch propellers selected for good overall performance will develop 3 to 4 lb of thrust per shp.

At flight conditions the performance may be obtained from generalized charts presenting $C_P$ as a function of the advance ratio $J$, with the propeller efficiency and blade pitch angle at 75 percent radius, $\theta_{3/4}$, as parameters. Figure 25 is such a chart.

**Reverse Thrust** By operating the propeller blades at large negative angles of attack, reversed thrust can be developed. In that condition, the blades are stalled, and as in the case of static thrust, the forces acting on the blades cannot be calculated. In practice, the reverse-pitch stop is set by trial and error so that the propeller can absorb the rated power at zero speed, i.e., at $V_0 = 0$. With fixed-pitch reversing, the power absorbed when there is forward speed will be smaller than the rated value, but the reversed thrust is usually substantially larger than the static forward thrust corresponding to the rated power. The landing roll of an aircraft with reversed thrust

plus brakes is approximately 45 percent of that with brakes alone.

**Compressibility Effects** As the relative velocity between an airfoil and the free-stream air approaches the sonic speed, the lift coefficient $C_L$ decreases rapidly, and simultaneously there is a large increase in the drag coefficient $C_D$. The propel-

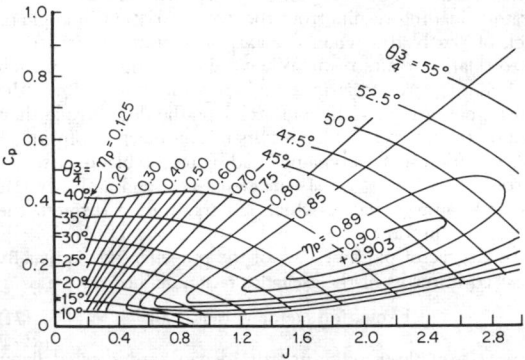

**Fig. 25** Propeller performance—efficiency in forward flight; four-bladed, 100 activity factor, 0.500 integrated design $C_L$ propeller. *(From Hamilton Standard Division, United Aircraft Corporation.)*

ler is subject to those effects. Since the acoustic speed decreases from sea level to the isothermal altitude, the effects appear at lower tip speeds at high altitude. Furthermore, the effect is more pronounced at higher tip speeds. Shock waves are formed at the leading edge of a propeller blade when its pitch-line velocity approaches the local sonic velocity of the air. As a consequence, the coefficient $C_F$ is reduced and $C_P$ is increased, thereby adversely affecting the propulsive efficiency of the propeller. Moreover, as the tip speed approaches the sonic speed, there is a large increase in the noise generated by the propeller.

Operation of a propeller in the transonic range requires that the blades be thin and have sweepback. Such designs present severe mechanical problems. Moreover, there is also the possibility of the propeller blades operating in the shock-stalled condition, which would cause serious losses if flow separation occurred. A propulsive efficiency of approximately 70 percent might be achieved with careful design for such a propeller.

**Noise** The noise produced by a propeller arises from the rotation of the pressure fields associated with the aerodynamic pressures acting on the blades. The fundamental frequency of the rotating pressure pattern produced by the propeller is, to a stationary observer, equal to the blade-passage frequency. The noise level associated with a given tip speed can be estimated by Gutin's method (see References under Aircraft Propellers). The noise intensity is determined by the number of blades, the tip speed, and the power input per unit of actuator-disk area. Increasing the number of blades, reducing the tip speed, and reducing the power loading reduce the noise. Reduction in noise level usually requires a compromise between performance and weight and is, therefore, not usually a factor in propeller selection. For high-subsonic-speed operation, see Garrick and Watkins (see References under Aircraft Propellers).

## Mechanical Design

**Blades** The principal blade loadings are (1) steady tensile, due to centrifugal forces; (2) steady bending, due to the aerodynamic thrust and torque forces; and (3) vibratory bending, due to cyclic variations in airloads and other excitations originating in the engine. The most serious and limiting stresses usually result from the vibratory loadings. The principal vibratory loading results from the cyclic variation in angle of attack of the blades when the axis of rotation is pitched or yawed relative to airstream. When the axis is pitched up, such as when the aircraft is in a high-angle-of-attack climb, the blades are at a higher angle of attack on the downstroke than when on the upstroke. This results in a once per revolution, $1 - P$, variation in aerodynamic loading on the blades, usually referred to as $1 - P$ **aerodynamic excitation,** and a steady **side force** and **yawing moment,** which are transmitted through the shaft to the aircraft.

The degree of pitch or yaw of the propeller axis is usually measured in terms of the **excitation factor,** defined as

$$\text{Excitation factor} = \psi(V_i/400)^2 \qquad (71)$$

where $\psi$ = angle of pitch or yaw, deg; $V_i$ = indicated flight velocity, mi/h. Another factor sometimes used is $Aq$, where $A = \psi^\circ$, $q = 12\rho V_i^2$, $Aq$ = excitation factor × 410. Ribner presents means of estimating the forces on inclined propellers.

Because of the restoring moment of the **centrifugal loads** on the blade elements when the blade deflects in bending, centrifugal forces have the effect of apparently stiffening the blade in **bending.** In Fig. 26, the bending natural frequency increases

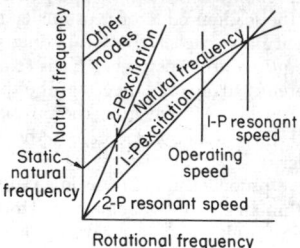

**Fig. 26** Relation between blade natural frequency and major excitations.

with rpm. At some rpm, the bending natural frequency will come into resonance with the $1 - P$ excitation, at which point small loadings will be greatly magnified (limited only by the damping present in the blade).

Propellers are designed so that the $1 - P$ resonant speed is above any expected operating speed. However, there will always be some magnification of the vibratory loads (due to proximity to resonant speed), resulting in disproportionately high vibratory stresses. As indicated in Fig. 26, blades normally pass through a $2 - P$ **resonance** in coming up to speed (usually not critical because it is a transient condition) and may pass through or be close to resonance with other modes.

The **natural frequencies** of the rotating propeller blade can be computed with good accuracy, but since the excitation varies from aircraft to aircraft and is usually not well defined, vibration surveys must be made of all new installations. These often result in the definition of ranges at which continuous

operation should be avoided. **Allowable vibratory stresses** are limited by the fatigue strength of the material. Unlike the steady-stress limits, **fatigue strength** is not a unique function of the material but is a function of surface finish as well. A sharp notch in the surface reduces the fatigue strength to a very low value. If propellers were designed to hold vibratory stresses below the lowest possible fatigue strength, the blades would be unacceptably heavy. Thus propellers are designed on the maintenance of a reasonably smooth surface finish. Since the blade surfaces are constantly subject to **store damage,** blades must be regularly inspected to make sure that the finish does not deteriorate below design standards.

Propeller blades are stalled when operating at low forward speeds, and when reversing (see Perfomance, above). Under these conditions **stall flutter** can occur. Stall flutter appears as a torsional oscillation caused by a lagging of the aerodynamic moment relative to the blade motion. This results in placing one component of the aerodynamic moment in phase with the twisting motions of the blade, thus feeding energy into said motions. This energy input increases with the magnitude of the aerodynamic forces i.e., approximately as the square of the velocity, and with the degree of lag, i.e., as the degree of stall. When this energy input exceeds that absorbed by mechanical and other damping, the motions become divergent. The ratio of aerodynamic to damping forces is, to a first approximation, proportional to $V/b\omega$, where $V$ = velocity at $0.75R$, $b$ = chord at $0.75R$, and $\omega$ = torsional frequency. Experience indicates that when this parameter is greater than 1.0, the blades will flutter when stalled. Increasing camber, chord, and rpm tend to unstall a blade at a given power. Increasing torsional stiffness increases $\omega$ and damping.

In general, $1 - P$ **considerations** control the design of the **inboard** portions of a blade, and **stall flutter** the **outboard** sections.

There are five types of **blade construction** in common use today: **solid aluminum; fabricated steel; one-piece steel; monocoque, fiber-glass construction,** supported on a steel shank; and a construction with a **hollow steel central spar and a lightweight fiber-glass cover** that forms the aerodynamic contour of the blade. As in any beam designed on the basis of bending loads, it is desirable to concentrate the structural material at the maximum radii, as in the flanges of a simple I beam, and omit it from the center, where it carries no load. This is accomplished, with steel blades, by making them hollow. The structural material is concentrated in the outer fibers, or surface. In general, steel blades weigh about 80 to 85 percent of the equivalent solid aluminum blade. The advantage of steel increases with size. The blades normally constitute from one-third to one-half of the total propeller weight.

The **hub** retains the blades and contains the pitch-change motor. A **pitch-change mechanism** is provided so that the blades can be positioned at the proper angle of attack $\alpha$ to absorb the desired power at the desired rpm, regardless of flight speed. If the pitch is unchanged, as in Fig. 27, the angle of attack is excessive at low speed. This prevents the engine from reaching rated speed and delivering rated power. The centrifugal loads on the blade elements tend to rotate the blades toward flat pitch (see Fig. 28). The pitch-change motor acts against this moment by applying a pitch-increasing moment. The addition of **counterweights,** shown dotted in Fig. 28, reduces the moment and the size of the pitch-change motor. Counterweights, however, add to the radial centrifugal load of the blades, necessitating a heavier retention, and consequently do not necessarily reduce the net weight of the propeller.

Most pitch-change mechanisms employ a **hydraulic piston,** mounted on the hub, with feed through the propeller shaft and with rotation of the blades by means of gears or links. **Electric motors** and **direct mechanical drives** have been used successfully. In the past, the weight of all systems, when fully developed, has turned out to be about the same.

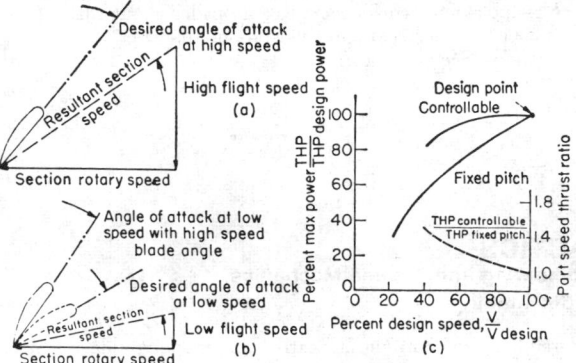

**Fig. 27** Effect of flight speed on the desired blade angle.

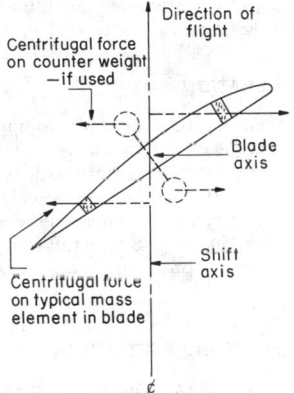

**Fig. 28** Twisting effect of centrifugal loads on blades.

Action of the pitch-change motor is limited by **low-** and **high-pitch stops.** These prevent the blades from assuming negative angles with reverse thrust. On reciprocating engines, the low-pitch stop is set to allow the propeller to absorb approximately rated power at zero forward speed. A lower setting results in higher windmilling drags in gliding flight, which could be dangerous if carried too far. In reversing propellers the low-pitch stop is automatically removed when the controls are set for reverse pitch. The high-pitch stop prevents the propeller from reversing through the positive range. In feathering propellers it is set at the angle which gives zero aerodynamic moment about the shaft.

**Propeller feathering** is provided in order to reduce the drag of a dead engine. To accomplish it, the blade-angle range is extended to 90°, approximately. Auxiliary pitch-change power is provided to complete the feathering as the engine slows down and to unfeather when the engine is stationary.

**Reversing** is provided in order to give propeller braking on the landing roll. It is accomplished by removing the flight low-pitch stop and driving the blades to a high negative **reverse-pitch angle.** The principal mechanical problem is to provide a mechanism which has the least possible chance of going into reverse inadvertently.

The **propeller control** is that portion of the system which regulates the blade angle. In light aircraft, this may consist merely of a mechanism which sets a given blade angle corresponding to a given position of the cockpit control. In such cases the propeller acts as a fixed-pitch propeller except when activated by the pilot. In most cases, the propeller control regulates blade angle so as to maintain a preset rotary rpm, irrespective of flight speed or shaft power. The **basic elements** of this system normally consist of a spring-balanced, engine-drive flyweight delivering an error signal which directs the servo, or pitch-change motor, to increase or decrease blade angle (Fig. 29). The control loop is closed by rpm feedback through the main engine. This basic system may be embel-

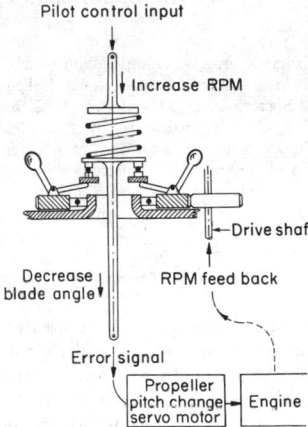

**Fig. 29** Propeller-speed governor.

lished with anticipating and delay devices to maintain speed in rapid throttle movements, synchronizing to coordinate two or more engines, and with overriding features to activate the reversing and feathering sequences. (See also Sec. 16.)

Propeller **controls** are fundamentally the same, with several added secondary features. First, because the engine is capable of a very rapid increase or decrease in torque and because of the high gear ratio between the propeller and engine shaft, anticipating features are necessary to prevent overspeeds in the propeller from producing large overspeeds and possible failure in the turbine. Second, because of the high idle rpm needed to maintain the pressure ratio in the compressor, the low-pitch stop of the turboprop engine propeller must be set substantially lower than for a piston engine. This is not true of a free-turbine drive and is only partially true of a split-compressor engine. In the event of an **engine failure,** the propeller, sensing the loss in rpm, will govern down to the low-pitch stop. This may lead to large, often uncontrollable, negative thrust. To avoid this possibility, an engine-shaft torque signal is put into the control, a negative torque automatically activating the feathering system. This portion of the system is known as the **ENT (emergency negative-thrust)** system. Third, because of the high idle rpm characteristic of the turboprop engine, and the resulting low blade angles at low power in the ground-taxi regime, provision must be made to cut out the governing system and substitute direct blade-angle regulation (known as **beta control**). This is required because the propeller will not govern under these conditions as the torque-to-blade-angle relationship goes to zero. The propeller control of a turboprop engine is normally mechanically coordinated directly to the engine throttle, leaving only a single control, with so-called power lever, in the aircraft.

The shaft on which the propeller is mounted is subjected to the steady-thrust load, vibratory loads due to the propeller side, and **gyroscopic forces** (the $1 - P$ moment, acting on the blades, is reactionless in propellers with three or more blades). In large aircraft, the shaft size is usually determined by the propeller side force (Ribner). In aircraft used for aerobatics, gyroscopic loads may become limiting. The **gyroscopic moment** on the propeller shaft is given by $2\omega_1 \cdot \omega_2(I_1 - I_2)$, where $\omega_1 =$ angular velocity of maneuver, $\omega_2 =$ angular velocity of shaft, $I_1 =$ polar moment of inertia of propeller about shaft, $I_2 =$ moment of inertia of propeller about diameter in plane of rotation. (See also Sec. 3.)

# ASTRONAUTICS
## by The Department of Aerospace Engineering and Applied Mechanics, University of Cincinnati

REFERENCES: "Progress in Astronautics and Aeronautics," AIAA series, Academic Press. "Advances in Astronautical Sciences," American Astronautical Society series, "Handbook of the British Astronautical Association." Koelle, "Handbook of Astronautical Engineering," McGraw-Hill. Purser, Faget, and Smith, "Manned Spacecraft: Engineering Design and Operation," Fairchild Publications. Herrick, "Astrodynamics," Vol. I, Van Nostrand Reinhold.

## SPACE FLIGHT
### by N. A. Armstrong

The science of astronautics deals with the design, construction, and operation of craft capable of flight through interplanetary or interstellar space. In addition to orbits around, and trajectories between, such bodies as stars, planets, and planetoids, the ascents from and descents to the surfaces of such bodies are considered to be part of the field.

The purposes of such craft are numerous and include the transportation of men and cargo, the transmission or relay of signals, and the carrying of instruments for the measurement of the characteristics of (1) the environment through which the craft flies, (2) the surface of the celestial body over which the craft flies, or (3) astronomical objects or phenomena.

### SATELLITES AND SPACE PROBES

An object that remains on a closed path (orbit) around some celestial body is called a **satellite.** A **space probe** is a craft flying along a nonclosed path (trajectory) for the purpose of exploring regions remote from the originating (launching) location. A number of significant flights of each type are listed in Table 1.

The wide variation in orbital characteristics available adds greatly to the flexibility and utility of various types of spacecraft. A **polar** orbit of each, for example, is one whose orbital plane includes the earth's axis and permits line-of-sight contact between the spacecraft and every point on earth on a periodic basis. The **synchronous** orbit has an angular rate equal to the rotational rate of the earth about its axis, providing a space-craft locus of position at nearly constant longitude. Orbital parameter variation provides a wide variety of other orbits, permitting the selection of characteristics most suitable for the specific flight objective, such as communications relays, meteorological observations, or earth surface monitoring.

### SPACECRAFT OPERATION

Trajectory adjustments and spacecraft equipment and payload operation may be performed by automatic control, remote control, human operators on board, or any combination of the three. As the time delays associated with commands radioed over very long distances increase, the difficulties of remote control increase. Spacecraft navigation is based on the mathematical processing of radar or optical measurements.

### SPACECRAFT LAUNCHING SYSTEMS

Spacecraft launched from earth have, in general, used launching systems powered by chemical rockets. Typical launch vehicles and their characteristics are listed in Fig. 1. Such launch vehicles have been abandoned after use. Recoverable launch vehicles are expected to produce significant economic advantages when the launch rate requirements are high. When the propulsion-system efficiency is sufficiently high and/or the gravitational field of the planetary body being departed is sufficiently low, the launching system may be an integral part of the spacecraft.

## ASTRONOMICAL CONSTANTS OF THE SOLAR SYSTEM
### by Helmut G. L. Krause

REFERENCES: Handbook of the British Astronautical Assoc. Blanco and McCuskey, "Basic Physics of the Solar System," Addison-Wesley. Clarke, Constants and Related Data for Use in Trajectory Calculations, *Tech. Rep.* 32-604, Jet Propulsion Lab., Pasadena. Francis, Constants

**Table 1. Significant Nonmilitary Spacecraft Flights**

| Name | Launch date | Weight, kg | Manned Unmanned | Results |
|---|---|---|---|---|
| Sputnik 1 | 10-57 (U.S.S.R.) | 84 | U | First artificial satellite |
| Explorer 1 | 2-58 (U.S.) | 14 | U | Radiation-belt monitor |
| Pioneer 1 | 10-58 (U.S.) | 38 | U | Lunar probe |
| Luna 2 | 9-59 (U.S.S.R.) | 390 | U | Lunar impact |
| Tiros 1 | 4-60 (U.S.) | 123 | U | Weather satellite |
| Echo 1 | 8-60 (U.S.) | 75 | U | Passive communications |
| Vostok 1 | 4-61 (U.S.S.R.) | 4,725 | M | First manned satellite |
| Mercury-Atlas 6 | 2-62 (U.S.) | 1,370 | M | 3 Orbits |
| Telstar | 8-62 (U.S.) | 78 | U | Active repeater, communications |
| Vostok 3 | 8-62 (U.S.S.R.) | 4,722 | M | Dual, simultaneous flight |
| Vostok 4 | 8-62 (U.S.S.R.) | 4,728 | M | |
| Mariner 2 | 8-62 (U.S.) | 206 | U | Venus probe |
| Atlas-Centaur | 12-63 (U.S.) | . . . . | U | Hydrogen stage |
| Ranger 7 | 7-64 (U.S.) | 362 | U | Lunar photographs |
| Early Bird | 4-65 (U.S.) | 39 | U | Communications, synchronous orbit |
| Luna 9 | 1-66 (U.S.S.R.) | 1,538 | U | Lunar landing |
| Gemini VI | 12-65 (U.S.) | 3,556 | M | Rendezvous |
| Gemini VII | 12-65 (U.S.) | 3,671 | M | |
| Gemini VIII | 3-66 (U.S.) | 3,795 | M | Docking |
| Agena | 3-66 (U.S.) | | U | |
| Apollo 8 | 12-68 (U.S.) | 129,000 (Earth orbit) | M | Circumlunar |
| Apollo II | 7-69 (U.S.) | | M | Lunar landing and return |
| Venera 7 | 8-70 (U.S.S.R.) | 1,108 | U | Venus landing |
| Salyut 1 | 4-71 (U.S.S.R.) | | U/M | Space laboratory |
| Soyus 10 | 4-71 (U.S.S.R.) | | M | |
| Soyus 11 | 6-71 (U.S.S.R.) | | M | |
| Pioneer 10 | 3-72 (U.S.) | 260 | U | Jupiter probe |
| Skylab 1 | 5-73 (U.S.) | 74,783 | U/M | Space laboratory |
| Skylab 2 | 5-73 (U.S.) | 19,979 | M | |
| Skylab 3 | 7-73 (U.S.) | 20,121 | M | |
| Skylab 4 | 11-73 (U.S.) | 20,847 | M | |
| Mariner 10 | 11-73 (U.S.) | 503 | U | Venus-Mercury flyby |

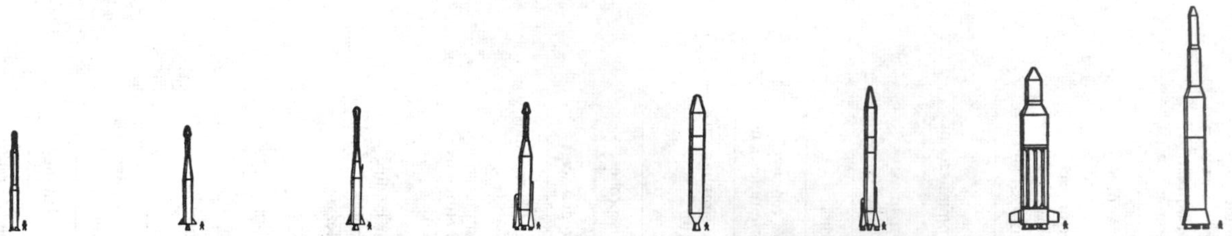

| | Scout | Delta | Thor-Agena B | Atlas-Agena B | Titan II | Centaur | Saturn 1B | Saturn V |
|---|---|---|---|---|---|---|---|---|
| Stages......<br>Propellants | 4<br>Solid | 3<br>1st stage, liquid oxygen and kerosine (LOX/RP); 2d stage, unsymmetrical dimethylhydrazine (UDMH) and inhibited red fuming nitric acid (IRFNA); 3d stage, solid | 2<br>1st stage, liquid oxygen and kerosine (LOX/RP); 2d stage, unsymmetrical dimethylhydrazine (UDMH) and inhibited red fuming nitric acid (IRFNA) | 2<br>1st stage, liquid oxygen and kerosine (LOX/RP); 2d stage, unsymmetrical dimethylhydrazine (UDMH) and inhibited red fuming nitric acid (IRFNA) | 2<br>Sterble [a blend of unsymmetrical dimethylhydrazine (UDMH) and hydrazine as fuel; nitrogen tetroxide as oxidizer] | 2<br>1st stage, liquid oxygen and kerosine (LOX/RP); 2d stage, liquid oxygen and liquid hydrogen (LOX/LM) | 2<br>1st stage, liquid oxygen and kerosine (LOX/RP); 2d stage, liquid oxygen and hydrogen (LOX/LM) | 3<br>1st stage, liq oxygen and ke sine (LOX/R 2d and 3d stages, liquid oxygen and liquid hydrog (LOX/LM) |
| Thrust.... | 1st stage (Algol II-A), 86,000 lb at sea level; 2d stage (Castor), 64,000 lb; 3d stage (Antares), 23,000 lb; 4th stage (Altair), 3,000 lb | 1st stage (Thor, DM-21), 170,000 lb at sea level; 2d stage (AJ10-118) 7,700 lb; 3d stage (Altair), 2,800 lb | 1st stage (Thor, DM-21), 170,000 lb at sea level; 2d stage (Agena B), 16,000 lb | Atlas D, 367,000 lb at sea level; Atlas D sustainer, 80,000 lb; Agena B, 16,000 lb | 1st stage (XLR-87), 430,000 lb at sea level; 2d stage (XLR-91), 100,000 lb | Atlas D, 367,000 lb at sea level; Atlas D sustainer, 80,000 lb; Centaur, 30,000 lb | 1st stage (S-1), 1,600,000 lb at sea level; 2d stage (S-IVB), 200,000 lb in a vacuum | 1st stage (S-I 7,500,000 lb sea level; 2d stage (S-11), 1,000,000 lb i vacuum; 3d stage (S-IVB) 200,000 lb in vacuum |
| Max diameter | 3.3 ft, excluding fins | 8 ft, excluding fins | 8 ft, excluding fins | 10 ft (16 ft at base) | 10 ft | 10 ft, excluding fins | 21.6 ft, excluding fins | 33 ft, excludi fins |
| Height.... | 65 ft, less spacecraft | 88 ft, less spacecraft | 76 ft, less spacecraft | 91 ft, less spacecraft | 90 ft, less spacecraft | 100 ft, less spacecraft | Approximately 150 ft, less spacecraft | 280 ft, less spa craft |
| Payload... | 220 lb in 300-nm orbit | 300 lb in 350-nm orbit; 130 lb escape | 1,600 lb in 300-nm orbit | 6,000 lb in 300-nm orbit; 750 lb escape; 425 lb to Mars or Venus | More than 6,000 lb in low earth orbit | 8,500 lb in 300-nm orbit; 2,300 lb escape; 1,300 lb to Mars or Venus | 34,000 lb in 100-nm orbit | 240,000 lb in 300-nm orbit; 90,000 lb esca 70,000 lb to Mars or Venu |
| Use........ | Launching of probes and satellites | Launching of scientific satellites and space probes | Launching of meteorological, communications, and scientific satellites | Lunar missions and launching of communications and scientific satellites | Launching of the two-man Gemini spacecraft | Launching of earth satellites and lunar and planetary exploration missions | Launching into earth orbit of the Apollo spacecraft with its Lunar Excursion Module (LEM), as well as the launching of unmanned payloads in support of the manned lunar-landing program | Launching of circumlunar a lunar leading flights |

**Fig. 1.** United States Launch Vehicles

for an Earth-Moon Transit, *Lockheed Rep. LAC/421571.* Krause, On a Consistent System of Astrodynamic Constants, *NASA-TN* D-1642. Makemson, Baker, and Westrom, Analysis and Standardization of Astrodynamic Constants, *Jour. Astro. Sci.*, Spring, 1961. Middlehurst and Kuiper, "The Solar System," Vols. 1–4, University of Chicago Press. Townsend, "Orbital Flight Handbook," Martin Co.

A greater number of astronomical constants is used in astrodynamics and astronautics than in classic astronomy because astronautics is a kind of experimental astronomy and its missions vary so greatly. A higher accuracy is also necessary because the astrodynamic missions include departure, landing, and flyby maneuvers. Some of the constants, given in relative and astronomical units in celestial mechanics, must be known in absolute units for astrodynamic purposes. Tables 2 to 4 provide the latest observed data for standardization of astrodynamic computations. See references for additional constants.

## SPACE VEHICLE ENVIRONMENTS
### by J. H. Farrow and R. E. Jewell

REFERENCES: Harris and Crede, "Shock and Vibration Handbook," McGraw-Hill. Beranek, "Noise Reduction," McGraw-Hill. Crandall, "Random Vibration," Technology Press, MIT. Neugebauer, The Space Environment, *Tech. Rep.* 34-229, Jet Propulsion Lab., Pasadena,

### Table 2. General, Terrestrial, and Lunar Constants

1. Ephemeris second: 1 sec$_E$ = 1/31,556,925.9747 of tropical year at 1900.0.
2. Mean solar day (culmination period of the mean sun):

$$1^d = 1^{d*}.002\ 737\ 909\ 265 + 0^{d*}.589 \times 10^{-10}T = 24^h03^m56^s.555\ 360\ 50 + 0^s.050\ 89 \times 10^{-4}T$$
$$= 1^{rot}.002\ 737\ 811\ 891 - 0^{rot}.001\ 4 \times 10^{-10}T = (1 \pm 10^{-8})^d{}_E$$

3. Mean sidereal day or mean equinoctial day (culmination period of the vernal equinox):

$$1^{d*} = 0^d.997\ 269\ 566\ 414 - 0^d.587 \times 10^{-10}T = 23^h56^m04^s.090\ 538\ 17 - 0^s.050\ 716\ 8 \times 10^{-4}T$$
$$= 0^{rot}.999\ 999\ 902\ 892 - 0^{rot}.589 \times 10^{-10}T$$

4. Mean stellar day or mean period of the earth's rotation (culmination period of an equatorial star without proper motion):

$$1^d{}_{st} = 1^{rot} = 1^{d*}.000\ 000\ 097\ 108 + 0^{d*}.589 \times 10^{-10}T$$
$$= 24^h*00^m*0^s*.008\ 390\ 13 + 0^s*.050\ 89 \times 10^{-4}T$$
$$= 0^d.997\ 269\ 663\ 257 + 0^d.001\ 4 \times 10^{-10}T = 23^h56^m04^s.098\ 905\ 40 + 0^s.001\ 21 \times 10^{-5}T$$

5. Tropical year (equinox to equinox):

$$P_{trop} = (365.242\ 198\ 78 - 0.000\ 006\ 138T)^d{}_E = 365^d{}_E05^h{}_E48^m{}_E45^s{}_E.530T$$

6. Siderial year (fixed star to fixed star):

$$P_{sid} = (365.256\ 360\ 42 + 0.000\ 000\ 11T)^d{}_E = 365^d{}_E06^h{}_E09^m09^s{}_E.54 + 0^s{}_E.010T$$

7. Synodic month (new moon to new moon):

$$P_{syn} = (29.530\ 588\ 2 - 0.000\ 000\ 2T)^d = 29^d12^h44^m02^s.78 - 0^s.017T$$

8. Tropical month (equinox to equinox):

$$P_{trop} = (27.321\ 581\ 7 - 0.000\ 000\ 2T)^d = 27^d07^h43^m04^s.7 - 0^s.017T$$

9. Siderial month (fixed star to fixed star):

$$P_{sid} = (27.321\ 661\ 0 - 0.000\ 000\ 2T)^d = 27^d07^h43^m11^s.47 - 0^s.017T$$

10. Astronomical unit (mean earth-sun distance): au = 149,598,700 ± 400 km.
11. Light year: ly = (9.460 530 ± 0.000 003) × $10^{12}$ km = 63,239.39 ± 0.15 au.
12. Parsec: pc = 206,264.806 247 au = (3.085 695 ± 0.000 008) × $10^{13}$ km.
13. Semimajor axis of the earth's orbit: $a\oplus$ = 1.000 000 236 au = 149,598,700 ± 400 km.
14. Mean orbital speed: $v\oplus$ = 29,784.90 ± 0.08 m/sec.
15. Mass of the earth: $M\oplus$ = (5.9761 ± 0.004 3) × $10^{24}$ kg.
16. Equatorial radius of the earth: $R_e\oplus$ = 6,378 170 ± 20 m.
17. Flattening (oblateness, ellipticity): $f\oplus = (R_e - R_p)/R_e$ = 0.003 352 55 = 1:(298.28 ± 0.05).
18. Acceleration of gravity at the earth's surface:

$$g = g_e(1 + \beta \sin^2 \phi + \gamma \sin^2 2\phi) = 9.780\ 315(1 + 0.005\ 302\ 74 \sin^2 \phi - 0.000\ 005\ 9 \sin^2 2\phi)\ \text{m/sec}^2$$

19. Moments of inertia of the earth:

$$A = (0.329\ 681_4 \pm 0.000\ 11)M\oplus R_e{}^2 = (0.801\ 50 \pm 0.000\ 85) \times 10^{38}\ \text{kg-m}^2$$
$$C = (0.330\ 763_9 \pm 0.000\ 11)M\oplus R_e{}^2 = (0.804\ 13 \pm 0.000\ 85) \times 10^{38}\ \text{kg-m}^2$$

20. Circular and escape velocities from the earth's surface at the equator:

$$v_{cir} = 7,905.39 \pm 0.06\ \text{m/sec} \qquad v_{esc} = v_{cir}\ 2^{1/2} = 11,179.91 \pm 0.08\ \text{m/sec}$$

21. Mean observed distance of the perturbed moon from the earth:

$$\bar{r}\mathbb{C} = 384,401.0 \pm 1.0\ \text{km} = (60.268\ 23 \pm 0.000\ 35)R_e\oplus = 0.002\ 569\ 548 \pm 0.000\ 000\ 014\ \text{au}$$

22. Semimajor axis of the moon's orbit: $a\mathbb{C}$ = 1.000 907 681$\bar{r}\mathbb{C}$ = 384 749.9 ± 1.0 km.
23. Mean orbital velocity: $v\mathbb{C}$ = 1,024.089 ± 0.003 m/sec.
24. Mass of the moon: $M\mathbb{C}$ = (7.353 4 ± 0.007 5) × $10^{22}$ kg = $M\oplus$: (81.270 ± 0.024).
25. Semiaxes of the moon ellipsoid: a = 1,738,780 ± 186 m; b = 1,738,452 ± 209 m; c = 1,737,688 ± 188 m.

The time T is in Julian centuries of 36,525 days from 1900 Jan. 0.5 U.T. (universal time). The constants are given with probable errors (pe).

### Table 3. Planetary Orbit Data
(Epoch 1960 Jan. 1.5)

| Planet | Number of satellites | Semimajor axis (au) | Sidereal period, year | Synodic period, days | Mean daily motion, s/day | Eccentricity, $e$ | Inclination to ecliptic, $i$, deg | Mean ascending node $\Omega$, deg | Longitude perihelion, $\omega$, deg | Orbital speed, km/s | Planetary escape speed, km/s | Gravity at surface, cm/s² |
|---|---|---|---|---|---|---|---|---|---|---|---|---|
| Mercury ☿ | . . . | 0.387 | 0.2411 | 115.88 | 14,732.42 | 0.206 | 7.004 | 47.857 | 76.833 | 47.8 | 4.3 | 400 |
| Venus ♀ | . . . | 0.723 | 0.6156 | 583.92 | 5767.67 | 0.007 | 3.394 | 76.320 | 131.008 | 35.0 | 10.3 | 875 |
| Earth ⊕ | 1 | 1.0000 | 1.0000 | | 3548.19 | 0.017 | . . . . | | 102.253 | 29.8 | 11.2 | 982 |
| Mars ♂ | 2 | 1.524 | 1.8822 | 779.94 | 1886.52 | 0.093 | 1.850 | 49.250 | 335.322 | 24.2 | 5.0 | 371 |
| Jupiter ♃ | 12 | 5.203 | 11.86 | 398.88 | 299.13 | 0.048 | 1.305 | 100.044 | 13.678 | 13.1 | 60.2 | 2599 |
| Saturn ♄ | 9 | 9.54 | 29.46 | 378.09 | 120.46 | 0.056 | 2.490 | 113.307 | 92.264 | 9.7 | 36.0 | 1108 |
| Uranus ♅ | 5 | 19.2 | 84.0 | 369.66 | 42.23 | 0.047 | 0.773 | 73.796 | 170.011 | 6.8 | 21.9 | 989 |
| Neptune ♆ | 3 | 30.1 | 164.8 | 367.49 | 21.53 | 0.009 | 1.774 | 131.340 | 44.274 | 5.4 | 23.4 | 1098 |
| Pluto ♇ | . . . | 39.4 | 247.7 | 366.74 | 14.29 | 0.249 | 17.170 | 109.886 | 224.160 | 4.7 | | |

### Table 4. Physical Data of Sun, Moon, and Planets

| Name | Radius $R$, km | Mass $M$ ($\oplus=1$) | Radius $R$ ($\oplus=1$) | Volume $V$ ($\oplus=1$) | Mean density $\bar{\mu}$ ($\oplus=1$) | Visual albedo $\bar{A}v$ | Rotational period at equator $P_{rot}$ | Inclination of equator to orbit $I$ | Rotational velocity at equator $V_{rot}$, m/sec | Gravitational parameter $\mu = GM$, km³/sec² |
|---|---|---|---|---|---|---|---|---|---|---|
| Sun | 695,995 | 332,948 | 109.121 | 1,303,735 | 0.2553 | | $25^d.035$ | 7°15' | 2,021.73 | 132,714.6 × 10⁶ |
| Moon | 1,738 | 0.0123 | 0.272 | 0.020 | 0.6058 | 0.067 | 27 .321 | 6°41' | 4.62 | 4,904.70 |
| Mercury | 2,422 | 0.0544 | 0.379 | 0.054 | 0.9896 | 0.056 | 87 .969 | 7°(?) | 2.00 | 21,684.1 |
| Venus | 6,114 | 0.8149 | 0.958 | 0.883 | 0.9221 | 0.76 | 30(?) | 32°(?) | 14.82 | 324,851 |
| Earth (e) | 6,378 | 1.0000 | 1.000 | 1.000 | 1.0000 | 0.36 | $23^h56^m4^s$ | 23°26'.7 | 465.10 | 398,604.6 |
| Earth (p) | 6,356 | | 0.996 | | | | | | | |
| Mars (e) | 3,416 | 0.1078 | 0.535 | 0.153 | 0.7031 | 0.16 | $24^h37^m22^s$ | 23°59' | 242.13 | 42,977.5 |
| Mars (p) | 3,398 | | 0.532 | | | | | | | |
| Jupiter (e) | 71,403 | 317.821 | 11.195 | 1,315.996 | 0.2415 | 0.73 | $9^h50^m$ | 3°6'.9 | 12,673.3 | 126,684,900 |
| Jupiter (p) | 66,749 | | 10.465 | | | | | | | |
| Saturn (e) | 60,502 | 95.112 | 9.485 | 772.552 | 0.1231 | 0.76 | $10^h14^m$ | 26°44.7' | 10,318.6 | 37,912,080 |
| Saturn (p) | 54,577 | | 8.556 | | | | | | | |
| Uranus (e) | 24,848 | 14.516 | 3.895 | 55.619 | 0.2610 | 0.93 | $10^h49^m$ | 97°53' | 4,009.5 | 5,786,140 |
| Uranus (p) | 23,295 | | 3.652 | | | | | | | |
| Neptune (e) | 25,000 | 17.140 | 3.919 | 59.217 | 0.2894 | 0.84 | $15^h40^m$ | 28°48' | 2,785 | 6,832,080 |
| Neptune (p) | 24,500 | | 3.841 | | | | | | | |
| Pluto | 2,966 | 0.1(?) | 0.465 | 0.10 | 0.991(?) | 0.14 | $6^d.390$ | (?) | 33.75 | 40,000(?) |

1960. Hart, Effects of Outer Space Environment Important to Simulation of Space Vehicles, *ASD Tech. Rep.* 61-201, Cornell Aeronautical Lab. Barrett, Techniques for Predicting Localized Vibratory Environments of Rocket Vehicles, *NASA-TN D-1836*. Wilhold, Guest, and Jones, A Technique for Predicting Far Field Acoustic Environments Due to a Moving Rocket Sound Source, *NASA-TN D-1832*. Eldred, Roberts, and White, Structural Vibration in Space Vehicles, *WADD Tech. Rep.* 61-62. Bolt, Beranek, and Newman, Exterior Sound and Vibration Fields of a Saturn Vehicle during Static Firing and during Launching, *U.S.A. Ord. Rep.* 764.

**Definitions** Space-vehicle environments are conditions or influences recognizable on and caused by the vehicle. Environments which are recognizable **on** the vehicle may be either **natural** or **induced**. Environments which are **caused by** the vehicle are **induced**. The natural environments exist in the absence of any operational influence of the space vehicle. The induced environments exist as a result of an outside agency, such as transportation, handling, and operation of the space vehicle.

At the time of launch, the total configuration which lifts from the launch pad is called the space vehicle. This configuration does not necessarily enter space in the sense of performing a space mission. The sequence of events is the burnout and separation of the various boost stages, followed by injection of a "spacecraft" into orbit or escape trajectory. The space-vehicle configuration differs for various phases of operation. The environments change because of a continuous interplay of natural and induced influences and changes in configurations.

The basic concepts concerning natural and induced environments are equally applicable to aircraft, but the environments of space vehicles cover a broader spectrum of severity. The aircraft experiences heat and cold, dry desert air, salt spray, and humidity but always remains within a radiation-protective convective atmosphere. The induced environment for aircraft is also relatively moderate due to moderate speeds, relatively low thrust levels, and unchanging configurations. The space-vehicle environments are much more severe. The forces required to propel a space vehicle from the launch pad are tremendous. The dynamic pressures generated in the atmosphere by large rocket engines are exceeded only by nuclear blasts. Slow initial ascent from the launch pad is followed by rapid acceleration and high $g$ loading, and acoustic and aerodynamic forces drive every point of the vehicle surface. The space-vehicle velocity quickly becomes supersonic and continues to hypersonic velocities. Gradually, as the space vehicle leaves the atmosphere, the aerodynamic forces recede. Then suddenly the rocket thrust decays and is followed by the ignition of another rocket, which quickly develops thrust and continues to accelerate the space vehicle. Finally, the space vehicle becomes weightless. While in space, the vehicle exists in a vacuum, is bombarded by solar radiation and micrometeorites, and experiences continuous cyclic variation of temperature. In completing a mission, a space vehicle is subjected to the additional environmental extremes of planetary landings, escape, and earth reentry (Fig. 2).

These environments represent basic criteria for space-vehicle design. For operational reliability, the most important are the shock, vibration, and acoustic environments which are present to varying magnitudes in all operational phases and which constitute a major engineering problem.

**Importance of Dynamic Environments** (See Fig. 3) The problem is selection of the most practical point to enter the inner cycle. It is not possible to have original environmental

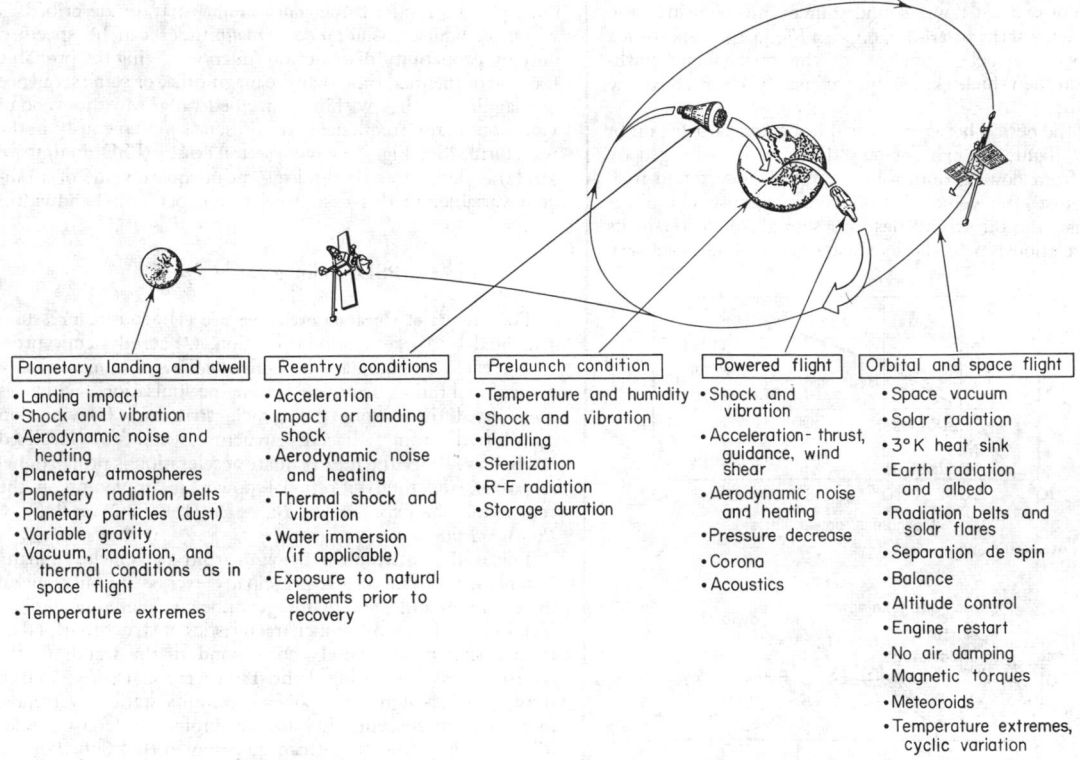

**Planetary landing and dwell**
- Landing impact
- Shock and vibration
- Aerodynamic noise and heating
- Planetary atmospheres
- Planetary radiation belts
- Planetary particles (dust)
- Variable gravity
- Vacuum, radiation, and thermal conditions as in space flight
- Temperature extremes

**Reentry conditions**
- Acceleration
- Impact or landing shock
- Aerodynamic noise and heating
- Thermal shock and vibration
- Water immersion (if applicable)
- Exposure to natural elements prior to recovery

**Prelaunch condition**
- Temperature and humidity
- Shock and vibration
- Handling
- Sterilization
- R-F radiation
- Storage duration

**Powered flight**
- Shock and vibration
- Acceleration- thrust, guidance, wind shear
- Aerodynamic noise and heating
- Pressure decrease
- Corona
- Acoustics

**Orbital and space flight**
- Space vacuum
- Solar radiation
- 3°K heat sink
- Earth radiation and albedo
- Radiation belts and solar flares
- Separation de spin
- Balance
- Altitude control
- Engine restart
- No air damping
- Magnetic torques
- Meteoroids
- Temperature extremes, cyclic variation

**Fig. 2**  Environmental conditions experienced by space vehicles.

data prior to fabrication and testing of a particular structure. Vibration, shock, and acoustic noise adversely affect structural integrity and vehicle reliability. These environments must be considered prior to design and fabrication and then again for design verification through ground testing. Precise vibration and acoustic environmental predictions are complicated because of structural nonlinearities, random forcing

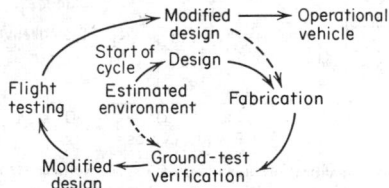

**Fig. 3**  Space-vehicle development.

functions, and multidegree-of-freedom systems. Current environmental-prediction techniques consist mainly of extrapolating measured data obtained from existing vehicles. Acoustic environments produced during ground operation are important factors in evaluating ground-facility locations and design, personnel safety, and public relations.

A **source of sound** common to all jet and rocket-engine propulsion systems is the turbulent exhaust flow. This high-velocity flow produces pressure fluctuations, referred to as **noise,** and has adverse effects on the vehicle and its operations. Also, far-field, uncontrolled areas may be subjected to intense

acoustic-sound-pressure levels which may require personnel protection or sound-suppression devices. (See also Sec. 12.)

The sound source is the exhaust flow, where the viscous atmospheric fluid is subjected to high-shear forces between the region of high-velocity flow and the ambient medium. These shear forces produce random eddy motions to propagate randomly fluctuating pressures, which are numerically described in standard statistical form.

The **sound-generation mechanism** is presented in empirical and analytical form, utilizing measured sound pressures and the known engine operational parameters. This combination of empirical terms and analytical reasoning provides a generalized relationship to predict sound-pressure levels for the vehicle structure and also for far-field areas. It has been shown by Lighthill (*Proc. Royal Soc.*, 1961) that **acoustic-power generation** of a subsonic jet is proportional to $\rho U^8 D^2/a_o^5$, where $\rho$ is the density of the ambient medium, $a_o$ is the speed of sound in the ambient medium, $U$ is the jet or rocket exit exhaust velocity, and $D$ is a characteristic dimension of the engine. Attempts have been made to extend this relationship to allow supersonic-jet acoustic-power predictions to be made, but no one scheme is accepted for broad use.

From acoustic-data-gathering programs utilizing rocket engines, various trends have been noted which facilitate acoustic predictions. For rocket engines up to $5 \times 10^9$ W of mechanical power, the **acoustic efficiency** shows a nonlinear rate of increase. For greater mechanical-power wattages, the acoustic efficiency approaches a value slightly higher than 0.5 percent (Fig. 4).

Turbulent-exhaust-flow sound-source mechanisms also exhibit **directional characteristics,** e.g., in Fig. 5 the data from a Saturn rocket vehicle launch with the microphone flush-mounted on the vehicle skin at approximately 100 ft above the nozzle plane.

During the period between on-pad firing and post-lift-off or main-stage flight, the exhaust-flow direction has changed by 90°, i.e., from flowing onto a bidirectional deflector to free-field flow just after vehicle lift-off. With exhaust-flow direction change, the directivity has also shifted 90° to retain its inherent relationship to the exhaust stream. The **sound-pres-**

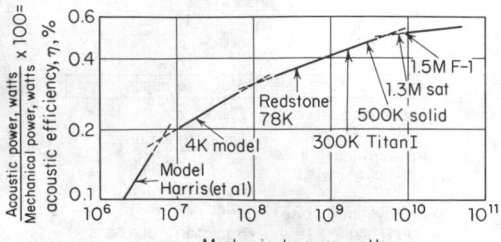

**Fig. 4**    Acoustic-efficiency trends.

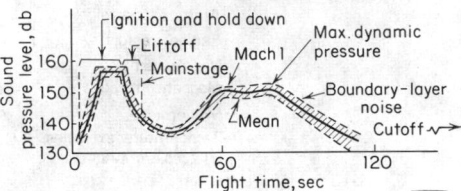

**Fig. 5**    Vehicle acoustic environment.

**sure-level** variation during a small time interval—between on-pad condition and shortly after lift-off—depends on the changes which have occurred in the directivity and the jet-exhaust-velocity vector (Fig. 5). This change in sound-pressure level is 20 dB approx in this case and is indicative of the significance of a sound source's directional properties, which depend on velocity and temperature of the jet.

The **acoustic environments** produced by a given conventional engine can be empirically described by that engine's flow parameters and geometry. It is difficult to predict the acoustic environments for nonconventional engines such as the plug-nozzle and expansion-deflection engines.

The **turbulent-boundary-layer** problem of high-velocity vehicle flights is difficult to evaluate in all but the simplest cases. Only for attached, homogeneous, subsonic flows can the boundary-layer noise be estimated with confidence.

**Space-vehicle Vibration** (See also Sec. 5.)   The vibratory environment on rocket vehicles consists of total-vehicle bending vibrations, in which the vehicle is considered a nonuniform beam, with low-frequency responses (0 to 20 Hz) and localized vibrations with frequencies up to thousands of hertz.

The response of a simple structure to a dynamic forcing function may be described by the universal equation of motion given as $M\ddot{X} + C\dot{X} + KX = F(t)$ or $F(t)/X(t) = (M - K/\omega^2) + j(C/\omega)$ and is a complex function containing real and imaginary quantities. In the case of rocket-vehicle vibrations, $F(t)$ is a **random** forcing phenomenon. Thus, the motion responses

must also be random. **Random vibration** may be described as vibration whose instantaneous magnitudes can be specified only by probability-distribution functions giving the probable fraction of the total time that the magnitude, or some sequence of magnitudes, lies within a specified range. In this type of vibration, many frequencies are present simultaneously in the waveform. (See Fig. 6.) **Power spectral density** (PSD; in units of $g^2$/Hz) is defined as the limiting mean-square value of a random variable, in this case acceleration per unit bandwidth; i.e.,

$$\text{PSD} = \lim_{\Delta f \to 0} \overline{g^2}/\Delta f = \overline{dg^2}/df$$

The **sources of vibration excitation** are (1) acoustic pressures generated by rocket-engine operation, (2) aerodynamic pressures created by boundary-layer fluctuations, (3) mechanically induced vibration from rocket-engine pulsation, which is transmitted throughout the vehicle structure, (4) vibration transmitted from adjacent structure, and (5) localized machinery. Since the mean-square acceleration is proportional to power, the total vibrational power at any point on the vehicle may be expressed as $Pv = Pv(1) + Pv(2) + Pv(3) + Pv(4) + Pv(5)$.

This is illustrated in Fig. 7 for an arbitrary vehicle. In many instances, the structure is principally excited by only one of these sources and the remaining can be considered negligible. (See Fig. 8.) The vibration characteristics of structure B, taken from a skin panel, closely correspond to the trends of the acoustic pressures during flight. However, structure A, taken from a rocket-engine component, exhibits stationary trends, indicating no susceptibility to the impinging acoustic field. Thus for this structure, it can be assumed that only $Pv(3)$ is significant, whereas for structure B, only $Pv(1)$ is the principal exciting source during the on-pad phase and only $Pv(2)$ is significant during the maximum dynamic-pressure phase. This also holds true for captive firings of the vehicle, in which only $Pv(1)$ is significant for structure B and $Pv(3)$ for structure A.

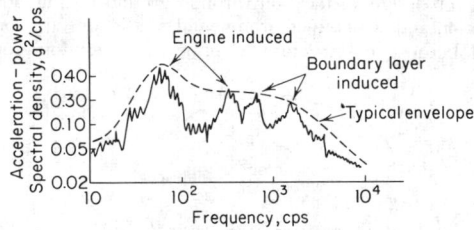

**Fig. 6**    Random-vibration spectrum of a space-vehicle structure.

## SPACE-VEHICLE TRAJECTORIES, FLIGHT MECHANICS, AND PERFORMANCE

by O. Elnan, W. R. Perry, J. W. Russell, A. G. Kromis, and D. W. Fellenz

REFERENCES: Russell, Dugan, and Steward, "Astronomy," Ginn. Sutton, "Rocket Propulsion Elements," Wiley. White, "Flight Performance Handbook for Powered Flight Operations," Wiley. "U.S. Standard Atmosphere, 1962," NASA, USAF, U.S. Weather Bureau. Space Flight Handbooks, *NASA SP*-33, *SP*-34, and *SP*-35. Gazley, Deceleration and Heating of a Body Entering a Planetary Atmosphere

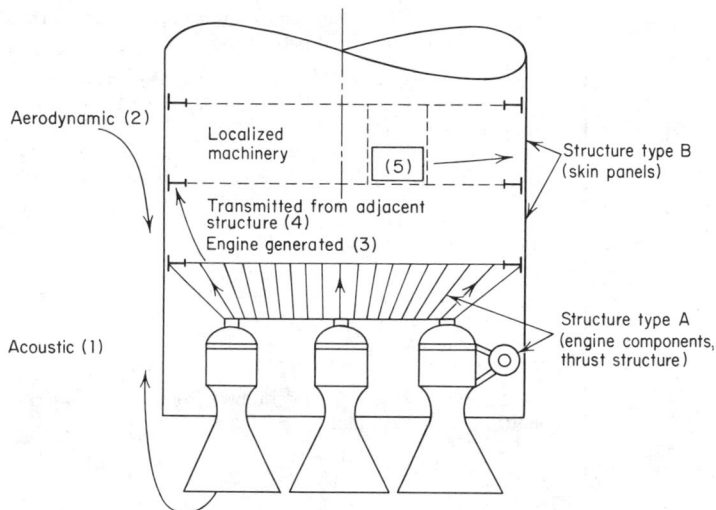

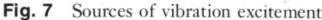

**Fig. 7**  Sources of vibration excitement.

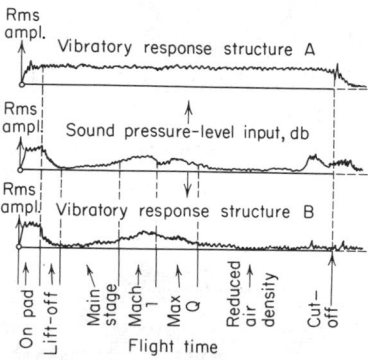

**Fig. 8**  Typical structural responses to a random-vibration environment.

from Space, "Visits in Astronautics," Vol. I, Pergamon. Chapman, An Approximate Analytical Method for Studying Entry into Planetary Atmospheres, *NACA-TN* 4276, May 1958. Chapman, An Analysis of the Corridor and Guidance Requirements for Super-circular Entry into Planetary Atmospheres, *NASA-TN D*-136, Sept 1959. Loh, "Dynamics and Thermodynamics of Planetary Entry," Prentice-Hall. Grant, Importance of the Variation of Drag with Lift in Minimization of Satellite Entry Acceleration, *NASA-TN D*-120, Oct 1959. Lees, Hartwig, and Cohen, Use of Aerodynamic Lift during Entry into the Earth's Atmosphere, *Jour. ARS*, 29 Sept 1959. Gervais and Johnson, Abort during Manned Ascent into Space, *AAS*, Jan. 1962.

## Notation

$A$ = reference area
$A_z$ = launch azimuth
$B$ = ballistic coefficient
$C_L$ = lift coefficent
$e$ = eccentricity
$f$ = stage-mass fraction ($WP/W_A$)
$H$ = total energy
$H_\infty$ = hyperbolic excess energy
$I$ = specific impulse
$\Delta_i$ = plane-change angle
$L$ = lift force
$m$ = mass
$n$ = mean motion ($2\pi/p_{sid}$)
$p$ = ambient atmospheric pressure
$q$ = dynamic pressure
$R$ = earth's equatorial radius
$S$ = wing area
$T$ = transfer time, time, absolute temperature
$t_{pp}$ = time after perigee passage
$V_c$ = injection velocity
$V$ = velocity
$V_g$ = velocity loss to gravity
$W$ = weight
$W_L$ = payload weight
$W_P$ = propellant weight
$Y$ = distance

$\alpha_\Omega$ = right ascension of the ascending node
$\gamma = \theta - 90°$ = flight-path angle used in reentry analysis
$\nu$ = true anomaly
$\omega$ = earth's rotational velocity
$\psi$ = central angle
$\sigma = \rho/\rho_0$ = relative atmospheric density
$\Upsilon$ = vernal equinox
$A_e$ = nozzle exit area
$a$ = semimajor axis
$C_D$ = drag coefficient
$D$ = drag force
$E$ = eccentric anomaly
$F$ = thrust
$g$ = acceleration due to gravity
$H_{esc}$ = escape energy
$b$ = altitude
$i$ = inclination
$J_2$ = oblateness coefficent of the earth's potential (second-zonal harmonic)
$M$ = molecular weight, mean anomaly
nm = nautical miles
$P$ = period of revolution
$P_e$ = nozzle exit pressure
$q_s$ = stagnation-point heat-transfer rate
$r$ = radius
$s$ = range
$t$ = time
$V_\infty$ = hyperbolic excess velocity
$V_{id}$ = ideal velocity
$\Delta V$ = impulsive velocity
$W_A$ = stage weight ($W_o - W_L$)
$W_o$ = gross weight
$X$ = distance
$\alpha$ = angle of attack, right ascension
$\beta$ = exponent of density-altitude function
$\lambda$ = longitude
$\tau$ = hour angle
$\omega$ = argument of perigee

$\phi'$ = latitude
$\rho$ = atmospheric density
$\theta$ = flight-path angle

**Subscripts**

$a$ = apogee
$e$ = entry
$f$ = final
$o$ = sea level conditions at 45° latitude
$s$ = space fixed
vac = vacuum
circ = circular orbital condition
esc = escape
$i$ = initial
$p$ = perigee
$M$ = mean
$SL$ = sea level
sid = sidereal

## ORBITAL MECHANICS
### by O. Elnan and W. R. Perry

The motion of the planets about the sun, as well as that of a satellite in its orbit about a planet, is governed by the inverse-square force law for attracting bodies. In those cases where the mass of the orbiting body is small relative to the central attracting body, it can be neglected. This simplifies the analysis of the orbital motion. The shape of the orbit is always a conic section, i.e., ellipse, parabola, or hyperbola, with the central attracting body at one of the foci. Parabolic and hyperbolic orbits are open, terminate at infinity, and represent cases where the orbiting body escapes the central force field of the attracting body.

The **laws of Kepler** for satellite orbits are (1) the radius vector to the satellite from the central body sweeps over equal areas in equal times; (2) the orbit is an ellipse, with the central attracting body at one of its foci; and (3) the square of the period of the satellite is proportional to the cube of the semimajor axis of the orbit.

Six **orbital elements** are required to describe the orbit in the orbit plane and the orientation of the plane in inertial space. The three elements that define the orbit are the semimajor axis $a$, eccentricity $e$, and period of revolution $P$. The orientation of the orbit (Fig. 9) is defined by the right ascension of the ascending node $\alpha_\Omega$, inclination of the orbit plane to the earth's equatorial plane $i$, and the argument of perigee $\omega$, which is the central angle measured in the orbit plane from ascending node

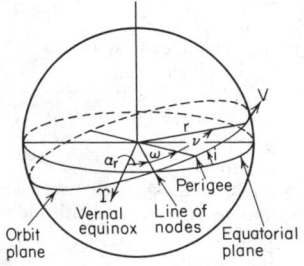

**Fig. 9** Elements of satellite orbit around a planet.

to perigee. The following equations compute these orbital elements and other orbital parameters.

$$a = (r_a + r_p)/2 = \mu/V_a V_p = r_a/(1 + e) = r_p/(1 - e)$$
$$e = (r_a/a) - 1 = (V_p - V_a)/(V_p + V_a) = (r_p V_p^2/\mu) - 1$$
$$r = a(1 - e^2)/(1 + e \cos \nu)$$
$$r_a = a(1 + e) = r_p(1 + e)/(1 - e)$$
$$r_p = a(1 - e) = r_a(1 - e)/(1 + e) \qquad V = [\mu(2/r - 1/a)]^{0.5}$$
$$P = 2\pi(a^3/\mu)^{0.5} \qquad \text{(For circular orbits, } r = a)$$
$$V_a = V_p(1 - e)/(1 + e) = [\mu(1 - e)/a(1 + e)]^{0.5}$$
$$V_p = V_a(1 + e)/(1 - e) = [\mu(1 + e))/a(1 - e)]^{0.5}$$
$$\nu = \cos^{-1}\{[2r_a r_p - r(r_a + r_p)]/r(r_a - r_p)\}$$
$$t_{pp} = (a^3/\mu)^{0.5}\{\cos^{-1}[(a - r)/ae] - e[1 - (a - r)^2/(ae)^2]^{0.5}\}$$
$$M = nt_{pp} = E - e \sin E$$
$$r = a(1 - e \cos E)$$
$$V_{esc} = (2\mu/r)^{0.5}$$

The elements may be determined from known injection conditions. For example,

$$1/a = V_i^2 \mu - 2/r_i$$
$$e^2 = \cos^2 V_i (r_i V_i^2/\mu - 1)^2 + \sin^2 V_i$$

General precession in longitude of vernal equinox:

$$x = 50''.2575 + 0''.0222T$$

where $T$ is in Julian centuries of 36,525 mean solar days reckoned from 1900 Jan. 0.5 UT.

### Measurement of Time

The concept of time is based on the position of celestial bodies with respect to the observer's meridian as measured along the geocentric celestial equator. The mean sun and the vernal equinox (♈) are used to define mean solar time and sidereal time. The mean sun, rather than the apparent (true) sun, is used since the latter's irregular motion from month to month gives a variation in the length of a day. The mean sun moves with uniform speed along the equator at a rate equal to the average of the true sun's angular motion during the year.

One sidereal day is the interval between two successive transits of the vernal equinox across the observer's meridian. A mean solar day is the interval between two successive transits of the mean sun. Local civil time (LCT) is defined to be the hour angle of the mean sun plus 12 h. Local sidereal time is the hour angle of the vernal equinox plus 12 h. For an observer at the Greenwich meridian, the local civil time is Greenwich mean time (GMT) or universal time (UT). The earth is divided into 24 time zones, each 15° of longitude wide. Each zone keeps the time of the standard meridian in the middle of the zone. (See Fig. 10.)

Time $T = \tau + 12$ h
Local sidereal time LST $= \tau_\Upsilon + 12$ h
Greenwich sidereal time GST $=$ LST $+ \lambda(w)$
Local civil time LCT $= \tau_M + 12$ h $=$ mean solar time
Equation of time ET $= \tau - \tau_M = \alpha_M - \alpha$
Greenwich mean time GMT $=$ UT $= \tau_M$(Greenwich)
$\quad + 12$ h $=$ universal time

$$\text{LCT} = \text{UT} - \lambda(w)$$
$$\text{GMT} = \text{ZT} \pm \lambda°/15$$

The "American Ephemeris and Nautical Almanac" gives the equation of time for every day of the year and conversions of mean solar time to and from sidereal time.

**Hohmann Transfer**  This is a maneuver between two coplanar orbits where the elliptical transfer orbit is tangent at its perigee to the lower orbit and tangent at its apogee to the higher orbit. To transfer from a low circular orbit to a higher one, the impulsive velocity required is the difference between

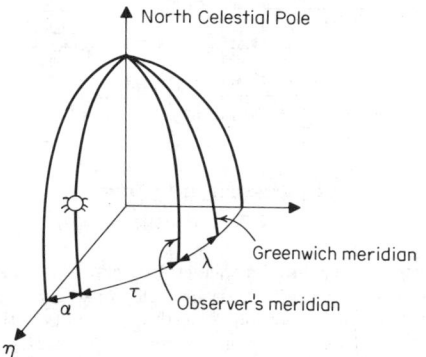

**Fig. 10**  Time-zone geometry.

the velocities in the circular orbits and the velocities in the corresponding points on the transfer orbit. Using the general velocity equation above, the transfer velocities are $\Delta V_i = \{\mu[2/r_i - 2/(r_i + r_f)]\}^{0.5} - (\mu/r_i)^{0.5}$ and $\Delta V_f = (\mu/r_f)^{0.5} - \{\mu[2/r_f - 2/(r_i + r_f)]\}^{0.5}$, where $V_i$ is the orbiting velocity in low orbit and $V_f$ is the orbiting velocity in the high orbit.

**Orbital Lifetime**  The satellite orbit will gradually decay to lower altitudes because of the drag effects of the atmosphere. This drag force has the form $D = \frac{1}{2}C_D A\rho V^2$, where values of the drag coefficient $C_D$, may range from 2.0 to 2.5. The atmospheric density, as a function of altitude, can be used for computing an estimated orbital lifetime of a satellite (Fig. 11.) (See NASA, "U.S. Standard Atmosphere" and "Space Flight Handbooks.")

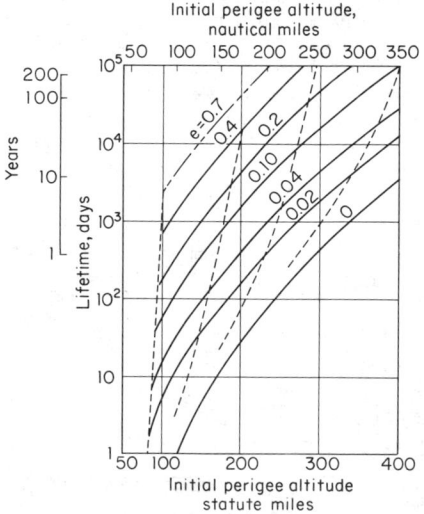

**Fig. 11**  Satellite lifetimes in elliptic orbits. $\rho$ = ARDC, 1959; $B = C_D A/2m = 1.0$ ft²/slug; $e$ = initial eccentricity; --- = decay histories.

**Perturbations of Satellite Orbits**  There are many secular and periodic perturbations of satellite orbits due to the effects of the sun, the moon, and some planets. The most significant perturbation on close-earth satellite orbits is caused by the oblateness of the earth. Its greatest effects are the precession of the orbit along the equator (nodal regression) and the rotation of the orbit in the orbit plane (advance of perigee). The nodal-regression rate is given to first-order approximation by

$$\dot{\alpha}_\Omega = -\tfrac{3}{2}(\mu/a^3)^{0.5} J_2[R/a(1 - e^2)]^2 \cos i$$

The mean motion of the perigee is

$$\dot{\omega} = J_2 n(2 - (5/2) \sin^2 i)/a^2(1 - e^2)^2$$

## LUNAR- AND INTERPLANETARY-FLIGHT MECHANICS
### by J. W. Russell

The extension of flight mechanics to the areas of lunar and interplanetary flight must include the effects of the sun, moon, and planets on the transfer trajectories. For most preliminary performance calculations, the **"sphere-of-influence,"** or **"patched-conic,"** method—whereby the multibody force field is treated as a series of central force fields—provides sufficient accuracy. By this method, the trajectories in the central force field are calculated to the "sphere of influence" of each body by standard Keplerian mechanics, and the velocity and position of the extremals are then matched to give a continuous trajectory. After the missions have been finalized, the precision trajectories are obtained by numerically integrating the equations of motion which include all the perturbative elements. It is necessary to know the exact positions of the sun, moon, and planets in their respective orbits.

**Lunar-flight Mechanics**  The moon moves about the earth in an orbit having an eccentricity of 0.055 and an inclination to the ecliptic of about 5.145°. The sun causes a precession of the lunar orbit about the ecliptic, making the inclination of the lunar plane to the earth's equatorial plane oscillate between 18.5 and 28.5° over a period of 18.5 years approx. To compute precision earth-moon trajectories, it is necessary to know the precise launch time to be able to include the perturbative effects of the sun, moon, and planets. The data presented are based on the moon being at its mean distance from earth and neglect the perturbation of the sun and planets; they are therefore only to be considered as representative data. The injection velocity (see Gazley) at 100 nm for earth-moon trajectories and the impulsive velocity for braking into a 100-nm orbit about the moon are shown in Fig. 12 as a function of

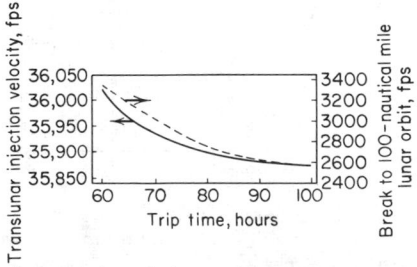

**Fig. 12**  Lunar-mission velocity requirements.

transfer time. Additional energy is required to offset losses due to gravity and aerodynamic drag as well as to provide plane-change and launch-window capabilities.

**Interplanetary-flight Mechanics** The fact that planetary orbits about the sun are not coplanar greatly restricts the number of feasible interplanetary trajectories. The plane of the interplanetary trajectory must include the position of the departure planet at departure, the sun, and the position of the target planet at arrival. The necessary plane change can be prohibitive even though the relative inclinations of the planetary orbits are small. The impulsive velocity required to effect a plane change for a spacecraft departing earth can be approximated by $\Delta V = 195,350 \sin(\Delta_i/2)$, where $\Delta_i$ is the amount of plane change required and $\Delta V$ is in ft/s.

For interplanetary flight, the "ideal" total energy that must be imparted to the spacecraft is the energy required to escape the gravitational field of the departure planet plus the energy required to change path about the sun so as to arrive at the target planet at the desired position and time. The energy required to escape the gravitational attraction of a planet can be determined by Keplerian mechanics to be $H_{esc} = \mu/r$, where $\mu$ is the gaussian gravitational constant and $r$ is the distance from the center of the planet. After escaping the departure planet, the velocity must be altered in both magnitude and direction—in order to arrive at the target planet at the chosen time—by supplying additional energy, $H_\infty = \frac{1}{2}V_\infty^2$.

For determination of vehicle size necessary to inject the spacecraft into the interplanetary trajectory, it is convenient to express the total energy $H = H_{esc} + H_\infty$ in terms of the required injection velocity

$$V_c = \sqrt{2(H_{esc} + H_\infty)} \quad \text{or} \quad V_c \sqrt{(2\mu/r) + V_\infty^2}$$

**Hyperbolic excess velocities,** $V_\infty$'s for **earth-to-Mars** missions are shown in Table 5. These velocities are near optimum for the trip time and year given but do not represent absolute-optimum trajectories. Mars has a cyclic period with respect to earth of 17 years approx; hence the energy requirements for Mars missions from earth also follow approximate 17-year cycles.

**Hyperbolic-excess velocity** requirements for **earth-Venus** missions are shown in Table 6. Venus has a cyclic period of about 8 years with respect to earth; therefore these velocities follow an 8-year cycle.

In Tables 5 and 6, the energy requirements are a function of

**Table 5. Hyperbolic-Excess-Velocity Requirements (ft/s) for Typical Mars Missions\***

| Trip time, days | Opposition year | | | | |
|---|---|---|---|---|---|
| | 1965 | 1967 | 1969 | 1971 | 1973 |
| 100 | 27,241 | 23,012 | 18,070 | 15,569 | 19,955 |
| 150 | 15,940 | 13,088 | 11,437 | 10,520 | 13,792 |
| 200 | 12,268 | 10,021 | 11,486 | 9,308 | 12,463 |
| 250 | 10,510 | 12,688 | 13,410 | 10,480 | 15,784 |

\*From Gammal, "Space Flight Handbooks."

date as well as of trip time. Figure 13 shows how the hyperbolic excess velocities of Tables 5 and 6 can be converted to stage-characteristics velocities for a vehicle leaving a 100-nm earth parking orbit.

**Table 6. Hyperbolic-Excess-Velocity Requirements (ft/s) for Typical Venus Missions\***

| Trip time, days | Conjunction year | | | | |
|---|---|---|---|---|---|
| | 1964 | 1966 | 1967 | 1969 | 1970 |
| 60 | 21,615 | 21,225 | 22,220 | 20,746 | 20,219 |
| 80 | 15,403 | 15,628 | 15,970 | 14,309 | 14,456 |
| 100 | 12,649 | 12,932 | 12,424 | 10,930 | 10,266 |
| 120 | 12,063 | 12,814 | 10,334 | 9,240 | 11,213 |

\*From Gammal, "Space Flight Handbooks."

## ATMOSPHERIC ENTRY
### by D. W. Fellenz

A vehicle approaching a planet possesses a considerable amount of energy. The entry vehicle must be designed to dissipate this energy without exceeding its limits with respect to maximum decelerations or heating.

The trajectory parameters of an entering vehicle are determined largely by its initial trajectory conditions (suborbital, orbital, superorbital), by the ratio of gas-dynamic forces acting upon it, and by its mass (ballistic factor, lift-drag ratio) and the type of atmosphere it is entering.

Planetary atmospheres (Table 7) can be assumed, as a first approximation, to have exponential density-altitude distributions: $\sigma = \rho/\rho_{SL} = e^{-\beta h}$, where

$$\beta = -(1/\rho)(d\rho/dh) = Mg/RT$$

For more exacting analyses, empirical atmospheric characteristics (e.g., U.S. Standard Atmosphere) have to be used.

The trajectory of the vehicle in flight-path fixed notation with the assumption of a nonrating atmosphere is described as $dV/dt = g \sin \gamma - (C_D A/m)(\rho/2)V^2$ and $(V/\cos \gamma)(d\gamma/dt) = g - (V^2/r) - (C_L A/m)(\rho/2)(V^2/\cos \gamma)$. Solutions exist for direct ($\gamma > 5°$) ballistic entry and for equilibrium glide-lifting entry for $L/D > 1$. General solutions of the equations of motion have been obtained for shallow entry of both ballistic and lifting bodies. For a survey of analytical methods available, see Chapman, Loh, Grant, Lees, and Gervais.

The **energy of an earth satellite** at 200 mi altitude is about 13,000 Btu/lb, and a vehicle entering at escape velocity possesses twice this energy. This energy is transformed, through the mechanism of gas-dynamic drag, into thermal energy in the air around the vehicle, of which only a fraction enters the

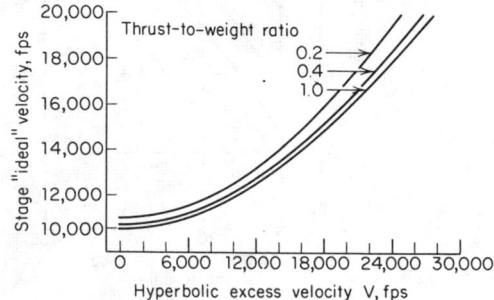

**Fig. 13** Stage-velocity requirements for earth departure from a 100-nm parking orbit (specific impulse = 450 s).

**Table 7. Planetary Atmospheres**

| Planet | $V_{esc}$, ft/s | Gases | $M$, gmol$^{-1}$ | $T$, K | $\beta^{-1}$, ft | $\beta r$ | $\rho_{st}$, lb/ft$^3$ |
|---|---|---|---|---|---|---|---|
| Venus......... | 34,300 | 10 % $N_2$ 90 % $CO_2$ | 40 | 250–350 | $2 \times 10^4$ | 1,006 | 1.0 |
| Earth.......... | 36,800 | 79 % $N_2$ 20 % $O_2$ | 29 | 240 | $2.35 \times 10^4$ | 880 | 0.0765 |
| Mars.......... | 16,900 | 95 % $N_2$ 5 % $CO_2$ | 30 | 200–300 | $6 \times 10^4$ | 132 | 0.0062 |
| Jupiter......... | 195,000 | $H_2$ $CH_4$ | 3 | 100–200 | $6 \times 10^4$ | 3,600 | |

vehicle surface as heat. This fraction depends on the characteristics of the boundary-layer flow, which is determined by shape, surface condition, and Reynolds number. Figure 14 illustrates the **energy conversion** where it is necessary to manage a given amount of energy in a way that minimizes structural and heat-protection weights, operational restraints, and cost.

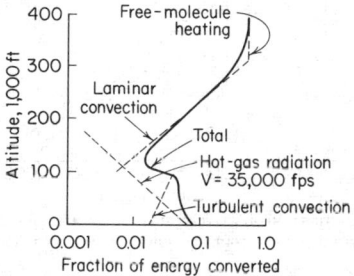

**Fig. 14**  Energy conversion during reentry. *(After Gazley in Koelle, "Handbook of Astronautical Engineering.")*

It would be most desirable if this energy could be dissipated at a constant rate, but with constant vehicle parameters, decelerations vary proportionally to $\rho V^2$ and heating rates proportionally to $\rho V^3$. The selection of a heat-protection system depends on the type of entry flown. Lifting entry from orbit results in relatively long flight times (in the order of ½ h) as compared with 10 min in the case of a steep ballistic entry. **Radiative-heat-transfer systems** favor low heating rates over long time periods. **Ablative systems** favor short heat pulses. In fact, longer soaking periods may melt the ablation coating without getting the benefit of heat absorption through multiple-phase changes. A more uniform dissipation of energy can be achieved by modulation of vehicle parameters.

Determination of an **entry-vehicle configuration** is a process of iteration. The entry-flight profile and the entry and recovery procedures are mainly determined by whether experiments or passengers are carried. The external shape is determined by requirements for hypersonic glide capability and subsonic handling and landing characteristics and also by the relations between aerodynamic shape, heat input, and structural-materials characteristics. Intermediate results are fed back into the evaluation of performance and operational effectiveness of the total transportation system.

If reentry capability from any point of the ascent trajectory

is desired for a manned vehicle, vehicle constraints and trajectory requirements must be compatible. **Performance-optimized trajectories** encompass combinations of relatively small velocities and large flight-path angles, which would result in considerable decelerations. In such cases, either reshaping the ascent trajectory or adding velocity at the time of abort, resulting in lower entry-flight-path angles, is effective (see Fig. 15). Lower flight-path angles during ascent depress the trajectory and increase drag losses. **Atmospheric entry is initiated** by changing the vehicle-velocity vector so that the virtual perigee of the descent ellipse comes to lie inside the atmosphere. The retrovelocity requirement for entry from low orbit is between 250 and 500 ft/s, depending on the range desired from deorbit to

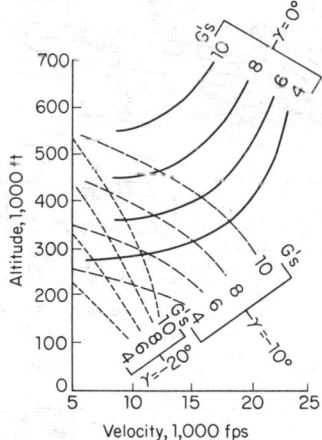

**Fig. 15**  Suborbital entry of a lifting vehicle. $W/C_D A = 28$ lb/ft$^2$; $L/D = 0.7$.

landing. For nonlifting entry, maximum deceleration and heat input into the vehicle are largely a function of entry angle, with the ballistic factor $W/C_D A$ determining the altitude at which the maximum decelerations and heating rates occur. Deceleration can be readily determined through the approximate relation $-(1/g)(dv/dt) \approx q/(W/C_D A)$. Decelerations and temperatures are drastically reduced with increasing lift-drag ratio (Fig. 16) and decreasing $W/C_D A$ (Figs. 17 and 18). This effect is particularly beneficial at steeper entry angles. The combination of longer flight times with lower heating rates, however, may actually result in a larger total heat input into the vehicle.

The **influence of lift** on the reduction of decelerations is greatest for the step up to $L/D \approx 1$. The influence of $W/C_D A$ on decelerations disappears beyond $L/D \approx 0.6$. Higher, hypersonic $L/D$ ratios serve to improve maneuverability. For entry from low orbit, the landing area selection (footprint) can be increased (see Table 8).

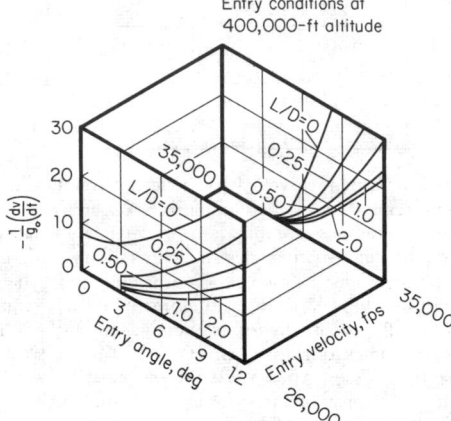

**Fig. 16** Peak decelerations for entry at constant $L/D$.

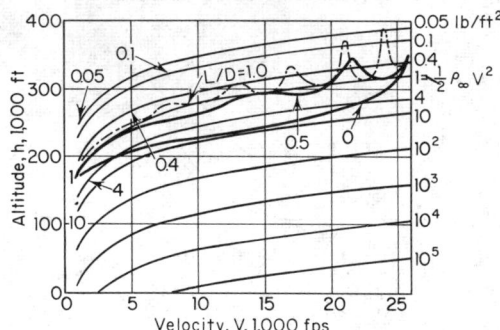

**Fig. 17** Glide reentry, dynamic pressure. $W/C_D A = 1$ lb/ft²; $\lambda = 2°$; $\frac{1}{2} \rho_\infty V^2 = (W/C_D A) (\frac{1}{2} \rho_\infty; V^2)_1$; $h = h_1 - 23,500 \ln (W/C_D A)$; $V = V_1$.

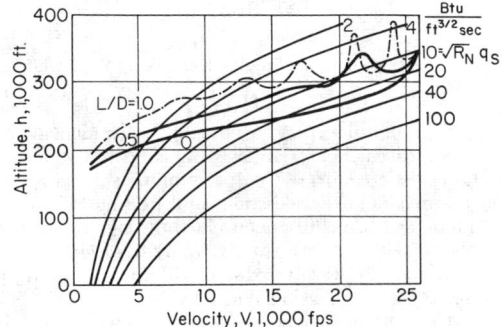

**Fig. 18** Glide reentry, stagnation-heating rate. $W/C_D A = 1$ lb/ft²; $\gamma = 2°$; $q_s = q_{s1} \sqrt{W/C_D A}$; $h = h_1 - 23,500 \ln (W/C_D A)$; $V = V_1$.

For entry at parabolic or hyperbolic speeds, it is necessary to dissipate sufficient energy so that the planet can capture the vehicle. Vehicle mass and aerodynamics characteristics determine the minimum allowable entry angle, the skip limit. The vehicle must be steered between the skip limit and the angle for maximum tolerable deceleration and heating. (See Figs. 19 and 20.)

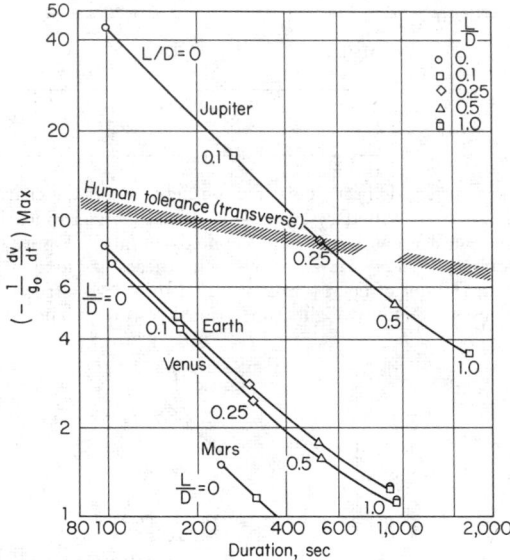

**Fig. 19** Comparison of decelerations and duration for entry into various planetary atmospheres from decaying orbits. (*After Chapman.*)

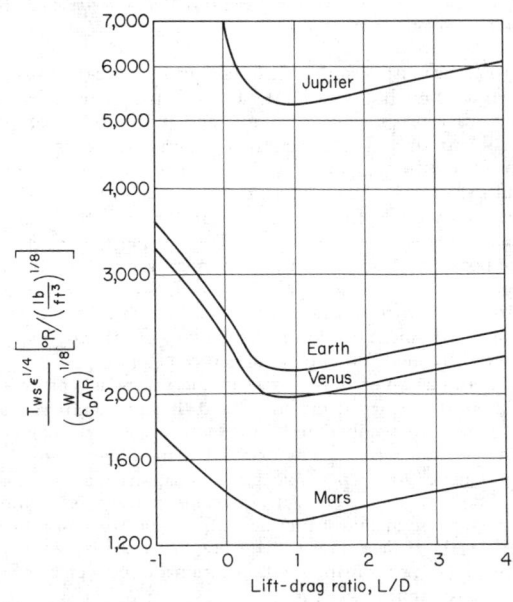

**Fig. 20** Maximum surface temperature for entry into various planets from decaying orbits. (*After Chapman.*)

**Table 8***

| $L/D$ | Lateral, nm | Longitudinal, nm |
|-------|-------------|------------------|
| 0.5 | $\approx$ 200 | $\approx 2,000$ |
| 1 | $\approx$ 600 | $\approx 4,000$ |
| 1.5 | $\approx 1,200$ | $\approx 6,000$ |
| 2 | $\approx 2,000$ | $\approx 9,000$ |

*Assumptions: $\gamma_e = 1°$; $b_e = 400,000$ ft; $V_e = 26,000$ ft/s.

## ATTITUDE DYNAMICS, STABILIZATION, AND CONTROL OF SPACECRAFT
### by M. R. M. Crespo da Silva

REFERENCES: Crespo da Silva, Attitude Stability of a Gravity-stabilized Gyrostat Satellite, *Celestial Mech.*, **2**, 1970; Non-linear Resonant Attitude Motions in Gravity-stabilized Gyrostat Satellites, *Int. Jour. Nonlinear Mech.*, **7**, 1972; On the Equivalence Between Two Types of Vehicles with Rotos, *Jour. Brit. Int. Soc.*, **25**, 1972. Thomson, Attitude Dynamics of Satellites, in Huang and Johnson (eds.), "Developments in Mechanics," Vol. 3, 1965. Kane, Attitude Stability of Earth Pointing Satellites, *AIAA Jour.*, **3**, 1965. Kane and Mingori, Effect of a Rotor on the Attitude Stability of a Satellite in a Circular Orbit, *AIAA Jour.*, **3**, 1965. Lange, The Drag-free Satellite, *AIAA Jour.*, **2**, 1964. Fleming and DeBra, Stability of Gravity-stabilized Drag-free Satellites, *AIAA Jour.*, **9**, 1971. Kendrick (ed.), TWR Space Book, 3d ed., 1967. Greensite, "Analysis and Design of Space Vehicle Flight Control Systems," Spartan Books, 1970.

A spacecraft is required to maintain a certain angular orientation, or attitude, in space in order to perform its mission adequately. For example, within certain tolerance, it may be required to point a face toward the earth for communications and observation purposes, as well as a solar panel toward the sun for power generation. A vehicle in space is subject to several external forces which produce a moment about its center of mass tending to change its attitude. The environmental moments that can act on a spacecraft can be due to solar radiation pressure, aerodynamic forces, magnetic forces, and the gravity-gradient effect. The relative importance of each of the above moments depends on the spacecraft design and on how close it is to a central attracting body. Most often, the effect on the vehicle's attitude of the first three moments mentioned above is undesirable, although they have been used occasionally for attitude control (e.g., Mariner IV, Tiros, OAO). Micrometeorite impacts can also produce a deleterious effect on the vehicle's attitude.

The most common ways to stabilize a vehicle's attitude in space are the gravity-gradient and the spin-stabilization methods. When a spacecraft is subject to the gravitational force of a central attracting body $E$, its mass element near the center of the gravity field will be subject to a greater force than that acting on the mass elements farther from $E$. This creates a moment about the center of mass, $C$, of the spacecraft, which depends on the inertia properties of the vehicle and also on the distance $r$ from $E$ to $C$. This gravity-gradient moment is given by

$$\mathbf{M} = (3GM_e/r^3)\,\hat{\mathbf{r}} \times (\mathbf{I} \cdot \hat{\mathbf{r}})$$

where $\mathbf{I}$ is the inertia dyadic of the spacecraft, referred to its mass center, and $\hat{\mathbf{r}}$ is the unit vector in the direction from $E$ to $C$. As a specific example, consider the elongated orbiting spacecraft shown in Fig. 21. The body-fixed $x$ axis (yaw) departs an angle $\theta_3$ from the local vertical, and the body-fixed

$z$ axis (pitch) remains normal to the orbital plane. The gravitational moment about $C$ is readily found to be as given in the equation below (it is assumed that the maximum dimension of the vehicle is much smaller than the distance $r$).

$$\mathbf{M} = (3GM_e/2r^3)\,(I_x - I_y)\,(\sin 2\theta_3)\,\hat{\mathbf{z}}$$

($I_x$ and $I_y$ are the spacecraft's moments of inertia about the $x$ and $y$ axes, respectively.) It is seen that unless the spacecraft's $x$ axis is vertical or horizontal, this moment is nonzero, and for $I_x < I_y$ it tends to force the $x$ axis of the vehicle to oscillate about the local vertical. Thus, the gravity-gradient moment provides a passive, and therefore very reliable, means for stabilizing the vehicle's attitude.

The orientation of a rigid body with respect to a reference frame is described by a set of three Euler angles. Figures 22 and 23 show a rigid body in orbit around a central attracting point $E$ and the coordinate systems used to describe its attitude. The unit vectors $\hat{\mathbf{a}}_1$, $\hat{\mathbf{a}}_2$, and $\hat{\mathbf{a}}_3$, with origin at the spacecraft's center of mass, define the orbital reference frame. The vector $\hat{\mathbf{a}}_1$ points in the direction of the vector from $E$ to $C$; $\hat{\mathbf{a}}_3$ in the direction of the orbital angular velocity (whose magnitude is denoted by $n$); and $\hat{\mathbf{a}}_2$ is such that $\hat{\mathbf{a}}_2 = \hat{\mathbf{a}}_3 \times \hat{\mathbf{a}}_1$. The unit vectors $\hat{\mathbf{x}}$, $\hat{\mathbf{y}}$, and $\hat{\mathbf{z}}$ in Figs. 22 and 23 are directed along the three principal axes of inertia of the spacecraft. The attitude of the vehicle with respect to the orbiting frame is defined by the three successive rotations $\theta_1$ (yaw), $\theta_2$ (roll), and $\theta_3$ (pitch) as shown in Fig. 23.

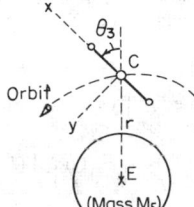

**Fig. 21**  Dumbbell satellite.

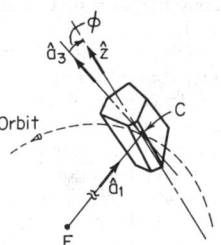

**Fig. 22**  Orbiting satellite.

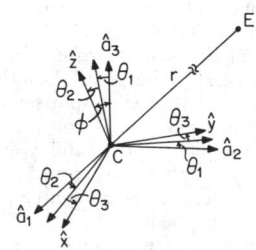

**Fig. 23**  Coordinate frames and Euler angles.

Using the above Euler angles, the gravity-gradient moment (about $C$) acting on the vehicle is given as

$$\mathbf{M} = \tfrac{3}{2}n^2[\hat{\mathbf{x}}(I_y - I_z)\sin 2\theta_2 \sin \theta_3 + \\ \hat{\mathbf{y}}(I_x - I_z)\sin 2\theta_2 \cos \theta_3 + \hat{\mathbf{z}}(I_x - I_y)\cos^2 \theta_2 \sin 2\theta_3]$$

If the spacecraft is subject to an external (other than the gravitational attraction of $E$) torque $\mathbf{T} = T_x\hat{\mathbf{x}} + T_y\hat{\mathbf{y}} + T_z\hat{\mathbf{z}}$ about its center of mass $C$, and if its attitude deviations $\theta_1$, $\theta_2$, and $\theta_3$ remain small, its linearized equations of motion are obtained as given in the following equations.

$$I_x\ddot{\theta}_1 + n\dot{\theta}_2(I_z - I_y - I_x) + n^2\theta_1(I_z - I_y) = T_x$$
$$I_y\ddot{\theta}_2 + n\dot{\theta}_1(I_x - I_z + I_y) - 4n^2\theta_2(I_x - I_z) = T_y$$
$$I_z\ddot{\theta}_3 + 3n^2(I_y - I_x)\theta_3 = T_z$$

Care must be taken in the design of a gravity-gradient stabilized satellite in order to guarantee that $I_x/I_z < I_y/I_z < 1$ for the attitude motions to be stable in the presence of the least amount of damping. For augmenting the gravity-gradient moments, booms are added to the satellite in order to make its inertia ellipsoid thinly shaped. It is interesting to note that theoretically stable motions can also be obtained when the axis of minimum moment of inertia is normal to the orbital plane. However, this orientation is unstable if damping is present, and this is what occurs in practice since all actual vehicles have parts that can move relative to each other, such as an antenna.

It is seen from the equations above that for infinitesimal oscillations the pitch motion of the satellite is uncoupled from its coupled roll-yaw motions. The natural frequency $\omega_p$ of the infinitesimal pitch oscillations is given by

$$\omega_p = n\sqrt{3(I_y - I_x)/I_z}$$

and those of the coupled roll-yaw oscillations are given by the roots of the polynomial $\omega^4 - a_2n^2\omega^2 - a_0n^4 = 0$, where

$$a_2 = 1 - [3 + (I_z - I_y)/I_x](I_x - I_z)/I_y$$
$$a_0 = 4(I_z - I_y)(I_x - I_z)/(I_xI_y)$$

For noninfinitesimal oscillations the pitch motion is coupled to the roll-yaw motion through nonlinear terms in the equations of motion. This coupling may give rise to an internal undesirable energy interchange between the modes of the oscillation, causing the roll-yaw motion, if uncontrolled, to oscillate slowly between two bounds. The upper bound of this nonlinear resonant roll-yaw motion may be much greater than (and independent of) the roll-yaw initial conditions. For given values of $I_x$, $I_y$, and $I_z$, it can be decreased only by reducing the initial conditions of the pitch motion. This phenomenon can be excited if $\omega_p \approx 2\omega_1$ or $\omega_p \approx 2\omega_2$, where $\omega_1$ and $\omega_2$ are the two natural frequencies of the roll-yaw oscillations. The same phenomenon can also be excited if $\omega_p \approx \omega_1$ (or $\omega_p \approx \omega_2$). However, the observation of this latter resonance requires a much finer "tuning" between the modes of oscillation, and therefore it is of lesser importance.

As the altitude of the orbit increases, the gravity-gradient moment decreases rapidly. For precise pointing systems, spin stabilization is used. Historically, spin stabilization was the first method used in space and is still the most commonly employed today. Simplicity and reliability are the advantages of this method when only two-axis stabilization is needed. By increasing the spin rate, the vehicle can be made very stiff in resisting disturbance moments.

A rigid body spins in a stable manner about either its axis of maximum or of minimum moment of inertia. However, in the presence of energy dissipation, even if it is infinitesimal, a spin about the axis of minimum moment of inertia leads to an unstable roll-yaw motion. Spinning satellites are built in such way as to achieve symmetry about the spin axis which, in practice, is the axis of maximum moment of inertia. A spin of a rigid body about its intermediate axis of inertia leads to an unstable roll-yaw motion. Since damping can destroy roll-yaw stability, an individual analysis of the equations of motion must be performed to guarantee the stability of the motions when the spacecraft houses a specific damper.

For planet-orbiting spinning satellites (for which the gravitational moment can now be viewed as a disturbance) the stability of the roll-yaw motion depends not only on the magnitude of the spin rate but also on the direction of its angular velocity (relative to the orbital reference frame) due to spin. Denoting the satellite's angular velocity due to spin by $\mathbf{s} = s\,\hat{\mathbf{z}}$, the following inequalities must be satisfied for the roll-yaw motions of a spinning symmetric ($I_x = I_y$) satellite in a circular orbit to be stable in the presence of an infinitesimal amount of damping:

$$(1 + s/n)\frac{I_z}{I_x} - 1 > 0$$
$$(4 + s/n)\frac{I_z}{I_x} - 4 > 0$$

Very often, a spinning pitch wheel is added to a satellite to provide additional stiffness for resisting motions out of the orbital plane. Also, the inclusion of such a wheel provides more freedom to the spacecraft designer when specifying the range of the inertia ratios to guarantee that the attitude motions of the vehicle are stable.

Let us assume that a pitch wheel with axial moment of inertia $I_w$ is connected to the spacecraft shown in Fig. 22. It is assumed that the wheel is driven by a motor at a constant angular velocity $\mathbf{s} = s\hat{\mathbf{z}}$ relative to the main body of the spacecraft. If the attitude deviations of the main body with respect to the orbiting reference frame remain small, the linearized equations for the attitude motion of the vehicle are now given as

$$I_x\ddot{\theta}_1 + n\dot{\theta}_2(I_z - I_y - I_x + \beta I_z) + n^2\theta_1(I_z - I_y + \beta I_z) = T_x$$
$$I_y\ddot{\theta}_2 + n\dot{\theta}_1(I_x - I_z + I_y - \beta I_z) \\ - n^2\theta_2[4(I_x - I_z) - \beta I_z] = T_y$$
$$I_z\ddot{\theta}_3 + 3n^2(I_y - I_x)\theta_3 = T_z$$

In these equations, the parameter $\beta$ is defined as $\beta = I_ws/I_zn$, and $I_x$, $I_y$, $I_z$ refer to the principal moments of inertia of the entire vehicle (rotor included).

Figure 24 shows a "stability chart" for this spacecraft when its center of mass, $C$, is describing a circular orbit around the central attracting point $E$. In this figure, a cut along the plane $\beta = 0$ produces the stability conditions for a gravity-gradient satellite, whereas a cut along the plane $K_1 + K_2 = 0$ (that is, $I_x = I_y$) gives rise to the inequalities for a spin-stabilized vehicle. Note that a rotor's spin rate with opposite sign to the orbital angular speed ($s < 0$) may destabilize the roll-yaw motions. Nonlinear resonances, as previously discussed, can also be excited in this vehicle. However, since the natural frequencies of the roll-yaw motions now also depend on the magnitude and sign of the internal angular momentum due to the rotor, the resonance lines are displaced in the region shown in Fig. 24 when the spin rate $s$ of the rotor is changed to a different constant. Therefore, this phenomenon, when excited, can be avoided simply by changing the parameter $\beta$.

For a spacecraft to perform its mission adequately during a long period of time, an attitude control system to maintain the vehicle's attitude within specified limits is necessary. Typically, this is done by means of a set of mass expulsion jets mounted on the vehicle, which are actuated by the output of a controller that receives information from the attitude sensors.

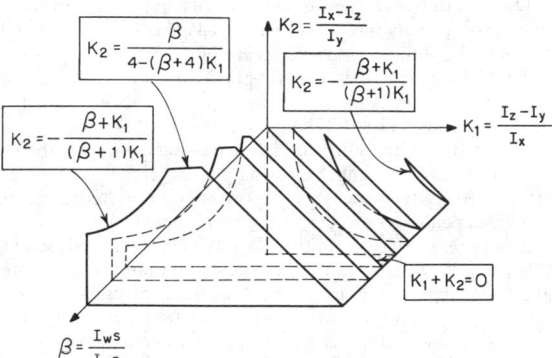

$$\beta = \frac{I_w s}{I_z n}$$

**Fig. 24**  Stable region of the vehicle parameter space.

**Drag-free Satellites**  For scientific applications that require the use of a satellite that follows a purely gravitational orbit, a translational control system is incorporated into the vehicle. Typical applications include satellite geodesy and navigation where accurate ephemeris prediction is needed. A drag-free satellite consists of a vehicle that contains a spherical cavity which houses an unsupported spherical proof mass. Sensors in the satellite detect the relative position of the proof mass inside the cavity and activate a set of thrusters placed in the vehicle, forcing it to follow the proof mass without touching it. Since the proof mass is shielded by the satellite from nongravitational forces, it follows a purely gravitational orbit.

## SPACE-VEHICLE STRUCTURES
### by Robert J. Kroll

The term **space-vehicle structures** is very broad. It includes launch vehicles, orbiting satellites, interplanetary probes, lunar landing vehicles, reentry vehicles, and the reusable space shuttle.

These structures are designed for minimum weight, because each point of structural weight can cost as much as 400 lb (181 kg) of system weight at launch. A typical space system can contain almost every type of structural element: monocoque, semimonocoque, truss, beam, frame, torque, tube, honeycomb, and filament-wound composite.

These structures must withstand the following loads during the transportation and handling, boost, space, reentry, and possibly landing phases of flight: acceleration, vibration, thermal shock, acoustic, wind gust and shear, fuel sloshing, control reactions, spin up, erection of structural components, radiation, micrometeorite impact, solar pressure, vacuum, gravity gradients, staging separation shocks, docking impacts, pull-up $g$'s, and landing loads.

Each structural component of the space vehicle must be designed by selecting the proper material and structural type for integrity relative to stresses, flexibility, stability, fatigue, vibrations, and service life. Several of these are discussed in the following subsections.

## MATERIALS FOR AEROSPACE APPLICATIONS
### by Stephen D. Antolovich

REFERENCES: Sessler and Weiss (eds.), "Aerospace Structural Metals Handbook," 4th Rev., Vol. I, Ferrous Alloys; Vol. II, Nonferrous Light Metal Alloys; Vol. IIA, Non-ferrous Heat Resistant Alloys, Syracuse University Press, 1967. (This handbook gives mechanical, physical, and chemical properties, fabrication, availability, etc.) Damage Tolerant Design Handbook MCIC-HB-01, Metals and Ceramics Information Center, Battelle Columbus Laboratories, 1972. Supplement to MCIC-HB-01, 1973. (These two documents provide the most up-to-date and complete compilation of fracture mechanics data for alloy steels, stainless steels, aluminum and titanium alloys.) "Metals Handbook," 8th ed., 8 vols., American Society for Metals: Vol. I, Properties and Selection of Metals, 1961; Vol. II, Heat Treating Cleaning and Finishing, 1964; Vol. III, Machining, 1967; Vol. IV, Forming, 1969; Vol. V, Forging and Casting, 1970; Vol. VI, Welding and Brazing, 1971; Vol. VII, Atlas of Microstructures, 1972; Vol. VIII, Metallography, Structures and Phase Diagrams, 1973. Titanium Alloys Handbook, MCIC-HB-02, Metals and Ceramics Information Center, Battelle Columbus Laboratories, 1972. Bibliography on Fibers and Composite Materials, 1969–1972, MCIC 72-09, Metals and Ceramics Information Center, Battelle Columbus Laboratories, 1972. 1974 Materials Selector, *Mater. Eng.*, **78**, no. 4, Sept. 1973.

A wide variety of materials are utilized in aerospace applications, and although many of the problems of the design engineer are quite conventional and easily handled, performance requirements place extreme demands on the materials. Some of the more important engineering properties that must be considered in the selection of materials are (1) strength-to-weight ratio, (2) density, (3) modulus of elasticity, (4) strength and toughness at operating temperature, (5) resistance to fatigue damage, and (6) environmental effects on strength, toughness, and fatigue properties. In addition, the engineer must consider factors such as weldability, formability, castability, quality control, and cost.

### LOW-TEMPERATURE APPLICATIONS

Although it is desirable to maximize the strength-to-weight ratio to achieve given design goals, materials with high strength-to-weight ratios generally are sensitive to the presence of small cracks and may exhibit catastrophic brittle failure at stresses below the nominal design stress. The problem of premature brittle fracture is accentuated as the temperature decreases and the yield and ultimate strength increase. For these reasons it is imperative that the tendency toward brittle failure be given major consideration for low-temperature applications. For example, liquid oxygen, a common oxidizer, boils at $-297°F$ ($-183°C$) whereas liquid hydrogen, an important fuel, boils at $-423°F$ ($-253°C$), and the problem of brittle failure is predominant. As a general rule, although the toughness of most important aerospace materials decreases with temperature, body-centered-cubic (BCC) and hexagonal close-packed (HCP) metals exhibit more drastic decreases in toughness with decreased temperature than do face-centered-cubic metals (FCC); for these reasons the FCC materials are preferred whenever practical and possible for use at low temperatures.

In addition to the more familiar common mechanical properties listed in Table 9 (which are extensively tabulated in "Aerospace Structural Metals Handbook," Vol. I, II, and IIA, Syracuse University Press, 1967), the fracture-mechanics approach has gained wide acceptance, and NDI characterization of flaws allows safe operating stresses to be calculated. A considerable amount of fracture-mechanics data has been collected and evaluated (Damage Tolerant Design Handbook, MCIC-HB-01, 1972, and its supplement, 1973, Metals and Ceramics Information Center, Battelle Columbus Laboratories). It should be emphasized that, although FCC alloys are generally superior to BCC for low-temperature applications, metallurgical factors such as melting and deoxidation practice, impurities, nonmetallic inclusions, grain size, processing variables, and alloy stability play important roles in determining the toughness of both FCC and BCC materials. For example, large-grain sizes in BCC promote brittleness whereas small inclusions in both FCC and BCC can be severely embrittling at low temperature. In addition, although most materials recover their original properties when returned to room temperature, some austenitic (FCC) stainless steels transform to brittle martensite upon low-temperature exposure.

## INTERMEDIATE-TEMPERATURE APPLICATIONS

At intermediate temperatures, encountered in repetitive-use components such as aircraft airframes, landing gear, fan shafts, fan and compressor blades, and disks, in addition to brittle fracture as discussed above, problems such as fatigue crack formation and propagation, stress corrosion cracking, and corrosion fatigue become important.

In general, because of the complexity of aerospace vehicles, the presence of flaws must be assumed (they may occur either during processing or fabrication) and design must be based on **fatigue crack propagation** (FCP) properties (Sec. 5). The FCP character of many materials follows an equation of the form

$$da/dN = R(\Delta K)^n \qquad (1)$$

where $a$ refers to crack length, $N$ is the number of cycles, $\Delta K$ is the fluctuation in the stress-intensity parameter, and $R$ and $n$ are material constants. (A considerable amount of FCP information is given in MCIC HB-01 and its supplement.) Equation (1) can be integrated to determine the cyclic life of a component provided that the stress-intensity parameter is known (Sec. 5). Although fatigue-crack initiation is generally retarded by having smooth surfaces, shot-peening, and other surface treatments, the FCP character of commercial aerospace alloys is relatively insensitive to metallurgical treatment, and to a reasonable approximation whole classes of materials (i.e., aluminum alloys, high-strength steels, etc.) can be represented within the same scatter band.

It should be noted that increases in the plane-strain fracture toughness, $K_{IC}$, do not contribute significantly to the life of a component if the design stress is maintained. This is due to the exponential nature of FCP. When the flaws are relatively small, the FCP rate is low, and the great majority of the useful life is spent in this regime. As the flaws become larger (and $\Delta K$

increases) the FCP rate is greatly accelerated and very few additional cycles can be accumulated before final brittle failure. Another pitfall for the design engineer to avoid where FCP is a potential problem is increasing the operating stress based on increases in $K_{IC}$. Increased stress, through its linear relationship to the stress-intensity parameter, causes an exponential increase in the FCP rate and may introduce a fatigue problem where previously there was none.

The **effect of environment** on materials used for aerospace problems is usually to aggravate the FCP rate. (Environmental effects on FCP for various alloys are given in MCIC HB-01.) It should be emphasized that fatigue loads have a synergistic effect on the environmental component of cracking and that the crack growth rate (CGR) during corrosion fatigue (CF) is rarely the sum of the FCP and stress corrosion cracking (SCC) rates. Furthermore, the SCC component of CF at a given stress-intensity level increases exponentially with temperature so that, depending on the system, small increases in temperature can greatly accelerate the CF rate. High-strength steels, 2000 and 7000 series aluminum alloys, and titanium alloys are quite sensitive to environmental factors. Stress corrosion effects may be minimized by proper heat-treatment below the recrystallization temperature to eliminate residual stresses after forming and machining. For aluminum alloys, special overaging treatments have been developed to minimize corrosion effects. Treatments that minimize grain boundary precipitation and promote uniform dispersions of precipitates in age-hardenable alloys usually provide superior resistance to corrosion.

**Welding** results in a cast structure in the fusion zone, larger grain size, concentration of impurities and defects such as porosity, modification of the microstructure in the adjacent heat-affected zones, and the formation of surface irregularities or notches which reduce ductility at room temperature. Low temperatures accentuate the problems associated with these factors. It is not acceptable to extrapolate the parent-metal behavior to that of the weldment. In weldments of castings or wrought material, tests are needed to simulate anticipated use conditions and to prove weld soundness.

## HIGH-TEMPERATURE APPLICATIONS

The turbine inlet temperatures of fan jet engines are in the vicinity of 2400°C or more, and the materials in the turbine section of the engine are subjected to deleterious gases such as $O_2$, $SO_2$, etc., high temperatures and stresses, and cyclic loads. Consequently the most significant metallurgical factors to consider are microstructural stability, resistance to oxidation and sulfidization, creep resistance, and resistance to low cycle fatigue (LCF). Nickel-base superalloys are generally used in the turbine section of the engine for both disks and blades. (See "Aerospace Structural Metals Handbook," Vols. II and IIA, for specific properties.) Although the Ni-base superalloys generally exhibit excellent strength and oxidation resistance at temperatures up to 1800°F (982.22°C), the interactive effects of sustained and cyclic loads are difficult to handle because at high temperatures plastic deformation of these materials is time-dependent. Recently an approach based on strain range partitioning has been developed (S. S. Manson, The Challenge to Unify Treatment of High Temperature Fatigue—A Partisan Proposal Based on Strain Range Partitioning, "Fatigue at Elevated Temperatures," ASTM

**Table 9. Mechanical Properties of Engineering Materials at Low Temperature**

| Alloy | 80 F Tensile strength, kpsi | 80 F Yield strength, kpsi | 80 F Percentage of elongation (2 in.) | 80 F Notched-unnotched ratio ($K_t = 10$) | −320 F Tensile strength, kpsi | −320 F Yield strength, kpsi | −320 F Percentage of elongation (2 in.) | −320 F Notched-unnotched ratio ($K_t = 10$) | −423 F Tensile strength, kpsi | −423 F Yield strength, kpsi | −423 F Percentage of elongation (2 in.) | −423 F Notched-unnotched ratio ($K_t = 10$) |
|---|---|---|---|---|---|---|---|---|---|---|---|---|
| **Aluminum** | | | | | | | | | | | | |
| 2014-T6 | 70 | 64 | 9.7 | 0.99 | 84 | 76 | 11.7 | 0.93 | 97 | 80 | 13.6 | 0.88 |
| 2020-T6 | 79 | 75 | 8.0 | 0.67 | 95 | 88 | 4.0 | 0.52 | 101 | 93 | 2.3 | 0.50 |
| 2219-T6 | 65 | 52 | 9.8 | 0.92 | 82 | 64 | 12.1 | 0.90 | 96 | 79 | 15.3 | 0.81 |
| 2219-T87 | 68 | 56 | 13.0 | 0.74 | 85 | 68 | 16.3 | 0.73 | 100 | 73 | 17.6 | 0.67 |
| 2119-T6 | 60 | 43 | 9.0 | 0.93 | 76 | 53 | 12.2 | 0.86 | 88 | 43 | 16.5 | 0.62 |
| 5052-H32 | 34 | 25 | 10.6 | 0.97 | 53 | 29 | 30.0 | 0.93 | 73 | 37 | 26.5 | 0.88 |
| 5086-H34 | 47 | 37 | 10.4 | 1.00 | 65 | 44 | 25.0 | 0.89 | 85 | 48 | 20.2 | 0.76 |
| 5456-H343 | 58 | 45 | 8.7 | 0.92 | 74 | 53 | 13.0 | 0.79 | 87 | 58 | 8.7 | 0.75 |
| 7002-T6 | 70 | 57 | 16.7 | 1.05 | 83 | 70 | 19.8 | 1.03 | 104 | 77 | 18.9 | 0.86 |
| 7075-T651 | 80 | 74 | 9.2 | 0.90 | 94 | 88 | 5.2 | 0.68 | 101 | 95 | 3.2 | 0.56 |
| 7079-T6 | 76 | 67 | 9.0 | 1.00 | 94 | 84 | 4.0 | 0.68 | 101 | 94 | 3.0 | 0.56 |
| 7178-T6 | 94 | 88 | 7.5 | 0.57 | 109 | 104 | 1.2 | 0.41 | 117 | 113 | 1.0 | 0.32 |
| Maraging steel—18 % Ni (annealed) | 192 | 175 | 7.0 | 1.16 | 268 | 250 | 8.0 | 1.01 | 313 | 283 | 8.7 | 0.59 |
| Maraging steel—18 % Ni (aged) | 254 | 245 | 2.8 | 1.09 | 321 | 309 | 2.5 | 0.90 | 365 | 355 | 3.2 | 0.41 |
| **Super alloys** | | | | | | | | | | | | |
| A-286 (annealed) | 93 | 42 | 37.3 | 0.86 | 144 | 68 | 71.0 | 0.80 | 161 | 81 | 47.3 | 0.82 |
| A-286 (age-hardened) | 140 | 94 | 22.0 | 0.94 | 191 | 122 | 40.7 | 0.82 | 218 | 137 | 28.5 | 0.83 |
| Inconel X (annealed) | 111 | 48 | 50.7 | 0.82 | 150 | 65 | 58.3 | 0.79 | 160 | 70 | 51.3 | 0.82 |
| Inconel X (age-hardened) | 180 | 126 | 25.3 | 0.90 | 220 | 139 | 32.0 | 0.79 | 224 | 140 | 28.0 | 0.82 |
| Waspaloy (annealed) | 144 | 80 | 48.0 | 0.85 | 200 | 108 | 53.0 | 0.77 | 214 | 118 | 50.4 | 0.80 |
| Waspaloy (age-hardened) | 177 | 116 | 26.3 | 0.81 | 205 | 142 | 15.0 | 0.80 | 197 | 154 | 10.2 | 0.85 |
| K-Monel (annealed) | 95 | 46 | 38.8 | 0.95 | 133 | 65 | 48.0 | 0.91 | 152 | 75 | 43.3 | 0.86 |
| K-Monel (age-hardened) | 148 | 106 | 22.7 | 0.92 | 177 | 128 | 30.7 | 0.92 | 192 | 137 | 28.3 | 0.90 |
| **Titanium** | | | | | | | | | | | | |
| Ti-6Al-4V | 139 | 133 | 11.0 | 1.02 | 218 | 214 | 13.0 | 0.82 | 240 | 240 | 1.7 | 0.61 |
| Ti-5Al-2.5 Sn | 134 | 128 | 12.8 | 1.20 | 213 | 207 | 14.0 | 0.81 | 234 | 234 | 5.0 | 0.66 |
| Ti-13V-11Cr-3Al | 137 | 137 | 13.3 | 1.20 | 285 | 282 | 2.5 | 0.54 | 289 | 289 | 0.7 | 0.40 |

Note: All tests were made on sheet.

STP 520, 1973, pp. 744–782). The utility of this approach is that frequency and temperature dependence are incorporated into the analysis by the way in which the strain is partitioned and reasonable envelopes of the expected life can be calculated.

## SPACE ENVIRONMENT
### by W. R. Lucas

The most adverse conditions in space for materials are vacuum, radiation, temperature extremes, meteoric bombardment, and possible synergistic effects of two or more environments. Where oxidation is involved as a degradation mechanism, e.g., in the irradiation of Teflon and the fatigue of oxide-forming metals, the space environment is less severe than the earth environment.

**Vacuum**  The **vacuum of space** may cause the evaporation of a material, or a volatile component of the material, and of the adsorbed gases on the surfaces of all materials. The evaporation rate of a pure material can be calculated by

$$G = \sqrt{\frac{M}{T}} \frac{P}{17.14}$$

where $G$ = evaporation rate, g/cm² (s); $M$ = molecular weight; $T$ = absolute temperature, K; and $P$ = vapor pressure, mmHg at temperature $T$. (See Dushman, "Scientific Foundations of Vacuum Technique," Wiley.) This simple formula is not applicable to heterogeneous materials or even to a pure substance, such as a plasticizer in an elastomer, which is removed from a matrix of another substance. In this case, other factors such as migration rate influence the rate of the reaction. Although the evaporation of a component of a material may not reduce the effectiveness of the material, e.g., the plasticizer in the insulation of an electrical conductor, the deposition of the vapor on a colder surface may be intolerable. Metals are not usually evaporated in space at modest temperatures, but organic materials, including elastomers, plastics, coatings, adhesives, and lubricants, must be of very high molecular weight to avoid evaporation.

**Radiation**  Spacecraft are exposed to electromagnetic and particle radiation and to the radiation environment of nuclear reactors. Metals and ionic compounds are relatively resistant to space-indigenous radiation. However, semiconductors are sensitive to permanent radiation damage, and other electrical materials are subject to permanent or transient damage. Organic materials are highly susceptible to degradation by both electromagnetic and particle radiation, especially in vacuum. Organic polymers of high molecular weight may have such low vapor pressure that their evaporation in vacuum at reasonable temperatures is not significant. However, radiation, which produces chain scission, yielding fragments of reduced molecular weight and increased vapor pressure, will result in the degradation of the mass at the same temperature which did not affect the nonirradiated material (Table 10). Note the almost complete degradation of the tensile strength of neoprene and Buna N as a result of irradiation in vacuum. On the other hand, Viton A appears to be satisfactory for this environment on the basis of increased tensile strength; however, the decreased elongation would be an important consideration in the application of this material as a seal, particularly as a dynamic seal. Furthermore, Viton A is not suitable for

very-low-temperature applications, another component sometimes encountered in the space environment.

**Lubrication**  (See also Secs. 6 and 8)  The better-known organic lubricants are subject to evaporation in space, either with or without degradation by radiation. Gases which are normally absorbed to surfaces and reduce friction between mating surfaces will be removed. Mating metal parts of the same material or of materials forming a solid solution in one another may weld together when clean surfaces are exposed as a result of the degrading influence of the space environment. Thus, lubrication is one of the most significant space-related problems. The most effective lubricants for space are those based upon the heavy metal sulfides, such as $MoS_2$. A lubricant known as MLF-5, consisting of $MoS_2$, graphite, and gold in the ratio of 10:1:5, has a coefficient of friction which is less in vacuum than at 1 atm. Moving electrical-contact surfaces, such as brushes, slip rings, and make-break switches, require either reliable isolation from the vacuum environment or special selection of materials, especially where long-time operation is involved. Composites of heavy metal sulfides with silver or copper are promising possibilities for this application.

**Meteoroids**  Meteoroids are metallic or stony bodies of various sizes which travel through space at velocities up to 300,000 ft/s. Large meteoroids are rare, but the great number of small ones increases the probability of impacts on a spacecraft. These small meteoroids may be only a few thousandths of an inch in diameter, but their tremendous velocities make them destructive. Experimental studies of hypervelocity impact phenomena have been conducted in velocity ranges up to 30,000 ft/s, but extrapolation is not advisable. However, there is widespread acceptance of the dual-wall technique of protecting a craft against meteoroid damage. The outer wall, called the **bumper,** is detached from the load-carrying structural wall. The bumper dissipates some of the kinetic energy of the particle, but primarily it serves to fragment the particle into a fine spray so that the impact energy is spread over a larger area of the structural wall.

## FINITE ELEMENT ANALYSIS OF STRUCTURES
### by Lawrence H. Sobel

REFERENCES: Clough, The Finite Element in Structural Mechanics, in Zienkiewicz and Hollister (eds.), "Stress Analysis," Wiley, 1965. Felippa and Clough, "The Finite Element Method in Solid Mechanics," *SIAM—AMS Proc.,* Vol. II, 1970. Gallagher, "Finite Element Analysis: Fundamentals," Prentice-Hall. Holand and Bell, "Finite Element Methods in Stress Analysis, Tapir. Marcal (ed.), On General Purpose Finite Element Computer Programs, *ASME Spec. Publ.,* 1970. NASTRAN, COSMIC, University of Georgia. Zienkiewicz, "The Finite Element Method in Engineering Science," McGraw-Hill.

The finite element method (henceforth abbreviated FEM) is a relatively recent numerical discretization technique that is widely used for the approximate solution of complex problems in structural mechanics. The popularity of this method lies primarily in its ability to treat complex structures of arbitrary shape, arbitrary boundary conditions, and variable geometrical and material properties. The FEM is an extension of the matrix structural analysis techniques originally employed in the analysis of skeleton structures, such as trusses and frames.

**Table 10. Effects of Vacuum and Radiation on Elastomers**

| Material | Test | Pressure, mm Hg | Temp, deg F | Radiation ergs, g⁻¹/ deg C | Tensile strength, psi | Elongation, percent |
|---|---|---|---|---|---|---|
| Neoprene | Air | 760 | 80 | 0 | 3,135 | 426 |
| | Vac | $1 \times 10^{-5}$ | 80 | 0 | 3,350 | 405 |
| | Air/Rad | 760 | 80 | $1.9 \times 10^{9}$ | 2,769 | 265 |
| | Vac/Rad | $5 \times 10^{-6}$ | 80 | $1.9 \times 10^{9}$ | 191 | 218 |
| Buna N | Air | 760 | 80 | 0 | 2,630 | 685 |
| | Vac | $1 \times 10^{-5}$ | 80 | 0 | 2,640 | 700 |
| | Air/Rad | 760 | 80 | $1.9 \times 10^{9}$ | 2,175 | 390 |
| | Vac/Rad | $5 \times 10^{-6}$ | 80 | $1.7 \times 10^{9}$ | 203 | 450 |
| Viton A | Air | 760 | 80 | 0 | 1,343 | 172 |
| | Vac | $1 \times 10^{-5}$ | 80 | 0 | 1,168 | 238 |
| | Air/Rad | 760 | 80 | $2 \times 10^{10}$ | 2,629 | 36 |
| | Vac/Rad | $5 \times 10^{-7}$ | 109 | $1.6 \times 10^{10}$ | 1,830 | 31 |

This extension has been made possible through the advent of large-scale, high-speed digital computers. The present availability of large-scale, general-purpose finite element computer programs (NASTRAN; also see Marcal) has helped bring to fruition the solution of many heretofore "unsolvable" practical complex structural problems. A brief description of the basic tenets of the FEM is presented herein. For comprehensive examples of the application of this method to complex problems, we must, of necessity, refer the reader to the cited references.

## BASIC CONCEPT

The basic concept of the FEM is that every structure may be considered to be an assemblage or collection of a finite number of structural components or building blocks, the so-called finite elements, which are interconnected at a finite number of points, called nodes or nodal points.

## SKELETON STRUCTURES

For skeleton structures, the assemblage occurs in a natural way. Each bar, beam, or column member may be considered to be a finite element, and the joints, which are the intersections of the elastic axes of the members, are considered to be the nodes. Skeleton structures, including space frames and space trusses, are considered mathematically to be one-dimensional structures, since only one spatial coordinate (which may be curvilinear) is needed to describe the state of elastic deformation of each of the individual members or elements. One way of describing this state of deformation is to obtain the so-called element stiffness equation $\{Q\}^e = [k]^e\{q\}^e$ that is written with respect to a conveniently chosen set of element-oriented axes, which are called local axes. This equation relates the generalized forces $\{Q\}^e$, exerted by the nodes on the ends of element $e$, to the generalized displacements $\{q\}^e$ of the ends of the element. It is common to call $\{Q\}^e$ and $\{q\}^e$ the nodal force and nodal displacement vectors, respectively. Corresponding elements of the displacement and force vectors are to be conjugate in the energy sense, which simply means that if one of the elements of $\{q\}^e$ is a rotation (for instance), then the corresponding element of $\{Q\}^e$ must be a moment, so that their product represents work or energy.

The nodal force vector is related to the nodal displacement vector by means of the element stiffness matrix $[k]^e$, whose elements are stiffness influence coefficients. For uniform one-dimensional elements, the element stiffness matrices can be determined exactly and are readily available in the references. An element stiffness equation in terms of local coordinates is obtained for each element of the structure. It is then necessary to express the element stiffness equation in terms of a common coordinate system, the so-called global coordinate system, which is used to describe the elastic behavior of the overall structure (or system).

After this transformation is completed, the elements are assembled together via the direct stiffness method. This assembly procedure furnishes the system stiffness equation $\{R\} = [K]\{r\}$, written in terms of global coordinates. In this equation $\{r\}$ is the vector of system nodal displacements and $\{R\}$ is the vector of conjugate system nodal forces. The elements of $\{R\}$ are either known applied nodal forces, or reactions, or applied forces required to produce a prescribed nodal displacement. $[K]$ is the stiffness matrix for the system.

The direct stiffness method for the assembly procedure can be interpreted from two different viewpoints (Felippa), one mathematical and the other physical. Here, only the physical viewpoint will be outlined. The elements of the system are joined together at their mutually common nodal points through the application of the simple requirements of nodal-point compatibility and nodal-point equilibrium. Nodal-point compatibility requires that all elements sharing the same node have the same system nodal displacement at that node. This therefore establishes a relation between system and element nodal displacements. Nodal-point equilibrium (method of joints in classic analysis of structures) requires that the nodal forces from all elements sharing the same node must be in equilibrium with any applied forces or reactions at that node. As a result, it is easy to show that the system stiffness matrix $[K]$ can be obtained from a direct superposition of the element stiffness matrices, expressed in terms of global coordinates.

The next step is to specify the prescribed applied forces and displacement boundary conditions and solve the system stiffness equation for the unknown nodal displacements and reactions. With the knowledge of all the nodal displacements, it is possible to determine, in a systematic way, all other quantities of interest, such as stresses, strains, or displacements at any point in any element.

## CONTINUUM STRUCTURES

Next we consider a different type of structure, the continuum, which is two- or three-dimensional in space and which is not blessed with a natural subdivision of structural components. Nevertheless, the continuum structure can also be considered to be composed of an assemblage of structural components or building blocks, i.e., finite elements (two- or three-continuum elements), and hence it can still be analyzed by essentially the same techniques employed for skeleton structures. However, for continuum structures, certain approximations and idealizations must be introduced.

Perhaps the basic concept (Gallagher) of the FEM, as applied to continuum structures, is best illustrated when it is discussed within the framework of a specific example, namely, the thin rectangular plate shown in Fig. 25. The plate, which is assumed to be in a state of plane stress, is fixed on three sides and is subjected to an in-plane distributed pressure on the fourth side (not shown in the figure). Next let us inscribe or paint some vertical and horizontal lines on the actual plate (Fig. 26). We can then consider a typical rectangular region

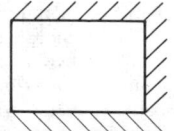

**Fig. 25** Actual plate.

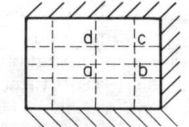

**Fig. 26** Actual plate with inscribed lines.

(*abcd*) bounded by these lines. Now, in the finite element approach, we introduce a structural idealization whereby we say that each typical rectangular region now defines a rectangular (plane stress) element. Thus, the original plate is modeled by an assemblage of rectangular elements that are interconnected at the nodes, which are the intersections of the vertical and horizontal edges of the elements. The (generalized) displacements of the nodes in the finite element model are taken to be the basic unknowns of the problem and they represent approximations to the corresponding displacements of the actual structure. Note that the behavior of such an assemblage of elements would represent a very crude approximation to the behavior of the continuous structure as gaps or overlaps would develop upon application of loads to the assemblage. That is, displacement continuity (or compatibility) conditions at the common-element interfaces would be violated except at the nodes where continuity conditions are explicitly enforced (in accord with the direct stiffness procedure). Therefore, to ensure that the behavior of the assemblage under load is a reasonable approximation to the behavior of the continuous structure, continuous displacement functions, with known shapes (so-called shape functions) but with unknown amplitudes (the nodal displacements), are selected to uniquely describe the displacements at any point within the element. These displacement functions are usually selected so as to satisfy displacement continuity at the element interfaces (however, interelement continuity of derivatives of displacements may not be satisfied, e.g., rotations for some plate bending elements). From the assumed displacement functions, the strains (from kinematic relations) and then the stresses (from constitutive equations) at any point in the element and along the edges of the element can be expressed in terms of the

nodal displacements of the element. Now, to employ the methods of analysis previously described for the skeleton structures, it is necessary to work with nodal displacements and nodal forces. Therefore, the distributed stresses along the edges of the element are replaced by a set of equivalent forces acting only at the nodes. These nodal forces are completely fictitious forces which do not exist in the actual structure. Nevertheless, they are still meaningful in the sense that they are statically equivalent to the distributed stresses on the edges of the element. Also, the work done by the nodal forces when subjected to the nodal displacements equals the work done by the edges stresses due to the corresponding edge displacements. On the basis of the assumed displacement functions, an approximate element stiffness matrix can be determined that relates the element nodal forces to the element nodal displacements. The analysis of the assemblage of continuum finite elements is then carried out by a procedure that is identical to the aforementioned one for skeleton structures.

Finally, it is noted that, although a plane stress rectangular element was chosen to illustrate the discretization process, the procedure outlined above is general and it applies to elements of other shapes, such as the popular triangular and quadrilateral elements, as well as to three-dimensional elements.

## STRESSES AND STABILITY
### by Robert J. Kroll

REFERENCES: Anderson, Brooks, Leonard, and Maltz, Shuttle: Structures—A Technology Overview, *Astronaut. Aeronaut.*, Feb. 1971, pp. 38–47. Love, Advanced Technology and the Space Shuttle, *Astronaut. Aeronaut.* Feb. 1973, pp. 30–66. Roark, "Formulas for Stress and Strain," McGraw-Hill. Peery, "Aircraft Structures," McGraw-Hill. Gatewood, "Thermal Stress," McGraw-Hill.

An intermediate design of the space shuttle will be used to illustrate the principles of stress and stability. As shown in Fig. 27, the space shuttle consists of two parts: the launch vehicle which is, in part, similar to previous launch vehicles such as the Saturn V, plus the orbiter which is similar to an airplane having internal propellant tanks.

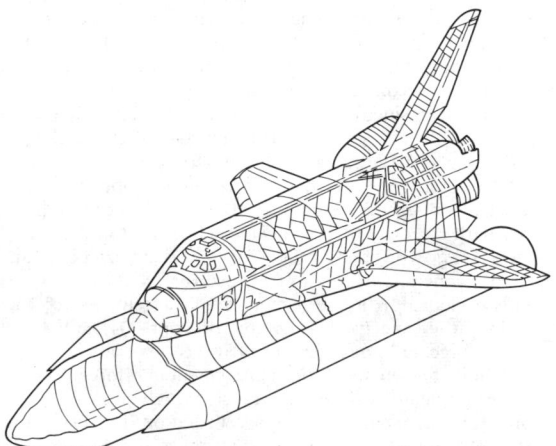

**Fig. 27** Space shuttle structure.

The launch vehicle is basically a stiffened pressure vessel which carries the propellants and rocket engines. Internal pressure creates hoop stress and meridional stress given by $pR/t$ and $pR/2t$, respectively. These are membrane stresses, but, in general, discontinuity stresses exist at the tank ends where flanges and end closures are located. The discontinuity bending stresses die out rapidly with distance from the end of the tank. The meridional discontinuity bending stress is given by

$$\sigma_b = \frac{6e^{-\lambda x}}{\lambda t^2}\left[V_0 \sin \lambda x + \lambda M_0 (\cos \lambda x + \sin \lambda x)\right]$$

where $M_0$ and $V_0$ are the moment and shear per unit length of circumference at the tank end, $x$ is the distance from the tank end, $R$ is the tank radius, $t$ is the wall thickness, and $\lambda = 1.28 (Rt)^{-0.5}$. The discontinuity stresses exist because the membrane stresses cannot satisfy the boundary conditions at the tank ends. The membrane and discontinuity stresses produce large combined stresses near the tank ends.

The launch vehicle also undergoes bending loads similar to a long beam. A pure monocoque cylinder would be inefficient in resisting bending. Stiffeners are added to the cylinder to provide a semimonocoque structure for resisting bending. The ring stiffeners help preserve the cross-sectional shape of the cylinder, the longitudinal stiffeners carry most of the longitudinal tension and compression bending stresses, and the cylinder skin carries the transverse bending shear.

A large monocoque cylinder would become structurally unstable at relatively low values of axial compression and bending. With the ring and longitudinal stiffeners, the cylinder has vastly improved stability. The longitudinal stiffeners that carry compression act as a column having a length equal to the distance between ring stiffeners. A conservative approximation would treat the column as having pinned ends giving a critical buckling load of $P_{cr} = \pi^2 EI/L^2$. The cylinder skin could become unstable because of transverse bending shear. The size of an individual skin panel is determined by the distance between adjacent ring and longitudinal stiffeners and is therefore relatively small and stable.

The orbiter airframe structure is semimonocoque, consisting of skin, bulkheads, and stringers. This structure is subject to a severe thermal environment during reentry. The most promising approach to thermal protection is exterior insulation. Even with this protection, the basic structure feels some elevated temperatures, which can create thermal stresses.

Material properties of the structural material are generally degraded at elevated temperature. In addition, thermal stresses are created when the strains due to thermal expansion are constrained. There are two types of constraints: mechanical and natural.

As an example of mechanical constraint, a bar which has a uniform temperature rise is stress-free if it is allowed to increase in length. If the bar is constrained between rigid walls and heated, then the bar has a compressive axial thermal stress ($E\alpha\Delta T$) because the walls impose an axial compressive force on the bar to suppress the thermal.

Natural constraints usually are associated with nonlinear temperature gradients which tend to create incompatible thermal strains that cannot exist because of continuity requirements. An example of natural constraint is a circular plate that has a uniform $T$ over a circular inner segment and no $T$ over the remaining outer annular segment. The inner disk has a tensile thermal expansion, and the outer ring has no thermal

expansion. But the disk and ring must be continuous and an internal radial compressive force must exist at their interface and thermal stresses are created.

Thermal stresses can be minimized by attempting to minimize and linearize the thermal gradients on the structure, plus designing expansion fitting into the structure to reduce the mechanically induced loads.

Most space-vehicle structures are sufficiently complex to prevent exact closed-form solutions for the stresses and stability. In Fig. 28 several of the computer methods for analyzing space structures are listed.

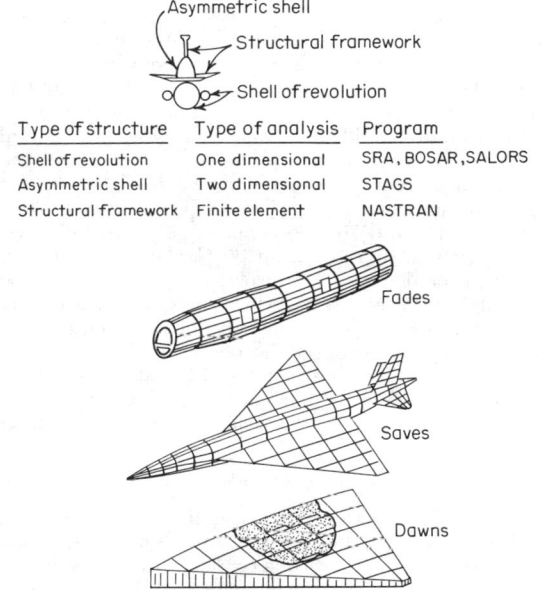

| Type of structure | Type of analysis | Program |
| --- | --- | --- |
| Shell of revolution | One dimensional | SRA, BOSAR, SALORS |
| Asymmetric shell | Two dimensional | STAGS |
| Structural framework | Finite element | NASTRAN |

**Fig. 28** Structural-design computer programs.

# VIBRATION OF STRUCTURES
### by Lawrence H. Sobel

REFERENCES: Crandall and McCalley, Numerical Methods of Analysis, chap. 28, in Harris and Crede (eds.), "Shock and Vibration Handbook," Vol. 2, McGraw-Hill. Hurty and Rubinstein, "Dynamics of Structures," Prentice-Hall. Meirovitch, "Analytical Methods in Vibrations," Macmillan. Przemienecki, "Theory of Matrix Structural Analysis," McGraw-Hill. Wilkinson, "The Algebraic Eigenvalue Problem," Oxford.

As a result of the vast improvements in large-scale, high-speed digital computers, numerical methods are being used at an exponentially increasing rate to analyze complex structural vibration problems. The numerical methods most widely used in structural mechanics are the finite element method and the finite difference method. In the finite element method, the actual structure is represented by a finite collection, or assemblage, of structural components whereas, in the finite difference method, spatial derivatives in the differential equations governing the motion of the structure are approximated by finite difference quotients. With either method, the continuous structure with an infinite number of degrees of freedom is, in effect, approximated by a discrete system with a finite number of degrees of freedom (unknowns). The linear, undamped free and forced vibration analysis of such a discrete system is discussed herein. The analysis is most conveniently carried out with the aid of matrix algebra, which is well suited for theoretical and computational purposes. The following discussion will be brief, and the reader is referred to the cited references for more comprehensive treatments and examples.

## EQUATION OF MOTION

Consider a conservative discrete system undergoing small motion about a state of equilibrium (neutral or stable). Let $q_1(t), \ldots, q_n(t)$, where $t$ is time, be the minimum number of independent coordinates (linear or rotational) that completely define the general dynamical configuration of the system and that are compatible with any geometrical constraints imposed on the system. Then the $n$ coordinates $q_1, \ldots, q_n$ are called generalized coordinates, and the system is said to have $n$ degrees of freedom. Let $\{\mathbf{Q}(t)\}$ be the vector of generalized forces. and let $\{\mathbf{q}(t)\}$ be the vector of generalized displacements, $\{\mathbf{q}(t)\} = \{q_1(t), \ldots, q_n(t)\}$. Corresponding elements of the generalized force and displacement vectors are to be conjugate in the energy sense, which simply means that if one of the elements of $\{\mathbf{q}\}$ is a rotation (for instance), then the corresponding element of $\{\mathbf{Q}\}$ must be a moment, so that their product represents work or energy. Methods of computing $\{\mathbf{Q}\}$ are discussed in the references (e.g., Meirovitch). The matrix equation governing the small undamped motion of the system is given by

$$[M]\{\ddot{\mathbf{q}}\} + [K]\{\mathbf{q}\} = \{\mathbf{Q}\} \qquad (1)$$

In this equation a dot denotes single differentiation in time, and $[M]$ and $[K]$ are the mass and stiffness matrices, respectively. $[M]$ and $[K]$ are real and constant matrices, which are assumed to be symmetric. Such symmetry will arise whenever an energy approach (e.g., Lagrange's equations) is employed to derive the equation of motion. The mass matrix is assumed always to be positive definite (and hence nonsingular, or can be made to be positive definite; see Przemieniecki).

## FREE-VIBRATION ANALYSIS

The following equation governing the free vibration motion of the conservative system is obtained from Eq. (1) with $\{\mathbf{Q}\} = \{\mathbf{0}\}$ : $[M]\{\ddot{\mathbf{q}}\} + [K]\{\mathbf{q}\} = \{\mathbf{0}\}$. To obtain a solution of this equation, we note that, for a conservative system, periodic motion may be possible. In particular, let us assume a harmonic solution for $\{\mathbf{q}\}$ in the form $\{\mathbf{q}\} = \{\mathbf{A}\} \cos (\omega t - \alpha)$. That is, we inquire whether it is possible for all coordinates to vibrate harmonically with the same angular frequency $\omega$ and the same phase angle $\alpha$. This means that all points of the system will reach their extreme positions at precisely the same instant of time and pass through the equilibrium position at the same time. Substitution of this trial solution into the free vibration equation yields the eigenvector equation $[K]\{\mathbf{A}\} = \omega^2[M]\{\mathbf{A}\}$. In this equation $\{\mathbf{A}\}$ is called the eigenvector and $\omega^2$ is the eigenvalue (the square of a natural frequency). This homogeneous equation always admits the trivial solution $\{\mathbf{A}\} = \{\mathbf{0}\}$.

Nontrivial solutions are possible provided that $\omega^2$ takes on certain discrete values, which are obtained from the requirement that the determinant of the coefficient matrix of $\{\mathbf{A}\}$ vanishes. This yields the eigenequation (frequency equation) $\Delta(\omega^2) = |[K] - \omega^2[M]| = 0$. It is an $n$th-degree polynomial in $\omega^2$, and its $n$ roots are denoted by $\omega_r^2$, $r = 1, \ldots, n$. The eigenvector $\{\mathbf{A}\}_r$, $r = 1, \ldots, n$, corresponding to each eigenvalue is determined from $[K]\{\mathbf{A}\}_r = \omega_r^2[M]\{\mathbf{A}\}_r$. However, the homogeneous nature of this equation precludes the possibility of obtaining the explicit values of all the elements of $\{\mathbf{A}\}_r$. It is possible to determine only their relative values or ratios in terms of one of the components. Thus, we see that for each eigenvalue $\omega_r^2$ there is a corresponding eigenvector $\{\mathbf{A}\}_r$, which has a unique shape (based on the relative values of its components) but which is determined only to within a scalar multiplicative factor that may be regarded as being the amplitude of the shape. This unique shape is called the mode shape, and it can be thought of as being the position of the system when it is in its extreme position, and hence momentarily stationary, as in a static problem. Each frequency along with its corresponding mode shape is said to define a so-called natural mode of vibration. Methods of determining these natural modes are presented in the references (e.g., Wilkinson). If the initial conditions are prescribed in just the right way, it is possible to excite just one of the natural modes. However, for arbitrary initial conditions all modes of vibration are excited. Thus, the general solution of the free-vibration problem is given by a linear combination of the natural modes, i.e.,

$$\{\mathbf{q}\} = \sum_{r=1}^{n} C_r \{\mathbf{A}\}_r \cos (\omega_r t - \alpha_r)$$

The $2n$ arbitrary constants $C_r$ and $\alpha_r$ are determined through specification of the $2n$ initial conditions $\{\mathbf{q}(0)\}$, and $\{\dot{\mathbf{q}}(0)\}$. An explicit representation for $\{\mathbf{q}\}$ in terms of $\{\mathbf{q}(0)\}$ and $\{\dot{\mathbf{q}}(0)\}$ is given later.

Some of the properties of the natural modes of vibration will now be listed. Unless stated otherwise, the following properties are all consequences of the assumptions that $[M]$ and $[K]$ are real and symmetric and that $[M]$ is positive definite: (1) the $n$ eigenvalues and eigenvectors are real. (2) If $[K]$ is positive definite (corresponding to vibration about a stable state of equilibrium), all eigenvalues are positive. (3) If $[K]$ is positive semidefinite (corresponding to vibration about a neutral state of equilibrium), there is at least one zero eigenvalue. All zero eigenvalues correspond to rigid-body modes. (4) Eigenvectors $\{\mathbf{A}\}_r$, $\{\mathbf{A}\}_s$ corresponding to different eigenvalues $\omega_r^2$, $\omega_s^2$ ($\omega_r^2 \neq \omega_s^2$; $r \neq s$; $r, s = 1, \ldots, n$) are orthogonal by pairs with respect to the mass and stiffness matrices (weighting matrices), that is, $\{\mathbf{A}\}_r^T[M]\{\mathbf{A}\}_s = 0$ and $\{\mathbf{A}\}_r^T[K]\{\mathbf{A}\}_s = 0$ for $r \neq s$; $r, s = 1, \ldots, n$. Since the eigenvectors are orthogonal, and hence independent, they form a basis of the $n$-dimensional vector space. Thus, any vector in the space, such as the solution vector $\{\mathbf{q}\}$, can be represented as a linear combination of the eigenvectors (base vectors). This theorem (expansion theorem) is of vital importance for the response problem to be considered next.

## RESPONSE ANALYSIS

The response of the system due to arbitrary deterministic excitations in the form of initial displacements $\{\mathbf{q}(0)\}$, initial velocities (impulses) $\{\dot{\mathbf{q}}(0)\}$, or forcing functions $\{\mathbf{Q}(t)\}$ may be obtained by means of the above expansion theorem. That is, the displacement solution of the response problem is expanded with respect to the eigenvectors of the corresponding free-vibration problem. The time-dependent coefficients in this expansion are the so-called normal coordinates of the system. The equations of motion when expressed in terms of the $n$ normal coordinates are uncoupled, and each one of the equations is mathematically identical in form to the equation governing the motion of a simple one-degree-of-freedom spring-mass system. Hence solutions of these equations are readily obtained for the normal coordinates. The normal coordinates are then inserted into the eigenvector expansion to obtain the following explicit solution for displacement response $\{q(t)\}$ (Przemieniecki) for undamped constrained or unconstrained (free) structures:

$$\{\mathbf{q}(t)\} = \sum_{r=1}^{m} \frac{\{\mathbf{A}\}_{ro}\{\mathbf{A}\}_{ro}^{T}}{\{\mathbf{A}\}_{ro}^{T}[M]\{\mathbf{A}\}_{ro}} \left( [M](\{\mathbf{q}(0)\} + t\,\{\dot{\mathbf{q}}(0)\}) + \int_{\tau_2=0}^{\tau_2=t} \int_{\tau_1=0}^{\tau_1=\tau_2} \{\mathbf{Q}\,(\tau_1)\}\, d\tau_1\, d\tau_2 \right)$$

$$+ \sum_{r=m+1}^{n} \frac{\{\mathbf{A}\}_{re}\{\mathbf{A}\}_{re}^{T}}{\{\mathbf{A}\}_{re}^{T}[M]\{\mathbf{A}\}_{re}} \left( [M](\{\mathbf{q}(0)\}\cos \omega_r t + \frac{1}{\omega_r}\{\dot{\mathbf{q}}(0)\}\sin \omega_r t + \frac{1}{\omega_r}\int_{\tau=0}^{\tau=t} \{\mathbf{Q}(\tau)\}\sin \omega_r(t - \tau)\, d\tau \right) \quad \textbf{(2)}$$

From this equation, it is seen that the displacement response is expressed directly in terms of the arbitrary excitations and the free-vibration frequencies and eigenvectors. The eigenvectors have been separated into two types denoted by $\{A\}_{ro}$ and $\{A\}_{re}$. The "rigid-body" eigenvectors $\{A\}_{ro}$ correspond to the $m$ rigid-body modes (if there are any) for which $\omega_r = 0$, $r = 1$, . . ., $m$. The "elastic" eigenvectors $\{A\}_{re}$ correspond to the remaining $n$-$m$ eigenvectors, for which $\omega_r \neq 0$, $r = m + 1$, . . ., $n$. The general solution given by Eq. **(2)** is very useful provided that the second integral (Duhamel's integral) can be evaluated. (The first integral is easier to evaluate.) Closed form solutions of this integral for some of the simpler types of elements of $\{Q(t)\}$, such as step functions, ramps, etc., are available (see Przemieniecki). Solutions of Duhamel's integral for more complicated forcing functions can be obtained through superposition of solutions for the simpler cases or, of course, from direct numerical integration.

The displacement response obtained from Eq. **(2)** may be employed to obtain other response variables of interest, such as velocities, accelerations, stresses, strains, etc. Velocities and accelerations are obtained from successive time differentiations of Eq. **(2)**. The method used to evaluate the stresses depends on the type of spatial discretization employed to approximate the structural continuum. For example, difference quotients are used in the finite difference approach. Note that the rigid-body components do not contribute to the stress or strain response and hence the first summation in Eq. **(2)** can be omitted for computation of these responses.

## SPACE-VEHICLE PROPULSION
### by W. Tabakoff

REFERENCES: Bussard and DeLauer, "Nuclear Rocket Propulsion," McGraw-Hill. Meghreblian and Holmes, "Reactor Analysis," McGraw-Hill. Glasstone and Sesonske, "Nuclear Reactor Engineering," Van Nostrand. Koelle, "Handbook of Aeronautical Engineering," McGraw-Hill. Hesse and Mumford, "Jet Propulsion," Pitman. Corliss, "Propulsion System for Space Flight," McGraw-Hill. Sutton, "Rocket Propulsion Elements," Wiley. Loh, "Jet, Rocket, Nuclear, Ion and Electric Propulsion," Springer-Verlag.

One of the central problems in space technology is the matching of the mission requirements with the capabilities of the multitude of space propulsion systems that have been proposed. It is well known that the mission and its demands upon the spaceship ultimately dominate the final selection and design of the engine. The major problem of the propulsion designer becomes the determination of the mission requirements and their translation into engine capabilities.

A rocket engine is the device which converts the energy into suitable form and ejects stored matter to derive momentum. Among many possible energy sources, four are considered to be useful in rocket propulsion.

1. Chemical combustion reaction
2. Nuclear reaction
3. Captured radiation energy from an emitter such as the sun
4. Electric energy which is stored or created in the vehicle

Accordingly, the various propulsion devices can be categorized into chemical propulsion, nuclear-energy propulsion, solar-energy propulsion, and electric-energy propulsion.

Table 11 lists several types of rocket engines with ranges of typical key parameters. All these, with the exception of the ion rocket, use the nozzle expansion and acceleration of a heated gas as the mechanism for imparting momentum to a vehicle. The definition of some of the parameters in Table 11 are as follows.

The specific impulse is one of the most important performance parameters used in rockets. It can be defined as the thrust that can be obtained from an equivalent rocket which has a propellant weight flow of unity. It is defined as

$$I_s = F/\dot{W} = c/g$$

where $I_s$ is the specific impulse in seconds, $F$ is the thrust, $\dot{W}$ is the weight flow rate, $c$ is the effective exhaust velocity, and $g$ is the gravitational constant.

In general, $I_s$ increases in engines using gas expansion as the gas temperature increases or the molecular weight of the reaction product gases decreases.

Specific power $P_s$ is a parameter indicating the utilization of mass in the propulsion system in producing a maximum of kinetic gas power of the ejected matter. It is defined as

$$P_s = \tfrac{1}{2}\dot{m}v^2/w_0 = FI_s g/2w_0$$

Here $w_0$ includes the weight of the energy source (such as a nuclear reactor) as well as the propellants, the rocket engine system, and the structure weight. For chemical energy systems which operate only for short periods, the values of $P_s$ can be several magnitudes higher than those for electrical propulsion devices which must carry a bulky, relatively inefficient source of electric power as part of their loaded engine weight. The thrust-to-weight ratio expresses the acceleration that the engine is capable of giving to its own loaded propulsion-system mass.

The chemical rockets, and to some extent also the nuclear-fission rockets, have relatively low values of specific impulse,

**Table 11. Ranges of Performance Parameters for Several Different Rocket Engine Types**

| Engine type | Specific impulse, s | Maximum temperature, K | Thrust-to-weight ratio | Duration | Typical working fluid | Status |
|---|---|---|---|---|---|---|
| Chemical (liquid) | 300–460 | 2,800–4,600 | $10^{-2}$ to 100 | Seconds to a few hours | $H_2$ to $O_2$ | In service use |
| Chemical (solid) | 200–310 | 2,800–4,400 | $10^{-2}$ to 100 | Seconds to minutes | Fuel and oxidizer | In service use |
| Chemical (hybrid) | 200–400 | 2,500–4,400 | $10^{-2}$ to 100 | Seconds to minutes | Fuel and oxidizer | Experimental flight |
| Nuclear fission | 600–1,100 | 3,100 | $10^{-2}$ to 30 | Seconds to a few hours | $H_2$ | In component development |
| Radioactive isotope decay | 400–700 | 1,500–2,000 | $10^{-5}$ to $10^{-3}$ | Days | $H_2$ | Experimental static devices have been tested |
| Arc heating | 400–2,000 | 6,000 | $10^{-4}$ to $10^{-2}$ | Days | $H_2$ | Experimental static devices have been tested |
| Ion | 5,000–25,000 | . . . . | $10^{-5}$ to $10^{-3}$ | Months | Cesium | Experimental engine has flown |
| Solar heating | 400–700 | 1,800 | $10^{-3}$ to $10^{-2}$ | Days | $H_2$ | Research and Analysis |

relatively light machinery, a very high thrust capability, and, therefore, high accelerations and high specific powers. At the other extreme, the ion propulsion devices have a very high specific impulse, but they must carry a heavy electric energy source with them to deliver the power necessary for high ejection velocities. The radioactive-isotope-decay type, the arc-heating rocket, and those using solar energy are intermediate in their values of specific impulse and thrust-to-weight ratio.

The very low acceleration potential for the electric propulsion units and those using solar radiation energy usually requires a long period for accelerating and thus is best used for those missions where the flight time is long. The low thrust values of electrical systems imply that they are not useful in fields of strong gravitational gradients (such as for takeoff or landing) but are best used in a true space-flight mission far away from the strong gravitational field of the earth, moon, or other heavenly bodies.

### NUCLEAR ROCKET PROPULSION

The major components of a nuclear rocket engine are shown in Fig. 29. The feed system forces the propellant through regenerative cooling passages in the nozzle, reflector, pressure shell, and shield into the reactor heat exchanger, where the propellant is heated and expanded through the rocket nozzle. The flow path of the propellant required to drive the turbine is shown in the figure for bleed and topping cycles. Liquid hydrogen is the best choice for the nuclear rocket's working fluid because its low molecular weight produces the highest exhaust velocity for a given nozzle-inlet temperature. The nuclear rocket engine is started by adjusting the reactor neutronic-control drums to increase neutron population. Propellant flow is initiated at a low reactor-power level and is increased in proportion to the increasing neutron population until the design steady-state reactor-power output is obtained.

For shutdown of the engine, control drums are adjusted to poison the core and decrease the neutron population.

**Nozzle Thermodynamics**  The most common type of nozzle used in nuclear rockets is the De Laval. Numerous factors like friction losses in the nozzle, nozzle divergence, and energy loss through radiation also detract from perfect performance. Texts on rockets and jet engines should be consulted for a more detailed discussion of these factors.

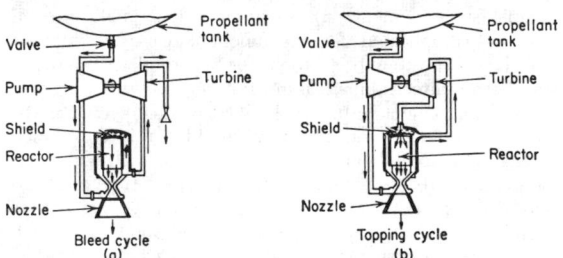

**Fig. 29**  Components of a nuclear rocket engine.

Past investigations also show that the more tractable propellants ($H_2O$, $NH_3$) seriously reduce the performance of nuclear rockets because of their high molecular weights. Hydrogen is the almost universal choice of propellant for the nuclear heat-transfer rocket. For example, for a given entrance nozzle pressure of 43 atm and temperatures of 1650, 3300, and 4950 K the following specific impulses are obtainable: 625, 890, and 1,216 s.

**Reactor Analysis**  Steady-state reactor thermal power, in megawatts, is determined by the relation $P_1 = K\dot{w}(b_{out} - b_{in})$, where $\dot{w}$ and $b$ are core flow rate and enthalpy, respectively, and $K$ is the appropriate conversion factor to megawatts.

The reactor is a high-power-density, self-energizing heat exchanger which elevates the temperature of the hydrogen propellant to the limit of component materials. Rigorous computations are necessary to obtain realistic solutions to the neutronic and thermodynamic equations for critical assembly data, uranium loadings, moderator and other material compositions, heat-transfer rates, and temperature and pressure gradients.

**Space-vehicle Applications** The gamma and neutron radiation environment emitted from a nuclear rocket requires (1) reactor shielding to attenuate the energy deposition in structure, propellant, and components and to obtain a certain separation distance and interface geometry between stage and engine; (2) careful selection of radiation-insensitive materials for critical components; (3) biological shielding for manned flight; and (4) special flight safety measures. The nuclear rocket is characterized by a high engine weight per pound of thrust produced, large overall engine dimensions, and bulky, low-density hydrogen tanks. The problems associated with nuclear space vehicles, however, must be weighed against the advantages derived from the very high nuclear performance potential. The high system specific impulse produces an increasingly advantageous pay-load fraction as the mission velocity requirement increases.

## ELECTRIC PROPULSION ENGINE

Electric propulsion engines generate exhaust velocities from about 4 to more than 100 times those of chemical rocket engines. The basic components for electric propulsion engines may be illustrated as follows:

1. Energy source (chemical, radioisotope, nuclear fission, nuclear fusion, and solar)
2. Electric power generator (batteries, chemical, radioisotope, solar, thermopile, turboelectric generator, and induction from moving plasma)
3. Thrust chamber (arc plasma, ion, magneto-plasma, and photon)

Any of the possible engines which may be considered will have the following characteristics in common.

The first of these characteristics is high specific impulse, which means low propellant weight.

$$\text{Propellant weight} = W_p = F(\Delta t)/I_{sp} \qquad (1)$$

where $\Delta t$ = propulsion time, s.

The second common characteristic is high power-plant weight, which results from heavy electric power generators.

$$\text{Specific weight} = \alpha = w_{pp}/P_j \qquad (2)$$

where $w_{pp}$ = power-plant weight, lbs
$P_j$ = jet power, kW

The jet power is equal to the product of the propellant flow and the kinetic energy of the exhaust gases. Hence,

$$P_j = \frac{(V_{ef})^2}{2g} G = \frac{(V_{ef})^2}{2g} \frac{W_p}{\Delta t} \qquad (3)$$

From combining Eqs. (1) and (3) one obtains

$$P_j = gFI_{sp}/2 = FI_{sp}/45.8 \qquad \text{kW} \qquad (4)$$

The total power-plant plus propellant weight is the parameter of real interest for space-vehicle applications.

Combining Eqs. (1), (2), and (4) one obtains

$$\text{Total propulsion-system weight} = F\left(\frac{\alpha I_{sp}}{45.8} + \frac{\Delta t}{I_{sp}}\right) \qquad (5)$$

The minimum propulsion-system weight is thus dependent on the thrust, specific weight, and propulsion time.

## SOLAR AND PHOTON PROPULSION

The radiant energy from the sun can be used for space propulsion, such as the solar-heating rocket and the solar sail. The solar-heating rocket uses the solar energy to heat a working fluid, which is then injected through a nozzle at a high velocity. The solar sail makes use of the pressure of the solar photons ($10^{-5}$ N/m$^2$) to create a propulsive force.

Figure 30 shows a schematic diagram of a possible solar-heating rocket system. The concentrated solar energy heats

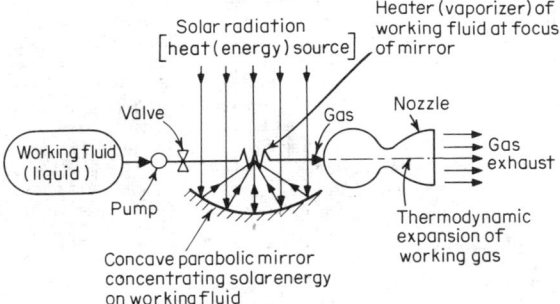

**Fig. 30** Solar-heating rocket.

the rocket propellant directly to about 1500 K. The heated propellant is then exhausted through the nozzle to produce thrust. Many design problems make the practical development of the solar-heating rocket quite difficult. The most important of these problems are:

1. The solar collector must be pointed toward the sun at all times.
2. The available solar energy varies inversely with the square of the distance from the sun. At the earth the incident radiation is about 1.36 kW/m$^2$. At the distance of Mars and Venus, the relative available solar energy is 43 and 190 percent, respectively.

## PHOTON ROCKET

This type of rocket has been referred to as the ultimate in jet propulsion. Basically, such an engine would consist of a high-intensity light source with collimating reflector. Small quantities of matter would produce large amounts of energy for the "ideal" photon rocket by conversion of mass to energy according to Einstein's equation ($E = mc^2$).

This concept is illustrated in Fig. 31.

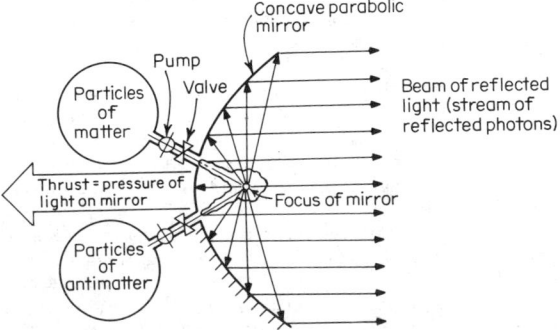

**Fig. 31** Photon rocket.

# PIPELINE TRANSMISSION
## by George H. Ewing

REFERENCES: AGA publication "Gas Facts." Huntington, "Natural Gas and Gasoline," McGraw-Hill. Leeston, Crichton, Jacobs, "The Dynamic Natural Gas Industry," University of Oklahoma Press. Lester, Hydraulics for Pipeliners, *Oildom*. Bell, "Petroleum Transportation Handbook," McGraw-Hill. Bureau of Mines, "Mineral Yearbook," Vol. II, Fuels. "The Transportation of Solids in Steel Pipelines," Colorado School of Mines Research Foundation, Inc. Ellis, Redberger, and Bolt, Transporting Solids By Pipe Line, *Ind. Eng. Chem.*, August 1963. Wasp, Regan, Withers, Cook, and Clancey, Cross Country Coal Pipe Line Hydraulics, *Pipe Line News*, July 1963.

## NATURAL GAS

**General** According to AGA estimates for 1971, approximately 80 percent of the U.S. natural-gas reserves occur in five states (TX, LA, OK, NM, and KS), and only about 40 percent of the total natural gas produced there is consumed within these states. In 1950, there were 314,000 mi (505,000 km) of gas pipelines in this country, and by 1971, with natural gas supplying 33 percent of the nation's energy requirements, this mileage had increased to 935,000 mi (1,504,000 km).

Transmission pipelines move the gas from the producing regions to consuming areas throughout the country. The trend for transmission lines has been toward the use of larger pipe diameters and pipe steel of greater tensile strength. Transmission lines of 36-in (0.91-m) diameter and pipe steel of 65,000 lb/in² (4,570 kgf/cm²) minimum specified yield are now in general use. Important mileage of 42-in (1.07-m) pipe has been installed. Larger-diameter pipe, along with increasing pressure capabilities and growing pipeline mileage, brought forth a considerable rise in prime-mover and compressor requirements. Natural-gas pipelines in the United States have experienced an increase from about 4 million installed horsepower (3 million kW) in 1953 to just under 11 million horsepower (8.2 million kW) by the end of 1971.

Gas engines, gas turbines, and electric motors are most generally employed as **prime movers on gas pipelines.** The gas engines generally drive reciprocating compressors; gas turbines and electric-motor drivers are usually connected to centrifugal compressors. The industry trend is toward the use of larger compressor units because of inherent economic advantages. Prime movers in general use have increased in size from the early 1950s to the early 1970s as follows: gas engines from 2,500 to 12,500 hp (1,865 to 9,325 kW), gas turbines from 5,000 to 32,500 hp (3,730 to 24,245 kW), and electric-motor drives from 2,500 to 20,000 hp (1,865 to 14,920 kW). Modified aircraft-type jet engines are also employed as prime movers driving centrifugal compressors.

Much progress has been made in **automated operations** of all types of pipelines. Valves, measuring and regulating stations, and other facilities are operated remotely or automatically. Factory-packaged compressor-station assemblies, most of which are highly automated, are now commonplace. Whole compressor stations with thousands of horsepower are operated unattended by the use of coded dispatching systems handled by high-speed communications.

**Flow** From the thermodynamic-energy-balance equations for the flow of compressible fluids (Sec. 4), the general flow formula is derived and expressed in *BuMines Monograph 6* (1935) as

$$Q = K \frac{T_0}{P_0} \left[ \frac{(P_1^2 - P_2^2)d^5}{GT_f L_f} \right]^{1/2}$$

The general flow equation for constant volume (steady state), isothermal, and horizontal flow of natural gas is given by *U.S. BuMines Monograph 9* (1956) as

$$Q = 38.7744 \frac{T_b}{P_b} \left[ \frac{(P_1^2 - P_2^2)D^5}{fZGTL} \right]^{0.5}$$

where $Q$ = rate of flow, ft³/day; $T_b$ = temperature base, °R; $P_b$ = pressure base, psia; $P_1$ = inlet pressure, psia; $P_2$ = outlet pressure, psia; $D$ = internal pipe diameter, in; $f$ = friction factor, dimensionless; $Z$ = average gas compressibility, dimensionless; $G$ = gas gravity (air = 1.00); $T$ = average flowing temperature, °R; $L$ = length of pipe, mi. *Monograph 9* also gives a good discussion of friction factors and elevation corrections.

Other flow equations may be derived from the general equation by substitution of the proper expression for the friction factor. This applies to the **Panhandle A equation** which has the form

$$Q = 435.87 \left( \frac{T_b}{P_b} \right)^{1.0788} \left( \frac{P_1^2 - P_2^2}{G^{0.8539} TL} \right)^{0.5394} D^{2.6182} E$$

where $E$ = pipeline flow efficiency, a dimensionless decimal fraction (design values of 0.88 to 0.96 are common). In the Panhandle A equation, the compressibility is included in the pipeline efficiency, and the other symbols are as previously defined. This formula may be solved graphically as described by C. W. Marvin, *Oil Gas Jour.*, Sept. 20, 1954.

An AGA paper Steady Flow in Gas Pipelines (1965) gives a good general review of constant-volume gas flow equations, both with and without elevation corrections. Equations and test data are included for both the partially turbulent (Reynolds-number-dependent) flow regime and the fully turbulent flow regime using effective internal pipe roughness. In this reference the drag factor is introduced to account for pressure losses for fittings such as bends and valves and is applied in the partially turbulent flow regime.

When the volume of gas flowing varies with time and position (transient or unsteady-state flow), the simultaneous momentum and mass balance equations are usually solved by use of numerical methods and with the aid of computers. An AGA computer program, **Pipetran,** is available for this purpose, as are several related texts on transient flow. A paper entitled Gas Transportation System Modeling appears in the 1972 *AGA Operating Proceedings*.

**Compression** The theoretical horsepower requirements for a station compressing natural gas may be calculated by polytropic compression formulas (see Sec. 4). The change of state that takes place in almost all reciprocating compressors is close to polytropic. Heat of compression is taken away by the jacket cooling and by radiation, and a small amount of heat is added by piston-ring friction. Compression by centrifugal compressors is even closer to polytropic (See Sec. 14).

For general design, the horsepower requirements for a station compressing natural gas with reciprocating compressors can be calculated as follows:

$$\text{Reciprocating station horsepower} = \text{hp}Z_s \frac{T_s}{520}$$

The value for hp can be taken from Fig. 1, typical manufacturer's curve; $T_s$ is suction temperature, °R; and $Z_s$ can be taken from Fig. 3.

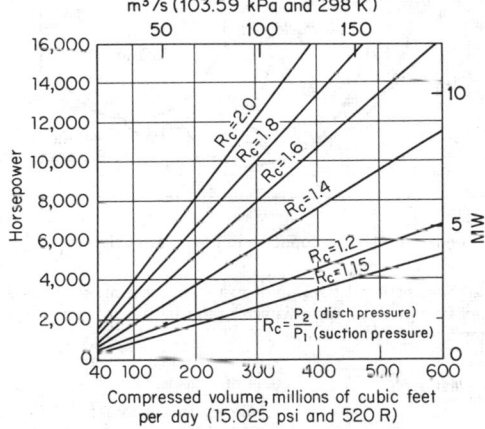

**Fig. 1** Reciprocating-compressor horsepower (MW) graph. $R_c = P_2/P_1 =$ discharge pressure, psia ÷ suction pressure, psia.

Similarly the horsepower requirements for a station compressing natural gas with centrifugal compressors can be calculated as follows:

$$\text{Centrifugal station horsepower} = \frac{\text{hp}Z_s}{E_c} \frac{T_s}{520}$$

The value for hp can be taken from Fig. 2, typical centrifugal horsepower curve; $E_c$ is the centrifugal-compressor shaft efficiency. (Design shaft efficiencies of 80 to 85 percent are common.)

**Design** The pipe and fittings for gas transmission lines are manufactured in accordance with the specifications of the API (see also Sec. 8). Interstate pipelines are installed and operated in conformance with regulations of the U.S. Department of Transportation (DOT). The ANSI B31.8 Code and the *Gas Measurement Committee Report* 3 of the AGA provide supplemental design guidelines. The equation for pipeline design pressure, as dictated by the DOT in Part 192 of Title 49, Code of Federal Regulations, is

$$P = \frac{2St}{D} \times F \times E \times T$$

where $P =$ design pressure, psig; $S =$ specified minimum yield strength, lb/in²; $t =$ nominal wall thickness, in; $D =$ nominal outside diameter, in; $F =$ design factor based on

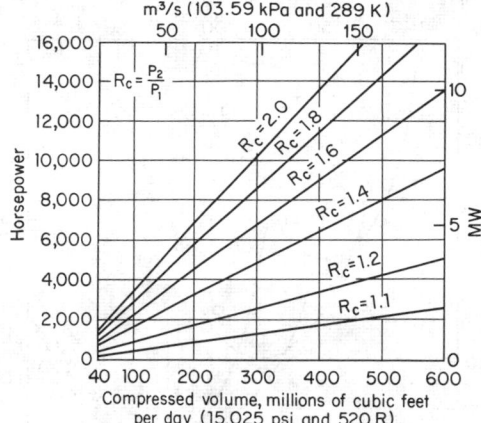

**Fig. 2** Centrifugal-pressure horsepower (MW) graph. $R_c = P_2/P_1 =$ discharge pressure, psia ÷ suction pressure, psia.

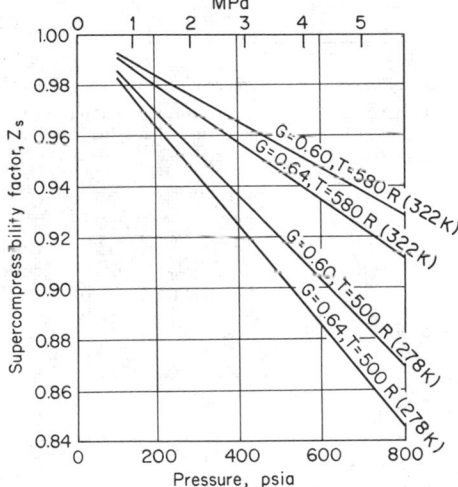

**Fig. 3** Supercompressibility factor Z for natural gas. (Natural gas, 1 percent N₂.)

population density in area through which the line passes, dimensionless; $E =$ longitudinal joint factor, dimensionless; and $T =$ temperature derating factor (applies above 250°F only), dimensionless. For cross-country pipelines in areas in which the population density is very low, the factor $F$ is 0.72, and assuming seamless or electric-resistance-weld pipe with an $E$ value of 1.00 and an operating temperature below 250°F, the equation can be written

$$P = 1.44St/D$$

The **selection** of the diameter, steel strength, and wall thickness of a line and the determination of the optimum station spacing and sizing are based on transportation economics. Figures 4 to 6 show schematically methods of analysis commonly used to relate design factors to the cost of transportation.

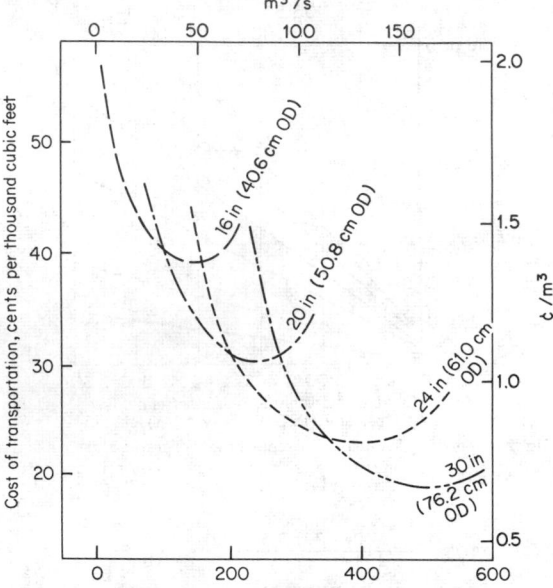

**Fig. 4**  Schematic relationship of cost of transportation versus design capacity for a given length of line with various pipeline diameters at equal design pressures.

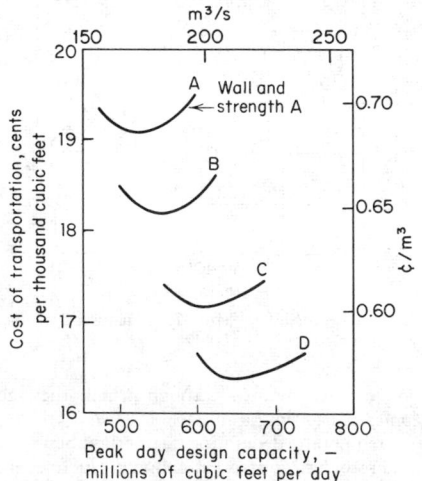

**Fig. 5**  Schematic relationship of cost of transportation versus pipeline steel strength and wall thickness. A = 30 in × 0.312-in, WT × 52,000 lb/in² min yield (779-lb/in² operating pressure). B = 30 in × 0.312-in WT × 56,000 lb/in² min yield (838-lb/in² operating pressure). C = 30 in × 0.375-in WT × 52,000-lb/in² min yield (935-lb/in² operating pressure). D = 30 in × 0.375-in WT × 56,000-lb/in² min yield (1,008-lb/in² operating pressure).

A perspective of various elements of the **cost** of transportation for a long-distance, large-diameter, fully developed pipeline system is presented in Fig. 7.

The cost of installed pipelines varies greatly with the location, the type of terrain, the design pressure, the total length to be constructed, and many other factors. In general, figures from $6,000 to $8,000 per inch (per 2.54 cm) of OD represent the range of approximate costs per mile (per 1.61 km) of installed pipeline. Compressor-station installed costs vary widely with the amount and type of horsepower to be installed and with the location, the type of construction, and weather conditions. The installed costs of gas-turbine stations range from $250 to $450 per installed hp ($335 to $603 per kW); gas-engine stations, from $375 to $475 per hp ($503 to $637 per kW); and electric-motor stations, from $200 to $300 per hp ($268 to $402 per kW).

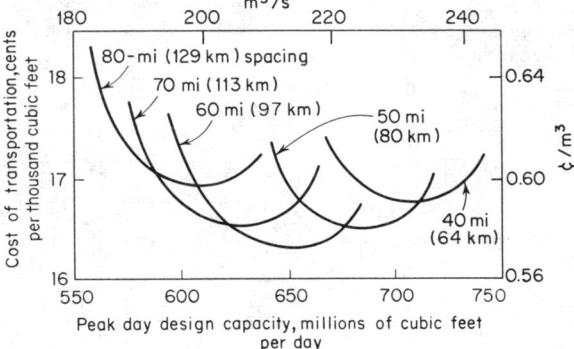

**Fig. 6**  Schematic relationship of cost of transportation versus design capacity for a given pipeline diameter and length, with various spacings of compressor stations.

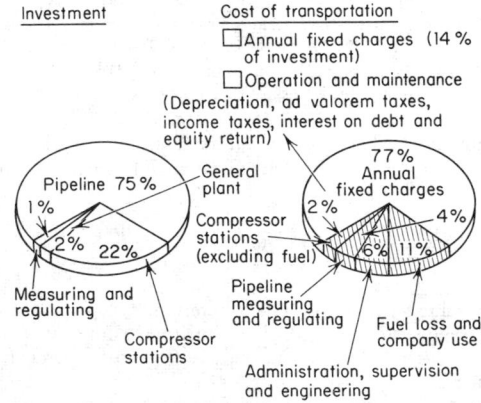

**Fig. 7**  Analysis of investment and cost of transportation.

After a pipeline system is constructed and fully utilized, expansion of delivery capacity can be accomplished by adding pipeline loops (connecting segments of line parallel to the original line), by adding additional horsepower at compressor stations, or by a combination of these two methods.

## CRUDE OIL AND OIL PRODUCTS

**General** Approximately 90 percent of the U.S. proved liquid hydrocarbon reserves are situated in the seven states of AK, CA, LA, KS, NM, OK and TX or in the marine areas off their coasts. With the exception of California, the markets in which petroleum products are consumed are remote from the proved reserves. One-fourth of all petroleum consumed in the United States is imported and must be moved from ports to the centers of consumption. Forty-seven percent of all movements of petroleum and petroleum products is made by pipelines. The interstate pipelines alone transport over 8 billion barrels (1.27 billion m³) of petroleum a year. There are about 147,500 mi (237,300 km) of liquid-petroleum trunk lines installed in the United States.

The trend in the industry is toward the use of larger-diameter pipe fabricated from stronger, tougher steels. From the mid-1960s to the mid-1970s 28-in-diameter (0.71-m) products pipeline system linked the Texas-Oklahoma refining center with Chicago, 36-in-diameter (0.91-m) products pipeline system linked the Texas-Louisiana refining area with New York, and a 40-in-diameter (1.02-m) crude line joined the Texas-Louisiana producing area with Chicago.

Centrifugal pumps have been used extensively. Generally these units are of one- or two-stage design, with several units placed in series (see Sec. 14). This arrangement provides greater stability, efficiency, and flexibility than is obtained with multistage centrifugals installed in parallel. Reciprocating pumps are installed in parallel; i.e., a common suction and a common discharge header are utilized by all pumps. There are instances where rotary, gear, and vane-type pumps have been employed to fulfill specific requirements.

Electrically driven centrifugal pumps are now widely employed for the liquid pipeline industry because of their compactness, low initial cost, and ease of control. Centrifugal pumps driven by gas-turbine engines have also come into routine use.

Positive-displacement and turbine meters are in general use on products systems (see Sec. 16). Totalizers for multimeter installations and remote transmission of readings are becoming widely accepted. Greater application of meters with crude-oil-gathering facilities is being made as automatic-custody units are installed.

**Flow** (See also Sec. 3) The **Darcy formula,**

$$h_f = fLV^2/D2g$$

[where $h_f$ = friction loss, ft; $f$ = friction factor (empirical values), dimensionless; $L$ = length of pipe, ft; $D$ = ID, ft; $V$ = velocity flow, ft/s; and $g$ = 32.2 ft/s²] is in general use by engineers making crude-oil pipeline calculations. The Darcy formula, when stated in a form utilizing conventional pipeline units, is

$$P = 34.87fB^2S/d^5$$

where $P$ = friction press drop, lb/(in²) (mi); $f$ = friction factor (empirical values), dimensionless; $B$ = flow rate, bbl/h (42 gal/bbl); $S$ = sp gr of the oil, dimensionless; $d$ = ID, in.

The **Williams and Hazen formula,** which has wide acceptance for product pipeline calculations, can be stated in a form employing conventional pipeline terms as follows:

$$P = \frac{2340B^{1.852}S}{C^{1.852}d^{4.870}}$$

where $P$ = friction press drop, lb/in² per 1,000 ft of pipe length; $B$ = flow rate, bbl/h; $S$ = sp gr of the oil product; $C$ = friction factor, dimensionless; and $d$ = ID, in. The friction factor $C$ includes the effect of viscosity and differs with each product ($C$ for gasoline is 150; for no. 2 furnace oil, 130; for kerosine, 134). A slide rule has been devised (Texas Eastern Transmission Corp.) which provides a scale to adjust the friction factor $C$ for viscosity.

The brake horsepower for pumping oil is $RSh/3960E$ or $BP/2450E$, where $R$ = flow rate, gal/min; $S$ = sp gr of the oil; $h$ = pump head, ft; $E$ = pump eff, decimal fraction; $B$ = flow rate, bbl/h; and $P$ = pump differential pressure, lb/in².

**Design** The pipe and fittings for oil transmission lines are manufactured in accordance with specifications of the API and are fabricated, installed, and operated in the United States in conformance with Federal Regulations, DOT, Pt. 195 (see also Sec. 8). Many factors influence the working pressure of an oil line, and the ANSI B 31.4 Code for pressure piping is used as a guide in these matters. A study of the hydraulic gradient, land profile, and static-head conditions is made in conjunction with the selection of the main-line pipe (see also Sec. 3).

Determining the pipe size and station spacing to transport a given oil or oil product at a specified rate of flow and to provide the lowest cost of transportation is a complex matter. Usually the basic approach is to prepare (1) a series of pipeline cost estimates covering a range of pipe diameters and wall thicknesses and (2) a series of station cost estimates for evaluation of the effect of station spacing. By applying capital charges to the system cost estimates and estimating the system operating costs, a series of transportation-cost curves similar to Fig. 8 can be drawn.

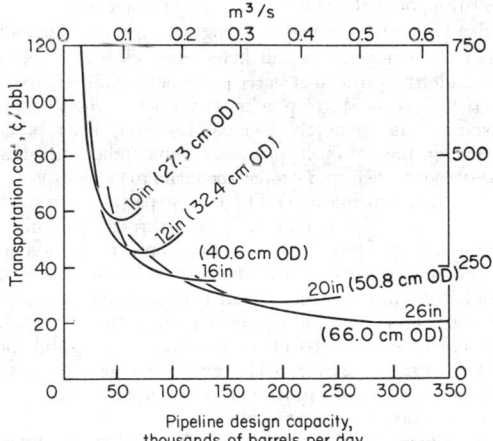

**Fig. 8** Schematic relationship of pipeline capacity to cost of transportation.

Estimates of $6,000 to $8,000 per in (per 2.54 cm) of OD represent the cost per mile (1.61 km) of installed trunk pipeline. The tabulation of pump-station investment costs for automated electric-motor–centrifugal-type installations presented in Table 1 can be used for preliminary evaluations.

Careful design and good operating practice must be followed to minimize contamination due to interface mixing in oil-product pipelines. When the throughput capacity of an oil

**Table 1**

| Pump-station installed hp, total | Cost per installed hp (1 hp = 0.746 kW) |
| --- | --- |
| 250–500 | $490–515 |
| 500–750 | $360–385 |
| 750–1,000 | $285–310 |
| 1,000–2,000 | $200–225 |
| Above 2,000 | $175 |

line becomes fully utilized, additional capacity can be obtained by installing a parallel pipeline or a partial-loop line. A partial loop is a parallel line which runs for only part of the distance between stations. In the case of product lines, careful attention must be paid to the design of the facilities where the loop line is tied into the original pipeline to prevent commingling of products.

## SOLIDS

**General**  Some of the **solids being transported** in slurry form considerable distances by pipeline are coal, coal refuse, gilsonite, phosphate rock, tin ore, nickel ore, copper ore, gold ore, kaolin, limestone, clay, borax, sand, and gravel. **Solids pipelines differ** from pipelines for oil, gas, and other true fluids in that the product to be transported must be designed and prepared for pipeline transportation.

Current trends in the construction of solids pipelines and those proposed for construction indicate that future transportation of minerals and other commodities will bear more consideration. Commercially successful long-distance pipelines throughout the world are listed in Table 2.

Advancements in slurry rheology studies and technology related to homogeneous and heterogeneous slurry transport have made the **potential of slurry pipelining** unlimited. Immediate advantages for slurry pipelines exist where the solids to be transported are remotely located; however, each potential slurry transport project possesses some other advantages. **Materials must satisfy the following conditions** to be transported by long-distance pipelines: (1) The largest particle size must be limited to that which can readily pass through commercially available pumps, pipes, and other equipment; (2) solids must mix and separate easily from the carrying fluids; (3) material degradation must be negligible or beneficial to pipelining and utilization; (4) material must not react with the carrying fluid or become contaminated in the pipeline; (5) the solids-liquid mixture must not be excessively corrosive to pipe, pumps, and equipment; (6) solids must not be so abrasive as to cause excessive wear at carrying velocities.

**Coal Slurry**  Uncleaned coal can be transported by pipeline, but a more stable and economic long-distance pipeline operation can be achieved with clean coal. The use of clean coal allows a variety of supply sources; produces a slurry with a lower, more uniform friction-head loss; reduces the pipe-wall wear; and increases the system capacity. Cleaning pipeline coal costs considerably less than cleaning coals for other

forms of transportation since drying can be eliminated after the cleaning operation.

**Coal-slurry preparation** normally consists of a wet grinding process for reducing clean coal to the proper particle size and size-range distribution. The final slurry-water concentration can be adjusted before it enters the pipeline with quality control of the slurry holding tanks. A typical size-range distribution (size consist) for a coal slurry with a design concentration of solids of 57 percent (percentage of coal by weight) would approximate the screen analysis shown in Table 3.

The **cost of slurry preparation** is only slightly affected by changes in concentration or size consist but is materially affected by plant capacity. For example, preparation-plant operating costs per ton (per 907 kg) of coal are nearly twice as high for a plant with a capacity of 1 million tons (907 million kg) per year as for a plant of 4 million tons (3.63 billion kg) per year.

The normal slurry-preparation-plant practice is to hold the concentration within 1 percent of design to provide a slurry

**Table 2. Summary of Selected Operating Long-Distance Slurry Pipelines**

| Location | Material transported | Length,* (mi) | Million tons/year† |
| --- | --- | --- | --- |
| U.S.A. | Coal | 273 | 4.8 |
| U.S.A. | Gilsonite | 72 | 0.38 |
| England | Limestone | 57 | 1.70 |
| Tasmania | Iron concentrates | 53 | 2.25 |
| Japan | Copper tailings | 40 | 1.0 |
| Japan | Ore tailings | 43 | 0.5 |

*1 mile = 1.61 km.
†1 ton = 907 kg.

with a uniform pressure drop. Different slurry designs have been used, ranging from 45 to 58 percent concentration. Continued research has made it possible accurately to predict the anticipated pressure drop in a pipeline by using laboratory and computer correlation.

**Optimum characteristics for a slurry** are (1) maximum concentration, (2) low rate of settling, and (3) minimum friction-head loss. Variables which can be manipulated to control and optimize slurry design are particle shape, size distribution, concentration, and ash content. Established computer techniques are used to evaluate these variables.

**Economics and Design**  Because of the many slurry variations, there is no single formula available by which the friction-head loss can be readily determined for all slurry applications. A general approach is described by Wasp, Thompson, and Snoek (The Era of Slurry Pipelines, *Chem. Tech.*, Sept. 1971, pp. 552–562), giving calculation methods which have been developed for predicting slurry pressure drops. The described methods are quite accurate and can be used for preliminary design. Slurry hydraulics should be verified with bench-scale tests and compared with commercial operating data reference points or pilot-plant operations before constructing a full-scale long-distance pipeline.

**Table 3**

| Mesh (Tyler wet screen) | 8 | 14 | 28 | 48 | 100 | 200 | 325 | −325 |
| --- | --- | --- | --- | --- | --- | --- | --- | --- |
| Percent solids by weight | 0.2 | 5.0 | 20.2 | 18.8 | 13.6 | 6.0 | 6.2 | 30.0 |

**Slurry velocities** are generally limited to a minimum critical velocity to prevent solids from settling to the bottom of the pipe and a maximum velocity to prevent excessive friction-head loss and pipe-wall wear. These velocity limitations and the throughput volume determine the diameter of the main-line pipe. The relationship of volume and velocity to the pipe diameter is expressed as $ID = 0.6392 \sqrt{(gal/min)/V}$, where the ID is in inches, gal/min is the volume rate of slurry flow (gallons per minute), and $V$ is flow velocity, ft/s.

**Pipe-wall thickness** in inches is computed by the modified Barlow formula as $[(PD/2SF) + tc]/tm$, where $P$ = maximum pressure, lb/in$^2$; $D$ = pipe OD, in; $S$ = minimum yield strength of the pipe, lb/in$^2$; $F$ = safety factor; $tm$ = pipe mill tolerance, usually 0.875; and $tc$ = internal erosion/corrosion allowances, in, for the life of the system. The internal metal loss results from the erosion of corrosion products from the pipe wall and thus can be limited with chemical inhibitors and velocity control. This metal loss occurs in all pipeline steels and, according to Cowper, Thompson, et al., Processing Steps: Keys to Successful Slurry-Pipeline Systems, *Chem. Eng.*, Feb. 7, 1972, pp. 58–67, must be evaluated on a case-by-case basis. A minimum pipe wall of 0.250 in (0.635 cm) is often used to provide mechanical strength for large lines even where pressures and wear permit the use of a thinner wall. Coal pipelines employ a graduated wall thickness to meet the demands of the hydraulic gradient, which saves pipe costs.

**Pump-station** hydraulic horsepower is calculated as $(P_d - P_s)$ (gal/min)/1,716, where $(P_d - P_s)$ = pressure rise across the station, lb/in$^2$; gal/min = maximum flow rate through the station; and 1,716 is a constant, gal/ft·in$^2$. Stations employ positive-displacement reciprocating pumps of the type used for oil-field mud pumps, as they attain high pressures with low slurry velocities. This allows fewer stations and lower overall costs than with centrifugal pumps. Stations are located where the friction-head loss and the effects of terrain have reduced the pressure head to 100 ft (30.48 m).

The annual tonnage **throughput** for any diameter line can be varied within the velocity limits or by intermittent operation near the low velocity limit. The optimum main-line size for high-load-factor systems is found by comparing the cost of transportation for a series of pipe sizes, as in Fig. 9.

When a main-line size is selected, the **cost of transportation** for that line size at various rates of throughput is weighed against the projected system growth and compared with other line sizes to determine the size which gives the best overall system economics. Coal pipelines cannot be expanded by line looping because of the velocity limitations. Some expansion can be attained by oversizing the line and operating it intermittently near the minimum velocity in the earlier years of life. The effect of system capacity on the cost of transportation per ton-mile and on the system investment cost is illustrated in Fig. 10.

The **cost of utilizing pipeline coal** includes (1) a cost of approximately 36 cents per ton (per 907 kg) to dewater it to about 17 percent moisture by centrifuge (see Hindenland, The Burning Question, *Coal Utilization*, Aug. 1962), or (2) an efficiency loss of about 4.5 percent when burned directly at 70 percent concentration (see Kelcec, Olivadoti, and Duzy, Coal Slurry Firing at Werner Station, *ASME Paper* 62-PWR-3). A savings

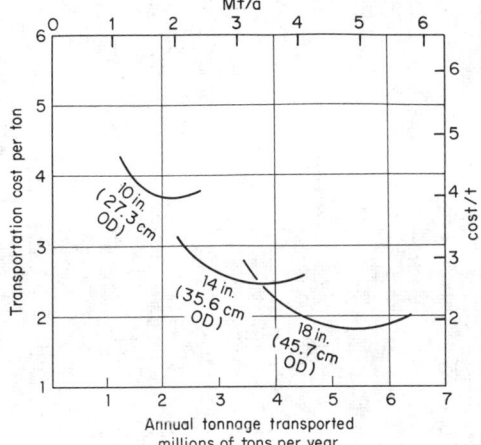

**Fig. 9**   System transportation-cost comparison for various main-line pipe sizes.

in new-plant investment of 5 percent or more can be realized at power stations by eliminating the usual coal-handling equipment when supplied by pipeline coal. Operating costs of conventional coal-handling equipment at either a new or an old plant can also be eliminated.

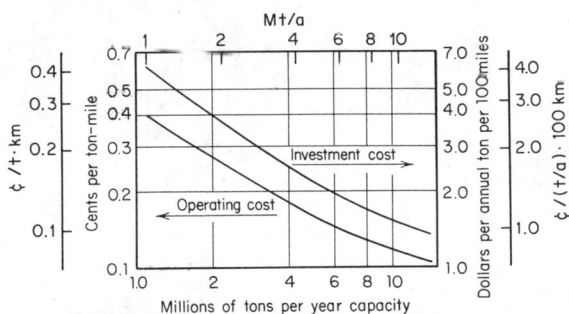

**Fig. 10**   Investment and operating costs of coal-slurry pipelines of various capacities. Basis: short tons of coal at 3 percent surface moisture; operating factor 90 percent. (*From Reichl and Thagard, National Power Conference Paper CP A 62-5113, Sept., 1962.*)

Section **12**

# Building Construction and Equipment

**BY**

**JOHN R. IMMER**  *President, Work Saving International, Washington, D.C.*
**JOHN A. BLUME**  *President, John A. Blume & Associates.*
**J. P. NICOLETTI**  *Vice President, John A. Blume & Associates.*
**WILLIAM L. GAMBLE**  *Professor of Civil Engineering, University of Illinois at Urbana-Champaign.*
**LEANDER ECONOMIDES**  *Consulting Engineer, New York, N.Y.*
**ABRAHAM ABRAMOWITZ**  *Consulting Engineer; Professor Emeritus of Electrical Engineering, The City College, The City University of New York.*
**BENSON CARLIN**  *President, O.E.M. Medical, Inc.*

### INDUSTRIAL PLANTS
#### by John R. Immer
General Planning of Industrial Plants . . . . . . . . . . . . . . . . . . . . 12-2
Processing . . . . . . . . . . . . . . . . . . . . . . . . . . . . . . . . . . . . . 12-3
Building Design and Construction . . . . . . . . . . . . . . . . . . . . . . 12-5
Contract Procedures . . . . . . . . . . . . . . . . . . . . . . . . . . . . . . 12-10

### STRUCTURAL DESIGN OF BUILDINGS
#### by John A. Blume and J. P. Nicoletti
##### with the assistance of Ray L. Holmes
Loads and Forces . . . . . . . . . . . . . . . . . . . . . . . . . . . . . . . . 12-11
Wind Pressure on Structures (BY WILLIAM BOLLAY) . . . . . . . . . 12-12
Design of Structural Members . . . . . . . . . . . . . . . . . . . . . . . . 12-14
Masonry Construction . . . . . . . . . . . . . . . . . . . . . . . . . . . . . 12-22
Timber Construction . . . . . . . . . . . . . . . . . . . . . . . . . . . . . . 12-23
Steel Construction . . . . . . . . . . . . . . . . . . . . . . . . . . . . . . . 12-28

### REINFORCED-CONCRETE DESIGN AND CONSTRUCTION
#### by William L. Gamble
Materials . . . . . . . . . . . . . . . . . . . . . . . . . . . . . . . . . . . . . . 12-48
Loads . . . . . . . . . . . . . . . . . . . . . . . . . . . . . . . . . . . . . . . . 12-49
Load Factors for Reinforced Concrete . . . . . . . . . . . . . . . . . . . 12-49
Reinforced-Concrete Beams . . . . . . . . . . . . . . . . . . . . . . . . . 12-50
Reinforced-Concrete Columns . . . . . . . . . . . . . . . . . . . . . . . . 12-53
Reinforced-Concrete Floor Systems . . . . . . . . . . . . . . . . . . . . 12-55
Footings . . . . . . . . . . . . . . . . . . . . . . . . . . . . . . . . . . . . . . 12-58
Walls and Partitions . . . . . . . . . . . . . . . . . . . . . . . . . . . . . . 12-59
Prestressed Concrete . . . . . . . . . . . . . . . . . . . . . . . . . . . . . . 12-59
Precast Concrete . . . . . . . . . . . . . . . . . . . . . . . . . . . . . . . . . 12-60

Joints . . . . . . . . . . . . . . . . . . . . . . . . . . . . . . . . . . . . . . . . 12-61
Forms . . . . . . . . . . . . . . . . . . . . . . . . . . . . . . . . . . . . . . . . 12-61

### HEATING, VENTILATION, AND AIR CONDITIONING
#### by Leander Economides
Heating . . . . . . . . . . . . . . . . . . . . . . . . . . . . . . . . . . . . . . 12-62
Cooling . . . . . . . . . . . . . . . . . . . . . . . . . . . . . . . . . . . . . . 12-76
Pumps . . . . . . . . . . . . . . . . . . . . . . . . . . . . . . . . . . . . . . . 12-101
Comfort Indexes . . . . . . . . . . . . . . . . . . . . . . . . . . . . . . . . . 12-104

### ILLUMINATION
#### by Abraham Abramowitz
Basic Units . . . . . . . . . . . . . . . . . . . . . . . . . . . . . . . . . . . . 12-116
Vision . . . . . . . . . . . . . . . . . . . . . . . . . . . . . . . . . . . . . . . . 12-117
Light Meters . . . . . . . . . . . . . . . . . . . . . . . . . . . . . . . . . . . 12-118
Light Sources . . . . . . . . . . . . . . . . . . . . . . . . . . . . . . . . . . . 12-118
Prescribing Illumination . . . . . . . . . . . . . . . . . . . . . . . . . . . . 12-120
Lighting Design . . . . . . . . . . . . . . . . . . . . . . . . . . . . . . . . . 12-126
The Economics of Lighting Installations . . . . . . . . . . . . . . . . . 12-135
Dimming Systems . . . . . . . . . . . . . . . . . . . . . . . . . . . . . . . . 12-135
Heat from Lighting . . . . . . . . . . . . . . . . . . . . . . . . . . . . . . . 12-135

### SOUND, NOISE, AND ULTRASONICS
#### by Benson Carlin
Definitions . . . . . . . . . . . . . . . . . . . . . . . . . . . . . . . . . . . . 12-136
The Production and Reception of Sounds . . . . . . . . . . . . . . . . 12-137
Noise Control . . . . . . . . . . . . . . . . . . . . . . . . . . . . . . . . . . 12-139
Applications . . . . . . . . . . . . . . . . . . . . . . . . . . . . . . . . . . . . 12-141
Safety . . . . . . . . . . . . . . . . . . . . . . . . . . . . . . . . . . . . . . . . 12-143

# INDUSTRIAL PLANTS

## by John R. Immer

REFERENCES: Muther, "Systematic Layout Planning," 2d ed., Cahners Books. Immer, "Materials Handling," McGraw-Hill. Apple, "Materials Handling Systems Design," Ronald. "Building Code," Building Officials Conference of America. Dagostino, "Methods and Materials of Commercial Construction," Reston Pub. Lepedes, et al., "McGraw-Hill Encyclopedia of Environmental Science," McGraw-Hill. Clough, "Construction Contracting," 3d ed., Wiley-Interscience.

## GENERAL PLANNING OF INDUSTRIAL PLANTS

**Top-management decisions** relative to the planning of new industrial plants or a major addition to an existing plant require a study of the present and future manufacturing requirements of the company, its financial position, general economic conditions and the condition of the capital market, changing marketing patterns and conditions, and a wide range of general policy matters affecting the most fundamental aspects of its operations. There will have to be considered in this order, the general location of the plant, site selection, processing and layout, the physical structure, and finally the contract procedures for the construction project.

**Computerized game plans** are almost a necessity on even medium-sized plants. Basic data concerning the major factors must be assembled, coordinated, and put in such form that all the factors can be evaluated. Many alternate locations and site plans should ordinarily be considered. Transport costs of raw materials and of finished products must be considered for each region and for each prospective site. With the requisite base data these alternate sites can be evaluated quickly by the computer. Varying labor, production, construction, and other cost items can also be included in the evaluating formula to provide a complete evaluation.

**Computerized scheduling** will expedite the preparation of schedules for various alternate regions, sites and construction proposals. The "critical path method" is effective in minimizing total time of the project through all the time requirements involved in site selection, preparation of environmental impact studies, design, construction, and the start of production processes (see Sec. 17).

A **construction progress chart** should be prepared showing the sequence and time required for each major item of construction and for the installation of principal equipment. This progress chart must take into account the time required to prepare detailed plans and specifications; to purchase, fabricate, and deliver major items of construction materials and process equipment; and to do the preliminary operation and make any necessary adjustment; the chart should be arranged to correlate and integrate each element of design and construction into an orderly sequence leading to a definite date for placing the plant in commercial operation. **Schedules** should then be prepared showing the sequence and date on which all drawings, structural, mechanical, and electrical, must be completed, specifications written, and contracts placed for the required materials and equipment, so that these elements may be fabricated and delivered to meet the requirements of the construction progress chart. Preliminary progress charts will be used as the basis for preliminary cost estimates. A complete cost estimate should be made as soon as the final location, site, and design of the buildings and the project permit.

## Factors Determining the General Location of a New Plant

The **dominant factor** in location must be considered first. Many types of manufacturing are basically tied into or oriented toward a certain factor such as raw materials (steel smelter), markets (concrete pipe), labor (clothing), or cheap electric power (aluminum). Some processes require a certain type or quantity of water or have problems of pollution and waste.

**Ancillary factors** may determine the selection of a region or site among a number of alternate sites providing the dominant factor. These may include existing communities with unusual recreational facilities which attract the type of employees that may be needed. State and local taxes vary from one state to another.

**Transportation** must be considered from a cost standpoint, time to markets, type of transport available, and the dependence of schedules.

**Environmental impact studies** will be required for most industries in most locations. Local, state, and federal environmental laws must be considered. These are proliferating rapidly at all three levels and must be checked to the current date.

## Factors Controlling Selection of Site

**Environmental impact studies** will have to be made of most sites. The effect of **hazardous or objectionable features** of plant operations on the surrounding countryside must be considered. **Size of site** must be sufficient for anticipated expansion and to afford protection against objectionable neighbors. **Disposal of plant waste** is becoming increasingly important in view of legislative trends. Do **zoning codes** permit all types of processes involved in operations of the proposed plant? **Building codes** may affect the layout and cost of buildings.

**Shipping facilities** and suitable rates, including rail, water, air, and truck, for raw materials and for finished products are important factors. Proximity to highway interchanges, air terminals, and major distribution terminals and freedom from congested areas may be important. **Public transportation** or **adequate parking** for employees may be a consideration.

**Topography and soil structure** are particularly important for large single-story plants and for those having special requirements such as large presses or other heavy-weight concentrations. Potential flood conditions, earthquake or fault zones, unstable soil or substructure, abandoned mines or underground caves, and the geology of the area must all be considered. Generally, one or more deep borings should be made on the construction site before acquisition. Potable and process water, waste elimination, light, power, gas, and sewers must be checked for availability.

**Power** may be purchased or generated. Process steam and

heating and air-conditioning requirements must be considered. Potentials of **solar energy** and **wind energy** and type of fuel upon which power and heat are based must be studied.

## PROCESSING

### Determine Processes and Space Requirements

**Develop space requirements** for each of the specific operations and processes to be employed. These will be based upon the output to be employed initially and at projected stages of expansion. State these **space requirements** in broad terms, as for production centers, cost centers, or major manufacturing departments. Figure 1 shows the relationship of these departments and the way in which materials-in-process flow through the plant. Summary of area requirements will provide total space requirements for plant.

**Special processes** should be noted. Large machines with special foundations requirements should be indicated. Processes with objectionable fumes or noises should be segregated. Mezzanines and upper floors may be utilized for processes involving small and lightweight machines and for storage and assembly areas.

**Occupational Safety and Health** standards and practices must be met. All operations, processes, layouts, machine arrangements, and industrial practices must meet these OSHA standards. Processes with special fumes, hazards, noises, chemicals, or vibrations should be carefully checked for their compliance.

### Flow of Materials

Flow of material through the plant in a straight line and with a minimum of backtracking and recrossing of lines is desirable.

The basic lines of flow may be a straight line, with the raw materials entering at one end of the plant and the finished product leaving at the other end, or they may be U-shaped, with both receiving and shipping departments at the same end of the plant. More complicated processes may require modification of these basic forms (see Figs. 2 and 3).

**Flow diagrams** establish the relationship of processes and working areas. They are of two basic types: (1) those employing **continuous** operations, such as paper mills and sugar refineries; (2) those employing **intermittent** operations, such as the manufacture and assembly of parts. As the layout of the former type is determined by the process, they are often referred to as "process" industries. Most manufacturing plants involve the intermittent type in which the "flow" of materials is more difficult to establish. Flow diagram of operations in a foundry is shown in Fig. 4.

**Lines of flow** of materials are shown on layouts, as in Fig. 1, for the flow of materials through various departments. The layout of the machine shop in Fig. 5 shows the flow of materials from the machines to inspection and to finish stores.

**Operation process charts** help to visualize the relationship of numerous assembly operations in a complicated process by using standard symbols. The flow process chart uses the same symbols but includes more detail as to movement and delay of materials in process.

### Materials Handling
(See also Sec. 10)

**Materials-handling equipment** must be selected to move materials into and out of machines, from one operation to the next, and from one department to another. In many industries the cost of handling exceeds the cost of machine and assembly operations. Usually, several types of equipment must be used in the

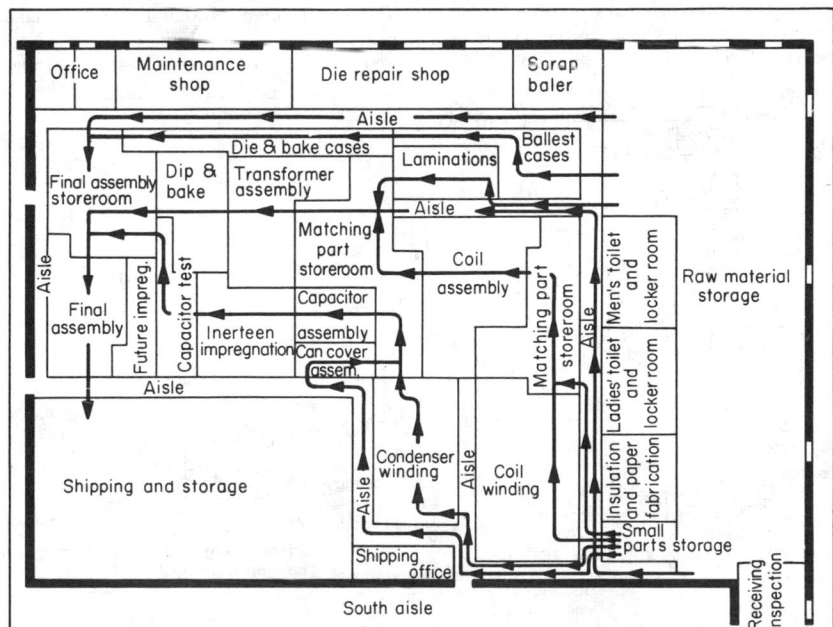

**Fig. 1**  Diagram of preliminary area allocation showing flow of materials.

same plant, and it is important that all types of handling equipment used in a plant fit into the overall plant handling system. In general, plants tend to rely on a system based predominantly on one type of equipment, such as conveyors or trucks, although other types will be used in connection with them.

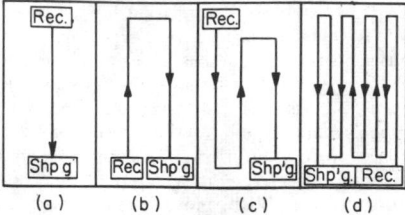

**Fig. 2**  Patterns of production lines. (*a*) Straight line; (*b*) U shape; (*c*) S shape; (*d*) convoluted.

**Overhead traveling cranes** are used for heavy lifts. They traverse production areas and provide for both horizontal and vertical movement of heavy items. As the weight of the crane and its load must be carried on elevated crane rails it is better

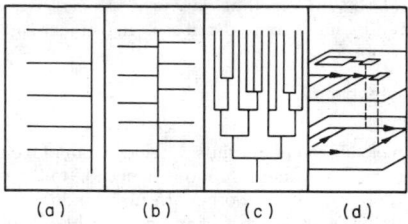

**Fig. 3**  Patterns of assembly lines. (*a*) Comb; (*b*) Tree; (*c*) Dendritic; (*d*) Overhead.

to plan these cranes as a part of the original structure. Swinging jib cranes of various types and single-leg gantry cranes may be operated underneath the overhead crane.

**Conveyors** are of the overhead or floor-level types. **Overhead types,** such as the continuous trolley conveyor or the monorail, on which individual units or items are moved, provide for overhead movement of materials. This relieves floor congestion. This type of equipment is planned around the process and is expensive to change once installed. Other conveyors are installed at floor level or at working height. These include **belt conveyors, roller conveyors** (gravity or powered), and **chain conveyors** (drag type for items, flight type for bulk materials). Smaller items are transported in **tote pans** or **boxes.**

**Industrial trucks** and **hand trucks** provide more flexibility of movement between processes. Two-wheel and four-wheel trailers are nonpowered. Powered trucks are used to pull trailers. Fork-lift trucks elevate loads and provide for high stacking of materials. This results in greater utilization of cubic space, particularly in high-ceilinged areas. **Pallets** and other means of unitizing materials, such as wire strapping, permit larger loads of small items to be moved at one time. **Scheduling of equipment** is important for greatest economy of operation.

### Layout Planning Techniques

Drawings of the layout showing the location of each machine, equipment, and physical features of the plan are used to translate generalized flow lines into the physical layout. Alternate layouts or other changes involve considerable expense in drafting.

**Templates** are two-dimensional representations of machines and equipment which are easily moved around and changed. If negatives of templates are used, the complete layout can be reproduced without any drafting. Scale commonly used is ¼ in to the foot. Larger layouts may use a scale of ⅛ in to the foot. Colored string or other lines may be used to indicate conveyors or to show flow of materials.

Scale models are used for better visualization of layout. Cost of models is often less than the cost of making drawings. Scale models provide the ultimate in visualization of layouts and frequently prevent serious mistakes in planning and location

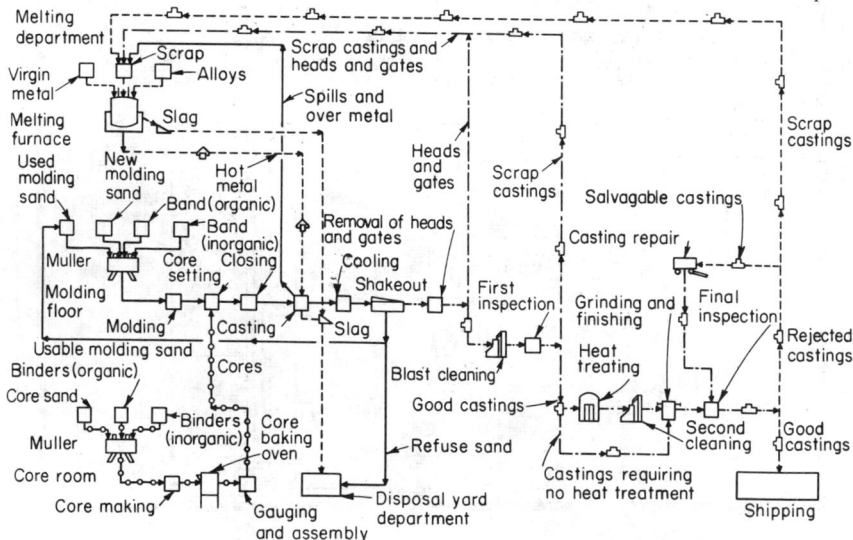

**Fig. 4**  Flow diagram of foundry operations.

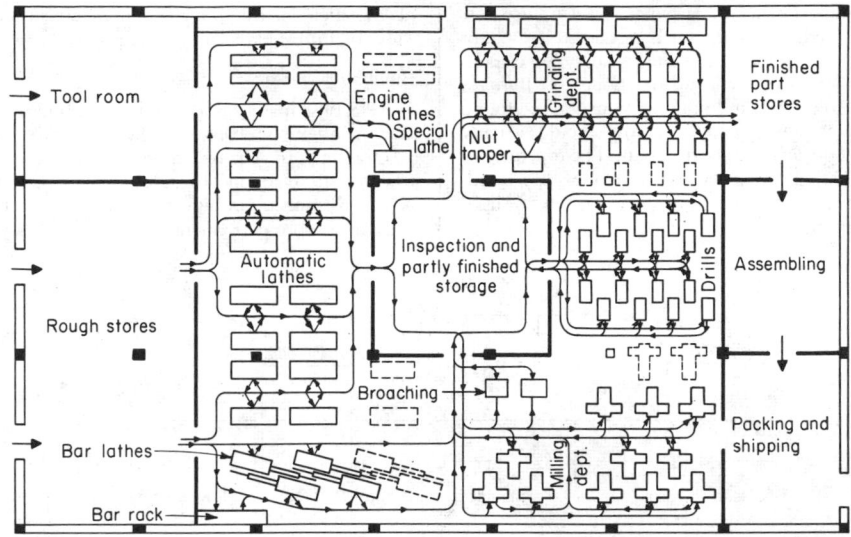

**Fig. 5**  Layout of a machine shop, providing for inspection of work after each operation, and showing the flow of materials in process.

of machines and other facilities. Models vary from plain wooden blocks to cast-metal models complete to minute detail. Both templates and models are available commercially.

## BUILDING DESIGN AND CONSTRUCTION

The selection of the proper type of building to house each step in the process has an important bearing on the economical performance of the completed industrial plant. **Space requirements** of the manufacturing equipment and the **functional relationships** of its various parts in the process flow will determine the basic features of the enclosing structures. The ideal arrangement will permit the work to be performed with practically as much freedom as if the building did not exist. The building units should then be assembled into a harmonious, pleasing group. This may require some rearrangement of the process equipment. Although the functioning of the equipment must remain the fundamental consideration, a proper overall solution of the problem must integrate a sound equipment layout with a suitable type and assemblage of enclosing structures.

**Site plan** must be developed for the overall economical relationship of structures and utilization of the site. Allowance must be made for future expansion of structures. Locate power plant as centrally as possible to demand, bearing in mind water supply, fuel delivery, and fuel storage space. Consider prevailing winds with respect to location of coal pile or other outside storage of loose materials from which dust may be blown.

### Type of Buildings

**Single-story buildings** are preferred where large uninterrupted floor areas are desired. The importance of the continuous flow of materials in process on the same level with a minimum of

obstruction favors single-story construction even in areas of high land value. Long-span construction can be adopted at reasonable cost, thus providing working spaces with a minimum of columns and, by means of monitors of various types, skylights, and ventilators, afford overhead light and ventilation. One-story structures are well adapted to processes requiring crane runways. In determining the spacing of columns, a longitudinal span of 20 to 25 ft provides an economical purlin design for most lightweight roof construction. For the transverse span, the advantage to the process layout of unobstructed floor space must be balanced against the cost of long-span girders or trusses. In some cases for buildings more than one bay in width, the use of longitudinal trusses to support one or more transverse trusses, thereby eliminating a number of interior columns, is warranted (see Figs. 10 and 11). The general trend is for wider spans and fewer columns in order to obtain greater flexibility of layout on the factory floor.

A variety of **roof profiles** for one-story buildings, with many types of monitors, skylights, and ventilators, offer a wide choice to meet the requirements of a particular problem. The single longitudinal monitor elevated well above the working area does not provide good interior lighting especially for the longer spans, and this type of roof is not well adapted to buildings more than one bay in width. Figures 6 to 9 illustrate a few of the many combinations of roof profile available for roofs having only a slight pitch. Although these types of trusses because of their practically constant depth are somewhat more expensive for a given span and load than the familiar pitched roof they offer advantages in adapting the building outline to the requirements of the process layout and in providing good interior light. Extension is practicable in more than one direction in combinations such as those illustrated in Figs. 8 and 9. Figure 10 shows a typical saw-tooth roof profile, which with the face of the saw tooth toward the north affords excellent natural interior lighting over large areas.

**Multiple-story buildings** are advantageous where a gravity flow of the process is desired, where the site is restricted in area, or where high land values prevail. For the usual types of construction and for floor live loads up to about 200 lb/ft², a column spacing of 20 to 25 ft in both directions will be found economical. For rectangular panels, beams should usually have the longer and girders the shorter span. The top story of a multiple-story building can be treated like a one-story building with longer spans and a roof profile providing interior light and natural ventilation.

arrangement of buildings and their construction should provide for future extensions as required.

### Building Materials

In selecting materials for industrial buildings, the engineer faces conflicting objectives, e.g., low first cost, with ready adaptability to change in arrangement and use, tending toward the selection of less permanent, less expensive materials and resulting in higher maintenance cost contrasted with the use of more permanent and more costly types of construc-

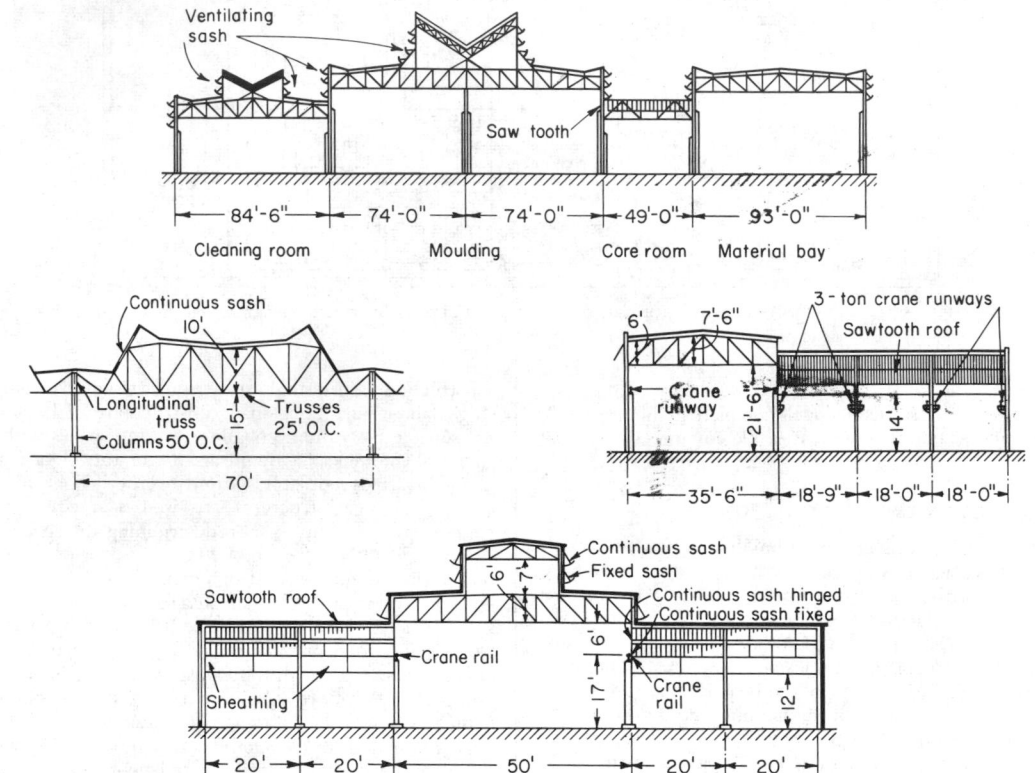

**Figs. 6 to 9**  Roof profiles with slight pitch.

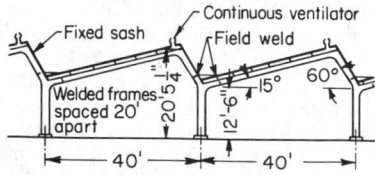

**Fig. 10**  Steel saw-tooth roof.

Combinations of multiple-story construction for those parts of a process requiring gravity flow and single-story structures for those parts of a process that require unobstructed floor space with overhead light and ventilation will often produce a desirable plant layout.

State and city codes and regulations frequently regulate number of floors, undivided floor areas, building population, exit provisions, service facilities, floor loading, ventilation, fire protection, hazards, and type of building construction. The

tion. **Speed of erection** may justify use of more expensive prefabricated materials even where the cost is greater than for other types of construction available. Comparative estimates of more than one type of construction are desirable. The possibility of a process becoming obsolete within a few years may not warrant more than a minimum-cost enclosing structure. Climatic conditions and the advertising value of the general appearance of the plant are important factors.

**Wood construction** is used only for temporary purposes and even then is often more costly than other materials. Because of its fire hazard, especially in the small sizes such as stud and joist construction, fire codes may preclude its use in these forms. As most machining operations require oil or grease, wood surfaces soon become impregnated and form a fire hazard.

The development of **laminated structural members** and of

improved types of fastening devices has led to the use of wood for roof structures. Spans of 40 to 240 ft are used. These are made of **impregnated timbers,** which increase their fire resistance. They are also used in hangar construction, as they will not melt in flash fires, and they retain their structural strength long enough to permit evacuation of personnel and planes.

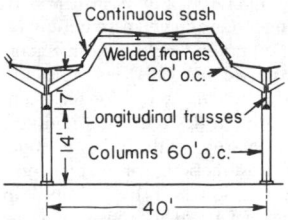

**Fig. 11**   Roof supported by longitudinal trusses.

**Steel-frame construction** is used almost exclusively for long-span single-story structures, and is generally economical for short-span single- and multistory construction where fireproof structures are not required. Unprotected steel begins to lose its strength at about 700°F and may warp, twist, and fail in the event of a serious fire in the contents of a building. An adequate sprinkler system is desirable. Where steel structures must be fireproofed, reinforced-concrete construction will generally be more economical except for long spans where steel framing, fireproofed with thin gunite or other light-weight material, may be lower in cost. Steel requires maintenance painting, the need for which depends on conditions of exposure (see Sec. 6).

**Masonry construction** is generally the most economical for single-story structures. Concrete block, with or without facing brick or other material, is constructed with supporting pillars built in at regular spaces to support roof girders. A waterproof surface can be provided by new silicon paints or by a plastered surface plain or sprayed with a quartz aggregate.

**Precast concrete** is used increasingly for structural members, walls, floors, and roofs. Structural sections are formed with conventional steel reinforcing, or wire or cable reinforcing, or may be prestressed or poststressed for additional strength and smaller sections. Units may be obtained commercially or cast at the site. In either case erection time is usually minimal.

**Reinforced-concrete construction** provides full fireproof construction. For approximately square panels and spans up to 25 ft, flat-slab construction is usually more economical than beam and girder construction, and its comparative economy increases with heavier live loads. Flat slabs are not well adapted to layouts requiring irregular column spacing or numerous floor openings after a building is completed. It is frequently desirable to provide inserts at regular intervals in all ceilings to allow for initial and future requirements in hanging piping, shafting, and the like. Concrete construction, because of its resistance to vibration, is especially adaptable to plants using high-speed machinery or having high floor-load requirements.

**Lift-slab construction** is used for multistory construction. Floor slabs are poured at ground level, and after reaching the full strength of the poured concrete, slabs are elevated on supporting columns to the desired height by means of hydraulic jacks. Frequently, major items of mechanical equipment are loaded on the slab before it is elevated into place.

**Walls** may be constructed of metal panels (galvanized iron, steel, or aluminum); asbestos cement or cinder blocks; hollow clay tile, brick, concrete (poured-in-place, tilt-up, or precast), or stone. All the corrugated or thin panel materials can be erected quickly and are light in weight, but they require girt framing on the inside. Increased use of air conditioning in summer and desire to reduce heat loss in winter have resulted in an increased use of insulating materials. Clay tile of various kinds and new epoxy paints provide sanitary, easily cleaned interior surfaces where needed.

**Windows** may consist of factory-type sash built of rolled sections of steel or aluminum; wood windows; double-hung metal windows; glass-block windows; and other special construction, the cost increasing roughly in the order listed. Standard steel sash is usually the economical window for industrial buildings. Pivoted ventilators are awkward where shades or screens are required, but the use of projected ventilators at small increased cost avoids this difficulty. An entire sidewall constructed of standard sash, located so as to run continuously past columns and floors, and glazed with steel plate at floor levels and glass elsewhere, is frequently a low-first-cost construction. **Glass** may be rough, clear, or ribbed in various forms for light diffusion. Corrugated glass is frequently used for skylights and to some extent for certain types of windows. For large lights, as in skylights, it is less likely to crack from expansion and contraction than flat glass. Wired glass frequently justifies its added cost as protection against the hazard of falling broken glass, particularly for high sidewalls and windows in monitors over working spaces. It is generally desirable for the psychological effect on employees to provide one or more lights of clear glass at eye level in working spaces. In contrast to the low first cost of standard steel sash, consideration must be given to the maintenance cost of painting to protect the light metal sections. Glass-block windows provide excellent diffused light and relatively good insulation, which is particularly important where condensation is a factor, and they require no maintenance painting. **Translucent plastic panels** also admit light and can be obtained in various colors to diffuse glare.

**Roofs** having a pitch of 4 in or more per foot can be economically constructed of corrugated materials, such as galvanized iron, zinc, asbestos-protected metal, and asbestos cement, the purlin span varying with the material selected but all light in weight and requiring no additional covering. Such roofs afford little insulation, and where conditions are such that condensation is likely to occur on the underside of the roof, an insulated roof deck is advisable. For roofs having a pitch less than 4 in/ft, a membrane roof covering is desirable. The corrugated materials listed above can be used by applying insulating board over the corrugations as a base for membrane roofing, and this construction can also be used on the steeper pitches where insulation is desired. Flat interlocking steel sheets formed with ribs for strength and stiffness provide a light-weight deck on which membrane roofing is laid with or without insulation. For relatively flat pitches, lightweight precast concrete slabs, either flat or channel shaped, provide a low-cost fire-resistant deck. Wood plank, gypsum plank, precast concrete plank, and poured-in-place reinforced concrete, covered with membrane roofing, are suitable for either pitched or flat decks, and insulation can be applied when desired. As between the many types of roof decks available, the relative first cost depends not only on the erected cost of the material itself but also on its effect on the cost of the supporting framing, particularly in long-span construction. Proper selection, in a given case, can best be based on comparative estimates of the cost of the entire roof system, weighing also the maintenance cost, resistance to heat loss and condensation, light-reflecting qualities, fire resistance, appearance, and other similar factors. For **roof insulation,** fiber insulating boards, cork, and processed glass products are excellent, the latter usually being slightly higher in cost. For membrane roofing, a variety of good factory-finished composition roofings ready to lay are available at low cost, but in general this type of roofing will not have so long a life as roofing consisting of several plies of high-grade roofing felts, each layer mopped down in hot pitch or asphalt. On relatively flat pitches, a surface coating of small gravel or slag embedded in the final hot mopping is desirable. **Flashings** against parapet walls and curbs around roof openings are important in tight roof construction. Flashings may be galvanized iron, membrane, or copper. The life of galvanized iron is limited, and, where metal flashing is used, the additional first cost of copper is generally warranted in avoiding maintenance expense.

**Floors** in industrial plants are usually of concrete poured in place as a concrete slab on grade or on concrete or steel beams. In multistory construction it may be monolithic reinforced concrete, either flat slab or beam and girder. Floors may also be of precast concrete floor slabs or of sections resting on beams, or they may be cast with an integral beam construction such as T- or inverted U-shaped sections. Finished floors may be exposed concrete, hardened concrete topping, or a vinyl floor finish. Cellular-type steel floors merit special consideration for offices or other areas where underfloor telephone or signal systems are required, since the wiring may be run through the cells of the steel floor deck. Relative costs depend on floor loadings, column spacing, number and size of openings required, and other similar factors. Comparative estimates of alternative designs, based on a typical cross section of the building, one bay in length, which include the effect on columns and foundations of the relative dead weight and the adaptability of the construction to conditions of service, offer the best basis for proper selection for a given case. An ideal **floor finish** should be resistant to wear and to the action of process spillage, sanitary, comfortable, nonslip, easily cleaned, and low in maintenance cost. A properly laid cement-finish floor using hard aggregate such as trap rock, granite, or quartzite chips laid over a concrete slab has excellent wearing qualities. Best results are obtained when the finish is at least 1 in thick, placed as a separate operation rather than monolithic with the slab. About 1 lb of carborundum grits per square foot worked into the cement finish provides a good nonslip surface and is especially desirable where floors are set during working operations. For aisles, areas at elevators, and loading platforms subject to heavy-duty iron-wheel trucking, grids of **cast iron** bedded flush with the cement finish, although high in first cost, give excellent service with low maintenance cost. For floors subject to strong acid or alkali action from process spillage, an **asphalt mastic** finish floor usually gives best results. Hardwood-finish floors such as maple provide an excellent working surface but are generally quite high in first cost and in maintenance cost, especially in areas subject to trucking, unless rubber tires are used. End-grain **wood-block** floors provide a good working surface at reasonable cost for dry locations and wear well under moderate trucking, but are subject to buckling if wet accidentally as from sprinkler operation. Creosoted wood-block and asphalt-block floors give good service in heavy machine shops, foundries, and the like, but are not attractive in appearance and are hard to keep clean.

**Partitions** fall into two general classes: those likely to be relatively permanent and those likely to require moving owing to future changes in layout. For **permanent** partitions in steel or concrete buildings, masonry construction of cinder blocks, hollow clay tile, common brick, concrete, salt-glazed tile, and ceramic-glazed tile are generally desirable, the cost increasing about in the order listed. Except for glazed materials, plaster can be applied readily if desired. Subdividing fire walls should be provided to comply with building-code and insurance requirements. Gypsum block or metal studs with wire lath and plaster provide low-cost lightweight fire-resistant construction in semipermanent locations. For **movable** partitions, wood studs faced with fiberboard or plywood sheets afford minimum-first-cost construction but present some fire hazard. Filling all spaces between studs with rock wool adds to the cost but lessens the fire risk and improves the sound-insulating quality. Woven-wire grille partitions are frequently used for tool- and stock-room enclosures where the principal requirement is to prevent unauthorized access. Sectional steel or steel-and-glass partitions are widely used for shop offices, inspection rooms, and other similar enclosures, as they can be readily enlarged, taken down, and rearranged, but they have little sound insulation in noisy locations. Double steel or asbestos cement sectional partitions with rock wool or other insulation and double glazing are readily moved, are effective against noise transmission, but are relatively high in first cost.

**Stair arrangement,** location, and construction is one of the most important problems in industrial-plant design, affecting, in the form of exits, the life and safety of all employees and, in the form of access to equipment, the safety and efficiency of operation. For exit stairways, a width of 22 in accommodates one person, and the total stair width should be in multiples of this unit, with 3 ft a desirable minimum where only a single unit is required. Up to 66 in width, a handrail should be provided on each side, and for wider stairs an additional center handrail is desirable. Handrails projecting not more than 3 in do not encroach on the effective width of a stair, but beams, columns, and other obstructions should not project into the effective stair width. Not more than 15 risers should be permitted in a single run, beyond which intermediate platforms should be provided. Winding treads at stair turns or straight runs of less than three risers should not be allowed. There should be at least two exists arranged so that, with a fire originating in any location, employees in any area can reach one of them. The maximum distance from any area to an exit should preferably not exceed 100 ft for hazardous occupancy or 150 ft in sprinklered fire-resistant construction. Stairs should be provided so that the total employee occupancy can be evacuated on one of three bases: (1) a safe total time depending on the hazards present based on 45 persons per 22-in stair width, walking down in 1 min; (2) all occupants of the most densely populated floor above, on the stairs in any story below that floor, assuming one person per 22-in width on every other tread and one person per 3½ ft² of landing area; (3) divide the gross square-foot area of any one floor above the ground by 6,000 for low-hazardous and by 3,000 for high-hazardous conditions, and provide a 22-in width of stair for each unit resulting from this division.

Exit stairways should be suitably enclosed with fire-resistant walls and doors, the latter equipped with self-closing devices and opening into the stair shaft without encroaching on the required effective landing width, except at ground level where doors should open out direct to safe open air or to a fireproof passage.

**Stairs** should be constructed of noncombustible materials with nonslip nosings. Steel stringers with pressed-metal treads filled with concrete, surfaced with carborundum grits, steel stringers with sheet steel risers and nonslip cast-iron treads, and grit-surfaced reinforced concrete are excellent constructions for exit stairs. Storage spaces under stairs should not be allowed.

The efficiency of convenient safe **access to working levels** around large equipment and to important operating valves justifies in most cases the cost of the necessary platforms and access stairs. Ship-ladder stairs, circular stairs, and ladders are inconvenient and somewhat hazardous. Grating platforms, with access stairs having a minimum width of 2 ft 2 in, grating treads with visible nonslip nosings and handrails are generally a desirable construction. **Nonskid surfaces** may also be obtained by the use of special tapes and by paints and other materials which prevent slipping even when the surface is wet or oily.

For stairs with open risers, plates about 2 in high at the back of the tread will prevent tools laid on the stair from being kicked off onto employees below; similarly curbs are advisable around floor openings and elevated platforms.

## Lighting

Natural lighting of building interiors is subject to wide variation in intensity due to fluctuations in ouside brightness. Although it is practically impossible to design a building so as to provide uniform distribution of daylight over an interior area of appreciable size, the arrangement of windows in sidewalls and monitors affects to a marked degree the uniformity of daylight lighting of the interior and warrants careful study in planning a new building.

For the average factory, the normal accumulation of dirt on the glass of vertical windows will, in a 6 months' period, reduce the light transmission about one-half as compared with clean glass. Although there is not much difference in the rate of dirt accumulation on clear, rough, and ribbed glass, the last is more difficult to clean. Much more dirt collects on the inside than on the outside; therefore, the inside should be washed twice as often as the outside. Windows set on a slope increase the amount of light transmitted to the interior as compared with those set vertical, but the former collect dirt more rapidly and with a 6 months' washing period there is little actual difference.

Assuming a desirable minimum lighting intensity of 10 foot-candles on a working plane 3 ft above the floor and overcast sky conditions, sidewall windows without monitors limit the working area to a distance from the wall of about three times the height from the floor to the top of the window with clean glass and only about twice the height to the window head, with no washing for 6 months. For the same conditions, the total window area in side walls and monitors should be not less than 30 percent of the floor area, and the heads of the windows should be set close to the ceiling; the width of a monitor should be not less than twice its height, and the total width of monitors should be not less than one-half the width of the building.

**Skylights** provide high-intensity lighting in the areas immediately beneath them, but nonuniform conditions immediately adjacent. **A-frame monitors** of the same width as skylights provide comparable lighting and, with movable sash, some gravity ventilation.

For multistoried buildings constructed in U, E, or other court form, the **width of court** should preferably be equal to the height of the building, court walls should be faced with light-colored material to reflect light into the building, and for equal story heights the intensity of interior light decreases from the top story to the ground so that, considering daylight lighting only, each lower story should have a greater story height than the story next above.

**Artificial lighting** is essential in most modern industrial plants during dull natural-light periods at any time and to permit two- or three-shift operation when desired. Its influence on labor productivity in many industries may justify continuous operation of artificial lighting during the entire working day. A **windowless industrial building** has many advantages, including freedom from the maintenance cost of painting and glazing windows and from the transmission of heat through them. The major drawback is the shut-in psychological reaction of most employees where no natural light is provided. Designs embodying low windows at working levels for employees' vision only with complete dependence on artificial light for proper working conditions merit consideration.

**Acoustics** Definite improvement in employee efficiency frequently results from reducing high-pitched objectionable sounds and creating an improved noise level by the application of sound-absorbing materials to the underside of roofs, ceilings, and frequently on walls and partitions.

**Heating and Ventilating** The heating and ventilating of industrial buildings are allied problems, each affecting the health and comfort of employees and, to a marked degree, their efficiency. Despite high initial and operating cost, complete air-conditioning systems, supplying an adequate quantity of clean properly humidified air at comfortable temperature throughout the entire seasonal cycle, have demonstrated their effectiveness in stores, offices, and to a limited extent in industry. The growing appreciation of the resulting efficiency, coupled with a continued downward trend in initial and operating costs, points the way to their wider application.

In many cases, the manufacturing equipment will furnish adequate heat during operation, and a heating system is necessary only to prevent freezing when the equipment is shut down. Quite frequently, the equipment will give off too much heat, and the problem then becomes one of ventilation.

## Plumbing

Codes in most states specify the minimum number of plumbing fixtures required. For modern plants, employees' comfort and convenience are a consideration, and in many cases it is desirable to provide a greater number of fixtures than the prescribed legal minimum. The following ratios of fixtures to employees are recommended:

For washrooms serving 25 or more, **circular wash fountains** are often economical in floor space required and cost. In the preceding schedule, one 36 in diam fountain is equivalent to four, and one 54 in diam fountain to seven individual lavatories. Frequently, **showers** also should be provided, one to serve 6 to 15 employees, depending on conditions. Where a number of showers are required, a circular arrangement consisting of

| | Employees per shift | | | |
|---|---|---|---|---|
| | 1–5 | 6–12 | 13–30 | 31 or more |
| Water closets, men | 1 | 2 | 2 | 1 per 15 |
| Water closets, women | 1 | 2 | 3 | 1 per 12 |
| Urinals | . | . | 1 | 1 per 20–30 |
| Individual lavatories | 1 | 2 | 1 per 7 | 1 per 8 |

five stalls arranged around a common supply and drain connection is inexpensive and will often save space. **Lockers** should be provided for all employees, and in general individual lockers are desirable. In cases where damp work clothing is customarily stored, exhaust-fan ventilation connected direct to each locker is effective. Centrally located locker rooms with adjacent sanitary and wash-up facilities can be properly ventilated, maintained in a clean condition, and supervised more easily and are generally more satisfactory than a number of smaller scattered rooms. If central locker, toilet, and washrooms are used, it is advisable to provide, in addition, small toilet rooms containing a few sanitary fixtures located throughout the factory within one story and 200 ft walking distance of any appreciable number of employees. Traveling distance between male and female toilet-room entrances should be preferably not less than 20 ft. **Drinking-water** facilities should be provided on every floor, spaced not over 300 ft, and closer for large employee groups. For most locations, refrigerated drinking water is desirable, and either a central refrigerating plant with pumped circulating water lines or unit-type coolers may be provided. Except where there are large employee groups, with drinking fountains closely spaced, unit coolers will generally be more economical to install.

## CONTRACT PROCEDURES

### Types of Estimates

Estimates of cost fall into three general classes:

1. Preliminary estimates made usually from sketch drawings and brief outline specifications to determine the approximate total cost of a project.

2. Comparative estimates made usually during the progress of design to determine the relative cost of two or more alternative arrangements of equipment, type of building, type of floor framing, and the like.

3. Detail estimates made from well-developed plans and specifications and based on a careful quantity survey of each component part of the work.

For **preliminary estimates** it is important to list all the major items, structural, mechanical, and electrical, required to complete the project since overlooking essential items that will be required later, when detailed designs are prepared, will generally affect the actual total cost to a much greater extent than inaccuracies in the amounts allowed for each of the various items included in the estimate. Fairly accurate comparative estimates can be made by preparing rough preliminary designs of two or more alternative building cross sections and then making quantity surveys for one typical bay in length, or from simple quantity surveys of alternative designs for a typical panel of floor framing.

In preliminary estimating for the cost of buildings, it is common practice to compute the **total cubic contents** of the

structure and then apply some arbitrary figure based on past experience for the cost per cubic foot. The results are likely to be quite misleading, since the cost per cubic foot is affected by variations in foundation conditions, shape, story height, floor loadings, type of construction, extent of subdividing partitions, and other similar factors. The one factor of shape, all others being identical, will affect the cubic-foot cost, a long narrow building costing more than an equal area, square in plan. Considering building cost only, the most economical shape generally is that containing the greatest volume for the least surface area, namely, a cube. Foundation conditions may have an important bearing on building cost; expert advice is frequently advisable. For a concrete or steel-frame multistory building, cubic-foot costs generally decrease up to about six stories, after which they increase slightly. Fireproof construction will add appreciably to the cost but is often an excellent investment. Providing earthquake resistance may add 5 percent to the building cost, and designs for high-wind conditions a like amount, but the two forces are not usually taken as acting together. The value of estimates based on cubic contents of the structure will depend upon the availability of figures for comparable structures in the same cost area.

**Preliminary quantity estimates** are usually more accurate where this comparison is lacking. Compute the overall square-foot area of exterior walls, interior partitions, floors, and roof, and multiply by known cost factors for each of these. Other items, such as number of electrical outlets required and number of plumbing fixtures and sprinklers, are estimated, and their cost computed. Equipment costs can be based on preliminary quotations from manufacturers.

The following items should be considered: land cost; fees to real-estate brokers, lawyers, architects, engineers, and contractors; interest during construction; building permits; taxes including sales or use taxes in force in some states; demolition of existing structures including removal of old foundations; yard work including leveling, drainage, fencing, roads, walks, landscaping, yard lighting, and parking spaces; transportation facilities including railroad tracks, wharves, and docks; power supply; water supply; sewer and industrial-waste disposal. A contingency item of 5 to 15 percent depending on the character of the estimate and the purpose for which it is to be used should be included to provide for unforeseen conditions that may arise during the development of the project.

### Working Drawings, Specifications, and Contracts

The technical staff of a given industry has special knowledge of trade practices, process requirements, operating conditions, and other fundamentals affecting successful operation in that field and is best fitted to determine the basic factors of process design and plant expansion. Unless the company is extremely large, it is unlikely that they will have the specialized staff or that their own staff will have the time available to undertake the complete layout. The efficient transformation of these requirements into a completed plant extension, or new plant ready to operate, usually requires experience of a different nature. Therefore, it is generally advisable and economical to employ engineers or architects who specialize in this particular field.

Three general procedures are in common use. First, the employment on a percentage or fixed-fee basis of an engineering and construction organization skilled in the industrial field to prepare the necessary working drawings and specifications, purchase equipment and materials, and execute the work. Such an organization becomes for the duration of the project a part of the owners' organization, working under their direction and in close cooperation with their technical staff in the development of the design, purchase of equipment and materials, and their installation. This procedure permits construction work to start as soon as basic factors of arrangement and cost have been determined but in advance of the time required to complete all working drawings and specifications. Such a program will result in the earliest possible completion consistent with economical construction.

A second procedure is to employ engineers or architects of wide experience in the industrial field to prepare working drawings and specifications and then to obtain **competitive lump-sum bids** and award separate contracts for each or a combination of two or more subdivisions of the work, such as foundations, structural steel, and brickwork. This method provides direct competition restricted to units of like character and permits intelligent consideration of the bids received.

Provision must be made for proper coordination of these separate contracts by experienced and skilled field supervision. This procedure requires more time than that first described, since all work of a given class should be completely designed before bids for that subdivision are taken. Where construction conditions are uncertain or hazardous, the first method is likely to be more economical, or combinations of the first method for uncertain conditions and the second method for the balance may prove most advantageous.

A third procedure is to employ engineers or architects to complete all plans and specifications and then **award lump-sum contracts** one for the entire work, or one for all building work, and one or more supplementary major contracts for equipment. This method is particularly useful where there are no serious complications or hazards affecting construction operations, but it requires considerably more time, since most of the working drawings and specifications must be complete before construction is started. It has the advantage, however, of fixing the total cost within narrow limits before the work starts, if the contracts cover the complete scope of work required and no major changes are required.

# STRUCTURAL DESIGN OF BUILDINGS

## by John A. Blume and J. P. Nicoletti

(With the assistance of Ray L. Holmes)

REFERENCES: "Steel Construction," American Institute of Steel Construction. "National Design Specification for Stress-Grade Lumber and Its Fastenings," National Forest Products Assoc. "Uniform Building Code," International Conference of Building Officials. "Analyses of Small Reinforced Concrete Buildings for Earthquake Forces," Portland Cement Assoc. Plummer and Blume, "Reinforced Brick Masonry and Lateral Force Design," Structural Clay Products Institute. Scofield and O'Brien, "Modern Timber Engineering," Southern Pine Assoc. "Timber Construction Manual," American Institute of Timber Construction. "Wind Forces on Structures," *Trans. ASCE,* 1961.

### LOADS AND FORCES

**Live loads on floors** are generally regulated by the building codes in cities. For areas not regulated, the following values will serve as a guide for live loads (pounds per square foot): rooms for habitation, 40 (195 kg/m²); offices, halls with fixed seats, 50 (244 kg/m²); corridors, halls, and other spaces where a crowd may assemble, 100; textile mills, 50 to 100 (244 to 488 kg/m²); machine shops, 50 to 200; foundries, warehouses, 200 to 300 (976 to 1,464 kg/m²). Floor decks and beams that support only a small floor area must also be designed for any local concentrations of load that may come upon them. Girders, columns, and members that support large floor areas, except in buildings such as warehouses where the full load may extend over the whole area, may often be designed for live loads progressively reduced as the supported area becomes greater. Where live loads, such as cranes and machinery, produce **impact** or **vibration**, 25 percent or more should be added to the static loads.

**Roof live loads,** for snow and workmen making repairs, should be taken at not less than 30 lb per horizontal ft² (146 kg/m²) of roof for slopes up to 15°, and 1 lb (0.45 kg) less for each additional degree up to 45°. In severe climates, this load should be increased; in mild climates, it may be reduced.

**Dead loads** are due to the weight of the structure, partitions, and all permanent equipment not included in the live load. The weights of common building materials used in floors and roofs are given below (see also Sec. 6).

| Material | Weight, lb per sq ft |
|---|---|
| Asphalt and felt, 4 ply . . . . . . . . . . . . | 2 |
| Corrugated asbestos board . . . . . . . . . | 5 |
| Glass, corrugated wire . . . . . . . . . . . . . | 5–6 |
| Glass, sheet, ⅛ in. thick . . . . . . . . . . . | 2 |
| Lead, ⅛ in. thick . . . . . . . . . . . . . . . . | 8 |
| Plaster ceiling (suspended) . . . . . . . . . | 10 |
| Sheet metal . . . . . . . . . . . . . . . . . . . . | 1–2 |
| Shingles, wood . . . . . . . . . . . . . . . . . . | 2 |
| Sheathing, 1 in., wood . . . . . . . . . . . . | 3 |
| Skylight, ³⁄₁₆ to ¼ in., glass and frame . . . . . . . . . . . . . . . . . . . . . . | 4–5 |
| Slate, ³⁄₁₆ to ½ in. thick . . . . . . . . . | 8–20 |
| Tar and gravel, 5 ply . . . . . . . . . . . . | 6 |
| Tar and slag, 5 ply . . . . . . . . . . . . . . | 5 |
| Tiles, plain, ⅝ in. thick . . . . . . . . . | 20 |

lb/ft² × 4.88 = kg/m².

**Earthquake** effects are usually represented, for design purposes, by static lateral forces. A common rule is to provide in the building frame for resistance to horizontal forces equal to one-tenth of the dead and live load supported. Roofs and floors should be designed as horizontal beams or diaphragms, or they should be provided with horizontal bracing between points of lateral support. Concrete and masonry walls can be made to act as resisting shear elements if properly connected to roof and floor bracing systems.

## WIND PRESSURE ON STRUCTURES

### by William Bollay

(See also Sec. 11)

REFERENCES: Flachsbart, Die Belastung von Bauwerken durch Luftkräfte, in Kaufmann, "Angewandte Hydromechanik," Vol. 2, 1934, p. 269. Rausch, Einwirkung von Windstössen auf Hohe Bauwerke, *ZVdI*, **77**, 1933. Dryden and Hill, Wind-Pressure of a Model of the Empire State Building, *NBS J. Res.*, **10**, 1933, p. 493. Nökkentved and Irminger, Windpressure on Buildings, *Ingeniorsvidenskab. Skrifter*, A23, 1930. Göttingen Ergebnisse, Vol. 3, p. 148.

**Wind** pressures on walls of buildings should be assumed at 15 lb/ft² (73 kg/m²) on surfaces less than 60 ft (18.3 m) above the ground and 20 lb/ft² (98 kg/m²) on higher surfaces. On the projected area of exposed structural-steel frames, provide for wind pressure 50 percent greater than on walls. Wind pressures normal to sloping roofs steeper than 4 in vertical in 1 ft horizontal (33 cm/m) should be taken at 1½ lb/ft² (7.33 kg/m²) for each inch vertical in 1 ft horizontal (8.5 cm/m), with a maximum of 20 lb/ft² (98 kg/m²). These pressures should be increased for buildings in exposed locations and in localities where extremely high wind velocity (over 70 mi/h or 112.5 km/h) may occur.

Natural wind is a highly turbulent flow of air over the surface of the earth. Near the ground, its velocity is reduced by friction so that at a height of 50 ft (15.2 m) it is about 90 percent as great as at 100 ft (30.5 m); the distribution of velocity is represented by a formula of the type $v \approx (\text{height})^n$, where $n$ varies with the roughness of the terrain. An average value is $n = 0.157$.

The weather-bureau records give **maximum** and **extreme wind velocities** observed in various cities. The maximum is an average velocity over 5 min, and the extreme is the average velocity over the time taken to cover a mile (1.61 km), i.e., over ½ to 1 min depending on the speed. The critical maximum velocity as far as the strength of any structure is concerned is the maximum velocity of gusts comparable in size with the structure. In a 60 mi/h (96.6 km/h) wind for a house about 90 ft (27.4 m) in linear dimensions, a gust of even a second's duration would thus be giving the effective maximum wind. The only way to record such short gusts is by means of a system having very little inertia, e.g., the hotwire anemometer. A British test made with such instruments on an airport tower 64 ft (19.5 m) high showed that eddies created by neighboring trees caused a sudden increase in velocity from a mean speed fo 38 mi/h (61.2 km/h) to a peak of 85 mi/h (136.8 km/h) and back again to the mean speed. The duration of this gust was 0.8 s, and its linear dimension was thus about 40 ft (12.2 m). The ratio of the maximum velocity to the mean velocity in this case was about 2¼. When using the maximum velocities of the weather bureau, it should be kept in mind that much higher velocities will be encountered locally. On a hill, the mean velocity at the peak may be as much as 50 percent greater than in level country.

The records of the U.S. Weather Bureau (see Sec. 9) indicate maximum wind velocities of the order of 70 to 90 mi/h (112.7 to 144.8 km/h) with an isolated maximum of 113 mi/h (181.9 km/h) at North Head, Wash., and a tropical hurricane at San Juan, Puerto Rico, of 149 mi/h (239.8 km/h). In choosing the maximum velocity for design purposes, it is not generally economical to design such structures as telephone lines and radio masts for the very highest wind velocities that might be encountered; the latter are so rare that the increased cost of construction is greater than the cost of insurance or an occasional repair.

Flachsbart (*loc. cit.*) suggests, for ordinary purposes, a design wind of 80 (90) mi/h (128.7 to 144.8 km/h) up to a height of 60 ft (18.3 m) and 90 (100) mi/h (144.8 to 160.9 km/h) for heights greater than 60 ft. The figures in parentheses refer to especially windy regions. If a structure is to be designed so strong that it will not fail under any possible winds, then, unless it is located on a mountain top, a design wind of 150 mi/h (241.4 km/h) should be sufficiently high.

The effect of the sudden application of **gust loads** has sometimes been blamed for peculiar failures due to wind. In most cases, these failures can be explained from the pressure distributions in a steady wind. If a relatively flexible structure such as a radio tower, chimney, or skyscraper with a natural period of 1 to 5 s is set into vibration, the stresses in the structure may be increased over those calculated from a static-load analysis. Rausch (*loc. cit.*), on the basis of a theoretical analysis, suggests that provision may be made for this effect by increasing the design wind by 5 to 25 percent, depending on the flexibility of the structure.

Even a steady wind may give rise to periodic forces which may build up into large vibrations and lead to failure of the structure when the frequency of the exciting force coincides with one of the natural frequencies of vibration of the structure (Den Hartog, "Mechanical Vibrations," 1940, p. 343). The periodic exciting force may be due to the separation of a system of **Kármán vortices** (Fig. 1) in the wake of the body. The exciting frequency $n$ in cycles per second is related to $d$, the dimension of the body normal to the wind velocity $V$, by the equation $nd/V = C$, where $C \approx 0.207$ for circular cylinders and $C \approx 0.18$ for rectangular plates (Blenk, Fuchs, and Liebers, Measurements of Vortex Frequencies, *Luftfahrt-Forsch.*, 1935, p. 38). Dangerous vibrations related to the "flutter" of airplane wings may arise on bridges and similar flat bodies. These self-induced vibrations may be caused (1) by a negative slope of the curve of lift against angle of attack (Den Hartog, *op. cit.*, pp. 343–354) or (2) by a dynamic instability which

**Fig. 1**  Kármán vortices.

arises when a body having two or more degrees of freedom (such as bending and torsion) moves in such a manner as to extract energy out of the air stream. The first of these vibrations will occur at one of the natural frequencies of the structure. The second type of vibration will occur at a frequency intermediate between the natural frequencies of the structure.

The wind forces and the **pressure distribution** over a structure corresponding to a design wind $V$ can be determined by **model** testing in a wind tunnel. Extrapolation from model to full scale is based on the fact that at every other point on the body, the pressure $p$ is proportional to the stagnation pressure $q$ (see Sec. 11) and thus the ratio $p/q$ = constant for a fixed point on the body, as the scale of the model or the velocity of the wind is changed. Since the principal component of the wind force is due to the pressures, the force $F$ acting on the surface $S$ is

$$F \approx p_{avg}S = (p/q)_{avg}qS$$

and denoting $(p/q)_{avg}$ by a normal force coefficient or **shape factor** $C_N$

$$F = C_N \tfrac{1}{2}\rho V^2 S = C_N qS$$

The shape factors so obtained apply to full scale for structures with sharp edges whose principal resistance is due to the pressure forces. For bodies that do not have any sharp edges perpendicular to the flow, such as spheres or streamlined bodies, the factor $C_N$ is not constant. It depends upon the Reynolds number (see Sec. 3). For such bodies, the law for variation of the shape factor $C_N$ must be determined experimentally before safe predictions of full-scale forces can be made from model measurements.

### Experimental Results

**Flat Plates in Free Air Perpendicular to the Flow**  The force $N$ normal to the plate depends upon the aspect ratio $A$ = (length)/(width) of the plate. Writing $N = C_{Nq}S$, the coefficient $C_N$ varies from about 1.18 to 2 as shown in Fig. 2. About 70 percent of normal force on the plate is due to the large underpressures existing over the rear surface.

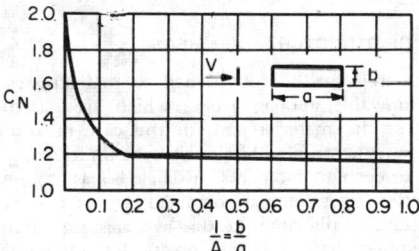

**Fig. 2**  Variation of "shape factor" with aspect ratio for rectangular plates perpendicular to the flow.

**Flat Plates in Free Air Inclined to the Flow**  The force $N$ normal to the plate depends both upon the inclination $\alpha$ of the air stream with respect to the plate and upon the aspect ratio $A = a/b$, as shown in Fig. 3.

**Inclined Roofs**  The largest normal forces occur on the windward side; and for relatively small inclinations they are suctions. Flachsbart gives the approximate curve of Fig. 4 for the ratio of the normal force coefficient $C_N$ of a roof inclined at an angle $\alpha$ to that inclined at 90° which was shown in Fig. 2. This curve was obtained from integrations of pressure distributions on roofs of aspect ratios 1:6. It is seen that this curve is in marked contrast with the building-code requirements which give a variation proportional to $\sin \alpha$ or $\sin^2 \alpha$.

**Forces on Closed Buildings**  Flachsbart (loc. cit.) gives measurements of pressure distributions on simplified building models as shown in Fig. 5. They show positive pressure only on the windward perpendicular faces. Suction pressures prevail over the side, top, and rear.

Figure 5a represents a building with a gabled roof at an

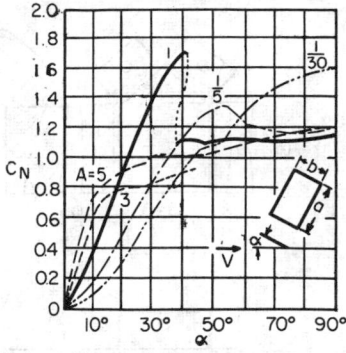

**Fig. 3**  Variation of "shape factor" with the inclination of the air stream for rectangular plates of various aspect ratios.

angle of 40° to the horizontal. The pressure distributions are plotted for cross sections I, II, and III for three different wind directions. When the wind blows against the end of the building ($\beta = 0$), the wind pressures are negative on the sides and on the roof, being largest at section I and decreasing toward II and III. The suction pressures at I are almost equal to the dynamic pressure $q$, i.e., $p/q \approx -1$. Suction pressures are indicated by curves drawn outside the sections, positive pressures by curves inside. When the wind blows at an angle of 30°, the maximum suction of $1.4q$ is found on the roof of section I. Even when the wind blows directly against the side of the model ($\beta = 90°$), the pressure on the roof is partly negative on the windward side and strongly negative on the lee side. Figure 5b represents a rectangular building under similar conditions. Again the suction forces are greater than the positive pressures, reaching a maximum of $p = -1.7q$ on the roof of the building when $\beta = 45°$. These high suction pressures explain the lifting off of roofs and similar phenomena.

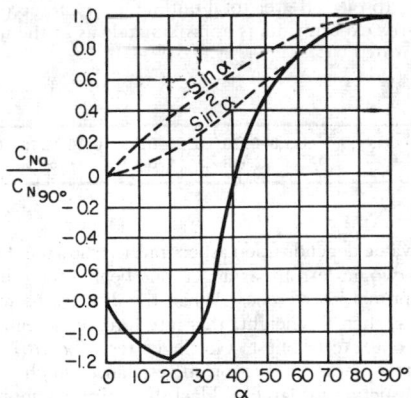

**Fig. 4**

Dryden (loc. cit.) gives pressure distributions and force measurements on a model of the Empire State Building. He finds a normal force coefficient $C_N = 1.5$.

Nökkentved (op. cit.) gives a very comprehensive series of pressure distributions on many different shapes of building models. He points out the fact that the internal pressure of a building is approximately an average of the external wind pressures. This is generally different from the external atmo-

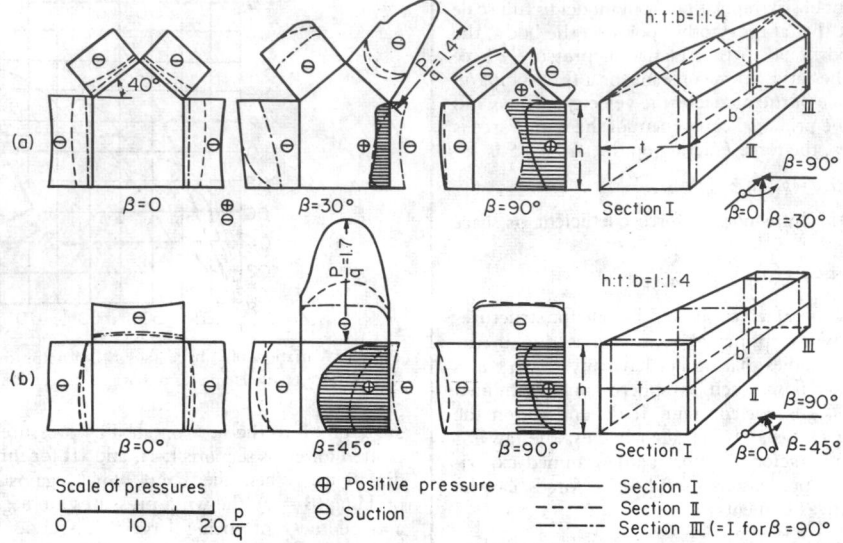

**Fig. 5**  Pressure distribution on two building models.

spheric pressure. For a building model with uniformly distributed holes, an internal underpressure of $-0.3q$ existed. For an actual building, they would probably not be quite so large, unless there are large openings. This underpressure will thus serve to increase the pressure differences on the windward face of the building and to decrease it on the other faces.

**Bridge Girders, Radio Towers, and Other Framed Structures**  For a bridge girder or a plane framework consisting of angles, I beams, and channels, the normal force coefficient $C_N$ is based on the projected area $S$ of the bridge members, i.e., $C_N = F/qS$. The value of $C_N$ depends principally upon the ratio $\varphi$ = (projected area/total outline area) and according to Flachsbart's experiments is approximately as in the following table:

| $\varphi$ | 0–0.20 | 0.20–0.30 | 0.30–0.90 | 0.90–1.0 |
|---|---|---|---|---|
| $C_N$ | 2.0 | 1.8 | 1.6 | 1.6–2.0 |

The last value depends upon aspect ratio as shown in Fig. 2.

When two girders are arranged one behind the other, they cause a mutual interference. When the distance between girders is less than the height, this interference is negligible on the front one. According to Göttingen tests *(loc. cit.)*, the rear girder has a marked reduction in resistance which may even become a suction for large $\varphi$. Flachsbart gives an approximate value for the case of girders spaced a distance equal to their height $C_{Nrear} = C_{Nfront} \cdot (1 - \varphi)^2$. If the rear girder is in the wind shadow, such that the members of the rear are behind the holes of the front one, this will have to be multiplied by about 1.2.

From these formulas, the wind forces on bridges or radio towers may be estimated. The maximum force on a radio tower occurs with the wind blowing at some angle to the side surfaces, generally about 45°. Streamlining of the framework is not possible, since the wind may blow from any direction.

By using round structural members instead of flat and angular sections, a substantial reduction in wind force can be effected.

For **cylinders, spheres, streamline bodies,** and **simple geometrical shapes,** see Sec. 11.

## DESIGN OF STRUCTURAL MEMBERS

Members are usually proportioned so that stresses do not exceed allowable **working stresses** which are based on the strength of the material and, in the case of compressive stresses, on the stiffness of the element under compression. Internal forces and moments in **simple beams,** columns, and pin-connected truss bars are obtained by means of the equations of static equilibrium. **Continuous beams,** rigid frames, and other members characterized by practically rigid joints require for analysis additional equations derived from consideration of deflections and rotations.

Design may also be on the basis of the **ultimate strength** of members, the factor of safety being embodied in stipulated increases in the design loads. In steel-frame construction, the procedures of **plastic design** determine points where the material may be allowed to yield, forming *plastic hinges,* and the resulting redistribution of internal forces permits a more efficient use of the material.

### Floors and Roofs

**Flat Framing**  Except in reinforced-concrete flat-slab construction, floors and roofs generally consist of flat decks supported upon beams, girders, or trusses. The decks may usually be considered a series of beamlike strips spanning between beams and themselves designed as beams. The design of a beam consists chiefly in proportioning its cross section to resist the maximum bending and shear and providing adequate connections at its supports, without exceeding the unit stresses allowed in the materials used (see Sec. 5).

Where the span of floor or roof exceeds 20 to 30 ft (6.1 to 9.1 m), **trusses** of some form may be more economical than beams of uniform section. For spans above 50 to 70 ft (15.2 to 21.3 m), trusses are usually economical. Between these limits, the line of economy is not definitely marked, but the conditions that favor the use of trusses are as follows: (1) identical trusses are repeated many times, (2) the height of the building need not be increased for the greater depth of the truss, (3) fire protection of wood or metal is not required.

**Roof trusses** usually have their top chords sloped with the roof. Common trusses for steeply pitched roofs are shown in Figs. 6 to 10, the top chord panels equal in each truss. The members shown by heavy lines are in compression under ordinary loads, those in light lines in tension. The trusses of Figs. 6 to 9 are adapted for either steel or wood. In wooden trusses, the tensile web members may be steel rods with plates, nuts, and threaded ends. The truss of Fig. 10 is usually made of steel.

The stresses in any member of these trusses under a vertical load uniformly distributed may be found by multiplying the coefficients in Tables 1 to 5 by the panel load $P$ on the truss. For other slopes, types of trusses, or loads, see below. Trusses for flat roofs are commonly of one of the types shown in Figs. 11 to 17 except that the top chords conform to the slope of the roof.

**Floor trusses** normally have parallel chords. Common types are shown in Figs. 11 to 17 in which heavy lines indicate members in compression, light lines in tension, and dash lines members with only nominal stress, under equal vertical panel loads. The panel lengths $l$ in each truss are equal. The stress in each member is written next to the member in the figure, in terms of the panel load and the lengths of members. For a truss like one of the figures turned upside down, the stresses in the chords and diagonals remain the same in magnitude but reversed in sign. Stresses in verticals must be computed (compare Figs. 11 and 12). For other loads and other types of trusses, see below.

**Weights of Trusses** The approximate weight in pounds of a wooden truss may be taken as $W = LS(L/25 + L^2/6{,}000)$, where $L$ is the span and $S$ the spacing of trusses in feet. The approximate weight in pounds of a steel roof truss may be taken as $W = \frac{1}{5}LS(\sqrt{L} + \frac{1}{8}L)$. For loads greater than 40 lb/ft², multiply by the load and divide by 40.

**Choice of Roof Trusses** WOODEN TRUSSES. For pitched roofs with spans up to 20 ft (6.1 m) the simple king-post truss, inversion of the trussed beam (Fig. 6), may be used. For spans up to 40 ft (60 ft) (12.2 to 18.3 m), the trusses of Figs. 6 and 8 (Figs. 7 and 9) are good. The number of panels rarely exceeds eight or the panel length 10 ft (3.0 m). For flat roofs, the Howe truss (Figs. 11, 14, and 16) is built of wood with steel rods for verticals; the depth is one-eighth to one-twelfth of the span. Wooden trusses are usually spaced 10 to 15 ft (3.0 to 4.6 m). Wood is rarely used in roof trusses over 60 ft (18.3 m) in span. Steel trusses for pitched roofs may well take the form of Figs. 7 to 10 for spans up to 100 ft (30.5 m). For flat roofs, the Warren truss (Figs. 13, 15, and 17) is common in steel; the depth of the truss ranges from one-eighth to one-twelfth of the span, the spacing from 15 to 25 ft (4.6 to 7.6 m).

### Stresses in Trusses

An ideal truss is a framework consisting of straight bars or members connected at their ends by frictionless ball-and-socket joints. The external forces are applied only at these ball-and-socket joints. Internal forces or stresses in such straight bars are axial, either tension or compression, without

bending. Since frictionless ball-and-socket joints are impossible, and the ends of bars are often riveted or welded, the ideal truss is never realized. For purposes of analysis, the primary stresses, which are always axial, are determined on the

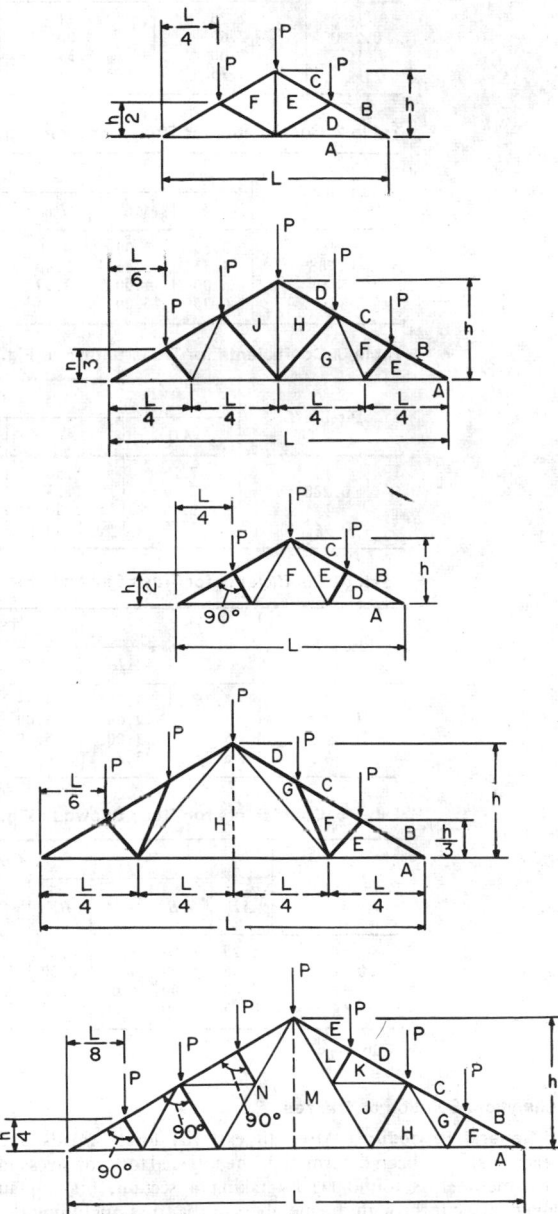

**Figs. 6 to 10** Types of steep-roof trusses.

assumption that the truss under consideration conforms to the ideal. Secondary stresses are additional stresses, generally flexural or bending, brought about by all the factors that make the actual truss different from the ideal. In the following discussion, primary stresses will be considered.

**Table 1. Coefficients for Truss Shown in Fig. 6**

| Pitch $h/L$ | Coefficients of $P$ for stress in | | | | |
|---|---|---|---|---|---|
| | AD | BD | CE | DE | EF |
| 1/8 | 2.25 | 2.71 | 1.80 | 0.90 | 1.00 |
| 0.288* | 2.60 | 3.00 | 2.00 | 1.00 | 1.00 |
| 1/4 | 3.00 | 3.35 | 2.24 | 1.12 | 1.00 |
| 1/3 | 3.75 | 4.03 | 2.69 | 1.35 | 1.00 |

**Table 2. Coefficients for Truss Shown in Fig. 7**

| Pitch $h/L$ | Coefficients of $P$ for stress in | | | | | | | | |
|---|---|---|---|---|---|---|---|---|---|
| | AE | AG | BE | CF | DH | EF | FG | GH | HJ |
| 1/8 | 3.75 | 3.00 | 4.50 | 3.90 | 3.60 | 0.83 | 0.72 | 1.25 | 2.00 |
| 0.288* | 4.33 | 3.46 | 5.00 | 4.33 | 4.00 | 0.88 | 0.73 | 1.32 | 2.00 |
| 1/4 | 5.00 | 4.00 | 5.59 | 4.84 | 4.47 | 0.94 | 0.75 | 1.42 | 2.00 |
| 1/3 | 6.25 | 5.00 | 6.74 | 5.84 | 5.39 | 1.07 | 0.79 | 1.60 | 2.00 |

**Table 3. Coefficients for Truss Shown in Fig. 8**

| Pitch $h/L$ | Coefficients of $P$ for stress in | | | | | |
|---|---|---|---|---|---|---|
| | AD | AF | BD | CE | DE | EF |
| 1/8 | 2.25 | 1.50 | 2.70 | 2.15 | 0.83 | 0.75 |
| 0.288* | 2.60 | 1.73 | 3.00 | 2.50 | 0.87 | 0.87 |
| 1/4 | 3.00 | 2.00 | 3.35 | 2.91 | 0.89 | 1.00 |
| 1/3 | 3.75 | 2.50 | 4.04 | 3.67 | 0.93 | 1.25 |

**Table 4. Coefficients for Truss Shown in Fig. 9**

| Pitch, $h/L$ | Coefficients of $P$ for stress in | | | | | | | |
|---|---|---|---|---|---|---|---|---|
| | AE | AH | BE | CF | DG | EF | FG | GH |
| 1/8 | 3.75 | 2.25 | 4.51 | 3.91 | 4.51 | 0.83 | 1.42 | 2.50 |
| 0.288* | 4.33 | 2.60 | 5.00 | 4.33 | 5.00 | 0.88 | 1.45 | 2.65 |
| 1/4 | 5.00 | 3.00 | 5.59 | 4.84 | 5.59 | 0.94 | 1.49 | 2.83 |
| 1/3 | 6.25 | 3.75 | 6.73 | 5.83 | 6.73 | 1.07 | 1.57 | 3.20 |

**Table 5. Coefficients for Truss Shown in Fig. 10**

| Pitch $h/L$ | Coefficients of $P$ for stress in | | | | | | | FG KL | GH JK | HJ | JM | LM |
|---|---|---|---|---|---|---|---|---|---|---|---|---|
| | AF | AH | AM | BF | CG | DK | EL | | | | | |
| 1/8 | 5.25 | 4.50 | 3.00 | 6.31 | 5.75 | 5.20 | 4.65 | 0.83 | 0.75 | 1.66 | 1.50 | 2.25 |
| 0.288* | 6.06 | 5.20 | 3.46 | 7.00 | 6.50 | 6.00 | 5.50 | 0.87 | 0.87 | 1.73 | 1.73 | 2.60 |
| 1/4 | 7.00 | 6.00 | 4.00 | 7.83 | 7.38 | 6.93 | 6.48 | 0.89 | 1.00 | 1.79 | 2.00 | 3.00 |
| 1/3 | 8.75 | 7.50 | 5.00 | 9.42 | 9.05 | 8.68 | 8.31 | 0.93 | 1.25 | 1.86 | 2.50 | 3.75 |

*30° slope.

## Analytical Solution of Trusses

**General Procedure**  After all external forces (loads and reactions) have been determined, the internal force or stress in any member is found (1) by taking a section, making an imaginary cut through the members of the truss, including the one whose stress is to be found, so as to separate the truss into two parts; (2) by isolating either of these parts; (3) by replacing each bar cut by a force, representing the stress in the bar or the force required to come to the part of the truss isolated from the other part removed; and (4) by applying the equations of statics to the part isolated.

The various ways in which sections can be taken and the equations used to determine the stresses are illustrated by a solution of the truss (Fig. 18).

A section 1–1 (Fig. 18) may be taken around a joint, $L_0$. Isolate the forces inside the section (Fig. 19a). Assume the unknown stresses to be tension. Since the forces are concurrent and coplanar, two independent equations of statics will establish equilibrium. These may be either $\Sigma x = 0$ and $\Sigma y = 0$ or $\Sigma M = 0$ taken about two axes perpendicular to the plane of the forces and passing through two points selected so that neither point is the intersection of the forces, or so that the line

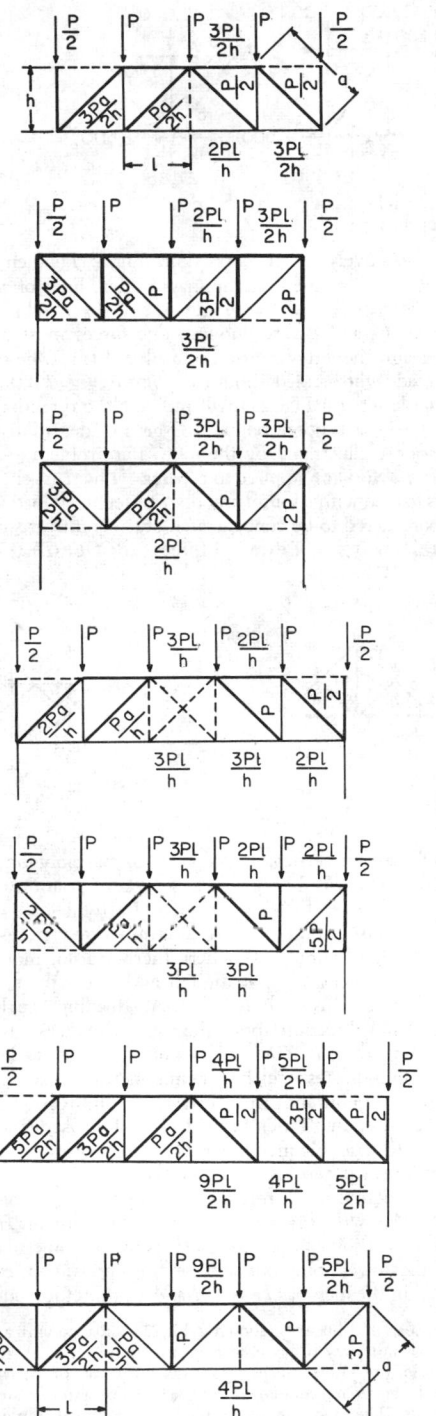

**Figs. 11 to 17** Types of floor trusses.

joining the two points is not coincident with either of the unknown forces.

Using the first set of equations and taking components of all forces along horizontal and vertical axes; e.g., the horizontal

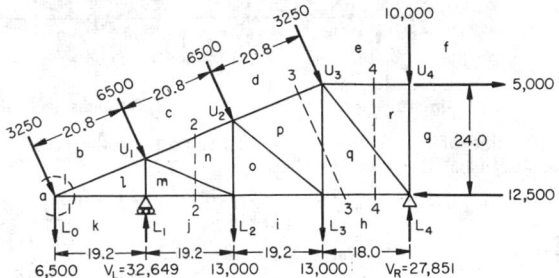

**Fig. 18**

component of the 3,250 lb force is 1,250 and the vertical component is 3,000 lb.

$$\Sigma x = 0 = s_{lk} + s_{bl}(19.2/20.8) + 1,250$$
$$\Sigma y = 0 = s_{bl}(8/20.8) - 9,500$$

From these equations, $s_{lk} = -24,050$ and $s_{bl} = 24,700$.

The minus sign indicates that the force acts opposite to the assumed direction. If all the unknown forces are assumed to be in tension, then a plus sign in the result indicates that the stress is tension and a minus sign indicates compression.

Hence, $s_{lk}$ is 24,050 lb compression and $s_{bl}$ is 24,700 lb tension.

Using the $\Sigma M = 0$ twice,

$$\Sigma M \text{ about } U_1 = 0 = -s_{lk} \times 8 - 1250 \times 8 - 9,500 \times 19.2$$

from which $S_{lk} = 24,050$ lb compression.

$$\Sigma M \text{ about } L_1 = 0 = s_{bl}(8/20.8)19.2 - 9,500 \times 19.2$$

from which $s_{bl} = 24,700$ tension.

Instead of taking a section around a joint, a cut may be made vertically or inclined, cutting a number of bars such as Fig. 19b or Fig. 19c. If only three members are cut, and they are neither concurrent nor parallel, the stresses can be found by taking moments of all forces on either side of the section about axes passing through the intersections of any two members.

Considering the part to the left of section 2–2, the stresses in the three members (Fig. 19b) may be determined by taking moments about $L_0$, $U_1$, and $L_2$ of all forces acting on the part on either side of the section, e.g., on the left of the section because it has the fewer forces:

$$\Sigma M \text{ about } L_0 = s_{mn}(8/20.8)38.4 + 2,500 \times 8 - (32,649 - 6,000)19.2$$
$$\Sigma M \text{ about } U_1 = 0 = -s_{mj} \times 8 - 1,250 \times 8 - 9,500 \times 19.2$$
$$\Sigma M \text{ about } L_2 = 0 = s_{cn}(19.2/20.8)16 + 2,500 \times 8 - 9,500 \times 38.4 + 26,649 \times 19.2$$

from which $s_{mn} = 33,290$ tension, $s_{mj} = 24,050$ compression, and $s_{cn} = 11,298$ compression.

This method is sometimes called the "method of moments."

Considering the part to the right of section 3–3, the stresses in the three members (Fig. 19c) may be determined by taking moments about $L_0$, $U_3$, $L_3$ of all the forces acting on part or

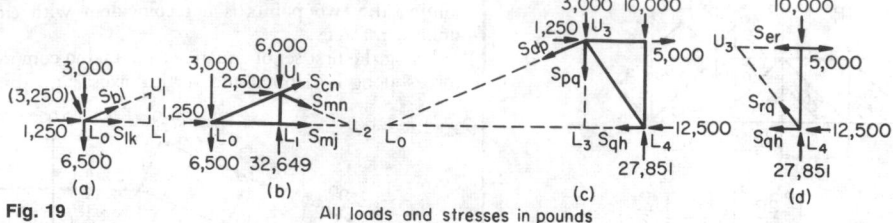

**Fig. 19**                All loads and stresses in pounds

either side of the section, e.g., on the right of the section because it has the fewer forces:

$$\Sigma M \text{ about } L_0 = 0 = s_{pq} \times 57.6 + 6{,}250 \times 24 + 3{,}000$$
$$\times 57.6 - (27{,}851 - 10{,}000) \times 75.6$$
$$\Sigma M \text{ about } L_3 = 0 = - s_{dp}(19.2/20.8)24 + 6{,}250 \times 24$$
$$- 17{,}851 \times 18$$
$$\Sigma M \text{ about } U_3 = 0 = s_{qh} \times 24 + 12{,}500 \times 24 - 17{,}851$$
$$\times 18$$

from which $s_{pq} = 17{,}825$ tension, $s_{dp} = 7{,}733$ compression, and $s_{qh} = 888$ tension.

Considering the part to the right of section 4–4, the stresses in the three members (Fig. 19d) may be found by taking moments about axes where two unknowns intersect, e.g., about $U_3$ and $L_4$. Since two unknown stresses are parallel, their lines of action do not intersect. However, the equation $\Sigma y = 0$ will enable one to find the stress in the member which is not parallel to the other two:

$$\Sigma M \text{ about } U_3 = 0 = s_{qh} \times 24 - (27{,}851 - 10{,}000)18$$
$$+ 12{,}500 \times 24$$
$$\Sigma M \text{ about } L_4 = 0 = -s_{er} \times 24 + 5{,}000 \times 24$$
$$\Sigma y = 0 = s_{rq}(24/30) + 27{,}851 - 10{,}000$$

from which $s_{qh} = 888$ tension, $s_{er} = 5{,}000$ tension, and $s_{rq} = 22{,}314$ compression.

### Graphical Solution of Trusses

When the external forces acting upon a truss are all in the same plane and may be assumed to act at the joints of the structure, it is often convenient to make use of graphical considerations to determine the stresses in the members. A series of stress polygons may be constructed for the forces acting at the different joints, using the methods shown in Sec. 3, but a combination diagram, using what is known as "Bow's notation," greatly simplifies the work.

The forces $F_1$, $F_2$, $F_3$, and $F_4$ (Fig. 20) are all in the same plane and form a system in equilibrium. Their lines of action all pass through the same point (o). Letters are so placed as to bring each force between two letters, these letters being usually read in a right-handed or clockwise direction around the joint. $F_1$, $F_2$, $F_3$, $F_4$ will then be designated $ab$, $bc$, $cd$, and

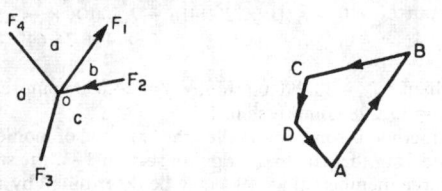

**Fig. 20**                **Fig. 21**

$da$, respectively. A stress polygon (Fig. 21) is then drawn for the forces. Assume that the magnitudes, lines of action, and directions of $F_1$ and $F_2$ are known and that the lines of action only of $F_3$ and $F_4$ are known. The forces must be taken in order and the letters must be so placed that when the forces are read right-handed about the point $o$ (Fig. 20) the sequence of the letters (in Fig. 21) will indicate the direction in which the forces act upon point $o$. The manner of using Bow's notation is illustrated by the following problem.

**Bow's Notation Applied to a Truss**   The truss shown in Fig. 22 is loaded with a uniformly distributed dead load which may be considered to be concentrated at the joints as shown. First plot the polygon of external forces $ABCDEFGA$ (Fig. 23). As

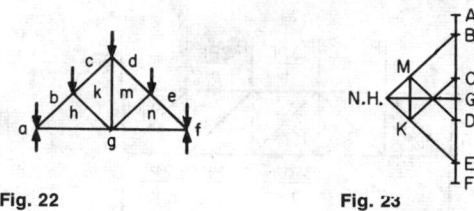

**Fig. 22**                **Fig. 23**

the forces act vertically, the sides of the polygon fall in the straight line $AF$. The supporting forces $GA$ and $FG$ are equal in this case, each being one-half the total load, so that no special construction for their determination is necessary. Start at any joint in the truss where there are not more than two unknown forces, e.g., at the left end of the truss. The stress polygon for this joint is $ABHGA$. Reading the letters in a right-hand direction about the joint, the stress in the upper chord member is $BH$. This sequence of letters in the stress polygon indicates that this member acts downward and to the left on the joint and, therefore, is in compression. The stress in the lower chord member is $HG$, and this sequence of letters indicates that this member acts to the right on the joint and therefore is in tension.

At the joint in the middle of the upper chord there are now two unknown stresses. Draw the stress polygon $HBCKH$ for the second joint. The stress $HB$ is of the same magnitude as for the lower joint but acts in the opposite direction. The stress in the member $kh$ is $KH$ and is, therefore, compression.

EXAMPLE.   The truss shown in Fig. 24 is loaded with a dead load of 21,760 uniformly distributed over the upper chord, together with a wind load of 13,600 on the right side. Both ends of the truss are fixed and the horizontal components of the supporting forces are assumed to be equal. The supporting forces may be determined graphically by first assuming a roller under one end of the truss, or by the funicular polygon construction (see Sec. 3). They are often calculated and the polygon of external forces plotted from the results of the calculation.

This is a desirable method, as it offers a check on this most important part of the graphical work.

The horizontal component of the wind load is (16/34)13,600 = 6,400. The $H$ component of each supporting force is assumed to be 6,400/2 = 3,200 lb. The vertical component of the wind load is (30/34)13,600 = 12,000 lb. Taking moments about the right end of the truss, $(21,760 \times 30) + (13,600 \times 17) = 60V_1$. $\therefore V_1 = 14,733$; $V_2 = 21,760 + 12,000 - 14,733 = 19,027$.

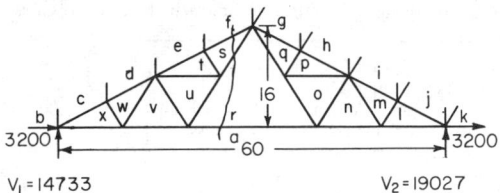

**Fig. 24**

The polygon of external forces can now be constructed, as in Fig. 25. The dotted part of the diagram is the combination of the dead and wind loads, assuming that they are each concentrated. The dotted line $BK$ is the resultant of these loads. The supporting forces $KA$ and $AB$ are determined by plotting to scale their horizontal and vertical components as calculated. The polygon of external forces for the truss is $BCDEFGHIJKAB$, and must check with the polygon shown dotted. The forces $GH$, $HI$, $IJ$, and $JK$ are the resultants of the forces acting at the joints on the right side of the truss.

When supporting forces are to be determined, the loads *may* be concentrated at the lines of action of their resultants, but when internal forces are to be determined, the loads *must* be distributed at the various joints. Start with some joint of the truss where there are only two unknown forces and draw the stress diagram, Fig. 25. The magnitudes of the stresses in the different members are determined directly from the lengths of the corresponding lines and the scale used in the construction of the diagram. The nature of the stress (tension or compression) is determined by the use of Bow's notation.

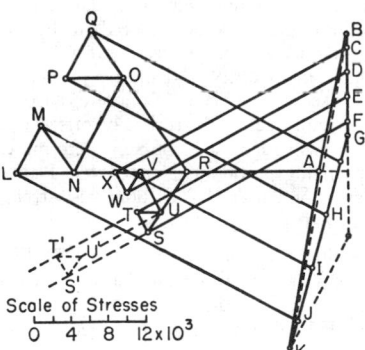

**Fig. 25**

A difficulty often arises that is illustrated by the truss of Fig. 24. It is impossible by the usual graphical procedure to complete the stress diagram, as it will be found that, after obtaining the stresses in the members meeting at the left end and at the next joint on both upper and lower chords, no joint with less than three unknown forces is available. To overcome this difficulty, some unknown stress may be calculated and placed on the diagram before proceeding with the graphical solution.

The stress which should be calculated in this case is that in the middle member of the lower chord. Taking moments about a point at the middle of the upper chord, $16RA = (14,733 \times 30) - (3,200 \times 16) - [2,720(22\frac{1}{2} + 15 + 7\frac{1}{2})] - (1,360 \times 30) = 227,600$. $\therefore RA = 14,225$ (tension). Place this stress in the diagram and proceed.

The solution may be entirely graphical by noting that $T$ must be on a line through $E$ parallel at $et$ and $S$ must be on a line through $f$ parallel to $fs$ and also that $TS$ must be parallel to $ts$. Hence the geometrical relationship may be indicated by any points $T'$ and $S'$ on the line through $E$ and $F$ as long as $T'S'$ is parallel to $ts$. Furthermore, $TU$ must be parallel to $tu$ and $SR$ parallel to $sr$. If $T'$ and $S'$ are arbitrarily selected as shown in dotted lines, then $U'$ is fixed. However, $U$ must lie on a line through $V$ parallel to $UV$. All conditions can be fulfilled by moving the triangle $T'S'U'$ so that the sides move parallel until $U'$ comes on the line through $V$ parallel to $uv$. This point will then be $U$, and $T$ and $S$ will be determined.

**Columns and Walls**

**Vertical elements** in building construction consist of columns, posts, or pilasters which transmit concentrated loads, walls or partitions which transmit linear loads from story to story, and rigid frames which transmit lateral as well as vertical loads.

**Columns** may be of timber, steel, or reinforced concrete and should be proportioned for the allowable stresses permitted for the material used and for the flexibility of the column. Care must be taken in the framing of beams and girders to avoid or to provide for the additional stresses due to connections which transfer loads to columns with large eccentricities. For instance, a beam framing to a flange of a steel column, with a seat or web connection, will produce an eccentric moment in the column equal to $Rd/2$, where $R$ is the beam reaction and $d$ the depth of the column section. Columns not adequately restrained against deflection of the top may be subject to considerable additional moment because of the resulting eccentricity of otherwise axial loads.

**Rigid frames** consist of columns and beams welded, riveted, bolted, or otherwise connected so as to produce continuity at the joints and permit the entire frame to behave as a unit. Advantages of rigid frames are the ease and simplicity of erection, increased headroom, and the resultant savings in wall heights. Rigid frames composed of rolled sections are commonly used for spans up to 100 ft (30.5 m) in length. Built-up members have been utilized on spans to 250 ft (76.2 m). Welded fabrication offers particular advantages for frames utilizing variable-depth members and on parabolic-shape roofs. The distribution of moments in the statically indeterminate rigid frame is effected by the relationship of the column height to span and roof rise to column height, as well as the relative stiffness of the various members. The solution for the moments in the frame is obtained from the usual equations of statics plus one or more additional equations pertaining to the elastic deformations of the frame under load.

**Walls** are designed as vertical elements of unit width transmitting vertical loads from story to story. Wind, earthquake, and other lateral loads are resisted by the walls spanning as a beam vertically between floors, horizontally between columns or pilasters, or as a two-way slab spanning in each direction.

**Stud walls** consist of wooden studs with one or more lines of horizontal bridging. The allowable vertical load on the wall is a function of the maximum permissible load on each stud as a column and the spacing of the studs (see Table 10).

**Corrugated or flat sheet steel, aluminum, or asbestos-cement sheets** are commonly employed for walls of industrial or mill buildings. These sheets are usually supported on steel girts framing horizontally between columns and supported from heavier eave struts by one or more lines of vertical steel sag rods.

**Reinforced-concrete or masonry walls** should be designed so that the allowable bending and/or axial stresses are not exceeded, but the minimum thicknesses of such walls should not be less than the following:

| Material | Max ratio, unsupported height or length to thickness | Nominal min thickness, in* |
|---|---|---|
| Reinforced concrete | 25 | 6 |
| Plain concrete | 22 | 7 |
| Reinforced brick masonry | 25 | 6 |
| Grouted brick masonry | 20 | 6 |
| Plain solid masonry | 20 | 8 |
| Hollow-unit masonry | 18 | 8 |
| Stone masonry (ashlar) | 14 | 16 |
| Interior nonbearing concrete or masonry (reinforced) | 48 | 2 |
| Interior nonbearing concrete or masonry (unreinforced) | 36 | 2 |

*× 25.4 = mm.

## Foundations

**Bearing Pressure of Soils**  The bearing pressure which may be allowed on soil may vary over a large range. For important structures, the nature of the underlying soil should be ascertained by borings or test pits. If the soil consists of medium or soft clay, a settlement analysis based on consolidation tests of undisturbed soil samples from the foundation strata is necessary. Structures founded upon mud, silt, peat, or artificial filling will almost certainly settle, and no foundation for a permanent structure should rest on or above such material without adequate provision for the resulting settlement. Table 6 gives a general classification of soils and the safe pressures which they may support.

**Table 6. Safe Bearing of Soils**

| Nature of soil | Safe bearing capacity tons/ft² | 1000 kg/m² |
|---|---|---|
| Solid ledge of hard rock, such as granite, trap, etc. | 25–100 | 245–975 |
| Sound shale and other medium rock, requiring blasting for removal | 10–15 | 98–146 |
| Hardpan, cemented sand, and gravel, difficult to remove by picking | 8–10 | 78–98 |
| Soft rock, disintegrated ledge; in natural ledge, difficult to remove by picking | 5–10 | 49–98 |
| Compact sand and gravel, requiring picking for removal | 4–6 | 39–59 |
| Hard clay, requiring picking for removal | 4–5 | 39–49 |
| Gravel, coarse sand, in natural thick beds | 4–5 | 39–49 |
| Loose, medium, and coarse sand; fine compact sand | 1.5–4 | 15–39 |
| Medium clay, stiff but capable of being spaded | 2–4 | 20–39 |
| Fine loose sand | 1–2 | 10–20 |
| Soft clay | 1 | 10 |

These values approximate the pressures allowed by the building law in most cities. Tests, when the cost thereof is warranted by the magnitude of the project, may show higher values to be safe. The foundation for a building housing heavy vibrating machinery such as steam hammers, heavy punches, and shears should receive some allowance for possible compression and rearrangement of soil due to the vibrations transmitted through it. The foundation for a tall chimney should be designed with a comparatively low pressure upon the soil, because of the disastrous results which might occur from local settlement.

**Footings**  The purpose of footings is to spread the concentrated loads of building walls and columns over an area of soil so that the unit pressure will come within allowable limits. Footings are usually constructed of concrete. Brick is occasionally used but is generally more expensive than concrete. Where the depth must be kept shallow, or for other reasons, a grillage of steel beams may be used. Under walls, a single layer of beams may be employed; under columns, two crossed layers with the upper layer no wider than the column base. The steel is embedded in concrete for protection. Stone, where available in quantity and of the proper quality, can sometimes be used economically. Concrete is used either plain or in mass, or reinforced with steel. The offset to each side of the footing for each successive course is determined by figuring the overhanging portion as a cantilever. Plain footings of 1:2½:5 concrete or better (1,800 lb/in² test strength) should have a thickness at least equal to the projection multiplied by $\sqrt{T}$, where $T$ is the soil pressure, tons/ft². The center of pressure in the wall or column should always pass through the center of the footing. Footings in ground exposed to freezing should be carried below the possible penetration of frost.

**Foundations** are ordinarily placed in open excavation, with or without sheeting and bracing. When a suitable bearing soil is more than 10 or 15 ft (3.0 to 4.6 m) below the surface, other methods are often desirable. Where the bearing soil is clay stiff enough to stand with undercutting, and the material immediately above it is peat or silt, the **open-caisson method** may be economical. In this method, steel cylinders 3 ft and more in diameter are sunk as excavation proceeds, the cylinders having successively smaller diameters. At the bottom of the shaft thus formed, the soil is undercut to obtain sufficient bearing area. The shaft and the enlargement at the base are then filled with concrete, the cylinders being withdrawn as the concrete is placed. The open-caisson method cannot be used where groundwater flows too freely into the excavation. Where large foundations under very heavy buildings must be carried to great depth to reach rock or hardpan, particularly where groundwater flows freely, the **pneumatic-caisson method** is used.

**Drilled-in piers** are formed by drilling with special power augers up to 5 ft (1.5 m) diameter or greater. The holes are drilled to the desired bearing level with or without metal casings, depending on the soil. Belling of the bottom may also be performed mechanically from the surface. In poor soils, or where groundwater is present, the hole may be retained with bentonite clay slurry which is displaced as concrete is placed in the caisson by the **tremie method.**

**Pile Foundations**  Where satisfactory bearing soil is not to be had at a reasonable depth, and the open-caisson method is for any reason impracticable, piles are often used. **Wood piles** are permanent when continually submerged in water. (Marine borers may attack wood in seawater above the mud line.) A subsequent lowering of the groundwater level, however, may be disastrous, because wood will rot when damp. **Concrete piles** are less destructible, and hence are adaptable to many conditions. Wood piles are usually designed to carry 15 to 20 tons (13,600 to 18,100 kg) per pile. They should be straight and not less than 6 in in diam under the bark at the tip. With a concrete pile, 25 to 60 tons (22,700 to 54,400 kg) or more are carried. Structural **steel H** columns and steel pipe with still higher capacities have been driven for piles where corrosion was not to be feared.

**Methods of Driving Piles**  The drop hammer and the steam hammer are usually employed in driving piles. The steam hammer, with its comparatively light blows delivered in rapid succession, is of advantage in a plastic soil, the speed with which the blows are delivered preventing the readjustment of the soil. It is also of advantage in soft soils where the driving is easy, but a light hammer may fail to drive a heavy pile satisfactorily. The water jet is sometimes used in sandy soils. Water supplied under pressure at the point of the pile through a pipe or hose run alongside it erodes the soil, allowing the pile to settle into place. To have full capacity, jetted piles should be driven after jetting is terminated.

**Determination of Safe Loads for Piles**  Piles may obtain their supporting power from friction on the sides or from bearing at the point. In the latter case, the bearing power may be limited by the strength of the pile, considered as a column, to which, however, the surrounding soil affords some lateral support. In the former case, no precise determination of the bearing power can be made. Many formulas have been developed for determining the safe bearing power in terms of the weight of the hammer, the fall, and the penetration of the pile per blow, the most generally accepted of which is that known as the *Engineering News* formula: $R = 2wh/(s + 1.0)$ for drop hammers, $R = 2wh/(s + 0.1)$ for single-acting steam hammers, $R = 2E/(s + 0.1)$ for double-acting steam hammers, where $R$ = safe load, lb; $w$ = weight of hammer, lb; $h$ = fall of hammer, ft; $s$ = penetration of last blow, in; $E$ = energy, ft·lb per blow transmitted to the pile. A factor of safety of 6 is assumed. This formula and similar ones are based on the determination of the energy in the falling hammer, and from this the pressure which it must exert on the top of the pile. It is a wise practice to drive test piles and determine their bearing value before proceeding with the final designs of important structures to be supported on piles. The designs then are

based upon the safe bearing power so determined, and the piles are driven to the penetration which the above formula, with constants modified to fit the results of the tested piles, shows necessary to give the bearing power for which they were designed.

**Concrete piles** may be divided into two classes: (1) those molded in place and (2) those precast, cured, and driven. Piles of both classes, longer than 100 ft (30.5 m), have been driven. Piles of the first class are made by driving a mandrel into the ground and filling the hole so formed with concrete. In one well-known pile of this type (Raymond), a thin steel sheet is fitted over a tapered mandrel before driving. This shell, which is left in the ground when the mandrel is withdrawn, is filled with concrete. Another well-known pile (Simplex) of the molded-in-place type uses a hollow cylindrical mandrel which is filled with concrete after having been driven to the desired depth and raised a few feet at a time, the concrete flowing out at the bottom and filling the hole in the earth. Precast and molded-in-place piles may be reinforced with steel. Prestressing of precast concrete piles is a recent innovation giving greater assistance to handling and driving stresses. Jetting is used extensively in placing precast concrete piles, but most are driven into place. Jetting is not often practical or safe in city work because of danger to the foundations of adjoining buildings. Driving by hammer necessitates a cushioned driving head. In underpinning buildings subject to settlement, steel pipes in sections have been jacked down and filled with concrete.

**Spacing of Piles**  Wood piles are preferably spaced not closer than 2½ ft (0.76 m), and concrete piles 3 ft (0.91 m) on centers. If driven closer than this, one pile is liable to force another up. Piles in a group must not cause excessive pressure in soil below their tips. The efficiency, or supporting value of friction piles when driven in groups, by the Converse-Labarre method, is

$$\frac{1 - d/s[(n - 1)m + (m - 1)n]}{90mn}$$

where $d$ = pile diameter, in (cm); $s$ = spacing center to center of piles, in (cm); $m$ = number of rows; $n$ = number of piles in a row.

**Capping of Piles**  Piles are usually capped with concrete; wood piles sometimes with timber. Timber capping to be permanent must be kept below groundwater level. Concrete is the most usual material and the most satisfactory for the reason that it gets a full bearing on all piles. The piles should be embedded 4 to 6 in (10 to 15 cm) in the concrete.

### Retaining Walls

A wall used to sustain the pressure of earth behind it is called a retaining wall. Retaining walls which depend for their stability upon the weight of the masonry are classed as gravity walls. Such walls built on firm soil will usually be stable when they have the following proportions: top of fill level, back vertical, base, 0.4 height; top of fill level, back battered, base, 0.5 height; top of fill steeply inclined, back vertical, base, 0.5 height; top of fill steeply inclined, back battered, base, 0.6 height. An additional factor of safety is obtained by building the face on a batter. Care should be taken in the design of a wall that the allowable soil pressure is not exceeded and that drainage is provided for the back of the wall. The foundations of retaining walls should be placed below the level of frost penetration. Retaining walls of reinforced concrete are made thin, with a broad base, and the wall either cantilevered from the base or braced with buttresses or counterforts.

It is impossible to derive formulas for the earth pressure on the back of the wall which will take account of all the actual conditions. Assuming the earth to be a loose, homogeneous, granular mass, and the coefficient of friction to be independent of the pressure, Rankine deduced the following formula for a wall with vertical back:

$$P = (\tfrac{1}{2}wb^2 + vb)\cos d(\cos d$$
$$- \sqrt{\cos^2 d - \cos^2 a})/(\cos d + \sqrt{\cos^2 d - \cos^2 a})$$

the center of pressure being at a height $\tfrac{1}{3}H(wb + 3v)/(wb + 2v)$ above the base, where $P$ = earth pressure per lin ft (m) of the wall, lb (kg); $b$ = height of the wall, ft (m); $w$ = weight of earth per ft³ (m³), lb (kg); $v$ = weight of superposed load per ft² (m²) of surface, lb (kg); $d$ = the angle with the horizontal of the earth surface behind the wall; and $a$ = angle of repose of the earth, deg. (For values of $a$, see Sec. 10.) The direction of the pressure is parallel to the earth's surface. The retaining wall should have sufficient thickness at the base so that the resultant of the earth pressure $P$ combined with the weight of the wall falls well within the base. If this resultant falls at the outside edge of the middle third, the maximum vertical pressure on the foundation (at the outer edge of the base) will be equal to $2W/T$ lb/ft² (kg/m²), where $W$ is the total vertical pressure on the base of 1 ft (m) length of wall and $T$ the thickness of the wall at the base, ft (m).

In the design of walls of buildings which must withstand earth pressure and low independent walls, where refinement is not necessary, the earth pressure is frequently assumed to be that of a fluid weighing 35 lb/ft³ (561 kg/m³).

## MASONRY CONSTRUCTION

**Brickwork**    The strength and durability of brick masonry

to which may be added slaked-lime paste or dry hydrated lime up to one-half the volume of the cement; for cement-lime mortar, 1 part cement, 1 part lime, and 6 parts sand. Portland-cement mortar should be used for all structural brickwork (see Sec. 6).

**Laying and Bonding**    Brick should be laid in a full bed of mortar and shoved laterally into place to secure solid bearing and a bed of even thickness, and to fill the vertical joints. Brick should be thoroughly wet before laying, except in freezing weather. Brick laid with long dimension parallel to the face of the work are called stretchers, perpendicular to the face, headers. Bats (half-brick) should not be used except where necessary to make corners or to form patterns on the face of the wall. Walls are bonded or tied together longitudinally by overlapping stretchers in successive courses. Transverse bond is obtained by making every sixth course headers, the headers themselves overlapping in successive courses in the interior of thick walls. Variations in the arrangement of headers are often used in the face of walls for appearance. The area of cross section of full-length headers should not be less than one-twelfth the face of the wall, in bonding each pair of transverse courses of brick. Three examples of bond are shown in Fig. 26.

**Arches** over windows and doorways are laid in concentric rings of headers on edge, with radial joints. The radius of the arch should be 1 to 1¼ times the width of the opening.

**Lateral Support**    Brick walls should be supported laterally by bonding to transverse walls or buttresses, or by anchoring to floors, at intervals not exceeding 20 times the thickness. Floors and anchors must be capable of transmitting wind pressure and earthquake forces, acting outward, to transverse walls or other adequate supports, and thus to the ground. The height of piers between lateral supports should not exceed 12 times their least dimension.

**Building Blocks**    Hollow blocks, of either burned clay or concrete, are used in the walls of buildings, often faced with brick or stone. Commercial sizes are designed for bonding

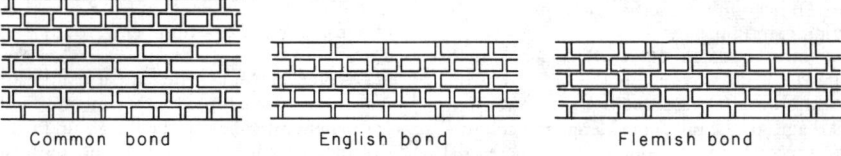

Common bond        English bond        Flemish bond

**Fig. 26**    Bonds used in bricklaying.

depend upon quality of the brick, quality of the mortar, and workmanship of laying. The strength depends also upon adequate bond and the shape of the structural unit.

**Brick**    Common red bricks are made of clay burned in a kiln. Quality characteristics are hardness and density. Light-colored brick are apt to be soft and porous. Brick for masonry exposed to the weather or where strength is desired should have a crushing strength of not less than 2,500 lb/in² (176 kg/cm²) and should absorb not over 20 percent of water by weight, after 5 h immersion (see Sec. 6).

**Mortar**    Mortars varying over a wide range of proportions are generally classified as (1) portland-cement mortar and (2) cement-lime mortar. Average proportions are, for portland-cement mortar, 1 part cement and 4½ parts sand, by volume,

with brick. For satisfactory quality, clay tile should have an absorption of not over 16 and concrete not over 10 percent. Clay tile tested with cells horizontal, or concrete blocks as laid, should have compressive strength not less than 700 lb/in² (49.2 kg/cm²) gross area. Walls of hollow blocks or hollow walls of brick bonded across the air space should have lateral support at intervals not exceeding 18 times their thickness.

**Faced Walls**    Stone or brick facing of walls, when adequately bonded in laying to a backing of brick or hollow blocks, may be considered structurally as part of the wall, working stresses to be those allowed for the backing material.

**Allowable working compressive stresses in masonry** in pounds per square inch and kilograms per square centimeter are given in the following table:

| Kind of masonry | Cement mortar | | Cement-lime mortar | | Lime mortar | |
|---|---|---|---|---|---|---|
| | lb/in² | kg/cm² | lb/in² | kg/cm² | lb/in² | kg/cm² |
| Stone ashlar | 400 | 28.1 | 320 | 22.5 | 250 | 17.6 |
| Rubble | 300 | 21.1 | 100 | 7.0 | 80 | 5.6 |
| Brickwork | 175 | 12.3 | 140 | 9.3 | 75 | 5.3 |
| Hollow building blocks | 80 | 5.6 | 70 | 4.9 | | |

The allowable stresses in brickwork may be increased 33 percent when the individual bricks have a compressive strength of 3,500 and a modulus of rupture of 600 lb/in². Local pressure under concentrations of load may exceed the stresses in the table by 25 percent.

**Plain Concrete** (see Sec. 6)   The allowable working stresses in plain concrete walls and piers are as follows:

| Approx proportions | Water, gal per sack of cement | Strength, 28 days | | Allowable stress | |
|---|---|---|---|---|---|
| | | lb/in² | kg/cm² | lb/in² | kg/cm² |
| 1-2½-5 | 8.00 | 1,500 | 105.5 | 375 | 26.3 |
| 1-2-4 | 7.50 | 2,000 | 140.6 | 500 | 35.2 |
| 1-1½-3 | 6.75 | 2,500 | 175.8 | 625 | 43.9 |
| 1-1-2 | 6.00 | 3,000 | 210.9 | 750 | 52.7 |

**Reinforced Masonry**   Bricks and building blocks may be used structurally when core spaces are partially or completely filled with concrete and reinforcing bars are embedded.

## TIMBER CONSTRUCTION

**Floors**   The framing of wooden floors may be divided into two general types: joist construction and solid, or mill, construction. The first consists of **joists** 2 to 6 in wide, of the necessary depth, and spaced about 12 to 16 in (30 to 40 cm) on centers. The wall ends should rest on and be anchored to walls and the interior ends carried by a line of girders on columns. These joists should be securely cross-bridged not over 8 ft (2.4 m) apart in each span to prevent twisting and to assist in distributing concentrated loads. Solid blocking should be provided at ends and at each point of support. The floor is formed of a thickness of rough boarding on which the finish flooring is laid. **Solid** or **mill-construction floors** are designed to do away with the small pockets which exist in joist construction and thus reduce the fire hazard. They are generally framed with beams spaced 8 to 12 ft (2.4 to 3.6 m) on centers and spanning 18 to 25 ft (5.5 to 7.6 m). The wall ends of beams rest on and are anchored to the wall, and the interior ends are carried on columns and tied together to form a continuous tie across the building. Ends of timbers in masonry walls should have metal bearing plates and ½ in space at sides and end for ventilation, to prevent rot. The ends should be beveled and the anchors placed low to avoid overturning the wall if the beams drop in a fire. In all cases, care should be taken to provide sufficient bearing at the points of support so that the allowable intensity of compression across the grain is not exceeded. In case it is desirable to omit columns, or the floor load requires a closer spacing of beams, girders are run lengthwise of the building over the columns to take the beams, the ends of which are hung in hangers or stirrup irons and tied together, over or through the girders. This is called **intermediate framing.** Steel beams are sometimes used in place of wooden beams in this type of construction, in which case a wooden strip is bolted to the top flange of the beam to take the nailing of the plank, or the plank is laid directly on top of the beam and secured by spikes driven from below and clinched over the flange. The floor is formed of 3 or 4 in (7.5 or 10 cm) plank grooved in each edge, put together with splines and securely spiked to beams. On top of the plank is laid flooring, with a layer of sheathing paper between. In case the floor loads require an excessive thickness of plank, or in localities where heavy plank is not easily obtainable, the floor is built up of 3 × 6 in (7.5 × 15 cm), or other sized pieces, placed on edge, and securely nailed together.

The roofs of buildings of joist and mill construction are framed in a manner similar to the floors of each type and should be securely anchored to the walls and columns. In case columns are not desired in the top story, steel beams or trusses of either steel or wood are used. For spans up to 35 ft (10.7 m), trussed beams can often be used to advantage.

For **unit stresses in timber,** see Sec. 6. For unit stresses in wooden columns, see Table 9. Table 7 gives the properties of mill floors made of dressed plank, and of laminated floors made of planks on edge, laid close.

### Timber Beams

**Properties of Timber Beams**   Table 8 presents those properties of wooden timbers most useful in computing their strength and deflection as beams. The "nominal size" of a timber is indicated by the breadth and depth of the section in inches. The "actual size" indicates the size of the dressed timber, according to National Lumber Manufacturers Assoc. The moment of inertia and section modulus are with the neutral axis perpendicular to the depth at the center. The **safe bending moment** in inch-pounds for a given beam is determined

Strength varies widely according to material, arrangement of reinforcement, and workmanship. Design procedures are similar to those for reinforced concrete.

**Reinforced Concrete**   **Efflorescence** on the face of brickwork can be reduced and sometimes avoided by waterproofing the mortar with an admixture of ammonium or calcium stearate, 2 percent by weight of the cement and lime.

**Table 7. Properties of Plank and Solid Laminated Floors**
($b$ = breadth = 12 in, $f$ = fiber stress)

| Nominal thickness or depth, in (1) | Actual thickness $d$, in (S4S) (2) | Area of section $A = bd$, in² (3) | Moment of inertia $I = bd^3/12$, in⁴ (4) | Section modulus $S = bd^2/6$, in³ (5) | Safe load, lb/ft² on 1-ft span* | | Coef of deflection, uniform load‡ | |
|---|---|---|---|---|---|---|---|---|
| | | | | | $f = 1,000$ lb/in²† (6) | $f = 1,600$ (7) | $E = 1,000,000$ (8) | $E = 1,760,000$ (9) |
| 1 | ¾ | 9.00 | 0.422 | 1.13 | 753 | 1,205 | 53.4 | 93.98 |
| 1½ | 1¼ | 15.00 | 1.95 | 3.13 | 2,085 | 3,336 | 11.51 | 20.26 |
| 2 | 1½ | 18.00 | 3.38 | 4.50 | 3,000 | 4,800 | 6.66 | 11.72 |
| 2½ | 2 | 24.00 | 8.00 | 8.00 | 5,334 | 8,534 | 2.82 | 4.96 |
| 3 | 2½ | 30.00 | 15.60 | 12.50 | 8,334 | 13,334 | 1.441 | 2.54 |
| 4 | 3½ | 42.00 | 42.9 | 24.5 | 16,334 | 26,134 | 0.524 | 0.922 |
| 5 | 4½ | 54.00 | 91.1 | 40.5 | 27,000 | 43,200 | 0.247 | 0.1404 |
| 6 | 5½ | 66.00 | 166.4 | 60.5 | 40,300 | 64,500 | 0.1348 | 0.0765 |
| 8 | 7½ | 90.00 | 422 | 112.5 | 75,000 | 120,000 | 0.0533 | 0.0303 |
| 10 | 9½ | 114.00 | 857 | 180.5 | 120,400 | 192,500 | 0.0263 | 0.0149 |
| 12 | 11½ | 138.00 | 1,521 | 264.5 | 176,400 | 282,000 | 0.0148 | 0.0084 |

*Divide tabular value by square of span in feet.
†For other fiber stress $f$, multiply tabular value by $f/1,000$.
‡For deflection in, multiply coefficient by load, lb/ft², and by fourth power of span in ft, and divide by 1,000,000. For other modulus of elasticity $E$, multiply coefficient of col. 8 by 1,000,000, and divide by $E$.

from the section modulus $S$ by multiplying the tabular value by the allowable fiber stress. To select a beam to withstand safely a given bending moment, divide the bending moment in inch-pounds by the allowable fiber stress, and choose a beam whose section modulus $S$ is equal to or larger than the quotient thus obtained. For formulas for computing bending moments, see Sec. 5.

**Maximum loads** in Table 8, cols. 7 and 8, are for uniform loading. Use half the values of col. 7 for a single load concentrated at midspan; for other loadings compute the bending moment and use the section modulus, col. 6. The values of col. 8 apply to all symmetrical loadings. For unsymmetrical loading, compute the maximum shear, which must not exceed one-half the tabular value.

The **coefficients of deflection** given are the deflections, in inches, of beams 1 ft in span, with a uniformly distributed load of 1,000,000 lb, the modulus of elasticity being taken at 1,000,000 lb/in². The deflection of a beam of a given span under uniformly distributed load is obtained by multiplying the coefficient of deflection of the beam by the cube of the span in feet and by the number of 1,000,000 lb units in the given load and by dividing by the number of 1,000,000 lb/in² units in the actual modulus of elasticity. Coefficients of deflection under concentrated loads applied at the middle of the span may be obtained by multiplying the values in the table by 1.6. The results are only approximate, as the modulus of elasticity varies with the moisture content of the wood.

The deflection of beams intended to carry plastered ceilings should not exceed $\frac{1}{360}$ of the span.

A convenient rule may be derived by assuming that the modulus of elasticity is 1,000 times the allowable fiber stress, which applies to all woods with sufficient accuracy for the purpose. Beams loaded uniformly to capacity in bending will then deflect $\frac{1}{360}$ of the span when the depth in inches is 0.90 times the span in feet; and beams with central concentration, when the depth is 0.72 times the span in the same units. For such beams, the deflection in inches is, for uniform load, $0.03L^2/d$; for central concentration, $0.024L^2/d$, where $L$ is the

span, ft and $d$ the depth, in. Variation in type of loading affects this result comparatively little.

**Timber Columns**

**Timber columns** may be either square or round and should have metal bases, usually galvanized steel, to cut off moisture and prevent lateral displacement. For supporting beams, they should have caps which, at roofs, may be of steel, or wood designed for bearing across the grain. At intermediate floors, caps should be of steel, although in some cases hardwood bolsters may be used. Except when caps or beams are of steel, columns should run down and rest directly on the baseplate. Table 9 gives **working unit stresses for wooden columns** recommended where the building laws do not prescribe lower stresses. Use actual, not nominal, dimension of timbers. The formula for columns on which Table 9 is based is $P/A = 0.30E/(l/d)^2$. The maximum unit stress should not exceed the allowable unit stress in compression parallel to grain $c$, as set forth in Sec. 6. When computing $l/d$, both axes of the column should be investigated.

**Stud Partitions** Table 10 gives the safe load in pounds per linear foot of partition, based on both capacity of the studs as columns and bearing across the grain of the plate or sill. Stresses used are those recommended above. Note that the capacity of the studs is generally more in low partitions than the values for cross bearing unless a hardwood is used for the sill and plate. It is well, however, to provide a margin of strength in the studs to cover cutting for wires, etc.

**Glued Laminated Timber** Structural glued laminated timber, commonly called glulam, refers to members which are fabricated by pressure gluing selected wood laminations of either ¾ or 1½ in (19 or 38 mm) surfaced thickness. The grain of all the laminations is approximately parallel longitudinally, with exterior laminations being of generally higher-quality wood since bending stresses are greater at the outer fibers. Curved and tapered structural members are available with the recommended minimum radii of curvature being 9 ft 4 in (2.84 m) for ¾-in laminations and 27 ft 6 in (8.4 m) for 1½-in

**Table 8. Properties of Wooden Beams (Surfaced Size)**

| Nominal size, in (1) | Actual size $b \times d$, in, dressed (S4S) size (2) | Area of section $bd$, in² (3) | Weight at 40 lb/ft³, lb/ft (4) | Moment of inertia $I = bd^3/12$, in⁴ (5) | Section modulus $S = bd^2/6$, in³ (6) | Max safe uniform load, lb, based on Bending on 1 ft span,* $f = 1,000$ lb/in² (7) | Shear at 100† lb/ in² (8) | Coef‡ of deflection, uniform load $E =$ 1,000,000 (9) |
|---|---|---|---|---|---|---|---|---|
| 2 × 4 | 1½ × 3½ | 5.25 | 1.46 | 5.36 | 3.06 | 2,040 | 700 | 4.20 |
| 3 × 4 | 2½ × 3½ | 8.75 | 2.43 | 8.93 | 5.10 | 3,400 | 1,166 | 2.52 |
| 4 × 4 | 3½ × 3½ | 12.25 | 3.40 | 12.51 | 7.15 | 4,760 | 1,632 | 1.80 |
| 2 × 6 | 1½ × 5½ | 8.25 | 2.29 | 20.8 | 7.56 | 5,040 | 1,100 | 1.082 |
| 3 × 6 | 2½ × 5½ | 13.75 | 3.82 | 34.7 | 12.60 | 8,390 | 1,835 | 0.648 |
| 4 × 6 | 3½ × 5½ | 19.25 | 5.35 | 48.5 | 17.65 | 11,760 | 2,570 | 0.464 |
| 6 × 6 | 5½ × 5½ | 30.3 | 8.40 | 76.3 | 27.7 | 18,490 | 4,040 | 0.295 |
| 2 × 8 | 1½ × 7¼ | 10.87 | 3.02 | 47.6 | 13.14 | 8,760 | 1,445 | 0.473 |
| 3 × 8 | 2½ × 7¼ | 18.12 | 5.04 | 79.4 | 21.9 | 14,600 | 2,410 | 0.284 |
| 4 × 8 | 3½ × 7¼ | 25.4 | 7.05 | 111.1 | 30.7 | 20,500 | 3,380 | 0.202 |
| 6 × 8 | 5½ × 7½ | 41.3 | 11.4 | 193 | 51.6 | 34,400 | 5,500 | 0.1162 |
| 8 × 8 | 7½ × 7½ | 56.3 | 15.6 | 264 | 70.3 | 46,900 | 7,500 | 0.0852 |
| 2 × 10 | 1½ × 9¼ | 13.87 | 3.85 | 98.9 | 21.4 | 14,290 | 1,850 | 0.227 |
| 3 × 10 | 2½ × 9¼ | 23.1 | 6.42 | 164.9 | 35.7 | 23,700 | 3,080 | 0.1364 |
| 4 × 10 | 3½ × 9¼ | 32.4 | 8.93 | 231 | 49.9 | 33,300 | 4,310 | 0.0974 |
| 6 × 10 | 5½ × 9½ | 52.3 | 14.5 | 393 | 82.7 | 55,200 | 6,970 | 0.0573 |
| 8 × 10 | 7½ × 9½ | 71.3 | 19.8 | 536 | 113 | 75,200 | 9,500 | 0.0421 |
| 10 × 10 | 9½ × 9½ | 90.3 | 25.0 | 679 | 143 | 95,300 | 12,030 | 0.0332 |
| 2 × 12 | 1½ × 11¼ | 16.87 | 4.69 | 178 | 31.6 | 21,100 | 2,250 | 0.1264 |
| 3 × 12 | 2½ × 11¼ | 28.1 | 7.81 | 297 | 52.7 | 35,100 | 3,750 | 0.0757 |
| 4 × 12 | 3½ × 11¼ | 39.4 | 10.94 | 415 | 73.9 | 49,300 | 5,250 | 0.0543 |
| 6 × 12 | 5½ × 11½ | 63.3 | 17.5 | 697 | 121 | 80,800 | 8,430 | 0.0323 |
| 8 × 12 | 7½ × 11½ | 86.3 | 23.9 | 951 | 165 | 110,200 | 11,510 | 0.0237 |
| 10 × 12 | 9½ × 11½ | 109.3 | 30.3 | 1,204 | 209 | 139,600 | 14,570 | 0.01864 |
| 12 × 12 | 11½ × 11½ | 132.3 | 36.7 | 1,458 | 253 | 169,000 | 17,620 | 0.01543 |
| 4 × 14 | 3½ × 13¼ | 46.4 | 12.88 | 678 | 102.4 | 68,300 | 6,180 | 0.0332 |
| 6 × 14 | 5½ × 13½ | 74.3 | 20.6 | 1,128 | 167 | 111,400 | 9,900 | 0.01987 |
| 8 × 14 | 7½ × 13½ | 101.3 | 28.0 | 1,538 | 228 | 152,000 | 13,500 | 0.01462 |
| 10 × 14 | 9½ × 13½ | 128.3 | 35.6 | 1,948 | 289 | 192,400 | 17,120 | 0.01153 |
| 12 × 14 | 11½ × 13½ | 155.3 | 43.1 | 2,360 | 349 | 233,000 | 20,700 | 0.00953 |
| 14 × 14 | 13½ × 13½ | 182.3 | 50.6 | 2,770 | 410 | 273,000 | 24,300 | 0.00812 |
| 6 × 16 | 5½ × 15½ | 85.3 | 23.6 | 1,707 | 220 | 146,800 | 11,380 | 0.01315 |
| 8 × 16 | 7½ × 15½ | 116.3 | 32.0 | 2,330 | 300 | 200,000 | 15,530 | 0.00967 |
| 10 × 16 | 9½ × 15½ | 147.3 | 40.9 | 2,950 | 380 | 254,000 | 19,610 | 0.00762 |
| 12 × 16 | 11½ × 15½ | 178.3 | 49.5 | 3,570 | 460 | 307,800 | 23,800 | 0.00630 |
| 14 × 16 | 13½ × 15½ | 209 | 58.1 | 4,190 | 541 | 360,000 | 27,900 | 0.00539 |
| 16 × 16 | 15½ × 15½ | 240 | 66.7 | 4,810 | 621 | 414,000 | 32,000 | 0.00468 |
| 8 × 18 | 7½ × 17½ | 131.3 | 36.4 | 3,350 | 383 | 255,000 | 17,500 | 0.00672 |
| 10 × 18 | 9½ × 17½ | 166.3 | 46.1 | 4,240 | 485 | 323,000 | 22,200 | 0.00531 |
| 12 × 18 | 11½ × 17½ | 201 | 55.9 | 5,140 | 587 | 391,000 | 26,800 | 0.00438 |
| 14 × 18 | 13½ × 17½ | 236 | 65.6 | 6,030 | 689 | 459,000 | 31,500 | 0.00373 |
| 16 × 18 | 15½ × 17½ | 271 | 75.3 | 6,920 | 791 | 528,000 | 36,200 | 0.00325 |
| 18 × 18 | 17½ × 17½ | 306 | 85.0 | 7,820 | 893 | 595,000 | 40,800 | 0.00288 |
| 12 × 20 | 11½ × 19½ | 224 | 62.3 | 7,110 | 729 | 485,000 | 29,900 | 0.00316 |
| 20 × 20 | 19½ × 19½ | 380 | 106 | 12,050 | 1,236 | 824,000 | 50,700 | 0.00187 |
| 24 × 24 | 23½ × 23½ | 552 | 153 | 25,400 | 2,160 | 1,440,000 | 73,400 | 0.000888 |
| 26 × 26 | 25½ × 25½ | 650 | 180.6 | 35,200 | 2,760 | 1,840,000 | 86,700 | 0.000639 |
| 28 × 28 | 27½ × 27½ | 756 | 210 | 47,700 | 3,470 | 2,320,000 | 100,600 | 0.000472 |
| 30 × 30 | 29½ × 29½ | 870 | 242 | 63,100 | 4,280 | 2,850,000 | 116,000 | 0.000356 |

*For total safe uniform load, pounds, on beam of span $L$, feet, divide tabular value by $L$. For fiber stress $f$ other than 1,000 lb/in² multiply by $f$ and divide by 1,000.
†For shearing stress other than 100 lb/in², multiply by stress and divide by 100.
‡For deflection, inches, multiply coefficient by total load, pounds, and by cube of span, feet, and divide by 1,000,000. For other modulus of elasticity $E$, multiply coefficient by 1,000,000 and divide by $E$.

lamination thickness. Laminations should be parallel to the tension face of members; sawn tapered cuts are permitted on the compression face.

Available net (surfaced) widths of members in inches are 2¼, 3⅛, 5⅛, 8¾, 10¾, 12¼, and 14¼; depths are determined by stress requirements. Economical spans (see "Timber Construction Manual," American Institute of Timber Construction) for roof framing range from 10 to 100 ft (3 to 30 m) for

**Table 9. Working Stresses for Square or Rectangular Timber Columns, lb/in²**
(Compression parallel to grain)

| $E$ | 10 or less | 15 | 20 | 25 | 30 | 35 | 40 | 45 | 50 | 55* | 60* | 70* | 80* |
|---|---|---|---|---|---|---|---|---|---|---|---|---|---|
| | | | | | | | $l/d$, in/in | | | | | | |
| 1,000,000 | 3,000 | 1,333 | 750 | 480 | 333 | 245 | 188 | 148 | 120 | 99 | 83 | 61 | 47 |
| 1,300,000 | 3,900 | 1,733 | 975 | 624 | 433 | 318 | 244 | 193 | 156 | 129 | 108 | 80 | 61 |
| 1,600,000 | 4,800 | 2,140 | 1,200 | 768 | 533 | 392 | 300 | 237 | 192 | 158 | 133 | 98 | 75 |
| 1,900,000 | 5,700 | 2,533 | 1,425 | 912 | 633 | 465 | 356 | 281 | 228 | 188 | 158 | 116 | 89 |

*Columns should be limited to $l/d = 50$, except for individual members in stud walls, which should be limited to $l/d = 80$.

simple spans. Floor framing, which is designed for much heavier live loads, economically spans from 6 to 40 ft (1.8 to 12 m) for simple beams and from 25 to 40 ft (7.5 to 12 m) for continuous beams.

Glued laminated members are generally fabricated from either Douglas fir and larch, Douglas fir (coast region), southern pine, or California redwood, depending on availability. Allowable design stresses depend on whether the condition of use is to be wet (moisture content in service of 16 percent or more) or dry (as in most covered structures), the species and grade of wood to be used, the manner of loading, and the number of laminations as well as the usual factors for duration

of loading. There are also cumulative reduction factors applicable to the allowable bending stress: (1) depth factor for beams greater than 12 in (0.3 m) is $C_d = 0.81 (d^2 + 143)/(d^2 + 88)$; and (2) slenderness reduction factor is $C_s = \sqrt{l_e d/b^2} \leqslant 50$, where $C_s$ = slenderness factor; $l_e$ = effective length of beam, in, taken conservatively as $1.92 \, l_u$; $l_u$ = laterally unsupported length of the compression face of beam, in; $d$ = depth of beam, in; $b$ = width of beam, in. When $C_s$ is less than 10, there is no adjustment to the bending stress $F_b$. When $C_s$ is greater than 10 but less than $C_k$, the allowable bending stress is determined by $F'_b = F_b [1\frac{1}{3}(C_s/C_k)^4]$ where $C_k =$

**Table 10. Safe Loads on Wood-Stud Bearing Walls**

| Nominal size of studs, in (net area in²) | Height, ft | $l/d$ max. | Based‡ on comp. ∥ to grain at $C =$ 1,000 lb/in² | $E =$ 1,000,000 lb/in² | Douglas fir $E =$ 1,600,000 | Southern pine $E =$ 1,900,000 | Redwood $C\perp =$ 305 | $C\perp =$ 390 | $C\perp =$ 455 |
|---|---|---|---|---|---|---|---|---|---|
| | | | | | Based on studs as columns† | | Based on bearing across the grain of the plate or sill ($C\perp$to grain) | Grades of southern pine and Douglas fir | |
| 2 × 3 | 8 | 38 | 3,750 | 760 | 1,220 | 1,440 | 1,145 | 1,445 | 1,705 |
| (3.75) | 10 | 48 | | 490 | 780 | 930 | | | |
| 2 × 4 | 8 | 32 | 5,250 | 1,540 | 2,460 | 2,930 | 1,600 | 2,020 | 2,390 |
| (5.25 | 10 | 40 | | 980 | 1,570 | 1,860 | | | |
| | 12 | 48 | | 680 | 1,090 | 1,290 | | | |
| 3 × 4 | 8 | 28 | 8,750 | 3,490 | 5,580 | 6,630 | 2,670 | 3,370 | 4,000 |
| (8.75) | 10 | 34 | | 2,230 | 3,570 | 4,240 | | | |
| | 12 | 41 | | 1,550 | 2,480 | 2,950 | | | |
| 4 × 4 | 8 | 28 | 12,250 | 4,880 | 7,810 | 9,270 | 3,740 | 4,700 | 5,575 |
| (12.25) | 10 | 34 | | 3,120 | 4,990 | 5,930 | | | |
| | 12 | 41 | | 2,170 | 3,470 | 4,120 | | | |
| 2 × 6 | 8 | 21 | 8,250 | 5,440 | 8,700 | 10,340 | 2,520 | 3,180 | 3,750 |
| (8.25) | 10 | 27 | | 3,480 | 5,570 | 6,610 | | | |
| | 12 | 32 | | 2,420 | 3,870 | 4,600 | | | |
| | 14 | 37 | | 1,770 | 2,830 | 3,360 | | | |
| | 16 | 43 | | 1,380 | 2,180 | 2,580 | | | |
| 3 × 6 | 8 | 19 | 13,750 | 11,190 | 17,900 | 21,260 | 4,190 | 5,300 | 6,250 |
| (13.75) | 10 | 24 | | 7,160 | 11,460 | 13,600 | | | |
| | 12 | 29 | | 4,970 | 7,950 | 9,440 | | | |
| | 14 | 34 | | 3,650 | 5,840 | 6,940 | | | |
| | 16 | 39 | | 2,800 | 4,480 | 5,320 | | | |
| 4 × 6 | 8 | 18 | 19,250 | 18,960 | 30,300 | 36,000 | 5,870 | 7,400 | 8,750 |
| (19.25) | 10 | 22 | | 12,130 | 19,410 | 23,050 | | | |
| | 12 | 26 | | 8,420 | 13,470 | 16,000 | | | |
| | 14 | 31 | | 6,190 | 9,900 | 11,760 | | | |
| | 16 | 35 | | 4,740 | 7,580 | 9,000 | | | |

Safe load, lb/ft of partition; studs at 12-in centers*

*For studs at 16 in on centers, multiply tabular value by 0.75.
†2 × 6 studs bridged twice in height at third points; others, once at midheight.
‡Values of allowable compression parallel to grain vary from 1,200 to 1,650 lb/in² for Douglas fir and from 1,800 to 2,350 lb/in² for southern pine, depending on grade.

$\sqrt{3E/5F_b}$, $E$ = modulus of elasticity. When the slenderness factor is greater than $C_k$ but less than 50, the allowable bending stress is determined by $F'_b = 0.4E/C_s^2$. For curved members the allowable bending stress is multiplied by a curvature factor $C_c = 1 - 2,000(t/R)^2$ where $t$ = thickness of laminations, in, and $R$ = radius at inside face of laminations, in. Maximum value of $t/R = \frac{1}{100}$ for southern pine and $\frac{1}{125}$ for other soft woods. There is also a limitation on radial stresses in curved members which is explained in paragraph 903-D of the National Design Specification.

A summary of allowable unit stresses may be found in Sec. 6 for glued laminated timber.

### Connections

**Bolted Joints** Compression may be transmitted by merely butting the timbers, with splice pieces bolted to the sides to keep alignment and resist incidental bending and shear. The same detail (Fig. 27) serves in tension, but the entire stress must then be transmitted through the bolts and splice pieces.

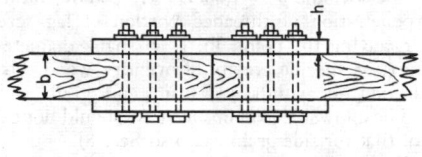

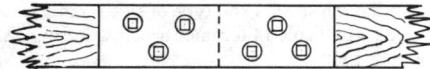

**Fig. 27** Bolted splice for timber framing.

If of wood, these should have a thickness $b$ equal to $\frac{1}{2}b$. In light, unimportant work, splice pieces may be spiked. Table 11 gives the allowable load in pounds for one bolt loaded at both ends (double shear) when $b$ is at least equal to $\frac{1}{2}b$. When steel side plates are used for side members, the tabulated loads may be increased 25 percent for parallel-to-grain loading, but no increase should be made for perpendicular-to-grain loads. When a joint consists of two members (single shear), one-half the tabulated load for a piece twice the thickness of the thinner member applies. The safe load for bolts loaded at an angle $\theta$ with the grain of the wood is given by the formula $N = PQ/(P \sin^2 \theta + Q \cos^2 \theta)$, where $N$ = allowable load per bolt in a direction at inclination $\theta$ with the direction of the grain, lb; $P$ = allowable load per bolt in compression parallel to the grain, lb; and $Q$ = allowable load per bolt in compression perpendicular to the grain, lb.

The size, arrangement, and **spacing of bolts** must be such that tension on the net section of the timber through the bolt holes and shear along the grain do not exceed allowable values. Bolts should be at least 7 diameters from the end of the timber for softwoods and 5 diameters for hardwoods and spaced at least 4 diameters on center parallel to the grain. Crossbolting, to prevent splitting the timber end, is sometimes desirable.

The efficiency of bolted timber connections may be greatly increased by the use of **ring connectors** (see Fig. 28). Split rings and shear plates are fitted into circular grooves, concentric with the bolt, in the contact surfaces, and transmit shear stresses across the joint. Grooves for split rings and shear plates are cut with a special tool, while toothed rings are usually seated by drawing together the timbers with high-strength bolts. Allowable loads for these various connectors are given in the "National Design Specification for Stress-Grade Lumber and Its Fastenings," published by the National Lumber Manufacturers Assoc. Selected values are given in Table 12.

The **holding power of wire nails** is as follows ("National Design Specification for Stress-Grade Lumber and Its Fastenings"): The resistance to withdrawal is proportional to the length of embedment, to the diameter of the nail (where the wood does not split), and to $G^{2.5}$, where $G$ is the ovendry specific gravity of the wood (see Sec. 6 for $G$ values of various species). The safe resistance to withdrawal of common wire nails driven into the side grain of seasoned wood is given by Table 13. Nails withdrawn from green wood have generally slightly higher resistance, but nails driven into green wood may lose much of

**Table 11. Allowable Load in Pounds on One Bolt Loaded at Both Ends (Double Shear)**
(For additional values and for conditions other than normal, see "National Design Specification for Stress-Grade Lumber and Its Fastenings")

| Length of bolt in main member, in | Diam of bolt, in | Douglas fir-larch, southern pine (dense) | | Douglas fir-larch, southern pine (med. grain), California redwood (close grain), southern cypress | | Oak, red and white | | Western hemlock, California redwood (open grain), eastern hemlock | |
|---|---|---|---|---|---|---|---|---|---|
| | | Parallel to grain | Perpendicular to grain | Parallel to grain | Perpendicular to grain | Parallel to grain | Perpendicular to grain | Parallel to grain | Perpendicular to grain |
| $1\frac{1}{2}$ | $\frac{1}{2}$ | 1,120 | 500 | 960 | 430 | 830 | 650 | 810 | 280 |
| | $\frac{3}{4}$ | 1,700 | 630 | 1,460 | 540 | 1,260 | 820 | 1,210 | 350 |
| | 1 | 2,270 | 760 | 1,940 | 650 | 1,690 | 980 | 1,610 | 420 |
| $2\frac{1}{2}$ | $\frac{1}{2}$ | 1,510 | 840 | 1,290 | 720 | 1,120 | 1,080 | 1,190 | 460 |
| | $\frac{3}{4}$ | 2,780 | 1,060 | 2,370 | 900 | 2,060 | 1,360 | 2,010 | 580 |
| | 1 | 3,770 | 1,270 | 3,230 | 1,080 | 2,800 | 1,640 | 2,690 | 690 |
| $3\frac{1}{2}$ | $\frac{1}{2}$ | 1,530 | 1,140 | 1,310 | 980 | 1,130 | 1,130 | 1,220 | 640 |
| | $\frac{3}{4}$ | 3,360 | 1,480 | 2,870 | 1,260 | 2,440 | 1,910 | 2,600 | 810 |
| | 1 | 5,120 | 1,770 | 4,380 | 1,520 | 3,800 | 2,290 | 3,740 | 970 |
| $5\frac{1}{2}$ | $\frac{3}{4}$ | 3,430 | 2,200 | 2,930 | 1,880 | 2,540 | 2,490 | 2,750 | 1,270 |
| | 1 | 6,080 | 2,790 | 5,200 | 2,380 | 4,510 | 3,560 | 4,860 | 1,520 |
| | $1\frac{1}{4}$ | 9,160 | 3,260 | 7,830 | 2,790 | 6,800 | 4,210 | 7,000 | 1,780 |

their resistance when the wood seasons; the allowable withdrawal load should be one-fourth of that given in Table 13. Cement and other coatings on nails may add materially to their resistance in softwoods. Drilling lead holes slightly smaller than the nail adds somewhat to the resistance and reduces danger of splitting. The structural design should be such that nails are not loaded in withdrawal from end grain. (See also Sec. 8.)

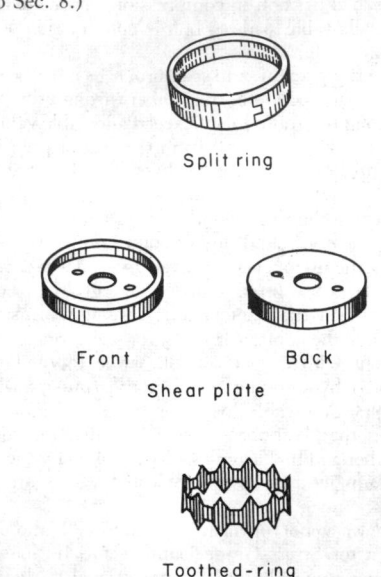

**Fig. 28**  Timber connectors.

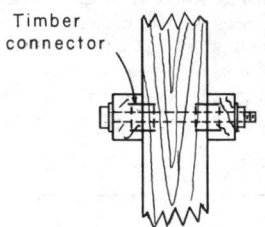

**Fig. 29**  Timber joint with connectors.

The safe **lateral resistance** of common wire nails driven in side grain to the specified penetrations is given in Table 13 and is proportional to $D^{1.5}$ where $D$ is the diameter, in. These values are for seasoned wood and should be reduced 25 percent for woods which will remain wet or will be loaded before seasoning. For nails driven into end grain, values should be reduced one-third.

Common wire **spikes** are larger for their lengths than nails. Their resistance to withdrawal and lateral resistance are given by the same formulas as for nails, but greater precautions need to be taken to avoid splitting. (See also Sec. 8.)

The **resistance of wood screws to withdrawal** from side grain of seasoned wood is given by the formula $P = 2{,}850G^2D$, where $P$ = the allowable load on the screw, lb/in penetration of the threaded portion; $G$ = specific gravity of ovendry wood; $D$ =

diameter of screw, in. Wood screws should not be designed to be loaded in withdrawal from end grain (see also Sec. 8).

The **allowable safe lateral resistance** of wood screws embedded 7 diameters in the side grain of seasoned wood is given by the formula $P = KD^2$, where $P$ is the lateral resistance per screw, lb; $D$ is the diameter, in; and $K$ is 4,800 for oak (red and white), 3,960 for Douglas fir (coast region) and southern pine, and 3,240 for cypress (southern) and Douglas fir (inland region).

The following rules should be observed: (1) the size of the lead hole in soft (hard) woods should be about 70 (90) percent of the core or root diameter of the screw; (2) lubricants such as soap may be used without great loss in holding power; (3) long, slender screws are preferable generally, but in hardwood too slender screws may reach the limit of their tensile strength; (4) in the screws themselves, holding power is favored by thin sharp threads, rough unpolished surface, full diameter under the head, and shallow slots.

The allowable **withdrawal load of lag screws** in side grain is given by the formula $p = 1{,}800D^{3/4}G^{3/2}$, allowable load per inch of penetration of threaded portion of lag screw into member receiving the point, lb; $D$ = shank diameter of lag screw, in; $G$ = specific gravity of ovendry wood. Use of lag screws loaded in withdrawal from end grain should be avoided. The allowable load in such case should not exceed 75 percent of that for side grain (see also Sec. 8).

The allowable **lateral resistance of lag screws** for parallel-to-grain loading with screws in side grain is proportional to $D^2$ and is dependent on species and type of side member. Selected values are given in Table 14 for one lag screw in single shear in a two-member joint.

**Lead holes** for lag screws (approximately 75 percent of shank diameter) should be prebored for the threaded portion. Lead holes for the shank should be of the same diameter and length as that of the unthreaded shank. Soap or other lubricant should be used to facilitate insertion and to prevent damage to the screw. Where steel-plate side pieces are used, the allowable loads given by the formula for parallel-to-grain loading may be increased by 25 percent.

The ultimate **withdrawal load per linear inch of penetration of a round drift bolt or pin** from side grain when driven into a prebored hole having a diameter ⅛ in less than that of the bolt diameter may be determined from the formula $p = 6{,}000G^2D$, where $p$ = ultimate withdrawal load of penetration, lb/lin in.; $G$ = specific gravity of ovendry wood; $D$ = diameter of drift bolt, in. A safety factor of about 5 is suggested for general use. The allowable load in lateral resistance for a drift bolt should ordinarily be taken as less than that for a common bolt.

## STEEL CONSTRUCTION

(**Note.** In the design of steel structures, 1,000 lb is frequently designated as a kilopound or "kip," and a stress of 1 kip per sq in is designated as 1 ksi.)

**Specifications**  The following are in part condensed excerpts from the Specifications of the American Institute of Steel Construction.

**Material**  Ordinary structural steel for rolled shapes, plates, and bars is specified by ASTM A36 with a yield stress of 36,000 lb/in² (2,530 kg/cm²). Steels of higher strength, A440, A441, and A242, are used in special applications and for highly stressed members of conventional structures.

**Table 12. Allowable Load in Pounds for One-Connector Unit in Single Shear***
(For additional values and for conditions other than normal, see "National Design Specification for Stress-Grade Lumber and Its Fastenings")

| Connector unit (diam) | Number of faces of piece with connectors on the same bolt | Net thickness of lumber, in | Min. edge distances, in | Group A Douglas fir-larch and southern pine (dense), oak, red and white | | Group B Douglas fir-larch, southern pine (med. grain) | | Group C California redwood (close grain), western hemlock, southern cypress | |
|---|---|---|---|---|---|---|---|---|---|
| | | | | ∥ to grain | ⊥ to grain | ∥ to grain | ⊥ to grain | ∥ to grain | ⊥ to grain |
| 2½-in split ring, ½-in bolt | 1 | 1 min, | 1¾ | 2,630 | 1,580 | 2,270 | 1,350 | 1,900 | 1,130 |
| | | 1½ or more | | 3,160 | 1,900 | 2,730 | 1,620 | 2,290 | 1,350 |
| | 2 | 1½ min, | | 2,430 | 1,460 | 2,100 | 1,250 | 1,760 | 1,040 |
| | | 2 or more | | 3,160 | 1,900 | 2,730 | 1,620 | 2,290 | 1,350 |
| 4-in split ring, ¾-in bolt | 1 | 1 min, | 2¾ | 4,090 | 2,370 | 3,510 | 2,030 | 2,920 | 1,700 |
| | | 1½ or more | | 6,020 | 3,490 | 5,160 | 2,990 | 4,280 | 2,490 |
| | 2 | 1½ min, | | 4,110 | 2,480 | 3,520 | 2,040 | 2,940 | 1,700 |
| | | 3 or more | | 6,140 | 3,560 | 5,260 | 3,050 | 4,380 | 2,540 |
| 2⅝-in shear plate, ¾-in bolt † | 1 | 1½ min | 1¾ | 3,110 | 1,810 | 2,670 | 1,550 | 2,220 | 1,290 |
| | 2 | 1½ min, | | 2,420 | 1,410 | 2,080 | 1,210 | 1,730 | 1,010 |
| | | 2½ or more | | 3,330 | 1,940 | 2,860 | 1,660 | 2,380 | 1,380 |
| 4-in shear plate, ¾-in or ⅞-in bolt † | 1 | 1½ min, | 2¾ | 4,370 | 2,540 | 3,750 | 2,180 | 3,130 | 1,810 |
| | | 1¾ or more | | 5,090 | 2,950 | 4,360 | 2,530 | 3,640 | 2,110 |
| | 2 | 1¾ min, | | 3,390 | 1,970 | 2,910 | 1,680 | 2,420 | 1,400 |
| | | 2½ | | 4,310 | 2,500 | 3,690 | 2,140 | 3,080 | 1,780 |
| | | 3½ or more | | 5,030 | 2,920 | 4,320 | 2,500 | 3,600 | 2,090 |

*One connector unit consists of one split ring with its bolt in single shear or two shear plates back to back in the contact faces of a timber-to-timber joint with their bolt in single shear.

†Allowable loads for all loadings, except wind, should not exceed 2,900 lb for 2⅝-in shear plates; 4,970 and 6,760 lb for 4-in shear plates with ¾- and ⅞-in bolts, respectively; multiply values by 1.33 for wind loading.

**Table 13. Allowable Loads in Pounds for Common Nails in Side Grain* of Seasoned Wood**

| Type of load | Specific gravity G | d | Size of nail | | | | | | | | | |
|---|---|---|---|---|---|---|---|---|---|---|---|---|
| | | | 6 | 8 | 10 | 12 | 16 | 20 | 30 | 40 | 50 | 60 |
| | | Length, in | 2 | 2½ | 3 | 3¼ | 3½ | 4 | 4½ | 5 | 5½ | 6 |
| | | Diam, in | 0.113 | 0.131 | 0.148 | 0.148 | 0.162 | 0.192 | 0.207 | 0.225 | 0.244 | 0.263 |
| Withdrawal load per in penetration | 0.31 | | 9 | 10 | 12 | 12 | 13 | 15 | 16 | 18 | 20 | 21 |
| | 0.40 | | 16 | 18 | 20 | 20 | 22 | 27 | 28 | 31 | 33 | 35 |
| | 0.44 | | 20 | 23 | 26 | 26 | 29 | 34 | 37 | 40 | 43 | 46 |
| | 0.47 | | 24 | 27 | 31 | 31 | 34 | 40 | 43 | 47 | 51 | 55 |
| | 0.51 | | 29 | 34 | 38 | 38 | 42 | 49 | 53 | 58 | 63 | 68 |
| | 0.55 | | 34 | 39 | 44 | 44 | 49 | 57 | 61 | 67 | 73 | 79 |
| | 0.67 | | 57 | 66 | 75 | 75 | 82 | 97 | 105 | 114 | 124 | 133 |
| Lateral load* † | 0.60–0.75 | | 78 | 97 | 116 | 116 | 132 | 171 | 191 | 218 | 249 | 276 |
| | 0.50–0.55 | | 63 | 78 | 94 | 94 | 107 | 139 | 154 | 176 | 202 | 223 |
| | 0.42–0.50 | | 51 | 64 | 77 | 77 | 88 | 113 | 126 | 144 | 165 | 182 |
| | 0.31–0.41 | | 41 | 51 | 62 | 62 | 70 | 91 | 101 | 116 | 132 | 146 |

*The allowable lateral load for nails driven in end grain is two-thirds the values shown above.

†The minimum penetration for full lateral resistance for the four groups listed is 10, 11, 13, and 14 diam from higher to lower specific gravities, respectively. Reduce by interpolation for lesser penetration; minimum penetration is one-third the above.

**Table 14. Allowable Lateral Loads in Pounds on Lag Bolts or Lag Screws**

| Side member | Length of bolt, in | Diam of bolt at shank, in | Ovendry specific gravity of species | | | | | | | |
| | | | 0.60–0.75 | | 0.51–0.55 | | 0.42–0.50 | | 0.31–0.41 | |
| | | | ∥ | ⊥ | ∥ | ⊥ | ∥ | ⊥ | ∥ | ⊥ |
| 1½-in wood | 4 | ¼ | 200 | 190 | 170 | 170 | 130 | 120 | 100 | 100 |
| | 4 | ½ | 390 | 250 | 290 | 190 | 210 | 140 | 170 | 110 |
| | 6 | ⅜ | 480 | 370 | 420 | 320 | 360 | 280 | 290 | 220 |
| | 6 | ⅝ | 860 | 510 | 710 | 430 | 510 | 310 | 410 | 250 |
| 2½-in wood | 6 | ½ | 620 | 410 | 470 | 310 | 340 | 220 | 270 | 180 |
| | 6 | 1 | 1,040 | 520 | 790 | 390 | 560 | 280 | 450 | 230 |
| | 8 | ¾ | 1,430 | 790 | 1,080 | 600 | 780 | 430 | 620 | 340 |
| | 8 | 1 | 1,800 | 900 | 1,360 | 680 | 970 | 490 | 780 | 390 |
| ½-in metal | 3 | ¼ | 240 | 185 | 210 | 160 | 155 | 120 | 125 | 100 |
| | 3 | ½ | 550 | 285 | 415 | 215 | 295 | 155 | 240 | 125 |
| | 6 | ½ | 1,100 | 570 | 945 | 490 | 770 | 400 | 615 | 320 |
| | 6 | ¾ | 1,970 | 865 | 1,480 | 650 | 1,060 | 460 | 850 | 370 |
| | 10 | ⅞ | 3,420 | 1,420 | 2,960 | 1,230 | 2,340 | 970 | 1,890 | 785 |
| | 12 | 1 | 4,520 | 1,810 | 3,900 | 1,560 | 3,290 | 1,320 | 2,630 | 1,050 |
| | 16 | 1¼ | 7,120 | 2,850 | 6,150 | 2,460 | 5,500 | 2,200 | 4,520 | 1,810 |

Ordinary unfinished machine bolts are specified by A307. The most common rivets are specified by A502, grade 1. Riveting is rapidly being supplanted, both in the shop and in the field, by bolting with high-strength bolts. The most common high-strength bolts are specified by A325. High-strength rivets and extra-high-strength bolts are available for use with high-strength structural steel.

## Allowable Stresses in A36 Steel

| | lb/in² | kg/cm² |
| --- | --- | --- |
| **Tension:** | | |
| On net section, except at pinholes | 22,000 | 1,547 |
| On net section, at pinholes | 16,000 | 1,139 |
| **Compression:** See Table 15 | | |
| **Bending** tension and compression on extreme fibers: | | |
| Basic stress, reduced in certain cases | 22,000 | 1,547 |
| Compact, adequately braced beams | 24,000 | 1,687 |
| Rectangular bearing plates | 27,000 | 1,891 |
| **Shear**—webs of beams, gross section | 14,500 | 1,019 |

Allowable stresses may be increased by one-third when produced by wind or seismic loading alone or when combined with design dead and live loads.

## Allowable Stresses in Connections

| | lb/in² | kg/cm² |
| --- | --- | --- |
| **Bearing:** A36 steel | | |
| Pins in reamed, drilled, or bored holes | 33,000 | 2,320 |
| Bolts and rivets | 48,500 | 3,417 |
| Rollers, lb/lin in (kg/lin cm) | 760 × diam | 53.5 × diam |
| **Shear:** bearing-type connections | | |
| A502, grade 1 hot-driven rivets | 15,000 | 1,055 |
| A307 bolts | 10,000 | 703 |
| A325 bolts, when threading is excluded from shear planes | 22,000 | 1,547 |
| A325 bolts, when threading is not excluded from shear planes | 15,000 | 1,055 |
| **Tension:** | | |
| A502, grade 1 hot-driven rivets | 20,000 | 1,406 |
| A307 bolts (net area) | 22,000 | 1,547 |
| A325 bolts (nominal area) | 40,000 | 2,812 |
| **Bending** in pins of A36 steel | 27,000 | 1,898 |

## Proportion of Parts

**Deflection** may govern in such members as cantilevers and lightly loaded roof beams. **Buckling**, rather than strength, may govern the design of compression members. The slenderness ratio $Kl/r$, where $Kl$ is the effective length of the member and $r$ is its radius of gyration, should be limited to 200 in compression members, 240 in main tension members, and 300 in bracing and other secondary tension members. $Kl$ should not be less than the actual unbraced length $l$ in columns of a frame which depends on its bending stiffness for lateral stiffness. **Width-thickness ratios** are specified for projecting elements under compression. **Reversal of stress** leading to fatigue may be a controlling factor. Rules are given for **combined stresses** of tension, compression, bending, and shear.

**Tension members** should be proportioned for the net section, deducting for bolt or rivet holes ⅛ in (0.3 cm) larger than the nominal diameter of the fastener.

**Columns** and other **compression members** subject to eccentric load or to axial load and bending are governed by special rules. A long-established rule is that $f_a/F_a + f_b/F_b$ should be equal to or less than unity, where $f_a$ is the axial stress, $f_b$ the bending stress, and $F_a$ and $F_b$ are the corresponding allowable stresses if axial or bending stress alone exist. This is still considered valid when $f_a/F_a$ is less than 0.15. Joints shall be fully spliced, except that where reversal of stress is not expected and the joint is laterally supported, the ends of the members may be milled to plane parallel surfaces normal to the stresses and abutted with sufficient splicing to hold the connected members accurately in place. Column bases should be planed on top for the column bearing, except for rolled-steel bearing plates 4 in (10 cm) or less in thickness.

**Beams and girders,** of rolled section or built-up, should in general be proportioned by the moment of inertia of the gross section. For A36 steel, flanges in compression should have a thickness of 1/16 the projecting half width, and webs should have a thickness of 1/320 the maximum clear distance between

Table 15. Allowable Stress, in Ksi, for Compression Members of A36 Steel

| Main and secondary members, $Kl/r$ not more than 120 | | | | | | Main members, $Kl/r$, 121–200 | | | | Secondary members, $l/r$, 121–200 | | | |
|---|---|---|---|---|---|---|---|---|---|---|---|---|---|
| $\dfrac{Kl}{r}$ | $F_a$ | $\dfrac{Kl}{r}$ | $F_a$ | $\dfrac{Kl}{r}$ | $F_a$ | $\dfrac{Kl}{r}$ | $F_a$ | $\dfrac{Kl}{r}$ | $F_a$ | $\dfrac{l}{r}$ | $F_{as}$ | $\dfrac{l}{r}$ | $F_{as}$ |
| 1 | 21.56 | 41 | 19.11 | 81 | 15.24 | 121 | 10.14 | 161 | 5.76 | 121 | 10.19 | 161 | 7.25 |
| 5 | 21.39 | 45 | 18.78 | 85 | 14.79 | 125 | 9.55 | 165 | 5.49 | 125 | 9.80 | 165 | 7.08 |
| 10 | 21.16 | 50 | 18.35 | 90 | 14.20 | 130 | 8.84 | 170 | 5.17 | 130 | 9.30 | 170 | 6.89 |
| 15 | 20.89 | 55 | 17.90 | 95 | 13.60 | 135 | 8.19 | 175 | 4.88 | 135 | 8.86 | 175 | 6.73 |
| 20 | 20.60 | 60 | 17.43 | 100 | 12.98 | 140 | 7.62 | 180 | 4.61 | 140 | 8.47 | 180 | 6.58 |
| 25 | 20.28 | 65 | 16.94 | 105 | 12.33 | 145 | 7.10 | 185 | 4.36 | 145 | 8.12 | 185 | 6.46 |
| 30 | 19.94 | 70 | 16.43 | 110 | 11.67 | 150 | 6.64 | 190 | 4.14 | 150 | 7.81 | 190 | 6.36 |
| 35 | 19.58 | 75 | 15.90 | 115 | 10.99 | 155 | 6.22 | 195 | 3.93 | 155 | 7.53 | 195 | 6.28 |
| 40 | 19.19 | 80 | 15.36 | 120 | 10.28 | 160 | 5.83 | 200 | 3.73 | 160 | 7.29 | 200 | 6.22 |

flanges. Web stiffeners should be provided at points of high concentrated loads; additional web stiffeners are required in plate girders. Splices in the webs of plate girders should be made by plates on both sides of the web. When two or more rolled beams or channels are used side by side to form a beam, they should be connected at separators spaced no more than 5 ft (1.52 m); beams deeper than 12 in (30 cm) are to have at least two bolts to each separator.

The **lateral force on crane runways** due to the effect of moving crane trolleys may be assumed as 20 percent of the sum of the weights of the lifted loads and of the crane trolley (but exclusive of the other parts of the crane) applied at the top of the rail, one-half on each side of the runway, and shall be considered as acting in either direction normal to the runway rail. The **longitudinal force** may be assumed as 10 percent of the maximum wheel reactions of the crane applied at the top of the rail.

**Riveted** or **bolted connections** carrying calculated stress, except lacing and sag bars, should be designed to support not less than 6,000 lb (2721 kg). Rivets or high-strength bolts are preferred in all places, and high-strength bolting is implied in these paragraphs wherever riveting is mentioned; unfinished bolts, A307, may be used in the shop or in field connections of small unimportant structures of secondary members, bracing, and beams.

Members in tension or compression, meeting at a joint, shall have their lines of center of gravity pass through a point, if practicable; if not, provision shall be made for the eccentricity. A group of rivets transmitting stress to a member shall have its center of gravity in the line of the stress, if practicable; if not, the group shall be designed for the resulting eccentricity. Pins and rivet groups may be so placed as to offset the effect of bending in the member due to dead load. Where stress is transmitted from one member to another by rivets through a loose filler, the filler shall be extended beyond the connected member and the extension secured by enough rivets to distribute the total stress in the member uniformly over the combined sections of the member and the filler.

The finished shank of *turned bolts* shall be long enough to provide full bearing, and washers shall be used under the nuts to grip the parts when nuts are turned tight. The holes shall be reamed, after assembly of the connected parts, to a clearance not more than $1/50$ in (0.05 cm) over the diameter of the bolt. (See also Sec. 8.)

The required strength of riveted connections shall be developed by the shearing and bearing values of the rivets, but

rivets in shelf angles and brackets, and in connections in so far as they give stiffness to the structure, may transmit stress by tension. Rivets connecting the web to a flange subject to transverse load shall be proportioned for the resultant of the longitudinal and transverse stress. Rivets connecting any portion of a girder flange to the remainder, and thus to the web, shall be proportioned for the increment of stress in that portion, between rivets. The pitch of such rivets is determined as follows: $p = nRI/Vm$, where $p$ is the pitch of rivets in any row, in or cm, $n$ is the number of rows of rivets serving the purpose in question, $R$ is the value of one rivet, lb or kg, whether determined by shear or bearing, $V$ is the transverse shear in the girder, $m$ is the product of the area of cross section of the flange portion in question, times the distance from its center of gravity to the gravity axis of the girder section, $in^3$ or $cm^3$. $I$ is the moment of inertia of the girder section, $in^4$ or $cm^4$. (See also Sec. 8.)

Rivets shall be proportioned by the nominal diameter. Rivets whose grip exceeds 5 diam shall be allowed 1 percent less safe stress for each $1/16$ in (0.16 cm) excess length. The minimum distance between centers of rivet holes shall be $2\frac{2}{3}$ diam of the rivet; but preferably not less than 3 diam.

The minimum distance from the center of any rivet hole to a sheared edge shall be $2\frac{1}{4}$ in (5.7 cm) for $1\frac{1}{4}$ in (32 mm) rivets, 2 in (5.1 cm) for $1\frac{1}{8}$ in (28 mm) rivets, $1\frac{3}{4}$ in (4.4 cm) for 1 in (25 mm) rivets, $1\frac{1}{2}$ in (3.8 cm) for $7/8$ in (22 mm) rivets, $1\frac{1}{4}$ in (3.2 cm) for $3/4$ in (19 mm) rivets, $1\frac{1}{8}$ in (2.8 cm) for $5/8$ in (16 mm) rivets, and 1 in (2.5 cm) for $1/2$ in (13 mm) rivets. The distance from any edge shall not exceed 12 times the thickness of the plate and shall not exceed 6 in (15 cm).

**Pins** may be used in heavy trusses of long span when riveted field connection would become unwieldy, or for other reasons. Pinholes in rolled members should be reinforced with plates if necessary, with sufficient rivets to transmit to the member their portion of the pin pressure. Members packed on pins should be held against lateral movement.

**Tie Plates.** The open sides of compression members built up from plates shall be provided with lacing having tie plates at each end and at intermediate points if the lattice is interrupted. Tie plates shall be as near the ends as practicable. In main members carrying calculated stresses, the end tie plates shall have a length of not less than the distance between the lines of rivets connecting them to the flanges, and intermediate ones of not less than one-half of this distance. The thickness of tie plates shall not be less than one-fiftieth of this distance, and the rivet pitch shall be not more than 6 diam. Tie plates for tension members shall have a length not less than two-thirds that required for compression members.

**Lacing bars** shall be proportioned to resist a shearing stress normal to the axis of the member equal to at least two percent of the total compression stress in the member. Lacing bars shall preferably be arranged in a single system for which the ratio $l/r$ shall not exceed 140. For double lacing this ratio shall not exceed 200. Double lacing bars shall be joined at their intersections.

The inclination of lacing bars to the axis of the members shall generally be not less than 60 deg for single lacing and 45 deg for double lacing. When the distance between the rivet lines in the flanges is more than 15 in (38.1 cm), the lacing shall be double and riveted at the intersection if bars are used, or else shall be made of angles.

Lacing bars shall be so spaced that the ratio $l/r$ of the flange included between their connections shall be not over three-fourths of that of the member as a whole.

**Bracing** provided in steel frames to resist wind and other lateral forces should preferably be of stiff members and should also serve to brace the frame during erection.

### Design of Members

**Properties of Standard Structural Shapes**  Tables 17 to 24 give the properties of American Standard channels and I beams, wide-flange beams and columns, angles, and tees. In these tables, $I$ = moment of inertia, $r$ = radius of gyration, $S$ = section modulus, $x$ = distance from gravity axis to face, $V$ = max safe shearing strength of the web in kips, and $R$ = max end reaction on $3\frac{1}{2}$-in (9-mm) seat, based on crippling of web, in kips. $R$ values are omitted where web crippling does not govern.

A great variety of tees is produced by shearing or gas-cutting standard beams or wide-flange sections (WF) lengthwise at midheight of the web, making two similar shapes of T section. Table 24 lists a selection of such tees.

For additional data regarding structural shapes, their strengths as beams and columns, and means of making connections, see "Steel Construction," American Institute of Steel Construction, New York.

**Standard Rivet Gages**  The standard gages for rivets in

angles are given in Table 26, and gages for rivets in the flanges of channels and I beams in Table 27.

**Welding** (see also Sec. 13).  Advantages of assembling steel frames by welding are quietness of erection compared with riveting and reduction in the amount of metal used. The saving in metal is made through (1) elimination of rivet holes which reduce the net section of tension members, (2) simplification of details, and (3) taking advantage of continuity in beams. For an example of simplification, web stiffeners are made of plates with their edges welded to the web.

The AWS Building Code specifies the following working unit stresses for use in designing steel building frames, in pounds per square inch through throat of weld:

Shear, 21,000 lb/in² (1,476 kg/cm²); tension, compression, and shear on section through throat of butt welds, same as corresponding allowable stress for base metal. The value of 21,000 lb/in² for shear gives the following values per lineal inch for fillets subjected to longitudinal shear: ¼-in fillets, 3,700 lb; ⅜-in fillets, 5,570 lb; ½-in fillets, 7,420 lb. Both gas and electric arc-welding processes are used.

**Safe Loads for Steel Beams**  To determine the safe load uniformly distributed, as limited by bending, for a structural steel beam on a given span, apply the formula $W = 8fS/l$, where $W$ is the total load, lb.; $f$ is the fiber stress (20,000 lb/in² or any other); $S$ is the section modulus for the beam in question, given in Tables 17 to 24; and $l$ is the span, in. (This formula may also be used with equivalent metric units.) The safe load concentrated at midspan is one-half this amount. For other safe loads, note that $fS$ is the safe resistance to bending in inch-pounds (or Kg·cm) afforded by the beam. Compute the load, of whatever type or distribution, which will produce a maximum bending moment equal to safe moment of resistance (see Sec. 5 for bending-moment formulas).

**To select a beam** to support a given load, compute the maximum bending moment in inch-pounds, divide by the allowable fiber stress, and refer to the table for a beam having a section modulus which is not smaller than the quotient.

Formulas for the safe loads and deflections of beams with various methods of support and of loading are given in Sec. 5.

**Short beams** should be investigated for crippling of the web. In the tables are given the safe end reactions for beams of A36

**Table 16. Coefficients of Deflection for Steel Beams under Uniformly Distributed Loads**

| Span, ft | Fiber stress, psi | | Span, ft | Fiber stress, psi | | Span, ft | Fiber stress, psi | | Span, ft | Fiber stress, psi | |
|---|---|---|---|---|---|---|---|---|---|---|---|
| | 24,000 | 10,000 | | 24,000 | 10,000 | | 24,000 | 10,000 | | 24,000 | 10,000 |
| 1 | 0.026 | 0.011 | 14 | 4.87 | 2.029 | 27 | 18.1 | 7.54 | 39 | 37.7 | 15.7 |
| 2 | 0.098 | 0.041 | 15 | 5.59 | 2.328 | 28 | 19.5 | 8.12 | 40 | 39.8 | 16.6 |
| 3 | 0.223 | 0.093 | 16 | 6.36 | 2.648 | 29 | 20.9 | 8.71 | 41 | 41.8 | 17.4 |
| 4 | 0.398 | 0.166 | 17 | 7.18 | 2.990 | 30 | 22.4 | 9.32 | 42 | 43.9 | 18.3 |
| 5 | 0.621 | 0.259 | 18 | 8.04 | 3.35 | 31 | 23.9 | 9.94 | 43 | 45.8 | 19.1 |
| 6 | 0.892 | 0.372 | 19 | 8.97 | 3.74 | 32 | 25.4 | 10.60 | 44 | 48.0 | 20.0 |
| 7 | 1.23 | 0.507 | 20 | 9.93 | 4.14 | 33 | 27.0 | 11.27 | 45 | 50.4 | 21.0 |
| 8 | 1.59 | 0.662 | 21 | 10.9 | 4.56 | 34 | 28.7 | 11.96 | 46 | 52.6 | 21.9 |
| 9 | 2.01 | 0.838 | 22 | 12.1 | 5.01 | 35 | 30.5 | 12.7 | 47 | 54.7 | 22.8 |
| 10 | 2.48 | 1.034 | 23 | 13.1 | 5.47 | 36 | 32.2 | 13.4 | 48 | 57.1 | 23.8 |
| 11 | 3.00 | 1.251 | 24 | 14.3 | 5.96 | 37 | 34.1 | 14.2 | 49 | 59.5 | 24.8 |
| 12 | 3.58 | 1.489 | 25 | 15.6 | 6.47 | 38 | 35.8 | 14.9 | 50 | 62.2 | 25.9 |
| 13 | 4.20 | 1.748 | 26 | 16.8 | 7.00 | | | | | | |

For a load concentrated at midspan, use ⅝ of the coefficient given.

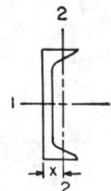

**Table 17. American Standard Channels**

| Depth of chan- nel, in. | Weight per ft, lb | Area of sec- tion, sq in. | Width of flange, in. | Thick- ness of web, in. | Axis 1–1 | | | Axis 2–2 | x, in. | V* | R* | Rows in std conn† |
|---|---|---|---|---|---|---|---|---|---|---|---|---|
| | | | | | I, in.⁴ | r, in. | S, in.³ | r, in. | | 1,000 lb | | |
| 15 | 50.0 | 14.64 | 3.716 | 0.716 | 401.4 | 5.24 | 53.6 | 0.87 | 0.80 | 156 | 93 | 4 |
| | 40.0 | 11.70 | 3.520 | 0.520 | 346.3 | 5.44 | 46.2 | 0.89 | 0.78 | 113 | 68 | |
| | 33.9 | 9.90 | 3.400 | 0.400 | 312.6 | 5.62 | 41.7 | 0.91 | 0.79 | 87 | 52 | |
| 12 | 30.0 | 8.79 | 3.170 | 0.510 | 161.2 | 4.28 | 26.9 | 0.77 | 0.68 | 89 | 63 | 3 |
| | 25.0 | 7.32 | 3.047 | 0.387 | 143.5 | 4.43 | 23.9 | 0.79 | 0.68 | 67 | 48 | |
| | 20.7 | 6.03 | 2.940 | 0.280 | 128.1 | 4.61 | 21.4 | 0.81 | 0.70 | 49 | 34 | |
| 10 | 30.0 | 8.80 | 3.033 | 0.673 | 103.0 | 3.42 | 20.6 | 0.67 | 0.65 | 98 | 81 | 2 |
| | 25.0 | 7.33 | 2.886 | 0.526 | 90.7 | 3.52 | 18.1 | 0.68 | 0.62 | 76 | 63 | |
| | 20.0 | 5.86 | 2.739 | 0.379 | 78.5 | 3.66 | 15.7 | 0.70 | 0.61 | 55 | 45 | |
| | 15.3 | 4.47 | 2.600 | 0.240 | 66.9 | 3.87 | 13.4 | 0.72 | 0.64 | 35 | 29 | |
| 9 | 20.0 | 5.86 | 2.648 | 0.448 | 60.6 | 3.22 | 13.5 | 0.65 | 0.59 | 58 | 53 | 2 |
| | 15.0 | 4.39 | 2.485 | 0.285 | 50.7 | 3.40 | 11.3 | 0.67 | 0.59 | 37 | 34 | |
| | 13.4 | 3.89 | 2.430 | 0.230 | 47.3 | 3.49 | 10.5 | 0.67 | 0.61 | 30 | 27 | |
| 8 | 18.75 | 5.49 | 2.527 | 0.487 | 43.7 | 2.82 | 10.9 | 0.60 | 0.57 | 56 | ... | 2 |
| | 13.75 | 4.02 | 2.343 | 0.303 | 35.8 | 2.99 | 9.0 | 0.62 | 0.56 | 35 | ... | |
| | 11.5 | 3.36 | 2.260 | 0.220 | 32.3 | 3.10 | 8.1 | 0.63 | 0.58 | 26 | ... | |
| 7 | 14.75 | 4.32 | 2.299 | 0.419 | 27.1 | 2.51 | 7.7 | 0.57 | 0.53 | 43 | ... | 1 |
| | 12.25 | 3.58 | 2.194 | 0.314 | 24.1 | 2.59 | 6.9 | 0.58 | 0.53 | 32 | ... | |
| | 9.8 | 2.85 | 2.090 | 0.210 | 21.1 | 2.72 | 6.0 | 0.59 | 0.55 | 21 | ... | |
| 6 | 13.0 | 3.81 | 2.157 | 0.437 | 17.3 | 2.13 | 5.8 | 0.53 | 0.52 | 38 | ... | 1 |
| | 10.5 | 3.07 | 2.034 | 0.314 | 15.1 | 2.22 | 5.0 | 0.53 | 0.50 | 27.3 | ... | |
| | 8.2 | 2.39 | 1.920 | 0.200 | 13.0 | 2.34 | 4.3 | 0.54 | 0.52 | 17.4 | ... | |
| 5 | 9.0 | 2.63 | 1.885 | 0.325 | 8.8 | 1.83 | 3.5 | 0.49 | 0.48 | 23.6 | ... | 1 |
| | 6.7 | 1.95 | 1.750 | 0.190 | 7.4 | 1.95 | 3.0 | 0.50 | 0.49 | 13.8 | ... | |
| 4 | 7.25 | 2.12 | 1.720 | 0.320 | 4.5 | 1.47 | 2.3 | 0.46 | 0.46 | 18.6 | | |
| | 5.4 | 1.56 | 1.500 | 0.100 | 3.8 | 1.56 | 1.9 | 0.45 | 0.46 | 10.4 | | |
| 3 | 6.0 | 1.75 | 1.596 | 0.356 | 2.1 | 1.08 | 1.4 | 0.42 | 0.46 | 15.5 | | |
| | 5.0 | 1.46 | 1.498 | 0.258 | 1.8 | 1.12 | 1.2 | 0.41 | 0.44 | 11.2 | | |
| | 4.1 | 1.19 | 1.410 | 0.170 | 1.6 | 1.17 | 1.1 | 0.41 | 0.44 | 7.4 | | |

*V and R values are for channels of A36 steel.       †For standard connections see Table 25.

steel resting on a seat 3½ in (9 cm) longitudinally of the beam, computed by the formula $R = 27,000(a + k)$, where $t$ = thickness of the web; $a$ = length of the seat; and $k$ = distance from outer face of flange to web toe of fillet; all in inches. Short beams, whose webs are stiffened against crippling, should be investigated for shear, by dividing the maximum shear, in pounds, by the area of the web, excluding the flanges.

**Single angles used as beams** and loaded in the plane of axis X-X or Y-Y tend to deflect laterally as well as in the plane of the loads. Unless this is prevented, as by pairing the angles back to back and securing them together, the unit fiber stress due to bending may be as much as 40 percent above that computed by dividing the bending moment by $S$ for the axis perpendicular to the plane of the loads. The relation $f = M/S$ does not hold for single angles, and Z bars, which are unsymmetrical about both axes.

**Deflection of I Beams and Other Structural Shapes**   Table 16 gives **coefficients of deflection** for steel shapes under uniformly distributed loads, and is based on the formula: deflection in inches = $30fL^2/Ed$, the table giving the values of $30fL^2/E$. ($f$ = fiber stress, lb/in²; $L$ = span, ft; $d$ = depth of section, in; $E$ = modulus of elasticity = 29,000,000 lb/in².)

To find the deflection in inches of a section symmetrical about the neutral axis, such as a beam, channel, etc., divide the coefficient in the table corresponding to given span and fiber stress by the depth of the section in inches.

To find the deflection in inches of a section which is not symmetrical about the neutral axis but which is symmetrical about an axis at right angles thereto, such as a tee or pair of

**Table 18. American Standard I Beams**

| Depth of beam, in. | Weight per ft, lb | Area of section, sq in. | Width of flange, in. | Thickness of web, in. | Neutral axis perpendicular to web at center | | | Neutral axis coincident with center line of web | | | V* | R* | Rows in std conn† |
|---|---|---|---|---|---|---|---|---|---|---|---|---|---|
| | | | | | I, in.⁴ | r, in. | S, in.³ | I, in.⁴ | r, in. | S, in.³ | 1,000 lb | | |
| 24 | 120.0 | 35.13 | 8.048 | 0.798 | 3010.8 | 9.26 | 250.9 | 84.9 | 1.56 | 21.1 | 278 | 117 | 4 |
| | 105.9 | 30.98 | 7.875 | 0.625 | 2811.5 | 9.53 | 234.3 | 78.9 | 1.60 | 20.0 | 218 | 92 | |
| | 100.0 | 29.25 | 7.247 | 0.747 | 2371.8 | 9.05 | 197.6 | 48.4 | 1.29 | 13.4 | 260 | 103 | |
| | 90.0 | 26.30 | 7.124 | 0.624 | 2230.1 | 9.21 | 185.8 | 45.5 | 1.32 | 12.8 | 217 | 86 | |
| | 79.9 | 23.33 | 7.000 | 0.500 | 2087.2 | 9.46 | 173.9 | 42.9 | 1.36 | 12.2 | 174 | 69 | |
| 20 | 95.0 | 27.74 | 7.200 | 0.800 | 1599.7 | 7.59 | 160.0 | 50.5 | 1.35 | 14.0 | 232 | 113 | 4 |
| | 85.0 | 24.80 | 7.053 | 0.653 | 1501.7 | 7.78 | 150.2 | 47.0 | 1.38 | 13.3 | 189 | 93 | |
| | 75.0 | 21.90 | 6.391 | 0.641 | 1263.5 | 7.60 | 126.3 | 30.1 | 1.17 | 9.4 | 186 | 88 | |
| | 65.4 | 19.08 | 6.250 | 0.500 | 1169.5 | 7.83 | 116.9 | 27.9 | 1.21 | 8.9 | 145 | 68 | |
| 18 | 70.0 | 20.46 | 6.251 | 0.711 | 917.5 | 6.70 | 101.9 | 24.5 | 1.09 | 7.8 | 186 | 94 | 4 |
| | 54.7 | 15.94 | 6.000 | 0.460 | 795.5 | 7.07 | 88.4 | 21.2 | 1.15 | 7.1 | 120 | 61 | |
| 15 | 50.0 | 14.59 | 5.640 | 0.550 | 481.1 | 5.74 | 64.2 | 16.0 | 1.05 | 5.7 | 120 | 71 | 4 |
| | 42.9 | 12.49 | 5.500 | 0.410 | 441.8 | 5.95 | 58.9 | 14.6 | 1.08 | 5.3 | 89 | 53 | |
| 12 | 50.0 | 14.57 | 5.477 | 0.687 | 301.6 | 4.55 | 50.3 | 16.0 | 1.05 | 5.8 | 120 | 89 | 3 |
| | 40.8 | 11.84 | 5.250 | 0.460 | 268.9 | 4.77 | 44.8 | 13.8 | 1.08 | 5.3 | 80 | 60 | |
| | 35.0 | 10.20 | 5.078 | 0.428 | 227.0 | 4.72 | 37.8 | 10.0 | 0.99 | 3.9 | 74 | 53 | |
| | 31.8 | 9.26 | 5.000 | 0.350 | 215.8 | 4.83 | 36.0 | 9.5 | 1.01 | 3.8 | 61 | 44 | |
| 10 | 35.0 | 10.22 | 4.944 | 0.594 | 145.8 | 3.78 | 29.2 | 8.5 | 0.91 | 3.4 | 86 | 72 | 2 |
| | 25.4 | 7.38 | 4.660 | 0.310 | 122.1 | 4.07 | 24.4 | 6.9 | 0.97 | 3.0 | 45 | 38 | |
| 8 | 23.0 | 6.71 | 4.171 | 0.441 | 64.2 | 3.09 | 16.0 | 4.4 | 0.81 | 2.1 | 51 | ... | 2 |
| | 18.4 | 5.34 | 4.000 | 0.270 | 56.9 | 3.26 | 14.2 | 3.8 | 0.84 | 1.9 | 31 | ... | |
| 7 | 20.0 | 5.83 | 3.860 | 0.450 | 41.9 | 2.68 | 12.0 | 3.1 | 0.74 | 1.6 | 46 | ... | 1 |
| | 15.3 | 4.43 | 3.660 | 0.250 | 36.2 | 2.86 | 10.4 | 2.7 | 0.78 | 1.5 | 25 | ... | |
| 6 | 17.25 | 5.02 | 3.565 | 0.465 | 26.0 | 2.28 | 8.7 | 2.3 | 0.68 | 1.3 | 40.5 | ... | 1 |
| | 12.5 | 3.61 | 3.330 | 0.230 | 21.8 | 2.46 | 7.3 | 1.8 | 0.72 | 1.1 | 20 | ... | |
| 5 | 14.75 | 4.29 | 3.284 | 0.494 | 15.0 | 1.87 | 6.0 | 1.7 | 0.63 | 1.0 | 35.8 | ... | 1 |
| | 10.0 | 2.87 | 3.000 | 0.210 | 12.1 | 2.05 | 4.8 | 1.2 | 0.65 | 0.82 | 15.2 | ... | |
| 4 | 9.5 | 2.76 | 2.796 | 0.326 | 6.7 | 1.56 | 3.3 | 0.91 | 0.58 | 0.65 | 18.9 | | |
| | 7.7 | 2.21 | 2.660 | 0.190 | 6.0 | 1.64 | 3.0 | 0.77 | 0.59 | 0.58 | 11.0 | | |
| 3 | 7.5 | 2.17 | 2.509 | 0.349 | 2.9 | 1.15 | 1.9 | 0.59 | 0.52 | 0.47 | 15.2 | | |
| | 5.7 | 1.64 | 2.330 | 0.170 | 2.5 | 1.23 | 1.7 | 0.46 | 0.53 | 0.40 | 7.4 | | |

Lightweight beams of each depth are usual stock sizes.
*V and R values are for beams of A36 steel.
†For standard connections see Table 25.

angles, divide the coefficient corresponding to given span and fiber stress by twice the distance of extreme fiber from neutral axis obtained from table of elements of sections.

To find the deflection in inches of a section for any other fiber stress than those given, multiply this fiber stress by either of the coefficients in the table for the given span and divide by the fiber stress corresponding to the coefficients used.

I beams and channels loaded to a fiber stress of 20,000 lb/in² will not deflect in excess of $\frac{1}{360}$ of the span (allowed for plastered ceilings) if the depth in inches is not less than 0.62 times the span in feet for uniform loads and 0.50 times the span for central concentration.

**Beam Supports** Steel beams are supported at the ends generally (1) by means of web connections to girders and columns, (2) by resting on structural-steel seats, or (3) by resting on masonry. Limiting values of end reactions of the second type, for seats 3½ in (9 cm) long, are given in Tables 17 to 19. Standard AISC web connections of the first type are called *framed beam* connections and are designated by the number of rows of rivets or bolts. These numbers are listed in Tables 17 to 19, and examples of connections are given in Fig. 30. These connections may be specified as "Standard 3 row, 4 row, etc., connections." Standard AISC connections for heavier loads are called *heavy framed beam* connections. Special connections must always be designated and detailed when the end reactions exceeds the capacity of the standard connection.

The capacity of web connections is governed by the shearing of the fastener, or the bearing of the fastener on the web or on the material to which the beam is connected, or by the strength of the connecting angles. The supporting values of standard framed beam connections, using ⅞ in fasteners in members of A36 material, are given in Table 25. For fasteners in webs thicker than 0.42 in use the values in the column

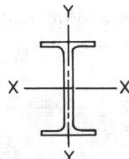

## Table 19. Properties of Wide-Flange Beams and Columns

| Nominal size, in. | Weight per ft, lb | Area of section, sq in. | Depth of section, in. | Flange Width, in. | Flange Thickness, in. | Web thickness, in. | Neutral axis perpendicular to web at center I, in.⁴ | S, in.³ | r, in. | Neutral axis parallel to web at center I, in.⁴ | S, in.³ | r, in. | V, 1,000 lb* | R, 1,000 lb* | Rows in std conn† |
|---|---|---|---|---|---|---|---|---|---|---|---|---|---|---|---|
| 36 × 16½ | 300 | 88.17 | 36.72 | 16.65 | 1.680 | 0.945 | 20,290 | 1,105 | 15.17 | 1,225 | 147.1 | 3.73 | 503 | 161 | 10 |
|  | 280 | 82.32 | 36.50 | 16.59 | 1.570 | 0.885 | 18,819 | 1,031 | 15.12 | 1,127 | 135.9 | 3.70 | 468 | 148 |  |
|  | 260 | 76.56 | 36.24 | 16.55 | 1.440 | 0.845 | 17,233 | 951 | 15.00 | 1,020 | 123.3 | 3.65 | 444 | 138 |  |
|  | 245 | 72.03 | 36.06 | 16.51 | 1.350 | 0.802 | 16,092 | 892 | 14.95 | 944 | 114.4 | 3.62 | 419 | 129 |  |
|  | 230 | 67.73 | 35.88 | 16.47 | 1.260 | 0.765 | 14,988 | 835 | 14.88 | 870 | 105.7 | 3.59 | 398 | 121 |  |
| 36 × 12 | 194 | 57.11 | 36.48 | 12.11 | 1.260 | 0.770 | 12,103 | 663 | 14.56 | 355 | 58.7 | 2.49 | 407 | 117 | 10 |
|  | 182 | 53.54 | 36.32 | 12.07 | 1.180 | 0.725 | 11,281 | 621 | 14.52 | 327 | 54.3 | 2.47 | 382 | 109 |  |
|  | 170 | 49.98 | 36.16 | 12.02 | 1.100 | 0.680 | 10,470 | 579 | 14.47 | 300 | 50.0 | 2.45 | 357 | 100 |  |
|  | 160 | 47.09 | 36.00 | 12.00 | 1.020 | 0.653 | 9,738 | 541 | 14.38 | 275 | 45.9 | 2.42 | 341 | 95 |  |
|  | 150 | 44.16 | 35.84 | 11.97 | 0.940 | 0.625 | 9,012 | 502 | 14.29 | 250 | 41.8 | 2.38 | 325 | 90 |  |
| 33 × 15¾ | 240 | 70.52 | 33.50 | 15.86 | 1.400 | 0.830 | 13,585 | 811 | 13.88 | 874 | 110.2 | 3.52 | 403 | 133 | 9 |
|  | 220 | 64.73 | 33.25 | 15.81 | 1.275 | 0.775 | 12,312 | 740 | 13.79 | 782 | 99.0 | 3.48 | 374 | 122 |  |
|  | 200 | 58.79 | 33.00 | 15.75 | 1.150 | 0.715 | 11,048 | 669 | 13.71 | 691 | 87.8 | 3.43 | 342 | 110 |  |
| 33 × 11½ | 152 | 44.71 | 33.50 | 11.56 | 1.055 | 0.635 | 8,147 | 486 | 13.50 | 256 | 44.3 | 2.39 | 308 | 92 | 9 |
|  | 141 | 41.51 | 33.31 | 11.53 | 0.960 | 0.605 | 7,442 | 446 | 13.39 | 229 | 39.8 | 2.35 | 292 | 86 |  |
|  | 130 | 38.26 | 33.10 | 11.51 | 0.855 | 0.580 | 6,699 | 404 | 13.23 | 201 | 35.0 | 2.29 | 278 | 81 |  |
| 30 × 15 | 210 | 61.78 | 30.38 | 15.10 | 1.315 | 0.775 | 9,872 | 649 | 12.64 | 707 | 93.7 | 3.38 | 341 | 122 | 8 |
|  | 190 | 55.90 | 30.12 | 15.04 | 1.185 | 0.710 | 8,825 | 586 | 12.57 | 624 | 83.1 | 3.34 | 310 | 109 |  |
|  | 172 | 50.65 | 29.88 | 14.98 | 1.065 | 0.655 | 7,891 | 528 | 12.48 | 550 | 73.4 | 3.30 | 284 | 98 |  |
| 30 × 10½ | 132 | 38.83 | 30.30 | 10.55 | 1.000 | 0.615 | 5,753 | 379 | 12.17 | 185 | 35.1 | 2.18 | 270 | 86 | 8 |
|  | 124 | 36.45 | 30.16 | 10.52 | 0.930 | 0.585 | 5,347 | 354 | 12.11 | 169 | 32.3 | 2.16 | 256 | 81 |  |
|  | 116 | 34.13 | 30.00 | 10.50 | 0.850 | 0.564 | 4,919 | 327 | 12.00 | 153 | 29.2 | 2.12 | 245 | 77 |  |
|  | 108 | 31.77 | 29.82 | 10.48 | 0.760 | 0.548 | 4,461 | 299 | 11.85 | 135 | 25.8 | 2.06 | 237 | 74 |  |
| 27 × 14 | 177 | 52.10 | 27.31 | 14.09 | 1.190 | 0.725 | 6,728 | 492.8 | 11.36 | 518.9 | 73.7 | 3.16 | 287 | 110 | 7 |
|  | 160 | 47.04 | 27.08 | 14.02 | 1.075 | 0.658 | 6,018 | 444.5 | 11.31 | 458.0 | 65.3 | 3.12 | 258 | 99 |  |
|  | 145 | 42.68 | 26.88 | 13.96 | 0.975 | 0.600 | 5,414 | 402.9 | 11.26 | 406.9 | 58.3 | 3.09 | 234 | 88 |  |
| 27 × 10 | 114 | 33.53 | 27.28 | 10.07 | 0.932 | 0.570 | 4,080 | 299.2 | 11.03 | 149.6 | 29.7 | 2.11 | 225 | 79 | 7 |
|  | 102 | 30.01 | 27.07 | 10.02 | 0.827 | 0.518 | 3,604 | 266.3 | 10.96 | 129.5 | 25.9 | 2.08 | 203 | 71 |  |
|  | 94 | 27.65 | 26.91 | 9.99 | 0.747 | 0.490 | 3,266 | 242.8 | 10.87 | 115.1 | 23.0 | 2.04 | 191 | 65 |  |
| 24 × 14 | 160 | 47.04 | 24.72 | 14.09 | 1.135 | 0.656 | 5,110 | 413.5 | 10.42 | 492.6 | 69.9 | 3.23 | 235 | 97 | 6 |
|  | 145 | 42.62 | 24.49 | 14.04 | 1.020 | 0.608 | 4,561 | 372.5 | 10.34 | 434.3 | 61.8 | 3.19 | 216 | 88 |  |
|  | 130 | 38.21 | 24.25 | 14.00 | 0.900 | 0.565 | 4,009 | 330.7 | 10.24 | 375.2 | 53.6 | 3.13 | 199 | 80 |  |
| 24 × 12 | 120 | 35.29 | 24.31 | 12.08 | 0.930 | 0.556 | 3,635 | 299.1 | 10.15 | 254.0 | 42.0 | 2.68 | 196 | 78 | 6 |
|  | 110 | 32.36 | 24.16 | 12.04 | 0.855 | 0.510 | 3,315 | 274.4 | 10.12 | 229.1 | 38.0 | 2.66 | 179 | 71 |  |
|  | 100 | 29.43 | 24.00 | 12.00 | 0.775 | 0.468 | 2,987 | 248.9 | 10.08 | 203.5 | 33.9 | 2.63 | 163 | 64 |  |
| 24 × 9 | 94 | 27.63 | 24.29 | 9.06 | 0.872 | 0.516 | 2,683 | 220.9 | 9.85 | 102.2 | 22.6 | 1.92 | 182 | 69 | 6 |
|  | 84 | 24.71 | 24.09 | 9.02 | 0.772 | 0.470 | 2,364 | 196.3 | 9.78 | 88.3 | 19.6 | 1.89 | 164 | 62 |  |
|  | 76 | 22.37 | 23.91 | 8.98 | 0.682 | 0.440 | 2,096 | 175.4 | 9.68 | 76.5 | 17.0 | 1.85 | 153 | 56 |  |
| 21 × 13 | 142 | 41.76 | 21.46 | 13.13 | 1.095 | 0.659 | 3,403 | 317.2 | 9.03 | 385.9 | 58.8 | 3.04 | 205 | 96 | 5 |
|  | 127 | 37.34 | 21.24 | 13.06 | 0.985 | 0.588 | 3,017 | 284.1 | 8.99 | 338.6 | 51.8 | 3.01 | 181 | 83 |  |
|  | 112 | 32.93 | 21.00 | 13.00 | 0.865 | 0.527 | 2,620 | 249.6 | 8.92 | 289.7 | 44.6 | 2.96 | 160 | 73 |  |

*V and R values are for beams of A36 steel.
†For standard connections see Table 25.

### Table 19. Properties of Wide-Flange Beams and Columns  (*Continued*)

| Nominal size, in. | Weight per ft, lb | Area of section, sq in. | Depth of section, in. | Flange Width, in. | Flange Thickness, in. | Web thickness, in. | Neutral axis perpendicular to web at center $I$, in.⁴ | $S$, in.³ | $r$, in. | Neutral axis parallel to web at center $I$, in.⁴ | $S$, in.³ | $r$, in. | $V$, 1,000 lb* | $R$, 1,000 lb* | Rows in std conn† |
|---|---|---|---|---|---|---|---|---|---|---|---|---|---|---|---|
| 21 × 9 | 96 | 28.21 | 21.14 | 9.03 | 0.935 | 0.575 | 2,088 | 197.6 | 8.60 | 109.3 | 24.2 | 1.97 | 176 | 79 | 5 |
|  | 82 | 24.10 | 20.86 | 8.96 | 0.795 | 0.499 | 1,752 | 168.0 | 8.53 | 89.6 | 20.0 | 1.93 | 151 | 67 |  |
| 21 × 8¼ | 73 | 21.46 | 21.24 | 8.29 | 0.740 | 0.455 | 1,600 | 150.7 | 8.64 | 66.2 | 16.0 | 1.76 | 140 | 59 | 5 |
|  | 68 | 20.02 | 21.13 | 8.27 | 0.685 | 0.430 | 1,478 | 139.9 | 8.59 | 60.4 | 14.6 | 1.74 | 132 | 55 |  |
|  | 62 | 18.23 | 20.99 | 8.24 | 0.615 | 0.400 | 1,326 | 126.4 | 8.53 | 53.1 | 12.9 | 1.71 | 122 | 51 |  |
| 18 × 11¾ | 114 | 33.51 | 18.48 | 11.83 | 0.991 | 0.595 | 2,033 | 220.1 | 7.79 | 255.6 | 43.2 | 2.76 | 159 | 83 | 4 |
|  | 105 | 30.86 | 18.32 | 11.79 | 0.911 | 0.554 | 1,852 | 202.2 | 7.75 | 231.0 | 39.2 | 2.73 | 147 | 77 |  |
|  | 96 | 28.22 | 18.16 | 11.75 | 0.831 | 0.512 | 1,674 | 184.4 | 7.70 | 206.8 | 35.2 | 2.71 | 135 | 69 |  |
| 18 × 8¾ | 85 | 24.97 | 18.32 | 8.83 | 0.911 | 0.526 | 1,429 | 156.1 | 7.57 | 99.4 | 22.5 | 2.00 | 140 | 71 | 4 |
|  | 77 | 22.63 | 18.16 | 8.78 | 0.831 | 0.475 | 1,286 | 141.7 | 7.54 | 88.6 | 20.2 | 1.98 | 125 | 63 |  |
|  | 70 | 20.56 | 18.00 | 8.75 | 0.751 | 0.438 | 1,153 | 128.2 | 7.49 | 78.5 | 17.9 | 1.95 | 114 | 57 |  |
|  | 64 | 18.80 | 17.87 | 8.71 | 0.686 | 0.403 | 1,045 | 117.0 | 7.46 | 70.3 | 16.1 | 1.93 | 104 | 52 |  |
| 18 × 7½ | 60 | 17.64 | 18.25 | 7.56 | 0.695 | 0.416 | 984 | 107.8 | 7.47 | 47.1 | 12.5 | 1.63 | 110 | 53 | 4 |
|  | 55 | 16.19 | 18.12 | 7.53 | 0.630 | 0.390 | 889 | 98.2 | 7.41 | 42.0 | 11.1 | 1.61 | 102 | 49 |  |
|  | 50 | 14.71 | 18.00 | 7.50 | 0.570 | 0.358 | 800 | 89.0 | 7.38 | 37.2 | 9.9 | 1.59 | 93 | 44 |  |
| 16 × 11½ | 96 | 28.22 | 16.32 | 11.53 | 0.875 | 0.535 | 1,355 | 166.1 | 6.93 | 207.2 | 35.9 | 2.71 | 127 | 74 | 4 |
|  | 88 | 25.86 | 16.16 | 11.50 | 0.795 | 0.504 | 1,222 | 151.3 | 6.87 | 185.2 | 32.2 | 2.67 | 118 | 68 |  |
| 16 × 8½ | 78 | 22.92 | 16.32 | 8.58 | 0.875 | 0.529 | 1,042 | 127.8 | 6.74 | 87.5 | 20.4 | 1.95 | 125 | 71 | 4 |
|  | 71 | 20.86 | 16.16 | 8.54 | 0.795 | 0.486 | 936 | 115.9 | 6.70 | 77.9 | 18.2 | 1.93 | 114 | 64 |  |
|  | 64 | 18.80 | 16.00 | 8.50 | 0.715 | 0.443 | 833 | 104.2 | 6.66 | 68.4 | 16.1 | 1.91 | 103 | 58 |  |
|  | 58 | 17.04 | 15.86 | 8.46 | 0.645 | 0.407 | 746 | 94.1 | 6.62 | 60.5 | 14.3 | 1.88 | 94 | 52 |  |
| 16 × 7 | 50 | 14.70 | 16.25 | 7.07 | 0.628 | 0.380 | 655 | 80.7 | 6.68 | 34.8 | 9.8 | 1.54 | 90 | 47 | 4 |
|  | 45 | 13.24 | 16.12 | 7.03 | 0.563 | 0.346 | 583 | 72.4 | 6.64 | 30.5 | 8.7 | 1.52 | 81 | 43 |  |
|  | 40 | 11.77 | 16.00 | 7.00 | 0.503 | 0.307 | 515 | 64.4 | 6.62 | 26.5 | 7.6 | 1.50 | 71 | 37 |  |
|  | 36 | 10.59 | 15.85 | 6.99 | 0.428 | 0.299 | 446 | 56.3 | 6.49 | 22.1 | 6.3 | 1.45 | 69 | 36 |  |
| 14 × 16 | 426 | 125.2 | 18.69 | 16.69 | 3.033 | 1.875 | 6,610 | 707.4 | 7.26 | 2,359 | 282.7 | 4.34 | ..... | ..... | 3 |
|  | 398 | 116.9 | 18.31 | 16.59 | 2.843 | 1.770 | 6,013 | 656.9 | 7.17 | 2,169 | 261.6 | 4.31 | ..... | ..... |  |
|  | 370 | 108.8 | 17.94 | 16.47 | 2.658 | 1.655 | 5,454 | 608.1 | 7.08 | 1,986 | 241.1 | 4.27 | ..... | ..... | 3 |
|  | 342 | 100.6 | 17.56 | 16.36 | 2.468 | 1.545 | 4,911 | 559.4 | 6.99 | 1,806 | 220.8 | 4.24 | ..... | ..... |  |
|  | 314 | 92.30 | 17.19 | 16.23 | 2.283 | 1.415 | 4,399 | 511.9 | 6.90 | 1,631 | 201.0 | 4.20 | ..... | ..... | 3 |
|  | 287 | 84.37 | 16.81 | 16.13 | 2.093 | 1.310 | 3,912 | 465.5 | 6.81 | 1,466 | 181.8 | 4.17 | ..... | ..... |  |
|  | 264 | 77.63 | 16.50 | 16.02 | 1.938 | 1.205 | 3,526 | 427.4 | 6.74 | 1,331 | 166.1 | 4.14 | ..... | ..... | 3 |
|  | 246 | 72.33 | 16.25 | 15.94 | 1.813 | 1.125 | 3,228 | 397.4 | 6.68 | 1,226 | 153.9 | 4.12 | ..... | ..... |  |
|  | 237 | 69.69 | 16.12 | 15.91 | 1.748 | 1.090 | 3,080 | 382.2 | 6.65 | 1,174 | 147.7 | 4.11 | ..... | ..... |  |
|  | 228 | 67.06 | 16.00 | 15.86 | 1.688 | 1.045 | 2,942 | 367.8 | 6.62 | 1,124 | 141.8 | 4.10 | ..... | ..... | 3 |
|  | 219 | 64.36 | 15.87 | 15.82 | 1.623 | 1.005 | 2,798 | 352.6 | 6.59 | 1,073 | 135.6 | 4.08 | ..... | ..... |  |
|  | 211 | 62.07 | 15.75 | 15.80 | 1.563 | 0.980 | 2,671 | 339.2 | 6.56 | 1,028 | 130.2 | 4.07 | ..... | ..... |  |
|  | 202 | 59.39 | 15.63 | 15.75 | 1.503 | 0.930 | 2,538 | 324.9 | 6.54 | 979 | 124.4 | 4.06 | ..... | ..... |  |
| 14 × 16 | 193 | 56.73 | 15.50 | 15.71 | 1.438 | 0.890 | 2,402 | 310.0 | 6.51 | 930 | 118.4 | 4.05 | ..... | ..... | 3 |
|  | 184 | 54.07 | 15.38 | 15.66 | 1.378 | 0.840 | 2,274 | 295.8 | 6.49 | 882 | 112.7 | 4.04 | ..... | ..... |  |
|  | 176 | 51.73 | 15.25 | 15.64 | 1.313 | 0.820 | 2,149 | 281.9 | 6.45 | 837 | 107.1 | 4.02 | ..... | ..... |  |
|  | 167 | 49.09 | 15.12 | 15.60 | 1.248 | 0.780 | 2,020 | 267.3 | 6.42 | 790 | 101.3 | 4.01 | ..... | ..... |  |
|  | 158 | 46.47 | 15.00 | 15.55 | 1.188 | 0.730 | 1,900 | 253.4 | 6.40 | 745 | 95.8 | 4.00 | ..... | ..... | 3 |
|  | 150 | 44.08 | 14.88 | 15.51 | 1.128 | 0.695 | 1,786 | 240.2 | 6.37 | 702 | 90.6 | 3.99 | ..... | ..... |  |
|  | 142 | 41.85 | 14.75 | 15.50 | 1.063 | 0.680 | 1,672 | 226.7 | 6.32 | 660 | 85.2 | 3.97 | ..... | ..... |  |
|  | 320‡ | 94.12 | 16.81 | 16.71 | 2.093 | 1.890 | 4,141 | 492.8 | 6.63 | 1,635 | 195.7 | 4.17 | ..... | ..... |  |

*$V$ and $R$ values are for beams of A36 steel.
†For standard connections see Table 25.
‡Column core section.

**Table 19. Properties of Wide-Flange Beams and Columns** *(Continued)*

| Nominal size, in. | Weight per ft., lb | Area of section, sq in. | Depth of section, in. | Flange Width, in. | Flange Thickness, in. | Web thickness, in. | Neutral axis perpendicular to web at center $I$, in.⁴ | $S$, in.³ | $r$, in. | Neutral axis parallel to web at center $I$, in.⁴ | $S$, in.³ | $r$, in. | $V$, 1,000 lb* | $R$, 1,000 lb* | Rows in std conn† |
|---|---|---|---|---|---|---|---|---|---|---|---|---|---|---|---|
| 14 × 14½ | 136 | 39.98 | 14.75 | 14.74 | 1.063 | 0.660 | 1,593 | 216.0 | 6.31 | 567 | 77.0 | 3.77 | .... | .... | 3 |
| | 127 | 37.33 | 14.62 | 14.69 | 0.998 | 0.610 | 1,476 | 202.0 | 6.29 | 527 | 71.8 | 3.76 | .... | .... | |
| | 119 | 34.99 | 14.50 | 14.65 | 0.938 | 0.570 | 1,373 | 189.4 | 6.26 | 491 | 67.1 | 3.75 | 120 | 78 | |
| | 111 | 32.65 | 14.37 | 14.62 | 0.873 | 0.540 | 1,266 | 176.3 | 6.23 | 454 | 62.2 | 3.73 | 113 | 73 | |
| | 103 | 30.26 | 14.25 | 14.57 | 0.813 | 0.495 | 1,165 | 163.6 | 6.21 | 419 | 57.6 | 3.72 | 102 | 66 | |
| | 95 | 27.94 | 14.12 | 14.54 | 0.748 | 0.465 | 1,063 | 150.6 | 6.17 | 383 | 52.8 | 3.71 | 95 | 61 | |
| | 87 | 25.56 | 14.00 | 14.50 | 0.688 | 0.420 | 966 | 138.1 | 6.15 | 349 | 48.2 | 3.70 | 85 | 55 | |
| 14 × 12 | 84 | 24.71 | 14.18 | 12.02 | 0.778 | 0.451 | 928 | 130.9 | 6.13 | 225.5 | 37.5 | 3.02 | 93 | 59 | 3 |
| | 78 | 22.94 | 14.06 | 12.00 | 0.718 | 0.428 | 851 | 121.1 | 6.09 | 206.9 | 34.5 | 3.00 | 87 | 56 | |
| 14 × 10 | 74 | 21.76 | 14.19 | 10.07 | 0.783 | 0.450 | 796 | 112.3 | 6.05 | 133.5 | 26.5 | 2.48 | 93 | 59 | 3 |
| | 68 | 20.00 | 14.06 | 10.04 | 0.718 | 0.418 | 724 | 103.0 | 6.02 | 121.2 | 24.1 | 2.46 | 85 | 54 | |
| | 61 | 17.94 | 13.91 | 10.00 | 0.643 | 0.378 | 641 | 92.2 | 5.98 | 107.3 | 21.5 | 2.45 | 76 | 48 | |
| 14 × 8 | 53 | 15.59 | 13.94 | 8.06 | 0.658 | 0.370 | 542 | 77.8 | 5.90 | 57.5 | 14.3 | 1.92 | 75 | 47 | 3 |
| | 48 | 14.11 | 13.81 | 8.03 | 0.593 | 0.339 | 484 | 70.2 | 5.86 | 51.3 | 12.8 | 1.91 | 68 | 43 | |
| | 43 | 12.65 | 13.68 | 8.00 | 0.528 | 0.308 | 429 | 62.7 | 5.82 | 45.1 | 11.3 | 1.89 | 61 | 38 | |
| 14 × 6¾ | 38 | 11.17 | 14.12 | 6.77 | 0.513 | 0.313 | 385 | 54.6 | 5.87 | 24.6 | 7.3 | 1.49 | 64 | 38 | 3 |
| | 34 | 10.00 | 14.00 | 6.75 | 0.453 | 0.287 | 339 | 48.5 | 5.83 | 21.3 | 6.3 | 1.46 | 58 | 34 | |
| | 30 | 8.81 | 13.86 | 6.73 | 0.383 | 0.270 | 289 | 41.8 | 5.73 | 17.5 | 5.2 | 1.41 | 54 | 32 | |
| 12 × 12 | 190 | 55.86 | 14.38 | 12.67 | 1.736 | 1.060 | 1,892 | 263.2 | 5.82 | 589.7 | 93.1 | 3.25 | .... | .... | 3 |
| | 161 | 47.38 | 13.88 | 12.51 | 1.486 | 0.905 | 1,541 | 222.2 | 5.70 | 486.2 | 77.7 | 3.20 | .... | .... | |
| | 133 | 39.11 | 13.38 | 12.36 | 1.236 | 0.755 | 1,221 | 182.5 | 5.59 | 389.9 | 63.1 | 3.16 | .... | .... | |
| | 120 | 35.31 | 13.12 | 12.32 | 1.106 | 0.710 | 1,071 | 163.4 | 5.51 | 345.1 | 56.0 | 3.13 | .... | .... | 3 |
| | 106 | 31.19 | 12.88 | 12.23 | 0.986 | 0.620 | 930 | 144.5 | 5.46 | 300.9 | 49.2 | 3.11 | .... | .... | |
| | 99 | 29.09 | 12.75 | 12.19 | 0.921 | 0.580 | 858 | 134.7 | 5.43 | 278.2 | 45.7 | 3.09 | .... | .... | |
| | 92 | 27.06 | 12.62 | 12.15 | 0.856 | 0.545 | 788 | 125.0 | 5.40 | 256.4 | 42.2 | 3.08 | .... | .... | |
| | 85 | 24.98 | 12.50 | 12.10 | 0.796 | 0.495 | 723 | 115.7 | 5.38 | 235.5 | 38.9 | 3.07 | 90 | 65 | 3 |
| | 79 | 23.22 | 12.38 | 12.08 | 0.736 | 0.470 | 663 | 107.1 | 5.34 | 216.4 | 35.8 | 3.05 | 84 | 61 | |
| | 72 | 21.16 | 12.25 | 12.04 | 0.671 | 0.430 | 597 | 97.5 | 5.31 | 195.3 | 32.4 | 3.04 | 76 | 55 | |
| | 65 | 19.11 | 12.12 | 12.00 | 0.606 | 0.390 | 533 | 88.0 | 5.28 | 174.6 | 29.1 | 3.02 | 69 | 49 | |
| 12 × 10 | 58 | 17.06 | 12.19 | 10.01 | 0.641 | 0.359 | 476 | 78.1 | 5.28 | 107.4 | 21.4 | 2.51 | 63 | 46 | 3 |
| | 53 | 15.59 | 12.06 | 10.00 | 0.576 | 0.345 | 426 | 70.7 | 5.23 | 96.1 | 19.2 | 2.48 | 60 | 44 | |
| 12 × 8 | 50 | 14.71 | 12.19 | 8.07 | 0.641 | 0.371 | 394 | 64.7 | 5.18 | 56.4 | 14.0 | 1.96 | 66 | 48 | 3 |
| | 45 | 13.24 | 12.06 | 8.04 | 0.576 | 0.336 | 350 | 58.2 | 5.15 | 50.0 | 12.4 | 1.94 | 59 | 43 | |
| | 40 | 11.77 | 11.94 | 8.00 | 0.516 | 0.294 | 310 | 51.9 | 5.13 | 44.1 | 11.0 | 1.94 | 51 | 37 | |
| 12 × 6½ | 36 | 10.59 | 12.24 | 6.56 | 0.540 | 0.305 | 280 | 45.9 | 5.15 | 23.7 | 7.2 | 1.50 | 54 | 37 | 3 |
| | 31 | 9.12 | 12.09 | 6.52 | 0.465 | 0.265 | 238 | 39.4 | 5.11 | 19.8 | 6.1 | 1.47 | 46 | 31 | |
| | 27 | 7.97 | 11.95 | 6.50 | 0.400 | 0.240 | 204 | 34.1 | 5.06 | 16.6 | 5.1 | 1.44 | 42 | 28 | |
| 10 × 10 | 112 | 32.92 | 11.38 | 10.41 | 1.248 | 0.755 | 718.7 | 126.3 | 4.67 | 235.4 | 45.2 | 2.67 | .... | .... | 2 |
| | 100 | 29.43 | 11.12 | 10.34 | 1.118 | 0.685 | 625.0 | 112.4 | 4.61 | 206.6 | 39.9 | 2.65 | .... | .... | |
| | 89 | 26.19 | 10.88 | 10.27 | 0.998 | 0.615 | 542.4 | 99.7 | 4.55 | 180.6 | 35.2 | 2.63 | .... | .... | |
| | 77 | 22.67 | 10.62 | 10.19 | 0.868 | 0.535 | 457.2 | 86.1 | 4.49 | 153.4 | 30.1 | 2.60 | .... | .... | |
| | 72 | 21.18 | 10.50 | 10.17 | 0.808 | 0.510 | 420.7 | 80.1 | 4.46 | 141.8 | 27.9 | 2.59 | .... | .... | |
| | 66 | 19.41 | 10.38 | 10.11 | 0.748 | 0.457 | 382.5 | 73.7 | 4.44 | 129.2 | 25.5 | 2.58 | 69 | 59 | |
| | 60 | 17.66 | 10.25 | 10.07 | 0.683 | 0.415 | 343.7 | 67.1 | 4.41 | 116.5 | 23.1 | 2.57 | 62 | 53 | |
| | 54 | 15.88 | 10.12 | 10.02 | 0.618 | 0.368 | 305.7 | 60.4 | 4.39 | 103.9 | 20.7 | 2.56 | 54 | 46 | |
| | 49 | 14.40 | 10.00 | 10.00 | 0.558 | 0.340 | 272.9 | 54.6 | 4.35 | 93.0 | 18.6 | 2.54 | 49 | 42 | |

*$V$ and $R$ values are for beams of A36 steel.
†For standard connections see Table 25.

**Table 19. Properties of Wide-Flange Beams and Columns**  *(Continued)*

| Nominal size, in. | Weight per ft, lb | Area of section, sq in. | Depth of section, in. | Flange Width, in. | Flange Thickness, in. | Web thickness, in. | Neutral axis perpendicular to web at center $I$, in.⁴ | $S$, in.³ | $r$, in. | Neutral axis parallel to web at center $I$, in.⁴ | $S$, in.³ | $r$, in. | $V$, 1,000 lb* | $R$, 1,000 lb* | Rows in std conn† |
|---|---|---|---|---|---|---|---|---|---|---|---|---|---|---|---|
| 10 × 8 | 45 | 13.24 | 10.12 | 8.02 | 0.618 | 0.350 | 248.6 | 49.1 | 4.33 | 53.2 | 13.3 | 2.00 | 51 | 44 | 2 |
|  | 39 | 11.48 | 9.94 | 7.99 | 0.528 | 0.318 | 209.7 | 42.2 | 4.27 | 44.9 | 11.2 | 1.98 | 46 | 39 |  |
|  | 33 | 9.71 | 9.75 | 7.96 | 0.433 | 0.292 | 170.9 | 35.0 | 4.20 | 36.5 | 9.2 | 1.94 | 41 | 35 |  |
| 10 × 5¾ | 29 | 8.53 | 10.22 | 5.79 | 0.500 | 0.289 | 157.3 | 30.8 | 4.29 | 15.2 | 5.2 | 1.34 | 43 | 34 | 2 |
|  | 25 | 7.35 | 10.08 | 5.76 | 0.430 | 0.252 | 133.2 | 26.4 | 4.26 | 12.7 | 4.4 | 1.31 | 37 | 29 |  |
|  | 21 | 6.19 | 9.90 | 5.75 | 0.340 | 0.240 | 106.3 | 21.5 | 4.14 | 9.7 | 3.4 | 1.25 | 34 | 27 |  |
| 8 × 8 | 67 | 19.70 | 9.00 | 8.28 | 0.933 | 0.575 | 271.8 | 60.4 | 3.71 | 88.6 | 21.4 | 2.12 | .... | .... | 2 |
|  | 58 | 17.06 | 8.75 | 8.22 | 0.808 | 0.510 | 227.3 | 52.0 | 3.65 | 74.9 | 18.2 | 2.10 | .... | .... |  |
|  | 48 | 14.11 | 8.50 | 8.11 | 0.683 | 0.405 | 183.7 | 43.2 | 3.61 | 60.9 | 15.0 | 2.08 | .... | .... |  |
|  | 40 | 11.76 | 8.25 | 8.07 | 0.558 | 0.365 | 146.3 | 35.5 | 3.53 | 49.0 | 12.1 | 2.04 | .... | .... |  |
|  | 35 | 10.30 | 8.12 | 8.02 | 0.493 | 0.315 | 126.5 | 31.1 | 3.50 | 42.5 | 10.6 | 2.03 | 37 | 37 |  |
|  | 31 | 9.12 | 8.00 | 8.00 | 0.433 | 0.228 | 109.7 | 27.4 | 3.47 | 37.0 | 9.2 | 2.01 | 33 | ... |  |
| 8 × 6½ | 28 | 8.23 | 8.06 | 6.54 | 0.463 | 0.285 | 97.8 | 24.3 | 3.45 | 21.6 | 6.6 | 1.62 | 33 | 33 | 2 |
|  | 24 | 7.06 | 7.93 | 6.50 | 0.398 | 0.245 | 82.5 | 20.8 | 3.42 | 18.2 | 5.6 | 1.61 | 28 | ... |  |
| 8 × 5¼ | 20 | 5.88 | 8.14 | 5.27 | 0.378 | 0.248 | 69.2 | 17.0 | 3.43 | 8.5 | 3.2 | 1.20 | 29 | 27 | 2 |
|  | 17 | 5.00 | 8.00 | 5.25 | 0.308 | 0.230 | 56.4 | 14.1 | 3.36 | 6.7 | 2.6 | 1.16 | 28 | 26 |  |

Flanges of wide-flange beams and columns are not tapered, have constant thickness.

Sections without values of *V* and *R* are used chiefly for columns.

Lightweight beams for each nominal size, and beams with depth in even inches, are most usually stocked.

Designation of wide-flange beams is made by giving nominal depth and weight, thus, 8 WF 40.

*V and R values are for beams of A36 steel.

†For standard connections see Table 25.

headed Double Shear; for thinner webs, bearing limits the value, and the coefficients for web bearing are to be used. For ¾-in fasteners, multiply tabular bearing values by ⁶⁄₇ and shear values by ³⁶⁄₄₉. Fasteners connecting the outstanding legs to the supporting metal are in double shear if two beams are framed opposite or in single shear if a beam is connected on one side only. If the supporting material of A36 steel is thinner than 0.42 in in double-shear connections or thinner than 0.21 in in single-shear connections, the capacity is limited by bearing. The value of any ⅞-in fastener in bearing on A36 material is 42,525*t*, where *t* is the thickness of the plate. The value of

⅞ in A502, grade 1 rivets or A325 HS bolts (friction-type connections) is 9,020 lb in single shear and 18,040 lb in double shear. The corresponding values for A307 unfinished bolts are 6,010 lb and 12,030 lb, respectively.

**Cast-iron columns** were often used in the past instead of wood, to save space, in the lower stories of heavy buildings. Their use is now obsolete, but they are occasionally encountered in repair and alteration work to older buildings. The ratio of length to least radius of gyration $l/r$ should not exceed 70, and the average unit stress under axial compression should not exceed $9,000 - 40l/r$ lb/in².

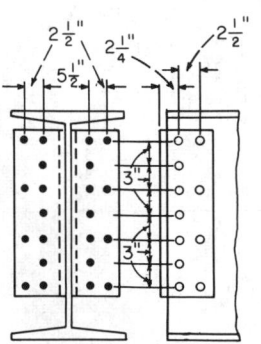
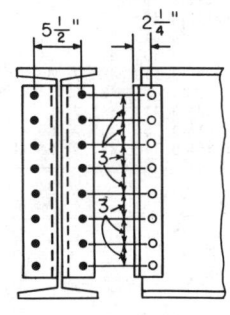
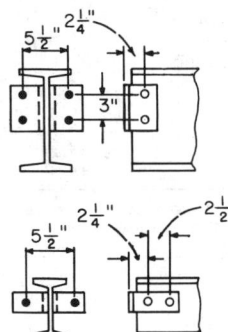

**Fig. 30**  Framed beam connections.

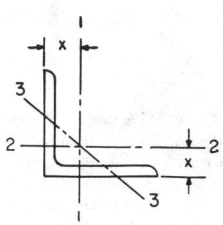

**Table 20. Selected Standard Angles, Equal Legs**

(One to three intermediate thicknesses in each size group are available, varying by $1/16$ in)

A single angle should never be used as a beam. Two angles, riveted at frequent intervals, may be used.

| Size, in. | | | Weight per ft, lb | Area of section, sq in. | Axis 1-1 and axis 2-2 | | | | Axis 3-3, r min, in. | Net areas after deducting holes for $7/8$ in. rivets | |
|---|---|---|---|---|---|---|---|---|---|---|---|
| | | | | | $I$, in.⁴ | $r$, in. | $S$, in.³ | $x$, in. | | 1 hole | 2 holes |
| 8 | × 8 | × 1⅛ | 56.9 | 16.73 | 98.0 | 2.42 | 17.5 | 2.41 | 1.56 | 15.60 | 14.48 |
| | | 1 | 51.0 | 15.00 | 89.0 | 2.44 | 15.8 | 2.37 | 1.56 | 14.00 | 13.00 |
| | | ⅞ | 45.0 | 13.23 | 79.6 | 2.45 | 14.0 | 2.32 | 1.57 | 12.36 | 11.48 |
| | | ¾ | 38.9 | 11.44 | 69.7 | 2.47 | 12.2 | 2.28 | 1.57 | 10.69 | 9.94 |
| | | ⅝ | 32.7 | 9.61 | 59.4 | 2.49 | 10.3 | 2.23 | 1.58 | 8.98 | 8.36 |
| | | ½ | 26.4 | 7.75 | 48.6 | 2.50 | 8.4 | 2.19 | 1.59 | 7.25 | 6.75 |
| 6 | × 6 | × 1 | 37.4 | 11.00 | 35.5 | 1.80 | 8.6 | 1.86 | 1.17 | 10.00 | 9.00 |
| | | ⅞ | 33.1 | 9.73 | 31.9 | 1.81 | 7.6 | 1.82 | 1.17 | 8.86 | 7.98 |
| | | ¾ | 28.7 | 8.44 | 28.2 | 1.83 | 6.7 | 1.78 | 1.17 | 7.69 | 6.94 |
| | | ⅝ | 24.2 | 7.11 | 24.2 | 1.84 | 5.7 | 1.73 | 1.18 | 6.48 | 5.86 |
| | | ½ | 19.6 | 5.75 | 19.9 | 1.86 | 4.6 | 1.68 | 1.18 | 5.25 | 4.75 |
| | | ⅜ | 14.9 | 4.36 | 15.4 | 1.88 | 3.5 | 1.64 | 1.19 | 3.98 | 3.61 |
| 5 | × 5 | × ⅞ | 27.2 | 7.98 | 17.8 | 1.49 | 5.2 | 1.57 | 0.97 | 7.10 | 6.23 |
| | | ¾ | 23.6 | 6.94 | 15.7 | 1.51 | 4.5 | 1.52 | 0.97 | 6.19 | 5.44 |
| | | ⅝ | 20.0 | 5.86 | 13.6 | 1.52 | 3.9 | 1.48 | 0.98 | 5.24 | 4.61 |
| | | ½ | 16.2 | 4.75 | 11.3 | 1.54 | 3.2 | 1.43 | 0.98 | 4.25 | 3.75 |
| | | ⅜ | 12.3 | 3.61 | 8.7 | 1.56 | 2.4 | 1.39 | 0.99 | 3.24 | 2.86 |
| 4 | × 4 | × ¾ | 18.5 | 5.44 | 7.7 | 1.19 | 2.8 | 1.27 | 0.78 | 4.69 | 3.94 |
| | | ⅝ | 15.7 | 4.61 | 6.7 | 1.20 | 2.4 | 1.23 | 0.78 | 3.98 | 3.36 |
| | | ½ | 12.8 | 3.75 | 5.6 | 1.22 | 2.0 | 1.18 | 0.78 | 3.25 | 2.75 |
| | | ⅜ | 9.8 | 2.86 | 4.4 | 1.23 | 1.5 | 1.14 | 0.79 | 2.48 | 2.11 |
| | | ¼ | 6.6 | 1.94 | 3.0 | 1.25 | 1.1 | 1.09 | 0.80 | 1.70 | 1.45 |
| 3½ | × 3½ | × ½ | 11.1 | 3.25 | 3.6 | 1.06 | 1.5 | 1.06 | 0.68 | 2.75 | 2.25 |
| | | ⅜ | 8.5 | 2.48 | 2.9 | 1.07 | 1.2 | 1.01 | 0.69 | 2.10 | 1.73 |
| | | ¼ | 5.8 | 1.69 | 2.0 | 1.09 | 0.79 | 0.97 | 0.69 | 1.44 | 1.19 |
| 3 | × 3 | × ½ | 9.4 | 2.75 | 2.2 | 0.90 | 1.1 | 0.93 | 0.58 | | |
| | | ⅜ | 7.2 | 2.11 | 1.8 | 0.91 | 0.83 | 0.89 | 0.58 | | |
| | | ¼ | 4.9 | 1.44 | 1.2 | 0.93 | 0.58 | 0.84 | 0.59 | | |
| 2½ | × 2½ | × ½ | 7.7 | 2.25 | 1.2 | 0.74 | 0.72 | 0.81 | 0.49 | | |
| | | ⅜ | 5.9 | 1.73 | 0.98 | 0.75 | 0.57 | 0.76 | 0.49 | | |
| | | ¼ | 4.1 | 1.19 | 0.70 | 0.77 | 0.39 | 0.72 | 0.49 | | |
| 2 | × 2 | × ⅜ | 4.7 | 1.36 | 0.48 | 0.59 | 0.35 | 0.64 | 0.39 | | |
| | | ¼ | 3.19 | 0.94 | 0.35 | 0.61 | 0.25 | 0.59 | 0.39 | | |
| | | ⅛ | 1.65 | 0.48 | 0.19 | 0.63 | 0.13 | 0.55 | 0.40 | | |
| 1¾ | × 1¾ | × ¼ | 2.77 | 0.81 | 0.23 | 0.53 | 0.19 | 0.53 | 0.34 | | |
| | | ⅛ | 1.44 | 0.42 | 0.13 | 0.55 | 0.10 | 0.48 | 0.35 | | |
| 1½ | × 1½ | × ¼ | 2.34 | 0.69 | 0.14 | 0.45 | 0.13 | 0.47 | 0.29 | | |
| | | ⅛ | 1.23 | 0.36 | 0.08 | 0.47 | 0.07 | 0.42 | 0.30 | | |
| 1¼ | × 1¼ | × ¼ | 1.92 | 0.56 | 0.08 | 0.37 | 0.09 | 0.40 | 0.24 | | |
| | | ⅛ | 1.01 | 0.30 | 0.04 | 0.38 | 0.05 | 0.36 | 0.25 | | |
| 1 | × 1 | × ¼ | 1.49 | 0.44 | 0.04 | 0.29 | 0.06 | 0.34 | 0.20 | | |
| | | ⅛ | 0.80 | 0.23 | 0.02 | 0.30 | 0.03 | 0.30 | 0.20 | | |

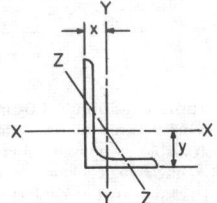

### Table 21. Selected Standard Angles, Unequal Legs

(Intermediate thicknesses are available in each size group, varying by $\frac{1}{16}$ in among the thinner angles)
A single angle should never be used as a beam. Two angles, riveted at frequent intervals, may be used.

| Size, in. | Thickness, in. | Weight per ft, lb | Area of section, sq in. | Axis X-X | | | | Axis Y-Y | | | | Axis Z-Z | Net areas after deducting holes for 7/8 in. rivets | |
|---|---|---|---|---|---|---|---|---|---|---|---|---|---|---|
| | | | | $I$, in.$^4$ | $S$, in.$^3$ | $r$, in. | $y$, in. | $I$, in.$^4$ | $S$, in.$^3$ | $r$, in. | $x$, in. | $r$, in. | 1 hole | 2 holes |
| 8 × 6 | 1 | 44.2 | 13.00 | 80.8 | 15.1 | 2.49 | 2.65 | 38.8 | 8.9 | 1.73 | 1.65 | 1.28 | 12.00 | 11.00 |
| | ¾ | 33.8 | 9.94 | 63.4 | 11.7 | 2.53 | 2.56 | 30.7 | 6.9 | 1.76 | 1.56 | 1.29 | 9.19 | 8.44 |
| | ½ | 23.0 | 6.75 | 44.3 | 8.0 | 2.56 | 2.47 | 21.7 | 4.8 | 1.79 | 1.47 | 1.30 | 6.25 | 5.75 |
| | ⁷⁄₁₆ | 20.2 | 5.93 | 39.2 | 7.1 | 2.57 | 2.45 | 19.3 | 4.2 | 1.80 | 1.45 | 1.31 | 5.49 | 5.06 |
| 8 × 4 | 1 | 37.4 | 11.00 | 69.6 | 14.1 | 2.52 | 3.05 | 11.6 | 3.9 | 1.03 | 1.05 | 0.85 | 10.00 | 9.00 |
| | ¾ | 28.7 | 8.44 | 54.9 | 10.9 | 2.55 | 2.95 | 9.4 | 3.1 | 1.05 | 0.95 | 0.85 | 7.69 | 6.94 |
| | ½ | 19.6 | 5.75 | 38.5 | 7.5 | 2.59 | 2.86 | 6.7 | 2.2 | 1.08 | 0.86 | 0.86 | 5.25 | 4.75 |
| | ⁷⁄₁₆ | 17.2 | 5.06 | 34.1 | 6.6 | 2.60 | 2.83 | 6.0 | 1.9 | 1.09 | 0.83 | 0.87 | 4.62 | 4.18 |
| 7 × 4 | ⅞ | 30.2 | 8.86 | 42.9 | 9.7 | 2.20 | 2.55 | 10.2 | 3.5 | 1.07 | 1.05 | 0.86 | 7.98 | 7.11 |
| | ¾ | 26.2 | 7.69 | 37.8 | 8.4 | 2.22 | 2.51 | 9.1 | 3.0 | 1.09 | 1.01 | 0.86 | 6.94 | 6.19 |
| | ½ | 17.9 | 5.25 | 26.7 | 5.8 | 2.25 | 2.42 | 6.5 | 2.1 | 1.11 | 0.92 | 0.87 | 4.75 | 4.25 |
| | ⅜ | 13.6 | 3.98 | 20.6 | 4.4 | 2.27 | 2.37 | 5.1 | 1.6 | 1.13 | 0.87 | 0.88 | 3.62 | 3.24 |
| 6 × 4 | ⅞ | 27.2 | 7.98 | 27.7 | 7.2 | 1.86 | 2.12 | 9.8 | 3.4 | 1.11 | 1.12 | 0.86 | 7.10 | 6.23 |
| | ¾ | 23.6 | 6.94 | 24.5 | 6.3 | 1.88 | 2.08 | 8.7 | 3.0 | 1.12 | 1.08 | 0.86 | 6.19 | 5.44 |
| | ½ | 16.2 | 4.75 | 17.4 | 4.3 | 1.91 | 1.99 | 6.3 | 2.1 | 1.15 | 0.99 | 0.87 | 4.25 | 3.75 |
| | ⅜ | 12.3 | 3.61 | 13.5 | 3.3 | 1.93 | 1.94 | 4.9 | 1.6 | 1.17 | 0.94 | 0.88 | 3.24 | 2.86 |
| | ⁵⁄₁₆ | 10.3 | 3.03 | 11.4 | 2.8 | 1.94 | 1.92 | 4.2 | 1.4 | 1.17 | 0.92 | 0.88 | 2.72 | 2.40 |
| 6 × 3½ | ½ | 15.3 | 4.50 | 16.6 | 4.2 | 1.92 | 2.08 | 4.3 | 1.6 | 0.97 | 0.83 | 0.76 | 4.00 | 3.50 |
| | ⅜ | 11.7 | 3.42 | 12.9 | 3.2 | 1.94 | 2.04 | 3.3 | 1.2 | 0.99 | 0.79 | 0.77 | 3.04 | 2.67 |
| | ¼ | 7.9 | 2.31 | 8.9 | 2.2 | 1.96 | 1.99 | 2.3 | 0.85 | 1.01 | 0.74 | 0.78 | 2.06 | 1.81 |
| 5 × 3½ | ¾ | 19.8 | 5.81 | 13.9 | 4.3 | 1.55 | 1.75 | 5.6 | 2.2 | 0.98 | 1.00 | 0.75 | 5.06 | 4.31 |
| | ½ | 13.6 | 4.00 | 10.0 | 3.0 | 1.58 | 1.66 | 4.1 | 1.6 | 1.01 | 0.91 | 0.75 | 3.50 | 3.00 |
| | ¼ | 7.0 | 2.06 | 5.4 | 1.6 | 1.61 | 1.56 | 2.2 | 0.83 | 1.04 | 0.81 | 0.76 | 1.81 | 1.56 |
| 5 × 3 | ½ | 12.8 | 3.75 | 9.5 | 2.9 | 1.59 | 1.75 | 2.6 | 1.1 | 0.83 | 0.75 | 0.65 | 3.25 | 2.75 |
| | ⅜ | 9.8 | 2.86 | 7.4 | 2.2 | 1.61 | 1.70 | 2.0 | 0.89 | 0.84 | 0.70 | 0.65 | 2.48 | 2.11 |
| | ¼ | 6.6 | 1.94 | 5.1 | 1.5 | 1.62 | 1.66 | 1.4 | 0.61 | 0.86 | 0.66 | 0.66 | 1.69 | 1.44 |
| 4 × 3½ | ⅝ | 14.7 | 4.30 | 6.4 | 2.4 | 1.22 | 1.29 | 4.5 | 1.8 | 1.03 | 1.04 | 0.72 | 3.68 | 3.05 |
| | ½ | 11.9 | 3.50 | 5.3 | 1.9 | 1.23 | 1.25 | 3.8 | 1.5 | 1.04 | 1.00 | 0.72 | 3.00 | 2.50 |
| | ⅜ | 9.1 | 2.67 | 4.2 | 1.5 | 1.25 | 1.21 | 3.0 | 1.2 | 1.06 | 0.96 | 0.73 | 2.30 | 1.92 |
| | ¼ | 6.2 | 1.81 | 2.9 | 1.0 | 1.27 | 1.16 | 2.1 | 0.81 | 1.07 | 0.91 | 0.73 | 1.56 | 1.31 |
| 4 × 3 | ⅝ | 13.6 | 3.98 | 6.0 | 2.3 | 1.23 | 1.37 | 2.9 | 1.4 | 0.85 | 0.87 | 0.64 | 3.36 | 2.73 |
| | ½ | 11.1 | 3.25 | 5.1 | 1.9 | 1.25 | 1.33 | 2.4 | 1.1 | 0.86 | 0.83 | 0.64 | 2.25 | 2.25 |
| | ¼ | 5.8 | 1.69 | 2.8 | 1.0 | 1.28 | 1.24 | 1.4 | 0.60 | 0.90 | 0.74 | 0.65 | 1.44 | 1.19 |
| 3½ × 3 | ½ | 10.2 | 3.00 | 3.5 | 1.5 | 1.07 | 1.13 | 2.3 | 1.1 | 0.88 | 0.88 | 0.62 | 2.50 | |
| | ¼ | 5.4 | 1.56 | 1.9 | 0.78 | 1.11 | 1.04 | 1.3 | 0.59 | 0.91 | 0.79 | 0.63 | 1.31 | |
| 3½ × 2½ | ½ | 9.4 | 2.75 | 3.2 | 1.4 | 1.09 | 1.20 | 1.4 | 0.76 | 0.70 | 0.70 | 0.53 | 2.25 | |
| | ¼ | 4.9 | 1.44 | 1.8 | 0.75 | 1.12 | 1.11 | 0.78 | 0.41 | 0.74 | 0.61 | 0.54 | 1.19 | |
| 3 × 2½ | ½ | 8.5 | 2.50 | 2.1 | 1.0 | 0.91 | 1.00 | 1.3 | 0.74 | 0.72 | 0.75 | 0.52 | | |
| | ⅜ | 6.6 | 1.92 | 1.7 | 0.81 | 0.93 | 0.96 | 1.0 | 0.58 | 0.74 | 0.71 | 0.52 | | |
| | ¼ | 4.5 | 1.31 | 1.2 | 0.56 | 0.95 | 0.91 | 0.74 | 0.40 | 0.75 | 0.66 | 0.53 | | |
| 3 × 2 | ½ | 7.7 | 2.25 | 1.9 | 1.0 | 0.92 | 1.08 | 0.67 | 0.47 | 0.55 | 0.58 | 0.43 | | |
| | ³⁄₁₆ | 3.07 | 0.90 | 0.84 | 0.41 | 0.97 | 0.97 | 0.31 | 0.20 | 0.58 | 0.47 | 0.44 | | |
| 2½ × 2 | ⅜ | 5.3 | 1.55 | 0.91 | 0.55 | 0.77 | 0.83 | 0.51 | 0.36 | 0.58 | 0.58 | 0.42 | | |
| | ³⁄₁₆ | 2.75 | 0.81 | 0.51 | 0.29 | 0.79 | 0.76 | 0.29 | 0.20 | 0.60 | 0.51 | 0.43 | | |
| 2 × 1½ | ¼ | 2.77 | 0.81 | 0.32 | 0.24 | 0.62 | 0.66 | 0.15 | 0.14 | 0.43 | 0.41 | 0.32 | | |
| | ⅛ | 1.44 | 0.42 | 0.17 | 0.13 | 0.64 | 0.62 | 0.09 | 0.08 | 0.45 | 0.37 | 0.33 | | |
| 1¾ × 1¼ | ¼ | 2.34 | 0.69 | 0.20 | 0.18 | 0.54 | 0.60 | 0.09 | 0.10 | 0.35 | 0.35 | 0.27 | | |
| | ⅛ | 1.23 | 0.36 | 0.11 | 0.09 | 0.56 | 0.56 | 0.05 | 0.05 | 0.37 | 0.31 | 0.27 | | |

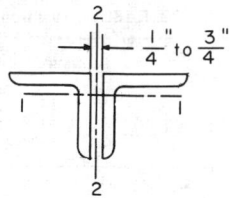

**Table 22. Radii of Gyration for Two Angles, Unequal Legs**

| Single angle | | Two angles | Radii of gyration, in. | | | | | | | | |
|---|---|---|---|---|---|---|---|---|---|---|---|
| | | | Long legs vertical | | | | Short legs vertical | | | | |
| Size, in. | Weight per ft, lb | Area, sq in. | Axis 1-1 | Axis 2-2 | | | Axis 1-1 | Axis 2-2 | | | |
| | | | | In contact | 3/8 in. apart | 3/4 in. apart | | In contact | 3/8 in. apart | 3/4 in. apart | |
| 8 × 6 × 1 | 44.2 | 26.00 | 2.49 | 2.39 | 2.52 | 2.66 | 1.73 | 3.64 | 3.78 | 3.92 | |
| 7/16 | 20.2 | 11.86 | 2.57 | 2.31 | 2.43 | 2.56 | 1.80 | 3.55 | 3.68 | 3.82 | |
| 8 × 4 × 1 | 37.4 | 22.00 | 2.52 | 1.47 | 1.61 | 1.76 | 1.03 | 3.95 | 4.10 | 4.25 | |
| 1/2 | 19.6 | 11.50 | 2.59 | 1.38 | 1.51 | 1.64 | 1.08 | 3.86 | 4.00 | 4.14 | |
| 7 × 4 × 7/8 | 30.2 | 17.72 | 2.20 | 1.50 | 1.64 | 1.78 | 1.07 | 3.37 | 3.51 | 3.66 | |
| 3/8 | 13.6 | 7.96 | 2.27 | 1.43 | 1.55 | 1.68 | 1.13 | 3.28 | 3.42 | 3.56 | |
| 6 × 4 × 7/8 | 27.2 | 15.96 | 1.86 | 1.57 | 1.71 | 1.86 | 1.11 | 2.82 | 2.97 | 3.11 | |
| 3/8 | 12.3 | 7.22 | 1.93 | 1.50 | 1.62 | 1.76 | 1.17 | 2.74 | 2.87 | 3.02 | |
| 5 × 3½ × 3/4 | 19.8 | 11.62 | 1.55 | 1.40 | 1.54 | 1.69 | 0.98 | 2.34 | 2.48 | 2.63 | |
| 5/16 | 8.7 | 5.12 | 1.61 | 1.33 | 1.45 | 1.59 | 1.03 | 2.26 | 2.38 | 2.53 | |
| 4 × 3½ × 5/8 | 14.7 | 8.60 | 1.22 | 1.46 | 1.60 | 1.75 | 1.03 | 1.77 | 1.91 | 2.06 | |
| 5/16 | 7.7 | 4.50 | 1.26 | 1.42 | 1.55 | 1.69 | 1.07 | 1.73 | 1.86 | 2.00 | |
| 4 × 3 × 5/8 | 13.6 | 7.96 | 1.23 | 1.22 | 1.36 | 1.51 | 0.85 | 1.84 | 1.99 | 2.14 | |
| 1/4 | 5.8 | 3.38 | 1.28 | 1.16 | 1.29 | 1.43 | 0.90 | 1.78 | 1.92 | 2.06 | |
| 3½ × 3 × 1/2 | 10.2 | 6.00 | 1.07 | 1.25 | 1.38 | 1.53 | 0.88 | 1.56 | 1.70 | 1.85 | |
| 1/4 | 5.4 | 3.12 | 1.11 | 1.21 | 1.34 | 1.48 | 0.91 | 1.52 | 1.65 | 1.80 | |
| 3 × 2½ × 3/8 | 6.6 | 3.84 | 0.93 | 1.02 | 1.16 | 1.31 | 0.74 | 1.34 | 1.48 | 1.63 | |
| 1/4 | 4.5 | 2.62 | 0.95 | 1.00 | 1.13 | 1.28 | 0.75 | 1.31 | 1.45 | 1.60 | |
| 2½ × 2 × 3/8 | 5.3 | 3.10 | 0.77 | 0.82 | 0.96 | 1.11 | 0.58 | 1.13 | 1.27 | 1.43 | |
| 1/4 | 3.6 | 2.12 | 0.78 | 0.80 | 0.94 | 1.09 | 0.59 | 1.11 | 1.25 | 1.40 | |

**Table 23. Radii of Gyration for Two Angles, Equal Legs**

| Single angle | | Two angles | Radii of gyration, in. | | | |
|---|---|---|---|---|---|---|
| | | | | Axis 2-2 | | |
| Size, in. | Weight per ft, lb | Area, sq in. | Axis 1-1 | In contact | ⅜ in. apart | ¾ in. apart |
| 8 × 8 × 1⅛ | 56.9 | 33.46 | 2.42 | 3.42 | 3.55 | 3.69 |
| ½ | 26.4 | 15.50 | 2.50 | 3.33 | 3.45 | 3.59 |
| 6 × 6 × 1 | 37.4 | 22.00 | 1.80 | 2.59 | 2.72 | 2.87 |
| ⅜ | 14.9 | 8.72 | 1.88 | 2.49 | 2.62 | 2.76 |
| 5 × 5 × ⅞ | 27.2 | 15.96 | 1.49 | 2.17 | 2.31 | 2.45 |
| ⅜ | 12.3 | 7.22 | 1.56 | 2.09 | 2.22 | 2.35 |
| 4 × 4 × ¾ | 18.5 | 10.88 | 1.19 | 1.74 | 1.88 | 2.03 |
| ¼ | 6.6 | 3.88 | 1.25 | 1.66 | 1.79 | 1.93 |
| 3½ × 3½ × ½ | 11.1 | 6.50 | 1.06 | 1.50 | 1.64 | 1.78 |
| ¼ | 5.8 | 3.38 | 1.09 | 1.46 | 1.59 | 1.73 |
| 3 × 3 × ½ | 9.4 | 5.50 | 0.90 | 1.29 | 1.43 | 1.58 |
| ¼ | 4.9 | 2.88 | 0.93 | 1.25 | 1.38 | 1.53 |
| 2½ × 2½ × ½ | 7.7 | 4.50 | 0.74 | 1.10 | 1.24 | 1.40 |
| ¼ | 4.1 | 2.38 | 0.77 | 1.05 | 1.19 | 1.34 |
| 2 × 2 × ⅜ | 4.7 | 2.72 | 0.59 | 0.87 | 1.02 | 1.18 |
| ¼ | 3.19 | 1.88 | 0.61 | 0.85 | 0.99 | 1.14 |

**Steel joists** consisting of lightweight rolled sections, thin for their height, or open-web trussed members fabricated by welding or otherwise, are used with economy in schools and buildings where spans are long and loads are light, and where a plaster ceiling affords sufficient fire protection. They are rarely used in industrial buildings, except for roof loads.

**Steel pipe** is often used for columns under light loads. Table 28 gives the safe loads on standard size pipes (ASTM A501) used as columns. For extra-strong and double extra-strong pipe used as columns, the safe loads will increase approximately in the same proportion as the weight per foot (see Sec. 8).

**Corrugated steel sheets** used for roofs and walls are subject to stress in bending. The section modulus of corrugated sheets is approximately $3td$ in$^3$ per ft wide, where $t$ is the thickness of the sheet and $d$ is the overall depth of the corrugations, in. The safe load, pounds per square foot, in bending is given by the formulas $w = 25,000td/L^2$ where $L$ is the span in feet. The spacing of purlins on roofs and girts on wall is usually 4 to 5 ft. Numbers 20 and 22, U.S. Standard gage, are generally used for roofing, No. 24 for siding.

**Fire Resistance**  The resistance to fire of building materials has been tested extensively by various agencies. Table 29 gives the fire-resistance rating of a few of the common building materials and methods of construction as established by the Uniform Building Code from standard fire tests.

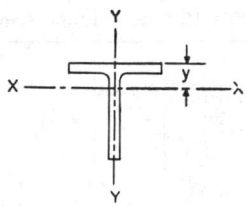

## Table 24. Tees Cut from Standard Sections

| Section No. | Weight per ft, lb | Area, sq in. | Depth of tee, in. | Flange Width, in. | Flange Avg thickness, in. | Stem thickness, in. | Axis $x$-$x$ $I$, in.$^4$ | $S$, in.$^3$ | $r$, in. | $y$, in. | Axis $y$-$y$ $I$, in.$^4$ | $S$, in.$^3$ | $r$, in. |
|---|---|---|---|---|---|---|---|---|---|---|---|---|---|
| ST 18 WF | 150 | 44.1 | 18.4 | 16.7 | 1.68 | 0.945 | 1,223 | 85.9 | 5.27 | 4.13 | 613 | 73.6 | 3.73 |
| | 115 | 33.9 | 17.9 | 16.5 | 1.26 | 0.765 | 936 | 67.2 | 5.26 | 4.02 | 436 | 52.9 | 3.59 |
| | 97 | 28.6 | 18.2 | 12.1 | 1.26 | 0.770 | 904 | 67.3 | 5.63 | 4.81 | 178 | 29.3 | 2.49 |
| | 75 | 22.1 | 17.9 | 12.0 | 0.940 | 0.625 | 697 | 53.0 | 5.62 | 4.79 | 125 | 20.9 | 2.38 |
| ST 16 WF | 120 | 35.3 | 16.8 | 15.9 | 1.40 | 0.830 | 822 | 63.2 | 4.83 | 3.73 | 437 | 55.1 | 3.52 |
| | 100 | 29.4 | 16.5 | 15.8 | 1.15 | 0.715 | 684 | 53.3 | 4.82 | 3.67 | 346 | 43.9 | 3.43 |
| | 76 | 22.4 | 16.8 | 11.6 | 1.06 | 0.635 | 592 | 47.4 | 5.15 | 4.26 | 128 | 22.1 | 2.39 |
| | 65 | 19.1 | 16.6 | 11.5 | 0.855 | 0.580 | 513 | 42.1 | 5.18 | 4.37 | 101 | 17.5 | 2.29 |
| ST 15 WF | 105 | 30.9 | 15.2 | 15.1 | 1.32 | 0.775 | 578 | 48.7 | 4.33 | 3.31 | 354 | 46.9 | 3.38 |
| | 86 | 25.3 | 14.9 | 15.0 | 1.06 | 0.655 | 471 | 40.2 | 4.31 | 3.23 | 275 | 36.7 | 3.30 |
| | 66 | 19.4 | 15.2 | 10.6 | 1.00 | 0.615 | 421 | 37.4 | 4.66 | 3.90 | 92.5 | 17.5 | 2.18 |
| | 54 | 15.9 | 14.9 | 10.5 | 0.760 | 0.548 | 350 | 32.1 | 4.69 | 4.03 | 67.6 | 12.9 | 2.06 |
| ST 13 WF | 88.5 | 26.0 | 13.7 | 14.1 | 1.19 | 0.725 | 392 | 36.7 | 3.88 | 2.97 | 259 | 36.8 | 3.16 |
| | 72.5 | 21.3 | 13.4 | 14.0 | 0.975 | 0.600 | 316 | 29.9 | 3.85 | 2.85 | 204 | 29.1 | 3.09 |
| | 57 | 16.8 | 13.6 | 10.1 | 0.932 | 0.570 | 289 | 28.3 | 4.15 | 3.42 | 74.8 | 14.9 | 2.11 |
| | 47 | 13.8 | 13.4 | 9.99 | 0.747 | 0.490 | 238 | 23.7 | 4.15 | 3.41 | 57.5 | 11.5 | 2.04 |
| ST 12 WF | 80 | 23.5 | 12.4 | 14.1 | 1.14 | 0.656 | 272 | 27.6 | 3.40 | 2.51 | 246 | 35.0 | 3.23 |
| | 60 | 17.6 | 12.2 | 12.1 | 0.930 | 0.556 | 214 | 22.4 | 3.48 | 2.62 | 127 | 21.0 | 2.68 |
| | 50 | 14.7 | 12.0 | 12.0 | 0.775 | 0.468 | 177 | 18.7 | 3.46 | 2.54 | 102 | 17.0 | 2.63 |
| | 38 | 11.2 | 12.0 | 8.98 | 0.682 | 0.440 | 151 | 16.9 | 3.68 | 3.00 | 38.3 | 8.5 | 1.85 |
| ST 10 WF | 71 | 20.9 | 10.7 | 13.1 | 1.10 | 0.659 | 177 | 20.8 | 2.91 | 2.18 | 193 | 29.4 | 3.04 |
| | 48 | 14.1 | 10.6 | 9.04 | 0.935 | 0.575 | 137 | 17.1 | 3.11 | 2.55 | 54.7 | 12.1 | 1.97 |
| | 41 | 12.0 | 10.4 | 8.96 | 0.795 | 0.499 | 115 | 14.5 | 3.09 | 2.48 | 44.8 | 10.0 | 1.93 |
| | 31 | 9.12 | 10.5 | 8.24 | 0.615 | 0.400 | 93.7 | 11.9 | 3.21 | 2.59 | 26.6 | 6.45 | 1.71 |
| ST 9 WF | 57 | 16.8 | 9.24 | 11.8 | 0.991 | 0.595 | 103 | 13.9 | 2.47 | 1.85 | 128 | 21.6 | 2.76 |
| | 42.5 | 12.5 | 9.16 | 8.84 | 0.911 | 0.526 | 84.4 | 11.9 | 2.60 | 2.05 | 49.7 | 11.3 | 2.00 |
| | 32 | 9.40 | 8.94 | 8.72 | 0.686 | 0.403 | 61.8 | 8.82 | 2.56 | 1.93 | 35.2 | 8.07 | 1.93 |
| | 25 | 7.35 | 9.00 | 7.50 | 0.570 | 0.358 | 53.9 | 7.85 | 2.71 | 2.14 | 18.6 | 4.96 | 1.59 |
| ST 8 WF | 48 | 14.1 | 8.16 | 11.5 | 0.875 | 0.535 | 64.7 | 9.82 | 2.14 | 1.57 | 104 | 18.0 | 2.71 |
| | 39 | 11.5 | 8.16 | 8.59 | 0.875 | 0.529 | 60.0 | 9.45 | 2.28 | 1.81 | 43.8 | 10.2 | 1.95 |
| | 29 | 8.52 | 7.93 | 8.46 | 0.645 | 0.407 | 43.6 | 7.00 | 2.26 | 1.70 | 30.2 | 7.14 | 1.88 |
| | 18 | 5.30 | 7.93 | 6.99 | 0.428 | 0.299 | 30.7 | 5.10 | 2.41 | 1.90 | 11.1 | 3.17 | 1.45 |
| ST 7 WF | 105.5 | 31.0 | 7.88 | 15.8 | 1.56 | 0.980 | 102.2 | 16.2 | 1.81 | 1.57 | 514 | 65.1 | 4.07 |
| | 71 | 20.9 | 7.38 | 15.5 | 1.06 | 0.680 | 62.1 | 10.2 | 1.72 | 1.29 | 330 | 42.6 | 3.97 |
| | 43.5 | 12.8 | 7.00 | 14.5 | 0.688 | 0.420 | 34.9 | 5.88 | 1.65 | 1.08 | 175 | 24.1 | 3.70 |
| | 39 | 11.5 | 7.03 | 12.0 | 0.718 | 0.428 | 34.8 | 5.96 | 1.74 | 1.19 | 104 | 17.2 | 3.00 |
| | 30.5 | 8.97 | 6.96 | 10.0 | 0.643 | 0.378 | 29.2 | 5.13 | 1.80 | 1.25 | 53.6 | 10.7 | 2.45 |
| | 21.5 | 6.32 | 6.84 | 8.00 | 0.528 | 0.308 | 22.2 | 4.02 | 1.87 | 1.33 | 22.6 | 5.64 | 1.89 |
| | 15 | 4.41 | 6.93 | 6.73 | 0.383 | 0.270 | 19.0 | 3.55 | 2.08 | 1.59 | 8.77 | 2.61 | 1.41 |
| ST 6 WF | 80.5 | 23.7 | 6.94 | 12.5 | 1.49 | 0.905 | 62.6 | 11.5 | 1.63 | 1.47 | 243 | 38.9 | 3.20 |
| | 32.5 | 9.55 | 6.06 | 12.0 | 0.606 | 0.390 | 20.6 | 4.06 | 1.47 | 0.98 | 87.3 | 14.6 | 3.02 |
| | 26.5 | 7.80 | 6.03 | 10.0 | 0.576 | 0.345 | 17.7 | 3.54 | 1.51 | 1.02 | 48.0 | 9.60 | 2.48 |
| | 20 | 5.89 | 5.97 | 8.00 | 0.516 | 0.294 | 14.4 | 2.94 | 1.56 | 1.08 | 22.0 | 5.50 | 1.94 |
| | 13.5 | 3.98 | 5.98 | 6.50 | 0.400 | 0.240 | 11.4 | 2.39 | 1.69 | 1.21 | 8.3 | 2.55 | 1.44 |

### Table 24. Tees Cut from Standard Sections (Continued)

| Section No. | Weight per ft, lb | Area, sq in. | Depth of tee, in. | Flange Width, in. | Flange Avg thickness, in. | Stem thickness, in. | Axis x-x I, in.⁴ | S, in.³ | r, in. | y, in. | Axis y-y I, in.⁴ | S, in.³ | r, in. |
|---|---|---|---|---|---|---|---|---|---|---|---|---|---|
| ST 5 WF | 56 | 16.5 | 5.69 | 10.4 | 1.25 | 0.755 | 28.8 | 6.42 | 1.32 | 1.21 | 118 | 22.6 | 2.67 |
|  | 24.5 | 7.20 | 5.00 | 10.0 | 0.558 | 0.340 | 10.1 | 2.40 | 1.18 | 0.81 | 46.5 | 9.30 | 2.54 |
|  | 16.5 | 4.85 | 4.88 | 7.96 | 0.433 | 0.292 | 7.80 | 1.95 | 1.27 | 0.88 | 18.2 | 4.58 | 1.94 |
|  | 10.5 | 3.10 | 4.95 | 5.75 | 0.340 | 0.240 | 6.31 | 1.62 | 1.43 | 1.06 | 4.87 | 1.69 | 1.25 |
| ST 4 WF | 33.5 | 9.85 | 4.50 | 8.29 | 0.933 | 0.575 | 10.9 | 3.07 | 1.05 | 0.94 | 44.3 | 10.7 | 2.12 |
|  | 15.5 | 4.56 | 4.00 | 8.00 | 0.433 | 0.288 | 4.31 | 1.30 | 0.97 | 0.67 | 18.5 | 4.60 | 2.01 |
|  | 12 | 3.53 | 3.97 | 6.50 | 0.398 | 0.245 | 3.53 | 1.08 | 1.00 | 0.70 | 9.10 | 2.80 | 1.61 |
|  | 8.5 | 2.50 | 4.00 | 5.25 | 0.308 | 0.230 | 3.21 | 1.01 | 1.13 | 0.84 | 3.36 | 1.28 | 1.16 |
| ST 6 I | 25 | 7.29 | 6.00 | 5.48 | 0.660 | 0.687 | 25.2 | 6.05 | 1.85 | 1.84 | 7.85 | 2.87 | 1.03 |
|  | 15.9 | 4.63 | 6.00 | 5.00 | 0.544 | 0.350 | 14.9 | 3.31 | 1.78 | 1.51 | 4.68 | 1.87 | 1.00 |
| ST 5 I | 12.7 | 3.69 | 5.00 | 4.66 | 0.491 | 0.310 | 7.81 | 2.05 | 1.45 | 1.20 | 3.39 | 1.46 | 0.95 |
| ST 4 I | 9.2 | 2.67 | 4.00 | 4.00 | 0.425 | 0.270 | 3.50 | 1.14 | 1.14 | 0.94 | 1.86 | 0.93 | 0.83 |
| ST 3.5 I | 7.65 | 2.22 | 3.50 | 3.66 | 0.392 | 0.250 | 2.18 | 0.81 | 0.99 | 0.81 | 1.32 | 0.72 | 0.77 |
| ST 3 I | 6.25 | 1.81 | 3.00 | 3.33 | 0.359 | 0.230 | 1.27 | 0.55 | 0.83 | 0.69 | 0.93 | 0.56 | 0.71 |

### Table 25. Values of Standard Framed-Beam Connections

(⅞-in A502, grade 1 rivets of ⅞-in A325 HS bolts,† A36 members)

| AISC designation | Two angles, size | Fasteners in outstanding legs No. | Single* shear, 1,000 lb | Fasteners in web legs No. | Double shear, 1,000 lb | Bearing, 1,000 lb |
|---|---|---|---|---|---|---|
| 10 rows........ | 4 × 3½ × ⅜ × 2'5½'' | 20 | 180 | 10 | 180 | 424t |
| 9 rows......... | 4 × 3½ × ⅜ × 2'2½'' | 18 | 164 | 9 | 164 | 382t |
| 8 rows......... | 4 × 3½ × ⅜ × 1'11½'' | 16 | 144 | 8 | 144 | 340t |
| 7 rows......... | 4 × 3½ × ⅜ × 1'8½'' | 14 | 126 | 7 | 126 | 297t |
| 6 rows......... | 4 × 3½ × ⅜ × 1'5½'' | 12 | 108 | 6 | 108 | 255t |
| 5 rows......... | 4 × 3½ × ⅜ × 1'2½'' | 10 | 90 | 5 | 90 | 212t |
| 4 rows......... | 4 × 3½ × ⅜ × 0'11½'' | 8 | 72 | 4 | 72 | 170t |
| 3 rows......... | 4 × 3½ × ⅜ × 0'8½'' | 6 | 54 | 3 | 54 | 127t |
| 2 rows......... | 4 × 3½ × ⅜ × 0'5½'' | 4 | 56 | 2 | 36 | 85t |
| 1 row.......... | 6 × 4 × ⅜ × 0'3'' | 2 | 18 | 2 | 36 | 85t |

†Values indicated are for friction-type connections or bearing-type where threads are not excluded from the shear plane. For bearing-type connections where threads are excluded from the shear plane, shear values may be increased by 22/15.

*If the web of the supporting beam is thinner than 0.21 in (0.42 in if beams frame on both sides) bearing must also be investigated.

### Table 26. Gages for Angles, Inches

| Leg | 8 | 7 | 6 | 5 | 4 | 3½ | 3 | 2½ | 2 | 1¾ | 1½ | 1⅜ | 1¼ | 1 |
|---|---|---|---|---|---|---|---|---|---|---|---|---|---|---|
| g¹ | 4½ | 4 | 3½ | 3 | 2½ | 2 | 1¾ | 1⅜ | 1⅛ | 1 | ⅞ | ⅞ | ¾ | ⅝ |
| g² | 3 | 2½ | 2¼ | 2 | | | | | | | | | | |
| g³ | 3 | 3 | 2½ | 1¾ | | | | | | | | | | |
| Max rivet | 1⅛ | 1 | ⅞ | ⅞ | ⅞ | ⅞ | ⅞ | ¾ | ⅝ | ½ | ⅜ | ⅜ | ⅜ | ¼ |

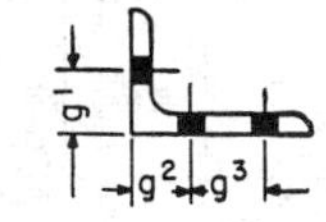

For diagonal angles, etc., gage in middle, where riveted leg equals or exceeds 3 in for ¾-in rivets, 3½ in for ⅞-in rivets. Use special gages to adapt work to multiply punch, or to secure desirable details. Rivet size may be increased in thin angles by adjusting gage lines.

**Table 27. Standard Gages in Flanges of Channels and I Beams; Maximum Rivets**

| Section | Depth, in. | Weight per ft, lb | Gage, in. | Max rivet, in. | Depth, in. | Weight per ft, lb | Gage, in. | Max rivet, in. | Depth, in. | Weight per ft, lb | Gage, in. | Max rivet, in. | Depth, in. | Weight per ft, lb | Gage, in. | Max rivet, in. |
|---|---|---|---|---|---|---|---|---|---|---|---|---|---|---|---|---|
| American Standard channels | 15 | 50 | 2¼ | 1 | 10 | 30 | 1¾ | ¾ | 8 | 18.75 | 1½ | ¾ | 6 | 13.0 | 1⅜ | ⅝ |
| | 15 | 40 | 2 | 1 | 10 | 20 | 1½ | ¾ | 8 | 13.75 | 1⅜ | ¾ | 6 | 10.5 | 1⅛ | ⅝ |
| | | | | | 9 | 20 | 1½ | ¾ | | | | | 5 | 9.0 | 1⅛ | ½ |
| | 12 | 30 | 1¾ | ⅞ | 9 | 15 | 1⅜ | ¾ | 7 | 14.75 | 1¼ | ⅝ | 4 | 7.25 | 1 | ½ |
| | | | | | | | | | | | | | 3 | 6 | ⅞ | ½ |
| American Standard I beams | 24 | 120 | 4 | 1 | 18 | 70 | 3½ | ⅞ | 10 | 35 | 2¾ | ¾ | 6 | 17.25 | 2 | ⅝ |
| | 20 | 95 | 4 | 1 | 15 | 50 | 3½ | ¾ | | | | | 5 | 14.75 | 1¾ | ½ |
| | 20 | 75 | 3½ | ⅞ | | | | | 8 | 23.0 | 2¼ | ¾ | 4 | 9.5 | 1½ | ½ |
| | | | | | 12 | 50 | 3 | ¾ | 7 | 20 | 2¼ | ⅝ | 3 | 7.5 | 1½ | ⅜ |
| Wide-flange beams | 36 | ... | 5½ | | 18 | 114 | 5½ | | 14 | ..... | 5½ | | 10 | 112 | 5½ | |
| | 33 | ... | 5½ | | | | | | | | | | | | | |
| | 30 | ... | 5½ | | 18 | 60 | 3½ | | 14 | 38 | 3½ | 1 | 10 | 29 | 2¾ | ⅞ |
| | 27 | ... | 5½ | | | | | | 12 | ..... | 5½ | | 8 | 67 | 5½ | 1 |
| | 24 | ... | 5½ | | 16 | 96 | 5½ | 1 | | | | | 8 | 28 | 3½ | 1 |
| | 21 | ... | 5½ | | 16 | 50 | 3½ | 1 | 12 | 36 | 3½ | 1 | 8 | 20 | 2¾ | ⅞ |

Gage in channels is the distance from center of rivet holes to flat side of the channel; in I beams it is the distance between lines of rivet holes in opposite flanges.

For weights lighter than those tabulated, the gage and maximum rivet are the same as for the next heavier tabulated section.

**Table 28. Safe Axial Loads for Standard Pipe Columns, Kips**
(Stress according to AISC Specification for A501 pipe*)

| Nominal pipe size, in | Outside diam, in | Wall thickness, in | Effective length of column K1, ft: 6 | 7 | 8 | 9 | 10 | 11 | 12 | 14 | 16 | 18 | 20 |
|---|---|---|---|---|---|---|---|---|---|---|---|---|---|
| 3 | 3.500 | 0.216 | 38 | 36 | 34 | 31 | 28 | 25 | 22 | 16 | 12 | 10 | |
| 3½ | 4.000 | 0.226 | 48 | 46 | 44 | 41 | 38 | 35 | 32 | 25 | 19 | 15 | 12 |
| 4 | 4.500 | 0.237 | 59 | 57 | 54 | 52 | 49 | 46 | 43 | 36 | 29 | 23 | 19 |
| 5 | 5.563 | 0.258 | 83 | 81 | 78 | 76 | 73 | 71 | 68 | 61 | 55 | 47 | 39 |
| 6 | 6.625 | 0.280 | 110 | 108 | 106 | 103 | 101 | 98 | 95 | 89 | 82 | 75 | 67 |
| 8 | 8.625 | 0.322 | 171 | 168 | 166 | 163 | 161 | 158 | 155 | 149 | 142 | 135 | 127 |
| 10 | 10.750 | 0.365 | 246 | 243 | 241 | 238 | 235 | 232 | 229 | 223 | 216 | 209 | 201 |
| 12 | 12.750 | 0.375 | 303 | 301 | 299 | 296 | 293 | 291 | 288 | 282 | 275 | 268 | 261 |

For dimensions of standard pipe see Sec. 8. Safe loads above underscore lines are for values of K1/r more than 120 but not over 200.

*Yield stress is 36 ksi. Pipe ordered to ASTM A53, type E or S grade B, or to API standard 5L grade B will have a yield point of 35 ksi and may be designed at stresses allowed for A501 pipe.

**Table 29. Selected Fire-Resistance Ratings**

| Type | Details of construction | Rating |
|---|---|---|
| Reinforced-concrete beams and girders | Grade A concrete, 1½ in clear to reinforcement | 4 h |
| | Grade B concrete, 1½ in clear to reinforcement | 3 h |
| Steel beams, girders, and trusses | 1½ in gypsum-perlite plaster on metal lath. 1¼ in clear of steel | 4 h |
| | 1 in gypsum-perlite plaster on metal lath, 1¼ in clear of steel | 3 h |
| | Ceiling of 1⅛ in gypsum-perlite plaster on metal lath with 2½ in min air space between lath and structural members | 4 h |
| Reinforced-concrete columns | Grade A concrete 1½ in clear to reinforcement; 12-in columns or larger | 4 h |
| | Grade B concrete 2 in clear to reinforcement; 12-in columns or larger | 4 h |
| Steel columns, 8 × 8 in or larger | Concrete (siliceous gravel): | |
| | 2½ in clear to reinforcement | 4 h |
| | 2 in clear to reinforcement | 3 h |
| | 1 in clear to reinforcement | 2 h |
| | 1½ in gypsum-perlite plaster on metal lath spaced from flanges with ¾-in steel furring channels | 4 h |
| | ⅞ in portland-cement plaster on metal lath over ¾-in channels | 1 h |
| Reinforced-concrete slabs | 5 in concrete (expanded clay, shale, slate, or slag) 1 in clear to reinforcement | 4 h |
| | 6½ in concrete (all other aggregate) 1 in clear to reinforcement | 4 h |
| Heavy-timber floors | 3 in tongue-and-groove plank floor with 1 in finish flooring | 1 h |

**Table 29. Selected Fire-Resistance Ratings** *(Continued)*

| Type | Details of construction | Rating |
|---|---|---|
| Wood joists | Wood floor; 1 in tongue-and-groove subfloor and 1 in finish floor with asbestos paper between. Ceiling of ⅝ in Underwriters' Laboratories listed wallboard | 1 h |
| Brick walls | Solid walls, unplastered, with no combustible members framed in wall: | |
| | 8 in nominal | 4 h |
| | 4 in nominal | 1 h |
| Concrete masonry units | 8 in Underwriters' Laboratories listed concrete blocks, laid as specified in Underwriters' Laboratories listing | 4 h |
| | 4 in Underwriters' Laboratories listed concrete blocks; cells filled with perlite mortar and laid as specified in Underwriters' Laboratories listing | 4 h |
| Steel-stud partitions | ¾ in gypsum-perlite plaster both sides on metal lath | 2 h |
| | ⅝ in gypsum wallboard on 3⅝-in steel studs; attached with 2 in cement-coated nails; joints taped and cemented | 1 h |
| Wooden-stud partitions | Exterior walls; one side covered with ½ in gypsum sheathing and wood siding; other side faced with ½ in gypsum-perlite plaster on ⅜ in perforated gypsum lath | 1 h |
| | ⅞ in exterior cement plaster on outside inside face same as above | 1 h |
| Plain or reinforced-concrete walls | Solid, unplastered: | 4 h |
| | 7½ in. thick | 3 h |
| | 6½ in. thick (6½) | 2 h |
| | 5½ in. thick Grade B (6) (5) Grade A | 1 h |
| | 4 in thick (3½) | |

Grade A concrete is made with aggregates such as limestone, calcareous gravel, trap rock, slag, expanded clay, shale, or slate or any other aggregates possessing equivalent fire-resistance properties.

Grade B concrete is all concrete other than Grade A concrete and includes concrete made with aggregates containing more than 40 percent quartz, cherts, or flint.

# REINFORCED-CONCRETE DESIGN AND CONSTRUCTION

## by William L. Gamble

REFERENCES: Ferguson, "Reinforced Concrete Fundamentals," Wiley. Winter and Nilson, "Design of Concrete Structures," McGraw-Hill. Lin, "Design of Prestressed Concrete Structures," Wiley. Khachaturian and Gurfinkel, "Prestressed Concrete," McGraw-Hill. "Building Code Requirements for Reinforced Concrete (318-77)," "Commentary on Building Code Requirements for Reinforced Concrete," "Formwork for Concrete, SP-4," and "Manual of Standard Practice for Detailing Reinforced Concrete Structures," American Concrete Institute. "Standard Specifications for Highway Bridges," American Association of State Highway Officials (ASSHO).

The design, theory, and notation of this chapter are in general accord with the 1977 Building Code Requirements for Reinforced Concrete of the American Concrete Institute, though many detailed provisions have been omitted.

## Standard Notation

### Load Factors

$D$ = dead load of structure or force caused by dead load

$E$ = earthquake load

$L$ = live load of structure or force caused by live load

$W$ = wind load

$U$ = required strength of structure to resist design ultimate loads

$\phi$ = understrength or capacity reduction factor

### Beams and General Notation

$a$ = depth of compression zone, using approximate method

$A_b$ = area of individual reinforcing bar, in$^2$

$A_{ps}$ = area of prestressing steel

$A_s$ = area of tension reinforcement

$A'_s$ = area of compression reinforcement

$A_v$ = area of steel in one stirrup

$b$ = width of compression face of beam

$b_w$ = width of stem of T beam

$c = k_u d$ = depth to neutral axis at ultimate, from compression face

$d$ = effective depth of beam, compression face to centroid of tension steel

$d'$ = depth of compression steel, from compression face

$d_b$ = diam of individual reinforcing bar, in

$E_c$ = Young's modulus of concrete

$E_s$ = Young's modulus of steel

$f'_c$ = compressive strength of concrete from tests of $6 \times 12$ in cylinders, lb/in$^2$

$\sqrt{f'_c}$ = measure of shear and tensile strength of concrete, lb/in$^2$, i.e., if $f'_c = 4{,}900$ lb/in$^2$, then $\sqrt{f'_c} = 70$ lb/in$^2$

$f_{pu}$ = ultimate stress of prestressing steel, lb/in$^2$

$f_y$ = yield stress of reinforcing steel, lb/in$^2$

$h$ = overall height or thickness of member

$k_u$ = ratio of ultimate neutral axis depth to effective depth

$l_d$ = development length of reinforcing bar, in

$s$ = spacing of shear reinforcement

$M_u$ = ultimate moment of section or required ultimate moment

$v_c$ = shear stress in concrete

$V_c$ = shear force resisted by concrete

$V'$ = shear force resisted by web reinforcement

$V_u$ = shear strength of section or required ultimate shear

$\beta_1$ = factor relating neutral axis position to depth of equivalent approximate stress block

$\epsilon_{su}$ = reinforcement strain at time of failure of member

$\epsilon_y$ = yield strain of reinforcement

$\rho$ = tension reinforcement ratio = $A_s/bd$

$\rho'$ = compression reinforcement ratio = $A'_s/bd$

$\rho_{bal}$ = balanced reinforcement ratio

### Columns

$A_c = A_g - A_s$ = net area of concrete in cross section

$A_{core}$ = area within spiral

$A_g$ = gross area of column

$A_s$ = area of steel in column

$e$ = eccentricity of axial load on column

$k$ = effective length factor

$l_u$ = unsupported length of column, from top of floor to bottom of beam, capital, or slab

$M = Pe$ = applied bending moment

$P$ = axial thrust

$P_L$ = design ultimate load on column

$P_0$ = failure load of short column under concentric load

$r$ = radius of gyration of gross concrete section

$r'$ = ratio of flexural stiffness of columns to restraining beam stiffness at joint

$R$ = capacity reduction factor due to length effects

$\rho_g$ = gross steel ratio = $A_s/A_g$

$\rho_s$ = spiral steel ratio = volume of spiral steel/volume of core

### Floor Systems

$b_0$ = effective shear perimeter around column in flat plate, flat slab, or footing

$c_1$ = width of supporting column or capital, in direction of span being considered

$c_2$ = width of supporting column, in transverse direction

$C$ = torsional constant

$E_{cs}$ = Young's modulus of concrete in torsional member

$K_b$ = flexural stiffness of beam, moment per unit rotation

$K_c$ = flexural stiffness of columns at joint, moment per unit rotation

$K_{ec}$ = flexural stiffness of equivalent column, moment per unit rotation

$K_s$ = flexural stiffness of slab of width $l_2$, moment per unit rotation

$K_t$ = rotational stiffness of torsional member, moment per unit rotation

$l_1$ = span, center to center of supports, in direction considered

$l_2$ = span, center to center of supports, in transverse direction

$l_n = l_1 - c_1$ = clear span in direction considered

$M_0$ = static moment

$w$ = distributed design load, including load factors

$x$ = shorter side of rectangular area of torsion member

$y$ = longer side of rectangular area of torison member

$\alpha_1$ = $EI$ of beam in direction $1/EI$ of slab of width $l_2$

$\alpha_c = K_c/\Sigma(K_b + K_s)$ = relative stiffness of column

$\alpha_{ec} = K_{ec}/\Sigma(K_b + K_s)$ = relative stiffness of equivalent column

$\beta_a$ = ratio of dead load/live load, without load factors

$\beta_t = (EC)_{beam}/2(EI)_{slab}$, where $I_s$ is for slab of width $l_2$

$\delta_c$ = factor to increase positive moments if column stiffness is inadequate

### Footings

$A_1$ = loaded area

$A_2$ = area of same shape and concentric with $A_1$

$d_p$ = diam of pile

$f_b$ = ultimate bearing stress on concrete

$\beta$ = ratio of long side/short side of footing

### Walls

$l_c$ = unbraced height of wall

**Prestressed Concrete**

$A_{ps}$ = area of prestressed reinforcement
$A_t$ = area of anchorage-zone reinforcement
$f'_{ci}$ = compressive strength of concrete at time of prestressing
$f_{ps}$ = stress in reinforcement at time of failure of member
$f_{pu}$ = ultimate stress of prestressing reinforcement
$f_{se} = f_{si} - \Delta f_s$ = stress in reinforcement at service load
$f_{si}$ = reinforcement stress at time of tensioning steel
$\Delta f_s$ = loss of prestress from initial tensioning value
$\rho_p = A_{ps}/bd$

## MATERIALS

**Reinforced concrete** is a combination of concrete and steel acting as a unit because of bond between the two materials. Concrete has a high compressive strength but a relatively low tensile strength. Beams of plain concrete fail by tension at very low stresses, but if properly reinforced by embedment of steel in their tensile regions, they may be loaded to utilize the much higher compressive strength of the concrete. Reinforced-concrete structures are practically monolithic, are more rigid than steel structures of the same strength, and are inherently fire-resistant. Reinforcement in the concrete also controls cracking caused by temperature changes and shrinkage.

**Prestressed concrete** is a form of reinforced concrete in which initial stresses opposite those caused by the applied loads are induced by tensioning high-strength steel embedded in the concrete. Members may be *pretensioned*, in which the steel is tensioned and the concrete then cast around it, or *posttensioned*, in which the concrete is cast and cured, after which steel placed in ducts through the concrete is tensioned.

**Concrete** For reinforced-concrete work, only high-quality portland cement concrete may be used, and the aggregates must be carefully selected. The proportions are governed by the required strength, durability, economy, and the quality of the aggregates. Concretes for building construction normally have compressive strengths of 3,000 to 5,000 lb/in² (20 to 35 MPa), except that concrete for columns may be considerably stronger. Most concrete for prestressed members will have 5,000 lb/in² or higher compressive strength. The higher-strength concretes require thorough quality control if the strengths are to be consistently obtained. Generally, lean, harsh mixes should be avoided because the bond with the steel will be poor, permeability will be high, and durability of the concrete may be poor. A consistency of concrete that will flow sluggishly but not so wet as to produce segregation of the materials when transported must be used for all reinforced-concrete work in order to embed the steel and completely fill the molds or forms. The use of vibration is almost mandatory, and enables the use of stiffer, more economical, concretes than would otherwise be possible (see also Sec. 6).

**Steel** Reinforcing steel may be deformed or plain bars, welded-wire fabric, or high-strength wire and strand for pre-stressed concrete. Bars with deformations on their surfaces are designed to produce mechanical bond and greater adhesion between the concrete and steel, and are used almost universally in the United States. Welded-wire fabric is suitable in many cases for slabs and walls, and may result in cost savings through labor savings. Fabric is made with both smooth and deformed wires, and the deformed fabric may have some advantage in terms of better crack control. Deformed reinforcing bars having minimum yield stresses from 40,000 to 60,000

lb/in² (275 to 414 MPa) are currently manufactured. The higher-strength steels have the advantage of allowing higher working stresses, but their ductility may be less and it may be difficult to cold-bend them successfully, especially in the larger sizes. The chemical makeup of most reinforcing steels is such that it is not readily weldable without special techniques, including careful preheating and controlled cooling. Furthermore, the variation of the material from batch to batch is so great that separate procedures must be devised for each batch. Tack welding in assembling bar cages can be particularly troublesome because of the stress raisers introduced. Prestressing steel is heat-treated high-carbon steel, and 7-wire strand will have a breaking stress of 250,000 or 270,000 lb/in² (1,720 or 1,860 MPa). The strength of solid wire for prestressing is slightly less. ASTM specifications cover the various steels, and all steel used as reinforcement should comply with the appropriate specification.

**Reinforcing Steel Sizes** The sizes of reinforcing bars have been standardized, and the designation numbers are approximately the bar diameter, in ⅛-in units. Table 1 gives the nominal diameter, cross-sectional area, and perimeter for each bar size. The areas of the four most common 7-wire prestressing strands are given below:

| Diam, in | Area, in² | |
|---|---|---|
| | Grade 250 | Grade 270 |
| 7/16 | 0.109 | 0.115 |
| 1/2 | 0.144 | 0.153 |

Plain and deformed wires are used as reinforcement, usually in the form of welded mats. Plain wires are made in sizes W0.5 through W31, where the number designates the cross-sectional area of the wire in hundredths of square inches. Deformed wires are made in sizes D1 through D31, and the number has the same meaning. Not all sizes are made by all manufacturers, and local suppliers should be consulted about availabilities of sizes. The formerly used wire gage numbers are no longer used for size specifications.

**Moduli of Elasticity** For concrete the modulus of elasticity $E_c$ may be taken as $w^{1.5}33 \sqrt{f'_c}$ in lb/in², where $w$ is the weight of the concrete between 90 and 155 lb/ft³. Normal weight concrete may be assumed to weigh 145 lb/ft³. For steel the modulus of elasticity may be taken as 29,000 lb/in² (200 GPa), except for prestressing steel for which the modulus shall be determined by tests or supplied by the manufacturer, but is usually about 28,000,000 lb/in².

**Table 1. Dimensions of Deformed Bars**

| No. | Diam, in. | Area, sq in. | Perimeter, in. |
|---|---|---|---|
| 2* | 0.250 | 0.05 | 0.786 |
| 3 | 0.375 | 0.11 | 1.178 |
| 4 | 0.500 | 0.20 | 1.571 |
| 5 | 0.625 | 0.31 | 1.963 |
| 6 | 0.750 | 0.44 | 2.356 |
| 7 | 0.875 | 0.60 | 2.749 |
| 8 | 1.000 | 0.79 | 3.142 |
| 9 | 1.128 | 1.00 | 3.544 |
| 10 | 1.270 | 1.27 | 3.990 |
| 11 | 1.410 | 1.56 | 4.430 |
| 14S | 1.693 | 2.25 | 5.32 |
| 18S | 2.257 | 4.00 | 7.09 |

*No. 2 bars are obtainable in plain rounds only.

The **modular ratio** $n = E_s/E_c$ is of importance in designing reinforced concrete. It may be taken as the nearest whole number but not less than six. The value of $n$ for lightweight concrete may be taken as the same for normal weight concrete of the same strength, except in calculations for deflections.

**Protection of Reinforcement** Reinforcement, for both regular and prestressed concrete, must be protected by the concrete so as to prevent corrosion. The amount of cover needed for various degrees of exposure is as follows:

| Member and Exposure | Cover, in |
|---|---|
| Concrete surface deposited against the ground | 3 |
| Concrete surface to come in contact with the ground after casting: | |
| Reinforcement larger than No. 5 | 2 |
| Reinforcement smaller than No. 5 | 1½ |
| Beams and girders not exposed to weather: | |
| Main steel | 1½ |
| Stirrups and ties | 1 |
| Joists, slabs, and walls not exposed to weather | ¾ |
| Column spirals and ties | 1½ |

Except for joists and slabs, the cover must be at least equal to the bar diameter, and for columns the cover must be 1½ times the maximum size of the coarse aggregate.

The amount of protection recommended is a minimum, and when corrosive environments or other severe exposure occurs, the cover should be increased. The concrete in the cover should be made as impermeable as possible. Fire-resistance requirements may also control the cover requirements.

## LOADS

The dead and live loads are combined in determining the cross sections of the members. The dead load includes the weight of the structure, all finishing materials, and usually the installed equipment. The live load includes the contents and ordinarily refers to the movable items.

The **live load** (pounds per square foot) to be used for design depends upon the loadings that will occur in the particular structure as well as on the requirements of the local building code. The Boston Building Code illustrates good practice and is as follows for floor loads:

Heavy manufacturing, sidewalks, heavy storage, truck garages, 250; public garages, intermediate manufacturing, and hangars, 150; stores, heavy merchandise, light storage, 125; armories, assembly halls, gymnasiums, grandstands, public portions of hotels, theaters, and public buildings, corridors and fire escapes from public assembly buildings, light merchandising stores, stairs, first and basement floors of office buildings, theater stages, 100; upper floors of public buildings, office portions of public buildings, stairs, corridors and fire escapes except from public assembly buildings, theater and assembly halls with fixed seats, light manufacturing, locker rooms, stables, 75; church auditoriums, 60; office buildings above first floor including corridors, classrooms with fixed seats, 50; residence buildings and residence portions of hotels, apartment houses, clubs, hospitals, educational and religious institutions, 40. [Note: 100 lb/ft² = 4.79 kPa.]

**Live loads** affecting structural members supporting considerable tributary floor areas are sometimes reduced in recognition of the low probability that the entire area will be loaded to the full design load at the same time. Roofs are commonly designed to support live loads of 20 to 30 lb/ft². Wind loads are commonly from 15 to 30 lb/ft² (higher in areas subject to hurricanes) of vertical projection. In the case of heavy moving loads, allowance should be made for impact by increasing the live load by 25 to 100 percent.

**Dead loads** include the weight of both structure and finishing materials, and the weights of some typical wall, floor, and ceiling materials are as follows:

| Description | Weight, lb/ft² |
|---|---|
| Granolithic finish, per in of thickness | 12 |
| ⅞ in hardwood, 1⅝ in plank intermediate floor, and tar base | 16 |
| 3-in wood block in coal-tar pitch | 10 |
| Lightweight concrete fill, 2 in thick | 14 |
| Plaster on concrete, tile, or concrete block, two coats | 5 |
| Plaster on lath | 10 |
| 6 in concrete block wall | 25 |
| Suspended ceiling | 12 |

## LOAD FACTORS FOR REINFORCED CONCRETE

The current ACI Building Code for Reinforced Concrete is based on a **strength design** concept, in which the strength of a member or cross section is the basis of design. This approach has been adopted because of the difficulty in assigning reasonable and consistent allowable stresses to the concrete and steel. Factors of safety are expressed in terms of *overload* and *understrength* factors. The overload factors, reflecting the uncertainity of the applied loads, are expressed as

No wind or earthquake loads:

$$U = 1.4D + 1.7L \tag{1}$$

With wind acting, use the larger of **(2)** or **(3)**:

$$U = 0.75(1.4D + 1.7L + 1.7W) \tag{2}$$
$$U = 0.9D + 1.3W \tag{3}$$

No section may be weaker than required by Eq. **(1)**. In case earthquake loadings are considered, $1.1E$ is substituted for $W$ in the above equations. Liquids are treated as dead loads, and other provisions are made for earth-pressure loadings. The understrength factors reflect the ductility of failure in the mode considered, and the consequences of a failure on the rest of the structure. These are expressed as $\phi$ factors with the following values:

| | |
|---|---|
| Bending and tension | $\phi = 0.90$ |
| Shear and torsion | $\phi = 0.85$ |
| Spiral columns | $\phi = 0.75$ |
| Tied columns and bearing | $\phi = 0.70$ |

These factors are used to reduce the computed ideal strengths of members, reflecting possible weaknesses related to materials, dimensions, and workmanship.

In addition to strength requirements, serviceability checks must be made to ensure freedom from excessive cracking, deflections, vibrations, etc., at working loads. Prestressed-concrete members must also satisfy a set of allowable stresses at this load level.

Deflections are usually controlled by specifying minimum member depths. The values of $l/16$ and $l/21$, for beams which are simply supported and continuous at both ends, respectively, are typical, unless an investigation is conducted to show that shallower members will result in acceptable deflections. Crack control is obtained by careful distribution of the steel through the tensile zone. In this regard, several small bars are superior to one large bar.

The approach to design is not a "limit design" or "plastic design" concept, as the ultimate forces are derived from elastic analyses of structures, using maximums for each critical section. The plastic collapse loads and mechanisms currently used in the "plastic design" of some steel structures in the United States are not considered.

Bridges may be designed using the same general concepts, but different overload and understrength factors are used, and the serviceability requirements include checks on the fatigue strength.

### REINFORCED-CONCRETE BEAMS

Concrete beams are reinforced to resist both flexural and shear forces. A typical reinforcement scheme for a simply supported beam is shown in Fig. 1.

As the load on a reinforced-concrete beam is increased, vertical tension cracks appear in the maximum moment regions, and gradually grow in length, width, and number. By

mum the member can sustain. Further attempts at loading produce large increases in deflection and crack widths with very small increases in load and a gradual reduction in the remaining concrete compression area. When a limiting concrete strain of about 0.003 is reached at the compression face of the beam, the concrete starts crushing and the capacity of the beam starts dropping. For static design purposes, the achievement of the 0.003 strain is usually regarded as the end of the useful life of the beam.

Within limits to be checked later, at the time of flexural failure of a beam the stress in the steel is equal to the yield stress, which greatly simplifies the analysis of the member. The strain and stress distributions in a beam are shown in Fig. 2, at failure.

It is assumed that plane sections remain plane, that adequate bond exists, that the stress-strain relationships for concrete and steel are known, and that tension in the concrete has a negligible influence.

The **flexural strength** of a cross section may be written as,

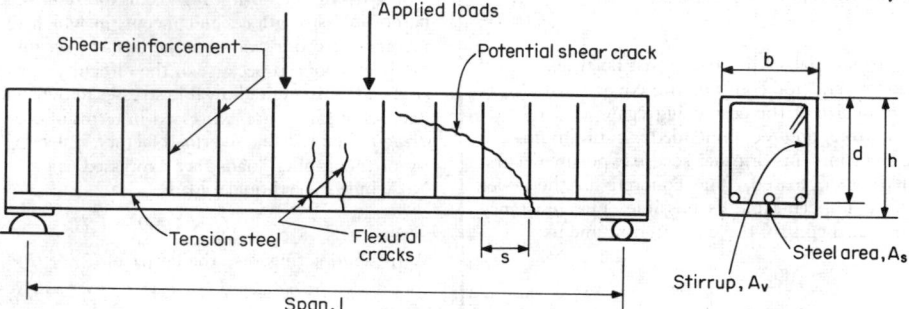

**Fig. 1**  Typical arrangement of reinforcement in beams.

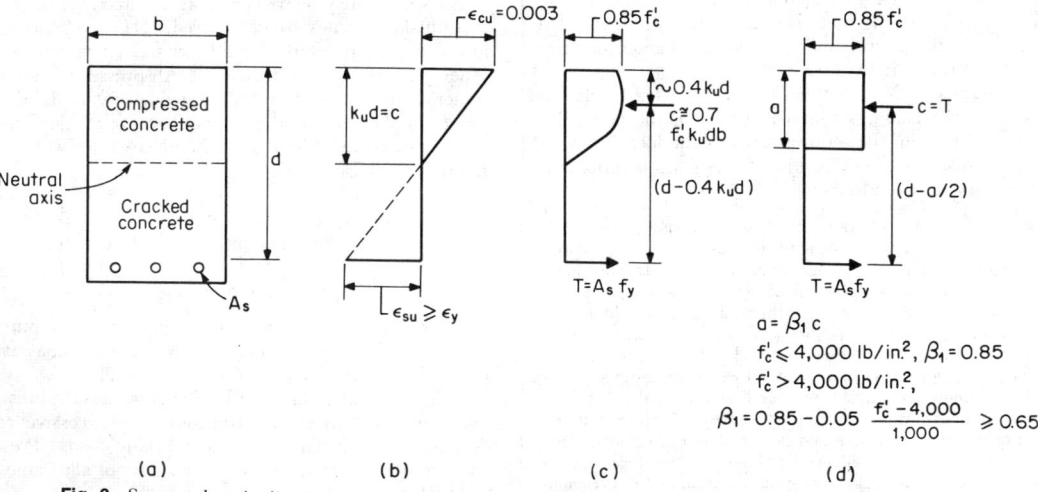

**Fig. 2**  Stress and strain distributions in reinforced-concrete beam at failure. (*a*) Section; (*b*) strains; (*c*) stresses; (*d*) approximate stresses.

the time the working load is reached, some of the cracks extend to the neutral axis and the contribution of the tensile strength of the concrete to the flexural capacity of the beam has become very small. As the load is increased further, the reinforcement eventually yields. This load is nearly the maxi-

including the understrength factor and rounding the terms in the parentheses to one significant figure:

$$M_u = \phi A_s f_y d(1 - 0.4 k_u)$$
$$= \phi A_s f_y d(1 - 0.6 \rho f_y / f'_c) \qquad (4)$$

where $\rho = A_s/bd$ = reinforcement ratio. The reinforcement ratio will ordinarily be between 0.005 and 0.02.

A satisfactory approximate stress distribution is shown in Fig. 2d. Since $T = C = 0.85 f_c' ba$, then $a = A_s f_y/0.85 f_c' b$ and the flexural capacity is

$$M_u = \phi A_s f_y(d - a/2) \tag{5}$$

It must be demonstrated for each case that the stress in the reinforcement has reached $f_y$, and the simplest approach is to show that $\epsilon_{su} \geqslant \epsilon_y$. From Fig. 2b, $\epsilon_{su} = 0.003 (1 - k_u)/k_u$. From equilibrium (see Fig. 2c), $k_u = A_s f_y/0.7 f_c' bd = \rho f_y/0.7 f_c'$. The limiting case for the validity of Eqs. (4) and (5) is a balanced condition in which the yield strain in the steel and the 0.003 compressive strain in the concrete are reached simultaneously, and this can be found using Fig. 2 and assuming $\epsilon_{su} = \epsilon_y$. It can then be shown that

$$\rho_{bal} = \frac{0.85 \beta_1 f_c'}{f_y} \frac{0.003}{0.003 + \epsilon_y} \tag{6}$$

Values of $\rho_{bal}$ are plotted vs. $f_c'$ in Fig. 3 for three values of $f_y$. The steel ratio should not exceed $0.75 \rho_{bal}$, in order to ensure that at least some yielding, with resultant large deflections, occurs before a member fails.

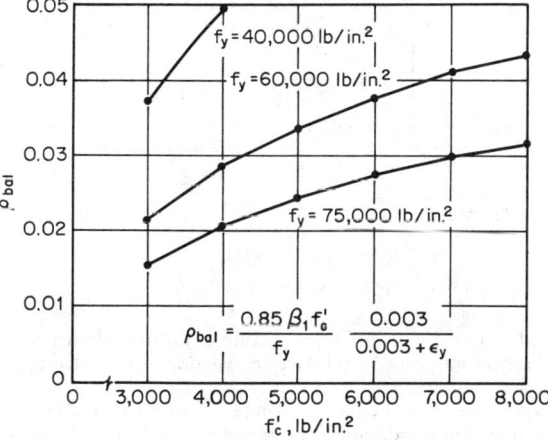

**Fig. 3**   Balanced steel ratio as a function of $f_y$ and $f_c'$.

Beams may contain both compression and tension reinforcement, especially when the beam must be kept as small as possible or when the long-term deflections must be minimized. The forces at ultimate are shown in Fig. 4, and the moment may be calculated as

$$M_u = \phi[A_s' f_y(d - d') + (A_s - A_s')f_y(d - a/2)] \tag{7}$$

where $a = (A_s - A_s')f_y/0.85 f_c' b$. The net steel ratio $\rho - \rho' = (A_s - A_s')/bd$ should not exceed 0.75 of the balanced value given by Eq. (6).

Many reinforced-concrete beams are flanged sections, or T beams, by virtue of having a slab cast monolithically with the beam, and such a cross section is shown in Fig. 5. It will be found that the neutral axis lies within the flange in most instances, and if $k_u d \leqslant t$, then the beam is treated as a rectangular beam of width $b$. This is checked using $k_u d =$

$A_s f_y/0.7 f_c' b$, developed from equilibrium (Fig. 2). If $k_u d > t$, the flexural capacity is computed by

$$M_u = \phi[(A_s - A_{sw})f_y(d - a/2) + A_{sw}f_y(d - t/2)] \tag{8}$$

where $A_{sw} = (b - b_w)t0.85 f_c'/f_y$

$$a = (A_s - A_{sw})f_y/0.85 f_c' b_w$$

Continuous T beams are treated as rectangular beams of width $b_w$ in the negative moment regions where the lower surface of the beam is in compression. Most continuous beams will have compression steel in the negative moment regions, as some bottom steel is always continued into the support regions.

Both T beams and beams with compression steel may be thought of in terms of dividing the beam into two components—one a rectangular beam containing part of the tension steel and the other a couple with the rest of the tension steel at the bottom and either the compression steel or the T-beam flanges at the top. Both components of the beam must satisfy horizontal equilibrium.

The reinforcement ratio should not be less than $\rho_{min} = 200/f_y$. This is to ensure that the ultimate moment is somewhat greater than the initial cracking moment. For T beams, $\rho = A_s/b_w d$ for purposes of the minimum steel requirement.

In the selection of a cross section, it is convenient to transform Eq. (4) by substituting $\rho bd$ for $A_s$ and rearranging to obtain

$$M_u/\phi bd^2 = \rho f_y(1 - 0.6 \rho f_y/f_c') \tag{9}$$

With given materials $f_y$ and $f_c'$ selection of a $\rho$ value enables calculation of the required $bd^2$, from which a cross section may be selected. For convenience, values of $M_u/\phi bd^2$ are plotted against $\rho f_y$ in Fig. 6, for several values of $f_c'$.

The strengths of prestressed-concrete beams are treated much the same as reinforced-concrete beams, and the differences will be noted later.

In addition to the tension stresses caused by bending forces, **shear forces** cause included tension stresses which may lead to inclined cracking such as is sketched in Fig. 1. Unless web reinforcement is present, the formation of such a crack usually leads directly to the complete collapse of the member at the load which caused the crack, or only slightly higher.

As a result of this undesirable behavior, shear reinforcement is required in all major beams, and the normal design procedure is to proportion the beam for the flexural requirements and then add shear steel, usually in the form of stirrups, to make the shear strength adequate.

The concrete can be assumed to resist a shear stress of $v_c = \phi 2\sqrt{f_c'}$, or a shear force of

$$V_c = \phi 2\sqrt{f_c'} \, bd \tag{10}$$

(or $V_c = \phi 2\sqrt{f_c'} \, b_w d$ for T beams). The sheer reinforcement must resist the force in excess of $V_c$, so that

$$V_u - V_c = V' \tag{11}$$

The area of shear reinforcement is then selected to satisfy

$$V' = \phi A_v f_y d/s \tag{12}$$

Shear reinforcement to satisfy this requirement is provided at every section of the beam, except that the region between the

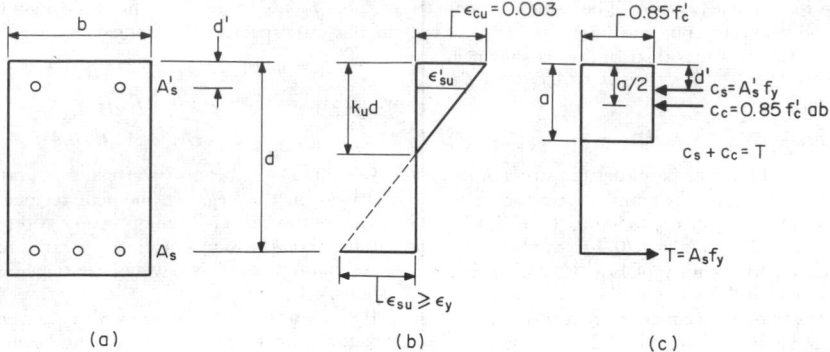

**Fig. 4** Strain and stress distributions in a doubly reinforced concrete beam. (*a*) Section; (*b*) strains; (*c*) approximate stresses.

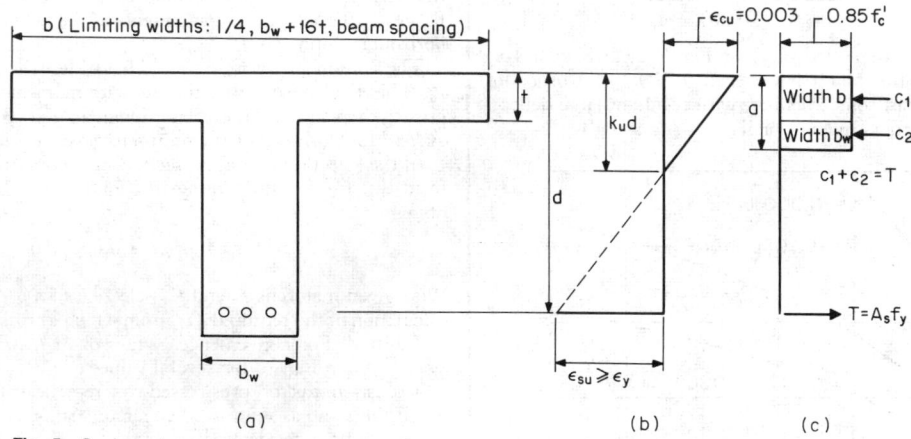

**Fig. 5** Strain and stress distributions in a T beam. (*a*) Section; (*b*) strains; (*c*) approximate stresses.

support and the section at the distance $d$ from the support is supplied with the steel required at $d$ from the support.

The stirrup spacing should not exceed $d/2$, and is reduced to a maximum of $d/4$ if $V_u > \phi 6\sqrt{f'_c}\, bd$. $V_u$ must not exceed $\phi 10\sqrt{f'_c}\, bd$. The minimum area of shear reinforcement

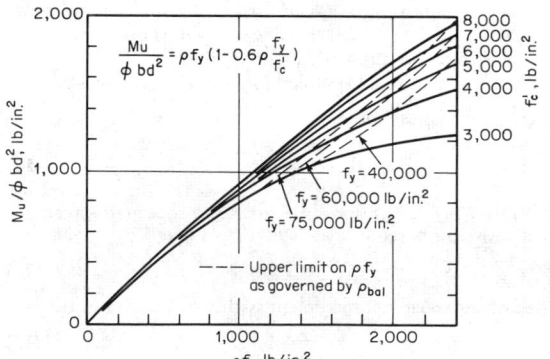

**Fig. 6** Design factors for singly reinforced concrete beams.

allowed is $A_v = 50 b_w s/f_y$. Closed stirrups of the form shown in Fig. 1 are recommended, and are essential in areas subjected to earthquake loadings.

Shear in prestressed-concrete members is handled in a similar manner, and the designer is referred to the ACI Code for details. However, it will be found that most prestressed members designed for buildings and which do not support major concentrated loads will have adequate shear strength if the minimum steel given by the following expression is supplied:

$$A_v = \frac{A_{ps}}{80} \frac{f_{pu}}{f_y} \frac{s}{d} \sqrt{\frac{d}{b_w}} \qquad \text{in}^2 \qquad (13)$$

The maximum stirrup spacing is 0.75 of the member depth, or 24 in, and for constant-depth members, $d$ is measured at the section of maximum moment.

At least the minimum area of shear reinforcement must be supplied over the full length of most reinforced and prestressed members unless it can be shown, by a test acceptable to the building official, that the members can sustain the required ultimate loads without the steel.

Occasionally inclined stirrups and bars bent up from the bottom to the top of the beam are used to provide part or all of the shear reinforcement, and the inclination must be taken into account in determining areas and spacings.

The adequacy of the **bond** between the concrete and steel must be checked, although this is not often a serious problem for static designs using modern deformed bars. The primary check is to ensure that there is sufficient length of embedded bar between the points of maximum stress and the ends of the bars.

The basic development lengths $l_d$ for bars in tension are given, in inches, as follows, with a 12 in minimum in all cases:

| | |
|---|---|
| No. 3 to No. 11 | $0.04\, A_b f_y / \sqrt{f'_c}$ |
| but not less than | $0.0004\, d_b f_y$ |
| No. 14 | $0.085\, f_y / \sqrt{f'_c}$ |
| No. 18 | $0.11\, f_y / \sqrt{f'_c}$ |
| Deformed wire | $0.03\, d_b f_y / \sqrt{f'_c}$ |

Horizontal bars with more than 12 in of concrete below them have development lengths 1.4 times those listed. Bars with yield stresses greater than 60,000 lb/in² must have the development lengths multiplied by $(2 - 60,000/f_y)$.

Bars No. 11 and smaller are commonly spliced by simply lapping them for a distance. These splices should be avoided in regions of high computed stress, and should be spread out so that not many bars are spliced near the same section. If the computed tensile stress is more than $0.5\, f_y$ at the design ultimate load and not more than half the bars are spliced at one section, the lap length is $1.3\, l_d$. For lower stress levels, a lap of $l_d$ is sufficient.

Development lengths in compression, important in columns and compression reinforcement, are somewhat shorter. The development length is $l_d = 0.02\, f_y d_b / \sqrt{f'_c}$, but not less than $0.0003\, f_y d_b$ or 8 in. The minimum lap splice length is $l_d$, but not less than $0.0005\, f_y d_b$ for $f_y = 60,000$ lb/in² or less.

Requirements for reinforcement of beams subjected to torsional moments are contained in the ACI Code. The requirements are too complex for discussion here, but the basic reinforcement scheme consists of closed stirrups with longitudinal bars in each corner of the stirrup. The most efficient method of dealing with torsion in many instances will be to rearrange the structure so as to reduce or eliminate the torsional moments.

## REINFORCED-CONCRETE COLUMNS

Reinforced-concrete compression members are proportioned taking into account the applied thrust, the bending moments, and the relationship between length and thickness for the member. The strength of a cross section or a short column can conveniently be shown with the aid of a *moment-thrust interaction diagram* such as is shown in Fig. 7a, which is drawn without considering the $\phi$ factor. The variation in $\phi$ with thrust is shown in Fig. 7b for the same column.

The load $P_0$ is the strength of a short column under a concentric load, and is expressed as

$$P_0 = 0.85\, A_c f'_c + A_s f_y \tag{14}$$

The contribution of the concrete is slightly less than the cylinder strength because of differences in workmanship, curing, and position of casting. $M_u$ is simply the strength in flexure, as was discussed for beams. The $M_b - P_b$ point is the "balance point," at which simultaneous crushing of the concrete and yielding of the reinforcement occur. Failure is initiated by crushing of the concrete at loads higher than $P_b$, and by yielding of the reinforcement in tension at lower loads. Only the reinforcement contributes to the tensile capacity of $T_u = A_s f_y$.

The balance-point moment and thrust are found using the strain and stress distributions shown in Fig. 8, for a symmetrical section with steel in two faces. From equilibrium,

$$P_b = C_s + C_c - T \tag{15}$$

Summing moments about the centroidal axis gives

$$M_b = A_s f_y (d - d')/2 + 0.7 f'_c k_u db(h/2 - 0.4\, k_u d) \tag{16}$$

This assumes that the compression steel has yielded, and this will be true except for very small members or members with exceptionally large cover over the steel.

The portion of the interaction diagram below $P_b$ may be constructed in a point-by-point manner. For any value of $\epsilon_s > \epsilon_y$, or for any value of $k_u d < k_u d_{bal}$, the strain and stress distributions are defined. Once the forces are defined, the moments and thrusts are computed using the same formulas as for the balance point. A modification will have to be made when the compression-steel stress is less than $f_y$.

Care must be used in the selection of the axis about which moments are to be summed, especially if the section is not symmetrical or symmetrically reinforced, since the force system is not a pure couple. The most important thing is to remain consistent, and to be certain that the interval and external moments are summed about the same axis.

In practice, either the moment-thrust curve can be reduced by the appropriate $\phi$ factor, or the computed ultimate $M$ and $P$ increased by dividing by $\phi$. If the required $M - P$ point lies on or slightly inside the interaction curve, the design is acceptable. In all cases a minimum moment requirement is imposed even if the structural analysis indicates that no bending exists. This is a moment of $M = 0.1\, hP$ for tied columns and $M = 0.05\, hP$ for spiral columns. The purpose of this minimum load eccentricity is to take account of accidental eccentricities of loading and construction. Column reinforcement consists of longitudinal bars and lateral ties or spiral bars. The ratio of longitudinal steel $\rho_g$ should be between 0.01 and 0.08. Ties are usually No. 3 or No. 4 bars which are bent to enclose the longitudinal bars, and are spaced at not more than 16 longitudinal bar diam, 48 tie diam, or the least thickness of the column. Ties are arranged to bind each corner bar and alternate intermediate bars. Several typical arrangements are shown in Fig. 9. Ties hold the longitudinal bars in place during construction, and may provide some shear resistance and improve the behavior of the column at loads near failure.

Spiral bars serve to confine the concrete core of the section as well as holding the longitudinal bars. The minimum amount of spiral steel, if advantage is to be taken of the higher $\phi$ factor for spiral columns, is

$$\rho_s = 0.45\, (A_g/A_{core} - 1) f'_c / f_y \tag{17}$$

where $f_y$ is the yield stress of the spiral, but not more than 60,000 lb/in². The clear spacing of the turns of the spiral must be between 1 and 3 in.

The strengths of compression members may be reduced

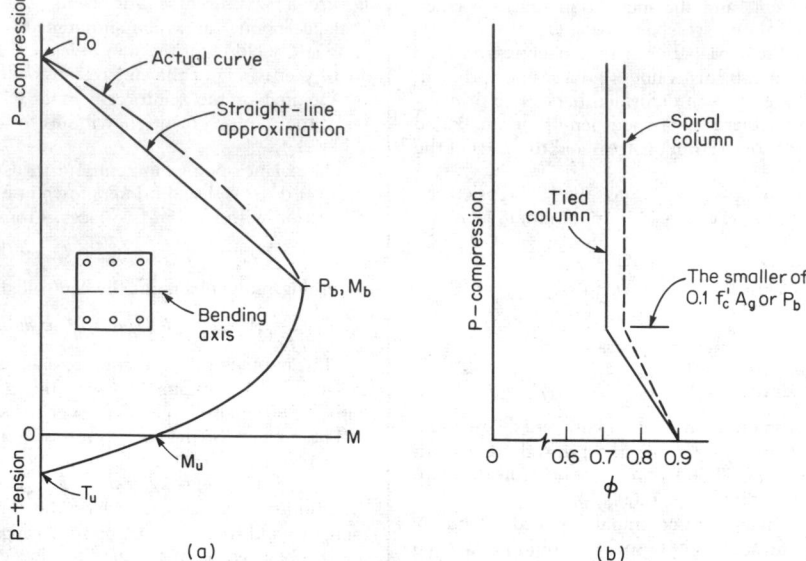

**Fig. 7** Column capacity diagrams. (*a*) Typical moment-thrust interaction diagram; (*b*) variation of $\phi$ with $P$.

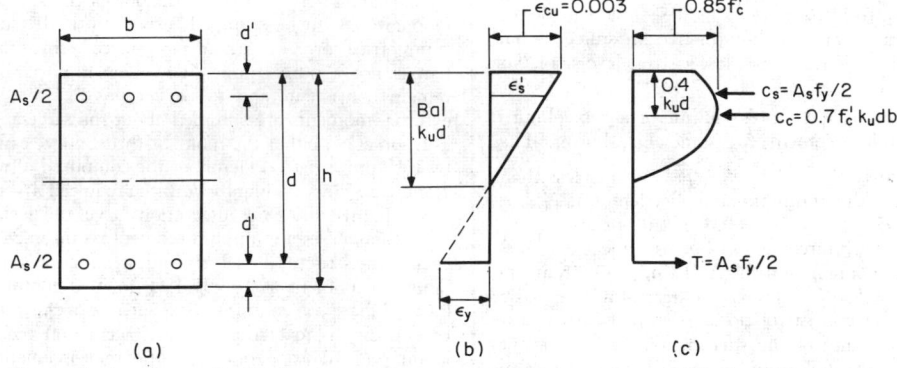

**Fig. 8**  Strain and stress distributions in a column at balanced moment and thrust. (*a*) Section; (*b*) strains; (*c*) stresses and forces.

below the cross-sectional strengths by length effects. Most columns in unbraced frames (frames in which the columns resist all horizontal forces) will have some strength reduction because of length effects, and most columns in braced frames will not, but this depends on the precise details of the length, width, restraint by other frame members, reinforcement, and amount of creep expected.

A comprehensive method of taking column length into account is contained in the 1977 ACI Code. A somewhat simpler method, based on the Commentary to that Code, is outlined below and may be used satisfactorily in most cases. In this method, a reduction factor $R$ is found to reduce the short column strength further below that found when including the $\phi$ factor. The $R$ factor is applied to both moment and thrust.

The effective length and the radius of gyration are the most important variables, and $r$ may be taken as $0.3h$ for a rectangular section, with $h$ measured in the direction being consi¹-

ered. For a circular section, $r = 0.25h$, and for other shapes $r$ is calculated for the gross concrete section. The provisions are as follows:

Braced frames:

Bending with point of contraflexure in member:
No reduction if $l_u/r \leq 54$
If $54 < l_u/r \leq 100$, then

$$R = 1.32 - 0.006\, l_u/r \leq 1.0 \tag{18}$$

Bending in single curvature:
$e/h < 0.1$

$$R = 1.23 - 0.008\, l_u/r \leq 1.0 \tag{19}$$

$e/h > 0.1$

$$R = 1.07 - 0.008\, l_u/r \leq 1.0 \tag{20}$$

Unbraced frames:
Restrictions: $k\, l_u/r \leq 40$

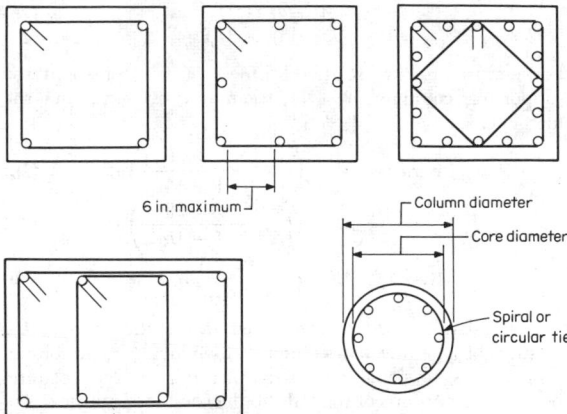

**Fig. 9** Typical arrangements of steel in columns.

Bracing beams to have $\rho \geq 0.01$ in negative moment regions.

$$k\,l_u = l_u\,(0.78 + 0.22\,r') \geq l_u \qquad (21)$$

where $r' = \Sigma K_{col}/\Sigma K_{beams}$, averaged at the two ends.

For loads of short duration

$$R = 1.07 - 0.008\,k\,l_u/r \leq 1.0 \qquad (22)$$

For loads of long duration

$$R = 0.97 - 0.008\,k\,l_u/r \leq 1.0 \qquad (23)$$

In addition to the reduction in the effective column strength, the beams framing into columns in an unbraced frame must be designed for negative moments larger than those from a first-order frame analysis. The design beam moments at a joint become

$$M = M_{nom}(1 - P_L/P_0)/(R - P_L/P_0) \qquad (24)$$

where $M_{nom}$ = nominal bending moment from frame analysis
$P_L$ = design ultimate load on column

These restrictions are stringent enough to force the designer to use the moment-magnifier approach given in the ACI Code for many unbraced-frame cases.

Columns are also made by encasing structural-steel sections in concrete, in which case the covering concrete must contain at least some steel in order to control cracks and maintain the integrity of the concrete in case of fire. Heavy steel pipes filled with concrete may also be used as columns. These are also used as piling, in which case the pipe is filled with concrete, usually after being driven to the final location.

## REINFORCED-CONCRETE FLOOR SYSTEMS

Several types of reinforced-concrete floors are used, with the choice depending on a number of factors such as span, live load, deflection limits, cost, story-height limitations, local custom, the nature of the rest of the structural frame or system, and the probability of future alterations.

The floor systems may be divided, somewhat arbitrarily, into one-way and two-way systems. One-way systems include solid and hollow slabs and joists spanning between parallel supporting beams or walls. Floors in which panels are subdivided into a grid by subbeams spanning between main girders have usually been designed as one-way slabs when the grid length is several times the width.

Two-way systems include slabs supported on all four sides on beams or walls, traditionally called *two-way slabs*. Slabs supported only on columns located at the corners of the panels also carry loads by developing stresses in the two major directions, and are usually termed *flat plates* if the slab is supported directly on the columns and *flat slabs* if there are capitals on the columns to increase the effective support size. A waffle slab is usually designed as a flat plate or flat slab, with pockets of concrete omitted from the lower surface, and the slab appears as a series of crossing joists.

One-way slabs and joists are designed as beams. A 12-in or other convenient width of slab is selected, analyzed as an isolated beam, and a depth and steel are picked. The main steel is perpendicular to the supports. Additional steel, parallel to the supports, is placed to control cracking and help distribute minor concentrated loads. This steel is usually one-fourth to one-third of the main steel, but not less than a gross steel ratio of 0.0018 for Grade 60 and 0.002 for lower-grade steels. These minimums govern in both directions, and in two-way systems as well.

One-way joists, such as that shown in Fig. 10, are usually cast with reusable sheet metal or fiberglass forms owned or rented by the contractor. Standard form sizes range from about 20 to 36 in wide and 8 to 20 in deep. The web thicknesses are made to suit the shear and fireproofing requirements, and special tapered end sections may be available to increase the web widths in the regions of high shear near the supports. Joists are exempted from the requirement that web steel be supplied regardless of the concrete shear stress. The allowable shear stress for the concrete is 1.1 times that for beams. The top slabs usually range from 2.5 to 4.5 in thick, and are reinforced to span from rib to rib. The joists are essentially designed as isolated T beams, and may be supported on girders or walls. Joist systems are suitable for reasonably long spans and heavy loads, and have low dead weights for the effective depths attainable.

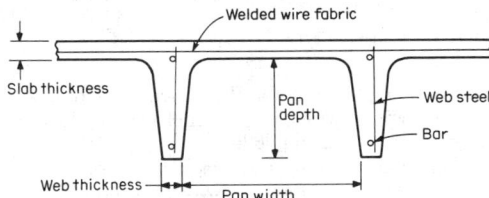

**Fig. 10** Cross section of concrete joist floor.

Slabs spanning in two directions are all designed taking into account the shape of the panel and the relative stiffnesses of the supporting beams, if any. The choice of types is a matter of loadings and economics. For residential and light office loadings, the flat plate is frequently the choice, as the very simple formwork may lead to substantial economy and the story height is minimized. For heavier loads or longer spans, punching shear around the columns becomes a limiting factor, and the flat slab, with its column capitals, may be the most suitable.

In case of extremely heavy floor loads or very stringent limits on deflections, slabs with beams on all four sides of each panel will be most satisfactory. The formwork is more complex than for the other slabs, but there will be some compensating savings in the amounts of steel because of the greater depths of the beams. In addition, the two-way slab may be much more efficient if the building is to resist major lateral loads by frame action alone, because of the difficulties in transferring large moments between columns and flat plates or slabs.

The design procedure is the same for slabs with and without beams. The basic steps, for each direction of span in each panel, are:

1. Compute static moment $M_0$.
2. Distribute $M_0$ to positive and negative moment sections.
3. Distribute section moments to column and middle strips and beams.

Most buildings slabs are designed for uniformly distributed loads, and the *static moment*, defined as the absolute sum of the positive plus average negative moments, is

$$M_0 = w\, l_2 (l_n)^2 / 8 \qquad (25)$$

The dimensions are illustrated in Fig. 11.

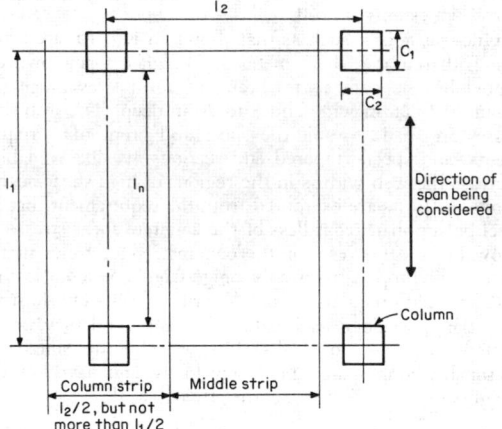

**Fig. 11** Arrangement of typical slab panel.

Relatively simple rules for the distribution of the static moment to various parts of the panel exist as long as the structure meets several simple limits:

1. Minimum of three spans in each direction.
2. Panel length no more than twice panel width.
3. Successive spans differ by not more than one-third the longer span.
4. Columns on a rectangular grid, or offset no more than 10 percent of the span.
5. Live load not more than three times the dead load.
6. If beams are used, they are used on all four sides of each panel and are approximately the same size, except that spandrel beams only are acceptable.

The following is for slabs meeting these restrictions. Information on other cases is contained in the ACI Code.

The positive-negative moment distribution for interior spans is

$$+M = 0.35\, M_0 \qquad (26)$$
$$-M = 0.65\, M_0 \qquad (27)$$

For end spans, the relative stiffness of the span and the supporting columns must be taken into account, and the moments are

$$\text{Interior} - M = \left(0.75 - \frac{0.10}{1 + 1/\alpha_{ec}}\right) M_0 \qquad (28)$$

$$+ M = \left(0.63 - \frac{0.28}{1 + 1/\alpha_{ec}}\right) M_0 \qquad (29)$$

$$\text{Exterior} - M = \left(\frac{0.65}{1 + 1/\alpha_{ec}}\right) M_0 \qquad (30)$$

The relative stiffness considered $\alpha_{ec}$ is the ratio of the stiffness of an equivalent column, which includes the column proper plus torsional members consisting of the spandrel beams plus portions of the slab, to the combined stiffnesses of the slabs and beams at the joint.

When there are substantial edge beams, with minimum widths and depths of about 1.5 and 2.5 times the slab thickness, respectively, moments for preliminary design purposes may be calculated neglecting the influence of the torsional members and using $\alpha_c$ in place of $\alpha_{ec}$. The exterior negative moments thus calculated should be multiplied by 0.9 if there are also beams perpendicular to the edge of the structure, and by 0.85 if there are only spandrel beams, before doing the initial proportioning. The final check on the design should be done using the equivalent column.

If there are no edge beams or only shallow ones, the use of the equivalent column is essential at all stages of the calculations, as the calculated exterior negative moments may be as much as 75 percent too large if the torsional members are neglected.

The stiffness of the "equivalent column" is calculated by adding the flexibilities of the column proper and the torsional members, as follows:

$$1/K_{ec} = 1/\Sigma K_c + 1/K_t \qquad (31)$$

The value of $K_t$ is calculated from

$$K_t = \Sigma \frac{9 E_{cs} C}{l_2 (1 - c_2/l_2)^3} \qquad (32)$$

where $l_2$ and $c_2$ refer to the transverse spans on each side of the column considered. If there are beams perpendicular to the edge of the structure, the value of $K_t$ from Eq. (32) is increased by multiplying it by the ratio of the moment of inertia of the slab plus beam (calculated for the entire unit and not the sum of the individual moments of inertia of slab and beam) to the moment of inertia of the slab, before calculating $K_{ec}$.

The torsional constant $C$ is calculated conservatively by the application of the following equation to each of the rectangular areas of the cross section, and summing:

$$C = \Sigma (1 - 0.63\, x/y) x^3 y / 3 \qquad (33)$$

A typical cross section, subdivided for the calculation of $C$, is shown in Fig. 12.

Once the section moments have been determined, they are distributed to the column and middle strips and beams, taking into account the panel shape and the beam stiffness. The beam relative stiffness coefficient $\alpha_1$ is calculated using an effective beam cross section as shown in Fig. 13, and the full width of the slab panel $l_2$. The locations of the column and middle strips are shown in Fig. 11.

The interior negative and positive moments are distributed to the column strips in the proportions shown in Fig. 14, with the remainder of the moment going to the middle strip. Linear

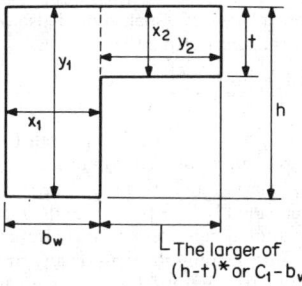

*But not more than 4t*

**Fig. 12** Subdivision of a beam for computation of $C$.

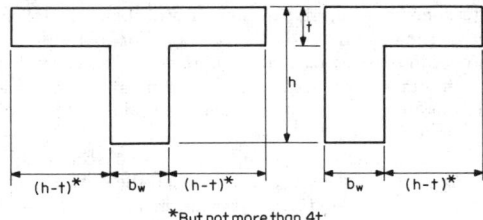

*But not more than 4t*

**Fig. 13** Beam sections for calculation of $I_b$ and $\alpha_1$.

interpolations are made for intermediate beam stiffnesses, but in most instances where there are beams, they will be found to have

$$\alpha_1 l_2 / l_1 \geq 1.0.$$

At the exterior negative moment sections, the distribution between column and middle strips is a function of both the torsional stiffness of the edge beam $\beta_t$ and the flexural stiffness

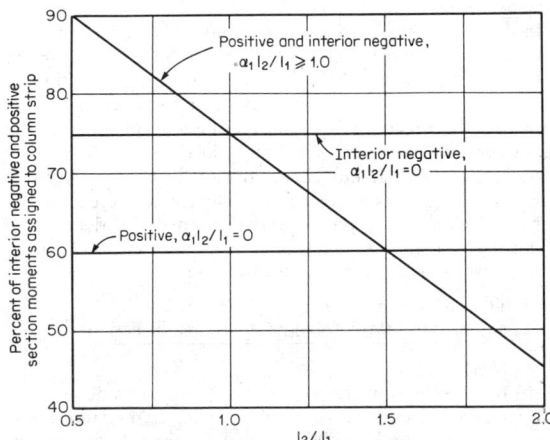

**Fig. 14** Percentage of interior negative and positive section moments assigned to a column strip.

$\alpha_1 l_2 / l_1$, of the beam perpendicular to the edge of the structure. The percentage of the section moment assigned to the column strip may be found by interpolation from Fig. 15. In most instances, 95 to 100 percent of the moment is assigned to the column strip, and minimum reinforcement governs in the middle strip.

Beam moments are found by dividing the column strip moment between beam and slab. If $\alpha_1 l_2 / l_1 \geq 1.0$, the beam moment is 85 percent of the column strip moment. This moment is reduced linearly to zero as $\alpha_1 l_2 / l_1$ approaches zero.

This design method assumes that all panels are loaded with the same uniformly distributed load at all times. Since this is obviously a gross simplification, two additional steps are taken to ensure that the structure will be adequate under partial loadings.

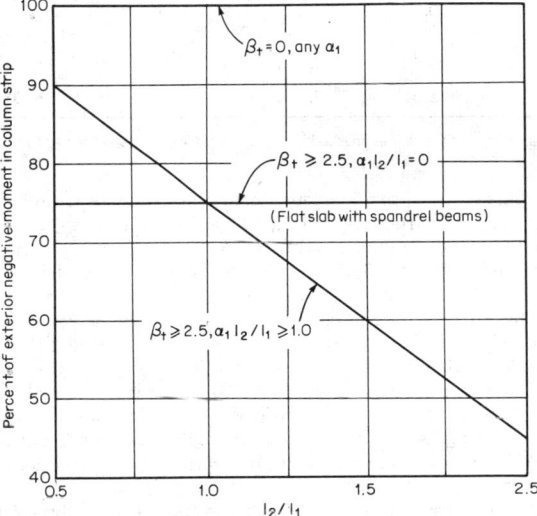

**Fig. 15** Percentage of exterior negative moment in a column strip.

First is a requirement for column design moments, and unless a more complete analysis is made, the following moment, divided between the columns above and below the slab in proportion to their stiffnesses, must be provided for:

$$M = 0.08 \frac{(w_d + w_l) \, l_2 \, (l_n)^2 - w'_d \, l'_2 \, (l'_n)^2}{1 + 1/\alpha_{ec}} \quad \text{(34)}$$

The loads $w_d$ and $w_l$ are the distributed dead and live loads including the overload factors. The terms $w'_d$, $l'_2$, and $l'_n$ are for the shorter of the two spans meeting at the column considered. As was discussed earlier, the term $\alpha_c$ could be used in place of $\alpha_{ec}$ when there are beams at the joint, and in this case this is always a conservative approximation.

The second check to ensure the serviceability of the structure under partial loadings is a column stiffness limit. If the interior column relative stiffness $\alpha_c$ is less than the value of $\alpha_{min}$ listed in Table 2, and the live load is more than half the dead load, the positive bending moments must be increased by multiplying by the coefficient

$$\delta_c = 1 + \frac{2 - \beta_a}{2 + \beta_a} \left( 1 - \frac{\alpha_c}{\alpha_{min}} \right) \quad \text{(35)}$$

**Table 2. Minimum $\alpha_{min}$**

| $\beta a$ | Aspect ratio $l_2/l_1$ | Relative beam stiffness $a$ | | | | |
|---|---|---|---|---|---|---|
| | | 0 | 0.5 | 1.0 | 2.0 | 4.0 |
| 2.0 | 0.5–2.0 | 0 | 0 | 0 | 0 | 0 |
| 1.0 | 0.5 | 0.6 | 0 | 0 | 0 | 0 |
| | 0.8 | 0.7 | 0 | 0 | 0 | 0 |
| | 1.0 | 0.7 | 0.1 | 0 | 0 | 0 |
| | 1.25 | 0.8 | 0.4 | 0 | 0 | 0 |
| | 2.0 | 1.2 | 0.5 | 0.2 | 0 | 0 |
| 0.5 | 0.5 | 1.3 | 0.3 | 0 | 0 | 0 |
| | 0.8 | 1.5 | 0.5 | 0.2 | 0 | 0 |
| | 1.0 | 1.6 | 0.6 | 0.2 | 0 | 0 |
| | 1.25 | 1.9 | 1.0 | 0.5 | 0 | 0 |
| | 2.0 | 4.9 | 1.6 | 0.8 | 0.3 | 0 |
| 0.33 | 0.5 | 1.8 | 0.5 | 0.1 | 0 | 0 |
| | 0.8 | 2.0 | 0.9 | 0.3 | 0 | 0 |
| | 1.0 | 2.3 | 0.9 | 0.4 | 0 | 0 |
| | 1.25 | 2.8 | 1.5 | 0.8 | 0.2 | 0 |
| | 2.0 | 13.0 | 2.6 | 1.2 | 0.5 | 0.3 |

From the 1977 ACI Building Code Requirements for Reinforced Concrete.

The intent is to provide enough column stiffness to isolate a panel partially from the effects of the loadings on adjacent panels, and the required stiffness depends strongly on the magnitude of the live load in addition to the structural characteristics of the system.

The shear strength of slab structures must always be checked, and shear stresses often govern the thickness of beamless slabs, especially flat plates.

If there are beams with $\alpha_1 l_2/l_1 \geq 1.0$, all shear is assigned to the beams, and stirrups are provided to make the shear capacity adequate, as was described in earlier coverage on beams. The beam shear is linearly reduced to zero as $\alpha_1 l_2/l_1$ is reduced to zero.

For the case of no beams, punching shear around the columns becomes a controlling factor. In this case the average shear stress, calculated as

$$v_u = V_u/b_0 d \qquad (36)$$

must not exceed $\phi 4\sqrt{f'_c}$. The critical shear perimeter $b_0$ is defined by a section located $d/2$ away from and extending all around the column, as shown in Fig. 16. It is very important that holes in the slab in the vicinity of the column be taken into account in reducing the value of $b_0$, and that no unauthorized

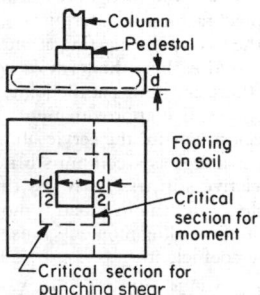

**Fig. 16** Two-way reinforced-concrete footing.

holes, such as for piping, be made either during or after construction.

It is possible to increase the shear resistance by the use of properly designed shear reinforcement, but this is not often done and is not recommended as a standard practice.

Transfer of moments between columns and slabs sets up shear and torsional stresses which should also be considered in the analysis of the shear strength.

## FOOTINGS

Footings (Fig. 16) may be classified as wall footings, isolated column footings, and combined column footings. The bending moments, shears, and bond stresses in such footings should be determined by the principles of statics on the basis of assumed or known soil-pressure distribution over the area of the footing. The bending moment on any projecting portion of a footing may be computed as the moment of the forces acting on the area to one side of a vertical plane through the critical section.

The critical section for bending in a concrete footing supporting a concrete column, pedestal, or wall should be taken at the face of the column, pedestal, or wall. For footings under metallic column bases or under masonry walls where bond with the footings is reduced to the friction value, the critical section is assumed midway between the middle and edge of the base or wall.

Shear stresses must be considered on two sections. The footing may act as a wide beam, for which the critical section is a vertical plane located $d$ away from the critical section for moment, and the stresses must satisfy those for a beam. Punching shear will often govern, and the critical section lies at a distance $d/2$ from the face of the column or other critical section for moment, as shown in Fig. 16, and as was the case for flat plates and slabs. If the footing is on piles, piles located $d_p/2$ outside the critical shear section are assumed to contribute shear, while those $d_p/2$ inside the section do not, with a linear variation between these extremes.

The critical section for bond should be taken at the same plane as for bending. Other vertical planes where abrupt changes of section occur should also be investigated for bond and shear stresses.

In sloped or stepped footings, sections other than the critical ones may require consideration. A square footing, reinforced in two directions, should have the reinforcement uniformly distributed across the entire width. Rectangular footings, reinforced in two directions, should have the reinforcement in the long direction uniformly distributed; in the short direction a portion, Eq. (37), should be uniformly distributed across a strip equal in width to the short side and centered on the structural element supported and the remainder distributed uniformly in the outer portions. The amount included in the center strip may be computed as follows:

Reinforcement in center strip
$$= \frac{2 \text{ (total reinforcement in short direction)}}{\beta + 1} \qquad (37)$$

where $\beta$ is the ratio of the long side to the short side.

**Combined Footings** Footings supporting two or more columns may be designed with sufficient accuracy by assuming uniform soil pressure and applying the laws of statics. The footing shape must be such that the center of gravity coincides with the center of gravity of the superimposed loads; otherwise unequal settlement may occur. The longitudinal and diagonal tension reinforcement should be designed by the ordinary rules of beam design. Lateral reinforcement should be designed as for isolated footings and should preferably be concentrated in bands under and near the columns proportionate in area to the column loads. The transverse reinforcement at each column should be uniformly distributed within a width centered on the column and should not be greater than the width of the column plus twice the effective depth of the footing.

**Spread or raft foundation,** consisting of a slab extending over the entire area under the columns or of a slab supported by beams, may be considered as loaded by a uniform upward reaction of the ground. The principle of design is exactly the same as that applied to a floor system, except that the load acts upward instead of downward.

**Concrete piles** of various types are widely used for foundations as they have larger carrying capacity and greater durability under many conditions of exposure than wooden piles. Precast piles are designed as columns with allowance for driving and handling stresses. Cast-in-place piles are constructed either by driving a steel shell and filling it with concrete or by filling the hole formed by a shell as it is withdrawn. Another method forms a bulb at the bottom by means of a ram which forces the concrete into the ground. The design load or capacity of cast-in-place concrete piles is largely empirical, being based on load-test data. The concrete for precast piles is usually over 3,500 lb/in² strength and for cast-in-place piles, over 3,000 lb/in² strength.

**Dowels and Bearing Plates** The stress in the longitudinal reinforcement of concrete columns should be transferred to the footing by means of dowels, equal in number and area to the column rods and of sufficient length to transfer the stress as in a lap splice in the column.

Bearing stresses in concrete, under design ultimate loads, should not exceed the following values:

Entire surface loaded: $f_b = 0.85 \, \phi f'_c$      **(38)**

Part of surface loaded: $f_b = 0.85 \, \phi f'_c \sqrt{A_2/A_1}$     **(39)**

where $A_1$ = loaded area
     $A_2$ = surface area of same shape and concentric with $A_1$

$\sqrt{A_2/A_1}$ should not exceed 2.0.

## WALLS AND PARTITIONS

Reinforced concrete is well suited to the construction of walls, especially where they have to withstand heavy pressures, such as the retaining walls of a cellar or basement, walls for coal pockets, silos, reservoirs, or grain elevators. Such walls must be designed for flexural shear and bond stresses as well as stability against overturning, sliding, and soil pressure. Drainage should be provided for by weep holes or drains. Partitions may be built of solid concrete 4 to 6 in thick, reinforced to control temperature and shrinkage cracks. Reinforced-concrete walls need to be anchored by reinforcement to adjacent structural members. All walls must be reinforced for temperature with steel placed horizontally and vertically.

The horizontal reinforcement shall not be less than 0.25 percent and the vertical reinforcement not less than 0.15 percent of the area of the reinforced section of the wall when bars are used and three-fourths of these amounts when welded fabric is used. Adequate reinforcement must be provided around all openings for windows and doors.

**Retaining walls** of reinforced concrete are used to resist the pressures of earth, water, and other retained materials and are usually of T or L shape. The base must be so proportioned that there is sufficient resistance to sliding and overturning and that the safe bearing strength of the soil is not exceeded. The dimensions of the concrete section and the position and amount of steel reinforcement are determined by the moments and shears at critical vertical and horizontal sections at the junction of the wall and the base. Particular attention should be given to drainage to prevent excessive water pressure behind walls retaining earth or other materials. Walls retaining water, such as tanks, should have steel tensile stresses limited to 12,000 lb/in² unless special consideration is given to controlling cracks and should have ample reinforcement to provide for effects caused by shrinkage of the concrete and temperature change.

**Bearing Walls** The allowable compressive force for reinforced-concrete bearing walls subject to concentric loads can be computed as follows:

$$P_u = 0.55 \, \phi f'_c A_g \left[ 1 - (1_c/40b)^2 \right] \qquad \textbf{(40)}$$

For the case of concentrated loads, the effective length for computational purposes can be considered as the width of the bearing plus four times the wall thickness but not greater than the distance between loads. The wall thickness should be at least $\frac{1}{25}$ of the unsupported height or width, whichever is smaller. For the upper 15 feet, bearing walls must be at least 6 in thick and increase at least 1 in in thickness for each successive 25 feet downward, except that walls of a two-story dwelling need to be only 6 in thick over the entire height, provided that the strength is adequate.

## PRESTRESSED CONCRETE

**Prestressed-concrete members** have initial internal stresses, set up by highly stressed steel tendons embedded in the concrete, which are generally opposite those caused by applied loads. Prestressed members are constructed in one of two ways: (1) *Pretensioned* members are factory precast products, made by tensioning steel tendons between abutments and then casting concrete directly around the steel. After the concrete has reached sufficient strength, the steel is cut and the force transferred to the concrete by bond. (2) *Posttensioned* members, either factory or site cast, contain steel in ducts cast in the concrete. After the concrete has cured, the steel is tensioned and mechanically anchored against the concrete. The ducts are preferably pumped full of grout after tensioning to provide bond and corrosion protection, or the tendons may be coated with corrosion inhibitors.

Very high strength steel is used for prestressing in order to overcome the losses of steel stress due to creep and shrinkage of the concrete, and as a result of the strength, relatively small amounts of steel are required. The concrete for pretensioned members will usually be at least 5,000 lb/in² compressive strength, and at least 4,000 lb/in² for posttensioned members.

Because of the initial stress conditions, prestressed-concrete members are generally crack-free at working loads and consequently are quite suitable for water-containing structures. Circular tanks and pipes are posttensioned by wrapping them with highly stressed wires, using specialized equipment.

Losses of prestress occur with time owing to creep and shrinkage of concrete and relaxation of steel stress. Pretensioned members also have an initial elastic shortening loss accompanying transfer of force to the concrete, and posttensioned members have losses due to friction between ducts and tendons and anchor set. These losses must be taken into account in the design of members.

Prestressed-concrete members are checked for both strength and stresses at working loads. Because of the built-in stresses, the condition of dead load only may also govern. The flexural strength is computed using the same equations as for reinforced concrete which were developed earlier. The steel does not have a well-defined yield stress, and the steel stress at failure can be predicted from the following expressions:
Bonded tendons:

$$f_{ps} = f_{pu} \left( 1 - 0.5 \, \rho_p \frac{f_{pu}}{f'_c} \right) \qquad (41)$$

Unbonded:

$$f_{ps} = f_{se} + 10,000 + f'_c/100 \, \rho_p \qquad (42)$$

but not more than $f_{se}$ or $f_{pu} + 60,000$.

The stresses are used directly in Eqs. (4), (7), or (8), as appropriate, substituting $f_{ps}$ for $f_y$ and $A_{ps}$ for $A_s$.

### Allowable Stresses at Working Load

Steel:
   Maximum jacking stress, but not to
    exceed recommendation by steel or
    anchorage manufacturer      $0.8 \, f_{pu}$
   Immediately after transfer or
    posttensioning      $0.7 \, f_{pu}$
Concrete:
   Temporary stresses immediately
    after prestressing
     Compression      $0.6 \, f'_{ci}$
   Tension in areas without
    reinforcement      $3\sqrt{f'_{ci}}$
   Design load stresses (after losses)
    Compression      $0.45 \, f'_c$
   Tension in precompressed tensile
    zones      $6\sqrt{f'_c}$

The allowable tension may be increased to $12\sqrt{f'_c}$ if it is demonstrated, by a comprehensive analysis taking cracking into account, that the short- and long-term deflections will be satisfactory.

The final steel stress, at working loads, will usually be 30,000 to 45,000 lb/in² less than the initial stress for pretensioned members. Posttensioned members will have slightly lower losses. Losses, from initial tensioning values, for pretensioned members may be predicted satisfactorily using the following expressions from the AASHO Bridge Specifications:

$$\Delta f_s = SH + ES$$
$$+ \, CR_c + CR_s = \text{prestress loss} \qquad \text{lb/in}^2 \qquad (43)$$

where $SH$ = shrinkage loss = $17,000 - 150 \, RH$
      $ES$ = elastic shortening loss = $(E_s/E_c)f_{cir}$

      $CR_c$ = creep loss = $12 \, f_{cir} - 7 \, f_{cds}$
      $CR_s$ = relaxation loss = $20,000 - 0.4ES - 0.2(SH + CR_c)$
      $f_{cir}$ = concrete stress at level of center of gravity of steel (cgs) at section considered, due to initial prestressing force and dead load
      $f_{cds}$ = change in concrete stress at cgs due to superimposed composite or noncomposite dead load
      $RH$ = relative humidity, percent

The loss calculations are carried out for each critical moment section. The average annual relative humidity of the service environment should be used in the $SH$ calculation.

The shear reinforcement requirements for simple cases are covered in an earlier section. In addition to the shear steel, a few stirrups or ties should be placed transverse to the member axis as close to the ends as possible, to control potential splitting cracking. The area of steel, from the AASHO Bridge Specification, should be $A_t = 0.04 \, f_{st}A_{ps}/20,000$. In posttensioned beams, end blocks will often have to be used to provide space for anchorage bearing plates.

The minimum clear spacing between strands in pretensioned members is three times the strand diameter near the ends of the beam, but many plants are set up to handle only 2-in spacings. Strands may be closer together in central positions of members, which will help in maximizing member effective depths and steel eccentricity.

Few precast, pretensioned members are solid rectangular sections, and single and double T beams and hollow floor slab units are used extensively in buildings. Hollow box beams and I-section beams are used extensively in bridges and in buildings with heavy design loads. Square piling with the prestressing strands arranged in a circular pattern is widely used. Because of the larger number of possible sections, it is necessary to check availability of any particular section with local producers before designing any precast structure.

### PRECAST CONCRETE

Precast slabs, beams, walls, and partitions as well as piles, retaining wall units, light standards, railroad crossings, and bridge slabs are being increasingly used because of the saving in time and labor cost. Such units vary in size from small slabs for use in floors of residences to large frames for industrial buildings. The small units, such as roof slabs, are cast in steel forms at central plants. Some of the larger units, such as bridge or highway slabs and wall units, are cast in wood forms at or near the place of use. A method by which wall or partition slabs are cast so that they are simply tilted into position has found wide use in housing and industrial construction. Another special adaptation is the method of casting complete floors on top of each other, then lifting into position vertically at the columns.

Particular attention must be given in the design of precast units to reduction in weight and to details to minimize the cost of erection and installation. Reduction in weight is obtained by the use of lightweight aggregates, high-strength concrete, and hollow units. Precast reinforced-concrete units are seldom designed for concrete strengths of less than 4,000 lb/in². They are often combined with cast-in-place concrete so as to obtain the advantages of continuity. The combination of precast beams with cast-in-place slabs gives the advantage of T-beam action. Wall units are tied together by interlocking joints or by bolts. Care must be taken in shipping and handling to avoid damage to the precast units, and the design must take care of the stresses that come from such causes. All lifting devices built into the units should be designed for 100 percent impact. All units must be identified as to proper location and orientation in the structure.

Because precast units are made under conditions which allow good control of dimensions, certain restrictions can be relaxed that must be observed for cast-in-place concrete. Cover over the reinforcement for members not exposed to freezing need not be more than the nominal diameter of the steel but not less than $\frac{5}{8}$ in. The maximum size of the coarse aggregate can be as large as one-third of the smaller dimension of the member. Precast wall panels are not limited to the minimum thickness requirements for cast-in-place walls.

To reduce the number of connections, precast units should generally be cast as large as can be properly handled. However, some joints will be needed to transfer moments, torsion, shear, and axial loads from one member to another. The integrity of the structure depends on the adequacy of the design of the various joints and connections. They may be made by use of bolts and pins or clips and keys, by welding the reinforcement or steel inserts, or by a number of other methods limited only by the ingenuity of the designer. The connections should not be the weak links in the structure. Thought as to their location will avoid many problems.

### JOINTS

**Contraction and expansion joints** may be needed at intervals in a structure to help care for movement due to temperature changes and shrinkage. Joints at 20- to 30-ft intervals provide good crack control. A weakened plane, formed in the tension side of the member by a slot $\frac{1}{4}$ in wide and $\frac{1}{2}$ in deep, will induce the formation of contraction cracks at selected points. Structures over 200 ft in length should have special consideration given to contraction provisions.

**Construction joints** are necessary in most structures because all sections cannot be cast continuously. They should be made at points of minimum shearing stress and reinforced across the joint with a steel area of not less than 0.5 percent of the area of the section cut. Provision must be made for the transfer of shear and other forces through the construction joint. Joints in columns should be made at the underside of the floor members, haunches, T beams, and column capitals.

The hardened concrete at a joint should be properly prepared for bonding with the new concrete by being cleaned, roughened, and wetted. On this surface, a coat of neat cement grout or other bonding agent should be applied just before depositing the new concrete.

### FORMS

Forms are usually built of wood or metal but in special cases may be made of plastic or fiberglass reinforced plastic. Wooden forms may be the most economical unless the construction allows for the repeated use of the same forms. Plywood and compressed wood fiber sheets, specially treated to make them waterproof, are frequently used for form faces where good surfaces are required. Forms must be designed so that they can be easily erected, removed, and reerected. The usual order of removing forms is (1) column sides, (2) joists, (3) girder and beam sides, (4) slab bottoms, and (5) girder and beam bottoms. Column forms are held together by clamps made of wood or steel, the spacing of which is smallest at the bottom and increases with the decrease in pressure. Beam forms consist of the bottom and two sides held together by clamps or cleats and supported by posts. Slab forms consist of boards or other form material supported by joists spaced 2 or 3 ft apart or other means. The joists either rest on a horizontal joist bearer fastened to the clamps of the beam or girder or are supported by stringers, or posts, or both.

Special consideration must be given to forms for prestressed-concrete members. For pretensioned members the form must be constructed such that it will permit movement of the member during release of the prestressing force. For posttensioned members, the form should provide a minimum of resistance to shortening of the member. It is also necessary to consider the deflection of the members due to the stressing force.

**Design in Formwork**    The formwork is an appreciable portion of the cost of most concrete structures. Any efforts, however, to reduce the cost of the forms must not go beyond the point of safe design to prevent failures which would in themselves raise the cost of construction.

All forms must conform to the dimensions and shape of the members and must be sufficiently tight to prevent leakage of the mortar. They must be properly braced and tied together to maintain their position and shape during the construction procedure.

The formwork must support all the vertical and lateral loads that may be applied until these loads can be carried by the concrete structure. Loads on the form include the weight of the forms, reinforcing steel, fresh concrete, and various construction live loads. The construction load varies with conditions but is often assumed to be 75 lb/ft² of floor area. The formwork should also be designed to resist lateral loads produced by wind and movement of construction equipment. Most frequently the steel and concrete will not be placed in a symmetrical pattern and frequently large impact loads will occur. Because of the many varied conditions, it is frequently impossible to determine with any great precision the loads which the form must carry. The designer must therefore make safe assumptions by which the forms can be designed such that failure will not result.

Lateral pressures in forms for walls and columns are influenced by a number of factors: weight of concrete, height of placing, vibration, temperature, size and shape of form, amount and distribution of reinforcing steel, and several other variables. Formulas have been suggested for computing safe lateral pressures to be used in form design. However, because of limited test data, they are not generally accepted by all engineers.

**Form Liners**    Absorbent form liners are occasionally applied to the surface of forms to extract the water from the surface of the concrete, eliminate air and water voids, and produce a concrete of uniform appearance with surfaces which are superior in durability and resistance to abrasion.

The vacuum process, whereby water is absorbed from the concrete through a special form liner made of two layers of screen or wire mesh covered by a layer of cloth, has a similar effect and if properly used reduces the water content of the concrete to a depth of several inches.

Neoprene and other types of rubber have been successfully used as liners in precasting work in which a number of units are made from one form. Rubber is particularly suited for patterned work.

Plastic form liners make it easy to obtain a textured surface or a glossy smooth surface. Generally speaking, plastic liners are easily cleaned and if not too thin are suitable for a number of reuses.

**Removal of Forms**    The time that forms should remain in place depends on the character of the members and weather conditions. The strength of concrete must be ascertained before removing the forms. Unless special precautions are taken, concrete should not be placed below 40°F. Fresh concrete should never be subjected to temperatures below freezing. As an approximate guide for the minimum time for form removal, the following rules, which assume moist curing at not less than 70°F for the first 24 h, may be observed.

WALLS IN MASS WORK.    In summer, 1 day; in cold weather, 3 days.

THIN WALLS.    In summer, 1 day; in cold weather, 5 days.

COLUMNS.    In summer, 1 day; in cold weather, 4 days, provided girders are shored to prevent appreciable weight reaching the columns.

SLABS UP TO 7 FT SPAN.    In summer, 4 days; in cold weather, 2 weeks.

BEAMS AND GIRDER SIDES.    In summer, 1 day; in cold weather, 5 days.

BEAMS AND GIRDERS AND LONG-SPAN SLABS.    In summer, 7 days; in cold weather, 2 weeks.

CONDUITS.    2 or 3 days, provided there is not a heavy fill upon them.

ARCHES.    If a small size, 1 week; large arches with a heavy dead load, 3 weeks.

Forms for prestressed members may be removed when sufficient prestressing has been applied to enable them to carry their dead loads and the expected construction loads.

# Heating, Ventilation, and Air Conditioning

## by Leander Economides

REFERENCES: ASHRAE "Handbook of Fundamentals." ASHRAE "Handbook & Product Directory—Systems and Applications." ASHRAE "Guide and Data Book—Equipment." Strock and Koral, "Handbook of Air Conditioning, Heating and Ventilating," Industrial Press.

This section presents standards, basic data, and physical laws for use in the design of heating, ventilating, and air-conditioning systems.

## HEATING

Heating load

$$Q_t = Q_{tr} + Q_{inf} + Q_{vent} \tag{1}$$

where $Q_t$ = total heating load, Btu/h
$\quad Q_{tr}$ = transmission load
$\qquad = AU(t_i - t_o)$, Btu/h
$\quad Q_{inf}$ = infiltration load
$\qquad = 1.08 \ (\text{ft}^3/\text{min})(t_i - t_o)$, Btu/h
$\quad Q_{vent}$ = ventilation load
$\qquad = 1.08 \ (\text{ft}^3/\text{min})(t_i - t_o)$, Btu/h
$\quad A$ = area through which heat flow occurs, ft²
$\quad U$ = overall heat-transfer coefficient
$\qquad = 1/R_t$ Btu/(h)(ft²)(°F)
$\quad t_o$ = outside-air design dry-bulb temperature, °F    (Fig. 1)
$\quad t_i$ = inside design dry-bulb temperature, °F (Table 1)
$\quad R_t$ = thermal resistance, °F/(Btu)(h)(ft²)
$\qquad = R_1 + R_2 + R_3 + \cdots + R_n$
$R_1, R_2, \ldots$
$\qquad$ = resistances to heat flow to the individual components of a composite construction

A comparable equation in SI units is

$$Q_t = Q_{tr} + Q_{inf} + Q_{vent}$$

where
$\quad Q_t$ = total heating load, kW
$\quad Q_{tr}$ = $AU \ (t_i - t_o)$, kW
$\quad Q_{inf}$ = $1.2 \ (\text{m}^3/\text{s})(t_i - t_o)$, kW
$\quad Q_{vent}$
$\qquad = 1.2 \ (\text{m}^3/\text{s})(t_i - t_o)$, kW
$\quad A$ = area through which heat flow occurs, m²
$\quad U$ = overall heat-transfer coefficient
$\qquad = 1/R_t$, W/(m² · K)
$\quad t_o$ = outside-air design dry-bulb temperature, °C
$\quad t_i$ = inside design dry-bulb temperature, °C
$\text{m}^3/\text{s}$ = cubic meters per second, air
$\quad R_t$ = thermal resistance, K/(W·m²)

### Thermal Expansion—Water

The volume of water in a system should be determined from the volumetric water contact of heating units, as given by manufacturers, and the volume of pipe or tube (Table 11 and Fig. 2).

### Design Water Temperatures

At any given average operating temperature of a system, an increase in the design temperature drop will result in a smaller expansion tank, since the lower circulating rate will reduce pipe sizes.

For example, if the design range is changed from 170 to 190°F (180°F average) (77 to 88°C, 82.5°C average) to 150 to 210°F (180°F average) (66 to 99°C, 82.5°C average), the water-circulation rate will be reduced to one-third of the original value. The actual reduction in water quantity, and therefore in water expansion, for any given change in water temperature

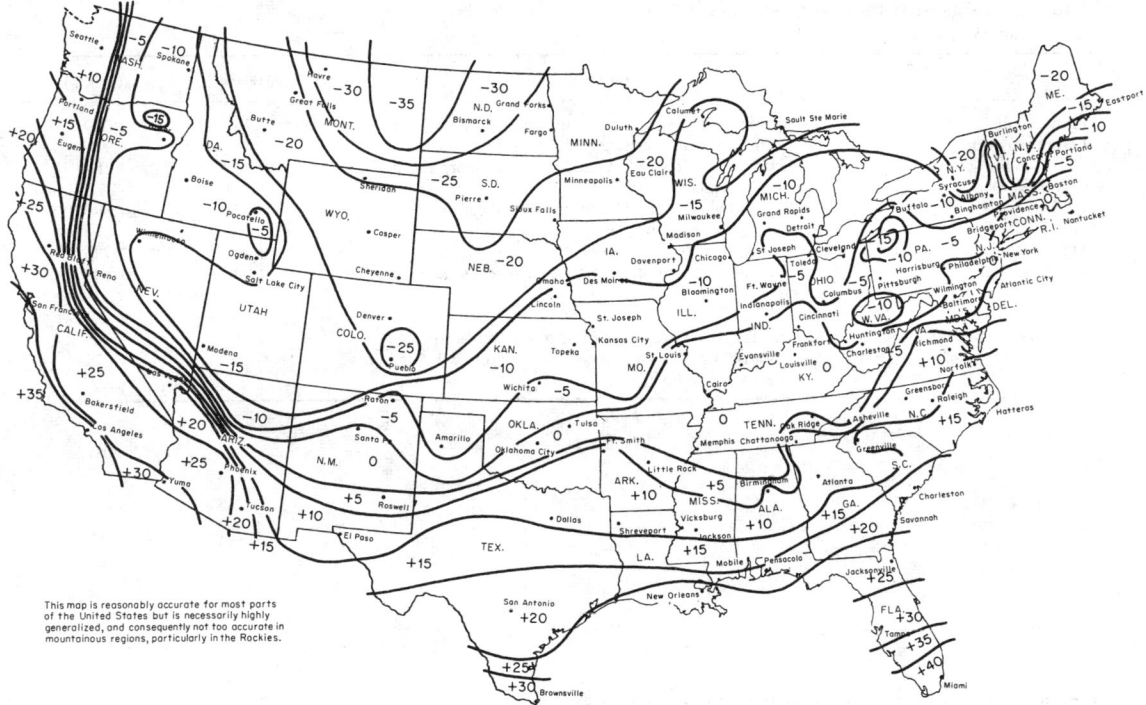

**Fig. 1**  Isotherms of winter outdoor temperatures. Given in Fahrenheit; to convert to Celsius: $t_c = \frac{5}{9}(t_f - 32)$. [*C. Strock and R. L. Koral (eds.),* "*Handbook of Air Conditioning, Heating and Ventilating,*" *2d ed., Industrial Press, New York, 1965.*]

## Table 1. Recommended Inside Design Conditions—Winter

| Type of application | With humidification | | | Without humidification | |
|---|---|---|---|---|---|
| | Dry-bulb, °F | Rel. hum., % | Temp. swing, °F* | Dry-bulb, °F | Temp. swing, °F* |
| General comfort (apartment, house, hotel, office, hospital, school, etc.) | 74–76 | 35–30 | −3 to −4 | 75–77 | −4 |
| Retail shops (short-term occupancy) (bank, barber or beauty shop, department store, supermarket, etc.) | 72–74 | 35–30† | −3 to −4 | 73–75 | −4 |
| Low-sensible-heat-factor applications (high latent load) (auditorium, church, bar, restaurant, kitchen, etc.) | 72–74 | 40–35 | −2 to −3 | 74–76 | −4 |
| Factory comfort (assembly areas, machining rooms, etc.) | 68–72 | 35–30 | −4 to −6 | 70–74 | −6 |

To convert to degrees Celsius: $t_c = 5/9(t_f - 32)$.
SOURCE: "Carrier Corporation System Design Manual," Part I, Load Estimating, 1970.
*Temperature swing is below the thermostat setting at peak winter load conditions (no lights, people, or solar-heat gain).
†Winter humidification in retail clothing shops is recommended to maintain the quality texture of goods.

**Table 2. Overall Heat-Transfer Coefficient** *U*

Outside air 15 mi/h wind, inside still air

| Example | Construction | Btu/(h)(ft²)(°F) | W/(m²)(°k) |
|---|---|---|---|
| Frame walls | Wood siding, insulation board, air space, gypsum board | 0.19 | 1.079 |
| Frame partition | Gypsum board, air space, gypsum board | 0.34 | 1.931 |
| Frame construction ceilings and floors | Linoleum or tile, felt, plywood, wood subfloor, air space, metal lath, plaster | 0.23 | 1.306 |
| Pitched roofs | Asphalt shingles, building paper, wood sheathing, air space, gypsum, lath, plaster | 0.28 | 1.590 |
| Masonry wall | Face brick 4 in, common brick 4 in | 0.48 | 2.725 |
|  | Face brick 4 in, common brick 4 in, air space, gypsum lath, plaster | 0.29 | 1.647 |
| Masonry partition | Cement block (cinder aggregate), plaster on both sides | 0.31 | 1.760 |
| Flat masonry roof | Built-up roofing, roof insulation 1 in, concrete slab 4 in, air space, metal lath, plaster | 0.18 | 1.022 |

**Table 3. Coefficient of Heat Transmission** *U* **for Windows and Skylights**

Air-to-air heat transfer, Btu/(h)(ft²)(°F) and W/(m²)(K); outside air 0°F, 15 mi/h (24 km/h) wind, no solar radiation; inside still air

| | Vertical glass sheets | | | | Horizontal glass sheets | | | |
|---|---|---|---|---|---|---|---|---|
| | Outdoor exposure | | Indoor exposure | | Outdoor exposure | | Indoor exposure | |
| | Btu/ (h)(ft²)(°F) | W/ (m²)(K) | Btu/ (h)(ft²)(°F) | W/ (m²)(K) | Btu/ (h)(ft²)(°F) | W/ (m²)(K) | Btu/ (h)(ft²)(°F) | W/ (m²)(K) |
| Common window glass, single sheet | 1.13 | 6.416 | 0.75 | 4.259 | 1.22 | 6.927 | 0.96 | 5.450 |
| Common window glass, two sheets, 1 in air space (2.54 cm) | 0.53 | 3.009 | 0.45 | 2.555 | 0.63 | 3.577 | 0.56 | 3.180 |

**Table 4. Coefficient of Heat Transmission** *U* **for Wood Doors**

| Construction | Single | | With glass storm door | |
|---|---|---|---|---|
| | Btu/(h)(ft²)(°F) | W/(m²)(K) | Btu/(h)(ft²)(°F) | W/(m²)(K) |
| 1 in-thick solid door (²⁵/₃₂ in)(2 cm) | 0.64 | 3.634 | 0.37 | 2.101 |
| 2 in-thick solid door (1⁵/₈ in)(4 cm) | 0.43 | 2.442 | 0.28 | 1.590 |
| Door containing wood or glass panels | 0.85 | 4.826 | 0.39 | 2.214 |

**Table 5. Infiltration through Double-Hung Wood Windows**
Expressed in ft³/(h)(ft of crack) and m³/(h)(m of crack)

| Type of window | in H₂O 0.10 / N/m² 24.9 | in H₂O 0.20 / N/m² 49.8 | in H₂O 0.30 / N/m² 74.7 | in H₂O 0.40 / N/m² 99.6 | in H₂O 0.50 / N/m² 124.5 |
|---|---|---|---|---|---|
| Wood double-hung window (locked); leakage expressed as ft³/(h)(ft of sash crack) and m³/(h)(m of sash crack); only leakage around sash and through frame given | | | | | |
| Nonweatherstripped, average fit* | 27 / 2.48 | 43 / 3.95 | 57 / 5.23 | 69 / 6.33 | 80 / 7.34 |
| Weatherstripped, average fit | 14 / 1.29 | 23 / 2.11 | 30 / 2.75 | 36 / 3.30 | 42 / 3.86 |
| Frame-wall leakage† (leakage is that passing between the frame of a wood double-hung window and the wall) | | | | | |
| Around frame in masonry wall, not calked | 17 / 1.56 | 26 / 2.39 | 34 / 3.12 | 41 / 3.76 | 48 / 4.41 |
| Around frame in masonry wall, calked | 3 / 0.28 | 5 / 0.46 | 6 / 0.55 | 7 / 0.64 | 8 / 0.73 |
| Around frame in wood frame wall | 13 / 1.19 | 21 / 1.93 | 29 / 2.66 | 35 / 3.21 | 42 / 3.86 |

SOURCE: Reprinted by permission from ASHRAE "Handbook of Fundamentals," 1972.

*The fit of the average double-hung wood window was determined as ¹/₁₆ in (0.16-cm) crack and ³/₆₄-in (0.12-cm) clearance by measurements on approximately 600 windows under heating-season condition.

†The values given for frame leakage are per foot of sash perimeter, as determined for double-hung wood windows. Some of the frame leakage in masonry walls originates in the brick wall itself, and cannot be prevented by calking. For the additional reason that calking is not done perfectly and deteriorates with time, it is considered advisable to choose the masonry frame leakage values for calked frames as the average determined by the calked and noncalked tests.

**Table 6. Infiltration through Walls**
Expressed in ft³/(h)(ft²) and m³/(h)(m²)

| Type of wall | in H₂O 0.05 / N/m² 12.45 | in H₂O 0.10 / N/m² 24.9 | in H₂O 0.20 / N/m² 49.8 | in H₂O 0.30 / N/m² 74.7 | in H₂O 0.40 / N/m² 99.6 |
|---|---|---|---|---|---|
| Brick wall* | | | | | |
| 8½ in (21.6 cm) | | | | | |
| Plain | 5 / 1.52 | 9 / 2.74 | 16 / 4.88 | 24 / 7.32 | 28 / 8.53 |
| Plastered† | 0.05 / 0.015 | 0.08 / 0.024 | 0.14 / 0.043 | 0.20 / 0.061 | 0.27 / 0.082 |
| 13 in (33 cm) | | | | | |
| Plain | 5 / 1.52 | 8 / 2.44 | 14 / 4.27 | 20 / 6.10 | 24 / 7.32 |
| Plastered† | 0.01 / 0.003 | 0.04 / 0.012 | 0.05 / 0.015 | 0.09 / 0.027 | 0.11 / 0.034 |
| Frame wall, lath and plaster‡ | 0.09 / 0.027 | 0.15 / 0.046 | 0.22 / 0.067 | 0.29 / 0.088 | 0.32 / 0.098 |

*Constructed of porous brick and lime mortar—workmanship poor.

†Two coats prepared gypsum plaster on brick.

‡Wall construction: bevel siding painted or cedar shingles, sheathing, building paper, wood lath, and three coats gypsum plaster.

**Table 7. Infiltration through Doors—Winter***

15 mi/h (24 km/h) wind velocity†; doors on one or adjacent windward sides‡

| | ft³/min per ft² area (m³/min per m²)§ | | | | | | | | | |
|---|---|---|---|---|---|---|---|---|---|---|
| | Infrequent use | | Average use | | | | | | | |
| | | | 1 and 2 story buildings | | Tall buildings, ft (m) | | | | | |
| | ft³/min per ft² | m³/min per m² | ft³/min per ft² | m³/min per m² | 50 | (15.25 m) | 100 | (30.5 m) | 200 | (61 m) |
| Revolving door | 1.6 | 0.48 | 10.5 | 3.15 | 12.6 | 3.78 | 14.2 | 4.26 | 17.3 | 5.19 |
| Glass door [3/16 in (5 mm) crack] | 9.0 | 2.70 | 30.0 | 9.00 | 36.0 | 10.80 | 40.5 | 12.15 | 49.5 | 14.85 |
| Wood door 3 × 7 ft (0.9 × 2 m) | 2.0 | 0.60 | 13.0 | 3.90 | 15.5 | 4.65 | 17.5 | 5.25 | 21.5 | 6.45 |
| Garage and shipping-room door | 4.0 | 1.2 | 9.0 | 2.70 | | | | | | |

SOURCE: Carrier Corporation "System Design Manual," Part I, Load Estimating, 1970.

*All values are based on the wind blowing directly at the window or door. When the prevailing wind direction is oblique to the window or doors, multiply the values by 0.60 and use the total window and door area on the windward side(s).

†Based on a wind velocity of 15 mi/h (24 km.p.a). For design wind velocities different from the base, multiply the table values by the ratio of velocities.

‡Stack effect in tall buildings may also cause infiltration on the leeward side. To evaluate this, determine the equivalent velocity $V_l$ and subtract the design velocity $V$. The equivalent velocity is

$$V_l = \sqrt{V^2 - 1.75a} \text{ (upper section)}$$
$$= \sqrt{V^2 - 1.75b} \text{ (lower section)}$$

where $a$ and $b$ are the distances above and below the midheight of the building, respectively, in feet.

Multiply the table values by the ratio $(V_l - V)/15$ for one-half of the windows and doors on the leeward side of the building. (Use values under one- and two-story building for doors on leeward side of tall buildings.)

§Doors on opposite sides increase values 25 percent.

**Table 8. Outdoor Air Requirements**

| Application | Smoking | ft³/min per person | | ft³/min per ft² of floor, min |
|---|---|---|---|---|
| | | Recommended | Min | |
| Apartment: | | | | |
|   Average | Some | 20 | 10 | |
|   Deluxe | Some | 20 | 10 | |
| Banking space | Occasional | 10 | 7½ | |
| Barber shops | Considerable | 15 | 10 | |
| Beauty parlors | Occasional | 10 | 7½ | |
| Brokers' board rooms | Very heavy | 50 | 20 | |
| Cocktail bars | | 40 | 25 | |
| Corridors (supply or exhaust) | | | | 0.25 |
| Department stores | None | 7½ | 5 | 0.05 |
| Directors' rooms | Extreme | 50 | 30 | |
| Drugstores | Considerable | 10 | 7½ | |
| Factories | None | 10 | 7½ | 0.10 |
| Five and ten cent stores | None | 7⅓ | 5 | |
| Funeral parlors | None | 10 | 7½ | |
| Garages | | | | 1.0 |
| Hospitals: | | | | |
|   Operating rooms | None | | | 2.0 |
|   Private rooms | None | 30 | 25 | 0.33 |
|   Wards | None | 20 | 10 | |
| Hotel rooms | Heavy | 30 | 25 | 0.33 |
| Kitchens: | | | | |
|   Restaurant | | | | 4.0 |
|   Residence | | | | 2.0 |
| Laboratories | Some | 20 | 15 | |
| Meeting rooms | Very heavy | 50 | 30 | 1.25 |
| Offices: | | | | |
|   General | Some | 15 | 10 | |
|   Private | None • | 25 | 15 | 0.25 |
| | Considerable | 30 | 25 | 0.25 |
| Restaurants: | | | | |
|   Cafeteria | Considerable | 12 | 10 | |
|   Dining room | Considerable | 15 | 12 | |
| Schoolrooms | None | | | |
| Shop, retail | None | 10 | 7½ | |
| Theater | None | 7½ | 5 | |
| | Some | 15 | 10 | |
| Toilets (exhaust) | | | | 2.0 |

To convert ft³/min to m²/s multiply by 0.00047.
SOURCE: Reprinted by permission from ASHRAE "Handbook of Fundamentals," 1972.

**Table 9. Weight Flow Rate of Steam in Schedule 40 Pipe\* at Initial Saturation Pressures of 3.5 and 12 psig†**
Weight flow rate, lb/h

| Nom. pipe size, in | Pressure drop, lb/in² per 100 ft in length | | | | | | | | | | | | | |
|---|---|---|---|---|---|---|---|---|---|---|---|---|---|---|
| | 1/16 lb/in² (1 oz) | | 1/8 lb/in² (2 oz) | | 1/4 lb/in² (4 oz) | | 1/2 lb/in² (8 oz) | | 3/4 lb/in² (12 oz) | | 1 lb/in² | | 2 lb/in² | |
| | Saturation pressure, psig | | | | | | | | | | | | | |
| | 3.5 | 12 | 3.5 | 12 | 3.5 | 12 | 3.5 | 12 | 3.5 | 12 | 3.5 | 12 | 3.5 | 12 |
| ¾ | 9 | 11 | 14 | 16 | 20 | 24 | 29 | 35 | 36 | 43 | 42 | 50 | 60 | 73 |
| 1 | 17 | 21 | 26 | 31 | 37 | 46 | 54 | 66 | 68 | 82 | 81 | 95 | 114 | 137 |
| 1¼ | 36 | 45 | 53 | 66 | 78 | 96 | 111 | 138 | 140 | 170 | 162 | 200 | 232 | 280 |
| 1½ | 56 | 70 | 84 | 100 | 120 | 147 | 174 | 210 | 218 | 260 | 246 | 304 | 360 | 430 |
| 2 | 108 | 134 | 162 | 194 | 234 | 285 | 336 | 410 | 420 | 510 | 480 | 590 | 710 | 850 |
| 2½ | 174 | 215 | 258 | 310 | 378 | 460 | 540 | 660 | 680 | 820 | 780 | 950 | 1,150 | 1,370 |
| 3 | 318 | 380 | 465 | 550 | 660 | 810 | 960 | 1,160 | 1,190 | 1,430 | 1,380 | 1,670 | 1,950 | 2,400 |
| 3½ | 462 | 550 | 670 | 800 | 990 | 1,218 | 1,410 | 1,700 | 1,740 | 2,100 | 2,000 | 2,420 | 2,950 | 3,450 |
| 4 | 640 | 800 | 950 | 1,160 | 1,410 | 1,690 | 1,980 | 2,400 | 2,450 | 3,000 | 2,880 | 3,460 | 4,200 | 4,900 |
| 5 | 1,200 | 1,430 | 1,680 | 2,100 | 2,440 | 3,000 | 3,570 | 4,250 | 4,380 | 5,250 | 5,100 | 6,100 | 7,500 | 8,600 |
| 6 | 1,920 | 2,300 | 2,820 | 3,350 | 3,960 | 4,850 | 5,700 | 7,000 | 7,200 | 8,600 | 8,400 | 10,000 | 11,900 | 14,200 |
| 8 | 3,900 | 4,800 | 5,570 | 7,000 | 8,100 | 10,000 | 11,400 | 14,300 | 14,500 | 17,700 | 16,500 | 20,500 | 24,000 | 29,500 |
| 10 | 7,200 | 8,800 | 10,200 | 12,600 | 15,000 | 18,200 | 21,000 | 26,000 | 26,000 | 32,000 | 30,000 | 37,000 | 42,700 | 52,000 |
| 12 | 11,400 | 13,700 | 16,500 | 19,500 | 23,400 | 28,400 | 33,000 | 40,000 | 41,000 | 49,500 | 48,000 | 57,500 | 67,800 | 81,000 |

To convert lb/h to kg/s, multiply by $1.2597 \times 10^{-4}$.
To convert lb/in² to kN/m², multiply by 6.894.
To convert in. to mm multiply by 25.4.
SOURCE: Reprinted by permission from ASHRAE "Handbook of Fundamentals," 1972.
\*Based on Moody friction factor, where flow of condensate does not inhibit the flow of steam.
†The weight flow rates at 3.5 psig can be used to cover saturation pressure from 1 to 6 psig, and the rates at 12 psig can be used to cover saturation pressure from 8 to 16 psig with an error not exceeding 8 percent.

**Table 10. Return Main and Riser Capacities for Low-Pressure Systems, lb/h (kg/s)**

| Pipe size, in | 1/32 lb/in² or 1/2 oz drop per 100 ft | | | 1/24 lb/in² or 2/3 oz drop per 100 ft | | | 1/16 lb/in² or 1 oz drop per 100 ft | | | 1/8 lb/in² or 2 oz drop per 100 ft | | | 1/4 lb/in² or 4 oz drop per 100 ft | | | 1/2 lb/in² or 8 oz drop per 100 ft | | |
|---|---|---|---|---|---|---|---|---|---|---|---|---|---|---|---|---|---|---|
| | Wet | Dry | Vac. | Wet | Dry | Vac. | Wet | Dry | Vac. | Wet | Dry | Vac. | Wet | Dry | Vac. | Wet | Dry | Vac. |
| G | H | I | J | K | L | M | N | O | P | Q | R | S | T | U | V | W | X | Y |
| **Mains** | | | | | | | | | | | | | | | | | | |
| 3/4 | | | | | | 42 | | | 100 | | | 142 | | | 200 | | | 283 |
| 1 | 125 | 62 | | 145 | 71 | 143 | 175 | 80 | 175 | 250 | 103 | 249 | 350 | 115 | 350 | | | 494 |
| 1 1/4 | 213 | 130 | | 248 | 149 | 244 | 300 | 168 | 300 | 425 | 217 | 426 | 600 | 241 | 600 | | | 848 |
| 1 1/2 | 338 | 206 | | 393 | 236 | 388 | 475 | 265 | 475 | 675 | 340 | 674 | 950 | 378 | 950 | | | 1,340 |
| 2 | 700 | 470 | | 810 | 535 | 815 | 1,000 | 575 | 1,000 | 1,400 | 740 | 1,420 | 2,000 | 825 | 2,000 | | | 2,830 |
| 2 1/2 | 1,180 | 760 | | 1,580 | 868 | 1,360 | 1,680 | 950 | 1,680 | 2,350 | 1,230 | 2,380 | 3,350 | 1,360 | 3,350 | | | 4,730 |
| 3 | 1,880 | 1,460 | | 2,130 | 1,560 | 2,180 | 2,680 | 1,750 | 2,680 | 3,750 | 2,250 | 3,800 | 5,350 | 2,500 | 5,350 | | | 7,560 |
| 3 1/2 | 2,750 | 1,970 | | 3,300 | 2,200 | 3,250 | 4,000 | 2,500 | 4,000 | 5,500 | 3,230 | 5,680 | 8,000 | 3,580 | 8,000 | | | 11,300 |
| 4 | 3,880 | 2,930 | | 4,580 | 3,350 | 4,500 | 5,500 | 3,750 | 5,500 | 7,750 | 4,830 | 7,810 | 11,000 | 5,380 | 11,000 | | | 15,500 |
| 5 | | | | | | 7,880 | | | 9,680 | | | 13,700 | | | 19,400 | | | 27,300 |
| 6 | | | | | | 12,600 | | | 15,500 | | | 22,000 | | | 31,000 | | | 43,800 |
| **Risers** | | | | | | | | | | | | | | | | | | |
| 3/4 | | 48 | | | 48 | 143 | | 48 | 175 | | 48 | 249 | | 48 | 350 | | | 494 |
| 1 | | 113 | | | 113 | 244 | | 113 | 300 | | 113 | 426 | | 113 | 600 | | | 848 |
| 1 1/4 | | 248 | | | 248 | 388 | | 248 | 475 | | 248 | 674 | | 248 | 950 | | | 1,340 |
| 1 1/2 | | 375 | | | 375 | 815 | | 375 | 1,000 | | 375 | 1,420 | | 375 | 2,000 | | | 2,830 |
| 2 | | 750 | | | 750 | 1,360 | | 750 | 1,680 | | 750 | 2,380 | | 750 | 3,350 | | | 4,730 |
| 2 1/2 | | | | | | 2,180 | | | 2,680 | | | 3,800 | | | 5,350 | | | 7,560 |
| 3 | | | | | | 3,250 | | | 4,000 | | | 5,680 | | | 8,000 | | | 11,300 |
| 3 1/2 | | | | | | 4,430 | | | 5,500 | | | 7,810 | | | 11,000 | | | 15,500 |
| 4 | | | | | | 7,880 | | | 9,680 | | | 13,700 | | | 19,400 | | | 27,300 |
| 5 | | | | | | 12,600 | | | 15,500 | | | 22,000 | | | 31,000 | | | 43,800 |

This table is based on pipe-size data developed through the research investigations of The American Society of Heating, Refrigerating and Air-Conditioning Engineers.

To convert lb/h to kg/s, multiply by $1.2597 \times 10^{-4}$.

To convert lb/in² to kN/m², multiply by 6.894.

SOURCE: Reprinted by permission from ASHRAE "Handbook of Fundamentals," 1972.

**Table 11. Volume of Water in Standard Pipe and Tube**

| Nominal pipe size | | Standard steel pipe | | | | | Type L copper tube | | | |
|---|---|---|---|---|---|---|---|---|---|---|
| | | Schedule No. | Inside diam | | Gal/lin ft | Litres/ lin m | Inside diam | | Gal/lin ft | Litres/ lin m |
| in | mm | | in | mm | | | in | mm | | |
| ⅜ | 10 | | | | | | 0.430 | 10.9 | 0.0075 | 0.0931 |
| ½ | 12 | 40 | 0.622 | 15.8 | 0.0157 | 0.1950 | 0.545 | 13.8 | 0.0121 | 0.1503 |
| ⅝ | 16 | | | | | | 0.666 | 16.9 | 0.0181 | 0.2248 |
| ¾ | 20 | 40 | 0.824 | 20.9 | 0.0277 | 0.3440 | 0.785 | 19.9 | 0.0251 | 0.3117 |
| 1 | 25 | 40 | 1.049 | 26.6 | 0.0449 | 0.5576 | 1.025 | 26.0 | 0.0429 | 0.5328 |
| 1¼ | 32 | 40 | 1.380 | 35 | 0.0779 | 0.9674 | 1.265 | 32.1 | 0.0653 | 0.8110 |
| 1½ | 40 | 40 | 1.610 | 40.9 | 0.106 | 1.3164 | 1.505 | 38.2 | 0.0924 | 1.1475 |
| 2 | 50 | 40 | 2.067 | 52.5 | 0.174 | 2.1609 | 1.985 | 50.4 | 0.161 | 1.9995 |
| 2½ | 60 | 40 | 2.469 | 62.7 | 0.249 | 3.0923 | 2.465 | 62.6 | 0.248 | 3.0799 |
| 3 | 80 | 40 | 3.068 | 77.9 | 0.384 | 4.7689 | 2.945 | 74.8 | 0.354 | 4.3963 |
| 3½ | 90 | 40 | 3.548 | 90.0 | 0.514 | 6.3834 | 3.425 | 87.0 | 0.479 | 5.9487 |
| 4 | 100 | 40 | 4.026 | 102.3 | 0.661 | 8.2090 | 3.905 | 99.2 | 0.622 | 7.7246 |
| 5 | 130 | 40 | 5.047 | 128.2 | 1.04 | 12.916 | 4.875 | 123.8 | 0.970 | 12.0464 |
| 6 | 150 | 40 | 6.065 | 154.0 | 1.50 | 18.629 | 5.845 | 148.5 | 1.39 | 17.2624 |
| 8 | 200 | 30 | 8.071 | 205.0 | 2.66 | 33.035 | 7.725 | 196.2 | 2.43 | 30.1782 |
| 10 | 250 | 30 | 10.136 | 257.5 | 4.19 | 52.036 | 9.625 | 244.5 | 3.78 | 46.9438 |
| 12 | 300 | 30 | 12.090 | 307.0 | 5.96 | 74.017 | 11.565 | 293.8 | 5.46 | 67.8077 |

SOURCE: Reprinted by permission from ASHRAE "Handbook & Product Directory," 1973.

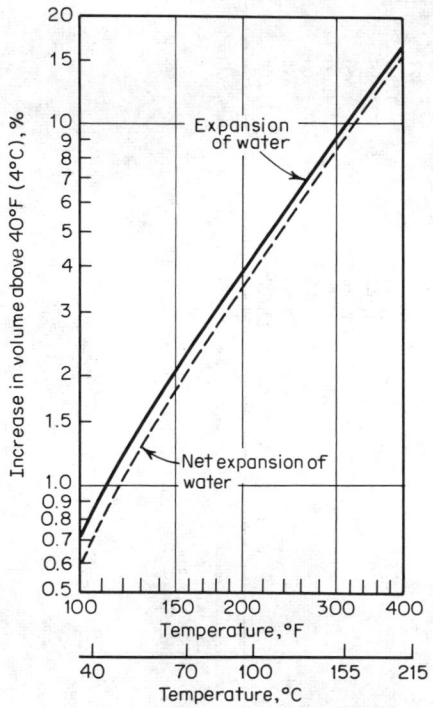

**Fig. 2** Expansion of water above 40°F (4°C). Net expansion of water equals actual expansion of water less volumetric expansion of metal system. (*Reprinted with permission from ASHRAE "Systems Handbook," 1973.*)

difference will vary with the amount of piping in the system. The water content of the hot-water generator, radiation, and other heat-exchange components may or may not change in size. On larger systems, an economic study of the effects of various operating-temperature ranges is often warranted.

**Expansion-Tank Design**

Design of expansion tanks is based on the total volume of system water, the hydrostatic head at the tank, the pumping head at the tank, and the water-temperature range over which the system operates. For operating temperatures between 160 and 280°F (70 and 140°C) expansion-tank sizes may be determined by the following ASME formula:

$$V_t = \frac{(0.00041t - 0.0466)V_s}{P_a/P_f - P_a/P_o} \tag{2}$$

where $V_t$ = minimum volume of the expansion tank, gal

$V_s$ = system volume, gal

$t$ = maximum average operating temperature, °F

$P_a$ = pressure in expansion tank when water first enters (usually atmospheric pressure), ft of water absolute

$P_f$ = initial fill or minimum pressure at tank, ft of water absolute

$P_o$ = maximum operating pressure (including pumping head) at tank, ft of water absolute

For operating temperatures below 160°F (70°C), use the following formula to determine the tank capacity:

$$V_t = \frac{EV_s}{P_a/P_f - P_a/P_o} \tag{3}$$

where $E$ = percent increase in the volume of water in the system (Fig. 2)

**Boiler Load** *Net load* is the sum of direct-connected load components. These include direct radiation, infiltration, air tempering, humidification, hot water, process steam, and snow melting.

*Design load* is the sum of the *net load* and the *piping tax*. Piping tax is the estimated heat emission in Btu per hour (watts) of the piping connecting the radiation and other apparatus to the boiler. In average heating systems it is common practice to consider the piping tax to be 20 percent of the net load.

*Gross, or maximum, load* is the sum of the *design load* and the *pickup allowance*. Pickup allowance is the estimated increase in the normal load in BTU per hour (watts), caused by the heating up of the cold system. For automatically fired boilers the sum of the piping tax and pickup allowance varies from 33.3 to 28.8 percent of the *net load*. The larger percentage should be applied to smaller boilers.

Information on *boiler performance* and the *heating value of various fuels* may be found in Sec. 8.

### Fuel Consumption

The general equation for calculating the probable fuel consumption by the degree-day method is

$$F = UN_bDC_f \qquad (4)$$

where $F$ = fuel consumption for the estimated period
$U$ = unit fuel consumption, or quantity of fuel used per (degree-day) (*building load unit*)
$N_b$ = number of *building load units* (when available, use calculated hourly heat loss instead of actual amount of radiation installed)
$D$ = number of degree-days for the estimated period
$C_f$ = temperature-correction factor from Table 12A

Values of $N_b$ depend on the particular residence for which the estimate is being prepared and must be found by surveying plans, by observation, or by measurement of the residence.

Values of $U$ for use in this equation are the unit fuel consumption per degree-day, obtained as a result of the collection of operating information and listed in Table 12B.

**Chimneys** Equations (5) to (8) are simplified equations for chimney sizes if the following typical values for boiler plants are assumed:

Average chimney-gas temperature:

$$T_c = 500°F \ (960°F \ abs)$$
$$= [260°C \ (533 \ K)]$$

Average atmospheric temperature:

$$T_0 = 62°F \ (522°F \ abs)$$
$$= [16.7°C \ (289.7 \ K)]$$

Average coefficient of friction:

$$f = 0.016$$

Average chimney-gas density at 0°F (−17.8°C) and 1 atm (101 kN/m²):

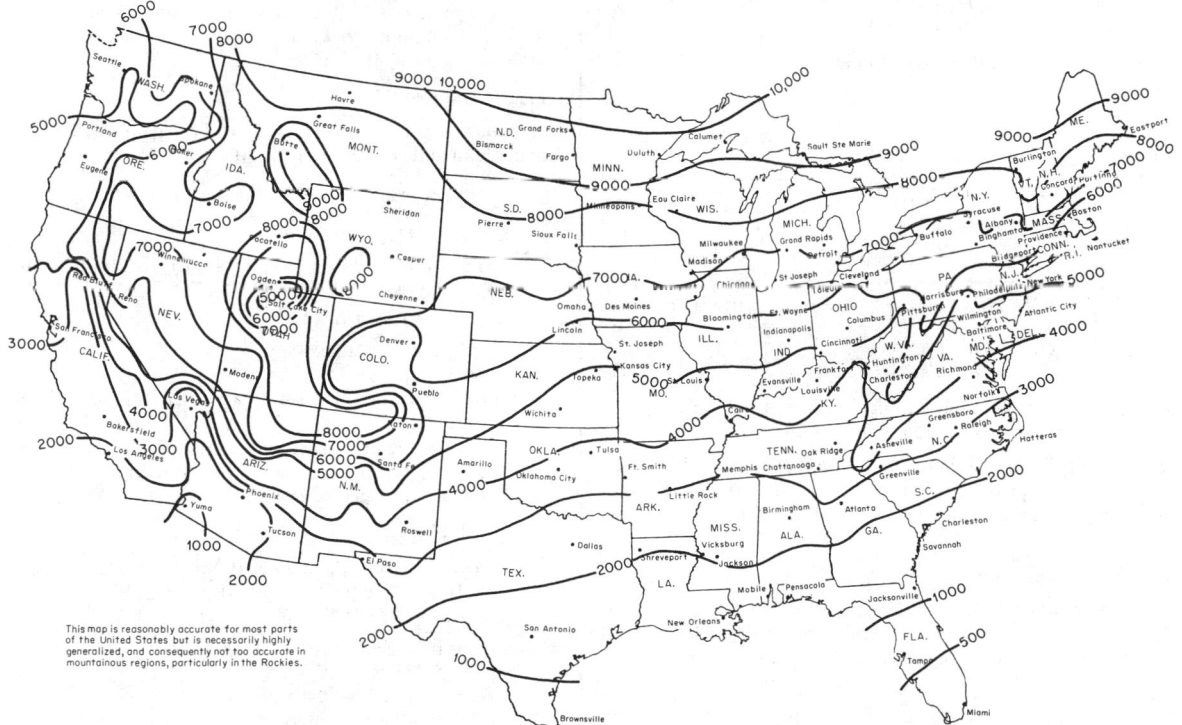

**Fig. 3** Number of degree-days in a normal heating season. To convert degree-days (K), multiply by ⅝. [*C. Strock and R. L. Koral (eds.), "Handbook of Air Conditioning, Heating and Ventilating," 2d ed., Industrial Press, New York, 1965.*]

**Table 12A. Correction Factors for Outdoor Design Temperatures***

| Outdoor design temp | $-20°F$ $(-29°C)$ | $-10°F$ $(-23.3°C)$ | $0°F$ $(-17.8°C)$ | $+10°F$ $(12.2°C)$ | $+20°F$ $(6.7°C)$ |
|---|---|---|---|---|---|
| Correction factor | 0.778 | 0.875 | 1.000 | 1.167 | 1.400 |

*The multipliers, which are high for mild climates and low for cold regions, are not in error as might appear. The unit figures are per square foot of radiator or 1,000 Btu heat loss per degree-day (1,000 W heat loss per °C day). For equivalent buildings and heating seasons, those in warm climates have lower design heat losses and smaller radiator quantities than those in cold cities. Consequently, the unit figure in quantity of fuel per (square foot of radiator) (degree-day) is larger for warm localities than for colder regions. Since northern cities have more radiator surface per given building and a higher seasonal degree-day total than cities in the south, the total fuel per season will be larger for the northern city.

**Table 12B. Efficiency of Utilization over the Heating System (Residential Systems)**

| Type of fuel-burning unit | Efficiency, % |
|---|---|
| Gas | 70–80 |
| Oil | 70–80 |

$$0.09 \text{ lb/ft}^3$$
$$(1.44 \text{ kg/m}^3)$$

Barometer reading, sea level:

$$B_0 = 29.92 \text{ inHg}$$
$$= (3.386 \text{ kN/m}^2)$$

Required height of chimney above inlet:

$$H = 190D_r \quad \text{ft}$$
$$H = 57.9D_r \quad \text{m} \tag{5}$$

Required minimum diameter of chimney:

$$d = 1.5 \ W^{2/5} \quad \text{ft}$$
$$d' = 0.33 \ W^{2/5} \quad \text{m} \tag{6}$$

Chimney gas velocity:

$$V_c = 13.7 \ W^{1/4} \quad \text{ft/s}$$
$$V'_c = 3.56 \ W^{1/5} \quad \text{m/s} \tag{7}$$

Stack draft:

$$D_r = 0.256 \ HB_0 \left( \frac{1}{T_0 \text{ abs}} = \frac{1}{T_c \text{ abs}} \right) \quad \text{in water}$$
$$D'_r = 0.1378 \ HB_0 \left( \frac{1}{T_0 \text{ abs}} - \frac{1}{T_c \text{ abs}} \right) \quad \text{kN/m}^2 \tag{8}$$

where $D_r$ = total required draft, in of water (kN/m²)
$W$ = flue-gas flow rate, lb/s (kg/h)

$T_0$ abs and $T_c$ abs are in kelvins, $H$ in meters, and $B_0$ in kN/m². Total required draft is the sum of draft loss through the breeching and through the boiler and the required draft in the firebox.

**Infiltration (Moisture)**

Infiltration of air and particularly moisture into a conditioned space is frequently a source of sizable heat gain or loss.

The moisture entering a building as water vapor may be expressed as

$$W_t = W_{trans} + W_{inf} + W_{vent} \tag{9}$$

where $W_t$ = total weight of vapor, grains (grams)

$W_{trans}$ = transmitted vapor
$$= MA\theta \ \Delta p \quad \text{grains (grams)} \tag{10}$$

$W_{inf}$ = air-infiltrated vapor = $W(M_o - M_i)$ grains (grams) **(11)**
$$= W(M_o - M_i) \ \Delta p \quad \text{grains (grams)}$$
$W_{vent}$ = ventilation air vapor **(12)**
$$= W(M_o - M_i) \quad \text{grains (grams)}$$
$M$ = permeance coefficient perms (kg/Ns), Table 14
$$= \bar{\mu}/l \quad \text{g/(ft}^2)(\text{h})(\text{in Hg } \Delta p)$$

where $A$ = area of low path, ft² (m²)
$\theta$ = time of transmission, h (s)
$\Delta p$ = vapor-pressure difference through flow path, in Hg (N/m²)
$W$ = weight of air, lb (kg)
$M_o$ = moisture content of outside air, g/lb (g/kg)
$M_i$ = moisture content of inside air, g/lb (g/kg)
1 perm = 1 g/(ft²)(h)(in Hg $\Delta p$)
$$= 5.738 \times 10^{-11} \text{ kg/Ns, in SI units}$$
$\mu$ = permeability, perm·in, g·in/(ft²)(h)(in Hg $\Delta p$)
$$= 1.457 \times 10^{-2} \text{ kg·m/Ns, in SI units}$$
$l$ = length of flow path, in
$$= 2.54 \times 10^{-2} \text{ m}$$

**Moisture Load**

Moisture as a cooling load is a latent-heat gain expressed in Btu per hour and is computed with the following equations:

$$Q_t = Q_{trans} + Q_{inf} + Q_{vent} \tag{13}$$
$$Q_{trans} = \frac{1,076}{7,000} \times MA\theta \ \Delta p \quad \text{Btu}$$

$$Q_{inf} = 0.68 \times \text{ft}^3/\text{min} \times (M_o - M_i) \quad \text{Btu/h} \tag{14}$$
$$Q_{vent} = 0.68 \times \text{ft}^3/\text{min} \times (M_o - M_i) \quad \text{Btu/h} \tag{15}$$

where $0.68 = \dfrac{60}{13.5} \times \dfrac{1,076}{7,000}$

60 = min/h
13.5 = specific volume of moist air at 70°F dB and 50% RH
1,076 = average heat removal required to condense 1 lb of water vapor from the room air
7,000 = grains per pound

A comparable equation in SI units is

$$Q_{trans} = 1,054.35 \times M'A'\theta \ \Delta p' \quad \text{joules}$$
$$Q_{inf} = 42.6 \text{ m}^3/\text{s} \ (M'_o - M'_i) \quad \text{kW}$$
$$Q_{vent} = 42.6 \text{ m}^3/\text{s} \ (M'_o - M'_i) \quad \text{kW}$$

where $M'$ = permeance coefficient, kg/N·s
$$= 5.738 \times 10^{-11} \text{ kg/N·s}$$
$A'$ = area of flow path, m²
$\theta$ = time of transmission, s
$\Delta p$ = vapor-pressure difference through flow path, N/m²
$M'_o$ = moisture content of outside air, g/kg
$M'_i$ = moisture content of inside air, g/kg

**Table 13. Unit Fuel-Consumption Constants**
Based on 0°F (−18°C) outdoor temperature, 70°F (23.3°C) indoor temperature

| | Utilization efficiency | | |
| --- | --- | --- | --- |
| | 60 | 70 | 80 |
| Fuel and units | Unit fuel consumption per degree-day per 1,000 Btu/h heat loss (consumption per °C day per 1,000 W) | | |
| Gas, therms* (m³) | 0.00572 (0.09947) | 0.00490 (0.08521) | 0.00429 (0.07460) |
| Oil, gal† (liters) | 0.00405 (0.09424) | 0.00347 (0.08076) | 0.00304 (0.07074) |
| Coal, lb‡ (kg) | 0.0476 (0.13269) | 0.0408 (0.11374) | 0.0357 (0.09952) |

*One therm is equal to 100,000 Btu (37,220 J/m³).
†Based on a heating value of 141,000 Btu/gal (39,240 J/l or 10.91 Wh/l.)
‡Based on a heating value of 12,000 Btu/lb (27,890 J/g or 7.748 Wh/kg).

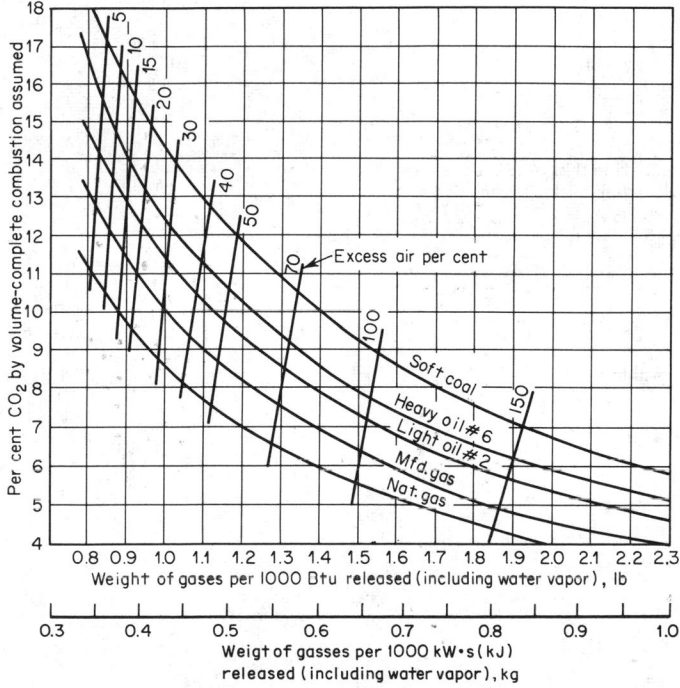

**Fig. 4** Graphical evaluation and rate of flue gas flow from percent $CO_2$ and fuel rate. *(Reprinted by permission from ASHRAE "Handbook and Product Directory," 1975.)*

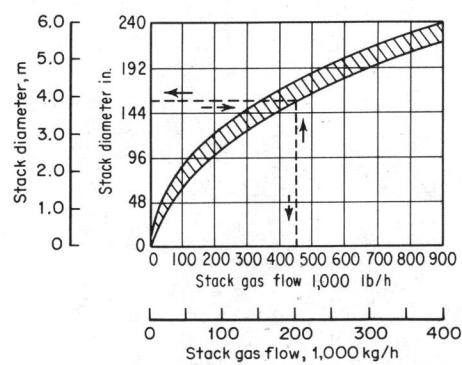

**Fig. 5** Economical stack size based on approximately 5 percent draft loss.

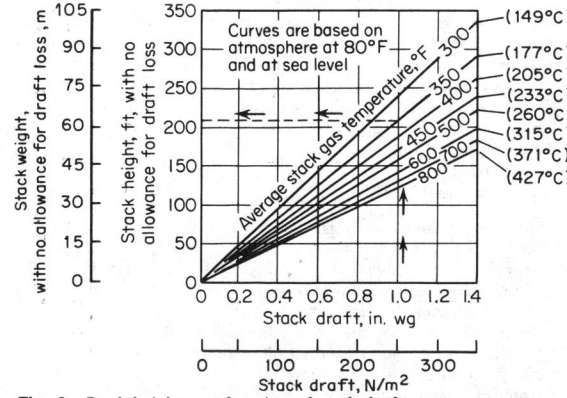

**Fig. 6** Stack height as a function of stack draft.

**Table 14. Permeance and Permeability of Materials to Water Vapor**[a]

| Material | Permeance (perm) | | | Permeability (perm-inch) | | |
|---|---|---|---|---|---|---|
| | Dry cup | Wet cup | Other | Dry cup | Wet cup | Other |
| **Materials used in construction:** | | | | | | |
| Concrete (1:2:4 mix) | | | | | 3.2 | |
| Brick masonry (4 in thick) (10 cm) | | | 0.8 | | | |
| Concrete block (8 in cored, limestone aggregate) | | | 2.4 | | | |
| Tile masonry, glazed (4 in thick) (10 cm) | | | 0.12 | | | |
| Asbestos-cement board (0.2 in thick) (0.5 cm) | 0.54 | | | | | |
| Plaster on metal lath (¾ in) (2 cm) | | | 15 | | | |
| Plaster on wood lath | | 11 | | | | |
| Plaster on plain gypsum lath (with studs) | | | 20 | | | |
| Gypsum wallboard (⅜ in plain) (2 cm) | | | 50 | | | |
| Gypsum sheathing (½ in asphalt impreg.) (1.5 cm) | | | | 20 | | |
| Structural insulating bd. (sheathing qual.) | | | | | | 20–50 |
| Structural insulating bd. (interior, uncoated, ½ in) (1.5 cm) | | | 50–90 | | | |
| Hardboard (⅛ in standard) (0.5 cm) | | | 11 | | | |
| Hardboard (⅛ in tempered) (0.5 cm) | | | 5 | | | |
| Built-up roofing (hot-mopped) | 0.0 | | | | | |
| Wood, sugar pine | | | | | | 0.4–5.4[b] |
| Plywood (Douglas fir, exterior, glue, ¼ in thick) (0.5 cm) | | | 0.7 | | | |
| Plywood (Douglas fir, interior, glue, ¼ in thick) (0.5 cm) | | | 1.9 | | | |
| Acrylic, glass-fiber-reinforced sheet, 56 mil | 0.12 | | | | | |
| Polyester, glass-fiber-reinforced sheet, 48 mil | 0.05 | | | | | |
| **Thermal insulations:** | | | | | | |
| Air (still) | | | | | | 120 |
| Cellular glass | | | | 0.0 | | |
| Corkboard | | | | 2.1–2.6 | 9.5 | |
| Mineral wool (unprotected) | | | | | 116 | |
| Expanded polyurethane (R-11 blown) board stock | | | | 0.4–1.6 | | |
| Expanded polystyrene—extruded | | | | 1.2 | | |
| Expanded polystyrene—bead | | | | 2.0–5.8 | | |
| Unicellular synthetic flexible rubber foam | | | | 0.02–0.15 | | |
| **Plastic and metal foils and films**[c] | | | | | | |
| Aluminum foil (1 mil) | 0.0 | | | | | |
| Aluminum foil (0.35 mil) | 0.05 | | | | | |
| Polyethylene (2 mil) | 0.16 | | | | | |
| Polyethylene (4 mil) | 0.08 | | | | | |
| Polyethylene (6 mil) | 0.06 | | | | | |
| Polyethylene (8 mil) | 0.04 | | | | | |
| Polyethylene (10 mil) | 0.03 | | | | | |
| Polyester (1 mil) | 0.7 | | | | | |
| Cellulose acetate (125 mil) | 0.4 | | | | | |
| Polyvinylchloride, unplasticized (2 mil) | 0.68 | | | | | |
| Polyvinylchloride, plasticized (4 mil) | 0.8–1.4 | | | | | |
| **Building papers, felts, roofing papers**[d] | | | | | | |
| Duplex sheet, asphalt laminated, aluminum foil one side (43)[e] | 0.002 | 0.176 | | | | |
| Saturated and coated roll roofing (326)[e] | 0.05 | 0.24 | | | | |
| Kraft paper and asphalt laminated, reinforced 30-120-30 (34)[e] | 0.3 | 1.8 | | | | |
| Blanket thermal insulation back-up paper, asphalt coated (31)[e] | 0.4 | 0.6–4.2 | | | | |
| Asphalt-saturated and coated vapor-barrier paper (43)[e] | 0.2–0.3 | 0.6 | | | | |
| Asphalt-saturated but not coated sheathing paper (22)[e] | 3.3 | 20.2 | | | | |
| 15-lb asphalt felt (70)[e] | 1.0 | 5.6 | | | | |
| 15-lb tar felt (70)[e] | 4.0 | 18.2 | | | | |
| Single-kraft, double infused (16)[e] | 31 | 42 | | | | |
| **Liquid-applied coating materials:** | | | | | | |
| Paint—2 coats | | | | | | |
| Asphalt paint on plywood | | 0.4 | | | | |
| Aluminum varnish on wood | 0.3–0.5 | | | | | |
| Enamels in smooth plaster | | | 0.5–1.5 | | | |
| Primers and sealers on interior insulation board | | | 0.9–2.1 | | | |
| Various primers plus 1 coat flat oil paint on plaster | | | 1.6–3.0 | | | |
| Flat paint on interior insulation bd | | | 4 | | | |
| Water emulsion on interior insul bd | | | 30–85 | | | |
| Paint—3 coats | | | | | | |
| Exterior paint, white lead and oil on wood siding | 0.3–1.0 | | | | | |
| Exterior paint, white lead-zinc oxide and oil on wood | 0.9 | | | | | |

**Table 14. Permeance and Permeability of Materials to Water Vapor** *(Continued)*

| Material | Permeance (perm) | | | Permeability (perm-inch) | | |
|---|---|---|---|---|---|---|
| | Dry cup | Wet cup | Other | Dry cup | Wet cup | Other |
| Styrene-butadiene latex coating, 2 oz/ft² | 11 | | | | | |
| Polyvinyl acetate latex coating, 4 oz/ft² | 5.5 | | | | | |
| Chlorosulfonated polyethylene mastic, 3.5 oz/ft² | 1.7 | | | | | |
| 7.0 oz/ft² | 0.06 | | | | | |
| Asphalt cut-back mastic, ¹⁄₁₆ in dry | 0.14 | | | | | |
| ³⁄₁₆ in dry | 0.0 | | | | | |
| Hot-melt asphalt, 2 oz/ft² | 0.5 | | | | | |
| 3.5 oz/ft² | 0.1 | | | | | |

To convert perm [grain/(hr)(ft²)(in Hg)] to kg/Ns, multiply by $5.733 \times 10^{-11}$.
To convert perm-inch [grain/(h)(ft²)(in Hg/in)] to kg/Ns, multiply by $1.457 \times 10^{-12}$.
[a]Table 4-14 gives the water-vapor transmission rates of some representative materials. The data are provided to permit comparisons of materials; but in the selection of vapor-barrier materials, exact values for permeance or permeability should be obtained from the manufacturer of the materials under consideration or secured as a result of laboratory tests. A range of values shown in the table indicated variations among mean values for materials that are similar but of different density, orientation, lot, or source. The values are intended for design guidance and should not be used as design or specification data. The compilation is from a number of sources; values from dry-cup and wet-cup methods were usually obtained from investigations using ASTM E96 and C355; values shown under Other were obtained from investigations using such techniques as *two-temperature, special cell,* and *air velocity.*
[b]Depending on construction and direction of vapor flow.
[c]Usually installed as vapor barriers, although sometimes used as exterior finish and elsewhere near cold side where special considerations are then required for warm-side-barrier effectiveness.
[d]Low-permeance sheets used as vapor barriers. High permeance used elsewhere in construction.
[e]Basis weight in lb per 500 ft².

**Table 15. Grains of Moisture per Pound of Dry Air (Grams of Moisture per kg of Dry Air) vs. Dew-Point Temperature, °F (°C)**

| DP °F (°C) | gr (g) | DP °F (°C) | gr (g) | DP °F (°C) | gr (g) | DP °F (°C) | gr (g) | DP °F (°C) | gr (g) |
|---|---|---|---|---|---|---|---|---|---|
| 0 (−17.8) | 5.50 (0.787) | 16 (−8.9) | 12.36 (1.767) | 32 ( 0) | 26.40 (3.775) | 48 ( 8.9) | 49.50 ( 7.078) | 64 (17.8) | 89.18 (12.753) |
| 1 (−17.2) | 5.79 (0.828) | 17 (−8.3) | 12.99 (1.857) | 33 (0.6) | 27.52 (3.935) | 49 ( 9.4) | 51.42 ( 7.353) | 65 (18.3) | 92.40 (13.213) |
| 2 (−16.7) | 6.10 (0.872) | 18 (−7.8) | 13.63 (1.949) | 34 (1.1) | 28.66 (4.098) | 50 (10.0) | 53.38 ( 7.633) | 66 (18.9) | 95.76 (13.694) |
| 3 (−16.1) | 6.43 (0.919) | 19 (−7.2) | 14.30 (2.045) | 35 (1.7) | 29.83 (4.265) | 51 (10.6) | 55.45 ( 7.929) | 67 (19.4) | 99.19 (14.184) |
| 4 (−15.6) | 6.77 (0.968) | 20 (−6.7) | 15.01 (2.146) | 36 (2.2) | 31.07 (4.443) | 52 (11.1) | 57.58 ( 8.234) | 68 (20.0) | 102.8 (14.700) |
| 5 (−15.0) | 7.12 (1.018) | 21 (−6.1) | 15.75 (2.252) | 37 (2.8) | 32.33 (4.623) | 53 (11.7) | 59.74 ( 8.543) | 69 (20.6) | 106.4 (15.215) |
| 6 (−14.4) | 7.50 (1.073) | 22 (−5.5) | 16.53 (2.364) | 38 (3.3) | 33.62 (4.807) | 54 (12.2) | 61.99 ( 8.865) | 70 (21.1) | 110.2 (15.759) |
| 7 (−13.9) | 7.89 (1.128) | 23 (−5.0) | 17.33 (2.478) | 39 (3.9) | 34.97 (5.000) | 55 (12.8) | 64.34 ( 9.200) | 71 (21.7) | 114.2 (16.330) |
| 8 (−13.3) | 8.30 (1.187) | 24 (−4.4) | 18.17 (2.598) | 40 (4.4) | 36.36 (5.199) | 56 (13.3) | 66.75 ( 9.545) | 72 (22.2) | 118.2 (16.903) |
| 9 ( 12.8) | 8.73 (1.248) | 25 (−3.9) | 19.05 (2.724) | 41 (5.0) | 37.80 (5.405) | 57 (13.9) | 69.23 ( 9.900) | 73 (22.8) | 122.4 (17.503) |
| 10 (−12.2) | 9.18 (1.313) | 26 (−3.3) | 19.97 (2.856) | 42 (5.5) | 39.31 (5.621) | 58 (14.4) | 71.82 (10.270) | 74 (23.3) | 126.6 (18.104) |
| 11 (−11.7) | 9.65 (1.380) | 27 (−2.8) | 20.94 (2.994) | 43 (6.1) | 40.88 (5.846) | 59 (15.0) | 74.48 (10.650) | 75 (23.9) | 131.1 (18.747) |
| 12 (−11.1) | 10.15 (1.451) | 28 (−2.2) | 21.93 (3.136) | 44 (6.7) | 42.48 (6.075) | 60 (15.6) | 77.21 (11.041) | 76 (24.4) | 135.7 (19.405) |
| 13 (−10.6) | 10.66 (1.524) | 29 (−1.7) | 22.99 (3.287) | 45 (7.2) | 44.14 (6.312) | 61 (16.1) | 80.08 (11.451) | 77 (25.0) | 140.4 (20.077) |
| 14 (−10.0) | 11.20 (1.602) | 20 (−1.1) | 24.07 (3.442) | 46 (7.8) | 45.87 (6.559) | 62 (16.7) | 83.02 (11.871) | 78 (25.6) | 145.3 (20.778) |
| 15 (−9.4) | 11.77 (1.683) | 31 (−0.6) | 25.21 (3.605) | 47 (8.3) | 47.66 (6.815) | 63 (17.2) | 86.03 (12.302) | 79 (26.1) | 150.3 (21.493) |

**Table 16. Vapor Pressure of Saturated Air, Inches of Hg (N/m²), vs. Dry-Bulb Temperature, °F (°C)**

| T °F (°C) | $P_{sat}$ in Hg (N/m²) | T °F (°C) | $P_{sat}$ in Hg (N/m²) | T °F (°C) | $P_{sat}$ in Hg (N/m²) | T °F (°C) | $P_{sat}$ in Hg (N/m²) | T °F (°C) | $P_{sat}$ in Hg (N/m²) |
|---|---|---|---|---|---|---|---|---|---|
| 0 (−17.8) | 0.03764 (127.08) | 16 (−8.9) | 0.08461 (285.68) | 32 ( 0) | 0.18035 ( 608.93) | 48 ( 8.9) | 0.33629 (1,135.45) | 64 (17.8) | 0.60073 (2,028.30) |
| 1 (−17.2) | 0.03966 (133.91) | 17 (−8.3) | 0.08884 (299.96) | 33 (0.6) | 0.18778 ( 634.02) | 49 ( 9.4) | 0.34913 (1,178.80) | 65 (18.3) | 0.62209 (2,100.42) |
| 2 (−16.7) | .04178 (141.07) | 18 (−7.8) | .09326 (314.88) | 34 (1.1) | 0.19546 ( 659.95) | 50 (10.0) | 0.36240 (1,223.60) | 66 (18.9) | 0.64411 (2,174.76) |
| 3 (−16.1) | 0.04400 (148.56) | 19 (−7.2) | 0.09789 (330.51) | 35 (1.7) | 0.20342 ( 686.83) | 51 (10.6) | 0.37611 (1,269.89) | 67 (19.4) | 0.66681 (2,251.41) |
| 4 (−15.6) | 0.04633 (156.43) | 20 (−6.7) | 0.10272 (346.82) | 36 (2.2) | 0.21166 ( 714.65) | 52 (11.1) | 0.39028 (1,317.74) | 68 (20.0) | 0.69019 (2,330.35) |
| 5 (−15.0) | 0.04877 (164.67) | 21 (−6.1) | 0.10777 (363.87) | 37 (2.8) | 0.22020 ( 743.48) | 53 (11.7) | 0.40492 (1,367.17) | 69 (20.6) | 0.71430 (2,411.75) |
| 6 (−14.4) | 0.05133 (173.31) | 22 (−5.5) | 0.11305 (381.70) | 38 (3.3) | 0.22904 ( 773.33) | 54 (12.2) | 0.42004 (1,418.22) | 70 (21.1) | 0.73915 (2,495.66) |
| 7 (−13.9) | 0.05402 (182.39) | 23 (−5.0) | 0.11856 (400.30) | 39 (3.9) | 0.23819 ( 804.22) | 55 (12.8) | 0.43565 (1,470.92) | 71 (21.7) | 0.76475 (2,582.09) |
| 8 (−13.3) | 0.05683 (191.88) | 24 (−4.4) | 0.12431 (419.72) | 40 (4.4) | 0.24767 ( 836.23) | 56 (13.3) | 0.45176 (1,525.32) | 72 (22.2) | 0.79112 (2,671.13) |
| 9 (−12.8) | 0.05977 (201.81) | 25 (−3.9) | 0.13032 (440.01) | 41 (5.0) | 0.25748 ( 869.35) | 57 (13.9) | 0.46480 (1,569.35) | 73 (22.8) | 0.81828 (2,762.83) |
| 10 (−12.2) | 0.06285 (212.21) | 26 (−3.3) | 0.13659 (461.18) | 42 (5.5) | 0.26763 ( 903.62) | 58 (14.4) | 0.48558 (1,639.51) | 74 (23.3) | 0.84624 (2,857.24) |
| 11 (−11.7) | 0.06608 (223.11) | 27 (−2.8) | 0.14313 (483.26) | 43 (6.1) | 0.27813 ( 939.07) | 59 (15.0) | 0.50330 (1,699.34) | 75 (23.9) | 0.87504 (2,954.47) |
| 12 (−11.1) | 0.06946 (234.52) | 28 (−2.2) | 0.14966 (505.31) | 44 (6.7) | 0.28889 ( 975.41) | 60 (15.6) | 0.52159 (1,761.09) | 76 (24.4) | 0.90470 (3,054.62) |
| 13 (−10.6) | 0.07299 (246.44) | 29 (−1.7) | 0.15707 (530.33) | 45 (7.2) | 0.30023 (1,013.69) | 61 (16.1) | 0.54047 (1,824.84) | 77 (25.0) | 0.93523 (3,157.70) |
| 14 (−10.0) | 0.07669 (258.94) | 30 (−1.1) | 0.16452 (555.48) | 46 (7.8) | 0.31185 (1,052.93) | 62 (16.7) | 0.55994 (1,890.57) | 78 (25.6) | 0.96665 (3,263.78) |
| 15 (− 9.4) | 0.80856 (272.00) | 31 (−0.6) | 0.17727 (581.65) | 47 (8.3) | 0.32386 (1,093.48) | 63 (17.2) | 0.58002 (1,958.37) | 79 (26.1) | 0.99899 (3,372.98) |

SOURCE: Reprinted by permission from ASHRAE "Handbook of Fundamentals," 1963.

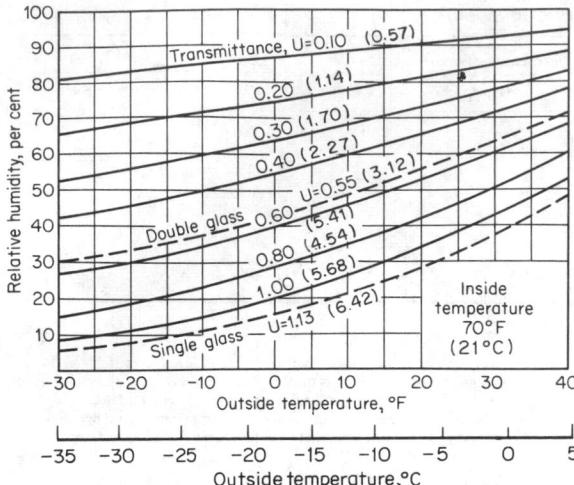

**Fig. 7** Relative humidity at which visible condensation will appear on inside surface. First value given for $U$ in Btu/(h·ft²·°F); values in parentheses given in W/(m² − K). (*Reprinted by permission from ASHRAE "Handbook of Fundamentals," 1972.*)

## COOLING

Cooling load $Q_t$, the total simultaneous cooling load, expressed in Btu/hour or (watts):

$$Q_t = Q_{ext} + Q_{int} + Q_{outside\ air} \qquad (16)$$
$$Q_{ext} = \text{external heat gains}$$
$$= Q_{transmission} + Q_{solar} \qquad (17)$$

For walls and roofs

$$Q_{tr-sol} = AU\ (Sa\ \Delta t) \qquad \text{Btu/h (watts)} \qquad (18)$$

where Sa $\Delta t$ = sol-air equivalent temperature differential, °F (°C), Table 17.

For glass

$$Q_{glass} = Q_{solar} + Q_{tr} \qquad (19)$$
$$= \text{SHGF (SF)}\ A_1 + A_2 U\ (t_o - t_i) \qquad (20)$$

where SHGF = solar-heat gain factor (Table 23)
SF = shading factor (Tables 24 to 27)
$A_1$ = area of sunlit glass, ft² (m²)
$A_2$ = area of total glass, ft² (m²)
$U$ = overall coefficient (Table 28)
$t_o$ = outside design temperature, °F (°C) (Figs. 10, 11)
$t_i$ = inside design temperature, °F (°C) (Table 32)

$Q_{int}$ = internal heat gains
$$= Q_{lights} + Q_{people} + Q_{equipment} + Q_{transmission} \qquad (21)$$
$Q_{lights}$ = 3.41 × wattage input to conditioned space
Btu/h (Figs. 8 and 9)
= wattage input to conditioned space      watts  **(22)**
$Q_{people}$ = number of people ($q_{sensible} + q_{latent}$)
Btu/h (watts) (Table 29)      **(23)**
$Q_{equipment} = q_{sensible} + q_{latent}$    Btu/h (watts) (Table 30, 31)
**(24)**

$$Q_{transmission} = AU(t_o - t_i) \qquad \text{Btu/h (watts)}$$
$$\text{(for partitions, floors, and ceilings)} \qquad (25)$$
$$Q_{outside\ air} = q_{sensible} + q_{latent} \qquad \text{Btu/h (watts)}$$
$$= AU(t_o - t_i)\ \text{ft}^3/\text{min}$$
$$+ 0.68\ (M_o - M_i)\ \text{ft}^3/\text{min} \qquad \text{Btu/h} \qquad (26)$$

where $M_o$ = outside moisture content at design wet bulb, °F, g/lb
$M_i$ = inside moisture content at design relative humidity, g/lb

A comparable equation in SI units is

$$Q_{outside\ air} = 1.2(t'_o - t'_i)\text{m}^2/\text{s}$$
$$+ 42.6\ (M'_o - M'_i) \qquad \text{m}^3/\text{s, kW}$$

where $M'_o$ = outside moisture content at design wet bulb, °C, g/kg
$M'_i$ = inside moisture content at design relative humidity, g/kg

$t'_o$, $t'_i$ are expressed in °C

**Supply-Air Temperature**   The room cooling load is the sum of external and internal sensible and latent heat gains plus the difference in enthalpy between outside and room air for that portion of outside air which does not contact the cooling-coil surfaces. The percentage of air that passes through a cooling coil untreated is the numerical value of the *coil bypass factor*; e.g., a bypass factor of 20 percent represents a cooling-coil saturation efficiency of 80 percent.

The ratio of room sensible heat gains to total room sensible and latent heat gains is the **room sensible heat ratio** (RSHR).

$$\text{RSHR} = \frac{Q_{rs}}{Q_{rs} + Q_{rl}} \qquad (27)$$

It represents the ratio of sensible cooling capacity to the total cooling capacity required of the supply air to satisfy room conditions. It is used to plot the slope of the *room-condition line* on a psychrometric chart (Fig. 12) for the determination of the *apparatus dew point* (ADP).

The actual supply-air temperature and off-coil wet-bulb temperature will depend on the bypass characteristic of the selected cooling coil (Fig. 14).

**Supply-Air Rate**   The rate of supply air required is expressed by

$$Q_{sa} = Q_{rs}/1.08(t_r - t_s) \qquad \text{ft}^3/\text{min} \qquad (28)$$

in SI units:

$$Q'_{sa} = Q'_{rs}/1.2(t_r - t_s) \qquad \text{m}^3/\text{s}$$

where $Q_{sa}$ = supply air, ft³/min (m³/s)
$Q'_{rs}$ = room sensible heat, kW
$t_r$ = room design temperature, °F (K)
$t_s$ = supply-air temperature, °F (K)

1.08 = (60 min)[0.244 Btu/(lb)(°F)](0.075 lb/ft³)  **(29)**

### Air Distribution

OUTLETS.  Purpose. Outlets are designed:

1. To control air motion, noise level, and temperature gradients caused by the introduction of air to and the removal of air from a space.

2. To counteract the natural convection and radiation effects within the room.

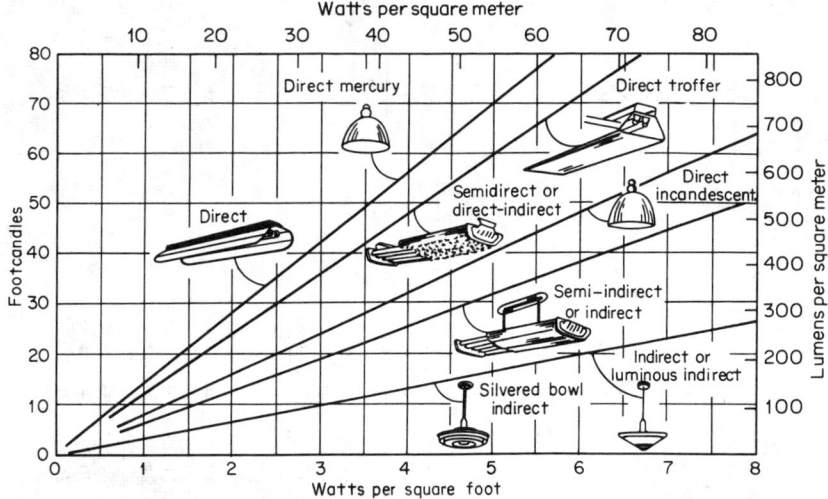

**Fig. 8** Typical heat-gain lighting fixtures. On the curve above, follow the horizontal line, beginning at the maintained footcandle (lm/lm²) value selected in (1), until it intersects the curve corresponding to the fixture type to be installed.

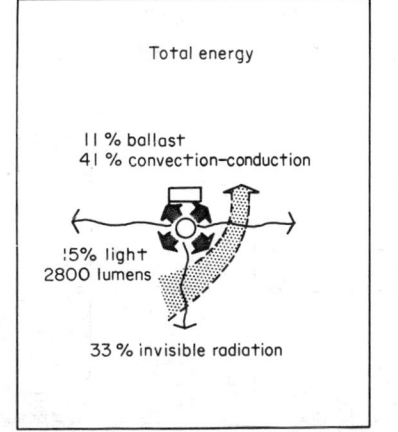

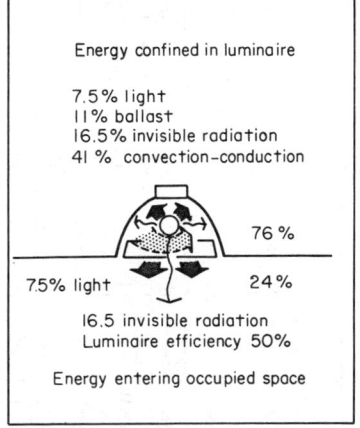

Energy output for 150-watt Incandescent lamp

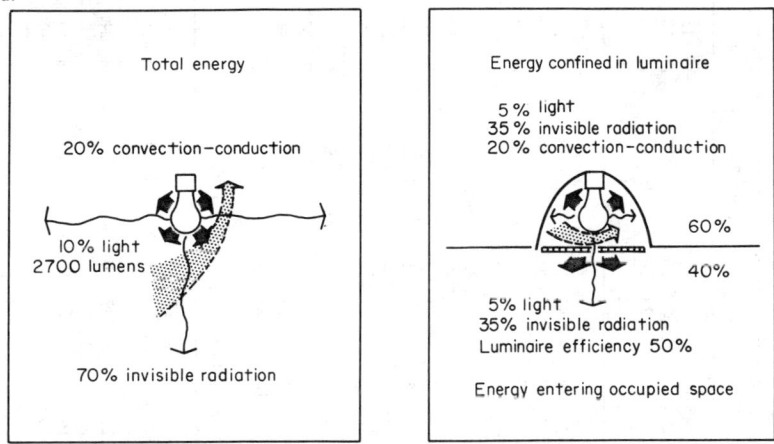

Energy output for 40-watt Fluorescent lamp and ballast

**Fig. 9** Distribution of energy output of incandescent and fluorescent lamps. (*Light Magazine, vol. 29, Fig. 1.*)

# Table 17. Total Equivalent Temperature Differentials for Calculating Heat Gain through Sunlit Walls

Exterior color of wall: D = dark, L = light

| North latitude wall facing | A.M. 8 D | 8 L | 10 D | 10 L | 12 D | 12 L | P.M. 2 D | 2 L | 4 D | 4 L | 6 D | 6 L | 8 D | 8 L | 10 D | 10 L | 12 D | 12 L | Amplitude decrement factor λ | time lag δ h | South latitude wall facing |
|---|---|---|---|---|---|---|---|---|---|---|---|---|---|---|---|---|---|---|---|---|---|
| **Group A** | | | | | | | | | | | | | | | | | | | | | |
| NE | 27 | 16 | 31 | 18 | 26 | 17 | 24 | 17 | 24 | 18 | 23 | 17 | 20 | 15 | 17 | 13 | 15 | 11 | | | SE |
| E | 32 | 18 | 41 | 24 | 37 | 22 | 29 | 20 | 28 | 20 | 26 | 19 | 23 | 16 | 20 | 14 | 18 | 13 | | | E |
| SE | 25 | 15 | 36 | 21 | 38 | 23 | 33 | 21 | 28 | 20 | 26 | 18 | 22 | 16 | 19 | 14 | 18 | 12 | 0.34 | 2 | NE |
| S | 14 | 9 | 20 | 13 | 28 | 18 | 33 | 22 | 31 | 21 | 25 | 18 | 20 | 15 | 17 | 13 | 15 | 11 | | | N |
| SW | 17 | 11 | 20 | 13 | 24 | 16 | 34 | 22 | 42 | 27 | 41 | 26 | 28 | 19 | 20 | 14 | 18 | 12 | | | NW |
| W | 17 | 11 | 20 | 13 | 24 | 16 | 30 | 20 | 42 | 27 | 48 | 30 | 33 | 22 | 22 | 15 | 19 | 13 | | | W |
| NW | 14 | 9 | 17 | 11 | 21 | 14 | 23 | 17 | 31 | 21 | 38 | 25 | 28 | 19 | 18 | 13 | 16 | 11 | | | SW |
| N | 14 | 9 | 15 | 10 | 18 | 12 | 20 | 15 | 21 | 16 | 25 | 16 | 18 | 14 | 14 | 11 | 12 | 9 | | | S |
| **Group B** | | | | | | | | | | | | | | | | | | | | | |
| NE | 12 | 7 | 27 | 14 | 31 | 17 | 30 | 19 | 31 | 21 | 30 | 22 | 27 | 20 | 21 | 17 | 16 | 13 | | | SE |
| E | 14 | 8 | 34 | 18 | 45 | 24 | 43 | 25 | 39 | 25 | 35 | 24 | 30 | 22 | 23 | 18 | 17 | 14 | | | E |
| SE | 9 | 5 | 25 | 13 | 39 | 21 | 44 | 26 | 41 | 26 | 37 | 25 | 31 | 23 | 24 | 18 | 17 | 14 | 0.51 | 3 | NE |
| S | 4 | 3 | 7 | 4 | 18 | 11 | 32 | 19 | 41 | 26 | 39 | 27 | 33 | 24 | 25 | 19 | 18 | 15 | | | N |
| SW | 5 | 3 | 7 | 4 | 11 | 7 | 23 | 15 | 41 | 26 | 54 | 34 | 51 | 33 | 38 | 25 | 26 | 19 | | | NW |
| W | 6 | 4 | 7 | 4 | 11 | 4 | 18 | 12 | 35 | 23 | 55 | 34 | 59 | 37 | 43 | 28 | 30 | 20 | | | W |
| NW | 5 | 3 | 6 | 4 | 11 | 7 | 17 | 12 | 26 | 18 | 41 | 27 | 47 | 31 | 36 | 24 | 25 | 18 | | | SW |
| N | 6 | 4 | 9 | 5 | 12 | 8 | 18 | 12 | 22 | 17 | 25 | 20 | 27 | 21 | 22 | 17 | 16 | 14 | | | S |
| **Group C** | | | | | | | | | | | | | | | | | | | | | |
| NE | 9 | 6 | 19 | 10 | 26 | 15 | 28 | 17 | 29 | 18 | 29 | 20 | 28 | 20 | 24 | 19 | 20 | 16 | | | SE |
| E | 10 | 7 | 22 | 12 | 36 | 19 | 40 | 23 | 39 | 23 | 36 | 24 | 33 | 23 | 28 | 20 | 22 | 17 | | | E |
| SE | 8 | 6 | 16 | 9 | 29 | 16 | 38 | 21 | 39 | 24 | 37 | 24 | 34 | 23 | 28 | 21 | 23 | 17 | 0.40 | 4 | NE |
| S | 7 | 5 | 7 | 4 | 12 | 7 | 22 | 14 | 32 | 20 | 36 | 24 | 34 | 24 | 29 | 21 | 23 | 17 | | | N |
| SW | 9 | 6 | 8 | 5 | 10 | 6 | 16 | 10 | 28 | 18 | 42 | 26 | 48 | 30 | 42 | 28 | 33 | 22 | | | NW |
| W | 10 | 7 | 9 | 5 | 10 | 6 | 14 | 9 | 24 | 16 | 40 | 25 | 52 | 32 | 47 | 30 | 37 | 24 | | | W |
| NW | 8 | 6 | 8 | 5 | 9 | 6 | 13 | 9 | 19 | 14 | 30 | 20 | 40 | 27 | 38 | 26 | 30 | 21 | | | SW |
| N | 7 | 5 | 8 | 5 | 10 | 7 | 14 | 9 | 18 | 13 | 22 | 16 | 25 | 19 | 23 | 18 | 19 | 16 | | | S |
| **Group D** | | | | | | | | | | | | | | | | | | | | | |
| NE | 8 | 5 | 19 | 10 | 28 | 15 | 29 | 17 | 30 | 19 | 30 | 21 | 28 | 21 | 24 | 19 | 19 | 16 | | | SE |
| E | 9 | 6 | 23 | 12 | 38 | 20 | 42 | 24 | 40 | 24 | 37 | 24 | 33 | 23 | 27 | 20 | 21 | 17 | | | E |
| SE | 7 | 5 | 16 | 9 | 30 | 16 | 40 | 22 | 41 | 25 | 38 | 25 | 34 | 24 | 28 | 21 | 22 | 17 | 0.45 | 4 | NE |
| S | 5 | 4 | 6 | 4 | 12 | 7 | 23 | 14 | 34 | 21 | 38 | 25 | 35 | 24 | 29 | 21 | 23 | 17 | | | N |

## Group E

| Dir | | | | | | | | | | | | | | | | | | | Dir |
|-----|--|--|--|--|--|--|--|--|--|--|--|--|--|--|--|--|--|--|-----|
| SW | 8 | 5 | 7 | 4 | 9 | 6 | 16 | 10 | 30 | 19 | 44 | 28 | 51 | 32 | 43 | 28 | 33 | 22 | NW |
| W | 8 | 6 | 7 | 5 | 9 | 6 | 14 | 9 | 25 | 16 | 42 | 27 | 55 | 34 | 49 | 31 | 37 | 25 | W |
| NW | 7 | 5 | 7 | 4 | 9 | 6 | 13 | 9 | 20 | 14 | 31 | 21 | 42 | 28 | 40 | 27 | 31 | 21 | SW |
| N | 6 | 4 | 8 | 5 | 10 | 6 | 14 | 10 | 19 | 14 | 23 | 17 | 25 | 19 | 24 | 19 | 19 | 16 | S |
| NE | 10 | 6 | 23 | 12 | 30 | 16 | 30 | 18 | 30 | 20 | 30 | 21 | 28 | 21 | 23 | 18 | 18 | 14 | SE |
| E | 11 | 6 | 28 | 15 | 42 | 22 | 43 | 24 | 39 | 24 | 36 | 23 | 32 | 23 | 25 | 19 | 19 | 15 | E |
| SE | 8 | 5 | 20 | 11 | 35 | 19 | 42 | 24 | 41 | 25 | 38 | 25 | 33 | 23 | 26 | 20 | 20 | 16 | NE |
| S | 4 | 3 | 6 | 4 | 15 | 9 | 28 | 17 | 38 | 24 | 39 | 26 | 34 | 24 | 27 | 20 | 20 | 16 | N |
| SW | 6 | 4 | 7 | 4 | 10 | 6 | 19 | 11 | 35 | 22 | 49 | 31 | 52 | 33 | 41 | 27 | 30 | 21 | NW |
| W | 7 | 5 | 7 | 4 | 10 | 6 | 16 | 11 | 30 | 20 | 48 | 31 | 57 | 36 | 47 | 30 | 34 | 23 | W |
| NW | 6 | 4 | 6 | 4 | 10 | 6 | 15 | 10 | 23 | 16 | 36 | 24 | 45 | 30 | 38 | 26 | 28 | 20 | SW |
| N | 6 | 4 | 8 | 5 | 11 | 7 | 16 | 11 | 21 | 15 | 24 | 18 | 26 | 20 | 23 | 18 | 18 | 15 | S |

(Group E parameters: 0.48, 4)

## Group F

| Dir | | | | | | | | | | | | | | | | | | | Dir |
|-----|--|--|--|--|--|--|--|--|--|--|--|--|--|--|--|--|--|--|-----|
| NE | 9 | 7 | 14 | 9 | 21 | 12 | 25 | 15 | 27 | 17 | 29 | 19 | 28 | 20 | 26 | 19 | 23 | 17 | SE |
| E | 10 | 8 | 17 | 10 | 28 | 15 | 35 | 19 | 37 | 22 | 37 | 23 | 35 | 23 | 31 | 22 | 26 | 19 | E |
| SE | 10 | 7 | 13 | 8 | 22 | 12 | 31 | 17 | 36 | 21 | 37 | 23 | 35 | 23 | 32 | 22 | 27 | 19 | NE |
| S | 9 | 7 | 7 | 5 | 10 | 6 | 17 | 10 | 26 | 16 | 32 | 20 | 33 | 22 | 31 | 22 | 27 | 19 | N |
| SW | 12 | 9 | 10 | 6 | 9 | 6 | 13 | 8 | 22 | 14 | 33 | 21 | 42 | 27 | 42 | 27 | 37 | 25 | NW |
| W | 14 | 9 | 11 | 7 | 10 | 6 | 12 | 8 | 19 | 12 | 31 | 20 | 43 | 27 | 46 | 29 | 41 | 27 | W |
| NW | 12 | 8 | 9 | 6 | 9 | 6 | 11 | 8 | 16 | 11 | 24 | 16 | 33 | 22 | 36 | 24 | 33 | 23 | SW |
| N | 8 | 7 | 8 | 6 | 9 | 6 | 12 | 8 | 15 | 11 | 19 | 14 | 22 | 17 | 23 | 18 | 21 | 17 | S |

(Group F parameters: 0.32, 6)

To convert temperature differentials in °F to °C, multiply by 5/9.

**Table 18. Description of Wall Construction**

| Group | Components | Weight lb/ft² | Weight kg/m² | U value Btu/(h)(ft²)(°F) | U value W/(m²)(K) |
|-------|-----------|---------------|--------------|--------------------------|-------------------|
| A | 1 in (2.5 cm) stucco + 4 in (10 cm) l.w. concrete block + air space | 28.6 | 139.63 | 0.267 | 0.047 |
|   | 1 in (2.5 cm) stucco + air space + 2 in (5 cm) insulation | 16.3 | 79.58 | 0.106 | 0.019 |
| B | 1 in (2.5 cm) stucco + 4 in (10 cm) common brick | 55.9 | 272.90 | 0.393 | 0.069 |
|   | 1 in (2.5 cm) stucco + 4 in (10 cm) h.w. concrete | 62.5 | 305.13 | 0.481 | 0.085 |
| C | 4 in (10 cm) face brick + 4 in (10 cm) l.w. concrete block + 1 in (2.5 cm) insulation | 62.5 | 305.13 | 0.158 | 0.028 |
|   | 1 in (2.5 cm) stucco + 4 in (10 cm) h.w. concrete + 2 in (5 cm) insulation | 62.9 | 307.08 | 0.114 | 0.020 |
| D | 1 in (2.5 cm) stucco + 8 in (20 cm) l.w. concrete block + 1 in (2.5 cm) insulation | 41.4 | 202.11 | 0.141 | 0.025 |
|   | 1 in (2.5 cm) stucco + 2 in (5 cm) insulation + 4 in (10 cm) h.w. concrete block | 36.6 | 178.68 | 0.111 | 0.019 |
| E | 4 in (10 cm) face brick + 3 in (10 cm) l.w. concrete block | 62.2 | 303.66 | 0.333 | 0.059 |
|   | 1 in (2.5 cm) stucco + 8 in (20 cm) h.w. concrete block | 56.6 | 276.32 | 0.349 | 0.061 |
| F | 4 in (10 cm) face brick + 4 in (10 cm) common brick | 89.5 | 436.94 | 0.360 | 0.063 |
|   | 4 in (10 cm) face brick + 2 in (5 cm) insulation + 4 in (10 cm) l.w. concrete block | 62.5 | 305.13 | 0.103 | 0.018 |

**Fig. 10** Summer outside dry-bulb design temperature. Given in Fahrenheit; to convert to Celsius: $t_C = \frac{5}{9}(t_F - 32)$. [*C. Strock and R. L. Koral (eds.), "Handbook of Air Conditioning, Heating and Ventilating," Industrial Press, New York, 1965.*]

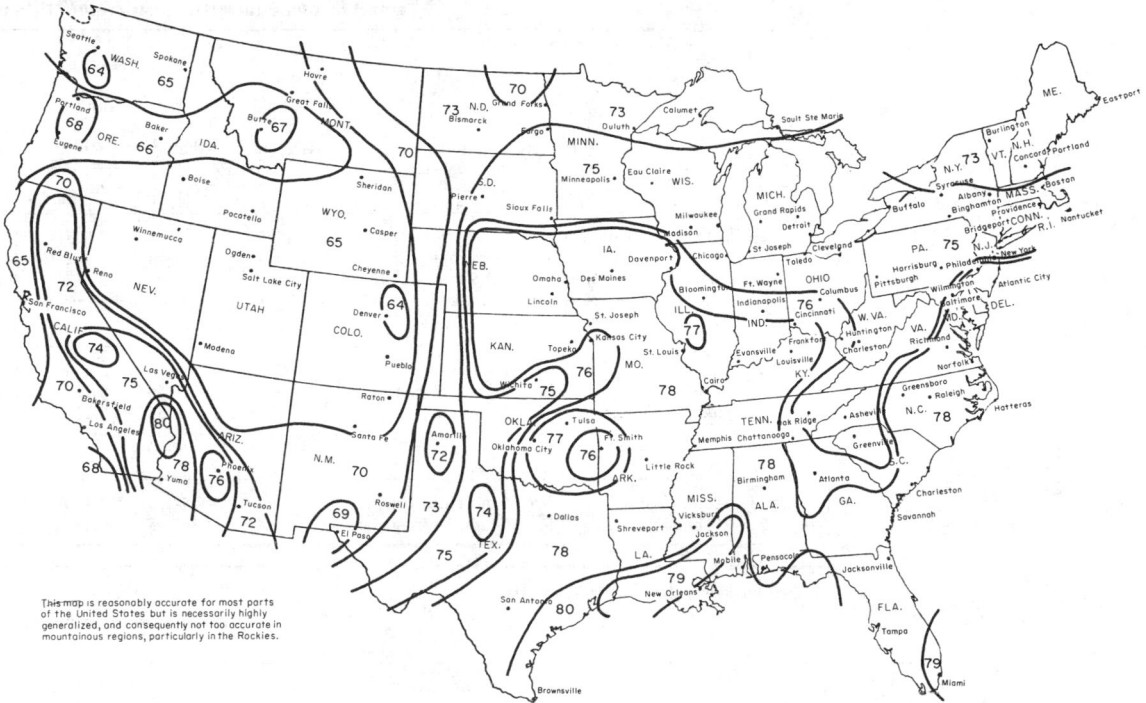

**Fig. 11** Summer outside wet-bulb design temperature. Given in Fahrenheit; to convert to Celsius: $t_C = {}^5/_9 (t_f - 32)$. [*C. Strock and R. L. Koral (eds.), "Handbook of Air Conditioning, Heating and Ventilating," 2d ed., Industrial Press, New York, 1965.*]

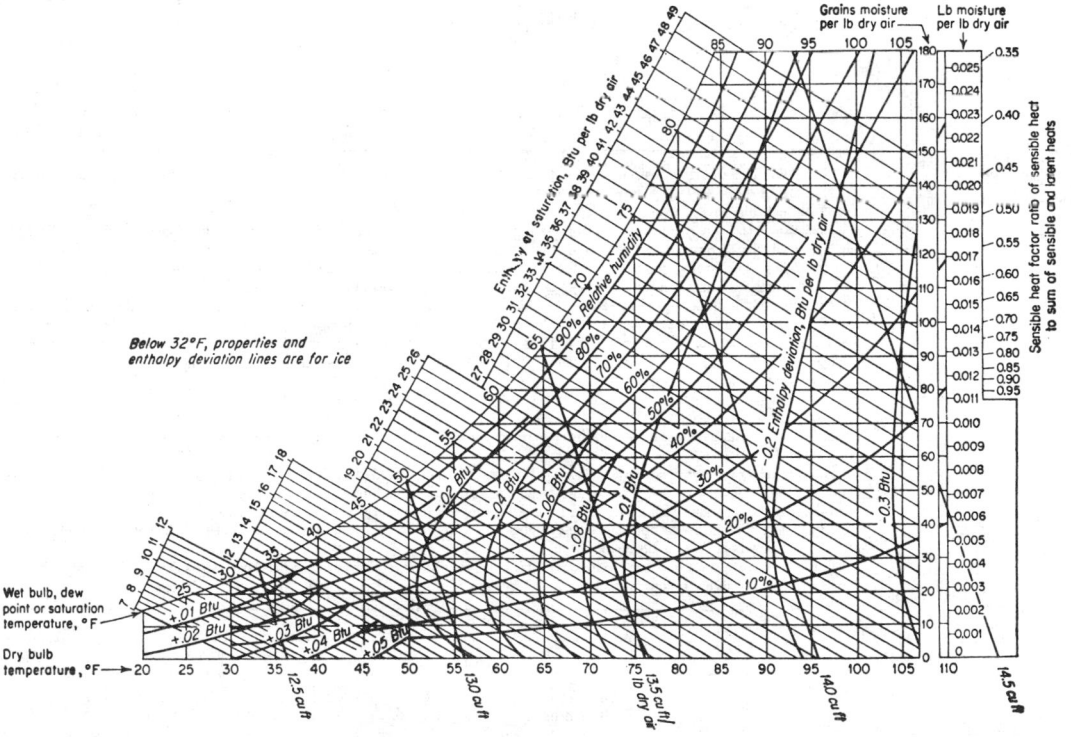

**Fig. 12** Psychrometric chart—normal temperatures.

**Table 19. Total Equivalent Temperature Differentials**

| Description of roof construction*·† | Wt., lb/ft² | kg/m² | U value Btu/(h)(ft²)(°F) | W/(m²)(K) |
|---|---|---|---|---|
| 2 in (50 mm) insulation + steel siding | 7.8 | 38.06 | 0.125 | 0.710 |
| 2 in (50 mm) insulation + 1 in (25 mm) wood‡ | 8.5 | 41.48 | 0.122 | 0.693 |
| 2 in (50 mm) insulation + 2.5 in (62.5 mm) wood‡ | 13.1 | 63.93 | 0.117 | 0.664 |
| 2 in (50 mm) insulation + 4 in (100 mm) wood‡ | 17.8 | 88.86 | 0.113 | 0.642 |
| 2 in (50 mm) insulation + 2 in (50 mm) h.w. concrete | 28.8 | 140.54 | 0.122 | 0.693 |
| 4 in (100 mm) l.w. concrete | 17.8 | 86.86 | 0.213 | 1.209 |
| 2 in (50 mm) insulation + 4 in (100 mm) h.w. concrete | 52.1 | 254.25 | 0.120 | 0.681 |
| 2 in (50 mm) insulation + 6 in (150 mm) h.w. concrete | 75.4 | 367.95 | 0.117 | 0.664 |

| | Outside-air dew point °F | °C | Water-layer thickness in | cm | Roofs covered with water—exposed to sun | | | | | | | | |
|---|---|---|---|---|---|---|---|---|---|---|---|---|---|
| Light construction | 60 | 15.5 | 6 | 15.0 | − 6 | −6 | −1 | 6 | 13 | 17 | 17 | 13 | 7 |
| | | | 1 | 2.5 | −12 | −6 | 4 | 15 | 21 | 22 | 17 | 8 | 0 |
| | | | 0 | 0 | −12 | −4 | 7 | 17 | 23 | 22 | 16 | 5 | −3 |
| | 70 | 21 | 6 | 15.0 | − 1 | 0 | 4 | 11 | 18 | 21 | 21 | 17 | 12 |
| | | | 1 | 2.5 | − 5 | 0 | 10 | 19 | 25 | 26 | 21 | 12 | 5 |
| | | | 0 | 0 | − 5 | 2 | 12 | 21 | 26 | 26 | 19 | 9 | 2 |
| Heavy construction | 60 | 15.5 | 6 | 15.0 | − 3 | −4 | −1 | 4 | 9 | 13 | 15 | 13 | 10 |
| | | | 1 | 2.5 | − 8 | −6 | 1 | 8 | 15 | 18 | 17 | 11 | 6 |
| | | | 0 | 0 | − 9 | −5 | 2 | 10 | 16 | 19 | 16 | 10 | 4 |
| | 70 | 21 | 6 | 15.0 | 2 | 2 | 4 | 9 | 14 | 18 | 20 | 18 | 15 |
| | | | 1 | 2.5 | − 2 | 0 | 6 | 14 | 20 | 23 | 21 | 16 | 11 |
| | | | 0 | 0 | − 2 | 1 | 8 | 16 | 21 | 23 | 21 | 15 | 9 |

*Includes outside surface resistance, ½ in (12.5 mm) membrane and ⅜ in (9 mm) felt on the top and inside surface resistance on the bottom.
†Dark roof (D) = 0.30; light roof (L) = 0.15.
‡Nominal thickness of wood.

*Explanation:* Total heat transmission from solar radiation and temperature difference between outdoor and room air, Btu/(h)(ft²)(W/m²) of roof area = equivalent temperature differential from above table × heat-transmission coefficient for summer, Btu/(h)(ft²) (°F) (W/m² K)

*Application.* These values may be used for all normal air conditioning estimated; usually without correction (except as noted below in latitude 0 to 50° north or south when the load is calculated for the hottest weather.

*Corrections.* The values in the table were calculated for an inside temperature of 75°F (24°C) and an outdoor maximum temperature of 95°F (35°C) with an outdoor daily range of 21°F (12°C). The table remains approximately correct for other outdoor maximums (93 to 102°F) (34 to 39°C) and other outdoor daily ranges (16 to 34°F) (9 to 19°C) provided the outdoor daily average temperature remains approximately 85°F (30°C). If the room air temperature is different from 75°F (24°C) and/or the outdoor daily average temperature is different from 85°F (30°C), the following rules can be applied:

1. For room air temperature less than 75°F (24°C) add the difference between 75°F (24°C) and room air temperature; if greater than 75°F (24°C), subtract the difference.

2. For outdoor daily average temperature less than 85°F (30°C) subtract the difference between 85°F (30°C) and the daily average temperature, if greater than 85°F (30°C), add the difference.

*Attics or other spaces between the roof and ceiling.* If the ceiling is insulated and a fan is used for positive ventilation in the space between the ceiling and roof, the total temperature differential for calculating the room load may be decreased by 25 percent. If the attic space contains a return duct or other air plenum, care should be taken in determining the portion of the heat gain that reaches the ceiling.

*Light colors.* Credit should not be taken for light-colored roofs except where the permanence of light color is established by experience as in rural areas or where there is little smoke.

For solar transmission in other months. The table values of temperature differentials that were calculated for July 21 will be approximately correct for a roof in the following months:

| | North latitude | | South latitude | |
|---|---|---|---|---|
| Latitude, deg | Months | | Latitude, deg | Months |
| 0 | All months | | 0 | All months |
| 10 | All months | | 10 | All months |
| 20 | All months except Nov., Dec., Jan. | | 20 | All months except May, June, July |
| 30 | Mar., Apr., May, June, July, Aug., Sept. | | 30 | Sept., Oct., Nov., Dec., Jan., Feb., March |
| 40 | April, May, June, July, Aug. | | 40 | Oct., Nov., Dec., Jan., Feb. |
| 50 | May, June, July | | 50 | Nov., Dec., Jan. |

| | | | | | | | | | | | | Sun time | | | | | | | | | |
|---|---|---|---|---|---|---|---|---|---|---|---|---|---|---|---|---|---|---|---|---|---|
| | A.M. | | | | | | P.M. | | | | | | | | | | | | | | |
| 8 | | 10 | | 12 | | 2 | | 4 | | 6 | | 8 | | 10 | | 12 | | | | | |
| D | L | D | L | D | L | D | L | D | L | D | L | D | L | D | L | D | L | λ | δ | | |
| | | | | | | Light construction roofs exposed to sun | | | | | | | | | | | | | | | |
| 24 | 8 | 61 | 29 | 88 | 46 | 96 | 53 | 81 | 46 | 48 | 30 | 10 | 8 | 2 | 2 | −3 | −3 | 0.99 | 1 | | |
| 8 | 0 | 41 | 18 | 72 | 36 | 90 | 48 | 88 | 49 | 65 | 38 | 30 | 19 | 9 | 7 | 1 | 0 | 0.93 | 2 | | |
| 1 | −2 | 19 | 6 | 43 | 20 | 65 | 33 | 76 | 41 | 72 | 40 | 53 | 31 | 33 | 20 | 18 | 11 | 0.68 | 4 | | |
| | | | | | | Medium construction roofs exposed to sun | | | | | | | | | | | | | | | |
| 6 | 1 | 13 | 4 | 28 | 12 | 45 | 22 | 58 | 30 | 63 | 34 | 56 | 31 | 43 | 25 | 32 | 18 | 0.48 | 5 | | |
| 2 | −2 | 23 | 9 | 49 | 23 | 70 | 36 | 79 | 43 | 71 | 40 | 49 | 29 | 28 | 17 | 15 | 9 | 0.73 | 3 | | |
| 1 | −3 | 28 | 11 | 59 | 28 | 82 | 43 | 88 | 48 | 74 | 42 | 44 | 27 | 19 | 12 | 6 | 4 | 0.87 | 3 | | |
| | | | | | | Heavy construction roofs exposed to sun | | | | | | | | | | | | | | | |
| 7 | 2 | 15 | 6 | 30 | 13 | 46 | 23 | 58 | 30 | 61 | 33 | 54 | 30 | 41 | 23 | 31 | 17 | 0.45 | 5 | | |
| 15 | 7 | 17 | 7 | 25 | 11 | 36 | 17 | 46 | 23 | 51 | 27 | 50 | 27 | 43 | 24 | 36 | 20 | 0.30 | 6 | | |

SUPPLY OUTLETS.  Supply outlets should be selected on the basis of manufacturers' data. Factors which usually affect the selection of supply outlets are (1) noise, (2) location of outlet, (3) temperature of supply air, and (4) area of diffusion.

RETURN OUTLETS.  Selection of return registers or grilles is usually governed by face velocity.

**Ductwork**  AIR VELOCITY.  Supply and return air ducts and apparatus are sized on the basis of air quantity, within the limitations of allowable friction losses, velocity, and noise.

**Pressure Loss in Duct Systems**  Pressure losses in duct systems are due to friction of the air in contact with the sides of the duct and dynamic losses caused by changes of duct shape or direction and by obstructions to flow.

$$\text{Friction } H_f = f \frac{L}{D}\left(\frac{V}{4,005}\right)^2 \qquad (30)$$

where $H_f$ = head loss due to friction, in $H_2O$
$L$ = length of duct, ft
$D$ = diameter of duct, ft
$V$ = velocity of air, ft/min
$f$ = nondimensional friction coefficient

A comparable equation of SI unit is

$$H_f = f \frac{L}{D} \frac{(V)^2}{2} \rho$$

where $H_f$ = head loss due to friction, N/m²
$L$ = length of duct, m
$D$ = diameter of duct, m
$V$ = velocity of air, m/s
$f$ = nondimensional friction coefficient
$\rho$ = density of air, kg/m³

$$\text{Dynamic losses } H_v = C\left(\frac{V}{4,005}\right)^2 \qquad (31)$$

where $H_v$ = velocity-head loss, in $H_2O$
$C$ = experimentally determined constant
$V = Q/A$
= air velocity, ft/min
$Q$ = airflow rate, ft³/min

$A$ = cross-sectional area of duct, ft²

A comparable equation in SI units is

$$H'_v = C' \frac{(V')^2}{2}$$

where $H'_v$ = velocity-head loss, N/m²
$V'$ = air velocity, m/s
= density of air, kg/m³

When the equation for $H'_v$ is written in the form above, $C' = C$.

**Design Methods**  The equal-friction method is applicable primarily to systems using low or moderate velocities where the velocity head is not an important factor. A friction drop per 100 ft of length is chosen, and the duct mains and branches are all sized on the basis of this friction drop. This will invariably result in higher velocities in the mains, where they can be tolerated, and low velocities in the branches, where they are desirable.

The static-regain method is used for both conventional and high-velocity systems. It is especially applicable in the latter, where the velocity head may be appreciable. In the static-regain method, the static pressure required to give proper airflow through the system outlets is determined, and this pressure is maintained by reducing the velocity at each branch or takeoff, so that the recovery in pressure due to reduction of velocity balances the friction loss in the preceding section of duct. This is possible because of the convertibility of static and velocity pressures. For practical applications it is usually assumed that 50 percent of the velocity pressure available will be converted to static pressure.

$$H_R = 0.5\left(\frac{V_1}{4,005}\right)^2 - \left(\frac{V_2}{4,005}\right)^2 \qquad (32)$$

where $H_R$ = head recovered, in $H_2O$
$V_1$ = system inlet velocity, ft/min
$V_2$ = system outlet velocity, ft/min

# Table 20. Conductivities, Conductances, and Resistances of Building and Insulating Materials (Design Values)

| Material | Description | Density lb/ft³ | Density kg/m³ | Conductivity Btu/(h)(ft²)(in)(°F) | Conductivity W/(m²/m) (K) | Conductance Btu/(ft²)(h)(°F) | Conductance W/(m²) (K) | R Per inch thickness 1/K | R Per m thickness 1/K' | R For thickness listed 1/c | R For thickness listed 1/c' |
|---|---|---|---|---|---|---|---|---|---|---|---|
| Building board, boards, panels, subflooring, sheathing, wood-based panel products | Asbestos-cement board | 120 | 1,922.16 | 4.0 | 0.576 | | | 0.25 | 1.736 | | |
| | Gypsum or plaster board, ½ in (12.5 mm) | 50 | 800.9 | | | 2.25 | 12.78 | | | 0.45 | 0.078 |
| | Plywood | 34 | 544.6 | 0.80 | 0.115 | | | 1.25 | 8.696 | | |
| | Insulating board, Sheathing, regular density, ½ in (12.5 mm) | 18 | 288.3 | | | 0.76 | 4.32 | | | 1.32 | 0.232 |
| | | | | | | 0.76 | 4.32 | | | 1.32 | 0.232 |
| | Sheathing, intermediate density, ½ in (12.5 mm) | 22 | 352.4 | | | 0.82 | 4.66 | | | 1.22 | 0.215 |
| | Tile and lay-in panels, plain or acoustic | 18 | 288.3 | 0.40 | 0.058 | | | 2.50 | 17.241 | | |
| | Hardboard, Medium-density siding, ⁷⁄₁₆ in (11 mm) | 40 | 640.7 | | | 1.49 | 8.46 | | | 0.67 | 0.118 |
| | High density, std. tempered | 63 | 1009.1 | 1.00 | 0.144 | | | 1.00 | 6.944 | | |
| | Particleboard, Medium density | 50 | 800.9 | 0.94 | 0.135 | | | 1.06 | 7.407 | | |
| | Underlayment, ⅝ in (15.5 mm) | 40 | 640.7 | | | 1.22 | 6.93 | | | 0.82 | 0.144 |
| | Wood subfloor, ¾ in (19 mm) | | | | | 1.06 | 6.02 | | | 0.94 | 0.166 |
| Building paper | Vapor-seal, 2 layers of mopped 15-lb felt | | | | | 8.35 | 47.41 | | | 0.12 | 0.021 |
| Finish flooring materials | Carpet and rubber pad | | | | | 0.81 | 4.60 | | | 1.23 | 0.217 |
| | Tile-asphalt, linoleum, vinyl, rubber | | | | | 20.00 | 113.56 | | | 0.05 | 0.009 |
| | Wood, hardwood, ¾ in (19 mm) | | | | | 1.47 | 8.35 | | | 0.68 | 0.120 |
| Board and slabs | Cellular glass | 9 | 144.16 | 0.40 | 0.058 | | | 2.50 | 17.24 | | |
| | Glass fiber, organic bonded | 4–9 | 64.01–144.16 | 0.25 | 0.036 | | | 4.00 | 27.78 | | |
| | Expanded polystyrene extruded, (R-12 exp.) | 2.2 | 35.24 | 0.20 | 0.029 | | | 5.00 | 34.48 | | |
| | Acoustical tile | 18 | 288.32 | 0.35 | 0.050 | | | 2.86 | 20.00 | | |
| | Acoustical tile | 23 | 368.41 | 0.42 | 0.061 | | | 2.38 | 16.39 | | |
| | Acoustical tile ½ in (12.5 mm) | | | | | 0.80 | 4.54 | | | 1.25 | 0.22 |
| | Insulating roof deck, approx. 2 in (50 mm) | | | | | 0.18 | 1.02 | | | 5.56 | 0.98 |
| Loose fill | Perlite, expanded | 5.0–8.0 | 80.09–128.14 | 0.37 | 0.053 | | | 2.70 | 18.87 | | |
| | | 7.0–8.2 | 112.13–131.35 | 0.47 | 0.068 | | | 2.13 | 14.71 | | |
| | Vermiculite (expanded) | 4.0–6.0 | 64.07–96.11 | 0.44 | 0.063 | | | 2.27 | 15.87 | | |
| Roof insulation | Preformed, for use above deck, approx. 2 in (50 mm) | | | | | 0.18 | 1.02 | | | 5.56 | 0.980 |
| Masonry materials, concretes | Cellular glass | 9 | 144.16 | 0.40 | 0.058 | | | 2.50 | 17.24 | | |
| | Lightweight aggregates including expanded shale, clay, slate, slags, cinders, pumice, vermiculite | 120 | 1,922.16 | 5.2 | 0.749 | | | 0.19 | 1.34 | | |
| | | 100 | 1,601.80 | 3.6 | 0.518 | | | 0.28 | 1.93 | | |
| | | 80 | 1,281.44 | 2.5 | 0.360 | | | 0.40 | 2.78 | | |
| | Perlite | 40 | 640.72 | 0.93 | 0.134 | | | 1.08 | 7.46 | | |
| | Sand and gravel or stone aggregate (not dried) | 140 | 2,242.52 | 12.0 | 1.728 | | | 0.08 | 0.58 | | |

| Category | Material | Density, lb/ft³ | Density, kg/m³ | $k$ | $k$ (SI) | $C$ | $C$ (SI) | $1/k$ | $1/k$ (SI) | $1/C$ | $1/C$ (SI) |
|---|---|---|---|---|---|---|---|---|---|---|---|
| Masonry units | Brick, common (150 mm) | 120 | 1,922.16 | 5.0 | 0.720 | | | 0.20 | 1.39 | | |
| | Brick, face | 130 | 2,082.34 | 9.0 | 1.296 | | | 0.11 | 0.77 | | |
| | Clay tile, hollow, 2 cells deep, 6 in | | | | | 0.66 | 3.75 | | | 1.52 | 0.267 |
| | Concrete blocks, three oval core: | | | | | | | | | | |
| |   Sand and gravel aggregate, 8 in (200 mm) | | | | | 0.90 | 5.11 | | | 1.11 | 0.196 |
| |   Cinder aggregate, 8 in (200 mm) | | | | | 0.58 | 3.29 | | | 1.72 | 0.304 |
| |   Lightweight aggregate, 8 in (200 mm) | | | | | 0.50 | 2.84 | | | 2.00 | 0.352 |
| | Concrete blocks, rectangular core, sand and gravel aggregate 2 core, 8 in (200 mm), 36 lb (16.4 kg) | | | | | 0.96 | 5.45 | | | 1.04 | 0.184 |
| | Same with filled cores | | | | | 0.52 | 2.95 | | | 1.93 | 0.339 |
| | Stone, lime, or sand | | | 12.50 | 1.80 | | | 0.08 | 0.56 | | |
| Plastering materials | Cement plaster, sand aggregate (19 mm) | 116 | 1,858.09 | 5.0 | 0.720 | | | 0.20 | 1.39 | | |
| | Gypsum plaster, lightweight aggregate on metal lath, ⅜ in (19 mm) | | | | | 2.13 | 12.09 | | | 0.47 | 0.083 |
| | Sand aggregate | 105 | 1,681.89 | | | 7.70 | 43.72 | | | 0.1 | 0.023 |
| | Sand aggregate on metal lath, ¾ in (19 mm) | | | 5.6 | 0.806 | | | 0.18 | 1.24 | | |
| Roofing | Asbestos-cement shingles | 120 | 1,922.2 | | | 4.76 | 27.03 | | | 0.21 | 0.037 |
| | Asphalt shingles | 70 | 1,121.3 | | | 2.27 | 12.89 | | | 0.44 | 0.078 |
| | Built-up roofing, ⅜ in (9.5 mm) | 70 | 1,121.3 | | | 3.00 | 17.03 | | | 0.33 | 0.059 |
| | Slate, ½ in (12.5 mm) | 70 | 1,121.3 | | | 20.00 | 113.56 | | | 0.05 | 0.009 |
| | Wood shingles, plain and plastic-film-faced | | | | | 1.06 | 6.02 | | | 0.94 | 0.166 |
| Siding materials (on flat surface) | Shingles | | | | | | | | | | |
| |   Asbestos-cement | 120 | 1,922.2 | | | 4.76 | 27.03 | | | 0.21 | 0.037 |
| |   Wood, double, 16 in (400 mm) 12 in (300 mm) exposure | | | | | 0.84 | 4.770 | | | 1.19 | 0.210 |
| |   Wood, plus insulating backer board, ⁵⁄₁₆ in (8 mm) | | | | | 0.71 | 4.031 | | | 1.40 | 0.248 |
| | Siding, wood, bevel, ¾ × 10 in (19 × 250 mm), lapped | | | | | 0.95 | 5.394 | | | 1.05 | 0.185 |
| Woods | Maple, oak, and similar hardwoods | 45 | 720.8 | 1.10 | 0.158 | | | 0.91 | 6.329 | | |
| | Fir, pine, and similar softwoods | 32 | 512.6 | 0.80 | 0.115 | | | 1.25 | 8.696 | | |

**Table 21A. Surface Conductances and Resistances for Air**

| Position of surface | Direction of heat flow | Nonreflective ε = 0.90 | | | | Surface emissivity, reflective ε = 0.20 | | | | Reflective ε = 0.05 | | | |
|---|---|---|---|---|---|---|---|---|---|---|---|---|---|
| | | C | C' | R | R' | C | C' | R | R' | C | C' | R | R' |
| Still air: | | | | | | | | | | | | | |
| Horizontal | Upward | 1.63 | 9.26 | 0.61 | 0.108 | 0.91 | 5.17 | 1.10 | 0.193 | 0.76 | 4.32 | 1.32 | 0.232 |
| Vertical | Horizontal | 1.46 | 8.29 | 0.68 | 0.121 | 0.74 | 4.20 | 1.35 | 0.238 | 0.59 | 3.35 | 1.70 | 0.299 |
| Horizontal | Downward | 1.08 | 6.76 | 0.92 | 0.148 | 0.37 | 2.10 | 2.70 | 0.476 | 0.22 | 1.25 | 4.55 | 0.800 |
| Moving air (any position): | | | | | | | | | | | | | |
| 15 mi/s (24 km/h) wind (for winter) | Any | 6.00 | 34.07 | 0.17 | 0.029 | | | | | | | | |
| 7½ mi/h (12 km/h) wind (for summer) | Any | 4.00 | 22.71 | 0.25 | 0.044 | | | | | | | | |

$C$ = conductance, Btu/(h)(ft²)(°F temp. diff.).
$C'$ = conductance, W/(m²)(K).
$R$ = resistance = $1/C$.
$R'$ = resistance = $1/C'$.

**Table 21B. Reflectivity and Emissivity Values of Various Surfaces and Effective Emissivities of Air Spaces**

| Surface | Reflectivity, % | Average emissivity $\epsilon$ | Effective emissivity E of air space | |
|---|---|---|---|---|
| | | | With one surface having emissivity $\epsilon$ and other 0.90 | With both surfaces of emissivity $\epsilon$ |
| Aluminum foil, bright | 92–97 | 0.05 | 0.05 | 0.03 |
| Aluminum sheet | 80–95 | 0.12 | 0.12 | 0.06 |
| Aluminum-coated paper, polished | 75–84 | 0.20 | 0.20 | 0.11 |
| Steel, galvanized, bright | 70–80 | 0.25 | 0.24 | 0.15 |
| Aluminum paint | 30–70 | 0.50 | 0.47 | 0.35 |
| Building materials: wood, paper, glass, masonry, nonmetallic paints | 5–15 | 0.90 | 0.82 | 0.82 |

SOURCE: Reprinted by permission from ASHRAE "Handbook of Fundamentals," 1972.
For ventilated attics or spaces above ceilings under summer conditions (heat flow down).
Conductances are for surfaces of the stated emissivity facing virtual blackbody surroundings at the same temperature as the ambient air. Values are based on a surface-air temperature difference of 10°F (5.5°C) and for surface temperature of 70°F (21°C).

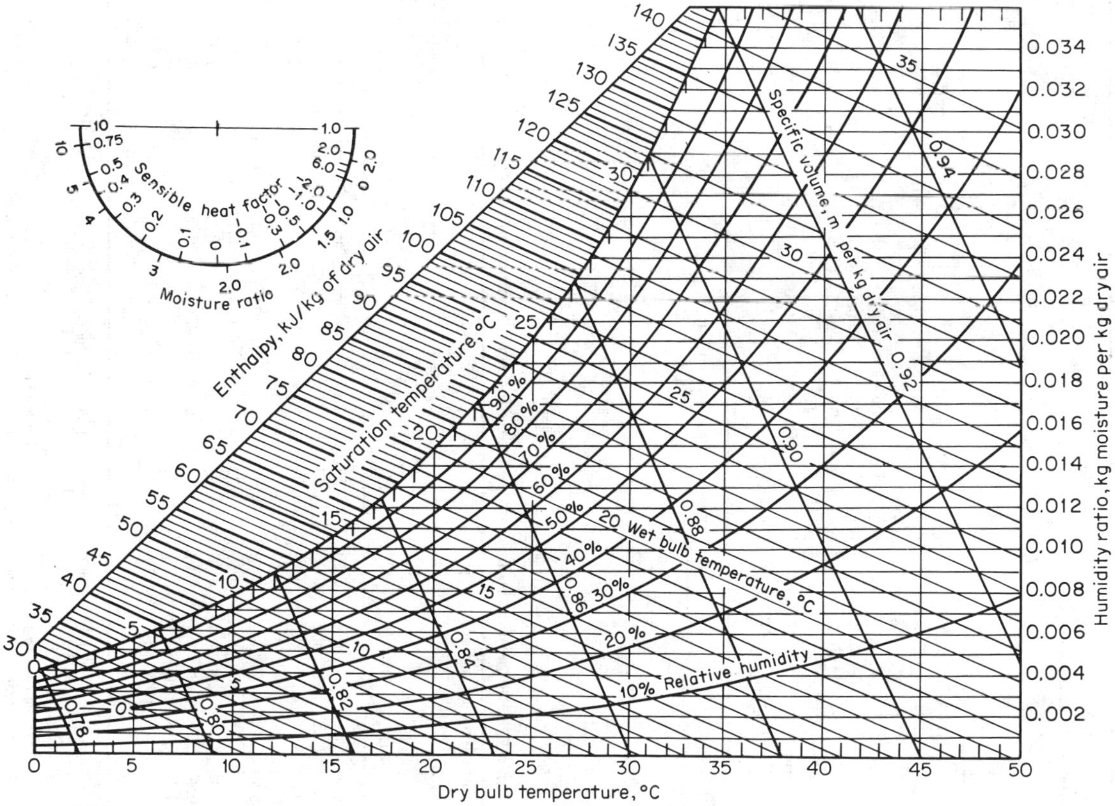

**Fig. 13**  Psychrometric chart in SI units. *(Adapted, by permission, from material copyrighted by Business News Publishing Company, 1975.)*

**Table 22A. Determination of U Value Resulting from Addition of Insulation to Uninsulated Building Sections**

| U value of roof without roof-deck insulation* | | Conductance C (C') of roof-deck insulation | | | | | | | | | | | |
|---|---|---|---|---|---|---|---|---|---|---|---|---|---|
| | | C = 0.12, C' = 0.68 | | C = 0.15, C' = 0.85 | | C = 0.19, C' = 1.08 | | C = 0.24, C' = 1.36 | | C = 0.36, C' = 2.04 | | C = 0.72, C' = 4.09 | |
| U | U' | U | U' | U | U' | U | U' | U | U' | U | U' | U | U' |
| 0.10 | 0.57 | 0.05 | 0.28 | 0.06 | 0.34 | 0.07 | 0.40 | 0.07 | 0.40 | 0.08 | 0.45 | 0.09 | 0.51 |
| 0.15 | 0.85 | 0.07 | 0.40 | 0.08 | 0.45 | 0.08 | 0.45 | 0.09 | 0.51 | 0.11 | 0.62 | 0.12 | 0.68 |
| 0.20 | 1.14 | 0.08 | 0.45 | 0.09 | 0.51 | 0.10 | 0.57 | 0.11 | 0.62 | 0.13 | 0.74 | 0.16 | 0.91 |
| 0.25 | 1.42 | 0.08 | 0.45 | 0.09 | 0.51 | 0.11 | 0.62 | 0.12 | 0.68 | 0.15 | 0.85 | 0.19 | 1.08 |
| 0.30 | 1.70 | 0.09 | 0.51 | 0.10 | 0.57 | 0.12 | 0.68 | 0.13 | 0.74 | 0.16 | 0.91 | 0.21 | 1.19 |
| 0.35 | 1.99 | 0.09 | 0.51 | 0.10 | 0.57 | 0.12 | 0.68 | 0.14 | 0.79 | 0.18 | 1.02 | 0.24 | 1.36 |
| 0.40 | 2.27 | 0.09 | 0.51 | 0.11 | 0.62 | 0.13 | 0.74 | 0.15 | 0.85 | 0.19 | 1.08 | 0.26 | 1.47 |
| 0.50 | 2.84 | 0.10 | 0.57 | 0.12 | 0.68 | 0.14 | 0.79 | 0.16 | 0.91 | 0.21 | 1.19 | 0.29 | 1.65 |
| 0.60 | 3.41 | 0.10 | 0.57 | 0.12 | 0.68 | 0.14 | 0.79 | 0.17 | 0.96 | 0.22 | 1.25 | 0.33 | 1.87 |
| 0.70 | 3.97 | 0.10 | 0.57 | 0.12 | 0.68 | 0.15 | 0.85 | 0.18 | 1.02 | 0.24 | 1.36 | 0.35 | 1.99 |

U and C expressed in U.S. customary units, U' and C' in SI units.
*Interpolation or mild extrapolation may be used.

**Table 22B. Determination of U Value Resulting from Addition of Insulation to Any Given Building Section**

| Given building section property*,† | | | | Added R‡§ (R') | | | | | | | | | | | | | |
|---|---|---|---|---|---|---|---|---|---|---|---|---|---|---|---|---|---|
| | | | | R = 4, R' = 0.70 | | R = 6, R' = 1.05 | | R = 8, R' = 1.41 | | R = 12, R' = 2.13 | | R = 16, R' = 2.86 | | R = 20, R' = 3.57 | | R = 24, R' = 4.17 | |
| U | U' | R | R' | U | U' | U | U' | U | U' | U | U' | U | U' | U | U' | U | U' |
| 1.00 | 5.68 | 1.00 | 0.18 | 0.20 | 1.13 | 0.14 | 0.79 | 0.11 | 0.62 | 0.08 | 0.45 | 0.06 | 0.34 | 0.05 | 0.28 | 0.04 | 0.23 |
| 0.90 | 5.11 | 1.11 | 0.20 | 0.20 | 1.13 | 0.14 | 0.79 | 0.11 | 0.62 | 0.08 | 0.45 | 0.06 | 0.34 | 0.05 | 0.28 | 0.04 | 0.23 |
| 0.80 | 4.54 | 1.25 | 0.22 | 0.19 | 1.08 | 0.14 | 0.79 | 0.11 | 0.62 | 0.08 | 0.45 | 0.06 | 0.34 | 0.05 | 0.28 | 0.04 | 0.23 |
| 0.70 | 3.97 | 1.43 | 0.25 | 0.19 | 1.08 | 0.13 | 0.74 | 0.11 | 0.62 | 0.07 | 0.40 | 0.06 | 0.34 | 0.05 | 0.28 | 0.04 | 0.23 |
| 0.60 | 3.41 | 1.67 | 0.29 | 0.19 | 1.08 | 0.13 | 0.74 | 0.10 | 0.57 | 0.07 | 0.40 | 0.06 | 0.34 | 0.05 | 0.28 | 0.04 | 0.23 |
| 0.50 | 2.84 | 2.00 | 0.35 | 0.18 | 1.02 | 0.13 | 0.74 | 0.10 | 0.57 | 0.07 | 0.40 | 0.06 | 0.34 | 0.05 | 0.28 | 0.04 | 0.23 |
| 0.40 | 2.27 | 2.50 | 0.44 | 0.16 | 0.91 | 0.12 | 0.68 | 0.10 | 0.57 | 0.07 | 0.40 | 0.05 | 0.28 | 0.05 | 0.28 | 0.04 | 0.23 |
| 0.30 | 1.70 | 3.33 | 0.59 | 0.14 | 0.79 | 0.11 | 0.62 | 0.09 | 0.51 | 0.07 | 0.40 | 0.05 | 0.28 | 0.04 | 0.23 | 0.04 | 0.23 |
| 0.20 | 1.14 | 5.00 | 0.88 | 0.11 | 0.62 | 0.09 | 0.51 | 0.08 | 0.45 | 0.06 | 0.34 | 0.05 | 0.28 | 0.04 | 0.23 | 0.03 | 0.17 |
| 0.10 | 0.57 | 10.00 | 1.75 | 0.06 | 0.34 | 0.06 | 0.34 | 0.06 | 0.34 | 0.05 | 0.28 | 0.04 | 0.23 | 0.04 | 0.23 | 0.03 | 0.17 |
| 0.08 | 0.45 | 12.50 | 2.22 | 0.06 | 0.34 | 0.05 | 0.28 | 0.05 | 0.28 | 0.04 | 0.23 | 0.04 | 0.23 | 0.03 | 0.17 | 0.03 | 0.17 |

U and R expressed in U.S. customary units, U' and R' in SI units.
SOURCE: Reprinted by permission from ASHRAE "Handbook of Fundamentals," 1972.
*For U or R values not shown in table, interpolate as necessary.
†Enter column 1 with U or R of the design building section.
‡Under appropriate column heading for Added R, find U value of resulting design section.
§If the insulation occupies a previously considered air space, an adjustment must be made in the given building section R value.

**Example of Calculating Coefficients of Transmission** *U*
*U* and *R* expressed in U.S. customary units, *U'* and *R'* in SI units

Example 1. Coefficients of transmission *U* of flat masonry roofs with built-up roofing

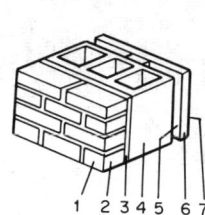

| Construction (*heat flow up*) | Resistance | |
|---|---|---|
| | *R* | *R'* |
| 1. Outside surface (15 mi/h wind (24 km/h) | 0.17 | 0.03 |
| 2. Built-up roofing ⅜ in (9 mm) | 0.33 | 0.06 |
| 3. Roof insulation (none) | | |
| 4. Concrete slab (lt. wt. agg.) (2 in) (50 mm) | 2.22 | 0.39 |
| 5. Corrugated metal | 0 | 0 |
| 6. Air space | 0.85 | 0.15 |
| 7. Metal lath and ¾ in (19 mm) plaster (lt. wt. agg.) | 0.47 | 0.08 |
| 8. Inside surface (still air) | 0.61 | 0.11 |
| Total resistance | 4.65 | 0.82 |

$U = 1/R = 1/4.65 = 0.22$
$U' = 1/R' = 1/0.82 = 1.22$

Example 2. Coefficients of transmission *U* of masonry walls

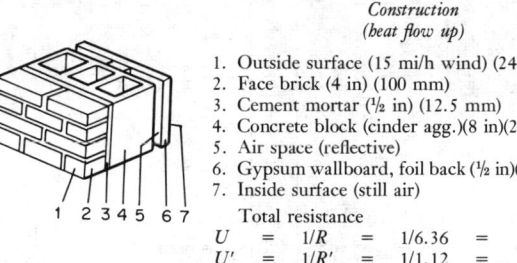

| Construction (*heat flow up*) | Resistance | |
|---|---|---|
| | *R* | *R'* |
| 1. Outside surface (15 mi/h wind) (24 km/h) | 0.17 | 0.03 |
| 2. Face brick (4 in) (100 mm) | 0.44 | 0.08 |
| 3. Cement mortar (½ in) (12.5 mm) | 0.10 | 0.02 |
| 4. Concrete block (cinder agg.)(8 in)(200 mm) | 1.72 | 0.30 |
| 5. Air space (reflective) | 2.80 | 0.49 |
| 6. Gypsum wallboard, foil back (½ in)(12.5 mm) | 0.45 | 0.08 |
| 7. Inside surface (still air) | 0.68 | 0.12 |
| Total resistance | 6.36 | 1.12 |

$U = 1/R = 1/6.36 = 0.16$
$U' = 1/R' = 1/1.12 = 0.89$

**Table 23. Solar Position and Intensity; Solar-Heat-Gain Factors* for 40° North Latitude**

| Date | Solar time A.M. | Solar position Alt. | Solar position Azimuth | Direct normal irradiation, Btu/(h)(ft²)† | N | NE | E | SE | S | SW | W | NW | Hor. | Solar time P.M. |
|---|---|---|---|---|---|---|---|---|---|---|---|---|---|---|
| Jan. 21 | 8 | 8.1 | 55.3 | 141 | 5 | 17 | 111 | 133 | 75 | 5 | 5 | 5 | 13 | 4 |
| | 9 | 16.8 | 44.0 | 238 | 11 | 12 | 154 | 224 | 160 | 13 | 11 | 11 | 54 | 3 |
| | 10 | 23.8 | 30.9 | 274 | 16 | 16 | 123 | 241 | 213 | 51 | 16 | 16 | 96 | 2 |
| | 11 | 28.4 | 16.0 | 289 | 18 | 18 | 61 | 222 | 244 | 118 | 18 | 18 | 123 | 1 |
| | 12 | 30.0 | 0.0 | 293 | 19 | 19 | 20 | 179 | 254 | 179 | 20 | 19 | 133 | 12 |
| | Half-day totals | | | | 59 | 68 | 449 | 903 | 815 | 271 | 59 | 59 | 353 | |
| Feb. 21 | 7 | 4.3 | 72.1 | 55 | 1 | 22 | 50 | 47 | 13 | 1 | 1 | 1 | 3 | 5 |
| | 8 | 14.8 | 61.6 | 219 | 10 | 50 | 183 | 199 | 94 | 10 | 10 | 10 | 43 | 4 |
| | 9 | 24.3 | 49.7 | 271 | 16 | 22 | 186 | 245 | 157 | 17 | 16 | 16 | 98 | 3 |
| | 10 | 32.1 | 35.4 | 293 | 20 | 21 | 142 | 247 | 203 | 38 | 20 | 20 | 143 | 2 |
| | 11 | 37.3 | 18.6 | 303 | 23 | 23 | 71 | 219 | 231 | 103 | 23 | 23 | 171 | 1 |
| | 12 | 39.2 | 0.0 | 306 | 24 | 24 | 25 | 170 | 241 | 170 | 25 | 24 | 180 | 12 |
| | Half-day totals | | | | 81 | 144 | 634 | 1035 | 813 | 250 | 81 | 81 | 546 | |
| Mar. 21 | 7 | 11.4 | 80.2 | 171 | 8 | 93 | 163 | 135 | 21 | 8 | 8 | 8 | 26 | 5 |
| | 8 | 22.5 | 69.6 | 250 | 15 | 91 | 218 | 211 | 73 | 15 | 15 | 15 | 85 | 4 |
| | 9 | 32.8 | 57.3 | 281 | 21 | 46 | 203 | 236 | 128 | 21 | 21 | 21 | 143 | 3 |
| | 10 | 41.6 | 41.9 | 297 | 25 | 26 | 153 | 229 | 171 | 28 | 25 | 25 | 186 | 2 |
| | 11 | 47.7 | 22.6 | 304 | 28 | 28 | 78 | 198 | 197 | 77 | 28 | 28 | 213 | 1 |
| | 12 | 50.0 | 0.0 | 306 | 28 | 28 | 30 | 145 | 206 | 145 | 30 | 28 | 223 | 12 |
| | Half-day totals | | | | 112 | 310 | 849 | 1100 | 692 | 218 | 112 | 112 | 764 | |
| Apr. 21 | 6 | 7.4 | 98.9 | 89 | 11 | 72 | 88 | 52 | 5 | 4 | 4 | 4 | 11 | 6 |
| | 7 | 18.9 | 89.5 | 207 | 16 | 141 | 201 | 143 | 16 | 14 | 14 | 14 | 61 | 5 |
| | 8 | 30.3 | 79.3 | 253 | 22 | 128 | 225 | 189 | 41 | 21 | 21 | 21 | 124 | 4 |
| | 9 | 41.3 | 67.2 | 275 | 26 | 80 | 203 | 204 | 83 | 26 | 26 | 26 | 177 | 3 |
| | 10 | 51.2 | 51.4 | 286 | 30 | 37 | 153 | 194 | 121 | 32 | 30 | 30 | 218 | 2 |
| | 11 | 58.7 | 29.2 | 292 | 33 | 34 | 81 | 161 | 146 | 52 | 33 | 33 | 244 | 1 |
| | 12 | 61.6 | 0.0 | 294 | 33 | 33 | 36 | 108 | 155 | 108 | 36 | 33 | 253 | 12 |
| | Half-day totals | | | | 153 | 509 | 969 | 1003 | 489 | 196 | 146 | 145 | 962 | |
| May 21 | 5 | 1.9 | 114.7 | 1 | 0 | 0 | 0 | 0 | 0 | 0 | 0 | 0 | 0 | 7 |
| | 6 | 12.7 | 105.6 | 143 | 35 | 128 | 141 | 71 | 10 | 10 | 10 | 10 | 30 | 6 |
| | 7 | 24.0 | 96.6 | 216 | 28 | 165 | 209 | 131 | 20 | 18 | 18 | 18 | 87 | 5 |
| | 8 | 35.4 | 87.2 | 249 | 27 | 149 | 220 | 164 | 29 | 25 | 25 | 25 | 146 | 4 |
| | 9 | 46.8 | 76.0 | 267 | 31 | 105 | 197 | 175 | 53 | 30 | 30 | 30 | 196 | 3 |
| | 10 | 57.5 | 60.9 | 277 | 34 | 54 | 148 | 163 | 83 | 35 | 34 | 34 | 234 | 2 |
| | 11 | 66.2 | 37.1 | 282 | 36 | 38 | 81 | 130 | 105 | 42 | 36 | 36 | 258 | 1 |
| | 12 | 70.0 | 0.0 | 284 | 37 | 37 | 40 | 82 | 112 | 82 | 40 | 37 | 265 | 12 |
| | Half-day totals | | | | 203 | 643 | 1002 | 874 | 356 | 194 | 171 | 170 | 1083 | |

**June 21**

|  |  |  |  |  |  |  |  |  |  |  |  |  |  |
|---|---|---|---|---|---|---|---|---|---|---|---|---|---|
| 5 | 117.3 | 4.2 | 21 | 10 | 21 | 20 | 6 | 1 | 1 | 1 | 1 | 2 | 7 |
| 6 | 108.4 | 14.8 | 154 | 47 | 142 | 151 | 70 | 12 | 12 | 12 | 12 | 39 | 6 |
| 7 | 99.7 | 26.0 | 215 | 37 | 172 | 207 | 122 | 21 | 20 | 20 | 20 | 97 | 5 |
| 8 | 90.7 | 37.4 | 246 | 29 | 156 | 215 | 152 | 29 | 26 | 26 | 26 | 153 | 4 |
| 9 | 80.2 | 48.8 | 262 | 33 | 113 | 192 | 161 | 45 | 31 | 31 | 31 | 201 | 3 |
| 10 | 65.8 | 59.8 | 272 | 35 | 62 | 145 | 148 | 69 | 36 | 35 | 35 | 237 | 2 |
| 11 | 41.9 | 69.2 | 276 | 37 | 40 | 80 | 116 | 88 | 42 | 37 | 37 | 260 | 1 |
| 12 | 0.0 | 73.5 | 278 | 38 | 38 | 41 | 71 | 95 | 80 | 41 | 38 | 267 | 12 |
| Half-day totals |  |  |  | 242 | 714 | 1019 | 810 | 311 | 197 | 181 | 180 | 1121 |  |

**July 21**

|  |  |  |  |  |  |  |  |  |  |  |  |  |  |
|---|---|---|---|---|---|---|---|---|---|---|---|---|---|
| 5 | 115.2 | 2.3 | 2 | 0 | 2 | 1 | 0 | 0 | 0 | 0 | 0 | 0 | 7 |
| 6 | 106.1 | 13.1 | 137 | 37 | 125 | 137 | 68 | 10 | 10 | 10 | 10 | 31 | 6 |
| 7 | 97.2 | 24.3 | 208 | 30 | 163 | 204 | 127 | 20 | 19 | 19 | 19 | 88 | 5 |
| 8 | 87.8 | 35.8 | 241 | 28 | 148 | 216 | 160 | 29 | 26 | 26 | 26 | 145 | 4 |
| 9 | 76.7 | 47.2 | 259 | 32 | 106 | 194 | 170 | 52 | 31 | 31 | 31 | 194 | 3 |
| 10 | 61.7 | 57.9 | 269 | 35 | 56 | 146 | 159 | 80 | 36 | 35 | 35 | 231 | 2 |
| 11 | 37.9 | 66.7 | 274 | 37 | 39 | 81 | 127 | 102 | 42 | 37 | 37 | 255 | 1 |
| 12 | 0.0 | 70.6 | 276 | 38 | 38 | 41 | 80 | 109 | 80 | 41 | 38 | 262 | 12 |
| Half-day totals |  |  |  | 211 | 645 | 986 | 850 | 347 | 197 | 177 | 176 | 1074 |  |

**Aug. 21**

|  |  |  |  |  |  |  |  |  |  |  |  |  |  |
|---|---|---|---|---|---|---|---|---|---|---|---|---|---|
| 6 | 99.5 | 7.9 | 80 | 12 | 67 | 82 | 48 | 5 | 5 | 5 | 5 | 11 | 6 |
| 7 | 90.0 | 19.3 | 195 | 17 | 135 | 191 | 135 | 17 | 15 | 15 | 15 | 62 | 5 |
| 8 | 79.9 | 30.7 | 236 | 23 | 126 | 216 | 180 | 40 | 22 | 22 | 22 | 122 | 4 |
| 9 | 67.9 | 41.8 | 259 | 28 | 82 | 197 | 196 | 79 | 28 | 28 | 28 | 174 | 3 |
| 10 | 52.1 | 51.7 | 271 | 32 | 40 | 149 | 187 | 116 | 34 | 32 | 32 | 213 | 2 |
| 11 | 29.7 | 59.3 | 277 | 34 | 35 | 81 | 156 | 140 | 52 | 34 | 34 | 238 | 1 |
| 12 | 0.0 | 62.3 | 279 | 35 | 35 | 38 | 105 | 149 | 105 | 38 | 35 | 247 | 12 |
| Half-day totals |  |  |  | 161 | 503 | 936 | 961 | 471 | 202 | 154 | 153 | 945 |  |

**Sept. 21**

|  |  |  |  |  |  |  |  |  |  |  |  |  |  |
|---|---|---|---|---|---|---|---|---|---|---|---|---|---|
| 7 | 80.2 | 11.4 | 149 | 8 | 84 | 146 | 121 | 21 | 8 | 8 | 8 | 25 | 5 |
| 8 | 69.6 | 22.5 | 230 | 16 | 87 | 205 | 199 | 71 | 16 | 16 | 16 | 82 | 4 |
| 9 | 57.3 | 32.8 | 263 | 22 | 47 | 195 | 226 | 124 | 23 | 22 | 22 | 138 | 3 |
| 10 | 41.9 | 41.6 | 279 | 26 | 28 | 148 | 221 | 165 | 30 | 26 | 26 | 180 | 2 |
| 11 | 22.6 | 47.7 | 287 | 29 | 29 | 77 | 192 | 191 | 77 | 29 | 29 | 206 | 1 |
| 12 | 0.0 | 50.0 | 290 | 30 | 30 | 32 | 141 | 200 | 141 | 32 | 30 | 215 | 12 |
| Half-day totals |  |  |  | 116 | 300 | 803 | 1045 | 672 | 221 | 117 | 116 | 738 |  |

**Oct. 21**

|  |  |  |  |  |  |  |  |  |  |  |  |  |  |
|---|---|---|---|---|---|---|---|---|---|---|---|---|---|
| 7 | 72.3 | 4.5 | 48 | 1 | 20 | 45 | 41 | 12 | 1 | 1 | 1 | 3 | 5 |
| 8 | 61.9 | 15.0 | 203 | 10 | 49 | 173 | 187 | 88 | 10 | 10 | 10 | 43 | 4 |
| 9 | 49.8 | 24.5 | 257 | 17 | 23 | 180 | 235 | 151 | 18 | 17 | 17 | 96 | 3 |
| 10 | 35.6 | 32.4 | 280 | 21 | 22 | 139 | 238 | 196 | 38 | 21 | 21 | 140 | 2 |
| 11 | 18.7 | 37.6 | 290 | 23 | 23 | 70 | 212 | 224 | 100 | 23 | 23 | 167 | 1 |
| 12 | 0.0 | 39.5 | 293 | 24 | 24 | 26 | 165 | 234 | 165 | 26 | 24 | 177 | 12 |
| Half-day totals |  |  |  | 83 | 143 | 610 | 989 | 783 | 245 | 84 | 83 | 535 |  |

**Table 23. Solar Position and Intensity; Solar-Heat-Gain Factors* for 40° North Latitude (Cont.)**

| Date | A.M. | Altitude | Azimuth | Direct Normal | N | NW | W | SW | S | SE | E | NE | Hor. | P.M. |
|---|---|---|---|---|---|---|---|---|---|---|---|---|---|---|
| Nov. 21 | 8 | 8.2 | 55.4 | 136 | 5 | 17 | 107 | 128 | 72 | 5 | 5 | 5 | 14 | 4 |
| | 9 | 17.0 | 44.1 | 232 | 12 | 13 | 151 | 219 | 156 | 13 | 12 | 12 | 54 | 3 |
| | 10 | 24.0 | 31.0 | 267 | 16 | 16 | 122 | 237 | 209 | 50 | 16 | 16 | 96 | 2 |
| | 11 | 28.6 | 16.1 | 283 | 19 | 19 | 61 | 218 | 240 | 116 | 19 | 19 | 123 | 1 |
| | 12 | 30.2 | 0.0 | 287 | 19 | 19 | 21 | 176 | 250 | 176 | 21 | 19 | 132 | 12 |
| | Half-day totals | | | | 61 | 71 | 442 | 884 | 798 | 267 | 62 | 61 | 353 | |
| Dec. 21 | 8 | 5.5 | 53.0 | 88 | 2 | 7 | 67 | 83 | 49 | 3 | 2 | 2 | 6 | 4 |
| | 9 | 14.0 | 41.9 | 217 | 9 | 10 | 135 | 205 | 151 | 12 | 9 | 9 | 39 | 3 |
| | 10 | 20.7 | 29.4 | 261 | 14 | 14 | 113 | 232 | 210 | 55 | 14 | 14 | 77 | 2 |
| | 11 | 25.0 | 15.2 | 279 | 16 | 16 | 56 | 217 | 242 | 120 | 16 | 16 | 103 | 1 |
| | 12 | 26.6 | 0.0 | 284 | 17 | 17 | 18 | 177 | 253 | 177 | 18 | 17 | 113 | 12 |
| | Half-day totals | | | | 49 | 54 | 380 | 831 | 781 | 273 | 50 | 49 | 282 | |
| | | | | | N | NW | W | SW | S | SE | E | NE | Hor. | ←P.M. |

SOURCE: Reprinted by permission from ASHRAE "Handbook of Fundamentals," 1972.
*Total solar heat gains for DS (⅛ in) (3 mm) sheet glass. Based on a ground reflectance of 0.20.
†To convert Btu/(h)(ft²) to W/m², multiply by 3.15.

**Table 24. Shading Coefficients for Single Glass and Insulating Glass***

| | | Single glass | | |
|---|---|---|---|---|
| | Nominal thickness,† in (mm) | | Shading coefficient | |
| Type of glass | | Solar trans.† | $b_0 = 4.0$ | $b_0 = 3.0$ |
| Regular sheet | ³⁄₃₂, ⅛ (2,3) | 0.87 | 1.00 | 1.00 |
| Regular plate/ | ¼ (6) | 0.80 | 0.95 | 0.97 |
| float | ⅜ (9) | 0.75 | 0.91 | 0.93 |
| | ½ (12) | 0.71 | 0.88 | 0.91 |
| Gray sheet | ⅛ (3) | 0.59 | 0.78 | 0.80 |
| | ³⁄₁₆ (5) | 0.74 | 0.90 | 0.92 |
| | ⁷⁄₃₂ (6) | 0.45 | 0.66 | 0.70 |
| | ⁷⁄₃₂ (6) | 0.71 | 0.88 | 0.90 |
| | ¼ | 0.67 | 0.86 | 0.88 |
| Heat-absorbing | ³⁄₁₆ (5) | 0.52 | 0.72 | 0.75 |
| plate/float§ | ¼ (6) | 0.47 | 0.70 | 0.74 |
| | ⅜ (9) | 0.33 | 0.56 | 0.61 |
| | ½ (12) | 0.24 | 0.50 | 0.57 |

| | | Insulating glass* | | | |
|---|---|---|---|---|---|
| | Nominal thickness,‡ in (mm) | Solar trans.† | | Shading coefficient | |
| Type of glass | | Outer pane | Inner pane | $b_0 = 4.0$ | $b_0 = 3.0$ |
| Regular sheet out, regular sheet in | ³⁄₃₂, ⅛ (2,3) | 0.87 | 0.87 | 0.90 | 0.90 |
| Regular plate/float out, regular plate/float in | ¼ (6) | 0.80 | 0.80 | 0.83 | 0.83 |
| Heat-absorbing plate/float out, regular plate/float in | ¼ (6) | 0.46 | 0.80 | 0.56 | 0.58 |

SOURCE: Reprinted by permission from ASHRAE "Handbook of Fundamentals," 1972.
*Refers to factory-fabricated units with ³⁄₁₆(5), ¼ (6), or ½ (12) in air space or to prime windows plus storm windows.
†Refer to manufacturer's literature for values.
‡Thickness of each pane of glass, not thickness of assembled unit.
§Refers to gray, bronze, and green tinted heat-absorbing plate/float glass.

**Table 25. Shading Coefficients for Single Glass with Indoor Shading by Venetian Blinds and Roller Shades**

| | | | Type of shading | | | | |
| | | | Venetian blinds | | Roller shade | | |
| | | | | | Opaque | | Translucent |
| Type of glass | Nominal thickness,* in (mm) | Solar trans.† | Medium | Light | Dark | White | Light |
|---|---|---|---|---|---|---|---|
| Regular sheet | ³⁄₃₂–¹⁄₄ (2–6) | 0.87–0.80 | | | | | |
| Regular plate/float | ¹⁄₄–¹⁄₂ (6–12) | 0.80–0.71 | | | | | |
| Regular pattern | ¹⁄₈–⁹⁄₃₂ (3–7) | 0.87–0.79 | 0.64 | 0.55 | 0.59 | 0.25 | 0.39 |
| Heat-absorbing pattern | ¹⁄₈ (3) | | | | | | |
| Gray sheet | ³⁄₁₆, ⁷⁄₃₂ | 0.74, 0.71 | | | | | |
| Heat-absorbing plate/float‡ | ³⁄₁₆, ¹⁄₄ (5,6) | 0.46 | | | | | |
| Heat-absorbing pattern | ³⁄₁₆, ¹⁄₄ (5,6) | | 0.57 | 0.53 | 0.45 | 0.30 | 0.36 |
| Gray sheet | ¹⁄₈, ⁷⁄₃₂ (3,5) | 0.59, 0.45 | | | | | |
| Heat-absorbing plate/float or pattern | | 0.44–0.30 | | | | | |
| Heat-absorbing plate/float‡ | ³⁄₈ (10) | 0.34 | 0.54 | 0.52 | 0.40 | 0.28 | 0.32 |
| Heat-absorbing plate or pattern | | 0.29–0.15 0.24 | 0.42 | 0.40 | 0.36 | 0.28 | 0.31 |
| Reflective coated glass SC§ = 0.30 | | | 0.25 | 0.23 | | | |
| 0.40 | | | 0.33 | 0.29 | | | |
| 0.50 | | | 0.42 | 0.38 | | | |
| 0.60 | | | 0.50 | 0.44 | | | |

SOURCE: Reprinted by permission from ASHRAE "Handbook of Fundamentals," 1972.
*Refer to manufacturer's literature for values.
†For vertical blinds with opaque white and beige louvers in the tightly closed position, SC is 0.25 and 0.29 when used with glass of 0.71 to 0.80 transmittance.
‡Refers to gray, bronze, and green tinted heat-absorbing plate/float glass.
§Shading coefficient for glass with no shading device.

**Table 26. Shading Coefficients for Insulating Glass* with Indoor Shading by Venetian Blinds and Roller Shades**

| | | | | Type of shading | | | | |
| | | | | Venetian blinds‡ | | Roller shade | | |
| | | Solar trans† | | | | Opaque | | Translucent |
| Type of glass | Nominal thickness, each light in (mm) | Outer pane | Inner pane | Medium | Light | Dark | White | Light |
|---|---|---|---|---|---|---|---|---|
| Regular sheet out, regular sheet in | ³⁄₃₂, ¹⁄₈ (2,3) | 0.87 | 0.87 | | | | | |
| Regular plate/float out, Regular plate/float in | ¹⁄₄ (6) | 0.80 | 0.80 | 0.57 | 0.51 | 0.60 | 0.25 | 0.37 |
| Heat-absorbing plate/float§ out, regular plate/float in | ¹⁄₄ (6) | 0.46 | 0.80 | 0.39 | 0.36 | 0.40 | 0.22 | 0.30 |
| Reflective coated glass SC¶ = 0.20 | | | | 0.19 | 0.18 | | | |
| 0.30 | | | | 0.27 | 0.26 | | | |
| 0.40 | | | | 0.34 | 0.33 | | | |

*Refers to factory-fabricated units with ³⁄₁₆ (2), ¹⁄₄ (3), or ¹⁄₂ (12) in air space, or to prime windows plus storm windows.
†Refer to manufacturer's literature for exact values.
‡For vertical blinds with opaque white or beige louvers, tightly closed, SC is approximately the same as for opaque white roller shades.
§Refers to bronze or green tinted heat-absorbing plate/float glass.
¶Shading coefficient for glass with no shading device.

**Table 27. Shading Coefficients for Double Glazing with Between-Glass Shading**

| Type of glass | Nominal thickness, each pane | Solar trans* | | Description of air space | Type of shading | | |
| | | Outer pane | Inner pane | | Venetian blinds | | Louvered sun screen |
| | | | | | Light | Medium | |
| Regular sheet out, regular sheet in | 3/32, 1/8 (2,3) | 0.87 | 0.87 | Shade in contact with glass or shade separated from glass by air space | 0.33 | 0.36 | 0.43 |
| Regular plate out, regular plate in | 1/4 (6) | 0.80 | 0.80 | Shade in contact with glass voids filled with plastic | | | 0.49 |
| Heat-absorbing, plate/float† out, | 1/4 (6) | 0.46 | 0.80 | Shade in contact with glass or shade separated from glass by air space | 0.28 | 0.30 | 0.37 |
| Regular plate in | | | | Shade in contact with glass voids filled with plastic | | | 0.41 |

SOURCE: Reprinted by permission from ASHRAE "Handbook of Fundamentals," 1972.
*Refer to manufacturer's literature for exact values.
†Refers to gray, bronze, and green tinted heat-absorbing plate/float glass.

**Table 28. Overall Coefficients of Heat Transmission (U Values) for Fenestration under Summer Conditions (7.5 mi/h (12 km/h) Wind Outdoors, Still Air Indoors)**

| Type of glass | U value | |
| | No shading | Internal shading* |
| --- | --- | --- |
| Any uncoated single glass‡ | 1.06 | 0.81 |
| Insulating glass,‡ uncoated   3/16 in (5 mm) air space | 0.66 | 0.54 |
| 1/4 in (6) air space | 0.65 | 0.52 |
| 3/8 in (10) air space | 0.61 | 0.50 |
| 1/2 in (12) air space | 0.59 | 0.48 |
| Prime window plus storm window, air space 1 in (25 mm) or more | 0.54† | 0.47† |

| | No supplementary shading |
| --- | --- |
| Double glazing with between-glass shading: | |
| Louvered sun screen separated by air space | 0.63 |
| Venetian blinds, closed, in air space | 0.44 |
| Glass-block panels§: | |
| Types I and II | 0.56 |
| Types II, III, and IIIA | 0.48 |

SOURCE: Reprinted by permission from ASHRAE "Handbook of Fundamentals," 1972.
*Values apply to tightly closed venetian and vertical blinds, draperies, and roller shades.
†Values apply to storm sash with a tight air space. Air leakage present in virtually all storm windows will, in effect, increase this value.
‡U values can be substantially reduced by low-emittance coatings applied to the inner surface of single or double glazing and to an air-space surface of insulating glass. Consult manufacturers for applicable U values.
§Values listed are for 7¾ × 7¾ × 3⅞ in (200 × 200 × 100 mm) block. For 11¾ × 11¾ × 3⅞ in (300 × 300 × 100 mm) block, reduce the listed value by 0.04, and for 5¾ × 5¾ × 3⅞ in (150 × 150 × 100 mm) block, increase the listed value by 0.04. See Table 4-33 for definition of types.

**Table 29. Rates of Heat Gain from Occupants of Conditioned Spaces***

| Degree of activity | Typical application | Total heat Adults, male, Btu/h (W) | Total heat adjusted,† Btu/h (W) | Sensible heat, Btu/h (W) | Latent heat, Btu/h (W) |
|---|---|---|---|---|---|
| Seated at rest | Theater—matinee | 390 (114.2) | 330 ( 96.5) | 225 ( 65.8) | 105 ( 30.7) |
| | Theater—evening | 390 (114.2) | 350 (102.4) | 245 ( 81.7) | 105 ( 30.7) |
| Seated, very light work | Offices, hotels, apartments | 450 (131.7) | 400 (117) | 245 ( 71.7) | 155 ( 45.35) |
| Moderately active office work | Offices, hotels, apartments | 475 (139) | 450 (131.7) | 250 ( 73.2) | 200 ( 58.5) |
| Standing, light work; or walking slowly | Department store, retail store, dime store | 550 (161) | 450 (131.7) | 250 ( 73.2) | 200 ( 58.5) |
| Walking, seated; standing; walking slowly | Drugstore, bank | 550 (161) | 500 (146.3) | 250 ( 73.2) | 250 ( 73.2) |
| Sedentary work | Restaurant‡ | 490 (143.4) | 550 (161) | 275 ( 80.5) | 275 ( 80.5) |
| Light bench work | Factory | 800 (234) | 750 (219.5) | 275 ( 80.5) | 475 (139) |
| Moderate dancing | Dance hall | 900 (264) | 850 (249) | 305 ( 89.3) | 545 (160) |
| Walking 3 mi/h; moderately heavy work | Factory | 1,000 (292.8) | 1,000 (292.8) | 375 (109.8) | 625 (183) |
| Bowling§ | Bowling alley | | | | |
| Heavy work | Factory | 1,500 (439) | 1,450 (425) | 580 (170) | 870 (252.5) |

SOURCE: Reprinted by permission from ASHRAE "Handbook of Fundamentals," 1972.

*Tabulated values are based on 75°F (24°C) room dry-bulb temperature. For 80°F (27°C) room dry-bulb, the total heat remains the same, but the sensible-heat values should be decreased by approximately 20 percent, and the latent-heat values increased accordingly.

†Adjusted total heat gain is based on normal percentage of men, women, and children for the application listed, with the postulate that the gain from an adult female is 85 percent of that for an adult male, and that the gain from a child is 75 percent of that for an adult male.

‡Adjusted total heat value for sedentary work, restaurant, includes 60 Btu/h (17.6 W/h) for food per individual [30 Btu (8.8 W) sensible and 30 Btu (8.8 W) latent].

§For bowling figure one person per alley actually bowling, and all others as sitting [400 Btu/h (117 W/h)] or standing [550 Btu/h (161 W/h)].

**Table 30. Electric Motor-Driven Appliances**
Motor and driven appliance both in same room

| Fans* [blade diameters, in (cm)] | | Btu/h | Watts | Appliances† | Btu/h | Watts |
|---|---|---|---|---|---|---|
| Ceilings | 32    80 | 340 | 100 | Clock | 7 | 2 |
| | 52    130 | 410 | 120 | Hair dryer | 1,900 | 556 |
| | 56    140 | 600 | 176 | Drink mixer | 240 | 70 |
| Desk or wall | 8    20 | 120 | 35 | Sewing machine (domestic) | 220 | 65 |
| | 10    25 | 140 | 40 | Vacuum cleaner (domestic) | 250 | 73 |
| | 12    30 | 200 | 59 | Hair clipper | 78 | 23 |
| | 16    40 | 300 | 88 | Vibrator (beauty) | 11 | 3 |

SOURCE: C. Strock and R. L. Koral (eds.), "Handbook of Air Conditioning, Heating and Ventilation," 2d ed., Industrial Press, New York 1965.

*Figures are thermal equivalent of nameplate rating corrected for motor efficiency.

†Figures are thermal equivalent of nameplate ratings.

## Table 31. Heat Gain from Electric Motors (Continuous Operation*)

| Nameplate† or brake horsepower | Full-load motor efficiency, % | Location of equipment with respect to conditioned space or air stream‡ | | | | | |
|---|---|---|---|---|---|---|---|
| | | Motor in, driven machines in $\dfrac{\text{hp} \times 2{,}545}{\% \text{ eff.}}$ | | Motor out, driven machine in hp × 2,545 | | Motor in, driven machine out $\dfrac{\text{hp} \times 2{,}545 \,(1 - \% \text{ eff.})}{\% \text{ eff.}}$ | |
| | | Btu/h | Watts | Btu/h | Watts | Btu/h | Watts |
| 1/8 | 55 | 580 | 170 | 320 | 95 | 260 | 75 |
| 1/2 | 70 | 1,820 | 535 | 1,280 | 375 | 540 | 160 |
| 1 | 79 | 3,220 | 945 | 2,540 | 745 | 680 | 200 |
| 5 | 82 | 15,600 | 4,570 | 12,800 | 3,750 | 2,800 | 820 |
| 10 | 85 | 30,000 | 8,785 | 25,500 | 7,465 | 4,500 | 1,320 |
| 25 | 88 | 72,400 | 21,200 | 63,600 | 18,625 | 8,800 | 2,575 |
| 50 | 89 | 143,000 | 41,870 | 127,000 | 37,185 | 16,000 | 4,685 |
| 100 | 90 | 284,000 | 83,155 | 255,000 | 74,665 | 29,000 | 8,490 |
| 200 | 91 | 560,000 | 163,970 | 510,000 | 149,330 | 50,000 | 15,640 |

SOURCE: Reprinted by permission from ASHRAE "Handbook of Fundamentals," 1972.
*For intermittent operation, an appropriate usage factor should be used, preferably measured.
†If motors are overloaded and amount of overloading is unknown, multiply the above heat-gain factors by the following maximum service factors:

Maximum service factors

| hp | 1/20–1/8 | 1/6–1/3 | 1/2–3/4 | 1 | 1 1/2–2 | 3–250 |
|---|---|---|---|---|---|---|
| ac open type | 1.4 | 1.35 | 1.25 | 1.25 | 1.20 | 1.15 |
| dc open type | | | | 1.15 | 1.15 | 1.15 |

No overload is allowable with enclosed motors.
‡For a fan or pump in air-conditioned space, exhausting air, and pumping fluid to outside of space, use values in last column.

## Table 32. Recommended Inside Design Conditions—Summer

| Type of application | Deluxe | | Commercial | | Temp. swing* °F (°C) |
|---|---|---|---|---|---|
| | Dry-bulb, °F (°C) | Rel. hum., % | Dry-bulb, °F (°C) | Rel. hum., % | |
| General comfort, apartment, house, hotel, office, hospital, school, etc. | 74–76 (23.3–24.4) | 50–45 | 77–79 (25.0–26.1) | 50–45 | 2–4 (1.1–2.2) |
| Retail shops (short-term occupancy), bank, barber, or beauty shop, department store, supermarket, etc. | 76–78 (24.4–25.6) | 50–45 | 78–80 (25.6–26.7) | 50–45 | 2–4 (1.1–2.2) |
| Low-sensible-heat-factor applications (high latent load), auditorium, church, bar, restaurant, kitchen, etc. | 76–78 (24.4–25.6) | 55–50 | 78–80 (25.6–26.7) | 60–50 | 1–2 (0.55–1.1) |
| Factory comfort, assembly areas, machining rooms, etc. | 77–80 (25–26.7) | 55–45 | 80–85 (26.7–29.4) | 60–50 | 3–6 (1.67–3.35) |

SOURCE: Carrier Corporation "System Design Manual," Part I, Load Estimating, 1970.
*Temperature swing is above the thermostat setting at peak summer load conditions.

## Table 33. Recommended Return-Intake Face Velocities

| Intake location | Velocity over gross area | |
|---|---|---|
| | ft/min | m/min |
| Above occupied zone | 800 up | 245 up |
| Within occupied zone, not near seats | 600–800 | 185–245 |
| Within occupied zone, near seats | 400–600 | 120–185 |
| Door or wall louvers | 200–300 | 60–90 |
| Undercutting of doors (through undercut area) | 200–300 | 60–90 |

SOURCE: Reprinted by permission from ASHRAE "Handbook of Fundamentals," 1972.

**Table 34. Recommended and Maximum Duct Velocities for Low-Velocity Systems**

| Designation | Residences | | Schools, theaters, public buildings | | Industrial buildings | |
|---|---|---|---|---|---|---|
| | Recommended velocities | | | | | |
| | ft/min | (m/min) | ft/min | (m/min) | ft/min | (m/min) |
| Outdoor-air intakes* | 500 | (150) | 500 | (150) | 500 | (150) |
| Filters* | 250 | (75) | 300 | (90) | 350 | (110) |
| Heating coils*,† | 450 | (140) | 500 | (150) | 600 | (180) |
| Cooling coils* | 450 | (140) | 500 | (150) | 600 | (180) |
| Air washers | 500 | (150) | 500 | (150) | 500 | (150) |
| Fan outlets | 1,000–1,600 | (300–490) | 1,300–2,000 | (400–610) | 1,600–2,400 | (500–730) |
| Main ducts† | 700–900 | (215–275) | 1,000–1,300 | (300–400) | 1,200–1,800 | (365–550) |
| Branch ducts | 600 | (180) | 600–900 | (180–275) | 800–1,000 | (245–300) |
| Branch risers† | 500 | (150) | 600–700 | (180–215) | 800 | (245) |

SOURCE: Reprinted by permission from ASHRAE "Handbook of Fundamentals," 1972.
*These velocities are for total face area, not the net free area; other velocities in the table are for net free area.
†For low-velocity systems only. For recommendations on high-velocity systems, see Table 46.

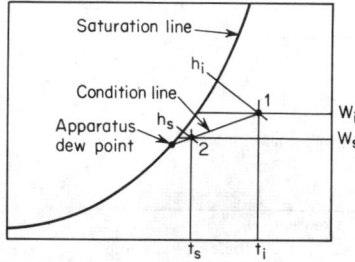

**Fig. 14**   Apparatus dew-point and condition line.

A comparable equation in SI units is

$$H'_R = \frac{0.5(V'_1)^2}{2} - \frac{V'_2{}^2}{2}$$

where $H_{R'}$ = head recovered, $N/m^2$
$V'_1$ = system inlet velocity, m/s
$V'_2$ = system outlet velocity, m/s

**Fans**   Fan laws are given in Table 36.
**Fan Equations**

Mechanical efficiency

$$= \frac{ft^3/min \times total\ pressure,\ inH_2O}{6{,}356 \times hp\ input} \quad (33)$$

A comparable equation in SI units is

Mechanical efficiency

$$= \frac{flow\ rate,\ m^3/s \times total\ pressure,\ N/m^2}{hp\ input,\ W}$$

These equations are applicable to fans operating with high outlet velocity pressure relative to static pressure.

$$Static\ efficiency = \frac{ft^3/min \times static\ pressure,\ in\ H_2O}{hp\ input} \quad (34)$$

A comparable equation in SI units is

$$Static\ efficiency = \frac{flow\ rate,\ m^3/s \times static\ pressure,\ N/m^2}{hp\ input,\ W}$$

These equations are more applicable to fans with high static pressure relative to velocity pressure.
See Table 37 for fan characteristics.

**Water Distribution, Chilled-Water Systems**
**Temperature Differential**

| Application | Temp rise | |
|---|---|---|
| | °C | °F |
| Close-coupled system on one floor | 5–8 | 3–4.5 |
| Two- or three-story building | 8–11 | 4.5–6 |
| Multistory building | 12–20 | 6.5–11 |

$$gal/min = \frac{total\ load,\ Btu/h + piping\ heat\ gains + pump\ heat}{500 \times temperature\ differential,\ °F} \quad (35)$$

A comparable equation in SI units is

Volume flow rate, $m^3/s$

$$= \frac{total\ load,\ kW + piping\ heat\ gains,\ kW + pump\ heat,\ kW}{4{,}190 \times temperature\ differential,\ °C}$$

**Condenser Water System**   For electrically driven refrigeration compressors a temperature differential of 10°F (5.5°C) may be assumed, and for steam-driven equipment a temperature differential of 20°F (11°C) is usual. In the latter case the refrigeration and steam condensers are piping in series with a temperature rise of approximately 10°F (5.5°C) each.

**Atmospheric Cooling Equipment**   The lowest temperature to which water can be cooled in atmospheric cooling equipment is the wet-bulb temperature of the ambient air.

**Water-Cooling Effectiveness in Percent**

$$E = \frac{(hot\text{-}water\ temp. - cold\text{-}water\ temp.) \times 100}{hot\text{-}water\ temp. - wet\text{-}bulb\ temp.\ entering} \quad (36)$$

The cold-water temperature must be chosen to place the requirement within the effectiveness range of the equipment used.

**Makeup Water for Atmospheric Cooling Equipment**   Makeup water is introduced to replace losses due to evaporation, drift, and blowdown. If all water were cooled by evaporation, the loss by evaporation for the usual 10°F (5.5°C) cooling range would be

$$Evaporation,\ \% = \frac{Q \times 100}{8.3 \times gal/min \times hfg} \quad (37)$$

where $Q$ = total heat rejected, Btu/h
gal/min = total condenser water circulated

hfg = evaporation heat of water, Btu/lb, at ambient design temperature

A comparable equation in SI units is

$$\text{Evaporation, \%} = \frac{Q' \times 100}{1{,}000 \times \text{m}^3/\text{s} \times \text{hfg}} \quad \textbf{(38)}$$

where $Q'$ = total heat rejected, kJ/min
$\quad$ m³/s = total condenser water circulated
$\quad$ hfg = evaporation heat of water, kJ/kg, at ambient design temperature

In practice, the loss of circulating water by evapoartion due to additional cooling by sensible heat transfer will vary from about 0.64 percent in winter to 0.88 percent in the summer for a water-cooling range of 10°F (5.5°C).

Drift losses depend on the tower design, but generally, from the cooling tower, they are limited to 0.2 percent of the circulated rate.

The makeup water replacing losses due to evaporation, drift, and blowdown introduces dissolved solids into the system.

To prevent excessive concentration, a portion of the circu-

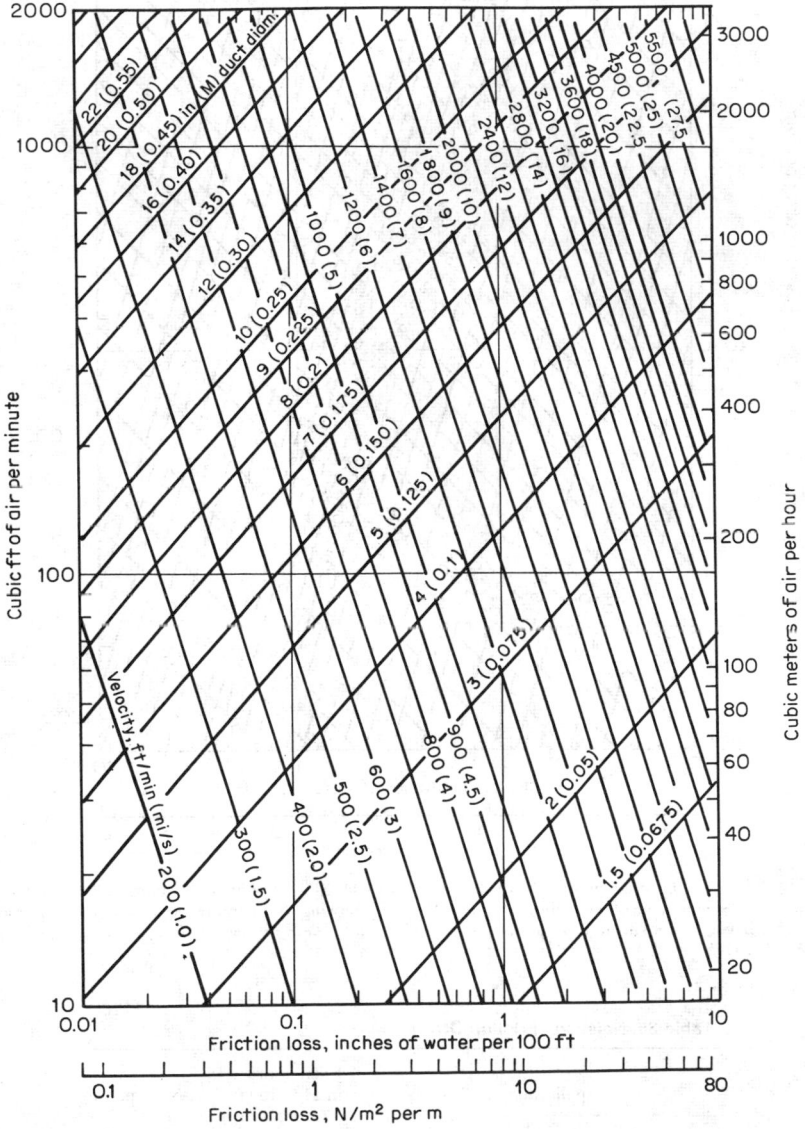

**Fig. 15** Friction of air in straight ducts for volumes of 10 to 2,000 ft³/min (20 to 3,000 m³/h). Based on standard air of 0.075 lb/ft³ (1.2 kg/m³) density flowing through average, clean, round, galvanized-metal ducts having approximately 40 joints per 100 ft (30 m). Do not extrapolate below chart. *(Reprinted by permission from ASHRAE "Handbook of Fundamentals," 1972.)*

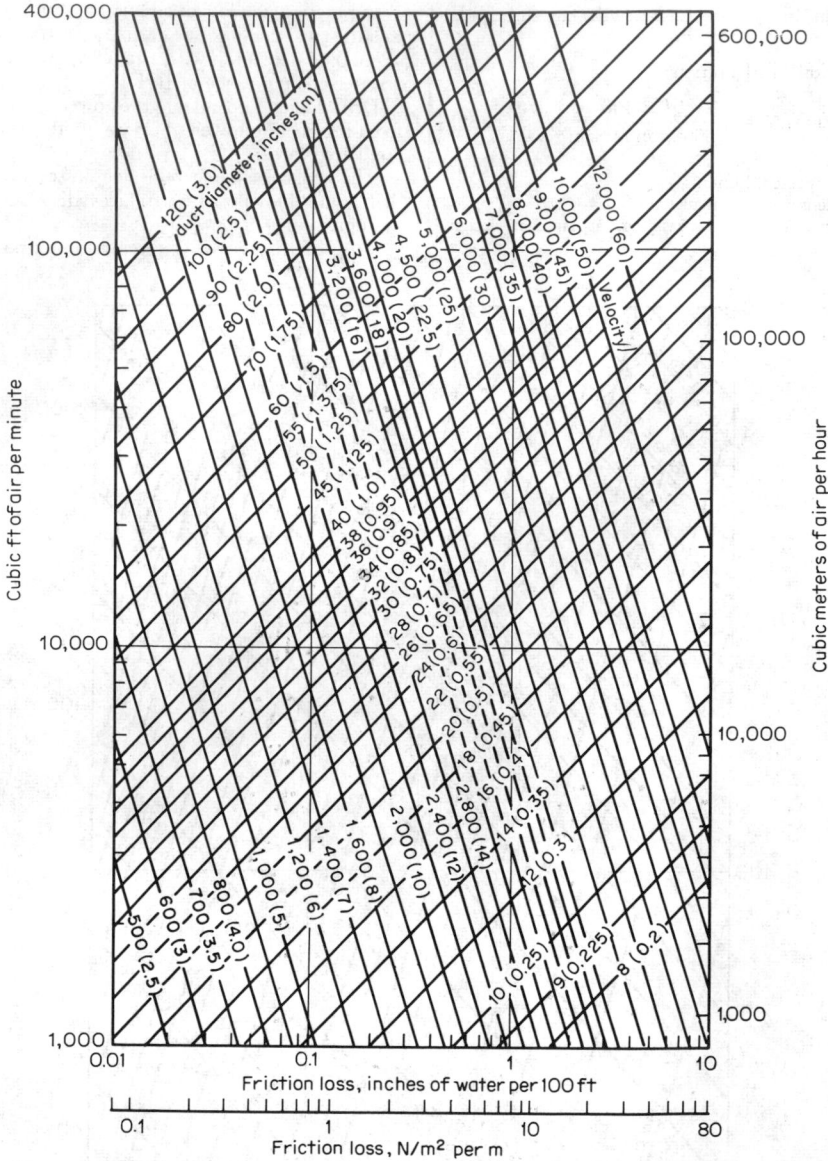

**Fig. 16**  Friction of air in straight ducts for volumes of 1,000 to 400,000 ft³/min (1,000 to 6,000,000 m³/h). Based on standard air of 0.075 lb/ft³ (1.2 kg/m³) density flowing through average, clean, round, galvanized-metal ducts having approximately 40 joints per 100 ft (30 m). *(Reprinted by permission from ASHRAE "Handbook of Fundamentals," 1972.)*

**Table 35. Noise vs. Friction Drops**

| Application | Friction drop, in H₂O/100 ft | N/m² per m |
|---|---|---|
| Noise critical, low velocity | 0.05–0.07 | 0.41–0.17 |
| Average application | 0.08–0.1 | 0.65–0.82 |
| Equipment rooms, industrial applications | 0.11–0.13 | 0.90–1.06 |

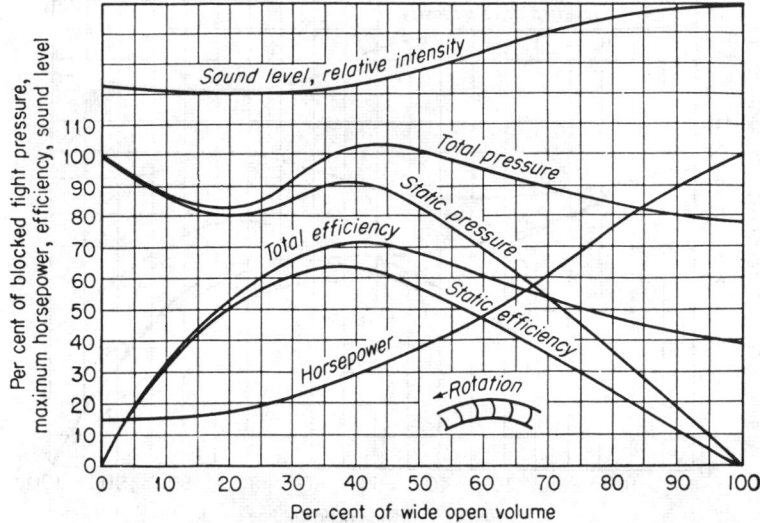

**Fig. 17** Percentage performance curves of a forward-blade centrifugal fan.

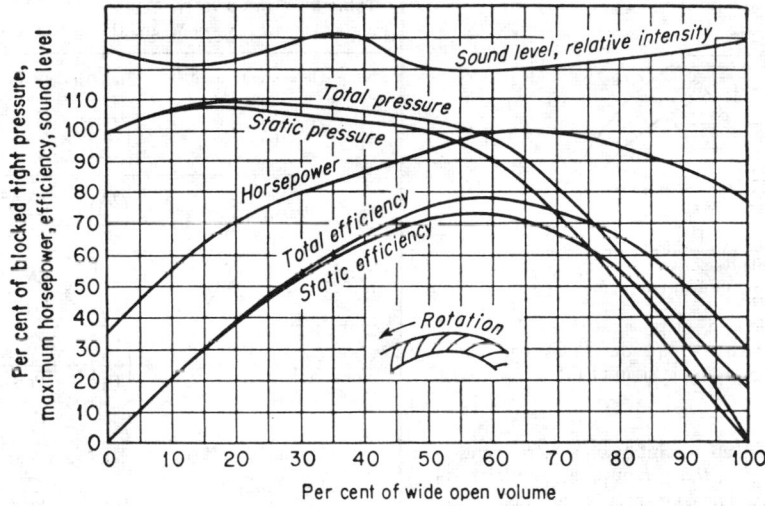

**Fig. 18** Percentage performance curves of a backward-curved-blade centrifugal fan.

lating water is wasted. The quantity of blowdown depends on the original quantity of dissolved solids in the makeup water and the permissible concentration.

For larger installation, chemical water-treatment processes are used, which also require a controlled blowdown rate.

### PUMPS

#### Water-Piping Design

There is a friction loss in any pipe through which water is flowing. This loss depends on the following factors:
  1. Water velocity
  2. Pipe diameter
  3. Interior surface roughness
  4. Pipe length

System pressure has no effect on the head loss of the equipment in the system. However, higher than normal system pressures may dictate the use of heavier pipe, fittings, and valves along with specially designed equipment.

To properly design a water-piping system, the engineer must evaluate not only the pipe friction loss but the loss through valves, fittings, and other equipment. In addition to these friction losses, the use of diversity in reducing the water quantity and pipe size should be considered in designing the water-piping system.

Most air-conditioning applications use either steel pipe or copper tubing in the piping system. For friction loss values in steel pipe or copper tubing refer to Figs. 21 and 22.

**Pipe Length** The friction loss in a water-piping system is the sum of:

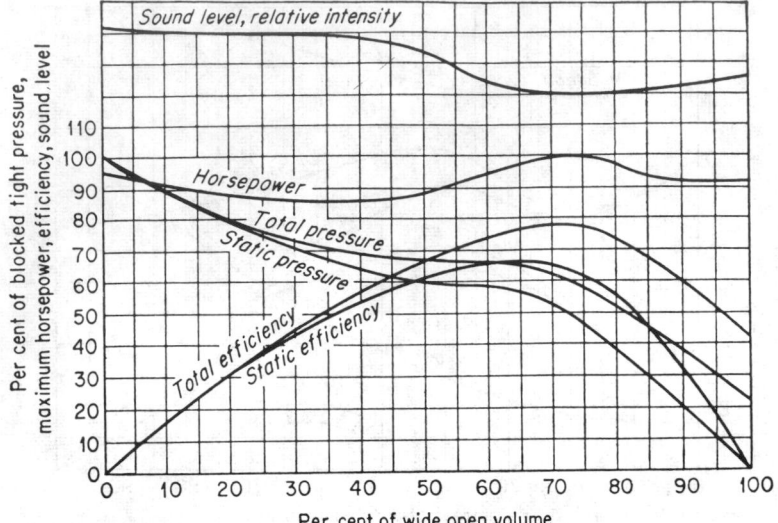

**Fig. 19**  Percentage performance curves of an axial-flow fan.

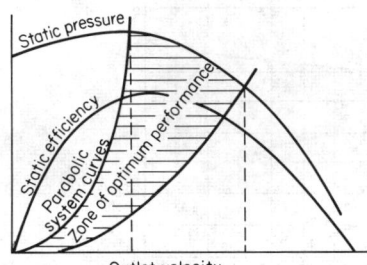

**Fig. 20**  Zone of optimum performance for fans.

1. The total straight lengths of pipe
2. The equivalent lengths of pipe due to fittings, valves, and other elements in the piping system (Tables 41 to 43)

### Heat Pumps

A heat pump is basically an assembly of a refrigerant gas compressor, heat exchangers, refrigerant piping, and controls and accessories which can provide heating or cooling to a building space. In the heating cycle heat is extracted from a natural or waste heat source and transferred to the space. In the cooling cycle heat is removed from the building space and discharged to a heat sink.

Heat pumps differ in the type of heat exchangers employed, depending on the heat sink and sources (Table 46) and the indoor heating and cooling medium used. The four basic types of heat pumps are air-to-air, water-to-air, water-to-water, and earth-to-air. Figures 23, 24, and 25 show cycle diagrams for the first three types. Earth-to-air heat pumps are similar to air-to-air heat pumps except that the outdoor heat exchanger is buried piping.

The efficiency of the perfect thermodynamic cycle relates to the operating-temperature levels:

$$\text{Eff.} = (T_2 - T_1)/T_2 \qquad (39)$$

**Table 36. Fan Laws\*†**

For all fan laws: $N_{t_1} = N_{t_2}$ and $(\text{Pt. of Rtg.})_1 = (\text{Pt. of Rtg.})_2$

| No. | Dependent variables | Independent variables | | |
|---|---|---|---|---|
| 1a | $Q_1 = Q_2$ | $\times \left(\dfrac{D_1}{D_2}\right)^3 \times \dfrac{N_1}{N_2}$ | | $\times 1$ |
| 1b | $\text{Press.}_1 = \text{Press.}_2{\ddagger}$ | $\times \left(\dfrac{D_1}{D_2}\right)^2 \times \left(\dfrac{N_1}{N_2}\right)^2$ | | $\times \dfrac{\rho_1}{\rho_2}$ |
| 1c | $H_1 = H_2$ | $\times \left(\dfrac{D_1}{D_2}\right)^5 \times \left(\dfrac{N_1}{N_2}\right)^3$ | | $\times \dfrac{\rho_1}{\rho_2}$ |
| 2a | $Q_1 = Q_2$ | $\times \left(\dfrac{D_1}{D_2}\right)^2 \times \left(\dfrac{\text{Press.}_1}{\text{Press.}_2}\right)^{1/2}$ | | $\times \left(\dfrac{\rho_2}{\rho_1}\right)^{1/2}$ |
| 2b | $N_1 = N_2$ | $\times \left(\dfrac{D_2}{D_1}\right) \times \left(\dfrac{\text{Press.}_1}{\text{Press.}_2}\right)^{1/2}$ | | $\times \left(\dfrac{\rho_2}{\rho_1}\right)^{1/2}$ |
| 2c | $H_1 = H_2$ | $\times \left(\dfrac{D_1}{D_2}\right)^2 \times \left(\dfrac{\text{Press.}_1}{\text{Press.}_2}\right)^{3/2}$ | | $\times \left(\dfrac{\rho_2}{\rho_1}\right)^{1/2}$ |
| 3a | $N_1 = N_2$ | $\times \left(\dfrac{D_2}{D_1}\right)^3 \times \dfrac{Q_1}{Q_2}$ | | $\times 1$ |
| 3b | $\text{Press.}_1 = \text{Press.}_2$ | $\times \left(\dfrac{D_2}{D_1}\right)^4 \times \left(\dfrac{Q_1}{Q_2}\right)^2$ | | $\times \dfrac{\rho_1}{\rho_2}$ |
| 3c | $H_1 = H_2$ | $\times \left(\dfrac{D_2}{D_1}\right)^4 \times \left(\dfrac{Q_1}{Q_2}\right)^3$ | | $\times \dfrac{\rho_1}{\rho_2}$ |

NOTE: $D$ = fan size, $N$ = revolutions per minute, $\rho$ = gas density, $Q$ = volume flow rate, $P$ = pressure, and $H$ = horsepower.
\*The subscript 1 denotes that the variable is for the fan under consideration.
†The subscript 2 denotes that the variable is for the tested fan.
‡$P_t$ or $P_s$.

**Table 37. Relative Characteristics of Centrifugal Fans**

| Characteristic | Backward | Radial | Forward |
|---|---|---|---|
| First cost | High | Medium | Low |
| Efficiency | High | Medium | Poor |
| Stability of operation | Good | Good | Poor |
| Space required | Medium | Medium | Small |
| Tip speed | High | Medium | Low |
| Resistance to abrasion | Medium | Good | Poor |

The heating coefficient of performance within the operating-temperature levels is the reciprocal of the efficiency

$$COP = T_2/(T_2 - T_1) \qquad (40)$$

where $T_1$ = absolute temperature of the evaporator
$T_2$ = absolute temperature of the condenser

In practical systems the heating and cooling coefficients of performance are

$$(COP)_h = g_h/g_t \qquad (41)$$

**Table 38. Correction Factor for Altitude and Temperature to Air Volume**

| Altitude, ft (m) above sea level | 0 | 1,000 (305) | 2,000 (610) | 3,000 (915) | 4,000 (1,220) | 5,000 (1,525) | 6,000 (1,880) | 7,000 (2,135) | 8,000 (2,440) |
|---|---|---|---|---|---|---|---|---|---|
| Barometric pressure, in Hg (kN/m²) | 29.92 (101.3) | 28.86 (97.7) | 27.82 (94.2) | 26.81 (90.8) | 25.84 (87.5) | 24.89 (84.3) | 23.98 (81.2) | 23.09 (78.2) | 22.2 (75.2) |
| Air temp, °F (°C) | | | | | Correction factors | | | | |
| 70 (21.1) | 1.040 | 1.003 | 0.967 | 0.932 | 0.898 | 0.865 | 0.833 | 0.803 | 0.772 |
| 100 (38) | 0.984 | 0.948 | 0.915 | 0.882 | 0.850 | 0.818 | 0.788 | 0.759 | 0.731 |
| 150 (66) | 0.904 | 0.872 | 0.840 | 0.801 | 0.781 | 0.752 | 0.724 | 0.698 | 0.672 |
| 200 (93) | 0.835 | 0.805 | 0.777 | 0.749 | 0.722 | 0.694 | 0.668 | 0.645 | 0.620 |
| 250 (121) | 0.777 | 0.749 | 0.722 | 0.696 | 0.671 | 0.647 | 0.622 | 0.599 | 0.577 |
| 300 (149) | 0.725 | 0.699 | 0.674 | 0.649 | 0.628 | 0.603 | 0.580 | 0.560 | 0.538 |
| 350 (177) | 0.680 | 0.656 | 0.632 | 0.609 | 0.588 | 0.566 | 0.545 | 0.525 | 0.505 |
| 400 (204) | 0.641 | 0.618 | 0.596 | 0.574 | 0.553 | 0.533 | 0.512 | 0.495 | 0.476 |
| 450 (232) | 0.605 | 0.583 | 0.564 | 0.543 | 0.523 | 0.503 | 0.485 | 0.467 | 0.450 |
| 500 (260) | 0.574 | 0.553 | 0.534 | 0.515 | 0.496 | 0.477 | 0.460 | 0.443 | 0.426 |
| 550 (288) | 0.546 | 0.526 | 0.508 | 0.490 | 0.472 | 0.454 | 0.438 | 0.421 | 0.406 |
| 600 (316) | 0.520 | 0.501 | 0.484 | 0.466 | 0.449 | 0.433 | 0.416 | 0.401 | 0.387 |
| 650 (343) | 0.496 | 0.478 | 0.462 | 0.444 | 0.428 | 0.413 | 0.397 | 0.383 | 0.368 |
| 700 (371) | 0.475 | 0.458 | 0.442 | 0.426 | 0.411 | 0.395 | 0.381 | 0.367 | 0.354 |

NOTE: Equivalent = $\dfrac{\text{ft}^3\text{/min at actual conditions}}{\text{correction factor}}$.

SOURCE: "Bulletin 3576-B, Correction Factors for Temperature and Altitude," Buffalo Forge Co., Buffalo, N.Y.

**Table 39. Heat Rejection of Typical Processes**

| Equipment | Btu/min/ton | kJ/min/ton | Btu/kWh | kJ/kWh | Btu/bhph | kJ/bhph |
|---|---|---|---|---|---|---|
| Refrigeration compressors, open-drive | 250 | 264 | | | | |
| Refrigeration compressors, hermetic | 300 | 317 | | | | |
| Refrigeration absorption system | 550 | 580 | | | | |
| Steam-jet refrigerating system | 550 | 580 | | | | |
| Steam-electric power plant, kW: | | | | | | |
| 500 | | | 11,210 | 11,827 | | |
| 1,000 | | | 10,750 | 11,342 | | |
| 5,000 | | | 8,150 | 8,599 | | |
| 7,500 | | | 7,700 | 8,124 | | |
| 10,000 | | | 7,020 | 7,407 | | |
| Diesel-engine jacket and lube oil: | | | | | | |
| Four-cycle, supercharged | | | | | 2,600 | 2,743 |
| Four-cycle, nonsupercharged | | | | | 3,000 | 3,165 |
| Two-cycle, crankcase compressor | | | | | 2,000 | 2,110 |
| Two-cycle, pump-scavenging (large unit) | | | | | 2,500 | 2,638 |
| Two-cycle, pump-scavenging (high-speed) | | | | | 2,200 | 2,321 |
| Natural-gas engine: | | | | | | |
| Four-cycle | | | | | 4,500 | 4,784 |
| Two-cycle | | | | | 4,000 | 4,220 |

SOURCE: "ASHRAE Guide and Data Book," Chap. 37, Table 4, ASHRAE, New York, 1961.

**Table 40. Effectiveness of Water-Cooling Equipment**

| Cooling equipment | Water-cooling effectiveness, % | | |
|---|---|---|---|
| | Minimum | Typical | Maximum |
| Spray ponds | 30 | 40–50 | 60 |
| Spray-filled atmospheric towers | 40 | 45–55 | 60 |
| Atmospheric deck towers | 50 | 50–60 | 90 |
| Mechanical draft towers | 50 | 55–75 | 93 |

SOURCE: "Heating, Ventilating, and Air Conditioning Guide," chap. 34, table 3, ASH-RAE, New York.

where $g_h$ = total heat output rate, Btu/h
$g_t$ = total heat input rate, Btu/h

$$(COP)_c = q_c/q_t \tag{42}$$

where $q_c$ = total cooling rate, Btu/h

## COMFORT INDEXES

### Effective Temperature Scales

*Effective temperature scales* are lines drawn on a psychrometric

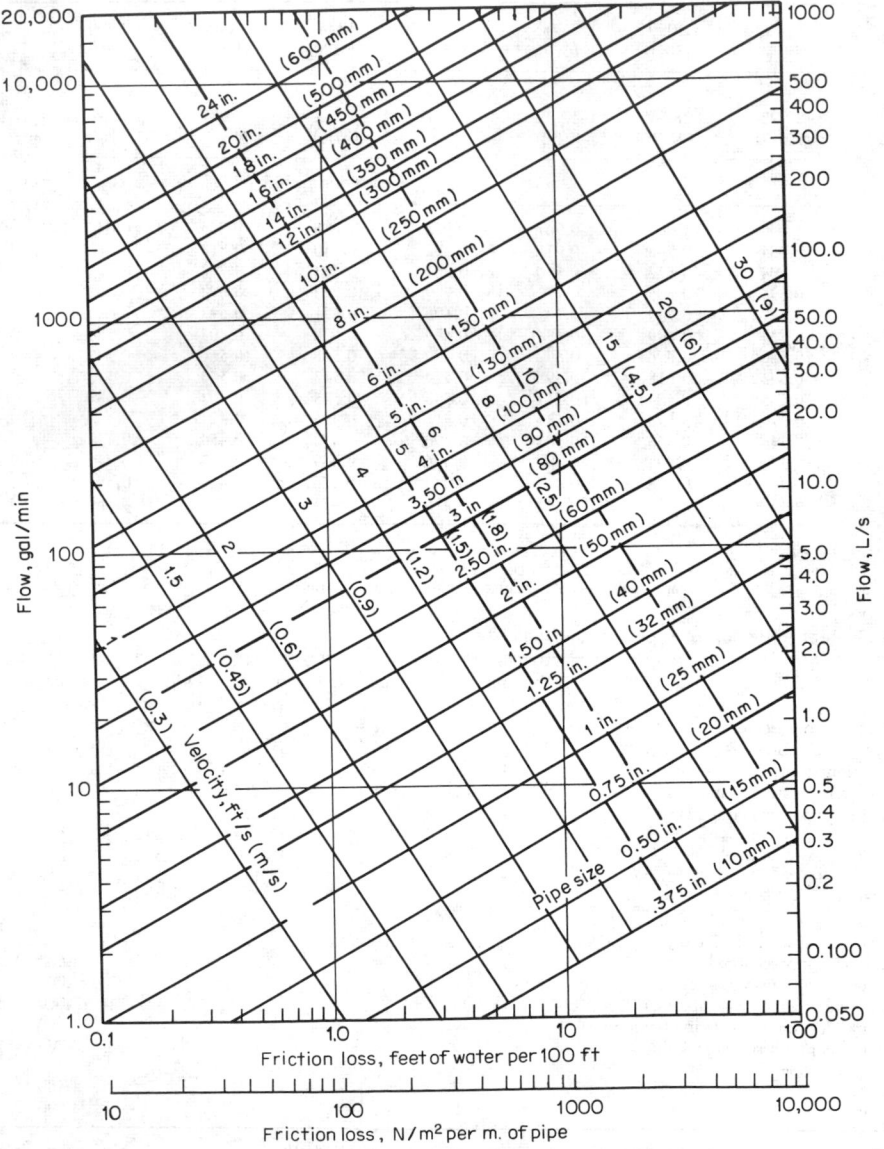

**Fig. 21** Friction loss for closed piping systems (schedule 40 pipe).

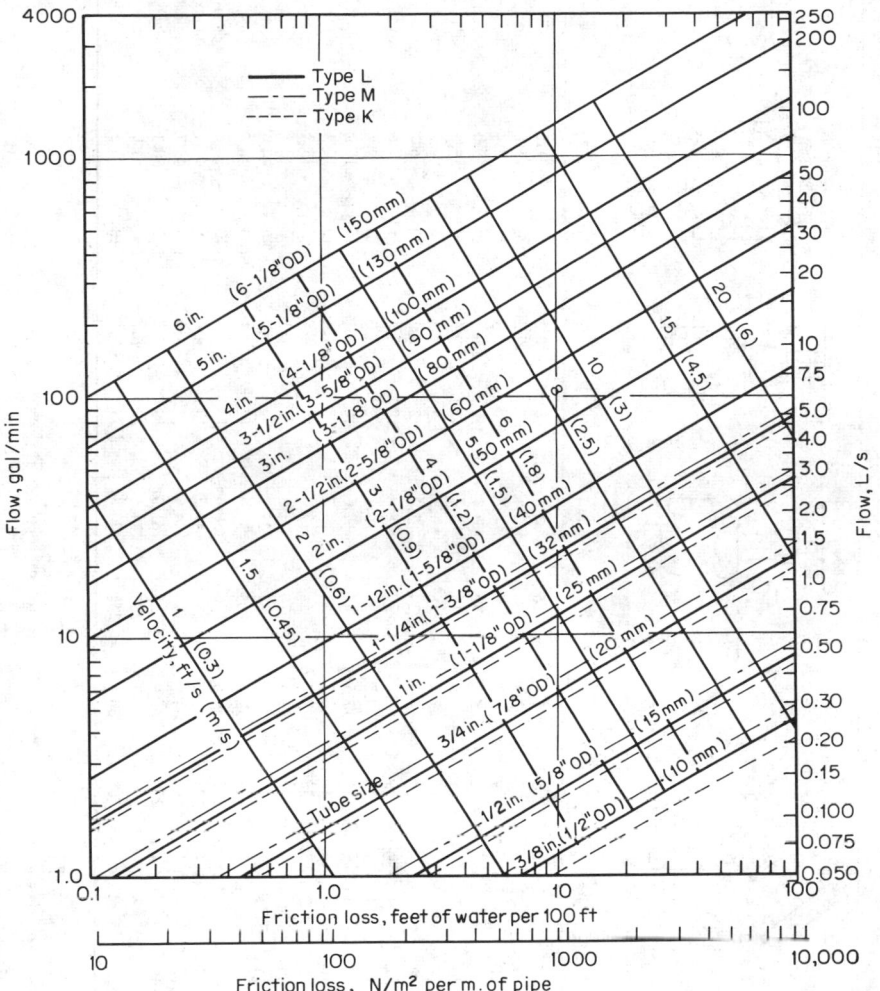

**Fig. 22** Friction loss for closed and open piping systems (copper tubing).

**Table 41. Maximum Water Velocity to Minimize Erosion**

| Normal operation, h | Water velocity | |
| --- | --- | --- |
| | ft/s | m/s |
| 1,500 | 12 | 0.060 |
| 2,000 | 11.5 | 0.058 |
| 3,000 | 11 | 0.055 |
| 4,000 | 10 | 0.050 |
| 6,000 | 9 | 0.045 |
| 8,000 | 8 | 0.040 |

SOURCE: Carrier "Design Manual," 1970.

chart on loci of constant degree of wettedness caused by regulatory sweating. Figure 26 relates to degrees of comfort experienced by a test group of clothed people in sedentary activity with air movement at 30 ft/min (0.15 m/s).

### ASHRAE Comfort Chart

The most recent comfort envelope obtained from a Kansas State University–ASHRAE study project is shown in Fig. 27 in the diamond-shaped envelope; the comfort envelope in ASHRAE Comfort Standard 55-56 is shown in the cross-hatched area. These apply to altitudes between sea level and 7,000 ft (2,134 m), a mean radiant temperature close to dry-bulb temperature, and air movement less than 45 ft/min (0.23 m/s).

### Temperature-Humidity Index

The term temperature-humidity index (THI) is used to describe the combined effects of temperature and humidity on comfort experienced by people. Since individual reactions can vary considerably from person to person, this quantity should be considered as a guide rather than as an absolute. However, relatively few people will feel discomfort when the THI is 70 or below. By the time it reaches 75, about half of the people

Table 42. Valve Losses in Equivalent Feet (meters) of Pipe*

Screwed, welded, flanged, and flared connections

| Nominal pipe or tube size | | Globe† | | 60° Y | | 45° Y | | Angle† | | Gate‡ | | Swing check§ | | Lift check |
|---|---|---|---|---|---|---|---|---|---|---|---|---|---|---|
| in | mm | ft | m | ft | m | ft | m | ft | m | ft | m | ft | m | |
| ⅜ | 10 | 17 | 5.0 | 8 | 2.4 | 6 | 1.8 | 6 | 1.8 | 0.6 | 0.2 | 5 | 1.5 | |
| ½ | 12 | 18 | 5.5 | 9 | 2.7 | 7 | 2.1 | 7 | 2.1 | 0.7 | 0.2 | 6 | 1.8 | Globe and |
| ¾ | 20 | 22 | 6.7 | 11 | 3.4 | 9 | 2.7 | 9 | 2.7 | 0.9 | 0.3 | 8 | 2.4 | vertical |
| 1 | 25 | 29 | 8.8 | 15 | 4.6 | 12 | 3.7 | 12 | 3.7 | 1.0 | 0.3 | 10 | 3.0 | lift |
| 1¼ | 32 | 38 | 11.6 | 20 | 6.1 | 15 | 4.6 | 15 | 4.6 | 1.5 | 0.5 | 14 | 4.3 | same as |
| 1½ | 40 | 43 | 13.1 | 24 | 7.3 | 18 | 5.5 | 18 | 5.5 | 1.8 | 0.5 | 16 | 4.9 | globe |
| 2 | 50 | 55 | 16.8 | 30 | 9.1 | 24 | 7.3 | 24 | 7.3 | 2.3 | 0.7 | 20 | 6.1 | valve¶ |
| 2½ | 60 | 69 | 21.0 | 35 | 10.7 | 29 | 8.8 | 29 | 8.8 | 2.8 | 0.9 | 25 | 7.6 | |
| 3 | 80 | 84 | 25.6 | 43 | 13.1 | 35 | 10.7 | 35 | 10.7 | 3.2 | 1.0 | 30 | 9.1 | |
| 3½ | 90 | 100 | 30.5 | 50 | 15.2 | 41 | 12.5 | 41 | 12.5 | 4.0 | 1.2 | 35 | 10.7 | |
| 4 | 100 | 120 | 36.6 | 58 | 17.7 | 47 | 14.3 | 47 | 14.3 | 4.5 | 1.4 | 40 | 12.2 | |
| 5 | 130 | 140 | 42.7 | 71 | 21.6 | 58 | 17.7 | 58 | 17.7 | 6 | 1.8 | 50 | 15.2 | |
| 6 | 150 | 170 | 51.8 | 88 | 26.8 | 70 | 21.3 | 70 | 21.3 | 7 | 2.1 | 60 | 18.3 | |
| 8 | 200 | 220 | 67.1 | 115 | 35.1 | 85 | 25.9 | 85 | 25.9 | 9 | 2.7 | 80 | 24.4 | |
| 10 | 250 | 280 | 85.3 | 145 | 44.2 | 105 | 32.0 | 105 | 32.0 | 12 | 3.7 | 100 | 30.5 | |
| 12 | 300 | 320 | 97.5 | 165 | 50.3 | 130 | 39.6 | 130 | 39.6 | 13 | 4.0 | 120 | 36.6 | Angle lift |
| 14 | 350 | 360 | 109.7 | 185 | 56.4 | 155 | 47.2 | 155 | 47.2 | 15 | 4.6 | 135 | 41.1 | same as |
| 16 | 400 | 410 | 125.0 | 210 | 64.0 | 180 | 54.9 | 180 | 54.9 | 17 | 5.0 | 150 | 45.7 | angle |
| 18 | 450 | 460 | 140.2 | 240 | 73.2 | 200 | 61.0 | 200 | 61.0 | 19 | 5.8 | 165 | 50.3 | valve |
| 20 | 500 | 520 | 158.5 | 275 | 83.8 | 235 | 71.6 | 235 | 71.6 | 22 | 6.7 | 200 | 61.0 | |
| 24 | 600 | 610 | 185.9 | 320 | 97.5 | 265 | 80.8 | 265 | 80.8 | 25 | 7.6 | 240 | 73.2 | |

*Losses are for all valves in fully open position.

†These losses do not apply to valves with needlepoint-type seats.

‡Regular and short pattern plug cock valves, when fully open, have same loss as gate valve. For valve losses of short-pattern plug cocks above 6 in, check manufacturer.

§Losses also apply to the in-line, ball-type check valve.

¶For Y pattern globe lift-check valve with seat approximately equal to the nominal pipe diameter, use values of 60° Y valve for loss.

will be uncomfortable. When the THI reaches 79, few people will not feel uncomfortable. An index of 80 in a work area may cause a decrease in efficiency of the workers. To compute THI, any one of the following formulas may be used:

$$\begin{aligned}
\text{THI} &= 0.4(t_d + t_w) + 15 \\
&= 0.55t_d + 0.2t_{dp} + 17.5 \\
&= t_d - (0.55 - 0.55 \text{ RH})(t_d - 58)
\end{aligned} \tag{43}$$

where $t_d$ = dry-bulb temperature, °F
$t_w$ = wet-bulb temperature, °F
$t_{dp}$ = dew-point temperature, °F
RH = relative humidity, expressed as a decimal

To convert to degrees Celsius, $t_c = \frac{5}{9}(t_f - 32)$

### Wind-Chill Index

The wind-chill index attempts to describe how much heat the body will lose under certain conditions of wind and temperature. It is determined empirically by an equation which is used to describe the rate of heat loss from a litre cylinder of water at 33°C (91.4°F) as a function of ambient temperature and wind velocity.

However, it is more useful and has become common practice to use the "equivalent wind-chill temperature." The formulas for the wind-chill index (WCI) are

$$\text{WCI} = (10.45 - V + 10\sqrt{V})(33 - t_a) \qquad \text{kcal/m}^2/\text{h} \tag{44}$$

where $V$ = meters per second, $t_a$ = °C

$$\text{WCI} = (10.45 - 0.447V - 6.6858\sqrt{V})(91.4 - t_a) \tag{45}$$

where $V$ = miles per hour, $t_a$ = °F

The more useful equivalent wind-chill temperature is expressed as

$$t_{eq} = -0.04544(\text{WCI}) + 33°\text{C} \tag{46}$$
$$t_{eq} = -0.04544(\text{WCI}) + 91.4°\text{F} \tag{47}$$

The National Weather Service has tabulated equivalent wind-chill temperature values from this formula as shown in Table 47.

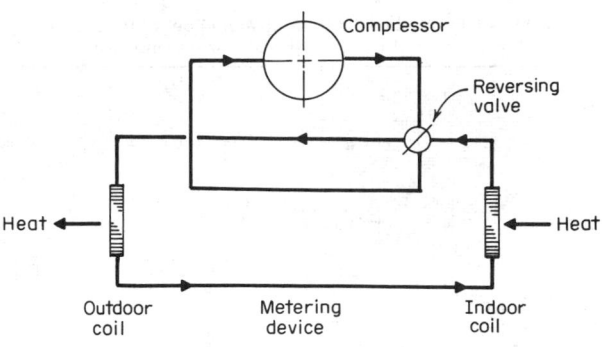

Cooling cycle

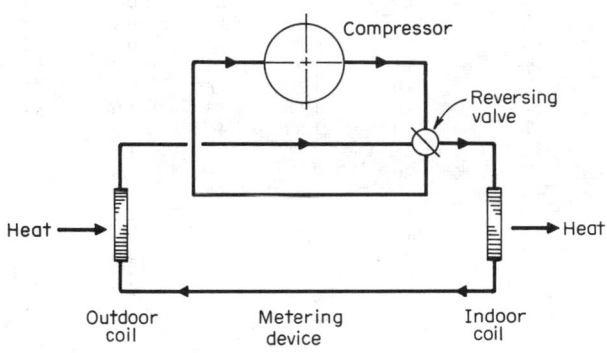

Heating cycle

**Fig. 23**  Air-to-air heat pump.

**Table 43. Fitting Losses in Equivalent Feet of Pipe**
Screwed, welded, flanged, flared, and brazed connections

| Nominal pipe or tube size, in | Smooth-bend elbows | | | | | | | | | | | |
|---|---|---|---|---|---|---|---|---|---|---|---|---|
| | 90° std* | | 90° long rad.† | | 90° street* | | 45° std* | | 45° street* | | 180° std* | |
| | ft | m | ft | m | ft | m | ft | m | ft | m | ft | m |
| 3/8 | 1.4 | 0.4 | 0.9 | 0.3 | 2.3 | 0.7 | 0.7 | 0.2 | 1.1 | 0.3 | 2.3 | 0.7 |
| 1/2 | 1.6 | 0.5 | 1.0 | 0.3 | 2.5 | 0.8 | 0.8 | 0.2 | 1.3 | 0.4 | 2.5 | 0.8 |
| 3/4 | 2.0 | 0.6 | 1.4 | 0.4 | 3.2 | 1.0 | 0.9 | 0.3 | 1.6 | 0.5 | 3.2 | 1.0 |
| 1 | 2.6 | 0.8 | 1.7 | 0.5 | 4.1 | 1.2 | 1.3 | 0.4 | 2.1 | 0.6 | 4.1 | 1.2 |
| 1 1/4 | 3.3 | 1.0 | 2.3 | 0.7 | 5.6 | 1.7 | 1.7 | 0.5 | 3.0 | 0.9 | 5.6 | 1.7 |
| 1 1/2 | 4.0 | 1.2 | 2.6 | 0.8 | 6.3 | 1.9 | 2.1 | 0.6 | 3.4 | 1.0 | 6.3 | 1.9 |
| 2 | 5.0 | 1.5 | 3.3 | 1.0 | 8.2 | 2.5 | 2.6 | 0.8 | 4.5 | 1.4 | 8.2 | 2.5 |
| 2 1/2 | 6.0 | 1.8 | 4.1 | 1.2 | 10 | 3.0 | 3.2 | 1.0 | 5.2 | 1.6 | 10 | 3.0 |
| 3 | 7.5 | 2.3 | 5.0 | 1.5 | 12 | 3.7 | 4.0 | 1.2 | 6.4 | 2.0 | 12 | 3.7 |
| 3 1/2 | 9.0 | 2.7 | 5.9 | 1.8 | 15 | 4.6 | 4.7 | 1.4 | 7.3 | 2.2 | 15 | 1.6 |
| 4 | 10 | 3.0 | 6.7 | 2.0 | 17 | 5.2 | 5.2 | 1.6 | 8.5 | 2.6 | 17 | 5.2 |
| 5 | 13 | 4.0 | 8.2 | 2.5 | 21 | 6.4 | 6.5 | 2.0 | 11 | 3.4 | 21 | 6.4 |
| 6 | 16 | 4.9 | 10 | 3.0 | 25 | 7.6 | 7.9 | 2.4 | 13 | 4.0 | 25 | 7.6 |
| 8 | 20 | 6.1 | 13 | 4.0 | | | 10 | 3.0 | | | 33 | 10.1 |
| 10 | 25 | 7.6 | 16 | 4.9 | | | 13 | 4.0 | | | 42 | 12.8 |
| 12 | 30 | 9.1 | 19 | 5.8 | | | 16 | 4.9 | | | 50 | 15.2 |
| 14 | 34 | 10.4 | 23 | 7.0 | | | 18 | 5.5 | | | 55 | 16.8 |
| 16 | 38 | 11.6 | 26 | 7.9 | | | 20 | 6.1 | | | 62 | 18.9 |
| 18 | 42 | 12.8 | 29 | 8.8 | | | 23 | 7.0 | | | 70 | 21.3 |
| 20 | 50 | 15.2 | 33 | 10.1 | | | 26 | 7.9 | | | 81 | 24.7 |
| 24 | 60 | 18.3 | 40 | 12.2 | | | 30 | 9.1 | | | 94 | 28.7 |

| Flow-through branch | | Smooth-bend tees Straight-through flow | | | | | | Mitre elbows 90° ell | | 60° ell | | 45° ell | | 30° ell | |
| | | No reduction | | Reduced ¼ | | Reduced ½ | | | | | | | | | |
| ft | m | ft | m | ft | m | ft | m | ft | m | ft | m | ft | m | ft | m |
|---|---|---|---|---|---|---|---|---|---|---|---|---|---|---|---|
| 2.7 | 0.8 | 0.9 | 0.3 | 1.2 | 0.4 | 1.4 | 0.4 | 2.7 | 0.8 | 1.1 | 0.3 | 0.6 | 0.2 | 0.3 | 0.1 |
| 3.0 | 0.9 | 1.0 | 0.3 | 1.4 | 0.4 | 1.6 | 0.5 | 3.0 | 0.9 | 1.3 | 0.4 | 0.7 | 0.2 | 0.4 | 0.1 |
| 4.0 | 1.2 | 1.4 | 0.4 | 1.9 | 0.6 | 2.0 | 0.6 | 4.0 | 1.2 | 1.6 | 0.5 | 0.9 | 0.3 | 0.5 | 0.2 |
| 5.0 | 1.5 | 1.7 | 0.5 | 2.3 | 0.7 | 2.6 | 0.8 | 5.0 | 1.5 | 2.1 | 0.6 | 1.0 | 0.3 | 0.7 | 0.2 |
| 7.0 | 2.1 | 2.3 | 0.7 | 3.1 | 0.9 | 3.3 | 1.0 | 7.0 | 2.1 | 3.0 | 0.9 | 1.5 | 0.5 | 0.9 | 0.3 |
| 8.0 | 2.4 | 2.6 | 0.8 | 3.7 | 1.1 | 4.0 | 1.2 | 8.0 | 2.4 | 3.4 | 1.0 | 1.8 | 0.5 | 1.1 | 0.3 |
| 10 | 3.0 | 3.3 | 1.0 | 4.7 | 1.4 | 5.0 | 1.5 | 10 | 3.0 | 4.5 | 1.4 | 2.3 | 0.7 | 1.3 | 0.4 |
| 12 | 3.7 | 4.1 | 1.2 | 5.6 | 1.7 | 6.0 | 1.8 | 12 | 3.7 | 5.2 | 1.6 | 2.8 | 0.9 | 1.7 | 0.5 |
| 15 | 1.6 | 5.0 | 1.5 | 7.0 | 2.1 | 7.5 | 2.3 | 15 | 4.6 | 6.4 | 2.0 | 3.2 | 1.0 | 2.0 | 0.6 |
| 18 | 5.5 | 5.9 | 1.8 | 8.0 | 2.4 | 9.0 | 2.7 | 18 | 5.5 | 7.3 | 2.2 | 4.0 | 1.2 | 2.4 | 0.7 |
| 21 | 6.4 | 6.7 | 2.0 | 9.0 | 2.7 | 10 | 3.0 | 21 | 6.4 | 8.5 | 2.6 | 4.5 | 1.4 | 2.7 | 0.8 |
| 25 | 7.6 | 8.2 | 2.5 | 12 | 3.7 | 13 | 4.0 | 25 | 7.6 | 11 | 3.4 | 6.0 | 1.8 | 3.2 | 1.0 |
| 30 | 9.1 | 10 | 3.0 | 14 | 4.3 | 16 | 4.9 | 30 | 9.1 | 13 | 4.0 | 7.0 | 2.1 | 4.0 | 1.2 |
| 40 | 12.2 | 13 | 4.0 | 18 | 5.5 | 20 | 6.1 | 40 | 12.2 | 17 | 5.2 | 9.0 | 2.7 | 5.1 | 1.6 |
| 50 | 15.2 | 16 | 4.9 | 23 | 7.0 | 25 | 7.6 | 50 | 15.2 | 21 | 6.4 | 12 | 3.7 | 7.2 | 2.2 |
| 60 | 18.3 | 19 | 5.8 | 26 | 7.9 | 30 | 9.1 | 60 | 18.3 | 25 | 7.6 | 13 | 4.0 | 8.0 | 2.4 |
| 68 | 20.7 | 23 | 7.0 | 30 | 9.1 | 34 | 10.4 | 68 | 20.7 | 29 | 8.8 | 15 | 1.6 | 9.0 | 2.7 |
| 78 | 23.8 | 26 | 7.9 | 35 | 10.7 | 38 | 11.6 | 78 | 23.8 | 31 | 9.4 | 17 | 5.2 | 10 | 3.0 |
| 85 | 25.8 | 29 | 8.8 | 40 | 12.2 | 42 | 12.8 | 85 | 25.9 | 37 | 11.3 | 19 | 5.8 | 11 | 3.4 |
| 100 | 30.5 | 33 | 10.1 | 44 | 13.4 | 50 | 15.2 | 100 | 30.5 | 41 | 12.5 | 22 | 6.7 | 13 | 4.0 |
| 115 | 35.1 | 40 | 12.2 | 50 | 15.2 | 60 | 18.3 | 115 | 35.1 | 49 | 14.9 | 25 | 7.6 | 16 | 4.9 |

*R/D approximately equal to 1.
†R/D approximately equal to 1.5.

## Table 44. Special Fitting Losses in Equivalent Feet of Pipe

| Nom. pipe or tube size, in | Sudden enlargement,* d/D ¼ ft | ¼ m | ½ ft | ½ m | ¾ ft | ¾ m | Sudden contraction,* d/D ¼ ft | ¼ m | ½ ft | ½ m | ¾ ft | ¾ m | Sharp edge* Entrance ft | m | Sharp edge* Exit ft | m | Pipe projection Entrance ft | m | Pipe projection Exit ft | m |
|---|---|---|---|---|---|---|---|---|---|---|---|---|---|---|---|---|---|---|---|---|
| ⅜ | 1.4 | 0.4 | 0.8 | 0.2 | 0.3 | 0.1 | 0.7 | 0.2 | 0.5 | 0.2 | 0.3 | 0.1 | 1.5 | 0.5 | .8 | 0.2 | 1.5 | 0.5 | 1.1 | 0.3 |
| ½ | 1.8 | 0.5 | 1.1 | 0.3 | 0.4 | 0.1 | 0.9 | 0.3 | 0.7 | 0.2 | 0.4 | 0.1 | 1.8 | 0.5 | 1.0 | 0.3 | 1.8 | 0.5 | 1.5 | 0.5 |
| ¾ | 2.5 | 0.8 | 1.5 | 0.5 | 0.5 | 0.2 | 1.2 | 0.4 | 1.0 | 0.3 | 0. | 0.2 | 2.8 | 0.9 | 1.4 | 0.4 | 2.8 | 0.9 | 2.2 | 0.7 |
| 1 | 3.2 | 1.0 | 2.0 | 0.6 | 0.7 | 0.2 | 1.6 | 0.5 | 1.2 | 0.4 | 0.7 | 0.2 | 3.7 | 1.1 | 1.8 | 0.5 | 3.7 | 1.1 | 2.7 | 0.8 |
| 1¼ | 4.7 | 1.4 | 3.0 | 0.9 | 1.0 | 0.3 | 2.3 | 0.7 | 1.8 | 0.5 | 1.0 | 0.3 | 5.3 | 1.6 | 2.6 | 0.8 | 5.3 | 1.6 | 4.2 | 1.3 |
| 1½ | 5.8 | 1.8 | 3.6 | 1.1 | 1.2 | 0.4 | 2.9 | 0.8 | 2.2 | 0.7 | 1.2 | 0.4 | 6.6 | 2.0 | 3.3 | 1.0 | 6.6 | 2.0 | 5.0 | 1.5 |
| 2 | 8.0 | 2.4 | 4.8 | 1.5 | 1.6 | 0.5 | 4.0 | 1.2 | 3.0 | 0.9 | 1.6 | 0.5 | 9.0 | 2.7 | 4.4 | 1.3 | 9.0 | 2.7 | 6.8 | 2.1 |
| 2½ | 10 | 3.0 | 6.1 | 1.9 | 2.0 | 0.6 | 5.0 | 1.5 | 3.8 | 1.1 | 2.0 | 0.6 | 12 | 3.7 | 5.6 | 1.7 | 12 | 3.7 | 8.7 | 2.7 |
| 3 | 13 | 4.0 | 8.0 | 2.4 | 2.6 | 0.8 | 6.5 | 2.0 | 4.9 | 1.5 | 2.6 | 0.8 | 14 | 4.3 | 7.2 | 2.2 | 14 | 4.3 | 11 | 3.4 |
| 3½ | 15 | 4.6 | 9.2 | 2.8 | 3.0 | 0.9 | 7.7 | 2.3 | 6.0 | 1.8 | 3.0 | 0.9 | 17 | 5.2 | 8.5 | 2.6 | 17 | 5.2 | 13 | 4.0 |
| 4 | 17 | 5.2 | 11 | 3.4 | 3.8 | 1.1 | 9.0 | 2.7 | 6.8 | 2.1 | 3.8 | 1.1 | 20 | 6.1 | 10 | 3.0 | 20 | 6.1 | 16 | 4.9 |
| 5 | 24 | 7.3 | 15 | 4.6 | 5.0 | 1.5 | 12 | 3.7 | 9.0 | 2.7 | 5.0 | 1.5 | 27 | 8.2 | 14 | 4.3 | 27 | 8.2 | 20 | 6.1 |
| 6 | 29 | 8.8 | 22 | 6.7 | 6.0 | 1.8 | 15 | 4.6 | 11 | 3.4 | 6.0 | 1.8 | 33 | 10.1 | 19 | 5.8 | 33 | 10.1 | 25 | 7.6 |
| 8 | | | 25 | 7.6 | 8.5 | 2.6 | | | 15 | 4.6 | 8.5 | 2.6 | 47 | 14.3 | 24 | 7.3 | 47 | 14.3 | 35 | 10.7 |
| 10 | | | 32 | 9.8 | 11 | 3.4 | | | 20 | 6.1 | 11 | 3.4 | 60 | 18.3 | 29 | 8.8 | 60 | 18.3 | 46 | 14.0 |
| 12 | | | 41 | 12.5 | 13 | 4.0 | | | 25 | 7.6 | 13 | 4.0 | 73 | 22.3 | 37 | 11.3 | 73 | 22.3 | 57 | 17.4 |
| 14 | | | | | 16 | 4.9 | | | | | 16 | 4.9 | 86 | 26.2 | 45 | 13.7 | 86 | 26.2 | 66 | 20.1 |
| 16 | | | | | 18 | 5.5 | | | | | 18 | 5.5 | 96 | 29.3 | 50 | 15.2 | 96 | 29.3 | 77 | 23.5 |
| 18 | | | | | 20 | 6.1 | | | | | 20 | 6.1 | 115 | 35.1 | 58 | 17.7 | 115 | 35.1 | 90 | 27.4 |
| 20 | | | | | | | | | | | | | 142 | 43.3 | 70 | 21.3 | 142 | 43.3 | 108 | 32.9 |
| 24 | | | | | | | | | | | | | 163 | 49.7 | 83 | 25.3 | 163 | 49.7 | 130 | 39.6 |

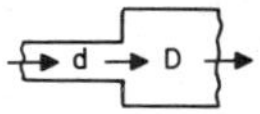

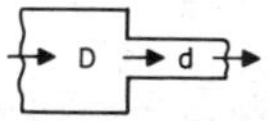

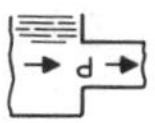

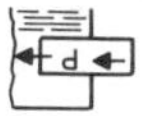

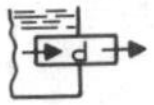

*Enter table for losses at smallest diameter d.

**Table 45. Approximate kW Input per 12,000 Btu/h Cooling Capacity**

Electric consumption, kW/ton cooling capacity

| Item | Type of heat rejection, kW/ton | |
|---|---|---|
| | Mechanical-draft cooling tower | Air-cooled condensers |
| Room air-conditioners | | 1.2–2.45* |
| Refrigeration compressors (motor-driven): | | |
|    Reciprocating (20 to 100 tons) | 1.03–0.86 | 1.27–1.09 |
|    Centrifugal-hermetic | 0.94–0.74 | |
|    Centrifugal-external drive (350 to 1,250 tons) | 1.02–0.81 | |
| Auxiliary Equipment: | | |
|    Pump, cooling water | 0.15 | |
|    Blower for air conditioner | 0.10 | 0.10 |
|    Blower for rejection air | 0.15 | 0.15 |

Steam-consumption rate lb/h/ton (kg/h/ton) cooling capacity

| Item | Steam rate | |
|---|---|---|
| | lb/h/ton | kg/h/ton |
| Centrifugal compressors (steam-driven): | | |
|    Condensing (27 in vacuum) | 14.5 | 6.6 |
|    Noncondensing (14.7 psig) | 30–35 | 13.6–15.9 |
| Absorption chillers (100 to 1,200 ton) | 18–18.5 | 8.2–8.4 |

Gas-consumption rate, Btu/ton (J/ton) cooling capacity

| Item | Gate rate | |
|---|---|---|
| | Btu/ton | kJ/ton |
| Refrigeration compressors (natural-gas-engine-driven) | 8,500 | 8,970 |

*Depending on the EER (energy efficiency ratio) rating for the different manufacturers.

**Table 46. Heat-Pump Heat Sources and Sinks**

| Heat source | Air | City water | Well water | Surface water | Waste water | Earth | Solar |
|---|---|---|---|---|---|---|---|
| Source classification | Primary | Primary or auxiliary | Primary | Primary | Primary or auxiliary | Primary or auxiliary | Auxiliary |
| Suitability as heat sink | Good | Good | Good | Good | Variable with source | Usually poor | May be used to dissipate heat to air |
| Availability (location) | Universal | Cities | Uncertain | Rare | Limited | Extensive | Universal |
| Availability (time) | Continuous | Continuous—except local shortages | Continuous—check water table | Continuous | Variable | Continuous, temperature drops as heat is removed, slowly rises when pump stops | Intermittent, unpredictable, except over extended time |
| Expense (original) | Low, less than earth and water sources except city | Usually lowest | Variable, depending on cost of drilling well | Low | Variable | High | High |
| Expense (operating) | Relatively low | High, usually prohibitive | Low to moderate | Relatively low | Low | Relatively moderate | Unexplored. Promising as auxiliary for reducing operating cost |
| Temperature (level) | Favorable 75–95% of time in most of United States | Usually satisfactory | Satisfactory | Satisfactory | Usually good | Initially good—drops with time and rate of heat withdrawal | Excellent |
| Temperature (variation) | Extreme | Variable with location (10 to 25°F) | Small | Moderate | Usually moderate | Large—less than for air, however | Extreme |
| Design information | Usually adequate | Usually adequate | Usually adequate | Usually adequate | Adequate if source is constant in supply and temperature | Inadequate | Practically available |
| Size of equipment | Moderate | Small | Small (except for well) | Small | Variable (usually moderate) | Small (except ground coils) | Available in some areas |
| Adaptability to standard product | Excellent, can be factory assembled and tested | Excellent | Excellent (except for well) | Excellent | Poor | Poor | Poor |
| Sources it may augment | | Air, earth | | | | | |
| Special problems | Least heat available when demand greatest. Coil frosting requires extra capacity, alternate source, or standby heat. May require ductwork | Scale on coils. Local use restrictions during shortages. Disposal. Water temperature may become too low to permit further heat removal | Corrosion, scale may form on heat-transfer surface. Disposal may require second well. Water location, temperature, composition usually unknown until well drilled. Well may run dry | Water may cause scale, corrosion, and algae fouling | Usually scale forming or corrosive. Often insufficient supply. Very limited application, hence required individual design. Freeze-up hazards | Limited by local geology and climate. Installation costs difficult to estimate. Requires considerable ground area, may damage lawns, gardens. Leaks difficult to repair | Probably will require heat-storage equipment at either evaporator or condenser side. |

## Table 47. Equivalent Wind-Chill Temperatures

NOTE: Wind speeds greater than 40 mi/h have little additional chilling effect.

| mi/h | Dry bulb temperature, °F | | | | | | | | | | | | | | | | |
|---|---|---|---|---|---|---|---|---|---|---|---|---|---|---|---|---|---|
| | 35 | 30 | 25 | 20 | 15 | 10 | 5 | 0 | −5 | −10 | −15 | −20 | −25 | −30 | −35 | −40 | −45 |
| | Wind chill index (Equivalent temperature) − Equivalent in cooling power on exposed flesh under calm conditions | | | | | | | | | | | | | | | | |
| Calm | 35 | 30 | 25 | 20 | 15 | 10 | 5 | 0 | −5 | −10 | −15 | −20 | −25 | −30 | −35 | −40 | −45 |
| 5 | 33 | 27 | 21 | 16 | 12 | 7 | 1 | −6 | −11 | −15 | −20 | −25 | −31 | −35 | −41 | −47 | −54 |
| 10 | 21 | 16 | 9 | 2 | −2 | −9 | −15 | −22 | −27 | −31 | −38 | −45 | −52 | −58 | −64 | −70 | −77 |
| 15 | 16 | 11 | 1 | −6 | −11 | −18 | −25 | −33 | −40 | −45 | −51 | −60 | −65 | −70 | −78 | −85 | −90 |
| 20 | 12 | 3 | −4 | −9 | −17 | −24 | −32 | −40 | −46 | −52 | −60 | −68 | −76 | −81 | −88 | −96 | −103 |
| 25 | 7 | 0 | −7 | −15 | −22 | −29 | −37 | −45 | −52 | −58 | −67 | −75 | −83 | −89 | −96 | −104 | −112 |
| 30 | 5 | −2 | −11 | −18 | −26 | −33 | −41 | −49 | −56 | −63 | −70 | −78 | −87 | −94 | −101 | −109 | −117 |
| 35 | 3 | −4 | −13 | −20 | −27 | −35 | −43 | −52 | −60 | −67 | −72 | −83 | −90 | −98 | −105 | −113 | −123 |
| 40 | 1 | −4 | −15 | −22 | −29 | −35 | −45 | −54 | −62 | −69 | −76 | −87 | −94 | −101 | −107 | −116 | −128 |
| 45 | 1 | −6 | −17 | −24 | −31 | −38 | −46 | −54 | −63 | −70 | −78 | −87 | −94 | −101 | −108 | −118 | −120 |
| 50 | 0 | −7 | −17 | −24 | −31 | −34 | −47 | −56 | −63 | −70 | −79 | −88 | −96 | −103 | −110 | −120 | −126 |

Very cold

Bitterly cold

Extreme cold

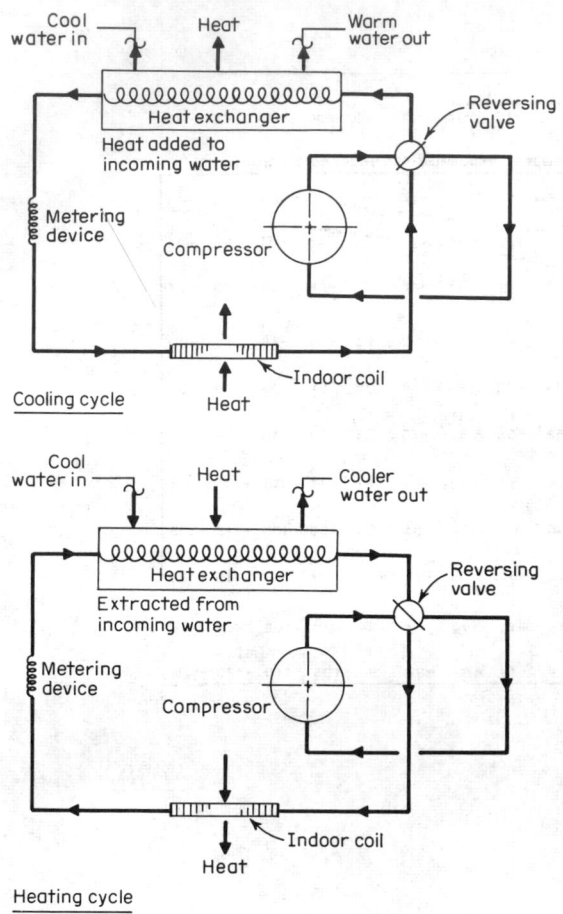

**Fig. 24** Water-to-air heat pump.

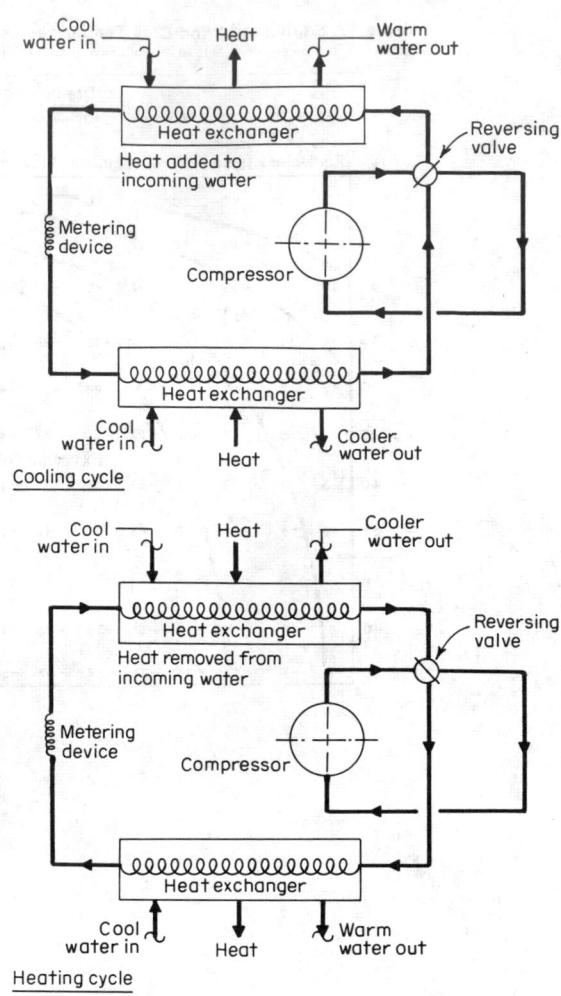

**Fig. 25** Water-to-water heat pump.

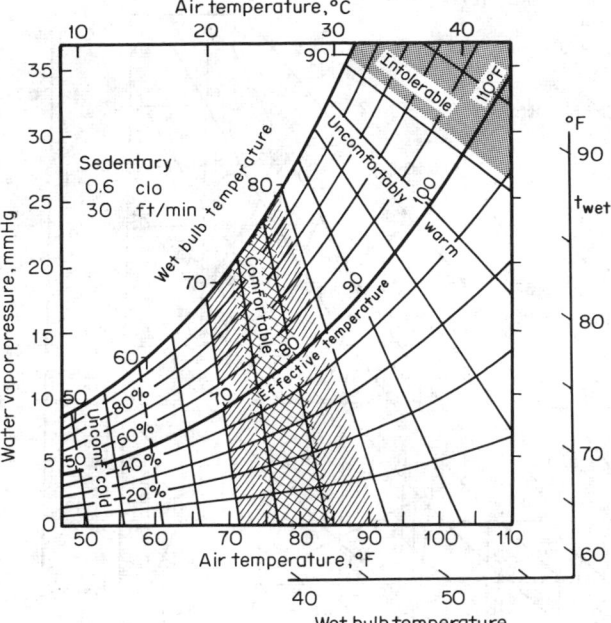

**Fig. 26** Effective temperature (ET) scale based on loci of constant wettedness caused by regulatory sweating.

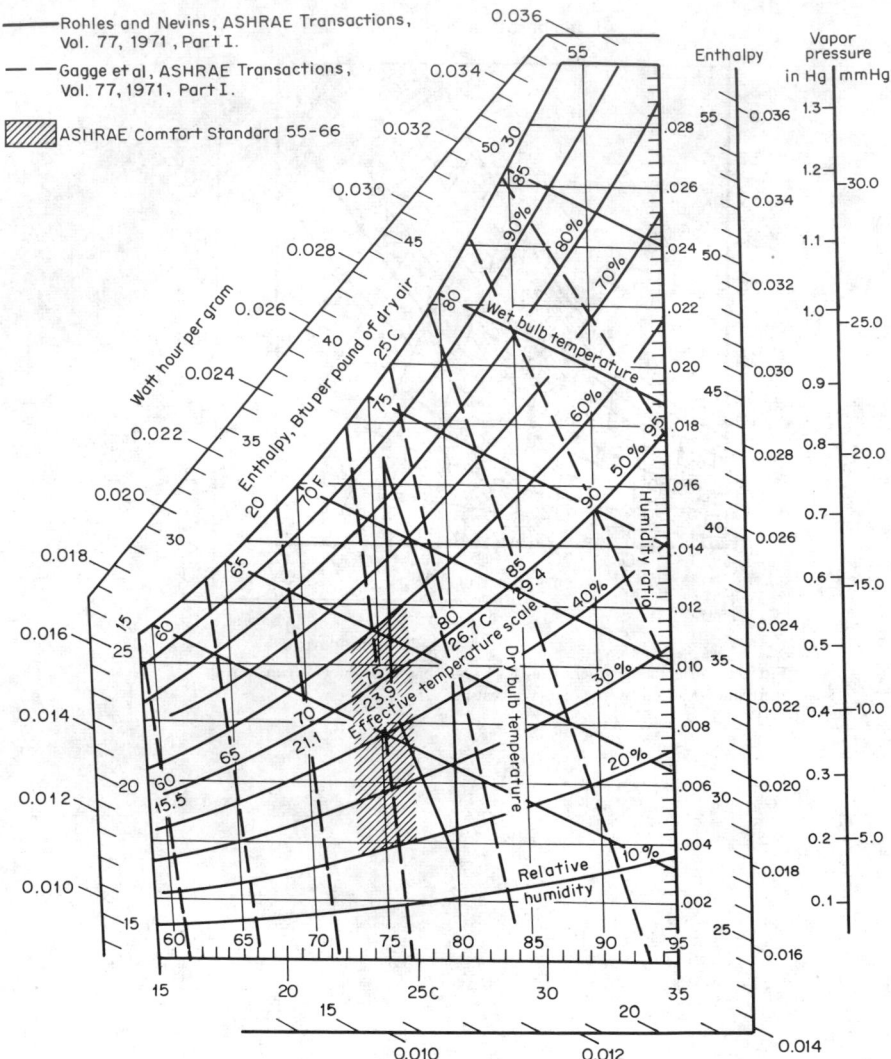

**Fig. 27**   The new ASHRAE comfort chart.

# ILLUMINATION

## by Abraham Abramowitz

REFERENCES: Amick, "Fluorescent Lighting Manual," McGraw-Hill. IES Lighting Handbook (5th ed.). Design publications of General Electric Co., Westinghouse Electric Co., and GTE-Sylvania.

## BASIC UNITS

**Candela, cd** (formerly candle) is the unit of luminous intensity of a light source. One candela is defined as the luminous intensity of $\frac{1}{6} \times 10^{-5}$ m² of projected area of a blackbody radiator operating at the temperature of solidification of platinum at standard pressure.

**Lumen, lm,** is the unit of luminous flux $\phi$. It is equal to the flux on a unit surface all points of which are one unit distant from a uniform point source of one candela. Such a point source emits $4\pi$ lumens. (For an additional definition of lumen, see the following material on vision.)

**Illuminance** $E$ is the density of luminous flux on a surface. If the foot is taken as the unit of length and the flux is uniformly distributed over the surface, the density in **lumens per square foot** is called **footcandles, fc**; in SI units **lumens per square meters lux (lx)**, are used. (One footcandle equals 10.76 lux.) In order to make the units comparable, dekalux (10 lux) is frequently used.

The term *illumination* is frequently used for the word *illuminance*. Modern practice reserves *illumination* for the process of lighting and *illuminance* for the result.

**Luminance** (formerly photometric brightness) is the luminance intensity of any surface in a given direction per unit of projected area of the surface as viewed from that direction. The unit of luminance is candela/in²; in SI units cd/m² is used. (1 cd/in² = 1,550 cd/m².) Another unit frequently used is cd/cm² (stilb). In general, a luminous surface will have a different luminance when viewed from different angles. An important exception is a **perfectly diffuse reflecting (lambertian) surface** which has a constant luminance regardless of the viewing angle. If such a surface has a luminance of 1 cd/in², it emits 452 lm/ft².

**Footlamberts**, fL, in lumens per square foot, is the unit of luminance applied in that case. While this conversion applies only to the perfectly diffuse case, it is frequently used in all cases. Thus a surface with a luminance of 1 cd/in² is said to have a luminance of 452 fL. In practice the average lumens emitted per square foot of surface is taken to be the footlamberts. The above conversion practices are deprecated.

**Subjective brightness** is the subjective attribute of any light sensation giving rise to the whole scale of qualities of becoming bright, light, brilliant, dim, or dark. Unfortunately, the term "brightness" often is used when referring to the measurable "photometric brightness" or luminance. (NOTE: The above definitions are adapted from the IES Lighting Handbook).

**Absorption, reflection, and transmission** are the general processes by which incident light flux interacts with a medium. **Absorption** is the process whereby incident flux is dissipated. **Reflection** is the process by which the incident flux leaves a surface or medium from the incident side. [NOTE: Reflection may occur as from a mirror (specular reflection), or it may be reflected at angles different from that of the incident flux to incident plane (diffuse reflection), or it may be a combination of the two types of reflection.] **Transmission** is the process by which incident flux leaves a surface or medium on a side other than the incident side. If the light ray is reduced only in intensity, the transmission is called *regular*. If the ray emerges in all directions, transmission is called *diffuse*. Both modes may exist in combination.

The incident flux $\phi_i$ equals the flux absorbed $\phi_a$, reflected $\phi_r$, and transmitted $\phi_t$. That is,

$$\phi_i = \phi_a + \phi_r + \phi_t$$

Dividing this equation by $\phi_i$, we obtain

$$1 = \phi_a/\phi_i + \phi_r/\phi_i + \phi_t/\phi_i$$
$$\text{or } 1 = \alpha + \rho + \tau$$

$\alpha$ is the *absorptance*, $\rho$ is the *reflectance*, and $\tau$ is the *transmittance*. In each case, the incident flux may be restricted to a single wavelength, a particular direction, and a given solid angle. These must be specified.

The **wavelength** of electromagnetic radiation is measured in meters. For the frequencies involved in illumination, the wavelength is given in nanometers, nm, equal to $10^{-9}$ m; micrometers, $\mu$m, equal to $10^{-6}$ m, and Angstroms, Å, equal to $10^{-10}$ m.

## VISION

Most engineering designs, (bridges, structures, roads etc.) are based on strength and are not concerned with the way the human organism reacts. The **response of the eye** is central to illuminating engineering. The **lens** of the **eye** focuses an image on the **retina**. Here a photochemical process takes place which sends nerve impulses to the brain via the optic nerve. The amount of light entering the eye is controlled by the **pupil**. The normal eye automatically accommodates itself to focus on an object, while the pupil adjusts itself to allow for a high or low level of object luminance. The sensors in the eye are known as **rods** and **cones**. The cones are clustered in a small central part of the retina called the **fovea**. They transmit a sharp image to the brain and give color response. Outside the fovea the rods predominate. They give neither a sharp image nor a color response. When the luminance of the visual field is 0.01 fL, as at night, seeing is due to the rods only and is called **scotopic vision**. At higher levels, with the cones primarily involved, seeing is called **photopic vision**. There is an intermediate region called **mesopic vision**.

The response of the eye to colors of different wavelengths is given in Fig. 1. Note the shift in maximum response at lower luminance levels called the "Purkinje shift." Note that these curves are relative ones, and that the two peaks do not correspond to the same levels of illumination. The **luminous efficacy** (lumen output per radiated watt) is 680 lm/W at the wavelength of maximum photopic response 555 nm. For white light, radiation which has the characteristic of an equal-energy spectrum with all the energy in the visual region, it is approximately 220 lm/W.

**Spectral Lumen**    If the response curve of the eye for pho-

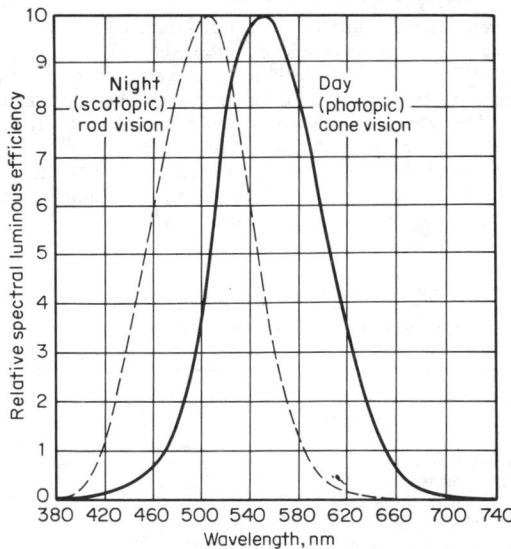

**Fig. 1**  Relative spectral luminous efficiency curves for photopic and scotopic vision, showing the Purkinje shift on the wavelength of maximum efficiency. Note the wavelength of the visual region of the electromagnetic spectrum. (*IES Lighting Handbook.*)

topic vision, versus λ in nanometers, is expressed as $k(\lambda)$, and the spectral power function of the source in watts per nanometer is taken to be $Q_e(\lambda)$, then the luminous flux is given by the equation

$$\phi_{\text{lumens}} = 680 \int_{380}^{780} k(\lambda)\, Q_e(\lambda)\, d(\lambda) \tag{1}$$

## LIGHT METERS

Early light meters compared the luminance of a diffuse highly reflecting surface with that obtained from a calibrated standard. The most common light meter in use today is similar to a photographic exposure meter. A photovoltaic cell is directly connected to a sensitive microammeter calibrated in footcandles (or decalux). The best meters (called color-corrected) have a response similar to that of the eye in photopic vision. Special shapes are used on the cover to avoid total reflection of light from the glass surface of the cell. Such meters are said to be cosine law corrected.

## LIGHT SOURCES

The original and still major source of light is the sun. Next came fire, derived from candles, oil, and gas lamps. With the discovery of electricity came arc lamps, gas-discharge lamps, and hot-filament lamps. "Flame" or hot sources give a continuous spectrum. Gas-discharge devices such as neon lamps and mercury-arc lamps give discrete, or line, spectra. The lines may be modified in various ways: by pressure broadening, use of phosphor coatings (to convert ultraviolet radiation into visible light), and using a mixture of gases. The continuous spectra of phosphors have colors which depend upon the mixture used. Light-emitting diodes, LED, consisting of a layer of two different semiconductors, are in use for display purposes.

### Color Temperature and Luminance

In general, three quantities are required to specify the color of a light and its luminous level. However, an approximate designation is used by specifying the temperature of a hot (blackbody) emitter whose color almost matches that of the light. The **color temperature** of daylight is about 6000 K and that of tungsten lamps about 2300 to 3300 K.

Different light sources have markedly different luminances as shown in Table 1. "Large" sources have low luminances, while small sources have high luminances.

### Lamps

Electric lamps are the principal source of artificial light in

**Table 1. Approximate Luminances of Various Light Sources (IES)**

| Light source | Approximate average luminance, cd/cm² |
|---|---|
| Clear sky | 0.8 |
| Candle flame (sperm) | 1.0 |
| 60-W inside frosted bulb | 60 |
| 60-W "white bulb" | 3 |
| High-intensity mercury-arc type H33, 35 atm | 150 |
| Clear glass neon tube 15 mm, 60 mA | 0.16 |
| Sodium-arc lamp, 10,000 lm size | 4,500 |

common use. They convert electrical energy into light or radiant energy.

An **incandescent-filament lamp** contains a filament which is heated by the current passing through it. The filament is enclosed in a glass bulb which has a base suitable to connect the lamp to an electrical socket. To prevent oxidation of the filament at elevated temperature, the bulb is evacuated of air or filled with an inert gas. The bulb also serves to control the light from the incandescent filament, which is essentially a point source. High luminance of the source is typically reduced by acid etching to frost the inside surface of the bulb. Silica coatings will also provide diffusion and can alter the color of the light emitted. Portions of the bulb's interior can be covered with reflecting material to give a predetermined direction to the emitted light. Chemical tinting of clear glass bulbs provides a variety of colors. Whenever the color that is normally produced by an incandescent filament is changed, the filtering process removes from the radiated light the energy of all wavelengths except those necessary to produce the desired color. This subtractive method of color alteration is less efficacious than the generation of light of varying colors by gaseous discharge.

**Sizes** and **shapes** of lamp bulbs are designated by a letter code followed by a numeral; the letter indicates the shape (Fig. 2), and the number indicates the diameter of the bulb in eighths of an inch. Thus a T-12 lamp has a tubular shape and is 1⅛ or 1½ in in diameter.

**Fig. 2** Typical filament lamp shapes: S, straight; F, flame; G, globe; A, general service; T, tubular; PS, pear shape; PAR, parabolic; R, reflector. (*Westinghouse.*)

**Incandescent lamps** are available with several **types of bases** (Fig. 3). Most general-service lamps have medium screw bases; larger or smaller screw bases are used depending on lamp wattage. Bipost and prefocus bases accurately position the filament, as in optical projection systems. Bipost lamps also serve where ruggedness and greater heat dissipation are required.

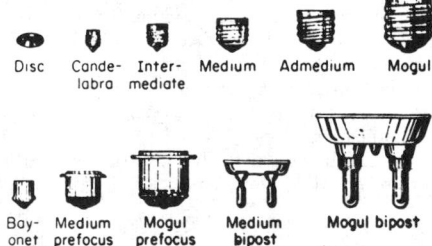

**Fig. 3** Typical incandescent-lamp bases. (*Westinghouse.*)

Incandescent-lamp filaments are generally constructed of tungsten. Tungsten has a high melting point and a low vapor pressure, which permits high operating temperatures without evaporation: the higher the operating temperature, the higher the efficacy (lumens per watt) and the shorter the life. Filament evaporation throughout the life of the lamp causes blackening of the bulb and thinning of the filament with consequent lower light output.

Tungsten filaments are also placed in compact quartz tubes filled with a halogen atmosphere where the tungsten halide lighting source continuously returns evaporated tungsten particles to the filament. The inside walls do not blacken, and light output remains fairly constant throughout the life of the lamp.

A **fluorescent lamp** consists of a glass tube coated on the inside with phosphor powders which fluoresce when excited by ultraviolet light; filament electrodes are mounted in end seals connected to base pins (Fig. 4). The tube is filled with a rare

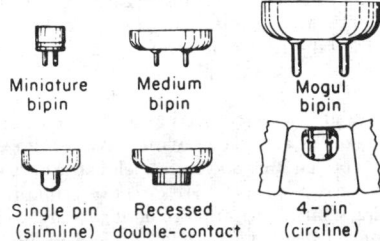

**Fig. 4**  Fluorescent-lamp bases. (*Amick.*)

gas (such as argon) and a drop of mercury (Fig. 5), and operates at a relatively low pressure. In operation, electrons are emitted from the hot electrodes. These electrons are accelerated by the voltage across the tube until they collide with mercury atoms, causing them to be ionized and excited. When the mercury atom returns to its normal state, mercury spectral lines in both the visible and the ultraviolet region are generated. The low pressure enhances ultraviolet radiation. The

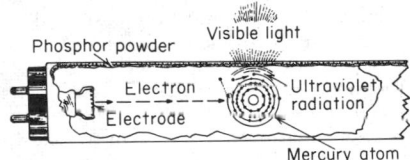

**Fig. 5**  Fluorescent-lamp operation. (*Westinghouse.*)

ultraviolet radiation excites the phosphor coating to luminance. The resulting light output is not only much higher than that obtained from the mercury lines alone but also results in a continuous spectrum with colors dependent upon the phosphors used.

As with all gas-discharge devices, these lamps have negative voltampere characteristics. Unless the voltage difference between the applied voltage and the lamp operating voltage is absorbed in some way, damaging currents will result. A reactor is used in series with the lamp. It may be capacitive or inductive (many turns of wire on an iron core). The supply voltage should be at least twice the lamp operating voltage. Where this is not the case, the supply voltage (up to 277 V is used) is stepped up by an autotransformer. The necessary reactance is frequently part of the transformer leakage inductance.

For starting purposes, voltages higher than twice lamp operating voltages may be used. For minimum line current a power-factor-correcting capacitor is used and assembled with

the autotransformer. A capacitor is put across the lamp to reduce radio-frequency interference with nearby radio receivers. All these elements are placed inside a case filled with a potting compound. The assembly is called a ballast. The object of the compound is to reduce noise from the case lamination vibrations and to improve heat dissipation. Built-in thermal protectors, which deenergize the ballast when dangerous temperatures are reached, are now required in the United States. Ballast manufacturers rate their units by noise levels.

To start a fluorescent lamp, electron emission from the electrodes must be induced. Two methods are generally employed: (1) the filament electrodes are heated by passing current through them; (2) a high voltage, sufficient to start an electric discharge in the lamp, is applied across it. Once a discharge starts, mercury-ion bombardment keeps the filaments at a hot electron-emitting temperature. Lamps are designed for either type of operation. The first group is further divided into preheat lamps and rapid-start lamps. Some lamps can be used for both types of circuits. The lamp current is carried primarily by electrons emitted from the filaments. The end or weakening of electron emission is an important cause of the end of lamp life.

**Preheat circuits** contain starters (Fig. 6) which are switches, closed when power is first applied, permitting current to flow and preheat the electrodes. After a predetermined period of time, the starter switch opens, throwing a potential across the lamp which starts the discharge. Lamps used for preheat circuits have bi-pin bases (Fig. 4).

**Instant-start circuits** have ballasts which apply sufficient voltage across the lamp to induce current flow without preheating the electrodes. **Slimline lamps** are the principal instant-start type. They have single-pin bases because no preheating is required, are available in sizes up to 8 ft in length, and can have varying lumen outputs dependent upon the ballast current rating and wattage. Slimline lamps are available in operating amperages of 200 and 425 mA. Because of their high starting voltage, they generally employ spring-loaded push-pull lampholders which disconnect the ballast circuit unless the lamps are properly seated in position.

Instant-start lamps are sometimes available with bi-pin

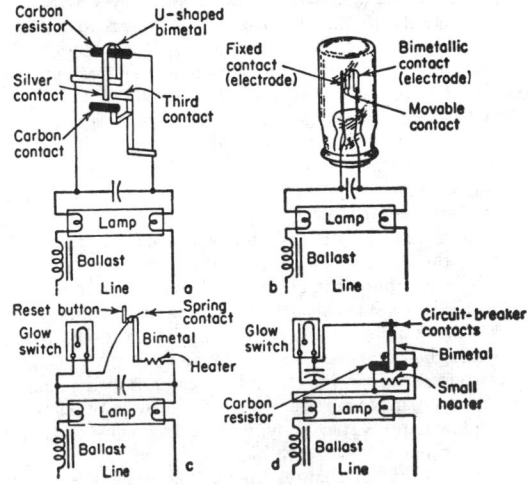

**Fig. 6**  Starter switches for preheat cathode circuits. (*IES.*)

bases similar to those used in preheat lamps. In these instances the lead wires from the pins are connected together inside the lamp. These lamps, marked "instant start," are not interchangeable with rapid-start equipment. Cold-cathode lamps employ the instant-start principle, have cylindrical iron electrodes, and tend to be less efficacious at shorter lengths because of high wattage losses at the electrodes. They are limited to low current densities because the electrodes operate at temperatures below that necessary for thermionic emission. Cold-cathode lamps, whose operation is not affected by dimming or flashing, have long life and are generally used for custom-built shapes and patterns that require bending, such as for electric signs.

**Rapid-start** circuit ballasts have separate windings for the electrodes which are immediately and continuously heated when the circuit is energized. This rapid heating causes sufficient ionization in the lamp for a discharge to start from the voltage of the main ballast windings. Two-lamp rapid-start ballasts are of the series sequence type, in which the lamps start in sequence and, when fully lighted, operate in series. Rapid-start lamps are available in operating amperages from 430 to 1,500 mA.

Typical fluorescent lamp circuits are shown in Fig. 7.

**High-intensity-discharge lamps** consist of tubes in which electric arcs in a variety of materials are produced. Outer glass jackets provide thermal insulation in order to maintain the arc tube temperature. The temperature and amount of material is controlled so that the discharge operates in a vapor pressure of several atmospheres. This results in enhancing the radiation in the visible region.

**Mercury-vapor lamps** consist of mercury-argon-filled quartz tubes surrounded by a nitrogen-filled glass jacket. Clear lamps radiate the visible mercury lines (bluish-green). Ultraviolet radiation is absorbed to some extent by the outer jackets. The color of the light and the lumen output is improved by coating the inside of the outer jackets with a phosphor. When excited by the ultraviolet radiation of the arc, the phosphors add light in the red part of the spectrum to the output. The resulting lamps are called white, color-improved, or de luxe white. The lamps start by a discharge in argon between an electrode and a starting electrode (see Fig. 8). As the mercury vaporizes, the pressure builds up and the discharge transfers to a mercury discharge. This takes several minutes. After shutdown, the lamps cannot be restarted until the inner tube pressure drops so that an argon discharge can start.

**Metal halide (multivapor) lamps** use small quantities of sodium, thallium, scandium, dysprosium, and indium iodides in addition to the usual mercury-argon mix. Color is improved and output substantially increased over high-intensity-discharge lamps using mercury alone. While the construction is similar to mercury lamps, a bimetal switch is built into the lamp to short out the starting resistor after the lamps start. A vacuum jacket is used around the quartz discharge tube (see Fig. 8).

**High-pressure sodium-vapor lamps** use metallic sodium sealed in translucent aluminum oxide tubes. This material is used to withstand the corrosive effect of hot sodium vapor. For starting purposes a xenon fill gas and a sodium-mercury amalgam is used. Arc temperatures are maintained by an outer vacuum jacket. The lamp is started by generating a high-voltage pulse for about a microsecond (see Fig. 8).

High-pressure discharge lamps, like fluorescent lamps, require ballasts. These provide the necessary voltage, react-ances, and power-factor-correcting capacitors. Typical circuits are shown in Fig. 7.

Comparative lamp efficacies (lumens/watt) are given in Table 2. Lamp data for commonly used incandescent, fluorescent, and high-intensity-discharge lamps are listed in Tables 3, 4, and 5.

### Luminaires

Luminaires are generally categorized as **industrial, commercial,** or **residential.** Use within these categories usually determines the quality and ruggedness of materials of construction. Generally speaking, style, ornament, and in most cases low cost are prime considerations for residential fixtures. Industrial fixtures require low maintenance, low operating cost, efficiency, and durability. Commercial fixtures combine the elements of all of these and place heavy emphasis on visual comfort.

Luminaires are classified by the International Commission on Illumination in accordance with the percentages of total luminaire output emitted above and below the horizontal (Fig. 9). Industrial fixtures usually are direct or semidirect.

Luminaires control the source of light so that it can be better used for a given seeing task. Materials used in luminaires are designed to reflect, refract, diffuse, or obscure light.

**Reflectors** are commonly made of specular alzak aluminum, glass, baked-enameled coated steel, and porcelain-enameled coated steel.

**Lenses** with prismatic patterns refract light sources to disperse the rays or to direct them most effectively. Lenses are of glass, acrylic plastic (Plexiglas), or polystyrene. **Glass** is generally superior for incandescent fixtures because of its heat resistance. **Acrylic** lenses, in fluorescent lamps, have a light life at least twice that of **polystyrene** because of their color stability. **Translucent** glass and plastic are used in bottom lenses, louvers, and side panels to diffuse light and to obscure light sources. Baked-enamel coated-steel **louvers,** in egg-crate, concentric-ring, or cellular configurations are widely used to shield light sources from normal viewing angles.

Fixture bodies, trims, and lens frames are commonly constructed of steel, electrogalvanized and/or treated with a rust-inhibiting coating, and painted with several coats of baked enamel. Stamped, spun, cast, and die-cast aluminum are also used.

### PRESCRIBING ILLUMINATION

The object of a **lighting design** is to provide sufficient illuminance for a given seeing task without introducing discomfort. Sufficient light is not difficult to obtain with modern light sources, but unless properly placed and controlled, uncomfortable, glaring light will result.

A given task has a size, luminance contrast, luminance, and color. The luminance of a perfectly diffusing reflecting surface is given by

$$L = E\rho \qquad (2)$$

where $E$ is illuminance in footcandles, $\rho$ its reflectance (ratio of reflected light flux to incident flux), and luminance $L$ is in footlamberts. The contrast $C$ between two adjacent areas is given by

$$C = (L_b - L_o)/L_b \qquad (3)$$

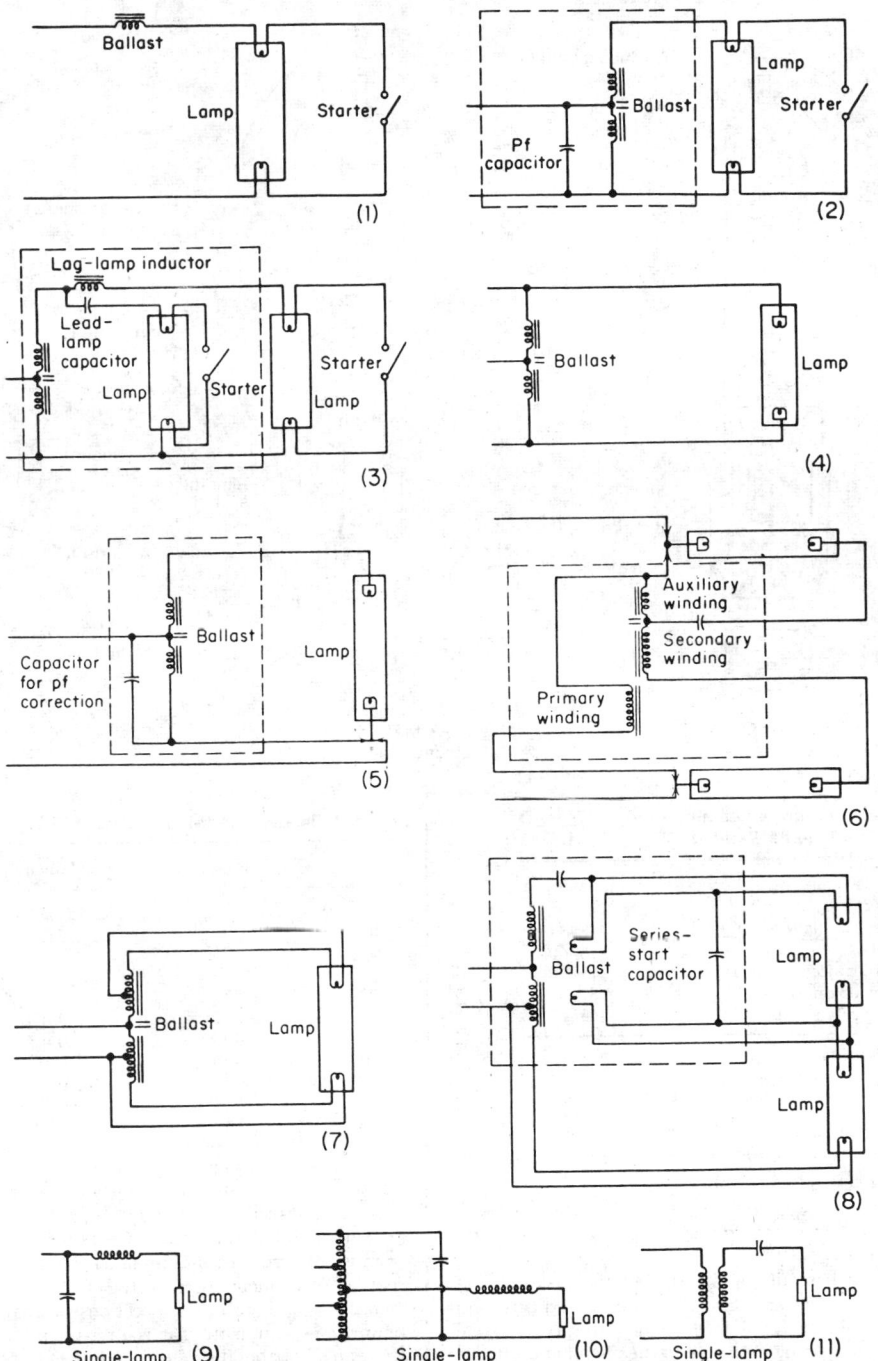

**Fig. 7**  Typical circuits for fluorescent and high-intensity discharge lamps. (1) Basic preheat circuit; (2) preheat circuit with autotransformer to step up voltage and capacitor to correct power factor; (3) headlag preheat; (4) basic instant-start circuit; (5) instant-start circuit showing disconnect lampholder; (6) typical series instant-start circuit; (7) basic rapid-start circuit; (8) two-lamp series lead circuit; (9) mercury-reactor ballast circuit; (10) mercury autotransformer circuit; (11) mercury stabilizing ballast circuit. *(GE.)*

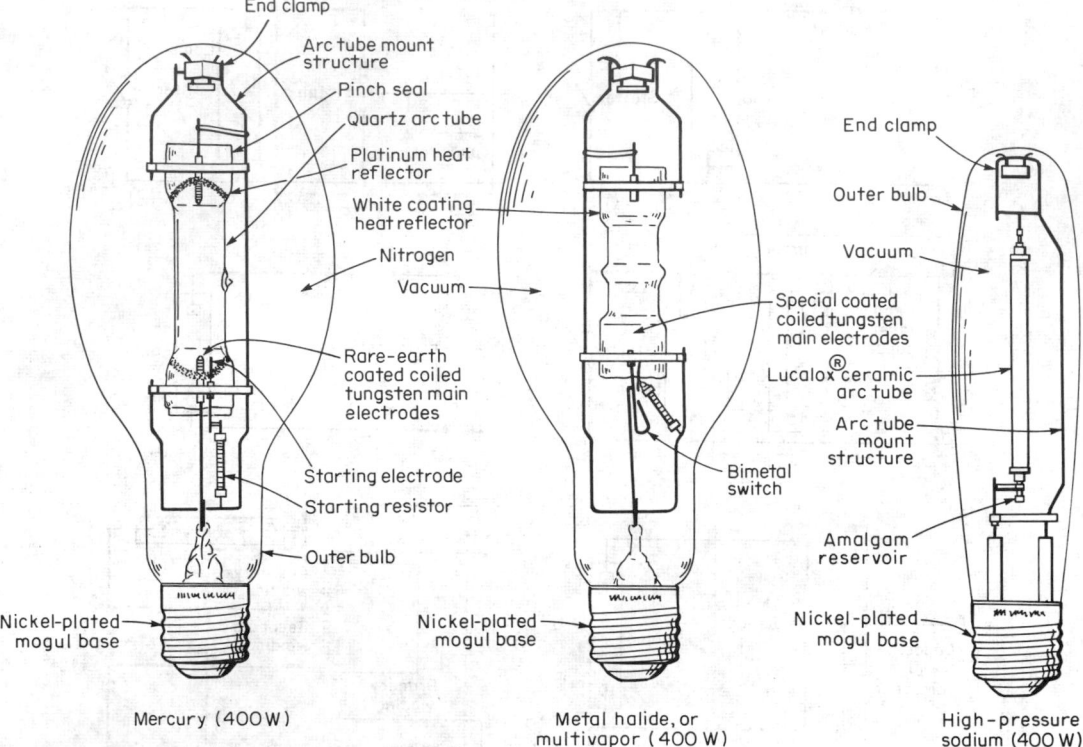

**Fig. 8** High-output lamps. *(GE Co.)*

**Table 2. Comparable Luminous Efficacies (lumens/watt)\* (IES)**

| Lamp | Lumens/watt |
|---|---|
| Tungsten incandescent | 8–33 |
| High-intensity mercury † | 28–60‡ |
| Fluorescent† | 13–90‡ |
| Metal halide (multivapor)† | 69–100 |
| High-pressure sodium† | 100–117 |

\*Constantly being improved.
†Ballast losses not included.
‡Depends upon lamp size, type, and color.

**Table 3. Incandescent-Lamp Data (IES Lighting Handbook)**

| Watts | Bulb size | Initial lumens | Rated life, h |
|---|---|---|---|
| 25 | A-19 | 230 | 2,500 |
| 40 | A-19 | 455 | 1,500 |
| 60 | A-19 | 860 | 1,000 |
| 75 | A-19 | 1,180 | 750 |
| 100 | A-19 | 1,740 | 750 |
| 150 | A-21 | 2,880 | 750 |
| 200 | A-23 | 4,000 | 750 |
| 300 | PS-30 | 6,100 | 750 |
| 500 | PS-35 | 10,600 | 1,000 |
| 750 | PS-52 | 17,000 | 1,000 |
| 1,000 | PS-52 | 23,600 | 1,000 |

For general-service lamps 115-, 120-, and 125-V service, inside frosted.
NOTE: Lamps are constantly being improved. The latest manufacturer's data should be used for accuracy.

where $L_b$ is the luminance of the larger or background area and $L_o$ is the object or task luminance for a given illuminance $E$. Substituting Eq. **(2)** into Eq. **(3)**,

$$C = \frac{E\rho_b - E\rho_o}{E\rho_b} = (\rho_b - \rho_o)/\rho_b \qquad \textbf{(4)}$$

Thus, contrast is basically a function of the task-to-background reflectance. Most surfaces are not perfectly diffuse and require the use of the luminance factor $\beta$ (ratio of actual luminance for a given viewing angle to the illumination under actual conditions) instead of $\rho$. The contrast which can just be seen or detected (minimum perceptible contrast) is a function of background luminance for a given task.

Unfortunately most visual tasks have mirrorlike (specular) or semimirrorlike surfaces. This results in reducing the work contrast. Where a highly luminous object such as an incandescent-lamp filament is reflected from a polished surface, "reflected glare" results. Reflections of a large luminous area by matte or semimatte material results in "veiling glare." For work on a horizontal surface, the line of sight from the eye for most people is from 0 to 40° from the vertical, with a peak angle of about 25°. If light from the source is directly reflected into the eyes, contrast and resulting visibility are greatly reduced. When a highly luminous source is directly reflected

**Table 4. Fluorescent-Lamp Data***

| Nominal lamp, watts† | Length, in | Bulb size (diam in 1/8 in) | Approximate lamp | | | Cool white lumens at 100 h | Related average life, h, 3 h burning/start |
|---|---|---|---|---|---|---|---|
| | | | Amperes | Volts | Watts | | |
| Preheat starting | | | | | | | |
| 15 (19.5)(19.5) | 18 | T12 | 0.33 | 47 | 13 | 775 | 7,500 |
| 20 (25)(25) | 24 | T12 | 0.38 | 57 | 20.5 | 1,250 | 7,500–9,000 |
| 30 (40.5)(38.5) | 36 | T8 | 9.36 | 98 | 30 | 2,190 | 7,500 |
| 40 (52)(48) | 48 | T12 | 0.43 | 103 | 39.8 | 3,200 | 18,000 |
| 90 (110)(102) | 60 | T17 | 1.52 | 63 | 87.5 | 6,350 | 9,000 |
| Rapid start (lightly loaded lamps) | | | | | | | |
| 30 (44)(36) | 36 | T12 | 0.43 | 81 | 32.4 | 2,300 | 18,000 |
| 40 (54)(46) | 48 | T12 | 0.43 | 100 | 40.7 | 3,200 | 1,800 |
| Rapid start (medium loaded lamps—high output) | | | | | | | |
| 60 (85)(70) | 48 | T12 | 0.8 | 76 | 61 | 4,150 | 12,000 |
| 85 (106)(100) | 72 | T12 | 0.8 | 112 | 86 | 6,450 | 12,000 |
| 110 (130)(125) | 96 | T12 | 0.8 | 149 | 111 | 9,150 | 12,000 |
| Rapid start (highly loaded lamps—super high output) | | | | | | | |
| 110 (140)(120) | 48 | T12 | 1.5 | 80 | 110 | 11,000 | 9,000 |
| 165 (210)(160) | 72 | T12 | 1.5 | 120 | 165 | 11,000 | 9,000 |
| 215 (260)(225) | 96 | T12 | 1.5 | 155 | 215 | 15,300 | 9,000 |
| Rapid start (highly loaded lamps—power grove§) | | | | | | | |
| 110 (140)(120) | 48 | PG17 | 1.5 | 85 | 110 | 6,800 | 9,000 |
| 165 (210)(160) | 72 | PG17 | 1.5 | 120 | 165 | 11,000 | 9,000 |
| 215 (260)(225) | 96 | PG17 | 1.5 | 175 | 215 | 16,000 | 9,000 |
| Instant start (slimline) | | | | | | | |
| 40 (-)(4/)‡ | 48 | T12 | 0.425 | 88 | 39 | 2,950 | 9,000–12,000 |
| 57 (-)(70)‡ | 72 | T12 | 0.425 | 147 | 57.5 | 4,500 | 9,000–12,000 |
| 75 (-)(86)‡ | 96 | T12 | 0.425 | 200 | 75 | 6,200 | 12,000 |
| Circline lamps | | | | | | | |
| 22 (34) | 8¼ OD | T9 | 0.39 | 62 | 22.5 | 980 | 7,500–12,000 |
| 32 (45)(45) | 12 OD | T10 | 0.435 | 80 | 33 | 1,750 | 7,500 |
| 40 (54)(54) | 16 OD | T10 | 0.42 | 109 | 41.5 | 2,450 | 7,500 |

Adapted from IES Lighting Handbook.
*Lamps are continuously being improved. For design purposes consult the latest manufacturers' data.
†Numerals in parentheses are the input wattages per lamp when single lamp and two-lamp 120-V ballasts are used.
‡For series ballast.
§General Electric Co. trademark.

into the eye, it is possible to completely obliterate a task. Veiling reflections are controlled by proper luminaire design and placement of fixtures. Some designs keep flux out of the 0 to 40° zone. As far as possible the work area should have a matte surface without shining details, and light should come from the side or behind the worker.

In addition to veiling reflectance, there is a reduction in contrast due to light directly entering the eye from the source. This is called the disability glare effect. It produces a light veil over the image of the task on the retina. It is not a serious problem in interior lighting, but it is important in roadway lighting and similar situations.

### The Illuminating Engineering Society (IES) Method for Prescribing Illumination

The criteria used by the IES for prescribing illumination are based upon both visual performance and visual comfort.

**Visual-Performance Criteria**    **Visual-performance potential** refers to the best completion of a task using the eyes in a fixed position for a given visibility. Actual performance will be less

### Table 5. High-Intensity-Discharge Lamp Data*

| Watt† | Nominal lamp | | Approx initial lumens‡ (100 h) | Life, h |
|-------|---------|---------|----------------|---------|
|       | Voltage | Amperes |                |         |
| *Mercury lamps* | | | | |
| 100   | 130 | 0.85–0.9 | 2,500–4,000 | 10,000–24,000+ |
| 175   | 130 | 1.5 | 6,000–8,500 | 16,000–24,000+ |
| 250   | 135 | 2.1 | 11,270–13,000 | 16,000–24,000+ |
| 400   | 135 | 3.2 | 12,00–23,000 | 12,000–24,000+ |
| 700   | 265 | 5.0 | 28,000–43,500 | 16,000–24,000+ |
| 1,000 | 265 | 4.0 | 40,000–60,000 | 16,000–24,000+ |
| *Metal-halide lamps* | | | | |
| 175   | 130 | 1.5 | 12,000 | 7,500 |
| 400   | 130 | 3.3 | 32,000–34,000 | 8,000–10,500 |
| 1,000 | 252 | 4.3 | 85,500–100,000 | 6,000–8,000 |
| *High-pressure sodium-vapor lamps* | | | | |
| 250   | 100 | 3.0 | 25,500 | 10,000 |
| 400   | 100 | 4.7 | 47,000 | 15,000 |

Abstracted from IES Lighting Handbook.
*Lamps are continuously being improved. For design purposes, consult the latest manufacturers' data.
†Necessary ballasts consume approximately an additional 10 percent energy.
‡Dependent upon ballast used, lamps may have outputs which change with burning position.

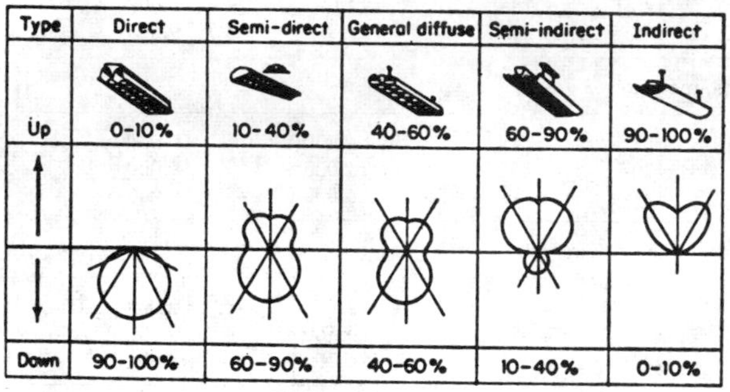

| Type | Direct | Semi-direct | General diffuse | Semi-indirect | Indirect |
|------|--------|-------------|-----------------|---------------|----------|
| Up | 0-10% | 10-40% | 40-60% | 60-90% | 90-100% |
| Down | 90-100% | 60-90% | 40-60% | 10-40% | 0-10% |

**Fig. 9** Classification of general lighting units by their percentage of upward and downward light. *(Amick.)*

than the potential. The required illuminance is determined by the task visibility. This, in turn, is determined by the level of illuminance, spatial distribution of the illumination reaching the task, and the resulting luminance of the task and its surroundings.

For a given task and background luminance, a luminance task contrast on the border of visibility and invisibility can be found. An average curve for a 4-minute (angle subtended at the eyes) disk exposed for 1/5 s, with the eyes in a fixed position, is shown in Fig. 10, and marked VL-1, the **visibility reference function.** The measurements were made under reference conditions of diffuse illumination of the task with uniform background luminance. The curve marked VL-8, the **visual performance criteria function,** has been scaled up eight times to allow for the more usual dynamic seeing tasks where the eyes move and the position of the object is not known.

Tasks other than the standard disk are evaluated by a contrast-reducing visibility meter known as a **visual task evalua-**tor, VTE. Each task of interest is placed under the reference diffuse illumination of a photometric sphere whose luminance is matched to the task background luminance. The sphere illumination simulates that available in the shade of a tree, or better, under an overcast sky. The VTE yields a value of equivalent contrast ($\tilde{C}$) equal to the luminance contrast of the 4-minute disk reference task equal in visibility to the task at the visibility threshold. From $\tilde{C}$, the required reference luminance $L_r$ is found from curve VL-8 of Fig. 10. The necessary illuminance under sphere illumination is equal to $L_r/\beta$.

**Determination of Design Illumination for Actual Conditions** The effect of veiling reflections has been discussed in reducing task contrast and visibility. The contrast $\tilde{C}$ measured under actual conditions will be higher than the effective contrast $\tilde{C}_e$ under the reference sphere illumination conditions.

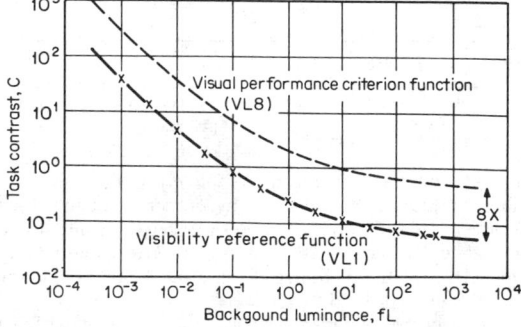

**Fig. 10** Plot of the visibility reference function (solid curve) representing task contrast required at different levels of task background luminance to achieve threshold visibility for a 4-min luminous disk exposed for 1/5 s. The dashed curve represents the visual-performance criterion function (values of the solid curve multiplied by 8). *(IES.)*

The ratio $\tilde{C}_e/\tilde{C}$ is called the **contrast rendition factor (CRF)**. For an actual installation and task both $\tilde{C}_e$ and $\tilde{C}$ are obtained by using the VTE. Fig. 11 shows how the effective illuminance is obtained from VL-8 of Fig. 10 after CRF has been measured.

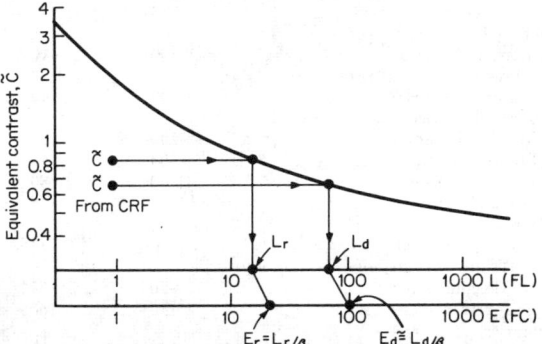

**Fig. 11** Illustration of the determination of the design illuminance $E_d$. The known task with a $\tilde{C}$ as shown requires an $L_r$ and $E_r$ as shown (under references lighting conditions—sphere illumination). Under the real luminous environment CRF has been used to determine $\tilde{C}_e$ (where $\tilde{C}_e = \tilde{C} \times$ CRF). $\tilde{C}_e$ is then used to determine $L_d$ and finally $E_d$. (IES.)

The **equivalent illuminance** $E_r$ found from the $L_r$ corresponding to $\tilde{C}_e$ is called the **equivalent sphere illumination ESI**. This is the reference illuminance necessary for the task, if there are no veiling reflections. If necessary, a similar factor, **disability glare factor DGF**, is used for disability glare. The ratio of ESI to the actual designed illuminance $E_d$, $E_s/E_d$, is the **lighting-effectiveness factor LEF**. For some conditions the CRF can be greater than 1, with the result that LEF is also greater than 1.

**Relative-Contrast Sensitivity (RCS)** In practice, instead of using the visual-performance criterion functions of Fig. 10, the relative-contrast sensitivity (RCS) function of luminance is used. It is essentially the normalized reciprocal of curve VL-8, and is shown in Fig. 12. The ordinates of these curves may be obtained from the equation

$$\text{RCS} = 45.9/\tilde{C} \qquad (5)$$

Typical values of RCS are given in Table 6. Veiling reflections reduce the effective RCS by the CRF, so that

$$\text{RCS}_e = \text{RCS} \times \text{CRF} \qquad (6)$$

The method of using the standard RCS function of luminance

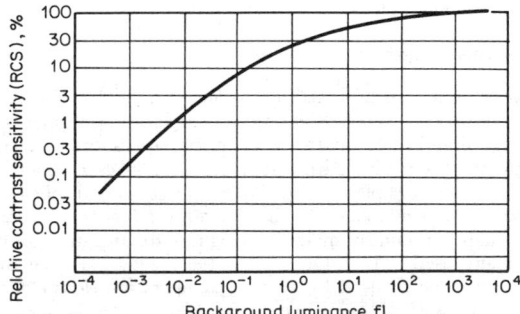

**Fig. 12** Standard RCS function of illuminance. (IES.)

**Table 6. Complete Range of Values of Relative-Contrast Sensitivity RCS as a Function of Task Luminance $L_1$ (IES.)**

| RCS, % | Task luminance, fL |
|---|---|
| 100 | 2,920 |
| 93.1 | 875 |
| 83.3 | 292 |
| 72.4 | 87.5 |
| 62.2 | 29.2 |
| 49.8 | 8.75 |
| 36.2 | 2.92 |
| 22.4 | 0.875 |
| 13.5 | 0.292 |
| 6.9 | 0.0875 |
| 3.3 | 0.0292 |
| 1.3 | 0.00875 |
| 0.46 | 0.00292 |
| 0.14 | 0.000875 |
| 0.047 | 0.000292 |

is shown in Fig. 13. From the measured task luminance $L_t$ the corresponding RCS is obtained and multiplied by the CRF to give the effective luminance $L_e$. The effective sphere illumination $E_s = L_e/\beta$ (Fig. 12).

**Visual-Comfort Criteria** High luminances directly or reflected in the field of view can cause discomfort without necessarily interfering with seeing even though visual performance may be impaired. This discomfort glare can be caused by direct glare from sources which have too high a luminance, are inadequately shielded, or have too great an area. Lighting systems are rated by a **visual-comfort probability, VCP**, expressed as a percentage of people who, if seated in the most undesirable location, will be expected to find it acceptable. (For a complete description of VCP, see the IES Handbook.) If the following conditions are met, direct glare will not be a problem in lighting installations:

1. The VCP is 70 or more.

2. The ratio of maximum-to-average luminaire luminance does not exceed 5:1 (preferably 3:1) at 45, 55, 65, 75, and 85° from the nadir crosswise and lengthwise.

3. Maximum luminaire luminances crosswise and lengthwise do not exceed the following values:

| Angle above nadir, deg | Maximum luminance, fL |
|---|---|
| 45 | 2,250 |
| 55 | 1,605 |
| 65 | 1,125 |
| 75 | 750 |
| 85 | 495 |

Visual-comfort probabilities are complicated to calculate. An approximate guide to VCP of about 70 is to plot the crosswise and endwise luminance on the plot of Fig. 14. The VCP criterion will be satisfied if the luminance plot falls between lines *A* and *B*. The dotted curve satisfies the criterion, while curve *C* does not.

**Illuminance Levels**

Table 7 lists levels of illuminance currently recommended. While sometimes locations are listed rather than tasks, the

**Table 7. Levels of Illuminance Currently Recommended**

| Area or task | Min maintained task | |
| --- | --- | --- |
| | Footcandles | Dekalux* |
| **Factories:** | | |
| Locker rooms, toilets, washrooms | 20 | 22 |
| Storage (inactive) | 5 | 54 |
| Storage (active) | | |
| Rough, bulky | 10 | 11 |
| Medium | 20 | 22 |
| Fine | 50 | 54 |
| Inspection: | | |
| Ordinary | 50 | 54 |
| Difficult | 100 | 110 |
| Highly difficult | 200 | 220 |
| Very difficult | 500 | 540 |
| Most difficult | 1,000† | 1,080† |
| Wrapping, packing, labeling | 50 | 54 |
| Drilling, riveting, screw fastening | 70 | 75 |
| Assembly: | | |
| Rough, easy seeing | 30 | 32 |
| Rough, difficult seeing | 50 | 54 |
| Medium | 100 | 110 |
| Fine | 500† | 540† |
| Extra-fine | 1,000† | 1,080† |
| Woodworking: | | |
| Rough sawing and bench work | 30 | 32 |
| Sizing, planing, rough sanding, medium-quality machine and bench work, gluing | 50 | 54 |
| Fine work, fine sanding and finishing | 100 | 110 |
| Printing: | | |
| Font assembly—sorting, electroplating | 50 | 54 |
| Composing room, machine composition; dressing type, casting, molding, masking | 100 | 100 |
| Proofreading, imposing stones | 150 | 160 |
| Color inspection and appraisal | 200† | 220† |
| Welding: | | |
| General illumination | 50 | 54 |
| Precision manual arc welding | 1,000† | 1,080† |
| Shoe manufacturing (cutting, stitching, marking, etc.) | 300† | 320† |
| **Offices:** | | |
| Stairways, elevators, escalators, corridors | 20‡ | 22‡ |
| Conference rooms | | |
| Critical seeing tasks | 100† | 110† |
| Conferring | 30 | 33 |
| Regular office work (reading, transcribing, active filing) | 70† | 75† |
| Accounting, auditing, tabulating, business-machine operating, etc. | 150+ | 160+ |
| Cartography, designing, detailed drawing | 200+ | 220+ |
| **Schools:** | | |
| Corridors, stairways | 20 | 22 |
| Auditoriums | | |
| Assembly only | 15 | 16 |
| Used for lectures | 70+ | 75+ |
| Gymnasiums: | | |
| General exercising | 30 | 32 |
| Exhibition games | 50 | 54 |
| Cafeterias | 50 | 54 |
| Classrooms: | | |
| Classroom work, study halls, libraries | 70+ | 75+ |
| Manual arts, drafting, shops, sewing rooms, sight-saving classes, lip-reading classes | 100+† | 110+† |
| Chalk boards (supplementary) | 150+† | 160+† |
| Stores circulation areas | 30 | 32 |
| **Stores (clerk-service types):** | | |
| Merchandising areas | 100 | 110 |
| Showcases and wallcases | 200 | 220 |
| Feature displays | 500 | 540 |
| **Stores (self-service types):** | | |
| Merchandising areas | 200 | 220 |
| Showcases and wallcases | 500 | 540 |
| Feature displays | 1,000 | 1,080 |
| **Public buildings:** | | |
| Banks | | |
| Lobby (general) | 50 | 54 |
| Writing areas in lobby, posting and keypunching | 70† | 75† |
| Tellers' stations | 150† | 160† |
| **Libraries:** | | |
| Stacks | 30† | 30 |
| Study, cataloging, circulation desk | 70† | 75 |
| Card files | 100† | 110† |
| **Art galleries:** | | |
| General | 30 | 32 |
| On paintings (supplementary) minimum | 30§ | 32§ |
| On statuary (some require higher values) | 100 | 100 |

Adapted from IES Lighting Handbook.

*Dekalux is an SI unit equal to 1.076 fc. 1 dekalux = 10 lux + equivalent sphere illumination.

†Requires supplementary lighting, in addition to general lighting. The supplementary lighting requires great care to avoid direct and reflected glare.

‡Or not less than one-fifth the level in adjacent areas.

§Dark paintings with fine detail should have two to three times higher illumination.

levels have been selected for the more difficult tasks in a given area. It should be noted that the levels are for the minimum on the task at any time for young adults with normal and better than 20/30 corrected vision. Initial measured values of a new installation will show higher readings, since the designer allows for a light loss factor. Values marked by a † are for equivalent sphere illumination $E_s$. The recommended levels may be obtained by using supplementary luminaires in combination with general lighting. The latter should be not less than 20 footcandles and should contribute at least one-tenth of the total illumination level.

## LIGHTING DESIGN

Interior lighting is designed by the **lumen** method. This takes into account the interreflections of light inside a room. The average illuminance on the work plane equals the incident luminous flux $\phi$ divided by the area, or $E = \phi/A$. Lumens reaching the work plane is equal to lamp lumens multiplied by the **coefficient of utilization** CU. This factor is a function of room size, shape, and finish, mounting height of fixture, and type of luminaire used. The lumens $\phi_L$ initially available from the lamps may be reduced by ambient temperature, lower voltage, and the ballast used. As time goes by the room surfaces and luminaires become dirty, which further reduces the illuminance. In addition, lamp output falls, and some of them burn out. The total effect of all these factors is expressed by the *light-loss factor* LLF. The maintained illuminance $E_m$ is the initial illumination times the LLF, or

$$E_m = (\phi_L \times CU \times LLF)/A \qquad (7)$$

The required maintained illuminance is selected from Table 7 or from the more extensive data in the IES Handbook. A fixture and lamp is selected, and Eq. (7) is solved for the necessary lamp flux $\phi_L$. The number of luminaires $N$ is found by dividing the total lamp lumens $\phi_L$ by the lumens per fixture $\phi_F$. A trial layout is then made. A simple layout keeps

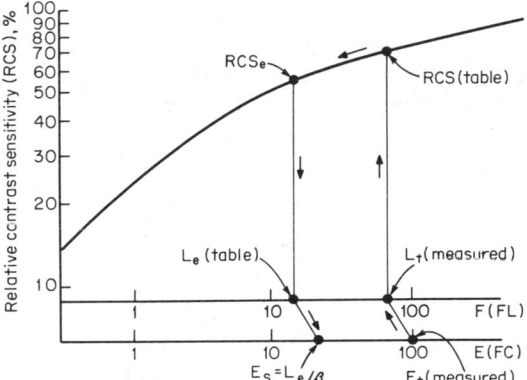

**Fig. 13** Determination of equivalent sphere illumination $E_s$.

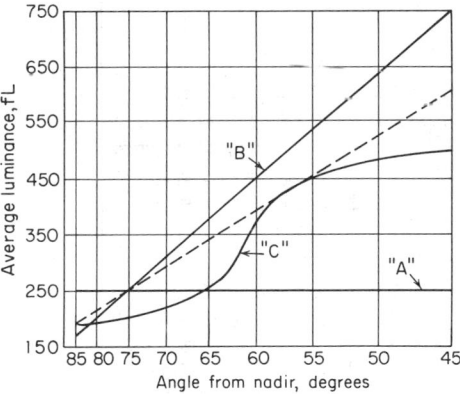

**Fig. 14** Sample luminaire average luminance, scissors curve. (*IES Lighting Handbook*, 4th ed.)

spacing between units equal to twice the distance between fixtures and wall. Other layouts are used with spacing closer to the walls (see the IES Handbook). Spacing is checked against the maximum allowable luminaire spacing from manufacturers' data to ensure uniform illumination. If equivalent sphere illumination (ESI) is required, a computer program is used to compute the ESI for the trial layout. The result should be equal to the recommended ESI. If not, a new design is undertaken by selecting another luminaire or repositioning the luminaires. Note that this procedure requires designing for a

higher illuminance than the required ESI. To ensure eye comfort, the visual-comfort probability (VCP) is investigated.

The **coefficient of utilization** is found by using the **zonal-cavity** method. In this method effects of the room proportion, luminaire suspension lengths, and work-plane height on the CU are found by dividing the room into three cavities as shown in Fig. 15. For each cavity a cavity ratio is calculated:

$$\text{Cavity ratio} = \frac{5h \, (\text{room length} + \text{room width})}{(\text{room length}) \times (\text{room width})} \qquad (8)$$

where $h = h_{RC}$ for the room cavity ratio RCR
$\quad\quad = h_{CC}$ for the ceiling cavity ratio CCR
$\quad\quad = h_{FC}$ for the floor cavity ratio FCR

Table 8 is used to obtain a single effective ceiling cavity reflectance $\rho_{CC}$ and a single effective floor cavity reflectance $\rho_{FC}$. For surface-mounted and recessed luminaires, CCR = 0 and the ceiling reflectance is used as $\rho_{CC}$. Figure 16 gives CU for selected fixtures. In using Fig. 16, interpolation may be necessary. Additional fixture data are given in the IES Hand-

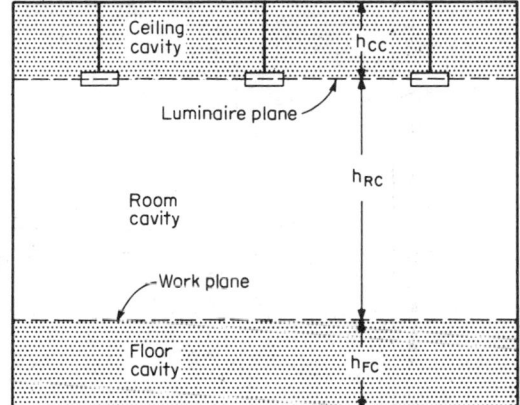

**Fig. 15** The three cavities used in the zonal-cavity method.

book. Fixture manufacturers furnish such data for their units. Those data should be used for the best accuracy. If the effective floor cavity reflectance $\rho_{FC}$ differs from 20 percent, an adjustment is made by using Table 9.

For simplicity in calculating the light-loss factor, the effects of ambient temperature, luminaire voltage variation, ballasts, and burnouts will be neglected. Room-surface dirt depreciation factors are shown in Fig. 17; luminaire dirt depreciation factors are in Fig. 18. The importance of frequent cleaning is evident. Categories are given for each fixture in Fig. 16. Lamp lumen depreciation (LLD) depends upon when lamps are replaced before complete burnout. If replacement at 30 percent rated life is used, the LLD for incandescent lamps varies from 78 to 90 percent, with an average about 87 percent. For fluorescent lamps the LLD varies from 67 to 91 percent, with an average about 82 percent. For better values consult the IES Handbook or manufacturers' data.

A design summary sheet is given in Fig. 19.

**Point (formerly Point-by-Point) Method of Design**

If uniformity of lighting is to be investigated, or if outdoor

**Table 8. Percent Effective Ceiling or Floor Cavity Reflectances for Various Reflectance Combinations (IES)**

| % base* reflectance → | 90 | | | | | | | | | | 80 | | | | | | | | | | 70 | | | | | | | | | | 60 | | | | | | | | | | 50 | | | | | | | | | |
|---|---|---|---|---|---|---|---|---|---|---|---|---|---|---|---|---|---|---|---|---|---|---|---|---|---|---|---|---|---|---|---|---|---|---|---|---|---|---|---|---|---|---|---|---|---|---|---|---|---|---|
| % wall reflectance → / Cavity ratio ↓ | 90 | 80 | 70 | 60 | 50 | 40 | 30 | 20 | 10 | 0 | 90 | 80 | 70 | 60 | 50 | 40 | 30 | 20 | 10 | 0 | 90 | 80 | 70 | 60 | 50 | 40 | 30 | 20 | 10 | 0 | 90 | 80 | 70 | 60 | 50 | 40 | 30 | 20 | 10 | 0 | 90 | 80 | 70 | 60 | 50 | 40 | 30 | 20 | 10 | 0 |
| 0.2 | 89 | 88 | 88 | 87 | 86 | 85 | 85 | 84 | 84 | 82 | 79 | 78 | 78 | 77 | 77 | 76 | 76 | 75 | 74 | 72 | 70 | 69 | 68 | 68 | 67 | 67 | 66 | 66 | 65 | 64 | 60 | 59 | 59 | 59 | 58 | 57 | 56 | 56 | 55 | 53 | 50 | 50 | 49 | 49 | 48 | 48 | 47 | 46 | 46 | 44 |
| 0.4 | 88 | 87 | 86 | 85 | 84 | 83 | 81 | 80 | 79 | 76 | 79 | 77 | 76 | 75 | 74 | 73 | 72 | 71 | 70 | 68 | 69 | 68 | 67 | 66 | 65 | 64 | 63 | 62 | 61 | 58 | 60 | 59 | 59 | 58 | 57 | 55 | 54 | 53 | 52 | 50 | 50 | 49 | 48 | 48 | 47 | 46 | 45 | 45 | 44 | 42 |
| 0.6 | 87 | 86 | 84 | 82 | 80 | 79 | 77 | 76 | 74 | 73 | 78 | 76 | 75 | 73 | 71 | 70 | 68 | 66 | 65 | 63 | 69 | 67 | 65 | 64 | 63 | 61 | 59 | 58 | 57 | 54 | 60 | 58 | 57 | 56 | 56 | 55 | 51 | 51 | 50 | 46 | 50 | 48 | 47 | 47 | 46 | 44 | 43 | 42 | 41 | 38 |
| 0.8 | 87 | 85 | 82 | 80 | 77 | 75 | 73 | 71 | 69 | 67 | 78 | 75 | 73 | 71 | 69 | 67 | 65 | 63 | 61 | 57 | 68 | 66 | 64 | 62 | 60 | 58 | 56 | 55 | 53 | 50 | 59 | 57 | 56 | 55 | 54 | 51 | 48 | 47 | 46 | 43 | 50 | 48 | 46 | 45 | 44 | 42 | 40 | 39 | 38 | 36 |
| 1.0 | 86 | 83 | 80 | 77 | 75 | 72 | 69 | 66 | 64 | 62 | 77 | 74 | 72 | 69 | 67 | 65 | 62 | 60 | 57 | 55 | 68 | 65 | 62 | 60 | 58 | 55 | 53 | 52 | 50 | 47 | 59 | 57 | 55 | 53 | 51 | 48 | 45 | 44 | 43 | 41 | 50 | 48 | 46 | 44 | 43 | 41 | 38 | 37 | 36 | 34 |
| 1.2 | 85 | 82 | 78 | 75 | 72 | 69 | 66 | 63 | 60 | 57 | 76 | 73 | 70 | 67 | 64 | 61 | 58 | 55 | 53 | 51 | 67 | 64 | 61 | 59 | 57 | 54 | 50 | 48 | 46 | 44 | 59 | 56 | 54 | 51 | 49 | 46 | 44 | 42 | 40 | 38 | 50 | 47 | 45 | 43 | 41 | 39 | 36 | 35 | 34 | 29 |
| 1.4 | 85 | 80 | 77 | 73 | 69 | 65 | 62 | 59 | 57 | 52 | 76 | 72 | 68 | 65 | 62 | 59 | 55 | 53 | 50 | 48 | 67 | 63 | 60 | 58 | 55 | 51 | 47 | 45 | 44 | 41 | 59 | 56 | 53 | 49 | 47 | 44 | 41 | 39 | 38 | 36 | 50 | 47 | 45 | 42 | 40 | 38 | 35 | 34 | 32 | 27 |
| 1.6 | 84 | 79 | 75 | 71 | 67 | 63 | 59 | 56 | 53 | 50 | 75 | 71 | 67 | 63 | 60 | 57 | 53 | 50 | 47 | 44 | 67 | 62 | 59 | 56 | 53 | 49 | 45 | 43 | 41 | 38 | 59 | 55 | 52 | 48 | 45 | 42 | 37 | 37 | 35 | 33 | 50 | 46 | 43 | 41 | 39 | 36 | 33 | 32 | 30 | 26 |
| 1.8 | 82 | 78 | 73 | 69 | 64 | 60 | 56 | 53 | 50 | 48 | 75 | 70 | 66 | 62 | 58 | 54 | 50 | 47 | 44 | 41 | 66 | 61 | 58 | 54 | 51 | 46 | 42 | 40 | 38 | 35 | 58 | 55 | 51 | 47 | 44 | 40 | 37 | 35 | 33 | 31 | 50 | 46 | 43 | 40 | 38 | 35 | 33 | 30 | 28 | 25 |
| 2.0 | 83 | 77 | 72 | 67 | 62 | 56 | 53 | 50 | 47 | 43 | 74 | 69 | 64 | 60 | 56 | 52 | 48 | 45 | 41 | 38 | 66 | 60 | 56 | 52 | 49 | 45 | 42 | 40 | 36 | 33 | 58 | 54 | 50 | 46 | 43 | 39 | 35 | 33 | 31 | 29 | 50 | 46 | 43 | 40 | 37 | 34 | 30 | 28 | 26 | 24 |
| 2.2 | 82 | 76 | 70 | 65 | 59 | 54 | 50 | 47 | 44 | 40 | 74 | 68 | 63 | 58 | 54 | 49 | 45 | 42 | 38 | 35 | 66 | 60 | 55 | 51 | 48 | 43 | 38 | 36 | 34 | 32 | 58 | 53 | 49 | 45 | 42 | 37 | 34 | 31 | 29 | 28 | 50 | 46 | 42 | 38 | 36 | 33 | 29 | 27 | 24 | 22 |
| 2.4 | 82 | 75 | 69 | 64 | 58 | 53 | 48 | 45 | 41 | 37 | 73 | 67 | 61 | 56 | 52 | 47 | 43 | 40 | 36 | 33 | 65 | 60 | 54 | 50 | 46 | 41 | 37 | 35 | 32 | 30 | 58 | 53 | 48 | 44 | 41 | 36 | 32 | 30 | 27 | 26 | 50 | 46 | 42 | 37 | 35 | 31 | 27 | 25 | 23 | 21 |
| 2.6 | 81 | 74 | 67 | 62 | 56 | 51 | 46 | 42 | 38 | 35 | 73 | 66 | 60 | 55 | 50 | 45 | 41 | 38 | 34 | 31 | 65 | 59 | 54 | 49 | 45 | 40 | 35 | 33 | 30 | 28 | 58 | 53 | 48 | 43 | 39 | 35 | 31 | 28 | 26 | 24 | 50 | 46 | 41 | 37 | 34 | 30 | 26 | 23 | 21 | 20 |
| 2.8 | 81 | 73 | 66 | 60 | 54 | 49 | 44 | 40 | 36 | 34 | 73 | 65 | 59 | 53 | 48 | 43 | 39 | 36 | 32 | 29 | 65 | 59 | 53 | 48 | 43 | 38 | 33 | 30 | 27 | 26 | 58 | 53 | 47 | 43 | 38 | 34 | 30 | 27 | 24 | 22 | 50 | 46 | 41 | 36 | 33 | 29 | 25 | 22 | 19 | 17 |
| 3.0 | 80 | 72 | 64 | 58 | 52 | 47 | 42 | 38 | 34 | 30 | 72 | 65 | 58 | 52 | 47 | 42 | 37 | 34 | 30 | 27 | 64 | 58 | 52 | 47 | 42 | 37 | 32 | 29 | 27 | 24 | 57 | 52 | 46 | 42 | 37 | 32 | 28 | 25 | 23 | 20 | 50 | 45 | 40 | 36 | 32 | 28 | 24 | 21 | 19 | 17 |
| 3.2 | 79 | 71 | 63 | 56 | 50 | 45 | 40 | 36 | 32 | 28 | 72 | 65 | 57 | 51 | 45 | 40 | 35 | 33 | 28 | 25 | 64 | 58 | 51 | 46 | 40 | 36 | 31 | 28 | 25 | 23 | 57 | 51 | 45 | 41 | 36 | 31 | 27 | 23 | 22 | 18 | 50 | 44 | 39 | 35 | 31 | 27 | 23 | 20 | 18 | 16 |
| 3.4 | 79 | 70 | 62 | 54 | 48 | 43 | 38 | 34 | 30 | 27 | 71 | 64 | 56 | 49 | 44 | 39 | 34 | 32 | 27 | 24 | 64 | 57 | 50 | 45 | 39 | 35 | 29 | 27 | 24 | 22 | 57 | 51 | 45 | 40 | 35 | 30 | 26 | 23 | 20 | 17 | 50 | 44 | 39 | 35 | 30 | 26 | 22 | 19 | 17 | 15 |
| 3.6 | 78 | 69 | 61 | 53 | 47 | 42 | 36 | 32 | 28 | 25 | 71 | 63 | 54 | 48 | 43 | 38 | 32 | 30 | 25 | 23 | 63 | 56 | 49 | 44 | 38 | 33 | 28 | 25 | 22 | 20 | 57 | 50 | 44 | 39 | 34 | 29 | 25 | 22 | 19 | 16 | 50 | 44 | 39 | 34 | 29 | 25 | 21 | 18 | 16 | 14 |
| 3.8 | 78 | 69 | 60 | 51 | 45 | 40 | 35 | 31 | 27 | 23 | 70 | 62 | 53 | 47 | 41 | 36 | 31 | 28 | 24 | 22 | 63 | 55 | 48 | 43 | 37 | 32 | 27 | 24 | 21 | 19 | 57 | 50 | 44 | 38 | 33 | 28 | 24 | 21 | 18 | 15 | 50 | 44 | 38 | 34 | 29 | 25 | 21 | 17 | 15 | 13 |
| 4.0 | 77 | 69 | 58 | 51 | 44 | 39 | 33 | 29 | 25 | 22 | 70 | 61 | 53 | 46 | 40 | 35 | 30 | 26 | 22 | 20 | 63 | 55 | 48 | 42 | 36 | 31 | 26 | 23 | 20 | 17 | 57 | 49 | 42 | 37 | 32 | 28 | 24 | 20 | 18 | 14 | 50 | 44 | 38 | 33 | 28 | 24 | 20 | 17 | 14 | 12 |
| 4.2 | 77 | 62 | 57 | 50 | 43 | 37 | 32 | 28 | 24 | 21 | 69 | 60 | 52 | 45 | 39 | 34 | 29 | 25 | 21 | 18 | 62 | 55 | 47 | 41 | 35 | 30 | 25 | 22 | 19 | 16 | 56 | 49 | 42 | 37 | 32 | 27 | 22 | 19 | 17 | 14 | 50 | 43 | 37 | 32 | 27 | 23 | 20 | 17 | 14 | 12 |
| 4.4 | 76 | 61 | 56 | 49 | 42 | 36 | 31 | 27 | 23 | 20 | 69 | 60 | 51 | 44 | 38 | 33 | 28 | 24 | 20 | 17 | 62 | 54 | 46 | 40 | 34 | 29 | 24 | 21 | 18 | 15 | 56 | 49 | 42 | 36 | 31 | 27 | 22 | 19 | 16 | 13 | 50 | 43 | 37 | 32 | 27 | 23 | 19 | 16 | 13 | 11 |
| 4.6 | 76 | 60 | 55 | 47 | 40 | 35 | 30 | 26 | 22 | 19 | 69 | 59 | 50 | 43 | 37 | 32 | 27 | 23 | 19 | 15 | 62 | 53 | 45 | 39 | 33 | 28 | 24 | 21 | 17 | 14 | 56 | 49 | 41 | 35 | 30 | 26 | 21 | 18 | 16 | 13 | 50 | 43 | 36 | 31 | 26 | 22 | 18 | 15 | 13 | 10 |
| 4.8 | 75 | 59 | 54 | 46 | 39 | 34 | 28 | 25 | 21 | 18 | 68 | 58 | 49 | 42 | 36 | 31 | 26 | 22 | 18 | 14 | 62 | 53 | 45 | 38 | 32 | 27 | 23 | 20 | 16 | 13 | 56 | 48 | 41 | 34 | 29 | 25 | 21 | 18 | 15 | 12 | 50 | 43 | 36 | 31 | 26 | 22 | 18 | 15 | 12 | 09 |
| 5.0 | 75 | 59 | 53 | 45 | 38 | 33 | 28 | 24 | 20 | 16 | 68 | 58 | 48 | 41 | 35 | 30 | 25 | 22 | 18 | 14 | 61 | 52 | 44 | 36 | 31 | 26 | 22 | 19 | 16 | 12 | 56 | 48 | 40 | 34 | 28 | 24 | 20 | 17 | 14 | 11 | 50 | 42 | 35 | 30 | 25 | 21 | 17 | 14 | 12 | 09 |
| 6.0 | 73 | 61 | 49 | 41 | 34 | 29 | 24 | 20 | 16 | 11 | 66 | 55 | 44 | 38 | 31 | 27 | 22 | 19 | 15 | 10 | 60 | 51 | 44 | 35 | 28 | 24 | 19 | 16 | 13 | 09 | 55 | 45 | 37 | 30 | 25 | 21 | 17 | 14 | 11 | 07 | 50 | 42 | 34 | 29 | 23 | 19 | 15 | 13 | 10 | 06 |
| 7.0 | 70 | 58 | 45 | 38 | 30 | 27 | 21 | 18 | 14 | 08 | 64 | 53 | 41 | 35 | 27 | 24 | 19 | 16 | 12 | 07 | 58 | 48 | 38 | 32 | 26 | 22 | 17 | 14 | 11 | 06 | 54 | 43 | 35 | 28 | 22 | 19 | 14 | 11 | 08 | 05 | 49 | 41 | 32 | 27 | 21 | 18 | 13 | 11 | 08 | 05 |
| 8.0 | 68 | 55 | 42 | 35 | 27 | 23 | 18 | 15 | 12 | 06 | 62 | 50 | 38 | 32 | 25 | 21 | 17 | 14 | 11 | 05 | 57 | 46 | 35 | 29 | 23 | 19 | 15 | 13 | 10 | 05 | 53 | 42 | 33 | 26 | 21 | 18 | 14 | 11 | 08 | 04 | 49 | 40 | 30 | 25 | 19 | 16 | 12 | 10 | 07 | 03 |
| 9.0 | 66 | 52 | 38 | 31 | 25 | 21 | 16 | 14 | 11 | 05 | 61 | 49 | 36 | 30 | 23 | 19 | 16 | 13 | 10 | 04 | 56 | 45 | 33 | 27 | 21 | 18 | 14 | 12 | 09 | 04 | 52 | 40 | 31 | 26 | 20 | 16 | 12 | 10 | 07 | 03 | 48 | 39 | 29 | 24 | 18 | 15 | 11 | 09 | 07 | 03 |
| 10.0 | 65 | 51 | 36 | 29 | 22 | 19 | 15 | 11 | 09 | 04 | 59 | 46 | 33 | 27 | 21 | 18 | 14 | 11 | 08 | 03 | 55 | 43 | 31 | 25 | 19 | 16 | 12 | 10 | 08 | 03 | 51 | 39 | 29 | 24 | 18 | 15 | 12 | 09 | 07 | 02 | 47 | 37 | 27 | 22 | 17 | 14 | 10 | 08 | 06 | 02 |

*Ceiling, floor, or floor of cavity.

| 40 | | | | | | | | | | 30 | | | | | | | | | | 20 | | | | | | | | | | 10 | | | | | | | | | | 0 | | | | | | | | | |
|---|---|---|---|---|---|---|---|---|---|---|---|---|---|---|---|---|---|---|---|---|---|---|---|---|---|---|---|---|---|---|---|---|---|---|---|---|---|---|---|---|---|---|---|---|---|---|---|---|---|
| 90 | 80 | 70 | 60 | 50 | 40 | 30 | 20 | 10 | 0 | 90 | 80 | 70 | 60 | 50 | 40 | 30 | 20 | 10 | 0 | 90 | 80 | 70 | 60 | 50 | 40 | 30 | 20 | 10 | 0 | 90 | 80 | 70 | 60 | 50 | 40 | 30 | 20 | 10 | 0 | 90 | 80 | 70 | 60 | 50 | 40 | 30 | 20 | 10 | 0 |
| 40 | 40 | 39 | 39 | 39 | 38 | 38 | 37 | 36 | 36 | 31 | 31 | 30 | 30 | 29 | 29 | 29 | 28 | 28 | 27 | 21 | 20 | 20 | 20 | 20 | 20 | 19 | 19 | 19 | 17 | 11 | 11 | 11 | 10 | 10 | 10 | 10 | 09 | 09 | 09 | 02 | 02 | 02 | 01 | 01 | 01 | 01 | 00 | 00 | 0 |
| 41 | 40 | 39 | 39 | 38 | 37 | 36 | 35 | 34 | 34 | 31 | 31 | 30 | 30 | 29 | 28 | 28 | 27 | 26 | 25 | 22 | 21 | 20 | 20 | 20 | 19 | 19 | 18 | 18 | 16 | 12 | 11 | 11 | 11 | 11 | 10 | 10 | 09 | 09 | 08 | 04 | 03 | 03 | 02 | 02 | 02 | 01 | 01 | 00 | 0 |
| 41 | 40 | 39 | 38 | 37 | 36 | 34 | 33 | 32 | 31 | 32 | 31 | 30 | 29 | 28 | 27 | 26 | 26 | 25 | 23 | 23 | 21 | 21 | 20 | 19 | 19 | 18 | 18 | 17 | 15 | 13 | 13 | 12 | 11 | 11 | 10 | 10 | 09 | 08 | 08 | 05 | 05 | 04 | 03 | 03 | 02 | 02 | 01 | 01 | 0 |
| 41 | 40 | 38 | 37 | 36 | 35 | 33 | 32 | 31 | 29 | 32 | 31 | 30 | 29 | 28 | 26 | 25 | 25 | 23 | 22 | 24 | 22 | 21 | 20 | 19 | 19 | 18 | 17 | 16 | 14 | 15 | 14 | 13 | 12 | 11 | 10 | 10 | 09 | 08 | 07 | 07 | 06 | 05 | 04 | 04 | 03 | 02 | 02 | 01 | 0 |
| 42 | 40 | 38 | 37 | 35 | 33 | 32 | 31 | 29 | 27 | 33 | 32 | 30 | 29 | 27 | 25 | 24 | 23 | 22 | 20 | 25 | 23 | 22 | 20 | 19 | 18 | 17 | 16 | 15 | 13 | 16 | 14 | 13 | 12 | 12 | 11 | 10 | 09 | 08 | 07 | 08 | 07 | 06 | 05 | 04 | 03 | 02 | 02 | 01 | 0 |
| 42 | 40 | 38 | 36 | 34 | 32 | 30 | 29 | 27 | 25 | 33 | 32 | 30 | 28 | 27 | 25 | 23 | 22 | 21 | 19 | 25 | 23 | 22 | 20 | 19 | 17 | 17 | 16 | 14 | 12 | 17 | 15 | 14 | 13 | 12 | 11 | 10 | 09 | 07 | 06 | 10 | 08 | 07 | 06 | 05 | 04 | 03 | 02 | 01 | 0 |
| 42 | 39 | 37 | 35 | 33 | 31 | 29 | 27 | 25 | 23 | 34 | 32 | 30 | 28 | 26 | 24 | 22 | 21 | 19 | 18 | 26 | 24 | 22 | 20 | 18 | 17 | 16 | 15 | 13 | 12 | 18 | 16 | 14 | 13 | 12 | 11 | 10 | 09 | 07 | 06 | 11 | 09 | 08 | 07 | 06 | 04 | 03 | 02 | 01 | 0 |
| 42 | 39 | 37 | 35 | 32 | 30 | 27 | 25 | 23 | 22 | 34 | 33 | 29 | 25 | 27 | 23 | 22 | 20 | 18 | 17 | 26 | 24 | 22 | 20 | 18 | 17 | 16 | 15 | 13 | 11 | 19 | 17 | 15 | 14 | 12 | 11 | 09 | 08 | 07 | 06 | 12 | 10 | 09 | 07 | 06 | 05 | 03 | 02 | 01 | 0 |
| 42 | 39 | 36 | 34 | 31 | 29 | 26 | 24 | 22 | 21 | 35 | 33 | 29 | 27 | 25 | 23 | 21 | 19 | 17 | 16 | 27 | 25 | 23 | 20 | 28 | 17 | 15 | 14 | 12 | 10 | 19 | 17 | 15 | 14 | 13 | 11 | 09 | 08 | 06 | 05 | 13 | 11 | 09 | 08 | 07 | 05 | 04 | 03 | 01 | 0 |
| 42 | 39 | 36 | 34 | 31 | 28 | 25 | 23 | 21 | 19 | 35 | 33 | 29 | 26 | 24 | 22 | 20 | 18 | 16 | 14 | 28 | 25 | 23 | 20 | 18 | 16 | 15 | 13 | 11 | 09 | 20 | 18 | 16 | 14 | 13 | 11 | 09 | 08 | 06 | 05 | 14 | 12 | 10 | 09 | 07 | 05 | 04 | 03 | 01 | 0 |
| 42 | 39 | 36 | 33 | 30 | 27 | 24 | 22 | 19 | 18 | 36 | 32 | 29 | 26 | 24 | 22 | 19 | 17 | 15 | 13 | 28 | 25 | 23 | 20 | 18 | 16 | 14 | 12 | 10 | 09 | 21 | 19 | 16 | 14 | 13 | 11 | 09 | 07 | 06 | 05 | 15 | 13 | 11 | 09 | 07 | 06 | 04 | 03 | 02 | 0 |
| 43 | 39 | 35 | 33 | 29 | 27 | 24 | 21 | 18 | 17 | 36 | 32 | 29 | 26 | 24 | 22 | 19 | 16 | 14 | 12 | 29 | 26 | 23 | 20 | 18 | 16 | 14 | 12 | 10 | 08 | 22 | 19 | 17 | 15 | 13 | 11 | 09 | 07 | 06 | 05 | 16 | 13 | 11 | 09 | 08 | 06 | 04 | 03 | 01 | 0 |
| 43 | 39 | 35 | 32 | 29 | 26 | 23 | 20 | 17 | 15 | 36 | 32 | 29 | 25 | 23 | 21 | 18 | 16 | 14 | 12 | 29 | 26 | 23 | 20 | 18 | 16 | 14 | 11 | 09 | 08 | 23 | 20 | 17 | 15 | 13 | 11 | 09 | 07 | 06 | 04 | 17 | 14 | 12 | 10 | 08 | 06 | 05 | 03 | 02 | 0 |
| 43 | 39 | 35 | 32 | 28 | 25 | 22 | 19 | 16 | 14 | 37 | 33 | 29 | 25 | 23 | 21 | 17 | 15 | 13 | 11 | 30 | 27 | 23 | 20 | 18 | 15 | 13 | 11 | 09 | 07 | 23 | 20 | 18 | 16 | 13 | 11 | 09 | 07 | 05 | 03 | 17 | 15 | 13 | 10 | 08 | 07 | 05 | 03 | 02 | 0 |
| 43 | 39 | 35 | 31 | 27 | 24 | 21 | 18 | 16 | 13 | 37 | 33 | 29 | 25 | 22 | 20 | 17 | 15 | 12 | 10 | 30 | 27 | 23 | 20 | 17 | 15 | 13 | 11 | 09 | 07 | 24 | 21 | 18 | 16 | 13 | 11 | 09 | 07 | 05 | 03 | 18 | 16 | 13 | 11 | 09 | 07 | 05 | 03 | 02 | 0 |
| 43 | 39 | 35 | 31 | 27 | 23 | 20 | 17 | 15 | 13 | 37 | 33 | 29 | 25 | 22 | 19 | 16 | 14 | 12 | 10 | 31 | 27 | 23 | 20 | 17 | 15 | 12 | 11 | 09 | 06 | 25 | 21 | 18 | 16 | 13 | 11 | 09 | 07 | 05 | 03 | 19 | 16 | 14 | 11 | 09 | 07 | 05 | 03 | 02 | 0 |
| 43 | 39 | 34 | 30 | 26 | 23 | 20 | 17 | 14 | 12 | 37 | 33 | 29 | 25 | 22 | 19 | 16 | 14 | 12 | 09 | 31 | 27 | 23 | 20 | 17 | 15 | 12 | 10 | 08 | 06 | 26 | 22 | 18 | 16 | 13 | 11 | 09 | 07 | 05 | 03 | 20 | 17 | 14 | 12 | 09 | 07 | 05 | 03 | 02 | 0 |
| 44 | 39 | 34 | 30 | 26 | 22 | 19 | 16 | 14 | 11 | 38 | 33 | 29 | 24 | 21 | 18 | 15 | 13 | 10 | 09 | 32 | 27 | 23 | 20 | 17 | 15 | 12 | 10 | 08 | 05 | 26 | 22 | 19 | 16 | 13 | 11 | 09 | 06 | 04 | 03 | 20 | 17 | 15 | 12 | 10 | 08 | 05 | 04 | 02 | 0 |
| 44 | 38 | 33 | 29 | 25 | 22 | 18 | 16 | 13 | 10 | 38 | 33 | 28 | 24 | 21 | 18 | 15 | 13 | 10 | 08 | 32 | 28 | 23 | 20 | 17 | 15 | 12 | 10 | 07 | 05 | 27 | 23 | 19 | 17 | 14 | 11 | 09 | 06 | 04 | 02 | 21 | 18 | 15 | 12 | 10 | 08 | 05 | 04 | 02 | 0 |
| 44 | 38 | 33 | 29 | 25 | 21 | 18 | 15 | 12 | 10 | 38 | 33 | 28 | 24 | 21 | 18 | 14 | 12 | 09 | 07 | 33 | 28 | 23 | 20 | 17 | 14 | 11 | 09 | 07 | 05 | 27 | 23 | 20 | 17 | 14 | 11 | 09 | 06 | 04 | 02 | 22 | 18 | 15 | 13 | 10 | 08 | 05 | 04 | 02 | 0 |
| 44 | 38 | 33 | 29 | 24 | 21 | 17 | 15 | 12 | 10 | 38 | 33 | 28 | 24 | 20 | 17 | 14 | 12 | 09 | 07 | 33 | 28 | 23 | 20 | 17 | 14 | 11 | 09 | 07 | 04 | 28 | 24 | 20 | 17 | 14 | 11 | 09 | 06 | 04 | 02 | 22 | 19 | 16 | 13 | 10 | 08 | 06 | 04 | 02 | 0 |
| 44 | 38 | 33 | 28 | 24 | 20 | 17 | 14 | 11 | 09 | 39 | 33 | 28 | 24 | 20 | 17 | 14 | 11 | 09 | 06 | 34 | 28 | 24 | 20 | 17 | 14 | 11 | 09 | 07 | 04 | 28 | 24 | 20 | 17 | 14 | 11 | 08 | 06 | 04 | 02 | 23 | 19 | 16 | 13 | 10 | 08 | 06 | 04 | 02 | 0 |
| 44 | 38 | 32 | 28 | 23 | 19 | 16 | 14 | 11 | 08 | 39 | 33 | 28 | 24 | 20 | 17 | 13 | 10 | 08 | 06 | 34 | 29 | 24 | 20 | 17 | 14 | 11 | 09 | 07 | 04 | 29 | 25 | 20 | 17 | 14 | 11 | 08 | 06 | 04 | 02 | 23 | 20 | 17 | 13 | 11 | 08 | 06 | 04 | 02 | 0 |
| 44 | 38 | 32 | 27 | 22 | 19 | 16 | 13 | 10 | 08 | 39 | 33 | 28 | 24 | 20 | 17 | 13 | 10 | 08 | 05 | 35 | 29 | 24 | 20 | 17 | 13 | 10 | 08 | 06 | 04 | 29 | 25 | 20 | 17 | 14 | 11 | 08 | 06 | 04 | 02 | 24 | 20 | 17 | 14 | 11 | 08 | 06 | 04 | 02 | 0 |
| 45 | 38 | 31 | 27 | 22 | 19 | 15 | 13 | 10 | 07 | 39 | 33 | 28 | 24 | 19 | 16 | 13 | 10 | 08 | 05 | 35 | 29 | 24 | 20 | 16 | 13 | 10 | 08 | 06 | 04 | 30 | 25 | 20 | 17 | 14 | 11 | 08 | 06 | 04 | 02 | 25 | 21 | 17 | 14 | 11 | 08 | 06 | 04 | 02 | 0 |
| 44 | 37 | 30 | 25 | 20 | 17 | 13 | 11 | 08 | 05 | 39 | 33 | 27 | 23 | 18 | 15 | 11 | 09 | 06 | 04 | 36 | 30 | 24 | 20 | 16 | 13 | 10 | 08 | 05 | 02 | 31 | 26 | 21 | 18 | 14 | 11 | 08 | 06 | 03 | 01 | 27 | 23 | 18 | 15 | 12 | 09 | 06 | 04 | 02 | 0 |
| 44 | 36 | 29 | 24 | 19 | 16 | 12 | 10 | 07 | 04 | 40 | 33 | 26 | 22 | 17 | 14 | 10 | 08 | 05 | 03 | 36 | 30 | 24 | 20 | 15 | 12 | 09 | 07 | 04 | 02 | 32 | 27 | 21 | 17 | 13 | 11 | 08 | 06 | 03 | 01 | 28 | 24 | 19 | 15 | 12 | 09 | 06 | 04 | 02 | 0 |
| 44 | 35 | 28 | 23 | 18 | 15 | 11 | 09 | 06 | 03 | 40 | 33 | 26 | 21 | 16 | 13 | 09 | 07 | 04 | 02 | 37 | 30 | 23 | 19 | 15 | 12 | 08 | 06 | 03 | 01 | 33 | 27 | 21 | 17 | 13 | 10 | 07 | 05 | 03 | 01 | 30 | 25 | 20 | 15 | 12 | 09 | 06 | 04 | 02 | 0 |
| 44 | 35 | 26 | 21 | 16 | 13 | 10 | 08 | 05 | 02 | 40 | 33 | 25 | 20 | 15 | 12 | 09 | 07 | 04 | 02 | 37 | 29 | 23 | 19 | 14 | 11 | 08 | 06 | 03 | 01 | 34 | 28 | 21 | 17 | 13 | 10 | 07 | 05 | 02 | 01 | 31 | 25 | 20 | 15 | 12 | 09 | 06 | 04 | 02 | 0 |
| 43 | 34 | 25 | 20 | 15 | 12 | 08 | 07 | 05 | 02 | 40 | 32 | 24 | 19 | 14 | 11 | 08 | 06 | 03 | 01 | 37 | 29 | 22 | 18 | 13 | 10 | 07 | 05 | 03 | 01 | 34 | 28 | 21 | 17 | 12 | 10 | 07 | 05 | 02 | 01 | 31 | 25 | 20 | 15 | 12 | 09 | 06 | 04 | 02 | 0 |

Fig. 16 table: Coefficients of utilization for typical luminaires.

| Typical luminaire | Maint. cat. | Max S/MH guide§ | RCR‡ ↓ | ρcc* → 80, ρw† → 50 | 30 | 10 | 70, 50 | 30 | 10 | 50, 50 | 30 | 10 | 30, 50 | 30 | 10 | 10, 50 | 30 | 10 | 0, 0 |
|---|---|---|---|---|---|---|---|---|---|---|---|---|---|---|---|---|---|---|---|
|   Porcelain-enameled ventilated standard dome with incandescent lamp (0%↑, 83½%↓) | IV | 1.3 | 0 | .99 | .99 | .99 | .97 | .97 | .97 | .92 | .92 | .92 | .88 | .88 | .88 | .85 | .85 | .85 | .83 |
| | | | 1 | .88 | .85 | .82 | .86 | .83 | .81 | .83 | .80 | .78 | .79 | .78 | .76 | .77 | .75 | .73 | .72 |
| | | | 2 | .78 | .73 | .68 | .76 | .72 | .67 | .73 | .69 | .66 | .71 | .67 | .64 | .68 | .65 | .63 | .61 |
| | | | 3 | .69 | .62 | .57 | .67 | .61 | .57 | .65 | .60 | .56 | .63 | .58 | .55 | .61 | .57 | .54 | .52 |
| | | | 4 | .61 | .54 | .49 | .60 | .53 | .48 | .58 | .52 | .48 | .56 | .51 | .47 | .54 | .50 | .46 | .45 |
| | | | 5 | .54 | .47 | .41 | .53 | .46 | .41 | .51 | .45 | .41 | .50 | .44 | .40 | .48 | .43 | .40 | .38 |
| | | | 6 | .48 | .41 | .35 | .47 | .40 | .35 | .46 | .39 | .35 | .44 | .39 | .34 | .43 | .38 | .34 | .32 |
| | | | 7 | .43 | .35 | .30 | .42 | .35 | .30 | .41 | .34 | .30 | .39 | .34 | .30 | .38 | .33 | .29 | .28 |
| | | | 8 | .38 | .31 | .26 | .38 | .31 | .26 | .37 | .30 | .26 | .36 | .30 | .26 | .35 | .30 | .26 | .24 |
| | | | 9 | .35 | .28 | .23 | .34 | .27 | .23 | .33 | .27 | .23 | .32 | .27 | .23 | .31 | .26 | .22 | .21 |
| | | | 10 | .31 | .25 | .20 | .31 | .24 | .20 | .30 | .24 | .20 | .29 | .24 | .20 | .29 | .23 | .20 | .18 |
| 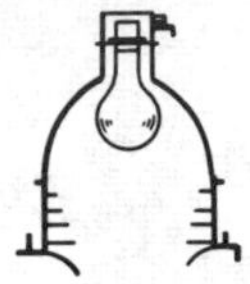  Reflector downlight with baffles and inside frosted lamp (0%↑, 44½%↓) | IV | 0.7 | 0 | .53 | .53 | .53 | .52 | .52 | .52 | .49 | .49 | .49 | .47 | .47 | .47 | .45 | .45 | .45 | .44 |
| | | | 1 | .51 | .50 | .49 | .50 | .49 | .48 | .48 | .47 | .47 | .46 | .46 | .45 | .45 | .44 | .44 | .43 |
| | | | 2 | .48 | .47 | .46 | .48 | .46 | .45 | .46 | .45 | .44 | .45 | .44 | .44 | .44 | .44 | .43 | .42 |
| | | | 3 | .47 | .45 | .44 | .46 | .45 | .43 | .45 | .44 | .43 | .44 | .43 | .42 | .43 | .42 | .41 | .41 |
| | | | 4 | .45 | .43 | .42 | .44 | .43 | .42 | .43 | .42 | .41 | .43 | .41 | .41 | .42 | .41 | .40 | .40 |
| | | | 5 | .43 | .41 | .40 | .43 | .41 | .40 | .42 | .40 | .39 | .41 | .40 | .39 | .41 | .40 | .39 | .38 |
| | | | 6 | .42 | .40 | .39 | .41 | .40 | .38 | .41 | .39 | .38 | .40 | .39 | .38 | .40 | .39 | .38 | .37 |
| | | | 7 | .40 | .38 | .37 | .40 | .38 | .37 | .39 | .38 | .37 | .39 | .38 | .37 | .38 | .37 | .36 | .36 |
| | | | 8 | .39 | .37 | .36 | .38 | .37 | .36 | .38 | .37 | .35 | .38 | .36 | .35 | .37 | .36 | .35 | .35 |
| | | | 9 | .37 | .36 | .34 | .37 | .35 | .34 | .37 | .35 | .34 | .36 | .35 | .34 | .36 | .35 | .34 | .33 |
| | | | 10 | .36 | .34 | .33 | .36 | .34 | .33 | .36 | .34 | .33 | .35 | .35 | .33 | .35 | .34 | .33 | .32 |
| 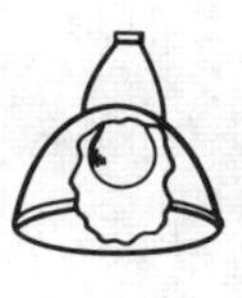  Enclosed reflector with an incandescent lamp (0%↑, 71½%↓) | V | 1.4 | 0 | .85 | .85 | .85 | .83 | .83 | .83 | .79 | .79 | .79 | .76 | .76 | .76 | .73 | .73 | .73 | .71 |
| | | | 1 | .78 | .76 | .74 | .76 | .74 | .73 | .73 | .72 | .70 | .71 | .69 | .68 | .68 | .67 | .66 | .65 |
| | | | 2 | .71 | .68 | .65 | .70 | .67 | .64 | .68 | .65 | .63 | .65 | .63 | .61 | .63 | .62 | .60 | .59 |
| | | | 3 | .65 | .61 | .57 | .64 | .60 | .57 | .62 | .59 | .56 | .60 | .57 | .55 | .59 | .56 | .54 | .53 |
| | | | 4 | .60 | .55 | .51 | .59 | .54 | .51 | .57 | .53 | .50 | .55 | .52 | .50 | .54 | .51 | .49 | .48 |
| | | | 5 | .54 | .49 | .45 | .54 | .49 | .45 | .52 | .48 | .45 | .51 | .47 | .44 | .50 | .46 | .44 | .43 |
| | | | 6 | .49 | .44 | .40 | .49 | .44 | .40 | .47 | .43 | .40 | .46 | .42 | .40 | .45 | .42 | .39 | .38 |
| | | | 7 | .44 | .39 | .35 | .44 | .39 | .35 | .43 | .38 | .35 | .42 | .38 | .35 | .41 | .37 | .35 | .33 |
| | | | 8 | .40 | .35 | .31 | .40 | .35 | .31 | .39 | .35 | .31 | .38 | .34 | .31 | .38 | .34 | .31 | .30 |
| | | | 9 | .37 | .31 | .28 | .36 | .31 | .28 | .36 | .31 | .28 | .35 | .31 | .28 | .34 | .30 | .27 | .26 |
| | | | 10 | .33 | .28 | .25 | .33 | .28 | .25 | .32 | .28 | .25 | .32 | .28 | .25 | .31 | .27 | .24 | .23 |
| 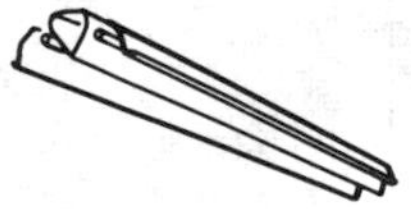 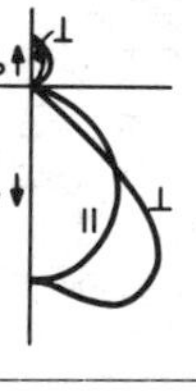 Diffuse aluminum reflector with 35°CW shielding (17%↑, 66%↓) | II | 1.5/1.3 | 0 | .94 | .94 | .94 | .90 | .90 | .90 | .82 | .82 | .82 | .75 | .75 | .75 | .69 | .69 | .69 | .66 |
| | | | 1 | .85 | .82 | .80 | .82 | .79 | .77 | .75 | .73 | .72 | .69 | .68 | .66 | .64 | .63 | .62 | .59 |
| | | | 2 | .76 | .72 | .68 | .74 | .70 | .66 | .68 | .65 | .62 | .63 | .61 | .58 | .58 | .56 | .55 | .52 |
| | | | 3 | .69 | .63 | .59 | .66 | .61 | .57 | .62 | .58 | .54 | .57 | .54 | .51 | .53 | .51 | .48 | .46 |
| | | | 4 | .62 | .56 | .51 | .60 | .54 | .50 | .56 | .51 | .47 | .52 | .48 | .45 | .48 | .45 | .43 | .41 |
| | | | 5 | .55 | .49 | .44 | .53 | .48 | .43 | .50 | .45 | .41 | .47 | .43 | .39 | .44 | .40 | .38 | .36 |
| | | | 6 | .50 | .43 | .39 | .48 | .42 | .38 | .45 | .40 | .36 | .42 | .38 | .35 | .40 | .36 | .33 | .31 |
| | | | 7 | .45 | .38 | .34 | .43 | .37 | .33 | .41 | .36 | .32 | .38 | .34 | .30 | .36 | .32 | .29 | .27 |
| | | | 8 | .40 | .34 | .29 | .39 | .33 | .29 | .37 | .31 | .28 | .34 | .30 | .26 | .32 | .28 | .25 | .24 |
| | | | 9 | .36 | .30 | .25 | .35 | .29 | .25 | .33 | .28 | .24 | .31 | .26 | .23 | .29 | .25 | .22 | .20 |
| | | | 10 | .33 | .26 | .22 | .32 | .26 | .22 | .30 | .25 | .21 | .28 | .23 | .20 | .26 | .22 | .19 | .18 |
|  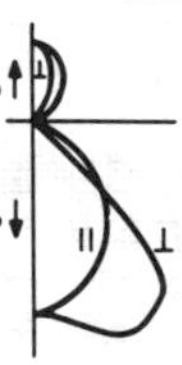 Diffuse aluminum reflector with 35°CW × 35°LW shielding (17%↑, 56½%↓) | II | 1.5/1.1 | 0 | .83 | .83 | .83 | .79 | .79 | .79 | .71 | .71 | .71 | .65 | .65 | .65 | .59 | .59 | .59 | .56 |
| | | | 1 | .75 | .72 | .70 | .72 | .69 | .68 | .65 | .64 | .62 | .60 | .59 | .58 | .55 | .54 | .53 | .50 |
| | | | 2 | .67 | .63 | .60 | .65 | .61 | .58 | .59 | .57 | .54 | .55 | .53 | .51 | .50 | .49 | .47 | .45 |
| | | | 3 | .61 | .56 | .52 | .58 | .54 | .51 | .54 | .50 | .48 | .50 | .47 | .45 | .46 | .44 | .42 | .40 |
| | | | 4 | .55 | .49 | .45 | .53 | .48 | .44 | .49 | .45 | .42 | .45 | .42 | .40 | .42 | .39 | .37 | .36 |
| | | | 5 | .49 | .44 | .40 | .47 | .42 | .39 | .44 | .40 | .37 | .41 | .38 | .35 | .38 | .35 | .33 | .31 |
| | | | 6 | .45 | .39 | .35 | .43 | .38 | .34 | .40 | .36 | .33 | .37 | .34 | .31 | .35 | .32 | .30 | .28 |
| | | | 7 | .40 | .35 | .31 | .39 | .34 | .30 | .36 | .32 | .29 | .34 | .30 | .27 | .32 | .29 | .26 | .25 |
| | | | 8 | .36 | .31 | .27 | .35 | .30 | .26 | .33 | .28 | .25 | .31 | .27 | .24 | .29 | .25 | .23 | .22 |
| | | | 9 | .33 | .27 | .23 | .32 | .26 | .23 | .29 | .25 | .22 | .28 | .24 | .21 | .26 | .22 | .20 | .19 |
| | | | 10 | .30 | .24 | .21 | .29 | .24 | .20 | .27 | .22 | .19 | .25 | .21 | .19 | .23 | .20 | .18 | .16 |

Coefficients of utilization for 20% effective floor cavity reflectance ($\rho FC = 20$)

**Fig. 16** Coefficients of utilization for typical luminaires. (*Abstracted from IES Lighting Handbook.*)

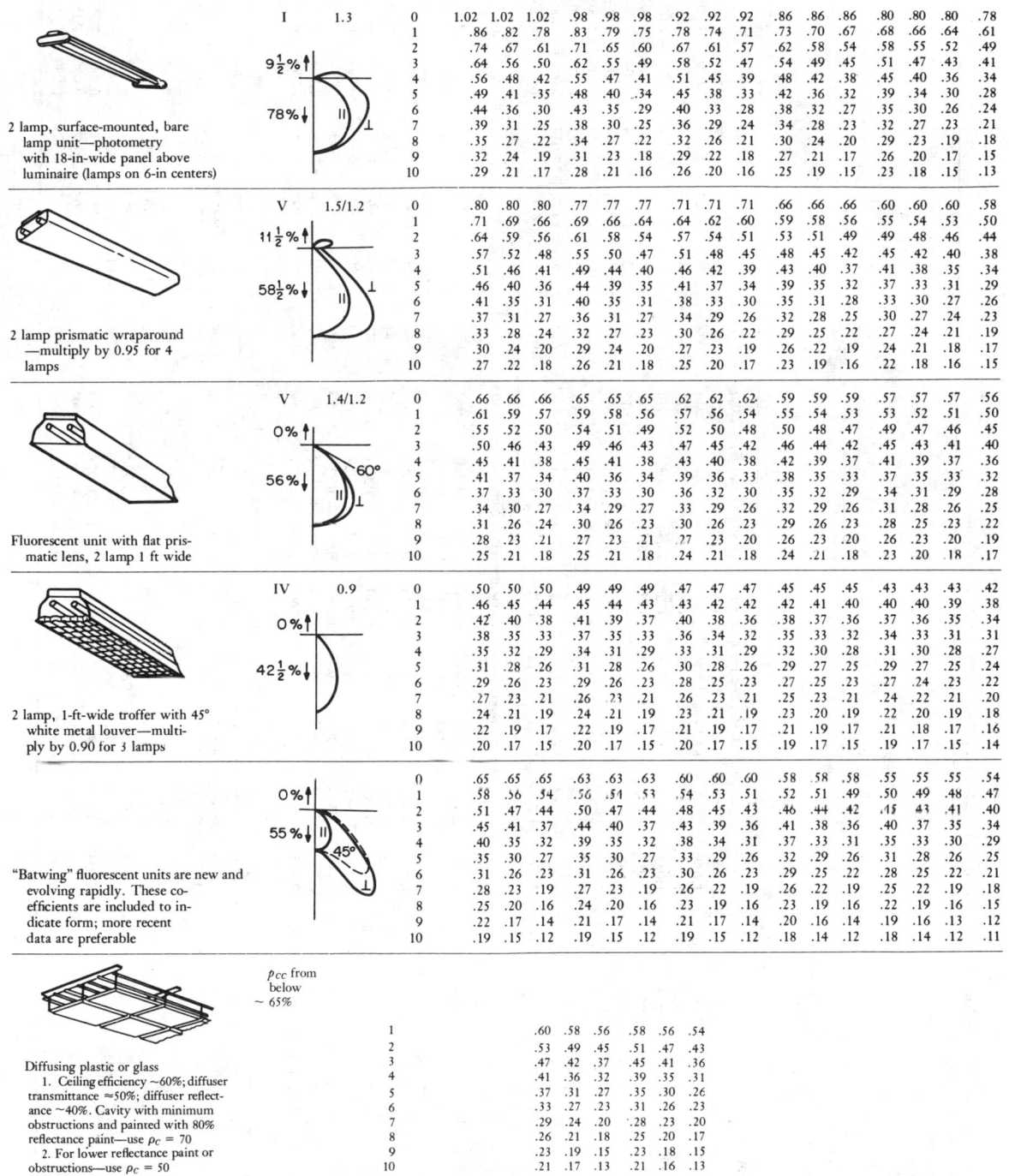

**2 lamp, surface-mounted, bare lamp unit—photometry with 18-in-wide panel above luminaire (lamps on 6-in centers)**

I  1.3  9½%↑  78%↓  ∥ ⊥

| RCR | | | | | | | | | | | | | | | | |
|---|---|---|---|---|---|---|---|---|---|---|---|---|---|---|---|---|
| 0 | 1.02 | 1.02 | 1.02 | .98 | .98 | .98 | .92 | .92 | .92 | .86 | .86 | .86 | .80 | .80 | .80 | .78 |
| 1 | .86 | .82 | .78 | .83 | .79 | .75 | .78 | .74 | .71 | .73 | .70 | .67 | .68 | .66 | .64 | .61 |
| 2 | .74 | .67 | .61 | .71 | .65 | .60 | .67 | .61 | .57 | .62 | .58 | .54 | .58 | .55 | .52 | .49 |
| 3 | .64 | .56 | .50 | .62 | .55 | .49 | .58 | .52 | .47 | .54 | .49 | .45 | .51 | .47 | .43 | .41 |
| 4 | .56 | .48 | .42 | .55 | .47 | .41 | .51 | .45 | .39 | .48 | .42 | .38 | .45 | .40 | .36 | .34 |
| 5 | .49 | .41 | .35 | .48 | .40 | .34 | .45 | .38 | .33 | .42 | .36 | .32 | .39 | .34 | .30 | .28 |
| 6 | .44 | .36 | .30 | .43 | .35 | .29 | .40 | .33 | .28 | .38 | .32 | .27 | .35 | .30 | .26 | .24 |
| 7 | .39 | .31 | .25 | .38 | .30 | .25 | .36 | .29 | .24 | .34 | .28 | .23 | .32 | .27 | .23 | .21 |
| 8 | .35 | .27 | .22 | .34 | .27 | .22 | .32 | .26 | .21 | .30 | .24 | .20 | .29 | .23 | .19 | .18 |
| 9 | .32 | .24 | .19 | .31 | .23 | .18 | .29 | .22 | .18 | .27 | .21 | .17 | .26 | .20 | .17 | .15 |
| 10 | .29 | .21 | .17 | .28 | .21 | .16 | .26 | .20 | .16 | .25 | .19 | .15 | .23 | .18 | .15 | .13 |

**2 lamp prismatic wraparound—multiply by 0.95 for 4 lamps**

V  1.5/1.2  11½%↑  58½%↓  ∥

| RCR | | | | | | | | | | | | | | | | |
|---|---|---|---|---|---|---|---|---|---|---|---|---|---|---|---|---|
| 0 | .80 | .80 | .80 | .77 | .77 | .77 | .71 | .71 | .71 | .66 | .66 | .66 | .60 | .60 | .60 | .58 |
| 1 | .71 | .69 | .66 | .69 | .66 | .64 | .64 | .62 | .60 | .59 | .58 | .56 | .55 | .54 | .53 | .50 |
| 2 | .64 | .59 | .56 | .61 | .58 | .54 | .57 | .54 | .51 | .53 | .51 | .49 | .49 | .48 | .46 | .44 |
| 3 | .57 | .52 | .48 | .55 | .50 | .47 | .51 | .48 | .45 | .48 | .45 | .42 | .45 | .42 | .40 | .38 |
| 4 | .51 | .46 | .41 | .49 | .44 | .40 | .46 | .42 | .39 | .43 | .40 | .37 | .41 | .38 | .35 | .34 |
| 5 | .46 | .40 | .36 | .44 | .39 | .35 | .41 | .37 | .34 | .39 | .35 | .32 | .37 | .33 | .31 | .29 |
| 6 | .41 | .35 | .31 | .40 | .35 | .31 | .38 | .33 | .30 | .35 | .31 | .28 | .33 | .30 | .27 | .26 |
| 7 | .37 | .31 | .27 | .36 | .31 | .27 | .34 | .29 | .26 | .32 | .28 | .25 | .30 | .27 | .24 | .23 |
| 8 | .33 | .28 | .24 | .32 | .27 | .23 | .30 | .26 | .22 | .29 | .25 | .22 | .27 | .24 | .21 | .19 |
| 9 | .30 | .24 | .20 | .29 | .24 | .20 | .27 | .23 | .19 | .26 | .22 | .19 | .24 | .21 | .18 | .17 |
| 10 | .27 | .22 | .18 | .26 | .21 | .18 | .25 | .20 | .17 | .23 | .19 | .16 | .22 | .18 | .16 | .15 |

**Fluorescent unit with flat prismatic lens, 2 lamp 1 ft wide**

V  1.4/1.2  0%↑  56%↓  60°  ∥ ⊥

| RCR | | | | | | | | | | | | | | | | |
|---|---|---|---|---|---|---|---|---|---|---|---|---|---|---|---|---|
| 0 | .66 | .66 | .66 | .65 | .65 | .65 | .62 | .62 | .62 | .59 | .59 | .59 | .57 | .57 | .57 | .56 |
| 1 | .61 | .59 | .57 | .59 | .58 | .56 | .57 | .56 | .54 | .55 | .54 | .53 | .53 | .52 | .51 | .50 |
| 2 | .55 | .52 | .50 | .54 | .51 | .49 | .52 | .50 | .48 | .50 | .48 | .47 | .49 | .47 | .46 | .45 |
| 3 | .50 | .46 | .43 | .49 | .46 | .43 | .47 | .45 | .42 | .46 | .44 | .42 | .45 | .43 | .41 | .40 |
| 4 | .45 | .41 | .38 | .45 | .41 | .38 | .43 | .40 | .38 | .42 | .39 | .37 | .41 | .39 | .37 | .36 |
| 5 | .41 | .37 | .34 | .40 | .36 | .34 | .39 | .36 | .33 | .38 | .35 | .33 | .37 | .35 | .32 | .32 |
| 6 | .37 | .33 | .30 | .37 | .33 | .30 | .36 | .32 | .30 | .35 | .32 | .29 | .34 | .31 | .29 | .28 |
| 7 | .34 | .30 | .27 | .34 | .29 | .27 | .33 | .29 | .26 | .32 | .29 | .26 | .31 | .28 | .26 | .25 |
| 8 | .31 | .26 | .24 | .30 | .26 | .23 | .30 | .26 | .23 | .29 | .26 | .23 | .28 | .25 | .23 | .22 |
| 9 | .28 | .23 | .21 | .27 | .23 | .21 | .27 | .23 | .20 | .26 | .23 | .20 | .26 | .23 | .20 | .19 |
| 10 | .25 | .21 | .18 | .25 | .21 | .18 | .24 | .21 | .18 | .24 | .21 | .18 | .23 | .20 | .18 | .17 |

**2 lamp, 1-ft-wide troffer with 45° white metal louver—multiply by 0.90 for 3 lamps**

IV  0.9  0%↑  42½%↓

| RCR | | | | | | | | | | | | | | | | |
|---|---|---|---|---|---|---|---|---|---|---|---|---|---|---|---|---|
| 0 | .50 | .50 | .50 | .49 | .49 | .49 | .47 | .47 | .47 | .45 | .45 | .45 | .43 | .43 | .43 | .42 |
| 1 | .46 | .45 | .44 | .45 | .44 | .43 | .43 | .42 | .42 | .42 | .41 | .40 | .40 | .40 | .39 | .38 |
| 2 | .42 | .40 | .38 | .41 | .39 | .37 | .40 | .38 | .36 | .38 | .37 | .36 | .37 | .36 | .35 | .34 |
| 3 | .38 | .35 | .33 | .37 | .35 | .33 | .36 | .34 | .32 | .35 | .33 | .32 | .34 | .33 | .31 | .31 |
| 4 | .35 | .32 | .29 | .34 | .31 | .29 | .33 | .31 | .29 | .32 | .30 | .28 | .31 | .30 | .28 | .27 |
| 5 | .31 | .28 | .26 | .31 | .28 | .26 | .30 | .28 | .26 | .29 | .27 | .25 | .29 | .27 | .25 | .24 |
| 6 | .29 | .26 | .23 | .29 | .26 | .23 | .28 | .25 | .23 | .27 | .25 | .23 | .27 | .24 | .23 | .22 |
| 7 | .27 | .23 | .21 | .26 | .23 | .21 | .26 | .23 | .21 | .25 | .23 | .21 | .24 | .22 | .21 | .20 |
| 8 | .24 | .21 | .19 | .24 | .21 | .19 | .23 | .21 | .19 | .23 | .20 | .19 | .22 | .20 | .19 | .18 |
| 9 | .22 | .19 | .17 | .22 | .19 | .17 | .21 | .19 | .17 | .21 | .19 | .17 | .21 | .18 | .17 | .16 |
| 10 | .20 | .17 | .15 | .20 | .17 | .15 | .20 | .17 | .15 | .19 | .17 | .15 | .19 | .17 | .15 | .14 |

**"Batwing" fluorescent units are new and evolving rapidly. These coefficients are included to indicate form; more recent data are preferable**

0%↑  55%↓  45°  ∥ ⊥

| RCR | | | | | | | | | | | | | | | | |
|---|---|---|---|---|---|---|---|---|---|---|---|---|---|---|---|---|
| 0 | .65 | .65 | .65 | .63 | .63 | .63 | .60 | .60 | .60 | .58 | .58 | .58 | .55 | .55 | .55 | .54 |
| 1 | .58 | .56 | .54 | .56 | .54 | .53 | .54 | .53 | .51 | .52 | .51 | .49 | .50 | .49 | .48 | .47 |
| 2 | .51 | .47 | .44 | .50 | .47 | .44 | .48 | .45 | .43 | .46 | .44 | .42 | .45 | .43 | .41 | .40 |
| 3 | .45 | .41 | .37 | .44 | .40 | .37 | .43 | .39 | .36 | .41 | .38 | .36 | .40 | .37 | .35 | .34 |
| 4 | .40 | .35 | .32 | .39 | .35 | .32 | .38 | .34 | .31 | .37 | .33 | .31 | .35 | .33 | .30 | .29 |
| 5 | .35 | .30 | .27 | .35 | .30 | .27 | .33 | .29 | .26 | .32 | .29 | .26 | .31 | .28 | .26 | .25 |
| 6 | .31 | .26 | .23 | .31 | .26 | .23 | .30 | .26 | .23 | .29 | .25 | .22 | .28 | .25 | .22 | .21 |
| 7 | .28 | .23 | .19 | .27 | .23 | .19 | .26 | .22 | .19 | .26 | .22 | .19 | .25 | .22 | .19 | .18 |
| 8 | .25 | .20 | .16 | .24 | .20 | .16 | .23 | .19 | .16 | .23 | .19 | .16 | .22 | .19 | .16 | .15 |
| 9 | .22 | .17 | .14 | .21 | .17 | .14 | .21 | .17 | .14 | .20 | .16 | .14 | .19 | .16 | .13 | .12 |
| 10 | .19 | .15 | .12 | .19 | .15 | .12 | .19 | .15 | .12 | .18 | .14 | .12 | .18 | .14 | .12 | .11 |

**Diffusing plastic or glass**

1. Ceiling efficiency ~60%; diffuser transmittance ≈50%; diffuser reflectance ~40%. Cavity with minimum obstructions and painted with 80% reflectance paint—use $\rho_C = 70$

2. For lower reflectance paint or obstructions—use $\rho_C = 50$

$\rho_{CC}$ from below ~65%

| RCR | | | | | | |
|---|---|---|---|---|---|---|
| 1 | .60 | .58 | .56 | .58 | .56 | .54 |
| 2 | .53 | .49 | .45 | .51 | .47 | .43 |
| 3 | .47 | .42 | .37 | .45 | .41 | .36 |
| 4 | .41 | .36 | .32 | .39 | .35 | .31 |
| 5 | .37 | .31 | .27 | .35 | .30 | .26 |
| 6 | .33 | .27 | .23 | .31 | .26 | .23 |
| 7 | .29 | .24 | .20 | .28 | .23 | .20 |
| 8 | .26 | .21 | .18 | .25 | .20 | .17 |
| 9 | .23 | .19 | .15 | .23 | .18 | .15 |
| 10 | .21 | .17 | .13 | .21 | .16 | .13 |

*$\rho_{CC}$ = percent effective ceiling cavity reflectance.

†$\rho_W$ = percent wall reflectance.

‡RCR = room cavity ratio.

§Maximum $S/MH$ guide—ratio of maximum luminaire spacing to mounting or ceiling height above work plane.

## Table 9. Multiplying Factors for Other than 20 Percent Effective Floor Cavity Reflectance (IES)

| % effective ceiling cavity reflectance $\rho_{CC}$ | 80 | | | | 70 | | | | 50 | | | 30 | | | 10 | | |
|---|---|---|---|---|---|---|---|---|---|---|---|---|---|---|---|---|---|
| % wall reflectance $\rho_W$ | 70 | 50 | 30 | 10 | 70 | 50 | 30 | 10 | 50 | 30 | 10 | 50 | 30 | 10 | 50 | 30 | 10 |
| Room cavity ratio | | | | | | | | | | | | | | | | | |

**For 30 % effective floor cavity reflectance (20 % = 1.00)**

| Room cavity ratio | 70 | 50 | 30 | 10 | 70 | 50 | 30 | 10 | 50 | 30 | 10 | 50 | 30 | 10 | 50 | 30 | 10 |
|---|---|---|---|---|---|---|---|---|---|---|---|---|---|---|---|---|---|
| 1 | 1.092 | 1.082 | 1.075 | 1.068 | 1.077 | 1.070 | 1.064 | 1.059 | 1.049 | 1.044 | 1.040 | 1.028 | 1.026 | 1.023 | 1.012 | 1.010 | 1.008 |
| 2 | 1.079 | 1.066 | 1.055 | 1.047 | 1.068 | 1.057 | 1.048 | 1.039 | 1.041 | 1.033 | 1.027 | 1.026 | 1.021 | 1.017 | 1.013 | 1.010 | 1.006 |
| 3 | 1.070 | 1.054 | 1.042 | 1.033 | 1.061 | 1.048 | 1.037 | 1.028 | 1.034 | 1.027 | 1.020 | 1.024 | 1.017 | 1.012 | 1.014 | 1.009 | 1.005 |
| 4 | 1.062 | 1.045 | 1.033 | 1.024 | 1.055 | 1.040 | 1.029 | 1.021 | 1.030 | 1.022 | 1.015 | 1.022 | 1.015 | 1.010 | 1.014 | 1.009 | 1.004 |
| 5 | 1.056 | 1.038 | 1.026 | 1.018 | 1.050 | 1.034 | 1.024 | 1.015 | 1.027 | 1.018 | 1.012 | 1.020 | 1.013 | 1.008 | 1.014 | 1.009 | 1.004 |
| 6 | 1.052 | 1.033 | 1.021 | 1.014 | 1.047 | 1.030 | 1.020 | 1.012 | 1.024 | 1.015 | 1.009 | 1.019 | 1.012 | 1.006 | 1.014 | 1.008 | 1.003 |
| 7 | 1.047 | 1.029 | 1.018 | 1.011 | 1.043 | 1.026 | 1.017 | 1.009 | 1.022 | 1.013 | 1.007 | 1.018 | 1.010 | 1.005 | 1.013 | 1.008 | 1.003 |
| 8 | 1.044 | 1.026 | 1.015 | 1.009 | 1.040 | 1.024 | 1.015 | 1.007 | 1.020 | 1.012 | 1.006 | 1.017 | 1.009 | 1.004 | 1.013 | 1.007 | 1.003 |
| 9 | 1.040 | 1.024 | 1.014 | 1.007 | 1.037 | 1.022 | 1.014 | 1.006 | 1.019 | 1.011 | 1.005 | 1.016 | 1.009 | 1.004 | 1.013 | 1.007 | 1.002 |
| 10 | 1.037 | 1.022 | 1.012 | 1.006 | 1.034 | 1.020 | 1.012 | 1.005 | 1.017 | 1.010 | 1.004 | 1.015 | 1.009 | 1.003 | 1.013 | 1.007 | 1.002 |

**For 10 % effective floor cavity reflectance (20 % = 1.00)**

| Room cavity ratio | 70 | 50 | 30 | 10 | 70 | 50 | 30 | 10 | 50 | 30 | 10 | 50 | 30 | 10 | 50 | 30 | 10 |
|---|---|---|---|---|---|---|---|---|---|---|---|---|---|---|---|---|---|
| 1 | 0.923 | 0.929 | 0.935 | 0.940 | 0.933 | 0.939 | 0.943 | 0.948 | 0.956 | 0.960 | 0.963 | 0.973 | 0.976 | 0.979 | 0.989 | 0.991 | 0.993 |
| 2 | 0.931 | 0.942 | 0.950 | 0.958 | 0.940 | 0.949 | 0.957 | 0.963 | 0.962 | 0.968 | 0.974 | 0.976 | 0.980 | 0.985 | 0.988 | 0.991 | 0.995 |
| 3 | 0.939 | 0.951 | 0.961 | 0.969 | 0.945 | 0.057 | 0.966 | 0.973 | 0.967 | 0.975 | 0.981 | 0.978 | 0.983 | 0.988 | 0.987 | 0.992 | 0.996 |
| 4 | 0.944 | 0.958 | 0.969 | 0.978 | 0.950 | 0.963 | 0.973 | 0.980 | 0.972 | 0.980 | 0.986 | 0.980 | 0.986 | 0.991 | 0.987 | 0.992 | 0.996 |
| 5 | 0.949 | 0.964 | 0.976 | 0.983 | 0.954 | 0.968 | 0.978 | 0.985 | 0.975 | 0.983 | 0.989 | 0.981 | 0.988 | 0.993 | 0.987 | 0.992 | 0.997 |
| 6 | 0.953 | 0.969 | 0.980 | 0.986 | 0.958 | 0.972 | 0.982 | 0.989 | 0.977 | 0.985 | 0.992 | 0.982 | 0.989 | 0.995 | 0.987 | 0.993 | 0.997 |
| 7 | 0.957 | 0.973 | 0.983 | 0.991 | 0.961 | 0.975 | 0.985 | 0.991 | 0.979 | 0.987 | 0.994 | 0.983 | 0.990 | 0.996 | 0.987 | 0.993 | 0.998 |
| 8 | 0.960 | 0.976 | 0.986 | 0.993 | 0.963 | 0.977 | 0.987 | 0.993 | 0.981 | 0.988 | 0.995 | 0.984 | 0.991 | 0.997 | 0.988 | 0.994 | 0.998 |
| 9 | 0.963 | 0.978 | 0.987 | 0.994 | 0.965 | 0.979 | 0.989 | 0.994 | 0.983 | 0.990 | 0.996 | 0.985 | 0.992 | 0.998 | 0.988 | 0.994 | 0.999 |
| 10 | 0.965 | 0.980 | 0.989 | 0.995 | 0.967 | 0.981 | 0.990 | 0.995 | 0.984 | 0.991 | 0.997 | 0.986 | 0.993 | 0.998 | 0.988 | 0.994 | 0.999 |

**For 0% effective floor cavity reflectance (20% = 1.00)**

| Room cavity ratio | 70 | 50 | 30 | 10 | 70 | 50 | 30 | 10 | 50 | 30 | 10 | 50 | 30 | 10 | 50 | 30 | 10 |
|---|---|---|---|---|---|---|---|---|---|---|---|---|---|---|---|---|---|
| 1 | 0.859 | 0.870 | 0.879 | 0.886 | 0.873 | 0.884 | 0.893 | 0.901 | 0.916 | 0.923 | 0.929 | 0.948 | 0.954 | 0.960 | 0.979 | 0.983 | 0.987 |
| 2 | 0.871 | 0.887 | 0.903 | 0.919 | 0.886 | 0.902 | 0.916 | 0.928 | 0.926 | 0.938 | 0.949 | 0.954 | 0.963 | 0.971 | 0.978 | 0.983 | 0.991 |
| 3 | 0.882 | 0.904 | 0.915 | 0.942 | 0.898 | 0.918 | 0.934 | 0.947 | 0.936 | 0.950 | 0.964 | 0.958 | 0.969 | 0.979 | 0.976 | 0.984 | 0.993 |
| 4 | 0.893 | 0.919 | 0.941 | 0.958 | 0.908 | 0.930 | 0.948 | 0.961 | 0.945 | 0.961 | 0.974 | 0.961 | 0.974 | 0.984 | 0.975 | 0.985 | 0.994 |
| 5 | 0.903 | 0.931 | 0.953 | 0.969 | 0.914 | 0.939 | 0.958 | 0.970 | 0.951 | 0.967 | 0.980 | 0.964 | 0.977 | 0.988 | 0.975 | 0.985 | 0.995 |
| 6 | 0.911 | 0.940 | 0.961 | 0.976 | 0.920 | 0.945 | 0.965 | 0.977 | 0.955 | 0.972 | 0.985 | 0.966 | 0.979 | 0.991 | 0.975 | 0.986 | 0.996 |
| 7 | 0.917 | 0.947 | 0.967 | 0.981 | 0.924 | 0.950 | 0.970 | 0.982 | 0.959 | 0.975 | 0.988 | 0.968 | 0.981 | 0.993 | 0.975 | 0.987 | 0.997 |
| 8 | 0.922 | 0.953 | 0.971 | 0.985 | 0.929 | 0.955 | 0.975 | 0.986 | 0.963 | 0.978 | 0.991 | 0.970 | 0.983 | 0.995 | 0.976 | 0.988 | 0.998 |
| 9 | 0.928 | 0.958 | 0.975 | 0.988 | 0.933 | 0.959 | 0.980 | 0.989 | 0.966 | 0.980 | 0.993 | 0.971 | 0.985 | 0.996 | 0.976 | 0.988 | 0.998 |
| 10 | 0.933 | 0.962 | 0.979 | 0.991 | 0.937 | 0.963 | 0.983 | 0.992 | 0.969 | 0.982 | 0.995 | 0.973 | 0.987 | 0.997 | 0.977 | 0.989 | 0.999 |

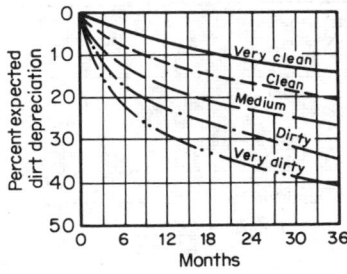

| % expected dirt depreciation | Luminaire distribution type | | | | | | | | | | | | | | | | | | | |
|---|---|---|---|---|---|---|---|---|---|---|---|---|---|---|---|---|---|---|---|---|
| | Direct | | | | Semidirect | | | | Direct-indirect | | | | Semi-indirect | | | | Indirect | | | |
| Room cavity ratio | 10 | 20 | 30 | 40 | 10 | 20 | 30 | 40 | 10 | 20 | 30 | 40 | 10 | 20 | 30 | 40 | 10 | 20 | 30 | 40 |
| 1 | .98 | .96 | .94 | .92 | .97 | .92 | .89 | .84 | .94 | .87 | .80 | .76 | .94 | .87 | .80 | .73 | .90 | .80 | .70 | .60 |
| 2 | .98 | .96 | .94 | .92 | .96 | .92 | .88 | .83 | .94 | .87 | .80 | .75 | .94 | .87 | .79 | .72 | .90 | .80 | .69 | .59 |
| 3 | .98 | .95 | .93 | .90 | .96 | .91 | .87 | .82 | .94 | .86 | .79 | .74 | .94 | .86 | .78 | .71 | .90 | .79 | .68 | .58 |
| 4 | .97 | .95 | .92 | .90 | .95 | .90 | .85 | .80 | .94 | .86 | .79 | .73 | .94 | .86 | .78 | .70 | .89 | .78 | .67 | .56 |
| 5 | .97 | .94 | .91 | .89 | .94 | .90 | .84 | .79 | .93 | .86 | .78 | .72 | .93 | .86 | .77 | .69 | .89 | .78 | .66 | .55 |
| 6 | .97 | .94 | .91 | .88 | .94 | .89 | .83 | .78 | .93 | .85 | .78 | .71 | .93 | .85 | .76 | .68 | .89 | .77 | .66 | .54 |
| 7 | .97 | .94 | .90 | .87 | .93 | .88 | .82 | .77 | .93 | .84 | .77 | .70 | .93 | .84 | .76 | .68 | .89 | .76 | .65 | .53 |
| 8 | .96 | .93 | .89 | .86 | .93 | .87 | .81 | .75 | .93 | .84 | .76 | .69 | .93 | .84 | .76 | .68 | .88 | .76 | .64 | .52 |
| 9 | .96 | .92 | .88 | .85 | .93 | .87 | .80 | .74 | .93 | .84 | .76 | .68 | .93 | .84 | .75 | .67 | .88 | .75 | .63 | .51 |
| 10 | .96 | .92 | .87 | .83 | .93 | .86 | .79 | .72 | .93 | .84 | .75 | .67 | .92 | .83 | .75 | .67 | .88 | .75 | .62 | .50 |

**Fig. 17**   Room-surface dirt depreciation factors. *(IES.)*

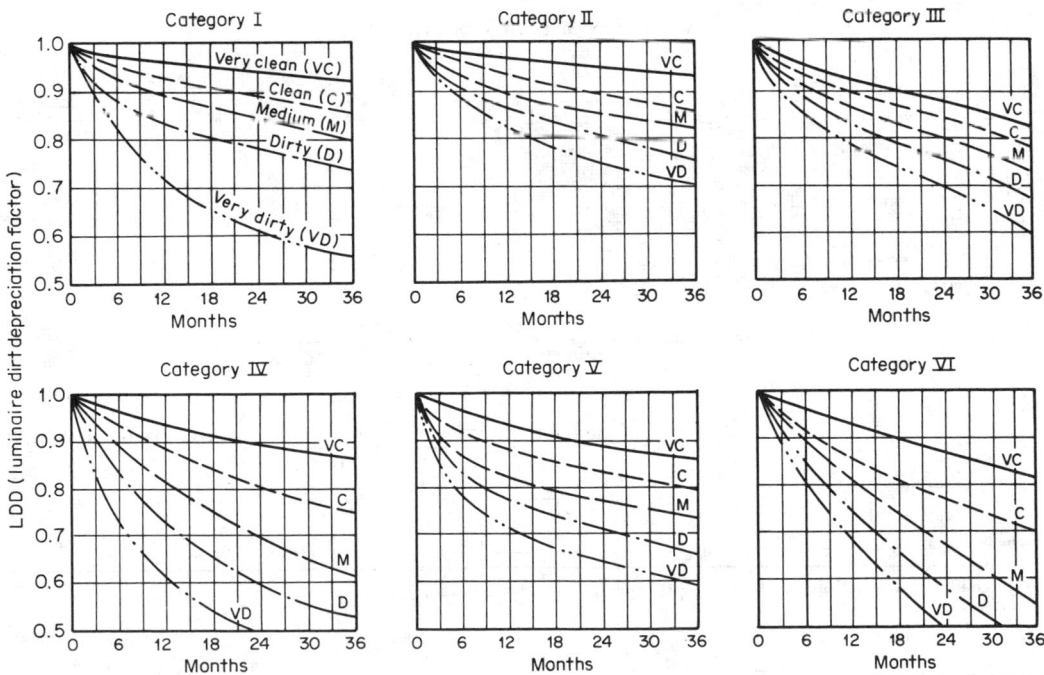

**Fig. 18**   Luminaire dirt depreciation (LDD) factors for six luminaire categories (I to VI) and for five degrees of dirtiness. *(IES.)*

## GENERAL INFORMATION

Project identification: _____
*(Give name of area and/or building and room number)*

Average maintained illumination for design: ____footcandles        Lamp data:

Luminaire data:                                                        Type and color: _____

  Manufacturer: _____        Number per luminaire: _____

  Catalog number: _____        Total lumens per luminaire: _____

### SELECTION OF COEFFICIENT OF UTILIZATION

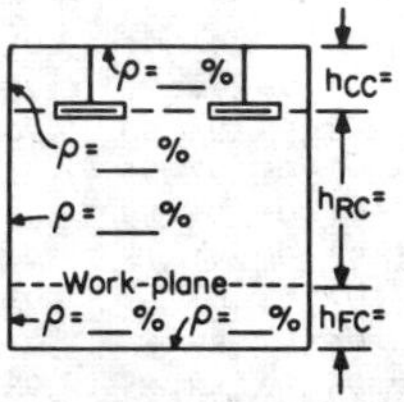

Step 1: Fill in sketch at right.

Step 2: Determine cavity ratios [Eq. **(8)**]

    Room cavity ratio RCR = _____
    Ceiling cavity ratio CCR = _____
    Floor cavity ratio FCR = _____

Step 3: Obtain effective ceiling cavity reflectance $\rho_{CC}$ from Table 8        $\rho_{CC}$ = _____

Step 4: Obtain effective floor cavity reflectance $\rho_{FC}$ from Table 8        $\rho_{FC}$ = _____

Step 5: Obtain coefficient of utilization CU from manufacturer's data (or Fig. 16 and Table 9)        CU = _____

### SELECTION OF LIGHT-LOSS FACTORS

Room-surface dirt depreciation
    RSDD (see Fig. 17)        _____
Lamp lumen depreciation LLD
Average values:        _____
  Incandescent = 0.87
  Fluorescent = 0.82
Luminaire dirt depreciation LLD (see Fig. 18)        _____
Total light-loss factor LLF (product of individual factors above): ____

### CALCULATIONS
(Average maintained illumination level)

$$\text{Number of luminaires} = \frac{(\text{footcandles}) \times (\text{area, ft}^2)}{(\text{lumens per luminaire}) \times (\text{CU}) \times (\text{LLF})}$$

$$= \underline{\hspace{10cm}} =$$

$$\text{Footcandles} = \frac{(\text{number of luminaires}) \times (\text{lumens per luminaire}) \times (\text{CU}) \times (\text{LLF})}{(\text{area, ft}^2)}$$

$$= \underline{\hspace{10cm}} =$$

Calculated by: _____ Date: _____

**Fig. 19**  Average-illumination calculation sheet. *(Adapted from IES Lighting Handbook.)*

lighting is to be designed, the point method is used. Manufacturers furnish candlepower distribution curves for their fixtures. An average curve is given for symmetrical fixtures while curves in various planes are given for asymmetrical ones. The basic equation for calculating the illumination from such curves is

$$E_h = (I_\theta \cos \theta)/D^2$$
$$= I_\theta\, H/D^3 \qquad (9)$$

where $E_h$ is the illumination on the horizontal plane, $I\theta$ the candlepower in the given direction, and $D$ the distance of the luminaire to the point $P$. See Fig. 20.

For vertical surfaces,

$$E_v = (I_\theta \sin \theta)/D^2$$
$$= I_\theta\, R/D^3 \qquad (10)$$

Nomograms and graphical solutions are available for Eqs. (9) and (10).

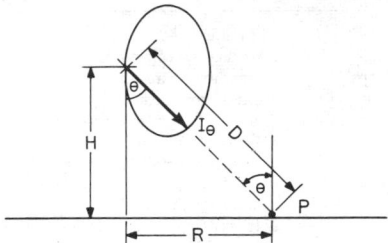

**Fig. 20**  Footcandle-calculation diagram.

## THE ECONOMICS OF LIGHTING INSTALLATIONS

The cost of lighting is computed by summing the annual cost of energy; relamping; cost of labor for cleaning, relamping, and servicing; interest; and depreciation.

Another way to compare installations is to compute the watts per square foot for each proposed installation. This is computed by either method:

$$\text{watts/ft}^2 = \frac{\text{total lamp lumens}}{\text{area, ft}^2} \times \frac{1}{\text{lumens/watt of lamp}} \qquad (11)$$

$$\text{watts/ft}^2 = \frac{\text{designed illuminance}}{\text{CU} \times \text{LLF}}$$
$$\times \frac{1}{\text{lumens/watt of lamp}} \qquad (12)$$

## DIMMING SYSTEMS

Dimming systems are used in theaters, auditoriums, ballrooms, etc. Originally, power-consuming rheostats were used. These have been replaced by continuously variable autotransformers, variable reactors, and silicon controlled rectifiers (SCR). The development of controlled solid-state rectifiers has resulted in small, reliable dimmers which can be readily programmed. Only incandescent and cold-cathode lamps can be dimmed easily. Fluorescent lamps require special ballasts which keep the electrodes hot at all times.

## HEAT FROM LIGHTING

Lighting installations are a substantial source of heat, have

long been a factor in the design of air-conditioning (cooling) systems, and are increasingly significant in the design of heating systems. The heating effect for 1 W is 3.413 Btu/h, and for 1 ton of refrigeration is 12,000 Btu/h. Approximate wattage data for lighting systems at various lighting levels can be calculated by using watts per square foot calculated from Eq. (11) or (12). The wattage load of an incandescent system is roughly 2½ times the load of an equivalent fluorescent system. Heat generated is delivered to surrounding areas in several ways, with energy distribution for fluorescent and incandescent lamps as illustrated in Fig. 21. With the prevalent high lighting intensities of modern buildings, it is essential to control the heat generated by a lighting system. Luminaires radiate approximately 25 percent of their energy supply into the room in the form of light and invisible radiation; the remaining 75 percent is emitted by conduction or convection. Substantial portions of the energy which is not radiated into the room may be conducted away from the luminaire by an air stream or by water flowing through a coil attached to the

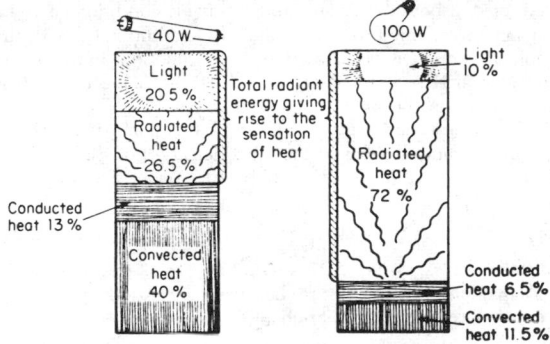

**Fig. 21**  Energy distribution of lamps. *(Westinghouse.)*

luminaire. In the heating season, this heat energy is delivered to the perimeter of the building for effective space warming. In the cooling season, the heat is rejected to the exterior, thus reducing the load on the cooling system. Air-handling luminaires (Fig. 22) are receiving wide acceptance.

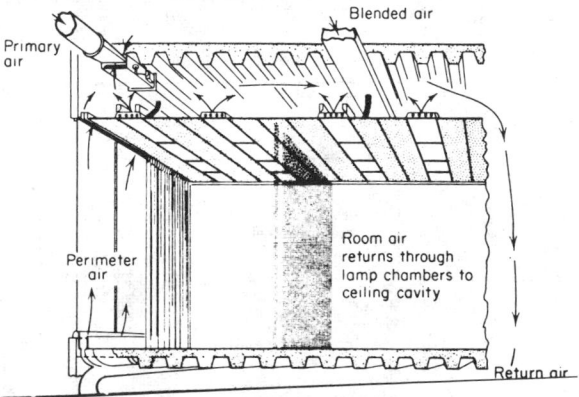

**Fig. 22**  Typical air-handling system. *(Barber-Coleman.)*

# SOUND, NOISE, AND ULTRASONICS

## by Benson Carlin

REFERENCES: ANSI 51.1, 1960, (R1971) Acoustical Terminology. *J. Acoust. Soc. Am.* 1929 et seq. Beranek, "Acoustics," McGraw-Hill. Carlin, "Ultrasonics," McGraw-Hill. Morse and Ingard, "Theoretical Acoustics." Harris, "Handbook of Noise Control," McGraw-Hill. Mason, "Physical Acoustics," Academic Press.

### DEFINITIONS

Sound is an alteration in pressure, stress, particle displacement, and particle velocity, which is propagated in an elastic material. It is longitudinal in gases but may also be transverse (shear) surface, or other types in elastic media which can support such energy. It may be reflected, diffracted, or refracted at boundaries and under suitable conditions may be changed from one form to another. In longitudinal waves, the molecules move in the direction of wave motion, in the others at right angles to it. Waves may also be plane or circular depending on the source.

The **velocity of propagation** $V$ is a function of the elastic (Young's) modulus $E$ and the mass density $\rho$ of the medium. For longitudinal waves, in solids: $V = \sqrt{(E/\rho)[(1-\mu)(1-2\mu)]}$, where $\mu$ is Poisson's ratio; for shear waves, $V = \sqrt{G/\rho}$, where $G$ is the modulus of rigidity; for surface waves, $V = 0.02\sqrt{G/\rho}$. In gases, the velocity is independent of the pressure, because the elasticity changes to compensate for the density changes; the general equation is $V = \sqrt{kp/\rho}$, where $k$ is the ratio of specific heats and $p$ is the pressure. The velocity in air at 68°F is 1,126 ft/s (33,160 cm/s) and increases by 0.1 percent per °F. In liquids, empirical formulas are easier to use than theoretical ones to predict actual velocities, since velocity varies in a complex way with temperature, pressure, and other factors. With sea water, a standard velocity of 5,100 ft/s (150,000 cm/s) may be used. The velocity of sound in liquids and solids is usually much higher than in gases (see Table 1 and Kinsler, "Fundamentals of Acoustics," Wiley).

The **frequency** of a sound is the number of periods (cycles) occurring in unit time, customarily expressed as cycles per sec (cps) or sometimes "hertz"; kilocycles per sec, kc = $10^3$ cps; megacycles per sec, Mc = $10^6$ cps. Sound frequencies are usually defined as 20 to 20,000 cps (**audible**), higher (**ultrasonic**), and lower (**infrasonic**). Frequencies as high as the thousand megacycle range ($10^9$ cps) are now being generated (see Table 2).

### Table 2. Sound Spectrum

| Frequency | Action |
|---|---|
| 20–40 cps | Thunder |
| 128 cps | Average speech (male) |
| 250–2,740 cps | Telephone bandwidth |
| 90–5,000 cps | Radio broadcast |
| 15 cps–15 kc | Limits of average human hearing |
| 10–90 kc | Ultrasonic cleaning |
| 15–50 kc | Ultrasonic depth sounding, sonar |
| 20 kc | Ultrasonic burglar alarm, control apparatus, door opening |
| 30 kc | Highest frequency obtained by friction |
| 40 kc | Highest frequency of Hartmann generator |
| 48 kc | Bat cries |
| 90 kc | Top limit of tuning fork |
| 100 kc | Highest frequency of Galton whistle |
| 500–15,000 kc | Ultrasonic pulse-echo testing |
| 1,000 kc | Medical therapy |
| 1,500–30,000 kc | Ultrasonic delay lines |
| 15,000 kc | Radar trainer |

The relation between frequency $f$ and wavelength $\lambda$ is $V = \lambda/f$. In air, at 1,126 cps, the wavelength is 1 ft. In nature, the waves may be simple sinusoidal, complex, or explosive (shock) depending on the source. The first is of course rare.

**Attenuation** of sound depends on the media of propagation and the frequency and is caused by absorption, spreading, and scattering. At audible frequencies in air attentuation is small

### Table 1. Velocity of Sound

| Material | Sound velocity, fps | Density, lb per cu ft | Density × velocity, lb per sq ft per sec |
|---|---|---|---|
| Aluminum | 16,740 | 168 | 2.82 × $10^6$ |
| Brass | 11,480 | 530 | 6.08 × $10^6$ |
| Copper | 11,670 | 555 | 6.47 × $10^6$ |
| Iron and soft steel | 16,410 | 486 | 7.98 × $10^6$ |
| Lead | 4,026 | 1125 | 4.54 × $10^6$ |
| Brick | 11,980 | 125 | 1.5 × $10^6$ |
| Cork | 1,640 | 15 | 0.025 × $10^6$ |
| Wood | 10,000–15,000 | 30–50 | 0.3 × $10^6$–0.75 × $10^6$ |
| Water | 4,794 | 62.4 | 0.299 × $10^6$ |
| Air, dry, $CO_2$ free, 32 F | 1,088.5 | 0.0808 | 88.0 |
| Hydrogen | 4,165 | 0.00560 | 23.3 |
| Water vapor, 212 F | 1,328 | 0.0372 | 49.4 |

Approximate values from Smithsonian Tables.

except for the spreading of the energy over wide areas as the sound waves are propagated. By this means the intensity drops according to the inverse square law. However, in other media, the absorption, scattering, or other characteristic may be predominant.

The sound **intensity** is the average rate of sound energy transmitted through a unit area normal to the wave direction at the point considered. This is a definition of power and may be expressed in watts per sq meter. It is usual however to express power in **decibels**, dB, which is a term used to give the relative magnitude of two powers by comparing the one under consideration to a standard. The sound-pressure level in decibels, dB, is defined as twenty times the logarithm to the base 10 of the ratio of sound pressure to the reference sound pressure. All values are for air at 20°C and atmospheric pressure. Pressure measurements in air use a pressure reference (rms) of 0.0002 dyne/cm²; 1 dyne/cm² is used underwater.

Intensity references for air are $10^{-16}$ w/cm² [equivalent to a pressure (rms) of 0.0002 dyne/cm², and 0.02 erg/cm²s, equivalent to a pressure of 1 dyne/cm²]. Since the references are equivalent (i.e., the reference pressure corresponds to the reference intensity in this particular case), numerical results are identical for plane waves using either expression $IL = 10 \log (I/I_0)$, or $PL = 20 \log (P_e/P_0)$, where $IL$ and $PL$ are the intensity and pressure levels, $I_0$ and $P_0$ are the reference intensity and pressure, $P_e$ is the effective pressure, and $I$ is the intensity in question. When making measurements with pressure or velocity microphones, it is the pressure level or velocity level which is measured and the relationship between the measurement and the intensity is unknown except in the special cases indicated.

Decibels do not add numerically as linear figures do; i.e., 70 dB + 70 dB = 73 dB since doubling power results in a 3-dB increase in sound pressure. Figure 1 shows how to add decibels within 14 dB of each other. If the difference is greater between two readings, ignore the weaker one.

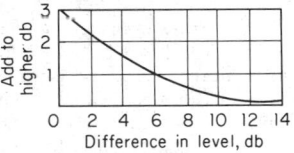

**Fig. 1** Chart for the addition of decibels.

**Specific Acoustic Impedance** The relationship between the pressure and the associated particle velocity at a point in a medium is called the specific acoustic impedance; its unit is the kilogram per meter second or mks rayl. The magnitude $\rho_c$ is called the characteristic impedance of the medium, or the radiation resistance. This applies in the case of plane waves. The $\rho_c$ of a material is one of its most useful acoustic characteristics, since by means of it the amount of energy reflected at boundaries may be computed, horns may be analyzed according to the acoustic resistance at throat, and other calculations analogous to those made in electrical design may be carried out.

## THE PRODUCTION AND RECEPTION OF SOUNDS

REFERENCES: Rinsler and Frey, "Fundamentals of Acoustics," Wiley. Mason, "Electromechanical Transducers and Wave Filters," Van Nostrand. Olsen, "Elements of Acoustical Engineering," Van Nostrand.

**Transducer** A device for converting energy from one form to another, e.g., from electrical to acoustic or vice versa, is called a transducer. Among these are loudspeakers, microphones, hydrophones, and piezoelectric and magnetostrictive transducers.

**Loudspeakers** are usually classified as direct-radiator or horn type. The direct-radiator type consists of a cone, a magnet, a voice coil moving in the magnetic field, a vibrating diaphragm coupled to the cone, and suitable supports. The attachment of a horn improves the impedance match between the speaker and the air since it is essentially an acoustic transformer. The dimensions and flare of the horn contribute to its matching ability.

One of the more common types of horn is exponential, although straight and other types are also possible. In a similar manner, mechanical transformers may be used to concentrate the energy of ultrasonic transducers. In such forms they operate to concentrate rather than to spread the energy. The operation of the speaker may be variously influenced by its enclosure, by the baffle which separates the front from the back radiation, or by its resonances.

**Microphones** (for gases) and **hydrophones** (for liquids) are transducers for converting mechanical to electrical energy. They may be piezoelectric, electromagnetic, magnetostrictive, or capacitive. The variation in electrical output is proportional to the effect of the acoustic field on the characteristics. Ultrasonic transducers may be any of the above types but are usually crystal (piezoelectric) or magnetostrictive. Among the common piezoelectric materials are quartz, barium titanate, lithium sulfate, ADP (ammonium dihydrogen phosphate), and rochelle salt. In sonar and high-power industrial systems, mosaics of crystals are used; in low-power, high-frequency systems, a single crystal is usual.

**Whistles** and **sirens** may also be used to produce intense sound fields in gases and liquids. These are devices which produce sound by passing a fluid over an obstacle, thereby creating turbulence in the fluid. When the obstacle is an edge, these are referred to as edge or E tones; when an orifice, as jet tones. Organ pipes, whistles, and nozzles for spraying are devices of this class. Frequencies up to 100,000 cps are possible, although 30,000 cps is the approximate limit at which appreciable power can be generated. Resonators may be placed in the sound field to reinforce it and to stabilize the frequency. These take the form of small pipes tuned to the approximate frequency. Common types of whistles are the Hartmann and Galton (for gases) and the jet edge (for liquids).

**Sirens** are devices in which a revolving disk with holes in it interrupts a jet from a nearby tube. Compressed air, steam, and water have been used. Frequencies up to 30 kc may be produced at efficiencies of 50 percent approximately; a 1-hp motor produces between 300 and 1,000 W (see also Jones, *J. Acoust. Soc. Am.*, 1946).

Transducers are generally driven by electronic generators, motor generators, or air compressors. As receivers, they activate amplifiers or indicating devices.

**The Perception of Sound** The average young observer perceives sound between 20 and 20,000 cps. High-frequency response deteriorates with advancing age. The ear responds to a wide range of intensities; e.g., between 500 and 5,000 cps, the ratio of tolerated intensities is about $10^{12}$. The minimum intensity perceived varies with frequency. Figure 2 shows the audible frequency and intensity range for a standard listener, where the lowest curve represents the threshold of hearing and the top one the beginning of sensation in the ear. These curves show the pressure levels required for a given tone to sound as loud as the corresponding reference tone of 1,000 cps (see also Fletcher and Munson, *J. Acoust. Soc. Am.*, 1933).

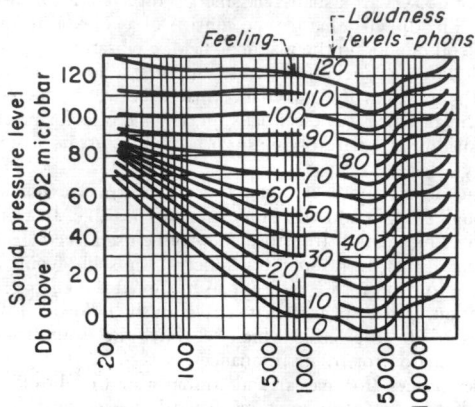

**Fig. 2** Loudness contours.

**Loudness** is a subjective rather than a purely physical attribute. To provide a qualitative basis, the loudness level in **phons** is defined as the pressure level in decibels of a pure 1,000 cps tone which a typical observer judges to sound as loud as the sound in question. Observers can experimentally judge the loudness of pure or complex tones. However, this does not mean that the apparent level is proportional to its level in phons; i.e., a level of 10 phons is not twice as loud as one of 5 phons. An additional expression, **sones**, defined as the loudness of a 1,000 cps tone at 40 dB intensity, is necessary to compare various loudness. The relationship between sones and phons is shown in Fig. 3.

**Quality** is a subjective attribute of sound in which equally loud sounds may be distinguished as to kind. Basically, differences in quality arise from differences in the distribution of

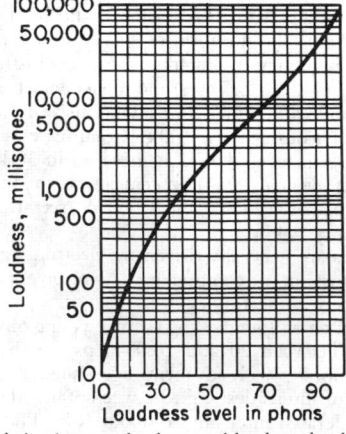

**Fig. 3** Relation between loudness and loudness level.

energy in different parts of the frequency spectrum. In *music* this takes the form of the energy relationship of fundamental and harmonics; in *noise* it is random. These differences affect the sensation of loudness of noise and the psychological annoyance it produces. Shrill, high-pitched, and irregular sounds are usually judged less pleasant than low-pitched and regular sounds.

Among terms used to define quality are **pitch,** determined by frequency (mostly), together with intensity and waveshape (unit is the mel); **timbre,** determined by wave shape (mostly), together with intensity and frequency (see Seashore, "Psychology of Music," Dover).

**Masking** describes the ability of one sound to make the ear incapable of perceiving a second one. ANSI 53.20 (1973), Psychoacoustical Terminology. It is measured by the shift in the threshold of audibility of the masked sound in decibels. The partial deafening of the ear by the masking effect of noise affords a direct quantitative measure of the interfering effect of the noise. If the masking is measured at several frequencies throughout the audible ranges, the overall pressure level of the sound can be computed. For any given frequency, the masking is expressed in decibels relative to the unmasked threshold for a **critical bandwidth** centered on the frequency of the masked tone. This critical bandwidth is that beyond which an increase in the passband has little effect on the masking of a pure tone at its center frequency. Critical bandwidths vary from 40 cycles at 100 cps to 200 cycles at 6,000 cps.

**Noise** is an undesired sound. It implies an unwanted disturbance in a useful band which interferes with the useful information. Any elastic structure may produce noise when set into vibration. Generally the motion is an unwanted concomitant of some desired function, e.g., the vibration of a machine tool.

The term has a connotation of unpleasantness in quality or loudness. Typical examples are gear noises, 60-cycle hum, motor traffic, hammer blows, pneumatic-tool operations, and hissing of gases in an orifice. Noise measurements are usually made with a **sound-level meter,** comprising a microphone, attenuator, amplifier, frequency-weighing networks, and an indicator or recorder. ANSI 51.4 (1971), Sound Level Meters. A **sound analyzer** indicates sound pressure as a function of frequency. In some cases electrical filters may be included which permit measurement in certain restricted frequency ranges. By using these instruments the components of the heterogeneous noise may be identified, and this helps to correlate it with its production. Once located, techniques for elimination flow, and the results may be used to compute masking. Individual frequencies may be identified by beating against a known source (**beat-frequency oscillator**) until a null is observed.

**Contact transducers (vibration pickups)** may be used to locate sources of noise, such as in partitions and machine parts. They may be piezoelectric or magnetic and may be used in place of the microphone or the sound-level meter; hydrophones may also be used.

Where the frequency distribution of the noise is significant, an analyzer may be used since the level meter tells nothing about frequency distribution. Analyzers are manufactured in various forms, e.g., the octave-band analyzer, the impact-noise analyzer, and the wave analyzer, each of which uses a different method of finding out which frequency components are present.

Typical sound levels are shown in Table 3.

### Table 3. Typical Sound Levels

|  | Decibels |  |
|---|---|---|
|  | 120 | Threshold of feeling |
|  |  | Thunder, artillery |
| Deafening | 110 | Nearby riveter |
|  |  | Elevated train |
|  | 100 | Boiler factory |
|  |  | Loud street noise |
| Very loud | 90 | Noisy factory |
|  |  | Truck unmuffled |
|  | 80 | Police whistle |
|  |  | Noisy office |
| Loud | 70 | Average street noise |
|  |  | Average radio |
|  | 60 | Average factory |
|  |  | Noisy home |
| Moderate | 50 | Average office |
|  |  | Average conversation |
|  | 40 | Quiet radio |
|  |  | Quiet home or private office |
| Faint | 30 | Average auditorium |
|  |  | Quiet conversation |
|  | 20 | Rustle of leaves |
|  |  | Whisper |
| Very faint | 10 | Soundproof room |
|  |  | Threshold of audibility |
|  | 0 |  |

## NOISE CONTROL

REFERENCES: General Radio Co., "Handbook of Noise Measurement." Harris, "Handbook of Noise Control," McGraw-Hill. Bolt, "Handbook of Acoustic Noise Control." Faulkner, "Handbook of Industrial Noise Control," Industrial Press.

Noise control may be carried out at several stages (1) at the source, by design changes or by quieting procedures, (2) during transmission, by attention to the path by which it is propagated to the listener, and (3) by quieting at the listening position. It may also be controlled architecturally as by the careful placement of necessarily noisy rooms in a building.

**The Source**  By inspection and test procedures, a noise is tracked to its source. In some cases, design procedures which attempt to reduce the vibration or to prevent its radiation may be used. This may require the redesign of elements, such as cams, gears, housings, or provisions for cushioning. Viscous damping materials, e.g., putty or tar, may be applied to the vibratory surfaces in the form of nonhardening plastic mixtures. A machine may be isolated by sections or shock mounts to prevent transmission of vibration from one section to another. Absorbing materials may be placed on walls to absorb sound after it has been radiated.

**Transmission Isolation**  If the vibrations of noisy machinery cannot be suppressed at the source, their transmission to the listener should be impeded. For the higher frequencies constituting noise, the most effective isolation method is the introduction of elastic discontinuities in the structure transmitting the noise (measured by the difference between density-velocity products as given in Table 1). The discontinuities may be obtained by the use of felt, cork, rubber, or springs in machinery mountings, or by the introduction of alternate lead and cork sheeting at masonry junctions. The isolation treatment should be applied as close to the source as possible in order to eliminate sound radiation from the structures transmitting the vibrations. Where this is not possible, the listening space itself may be isolated. Thus **quiet rooms,**

constructed especially for noise measurements, are usually built as separate structures isolated from the main building.

**Filtration**  Some problems of noise transmission through air lend themselves to solution by methods of filtration. Typical examples are the transmission of sound in ventilating ducts and the noise production at engine exhaust pipes. In each of these cases, the steady flow of gas must not be impeded, but the alternating flow, representing sound transmission, must be effectively suppressed.

For ventilating ducts, an acceptable degree of noise suppression may be obtained by lining the ducts (on at least two nonopposite walls) with an efficient sound absorbent for a distance of 10 to 15 ft from both the inlet and the outlet. Where the length of duct available is insufficient, or where additional noise suppression is required, baffles, covered with absorbing material, may be introduced in the duct. A plenum chamber, used to serve several ducts, should be lined with sound absorbents. If the air velocities are high, it may be necessary to introduce additional baffles at bends in the ducts to avoid noise production through turbulence.

Exhaust mufflers are usually modifications of the elementary low-pass acoustical filter, comprising a through tube to which closed cavities are coupled through small holes at intervals along the tube. Typical structures of this type (Fig. 4) produce little increase in back pressure and considerable attenuation of sound waves having frequencies above a cutoff frequency determined by the size of the holes and cavities.

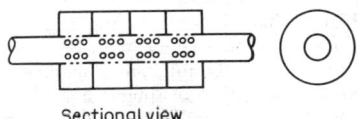

Sectional view

**Fig. 4**  Straight-through exhaust muffler.

Porous packing, such as steel wool, in the side cavities or studied irregularity in the size and spacing of the cavities will increase the uniformity of noise suppression, whereas increasing the number of side cavities and the length of the muffler will increase the amount of suppression. Baffles in the tailpipe or irregular obstructions producing devious flow paths, e.g., the stone-filled pit for stationary engine exhausts, produce muffling action at the expense of appreciable increase in exhaust back pressure.

**Shielding** of airborne noise must be done by sound-opaque screens large in comparison with the wavelength of the sounds whose transmission they are to impede. This is seldom possible in building interiors except by utilization of building partitions as screens. Sound is transmitted through such partitions principally by minute flexure of the wall as a whole in response to the incident sound pressure on the noisy side, with consequent reradiation on the quiet side. Reduction of sound transmission is obtained by increasing the mass per unit area of the partition, by constructing the partition of material having large viscosity for bending, such as Thermax, or by the use of double partitions, vibrationally isolated.

Sound-transmission loss is usually greater for high frequencies than for low and is measured by comparing the average sound level on each side of the partition under standardized conditions. Average values of transmission loss, for frequencies from 125 to 4,000 cps, for typical partitions are shown in

Table 4. In any specific case, a more exact measure of the effectiveness of an insulating partition can be obtained by direct comparison of the transmission-loss vs. frequency curve for the partition and the intensity vs. frequency curve for the noise. Additional data on transmission loss for a wide variety of building matierals and structures are available in the NBS publications TRBM-44; BMS 17 and Supplements 1 and 2.

In general, double partitions (including floated floor constructions) provide greater transmission loss than equally heavy concrete, masonry, or brick walls but, except for special designs, less transmission loss than equally thick masonry walls. Double walls must be constructed carefully to avoid loss of vibration isolation through mechanical bridging between the opposite surfaces. Sound-absorbing fillers (e.g., mineral wool) are usually detrimental to sound insulation if in contact with both interior surfaces, and a single bridging nail may alter significantly the insulating efficiency. For maximum effectiveness, one of the wall surfaces should be hung structurally free at all four edges with the boundary cracks sealed, with felt or asphalt compounds, against sound leakage. Through piping should be made vibrationally discontinuous by introducing canvas or metallic sylphon sections, and clearance holes at the walls should be sealed. Sound leakage through small clearance cracks contributes to the low transmission loss of ordinary doors. Special self-sealing "soundproof" doors are required to maintain the effectiveness of an efficient sound-insulating partition.

**Quieting**   The sound level established in a room by a noise source is higher than that which the same source would produce in free space on account of successive reflections of sound at the walls. It is the function of **quieting** to avoid such enhancement of noise by providing a high degree of sound absorption at all interior reflecting surfaces exposed to the noise. Commercially available sound-absorbing materials may be cemented to flat surfaces or secured to wood or metal furring strips. They derive their absorbing property either from capillary porosity of the surface or from the dissipative vibration of surface layers. Hanging "functional absorber" units comprising vibratile matte surfaces, enclosing a volume of about 1 ft³, can be used where surface absorbents cannot be installed conveniently. The effectiveness of sound absorbents varies with frequency, usually being greater for high and intermediate than for low frequencies. It may be measured by determining the **absorption coefficient,** defined as the fraction of sound energy diffusely incident on the material that is not reflected, or by determining the **specific acoustic impedance** of the material. The measured absorption coefficient is not a property of the material alone, but depends partly on the size and mounting of the test sample and the size and shape of the test chamber; thus comparison of the coefficients for different materials should be based only on measurements made under identical conditions. Such measurements on a wide variety of materials have been made available by the Acoustical Materials Assoc. (Chicago, Ill.), although it is to be expected that the absorption coefficients effective in various practical applications may differ somewhat from the published values.

For ordinary noise quieting, the average of absorption coefficients measured at frequencies of 250, 500, 1,000, and 2,000 cps, called the **noise-reduction coefficient,** may be used. Typical values of this coefficient for representative materials are given in Table 5. In making quantitative estimates of noise reduction, the **total sound absorption** of the room boundaries may be computed by multiplying the noise-reduction coefficient of each different material present by the total exposed area of that material and summing up the resulting products. The noise reduction is then given by

$$\text{Noise reduction in decibels} = 10 \log \frac{\text{total absorption after treatment}}{\text{total absorption before treatment}}$$

When the frequency spectrum of the offending noise is known, greater precision in calculation of total absorption is obtained by replacing the noise-reduction coefficient by the absorption coefficient measured at the frequency of maximum loudness level from the noise source. Subjective judgments of the loudness reduction obtained by quieting can be estimated by using the noise reduction in decibels in connection with the loudness chart of Fig. 3.

In general, the larger the area of absorbing material introduced and the higher its noise-reduction coefficient, the more effectively the noise is reduced. No amount of quieting treatment can reduce the level of the noise received directly from the source. If full coverage of walls and ceiling is not possible, distribution of the material in several small patches is more effective than the same total area of material concentrated in one location. Similarly, the same area of material is more effective when applied to nonopposite walls and ceiling than when concentrated on either of these areas, and more effective when located near the edges and corners of a given area than when located in the center.

**Table 4. Sound-Transmission Loss in Building Partitions**

| Wall | Thickness, in. | Weight, lb per sq ft | Transmission loss, db |
|---|---|---|---|
| Wood | 0.2 | 0.45 | 18.5 |
| Plate glass | 0.25 | 3.2 | 27.0 |
| Hollow gypsum tile, unplastered | 3 | 11.1 | 27.2 |
| Brick wall, unplastered | .... | 22.0 | 33 |
| Brick wall, plastered | 6 | 46 | 43 |
| Brick wall, plastered | 10.5 | 93 | 49 |
| Double wall; metal lath, ½ in. gypsum plaster, on staggered 2 × 4 in. wood studs | 7.5 | 19.8 | 44 |
| Double 3 in. hollow gypsum tile, unplastered, 3 in. air space... | 9 | 22.0 | 42.6 |
| 1 in. Thermax nailed over building paper to 3 in. Thermax laid up in mortar, ½ in. plaster on both sides | 5 | 15 | 47 |
| Double 2 in. solid-gypsum tile, unplastered, completely isolated structurally by separate foundations, 4 in. air space | 8 | 20.4 | 59 |

Based on Sabine, "Acoustics and Architecture," McGraw-Hill.

**Table 5. Sound-Absorption Coefficients**

| Maker, material thickness | Absorption coef at indicated frequencies | | | | | | Noise-reduction coef | Weight, lb per sq ft | Surface | AMA test No. |
|---|---|---|---|---|---|---|---|---|---|---|
| | 128 | 256 | 512 | 1,024 | 2,048 | 4,096 | | | | |
| Armstrong Cork Co. Cushiontone A, ¾ in... | 0.10 | 0.28 | 0.66 | 0.91 | 0.82 | 0.69 | 0.65 | 1.05 | 484 holes per sq ft, ³⁄₁₆ in. diam, ⅝ in. deep. Painted by mfr | 47–28 |
| Travertone, ¾ in...... | 0.06 | 0.23 | 0.78 | 0.97 | 0.84 | 0.80 | 0.70 | 1.20 | Fissured, painted | 48–58 |
| The Celotex Corp. Acousti-Celotex Type C-9, ¾ in..... | 0.11 | 0.23 | 0.80 | 0.93 | 0.58 | 0.50 | 0.65 | 0.96 | 441 holes per sq ft, ³⁄₁₆ in. diam. Any paint | 46–132 |
| Type M-1, ⅝ in..... | 0.07 | 0.21 | 0.64 | 0.86 | 0.93 | 0.83 | 0.65 | 1.31 | 676 holes per sq ft. ⁵⁄₃₂ in. diam. Any paint | 46–12 |
| Q-T Ductliner, . in.... | 0.21 | 0.42 | 0.71 | 0.86 | 0.79 | 0.75 | 0.70 | 1.3 | Unpainted | A 48–10 |
| Johns-Manville Corp. Sanacoustic, KK, pad plus metal facing and pad supports 1³⁄₁₆ in............. | 0.25 | 0.58 | 0.96 | 0.97 | 0.85 | 0.72 | 0.85 | Pad 1.28 | 4,608 holes per sq ft, 0.068 in. diam. Enameled metal pan backed with wool pad | 46–88 |
| Fibretone, 1³⁄₁₆ in..... | 0.14 | 0.37 | 0.69 | 0.80 | 0.76 | 0.73 | 0.65 | 1.17 | 484 holes per sq ft, ³⁄₁₆ in. diam. Any paint | 46–124 |
| Airacoustic, 1 in....... | 0.29 | 0.31 | 0.70 | 0.82 | 0.79 | 0.80 | 0.70 | 1.50 | Unpainted | 46–71 |
| National Gypsum Co. Acoustifibre, ⅝ in..... | 0.10 | 0.16 | 0.62 | 0.97 | 0.81 | 0.73 | 0.65 | 0.56 | 441 holes per sq ft, ³⁄₁₆ in. diam, ³⁄₁₆ in. deep. Painted | 46–137 |
| Owens-Corning Fiberglas Corp. Fiberglas Acoustical Tile, plain type, ¾ in........ | 0.04 | 0.20 | 0.63 | 0.91 | 0.82 | 0.82 | 0.65 | 0.69 | Painted | A 48–99 |
| United States Gypsum Co., Acoustone F, 1¹⁄₁₆ in.. | 0.08 | 0.25 | 0.76 | 0.84 | 0.78 | 0.73 | 0.65 | 1.35 | Fissured, painted | 46–50 |
| Brick wall, painted...... | 0.012 | .... | 0.017 | .... | 0.023 | .... | 0.02 | | | |
| Concrete wall or floor.... | 0.01 | .... | 0.015 | .... | 0.02 | .... | 0.02 | | | |
| Wood floor............. | 0.05 | .... | 0.03 | .... | 0.03 | .... | 0.03 | | | |
| Cork or rubber tile on concrete | .... | .... | 0.03–0.08 | .... | .... | .... | 0.05 | | | |
| Glass................. | 0.035 | .... | 0.027 | .... | 0.02 | .... | 0.02 | | | |

This tabulation is based on *Bull.* XI (1949), Acoustical Materials Assoc. All samples were cemented to plasterboard for test, except that the Sanacoustical unit is attached to wood furring with special clips, and the duct linings are laid on 24 gage sheet iron, nailed to 1 × 3-in wood furring, 24-in O.C.

## APPLICATIONS

### Industrial Applications

(See also Sec. 5)

REFERENCES: Carlin, "Ultrasonics," McGraw-Hill. Bergmann, "Ultrasonics," S. Hirzel Verlag. ANSI Z24.18 (1957) (R1971), Ultrasonic Therapeutic Equipment.

Acoustic waves of high powers used in industrial applications are generally called **sonic** (or **ultrasonic,** when greater than about 20,000 cps). High-intensity sonic waves, in the 10,000 cps to the megacycle range ($10^6$ cps), are applied to many industrial processes. The effects seem to be a function of cavitation, heating, particle acceleration, short wavelength, and other characteristics of the waves. Application categories are (1) high amplitude, and (2) low amplitude.

**High-amplitude waves** are used in operations such as cleaning, welding, drilling, emulsification, soldering, chemical and biological applications, medical therapy, and sonar. The energy may be continuous, pulsed, or modulated, in various ways.

**Low-amplitude waves** are used in operations such as material testing, burglar alarms, delay lines, or medical diagnoses. Any waveshape may be used; basically, a physical characteristic of the waves, such as velocity, is measured.

**Cavitation** may be defined as the formation and collapse of gas- or vapor-filled bubbles. Most significant industrial applications, e.g., cleaning, take place during the vaporous phase. The amount and force of cavitation are affected by the character of the liquid and the gas in it. The bubble collapse generates powerful local forces which cause the desired action. Levels of power in the liquid of approximately 3 W/cm² are required for intense cavitation and are dependent upon factors such as the liquid, temperature, and external pressure; powers as low as 0.3 W/cm² in water will produce a threshold cavitation (see also Briggs et al., *J. Acoust. Soc. Am.*, 1947).

### High-Amplitude Applications

(See Fig. 5)

**Cleaning**  An ultrasonically agitated bath will erosively clean dirt from immersed articles. A high-power generator, usually electronic, produces sonic energy which is impressed on a transducer to drive the bath. Barium titanate transducers prevail, but magnetostrictive designs may be used. 10,000 to 90,000 cps are commonly used, with the lower frequencies generally more effective. Generator tuning may be manual or by feedback from the transducer controls. Power levels of 5 ± W/cm² are commonly used; size (depth) of the tank may affect the output; 50 ± W/gal is a rough empirical relationship; cavitation must exist for effective, speedy cleaning. A proper cleaning solution must be used, i.e., one which supports intense cavitation and also clean (e.g., water solutions of alkalines or acids, or solvents like trichlorethylene); temperatures between 120 and 160°F prevail. Among items commonly cleaned are jewelry (lost-wax castings), eyeglass frames, lenses, metal parts, and watches and clocks.

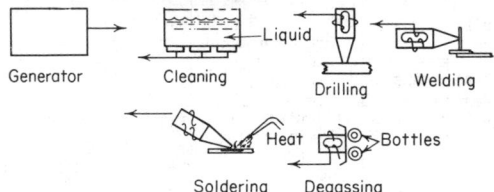

**Fig. 5**  High-amplitude systems.

**Foaming of Beverages**  Air content, which determines the life of a carbonated beverage, is reduced by foaming the bottles or cans before capping. Magnetostrictive transducers at 20,000 cps and 250 W, in contact with the containers, produce enough foam from the $CO_2$ to expel the air. The ultrasonic power requirement is basically small, but the losses in the coupling dictate the use of large generators. Similar techniques and apparatus may be used for the removal of gases from materials and for chemical effects such as the acceleration of iodine reactions and oxidation.

**Soldering**  Some materials, such as aluminum, oxidize when exposed to air so that soldering is not possible. However, an ultrasonically driven solder bath will cause wetting of the material with solder and tinning of the surface. Magnetostrictive units are indicated because of the temperature requirements, with external heating and high tin-content solders; pots, as well as irons, may be constructed. Applications include aluminum wire, foil capacitors, and the filling of holes in castings (see also Sec. 13).

**Welding**  Similar equipment (between 100 and 5,000 W output) may be used to weld thin metal or thin-to-thick sections. Such units apply the ultrasonics in a shear direction with respect to the parts to be welded. The process depends primarily on the sonic energy, the clamping force, and the amount of external heating. Either spot or lap welds are possible, but in all cases one of the sections must be thin (see also Sec. 13 and AWS Welding Handbook).

**Drilling** is effected with the same sort of apparatus, but the force is longitudinal rather than shear. An abrasive is flowed over the tool head and is driven by the cavitation against the part to be drilled, causing the material to erode away. Any shape may be obtained in this manner. The head is usually mounted on a milling machine or similar apparatus. Tolerances depend principally on the physical rigidity of the system and on grit size. Applications include hard materials such as ceramics, jewels, and glasses. The same apparatus has been applied to dental drilling, but the time required and the necessity for use of a slurry has kept it from greater acceptance. However, the method is widely applied to cleaning the surface of the teeth. The technique has been applied to forming lesions in the brain and spinal column of human beings, using a focused beam of sound at 3 ± Mc rather than a velocity transformer (see also Sec. 13).

A **whistle** may be operated in a gas for agglomeration or foam settling and in a liquid for emulsification. Ultrasonic fields introduce additional forces on particles suspended in a gas, causing them to come together. Materials such as smoke, dust, and fog have been experimentally treated in this way. Generally, the whistle drives a resonant cavity, generating about 150 dB intensity (power output of 150 ± W) at 10 to 20 kc. Atomization of a liquid may be effected by introducing the liquid into a strong sonic field, either passing it directly over or through the whistle. The use of waves has been reported for drying solids, such as sugar, for atomically driven sound beacons under the sea, and for emulsification of liquid rubber and other materials.

**Sonar**  Underwater signaling and detection are among the older applications of ultrasonics and comprise the active (pulse-echo, Doppler) and passive (listening) systems. The principles of operation are similar to those of pulse-echo testing in the active case (see Albers, "Underwater Acoustics Handbook," Penn. State Univ. Press).

Among the **miscellaneous applications** of high-power ultrasonics is metal treatment, atomization of oil for burners, ultrasonic diathermy for bursitis, humidification, and ultrasonic neurosonic surgery.

**Testing Materials (see Sec. 5 and McMaster, "Nondestructive Testing Handbook," Ronald).**  The most common industrial use of low-power ultrasonic waves is for testing materials. The technique basically depends on the ability of a discontinuity in a material to reflect part of the energy hitting it. Various types of ultrasonic waves such as longitudinal, shear, or surface may be used. Transducers are usually crystal, such as barium titanate or quartz, and measure from ¼ to 1 in. diam (Fig. 6).

The basic types of ultrasonic systems are (1) pulse-echo, (2) through-transmission, and (3) resonance. The **pulse-echo system** uses a pulse ranging in length (time) from a fraction of a microsecond to several microseconds and an amplitude from 50 to 250 V radio frequency across the transducer. The pulse travels in the material and is reflected at an interface; the time of travel is measured. The pulse-echo method has also been applied to medical mapping of the body interior and for tracing brain centerline displacement and heart valve action.

The **through-transmission system** places continuous pulsed or modulated waves on a transducer coupled to one side of a part with pickup on the other side. If a flaw interrupts, the waves do not penetrate the part.

The **resonance system** uses a single transducer and varies the frequency applied to it. Within the applied frequencies is one whose wavelength is related to the thickness of the part in such a way that less power is required of the driving system. This condition is indicated, and since the wavelength within the material is known, the thickness is determined. The most common applications of the resonance system are (1) for thickness measurement when one side only is available, and (2) for finding laminations in thin sections and lack of bond (see also Sec. 5).

The **sonic burglar alarm** depends on the Doppler shift caused in a sonic field by a moving object. The unit operates at 20 kc and will find minimum objects of 0.03 ft² in an enclosure of 100 ft². Magnetostrictive transducers are used coupled to

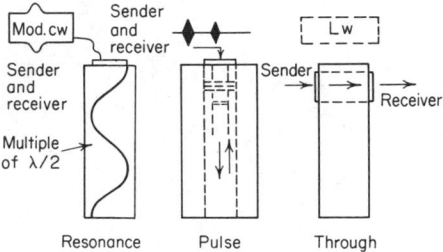

**Fig. 6** Low-amplitude systems.

diaphragms; they are not unlike radio speaker systems in appearance.

**Liquid-Level Sensors** Ultrasonic devices may be used to measure the level of liquid in a tank either by the pulse-echo technique or by indicating the transducer lead, i.e., a transducer driven by an oscillator where the reaction of the liquid load causes a change in the driving current.

**Sonic microscopes** are devices in which a beam of sound illuminates a part; the shadows of the field are scanned on a cathode-ray tube (usually constructed of barium titanate) by a flying spot.

## SAFETY

Under the Occupational Safety and Health Act (OSHA) (Federal Register, vol. 37, no. 202, Oct. 18, 1972), definitions have been made which legally define levels either as safe or as hazardous. Moreover, noise is now beginning to be recognized as a pollutant, both as a nuisance and as the cause of hearing impairment.

General noise levels in the environment have already been defined (Fig. 2). There is some evidence that noise may cause ailments such as anxiety and heart disorders.

Protection from noise is required when sound levels exceed those in Table 6 (Table G-16 of the Act), when measured on the A scale at slow response on a standard sound-level meter (except for certain alarms, etc., as provided in the Act). When several successive exposures occur, they are combined (see paragraph 1910.95, reference above).

Conversion from octave-band analysis levels to A-weighted levels may be made from the figure in the Act. This figure has been also adopted by some local or state laws.

**Table 6. Permissible Noise Exposures**

| Duration per day, h | Sound level, dBA slow response |
|---|---|
| 8 | 90 |
| 6 | 92 |
| 4 | 95 |
| 3 | 97 |
| 2 | 100 |
| 1½ | 102 |
| 1 | 105 |
| ½ | 110 |
| ¼ or less | 115 |

When the environmental noise is greater than specified in the Act, protection must be provided.

However, when the noise is intermittent, if the peaks occur within 1 s or less, it is considered continuous. When protective

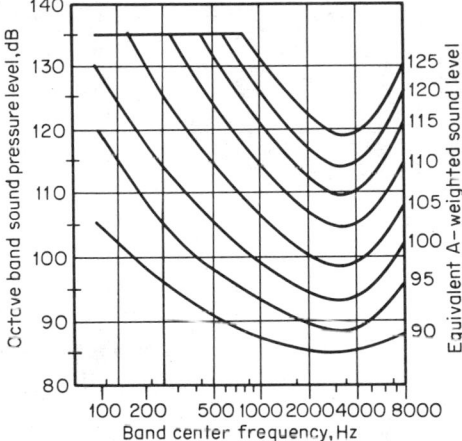

**Fig. 7** Equivalent sound-level contours. Octave-band sound-pressure levels may be converted to the equivalent A-weighted sound level by plotting them on this graph and noting the A-weighted sound level corresponding to the point of highest penetration into the sound-level contours. This equivalent A-weighted sound level, which may differ from the actual A-weighted sound level of the noise, is used to determine exposure limits.

equipment is required, it must be provided by a trained person and periodic checks made of the effectiveness.

In addition to the levels, time of exposure is also involved, as shown in Table 6.

OSHA also requires, when 90 dBA is exceeded, "a continuing effective hearing conservation program shall be administered," i.e., consisting of periodic hearing checks and noise surveys.

The type of facility required for these tests is spelled out in the reference above.

# 13

# Shop Processes

BY

**CARL R. LOPER, JR.,** *Professor, Minerals and Metals Engineering, University of Wisconsin.*
**E. V. CRANE,** *Late Chief Engineer and Research Director, E. W. Bliss Co.*
**EDWARD A. FENTON,** *Consulting Engineer, Miami, Fl.*
**SEROPE KALPAKJIAN,** *Professor of Mechanical Engineering, Illinois Institute of Technology.*
**JAMES A. BROADSTON,** *Chief Engineer, Surface Checking Gage Co.*
**RICHARD W. PERKINS,** *Professor of Mechanical and Aerospace Engineering, Syracuse University*

### FOUNDRY PRACTICE AND EQUIPMENT
#### by Carl R. Loper, Jr.

Basic Steps in Making Sand Castings ..................... 13-2
Patterns ................................................ 13-2
Molding Processes and Materials ....................... 13-3
Molding Equipment and Mechanization .................. 13-4
Molding Sand .......................................... 13-6
Casting Alloys ......................................... 13-7
Melting Furnaces ...................................... 13-8
Cleaning and Inspection ............................... 13-8
Casting Design ........................................ 13-9

### WORKING METALS AND PLASTICS
#### by E. V. Crane

Structure .............................................. 13-9
Plasticity ............................................. 13-10
Plastic-Working Techniques ............................ 13-15
Rolling Operations .................................... 13-16
Shearing .............................................. 13-17
Bending ............................................... 13-19
Drawing ............................................... 13-19
Squeezing ............................................. 13-21
Equipment for Working Metals and Plastics ............. 13-23

### WELDING
#### by Edward A. Fenton

Fundamentals of Welding .............................. 13-27
Design Strengths ...................................... 13-27
Arc Welding ........................................... 13-28
Gas Welding ........................................... 13-33
Thermal Cutting of Metals ............................. 13-36
Resistance Welding .................................... 13-38
Surfacing by Welding .................................. 13-42
Brazing ............................................... 13-42
Welding Procedures ................................... 13-43

### MATERIAL-REMOVAL PROCESSES AND EQUIPMENT
#### by Serope Kalpakjian

Metal-Cutting Principles .............................. 13-49
Cutting-Tool Materials ................................ 13-51
Preparation of Tool Steels ............................ 13-54
Cutting Fluids ........................................ 13-54
Machining Processes and Equipment .................... 13-55
Lathes ................................................ 13-56
Tool Shapes ........................................... 13-57
Turret Lathes ......................................... 13-57
Screw Machines ....................................... 13-59
Boring Machines ...................................... 13-59
Drilling Machines ..................................... 13-60
Reaming Machines ..................................... 13-60
Threading Machines ................................... 13-61
Milling Machines ...................................... 13-61
Gear Manufacturing ................................... 13-62
Planers and Shapers ................................... 13-64
Broaching ............................................. 13-65
Cutting-Off Machines .................................. 13-66
Grinding Machines ..................................... 13-66
Polishing ............................................. 13-69
Machining of Plastics ................................. 13-70
Alternative Material-Removal Processes ................ 13-71

### SURFACE-TEXTURE DESIGNATION, PRODUCTION, AND CONTROL
#### by James A. Broadston

Design Criteria ....................................... 13-74
Designation Standards, Symbols, and Conventions ....... 13-75
Measurement and Production ........................... 13-77
Surface Quality versus Tolerances ..................... 13-77

### WOODCUTTING TOOLS AND MACHINES
#### by Richard W. Perkins

Sawing ................................................ 13-80
Planing and Molding .................................. 13-81
Boring ................................................ 13-82
Sanding ............................................... 13-83

# FOUNDRY PRACTICE AND EQUIPMENT

## by Carl R. Loper, Jr.

REFERENCES: Publications of the American Foundrymen's Society: "Cast Metals Handbook," "Alloy Cast Irons Handbook," "The Cupola and Its Operation," "Copper-base Alloys Foundry Practice," "Foundry Sand Handbook," "Statistical Quality Control for Foundries," "Symposium on Principles of Gating," "Time and Motion Study for the Foundry," "Patternmaker's Manual"; Caine, "Design of Ferrous Castings." "Ferrous Foundry Process Control," SAE. "Steel Castings Handbook," Steel Founders' Society of America. "American Malleable Iron Handbook," Malleable Founders' Society. Heine and Rosenthal, "Principles of Metal Casting," McGraw-Hill. Cady, "Precision Investment Castings," Reinhold. Marek, "Fundamentals in the Production and Design of Castings," Wiley. Cook, "Engineered Castings," McGraw-Hill. "Casting Design Handbook," ASM.

## BASIC STEPS IN MAKING SAND CASTINGS

The basic steps involved in making sand castings are:

1. **Patternmaking.** Patterns are required to make molds. The mold is made by packing molding sand around the pattern. The mold is usually made in two parts so that the pattern can be withdrawn, the top half being referred to as the **cope,** and the bottom half as the **drag.** When the pattern is withdrawn, the cope is replaced on the drag, the imprint of the pattern providing the cavity which ultimately is filled with metal to form the casting.

2. If the casting is to be hollow, additional patterns, referred to as **core boxes,** are needed to shape the sand forms, or **cores,** that are placed in the mold cavity to form the interior surfaces of castings. Thus the void between the mold and core eventually becomes the casting.

3. **Molding** is the operation necessary to prepare a mold for receiving the metal. It consists of ramming sand around the pattern placed in a support, or **flask,** removing the pattern, setting cores in place, cutting the **feeding system** to direct the metal if this feeding system is not a part of the pattern, removing the pattern, and closing the mold.

4. **Melting** and **pouring** are the processes of preparing molten metal of the proper composition and temperature and pouring this into the mold from transfer **ladles.**

5. **Cleaning** is all the operations required to remove the **gates** and **risers** that constitute the feeding system and to remove the adhering sand, scale, and other foreign material that must be removed before the casting is ready for shipment or other processing. Inspection for defects follows, and additional processing, such as heat-treatment, surface finishing, or machining may be necessary.

## PATTERNS

Since patterns are the forms for the castings, the casting can be no better than the patterns from which they are made. Where close tolerances or smooth casting finishes are desired, it is particularly important that patterns be carefully designed, constructed, and finished.

Patterns serve a variety of functions, the more important being (1) to shape the mold cavity to produce castings, (2) to make corrections for the characteristics of the metal cast, (3) to provide accurate dimensions, (4) to provide a means of getting liquid metal into the mold (gating system), and (5) to provide a means to support cores.

Usual allowances built into the pattern to ensure dimensional accuracy include the following: (1) **Draft,** the taper on the vertical walls of the casting which is necessary to extract the pattern from the mold without disturbing the mold walls. (2) **Shrinkage allowance,** a correction to compensate for the solidification shrinkage of the metal and its contraction during cooling. These allowances vary with the type of metal and size of casting. Typical allowances for cast iron are $\frac{1}{8}$ to $\frac{1}{12}$ in/ft; for steel, $\frac{1}{8}$ to $\frac{1}{4}$ in/ft; and for aluminum, $\frac{1}{16}$ to $\frac{5}{32}$ in/ft; however, a designer should consult appropriate references (AFS, "Cast Metals Handbook"; ASM, "Casting Design Handbook"; "Design of Ferrous Castings") or the foundry. These allowances also include a size tolerance for the process so that the casting is dimensionally correct. (See also Sec 6.) (3) **Machine-finish allowance** is necessary if machining operations are to be used in order that stock is provided for machining. Tabulated data are available in the references cited for shrinkage allowances. (4) If a casting is prone to distortion, a pattern may be intentionally distorted to compensate. This is a **distortion allowance.**

Patterns vary in complexity, depending on the size and number of castings required. **Loose patterns** are single prototypes of the casting and are used only when a few castings are needed. They are usually constructed of wood, but metal, plaster, plastics, or other suitable material may be used. The gating system for feeding the the casting is cut into the sand by hand. Some loose patterns may be split into two parts to facilitate molding.

**Gated patterns** incorporate a gating system along with the pattern to eliminate hand cutting.

**Match-plate patterns** have the cope and drag portions of the pattern mounted on opposite sides of a wooden or metal plate, and are designed to speed up the molding process. Gating systems are also usually attached. These patterns are generally used with some type of molding machine and are recommended where a large number of castings are required.

For fairly large castings or where an increase in production rate is desired, the cope and drag portions of the pattern may be mounted on separate plates, referred to as **cope- and drag-pattern** plates. These may thus be used on separate molding machines by separate workers and combined on the molding floor prior to pouring.

**Special Patterns and Devices** For extremely large castings, **skeleton** patterns may be employed. Large molds of a symmetrical nature may be made for forming the sand mold by **sweeps,** which provide the contour of the casting through the movement of a template around an axis.

**Follow boards** are used to support irregularly shaped patterns which require an irregular parting line between cope and drag. A **master pattern** is used as an original to make up a number of similar patterns that will be used directly in the foundry.

## MOLDING PROCESSES AND MATERIALS

### Molding Processes

**Green Sand** Most castings are made in **green sand,** i.e., sand bonded with clay or bentonite and properly tempered with water to give it **green strength.** Miscellaneous additions may be used for special properties. This method is adaptable to high production of small- or medium-sized castings because the mold can be poured immediately after forming, and the sand can be reused and reprocessed after the casting has solidified.

**Dry-sand Molds** These molds are made with green sand but are baked prior to use. The surface is usually given a refractory wash before baking to prevent erosion and to produce a better surface finish. Somewhat the same effects are obtained if the mold is allowed to **air-dry** by leaving it open for a period of time before pouring, or if it is **skin-dried** by using a torch, infrared lamps, or heating elements directed at the mold-cavity surface.

**Core molding** makes use of assembled cores to construct the mold. The sand is prepared by mixing with oil, or cereal, forming in core boxes, and baking. This process is used where the intricacy of the casting justifies the extra costs involved.

**Carbon Dioxide Process Molds** These molds are made in a manner similar to the green-sand process but use sand bonded with sodium silicate. When the mold is finished, carbon dioxide gas is passed through the sand to produce a very hard mold with many of the advantages of dry-sand and core molds but requiring no baking.

**Floor and Pit Molding** When large castings are to be produced, these may be either directly on the floor of the foundry or formed in pits in the floor which serve as the flask. **Loam molding** is a variation of floor molding in which molding material composed of 50 percent sand and 50 percent clay (approx) is troweled onto a brickwork surface and brought to dimension by use of patterns, sweeps, or templates.

**Shell Molding (C Process)** Sand castings having close dimensional tolerances, and smooth finish can be produced by a process using a synthetic-resin binder. The sand and resin mixture is dumped onto a preheated metal pattern, which causes the resin in the mixture to set as a thin shell over the pattern. When the shell has reached the proper thickness, the excess sand is removed by rotating the pattern to dump out the sand. The remaining shell is then cured on the pattern and subsequently removed by stripping it off, using mold-release pins which have been properly spaced and that are mechanically or hydraulically made to protrude through the pattern. Mating shell halves are then bonded together with a cement or by a suitable backing, and poured.

**Plaster Molds** Plaster or plaster-bonded molds are used for casting certain aluminum or copper-base alloys. Dimensional accuracy and excellent surface finish make this a useful process for making rubber-tire molds, match plates, etc.

A variation of this method of molding is the **Antioch process,** using mixtures of 50 percent silica sand, 40 percent gypsum cement, 8 percent talc, and small amounts of sodium silicate,

portland cement, and magnesium oxide. These dry ingredients are mixed with water and poured over the pattern. After the mixture is poured, the mold is steam-treated in an autoclave and then allowed to set in air before drying in an oven. When the mold has cooled it is ready for pouring. Tolerances of ±0.005 in (±0.13 mm) on small castings and ±0.015 in (±0.38 mm) on large castings are obtained by this process.

The **lost-wax, investment, or precision-casting, process** permits the accurate casting of highly alloyed steels and of nonferrous alloys which are impossible to forge and difficult to machine. The procedure consists of making an accurate metal die into which the wax or plastic patterns are cast. The patterns are assembled on a sprue and the assembly sprayed, brushed, or dipped in a slurry of a fine-grained, highly refractory aggregate, and a proprietary bonding agent composed chiefly of ethyl silicate. This mixture is then allowed to set. The pattern is coated repeatedly with coarser slurries until a shell of the aggregate is produced around the pattern. The molds are allowed to stand until the aggregate has set, after which they are heated in an oven in an inverted position so that the wax will run out. After the wax is removed, the molds are baked in a preheat furnace. The molds may then be supported with loose sand and poured in any conventional manner.

All dimensions can be held to a tolerance of ±0.005 in (±0.13 mm) with some critical dimensions held to 0.002 in. (0.05 mm) Most castings produced by this process are small.

Faithful reproduction and accurate tolerances can also be attained by the **Shaw process,** which combines the advantages of the dimensional control of precision molds with the ease of production of conventional molding. The process makes use of wood or metal patterns and a refractory mold bonded with an ethyl silicate base material. Since the mold is rubbery when stripped from the pattern, some back draft is permissible.

In the **cement-sand process** portland cement is used as the sand binder. A typical mixture has 11 percent portland cement, 89 percent silica sand, and water 4½ to 7 percent of the total sand and cement. New sand is used for facing the mold and is backed with ground-up sand which has been rebonded. Cores are made of the same material. The molds and cores must air-dry 24 to 72 h before pouring. The process can be used for either ferrous or nonferrous castings. This molding mixture practically eliminates the generation of gases, forms a hard surface which resists the erosive action of the metal, and produces castings with good surfaces and accurate dimensions.

### Permanent-mold Casting Methods

In the **permanent-mold casting method,** fluid metal is poured by hand into metal molds and around metal cores without external pressure. The molds are held together by C clamps or by a screw or toggle.

Metals suitable for this type of casting are lead, zinc, aluminum and magnesium alloys, certain bronzes, and cast iron.

For making iron castings of this type, a number of metal-mold units are usually mounted on a turntable. The individual operations, such as coating the mold, placing the cores, closing the mold, pouring, opening the mold, and ejection of the casting, are performed as each mold passes certain stations. The molds are preheated before the first casting is poured. The process produces castings having a dense, fine-grained structure, free from shrink holes or blowholes. The tool changes are relatively low, and better surface and closer tolerances are obtained than with the sand-cast method. It does not maintain tolerances as close or sections as thin as the die-casting or the plaster-casting methods.

Yellow brasses, which are high in zinc, should not be cast by the permanent-mold process because the zinc oxide fouls the molds or dies.

The **semipermanent-mold casting method** differs from the permanent-mold casting in that sand cores are used, in some places, instead of metal cores. The same metals may be cast by this method. This process is used where cored openings are so irregular in shape, or so undercut, that metal cores would be too costly or too difficult to handle. The structure of the metal cast around the sand cores is like that of a sand casting. The advantages of permanent-mold casting in tolerances, density, appearance, etc., exist only in the section cast against the metal mold.

**Graphite molds** may be used as short-run permanent molds since they are easier to machine to shape and can be used for higher-melting-point alloys, e.g., steel. The molds are softer, however, and more susceptible to erosive damage. Steel railroad-car wheels may be made in these molds and can be cast by filling the mold by **low-pressure casting** methods.

In the **slush casting** process, the cast metal is allowed partially to solidify next to the mold walls to produce a thin-walled hollow casting where the excess liquid metal is poured out of the permanent mold.

In **centrifugal casting** the metal is under centrifugal force, developed by rotating the mold at high speed. This process, used in the manufacture of bronze, steel, and iron castings, has the advantage of producing sound castings with a minimum of risers. In **true centrifugal castings** the metal is poured directly into a mold which is rotated around its own axis.

In **pressure casting,** for asymmetrical castings which cannot be spun around their own axes, the mold cavities are arranged around a common sprue located on the neutral axis of the mold. The molds used in the centrifugal-casting process may be metal cores or dry sand, depending on the type of casting and the metal cast.

Most of the cast-iron pipe produced in the United States is centrifugally cast by the **de Lavaud process,** which employs water-cooled metal molds. The U.S. Army produces gun barrels at the Watertown Arsenal by the centrifugal process. Other types of castings currently being produced are gear blanks, bushings, pistons, piston rings, alloy-steel tubes, liners, valve bodies, and sheaves.

**Die-casting machines** consist of a basin holding molten metal, a metallic mold or die, and a metal-transferring device which automatically withdraws molten metal from the basin and forces it under pressure into the die. Two forms of die-casting machines are in general use. Lead, tin, and zinc alloys containing aluminum are handled in **piston machines.** Aluminum alloys and pure zinc, or zinc alloys free from aluminum, rapidly attack the iron in the piston and cylinder and require a different type of casting machine. The pressures in a piston machine range from a few hundred to 5,000 lb/in² (1 to 35 MN/m²). Zinc-aluminum alloys use about 1,000 lb/in². (See also Sec. 6.)

The **gooseneck machine** has a cast-iron gooseneck which dips the molten metal out of the melting pot and transfers it to the die. The pressure is applied to the molten metal by compressed air after the gooseneck is brought in contact with the die. The operation pressures average 350 to 500 lb/in² (2.5 to 3.5 MN/m²), with a maximum of about 700. This machine, developed primarily for aluminum alloys, is sometimes used for zinc-aluminum alloys, especially for large castings, but, owing to the lower pressure, the casting is likely to be less dense than when made in the piston machine. It is seldom used for magnesium alloys.

In **cold-chamber machines** the molten-metal reservoir is separated from the casting machine, and just enough metal for one casting is ladled by hand into a small chamber, from which it is forced into the die under high pressure. The pressures, ranging from a few thousand to 10,000 lb/in² (13 to 70 MN/m²) and more, are produced by a hydraulic system connected to the piston in the hot metal chamber. The alloy is kept so close to its melting temperature that it is in a slushlike condition. The process is applicable to aluminum alloys, magnesium alloys, zinc alloys, and even high-melting-point alloys like brasses and bronzes, since the pouring well, cylinder, and piston are exposed to the high temperature for only a short time. The production rates are lower than those obtained with the piston or gooseneck machines.

All metal-mold external-pressure castings have close tolerances, sharp outlines and contours, fine smooth surface, and high rate of production, with low labor cost. They have a hard skin, owing to the rapid chilling of the surface of the casting from the metal mold and a softer core.

The dies usually consist of two blocks of steel, each containing a part of the cavity, which are locked together while the casting is being made and drawn apart when it is ready for ejection. One-half of the die (next to the ejector nozzle) is stationary; the other half moves on a carriage. The dies are preheated before using and are either air- or water-cooled to maintain the desired operating temperature. Die life varies with the alloy and dimensional tolerances required. Retractable and removable metal cores are used to form internal surfaces. Inserts can be cast into the piece by placing them on locating pins in the die.

A wide range of sizes and shapes can be made by these processes, including threaded pieces and gears. Holes can be accurately located. The process is best suited to large-quantity production.

## MOLDING EQUIPMENT AND MECHANIZATION

Flasks may be filled with sand by hand shoveling, gravity feed from overhead hoppers, continuous belt feeding from a bin, sand slingers, and, for large molds, by an overhead crane equipped with a grab bucket.

**Hand ramming** is the simplest method of compacting sand. To increase the rate, pneumatic rammers are used. The method is slow, the sand is rammed in layers, and it is difficult to gain uniform density.

More uniform results and higher production rates are obtained by **squeezing machines.** Hand-operated squeezers are limited to small molds and are becoming obsolete; **air-operated machines,** using air at 80 to 100 lb/in², permit an increase in the allowable size of molds as well as in the production rate. These machines are suitable for shallow molds. Squeezer molding machines produce greatest sand density at the top of the flask and softest near the parting line of the pattern.

In **jolt molding machines** the pattern is placed on a platen

**Table 1. Design and Cost Features of Basic Casting Methods**

| Design and cost features | Process | | | | | |
|---|---|---|---|---|---|---|
| | Sand casting | Shell-mold casting | Permanent-mold casting | Plaster-mold casting | Investment casting | Die casting |
| Choice of materials | Wide—ferrous and non-ferfous | Wide—except for low-carbon steels | Restricted—brass, bronze, aluminum, some gray iron | Narrow—brass, bronze, aluminum | Wide—includes hard-to-forge and machine materials | Narrow—zinc, aluminum, brass, magnesium |
| Complexity | Considerable | Moderate | Moderate | Considerable | Greatest | Considerable |
| Size range | Great | Limited | Moderate | Moderate | Moderate | Moderate |
| Minimum section, in. | 3/32 | 1/16 | 0.100 | 0.010 | 0.010 | 0.025 |
| Tolerances, in. per ft* | 1/16–1/8 | 1/32–3/32 | 1/32–7/64 | 1/32–5/64 | 0.003–0.006 | 1/32–1/16 |
| Surface smoothness, microin., rms | 250–300 | 150–200 | 90–125 | 90–125 | 90–125 | 60–125 |
| Design feature remarks | Basic casting method of the industry | Considered to be a good low-cost casting method | Production economics with substantial quantities | Little finishing required | Best for parts too complicated for other casting methods | Most economical where applicable |
| Tool and die costs | Low | Low to moderate | Medium | Medium | Low to moderate | High |
| Optimum lot size | Wide—range from few pieces to huge quantities | More required than sand castings | Best when requirements are in thousands | From one to several hundred | Wide—but best for small quantities | Substantial quantities required |
| Direct labor costs | High | Moderate | Moderate | High | Very high | Low to medium |
| Finishing costs | High | Low | Low to moderate | Low | Low | Low |
| Scrap costs | Moderate | Low | Low | Low | Low | Low |

*Source:* Cook, "Engineered Castings," McGraw-Hill.
*Closer at extra cost.

attached to the top of an air cylinder. After the table is raised, a quick-release port opens, and the piston, platen, and mold drop free against the top of the cylinder or striking pads. The impact packs the sand. The densities produced by this machine are greatest next to the parting line of the pattern and softest near the top of the flask. This procedure can be used for any flask that can be rammed on a molding machine. As a separate unit, it is used primarily for medium and large work. The largest units will jolt a combined weight of sand, flask, and pattern of 100,000 lb (445 kN). Where plain jolt machines are used on large work, it is usual to ram the top of the flask manually with an air hammer.

**Jolt squeeze machines** use both the jolt and the squeeze procedures. The platen is mounted on two air cylinders: a small cylinder to jolt and a large one to squeeze the mold. They are widely used for small and medium work, and with matchplate or gated patterns. Pattern-stripping devices can be incorporated with jolt or squeezer machines to permit mechanical removal of the pattern. Pattern removal can also be accomplished by using jolt-rockover-draw or jolt-squeeze-rollover-draw machines.

The **sand slinger** is the most widely applicable type of ramming machine. It consists of an impeller mounted on the end of a double-jointed arm which is fed with sand by belt conveyors mounted on the arm. The impeller rotating at high speed gives sufficient velocity to the sand to ram it in the mold by impact. The head may be directed to all parts of the flask manually on the larger machines and may be automatically controlled on smaller units used for the high-speed production of small molds.

**Vibrators** are used on all pattern-drawing machines to free the pattern from the grip of the sand before drawing. Their use reduces mold damage to a minimum when the pattern is removed, and has the additional advantage of producing castings of more uniform size than can be secured by hand-rapping the pattern. Pattern damage is also kept to a minimum. Vibrators are usually air-operated, but some electrically operated types are in use.

**Flasks** generally consist of two parts: the upper section, called the **cope**, and the bottom section, the **drag**. When more than two parts are used, the intermediate sections are called **cheeks**. Flasks are classified as tight, snap, and slip. **Tight flasks** are those in which the flask remains until the metal is poured. **Snap flasks** are hinged on one corner and have a locking device on the diagonally opposite corner. In use, these flasks are removed as soon as the mold is closed. **Slip flasks** are of solid construction tapered from top to bottom on all four sides so that they can be removed as soon as the mold is closed. Snap or slip flasks permit the molder to make any number of molds with one flask. Before pouring snap- or slip-flask molds, a wood or metal pouring jacket is placed around the mold and a weight set on the top to keep the cope from lifting. The cope and drag sections on all flasks are maintained in proper alignment by flask pins and guides.

Tight flasks can be made in any size and are fabricated of wood, rolled steel, cast steel, cast iron, magnesium, or aluminum. Wood, aluminum, and magnesium are used only for small- and medium-sized flasks. Snap and slip flasks are made of wood, aluminum, or magnesium, and are generally used for molds not over 20 by 20 in (500 mm by 500 mm).

### Mechanization of Sand Preparation

In addition to the various types of molding machines, the modern foundry makes use of a variety of equipment to handle the sand and castings.

**Sand Preparation and Handling**    Sand is prepared in **mullers,** which serve to mix the sand, bonding agent, and water. **Aerators** are used in conjunction to loosen the sand to make it more amenable to molding. **Sand cutters** that operate over a heap on the foundry floor may be used instead of mullers. Delivery of the sand to the molding floor may be by means of dump or scoop trucks or by belt conveyors. At the molding floor the molds may be placed on the floor or delivered by conveyors to a pouring station. After pouring, the castings are removed from the flasks and adhering sand at a **shakeout** station. This may be a mechanically operated jolting device that shakes the loose sand from flask and casting. The used sand, in turn, is returned to the storage bins by belt conveyor or other means. Small castings may be poured by using **stack-molding** methods. In this case, each flask has a drag cavity molded in its upper surface and a cope cavity in its lower surface. These are stacked one on the other to a suitable height and poured from a common sprue.

There is an almost infinite variety of equipment and methods available to the foundry, ranging from simple, work-saving devices to completely mechanized units, including completely automatic molding machines. Because of this wide selection available, the degree to which a foundry can be mechanized depends almost entirely on the economics of the operations, rather than the availability or lack of availability of a particular piece of equipment.

## MOLDING SAND

Molding sand consists of silica grains held together by some bonding material, usually clay or bentonite.

**Grain size** greatly influences the surface finish of a casting. The proper grain size is determined by the size of the casting, the quality of surface required, and the surface tension of the molten metal. The grain size should be approximately uniform when maximum permeability is desired.

**Naturally bonded sands** are mixtures of silica and clay as taken from the pits. Modification may be necessary to produce a satisfactory mixture. This type of sand is used in gray iron, malleable iron, and nonferrous foundries (except magnesium).

**Synthetically bonded sands** are produced by combining clay-free silica sand with clay or bentonite. These sands can be compounded to suit foundry requirements. They are more uniform than naturally bonded sands but require more careful mixing and control. Steel foundries, gray-iron and malleable-iron foundries, and magnesium foundries use this type of sand.

Special additives may be used in addition to the basic sand, clay, and water. These include cereals, ground pitch, sea coal, gilsonite, fuel oil, wood flour, silica flour, iron oxide, perlite, molasses, dextrin, and proprietary materials. These all serve the purpose of altering specific properties of the sand to give better results.

The properties of the sand that are of major interest to the foundryman are **permeability,** or the venting power, of the sand; **green compressive strength; green shear strength; deformation,** or the sand movement under a given load; **dry compressive strength;** and **hot strength,** i.e., strength at elevated temperatures. Several auxiliary tests are often made, including moisture content, clay content, and grain-size determination.

The foundry engineer or metallurgist who usually is entrusted with the control of the sand properties makes the adjustments required to keep it in good condition.

**Facing sands,** for giving better surface to the casting, are used for gray-iron, malleable-iron, steel, and magnesium castings. The iron sands usually contain **sea coal,** a finely ground coal which keeps the sand from adhering to the casting by generating a gas film when in contact with the hot metal. Steel facings contain silica flour or other very fine highly refractory material to form a dense surface which the metal cannot readily penetrate.

**Mold washes** are coatings applied to the mold or core surface to improve the finish of the casting. They are applied either wet or dry. The usual practice is to brush or spray the wet mold washes and to brush or rub on the dry ones. Graphite or silica flour mixed with clay and molasses water is frequently used.

### Core Sands and Core Binders

**Green-sand cores** are made from standard molding-sand mixtures, sometimes strengthened by adding a binder, such as dextrin, which hardens the surface. Cores of this type are very fragile and are usually made with an arbor or wires on the inside to facilitate handling. Their collapsibility is useful to prevent hot tearing of the casting.

**Dry-sand cores** are made from silica sand and a binder (usually oil) which hardens under the action of heat. The amount of oil used should be the minimum which will produce the necessary core strength.

Core binders are either organic, such as core oil, which are destroyed under heat, or inorganic, which are not destroyed.

**Organic Binders**    The main organic binder is **core oil.** Pure linseed oil is used extensively as one of the basic ingredients in blended-oil core binders. These consist primarily of linseed oil, resin, and a thinner, such as high-grade kerosene. They have good wetting properties, good workability, and better oxidation characteristics than straight linseed oil.

**Corn flour** produces good green strength and dry strength when used in conjunction with oil. Cores made with this binder are quick drying in the oven and burn out rapidly and completely in the mold.

**Dextrin** produces a hard surface and weak center because of the migration of dextrin and water to the surface. Used with oil, it produces a hard smooth surface but does not produce a green bond as good as that with corn flour.

Commercial **protein binders,** such as gelatin, casein, and glues, improve flowability of the sand, have high binding power, rapid drying, fair resistance to moisture, and low burning-out point, with only a small volume of gas evolved on burning. They are used where high collapsibility of the core is essential.

Other binders include paper-mill by-products, which absorb moisture readily, have high dry strength, low green strength, high gas ratio, and high binding power for clay materials.

**Coal tar pitch** and **petroleum pitch** flow with heat and freeze around the grains on cooling. These compounds have low moisture-absorption rates and are used extensively for large iron cores. They can be used effectively with impure sands.

**Wood** and **gum rosin, plastic resins,** and rosin by-products are used to produce collapsibility in cores. They must be well ground. They tend to cake in hot weather, and large amounts are required to get desired strength.

Plastics of the **urea-** and **phenol-formaldehyde** groups and **furan resins** are being used for core binders. They have the advantage of low-temperature baking, collapse readily, and produce only small amounts of gas. These can be used in **dielectric baking ovens** or in the **shell molding, hot box, or air setting** processes for making cores.

**Inorganic binders** include fire clay, southern bentonite, western bentonite, and iron oxide.

Cores can also be made by mixing sand with sodium silicate. When this mixture is in the core box, it is infiltrated with $CO_2$, which causes the core to harden. This is called the **$CO_2$ process.**

### Core-making Methods

Cores are made by the methods employed for sand molds. In addition, **core blowers** and **extrusion machines** are used.

**Core blowers** force sand into the core box by compressed air at about 100 lb/in². They can be used for making all types of small- and medium-sized cores. The cores produced are very uniform, and production rates up to 200 boxes per hour per machine are achieved.

**Screw-feed machines** are used largely for plain cylindrical cores of uniform cross section. The core sand is extruded through a die onto a core plate. The use of these machines is limited to the production of stock cores, which are cut to the desired length after baking.

**Core Ovens** Core-oven walls are constructed of inner and outer layers of sheet metal separated by rock wool or Fiberglas insulation and with interlocked joints. Combustion chambers are refractory-lined, and the hot gases are circulated by fans. They are designed for operating at temperatures ranging from 300 to 650°F (149 to 343°C). Baking temperatures are approximately: urea-formaldehyde binder, 300 to 350°F (149 to 177°C); flour, starch, or dextrin, 350 to 375°F (177 to 191°C) with a maximum of 410°F (210°C); pitch and resin, 350 to 400°F (177 to 205°C); dextrin, below 400°F (205°C); phenol formaldehyde, 425 to 450°F (219 to 232°C); cores bonded with oil, 400 to 450°F (205 to 232°C), but 500°F (260°C) will not injure.

**Core driers** are light skeleton cast-iron or aluminum boxes, the internal shape of which conforms closely to the cope portion of the core. They are used to support, during baking, cores which cannot be placed on a flat plate.

## CASTING ALLOYS

In general, the types of alloys that can be produced as wrought metals can also be prepared as castings. Certain alloys, however, cannot be forged or rolled and can only be produced as castings.

A very general subdivision is into **ferrous,** or iron-base alloys, and **nonferrous** alloys. The nonferrous group includes aluminum-base, copper-base, magnesium-base, nickel-base, and other miscellaneous alloys.

### Ferrous Alloys

**Steel Castings** (See Sec. 6) Steel castings may be classified as:

1. Low carbon (C < 0.20 percent). These are relatively soft and not readily heat-treatable.

2. Medium carbon (0.20 percent < C < 0.50 percent). These castings are somewhat harder and amenable to strengthening by heat-treatment.

3. High carbon (C > 0.50 percent). These steels are used where maximum hardness and wear resistance are desired.

In addition to the classification based on carbon content, which determines the maximum hardness obtainable in steel, the castings can be also classified as **low-alloy** content (< 8 percent) or **high-alloy** content (> 8 percent).

The low-alloy steels behave essentially as plain-carbon steels but have a higher **hardenability,** which is a measure of ability to be hardened by heat-treatment. The high-alloy steels are designed to produce some specific property, like corrosion resistance, heat resistance, wear resistance, or some other special property.

**Malleable-Iron Castings** The carbon content of malleable iron ranges from about 2.00 to 2.80 percent and may reach as high as 3.30 percent if the iron is melted in a cupola. Silicon ranging from 0.90 to 1.80 percent is an additional alloying element required to aid the annealing of the iron. As cast, this iron is hard and brittle and is rendered soft and malleable by a long heat-treating or annealing cycle. This product is usually marketed to meet one of two ASTM classes, the 32510 grade and the 35018 grade. The former has minimum standards of 50,000 lb/in² tensile strength (345 MN/m²), 32,500 lb/in² yield strength (225 MN/m²), and 10 percent minimum elongation, while standards for the latter are 53,000 lb/in² (365 MN/m²), 35,000 lb/in² (240 MN/m²), and 18 percent elongation, respectively, for these properties. (See also Sec. 6.)

Malleable iron can also be heated in a number of ways to produce **pearlitic malleable** that may reach 90,000 lb/in² (620 MN/m²) in tensile strength, 70,000 lb/in² (485 MN/m²) in yield strength, and a minimum elongation of 2 percent in 2 in.

**Gray-Iron Castings** Gray iron is an alloy of iron, carbon, and silicon, containing a higher percentage of these last two elements than found in malleable iron. Much of the carbon is present in the elemental form as graphite. Other elements present include manganese, phosphorus, and sulfur. Because the properties are controlled by proper proportioning of the carbon and silicon and by the cooling rate of the casting, it is usually sold on the basis of specified properties rather than composition. The carbon content will usually range between 3.00 and 4.00 percent and the silicon will be between 1.00 and 3.00 percent, the higher values of carbon being used with the lower silicon values (usually), and vice versa. As evidence of the fact that gray iron should not be considered as a material having a single set of properties, the ASTM and AFS recognize seven classes, ranging in tensile strength from 20,000 to 60,000 lb/in² (138 to 414 MN/m²) or over. The high strengths are obtained by proper adjustment of the carbon and silicon contents or by alloying. (See also Sec. 6.)

An important variation of gray iron is **nodular iron,** in which the graphite appears as nodules rather than as flakes. This iron is prepared by treating the metal in the ladle with additives that usually include magnesium in alloy form. Nodular iron can exceed 100,000 lb/in² (690 MN/m²) as cast and is much more ductile than gray iron, measuring about 2 to 5 percent elongation at these higher strengths, and even higher percentages if the strength is lower. (See also Sec. 6.)

### Nonferrous Alloys

**Aluminum-Base Castings** Aluminum is alloyed with copper, silicon, magnesium, zinc, nickel, and other elements to produce a wide variety of casting alloys having specific characteristics of foundry properties, mechanical properties, machinability, and/or corrosion resistance. Alloys are produced for use in sand casting, permanent-mold casting, or die casting. Some alloys are heat-treatable using **solution** and **age-hardening** treatments. These alloys range in strength from about 17,000 lb/in² (117 MN/m²) to as high as 43,000 lb/in² (297 MN/m²) in some die-cast alloys. (See also Sec. 6.)

**Copper-Base Alloys** The alloying elements used with copper include zinc (brasses), tin (bronzes), nickel (nickel bronze), aluminum (aluminum bronze), silicon (silicon bronze), and beryllium (beryllium bronze). The brasses and tin bronzes may contain lead for machinability. Various combinations of zinc and tin, or of tin or zinc with other elements, are also available. With the exception of some of the aluminum bronzes and beryllium bronze, most of the copper-base alloys cannot be hardened by heat-treatment. Certain copper-base alloys may exceed 100,000 lb/in² (690 MN/m²) in tensile strength. (See also Sec. 6.)

**Special Casting Alloys** Other metals cast in the foundry include **magnesium-base alloys** for light weight, **nickel-base alloys** for high-temperature applications, **titanium-base alloys** for strength-to-weight ratio, etc. The magnesium-base alloys require special precautions during melting and pouring to avoid burning. They can be heat-treated and may exceed 40,000 lb/in² (275 MN/m²) in tensile strength. Most magnesium alloys have excellent machinability. (See also Sec. 6.)

## MELTING FURNACES

The melting furnaces used for castings, arranged for each metal in the approximate order of their general use, are as follows:

**Steel.** Electric (direct-arc, acid), open-hearth (acid), open-hearth (basic), electric (direct-arc, basic), Bessemer, induction (high-frequency).

**Cast iron.** Cupola, air furnace (reverberatory), electric (direct-arc).

**Malleable iron.** Duplexing (cupola and air furnace or electric furnace), air furnace, cupola, rotary furnace (Brackelsberg), open-hearth (acid).

**Brass and bronze.** Crucible, electric (indirect-arc), induction (low-frequency), air furnace, induction (high-frequency).

**Aluminum.** Crucible furnace with metal or refractory crucibles, induction (low-frequency).

**Magnesium.** Crucible furnace with steel crucibles, induction (high-frequency).

**Annealing furnaces.** (See Sec. 7.)

## CLEANING AND INSPECTION

**Tumbling barrels** consist of a power-driven drum in which the castings are tumbled in contact with hard-iron stars or balls. Their impact removes the sand and scale.

In **air-blast cleaning units,** compressed air forces silica sand or

chilled-iron shot into violent contact with the castings, which are tumbled in a barrel, rotated on a table, or passed between multiple orifices on a conveyor. Large rooms are sometimes utilized, with an operator directing the nozzle. These machines are equipped with hoppers and elevators to return the sand or shot to the magazine. Dust-collecting systems are required.

In **centrifugal-blast cleaning units,** a rotating impeller is used to impart the necessary velocity to the chilled-iron shot or grit abrasive. The velocities are not so high as with air, but the volume of abrasive is much greater. The construction is otherwise similar to the air-blast machine.

**Water** in large volume at pressures of 250 to 600 lb/in² is used to remove sand and cores from medium and large castings. The stream is directed by an operator located outside the cleaning chamber or room.

**High-pressure water and sand cleaning (Hydroblast)** employs water at 1,200 to 1,900 lb/in² (8.3 to 13.1 MN/m²) mixed with molding sand which has been washed off the casting. A sand classifier is incorporated in the sand-reclamation system.

**Pneumatic chipping hammers** may be used to clean large castings where the sand is badly burned on and for deep pockets.

**Removal of Gates and Risers and Finishing Castings** The following tabulation shows the most generally used methods for removing gates and risers (marked R) and for finishing (marked F).

**Steel.** Oxyacetylene (R), hand hammer or sledge (R), grinders (F), chipping hammer (F), and machining (F).

**Cast iron.** Chipping hammer (R, F), hand hammer or sledge (R), abrasive cutoff (R), power saw (R), and grinders (F).

**Malleable iron.** Hand hammer or sledge (R), grinders (F), shear (F), and machining (F).

**Brass and bronze.** Chipping hammer (R, F), shear (R, F), hand hammer or sledge (R), abrasive cutoff (R), power saw (R), belt sanders (F), grinders (F), and machining (F).

**Aluminum.** Chipping hammer (R), shear (R), hand hammer or sledge (R), power saw (R), grinders (F), and belt sander (F).

**Magnesium.** Band saw (R), machining (F), and flexible-shaft machines with steel burr cutters (F).

### Casting Inspection

Castings are inspected for dimensional accuracy, hardness, surface finish, physical properties, internal soundness, and cracks. For **hardness** and for **physical properties,** see Sec. 5.

**Internal soundness** is checked by cutting or breaking up pilot castings or by nondestructive testing using X-ray, gamma ray, and fluoroscope.

**Destructive testing** tells only the condition of the piece tested and does not ensure that other pieces not tested will be sound. It is the most commonly used procedure at the present time.

**X-ray, gamma ray, and the fluoroscope** have made possible the nondestructive checking of castings to determine internal soundness on all castings produced. Shrinks, cracks, tears, and gas holes can be determined and repairs made before the castings are shipped.

**Magnetic-powder tests (Magnaflux)** are used to locate structural discontinuities in iron and steel except austenitic steels, but they are not applicable to most nonferrous metals or their alloys. The method is most useful for the location of surface discontinuities, but it may indicate deep-seated defects if the magnetizing force is sufficient to produce a leakage field at the surface.

In this test a magnetic flux is induced in ferromagnetic material. Any abrupt discontinuity in its path results in a local flux-leakage field. If finely divided particles of ferromagnetic material are brought into the vicinity, they offer a low reluctance path to the leakage field and take a position that outlines approximately its effective boundaries. The casting to be inspected is magnetized and its surface dusted with the magnetic powder. A low-velocity air stream blows the excess powder off and leaves the defect outlined by the powder particles. The powder may be applied while the magnetizing current is flowing (**continuous method**) or after the current is off (**residual method**). It may be applied dry or suspended in a light petroleum distillate similar to kerosene. The dry-powder method using dc magnetization is recommended. Expert interpretation of the tests is necessary for satisfactory results.

## CASTING DESIGN

Design for the best utilization of metal in the cast form requires a knowledge of metal solidification characteristics, foundry practices, and the metallurgy of the metal being used. Metals exhibit certain peculiarities in the formation of solid metal during freezing and also undergo shrinkage in the liquid state during the freezing process and after freezing, and the casting must be designed to take these factors into consideration. Knowledge concerning the freezing process will also be of assistance in determining the fluidity of the metal, its resistance to **hot tearing,** and its tendency to evolve dissolved gases. For economy in production, casting design should take into consideration those factors in molding and coring that will lead to the simplest procedures. Elimination of expensive cores, irregular parting lines, and deep drafts in the casting can often be accomplished with a slight modification of the original design. Combination of the foregoing factors with the selection of the right metal for the job is an important facet of casting design. Consultation with the plant metallurgist or foundry engineer is recommended as a first step in the development of a well-designed casting. The references, ASM "Casting Design Handbook," and "Design of Ferrous Castings" are recommended.

# WORKING METALS AND PLASTICS

## by E. V. Crane

(See also Secs. 5 and 6.)

REFERENCES: Crane, "Plastic Working of Metals and Power Press Operations," Wiley. Woodworth, "Punches, Dies and Tools for Manufacturing in Presses," Henley. Jones, "Die Design and Die Making Practice," Industrial Press. Stanley, "Punches and Dies," McGraw-Hill. Woodworth, "Dies—Their Construction and Use," Henley. Dowd and Curtis, "Punches, Dies and Gages," McGraw-Hill. DeGarmo, "Materials and Processes in Manufacturing," Macmillan. "Modern Plastics Encyclopedia and Engineers Handbook" Plastics Catalogue Corp., New York. Jevons, "The Metallurgy of Deep Drawing and Pressing," Wiley. "The Tool Engineers Handbook," McGraw-Hill. Bridgman, "Large Plastic Flow and Fracture," McGraw-Hill. "Cold Working of Metals," ASM. Pearson, "The Extrusion of Metals," Wiley. Shockley, "Imperfections in Nearly Perfect Crystals," Wiley.

## STRUCTURE

**Yieldable structural forces** between the particles composing a material to be worked are the key to its behavior. Simple internal structures contain only a single element, as pure copper, silver, or iron. Relatively more difficult to work are the solid solutions in which one element tends to distribute uniformly in the structural pattern of another. Thus silver and gold form a continuous series of solid-solution alloys as their proportions vary. Next are alloys in which strongly bonded molecular groups dispersed through or along the grain boundaries of softer metals offer increasing resistance to working, as does iron carbide ($Fe_3C$) in solution in iron. Molecules may become complex groups of many atoms, as in the plastics. Such molecules may bond strongly together in chain relationships, as in polymers and rubbery elastomers. In other cases, mixtures may be used in which filling materials supply bulk, hardness, or other characteristics and are held together in a matrix of metallic, plastic, or ceramic bonding materials.

**Bonding forces** are supplied by electric fields characteristic of individual atoms. These forces in turn are subject to modification by temperature as energy is added, increasing electron activity. The atom of each individual element includes a closely associated group of neutrons and positively charged protons at its core and a corresponding number of smaller negatively charged electrons, grouped in orbits around that core. The atomic number of the element also designates the normal number of electrons in its orbits. (See Sec. 6.)

The **particles** which constitute an atom are so small that most of its volume is empty space. For a similar energy state, there is some rough uniformity in the outside size of atoms. In general, therefore, the more complex elements have their larger number of particles more densely packed and so are heavier. For each element, the energy pattern of its electric charges in motion determines the field characteristics of that atom and which of the orderly arrangements it will seek to assume with relation to others like it in the orderly crystalline form. For typical crystal patterns, see Sec. 6.

**Space lattice** is the term used to describe the orderly arrangement of rows and layers of atoms in the crystalline form. This orderly state is also described as balanced, unstrained, or **annealed.** The working or deforming of materials distorts the orderly arrangement, unbalancing the forces between atoms. Cubic patterns or space lattices characterize the more ductile or workable materials. Hexagonal and more complex patterns tend to be more brittle or more rigid. Flaws, irregularities, or distortions, with corresponding unbalanced strains among adjacent atoms, may occur in the pattern or along grain boundaries. **Slip-plane** movements in working to new shapes tend to slide the once orderly layers of atoms within the grain-boundary limitations of individual crystals. Such sliding movement tends to take place at 45° to the direction of the applied load because much higher stresses are required to pull atoms directly apart or to push them straight together.

**Chemical combinations,** in liquid or solid solutions, or molecular compounds depend upon relative field patterns of elements or upon actual displacement of one or more electrons from the outer orbit of a donor element to the outer orbit of a receptor element. Thus the molecules of hard iron carbide, $Fe_3C$, may be held in solid solution in soft pure iron (ferrite) in increasing proportions up to 0.83 percent of carbon in iron, which is described as **pearlite.** Zinc may occupy solid-solution positions in the copper space lattice up to about 45 percent, the range of the ductile red and yellow brasses. Typical of the plastics, cellulose triacetate, $C_6H_7O_2(O_2CCH_3)_3$, combines 12 carbon atoms, 16 hydrogen atoms, and 8 oxygen atoms into a complex molecule having a molecular weight nearly six times that of an iron atom. But since the mass is distributed among the 36 lighter atoms of carbon, hydrogen, and oxygen, the plastic weighs less than one-sixth of the same volume of iron.

**Thermal Changes** Adding heat (energy) increases electron activity and therefore also the mobility of the atom. Probability of brittle failure at low temperatures usually becomes less as temperature increases. Transition temperatures from one state to another differ for different elements. Thermal transitions therefore become more complex as such differing elements are combined in alloys and compounds. As temperatures rise, a **stress-relieving** range is reached at which the most severely strained atoms are able to ease themselves around into less strained positions. At somewhat higher temperatures, **annealing** or **recrystallization** of worked or distorted structure takes place. Old grain boundaries disappear and small new grains begin to grow, aligning nearby atoms into their orderly lattice pattern. The more severely the material has been worked, the lower is the temperature at which recrystallization begins. Grain growth is more rapid at higher temperatures. In working materials above their recrystallization range, as in forging, the relief of interatomic strains becomes more nearly spontaneous as the temperature is increased. **Creep** takes place when materials are under some stress above the recrystallization range, and the thermal mobility permits individual atoms to ease around to relieve that stress, with an accompanying gradual change of shape. Thus a wax candle droops due to gravity on a hot day. Lead, which recrystallizes below room temperature, will creep when used for roofing or spouting. Steels in rockets and jet engines begin to creep around 1300 to 1500°F (704 to 815°C). Creep is more rapid as the temperature rises farther above the recrystallization range.

## PLASTICITY

**Plasticity** is that property of materials which commends them to the mass-production techniques of pressure-forming desired shapes. It is understood more easily if several types of plasticity are considered.

**Solvents** —water, oils, acetone, and waxes—contribute to the workability of clays, paper, wood, plaster, concrete, and some of the synthetic plastics. The term **soluplastic** applies when a material can be restored to a workable state by replen-

ishing the plasticizing solvent. Thus paper is stored in a proper humidity and wood is placed in a steam chamber to render them formable. The term **solusetting** describes forming processes in which chemical change takes place such that plasticity cannot be restored by reapplication of the solvent. Thus plaster, concrete, and baked ceramics take a chemical set. Glass fabric and epoxy set quickly to a master form.

**Crystoplastic** describes materials, notably metals, which can be worked in the stable crystalline state, below the recrystallization range. Metals which crystallize in the cubic patterns have a wider plastic range than those of hexagonal pattern. Alloying narrows the range and increases the resistance to working. Tensile or compressive testing of an annealed specimen can be used to show the plastic range which lies between the initial yield point and the point of ultimate tensile or compressive failure.

The **plastic range,** as of an annealed metal, is illustrated in Fig. 1. Changing values of **true stress** are determined by divid-

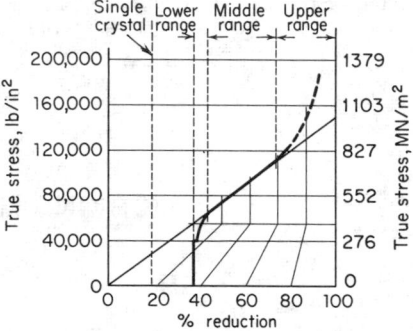

**Fig. 1** Three ranges of crystoplastic work hardening of a low-carbon steel. *(ASME, 1954, W. S. Wagner, E. W. Bliss Co.)*

ing the applied load at any instant by the cross-section area at that instant. As material is worked, a progressive increase in elastic limit and yield point registers the slip-plane movement or work hardening which has taken place and the consequent reduction in residual plasticity. This changing yield point or resistance, shown in Fig. 1, is divided roughly into three characteristic ranges. The contour of the lower range can be varied by nonuniformity of grain sizes or by small displacements resulting from prior direction of working. Random large, soft grains yield locally under slight displacement, with resulting **surface markings,** described as *orange peel, alligator skin,* or *stretcher strain markings.* These can be prevented by preparatory roller leveling, which gives protection in the case of steel for perhaps a day, or by a 3 to 5 percent temper pass of cold-rolling, which may stress relieve in perhaps 3 months, permitting recurrent trouble. The middle range covers most drawing and forming operations. Its upper limit is the point of normal tensile failure. The upper range requires that metal be worked primarily in compression to inhibit the start of tensile fracture. Severe extrusion, spring-temper rolling, and music-wire drawing use this range.

**Dispersion hardening** of metal alloys by heat treatment (see Fig. 2) reduces the plastic range and increases the resistance to work hardening. Figure 2 also shows the common methods of plotting change of true stress against **percentage of reduction** —

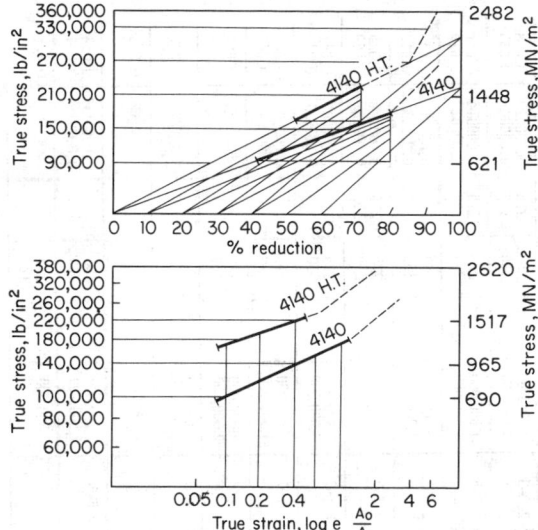

**Fig. 2** High-range plasticity (dotted) of 4140 steel, showing also the effect of dispersion hardening. Two plotting methods. (*ASME*, 1958, *Crane and Wagner, E. W. Bliss Co.*)

e.g., reduction of thickness in rolling or compressive working, of area in wire drawing, ironing, or tensile testing, or of diameter in cup drawing or reducing operations—and against **true strain,** which is the natural logarithm of change of area, for convenience in higher mathematics.

The term **thermoplastic** describes the working range of temperatures between recrystallization and fluidity. As the temperature increases, the material offers less resistance to flow and can be worked more severely; or at higher temperatures, it can be forced through more difficult passageways. The thermoplastic materials are not subject to chemical change and will soften when reheated.

For metals, thermoplastic working is usually described as **hot forging,** except for tin and lead, which recrystallize below room temperature. Forgings may be etched to show **flow lines,** which are usually made up of old-grain boundaries. Where these show, recrystallization has not yet taken place, and some work hardening is retained to improve physical properties. Zinc and magnesium, which are typical of the hexagonal-structure metals, take only small amounts of cold working but can be drawn or otherwise worked severely at rather moderate temperatures [Zn, 200 to 400°F (90 to 200°C); Mg, 500 to 700°F (260 to 400°C)]. Note that, although hexagonal-pattern metals are less easily worked than cubic-pattern metals, they are for that same reason structurally more rigid for a similar relative weight. Advantageous forging temperatures change with alloy composition: copper, 1800 to 1900°F (980 to 1040°C); red brass, Cu 70, Zn 30, 1600 to 1700°F (870 to 930°C); yellow brass, Cu 60, Zn 40, 1200 to 1500°F (650 to 815°C). See Sec. 6 for general physical properties of metals.

**Thermoplastic synthetic plastics** usually recrystallize a little above or below room temperature. They are usually too brittle for working purposes at lower temperatures. Plasticizing waxes may be used in a solid-solution relationship to improve flexibility in wrapping foils. Such plasticizers tend to leach out

as moisture is taken up or lost on damp or dry days. Moisture sensitivity varies considerably among the plastics. Extrusion-, injection-, and compression-molding temperatures are usually in the range of 200 to 500°F (90 to 260°C), depending upon the plastic, the production rate, and the die passageways.

**Physical properties** of the plastics (Table 1) and of the metals (Secs. 5 and 6) are similar in principle though varying widely with materials, alloys, and their temperature or degree of work hardening. Figure 3 shows a stress-strain curve for a typical thermoplastic. Note the increasing elastic limit and decreasing plasticity at lower and lower temperatures. Substantially pure iron shows an increasing elastic limit and decreasing plasticity with increasing amounts of work hardening by cold-rolling. The rate at which such work hardening takes place is greatly increased, and the remaining plasticity reduced, as alloying becomes more complex.

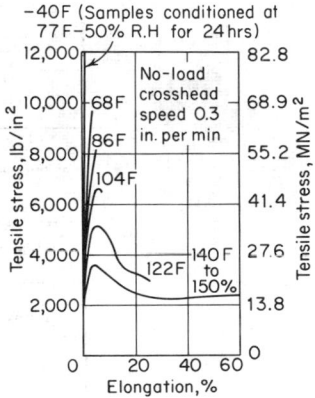

**Fig. 3** Tensile stress-strain curves for thermoplastic Lucite. (*E. I. Du Pont de Nemours & Co.*)

**Thermosetting** plastics are those in which a chemical change takes place with heat, so that the process is not reversible and reheating does not restore plasticity. Such thermosetting resins as the phenolics, ureas, melamines, and shellac are used in many or most cases with a variety of fillers for bulk, reinforcement, abrasive grinding, appearance, or resistance to heat, wear, water, or chemicals. Although there is some injection molding, most working of thermosetting mixtures is in compression molding, transfer molding, and multiplaten laminating.

**Elastomers,** synthetic and natural rubbers, should probably be included under the thermosetting molding and forming techniques, although such work as rubber-glove production using molds dipped in solution and then heat-cured would be solusetting.

**Ceramics** industries, with roots farthest in the past, remain more of an art by reason of the chance variety of metallic oxides in the natural clays. In the forming stage, the process is soluplastic and control of the plasticizing water is critical. The subsequent firing is an irreversible solusetting process, and for production purposes, there are two stages where time and temperature are critical with respect to shrinkage and cracking: In the first, water which was holding clay platelettes in solution is evaporated; and in the second, water which was in

**Table 1. Properties of Plastics***

THERMOSETTING MIXTURES

| | Yield point at 77 F, approx psi | Hot-shearing temp, approx deg F | Knife-edge cutting load at 77 F, lb lineal in. | Shearing strength at 77 F, psi | Compression-molding pressure, psi | Injection-molding pressure, psi | Compressive strength, psi | Tensile strength, psi | Elongation in 2 in., % | Powder-to-solid compression ratio | Modulus of elasticity, psi | Lowest recrystallization temp, approx deg F | Forming compression-molding temp, approx deg F | Injection-molding temp, approx deg F | Thermal coef of expansion, in./in., deg F | Weight, lb/cu in. |
|---|---|---|---|---|---|---|---|---|---|---|---|---|---|---|---|---|
| Phenol-formaldehyde (no filler) | | 200 / 250 | | 10,000 | 2,000 / 5,000 | | 10,000 / 30,000 | 7,000 / 12,000 | 1.1 | 2–2.6 | 700,000 / 1,000,000 | | 300 / 340 | | 0.000013 / 0.00003 | 0.046 |
| Phenol-formaldehyde (wood-flour filler) | | 200 / 250 | | | 2,000 / 4,500 | 2,000 / 10,000 | 16,000 / 36,000 | 6,000 / 11,000 | 0.6 | 2.2–3 | 1,000,000 / 1,500,000 | | 280 / 360 | 275 / 375 | 0.000020 / 0.000041 | 0.0481 / 0.055 |
| Phenol-formaldehyde (mineral filler) | | 200 / 250 | | | 2,000 / 6,000 | 2,000 / 15,000 | 18,000 / 36,000 | 4,000 / 8,000 | 0.6 | 2–8 | 1,000,000 / 4,500,000 | | 270 / 350 | 275 / 350 | 0.000014 / 0.00002 | 0.061 / 0.075 |
| Phenol-formaldehyde (macerated-fabric filler) | | 200 / 250 | | | 2,000 / 8,000 | | 20,000 / 32,000 | 5,500 / 8,000 | 0.7 | 2.5–15 | 700,000 / 1,200,000 | | 270 / 350 | | 0.00001 / 0.00003 | 0.049 / 0.053 |
| Phenol-formaldehyde (sisal-felt filler) | | 200 / 250 | | | 300 / 3,000 | 10,000 / 20,000 | 10,000 / 35,000 | 7,000 / 12,000 | | 2–5 | | | 275 / 350 | 275 / 350 | 0.0001 / 0.00003 | 0.025 / 0.050 |
| Phenol-formaldehyde (paper laminate) | 4,000 / 18,000 | 200 / 250 | | 12,500 | 1,000 / 3,000 | | 20,000 / 40,000 | 7,000 / 18,000 | 1.5 | 1.5–3 | 400,000 / 3,000,000 | | 275 / 350 | | 0.000009 / 0.000013 | 0.047 / 0.049 |
| Phenol-formaldehyde (cotton-fabric laminate) | 3,000 / 8,000 | 200 / 250 | | 15,000 | 1,000 / 3,000 | | 30,000 / 44,000 | 8,000 / 12,000 | 1–2 | 1.5–3 | 350,000 / 1,500,000 | | 275 / 350 | | 0.000009 / 0.00001 | 0.047 / 0.049 |
| Phenol-formaldehyde (glass-fabric laminate) | 4,500 / 28,000 | 200 / 250 | | | 1,000 / 3,000 | | 42,000 / 47,000 | 14,000 / 20,000 | 2 | | 1,000,000 / 2,000,000 | | 275 / 350 | | | 0.050 / 0.057 |
| Phenol-formaldehyde (asbestos-cloth laminate) | | 200 / 250 | | | 1,000 / 3,000 | | 18,000 / 45,000 | 7,000 / 12,000 | | 1.6 | 350,000 / 1,500,000 | | 300 / 350 | | 0.000009 / 0.000013 | 0.056 / 0.086 |
| Birch plywood (phenolic binder) | | | | 13,000 | 200 / 2,000 | | 5,700 | 13,100 | | | 1,400,000 | | | | | 0.0288 |
| Urea-formaldehyde (alpha-cellulose filler) | | | | | 1,500 / 6,000 | | 20,000 / 24,000 | 5,500 / 7,000 | | 2.5–3 | 1,200,000 / 1,500,000 | | 290 / 325 | | 0.0000138 / 0.000016 | 0.052 / 0.054 |
| Urea-formaldehyde (cotton-fabric laminate) | 600 / 2,500 | | | | 1,500 / 6,000 | | 7,000 / 8,700 | 5,100 / 6,900 | 8–22 | | 560,000 | | 290 / 325 | | | 0.044 |

*From "Plastics and Plasticity," E. W. Bliss Co.

| Material | | | | | | | | | | Temp. | | | |
|---|---|---|---|---|---|---|---|---|---|---|---|---|---|
| Melamine-formaldehyde (alpha-cellulose filler) | | 1,500 / 6,000 | | | | | 2.0–2.3 | | | 280 / 340 | | | 0.0537 |
| Melamine-formaldehyde (asbestos filler) | | 1,000 / 4,000 | | 30,000 | 5,500 / 7,000 | 0.30–0.45 | 2.1–2.5 | 1,600,000 | | 280 / 330 | | 0.00001 / 0.000025 | 0.0612 / 0.087 |
| Aniline-formaldehyde (no filler) | | 1,500 / 6,000 | | 20,000 / 23,000 | 8,500 / 10,000 | | 2.5–3 | 500,000 / 600,000 | | 300 / 340 | | 0.00002 / 0.00003 | 0.044 / 0.045 |
| Casein-formaldehyde | | 2,000 / 2,500 | | 5,300 / 27,000 | 10,000 | 2.5 | | 510,000 / 570,000 | | 200 / 225 | | 0.00004 | 0.0487 |
| Phenol-furfural (wood-floor filler) | | 1,000 / 4,000 | 300 / 10,000 | 28,000 / 36,000 | 6,000 / 11,000 | | 2.5–3 | 1,000,000 / 2,500,000 | | 330 / 400 | 250 / 375 | 0.0001 | 0.047 / 0.050 |
| Phenol-furfural (mineral filler) | | 1,000 / 4,000 | 300 / 15,000 | 24,000 / 36,000 | 5,000 / 10,000 | | 2.5–6 | 1,000,000 / 4,500,000 | | 330 / 360 | 250 / 375 | 0.0001 | 0.057 / 0.072 |
| Phenol-furfural (fabric filler) | | 1,000 / 8,000 | 300 / 30,000 | 26,000 / 30,000 | 6,500 / 8,000 | | 4–15 | 700,000 / 1,200,000 | 2 | 300 / 360 | 250 / 375 | 0.00025 | 0.047 / 0.050 |
| Phenol-lignin (laminate) | | 1,500 / 2,000 | | 25,000 / 30,000 | 7,500 / 12,000 | | 2–3 | 800,000 / 2,000,000 | | 365 | | 0.000011 / 0.000013 | 0.049 / 0.051 |
| Columbia allyl resin 39 (paper laminate) | | | | 15,000 / 31,000 | 10,000 / 21,000 | | | | | | | | 0.048 |
| Columbia allyl resin 39 (fabric laminate) | | | | 29,000 | 1,000 / 7,500 | 4 | | 550,000 | | 160 / 240 | | 0.000013 / 0.000023 | 0.049 |
| Columbia allyl resin 39 (glass-cloth laminate) | 12,700 | | | 52,000 / 60,000 | 30,400 / 39,000 | | | 1,700,000 | | 160 / 240 | | 0.00001 | 0.062 |
| Columbia allyl resin 39 (sheet) | | | | | 6,000 | | | 350,000 | 140 | 160 / 240 | | 0.000049 | 0.047 |
| Shellac | | 1,000 / 2,500 | 1,000 / 1,200 | 10,000 / 17,000 | 900 / 2,000 | | 2–3 | 500,000 / 600,000 | | 240 | 180 / 260 | | 0.039 / 0.098 |
| Rubber-sulphur | | | | | 1,000 / 4,000 | 600 | | 400 | | | | | 0.035 / 0.045 |
| Ceramic "prestite" (flint feldspar, ball clay, china clay) | | | | 48,000 | 5,000 | | | 10,000,000 | 1200 | | | 0.000002 / 0.000006 | 0.088 |

**Table 1. Properties of Plastics** *(Continued)*

SYNTHETIC THERMOPLASTICS

| | Yield point at 77 F, approx psi | Hot-shearing temp., approx deg F | Knife-edge cutting load at 77 F, lb lineal in. | Shearing strength at 77 F, psi | Compression-molding pressure, psi | Injection-molding pressure, psi | Compressive strength, psi | Tensile strength, psi | Elongation in 2 in., % | Powder-to-solid compression ratio | Modulus of elasticity, psi | Lowest recrystallization temp, approx deg F | Forming compression-molding temp, approx deg F | Injection-molding temp, approx deg F | Thermal coef of expansion, in./in./deg F | Weight, lb/cu in. |
|---|---|---|---|---|---|---|---|---|---|---|---|---|---|---|---|---|
| Cellulose acetate | 5,000 | 77+ | 400 / 500 | 6,000 / 10,000 | 500 / 5,000 | 3,000 / 30,000 | 5,000 / 27,000 | 2,200 / 14,600 | 7.3-43 | 2-2.6 | 300,000 | 30-40 | 250 / 350 | 300 / 420 | 0.000044 / 0.000088 | 0.045 / 0.050 |
| Cellulose nitrate (celluloid) | | 120 / 150 | 300 / 400 | | 2,000 / 5,000 | | 20,000 / 30,000 | 5,000 / 12,000 | 10-50 | | 200,000 / 400,000 | | 185 / 250 | | 0.000066 / 0.000088 | 0.049 / 0.050 |
| Cellulose acetate butyrate | | | | 6,000 | 500 / 5,000 | 8,000 / 30,000 | 7,500 / 30,400 | 2,500 / 7,500 | 5 / 90 | 2-2.8 | 200,000 / 350,000 | | 260 / 370 | 340 / 420 | | 0.041 |
| Ethyl cellulose | 1,500 / 3,500 | | | 7,500 | 1,500 / 5,000 | 3,000 / 30,000 | 10,000 / 12,000 | 2,000 / 9,000 | 10-40 | 2.2-2.5 | 100,000 / 500,000 | | 300 / 360 | 350 / 425 | 0.000055 / 0.000077 | |
| Methyl methacrylate | 7,000 / 9,000 | 200 | | 11,500 | 2,000 / 7,500 | 10,000 / 35,000 | 9,000 / 12,500 | 6,000 / 9,000 | 1-5 | 1.7-2.2 | 300,000 / 500,000 | 100-110 | 280 / 350 | 325 / 475 | 0.000036 / 0.000052 | 0.0417 / 0.0435 |
| Nylon, molded | | | | | | | | 5,000 / 10,500 | 30-60 | | | | 450 | 300 / 500 | | |
| Vinyl chloride | 26,000 | | | | | | | 1,000 / 10,000 | | | 350,000 / 400,000 | | | | 0.000038 | 0.0486 |
| Vinylidene chloride | | | | 8,000 / 10,500 | 250 / 5,000 | 10,000 / 30,000 | 7,500 / 8,500 | 4,000 / 8,000 | 15-25 | | 70,000 / 200,000 | 30-40 | 250 / 350 | 300 / 400 | 0.000087 | 0.0595 / 0.0633 |
| Polystyrene (vinyl benzene) | | | | 8,000 | 1,000 / 5,000 | 10,000 / 30,000 | 11,500 / 15,000 | 5,000 / 9,000 | 2-5 | 2-2.3 | 170,000 / 470,000 | -78 | 275 / 350 | 300 / 500 | 0.000033 / 0.000044 | 0.038 / 0.0385 |
| Soluplastics: Vulcanized fiber | | 180 | 400 | 4,000 | | | 20,000 / 32,000 | 5,000 / 12,000 | | | | | | | | 0.036 / 0.054 |
| Aircraft spruce (douglas fir) | | | | | | | 5,000 | 10,000 | | | 1,300,000 | | | | | 0.017 |

chemical combination is driven off. This is prior to decorative glazing. Fillers are used with clay binders for specialized purposes, such as iron powders for electronic items.

## PLASTIC-WORKING TECHNIQUES

Reference should be made to a vast file of papers on plasticity, flow research, processes, techniques, and materials and to applicable bibliographies, especially as indexed in the Engineering Societies Library, 345 E. 47 St., New York 10017. The library provides a research service by mail at a moderate fee.

In the **metalworking** operations, as distinguished from metal cutting, material is forced to move into new shapes by plastic flow. **Hot-working** is carried on above the recovery temperature, and spontaneous recovery, or annealing, occurs about as fast as the properties of the material are altered by the deformation. This process is limited by the chilling of the material in the tools, scaling of the material, and the life of the tools at the required temperatures. **Cold-working** is carried on at room temperature and may be applied to most of the common metals. Since, in most cases, no recovery occurs at this temperature, the properties of the metal are altered in the direction of increasing strength and brittleness throughout the working process, and there is consequently a limit to which cold-working may be carried without danger of fracture.

A convenient way of representing the action of the common metals when cold-worked consists of plotting the actual stress in the material against the percentage reduction in thickness. Within the accuracy required for shop use, the relationship is linear, as in Fig. 4. The lower limit of stress shown is the yield point at the softest temper, or anneal, commercially available, and the upper limit is the limit of tensile action, or the stress at which fracture, rather than flow, occurs. This latter value does not correspond to the commercially quoted "tensile strength" of the metal, but rather to the "true tensile strength," which is the stress that exists at the reduced section of a tensile specimen at fracture and which is higher than the nominal value in inverse proportion to the reduction of area of the material.

As an example of the construction and use of the cold-working plots shown in Fig. 4, the action of a very-low-carbon

deep-drawing steel has been shown in Fig. 5. Starting with the annealed material with a yield point of 35,000 lb/in² (240 MN/m²), the steel was drawn to successive reductions of thickness up to about 58 percent, and the corresponding stresses plotted as the heavy straight line. The entire graph was then extrapolated to 100 percent reduction, giving the **modulus of strain hardening** as indicated, and to zero stress so that all materials might be plotted on the same graph. Lines of equal reduction

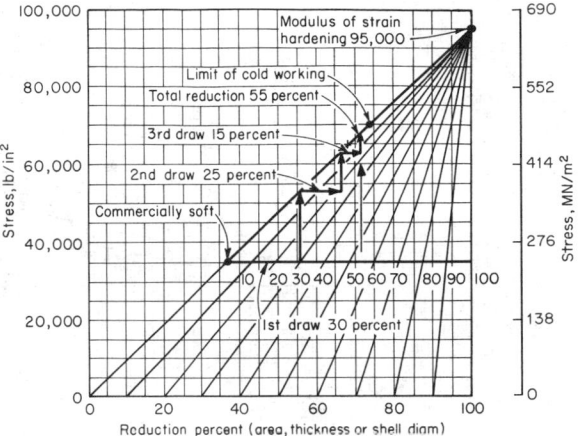

**Fig. 5** Graphical solution of a metalworking problem.

are slanting lines through the point marking the modulus of strain hardening at theoretical 100 percent reduction. Starting at any initial condition of previous cold work on the heavy line, a percentage reduction from this condition will be indicated by a horizontal traverse to the slanting reduction line of corresponding magnitude and the resulting increase in stress by the vertical traverse from this point to the heavy line.

The traverse shown involved three draws from the annealed condition of 30, 25, and 15 percent each, and resulting stresses of 53,000, 63,000, and 68,000 lb/in² (365, 434, and 469 MN/m²). After the initial 30 percent reduction, the next 25 percent uses $(1.00 - 0.30) \times 0.25$, or 17.5 percent more of the cold-working range; the next 15 percent reduction uses $(1.00 - 0.30 - 0.175) \times 0.15$, or about 8 percent of the original range, totaling $30 + 17.5 + 8 = 55.5$ percent. This may be compared with the test value percent reduction in area for the particular material. The same result might have been obtained, die operation permitting, by a single reduction of 55 percent, as shown. Any appreciable reduction beyond this point would come dangerously close to the limit of plastic flow, and consequently an annealing is called for before any further work is done on the piece.

Figure 6 shows the approximate **true-stress–true-strain** plotting of common plastic range values, for comparison with Fig. 4.

A practical manufacturing method of judging relative plasticity is to compute the ratio of initial yield point to the ultimate tensile strength as developed in the tensile test. Thus a General Motors research memo in 1955 listed steel with an 0.51 **yield-tensile ratio** [22,000 lb/in² (152 MN/m²) yp/43,000 lb/in² (296 MN/m²) U.T.S.] as being suitable for really severe draws of exposed parts. When the ratio reaches about 0.75, the steel should be used only for flat parts or possibly those

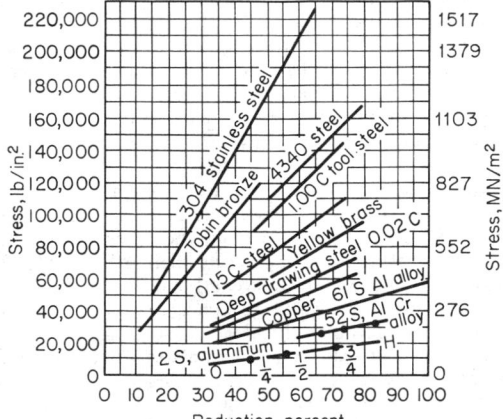

**Fig. 4** Plastic range chart of commonly worked metals.

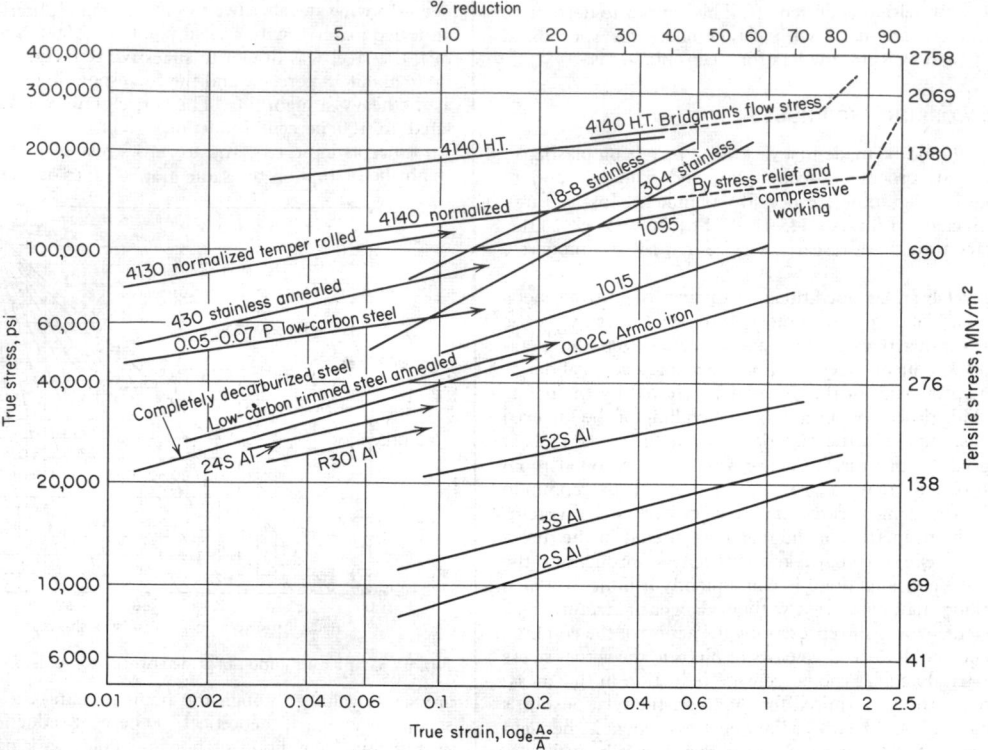

**Fig. 6**  True-stress–true-strain curves for typical metals. *(Crane and Hauf, E. W. Bliss Co.)*

with a bend of not more than 90°. The higher ratios obviously represent a narrowing range of workability or residual plasticity.

### Rolling Operations

Rolling of sheets, coils, bars, and shapes is a primary process using plastic ranges both above and below recrystallization to prepare metals for further working or for fabrication. Metal squeezed in the bite area of the rolls moves out lengthwise with very little spreading in width. This compressive working above the yield point of the metal may be aided in some cases by maintaining a substantial tensile strain in the direction of rolling.

A cast or forged billet or slab is usually preheated for the preliminary breakdown stage of rolling, although considerable progress is being made with pouring some metals directly into the first pair of rolls. A reversing hot mill may achieve 5,000 percent elongation of an original billet in a series of manual or automatic passes. Alternatively, the billet may pass progressively through, say, 10 hot mills in rapid succession. Such a production setup requires precise control so that each mill stand will run enough faster than the previous one to make up for the elongation of the metal that has taken place. Hot-rolled steel may be sold for many purposes with the black mill scale on it. Alternatively, it may be acid-pickled to remove the scale and treated with oil or lime for corrosion protection. To prevent scale from forming in hot-rolling, a nonoxidizing

atmosphere may be maintained in the mill area, a highly special plant design.

**Pack rolling** of a number of sheets stacked together provides means of retaining enough heat to hot-roll thin sheets, as for high-silicon electric steels.

**Cold-rolling** is practical in production of thin coil stock with the more ductile metals. The number of passes or amount of reduction between anneals is determined by the rate of work hardening of the metal. Successive stands of cold-rolling help to retain heat generated in working. Tension provided by mill reels and between stands helps to increase the practical reduction per step. Bright annealing in a controlled atmosphere avoids surface pockmarks, which are difficult to get out. For high-finish stock, the rolls must be maintained with equal finish.

**Protective coating** is best exemplified by high-speed tinplate mills in which coil stock passes continuously through the necessary series of cleaning, plating, and heating steps. Zinc and other metals are also applied by plating but not on the same scale. **Clad** sheets (high-strength aluminum alloys with pure aluminum surface for protection against electrolytic oxidation) are produced by rolling together; e.g., an alloy-aluminum billet is hot-rolled together with plates of pure aluminum above and below it through a series of reducing passes, with precautions to ensure clean adhesion.

On the other hand, prevention of adhesion, as by a separating film, is essential in the final stages of **foil rolling**, where two coils may have to be rolled together. Such foil may then be **laminated** with suitable adhesive to paper backing materials for wrapping purposes. (See also Sec. 6.)

**Shape rolling** of structural shapes and rails is usually a **hot** operation with roll-pass contours designed to distribute the displacement of metal in a series of steps dictated largely by experience. **Contour rolling** of relatively thin stock into tubular, channel, interlocking, or varied special cross sections is usually done cold in a series of roll stands for lengthwise bending and setting operations. There is also a wide range of simple bead-rolling, flange-rolling, and seam-rolling operations in relatively thin materials, especially in connection with the production of barrels, drums, and other containers.

**Oscillating** or **segmental rolling** probably developed first in the manually fed contour rolling of agricultural implements. In some cases, the suitably contoured pair of roll inserts or roll dies oscillates before the operator, to form hot or cold metal. In other cases, the rolls rotate constantly, toward the operator. The working contour takes only a portion of the circumference, so that a substantial clearance angle leaves a space between the rolls. This permits the operator to insert the blank to the tong grip between the rolls and against a fixed gage at the back. Then, as rotation continues, the roll dies grip and form the blank, moving it back to the operator. This process is sometimes automated; such units as **tube-reducing mills** oscillate an entire rolling-mill assembly and feed the work over a mandrel and into the contoured rolls, advancing it and possibly turning it between reciprocating strokes of the roll stands for cold reduction, improved concentricity, and (possibly) the tapering or forming of special sections.

**Spinning** operations (Fig. 7) apply a rolling-point pressure to relatively limited-lot production of cup, cone, and disk shapes, from floor lamps and TV tube housings to car wheels and large tank ends. Where substantial metal thickness is required, powerful machines and hydraulic servo controls may be used. Some of the large, heavy sections and difficult metals are spun hot.

Rolling operations are distinguished by the relatively rapid and continuous application of working pressure along a limited line of contact. In determining the working area, consider

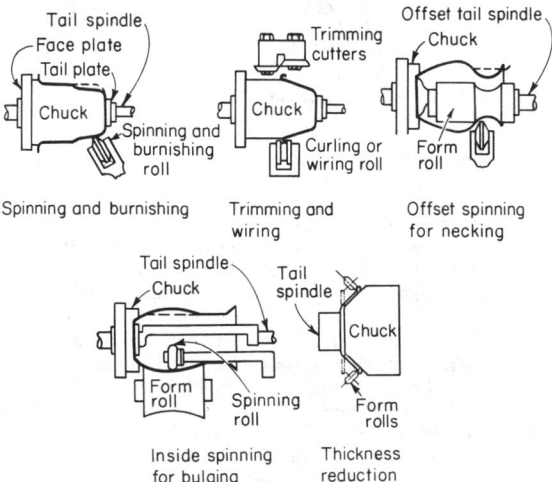

**Fig. 7**  Spinning operations.

the lineal dimension (width of coil), the bit or reduction in thickness, and the roll-face deflection, which tends to increase the contact area. Approximations of rolling-mill load and power requirements have been worked out in literature of the AISE and ASME.

### Shearing

The shearing group of operations includes such **power-press operations** as blanking, piercing, perforating, shaving, broaching, trimming, slitting, and parting. Shearing operations traverse the entire plastic range of metals to the point of failure.

The maximum pressure $P$, in pounds, required in shearing operations is given by the equation $P = \pi DtS = Lts$, where $s$ is the resistance of the material to shearing, lb/in$^2$; $t$ is the thickness of the material, in; $L$ is the length of cut, in, which is the circumference of a round blank $\pi D$ or the periphery of a rectangular or irregular blank. Approximate values of $s$ are given in Table 2.

**Shear** (Fig. 8) is the advance of that portion of the shearing edge which first comes in contact with the material to be sheared over the last portion to establish contact, measured in the direction of motion. It should be a function of the thickness $t$. Shear reduces the maximum pressure because, instead

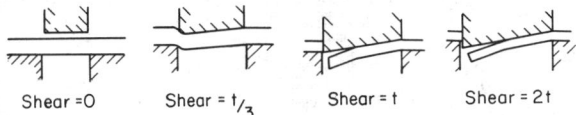

Shear =0        Shear = $t/3$        Shear = $t$        Shear = $2t$

**Fig. 8**  Diagrams illustrating shear.

of shearing the whole length of cut at once, the shearing action takes place progressively, shearing on only a portion of the length at any instant. This is illustrated in Figs. 9 and 10, which show the relationships of metal thickness, instantaneous pressure, clearance, and shear for conditions of adequate and of insufficient clearance, respectively. The maximum pressure for any case where the shear is equal to or greater than $t$ is given by the equation $P_{max} = P_{av} \times t/\text{shear}$, where $P_{av}$ is the average value of the pressure on a punch, with shear = $t$, from the time it strikes the metal to the time it leaves.

**Distortion** results from shearing at an angle (Fig. 8) and accordingly, in blanking, where the blank should be flat, the punch should be flat and the shear should be on the die. Conversely, in hole punching, where the scrap is punched out, the die should be flat and the shear should be on the punch. Where there are a number of punches, the effect of shear may be obtained by stepping the punches.

**Crowding** results during the plastic deformation period, before the fracture occurs, in any shearing operation. Accordingly, when small delicate punches are close to a large punch, they should be stepped shorter than the large punch by at least a third of the metal thickness.

**Clearance** between the punch and die is required for a clean cut and durability. An old rule of thumb places the clearance all around the punch at 8 to 10 percent of the metal thickness for soft metal and up to 12 percent for hard metal. Actually, hard metal requires less clearance for a clean fracture than soft, but it will stand more. In some cases, with delicate punches, clearance is as high as 25 percent. Where the hole diameter is important, the punch should be the desired diame-

### Table 2. Approximate Resistance to Shearing in Dies

| Material | Annealed state | | Hard, cold-worked | |
|---|---|---|---|---|
| | Resistance to shearing, lb/in²* | Penetration to fracture, percent | Resistance to shearing, lb/in²* | Penetration to fracture, percent |
| Lead | 3,500 | 50 | Anneals at room temperature | |
| Tin | 5,000 | 40 | Anneals at room temperature | |
| Aluminum 2S, 3S | 9–11,000 | 60 | 13–16,000 | 30 |
| Aluminum 52S, 61S, 62S | 12–18,000 | . . . | 24–30,000 | |
| Aluminum 75S | 22,000 | . . . | 46,000 | |
| Zinc | 14,000 | 50 | 19,000 | 25 |
| Copper | 22,000 | 55 | 28,000 | 30 |
| Brass | 33–35,000 | 50–55 | 52,000 | 25–30 |
| Bronze 90-10 | . . . | . . . | 40,000 | |
| Tobin bronze | 36,000 | 25 | 42,000 | |
| Steel 0.10C | 35,000 | 50 | 43,000 | 38 |
| Steel 0.20C | 44,000 | 40 | 55,000 | 28 |
| Steel 0.30C | 52,000 | 33 | 67,000 | 22 |
| Steel 0.40C | 62,000 | 27 | 78,000 | 17 |
| Steel 0.60C | 80,000 | 20 | 102,000 | 9 |
| Steel 0.80C | 97,000 | 15 | 127,000 | 5 |
| Steel 1.00C | 115,000 | 10 | 150,000 | 2 |
| Stainless steel | 57,000 | 39 | | |
| Silicon steel | 65,000 | 30 | | |
| Nickel | 35,000 | 55 | | |

*1,000 lb/in² = 6.895 MN/m².

Available test data do not agree closely. The above table is subject to verification with closer control of metal analysis, rolling and annealing conditions, die clearances. In dinking dies, steel-rule dies, hollow cutters, etc., cutting-edge resistance is substantially independent of thickness: cotton glove cloth (stack, 2 or 3 in thick), 240 lb/in; kraft paper (stack tested, 0.20 in thick), 385 lb/in, celluloid [¹¹/₃₂ in thick, warmed in water to 120 to 150°F (49 to 66°C)], 300 lb/in.

ter and the clearance should be added to the die diameter. Conversely, where the blank size is important, the die and blank dimensions are the same and clearance is deducted from the punch dimensions.

The **work per stroke,** as measured by the area under the pressure-distance curve (Fig. 9 or 10), is not affected by shear; thus the area under each of the curves in Fig. 9 is the same. It may be approximated as the product of the maximum pressure and the metal thickness, although it is only about 20 to 80 percent of that product, depending upon the clearance and the ductility of the metal. Reducing the clearance causes secondary fractures and increases the work done. With sufficient clearance for a clean fracture, the work is a little less than the product of the maximum pressure, the metal thickness, and the percentage reduction in thickness at which the fracture occurs. Approximate values for this are given in Table 2. The **power required** may be obtained from the work per stroke plus a 10 to 20 percent friction allowance.

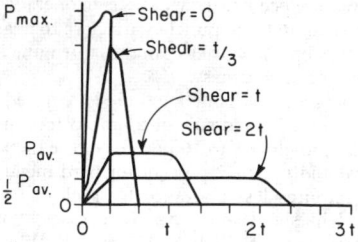

**Fig. 9** Influence of shear with clearance adequate for a clean fracture.

**Shaving**  A sheared edge may be squared up roughly by shaving once, allowing for the shaving of mild sheet steel about 10 percent of the metal thickness. This allowance may be increased somewhat for thinner material and should be

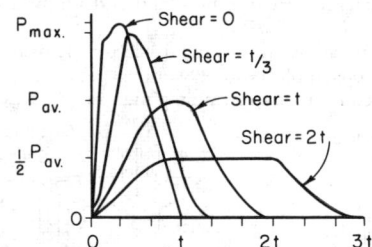

**Fig. 10**  Influence of shear with insufficient clearance.

decreased for thicker and softer material. In making several cuts, the amount removed is reduced each time. For extremely fine finish a round-edged burnishing die or punch, say 0.001 or 0.0015 in tight, may be used. Aluminum parts may be blanked (as for impact extrusion) with a fine finish by putting a 30° bevel, approx one-third the metal thickness on the die opening, with a near metal-to-metal fit on the punch and die, and pushing the blank through the highly polished die.

**Squaring shears** for sheet or plate may have their blades arranged in either of the ways shown in Fig. 11. The square-edged blades in Fig. 11a may be reversed to give four cutting edges before they are reground. Single-edged blades, as shown in Fig. 11b, may have a clearance angle on the side where the blades pass, to reduce the working friction. They may also be ground at an angle or rake, on the face which comes in contact with the metal. This reduces the bending and consequent distortion at the edge. Either type of blade distorts also in the other direction owing to the angle of shear on the length of the blades (see Fig. 8). Cutting speed is 3 to 30 ft/min

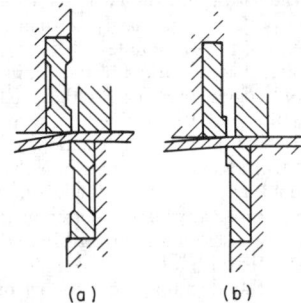

(a)        (b)

**Fig. 11** Squaring shears.

(1 to 10 m/min), depending in part upon the thickness of the material.

**Circular cutters** for slitters and circle shears may also be square-edged (on most slitters) or knife-edged (on circle shears). According to one rule, their diameter should be not less than 70 times the metal thickness. Cutting speeds vary from 50 to 200 ft/min (17 to 65 m/min), depending largely upon metal thickness (inverse proportion).

**Knife-edge hollow cutters** working against end-grain maple blocks represent an old practice in cutting leather, rubber, and cloth in multiple thicknesses. **Steel-rule dies,** made up of knife-edge hard-steel strip economically mounted against a steel plate in a wood matrix with rubber strippers and cutting against hard saw-steel plates, extend the practice to corrugated-carton production and even some limited-lot metal cutting.

### Bending

The bending group of operations is performed in **presses** (variety), **brakes** (metal furniture, cornices, roofing), **bulldozers** (heavy rolled sections), **multiple-roll forming machines** (molding, etc.), **draw benches** (door trim, molding, etc.), **forming rolls** (cylinders), and **roll straighteners** (strips, sheets, plates).

**Spring back,** due to the elasticity of the metal and amount of the bend, may be compensated for by overbending or largely prevented by striking the metal at the radius with a **coining** (i.e., squeezing, as in production of coins) pressure sufficient to set up compressive stresses to counterbalance surface tensile stresses. A very narrow bead may be used to localize the pinch where needed and minimize danger to the press in squeezing on a large area. Under such conditions, good sharp bends in V dies have been obtained with two to four times the pressure required to shear the metal across the same section.

**Bending Allowance** The thickness of the metal over a small radius or a sharp corner is 10 or 15 percent less than before bending because the metal moves more easily in tension than in compression. For the same reason the neutral axis of the metal moves in toward the center of the corner radius. Therefore, in figuring the length of blank $L$ to be allowed for the bend up to an inside radius $r$ of two or three times the

metal thickness $t$, the length may be figured closely as along a neutral line at $0.4t$ out from the inside radius. Thus, with reference to Fig. 12, for any angle $a$ in deg and other dimensions in inches, $L = (r + 0.4t)2\pi a/360 = (r + 0.4t)a/57.3$.

The factor $0.4t$, which locates the neutral axis, is subject to some variation (say $0.35$ to $0.45t$) according to radius, condition of metal, and angle. In figuring allowances for sharp bends, note that the metal builds up on the compression side of the corner. Therefore, in locating the neutral axis, consider an inside radius $r$ of about $0.05t$ as a minimum.

**Roll straighteners** work on the principle of bending the metal beyond its elastic limit in one direction over rolls small enough in diameter, in proportion to the metal thickness, to give a permanent set, and then taking that bend out by repeatedly reversing it in direction and reducing it in amount. Metal is also straightened by gripping and stretching it beyond its elastic limit and also by hammering; the results of the latter operation depend entirely upon the skill of the operator.

For approximating **bending loads,** the beam formula may be used but must be very materially increased because of the short spans. Thus, for a span of about four times the depth of section, the bending load is about 50 percent more than that indicated by the beam formula. It increases from this to nearly

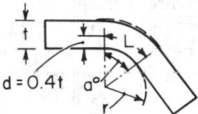

**Fig. 12** Bending allowance.

the shearing resistance of the section where some **ironing** (i.e., the thinning of the metal when clearance between punch and die is less than the metal thickness) occurs. Where hit-home dies do a little coining to "set" the bend, the pressure may range from two or three times the shearing resistance, with striking beads and proper care, up to very much higher figures.

The work to roll-bend a sheet or plate $t$ in thick with a volume of $V$ in$^3$, into curved shape of radius $r$ in, has been given as $W = CS(t/r)V/48$ ft·lb, in which $S$ is the tensile strength and $C$ is an experience factor between $1.4$ and $2$.

### Drawing

**Drawing** includes operations in which metal is pulled or drawn, in suitable containing tools, from flat sheets or blanks into cylindrical cups or rectangular or irregular shapes, deep or shallow. It also includes reducing operations on shells, tube, wire, etc., in which the metal being drawn is pulled through dies to reduce the diameter or size of the shape. All drawing and reducing operations, by an applied tensile stress in the material, set up circumferential compressive stresses which crowd the metal into the desired shape. The relation of the shape or diameter before drawing to the shape or diameter after drawing determines the magnitude of the stresses. Excessive draws or reductions cause thinning or tearing out near the bottom of a shell. Severe cold-drawing operations require very ductile material and, in consequence of the amount of plastic deformation, harden the metal rapidly and necessitate annealing to restore the ductility for further working.

The **pressure used in drawing** is limited to the load to shear the bottom of the shell out, except in cases where the side wall is ironed thinner, when wall friction makes somewhat higher loads possible. It is less than this limit for round shells which are shorter than the limiting height and also for rectangular shells. Drawing occurs only around the corner radii of rectangular shells, the straight sides being merely free bending.

A **holding pressure** is required in most initial drawing and some redrawing, to prevent the formation of wrinkles due to the circumferential compressive stresses. Where the blank is relatively thin compared with its diameter, the blank-holding pressure for round work is likely to vary up to about one-third of the drawing pressure. For material heavy enough to provide sufficient internal resistance to wrinkling, no pressure is required. Where a drawn shape is very shallow, the metal must be stretched beyond its elastic limit in order to hold its shape, making it necessary to use higher blank-holding loads, often in excess of the drawing pressure. To grip the edges sufficiently to do this, it is often advisable to use **draw beads** on the blank-holding surfaces if sufficient pressure is available to form these beads.

Some shells, which are very thick or very shallow compared with their diameter, do not require a blank holder. Blank-holding pressure may be obtained through toggle, crank, or cam mechanisms built into the machine or by means of air cylinders, spring-pressure attachments, or rubber bumpers under the bolster plate. The length of car springs should be about 18 in/in (18 cm/cm) of draw to give a fairly uniform drawing pressure and long life. The use of car springs has been largely superseded by hydraulic and pneumatic cushions. Rubber bumpers may be figured on a basis of about 7.5 lb/in² (50 kN/m²) of cross-sectional area per 1 percent of compression. In practice they should never be loaded beyond 20 percent compression, and as with springs, the greater the length relative to the working stroke, the more uniform is the pressure.

**Dimensions of Drawn Shells**   The smallest and deepest round shell that can be drawn from any given blank has a diameter $d$ of 65 to 50 percent of the blank diameter $D$. The height of these shells is $h = 0.35d$ to $0.75d$, approximately. Higher shells have occasionally been drawn with ductile material and large punch and die radii. Greater thickness of material relative to the diameter also favors deeper drawing.

The area of the bottom and of the side walls added together may be considered as equal to the area of the blank for approximations. If the punch radius is appreciable, the area of a neutral surface about $0.4t$ out from the inside of the shell may be taken for approximations. Accurate blank sizes may be obtained only by trial, as the metal tends to thicken toward the top edge and to get thinner toward the bottom of the shell wall in drawing.

Approximate diameters of blanks for shells are given by the expression $\sqrt{d^2 + 4dh}$, where $d$ is the diameter and $h$ the height of the shell.

In **redrawing** to smaller diameters and greater depths the amount of reduction is usually decreased in each step. Thus in double-action redrawing with a blank holder, the successive reductions may be 25, 20, 16, 13, 10 percent, etc. This progression is modified by the relative thickness and ductility of the metal. Single-action redrawing without a blank holder necessitates smaller steps and depends upon the shape of the dies and punches. The steps may be 19, 15, 12, 10 percent, etc. Smaller reductions per operation seem to make possible greater total reductions between annealings.

**Rectangular shells** may be drawn to a depth of four to six times their corner radius. It is sometimes desirable, where the sheet is relatively thin, to use draw beads at the corners of the shell or near reverse bends in irregular shapes to hold back the metal and assist in the prevention of wrinkles.

**Work** in drawing is approximately the product of the length of the draw, and the maximum punch pressure, as the load rises quickly to the peak, remains fairly constant, and drops off sharply at the end of the draw unless there is stamping or wall friction. To this, add the work of blank holding which, in the case of cam and toggle pressure, is the product of the blank-holding pressure and the spring of the press at the pressure (which is small). For single-action presses with spring, rubber, or air-drawing attachments it is the product of the average blank-holding pressure and the length of draw.

**Rubber-die forming,** especially of the softer metals and for limited-lot production, uses one relatively hard member of metal, plaster, or plastic with a hard powder filler to control contour. The mating member may be a rubber or neoprene mattress or a hydraulically inflatable bag, confined and at 3,000 to 7,000 lb/in² (20 to 48 MN/m²). Babbitt, oil, and water have also been used directly as the mobile member. A large hydraulic press, often with a sliding table or tables, and even static containers with adequate pumping systems are used.

**Hot drawing** above the recrystallization range applies single- and double-action drawing principles. For light gages of plastics, paper, and hexagonal-lattice metals such as magnesium, dies and punches may be heated by gas or electricity. For thick steel plate and heat-treatable alloys, the mass of the blank may be sufficient to hold the heat required.

**Lubricants for Presswork**   Many jobs may be done dry, but better results and longer life of dies are obtained by the use of a lubricant. Lard or sperm oil is used when punching iron, steel, or copper. Petroleum jelly is used for drawing aluminum. A soap solution is commonly used for drawing brass, copper, or steel. One manufacturer uses 90 percent mineral oil, 5 percent rosin, and 5 percent oleic acid for light work and an emulsion of a mineral oil, degras, and a pigment consisting of chalk, sulfur, or lithopone for heavy work. (See also Sec. 6.)

For heavy drawing operations and extrusion, steels may have a zinc phosphate coating bonded on, and a zinc or sodium stearate bonded to that, to withstand pressures over 300,000 lb/in² (2,070 MN/m²). An anodized coating for aluminum may be used as a host for the lubricant. It is reported that such a chemical treatment plus a lacquer or plastic coating and plus a lubricant is effective for severe ironing operations.

The problem is to prevent local pressure welding from starting as galling or pickup, with resulting scratching, by maintaining a fluid film separation between metal surfaces. At moderate pressures, almost any viscous liquid lubricant will do the job. Rust protection and easy removal of the lubricant are often major factors in the choice.

**Shock-wave forming** for limited lots is developing in several ways. **Explosive forming,** especially for large-area drawn or formed shapes, usually requires one metal contour-control die immersed in fluid as in a lake. Explosives manufacturers have developed means of computing the charge and the distance that it should be suspended above the blank to be formed. The space back of the blank in the die has to be evacuated. A blank-holding ring to minimize wrinkle formation in the flange area is bolted very tightly to the die, with an O-ring seal to prevent leakage.

**Electrohydraulic forming** is similar to explosive forming except that the shock wave is imparted electrically from a large battery of capacitors. ASTM Papers SP-1963 184, 185, and !86, by engineers of Republic Aviation Corp., describe the electrodes, tool buildup, and impact distribution demonstrating this method at 10,000 V and 155,000 J capacity. **Magnetic forming** uses the same source of power but does not require a fluid medium. A flexible pancake coil under a Republic patent delivers the magnetic shock pulse.

### Squeezing

The squeezing group of operations are those in which the metal is worked in compression. Resultant tensile strains occur, however; in cases where the metal is thin compared with its area and there is an appreciable movement of the metal, there results a pyramiding of pressure toward the center of the die which may prove serious. The metal is incompressible (beyond about 1 percent), and consequently, to reduce the thickness of any volume of metal in the center of the blank, its area must be increased which involves spreading or stretching all the metal around it. The surrounding metal acts like shrunk bands and offers a resistance increasing toward the center and often many times the mere compressive resistance of the material.

Squeezing operations and particularly the squeezing of steel are practically the severest of all press operations. They may be divided into four general classifications according to severity, although in every group there will be found examples of working to the limit of what the die steels will stand, which may be taken at about 100 tons/in². The severer operations, such as cold bottom extrusion and wall extrusion, are limited to the softer metals. Squeezing operations ordinarily require pressure through a very short distance, the pressure starting at the compressive yield strength of the material over the surface being squeezed and rising to a maximum at bottom stroke. This maximum is greatest when the metal is thin compared with its area or when the die is entirely closed as for coining. Care must be taken, on all squeezing operations, in the setting of presses and avoidance of double blanks or extra-heavy

blanks as the presses must be stiff. In squeezing solidly across bottom center the mechanical advantage is such that a small difference in thickness or setup can make a very large difference in pressure exerted. For this reason high-speed self-contained hydraulic presses, with automatic pressure-control and size blocks, are now finding favor for some of this work.

**Sizing,** or the flattening or surfacing of parts of forgings or castings, is usually the least severe of the squeezing group. Tolerances are ordinarily closer than for the milling operations which are supplanted. When extremely close tolerances are required, say plus or minus one-thousandth, arrange substantial size blocks to take half or two-thirds of the total load. These take up uniformly the bearing-oil films and any slight deflection of the bed and bolster and minimize the error in spring back due to variation in thickness, hardness, and area of the rough forging or casting. The usual amount left for squeezing is $\frac{1}{32}$ to $\frac{1}{16}$ in (0.8 to 1.6 mm). Presses may be selected for this service on a basis of 60 to 80 tons/in² (830 to 1,100 MN/m²), although 100 tons/in² (1,380 MN/m²) is more often used in the automobile trade for reserve capacities. When figuring from experimental results obtained in testing machines, the recorded loads are usually doubled in selecting a press, in order to allow for the difference among the speed of the machines, the positive action, and a safety margin.

**Swaging** or **cold forging** involves squeezing of the blank to an appreciably different shape. Success in performing such operations on steel usually depends upon squeezing a relatively small area with freedom to flow without restraint. Dies for this work must usually be substantially backed up with hardened steel plates. The edge of the blank after coining is usually ragged and must be trimmed for appearance.

**Hot forging** is similar in certain respects to the above but permits much greater movement of metal. Hot forging may be done in drop hammers, percussion presses, power presses, or forging machines, when dies are used, or in steam hammers, helve hammers, or hydraulic presses, on plain anvils.

The pressure exerted by hydraulic presses and steam hammer for jobbing work should be about as follows:

| Ingot diam, in* | 5 | 8 | 12 | 16 | 24 | 36 | 48 | 60 | 72 |
|---|---|---|---|---|---|---|---|---|---|
| Press, tons† | 100 | 200 | 400 | 600 | 1,000 | 1,500 | 2,000 | 3,000 | 4,000 |
| Hammer, tons† | ½ | 1 | 3 | 5 | 10 | 20 | 40 | 80 | 120 |

*1 in = 2.54 cm.
†1 ton = 8.9 kN.

**Drop hammers** are rated according to weight of ram. For carbon steel they may be selected on a basis of 50 to 55 lb of ram weight per square inch of projected area (3.5 to 3.9 kg/cm²) of the forging, including as much of the flash as is squeezed. This allowance should be increased to 60 lb/in² (4.2 kg/cm²) for 0.20 carbon steel, 70 lb/in² (4.9 kg/cm²) for 0.30 carbon steel, and up to about 130 lb/in² (9.2 kg/cm²) for tungsten steel.

In figuring the **forging pressure,** multiply the projected area of the forging, including the portion of the flash that is squeezed, by approximately one-third of the cold compressive strength of the material. Another method gives the forging pressure at three to four times the compressive strength of the material at forging temperatures times the projected area, for presses; or at ten times the compressive strength at forging temperatures times the projected area, for hammers. The pressure builds up to a rather high figure at bottom stroke owing to

the cooling of the metal particularly in the flash and to the small amount of relief for excess metal which the flash allows.

For **brass press forgings** a good mixture is about Cu, 59; Zn, 39; Pb, 2 percent forged at 1300 to 1400°F (700 to 760°C). The power curve in press forging rises sharply, from the compressive strength at forging heat times the projected area of the slug to three or more times that quantity at bottom stroke. A large flash area assists in driving the metal into deep die recess.

In **heading operations,** hot or cold, the length of wire or rod that can be gathered into a head, without side restraint, in a single operation, is limited to three times the diameter. In coining and then heading large heads, cold, wire of about 0.08 carbon must be used to avoid excessive strain-hardening.

**Forging Dies** Drop-forge dies are usually of steel or steel castings. A good all-round grade of **steel** is a 0.60 percent carbon open-hearth. Dies of this steel will forge mild steel, copper, and tool steel satisfactorily if the number of forgings required is not too large. For a large number of tool-steel forgings, tool-steel dies of 0.80 to 0.90 percent carbon may be used and for extreme conditions, 3½ percent nickel steel.

Die blocks of alloy steels have special value for the production of drop forgings in large quantities. Metals Handbook, 1961, recommends the following alloy steel compositions: for hot forgings, 0.60 C, 0.25 Cr, or 0.70 C, 0.20 V; for cold forgings, 0.85 C, 0.20 V; for drop-hammer work, 0.55 C, 1.00 Cr, 0.45 Mo, or 0.50 C, 1.60 Ni, 1.00 Cr, 0.30 Mo; for forging presses, 0.55 C, 2.00 Ni, 1.00 Cr, 0.75 Mo, or 0.35 C, 5.00 Cr, 1.50 Mo. Steels recommended for hot die inserts include the above and also 0.55 C, 1.00 Si, 1.25 Cr, 2.50 W, as well as tungsten steels of 9.00, 12.00, and 15.00 W.

For large massive dies or for intermittent service, chrome-nickel-molybdenum alloys are preferred. In closed die work, where the dies must dissipate considerable heat, the tungsten steels are preferred, with resulting increase in wear resistance but decrease in toughness.

For very large pieces with deep impressions, **cast-steel** dies are sometimes used. For large dies liable to spring in hardening, 0.85 **carbon steel** high in manganese is sometimes used unhardened.

Good **die-block proportions** for width and depth are as follows:

| Width, in (cm) | 8 (20) | 10 (24) | 12 (30) | 14 (36) |
| Depth, in (cm) | 6 (15) | 7 (18) | 7 (18) | 7 or 8 (18 or 20) |

For ordinary work, 1½ in (4 cm) of metal between impression and edge of block is sufficient.

Dimensions of **dovetailed die shanks:** for hammers up to 1,200 lb (550 kg) 4 in (10 cm) wide and 1⅛ in (3 cm) deep, with sides dovetailed at angles of 6° with the vertical; for hammers from 1,200 to 3,000 lb (550 to 1,360 kg) in size, 6 in (15 cm) wide and 1½ in (4 cm) deep, with 6° angles.

The **minimum draft for the impressions** is 7°, although for parts difficult to draw this may be increased up to 15°. It is not uncommon to have several drafts in the same impression.

Open-hearth and tool-steel **dies** are **hardened** by heating in a carbonizing box packed in charcoal and dipping face downward over a jet of brine. The jet is allowed to strike into the impression, thus freeing the face of steam and producing uniform hardness. After hardening, they are drawn in an oil bath to a temperature of 500 to 550°F (260 to 290°C).

The forging production per pair of dies is largely affected by the size and shape of the impression, the material forged, the material in the dies, the quality of heating of stock to be forged, and the care exercised in use. It may vary from a few hundred pieces to 50,000 or more. A normal **life** for a pair of dies under average conditions may be 20,000 pieces.

**Coining, Stamping, and Embossing** The metal is well confined in closed dies in which it is forced to flow to fill the shape. The government gives the following pressures: silver quarter, 100 tons/in² (1,380 MN/m²); nickel (0.25 Ni, 0.75 Cu), 90 tons/in² (1,240 MN/m²); copper cent, 40 tons/in² (550 MN/m²). In stamping designs, lettering, etc., in sheet metal the thickness is so little compared with the area that there is practically no relief for excess pressure. Where sharp designs are required, as in stamping panels, the dies should be arranged to strike on a narrow line [say ¹⁄₃₂ in (0.8 mm)] around the outline. If a sharp design is not obtained, it is often best to correct deflection in the machine by shimming or more substantial backing. Increasing the pressure only aggravates the condition and may break the press. General practice for light overall stamping is to allow 5 to 10 tons/in² (68 to 132 MN/m²) of area that is to be stamped, except in areas where the yield point must be exceeded.

**Extrusion** is the severest of the squeezing processes. The metal is forced to flow rapidly through an orifice, being otherwise confined and subject largely to the laws of hydraulics, with allowances for restraint of flow and for work hardening. Power-press **impact extrusion** began with tin and lead collapsible tubes. It has been extended to the backward and forward extrusion of aluminum, brass, and copper in pressure ranges of 30 to 60 or more tons/in² (413 to 825 MN/m²), and mild steel at pressures up to 165 tons/in² (2,275 MN/m²). Hot impact extrusion of steel, as in projectile piercing, ranges from about 25 to 50 tons/in² (345 to 690 MN/m²). **Forward extrusion** of long tubes, rods, and shapes usually performed hot in hydraulic presses has been extended from the softer metals to the extrusion of steels. Most work is done horizontally because of the lengths of the extrusions. Some vertical mechanical-press equipment is used in hot extrusion of steel tubing. **Screw extrusion** of the plastics requires closely controlled heating of the barrel around the feeding screw, which is between the feeding hopper and the contour die. Preheating for extrusion is in approximately the same range as temperatures for injection molding (Table 1).

**Powder compacting** apparently began with drug pills and carbon electrodes. A reasonable uniformity of particle size, fairly free of fines, is desirable for automatic feeding. The powders tend to arch and clog, particularly in narrow sections, so that vibrating devices are sometimes added. Ejectors are often timed so that they help to draw the powder into the die in which it is to be compacted. The volume of the powder fill is usually about three times the volume (depth) of the finished compact. Compacting pressures vary from as little as 1 to about 20 tons/in² (14 to 280 MN/m²) for the metals. Humidity, oxidation, and even static electricity are often problems in the storage and use of powders. Metal powders usually require careful blending with ½ to 1 percent of metallic soap (e.g., zinc stearate) to lubricate ejection from the compacting die and perhaps to provide a temporary bonding. Die or mold surfaces must be polished in the direction of ejection to a high finish to facilitate stripping. Ejecting pressures may range from 10 to 100 percent of the compacting pressure, depending on the contour. Thus a gear is much harder to strip than a rectangular contour.

**Sintering** of metal-powder compacts at temperatures below the melting point provides the time and electron activity needed to establish bonded strength between particles. An initial lower temperature permits volatilization of much of the lubricant. A reducing atmosphere is often needed to eliminate oxides. Both batch and conveyor-type continuous furnaces are used. For some sintered carbides, the compacting pressure and sintering action are combined in a hydraulic press using carbon dies and induction heating.

**Molding** under pressure of synthetic plastics or rubbers requires control of temperature, heating time, and in some cases cooling time to suit the material and volume being handled. **Compression molding** usually requires the loading of preformed compacts or blanks in heated dies. Preforms may also be preheated electronically to save press time. **Transfer**

molding usually first closes the die or mold with a large hydraulic cylinder to hold the joint line tight against flash and then forces the charge into the cavity by means of a smaller cylinder. **Injection molding** makes this process automatic by adding a hopper to maintain a supply of powder, a controlled injection stroke to meter the correct amount of powder, a controlled heating zone to bring the charge up to temperature before it is injected into the mold cavities, an ejecting means, and automatic cycle timing.

**Laminating** of plastic bonded sheet materials uses steam or electrically heated plates or platens in hydraulic presses. Where more than one set of hot plates is used, the method of suspension opens the plate spacing for removal and reloading. Thus a dozen or more sheets of plywood can be cured at once, using sheet Tago glue interleaved between wood plies. Layers of fabric or other filler material are also cured in sheet or other forms, the curing times depending upon thermal conductivity and reaction time and temperatures required. Different materials may be bonded together, e.g., plywood, metal foil, and patterned plastics for table tops.

## EQUIPMENT FOR WORKING METALS AND PLASTICS

REFERENCES: "Thomas Register" (of machinery builders, etc.), Thomas Publishing Co., New York.

The **mass production** industries use an extremely wide variety of machines to force materials to flow plastically into desired shapes (as compared with the more gradual methods of obtaining shapes by cutting away surplus material in machine tools). The application of working pressure may use hydraulic, pneumatic, mechanical, or electrical means to apply pressing, hammering, or rolling forces. Mechanically and hydraulically actuated devices cover much the same range. In general, the mechanical equipment is faster, easier to maintain, and more efficient to operate by reason of energy-storing flywheels. The hydraulic equipment is more flexible and more easily adjusted to limited lots in pressure, positions, and strokes. Mechanical handling or feeding devices incorporated in or serving many of these more or less specialized machines further extend their productivity.

**Power presses** consist of a frame of substantial construction with devices for holding the dies or tools and a moving member or slide for actuating one portion of the dies. This slide usually receives its movement from a crankshaft furnished with a clutch for intermittent operation and a flywheel to supply the sudden power requirement. **Hydraulic presses** have no crankshaft, clutch, or flywheel but employ rams actuated by pumps.

The crankshaft is ordinarily the limiting factor in the pressure capacity of the machine and accordingly is often taken as the basis for tonnage ratings. There is no uniform basis for this rating, owing to variations in shaft proportions and materials and in the different relative severity of various press operations. The following valuation is tentative and is based on the shaft diameter in the main bearings. The bending strength is figured at a section through the center of the crankpin and the combined bending and torsional strength at the inside ends of the main bearings, taking the bending fulcrum at a distance out from these points equal to one-third the length of the main bearings. In the case of double-crank presses and twin-drive

arrangements, the relative proportion of the torsional load must be varied to suit, but, except in the cases of long strokes, it is usually small. The working strength is based upon a stress in the extreme fibers of 28,000 lb/in² (193 MN/m²). The limit bearing capacities are taken approximately at 5,000 lb/in² (35 MN/m²) on the crankpins and 2,500 lb/in² (18 MN/m²) average over the main bearings for ordinary steel on cast-iron press bearings with proper grooving. On the knuckle-joint-type presses with hardened tool-steel bearing surfaces and flood lubrication, the bearings will take up to about 30,000 lb/in² (207 MN/m²). On eccentric-type shafts where the main bearings support right up to the oversize pin on each side, the limiting factor is the bearing load. The shaft is practically in shear, so that it has a considerable overload capacity (about $7d^2$ tons). In Table 3, uniform-diameter single crankshafts are those in which for manufacturing reasons the diameter is the same at the crankpin and at the main bearings. Other crankshafts have an oversize crankpin to balance the bending load at the center with the combined bending and torsional load at the side. The strength of the shaft is figured at mid-stroke, and the stroke and tonnage capacity are given in terms of the diameter $d$ at the main bearings. Where the working load comes on only near the bottom stroke, the shaft press capacity may be figured as if the stroke were shorter in proportion.

Table 3 gives the rated capacities of a series of power presses as a function of the shaft diameter.

The speed of operation of the press depends upon the energy requirement and the crankpin velocity. The latter determines the velocity of impact on the tools. In blanking, the blow varies directly with the contact speed and the thickness and hardness of the material. On drawing operations the variation depends upon contact speed, ductility of material, lubricant, etc.

The energy required per stroke is practically the product of the average load and the working distance, plus friction allowance, assumed at about 16 percent. On short-stroke operations, such as blanking, the working energy is supplied almost entirely by slowing down the flywheel; motor or belt pull serves merely to return the flywheel to speed during the large part of the cycle in which no work is done. In drawing operations, the working period is considerable, and in many cases the belt takes the largest part of the working load. In this case, add to the available flywheel energy, the work done by the belt. This amounts, for example, to 70 lb/in of width of the belt (123 N/cm), multiplied by the ratio of the belt velocity to the crankpin velocity, multiplied by the length of the working stroke on the crank circle in feet. The maximum flywheel slowdown has been assumed as up to 10 percent for continuous operation and up to 20 percent for intermittent operation. The following formula is based upon average press-flywheel proportions and a slowdown of 10 percent. The result may be doubled for 20 percent slowdown.

The flywheel capacity per stroke at 10 percent slowdown in inch-tons equals $WD^2N^2/5,260,000,000$, where $W$ is the weight of the flywheel, $D$ is the diameter, in, and $N$ is r/min.

The difference between nongeared and geared presses is only in speed of operation and the relatively greater flywheel capacity.

**Press frames** are designed for stiffness and usually have a considerable excess strength. Good practice is to figure cast-iron sections on a basis of about 2,000 to 3,000 lb/in² (13.8 to 20.6 MN/m²). **C-frame presses** are subjected to an appreciable

Table 3. Power-Press Shaft Capacities*

| Type of press crankshaft | Max stroke, in† | | Capacity, tons‡ |
|---|---|---|---|
| Single crank, single drive, uniform diameter............................ | | $d$ | $2.8d^2$ |
| Single crank, single drive, oversize crankpin.......................... | | $d$ | $3.5d^2$ |
| Single crank, single drive, oversize crankpin.......................... | 2 | $d$ | $2.2d^2$ |
| Single crank, single drive, oversize crankpin.......................... | 3 | $d$ | $1.6d^2$ |
| Single crank, twin drive, oversize crankpin........................... | 2 | $d$ | $3.5d^2$ |
| Single crank, twin drive, oversize crankpin........................... | 3 | $d$ | $2.7d^2$ |
| Double crank, single drive, oversize crankpin......................... | | $0.75d$ | $5.5d^2$ |
| Double crank, single drive, oversize crankpin......................... | | $d$ | $4.4d^2$ |
| Double crank, single drive, oversize crankpin......................... | 2 | $d$ | $2.5d^2$ |
| Double crank, single drive, oversize crankpin......................... | 3 | $d$ | $1.7d^2$ |
| Double crank, twin drive, oversize crankpin.......................... | 1.5 | $d$ | $5.5d^2$ |
| Double crank, twin drive, oversize crankpin.......................... | 3 | $d$ | $3.2d^2$ |
| Single eccentric, single or twin drive................................ | 0.5 | $d$ | $4.3d^2$ |

*E. W. Bliss Co.
† 1 in = 2.54 cm.
‡ 1 ton = 8.9 kN.

arc spring amount ordinarily to between 0.0005 and 0.002 in/ton (1.5 to 6 mm/MN), because the center of gravity of the frame section is a considerable distance back of the working centerline of the press. **Straight-sided presses** eliminate that portion of the spring or deflection which is on an arc. **Built-up frame presses** are held together with steel tie rods shrunk in under an initial tension in excess of the working load so that they minimize stretch in that portion of the press.

**Power presses** are built in a very wide variety of styles and sizes with shafts ranging from 1 to 21 in (2.5 to 53 cm) diam. Over a large part of this range they are built with C frames for convenience, straight-sided frames for heavier and thinner work, eccentric shafts for heavy forgings and stampings, double crankshafts for wide jobs, four-point presses for large panel work, underdrive presses in high-production plants where repairs to presses would interfere with flow of production, and knuckle-joint presses for intensely high pressures at the very bottom of the stroke. All these are classified as single-action presses and are used for most of the operations previously discussed.

**Double-action presses** combine the functions of blank holding with drawing. In the smaller sizes, such presses have cams mounted on the cheeks of the crankshaft to actuate the outer or blank-holding slide. In larger machines, toggle mechanisms are provided to actuate the outer slide, with the advantage that the blank-holding load is taken on the frame instead of the crankshaft. Both of these types afford a considerable power saving over single-action presses equipped with drawing attachments, because the latter must add the blank-holding pressure to the working load for the full depth of the draw.

Types of presses include **foot presses,** in which the pendulum type has the lowest mechanical advantage and the longest stroke; the **lever type,** which has higher mechanical advantage and shorter strokes; **toggle** or **knuckle type,** which has the highest mechanical advantage and works through the shortest stroke with considerable advantage obtainable from the use of tie rods on fine stamping or embossing work; long-stroke rack and pinion-driven presses; triple-action drawing presses; cam-actuated presses; etc.

**Screw presses** consist of a conventional frame and a slide which is forced down by a steep pitch screw on the upper end of which is a flywheel or weight bar. Hand-operated machines

are used for die testing and for small production stamping, embossing, forming, and other work requiring more power than foot presses. Power-driven screw presses are built with a friction drive for the flywheel and automatic control to limit the stroke. Such presses are built in comparatively large sizes and used to a considerable extent for press forging. They lack the accuracy and speed of power presses built for this work but have a safety factor which power presses have not, in that their action is not positive. In this they closely resemble a drop hammer, although their motion is slower. The energy available for work in these presses is $\frac{1}{2}I_f v^2 + \frac{1}{2}I_s v^2$, in which $I_f$ is the moment of inertia of the flywheel, $I_s$ is the moment of inertia of the spindle, and $v$ is the angular velocity of both.

Self-contained fast-acting **hydraulic presses** are being increasingly used. Equipped with motor-driven variable-displacement oil hydraulic pumps, the speed and pressure of the operating ram or rams are under instant and automatic control; this is particularly advantageous for deep drawing operations. The punch can be brought into initial contact with the work without shock and moved with a uniform controlled velocity through the drawing portion of the cycle. The drawing of stainless steels and alloy aluminums (in which the control of drawing speed is vital), as well as the hot drawing of magnesium, is best done on hydraulic presses.

The hydraulic press is used in the rubber pad, or *Guerin*, process of blanking or forming metals, in which a laminated-rubber pad replaces one half—usually the female half—of a die. In forming aluminum the practice has developed of using inexpensive dies of soft metal, vulcanized fiber, plastic, wood, or plaster; and cast dies in industries which, like the aircraft industry, require short runs on many different sizes of shapes and parts.

The older accumulator type of hydraulic-press construction is still used for hot extrusion and some forging work.

**Drop Hammers or Presses** Small belt-lift and board-lift drop presses are used for variety of sheet-metal operations on hardware, cutlery, silverware, etc. Large-area rope-lift drop hammers are used for stamping metal ceilings and the like, usually having a hard die and a soft babbitt or steel punch. Board-lift drop hammers with heads weighing up to 5,000 lb (22 kN) are widely used in the production of steel drop forgings. The energy available for work is the product of the weight of the ram and the length of fall. The following table gives the shaft diameters of trimming presses (for removing the flash) ordinarily used with drop hammers for trimming the same range of forgings:

**Drop-hammer Ram Weights and Suitable Trimming-press Shaft Diameters\***

| Ram weight, lb† | 600 | 800 | 1,000 | 1,200 | 1,500 | 2,000 | 2,500 | 3,000 | 5,000 | 8,000 |
|---|---|---|---|---|---|---|---|---|---|---|
| Press shaft diam, in‡ | $3\frac{1}{2}$ | 4 | $4\frac{1}{2}$ | 5 | $5\frac{1}{2}$ | 6 | $6\frac{1}{2}$ | 7 | 8 | 9 |

\*E. W. Bliss Co.
†1,000 lb = 4,450 N.
‡1 in = 2.54 cm.

the blows. They are **used for general** and duplicate **forging,** welding, plating drawing, swaging, collaring, spindle making, etc. **Commercial sizes** are 15, 25, 40, 60, 80, 100, 200, 300, and 500 lb (66 to 2,200 N). The 200-lb (890-N) size requires approximately 2 hp (1.5 kW) to operate.

**Strap hammers** carry the hammer slung from a strap, usually of leather. The control and operation are the same as for the helve type. They are **adapted for general work,** tool dressing, and the like. **Commercial sizes** are 15, 30, 50, 75, 100, 125, 150, and 200 lb (66 to 890 N).

**Board Drop Hammer**    Let $W$ = work of blow, ft·lb; $H$ = weight of hammer and die, lb; $g$ = acceleration due to gravity [= 32.2 ft/(s)(s)]; $b$ – actual hammer stroke, ft; and $v$ = terminal velocity of hammer, ft/s. Then, if the hammer and die fall of their own weight, $W = (H/g)v^2/2 = Hb = 0.015\,Hv^2$ (approx). The **lifting power** $L = fF$ when the board is lifted by one roller (when both rollers are driven $L = 2fF$), where $f$ = coefficient of friction of roll and board and $F$ = force with which rolls are pressed together. Since $f = 0.25$ (approx), $L = 0.25F$ and $0.5F$ (approx), respectively.

Part of the lifting power must at first accelerate the motion of the hammer head. Consequently, $L = bH$, where $b$ is a constant which is always $> 1$.

The motion of the hammer is accelerated with the effort $(b - 1)H$ until the speed $v$ has been attained. The friction driving wheels or belt must consequently slip at first. Disregarding frictional resistances, **time** (in seconds) **required to lift hammer** = $t_1 = [0.016v/(b - 1)] + b/v$, and the **duration of drop** = $t_2 = 0.124\sqrt{b}$. Generally $v = 2.5$ to 3.5 ft/s; $b = 1.2$ to 2; $b$ = 3 to 6 ft; and $H = 100$ to 3,000 lb.

| Weight of hammer, lb\* | 250 | 400 | 600 | 800 | 1,000 | 1,500 |
|---|---|---|---|---|---|---|
| Approximate hp required† | 2 | 2.5 | 3 | 3.5 | 4 | 5 |

\*1,000 lb = 4,450 N.
†1 hp = 0.746 kN.

**Steam hammers** may be divided into **three classes.** In the first, the hammer is lifted by steam and drops of its own weight; in the second, steam is admitted above the piston and through its expansion increases the force of the blow; in the third, live steam is admitted above the piston throughout the stroke and the force of the blow is from combined weight of the falling hammer and the pressure of the steam.

In the **first class,** that of the single-acting steam hammer, the lifting power $L = bH$. That part of the force of the steam equal to $(b - 1)H$ is used for accelerating the lifting motion, continu-

**Helve hammers** (Bradley type) are usually belt-driven and carry the hammer face or swage on the end of a beam. The belt is provided with a tightening device, treadle-controlled, permitting the operator to regulate the number and speed of ing after the steam has been cut off as long as the pressure under the piston $> H$. These hammers are **used** only **for very large work.** The **weight of the hammer** ranges from 25 to 125 tons (220 to 1,100 kN), and the value of $b$ ranges from 1.5 to 1.2. The disadvantage of this type lies in the fact that the height of the clearance under the piston is directly dependent upon the thickness of the piece of work.

In the **second class,** in which steam is admitted above the piston and allowed to expand, there is an economy in steam consumption, a greater acceleration of the hammer head, and a large number of blows per unit of time. The reliability of operation is not of the best.

To overcome this defect in operation, the **third class** was developed, in which live steam is admitted above the piston. Here the **weight of the hammer head** varies from 1 to 25 tons (9 to 1,100 kN). The control is such that the weight of the hammer itself is available for light blows, and for heavier blows steam is admitted above the piston.

In comparatively small hammers where the head weighs 150 to 2,000 lb (670 to 8,900 N) $b = 2$ to 3.5, and the diameter of the piston rod equals 0.5 to 0.65 of the diameter of the piston. The area of the upper piston face is, consequently, 1.3 to 1.7 times that of the lower face. Up to 350 **blows per min** can be obtained, depending on the length of the stroke and the tightness of the stuffing box. These hammers are provided with an automatic reversing gear. The number and force of the blows can be regulated by throttling the steam, changing the center position of the operating valve, and changing the backlash in the operating gear. The frame of these hammers is usually C-shaped.

The **anvil** in small hammers is usually a single casting. In large hammers it is usually divided into upper and lower anvil blocks. In smaller hammers the anvil is connected with the shears and upper part of the machine. In the larger hammers, however, this is not the case, as the concussions tend to injure the hammer mechanism. For good practice the **weight of the anvil** (= $Q$) for hammers used in forging iron is at least eight times the weight of the hammer head; for forging steel, at least twelve times.

The pressure $Q_1$ exerted by the anvil block on the surface which it supports is assumed to be as follows: for blooming hammers, $Q_1 = (30$ to $60)bH + Q$; for billet-forging hammers, $Q_1 = (60$ to $95)bH + Q$; for hammers for steel forging, $Q_1 = (95$ to $125)bH + Q$.

**Commercial Sizes**    From mechanical arrangement, steam hammers are called steam drops and single-frame and double-frame steam hammers. Sizes of steam drops range from 400 to

| $c$ (in)* | 0.2 | 0.4 | 0.6 | 0.8 | 1.0 | 1.2 | 1.25 |
|---|---|---|---|---|---|---|---|
| $E$ (in·lb)† @ 0.005 ft/s‡ | 600 | 2,100 | 4,000 | 7,000 | 11,000 | 19,200 | 23,500 |
| $E$ (in·lb)† @ 0.0007 ft/s‡ | 400 | 1,600 | 3,000 | 5,000 | 8,000 | 14,800 | 18,000 |
| $E$ (in·lb)† @ static pressure (Purdue) | 400 | 1,400 | 2,600 | 4,100 | 6,600 | 12,000 | 15,000 |

*1 in = 2.54 cm.
†1 in·lb = 0.06 J.
‡1 ft/s = 30.5 cm/s.

30,000 lb (1.8 to 134kN). Sizes of single- and double-frame hammers range from 250 to 4,000 lb (1.1 to 36 kN), and 1,500 to 30,000 lb (6.7 to 267 kN), respectively. A series of small double-frame steam hammers, known as *tilting hammers*, is also made. They are used chiefly by forges producing the smaller sizes of steel bars, as squares and octagons of tool steel. Uniformity in force of blow and length of stroke are the important features. Commercial sizes range from 500 to 2,500 lb (2.2 to 11 kN).

**Pneumatic Hammers**  A self-contained type of pneumatic forge hammer (the Bêché) has an air-operated ram with an air-compressing cylinder integral with the frame. The ram is raised by admitting compressed air beneath the ram piston; at the same time a partial vacuum is caused above it. The ram is forced down by a reversal of this action.

This type is made in a number of different models by the Nazel Engineering and Machine Works, Philadelphia. Tests by the maker give the force of the blows for the various sizes and the capacity in maximum sizes of stock, as follows:

| Size, lb* | 66 | 165 | 250 | 350 | 500 | 770 |
|---|---|---|---|---|---|---|
| Blow, ft·lb† | 268 | 948 | 1,338 | 2,351 | 4,116 | 6,452 |
| Max diam of stock, forged, in‡ | 2 | 4 | 5 | 7 | 8 | 9 |

*1,000 lb = 4,450 N.
†1 ft·lb = 0.737 J.
‡1 in = 2.54 cm.

**Energy of Hammer Blows**  Where the striking velocity of a hammer ram is greater than 10 ft/s, the energy $E$ of the blow (in in·lb) may be determined from the compression of lead cylinders. From experiments made by W. T. Sears, of the Niles-Bement-Pond Co. (*Am. Mach.*, Mar. 10, 1910), using $1\frac{1}{2} \times 1\frac{1}{2}$ in cylinders, the following results were obtained ($c$ = compression or shortening of cylinder, in):

| $c$, in* | 0.2 | 0.4 | 0.6 | 0.8 | 1.0 | 1.2 | 1.3 |
|---|---|---|---|---|---|---|---|
| $E$, in·lb† | 900 | 3,000 | 6,000 | 10,500 | 18,000 | 34,000 | 54,000 |

*1 in = 2.54 cm.
†1 in·lb = 0.06 J.

For any value of $c$, close results may be obtained from the formula $E = (10,800c - 870)/(1.55 - c)$. Where the speed of compression is slow, as in **presses**, it is necessary to know the speed to estimate the energy. Sears gives values at two rates of speed and also quotes static-pressure results obtained at Purdue University (see table above).

**Rotary motion** is used **for working sheet metal** in a variety of machines, including bending rolls (three rolls); rolling straighteners with five, seven, or more rolls; roll forming machines, in which a series of rolls in successive pairs are used to bend the strip material step by step to some desired shape; a series of two-spindle and multiple-spindle machines used for rolling beads, threads, knurls, flanges, and trimming or curling the edges of drawn shells of cylinders; seaming machines for double seaming, crimping, curling, and other operations in the production of tin cans, pieced tinware, etc.; and spinning machines for spinning, burnishing, trimming, curling, shape forming, and thickness reduction. Various production spinning operations and tool arrangements are shown in Fig. 7.

**Plate-straightening Machines**  The **horsepower required** for plate-straightening machines operating on steel plate is shown in the table at the bottom of this page.

Power required for angle-iron-straightening machines: for 4-in angles, 12 hp; for 6-in, 18 hp; for 8-in, 25 hp.

Power or hydraulic presses are used to straighten **large rolled sections.** The presses make 20 to 30 strokes per min, and the amount of flexure is regulated by inserting wedges or pieces of flat iron. The beams are supported on rolls so they can be easily handled. The **power required for presses** of this kind is as follows:

| Depth of girder, in* | 4 | 6 | 8 | 10 | 12 | 16 | 20 | 24 |
|---|---|---|---|---|---|---|---|---|
| Horsepower (approx)† | 3 | 4 | 7 | 11 | 13 | 19 | 23 | 35 |

*1 in = 2.54 cm.
†1 hp = 0.746 kW.

**Horizontal plate-bending machines** consist of two stationary rolls and a third vertical adjustable upper roll which can be fitted obliquely for taper bending and is held in bearings with spherical seats. The diameter of the rolls can be determined approximately from the equation $r^2 = bt$, in which $r$ is the radius of the roll, $b$ the width of the plate or sheet, and $t$ its thickness, all in inches.

| Thickness of plate, in* | 0.25 | 0.4 | 0.6 | 0.8 | 1.0 | 1.2 | 1.4 | 1.6 |
|---|---|---|---|---|---|---|---|---|
| Width of plate, in* | 48.0 | 52.0 | 60.0 | 72.0 | 88.0 | 102.0 | 120.0 | 140.0 |
| Diameter of rolls, in* | 5.0 | 8.0 | 10.0 | 12.0 | 13.0 | 14.0 | 15.0 | 16.0 |
| Horsepower (approx)† | 6.0 | 8.0 | 12.0 | 20.0 | 30.0 | 55.0 | 90.0 | 130.0 |

*1 in = 2.54 cm.
†1 hp = 0.746 kW.

| Thickness of plate, in* | 0.5 | 0.6 | 0.8 | 1 | 1.2 |
|---|---|---|---|---|---|
| Horsepower for plate 120 in wide† | 10.0 | 12.0 | 18.0 | 27.0 | 40.0 |
| Horsepower for plate 240 in wide† | 30.0 | 30.0 | 40.0 | 55.0 | 75.0 |

*1 in = 2.54 cm.
†1 hp = 0.746 kW.

The **power requirements** of horizontal plate- and sheet-bending machines are shown in the table above.

**Vertical plate-bending machines** have a hydraulically operated piston which moves an upper and a lower pair of rolls between inclined surfaces of the stationary upright and the crosshead. The bending is done piece by piece against a second stationary upright. Heavy ship plates are rigidly clamped down and bent by a roll operated by two hydraulic pistons. For angular bends or for the production of warped surfaces, the pistons can be operated independently or together. In vertical machines, angles and other rolled shapes are bent between suitably shaped rolls. Pipes are filled with sand to prevent flattening when being bent. For some work, pipes are bent hot between suitable forms operated by hydraulic pressure.

The **rotary swaging machine** for tapering, closing in, and reducing tubes, rods, and hollow articles is essentially a cage carrying a number of rollers and revolving at high speed; e.g., 14 rolls in a cage revolving at 600 r/min will strike 8,400 blows per min on the work.

A rapid succession of light blows is applied to a considerable variety of commercial **riveting** operations such as pneumatic riveting. Another method of riveting, described as spinning, involves rotating small rollers rapidly over the top of the rivet and at the same time applying pressure. Neither of these methods involves as intense pressures as are used in riveting by direct pressure either hot or cold. Power presses and C-frame riveters, employing hydraulic pressure or air pressure of 80 to 100 lb/in² (550 to 690 kN/m²), are figured to apply 150,000 lb/in² (1,035 MN/m²) of the cross section of the body of the rivet for hot-working and 300,000 lb/in² (2,070 MN/m²) for cold-working. The plates should be pressed together by a pressure of 0.3 to 0.4 times that used in riveting.

# WELDING

## by Edward A. Fenton

REFERENCES: "Welding Handbook" (6 vols.), AWS. "Recommended Practices for Resistance Welding," AWS. "Resistance Welding Manual," Resistance Welder Manufacturers' Assoc. "Resistance Welding—Theory and Use," AWS. Grover, "Manual of Design for Arc Welded Steel Structures," Air Reduction Co. "Procedure Handbook of Arc Welding Design and Practice," The Lincoln Electric Co. "The Oxyacetylene Handbook," The Linde Div. Union Carbide Corp. "Brazing Manual," AWS. "Safety in Welding and Cutting," ANSI. Slottman and Roper, "Oxygen Cutting," McGraw-Hill.

### FUNDAMENTALS OF WELDING

A weld is defined as a localized coalescence of metal wherein coalescence is produced by heating to suitable temperatures, with or without the application of pressure, and with or without the use of filler metal. The filler metal may have a melting point approximately the same as the base metals (as in arc or gas welding), or it may have a lower melting point but above 800°F (427°C) (as in brazing). This definition distinguishes welding from mechanical joining and adhesive bonding. The 800°F limit distinguishes brazing from soldering, which is excluded. The definition includes some 34 different processes falling under six general categories: arc welding, gas welding, resistance welding, brazing, solid-state welding, and "other" processes. Of these, the first four are of wide industrial significance.

Basic **types of arc and gas welds** are shown in cross section in Fig. 1, top; at bottom are views of joints commonly used in plate and sheet fabrication. Figure 2 gives typical **dimensions** for plate-edge preparation for making butt joints with six types of groove welds deposited by the manual shielded metal-arc process. This process is still the most widely used process in the industry.

The basis for weld-groove design is to provide a shape and size of opening that will enable a sound deposition of filler metal, under given conditions, with maximum economy. For other welding processes that employ filler metals, groove dimensions may vary from those shown in Fig. 2. For instance, in submerged arc welding, greater power input is available and deeper penetration is possible. Therefore, smaller weld grooves are required.

To establish uniform and simple drafting practice, welding symbols have been standardized by AWS and adopted by ANSI. Figure 3 shows several basic symbols.

### DESIGN STRENGTHS

The AWS Structural Welding Code (D1.1-72) permits, in general, the same values of stress for butt welds as are applicable to the base metal, except that for fatigue loading, special formulas are provided. Use of these weld stresses is predicated

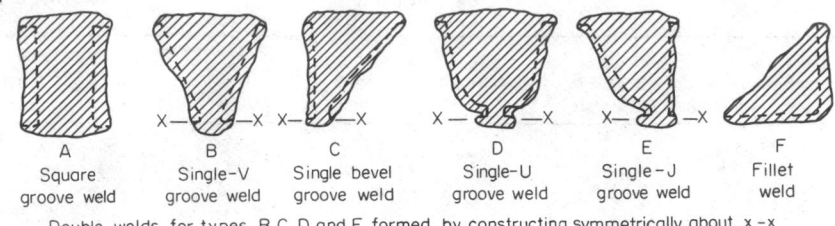

A
Square
groove weld

B
Single-V
groove weld

C
Single bevel
groove weld

D
Single-U
groove weld

E
Single-J
groove weld

F
Fillet
weld

Double welds for types B, C, D and E formed by constructing symmetrically about x-x

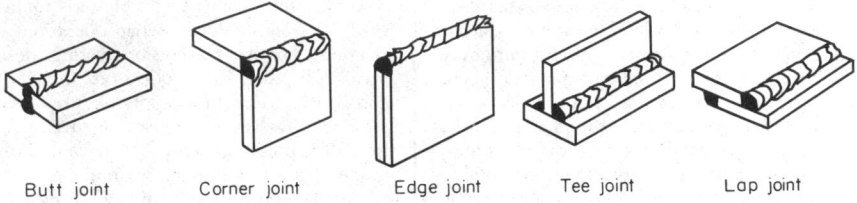

Butt joint        Corner joint        Edge joint        Tee joint        Lap joint

**Fig. 1**  Some types of welds (upper) and joints (lower).

upon the employment of base metals and filler metals prescribed in the Code. Stress on fillet welds is considered as shear on the throat of the weld and limited to 16,500 to 30,000 lb/in² (114 to 207 MPa) for bridges, depending on filler metal and base metal. In building construction, such shear stresses are similarly limited to 18,000 to 33,000 lb/in² (124 to 228 MPa) for buildings. For both bridges and buildings, provisions are based on the use of ASTM steels (about 15 grades are listed). The bridge specification further provides extensive formulas based upon the work of the Welding Research Council, for computing the required amounts of welding and the required areas of members connected by fillet welding, when subjected to various conditions of fatigue loading. In most respects the specifications of the AISC are in substantial agreement with the AWS Structural Welding Code.

In operation, this AWS Code establishes certain welding

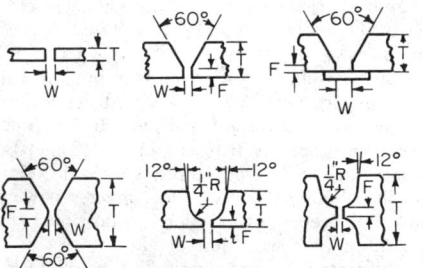

**Fig. 2**  Weld grooves for manual shielded metal-arc welding. Dimensions in inches are shown in the table immediately below.

procedures and joints as acceptable standards, and if used, no further qualification is necessary. Typical butt joints in this category include single- and double-V and U grooves, single- and double-bevel and J grooves, all of controlled dimensions and welded in such a manner (root chipping or gouging required if not welded on backing material) as to ensure fusion of the weld and base metal throughout the entire depth of the joint. In a similar manner controlled types of fillet welds may be used without special qualification.

## ARC WELDING

Arc welding is defined as a group of welding processes wherein coalescence is produced by heating with an electric arc or arcs, with or without the application of pressure and with or without the use of filler metal. The most significant processes are shielded metal-arc welding, submerged arc welding, gas tungsten-arc welding, gas metal-arc welding, and arcspot welding. Other processes exist; some have several variations. In all these cases, an arc is maintained between an electrode and the work (or between two electrodes), which form the terminals of an electric circuit. Whether direct or alternating current is employed may depend on the process, the filler metal, the type of shielding, the base metal, or other factors. Welding processes may be manual, semiautomatic (partly mechanized), or automatic (fully mechanized).

**Welding Arc**  More than a heat source, the arc is a complex mixture of ionized gas particles accelerated through an electric field constricted by its magnetic field, and exerting a profound effect on the transfer of filler metal. In turn, arc behavior is dependent on filler metal, base metal, type of shielding, circuit characteristics, and other factors.

| Groove | 1 | 2 | 3 | 4 | 5 | 6 |
|---|---|---|---|---|---|---|
| T | $\frac{1}{8} - \frac{3}{16}$ | $\frac{1}{4}$ up | $\frac{1}{4}$ up | $\frac{1}{2}$ up | $\frac{3}{8}$ | $\frac{3}{4}$ up |
| W | $0 - \frac{1}{16}$ | $0 - \frac{3}{16}$ | $\frac{3}{16} - \frac{1}{4}$ | $0 - \frac{3}{16}$ | $0 - \frac{1}{16}$ | $0 - \frac{1}{16}$ |
| F | .......... | $0 - \frac{1}{16}$ | $0 - \frac{1}{16}$ | $0 - \frac{1}{16}$ | $\frac{1}{8} \pm \frac{1}{32}$ | $\frac{1}{8} \pm \frac{1}{32}$ |

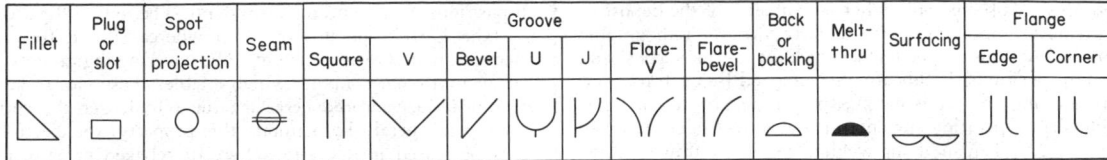

| Fillet | Plug or slot | Spot or projection | Seam | Groove | | | | | | | Back or backing | Melt-thru | Surfacing | Flange | |
|---|---|---|---|---|---|---|---|---|---|---|---|---|---|---|---|
| | | | | Square | V | Bevel | U | J | Flare-V | Flare-bevel | | | | Edge | Corner |

**Fig. 3** Basic arc- and gas-weld symbols. *(WES.)*

In a bare-electrode dc arc, more heat is liberated at the positive terminal (anode) than at the negative terminal (cathode); in an ac arc, the heat is about the same at both terminals. Electrode polarity affects not only heat input but weld penetration, fluidity, and metal transfer. Through their action in the arc, shielding fluxes and gases enhance, retard, or control these effects. In dc arc welding, the term **straight polarity** signifies that the electrode is the negative terminal; **reverse polarity,** that it is the positive terminal. The other terminal, in each case, is connected to the work.

There is a direct relation between the length of the arc and the arc voltage—the longer the arc, the higher the voltage. The precise relationship depends on arc conditions, i.e., bare-metal arcs, arcs with covered electrodes, submerged arcs, or arcs in gaseous atmospheres. Under each of these conditions there is a range of arc lengths to achieve optimum welding conditions.

**Arc-welding machines** may be motor-generator sets, rectifier sets, transformers, or generators driven by any of the common prime movers. Motor-generator sets consist of a dc welding generator connected to an ac or dc motor. The rectifier sets employ a single- or three-phase step-down transformer, the output of which is fed through rectifiers to produce direct current. Transformer welding machines are single-phase step-down transformers which supply ac welding power. Motor-generator, rectifier, and transformer units are used in shops and field locations where power lines are available. Engine-driven machines are used on field work where power is not available.

**Shielding** serves to exclude oxygen and nitrogen of the air from the arc and metal, thus eliminating the formation of oxides and nitrides, which decrease weld-metal ductility and, sometimes, strength. Shielding may be accomplished by (1) a flux covering applied (usually extruded) on a core wire, (2) powdered flux heaped over the weld area and the terminal end of the electrode, (3) a flow of inert or active gases, or mixtures of gases, projected around the arc and molten metal.

**Shielded metal-arc welding** employs covered electrodes and can be performed with ac or dc power sources. The most widely used of all processes, it finds application on mild and alloy steels, stainless steels, and, to a lesser extent, nonferrous metals. The electrode is clamped in an electrode holder which has a cable leading to the power source. The work is electri-

cally grounded. The electrode tip is touched to the work to establish the circuit and then retracted slightly, initiating the arc. In general, with mild- and low-alloy-steel electrodes, welding in the flat and horizontal positions may be performed with electrodes 5/16 in (4.8 mm) in diameter and smaller. In the vertical and overhead positions, electrode diameters of 3/16 in or less are usual. Arc voltage ranges from 20 to 40 V approx; current from 20 to 500 A approx, depending mainly on size of electrode.

**Electrode coverings** serve several purposes: (1) to facilitate the establishment and maintenance of the arc; (2) to protect the molten metal from the air; (3) to provide fluxing of the molten metals, particularly with the nonferrous metals; (4) to provide a means of introducing alloying ingredients not contained in the core wires. A series of standard specifications covering these electrodes has been issued by AWS.

The Specification for Mild Steel Covered Arc-welding Electrodes (AWS A5.1-69) provides 12 classifications based on chemical requirements, mechanical properties, types of electrode covering, usability, and soundness. Usability is taken as the capability of an electrode to pass the fillet-weld test when used in the positions and with the type of current for which it is intended. Soundness is based on comparison of weld-metal radiographs with standard porosity charts. Type of covering and its condition when used are the fundamental factors determining usability and soundness. The 12 classifications are subdivided into two groups: the E60XX series and the E70XX series. In a classification designation such as E6010, "E" designates an electrode, the first two digits designate the minimum tensile strength in ksi of the deposited metal in the as-welded condition, the third digit indicates the position in which the electrode is capable of making satisfactory welds, and the last digit designates the type of covering on the electrode and the type of current with which it is to be used. The main differences between the E60XX and the E70XX series are the mechanical properties shown in Table 1 and the absence of chemical requirements for the E60XX series. In many applications, electrodes of either series may be used.

**E6010** and **E6011** electrodes are for all-position welding in sizes up to 3/16 in and for flat and horizontal positions in larger sizes. E6010 is used with direct current, reverse polarity (dcrp); E6011 is designed for alternating current but may be used with dcrp to lesser advantage. Other than this, the

### Table 1. Typical Mechanical Properties of Mild Steel Electrodes*

| | E6010 E6011 | E6012 E6013 | E6020 E6027 | E7014 E7024 | E7015, E7016 E7018, E7028 |
|---|---|---|---|---|---|
| Tensile strength, min psi | 62,000 | 67,000 | 62,000 | 72,000 | 72,000 |
| Yield point, min psi | 50,000 | 55,000 | 50,000 | 60,000 | 60,000 |
| Elongation in 2 in., min % | 22 | 17 | 25 | 17 | 22 |

*All-weld-metal tension test in the as-welded condition.

electrodes are used equally where the quality of the deposit is of greatest importance, particularly where radiographic requirements must be met when welding in the vertical and overhead positions. Welds are characterized by deep penetration resulting from the harsh, spray-type arc and a light, friable slag. Operating currents for these electrodes are given in Table 2. When used for welding vertically upward, currents near the lower limit are generally used. With the larger sizes of electrodes, maximum currents are somewhat limited as compared with other classifications because of spatter loss that occurs with high currents.

**E6012** electrodes are designed for dc use but are usable with alternating current. In sizes up to $\frac{3}{16}$ in, they are all-position electrodes; however, they are used mostly in the flat and horizontal positions. They are especially recommended for single-pass, high-speed, high-current, horizontal fillet welds. Ease of handling, good fillet-weld profile, and ability to withstand high current and to bridge gaps resulting from poor fit-up make them well suited to this type of work. Medium penetration, quiet arc, slight spatter, and dense slag are characteristics of E6012. Single-pass welds may meet radiographic standards; multipass welds do not.

**E6013** electrodes, although similar to E6012, have several differences. Used with alternating or direct current, either polarity, the arc is softer, the penetration less, the slag removal easier, and the arc more easily started and maintained, especially with the smaller diameters. Originally designed for sheet-metal work with lower open-circuit voltage, the E6013 electrodes, especially in larger sizes, are used in many cases instead of the E6012. Fillet welds are flatter; radiographic quality is better.

**E6020** electrodes operate with alternating or with direct current, straight polarity (dcsp). They are designed to produce flat or slightly concave horizontal fillet welds and will produce satisfactory fillet and groove welds in the flat position. Penetration is medium when normal welding currents and techniques are used. However, this classification is considered to be the best for operating at high currents to obtain deep-penetration fillet welds. High deposition rates are obtained at the higher currents in heavy plate. The heavy, easily removed slag is honeycombed on the underside. The welds meet radiographic standards.

**E6027** electrodes are the heavily covered, **iron-powder** electrodes designed for fillet or groove welds in the flat position with alternating or direct current, either polarity. They will produce flat or slightly concave horizontal fillet welds equally with alternating current or dcsp. High currents with spray-type metal transfer produce high deposition rates with medium penetration and low spatter loss. The heavy, honeycombed slag is easily removed. Weld metal varies in radiographic quality and tends to be inferior to that from E6020 electrodes.

**E7014** electrodes are used with alternating or direct current, either polarity, and are designed for all-position welding in sizes up to $\frac{5}{32}$ in (4.0 mm) diam. Operating characteristics of this iron-powder electrode are a compromise between the E6013 and the E7024 types. Although deposition rate falls midway between the two, penetration is the same as with E6012. This is an advantage when welding over poor fit-up. Fillet welds are flat to slightly convex, and the slag is often self-cleaning.

**E7015** and **E7016** are the low-hydrogen electrodes usable in

all positions in sizes up to $\frac{5}{32}$-in diam. The only difference is that the E7015 operates on dcrp whereas the E7016 also operates on alternating current. The covering ingredients are low in hydrogen content, which enables these electrodes to avoid or reduce **underbead cracking** due to hydrogen absorption in the weld metal. To maintain this property, the electrodes must be stored in dry atmosphere or rebaked prior to use. Underbead cracking does not occur in mild steels. These electrodes were originally developed for use on high-carbon, alloy, and high sulfur steels and have been found useful on malleable iron, spring steels, and the mild-steel side of clad plates. They are also useful in making small welds on heavy plate since they are less susceptible to cracking. For best results, the arc should be maintained as short as possible.

**E7018** electrodes are low-hydrogen with high percentages of **iron powder** in the covering. Usable with alternating current and dcrp, they are designed for the same applications as the E7015 and E7016 electrodes. They are also well suited for fillet welds in high-carbon or alloy steels. Operation is characterized by a smooth, quiet arc, very low spatter, low penetration, and high linear speeds.

**E7024** electrodes are used with alternating or direct current, either polarity, and are well suited for making fillet welds in mild steel. The **iron-powder** covering amounts to about 50 percent of the total weight of the electrode. The welds are slightly convex in profile, with a very smooth surface. High linear speeds are possible, with low penetration, very low spatter, and a smooth, quiet arc.

**E7028** electrodes are usable in the horizontal-fillet and flat positions. The low-hydrogen **iron-powder** coverings represent about 50 percent of the weight of the electrode. Deposition rate with spray-type transfer is higher than for the E7018 electrodes with the globular-type transfer. E7028 also produces more weld metal per unit weight of core wire consumed. Except for these differences, the characteristics of the E7028 electrodes are the same as for the E7018 classification.

Specifications for Low-alloy Steel Arc-welding Electrodes (AWS A5.5-69) provide for six groups of classifications, at strength levels ranging from 70,000 to 120,000 lb/in² (483 to 830 MPa), in increments of 10,000 lb/in² (69 MPa). The system of classification follows that described for mild-steel electrodes above: the first two digits indicate the tensile strength in ksi, 70, 80, etc.; the last two digits indicate the usability and have the same significance as for the mild-steel electrodes. Electrodes may be further identified by supplementary classification to show the chemistry of the deposited metal. This is accomplished by means of a letter and digits following the normal classification. Thus E7010-A1 signifies a cellulose-type coating, usable with direct current, reverse polarity only, depositing weld metal, in all positions and of the nominal 0.50 molybdenum (carbon-molybdenum) analysis.

Specifications for Corrosion-resisting Chromium and Chromium-Nickel Steel Covered Welding Electrodes (AWS 5.4-69) provide for the classification of stainless and straight chrome-steel electrodes on the basis of AISI type numbers. Usability is covered by two terminal digits, 15 and 16, which have the same significance as in the mild- and low-alloy-steel-electrode specifications for the low-hydrogen electrodes. The classification E308-15 means E, electrode; 308, the AISI type number for 19 Cr-9 Ni stainless steel; and 15 indicates the electrode usable on direct current, reverse polarity only; terminal digits 16 indicate usability on alternating or direct

**Table 2. Typical Current Ranges in Amperes for Mild steel Electrodes**

| Electrode diameter, in. | E6010 and E6011 | E6012 | E6013 | E6020 | E6027 | E7014 | E7015 and E7016 | E7018 | E7024 and E7028 |
|---|---|---|---|---|---|---|---|---|---|
| 1/16 | . . . . . . . | 20 to 40 | 20 to 40 | . . . . . . . | . . . . . . . | | | | |
| 5/64 | . . . . . . . | 25 to 60 | 25 to 60 | . . . . . . . | . . . . . . . | | | | |
| 3/32 | 40 to 80 | 35 to 85 | 45 to 90 | . . . . . . . | . . . . . . . | 80 to 125 | 65 to 110 | 70 to 100 | 100 to 145* |
| 1/8 | 75 to 125 | 80 to 140 | 80 to 130 | 100 to 150 | 125 to 185 | 110 to 160 | 100 to 150 | 115 to 165 | 140 to 190 |
| 5/32 | 110 to 170 | 110 to 190 | 105 to 180 | 130 to 190 | 160 to 240 | 150 to 210 | 140 to 200 | 150 to 220 | 180 to 250 |
| 3/16 | 140 to 215 | 140 to 240 | 150 to 230 | 175 to 250 | 210 to 300 | 200 to 275 | 180 to 255 | 200 to 275 | 230 to 305 |
| 7/32 | 170 to 250 | 200 to 320 | 210 to 300 | 225 to 310 | 250 to 350 | 260 to 340 | 240 to 320 | 260 to 340 | 275 to 365 |
| 1/4 | 210 to 320 | 250 to 400 | 250 to 350 | 275 to 375 | 300 to 420 | 330 to 415 | 300 to 390 | 315 to 400 | 335 to 430 |
| 5/16 | 275 to 425 | 300 to 500 | 320 to 430 | 340 to 450 | 375 to 475 | 390 to 500 | 375 to 475 | 375 to 470 | 400 to 525* |

*These values do not apply to the E7028 classification.

current, reverse polarity. The following AISI types are covered as to chemistry and mechanical properties of deposited metal: 308, 308L, 309, 309Cb, 309Mo, 310, 310Cb, 310Mo, 312, 16-8-2, 316, 316L, 317, 318, 330, 347, 349, 410, 430, 502, 505, and 7Cr.

Specifications for Copper and Copper-alloy Arc-welding Electrodes (AWS A5.6-69) classify electrodes according to principal alloying ingredients to identify the composition. Where further classification under a given alloy group is necessary, supplementary letters and numerals are employed to differentiate the several analyses. The classification ECuAl-A2 indicates E, electrode; CuAl, a copper-aluminum alloy; A2, a specific composition, that is, 9-11 percent Al, 1.5 percent Fe. The basic groups covered are copper, copper-tin (phosphor bronze), copper-nickel, copper-silicon, and copper-aluminum. The requirements cover chemical composition and mechanical properties, together with usability.

Specifications for Nickel and Nickel-base Alloy Covered Welding Electrodes (AWS A5.11-69) employ chemical composition as the basis of classification. The system of classification uses a series of letters and numbers preceded by the usual letter E to indicate electrodes.

The electrodes are classified on the basis of chemical composition; the chemical symbol Ni is used to identify the electrodes as nickel-base alloys, and the additional chemical symbols (that is, Cr, Cu, Fe, and Mo) in the designations indicate the principal alloying elements in each classification or series. The individual classifications within a series are identified by the number at the end of the designation (that is, ENiCu-1, ENiCu-2, etc.). The chemical analysis serves to identify and control the electrode through analysis of weld deposit.

Specifications for Aluminum and Aluminum-alloy Arc-welding Electrodes (AWS A5.3-69) provide for two classifications of covered aluminum electrodes. Classification Al-2 covers the commercially pure grade of aluminum, and Al-43 covers the 5 percent silicon-aluminum alloy.

**Submerged arc welding** employs the heat of an arc between a mechanically fed bare-metal electrode and the work. The weld and arc are shielded by a blanket of powdered material called a *flux*. The arc therefore is "submerged," i.e., not visible. Welding power may be alternating or direct current, either polarity, and current densities are high. Welding currents range from 250 A for $\frac{3}{32}$ in (2.4 mm) wire to 2,000 A for $\frac{3}{8}$ in (9.5 mm) wire. As a result of these high currents, a submerged arc weld is characterized by deep melting of the base metal and high welding speeds. Weld quality is high, readily meeting ASME Boiler and Pressure Vessel Code and Piping Code requirements. Submerged arc welding is used for welding mild and low-alloy steels, stainless steels, copper, and nickel, and alloys of these metals. The process is automatic, and either the work or the arc may be moved. Welding is performed in the flat and horizontal positions. A manual version retains all the essential features but provides for manually guiding the arc. Typical welding speeds and other production data are given in Table 3.

**Gas tungsten-arc welding** involves an arc between a single tungsten electrode and the work to be welded. A shield of monatomic inert gas, argon, helium, or mixtures of these gases is projected around the electrode. Welding may be performed with or without filler metal. When filler metal is used it is introduced separately into the arc: manually from a rod, as in gas welding; mechanically from a power-driven wire-feed reel.

## Table 3. Production Data—Submerged Arc Welding

| Thickness | Weld or joint type | Root opening, in | Root face, in | Groove angle, deg | Current, amp | Arc, volts | Speed, ipm | Electrode Diam, in | Electrode Lb per ft of weld |
|---|---|---|---|---|---|---|---|---|---|
| 16 gage | Square groove | 0-$\frac{1}{32}$ | .... | ..... | 250 | 22 | 100 | $\frac{3}{32}$ | 0.015 |
|  |  |  |  |  | 350 | 24 | 150 | $\frac{1}{8}$ | 0.020 |
| 14 gage | Square groove | 0-$\frac{1}{32}$ | .... | ..... | 325 | 24 | 100 | $\frac{3}{32}$ | 0.020 |
|  |  |  |  |  | 400 | 26 | 120 | $\frac{1}{8}$ | 0.025 |
| 12 gage | Square groove | 0-$\frac{1}{32}$ | .... | ..... | 350 | 24 | 75 | $\frac{1}{8}$ | 0.027 |
|  |  |  |  |  | 500 | 30 | 120 |  |  |
| 10 gage | Square groove | 0-$\frac{1}{16}$ | .... | ..... | 575 | 24 | 60 | $\frac{1}{8}$ | 0.05 |
|  |  |  |  |  | 650 | 31 | 100 |  |  |
| $\frac{3}{16}$ in. | Square groove | 0-$\frac{1}{16}$ | .... | ..... | 575 | 25 | 40 | $\frac{5}{32}$ | 0.07 |
|  |  |  |  |  | 700 | 31 | 65 |  |  |
| $\frac{1}{4}$ in. | Square groove | 0-$\frac{3}{32}$ | .... | ..... | 750 | 25 | 30 | $\frac{3}{16}$ | 0.10 |
|  |  |  |  |  | 850 | 35 | 40 | $\frac{7}{32}$ |  |
| $\frac{5}{16}$ in. | Square groove | 0-$\frac{3}{32}$ | .... | ..... | 800 | 26 | 26 | $\frac{3}{16}$ | 0.25 |
|  |  |  |  |  | 900 | 36 | 30 | $\frac{7}{32}$ |  |
| $\frac{1}{4}$ in. | Single vee | 0 | 0-$\frac{1}{8}$ | 50-60 | 625 | 25 | 28 | $\frac{5}{32}$ | 0.10 |
|  |  |  |  |  | 825 | 32 | 50 | $\frac{3}{16}$ | 0.23 |
| $\frac{3}{8}$ in. | Single vee | 0 | 0-$\frac{1}{8}$ | 40-60 | 900 | 28 | 24 | $\frac{1}{4}$ | 0.16 |
|  |  |  |  |  | 1,100 | 36 | 47 |  | 0.35 |
| $\frac{1}{2}$ in. | Single vee | 0 | $\frac{3}{16}$ | 60 | 1,075 | 30 | 19 | $\frac{1}{4}$ | 0.38 |
|  |  |  |  |  | 1,175 | 37 | 23 |  | 0.45 |
| $\frac{3}{4}$ in. | Single vee | 0 | $\frac{3}{16}$ | 45 | 1,200 | 32 | 12 | $\frac{1}{4}$ | 0.60 |
|  |  |  |  |  | 1,300 | 39 | 14 |  | 0.75 |
| 1 in. | Single vee | 0 | $\frac{1}{8}$ | 35 | 1,500 | 35 | 10 | $\frac{5}{16}$ | 0.90 |
|  |  |  |  |  | 1,600 | 41 | 12 |  | 1.10 |
| 1$\frac{1}{4}$ in. | Single vee | 0 | $\frac{1}{8}$ | 30 | 1,600 | 37 | 8 | $\frac{5}{16}$ | 1.25 |
|  |  |  |  |  | 1,700 | 41 | 9 |  | 1.60 |
| 1$\frac{1}{2}$ in. | Single vee | 0 | $\frac{1}{8}$ | 30 | 1,900 | 38 | 7 | $\frac{3}{8}$ | 1.85 |
|  |  |  |  |  | 2,000 | 43 | 8 |  | 2.00 |

The inert-gas envelope permits welding such metals as aluminum, magnesium, nickel alloys, and stainless steels without flux. Welding power sources may be either alternating or direct current, either polarity. Selection of power sources depends upon metal being welded, whether manual or automatic welding, and desired results. With ac power sources, especially on aluminum and magnesium, it is necssary to stabilize the arc by using either a high-frequency, high-voltage pilot circuit superimposed on the welding circuit, or high open-circuit voltage.

**Semiautomatic gas tungsten-arc welding** involves either a fixed tungsten electrode holder with work moved under it, or else the holder is mounted on a motorized torch carriage and traverses the joint; control of arc length is manual. Automatic gas tungsten-arc welding may be employed in a similar manner, except that arc voltage, and hence arc length, is electronically controlled. Typical production data and welding conditions are provided in Table 4.

Filler metals, in the form of welding rods, for gas tungsten-arc welding are covered by a series of five AWS specifications. The classification numbers start out with the first letter R, which indicates welding rods. The combination of numbers and letters following the R identifies the metal composition by means of chemical symbols or other standard identification, such as AISI type numbers for stainless steels. The following are the applicable specifications: Specifications for Copper and Copper-alloy Welding Rods (AWS A5.7); Specifications for Corrosion-resisting Chromium and Chromium-Nickel Steel Welding Rods and Bare Electrodes (AWS A5.9); Specifications for Aluminum and Aluminum-alloy Welding Rods and Bare Electrodes (AWS A5.10); Specifications for Nickel and Nickel-base-alloy Bare Welding Filler Metals (AWS A5.14).

**Tungsten electrodes** used in the gas-shielded tungsten-arc-welding process are not filler-metal electrodes since they are consumed very slowly and are not deposited in the weld. Tungsten electrodes for this process are covered by Specifications for Tungsten Arc-welding Electrodes (AWS A5.12). Tungsten electrodes with small additions of thoria are also used for their improvement in arc starting and stabilization.

**Gas metal-arc welding** involves an arc between a continuously fed filler-metal wire and the work to be welded. A gas shield is projected around the arc and the filler wire. The gas shield may be helium or argon or mixtures of these gases; small percentages of oxygen may be introduced into these inert-gas shields; carbon dioxide may also be used as the shielding gas. Welding power sources may be direct current, either polarity, or alternating current. The filler-metal wires, of small size, are fed into the arc at high speeds [175 to 225 in/min (4.45 to 5.72 m/min)] and high current densities, usually above 50,000 A psi. This combination of welding conditions produces high deposition rates and high welding speeds. With the smaller sizes of filler metal it is possible to weld in all positions. The use of inert-gas envelopes permits the welding of aluminum, magnesium, stainless steels, copper, nickel, and alloys without flux. The inert-gas shield protects the metal in transfer across the arc, thus providing weld metal of substantially the same composition as the filler metal. Typical production data are given in Table 5. The manually operated equipment uses wire sizes from 0.020 to ⅛ in (0.5 to 3.2 mm). Automatic equipment with this process uses wire up to ⅛ in.

An important version of this process is the short-circuiting type of metal transfer which takes place when small-diameter electrode wire [ $\frac{1}{32}$ to $\frac{3}{64}$ in (0.8 to 1.2 mm)] is used to form a low-current arc (50 to 225 A). The resulting globular transfer causes a short circuit. However, by adding an inductive reactance in the power supply, a stable rate of transfer (roughly 100 times a second) can be achieved. The relatively low heat input and the strong directional transfer are desirable for welding thin sheets in any position or heavy sections in the vertical or overhead position.

**Filler metals** for gas metal-arc welding are supplied, level-wound on reels, spools, or coils. The standard specifications covering these filler metals are combined with other filler-metal specifications covered under welding rods or electrodes. The method of identification involves the letter E to denote an electrode wire. Following this is either the chemical symbols of the principal elements or a type number, such as is used for stainless steels. The following specifications provide for these types of filler metals: Specifications for Copper and Copper-alloy Welding Electrodes (AWS A5.6); Specifications for Corrosion-resisting Chromium and Chromium-Nickel Steel Welding Rods and Bare Electrodes (AWS A5.9); Specifications for Aluminum and Aluminum-alloy Welding Rods and Bare Electrodes (AWS A5.10); Specifications for Nickel and Nickel-base-alloy Bare Welding Filler Metals (AWS A5.14).

**Arc Spot Welding** In arc spot welding, small circular welds are produced in lap joints, using a manual tool employing the gas tungsten-arc-welding process; generally no filler metal is added. In another version of arc spot welding, similar welds are produced employing the gas metal-arc-welding process; in this case filler metal is added. This method may be used with either manual or automatic equipment. Equipment includes controls for regulating rate of heat input. These two methods may also be used for tack-welding members in butt, lap, T, and corner joints.

## GAS WELDING

The heat for gas welding is supplied by burning a mixture of oxygen and a suitable combustible gas. The gases are mixed in a torch which gives control of the welding flame.

**Acetylene** is almost universally used as the combustible gas because of its high flame temperature. This temperature, estimated to be about 6000°F (3315°C), is so far above the melting point of all commercial metals that it provides a means for the rapid localized melting essential in welding. The oxyacetylene flame is also used in cutting ferrous metals.

The **oxyhydrogen** flame is used in welding metals that have low melting points, such as lead, and in welding thin aluminum sheet.

With a 1:1 mixture of oxygen and acetylene the resulting flame is neutral. The neutral flame has an inside portion, consisting of a brilliant cone $\frac{1}{16}$ to $\frac{3}{4}$ in (1.6 to 19.1 mm) long, surrounded by a faintly luminous envelope flame. When acetylene is in excess, the flame consists of three easily recognizable zones: a sharply defined inner cone, an intermediate cone of whitish color, and the bluish outer envelope. The length of the intermediate cone is a measure of the amount of excess acetylene. This flame is reducing, or carburizing.

When oxygen is in excess in the mixture, the flame resembles the neutral flame, but the inner cone is shorter, is "necked in" on the sides, is not so sharply defined, and acquires a purplish tinge. A slightly oxidizing flame may be used in braze welding and bronze surfacing, and a more strongly oxidizing

Table 4. Typical Production Data for Gas Tungsten-Arc Welding—Flat Position

| Metal | Thickness | Weld or joint design | | | Number of passes | Filler metal size, in | Filler metal speed, ipm | Welding power* | | | | Shielding gas | | Tungsten electrode | | Type§ of welding |
|---|---|---|---|---|---|---|---|---|---|---|---|---|---|---|---|---|
| | | Type | Spacing, in. | Backing | | | | Arc,† volts, or length, in | Current, amp | High-frequency | Welding speed, ipm | Type | Flow, cfh | Type‡ | Size, in. | |
| Aluminum, type 3003 | 0.030 in. | Square groove | 0 | Yes | 1 | None | ...... | 0.04 | 80 | No | 160 | He | 30 | Th | 0.040 | ME |
| Aluminum, type 3003 | 0.096 in. | Square groove | 0 | Yes | 1 | None | ...... | 15 | 135 | No | 38 | He | 30 | Th | 3⁄32 | AU |
| Stainless steel, type 304 | 12 gage | Square groove | 3⁄32 | Copper | 1 | 0.045 | 82 | 12 | 200 | No | 13 | He | 40 | Th | 3⁄32 | AU |
| Stainless steel, type 304 | 12–20 gage | Corner | 0 | Steel angle | 1 | None | ...... | 12 | 155 | No | 25 | He | 40 | Th | 3⁄32 | AU |
| Stainless steel, type 321 | 0.029 in. | Square groove | 0 | Copper | 1 | None | ...... | 3⁄32 | 70 | Start only | 72 | He | 20 | Th | 3⁄32 | ME |
| Stainless steel, type 347 | 0.020 in. | Square groove | 0 | Copper | 1 | None | ...... | 3⁄32 | 55 | Start only | 120 | He | 20 | Th | 3⁄32 | ME |
| Stainless steel, type 410 | 0.010 in. | Square groove | 0 | Yes | 1 | None | ...... | 3⁄32 | 40 | Start only | 130 | Ar | 12 | Th | 0.040 | ME |
| Killed steel | 0.075 in. | Square groove | 0 | Yes | 1 | None | ...... | 14 | 155 | No | 30 | He | 30 | Th | 3⁄32 | AU |
| Rimmed steel | 0.062 in. | Square groove | 0 | Yes | 1 | 0.045 | 50 | 14 | 160 | No | 34 | He | 30 | Th | 3⁄32 | AU |
| Monel | 0.084 in. | Square groove | 0.038 | Copper | 1 | 0.045 | 77 | 11 | 270 | No | 30 | He | 40 | Th | 3⁄32 | AU |
| Titanium, type AMS 4901B | 0.063 in. | Square groove | 0 | Gas¶ | 1 | None | ...... | 14 | 115 | Start only | 36 | He | 20¶ | Th | 3⁄32 | ME |

* Welding power direct current, straight polarity.
† Dimensional values are arc length.
‡ Th = thoriated tungsten electrode.
§ ME = Mechanized welding—mechanical arc length control.
  AU = Automatic welding—automatic arc voltage control.
¶ He backing at 15 cfh (ft³/h); He trailing shield 25 cfh (ft³/h).

**Table 5. Typical Production Data for Gas Metal-Arc Welding—Flat Position**

| Metal | Thickness, in | Weld or joint design | | | | | Number of passes | Filler metal diam, in | Welding power | | | Welding speed per pass, ipm | Shielding Gas | |
|---|---|---|---|---|---|---|---|---|---|---|---|---|---|---|
| | | Type | Bevel angle, deg | Root face, in | Spacing, in | Backing | | | Type | Arc, volts | Current, amp | | Type | Flow, cfh |
| Aluminum | 1/8 | Square groove | ..... | ..... | 0-1/8 | ....... | 1 | 3/64 | d-c rp | 30 | 110 | 24 | Ar | 30 |
| Aluminum | 1/4 | Single vee | 60 | 1/16 | 0-1/16 | Grooved steel | 1 | 1/16 | d-c rp | 27 | 200 | 24 | Ar | 40 |
| Aluminum | 1/2 | Single vee | 60 | 1/16 | 0-1/8 | Grooved steel | 2 | 3/32 | d-c rp | 28 | 320 | 16 | Ar | 50 |
| Stainless steel | 1/2 | Double vee | 90 | 1/8 | 0 | Stainless steel | 2 | 1/16 | d-c rp | 25 | 300 | 12 | Ar | 50 |
| Stainless steel | 1/2 | Single vee | 70 | 0 | 1/4 | Stainless steel | 6 | 1/16 | d-c rp | 26 | 300 | 11 | Ar | 50 |
| Deoxidized copper | 1/4 | Single vee | 60 | 0 | 0 | Steel | 1 | 3/32 | d-c rp | 28 | 450 | 17 | Ar | 45 |
| Cu-Al (8 Al, 2 Fe) | 1/2 | Single vee | 60 | 0 | 0 | Steel | 3 | 1/16 | d-c rp | 29 | 300 | 4.1* | Ar | 35 |
| Cu-Si | 3/8 | Single vee | 60 | 0 | 0 | ....... | 3 | 1/16 | d-c rp | 26 | 265 | 4.6* | Ar | 35 |
| Magnesium | 1/4 | Square butt | ..... | 1/4 | 0 | Grooved steel | 1 | 1/16 | d-c rp | 26 | 195 | 19.2 | He | 60 |
| Carbon steel | 1/4 | Square butt | 0 | 1/4 | 1/16 | Grooved copper | 1 | 1/16 | d-c rp | 26 | 340 | 9.0 | Ar + 1% O2 | 40 |
| Carbon steel | 1/2 | Double vee | 60 | 1/16 | 1/16 | ....... | 2 | 1/16 | d-c rp | 26 | 340 | 5.0 | Ar + 1% O2 | 40 |
| Monel | 5/16 | Single vee | 60 | 3/16 | 0 | Steel | 2 | 1/16 | d-c rp | 28 | 325 | 27 | Ar | 60 |

*Speed of progression of total joint of number of passes indicated.

flame is sometimes used in gas-welding brass, bronze, and copper.

**Forehand** and **backhand welding** are employed in oxyacetylene welding. In forehand welding the torch flame is pointed ahead in the direction of welding, and the welding rod precedes the flame. Distribution of the heat and the molten metal is obtained by imparting to the torch and rod opposite oscillating motions. In backhand welding the torch flame is pointed back in the direction of welding, and the rod is interposed between the flame and the molten pool. Distribution of the heat and molten metal is secured by a slight motion of the flame in a vertical plane in the line of the joint; the rod is rolled in the joint. Because of simplified technique, backhand welding permits the use of narrower V's (60 to 75°). Weld quality is generally higher with backhand welding since the molten metal has less chance to be exposed to atmosphere. Flame adjustment for backhand welding usually employs a slight excess of acetylene (carburizing) for the purpose of ensuring absence of an oxidizing condition, which is particularly harmful with welding rods containing small amounts of alloying ingredients.

It is essential that the weld should penetrate entirely through the metal. Joint designs suitable for gas welding are shown in Figs. 1 and 2. To build the weld up to the original surface, welding rods must be used to fill the V.

In **braze welding,** coalescence is produced by heating above 800° F (427°C) and by using a nonferrous filler metal having a melting point below that of the base metals. Braze welding with brass (bronze) rods is used extensively on cast iron, steel, copper, brass, etc. Since it operates at temperatures lower than base-metal melting points, it is used where control of distortion is necessary. Braze-welded joints on mild steel, made with rods of classifications RCuZn-B and RBCuZn-D, will show transverse tensile values of 60,000 to 70,000 lb/in² (414 to 483 MPa). Joint designs for braze welding are similar to those used for gas and arc welding (Figs. 1 and 2).

In braze welding it is necessary to remove rust, grease, scale, etc., and to use a suitable flux to dissolve oxides and clean the metal. The parts are heated to red heat [1150 to 1350°F (621 to 732°C)], and the rod is introduced into the heated zone. The brass rod melts first and "tins" the surfaces, following which additional brass is added by welding.

Welding rods for use in oxyacetylene braze welding are usually of the copper-zinc (60 Cu-40 Zn) analysis. Additions of tin, manganese, iron, nickel, and silicon are made to improve the mechanical properties and usability of the rods. These rods are covered by Specifications for Copper and Copper-alloy Welding Rods (AWS A5.7-69).

**Oxyacetylene welding** of steel is performed without any flux. Fluxes are necessary for welding practically all other metals. The flux used depends on the metal to be welded, although fluxes for braze welding with brass rods may be used for welding brasses and copper. Fluxes are available for welding stainless steel, aluminum and its alloys, nickel and its alloys, magnesium and its alloys, and cast iron. Some nonferrous welding fluxes (aluminum, chromium-nickel, magnesium, etc.) contain fluorides and require adequate ventilation to remove fumes. Ventilation is also necessary in the welding or brazing of metals containing or coated with zinc, lead, or cadmium.

Welding rods for oxyacetylene welding are covered by Specifications for Iron and Steel Gas-welding Rods (AWS A5.2). Three classifications, based on strength and ductility of all-weld-metal tensile-strength specimens, are provided. The significance of the classification numbers, such as RG65, is as follows: The R indicates a welding rod, and the G indicates that it is for use expressly with gas welding; 60 gives the first two digits of the tensile strength. Classification RG65 is commonly used for welding where high quality is required, as in pressure piping. Weld metal with this classification of rods is required to show 65,000 lb/in² (448 MPa) minimum tensile strength; 25 percent elongation in 2 in (50 mm).

Oxyacetylene welding may be employed for the welding of practically all metals. In addition, it has the advantage of permitting welding in all positions. Welding rods are available for most commercial metals and alloys. Where rods are not available, satisfactory results may be obtained by using strips sheared from the base metal for filler metal.

**Mechanized Oxyacetylene Welding**   Oxyacetylene welding by machines is applied in the manufacture of steel barrels, tubing, irrigation pipes, and special forms fabricated of steel and many alloys. Multiple-flame water-cooled machine torches make welds at rates several times that possible by the most expert hand welders. Machine welding is done with and without filler metal, mostly without, on tubing and pipe, the welds being compressed in the process to make a flush or reinforced joint.

## THERMAL CUTTING OF METALS

Two types of metal cutting closely allied to welding are important in the metal-fabricating industry and elsewhere: The most important is oxygen cutting; the other is arc cutting.

**Oxygen cutting** is based on the rapid, exothermic oxidation of iron when heated to about 1500°F (816°C) in the presence of oxygen. The process is therefore usable only with the ferrous metals, especially such steel products as sheet, plate, bars, shapes, piping, tubing, forgings, and castings, and with wrought-iron products. Instead of a welding torch, a cutting torch is coupled to the gas hoses, and the oxygen-supply pressure is increased. The tip of the cutting torch contains a ring of small orifices for the preheating flame and a central orifice for the oxygen jet. When the steel is suitably heated, the oxygen is turned on, "burning" a clean, narrow cut as the torch advances. For preheating, oxygen and acetylene are commonly used; other fuel gases are hydrogen, propane, natural gas, city gas, or proprietary gases. The oxygen-cutting process finds wide application in severing, trimming, and plate-edge preparation for welding, metal removal, and gouging. Table 6 provides operating data for oxyacetylene cutting.

The **volume of oxygen** (in cubic feet per hour at 1 atm and 70°F) required to cut a heavy section ranges from 80 to 120 times the thickness, measured in inches. The supply pressure depends on the size of the cutting orifice, with higher pressures for small orifices. The volume of acetylene for preheating ranges from 10 to 15 percent of the cutting oxygen flow. In cutting heavy sections the principal object is to secure good, clean "drop" cuts; speed of cutting is secondary. For thicknesses from 12 to 48 in (0.31 to 1.22 m) the **cutting speed** varies from 6 to 2 in/min (150 to 50 mm/min). High speeds may cause loss of cut and spoilage of metal. Production cuts have been made on steel up to 62 in (1.57 m) thick, using oxyacetylene cutting.

**Influence of Alloying Elements on Cutting**   Alloying ele-

**Table 6. Oxyacetylene Cutting**
(Manual and machine)

| Thickness, in | ¼ | ½ | 1 | 2 | 4 | 6 | 8 | 10 |
|---|---|---|---|---|---|---|---|---|
| Speed, in/min, manual | 16–18 | 12–14.5 | 8–12 | 5–7 | 4–5 | 3–4 | 2.5–3.5 | 2–3 |
| Speed, in/min, machine | 20–26 | 17–22 | 14–18 | 10–13 | 7–9 | 5–7 | 4–6 | 3–4 |
| Oxygen consumption, ft³/h | 50–90 | 90–125 | 130–200 | 200–300 | 300–400 | 400–500 | 500–650 | 700–1,000 |

ments have two possible effects: They may increase the resistance of steel to cutting, and they may give rise to harder cut surfaces. Where an alloy steel is difficult to cut by the normal oxygen jet, an improvement is effected by tightly clamping a "waster" plate to the upper surface and cutting through both thicknesses. Another method is to weld a heavy bead on the upper surface along the proposed line of cut. **Chemical-flux cutting** provides an effective means for cutting high-alloy compositions.

Steels up to 0.30 percent **carbon** can be cut without difficulty. Higher-carbon steels should be preheated to about 600°F (316°C) to prevent hardening and cracking. Cast irons containing up to 4 percent **carbon** may be cut by special technique. Steels up to 14 percent **manganese** and 1.5 percent **carbon** are cut with difficulty and for best results should be preheated.

**Silicon** in amounts usually present has no effect. Silicon steel containing considerable amounts of carbon and manganese must be carefully preheated and postannealed for best mechanical properties.

Steels up to 5 percent **chromium** are cut without much difficulty when the surfaces are clean. If the steel is of an air-hardening type, preheat and postheat should be employed. Higher chromium and chromium-nickel steels should be cut by the chemical-flux and metal-powder methods.

**Nickel** up to 20 or 30 percent (if the carbon is not too high) may be cut. Up to about 7 percent nickel content, cuts are very satisfactory.

Aircraft-quality **chrome-molybdenum** steel offers no difficulties. High-**molybdenum-tungsten** steels, however, may be cut only by means of special techniques.

**Tungsten** alloys up to 12 or 14 percent may be cut very readily, but with a higher percentage of tungsten, cutting is difficult.

**Vanadium** in small amounts may improve cutting. The other common alloying elements have no appreciable influence.

**Oxygen-cutting machines** are capable of making oxygen cuts of intricate shape and of such high quality and accuracy as to require no further finishing (see Table 6). Cutting machines are frequently equipped with more than one cutting torch, centrally controlled and guided, and will oxygen-cut a number of identical shapes simultaneously, thereby effecting marked economies where a high production rate prevails. Oxygen-cutting machines may be equipped with photoelectric cells (electric eyes) for following black-and-white drawings of parts to be cut. Other machines crawl around pipes and make one or two square or beveled cuts, as desired.

**Oxygen gouging (flame gouging)** is a form of oxygen cutting. Instead of a straight tip, a slightly curved tip is attached to the cutting torch. This enables the flame to strike the metal surface at a low angle, making shallow cuts possible. It is used for cutting weld grooves, especially to clean out the underside of root passes to make a sound back weld. It is also used for removing defective welds and surface defects.

**Cutting metals under water** can be accomplished with oxygen, using either a fuel gas or the electric arc to preheat the metals. Since acetylene cannot be used at pressures above 15 psig (103 kPa) for safety reasons, its use in underwater cutting is limited to shallow depths; hydrogen is employed for greater depths. **Oxygen-arc** underwater cutting is performed with a special hollow, shielded, and insulated electrode. A fully insulated electrode holder conducts current and oxygen to the electrode. Striking of the arc preheats the metal instantaneously, and the oxygen makes the cut.

The torch method consists of an oxyacetylene cutting torch surrounded by a protecting bell through which is forced compressed air. Special methods are used to permit the lighting of the torch under water.

**Stainless-steel Cutting** Stainless steels (chrome-nickel and straight chrome) are virtually impossible to cut by normal oxygen-cutting procedures because chromium oxide is formed and prevents continuation of the cut. Two methods have been developed to overcome this condition. One employs a chemical flux to react with the chromium oxide. The other method uses iron powder to flux the oxide. Use of these methods permits cutting stainless steels and some nonferrous metals with almost the same ease as oxygen cutting of mild steel. The processes have been adapted to machine gas cutting, and the results equal those with mild steel. Unstabilized varieties of stainless steel will show carbide precipitation in a zone about ⅛ to ¼ in (3.2 to 6.4 mm) deep, which may be removed by grinding. Steels with alloy contents up to about 45 percent have been cut successfully. Oxygen-arc cutting using a hollow electrode, similar to that employed for underwater cutting, may also be employed for manual cutting of stainless steel.

**Arc-cutting** processes rely on arc heat to melt a path through metal and are therefore capable of cutting nonferrous as well as ferrous metals. As in welding, the arc is established between an electrode forming one terminal of an electric circuit and the workpiece forming the other terminal. Various means are employed to flush the molten metal from the cut. The selection of the process to be used depends on the type of cutting required.

For cutting weld grooves, for back-gouging roots of welds, or for simple removal of metal, **air carbon-arc cutting** is fast and efficient. The electrode holder is equipped to direct a jet of compressed air in line with the electrode so as to blow away the molten metal. In plate work, grooves may be cut having a width roughly equal to the size of the electrode and a depth depending on the angle of approach and speed of travel.

For piercing, severing, and other types of rough cutting, **oxygen-arc cutting** may be used. Its use in underwater cutting is described above. As a result of both chemical and mechanical interactions of the flux covering, the mild-steel electrode, and the oxygen, the arc action is greatly increased. This process combines deep penetration with speed.

One of the most important cutting operations in the welding industry is the trimming and edge preparation of plates prior to forming and welding. The development of the plasma arc and its application to this type of cutting has displaced all former methods of arc cutting. **Plasma-arc cutting** employs the heavy-duty equipment needed to optimize the basic demand for a hot, thin jet capable of rapidly melting a narrow kerf in any metal and of flushing away the melt. In this process, a tungsten electrode (cathode) is centered in a tubular, water-cooled nozzle constricted at the tip to form a narrow passage about ⅛ in (3.2 mm) or smaller in diameter. Gas is introduced into the nozzle, where it flows past the electrode, through the constricted passage, and is directed at the point to be cut in the workpiece. This point forms the anode of the circuit when a transferred arc is employed. In operation, the arc extends from the tip of the electrode, through the passage, and across the intervening distance to the workpiece. Open-circuit voltage for this arc may be 250 V or more. However, a pilot arc drawn between the electrode and nozzle wall is needed to initiate the main arc. The heated gas quickly expands into a high-speed plasma jet with a temperature said to be in the region of 50,000°F (27,760°C). Mixtures of argon and hydrogen or nitrogen and hydrogen are used in cutting nonferrous metals and stainless steels. Air or oxygen is more economical for cutting carbon steels.

An alternative mode of operation makes use of the nontransferred arc. In this case, the arc is drawn between the electrode and the outer nozzle wall, the plasma jet issuing in the same manner as with the transferred arc but of shorter length.

Safety precautions are necessary to protect the user from high-voltage power, ultraviolet radiation, and fumes.

## RESISTANCE WELDING

In resistance welding, coalescence is produced by the heat obtained from the resistance offered by the work to the flow of electric current in a circuit of which the work is a part, and by the application of pressure. The specific processes include **resistance-spot welding, resistance-seam welding, projection welding, upset welding,** and **flash welding.** Figure 4 shows diagrammatic outlines of the processes; Fig. 5 shows resistance-welding symbols.

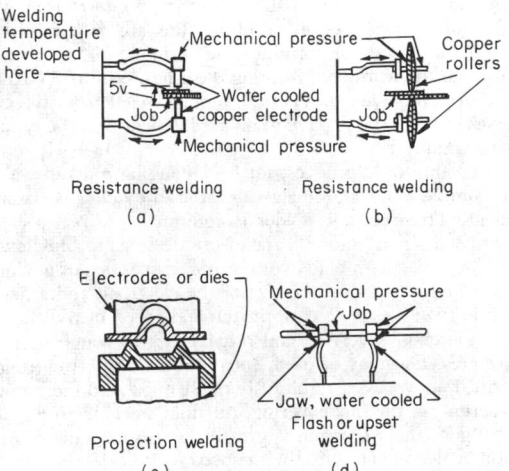

**Fig. 4** (*a*) Resistance-spot, (*b*) resistance-seam, (*c*) projection, (*d*) flash, or upset, welding.

The resistance of the welding circuit should be a maximum at the interface of the parts to be joined, and the heat generated there must reach a value high enough to cause a localized fusion under pressure. There is an exception to this principle in flash butt welding, where a portion of the heat is derived from the flashing and combustion of the metal at the interface. Even in this instance a part of the heat is generated in the work. The duration of the application of the current must be short so as to limit the zone of melting; otherwise an inferior weld is produced.

**Electrodes** are of copper alloyed with such metals as molybdenum and tungsten, with high electrical conductivity, good thermal conductivity, and sufficient mechanical strength to withstand the high pressures to which they are subjected. The electrodes should be water-cooled. The resistance at the surfaces of contact between the work and the electrodes must be kept low. This may be accomplished by using smooth, clean work surfaces and a high electrode pressure.

In **resistance-spot welding** (Fig. 4), the parts are lapped and held in place under pressure. The size and shape of the electrodes control the size and shape of the welds, which are usually circular.

| Type of weld | | |
|:---:|:---:|:---:|
| Spot or projection | Seam | Flash or upset butt |
| ◯ | ⊖ | ❘ |

**Fig. 5** Resistance-welding symbols.

Designing for spot welding involves six elements: tip size, edge distance, contacting overlap, spot spacing, spot-weld shear strength, and electrode clearance. For mild steel, the diameter of the tip face, in terms of sheet thickness $t$, may be taken as $0.1 + 2t$ for thin material, and as $\sqrt{t}$ for thicker material, dimensions in inches. Edge distance should be sufficient to provide enough metal around the weld to retain it when in molten condition. Contacting overlap is generally taken as the diameter of the weld nugget plus twice the minimum edge distance. Spot spacing must be sufficient to ensure that the welding current will not shunt through the previously made weld. Table 7 gives values ot the above dimensions and also spot-weld shear-strength values based on base-metal tensile strength up to 70,000 lb/in² (483 MPa).

**Resistance-spot welding machines** vary from small, manually operated units to large, elaborately instrumented units designed to produce high-quality welds, as on aircraft parts. Portable gun-type machines are available for use where the assemblies are too large to be transported to a fixed machine. Spot welds may be made singly or in multiple, the latter being generally by special-purpose machines. Spacing of electrodes is important to avoid excessive shunting of welding current.

The **resistance-seam welding process** (Fig. 4) produces a series of spot welds made by circular or wheel-type electrodes. The weld may be a series of closely spaced individual spot welds, overlapping spot welds, or a continuous weld nugget. The weld shape for individual welds is rectangular; continuous welds are about 80 percent of the width of the roll electrode face.

Machines for seam welding may provide for the rotation of one or both rollers. Three general types are available: **longitudi-**

**Table 7. Spot-Welding SAE 1010 Steel**
(Free from Scale, Oxides, Paint, Grease, and Oil)

| Thickness $t$ of thinnest outside piece, in | Electrode diam and shape | | Net electrode force, lb | Weld time, (60 Hz) | Min shear strength, lb | Welding current (approx), A | Diam (approx) of fused zone, in | Min weld spacing, in | Min contacting overlap $L$, in |
|---|---|---|---|---|---|---|---|---|---|
| | $D$, in, min | $d$, in, max | | | | | | | |
| 0.010 | 3/8 | 1/8 | 200 | 4 | 130 | 4,000 | 0.10 | 1/4 | 3/8 |
| 0.021 | 3/8 | 3/16 | 300 | 6 | 320 | 6,500 | 0.13 | 3/8 | 7/16 |
| 0.031 | 3/8 | 3/16 | 400 | 8 | 570 | 8,000 | 0.16 | 1/2 | 7/16 |
| 0.040 | 1/2 | 1/4 | 500 | 10 | 920 | 9,500 | 0.19 | 3/4 | 1/2 |
| 0.050 | 1/2 | 1/4 | 650 | 12 | 1,350 | 10,500 | 0.22 | 7/8 | 9/16 |
| 0.062 | 1/2 | 1/4 | 800 | 14 | 1,850 | 12,000 | 0.25 | 1 | 5/8 |
| 0.078 | 5/8 | 5/16 | 1,100 | 17 | 2,700 | 14,000 | 0.29 | 1 1/4 | 11/16 |
| 0.094 | 5/8 | 5/16 | 1,300 | 20 | 3,450 | 15,500 | 0.31 | 1 1/2 | 3/4 |
| 0.109 | 5/8 | 3/8 | 1,600 | 23 | 4,150 | 17,500 | 0.32 | 1 5/8 | 13/16 |
| 0.125 | 7/8 | 3/8 | 1,800 | 26 | 5,000 | 19,000 | 0.33 | 1 3/4 | 7/8 |

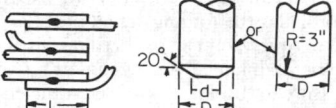

The welding conditions are governed by $t$ of thinnest outside piece. The thickness of the assembly of sheets must not exceed $4t$: max ratio of any two thicknesses not to exceed 3:1. Electrode material conductivity min, 75 percent of copper, hardness min, −5 Rockwell B. Dimensions and contacting overlap $L$, as shown.

nal seam, circular seam, and **universal welding machines.** Interrupters are necessary to provide control of the heat effect and allow cooling, under pressure.

**Electrodes** (rollers or wheels) are generally of the shape shown in Table 8. In joints with limited space, as in welding flanges to containers, one face is flat. The axes of rollers may be inclined up to 6° for clearance. Either roller may be driven, with the other idling, or both may be driven. Drives may be through friction rolls or knurled rollers. Wheels may range from 2 to 24 in (50 to 610 mm) diam, with the average 7 to 10 in (178 to 254 mm).

A **mash weld** is a seam weld in which the finished weld is only slightly thicker than the sheets, and the lap disappears. It is limited to thicknesses of about 16 gage and an overlap of 1 1/2 times the sheet thickness. Operating the machine at reduced speed, with increased pressure and noninterrupted current, a strong quality weld may be secured that will be 10 to 25 percent greater in thickness that the sheets. The process is applicable to mild steel but has limited use on stainless steel; it cannot be used on nonferrous metals. A modification of this technique employs a straight butt joint. This produces a slight depression at the weld, but the strength is satisfactory on some applications, e.g., for the production of some electric-welded pipe and tubing.

Cleanliness of sheets is of even more importance in seam welding than in spot welding. Best results are secured with cold-rolled steel, wiped clean of oil; the next best with pickled hot-rolled steel. Grinding or polishing is sometimes satisfactory, but not sand- or shot-blasting.

Typical production data for seam welding for use with clean SAE 1010 steel are given in Table 8.

In **projection welding** (Fig. 4), the heat for welding is derived

**Table 8. Seam-Welding SAE 1010 Steel**
(Free from Scale, Oxides, Paint, Grease, and Oil)

| Thickness $t$ of thinnest outside piece, in | Electrode width and shape | | Net electrode force, lb | On-time, (60 Hz) | Off-time (pressure-tight), cycles | Weld speed, in/min | Welding current (approx), A | Welds per in | Min contacting overlap $L$, in |
|---|---|---|---|---|---|---|---|---|---|
| | $W$, in, min | $w$, in, max | | | | | | | |
| 0.010 | 3/8 | 3/16 | 400 | 2 | 1 | 80 | 8,000 | 15 | 3/8 |
| 0.021 | 3/8 | 3/16 | 550 | 2 | 2 | 75 | 11,000 | 12 | 7/16 |
| 0.031 | 1/2 | 1/4 | 700 | 3 | 2 | 72 | 13,000 | 10 | 1/2 |
| 0.040 | 1/2 | 1/4 | 900 | 3 | 3 | 67 | 15,000 | 9 | 1/2 |
| 0.050 | 1/2 | 5/16 | 1,050 | 4 | 3 | 65 | 16,500 | 8 | 9/16 |
| 0.062 | 1/2 | 5/16 | 1,200 | 4 | 4 | 63 | 17,500 | 7 | 5/8 |
| 0.078 | 5/8 | 3/8 | 1,500 | 6 | 5 | 55 | 19,000 | 6 | 11/16 |
| 0.094 | 5/8 | 7/16 | 1,700 | 7 | 6 | 50 | 20,000 | 5.5 | 3/4 |
| 0.109 | 3/4 | 1/2 | 1,950 | 9 | 6 | 48 | 21,000 | 5 | 13/16 |
| 0.125 | 3/4 | 1/2 | 2,200 | 11 | 7 | 45 | 22,000 | 4.5 | 7/8 |

See footnotes for Table 7.

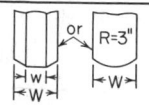

Wheel type electrodes

from the localization of resistance at predetermined points by means of projections, embossments, or the intersections of elements of the assembly. The projections may be made by stamping or machining. The process is essentially the same as spot welding, and the projections seem to concentrate the current and pressure. Welds may be made singly or in multiple with somewhat less difficulty than is encountered in spot welding. When made in multiple, all welds may be made simultaneously. The advantages of projection welding are (1) the heat balance for difficult assemblies is readily secured, (2) the results are generally more uniform, (3) a closer spacing of welds is possible, and (4) electrode life is increased. Sometimes it is possible to projection-weld joints that could not be welded by other means.

The design and forming of the projections are important. For sheet metal the dies must be so designed as to prevent shearing of the projection. This would weaken the weld. The weld growing from the center outward will tend to a circular shape. When projections must be formed on curved or irregular surfaces, it is preferable to elongate them to ensure that the weld initiates at a point contact. Heavy metal sections should be provided with raised V-shaped projections located so far in from the edges that the metal will not squeeze out when molten. Studs, bolts, etc., should be domed to a radius 1½ times the diameter of the stud to provide proper contact. Crossed wires may be projection-welded without any special preparation.

The **welding pressure** should be set at a value which will ensure complete flattening of the projection; too much pressure causes premature collapse of the projection. **Welding currents** are high, particularly on multiple-projection assemblies; the maximum current which does not cause splashing is correct.

Projection welding may be applied to most metals except copper and red brasses. Other brasses may prove difficult to handle. Galvanized iron, terneplate, and tin plate are successfully handled. Dissimilar metals, such as steel to brass, bronze, or aluminum, may be welded with special techniques. Heavy steel sections may also be welded by projection welding.

**Upset welding** (Fig. 4), an early form of resistance welding, is limited to joining members of approximately equal cross section. The parts to be welded are brought together under pressure, and current is passed through the contact area. This results in creating a forge weld of symmetrical shape. Heat generated by contact resistance is a function of surface resistivity, the nature of the surfaces, the unit pressure, and other factors; the resistance is approximately inversely proportional to pressure, other factors remaining constant. Pressure and current are maintained throughout the welding cycle, although pressure is initiated at a low value (to raise the initial contact resistance) and subsequently raised to that necessary for forging. When the required upset is achieved, welding current is cut off.

In **flash welding,** the parts are brought together lightly, with current flowing, and then separated slightly; a flashing action is created at the interface. The flashing generates the greater part of the heat, the balance coming from the resistance at the joint. When the metal at the joint is molten, heavy pressure is quickly applied, and this forces out the molten metal and makes the weld in the plastic metal just in back of the molten metal. Flash welding is used more than upset welding because of greater weld strength; no need for special preparation of weld surface; lower power demand with less power consumption; faster speed and smaller upset; and less heat in the work since most heat appears at the interface. It is possible to weld dissimilar metals of widely differing melting points since flashing may be continued until both metals have reached their individual fusing temperatures.

In flashing it is necessary to move the ends together by means of the platen. The **flashing rate** must be maintained at a correct value to assure continuous flashing. If too slow, the flashing is intermittent and it is difficult to build up heat. If too fast, there is a possibility of freezing the pieces. Too short flashing time results in lowered heat and renders it impossible to secure proper upset. Too long heating time creates such a mass of plastic metal that it cannot be squeezed out properly in the upsetting cycle, resulting in improper upset.

**Upsetting pressure** must be applied quickly to create the weld after the flashing action has established a zone of plastic metal. The upsetting pressure provides the forging action necessary to make the weld and also to squeeze out all oxidized metal and slag. The current must be maintained during upsetting to prevent chilling. The pressure for upsetting must be sufficient to ensure complete extrusion of molten metal at the weld line beyond the original cross section. Upset pressures for welding without preheat (pressures indicated for cross section of weld area) are low-carbon steel, 10,000 lb/in² (69 MPa); medium-carbon and low-alloy steels, 15,000 lb/in² (103 MPa); special stainless alloys with high hot compressive strengths, 35,000 lb/in² (241 MPa).

After welding, the flash and upset may be removed by chipping, grinding, machining, flame machining, die trimming, or with a high-speed sander. When the flash has been removed, the weld line should not be visible. The joint efficiency should approach 100 percent. Welds in hardenable steels may require preheating, either outside the machine or by means of the machine, to slow down the cooling rate after welding. Postheating may also be accomplished in the machine.

**Multiple-impulse welding** is applicable to spot-, seam-, and projection-welding processes. It consists of applying the current in a series of impulses, which may be a fraction of a cycle or a number of cycles. The welding machine must be capable of maintaining the pressure while the cyclic operations of current "on" and current "off" are in progress. The advantages of **pulsation welding** are (1) increased electrode life; (2) the welding of thicker material, frequently with the same equipment; (3) better welds in some cases; (4) reduction of "spitting" of weld metal; and (5) the successful spot welding of many thin sections stacked at greater heights. Typical production data for pulsation welding are given in Table 9.

A continuous resistance-welding process is used in the manufacture of tubular goods. The finished weld resembles a flash weld with a burr of flash ejected at the top and bottom of the weld. Steel strip is fed into the machine and is formed into a circular shape as it passes through several rolls. After the circular shape is secured, two copper rollers, one on each side of the joint, apply the welding current while side rollers apply pressure effecting the weld. Tubing has been produced by this process in low-carbon steels, low-alloy steels, and stainless steels.

In **percussion welding,** the heat for welding is secured simultaneously over the entire area of the abutting surfaces from an

**Table 9. Pulsation-Welding SAE 1010 Steel**
(Free from Scale, Oxide, Paint, Grease, and Oil)

| Thicknesses of the two plates welded, in | Electrode diam and shape | | Net electrode force, lb | Weld time, 20 cycles on, 5 cycles off (60 Hz), number of pulsations | | | Min shear strength, lb | Welding current (approx), A | Diam (approx) of fused zone, in | Min contacting overlap $L$, in |
|---|---|---|---|---|---|---|---|---|---|---|
| | $D$, in min | $d$, in max | | Single weld | Adjacent welds | | | | | |
| | | | | | 1–2 in | 2–4 in | | | | |
| 1/8, 1/8 | 1 | 7/16 | 1,800 | 3 | 5 | 4 | 5,000 | 18,000 | 3/8 | 7/8 |
| 1/8, 3/16 | 1 | 7/16 | 1,800 | 3 | 5 | 4 | 5,000 | 18,000 | 3/8 | 7/8 |
| 1/8, 1/4 | 1 | 7/16 | 1,800 | 3 | 5 | 4 | 5,000 | 18,000 | 3/8 | 7/8 |
| 3/16, 3/16 | 1 1/4 | 1/2 | 1,950 | 6 | 20 | 14 | 10,000 | 19,500 | 9/16 | 1 1/8 |
| 3/16, 1/4 | 1 1/4 | 1/2 | 1,950 | 6 | 20 | 14 | 10,000 | 19,500 | 9/16 | 1 1/8 |
| 3/16, 5/16 | 1 1/4 | 1/2 | 1,950 | 6 | 20 | 14 | 10,000 | 19,500 | 9/16 | 1 1/8 |
| 1/4, 1/4 | 1 1/4 | 9/16 | 2,150 | 12 | 24 | 18 | 15,000 | 21,500 | 3/4 | 1 3/8 |
| 1/4, 5/16 | 1 1/4 | 9/16 | 2,150 | 12 | 24 | 18 | 15,000 | 21,500 | 3/4 | 1 3/8 |
| 5/16, 5/16 | 1 1/2 | 5/8 | 2,400 | 15 | 30 | 23 | 20,000 | 24,000 | 7/8 | 1 1/2 |

Same footnote as for Table 7, except bevel angle 10° (instead of 20°) and $d$ min = 1/4 in.

arc produced by a rapid discharge of stored electric energy, followed immediately by the application of pressure. The process is used only to a limited extent today.

## SURFACING BY WELDING

**Surfacing** by welding is a method of applying an alloy material to a metal part so as to form a protective surface to resist abrasion, corrosion, heat, impact, or any combination of these factors. The alloy may be applied by oxyacetylene, metal arc, submerged arc, or gas-shielded arc welding, or by the metal-spray method. New parts may be surfaced before use, or worn parts may be built up to the original size and reclaimed. The most important economy derived from surfacing results from prolonged life of parts. Surfaced parts may outwear plain or unfaced ones many times, depending on the type of hard metal used and the service to which they are subjected. To meet the various requirements of hardness, toughness, impact, abrasion-corrosion, heat resistance, and other qualities, numerous surfacing alloys are available. Practically all ferrous metals can be surfaced. Welding rods and electrodes for this work are covered by Specifications for Surfacing Welding Rods and Electrodes (AWS A5.13-69). Nonferrous alloys with low melting points cannot generally be surfaced with the alloys designed for ferrous application. However, Cu-Al covered electrodes of hard grades are available and included in the above specification. In addition, a commentary is appended to guide the user in the application and expected performance of six groups of surfacing filler metals, namely, high-speed steels, austenitic manganese steels, austenitic high-chromium iron, cobalt-base alloys, copper-base alloys, and nickel-chromium-boron alloys.

## BRAZING

**Brazing** is another one of the general groups of welding processes, consisting of the torch, furnace, induction, dip, resistance, twin carbon arc, flow, and block-brazing processes; the first five processes are of industrial significance. Brazing may be used for joining virtually all metals and dissimilar combinations of metals, although not all combinations of dissimilar metals are satisfactory (e.g., aluminum or magnesium to other metals). In brazing, coalescence is produced by heating above 800°F (427°C) but below the melting point of the metals being joined. The nonferrous filler metal used has a melting point below that of the base metal, and the filler metal is distributed in the closely fitted lap or butt joints by capillary attraction. Cleaning of the joints is essential for satisfactory brazing. The use of a flux or atmosphere to control surface cleanliness is usually necessary. Filler metal may be hand-held and fed into the joint (face feeding), or preplaced as rings, washers, shims, slugs, etc.

Brazing with the silver-alloy types of filler metals has previously been known as **silver soldering** and **hard soldering**. Similarly, brazing with **spelter solders** has been known as **spelter brazing**. These terms are now considered obsolete, as the term *brazing* adequately covers joints made by the flow of molten filler metal by capillary attraction. *Braze welding* should not be confused with brazing. Braze welding is a method of welding employing a filler metal which melts below the melting points of the base metals jointed, but the filler metal *is not* distributed in the joint by capillary attraction. (See also Sec. 6.)

Specifications for Brazing Filler Metal (AWS A5.8) provide for seven basic classifications. The classification numbers use the letter B to signify brazing. Following this letter there appear the chemical symbols representing the principal alloying ingredients. The final numerals differentiate between the several analyses in a group. The BAlSi, aluminum-silicon group operates in the range of 1060 to 1150°F (571 to 621°C); the copper-phosphorus group, BCuP, operates between 1300 and 1700°F (704 to 927°C); the group carrying the symbols BAg are the widely used silver alloys which operate in a range of 1145 to 1800°F (618 to 982°C); the symbols BCu and BCuZn cover the copper and copper-zinc groups, brazing in the range of 1670 to 2100°F (910 to 1149°C); a copper-gold group, BAu, operates between 1635 to 2000°F (891 to 1093°C); BMg represents the magnesium group operating between 1080 to 1160°F (582 to 627°C); finally, there is the heat-resistant group, BNi, operating between 1700 and 2200°F (927 to 1204°C).

**Torch brazing** uses acetylene, propane, or other fuel gas, burned with oxygen or air. The combination employed is governed by the brazing-temperature range of the filler metal, which is usually above the liquidus. Flux with a melting point appropriate to the brazing-temperature range and the filler metal is essential. Torch brazing may be manual or mechanized.

**Furnace brazing** employs the heat of a gas-fired, electric, or other type of furnace to raise the parts to brazing temperature. Fluxes may be used, although reducing or inert atmospheres are more common since they eliminate postbraze cleaning necessary with fluxes.

**Induction brazing** utilizes a high-frequency current to generate the necessary heat in the part by induction. Distortion in the brazed joint can be controlled by current frequency and other factors. Fluxes or gaseous atmospheres must be used in induction brazing.

**Dip brazing** involves the immersion of the parts in a molten bath. The bath may be either molten brazing filler metal or molten salts, usually brazing flux. The former is limited to small parts such as electrical connections; the latter is capable of handling assemblies weighing several hundred pounds. The particular merit of dip brazing is that the joint is virtually completed all over the assembly at one time.

**Resistance brazing** utilizes standard resistance-welding machines to supply the heat. Although theoretically any of the filler metals may be used, those of the BAg, BCuP, and BCuZn groups are most common. Either replacement or face feeding of the filler metal may be used. Fluxes or atmospheres must be used, with flux predominating. Standard spot or projection welders may be used. Pressures are lower than those for conventional resistance welding. As currents are large, water cooling of electrodes is essential.

**Step brazing** is a technique that has found increasing use through the development of new brazing filler metals. This technique makes possible the brazing of several joints in succession by employing filler metals having successively lower melting points. After the first joint is brazed, any type of suitable machining or forming operation can be performed. The next joint is brazed at a lower temperature so as not to affect the first joint, etc. Another effective way of using this technique is to select a filler metal that, upon melting, combines with the base metal to form a new alloy having a melting point higher than that of the original. Precise control over filler-metal composition, brazing temperatures, and other factors is usually required.

## WELDING PROCEDURES

### Steel

**Low-carbon Steels**  (Carbon up to 0.30 by ladle analysis) Steels in this class are readily welded by most arc and gas processes. Preheating is unnecessary unless parts are very heavy or welding is performed below 32°F (0°C). Torch-heating steel in the vicinity of welding to 70°F (21°C) offsets low temperatures. Postheating is necessary only for important structures such as boilers, pressure vessels, and piping. Gas tungsten-arc welding is usable only on killed steels; rimmed steels produce porous, weak welds. Resistance welding is readily accomplished if carbon is below 0.20 percent; higher carbon requires heat-treatment to slow the cooling rate and avoid hardness. Brazing with BAg, BCu, and BCuZn filler metals is very successful.

**Medium-carbon Steels**  (Carbon from 0.30 to 0.45 by ladle analysis) This class of steel may be welded by the arc, resistance, and gas processes. As the rapid cooling of the metal in the welded zone produces a harder structure, it is desirable to hold the carbon as near 0.30 percent as possible. These hard areas are proportionately more brittle and difficult to machine. The cooling rate may be diminished and hardness decreased by preheating the metal to be welded above 300°F (149°C), preferably to 500°F (260°C). The degree of preheating depends somewhat on the thickness of the section. Subsequent heating of the welded zone to 1100 to 1200°F (593 or 649°C) will restore ductility and relieve strain. Brazing may also be used as noted for low-carbon steels.

**High-carbon Steels**  (Carbon from 0.45 to 0.80 by ladle analysis) These steels are rarely welded except in special cases. The tendency for the metal heated above the critical range to become brittle is more pronounced than with lower- or medium-carbon steels. Thorough preheating of metal, in and near the welded zone, to a minimum of 500°F is essential. Subsequent annealing at 1350 to 1450°F (732 to 733°C) is also desirable. Brazing is used with these steels, in the form of tools, and is combined with the heat-treatment cycle.

**Low-Alloy Steels**  The weldability of low-alloy steels is dependent upon the analysis and the hardenability, those exhibiting low hardenability being welded with relative ease, whereas those of high hardenability requiring preheating and postheating. Sections of ¼ in (6.4 mm) or less may be welded with mild-steel filler metal and may secure joint strengths approximating base-metal strength by virtue of alloy pickup in the weld metal, and weld reinforcement. Alloys of higher strength require filler metals of mechanical properties matching the base metal. Special alloys with creep-resistant or corrosion-resistant properties must be welded with filler metals of the same chemical analysis. Austenitic chrome-nickel shielded metal-arc electrodes are helpful for welding hardenable alloys when preheating and postheating are not practicable. Low-hydrogen-type electrodes (either mild- or alloy-steel analyses) permit the welding of alloy steels, minimizing the occurrence of underbead cracking. No preheating is usually necessary on sections up to about ½ in (12.7 mm); low preheat of about 300°F (149°C) is necessary for heavier sections. Brazing may also be used with these steels.

**Chrome-nickel austenitic steels** are excellent for welding and,

under satisfactory conditions, produce strong, tough, and reasonably ductile welds. Alloys containing less than 0.03 percent carbon can be welded without any subsequent heat-treatment. Alloys containing up to 0.10 percent carbon can be welded, provided the structures welded will not be used for service where high corrosion resistance is required.

Addition of titanium, columbium, or some of the other rare metals stabilizes the alloy by formation of titanium or columbium carbide in preference to chromium carbide. In general, titanium or columbium is added to the extent of about 10 times the carbon content present. Such **stabilized** steel is not entirely proof against dangerous carbide precipitation. For the welding of the stabilized analyses only stabilized filler metals should be used. In the gas tungsten-arc and gas metal-arc welding processes the stabilizer is incorporated in the filler metal. In stainless-steel covered electrodes, the columbium is frequently introduced through the coating. Since titanium cannot be successfully transferred across the metal arc with covered electrodes, columbium is the principal stabilizer employed.

**Unstabilized alloys** containing more than 0.07 percent carbon can generally be spot- or seam-welded without impairing their corrosion-resistant properties, provided that the welding time is made very short. The maximum safe duration of the welding time will vary inversely with the carbon content of the alloy and somewhat with the welding technique used. With alloys with 0.10 percent carbon no dangerous carbide precipitation will occur if the duration of the application of the welding current does not exceed 0.20 s. With alloys having a carbon content of around 0.15 percent this figure should be reduced to 0.10 or 0.15 s max. All welding will produce some carbide precipitation, and the line of demarcation between the alloys that can and cannot be welded is a compromise between the extent of the carbide precipitation and the severity of the corrosion attack to which the welded product will be subjected. No welding should be attempted without first thoroughly cleaning the surfaces to be welded. Flash welding is, in general, quite satisfactory, as most of the fused metal produced by the arc of the flash which could be expected to contain chromium oxides and nitrides is squeezed out by the upsetting operation, and the weld obtained is therefore fairly ductile. Because of the amount of heating involved, the upset welding process should be avoided. Gas tungsten-arc welding is particularly useful on stainless steels up to about ³⁄₁₆ in (4.8 mm). Manual welding with this process is performed in all positions, with and without filler rods. Semiautomatic and automatic welding are performed at high speeds and yield welds of high quality. As flux is not needed, postweld cleaning is greatly simplified. Gas metal-arc welding is especially suitable for stainless steels above ³⁄₁₆ in (4.8 mm) thick. Welds of highest quality are readily produced.

**Chromium Irons and Steels**  Welding of the chromium irons and steels can be divided into two classes: (1) welding in which the filler metal deposited has essentially the same chemical analysis as the base metal; (2) welding in which the filler metal deposited differs in analysis and characteristics from the base metal, the filler metal commonly employed being an austenitic chromium-nickel steel of the 18 percent chromium, 8 percent nickel type, or higher-alloy analyses. An austenitic chromium-nickel filler metal is used for all welds that cannot be annealed, including field and repair welds, and a filler metal of the same analysis as the base metal for welds that can be annealed. For operation at high temperatures the expansion of the austenitic filler metal as related to the base metal must be considered. Operation at high temperature will result in warpage or high stresses and, under repeated heating and cooling, may result in failure through fatigue.

### Cast Steel

In good grades of cast steel with a carbon content below 0.25 percent, welding procedures are approximately the same as for wrought steel. With a carbon content above 0.25 percent or with special alloy compositions, precautions and sometimes special procedures are necessary. Segregations of phosphorus and sulfur should be removed from areas to be welded by oxygen gouging or chipping. The problem of overcoming shrinkage in the repair of castings requires special care. Stress relieving is desirable, especially in repair work. In higher-carbon and alloy steels full annealing may be necessary.

For arc welding, a high-grade heavy-covered electrode is essential. A tough, general-purpose electrode (arc) or welding rod (gas) is used for plain-carbon castings, but filler metal of the same composition as the casting is used for alloy cast steel. Electrodes with low-hydrogen coverings are particularly useful in this work.

Multilayer welding and high currents should be employed for low-carbon castings. Low currents and small-diameter electrodes (and, in some cases, preheating and slow cooling) are necessary for high-carbon and special-alloy castings. For gas welding, only large sections are preheated, locally or generally, to a bright red.

### Cast Iron

Even though cast iron has a high carbon content and is a relatively brittle and rigid material, welding can be performed successfully if proper precautions are taken. Optimum conditions for welding include the following: (1) A weld groove large enough to permit manipulation of the electrode or the welding torch and rod. The groove must be clean and free of oil, grease, and any foreign material. (2) Adequate preheat, depending on the welding process used, the type of cast iron, and the size and shape of the casting. Preheat temperature must be maintained throughout the welding operation. (3) Welding-heat input sufficient for a good weld but not enough to superheat the weld metal; i.e., welding temperature should be kept as low as practicable. (4) Slow cooling after welding. Gray iron may be enclosed in asbestos, lime, or vermiculite. Other irons may require postheat treatment immediately after welding to restore mechanical properties.

Most cast-iron welding takes place in foundries, where small defects in new castings are repaired by welding. A small but growing application exists in the production of cast-iron weldments. In these cases, optimum conditions for welding usually can be realized. This is not always the case when castings that have cracked or broken in service require repair at the site. However, if optimum conditions are not realized, the overall quality of the result will be lower. Preheating has two purposes: (1) to reduce the rate of cooling in the vicinity of the weld in order to reduce embrittlement, which reacts unfavorably to weld-shrinkage stress and to machinability; and (2) to reduce unequal expansion stresses in the relatively brittle material during heating, as well as unequal contraction stress during cooling. If the parts being welded are free of restraint, local preheat may be satisfactory. If the parts are under restraint, the effect of heating and cooling stresses may be difficult to assess. Unless a special preheat program can be devised, the best general rule is to uniformly preheat the entire casting.

**Welding rods** and **electrodes** for welding cast iron are described in Specifications for Welding Rods and Covered Electrodes for Welding Cast Iron (AWS 5.15). In this specification, the initial letter R identifies a welding rod for gas welding and the initial letter E an electrode for arc welding. Other letters identify the composition. Thus RCI and RCuZn-A identify welding rods of gray iron and a copper-zinc alloy, respectively. ESt and ENiFe similarly identify electrodes of steel and of a nickel-iron alloy. Many different welding processes have been used to weld cast iron, the most common being manual shielded metal-arc welding, gas welding, and braze welding.

**Manual Shielded Metal-Arc Welding** Of the classified electrodes, the most successful to date has been the nickel-iron electrode (ENiFe). Machinability and color match are good. Weld deposits have some ductility, which is important. Good results have been obtained in welding gray, malleable, and nodular iron. Weld grooves of about 60° are used, as in Fig. 2. For gray iron, the preheat is 400 to 500°F (204 to 260°C). Small-diameter electrodes ($\frac{5}{32}$ in) (4.0 mm) are used with dcrp at about 130 A. Beads should be kept well under $\frac{3}{8}$ in (9.5 mm) wide and should be staggered to avoid high temperature buildup. After welding, the casting is cooled slowly under insulation. Malleable iron may be welded in a similar manner, except that a preheat up to 300°F (149°C) is used and the casting must be specially heat-treated after welding if most of the original mechanical properties are to be restored. For nodular iron using a $\frac{3}{16}$-in (4.8 mm) ENiFe electrode with dcrp at 185 A, a normal bead-on-bead deposition may be used. Again, to restore some of the properties of the base metal, a special heat-treatment is required. The ENiFe electrode may also be used to join various metals to cast iron. The cast-iron electrode (ECI) has a limited use, mainly because better results are obtained when gas welding with the cast-iron welding rod. The steel electrode (ESt) is no longer widely used in large repairs because of the elaborate procedure, known as **studding,** which is required. For small repairs, it yields a hard, unmachinable deposit.

**Braze welding** with copper-alloy or "bronze" electrodes such as ECuSn-A, ECuSn-B, and ECuAl-A2 has the advantage of being faster than with the so-called *brasses* used with gas welding. The latter process offers better heat control, however, and is generally preferred unless time is an overriding factor. In arc welding, striking the arc outside the weld groove must be avoided, as these **arc strikes** cause unmachinable hard spots.

**Gas welding** with the cast-iron welding rod (RCI) is slower than arc welding but affords better heat control because of the lower temperature of the gas flame. The weld groove is chipped or ground to 75 to 90° to permit manipulation. Preheat is in the range of 900 to 1250°F (482 to 677°C). A neutral flame is generally employed. A little flux is spread over the starting point, and the flame is played on the groove until the walls begin to melt, starting at the root. The welding rod is dipped in flux, brought to a red heat, and then rubbed into the molten metal. The tip of the inner cone of the flame should be kept $\frac{3}{16}$ to $\frac{7}{8}$ in (4.8 to 22.2 mm) away from the molten puddle to avoid overheating. The casting should be cooled slowly under insulation.

**Braze welding** with the copper-alloy gas welding rods

(RBCuZn-A,-D, RCuZn-B,-C) requires that the weld groove first be cleaned of all foreign matter and of graphite by heating to red heat with an oxidizing flame or by grit blasting. Preheat is in the range of 900 to 1100°F (482 to 593°C). A neutral-to-slightly-oxidizing flame is recommended. If the temperature of the base metal is too low, the filler metal does not spread out; if too high, it collects in little balls, which are driven away by the force of the flame. To avoid overheating, the torch should be held at a smaller angle to the deposited metal than in welding. The flux should be oxidizing in order to remove graphite, and it should also remove oxide films from the base metal; otherwise the flow of the filler metal will not occur without overheating.

### Wrought Iron

In **metal-arc welding** of wrought iron, slightly slower welding speeds should be used than for mild steel of the same thickness. In this way the metal is kept fluid longer, and gases and slag are eliminated. It may be necessary to use a slightly lower current than for the same thickness of mild steel, especially in thin sections where burning through is possible. Excessive penetration into the face of the metal should be avoided. Electrodes of classification E6010 have proved best and should be used.

**Resistance welding** of wrought iron is the same as that of mild steel.

### Nickel Alloys

**Filler metals** for welding nickel and nickel alloys are covered by the following specifications: Specifications for Nickel and Nickel-alloy Covered Welding Electrodes (AWS A5.11); Specifications for Nickel and Nickel-alloy Bare Welding Filler Metals (AWS A5.14).

**Gas Welding**   The oxyacetylene flame used should be very slightly reducing, with only a small feather, no longer than ⅛ in (3.2 mm), showing beyond the tip of the luminous cone. The end of the welding rod should be kept well within the flame, so as to prevent its oxidation. Besides being reducing, the flame should be soft, rather than harsh, as is the case when too small a tip is used.

A **flux** is always used for Monel (except for gas-welding Monel with the silicon-Monel gas-welding rod for pickling service), but none should be used with pure nickel. A flux recommended by manufacturers should be used. The silicon-Monel gas-welding rod requiring no gas-welding flux was developed particularly for the welding of equipment exposed to sulfuric acid service and specifically for Monel equipment for the pickling of steel. After the flux is painted on both sides of the joint, the flame adjusted to slightly reducing conditions, and the welding begun, Monel flows freely. Nickel, which is not fluxed, flows a little sluggishly. The appearance of properly made gas welds is quite similar to that of good steel welds. For the best results on high-nickel materials there should be little puddling; the molten pool should be kept quiet, with the tip of the luminous flame just touching its surface.

**Welding rods** are of the same composition as the alloy being welded if uniform corrosion resistance, with lack of galvanic effects, is desired. Some leeway is possible with the deoxidizing additions. The manufacturers of nickel furnish proper welding rods where these are specified. Rods should be bright-annealed and free from oxide. Only rods tested for their usability should be used for Monel or nickel welding.

**Arc Welding**   Covered electrodes are used for metal arc welding of Monel and nickel. As the presence of aluminum is desirable in, or near the molten pool of Monel weld metal, a small amount of aluminum is included in the Monel-core wire. The use of a slightly alloyed Monel electrode results in developing strengths of 70,000 to 80,000 lb/in² (483 to 552 MPa) in single-bead and multiple-bead butt joints, metal-arc-welded in flat, vertical, or overhead positions. This is the range of tensile strengths obtained in plate material. Monel and nickel electrodes carrying relatively heavy flux coatings require reverse polarity.

In the **metal arc-welding** of sheets of light to medium gage between 0.037 and 0.125 in (0.9 and 3.2 mm) thick, it is desirable to clamp the sheets to restrain buckling. For lighter gages, it has been found that beads made without weaving are entirely satisfactory. Electrode sizes of ³⁄₃₂ (⅛) [⁵⁄₃₂] in (2.4, 3.2, 4.0 mm) with 50 to 70 (70 to 80) [80 to 140] A are used for 12 (9) [9] gage sheets.

**Gas tungsten-arc** and **gas metal-arc** welding are successfully used on nickel and nickel alloys. No flux is necessary, and welds of high quality are consistently obtainable with both manual and automatic operation.

Nickel and high-nickel alloys may be **resistance-welded** by the spot- and flash-welding processes; seam and projection welding may also be used. Since ease of resistance welding varies as the resistivity of the material, alloys such as Inconel X may be welded more easily than pure nickel. Cleanliness is of extreme importance, and sulfur-bearing substances, such as some oils and greases, must be thoroughly removed by cleaning as they can lead to sulfur embrittlement. High unit pressures are necessary to permit adequate forging of these alloys.

**Brazing** with the BAg, BCu, or BNi filler metals is widely used. Because nickel and high-nickel alloys are usually employed under severe service conditions, careful attention must be given to select the proper brazing procedure and to follow procedural details. The presence, in any form, of low-melting-point elements (such as lead or sulfur) must be avoided. The same is true of oxides. Annealing or stress relieving is advisable before brazing if residual or applied stresses may be present, since the molten filler metal may cause stress cracking. The age-hardenable alloys are particularly susceptible to stress cracking and to the formation of detrimental oxides from oxygen-containing atmospheres. The most common processes are torch, furnace, induction, and resistance brazing. Generally, BAg filler metals may be used in torch brazing, and BCu and BNi filler metals are usually used in controlled-atmosphere brazing.

### Aluminum and Aluminum Alloys

The **properties** that distinguish the **aluminum alloys** from other metals determine which welding processes can be used and which particular procedures must be followed for best results. Among the welding processes that can be used, choice is further dictated by the requirements of the end product and by economic considerations.

Physical properties of aluminum alloys that most significantly affect all welding procedures include low melting-point range (approx 900 to 1215°F) (482 to 657°C), high thermal conductivity (about two to four times that of mild steel), high rate of thermal expansion (about twice that of mild steel), and high electrical conductivity (about three to five times that of mild steel). Interpreted in terms of welding, this means that, as compared with mild steel, much higher welding speeds are demanded, greater care must be exercised to avoid distortion, and for arc and resistance welding, much higher current densities are required.

It should be noted that the aluminum alloys are not quench-hardenable. However, weld cracking may result from excessive shrinkage stresses due to the high rate of thermal contraction. To offset this tendency, welding procedures, where possible, require a fast weld cycle and a narrow-weld zone, e.g., a highly concentrated heat source with deep penetration, moving at a high rate of speed. Shrinkage stresses can also be reduced by using a filler metal of lower melting point than the base metal. The filler metal ER4043 is often used for this purpose.

**Welding procedures** also call for the removal of the thin, tough, transparent film of aluminum oxide that forms on and protects the surface of these alloys. The oxide has a melting point of about 3700°F (2038°C) and can therefore exist as a solid in the molten weld. Removal may be by chemical reduction or by mechanical means such as machining, filing, rubbing with steel wool, or brushing with a stainless-steel wire brush. The aluminum alloys, when molten, can dissolve relatively large amounts of hydrogen, which is given up upon cooling, causing porosity in the weld. To overcome this, all hydrogen-containing compounds, such as moisture, water vapor, oil, and grease, must be removed from the weld zone by degreasing and drying.

As classified by the Aluminum Association, the commercial wrought alloys fall into seven general types according to the principal alloying element. The **non-heat-treatable types** are the 1XXX, 3XXX, and 5XXX series—denoting, respectively, the 99 min percent pure Al, the Al-Mn series, and the Al-Mg series. Second, third, and fourth digits denote alloy modification and specific alloy. These three alloy series are supplied in the annealed form as well as in several work-hardened tempers. Since the heat of welding removes any work-hardened structure in the area near the weld, the strength of the welded joint is based on the annealed strength of the base metal, regardless of the original temper.

The **heat-treatable types** are the 2XXX, 4XXX, 6XXX, and 7XXX series—denoting, respectively, the Al-Cu, the Al-Si, the Al-Si-Mg, and the Al-Zn series. These types can achieve very high mechanical strength through solution heat-treatment combined with age hardening and/or work hardening. The 6XXX series is commonly welded. Although welding will remove the work hardening in the heat-affected zone, the effect on the heat-treated strength is to lower it to somewhere between the original temper and the fully annealed state. Postweld heat-treatment may be effective in restoring strength, depending on the alloy and other conditions. In general, the 2XXX, 4XXX, and 7XXX series are somewhat specialized in application and weldability. Before welding any of the heat-treatable alloys, consultation with the aluminum producer is advisable.

**Filler metals** for welding aluminum and aluminum alloys are covered by Specifications for Aluminum and Aluminum-alloy Arc-welding Electrodes (AWS A5.3-69) and Specifications for Aluminum and Aluminum-alloy Welding Rods and Bare Electrodes (AWS A5.10-69).

**Gas Tungsten-Arc Welding** Most of the advances made in welding aluminum are due to this process and to the one immediately following. Welding is performed in an inert atmosphere and may be manual or mechanized. High current densities and the inherently stable tungsten arc make possible precision and high speed as well as a narrow, deeply penetrated weld zone. These are the factors that minimize distortion, reduce shrinkage stresses, and inhibit microstructural changes in the heat-treated alloys. Economically best suited for welding square-groove welds in thin gages without added filler metal, this process can also be used for welding V grooves in thicker material with cold filler metal added. Welding is effected in all positions. With ac welding, argon is used as the shielding gas and the power supply must deliver high current with balanced wave characteristics or superimposed high-frequency current. With dcsp, helium is used as the shielding gas. Deeper penetration, a narrower weld zone, and faster speed are possible with this method. Welding with either ac or dc power is predicated on absolutely clean weld grooves, clean filler metal, and properly functioning equipment.

**Gas metal-arc welding** employs dcrp in a shielding gas that may be argon, helium, or a mixture of the two. In this process, the welding arc is formed by the filler metal, which serves as the electrode. Since the filler metal is fed from a coil as it melts in the arc, some arc instability may arise. For this reason, the process does not have the same precision as the gas tungsten-arc process for welding very thin gages. However, it is more economical for welding thicker sections because of its higher deposition rates. With a special power supply, a short-circuiting technique is possible which results in a globular transfer of weld metal with low heat input and low penetration. Thinner sections can be welded in any position. Proper functioning of equipment is mandatory, as are clean weld grooves and filler metal.

**Resistance-spot welding** is the most widely used of the aluminum resistance-welding processes. The metal properties indicated above impose specific demands upon the three basic variables—current, force, and time—which define the operating capacity of resistance-welding equipment. Welding currents are roughly three to four times those used for equal thicknesses of mild steel, but weld times are shorter. Special power supplies and timing controls have been developed for this purpose. In addition, both the magnitude and the timing of the electrode force must be closely coordinated with the current flow. Aluminum alloys soften quickly upon heating and undergo considerable shrinkage upon solidifying. To meet these conditions, low-inertia pressure systems have been devised. They afford rapid electrode follow-up to maintain proper electrical contact during heating and proper forging force during cooling. Resistance-spot welding of the aluminum alloys in the soft or annealed condition is not recommended.

The operational limits for current, force, and time are determined somewhat by the service requirements of the end product, by the particular alloy being welded, and by the capabilities of the welding machine and its controls. Nearly all the aluminum alloys can be resistance-spot-welded with excellent results. The non-heat-treatable alloys (1000, 3000, 5000 series) are the easiest to weld; most of the heat-treatable alloys (2000, 6000, 7000 series) are weldable over a narrow range of settings. Except for low-stress applications employing the 1100 or 3003 alloys, cleaning and surface preparation are highly important. Emphasis is placed on obtaining low and uniform surface-contact resistance.

**Resistance-seam welding** makes use of electrodes in the form of wheels, which can produce a series of overlapping welds that result in a pressure-tight seam. In other respects, this process is similar to resistance-spot welding.

**Flash welding** finds an application in welding butt and miter joints in aluminum extrusions and butt joints in aluminum rod and bar stock as well as in dissimilar metals, such as copper to aluminum.

**Aluminum brazing** requires close temperature control because of the narrow range between the melting points of filler metals and base metal. A flux is required to remove the aluminum oxide film and to prevent it from re-forming. After brazing, the flux must be removed completely to avoid corrosion of the aluminum. The commonly brazed alloys are 1100, 3003, 3004, 5050, the 6000 series, and the A612, C612 casting alloys. The common processes used are torch brazing, furnace brazing, and dip brazing, the latter two being preferred because closer temperature control is possible. The filler metals are Al-Si alloys and are supplied as wire, as shims, as a paste mixture of flux and alloy powder, or as brazing sheet. The latter consists of a core alloy, such as 3003 or 6951, clad on one or both sides with the filler metal. Prebraze cleaning is essential for strong joints.

**Gas welding** was the first process used for welding the aluminum alloys and is still used for inexpensive repairs and small jobs. A flux is required; as always, it must be removed after welding. Filler metals are covered by the welding-rod specification indicated above. Smooth, satisfactory joints can be produced in the flat position. Out-of-position welding is very difficult. The high total heat input abets distortion, which may be overcome by suitable preheat and strategic location of weld joints.

**Manual shielded metal-arc welding** using dcrp and flux-covered electrodes produces welded joints that are satisfactory in some applications. Thicknesses ranging from ⅛ to 1 in (3.2 to 25.4 mm) are easily welded. Deep penetration makes possible the welding of ¼-in (6.4 mm) square-groove welds in a single pass with a backing strip. The fluxing slag must be removed between passes and after welding. Filler-metal specifications provide two classifications having core wires equal to the ER1100 and ER4043 welding rods and bare electrodes.

### Copper and Copper Alloys

In welding **commercially pure copper** it is important to select the correct type. Electrolytic, or "tough-pitch," copper contains a small percentage of copper oxide, which at welding heat leads to oxide embrittlement. For welded assemblies it is recommended that deoxidized, or oxygen-free, copper be used and that welding rods, when needed, be of the same analysis. The preferred processes for welding copper are **gas tungsten-arc** and **gas metal-arc welding;** manual **shielded metal-arc** welding can also be used. It is also welded by oxyacetylene method and braze-welded; brazing with brazing filler metals conforming to BAg, BCuP, and RBCuZn-A classifications is also employed. The high heat conductivity of copper requires special consideration in welding; generally higher welding heats are necessary together with concurrent supplementary heating.

**Copper alloys** are extensively welded in industry. The specific procedures employed are dependent upon the analysis, and reference should be made to the AWS Welding Handbook. Filler metals for welding copper and its alloys are covered in the following specifications: Specifications for Copper and Copper-alloy Welding Electrodes (AWS A5.6); Speci-

fications for Copper and Copper-alloy Welding Rods (AWS A5.7).

### Other Metals

**Magnesium** and **magnesium alloys** can be welded in the work-hardened and heat-treated condition with joint strengths approaching that of the base metal. **Gas tungsten-arc** welding and **gas metal-arc** welding are preferred to the **gas welding** process. Resistance-spot welding, seam welding, and flash welding are commonly performed on a number of alloys. Brazing methods are also employed on some alloys. Magnesium alloys are not generally welded to dissimilar metals or alloys.

The **reactive metals,** or so-called **exotic metals,** include titanium, zirconium, columbium (niobium), molybdenum, and tantalum. Their unusual properties make them uniquely suited to specialized applications in the aerospace and nuclear-energy fields. However, their prompt reaction to minute quantities of oxygen, nitrogen, hydrogen, and other impurities at elevated temperatures results in embrittlement. Welding operations, therefore, must be carried out under a shield of high-purity inert gases, such as argon or helium. Gas tungsten-arc welding with rigid control over cleanliness and with careful joint fixturing has been generally most successful. Electron-beam welding is unequalled but is limited by the size of the vacuum chamber. In some instances, brazing under a vacuum or in an inert atmosphere gives excellent results. Resistance-spot and seam welding also have been found useful. As welding procedures have improved and welding costs reduced, many of the reactive metals have begun to find wider industrial applications.

### Other Processes

A number of the welding processes of lesser industrial significance have considerable importance from a technical point of view. Most of these processes are uniquely successful in performing difficult tasks in specialized fields. Many of them promise wider use in industry, especially if present applied research-and-development programs are successful. The following are a few important examples.

**Electron beam welding** makes use of a concentrated beam of electrons accelerated under a high-voltage potential and focused on a small area of the workpiece. The entire process is under high vacuum ($10^{-4}$ mmHg pressure) so that it avoids electron scattering, while also providing an ideal welding environment. Energy transfer is highly efficient, resulting in narrow, deeply penetrating single-pass welds, narrow heat-affected zones, and high welding speeds. Advantages are sound welds completely free of gaseous contamination, freedom from distortion, and high joint efficiencies. Major applications are the metals and alloys which are highly reactive to gases of the atmosphere as well as to gases formed from volatilized impurities. Limitations are imposed by the size of the vacuum chamber, the difficulty in maintaining the vacuum for high production rates, and cost and maintenance of equipment. Present high-voltage equipment operating at about 150 kV and about 6 kW produces single-pass welds in most metals up to about 1½ in (38 mm) thick. Present development work is aimed at higher power input, at greater thicknesses, and at the possibility of welding with the workpiece outside the vacuum chamber.

**Plasma-arc welding,** a more recent application of the cutting process described earlier, is a machine-welding operation. The plasma arc melts completely through the joint, but by precise control of welding speed and other parameters, the molten metal re-forms behind the advancing arc, resulting in a solid weld joint. The advantage is single-pass welding at high speed for limited thicknesses.

**Flame spraying** is a process whereby a metal or a metal compound may be sprayed on a workpiece to provide a protective coating or to build up a worn or undersized area. Metals are introduced into an oxyacetylene flame, either as a powder or as a wire that is melted in the flame, and are impelled toward the workpiece. The bond is mechanical and somewhat porous, although subsequent melting can be used to increase the density of the deposit. The workpiece is usually scored to facilitate the bond. The plasma arc is effectively used to apply the more refractory metals, carbides, and ceramics. Because of its greater heat and velocity, the bond more nearly resembles a weld.

**Electroslag welding** is a vertical welding process designed for welding sections of about 1 to 14 in (25 to 360 mm) thick, or more, in a single pass. When welding heavy steel plates, the plates are positioned vertically about 1 in (25 mm) apart. A starting plate is welded to the underside of the joint, and water-cooled copper shoes are positioned at either side to form the starting weld. Electrodes are conventionally introduced from the top or side through a wire-feeding device which curves down into the mold. Flux is introduced at the start of the weld and is fed continuously as the welding proceeds in order to maintain a molten slag blanket that covers the pool of molten weld metal. Vertical motion is accomplished by means of a relay actuated by a sensing circuit through one of the copper shoes, the essential equipment being mounted, usually, on a vertical mast. The electrically conductive slag remains molten throughout the operation, serving as the heat source and otherwise acting in the same capacity as the slag covering an open-hearth melt. The resulting weld has a dendritic structure resembling an ingot but without segregations or piping. The process is sometimes used to produce ingots of refined metal.

**Solid-state welding** encompasses a group of processes in which the weld is effected by bringing clean metal surfaces into intimate contact under certain specific conditions. In **friction welding,** one part is rotated at high speed with respect to the other, under pressure. The parts are heated, but not to the melting point of the metal. Rotation is stopped at the critical moment of welding. Base-metal properties across the joint show little change because the process is so rapid. **Ultrasonic welding** employs mechanical vibrations at ultrasonic frequencies plus pressure to effect the intimate contact between faying surfaces needed to produce a weld. (See also Sec. 12.) The welding tool is essentially a transducer that converts electric frequencies to mechanical frequencies. By applying the tip of the tool, or anvil, to a small area in the external surface of two lapped parts, the vibrations and pressure are transmitted to the faying surfaces. Foils, thin-gage sheets, or fine wires can be spot- or seam-welded to each other or to heavier parts. In **cold welding,** the parts to be welded are permanently deformed by dies in such a way as to cause some relative displacement between the faying surfaces. Aluminum-to-copper electrical joints are a common application. **Explosive welding** is a relatively new process used to clad a metal substrate, such as steel, with a protective layer of dissimilar metal, such as aluminum. The force and speed of the explosion are directed in such a way as to cause a series of progressive shock waves that deform the faying surfaces at the moment of impact. A magnified section of the joint reveals a true weld with an interlocking waveshape and usually some alloying.

**Laser-beam welding,** as a process, is in the development-application stage. A small area of intense heat can be created by focusing the beam with a condensing lens. Small parts have been spot-welded with high-energy pulses in a fraction of a second. The development of the continuous-wave $CO_2$ gas laser equipment has expanded the application of laser-beam welding considerably. The $CO_2$ laser also is applicable to plastics and ceramics, and both welding and cutting operations are possible.

# MATERIAL-REMOVAL PROCESSES AND EQUIPMENT
## by Serope Kalpakjian

REFERENCES: "Machining Data Handbook," Machinability Data Center, Metcut Research Associates Inc. "Machining," ASM. Armarego and Brown, "The Machining of Metals," Prentice-Hall. Kronenberg, "Machining Science and Application," Pergamon. Boothroyd, "Fundamentals of Metal Machining," Edward Arnold. "Tool and Manufacturing Engineers Handbook," Third ed., McGraw-Hill. Brierly and Siekmann, "Machining Principles and Cost Control," McGraw-Hill. "Machining with Carbides and Oxides," SME. Woldman and Gibbons, "Machinability and Machining of Metals," McGraw-Hill. Roberts, Hamaker, and Johnson, "Tool Steels," ASM. "American Machinists' Handbook," McGraw-Hill. Zorev, "Metal Cutting Mechanics," Pergamon. "Machinery's Handbook," Industrial Press. "Cutting Tool Material Selection," SME. Kobayashi, "Machining of Plastics," McGraw-Hill. Kalpakjian, "Mechanical Processing of Materials," Van Nostrand. King and Wheildon, "Ceramics in Machining Processes," Academic Press. Ham and Bhattacharyya, "Design of Cutting Tools,"

SME. Shaw, "New Developments in Grinding," Carnegie Press. Schlesinger, "Testing Machine Tools," Industrial Press. "Non-traditional Machining Processes," SME. "Cutting and Grinding Fluids: Selection and Application," SME. Koenigsberger, "Design Principles of Metal-cutting Machine Tools." Pergamon. Koenigsberger and Tlusty, "Machine Tool Structures," Pergamon. Tobias, "Machine Tool Vibration," Blackie & Son. DeBarr and Oliver, "Electrochemical Machining," American Elsevier. Wilson, "Practice and Theory of Electrochemical Machining," Wiley-Interscience. Welbourn and Smith, "Machine-tool Dynamics: An Introduction," Cambridge. Koch, "Wood Machining Processes," Ronald. Morse and Cox, "Numerically Controlled Machine Tools," American Data Processing. "Introduction to Numerical Control in Manufacturing," SME. Childs, "Principles of Numerical Control," Industrial Press. "N/C Handbook," Bendix. "International Conference on Manufacturing Technology," SME. "N/C Machinability Data Systems," SME. "Manufacturing

Engineers' Manual," McGraw-Hill. Boothroyd, "Fundamentals of Metal Machining and Machine Tools," McGraw-Hill. "Proceedings of the North American Metalworking Research Conference," published annually since 1973.

## METAL-CUTTING PRINCIPLES

The proper understanding and successful solution of machining problems require a knowledge of several fields, such as mechanics, plasticity, surface phenomena, chemistry, metallurgy, and heat transfer. Investigations in machining making use of these different technical areas already have resulted in a better understanding of the problems involved and have led to improved machining conditions and equipment. The following is a brief summary of the present state of knowledge of the more practical aspects in this field.

### Basic Mechanics of Metal Cutting

The basic mechanics of chip-type machining processes (Fig. 1) are shown, in their simplest form, in Fig. 2. A tool of a certain **rake angle** $\alpha$ (positive as shown) and **relief angle** moves along the surface of the workpiece at a depth of $t_1$. The material ahead of the tool is sheared continuously along the **shear plane,** which makes an angle of $\phi$ with the surface of the workpiece. This angle is called the **shear angle** and, together with the rake angle, determines the chip thickness $t_2$. The ratio of $t_1$ to $t_2$ is called the **cutting ratio** $r$. The relationship between the shear angle, the rake angle, and the cutting ratio is given by the equation

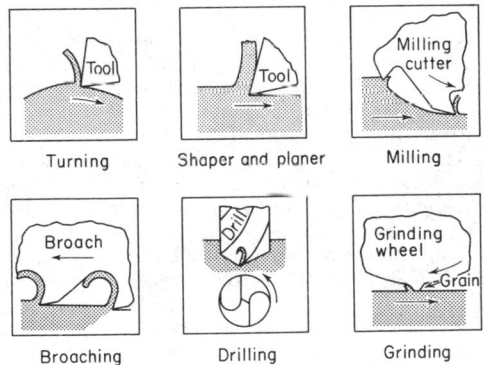

**Fig. 1**  Examples of chip-type machining operations.

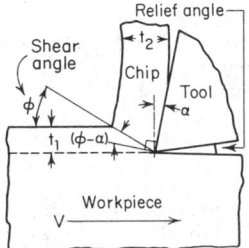

**Fig. 2**  Basic mechanics of the metal cutting process.

$\tan \phi = r \cos \alpha/(1 - r \sin \alpha)$. It can readily be seen that the shear angle is important in that it controls the thickness of the chip. This, in turn, has great influence on cutting performance. The **shear strain** that the material undergoes is given by the equation $\gamma = \cot \phi + \tan (\phi - \alpha)$. Shear strains in metal cutting are usually less than 5.

Investigations have shown that the shear plane may be neither a plane nor a narrow zone, as assumed in simple analysis. Various formulas have been developed which define the shear angle in terms of such factors as the rake angle and the friction angle $\beta$.

Because of the large shear strains that the chip undergoes, it becomes hard and brittle. In most cases, the chip curls away from the tool. Among possible factors contributing to chip curl is nonuniform normal stress distribution on the shear plane.

Regardless of the type of machining operation, four basic types of chips or combinations of these are found:

**Continuous chips** are formed by continuous deformation of the workpiece material ahead of the tool, followed by smooth flow of the chip along the tool face. These chips ordinarily are obtained in cutting ductile materials at high speeds.

**Discontinuous chips** consist of segments which are produced by fracture of the metal ahead of the tool. The segments may be either loosely connected to each other or unconnected. Such chips are most often found in the machining of brittle materials or in cutting ductile materials at very low speeds.

**Inhomogeneous chips** consist of regions of large and small strain. Such chips are characteristic of metals with low thermal conductivity or metals whose yield strength decreases sharply with temperature. Chips from titanium alloys frequently are of this type.

**Built-up edge chips** consist of a mass of metal which adheres to the tool face while the chip itself flows continuously along the face. This type of chip is often encountered in machining operations and is associated with high friction between chip and tool and with poor surface finish.

The **forces** acting on the cutting tool are shown in Fig. 3. The resultant force $R$ has two components, $F_c$ and $F_t$. The cutting force $F_c$ in the direction of tool travel determines the amount of work done in cutting. The thrust force $F_t$ does no work but, together with $F_c$, produces deflections of the tool. The resultant force also has two components on the shear plane: $F_s$ is the force required to shear the metal along the shear plane, and $F_n$ is the normal force on this plane. Two force components also exist on the face of the tool: the friction force $F$ and the normal force $N$.

Whereas the cutting force $F_c$ is always in the direction shown in Fig. 3, the thrust force $F_t$ may be in the opposite direction to that shown in the figure. This occurs when both the rake angle and the depth of cut are large.

From the geometry of Fig. 3, the following relationships can be derived: The **coefficient of friction** on the face of the tool is given by $\mu = (F_t + F_c \tan \alpha)/(F_c - F_t \tan \alpha)$. The **friction force** along the tool is $F = F_t \cos \alpha + F_c \sin \alpha$. The **shear stress** in the shear plane is $\tau = (F_c \sin \phi \cos \phi - F_t \sin^2 \phi)/A_0$, where $A_0$ is the cross-sectional area that is being cut from the workpiece.

The coefficient of friction on the tool face is a complex but important factor in cutting performance and can be reduced by such means as the use of an effective cutting fluid, higher cutting speed, improved tool material and condition, or chemical additives in the workpiece material.

The net **power** consumed at the tool is calculated from the equation $P = F_c V/33,000$. Since $F_c$ is a function of tool geometry, workpiece material, and process variables, it is difficult reliably to calculate its value in a particular machining operation. Depending on workpiece material and the condition of the tool, **unit power** requirements in machining range between 0.2 hp/(in³)(min) [9 W/(cm³)(min)] of metal removal for aluminum and magnesium alloys, to 3.5 (160) for high-strength alloys. The power consumed is the product of unit power and rate of metal removal: $P =$ (unit power)(vol/min).

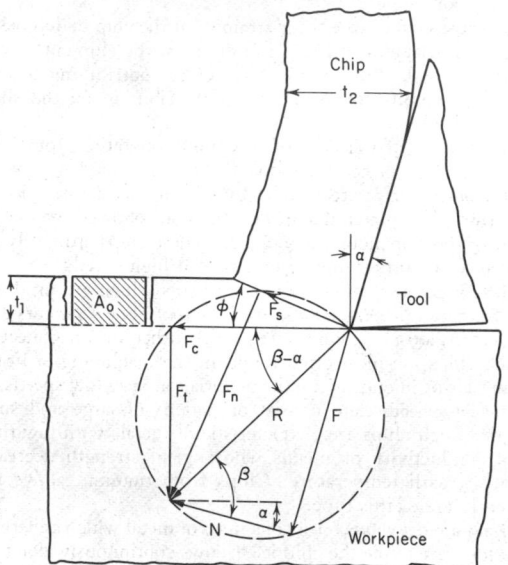

**Fig. 3**  Force system in the metal cutting process.

Most of the power consumed in cutting is transformed into **heat.** Most of the heat is carried away by the chip, and the remainder is divided between the tool and the workpiece. An increase in cutting speed or feed will increase the proportion of the heat transferred to the chip. It has been observed that, in turning, the average interface **temperature** between the tool and the chip increases with cutting speed and feed, while the influence of the depth of cut on temperature has been found to be limited. Interface temperatures to the range of 1500 to 2000°F (800 to 1100°C) have been measured in metal cutting. The effect of a cutting fluid on temperature at speeds normally employed with carbides or ceramics has been found to be negligible. However, the cutting fluid removes heat and thus avoids temperature buildup on the cutting edge.

A factor of great significance in metal cutting is **tool wear.** Many factors determine the type and rate at which wear occurs on the tool. The major critical variables that affect wear are tool temperature, hardness and type of tool material, grade and condition of workpiece, abrasiveness of the microconstituents in the workpiece material, tool geometry, feed, cutting fluid, and surface finish on the tool. The type of wear pattern that develops depends on the relative role of these variables.

Tool wear can be classified as (1) uniform abrasive wear, such as that resulting in a flank wear land (Fig. 4); (2) crater wear on the tool face; (3) localized wear, such as the rounding

of the cutting edge; (4) chipping of the cutting edge; (5) concentrated wear resulting in a deep groove at the edge of a turning tool.

In general, the wear on the flank or relief side of the tool is the most dependable guide for **tool life.** A wear land of 0.060 in (1.5 mm) on high-speed steel tools and 0.015 or 0.030 in (0.4 or

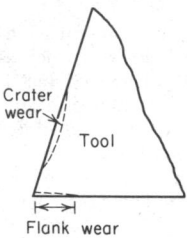

**Fig. 4**  Types of tool wear in cutting

0.8 mm) for carbide tools is usually used as the end point. The cutting speed is the variable which has the greatest influence on tool life. The relationship between tool life and cutting speed is given by the well-known Taylor equation $VT^n = C$, where $V$ is the cutting speed, ft/min; $T$ is the actual cutting time between resharpenings, min; $C$ is a constant whose value depends on workpiece material and machine variables, numerically equal to the cutting speed that gives a tool life of 1 min; and $n$ is the exponent whose value depends on workpiece material and other machine variables. The recommended cutting speed for a high-speed steel tool is generally the one which produces a 60- to 120-min tool life. With carbide tools, a 30- to 60-min tool life may be satisfactory.

For many years, numerous investigations have been carried out modifying the Taylor equation to obtain relationships between cutting speed, feed, depth of cut, workpiece material, tool material, etc. Some of the modified equations for turning are in the form $V_1 d^x f^y = C_1$, where $V_1$ is the equivalent cutting speed, i.e., speed at which a certain cutting time, such as 60 min, is obtained under a given set of cutting conditions between resharpenings; $d$ is the depth of cut, in; $f$ is the feed per revolution, in; $x$ and $y$ are exponents determined experimentally—their average value is less than 1. Two observations can be made from this equation: (1) As the feed or the depth of cut is increased, the cutting speed must be decreased in order to keep the tool life constant. (2) In doing this, the amount of metal removed for the same tool life is increased considerably. Thus, for a given tool life, a large amount of metal can be removed as a result of large depth of cut and feed with a low cutting speed. There are, however, some exceptions to this rule.

When using tool-life equations similar to the ones given above, caution should be exercised in extrapolation of the curves beyond the operating region in which they are derived. In a log-log plot, tool life curves may be linear over a short cutting-speed range but are rarely linear over a wide range of cutting speeds. In spite of the considerable data obtained to date, no simple formulas can be given for quantitative relationships between tool life and process variables for a wide range of materials and conditions.

A term commonly used in machining and comprising most of the items discussed above is **machinability.** This is best defined in terms of (1) tool life, (2) power requirement, and (3) surface integrity. Thus, a good machinability rating would indicate a combination of long tool life, low power requirement, and a good surface. However, it is difficult to develop quantitative relationships between these variables. Tool life is considered as the important factor and, in production, is usually expressed as the number of pieces machined between tool resharpenings. Various tables are available in literature that show the machinability rating for different materials;

however, these ratings are relative. To determine the proper machining conditions for a given material, refer to the machining recommendations given later in this section. These recommendations are the result of extensive tests and data collection from industry.

Metals which machine with great difficulty at room temperature because of their high strength or high rate of work hardening may be machined at elevated temperatures. This technique is called **hot machining** and may result in lower cutting force and longer tool life. However, since optimum conditions of machining exist for each material, the best cutting condition with regard to temperature and cutting speed for maximum tool life must be determined experimentally. Investigations have shown that, except in isolated cases where limited success was obtained, elevated temperature machining of refractory metals and titanium alloys offers no advantages over room temperature machining with appropriate cutting fluids.

The major factors influencing **surface finish** (see later in this section) in machining are (1) the outline of the cutting tool in contact with the workpiece, (2) fragments of built-up edge left on the workpiece during cutting, and (3) vibration. Little quantitative data are available to show relationships between surface roughness and material and process variables. Improvement in surface finish may be obtained to various degrees by increasing the cutting speed and decreasing the feed and depth of cut. Changes in cutting fluid, tool geometry, and material are also important; the microstructure and chemical composition of the material have great influence on surface finish.

Because of plastic deformation, thermal effects, and chemical reactions during material-removal processes, alterations of machined surfaces may take place which can seriously affect the **surface integrity** of a part. Typical detrimental effects may be lowering of the fatigue strength of the part, distortion, changes in stress-corrosion properties, burns, cracks, and residual stresses. Because of its great importance to manufacturing technology, particularly in critical aerospace components, surface integrity is now a recognized and rapidly developing subject. Guidelines for surface-integrity control are being established; a comprehensive summary of the subject is given in "Machining Data Handbook." Surface control generally results in decreasing production rate and increasing costs. Improvements in surface integrity may be obtained by postprocessing techniques such as polishing, sanding, peening, finish machining, and fine grinding.

**Vibration** in machine tools, a very complex behavior, is often the cause of premature tool failure or short tool life, poor surface finish, damage to the workpiece, and damage to the machine itself. Vibration may be **forced** or **self-excited.** The term **chatter** is commonly used to designate self-excited vibrations in metal-cutting machines. The excited amplitudes are usually very high and may cause damage to the machine. Although there is no complete solution to all types of vibration problems, certain measures may be taken. If the vibration is being forced, it may be possible to remove or isolate the forcing element from the machine. In cases where the forcing frequency is near a natural frequency, either the forcing frequency or the natural frequency may be raised or lowered. Damping will also greatly reduce the amplitude. Self-excited vibrations are generally controlled by increasing the rigidity and damping. (See also Sec. 5.)

General experience indicates that good machining practice requires a rigid setup. The machine tool must be capable of providing the **rigidity** required for the machining conditions used. If a rigid setup is not available, the depth of cut must be reduced or the feed must be changed. Excessive tool overhang should be avoided, and in milling, cutters should be mounted as close to the spindle as possible. The length of end mills and drills should be kept to a minimum. Tools with large nose radius or with a long, straight cutting edge increase the possibility of chatter.

As a result of mechanical working and thermal effects, **residual stresses** exist near the surfaces of metals that have been machined or ground. These stresses may cause warping of the workpiece as well as affect the resistance to fatigue and stress corrosion. To minimize residual stresses, sharp tools, medium feeds, and medium depths of cut are recommended. In grinding, high values of the more harmful tensile stresses often result. To minimize residual stresses in grinding, the following is recommended: soft wheels dressed frequently, low wheel speeds, high work speeds, low feeds, and, whenever possible, sulfurized oil. Stress relieving or shot-peening often follows grinding when the endurance limit is critical.

In cutting metal at high speeds, the chips may become very hot and cause safety hazards because of long spirals which whirl around. In such cases, **chip breakers** are introduced on the tool geometry, which curl the chips and cause them to break into short sections. Chip breakers can be produced on the face of the cutting tool, or they are separate pieces clamped on top of the tool.

## CUTTING-TOOL MATERIALS

As a result of research and development over many years, a wide variety of **cutting-tool materials** is now available. The selection of a proper material depends on such factors as the cutting operation involved, the machine to be used, the workpiece material, production requirements, cost, and surface finish and accuracy desired. The major qualities required in a cutting tool are (1) hot hardness, (2) impact toughness or mechanical shock resistance, and (3) wear resistance. (See Table 1.)

Materials for cutting tools include carbon steels, medium-alloy steels, high-speed steels, cast alloys, cemented carbides, ceramics or oxides, and diamonds. Understanding the different types of **tool steels** (see Sec. 6) requires a knowledge of the role of different alloying elements. These elements are added to (1) obtain greater hardness and wear resistance, (2) obtain greater impact toughness, (3) impart hot hardness to the steel such that its hardness is maintained at high cutting temperatures, and (4) decrease distortion and warpage during hardening.

**Carbon** forms a carbide with iron, making it respond to hardening and thus increasing the hardness, strength, and wear resistance. The carbon content of tool steels ranges from 0.6 to 1.4 percent. **Chromium** is added to increase wear resistance and toughness; the content ranges from 0.25 to 4.5 percent. **Cobalt** is commonly used in high-speed steels to increase hot hardness so that tools may be used at higher cutting speeds and temperatures and still maintain hardness and keen cutting edges; the content ranges from 5 to 12 percent. **Molybdenum** is a strong carbide-forming element and increases strength, wear resistance, and hot hardness. It is

### Table 1. General Properties of Tool Materials*

| | Mechanical shock resistance Transverse rupture strength, ksi | Abrasion resistance hardness, Rockwell A | Crater resistance in machining ferrous metals (Arbitrary scale) |
|---|---|---|---|
| High-speed steel | 500–700 (3450–4850 N/mm²) | 81.0–86.5 (60–70 Rc) | Poor |
| Cast alloy | 300–350 (2100–2400) | 81–84 (60–65 Rc) | |
| High-strength carbide | 400–540 (2750–3700) | 91.5–92.0 | |
| C-1 and C-2 carbide | 240–325 (1650–2250) | 90.0–92.0 | Fair |
| C-3 and C-4 carbide | 175–260 (1200–1800) | 92.0–93.0 | |
| C-5 and C-6 carbide | 200–300 (1400–2100) | 90.0–92.5 | Good |
| C-7 and C-8 carbide | 100–250 (700–1700) | 92.0–95.0 | |
| Ceramic | 80–110 (550–750) | 93.5–94.0 | Excellent |

*"Machining Data Handbook," published by the Machinability Data Center, Metcut Research Associates, Inc.

always used in conjunction with other alloying elements, and its content ranges to 10 percent. **Tungsten** promotes hot hardness and strength; content ranges from 1.25 to 20 percent. **Vanadium** increases hot hardness and abrasion resistance; in carbon tool steels, the content is of the order of 0.20 to 0.50 percent, and in high-speed steels, it ranges from 1 to 5 percent.

**Carbon tool steels** (Table 2) are the oldest type of steel used for cutting metals. The steel is inexpensive, has shock resistance, can be heat-treated to obtain a wide range of hardness, is easily formed and ground, and holds a keen cutting edge where excessive abrasion and high heat are absent. Carbon tool steels are used for drills which are to be run at relatively low speeds, taps, broaches, and reamers.

**Medium-alloy tool steels** have greater wear resistance than carbon steels. Alloying elements are introduced to improve

### Table 2. Nominal Composition of Some Carbon Tool Steels
(See also Sec. 6)

| Type | Percentage | | | |
|---|---|---|---|---|
| | C | Cr | V | W |
| W1 | 0.60–1.40 | .... | .... | .... |
| W2 | 0.60–1.40 | .... | 0.25 | .... |
| W3 | 0.60–1.40 | .... | 0.50 | .... |
| W4 | 0.60–1.40 | 0.25 | .... | .... |
| W5 | 0.60–1.40 | 0.50 | .... | .... |
| W6 | 0.60–1.40 | 0.25 | 0.25 | .... |
| W7 | 0.60–1.40 | 0.50 | 0.20 | .... |
| F1 | 1.00 | .... | .... | 1.25 |
| F2 | 1.25 | .... | .... | 3.50 |
| F3 | 1.25 | 0.75 | .... | 3.50 |

hardenability with less distortion. Manganese is the chief alloying element. These steels have been used for drills, taps and reamers but do not have sufficient hot hardness to be used in high-speed turning or milling.

**High-speed steels** (Table 3) are the most highly alloyed group among tool steels and maintain their hardness, strength, and keen cutting edge at high operating temperatures. They are oil-hardening and are little subject to warping. With suitable procedures and equipment, they can be fully hardened with little danger of distortion or cracking. High-speed steel tools are widely used in operations using form tools, drilling, reaming, end-milling, broaching, tapping, and screw machines.

**Cast alloys** (Table 4) maintain high hardness at high temperatures and have good wear resistance. Cast-alloy tools, which are cast and ground into any desired shape, are not as tough as high-speed steels and are sensitive to shock loading. These alloys are recommended for deep roughing operations at relatively high speeds and feeds. Cutting fluids are not necessary and are usually used only to obtain a special surface finish.

**Cemented carbides** have metal carbides as key ingredients and are manufactured by powder-metallurgy techniques. They have the following properties which make them good cutting-tool materials: (1) high hardness over a wide range of temperatures; (2) high Young's modulus, two to three times that of steel; (3) no plastic flow even at very high stresses; (4) low thermal expansion; and (5) high thermal conductivity. Cemented carbides are used in the form of inserts or tips which are brazed or clamped to a steel shank. Because of the difference in coefficients of expansion, brazing should be done carefully. The mechanically fastened tool tips are called **throwaway tips;** they are available in different shapes, such as square, triangular, circular, and special forms.

**Table 3. Nominal Composition of Some High-speed Steels**
(See also Sec. 6)

| Type | Percentage | | | | | Application |
| | W | Cr | V | Mo | Co | |
|------|------|------|------|------|------|-------------|
| T1 | 18 | 4 | 1 | . . . . | . . . . | General purpose |
| T2 | 18 | 4 | 2 | . . . . | . . . . | General purpose, higher strength |
| T3 | 18 | 4 | 3 | . . . . | . . . . | Light cuts, high speed |
| T4 | 18 | 4 | 1 | . . . . | 5 | Heavy cuts |
| T5 | 18 | 4 | 2 | . . . . | 8 | Heavy cuts, abrasion resistant |
| T6 | 20 | 4.5 | 1.5 | . . . . | 12 | Heavy cuts, hard material |
| T7 | 14 | 4 | 2 | . . . . | . . . . | Planer tools |
| T8 | 14 | 4 | 2 | . . . . | 5 | General purpose, hard material |
| T15 | 12 | 4 | 5 | . . . . | 5 | Extreme abrasion resistant |
| M1 | 1.5 | 4 | 1 | 8 | . . . . | General purpose |
| M2 | 6 | 4 | 2 | 5 | . . . . | General purpose |
| M3 | 6 | 4 | 2.75 | 6 | . . . . | Fine-edge tools |
| M4 | 5.5 | 4 | 4 | 4.5 | . . . . | Abrasion resistant |
| M6 | 4 | 4 | 1.5 | 5 | 12 | Heavy cuts, abrasion resistant |
| M7 | 1.75 | 4 | 2 | 8.75 | . . . . | Fine-edge tools, abrasion resistant |
| M8 | 5 | 4 | 1.5 | 5 | * | General purpose, abrasion resistant |
| M10 | . . . . | 4 | 2 | 8 | . . . . | General purpose, high strength |
| M15 | 6.5 | 4 | 5 | 3.5 | 5 | Heavy cuts, abrasion resistant |
| M33 | 1.75 | 3.75 | 1 | 9.75 | 8.25 | Heavy cuts, abrasion resistant |
| M34 | 2 | 4 | 2 | 8 | 8 | Heavy cuts, abrasion resistant |
| M36 | 6 | 4 | 2 | 5 | 8 | Heavy cuts, abrasion resistant |
| M41 | 6.75 | 4.25 | 2 | 3.75 | 5 | Heavy cuts, abrasion resistant |
| M42 | 1.5 | 3.75 | 1.15 | 9.5 | 8 | Heavy cuts, abrasion resistant |
| M43 | 2.75 | 3.75 | 1.60 | 8 | 8.25 | Heavy cuts, abrasion resistant |
| M44 | 5.25 | 4.25 | 2.25 | 6.25 | 12 | Heavy cuts, abrasion resistant |
| M46 | 2 | 4 | 3.2 | 8.25 | 8.25 | Heavy cuts, abrasion resistant |
| M47 | 1.5 | 3.75 | 1.25 | 9.5 | 5 | Heavy cuts, abrasion resistant |

*1.25 Cb.

There are three general groups of cemented carbides in use: (1) tungsten carbide with cobalt as a binder, used in machining cast irons and nonferrous abrasive metals; (2) tungsten carbide with cobalt as a binder, plus a solid solution of WC-TiC-TaC-NbC, for use in machining steels; and (3) titanium carbide with nickel and molybdenum as a binder, for use where cutting temperatures are high because of high cutting speeds or the high strength of the workpiece material.

Carbides are identified as Grade C-1, C-2, etc. (see Table 1). Grades 1 to 4 are recommended for machining cast iron and nonmetallic materials, and grades 5 to 8 are for machining steel and steel alloys. Grades 1 and 5 are for

**Table 4. Nominal Composition of Some Cast-Alloy Tool Materials**

| Trade name | Percentage | | |
| | Co | Cr | W |
|------------|------|------|------|
| Blackalloy 525 | 44 | 24 | 20 |
| Blackalloy T.X. 90 | 42 | 24 | 22 |
| Crobalt 1 | 48 | 30 | 14 |
| Crobalt 2 | 40 | 33 | 18 |
| Crobalt 3 | 40 | 33 | 20 |
| Stellite 98M2 | 38 | 30 | 18 |
| Stellite Star J | 41 | 32 | 17 |
| Stellite No. 3 | 52 | 30 | 11 |
| Stellite No. 19 | 53 | 31 | 10 |
| Tantung G | 47 | 30 | 15 |
| Tantung 144 | 45 | 32 | 18 |

roughing operations, 2 and 6 are for general purpose, 3 and 7 are for finishing, and 4 and 8 are for precision finishing operations. Other grades also are available for a variety of applications and severity of the machining operation.

A new group of carbide tool materials, **coated carbides**, consists of conventional carbide inserts that are coated with a thin layer of titanium carbide or other materials. The coating provides additional wear resistance while maintaining the strength and toughness of the carbide tool.

Rigidity is of great importance when using carbide tools. Light feeds, low speeds, and chatter are deleterious. No cutting fluid is needed, but if one is used for cooling, it should be applied in large quantities and continuously to prevent heating and quenching. To take full advantage of carbide cutting tools, rigid, higher-speed, and more powerful machine tools have been developed.

**Ceramic**, or **oxide**, tool tips consist primarily of fine aluminum oxide grains which have been bonded together. Minor additions of other elements help to obtain optimum properties. Ceramic tools have very high abrasion resistance, are harder than cemented carbides, and have less tendency to weld to metals during cutting. However, they lack impact toughness, and premature tool failure can result by chipping or general breakage (Table 1). Ceramic tools have been found to be effective for high-speed, uninterrupted turning operations. Tool and setup geometry is important. Tool failures can be reduced by the use of rigid tool mountings and rigid machine tools. Included in oxide cutting-tool materials are also borides and cermets.

**Diamonds,** known as **bort,** are used where good surface finish and dimensional accuracy are desired, particularly on soft materials that are difficult to machine. The general properties of diamonds are extreme hardness, low thermal expansion, high heat conductivity, and a very low coefficient of friction. Because of brittleness, tool shape is important in the use of diamonds; special care is given to the proper mounting and crystal orientation of diamonds in order to obtain optimum use. Low values of rake angles are normally used to provide a stronger cutting edge. Diamond tools work satisfactorily at almost any speed.

### Preparation of Tool Steels

As cutting tools, tool steels undergo a number of processes such as forging, machining, normalizing, annealing, hardening, tempering, and grinding and lapping. Each type of tool steel must be given its own special heat-treatment to secure optimum performance. Because of the large variety of tool steels available in industry, it is advisable to consult the literature for specific heat-treating and grinding procedures. The major sources of information are the following two references: "Tool Steels," by Roberts, Hamaker, and Johnson; and "Heat Treating, Cleaning and Finishing," Vol. 2 of "Metals Handbook," published by the American Society for Metals. For additional literature on grinding tool steels, it is advisable to consult the literature of grinding-wheel manufacturers.

### CUTTING FLUIDS
(See also Sec. 6)

**Definition**  Cutting fluids, frequently referred to as lubricants or coolants, comprise those liquids and gases which are applied to the material being cut and to the tool in order to facilitate the cutting operation.

**Purpose**  A cutting fluid is used (1) to keep the tool cool and prevent it from being heated to a temperature at which the hardness and resistance to abrasion are reduced; (2) to keep the work cool, thus preventing it from being machined in a warped shape to inaccurate final dimensions; (3) possibly through lubrication to reduce the power consumption, wear on the tool, and the generation of heat; (4) to provide a good finish on the work; (5) to aid in providing a satisfactory chip formation; (6) to wash away the chips (this is particularly desirable in deep-hole drilling, hacksawing, milling, and grinding); (7) to prevent corrosion of the work and machine; and (8) to lubricate moving machine parts close to the cutting tool.

**Classification**  Cutting fluids may be classified as follows: (1) air used as suction or blast; (2) emulsifiable oils; (3) oils; and (4) chemicals and synthetics. Cutting fluids are also classified as light, medium, and heavy-duty; light-duty fluids are for general-purpose machining.

Induced **air draft** is often used with internal and surface grinding and polishing operations or on grinding and boring operations on gray iron. Its main purpose is to remove the small chips or dust, although some cooling is also obtained.

**Emulsions** consist of a soluble oil or paste emulsified with water in the ratio of 1 part oil to 10 to 100 parts water, depending upon the type of product and the operation. This is a low-cost cutting fluid and is used for practically all types of cutting and grinding when machining all types of metals. The richer mixtures of oil and water, such as 1:10, are used for broaching, threading, and gear cutting, where an oil is not required to secure the desired surface finish or machine lubrication. For most operations, a solution of 1 part soluble oil to 20 parts water will be satisfactory for turret-lathe work, some screw-machine work, gear hobbing, milling, and drilling. The mix of the emulsion is often determined by the rust-prevention requirements of the metal being machined or the lubrication requirements of the machine, and not by the actual machining operation.

A variety of **oils** are used for metal cutting. They are used where lubrication rather than cooling is essential or on high-grade finishing cuts, although sometimes superior finishes are obtained with the emulsions.

**Sulfurized oils** are used generally as cutting fluids for rapid production involving good surface finish and close tolerances on metals difficult to machine. These sulfurized oils may be classified into four groups: (1) sulfurized mineral oils containing up to 4 percent active sulfur in straight mineral oil; (2) sulfurized mixed oils—combinations of fatty oils, such as lard, rapeseed, degras, and petroleum oils to which sulfur is added; (3) sulfurized-base oils—fatty oils containing 8 to 12 percent sulfur, which the user blends with mineral oil for individual requirements; (4) **chlorinated** and sulfurized, and chlorinated mineral oils.

Sulfurized oils are useful in machining low-carbon steel, hot-rolled steels, stainless steel, high-nickel alloys, and Monel. They are used extensively in broaching, gear cutting, and in automatic screw-machine work involving threading and tapping.

**Chemicals** and **synthetics** are a family of cutting fluids that are a blend of water and various chemical agents such as amines, nitrites, nitrates, phosphates, chlorine, and sulfur compounds. These agents are added for purposes of rust prevention, water softening, lubrication, and reduction of surface tension. Most of these chemical fluids are coolants but some are lubricants.

The chief advantage of petroleum-based fluids is their low surface tension; the advantage for water-miscible fluids lies in the high specific heat of water. The choice of a cutting fluid depends on many factors such as speed, ease of cleaning, distortion by heat, and susceptibility of the workpiece material to stress corrosion. As a general rule, if the cutting speed is below 75 ft/min (23 m/min) oils should be used, and for speeds above 75 to 100 ft/min (30 m/min) water-base fluids are recommended.

**Methods of Applying Cutting Fluids**  The most common method is **flood cooling** in quantities such as 3 to 5 gal/min (about 10 to 20 l/min) for single-point tools and up to 60 gal/min (230 l/min) per cutter for multiple-tooth cutters. Whenever possible, multiple nozzles should be used. In **mist cooling** a small jet equipment is used to disperse water-base fluids as very fine droplets in a carrier that is generally air at pressures 10 to 80 lb/in² (0.07 to 0.55 N/mm²). Mist cooling has a number of advantages, such as providing high-velocity fluids to the working areas, better visibility, and improving tool life in certain instances. The disadvantages are that venting is required and also the cooling capability is rather limited. In tapping operations **solid** lubricants such as graphite, molybdenum disulfide, soaps, and waxes may be applied by a brush. In operations such as gun drilling, counterboring, and trepanning a **high-pressure** system may be used where cutting fluids are applied at pressures from 100 to 2,000 lb/in² (0.7 to 14 N/mm²).

Consideration should be given to the proper **disposal** of used cutting fluids in relation to local pollution regulations.

## MACHINING PROCESSES AND EQUIPMENT

The general types of **machine tools** are lathes; turret lathes; screw, boring, drilling, reaming, threading, milling, and gear-cutting machines; planers and shapers; broaching, cutting-off, grinding, and polishing machines. Each of these is subdivided into many types and sizes.

"American Standards for Small Tools and Machine Tool Elements" is published by and may be obtained from the ASME. Names and addresses of machine-tool manufacturers may be obtained from various sources such as the "Mechanical Engineers' Catalog" of the ASME and the special publications of the SME and NMTBA (National Machine Tool Builders' Association). A growing number of established and tentative standards on many aspects of machining processes, equipment, tooling, and safety are available from organizations such as the ASME, ANSI, ASTM, NSA, and NMTBA. Standards on specific topics such as cemented carbides and numerical control are also being established by organizations such as the Cemented Carbide Producers Association and Electronic Industries Association, respectively. Extensive safety and health standards pertaining to machinery and machine guarding, materials handling, and storage, tools, etc., have been promulgated in the Occupational Safety and Health Act (OSHA) of 1970. The major requirements of this Act are that employers provide a place of employment free from recognized hazards that are likely to cause death or serious injury to employees, and that they comply with the standards described in various sections of the Act.

General items common to all machine tools are discussed first, and individual machining processes are treated later in this section.

**Automation** is the application of special equipment to control and perform manufacturing processes with little or no manual effort. It is applied to the manufacturing of all types of goods and processes, from the raw material to the finished product. Automation involves many items, such as handling, processing, assembly, inspecting, and packaging. Its primary objective is to lower manufacturing cost through controlled production and quality, lower labor cost, reduced damage to work by handling, higher degree of safety for personnel, and economy of floor space. Automation may be partial, such as gaging in cylindrical grinding, or it may be complete.

The conditions which play a role in decisions concerning automation are rising production costs, high percentage of rejects, lagging output, scarcity of human labor, hazardous working conditions, and work requiring repetitive operation. Factors which must be carefully studied before deciding on automation are high initial cost of equipment, maintenance problems, and type of product. (See also Sec. 16.)

**Numerical control (N/C)**, which is a method of controlling the motions of machine components by numbers, was first applied to machine tools in the 1950s. Numerically controlled machine tools are now at work in a growing number of plants. Such machines are classified according to the type of cutting process. For instance, in drilling and boring machines, the positioning and the cutting take place sequentially, whereas in die-sinking machines, positioning and cutting take place simultaneously. The latter are often described as **continuous-path** machines, and since they require more exacting specifications, they give rise to more numerous and complex problems. Great efforts are being made to make the machines perform over a very wide range of cutting conditions without requiring adjustment, to eliminate chatter, and to improve accuracy. Information is stored on punch cards or tapes; the basic concept is that holes, representing information, are read by sensing devices which then actuate relays and other devices to control various electrical or mechanical systems. In this way, great savings can be achieved by eliminating templates; complex contours can be machined which would be almost impossible by any other method. A large variety of programming systems has been developed.

Numerical control, however, is not the answer to all manufacturing problems. Selection of parts should be based on economics, involving considerations such as tooling costs, lot size, number and complexity of operations required, and anticipated design changes.

**Direct numerical control (DNC)**, also known as computer-directed numerical control, is a system in which the machine tool is operated through direct linkage with a data-processing computer. Punched tapes and tape readers are eliminated, except for possible use as a backup in case the computer fails.

The proper design of **machine-tool structures** requires analysis of such factors as form and materials of structures, stresses, weight, and manufacturing and performance considerations. Current thinking is that the best approach to obtain the ultimate in machine-tool accuracy is to employ both improvements in structural rigidity and compensation of deflections by use of special controls. The C-frame structure has been used extensively in the past because it provides ready accessibility to the working area of the machine. With the advent of numerical control, the box-type frame with its considerably improved static stiffness becomes practical since the need for manual access to the working area is greatly reduced. The use of a box-type structure with thin walls can provide low weight for a given stiffness. The lightweight-design principle offers high dynamic stiffness by providing a high natural frequency of the structure through combining high static stiffness with low weight rather than through the use of large mass. (Dynamic stiffness is the stiffness exhibited by the system when subjected to dynamic excitation where the elastic, the damping, and the inertia properties of the structure are involved; it is a frequency-dependent quantity.)

Manufacturing methods for structures should make the most efficient use of material. Lightweight designs can best be obtained by fabrication processes such as welding. This requires the use of steel instead of cast iron and raises the question of the lower damping capacity of steel, but the problem can be overcome by the introduction of preloaded frictional surfaces. Against the advantages of lightweight construction must be held the higher cost of labor required. New approaches to correcting lateral deviation of moving parts from their intended paths, caused by deformation under load, tolerances, wear, etc., consist of introducing additional control to sense these deviations and automatically reposition the machine element to compensate for them.

Mass production with modern machine tools has been achieved through the development of self-contained **power-head** production units and the development of **transfer** mechanisms. Power-head units, consisting of a frame, electric driving motor, gearbox, tool spindles, etc., are available for many types of machining operations. Transfer mechanisms move the workpieces from station to station by various methods. Transfer-type machines can be arranged in several configurations, such as a straight line or a U pattern. Various types of machine tools for mass production can be built from components; this is known as the **building-block** principle. Such a system combines flexibility and adaptability with high productivity.

A new approach to optimize machining operations is **adaptive control (A/C).** While the material is being machined, the system senses operating conditions such as forces, tool-tip temperature, rate of tool wear, and surface finish and converts these data into feed and speed control that enables the machine to cut under optimum conditions for maximum productivity. Combined with numerical controls and computers, adaptive controls are expected to result in increased efficiency of metal-working operations.

**Machining Recommendations** Extensive data are being compiled from a large variety of sources such as machine tool manufacturers, material producers, cutting-tool and cutting-fluid manufacturers, and miscellaneous sources such as machinability reports, handbooks, etc. The most comprehensive source of information is "Machining Data Handbook" (2d ed., 1972) published by the Machinability Data Center, Metcut Research Associates Inc., Cincinnati, Ohio. Specific recommendations are made for each material and each machining process, based on a critical evaluation of all available data. The recommendations are, of course, nominal and should be considered good starting points.

In addition to parameters such as speeds and feeds, optimum machining conditions depend on factors such as the surface finish desired, the condition of the machine tool, tolerances, and part configuration. For most common materials, machining recommendations are based on a tool life of 1 to 2 h for high-speed steel or brazed-carbide tools, and 30 to 60 min for throwaway carbide tools. A tool life in excess of 2 h actual cutting time would generally indicate that the speeds and feeds are too low.

Specific information on machining problems may be obtained from the Machinability Data Center. Sources of literature on all aspects of material-removal processes may be found in references such as *Metals Abstracts* and *The Engineering Index*. Technical papers and articles on machining appear regularly in periodicals such as *American Machinist, International Journal of Machine Tool Design and Research, C.I.R.P. Annals, Manufacturing Engineering Transactions, Journal of Engineering Materials and Technology*, and *Journal of Engineering for Industry*. Individual technical papers are also available from technical societies such as the SME, ASME, IME, and SAE.

## LATHES

Lathes are generally considered to be the oldest member of machine tools, having been first developed in the late eighteenth century. The most common lathe is called an engine lathe because it was one of the first machines driven by Watt's steam engine. The basic lathe has the following main parts: bed, headstock, tailstock, and carriage. The types of lathes available for a variety of applications may be listed as follows: engine lathes, speed lathes, horizontal turret lathes, vertical lathes, and automatics. A great variety of lathes and attachments are available within each category, also depending on the production rate required.

**Size of Lathe** It is common practice to specify the size of an engine lathe by giving the **swing** (diameter) and the **distance between centers** when the tailstock is flush with the end of the bed. The maximum swing over the ways is usually greater than the nominal swing. The **length of the bed** is given frequently to specify the overall length of the bed. A lathe size is indicated thus: 14 in (35.6 cm) (swing) by 30 in (76.2 cm) (between centers) by 6 ft (183 cm) (length of bed). Lathes are made for light, medium, or heavy-duty work.

**Lathe Centers** The tailstock center of a lathe is generally made of high-carbon steel ground to an included point angle of 60°. For production or high-speed work, dead centers pointed with Stellite or sintered-carbide tips are used. All centers of this type should be lubricated with a paste such as powdered lead oxide mixed with machine oil. Live centers are used generally. The point, which rotates with the work, is mounted on the shank by means of antifriction bearings.

**Feed and Speed Ratios for Lathes** Geometrical progression is used extensively in designing machine-tool feeds and speeds. Feeds in geometrical progression are used on cylindrical grinders, boring mills, milling machines, drilling machines, etc.; but for screw-cutting lathes, the power feeds for thread cutting and turning must be in proportion to the threads per inch to be cut.

The carriage may be fed along the bed by means of a lead screw for thread cutting or through a feed rod driving through a friction clutch in the apron for general turning and facing. The lead screw is driven positively by a train of gears from the spindle. The feed rod may be driven from the spindle by a belt and change gears, pick-off change gears, or built-in easily selected quick-change gears. Most screw-cutting lathes will cut 36 differently pitched screw threads, ranging from 1½ to 80 per in. By means of gearing and a friction clutch in the apron, the feed rod (or single-splined lead screw if used for turning feeds) provides turning feeds reduced from one-half to one-tenth of these. A reduction of one-tenth the above threads gives 36 turning feeds from 15 to 800 r/min of the work per inch of tool travel, or 0.0666 to 0.00125 in feed per revolution. The power cross feeds are often equivalent to the turning feeds.

To select the pick-off change gears for **thread cutting** if the lathe is simple-geared and the stud runs at the same speed as the spindle, choose some convenient gear for the screw and multiply its number of teeth by the number of threads per inch of the lead screw. Divide this product by the number of threads per inch to be cut. The quotient will be the number of teeth in the gear for the stud.

If the lathe is compound-geared, select at random a set of driving gears, multiply all the numbers of teeth, and multiply this product by the number of threads per inch to be cut. Then select at random all the driven gears except one. Multiply the numbers of teeth of these gears and this product by the number of threads per inch on the lead screw. Divide the first result by the second. The quotient is the number of teeth in the remaining driven gear. **Metric screw threads** may be cut on lathes having lead screws cut to a number of threads per inch by using change gears for each pitch in combination with a pair of transposing gears having 50 teeth in the driver and 127 teeth in the driven. (This is because 127 cm = 50 in.)

**Multispeeds for Lathes**   All geared-head lathes, which are single-pulley belt-driven or arranged for direct-motor drive through short, flat, or V belts, gears, or silent chain, increase the power of the drive and provide a means for obtaining 8, 12, 16, or 24 spindle speeds. The teeth may be of the spur, helical, or herringbone type and may be ground or lapped after hardening.

**Variable speeds** are obtained by driving with adjustable-speed dc shunt-wound motors with stepped field-resistance control or by electronics or motor-generator system to give speed variation in infinite steps. AC motors driving through infinitely variable-speed transmissions of the mechanical or hydraulic type are also in general use.

Modern lathes are built with the speed capacity, rigidity, and strength capable of taking full advantage of new and stronger tool materials. The main drive-motor capacity of lathes ranges from fractional to more than 200 hp (150 kW). Speed preselectors, which give speed as a function of work diameter, are introduced, and variable-speed drives using dc motors with panel control are standard on many lathes. Lathes with contour facing, turning, and boring attachments are also available. They may be operated electrically, hydraulically, and pneumatically. Punched cards and tape are used for programming.

### Tool Shapes

The standard **nomenclature** for single-point tools, such as those used on lathes, planers, and shapers, is shown in Fig. 5. Each tool consists of a shank and point. The point of a single-point tool may be formed by grinding on the end of the shank; it may be forged on the end of the shank and subsequently ground; a tip or insert may be clamped or brazed to the end of the shank. The **best tool shape** for each material and each operation depends on many factors. For specific information and recommendations, the various sources listed in the References should be consulted. See also Table 5.

Positive **rake angles** improve the cutting operation with regard to forces and deflection; however, a high positive rake angle may result in early failure of the cutting edge. Positive rake angles are generally used on lower-strength materials. For higher-strength materials, negative rake angles are used. **Back rake** usually controls the direction of chip flow and is of less importance than the **side rake**. The purpose of **relief angles** is to avoid interference and rubbing between the workpiece and tool flank surfaces. In general, they should be small for high-strength materials and larger for softer materials. Excessive relief angles may weaken the tool. The **side cutting-edge angle** influences the length of chip contact and the true feed. This angle is often limited by the workpiece geometry, e.g., the shoulder contour. Large angles are apt to cause tool chatter. Small **end cutting-edge angles** may create excessive force normal to the workpiece, and large angles may weaken the tool point. The purpose of the **nose radius** is to give a smooth surface finish and to obtain longer tool life by increasing the strength of the cutting edge. The nose radius should be tangent to the cutting-edge angles. A large nose radius gives a stronger tool and

may be used for roughing cuts; however, large radii may lead to tool chatter. A small nose radius reduces forces and is therefore preferred on thin or slender workpieces.

**Turning Recommendations**   The most common tool materials for turning are M2 and M3 high-speed steels and C-6 and C-7 carbides. Ceramic tools may also be used for finish and

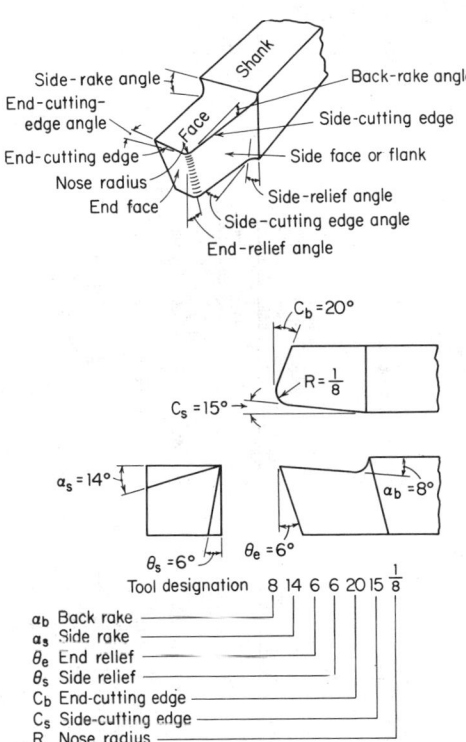

**Fig. 5**   Standard nomenclature for single-point cutting tools.

semifinish turning. Optimum turning speeds depend on factors such as the material and its condition, depth of cut, and type of cutting tool used. Speeds may be in the low range of 10 to 20 ft/min (3 to 6 m/min) for high-temperature alloys, and range up to 2,500 ft/min (750 m/min) and even higher for aluminum and magnesium alloys. Under otherwise similar conditions, the cutting speed should be increased in the following order of tool materials: high-speed steel, cast alloy, brazed carbide, throwaway carbide. Thus, the highest speeds are used with throwaway carbide tools. Also, as the depth of cut increases, the speed should be reduced. In general, the harder the condition of the material, the lower the cutting speed should be in order to maintain the same tool life. For specific, detailed information see "Machining Data Handbook."

### Turret Lathes

Turret lathes are used for the production of parts in moderate quantities and produce interchangeable parts at low production cost. Turret lathes may be chucking, screw machine, or

**Table 5. Tool Geometry for Turning\***

| Material | High-speed steel and cast-alloy tools | | | | | Carbide tools (throwaway) | | | | |
|---|---|---|---|---|---|---|---|---|---|---|
| | Back rake | Side rake | End relief | Side relief | Side and end cutting edge | Back rake | Side rake | End relief | Side relief | Side and end cutting edge |
| Aluminum alloys | 20 | 15 | 12 | 10 | 5 | 0 | 5 | 5 | 5 | 15 |
| Magnesium alloys | 20 | 15 | 12 | 10 | .5 | 0 | 5 | 5 | 5 | 15 |
| Copper alloys | 5 | 10 | 8 | 8 | 5 | 0 | 5 | 5 | 5 | 15 |
| Steels | 10 | 12 | 5 | 5 | 15 | −5 | −5 | 5 | 5 | 15 |
| Stainless steels, ferritic | 5 | 8 | 5 | 5 | 15 | 0 | 5 | 5 | 5 | 15 |
| Stainless steels, austenitic | 0 | 10 | 5 | 5 | 15 | 0 | 5 | 5 | 5 | 15 |
| Stainless steels, martensitic | 0 | 10 | 5 | 5 | 15 | −5 | −5 | 5 | 5 | 15 |
| High-temp. alloys | 0 | 10 | 5 | 5 | 15 | 5 | 0 | 5 | 5 | 45 |
| Refractory alloys | 0 | 20 | 5 | 5 | 5 | : | : | 5 | 5 | 15 |
| Titanium alloys | 0 | 5 | 5 | 5 | 15 | −5 | −5 | 5 | 5 | 5 |
| Cast irons | 5 | 10 | 5 | 5 | 15 | −5 | −5 | 5 | 5 | 15 |
| Thermoplastics | 0 | 0 | 20–30 | 15–20 | 10 | 0 | 0 | 20–30 | 15–20 | 10 |
| Thermosetting plastics | 0 | 0 | 20–30 | 15–20 | 10 | 0 | 15 | 5 | 5 | 15 |

\*"Machining Data Handbook," published by the Machinability Data Center, Metcut Research Associates Inc.

universal. The universal machine may be set up to machine bar stock as a screw machine or have the work held in a chuck. These machines may be semiautomatic, i.e., so arranged that, after a piece is chucked and the machine started, it will complete the machining cycle automatically and come to a stop. They may be horizontal or vertical and single- or multiple-spindle.

The basic principle of the turret lathe is that, with standard tools, setups can be made quickly so that combined, multiple, and successive cuts can be made on a part. By **combined** cuts, tools on the cross slide operate simultaneously with those on the turret, e.g., facing from the cross slide and boring from the turret. **Multiple** cuts permit two or more tools to operate from either or both the cross slide or turret. By **successive** cuts, one tool may follow another to rough or finish a surface; e.g., a hole may be drilled, bored, and reamed at one chucking. In the tool-slide machine only roughing cuts, such as turn and face, can be made in one machine. A second machine similarly tooled must be available to make the finishing cuts.

**Ram-type turret lathes** have the turret mounted on a ram which slides in a separate base. This base can be clamped at any position along the bed to suit long or short work. A cross slide can be used so that combined cuts can be taken from the turret and the cross slide at the same time. Turret and cross slide can be equipped with manual or power feed. The short stroke of the turret slide limits this machine to comparatively short light work, in both small and quantity-lot production.

**Saddle-type turret lathes** have the turret mounted on a saddle which slides directly on the bed. Hence, the length of stroke is limited only by the length of bed. A separate square-turret carriage with longitudinal and transverse movement can be mounted between the head and the hex-turret saddle so that combined cuts from both stations at one time are possible. The saddle type of turret lathe generally has a large hollow vertically faced turret for accurate alignment of the tools.

Turret lathes of a wide range of sizes of the ram type and, in still larger sizes, of saddle-type machines are built. The newer machines are made more rigid and powerful and embody more automatic features. Some have speed and feed calculators, rapid traverse on longitudinal and cross-feed, higher turning speeds, and automatic pressure lubrication.

### Screw Machines

When turret lathes are set up for bar stock, they are often called **screw machines.** Turret lathes that are adaptable only to bar-stock work are constructed for light work. As with turret lathes, they have spring collets for holding the bars during machining and friction fingers or rolls to feed the bar stock forward. Some bar-feeding devices are operated by hand and others semiautomatically. Some screw machines are provided with power feed, but more often both turret and cross slide are operated manually.

**Automatic screw machines** may be classified as single-spindle or multiple-spindle. Single-spindle machines rotate the bar stock from which the part is to be made. The tools are carried on a turret and on cross slides or on a circular drum and on cross slides. Multiple-spindle machines have four, five, six, or eight spindles, each carrying a bar of the material from which the piece is to be made. Capacities range from ⅛ to 6 in (3 to 150 mm) diam of bar stock.

**Feeds** of forming tools vary with the width of the cut. The wider the forming tool and the smaller the diameter of stock, the smaller should be the feed. On multiple-spindle machines, where many tools are working simultaneously, the feeds should be such as to reduce the actual cutting time to a minimum. Often only one or two tools in a set are working up to capacity, as far as actual speed and feed are concerned.

A cutting fluid should always be used to increase tool life and to provide a better finish on the product. A rich emulsion is often satisfactory for general work, provided that it does not interfere with the lubrication of the machine. A light paraffin oil is satisfactory for brass. For aluminum, a paraffin oil plus 5 to 10 percent lard oil is very satisfactory. On most steels, a sulfurized mineral oil or a sulfurized-chlorinated oil with some fatty oil addition is best for general work.

### BORING MACHINES

**Boring machines** are of two general types, horizontal and vertical, and are frequently referred to as horizontal boring machines and vertical boring and turning mills. A classification of boring machines comprises horizontal boring, drilling, and milling machines; vertical boring and turning mills; vertical multispindle cylinder boring mills; vertical cylinder boring mills; vertical turret boring mills (vertical turret lathes); car-wheel boring mills; diamond or precision boring machines (vertical and horizontal); and jig borers.

The **horizontal type** is made for both precision work and general manufacturing. It is particularly adapted for work not conveniently revolved, for milling, slotting, drilling, tapping, boring, and reaming long holes, and for making interchangeable parts that must be produced without jigs and fixtures. The machine is universal and has a wide range of speeds and feeds, for a face-mill operation may be followed by one with a small-diameter drill or end mill.

**Vertical boring mills** are adapted to a wide range of faceplate work that can be revolved. The advantage lies in the ease with which work is fastened to the horizontal table, which resembles a four-jaw independent chuck with extra radial T slots, and in the lessened effect of centrifugal forces arising from unsymmetrically balanced work. Pulleys, flywheels, gear blanks, piston heads, and electric-motor and electric-generator frames and spiders are commonly finished on these machines.

A **jig-boring machine** has a single-spindle sliding head mounted over a table adjustable longitudinally and transversely by lead screws which roughly locate the work under the spindle. Precision setting of the table may be obtained with end measuring rods, or it may depend only on the accuracy of the lead screw. These machines, made in various sizes, are used for accurately finishing holes and surfaces in definite relation to one another. They may use drills, rose or fluted reamers, or single-point boring tools. The latter are held in an adjustable **boring head** by which the tool can be moved eccentrically to change the diameter of the hole.

**Precision-boring machines** may have one or more spindles operating at high speeds for the purpose of boring to accurate dimensions such surfaces as wrist-pin holes in pistons and connecting-rod bushings. Diamonds or cemented carbide are used as the cutters.

**Tools for Boring Mills** The tools used are generally the same as those for lathes. In addition, many special tools are made for particular jobs, and these are frequently furnished by the machine builder. Most common tool materials are M2 and M3 high-speed steel and C-7 and C-8 carbides; ceramic tools may also be used for finish and semifinish operations. Optimum cutting speed depends on the tool material used and the depth of cut. Speeds are generally of the same order of magnitude as in turning: they should be reduced as the depth of cut increases. With carbide tools the speed is generally two to three times that for high-speed steel tools.

## DRILLING MACHINES

**Drilling machines** are intended for drilling holes, tapping, counterboring, reaming, and general boring operations. They may be classified into a large variety of types.

**Sizes of Drilling Machines**   **Vertical drilling machines** are usually designated by a dimension which roughly indicates the diameter of the largest circle that can be drilled at its center under the machine. This dimensioning, however, does not hold for all makes of machines. The sizes begin with about 6 in and continue to 50 in. Heavy-duty presses of the vertical type, with all-geared speed and feed drive, are constructed with a box-type column instead of the older cylindrical column.

The size of a **radial drill** is designated by the length of the arm. This represents the radius of a piece which can be drilled in the center.

**Twist drills** are the most common tools used in drilling and are made in many sizes and lengths. For years they have been grouped according to numbered sizes, 1 to 80, inclusive, corresponding approximately to Stubbs steel wire gage; some by lettered sizes A to Z, inclusive; some by fractional inches from $\frac{1}{64}$ up, and the group of millimeter sizes.

**Straight-shank twist drills** of fractional size and various lengths range from $\frac{1}{64}$ in diam to $1\frac{1}{4}$ in by $\frac{1}{64}$ in increments; to $1\frac{1}{2}$ in by $\frac{1}{32}$ in; and to 2 in by $\frac{1}{16}$ in. **Taper-shank drills** range from $\frac{1}{8}$ in diam to $1\frac{3}{4}$ in by $\frac{1}{64}$ increments; to $2\frac{1}{4}$ in by $\frac{1}{32}$ in; and to $3\frac{1}{2}$ in by $\frac{1}{16}$ in. Larger drills are made by various drill manufacturers.

Tolerances have been set on the various features of all drills so that the products of diffferent manufacturers will be interchangeable in the user's plants.

Twist drills (Fig. 6) are decreased in diameter from point to shank (back taper) to prevent binding. If the web is increased gradually in thickness from point to shank to increase the strength, it is customary to reduce the helix angle as it approaches the shank. The shape of the groove is important, the one that gives a straight cutting edge and allows a full curl to the chip being the best. The **helix angles** of the flutes vary from 10 to 45°. The standard **point angle** is 118°. There are a number of **drill grinders** on the market designed to give the proper angles. The point may be ground either in the **standard** or the **crankshaft** geometry. The drill geometry for high-speed-steel twist drills for a variety of workpiece materials is given in Table 6.

Among the common **types of drills** are the combined drill and countersink or **center drill,** a short drill used to center shafts before squaring and turning; the **step drill,** with two or more diameters; the **spade drill** which has a removable tip or bit clamped in a holder on the drill shank, used for large and deep holes; the **trepanning tool** used to cut a core from a piece of metal instead of reducing all the metal removed to chips; the **gun drill,** run at a high speed under a light feed, and used to drill small long holes; the **core drill** used to bore out cored holes; the **oil-hole drill,** having holes or tubes in its body through which oil is forced to the cutting lips; **three-** and **four-fluted drills,** used to enlarge holes after a leader hole has been cored, punched, or drilled with a two-fluted drill; **twisted drills** made from flat high-speed steel or drop-forged to desired shape and then twisted. Drills are also made of cemented carbide or of high-speed steel with an insert of carbide to form the chisel edge and both cutting edges. They are used primarily for drilling abrasive or very hard materials.

**Drilling Recommendations**   The most common tool material for drills is high-speed steel M1, M7, and M10. Speeds may be of the order of 15 to 20 ft/min (5 to 6 m/min) for high-temperature alloys and range up to 350 ft/min (110 m/min) for aluminum and magnesium alloys. The harder the condition of the material, the lower the speed should be. As the drill diameter increases, the feed, which may be as low as 0.001 in/rev (0.025 mm/rev) for small-diameter drills, should be increased. For specific, detailed information see "Machining Data Handbook."

## REAMING MACHINES

A **reamer** is a multiple-cutting-edge tool used to enlarge or finish holes, to give accurate dimensions as well as good finish. Reamers are of two types: (1) rose and (2) fluted.

The **rose reamer** is a heavy-bodied tool with end cutting edges. It is used to remove considerable metal and to true up a hole preparatory to flute reaming. It is similar to the three- and four-fluted drills. Wide cylindrical lands are provided back of the flute edges.

**Fluted reamers** cut principally on the periphery and remove only 0.004 to 0.008 in (0.1 to 0.2 mm) on the bore. Very narrow cylindrical margins are provided back of the flute edges, 0.012 to 0.015 in (0.3 to 0.4 mm) wide for machine-finish reaming and 0.004 to 0.006 in (0.1 to 0.15 mm) for hand reaming, to provide free cutting of the edges due to the slight body taper and also to pilot the reamer in the hole. The hole to be flute- or finish-reamed should be true. A rake of 5° is recommended for most reaming operations. A reamer may be straight or helically fluted. The latter provide much smoother cutting and give a better finish.

**Expansion reamers** permit a slight expansion by a wedge so that the reamer may be resharpened to its normal size or for job shop use; they provide slight variations in size. **Adjustable reamers** have means of adjusting inserted blades so that a definite size can be maintained through numerous grindings and used-up blades can be replaced with new ones. **Shell**

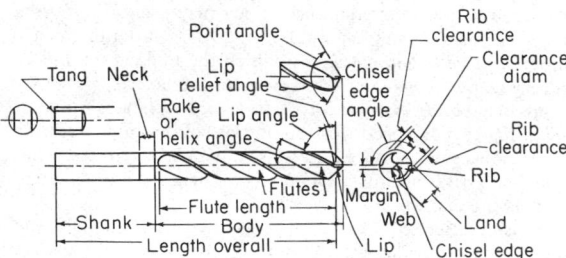

**Fig. 6**   Straight-shank twist drill.

**Table 6. Drill Geometry for High-speed Steel Twist Drills***

| Material | Point angle | Lip relief angle | Chisel edge angle | Helix angle | Point grind |
|---|---|---|---|---|---|
| Aluminum alloys | 90–118 | 12–15 | 125–135 | 24–48 | Standard |
| Magnesium alloys | 70–118 | 12–15 | 120–135 | 30–45 | Standard |
| Copper alloys | 118 | 12–15 | 125–135 | 10–30 | Standard |
| Steels | 118 | 10–15 | 125–135 | 24–32 | Standard |
| High-strength steels | 118–135 | 7–10 | 125–135 | 24–32 | Crankshaft |
| Stainless steels, low-strength | 118 | 10–12 | 125–135 | 24–32 | Standard |
| Stainless steels, high-strength | 118–135 | 7–10 | 120–130 | 24–32 | Crankshaft |
| High-temp. alloys | 118–135 | 9–12 | 125–135 | 15–30 | Crankshaft |
| Refractory alloys | 118 | 7–10 | 125–135 | 24–32 | Standard |
| Titanium alloys | 118–135 | 7–10 | 125–135 | 15–32 | Crankshaft |
| Cast irons | 118 | 8–12 | 125–135 | 24–32 | Standard |
| Plastics | 60–90 | 7 | 120–135 | 29 | Standard |

*"Machining Data Handbook," published by the Machinability Data Center, Metcut Research Associates Inc.

**reamers** constitute the cutting portion of the tool which fits interchangeably on arbors to make many sizes available or to make replacement of worn-out shells less costly. Reamers float in their holding fixtures to ensure alignment, or they should be piloted in guide bushings above and below the work. They may also be held rigidly, such as in the tailstock of a lathe.

The **speed** of high-speed steel reamers should be two-thirds to three-quarters and **feeds** usually are two or three times that of the corresponding drill size. The most common tool materials for reamers are M1, M2, and M7 high-speed steels and C-2 carbide.

## THREADING MACHINES

Threads may be formed on the inside or outside of a cylinder or cone (1) with single-point threading tools, (2) with threading chasers, (3) with taps, (4) with dies, (5) by thread milling, (6) by thread rolling, (7) by grinding. There are numerous types of taps, such as hand, machine screw, tapper, nut, pulley, boiler, mud or washout, staybolt, pipe, combined pipe tap and drill. Small taps usually have no radial relief. They may be made in two, three, or four flutes. Large taps may have still more flutes.

The **feed** of a tap depends upon the lead of the screw thread. The **cutting speed** depends upon numerous factors: Hard tough materials, great length of hole, taper taps, and full-depth thread reduce the speed; long chamfer, fine pitches, and a good cutting fluid applied in quantity increase the speed. Taps are cut or formed by grinding. The ground-thread taps may operate at much higher speeds than the cut taps. Speeds may range from 3 ft/min (1 m/min) for high-strength steels to 150 (45 m/min) for aluminum and magnesium alloys. Common high-speed steel tool materials for taps are M1, M7, and M10.

**Threading dies,** used to produce external threads, may be solid, adjustable, spring-adjustable, or self-opening die heads. Replacement chasers are used in die heads and may be of the fixed or self-opening type. These chasers may be of the radial type, hobbed or milled; of the tangential type; or of the circular type. Emulsions may prove satisfactory for many threading operations, such as on brass, bronze, cast iron,

malleable iron, and some steels. When threading low-carbon steel, alloy steel, and Monel metal, sulfurized oils have proved satisfactory.

## MILLING MACHINES

**Milling machines** use cutters with multiple teeth in contrast with the single-point tools of the lathe and planer. The work is generally fed past the cutter perpendicular to the cutter axis. Milling usually is face or peripheral cutting. The cutting edge is cooled intermittently, as the cuts are not continuous.

Milling machines have shown marked improvement. The machines are heavier, provide more power to the spindle, and give greater accuracy and convenience of operation for the effective use of carbide milling cutters at optimum cutting speeds and feeds.

**Standard spindle noses and arbors for milling machines** provide interchangeability of arbors and face-milling cutters, regardless of make or size of machine. The taper of the spindle end and arbor is 3½ in/ft, to make them **self-releasing.** The retention of the shank is dependent upon a positive locking device, such as screws or draw-in bolt. When unlocked, these tapers release themselves.

Milling-machine classification is based on design, operation, or purpose. **Knee-and-column** type milling machines have the table and saddle supported on the vertically adjustable knee gibbed to the face of the column. The table is fed longitudinally on the saddle, and the latter transversely on the knee to give three feeding motions.

Knee-type machines are made with horizontal or vertical spindles. The **horizontal** universal machines have a swiveling table for cutting helices. The plain machines are used for jobbing or manufacturing work, the universal for toolroom work. **Vertical** milling machines with fixed or sliding heads are otherwise similar to the horizontal type. They are used for face or end mill work and are frequently provided with a rotary table for making cylindrical surfaces.

**Hand millers,** with feeding movements hand-controlled, are sensitive, as the cuts can be felt. They are used only on light work.

The **fixed-bed** machines have a spindle mounted in a head dovetailed to and sliding on the face of the column. The table rests directly on the bed. They are simple and rigidly built and are used primarily for high-production work. These machines are usually provided with work-holding fixtures and may be constructed as plain or multiple-spindle machines, simple or duplex.

**Rotary-type** millers usually have a rotating table on which fixtures carrying the work are mounted. The cutter spindles are mounted over the edge of the table past which the work is fed.

**Drum-type** millers consist of a drum carrying the work and rotating on a horizontal axis. Both rotary and drum types are mass-production machines, there being no idle time, for the drums rotate continuously while the parts, loaded on one side of the machine, pass first the roughing and then the finishing cutters and are replaced when they return to the loading position.

In **planetary** milling machines the work is stationary on the bed or clamped to the tailstock while the cutter rotates. Plain and formed internal or external surfaces and threads are produced by inserting the cutter into the bore to be milled, feeding it to depth radially, making a sweeping cut about the bore, and withdrawing first radially and then axially.

**Planer-type** millers are used only on the heaviest work. They are used to machine a number of surfaces on a particular part or group of parts arranged in series in fixtures on the table.

**Thread millers** are used to cut threads and worms. A single formed cutter may be used or all the threads may be cut at one time by a multiple-thread cutter.

### Milling Cutters

**Milling cutters** are made in a wide variety of shapes and sizes. The nomenclature of tooth parts and angles is standardized as in Fig. 7. Milling cutters may be classified in various ways, such as purpose or use of the cutters (Woodruff keyseat

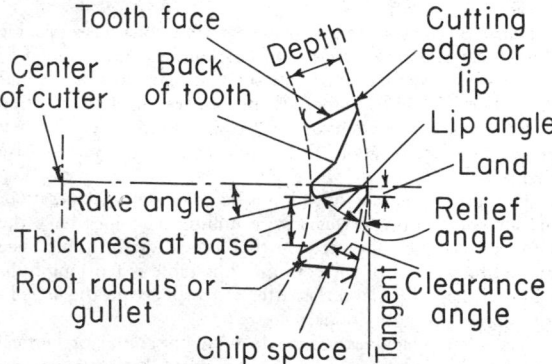

**Fig. 7**  Plain milling-cutter teeth.

cutters, T-slot cutters, gear cutters, etc.); construction characteristics (solid cutters, carbide-tipped cutters, inserted-blade cutters, etc.); method of mounting (arbor type, shank type, etc.); and relief of teeth. The latter has two categories: profile cutters which produce flat, curved, or irregular surfaces, with the cutter teeth sharpened on the land; and formed cutters which are sharpened on the face to retain true cross-sectional form of the cutter.

Two kinds of milling are generally considered to represent all forms of milling processes: **peripheral** and **face** milling. In the peripheral-milling process the axis of the cutter is parallel to the surface milled, whereas in face milling, or slab milling, the cutter axis is generally at a right angle to the surface. The peripheral-milling process is also divided into two types: **up** (conventional) milling and **down** (climb) milling. Each has its advantages, and the choice depends on a number of factors such as the type and condition of the equipment, tool life, surface finish, and machining parameters.

**Milling Recommendations**  The most common tool materials for milling cutters are M2 and M7 high-speed steel and, for face milling, C-6 and C-7 carbide. Speeds recommended are similar to those in turning operations. The feed may range from 0.003 in per tooth (0.076 mm per tooth) for high-temperature alloys to 0.022 for aluminum and magnesium alloys (0.076 mm per tooth to 0.56 mm per tooth). For specific information, see "Machining Data Handbook."

## GEAR MANUFACTURING
(See also Sec. 8)

**Gear Cutting**  Most gear-cutting processes can be classified as either **forming** or **generating**. In a forming process, the shape of the tool is reproduced on the work; in a generating process, the shape produced on the work depends on both the shape of the tool and the relative motion between the tool and the work during the cutting operation. A soft live center on a lathe, for example, can be formed by means of a broad, flat form tool fed at right angles to the lathe spindle, or generated by a single-point tool fed at the point angle in the compound rest. In general, a generating process is more accurate than a forming process.

In the form cutting of gears, the tool has the shape of the space between the teeth. For this reason, form cutting will produce precise tooth profiles only when the cutter is accurately made and the tooth space is of constant width, such as on spur and helical gears. A form cutter may cut or finish one or all of the spaces in one pass. Single-space cutters may be disk-type or end-mill-type milling cutters. In all single-space operations, the gear blank must be retracted and indexed, i.e., rotated one tooth space, between each pass.

Single-space form milling with disk-type cutters is particularly suitable for gears with large teeth, because, as far as metal removal is concerned, the cutting action of a milling cutter is more efficient than that of the tools used for generating. In gear manufacturing plants the form milling of spur gears is done on machines that retract and index the gear blank automatically. In job and repair shops, form cutting is done on a milling machine equipped with a **dividing head** that is indexed manually; the cutter is the only special equipment required.

For the same tooth size (pitch), the shape (profile) of the teeth on an involute gear depends on the number of teeth on the gear. Most gears have active profiles that are wholly, partially, or approximately involute, and, consequently, accurate form cutting would require a different cutter for each number of teeth. In most cases, satisfactory results can be obtained by using the eight cutters for each pitch that are commercially available. Each cutter is designed to cut a range of tooth numbers; the no. 1 cutter, for example, cuts from 135 teeth to a rack, and the no. 8 cuts 12 and 13 teeth.

| No. of cutter | 1 | 2 | 3 | 4 | 5 | 6 | 7 | 8 |
|---|---|---|---|---|---|---|---|---|
| No. of teeth | 135–∞ | 55–134 | 35–54 | 27–34 | 21–26 | 17–20 | 14–16 | 12 and 13 |

For more accurate gears, 15 cutters are available.

| No. of cutter | 1 | 1½ | 2 | 2½ | 3 | 3½ | 4 | 4½ |
|---|---|---|---|---|---|---|---|---|
| No. of teeth | 135–∞ | 80–134 | 55–79 | 42–54 | 35–41 | 30–34 | 26–29 | 23–25 |
| No. of cutter | 5 | 5½ | 6 | 6½ | 7 | 7½ | 8 | |
| No. of teeth | 21 and 22 | 19 and 20 | 17 and 18 | 15 and 16 | 14 | 13 | 12 | |

Multiple-space form cutting is done with a broach or with a patented Shear-Speed toolhead. A broach resembles a long external or internal gear, with teeth of increasing depth and provided with cutting clearances. The broach is pushed down into or over the gear blank, usually on a hydraulic press, and all the spaces can be cut in one pass.

The Shear-Speed toolhead contains three main parts: a housing, a member with radial slots in which a tool for each of the tooth spaces on the gear being cut can slide, and a movable, double cone-shaped guiding unit that controls the radial movement (feed) of the tools. With the toolhead stationary, the part is reciprocated past the cutting tools; each tool is fed radially a predetermined amount each stroke until the full tooth depth is cut. To avoid drag, the tools are retracted on each return (noncutting) stroke of the gear.

Gears always operate in pairs and are usually required to have contacting profiles of such a shape that the ratio of the speeds of the pair remains constant at all times; such gears are said to be **conjugate** to one another.

In a gear generating machine, the generating tool can be considered as one of the gears in a conjugate pair and the gear blank as the other gear. The correct relative motion between the tool arbor and the blank arbor is obtained by means of a train of indexing gears within the machine.

One of the most valuable properties of the involute as a gear-tooth profile is that if a cutter is made in the form of an involute gear of a given pitch and any number of teeth, it can generate all gears of all tooth numbers of the same pitch and they will all be conjugate to one another. The generating tool may be a pinion-shaped cutter, a rack-shaped (straight) cutter, or a hob, which is essentially a series of racks wrapped around a cylinder in a helical, screwlike, form.

On a gear shaper, the generating tool is a pinion-shaped cutter that rotates slowly at the proper speed as if in mesh with the blank; the cutting action is produced by a reciprocation of the cutter parallel to the work axis. These machines can cut spur and helical gears, both internal and external; they can also cut continuous-tooth helical (herringbone) gears and are particularly suitable for cluster gears, or gears that are close to a shoulder.

On a rack shaper the generating tool is a segment of a rack that moves perpendicular to the axis of the blank while the blank rotates about a fixed axis at the speed corresponding to conjugate action between the rack and the blank; the cutting action is produced by a reciprocation of the cutter parallel to the axis of the blank. Since it is impracticable to have more than 6 to 12 teeth on a rack cutter, the cutter must be disengaged from the blank at suitable intervals and returned to the starting point, the blank meanwhile remaining fixed. These machines can cut both spur and helical external gears.

A gear-cutting hob is basically a worm, or screw, made into a generating tool by cutting a series of longitudinal slots or "gashes" to form teeth; to form cutting edges, the teeth are "backed-off," or relieved, in a lathe equipped with a backing-off attachment. A hob may have one, two, or three threads; on involute hobs with a single thread, the generating portion of the hob-tooth profile usually has straight sides (like an involute rack tooth) in a section taken at right angles to the thread.

When hobbing a spur gear, the acute angle between the hob and work spindles is 90° minus the lead angle at the hob threads. If the gear is to have 60 teeth, for example, and the hob has a single thread, the hob and work spindles must be geared together by the indexing gears so that the hob makes 60 revolutions while the work makes one; if a double-threaded hob were used, the hob would make 30 revolutions to one of the work.

In addition to the conjugate rotary motions of the hob and work, the hob must be fed parallel to the work axis for a distance greater than the face width of the gear. The feed, per revolution of the work, is produced by the feed gears, and its magnitude depends on the material, pitch, and finish desired; the feed gears are independent of the indexing gears. The hobbing process is continuous until all the teeth are cut.

The same machines and the same hobs that are used for cutting spur gears can be used for helical gears; it is only necessary to tip the hob axis so that the hob and gear pitch helices are tangent to one another and to correlate the indexing and feed gears so that the blank and the hob are advanced or retarded with respect to each other by the amount required to produce the helical teeth. Some hobbing machines have a differential gear mechanism that permits the indexing gears to be selected as for spur gears and the feed gearing to be chosen independently.

The threads of worms are usually cut with a disk-type milling cutter on a thread-milling machine and finished, after hardening, by grinding. Worm gears are usually cut with a hob on the machines used for hobbing spur and helical gears. Except for the gashes, the relief on the teeth, and an allowance for grinding, the hob is a counterpart of the worm. The hob and work axes are inclined to one another at the shaft angle of the worm and gear set, usually 90°. The hob may be fed in to full depth in a radial (to the blank) direction or parallel to the hob axis.

Although it is possible to approximate the true shape of the teeth on a straight bevel gear by taking two or three cuts with a form cutter on a milling machine, this method, on account of the taper of the teeth, is obviously unsuited for the rapid production of accurate teeth. Most straight bevel gears are roughed out in one cut with a form cutter on machines that index automatically and then finished to the proper shape on a generator.

The generating method used for straight bevel gears is analogous to the rack-generating method used for spur gears. Instead of using a rack with several complete teeth, however, the cutter has only one straight cutting edge that moves, during generation, in the plane of the tooth of a basic crown gear conjugate to the gear being generated. A crown gear is the rack among bevel gears; its pitch surface is a plane, and its teeth have straight sides.

The generating cutter moves back and forth across the face of the bevel gear like the tool on a shaper; the "generating roll" is obtained by rotating the gear slowly relative to the tool. In practice two tools are used, one for each side of a tooth; after each tooth has been generated, the gear must be retracted and indexed to the next tooth.

The machines used for cutting spiral bevel gears operate on essentially the same principle as those used for straight bevel gears; only the cutter is different. The spiral cutter is basically a disk that has a number of straight-sided cutting blades protruding from its periphery on one side to form the rim of a cup. The machines have means for indexing, retracting, and producing a generating roll; by disconnecting the roll gears, spiral bevel gears can be form cut.

**Gear Shaving**　For improving the surface finish and profile accuracy of cut spur and helical gears (internal and external), gear shaving, a free-cutting gear finishing operation that removes small amounts of metal from the working surfaces of the teeth, is employed. The teeth on the shaving cutter, which may be in the form of a pinion (spur or helical) or a rack, have a series of sharp-edged rectangular grooves running from tip to root. The intersection of the grooves with the tooth profiles creates cutting edges, and when the cutter and the work, in tight mesh, are caused to move relative to one another along the tooth, the cutting edges scrape metal from the teeth of the work gear. Usually the cutter drives the work, which is free to rotate and is traversed past the cutter parallel to the work axis. Shaving requires less time than grinding, but ordinarily it cannot be used on gears harder than approximately 400 Brinell (42 Rockwell C).

**Gear Grinding**　Machines for the grinding of spur and helical gears utilize either a forming or a generating process. For form grinding, a disk-type grinding wheel is dressed to the proper shape by a diamond held on a special dressing attachment; for each number of teeth a special index plate, with V-type notches on its periphery, is required. When grinding helical gears, means for producing a helical motion of the blank must be provided.

For grinding-generating, the grinding wheel may be a disk-type, double-conical wheel with an axial section equivalent to the basic rack of the gear system. A master gear, similar to the gear being ground, is attached to the work arbor and meshes with a master rack; the generating roll is created by rolling the master gear in the stationary rack.

In the 1940s an interesting extension of the hobbing principle to the finish-grinding of gears was introduced in the United States by a Swiss machine-tool builder. The machine employs a large-diameter grinding wheel that is dressed, on its periphery, to the form of an ungashed involute hob. An unusual feature is the use of separate synchronous motors to drive the gear blank and the wheel at the proper relative speeds. These machines can produce precision spur and helical gears at high production rates.

Spiral bevel and hypoid gears can be ground on the machines on which they are generated. The grinding wheel has the shape of a flaring cup with a double-conical rim having a cross section equivalent to the surface that is the envelope of the rotary cutter blades.

**Gear Rolling**　The cold-rolling process for gear production, which was introduced in the 1960s, is now firmly established as a production process for the finishing of spur and helical gears for automatic transmissions and power tools; in some cases it has replaced gear shaving. It differs from cutting in that the metal is not removed in the form of chips but is displaced under heavy pressure.

There are two main types of cold-rolling machines on the market, namely, those employing dies in the form of racks or gears that operate in a parallel-axis relationship with the blank and those employing worm-type dies that operate on axes at approximately 90° to the work axis. The dies, under pressure, create the tooth profiles by the plastic deformation of the blank.

When racks are used, the process resembles thread rolling; with gear-type dies the blank can turn freely on a shaft between two dies, one mounted on a fixed head and the other on a movable head. The dies have the same number of teeth and are connected by gears to run in the same direction at the same speed. In operation, the movable die presses the blank into contact with the fixed die, and a conjugate profile is generated on the blank. On some of these machines the blank can be fed axially, and gears can be rolled in bar form to any convenient length.

On machines employing worm-type dies, the two dies are diametrically opposed on the blank and rotate in opposite directions. The speeds of the blank and the dies are synchronized by change gears, like the blank and the hob on a hobbing machine; the blank is fed axially between the dies.

Although gears have been cold-rolled from solid blanks, most commercial rolling is done as a finishing operation on fine-pitch, precut gears.

## PLANERS AND SHAPERS

**Planers** are used to rough and finish large flat surfaces, although arcs and special forms can be made with proper tools and attachments. Surfaces to be finished by scraping, such as ways and long dovetails and, particularly, parts of machine tools, are, with few exceptions planed. With fixtures to arrange parts in parallel and series, quantities of small parts can be produced economically on planers. Milling planers will plane, mill, drill, and bore work at one clamping.

Planer tools are clamped to a head, comprising a clapper attached to a ram mounted in a saddle which, in turn, is mounted on the horizontal rail or attached to the face of the vertical housing. The ram is fed vertically and the whole saddle horizontally. The clapper, hinged at its upper end, permits the tool point to be swung by dragging or by positive means away from the machined surface on the noncutting stroke; the latter is necessary for cemented-carbide tools. Large machines may have two or more tool heads on the cross rail and frequently one on each housing below the rail, known as side heads. The latter are used to plane down the sides of work and for undercutting.

The **size** of a planer is indicated by the width of the table, or distance between housings, the maximum height cleared under the rail (these dimensions usually being equal), and the length of the table between pockets. In modern planers, the length of the bed is twice the length of the table.

Cutting speeds for planers range up to 400 ft/min (120 m/min) with capacities up to 150 hp (110 kW). Recommended speeds range from 10 to 20 ft/min (3 to 6 m/min) for cast irons and stainless steels, up to 300 ft/min (90 m/min) for aluminum and magnesium alloys. Feeds range between 0.020 and 0.125 in per stroke (0.5 and 0.32 mm per stroke). The feed is such as to allow a small overlap. The most common tool materials for planers are M2 and M3 high-speed steels and C-2 and C-6 carbides. Carbide tools must be lifted from the work on the return stroke to avoid chipping. This is accomplished hydraulically or with electromagnets.

**Shapers** are used for miscellaneous planing, surfacing, notching, key seating, and production of flat surfaces on flat parts. They are essentially toolroom, repair-job, or job-shop machines. The tool is held in a holder supported on a clapper on the end of a ram which is reciprocated hydraulically or by crank and rocker arm, in a straight line. A table carrying the vise and the work feeds transversely on each return stroke.

**Horizontal shapers** have cutting speeds usually ranging up to 70 ft/min (20 m/min) with return cutting speeds nearly twice that amount. Ram reciprocations vary up to 150 per min. In the hydraulic shaper a motor drives an oil pump which provides uniform cutting speed and pressure throughout the stroke. Commercial sizes of shapers are up to 36-in (91.4-cm) length of stroke. A 48-in (122-cm) all-steel shaper with a triangular ram provides speeds up to 400 ft/min (120 m/min).

**Vertical shapers** or **slotters** reciprocate the ram vertically. They are made in a wide range of sizes and are used on general work for key seating, notching, facing, die making, and surfacing. A rotary table with indexing attachment may be mounted on the knee which may be fed mechanically along and toward the rail.

## BROACHING

**Broaching** is a production process whereby a cutter, called a **broach,** is used to finish internal or external surfaces such as holes of circular, square, or irregular section, keyways, the teeth of internal gears, multiple spline holes, and flat surfaces. In broaching, the action of the broach itself serves as a clamping medium so that in many cases the operation may be completed in the time ordinarily taken to chuck the piece. The expense of making a broach may be large. However, the cost of maintaining its size is usually not excessive, and very little scrap work results. Broaching round holes gives greater accuracy and better finish than reaming, but, as the broach may be guided only by the work it is cutting, the hole may not be accurate with respect to previously machined surfaces. Where such accuracy is required, it is better practice to broach first and then turn other surfaces with the work mounted on a mandrel. The cutting tool or broach is usually long and is provided with many teeth so graded in size that each takes a small chip when the tool is pulled or pushed through the previously prepared leader hole or past the surface.

The main features of the broach are the pitch, the degree of taper or increase in height of each successive tooth, relief, tooth depth, and rake.

The **pitch** of the teeth, i.e., the distance from one tooth to the next, depends upon tooth strength, length of cut, shape and size of chips, etc. The pitch should be as coarse as possible to provide ample chip clearance, but at least two teeth should be in contact with the work at all times. The formula $p = 0.35 \sqrt{l}$ may be used, where $p$ is pitch of the roughing teeth and $l$ the length of hole or surface, in. An average pitch for small broaches is $\frac{1}{8}$ to $\frac{1}{4}$ in (3.175 to 6.35 mm) and for large ones $\frac{1}{2}$ to 1 in (12.7 to 25.4 mm). Where the hole or other surface to be broached is short, the teeth are often cut on an angle or helix, so as to give more continuous cutting action by having at least two teeth cutting simultaneously.

The degree of **taper,** or increase in size per tooth, depends largely on the hardness or toughness of the material to be broached and the finish desired. The degree of taper or feed for broaching cast iron is approximately double that for steel. Usually the first few teeth coming in contact with the work are undersize but of uniform taper to take the greatest feeds per tooth, but as the finished size is approached, the teeth take smaller and smaller feeds with several teeth at the finishing end of nearly zero taper. In some cases, for soft metals and even cast iron, the large end is left plain or with rounded lands a trifle larger than the last cutting tooth so as to burnish the surface. For medium-sized broaches, the taper per tooth is 0.001 to 0.003 in (0.025 to 0.076 mm). Large broaches remove 0.005 to 0.010 in (0.127 to 0.254 mm) per tooth or even more. The teeth are given a **front rake** angle of 5 to 15° to give a curl to the chip, provide a cleaner cut surface, and reduce the power consumption. The **land** back of the cutting edge, which may be $\frac{1}{64}$ to $\frac{1}{16}$ in (0.4 to 1.6 mm) wide, usually is provided with a land relief varying from 1 to 3° with a clearance of 30 to 45°.

The heavier the feed per tooth or the longer the surface being broached, the greater must be the **chip clearance** or space between successive teeth for the chips to accumulate. The root should be a smooth curve.

Broaches are generally made of M2 or M7 high-speed steel. Broaches of complicated shape are apt to warp during the heat-treating process. For this reason, in hardening, they are often heated in a vertical cylindrical furnace and quenched by being hung in an air blast furnished from small holes along the side of pipes placed vertically about the broach. Cemented carbide is also used for the teeth of large broaches or for burnishing buttons.

**Push broaches** are usually shorter than pull broaches, being 6 to 14 in (15 to 35 cm) long, depending on their diameter and the amount of metal to be removed. In many cases, for accuracy, four to six broaches of the push type constitute a set used in sequence to finish the surface being broached. Push broaches usually have a large cross-sectional area so as to be sufficiently rigid. With **pull broaches,** pulling tends to straighten the hole, whereas pushing permits the broaches to follow any irregularity of the leader hole. Push broaching permits the use of cheaper broaches. Pull broaches are attached to the cross head of the broaching machine by means of a key slot and key, by a threaded connection, or by a head that fits into an automatic broach puller. The threaded connection is used where the broach is not removed from the drawing head while the work is placed over the cutter, as in cutting a keyway. In enlarging holes, however, the small end of the pull broach must first be extended through the reamed, drilled, or cored hole and then fixed in the drawing head before being pulled through the work.

**Broaching Machines**  Push broaching is done on machines of the press type with a sort of fixture for holding the work and broach or on presses operated by power. They are usually vertical and may be driven hydraulically or by screw, rack, or crank. The pull type of broach may be either vertical or horizontal. The ram may be driven hydraulically or by screw, rack, or crank. Both are made in the duplex- and multiple-head type.

**Speeds and Feeds for Broaching**  Cutting speeds for broaching may range from 5 ft/min (1.5 m/min) for high-strength materials to as high as 30 to 50 ft/min (9 to 15 m/min) for aluminum and magnesium alloys. The most common tool materials are M2 and M7 high-speed steels. An emulsion of soluble oil is often used for broaching for general work, but some grades of steel forgings require an oil either of the compounded type or of the sulfur-base mineral or animal-oil type. In broaching cast iron where a polished or burnished finish is required, it is advisable to use a cutting compound.

## CUTTING-OFF MACHINES

These machines, used for cutting off bars, extruded and rolled shapes, etc., are made in six types: the lathe type using single-point cutoff tools as in screw machines, hack saws, band saws, friction-wheel saws, abrasive-wheel saws, and circular saws.

In **power hack saws,** the frame in which the blade is strained is reciprocated above the work which is held in a vise on the bed. The cutting feed is effected by weighting the frame, with 12 to 50 lb (5 to 22 kg) from small to large machines; adding weights or spring tension giving up to 180 lb (82 kg); providing a positive screw feed or a friction screw feed; and by hydraulic feed mechanism giving forces up to 300 lb (136 kg) between the blade and work. With high-speed steel blades, cutting speeds range from about 30 strokes per minute for high-strength materials to 180 strokes per minute for carbon steels.

Hack-saw **blades** for hand frames are made 8, 10, and 12 in long, 7/16 to 9/16 in wide, and 0.025 in thick. Number of teeth per inch for cutting soft steel or cast iron, 14; tool steel and angle iron, 18; brass, copper, and heavy tubing, 24; sheet metal and thin tube, 32.

Blades for power hack saws are made of alloy steels and of high-speed steels. Each length is made in two or more widths. The coarsest teeth should be used on large work and with heavy feeds.

**Band saws,** vertical, horizontal, and universal, are used for cutting-off work. They are particularly useful for cutting off gates and sprues of brass, bronze, and iron castings and for cutting off high-speed and stainless steel, especially for disks of a thickness that is small compared with the diameter. The kerf or width of cut is small with a consequently small loss in expensive material. The teeth of band saws, like those of hack saws, are set with the regular alternate type, one bent to the right and the next to the left, or with the alternate and center set, in which one tooth is bent to the right, the second to the left, and the third straight in the center. With high-speed steel saw blades, band speeds range from about 30 ft/min (10 m/min) for high-temperature alloys to about 400 (120) for carbon steels. For aluminum and magnesium alloys the speed ranges up to about 1,300 ft/min (400 m/min) with high-carbon blades.

The band speed should be decreased as the work thickness increases.

**Friction sawing machines** are used largely for cutting off structural shapes. Peripheral speeds of about 20,000 ft/min (6,000 m/min) are used. The wheels may be plain on the periphery, V-notched, or with milled square notches.

**Abrasive cutoff saws,** which are similar to the steel friction saw, are made of thin rubber or resinoid bonded abrasive wheels. The wheels operate at speeds of 12,000 to 16,000 ft/min (3,600 to 4,800 m/min). These wheels are excellent for cutting off tubes, shapes, hardened high-speed steel, Stellite, etc.

**Circular saws** are made in a wide variety of styles and sizes. Some makers list them in numbered sizes, others according to the diameter of the saw blade that the machine will carry. Circular saws may have teeth of several shapes, as radial face teeth for small fine-tooth saws; radial face teeth with a land for fine-tooth saws; alternate bevel-edged teeth to break up the chips, with every other tooth beveled 45° on each side with the next tooth plain; alternate side-beveled teeth; and the Simonds patented tooth, having one tooth beveled on the right, the next on the left, and the third beveled on both sides, each leaving slightly overlapping flats on the periphery of the teeth. The most common high-speed steels for circular saws are M2 and M7.

## GRINDING MACHINES

**Grinding machines** may be classified as to purpose and type as follows: for **rough removal of stock,** the swinging-frame, portable, flexible shaft, two-wheel stand, and disk; **cutting off** or parting, the cutting-off machine; **surface finishing,** band polisher, two-wheel combination, two-wheel polishing machine, two-wheel buffing machine, and semiautomatic polishing and buffing machine; **precision grinding,** tool post, cylindrical (plain and universal), crankshaft, centerless, internal, and surface (reciprocating table with horizontal or vertical wheel spindle, and rotary table with horizontal or vertical wheel spindle); **special form** grinders, gear or worm, ball-bearing balls, cams, and threads; **tool and cutter** grinders for single-point tools, drills, and milling cutters, reamers, taps, dies, knives, etc; **finishing and sizing,** as cylinder honing machines and lapping machines; and **pulpwood** grinders.

All these machines, except the polishing, buffing, disk, lapping, and honing types, use **grinding wheels** made by fixing an abrasive in a binder which is usually hardened by baking. The disk type uses an abrasive disk or abrasive cloth pasted to the surface of a metal disk or a thin abrasive wheel anchored or glued to the steel wheel face. **Lapping machines** use a lapping plate charged with diamond grit or substitute or a plain soft lap formed to fit the work with an oil which contains a fine abrasive, while **honing machines,** such as those used to finish automotive cylinders, use several honing stones which are fixed in a body adjustable to any diameter. **Centerless grinders** are used to good advantage where large numbers of relatively small pieces must be ground and where the ground surface has no exact relation to any other surface except as a whole; the work is carried on a support between two abrasive wheels, one a normal grinding wheel, the second a rubber-bonded wheel, rotating at about 1/20th the grinding speed, and is tilted 3 to 8° to cause the work to rotate and feed past the grinding wheel (see also Sec. 6). **Cutter grinders** are designed to grind all kinds of milling and similar toothed cutters.

**Cylinder grinders** are a special type for grinding the inside of cylinders of engines; one form has a planetary motion for the grinding spindle. The **cylindrical grinder** is a companion machine to the engine lathe; shafts, cylinders, rods, studs, and a wide variety of other cylindrical parts are first roughed out on the lathe, then finished accurately to size by the cylindrical grinder. The work is carried on centers, rotated slowly, and traversed past the face of a grinding wheel.

Grinding equipment of all types has been improved during the past few years so as to be more rigid, provide more power to the grinding wheel, and provide automatic cycling, loading, clamping, wheel dressing, and, in some instances, automatic feedback.

**Universal grinders** are cylindrical machines arranged with a swiveling table so that both straight and taper internal and external work can be ground; they are used on tool work and in refined manufacturing. **Drill grinders** are provided with rests so mounted that by a simple swinging motion, correct cutting angles are produced automatically on the lips of drills; a cupped wheel is usually employed. **Internal grinders** are used for finishing the holes in bushings, rolls, sleeves, cutters, and the like; speeds from 15,000 to 30,000 r/min are common.

**Horizontal surface grinders** range from small capacity, used mainly in tool making or small production work, to large sizes used for production work.

**Vertical surface grinders** are used for producing flat surfaces on production work. **Vertical** and **horizontal disk grinders** are used for surfacing. Grinding machines are used for **cutting off** steel, especially tubes, structural shapes, and hard metals. A thin resinoid or rubber-bonded wheel is used, with aluminum oxide abrasive for all types of steel, aluminum, brass, bronze, nickel, Monel, and Stellite; silicon carbide for cast iron, copper, carbon, glass, stone, plastics, and other nonmetallic materials; and crushed diamonds for cemented carbides and quartz.

**Belt grinders** use a coated abrasive belt running between pulleys. Belt grinding is generally considered to be a roughing process, but finer finishes may be obtained by using finer grain size. Belt speeds generally range from 2,000 to 10,000 ft/min (600 to 3,000 m/min) with grain sizes ranging between 24 and

320, depending on the workpiece material and the surface finish desired. The process has the advantage of high-speed material removal and is applied to flat as well as irregular surfaces.

Although grinding is primarily regarded as a finishing operation, it is possible to increase the rate of stock removal whereby the process becomes, in certain instances, competitive with milling. This type of grinding operation is usually called **abrasive machining.** It uses equipment such as reciprocating table or vertical-spindle rotary table surface grinders with capacities up to 300 hp (220 kW). The normal stock removal may range up to $\frac{1}{4}$ in (6.4 mm) with wheel speeds between 3,400 and 5,000 surface ft/min (1,000 to 1,500 m/min).

A good ground surface with **surface quality** of 50 to 200 $\mu$in rms (0.0012 to 0.005 mm) is sufficient for many purposes and is a basic requirement for further finishing operations, such as lapping, honing, and superfinishing. The size and depth of the scratches can be varied considerably by the selection of the wheel.

**Grinding Wheels**
(See also Sec. 6)

**Grinding wheels** have characteristics influenced by (1) type of abrasive; (2) grain size, sometimes called the grit size; (3) grade; (4) structure; and (5) type of bond. (See Fig. 8.)

**Selection of Abrasive**   Although a number of natural abrasives are available, such as emery, corundum, quartz, garnet, and diamond, the most commonly used abrasives in grinding wheels are **aluminum oxide** and **silicon carbide,** the former being more commonly used than the latter. Aluminum oxide is softer than silicon carbide, and because of its friability and low attritious wear it is suitable for most applications. Silicon carbide is used for grinding aluminum, magnesium, titanium, copper, tungsten, rubber, Teflon, silverplate, and porcelain enamel. It is also used for grinding very hard and brittle materials such as carbides, ceramics, and stones. **Diamond** grains are used to grind very hard materials.

Selection of **grain size** depends on the rate of material removal desired and the surface finish. Coarse grains are used

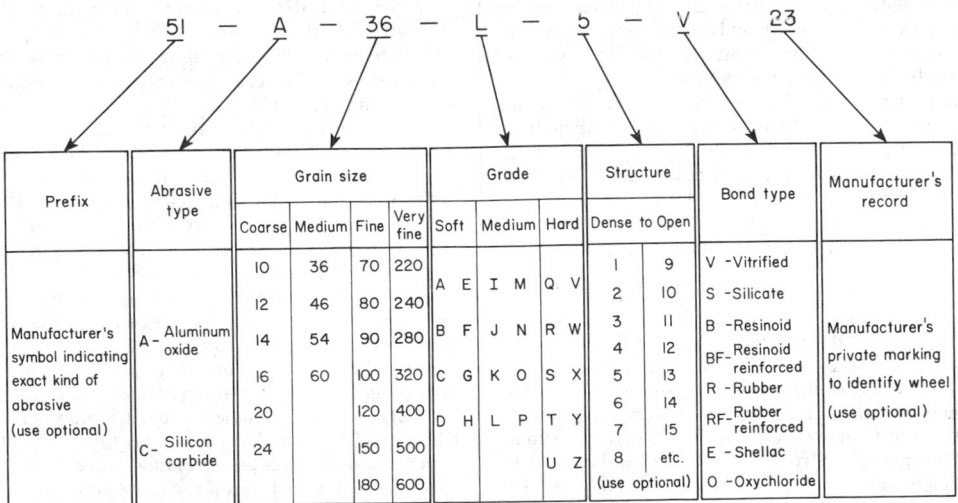

51 — A — 36 — L — 5 — V   23

| Prefix | Abrasive type | Grain size | | | | Grade | | | Structure | | Bond type | | Manufacturer's record |
|---|---|---|---|---|---|---|---|---|---|---|---|---|---|
| | | Coarse | Medium | Fine | Very fine | Soft | Medium | Hard | Dense to Open | | | | |
| | | 10 | 36 | 70 | 220 | | | | 1 | 9 | V -Vitrified | | |
| | | 12 | 46 | 80 | 240 | A  E | I  M | Q. V | 2 | 10 | S -Silicate | | |
| | A - Aluminum oxide | 14 | 54 | 90 | 280 | B  F | J  N | R W | 3 | 11 | B -Resinoid | | Manufacturer's |
| Manufacturer's symbol indicating exact kind of abrasive (use optional) | | 16 | 60 | 100 | 320 | C  G | K  O | S  X | 4 | 12 | BF- Resinoid reinforced | | private marking to identify wheel (use optional) |
| | | 20 | | 120 | 400 | D  H | L  P | T  Y | 5 | 13 | R -Rubber | | |
| | C - Silicon carbide | 24 | | 150 | 500 | | | U  Z | 6 | 14 | RF- Rubber reinforced | | |
| | | | | 180 | 600 | | | | 7 | 15 | E - Shellac | | |
| | | | | | | | | | 8 (use optional) | etc. | O -Oxychloride | | |

**Fig. 8**   Standard-marking-system chart for grinding wheels and other bonded abrasives.

for fast removal of stock; fine grain for small removal rates and for fine finish. Coarse grains are also used for ductile materials and a finer grain for hard and brittle materials.

The **grade** of a grinding wheel is a measure of the strength of its bond. The force that acts on the grain in grinding depends on process variables (such as speeds, depth of cut, etc.) and the strength of the work material. Thus a greater force on the grain will increase the possibility of dislodging the grain; if the bond is too strong, the grain will tend to get dull, and if it is too weak then wheel wear will be great. If glazing occurs, the wheel is **acting hard;** reducing the wheel speed or increasing the work speed or the depth of cut causes the wheel to **act softer.** If the wheel breaks down too rapidly, reversing this procedure will make the wheel act harder. Harder wheels are generally recommended for soft work materials, and vice versa.

A variety of **bond** types are used in grinding wheels; these are generally categorized as **organic** and **inorganic.** Organic bonds are materials such as resin, rubber, shellac, and other similar bonding agents. Inorganic materials are glass, clay, porcelain, sodium silicate, magnesium oxychloride, and metal. The most common bond type is the vitrified bond which is composed of clay, glass, porcelain, or related ceramic materials. This type of bond is brittle and produces wheels that are rigid, porous, and resistant to oil and water. The most flexible bond is rubber which is used in making very thin, flexible wheels. Wheels subjected to bending strains should be made with organic bonds. In selecting a bonding agent, attention should be paid to its sensitivity to temperature, stresses, and grinding fluids, particularly over a period of time. The term **reinforced** as applied to grinding wheels indicates a class of organic wheels which contain strengthening fabric or filament, such as fiber glass. This term does not cover wheels with reinforcing elements such as steel rings, steel cup backs, or wire or tape winding, Fiber glass and filament reinforcing increases the ability of wheels to withstand operational forces when cracked.

The **structure** of a wheel is important in two aspects: It supplies a clearance for the chip, and it determines the number of cutting points on the wheel.

In addition to wheel characteristics, grinding wheels come in a very large variety of shapes and dimensions. They are classified as **types,** such as Type 1-straight wheels, Type 4-taper side wheels, Type 12-dish wheels, etc.

The **grinding ratio** is defined as the ratio of the volume of material removed to the volume of wheel wear. The grinding ratio depends on parameters such as the type of wheel, work speed, wheel speed, cross-feed, down-feed, and the grinding fluid used. Values ranging from a low of 2 to over 200 have been observed in practice. A high grinding ratio, however, may not necessarily result in the best surface integrity of the part.

**Wheel Speeds**  Depending on the type of wheel and the type and strength of bond, wheel speeds for standard applications range between 4,500 and 16,000 surface ft/min (1,400 and 4,800 m/min). The lowest speeds are for low-strength, inorganic bonds whereas the highest speeds are for high-strength organic bonds. The majority of surface grinding operations are carried out at speeds from 5,500 to 6,500 ft/min (1,750 to 2,000 m/min). A recent trend is toward high-efficiency grinding where wheel speeds from 12,000 to 18,000 ft/min (3,600 to 5,500 m/min) are employed. It has been found that, by increasing the wheel speed, the rate of material removal can be increased, thus making the process more economical. This, of course, requires special grinding wheels to withstand the high stresses. Design changes or improvements involve items such as a composite wheel with a vitrified bond on the outside and a resinoid bond toward the center of the wheel; elimination of the central hole of the wheel by providing small bolt holes; and clamping of wheel segments instead of using a one-piece wheel. Grinding machines for such high-speed applications have requirements such as rigidity, high work and wheel speeds, high power, and special provisions for safety.

**Work speeds** depend on the size and type of work material and on whether it is rigid enough to hold its shape. In surface grinding, table speeds generally range from 50 to 100 ft/min (15 to 30 m/min); for cylindrical grinding, work speeds from 70 to 100 ft/min (20 to 30 m/min), and for internal grinding they generally range from 75 to 200 ft/min (20 to 60 m/min).

**Cross-feed** depends entirely on the width of the wheel. In roughing, the work should travel past the wheel $\frac{3}{4}$ to $\frac{7}{8}$ of the width of the wheel for each revolution of the work. As the work travels past the wheel with a helical motion, the preceding rule allows a slight overlap. In finishing, a finer feed is used, generally $\frac{1}{10}$ to $\frac{1}{4}$ of the width of the wheel for each revolution of the work.

**Depth of Cut**  In the roughing operation the depth of cut should be all the wheel will stand without crowding. This varies with the hardness of the material and the diameter of the work; the operator's experience is the only guide. In the finishing operation the depth of cut is always slight, 0.0005 to 0.001 in (0.013 to 0.025 mm). Excellent results as regards finish are obtained by letting the wheel run over the work several times without cross-feeding. This practice of letting the wheel "grind out" has been found by the majority of expert operators to give satisfactory results even with a comparatively coarse wheel.

**Grinding Allowances**  From 0.005 to 0.040 in (0.13 to 1 mm) is generally removed from the diameter in rough grinding in a cylindrical machine. For finishing, 0.002 to 0.010 in (0.05 to 0.25 mm) is common. Work can be finished by grinding to a tolerance of 0.0002 in (0.005 mm) and a surface roughness of 50 + $\mu$in rms (0.0012 mm).

In situations where grinding leaves unfavorable surface residual stresses, the technique of **gentle** or **low-stress** grinding may be employed. This generally consists of removing a layer of about 0.010 in (0.25 mm) at depths of cut of 0.0002 to 0.0005 in (0.005 to 0.013 mm) with wheel speeds that are lower than the conventional 5,500 to 6,500 ft/min.

**Truing and Dressing**  The wheel face should be sharp, i.e., present newly fractured crystals to the face to act as cutting tools. In **truing,** a diamond supported in the end of a soft steel rod held rigidly in the machine is passed over the face of the wheel two or three times to remove just enough material to give the wheel its true geometric shape. **Dressing** is a more severe operation of removing the dull or loaded surface of the wheel. Abrasive sticks or wheels or steel star wheels are pressed against and moved over the wheel face.

**Safety**  If stored, handled, and used improperly, a grinding wheel can be a very dangerous tool. Because of its mass and high rotational speed, a grinding wheel has considerable energy and, if it fractures, it can cause serious injury and even death to the operator or to personnel nearby. A safety code B7.1 entitled "The Use, Care, and Protection of Abrasive

Wheels" is available from ANSI; other safety literature is available from the Grinding Wheel Institute, Cleveland, Ohio, and from the National Safety Council, Chicago.

The salient features of safety in the use of grinding wheels may be listed as follows: Wheels should be stored and handled carefully; a wheel that has been dropped should not be used. Before it is mounted on the machine, a wheel should be visually inspected for possible cracks; a simple "ring" test may be employed whereby the wheel is tapped gently with a light nonmetallic implement and if it sounds cracked it should not be used. The wheel should be mounted properly with the required blotters and flanges, and the mounting nut tightened not too excessively. The label on the wheel should be read carefully for maximum operating speed. An appropriate guard should always be used with the machine, whether portable or stationary. Newly mounted wheels should be allowed to run idle at the operating speed for at least 1 min before grinding. The operator should always wear safety glasses and should not stand directly in front of a grinding wheel when a grinder is started. If a grinding fluid is used, it should be turned off first before stopping the wheel to avoid creating an out-of-balance condition. Because for each type of job and workpiece material there usually is a specific type of wheel recommended, the operator must make sure that the appropriate wheel has been selected.

## POLISHING

**Polishing** is an operation by which scratches or tool marks or, in some instances, rough surfaces left after forging, rolling, or similar operations are removed. It is not a precision operation. Polishing wheels are built-up wheels of wood, leather, cloth, felt, etc., with the abrasive glued to the surface.

The nature of the polishing process has been debated for a long time. Two mechanisms appear to play a role: One is fine-scale abrasion, and the other is melting of surface layers. In addition to removal of material by the abrasive particles, the high temperatures generated because of friction soften the asperities of the surface of the workpiece, resulting in a smeared surface layer. Furthermore, chemical reactions may also take place in polishing whereby surface irregularities are removed by chemical attack.

Polishing may be divided into three steps: **rough polishing, dry fining,** and **finishing** or **oiling.** The abrasive grain used for roughing usually runs from no. 20 to 80, for dry fining from no. 90 to 120, and for finishing or oiling from no. 150 to the fine flours. For the first two steps the polishing wheels are used dry. For finishing, the wheels are first worn down a little and then coated with tallow, oil, beeswax, or similar substances. This step is partly polishing and partly buffing, as additional abrasive is often added in cake form with the grease. The cutting action is freer and the life of the wheel is prolonged by making the wheel surface flexible. Buffing wheels are also used for the finishing step when tallow, etc., containing coarse or fine abrasive grains is periodically rubbed against the wheel.

To prepare a rough forging for nickel-plating may require all three steps of polishing. The first or even second polishing step may be omitted on some smooth objects or soft materials. Many steel or nonferrous metal parts are given a high luster without being plated. When nickel-plated, the final luster or "color" is secured by a buffing operation.

**Polishing wheels** consisting of wooden disks faced with leather, turned to fit the form of the piece to be polished, are used for flat surfaces or on work where it is necessary to maintain square edges. A large variety of other types of wheels are in common use. Compress wheels are used extensively and are strong, durable, and easily kept in balance. They consist of a steel center the rim of which holds a laminated surface of leather, canvas, linen, felt, rubber, etc., of various degrees of pliability. Wheels of solid leather disks of walrus hide, buffalo hide, sheepskin, or bull's neck hide, or of soft materials such as felt, canvas, and muslin, built up of disks either loose, stitched, or glued, depending on the resiliency or pliability required, are used extensively for polishing as well as buffing. Belts of cloth or leather are often charged with abrasive for polishing flat or other classes of work. Wire brushes made with coarse brass, steel, or German silver wires may be used with no abrasive for a final operation to give a satin finish to nonferrous metals.

For most polishing operations speeds range from 5,000 to 7,500 surface ft/min (1,500 to 2,250 m/min). The higher range is for high-strength steels and stainless steels. Excessively high speeds may cause burning of the work and glazing.

**Buffing** is a form of finish polishing in which the surface finish is improved; very little material is removed. The powdered abrasives, usually the fine flours, are applied to the surface of the wheel by pressing a mixture of abrasive and tallow or wax against the face for a few seconds. The abrasive is replenished periodically. The wheels are made of a soft pliable material, such as soft leather, felt, linen, or muslin. Rotated at a high speed, they present a flexible face which resists pressure.

Buffing generally comprises two stages: **cutting down** and **coloring.** Cutting down makes a surface smoother by removing scratches and other marks from previous operations. Coloring further refines the surface and brings out maximum luster. A variety of buffing compounds are available: Tripoli (an amorphous silica), aluminum oxide, chromium oxide, soft silica, rouge (iron oxide), pumice, lime compounds, emery, and crocus. In cutting down nonferrous metals Tripoli is used, and for steels and stainless steels aluminum oxide is the common abrasive. For coloring, soft silica, rouge, and chrome oxide are the more common compounds used. Buffing speeds range from 6,000 to 10,000 surface ft/min (1,800 to 3,000 m/min); the higher speeds are for steels, although the speed may be as high as 12,000 surface ft/min (3,600 m/min) for coloring brass and copper.

**Lapping** is a process of producing extremely smooth and accurate surfaces by rubbing the surface which is to be lapped against a mating form which is called a **lap.** The lap may either be charged with a fine abrasive and moistened with oil or grease, or the fine abrasive may be introduced with the oil. If a part is to be lapped to a final accurate dimension, a mating form of a softer material such as soft close-grained cast iron, copper, brass, or lead is made up. Aluminum oxide, silicon carbide, and diamond grits are used for lapping. Lapping requires considerable time. No more than 0.0002 to 0.0005 in (0.005 to 0.013 mm) should be left for removal by this method. Surface plates, rings, and plugs are common forms of laps. Seating valves in a gas engine is one common illustration of lapping, the valve itself serving as the lap. For most applications grit sizes range between 100 and 800, depending on the finish desired. For most efficient lapping, speeds generally range from 300 to 800 surface ft/min (150 to 240 m/min) with pressures of 1 to 3 lb/in² (0.007 to 0.021 N/mm²) for soft materials and up to 10 lb/in² (0.07 N/mm²) for harder materials.

**Honing** is an operation similar to lapping. Instead of a metal lap charged with fine abrasive grains, a honing stone made of fine abrasives is used. Small stones of various cross-sectional shapes and lengths are manufactured for honing the edges of cutting tools. Automobile cylinders are honed to secure a fine finish and accurate dimensions. This honing usually follows a light-finish reaming operation or a precision-boring operation using diamonds or carbide tools. The tool consists of several honing stones adjustable at a given radius or forced outward by springs or a wedge forced mechanically or hydraulically and is given a reciprocating (25 to 40 per min) and a rotating motion (about 300 r/min) in the cylinder which is flooded with kerosene.

Hones operate at speeds of 50 to 200 surface ft/min (15 to 60 m/min) and use universal joints to allow the tool to center itself in the work. The automatic pressure-cycle control of hone expansion, in which the pressure is reduced in steps as the final finish is reached, removes metal 10 times as fast as with the spring-expanded hone. Rotational and reciprocating movements are provided to give an uneven ratio and thus prevent an abrasive grain from ever traversing its own path twice.

**Superfinish** is a honing process. Formed honing stones bear against the work previously finished to 0.0005 in (0.013 mm) or at the most to 0.001 in (0.025 mm) by a pressure of a few ounces (1 oz = 28g) which gradually increases to several pounds per square inch ( 1 lb/in² = 0.0069 N/mm²) of stone area in proportion to the development of the increased area of contact between the work and stone. The work or tool rotates for speed and where possible is reciprocated slowly over the surface which may be finished in a matter of 20 s to a surface quality of 1 to 3 $\mu$in (0.025 to 0.075 $\mu$m). Superfinishing is applied to many types of work such as crankshaft pins and bearings, cylinder bores, pistons, valve stems, cams, and other metallic moving parts.

The **sand blast** consists of particles of sand, powdered quartz, chilled-iron globules, emery, or other hard granular material blown by a jet of compressed air or of steam against a hard surface which it is desired to abrade. It is commonly used for cleaning metal castings, frosting smooth surfaces, etc. Portions of the surface which are not to be abraded can be protected by coating with a soft material such as wax, lead, or rubber.

## MACHINING OF PLASTICS

The low shear strength of plastics permits high cutting speeds and feeds but the low heat conductivity and greater resilience require increased reliefs and less rake in order to avoid undersize cutting. Hard and sharp tools should be used. Nitrided, chromium-plated, cast nonferrous, or even cemented-carbide tools are better, where production justifies the preparation of these tools, than the conventional high-speed steel tools. Plastics are usually abrasive and cause the tools to wear or become dull rapidly. Dull tools generate heat and cause the tools to cut to shallow depths. The depth of cut should be small. When high production justifies the cost, diamond turning and boring tools are justified. Diamond tools maintain sharp cutting edges and produce an excellent machined surface. They are particularly advantageous when a more abrasive plastic such as asbestos-filled plastic or rubber is machined.

A cutting fluid, such as a small blast of air, a stream of water, or an emulsion, improves the **turning** and cutting of plastics as it prevents the heating of the tool and causes the chips to remain brittle and to break rather than become sticky and gummy. A zero or slightly negative back rake and a relief angle of 8 to 12° should be used. For thermoplastics cutting speeds generally range from 250 to 400 ft/min (75 to 120 m/min) and for thermosetting plastics from 400 to 1,000 ft/min (120 to 300 m/min). Recommended tool materials are M2 and T5 high-speed steels and C-2 carbide.

In **milling** plastics, speeds of 400 to 1,000 ft/min (120 to 300 m/min) should be used with angles similar to those on a single-point tool. From 0 to 10° negative rake may be used. Excellent results have been obtained by hobbing plastic gears with carbide-tipped hobs. Recommended tool materials are M2 and M7 high-speed steels and C-2 carbide.

In **drilling,** speeds range from 150 to 400 ft/min (45 to 120 m/min), and the recommended drill geometry is given in Table 6. Tool materials are M1, M7, and M10 high-speed steel. Usually the drill cuts under size; drills 0.002 to 0.003 in (0.05 to 0.075 mm) oversize should be used.

In **sawing** plastics, either precision or buttress tooth forms may be used, with a pitch ranging from 3 to 14 teeth/in (1.2 to 5.5 teeth/cm), the thicker the material the lower the number of teeth per unit length of saw. Cutting speeds for thermoplastics range from 1,000 to 4,000 ft/min (300 to 1,200 m/min) and for thermosetting plastics from about 3,000 to 5,500 ft/min (900 to 1,700 m/min), with the higher speeds for thinner stock. High-carbon-steel blades are recommended. An air blast is helpful in preventing the chips from sticking to the saw. Plastics are not machined by friction sawing although razor-edge or sharp scalloped-edge blades work well on resilient materials and soft rubber, cork, etc. Abrasive saws operating at 3,500 to 6,000 ft/min (1000 to 1800 m/min) are used for cutting off bars and forms.

The cast phenolics are tapped and threaded with standard tools. Ground M1, M7, or M10 high-speed steel **taps** with large polished flutes are recommended. **Tapping speeds** are usually from 25 to 50 ft/min (8 to 15 m/min); water serves as a good cutting fluid as it keeps the material brittle and prevents sticking in the flutes. As taps become worn through the abrasive action of the plastics, they will cut under size. **Thread cutting** is generally accomplished with tools similar to those used on brass.

**Reaming** is best accomplished in production by using tools of the expansion or adjustable type with relatively low speeds but high feeds. Less material should be removed in reaming plastics than in reaming other materials.

**Polishing, buffing,** and **ashing** are done on many types of plastics. Polishing is done with special compounds containing wax or a fine abrasive. Buffing wheels for plastics should have loose stitching. Ashing is done with wet pumice. Vinyl plastics can be buffed and polished with fabric wheels of standard types, using light pressures.

For specific information, refer to the literature of the manufacturer or to publications such as the yearly "Modern Plastics Encyclopedia."

## ALTERNATIVE MATERIAL-REMOVAL PROCESSES

In addition to the mechanical methods of material removal described above, there are a number of unconventional processes which may be preferred over conventional methods. Among the important factors to be considered are the hardness of the workpiece material, the shape of the part, its sensitivity to thermal damage, residual stresses, tolerances, and economics. Some of these processes produce a heat-affected layer on the surface; improvements in surface integrity may be obtained by postprocessing techniques such as polishing or peening.

**Electric-discharge machining (EDM)** is based on the principle of erosion of metals by spark discharges. Figure 9 gives a schematic diagram of this process. The spark is a transient electric discharge through the space between two charged electrodes, which are the tool and the workpiece. The discharge occurs when the potential difference between the tool and the workpiece is large enough to cause a breakdown in the medium (which is called the **dielectric fluid** and is usually a hydrocarbon) and to procure an electrically conductive spark channel. The breakdown potential is usually established by connecting the two electrodes to the terminals of a capacitor charged from a power source. The spacing between the tool and workpiece is critical; therefore, the feed is controlled by servomechanisms. The dielectric fluid has the additional functions of providing a cooling medium and carrying away particles produced by the electric discharge. The discharge can be repeated rapidly, and each time a minute amount of workpiece material is removed.

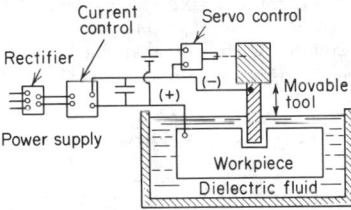

**Fig. 9**  Schematic diagram of the electric-discharge machining process.

The rate of metal removal depends mostly on the average current in the discharge circuit; it is also a function of the electrode characteristics, the electrical parameters, and the nature of the dielectric fluid. In practice, this rate is normally varied by changing the number of discharges per second or the energy per discharge. Rates of metal removal may range from 0.01 to 25 in³/h (0.17 to 410 cm³/h), depending on surface finish and tolerance requirements. In general, higher rates produce rougher surfaces. Surface finishes may range from 1,000 $\mu$in rms (0.025 mm) in roughing cuts to less than 25 $\mu$in (0.0006 mm) in finishing cuts.

The response of materials to this process depends mostly on their thermal properties. Thermal capacity and conductivity, latent heats of melting and vaporization are important. Hardness and strength do not necessarily have significant effect on metal-removal rates. The process is applicable to all materials which are sufficiently good conductors of electricity. The tool has great influence on permissible removal rates. It is usually

made of brass, although other materials have been used successfully. Tools have been made by casting, extruding, machining, powder metallurgy, and other techniques and are made in any desired shape. Tool wear is an important consideration, and in order to control tolerances and minimize cost, the ratio of tool material removed to workpiece material removed should be low. This ratio varies with different tool and workpiece material combinations and with operating conditions. Therefore, a particular tool material may not be best for all workpieces. Tolerances as low as 0.0001 to 0.0005 in (0.0025 to 0.0127 mm) can be held with slow metal-removal rates.

The electric-discharge machining process has numerous applications, such as machining cavities and dies, cutting small-diameter holes, blanking parts from sheets, cutting off rods of materials with poor machinability, and flat or form grinding. It is also applied to sharpening tools, cutters, and broaches. The process can be used to generate almost any geometry if a suitable tool can be fabricated and brought into close proximity to the workpiece.

**Electric-discharge grinding (EDG)** is similar to the electric-discharge machining process with the exception that the electrode is in the form of a grinding wheel. Removal rates are up to 1.5 in³/h (25 cm³/h) with practical tolerances of the order of 0.001 in (0.025 mm). A graphite electrode wheel is operated around 100 to 600 surface ft/min (30 to 180 m/min) to minimize splashing of the dielectric fluid. Typical applications of this process are in grinding of carbide tools and dies, thin slots in hard materials, and production grinding of intricate forms.

The **electrochemical machining (ECM)** process (Fig. 10) uses electrolytes which dissolve the reaction products formed on the workpiece by electrochemical action; it is similar to a reverse electroplating process. The electrolyte is pumped at high velocities through the tool. A gap of 0.005 to 0.020 in (0.13 to 0.5 mm) is maintained. A dc power supply maintains very high current densities between the tool and the workpiece. In most applications, a current density of 1,000 to 5,000 A is required per in² of active cutting area. The rate of metal removal is proportional to the amount of current passing between the tool and the workpiece. Removal rates up to 1 in³/min (16 cm³/min) can be obtained with a 10,000-A power supply. The penetration rate is proportional to the current density for a given workpiece material.

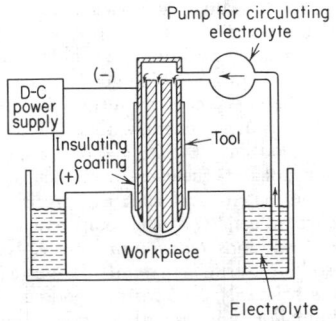

**Fig. 10**  Schematic diagram of the electrochemical machining process.

The process leaves a burr-free surface. It is also a cold machining process and does no thermal damage to the surface of the workpiece. Electrodes are normally made of brass or copper; stainless steel, titanium, sintered copper-tungsten, aluminum, and graphite have also been used. The electrolyte is usually a sodium chloride solution up to 2.5 lb/gal (300 g/l); other solutions and proprietary mixtures are also available. The amount of overcut, defined as the difference between hole diameter and tool diameter, depends upon cutting conditions. For production applications, the average overcut is around 0.015 in (0.4 mm). The rate of penetration is up to 0.750 in/min (20 mm/min). Very good surface finishes may be obtained with this process. However, sharp square corners or sharp corners and flat bottoms cannot be machined to high accuracies. The process is applied mainly to round or odd-shaped holes with straight parallel sides. It is also applied to cases where conventional methods produce burrs which are costly to remove. The process is particularly economical for materials with a hardness above 400 BHN.

The **electrochemical grinding (ECG)** process (Fig. 11) is a combination of electrochemical machining and abrasive grinding where most of the metal removal results from the electrolytic action. The process consists of a rotating cathode, a neutral electrolyte, and abrasive particles in contact with the workpiece. The equipment is similar to a conventional grinding

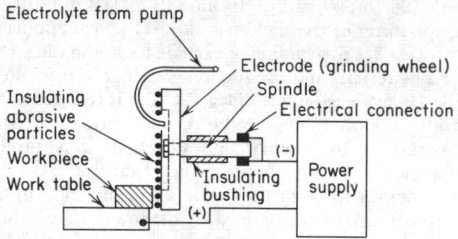

**Fig. 11** Schematic diagram of the electrochemical grinding process.

machine except for the electrical accessories. The cathode usually consists of a metal-bonded diamond or aluminum oxide wheel. An important function of the abrasive grains is to maintain a space for the electrolyte between the wheel and workpiece. The wheel runs at normal grinding speeds of 3,000 to 6,000 surface ft/min (900 to 1,800 m/min). The abrasive particles rub against the workpiece, scrubbing off the products of the composition, thus allowing good dimensional control. In some cases, the heavy cuts are made with full electrolytic action and the final cuts are made with little electrolytic action for purposes of good surface finish and high accuracy. For best dimensional control and efficiency, the noncutting areas of the tool should be insulated. Surface finish, precision, and metal-removal rate are influenced by the composition of the electrolyte. Aqueous solutions of sodium silicate, borax, sodium nitrate, and sodium nitrite are commonly used as electrolytes. The process is primarily used for tool and cutter sharpening and for machining of high-strength materials. It is not adapted to cavity-sinking operations. It has been successfully applied to refractory metals, high-strength steels, and nickel- and cobalt-base alloys and carbides.

A combination of the electric-discharge and electrochemical

methods of material removal is known as **electric-chemical discharge machining (ECDM)**. The electrode is a pure graphite rotating wheel which electrochemically grinds the workpiece. The intermittent spark discharges remove oxide films that form as a result of electrolytic action. The equipment is similar to that for electrochemical grinding. Typical applications include machining of fragile parts and resharpening or form grinding of carbides and tools such as milling cutters.

In **chemical machining (CHM)** material is removed by chemical or electrochemical dissolution of preferentially exposed surfaces of the workpiece. Selective attack on different areas is controlled by masking or by partial immersion. There are two processes involved: chemical **milling** and chemical **blanking.** Milling applications are in producing shallow cavities for overall weight reduction, and in making tapered sheets, plates, or extrusions. Masking with paint or tapes is common. Masking materials may be elastomers (such as butyl rubber, neoprene, and styrene-butadiene) or plastics (such as polyvinyl chloride, polystyrene, and polyethylene). Typical blanking applications are decorative panels, printed-circuit etching, and thin stampings. Etchants are solutions of sodium hydroxide for aluminum, and solutions of hydrochloric and nitric acids for steel.

**Ultrasonic machining (USM)** is a process in which a tool is given a high-frequency, low-amplitude oscillation, which, in turn, transmits a high velocity to fine abrasive particles that are present between the tool and the workpiece (See also p. 12-185 et seq.) Minute particles of the workpiece are chipped away on each stroke. Aluminum oxide, boron carbide, or silicone carbide grains are used in a water slurry (usually 50 percent by volume), which also carries away the debris. Grain size ranges from 200 to 1,000 (see Sec. 6 and Fig. 8). The equipment consists of an electronic oscillator, a transducer (Fig. 12), a connecting cone or toolholder, and the tool. The oscillatory motion is obtained most conveniently by magneto-

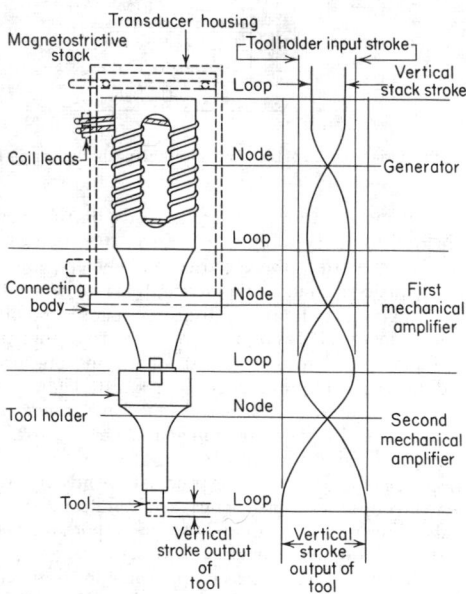

**Fig. 12** Schematic diagram of a transducer used in the ultrasonic machining process. *(The Sheffield Corp.)*

striction, at approximately 20,000 Hz and a stroke of 0.002 to 0.005 in (0.05 to 0.13 mm). The tool material is normally cold-rolled steel or stainless steel and is braised, soldered, or fastened mechanically to the transducer through a toolholder. The tool is ordinarily 0.003 to 0.004 in (0.075 to 0.1 mm) smaller than the cavity it produces. Tolerances of 0.0005 in (0.013 mm) or better can be obtained with fine abrasives. For best results, roughing cuts should be followed with one or more finishing operations with finer grits. The ultrasonic machining process is used in drilling holes, engraving, cavity sinking, slicing, broaching, etc. It is best suited to materials which are hard and brittle, such as ceramics, carbides, borides, ferrites, glass, precious stones, and hardened steels.

In **abrasive-jet machining (AJM),** material is removed by fine abrasive particles (aluminum oxide or silicon carbide) carried in a high-velocity stream of air, nitrogen, or carbon dioxide. The gas pressure ranges up to 120 lb/in² (0.83 N/mm²), providing a nozzle velocity of up to 1,000 ft/s (300 m/s). Nozzles are made of tungsten carbide or sapphire. Typical applications are in drilling, sawing, slotting, and deburring of hard, brittle materials such as glass.

In **laser-beam machining (LBM),** material is removed by converting electric energy into a single-wavelength, narrow beam of light and focusing it on the workpiece. The high energy density of the beam is capable of melting and vaporizing all materials. Typical applications are in drilling small holes in all types of materials, as small as 0.0002 in (0.005 mm) in diameter, and cutting titanium and nonmetallic materials such as fabric, wood, cardboard, and plastics. It is desirable for the workpiece material to have low thermal conductivity and low reflectivity.

The **electron-beam machining (EBM)** process removes material by focusing high-velocity electrons on the workpiece. Unlike lasers, this process is carried out in a vacuum chamber and is used for drilling small holes in all materials including ceramics, scribing, and cutting slots.

# SURFACE-TEXTURE DESIGNATION, PRODUCTION, AND CONTROL

## by James A. Broadston

REFERENCES: American National Standards Institute, "Surface Texture," ANSI B46.1, 1976. British Standard BS 1134, 1972; Canadian Standard CSA B-95, 1962; International Standards ISO R-468, 1974; ISO 1880, 1974; ISO 3274, 1975. *Proceedings of Institution of Mechanical Engineers on Properties and Metrology of Surfaces,* Oxford, April, 1968. *Proceedings, International Conference on Surface Technology* at Dearborn, MI, May, 1973, Carnegie Mellon University, SME, Pittsburgh. *Proceedings, International Production Engineering Research Conference,* Carnegie Institute of Technology, Pittsburgh, Sept., 1963. Reason, The Measurement of Surface Texture, in "Modern Workshop Technology," Pt. 2, Macmillan, 1970. Broadston, "Control of Surface Quality," Surface Checking Gage Co., Hollywood, CA, 1977. ASME "Metals Engineering Design Handbook," McGraw-Hill, 1965. SME "Tool Engineers Handbook," 1959, McGraw-Hill.

Rapid changes in the complexity and precision requirements of mechanical products since 1945 have created a need for improved methods of determining, designating, producing, and controlling the surface texture of manufactured parts. Although standards are aimed at standardizing methods for measuring by using stylus probes and electronic transducers for surface quality control, other descriptive specifications are sometimes required, i.e., interferometric light bands, peak-to-valley by optical sectioning, light reflectance by commercial glossmeters, etc. Other parameters are used by highly industrialized foreign countries to solve their surface specification problems. These include the high spot counter and bearing area meter of England (Talysurf); the total peak-to-valley, or $R_t$, of Germany (Perthen); and the $R$ or average amplitude of surface deviations of France. In the United States peak counting is used in the sheet-steel industry, instrumentation is available (Bendix), and a standard for specification, SAE J-911, exists.

Surface texture control should be considered for many reasons, among them being the following:

1. Advancements in the technology of metal-cutting tools and machinery have made the production of higher-quality surfaces possible.

2. Products are now being designed that depend upon proper quality control of critical surfaces for their successful operation as well as for long, trouble-free performance in service.

3. Craftsmen who knew the function and finish requirements for all the parts they made are gradually being replaced by machine operators who are not qualified to determine the proper texture requirements for critical surfaces.

4. Remote manufacture and the necessity for controlling costs have made it preferable that finish requirements for all the critical surfaces of a part be specified on the drawing.

5. The design engineer, who best understands the overall function of a part and all its surfaces, should be able to determine the requirement for surface-texture control where applicable and to use a satisfactory standardized method for providing this information on the drawing for use by manufacturing departments.

6. Manufacturing personnel should know what processes are able to produce surfaces within specifications and should be able to verify that the production techniques in use are under control.

7. Quality-control personnel should be able to check conformance if product quality is to be maintained and product performance and reputation ensured.

8. Test personnel should be able to operate completed products, as well as detail components, under simulated environmental conditions to determine shortcomings in design that may prevent satisfactory and trouble-free performance of the product in service.

9. The design engineer should be fully cognizant of prod-

uct performance and/or failure and of the reasons therefor, both in test and during customer operation, and should be able to apply such information toward the improvement of future designs.

10. Too much control may be worse than too little; hence, overuse of available techniques may hinder rather than assist, there being no payoff in producing surfaces that are more expensive than required to ensure product performance to established standards.

### DESIGN CRITERIA

Surfaces produced by various processes exhibit distinct differences in texture. These differences make it possible for honed, lapped, polished, turned, milled, or ground surfaces to be easily identified. As a result of its unique character, the surface texture produced by any given process can be readily compared with other surfaces produced by the same process through the simple means of comparing the average size of its irregularities, using applicable standards and modern measurement methods. It is then possible to predict and control its performance with considerable certainty by limiting the range of the average size of its characteristic surface irregularities. Surface-texture standards make this control possible.

Variations in the texture of a critical surface of a part influence its ability to resist wear and fatigue; to assist or destroy effective lubrication; to increase or decrease its friction and/or abrasive action on other parts, and to resist corrosion, as well as affect many other properties that may be critical under certain conditions.

Clay has shown that the load-carrying capacity of nitrided shafts of varying degrees of roughness, all running at 1,500 r/min in diamond-turned lead-bronze bushings finished to 20 $\mu$in, (0.50 $\mu$m), varies as shown in Fig. 1. The effects of roughness values on the friction between a flat slider on a well-lubricated rotating disk are shown in Fig. 2.

Surface-texture control should be a normal design consideration under the following conditions:

1. For those parts whose roughness must be held within closely controlled limits for optimum performance. In such cases, even the process may have to be specified. Automobile-

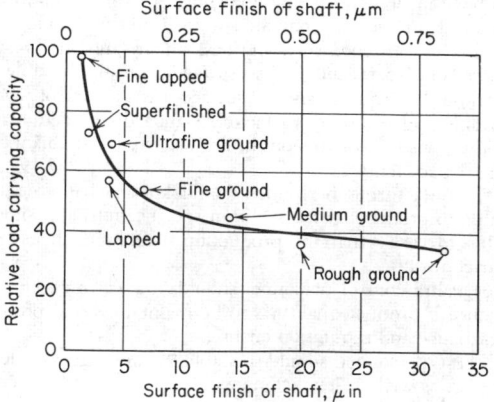

**Fig. 1**  Load-carrying capacity of journal bearings as related to the surface roughness of a shaft. *(Clay, ASM Metal Progress, Aug. 15, 1955.)*

engine cylinder walls, which should be finished to about 13 $\mu$in (0.32 $\mu$m) and have a circumferential (ground) or an angular (honed) lay, are an example. If too rough, excessive wear occurs; if too smooth, piston rings will not seat properly, lubrication is poor, and surfaces will seize or gall.

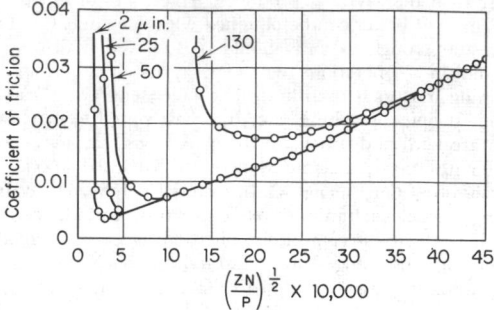

**Fig. 2**  Effect of surface texture on friction with hydrodynamic lubrication using a flat slider on a rotating disk. $Z$ is oil viscosity, P; $N$ is rubbing speed, ft/min; $P$ is load, lb/in².

2. Some parts, such as antifriction bearings, cannot be made too smooth for their function. In these cases, the designer must optimize the trade-off between the added costs of production and the market value of the added performance.

3. There are some parts where surfaces must be made as smooth as possible for optimum performance regardless of cost, such as gages, gage blocks, lenses, and carbon pressure seals.

4. In some cases, the nature of the most satisfactory finishing process may dictate the surface-texture requirements to attain production efficiency, uniformity, and control even though the individual performance of the part itself may not be dependent on the quality of the controlled surface. Hardened steel bushings, for example, which must be ground to close tolerance for press fit into housings, could have outside surfaces well beyond the roughness range specified and still perform their function satisfactorily.

5. For parts which the shop, with unjustified pride, has traditionally finished to greater perfection than is necessary, the use of proper surface-texture designations will encourage rougher surfaces on exterior and other surfaces that do not need to be finely finished.

It is the designer's responsibility to decide which surfaces of a given part are critical to its design function and which are not. This decision should be based upon a full knowledge of the part's function as well as of the performance of various surface textures that might be specified. From both a design and an economic standpoint, it may be just as unsound to specify too smooth a surface as to make it too rough—or to control it at all if not necessary. Wherever normal shop practice will produce acceptable surfaces, as in drilling, tapping, and threading, or in keyways, slots, and other purely functional surfaces, unnecessary surface-texture control will add costs which should be avoided.

Whereas each specialized field of endeavor has its own traditional criteria for determining which surface finishes are optimum for adequate performance, Table 1 provides some common examples for design review, and Table 6 provides

**Table 1. Typical Surface Texture Design Requirements**

| | |
|---|---|
| (250 μin.) 6.3 ✓ | Clearance surfaces<br>Rough machine parts |
| (125 μin.) 3.2 ✓ | Mating surfaces (static)<br>Chased and cut threads<br>Clutch-disk faces<br>Surfaces for soft gaskets |
| (63 μin.) 1.60 ✓ | Piston-pin bores<br>Brake drums<br>Cylinder block, top<br>Gear locating faces<br>Gear shafts and bores<br>Ratchet and pawl teeth<br>Milled threads<br>Rolling surfaces<br>Gearbox faces<br>Piston crowns<br>Turbine-blade dovetails |
| (32 μin.) 0.80 ✓ | Broached holes<br>Bronze journal bearings |
| | Gear teeth<br>Slideways and gibs<br>Press-fit parts<br>Piston-rod bushings<br>Antifriction-bearing seats<br>Sealing surfaces for hydraulic tube fittings |
| (16 μin.) 0.40 ✓ | Motor shafts<br>Gear teeth (heavy loads)<br>Spline shafts<br>O-ring grooves (static)<br>Antifriction-bearing bores and faces<br>Camshaft lobes<br>Compressor-blade airfoils<br>Journals for elastomer lip seals |
| (13 μin.) 0.32 ✓ | Engine cylinder bores<br>Piston outside diameters<br>Crankshaft bearings |
| (8 μin.) 0.20 ✓ | Jet-engine stator blades<br>Valve-tappet cam faces<br>Hydraulic-cylinder bores<br>Lapped antifriction bearings |
| (4 μin.) 0.10 ✓ | Ball-bearing races<br>Piston pins<br>Hydraulic piston rods<br>Carbon-seal mating surfaces |
| (2 μin.) 0.050 ✓ | Shop-gage faces<br>Comparator anvils |
| (1 μin.) 0.025 ✓ | Bearing balls<br>Gages and mirrors<br>Micrometer anvils |

data on the surface-texture ranges that can be obtained from normal production processes.

## DESIGNATION STANDARDS, SYMBOLS, AND CONVENTIONS

The precise definition and measurement of surface-texture irregularities of machined surfaces are almost impossible because the irregularities are very complex in shape and character and, being so small, do not lend themselves to direct measurement. Although both their shape and length may affect their properties, control of their average height and direction usually provides sufficient control of their performance. The standards do not specify the surface texture suitable for any particular application, nor the means by which it may be produced or measured. Neither are the standards concerned with other surface qualities such as appearance, luster, color, hardness, microstructure, or corrosion and wear resistance, any of which may be a governing design consideration.

The standards provide definitions of the terms used in delineating critical surface-texture qualities and a series of symbols and conventions suitable for their designation and control. The ANSI B46.1, 1976, used in this section, has replaced all other domestic standards and conforms in all essential elements with the British, Canadian, and most ISO international standards, even though different terms are used; i.e., the $R_a$, the AA (arithmetical average), and the CLA (centerline average) are identical with the internationally adopted symbol $R_a$ of ISO R468.

The basic ANSI symbol for designating surface texture is the checkmark with horizontal extension shown in Fig. 3. The symbol with the triangle at the base indicates a requirement for a machining allowance, in preference to the old $f$ symbol. Another, with the small circle in the base, prohibits machining; hence surfaces must be produced without the removal of material by processes such as cast, forged, hot- or cold-finished, die-cast, sintered- or injection-molded, to name a few. The surface-texture requirement may be shown at A; the machining allowance at B; the process may be indicated above

**Table 2. Application of Surface Texture Values to Surface Symbols**

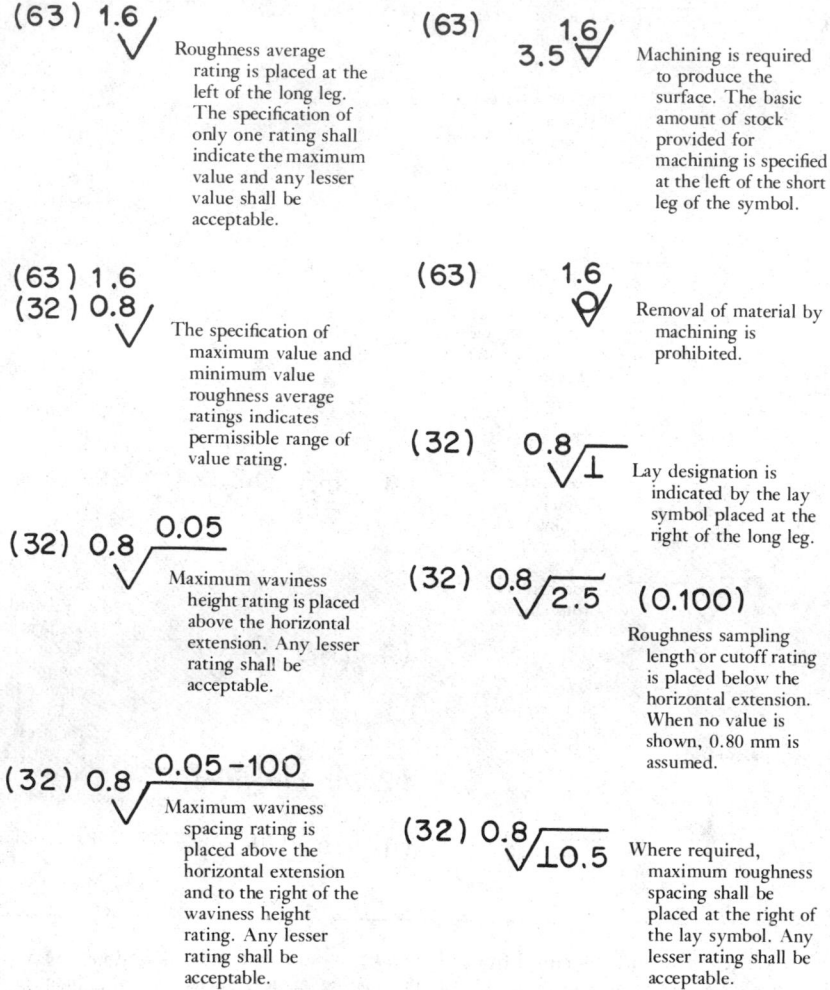

(63) 1.6

Roughness average rating is placed at the left of the long leg. The specification of only one rating shall indicate the maximum value and any lesser value shall be acceptable.

(63) 1.6
(32) 0.8

The specification of maximum value and minimum value roughness average ratings indicates permissible range of value rating.

(32) 0.8   0.05

Maximum waviness height rating is placed above the horizontal extension. Any lesser rating shall be acceptable.

(32) 0.8   0.05–100

Maximum waviness spacing rating is placed above the horizontal extension and to the right of the waviness height rating. Any lesser rating shall be acceptable.

(63)   1.6
       3.5

Machining is required to produce the surface. The basic amount of stock provided for machining is specified at the left of the short leg of the symbol.

(63)   1.6

Removal of material by machining is prohibited.

(32)   0.8 ⊥

Lay designation is indicated by the lay symbol placed at the right of the long leg.

(32) 0.8 / 2.5   (0.100)

Roughness sampling length or cutoff rating is placed below the horizontal extension. When no value is shown, 0.80 mm is assumed.

(32) 0.8 / ⊥0.5

Where required, maximum roughness spacing shall be placed at the right of the lay symbol. Any lesser rating shall be acceptable.

the line at C; the roughness width cutoff (sampling length) at D, and the lay at E. The ANSI symbol provides places for the insertion of numbers to specify a wide variety of texture characteristics, as shown in Table 2.

Control of **roughness,** the finely spaced surface-texture irregularities resulting from the manufacturing process or the cut-

ting action of tools or abrasive grains, is the most important function accomplished through the use of these standards, because roughness, in general, has a greater effect on performance than any other surface quality. The roughness-height index value is a number which equals the arithmetical average deviation of the minute surface irregularities from a hypotheti-

**Table 3. Preferred Series Roughness Average Values ($R_a$) Micrometres ($\mu$m); Microinches ($\mu$in)**

| $\mu$m | $\mu$in | $\mu$m | $\mu$in | $\mu$m | $\mu$in | $\mu$m | $\mu$in | $\mu$m | $\mu$in |
|---|---|---|---|---|---|---|---|---|---|
| 0.012 | 0.5 | 0.125 | 5 | 0.50 | 20 | 2.00 | 80 | 8.0 | 320 |
| 0.025 | 1 | 0.15 | 6 | 0.63 | 25 | 2.50 | 100 | 10.0 | 400 |
| **0.050** | **2** | **0.20** | **8** | **0.80** | **32** | **3.20** | **125** | **12.5** | **500** |
| 0.075 | 3 | 0.25 | 10 | 1.00 | 40 | 4.0 | 160 | 15.0 | 600 |
| **0.10** | **4** | 0.32 | 13 | 1.25 | 50 | 5.0 | 200 | 20.0 | 800 |
| | | **0.40** | **16** | **1.60** | **63** | **6.3** | **250** | **25.0** | **1000** |

**Table 4. Preferred Series Maximum Waviness Height Values**

| mm | in | mm | in | mm | in |
|---|---|---|---|---|---|
| 0.0005 | 0.00002 | 0.008 | 0.0003 | 0.12 | 0.005 |
| 0.0008 | 0.00003 | 0.012 | 0.0005 | 0.20 | 0.008 |
| 0.0012 | 0.00005 | 0.020 | 0.0008 | 0.25 | 0.010 |
| 0.0020 | 0.00008 | 0.025 | 0.001 | 0.38 | 0.015 |
| 0.0025 | 0.0001 | 0.05 | 0.002 | 0.50 | 0.020 |
| 0.005 | 0.0002 | 0.08 | 0.003 | 0.80 | 0.030 |

cal perfect surface, expressed in either millionths of an inch (microinches, $\mu$in, 0.000001 in) or in micrometres, $\mu$m, if drawing dimensions are in metric, SI units. For control purposes, roughness-height values are taken from Table 3, with those in boldface given preference.

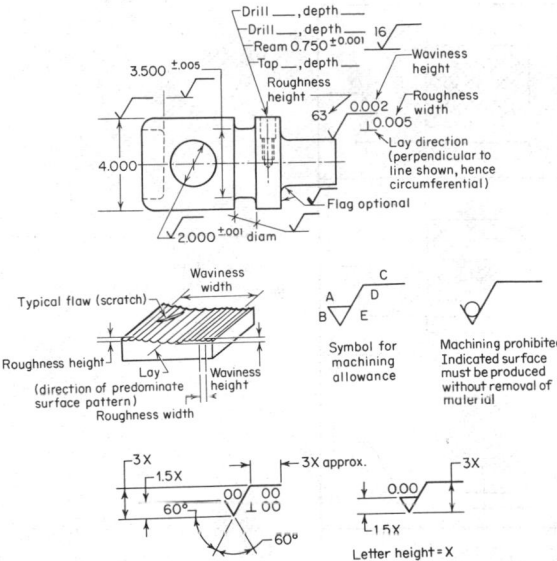

**Fig. 3**  Application and use of surface-texture symbols.

The term *roughness cutoff*, a characteristic of tracer-point measuring instruments, is used to limit the length of trace within which the asperities of the surface must lie for consideration as roughness. Asperity spacings greater than roughness cutoff are then considered as waviness.

**Waviness** refers to the secondary irregularities upon which roughness is superimposed, which are of significantly longer wavelength and are usually caused by machine or work deflections, tool or workpiece vibration, heat treatment, or warping. Waviness can be measured by a dial indicator or a profile recording instrument from which roughness has been filtered out. It is rated as maximum peak-to-valley distance and is indicated by the preferred values of Table 4. For fine waviness control, techniques involving contact-area determination in percent (90, 75, 50 percent preferred) may be required. Waviness control by interferometric methods is also common, where notes, such as "Flat within XX helium light Bands,"

may be used. Dimensions may be determined from the precision length table (see Sec. 1).

**Lay** refers to the direction of the predominant visible surface-roughness pattern. It can be controlled by use of the approved symbols given in Table 5, which indicate desired lay direction with respect to the boundary line of the surface upon which the symbol is placed.

**Flaws** are imperfections in a surface that occur only at infrequent intervals. They are usually caused by nonuniformity of the material, or they result from damage to the surface subsequent to processing, such as scratches, dents, pits, and cracks. Flaws shoud not be considered in surface-texture measurements, as the standards do not consider or classify them. Acceptance or rejection of parts having flaws is strictly a matter of judgment based upon whether the flaw will compromise the intended function of the part.

To call attention to the fact that surface-texture values are specified on any given drawing, a note and typical symbol may be used as follows:

$\sqrt{}$   Surface texture per ANSI B46.1, 1976.

Values for nondesignated surfaces can be limited by the note

$\overset{xx}{\sqrt{}}$   All machined surfaces except as noted

### MEASUREMENT AND PRODUCTION

Tracer-point analyzers provide an effective and rapid means for determining roughness values. Optical straightedge shadow and interference microscopes provide for measurement and comparison. Standard replicas of typical machined surfaces provide less accurate but adequate reference and control of rougher surfaces over 16 $\mu$in.

Various production processes can produce surfaces within the ranges shown in Table 6. For production efficiency, it is best that critical areas requiring surface-texture control be clearly designated on drawings so that proper machining and adequate protection from damage during processing will be ensured.

### SURFACE QUALITY VERSUS TOLERANCES

It should be remembered that surface quality and tolerances are distinctly different attributes that are controlled for completely separate purposes. **Tolerances** are established to limit the range of the size of a part at the time of manufacture, as measured with gages, micrometers, or other traditional measuring devices having anvils that make contact with the part. **Surface quality** controls, on the other hand, serve to limit the minute surface irregularities or asperities that are formed by the manufacturing process. These lie under the gage anvils during measurement and **do not use up tolerances.**

**Table 5. Lay Symbols**

| Lay symbol | Interpretation | Example showing direction of tool marks |
|---|---|---|
| = | Lay parallel to the line representing the surface to which the symbol is applied | |
| ⊥ | Lay perpendicular to the line representing the surface to which the symbol is applied | |
| X | Lay angular in both directions to line representing the surface to which symbol is applied | |
| M | Lay multidirectional | |
| C | Lay approximately circular relative to the center of the surface to which the symbol is applied | |
| R | Lay approximately radial relative to the center of the surface to which the symbol is applied | |
| P | Pitted, protuberant, porous, or particulate nondirectional lay | |

**Table 6. Surface-Roughness Ranges of Production Processes**

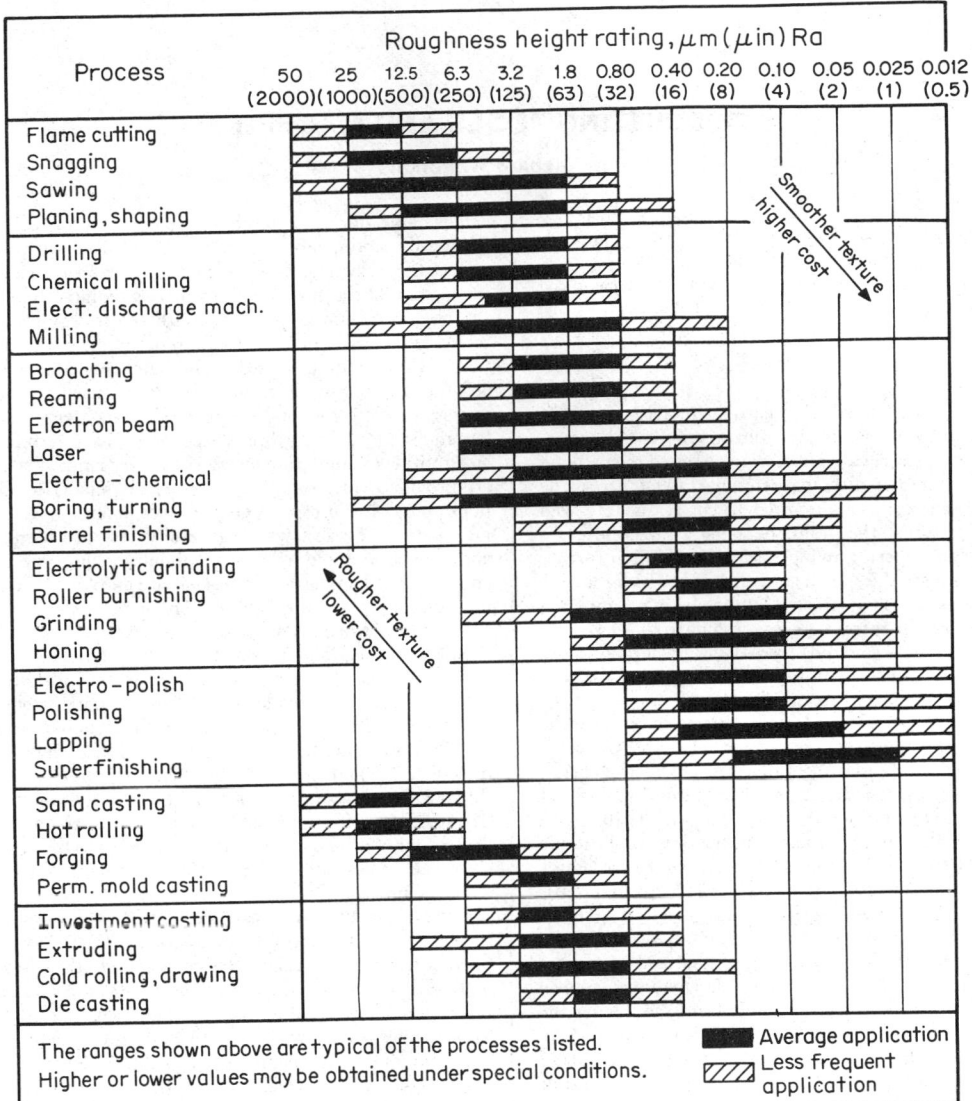

The ranges shown above are typical of the processes listed.
Higher or lower values may be obtained under special conditions.

Average application
Less frequent application

# WOODCUTTING TOOLS AND MACHINES

## by Richard W. Perkins

REFERENCES: Davis, Machining and Related Characteristics of United States Hardwoods, *USDA Tech. Bull.* 1267. Harris, "A Handbook of Woodcutting," Her Majesty's Stationery Office, London, 1946. Koch, "Wood Machining Processes," Ronald Press, 1964. Kollmann, Wood Machining, in Kollmann and Côté, "Principles of Wood Science and Technology," chap. 9, Springer-Verlag, 1968.

## SAWING

Sawing machines are classified according to basic machine design, i.e., band saw, gang saw, chain saw, circular saw. Saws are designated as **ripsaws** if they are designed to cut along the grain or **crosscut** saws if they are designed to cut across the grain. A **combination** saw is designed to cut reasonably well along the grain, across the grain, or along a direction at an angle to the grain (**miter**). Sawing machines are often further classified according to the specific operation for which they are used, e.g., **headsaw** (the primary log-breakdown saw in a sawmill), **resaw** (saw for ripping cants into boards), **edger** (saw for edging boards in a sawmill), **variety saw** (general-purpose saw for use in furniture plants), **scroll saw** (general-purpose narrow-band saw for use in furniture plants).

The thickness of the saw blade is designated in terms of the Birmingham wire gage (BWG) (see Sec. 6). Large-diameter [40 to 60 in (1.02 to 1.52 mm)]) circular-saw blades are tapered so that they are thicker at the center than at the rim. Typical headsaw blades range in thickness from 5 to 6 BWG [0.203 to 0.220 in (5.16 to 5.59 mm)] for use in heavy-duty applications to 8 to 9 BWG [0.148 to 0.165 in (3.76 to 4.19 mm)] for lighter operations. Small-diameter [6 to 30 in (152 to 762 mm)] circular saws are generally flat-ground and range from 10 to 18 BWG [0.049 to 0.134 in (1.24 to 3.40 mm)] in thickness. Band-saw and gang-saw blades are flat-ground and are generally thinner than circular-saw blades designed for similar applications. For example, typical wide-band-saw blades for sawmill use range from 11 to 16 BWG [0.065 to 0.120 in (1.65 to 3.05 mm)] in thickness. The thickness of a band-saw blade is determined by the cutting load and the diameter of the band wheel. Gang-saw blades are generally somewhat thicker than band-saw blades for similar operations. Narrow-band-saw blades for use on scroll band saws range in thickness from 20 to 25 BWG [0.020 to 0.035 in (0.51 to 0.89 mm)] and range in width from $\frac{1}{8}$ to about $1\frac{3}{4}$ in (3.17 to 44.5 mm) depending upon the curvature of cuts to be made.

The considerable amount of heat generated at the cutting edge results in compressive stresses in the rim of the saw blade of sufficient magnitude to cause mechanical instability of the saw blades. Circular-saw blades and wide-band-saw blades are commonly prestressed (or **tensioned**) to reduce the possibility of buckling. Small circular-saw blades for use on power-feed ripsaws and crosscut saws are frequently provided with **expansion slots** for the same purpose.

The **shape of the cutting portion of the sawtooth** is determined by specifying the hook, face bevel, top bevel, and clearance angles. The optimum tooth shape depends primarily upon cutting direction, moisture content, and density of the work material. Sawteeth are, in general, designed in such a way that the portion of the cutting edge which is required to cut across the fiber direction is provided with the maximum effective rake angle consistent with tool strength and wear considerations. Ripsaws are designed with a hook angle between some 46° for inserted-tooth circular headsaws used to cut green material and 10° for solid-tooth saws cutting dense material at low moisture content. Ripsaws generally have zero face bevel and top bevel angle; however, spring-set ripsaws sometimes are provided with a moderate top bevel angle (5 to 15°). The hook angle for crosscut saws ranges from positive 10° to negative 30°. These saws are generally designed with both top and face bevel angles of 5 to 15°; however, in some cases top and face bevel angles as high as 45° are employed. A compromise design is used for combination saws which embodies the features of both ripsaws and crosscut saws in order to provide a tool which can cut reasonably well in all directions. The clearance angle should be maintained at the smallest possible value in order to provide for maximum tooth strength. For ripsawing applications, the clearance angle should be about 12 to 15°. The minimum satisfactory clearance angle is determined by the nature of the work material, not from kinematical considerations of the motion of the tool through the work. In some cases of cutoff, combination, and narrow-band-saw designs where the tooth pitch is relatively small, much larger clearance angles are used in order to provide the necessary gullet volume.

A certain amount of clearance between the saw blade and the generated surface (**side clearance** or **set**) is necessary to prevent frictional heating of the saw blade. In the case of solid-tooth circular saws and band or gang saws, the side clearance is generally provided either by deflecting alternate teeth (**spring-setting**) or by spreading the cutting edge (**swage-setting**). The amount of side clearance depends upon density, moisture content, and size of the saw blade. In most cases, satisfactory results are obtained if the side clearance $S$ is determined from the formula $S$ [in (mm)] $= A\frac{1}{2}[f(g - 5) - f(g)]$, where $g =$ gage number (BWG) of the saw blade, $f(n) =$ dimension in inches (mm) corresponding to the gage number $n$, and $A$ has values from Table 1. Certain specialty circular saws such as planer, smooth-trimmer, and miter saws are hollow-ground to provide side clearance. Inserted-tooth saws, carbide-tipped saws, and chain-saw teeth are designed so that sufficient side clearance is provided for the life of the tool; consequently, the setting of such saws is unnecessary.

The **tooth speed** for sawing operations ranges from 3,000 to 17,000 ft/min (15 to 86 m/s) approx. Large tooth speeds are in general desirable in order to permit maximum work rates. The upper limit of permissible tooth speed depends in most cases on machine design considerations and not on considerations of wear or surface quality as in the case of metal cutting. Excep-

**Table 1. Values of $A$ for Computing Side Clearance**

| Saw type | Work material | | | |
|---|---|---|---|---|
| | Specific gravity less than 0.45 | | Specific gravity greater than 0.55 | |
| | Air dry | Green | Air dry | Green |
| Circular rip and combination | 0.90 | 1.00 | 0.85 | 0.95 |
| Glue-joint ripsaw | 0.80 | .... | 0.60 | |
| Circular crosscut | 0.95 | 1.05 | 0.90 | 1.00 |
| Wide-band saw | 0.55 | 0.65 | 0.30 | 0.40 |
| Narrow-band saw | 0.65 | .... | 0.55 | |

tionally high tooth speeds may result in charring of the work material, which is machined at slow feed rates.

In many sawing applications, **surface quality** is not of prime importance since the sawed surfaces are subsequently machined, e.g., by planing, shaping, sanding; therefore, it is desirable to operate the saw at the largest feed per tooth consistent with gullet overloading. Large values of feed per tooth result in lower amounts of work required per unit volume of material cut and in lower amounts of wear per unit tool travel. Large-diameter circular saws, wide-band saws, and gang saws for ripping green material are generally designed so that the feed per tooth should be about 0.08 to 0.12 in (2.03 to 3.05 mm). Small-diameter circular saws are designed so that the feed per tooth ranges from 0.03 in (0.76 mm) for dense hardwoods to 0.05 in (1.27 mm) for low-density softwoods. Narrow-band saws are generally operated at somewhat smaller values of feed per tooth, e.g., 0.005 to 0.04 in (0.13 to 1.02 mm). Smaller values of feed per tooth are necessary for applications where surface quality is of prime importance, e.g., glue-joint ripsawing and variety-saw operations. The degree of gullet loading is measured by the **gullet-feed index (GFI)**, which is computed as the feed per tooth times the depth of face divided by the gullet area. The maximum GFI depends primarily upon species, moisture content, and cutting direction. It is generally conceded that the maximum GFI for ripsawing lies between 0.3 for high density, low-moisture-content material and 0.4 for low density, high-moisture-content material. For specific information, see Telford, *For. Prod. Res. Soc. Proc.*, 1949.

Saws vary considerably in design of the **gullet shape**. The primary design considerations are gullet area and tooth strength; however, special design shapes are often required for certain classes of work material, e.g., for ripping frozen wood.

**Materials** Saw blades and the sawteeth of solid-tooth saws are generally made of a nickel tool steel. The bits for inserted-tooth saws are usually a plain carbon tool steel; however, high-speed-steel bits or bits with a cast-alloy inlay (e.g., Stellite) are sometimes used in applications where metal or gravel will not be encountered. Small-diameter circular saws of virtually all designs are made with cemented-carbide tips. This design type is almost imperative in applications where highly abrasive material is cut, namely, in plywood and particle-board operations.

**Sawing Power** References: Endersby, The Performance of Circular Plate Ripsaws, *For. Prod. Res. Bull.* 27, Her Majesty's Stationery Office, London, 1953. Johnston, Experimental Cut-off Saw, *For. Prod. Jour.*, June, 1962. Oehrli, Research in Cross-cutting with Power Saw Chain Teeth, *For. Prod. Jour.*,

Jan., 1960. Telford, Energy Requirements for Insert-point Circular Headsaws, *Proc. For. Prod. Res. Soc.*, 1949.

An approximate relation for computing the power $P$, ft·lb/min (W) required to saw is

$$P = kvb(A + Bt_a)/p$$

where $k$ is the kerf, in (m), $v$ is the tooth speed, ft/min (m/s); $p$ is the tooth pitch, in (m); $A$ and $B$ are constants for a given sawing operation, lb/in (N/m) and lb/in² (N/m²), respectively; and $t_a$ is the average chip thickness, in (m). The average chip thickness is computed from the relation $t_a = \gamma f_t \times d/b$, where $f_t$ is the feed per tooth; $d$ is the depth of face; $b$ is the length of the tool path through the workpiece; and $\gamma$ has the value *unity* except for saws with spring-set or offset teeth, in which case $\gamma$ has the value 2. The constants $A$ and $B$ depend primarily upon cutting direction (ripsawing, crosscutting), moisture content below the fiber-saturation point and specific gravity of the work material, and tooth shape. The values of $A$ and $B$ (see Table 2) depend to some degree upon the depth of face, saw diameter, gullet shape, gullet-feed index, saw speed, and whether the tool motion is linear or rotary; however, the effect of these variables can generally be neglected for purposes of approximation.

### PLANING AND MOLDING

**Machinery** Planing and molding machines employ a rotating cutterhead to generate a smooth, defect-free surface by cutting in a direction approximately along the grain. A **surfacer** (or **planer**) is designed to machine boards or panels to uniform thickness. A **facer** (or **facing planer**) is designed to generate a flat (plane) surface on the wide faces of boards. The **edge jointer** is intended to perform the same task on the edges of boards in preparation for edge-gluing into panels. A **planer-matcher** is a heavy-duty machine designed to plane rough boards to uniform width and thickness in one operation. This machine is commonly used for dressing dimension lumber and producing millwork. The **molder** is a high-production machine for use in furniture plants to generate parts of uniform cross-sectional shape.

**Recommended Operating Conditions** It is of prime importance to adjust the operating conditions and knife geometry so that the machining defects are reduced to a satisfactory level. The most commonly encountered defects are torn (chipped) grain, fuzzy grain, raised and loosened grain, and chip marks. **Torn grain** is caused by the wood splitting ahead of the cutting edge and below the generated surface. It is generally associated with large cutting angle, large chip thickness, low mois-

**Table 2. Constants for Sawing-Power Estimation**

| Material | | | Tool | | Constants | | | |
|---|---|---|---|---|---|---|---|---|
| Species | Specific gravity | Moisture content, % | Angles[e] | Sawing situation | A, lb/in | A, N/m | B, lb/in² × 10⁻³ | B, N/m² × 10⁻⁶ |
| Beech, European[a] | 0.72 | 12 | 20, 0, 12 | SS, R | 27.8 | 4,869 | 5.760 | 39.71 |
| Birch, yellow[b] | 0.55 | FSP | − 30, 10, 10 | SS, CC | 19.7 | 3,450 | 4.100 | 28.27 |
| Elm, wych[a] | 0.67 | 12 | 20, 0, 12 | SS, R | 23.2 | 4,063 | 4.840 | 33.37 |
| Maple, sugar[c] | 0.63 | FSP | 41, 0, 0 | IT, R | 85.6 | 15,991 | 2.995 | 20.65 |
| Maple, sugar[c,f] | 0.63 | FSP | 41, 0, 0 | IT, R | 48.0 | 8,406 | 4.400 | 30.34 |
| Pine, northern white[c] | 0.34 | FSP | 41, 0, 0 | IT, R | 27.1 | 4,746 | 1.675 | 11.55 |
| Pine, northern white[c,f] | 0.34 | FSP | 41, 0, 0 | IT, R | 28.2 | 4,939 | 2.085 | 14.38 |
| Pine, northern white[b] | 0.34 | FSP | − 30, 10, 10 | SS, CC | 0.0 | 0 | 3.300 | 22.75 |
| Pine, ponderosa[d] | 0.38–0.40 | 15–40 | 28, 25, 0 | OFT, R | 29.3 | 5,131 | 1.700 | 11.72 |
| Pine, ponderosa[d] | 0.38–0.40 | 15–40 | 28, 25, 0 | OFT, CC | 0.0 | 0 | 2.120 | 14.62 |
| Poplar (P. serotina)[a] | 0.48 | 12 | 20, 0, 12 | SS, R | 18.6 | 3,257 | 3.290 | 22.68 |
| Redwood, California[a] | 0.37 | 12 | 20, 0, 12 | SS, R | 15.0 | 2,627 | 2.260 | 15.58 |
| Spruce, white[b] | 0.32 | FSP | − 30, 10, 10 | SS, CC | 0.0 | 0 | 4.680 | 32.27 |

[a] Endersby.
[b] Johnston.
[c] Hoyle, unpublished report, N.Y. State College of Forestry, Syracuse, N.Y., 1958.
[d] Oehrli.
[e] The numbers represent hook angle, face bevel angle, and top bevel angle in degrees.
[f] Cutting performed on frozen material.
Note: FSP = moisture content greater than the fiber-saturation point; CC = crosscut; IT = insert-tooth: OFT = offset-tooth; R = rip; SS = spring-set.

ture content, and low work-material density. The **fuzzy-grain** defect is characterized by small groups of wood fibers which stand up above the generated surface. This defect is caused by incomplete severing of the wood by the cutting edge and is generally associated with small cutting angles, dull knives, low-density species, high moisture content, and (often) the presence of reaction wood. The **raised-grain** defect is characterized by an uneven surface where one portion of the annual ring is raised above the remaining part. **Loosened grain** is similar to raised grain; however, loosened grain is characterized by a separation of the early wood from the late wood which is readily discernible to the naked eye. The raised- and loosened-grain defects are attributed to the crushing of spring-wood cells as the knife passes over the surface. (Edge-grain material may exhibit a defect similar to the raised-grain defect if machining is performed at a markedly different moisture content from that encountered at some later time.) Raised and loosened grains are associated with dull knives, excessive jointing of knives [the jointing land should not exceed ¹⁄₃₂ in (0.79 mm)], and high moisture content of work material. **Chip marks** are caused by chips which are forced by the knife into the generated surface as the knife enters the work material. Chip marks are associated with inadequate exhaust, low moisture content, and species (e.g., birch, Douglas fir, and maple have a marked propensity toward the chip-mark defect).

**Depth of cut** is an important variable with respect to surface quality, particularly in the case of species which are quite prone to the torn-grain defect (e.g., hard maple, Douglas fir, southern yellow pine). In most cases, the depth of cut should be less than ¹⁄₁₆ in (1.59 mm). The number of **marks per inch (marks per meter)** (reciprocal of the feed per cutter) is an important variable in all cases; however, it is most important in those cases for which the torn-grain defect is highly probable. The marks per inch (marks per meter) should be between

8 and 12 (315 and 472) for rough planing operations and from 12 to 16 (472 to 630) for finishing cuts. Slightly higher values may be necessary for refractory species or for situations where knots or curly grain are present. It is seldom necessary to exceed a value of 20 marks per inch (787 marks per meter). The **clearance angle** should in all cases exceed a value of 10°. When it is desired to hone or joint the knives between sharpenings, a value of about 20° should be used. The optimum **cutting angle** lies between 20 and 30° for most planing situations; however, in the case of interlocked or wavy grain, low moisture content, or species with a marked tendency toward the torn-grain defect, it may be necessary to reduce the cutting angle to 10 or 15°.

## BORING

**Machinery** The typical general-purpose wood-boring machine has a single vertical spindle and is a hand-feed machine. Production machines are often of the vertical, multiple-spindle, adjustable-gang type or the horizontal type with two adjustable, independently driven spindles. The former type is commonly employed in furniture plants for boring holes in the faces of parts, and the latter type is commonly used for boring dowel holes in the edges and ends of parts.

**Tool Design** A wide variety of tool designs is available for specialized boring tasks; however, the most commonly used tools are the taper-head drill, the spur machine drill, and the machine bit. The **taper-head** drill is a twist drill with a point angle of 60 to 90°, lip clearance angle of 15 to 20°, chisel-edge angle of 125 to 135°, and helix angle of 20 to 40°. Taper-head drills are used for drilling screw holes and for boring dowel holes along the grain. The **spur machine** drill is equivalent to a twist drill having a point angle of 180° with the addition of a pyramidal point (instead of a web) and spurs at the circumference. These drills are designed with a helix angle of 20 to 40° and a clearance angle of 15 to 20°. The **machine bit** has a specially formed head which determines the configuration of the spurs. It also has a point. Machine bits are designed with a helix angle of 40 to 60°, cutting angle of 20 to 40°, and clearance angle of 15 to 20°. Machine bits are designed with spurs contiguous to the cutting edges (**double-spur machine bit**), with spurs removed from the vicinity of the cutting edges (**extension-lip machine bit**), and with the outlining portion of the spurs removed (**flat-cut machine bit**).

The purpose of the spurs is to aid in severing wood fibers across their axes, thereby increasing hole-wall smoothness when boring across the grain. Therefore, drills or bits with spurs (double-spur machine drill and bit) are intended for boring across the grain, whereas drills or bits without spurs (taper-head drill, flat-cut machine bit) are intended for boring along the grain or at an angle to the grain.

Taper-head and spur machine drills can be sharpened until they become too short for further use; however, machine bits and other bit styles which have specially formed heads can only be sharpened a limited number of times before the spur and cutting-face configuration is significantly altered. Since most wood-boring tools are sharpened by filing the clearance face, it is important to ensure that sufficient clearance is maintained. The clearance angle should be at least 5° greater than the angle whose tangent (function) is the feed per revolution divided by the circumference of the drill point.

**Recommended Operating Conditions** The most common defects are tearing of fibers from the end-grain portions of the hole surface and charring of hole surfaces. Rough hole surfaces are most often encountered in low-density and ring-porous species. This defect can generally be reduced to a satisfactory level by controlling the chip thickness. Charring is commonly a problem in high-density species. It can be avoided by maintaining the peripheral speed of the tool below a level which depends upon density and moisture content and by maintaining the chip thickness at a satisfactory level. Large chip thickness may result in excessive tool temperature and therefore rapid tool wear; however, large chip thickness is seldom a cause of hole charring. The following recommendations pertain to the use of spur-type drills or bits for boring material at about 6 percent moisture content across the grain. For species having a specific gravity less than 0.45, the chip thickness should be between 0.015 and 0.030 in (0.38 and 0.76 mm), and the peripheral speed of the tool should not exceed 900 ft/min (4.57 m/s). For material of specific gravity between 0.45 and 0.65, satisfactory results can be obtained with values of chip thickness between 0.015 and 0.045 in (0.38 and 1.14 mm) and with peripheral speeds less than 700 ft/min (3.56 m/s). For material of specific gravity greater than 0.65, the chip thickness should lie between 0.015 and 0.030 in (0.38 and 0.76 mm) and the peripheral speed should not exceed 500 ft/min (2.54 m/s). Somewhat higher values of chip thickness and peripheral speed can be employed when the moisture content of the material is higher.

## SANDING

(See Sec. 6)

**Machinery** Machines for production sanding of parts having flat surfaces are multiple-drum sanders, automatic-stroke sanders, and wide-belt sanders. **Multiple-drum** sanders are of the endless-bed or roll-feed type and have from two to six drums. The drum at the infeed end is fitted with a relatively coarse abrasive (40 to 100 grit), takes a relatively heavy cut [0.010 to 0.015 in (0.25 to 0.38 mm)], and operates at a relatively slow surface speed [3,000 to 3,500 ft/min (15.24 to 17.78 m/s)]. The drum at the outfeed end has a relatively fine abrasive paper (60 to 150 grit), takes a relatively light cut [about 0.005 in (0.13 mm)], and operates at a somewhat higher surface speed [4,000 to 5,000 ft/min (20.3 to 25.4 m/s)]. **Automatic-stroke** sanders employ a narrow abrasive belt and a reciprocating shoe which forces the abrasive belt against the work material. This machine is commonly employed in furniture plants for the final white-sanding operation prior to finish coating. The automatic-stroke sander has a relatively low rate of material removal (about one-tenth to one-third of the rate for the final drum of a multiple-drum sander) and is operated with a belt speed of 3,000 to 7,500 ft/min (15.2 to 38.1 m/s). **Wide-belt** sanders are commonly used in board plants (plywood, particle board, hardboard). They have the advantage of higher production rates and somewhat greater accuracy than multiple-drum sanders [e.g., feed rates up to 100 ft/min (0.51 m/s) as opposed to about 35 ft/min (0.18 m/s)]. Wide-belt sanders operate at surface speeds of approximately 5,000 ft/min (25.4 m/s) and are capable of operating at depths of cut of 0.006 to 0.020 in (0.15 to 0.51 mm) depending upon work-material density.

**Abrasive Tools** The abrasive tool consists of a **backing** to carry the **abrasive** and an **adhesive** coat to fix the abrasive to the backing. Backings are constructed of paper, cloth, or vulcanized fiber or consist of a cloth-paper combination. The adhesive coating (see also Sec. 6) is made up of two coatings; the first coat (**make coat**) acts to join the abrasive material to the backing, and the second coat (**size coat**) acts to provide the necessary support for the abrasive particles. Coating materials are generally animal glues, urea resins, or phenolic resins. The choice of material for the make and size coats depends upon the required flexibility of the tool and the work rate required of the tool. Abrasive materials (see also Sec. 6) for woodworking applications are garnet, aluminum oxide, and silicon carbide. **Garnet** is the most commonly used abrasive mineral because of its low cost and acceptable working qualities for low-work-rate situations. It is generally used for sheet goods, for sanding softwoods with all types of machines, and for sanding where the belt is loaded up (as opposed to worn out). **Aluminum oxide** abrasive is used extensively for sanding hardwoods, particle board, and hardboard. **Silicon carbide** abrasive is used for sanding and polishing between coating operations and for machine sanding of particle board and hardboard. Silicon carbide is also frequently used for the sanding of softwoods where the removal of raised fibers is a problem. The size of the abrasive particles is specified by the mesh number (the approximate number of openings per inch in the screen through which the particles will pass). (See Commercial Std CS217-59, "Grading of Abrasive Grain on Coated Abrasive Products," U.S. Government Printing Office.) Mesh numbers range from about 600 to 12. Size may also be designated by an older system of symbols which range from 10/0 (mesh no. 400) through 0 (mesh no. 80) to 4½ (mesh no. 12). Some general recommendations for common whitewood sanding operations are presented in Table 3.

**Table 3. Recommendations for Common Whitewood Sanding Operations***

| | | Abrasive | | | | | | | | | |
| | | Backing | | Adhesive | | Mineral | | | Grit size | | |
| | | Material | Weight | Make coat | Size coat | 1st drum | 2d drum | 3d drum | 1st drum | 2d drum | 3d drum |
|---|---|---|---|---|---|---|---|---|---|---|---|
| Multiple-drum | Softwood | Paper | E | Glue | Resin | G | G | G or S | 50 | 80 | 100 |
| | Hardwood | Paper | E | Glue | Resin | A | A | A | 60 | 100 | 120 |
| | Particle board | Paper | E | Glue | Resin | G | G | G or A | 40 | 60 | 80 |
| | | Fiber | 0.020 | Resin | Resin | S | S | S | 40 | 60 | 80 |
| | Hardboard | Paper | E | Glue | Resin | S | S | S | 60 | 80 | 120 |
| | | Fiber | 0.020 | Resin | Resin | S | S | S | 60 | 80 | 120 |
| Wide-belt | Softwood | Paper | E | Glue | Resin | G | or | S | | 80–220[a] | |
| | | Cloth | X | Resin | Resin | G | or | S | | 80–220[a] | |
| | Hardwood | Paper | E | Glue | Resin | | A | | | 80–220[a] | |
| | | Cloth | X | Resin | Resin | | A | | | 80–220[a] | |
| | Burnishing | Paper | E | Glue | Glue | | A | | | 280–400 | |
| | Particle board | Cloth | X | Resin | Resin | | S | | | 24–150[a] | |
| | Hardboard | Cloth | X | Resin | Resin | | S | | | 100–150 | |
| Stroke sanding | Softwood | Paper | E | Glue | Resin | | G | | | 80;120[b] | |
| | Hardwood | Paper | E | Glue | Resin | | A | | | 100;150–180[b] | |
| | | Cloth | X | Glue | Resin | | A | | | 100;150–180[b] | |
| | Particle board | Paper | E | Glue | Resin | A | or | S | | 80;120[b] | |
| | | Cloth | X | Resin | Resin | A | or | S | | 80;120[b] | |
| | Hardboard | Paper | E | Glue | Resin | A | or | S | | 100;150[b] | |
| | | Cloth | X | Resin | Resin | A | or | S | | 100;150[b] | |
| Edge sanding | Softwood | Cloth | X | Glue | Resin | | G | | | 60;100[b] | |
| | Hardwood | Cloth | X | Glue | Resin | | A | | | 60–150[a] | |
| | | Cloth | X | Resin | Resin | | A | | | 60–150[a] | |
| Mold sanding | Softwood | Cloth | J | Glue | Glue | | G | | | 80–120[a] | |
| | | Cloth | J | Glue | Resin | | G | | | 80–120[a] | |
| | Hardwood | Cloth | J | Glue | Glue | G | or | A | | 80–120[a] | |
| | | Cloth | J | Glue | Resin | G | or | A | | 80–120[a] | |

*Source: Graham, *Furniture Production,* July and Aug., 1961; and Martin, *Wood Working Digest,* Sept., 1961.
Note: G = garnet; A = aluminum oxide; S = silicon carbide.
[a] May be single- or multiple-grit operation.
[b] First number for cutting-down operations, second number for finishing operations.

Section **14**

# Pumps and Compressors

BY

**F. W. BUSE,** *Chief Engineer, Standard Pump-Aldrich Division, Ingersoll-Rand Co.*

**IGOR J. KARASSIK,** *Vice President and Chief Consulting Engineer, Worthington Pump Corp.*

**WILLIAM C. KRUTZSCH,** *Manager, Product Development, Engineered Pump Division, Worthington Pump Corp. (U.S.A.).*

**A. R. WORSTER,** *Manager, Engineering, Air Power Division, Ingersoll-Rand Company.*

**B. B. DAYTON,** *Consulting Engineer, East Flat Rock, NC.*

**ROBERT JORGENSEN,** *Chief Engineer, Air Handling, Division Buffalo Forge Company.*

## PUMPS
### by F. W. Buse

Reciprocating Power Pumps ............................... 14-2
Pulsation Dampeners (Cushion Chambers) ................ 14-7
Power-Pump Speeds ...................................... 14-7
Direct-Acting Steam Pumps ............................. 14-7
Pump Ends.............................................. 14-7
Power Requirements..................................... 14-7
Plunger Load .......................................... 14-8
Pump Valves ........................................... 14-8
Rotary Pumps .......................................... 14-10
Single Rotor .......................................... 14-11
Pistonless Pumps ...................................... 14-14
Air-Lift Pumps ........................................ 14-14

## CENTRIFUGAL AND AXIAL PUMPS
### by Igor J. Karassik and William C. Krutzsch

Nomenclature and Mechanical Design .................... 14-15
Materials of Construction ............................. 14-22
Pump Performance ...................................... 14-22
Installation, Operation, Maintenance .................. 14-28

## COMPRESSORS
### by A. R. Worster

Air Compressors ....................................... 14-31
Power Requirement ..................................... 14-32
Lubrication ........................................... 14-37
Compressor Accessories ................................ 14-38
Turbocompressors ...................................... 14-39
Helical-Screw Compressors ............................. 14-42
Other Rotary Compressors .............................. 14-43

## HIGH-VACUUM PUMPS
### by B. B. Dayton

Selection of Pumps .................................... 14-44
Types and Sizes ....................................... 14-45
Installation .......................................... 14-47

## FANS
### by Robert Jorgensen

Fan Types and Nomenclature ............................ 14-49
Fan Performance and Testing ........................... 14-50
Fan and System Performance Characteristics ............ 14-52

# PUMPS
## by F. W. BUSE

(For centrifugal and axial pumps, see following subsection)

REFERENCES: Greene, "Pumping Machinery," Wiley. Nickel, "Direct-acting Steam Pumps," McGraw-Hill. Butler, "Modern Pumping and Hydraulic Machinery," Griffin. Kristal and Annett, "Pumps," McGraw-Hill. Hydraulic Institute Standards.

**Displacement pumps** are usually divided into four general classes: (1) reciprocating power, (2) steam, (3) rotary, and (4) pistonless. A power pump is a reciprocating pump driven by power from an outside source applied to the crankshaft of the pump. A steam pump is a reciprocating pump and a steam engine built together as a unit. The power to drive the pump is furnished by the steam engine. A rotary pump is a positive-displacement pump, consisting of a fixed casing containing gears, cams, screws, vanes, plungers, or similar elements, actuated by rotation of the drive shaft. These pumps are characterized by their close running clearances and the absence of suction and discharge valves. Rotary pumps are frequently lubricated only by the fluid being pumped. Pistonless pumps utilize the direct pressure of air, gas, or steam on the fluid pumped.

## RECIPROCATING POWER PUMPS

**Power pumps** are positive-displacement machines which, at constant speed, deliver essentially the same capacity at any pressure within the capability of the driver and the strength of the pump. The inherently high efficiency of a power pump is almost independent of pressure and capacity and is only slightly lower for a small pump than for a large pump. Thus the power pump is most useful in the field of high pressure and low capacity, where its high efficiency more than offsets the high initial cost. In some applications the constant delivery at varying pressure is a definite advantage, with the power pump acting also as a metering device. In other applications this creates a control problem to be met by varying the speed, bypassing at constant speed, or intermittently loading and unloading the pump. Many power pumps are arranged so that the size of the piston or plunger can be changed easily, thus providing a pump adaptable to a considerable range of pressure, with capacity varying inversely with pressure at constant hydraulic-power output. Power pumps are supplied with two, three, five, seven, and nine plungers termed duplex, triplex, quintuplex, septuplex, and nonuplex, respectively.

**High-Speed Power Pumps** Modern power pumps are built with totally enclosed, self-lubricating power ends effectively protected from damage by any leakage of the fluid pumped or from dirt in the surrounding atmosphere. Plunger speeds up to 400 ft/min (20 m/s) in short-stroke pumps with rotative speeds of 300 to 720 r/min permit direct connection to the driver or a single-reduction drive instead of the double-reduction drive usually required with older slow-speed pumps. Figure 1 shows a typical inverted vertical pump of 50 to 1,500 bhp (37 to 1,100 kW). The suction manifold (1) communicates

with three suction-valve assemblies (2) set in the individual plunger chambers in the cylinder (3). Similar discharge-valve assemblies (4) are contained partly in the discharge manifold (5) on the opposite side of the cylinder. The stuffing-box barrel (6) extends upward from the cylinder, where it is held by the stuffing-box nuts (7) and studs (8). The plunger (9) and upper crosshead (10) are connected to the power-end crosshead by pull rods, which pass through vertical holes drilled through the cylinder forging. The power end is protected from dirt by the telescopic tubes (11) surrounding the pull rods. The crankshaft is carried by roller bearings or sleeve bearings in the pump power frame, and all parts are pressure-lubricated through drilled passages in the crankshaft, the connecting rods, and the crossheads.

These pumps are offered with different diameter and number of plungers to cover a range of displacement with strokes from 2½ to 9 in (64 to 229 mm). Power pumps can handle slurries consisting of 65 percent weight of solids.

Conventional displacement-pump design involves intersecting bores, as in Fig. 1, where the plunger chamber enters the passage between the suction and discharge valves. This leads to stress concentration in the pulsation chamber which limits

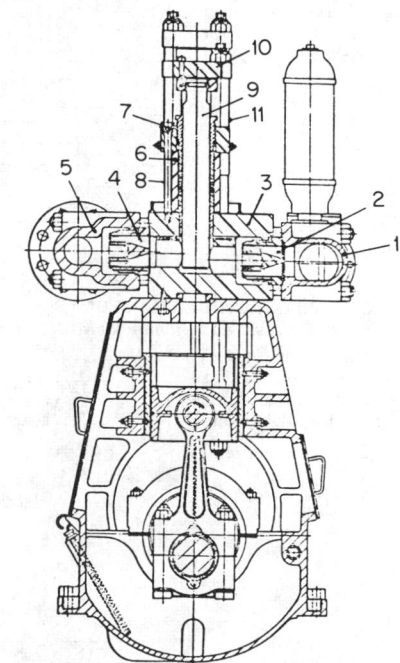

**Fig. 1** High-speed triplex pump. *(Ingersoll-Rand Company.)*

the maximum pressure, even with high-strength forgings, to about 20,000 lb/in². Figure 2 shows a valve-in-line fluid end for pressures up to 32,000 lb/in² (220 kN/m²). Here a one-piece valve seat (1) receives both the discharge ball valve (2) and the ring-type suction valve (3). This valve seat, with its inlet passages intersecting at a right angle, is reinforced by the taper shrink fit of the valve-seat body (4). This shrink fit also seals inlet passages from discharge pressure at the top and pulsation-chamber pressure at the bottom. The suction-port ring (5) connects the suction manifolds to the valve-seat body.

The stuffing-box barrel (6), the cylinder head (7), and the lower portion of the valve-seat body are all simple cylindrical forms, well adapted to high internal pressures because of favorable stress distribution. This design also keeps the clearance volume down to a very low value, which is important at pressures where even water is appreciably compressed.

**Horizontal power pumps** are also made in a similar range of sizes for approximately the same speed with both cast and forged fluid cylinders (Fig. 3). They are offered as triplex and quintuplex pumps with single-acting plungers. The plungers are connected directly to the power-end crossheads with the stuffing box between the cylinder and the frame, thus eliminating the upper crosshead and the pull rods characteristic of the inverted vertical pump. Pumps of this type are used for water flooding and saltwater disposal in the oil fields, for gathering and product pipelines, and for small hydraulic systems and hydraulic testing. (See Table 1.) Figure 4 shows a quintuplex vertical pump equipped with a synchronized suction-valve unloading system for use in an accumulator-type hydraulic system. Because of the heavy loading the crankshaft is carried in main bearings on either side of each crankpin with the bearing journals formed on the cheeks or webs of the crankshaft.

The fluid end of this pump is arranged with the suction and discharge manifolds bored in the cylinder forging and the valve assemblies held in place with individual covers and spacers. Separate plunger barrels are used to permit the use of a range of plunger sizes for each size cylinder; for higher pressures the forged cylinders are sectionalized with one, two, or three plunger chambers in each cylinder forging. It should be noted in the inverted construction that the only parts of the power end which are loaded in tension are the main-bearing bolts, the connecting rods, and the pull rods. The plunger force presses the cylinders against the upper half of the frame,

which is loaded in compression against the lower half of the frame.

The synchronized suction-valve unloading system, as shown in Fig. 5, permits this pump to operate continuously at constant speed, with the delivery starting and stopping to meet the requirements of the hydraulic system. The double-

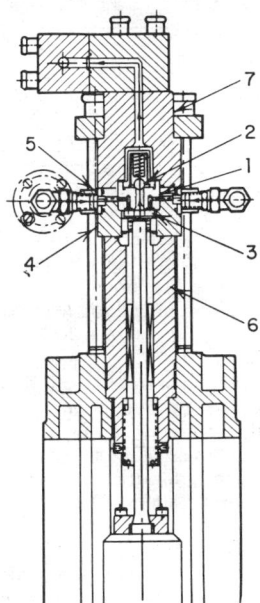

**Fig. 2**   Valve-in-line fluid end.

ported, or ring-type, suction valve (1) is made with several lugs extending radially inward which are engaged by the mushroom head of the valve lifter tappet (2) to hold the valve wide open when the pump is unloaded, thus permitting free flow in and out of the cylinder. This tappet is pushed to the unload position by the spring engaging the piston (3) on the valve lifter rod. When delivery is required by the system, an electrical timer energizes, in proper sequence, solenoid air valves which control the application of compressed air to the

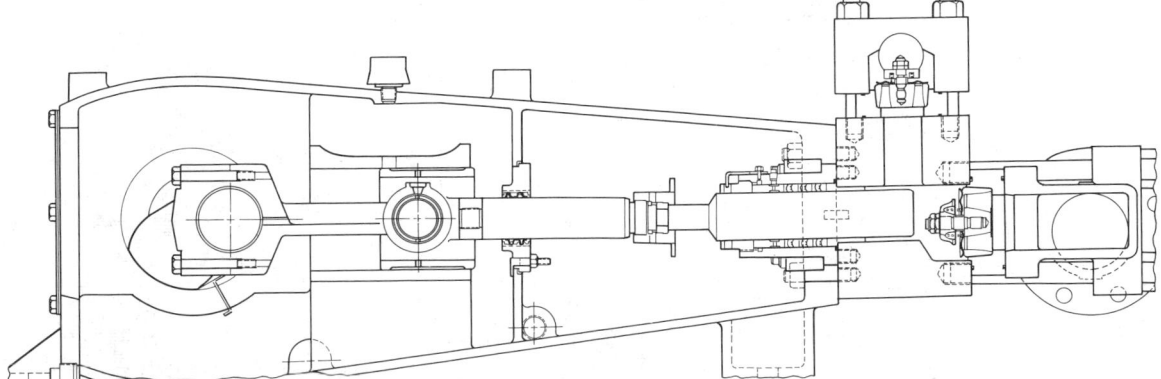

**Fig. 3**   Horizontal triplex pump. *(Ingersoll-Rand Company.)*

**Table 1. Typical Ratings of Triplex Horizontal Power (Spec) Pumps**

| Stroke, in | 2½ | 3 | 4 | 5 | 6 |
|---|---|---|---|---|---|
| mm | 64 | 76 | 102 | 127 | 152 |
| r/min | 690 | 530 | 420 | 360 | 300 |
| bhp | 28 | 60 | 100 | 150 | 300 |
| kW | 21 | 44 | 74 | 110 | 220 |
| Efficiency, percent | 85 | 86 | 87 | 88 | 90 |

pistons on the lifting rods, thus compressing the spring and withdrawing the lifter rods to the position shown so as to permit the suction valve to open and close for normal operation. The motion of the valve lifter rod and the tappet for both loading and unloading is synchronized to take place during the early part of the suction stroke of each plunger so that the suction valve is either free to close during the latter part of the suction stroke when delivery is required, or held firmly against the stop at the start of the next discharge stroke to prevent delivery. The mushroom head of the tappet has the additional function of sealing the hole in the valve seat through which it extends when pressure is developed by the discharge stroke of the plunger.

With this system of control the rate of flow from the pump changes from zero to maximum and from maximum to zero during one-half revolution of the pump, with the rate of change following the sum of the sine-curve flow patterns of the individual plungers in operation at each instant.

**Low-Speed Power Pumps** Another important type of power pump, illustrated by Fig. 6 usually operates in the speed range of 50 to 100 r/min and is built with a single reduction gear in the enclosed crankcase. Both the pinion shaft and the crankshaft are mounted in antifriction bearings, and some designs use roller bearings in both ends of the connecting rod. The horizontal power end is usually self-oiling, with the gear dipping in oil and carrying it up to a distribution system from which it flows by gravity to all moving parts. The side-pot piston liquid cylinder is fitted with easily removable liners so that a range of capacity and pressure is covered by using different sizes of pistons and liners. They use single- and double-acting pistons.

For general service the horizontal piston-pattern pump is used in sizes from 10 to 100 hp (7.4 to 74 kW) and pressures up to 1,000 lb/in² (6.9 MN/m²). In oil-field use the pressure range is extended to 1,500 lb/in², and specially fitted slush pumps of this type for oil-well drilling are built in sizes from 100 to 1,750 hp (74 to 1310 kW).

**Pumps for Very High Pressure** As pressures rise substantially above 15,000 to 20,000 lb/in² (104 to 138 MN/m²), piston or plunger speeds must be reduced drastically to obtain acceptable packing life and to reduce the number of pressure reversals and cyclical stresses which contribute to fatigue failures. This leads to plunger sizes and loads which make crank-driven pumps impractical because of the large size, high torque, low speed, and side thrust on the crossheads. The single-acting intensifier has long been used to produce extremely high pressures for research on a laboratory scale. Based on the principle of the intensifier, two- and four-

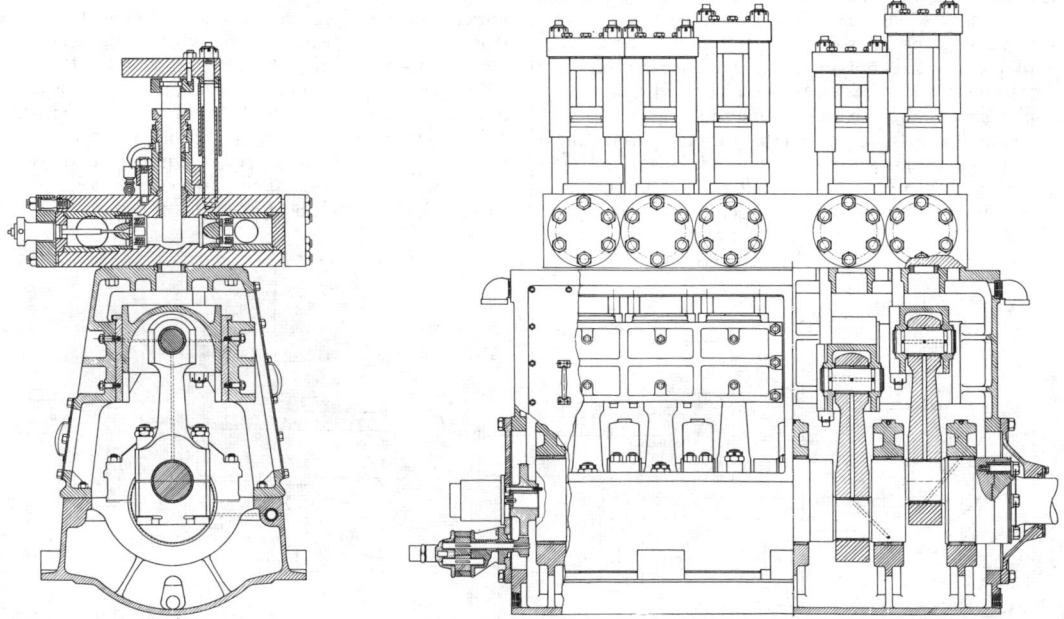

**Fig. 4** Quintuplex vertical pump with suction-valve unloading.

plunger pumps can be driven by double-acting oil hydraulic cylinders for pressures from 10,000 to 200,000 lb/in² (69 to 1,380 MN/m²). Figure 7 shows a section through one end of a pump of this type with the compound construction used for pressures over 50,000 lb/in² (345 MN/m²). The inner cylinder (1) is reinforced by the taper shrink fit of the outer cylinder (2). High pressure is confined to the space between the packed plunger (3) and the end closure (4), which is drilled to communicate with the external check valves. Because of the slow plunger speed of not more than 180 in/min (0.076 m/s), the plunger chamber can be connected to the valve body with small-bore, high-pressure tubing using double-cone high-pressure fittings at each end of the tubing. The special packing (5) is located on the end of the plunger, but when it is desirable for specific applications, the packing can be in a stuffing-box surrounding the plunger.

The low-pressure or driving cylinder is a conventional dou-

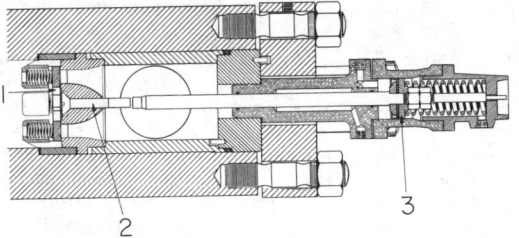

**Fig. 5** Suction-valve unloader.

ble-acting hydraulic cylinder sized to develop the desired thrust with not more than 2,000 lb/in² (13.9 kN/m²) oil pressure. Four different sizes of oil cylinders are used to cover a range of 15 to 100 hp (11 to 75 kW), and each is fitted with suitable plunger sizes and fluid cylinders to cover the desired range of pressure with corresponding capacities. Electric limit switches or fluidic control are used to control the hydraulic reversing valves connected to the oil cylinders. These pumps

may be arranged with a single power cylinder between two pumping cylinders or as a duplex unit with two oil cylinders and four pumping cylinders. Where continuous delivery of a relatively compressible fluid is desired, a special valving control is available which advances the next operating plunger to compress the charge almost to pumping pressure while the operating plunger is completing its stroke.

The in-line arrangement of these pumps minimizes side thrust, and the symmetrical, simple cylinders ensure the favorable stress distribution so important at high working loads. Hydraulic shock is practically eliminated by the controlled buildup of pressure possible with the hydraulic drive, thus contributing to long life even at very high pressures. Variable-capacity primary oil pumps may be used, or oil may be bypassed to control the high-pressure pump. Pressure-control or relief valves in the oil circuits are used to limit the maximum pressure, and other safety controls are readily adaptable to the primary pump control. The almost complete lack of inertia effects in the high-pressure pump make the unit very responsive to controls and contribute to the operating safety.

**Suction Lift, or Head**    Liquid flows into the suction side of a pump as a result of pressure exerted on the liquid. If the liquid is exposed to the atmosphere, this pressure will be atmospheric; if the liquid is in a closed vessel (as with water above 212°F (100°C) or volatile liquids such as ammonia or butane), the pressure will be the saturation pressure corresponding to the temperature of the liquid. To these pressures there will be added a positive hydrostatic head whenever the free surface of the liquid is at a higher level than that of the pump discharge valve.

The pressure exerted must at least equal the sum of the resistances to flow: (1) the vapor pressure of the liquid in the pump chamber; (2) the suction lift when the liquid level is below the pump level; (3) the pressure required to lift the suction valve and overcome the resistance of its spring; (4) the liquid friction in the suction pipeline; (5) the forces required to accelerate the liquid in the suction pipeline; and (6) hydraulic

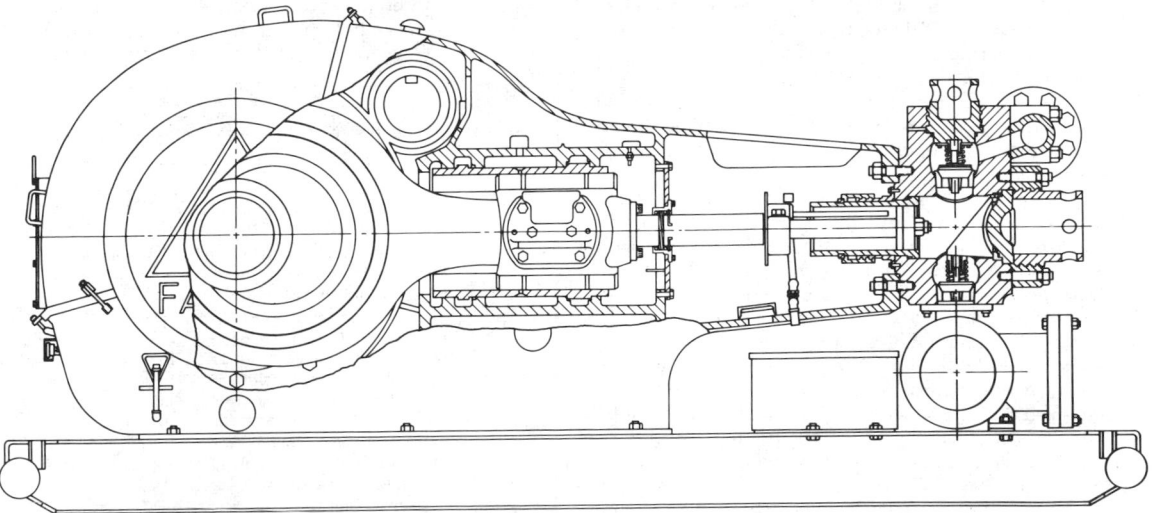

**Fig. 6** Horizontal piston pump. *(Continental-Ensco Co.)*

losses in the pump. All these quantities are conveniently expressed in feet of the liquid.

Low-speed piston power pumps and steam pumps will usually operate satisfactorily at a considerable suction lift. Figure 8 gives the variation with temperature of permissible suction lifts for water. $T$, $N$, and $M$ are, respectively, the theoretical suction lift, the normal suction lift, and the maximum possible under favorable conditions, all at sea level. The horizontal distance between $T$ and $M$ represents 8 ft (2.45 m) head and between $T$ and $N$ represents 12 ft (3.7 m) head, which are allowances to cover items 3 to 6 above. If the total of items 3 to 6 exceeds 12 ft (3.7 m), deduct the excess from the possible suction lift. The static suction lift is the vertical distance from the water in the suction supply to the top of the pump discharge deck. For water temperatures above 212°F (100°C), the same differential of 12 ft (3.7 m) between $T$ and $N$ must be added. The broken lines show normal static suction lift at the stated altitudes.

For a liquid, such as anhydrous ammonia, whose temperature is much below room temperature, it is important to have heat insulation on the suction piping. With the ammonia at 32°F (0°C), a rise in temperature of 1°F (0.56°C) would require 4.75 ft (1.45 m) **additional static** head to compensate for the increased vapor pressure at the pump. Similarly, when pump-

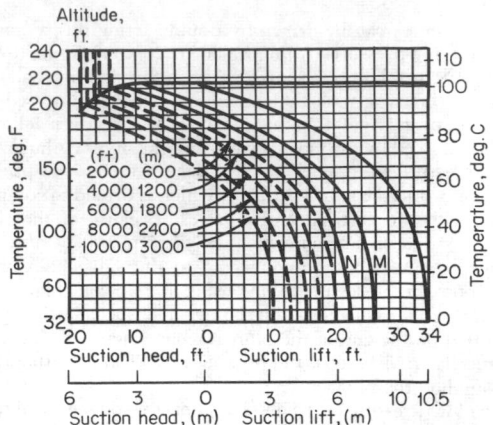

**Fig. 8** Suction lifts for low-speed reciprocating pumps.

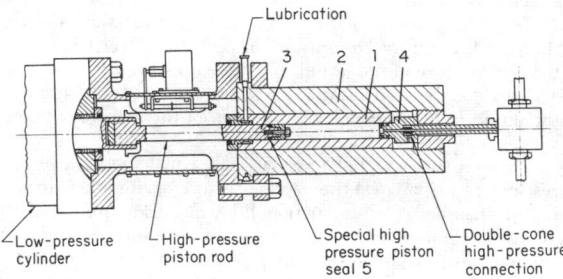

**Fig. 7** Very-high-pressure pump.

ing from a closed storage tank, the evaporation in the tank will lower the temperature there, thus reducing the saturation pressure and demanding an increased static head at the pump to maintain the effective pressure at the pump.

In water pumps where the water carries considerable air, there should be a large receiver in the suction line near the pump. Air should be removed from the top of the receiver by an independent vacuum pump.

The suction line should be short, with a minimum number of elbows or fittings, and any valve in it should be a gate valve. The suction pipe should slope up toward the pump with a uniform grade of at least 6 in in 100 ft (15 cm in 30 m) to avoid air or gas pockets. If the suction pipe is from an elevated tank, it should first drop below the pump level and then grade upward to the pump. Where pipe size is changed, eccentric reducers should be used so that the top layer of the pipe is continuous without air or gas pockets.

**High-speed plunger pumps** require considerably more net positive suction head than low-speed pumps and frequently will not operate with any suction lift, even on cold water. In addition, item 5, the head required to accelerate the liquid in the suction pipeline, is a much greater factor at higher rotative speeds. An empirical approximation of this quantity for crank-driven pumps is given by the equation $H = LVnC/gK$, where $L$ is the length of the pipe, ft (m); $n$ is the revolutions per minute

of the pump crankshaft; $H$ is the head of the liquid pumped to produce the required acceleration, ft (m); $V$ is the mean velocity of flow in the suction line, ft/s (m/s); $g$ is gravitational acceleration, ft/s² (m/s²), and $C$ is a factor for the type of pump. $C$ has the following values: simplex double-acting, 0.20; duplex double-acting, 0.115; triplex double-acting or single-acting, 0.066; quintuplex single-acting, 0.040; septuplex single-acting, 0.028; nonuplex single-acting, 0.022. An increase in pump speed with an existing suction line increases $H$ as the square of the speed because both $V$ and $n$ increase proportionally to the speed.

The value of $K$ is 2.5 for hot oil, 2.0 for most hydrocarbons, 1.5 for amine, glycol, and water, 1.4 for deaerated water, and 1.0 for urea and liquids with a small amount of entrained gas. These values assume short, nonelastic suction lines.

**Flow and Acceleration** Figure 9 shows the theoretical flow variation for various types of pumps as listed. Pumps with an odd number of cranks have the same flow variation for both single-acting and double-acting liquid ends. Flow curves for septuplex and nonuplex pumps are similar to the quintuplex curve, with more but smaller variations as indicated in the tabulation. The values tabulated will vary slightly with

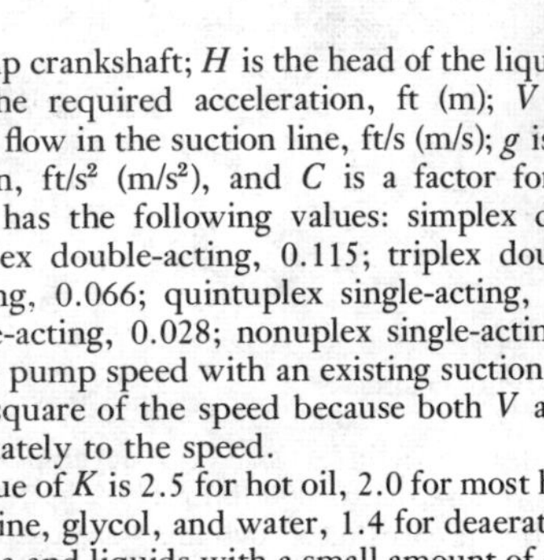

A = Simplex - double acting
Max. rate of flow above mean    59.96%
Min. rate of flow below mean    100.0 %

B = Duplex - double acting - cranks 90° apart
Max. rate of flow above mean    26.72%
Min. rate of flow below mean    21.56%

C = Triplex - single acting - cranks 120° apart
Max. rate of flow above mean    6.64%
Min. rate of flow below mean    18.42%

D = Quintuplex - single acting - cranks 72° apart
Max. rate of flow above mean    1.88%
Min. rate of flow below mean    5.63%

Septuplex - single acting - cranks 51⁴/₇° apart
Max. rate of flow above mean    1.2%
Min. rate of flow below mean    2.8%

Nonuplex - single acting - cranks 40° apart
Max. rate of flow above mean    0.7%
Min. rate of flow below mean    1.5%

**Fig. 9** Flow-rate variation in reciprocating power pumps.

changes in crank-to-connecting-rod ratio. The curves are theoretical, being based on crosshead pin velocity for constant rotational speed, and do not represent actual flow rate. This will vary considerably from the ideal, particularly in high-speed pumps, where the valve closing and opening may lag behind the crankshaft rotation as much as 10 to 15° (0.018 to 0.027 rad), and at high pressures, where liquid compression may cause a further lag to 5 to 10° (0.009 to 0.018 rad), or even more if the clearance volume is relatively large.

These theoretical flow curves are frequently taken as being directly indicative of the **pressure pulsation** which can be expected in the operation of pumps of various types. However, flow curves indicate only the percentage change in flow rate at various positions of the crankshaft and in no way take into account the rate of change or departures from the ideal curve. For most operating conditions the rate of change in flow, or acceleration, is the primary factor in pressure pulsation. The constant $C$ in the acceleration-head formula for various types of pumps is an index of pumps' relative pressure pulsation at the same rotative speed.

### PULSATION DAMPENERS (CUSHION CHAMBERS)

To take up irregularities and induce a uniform flow in suction and discharge lines, pulsation dampeners must often be used. Their use is particularly desirable for single pumps and high-speed pumps. The volume of an air dampener for single pumps should be six to eight times the displacement of one plunger per stroke; for duplex and triplex pumps, three to four times. For high-speed multiplex pumps the size of dampener may be less than that indicated for a triplex pump of the same plunger size, but it is of prime importance to make the connection between the dampener and the pump as short and as large as practicable. A simple air dampener is usually satisfactory on the inlet of the suction side of pumps, but at higher delivery pressures the air or gas charge is soon lost by dissolving into the fluid pumped. Here a dampener with some form of diaphragm or bladder, to hold the charge, is needed. The charge is approximately two-thirds of the system pressure.

With deaerated water for boiler feeding, a suction chamber using steam as the cushioning medium is simple and effective. This steam cushion is maintained by jacketing the chamber with steam at a somewhat higher pressure than the maximum suction pressure. Suction cushion chambers are particularly desirable where the liquid comes to the pump under a static head or the suction line is relatively long.

### POWER-PUMP SPEEDS

Table 2 gives data on standard pumps of conventional design. Slower speeds are used for viscous hot liquids, oil-refinery service, slurries, and urea service. Higher speeds are used where weight is a consideration, as in marine service, and also in high-pressure hydraulic-press service, where the volume of liquid pumped is comparatively small, so that inertial effects are not so important. Speed is a limiting factor in liquid separation from the plunger. Low-speed limits with sleeve bearings have to be considered because of the lack of lubrication from proper oil film.

### DIRECT-ACTING STEAM PUMPS

In the direct-acting steam pump the steam piston is connected to the pump piston by a rod, without crank motion. There is no cutoff or expansion of the steam, since it is admitted at a constant rate throughout the stroke. The moving parts are cushioned and brought to rest by steam trapped in the end of the steam cylinder at the end of each stroke, with full steam pressure on the opposite side of the piston, The actual velocity of the moving piston is practically constant during 80 to 90 percent of the stroke. In single and duplex pumps there is a definite pause at the end of each stroke, which is important for closing the fluid-end valves. The steam end of a duplex pump is shown in Fig. 10. The steam piston on each side is mechani-

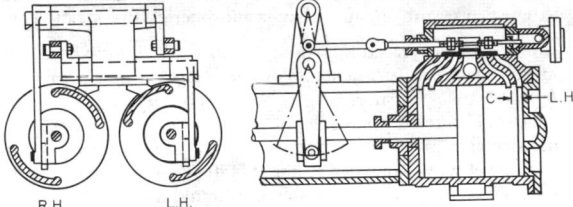

**Fig. 10** Steam end of a duplex pump.

cally connected to the steam valve of the opposite side. Since there is some overlapping of the strokes, a duplex pump will deliver a continuous flow of fluid without marked pressure fluctuation. A duplex pump will short-stroke badly in pumping a volatile fluid if the suction head is not sufficient to prevent flashing in the fluid cylinder.

### PUMP ENDS

Double-acting fluid ends are used on all direct-acting steam pumps. A typical cap-and-valve-plate piston fluid cylinder is shown in Fig. 11. This style is used for pressures up to 250 or 350 lb/in² (171 or 240 kN/m²); for higher pressures side-pot fluid cylinders are usually preferred.

### POWER REQUIREMENTS

The steam consumption of direct-acting pumps will vary from

**Table 2. Approximate Speeds of Power Pumps**

| Stroke | | | Stroke | | | Stroke | | |
|---|---|---|---|---|---|---|---|---|
| in | mm | r/min | in | mm | r/min | in | mm | r/min |
| 2½ | 64 | 800 | 6 | 152 | 400 | 12 | 305 | 100 |
| 3 | 76 | 500 | 7 | 178 | 330 | 18 | 457 | 70 |
| 5 | 127 | 400 | 9 | 229 | 225 | 24 | 610 | 60 |

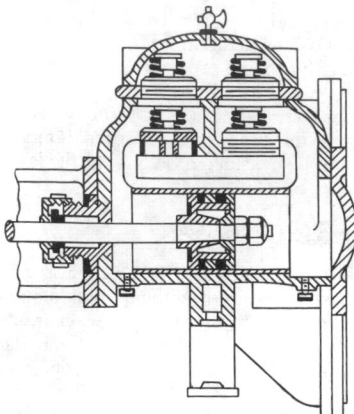

**Fig. 11**  Submerged piston pump.

200 lb (90 kg) of steam per water hp·h (kWh) for small pumps at light loads to as little as 50 lb (23 kg) of steam per water hp·h (kWh) for large pumps operating at 350 to 400 lb/in² (240 to 275 kN/m²) steam pressure. For pumps operating with an exhaust pressure greater than atmospheric, the steam consumption is increased in proportion to $\sqrt{p/(p-b)}$, where $p$ = initial steam pressure at steam cylinder inlet, psig (kN/m²) and $b$ = exhaust pressure, psig (kN/m²). For superheated steam deduct 1 percent for each 10°F (5.6°C) of superheat. The performance of a steam pump is often expressed as duty in ft·lb (J) of work done per 1,000 lb (450 kg) of dry steam. **Brake horsepower** for power pumps can be found using

$$bhp = Qd \times \text{lb/in}^2/1714 \times ME$$
$$(kW = Qd \times \text{kN/m}^2/60 \times ME)$$

where $Qd$ is delivered U.S. gal/min (m³/min) and lb/in² (kN/m²) is differential pressure. ME is mechanical efficiency. Figure 12 shows a typical power-pump performance curve. Capacity $Qd$ is the total volume of fluid delivered per unit of time. The fluid includes liquids and entrained gases and solids at specified conditions. Capacity is equal to volume displaced $D$ less slip $S$. $Qd = D(1 - S)$, where displacement $D$ is the calculated capacity with no loss due to slip. $D = Amns/231$ gal ($D = Amns/10^9$ m³) for single-acting piston or plunger; $D = (2A - a)mns/231$ ($D = (2A - a)mns$) for double-acting, where $A$ = area of piston or plunger, in² (mm²); $a$ = area of piston rod, in² (mm²); $m$ = number of plungers or pistons; $n$ = r/min; $s$ = stroke, in (mm), i.e., half the linear distance the plunger or piston moves linearly in one revolution, and 231 in³/gal ($10^9$ mm³/m³).

**Slip** is the capacity loss as a percentage of suction capacity, $S = Ve_1 + B_1 + V_1$, where $Ve_1$ = volumetric efficiency loss, $B_1$ = stuffing-box loss (negligible), $V_1$ = the valve loss (2 to 10 percent depending on valve design and conditions). $Ve_1 = 1 - Ve$, where **volumetric efficiency** $Ve$ is the ratio of discharge volume to suction volume as a percentage proportional to ratio $r$ and differential pressure (Fig. 13). $r = (C + D)/D$ is the ratio of internal volume of fluid between valves when the plunger or piston is at the top of its back stroke $(C + D)$ to plunger displacement $D$ (Fig. 14). Since volume cannot readily be measured at discharge pressure, it is taken at suction pressure. This will result in a higher $Ve$ due to fluid compressibility not

accounted for; compressibility loss becomes important for water over 6,000 lb/in² (41,000 kN/m²). Approximate percentage loss for 1,000 lb/in² (7000 kN/m²) at 68°F (22°C) is water, 0.29; lubricating oil, 0.44; kerosene, 0.51; gasoline, 0.70; ammonia, 0.98; $n$-butane, 1.9; isobutane, 2.4; propane, 3.0. Refer to Hydraulic Institute Standards for calculations at higher pressures.

Mechanical efficiency $e_m$ = whp/bhp, where whp = water-horsepower input to a power pump. This efficiency gives information about the mechanical friction in the mechanism transmitting the power to the liquid end. It can be determined only by actual experiment with different types of pumps and depends upon the size of the pump and the service. $e_m$ with motor drive is 90 to 92 percent; with single-reduction gear or V belt it is 88 to 90 percent.

## PLUNGER LOAD

A power pump is designed for a **plunger-load limit**. This is the load in pounds that is applied to the plunger or piston and the bearing system.

$PL = A \times$ lb/in² (kN/m²), where lb/in² (kN/m²) is differential pressure and $A$ is area of plunger or piston, in² (mm²) on the double-acting pistons; the piston rod area $a$ is subtracted on the forward stroke.

The bearing system is rated for a specific load at design speed. Higher than rated plunger loads will result in short bearing life. With the sleeve bearing, high plunger loads at slow speed will destroy the lubrication film.

The additional load on the bearings from the moment of inertia of unbalanced reciprocating and unbalanced rotating parts is approximately equal to 10 to 25 percent of the rated plunger load.

## PUMP VALVES

**Disk Valves**  For pressures of 6,000 lb/in² (41,200 kN/m²) disk valves on ground seats, as shown in Figs. 15 and 16, are generally used. The seats are metallic with smooth-taper fit in the valve deck. **Conical-faced wing** and ball valves are used for pressures of 10,000 and 30,000 lb/in² (69,000 and 280,000 kN/m²), respectively, as they can be ground pressure-tight more

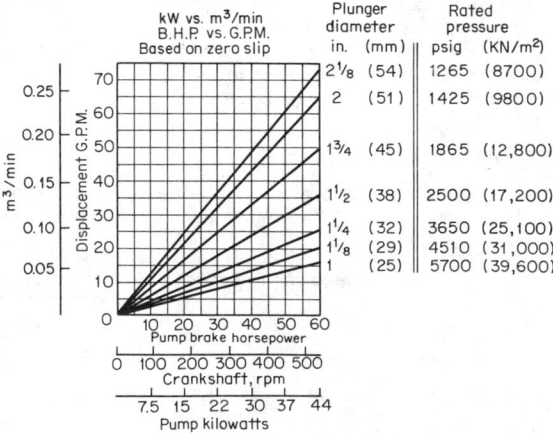

**Fig. 12**  Power-pump performance curve. (*Ingersoll-Rand Company.*)

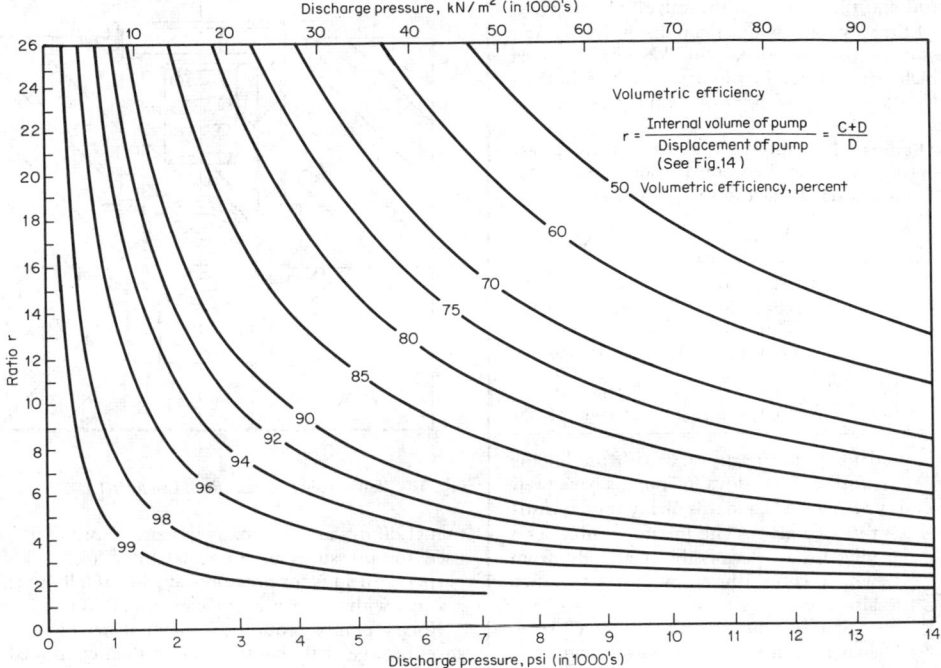

**Fig. 13** Volumetric effiency related to *r* and discharge pressure. (*Contained volume including plunger displacement.) *(Ingersoll-Rand Company.)*

readily than flat valves. Figures 17 and 18 show this type as applied to a forged steel cylinder. **Ball and elastomer insert** valves (Figs. 18 and 19) are used for viscous liquids and slurry service. Figure 20 shows a double-ported, or ring-type, valve. In closing, a single-ported valve must displace a quantity of fluid proportional to the square of the valve diameter through the outlet area, which is proportional to the first power of the diameter. In order to obtain a greater outlet area with the

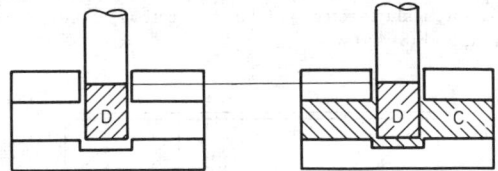

**Fig. 14** Volumes *C* and *D* in Fig. 13. *(Ingersoll-Rand Company.)*

small lift permissible at high pump speeds and to reduce the quantity of fluid which the valve must displace in closing, the ring-type, or double-ported, valve is used except for very small capacities. Flow from the annular seat is both radially outward around the valve and radially inward and through the center hole in the valve.

**Velocity through Valve Seats** At conventional pump speeds, as shown in Table 2, the velocity of flow for cold water through the lifted inlet valve seat, is usually 3 to 8 ft/s (0.91 to 2 m/s). For viscous liquids, the velocity may be 1.5 ft/s (0.46 m/s) or less; for slurry pumps, it is 6 to 10 ft/s (2.83 to 3.05 m/s). The velocity through the outlet port between the lifted valve and its seat is much higher and depends on the sum

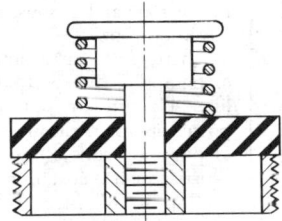

**Fig. 15** Nonmetallic disk valve.

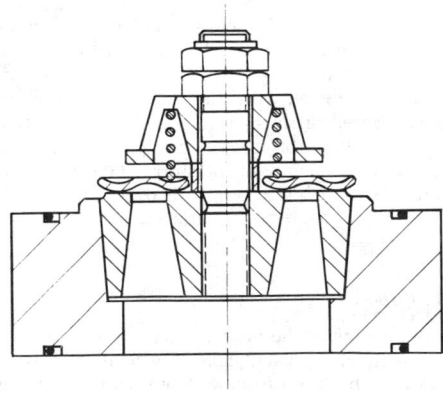

**Fig. 16** Stainless steel disk valve. *(Durabla Mfg. Company.)*

of the spring load and the weight of the valve. The required area of contact of the valve seat with a plain disk valve is $pA/b$, where $p$ = pressure on back of valve, lb/in² (kN/m²); $A$ = area of valve on which $p$ is acting, in² (mm²); and $b$ = bearing pressure of valve on seat, lb/in² (kN/m²). Unit pressures are given in Table 3.

**Materials for Pumps**  The allowable stresses in pumps are lower than in most machinery because of shocks and water hammer. The following tensile stresses may be used:

| Metal | lb/in² | kN/m² |
| --- | --- | --- |
| Cast iron | 1,500–1,800 | 10,000–12,400 |
| High-test alloy | 2,000–2,500 | 14,000–17,000 |
| Malleable iron | 3,000 | 20,000 |
| Steel castings | 6,000 | 41,000 |
| Cast bronze | 3,000 | 20,000 |
| Rolled Ni-Al bronze | 10,000 | 69,000 |
| Steel forgings | 10,000–16,000 | 69,000–111,000 |

The materials used for pump parts vary with the liquids handled. Corrosion-resisting steel alloys for pumps have been classified in several types in the Standards of Hydraulic Institute (Table 4). The Institute publishes a long list of permissible materials, from which Table 5 has been abstracted. (See also Sec. 6.)

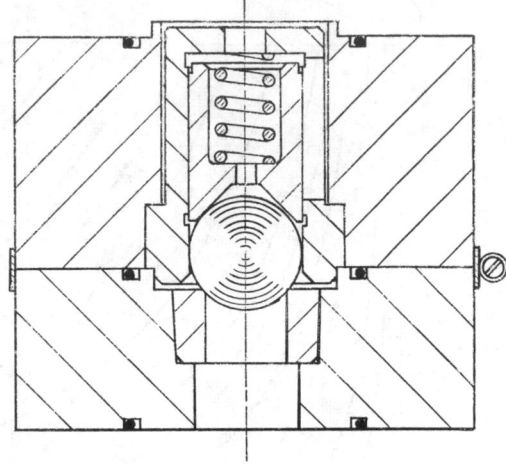

Standard-fitted pumps have steel-coated plungers in a brass bushing, steel rods, plastic or steel valves, stainless-steel valve seats, stems, and springs. Bronze-fitted pumps differ only in having bronze piston rods and pistons, All-iron pumps have no bronze.

There is now great interest in the use of pumps to transport slurries via pipeline. The present state of the art uses power pumps as the most efficient means of pumping slurries (80 to 92 percent). Slurries such as coal, iron ore, and copper ore up to 65 percent by weight are pumped hundreds of miles using power pumps. The technology of pumping slurries is relatively new and being improved with experience. Nomenclature and standards for this technology are being established.

**Fig. 17**  Conical-faced wing valve.

The Miller number is being referenced as an index of abrasivity for slurries. The relation of wear rates of parts relative to velocity varies with the slurry and is not firmly established. In general, wear rate varies as the square to cube of velocity. Speed of slurry pumps is from 60 to 120 r/min. Velocity of the fluid through the pumps is usually 2 or 3 ft/s above the critical velocity of the slurry. Horsepower of these pumps ranges from 200 to proposals over 2,000.

## ROTARY PUMPS

Rotary pumps are of the positive-displacement type, usually valveless, simple, compact, light in weight, and low in first cost. They are built in capacities from a fraction of a gal/min (m³) (as in domestic oil burners and refrigerators) to 5,000 gal/

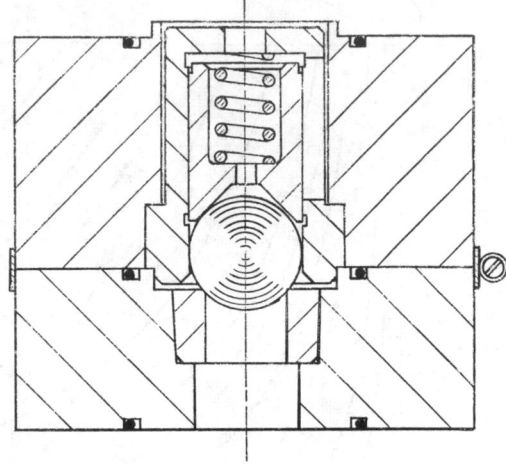

**Fig. 18**  Ball valve. *(Ingersoll-Rand Company.)*

min (19.0 m³) and above, as in marine cargo service. Though used for pressures up to 1,000 lb/in² (6,900 kN/m²), their particular field is for pressures of 25 to 500 lb/in² (170 to 3,500 kN/m²), with mechanical efficiencies of 80 to 85 percent.

Rotary pumps require the maintenance of very close clearances between rubbing surfaces for their continued volumetric efficiency. No satisfactory method of packing the moving surfaces to compensate for wear has been developed; consequently, although some rotary pumps are used successfully for clean water, their great field of application is in pumping oils or other liquids having lubricating value and sufficient viscosity to prevent excessive leakage. Rotary pumps are being used in the oil industry in increasing numbers. They are also used for liquids of high viscosities, up to 500,000 SSU.

The diameter of the discharge opening of a rotary pump designates the nominal size but does not fix the displacement. The Hydraulic Institute classifies rotary pumps in the following groups: **single-rotor** (1) valve, (2) piston, (3) flexible-valve, (4) screw; **multiple-rotor** (5) lobe, (6) gear, (7) circumferential piston, and (8) screw.

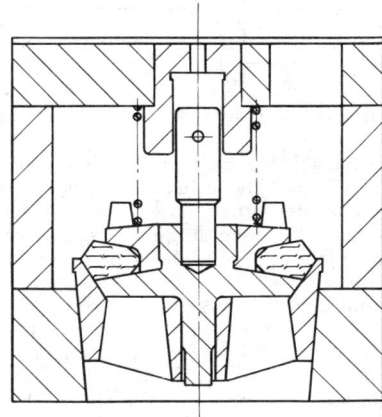

**Fig. 19**  Elastomer insert valve. *(TRW-Mission Mfg. Co.)*

**Table 3. Bearing Pressures for Valve Materials***

| Material | Brinell hardness | Allowable bearing pressure | |
|---|---|---|---|
| | | lb/in² | kN/m² |
| Bronze, regular | 60–80 | 3,000 | 21,000 |
| Hard | 175–200 | 8,000 | 55,000 |
| Cast iron | 200 | 10,000 | 69,000 |
| Steel, not heat-treated | 180 | 9,000 | 62,000 |
| Hardened alloy | 500–600 | 25,000 | 170,000 |
| 11–13 chrome stainless steel (hardened) | 350 | 17,500 | 120,000 |
| 18–8 stainless steel | 180 | 9,000 | 62,000 |
| Monel (hardened) | 260–300 | 13,000 | 90,000 |
| Bakelite | ....... | 2,000 | 14,000 |
| Leather | ....... | 750 | 5,200 |
| Rubber, Hard | ....... | 1,000 | 7,000 |
| Soft | ....... | 375 | 2,600 |

*For high-speed pumps, these values should be reduced to compensate for increased shock.

Rotary pumps up to 100 lb/in² (690 kN/m²) may be considered low pressure, from 100 to 500 lb/in² (690 to 3430 kN/m²) moderate pressure, and above 500 lb/in² (3430 kN/m²) high pressure; fractional [to 50 gal/min (0.2 m³)] are small-volume pumps, 50 to 500 gal (1.2 to 1.9 m³) moderate-volume, and above 500 gal (1.9 m³) large-volume.

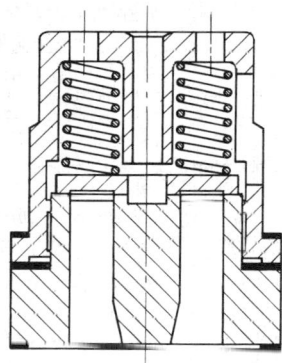

**Fig. 20**   Double-ported valve.

## SINGLE ROTOR

**Vane Pumps**   Leakage in vane-type pumps is across the tips and sides of the vanes. This leakage occurs when the vanes are under the two abutments. Since the vane tips cannot be made to fit the bore of the housing in all positions, there is line contact and low resistance to leakage. Wear may also be serious at the higher speeds unless the vanes are restrained against centrifugal force. Increasing the number of vanes materially decreases leakage.

Figure 21 is typical of the **guided-vane type.** A single rotor revolves in a case. The pumping element consists of multiple blades sliding in slots in the rotor. Impeller and case are eccentric. Centrifugal force or pressure maintains the outer end of the blades in contact with the casing bore. The blades are of hardened steel, bronze, or Bakelite. This type is useful for small and moderate capacities and low pressure. Rapid wear on the points of the sliding blades and in the casing occurs where speed is high or where the liquid pumped has low lubricating value. In some constructions, the blades are made with end trunnions operating in grooves in the side plate.

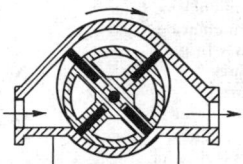

**Fig. 21**   Rotary pump with guided vanes.

Figure 22 illustrates a pump of the **swinging-vane type,** with vanes hinged or articulated. The hinge joints are subject to wear, and the comparatively small number of vanes or blades possible with this construction give a less satisfactory seal than the multiple blades in the sliding-vane type. Swinging-vane pumps are used for moderate volume, for low pressure and vacuum, and for low speeds.

**Eccentric-Piston Pumps**   Many pumps of this type are in

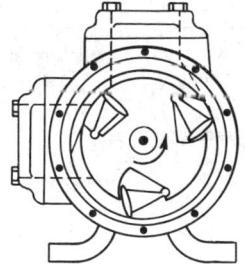

**Fig. 22**   Rotary pump with swinging vanes.

**Table 4. Approximate Analysis of Stainless Steels**

| Type no. | Approximate analysis, percent | | | | | | |
|---|---|---|---|---|---|---|---|
| | C*, max | Cr | Ni | Mo | Cu, max | Mn, max | Si, max |
| 4 | 0.20 | 5.00 | ........ | 0.50 | | | |
| 5 | 0.15 | 13.00 | | | | | |
| 6 | 0.30 | 20.00 | 1.00 | | | | |
| 7 | 0.50 | 28.00 | 2.00 | | | | |
| 8 | 0.08 | 18.0–21.0 | 8.0–11.0 | ....... | .... | 1.50 | 2.00 |
| 9 | 0.08 | 18.0–21.0 | 9.0–12.0 | 2.0–3.0 | .... | 1.50 | 2.00 |
| 10 | 0.07 | 18.0–22.0 | 20.0–30.0 | 3.5 max | 4.5 | 1.50 | 4.00 |

*Low-carbon alloys, only, are recommended to ensure protection from such failures as intercrystalline corrosion or intergranular attack.

**Table 5. Permissible Materials for Pumps**

| Liquid | Materials permissible | Liquid | Materials permissible |
|---|---|---|---|
| Fatty acids | A, 8, 9, 10, 11 | Glue | B, C |
| Fruit acids | A, 8, 9, 10, 11, 14 | Glycerol (glycerin) | A, B, C |
| Hydrochloric acid | 11, 12 | Lard | B, C |
| Nitric acid | 5, 6, 7, 8, 9, 10, 12 | Limewater (milk of lime) | C |
| Sulfuric acid (to 65%) | 10, 11, 12 | Magnesium chloride | A, 8, 9, 10, 11, 12 |
| (10%) | A, 10, 11, 12, 14 | Milk | 8 |
| Tannic acid | A, 8, 9, 10, 11, 14 | Molasses | A, B |
| Alcohol | A, B | Naphtha | B, C |
| Aqua ammonia | C | Fuel oil | B, C |
| Ammonium chloride | 9, 10, 11, 12, 14 | Vegetable oil | A, B, C, 8, 9, 10, 11, 14 |
| Asphaltum | C, 5 | Turpentine oil | B, C |
| Beer | A, 8 | Potassium carbonate | C |
| Benzene (benzol) | B, C | Potassium chloride | A, 8, 9, 10, 11, 14 |
| Calcium chloride brine | C | Soap liquor | C |
| Sodium chloride brine | A, C, 8 | Soda ash (sodium carbonate) | C |
| Seawater brine | A, B, C | Sugar | A, 8, 9, 10, 11, 13 |
| Cane juice | A, B, 13 | Sirup | A, 8, 9, 10, 11, 13 |
| Creosote | B, C | Tanning liquors (veg.) | A, 8, 9, 10, 11, 12 |
| Diphenyl | C, 3 | Tar | C, 3 |
| Ethyl acetate | C, 9, 10 | Toluene (toluol) | B, C |
| Ethylene chloride | A, 8, 9, 10, 11, 14 | Varnish | A, B, C, 8, 14 |
| Fruit juices | A, 8, 9, 10, 11, 14 | Vinegar | A, 8, 9, 10, 11, 12 |
| Gasoline | B, C | Wood pulp | A, B, C |

The letters are abbreviations of materials, as follows: A, all bronze; B, bronze-fitted; C, all iron. The numbers, other than those indicating the stainless steels, represent the following materials: 1, gray iron; 2, tin bronze; 3, carbon steel; 11, a series of nickel-base alloys; 12, high-silicon cast iron; 13, austenitic cast iron; 14, Monel metal.

service. The pump shown in Fig. 23 is of the single-shaft type with cylindrical body and with oscillating eccentric and strap. The contact between the strap and the body approximates single-line contact. Leakage therefore becomes excessive as wear progresses. This type is useful for small and medium capacities, low pressure, and limited speed. A jacketed construction is shown. Practically all types of rotary pumps may be jacketed for viscous materials which must be heated to be pumped or for water cooling.

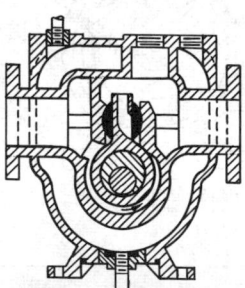

**Fig. 23**   Rotary pump, eccentric-piston type.

**Radial-Plunger and Swash-Plate Pumps**   The rotation of the body carrying the plungers connects each plunger flow periodically to the suction port on the plunger's suction stroke and to the discharge port on its discharge stroke. These groups can be adapted for variable capacity. In Fig. 24 this is done by varying the eccentricity between the plunger-carrying body and the ring that drives the plungers; in Fig. 25 by varying the angle between the drive shaft and the plunger-carrying body. The actual machines are complicated.

In a **flexible-member** pump the sealing and pumping actions depend on the elasticity of the flexible members, which may be a tube or vanes (Fig. 26). In **single-screw** pumps the fluid is pumped axially between the meshing internal stator screw threads (Fig. 27).

The **lobar type** (Fig. 28) is one of the earliest constructions used for rotary pumps and blowers. These machines are suitable for medium and large capacities and low pressures. As in the oscillating-piston type, there is line contact between the impeller and the body, and the leakage is excessive at higher pressures. The impellers are not self-actuating. Such pumps must therefore be built with external pilot gears capable of transmitting half the power utilized from the driving to the driven shaft.

**Gear pumps** (Fig. 29) are of the two-shaft type and cover a wide variety of constructions. They are used for practically all capacities and pressures. In many types, the impeller gears are self-actuating, requiring no pilot gears. The simplest form uses spur gears. The large number of teeth in contact with the casing minimizes leakages around the periphery. The utility of the straight spur-gear type is limited by trapping of liquid, which occurs on the discharge side at the point of gear inter-

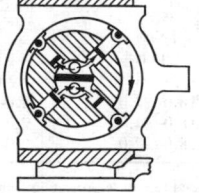
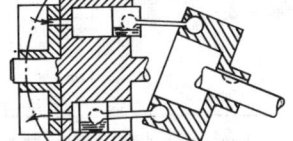

**Fig. 24**   Radial-plunger pump.   **Fig. 25**   Swash-plate pump.

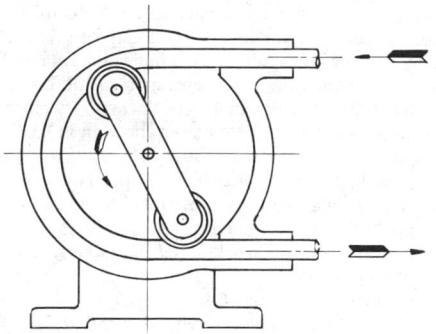

**Fig. 26**  Flexible tube pump. *(Hydraulic Institute.)*

mesh, resulting in noisy operation and low mechanical efficiency, particularly at high rotative speed. Discharge pockets in the side plates may be provided to reduce the effects of trapping. Impellers in other pumps of this type are of single-helical or double-helical construction with angles from 15 to 30° (0.26 to 0.52 rad) or more. With gears of single-helical type on higher pressures, considerable end thrust of the impeller gears on the pump side plates results. Either helical or herringbone gear construction largely eliminates the effects

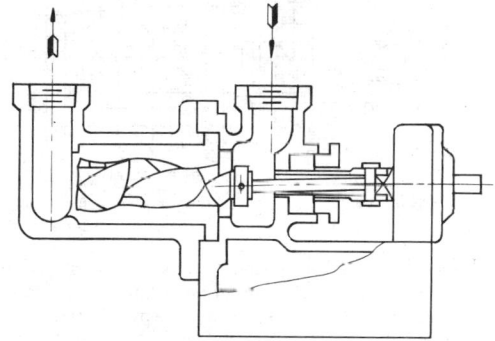

**Fig. 27**  Single-screw pump. *(Hydraulic Institute.)*

of trapping but introduces leakage losses between the teeth at the meshing point unless the teeth are cut without root clearance.

**Internal-Gear Pumps**  ONE-TOOTH DIFFERENCE (Fig. 30). In pumps of this type, an impeller mounted eccentrically with the body actuates an internal gear rotating in the body or in bearings carried in the end plates. Flow is practically continu-

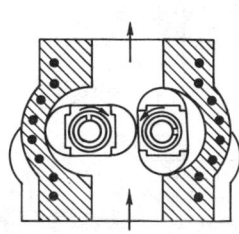

**Fig. 28**  Roots rotary pump, lobar type.

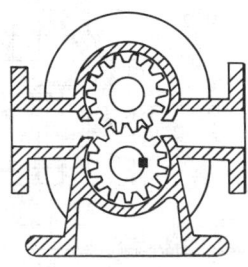

**Fig. 29**  Gear pump.

ous and without reversals. High rotative speeds may be used. In such pumps, leakage occurs around the periphery of the ring gear, over the tips of the gear teeth at open mesh, and through the contact line at full mesh. This type is particularly adaptable for high pressures and high speeds, for oils with lubricating value and considerable viscosity.

TWO-TEETH DIFFERENCE (Fig. 31). In this construction, an abutment on one side plate is used to fill the clearance between the external and internal gear. Such construction reduces leakage but involves the use of an overhung internal gear, which restricts the application to pumps for small and medium capacity and pressure.

CIRCUMFERENTIAL PISTON PUMPS. The fluid is pumped between the spaces of the piston surfaces. There is no real contact between piston surfaces (Fig. 32).

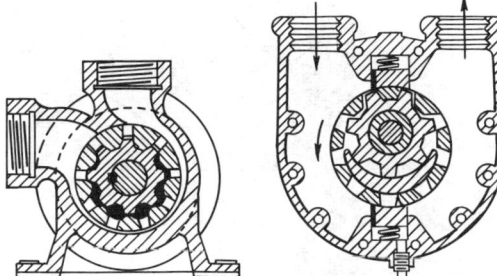

**Fig. 30**  Internal gear pump, one-tooth difference.

**Fig. 31**  Internal gear pump, two-tooth difference

In **screw pumps** (Fig. 33), a long single helical impeller of small diameter and special form actuates one or more idler impellers contained in a casing so that the pumped liquid is axially displaced. Multiple surface rather than line contacts between screws and case minimizes leakage. This construction

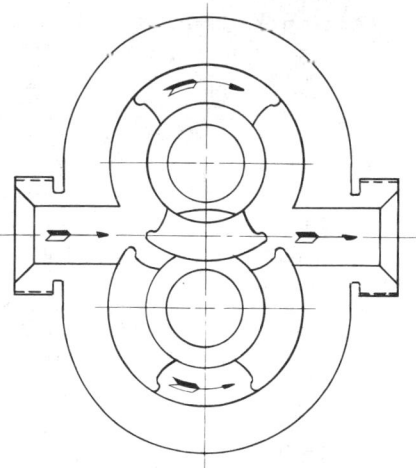

**Fig. 32**  Circumferential piston pump. *(Hydraulic Institute.)*

permits operation at very high speed. Where, as in the illustration, right- and left-hand helices are used, the pumping load is balanced and thrust is eliminated. No shaft bearings or timing gears are required thanks to the form of the impellers. Wear of rotating elements may be rapid with liquids of low lubricating value.

**Double-Screw Pumps** (Fig. 34) The screw-pump construction shown incorporates right- and left-hand intermeshing helices on parallel shafts with timing gears. These pumps have been extensively used for medium and large capacities and moderate to high pressures. There is some leakage axially at the impeller contact. Impellers are carried in bearings so that wear on impellers and casing is reduced. Flow is practically continuous.

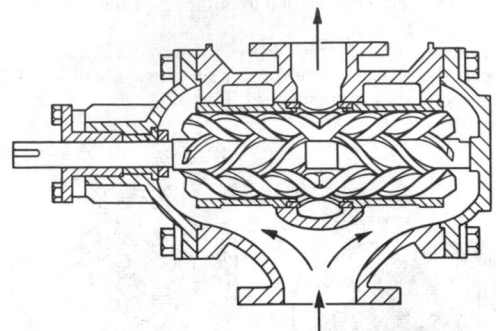

**Fig. 33** Screw pump.

The mechanical efficiency of the better types of rotary pumps in handling oils or other liquids with lubricating value is good. Figures 35 and 36 show the characteristics of pumps of the herringbone-gear type and the internal-gear one-tooth-difference type, respectively.

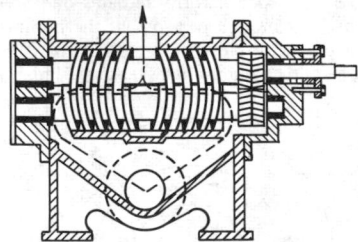

**Fig. 34** Double-screw pump.

## PISTONLESS PUMPS

Liquids may be pumped by the direct pressure of air, gas, or steam on the surface of the liquid.

**Injectors** An injector consists of a **steam nozzle** $a$ (Fig. 37), in which the steam acquires kinetic energy; a **combining tube** $b$, at the entrance to which the steam impinges on the feedwater entering at $c$; and a **delivery tube** $d$, in which the velocity head of the feedwater is reduced and its static pressure increased.

The velocity acquired by the steam in the steam nozzle can be calculated as indicated in Sec. 4. The pressure at the entrance to the combining tube is found to be 3 to 4 psia (20 to 28 kN/m²) for usual feed temperatures. The quantity of feedwater that can be handled is obtainable from the equation

$$(W + 1) V_m = C(V_s + W V_w)$$

where $W$ = weight, lb (kg), of water discharged per lb (kg);

$V_s$ = velocity of steam before impact; $V_w$ = velocity of feedwater before impact, ft/s (m/s); $V_m$ = velocity of mixture after impact. The value of $V_w$ is calculable when the head and lift of the feedwater, the vacuum at the entrance to the combining tube, and the frictional resistances are known. Its amount is generally negligible, so that the equation becomes $(W + 1) V_m = C V_s$. The value of $C$ averages about 0.5.

The actual weight of feedwater handled per pound of steam $W$ is easily determined in an injector from a heat balance. The heat given up by the steam is equal to the heat given to the feedwater plus the external work done. The external work is usually about 2 percent of the heat given up by the steam. The heat-balance equation becomes $0.98 (H - h_2) = W(h_2 - h_1)$, where $H$, $h_1$, and $h_2$ are, respectively, the enthalpies of 1 lb (0.45 kg) of steam, 1 lb (0.45 kg) of feedwater, and 1 lb of the mixture.

## AIR-LIFT PUMPS

The air lift consists of a drop pipe placed in a well with its lower end submerged. The depth of submergence measured

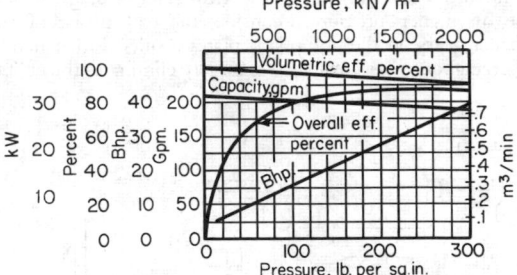

**Fig. 35** Performance curves for 600 r/min, 400 SSU, herringbone-gear pump.

from the level at which the water stands during operation to the air entrance is called the **submergence**. An air pipe delivers air at the bottom of the drop pipe and forms a mixture of air and water which is lighter than the column of solid water in the well; consequently, the mixture rises above the surrounding water. The necessary **percentage of submergence** (the per-

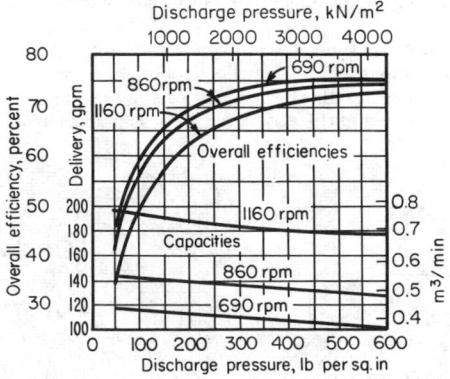

**Fig. 36** Performance curves for an internal gear, one-tooth difference, rotary pump.

centage of the total length of pipe which is submerged in "solid" water when pumping) decreases as the lift increases. Purchas (*Proc. Inst. Mech Eng.*, 1917) gives the following values:

| Lift | | Submergence/ | Submergence | |
|---|---|---|---|---|
| ft | m | Lift | ft | m |
| 25 | 7.6 | 4.0 | 100 | 30.3 |
| 50 | 15.2 | 3.0 | 150 | 45.7 |
| 75 | 22.9 | 2.33 | 175 | 53.3 |
| 100 | 30.5 | 2.0 | 200 | 61.0 |
| 150 | 45.7 | 1.7 | 255 | 77.7 |
| 200 | 61.0 | 1.38 | 275 | 83.8 |
| 250 | 76.2 | 1.22 | 305 | 93.0 |
| 300 | 91.4 | 1.0 | 300 | 91.4 |

The absolute air pressure required, measured in ft (m) of water, is $B + s$, where $B$ = barometric pressure, ft (m) of water, and $s$ = submergence, ft (m). The efficiency is about 50 percent for lifts up to 300 ft (91.4 m) and as low as 25 percent for high lifts. The whole operating mechanism is above ground, and the pump is able to handle dirty, gritty, or acidulous mine water, slimes from reduction plants, sewage etc. The theoretical consumption $v$ of free air in ft³/ft³ (m³/m³) of water pumped is given by Goodman (1899) and Purchas, (*Proc. Inst. Mech Eng.*, 1917) as follows: $v - l/B \ln(1 + s/B)$, where $l$ is the

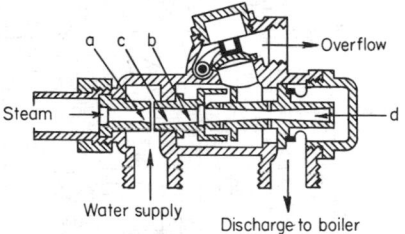

**Fig. 37** Injector.

lift, ft (m). Multiply by from 2 to 4 for efficiency, and make the air-compressor displacement twice the quantity so found.

The air pressure required for starting is higher than that for operating and is equivalent to the height of the water level at rest above the end of the air pipe. The actual pumping level

and yield of a well are seldom known in advance. After the piping is installed, the submergence is altered to suit by raising or lowering the pipe in the well. The quantity of water a well will yield depends upon its diameter. The water velocity should be 4 to 5 ft/s (1.2 to 1.5 m/s). Too large a pipe lets the air slip by, and too small a pipe means excessive friction and inefficient expansion of the air bubbles. The air should be injected into the water in small bubbles by the use of a foot piece provided with a large number of small holes, say about ¼ in (6 mm) diam. The slip increases rapidly with increase of bubble size.

The arrangement shown in Fig. 38*a* and *b* is the **Pohlé**, or side-inlet, **method**, where the discharge and air pipes are placed

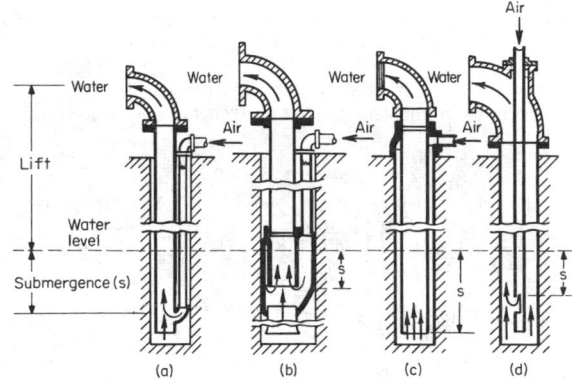

**Fig. 38** Arrangements of air lifts.

side by side in the well and joined by a suitable foot piece. In Fig. 38*b* compressed air fills the annular space or ring surrounding the uptake pipe and is free to enter the rising column at all points of its periphery. The **Saunders system** is shown in Fig. 38*c*. A central discharge pipe is suspended in the well, the air passing down between it and the well casing. If the well is not cased, a second pipe must be used outside the main discharge pipe, the air, as before, filling the annular space between the two pipes. The **central air-pipe system** shown in Fig. 38*d* is used where the lift is low. The air pipe is suspended in the well without the usual discharge pipe, making the well its own discharge pipe.

# CENTRIFUGAL AND AXIAL PUMPS

## by Igor J. Karassik and William C. Krutzsch

REFERENCES: (1) Theory: Stepanoff, "Centrifugal and Axial Flow Pumps," Wiley. Wislicenus, "Fluid Mechanics of Turbomachinery," McGraw-Hill. Pfleiderer, "Die Kreiselpumpen," Springer. Spannhake, "Centrifugal Pumps, Turbines and Propellers," Technology Press. (2) Practice: Karassik and Carter, "Centrifugal Pumps," F. W. Dodge. Hicks, "Pump Selection and Application," and "Pump Operation and Maintenance," McGraw-Hill. Standards of the Hydraulic Institute.

## NOMENCLATURE AND MECHANICAL DESIGN

Pumps covered in this section fall into three general classes: (1) **centrifugal** or **radial-flow**, (2) **mixed-flow**, and (3) **axial-flow** or propeller pumps. The essential elements of a centrifugal pump are (1) the **rotating element**, consisting of the shaft and the impeller, and (2) the **stationary element**, consisting of the casing, stuffing boxes, and bearings (see Fig. 1). Other parts, such as wearing rings and shaft sleeves, are generally added to produce better-operating, more economical machines as warranted by the various services on which the pumps are to be used. Names recommended by the Hydraulic Institute for various parts are given in Table 1.

In a centrifugal pump the liquid is forced, by atmospheric or other pressure, into a set of rotating vanes which constitute an impeller discharging the liquid at a higher pressure and a higher velocity at its periphery. The major portion of the velocity energy is then converted into pressure energy by means of a **volute** (Fig. 1) or by a set of stationary **diffusion vanes** (Fig. 3) surrounding the impeller periphery. Pumps with volute casings are called **volute pumps;** those with diffusion vanes are called **diffuser pumps.** The latter were once commonly called turbine pumps, but this term has recently been more selectively applied to vertical deep-well centrifugal diffuser pumps, now called **vertical turbine pumps.**

Centrifugal pumps are divided into other categories, several of which relate to the impeller. First, impellers are classified according to the major direction of flow in reference to the axis of rotation. Centrifugal pumps may have (1) **radial-flow** impellers (Figs. 1, 2, and 5), (2) **axial-flow** impellers (Fig. 4), and (3) **mixed-flow** impellers, which combine radial- and axial-flow principles (Fig. 7).

Impellers are further classified according to the flow

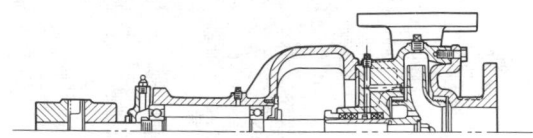

**Fig. 2** Single-stage end-suction volute pump.

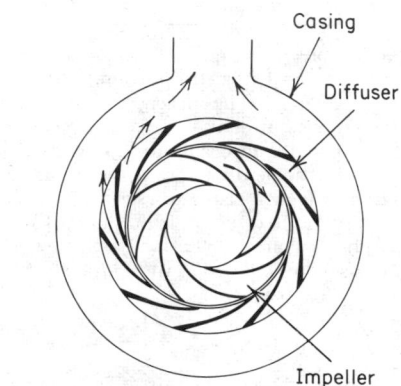

**Fig. 3** Diffuser-type pump.

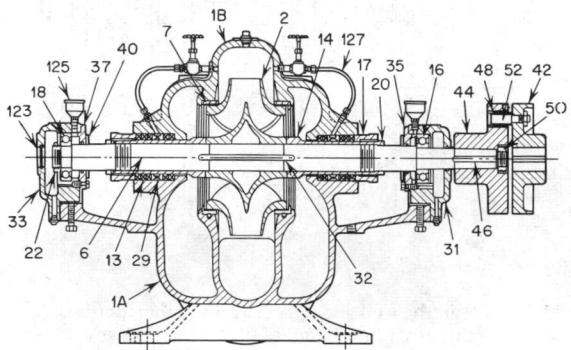

**Fig. 1** Horizontal single-stage double-suction volute pump. (Numbers refer to parts listed in Table 1.)

**Table 1. Recommended Names of Centrifugal-pump Parts**
(These parts are called out in Figs. 1, 4, and 5)

| Item no. | Name of part | Item no. | Name of part |
|---|---|---|---|
| 1 | Casing | 33 | Bearing housing (outboard) |
| 1A | Casing (lower half) | 35 | Bearing cover (inboard) |
| 1B | Casing (upper half) | 36 | Propeller key |
| 2 | Impeller | 37 | Bearing cover (outboard) |
| 4 | Propeller | 39 | Bearing bushing |
| 6 | Pump shaft | 40 | Deflector |
| 7 | Casing ring | 42 | Coupling (driver half) |
| 8 | Impeller ring | 44 | Coupling (pump half) |
| 9 | Suction cover | 46 | Coupling key |
| 11 | Stuffing-box cover | 48 | Coupling bushing |
| 13 | Packing | 50 | Coupling lock nut |
| 14 | Shaft sleeve | 52 | Coupling pin |
| 15 | Discharge bowl | 59 | Handhole cover |
| 16 | Bearing (inboard) | 68 | Shaft collar |
| 17 | Gland | 72 | Thrust collar |
| 18 | Bearing (outboard) | 78 | Bearing spacer |
| 19 | Frame | 85 | Shaft-enclosing tube |
| 20 | Shaft-sleeve nut | 89 | Seal |
| 22 | Bearing lock nut | 91 | Suction bowl |
| 24 | Impeller nut | 101 | Column pipe |
| 25 | Suction-head ring | 103 | Connector bearing |
| 27 | Stuffing-box-cover ring | 123 | Bearing end cover |
| 29 | Lantern ring | 125 | Grease (oil) cup |
| 31 | Bearing housing (inboard) | 127 | Seal piping (tubing) |
| 32 | Impeller key | | |

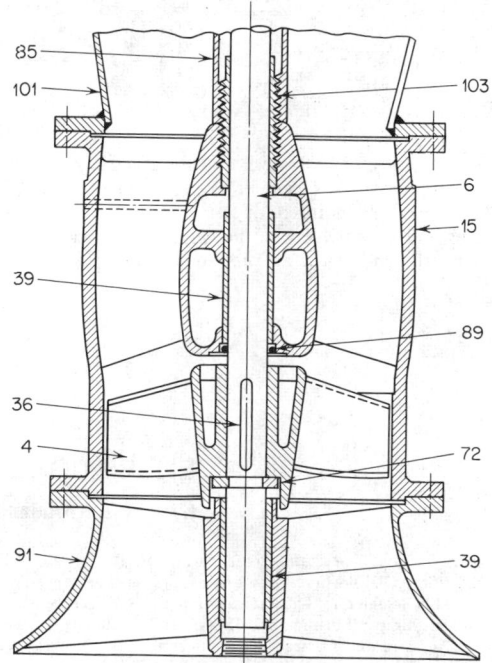

**Fig. 4** Vertical wet-pit diffuser pump bowl. (Numbers refer to parts listed in Table 1.)

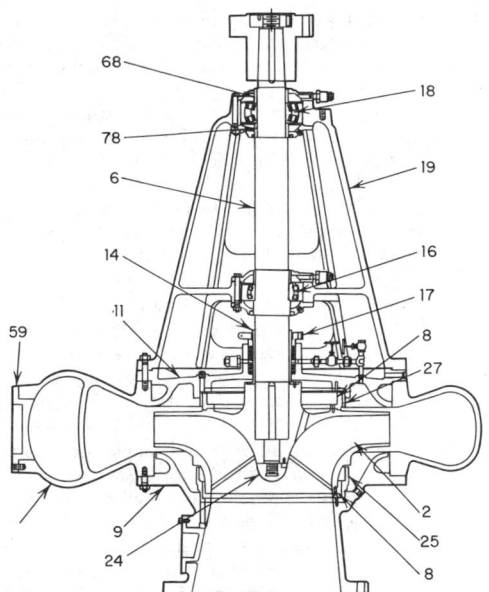

**Fig. 5** Vertical end-suction pump with a double-volute casing. (Numbers refer to parts listed in Table 1.)

arrangement into (1) **single-suction,** with a single inlet on one side, and (2) **double-suction,** with water flowing to the impeller symmetrically from both sides. They are categorized according to their mechanical construction into (1) **closed,** with shrouds or sidewalls enclosing the waterways, (2) **open,** with no shrouds, and (3) **semiopen** or semiclosed.

If the pump is one in which the head is developed by a single impeller, it is called a **single-stage pump.** When two or more impellers operating in series are used, the unit is called a **multistage pump.** The mechanical design of the casing provides the added classification of **axially split** (Fig. 1) or **radially split** (Figs. 2 and 5), and the axis of rotation determines whether the pump is horizontal-shaft, vertical-shaft, or (occasionally) inclined-shaft. Usually these are referred to simply as **horizontal** or **vertical** units.

Horizontal centrifugal pumps are classified still further according to suction-nozzle location into (1) **end-suction,** (2) **side-suction,** (3) **bottom-suction,** and (4) **top-suction.**

Some pumps operate with the liquid conducted to and from

the unit by piping. Other pumps, usually vertical types, are submerged in their suction supply. Vertical pumps are therefore called either **dry-pit** or **wet-pit** types. If the wet-pit pumps are axial-flow, mixed-flow, or vertical-turbine types, the liquid is discharged up through the supporting drop or column pipe to a discharge point either above or below the supporting floor. These pumps are consequently designated as **above-ground discharge** or **below-ground discharge** units.

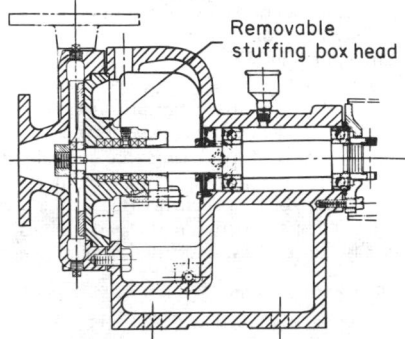

**Fig. 6** End-suction pump with removable stuffing-box head.

### Casings

The pressure acting on the impeller in a single-volute pump-casing design is nearly uniform when the pump is operated at or near its design capacity. At other capacities, the pressures around the impeller are not uniform, causing a radial reaction (or radial thrust) which can substantially increase the pump-shaft deflection. When it becomes impractical to counteract this radial thrust through the use of a heavier shaft and heavier bearings, a **double-** or **twin-volute** design (Fig. 5) may be used.

End-suction single-stage pumps are made of one-piece solid casings. At least one side of the casing must have an opening with a cover so that the impeller can be assembled in the pump. If the cover is on the suction side, it becomes the casing sidewall and contains the suction opening (Fig. 2). This is called the **suction cover** or **casing suction head.** Other designs are made with stuffing-box covers (Fig. 6), and still others have both casing suction covers and stuffing-box covers (Figs. 5 and 6).

In the inexpensive open-impeller pump, the impeller rotates within close clearance of the pump casing (Fig. 6). If the

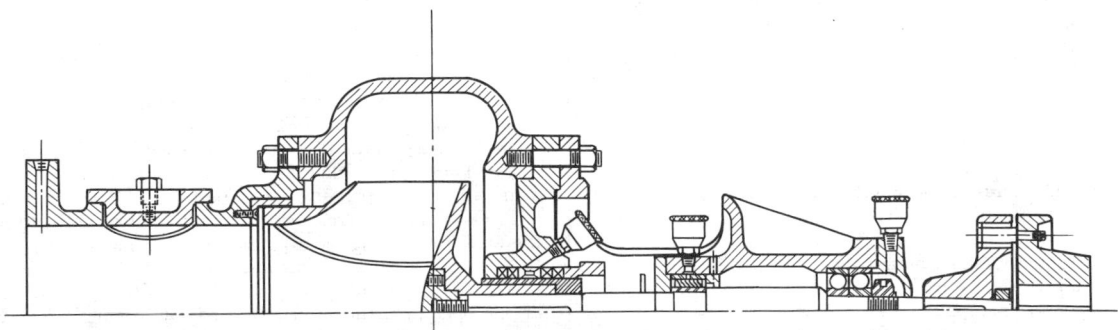

**Fig. 7** End-suction pump with removable suction and stuffing-box heads.

intended service is more severe, a side plate is mounted within the casing to provide a renewable close-clearance guide to the liquid flowing through the open impeller.

Thd discharge nozzle of end-suction single-stage horizontal pumps is usually in a top-vertical position (Fig. 2). However, other nozzle positions may be obtained, such as top-horizontal, bottom-horizontal, or bottom-vertical discharge. Practically all double-suction axially split casing pumps have a side-discharge nozzle and either a side- or a bottom-suction nozzle. Single-stage bottom-suction pumps are rarely made in sizes below 10 in discharge-nozzle diameter.

Both axially split (Fig. 8) and radially split (Fig. 9) casings are used for multistage pumps. The choice between the two designs is dictated by the discharge pressure, with 1,300 to 2,000 lb/in² (90 to 140 kg/cm²) forming the approximate limit between the two. Radially split casings are normally designed as double casings; the working parts of the pump are enclosed in an inner casing, which is then inserted into a second, or outer, casing. The space between the two casings is maintained at the discharge pressure of the last pump stage.

### Impellers and Wearing Rings

In addition to being classified with reference to the suction flow into the impeller, the basic flow component, and their mechanical features, impellers are also classified with reference to their profile and to their head-capacity characteristics at a given speed. This last relationship will be described later, in the discussion of **specific speed**.

Many impellers are designed for specific applications. Special nonclogging impellers with blunt edges and large waterways are used for sewage which ordinarily contains rags or stringy material. Impellers designed for handling paper-pulp stock are fully open and nonclogging and have screw conveyor vanes which project into the suction nozzle.

**Wearing rings** provide an easily and economically renewable leakage joint between the impeller and casing. A leakage joint without renewable parts is used only in very small, inexpensive pumps. The stationary ring is called (1) **casing ring** if mounted in the casing, (2) **suction-cover ring** or **suction-head ring** if mounted in a suction cover or head, and (3) **stuffing-box cover ring** if mounted in the stuffing-box cover. A renewable part for the impeller wearing surface is called the **impeller ring**. Pumps with both stationary and rotating rings are said to have **double-ring** construction.

There are various types of wearing-ring designs, and selection of the most desirable type depends on the liquid being handled, the pressure differential across the leakage joint, the surface speed, and the particular pump design. In general, centrifugal-pump designers use the ring construction they have found suitable for each particular pump service. The most common ring constructions are the **flat type** (Fig. 2) and the **L type** (Figs. 1 and 7).

### Axial Thrust in Single-Stage and Multistage Pumps

**Axial hydraulic thrust** is the summation of unbalanced impeller forces acting in the axial direction. Theoretically, a double-suction impeller is in hydraulic balance, with the pressures on one side equal to and counterbalancing the pressures on the other. In practice, some slight unbalance may exist, and even double-suction pumps are provided with thrust bearings.

The single-suction radial-flow impeller is subject to axial thrust because a portion of the front wall is exposed to suction pressure, with a greater back-wall surface subject to discharge pressure. In addition, an overhung single-suction impeller with a single stuffing box is subject to an axial force equivalent to the product of the shaft area through the stuffing box and the difference between suction and atmospheric pressure. This force acts toward the impeller suction when the suction pressure is less than the atmospheric and in the opposite direction when it is higher than the atmospheric.

To eliminate the axial thrust of a single-suction impeller, a pump can be provided with both front and back wearing rings (Figs. 2 and 5). Pressure approximately equal to the suction pressure is maintained in a chamber located on the inner side of the back wearing ring by providing so-called **balancing holes** through the impeller. Leakage past the back wearing ring is returned into the suction area through these holes. In large pumps, a piped connection usually replaces the balancing holes.

Most multistage pumps are built with single-suction impellers. To balance the axial thrust of these impellers, two arrangements are used: (1) The impellers all face in the same direction and are mounted in the ascending order of the stages. The axial thrust is balanced by a hydraulic balancing device (Figs. 8 and 9). (2) An even number of single-suction impellers is used, one half of these facing in a direction opposite to the second half (Fig. 10). This mounting of single-suction impellers back to back is frequently called **opposed impellers**.

**Hydraulic balancing devices** may take the form of (1) a **balancing drum**, (2) a **balancing disk**, or (3) a combination of these two. The **balancing drum** is illustrated in Fig. 11. The balancing chamber at the back of the last-stage impeller is separated from the pump interior by a drum mounted on the shaft. The drum is separated by a small radial clearance from the stationary portion of the balancing device, called the **balancing drum head**, which is fixed to the pump casing. The balancing chamber is connected either to the pump suction or to the vessel from which the pump takes its suction. The forces acting on the balancing drum are (1) toward the discharge end—the discharge pressure multiplied by the front balancing area (area $B$) of the drum; (2) toward the suction end—the back pressure in the balancing chamber multiplied by the back balancing area (area $C$) of the drum. The first force is greater than the second, thereby counterbalancing the axial thrust exerted upon the single-suction impellers. The drum diameter can be selected to balance the axial thrust completely or to balance 90 to 95 percent of this thrust, depending on whether a slight thrust load in a specific direction on the thrust bearing is desirable.

The operation of the simple **balancing disk** is illustrated in Fig. 12. The rotating disk is separated from the balancing-disk head by a small axial clearance. The leakage through this clearance flows into the balancing chamber and from there either to the pump suction or to the suction vessel. The back of the balancing disk is subject to the balancing-chamber back pressure, whereas the disk face experiences a range of pressures. These vary from discharge pressure at its smallest diameter to back pressure at its periphery. The inner and outer disk diameters are chosen so that the difference between the total force acting on the disk face and that acting on its back will balance the impeller axial thrust. If the axial thrust of the impellers should exceed the thrust acting on the disk during operation, the latter is moved toward the disk head, reducing the axial clearance. The amount of leakage through

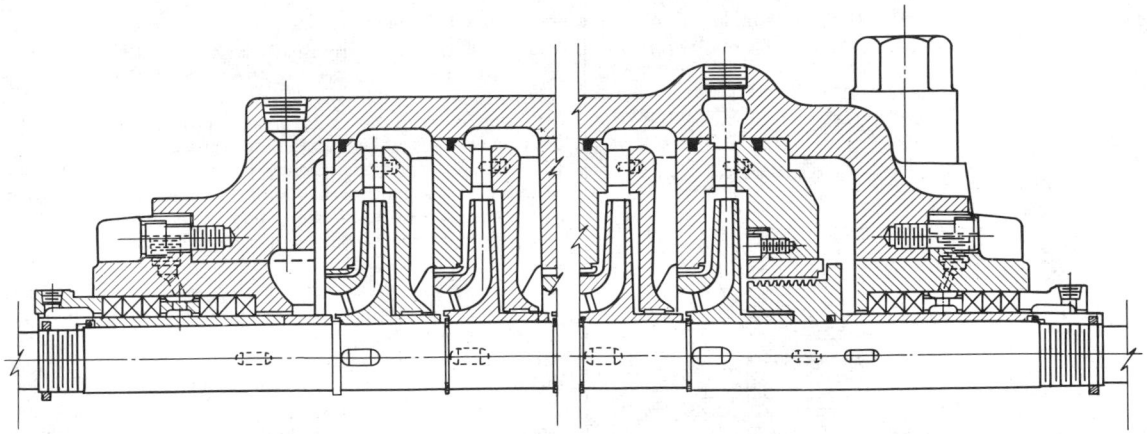

**Fig. 8** Multistage pump with single-suction impellers facing in one direction and with hydraulic balancing device.

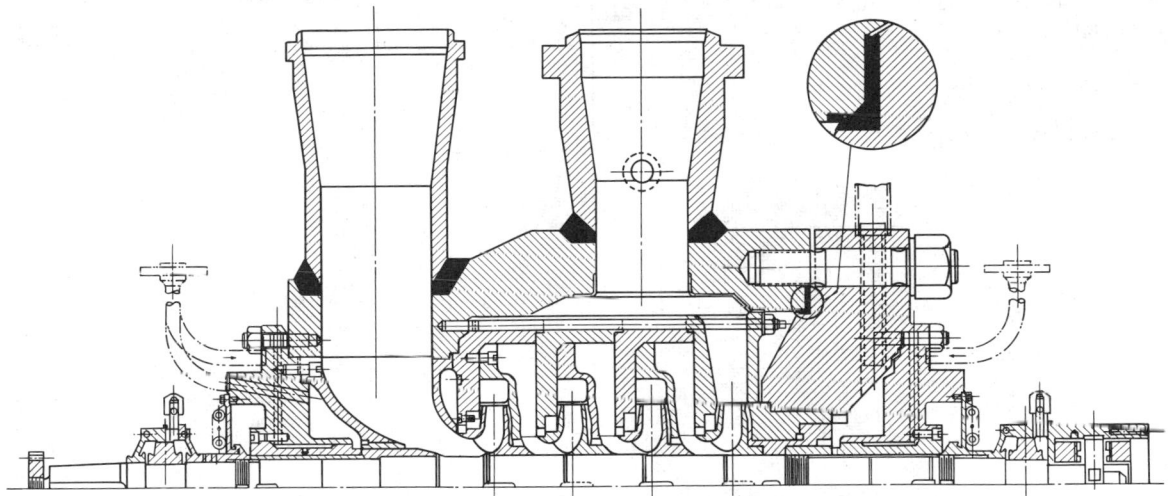

**Fig. 9** Double-casing multistage pump with radially split inner casing.

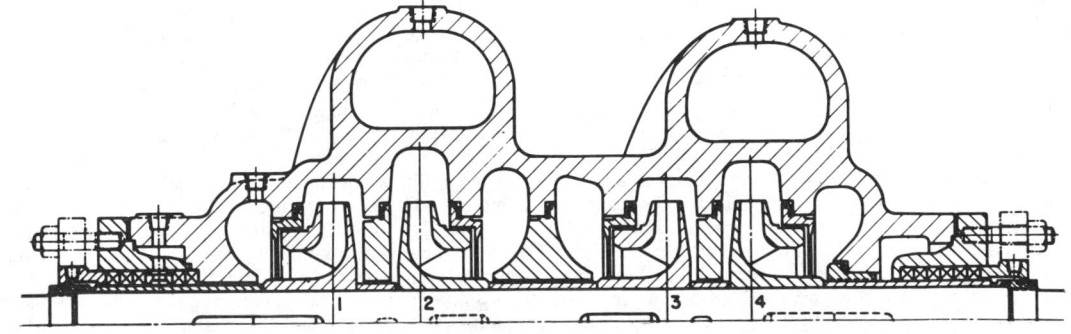

**Fig. 10** Four-stage pump with opposed impellers.

this clearance is reduced so that the friction losses in the leakage return line are also reduced, lowering the back pressure in the balancing chamber. This automatically increases the pressure difference acting on the disk and moves it away from the disk head, increasing the clearance. Now the pressure builds up in the balancing chamber, and the disk is again

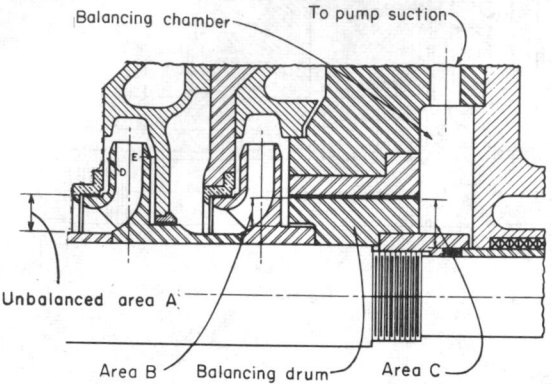

**Fig. 11**  Balancing drum.

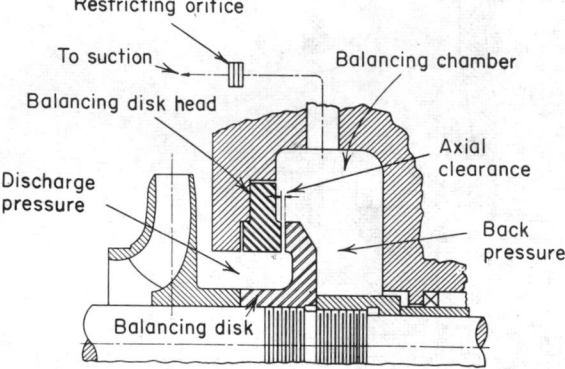

**Fig. 12**  Simple balancing disk.

moved toward the disk head until an equilibrium is reached. To ensure proper balancing-disk operation, the change in back pressure must be of an appreciable magnitude. This is accomplished by introducing a restricting orifice in the leakage return line.

The **combination disk and drum** (Fig. 9) is the most commonly used hydraulic balancing device. It incorporates portions rotating within radial clearances of stationary portions and a disk face rotating within an axial clearance of another portion of the stationary part. The radial clearance remains constant regardless of any axial displacement of the rotor within the casing. Such displacement, however, changes the axial clearance within the balancing device. These changes cause changes in the leakage, which in turn change the pressure drop across the radial clearances and thus increase or decrease the average value of the pressure acting on the disk face. These changes in the intermediate pressure on the disk face act to move the balancing device in whichever direction is required to restore equilibrium and axial balance.

## Shafts and Shaft Sleeves

Pump-shaft diameters are usually larger than actually needed to transmit the torque because their size is dictated by the maximum permissible or desirable shaft deflection. This deflection is itself chosen to prevent possible contact at the wearing surfaces while maintaining reasonable clearances that will not affect pump efficiency too unfavorably. The first **critical speed** of a shaft is related to its deflection. It follows that a shaft design permitting a deflection of, for instance, 0.005 to 0.006 in (0.13 to 0.15 mm) will have a first critical speed of 2,400 to 2,650 r/min. This is the reason for using rigid shafts (operating below their first critical speed) for pumps that operate at 1,750 r/min or lower. Multistage pumps operating at 3,600 r/min or higher use shafts of equal stiffness (for the same purpose of avoiding wearing-ring contact). However, their corresponding critical speed is about 25 to 40 percent less than their operating speed. This margin is sufficient to avoid any danger to the operation caused by critical-speed effect.

Pump shafts are usually protected from erosion, corrosion, and wear at the stuffing boxes and leakage joints and in the waterways by renewable **sleeves**. The most common shaft-sleeve function is that of protecting the shaft from wear at a stuffing box. Shaft sleeves serving other functions are given specific names to indicate their purpose. For example, a shaft sleeve used between two multistage impellers in conjunction with an interstage bushing to form an interstage leakage joint is called an **interstage** or **distance** sleeve.

## Stuffing Boxes

**Stuffing boxes** have the primary function of protecting the pump against leakage at the point where the shaft passes out through the pump casing. If the pump handles a suction lift and the pressure at the interior stuffing-box end is below atmospheric, the stuffing-box function is to prevent air leakage into the pump. If this pressure is above atmospheric, the function is to prevent liquid leakage out of the pump. The stuffing box takes the form of a cylindrical recess that accommodates a number of rings of **packing** around the shaft or shaft sleeve. If sealing the box is desired, a **lantern ring** or **seal cage** is used to separate the rings of packing into approximately equal sections. The packing is compressed to give the desired fit on the shaft or sleeve by a gland that can be adjusted in an axial direction.

Water or some other sealing fluid can be introduced under pressure into the space provided by the seal cage, causing flow of sealing fluid in both axial directions. This is useful for pumps handling flammable or chemically active and dangerous liquids since it prevents outflow of the pumped liquid.

When a pump handles clean, cool water, stuffing-box seals are usually connected to the pump discharge or, in multistage pumps, to an intermediate stage. An **independent supply of sealing water** should be provided if any of the following conditions exist: (1) a suction lift in excess of 15 ft (4.5 m); (2) a discharge pressure under 10 lb/in² (0.7 kg/cm²); (3) hot water handled without adequate cooling (except for boiler feed pumps, in which seal cages are not used); (4) muddy, sandy, or gritty water handled; (5) for all hot-well pumps; (6) no leakage to atmosphere permitted of the liquid handled. When sealing water is taken from the pump discharge, an external connection is generally made to the seal cage through small-diameter piping (Fig. 1), or an internal-passage connection is made within the pump itself (Fig. 2).

High temperatures or pressures complicate the problem of maintaining stuffing-box packing. Pumps in these more difficult services are usually provided with jacketed, **water-cooled stuffing boxes.** If the pressure ahead of the stuffing box makes it impractical to pack the stuffing box satisfactorily, a pressure-reducing breakdown or **labyrinth** may be located ahead of the box, with the leakage past the pressure-reducing breakdown being returned to some point of lower pressure in the pumping cycle.

Basically, **stuffing-box packing** is a pressure-breakdown device that is sufficiently plastic to be adjusted for proper operation. The most common types are asbestos packing and metallic packing, the latter being composed of flexible metallic strands or foil with graphite or oil lubricant and with either an asbestos or a plastic core. Other types of packing used may be hemp, cord, braided, duck fabric, chevron, etc. Packing is supplied either in continuous coils of square cross section or in preformed, die-molded rings.

**Mechanical seals** are used in centrifugal pumps when it becomes impractical to use conventional packing with radial sealing surfaces. The sealing surfaces of a mechanical seal are located in a plane perpendicular to the pump shaft and consist of two highly polished surfaces running adjacently, one surface being connected to the shaft and the other to the stationary portion of the pump. These surfaces are held essentially in contact by a spring, the axial clearance between the surfaces being provided by a thin film of liquid. The flow of liquid may be only a drop every few minutes or even a haze of escaping vapor.

There are two basic seal arrangements: (1) internal assembly and (2) external assembly. Two mechanical seals may be mounted inside a stuffing box to make a **double seal assembly** (Fig. 13). Such an arrangement is used for pumps handling toxic or highly inflammable liquids. A clear, filtered, and generally inert sealing liquid is injected between two seals at a pressure slightly in excess of the pressure in the pump ahead of the seal.

For some power-plant services, **condensate-injection sealing** is superior to either conventional packing or mechanical seals. A labyrinth breakdown bushing is substituted for the conventional packing, and the pump-shaft sleeve runs within this bushing with a small radial clearance. Cold condensate at a pressure in excess of the internal pump pressure is introduced centrally in this breakdown bushing (Fig. 9). A small portion of the injection water flows inwardly into the pump proper; the remainder flows out into a collecting chamber vented to the atmosphere and is piped back to the condenser.

### Bearings
(See also Sec. 8)

All types of **bearings** are used in centrifugal pumps. Even the same basic design of pump is often made with two or more different bearings, required by varying service conditions. Two external bearings are usually used for the double-suction single-stage general-service pump, one on either side of the casing. In horizontal pumps with bearings on each end, the **inboard bearing** is the one between the casing and the coupling and the **outboard bearing** is located at the opposite end. Pumps with overhung impellers have both bearings on the same side of the casing; the bearing nearest the impeller is the inboard bearing, and the one farthest away the outboard bearing.

**Ball bearings** are the most common antifriction bearings used on centrifugal pumps. Roller bearings are used less often, although the spherical roller bearing is used frequently for large shaft sizes. Ball bearings used in centrifugal pumps are usually grease-lubricated, although some services use oil lubrication.

**Sleeve bearings** are used for large, heavy-duty pumps with shaft diameters of such proportions that the necessary antifriction bearings are not commonly available. Another application is for high-pressure multistage pumps operating at speeds of 3,600 to 9,000 r/min. Still another application is in vertical submerged pumps, such as vertical turbine pumps, in which the bearings are subject to a water contact. Most sleeve bearings are oil-lubricated.

**Thrust bearings** used in combination with sleeve bearings are generally Kingsbury or Kingsbury-type bearings.

### Couplings

Centrifugal pumps are connected to their drivers through **couplings** of one sort or another, except for close-coupled units (Fig. 14), in which the impeller is mounted on an extension of the shaft of the driver. Couplings used with centrifugal pumps can be either rigid (of the clamp or compressor type) or **flexible** (pin-and-buffer, gear, grid, or flexible-disk type). (See also Sec. 8.)

### Pump Mounting

It is desirable that pumps and their drivers be removable from their mountings. Consequently, they are usually bolted and doweled to machined surfaces that, in turn, are firmly connected to the foundations. These machined surfaces are usually part of a common **bedplate** on which the pump and its driver have been prealigned. Bedplates are made of either cast iron or structural steel. Cast-iron or steel **soleplates** are customarily used for vertical dry-pit pumps and also for some of the larger horizontal units.

### Vertical Pumps

**Dry-pit pumps** with external bearings include most sewage pumps, most medium and large drainage and irrigation pumps for medium and high head, many large condenser-circulating and water-supply pumps, and many marine pumps. Some vertical dry-pit pumps are basically horizontal designs with minor modifications to adapt them for vertical-shaft drive. Other applications, such as small- and medium-sized sewage pumps, employ a purely vertical design. Most of these sewage

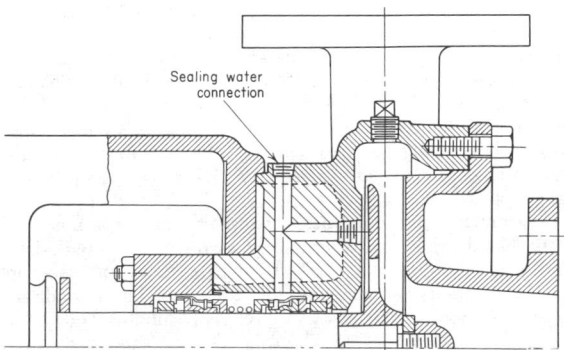

**Fig. 13**  Single-stage end-suction pump with double mechanical seal.

Sealing water connection

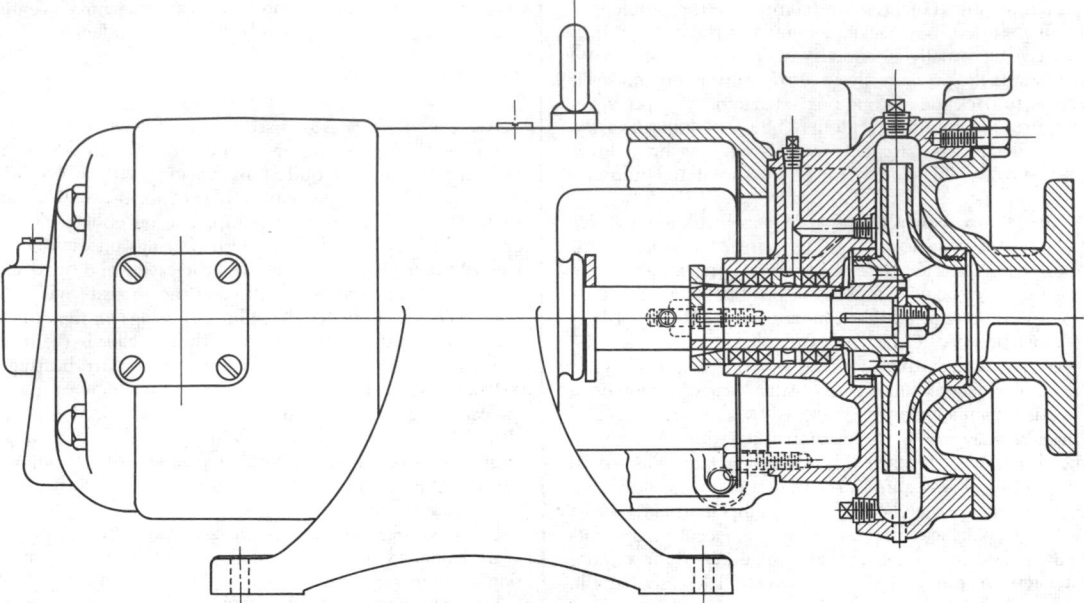

**Fig. 14**  Close-coupled (motor-mounted) pump.

pumps have elbow suction nozzles (Fig. 15) containing a handhole to provide easy access to the impeller. Although the driving motors are frequently mounted right on top of the pump casing, the use of vertical-shaft design permits mounting the motor at an elevation sufficiently above the pump to prevent accidental flooding. For such applications, the pump and its driver are separated by a length of shafting, which may require steady bearings between the two units.

**Vertical wet-pit centrifugal pumps** can be classified as (1) vertical turbine pumps, (2) propeller or modified propeller pumps, (3) sewage pumps, (4) volute pumps, and (5) sump pumps. The first of these is the most common type. **Vertical turbine pumps** (Fig. 16) are built with either closed or semiopen impellers and with either enclosed or open line shafting. The bowl assembly consists of the suction head, the impeller or impellers, the discharge bowl, the intermediate bowl or bowls, the discharge case, the various bearings, the shaft, and miscellaneous parts such as keys and impeller locking devices. The column-pipe assembly consists of the column pipe itself, the shafting above the bowl assembly, the shaft bearings, and the cover pipe or bearing retainers. The pump is suspended from the driving head, which consists of the discharge elbow, the motor or driver support, and either the stuffing box (in open-shaft construction) or the assembly for providing tension on and the introduction of lubricant to the cover pipe.

## MATERIALS OF CONSTRUCTION

Centrifugal pumps can be fabricated of almost any of the known common metals or metal alloys, as well as of porcelain, glass, and even synthetics. A listing of materials commonly recommended for various liquids can be found in the Standards of the Hydraulic Institute. Table 2 indicates the materials most commonly used for various pump parts.

## PUMP PERFORMANCE

The performance of a centrifugal pump is generally described in terms of the following of its **characteristics:** (1) rate of flow, or **capacity** $Q$, expressed in units of volume per unit of time, most frequently ft³/s, gal/min, or m³/h (1 ft³/s = 449 gal/min; 1 m³/h = 4.403 gal/min); (2) increase of energy content in the fluid pumped, or **head** $H$, expressed in units of energy per unit of mass, usually ft·lb/lb or, more simply, ft, or m; (3) input **power** $P$, expressed in units of work per unit of time, bhp; (4) **efficiency** $\eta$, the ratio of useful work performed to power input, (5) **rotative speed** $N$, in r/min.

Since the parameters indicated are all mutually interdependent, it is customary to represent the performance of a centrifugal pump by means of **characteristic curves** similar to that shown in Fig. 17. While it is possible, within certain limits, for the pump designer to regulate the shape of these curves to suit the needs of a particular application, this is essentially a function of design capacity, head, and speed and hence for a standard production unit is determined by experiment and is not subject to modification. For any given capacity on such a characteristic curve, the relationship between performance characteristics is expressed by the equation $\eta = \gamma QH/3{,}960P$, where $\eta$ is efficiency expressed as a decimal, $\gamma$ is the specific gravity of the fluid pumped, $Q$ is in gal/min, $H$ in ft, and $P$ in bhp. In metric units, $\eta = \gamma QH/270P$, where $Q$ is in m³/h and $H$ in meters.

Inasmuch as the actual performance of centrifugal pumps is determined largely by experimental means, it is highly desirable to be able to use the results of past tests as a basis for predicting the performance of future designs. To this end, an interesting and widely used characteristic number known as **specific speed,** $N_s = N\sqrt{Q}/H^{3/4}$, has been developed. In this expression, values of $N$, $Q$, and $H$ are all for the point of best

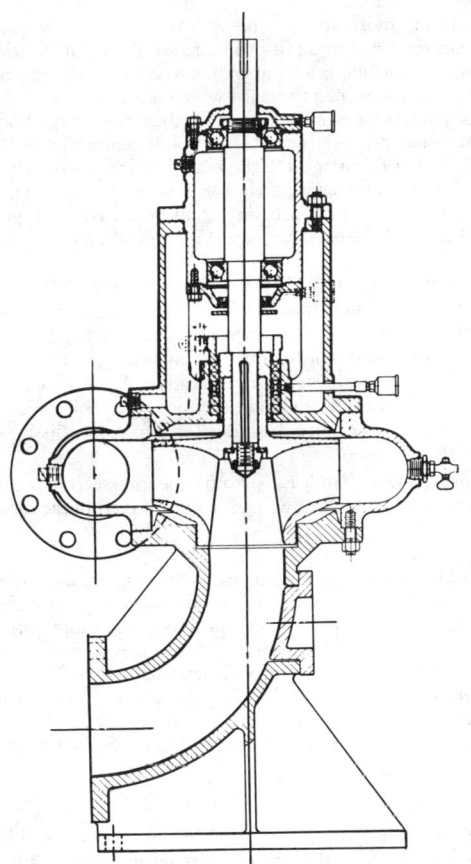

**Fig 15**  Small vertical sewage pump with intermediate shafting.

efficiency. While not dimensionless, $N_s$ is generally expressed simply as a number since its practical application is such that units are of no consequence except for their influence on the absolute magnitude of the number itself. Specific speed is of interest to both the pump designer and the pump user essentially in two ways: (1) All geometrically similar pumps,

regardless of their size, will have identical specific speeds (but all pumps of the same specific speed will not necessarily be geometrically similar). (2) Within reasonable limits, pump geometry and performance can be predicted as a function of $N_s$ and $Q$. For $N$ in r/min, $Q$ in gal/min, and $H$ in ft, the practical range of $N_s$ is approximately 500 to 15,000. In metric units $Q$ is in m³/s and $H$ is in meters, so that this range becomes approximately 10 to 300 ($N_{s_m} = N_s/51.66$). How pump efficiency and impeller design vary over this range is shown in Fig. 18. In addition, the general shape of the pump-characteristic curves will vary widely from one end of this range to the other, as illustrated by Figs. 23 and 24.

**Pump Theory**

The basic purpose of a centrifugal pump in any fluid-handling system is to add energy to the fluid, and since it is a dynamic machine, the pump depends entirely on changes in velocity relationships to provide the energy. While the measurable evidence of energy addition is in most cases largely in the form of static pressure, this is partially the result of velocity reductions and constraints occurring in the diffuser or casing and to this extent represents a conversion from the velocity energy produced by the impeller. Thus, any discussion of centrifugal-pump theory generally becomes a discussion of velocities occurring at various points within the pump.

The true velocity relationships existing within a pump are extremely complex and, to a substantial degree, still unknown in their ultimate detail; but for practical purposes, a one-dimensional analysis serves to illustrate the basic concepts and, indeed, has served as the basis of design for virtually all centrifugal pumps ever built.

**For radial and mixed-flow impellers** ($500 \leqslant N_s \leqslant 7,500$ or $10 \leqslant N_{s_m} \leqslant 150$) the velocities at the inlet and outlet of the impeller are shown by the vector diagrams in Fig. 19. The head produced by such an impeller is represented by

$$H = \eta_H(U_2 V_{u_2} - U_1 V_{u_1})/g \qquad (1)$$

where $H$ = head, ft (m); $U$ = circumferential velocity of impeller at radius being considered, ft/s (m/s), $V_u$ = average value of the circumferential component of absolute fluid velocity, ft/s (m/s), and $g$ = 32.174 ft/s² (9.80 m/s²). Subscript 1 refers to the impeller inlet section, and subscript 2 to the impeller discharge section. The coefficient $\eta_H$ is the hydraulic efficiency of the rotating-vane system and, for the range of

**Table 2. Materials for Various Fittings**
(Materials for bearing housings, bearings, and other parts are not usually affected by the liquid handled)

| Part | Standard fitting | All-iron fitting | All-bronze fitting |
|---|---|---|---|
| Casing | Cast iron | Cast iron | Bronze |
| Suction head | Cast iron | Cast iron | Bronze |
| Impeller | Bronze | Cast iron | Bronze |
| Impeller ring | Bronze | Cast iron or steel | Bronze |
| Casing ring | Bronze | Cast iron | Bronze |
| Diffuser | Cast iron or bronze | Cast iron | Bronze |
| Stage piece | Cast iron or bronze | Cast iron | Bronze |
| Shaft, with sleeve | Steel | Steel | Steel, bronze, or Monel |
|     Without sleeve | Stainless steel or steel | Stainless steel or steel | Bronze or Monel |
| Shaft sleeve | Bronze | Steel or stainless steel | Bronze |
| Gland | Bronze | Cast iron | Bronze |

specific speeds indicated above, will generally fall between 0.85 and 0.95. This hydraulic efficiency is considerably higher than pump efficiency $\eta$ since it does not include mechanical losses due to bearing or packing friction, volumetric losses due to internal wearing-ring clearances, impeller-disk friction, or fluid-friction losses due to velocity conversion or boundary-layer considerations ahead of or following the impeller. For pumps in this specific-speed range, the hydraulic losses $1 - \eta_H$ will generally be between one-quarter and three-quarters of total pump losses $1 - \eta$.

For pumps arranged with an axial inlet to the impeller (such as that shown in Fig. 5), it is generally assumed that the entering flow will have no rotational component, and $V_{u_1}$ is therefore zero. Equation (1) can thus be reduced to

$$H = \eta_H U_2 V_{u_2} /g \qquad (2)$$

In practice, Eq. (2) will provide a close approximation to total head for any pump up to a specific speed of 2,000 ($N_{s_m} \approx 40$) since the term $U_1 V_{u_1}$ in Eq. (1) is very small compared with $U_2 V_{u_2}$.

It should also be noted in Fig. 19 that the relative velocity $v_2$ does not coincide in direction with the vane angle at the impeller discharge. The angular difference between $v_2$ and the direction of the vane is due to the irrotational nature of the flow between the vanes. The effect of this difference can be taken into account as the vector difference $V_{u_2}^* - V_{u_2} = Na/229$, where $a$ is the shortest distance in inches taken in the radial plane between the discharge tip of any vane and the upper surface of the following vane. ($V_{u_2}^* - V_{u_2} = Na/19,100$, where $a$ is in mm and velocities are in m/s.)

The foregoing relationships provide the basis for determining impeller diameters and vane angles required to produce a given total head requirement at a specified rotative speed. It is equally necessary, of course, to provide in the design for handling a specified flow volume, and this is readily accomplished by providing the necessary cross-sectional area $A$ between vanes to pass the required flow at velocities previously determined. A useful relationship for this purpose, in light of the units commonly employed in pump design, is $Q = AV/0.321$, where $Q$ is in gal/min, $V$ in ft/s, and $A$ in in² ($Q = AV/278$, where $Q$ is in m³/h, $V$ in m/s, and $A$ in mm²).

Upon leaving the impeller, the liquid pumped enters either (1) a system of diffusing vanes surrounded by an outer casing or (2) directly into a casing designed to contain the fluid and control its velocities. Where diffusing vanes are used, they are designed on the basis of velocity relationships very similar to those employed in impeller design but with the objective being to slow the fluid down to convert velocity energy to pressure energy and, further, to reduce frictional losses in the discharge system following the diffuser. Where the impeller discharges directly into the casing, this component of the machine is most frequently designed in the form of a volute to provide constant velocity all around the impeller periphery up to the point of entry into the discharge nozzle. From this point, commonly called the **casing throat,** to the discharge flange or to the inlet of a succeeding stage, the velocity is gradually reduced. Special circumstances related to pump design or application often result in modifications to the constant-velocity design of such a casing, and variations may be found covering the entire range from constant velocity to constant area. In addition, many casings are now designed

with one or more spiral vanes placed in such a way as to approximate a condition of geometric similarity in relation to the impeller, which is advantageous when a pump is operated at capacities other than those for which it is designed. A casing of this nature represents an effort by the designer to obtain an optimum balance between the desirable geometric similarity of the diffuser discharge and the manufacturing simplicity and generally high efficiency of the volute-type casing. The most common form of such a casing is the twin-volute type discussed under Nomenclature and Mechanical Design (see Fig. 1 and Table 1).

**For axial-flow impellers** ($7,500 \leqslant N_s \leqslant 15,000$ or $150 \leqslant N_{s_m} \leqslant 300$) velocity relationships can be approximated in a manner similar to that used for lower-specific-speed pumps, but refinement of these approximations is approached in a somewhat different manner, largely because of the considerable body of knowledge available in the form of airfoil data which can be applied. Velocity diagrams for pumps of this type are shown in Fig. 20.

Considering a cylindrical stream tube intersecting the vanes of an axial-flow impeller, we can rewrite Eq. (1) in the form

$$H = \eta_H U \, \Delta V_u /g \qquad (3)$$

where $\Delta V_u$ represents the increase in the tangential component of the absolute velocity as the fluid passes through the impeller. For pumps in this specific-speed range, $\eta_H$ will generally fall between 0.80 and 0.90 and $1 - \eta_H$ will generally be between one-half and three-quarters of $1 - \eta$.

As in the case of radial-flow impellers, the relative velocity $v_2$ at the impeller exit does not coincide with the vane angle. In this case, the necessary correction can be applied by means of the expression

$$\Delta V_u = 2\Delta \, V_u^* \, /[(t/l)(2/\pi K)(1/\sin \beta) + 1] \qquad (4)$$

where $t$ is vane spacing, $l$ is vane length, $\beta$ is the discharge vane angle, and $K$ is the coefficient determined from Fig. 21, which provides for the fact that the impeller blades are, in effect, arranged in a continuous lattice.

In the design of axial-flow pumps, it is generally assumed that the head developed by the blade elements within all the cylindrical stream tubes between the impeller hub and its outer diameter will be the same. For this condition to be achieved, it will be evident from Eq. (3) that since $U$ will vary directly with the radius, $\Delta V_u$ must vary inversely with the radius. Thus vane camber (or curvature) will be greater near the hub than at the periphery. It is further generally assumed that the axial velocity is constant throughout the impeller, and to satisfy this condition, the blade angles will be greater at the hub than at the periphery, giving rise to the twist of the vane.

By designing the vane element at the hub, the periphery, and any reasonable number of radial stations between, it is possible to define the vane over its entire surface to produce the total head desired. The design capacity can be simply established from the constant axial velocity and the annular area between the hub and the periphery.

Axial-flow impellers will almost invariably discharge into a vaned diffuser designed primarily to convert into pressure the tangential component of the absolute velocity leaving the impeller, thus producing a uniform, nonrotating velocity profile at the pump discharge or at the entrance to any succeeding stage.

## Pump Application

In applying any centrifugal pump to a practical fluid-handling system, the engineer generally has two variables in the pump which may be used to effect a match between it and the system, namely, speed of operation and impeller diameter. (In axial-flow pumps, impeller diameter cannot conveniently be changed, but similar results can be accomplished within a limited range by reducing vane length.) To take advantage of these variables, it is necessary to understand the similarity relationships which govern pump behavior.

The effect of change in speed can be most readily explained by referring back to Eq. (1). For a fixed impeller diameter, it will immediately be evident that any increase in rotative speed will result in a directly proportional increase in the peripheral velocities of the impeller at both inlet and outlet, $U_1$ and $U_2$, respectively. Furthermore, since the directions of both the relative and absolute velocities $v$ and $V$, respectively, are controlled by the vane angles, it follows that both the inlet and outlet velocity diagrams remain geometrically similar; in other words, as pump speed changes, all fluid velocities change in direct proportion. Thus the effect of a speed change on pump capacity can be represented for two speeds $N_1$ and $N_2$ by

$$Q_1/Q_2 = N_1/N_2$$

Referring again to Eq. (1), it will be evident that the change in head will be proportional to the square of the change in speed, resulting in

$$H_1/H_2 = N_1^2/N_2^2$$

Also, since $P$ is proportional to $QH$, then

$$P_1/P_2 = Q_1 H_1/Q_2 H_2 = N_1^3/N_2^3$$

The effect of a reduction in impeller diameter is for practical purposes, identical to that of a reduction in speed, and the above equations can all be rewritten in the same form, substituting $D_1$ and $D_2$ for $N_1$ and $N_2$. In this case, however, there is somewhat less latitude for change, the practical limit for cutdown being approximately 25 to 30 percent in the case of low-specific-speed impellers and decreasing from this value as specific speed increases.

From the foregoing it should be noted that running a pump at speeds far below its normal rated speed would be uneconomical since the pump would obviously be overdesigned for the service. Conversely, the same pump could not be run at speeds far above its rating since it would soon exceed its horsepower capability. Thus the normal approach to pump application calls for selection of a unit which will meet the system requirements at or near the pump's maximum rated speed and preferably as close to maximum impeller diameter as possible.

A further aspect of the similarity laws, of interest primarily to pump designers, is the matter of geometrically similar pumps, or as they are commonly called, **factors of each other.** Assuming two such pumps with a size ratio $f_1/f_2$, then **at the same rotative speed** $H_1/H_2 = f_1^2/f_2^2$, which is the same as for a change in speed of the same ratio. In the case of capacity, however, we have, in addition to the linear increase due to velocity change, a further increase due to the change in the cross-sectional area of all fluid channels which is proportional to the square of the size factor, resulting in the relationship $Q_1/Q_2 = f_1^3/f_2^3$. The ratio of horsepower requirements therefore becomes $P_1/P_2 = f_1^5/f_2^5$.

Lines of geometrically similar pumps are more frequently designed, however, for constant head than for constant speed, and to have this condition apply for the size ratio indicated in the preceding paragraph, it is necessary that the speeds vary in exact inverse proportion to size ratio. Thus, **for constant head,** $N_2/N_1 = f_1/f_2$ and

$$H_1/H_2 = 1 = (f_1/f_2)^2 (N_2/N_1)^2$$
$$Q_1/Q_2 = (f_1/f_2)^3 (N_2/N_1) = (f_1/f_2)^2$$
$$P_1/P_2 = (H_1/H_2)(Q_1/Q_2) = (f_1/f_2)^2$$

To simplify selection of pumps, it is customary to plot, for any given speed, performance curves to show pump characteristics over the available range of impeller diameters rather than at the single diameter which would be implicit in a curve of the type shown in Fig. 17. A typical rating curve of this nature is shown in Fig. 22. Such rating curves will vary widely in their nature with changes in $N_s$, and an understanding of these variations is essential to the selection of properly sized pump drivers and, in many cases, proper design of discharge piping and/or the determination of limiting ranges of pump operation. Some idea of the diversity of characteristics which is available can be obtained by a comparison between Figs. 23 and 24, both of which are plotted entirely in percentages of design values occurring at the point of best efficiency.

In the case of the lower-specific-speed pump shown in Fig. 23, a driving motor selected for the rated condition would be more than adequate at lower capacities, and discharge piping would be subjected to only modest overpressures. On the other hand, if the pump shown in Fig. 24 is to be operated at reduced capacities, then both motor size and discharge piping must be designed for the minimum flows to be encountered.

Selection of drivers, as well as pumps, is also influenced by **properties of the fluid** handled. Pump horsepower varies directly with the specific gravity of the liquid and is influenced in a more complex manner by viscosity. The effect of the latter is illustrated in Fig. 25. For the example shown on the chart, the pump in question is rated for 750 gal/min (170 m³/h) at 100-ft (30-m) head. When handling a fluid having a viscosity of 1,000 SSU (220 cS), its capacity is reduced to 95 percent of that on water over its complete range of operation. Its head is reduced to 96, 94, 92, and 89 percent of its head on cold water at 60, 80, 100, and 120 percent of these reduced capacities, and its efficiency is reduced to 63.5 percent of its efficiency on water over its reduced capacity range. Thus, if the pump had a rated efficiency on cold water of 70 percent, it would now deliver 712.5 gal/min (161.8 m³/h) at 92-ft (28.0-m), head with an efficiency of 44.5 percent.

In order for a pump to deliver its rated output, it is obviously necessary that it be supplied with fluid at its inlet at the same rate. It is further necessary that the absolute pressure (including velocity head, $V^2/2g$) of the fluid at the inlet exceed the vapor pressure by an amount sufficient to overcome (1) any entrance or frictional losses between the point of entry into the pump and the impeller, and (2) the shock losses occurring at the impeller inlet. This gives rise to the definition of **net positive suction head (NPSH)** which is the absolute pressure at the pump inlet expressed in feet of liquid, plus velocity head, minus the vapor pressure of the fluid at pumping temperature, and corrected to the elevation of the pump centerline in the case of horizontal pumps or to the entrance to the first-stage impeller for vertical pumps.

**NPSH required** is determined by the pump manufacturer and is a function of both pump speed and pump capacity. **NPSH available** represents the energy level of the fluid over the vapor pressure at the pump inlet and is determined entirely by the system preceding the pump. Unless NPSH available at least equals NPSH required at any condition of operation, some of the fluid will vaporize in the pump inlet and bubbles of vapor will be carried into the impeller. These bubbles will collapse violently at some point downstream of the pump inlet (usually at some point within the impeller) and produce very sharp, crackling noises, frequently accompanied by physical damage of adjacent metal surfaces. This phenomenon is known as **cavitation** and is generally highly undesirable.

A term similar to that used for pump specific speed has been developed for pump inlet characteristics and has been identified as **suction specific speed,** $S = N \sqrt{Q}/H_{sv}^{3/4}$, where $H_{sv}$ is NPSH. For double-suction impellers $S = N \sqrt{Q/2}/H_{sv}^{3/4}$. The value $S$, like $N_s$, is expressed simply as a number and, for general-purpose pumps, does not generally exceed 10,000, where $N$ is in r/min, $Q$ is in gal/min, and $H_{sv}$ is in ft. When $Q$ is in m³/s and $H_{sv}$ is in meters, the equivalent value $S_m$ is 194.

Within the limits of NPSH capabilities, it is desirable to select a pump of the highest specific speed since this will produce the highest pump speed and consequently use the smallest pump. As a convenience in accomplishing this for the commonly occurring case of cold water, the Hydraulic Institute has published a series of charts defining upper limits of specific speed as a function of total head and suction lift or suction head. These charts provide conservative limits for water temperatures to 85°F (29°C). The chart for double-suction pumps is reproduced in Fig. 26.

Just as the head-capacity characteristic of a pump is represented by a curve indicating reducing head developed with increasing capacity, the head-capacity requirements of the system served by the pump can be depicted by a curve showing increased head requirements for increased flow, i.e., a **system-head curve** (Fig. 27). The capacity at which a pump will operate in this system is that at which the pump head-capacity curve intersects the system-head curve.

Every system-head curve consists of (1) a total static head (which may in some cases be equal to zero) plus (2) friction losses. The total static head $H_{stat} = H_d - H_s + P_d - P_s$,

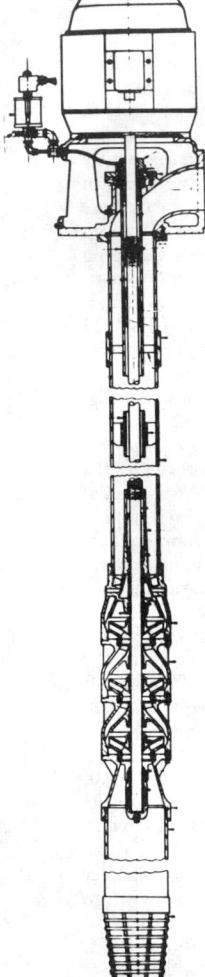

**Fig. 16** Vertical turbine pump with closed impellers and oil-lubricated enclosed shafting.

where $H_d$ and $H_s$ are the elevations at the system's termination and origin, respectively, and $P_d$ and $P_s$ are the gage pressures at the same points. In the special case of a continuous freshwater system where $P_d = P_s =$ zero, i.e., atmospheric, it must be remembered that siphon recovery cannot exceed 1 atm (34 ft less vapor pressure at sea level) and that recovery will not occur if the siphon leg is broken. Thus, any system over this height will have a total static-head component at least equal to the difference between its highest elevation and $H_s$ minus the static-head equivalent of 1 atm plus the vapor pressure, even where $H_d = H_s$.

The **friction losses** $H_{lr}$ are the sum of the line losses $H_1$, fitting losses $H_f$, and changes in velocity head across the system, $H_v$. (See Sec. 3 for calculation methods.) $H_v$ is simply the velocity-head difference between the system's termination and origin; $H_v = V_d^2/2g - V_s^2/2g$.

In complex pumping systems, such as municipal water-supply operations, both the total static-head and the friction-loss components may be variable, e.g., the former by changes in reservoir levels and the latter by the particular combination of lines being served at any given moment. This results in upper and lower limits of system-head requirements, and the intersection of these limiting curves with the pump head-capacity curve will define the capacity range over which the pump will be required to operate.

Since the capacity at which a pump will operate corresponds to the intersection of the system-head curve with the pump head-capacity curve, any changes in pump capacity can only be obtained by varying one or the other of these two curves (Fig. 28). Thus the capacity of a centrifugal pump operating in a system can be regulated by (1) changing the pump speed or (2) throttling in the discharge piping. The former method is preferable whenever the driver permits it, because throttling always involves an appreciable waste of power.

Pump capacity should never be permitted to be reduced to zero because the fluid within the pump would have to absorb the entire power input and therefore would heat up rapidly with injurious effects on the pump. A small bypass line should be provided in the pump discharge line to permit a predetermined amount of liquid to flow through the pump if the discharge valve is closed entirely. This bypass may be operated manually or automatically.

Centrifugal pumps may sometimes be operated in **parallel** or in **series.** To construct the head-capacity curve of two pumps in parallel, it is merely necessary to add the capacities of the individual pumps for various total heads. The head-capacity curve of two pumps in series is constructed by adding the individual pump heads for various capacities. Figure 29 shows series and parallel operation of two centrifugal pumps with both flat and steep system-head curves.

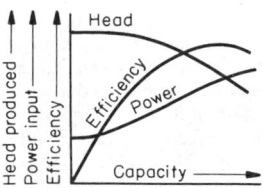

**Fig. 17** Typical characteristic curves at constant speed.

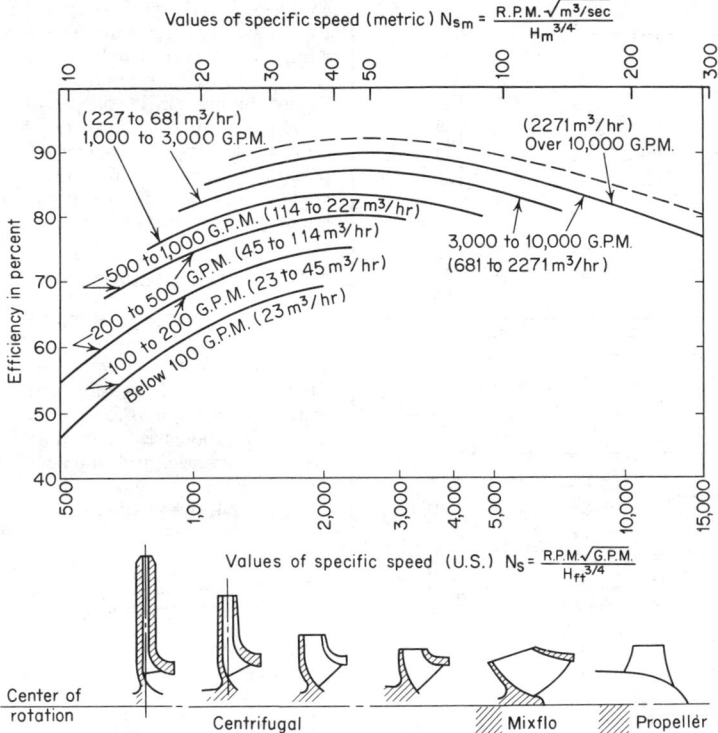

**Fig. 18** Approximate relative impeller shapes and efficiency variations with specific speed.

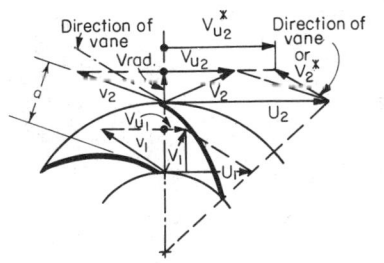

**Fig. 19** Velocity diagram of a radial-flow impeller.

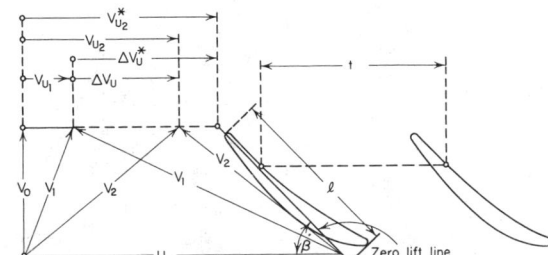

**Fig. 20** Velocity diagram of an axial-flow vane system.

## Priming

A centrifugal pump is primed when the waterways of the pump are filled with the liquid to be pumped. When first put into service, the waterways are filled with air. If the suction supply is above atmospheric pressure, priming is accomplished by venting the entrapped air out of the pump through

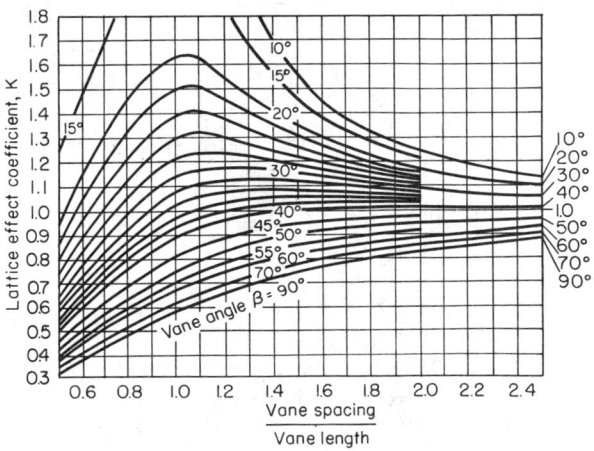

**Fig. 21** Lattice-effect coefficient. *(Weinig.)*

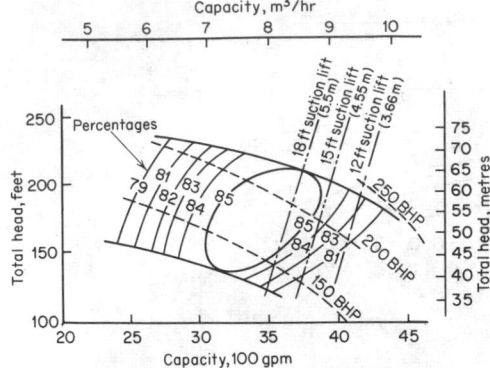

**Fig. 22**  Rating curve of 10-in double-suction single-stage pump.

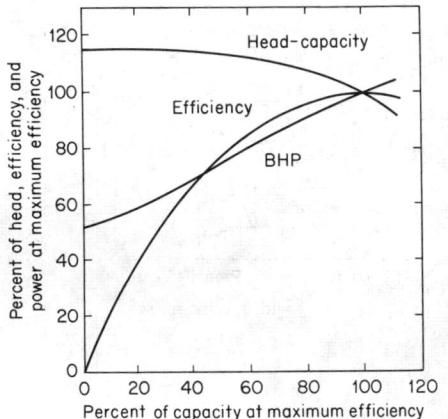

**Fig. 23**  Type characteristics for $N_s = 1,550$ single-suction impeller ($N_{sm} = 30$).

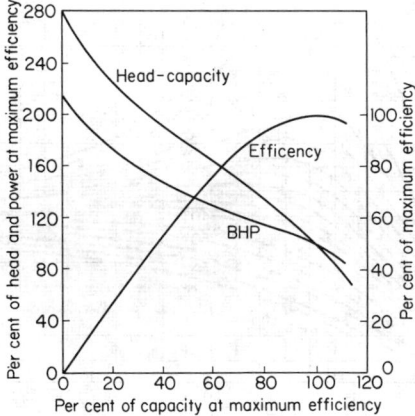

**Fig. 24**  Type characteristics for $N_s = 10,000$, single-suction impeller ($N_{sm} = 194$).

a valve provided for this purpose. If the pump takes its suction from a supply located below the pump itself, the air in the pump must be evacuated by some vacuum-producing device, by placing a foot valve in the suction line so that the pump and suction piping can be filled with liquid, or by providing a priming chamber in the suction line. Almost every commercially made vacuum-producing device can be used to prime pumps. Formerly, water- and steam-jet primers had wide application, but today electric-motor-driven vacuum pumps are most frequently used.

### INSTALLATION, OPERATION, MAINTENANCE

Proper installation, operation, and maintenance of centrifugal pumps will vary widely over the complete range of services to which the pumps may be applied, and satisfactory results in these areas can only be fully achieved by following the manufacturer's instructions for the size and type of unit involved. There are, however, certain general considerations which should be observed and which will seldom need to be modified under any circumstances.

In general, the location selected for installation should be as

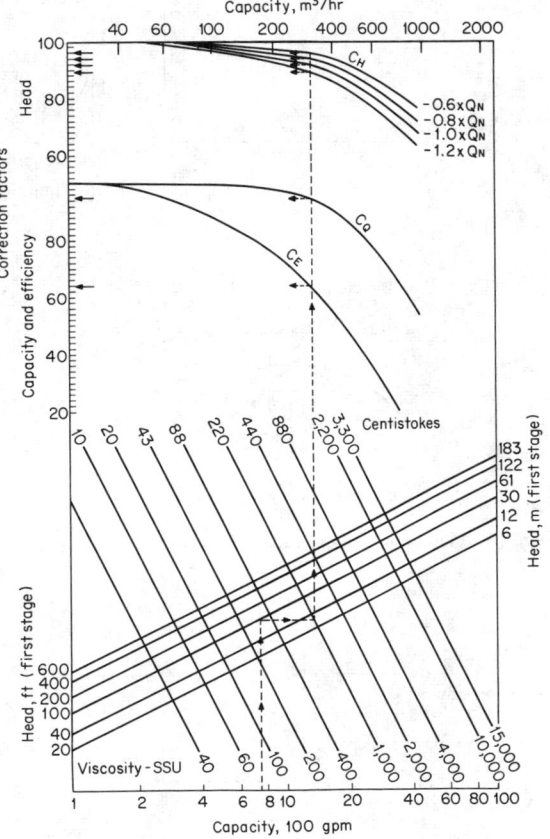

**Fig. 25**  Performance-correction chart for effect of viscosity. *(Hydraulic Institute.)*

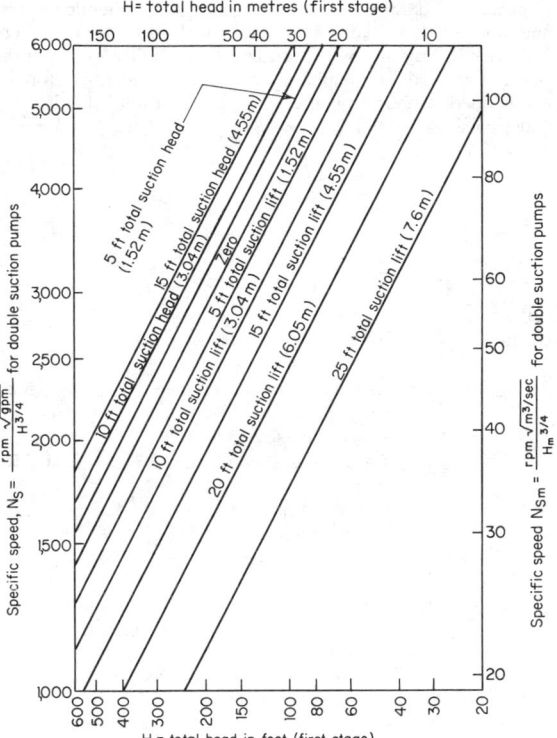

**Fig. 26** Upper limits of specific speed for single-stage double-suction pumps. *(Hydraulic Institute.)*

friction losses and to present a uniform velocity profile at the pump inlet. Suction and discharge (and/or check) valves must be suitable for the pressures involved and, in the case of large units, may also require independent support. If the pump will be required to operate against a suction lift, a suitable priming system must be installed, and where it is to be provided with a head-on suction, it will often be necessary to provide a venting arrangement. Care must also be exercised to ensure that all auxiliary connections for sealing, cooling, flushing, and drainage are made as required for the particular unit being installed.

Prior to initial operation of any centrifugal pump, it is necessary to make sure that the driver is connected to provide proper direction of rotation, that any shaft couplings between separate components of the entire unit are aligned within the manufacturer's stated limits, and that all bearings are provided with the proper amounts and grades of lubricants. The normal starting sequence will then be as follows: (1) open valves in all auxiliary sealing, cooling, flushing, and bypass lines; (2) open suction valve; (3) close discharge valve for low-specific-speed pumps where no check valve is installed after the pump, or open discharge valve for high-specific-speed pumps or wherever a discharge check valve has been provided; (4) prime or vent as necessary; (5) energize the driver; and (6) open discharge valve if it was previously closed in step 3.

Following start-up and until proper operation has been adequately established, it is desirable to monitor bearing temperature, stuffing-box leakage, and other outward symptoms

close to the source of the fluid as possible, consistent with the requirement that adequate space be made available to provide accessibility for operation, inspection, and maintenance. The pumping unit should be mounted on a foundation of sufficient size and rigidity to support the unit itself plus the weight of the fluid it will contain during operation and to maintain accurate alignment. Piping should be independently supported and anchored to avoid imposing stresses on the pump, and suction piping in particular must be designed to minimize

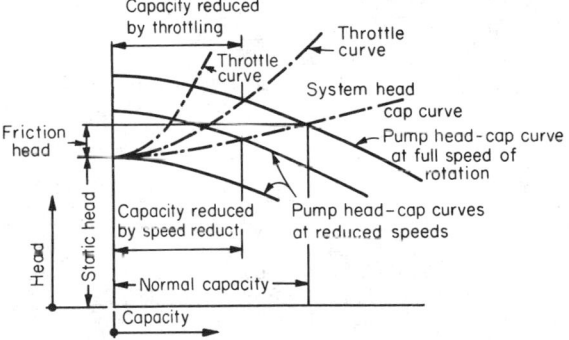

**Fig. 28** Pump operation in a system.

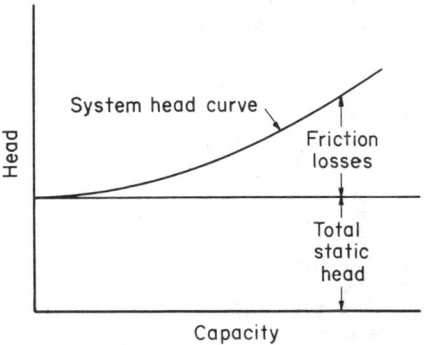

**Fig. 27** System-head curve.

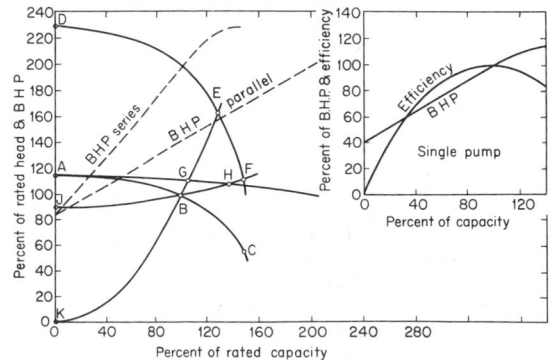

**Fig. 29** Series and parallel operation of centrifugal pumps.

of the unit's behavior. Securing of the pump is accomplished by a reversal of the start-up sequence, encompassing steps 6, 5, 3, and 1, in that order.

In the matter of pump maintenance, a generally accepted cardinal rule is that as long as operation continues normal, the unit should be left alone. Thus, except in special circumstances, periodic overhauls are not recommended. The amount and degree of maintenance likely to be required are influenced primarily by the nature of the service to which the pump is applied, and maintenance practices must therefore be determined largely by the user as a result of his own experience.

# COMPRESSORS
## by A. R. Worster

REFERENCES: API Standards 617 and 618. Balje, ASME Paper 60- WA-231. CAGI, "Compressed Air Handbook," McGraw-Hill. Scheel, "Gas and Air Compression Machinery," McGraw-Hill. Stepanoff, "Turboblowers," Wiley. Shepherd, "Principles of Turbomachinery," Macmillan. Gibbs, "Compressed Air and Gas Data," Ingersoll-Rand.

## Notation

$a$ = piston-rod area, in$^2$

$A$ = piston or impeller area, in$^2$

ACFM = actual volume, ft$^3$/min

atm abs (ata) = atmospheres of pressure absolute, 14.22 lb/in$^2$ (1 kg/cm$^2$)

atm gage (atu) = atmospheres of pressure gage, 14.22 lb/in$^2$ (1 kg/cm$^2$)

$b$ = impeller-tip width, in

$c$ = specific heat

$C$ = cylinder clearance, decimal fraction

$c_p$ = specific heat at constant pressure

$c_v$ = specific heat at constant volume

cfm = (ft$^3$/min)

cfs = (ft$^3$/s)

$d$ = diameter, in, or as referenced

$D$ = diameter, ft, or as referenced for specific equation

$D_s$ = specific diameter

$E_v$ = volumetric efficiency, decimal fraction

$f$ = valve resistance in velocity heads

fpm = piston speed, ft/min

fps = tip speed, ft/s

$g$ = acceleration due to gravity, 32.2 ft/s$^2$

$H_j$ = heat rejection, Btu/(bhp)(h)

$k$ = ratio of specific heats, $c_p/c_v$

$\mathbf{K}$ = valve-loss correction factor

$L$ = length of piston stroke, in, or as specified for reference equation

$L_{ad}$ = adiabatic head, ft of gas

$L_w''$ = head, in of water

$\mathrm{m}$ = metre, metric system

$\overline{m}$ = mole weight

M* = Mach number

Mcfd = thousands of cu ft per day (10$^3$ ft$^3$/day)

MMcfd = millions of cu ft per day (10$^6$ ft$^3$/day)

MMscfd = millions of std cu ft per day

$N$ = rpm; r/min (preferred abbreviation)

$N_s$ = specific speed

$P$ = pressure, psia

$PS$ = European unit of power = 0.986 hp

$Q$ = capacity or flow, as specified by equation involved

$q_{ad}$ = pressure coefficient

$\mathbf{R}$ = universal gas constant, 1,545.4 ft·lb/(°R)/(mole)

$R_c$ = ratio of compression

Re = Reynolds number

scfm = std cu ft per min

$T$ = absolute temperature, °R

$\mathbf{T}$ = thrust, lb

$U$ = piston speed, tip speed, or pitch-line velocity, ft/s

$v_s$ = specific volume, ft$^3$/lb

$w$ = weight flow, lb/s

$W$ = weight flow, lb/h

$X$ = ratio factor, as designated for specific equation

$Y$ = pressure drop, decimal fraction

$Z$ = compressibility factor, decimal fraction

## Greek

$\alpha$ = ratio of piston area to minimum valve area

$\Delta$(lb/in$^2$) = pressure loss or differential pressure

$\Delta t$ = isentropic temperature rise, °F

$\eta$ = efficiency, decimal fraction

$\Theta$ = compressor-valve resistance factor (Table 1)

$\Lambda$ = valve leakage factor, decimal fraction

$\rho$ = density, lb/ft$^3$

$\sigma$ = exponential function, $(k-1)/k$

$\Sigma$ = hp per MMcfd

$\tau$ = sonic velocity, ft/s

$\phi$ = flow coefficient

## Subscripts

1 = suction conditions

2 = discharge conditions

$ad$ = adiabatic

$ag$ = average gas

$aw$ = average water

$b$ = mixture

$c$ = compression or compressor

$d$ = disk

dng = dry natural gas

$m$ = mechanical

$o$ = orifice

$s$ = specific

$v$ = vapor
$w$ = water

For metric pressure, kilograms per square centimeter (kg/cm²) is used. SI units for pressure are pascals (Pa) or kilopascals (kPa). To convert kg/cm² to kPa, multiply by 98.81.

### Air Compressors

Compressed air is a form of energy used extensively for such operations as automatic machines, tools, material handling, and food processing. Most plant air systems are maintained at 90 to 110 psig (6.3 to 7.7 kg/cm²) to operate the tools and machines, which require 70 to 90 psig (4.9 to 6.3 kg/cm²). Instrument-control devices require 25 psig (1.8 kg/cm²), sometimes supplied from intermediate 50-psig (3.5 kg/cm²) feeders. Other applications that require large quantities of lower-pressure air are usually installed adjacent to the operation to avoid extensive pipe systems. In addition to these permanent air systems, portable air compressors used for construction, mining, road building, and painting, range from 1 ft³/min (1.7 m³/h) and ½ hp to 2000 ft³/min (3400 m³/h) and 500 hp. Most portable compressors are now of the rotary sliding-vane (Fig. 28) and the rotary-screw (Fig. 25) types. Most permanent installations use piston compressors available as stock items in sizes ranging from 30 to 2500 ft³/min (50 to 4,250 m³/h), with pressures of 60, 100, and 150 lb/in² (4.2, 7.0, and 10.5 kg/cm²) and to 500 hp. Engine-starting service requires 250 psig (17.6 kg/cm²). Oil-well drilling, soot blowing, and in situ operations require pressures of 350 to 5,000 psig (24.6 to 350 kg/cm²). The "free-air," or ambient, capacity is less than the piston displacement by the amount of the volumetric efficiency (see Fig. 6). The power load varies directly as the barometric pressure at a constant ratio of compression $R_c$. The inlet pressure is depressed roughly 0.48 psia (0.034 kg/cm²) per 1,000 ft (305 m) of elevation [the average barometric pressure, in mile-high Denver, is 12.2 psia (0.86 kg/cm²)].

### Gas Measurement Units

Gas must be defined to have identity. The most widely accepted unit of measurement in the United States is the standard cubic foot, fixed at 60°F (519.7°R) and 14.696 psia (760 mm, or 29.92 inHg). A dry cubic foot of air at these conditions weighs 0.0763 lb and has a specific gravity of 1.000. It is common practice to accept the basic condition as 14.70 psia and 60°F. Dry air has a molecular volume of 379.5 ft³ at standard conditions. The universal gas constant **R** is 1545.4 ft·lb/(°R)/(mol). Volume flow in the gas industry is measured in millions of standard cubic feet per day. However, temperature and pressure standards vary between user. The metric standard unit of volume is the normal m³/h; based on 0°C and 760 mm (1.0333 kg/cm²), it is equal to 0.622 std ft³/m. The British use std m³/h based on 60°F and 30.00 inHg. A unit often used in Europe for absolute pressure is an **ata**, equal to 14.22 psia, or 1 kg/cm². The term **atu** refers to the gage pressure in excess of ambient in kg/cm². A **bar** is a French unit equal to 14.5 lb/in² (1.02 kg/cm²). The metric unit of power PS is 0.986 hp.

### Types of Compressors

Compressors fall into two broad categories: positive-displacement and dynamic machines. The reciprocating-piston type (Figs. 2 and 3) prevails in the former category, with capacities ranging from 30 to 15,000 ft³/min (50 to 25,000 m³/h), suitable

for pressures to 50,000 lb/in² (3500 kg/cm²). The cycle is portrayed in Fig. 1, with the suction from $D$ to $A$, compression from $A$ to $B$, discharge from $B$ to $C$, and expansion of the trapped clearance gas from $C$ to $D$. Admission and delivery to the cylinder are through valves (Figs. 8 to 10), where the seating elements are motivated by the differential pressures $P_1 - P_3$ and $P_4 - P_2$. The positive-displacement category includes several types of rotary compressors, such as the multiple-lobe, sliding-vane, and liquid ring-seal compressors shown in Figs. 25 to 29. The inlet gas passes through a port, the rotor effects a cutoff, and the gas is compressed by the action of the rotor; sizes range from 40 to 20,000 ft³/min (70 to 34,000 m³/h) for low-pressure and vacuum services. The rotary-screw compressor can handle 20 to 13,000 ft³/min (35 to 22,000 m³/h) and in series they are normally limited to 250 psig (18 kg/cm²) and 450°F.

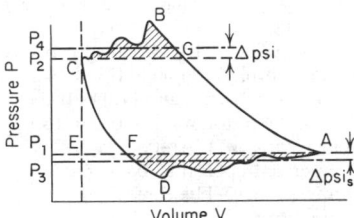

**Fig. 1** Typical gas-compression-indicator diagram.

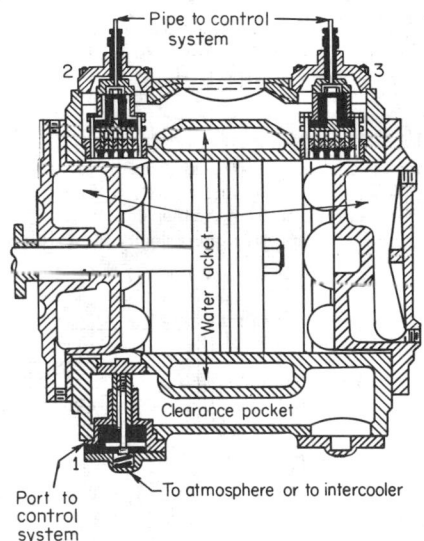

**Fig. 2** Cross section of gas-compressor cylinder, showing water jackets, clearance pocket with pneumatic control, and suction-valve lift devices.

The rotodynamic machine depends upon its high impeller velocity (1,000 ft/s, or 305 m/s) to impart sufficient momentum to the gas to effect the necessary rise in pressure. The centrifugal compressor dominates this category with sizes ranging from 100-hp stock refrigeration units to 20,000-hp units for heavy chemicals and volumes up to 200,000 actual ft³/min (340,000 m³/h). Figure 24 illustrates a horizontally split-case model used for pressures to 800 lb/in² (56 kg/cm²); Fig. 14

shows the vertically split-case barrel-type design for pressures to 5000 lb/in² (350 kg/cm²). For maximum volumes, the simpler steady flow axial compressor is used. Industrial models (Fig. 15) range from 8,000 actual ft³/min to 13 × 10⁶ actual ft³/min (14,000 m³/h to 22 Mm³/h), with ratio of compression $R_c$ limited to 4. The aircraft jet engine uses this type of compressor for volumes of 1,700 to 350 × 10⁶ actual ft³/min (2,900 m³/h to 600 Mm³/h) and to $R_c = 16$. There is a mixed-flow model where the radial outflow pattern lies 45° with respect to the axis. Still other designs attempt to handle smaller flow rates at higher heads and lower $N_s$, such as drag or regenerative compressors and partial-emission and pitot pumps (see Balje chart, Fig. 19).

### Piston-Rod Load

The piston-rod thrust is one of the limits of a compressor frame. The heaviest frames have a piston-rod capacity of 150,000 lb (70,000 kg); the medium sizes, 14- to 18-in (356- to 457-mm) stroke, have 95,000 lb (43,000 kg) capacity; 10- to 12-in (254- to 305-mm) stroke, 50,000 lb (23,000 kg); and the 5-in (127-mm), balanced-opposed units, 25,000 lb (11,400 kg). The specific loading on the piston rod is 9,000 to 16,000 lb/in² (633 to 1,125 kg/cm²) with SAE 4140. Bearing sizes and frame structure are predicted upon this loading. (Consult manufacturer for definitive limits.) The controlling piston-rod tension load is determined from

$$\text{Tension, lb} = A\,(P_2 - P_1) - aP_2 \qquad (1)$$

where $A$ = piston area, in², and $a$ = rod area, in²; subscripts 1 and 2, odd and even numbers, refer to the absolute suction and discharge pressures, respectively.

### Ratio of Compression

Performance is customarily evaluated on the ratio of compression $R_c = P_2/P_1$ (see Fig. 1). These pressures represent stagnation-line-connection conditions, which by strict interpretation should include the respective line-velocity head. These losses are usually negligible where line velocities are less than 50 ft/s (15 m/s) and densities less than 1,000 lb/in² (70 kg/cm²). The full area of the indicator diagram is obtained by integrating $P_4/P_3$, which includes the valve losses. (See Compression Efficiency below.) $R_c$ is usually held between 2.5 and 4 per stage for systems below 500 lb/in² (35 kg/cm²) and below 2.5 for 1,000 lb/in² (70 kg/cm²) systems; "big-inch" gas-transmission lines are operated at an average of $1.3R_c$. Allowing a 9 percent valve loss, 5 percent mechanical-gear loss, and 3 percent intercooler loss, single-stage operation can be economically justified to 6 $R_c$, two-stage to 20 $R_c$, and three-stage to 60 $R_c$. However, service continuity and plant and personnel safety are more important than power economy in limiting the $R_c$ to the 2.7 proximity. There was a time when large single-stage air compressors discharging at 100 psig (7 kg/cm²) and even 200 psig (14 kg/cm²) were commonplace and were responsible for numerous fires and explosions. Unsaturated gases from refinery processes are unstable above 220°F and form heavy gums and coke. The maintenance and service interruption resulting from such high $R_c$ operation and subsequent temperatures are prohibitive.

### Power Requirement

The power required for a given pressure rise is determined by the product of the volume and $\Sigma$, a function of $R_c$; that is,

$$\Sigma = [(\mathbf{K}R_c)^\sigma - 1]46/\sigma \qquad (2)$$

where $\sigma$ is $(k - 1)/k$ and $k$ is the ratio of specific heats. The equation includes a mechanical-gearing loss of 5 percent and is at a pressure base of 14.4 psia (1.01 kg/cm²) and 60°F. $\mathbf{K}$ is a valve-loss correction factor [see Table 2 and Eq. **(8)**]. For air, where $k = 1.40$ and $\mathbf{K} = 1.13$ for 6 percent loss in pressure through each set of valves,

$$\Sigma_{air} = 161[(1.13R_c)^{0.286} - 1] \qquad (3)$$

For dry natural gas, where $k = 1.255$ and $\bar{m} = 19$ and with the same valve loss, $\mathbf{K}$ becomes 1.08 and

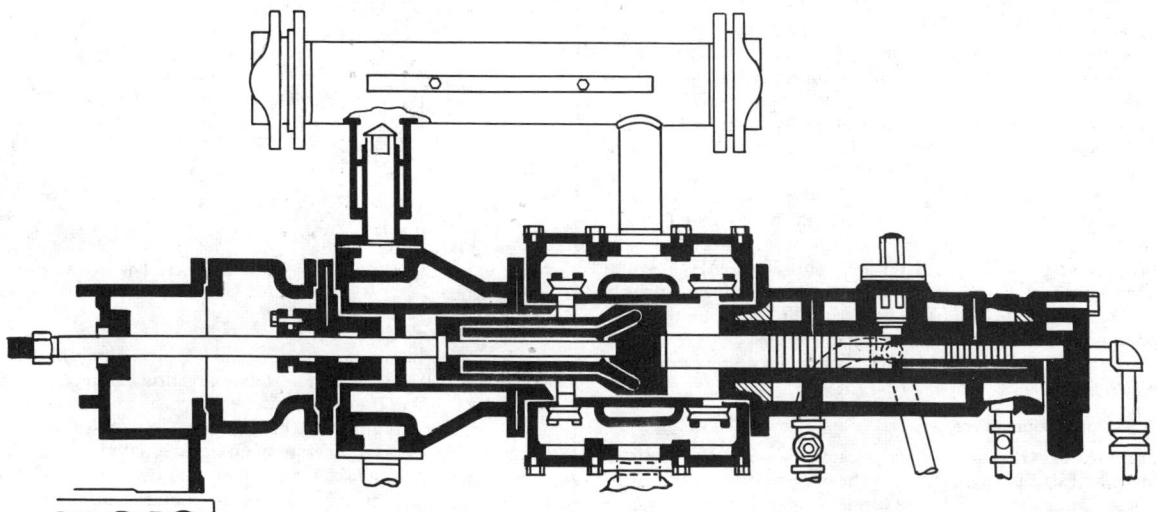

**Fig. 3**  Four-stage truncated tandem 4,000 lb/in² compressor.

$$\Sigma_{dng} = 227[(1.08R_c)^{0.203} - 1] \qquad (4)$$

Also $\Sigma_{dng} = 0.0456 \ \overline{m}c_v \ \Delta t$, where $c_v$ is the specific heat at constant volume and $\Delta t$ is the isentropic temperature rise. A multiplier of 0.147 applied to the above $\Sigma$ values gives the bhp per 100 ft³/min at 14.4 psia and 60°F. Figure 4 illustrates a popular empirical solution for $\Sigma$, wherein an arbitrary compression efficiency of 68 percent is applied at $1.5R_c$, 78 percent at $2R_c$, and 87 percent at $4R_c$. Mechanical efficiency of 95 percent is widely accepted, which includes loss allowances of 1 percent for piston-ring friction and piston-rod packing and 3 percent for gearing friction of the crosshead, slipper guides, connecting-rod pins, and crankshaft bearings. The latter losses are dissipated by convectional air circulation in frame sizes under 300 hp and into the lubricant system in larger sizes. The ring and packing losses are mostly absorbed by the jacket-water system. Where the cylinder power is less than 100 hp, these losses should be doubled.

## Temperature Rise

Piston compression is essentially an adiabatic function, especially when referred to the internal cylinder conditions. The compression-temperature rise follows the equation

$$T_2 = T_1(\mathbf{K}R_c)^{\sigma/m} \qquad (5)$$

where $\eta$ represents the heat leak factor applied in a manner consistent with the thermal efficiency. These factors are less than 1.05 for normal water-jacket cylinders, 1.09 for dry-jacket cylinders, 1.11 for forced-air-cooled cylinders with fins, and 1.15 for high-velocity water-jacket cooling and the expansion cycle, curve CFD on Fig. 1. There was a time when water was injected into the suction of air compressors to reduce the discharge temperature; when the speed of machinery was increased and the clearance volume reduced, the practice was abandoned as hazardous. The temperature drop was substantial; $\eta$ was 1.75. The scheme is still applied in chemical processes to wash out unsaturated gums and to suppress the discharge temperature of exothermic gases. The liquid is usually a light solvent of the same character as the gas and is atomized into the suction line. A short, 10-s blast of steam every 2 or 4 h can usually clear the gums from a cylinder.

The temperature behavior is only consistent below $4R_c$; beyond this, the cylinder cooling effect is perceptible because of the reduced mass flow at higher $R_c$ operation. European practice of process sizing includes a **warm-up** factor, which presumes the gas is heated 20 to 40°F in passing through the hot cylinder and suction valves. Such a correction complements the volumetric efficiency by a judgment factor of 0.95 to 0.90. Thermocouple probes in the suction valve and in the flow stream show no such evidence at the ambient-temperature range. American practice has always disregarded such corrections. The warm-up factor also allows for valve and piston-ring leakage. If such leakage is perceptible, the temperature rise is usually cumulative and readily detectable by thermometry.

## Compression Efficiency

Compression efficiency is an approximate method of accounting for all the power losses that occur between stagnant suction and discharge pressures. It presumes that all valve areas and gas channels offer equal resistance and that the compressor speed and character of the gas are inconsequential.

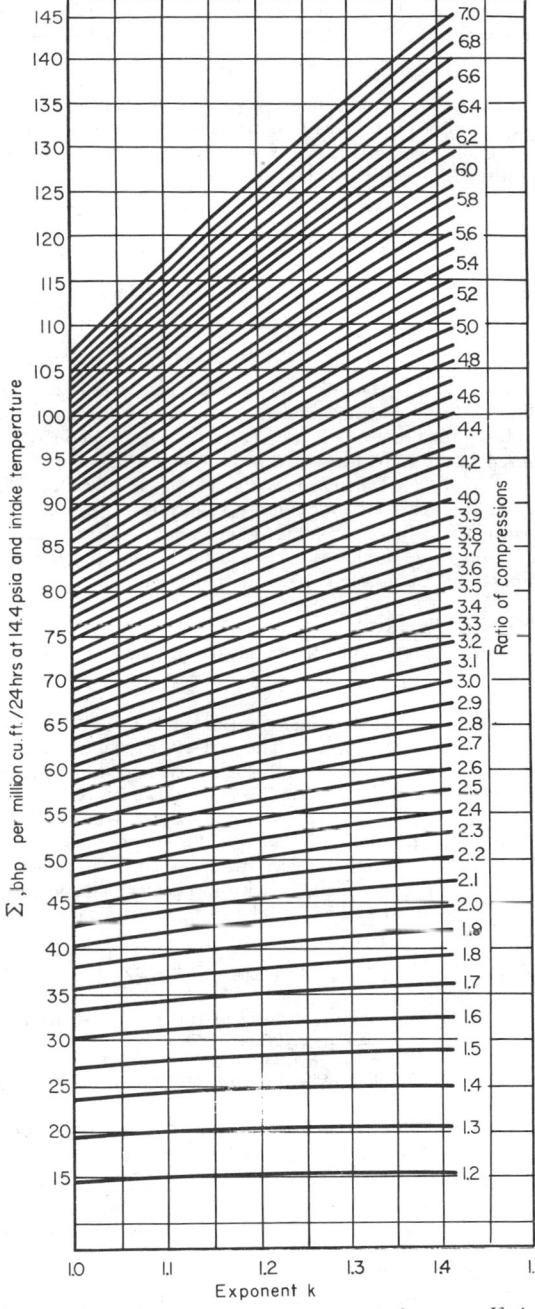

**Fig. 4** Approximate horsepower to compress air or gas. If single-stage, multiply cubic feet actual capacity of free gas per minute by 1,440 to obtain capacity in millions of cubic feet per 24 h. Then capacity in 24 h times horsepower per million as obtained from the chart will give the total horsepower. If two-stage, take the square root of the total number of compressions. Read the horsepower from the chart for this ratio, multiplying the same for the two stages, to which add 3 percent for cooler loss. Note that horsepower is for 14.4 psia intake. If horsepower based on capacity at 14.7 psia, add 2 percent to horsepower.

The height of the crosshatched area BCG and AFD (Fig. 1) represents the Δ(lb/in²) valve loss and can be evaluated from

$$\Delta(lb/in^2) = 1.26 \times 10^{-6}\alpha^2 fU^2 \text{ sp gr } P/Z \qquad (6)$$

where f is the valve resistance in velocity heads (usually 4), $\alpha$ is the ratio of piston area to minimum valve area (should be about 10), U is the average piston speed (13.3 ft/s). Introducing these values, the equation becomes

$$\Delta(lb/in^2) = 0.09P \text{ sp gr}/Z = \mathbf{K}P \text{ sp gr}/Z \qquad (7)$$

and

$$\mathbf{K}R_c = \frac{P_2}{P_1}\frac{1 + \Theta \text{ sp gr } Z_1}{1 - \Theta \text{ sp gr } Z_2} \qquad (8)$$

See Tables 1 and 2 for $\Theta$ and $\mathbf{K}$ correction values.

### Performance Definitions

The power required and the temperature are determined by the intrinsic corrected $\mathbf{K}R_c$ value and Eqs. (2) and (5). The cylinder charge pressure and the suction-line stagnation or snubbed pressure are one and the same and are depicted as point A in Fig. 1. A horizontal projection of this point to the expansion curve CD determines the volumetric efficiency AF/AE, which is corrected for compressibility by $Z_1/Z_2$, and the expansion is a function of the stagnation $R_c$, without the $\mathbf{K}$ correction. The k value is corrected for the mean adiabatic-temperature rise in reference to the mole fraction.

### Cylinder Sizing

The exact displacement of a double-acting (DA) cylinder is

$$ft^3/min = 9.1[D^2 - (d/2)^2]LN \ 10^{-4} \qquad (9)$$

where D = cylinder bore, in; d = piston-rod diameter, in; L = stroke, in; and N = r/min. The rod effect is negligible in cylinders larger than 10 in, and at the average piston speed of 13.3 ft/s (4.05 m/s) Eq. (9) simplifies to $4.23D^2$. Piston speed is established by effective response of the floating elements in the suction valve and the bearing design. Piston speeds of 45 ft/s (13.7 m/s) are common for aircraft engines and experience an excellent volumetric efficiency with mechanical suction-valve gearing, as shown in Fig. 5. The capacity of a process-gas cylinder is

$$10^6 \ ft^3/day = 10^{-4} \ ft/min \ P_1 E_v/Z \qquad (10)$$

EXAMPLE. Size a 50-hp air cylinder to handle 24.7 psia, 100°F to 74.0 psia, where Z = 1.0, $E_v$ = 0.70, and $\theta$ = 0.09. $R_c$ = 74/24.7 = 3.0; $\mathbf{K}R_c$ = 3(1.09/0.91) = 3.6; $(\mathbf{K}R_c)^{0.286}$ = 1.442. From Eq. (3), $\Sigma$ = 161(1.442 − 1) = 71 bhp/(10⁶ ft³)/(day) at 60°F and 76.3 at 100°F. The

air capacity is 50/76.5, or 654,000 ft³/day, and the cylinder displacement by Eq. (10) is 379 ft³/min which requires a (379/4.2)⁰·⁵, or 9.5-in bore. A 15-in-stroke cylinder would run; (13.3 × 60)/(1.25 × 2) = 320 r/min. The standard capacity is 654 × (14.4/14.7) × (520/560) = 595,000 ft³/day.

### Volumetric Efficiency

The clearance space in a cylinder is that volume ($CP_2$ in Fig. 1) not displaced by the piston sweep. A nominal clearance of 0.100 in (2.5 mm) is usually allowed between the extended piston and the cylinder head. This space plus the internal access passage to the valve seat makes the total clearance 10 to

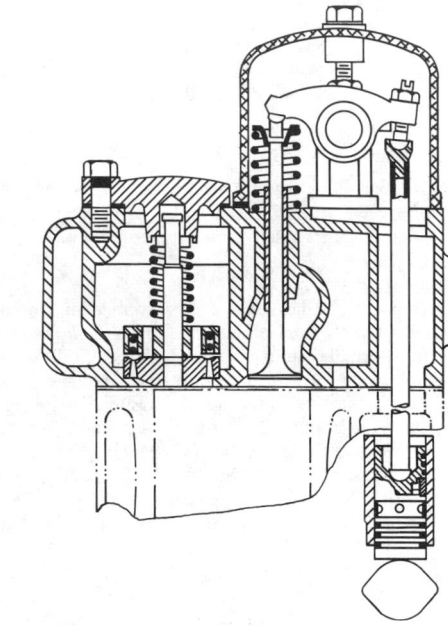

**Fig. 5** Compressor cylinder head with mechanically operated suction valve and spring-loaded poppet discharge valve. (*Schramm.*)

15 percent of the 10- to 20-in (254- to 508-mm) piston displacement. High-pressure cylinders smaller than 10 in (254 mm) usually have a clearance less than 8 percent so as to increase the volumetric efficiency and thereby reduce the rod loading

**Table 1. Compressor-Valve Resistance Factor $\Theta$***
(Showing influence of valve area and piston-speed resistance entering and leaving the cylinder)

| Piston-valve ratio $\alpha$ | Piston speed, ft/min and ft/s | | | | | | |
|---|---|---|---|---|---|---|---|
| | 600<br>10.0 | 700<br>11.7 | 800<br>13.3 | 900<br>15.0 | 1,000<br>16.7 | 1,100<br>18.3 | 1,200<br>20.0 |
| 20 | 0.202 | 0.275 | 0.360 | 0.457 | | | |
| 15 | 0.113 | 0.155 | 0.202 | 0.257 | 0.332 | 0.382 | 0.454 |
| 12 | 0.073 | 0.100 | 0.130 | 0.164 | 0.203 | 0.246 | 0.292 |
| 10 | 0.051 | 0.069 | 0.090 | 0.114 | 0.140 | 0.170 | 0.202 |
| 8 | 0.033 | 0.044 | 0.058 | 0.074 | 0.090 | 0.110 | 0.131 |
| 6 | 0.018 | 0.025 | 0.032 | 0.041 | 0.050 | 0.060 | 0.071 |
| 4 | 0.008 | 0.011 | 0.014 | 0.018 | 0.022 | 0.027 | 0.032 |

*$\Theta = 1.26 \times 10^{-6}\alpha^2 fU^2$, and $\Delta(lb/in^2) = \Theta$ sp gr P/Z.

**Table 2. Valve-Loss Correction Factors K***
(As applied to ratio of stagnation pressures for the intrinsic power of compression determination)

| Piston-valve ratio $\alpha$ | Piston speed, ft/min and ft/s | | | | | | | Specific gravity |
| --- | --- | --- | --- | --- | --- | --- | --- | --- |
| | 600 10.0 | 700 11.7 | 800 13.3 | 900 15.0 | 1,000 16.7 | 1,100 18.3 | 1,200 20 | |
| 12 | 1.34 | 1.50 | 1.70 | 1.97 | 2.37 | 2.93 | 3.72 | 2.0 |
| 12 | 1.16 | 1.22 | 1.30 | 1.39 | 1.51 | 1.66 | 1.83 | 1.0 |
| 12 | 1.09 | 1.13 | 1.17 | 1.22 | 1.27 | 1.35 | 1.43 | 0.6 |
| 10 | 1.23 | 1.32 | 1.44 | 1.59 | 1.78 | 2.09 | 2.34 | 2.0 |
| 10 | 1.11 | 1.15 | 1.19 | 1.26 | 1.33 | 1.41 | 1.51 | 1.0 |
| 10 | 1.06 | 1.09 | 1.12 | 1.15 | 1.18 | 1.23 | 1.28 | 0.6 |
| 8 | 1.14 | 1.19 | 1.26 | 1.35 | 1.44 | 1.57 | 1.71 | 2.0 |
| 8 | 1.07 | 1.10 | 1.12 | 1.16 | 1.20 | 1.25 | 1.30 | 1.0 |
| 8 | 1.04 | 1.06 | 1.07 | 1.09 | 1.11 | 1.13 | 1.15 | 0.6 |
| 6 | 1.06 | 1.11 | 1.14 | 1.17 | 1.22 | 1.27 | 1.33 | 2.0 |
| 6 | 1.04 | 1.05 | 1.07 | 1.09 | 1.11 | 1.13 | 1.20 | 1.0 |
| 6 | 1.03 | 1.03 | 1.04 | 1.05 | 1.06 | 1.07 | 1.09 | 0.6 |

*$K = (1 + \Theta \text{ sp gr})(1 - \Theta \text{ sp gr})$. **K** applied to the stagnation $R_c$ produces the intrinsic $R_c$ experienced with the cylinder.

with a smaller bore. Pipeline cylinders are usually provided with 100 percent clearances because they have an optimum loading at 1.2 to $1.6R_c$ with a constant discharge and variable suction pressure. The volumetric efficiency is

$$E_v = 1 + C - C \Lambda R_c^{1/k} Z_1/Z_2 \qquad (11)$$

where $C$ is clearance, decimal fraction, and $\Lambda$ is a leakage factor, usually 1.1 (or 10 percent) at $R_c$ below 7. (See Figs. 6 and 7.)

## Multistage Sizing

The diameter of a secondary cylinder can be estimated by dividing the preceding cylinder diameter by 1.41; i.e., if the primary cylinder bore is 20 in (508 mm), the second stage would be 20/1.41 = 14.2 in (361 mm), the third stage would be 14.2/1.41 = 10 in (254 mm), and the fourth, 7 in (178 mm). The intermediate pressures are determined by multiplying the preceding suction pressure by the average ratio of compression per stage. This is obtained by reducing the total $R_c$ by the

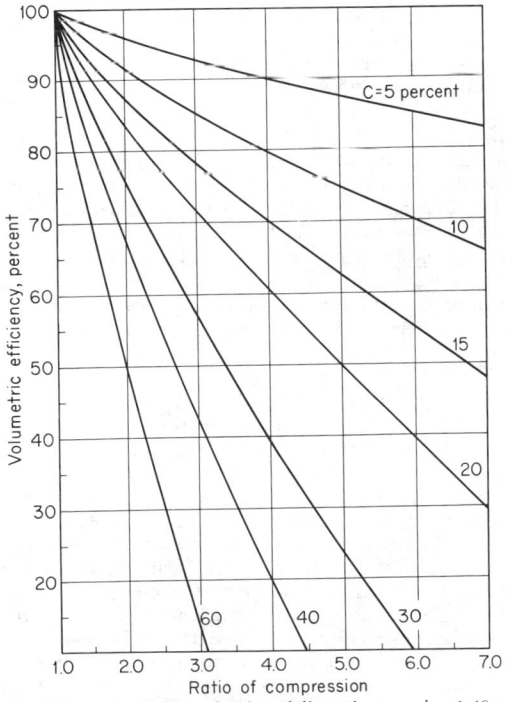

**Fig. 6** Volumetric efficiency for air and diatomic gases; $k = 1.40$, $n = 1.38$, $\eta = 1.05$, $A = 1.10$ [see Eq. (11)].

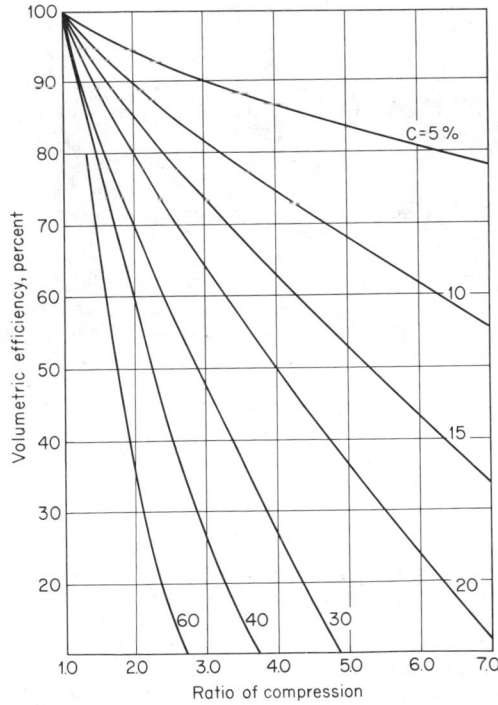

**Fig. 7** Volumetric efficiency for natural gas; $\bar{m} = 17.9$, $k = 1.28$, $n = 1.22$, $\eta = 1.10$, $A = 1.10$ [see Eq. (11)].

negative exponential power of the number of stages involved; i.e., if the operation is from 20 to 1,620 psia, $R_c = 81$, and $R_c^{-4} = 3$, intermediate pressures are 3(9)[27] × 1.03 (allowance for intercooler and piping losses) × 20 = 62(185) and [555] psia.

### Compressor Valves

The effective compressor-valve area is fixed by the passage under the raised valve elements and is defined as the product of the element lift and the sum of the valve-seat peripheries or strip edges, less the guide and end contacting surfaces. The plate- and strip-type valves (Figs. 8 and 9) are generally used in air and low-pressure service. The concentric-disk type (Fig. 10) is used in high-pressure chemical and natural-gas opera-

**Fig. 8** Plate valve, illustrating floating element, cushion spring, and seat. *(Ingersoll-Rand.)*

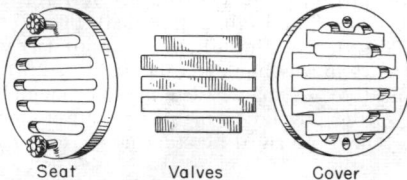

**Fig. 9** Feather valve, illustrating seat, flexing elements, and valve cover. *(Worthington.)*

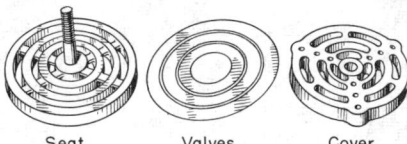

**Fig. 10** Concentric-disk valve, illustrating seat, floating elements, and valve cover. *(Chicago Pneumatic.)*

tions. The lift varies from 0.035 in (0.9 mm) for high-pressure, high-speed operation to 0.180 (4.6 mm) for low-pressure, low-speed operation, with 0.100 (2.5 mm) for general purposes. Various springs are used to dampen the opening impact load. The thickness of the element ranges from 0.050 to 0.125 in (1.3 to 3.2 mm) for high-pressure service. The optimum thickness, lift, and spring tension are matters of trial and error. The element in the feather valve (Fig. 9) functions as a spring and sealing element. It flexes in an arced segmental recess. An 8-in (203-mm) strip has an average lift of 0.100 in (2.5 mm). The strip thickness varies from 0.02 to 0.08 in (0.5 to 2.0 mm). Multiple nylon poppet valves (lift = 0.25 in) (6.3 mm) have been successfully applied to pipeline and other services where the discharge temperature does not exceed 250°F.

### Cylinder Wear

The wear pattern of horizontal compressors takes the form of an hourglass, with the emphasis in the vertical plane and on the head end. Malalignment of the cylinder or the piston rod is manifest by gray filing discolorations in the lubricant sampled

from the bottom connection. Severe cases produce abnormal heating, even at idling speeds. These conditions are corrected by meticulously aligning all components normal to a center-line-drawn piano wire. Cylinder-wear tolerance is between 0.5 and 1.0 percent of the cylinder diameter. Blow-by becomes excessive with greater wear, capacity is reduced, and the discharge temperature rises. The reciprocating action of the piston rings wears the radial face of the ring lands. It is possible to square the lands by machining and fitting the grooves with oversize rings or to rebuild the lands by metal spray. The latter repair can be applied to cylinder and piston surface wear, thereby maintaining the original bore. Other operators prefer to use the portable boring bar to true the cylinders to one of several regular oversized bores. When the wear exceeds 0.25 in (6.3 mm), a liner can sometimes be inserted to restore the original bore. The cylinder bore is finished within +0.003 and −0.001 in (+0.08 and −0.03 mm).

### Piston Rings

Pistons are finished with a nominal clearance of 0.01 in per in (mm per mm) of bore for sizes above 10 in (250 mm) and with an 0.0015 tolerance for smaller cylinders. Piston rings in a clean, well-lubricated, cylinder may serve effectively for more than 20 years. Such wear as occurs is observed at the open end, where it continues until the section no longer sustains the pressure differential and the ring breaks into small segments. This pressure loading can be relieved by fluting the radial edge to balance the peripheral pressure. Pistons operating below 300 lb/in² (21 kg/cm²) have 2 to 4 cast-iron rings; 8 to 16 rings are used for high pressures. A common one-piece snap cast piston ring has a proximate section 0.500 in (13 mm) wide by 0.375 in (9.5 mm) thick. Smaller rings are more fragile and give less service. The rings are finished to fit the cylinder and leave a 0.100-in (2.5-mm) gap at the step-cut or diagonal opening. The inherent spring and tension of the ring effect the seal.

Segmental bronze rings are used for high pressures and where snap-ring service is inadequate. Where bronze rings are inadequate, laminated-phenolic, dense-carbon, nylon, or Teflon rings are used. These materials are reinforced with asbestos, cotton, glass, and shredded-bronze fibers and are filled with graphite and molybdenum disulfide lubricants. Teflon rings reinforced with glass and bronze fibers and disulfide are rendering satisfactory NL service to 400°F.

### Piston-Rod Packing

Piston-rod packing includes the same range of materials plus soft rubberized asbestos rings for low-pressure services. Metallic packing consists of spacer cups containing a pair of three-piece rings, alternately cut radially and tangentially. The case for pressures less than 200 lb/in² (14 kg/cm²) contains three cups, and for 1,200 lb/in² (84 kg/cm²) eight cups. The pressure is presumed to be progressively reduced through each cup seal; actually the first cup facing the pressure does most of the reduction, and the subsequent rings are for reserve. The case is fed 2 to 5 drops/min of oil. New packing start-up operations require maximum lubricant, and in severe instances, bright-stock oil often provides the solution. The lateral clearance of the rings in the cup must permit free-floating action for the maximum radial runout of the rod. A runout of 0.002 in per ft (0.2 mm/m) of stroke is acceptable.

Metallic packing can tolerate rod wear of 0.15 percent of the rod diameter, with less tolerance above 2,000 lb/in² (140 kg/cm²). The rod should be relatively hard and ground to a 10-rms finish.

### Nonlubricated Cylinders

Industries such as the cryogenic and food and drug require gas compression absolutely void of contaminating lubricants. The NL-1 class does not permit any direct lubrication and tolerates the minute contamination introduced along the piston rod.

This can be eliminated by an additional spacer and rod wiper, qualifying the cylinder as a class NL-2. The pistons for this service use various Teflon rings and riders. The life of carbon rings is reduced by low humidities. The cylinder should have a ground 10-rms finish. Custom adds 5 percent friction for such NL service, but experience indicates this is overgenerous.

### Lubrication

Compressor-piston lubricants serve to (1) prevent wear by

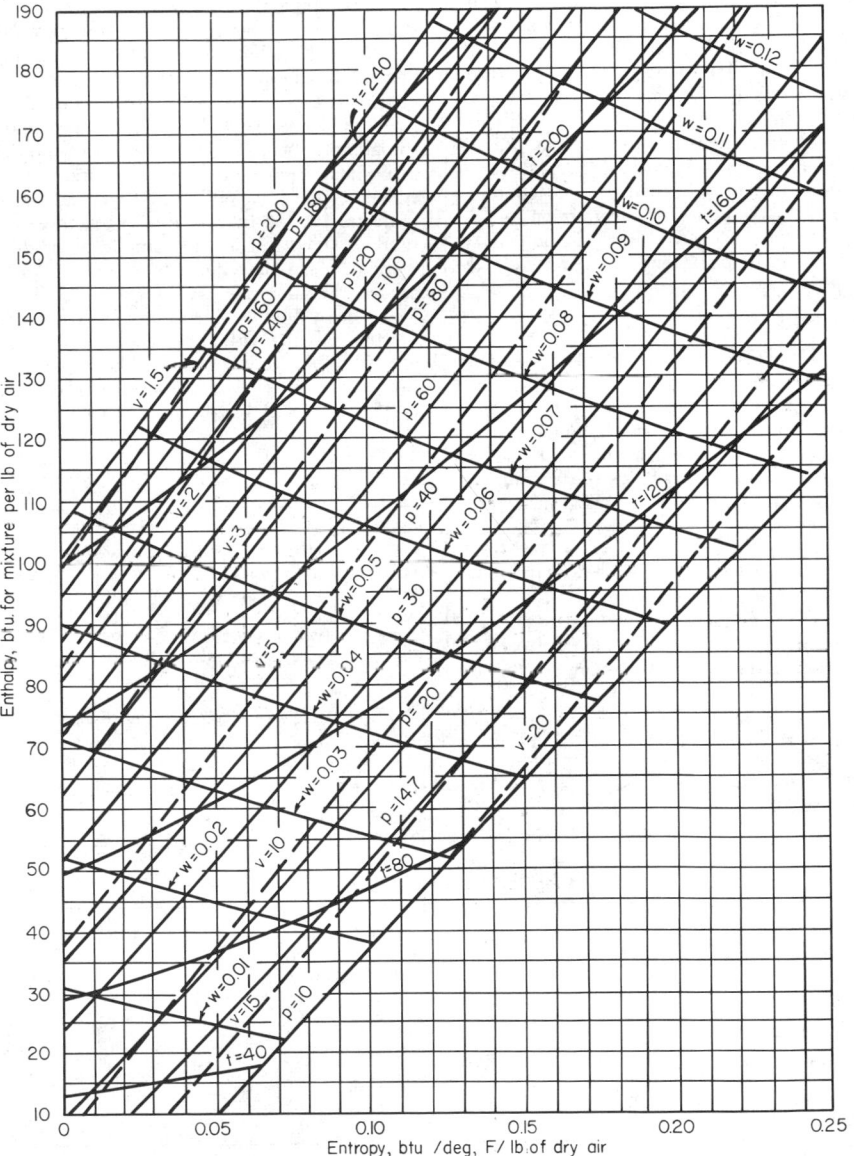

**Fig. 11** Enthalpy-entropy diagram for adiabatic compression of air saturated with water vapor (neglecting all consideration of the liquid water present); $w$ = weight of water vapor, lb per lb of dry air; $v$ = specific volume of air-vapor mixture, ft³ per lb of dry air; $p$ = total pressure of mixture, psia; $t$ = temperature of mixture, °F.

providing a low-frictional supporting film between the rubbing surfaces, (2) seal close clearances, (3) protect against corrosion, and (4) transmit heat of friction and minute wear particles away from points of contact. Industrial cylinders and packers require force-feed lubricators. Viscosity is the best index of suitability: general service at pressures below 500 lb/in² requires 400 SSU at 100°F (SAE 40), 700 SSU for 2,000 lb/in², and 1,000 SSU (SAE 50) for 8,000 lb/in² operations. These oils have very high viscosity at 40°F, which is difficult to roll for start-up and will result in poor lubrication. Good winterizing practice provides continuous circulation of warm jacket water and an immersed electric heater in the oil reservoir. Crankcase oils should include a foam inhibitor, sludge dispersant, and rust inhibitor. Phosphorus extreme-pressure agents are added to high-pressure lubricants to avoid wear and scuff damage. Castor-bean and rapeseed oils are additives resistant to solvent action of condensing hydrocarbons. Phosphate esters, Cellulube, Houghto-safe, Pydraul AC, and Fluorolubes are used to resist the hazard of exothermic reaction in air compression. The crankcase lubricant consumption for 10 plants over a score of years averaged 20,000 bhp/(h)(gal). Table 3 shows lubricant requirements for general process- and natural-gas operations.

**Compressor Accessories**

Most state laws and safe practice require a relief valve ahead of the first stop valve in every positive-displacement compressor. It is set to release at 1.25 times normal discharge or at the maximum working pressure of the cylinder, whichever is lower. The relief-valve piping system sometimes includes a manual vent valve and/or a bypass valve to the suction to facilitate start-up and shutdown operations. Quick line-sizing equations are (1) line connection, $d/1.75$; (2) bypass, $d/4.5$; (3) vent, $d/6.3$; (4) relief-valve port, $d/9$. For an 18-in (460-mm) diam cylinder, these would evaluate, respectively, to 10, 4, 3, and 2 in (250, 100, 75, and 50 mm) diameter.

The line connection to the cylinder involves two complications: (1) to provide attenuation or adequate reservoir capacity to minimize the variation in density at the instant of charge and dispel; (2) to damper undesirable pressure waves. The former is a matter of adequate volume (six to eight single displacements) adjacent to the cylinder. A pulse bottle equal to the cylinder bore or of the next larger nominal pipe size and between 3 and 7 ft (1 and 2 m) long ($^1/_{19}$ wavelength) is usually adequate. The wavelength is the quotient of the sonic velocity and the compressor action frequency; where $\tau$ is 1,200 ft/s (365 m/s) a DA cylinder at 360 r/min has a wavelength of 100 ft (30.4 m). A pulse-bottle length should avoid even quarter wavelengths to dampen the offending pressure wave. These unabated pressure waves can be destructive to piping and can make meter and control conditions difficult. Numerous baffles and chokes have been installed in the pulse bottle in an effort to diminish the pressure wave. A more logical and effective method is to install a Helmholtz resonator in the bottle outside-line connection. A simplified resonator consists of a straight tube with a bore one-third of the cylinder diameter and a length equal to the tube bore. This bore would be adequate for air at 60°F and 1 percent pressure drop. The frequency in the tube is accelerated three to four times the source, which restricts the amplitude of the sound-pressure wave that can pass through the filter. An orifice plate can achieve the same effect with slightly more resistance.

A rule of thumb which normally will reduce pulsations to acceptable limits with a reasonable pressure drop is to install a one-half pipe diameter orifice at the open end of the pipe. On a simple system, this is at the inlet filter for inlet-line pulsations or at the receiver for discharge-line pulsations. Complicated

| Item | Full load | ¾ load | ½ load |
|---|---|---|---|
| Displacement, ft³/min | 1,546 | | |
| Volumetric efficiency, percent | 86.0 | | |
| Actual capacity, ft³/min | 1,329 | 996 | 664 |
| bhp | 242 | 188 | 135 |
| bhp per 100 ft³/min | 18.25 | 18.9 | 20.3 |
| Motor efficiency, percent | 92.5 | 92.2 | 90.9 |
| Electrical hp per 100 ft³/mi | 19.73 | 20.5 | 20.9 |

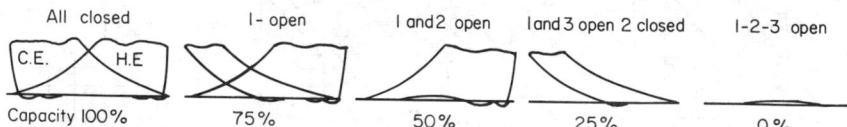

| All closed | 1- open | 1 and 2 open | 1 and 3 open 2 closed | 1-2-3 open |
|---|---|---|---|---|
| C.E.  H.E. | | | | |
| Capacity 100% | 75% | 50% | 25% | 0% |

**Fig. 12** Cards from compressor with combination clearance and suction-valve bypass control.

**Table 3. Lubrication Required for Compressor Cylinders**

| Range of cylinder diam, in | Film thickness, μm | 10³ hp/(h) per gal | Pints per cylinder per day per 100 hp of load | Oil drops/ min |
|---|---|---|---|---|
| 24–36 | 0.40 | 13 | 1.5 | 13 |
| 15–23 | 0.45 | 19 | 1.0 | 9 |
| 10–14 | 0.55 | 24 | 0.8 | 7 |
| 7–9 | 0.60 | 32 | 0.6 | 5 |
| 4–7 | 1.20 | 24 | 0.8 | 7 |
| 3–5 | 1.40 | 27 | 0.7 | 6 |

Based on 12,500 drops per pint of SAE 40 oil at 75°F with vacuum-type lubricator. Reduced drop count to roughly one-half for pressure-type lubricator. Example: 500 bhp, 20-in cylinder, requires 5 pint/day and feed of 45 drops/min. Each installation will vary on lubrication requirements. Manufacturers recommendations on oil type and quantity should be followed.

systems can best be analyzed by use of a specialized electrical analog. Most large compressor manufacturers have these available.

Control of volume flow is effected by variable-speed drivers; steam engines can operate at 20 percent of rated speed and gas engines at 60 percent, and electric-motor speed can be varied by means of eddy-current and hydraulic couplings and special wound rotors with rheostats, which are both costly and inefficient. Unloading can be applied by means of valve-lift and clearance pockets, as shown in Figs. 2 and 12. Utility air plants are throttled with suction unloaders, as shown in Fig. 13. Suction throttling and bypass controls are used in process operations.

### Cylinder Cooling

The principal purpose of a jacket-water system is to normalize cylinder-casting strains. Some cylinders have been operated **dry** for over 12 years without undue maintenance. The heat rejection to the jacket water is

$$H_j = 4(t_{ag} - t_{aw}) + 100 \quad \text{Btu/(bhp)(h)} \qquad (12)$$

where $t_{ag}$ and $t_{aw}$ represent the average gas and water temperature, respectively; good practice would offer 155 and 140°F, respectively, and a rejection of 160 Btu/(bhp)(h). Where the temperature rise is between 2 and 5°F, the jacket-water requirement is 160 bhp/(500 × 3.2$\Delta t_w$), or 0.1 gal/(min)/(bhp). Efforts to run cold jacket-water systems only produce liner sweating, washing of lubricant, and excessive ring wear.

### Turbocompressors

Typical turbocompressor types and performances are illustrated in Figs 14 to 20. These dynamic compressors operate on the same analytical principles as centrifugal pumps and have similar components and a similar terminology. Performance is

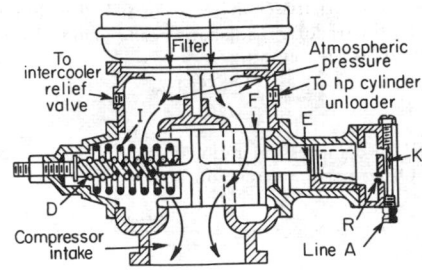

**Fig. 13**  Air suction unloader. Piston E moves valve to closed position as line A pressure reaches design point.

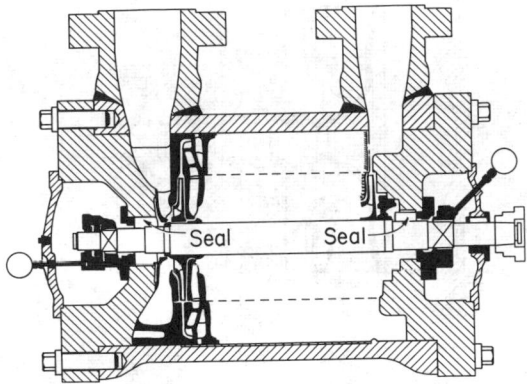

**Fig. 14**  Vertically split-case multistage 900 to 5,000 lb/in² centrifugal compressor.

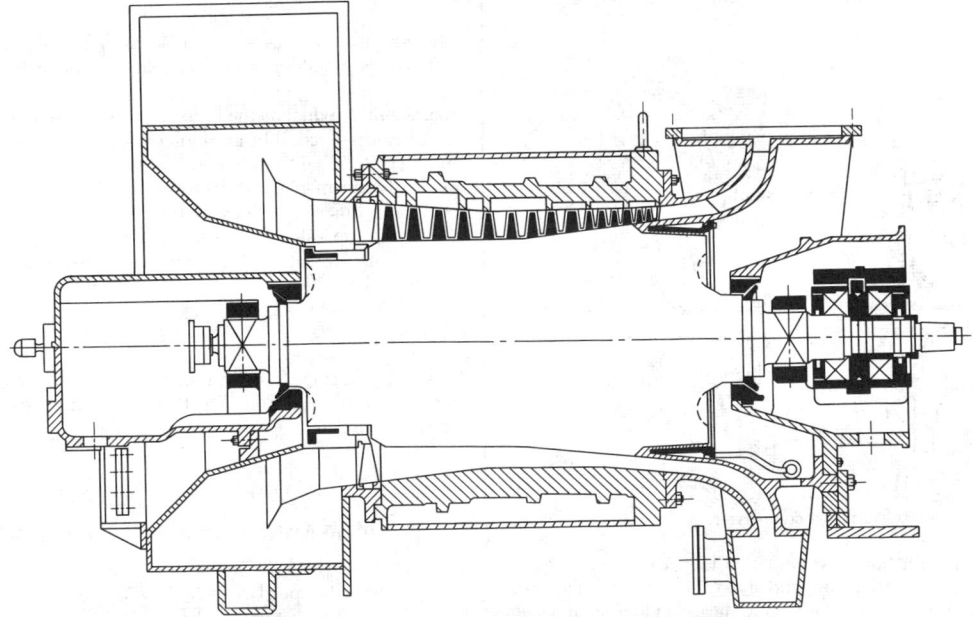

**Fig. 15**  Multistage axial compressor.

described by capacity $Q$ in terms of ft³/s (by designers) or ACFM (by the trade) and adiabatic head $L_{ad}$ in ft. Equation (2) can be modified as follows:

$$L_{ad} = \frac{1545\,T_1}{\overline{m}\sigma}(R_c^\sigma - 1) \qquad (13)$$

where $R_c$ = compression ratio; $\sigma = (k-1)/k$; $\overline{m}$ = mole weight. Other pertinent relationships include

$$\text{Power (hp)} = \frac{w\,L_{ad}}{550\eta_{ad}\eta_d\eta_m} \qquad (14)$$

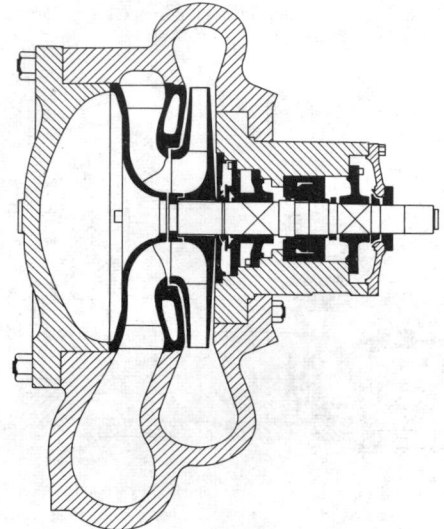

**Fig. 16** Single-stage cantilever-design centrifugal compressor.

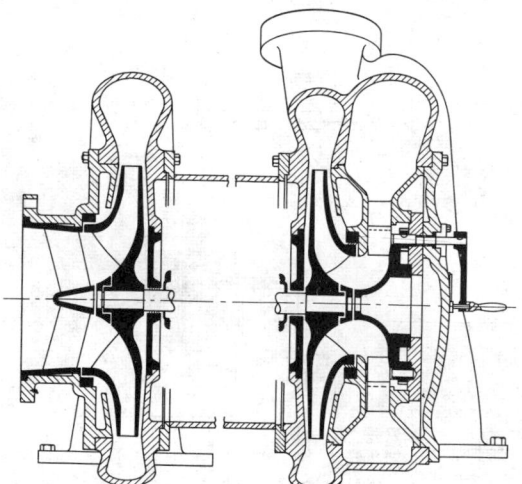

**Fig. 17** Hermetic-refrigeration compressor.

where $w$ = weight flow, lb/s; $\eta_{ad}$ = adiabatic efficiency; $\eta_d$ = disk and labyrinth efficiency, and $\eta_m$ = mechanical efficiency. It is important that the adiabatic head and efficiency be applied consistently and not interchanged with polytropic or hydraulic terms. The $k$ value should be determined from the gas analysis for the mean temperature. The adiabatic efficiency is the ratio of the adiabatic head power to the dynamic power. Heating-loss determination should include a detailed analysis of the frictional flow losses which occur throughout the entire flowpath. The adiabatic efficiency can also be determined by the ratio of the adiabatic power to the dynamometer power less the mechanical, disk, and labyrinth losses. The mechanical losses are assumed to be 5 percent for units smaller than 500 hp, 2 percent for 5,000 hp, and 1 percent for units larger than 10,000 hp. An idealized performance chart for a radial and several backward-leaning impellers is shown in Fig. 18. The abscissa represents the capacity in terms of the flow

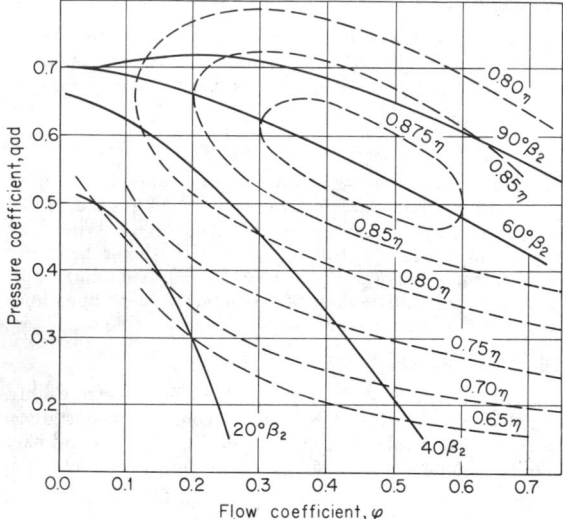

**Fig. 18** Basic characteristics of centrifugal compressor, showing effects of backward-leaning impeller blades. *(After Balje, ASME 51-F-12.)*

coefficient $\phi$, which is the ratio of the axial inlet velocity to the impeller-tip speed. The axial inlet velocity is usually at a Mach number of 0.2 and rarely exceeds Mach 0.4. The tip speed is generally less than 1,100 ft/s for industrial compressors. Aerospace machinery operates at between 1,400 and 1,800 ft/s. The optimum value of $\phi$ is 0.3 with 0.15 to 0.4 the practical range. The ordinate of Fig. 18 is the pressure coefficient $q_{ad}$, which is defined as

$$q_{ad} = \frac{gL_{ad}}{U^2} \qquad (15)$$

The negative slope lines on Fig. 19 are constant $q_{ad}$ values and can be determined from Eq. (16). This and other applicable equations are

$$q_{ad} = 11{,}750/N_s^2 D_s^2 \qquad (16)$$
$$q_{ad} = 1.7 \times 10^{-6} L_{ad}/d^2 N^2 \qquad (17)$$

$$10^6 \text{ std ft}^3/\text{day} = 32.8w/\overline{m} = 0.0091W/\overline{m} \qquad (18)$$

$$v_s = 10.73TZ/\overline{m}\,P \qquad (19)$$
Specific speed $N_s = NQ^{0.5}/L_{ad}^{0.75}$ $\qquad (20)$
Specific diameter $= DL_{ad}0.25/Q^{0.5}$ $\qquad (21)$
Sonic velocity $= 224(kT/m)^{0.5}$ $\qquad (22)$

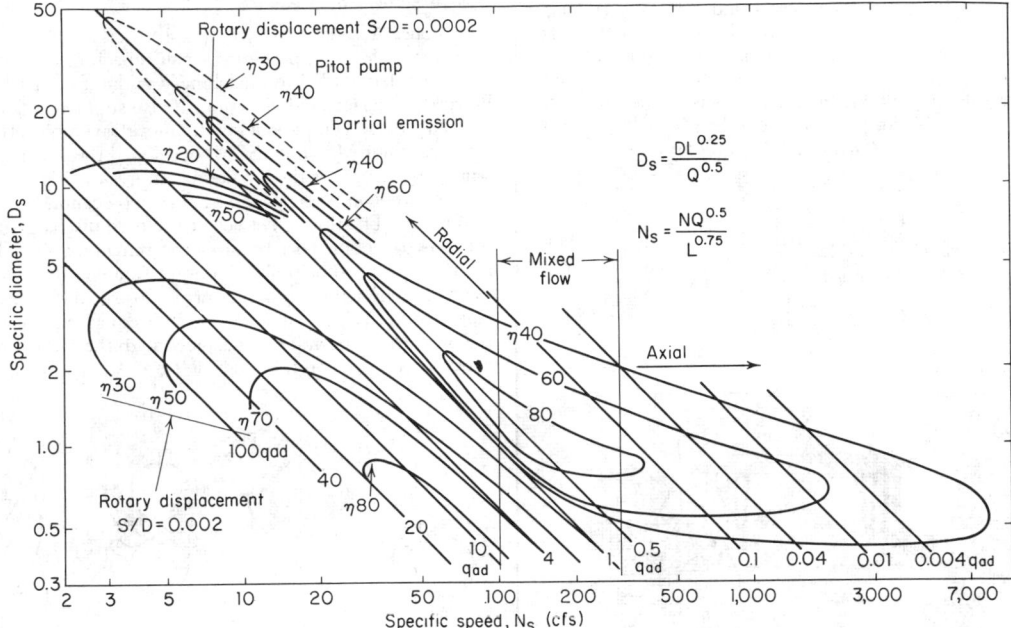

**Fig. 19** Turbocompressor performance chart. *(After Balje, ASME 60-WA-231.)*

where $Q$ = ft³/s, $d$ = impeller diameter, in; $D$ is in ft, $w$ is in lb/s and $W$ is lb/h.

The optimum performance range of the centrifugal compressor lies in the region between $q_{ad}$ values of 0.7 and 0.4 (Fig. 19). Where the volume flow exceeds this region, an axial compressor is required. The axial machine is further identified as a low-head machine, having a pressure coefficient less than 0.3 and a specific speed greater than 300. The two coordinates of Fig. 19, the specific speed $N_s$ [Eq. (20)] and the specific diameter $D_s$ [Eq. (21)], are useful for optimizing the geometry of any turbomachine.

Where the flow rate and head values are known, Eqs. (20) and (21) can be used in conjunction with Fig. 19 to determine practical impeller speeds.

The limiting head per stage is about 15,000 ft for industrial machines, but this value may be exceeded more than twofold. Aerospace machines generally exceed this limit by more than threefold. The efficiencies attained from Fig. 19 are more reliable than those arbitrarily associated with a given rotor diameter. A correction factor of 0.9 should be applied to these efficiencies to correct for multistage and off-design characteristics found in standard designs. These efficiencies are not applicable to off-design performance. They refer to the static pressure and the stagnation inlet pressure and assume that the inlet-line velocity $V_1$ equals the outlet-line velocity $V_2$. The data in Fig. 19 were obtained from calculated loss analysis using simplified assumptions and from the available performance data of high $N_S$ compressors and pumps. The efficiencies shown are attainable or can be bettered with precision production. These relationships are also dependent on a constant M* (Mach) and Re (Reynolds number). The effect of the M* (Mach number) is noted at $N_S$ values below 60 and where the relative inlet velocity exceeds the sonic velocity; the Reynolds number must be less than $10^6$ before it affects the performance; neither condition is likely in industrial machinery.

**Shaft Seals**

The restrictive edges of the labyrinth-type shaft seal are made of a soft expendable metal, aluminum or bronze, and finished to slight-pressure interference (j5) fit. Some applications use a labyrinth seal to break down the high pressure to a suction-pressure equalizer. Another oil-flushed labyrinth seal protects the shaft projection to the atmosphere (mechanical contact seals are also used). Figure 21 shows this seal with a rotating Stellite ring and a dense-carbon stationary ring supported on a convolution or coil-spring carrier. The spring tension and the effective contact area are varied to produce a satisfactory, balanced sealing pressure. The seal is cooled and lubricated with high-pressure seal oil. Segmental carbon rings (Fig. 22)

**Fig. 20** Labyrinth shaft seal.

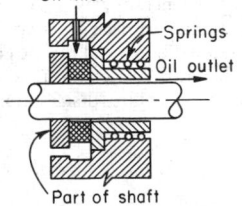

**Fig. 21** Mechanical-contact shaft seal.

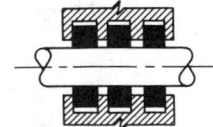

**Fig. 22** Carbon-ring shaft or interstage seal.

are used for interstage and end seals under 50 psig (3.5 kg/cm²). The oil-film seal (Fig. 23) is applied to shafts operating at 2,500 psig (175 kg/cm²). It depends upon the close clearance of the two floating bushings on the shaft and their lapped L-hub shoulder radial seals. The interior annulus is flooded with oil at 5 lb/in² (0.35 kg/cm²) above the high-pressure gas; the interior leakage is 2 to 10 gal/day (8 to 40 l/day).

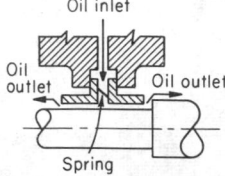

**Fig. 23**  Oil-film shaft seal.

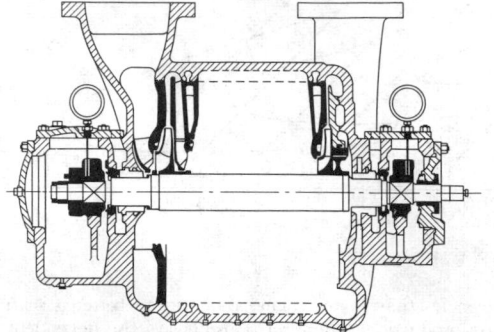

**Fig. 24**  Horizontally split-case multistage 800 lb/in² centrifugal compressor.

### Thrust Pressures

A double-inlet and an open impeller have no axial thrust; a semienclosed impeller, without a frontal shroud, can impose substantial thrust, evaluated by

$$T_A = Y(A_1 - A_2) \, \Delta(\text{lb/in}^2)$$
$$+ \, 0.8(A_2 - A_3) \, \Delta(\text{lb/in}^2) - P_1 A_3 \quad (23)$$

where $A_1$ = full impeller area, $A_2$ = impeller area less the area of the inlet-eye seal, $A_3$ = shaft area through the seal, all in in²; $\Delta(\text{lb/in}^2)$ = pressure differential per stage. $Y$ is the percentage of $\Delta(\text{lb/in}^2)$ acting in back of a single-stage disk: (1) without a frontal shroud and a plain back, $Y = 0.35$; (2) with 0.060-in back ribs, $Y = 0.28$; (3) with ¼-in equalizer holes at half radius and about 2 in apart, $Y = 0.22$; and (4) with deep scallops, $Y = 0.18$. For fully shrouded impellers, as used in multiple stage, $Y = 1.00$. These thrusts are balanced by opposing impeller inlet ends or, more generally, by a balancing drum wherein an equal and opposite thrust is created, equal to the product of the compressor $\Delta(\text{lb/in}^2)$ and the area of the drum. Stepanoff shows that radial impellers create a radial thrust, caused by the uneven volute-pressure distribution, which is

$$\mathbf{T}_R = 0.36 \, \Delta(\text{lb/in}^2) \, db$$

where $d$ is the impeller diameter and $b$ is the width, including the shrouds, both in in.

### Helical-Screw Compressors

The helical-screw rotary compressor has a relatively high efficiency at low $N_s$, providing small capacities at high heads. This complements piston machinery without the problems of lubrication, space, and vibration. It also supplements the low-$N_s$ centrifugal compressors with more favorable efficiency. Svenska Rotor Maskiner (SRM) developed the **built-in compression-space** feature for the Lysholm compressor (Fig. 25), where the compression phase of the axial sweep occupies about 300° of rotation. The timing between the closed discharge port and the closed suction port requires the remaining 60°. The gas retained in the rotor built-in compression space is released at the point where the discharge port is closed, to the unwrapped volute space. When the operating $R_c$ is equal to the **built-in** $R_c$, the pressure in the volute is balanced with the suction pressure and the compressor functions with 100 percent volumetric

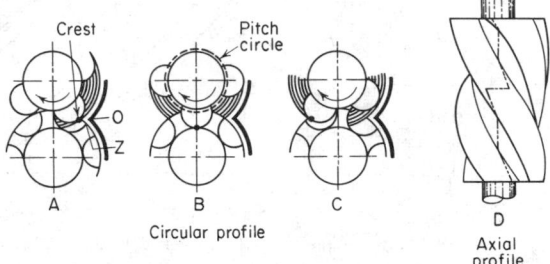

**Fig. 25**  Profile of SRM rotor in sequence of circular enclosure and axial sealing line.

efficiency. If the operating $R_c$ is greater than the built-in ratio, the capacity and the compression efficiency are reduced. If the operating $R_c$ is less than the design $R_c$, the compression efficiency is reduced but the capacity is slightly improved. Figure 25 illustrates the helical form and operation sequence. The unit is sized on the diameter of the male rotor, which drives the female gear via timing gears. Some manufacturers drive the female element directly with the aid of a generous oil flood and seal. Industrial sizes range from 4- to 25-in (100- to 630-mm) rotors, operating between 2,000 and 12,000 r/min and with capacities up to 20,000 ft³/min (34,000 m³/h). Industrial models are limited to rotor-tip velocities of 330 ft/s (100 m/s) on air, and higher tip speeds will give optimum efficiencies on gases lighter than air. The theoretical displacement of the helical-screw-type compressor is

$$Q_{ft^3/min} = dLUX \quad (24)$$

The $X$ factor is the ratio of the free cross-sectional area to the square of the rotor diameter $d$. The average functional $X$ factor, including gap losses at the **built-in** $R_c$ operation, is 0.059 for the most popular four-male-by-six-female lobe configuration. The power of compression for helical-screw-type compressors is

$$\Sigma = 50(R_c^\sigma - 1)/\sigma \eta_m \quad (25)$$

The mechanical efficiency is a function of speeds, bearing losses, seal losses, and gear losses. The normal range is from 90 percent on small units to 97 percent on large units. The slip losses must be added to the mechanical losses for overall efficiency. The geometric form varies from square, with $L =$

$d$, to $L = 1.65d$. Maximum differential pressures vary from 50 to 250 lb/in² (3.5 to 18 kg/cm²). The optimum range of application for this type of compressor is shown on the left margin of Fig. 19. The lower family of curves are for rotors having a nominal clearance of $0.002d$; the upper curves have a clearance of $0.0002d$.

### Other Rotary Compressors

Figure 26 illustrates an **axial-sweep** type of **screw compressor** which has an $X$ factor of 0.165 and a commercial top speed of 140 ft/s (43 m/s) in air service. The rotor sizes range from 6 to 16 in (150 to 400 mm) in diameter, with capacities up to 12,500 ft³/min (21,000 m³/h). A quick ft³/min capacity check is $22dL$. The $R_c$ is limited to 3, and the air power of compression is

$$\Sigma_{air} = 164(R_c^{0.375} - 1) \qquad (26)$$

The **straight two-lobe cycloidal rotor units** (Fig. 27) are used extensively for high vacuum and low (1.7 $R_c$) compression. A single-stage unit is effective from 1 mmHg to a blank-off at 2 $\mu$m. A two-stage unit or one operating in series with the sliding-lobe vacuum pump shown in Fig. 30 is effective from 5 to 0.05 $\mu$m, or $100R_c$. The sizes range from 6 by 6 to 28 by 70 in (150 by 150 to 700 by 1800 mm) with capacities up to 33,000 ft³/min (56,000 m³/h) at 55-ft tip speed. The average

displacement factor $X$ is 0.30 for dry-gas and 0.26 for moisture-passage models. The static slip for this class of machinery is determined by the speed required to sustain the desired pressure at zero flow. The slippage varies directly as the square root of the change in absolute pressure and inversely as the square root of the absolute temperature and specific gravity. The speed of heavy-duty two-lobe compressors is limited to 57 ft/s (17 m/s); light models range from 25 to 35 ft/s (8 to 11 m/s).

The **sliding-vane compressor** (Fig. 28) is limited to relatively small capacities and $R_c$ of 9.0 and has been most successful as a construction air compressor. The unit is flooded with 0.05 gal/ft³ (6.7 l/m³) of displacement. The oil picks up about 20°F (11°C) and holds the discharge temperature within 100°F (55°C) of ambient. The $X$ factor is 0.060, and the limiting vane-tip speed is 60 ft/s (18 m/s). The air power is

$$\Sigma_{air} = 164[(1.36R_c)^{0.286} - 1] \qquad (27)$$

The **rotary liquid-ring compressor** (Fig. 29) is used for handling highly saturated vapors, wet vacuums, and corrosive and exothermal gases. Compressor efficiency is less than 50 percent. Heat rejection from the liquid sealant is about 6,000 Btu/(hp)(h) of useful compression. Units are limited to capacities of 5,000 ft³/min (8,500 m³/h); pressures of 75 lb/in² (5 kg/cm²);

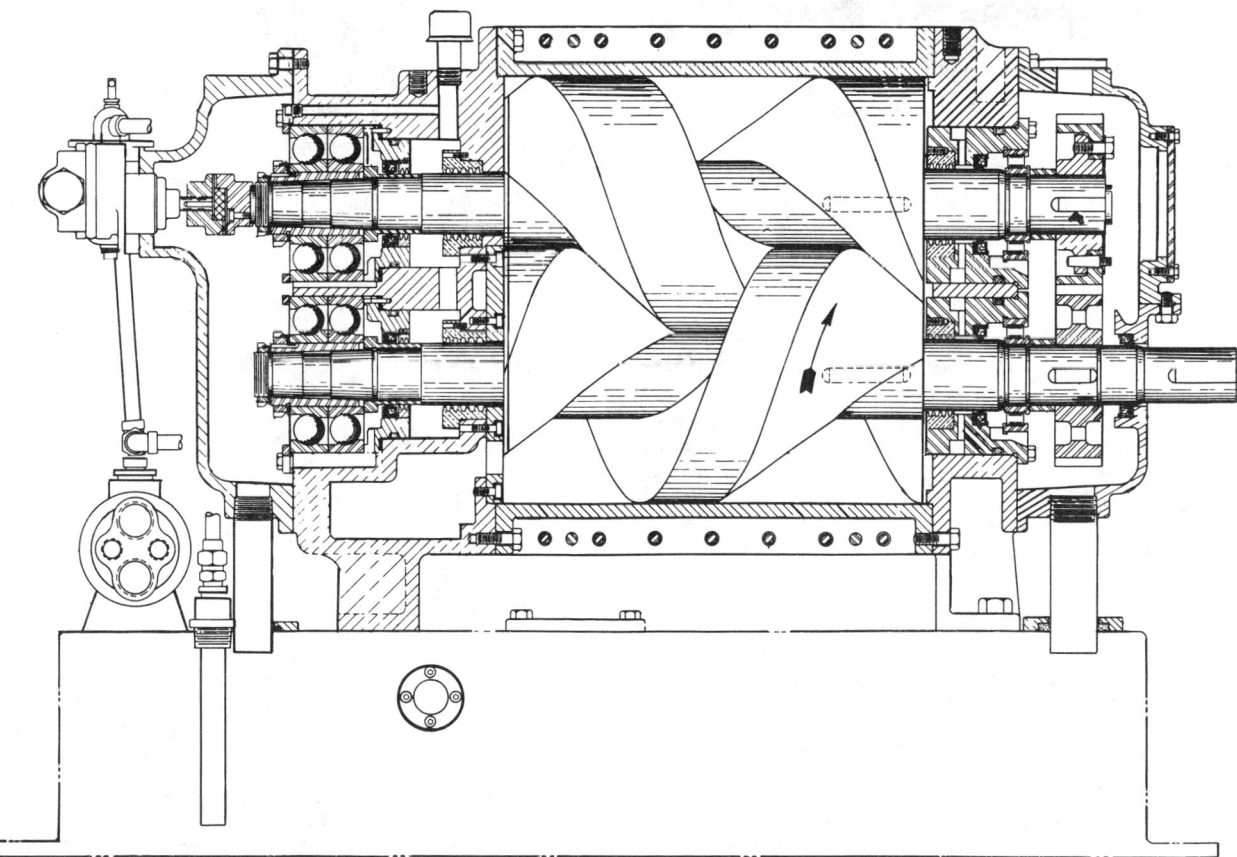

**Fig. 26**  Side section of spiral-axial compressor. (*Roots-Connersville, Spiraxial.*)

and vacuums to 25 in (635 mm) of mercury, single-stage, and 29 in (737 mm), two-stage. Figure 30 illustrates a single-lobe

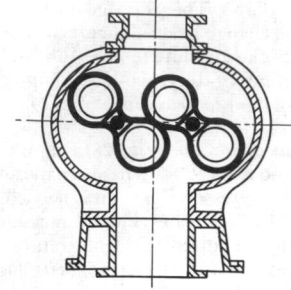

**Fig. 27** End view of straight two-lobe rotary compressor. *(Roots.)*

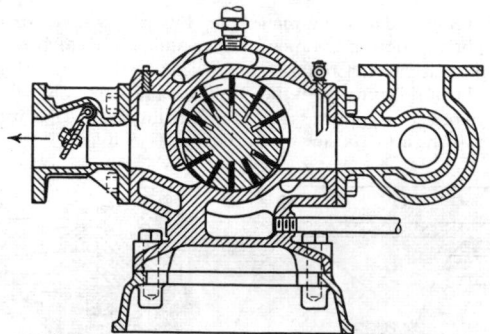

**Fig. 28** Sliding-vane-type rotary compressor.

rotary compressor used extensively as a low-micrometer vacuum pump.

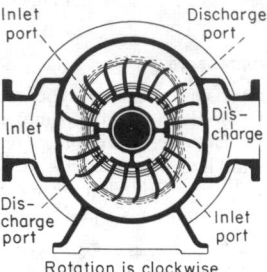

**Fig. 29** Rotary compressor with liquid-ring seal. *(Nash.®)*

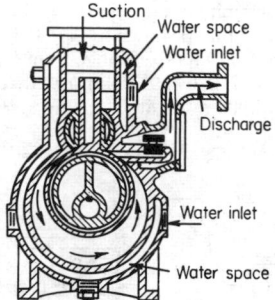

**Fig. 30** Single-lobe-type low-micrometer vacuum pump. *(Kinney.)*

# HIGH-VACUUM PUMPS
## by B. B. Dayton

REFERENCES. Dushman, "Scientific Foundations of High Vacuum Technique," Wiley. Pirani and Yarwood, "Principles of Vacuum Engineering," Reinhold. Guthrie, "Vacuum Technology," Wiley. Steinherz, "Handbook of High Vacuum Engineering," Reinhold. Guthrie and Wakerling, "Vacuum Equipment and Techniques," McGraw-Hill. Power, "High Vacuum Pumping Equipment," Reinhold. Van Atta, "Vacuum Science and Engineering," McGraw-Hill.

The pressure of gas in a chamber at a given temperature can be reduced by allowing the gas to escape through a port into a vacuum pump, or by a pumping system comprising two or more pumps in series, which compresses the gas and discharges it into the atmosphere, or by allowing the gas (or vapor) to condense on a sufficiently cold surface or to react with a chemically active surface exposed within the chamber or within an appendage to the chamber. The movement of the gas toward the pump or trapping surface can be accelerated by heating the walls of the chamber or by ionizing the gas and applying an electric field. **Rotary mechanical pumps** and **steam ejectors** which can compress the gas from about $10^{-3}$ mmHg up to atmospheric pressure are described earlier in Sections 9 and 14. Pumps and trapping techniques required to produce pressures in the "high-vacuum region" (below $10^{-3}$ mmHg) are described below.

**Units** The pressure unit *mmHg* formerly used in high-vacuum technology is difficult to define precisely and has been widely replaced by the **torr** defined as $1/760$ atm or exactly $(1,013,250/760)$ dyn/cm² or $(101,325/760)$ N/m². The international standard mmHg is equal to $1.000,000,14$ torr. One *micrometer Hg* (1 $\mu$m Hg) is approximately equal to one *millitorr* (1 mtorr) or $10^{-3}$ torr. **Pumping speed** is normally expressed in litres per second or cubic feet per minute (l/s or ft³/min); 1 l/s equals 2.12 ft³/min.

**Selection of Pumps** The type of vacuum pump selected depends primarily on the lowest pressure to be attained and the possible effects of vapor contamination. The lowest pressure that the pump itself can attain without vapor traps is called the **ultimate pressure** of the pump. The *size* of the pump depends either on the time to "pump-down" from atmospheric pressure or on the pressure to be maintained in the presence of a given gas load during the process cycle and the number of pumps which can be installed in parallel. The gas load is usually expressed in terms of **throughput** defined as the product of the static pressure and the volumetric flow rate across a given section at a given temperature. Size of pump is rated in terms of volumetric pumping speed (l/s or ft³/min) at the inlet pressure for which the speed is a maximum. As shown in Fig. 1, for rotary-piston mechanical pumps this maximum speed

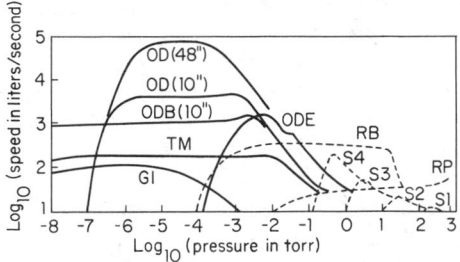

**Fig. 1**  Variation of pumping speed with inlet pressure for typical vacuum pumps. OD(48 in) = oil diffusion pump with nominal 48-in-diam inlet flange; OD(10 in) = 10-in oil diffusion pump; ODB(10 in) = 10-in oil diffusion pump with baffle; ODE = oil diffusion-ejector pump; TM = turbomolecular pump; GI = getter-ion pump (sputter type); RB = Roots-type blower; RP = rotary-piston oil-sealed mechanical pump; S1, S2, S3, S4 = matching stages in series of four-stage steam ejector.

occurs at atmospheric pressure, but for steam ejectors and oil vapor ejectors (similar in principle and design to steam ejectors but employing water cooling on the venturi section or diffuser) the maximum speed occurs at the top of a narrow peak on the speed-pressure curve. The **diffusion pump,** Fig. 2, which employs one or more jets of vapor into which molecules from

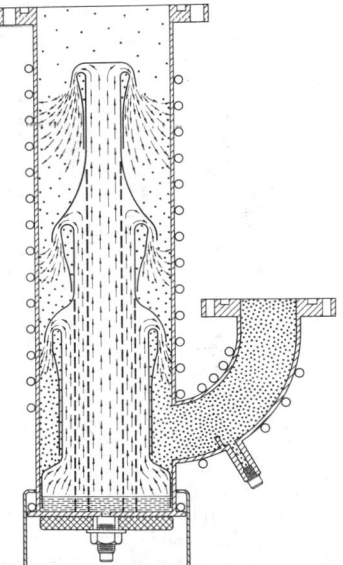

**Fig. 2**  Operating principle of oil diffusion pump. Arrows show vapor stream; dots represent gas molecules. Dashed lines show position of cylindrical vapor chimneys in fractionating pumps. *(Consolidated Vacuum Corporation and the Encyclopedia of Chemical Technology.)*

the chamber can diffuse and be carried forward into a region of higher gas pressure, has a broad plateau of maximum speed (see Fig. 1) from about $10^{-3}$ torr (1 mtorr) to pressures 20 times the ultimate pressure, the latter being approximately equal to the vapor pressure at room temperature of the fluid used to form the vapor jet.

The number, types, and sizes of vacuum pumps in series required to compress gas from the chamber up to atmospheric pressure usually depends on the compression ratio per pumping stage at the maximum throughput during the process cycle. For **ultrahigh vacuum** systems (pressures less than $10^{-9}$ torr) the number of pumps or stages required to reach the lowest pressures may depend on the rate of back leakage or back diffusion through the pumps at the minimum throughput. **Compression ratio** is defined as the ratio of the outlet (exhaust or discharge) pressure to the inlet (or intake) pressure at a given throughput when pumping a gas or vapor that is not absorbed or condensed within the pump. This ratio is a variable quantity which depends in general on the ratio of the pumping speed of a given stage to the net speed of the fore pump, or next stage of compression. However, for a given throughput there is an upper limit to the compression ratio which can be maintained across a vapor jet pumping stage at a given power input. At maximum throughput under normal operation the limiting compression ratio for single-stage ejector pumps is about 10, and for the individual stages of multistage diffusion pumps the limiting ratio is about 4. Thus, a three-stage steam ejector is required to compress air from 0.8 to 760 torr, a ratio of about $10^3$. Oil diffusion pumps usually have from three to five stages, and a four-stage diffusion pump can compress air from 1.5 mtorr to 0.4 torr at maximum full-speed throughput, corresponding to a total compression ratio of about $4^4$ or 256. The overall compression ratio can be much higher at very low throughputs, but the **forepressure** (pressure at the outlet) should not be allowed to exceed a value known as the **limiting forepressure** or **tolerable forepressure,** which is usually in the range of 0.2 to 0.6 torr for a four-stage diffusion pump.

Oil-sealed mechanical piston pumps can be operated with high compression ratios over a pressure range from atmospheric pressure (or above) to about 1 mtorr. Molecular drag, centrifugal, rotary blower, e.g., Roots-type, pumps, and turbomolecular pumps are limited by back flow or leakage of gas from the forepressure side through the clearances between the rotor and stator, and therefore require backing pumps to reduce the forepressure from atmospheric to a sufficiently low value to permit a high vacuum to be obtained on the inlet side. Oil vapor-jet pumps cannot discharge gas directly to the atmosphere because of limitations on the vapor pressure which can be generated without excessive thermal decomposition of the oil, and backing pumps must therefore be used to reduce their forepressures to values less than the vapor pressure of the oil at a safe temperature in the boiler. Some gas molecules (particularly hydrogen and helium) may succeed in diffusing back through a vapor jet from the forevacuum to the high vacuum, and for ultrahigh vacuum systems this back diffusion must be reduced by using a sufficient number of pumping stages and vapor jets with high vapor density and velocity.

Table 1 lists the types, sizes, and operating characteristics of typical commercially available pumps. The maximum (standard measured) speed (in liters per second) of oil diffusion pumps is about $18D^2$ for three-stage models and $28D^2$ for four-stage models, where $D$ is the true inlet diameter (in inches).

**Table 1. Types and Available Sizes of High-Vacuum Pumps**

| Type | Available sizes | | Operating range | | Power input, kW |
|---|---|---|---|---|---|
| | Nom. inlet diam, in. | Max. air speed, liters/sec | Ult. press, torr | Limiting forepress, torr | |
| Mechanical: | | | | | |
| rotary piston or plunger.... | 0.5, 1.0, 1.5, 2, 3, 4, 6, 8 | 0.1–400 | ~$10^{-3}$ | (atm.) | 0.2–30 |
| rotary blower (Roots)...... | 2, 4, 8, 16 | 40–2000 | ~$10^{-3}$ | 30 | 0.5–20 |
| turbomolecular........... | 6, 10 | 140–1,600 | <$10^{-9}$ | ~$10^{-1}$ | 0.4–1 |
| Ejectors: | | | | | |
| steam, six-stage......... | 48 | 50,000 | ~$10^{-3}$ | (atm.) | |
| mercury vapor.......... | 1 | 15 | <$10^{-4}$* | ~200 | 7 |
| Diffusion: | | | | | |
| oil, multistage...... ..... | 2, 3, 4, 6, 10, 12, 14, 16, 20, 24, 32, 35, 48 | 100–90,000 | <$10^{-7}$ | ~0.5 | 0.3–30 |
| mercury, multistage....... | 2, 4, 6, 10, 24, 32, 48 | 50–50,000 | <$10^{-10}$* | ~0.5 | 0.7–20 |
| Diffusion-ejector, oil........ | 4, 8, 12, 24 | 300–16,000 | ~$10^{-3}$ | 2.0 | 2–20 |
| Getter-ion: | | | | | |
| sublimation-ion........... | 0.3–36 | 1–$10^{5}$ | <$10^{-10}$ | ~$10^{-3}$ | |
| sputter-ion............... | 0.5–36 | 1–50,000 | <$10^{-10}$ | 0.02 | 0.4–20 |
| Cryosorption.............. | 1, 2 | 1–$10^{7}$† | <$10^{-3}$ | (atm.) | 0.1–50 |
| Cryopanel................ | (Sized to fit inside chamber) | $10^{3}$–$10^{6}$† | <$10^{-10}$ | | |

* With trap.
† $N_2$ speed.

The larger pumps are usally water-cooled, and about 80 percent of the heat generated must be carried away by the cooling water, the remainder being lost by convection, radiation, and conduction to the surroundings. The exit temperature of the cooling water should usually not exceed 40°C, and the required rate of flow in cubic centimeters per minute of water having an inlet temperature of 20°C and a exit temperature of 30°C is about numerically equal to the power input in watts. This rule of thumb applies to either diffusion pumps or water-cooled mechanical pumps where the horsepower of the motor is converted to watts by multiplying by 746.

**Diffusion Pumps**  Figure 2 shows a typical diffusion pump constructed of metal. Gas molecules wandering into the top of the pump are able to penetrate the diffuse boundary of the vapor jet (downward arrows) and reach the denser forward-moving core of vapors where they are driven at an acute angle toward the pump wall and on to the next stage of pumping. The vapor condenses on the wall and drains as a liquid back to the bottom, or boiler region, where the liquid is reheated to about 200° C to create a fresh vapor supply which rises up the chimneys to feed the nozzles. The operating fluid may be mercury or a low-vapor-pressure oil. The oils are sold under various trade names and usually consist of organic or silicone compounds of molecular weight in the range from 350 to 500. **Mercury-vapor pumps** require efficient refrigerated traps to keep mercury vapor out of the chamber being evacuated. Oil diffusion pumps require some form of purging or purification of the oil during use to eliminate dissolved gases and volatile decomposition products. **Fractionating oil diffusion pumps** purify the oil by circulating it through a series of boilers, or boiler compartments, feeding vapor through separate chimneys to the various nozzles in a multistage pump. The volatile impurities are ejected with the vapor feeding the stages nearest the fore vacuum, and the purged oil of lowest vapor pressure supplies the top nozzle from which vapor molecules scattered back out

of the inlet port create a partial pressure of oil vapor in the high vacuum which limits the ultimate pressure obtainable without cold traps [Hickman, *J. Appl. Phys.* **11**, 303 (1940)].

**Getter-ion** or **sputter-ion pumps** (Fig. 3) employ chemically active metal layers which are continuously or intermittently deposited on the wall of the pump by either thermal evaporation or sputtering and which chemisorb oxygen, nitrogen, water vapor, and other active gases while the inert gases such as helium, neon and argon are "cleaned up" by ionizing them

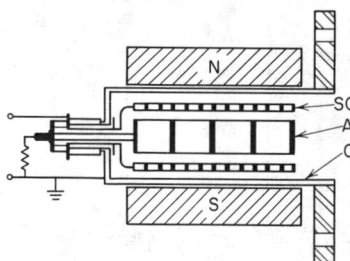

**Fig. 3**  Sputter-ion pump, triode type. SC = sputter cathode; A = anode; C = ion collector; N,S = magnet poles.

in an electric discharge and drawing the positive ions to the wall where the neutralized ions are buried by fresh deposits of metal. These pumps require a roughing pump to reduce the pressure to less than about 20 mtorr at which point the active metal (usually titanium) can be evaporated or sputtered at the required rate, but after they begin operation no backing pump is required since all of the gas is trapped at the wall. When isolated by valves from the roughing pump, the getter-ion pumps form an enclosure sealed to the vacuum system so that a power failure cannot result in leakage of atmospheric air or vapors from a forepump into the system. Sputter-ion pumps can operate continuously for more than 1 year at pressures below $10^{-6}$ torr.

**Cryopumps** consist of one or more exposed surfaces refrigerated to a temperature usually below 100 K, at which certain gases will be condensed and form a layer having an equilibrium vapor pressure below a specified limit. A plate cooled to 20 K, by circulating helium gas from a refrigeration unit through coils attached to the plate, will condense $N_2$, $CO_2$, CO, $H_2O$, $O_2$, Ar, and Xe to maintain partial pressures of these gases less than $10^{-10}$ torr. Hydrogen, helium, and neon are not adequately condensed at 20 K but may be "cryotrapped" in a deposit of $H_2O$ and other gases condensing on the plate. Cryopumping is used in space simulation chambers to create the necessary low pressures and to act as a heat sink comparable to "cold black space." In this application the 20-K plates are shielded by liquid-nitrogen-cooled panels, and the whole array is shaped to cover the inside wall of the vacuum chamber (see also Sec. 19).

**Cryosorption pumps** employ a sorbent such as activated charcoal or synthetic zeolite (Molecular Sieve) cooled by liquid nitrogen or other refrigerant. They can be used to rough a system down from atmospheric pressure to a few millitorr at which getter-ion pumps may begin operation, or an additional preconditioned cryosorption pump can be valved in to reduce the pressure to $10^{-5}$ torr or less.

**Turbomolecular pumps** (U.S. Patent 2,918,208) employ a system of alternate high-speed rotors with inclined blades and stators with inclined slots to impel the gas molecules from the high vacuum to a forepump. They have a broad plateau of pumping speed from about $10^{-2}$ to $10^{-8}$ torr. When water and hydrogen are removed from the walls of the system by baking, these pumps can reach ultimate pressures of less than $10^{-9}$ torr.

**Installation of Pumps**    In a typical assembly of pumps for a high vacuum system (Fig. 4) the diffusion pump A must be located close to the chamber B and should always have a water-cooled or refrigerated baffle C over the pump inlet so that vapor scattered back from the jets (backstreaming) will be condensed and returned as liquid to the pump. The pipe or manifold, which may include a valve D, connecting the chamber and the baffled pump should be of diameter equal to or larger than the inlet of the pump, and the length of the passage between the baffled pump and the chamber should preferably

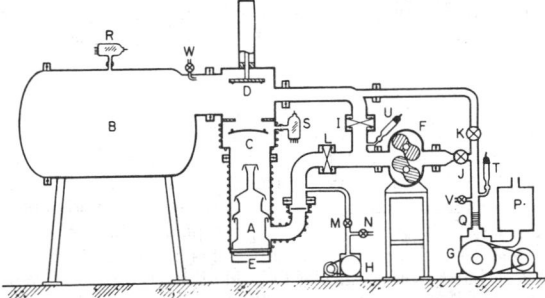

**Fig. 4**   Typical high-vacuum system. A = diffusion pump; B = chamber; C = baffle; D = valve; E = pump heater; F = Roots-type blower; G = roughing (mechanical) pump; H = holding (mechanical) pump; I to N = valves; P = oil reservoir and separator; Q = bellows; R, S = ion-gage tubes; T, U = Pirani-gage tubes; V, W = air-inlet valves.

be not more than about 3 times the mean diameter of the passage. A space at least 6 in high should be allowed below the diffusion pump boiler E for easy servicing of the heaters, draining the pump fluid, or removal of the pump from the system.

Some mechanical pumps have appreciable vibration so that they should be firmly anchored or mounted on vibration damping pads, and flexible metal bellows Q should be installed in the forevacuum line. These bellows also aid in aligning the pumps during assembly. It is not necessary to have the mechanical pumps close to the chamber because the resistance to gas flow of moderate lengths of pipe having the same diameter as the inlet of these pumps is not sufficiently large to create a serious pressure drop at mean pressures above 200 mtorr. Below 200 mtorr the pressure drop along the pipe is larger, but usually the diffusion pump operation is not affected unless the limiting forepressure (about 100 to 600 mtorr) is exceeded at the pump outlet.

**Vapor Contamination**    The vapors evolved from vacuum systems may be condensed within mechanical pumps or the intercondensers of steam-ejector systems. If the condensed vapor is the same as the working fluid in the pump, such as water vapor condensed in "wet" vacuum pumps, no harm is done, provided that the fluid level is maintained at the optimum working value. If the condensed vapor can be separated from the working fluid by centrifuging, evaporation, filtering, or settling, suitable separating means can be installed. In certain types of rotary pumps the Gaede **gas-ballast** principle may be employed to avoid condensation by admitting air at a certain point in the compression cycle. If separation cannot be satisfactorily accomplished during operation of the pump, the vapor should be condensed in cold traps or liquid absorption columns before reaching the pump.

**Baffles** and **traps** are often required to prevent the backstreaming or migration of the pump operating fluid or sealing fluid into the high-vacuum chamber. Unless cooled baffles and traps or sorption traps for oil vapor are included in the connecting line, the ultimate pressure in the vacuum chamber will usually not be less than the vapor pressure of the pump fluid at ambient temperature. While a water cooled "optically tight" baffle installed above the inlet of a diffusion pump will condense the backstreaming pump-fluid vapor and return most of the fluid to the pump, some fluid will reevaporate from such a baffle and migrate back into the vacuum chamber. To reduce the partial pressure of pump-fluid vapor below the vapor pressure at the water cooling temperature it is necessary to add a cold trap or baffle refrigerated to temperatures usually below the pour point of the fluid. These cold traps must be periodically warmed and the condensate removed to avoid inefficient cooling of the exposed surfaces during operation. The addition of an "optically tight" baffle and a cold trap over the inlet to a diffusion pump usually reduces the net pumping speed for permanent gases to less than 50 percent of the diffusion pump speed at the inlet.

**Flow of Gases at Low Pressure**    The pipe line between the high-vacuum pump and the vacuum chamber limits the volumetric flow so that the **net pumping speed** as measured by a vacuum gage located in the chamber is given by $S_n = S_0 U/(S_0 + U)$, where $S_0$ is the measured speed of the pump at its inlet and $U$ is the **conductance** of the pipe defined by $U = Q/(P_n - P_0)$, where $Q$ is the throughput, $P_n$ is the pressure in the chamber, and $P_0$ is the pressure near the inlet of the pump as measured by a gage installed in a similar manner to that used to determine the pump speed $S_0$. When there is no loss or gain of gas within the pipe line, $Q = S_n P_n = S_0 P_0 = U(P_n - P_0)$.

The conductance of a pipe depends on the geometry and the **Knudsen number** $K$ (defined as the ratio of the mean free path of the gas molecules to the mean diameter of the cross section) as well as the direction and velocity of the molecules entering the pipe. For $K > 1$ the conductance for air in liters per second at 25°C of a circular pipe of length $L$ (feet) and inside diameter $D$ (inches) connecting a high-vacuum pump to a chamber of diameter greater than $3D$, including the "entrance correction" at the chamber, but neglecting the "exit correction" at the pump which depends on the inlet diameter and other factors, may be calculated from $U = 6.6D^3/(L + 0.11D)$. The effect of right-angle bends in the pipe for "molecular flow" $K > 1$ and for $L > D/3$ may be approximated by adding $0.05D$ to $L$ for each bend (where $L$ is in feet and $D$ in inches). A single right-angle bend in a short pipe ($L < D/3$) has practically no effect on the conductance as computed for a straight pipe of the same length along the centerline [Davis, *J. Appl. Phys.* **5**, 358 (1954)].

When $K < 0.01$, the conductance in litres per second for air at 20°C of a long circular pipe of length $L$ (feet) and diameter $D$ (inches) may be calculated from $U = 0.25D^4\overline{P}/L$, where $\overline{P}$ is the mean pressure in the pipe in millitorr (micrometers of Hg). For $0.01 < K < 1$ the conductance for air at 20°C of a long tube can be estimated from

$$U = \frac{0.25D^4\overline{P}}{L} + \left(\frac{1 + 0.65D\overline{P}}{1 + 0.80\,D\overline{P}}\right)\left(\frac{6.6D^3}{L}\right)$$

The size of the primary pump, which acts as a "roughing pump" to pump the chamber down from atmospheric pressure to a pressure at which a diffusion pump or other high-vacuum pump can operate, may depend on the peak gas load during the process as well as the desired pump-down time. However, the size indicated by the peak load condition is frequently much smaller than the size required to meet the specified pump-down or roughing time. In this case, if the process cycle is much longer than the pump-down time, it is advisable to use two forepumps (primary pumps), a large one (G in Fig. 4) for roughing down and a smaller one (H in Fig. 4) for holding the vapor pumps during the roughing period and backing them during the processing period. Roots-type blowers are useful for shortening the roughing time for large chambers in the range below 20 torr and for handling unusually large bursts of gas which occur in some processes (F in Fig. 4).

An oil-sealed rotary mechanical pump is normally used as the primary pump, and the roughing time $t_r$ (in minutes) required to evacuate a chamber of volume $V$ (cubic feet) from atmospheric pressure to about 0.7 torr at which high-vacuum pumps begin to operate can be estimated from $t_r = 10V/C$, where $C$ is the rated speed (cubic feet per minute) at atmospheric pressure of the rotary pump. Since the speed of the high-vacuum pumps is usually of the order of 100 times that of the forepump, the pressure should drop quickly as soon as the high-vacuum pumps "take hold," but below $10^{-4}$ torr the pressure may begin to decrease more slowly because of the outgassing of the materials exposed inside the vacuum system.

For most of the materials exposed in high-vacuum systems the **outgassing rate** can be assumed proportional to the exposed area $A_m$, although for some very porous materials the rate is more proportional to the bulk or mass. Except for evaporation from pure liquid or solid phases, the outgassing rate normally decreases with time when the temperature is constant. For many industrial-type vacuum systems it has been found that the pressure $P$ in the chamber decreases approximately according to $P = P_u + K_1 A_m/S_n t^\alpha$ where $P_u$ is the ultimate pressure, $K_1$ is a constant which may be considered equal to the average outgassing rate per unit area after 1 h of pumping, $A_m$ is the exposed area, $t$ is the total pumping time in hours, and $\alpha$ is an exponent which is usually nearly constant for the first few hours. For rough calculations in typical metal vacuum systems it may be assumed that $\alpha = 1$, but for systems containing large amounts of elastomers or plastics $\alpha$ may be closer to 0.5 [Dayton, *Trans. 6th Natl. Vacuum Symp.*, pp. 101–119, Pergamon Press, Oxford, 1960; Kraus, ibid., pp. 204–205].

When the system is first evacuated after prolonged exposure to the atmosphere, most of the outgassing load for the first 10 to 100 h is usually water vapor, and for unbaked metal systems the numerical value of $K_1$ is of the order of $10^{-4}$ when $A_m$ is in square feet and $S_n$ in litres per second. For systems containing large amounts of elastomers or plastics $K_1$ is more of the order of $10^{-3}$ to $10^{-2}$. Systems constructed entirely of metal and glass may be heated to temperatures as high as 500°C to accelerate the outgassing, the average rate increasing by approximately a factor of 10 for each 100°C increase in temperature. In order to reach $10^{-6}$ torr in 10 h in an unconditioned vacuum chamber, the net pumping speed (in litres per second) should be 10 to 100 times the exposed area (in square feet).

The time $t_v$ (in minutes) to vent a vacuum chamber of volume $V$ (cubic feet) from pressures less than 1 mtorr to atmospheric pressure through a standard globe valve of nominal diameter $D$ (inches) is given approximately by $t_v = V/100D^2$. The oil in the boiler of oil diffusion pumps should be cooled to below 100°C before opening the pump to atmospheric pressure.

**Applications of High-Vacuum Pumps**  Steam ejectors, Roots-type blowers, diffusion-ejector oil vapor pumps, and rotary piston pumps are used to produce pressures in the $10^{-1}$ to $10^{-4}$ torr range for distillation of plasticizers, fat-soluble vitamins, and certain other organic chemicals; for dehydration of frozen foods, animal tissue, blood plasma, serum and antibiotics; for refining, degassing, and casting metals in vacuum furnaces; for vacuum sintering of cermets and powder metallurgy parts; for vacuum annealing of special alloys; and for pumping down vacuum chambers to a pressure at which high-vacuum pumps can begin operation. Diffusion pumps, getterion pumps, and turbomolecular pumps are used to evacuate radio, television, and X-ray tubes and vacuum-insulated containers before sealing off; to produce pressures in the $10^{-3}$ to $10^{-6}$ torr range in chambers for coating injection-molded plastic parts, rolls of plastic sheet, glass plates, and other substrates with aluminum and other metal films; for coating optical parts with antireflection films or layers producing dichroic mirrors and interference filters; for depositing metal and semiconductor films to produce microelectronic circuits; to maintain pressures of $10^{-4}$ to $10^{-7}$ torr in mass spectrometers, electron microscopes, synchrocyclotrons, betatrons, and other devices for accelerating or separating charged particles.

# FANS
## by Robert Jorgensen

REFERENCES. Baumeister, "Fans," McGraw-Hill. Jorgensen, "Fan Engineering," Buffalo Forge Company. "Laboratory Methods of Testing Fans for Rating Purposes," AMCA (Air Moving and Conditioning Association) Standard 210-75; also ASHRAE Standard 51 in preparation. "Fans and Systems," *AMCA Publ.* 201. "Test Code for Sound Rating," AMCA Standard 300-67. "Standards Handbook," *AMCA Publ.* 99.

## Symbols

$A$ = area, ft$^2$
$b$ = barometric pressure, in Hg
$D$ = diameter, ft
$D_s$ = specific diameter, ft
$H$ = fan power input, hp
$H_o$ = fan power output, hp
$K_p$ = compressibility factor, dimensionless
$L_p$ = sound pressure level, dB
$L_w$ = sound power level, dB
$L_{ws}$ = specific sound power level, dB
$\log$ = logarithm to base 10
$h$ = head, ft·lb/lbm
$M$ = molecular weight, dimensionless
$N$ = speed of rotation, r/min
$N_s$ = specific speed, r/min
$n$ = number or polytropic exponent
$p_s$ = fan static pressure, in wg
$p_t$ = fan total pressure, in wg
$p_v$ = fan velocity pressure, in wg
$p_{sx}$ = static pressure at plane $x$, in wg
$p_{tx}$ = total pressure at plane $x$, in wg
$p_{vx}$ = velocity pressure at plane $x$, in wg
$Q$ = fan capacity, ft$^3$/min
$Q_x$ = volumetric flow rate at plane $x$, ft$^3$/min
$T$ = absolute temperature, °R
$W$ = power, W
$x$ = factor used to determine $K_p$, dimensionless
$z$ = factor used to determine $K_p$, dimensionless
$\gamma$ = isentropic exponent, dimensionless
$\eta_s$ = fan static efficiency, per unit
$\eta_t$ = fan total efficiency, per unit
$\rho$ = fan air density, lbm/ft$^3$
$\rho_x$ = air density at plane $x$, lbm/ft$^3$

## Subscripts

1 = fan inlet plane
2 = fan outlet plane
$a$ = absolute
$b$ = basic known conditions
$c$ = calculated condition
$r$ = reading
$x$ = plane 1, 2, 3, or other

## FAN TYPES AND NOMENCLATURE

Any device which produces a current of air may be called a fan. This discussion will be limited to fans which have a rotating impeller to produce the flow and a stationary casing to guide the flow into and out of the impeller (see Fig. 1). The form of the casing or the impeller may vary widely.

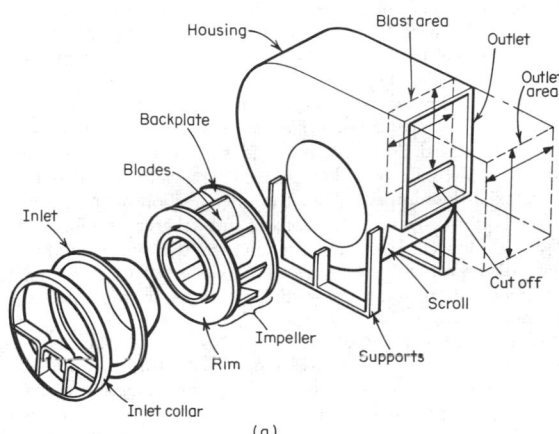

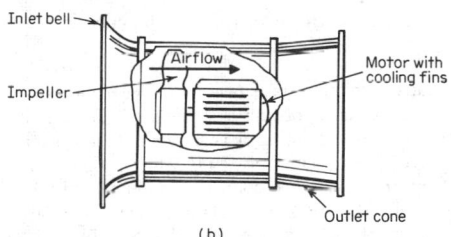

**Fig. 1** Elements of fans and preferred terminology: *(a)* centrifugal fans; *(b)* axial fans. *(Based on AMCA Publ. 201.)*

### Fan Classifications

One of the characteristics by which fans are classified is the nature of the flow through the blade passages of the impeller. Axial flow, radial flow, mixed flow, and cross flow are all possible in fan impellers. Certain fan names result from these classifications. Other fan names derive from other characteristics.

Propeller fans, tube-axial fans, and vane-axial fans all utilize **axial-flow impellers,** but their casings differ. Propeller fans may be mounted in a ring or panel. Tube-axial fans and vane-axial fans both use tubular casings, but for vane-axial fans they are equipped with stationary guide vanes. A great deal of the energy transferred to the air in axial-flow machines is in kinetic form. Some of this kinetic energy can be transformed into potential energy by straightening the swirl, e.g., with vanes, or by reducing the exit velocity, e.g., with a diffuser. Propeller fans effect very little transformation and hence have very low pressure-producing capability. Vane-axial fans can be equipped for maximum transformation as well as high transfer of energy and hence have high pressure-producing potential depending on tip speed and blade angles. High hub ratios promote high energy transfer.

Centrifugal fans and tubular centrifugal fans both utilize **radial-flow impellers.** Centrifugal fans usually employ a volute or scroll-type casing, the flow entering the casing axially and leaving tangentially. Tubular centrifugals use tubular casings so that the flow both entering and leaving the casing is axial. A considerable portion of the energy transferred to the air in a radial-flow machine may be in potential form due to centrifugal action; hence the name centrifugal fan. Since centrifugal action with blade depth, the pressure-producing capability of radial-flow fans will vary with this factor as well as tip speed and blade angles.

**Mixed-flow impellers** can be used in either axial or scroll-type casings. They are characterized as mixed flow because both axial and radial flow take place in the blading. Mixed-flow impellers used in axial-flow casings have a hub similar to a pure axial-flow impeller, but the inlet portion of the blading extends down over the face of the hub, thereby giving some radial guidance. Mixed-flow impellers used in scroll-cased fans have blades which give most of the axial guidance in the inlet portion and most of the radial guidance in the discharge portion.

In a **cross-flow impeller,** the air passes through the blading twice, entering more or less tangentially through the tip, passing across the impeller and out the other side. The casings are designed to provide this transverse flow. Cross-flow fans are also known as tangential fans or transverse-flow fans. Pressure-producing potential is low and depends on the formation of a vortex as the air leaves the impeller.

### Fan Details

Propeller fans and other **axial fans** may use blades shaped to airfoil sections or blades of uniform thickness. Blading may be fixed, adjustable at standstill, or variable in operation. Propeller fans have very small hubs. Hub-to-tip diameter ratios ranging from 0.4 to 0.7 are common in vane-axial fans. The larger the hub, the more important it is to have an inner cylinder approximately the hub size located downstream of the impeller. The guide vanes of a vane-axial fan are located in the annular space between the tubular casing and the inner cylinder. Diffusers are generally used between the fan and the discharge ductwork.

**Centrifugal fans** use various types of blading. Forward-curved blades are shallow and curved so that both the tip and the heel point in the direction of rotation. Radial and radial-tip blades both are radial at the tip, but the latter are curved at the heel to point in the direction of rotation. Backward-curved and backward-inclined blades point in the direction opposite rotation at the tip and in the direction of rotation at the heel. All the above blades are of uniform thickness and are designed for radial flow. Airfoil blades have backward-curved chord lines so that the leading edge of the airfoil is at the heel pointing forward and the trailing edge at the tip pointing backward with respect to rotation. Impellers for all blade shapes are usually shrouded and may have single or double inlets. Blade widths are related to the inlet-to-tip-diameter ratio. Tip angles may vary widely, but heel angles should be set to minimize

entrance losses. Scroll casings may be fitted with a streamlined inlet bell, an inlet cone, or simply a collar.

**Tubular centrifugals** may be designed for backward-curve, airfoil or mixed-flow impellers. An inlet bell and discharge guide vanes are required for good performance.

**Cross-flow fans** utilize impellers with blading similar to that of a forward-curved centrifugal, but the end shrouds have no inlet holes. Blade-length-to-tip-diameter ratios are limited only by structural considerations.

**Power roof ventilators** may use either axial- or radial-flow impellers. The casings will include either a propeller fan mounting ring or a tubular casing to guide the flow for an axial-flow impeller. If a radial-flow impeller is used, an inlet bell is required, but the scroll case may be replaced by the ventilator hood.

All fan types may have direct or indirect drive arrangements. Various standards including arrangement numbers have been adopted by AMCA. Arr. 1 signifies an overhung impeller on a fan shaft with two bearings on a base; Arr. 3 signifies an impeller on a fan shaft between bearings; Arr. 4 is for overhung impeller on a motor shaft; Arr. 7 is Arr. 3 with a motor base; Arr. 8 is Arr. 1 with an extended base for motor; Arr. 9 is Arr. 1 with motor mounted on side of unit; Arr. 10 is Arr. 1 with motor mounted inside of base. Rotation is specified as cw or ccw when viewed from the drive side. All fans can be equipped with variable-speed drives, variable inlet vanes, or dampers, but controllable-pitch axials do not need speed or vane control.

### FAN PERFORMANCE AND TESTING

The conventional terms used to describe fan performance in the United States are defined below.

**Fan Air Density**   Air density is the mass per unit volume of the air. The density of a perfect gas is a function of its molecular weight, temperature, and pressure as indicated by

$$\rho_x = \frac{M}{386.7} \frac{529.7}{T_x} \frac{b_x}{29.92}$$

where the subscript $x$ indicates the plane of the measurements. For dry air this reduces to

$$\rho_x = 1.325 b_x / T_x$$

This expression will usually be accurate enough even when moist air is involved. Fan air density is the density of the air corresponding to the total pressure and total temperature at the fan inlet

$$\rho = \rho_1$$

**Fan Capacity**   Volume flow rate is usually determined from pressure measurements, e.g., a velocity-pressure traverse taken with a Pitot static tube or a pressure drop across a flowmeter. The average velocity pressure for a Pitot traverse is

$$p_{rx} = (\Sigma \sqrt{p_{rxn}}/n)^2$$

where subscript $x$ indicates the plane of the measurements, subscript $r$ indicates a reading at one station, $n$ is the number of stations, and $\Sigma$ is the summation sign. The corresponding capacity is:

$$\dot{Q}_x = 1{,}097 A_x \sqrt{p_{rx}/\rho_x}$$

For a flowmeter

$$\dot{Q}_x = 1,097\,CA_xY\,\sqrt{\Delta p/\rho_x}/F$$

where $C$ = coefficient of discharge of meter, $Y$ = expansion factor for gas, $\Delta p$ = measured pressure drop, and $F$ = velocity-of-approach factor for the meter installation. Fan capacity is the volumetric flow rate at fan air density

$$\dot{Q} = \dot{Q}_x\rho_x/\rho$$

**Fan Total Pressure**  Fan total pressure is the difference between the total pressure at the fan outlet and the total pressure at the fan inlet

$$p_t = p_{t2} - p_{t1}$$

When the fan draws directly from the atmosphere,

$$p_{t1} = 0$$

When the fan discharges directly to the atmosphere,

$$p_{t2} = p_{r2}$$

If either side of the fan is connected to ductwork, etc., and the measuring plane is remote, the measured values should be corrected for the appropriate pressure drop

$$p_{t2} = p_{tx} + \Delta p_{2-x} \qquad p_{t1} = p_{tx} + \Delta p_{x-1}$$

**Fan Velocity Pressure**  Fan velocity pressure is the pressure corresponding to the average velocity at the fan outlet

$$p_v = \left(\frac{\dot{Q}_2/A_2}{1,097}\right)^2 \rho_2$$

**Fan Static Pressure**  Fan static pressure is the difference between the fan total pressure and the fan velocity pressure. Therefore, fan static pressure is the difference between the static pressure at the fan outlet and the total pressure at the fan inlet

$$p_s = p_t - p_v = p_{s2} - p_{t1}$$

**Fan Speed**  Fan speed is the rotative speed of the impeller.

**Compressibility Factor**  The compressibility factor is the ratio of the fan total pressure $p_t{}'$ that would be developed with an incompressible fluid to the fan total pressure $p_t$ that is developed with a compressible fluid, all other conditions being equal:

$$K_p = \frac{p'_t}{p_t} = \frac{[n/(n-1)]\,[(p_{t2a}/p_{t1a})^{(n-1)/n} - 1]}{(p_{t2a}/p_{t1a}) - 1}$$

Compressibility factor can be determined from test measurements using

$$x = p_t/p_{t1a}$$
$$z = [(\gamma - 1)/\gamma]\,6356H/\dot{Q}p_t$$
$$K_p = [z \log(1 + x)]/[x \log(1 + z)]$$

**Fan Power Output**  Fan power output is the product of fan capacity and fan total pressure and compressibility factor

$$H_o = \dot{Q}p_tK_p/6,356$$

**Fan Power Input**  Fan power input is the power required to drive the fan and any elements in the drive train which are considered a part of the fan. Power input can be calculated from appropriate measurements for a dynamometer, torque meter, or calibrated motor.

**Fan Total Efficiency**  Fan total efficiency is the ratio of the fan power output to the fan power input

$$\eta_t = \dot{Q}p_tK_p/6,356H$$

**Fan Static Efficiency**  Fan static efficiency is the fan total efficiency multiplied by the ratio of fan static pressure to fan total pressure

$$\eta_s = \eta_t p_s/p_t$$

**Fan Sound Power Level**  Fan sound power level is 10 times the logarithm (base 10) of the ratio of the actual sound power in watts to $10^{-12}$ watts,

$$L_w = 10 \log\,(W/10^{-12})$$

Fan sound power level can be calculated from sound pressure-level measurements in a known acoustical environment. The standard laboratory method of testing is to use a calibrated sound source to calibrate a semireverberant room. In-duct test methods are being developed. The total sound power level of a fan is usually assumed to be 3 dB higher than either the inlet or outlet component. The casing component varies with construction but will usually range from 15 to 30 dB less than the total.

**Head**  The difference between head and pressure is important in fan engineering. Both are measures of the energy in the air. Head is energy per unit mass and can be expressed in ft·lb/lbm, which is often abbreviated to ft (of fluid flowing). Pressure is energy per unit volume and can be expressed in ft·lb/ft³, which simplifies to lb/ft² or force per unit area. The use of the inch water gage (in wg) is a convenience in fan engineering reflecting the usual methods of measurement. It sounds like a head measurement but is actually a pressure measurement corresponding to 5.192 lb/ft², the pressure exerted by a column of water 1 inch high. Pressures can be converted into heads and vice versa:

$$p = \rho h/5.192 \qquad h = 5.192p/\rho$$

For instance, for air at 0.075 lbm/ft³ density, 1 in wg corresponds to 69.4 ft of air. That is, a column of air at that density would exert a pressure of 5.192 lb/ft². Lighter air would exert less pressure for a given head. For a given pressure, the head would be higher with lighter air. Although it is not widely used in the United States, head is commonly used in a number of European countries.

Fans, like other turbomachines, can be considered constant-head–constant-capacity (volumetric) machines. This means that a fan will develop the same head at a given capacity regardless of the fluid handled, all other conditions being equal. Of course, this also means that a fan will develop a pressure proportional to the density at a given capacity, all other conditions being equal.

All the preceding equations are based on the units listed under Symbols. If **SI units** are to be used, certain numerical coefficients will have to be modified. For instance, substitute $\sqrt{2}$ for 1,097 in the flow equations when using N/m² for pressure, kg/m³ for density, and m³/s for capacity. Similarly, substitute 1.0 for 6,356 in the power equations when using N/m² for pressure, m³/s for capacity, and W for power.

Common metric practice in fan engineering leads to other numerical coefficients. When mm wg is used for pressure, m³/s for capacity, kW for power, and kg/m³ for density, substitute

4.424 for 1,097 in the flow equations and 102.2 for 6,356 in the power equations.

## FAN AND SYSTEM PERFORMANCE CHARACTERISTICS

The performance characteristics of a fan are best described by a graph. The conventional method of graphing fan performance is to plot a series of curves with capacity as abscissa and all other variables as ordinates. System characteristics can be plotted in a similar manner.

### System Characteristics

Most systems served by a fan have characteristics which can be described by a parabola passing through the origin; i.e., the energy required to produce flow through the system (which can be expressed as pressure or head) varies approximately as the square of the flow. In some cases, the system characteristics will not pass through the origin because the energy required to produce flow through an element of the system may be controlled at a particular value, e.g., with venturi scrubbers. In some cases, the system characteristic will not be parabolic because the flow through an element of the system is laminar rather than turbulent, e.g., in some types of filters. Whatever the case, the system designer should establish the characteristics by determining the energy requirements at various flow rates. The energy requirement (pressure drops or head losses) for each element can be determined by reference to handbook or manufacturer's literature or by test.

The true measure of the energy requirement for a system element is the total pressure drop or the total head loss. Only if the entrance velocity for the element equals the exit velocity will the change in static pressure equal the total pressure drop. There are some advantages in using static pressure change, but the system designer is usually well advised to use total pressure drops to avoid errors in fan selection. The sum of the total pressure losses for elements on the inlet side of the fan will equal $-p_{t1}$. This should include energy losses at the entrance to the system but not the energy to accelerate the air to the velocity at the fan inlet, which is chargeable to the fan. The sum of the total pressure losses for elements on the discharge side of the fan will equal $p_{t2}$. This should include the kinetic energy of the stream issuing from the system. The total system requirement will be the arithmetic sum of all the appropriate losses or the algebraic difference $p_{t2} - p_{t1}$, which is also $p_t$.

A system characteristic curve based on total pressure is plotted in Fig. 2. A characteristic based on static pressure is also shown. The latter recognizes the definition of fan static pressure so that the only difference is the velocity pressure corresponding to the fan outlet velocity. This system could operate at any capacity provided a fan delivered the exact pressure to match the energy requirements shown on the system curve for that capacity. The advantages of plotting the system characteristics on the fan graph will become evident in the following discussions.

### Fan Characteristics

The constant-speed performance characteristics of a fan are illustrated in Fig. 2. These characteristics are for a particular size and type of fan operating at a particular speed and handling air of a particular density. The fan can operate at any capacity from zero to the maximum shown, but when applied on a particular system the fan will operate only at the intersection of the system characteristics with the appropriate fan pressure characteristic. For the case illustrated, the fan will operate at $\dot{Q} = 27,300$ ft³/min and $p_t = 3.4$ in wg or $p_s = 3$ in wg, requiring $H = 18.3$ hp at the speed and density for which the curves were drawn. The static efficiency at this point of operation is 73 percent, and the total efficiency is 80 percent. If the system characteristic had been lower, it would have intersected the fan characteristic at a higher capacity and the fan would have delivered more air. Contrariwise, if the system characteristic had been higher, the capacity of the fan would have been less. Capacity reduction can be accomplished, in fact, by creating additional resistance, as with an outlet damper.

Figure 3 illustrates the characteristics of a fan with damper control, variable inlet vane control, and variable speed control. These particular characteristics are for a backward-curved centrifugal fan, but the general principles apply to all fans. Outlet dampers do not affect the flow to the fan and therefore can alter fan performance only by adding resistance to the system and producing a new intersection. Point 1 is for wide-open dampers. Points 2 and 3 are for progressively closed dampers. Note that some power reduction accompanies the capacity reduction for this particular fan. Variable-inlet vanes produce inlet whirl, which reduces pressure-producing capability. Point 4 is for wide-open inlet vanes and corresponds to point 1. Points 5 and 6 are for progressively closed vanes. Note that the power reduction at reduced capacity is better for vanes than for dampers. Variable speed is the most efficient means of capacity control. Point 7 is for full speed and points 8 and 9 for progressively reduced speed. Note the improved power savings over other methods. Variable speed also has advantages in terms of lower noise and reduced erosion potential but is generally at a disadvantage regarding first cost.

Figure 4 illustrates the characteristics of a fan with variable pitch control. These characteristics are for an axial-flow fan, but comparisons to the centrifugal-fan ratings in Fig. 3 can be made. Point 10 corresponds to points 1, 4, or 7. Reduced ratings are shown at points 11 and 12 obtained by reducing the pitch. Power savings can be almost as good as with speed control. Notice that in all methods of capacity control, operation is at the intersection of a system characteristic with a fan characteristic. With variable vanes, variable speed, and variable pitch, the fan characteristic is modified. With damper control, the effect could be considered a change in fan characteristics, but it generally makes more sense to consider it a change in system characteristics.

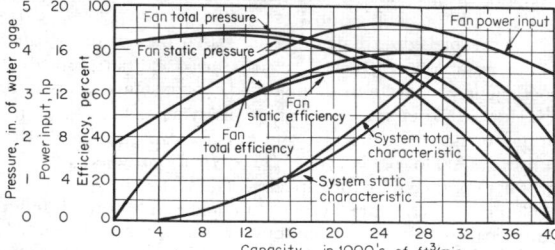

**Fig. 2**  Fan and system characteristics.

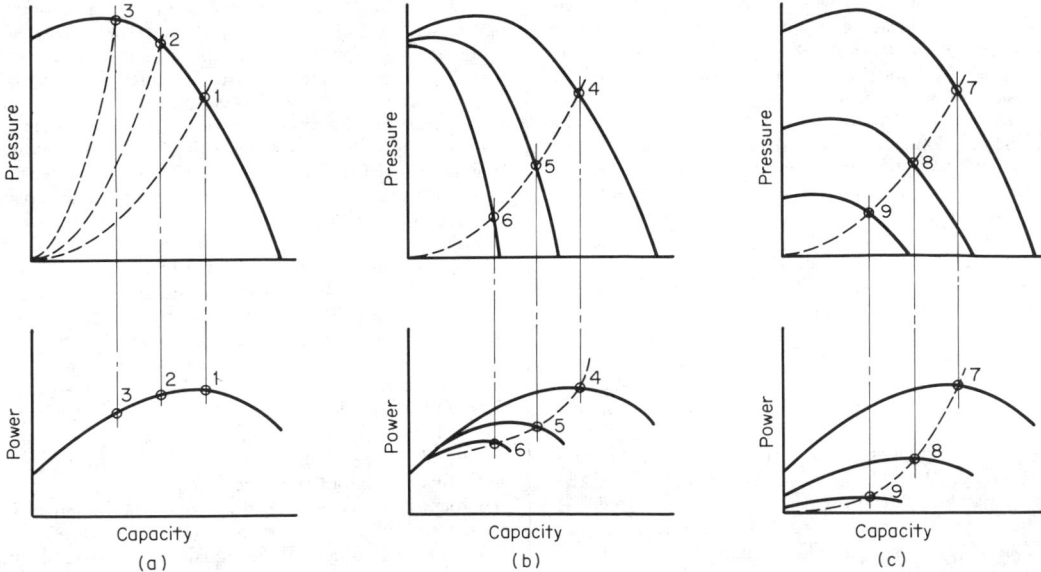

**Fig. 3** Fan characteristics with *(a)* damper control, *(b)* variable-inlet vane control, *(c)* variable speed control.

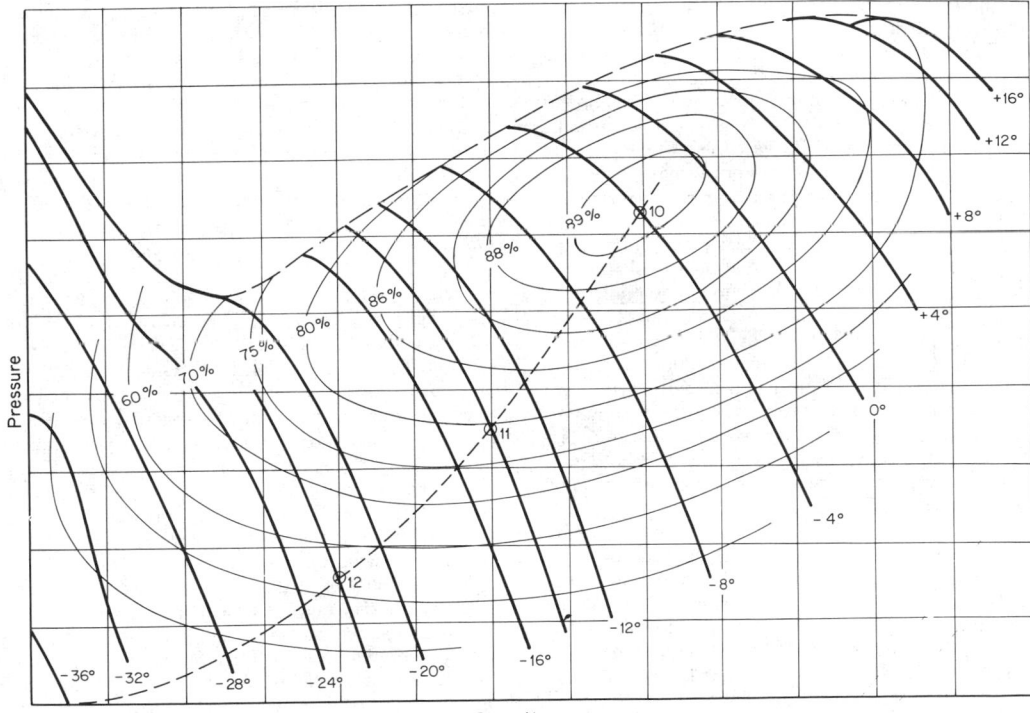

**Fig. 4** Characteristics of an axial-flow fan with variable pitch control.

## Fan-System Matching

It has already been observed in the discussions of both system and fan characteristics that the point of operation for **one fan** on a particular system will be at the intersection of their charac-teristics. Stated another way, the energy required by the system must be provided by the fan exactly. If the fan delivers too much or too little energy, the capacity will be more or less than desired. The effects of utilizing dampers, variable speed, variable-inlet vanes, and variable pitch were illustrated, but in all cases the fan was preselected. The more general case is the one involving the selection or design of a fan to do a particular job. The design of a fan from an aerodynamicist's point of view is beyond the scope of this discussion. Fortunately, most fan problems can be solved by selecting a fan from the many standard lines available commercially. Again, the crux of the matter is to match fan capability with system requirements.

Most fan manufacturers have several standard lines of fans. Each line may consist of various sizes all resembling each other. If the blade angles and proportions are the same, the line is said to be homologous. There is only one fan size in a line of homologous fans that will operate at the maximum efficiency point on a given system. If the fan is too big, operation will be at some point to the left of peak efficiency on a standard characteristic plot. If the fan is too small, operation will be to the right of peak. In either case it will be off peak. A slightly undersized fan is often preferred for reasons of cost and stability.

Selecting and rating a fan from a catalog is a matter of fan-system matching. When it is recognized that many of the catalog sizes may be able to provide the capacity and pressure, selection becomes a matter of trying various sizes and comparing speed, power, cost, etc. The choice is a matter of evaluation.

There are various reasons for using **more than one fan** on a system. Supply and exhaust fans are used in ventilation to avoid excessive pressure build-up in the space being served. Forced and induced-draft fans are used to maintain a specified draft over the fire. Two fans may fit the available space better than one larger fan. Capacity control by various fan combinations may be more economical than other control methods. Multistage arrangements may be necessary when pressure requirements exceed the capabilities of a single-stage fan. Standby fans are frequently required to ensure continuous operation.

When two fans are used, they may be located quite remote from each other or they may be close enough to share shaft and bearings or even casings. Double-width, double-inlet fans are essentially two fans in parallel in a common housing. Multistage blowers are, in effect, two or more fans in series in the same casing. Fans may also be in series but at opposite ends of the system. Parallel-arrangement fans may have almost any amount of their operating resistance in common. At one extreme, the fans may have common inlet and discharge plenums. At the other extreme, the fans may both have considerable individual ductwork of equal or unequal resistance.

Fans in **series** must all handle the same amount of gas by weight measurements, assuming no losses or gains between stages. The combined total pressure will be the sum of the individual fan total pressures. The velocity pressure of the combination can be defined as the pressure corresponding to the velocity through the outlet of the last stage. The static pressure for the combination is the difference between its total and velocity pressures and is therefore not equal to the sum of the individual fan static pressures. The volumetric capacities will differ whenever the inlet densities vary from stage to stage. Compression in one stage will reduce the volume entering the next if there is no reexpansion between the two. As with any fan, the pressure capabilities are also influenced by density.

The combined total pressure-capacity characteristic for two fans in series can be drawn by using the volumetric capacities of the first stage for the abscissa and the sum of the appropriate total pressures for the ordinate. Because of compressibility, the volumetric capacities of the second stage will not equal the volumetric capacities of the first stage. The individual total pressures must be chosen accordingly before they are combined. If the gas can be considered incompressible, the pressures for the two stages may be read at the same capacity. In the area near free delivery, it may be necessary to estimate the negative pressure characteristics of one of the fans in order to combine values at the appropriate capacity.

Fans in **parallel** must all develop sufficient pressure to overcome the losses in any individual ductwork, etc., as well as the losses in the common portions of the system. When such fans have no individual ductwork but discharge into a common plenum, their individual velocity pressures are lost and the fans should be selected to produce the same fan static pressures. If fan velocity pressures are equal, the fan total pressures will be equal in such cases. When the fans do have individual ducts but they are of equal resistance and joined together at equal velocities, the fans should be selected for the same fan total pressures. If fan velocity pressures are equal, fan static pressures will be equal in this case. If the two streams join together at unequal velocities, there will be a transfer of energy from the higher-velocity stream to the lower-velocity stream. The fans serving the lower-velocity branch can be selected for a correspondingly lower total pressure. The other fan must be selected for a correspondingly higher total pressure than if velocities were equal.

The combined pressure-capacity curves for two fans in parallel can be plotted by using the appropriate pressures for ordinates and the sum of the corresponding capacities for abscissa. Such curves are meaningful only when a combined-system curve can be drawn. In the area near shutoff, it may be necessary to estimate the negative capacity characteristics of one of the fans in order to combine values at the appropriate pressure.

Figure 5 illustrates the combined characteristics of two fans with slightly different individual characteristics (A-A and B-B). The combined characteristics are shown for the two fans in series (C-C) and in parallel (D-D). Only total-pressure curves are shown. This is always correct for series arrangements but may introduce slight errors for parallel arrangements. An incompressible gas has been assumed. The questionable areas near shutoff or free delivery have been omitted. Two different system characteristics (E-E and F-F) have been drawn on the chart. With the two fans in series, operation will be at point EC or FC if the fan is on system E or F, respectively. Parallel arrangement will lead to operation at ED or FD. Single-fan operation would be at the point indicated by the intersection of the appropriate fan and system curves provided the effect of an inoperative second fan is negligible. Some sort of bypass is required around an inoperative fan in series whereas an inoperative fan in parallel need only be dampered shut. Parallel operation yields a higher capacity than series operation on system F, but the reverse is true on system E. For the type of fan and system characteristics drawn there is only one possible point of operation for any arrangement.

### Fan Laws

The fan laws are based on the experimentally demonstrable

fact that any two members of a homologous series of fans have performance curves which are homologous. At the same point of rating, i.e., at similarly situated points of operation on their characteristic curves, efficiencies are equal and other variables are interrelated according to the fan laws. If size and speed are

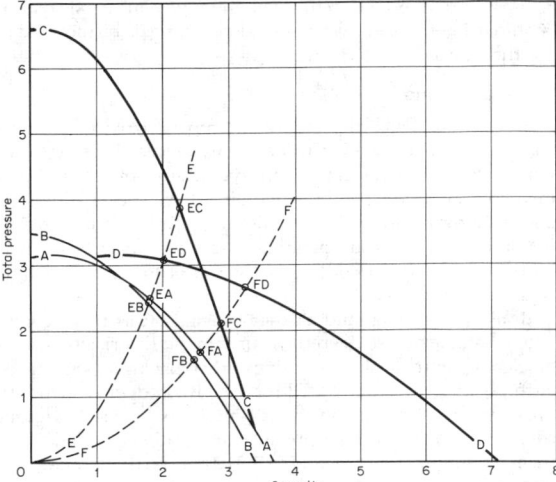

**Fig. 5** Combined characteristics of fans in series and parallel.

considered independent variables, the fan laws can be written as follows:

$$\eta_{tc} = \eta_{tb}$$
$$\dot{Q}_c = \dot{Q}_b (D_c/D_b)^3 (N_c/N_b)$$
$$p_{tc} = p_{tb} (D_c/D_b)^2 (N_c/N_b)^2 (\rho_c/\rho_b)$$
$$H_c = H_b (D_c/D_b)^5 (N_c/N_b)^3 (\rho_c/\rho_b)$$
$$L_{wc} = L_{wb} + 70 \log (D_c/D_b) + 50 \log (N_c/N_h) + 20 \log (\rho_c/\rho_b)$$

The above laws are useful, but they are dangerous if misapplied. The calculated fan must have the same point of rating as the known fan. When in doubt, it is best to reselect the fan rather than attempt to use the fan laws.

The fan designer utilizes the fan laws in various ways. Some of the more useful relationships in addition to those above derive from considering fan capacity and fan total pressure as the independent variables. This leads to specific diameter, specific speed, and specific sound power level

$$D_s = D(p_t/\rho)^{1/4}/\dot{Q}^{1/2}$$
$$N_s = N\dot{Q}^{1/2}/(p_t/\rho)^{3/4}$$
$$L_{ws} = L_w - 10 \log \dot{Q} - 20 \log p_t$$

$D_s$, $N_s$, and $L_{ws}$ are the diameter, speed, and sound power level of a homologous fan which will deliver 1 ft³/min at 1 in wg at the same point of rating as $Q$ and $p_t$ for $D$ and $N$. $D_s$ and $N_s$ can be used to advantage in fan selection. They can also be used by a designer to determine how well a line will fit in with other lines. This is illustrated in Fig. 6. Each segment represents a particular fan line. Note the trends for each kind of fan. Incidentally, some fan engineers utilize different formulas for specific diameter and specific speed

$$D_{se} = Dp_{te}^{1/4}/\dot{Q}^{1/2} \qquad N_{se} = N\dot{Q}^{1/2}/p_{te}^{3/4}$$

where $p_{te}$ is the equivalent total pressure based on standard air. This makes $D_{se} = D_s \times 1.911$ and $N_{se} = N_s \times 6.978$.

Specific sound power level is useful in predicting noise levels as well as in comparing fan designs. There appears to be a lower limit of $L_{ws}$ in the vicinity of 45 dB for the more efficient types ranging to 70 dB or more for cruder designs. Actual sound power levels can be figured from

$$L_w = L_{ws} + 10 \log \dot{Q} + 20 \log p_t$$

Another useful parameter which derives from the fan laws is orifice ratio

$$R_o = \dot{Q}/D^2 (p_t/\rho)^{1/2}$$

This ratio can be plotted on a characteristic curve for a known fan. If the ratio is determined for a calculated homologous fan, the point of rating can be established by inspection. Other ratios can be used in the same manner including $p_v/p_t$, $p_t/Q^2$, $D_s$, and $N_s$.

**Stability Considerations**

The flow through a system and its fan will normally be steady. If the fluctuations occasioned by a temporary disturbance are quickly damped out, the fan system may be described as having a **stable** operating characteristic. If the unsteady flow continues after the disturbance is removed, the operating characteristic is **unstable.**

To ensure stable operation the slopes of the pressure-capacity curves for the fan and system should be of opposite sign. Almost all systems have a positive slope; i.e., the pressure requirement or resistance increases with capacity. Therefore, for stable operation the fan curve should have a negative slope. Such is the case at or above the design capacity.

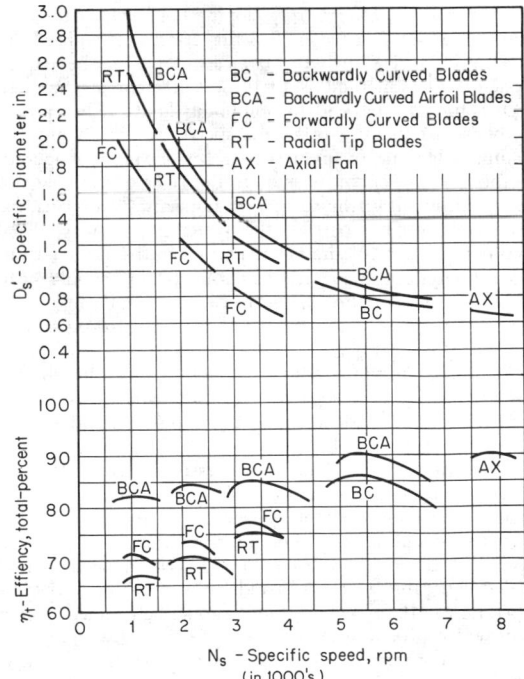

**Fig. 6** Specific diameter and efficiency vs. specific speed for single-inlet fan types.

When the slopes of the fan and system characteristics are of opposite sign, any system disturbance tending to produce a temporary decrease in flow is nullified by the increase in fan pressure. When the slopes are of the same sign, any tendency to decrease flow is strengthened by the resulting decrease in fan pressure. When fan and system curves coincide over a range of capacities, the operating characteristics are extremely unstable. Even if the curves exactly coincide at only one point, the flow may vary over a considerable range.

There may or may not be any obvious indication of unstable operation. The pressure and power fluctuations that accompany unsteady flow may be so small and rapid that they cannot be detected by any but the most sensitive instruments. Less rapid fluctuations may be detected on the ordinary instruments used in fan testing. The changes in noise which occur with each change in flow rate are easily detected by ear as individual beats if the beat frequency is below about 10 Hz. In any event, the overall noise level will be higher with unsteady flow than with steady flow.

The conditions which accompany unsteady flow are variously described as pulsations, hunting, surging, or pumping. Since these conditions occur only when the operating point is to the left of maximum pressure on the fan curve, this peak is frequently referred to as the **surge point** or **pumping limit.**

Pulsation can be prevented by rating the fan to the right of the surge point. Fans are usually selected on this basis, but it is sometimes necessary to control the volume delivered to the value below that at the surge point. This may lead to pulsation, particularly if the fan pressure exceeds 10 in wg.

If the required capacity is less than that at the pumping limit, pulsation can be prevented in various ways, all of which in effect provide a negatively sloping fan curve at the actual operating point. To accomplish this effect the required pressure must be less than the fan capabilities at the required capacity. One method is to bleed sufficient air for actual operation to be beyond the pumping limit. Other possible methods are the use of pitch, speed, or vane control for volume reduction. In any of these cases, the point of operation on the new fan curve must be to the right of the new surge point. Although in the section on Capacity Control dampers were considered a part of the system, they may also be considered a part of the fan if located in the right position. Accordingly, pulsations may be eliminated in a supply system if the damper is on the inlet of the blower. Similarly, dampering at the outlet of the exhauster may control pulsation in exhaust systems.

Another condition frequently referred to as instability is associated with flow separation in the blade passages of an impeller and is evidenced by slight discontinuities in the performance curve. There may be a small range of capacities at which two distinctly different pressures may be developed depending on which of the two flow patterns exists. Such a condition usually occurs at capacities just to the left of peak efficiency. Still another condition of unsteady flow may develop at extremely low capacities. This is known as **blowback** or **puffing** because air puffs in and out of a portion of the inlet. Operation in the blowback range should be avoided, particularly with high-energy fans.

A different type of unsteady flow may occur when two or more fans are used in parallel. If the individual fan characteristics exhibit a dip in pressure between shutoff and design, the combined characteristic will contain points where the point of operation of the individual fans may be widely separated even for identical fans. If the system characteristic intersects the combined-fan characteristic at such a point, the individual fans may suddenly exchange loads. That is, the fan operating at high capacity may become the one operating at low capacity and vice versa. This can produce undesirable shocks on motors and ducts. Careful matching of fan to system is required for either forward-curved centrifugals or most axials for this reason.

### Fan Applications

The selection of a particular size and type of fan for a particular application involves considerations of aerodynamic, economic, and functional suitability. Many of the factors involved in aerodynamic suitability have been discussed above. Determination of economic suitability requires an evaluation of first cost and operating costs. The functional suitability of various types of fans with respect to certain applications is discussed below.

**Heating, ventilating,** and **air-conditioning systems** may require supply and exhaust or return air fans. Historically, high-efficiency centrifugal fans, using either backward-curved or airfoil blades, have been used for supply on duct systems. In low-pressure applications these types can be used without sound treatment. In high-pressure applications, sound treatment is almost always required. Both centrifugal and axial fans have been used. Axials have long been used for shipboard ventilation because they generally can be made smaller than centrifugals. Both adjustable axials and tubular centrifugals have proved popular on duct systems for exhaust service in building ventilation. Centrifugals with variable inlet vanes and axials with pitch control are being used for supply in variable-air-volume systems. Propeller fans and power roof ventilators are used for either supply or exhaust systems when there is little or no ductwork. Heating, ventilating, and air-conditioning applications are considered clean-air service. Various classes of construction are available in standard lines for different pressure ranges. Both direct and indirect drive are used, the latter being most common.

**Industrial exhaust systems** generally require fans that are less susceptible to the unbalance that may result from dirty-gas applications than the clean-air fans used for heating, ventilating, and air conditioning. Simple, rugged, industrial exhausters are favored for applications up to 200 hp. They have a few radial blades and relatively low efficiency. Most are V-belt-driven. Extra-heavy construction may be required where significant material passes through the fan.

**Process air requirements** can be met with either centrifugal or axial fans; the latter may be used in single or double stage. The higher pressure ratings are usually provided by a single-stage centrifugal fan with radial blades, known as a pressure blower. These units are generally direct connected to the driver. They not only compete with centrifugal compressors but resemble them.

**Large industial process** and **pollution-control systems** involving more than about 200 hp are generally satisfied with a somewhat more sophisticated fan than an industrial exhauster or pressure blower. Centrifugal fans with radial-tip blades are frequently used on the more severe service. For the less severe requirements, backward-curved or airfoil blades may be used. Rugged fixed-pitch axials have also been used. Industrial fans are usually equipped with inlet boxes and independently mounted bearings and are usually direct-driven. Journal bearings are usually preferred. Inlet-box damper control can be used to approximate the power saving available from variable-inlet vane control. Variable-speed hydraulic couplings may be economically justified in some cases. Special methods or special construction may be required to provide protection against corrosion or erosion. These fans tend to take on the name of the application such as sintering fan, scrubber exhaust fan, etc.

**Mechanical draft systems** may utilize any of the fan types described in connection with the above applications. Ventilating fans, industrial exhausters, and pressure blowers have been used for forced draft, induced draft, and primary air service on small steam-generating units. The large generating units are generally equipped with the most efficient fans available consistent with the erosion-corrosion potential of the gas being handled.

Both centrifugals and axials are used for forced and induced draft. Axials have predominated in Europe, and there is a growing trend throughout the rest of the world toward axials. Centrifugals have been used almost exclusively in the United States until very recently.

Forced-draft centrifugals invariably have airfoil blade impellers. Induced-draft centrifugals may have airfoil-blade impellers, but for scrubber exhaust, radial-tip blades are more common. This is due, in part, to the high pressures required for scrubber operation and in part to the erosion-corrosion potential downstream of a scrubber. Gas recirculating fans are usually of the radial-tip design. Forced-draft control may be by variable speed but is more likely to be variable vanes on centrifugal fans. Variable inlet vanes or inlet-box dampers may be used to control induced-draft fans. All large centrifugals are direct-connected and have independent pedestal-mounted bearings. Journal bearings are almost always used.

Forced-draft axials are likely to be of the variable-pitch full-airfoil-section design. Hydraulic systems are almost always used for pitch control, but pneumatic and mechanical systems have been tried. Bearings are usually of the antifriction type. Fixed-pitch axials are usually used for induced-draft duty. Control is by variable-inlet vanes. Either journal or antifriction bearings may be used.

Fans are incorporated in many different kinds of machines. Electronic equipment may require cooling fans to prevent hot spots. Driers use fans to circulate air to carry heat to, and moisture away from, the product. Air-support structures require fans to inflate them and maintain the supporting pressure. Ground-effect machines use fans to provide the lift pressure. Air conditioners and other heat exchangers incorporate fans. Aerodynamic, economic, and functional considerations will dictate the type and size of fan to be used.

# Electrical and Electronics Engineering

BY

**WALTER A. LA PIERRE,** *Consulting Engineer in Private Practice, Westfield, N.J.*
**BYRON M. JONES,** *Manager, Analysis and Test Department, Bucyrys-Erie Company.*

### ELECTRICAL ENGINEERING
#### by Walter A. La Pierre

Magnetic and Electrical Units .......................... 15-2
Conductors and Resistance ............................ 15-5
Electrical Circuits ................................... 15-8
Batteries ........................................... 15-10
Magnetism .......................................... 15-14
Dielectric Circuit ................................... 15-18
Transients .......................................... 15-19
Alternating Currents ................................. 15-21
Electrical Instruments and Measurements ................ 15-24
DC Generators ...................................... 15-30
DC Motors .......................................... 15-32
Synchronous Generators .............................. 15-36
Transformers ....................................... 15-40
AC Motors .......................................... 15-43
Synchronous Converters .............................. 15-49
Rating of Electrical Apparatus ........................ 15-50

Electric Drives ...................................... 15-52
Switchboards ........................................ 15-53
Power Transmission .................................. 15-55
Power Distribution ................................... 15-61
Wiring Calculations .................................. 15-64
Interior Wiring ...................................... 15-66
Resistor Materials ................................... 15-69
Magnets ............................................ 15-73
Automobile Ignition Systems .......................... 15-80
Automobile Lighting and Starting Systems .............. 15-82

### ELECTRONICS
#### by Byron M. Jones

Components .......................................... 15-84
Discrete-Component Circuits .......................... 15-87
Integrated Circuits .................................. 15-92
Industrial Electronics ................................ 15-93
Communications ..................................... 15-94

# ELECTRICAL ENGINEERING

## Walter A. La Pierre

REFERENCES: Knowlton, "Standard Handbook for Electrical Engineers," McGraw-Hill. Pender and Del Mar, "Electrical Engineers' Handbook," Wiley. Dawes, "Course in Electrical Engineering," Vols. I and II, McGraw-Hill. Gray, "Principles and Practice of Electrical Engineering," McGraw-Hill. Laws, "Electrical Measurements," McGraw-Hill. Karapetoff-Dennison, "Experimental Electrical Engineering and Manual for Electrical Testing," Wiley. Langsdorf, "Principles of Direct-current Machines," McGraw-Hill. Hehre and Harness, "Electric Circuits and Machinery," Vols. I and II, Wiley. Timbie-Higbie, "Alternating Current Electricity and Its Application to Industry," Wiley. Lawrence, "Principles of Alternating-current Machinery," McGraw-Hill. Puchstein and Lloyd, "Alternating-current Machinery," Wiley. Lovell, "Generating Stations," McGraw-Hill. Underhill, "Coils and Magnet Wire" and "Magnets," McGraw-Hill. Abbott, "National Electrical Code Handbook," McGraw-Hill. Dyke, "Automobile and Gasoline Engine Encyclopedia," The Goodheart-Wilcox Co., Inc.

## MAGNETIC AND ELECTRICAL UNITS

**Systems of Units** There are three fundamental systems of electrical units of which two, the cgs electrostatic and the cgs electromagnetic, are based dimensionally on the centimetre, gram, and second; the third, the mks, is based on the metre, kilogram, and second. The **cgs electrostatic system** is derived from the force exerted between two unit charges of electricity concentrated at points 1 cm apart in a medium of unit capacitivity, or dielectric constant. The **cgs electromagnetic system** is derived from the force exerted between two unit magnetic poles concentrated at points 1 cm apart in a medium of unit magnetic permeability.

The **mks system,** based dimensionally on the metre, kilogram, and second, was adopted as a standard in 1935 by the International Electrotechnical Commission. There are two mks systems: the **unrationalized,** in which the mmf is equal to $4\pi NI$, where $N$ is the number of turns and $I$ the current in amperes; and the **rationalized** system, in which the mmf is equal to $NI$. In the unrationalized system the permeability of free space $\mu_v$ is equal to $10^{-7}$; in the rationalized system it is equal to $4\pi\mu_v$, or $4\pi10^{-7} = \mu_v'$. Similarly in the unrationalized system the capacitivity of free space $\epsilon_v$ is equal to $8.854 \times 10^{-12}$ and in the rationalized system is equal to $4\pi\epsilon_v = \epsilon_v' = 1.113 \times 10^{-10}$. The practical electrical units such as the V, A, W, F, H are mks units.

When the cgs systems are used, magnetic calculations are usually made in the electromagnetic system and capacitance calculations in the electrostatic system. Table 1 gives the relations between the units of the three systems.

Both the cgs and the mks systems have been replaced by the SI system. Basic SI units are metre, kilogram (mass), second, ampere, kelvin, and cd (luminous intensity).

### Magnetic Units

(See Table 6 for relations of magnetic units)

A **unit cgs magnetic pole** is one which is concentrated at a point and which has such strength that when it is placed at a unit distance 1 cm from an exactly similar pole in a medium of unit permeability, the two poles will repel each other with a unit force (dyne). One unrationalized mks unit pole = $10^8$ cgs unit poles; one rationalized mks unit pole = $10^8/4\pi = 7.958 \times 10^6$ cgs unit poles. Unit mks unit poles, one metre apart in vacuum, repel or attract each other with a force of one **newton** = $10^5$ dynes.

**Magnetic potential difference** $(M, \mathcal{F})$ between two points is measured by the work involved in moving a unit magnet pole between the two points.

**Magnetic field intensity** $(H)$ at a point is defined as the vector quantity which is measured by the force (mechanical) which is exerted on a unit magnetic pole placed at the point, when the point under consideration is in a vacuum.

In mediums whose permeability is unity, the cgs field intensity is given by the number of lines of force per square centimetre taken normal to their direction. The cgs unit of field intensity is the **oersted.**

**Magnetic flux** $(\Phi, \phi)$ is the magnetic flow that exists in any magnetic circuit. The cgs unit is the **maxwell,** the mks unit is the **weber;** 1 Wb = $10^8$ Mx.

**Magnetic flux density** $(B)$ is the ratio of the flux in any cross section to the area of that cross section, the cross section being taken normal to the direction of flux. In the cgs system the unit of flux density is the **gauss** = 1 Mx/cm²; in the two mks systems the unit of flux density is the Wb/m² = $10^4$ G.

The force on a unit cgs magnetic pole in a field intensity of one gauss is one dyne. To produce a field intensity of one gauss requires a mmf of one gilbert per centimetre or 0.896 ampere-turn.

**Magnetomotive force** ( $\mathcal{F}$ and mmf) tends to produce magnetic flux and corresponds to emf in the electric circuit. The cgs unit is the **gilbert** = $0.4\pi(ni)$, where $ni$ is an ampere-turn. The unrationalized mks unit is $4\pi(ni)$, and the rationalized unit is the **ampere-turn** $(ni)$. The respective mmfs acting on a magnetic circuit are $0.4\pi NI$, $4\pi NI$, and $NI$, where $NI$ is the ampere-turns.

**Relative permeability** $(\mu_r)$ is the ratio of the cgs magnetic-flux density to the magnetizing force $(B/H)$. In the unrationalized and rationalized mks systems $\mu_r = B/\mu_v H$, and $\mu_r = B/\mu_v' H$, respectively. Actually, relative permeability is the ratio of the magnetic flux in any element of a medium to the flux that would exist if that element were replaced with air, the mmf acting on the element remaining unchanged. The term **permeability** = $\mu$ is commonly used to denote relative permeability.

**Permeance** ($\mathcal{P}$) of a portion of a magnetic circuit bounded by two equipotential surfaces, and by a third surface at every point of which there is a tangent having the direction of the magnetic induction, is the ratio of the flux through any cross section to the magnetic potential difference between the surfaces when taken within the portion under consideration. The equation for the permeance of the medium as defined above is $\mathcal{P} = \phi/\mathcal{F}$. Permeance is the reciprocal of reluctance.

**Reluctivity** $(\nu)$ of a medium is the reciprocal of its permeabil-

**Table 1. Electrical Units**

| Quantity | Symbol | Equation (mks) | Cgs unit | Mks and SI unit | Ratio of magnitude of SI to cgs unit |
|---|---|---|---|---|---|
| Current......... | $I, i$ | $I = E/R$; $I = E/Z$; $I = Q/t$ | Abamp | Amp | $10^{-1}$ |
| Quantity........ | $Q, q$ | $Q = it$; $Q = CE$ | Abcoulomb | Coulomb | $10^{-1}$ |
| Electromotive force.......... | $E, e$ | $E = IR$; $E = W/Q$ | Abvolt | Volt | $10^8$ |
| Resistance....... | $R, r$ | $R = E/I$; $R = \rho l/A$ | Abohm | Ohm | $10^9$ |
| Resistivity....... | $\rho$ | $\rho = RA/l$ | Abohm-cm | Ohm-cm | $10^{11}$ |
| Conductance..... | $G, g$ | $G = \gamma A/l$ | Abmho | Mho, siemens | $10^{-9}$ |
| Conductivity..... | $\gamma$ | $\gamma = 1/\rho = l/RA$ | Abmho per cm | Mho per cm | $10^{-11}$ |
| Capacitance...... | $C$ | $C = Q/E$ | Abfarad | Farad* | $10^{-9}$ |
| Relative capacitivity (dielectric constant).. | $\epsilon_r$ | Numerical | ........... | ........... | 1 |
| Self-inductance... | $L$ | $L = -N\dfrac{d\phi}{di}$ | Abhenry | Henry | $10^9$ |
| Mutual inductance......... | $M$ | $M = K\sqrt{L_1 L_2}$ | Abhenry | Henry | $10^9$ |
| Energy.......... | $W$ | $W = eit$ | Erg | Joule | $10^7$ |
| | Wh | Wh = W/3600 | ........... | Watthour | $36 \times 10^9$ |
| | kWh | kWh = Wh/1,000 | ........... | Kilowatt-hour | $36 \times 10^{12}$ |
| Apparent power.. | ...... | $P = EI$ | ........... | Volt-amp | $10^7$ |
| Active power..... | $P, p$ | $P = \dfrac{dw}{dt} = ei$; $P = EI \cos\theta$ | Abwatt | Watt | $10^7$ |
| Reactive power... | $jQ$ | $Q = EI \sin\theta$ | Abvar | Var | $10^7$ |
| Power factor..... | pf | $\text{pf} = \dfrac{P}{EI} = \dfrac{P}{\sqrt{P^2 + Q^2}}$ | ........... | ........... | 1 |
| Time constant.... | .... | $L/R$ | Sec | Sec | 1 |
| Frequency....... | $f$ | $f = 1/T$ | Cps, Hz† | Cps, Hz† | 1 |
| Period.......... | $T$ | $T = 1/f$ | Sec | Sec | 1 |
| Angular velocity.. | $\omega$ | $\omega = 2\pi f$ | Radians per sec | Radians per sec | 1 |
| Reactance, inductive......... | $X_L$ | $X_L = 2\pi f L$ | Abohm | Ohm | $10^9$ |
| Reactance, capacitive....... | $X_c$ | $X_c = 1/(2\pi f C)$ | Abohm | Ohm | $10^9$ |
| Impedance....... | $Z$ | $Z = E/I = \sqrt{R^2 + (X_L - X_c)^2}$ | Abohm | Ohm | $10^9$ |
| Conductance..... | $G$ | $G = R/Z^2$ | Abmho | S | $10^{-9}$ |
| Susceptance..... | $B$ | $B = X/Z^2$ | Abmho | S | $10^{-9}$ |
| Admittance...... | $Y$ | $Y = I/E = \sqrt{G^2 + B^2}$ | Abmho | S | $10^{-9}$ |

The unit of force in the mks and SI system is joules per metre $= 10^5$ dynes $= 0.10197$ kgf and is called the *newton*.
The relations of magnetic units are given in Table 6.
*1 F $= 9 \times 10^{11}$ statfarads (cgs electrostatic units).
†1 cps = 1 hertz (Hz). This applies equally to cgs, mks, mksa, and SI units.

ity. In the cgs system it is the reluctance between any two parallel faces of a one-centimetre cube of the medium.

**Reluctance** ($\mathcal{R}$) is the reciprocal of permeance. It is the resistance to magnetic flow. In a homogeneous medium of uniform cross section, cgs reluctance is equal to the length divided by the product of the area and permeability, the length and area being expressed in centimetre units ($\mathcal{R} = L/A\mu_r$). In the mks unrationalized system the reluctance $\mathcal{R} = L/A\mu_r\mu_v$, where the length $L$ and the area $A$ are in metre units, $\mu_r$ is relative permeability, and $\mu_v$ is $10^{-7}$. In the rationalized system, $\mathcal{R} = L/A\mu_r(4\pi\mu_v) = L/A\mu_r\mu_v'$, where $\mu_v' = 4\pi 10^{-7}$.

**Electrical Units**

(See Table 1)

**Current** ($I, i$) The practical unit of current is the **ampere,** which is equal to one-tenth the absolute unit of current and is the current in a conductor having a resistance of one ohm and a difference of potential of one volt between its ends. The cgs absolute unit of current is defined as follows: if one centimetre

of a circuit is bent into an arc of one centimetre radius, the current is one cgs **abampere** if the magnetic field intensity at the center is one oersted, provided the remainder of the circuit produces no magnetic effect at the center of the arc. One international ampere (dc) will deposit 0.001118 g/s of silver from a standard silver solution. The **international ampere** equals 0.999835 absolute ampere. The absolute units, legalized by Congress, went into effect Jan. 1, 1948.

**Quantity** ($Q$) The practical unit of quantity is the **coulomb.** An **international coulomb** is the quantity of electricity which passes any section of an electric circuit in one second, when the current in the circuit is one international ampere. One international coulomb equals 0.999835 absolute coulomb.

**Potential Difference or Electromotive Force** ($E$, $V$, emf) The practical unit of electromotive force is the **volt.** The **international volt** is the voltage which will produce a current of one international ampere through a resistance of one international ohm. One international volt equals 1.000330 absolute volts. The **absolute volt** equals $10^8$ cgs absolute volts. A conductor one centimetre long cutting flux at the rate of one maxwell per square centimetre (one gauss) per second has induced in it one cgs abvolt. A conductor one metre long cutting flux at the rate of one weber per square metre per second has induced in it one mks (practical) volt. Emf tends to cause flow of electricity.

**Resistance** $(R, r)$ The practical unit of resistance is the **ohm** $(\Omega)$ and is that resistance through which the fall of potential is one volt when the current is one ampere. The **international ohm** is defined as the resistance at 0°C of a column of mercury of uniform cross section, having a length of 106.300 cm and a mass of 14.521 g. One international ohm equals 1.000495 **absolute ohms**. The absolute ohm equals $10^9$ cgs abohms.

**Resistivity** $(\rho)$ of a material is the dc resistance between the opposite parallel faces of a portion of the material having unit length and unit cross section. Common portions of the material are 1 cm³ and 1 cir mil·ft.

**Conductance** $(G, g)$ Conductance is the reciprocal of resistance and is expressed in reciprocal ohms or **mhos** (℧), now called siemens (S).

**Conductivity** $(\gamma)$ of a material is the dc conductance between the opposite parallel faces of a portion of the material having unit length and unit cross section.

**Capacitance** $(C)$ is that property of a system of conductors and dielectrics which permits the storage of electricity when potential difference exists between the conductors. Its value is expressed as a ratio of a quantity of electricity to a potential difference. A capacitance value is always positive.

The practical unit of capacitance is the **farad** (F) and is that capacitance the potential of which will be raised one volt by the addition of a charge of one coulomb. One **international farad** equals 0.999505 absolute farad. As the farad is too large a unit for practical purposes, the **microfarad** ($\mu$F) is generally used. For capacitors used for radio purposes the **nanofarad** (nF) or **picofarad** (pF) is a more suitable unit. The magnitude of the microfarad is $9 \times 10^5$ that of the cgs electrostatic unit (statfarad).

**Relative capacitivity,** or **dielectric constant** ($\epsilon_r$), of a dielectric is that property which determines the electrostatic energy stored per unit volume for unit potential gradient. In the electrostatic system of units, the capacitivity of a vacuum is unity, so that the relative capacitivity of a dielectric is the ratio of capacitance with the dielectric to the capacitance with a vacuum.

In the unrationalized mks system the capacitivity $\epsilon_v$ of free space, or of a vacuum, is $1.11279 \times 10^{-10}$; in the rationalized mks system the capacitivity $\epsilon_v'$ of free space is $\epsilon_v/4\pi = 8.854 \times 10^{-12}$.

**Self-inductance** $(L)$ is the property of an electric circuit which determines, for a given rate of change of current in the circuit, the emf induced in the same circuit. Thus $e_1 = -L \, di_1/dt$, where $e_1$ and $i_1$ are in the same circuit and $L$ is the coefficient of self-inductance.

The practical unit of self-inductance is the **henry.** An electric circuit has an inductance of one henry when a rate of change of one ampere per second will induce an emf of one volt. It also follows that in such a circuit one ampere will produce $10^8$ cgs linkages of magnetic lines (product of turns and flux) in the circuit, since a change of $10^8$ linkages per second is required to induce one volt. One henry is equal to $10^9$ cgs absolute units of self-inductance; one international henry is equal to 1.000495 absolute henrys. If the permeability is constant, $L = n\phi 10^{-8}/I$

henry, where $n\phi$ is the cgs linkages (product of turns and maxwells) and $I$ is the amperes; in both mks systems $L = n\phi/I$ henry, where $n\phi$ is the turn-weber linkages, and $I$ is the amperes.

**Mutual inductance** $(M)$ is the common property of two associated electric circuits which determines for a given rate of change of current in one of the circuits, the emf induced in the other. Thus $e_1 = -M di_2/dt$ and $e_2 = -M di_1/dt$, where $e_1$ and $i_1$ are in circuit 1, $e_2$ and $i_2$ are in circuit 2, and $M$ is the mutual inductance.

The unit of mutual inductance is the **henry.** When a change of current of 1 A/s in either of the two separate circuits induces an emf of 1 V in the other circuit, their mutual inductance is 1 H. If $M$ is the mutual inductance of two circuits and $k$ is the coefficient of coupling, i.e., the proportion of flux produced by one circuit which links the other, then $M = k \sqrt{L_1 L_2}$, where $L_1$ and $L_2$ are the respective self-inductances of the two circuits.

**Energy** $(W)$ in a system is measured by the amount of work which the system is capable of doing. The **joule,** or **watt-second,** is the practical unit of electrical energy. One **international joule** equals 1.000165 absolute joules. **Watthours** and **kilowatthours** are commonly used in practice. $1 \text{ J} = 10^7 \text{ ergs} = 0.2389 \text{ gcal}$. $1$ Wh $= 3,600 \text{ J} = 2,655.4 \text{ ft·lb} = 8,605 \text{ gcal} = 3.413 \text{ Btu} = 0.001341 \text{ hp·h}$.

**Power** $(p)$ is the time rate of transferring or transforming energy. The practical unit of power is the **watt.** One **international watt** equals 1.000165 absolute watts. One watt is produced when one ampere flows at an emf of one volt. One watt equals $10^7$ ergs per second. One **kilowatt** equals 1,000 watts.

$1 \text{ watt} = 0.00134 \text{ hp} = 44.25 \text{ ft·lb/min} = 0.2389 \text{ gcal/s} = 0.737 \text{ ft·lb/s} = 0.0569 \text{ Btu/min}$.

**Active power** $(P)$ at the points of entry of a single-phase two-wire circuit or of a polyphase circuit is the time average of the values of the instantaneous power at the points of entry, the average being taken over a complete cycle of the alternating current. The value of active power is given in **watts** when the rms currents are in amperes and the rms potential differences are in volts. For sinusoidal emf and current, $P = EI \cos \theta$, where $E$ and $I$ are the rms values of volts and currents and $\theta$ is the phase difference of $E$ and $I$.

**Reactive power** $(Q)$ at the points of entry of a single-phase two-wire circuit, or for the special case of a sinusoidal current and sinusoidal potential difference of the same frequency, is equal to the product obtained by multiplying the rms value of the current by the rms value of the potential difference and by the sine of the angular phase difference by which the current leads or lags the potential difference. $Q = EI \sin \theta$. The unit of $Q$ is the **var** (volt-ampere-reactive). 1 **kilovar** $= 10^3$ **vars.**

**Apparent power** $(EI)$ at the points of entry of a single-phase two-wire circuit is equal to the product of the rms current in one conductor multiplied by the rms potential difference between the two points of entry. Apparent power $= EI$.

**Power factor** (pf) is the ratio of power to apparent power. Pf $= P/EI = \cos \theta$, where $\theta$ is the phase difference between $E$ and $I$, both assumed to be sinusoidal.

The **reactance** $(X)$ of a portion of a circuit for a sinusoidal current and potential difference of the same frequency is the product of the sine of the angular phase difference between the current and potential difference times the ratio of the rms potential difference to the rms current, there being no source of power in the portion of the circuit under consideration. $X =$

$(E/I) \sin \theta = 2\pi fL$ ohms, where $f$ is the frequency and $L$ the inductance, H; or $X = 1/2\pi fC$ ohms, where $C$ is the capacitance, F.

The **impedance** $(Z)$ of a portion of an electric circuit to a completely specified periodic current and potential difference is the ratio of the rms value of the potential difference between the terminals to the rms value of the current, there being no source of power in the portion under consideration. $Z = E/I \, \Omega$.

**Admittance** $(Y)$ is the reciprocal of impedance. $Y = I/E$ S.

The **susceptance** $(B)$ of a portion of a circuit for a sinusoidal current and potential difference of the same frequency is the product of the sine of the angular phase difference between the current and the potential difference times the ratio of the rms current to the rms potential difference, there being no source of power in the portion of the circuit under consideration. $B = (I/E) \sin \theta$.

## CONDUCTORS AND RESISTANCE

**Resistivity,** or **specific resistance,** is the resistance of a sample of the material having both a length and cross section of unity. The two most common resistivity samples are the centimetre cube and the cir mil·ft. If $l$ is the length of a conductor of uniform cross section $a$, then its resistance is

$$R = \rho l/a \tag{1}$$

where $\rho$ is the resistivity. With a cir mil·ft $\rho$ is the resistance of a cir mil·ft and $a$ is the cross section, cir mils. Since $v = la$ is the volume of a conductor,

$$R = \rho l^2/v = \rho v/a^2 \tag{2}$$

A **circular mil** is a unit of area equal to that of a circle whose diameter is 1 mil (0.001 in). It is the unit of area which is used almost entirely in this country for wires and cables. To obtain the cir mils of a solid cylindrical conductor, square its diameter expressed in mils. For example, the diameter of 000 AWG solid copper wire is 410 mils and its cross section is $(410)^2 = 168,100$ cir mils. The diameter in mils of a solid cylindrical conductor is the square root of its cross section expressed in cir mils.

A **cir mil · ft** is a conductor having a length of 1 ft and a uniform cross section of 1 cir mil. In terms of the copper standard the resistance of a cir mil·ft of copper at 20°C is 10.371 $\Omega$. As a first approximation 10 $\Omega$ may frequently be used.

At 60°C a cir mil·in of copper has a resistance of 1.0 $\Omega$. This is a very convenient unit of resistivity for magnet coils since the resistance is merely the length of copper in inches divided by its cross section in cir mils.

**Temperature Coefficient of Resistance**   The resistance of the pure metals increases with temperature. The resistance at any temperature $t$°C is

$$R = R_0(1 + \alpha t) \tag{3}$$

where $R_0$ is the resistance at 0°C and $\alpha$ is the **temperature coefficient of resistance.** For copper, $\alpha = 0.00427$.

With any initial temperature $t_1$, the resistance at temperature $t$°C is

$$R = R_1[1 + \alpha_1(t - t_1)] \tag{4}$$

where $R_1$ is the resistance at temperature $t_1$ °C and $\alpha_1$ is the temperature coefficient of resistance at temperature $t_1$ [see Eq. (5)].

For any initial temperature $t_1$ the value of $\alpha_1$ is

$$\alpha_1 = 1/(234.5 + t_1) \tag{5}$$

**Inferred Absolute Zero**   Between 100 and 0°C the resistance of copper decreases at a rate which is practically uniform and which if continued would give a resistance of zero at −234.5°C (an easy number to remember). If the resistance at $t_1$°C is $R_1$ and the resistance at $t_2$°C is $R_2$, then

$$R_2/R_1 = (234.5 + t_2)/(234.5 + t_1) \tag{6}$$

EXAMPLE.   The resistance of a copper coil at 25°C is 4.26 $\Omega$. Determine is resistance at 45°C. Using Eq. (4) and $\alpha_1 = 1/(234.5 + 25) = 0.00385$, $R = 4.26 [1 + 0.00385 (45 − 25)] = 4.59 \,\Omega$. Using Eq. (6) $R = 4.26 (234.5 + 45)/(234.5 + 25) = 4.26 \times 1.077 = 4.59 \,\Omega$.

The inferred absolute zero for aluminum is −228.

In Table 2 are given values of $a_1$ for copper and aluminum at several initial temperatures.

The **international copper standard** of annealed copper (density, 8.99 g/cm$^3$ or 0.321 lb/in$^3$) at 100 percent conductivity and 20°C has a **resistivity,** for dimensions of conductors in the stated units, as follows: 0.15328 $\Omega$ (m, g); 875.20 $\Omega$ (mile, lb); 1.7241 $\mu\Omega$ (cm$^2$/cm); 0.67879 $\mu\Omega$ (in$^2$/in); 10.371 $\Omega$ (mil, ft); 0.017241 $\Omega$ (m, mm$^2$).

ASTM specifications for minimum conductivities of copper wire are as follows: soft or annealed, 98.16; medium-hard-drawn, 0.460 to 0.325 in diam, 97.66 percent; 0.324 to 0.040 in diam, 96.60 percent; hard-drawn 0.460 to 0.325 in diam, 97.16 percent; 0.324 to 0.040 in diam, 96.16 percent.

The **international aluminum standard** (density, 2.70 g/cm$^3$ or 0.0976 lb/in$^3$) at 100 percent conductivity and 20°C has a **resistivity** for dimensions of conductors in the units stated, as follows: 0.0764 $\Omega$ (m, g); 436.0 $\Omega$ (mile, lb); 2.828 $\mu\Omega$ (cm$^2$/cm); 1.113 $\mu\Omega$ (in$^2$/in); 17.01 $\Omega$ (mil, ft).

**Materials**   The materials generally used for the transmission and distribution of electrical energy are copper, alumi-

**Table 2. Temperature Coefficients of Resistance**

| Initial temperature, deg C | Increase in resistance per deg C | | Initial temperature, deg C | Increase in resistance per deg C | |
|---|---|---|---|---|---|
| | Copper | Aluminum | | Copper | Aluminum |
| 0 | 0.00427 | 0.00439 | 25 | 0.00385 | 0.00396 |
| 5 | 0.00418 | 0.00429 | 30 | 0.00378 | 0.00388 |
| 10 | 0.00409 | 0.00420 | 40 | 0.00364 | 0.00373 |
| 15 | 0.00401 | 0.00411 | 50 | 0.00352 | 0.00360 |
| 20 | 0.00393 | 0.00403 | | | |

num, and sometimes iron and steel. For resistors and heaters, iron, steel, commercial alloys, and carbon are most used.

**Copper** is the most widely used electrical conductor. It has high conductivity, relatively low cost, good resistance to oxidation, is readily soldered, and has good mechanical characteristics such as tensile strength, toughness, and ductility. Its tensile strength together with its low linear temperature coefficient of expansion are desirable characteristics in its use for overhead transmission lines.

**Aluminum** is used to considerable extent for high-voltage transmission lines, because its weight is one-half that of copper for the same conductance. Moreover, the greater diameter reduces corona loss. As it has 1.4 times the linear temperature coefficient of expansion, changes in sag with temperature are greater. Because of its lower melting point, spans may fail more readily with arc-overs. In aluminum cable steel-reinforced (ACSR), the center strand is a steel cable, which gives added tensile strength. Aluminum is used occasionally for bus bars because of its large heat-dissipating surface for a given conductance. The greater cross section for a given conductance requires a greater volume of insulation for a given voltage. When the ratio of the cost of aluminum to the cost of copper becomes economically favorable, aluminum is often used for insulated wires and cables.

**Steel,** either **galvanized** or **copper-covered** ("copperweld"), is used for high-voltage transmission spans where tensile strength is more important than high conductance. Steel is also used for third rails.

**Copper alloys** and **bronzes** are of increasing importance as electrical conductors. They have lower electrical conductivity but greater tensile strength and are resistant to corrosion. **Hitenso, Calsum bronzes, Signal bronze, Phono-electric,** and **Everdur** are bronzes containing phosphorus, silicon, manganese, or zinc. Their conductivities vary from 20 to 85 percent of 100 percent conductivity copper, and they have tensile strengths up to 130,000 lb/in², about twice that of hard-drawn copper. Such alloys were frequently used for trolley wires. Copper alloys having lower conductivity are usually classified as resistor materials.

In Table 3 are given the electrical properties of some of the pure metals and alloys.

**American Wire Gage (AWG)**   The AWG (formerly Brown & Sharpe gage) is based on a constant ratio between diameters of successive gage numbers. The ratio of any diameter to the next smaller is 1.123, and the corresponding ratio of cross sections is $(1.123)^2 = 1.261$, or $1\frac{1}{4}$ approximately. $(1.123)^6$ is 2.0050, so that diameters differing by 6 gage numbers have a ratio of approximately 2; cross sections differing by 3 gage numbers also have a ratio of approximately 2. The ratio of cross sections differing by 2 numbers is $(1.261)^2 = 1.590$, or 1.6 approximately. The ratio of cross sections differing by 10 numbers is approximately 10. The gage ordinarily extends from no. 40 to 0000 (4/0). Wires larger than 0000 must be stranded, and their cross section is given in cir mils.

The diameter of no. 10 wire is 102.0 mils. As an approximation this may be considered as being 100 mils; the cross section is 10,000 cir mils; the resistance is 1 $\Omega$ per 1,000 ft; and the weight of 1,000 ft is $31.4(10\pi)$ lb. Also the weight of 1,000 ft of no. 2 is 200 lb. These facts give many short cuts in estimating resistances and weights of various gage numbers.

**Table 3. Properties of Metals and Alloys**
(See Table 27 for properties of resistor alloys)

| Metals | Resistivity, 20°C | | Temperature coefficient of resistance at 20°C |
|---|---|---|---|
| | $\mu\Omega$ (cm²/cm) | $\Omega$ (mil: ft) | |
| Aluminum | 2.828 | 17.01 | 0.00403 |
| Antimony | 42.1 | 251.0 | 0.0036 |
| Bismuth | 111.0 | 668.0 | 0.004 |
| Brass | 6.21 | 37.0 | 0.0015 |
| Carbon: amorphous | 3,800–4,100 | ...... | (−) |
| Retort (graphite) | 720–812* | ...... | (−) |
| Copper (drawn) | 1.724 | 10.37 | 0.00393 |
| Gold | 2.44 | 14.7 | 0.0034 |
| Iron: electrolytic | 10.1 | 59.9 | 0.0064 |
| Cast | 75.2–98.8 | 448–588 | |
| Wire | 97.8 | 588 | |
| Lead | 22.0 | 132 | 0.00387 |
| Molybdenum | 5.78 | 34.8 | |
| Monel metal | 43.5 | 262 | 0.0019 |
| Mercury | 96.8 | 576 | 0.00089 |
| Nickel | 8.54 | 50.8 | 0.0041 |
| Platinum | 10.72 | 63.8 | 0.003 |
| Platinum silver, 2Ag + 1Pt | 24.6† | 148.0 | 0.00031 |
| Silver | 1.628 | 9.8 | 0.0038 |
| Steel: soft | 15.9 | 95.8 | 0.0016 |
| Glass hard | 45.7 | 275 | |
| Silicon (4 percent) | 51.18 | 308 | |
| Transformer | 11.09 | 66.7 | |
| Trolley wire | 12.7 | 76.4 | |
| Tin | 11.63 | 70 | 0.0042 |
| Tungsten | 5.51 | 33.2 | 0.005 |
| Zinc | 5.97 | 35.58 | 0.0037 |

Max working temperature: Cu, 260°C; Ni, 600°C; Pt, 1500°C.
*Furnace electrodes, 3000°C.
†0°C.

**Table 4. Working Table, Standard Annealed Copper Wire, Solid**
[American Wire Gage (B & S)]

| Gage no. | Diam, mils | Cross section | | Ω per 1,000 ft | | Ω/mi at 25°C (=77°F) | Weight per 1,000 ft, lb |
| --- | --- | --- | --- | --- | --- | --- | --- |
| | | cir mils | in² | 25°C (=77°F) | 65°C (=149°F) | | |
| 0000 | 460.0 | 212,000 | 0.166 | 0.0500 | 0.0577 | 0.264 | 641.0 |
| 000 | 410.0 | 168,000 | 0.132 | 0.0630 | 0.0727 | 0.333 | 508.0 |
| 00 | 365.0 | 133,000 | 0.105 | 0.0795 | 0.0917 | 0.420 | 403.0 |
| 0 | 325.0 | 106,000 | 0.0829 | 0.100 | 0.116 | 0.528 | 319.0 |
| 1 | 289.0 | 83,700 | 0.0657 | 0.126 | 0.146 | 0.665 | 253.0 |
| 2 | 258.0 | 66,400 | 0.0521 | 0.159 | 0.184 | 0.839 | 201.0 |
| 3 | 229.0 | 52,600 | 0.0413 | 0.201 | 0.232 | 1.061 | 159.0 |
| 4 | 204.0 | 41,700 | 0.0328 | 0.253 | 0.292 | 1.335 | 126.0 |
| 5 | 182.0 | 33,100 | 0.0260 | 0.319 | 0.369 | 1.685 | 100.0 |
| 6 | 162.0 | 26,300 | 0.0206 | 0.403 | 0.465 | 2.13 | 79.5 |
| 7 | 144.0 | 20,800 | 0.0164 | 0.508 | 0.586 | 2.68 | 63.0 |
| 8 | 128.0 | 16,500 | 0.0130 | 0.641 | 0.739 | 3.38 | 50.0 |
| 9 | 114.0 | 13,100 | 0.0103 | 0.808 | 0.932 | 4.27 | 39.6 |
| 10 | 102.0 | 10,400 | 0.00815 | 1.02 | 1.18 | 5.38 | 31.4 |
| 11 | 91.0 | 8,230 | 0.00647 | 1.28 | 1.48 | 6.75 | 24.9 |
| 12 | 81.0 | 6,530 | 0.00513 | 1.62 | 1.87 | 8.55 | 19.8 |
| 13 | 72.0 | 5,180 | 0.00407 | 2.04 | 2.36 | 10.77 | 15.7 |
| 14 | 64.0 | 4,110 | 0.00323 | 2.58 | 2.97 | 13.62 | 12.4 |
| 15 | 57.0 | 3,260 | 0.00256 | 3.25 | 3.75 | 17.16 | 9.86 |
| 16 | 51.0 | 2,580 | 0.00203 | 4.09 | 4.73 | 21.6 | 7.82 |
| 17 | 45.0 | 2,050 | 0.00161 | 5.16 | 5.96 | 27.2 | 6.20 |
| 18 | 40.0 | 1,620 | 0.00128 | 6.51 | 7.51 | 34.4 | 4.92 |
| 19 | 36.0 | 1,290 | 0.00101 | 8.21 | 9.48 | 43.3 | 3.90 |
| 20 | 32.0 | 1,020 | 0.000802 | 10.4 | 11.9 | 54.9 | 3.09 |
| 21 | 28.5 | 810 | 0.000636 | 13.1 | 15.1 | 69.1 | 2.45 |
| 22 | 25.3 | 642 | 0.000505 | 16.5 | 19.0 | 87.1 | 1.94 |
| 23 | 22.6 | 509 | 0.000400 | 20.8 | 24.0 | 109.8 | 1.54 |
| 24 | 20.1 | 404 | 0.000317 | 26.2 | 30.2 | 138.3 | 1.22 |
| 25 | 17.9 | 320 | 0.000252 | 33.0 | 38.1 | 174.1 | 0.970 |
| 26 | 15.9 | 254 | 0.000200 | 41.6 | 48.0 | 220 | 0.769 |
| 27 | 14.2 | 202 | 0.000158 | 52.5 | 60.6 | 277 | 0.610 |
| 28 | 12.6 | 160 | 0.000126 | 66.2 | 76.4 | 350 | 0.484 |
| 29 | 11.3 | 127 | 0.0000995 | 83.4 | 96.3 | 440 | 0.384 |
| 30 | 10.0 | 101 | 0.0000789 | 105 | 121 | 554 | 0.304 |
| 31 | 8.9 | 79.7 | 0.0000626 | 133 | 153 | 702 | 0.241 |
| 32 | 8.0 | 63.2 | 0.0000496 | 167 | 193 | 882 | 0.191 |
| 33 | 7.1 | 50.1 | 0.0000394 | 211 | 243 | 1,114 | 0.152 |
| 34 | 6.3 | 39.8 | 0.0000312 | 266 | 307 | 1,404 | 0.120 |
| 35 | 5.6 | 31.5 | 0.0000248 | 335 | 387 | 1,769 | 0.0954 |
| 36 | 5.0 | 25.0 | 0.0000196 | 423 | 488 | 2,230 | 0.0757 |
| 37 | 4.5 | 19.8 | 0.0000156 | 533 | 616 | 2,810 | 0.0600 |
| 38 | 4.0 | 15.7 | 0.0000123 | 673 | 776 | 3,550 | 0.0476 |
| 39 | 3.5 | 12.5 | 0.0000098 | 848 | 979 | 4,480 | 0.0377 |
| 40 | 3.1 | 9.9 | 0.0000078 | 1,070 | 1,230 | 5,650 | 0.0200 |

**Lay Cables** In order to obtain sufficient flexibility, wires larger than 0000 are stranded, and they are designated by their circular mils. Smaller wires may be stranded also since sizes as small as no. 4 when insulated are usually too stiff for easy handling. Lay cables are made up geometrically as shown in Fig. 1. Six strands will just fit around the single central conductor; the number of strands in each succeeding layer increases by 6. The number of strands that can thus be layed up are 1, 7, 19, 37, 61, 91, 127, etc. In order to obtain sufficient flexibility with large cables, the strands themselves frequently consist of stranded cable.

The **resistance of cables** is readily computed from Eq. **(1)**, using the cir mil ft as the unit of resistivity.

EXAMPLE. Determine the resistance of 3,500 ft of 800,000 cir mil cable at 20°C. *Answer:* ρ (of a cir mil·ft) = 10.37. $R = 10.37 \times 3,500/800,000 = 0.0454\ \Omega$.

ρ = 10 Ω/cir mil·ft is often sufficiently accurate for practical purposes.

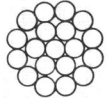

**Fig. 1** Makeup of a 19-strand cable.

## ELECTRICAL CIRCUITS

**Ohm's law** states that, with a steady current, the current in a circuit is **directly** proportional to the **total** emf acting in the circuit and is **inversely** proportional to the total resistance of the circuit. The law may be expressed by the following three equations:

$$I = E/R \qquad (7)$$
$$E = IR \qquad (8)$$
$$R = E/I \qquad (9)$$

where $E$ is the emf, V; $R$ the resistance, $\Omega$; and $I$ the current, A.

**Series Circuits**    The combined resistance of a number of series-connected resistors is the sum of their separate resistances. When batteries or other sources of emf are connected in series, the total emf of the combination is the sum of the separate emfs. The open-circuit emf of a battery is the total generated emf and can be measured at the battery terminals only when no current is being delivered by the battery. The internal resistance is the resistance of the battery alone. The current in a circuit connected in series with a source of emf is $I = E/(R + r)$, where $E$ is the open-circuit emf, $R$ the external resistance, and $r$ the internal resistance of the source of emf.

**Parallel Circuits**    The combined conductance of a number of parallel-connected resistors is the sum of their separate conductances.

$$G = G_1 + G_2 + G_3 + \cdots \qquad (10)$$
$$\frac{1}{R} = \frac{1}{R_1} + \frac{1}{R_2} + \frac{1}{R_3} + \cdots \qquad (11)$$

The equivalent resistance for two parallel resistors having resistances $R_1$, $R_2$ is

$$R = R_1 R_2 / (R_1 + R_2) \qquad (12)$$

The equivalent resistance for three parallel resistors having resistances $R_1$, $R_2$, $R_3$ is

$$R = \frac{R_1 R_2 R_3}{R_1 R_2 + R_2 R_3 + R_3 R_1} \qquad (13)$$

and for four parallel resistors having resistances $R_1$, $R_2$, $R_3$, $R_4$

$$R = \frac{R_1 R_2 R_3 R_4}{R_1 R_2 R_3 + R_2 R_3 R_4 + R_3 R_4 R_1 + R_4 R_1 R_2} \qquad (14)$$

To obtain the resistance of combined series and parallel resistors, the equivalent resistance of each parallel portion is obtained separately and then these equivalent resistances are added to the series resistances according to the principles stated above.

**Kirchhoff's laws** (derived from Ohm's law) make it possible to solve many circuit networks that would otherwise be difficult to solve. The first law states that: *In any branching network of wires the algebraic sum of the currents in all the wires that meet at a point is zero.* The second law states that: *The sum of all the electromotive forces acting around a complete circuit is equal to the sum of the resistances of its separate parts multiplied each by the strength of the current in it, or the total change of potential around any closed circuit is zero.*

In applying Kirchhoff's laws the following rules should be observed. Currents going toward a junction should be pre-

### Table 5. Bare Concentric Lay Cables of Standard Annealed Copper
From *NBS Circ.* 31. See Table 21 for the carrying capacity of wires.

| AWG no. | cir mils | $\Omega$ per 1,000 ft | | Weight per 1,000 ft, lb | Standard concentric standing | | |
|---|---|---|---|---|---|---|---|
| | | 25°C (=77°F) | 65°C (=149°F) | | No. of wires | Diam of wires, mils | Outside diam, mils |
| | 2,000,000 | 0.00539 | 0.00622 | 6,180 | 127 | 125.5 | 1,631 |
| | 1,700,000 | 0.00634 | 0.00732 | 5,250 | 127 | 115.7 | 1,504 |
| | 1,500,000 | 0.00719 | 0.00830 | 4,630 | 91 | 128.4 | 1,412 |
| | 1,200,000 | 0.00899 | 0.0104 | 3,710 | 91 | 114.8 | 1,263 |
| | 1,000,000 | 0.0108 | 0.0124 | 3,090 | 61 | 128.0 | 1,152 |
| | 900,000 | 0.0120 | 0.0138 | 2,780 | 61 | 121.5 | 1,093 |
| | 850,000 | 0.0127 | 0.0146 | 2,620 | 61 | 118.0 | 1,062 |
| | 750,000 | 0.0144 | 0.0166 | 2,320 | 61 | 110.9 | 998 |
| | 650,000 | 0.0166 | 0.0192 | 2,010 | 61 | 103.2 | 929 |
| | 600,000 | 0.0180 | 0.0207 | 1,850 | 61 | 99.2 | 893 |
| | 550,000 | 0.0196 | 0.0226 | 1,700 | 61 | 95.0 | 855 |
| | 500,000 | 0.0216 | 0.0249 | 1,540 | 37 | 116.2 | 814 |
| | 450,000 | 0.0240 | 0.0277 | 1,390 | 37 | 110.3 | 772 |
| | 400,000 | 0.0270 | 0.0311 | 1,240 | 37 | 104.0 | 728 |
| | 350,000 | 0.0308 | 0.0356 | 1,080 | 37 | 97.3 | 681 |
| | 300,000 | 0.0360 | 0.0415 | 926 | 37 | 90.0 | 630 |
| | 250,000 | 0.0431 | 0.0498 | 772 | 37 | 82.2 | 575 |
| 0000 | 212,000 | 0.0509 | 0.0587 | 653 | 19 | 105.5 | 528 |
| 000 | 168,000 | 0.0642 | 0.0741 | 518 | 19 | 94.0 | 470 |
| 00 | 133,000 | 0.0811 | 0.0936 | 411 | 19 | 83.7 | 418 |
| 0 | 106,000 | 0.102 | 0.117 | 326 | 19 | 74.5 | 373 |
| 1 | 83,700 | 0.129 | 0.149 | 258 | 19 | 66.4 | 332 |
| 2 | 66,400 | 0.162 | 0.187 | 205 | 7 | 97.4 | 292 |
| 3 | 52,600 | 0.205 | 0.237 | 163 | 7 | 86.7 | 260 |
| 4 | 41,700 | 0.259 | 0.299 | 129 | 7 | 77.2 | 232 |

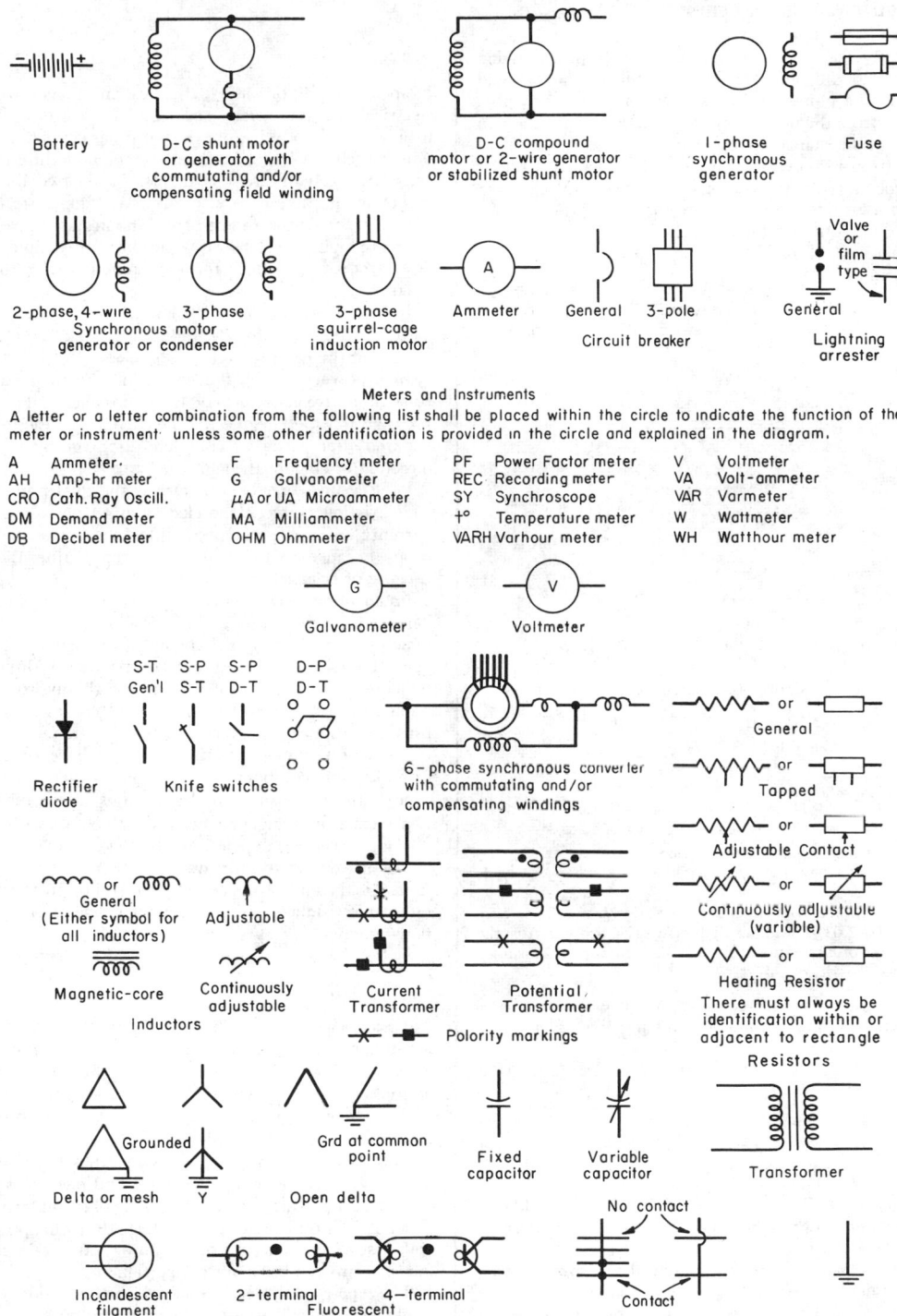

Battery

D-C shunt motor
or generator with
commutating and/or
compensating field winding

D-C compound
motor or 2-wire generator
or stabilized shunt motor

1-phase
synchronous
generator

Fuse

2-phase, 4-wire
Synchronous motor
generator or condenser

3-phase

3-phase
squirrel-cage
induction motor

Ammeter

General   3-pole

Circuit breaker

Valve
or
film
type

General

Lightning
arrester

Meters and Instruments

A letter or a letter combination from the following list shall be placed within the circle to indicate the function of the meter or instrument: unless some other identification is provided in the circle and explained in the diagram.

| A | Ammeter· | F | Frequency meter | PF | Power Factor meter | V | Voltmeter |
|---|---|---|---|---|---|---|---|
| AH | Amp-hr meter | G | Galvanometer | REC | Recording meter | VA | Volt-ammeter |
| CRO | Cath. Ray Oscill. | $\mu$A or UA | Microammeter | SY | Synchroscope | VAR | Varmeter |
| DM | Demand meter | MA | Milliammeter | t° | Temperature meter | W | Wattmeter |
| DB | Decibel meter | OHM | Ohmmeter | VARH | Varhour meter | WH | Watthour meter |

Galvanometer

Voltmeter

S-T    S-P    S-P    D-P
Gen'l  S-T    D-T    D-T

Rectifier
diode

Knife switches

6-phase synchronous converter
with commutating and/or
compensating windings

General

Tapped

Adjustable Contact

Continuously adjustable
(variable)

Heating Resistor
There must always be
identification within or
adjacent to rectangle

Resistors

or

General
(Either symbol for
all inductors)

Magnetic-core

Adjustable

Continuously
adjustable

Inductors

Current
Transformer

Potential
Transformer

✕ ■ Polarity markings

Grounded

Delta or mesh      Y

Grd at common
point

Open delta

Fixed
capacitor

Variable
capacitor

Transformer

Incandescent
filament

2-terminal      4-terminal
Fluorescent

Lamps

No contact

Contact

Crossings

Ground

**Fig. 2** Diagrammatic symbols for electrical machinery and apparatus. (*American Standard, "Graphic Symbols for Electrical Diagrams," ANSI Y 32.2, 1970.*)

ceded by a plus sign. Currents going away from a junction should be preceded by a minus sign. A rise in potential should be preceded by a plus sign. (This occurs in going through a source of emf from the negative to the positive terminal, and in going through resistance in opposition to the direction of current.) A drop in potential should be preceded by a minus sign. (This occurs in going through a source of emf from the positive to the negative terminal and in going through resistance in conjunction with the current.)

The application of Kirchhoff's laws is illustrated by the following example.

EXAMPLE.    Determine the three currents $I_1$, $I_2$, and $I_3$ in the circuit network (Fig. 3). The arrows show the assumed directions of the three currents.

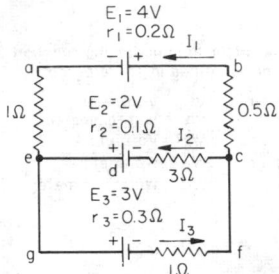

**Fig. 3**  Electric network and Kirchhoff's laws.

Applying Kirchhoff's second law to circuit *abcdea*,

$$+4 + 0.2I_1 + 0.5I_1 - 3I_2 + 2 - 0.1I_2 + I_1 = 0$$

or

$$+6 + 1.7I_1 - 3.1I_2 = 0 \qquad \textbf{(I)}$$

and for *edcfge*,

$$-2 + 0.1I_2 + 3I_2 + I_3 + 3 + 0.3I_3 = 0$$

or

$$+1 + 3.1I_2 + 1.3I_3 = 0 \qquad \textbf{(II)}$$

Applying Kirchhoff's first law to junction *c*,

$$-I_1 - I_2 + I_3 = 0 \qquad \textbf{(III)}$$

Solving **(I)**, **(II)**, and **(III)** simultaneously gives $I_1 = -2.56$, $I_2 = +0.53$, and $I_3 = -2.03$. The minus signs before $I_1$ and $I_2$ show that the actual directions of these two currents are opposite the assumed directions.

**Electrical Power**  With direct currents the electrical power is given by the product of the volts and amperes. That is,

$$P = EI \qquad \text{W} \qquad \textbf{(15)}$$

Also, by substituting for $E$ and $I$ Eqs. **(8)** and **(7)**,

$$P = I^2R \qquad \text{W} \qquad \textbf{(16)}$$
$$P = E^2/R \qquad \text{W} \qquad \textbf{(17)}$$

The watt is too small a unit for many purposes. Hence, the **kilowatt** (kW) is used. 746 watts = 1 hp = 0.746 kW; 1 kW = 1.340 hp. The **kilowatthour (kWh)** is the common engineering unit of electrical energy.

**Joule's Law**  When an electric current flows through resistance, the number of heat units developed is proportional to the square of the current, directly proportional to the resistance, and directly proportional to the time that the current flows. $h = 0.2389i^2rt$, where $h$ = number of gram calories; $i$ = current, A; $r$ = resistance, $\Omega$; and $t$ = time, s. $h$ (in Btu) = $0.0009478i^2rt$.

## BATTERIES

In an **electric cell,** or **battery,** chemical energy is converted into electrical energy. Strictly speaking, the word battery applies to an assembly of cells, but the word has come to mean single units or cells. A battery utilizes the potential difference which exists between different elements. When two different elements are immersed in electrolyte an emf exists tending to send current within the cell from the negative pole, which is the more highly electropositive, to the positive pole. The **poles,** or **electrodes** of a battery form the junction with the external circuit.

If the external circuit is closed, current flows from the battery at the **positive electrode,** or **cathode,** and enters the battery at the **negative electrode,** or **anode.**

In a **primary battery** the chemically reacting parts **require renewal;** in a **secondary battery,** the electrochemical processes **are reversible** to a high degree and the chemically reacting parts are restored after partial or complete discharge by reversing the direction of current through the battery.

**Electromotive force** of a battery is the total potential difference existing between the electrodes on open circuit. When current flows, the potential difference across the terminal drops because of the resistance drop within the cell and because of **polarization.**

**Polarization**  When current flows in a battery, hydrogen is deposited on the cathode. This produces two effects, both of which reduce the terminal voltage of the battery. The hydrogen in contact with the cathode constitutes a hydrogen battery which opposes the emf of the battery; the hydrogen bubbles reduce the contact area of the electrolyte with the cathode, thus increasing the battery resistance. The most satisfactory method of reducing polarization is to have present at the cathode some compound that supplies negative ions to combine with the positive hydrogen ions at the plate. In the Leclanché cell, manganese peroxide in contact with the carbon cathode serves as a depolarizer, its oxygen ion combining with the hydrogen ion to form water.

If $E$ is the emf of the cell, $E_p$ the emf of polarization, $r$ the internal resistance, $V$ the terminal voltage, when current $I$ flows, then

$$V = (E - E_p) - Ir \qquad \textbf{(18)}$$

### Primary Batteries

The **Leclanché cell** has carbon for the positive and zinc for the negative electrode, with sal ammoniac as solution. It is used where only low values of current and intermittent service are desired. In the improved type the cathode is a porous carbon cup in which are packed lumps of manganese dioxide to serve as an insoluble oxidizing agent.

The rather high internal resistance has been reduced by using a zinc cylinder for the anode and placing it about the carbon cup and as near as possible. The emf of this cell is about 1.5 V, but the terminal voltage drops to approximately 1 V in service. This type of cell now is only of importance in that it forms the basis of the dry cell.

The **copper-oxide, zinc, caustic-soda** battery is the most widely used wet primary battery at the present time as it is suited for both open- and closed-circuit work, provided the open-circuit periods are not of too long duration. The cell is made in two forms: one with multiple flat plates and the other with concentric cylindrical plates. The positive electrode is compressed cupric oxide the surface of which is reduced to metallic copper, the oxide serving as a depolarizer. The negative electrode is zinc and the electrolyte a strong caustic-soda solution (NaOH), about 1 part caustic to 4 of water. The surface is covered with mineral oil to minimize evaporation. A typical design is shown in Fig. 4. The jar is of heat-resisting glass to avoid breakage due to the heat evolved when the caustic-soda solution is mixed. The open-circuit emf is about 1 V and the terminal voltage under load is about 0.65 V. These batteries are extensively used in railroad service and have ratings from 250 to 1,000 A·h. On open-circuit work it is recommended that they be discharged for 10-min periods at least once a month.

**Dry Cells**  A dry cell is one in which the electrolyte exists in the form of a jelly, is absorbed in a porous medium, or is otherwise restrained from flowing from its intended position, such a cell being completely portable and the electrolyte nonspillable. There are only two forms of dry cell which are practicable, the common type developed from the Leclanché cell and the Ruben, or RM, cell. The most common type of dry cell consists of a cylindrical zinc container which serves as the negative electrode and is lined with specially prepared paper, or some similar absorbent material, to prevent the mixture of carbon and manganese dioxide, which is tamped tightly around the positive carbon electrode, from coming in contact with the zinc.

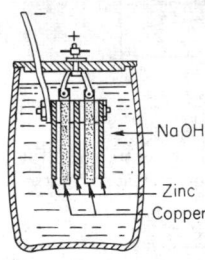

**Fig. 4**  Copper-oxide, zinc, caustic-soda cell.

The absorbent lining and the mixture are moistened with a solution of zinc chloride and sal ammoniac. In smaller cells the manganese-carbon mixture is often molded into a cylinder around the carbon electrode, the whole is then set into the zinc cup, and the space between the molded mixture and the zinc is filled with electrolyte made into a paste in such a manner that it can be solidified by either standing or heating. The top of the cell is closed with a sealing compound, and the cell is placed in a cardboard container. The emf of a dry cell when new is 1.4 to 1.6 V.

In **block assembly** the dry cells, especially in the smaller sizes, are assembled in series and sealed in blocks of insulating compound with only two terminals and, sometimes, intermediate taps brought out. This type of battery is used for radio B and C batteries. Another construction is to build the battery up of layers in somewhat the manner of the old voltaic pile. Each cell consists of a layer of zinc, a layer of treated paper, and a flat cake of the manganese-carbon mixture. The cells are separated by layers of a special material which conducts electricity but which is impervious to electrolyte. A sufficient number of such cells are built up to give the required voltage and the whole battery is sealed into the carton.

Dry cells and batteries fall generally into three classes: (1) the **large-size dry cells,** the no. 6, which are usually approximately 2½ in diam by 6 in high and have a capacity of about 30 A·h; (2) **flashlight batteries,** which are of small size, usually 1¼ in diam by 2½ in high or smaller, with a capacity of about 3 A·h; and (3) **radio B batteries,** which consist usually of 15 or 30 cells permanently connected into a battery which is used chiefly to supply the B-battery current for radio receiving sets.

The **efficiency** of a standard-size dry battery depends on the rate at which it is discharged. Up to a certain rate the lower the discharge rate, the greater the efficiency. Above this rate the efficiency decreases (see *Natl. Bur. Stand. Circ.* 79, p. 39).

When used efficiently, a 6-in dry cell will give over 30 A·h of service. As ordinarily used, however, the dry cell gives no more than 8 to 10 A·h of service and at times even less. The 1¼ by 2¼ in flashlight battery is usually employed with a lamp taking 0.25 to 0.35 A. Under these conditions 3 A·h or thereabouts may be expected if the battery is used for not more than an hour or so a day. The so-called "heavy-duty" radio battery will give about 8 to 10 A·h when efficiently used.

For the best results 6-in dry cells should not be used for current drains of over 0.5 A except for very short periods of time. Flashlight batteries should not be used for higher than the preceding current drain, and heavy-duty radio batteries will give best results if the current drain is kept below 25 mA.

Dry cells should be stored in a cool, dry place. Extreme heat during storage will shorten their life. The cell will not be injured by being frozen but will be as good as new after being brought back to normal temperature. In extreme cold weather dry cells may not give more than half of their normal service. At a temperature of about $-30°F$ they freeze solid and give neither voltage nor current.

The amperage of a dry cell by definition is the current that it will give when it is short-circuited (at about 70°F) through an ammeter which with its leads has a resistance of 0.01 Ω.

The **Ruben cell** (Ruben, Balanced Alkaline Dry Cells, *Trans. Electrochem. Soc.*, **92**, 1947) was developed jointly by the Ruben Laboratories and P. R. Mallory & Company during World War II for the operation of radar equipment and other electronic devices which require a high ratio of ampere-hour capacity to the volume of the cell at higher current densities than were considered practicable for the Leclanché type. The anode is of amalgamated zinc, and the cathode is a mercuric oxide depolarizing material intimately mixed with graphite in order to reduce its electrical resistivity. The electrolyte is a solution of potassium hydroxide (KOH) containing potassium zincate. The cell is made in two forms. In one form, the anode consists of a spirally wound corrugated strip of zinc (Fig. 5a) 0.002 to 0.005 in (0.051 to 0.13 mm) thick (1), which is

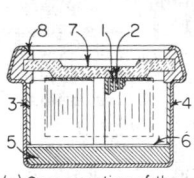

(a) Cross-section of the "roll anode" Ruben cell.

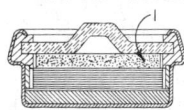

(b) Cross-section of the "pressed powder anode" Ruben cell.

**Fig. 5**  Ruben cell.

amalgamated after assembly. Two strips of alkali-resistant absorbent paper (2) are interwound with the zinc foil so that the zinc protrudes at the upper side and paper at the lower side. The anode is insulated from the steel container (4) by a polystyrene sleeve (3). The cathode depolarizer is shown at (5) separated from the anode by a barrier of alkali-resistant paper (6). The cell top (7) is copper and contacts the zinc strip to form the negative terminal of the cell. The cell is sealed by an insulating grommet of neoprene (8). The cell container (4), which is inert chemically to the cell ingredients, forms the positive electrode.

In the second or "button" type of cell (Fig. 5b) the anode is a pressed powdered-zinc amalgam disk (1). The other elements of the cell are identical with those in (a). The button type of cell has higher volumetric efficiency than the roll type.

The no-load emf of the cell is 1.34 V and remains essentially constant irrespective of time and temperature. Advantages of the cell are long shelf life, which enables them to be stored indefinitely; long service life, about four times that of the Leclanché dry cell of equivalent volume; small weight; a flat voltage characteristic which is advantageous for electronic uses in which the characteristics of tubes vary widely with voltage; adaptability to operating at high temperatures without deterioration; high resistance to shock.

The **Weston cell** is a primary cell used as a standard of emf. It consists of a glass H tube in the bottom of one leg of which is mercury which forms the positive or cathode; in the bottom of the other leg is cadmium amalgam forming the anode. The electrolytes consist of mercurous sulfate and cadmium sulfate. There are two forms of the Weston cell: the saturated or normal cell, and the unsaturated cell. In the normal cell the electrolyte is saturated. This is the official standard since it is more permanent than the unsaturated type and can be reproduced with far greater accuracy. When carefully made, the emfs of cells agree within a few parts per million. There is, however, a small temperature coefficient. Although the unsaturated cell is not so reliable as the normal cell and must be standardized, it has a negligible temperature coefficient and is more convenient for general use. The manufacturers recommend that the temperature be not less than 4°C and not more than 40°C and the current should not exceed 0.0001 A. The emf is between 1.0185 and 1.0190 V. Since no appreciable current can be taken from the cell, a null method must be used to utilize its emf.

### Storage Batteries

In a **storage battery** the electrolytic action must be **reversible** to a high degree. There are three types of storage batteries; the lead-lead-acid type, the nickel-iron-alkaline type (**Edison** battery), and the nickel-cadmium-alkali type (**Nicad**).

In the manufacture of the **lead-lead-acid cells** there are two general types of plates, or electrodes. In the **Planté type** the active material is electrically formed of pure lead by repeated reversals of the charging current. In the **Faure,** or **pasted-plate,** type, the positive and negative plates are formed by applying a paste, largely of lead oxides ($PbO_2$, $Pb_3O_4$), to lead-antimony supporting grids. A current is passed through the plates while they are immersed in weak sulfuric acid, the positive plates being connected as anodes and the negative ones as cathodes. The paste on the positive plates is converted into lead peroxide while that on the negative plate is reduced to spongy lead.

In order to obtain high capacity per unit weight it is necessary to expose a large plate area to the action of the acid. This is done in the Planté plate by "ploughing" with sharp steel disks, and by using corrugated helical inserts as active positive material (Manchester plate). In the pasted plate a large area of the material is necessarily exposed to the action of the acid.

The chemical reactions in a lead cell may be expressed by the following equation, based on the double sulfation theory:

$$\underset{\substack{\text{positive} \\ \text{plate}}}{PbO_2} + \underset{\substack{\text{negative} \\ \text{plate}}}{Pb} + \underset{\substack{\text{sulfuric} \\ \text{acid}}}{2H_2SO_4} \xrightleftharpoons[\text{discharge}]{\text{charge}} \underset{\substack{\text{positive and} \\ \text{negative plates}}}{2PbSO_4} + \underset{\text{water}}{2H_2O}$$

Between the extremes of complete charge and discharge, complex combinations of lead and sulfate are formed. After complete discharge a hard insoluble sulfate forms slowly on the plates, and this is reducible only by slow charging. This sulfation is objectionable and should be avoided.

**Specific Gravity** Water is formed with discharge and sulfuric acid is formed on charge, consequently the specific gravity must decrease on discharge and increase on charge. The variation of the specific gravity for a stationary battery is shown in Fig. 6. With starting and vehicle batteries it is necessary to operate the electrolyte from between 1.280 to 1.300 when fully charged to as low as 1.100 when completely discharged. The condition of charge of a battery can be determined by its specific gravity.

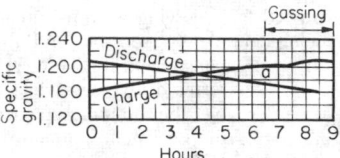

**Fig. 6** Variations of specific gravity in a stationary battery.

Battery electrolyte may be made from concentrated sulfuric acid (oil of vitriol, sp gr 1.84) by **pouring the acid into the water** in the following proportions:

**Parts Water to 1 Part Acid**

| Specific gravity | 1.200 | 1.210 | 1.240 | 1.280 |
|---|---|---|---|---|
| Volume | 4.3 | 4.0 | 3.4 | 2.75 |
| Weight | 2.4 | 2.2 | 1.9 | 1.5 |

**Freezing Temperatures of Sulfuric Acid**

| Specific gravity | 1.180 | 1.200 | 1.240 | 1.280 |
|---|---|---|---|---|
| Freezing temp, °F | −6 | −16 | −51 | −90 |

**Voltage** The emf of a lead cell when fully charged and idle is 2.05 to 2.10 V. Discharge lowers the voltage in proportion to the current. When charging at constant current and normal rate, the terminal voltage gradually increases from 2.14 to 2.3 V, then increases rapidly to between 2.5 and 2.6 V (Fig. 7). This latter interval is known as the **gassing period.** When this period is reached, the charging rate should be reduced in order to avoid waste of power and unnecessary erosion of the plates.

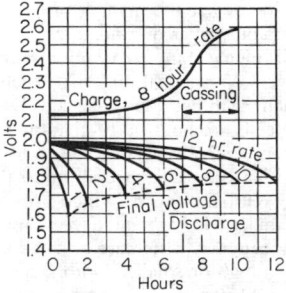

**Fig. 7** Voltage curves on charge and discharge for lead cell.

Practically all batteries have a **normal rating** based on the 8-h rate of discharge. Thus a 320 A·h battery would have a normal rate of 40 A. The ampere-hour capacity of batteries falls off rapidly with increase in discharge rate.

**Effect of Discharge Rate on Battery Capacity**

| Discharge rate, h | 8 | 5 | 3 | 1 | $\frac{1}{3}$ | $\frac{1}{10}$ |
|---|---|---|---|---|---|---|
| Percentage of rated capacity, | | | | | | |
| Planté type | 100 | 88 | 75 | 55.8 | 37 | 19.5 |
| Pasted type | 100 | 93 | 83 | 63 | 41 | 25.5 |

The following rule may be observed in **charging a lead battery**. The charging rate in amperes should be less than the number of ampere-hours out of the battery. For example, if 200 A·h are out of a battery, a charging rate of 200 A may be used until the ampere-hours out of the battery are reduced appreciably.

There are two common methods of charging: the **constant-current method** and the **constant-potential method**. Figure 8a shows a common method of charging with constant current, provided a low-voltage dc power supply is available. The resistor connected in series may be adjusted to give the required current. Several batteries may be connected in series. Figure 8b shows a more common method, using a copper oxide or silicon rectifier, since ac power supply is more common than dc. The rectifier disks, mounted in a stack, are bridge-connected, the directions of rectification being indicated. The polarity of the two wires can readily be determined by means of a dc voltmeter.

The constant-potential method is to be preferred since the rate automatically tapers off as the cell approaches the charged condition. Without resistance the terminal voltage should be 2.3 V per cell, but it is preferable to use 2.4 to 2.5 V per cell with low resistance in series.

When a battery is being charged, its terminal voltage

$$V = E + Ir \qquad (19)$$

Compare with Eq. **(18)**.

When a battery is fully charged, any rate will produce gassing, but the rate may be reduced to such a low value that gassing is practically harmless. This is called the **finishing rate**.

**Portable batteries** for automobile starting and lighting, airplanes, industrial trucks, electric locomotives, train lighting, and power boats employ the pasted-type plates because of their high discharge rates for a given weight and size. The separators are either of treated grooved wood; perforated hard rubber; glass-wool mats; perforated rubber, and grooved wood; ribbed microporous rubber. In low-priced short-lived batteries for automobiles, grooved wood alone is used; in the better types, the wood is reinforced with perforated hard rubber. Containers for the low-priced short-lived automobile-type starting batteries are of asphaltic compound; for other portable types they are usually of hard rubber.

The **Exide ironclad** battery is a portable type designed for propelling electric vehicles. The positive plate consists of a lead-antimony frame supporting perforated hard-rubber tubes. An irregular lead-antimony core runs down the center of each tube, and the lead peroxide paste is packed into these tubes so that shedding of active material from the positive plate cannot occur. Pasted negative plates are used. The separators are flat microporous rubber.

**Stationary Batteries**  The tanks of stationary batteries are made of hard rubber, a size of $25\frac{3}{16}$ by $18\frac{9}{16}$ and $52\frac{1}{8}$ in already having been attained. (Formerly glass tanks and lead-lined wooden tanks were used, but they have been superseded by hard rubber.) When the battery is used for regulating or cycling duty, the positive plates may be of the Planté type because of their long life. However, in most modern installations thick pasted plates are used. Because of the tight fit of the plate assembly within the container and the resulting pressure of the separator against the plate surfaces, shedding of active material is reduced to a minimum and long life is obtained. Pasted negative plates are used in almost all batteries.

With Planté plates, wood veneer, frequently in combination with wooden or hard-rubber dowels, is used for separation. With pasted plates, glass-wool mats are used since their pressure against the positive plates minimizes shedding of the active material as stated above.

A lead storage battery **removed from service** for less than 9 months should be charged once a month if possible; if not, it should be given a heavy overcharge before discontinuing service. If removed for a longer period, siphon off acid (which may be used again) and fill with fresh water. Allow to stand 15 h and siphon off water. Remove and throw away the wood separators. The battery will now stand indefinitely. To put in service again, install new separators, fill with acid (sp gr 1.210) and charge at normal rate 35 h or until gravity has ceased to rise over a period of 5 h. Charge at a low rate a few hours longer.

The **ampere-hour efficiency** of lead batteries is 85 to 90 percent. The **watthour efficiency** obtained from full charge to discharge at the normal rate and at rated amp-hour is 75 to 80 percent. Batteries which do regulating duty only may have a much higher watthour efficiency.

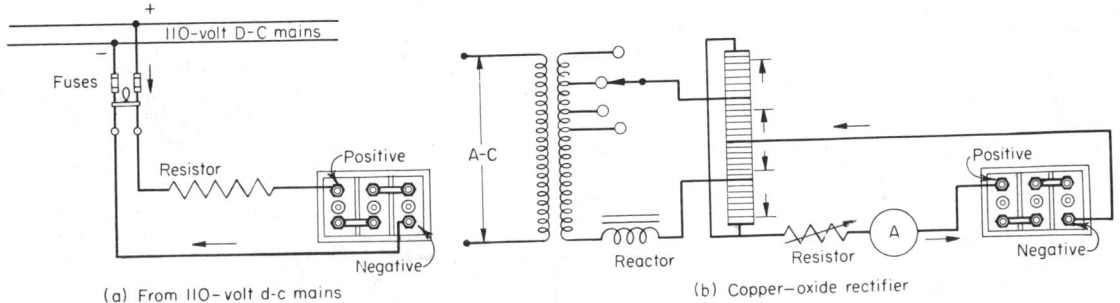

(a) From 110-volt d-c mains

(b) Copper-oxide rectifier

**Fig. 8**  Connections for charging storage battery.

The **Edison storage cell** when fully charged has a positive plate of nickel pencils filled with a higher nickel oxide and a negative plate of flat nickel-plated-steel stampings containing metallic iron in finely divided form. The active material for the positive plate is nickel hydrate and for the negative plate, iron oxide. The electrolyte is a 21 percent solution of potassium hydrate with lithium hydroxides. The initial emf is about 1.4 V and the average emf about 1.1 V throughout discharge. In Fig. 9 are shown typical voltage characteristics on charge and discharge for an Edison cell. On account of the higher internal resistance of the cell the battery is not so efficient from the energy standpoint as the lead cell. The jar is welded nickel-plated steel. The battery is compact and extremely light and strong and for these reasons is particularly adapted for propelling electric vehicles and for boat- and train-lighting systems. The battery is rugged, and since there is no opportunity for the growth of active material on the plates or flaking of active material, the battery has long life.

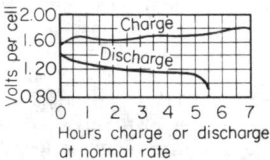

**Fig. 9**  Voltage during charge and discharge of Edison cell.

**Nickel-Cadmium-Alkali (Nicad) Battery**  Although the nickel-cadmium-alkali battery has been used in Europe for years, it has only recently been introduced into the United States. The positive active material is nickelic (black) hydroxide mixed with graphite to give it high conductivity. The negative active material is cadmium oxide. Both materials are used in powdered form and are contained within flat perforated pockets. These pockets are locked into steel plates, the positive and negative being alike in construction. All steel parts are nickel-plated. A complete plate group consists of a number of positive and negative plates assembled on bolts and terminal posts common to plates of the same polarity. The separators are thin strips of polystyrene, and all other battery insulation is also polystyrene. The entire plate assembly is contained within a welded-steel tank. The electrolyte is potassium hydroxide (KOH), specific gravity 1.210 at 72°F (22°C); it does not enter into any chemical reactions with the electrode materials, and its specific gravity remains constant during charge and discharge, neglecting any slight change due to the small amount of gassing. On charge, the voltage is 1.4 to 1.5 V until near the end when it rises to 1.8 V. On discharge, the voltage is nearly constant at 1.2 V.

Nicad batteries are strong mechanically and are not damaged by overcharge; they hold their charge over long periods of idleness, the active material cannot flake off, the internal resistance is low, there is no corrosion, and the battery has an indefinitely long life. It is a general-purpose battery.

In the **Sonotone** nickel-cadmium battery the positive plates are nickel oxide when the battery is charged, and the negative plates are metallic cadmium. On discharge the positive plates are reduced to a state of lower oxidation, and the negative plates regain oxygen. The electrolyte is a 30 percent solution of potassium hydroxide, the specific gravity of which is 1.29 at room temperature. The case is a transparent plastic. The terminal voltage at the normal discharge rate is 1.2 V per cell.

**Rechargeable batteries,** exemplified by Gould Nicad cells (Alkaline Battery Division, Gould National Batteries, Inc.), are hermetically sealed nickel-cadmium cells that contain no free alkaline electrolyte. Since there is no spillage or leakage, they can operate in any position, have long life, and require no maintenance or servicing, and their weight is small for their output. They are thus well adapted to power many types of cordless appliances such as tools, hedge shears, cameras, dictating equipment, electric razors, radios, and television sets. The electrodes consist of a plaque of microporous sintered nickel having an extremely high surface area. The electrochemical reactions differ from those of the conventional vented-type alkaline battery, a type which at the end of a charge liberates both oxygen and hydrogen gases as well as electrolytic fumes that must be vented through a valve in the top of the cell. In the sealed nickel-cadmium cell, the negative electrode (at the time that the cell is sealed) never becomes fully charged, and the evolution of hydrogen is completely suppressed. On charging, when the positive electrode has reached its full capacity, the oxygen which has evolved is channeled through the porous separator to the negative electrode and oxidizes the finely divided cadmium of the microporous plate to cadmium hydroxide, which at the same time is reduced to metallic cadmium. The cells are constructed in three different forms: the button type, the cylindrical type, and the prismatic type. Their ratings range from 20 mA·h to 23 A·h. Their average discharge voltage is 1.22 V, and they require 14 h of charge at the normal rate (one-tenth A·h rating), which for a 3.5 A·h cell is 0.35A.

Precautions in the **care of storage batteries:** An ammeter should not be connected directly across the terminals to test the condition of a cell; a battery should not be left to stand in a discharged condition; a flame should not be brought in the vicinity of a battery that is being charged; the battery should not be allowed to become heated when charging; water should never be added to the concentrated acid—always acid to the water; acid should never be equalized except when the battery is in a charged condition; a battery should never be exposed to the influence of external heat; voltmeter tests should be made when the current is flowing; batteries should always be kept clean. To replace acid lost through slopping, use a solution of 2 parts concentrated sulfuric acid in 5 parts water by weight, unless a hydrometer is at hand to enable the solution to be made up according to the specifications of the makers of the cell.

## MAGNETISM

### Magnetic Circuit

The magnetic circuit is analogous to the electric circuit in that the flux $\Phi$ is proportional to the magnetomotive force $\mathscr{F}$ and inversely proportional to the reluctance $\mathscr{R}$ or magnetic resistance. Thus

$$\Phi = \mathscr{F} / \mathscr{R} \qquad (20)$$

Compare with Eq. (**7**). $\Phi$ is in maxwells, where the maxwell is the cgs unit of flux, $\mathscr{F}$ in gilberts, and $B$ in cgs reluctance units. In the rationalized mks system, $\phi$ is in webers, $\mathscr{F}$ is in ampere-turns, and $\mathscr{R}$ is in mks reluctance units.

$$\mathscr{R} = l/\mu_r\mu_v A \qquad (21)$$

where $\mu_r$ is **relative permeability** (commonly called permeability, $\mu$), a property of the magnetic material, and $\mu_v$ is the permeability of evacuated space = unity in the cgs system; in cgs system $A$ is in square centimetres. In the rationalized mks system $\mu_v = 4\pi \times 10^{-7}$, so that

$$\mathcal{R} = \frac{l}{\mu_r(4\pi \times 10^{-7})A} = \frac{l}{\mu_r(1.257 \times 10^{-6})A} \qquad (22)$$

$l$ is in metres and $A$ in square metres.

The unit of **flux density** in the cgs system is the **gauss**, which is equal to the number of maxwells per square centimetre taken perpendicular to their direction. One gilbert between opposite faces of a centimetre cube of a magnetic medium produces $\mu_r$ gauss. For air, $\mu_r = 1$. In the mks system the unit of flux density is **webers per square metre** $= 10^4 \mathrm{G}$ (see Table 6), which equals 1 T in SI units.

Magnetic-circuit calculations cannot be made with the same degree of accuracy as electric-circuit calculations because of several factors. The cross-sectional dimensions of the magnetic circuit are large relative to its length; magnetic paths are irregular, and their geometry can only be approximated as with the air gap of electric machines, which usually have slots on one or both sides of the gap.

Magnetic flux cannot be confined to definite magnetic paths, but a considerable proportion usually takes paths external to the circuit giving magnetic leakage (see Fig. 13). The relative permeability of iron varies over wide ranges with the flux density and with the previous magnetic condition (see Fig. 11). These variations of relative permeability cannot be expressed by any simple equation. Although the foregoing factors prevent the obtaining of extremely high accuracy in magnetic calculations, yet, with experience, it is possible to design magnetic circuits with a precision that is satisfactory for all practical purposes.

The **magnetomotive force** $\mathcal{F}$ in Eq. (20) is expressed in **gilberts** $= 0.4\pi NI$, where $N$ is the number of turns linked with the circuit and $I$ is the current, A. The unit of reluctance is the reluctance of a 1-cm cube of air. The total reluctance is proportional to the length and inversely proportional to the cross-sectional area of the magnetic circuit, which is analogous to electrical resistance. Hence the reluctance of any given path of uniform cross section $A$ is $l/A\mu$, where $l = $ length of path, cm; $A = $ its cross section, cm², and $\mu = $ permeability. Reluctances in series are added to obtain their combined reluctance. Ohm's law of the magnetic circuit becomes

$$\Phi = \frac{0.4\pi NI}{l_1/A_1\mu_1 + l_2/A_2\mu_2 + l_3/A_3\mu_3 \ldots} \qquad \mathrm{Mx} \quad (23)$$

where $l_1$, $A_1$, $\mu_1$, etc., are the lengths, cross sections, and relative permeabilities of each series part of the circuit.

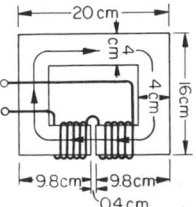

**Fig. 10** Magnetic circuit.

EXAMPLE. In Fig. 10 is shown a magnetic circuit of cast steel with a 0.4-cm air gap. The cross section of the core is 4 cm square. There are 425 turns wound on the core and the current is 10 A. The relative permeability of the steel at the operating flux density is 1,100. Assume that the path of the flux is as shown, the average path at the corners being quarter circles. Neglect fringing at the air gap and any leakage. Determine the flux and the flux density.

Using the cgs system, the length of the magnetic path in the iron $= 12 + 8 + 8 + 5.8 + 5.8 + 4\pi = 52.2$ cm. From Eq. (23),

$$\Phi = \frac{0.4\pi \times 425 \times 10}{[52.2/(16 \times 1,100)] + (0.4/16)} = 191,000 \text{ Mx}$$

$$B = \frac{191,000}{16} = 11,940 \text{ G}$$

Using the rationalized SI system, the length of the iron is 0.522 m, the length of the air gap is 0.004 m, and the cross section of the iron and air gap is 0.0016 m².

$$\Phi = \frac{425 \times 10}{\dfrac{0.522}{1,100 \times 4\pi \times 10^{-7} \times 0.0016} + \dfrac{0.004}{4\pi \times 10^{-7} \times 0.0016}}$$

$$= 0.00191 \text{ Wb}$$

**Magnetization and Permeability Curves** The magnetic permeability of air is a constant and is taken as unity. The relative permeability of iron and other magnetic substances varies with the flux density. In Fig. 11 is shown a magnetization curve for cast steel in which the flux density $B$ in gauss is plotted as a function of the field intensity, or Gb/cm, $H$. Also the relative permeability $\mu_r = B/H$ is plotted as a function of the flux density $B$. Note the wide range over which the relative permeability varies. No satisfactory equation has been found to express the relation between magnetizing force and flux density and between relative permeability and flux density. If an attempt is made to solve Eq. (23) for flux, the factors $\mu_1$, $\mu_2$, etc., are unknown since they are functions of the flux density, which is being determined. The simplest method is

### Table 6. Magnetic Units

| Quantity | Symbol | Equation,* cgs | Cgs unit | SI unit | Ratio of magnitude of SI to cgs unit |
|---|---|---|---|---|---|
| Pole strength | $m$ | $F = mm'/\mu_r l^2$ | Unit pole | $mm'/\mu_r(4\pi)^2 10^{-7}$ | $0.7958 \times 10^7$ |
| Magnetomotive force | $\mathcal{F}$ | $\mathcal{F} = 0.4\pi NI$ | Gb | A-turns, $NI$ | 1.257 |
| Magnetic field intensity | $H$ | $H = \mathcal{F}/l = F/m$ | Oe | A-turns/m | 0.01257 |
| Magnetic flux | $\Phi, \phi$ | $\Phi = \mathcal{F}/\mathcal{R}$ | Mx | Wb | $10^8$ |
| Magnetic flux density | $B$ | $B = \Phi/A$ | G | T, Wb/m² | $10^4$ |
| Relative permeability | $\mu_r$ | $\mu_r = B/H$ | Numeric | Numeric | 1 |
| Reluctivity | $\gamma$ | $\gamma = H/B = 1/\mu_r$ | Numeric | Numeric | 1 |
| Permeance | $\mathcal{P}$ | $\mathcal{P} = \mu_r A/l$ | ........ | ................ | $7.96 \times 10^7$ |
| Reluctance | $\mathcal{R}$ | $\mathcal{R} = l/A\mu_r$ | ........ | ................ | $1.257 \times 10^{-8}$ |

*$l = $ length; $A = $ sectional area in cm²; F $= $ force, dyn; $N = $ number of turns.

one of trial and error, i.e., a value of flux, and the corresponding permeability, is first assumed, the equation solved for the flux, and if the computed flux differs widely from the assumed flux, a second approximation is made, etc. In nearly all magnetic designs either the flux or flux density is the independent

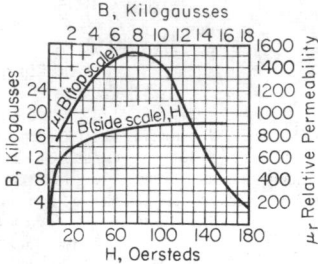

**Fig. 11**  Magnetization and relative-permeability curves for cast steel.

variable, and it is required to find the necessary ampere-turns to produce them. Let the flux $\Phi = BA$ where $B$ is the flux density, G. Then

$$\Phi = BA = 0.4\pi NI/(l/A\mu_r)$$
$$NI = Bl/0.4\pi\mu_r = 0.796Bl/\mu_r \qquad (24)$$

and

Equation (24) shows that the necessary ampere-turns are proportional to the **flux density** and the length of path and are inversely proportional to the relative permeability.

With air and nonmagnetic substances $\mu_r$ [Eq. (24)] becomes unity, and

$$NI = 0.796Bl \qquad (25)$$

in centimeter units. With inch units

$$NI = 0.313B'l' \qquad (26)$$

where $B'$ is the flux density, Mx/in$^2$; and $l'$ the length of the magnetic path, in.

EXAMPLE. The average flux density in the air gap of a generator is 40,000 Mx/in$^2$, and the effective length of the gap is 0.2 in. How many ampere-turns per pole are necessary for the gap?

$$NI = 0.313 \times 40,000 \times 0.2 = 2,500$$

Since the relation of $\mu_r$ to flux density $B$ in Eq. (24) is not simple, the relation of ampere-turns per unit length of magnetic circuit to flux density is ordinarily shown graphically. Typical curves of this character are shown in Fig. 12, inch

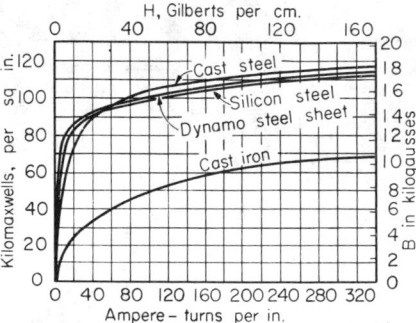

**Fig. 12**  Typical magnetization curves.

units being used although scales of kG and Gb/cm are also given. To find the kMx/cm$^2$ (kG), divide the ordinate scale by 6.45. To determine the number of ampere-turns necessary to produce a given total flux in a magnetic circuit composed of several parts in series having various lengths, cross sections, and relative permeabilities, determine the flux density if the cross section is fixed, or otherwise choose a cross section to give a suitable flux density. From the magnetization curve obtain the ampere-turns necessary to drive this **flux density** through a unit length of the portion of the circuit considered and multiply by the length. Add together the ampere-turns required for each series part of the magnetic circuit to obtain the total ampere-turns necessary to give the assumed flux.

It is desirable to operate magnetic circuits at as high flux densities as is practicable in order to reduce the amount of iron and copper. The air gaps of dynamos are operated at average densities of 40,000 to 50,000 Mx/in$^2$. Higher densities increase the exciting ampere-turns and tooth losses. At 45,000 Mx/in$^2$ the flux density in the teeth may be as high as 120,000 to 130,000 Mx/in$^2$. The flux densities in transformer cores are limited as a rule by the permissible losses. At 60 Hz and with silicon steel the maximum density is 60,000 to 70,000 Mx/in$^2$, at 25 Hz the density may run as high as 75,000 to 90,000 Mx/in$^2$. With laminated cores, the net iron is approximately 0.9 the gross cross section.

**Magnetic Leakage**  It is impossible to confine all magnetic flux to any desired path since there is no known insulator of magnetic flux. Figure 13 shows the magnetic circuit of a

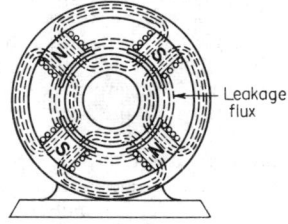

**Fig. 13**  Magnetic circuit of four-pole dynamo with leakage flux.

modern four-pole dynamo. A considerable proportion of the useful magnetic flux leaks between the pole shoes and cores, rather than across the air gap. The ratio of the maximum flux, which exists in the field cores, to the useful flux, i.e., the flux that crosses the air gap, is the **coefficient of leakage**. This coefficient must always be greater than unity and in carefully designed dynamos may be as low as 1.15. It is frequently as high as 1.30. Although the geometry of the leakage-flux paths is not simple, the leakage flux may be determined by approximations with a fair degree of accuracy.

**Magnetic Hysteresis**  The magnetization curves shown in Figs. 11 and 12 are called **normal curves**. They are taken with the magnetizing force continuously increased from zero. If at any point the magnetizing force be decreased, a greater value of flux density for any given magnetizing force will result. The effect of carrying iron through a complete cycle of magnetization, both positive and negative, is shown in Fig. 14.

The curve *OKB*, taken with increasing values of magnetizing force per centimeter $H$, is the **normal induction** curve. If after the magnetizing force has reached the value *OA*, it is decreased, the magnetic flux density $B$ will decrease in accordance with curve *BCD*, between $A$ and $O$ the values being much greater than those given by the normal curve, i.e., the flux density lags the magnetizing force. At zero magnetizing force, the flux density is $OC$, called the **remanence.** A negative magnetizing force $OD$, called the **coercive force,** is required to bring the flux density to zero. If the magnetizing force is

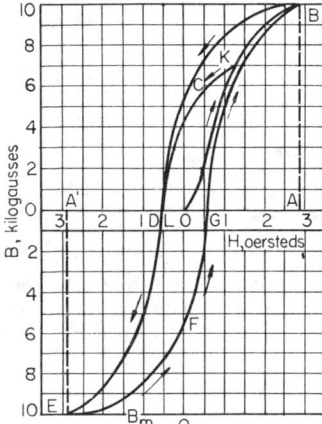

**Fig. 14**  Hysteresis loop for dynamo steel.

increased negatively to $OA'$, the flux density will be given by the curve $DE$. If the magnetizing force is increased positively from $A'$ to $A$, the flux density will be given by the curve $EFGB$, which is similar to the curve $BCDE$. $OF$ is the negative remanence and $OG$ again is the coercive force. The complete curve is called a **hysteresis loop.** When the normal curve reaches the point $K$, if the magnetizing force is then decreased, another hysteresis loop, a portion of which is shown at $KL$, will be obtained. It is seen that the flux density lags the magnetizing force throughout.

The energy dissipated per cycle is proportional to the area of the loop and is equal to $(1/4\pi)\int H\,dB$ ergs/(Hz)(cm³). For moderately high densities the energy loss per cycle varies according to the **Steinmetz law**

$$W = \eta B_m^{1.6} \text{ ergs/cm}^3 \qquad (27)$$

where $B_m$ is the maximum value of the flux density, G (Fig. 14). Table 7 gives values of the Steinmetz coefficient $\eta$, for common magnetic steels.

A permanent increase in the hysteresis constant occurs if the temperature of operation remains for some time above 80°C. This phenomenon is known as **aging** and may be much reduced by proper annealing of the iron. Silicon steels containing about 3 percent silicon have a lower hysteresis loss, somewhat larger eddy-current loss, and are practically nonaging.

**Eddy-current losses,** also known as Foucault-current losses, occur in iron subjected to cyclic magnetization. Eddy-current losses are reduced by laminating the iron, which subdivides the emf and increases greatly the length of path of the parasitic currents. Eddy currents have also a screening effect, which tends to prevent the flux penetrating the iron. Hence laminating also allows the full cross section of the iron to be utilized unless the frequency is too high.

Eddy-current loss in sheets is given by

$$P_e = (\pi t f B_m)^2 /6\rho 10^{16} \qquad \text{W/cm}^3 \qquad (28)$$

where $t$ = thickness, cm; $f$ = frequency, Hz; $B_m$ = the maximum flux density, G; $\rho$ = the resistivity, $\Omega\cdot$cm.

**Relations of Direction of Magnetic Flux to Current Direction**  The direction of the magnetizing force of a current is at right angles to its direction of flow. Magnetic lines about a cylindrical conductor carrying current exist in circular planes concentric with and normal to the conductor. This is illustrated in Fig. 15a. The $\oplus$ sign, corresponding to the feathered end of the arrow, indicates a direction of current away from the observer; a $\odot$ sign, corresponding to the tip of an arrow, indicates a direction of current toward the observer.

**Corkscrew Rule**  The direction of the current and that of the resulting magnetic field are related to each other as the forward travel of a corkscrew and the direction in which it is rotated.

**Hand Rule**  Grasp the conductor in the right hand with the thumb pointing in the direction of the current. The fingers will then point in the direction of the lines of flux.

The applications of these rules are illustrated in Fig. 15. If the currents in parallel conductors are in opposite directions (Fig. 15a), the conductors tend to move apart; if the currents in parallel conductors are in the same direction (Fig. 15b), the conductors tend to come together. The magnetic lines act like stretched rubber bands and, in attempting to contract, tend to pull the two conductors together.

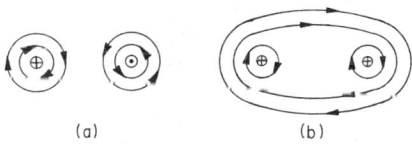

(a)                    (b)

**Fig. 15**  Currents in (a) opposite directions and (b) in the same direction.

The relation of the direction of current in a solenoid helix to the direction of flux is shown in Fig. 16. Figure 17 shows the effect on a uniform field of placing a conductor carrying current in that field and normal to it. In (a) the direction of the current is toward the observer. By applying the corkscrew rule it is seen that the current weakens the field immediately above it and strengthens the field immediately below it. The reverse is true in (b), where the direction of the current is away from the observer.

### Table 7. Steinmetz Coefficient

| | | | |
|---|---|---|---|
| Hard tungsten steel | 0.058 | Annealed cast steel | 0.008 |
| Hard cast steel | 0.025 | Ordinary sheet iron | 0.004 |
| Forged steel | 0.020 | Pure iron | 0.003 |
| Cast iron | 0.013 | Annealed iron sheet | 0.002 |
| Electrolytic iron | 0.009 | Best annealed sheet | 0.001 |
| Soft machine steel | 0.009 | Silicon steel sheet | 0.00046 |
| | | Permalloy | 0.0001 |

Figure 17 is illustrative of the force developed on a conductor carrying current in a magnetic field. In (a) the conductor will tend to move upward owing to the stretching of the magnetic lines beneath it. Similarly, the conductor in (b) will tend to move downward. This principle is the basis of motor action. (See also Magnets.)

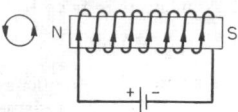

**Fig. 16**   Direction of current and poles in solenoid.

## DIELECTRIC CIRCUIT

**Dynamic and Static Electricity**   Electricity in motion such as an electric current is dynamic electricity; electricity at rest is static electricity. The two are identical physically. Since static electricity is frequently produced at high voltage and small quantity, the two are frequently considered as being two different types of electricity.

### Capacitors

**Capacitors (formerly condensers)**   Two conducting bodies, or electrodes, separated by a dielectric constitute a capacitor. If a positive charge is placed on one electrode of a capacitor, an equal negative charge is induced on the other. The medium between the capacitor plates is called a **dielectric**. The dielectric properties of a medium relate to its ability to conduct **dielectric lines**. This is in distinction to its **insulating** properties which relate to its property to conduct **electric current**. For example, air is an excellent insulator but ruptures dielectrically at low volatge. It is not a good dielectric so far as breakdown strength is concerned.

With capacitors

$$Q = CE \qquad (29)$$
$$C = Q/E \qquad (30)$$
$$E = Q/C \qquad (31)$$

where $Q$ = quantity, C; $C$ = capacitance, F; and $E$ = voltage. The unit of capacitance in the practical system is the **farad**. The farad is too large a unit for practical purposes, so that either the **microfarad** ($\mu$F) or the **picofarad** (pF), are used. However, in voltage, current, and energy relations the capacitance must be expressed in farads.

The energy stored in a capacitor

$$W = \tfrac{1}{2}QE = \tfrac{1}{2}CE^2 = \tfrac{1}{2}Q^2/C \qquad J \qquad (32)$$

**Capacitance of Capacitors**   The capacitance of a **parallel-electrode** capacitor (Fig. 18) is

$$C = \epsilon_r A/(4\pi d \times 9 \times 10^5) \qquad \mu F \qquad (33)$$

where $\epsilon_r$ = relative capacitivity; $A$ = area of one electrode, cm²; and $d$ = distance between electrodes, cm.

**Fig. 17**   Effect of current on uniform magnetic field.

The capacitance of **coaxial cylindrical** capacitors (Fig. 19) is

$$C = 0.2171\epsilon_r l \, / \, [9 \times 10^5 \log (R_2/R_1)] \qquad \mu F \qquad (34)$$

where $\epsilon_r$ is the relative capacitivity and $l$ the length, cm. Also

$$C = 0.03882\epsilon_r/\log (R_2/R_1) \qquad \mu F/mi \qquad (35)$$

Equation (35) is useful in that it is applicable to cables.

**Fig. 18**   Parallel-electrode capacitor.

The capacitance of two **parallel cylindrical conductors** $D$ cm between centers and having radii of $r$ cm is

$$C = 0.01941/\log (D/r) \qquad \mu F/mi \qquad (36)$$

In practice, the capacitance to neutral or to an infinite conducting plane midway between the conductors and perpendicular to their plane is usually used. The capacitance to neutral is

$$C = 0.03882/\log (D/r) \qquad \mu F/mi \qquad (37)$$

Equations (36) and (37) are used for calculating the capacitance of overhead transmission lines. When computing charging current, use voltage between lines in (36) and to neutral in (37).

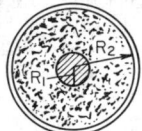

**Fig. 19**   Coaxial-cylinder capacitor.

**Capacitances in Parallel**   The equivalent capacitance of capacitances in parallel (Fig. 20)

$$C = C_1 + C_2 + C_3 \qquad (38)$$

Capacitances in parallel are all across the same voltage. If the voltage is $E$, then the total quantity $Q = CE$ and $Q_1 = C_1E$, etc.

**Capacitances in Series**   The equivalent capacitance $C$ of capacitances in series (Fig. 21) is found as follows:

$$1/C = 1/C_1 + 1/C_2 + 1/C_3 \qquad (39)$$

If the capacitances are not leaky, the charge $Q$ is the same on each. $Q = CE$, $E_1 = Q/C_1$, $E_2 = Q/C_2$, etc.

**Insulators and Dielectrics**   Insulating materials are applied to electric circuits to prevent the leakage of current. Insulating materials used with high voltage must not only have a high resistance to leakage current, but must also be able to resist dielectric puncture; i.e., in addition to being a good insulator, the material must be a good dielectric. Insulation resistance is usually expressed in M$\Omega$ and the resistivity given in M$\Omega$·cm. The dielectric strength is usually given in terms of voltage gradient, common units being V/mil, V/mm, and kV/cm.

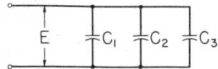

**Fig. 20**   Capacitances in parallel.

Insulation resistance decreases very rapidly with increase in temperature. Absorbed moisture reduces the insulation resistance, and moisture and humidity have a large effect on surface leakage. In Table 8 are given the insulating and dielectric properties of several common insulating materials (see also Sec. 6).

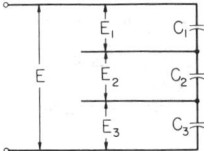

**Fig. 21**  Capacitances in series.

## TRANSIENTS

**Induced EMF**  If a flux $\phi$ maxwells linking $N$ turns of conductor changes, an emf

$$e = -N(d\phi/dt)10^{-8} \quad \text{V} \quad \text{(40)}$$

is induced.

**Self-inductance**  Let a flux $\phi$ link $N$ turns. The linkages of the circuit are $N\phi$ maxwell-turns. If the permeability of the circuit is assumed constant, the number of these linkages per ampere ($\times 10^{-8}$) is the **self-inductance** or **inductance** of the circuit. The unit of inductance is the **henry**. The inductance is

$$L = N\phi/(i \times 10^{8}) \quad \text{H} \quad \text{(41)}$$

If the permeability changes with the current

$$L = N(d\phi/di)10^{-8} \quad \text{H} \quad \text{(42)}$$

The energy stored in the magnetic field

$$W = \tfrac{1}{2}Li^2 \quad \text{J} \quad \text{(43)}$$

**EMF of Self-induction**  If Eq. (41) be written $Li = N\phi10^{-8}$ and differentiated with respect to the time $t$, $L(di/dt) = N(d\phi/dt)10^{-8}$ and from Eq. (40)

$$e = -L(di/dt) \quad \text{V} \quad \text{(44)}$$

$e$ is the emf of self-induction. If a rate of change of current of 1 A/s induces an emf of 1 V, the inductance is then 1 H. If the flux $\phi$ is given in terms of the SI unit, the Wb, the factor $10^{-8}$ should be omitted.

**Current in Inductive Circuit**  If a circuit containing resistance $R$ and inductance $L$ in series is connected across a steady voltage $E$, the voltage $E$ must supply the $iR$ drop in the circuit and at the same time overcome the emf of self-induction. That

### Table 8. Insulating and Dielectric Properties of Insulating Materials

| | Insulation resistance, MΩ·cm | Relative capacitivity or dielectric constant | Rupturing strength | |
|---|---|---|---|---|
| | | | V/mil | kV/cm |
| Asbestos board (ebonized) | $1.0 \times 10^7$ | ......... | 55 | 22 |
| Bakelite | $5$–$30 \times 10^{11}$ | 4.5–5.5 | 450–1,400 | 180–550 |
| Ebonite | $10^9$–$10^{12}$ | 1.9–3.5 | 1,000–2,000 | 390–780 |
| Empire cloth | ............ | 3.5–5.5 | 200–750 | 80–300 |
| Empire paper | ............ | ......... | 1,140 | 450 |
| Fiber | $5 \times 10^3$ | 2.5–5 | 50 | 20 |
| Fuller board | $11 \times 10^9$ | ......... | 120–760 | 47–300 |
| Glass | $17 \times 10^9$ | 5 4–9.9 | 760–3,800 | 300–1,500 |
| Flint | ............ | 6.61–9.90 | | |
| Jena-boron | ............ | 5.5–8.1 | | |
| Gutta percha | $25 \times 10^8$ | 2.9–4.9 | 200–510 | 80–200 |
| Linen, varnished (see Empire cloth) | | | | |
| Marble | $10^2$–$10^5$ | 8.3 | 50–100 | 20–39 |
| Mica: | | | | |
| India | $10^9$ | 7.07–7.90 | 4,000 | 1,580 |
| Canada | $0.44$–$22 \times 10^6$ | 2.9–3.0 | 1,270–3,800 | 500–1,500 |
| South America | $39 \times 10^6$ | 5.9 | 3,800 | 1,500 |
| Mica segment plate | ............ | ......... | 900–1,200 | 350–470 |
| Mica high-heat plate | ............ | ......... | 1,000 | 390 |
| Oils: | | | | |
| Castor | $6.6 \times 10^4$ | 4.7 | 330–480 | 130–190 |
| Cottonseed | $10^4$ | 3.2 | 300–400 | 120–160 |
| Lard | ............ | ......... | 102–355 | 40–140 |
| Linseed | $6 \times 10^5$ | 3.3 | 300–470 | 120–185 |
| Mineral | $21 \times 10^6$ | 2.0–4.7 | 300–400 | 120–160 |
| Paraffin | $1,000 \times 10^{12}$ | 2.41 | 410–550 | 160–215 |
| Paper | ............ | 1.7–2.6 | 110–230 | 43–90 |
| Paper, treated | ............ | 2.5–4.0 | 500–750 | 20–300 |
| Paraffin, solid | $1 \times 10^{10}$ | 1.9–2.5 | 580 | 230 |
| Polyethylene | ............ | 2.3 | 500–700 | 197–276 |
| Polyvinyl chloride | ............ | 6.5–12 | 250–450 | 98.5–177 |
| Porcelain | $3 \times 10^8$ | 5.7–6.8 | 240–300 | 95–120 |
| Rubber (vulcanized) | $10^{14}$–$10^{16}$ | 2.0–3.5 | | |
| Rubber (compounds) | $10^{14}$–$10^{16}$ | 2.5–6 | 300–500 | 120–200 |
| Slate | $10^2$–$10^4$ | 6–7.4 | 5–10 | 2–3.9 |
| Transil oil | ............ | 2.4–2.6 | 300 | 120 |
| Petroleum jelly | ............ | 2.16–2.2 | 230–330 | 90–130 |

is $E = Ri + L \, di/dt$. A solution of this differential equation gives

$$i = (E/R)(1 - \epsilon^{-Rt/L}) \quad \text{A} \qquad (45)$$

where $\epsilon$ is the base of the natural system of logarithms.

Figure 22 shows this equation plotted when $E = 10$ V, $R = 20 \, \Omega$, $L = 0.6$ H. It is to be noted that inductance causes the current to rise slowly to its Ohm's law value, $I_0 = E/R = {}^{10}\!/_{20} = 0.5$ A. When $t = L/R$, the current has reached 63.2 percent of its Ohm's law value. $L/R$ is the **time constant** of the circuit. In the foregoing circuit, the time constant $L/R = 0.6/20 = 0.03$ s. The initial rate of rise of current is $\tan \alpha = E/L$. If current continued at this rate, it would reach $a = E/R$ in $L/R$ s $[(E/L) \times (L/R) = E/R]$.

If a circuit containing inductance and resistance in series is short-circuited when the current is $I_0$, the equation of current becomes

$$i = I_0 \epsilon^{-Rt/L} \quad \text{A} \qquad (46)$$

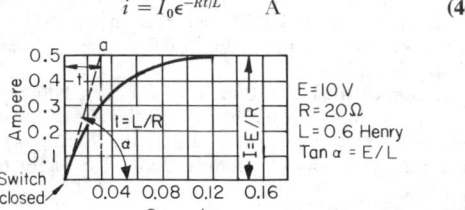

**Fig. 22**   Rise of current in inductive circuit.

Figure 23 shows this equation plotted when $I_0 = 0.5$ A, $R = 20\Omega$, $L = 0.6$ H. It is seen that inductance opposes the decay of current. Inductance always opposes change of current.

**Mutual Inductance**   If two circuits having inductances $L_1$ and $L_2$ henrys are so related to each other geometrically that any portion of the flux produced by the current in one circuit links the other circuit, the two circuits possess **mutual inductance**. It follows that a change of current in one circuit causes an emf to be induced in the other. Let $e_2$ be induced in circuit 2 by a change $di_1/dt$ in circuit 1. Then

$$e_2 = -M \, di_1/dt \quad \text{V} \qquad (47)$$

$M$ is the mutual inductance of the two circuits.

$$M = k \sqrt{L_1 L_2} \qquad (48)$$

where $k$ is the **coefficient of coupling** of the two circuits, or the proportion of the flux in one circuit which links the other. Also a change of current $di_2/dt$ in circuit 2 induces an emf $e_1$ in circuit 1, $e_1 = -M \, di_2/dt$.

The stored energy is

$$W = \tfrac{1}{2}L_1 I_1^2 + \tfrac{1}{2}L_2 I_2^2 + M I_1 I_2 \quad \text{J} \qquad (49)$$

where $I_1$ and $I_2$ are the currents in circuits 1 and 2.

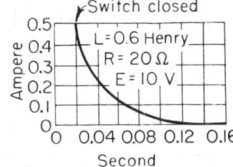

**Fig. 23**   Decay of current in inductive circuit.

**Current in Capacitive Circuit**   If capacitance $C$ farads and resistance $R$ ohms are connected in series across the steady voltage $E$, the current is

$$i = (E/R)\epsilon^{-t/CR} \quad \text{A} \qquad (50)$$

If a capacitor charged to voltage $E$ is discharged through resistance $R$, the current is

$$i = -(E/R)\epsilon^{-t/CR} \quad \text{A} \qquad (51)$$

Except for sign, these two equations are identical and are of the same form as Eq. (46).

In Fig. 24 is shown the transient current to a capacitor in series with a resistor when $E = 200$ V, $C = 4.0 \, \mu\text{F}$, $R = 2$ k$\Omega$. When $t = CR$, the current has reached $1/\epsilon = 0.368$ its initial value. $CR$ is the **time constant** of the circuit. The initial rate of decrease of current is $\tan \alpha = -E/CR^2$. If the current continued at this rate it would reach zero when the time is $CR$ s. If, in its fully charged condition, the capacitor of Fig. 24 is discharged through the resistor $R$, the curve will be the negative of that shown in Fig. 24.

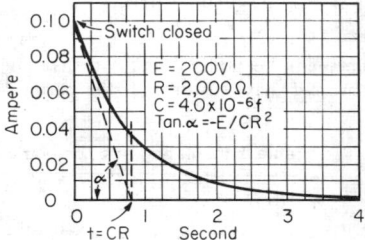

**Fig. 24**   Transient current to capacitor.

**Resistance, Inductance, and Capacitance in Series**   If a circuit having resistance, inductance, and capacitance in series is connected across a source of steady voltage, a transient condition results. If $R > \sqrt{4L/C}$, the circuit is nonoscillatory or overdamped.

The current is

$$i = \frac{EC}{\sqrt{R^2 C^2 - 4LC}} \left( \epsilon^{(-\alpha+\beta)t} - \epsilon^{(-\alpha-\beta)t} \right) \quad \text{A} \qquad (52)$$

where $\alpha = R/2L$ and $\beta = (\sqrt{R^2 C^2 - 4LC})/2LC$.

In Fig. 25 is shown the curve corresponding to Eq. (52). When $R = \sqrt{4L/C}$, the system is **critically damped** and the transient dies out rapidly without oscillation. The current is

$$i = (E/L)t\epsilon^{-Rt/2L} \quad \text{A} \qquad (53)$$

Figure 25 shows also the curve corresponding to Eq. (53).

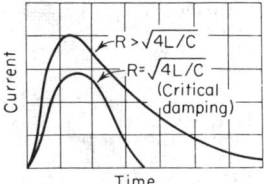

**Fig. 25**   Transient current in nonoscillatory circuits.

If $R < \sqrt{4L/C}$, the transient is oscillatory, being a logarithmically damped sine wave. The current is

$$i = \frac{2EC}{\sqrt{4LC - R^2C^2}} \, \epsilon^{-Rt/2L} \sin \frac{\sqrt{4LC - R^2C^2}}{2LC} t \qquad A \qquad (54)$$

The transient oscillates at a frequency very nearly equal to $1/(2\pi \sqrt{LC})$ Hz. This is the **natural frequency** of the circuit.

In Fig. 26 is shown the curve corresponding to Eq. (54). If the capacitor, after being charged to $E$ V, is discharged into the foregoing series circuits, the currents are given by Eqs. (52) to (54) multiplied by $-1$. Equations (52) to (54) are the same types obtained with dynamic mechanical systems with friction, mass, and elasticity.

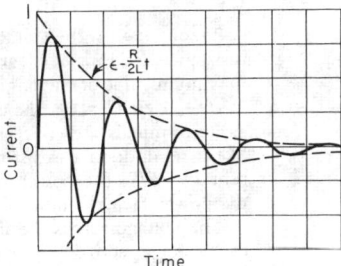

**Fig. 26**  Transient current in oscillatory circuit.

## ALTERNATING CURRENTS

**Sine Waves**  In the following discussion of alternating currents, sine waves of voltage and current will be assumed. That is, $e = E_m \sin \omega t$ and $i = I_m \sin (\omega t - \theta)$, where $E_m$ and $I_m$ are maximum values of voltage and current; $\omega$, the angular velocity, in rad/s, is equal to $2\pi f$, where $f$ is the frequency; $\theta$ is the angle of phase difference.

**Cycle; Frequency**  When any given armature coil has passed a pair of poles, the emf or current has gone through 360 electrical degrees, or 1 **cycle**. An **alternation** is one-half a cycle. The **frequency** of a synchronous machine in cycles per second (hertz)

$$f = NP/120 \qquad Hz \qquad (55)$$

where $N$ is the speed in r/min and $P$ the number of poles. In the United States and Canada the frequency of 60 Hz is almost universal for general lighting and power. For the ac power supply to dc transit systems, and for railroad electrification, a frequency of 25 Hz is used in many installations. In most of Europe and Latin America the frequency of 50 Hz is in general use. In order to obtain synchronous speeds up to 10,800 r/min so that direct drive can be used for planers and molders in woodworking, frequencies of from 90 to 180 Hz, obtained from frequency changers, are used. In aircraft the frequency of 400 Hz has become standard.

The **root-mean-square (rms)**, or **effective, value of a current wave** produces the same heating in a given resistance as a direct current of the same ampere value. Since the heating effect of a current is proportional to $i^2r$, the rms value is obtained by squaring the ordinates, finding their average value, and extracting the square root, i.e., the rms value is

$$I = \sqrt{1/T \int_0^T i^2 \, dt} \qquad A$$

where $T$ is the time of a cycle. The rms value $I$ of a sine wave equals $(1/\sqrt{2}) I_m = 0.707 I_m$.

**Average Value of a Wave**  The average value of a sine wave over a complete cycle is zero. For a half cycle the average is $(2/\pi)I_m$, or $0.637 \, I_m$, where $I_m$ is the maximum value of the sine wave. The average value is of importance only occasionally. A dc measuring instrument gives the average value of a pulsating wave. The average value is of use (1) when the effects of the current are proportional to the number of coulombs, as in electrolytic work and (2) when converting alternating to direct current.

**Form Factor**  The form factor of a wave is the ratio of rms value to average value. For a sine wave this is $\pi/(2\sqrt{2}) = 1.11$. This factor is important in that it enters equations for induced emf.

**Inductive reactance,** $2\pi fL$ or $\omega L$, opposes an alternating current in inductance $L$. It is expressed in $\Omega$. Reactance is usually denoted by the symbol $X$. Inductive reactance is denoted by $X_L$.

The **current** in an inductive reactance $X_L$ when connected across the voltage $E$ is

$$I = E/X_L = E/(2\pi fL) \qquad A \qquad (56)$$

This current lags the voltage by 90 electrical degrees. Inductance absorbs no energy. The energy stored in the magnetic field during each half cycle is returned to the source during the same half cycle.

**Capacitive reactance** is $1/(2\pi fC) = 1/\omega C$ and is denoted by $X_C$, where $C$ is in F. If $C$ is given in $\mu$F, $X_C = 10^6/2\pi fC$. The current in a capacitive reactance $X_C$ when connected across voltage $E$ is

$$I = E/X_C = 2\pi fCE \qquad A \qquad (57)$$

This current leads the voltage by 90 electrical degrees. Pure capacitance absorbs no energy. The energy stored in the dielectric field during each half cycle is returned to the source during the same half cycle.

**Impedance** opposes the flow of alternating current and is expressed in $\Omega$. It is denoted by $Z$. With resistance and inductance in series

$$Z = \sqrt{R^2 + X_L^2} = \sqrt{R^2 + (2\pi fL)^2} \qquad \Omega \qquad (58)$$

With resistance and capacitance in series

$$Z = \sqrt{R^2 + X_C^2} = \sqrt{R^2 + [1/(2\pi fC)]^2} \qquad \Omega \qquad (59)$$

With resistance, inductance, and capacitance in series

$$Z = \sqrt{R^2 + (X_L - X_C)^2} =$$
$$\sqrt{R^2 + [2\pi fL - 1/(2\pi fC)]^2} \qquad \Omega \qquad (60)$$

The current is

$$I = E/\sqrt{R^2 + [2\pi fL - 1/(2\pi fC)]^2} \qquad A \qquad (61)$$

**Phasor or Vector Representation**  Sine waves of voltage and current can be represented by phasors, these phasors being proportional in magnitude to the waves that they represent. The angle between two phasors is also equal to the time angle existing between the two waves that they represent.

Phasors may be combined as forces are combined in mechanics. Both graphical methods and the methods of complex algebra are used. Impedances and also admittances may be similarly combined, either graphically or symbolically. The usual method is to resolve series impedances into their component resistances and reactances, then combine all resistances and all reactances, from which the resultant impedance is obtained. Thus $Z_1 + Z_2 = \sqrt{(r_1 + r_2)^2 + (x_1 + x_2)^2}$, where $r_1$ and $x_1$ are the components of $Z_1$, etc.

**Phase Difference**   With resistance only in the circuit, the current and the voltage are in phase with each other; with inductance only in the circuit, the current lags the voltage by 90 electrical degrees; with capacitance only in the circuit, the current leads the voltage by 90 electrical degrees.

With resistance and inductance in series, the voltage leads the current by angle $\theta$ where $\tan \theta = X_L/R$. With resistance and capacitance in series, the voltage lags the current by angle $\theta$ where $\tan \theta = -X_C/R$.

With resistance, inductance, and capacitance in series, the voltage may lag, lead, or be in phase with the current.

$$\tan \theta = (X_L - X_C)/R = (2\pi fL - 1/2\pi fC)/R \quad \textbf{(62)}$$

If $X_L > X_C$ the voltage leads; if $X_L < X_C$ the voltage lags; if $X_L = X_C$ the current and voltage are in phase and the circuit is in resonance.

**Power Factor**   In ac circuits the power $P = I^2 R$ where $I$ is the current and $R$ the effective resistance (see below). Also the power

$$P = EI \cos \theta \quad \text{W} \quad \textbf{(63)}$$

where $\theta$ is the phase angle between $E$ and $I$. Cos $\theta$ is the **power factor** (pf) of the circuit. It can never exceed unity and is usually less than unity.

$$\cos \theta = P/EI \quad \textbf{(64)}$$

$P$ is often called the true power. The product $EI$ is the volt-amp (V·A) and is often called the apparent power.

**Active** or **energy** current is the projection of the total current on the voltage phasor. $I_e = I \cos \theta$. Power $= EI_e$.

**Reactive, quadrature,** or **wattless current** $I_q = I \sin \theta$ and is the component of the current that contributes no power but increases the $I^2 R$ losses of the system. In power systems it should ordinarily be made low.

The **vars** (volt-amp-reactive) are equal to the product of the voltage and reactive current. **Vars** $= EI_q$. **Kilovars** $= EI_q/1,000$.

**Effective Resistance**   When alternating current flows in a circuit, the losses are ordinarily greater than are given by the losses in the ohmic resistance alone. For example, alternating current tends to flow near the surface of conductors (skin effect). If iron is associated with the circuit, eddy-current and hysteresis losses result. These power losses may be accounted for by increasing the ohmic resistance to a value $R$, where $R$ is the **effective resistance,** $R = P/I^2$. Since the iron losses vary as $I^{1.8}$ to $I^2$, little error results from this assumption.

SOLUTION OF SERIES-CIRCUIT PROBLEM. Let a resistor $R$ of 10 $\Omega$, an inductor $L$ of 0.06 H, and a capacitor $C$ of 60 $\mu$F be connected in series across 120-V, 60-Hz mains (Fig. 27). Determine (1) the impedance, (2) the current, (3) the voltage across the resistance, the inductance, the capacitance, (4) the power factor, (5) the power, (6) the angle of phase difference.

(1) $\omega = 2\pi 60 = 377$. $X_L = 0.06 \times 377 = 22.6\,\Omega$; $X_C = 1/(377 \times 0.000060) = 44.2\ \Omega$; $Z = \sqrt{(10)^2 + (22.6 - 44.2)^2} = 23.8\ \Omega$; (2) $I = 120/23.8 = 5.04$ A; (3) $E_R = IR = 5.04 \times 10 = 50.4$ V; $E_L = IX_L = 5.04 \times 22.6 = 114.0$ V; $E_C = IX_C = 5.04 \times 44.2 = 223$ V; (4) $\tan \theta = (X_L - X_C)/R = -21.6/10 = -2.16$, $\theta = -65.2°$, $\cos \theta = \text{pf} = 0.420$; (5) $P =$

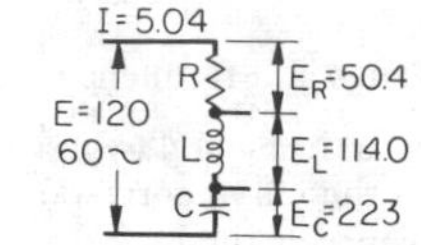

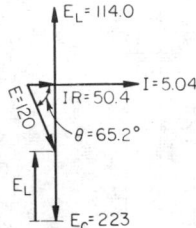

**Fig. 27**   Resistor, inductor, and capacitor in series.

$120 \times 5.04 \times 0.420 = 254$ W; $P = I^2 R = (5.04)^2 \times 10 = 254$ W (check); (6) From (4) $\theta = -65.2°$. Voltage lags. The phasor diagram to scale of this circuit is shown in Fig. 28. Since the current is common for all elements of the circuit, its phasor is laid horizontally along the axis of reference.

**Fig. 28** Phasor diagram for series circuit.

**Resonance**   If the voltage $E$ and the resistance $R$ [Eq. (61) ] are fixed, the maximum value of current occurs when $2\pi fL - 1/2\pi fC = 0$. The circuit so far as its terminals are concerned behaves like a noninductive resistor. The current $I = E/R$, the power $P = EI$, and the power factor is unity.

The voltage across the inductor and the voltage across the capacitor are opposite and equal and may be many times greater than the circuit voltage. The frequency

$$f = 1/(2\pi \sqrt{LC}) \quad \text{Hz} \quad \textbf{(65)}$$

is the **natural frequency** of the circuit and is the frequency at which it will oscillate if the circuit is not acted upon by some external frequency. This is the principle of radio sending and receiving circuits. Resonant conditions of this type should be avoided in power circuits, as the piling up of voltage may endanger apparatus and insulation.

EXAMPLE.   For what value of the inductance in the circuit (Fig. 27) will the circuit be in resonance, and what is the voltage across the inductor and capacitor under these conditions?

From Eq. **(65)** $L = 1/(2\pi f)^2 C = 0.1173$ H. $I = E/R = 120/10 = 12$ A. $L\omega I = I/C\omega = 0.1173 \times 377 \times 12 = 530$ V. This voltage is over four times the line voltage.

**Parallel Circuits**

Parallel circuits are used for nearly all power distribution. With several series circuits in parallel it is merely necessary to find the current in each and add all the current phasors vectorially to find the total current. Parallel circuits may be solved analytically.

A series circuit has resistance $r_1$ and **inductive reactance** $x_1$. The **conductance** is

$$g_1 = r_1/(r_1^2 + x_1^2) = r_1/Z_1^2 \quad \text{S} \quad \textbf{(66)}$$

and the **susceptance** is

$$b_1 = x_1/(r_1^2 + x_1^2) = x_1/Z_1^2 \quad \text{S} \quad \textbf{(67)}$$

Conductance is not the reciprocal of resistance unless the reactance is zero; susceptance is not the reciprocal of reactance unless the resistance is zero. With inductive reactance the susceptance is **negative;** with capacitive reactance the susceptance is **positive.**

If a second circuit has resistance $r_2$ and **capacitive reactance** $x_2$ in series, $g_2 = r_2/(r_2^2 + x_2^2) = r_2/Z_2^2$; $b_2 = x_2/(r_2^2 + x_2^2) = x_2/Z_2^2$. The total conductance $G = g_1 + g_2$; the total susceptance $B = -b_1 + b_2$. The **admittance** is

$$Y = \sqrt{G^2 + B^2} = 1/Z \qquad S \qquad (68)$$

The energy current is $EG$; the reactive current is $EB$; the power is

$$P = E^2 G \qquad W \qquad (69)$$
$$\text{vars} = E^2 B \qquad W \qquad (70)$$

The power factor is

$$pf = G/Y \qquad (71)$$

Also the following relations hold:

$$r = g/(g^2 + b^2) = g/Y^2 \qquad \Omega \qquad (72)$$
$$x = b/(g^2 + b^2) = b/Y^2 \qquad \Omega \qquad (73)$$

SOLUTION OF A PARALLEL-CIRCUIT PROBLEM. In the parallel circuit of Fig. 29 it is desired to find the joint impedance, the total current, the power in each branch, the total power, and the power factor, when $E = 100$, $f = 60$, $R_1 = 2 \ \Omega$, $R_2 = 4 \ \Omega$, $L_1 = 0.00795$ H, $X_1 = 2\pi f L_1 = 3 \ \Omega$, $C_2 = 1{,}326 \ \mu F$, $X_2 = 1/2\pi f C_2 = 2 \ \Omega$, $Z_1 = \sqrt{2^2 + 3^2} = 3.6 \ \Omega$, and $Y_1 = 1/3.6 = 0.278$ S. *Solution:* $g_1 = R_1/(R_1^2 + X_1^2) = 2/13 = 0.154$; $b_1 = -3/13 = -0.231$; $Z_2 = \sqrt{16 + 4} = 4.47$; $Y_2 = 1/4.47 = 0.224$; $g_2 = R_2/(R_2^2 + X_2^2) = 4/(16 + 4) = 0.2$ S; $b_2 = 2/20 = 0.1$ S; $G = g_1 + g_2 = 0.154 + 0.2 = 0.354$ S; $B = b_1 + b_2 = -0.231 + 0.1 = -0.131$ S; $Y = \sqrt{G^2 + B^2} = \sqrt{0.354^2 + (-0.131)^2} = 0.377$ S, and joint impedance $Z = 1/0.377 = 2.65 \ \Omega$. Phase angle $\theta = \tan^{-1}(-0.131/0.354) = -20.3°$. $I = EY = 100 \times 0.377 = 37.7$ A; $P_1 = E^2 g_1 = 100^2 \times 0.154 = 1{,}540$ W; $P_2 = E^2 g_2 = 100^2 \times 0.2 = 2{,}000$ W; total power $= E^2 G = 100^2 \times 0.354 = 3{,}540$ W. Power factor $= \cos \theta = 3{,}540/(100 \times 37.7) = 93.8$ percent.

With parallel circuits, unity power factor is obtained when the algebraic sum of the quadrature currents is zero. That is, $b_1 + b_2 + b_3 \cdots = 0$.

**Three-Phase Circuits** Ac generators are usually wound with three armature circuits which are spaced 120 electrical degrees apart on the armature. Hence these coils generate emfs 120 electrical degrees apart. The coils are connected either in $Y$ (star) or in $\Delta$ (mesh) as shown in Fig. 30. Whether $Y$- or $\Delta$-connected, with a balanced load, the three coil emfs $E_c$ and the three coil currents $I_c$ are equal. In the Y connection the line and coil currents are equal, but the line emfs $E_{AB}$, $E_{BC}$, $E_{CA}$ are $\sqrt{3}$ times in magnitude the coil emfs $E_{OA}$, $E_{OB}$, $E_{OC}$, since each is the phasor difference of two coil emfs. In the delta connection the line and coil emfs are equal, but $I$, the line current, is $\sqrt{3} \ I_c$, the coil current, i.e., it is the phasor difference of the currents in the two coils connected to the line. The power of a coil is $E_c I_c \cos \theta$, so that the total power is $3E_c I_c \cos \theta$. If $\theta$ is the angle between coil current and coil

voltage, the angle between line current and line voltage will be $30° \pm \theta$. In terms of line current and emf, the power is $\sqrt{3} \ EI \cos \theta$. A fourth or neutral conductor connected to $O$ is frequently used with the Y connection. The neutral point $O$ is frequently grounded in transmission and distribution circuits. The coil emfs are assumed to be sine waves. Under these conditions they balance, so that in the delta connection the sum of the two coil emfs at each instant is balanced by the third coil emf. Even though the third, ninth, fifteenth . . . harmonics, $3(2n + 1)f$, where $n = 0$ or an integer, exist in the coil emfs, they cannot appear between the three external line conductors of the three-phase Y-connected circuit. In the delta circuit, the same harmonics $3(2n + 1)f$ cause local currents to circulate around the mesh. This may cause a very appreciable heating. In a three-phase system the power

$$P = \sqrt{3} \ EI \cos \theta \qquad W \qquad (74)$$

the power factor is

$$P/\sqrt{3} \ EI \qquad (75)$$

and the kV·A

$$\sqrt{3} \ EI/1{,}000 \qquad (76)$$

where $E$ and $I$ are line voltages and currents.

**Two-Phase Circuits** Two-phase generators have two windings spaced 90 electrical degrees apart on the armature. These windings generate emfs differing in time phase by 90°. The two windings may be independent and power transmitted to the receiver though the two single-phase circuits are entirely insulated from each other. The two circuits may be combined into a two-phase three-wire circuit such as is shown in Fig. 31, where $OA$ and $OB$ are the generator circuits (or transformer secondaries) and $A'O'$ and $B'O'$ are the load circuits. The wire $OO'$ is the common wire and under balanced conditions carries a current $\sqrt{2}$ times the current wires $AA'$ and $BB'$. For example, if $I_c$ is the coil current, $\sqrt{2} \ I_c$ will be the value of the current in the common conductor $OO'$. If $E_c$ is the voltage across $OA$ or $OB$, $\sqrt{2} \ E_c$ will be the voltage across $AB$. The power of a two-phase circuit is twice the power in either coil if the load is balanced. Normally, the voltages $OA$ and $OB$ are equal, and the current is the same in both coils. Owing to nonsymmetry and the high degree of unbalancing of this system even under balanced loads, it is not used at the present time for transmission and is little used for distribution.

**Four-Phase Circuit** A four-phase or quarter-phase circuit is shown in Fig. 32. The windings $AC$ and $BD$ may be independent or connected at $O$. The voltages $AC$ and $BD$ are 90 electrical degrees apart as in two-phase circuits. If a neutral wire $O$-$O'$ is added, three different voltages can be obtained.

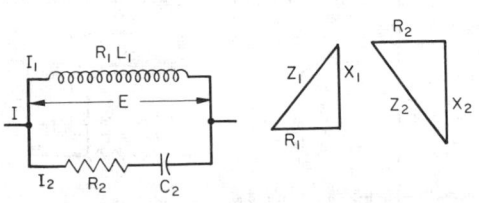

**Fig. 29** Parallel circuit and phasor diagram.

Let $E_1$ = voltage between $O$-$A$, $O$-$B$, $O$-$C$, $O$-$D$. Voltages between $A$-$B$, $B$-$C$, $C$-$D$, $D$-$A$ = $\sqrt{2}\,E_1$. Voltages between $A$-$C$, $B$-$D$ = $2E_1$. Because of this multiplicity of voltages and the fact that polyphase power apparatus and lamps may be connected at the same time, this system is still used to some extent in distribution.

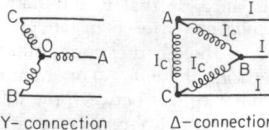

**Fig. 30** Three-phase connections.

**Advantages of Polyphase Power** The advantages of polyphase power over single-phase power are as follows. The output of synchronous generators and most other rotating machinery is from 60 to 90 percent greater when operated polyphase than when operated single phase; pulsating fluxes and corresponding iron losses which occur in many common types of machinery when operated single phase are negligible when operated polyphase; with balanced polyphase loads polyphase power is constant whereas with single phase the power fluctuates over wide limits during the cycle. Because of its minimum number of wires and the fact that it is not easily unbalanced, the three-phase system has for the most part superseded other polyphase systems.

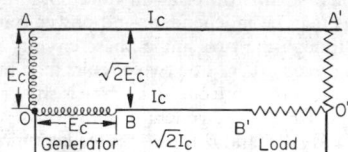

**Fig. 31** Two-phase three-wire circuit.

## ELECTRICAL INSTRUMENTS AND MEASUREMENTS

Electrical measuring devices that merely indicate, such as ammeters and voltmeters, are called **instruments;** devices that totalize with time such as watthour meters and ampere-hour meters are called **meters.** (See also Sec. 16.) Most types of electrical instruments are available with digital read out.

**DC Instruments** Direct current and voltage are both measured with an indicating instrument based on the principle of the D'Arsonval galvanometer. A coil with steel pivots and turning in jewel bearings is mounted in a magnetic field produced by permanent magnets. The motion is restrained by two small flat coiled springs, which also serve to conduct the current to the coil. The deflections of the coil are read with a

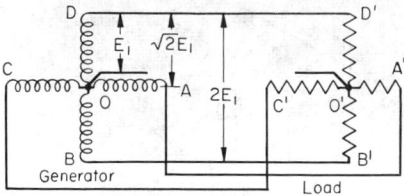

**Fig. 32** Four-phase or quarter-phase circuit.

light aluminum pointer attached to the coil and moving over a graduated scale. The same instrument may be used for either current or voltage, but the method of connecting in circuit is different in the two cases. Usually, however, the coil of an instrument to be used as an ammeter is wound with fewer

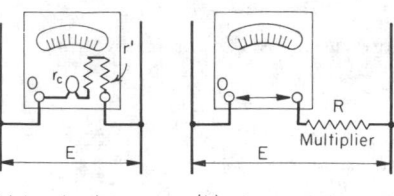

(a) Internal resistance    (b) With multiplier

**Fig. 33** Voltmeter.

turns of coarser wire than an instrument to be used as a voltmeter and so has lower resistance. The instrument itself is frequently called a **millivoltmeter.** It cannot be used alone to measure voltage of any magnitude since its resistance is so low that it would be burned out if connected across the line. Hence a resistance $r'$ in series with the coil is necessary as indicated in Fig. 33$a$ in which $r_c$ is the resistance of the coil. From 0.2 to 750 V this resistance is usually within the instrument. For higher voltages an external resistance $R$ called an extension coil or multiplier (Fig. 33$b$), is necessary. Let $e$ be the reading of the instrument, in volts (Fig. 33$b$), $r$ the internal resistance of the instrument, including $r'$ and $r_c$ in $(a)$, $R$ the resistance of the multiplier. Then the total voltage is

$$E = e(R + r)/r \qquad (77)$$

It is clear that by using suitable values of $R$ a voltmeter can be made to have several scales.

Instruments themselves can only carry currents of the magnitudes of 0.01 to 0.06 A. To measure larger values of current the instrument is provided with a shunt $R$ (Fig. 34). The current divides inversely as the resistances $r$ and $R$ of the instrument and the shunt. A low resistance $r'$ within the instrument is connected in series with the coil. This permits some adjustment to the deflection so that the instrument can be adapted to its shunt. Usually most of the current flows through the shunt, and the current in the instrument is negligible in comparison. Up to 50 and 75 A the shunt can be incorporated within the instrument. For larger currents it is usually necessary to have the shunt external to the instrument and connect the instrument to the potential terminals of the shunt by means of leads. Any given instrument may have any number of ranges by providing it with a sufficient number of shunts. The range of the usual instrument of this type is approximately 50 mV. Although the same instrument may be used for voltmeter or ammeter, the moving coils of voltmeters are usually wound with more turns of finer wire. They take approximately 0.01 A so that their resistance is approximately

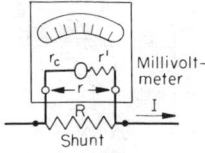

**Fig. 34** Millivoltmeter with shunt.

100 Ω/V. Instruments used as ammeters alone operate with 0.01 to 0.06 A.

Permanent-magnet moving-coil instruments may be used to measure unidirectional pulsating currents or voltages and in such cases will indicate the average value of the periodically varying current or voltage.

**AC Instruments** Instruments generally used for alternating currents may be divided into five types: **electrodynamometer, iron-vane, thermocouple, rectifier,** and **electronic.** Instruments of the **electrodynamometer** type, the most precise, operate on the principle of one coil carrying current, turning in the magnetic field produced by a second coil carrying current taken from the same circuit. If these circuits or coils are connected in series, the torque exerted on the moving system for a given relative position of the coil system is proportional to the square of the current and is not dependent on the direction of the current. Consequently, the instrument will have a compressed scale at the lower end and will usually have only the upper two-thirds of the scale range useful for accurate measurement. Instruments of this type ordinarily require 0.04 to 0.08 A or more in the moving-coil circuit for full-scale deflection. They read the rms value of the alternating or pulsating current. The wattmeter operates on the electrodynamometer principle. The fixed coil, however, is energized by the current of the circuit, and the moving coil is connected across the potential in series with high resistance. Unless shielded magnetically the foregoing instruments will not, in general, indicate so accurately on direct as on alternating current because of the effects of external stray magnetic fields. Also reversed readings should be taken. **Iron-vane** instruments consist of a fixed coil which actuates magnetically a light movable iron vane mounted on a spindle; they are rugged, inexpensive, and may be had in ranges of 30 to 750 V and 0.05 to 100 A. They measure rms values and tend to have compressed scales as in the case of electrodynamometer instruments.

The compressed part of the scale may, however, be extended by changing the shape of the vanes. Such instruments operate with direct current and are accurate to within 1 percent or so. AC instruments of the **induction type** (Westinghouse Electric Corp.) must be used on ac circuits of the frequency for which they have been designed. They are rugged and relatively inexpensive and are used principally for switchboards where a long-scale range and a strong deflecting torque are of particular advantage. **Thermocouple instruments** operate on the **Seebeck effect.** The current to be measured is conducted through a heater wire, and a thermojunction is either in thermal contact with the heater or is very close to it. The emf developed in the thermojunction is measured by a permanent-magnet dc type of instrument. By controlling the shape of the air gap, a nearly uniform scale is obtained. This type of instrument is well adapted to the measurement of high-frequency currents or voltages, and since it operates on the heating effect of current, it is convenient as a transfer instrument between direct current and alternating current.

In the **rectifier-type instrument** the ac voltage or current is rectified, usually by means of a small copper oxide or a selenium-type rectifier, connected in a bridge circuit to give full-wave rectification (Fig. 35). The rectified current is measured with a dc permanent-magnet-type instrument $M$. The instrument measures the **average** value of the half waves that have been rectified, and with the sine waves, the average value is 0.9 the rms value. The scale is calibrated to indicate rms

values. With nonsinusoidal waves the ratio of average to rms may vary considerably from 0.9 so that the instrument may be in error up to ±5 percent from this cause. This type of instrument is widely used in the measurement of high-frequency voltages and currents. **Electronic (vacuum-tube) voltme-**

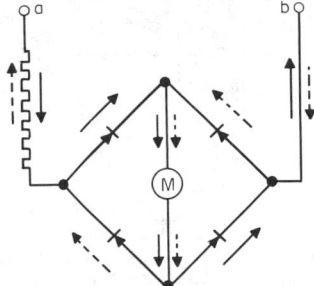

**Fig. 35** Rectifier-type instrument.

**ters** operate on the principle of the amplification which can be obtained with a triode, or three-element vacuum tube. Since the emf to be measured is applied to the grid, the instruments take practically no current and hence are adapted to measure potential differences which would change radically were any appreciable current taken by the measuring device. This type of instrument can measure voltages from a few tenths of a volt to several hundred volts, and with a potential divider, up to thousands of volts. They are also adapted to frequencies up to 100 MHz.

**Power Measurement in Single-Phase Circuits** Wattmeters are not rated primarily in W, but in A and V. For example, with a low power factor the current and voltage coils may be overloaded and yet the needle be well on the scale. The current coil may be carrying several times its rated current, and yet the instrument reads zero because the potential circuit is not closed, etc. Hence it is desirable to use both an ammeter and a voltmeter in conjunction with a wattmeter when measuring power (Fig. 36a). The instruments themselves consume appreciable power, and correction is often necessary unless these losses are negligible compared with the power being measured. For example, in Fig. 36a, the wattmeter measures the $I^2R$ loss in its own current coil and in the ammeter (1 to 2 W each), as well as the loss in the voltmeter (= $E^2/R$ where $R$ is the resistance of the voltmeter). The losses in the ammeter and voltmeter may be eliminated by short-circuiting the ammeter and disconnecting the voltmeter when reading the wattmeter. If the wattmeter is connected as shown in Fig. 36b, it measures the power taken by its own potential coil ($E^2/R_p$) which at 110 V is 5 to 7 W. ($R_p$ is the resistance of the potential circuit.) Frequently correction must be made for this power.

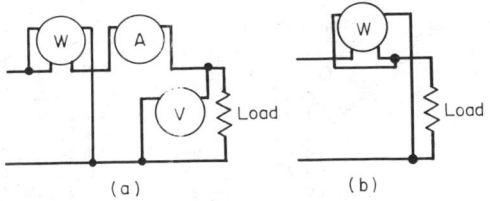

**Fig. 36** Connections of instruments to single-phase load.

**Power Measurement in Polyphase Circuits; Three-Wattmeter Method**  Let $ao$, $bo$, and $co$ be any Y-connected three-phase load (Fig. 37). Three wattmeters with their current coils in each line and their potential circuits connected to neutral measure the total power, since the power in each load is

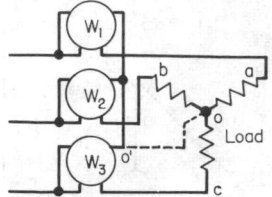

**Fig. 37**  Three-wattmeter method.

measured by one of the wattmeters. The connection $oo'$ may, however, be broken, and the total power is still the sum of the three readings; i.e., the power $P = P_1 + P_2 + P_3$. This method is applicable to any system of $n$ wires. The current coil of one wattmeter is connected in each of the $n$ wires. The potential circuit of each wattmeter is connected between its own phase wire and a junction in common with all the other potential circuits. The wattmeters must be connected symmetrically, and the readings of any that read negative must be given the negative sign.

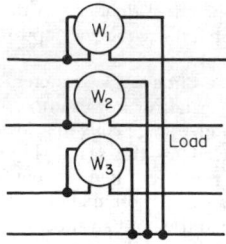

**Fig. 38**  Power measurement in $n$-wire system.

In the general case any system of $n$ wires requires at least $n - 1$ wattmeters to measure the power correctly. The $n - 1$ wattmeters are connected in series with $n - 1$ wires. The potential circuit of each is connected between its own phase wire and the wire in which no wattmeter is connected (Fig. 38).

The thermal watt converter is also used to measure power. This instrument produces a dc voltage proportional to three-phase ac power.

**Three-Phase Systems**  The three-wattmeter method (Fig. 37) is applicable to any three-phase system. It is commonly used with the three-phase four-wire system. If the loads are balanced, $P_1 = P_2 = P_3$ and the power $P = 3P_1$.

The **two-wattmeter** method is most commonly used with three-phase three-wire systems (Fig. 39). The current coils may be connected in *any* two wires, the potential circuits being connected to the third. It will be recognized that this is adapting the method of Fig. 38 to three wires. With balanced loads the readings of the wattmeters are $P_1 = Ei \cos (30° + \theta)$, $P_2 = Ei \cos (30° - \theta)$, and $P = P_2 \pm P_1$. $\theta$ is the angle of phase difference between coil voltage and current. Since

$$P_1/P_2 = \cos (30° + \theta)/\cos (30° - \theta) \qquad (78)$$

the power factor is a function of $P_1/P_2$. Table 9 gives values of power factor for different ratios of $P_1/P_2$.

$P = P_2 + P_1$ when $\theta < 60°$.

When $\theta = 60°$, pf $= \cos 60° = 0.5$, $P_1 = \cos (30° + 60°) = 0$, $P = P_2$. When $\theta > 60°$, pf $< 0.5$, $P = P_2 - P_1$. Also,

$$\tan \theta = \sqrt{3} \, (P_2 - P_1)/(P_2 + P_1) \qquad (79)$$

In a **polyphase wattmeter** the two single-phase wattmeter elements are combined to act on a single spindle. Hence the

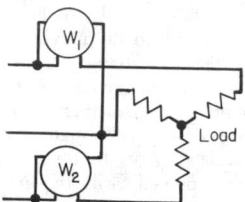

**Fig. 39**  Two-wattmeter method.

adding and subtracting of the individual readings are done automatically. The total power is indicated on one scale. This type of instrument is almost always used on switchboards. The connections of a portable type are shown in Fig. 40.

In the foregoing instrument connections, Y-connected loads are shown. These methods are equally applicable to delta-connected loads. The two-wattmeter method (Fig. 39) is obviously adapted to the two-phase three-wire system (Fig. 31).

**Measurement of Energy**

**Watthour meters** record the energy taken by a circuit over some interval of time. Correct registration occurs if the angular velocity of the rotating element at every instant is proportional to the power. The method of accomplishing this with dc meters is illustrated in Fig. 41. The meter is in reality a small motor. The field coils $FF$ are in series with the line. The armature $A$ is connected across the line, usually in series with a resistor $R$. The movable field coil $F'$ is in series with the armature $A$ and serves to compensate for friction. $C$ is a small commutator, either of copper or of silver, and the two small brushes are usually of silver. An aluminum disk, rotating between the poles of permanent magnets $M$, acts as a magnetic brake the retarding torque of which is proportional to the angular velocity of the disk. A small worm and the gears $G$ actuate the recording dials.

The following relation, or an equivalent, holds with most types of meter. With each revolution of the disk, $K$ Wh are

**Table 9. Ratio $P_1/P_2$ and Power Factor**

| $P_1/P_2$ | Power factor | $P_1/P_2$ | Power factor | $P_1/P_2$ | Power factor | $P_1/P_2$ | Power factor |
|---|---|---|---|---|---|---|---|
| +1.0 | 1.000 | +0.4 | 0.804 | −0.1 | 0.427 | −0.6 | 0.142 |
| +0.9 | 0.996 | +0.3 | 0.732 | −0.2 | 0.360 | −0.7 | 0.102 |
| +0.8 | 0.982 | +0.2 | 0.656 | −0.3 | 0.296 | −0.8 | 0.064 |
| +0.7 | 0.956 | +0.1 | 0.576 | −0.4 | 0.240 | −0.9 | 0.030 |
| +0.6 | 0.918 | 0.0 | 0.50 | −0.5 | 0.188 | −1.0 | 0.000 |
| +0.5 | 0.866 | | | | | | |

recorded, where $K$ is the **meter constant** found usually on the disk. It follows that the average watts $P$ over any period of time $t$ sec is

$$P = 3{,}600KN/t \qquad (80)$$

where $N$ is the revolutions of the disk during that period. Hence, the meter may be calibrated by connecting standard-

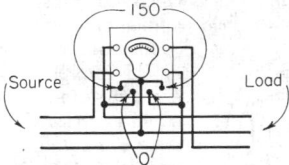

**Fig. 40**   Connections for polyphase wattmeter in three-phase circuit.

ized instruments to measure the average power taken by the load and by counting the revolutions $N$ for $t$ s. Near full load, if the meter registers fast, the magnets $M$ should be moved outward radially; if it registers slow, the magnets should be moved inward. If the meter registers fast at light (5 to 10 percent) load, the starting coil $F'$ should be moved further away from the armature; if it registers slow, $F'$ should be moved nearer the armature. A meter should not register more than 1.5 percent fast or slow, and with calibrated standards it can be made to register to within 1 percent of correct.

The **induction watthour meter** is used with alternating current. Although the dc meter registers correctly with alternating current, it is more expensive than the induction type, the commutator and brushes may cause trouble, and at low power factors compensation is necessary. In the induction watthour meter the driving torque is developed in the aluminum disk by the joint action of the alternating magnetic flux produced by the potential circuit and by the load current. The driving torque and the retarding torque are both developed in the same aluminum disk, hence no commutator and brushes are necessary. The rotating element is very light, and hence the friction torque is small. Equation (80) applies to this type of meter. When calibrating, the average power $W$ for $t$ s is determined with a calibrated wattmeter. The friction compensation is made at light loads by changing the position of a small hollow stamping with respect to the potential lug. The meter should also be adjusted at low power factor (0.5 is customary). If the meter is slow with lagging current, resistance should be cut out of the compensating circuit; if slow with leading current, resistance should be inserted.

**Power-Factor Measurement**   The usual method of determining power factor is by the use of voltmeter, ammeter, and wattmeter. The wattmeter gives the watts of the circuit, and the product of the voltmeter reading and the ammeter reading gives the volt-amperes. The power factor is the ratio of the two [see Eqs. (64) and (75)]. Also single-phase and three-phase

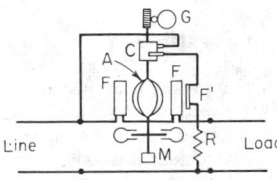

**Fig. 41**   DC watthour meter.

power-factor indicators, which can be connected directly in circuit, are on the market.

**Instrument Transformers**

With voltages higher than 600 V, and even at 600 V, it becomes dangerous and inaccurate to connect instruments and meters directly into power lines. It is also difficult to make

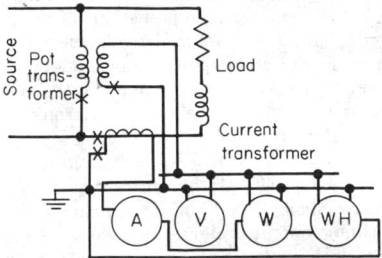

**Fig. 42**   Single-phase connections of instruments with transformers.

potential instruments for voltages in excess of 600 V and ammeters in excess of 60-A ratings. To insulate such instruments from high voltage and at the same time to permit the use of low-range instruments, instrument transformers are used. **Potential transformers** are identical with power transformers except that their volt-ampere rating is low, being 40 to 500 W. Their primaries are wound for line voltage and their secondaries for 110 V. **Current transformers** are designed to go in series with the line, and the rated secondary current is 5 A. The secondary of a current transformer **should always be closed** when current is flowing; it should never be allowed to become open circuited under these conditions. When open-circuited the voltage across the secondary becomes so high as to be dangerous and the flux becomes so large in magnitude that the transformer overheats. The secondaries of both potential and current transformers should be well grounded at one point (Figs. 42 and 43). Instrument transformers introduce slight errors because of small variations in their ratio with load. Also there is slight phase displacement in both current and potential transformers. The readings of the instruments must be multiplied by the instrument transformer ratios. The scales of switchboard instruments are usually calibrated to take these ratios into account.

Figure 42 shows the use of instrument transformers to measure the voltage, current, power, and kilowatthours of a single-phase load. Figure 43 shows the connections that would be used to measure the voltage, current, and power of a 26,400-V 600-A three-phase load.

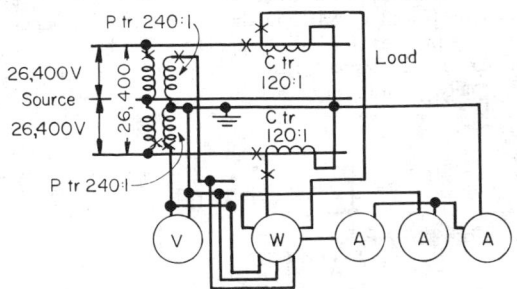

**Fig. 43**   Three-phase connections of instruments and instrument transformers.

**Measurement of High Voltages** Potential transformers such as those shown in Figs. 42 and 43 may be used even for very high voltages, but for voltages above 132 kV they become so large and expensive that they are used only sparingly. A convenient method used with testing transformers is the employment of a voltmeter coil, which consists of a coil of a few turns interwoven in the high-voltage winding and insulated from it. The voltage ratio is the ratio of the turns in the high-voltage winding to those in the voltmeter coil. A capacitance voltage divider consists of two or more capacitors connected in series across the high voltage to be measured. A high-impedance voltmeter, such as an electronic one, is connected across the capacitor at the grounded end. The high voltage $V = V_m C_m/C$ V, where $V_m$ is the voltmeter reading, $C$ the capacitance (in $\mu$F) of the entire divider, and $C_m$ the equivalent capacitance (in $\mu$F) of the capacitor at the grounded end. A bushing potential device consists of a high-voltage-transformer bushing having a capacitance tap brought out from one of the metallic electrodes within the bushing which is near ground potential. This device is obviously a capacitance voltage divider. For testing, **sphere gaps** are used for the very high voltages. Calibration data for sphere gaps are given in the ASA standard C68.1, 1953, Measurement of Voltage in Dielectric Tests. Even when it is not being used for the measurement of voltage, it is frequently advisable to connect a sphere gap in parallel with the specimen being tested so as to prevent overvoltages. The gap is set to a slightly higher voltage than that which is desired.

### Measurement of Resistance

**Voltmeter-Ammeter Method** A common method of measuring resistance, known as the voltmeter-ammeter or fall-in-potential method, makes use of an ammeter and a voltmeter. In Fig. 44, the resistance to be measured is $R$. The current in the resistor $R$ is $I$ A, which is measured by the ammeter $A$ in series. The drop in potential across the resistor $R$ is measured by the voltmeter $V$. The current shunted by the voltmeter is so small that it may generally be neglected. A correction may be applied if necessary, for the resistance of the voltmeter is generally given with the instrument. The potential difference divided by the current gives the resistance included between the voltmeter leads. As a check, determinations are generally made with several values of current, which may be varied by means of the controlling resistor $r$. If the resistance to be measured is that of the armature of a dc machine and the voltmeter leads are placed on the brush holders, the resistance determined will include that of the brush contacts. To measure the resistance of the armature alone, the voltmeter leads should be placed directly on the commutator segments on which the brushes rest but not under the brushes.

**Insulation Resistance** Insulation resistance is so high that it is usually given in megohms ($10^6$ ohms, M$\Omega$) rather than in

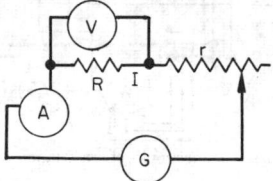

**Fig. 44** Voltmeter-ammeter method for resistance measurement.

ohms. Insulation resistance tests are important, for although they may not be conclusive they frequently reveal flaws in insulation, poor insulating material, presence of moisture, etc. Such tests are applied to the insulation of electrical machinery from the windings to the frame, to underground cables, to insulators, capacitors, etc.

For moderately low resistances, 1 to 10 M$\Omega$, the voltmeter method given in Fig. 45, which shows insulation measurement to the frame of the field winding of a generator, may be used.

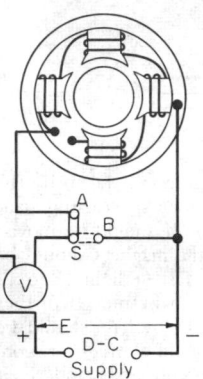

**Fig. 45** Voltmeter method for insulation resistance measurement.

To measure the current when a voltage $E$ is impressed across the resistor $R$, a high-reading voltmeter $V$ is connected in series with $R$. The current under this condition with the switch connecting $S$ and $A$ is $E/(R + r)$, where $r$ is the resistance of the voltmeter. A high-resistance voltmeter is necessary, since the method is in reality a comparison of the unknown insulation resistance $R$ with the known resistance $r$ of the voltmeter. Hence, the resistance of the voltmeter must be comparable with the unknown resistance, or the deflection of the instrument will be so small that the results will be inaccurate. To determine the impressed voltage $E$, the same voltmeter is used. The switch $S$ connects $S$ and $B$ for this purpose. With these two readings, the unknown resistance is

$$R = r(E - e)/e \qquad (81)$$

where $e$ is the deflection of the voltmeter when in series with the resistance to be measured as when $S$ is at $A$. If a special voltmeter, having a resistance of 100 k$\Omega$ per 150 V, is available, a resistance of the order of 2 to 3 M$\Omega$ may be measured very accurately.

When the insulation resistance is too high to be measured with a voltmeter, a sensitive galvanometer may be used. The connections for measuring the insulation resistance of a cable are shown in Fig. 46. The battery should have an emf of at least 100 V. Radio B batteries are convenient for this purpose. The method involves comparing the unknown resistance with a standard 0.1 M$\Omega$. To calibrate the galvanometer the cable is short-circuited (dotted line) and the switch $S$ is thrown to position (a). Let the galvanometer deflection be $D_1$ and the reading of the Ayrton shunt $S_1$. The short circuit is then removed. The 0.1 M$\Omega$ is left in circuit since it is usually negligible in comparison with the unknown resistance $X$. Let the reading of the galvanometer now be $D_2$ and the reading of the shunt $S_2$. Then

$$X = 0.1 \, S_2 D_1/S_1 D_2 \qquad M\Omega \qquad (82)$$

When the switch $S$ is thrown to position (b), the cable is short-circuited through the 0.1 M$\Omega$ and becomes discharged.

The **Megger** insulation tester is an instrument that indicates insulation resistance directly on a scale. It consists of a small hand-driven generator which generates approximately 500 V and sometimes 1 kV. A clutch slips when the voltage exceeds the rated value. The current through the unknown resistance flows through a moving element consisting of two coils fastened rigidly together, but which move in different portions of the magnetic field. A pointer attached to the spindle of the moving element indicates the insulation resistance directly. These instruments have a range up to 2,000 MΩ and are very convenient where portability and convenience are desirable.

The insulation resistance of electrical machinery may be of doubtful significance as far as dielectric strength is concerned.

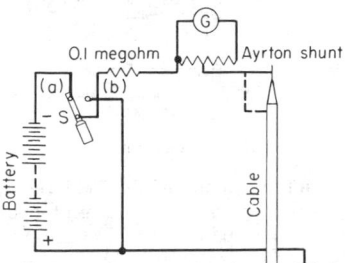

**Fig. 46**   Measurement of insulation resistance with galvanometer.

It varies widely with temperature, humidity, and cleanliness of the parts. When the insulation resistance falls below the prescribed value, it can (in most cases of good design) be brought to the required standard by cleaning and drying the machine. Hence it may be useful in determining whether or not the insulation is in proper condition for a dielectric test. The AIEE standards (5-451) specify minimum value of insulation resistance in MΩ = (rated voltage)/(rating in kW + 1,000). If the operating voltage is higher than the rated voltage, the operating voltage should be used. The rule specifies that a dc voltage of 500 be used in testing. If not, the voltage should be specified.

**Wheatstone Bridge**   Resistors from a fraction of an Ω to 100 kΩ and more may be measured with a high degree of precision with the Wheatstone bridge (Fig. 47). The bridge consists of four resistors *ABCX* connected as shown. *X* is the unknown resistance; *A* and *B* are ratio arms, the resistance units of which are in even decimal Ω as 1, 10, 100, etc. *C* is the rheostat arm. A battery or low-voltage source of direct current is connected across *ab*. A galvanometer *G* of moderate sensitivity is connected across *cd*. The values of *A* and *B* are so chosen that three or four significant figures in the value of *C* are obtained.   As a first approximation it is well to make *A* and *B* equal. When the bridge is in balance,

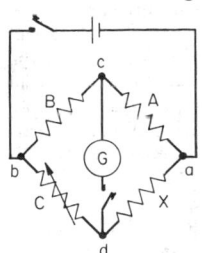

**Fig. 47**   Wheatstone bridge.

$$X/C = A/B \qquad (83)$$

The positions of the battery and galvanometer are interchangeable. There are many modifications of the bridge which adapt it to measurements of very low resistances and also to ac measurements.

**Kelvin Double Bridge**   The simple Wheatstone bridge is not adapted to measuring very low resistances since the contact resistances of the test specimen become comparable with the specimen resistance. This error is avoided in the Kelvin double bridge, the diagram of which is shown in Fig. 48. The specimen *X*, which may be a short length of copper wire or bus bar, is connected in series with an adjustable calibrated resistor *R* whose resistance is comparable with that of the

specimen. The arms *A* and *B* of the bridge are ratio arms usually with decimal values of 1, 10, 100 Ω. One terminal of the galvanometer is connected to *X* and *R* by means of two resistors *a* and *b*. If these resistors are set so that *a/b = A/B*, the contact resistance *r* between *X* and *R* is eliminated in the

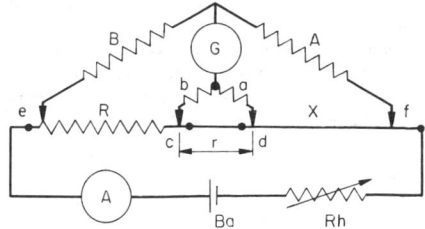

**Fig. 48**   Kelvin double bridge.

measurement. The contact resistances at *c* and *d* have no effect since at balance the galvanometer current is zero. The contact resistances at *f* and *e* need only be negligible compared with the resistances of arms *A* and *B* both of which are reasonably high. By means of the variable resistor *Rh* the value of current, as indicated by ammeter *A*, may be adjusted to give the necessary sensitivity. When the bridge is in balance,

$$X/R = A/B \qquad (84)$$

**Potentiometer**   The principle of the potentiometer is shown in Fig. 49. *ab* is a slide wire, and *bc* consists of a number of equal individual resistors between contacts. A battery *Ba* the emf of which is approximately 2 V supplies current to this wire through the adjustable rheostat *R*. A slider *m* makes contact with *ab*, and a contactor *m′* connects with the contacts in *bc*. A galvanometer *G* is in series with the wire connecting to *m*. By means of the double-throw double-pole switch *Sw*, either the standard cell or the unknown emf (*EMF*) may be connected to *mm′* through the galvanometer *G*. The potentiometer is standardized by throwing *Sw* to the standard-cell side, setting *mm′* so that their positions on *ab* and *bc* correspond to the emf of the standard cell. The rheostat *R* is then adjusted until *G* reads zero. (In commercial potentiometers a dial which may be set directly to the emf of the standard cell is usually provided.) The unknown emf is measured by throwing *Sw* to *EMF* and adjusting *m* and *m′* until *G* reads zero. The advantage of this method of measuring emf is that when the potentiometer is in balance no current is taken from either the standard cell or the source of emf. Potentiome-

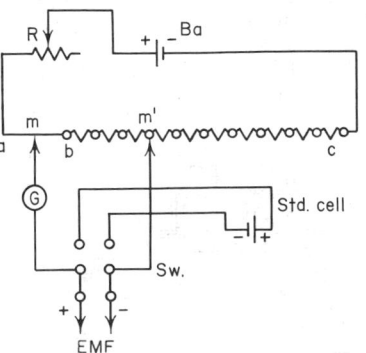

**Fig. 49**   Potentiometer principle.

ters seldom exceed 1.6 V in range. To measure voltage in excess of this, a **volt box** which acts as a multiplier is used. To measure current, the voltage drop across a standard resistor of suitable value is measured with the potentiometer. For example, with 50 A a 0.01-Ω standard resistance gives a voltage drop of 0.5 V which is well within the range of the potentiometer.

**Potentiometers of low range** are used extensively with thermocouple pyrometers. Figure 49 merely illustrates the principle of the potentiometer. There are many modifications, conveniences, etc., not shown in Fig. 49.

## DC GENERATORS

All electrical machines are comprised of a magnetic circuit of iron (or steel) and an electric circuit of copper. In a generator the armature conductors are rotated so that they cut the magnetic flux coming from and entering the field poles. In the dc generator (except the unipolar type) the emf induced in the individual conductors is alternating, but this is rectified by the commutator and brushes, so that the current to the external circuit is unidirectional.

The induced emf in a generator (or motor)

$$E = \phi ZNP/60P'10^8 \quad \text{V} \qquad \textbf{(85)}$$

where $\phi$ = flux in maxwells entering the armature from one north pole; $Z$ = total number of conductors on the armature; $N$ = speed, r/min; $P$ = number of poles; and $P'$ = number of parallel paths through the armature.

In the SI system $\phi$ is in Wb and $10^8$ is omitted. Since with a given generator, $Z$, $P$, $P'$ are fixed, the induced emf

$$E = K\phi N \quad \text{V} \qquad \textbf{(86)}$$

where $K$ is a constant. When the armature delivers current, the terminal volts are

$$V = E - I_a R_a \qquad \textbf{(87)}$$

where $I_a$ is the armature current and $R_a$ the armature resistance including the brush and contact resistance, which vary somewhat.

There are three standard types of dc generators: the **shunt generator,** the **series generator,** and the **compound generator.**

**Shunt Generator**  The field of the shunt generator in series with its rheostat is connected directly across the armature as shown in Fig. 50. This machine maintains approximately constant terminal voltage over its working range of load. An external characteristic of the generator is shown in Fig. 51. As load is applied the terminal voltage drops owing to the armature-resistance drop [Eq. (87)] and armature reaction which decreases the flux. The drop in terminal voltage reduces the field current which in turn reduces the flux, hence the induced emf, etc. At some point $B$, usually well above rated current,

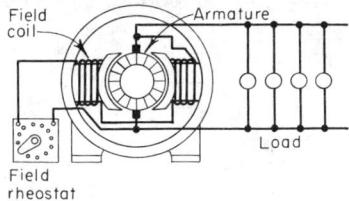

**Fig. 50**  Shunt generator.

the foregoing reactions become cumulative and the generator starts to break down. The current reaches a maximum value and then decreases to nearly zero at short circuit. With large machines, point $B$ is well above rated current, the operating range being between $O$ and $A$. The voltage may be main-

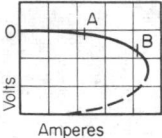

**Fig. 51**  Shunt generator characteristic.

tained constant by means of the field rheostat. Automatic regulators which operate through field resistance are frequently used to maintain constant voltage.

Shunt generators are commonly used in city substations which are all tied together through the network of feeders and mains. Their stability when in parallel is a distinct advantage for this service. If a generator fails to build up (1) the load may be connected; (2) the field resistance may be too high; (3) the field circuit may be open; (4) the residual magnetism may be insufficient; (5) the field connection may be reversed.

**Series Generator**  In the series generator (Fig. 52) the entire load current flows through the field winding, which

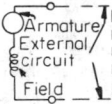

**Fig. 52**  Series generator.

consists of relatively few turns of wire of sufficient size to carry the entire load current without undue heating. The field excitation, and hence the terminal voltage, depends on the magnitude of the load current. The generator supplies an essentially constant current and for years was used to supply series arc lamps for street lighting requiring direct current. Except for some special applications, the series generator is now obsolete.

**Compound-wound Generators**  By the addition of a series winding to a shunt generator the terminal voltage may be automatically maintained very nearly constant, or, by properly proportioning the series turns, the terminal voltage may be made to increase with load to compensate for loss of voltage in the line, so that approximately constant voltage is maintained at the load. If the shunt field is connected outside the series field (Fig. 53), the machine is **long shunt;** if the shunt field is connected inside the series field, i.e., directly to the armature terminals, it is **short shunt.** So far as the operating characteristic is concerned, it makes little difference which way a machine is connected.

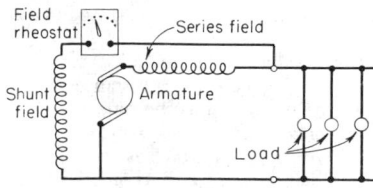

**Fig. 53**  Compound-wound dc generator.

Compound-wound generators are chiefly used for small isolated plants and for generators supplying a purely motor load subject to rapid fluctuations such as in railway work. When first putting a compound generator in service, the shunt field must be so connected that the machine builds up. The series field is then connected so that it aids the shunt field. Figure 54 gives the characteristics of an overcompounded 200-kW 600-V compound-wound generator.

**Table 10. Approximate Test Performance of Compound-wound DC Generators with Commutating Poles**

| kW | Rpm | Volts | Amperes | Efficiencies, percent | | |
|---|---|---|---|---|---|---|
| | | | | $^1/_4$ load | $^1/_2$ load | Full load |
| 5 | 1,750 | 125 | 40 | 77.0 | 80.5 | 82.0 |
| 10 | 1,750 | 125 | 80 | 80.0 | 83.0 | 85.0 |
| 25 | 1,750 | 125 | 200 | 84.0 | 86.5 | 88.0 |
| 50 | 1,750 | 125 | 400 | 83.0 | 86.0 | 88.0 |
| 100 | 1,750 | 125 | 800 | 87.0 | 88.5 | 90.0 |
| 200 | 1,750 | 125 | 1,600 | 88.0 | 90.5 | 91.0 |
| 400 | 1,750 | 250 | 1,600 | 91.7 | 91.9 | 91.7 |
| 1,000 | 1,750 | 250 | 4,000 | 92.1 | 92.6 | 92.1 |

Source: Westinghouse Electric Corp.

**Amplidynes**  The amplidyne is a dc generator in which a small amount of power supplied to a control field controls the generator output, the response being nearly proportional to the control field input. The amplidyne is a dc amplifier which can supply large amounts of power. The amplifier operates on the principle of armature reaction. In Fig. 55, NN and SS are the conventional north and south poles of a dc generator with central cavities. BB are the usual brushes placed at right angles to the pole axes of NN and SS. A control winding CC of small rating, as low as 100 W, is wound on the field poles. In Fig. 55, for simplicity, the control winding is shown as being wound on one pole only. The brushes BB are short-circuited, so that a small excitation mmf in the control field produces a large short-circuit current along the brush axis BB. This large short-circuit current produces a large armature-reaction flux AA along brush axis BB. The armature rotating in this field produces a large voltage along the brush axis B'B'. The load or working current is taken from brushes B'B' as shown. In Fig. 55 the working current only is shown by the crosses and dots in the circles. The short-circuit current would be shown by crosses in the conductors to the left of brushes BB and by dots in the conductors to the right of brushes BB.

A small current in the control winding produces a high output voltage and current as a result of the large short-circuit current in brushes BB.

In order that the brushes B'B' shall not be short-circuiting conductors which are cutting the flux of poles NN and SS, cavities are cut in these poles. Also the load current from brushes B'B' produces an armature reaction mmf in opposition to flux A'A' produced by the control field CC. Were this mmf not compensated, the flux A'A' and the output of the machine would no longer be determined entirely by the control field. Hence there is a compensating field FF' in series with the armature, which neutralizes the armature-reaction mmf which the load current produces. For simplicity the compensating field is shown on one field pole only.

The amplidyne is capable of controlling and regulating speed, voltage, current, and power with accurate and rapid response. The amplification is from 10,000 to 250,000 times in machines rated from 1 to 50 kW. Amplidynes are frequently used in connection with selsyns and are employed for gun and turret control and for accurate controls in many industrial power applications.

**Parallel Operation of Shunt Generators**  It is desirable to operate generators in parallel so that the station capacity can be adapted to the load. Shunt generators, because of their drooping characteristics (Fig. 51), are inherently stable when in parallel. To connect shunt generators in parallel it is neces-

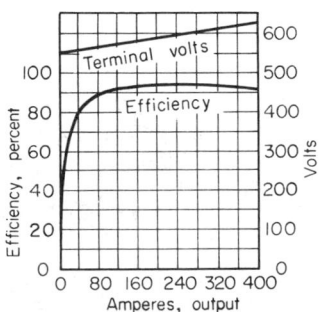

**Fig. 54**  Characteristics of a 200-kW compound-wound dc generator.

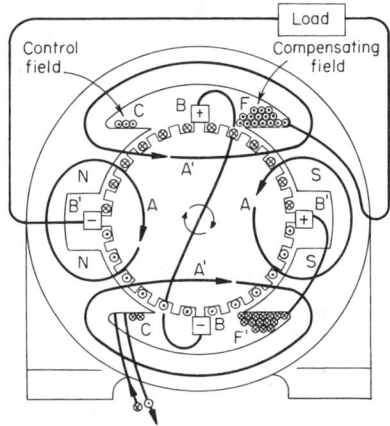

**Fig. 55**  Amplidyne.

sary that the switches be so connected that like poles are connected to the same bus bars when the switches are closed. Assume one generator to be in operation; to connect another generator in parallel with it, the incoming generator is first brought up to speed and its terminal voltage adjusted to a

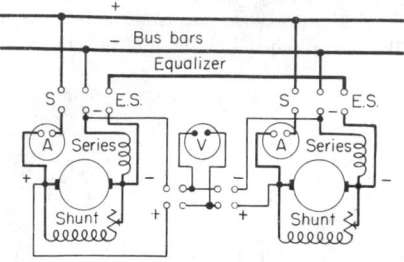

**Fig. 56** Connections for compound-wound generators operating in parallel.

value slightly greater than the bus-bar voltage. This generator may then be connected in parallel with the other without difficulty. The proper division of load between them is adjusted by means of the field rheostats and is maintained automatically if the machines have similar voltage-regulation characteristics.

**Parallel Operation of Compound Generators** As a rule, compound generators have either flat or rising voltage characteristics. Therefore, when connected in parallel, they are inherently unstable. Stability may, however, be obtained by using an equalizer connection, Fig. 56, which connects the terminals of the generator at the junctions of the series fields. This connection is of low resistance so that any increase of current divides proportionally between the series fields of the two machines. The equalizer switch (E.S.) should be closed first and opened last, if possible. In practice, the equalizer switch is often one blade of a three-pole switch, the other two being the bus switch S, as in Fig. 56. When compound generators are used on a three-wire system, two series fields—one at each armature terminal—and two equalizers are necessary. It is possible to operate any number of compound generators in parallel provided their characteristics are not too different and the equalizer connection is used.

## DC MOTORS

Motors operate on the principle that a conductor carrying current in a magnetic field tends to move at right angles to that field (see Fig. 17). The ordinary dc generator will operate entirely satisfactorily as a motor and will have the same rating. The conductors of the motor rotate in a magnetic field and therefore must generate an emf just as does the generator. The induced emf

$$E = K\phi N \tag{88}$$

where $K$ = constant $\phi$ flux entering the armature from one north pole and $N$ = r/min [see Eq. (**86**), ]. This emf is in opposition to the terminal voltage and tends to oppose current entering the armature. Its value is

$$E = V - I_a R_a \tag{89}$$

where $V$ = terminal voltage, $I_a$ = armature current, and $R_a$ = armature resistance [compare with Eq. (**87**) ]. From Eq. (**88**) it is seen that the speed

$$N = K_s E/\phi \tag{90}$$

when $K_s = 1/K$. This is the fundamental speed equation for a motor. By substituting in Eq. (**89**)

$$N = K_s(V - I_a R_a)/\phi \tag{91}$$

which is the general equation for the speed of a motor.

The internal or electromagnetic torque developed by an armature is proportional to the flux and to the armature current; i.e.,

$$T_t = K_t \phi I_a \tag{92}$$

when $K_t$ is a constant. The torque at the pulley is slightly less than the internal torque by the torque necessary to overcome the rotational losses, such as friction, windage, eddy-current and hysteresis losses in the armature iron and in the pole faces.

The total mechanical power developed internally

$$P_m = EI_a \quad \text{W} \quad = \quad EI_a/746 \quad \text{hp} \tag{93}$$

The internal torque thus becomes

$$T = EI_a 33,000/(2\pi \times 746N) = 7.04 EI_a/N \tag{94}$$

Let $VI$ be the motor input. The output is $VI\eta$ where $\eta$ is the efficiency. The horsepower is

$$P_H = VI\eta/746 \tag{95}$$

and the torque is

$$T = 33,000 P_H/2\pi N = 5,260 P_H/N \quad \text{lb·ft} \tag{96}$$

where $N$ is r/min.

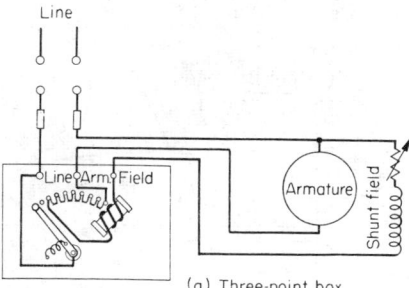

(a) Three-point box

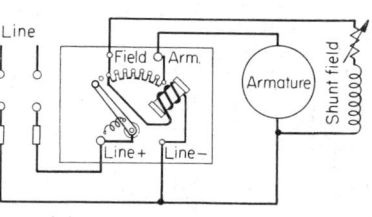

(b) Four-point box

**Fig. 57** Connections for shunt dc motors and starters.

**Shunt Motor** In the shunt motor (Fig. 57) the flux is substantially constant and $I_aR_a$ is 2 to 6 percent of $V$. Hence from Eq. (91), the speed varies only slightly with load (Fig. 58), so that the motor is adapted to work requiring constant speed. The speed regulation of constant-speed motors is defined by the American Definitions of Electrical Terms ANSI C42.100-1972 as follows:

The speed regulation of a constant-speed direct-current motor is the change in speed when the load is reduced gradually from the rated value to zero with constant applied voltage and field rheostat setting expressed as a percent of speed at rated load.

In Fig. 61 the speed regulation under each condition is $100(ac - bc)/bc$ (see Fig. 58a). Also from Eq. (92) it is seen that the torque is practically proportional to the armature current (see Fig. 58b). The motor is able to develop full-load torque and more on starting, but the ordinary starter is not designed to carry the current necessary for starting under load. If a motor is to be started under load, the starter should be provided with resistors adapted to carry the required current without overheating. A controller is also adapted for starting duty under load.

**Commutating poles** have so improved commutation in dc machines that it is possible to use a much shorter air gap than formerly. Since, with the shorter air gap, fewer field ampere-turns are required, the armature becomes magnetically strong with respect to the field. Hence, a sudden overload might weaken the field through armature reaction, thus causing an increase in speed; the effect may become cumulative and the motor run away. To prevent this, modern shunt motors are usually provided with a **stabilizing winding,** consisting of a few turns of the field in series with the armature and aiding the shunt field. The resulting increase of field ampere-turns with load will more than compensate for any weakening of the field through armature reaction. The series turns are so few that they have no appreciable compounding effect. The shunt motor is used to drive constant-speed line shafting, for machine tools, etc. Since its speed may be efficiently varied, it is very useful when **adjustable speeds** are necessary, such as individual drive for machine tools.

**Shunt-Motor Starters** At standstill the counter emf of the motor is zero and the armature resistance is very low. Hence, except in motors of very small size, series resistance in the armature circuit is necessary on starting. The field must, however, be connected across the line so that it may obtain full excitation.

Figure 57 shows the two common types of starting boxes used for starting shunt motors. The armature resistance remains in circuit only during starting. In the **three-point box** (Fig. 57a) the starting lever is held, against the force of a spring, in the running position, by an electromagnet in series with the field circuit, so that, if the field circuit is interrupted or the line voltage becomes too low, the lever is released and the armature circuit is opened automatically. In the **four-point starting box** the electromagnet is connected directly across the line, as shown in Fig. 57b. In this type the arm is released instantly upon failure of the line voltage. In the three-point type some time elapses before the field current drops enough to effect the release. Some starting rheostats are provided with an overload device so that the circuit is automatically interrupted if too large a current is taken by the armature. The four-point box is used where a wide speed range is obtained by

means of the field rheostat. The electromagnet is not then affected by changes in field current.

In large motors and in many small motors, automatic starters are widely used. The advantages of the automatic starter are that the current is held between certain maximum and minimum values so that the circuit does not become opened by too rapid starting as may occur with manual operation; the acceleration is smooth and nearly uniform. Since workmen can stop and start a motor merely by the pushing of a button, there results considerable saving by the shutting down of the motor when it is not needed. Automatic controls are essential to elevator motors so that smooth rapid acceleration with frequent starting and stopping may be obtained. Also automatic starting is very necessary with multiple-unit operation of electric-railway cars and with rolling-mill motors which are continually subjected to rapid acceleration, stopping, and reversing.

**Series Motor** In the series motor the armature and field are in series. Hence, if saturation is neglected, the flux is proportional to the current and the torque [Eq. (92)] varies as the current squared. Therefore any increase in current will produce a much greater proportionate increase in torque (see Fig. 58b). This makes the motor particularly well adapted to traction work, cranes, hoists, fork-lift trucks, and other types of work which require large starting torques. A study of Eq. (91) shows that with increase in current the numerator changes only slightly, whereas the change in the denominator is nearly proportional to the change in current. Hence the speed of the series motor is practically inversely proportional to the current. With overloads the speed drops to very low values (see Fig. 58a). With decrease in load the speed approaches infinity, theoretically. Hence the series motor should always be connected to its load by a direct drive, such as gears, so that it cannot reach unsafe speeds (see Speed Control of Motors). A series-motor starting box with no-voltage release is shown in Fig. 59.

**Differential Compound Motors** The cumulative compound winding of a generator becomes a differential compound winding when the machine is used as a motor. Its speed may be made more nearly constant than that of a shunt motor, or, if desired, it may be adjusted to increase with increasing load.

The speed as a function of armature current is shown in Fig. 58a and the torque as a function of armature current is shown in Fig. 58b.

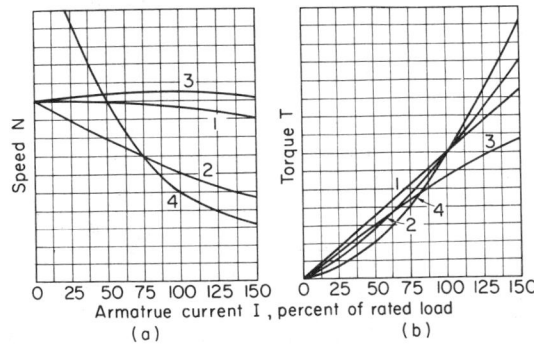

**Fig. 58** Speed and torque characteristics of dc motors: (1) shunt motor; (2) cumulative compound motor; (3) differential compound motor; (4) series motor.

Since the speed of the shunt motor is sufficiently constant for most purposes and the differential motor tends toward instability, particularly in starting and on overloads, the differential motor is little used.

**Cumulative compound motors** develop a more rapid increase in torque with load than shunt motors (Fig. 58b); on the other hand, they have much poorer speed regulation (Fig. 58a).

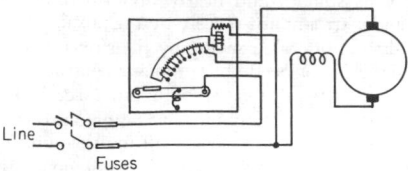

**Fig. 59**  Series motor starter, no-voltage release.

Hence they are used where larger starting torque than that developed by the shunt motor is necessary, as in some industrial drives. They are particularly useful where large and intermittent increases of torque occur as in drives for shears, punches, rolling mills, etc. In addition to the sudden increase in torque which the motor develops with sudden applications of load, the fact that it slows down rapidly and hence causes the rotating parts to give up some of their kinetic energy is another important advantage in that it reduces the peaks on the power plant. Performance data for compound motors are given in Tables 11 and 22.

**Commutation**  The brushes on the commutator of either a motor or generator should be set in such a position that the induced emf in the armature coils undergoing commutation and hence short-circuited by the brushes, is zero. In practice, this condition can at best be only approximately realized. Frequently conditions are such that it is far from being realized. At no load, the brushes should be set in a position corresponding to the geometrical neutral of the machine, for under these conditions the induced emf in the coils short-circuited by the brushes is zero. As load is applied, two factors cause sparking under the brushes. The mmf of the armature, or **armature reaction,** distorts the flux; when the current in the coils undergoing commutation reverses, an emf of self-induction $- L \, di/dt$ tends to prolong the current flow which produces sparking. In a generator, armature reaction distorts the flux in the direction of rotation and the brushes should be advanced. In order to neutralize the emf of self-induction the brushes should be set a little ahead of the neutral plane so that the emf induced in the short-circuited coils by the cutting of the flux at the fringe of the next pole is opposite to this emf of self-

induction. In a motor the brushes are correspondingly moved backward in the direction opposite rotation.

Theoretically, the brushes should be shifted with every change in load. However, practically all dc generators and motors now have **commutating poles** (or **interpoles**) and with these the brushes can remain in the no-load neutral plane, and good commutation can be obtained over the entire range of load. Commutating poles are small poles between the main poles (Fig. 60) and are excited by a winding in series with the armature. Their function is to neutralize the flux distortion in the **neutral plane** caused by armature reaction and also to supply a flux that will cause an emf to be induced in the conductors undergoing commutation, opposite and equal to the emf of self-induction. Since armature reaction and the emf of self-induction are both proportional to the armature current, saturation being neglected, they are neutralized theoretically at every load. Commutating poles have made possible dc generators and motors of very much higher voltage, greater speeds, and larger kW ratings than would otherwise be possible.

Occasionally, the commutating poles may be connected incorrectly. In a motor, passing from an N main pole in the direction of rotation of the armature, an N commutating pole should be encountered as shown in Fig. 60. In a generator

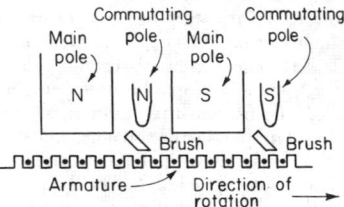

**Fig. 60**  Commutating poles in motor.

under these conditions an S commutating pole should be encountered. The test can easily be made with a compass. If poor commutation is caused by too strong interpoles, the winding may be shunted. If the poles are too weak and the shunting cannot be reduced, they may be strengthened by inserting sheet-iron shims between the pole and the yoke thus reducing the air gap.

Although the emfs induced in the coils undergoing commutation are relatively small, the resistance of the coils themselves is low so that unless further resistance is introduced, the short-circuit currents would be large. Hence, with the exception of certain low-voltage generators, carbon brushes that

**Table 11. Test Performance of Compound-wound DC Motors**

| Hp | Rpm | 115 volts | | 230 volts | | 550 volts | |
|---|---|---|---|---|---|---|---|
| | | Amp | Full-load eff, percent | Amp | Full-load eff, percent | Amp | Full-load eff, percent |
| 1 | 1,750 | 8.4 | 78 | 4.3 | 79 | 1.86 | 73.0 |
| 2 | 1,750 | 16.0 | 80 | 8.0 | 81 | 3.21 | 82.0 |
| 5 | 1,750 | 40.0 | 82 | 20.0 | 83 | 8.40 | 81.0 |
| 10 | 1,750 | 75.0 | 85.6 | 37.5 | 85 | 15.4 | 86.5 |
| 25 | 1,750 | 182.0 | 87.3 | 91.7 | 87.5 | 38.1 | 88.5 |
| 50 | 850 | ..... | .... | 180.0 | 89 | 73.1 | 90.0 |
| 100 | 850 | ..... | .... | 350.0 | 90.5 | 149.0 | 91.0 |
| 200 | 1,750 | ..... | .... | 700.0 | 91 | 295.0 | 92.0 |

SOURCE: Westinghouse Electric Corp.

have relatively large contact resistance are almost always used. Moreover, the graphite in the brushes has a lubricating action, and the usual carbon brush does not score the commutator.

### Speed Control of Motors

**Shunt Motors**  In Eq. **(90)** the speed of a shunt motor $N = K_s E/\phi$, where $K_s$ is a constant involving the design of the motor such as conductors on armature surface and number of poles. Obviously, in order to change the speed of a motor, without changing its construction, two factors may be varied, the counter emf $E$ and the flux $\phi$.

**Armature-Resistance Control**  The counter emf $E = V - I_a R_a$, where $V$ is the terminal voltage, assumed constant. $R_a$ must be small so that the armature heating can be maintained within permissible limits. Under these conditions the speed change with load is small. By inserting an external resistor, however, into the armature circuit the counter emf $E$ may be made to decrease rapidly with increase in load; that is, $E = V - I_a (R_a + R)$ [see Eq. **(89)**] where $R$ is the resistance of the external resistor. The resistor $R$ must be inserted in the **armature** circuit only. The advantages of this method are its simplicity, the full torque of the motor is developed at any speed, and the method introduces no commutating difficulties. Its disadvantages are the increased speed regulation with change of load (Fig. 61), the low efficiency, particularly at the

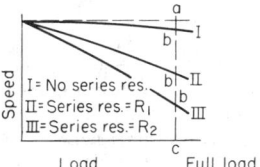

**Fig. 61**  Speed-load characteristics with armature resistance control.

lower speeds, and the fact that provision must be made to dissipate the comparatively large power losses in the series resistor. Figure 61 shows typical speed-load curves without and with series resistors in the armature circuit. The armature efficiency is nearly equal to the ratio of the operating speed to the no-load speed. Hence at 25 percent speed the armature efficiency is practically 25 percent. Frequently the controlling and starting resistors are one, and the device is called a **controller.** Starting rheostats themselves are not designed to carry the armature current continuously and must not be used as controllers. The armature-resistance method of speed control is frequently used to regulate the speed of ventilating fans where the power demand diminishes rapidly with decrease in speed.

**Control by Changing Impressed Voltage**  From Eq. **(91)** it is evident that the speed of a motor may be changed if $V$ is changed by connecting the armature across different voltages. Speed control by this method is accomplished by having mains (usually four), which are maintained at different voltages, available at the motor.

The shunt field of the motor is generally permanently connected to one pair of mains, and the armature circuit is provided with a controller by means of which the operator can readily connect the armature to any pair of mains. Such a system gives a series of distinct and widely separated speeds and generally necessitates the use of field-resistance control, in combination, to obtain intermediate speeds. This method, known as the **multivoltage method,** has the disadvantage that the

system is expensive, for it requires several generating machines, a somewhat complicated switchboard, and a number of service wires. The system is used somewhat in machine shops and is extensively used for dc elevator starting and speed control.

In the **Ward Leonard system,** the variable voltage is obtained from a separately excited generator whose armature terminals are connected directly to the armature terminals of the working motor. The generator is driven at essentially constant speed by a dc shunt motor if the power supply is direct current, or by an induction motor or a synchronous motor if the power supply is alternating current. The field circuit of the generator and that of the motor are connected across a constant-voltage dc supply. The terminal voltage of the generator, and hence the voltage applied to the armature of the motor, is varied by changing the generator-field current with a field rheostat. The rheostat has a wide range of resistance so that the speed of the motor may be varied smoothly from 0 to 100 percent. Since three machines are involved the system is costly, somewhat complicated, and has low power efficiency. However, because the system is flexible and the speed can be smoothly varied over wide ranges, it is used in many applications, such as elevators, mine hoists, large printing presses, paper machines, and electric locomotives.

The **Ilgner system** is quite similar to the Ward Leonard system, except that a heavy flywheel is provided for the ac motor—dc generator set and the ac motor is a wound-rotor induction motor with external variable resistance in the rotor circuit. As in the Ward Leonard system, the armature of the work motor is connected directly to the armature of the dc generator and the field of each machine is excited from a dc source, usually provided by a small ac motor—dc generator set. The system is particularly designed for sharply varying loads at widely varying speeds, such as for mine hoists and reversible rolling mills. For example, when the motor begins a hoisting operation, the field of the generator is strengthened and much of the power demanded by the motor comes from the kinetic energy stored in the flywheel, whose speed then drops only a few percent. During the descent the motor will accelerate the flywheel and frequently, through regeneration, will restore energy to it. Automatic insertion of resistance in the rotor circuit controls the rate and extent of the delivery of the kinetic energy of the flywheel.

**Control by Changing Field Flux**  Equation **(90)** shows that the speed of a motor is inversely proportional to the flux $\phi$. The flux can be changed either by varying the shunt-field current or by varying the reluctance of the magnetic circuit. The variation of the **shunt-field current** is the simplest and most efficient of all the methods of speed control.

With the ordinary motor, speed variation of 1.5 to 1.0 is obtainable with this method. If attempt is made to obtain greater ratios, severe sparking at the brushes results, owing to the field distortion caused by the armature mmf becoming large in comparison with the weakened field of the motor. Speed ratios of 5:1 and higher are, however, obtainable with motors which have commutating poles. Since the field current is a small proportion of the total current (1 to 3 percent), the rheostat losses in the field circuit are always small. This method is efficient. Also for any given speed adjustment the speed regulation is excellent, which is another advantage. Because of its simplicity, efficiency, and excellent speed regulation, the control of speed by means of the field current is by far the most common method.

**Speed Control of Series Motors** The series motor is fundamentally a variable-speed motor, the speed varying widely from light load to full load and more (see Fig. 58a). From Eq. **(91)** the speed for any value of $\phi$, or current, can be changed by varying the impressed voltage. Hence the speed can be controlled by inserting resistance in series with the motor. This method, which is practically the same as the armature-resistance control method for shunt motors, has the same objections of low efficiency and poor regulation with fluctuating loads. It is extensively used in controlling the speed of hoist and crane motors.

The **series-parallel** system of series-motor speed control is almost universally used in electric traction. At least two motors are necessary. The two motors are first connected in series with each other and with the starting resistor. The starting resistor is gradually cut out and, since each motor then operates at half line voltage, the speed of each is approximately half speed. Both motors take the same current, and each can develop full torque. This condition of operation is efficient since there is no external resistance in circuit. When the controller is moved to the next position, the motors are connected in parallel with each other and each in series with starting resistors. Full speed of the motors is obtained by gradually cutting out these resistors. Connecting the two motors in series on starting reduces the current to one-half the value that would be required for a given torque were both motors connected in parallel on starting. The power taken from the trolley is halved, and an intermediate running speed is efficiently obtained.

In the **multiple-unit** method of speed control which is used for electric railway trains, the starting contactors, reverser, etc., for each car are located under that car. The relays operating these control devices are actuated by energy taken from the train line consisting usually of seven wires. The train line runs the entire length of the train, the connections between the individual cars being made through the couplers. The train line is energized by the action of the motorman operating any one of the small master controllers which are located in each car. Hence corresponding relays, contactors, etc., in every car all operate simultaneously. High accelerations may be reached with this system because of the large tractive effort exerted by the wheels on every car.

## SYNCHRONOUS GENERATORS

The synchronous generator is the only type of ac generator now in general use.

**Construction** In the usual synchronous generator the armature or stator, is the stationary member. This construction has many advantages. It is possible to make the slots any reasonable depth, since the tooth necks increase in cross section with increase in depth of slot; this is not true of the rotor. The large slot section which is thus obtainable gives ample space for copper and insulation. The conductors from the armature to the bus bars can be insulated throughout their entire lengths, since no rotating or sliding contacts are necessary. The insulation in a stationary member does not deteriorate as rapidly as that on a rotating member, for it is not subjected to centrifugal force or to any considerable vibration.

The **rotating member** is ordinarily the field. There are two general types of field construction: the **salient-pole type** and the **cylindrical**, or **nonsalient-pole, type.** The salient-pole type is used almost entirely for slow and moderate-speed generators since this construction is the least expensive and permits ample space for the field ampere-turns.

It is not practicable to employ salient poles in high-speed turboalternators because of the excessive windage and the difficulty of obtaining sufficient mechanical strength. The **cylindrical type** consists of a cylindrical steel forging with radial slots in which the field copper, usually in strip form, is placed. The fields are ordinarily excited at low voltage, 125 and 250 V, the current being conducted to the rotating member by means of slip rings and brushes. The field power is ordinarily only 1.5 percent and less of the rated power of the machine (see Table 12).

**Classes of Synchronous Generators** Synchronous generators may be divided into three general classes: (1) the slow-speed engine-driven type; (2) the moderate-speed water-wheel-driven type; and (3) the high-speed turbine-driven type. In (1) a hollow box frame is used as the stator support, and the field consists of a spider to which a large number of salient poles are attached, usually bolted. The speed seldom exceeds 75 to 90 r/min, although it may run as high as 150 r/min. Water-wheel generators also have salient poles which are usually dovetailed to a cylindrical spider consisting of steel plates riveted together. Their speeds range from 80 to 900 r/min and sometimes higher, although the 9,000-kVA Keokuk synchronous generators rotate at only 58 r/min, operating at a very low head. The speed rating of direct-connected water-wheel generators decreases with decrease in head. It is desirable to operate synchronous generators at the highest permissible speed since the weight and costs diminish with increase in speed. Waterwheel-driven generators must be able to run at double speed, as a precaution against accident, should the governor fail to shut the gate sufficiently rapidly in case the circuit breakers open or should the governing mechanism become inoperative.

Turbine-driven generators operate at speeds of 720 to 3,600 r/min. Direct-connected exciters, belt-driven exciters from the generator shaft, and separately driven exciters are used. In large stations separately driven (usually motor) exciters may supply the excitation energy to excitation bus bars. Steam-driven exciters and storage batteries are frequently held in reserve. With slow-speed synchronous generators, the belt-driven exciter is frequently used because it can be driven at higher speed, thus reducing the cost.

**Synchronous-Generator Design** At the present time single-phase generators are seldom built. For single-phase service two phases of a standard three-phase Y-connected generator are used. A single-phase load or unbalanced three-phase load produces flux pulsations in the magnetic circuits of synchronous generators, which increase the iron losses and introduce harmonics into the emf wave. Two-phase windings consist of two similar single-phase windings displaced 90 electrical space degrees on the armature and ordinarily occupying all the slots on the armature. The most common type of winding is the three-phase lap-wound two-layer type of winding. In three-phase windings three windings are spaced 120 electrical space degrees apart, the individual phase belts being spaced 60° apart. Usually, all the slots on the armature are occupied. Standard voltages are 550, 1,100, 2,200, 6,600, 13,200, and 20,000 V. It is much more difficult to insulate for 20,000 V than it is for the lower voltages. However, if the power is to be transmitted at this voltage, its use would be justified by the saving of transformers. In machines of moderate and larger ratings it is common to generate at 6,600 and 13,200 V if transformers must be used. The higher voltage is preferable, particularly for the higher ratings, because it reduces the cross section of the connecting leads and bus bars.

The **standard frequency** in the United States for lighting and power systems is 60 Hz; the few former 50-Hz systems have practically all been converted to 60 Hz. The frequency of 25 Hz is commonly used in street-railway and subway systems to supply power to the synchronous converters and other ac-dc conversion apparatus; it is also commonly used in railroad electrification, particularly for single-phase series-motor locomotives (see Sec. 11). At 25 Hz incandescent lamps have noticeable flicker. In European (and most other) countries 50 Hz is standard. The frequency of a synchronous machine

$$f = P \times \text{r/min}/120 \quad \text{Hz} \qquad (97)$$

where $P$ is the number of poles. Synchronous generators are rated in kVA rather than in kW, since heating, which determines the rating, is dependent only on the current and is independent of power factor. If the kilowatt rating is specified, the power factor should also be specified.

**Induced EMF**   The induced emf per phase in synchronous generator is

$$E = 2.22 k_b k_p \Phi f Z 10^{-8} \quad \text{V/phase} \qquad (98)$$

where $k_b$ = breadth factor or belt factor (usually 0.9 to 1.0), which depends on the number of slots per pole per phase, 0.958 for three-phase, four slots per pole per phase; $k_p$ = pitch factor = 1.0 for full pitch, 0.966 for ⅚ pitch; $\Phi$ = total flux Mx, entering armature from one north pole and is assumed to be sinusoidally distributed along the air gap; $f$ = frequency; and $Z$ = number of series conductors per phase.

Synchronous generators usually are Y-connected. The advantages are that for a given line voltage the voltage per phase is $1/\sqrt{3}$ that of the delta-connected winding; third-harmonic currents and their multiples cannot circulate in the winding as with a delta-connected winding; third-harmonic emfs and their multiples cannot exist in the line emfs; a neutral point is available for grounding.

**Regulation**   The terminal voltage of synchronous generator at constant frequency and field excitation depends not only on the current load but on the power factor as well. This is illustrated in Fig. 62, which shows the voltage-current characteristics of a synchronous generator with lagging current, leading current and in-phase current (pf = 1.00). With leading current the voltage may actually rise with increase in load; the rate of voltage decrease with load becomes greater as the lag of the current increases. The regulation of a synchronous generator is defined by the American Definitions of Electrical Terms ANSI C42.100-1972 as follows:

The voltage regulation of a synchronous generator is the rise in voltage with constant field current, when, with the synchronous generator operated at rated voltage and rated speed, the specified load at the specified power factor is reduced to zero, expressed as a percent of rated voltage.

For example, in Fig. 62 the regulation under each condition is

$$100(ac - bc)/bc \qquad (99)$$

With leading current the regulation may be negative.

Three factors affect the regulation of synchronous generators; the **effective armature resistance,** the **armature leakage reactance,** and the **armature reaction.** With alternating current the armature loss is greater than the value obtained by multiplying the square of the armature current by the ohmic resistance. This is due to hysteresis and eddy-current losses in the iron adjacent to the conductor and to the alternating flux producing losses in the conductors themselves. Also the current is not distributed uniformly over conductors in the slot, but the current density tends to be greatest in the top of the slot. These factors all have the effect of increasing the resistance. The ratio of effective to ohmic resistance varies from 1.2 to 1.5. The **armature leakage reactance** is due to the flux produced by the armature current linking the conductors in the slots and also the end connections.

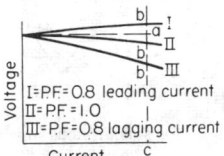

**Fig. 62**   Synchronous generator characteristics.

The armature mmf reacts on the field to change the value of the flux. With a single-phase generator and with an unbalanced load on a polyphase generator, the armature mmf is pulsating and causes iron losses in the field structure. With polyphase machines under a constant balanced load, the armature mmf is practically constant in magnitude and fixed in its relation to the field poles. Its direction with relation to the field-pole axis is determined by the power factor of the load.

A component of current in phase with the no-load induced emf, or the excitation emf, merely distorts the field by strengthening the trailing pole tip and weakening the leading pole tip. A component of current lagging the excitation emf by 90° weakens the field without distortion. A component of current leading the excitation emf by 90° strengthens the field without distortion. Ordinarily, both cross magnetization and one of the other components are acting simultaneously.

The foregoing effects are called **armature reaction.** Frequently the effects of armature reactance and armature reaction can be combined into a single quantity.

It is difficult to determine the regulation of synchronous generator by actual loading, even when in service, owing to the difficulty of obtaining, controlling, and absorbing the large balanced loads. Hence methods of **predetermining regulation without actually loading** the machine are used.

**Synchronous Impedance Method**   Both armature reactance and armature reaction have the same effect on the terminal voltage. In the synchronous impedance method the generator is considered as having no armature reaction, but the armature reactance is increased a sufficient amount to account for the effect of armature reaction. The phasor diagram for a current $I$ lagging the terminal voltage $V$ by an angle $\theta$ is shown in Fig. 63. In a polyphase generator the phasor diagram is applicable to one phase, a balanced load almost always being assumed.

The power factor of the load is cos $\theta$; $IR$ is the effective armature resistance drop and is parallel to $I$; $IX_s$ is the synchronous reactance drop and is at right angles to $I$ and leading it by 90°. $IX_s$ includes both the reactance drop and the drop in voltage due to armature reaction. That part of $IX_s$ which replaces armature reaction is in reality a fictitious quantity. The synchronous impedance drop is given by $IZ_s$. The no-load or open-circuit (excitation) voltage

$$E = \sqrt{(V \cos \theta + IR)^2 + (V \sin \theta \pm IX_s)^2} \quad \text{V} \quad \textbf{(100)}$$

All quantities are per phase. The negative sign is used with leading current.

$$\text{Regulation} = 100(E - V)/V \quad \textbf{(101)}$$

With leading current $E$ may be less than $V$ and a negative regulation results.

The synchronous impedance is determined from an open-circuit and a short-circuit test, made with a weak field. The voltage $E'$ on open circuit is divided by the current $I'$ on short circuit for the same value of field current.

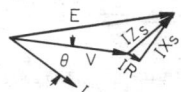

**Fig. 63** Phasor diagram for synchronous impedance method.

$$Z_s = E'/I' \qquad X_s = \sqrt{Z_s^2 - R^2} \qquad \Omega \quad \textbf{(102)}$$

$R$ is so small compared with $X_s$ that for all practical purposes $X_s = Z_s$. $R$ may be determined by measuring the ohmic resistance per phase and multiplying by 1.4 to 1.5 to obtain the effective resistance. This value of $R$ and the value of $X_s$ obtained from Eq. (**102**) may then be substituted in Eq. (**100**) to obtain $E$ at the specified load and power factor.

Since the synchronous reactance is determined at low saturation of the iron and used at high saturation, the method gives regulations that are too large; hence it is called the **pessimistic method.**

**MMF Method** In the mmf method the generator is considered as having no armature reactance but the armature reaction is increased by an amount sufficient to include the effect of reactance. That part of armature reaction which replaces the effect of armature reactance is in reality a fictitious quantity. To obtain the data necessary for computing the regulation, the generator is short-circuited and the field adjusted to give rated current in the armature. The corresponding value of field current $I_a$ is read. The field is then adjusted to give voltage $E'$ equal to rated terminal voltage + $IR$ drop (= $V$ + $IR$, as phasors, Fig. 64) on open circuit and the field current $I'$ read.

$I_a$ is 180° from the current phasor $I$, and $I'$ leads $E'$ by 90° (Fig. 64). The angle between $I'$ and $I_a$ is $90 - \theta + \phi$, but since $\phi$ is small, it can usually be neglected. The phasor sum of $I_a$ and $I'$ is $I_o$. The open-circuit voltage $E$ corresponding to $I_o$ is the no-load voltage and can be found on the saturation curve. The regulation is then found from Eq. (**101**). This method gives a value of regulation less than the actual value and hence is called the **optimistic method.** The actual regulation lies somewhere between the values obtained by the two methods but is more nearly equal to the value obtained by the mmf method.

**ANSI Method** The ANSI method (American Standard 50, Rotating Electrical Machinery) which has become the accepted standard for the predetermination of synchronous generator operation, eliminates in large measure the errors due to saturation which are inherent in the synchronous impedance and mmf methods. In Fig. 65*a* is shown the saturation

curve *OAF* of the generator. The axis *OP* is not only the field-current axis but also the axis of the current phasor $I$ as well. $V$ the terminal voltage is drawn $\theta$ deg from $I$ or *OP*, where $\theta$ is the power factor angle. The effective resistance drop $IR$ and the **leakage reactance** drop $IX$ are drawn parallel and perpendicular to the current phasor. $E_a$, the phasor sum of $V$, $IR$, and $IX$, is the internal **induced** emf. Arcs are swung with $O$ as the center and $V$ and $E_a$ as radii to intercept the axis of ordinates at $B$ and $C$. $OK$, tangent to the straight portion of the saturation curve, is the **air-gap line.** If there is no saturation, $I_v$ is the field current necessary to produce $V$, and $CK$ is the field current necessary to produce $E_a$. The field current $I_s$ is the increase in field current necessary to take into account the saturation corresponding to $E_a$.

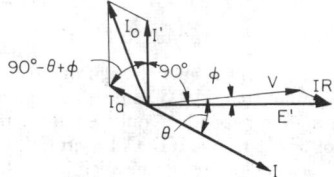

**Fig. 64** Phasor diagram for mmf method.

The corresponding phasor diagram to a larger scale is shown in Fig. 65*b*. $I_f'$, the field current necessary to produce rated current at short circuit, corresponding to $I_a$ (Fig. 64), is drawn horizontally. The field current $I_v$ is drawn at an angle $\theta$ to the right of a perpendicular erected at the right-hand end of $I_f'$. $I_r$ is the resultant of $I_f'$, and $I_v$. $I_s$ is added to $I_r$ giving $I_f$ the resultant field current. The no-load emf $E$ is found on the saturation curve, Fig. 65*a*, corresponding to $I_f = OD$.

**Excitation** is commonly supplied by a small dc generator driven from the generator shaft. On account of commutation, except in the smaller sizes, the dc generator cannot be driven at 3,600 r/min, the usual speed for turbine generators, and belt or gear drives are necessary. The use of the silicon rectifier has made possible simpler means of excitation as well as voltage regulation. In one system the exciter consists of a small rotating-armature synchronous generator (which can run at high speed) mounted directly on the main generator shaft. The three-phase armature current is rectified by three silicon rectifiers and is conducted directly to the main generator field without any sliding contacts. The main generator field current is controlled by the current to the stationary field of the exciter generator. In another system there is no rotating exciter, the generator excitation being supplied directly from the generator terminals, the 13,800 V, three-phase, being stepped down to 115 V, three-phase, by small transformers and rectified by silicon rectifiers. Voltage regulation is obtained by saturable reactors actuated by potential transformers connected across the generator terminals.

Most regulators such as the following operate through the field of the exciter. In the **Tirrill regulator** the field resistance of the exciter is short-circuited temporarily by contacts when the bus-bar voltage drops. Actually, the contacts are vibrating continuously, the time that they are closed depending on the value of the bus-bar voltage. The General Electric Co. manufactures a direct-acting regulator in which the regulating rheostat is part of the regulator itself. The rheostat consists of stacks of graphite plates, each plate being pivoted at the center. Tilting the plates changes the path of the current through the rheostat and thus changes the resistance. The plates are tilted by a sensitive torque armature which is actuated by variations of voltage from the normal value (for regulators employing silicon rectifiers).

**Parallel Operation of Synchronous Generators** The kilowatt division of load between synchronous generators in parallel is determined entirely by the speed-load characteristics of

circulating current to flow between machines. This current puts more electrical load on the machine whose driving torque is increased and tends to produce motor action in the other machines. In an extreme case, the driving torque of one prime mover may be removed entirely, and its generator will operate as a synchronous motor, driving the prime mover mechanically.

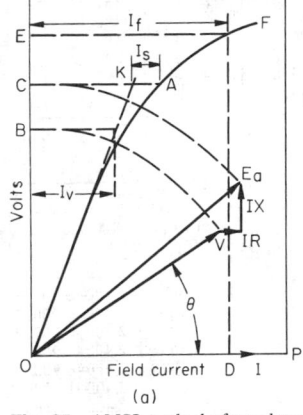

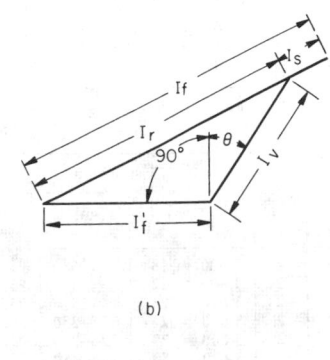

(b)

**Fig. 65** ANSI method of synchronous generator regulation.

their prime movers and not by the characteristics of the generators themselves. No appreciable adjustment of kilowatt load between synchronous generators in parallel can be made by means of their field rheostats, as with dc generators. Consider Fig. 66, which gives the speed-load characteristics in terms of frequency, of two synchronous generators, no. 1 and no. 2, these characteristics being the speed-load characteristics of their prime movers. These speed-load characteristics are drooping, which is necessary for stable parallel operation. The total load on the two machines is $P_1 + P_2$ kW. Both machines must be operating at the same frequency $F_1$. Hence generator 1 must be delivering $P_1$ kW, and generator 2 must be delivering $P_2$ kW (the small generator losses being neglected). If, under the foregoing conditions, the field of either machine is strengthened, it cannot deliver a greater kilowatt load, for its prime mover can deliver more power only by dropping its speed. This is impossible, for both generators must operate always at the same frequency $f_1$. For any fixed total power load, the division of kilowatt load between synchronous generators can be changed only by modifying in some manner the speed-load characteristics of their prime movers, such, for example, as changing the tension in the governor spring. Synchronous generators in parallel are of themselves in stable equilibrium. If the driving torque of one machine is increased, the resulting electrical reactions between the machines cause a

Variations in driving torques cause currents to circulate between synchronous generators, transferring power which tends to keep the generators in synchronism. If the power transfer takes the form of recurring pulsations, it is called **hunting**, which may be reduced by building heavy copper grids called **amortisseurs,** or **damper windings,** into the pole faces. Turbine- and water-wheel-driven synchronous generators are much better adapted to parallel operation than are synchronous generators which are driven by reciprocating engines, because of their uniformity of torque.

Increasing the field current of synchronous generators in parallel with others causes it to deliver a greater lagging component of current. Since the character of the load determines the total current delivered by the system, the lagging components of current delivered by the other generators must decrease and may even become leading components. Likewise if the field of one generator is weakened, it delivers a greater leading component of current and the other machines deliver components of current which are more lagging. These leading and lagging currents do not affect appreciably the division of **kilowatt** load between the synchronous generators. They do, however, cause unnecessary heating in their armatures. The fields of all synchronous generators should be so adjusted that the heating due to the quadrature components of currents is a minimum. With two generators having equal armature resistances, this occurs when both deliver equal quadrature currents.

Armature reactance in the armature of machines in parallel is desirable. If not too great, it stabilizes their operation by producing the synchronizing action. Synchronous generators with too little reactance are sensitive, and if connected in parallel with slight phase displacement or inequality of voltage, considerable disturbance results. Armature reactance also reduces the current on short circuit, particularly during the first few cycles when the short-circuit current is a maximum.

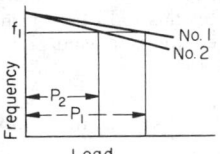

**Fig. 66** Speed-load characteristics of synchronous generators in parallel.

#### Table 12.  Performance Data for Synchronous Generators

80 Percent pf, 3 Phase, 60 Cycle, 240 to 2,400 Volts, Horizontal-coupled or Belted-type Engine-

| kVA | Poles | Rpm | Excitation, kW | Efficiencies, percent | | | Approx net weight lb |
|-----|-------|-----|----------------|-----------|-----------|-----------|---------------------|
| | | | | 1/2 load | 3/4 load | Full load | |
| 25 | 4 | 1,800 | 0.8 | 81.5 | 85.7 | 87.6 | 900 |
| 93.8 | 8 | 900 | 2 | 87 | 89.5 | 90.9 | 2,700 |
| 250 | 12 | 600 | 5 | 90 | 91.3 | 92.2 | 6,000 |
| 500 | 18 | 400 | 8 | 91.7 | 92.6 | 93.2 | 10,000 |
| 1,000 | 24 | 300 | 14.5 | 92.6 | 93.4 | 93.9 | 16,100 |
| 3,125 | 48 | 150 | 40 | 93.4 | 94.2 | 94.6 | 52,000 |

Industrial-size Turbine Generators, Direct-connected Type, 80 Percent pf, 3 Phase, 60 Cycle, Air-cooled

| kVA | Poles | Rpm | Excitation | | Efficiency, percent | | | Volume of air, cfm | Voltage | Approx wt, including exciter, lb |
|-----|-------|-----|-----|-------|------|------|------|--------|---------|------------------------|
| | | | kW | Volts | 1/2 load | 3/4 load | Full load | | | |
| 1,875 | 2 | 3,600 | 18 | 125 | 95.3 | 96.1 | 96.3 | 3,500 | 480–6,900 | 21,900 |
| 2,500 | 2 | 3,600 | 22 | 125 | 95.3 | 96.1 | 96.3 | 5,000 | 2,400–6,900 | 22,600 |
| 3,125 | 2 | 3,600 | 24 | 125 | 95.3 | 96.3 | 96.5 | 5,500 | 2,400–6,900 | 25,100 |
| 3,750 | 2 | 3,600 | 24 | 125–250 | 95.3 | 96.3 | 96.6 | 6,500 | 2,400–6,900 | 27,900 |
| 5,000 | 2 | 3,600 | 29 | 125–250 | 95.3 | 96.3 | 96.6 | 11,000 | 2,400–6,900 | 40,100 |
| 6,250 | 2 | 3,600 | 38 | 125–250 | 95.3 | 96.3 | 96.7 | 12,000 | 2,400–13,800 | 43,300 |
| 7,500 | 2 | 3,600 | 42 | 125–250 | 95.5 | 96.5 | 96.9 | 15,000 | 2,400–13,800 | 45,000 |
| 9,375 | 2 | 3,600 | 47 | 125–250 | 95.5 | 96.5 | 96.9 | 16,500 | 2,400–13,800 | 61,200 |

Central-station-size Turbine Generators, Direct-connected Type, 85 Percent pf, 3 Phase, 60 Cycle, 11,500 to 14,400 Volts

| kVA | Poles | Rpm | Excitation | | Efficiency, percent | | | Volume, of air, cfm | Ventilation | Approx wt, including exciter, lb |
|-----|-------|-----|-----|-------|------|------|------|--------|-------------|------------------------|
| | | | kW | Volts | 1/2 load | 3/4 load | Full load | | | |
| 13,529 | 2 | 3,600 | 70 | 250 | 96.3 | 97.1 | 97.3 | 22,000 | Air-cooled | 116,700 |
| 17,647 | 2 | 3,600 | 100 | 250 | 97.7 | 97.9 | 97.9 | 22,000 | H₂-cooled | 115,700 |
| 23,529 | 2 | 3,600 | 115 | 250 | 98.0 | 98.2 | 98.2 | 25,000 | H₂-cooled | 143,600 |
| 35,294 | 2 | 3,600 | 145 | 250 | 98.1 | 98.3 | 98.3 | 34,000 | H₂-cooled | 194,800 |
| 47,058 | 2 | 3,600 | 155 | 250 | 98.3 | 98.5 | 98.5 | 42,000 | H₂-cooled | 237,200 |
| 70,588 | 2 | 3,600 | 200 | 250 | 98.4 | 98.7 | 98.7 | 50,000 | H₂-cooled | 302,500 |

Source; Westinghouse Electric Corp.

Frequently, external power-limiting reactances are connected in series to protect the generators and equipment from injury that would result from the tremendous short-circuit currents. For these reasons, poor regulation in large synchronous generators is frequently considered to be an advantage rather than a disadvantage.

**Ground Resistors**  Most power systems operate with a grounded neutral. When the station generators deliver current directly to the system (without intervening transformers), it is customary to ground the neutral (of the Y-connected windings) of one generator in a station; this is usually done through a grounding resistor of from 2 to 6 Ω. If the neutral of more than one generator is grounded, third-harmonic (and multiples thereof) currents can circulate between the generators. The ground resistor reduces the short-circuit currents when faults to ground occur, and hence reduces the violence of the short circuit as well as the duty of the circuit breakers. Grounding reactors are sometimes used but have limited application owing to the danger of high voltages resulting from resonant conditions.

## TRANSFORMERS

**Transformer Theory**  The transformer is a device that transfers energy from one electric circuit to another without change of frequency and usually, but not always, with a change in voltage. The energy is transferred through the medium of a magnetic field: it is supplied to the transformer through a primary winding and is delivered by means of a secondary winding. Both windings link the same magnetic circuit. With no load on the secondary, a small current, called the exciting current, flows in the primary and produces the alternating flux. This flux links both primary and secondary windings and induces the same volts per turn in each. With a sine wave the emf is

$$E = 4.44\Phi_m nf 10^{-8} \quad \text{V} \qquad \textbf{(103)}$$

where $\Phi_m$ = maximum instantaneous flux in maxwells, $n$ = turns on either winding, and $f$ = frequency. Equation **(103)** may also be written

$$E = 4.44 B_m A n f 10^{-8} \quad \text{V} \quad \quad \textbf{(104)}$$

$B_M$ = maximum instantaneous flux density in iron and $A$ = net cross section of iron. If $B_m$ is in G, $A$ is in cm²; if $B_m$ is in Mx/in², $A$ is in in².

In SI units, Eq. (103) becomes

$$E = 4.44 \Phi_m n f \quad \text{V} \quad \quad \textbf{(103a)}$$

where $\Phi_M$ is in webers; Eq. (104) becomes

$$E = 4.44 B_m A n f \quad \text{V} \quad \quad \textbf{(104a)}$$

where $B_m$ is in Wb/m² and $A$ is in m².

$B_m$ is practically fixed. In large transformers with silicon steel it varies between 60,000 and 75,000 Mx/in² at 60 Hz and between 75,000 and 90,000 Mx/in² at 25 Hz. It is desirable to operate the iron at as high density as possible in order to minimize the weight of iron and copper. On the other hand, with too high densities the eddy-current and hysteresis losses become too great, and with low frequency the exciting current may become excessive. It follows from Eq. (103) that

$$E_1/E_2 = n_1/n_2 \quad \quad \textbf{(105)}$$

where $E_1$ and $E_2$ are the primary and secondary emfs and $n_1$ and $n_2$ are the primary and secondary turns. Since the impedance drops in ordinary transformers are small, the terminal voltages of primary and secondary are also practically proportional to their number of turns. As the change in secondary terminal voltage in the ordinary constant-potential transformer over its range of operation is small (1.5 to 3 percent), the flux must remain substantially constant. Therefore, the added ampere-turns produced by any secondary load must be balanced by opposite and equal primary ampere-turns. Since the exciting current is small compared with the load current (1.5 to 5 percent) and the two are usually out of phase, the exciting current may ordinarily be neglected. Hence,

$$n_1 I_1 = n_2 I_2 \quad \quad \textbf{(106)}$$
$$I_1/I_2 = n_2/n_1 \quad \quad \textbf{(107)}$$

where $I_1$ and $I_2$ are the primary and secondary currents.

When load is applied to the secondary of a transformer, the secondary ampere-turns reduce the flux slightly. This reduces the counter emf of the primary, permitting more current to enter and thus supply the increased power demanded by the secondary.

Both primary and secondary windings must necessarily have resistance. All the flux produced by the primary does not link the secondary; the counter ampere-turns of the secondary produce some flux which does not link the primary. These **leakage fluxes** produce reactance in each winding. The combined effect of the resistance and reactance produces an impedance drop in each winding when current flows. These impedance drops produce a slight drop in the secondary terminal voltage with load.

**Transformer Testing**  Transformer regulation and losses are so small that it is far more accurate to compute the regulation and efficiency than to determine them by actual measurement. The necessary measurements and computations are comparatively simple, and little power is involved in making the tests. In the **open-circuit** test, the power input to either winding is measured at its rated voltage. Usually it is

more convenient to make this test on the low-voltage winding, particularly if it is rated at 110, 220, or 550 V. The open-circuit power practically all goes to supply the core losses, consisting of eddy-current and hysteresis losses. Let this value of power be $P_0$. The eddy-current loss varies as the square of the voltage and frequency; the hysteresis loss varies as the 1.6 power of the voltage, and directly as the frequency. In the **short-circuit** test one winding is short-circuited, and the current in the other is adjusted to near its rated value. The voltage $V_c$, the current $I_1$, and the power input $P_c$ are measured. When one winding of a transformer is short-circuited, the voltage across the other winding is 3 to 4 percent of rated value when rated current flows. Since a voltage range of from 110 to 250 V is best adapted to measuring instruments, that winding whose rated voltage, multiplied by 0.03 or 0.04, is closest to this voltage range should be used for making the short-circuit test, the other winding being short-circuited. Practically all the power on short circuit goes to supply the copper loss of primary and secondary. If the measurements are made on the primary,

$$R_{01} = P_c/I_1^2 \quad \quad \textbf{(108)}$$
$$Z_{01} = V_c/I_1 \quad \quad \textbf{(109)}$$
$$X_{01} = \sqrt{Z_{01}^2 - R_{01}^2} \quad \quad \textbf{(110)}$$

where $R_{01}$, $Z_{01}$, and $X_{01}$ are the equivalent resistance, impedance, and reactance referred to the primary. Also $R_{02} = R_{01}(n_2/n_1)^2$; $Z_{02} = Z_{01}(n_2/n_1)^2$; $X_{02} = X_{01}(n_2/n_1)^2$, these quantities being the equivalent resistance, impedance, and reactance referred to the secondary. If the dc resistances $R_1$ and $R_2$ of the primary and secondary are measured,

$$R_{01} = R_1 + (n_1/n_2)^2 R_2 \quad \quad \textbf{(111)}$$
$$R_{02} = R_2 + (n_2/n_1)^2 R_1 \quad \quad \textbf{(112)}$$

The ac or effective resistances, determined from Eq. (108), are usually 10 to 15 percent greater than these values.

**Regulation**  The regulation may be computed from the foregoing data as follows:

$$V_1' = \sqrt{(V_1 \cos \theta + I_1 R_{01})^2 + (V_1 \sin \theta \pm I_1 X_{01})^2} \quad \textbf{(113)}$$
$$\text{Regulation} = 100(V_1' - V_1)/V_1 \quad \quad \textbf{(114)}$$

$V_1$ = rated primary terminal voltage; $\cos \theta$ = load power factor; $I_1$ = rated primary current; $R_{01}$ = equivalent resistance referred to primary [from Eq. (108)]; $X_{01}$ = equivalent reactance referred to primary. The + sign is used with lagging current and the − sign with leading current. Equations (113) and (114) are equally applicable to the secondary if the subscripts are changed.

**Efficiency**  The only two losses in a constant-potential transformer are the core loss in W, $P_0$, which is practically independent of load, and $P_c$ the copper loss in W, which varies as the load current squared. The efficiency for any current $I_1$ is

$$\eta = V_1 I_1 \cos \theta / (V_1 I_1 \cos \theta + P_0 + I_1^2 R_{01}) \quad \textbf{(115)}$$

Equation (115) applies equally well to the secondary if the subscripts are changed. The maximum efficiency occurs when the core and copper losses are equal.

**All-Day Efficiency**  Since transformers must usually be on the line 24 h/day, part of which time the load may be very light, the all-day efficiency is important. This is equal to the total energy or watthour output divided by the total energy or watthour input for the 24 h. That is,

$$\eta = \frac{(V_1 I_1 \cos \theta_1) t_1 + \cdots}{(V_1 I_1 \cos \theta_1) t_1 + \cdots + (I_1^2 R_{01}) t_1 + \cdots + 24 P_0} \quad (116)$$

where $t_1$ = time in hours that load $V_1 I_1 \cos \theta_1$ is being delivered, etc.

**Polyphase Transformer Connections** Three-phase transformer banks may be connected $\Delta$-$\Delta$, $\Delta$-Y, Y-Y, and Y-$\Delta$. The $\Delta$-$\Delta$ connection is very common, particularly at the lower voltages, and has the important advantage that the bank will operate V-connected if one transformer is disabled. The $\Delta$-Y connection is advantageous for stepping up to high voltages since the secondary of the transformers need be wound only for 58 percent ($1/\sqrt{3}$) of the line voltage; it is also necessary when a four-wire three-phase system is obtained from a three-wire three-phase system since "a floating neutral" on the secondary cannot occur. The Y-Y system may be used for stepping up voltage. It should not be used for obtaining a three-phase four-wire system from a three-phase three-wire system, because of the "floating neutral" on the secondary and the resulting high degree of unbalance of the secondary voltages. The Y-$\Delta$ system may be used to step down high voltages, the reverse of the $\Delta$-Y connection. In the $\Delta$-Y and Y-$\Delta$ systems the ratio of line voltage is obviously not that of the individual transformers. Because of different phase displacement between primaries and secondaries, a $\Delta$-$\Delta$ bank cannot be connected in parallel (on both sides) with a $\Delta$-Y bank, etc., even if they both have the correct voltage ratios between lines.

**Three-phase transformers** combine the magnetic circuits of three single-phase transformers so that they have parts in common. A material saving in cost, in weight, and in space results, the greatest saving occurring in the core and oil. The advantages of three-phase transformers are often outweighed by their lack of flexibility. The failure of a single phase shuts down the entire transformer. With three single units, one unit may be readily replaced with a single spare. The primaries of single-phase transformers may be connected in Y or $\Delta$ at will and the secondaries properly phased. The primaries, as well as the secondaries of three-phase transformers, must be phased.

For the transformation of moderate amounts of power from three-phase to three-phase, two transformers employing either the **V** or **T connection** (Fig. 67) may be used. With each connection the ratio of line voltages is the same as the transformer ratios. In the figure, ratios 10:1 are shown. In the T connection the primary and the secondary of the main transformer must be provided with a center tap to which one end of the teaser transformer is connected. The ratings of these systems

are only 58 percent of the rating of the system using three similar transformers, one for each phase. Owing to dissymmetry, the terminal voltages become somewhat unbalanced even with a balanced load.

To transform from two- to three-phase or the reverse, the **T connection** of Fig. 68 is used. To make the secondary voltages symmetrical a tap (called a **Scott tap**) is brought out at 86.6 percent ($\sqrt{3}/2$) of the primary winding of the teaser transformer as shown in Fig. 68. With balanced no-load voltages the voltages become slightly unbalanced even under a symmetrical load, owing to unequal phase differences in the individual coils. The three-phase neutral $O$ is one-third of the winding of the teaser transformer from the junction. In Fig. 68a the transformation is from three-phase to a two-phase three-wire system. In Fig. 68b the transformation is from three-phase to a four-phase, five-wire system. The voltages are given on the basis of 100-V primaries with 1:1 transformer ratios.

An **autotransformer**, also called **compensator**, consists essentially of a single winding linking a magnetic circuit. Part of the energy is transformed, and the remainder flows through conductively. Suitable taps are provided so that, if the primary voltage is applied to two of the taps, a voltage may be taken from any other two taps. The ratio of voltages is equal practically to the ratio of the turns between their taps. An autotransformer should be installed only when the ratio of transformation is not large. The ratio of power transformed to total power is $1 - n$, where $n$ is the ratio of low-voltage to high-voltage emf. This gives the saving over the ordinary transformer and is greatest when the ratio is not far from unity. Figure 69a shows 100 kW being changed from 3,300 to 2,300 V; 30.3 kW only are being actually transformed, and the remainder of the power flows through conductively. Figure 69b shows how an ordinary 10:1, 10-kW lighting transformer may be connected to boost 110 kW 10 percent in voltage. In Fig. 69b, however, the 230-V secondary must be insulated for 2,300 V to the core and ground. The voltage may likewise be reduced by reversing the 230-V coil. An autotransformer should never be used when it is desired to keep dangerous primary potentials from the secondary. It is used for starting induction motors (Fig. 71) and for a number of similar purposes.

**Data on Transformers** Single-phase 55° self-cooled oil-insulated transformers for 2,300-V primaries, 230/115-V secondaries, and in sizes from 5 to 200 kVA for 60(25) Hz have efficiencies from one-half to full load of about 98 (97 to 98.7) percent and regulation of 1.5 (1.1 to 2.1) percent with pf = 1, and 3.5 (2.7 to 4.1) percent with pf = 0.8. Power transformers with 13,200-V primaries and 2,300-V secondaries in sizes from 667 to 5,000 kVA and for both 60 and 25 Hz have

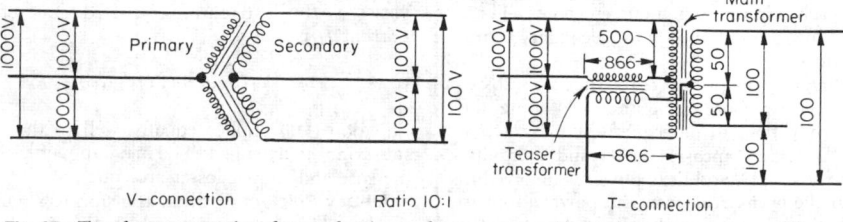

**Fig. 67** Transformer connections for transforming moderate amounts of three-phase power.

efficiencies from one-half to full load of about 99.0 percent and regulation of about 1.0 (4.2) percent with pf = 1 (0.8).

## AC MOTORS

**Polyphase Induction Motor** The polyphase induction motor is the most common type of motor used. It consists ordinarily of a stator which is wound in the same manner as

currents tend to lag the emfs producing them, because of the rotor-leakage reactance. From Eq. **(119)** the rotor frequency and hence the rotor reactance ($x_2 = 2\pi f_2 L_2$) are low when the motor is running near synchronous speed, so that there is a large component of rotor current in space phase with the flux. With large values of slip the increased rotor frequency increases the rotor reactance and hence the lag of the rotor currents behind their emfs, and therefore considerable space-

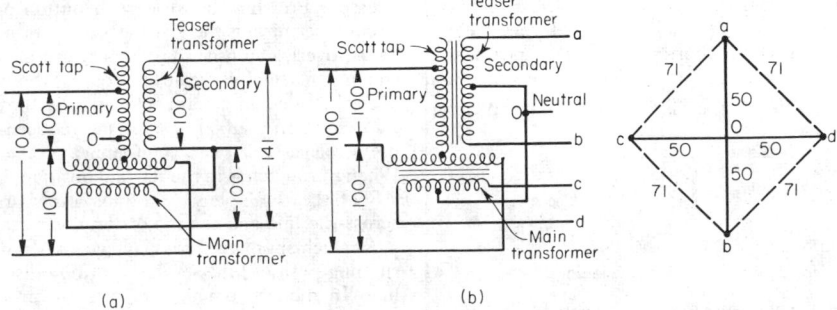

**Fig. 68** Connections for transforming from three-phase to two- and four-phase power.

the synchronous-generator stator. If two-phase current is supplied to a two-phase winding or three-phase current to a three-phase winding, a rotating magnetic field is produced in the air gap. The number of poles which this field has is the same as the number of poles that a synchronous generator employing the same stator winding would have. The speed of the rotating field, or the **synchronous speed,**

$$N = 120f/P \qquad \text{r/min} \qquad (117)$$

where $f$ = frequency and $P$ = number of poles.

There are two general types of rotors. The **squirrel-cage type** consists of heavy copper bars short-circuited by end rings, or the bars and end rings may be an integral aluminum casting. The **wound rotor** has a polyphase winding of the same number of poles as the stator, and the terminals are brought out to slip rings so that external resistance may be introduced. The rotor conductors must be cut by the rotating field, hence the rotor cannot run at synchronous speed but must slip. The **slip** is

$$s = (N - N_2)/N \qquad (118)$$

where $N_2$ = the rotor speed, r/min. The rotor frequency

$$f_2 = sf \qquad (119)$$

The torque is proportional to the air-gap flux and the components of rotor current in space phase with it. The rotor

phase difference between these currents and the flux develops. Consequently even with large values of current the torque may be small. The torque of the induction motor increases with slip until it reaches a maximum value called the **breakdown torque,** after which the torque decreases (see Fig. 72). The breakdown torque varies as the square of the voltage, inversely as the stator impedance and rotor reactance, and is independent of the rotor resistance.

The squirrel-cage motor develops moderate torque on starting ($s = 1.0$) even though the current may be three to seven times rated current. For any value of slip the torque of the induction motor varies as the square of the voltage. The torque of the squirrel-cage motor which, on starting, is only moderate is reduced in the larger motors because of the necessity for applying reduced voltage.

**Polyphase squirrel-cage motors** are used for constant-speed work. They are used widely on account of their rugged construction and the absence of moving electrical contacts, which makes them suitable for operation when exposed to flammable dust or gas. General-purpose squirrel-cage motors have starting torques of about 1.5 times full load torque at rated voltage. The highest torques occur at the higher rated speeds. The locked rotor currents vary between four and seven times full-load current. In the **double-squirrel-cage** type of motor there is a high-resistance winding in the top of the rotor slots and a low-resistance winding in the bottom of the slots. The low-resistance winding is made to have a high leakage reactance, either by separating the windings with a magnetic bridge, Fig. 70a, or by making the slot very narrow in the area between the two windings, Fig. 70b. On starting, because of the high reactance of the low-resistance winding, most of the rotor current will flow in the high-resistance winding, giving the motor a large starting torque. As the rotor approaches the low value of slip at which it normally operates, the rotor frequency and hence the rotor reactance become low and most of the rotor current now flows in the low-resistance winding. The rotor operates with a low value of slip. The high starting

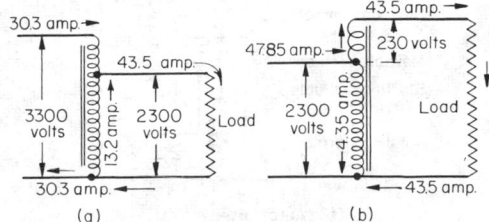

**Fig. 69** Autotransformer.

torque of the high-resistance motor and the excellent constant-speed operating characteristics of the low-resistance squirrel-cage rotor are combined in one motor.

When the bars of both squirrel cages are brazed to a single end ring at each end of the rotor, differential thermal expansion between the shallower bars and the deeper bars may cause cracks to develop in the bars in service. During starting, the temperature of the outer cage rises much more rapidly than the temperature of the inner cage. The larger bars of the inner cage clamp the end rings axially. The smaller bars of the outer cage are prevented from expanding axially. If they yield in compression, they become shorter than the deeper bars when cold and thus develop stress in tension. For this reason cracks may start and gradually progress completely through

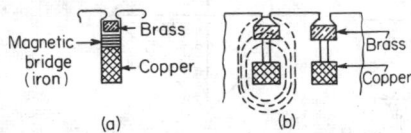

**Fig. 70**  Types of slot for double-squirrel-cage windings.

the bars. This sequence of events may be prevented by the use of a single squirrel cage having bars of variable cross section, such as the inverted T bar or other shapes to enhance skin effect. During starting, high current flows in the stem of the T, but the bar, being one continuous mass, tends to heat up uniformly enough to avoid bar breakage.

Nameplates of polyphase integral hp squirrel cage induction motors carry a *code* letter and a *design* letter. These provide information about motor characteristics, the former on locked rotor or starting inrush current (see Table 26) and the latter on torque characteristics. National Electrical Manufacturers Association standards publication No. MG1-1972 defines four design letters: A, B, C, and D. In all cases the motors are designed for full voltage starting. Locked rotor current and torque, pull-up torque and breakdown torque are tabulated according to horsepower and speed. Designs A, B, and C have full load slips less than 5 percent and design D more than 5 percent. The nature of the various designs can be understood by reference to the full voltage values for a 100-hp, 1800-r/min motor which follow:

|  | Design | | | |
|---|---|---|---|---|
|  | A | B | C | D |
| Locked rotor torque | 125* | 125* | 200* | 275* |
| Pull-up torque | 100† | 100† | 140† | . . . |
| Breakdown torque | 200‡ | 200* | 190* | . . . |
| Locked rotor current | . . . | 690* | 690* | 690* |
| Full load slip (%) | 5§ | 5§ | 5§ | 5‡ |

*Upper limit.
†Not less than.
‡Greater than.
§Less than.
NOTE: All quantities, except slip, are % of full load value.

**Starting**  It is desirable to start induction motors by direct connection across the line, since reduced voltage starters are expensive and almost always reduce the starting torque. In the larger sizes, reduced voltage may be necessary to meet starting-current restrictions, but for the most part full-voltage start

predominates up to 200 hp. For larger ratings it may be necessary to employ reduced-voltage starting.

In Fig. 71a is shown an **"across-the-line" starter** which may be operated from different push-button stations. The START push button closes the solenoid circuit between phases C and A through two bimetallic strips in series. This energizes solenoid S, which attracts armature D, which in turn closes the starting switch and the auxiliary blade G. This blade keeps the solenoid circuit closed when the START push button is released. Pressing the STOP push button opens the solenoid circuit, permitting the starting switch to open. A prolonged heavy overload raises the temperature of the heaters by an amount that will cause at least one of the bimetallic strips to open the solenoid circuit, releasing the starting switch.

A common method of applying reduced-voltage start is to use a **compensator** or **autotransformer** or **autostarter** (Fig. 71b). When the switch is in the starting position, the three windings AB of the three-phase autotransformer are connected in Y across the line and the motor terminals are connected to the taps which supply reduced voltage. When the switch is in the running position, the starter is entirely disconnected from the line. In modern practice, motors are protected by thermal overload relays (Fig. 71) which operate to trip the circuit breaker. Since a time element is involved in the operation of such relays, they do not respond to large starting currents,

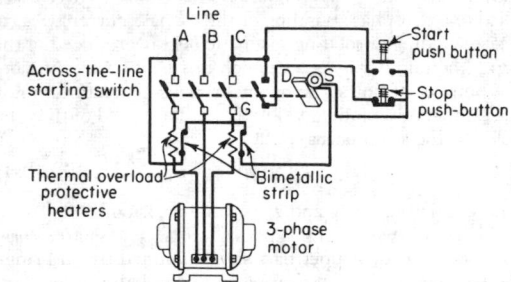

(a) Across-the-line-starter

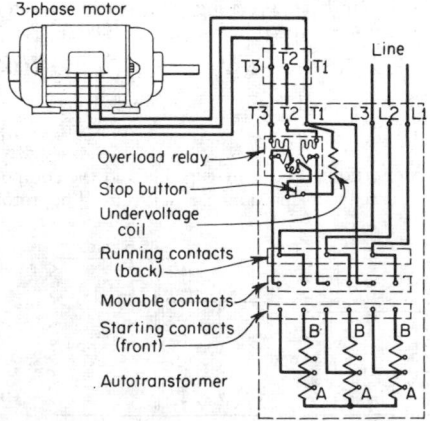

(b) Autostarter

**Fig. 71**  Starters for squirrel-cage induction motor.

because of their short duration. To limit the current to as low a value as possible, the lowest taps that will give the motor sufficient voltage to supply the required starting torque should be used. As the torque of an induction motor varies as the square of the voltage, the compensator produces a very low starting torque.

**Resistors in series with the stator** may also be used to start squirrel-cage motors. They are inserted in each phase and are gradually cut out as the motor comes up to speed. The resistors are generally made of wire-type resistor units or of graphite disks enclosed within heat-resisting porcelain-lined iron tubes. The disadvantage of resistors is that if the motor is started slowly the resistor becomes very hot and may burn out. Resistor starters are less expensive than autotransformers. Their application is to motors that start with light loads at infrequent intervals.

By **introducing resistance into the rotor circuit** through slip rings, the rotor currents may be brought nearly into phase with the air-gap flux and, at the same time, any value of torque up to maximum torque obtained. As the rotor develops speed, resistance may be cut out until there is no external resistance in the rotor circuit. The speed may also be controlled by inserting resistance in the rotor circuit. However, like the armature-resistance method of speed control with shunt motors, this method is also inefficient and gives poor speed regulation. Figure 72 shows graphically the effect on the torque of applying reduced voltage (curves *b*, *c*) and of inserting resistance in the rotor circuit (curve *d*). As shown by curves *b* and *c*, the torque for any given slip is proportional to the square of the line voltage. The effect of introducing resistance into the rotor circuit is shown by curve *d*. The point of maximum torque is shifted toward higher values of slip. The maximum torque at starting (slip − 1.0) occurs when the rotor resistance is equal to the rotor reactance at standstill. The wound-rotor motor is used where large starting torque is necessary as in railway work, hoists, and cranes. It has better starting characteristics than the squirrel-cage motor, but, because of the necessarily higher resistance of the rotor, it has greater slip even with the rotor resistance all cut out. Obviously, the wound rotor, controller, and external resistance make it more expensive than the squirrel-cage type.

One **disadvantage of induction motors** is that they take lagging current, and the power factor at half load and less is low. The speed-and torque-load characteristics of induction motors are almost identical with those of the shunt motor. The speed decreases slightly to full load, the slip being from 10 percent in small motors to 2 percent in very large motors. The torque is almost proportional to the load nearly up to the breakdown torque. The power factor is 0.8 to 0.9 at full load. The direction of rotation of any three-phase motor may be reversed by interchanging any two stator wires.

**Speed Control of Induction Motors** The induction motor inherently is a constant-speed motor. From Eqs. **(117)** and **(118)** the rotor speed is

$$N_2 = 120f(1 - s)/P \qquad (120)$$

The speed can be changed only by changing the frequency, poles, or slip. In some applications where the motors constitute the only load on the generators, as with electric propulsion of ships, their speed may be changed by changing the frequency. Even then the range is limited, for both turbines and generators must operate near their rated speeds for good

efficiency. By employing two distinct windings or by reconnecting a single winding by switching it is possible to change the number of poles. Complications prevent more than two speeds being readily obtained in this manner. Elevator motors frequently have two distinct windings. Another objection to changing the number of poles is the fact that the design is a compromise, and sacrifices of desirable characteristics usually are necessary at both speeds.

The change of slip by introducing resistance into the rotor circuit has been discussed under the wound-rotor motor. It is possible to introduce **counter** emfs into the rotor circuit at slip frequency, by means of special commutator machines. The power, which in resistance control is dissipated as heat, is converted into mechanical power and a portion returned to the line. This method of control is only practicable in very large units and its applications are limited. In the **concatenation method** of speed control the rotors of two wound-rotor motors are mechanically coupled together. The stator of the second motor is supplied with power from the rotor slip rings of the first motor. If both motors have the same number of poles and equal number of turns in both stator and rotor, the set will operate efficiently at a little less than half synchronous speed. The slip of the first motor is therefore slightly greater than 50 percent, and accordingly the frequency of its rotor currents is about half the line frequency. This makes the synchronous speed of the second motor about half that of the first motor. Each motor can operate at rated flux and current so that each can develop rated torque. Because of its complications this method is little used in this country.

**Induction Generator** If an induction motor, while connected to a source of power, is driven above synchronous speed, it becomes an **induction**, or **asynchronous**, **generator** and returns electrical energy to the line without any change in connections. When driven above synchronous speed, the rotor conductors cut the rotating field in a direction opposite to that when operating as a motor, and hence the mechanical power applied to the shaft is converted into electrical power. The load increases with the negative slip; this permits induction generators to be driven by prime movers without governor control. On the short circuit, the induction generator has the desirable characteristic that after the first few cycles it does not deliver any power. It must always be used in parallel with some synchronous apparatus or a capacitor bank. Since it must take lagging current from the line for its own excitation, and in addition cannot deliver any lagging current to the

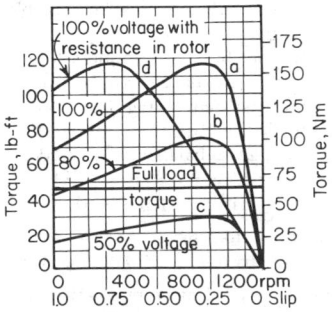

**Fig. 72** Speed-torque curve for 10-hp, 60-Hz, 1140-r/min induction motor.

**Table 13. Performance Data for Induction Motors**

3 PHASE, 220 VOLTS, 60 Hz, 1,750 RPM, SQUIRREL-CAGE TYPE

| Hp | Weight, lb | Amp | Power factor, percent | | | Efficiency, percent | | |
|---|---|---|---|---|---|---|---|---|
| | | | ½ load | ¾ load | Full load | ½ load | ¾ load | Full load |
| 1 | 65 | 3.1 | 59 | 72.6 | 80.4 | 74.5 | 78.2 | 78.8 |
| 2 | 90 | 6.06 | 62.4 | 73.3 | 80.3 | 77.8 | 80.5 | 80.5 |
| 5 | 145 | 13.84 | 69.5 | 79.4 | 83.6 | 83 | 84.3 | 84.7 |
| 10 | 225 | 26.1 | 78.7 | 85.6 | 87.6 | 86.5 | 86.7 | 85.6 |
| 20 | 380 | 50 | 86.7 | 89 | 90 | 89 | 89 | 87.3 |
| 40 | 515 | 98 | 80.4 | 86.4 | 89 | 90.8 | 90.8 | 90.4 |
| 100 | 1,100 | 233 | 89.0 | 91 | 91 | 90 | 90.5 | 91 |
| 200 | 2,500 | 465 | 89.0 | 91 | 91 | 90 | 91 | 91.5 |

3 PHASE, 220 VOLT, 60 Hz, 1,750 RPM, WOUND-ROTOR TYPE

| | | | | | | | | |
|---|---|---|---|---|---|---|---|---|
| 5 | 220 | 14.3 | 72.5 | 80 | 82.5 | 78 | 79 | 79.5 |
| 10 | 336 | 26.6 | 69 | 79 | 83 | 83 | 84.5 | 85 |
| 25 | 480 | 32 | 76.5 | 84.3 | 87.5 | 87 | 88.7 | 88.9 |
| 50 | 850 | 121 | 78.8 | 86.5 | 89.7 | 90.9 | 90.3 | 89.9 |
| 100 | 2,618 | 233 | 88 | 90.5 | 89.5 | 86 | 88 | 88 |
| 200 | 3,900 | 473 | 89 | 91 | 92 | 87 | 89 | 90 |

3 PHASE, 2,300 VOLT, 60 Hz, 1,750 RPM, SQUIRREL-CAGE TYPE

| | | | | | | | | |
|---|---|---|---|---|---|---|---|---|
| 300 | 3,200 | 67 | 84.4 | 88.2 | 90.9 | 90.8 | 92.5 | 92.8 |
| 700 | 5,200 | 153 | 85.7 | 89.7 | 91.9 | 91.7 | 93.4 | 93.7 |
| 1,000 | 7,700 | 216 | 85.9 | 89.9 | 92.1 | 92.1 | 93.8 | 94.1 |

3-PHASE, 2,300 VOLT, 60 Hz, 1,750 RPM, WOUND-ROTOR TYPE

| | | | | | | | | |
|---|---|---|---|---|---|---|---|---|
| 300 | 3,900 | 68 | 84.4 | 86.2 | 89.9 | 90.8 | 92.5 | 92.8 |
| 700 | 5,750 | 154 | 82.7 | 87.7 | 90.9 | 91.7 | 93.4 | 93.7 |
| 1,000 | 8,450 | 218 | 82.9 | 87.9 | 91.1 | 92.1 | 93.8 | 94.1 |

SOURCE: Westinghouse Electric Corp.

system, the induction generator is little used for power supply. Induction motors are frequently used in railway work, especially in mountain systems where it is advantageous on the downgrades to permit the motors to operate as induction generators, thus acting as brakes and in addition returning energy to the line (regenerative braking). (See also Sec. 11.)

If one supply line to a polyphase induction motor of reasonable industrial rating, say over 10 hp, were opened while the motor was running, the motor would continue to run, provided the load demanded was low enough. It might soon burn out, however. Such a motor could not start with one line open. The low rotor resistance of integral horsepower motors generally cannot limit the current to acceptable levels when operated in the single-phase mode.

Opening one supply line to an induction motor constitutes the limit of voltage unbalance. In this case one voltage becomes zero. Power system voltages do become unbalanced, but normally not for this extreme.

The percent voltage unbalance is defined in the National Electrical Manufacturers Association (NEMA) Publication No. MG-1, Art. 14.34B (rev. Jan. 1974), as follows:

Percent Voltage Unbalance = 100 × (Maximum voltage deviation from average voltage) ÷ Average voltage

Approximately 3.5 percent voltage unbalance would cause about 25 percent increase in temperature rise at full load. The input current unbalance at full load would probably be 6 to 10 times the input voltage unbalance.

**Single-Phase Induction Motor** Single-phase induction motors are usually made in fractional horsepower ratings, but they are listed by NEMA in integral ratings up to 20 hp. They have relatively high rotor resistances and can operate in the single-phase mode without overheating. Single-phase induction motors are not self-starting.

However, the single-phase motor runs in the direction in which it is started. There are several methods of starting single-phase induction motors. Short-circuited turns, or **shading coils,** may be placed around the pole tips which retard the time phase of the flux in the pole tip, and thus a weak torque in the direction of rotation is produced. A high-resistance starting winding, displaced 90 electrical degrees from the main winding, produces poles between the main poles and so provides a rotating field which is weak but is sufficient to start the motor. This is called the **split-phase** method. In order to minimize overheating this winding is ordinarily cut out by a centrifugal device when the armature reaches speed. In the larger motors a repulsion-motor start is used. The rotor is wound like an ordinary dc armature with a commutator, but with short-circuited brushes pressing on it axially rather than radially. The motor starts as a repulsion motor, developing high torque. When it nears its synchronous speed, a centrifugal device pushes the brushes away from the commutator, and at the same time causes the segments to be short-circuited, thus converting the motor into a single-phase induction motor.

**Capacitor Motors** Instead of splitting the phase by means of a high-resistance winding, it has become almost universal practice to connect a capacitor in series with the auxiliary winding (which is displaced 90 electrical degrees from the main winding). With capacitance, it is possible to make the flux produced by the auxiliary winding lead that produced by the main field winding by 90° so that a true two-phase rotating field results and good starting torque develops. However, the 90° phase relation between the two fields is obtainable at only one value of speed (as at starting), and the phase relation changes as the motor comes up to speed. Frequently the auxiliary winding is disconnected either by a centrifugal switch or a relay as the motor approaches full speed, in which case the motor is called a **capacitor-start** motor. With proper design the auxiliary winding may be left in circuit permanently (frequently with additional capacitance introduced). This improves both the power factor and torque characteristics. Such a motor is called **permanent-split capacitor motor.**

**Phase Converter**  If a polyphase induction motor is operating single-phase, polyphase emfs are generated in its stator by the combination of stator and rotor fluxes. Such a machine can be utilized, therefore, for converting single-phase power into polyphase power and, when so used, is called a **phase converter.** Unless corrective means are utilized, the polyphase emfs at the machine terminals are somewhat unbalanced. The power input, being single-phase and at a power factor less than unity, not only fluctuates but is negative for two periods during each cycle. The power output being polyphase is steady, or nearly so. The cyclic differences between the power output and the power input are accounted for in the kinetic energy stored in the rotating mass of the armature. The armature accelerates and decelerates, but only slightly, in accordance with the difference between output and input. The phase converter is used principally on railway locomotives, since a single trolley wire can be used to deliver single-phase power to the locomotive, and the converter can deliver three-phase power to the three-phase wound-rotor driving motors.

**AC Commutator Motors**  Inherently simple ac motors are not adapted to high starting torques and variable speed. There are a large number of types of commutating motor that have been developed to meet the requirement of high starting torque and adjustable speed, particularly with single phase. These usually have been accompanied by compensating windings, centrifugal switches, etc., in order to overcome low power factors and commutation difficulties. With proper compensation, commutator motors may be designed to operate at a power factor of nearly unity or even to take leading current.

One of the simplest of the **single-phase commutator motors** is the ac series railway motor such as is used on the erstwhile New York, New Haven, and Hartford Railroad. It is based on the principle that the torque of the dc series motor is in the same direction irrespective of the polarity of its line terminals. This type of motor must be used on low frequency, not over 25 Hz, and is much heavier and more costly than an equivalent dc motor. The torque and speed curves are almost identical with those of the dc series motor. Unlike most ac apparatus the power factor is highest at light load and decreased with increasing load. Such motors operate with direct current even better than with alternating current. For example, the New Haven locomotives also operate from the 600-V dc third-rail system (two motors in series) from the New York City line (238th St.) into Grand Central Station. (See also Sec. 11.)

On account of difficulties inherent in ac operation such as commutation and high reactance drops in the windings, it is economical to construct and operate such motors only in sizes adaptable to locomotives, the ratings being of the order of 300 to 400 hp. **Universal motors** are small simple series motors, usually of fractional horsepower, and will operate on either direct or alternating current, even at 60 Hz. They are used for vacuum cleaners, electric drills, and small utility purposes.

**Synchronous Motor**  Just as dc shunt generators operate as motors, a synchronous generator, connected across a suitable ac power supply, will operate as a motor and deliver mechanical power. Each conductor on the stator must be passed by a pole of alternate polarity every half cycle so that at constant frequency the rpm of the motor is constant and is equal to

$$N = 120f/P \quad \text{r/min} \quad \text{(121)}$$

and the speed is independent of the load.

The synchronous motor has the desirable characteristic that its power factor can be varied over a wide range merely by changing the field excitation. With a weak field the motor takes a lagging current. If the load is kept constant and the excitation increased, the current decreases (Fig. 73) and the phase difference between voltage and current becomes less until the current is in phase with the voltage and the power factor is unity. The current is then at its minimum value such as $I_0$, and the corresponding field current is called the **normal excitation.** Further increase in field current causes the armature current to lead and the power factor to decrease. Thus **underexcitation** causes the current to **lag; overexcitation** causes the current to **lead.** The effect of varying the field current at constant values of load is shown by the V-curves (Fig. 73). Unity power factor occurs at the minimum value of armature current, corresponding to normal excitation. The power factor for any point such as $P$ is $I_0/I_1$, leading current. Because of its adjustable power factor, the motor is frequently run light merely to improve power factor or to control the voltage at some part of a power system. When so used the motor is called a **synchronous condenser.** The motor may, however, deliver mechanical power and at the same time take either leading or lagging current. Its common applications are drives for motor-generating sets, ammonia compressors in refrigerating plants, rubber mills, and air compressors. The motor should not be used where fluctuations of torque are violent. As a rule, it

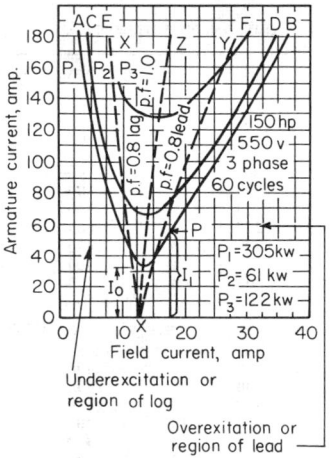

**Fig. 73**  V curves of synchronous motor.

should not be used in small sizes (under 50 hp) since it requires dc excitation, is more difficult to start than induction motors, and falls out of step quite readily when system disturbances occur.

If situated near an inductive load the motor may be overexcited, and its leading current will neutralize entirely or in part the lagging quadrature current of the load. This reduces the $I^2R$ loss in the transmission lines and also increases the kilowatt ratings of the system apparatus. The **synchronous condenser and motor** can also be used to control voltage and to stabilize power lines. If the condenser or motor is overexcited, its leading current flowing through the line reactance causes a rise in voltage at the motor; if it is underexcited, the lagging current flowing through the line reactance causes a drop in voltage at the motor. Thus within limits it becomes possible to control the voltage at the end of a transmission line by regulating the fields of synchronous condensers or motors. Long 220-kV lines and the 287-kV Hoover Dam–Los Angeles line require several thousand kVa in synchronous condensers floating at their load ends merely for voltage control. If the load becomes small, the voltage would rise to very high values if the synchronous condensers were not underexcited, thus maintaining nearly constant voltage.

The synchronous motor is started as an induction motor through the action of a starting or damper winding similar to the squirrel-cage rotor winding of an induction motor. Copper or alloy bars are inserted in the faces of the salient dc field poles, and their ends are brazed to copper segmental end rings bolted together to form, usually, a continuous ring. The process of starting a synchronous motor is simply one of accelerating the motor to as high a speed as it will reach as an induction motor with its damper winding and then applying field excitation in order to pull the rotor into synchronism. Because of its salient poles, the synchronous motor usually pulls into synchronism without dc field excitation.

As with the larger sizes of induction motors, synchronous motors are usually started at reduced voltage, a compensator (Fig. 71b) ordinarily being used. Sometimes the stator winding is connected in Y at starting and in Δ when running. In order to minimize line disturbances, the field is ordinarily

connected while reduced voltage is being applied to the stator, and the connection to the running position is made quickly so that the motor does not have opportunity to drop out of step. All starting functions may be automatically performed by the operation of relays.

It is possible to design synchronous motors for any required values of starting and pull-in torque and, by special methods, meet the low-starting inrush current limitations imposed by power companies.

For example, high starting torque can be obtained by using high-resistance dampers, made of brass or some high-resistivity alloy. The fact that the slip may be too high for synchronizing is overcome by short-circuiting the field winding when the rotor is near synchronism. Synchronous motors may also be provided with phase-wound dampers which are connected to external resistors through slip rings. Just as with the wound-rotor induction motor, high resistance is introduced on starting, and this is cut out as the speed increases.

The **synchronous-induction motor** is fundamentally a wound-rotor slip-ring induction motor with an air gap greater than normal, and the rotor slots are larger and fewer. On starting, resistance is inserted in the rotor circuit to produce high torque, and this is cut out as the speed increases. As synchronism is approached, the rotor windings are connected to a dc power source and the motor operates synchronously.

**Timing** or **clock motors** operate synchronously from ac power systems. Figure 74a illustrates the Warren Telechron motor which operates on the hysteresis principle. The stator consists of a laminated element with an exciting coil, and each pole piece is divided, a short-circuited shading turn being placed on each of the half poles so formed. The rotor consists of two or more hard-steel disks of the shape shown, mounted on a small shaft. The shaded poles produce a 3,600 r/min rotating magnetic field (at 60 Hz), and because of hysteresis loss, the disk follows the field just as the rotor of an induction motor does. When the rotor approaches the synchronous speed of 3,600 r/min, the rotating magnetic field takes a path along the two rotor bars and locks the rotor in with it. The rotor and the necessary train of reducing gears rotate in oil sealed in a small metal can. Figure 74b shows a subsynchronous motor. Six

**Table 14. Performance Data for Coupled Synchronous Motors**

| | | | | | Efficiencies, percent | | | |
|---|---|---|---|---|---|---|---|---|
| hp | Poles | r/min | A | Excitation, kW | $1/2$ load | $3/4$ load | Full load | Weight, lb |
| | | | | UNITY POWER FACTOR, 3 PHASE, 60 Hz, 2,300 V | | | | |
| 50 | 4 | 1,800 | 10.3 | 0.8 | 86.5 | 89.6 | 91 | 1,200 |
| 100 | 8 | 900 | 20.4 | 1.5 | 88.5 | 91 | 92.1 | 2,400 |
| 250 | 12 | 600 | 50.2 | 2.5 | 90.7 | 92.5 | 93.4 | 4,600 |
| 500 | 18 | 400 | 99.3 | 5 | 92.9 | 93.9 | 94.3 | 7,150 |
| 1,000 | 24 | 300 | 197 | 8.4 | 93.7 | 94.6 | 95 | 15,650 |
| 4,000 | 48 | 150 | 781 | 25 | 94.9 | 95.6 | 95 | 54,000 |
| | | | | 80% POWER FACTOR, 3 PHASE, 60 Hz, 2,300 V | | | | |
| 50 | 4 | 1,800 | 13.2 | 1.1 | 84 | 87.8 | 88.8 | 2,100 |
| 100 | 8 | 900 | 25.8 | 2 | 87 | 89.5 | 90.6 | 3,000 |
| 250 | 12 | 600 | 63.6 | 2.8 | 89.5 | 91.2 | 92.1 | 6,100 |
| 500 | 18 | 400 | 125 | 7.2 | 92.4 | 93.4 | 93.6 | 9,500 |
| 1,000 | 24 | 300 | 248 | 11.6 | 93.3 | 94.2 | 94.4 | 17,500 |
| 4,000 | 48 | 150 | 982 | 40 | 94.6 | 95.3 | 95.5 | 115,000 |

SOURCE: Westinghouse Electric Corp.

squirrel-cage bars are inserted in six slots of a solid cylindrical iron rotor, and the spaces between the slots form six salient poles. The motor, because of the squirrel cage, starts as an induction motor, attempting to attain the speed of the rotating field, or 3,600 r/min (at 60 Hz). However, when the rotor reaches 1,200 r/min, one-third synchronous speed, the salient poles of the rotor lock in with the poles of the stator and hold the rotor at 1,200 r/min.

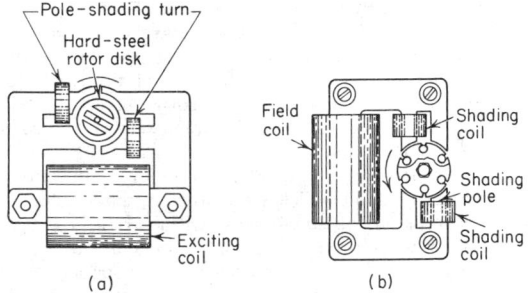

**Fig. 74**  Synchronous motors for timing: (*a*) Warren Telechron motor; (*b*) Holtz induction-reluctance subsynchronous motor.

## SYNCHRONOUS CONVERTERS

The synchronous converter is essentially a dc generator with slip rings connected by taps to equidistant points in the armature winding. Alternating current may also be taken from and delivered to the armature. The machine may be single-phase, in which case there are two slip rings and two slip-ring taps per pair of poles; it may be three-phase, in which case there are three slip rings and three slip-ring taps per pair of poles, etc. Converters are usually used to convert alternating to direct current, in which case they are said to be operating **direct**; they may equally well convert direct to alternating current, in which case they are said to be operating **inverted**. A converter will operate satisfactorily as a dc motor, a synchronous motor, a dc generator, a synchronous generator, or it may deliver direct and alternating current simultaneously, when it is called a **double-current generator.**

The rating of a converter increases very rapidly with increase in the number of phases owing, in part, to better utilization of the armature copper and also because of more uniform distribution of armature heating.

Because of the materially increased rating, converters are nearly all operated six-phase. The rating decreases rapidly with decrease in power factor, and hence the converter should operate near unity power factor (see Table 15). The diametrical ac voltage is the ac voltage between two slip-ring taps 180 electrical degrees apart. With a two-pole closed winding, i.e., a winding that closes on itself when the winding is completed,

the diametrical ac voltage is the voltage between any two slip-ring taps diametrically opposite each other.

With a sine-voltage wave, the dc voltage is the peak of the diametrical ac voltage wave. The voltage relations for sine waves are as follows: dc volts, 141; single phase, diametrical, 100; three-phase, 87; four-phase, diametrical, 100; four phase, adjacent taps, 71; six phase, diametrical, 100; six phase, adjacent taps, 50. These relations are obtained from the sides of polygons inscribed in a circle having a diameter of 100 V, as shown in Fig. 75.

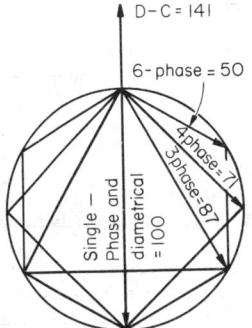

**Fig. 75**  EMF relations in converter.

At unity power factor and 100 percent efficiency, the ratio of alternating to direct current is as follows: two slip rings, 1.41; three slip rings, 0.94; four slip rings, 0.71; six slip rings, 0.47. At other efficiencies and power factors, divide the ratios as given by the product of the efficiency and power factor.

At unity power factor and 0.94 efficiency (a normal value) with three slip rings, the direct and alternating currents are equal; with six slip rings the alternating current is one-half the direct current; 25-Hz converters are slightly more efficient than 60-Hz converters.

The dc voltage of converters can be controlled a limited amount by varying the field current. As with the synchronous motor, decreasing the field current from its normal value causes the armature current to increase and to lag (Fig. 73); increasing the field current from its normal value causes the armature current to increase and to lead. The lagging and leading currents flowing through the transformer, armature, and any other system reactance, lower and raise the voltage at the commutator. The range of dc voltage change obtainable by this method is about 10 percent above and below normal and the power factor is changed simultaneously. The same effect may be obtained by compounding with dc series turns.

With large units, the most satisfactory method of controlling the dc voltage of converters is to use a synchronous generator of smaller rating and of the same number of poles mounted on the same shaft. This generator, called a **booster,** may boost or buck the converter voltage. Converters operate

### Table 15. Relative Outputs of Converters

| Power factor, percent | Continuous-current generator | Single-phase converter | Three-phase converter | Four-phase converter | Six-phase converter |
|---|---|---|---|---|---|
| 100 | 100 | 85 | 132 | 161 | 194 |
| 95.5 | 100 | 78 | 120 | 145 | 170 |
| 90 | 100 | 74 | 109 | 128 | 145 |

satisfactorily in parallel. When used to convert alternating to direct current, the machine must be in synchronism with the alternating supply. The converter may be started from the ac end in much the same manner that synchronous motors are started. Occasionally the machine is brought to speed as a dc motor and synchronized. When operated inverted (direct current to alternating current), some centrifugal or electrical device must be employed to prevent the converter from running away, since a highly inductive load weakens the field through armature reaction and causes the speed to increase.

Converters are cheaper, more efficient, and occupy less floor space than motor-generator sets. They are much less flexible in the matter of voltage and power-factor control. Where they cannot operate near unity power factor and where otherwise transformers are not necessary, their advantage over a motor-generator set diminishes.

Synchronous converters have efficiencies at one-half (full) load from 90 to 92.5 (92.4 to 94.3) percent, the larger sizes having the higher efficiencies. Synchronous booster efficiencies at one-half (full) load vary similarly from 91.3 to 93 (93.8 to 94.4) percent.

For the conversion of alternating to direct current at 600 volts (direct current) and higher, mercury-arc metal-tank rectifiers, particularly ignitrons, rather than motor-generator sets and synchronous converters, are in use. However, because of their small size and weight, low cost, and high efficiency, silicon rectifiers are being used almost exclusively in the newer installations.

**Selsyns**   The word **selsyn** is an abbreviation of self-synchronizing and is applied to devices which are connected electrically, and in which an angular displacement of the rotating member of one device produces an equal angular displacement in the rotating member of the second device. There are several types of selsyns and they may be dc or ac, single-phase or polyphase. A simple and common type is shown in Fig. 76. The two stators $S_1$, $S_2$ are phase-wound stators, identical electrically with synchronous-generator or induction-motor stators. For simplicity Gramme-ring windings are shown in Fig. 76. The two stators are connected three-phase and in parallel. There are also two bobbin-type rotors $R_1$, $R_2$, with single-phase windings, each connected to a single-phase supply such as 115 Volts, 60 Hz. When $R_1$ and $R_2$ are in the same angular positions, the emfs induced in the two stators by the ac flux of the rotors are equal and opposite, there are no interchange currents between stators, and the

system is in equilibrium. However, if the angular displacement of $R_1$, for example, is changed, the magnitudes of the emfs induced in the stator winding of $S_1$ are correspondingly changed. The emfs of the two stators then become unbalanced, currents flow from $S_1$ to $S_2$, producing torque on $R_2$. When $R_2$ attains the same angular position as $R_1$, the emfs in the two rotors again become equal and opposite, and the system is again in equilibrium.

If there is torque load on either rotor, a resultant current is necessary to sustain the torque, so that there must be an angular displacement between rotors. However, by the use of an auxiliary selsyn a current may be fed into the system which is proportional to the angular difference of the two rotors. This current will continue until the error is corrected. This is called **feedback.** There may be a master selsyn, controlling several secondary units.

Two similar wound-rotor induction motors with stators in parallel and rotors in parallel, with either system excited with single-phase or polyphase currents, will operate as a selsyn. Two similar wound-rotor induction motors in operation as motors will keep in exact synchronism if the stators are connected to the same supply and the rotors are connected in parallel. This system is used to maintain constant speed among different parts of a machine, which are driven by different motors.

Selsyns are used for position indicators, e.g., in bridge–engine-room signal systems. They are also widely used for fire control so that from any desired position all the turrets and guns on battleships can be turned and elevated simultaneously through any desired angle with a high degree of accuracy. The selsyn itself rarely has sufficient power to perform these operations, but it actuates control through power multipliers such as amplidynes.

### RATING OF ELECTRICAL APPARATUS

The **rating** of electrical apparatus is almost always determined by the maximum temperature at which the materials in the machine, especially the insulation and lubricant, may be operated for long periods without deterioration. It is permissible, as far as temperature is concerned, to overload the apparatus so long as the safe temperature is not exceeded. The IEEE Standard 100-1972 (ANSI C42.100-1972) classifies **insulating materials** in seven different **classes:**

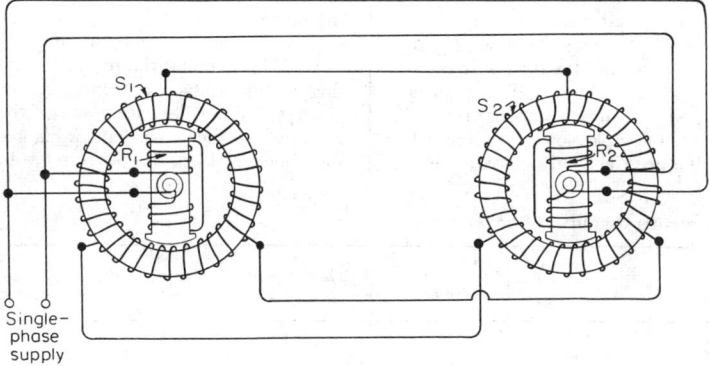

Single-phase supply

**Fig. 76**   Selsyn system.

1. **Class 90 insulation:** materials or combinations of materials such as cotton, silk, and paper without impregnation which will have suitable thermal endurance if operated continually at 90°C.

2. **Class 105 insulation:** materials or combinations of materials such as cotton, silk, and paper when suitably impregnated or coated or when immersed in a dielectric liquid. This class has sufficient thermal endurance at 105°C.

3. **Class 130 insulation:** materials or combinations of materials such as mica, glass fiber, asbestos, etc., with suitable bonding substances. This class has sufficient thermal endurance at 130°C.

4. **Class 155 insulation:** same materials as class 130 but with bonding substances suitable for continuous operation at 155°C.

5. **Class 180 insulation:** materials or combinations of materials such as silicone elastomer, mica, glass fiber, asbestos, etc., with suitable bonding substances such as appropriate silicone resins. This class has sufficient thermal life at 180°C.

6. **Class 220 insulation:** materials suitable for continuous operation at 220°C.

7. **Class over-220 insulation:** materials consisting entirely of mica, porcelain, glass, quartz, and similar inorganic materials which have suitable thermal life at temperatures over 220°C.

**Note:** In all cases, other materials or combinations of materials other than those mentioned above may be used in a given class if from experience or accepted tests they can be shown to have comparable thermal life. It is common practice also to specify **insulation systems in electrical machinery** by letter. For example, integral horsepower ac motors may have a maximum temperature rise in the winding (determined by winding resistance) of 60°C for **class A** insulation, 80°C for **class B**, 105°C for **class F**, and 125°C for **class H**—all based on a 40°C ambient.

The recommended methods of measurement are: (1) the thermometer method is preferred for uninsulated windings, exposed metal parts, gases and liquids, or surface methods generally; thermocouples are preferred for rapidly changing surface temperatures; (2) the applied-thermocouple method is suitable for making surface temperature measurements when it is desired to measure the temperature of surfaces that are accessible to thermocouples but not to liquid-in-glass thermometers; (3) the contact-thermocouple method is suitable for measuring temperatures of bare metal surfaces such as those of commutator bars and slip rings; (4) the resistance method is suitable for insulated windings, except for windings of such low resistance that measurements cannot be accurately made due to uncontrollable resistance in contacts or where it is impracticable to make connections to obtain measurements before an undesirable drop in temperature occurs; (5) the embedded-detector method is suitable for interior measurements at designated locations as specified in the standards for certain kinds of equipment, such as large rotating machines.

### Efficiency of Electrical Apparatus

The losses in dc machinery are classified as follows (IEEE Standard Test Code for Direct Current Machines, Nov. 1, 1973, IEEE Std. 113): (1) shunt-field loss; (2) shunt-field-rheostat loss; (3) exciter losses; (4) friction and windage; (5) brush-friction loss; (6) ventilating loss; (7) core loss; (8) armature $I^2R$ loss; (9) series-field loss; (10) brush contact loss; (11)

stray-load losses. (1), (8), and (9) are determined by measuring the resistance of the windings and correcting to a proper operating temperature by Eq. (6). The rotational losses, i.e., core loss, brush friction, and friction and windage, can be determined by measuring the power required to drive the machine at no load with specified conditions. The determination can be made by electrical input, mechanical input, the retardation method, or a calorimetric method. When the losses are evaluated separately, the calculation is labeled the conventional method. The efficiency is often obtained by direct measurement using torque meters, dynamometers, brakes, loading generators, or calibrated motors. The use of a dc source consisting of rectified ac increases the losses because the applied voltage usually has a large ac component (ripple). In that case, care must be exercised in the choice of measuring instruments to assure, for example, that the rms value of armature current is measured, rather than just the dc component. Stray load loss is taken as 1 percent of the output. Brush drop is taken as 2 V for carbon or graphite brushes with pigtails attached, and 3 V without pigtails.

With a motor,

$$\text{Output} = \text{input} - \text{losses} \qquad \textbf{(122)}$$
$$= VI - [(1) + (8) + (9) + (10) + P_s + (11)]$$

With a generator,

$$\text{Input} = \text{output} + \text{losses}$$
$$\text{Brush loss} = \text{Brush drop} \times I \qquad \textbf{(123)}$$

The motor efficiency is

$$\eta = (\text{input} - \text{losses})/\text{input} \qquad \textbf{(124)}$$

The generator efficiency is

$$\eta = \text{output}/(\text{output} + \text{losses}) \qquad \textbf{(125)}$$

Similar losses occur in synchronous machines.

Temperature rise under full-load conditions can be determined by a full-load heat run or by the pump-back method when another machine of appropriate capacity and similar voltage and speed rating is available. In the pump-back method, the shafts of the machines are mechanically coupled together. One runs as a motor, drawing power from the electric power system, while the other runs as a generator, delivering power to the electric power system. The net power input is the sum of the losses of the two machines.

### Industrial Applications of Motors

**Alternating or Direct Current**  The induction motor, particularly the squirrel-cage type, is preferable to the dc motor for constant-speed work, for the initial cost is less and the absence of a commutator reduces maintenance. Also there is less fire hazard in many industries, such as sawmills, flour mills, textile mills, and powder mills. The use of the induction motor in such places as cement mills is advantageous since with dc motors the grit makes the maintenance of commutators difficult.

For variable-speed work like cranes, hoists, elevators, and for adjustable speeds, the dc motor characteristics are superior to induction-motor characteristics. Even then, it may be desirable to use induction motors since their less desirable characteristics are more than balanced by their simplicity and the fact that ac power is available, and to obtain dc power conversion apparatus is usually necessary. DC power is supplied at 115 and 230 V, 230 V being preferable because of the saving in copper. In certain railway shops where 550 V is available, 550 V motors may be used, but their use, particularly in small sizes, is undesirable because of commutator difficulties. Alternating current is almost always 60 Hz, three-phase, and 220, 440, and 550 V are all used for smaller motors. For larger motors, 1,150, 2,300, and even 6,660 V may be used. Where both lights and motors are to be supplied from the same ac system, the 208/120-V four-wire three-phase system is now in common use. This gives 208 V three-phase for the motors, and 120 V to neutral for the lights.

Full-load speed, temperature rise, efficiency, and power factor as well as breakdown torque and starting torque have long been parameters of concern in the application and purchase of motors. Another qualification is service factor. The service factor of an alternating current motor is a multiplier applicable to the horsepower rating. When so applied, the result is a permissible horsepower loading under the conditions specified for the service factor. When operated at service factor load with 1.15 or higher service factor, the permissible temperature rise by resistance is as follows; class A insulation 70°C, class B, 90°C, and class F, 115°C.

Special enclosures, fittings, seals, ventilation systems, electromagnetic design, etc., are required when the motor is to be operated under unusual service conditions, such as exposure to (1) combustible, explosive, abrasive, or conducting dusts, (2) lint or very dirty conditions where the accumulation of dirt might impede the ventilation, (3) chemical fumes or flammable or explosive gases, (4) nuclear radiation, (5) steam, salt laden air, or oil vapor, (6) damp or very dry locations, radiant heat, vermin infestation, or atmosphere conducive to the growth of fungus, (7) abnormal shock, vibration, or external mechanical loading, (8) abnormal axial thrust or side forces on the motor shaft, (9) excessive departure from rated voltage, (10) deviation factors of the line voltage exceeding 10 percent (11) line voltage unbalance exceeding 1 percent, (12) situations where low noise levels are required, (13) speeds higher than the highest rated speed, (14) operation in a poorly ventilated room, in a pit, or in an inclined attitude, (15) torsional impact loads, repeated abnormal overloads, reversing or electric braking, (16) operation at standstill with any winding continuously energized, and (17) operation with extremely low structureborne and airborne noise. For dc machines, a further unusual service condition occurs when the average load is less than 50 percent over a 24-h period or the continuous load is less than 50 percent over a 4-h period.

The standard direction of rotation for all nonreversing dc motors, ac single-phase motors, synchronous motors, and universal motors is counterclockwise when facing the end of the machine opposite the drive end. For dc and ac generators, the rotation is clockwise.

Further information may be found in Publication No. MG-1 of the National Electrical Manufacturers Association.

It must be recognized that heat is conducted by electrical conductors. Windings in motors operating in a 40°C ambient at class F temperature rises are running at temperatures 90°C higher than the maximum allowable temperature (75°C) of cable ordinarily used in interior wiring. Heat conducted by the motor leads in such a situation could cause a failure of the branch circuit cable in the terminal box. See Tables 21 and 22.

## ELECTRIC DRIVES

**Cranes and Hoists**   The dc series motor is best adapted to cranes and hoists. When the load is heavy the motor slows down automatically and develops increased torque thus reducing the peaks on the electrical system. With light loads, the speed increases rapidly, thus giving a lively crane. The series motor is also well adapted to moving the bridge itself and also the trolley along the bridge. Where alternating current only is available and it is not economical to convert it, the slip-ring type of induction motor, with external-resistance speed control, is the best type of ac motor. Squirrel-cage motors with high resistance end rings to give high starting torque are used (Class D motors; also see Ilgner system).

**Woodworking Machinery**   The log chain which hauls logs from a pond up the inclined slip to the log deck is usually driven by 25 to 75 hp wound-rotor induction motors, the rating depending on the height, the size of logs, and the speed. **Circular saws** are usually driven by a belted squirrel-cage induction motor running at 1,140 and 1,720 r/min, the speeds of the saws being much greater than those of the motors. Squirrel-cage motors operating at 3,600 r/min (synchronous speed) are designed for direct connection. The ratings for woodworking shops vary from 3 to 5 hp for cutoff, 7½ to 15 hp for rip, and about 25 hp for heavy duty. **Band saws** are the first saws to operate on the logs, the head saws having band wheels from 8 to 12 ft in diameter. Such wheels have high inertia and require motors with large starting torque to bring them up to speed rapidly. For this reason wound-rotor induction motors are frequently used. Another advantage is that their higher values of slip permit them to slow down under severe cutting conditions, thus causing the rotating system to give up some of its stored kinetic energy. Synchronous motors are now commonly used, their constant speed producing a straighter cut. Also, electrically, they improve the system power factor. However, they must be provided with damper windings designed to produce high starting torque, and at the same time provide high transient torque when severe cutting conditions are suddenly encountered. In woodworking shops where wheels from 20 to 42 in are employed, squirrel-cage induction motors are used, the horsepowers varying from 1 to 7½ hp and the speeds from 900 to 660 r/min. **Planers** may be driven at high speed by belted squirrel-cage induction motors. In modern practice the motor is an integral part of the machine, using direct drive. Since speeds of 6,000 to 10,000 r/min are necessary and at 60 Hz the maximum speed obtainable is 3,600 r/min, the two-pole motors are supplied at higher frequencies from frequency changers. This is economical only with a number of planers. (See Sec. 13.)

**Pumps**   Single-acting **reciprocating pumps** should be driven with compound motors and duplex and triplex pumps with shunt motors, if direct current is used. Squirrel-cage and slip-ring motors are satisfactory with ac supply. To reduce starting torque a by-pass in the pump is frequently opened until the motor comes up to speed. Constant head requires constant torque, and variable capacity under these conditions necessitates variable speed. For efficient operation, field control should be used with dc motors and pole changing with induction motors. (See Sec. 14.)

**Centrifugal pumps** may be driven by shunt, compound, squirrel-cage, and slip-ring motors. Since such pumps require very small starting torque and operate at high speed, up to 3,600 r/min, both general-purpose squirrel-cage induction motors and synchronous motors make ideal drives. In the larger ratings the synchronous motor is favored because of its ability to improve power factor. (See Sec. 14.)

**Compressors** Shunt motors ordinarily are used where dc power is already available as in mines or with dc electric railways. For light intermittent duty and heavy industrial medium duty and up to 30 hp, low-starting-current, normal-torque squirrel-cage induction motors (Class B motors) with full voltage magnetic starters are used. Above 30 hp, the usual installation will be a normal-starting-current, normal-torque motor (Class A) with a reduced-voltage starter. Above 75 hp and for large heavy duty, the direct-connected synchronous motor is now almost universally used. It is not only efficient but operates to improve the system power factor. With many compressors, a large flywheel effect is desirable, to minimize the pulsations of torque which are reflected in the electric system. (See Sec. 14.)

## SWITCHBOARDS

**Switchboards** may, in general, be divided into four classes: direct-control panel type; remote mechanical-control panel type; direct-control truck type; electrically operated. With **direct-control panel-type boards** the switches, rheostats, bus bars, meters, and other apparatus are mounted on or near the board and the switches and rheostats are operated directly, or by operating handles if they are mounted in back of the board. The voltages, for both direct current and alternating current, are usually limited to 600 V and less but may operate up to 2,500 V ac if oil circuit breakers are used. Such panels are not recommended for capacities greater than 3,000 kVA. **Remote mechanical-control panel-type boards** are ac switchboards with the bus bars and connections removed from the panels and mounted separately away from the load. The oil circuit breakers are operated by levers and rods. This type of board is designed for heavier duty than the direct-control type and is used up to 25,000 kVA. **Direct-control truck-type switchboards** for 15,000 V or less consist of equipment enclosed in steel compartments completely assembled by the manufacturers. The high-voltage parts are enclosed, and the equipment is interlocked to prevent mistakes in operation. This equipment is designed for low- and medium-capacity plants and auxiliary power in large generating stations. **Electrically operated switch-**

boards employ solenoid or motor-operated circuit breakers, rheostats, etc., controlled by small switches mounted on the panels. This makes it possible to locate the high-voltage and other equipment independently of the location of switchboard.

In all large stations the switching equipment and buses are always mounted entirely either in separate buildings or in outdoor enclosures. Such equipment is termed **bus structures** and is electrically operated from the main control board.

Marble has high dielectric qualities and was formerly used exclusively for the panels. It is now used occasionally where its appearance is desired for architectural purposes. Slate is used extensively and is finished in black enamel, marine, and natural black. Ebony asbestos is also used frequently, is lighter than marble or slate, has high dielectric strength and insulation resistivity, and can be readily cut, drilled, and machined. Steel panels, usually ⅛ in thick, are light, economical in construction and erection, and at the present time are favored over other types.

Switchboards should be erected at least 3 or 4 ft from the wall. Switchboard frames and structures should be grounded. The only exceptions are effectively insulated frames of single-polarity dc switchboards. For low-potential work, the conductors on the rear of the switchboard are usually made up of flat copper strip, known as **bus-bar** copper. The size required is based upon a current density of about 1,000 A/in². Figure 77 gives the approximate continuous dc carrying capacity of copper bus bars for different arrangements and spacings for 35°C temperature rise.

Switchboards must be individually adapted for each specific electrical system. Space permits the showing of the diagrams of only three boards each for a typical electrical system (Fig. 78). Aluminum busbars are also frequently used.

**Equipment of Standard Panels** Following are enumerated the various parts required in the equipment of standard panels for varying services:

**Generator or synchronous-converter panel, dc two-wire system:** 1 circuit breaker; 1 ammeter; 1 handwheel for rheostat; 1 voltmeter; 1 main switch (three-pole single throw or double throw) or 2 single-pole switches.

**Generator or synchronous-converter panel, dc three-wire system:** 2 circuit breakers; 2 ammeters; 2 handwheels for field rheostats; 2 field switches; 2 potential receptacles for use with voltmeter; 3 switches; 1 four-point starting switch.

**Generator or synchronous-motor panel, three-phase three-wire sys-**

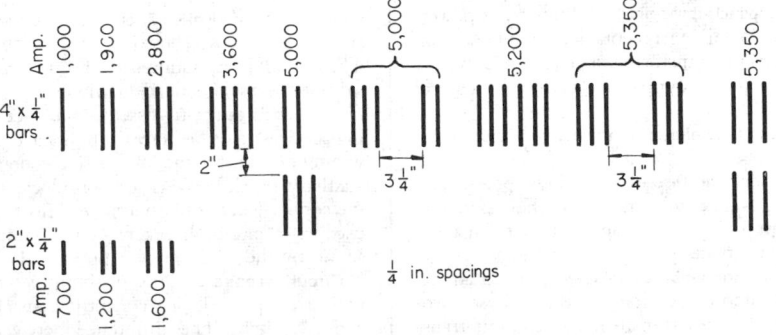

**Fig. 77** Current-carrying capacity of copper bus bars.

**tem:** 3 ammeters; 1 three-phase wattmeter; 1 voltmeter; 1 field ammeter; 1 double-pole field switch; 1 handwheel for field rheostats: 1 synchronizing receptacle (four-point); 1 potential receptacle (eight-point); 1 field rheostat; 1 triple-pole oil switch; 1 power-factor indicator; 1 synchronizer; 2 series transformers; 1 governor control switch.

**Exciter panel (for 1 or 2 exciters):** 1 ammeter (2 for 2 exciters); 1 field rheostat (2 for 2 exciters); 1 four-point receptacle (2 for 2 exciters); 1 equalizing rheostat for regulator.

**Switches**   The current-carrying parts of switches are usually designed for a current density of 1,000 A/in². At contact surfaces, the current density should be kept down to

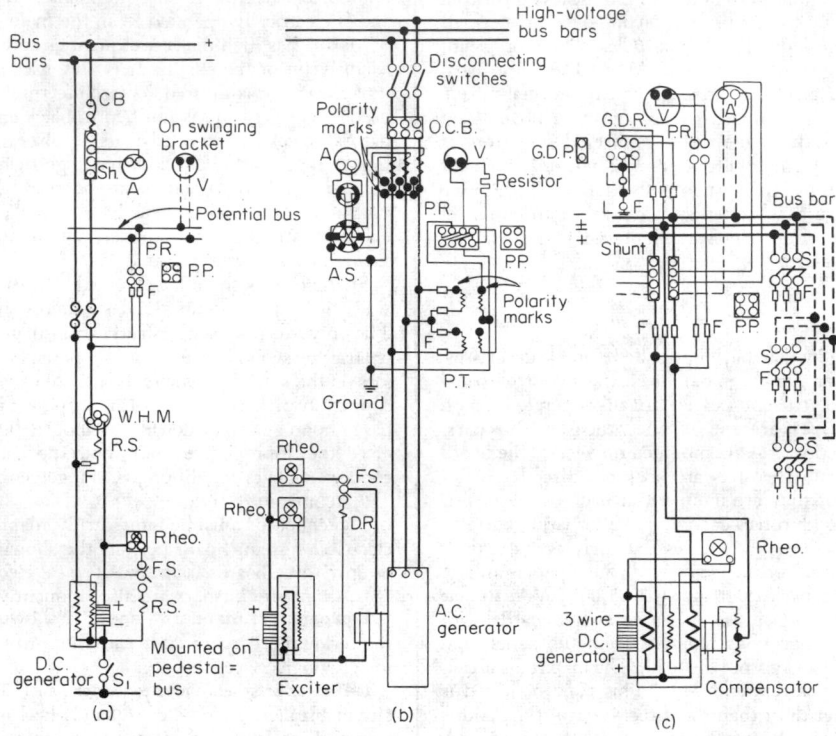

**Fig. 78**   Switchboard wiring diagrams for generators: (*a*) 125- or 250-V dc generator; (*b*) three-phase, synchronous generator and exciter for small or isolated plant; (*c*) three-wire dc generator for small or isolated plant. A., ammeter; A.S., three-way ammeter switch; C.B., circuit breaker; C.T., current transformer; D.R., ground detector receptacle; L., ground detector lamp; O.C., overload coil; O.C.B., oil circuit breaker; P.P., potential ring; P.R., potential receptacle; P.T., potential transformer; Rheo., rheostat; R.S., resistor; S, switch; Sh., shunt; V., voltmeter; W.H.M., watthour meter.

**Synchronous-converter panel, three-phase:** 1 ammeter; 1 power-factor indicator; 1 synchronizing receptacle; 1 triple-pole oil circuit breaker; 2 current transformers; 1 potential transformer; 1 watthour meter (polyphase); 1 governor control switch.

**Induction motor panel, three-phase:** 1 ammeter; series transformers; 1 oil switch.

**Feeder panel, dc, two-wire and three-wire:** 1 single-pole circuit breaker; 1 ammeter; 2 single-pole main switches; potential receptacles (1 four-point for two-wire panel; 1 four-point and 1 eight-point for three-wire panel).

**Feeder panel, three-wire, three-phase and single-phase:** 3 ammeters; 1 automatic oil switch (three-pole for three-phase, two-pole for single-phase); 2 series transformers; 1 shunt transformer; 1 wattmeter; 1 voltmeter; 1 watthour meter; 1 handwheel for control of potential regulator.

about 50 A/in². **Knife switches** are used on low-tension circuits and, in most cases, are mounted on the front of the board. They should be mounted to throw vertically, with the blade side of the switch dead or disconnected from the source of power when open, to lessen the danger of accidental contact. **Copper-brush switches** substitute a leaved copper brush with a wiping contact for the knife-blade contact and make use of an auxiliary break between carbon blocks to prevent burning of the copper leaves due to arcing. This type of switch has been used as a circuit breaker, being rendered automatic in its action by the addition of tripping coils.

**Circuit Breakers**   Any of the foregoing switches equipped with a tripping device constitutes an elementary load interrupter switch. The difference between a load interrupter switch and a circuit breaker lies in the interrupting capacity. A circuit breaker must open the circuit successfully under short circuit conditions when the current through the contacts may be several orders of magnitude greater than the rated current. As the circuit is being opened, the device must withstand the accompanying mechanical forces and the heat of the ensuing arc until the current is permanently reduced to zero.

The opening of a metallic circuit while carrying electric current causes an electric arc to form between the parting contacts. If the action takes place in air, the air is ionized (a plasma is formed) by the passage of current. When ionized, air becomes an electric conductor. The space between the parting contacts thus has relatively low voltage drop and the region close to the surface of the contacts has relatively high voltage drop. The thermal input to the contact surfaces (VI) is therefore relatively large and can be highly destructive. A major aim in circuit breaker design is to quench the arc rapidly enough to keep the contacts in a reusable state. This is done in several ways; (1) lengthening the arc mechanically, (2) lengthening the arc magnetically by driving the current-carrying plasma sideways with a magnetic field, (3) placing barriers in the arc path to cool the plasma and increase its length, (4) displacing and cooling the plasma by means of a jet of compressed air or inert gas, and (5) separating the contacts in a vacuum chamber.

By a combination of shunt and series coils the circuit breaker can be made to trip when the energy reverses. Circuit breakers may trip unnecessarily when the difficulty has been immediately cleared by a local breaker or fuse. In order that service shall not be thus interrupted unnecessarily, **automatically reclosing breakers** are used. After tripping, an automatic mechanism operates to reclose the breaker. If the short circuit still exists, the breaker cannot reclose. The breaker attempts to reclose two or three times and then if the short circuit still exists it remains permanently locked out.

**Metal-clad switch gears** are highly developed pieces of equipment that combine buses, circuit breakers, disconnecting devices, controlling devices, current and potential transformers, instruments, meters, and interlocking devices, all assembled at the factory as a single unit in a compact steel enclosing structure. Such equipment may comprise truck type circuit breakers, assembled as a unit, each housed in a separate steel compartment and mounted on a small truck to facilitate removal for inspection and servicing. The equipment is interlocked to prevent mistakes in operation and in the removal of the unit; the removal of the unit breaks all electrical connections by suitable disconnecting switches in the rear of the compartment, and all metal parts are grounded. This design provides compactness, simplicity, ease of inspection, and safety to the operator.

**High-voltage circuit breakers** are always either of the oil type, in which the contacts open under **oil,** or of the **air-blast** (or oilless) type, in which the arc is extinguished by a powerful blast of air directed through an orifice across the arc and into an arc chute. The tripping of high-voltage circuit breakers is initiated by an abnormal current acting through the secondary of a current transformer on an inverse-time relay in which the time of closing the relay contacts is an inverse time function of the current; i.e., the greater the current the shorter the time of closing. The breaker is tripped by a dc tripping coil, the dc circuit being closed by the relay contacts. Modern circuit breakers should open the circuit within 6 cycles from the time of the closing of the relay contacts.

Air-blast circuit breakers have received wide acceptance in all fields in recent years, both for indoor work and for outdoor applications. Indoor breakers are available up to 40 kV and interrupting capacities up to 2.5 GVA. Outdoor breakers are available in ratings up to that of the EHV (extra-high voltage) 765-kV three-pole breaker capable of interrupting 55 GVA, or 40,000-A symmetrical current. Its operating rating is 3,000 A, 765 kV. The arc is extinguished by a blast of air. Switching stations, gas-insulated and operating at 550 kV, are also in use.

## POWER TRANSMISSION

Power for long-distance transmission is usually generated at 6,600, 13,200 and 18,000 V and is stepped up to the transmission voltage by Δ-Y-connected transformers. The transmission voltage is roughly 1,000 V/mi. Preferred or standard transmission voltages are 22, 33, 44, 66, 110, 132, 154, 220, 287, 330, 500, and 765 kV. High-voltage lines across country are located on private rights of way. When they reach urban areas, the power must be carried underground to the substations which must be located near the load centers in the thickly settled districts. In many cases it is possible to go directly to underground cables since these are now practicable up to 345 kV between three-phase line conductors (200 kV to ground). High-voltage cables are expensive in both first cost and maintenance, and it may be more economical to step down the voltage before transmitting the power by underground cables. Within a city, alternating current may be distributed from a substation at 13,200, 6,600, or 2,300 V, being stepped down to 600, 480, and 240 V, three-phase for power and 240 to 120 V single-phase three-wire for lights, by transformers at the consumers' premises. **Direct current** at 1,200 or 600 V for railways, 230 to 115 V for lighting and power, is supplied by motor-generator sets, synchronous converters, and rectifiers. **Constant current** for series street-lighting systems is obtained through constant-current transformers.

### Transmission Systems

Power is almost always transmitted three-phase. The following fundamental relations apply to any transmission system. The weight of conductor required to transmit power by any given system with a given percentage power loss varies directly with the power, directly as the square of the distance, and inversely as the square of the voltage. The cross-sectional area of the conductors with a given percentage power loss varies directly with the power, directly with the distance, and inversely as the square of the voltage.

For two systems of the same length transmitting the same power at different voltages and with the same power loss for both systems, the cross-sectional area and weight of the conductors will vary inversely as the square of the voltages. The foregoing relations between the cross section or weight of the conductor and transmission distance and voltage hold for all systems, whether dc, single-phase, three-phase, or four-phase. With the power, distance, and power loss fixed, all symmetrical systems having equal voltages to neutral require equal weights of conductor. Thus, the three symmetrical systems shown in Fig. 79 all deliver the same power, have the same power loss and equal voltages to neutral, and the transmission distances are all assumed to be equal. They all require the same weight of conductor since the weights are inversely proportional to all resistances. (No actual neutral conductor is used.) The respective power losses are (1) $2I^2R$ W; (2) $3(2I/3)^2(3R/2) = 2I^2R$ W; (3) $4(I/2)^2(2R) = 2I^2R$ W, which are all equal.

**Size of Transmission Conductor**  Kelvin's law states, "The most economical area of conductor is that for which the annual cost of energy wasted is equal to the interest on that portion of

reactance per mile at 60 Hz and the resistance of stranded and solid copper conductor. (See Table 20.)

Any **symmetrical system** having $n$ conductors can be divided into $n$ equal single-phase systems, each consisting of one wire and a return circuit of zero impedance and each having as its voltage the system voltage to neutral.

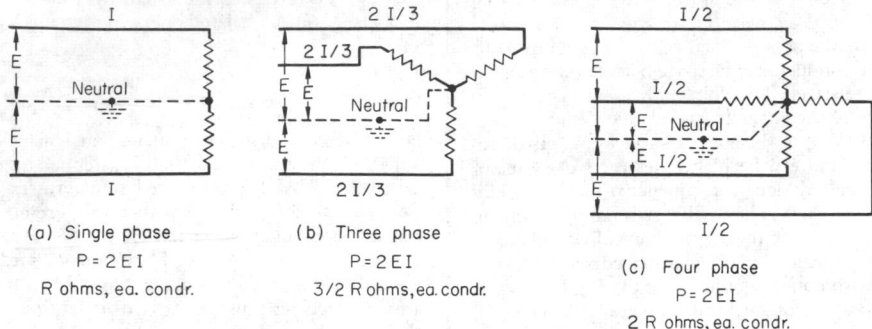

(a)  Single phase

P = 2 E I

R ohms, ea. condr.

(b)  Three phase

P = 2 E I

3/2 R ohms, ea. condr.

(c)  Four phase

P = 2 E I

2 R ohms, ea. condr.

**Fig. 79**  Three equivalent symmetrical transmission, or distribution, systems.

the capital outlay which can be considered proportional to the weight of copper used." In Fig. 80 are shown the annual interest cost, the annual cost of $I^2R$ loss, and the total cost as functions of circular mils cross section for both typical overhead conductors and three-conductor cables. Note that the total-cost curves have very flat minimums, and usually other factors such as the character of the load and the voltage regulation, are taken into consideration.

In addition to resistance, overhead power lines have inductive reactance to alternating currents. The inductive reactance

$$X = 2\pi f \{80 + 741.1 \log [ (D - r)/r ] \} 10^{-6} \quad \Omega/\text{conductor mile} \quad \textbf{(126)}$$

where $f$ = frequency, $D$ = distance between centers of conductors, and $r$ their radius. Table 16 gives the inductive

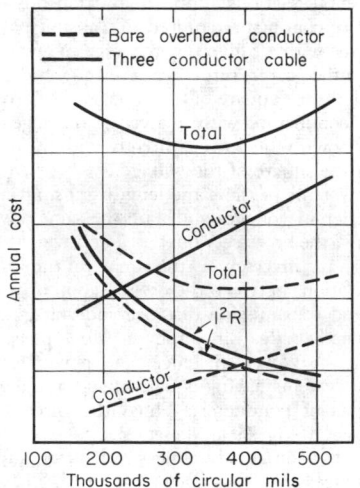

**Fig. 80**  Most economical sizes of overhead and underground conductors.

Figure 81 shows a symmetrical three-phase system, with one phase detached. The load or received voltage between line conductors is $E'_R$ so that the receiver voltage to *neutral* is $E_R = E'_R/\sqrt{3}$ V The current is $I$ A, the load power factor is $\cos \theta$, and the line resistance and reactance are $R$ and $X$ $\Omega$ per wire, and the sending-end voltage is $E_S$. The phasor diagram is shown in Fig. 82 (compare with Fig. 63). Its solution is

$$E_S = \sqrt{(E_R \cos \theta + IR)^2 + (E_R \sin \theta + IX)^2} \quad \textbf{(127)}$$

[see Eq. **(100)**].

Figure 83 (Mershon diagram) shows the right-hand portion of Fig. 82 plotted to large scale, the arc 00 corresponding to the arc $ab$ (Fig. 82). The abscissa 0 (Fig. 83) corresponds to point $b$ (Fig. 82) and is the load voltage $E_R$ taken as 100 percent. The concentric circular arcs 0–40 are given in percentage of $E_R$. To find the sending-end voltage $E_S$ for any power factor $\cos \theta$, compute first the resistance drop $IR$ and the reactance drop $IX$ in percentage of $E_R$. Then follow the ordinate corresponding to the load power factor to the inner arc 00 ($a$, Fig. 82). Lay off the percentage $IR$ drop horizontally to the right, and the percentage $IX$ drop vertically upward. The arc at which the $IX$ drop terminates ($c$, Fig. 82) when added to 100 percent gives the sending-end voltage $E_S$ in percent of the load voltage $E_R$.

EXAMPLE.  Let it be desired to transmit 20,000 kW three-phase 80 percent power factor lagging current, a distance of 60 mi. The voltage

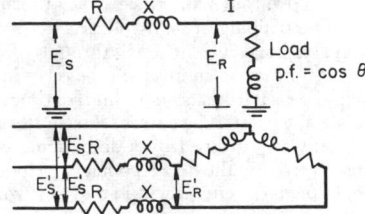

**Fig. 81**  Three-phase power system.

**Table 16. Resistance and Inductive Reactance per Single Conductor**

HARD-DRAWN COPPER, STRANDED

| Size, cir mils or AWG | No. of strands | O D, in | Ohms per mile | 60 Hz — Spacing, ft | | | | | | | | | | | | |
|---|---|---|---|---|---|---|---|---|---|---|---|---|---|---|---|---|
| | | | | 1 | 2 | 3 | 4 | 5 | 6 | 7 | 8 | 10 | 12 | 15 | 20 | 30 |
| 500,000 | 37 | 0.814 | 0.1130 | 0.443 | 0.527 | 0.576 | 0.611 | 0.638 | 0.660 | 0.679 | 0.695 | 0.722 | 0.745 | 0.772 | 0.807 | 0.856 |
| 400,000 | 19 | 0.725 | 0.1426 | 0.458 | 0.542 | 0.591 | 0.626 | 0.653 | 0.675 | 0.694 | 0.710 | 0.737 | 0.760 | 0.787 | 0.822 | 0.871 |
| 300,000 | 19 | 0.628 | 0.1900 | 0.476 | 0.560 | 0.609 | 0.644 | 0.671 | 0.693 | 0.712 | 0.728 | 0.755 | 0.778 | 0.805 | 0.840 | 0.889 |
| 250,000 | 19 | 0.574 | 0.2278 | 0.487 | 0.571 | 0.620 | 0.655 | 0.682 | 0.704 | 0.723 | 0.739 | 0.766 | 0.789 | 0.816 | 0.851 | 0.900 |
| 0000 | 19 | 0.528 | 0.2690 | 0.497 | 0.581 | 0.630 | 0.665 | 0.692 | 0.714 | 0.733 | 0.749 | 0.776 | 0.799 | 0.826 | 0.861 | 0.917 |
| 000 | 7 | 0.464 | 0.339 | 0.518 | 0.602 | 0.651 | 0.686 | 0.713 | 0.735 | 0.754 | 0.770 | 0.797 | 0.820 | 0.847 | 0.882 | 0.931 |
| 00 | 7 | 0.414 | 0.428 | 0.532 | 0.616 | 0.665 | 0.700 | 0.727 | 0.749 | 0.768 | 0.784 | 0.811 | 0.834 | 0.861 | 0.896 | 0.945 |
| 0 | 7 | 0.368 | 0.538 | 0.546 | 0.630 | 0.679 | 0.714 | 0.741 | 0.763 | 0.782 | 0.798 | 0.825 | 0.848 | 0.875 | 0.910 | 0.959 |

HARD-DRAWN COPPER, SOLID

| Size, cir mils or AWG | No. of strands | O D, in | Ohms per mile | 1 | 2 | 3 | 4 | 5 | 6 | 7 | 8 | 10 | 12 | 15 | 20 | 30 |
|---|---|---|---|---|---|---|---|---|---|---|---|---|---|---|---|---|
| 0000 | ... | 0.4600 | 0.264 | 0.510 | 0.594 | 0.643 | 0.678 | 0.705 | 0.727 | 0.746 | 0.762 | 0.789 | 0.812 | 0.839 | 0.874 | 0.923 |
| 000 | ... | 0.4096 | 0.333 | 0.524 | 0.608 | 0.657 | 0.692 | 0.719 | 0.741 | 0.760 | 0.776 | 0.803 | 0.826 | 0.853 | 0.888 | 0.937 |
| 00 | ... | 0.3648 | 0.420 | 0.538 | 0.622 | 0.671 | 0.706 | 0.733 | 0.755 | 0.774 | 0.790 | 0.817 | 0.840 | 0.867 | 0.902 | 0.951 |
| 0 | ... | 0.3249 | 0.528 | 0.552 | 0.636 | 0.685 | 0.720 | 0.747 | 0.769 | 0.788 | 0.804 | 0.831 | 0.854 | 0.881 | 0.916 | 0.965 |
| 1 | ... | 0.2893 | 0.665 | 0.566 | 0.650 | 0.699 | 0.734 | 0.761 | 0.783 | 0.802 | 0.818 | 0.845 | 0.868 | 0.895 | 0.930 | 0.979 |

at the receiving end is 66,000 V, 60 Hz and the line loss must not exceed 10 percent of the power delivered. The conductor spacing must be 7 ft (84 in). Determine the sending-end voltage and the actual efficiency. $I = 20,000,000/(66,000 \times 0.80 \times \sqrt{3}) = 218.8$ A. $3 \times 218.8^2 \times R' = 0.10 \times 20,000,000$. $R' = 13.9 \ \Omega = 0.232 \ \Omega/\text{mi}$. By referring to Table 16, 250,000 cir mils copper having a resistance of $0.2278 \ \Omega/\text{mi}$ may be used. The total resistance $R = 60 \times 0.2278 = 13.67 \ \Omega$. The reactance $X = 60 \times 0.723 = 43.38 \ \Omega$. The volts to

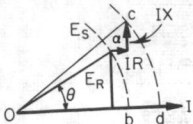

**Fig. 82**  Phasor diagram for power line.

neutral at the load, $E_R = 66,000/\sqrt{3} = 38,100$ V. $\cos \theta = 0.80$; $\sin \theta = 0.60$. Using Eq. **(127)**, $E_S = \{[(38,100 \times 0.80) + (218.8 \times 13.67)]^2 + [(38,100 \times 0.60) + (218.8 \times 43.38)]^2\}^{1/2} = 46,500$ V to neutral or $\sqrt{3} \times 46,500 = 80,500$ between lines at the sending end. The line loss is $3(218.8)^2 \times 13.67 = 1963$ kW. The efficiency $\eta = 20,000/(20,000 + 1963) = 0.911$, or 91.1 percent. This same line is solved by means of the Mershon diagram as follows. Let $E_R = 38,100$ V = 100 percent. $IR = 218.8 \times 13.67 = 2,991$ V = 7.85 percent. $IX = 218.8 \times 43.38 = 9,490$ V = 24.9 percent. Follow the 0.80 power-factor ordinate (Fig. 83) to its intersection with the arc 00; from this point go 7.85 percent horizontally to the right and then 24.9 percent vertically. (These percentages are measured on the horizontal scale.) This last distance terminates on the 22.5 percent arc. The sending-end voltage to neutral is then $1.225 \times 38,100 = 46,500$ V, so that the sending-end voltage between line conductors is $E'_S = 46,500 \sqrt{3} = 80,530$ V.

In Table 16 the spacing is the distance between the centers of the two conductors of a single-phase system or the distance between the centers of each pair of conductors of a three-phase system if they are equally spaced. If they are not equally

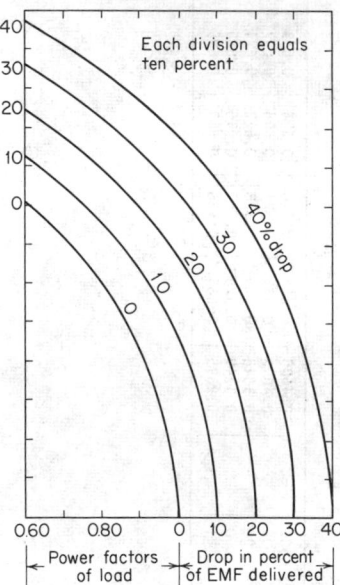

**Fig. 83**  Mershon diagram for determining voltage drop in ac power lines.

spaced, the geometric mean distance GMD is used, where $GMD = \sqrt[3]{D_1 D_2 D_3}$ (Fig. 84a.) With the flat horizontal spacing shown in Fig. 84b, $GMD = \sqrt[3]{2D^3} = 1.26\ D$.

In addition to copper, aluminum cable steel-reinforced (ACSR), Table 17, is used for transmission conductor. For the same resistance it is lighter than copper, and with high voltages the larger diameter reduces corona loss.

Until 1966, 345 kV was the highest operating voltage in the United States. The first 500 kV system put into operation (1966) was a 350-mi transmission loop of the Virginia Electric and Power Company; the longest transmission distance was 170 mi. The towers, about 94 ft high, are of corrosion-resistant steel, and the conductors are 61-strand cables of aluminum alloy, rather than the usual aluminum cable with a steel core (ACSR). The conductor diameter is 1.65 in with two "bundled" conductors per phase and 18-in spacing. The

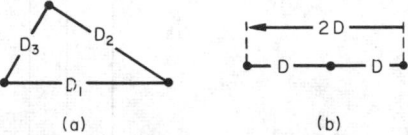

**Fig. 84**  Unequal spacing of three-phase conductors: (a) GMD = $\sqrt[3]{D_1 D_2 D_3}$; (b) flat horizontal spacing; GMD = 1.26 D.

standard span is 1,600 ft, the conductor spacing is flat with 30-ft spacing between phase-conductor centers, and the minimum clearance to ground is 34 to 39 ft. To maintain a minimum clearance of 11 ft to the towers and 30 ft spacing between phases, vee insulator strings, each consisting of twenty-four 10-in disks, are used with each phase. The highest EHV system in North America is the 765-kV system in the midwest region of the United States. A dc transmission line on the west coast of the United States operating at ±450 kV is transmitting power in bulk more than 800 miles.

**High-voltage dc transmission** has a greater potential for savings and a greater ability to transmit large blocks of power longer distances than has three-phase transmission. For the same crest voltage there is a saving of 50 percent in the weight of the conductor. Because of the power stability limit due to inductive and capacitive effects (inherent with ac transmission), the ability to transmit large blocks of power long distances has not kept pace with power developments, even at the present highest ac transmission voltage of 765 kV. With direct current there is no such power stability limit.

Where cables are necessary, as under water, the capacitive charging current may, with alternating current become so large that it absorbs a large proportion, if not all, of the cable-carrying capability. For example, at 132 kV, three-phase (76 kV to ground), with a 500 MCM cable, at 36 mi, the charging current at 60 Hz is equal to the entire cable capability so that no capability remains for the load current. With direct current there is no charging current, only the negligible leakage current, and there are no ac dielectric losses. Furthermore, the dc voltage at which a given cable can operate is twice the ac voltage.

The high dc transmission voltage is obtained by converting the ac power voltage to direct current by means of mercury-arc rectifiers; at the receiving end of the line the dc voltage is inverted back to a power-frequency voltage by means of mercury-arc inverters.

**Table 17. Properties of Aluminum Cable Steel-reinforced (ACSR)**

| Cir mils or AWG | | No. of wires | | OD, in | Cross section, in² | | Total lb/mi | Ω/mi of single conductor at 25°C | | | | |
|---|---|---|---|---|---|---|---|---|---|---|---|---|
| Aluminum | Copper equivalent | Aluminum | Steel | | Aluminum | Total | | 0 amp dc | 200 A | | 600 A | |
| | | | | | | | | | 25 Hz | 60 Hz | 25 Hz | 60 Hz |
| 1,590,000 | 1,000,000 | 54 | 19 | 1.545 | 1.249 | 1.4071 | 10,777 | 0.0587 | 0.0589 | 0.0594 | 0.0592 | 0.0607 |
| 1,431,000 | 900,000 | 54 | 19 | 1.465 | 1.124 | 1.2664 | 9,699 | 0.0652 | 0.0654 | 0.0659 | 0.0657 | 0.0671 |
| 1,272,000 | 800,000 | 54 | 19 | 1.382 | 0.9990 | 1.1256 | 8,621 | 0.0734 | 0.0736 | 0.0742 | 0.0738 | 0.0752 |
| 1,192,500 | 750,000 | 54 | 19 | 1.338 | 0.9366 | 1.0553 | 8,082 | 0.0783 | 0.0785 | 0.0791 | 0.0787 | 0.0801 |
| 1,113,000 | 700,000 | 54 | 19 | 1.293 | 0.8741 | 0.9850 | 7,544 | 0.0839 | 0.0841 | 0.0848 | 0.0843 | 0.0857 |
| 1,033,500 | 650,000 | 54 | 7 | 1.246 | 0.8117 | 0.9170 | 7,019 | 0.0903 | 0.0906 | 0.0913 | 0.0908 | 0.0922 |
| 954,000 | 600,000 | 54 | 7 | 1.196 | 0.7493 | 0.8464 | 6,479 | 0.0979 | 0.0980 | 0.0985 | 0.0983 | 0.0997 |
| 874,500 | 550,000 | 54 | 7 | 1.146 | 0.6868 | 0.7759 | 5,940 | 0.107 | 0.107 | 0.108 | 0.107 | 0.109 |
| 795,000 | 500,000 | 26 | 7 | 1.108 | 0.6244 | 0.7261 | 5,770 | 0.117 | 0.117 | 0.117 | 0.117 | 0.117 |
| 715,500 | 450,000 | 54 | 7 | 1.036 | 0.5620 | 0.6348 | 4,859 | 0.131 | 0.131 | 0.133 | 0.131 | 0.133 |
| 636,000 | 400,000 | 54 | 7 | 0.977 | 0.4995 | 0.5642 | 4,319 | 0.147 | 0.147 | 0.149 | 0.147 | 0.149 |
| 556,500 | 350,000 | 26 | 7 | 0.927 | 0.4371 | 0.5083 | 4,039 | 0.168 | 0.168 | 0.168 | 0.168 | 0.168 |
| 477,000 | 300,000 | 26 | 7 | 0.858 | 0.3746 | 0.4357 | 3,462 | 0.196 | 0.196 | 0.196 | 0.196 | 0.196 |
| 397,500 | 250,000 | 26 | 7 | 0.783 | 0.3122 | 0.3630 | 2,885 | 0.235 | 0.235 | 0.235 | 0.235 | 0.235 |
| 336,400 | 0000 | 26 | 7 | 0.721 | 0.2642 | 0.3073 | 2,442 | 0.278 | 0.278 | 0.278 | 0.278 | 0.278 |
| 266,800 | 000 | 26 | 7 | 0.642 | 0.2095 | 0.2367 | 1,936 | 0.350 | 0.350 | 0.350 | 0.350 | 0.350 |
| 0000 | 00 | 6 | 1 | 0.563 | 0.1662 | 0.1939 | 1,542 | 0.441 | 0.443 | 0.446 | 0.447 | 0.464 |
| 000 | 0 | 6 | 1 | 0.502 | 0.1318 | 0.1537 | 1,223 | 0.556 | 0.557 | 0.561 | 0.562 | 0.579 |
| 00 | 1 | 6 | 1 | 0.447 | 0.1045 | 0.1219 | 970 | 0.702 | 0.703 | 0.707 | 0.706 | 0.718 |
| 0 | 2 | 6 | 1 | 0.398 | 0.0829 | 0.0967 | 769 | 0.885 | 0.885 | 0.889 | 0.887 | 0.893 |

SOURCE: Aluminum Co. of America.

Alternating to direct to alternating current is nonsynchronous transmission of electric power. It can be overhead or under the surface. In the early 1970s the problem of bulk electric-power transmission over high-voltage transmission lines above the surface developed the insistent discussion of land use and environmental cost.

While the maximum ac transmission voltage in use in the United States (1977) is 765 kV, ultra-high voltage (UHV) is under consideration. Transmission voltages of 1200 to 2550 kV are being studied. The right-of-way requirements for power transmission are significantly reduced at higher voltages. For example, in one study the transmission of 7500 MVA at 345 kV ac was found to require 14 circuits on a corridor 725 ft (221.5 m) wide, whereas a single 1200-kV ac circuit of 7500-MVA capacity would need a corridor 310 ft (91.5 m) wide.

Continuing research and development efforts may result in the development of more economic high-voltage underground transmission links. Sufficient bulk power transmission capability would permit **power wheeling,** i.e., the use of generating capacity to the east and west to serve a given locality as the earth revolves and the area of peak demand glides across the countryside.

**Corona** is a reddish-blue electrical discharge which occurs when the voltage-gradient in air exceeds 30 kV peak, 21.1 kV rms, at 76 cm pressure. This electrical discharge is caused by ionization of the air and becomes more or less concentrated at irregularities on the conductor surface and on the outer conductors of stranded conductors. Corona is accompanied by a hissing sound; it produces ozone and, in the presence of moisture, nitrous acid. On high-voltage lines corona produces a substantial power loss, corrosion of the conductors, and

radio and television interference. The fair-weather loss increases as the square of the voltage above a critical value $e_0$ and is greatly increased by fog, smoke, rainstorms, sleet, and snow (see Fig. 85). To reduce corona, the diameter of high-voltage conductors is increased to values much greater than would be required for the necessary conductance cross section. This is accomplished by the use of hollow, segmented conductors and by the use of aluminum cable, steel-reinforced (ACSR), which often has inner layers of jute to increase the diameter. In extra-high-voltage lines (400 kV and greater), corona is reduced by the use of bundled conductors in which each phase consists of two or three conductors spaced about 16 in (0.41 m) from one another.

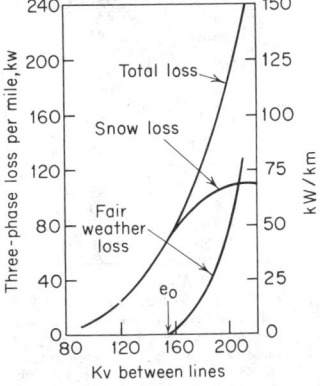

**Fig. 85** Corona loss with snowstorm.

### Underground Power Cables

Insulations for power cables include heat-resisting, low-water-absorptive synthetic rubber compounds, varnished cloth, impregnated paper, cross-linked polyethylene thermosetting compounds, and thermoplastics such as polyvinyl chloride (PVC) and polyethylene (PE) compounds (see Sec. 6).

Properly chosen **rubber-insulated cables** may be used in wet locations with a nonmetallic jacket for protective covering instead of a metallic sheath. Commonly used jackets are flame-resisting, such as neoprene and PVC. Such cables are relatively light in weight, easy to train in ducts and manholes, and easily spliced. When distribution voltages exceed 2,000 V phase to phase, an ozone-resisting type of compound is required. Such rubber insulation may be used in cables carrying up to 28,000 V between lines in three-phase grounded systems. The insulation wall will be thicker than with varnished cloth, polyethylene, or paper.

**Varnished-cloth cables** are made by applying varnish-treated closely woven cloth in the form of tapes, helically, to the metallic conductor. Simultaneously a viscous compound is applied between layers which fills in any voids at laps in the taping and imparts flexibility when the cable is bent by per-mitting movement of one tape upon another. This type of insulation has higher dielectric loss than impregnated paper but is suitable for the transmission of power up to 28,000 V between phases over short distances. Such insulated cables may be used in dry locations with flame-resisting fibrous braid, reinforced neoprene tape, or PVC jacket and are often further protected with an interlocked metallic tape armor; but in wet locations these cables should be protected by a continuous metallic sheath such as lead or aluminum. Since varnish-cloth-insulated cable has high ozone resistance, heat resistance, and impulse strength, it is well adapted for station or powerhouse wiring or for any service where the temperature is high or where there are sudden increases in voltage for short periods. Since the varnish is not affected by mineral oils, such cables make excellent leads for transformers and oil switches.

**PVC** is readily available in several fast, bright colors and is often chosen for color-coded multiconductor control cables. It has inherent flame and oil resistance, and as single conductor wire and cable with the proper wall thickness for a particular application, it usually does not need any outside protective covering. On account of its high dielectric constant and high power factor, its use is limited to low voltages, i.e., under 1,000 V, except for series lighting circuits.

### Table 18. Carrying Capacity of Power Cables in Underground Ducts, A*

(5,000 V, 75 percent load factor, 20°C ambient temp, three loaded ducts per bank)

| Size, AWG or MCM | Rubber or PVC | | | | Varnished cambric | | Impregnated paper | |
|---|---|---|---|---|---|---|---|---|
| | 60°C | | 75°C | | 85°C | | 85°C | |
| | Single conductor | Three conductors | Single conductor† | Three conductors | Single conductor† | Three conductors | Single conductor† | Three conductors |
| 8 | 67 | 48 | 76 | 55 | 83 | 56 | 83 | 56 |
| 6 | 89 | 65 | 100 | 71 | 109 | 73 | 109 | 73 |
| 4 | 116 | 84 | 135 | 93 | 142 | 96 | 142 | 96 |
| 2 | 151 | 108 | 173 | 119 | 186 | 124 | 186 | 124 |
| 1 | 172 | 123 | 199 | 137 | 214 | 141 | 214 | 141 |
| 0 | 197 | 140 | 230 | 154 | 245 | 162 | 245 | 162 |
| 00 | 225 | 159 | 264 | 174 | 283 | 185 | 283 | 185 |
| 000 | 258 | 180 | 303 | 197 | 324 | 211 | 324 | 211 |
| 0,000 | 295 | 207 | 348 | 226 | 371 | 240 | 371 | 240 |
| 250 | 325 | 224 | 385 | 252 | 409 | 263 | 409 | 263 |
| 300 | 372 | 240 | 424 | 275 | 459 | 292 | 459 | 292 |
| 350 | 395 | 267 | 546 | 302 | 500 | 319 | 500 | 320 |
| 400 | 435 | 282 | 502 | 323 | 540 | 343 | 540 | 344 |
| 500 | 485 | 319 | 571 | 364 | 611 | 386 | 611 | 390 |
| 600 | 555 | 350 | 635 | 400 | 679 | 420 | 679 | 430 |
| 700 | 608 | 378 | 691 | 434 | 742 | 458 | 741 | 466 |
| 750 | 628 | 383 | 718 | 448 | 771 | 475 | 771 | 482 |
| 800 | 653 | ... | 744 | ... | 797 | ... | 797 | |
| 1,000 | 709 | 425 | 840 | ... | 898 | ... | 898 | |
| 1,250 | 793 | ... | 940 | | 1,012 | ... | 1,012 | |
| 1,500 | 865 | ... | 1,032 | | 1,110 | ... | 1,110 | |
| 1,750 | 927 | ... | 1,112 | | 1,204 | ... | 1,204 | |
| 2,000 | 978 | ... | 1,187 | | 1,290 | ... | 1,300 | |

*Note: This is a general table. For complete and up-to-date ampacity (ampere-carrying capacity) values for various conductor temperatures and installation conditions, refer to IEEE Publication S-135, "AIEE-IPCEA Power Cable Ampacities," vols. I and II. Each volume contains 317 pages of ampacity tables and cable constants covering all conductor sizes, voltages, and operating temperatures with paper, rubber, thermoplastic, varnished-cloth, and asbestos insulations in ducts, directly buried conduits, ladders, raceways, trays, and in air.

†Sheaths assumed to be open-circuited. If sheaths are bonded together, the sheath currents will reduce the carrying capacity from 25 to 40 percent. The foregoing values are based on three loaded ducts. Values are reduced as number of loaded ducts increases.

**Polyethylene,** because of its excellent electrical characteristics, first found use when it was adapted especially for high-frequency cables used in radio and radar circuits; for certain telephone, communication, and signal cables; and for submarine cables. Submarine telephone cables with built-in repeaters laid first in the Atlantic Ocean and then in the Pacific are insulated with polyethylene. Because of polyethylene's thermal characteristics, the standard maximum conductor operating temperature is 75°C. It is commonly used for power cables (including large use for underground residential distribution), with transmissions up to 15,000 V. Successful installations have been in service at 46 kV and some at 69 kV. The upper limit has not been reached, inasmuch as work is in progress on higher-voltage polyethylene power cables as a result of advancements in the art of compounding.

**Cross-linked polyethylene** is another insulation which is gaining in favor in the power field. For power cable insulations, the cross-linking process is most commonly obtained chemically. It converts polyethylene from a thermoplastic into a thermosetting material; the result is a compound with a unique combination of properties, including resistance to heat and oxidation, thus permitting an increase in maximum conductor operating temperature to 90°C. Since the service record with this compound has been good at voltages which have been gradually increased to 15 kV, it is expected that its voltage range also will be extended in the future.

**Impregnated-paper insulation** is used for very-high-voltage cables whose range has been extended to 345 kV. To eliminate the detrimental effects of moisture and to maintain proper impregnation of the paper, such cables must have a continuous metallic sheath such as lead or aluminum or be enclosed within a steel pipe; the operation of the cable depends absolutely on the integrity of that enclosure. In three-conductor belted-type cables the individual insulated conductors are surrounded by a belt or wall of impregnated paper over which the lead sheath is applied. When all three conductors are within one sheath, their inductive effects practically neutralize one another and eddy-current loss in the sheath is negligible. In the type-H cable, each of the individual conductors is surrounded with a perforated metallic covering, either aluminum foil backed with a paper tape or thin perforated metal tapes wound over the paper. All three conductors are then enclosed within the metal sheath. The metallic coverings being grounded electrically, each conductor acts as a single-conductor cable. This construction eliminates "tangential" stresses within the insulation and reduces pockets or voids. When paper tapes are wound on the conductor, impregnated with an oil or a petrolatum compound, and covered with a lead sheath, they are called **solid type.**

Three-conductor cables are now operating at 33,000 V, and single-conductor cables at 66,000 V between phases (38,000 V to ground). In New York and Chicago, special hollow-conductor oil-filled single-conductor cables are operating successfully at 132,000 V (76,000 V to ground). In France, cables are operating at 345 kV between conductors.

Other methods of installing underground cables are to draw them into steel pipes, usually without the sheaths, and to fill the pipes with oil under pressure (**oilstatic**) or **nitrogen** under 200 lb pressure. The ordinary medium-high-voltage underground cables are usually drawn into duct lines. With a straight run and ample clearance the length of cable between manholes may reach 600 to 1,000 ft. Ordinarily, the distance is more nearly 400 to 500 ft. With bends of small radius the distance must be further reduced.

**Cable ratings** are based on the permissible operating temperatures of the insulation and environmental installation conditions.

## POWER DISTRIBUTION

**Distribution Systems** The choice of the system of power distribution is determined by the type of power that is available and by the nature of the load. To transmit a given power over a given distance with a given power loss ($I^2R$), the weight of conductor varies inversely as the square of the voltage. Incandescent lamps will not operate economically at voltages much higher than 120 V, the most suitable voltages for dc motors are 230 and 550 V, although 550 V is practically obsolete, except for railway motors; for ac motors, standard voltages are 230, 460, and 575 V, three-phase. When power for lighting is to be distributed in a district where the consumers are relatively far apart, alternating current is used, being distributed at high voltage (2,400, 4,160, 4,800, 6,900, and 13,800 V) and transformed at the consumer's premises, or by transformers on poles or located in manholes or vaults under the street or sidewalks, to 240/120 V three-wire for lighting and domestic customers, and to 208, 240, 480, and 600 volts, three-phase, for power.

The first central station power systems were built with dc generation and distribution. The economical transmission distance was short. Densely populated, downtown areas of cities were therefore the first sections to be served. Growth of electric service in the United States was phenomenal in the last two decades of the nineteenth century. After 1895 when ac generation was selected for the development of power from Niagara Falls, the expansion of dc distribution diminished. The economics overwhelmingly favored the new ac system.

Direct current service is still available in small pockets in some cities. In those cases, ac power is generated, transmitted, and distributed. The conversion to dc takes place in rectifiers installed in manholes near the load or in the building to be served. Some dc customers resist the change to ac service because of their need for motor speed control. Elevator and printing press drives and some cloth-cutting knives are examples of such needs (See Low-Voltage AC Network.)

**Series Circuits** Where the devices to be supplied with power are nearly of the same current rating, are located relatively far apart, and are ordinarily used simultaneously, it is often more economical to supply **power at constant current** than at constant potential. To operate at constant current, the power-consuming devices and the power source are all in series with the same current in all. In cutting any individual device out of service, an equivalent resistance must be inserted in its place to maintain the same current in the circuit, or else some means must be provided at the power source to adjust automatically the total voltage so as to maintain constant current. If, in a series circuit operating with direct current, the resistance of each of the power-consuming devices is $R_d$, the resistance of the lines is $R$, the current is $I$, the generator voltage when all the devices are operating is $E$, and the number of the receiving devices is $n$, $E = nIR_d + IR$, where $IR = e$ is the resistance drop in the line. Now $e = \rho Il/A$, where $\rho$ is the resistivity of the material in $\Omega$/cir mil·ft, $l$ = length of the conductor, ft, and $A$ = sectional area of the conductor, cir mils. If the permissible voltage drop $e$ has been decided upon, the proper cross section of conductor $A = \rho II/e$. For mechanical reasons, conductors smaller than no. 6 AWG are not generally used. This method of distribution is used almost exclusively for **street-lighting systems** in which the lamps are located over considerable areas. The power is supplied by a constant-current transformer which maintains constant current irrespective of the number of lamps in the circuit. The necessarily high voltage is not objectionable, as circuits placed on poles or underground can be safely and satisfactorily operated at voltages of 1,100 to 3,000 V per circuit. Standard currents are 4.4, 5.5, and 6.6 A. In some systems it is more economical to supply street-lighting power from the low-voltage parallel system, time switches or photoelectric switches being used to connnect and disconnect the lamps.

**Parallel Circuits**    **Power** is usually distributed at **constant potential,** and all the devices or receivers in the circuit are connected in parallel, giving a constant-potential system, Fig. 86$a$. If conductors of constant cross section are used and all

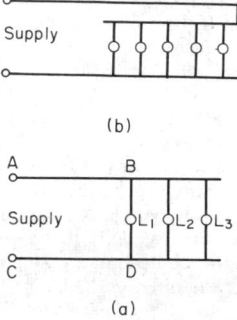

Fig. 86    ($a$) Parallel circuit. ($b$) Loop circuit.

the loads, $L_1$, $L_2$ etc., are operating, there will be a greater voltage $IR$ drop per unit length of wire in the portion of the circuit $AB$ and $CD$ than in the other portions; also the voltage will not be the same for the different lamps but will decrease along the mains with distance from the generating end.

**Loop Circuits**    A more nearly equal voltage for each load is obtained in the loop system, Fig. 86$b$. The electrical distance from one generator terminal to the other through any receiver is the same as that through any other receiver, and the voltage at the receivers may be maintained more nearly equal, but at the expense of additional conductor material.

**Series-Parallel Circuit**    For incandescent lamps the power must be at low voltage (115 V) and the voltage variations must be small. If the transmission distance is considerable or the loads are large, a large or perhaps prohibitive investment in conductor material would be necessary. In some special cases, lamps may be operated in groups of two in series as shown in Fig. 87. The transmitting voltage is thus doubled, and, for a given number of lamps, the current is halved, the permissible voltage drop $(IR)$ in conductors doubled, the conductor resistance quadrupled, the weight of conductor material thus being reduced to 25 percent of that necessary for simple parallel operation.

**Three-wire System**    In the series-parallel system the loads must be used in pairs and both units of the pair must have the same power rating. To overcome these objections and at the same time to obtain the economy in conductor material of operating at higher voltage, the three-wire system is used. It

consists merely of adding a third wire or neutral to the system of Fig. 87 as shown in Fig. 88.

If the neutral wire is of the same cross section as the two outer wires, this system requires only 37.5 percent the copper required by an equivalent two-wire system. Since the neutral

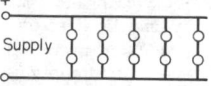

Fig. 87    Series-parallel dc system.

ordinarily carries less current than the outers, it is usually smaller and the ratio of copper to that of the two-wire system is even less than 37.5 percent (see Table 19).

When the loads on each half of the system are equal, there will be no current in the middle or neutral wire, and the condition is the same as that shown in Fig. 87. When the loads on the two sides are unequal, there will be a current in the neutral wire equal to the difference of the currents in the outside wires. For example, if each of the loads in the system

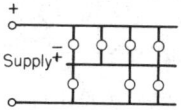

Fig. 88    Three-wire dc system.

shown in Fig. 89 takes 1 A, the current in each part of the system will be given by the numbers on the ammeters shown connected in the system.

The three-wire system shown in Fig. 88 is not practicable because no means are provided for holding the neutral at its correct potential. If, for example, four loads are in operation on one side of the system and three on the other, as shown, the voltages on the two sides of the system become seriously unbalanced and the three loads, which may be lamps, are subjected to overvoltage. One method of supplying the neutral in a dc system is shown in Fig. 89 where each side of the system is supplied by a separate generator. This is open to the objection of the greater complications of two machines, greater cost, more floor space, and the lesser efficiency of two machines.

**Balancer Set**    Another method of obtaining the neutral is to use a balancer set. This consists of two similar shunt or compound machines coupled together with the armatures connected in series across the outer lines as shown in Fig. 90. When the loads are balanced, there is no neutral current and the two machines merely run idle as motors, being connected in series across the outer conductors. If the load on one side of the system becomes greater than that on the other side, the machine on the more heavily loaded side operates as a generator and pumps some of the neutral current to its side of the

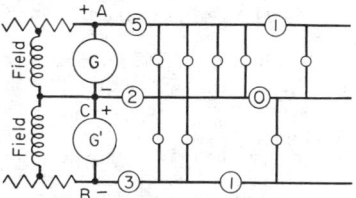

Fig. 89    Three-wire dc system with two generators.

**Table 19. Resistance and 60-Hz Reactance for Wires with Small Spacings, Ω, at 20°C**
(See also Table 16)

| AWG and size of wire, cir mils | Resistance in 1,000 ft of line (2,000 ft of wire), copper | Reactance in 1,000 ft of line (2,000 ft of wire) at 60 Hz for the distance given in inches between centers of conductors | | | | | | | | | | |
|---|---|---|---|---|---|---|---|---|---|---|---|---|
| | | $^1/_2$ | 1 | 2 | 3 | 4 | 5 | 6 | 9 | 12 | 18 | 24 |
| 14– 4,107 | 5.06 | 0.138 | 0.178 | 0.218 | 0.220 | 0.233 | 0.244 | 0.252 | 0.271 | 0.284 | 0.302 | |
| 12– 6,530 | 3.18 | 0.127 | 0.159 | 0.190 | 0.210 | 0.223 | 0.233 | 0.241 | 0.260 | 0.273 | 0.292 | |
| 10– 10,380 | 2.00 | 0.116 | 0.148 | 0.180 | 0.199 | 0.212 | 0.223 | 0.231 | 0.249 | 0.262 | 0.281 | |
| 8– 16,510 | 1.26 | 0.106 | 0.138 | 0.169 | 0.188 | 0.201 | 0.212 | 0.220 | 0.238 | 0.252 | 0.270 | 0.284 |
| 6– 26,250 | 0.790 | 0.095 | 0.127 | 0.158 | 0.178 | 0.190 | 0.201 | 0.209 | 0.228 | 0.241 | 0.260 | 0.272 |
| 4– 41,740 | 0.498 | 0.085 | 0.117 | 0.149 | 0.167 | 0.180 | 0.190 | 0.199 | 0.217 | 0.230 | 0.249 | 0.262 |
| 2– 66,370 | 0.312 | 0.074 | 0.106 | 0.138 | 0.156 | 0.169 | 0.180 | 0.188 | 0.206 | 0.220 | 0.238 | 0.252 |
| 1– 83,690 | 0.248 | 0.068 | 0.100 | 0.132 | 0.151 | 0.164 | 0.174 | 0.183 | 0.201 | 0.214 | 0.233 | 0.246 |
| 0–105,500 | 0.196 | 0.063 | 0.095 | 0.127 | 0.145 | 0.159 | 0.169 | 0.177 | 0.196 | 0.209 | 0.228 | 0.241 |
| 00–133,100 | 0.156 | 0.057 | 0.090 | 0.121 | 0.140 | 0.153 | 0.164 | 0.172 | 0.190 | 0.204 | 0.222 | 0.236 |
| 000–167,800 | 0.122 | 0.052 | 0.085 | 0.116 | 0.135 | 0.148 | 0.158 | 0.167 | 0.185 | 0.199 | 0.217 | 0.230 |
| 0000–211,600 | 0.098 | 0.046 | 0.079 | 0.111 | 0.130 | 0.143 | 0.153 | 0.161 | 0.180 | 0.193 | 0.212 | 0.225 |
| 250,000 | 0.085 | ..... | 0.075 | 0.106 | 0.125 | 0.139 | 0.148 | 0.157 | 0.175 | 0.189 | 0.207 | 0.220 |
| 300,000 | 0.075 | ..... | 0.071 | 0.103 | 0.120 | 0.134 | 0.144 | 0.153 | 0.171 | 0.185 | 0.203 | 0.217 |
| 350,000 | 0.061 | ..... | 0.067 | 0.099 | 0.188 | 0.128 | 0.141 | 0.149 | 0.168 | 0.182 | 0.200 | 0.213 |
| 400,000 | 0.052 | ..... | 0.064 | 0.096 | 0.114 | 0.127 | 0.138 | 0.146 | 0.165 | 0.178 | 0.197 | 0.209 |
| 500,000 | 0.042 | ..... | ..... | 0.090 | 0.109 | 0.122 | 0.133 | 0.141 | 0.160 | 0.172 | 0.192 | 0.202 |
| 600,000 | 0.035 | ..... | ..... | 0.087 | 0.106 | 0.118 | 0.128 | 0.137 | 0.155 | 0.169 | 0.187 | 0.200 |
| 700,000 | 0.030 | ..... | ..... | 0.083 | 0.102 | 0.114 | 0.125 | 0.133 | 0.152 | 0.165 | 0.184 | 0.197 |
| 800,000 | 0.026 | ..... | ..... | 0.080 | 0.099 | 0.112 | 0.122 | 0.130 | 0.148 | 0.162 | 0.181 | 0.194 |
| 900,000 | 0.024 | ..... | ..... | 0.077 | 0.096 | 0.109 | 0.119 | 0.127 | 0.146 | 0.159 | 0.178 | 0.191 |
| 1,000,000 | 0.022 | ..... | ..... | 0.075 | 0.094 | 0.106 | 0.117 | 0.125 | 0.144 | 0.158 | 0.176 | 0.188 |

For other frequencies the reactance will be in direct proportion to the frequency.

line. The remainder of the neutral current goes through the other machine supplying it with the power that enables it to operate as a motor and drive the generator. For example, in Fig. 90, the load on the positive side of the system is greater than that on the negative side. Hence, the machine G on the positive side is operating as a generator and the machine M is operating as a motor. If the machines are compounded so that when operating as a generator the winding is cumulative and hence is differential when a motor, the voltage unbalance with change in load can be made practically zero.

Since balancer sets take power continuously, they are used for the most part on large systems of high diversity where the percentage unbalance is small, so that the power rating of the set is but a small percentage of the power of the system.

**Three-Wire Generator** The three-wire generator is the most common and efficient method of obtaining a neutral in a dc system. It is a conventional generator that would ordinarily be used to supply the outer conductors with power. Two or more taps *a* and *b* are, however, brought out from the armature winding to two slip rings (Fig. 91). A compensator or reactance coil of low resistance called a **balance coil** is connected across the slip rings. The neutral of the three-wire system is connected to its center point. The voltage across the slip rings is alternating. Because of its choking action, but little alternating current flows in the balance coil. The unbalanced direct

current in the neutral flows back through the balance coil to the armature. The inductance of the balance coil has no effect on the steady direct current, and the resistance of the balance coil is low so that there is little voltage drop due to the direct current. Two or more balance coils with their neutrals connected may be used. With two such balance coils, the second balance coil is connected to slip rings that are tapped to the armature winding at points 90 electrical degrees from the first.

**Feeders and Mains** Where dc power is supplied to a large district, improved voltage regulation is obtained by having centers of distribution. Power is supplied from the station bus bars directly to the **centers of distribution** by large cables known as **feeders**. Power is distributed from the distribution centers to the consumers through the **mains**. As there are no loads connected to the feeders between the generating station and the centers of distribution, the voltage at the latter points may be maintained constant. Pilot wires from the centers of distribution often run back to the station voltmeter, assisting the operator in maintaining the potential constant at the centers. This system provides a means of maintaining very close voltage regulation at the consumer's premises.

**AC Three-Wire Distribution** Practically all energy for lighting and small motor work is distributed at 1,150, 2,300, or 4,160 V ac to transformers which step down the voltage to 240

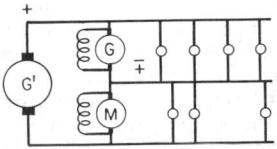

**Fig. 90** Motor-generator balancer.

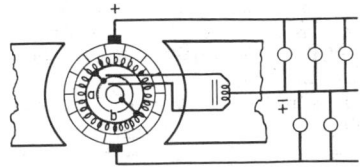

**Fig. 91** Three-wire generator.

and 120 V for three-wire domestic and lighting systems as well as 240, 480, and 600 volts, three-phase, for power. For the three-wire systems the transformers are so designed that the secondary or low-voltage winding will deliver power at 240 V, and the middle or neutral wire is obtained by connecting to the center or midpoint of this winding (see Fig. 92).

**Grounding**  The neutral wire of the secondary circuit of the transformer should be grounded on the pole (or in the manhole) and at the service switch in the building supplied. If, as a result of a lightning stroke or a fault in the transformer insulation, the transformer primary circuit becomes grounded at *a* (Fig. 92) and the transformer insulation between primary and secondary windings is broken down at *b* and if there were no permanent ground connection in the secondary neutral wire, the potential of wire 1 would be raised 2,300 V above ground potential. This constitutes a very serious hazard to life for persons coming in contact with the 120 V system. The National Electrical Code requires the use of a ground wire not smaller than 8 AWG copper; on secondary circuits, grounds should be provided at least every 500 ft. With the neutral grounded (Fig. 92), voltages to ground on the secondary system cannot exceed 120 V. (See National Electrical Code 1975, Art. 250.)

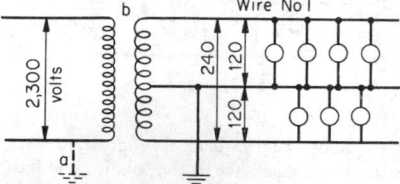

**Fig. 92**  Three-wire 230/115-V ac system.

A common and economical method of supplying business and thickly settled districts with high load densities is to employ a 208/120-V, three-phase, four-wire **low-voltage ac network.** The network operates with 208 V between outer wires giving 120 V to neutral (Fig. 93). Motors are connected across the three outer wires operating 208 V, three-phase. Lamp loads are connected between outer wires and the grounded neutral. The network is supplied directly from 13,800-V feeders by 13,800/208-V three-phase transformer units, usually located in manholes, vaults, or outdoor enclosures. This system thus eliminates the necessity for transformation in the substation. A large number of such units feed the network, so that the secondaries are all in parallel. Each transformer is provided with an overload reverse-energy circuit breaker (network protector), so that a feeder and its transformer are isolated if trouble develops in either. This system is flexible since units can be easily added or removed in accordance with the rapid changes in local loads that occur particularly in downtown business districts.

**Voltage Drops**  In ac distribution systems the voltage drop from transformer to consumer in lighting mains should not exceed 2 percent in first-class systems, so that the lamps along the mains can all operate at nearly the same voltage and the annoying flicker of lamps may not occur with the switching of appliances. This may require a much larger conductor than the most economical size. In transmission lines and in feeders where there are no intermediate loads and where means of regulating the voltage are provided, the drop is not limited to

the low values that are necessary with mains and the matter of economy may be given consideration.

## WIRING CALCULATIONS

**Wiring Calculations for DC Circuits**  (These same calculations can be used for ac if the reactance can be neglected.) The determination of the proper size of conductor is influenced by a number of factors. Except for short distances, the minimum size of conductor recommended by the National Board of Fire Underwriters in Table 21, which is based on the maximum permissible current for each type of insulation, cannot be used; the size of conductor must be larger so that the voltage drop *IR* shall not be too great. With branch circuits supplying an incandescent-lamp load, this drop should not be more than a small percentage of the voltage between wires. The National Electrical Code 1975 requires that conductors for feeders, i.e., from the service equipment to the final branch circuit overcurrent device, be sized to prevent (1) a voltage drop of more than 3 percent at the farthest outlet of power, heating, and lighting loads or combinations thereof, and (2) a maximum voltage drop on combined feeders and branch circuits to the farthest outlet of more than 5 percent.

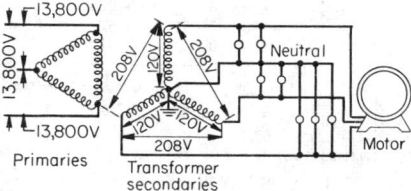

**Fig. 93**  208/120-V secondary network (single unit) showing voltages.

The resistance of 1 cir mil·ft of commercial copper may be taken as 10.8 Ω. The resistance of a copper conductor may be expressed as $R = 10.8l/A$, where $l$ = length, ft and $A$ = area, cir mils. If the length is expressed in terms of the transmission distance $d$ (since the two wires are usually run parallel), the voltage drop *IR* to the end of the circuit is

$$e = 21.6Id/A \qquad (128)$$

and the size of conductor in circular mils necessary to give the permissible voltage drop $e$ is

$$A = 21.6Id/e \qquad (129)$$

If $e$ is expressed as a percentage $x$ of the voltage $E$ between conductors, then

$$A = 2,160Id/xE \qquad (130)$$

EXAMPLE.  Determine the size of conductor to supply power to a 10-hp, 220-V dc motor 500 ft from the switchboard with 5 V drop. Assume a motor efficiency of 86 percent. The motor will then require a current of $(10 \times 746)/(0.86 \times 220) = 39.4$ A. From Eq. (**129**), $A = 21.6 \times 39.4 \times 500/5 = 85,100$ cir mils. The next largest wire is no. 0 AWG.

The calculation of the size conductor for dc three-wire circuits is made in practically the same manner. With a balanced circuit there is no current in the neutral wire, and the current in each outside wire will be equal to one-half the sum of the currents taken by all the receiving devices connected between neutral and outside wires plus the sum of the currents taken by the receivers connected between the outside wires.

Using this total current and neglecting the neutral wire, make calculations for the size of the outside wires by means of Eq. **(129)**. The neutral wire should have the same cross section as the outside wires in interior wiring.

EXAMPLE.   Determine the size wire which should be used for the three-wire main of Fig. 94. Allowable drop is 3 V and the distance to the load center 40 ft; circuit loaded with two groups of receivers each taking 60 A connected between the neutral and the outside wires, and one group of receivers taking 20 A connected across the outside wires.

*Solution:*   load = $(60 + 60)/2 + 20 = 80$ A. Substituting in Eq. **(129)**, cir mils = $21.6Id/e = 21.6 \times 80 \times 40/3 = 23,030$ cir mils.

From Tables 19 and 21, no. 6 wire, which has a cross section of 26,250 cir mils, is the next size larger. This size of wire would satisfy the voltage-drop requirements, but rubber-insulated no. 6 has a safe carrying capacity of but 55 A. The current in the circuit is 80 A. Therefore, rubber-insulated wire no. 3, which has a carrying capacity of 80 A, should be used. The neutral wire should be the same size as the outside wires.

See also examples in the National Electrical Code 1975.

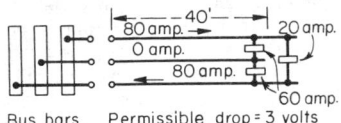

**Fig. 94**   Three-wire 230/115-V main.

**Wiring calculations for ac circuits** are essentially the same as for dc circuits, but other factors such as power factor, reactance, and skin effect may require consideration. Skin effect becomes pronounced only when very large conductors are used for alternating current. For interior wiring, conductors larger than 700,000 cir mils should not be used, and many prefer not to use conductors larger than 300,000 cir mils. Should the required copper cross section exceed these values, a number of conductors may be operated in parallel.

For voltages under 5,000 the effect of line capacitance may be neglected. With ordinary single-phase interior wiring, where the effect of the line reactance may be neglected and where the power factor of the load (incandescent lamps) is nearly 100 percent, the calculations are made the same as for dc circuits. Three-wire ac circuits of ordinary length with incandescent lamp loads are also determined in the same manner. When the load is other than incandescent lamps, it is necessary to know the power factor of the load in order to make calculations. When the exact power factor cannot be accurately determined, the following approximate values may be used: incandescent lamps, 0.95 to 1.00; lamps and motors, 0.75 to 0.85; motors 0.5 to 0.80. Equation **(131)** gives the value of current in a single-phase circuit. See also Table 25.

$$I = (P \times 1,000)/(E \times pf) \qquad (131)$$

where $I$ = current, A; $P$ = kW; $E$ = load voltage; and $pf$ = power factor of the load. The size of conductor is then determined by substituting this value of $I$ in Eq. **(129)** or **(130)**.

**For three-phase three-wire ac circuits** the current per wire

$$I = 1,000P/\sqrt{3}\ Epf = 580P/Epf \qquad (132)$$

Computations are usually made of voltage drop **per wire** (see Fig. 95). Hence, if reactance can be neglected, the conductor cross section in cir mils is one-half that given by Eq. **(129)**. That is,

$$A = 10.8Id/e \qquad \text{cir mils} \qquad (133)$$

where $e$ in Eq. **(133)** is the voltage drop per wire. The voltage drop between any two wires is $\sqrt{3}e$. The percent voltage drop should be in terms of the voltage to **neutral**. That is, percent drop = $[e/(E/\sqrt{3})]100 = [\sqrt{3}\ e/E]100$ (see Fig. 81).

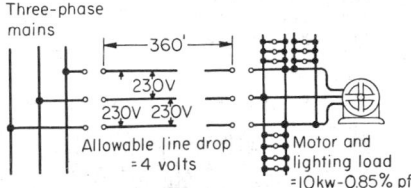

**Fig. 95**   Three-phase lamp and induction-motor load.

EXAMPLE.   In Fig. 95, load 10 kW; voltage of circuit 230; power factor 0.85; distance 360 ft; allowable drop per wire 4 V. Substituting in Eq. **(132)** $I = (580 \times 10)/(230 \times 0.85) = 29.7$ A. Substituting in Eq. **(133)**, $A = 10.8 \times 29.7 \times 360/4 = 28,900$ cir mils.

The next larger commercially available standard-size wire (see Table 19) is 41,700 cir mils corresponding to AWG no. 4. From Table 21 this will carry 70 A with rubber insulation, and is therefore ample in section for 29.7 A. Three no. 4 wires would be used for this circuit.

From Table 19 the resistance of 1,000 ft of no. 4 copper wire is 0.249 $\Omega$. Hence. the voltage drop per conductor, $e = 29.7 \times (360/1,000)0.249 = 2.66$ V. Percent voltage drop = $\sqrt{3} \times 2.66/230 = 2.00$ percent.

Where all the wires of a circuit, two wires for a single-phase circuit, four wires for a four-phase circuit (see Fig. 32 and Fig. 79c), and three wires for a three-phase circuit, are carried in the same conduit or where the wires are separated less than 1 in between centers, the effect of line (inductive) reactance may ordinarily be neglected. Where circuit conductors are large and widely separated from one another and the circuits are long, the inductive reactance may increase the voltage drop by a considerable amount over that due to resistance alone. Such problems are treated using $IR$ and $IX$ phasors. Line reactance decreases somewhat as the size of wire increases and decreases as the distance between wires decreases.

EXAMPLE.   Determine the size of wire necessary for the branch to the 50-hp, 60-Hz, 250-V single-phase induction motor of Fig. 96. The name-plate rating of the motor is 195 A, and its full-load power factor is 0.85. The wires are run open and separated 4 in; length of circuit, 600 ft. Assume the line drop must not exceed 7 percent, or $0.07 \times 250 = 17.5$ V. The point made by this example is emphasized by the assumption of an outsize motor.

*Solution.*   To ascertain approximately the size of conductor, substitute in Eq. **(129)** giving cir mils = $21.6 \times 195 \times 600/17.5 = 144,400$. Referring to Table 19, the next larger size wire is no. 000 or 167,800 cir mils. This size would be ample if there were no line reactance. In order to allow for reactance drop, a larger conductor is selected and the corresponding voltage drop determined. Inasmuch as this is a motor branch, the code rules require that the carrying capacity be sufficient for a 25 percent overload. Therefore the conductor should be capable of

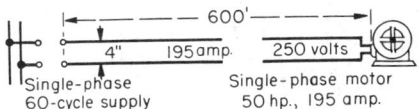

**Fig. 96**   Single-phase induction motor load on branch circuit.

carrying $195 \times 1.25 = 244$ A. From Table 21, a 350,000-cir mil conductor rubber-insulated cable would be required to carry 244 A. Resistance drop (see Table 19), $IR = 195 \times 0.061 \times 0.6 = 7.14$ V. $7.14/250 = 2.86$ percent. From Table 19, $X = 0.128 \times 0.6 = 0.0768 \ \Omega$. $IX = 195 \times 0.768 = 14.98$ V. $14.98/250 = 5.99$ percent. Using the Mershon diagram (Fig. 83), follow the ordinate corresponding to power factor, 0.85, until it intersects the smallest circle. From this point, lay off horizontally the percentage resistance drop, 2.86. From this last point, lay off vertically the percentage reactance drop 5.99. This last point lies about on the 6.0 percent circle, showing that with 195 amp the difference between the sending-end and receiving-end voltages is $0.06 \times 250 = 15.0$ V, which is within the specified limits.

Also Eq. **(127)**, may be used. $\cos \theta = 0.85$; $\sin \theta = 0.527$.

$$E_s = \{[(250 \times 0.85) + 7.14]^2 + [(250 \times 0.527) + 14.98]^2\}^{1/2} = 264.3 \text{ V}$$
$$264.3/250 = 105.7 \text{ percent}$$

In the calculation of three-phase three-wire circuits where line reactance must be considered, the method given on p. 15-56 may be used. The system is considered as being three single-phase systems having a ground return the resistance and inductance of which are zero, and the voltages are equal to the line voltages divided by $\sqrt{3}$. When the three conductors are spaced unequally, the value of GMD given in Fig. 84 should be used in Tables 16 and 19. (When the value of resistance or reactance per 1,000 ft of **conductor** is desired, the values in Table 19 should be divided by 2.)

The National Electrical Code of 1975 specifies that the size of conductors for branch circuits should be such that the voltage drop will not exceed 3 percent to the farthest outlet for power, heating, lighting, or combination thereof, requiring further that the total voltage drop for feeders and branch circuits should not exceed 5 percent overall. For examples of calculations for interior wiring, see National Electrical Code of 1975 (ANSI C1-1975, Chap. 9).

## INTERIOR WIRING

**Interior wiring requirements are based,** for the most part, on the **National Electrical Code** (NEC), which has been adopted by the National Fire Protection Association, American National Standards Institute (ANSI), and the Occupational Safety and Health Act (OSHA).

The Occupational Safety and Health Act of 1970 (OSHA) made the National Electrical Code a national standard. Conformance with the NEC became a requirement in most commercial, industrial, agricultural, etc., establishments in the United States. Some localities may not accept NEC standards. In those cases, local rules must be followed.

NEC authority starts at the point where the connections are made to the conductor of the **service drop** (overhead) or **lateral** (underground) from the electricity supply system. The **service equipment** must have a rating not less than the load to be carried (computed according to NEC methods). Service equipment is defined as the necessary equipment, such as circuit breakers or fused switches and accompanying accessories. This equipment must be located near the point of entrance of supply conductors to a building or other structure or an otherwise defined area. Service equipment is intended to be the main control and means of cutoff of the supply.

**Service-entrance conductors** connect the electricity supply to the service equipment. Service-entrance conductors running along the exterior or entering a building or other structure may be installed (1) as separate conductors, (2) in approved cables, (3) as cable bus, or (4) enclosed in rigid conduit. Also, for voltages less than 600 V, the conductors may be installed in electrical metallic tubing, wireways, auxiliary gutters, or busways. Service-entrance cables which are exposed to physical damage from awnings, swinging signs, coal chutes, etc., must be of the protected type or be protected by conduit, electrical metallic tubing, etc. Service heads must be raintight. Thermoplastic or rubber insulation is required in overhead services. A grounded conductor may be bare. If exposed to the weather or embedded in masonry, raceways must be raintight and arranged to drain. Underground service raceway or duct entering from an underground distribution system must be sealed with a suitable compound (spare ducts, also).

NEC rules permit **multiple services** to a building for various reasons, such as: (1) fire pumps, (2) emergency light and power, (3) multiple occupancy, (4) when the calculated load is greater than 3,000 A, (5) when the building extends over a large area, and (6) where different voltages, frequencies, number of phases, or classes of use are required.

Ordinary service drops (overhead) and lateral (underground) must be large enough to carry the load but not smaller than no. 8 copper or no. 6 aluminum. As an exception, for installations to supply only limited loads of a single branch

### Table 20. Wire Table for Standard Annealed Copper at 20°C in MKS and SI Units

| AWG size | Diameter, mm | Kgf/km | N/km | M/$\Omega$ | Area, mm² |
|---|---|---|---|---|---|
| 14 | 1.628 | 18.50 | 181.4 | 120.7 | 2.08 |
| 12 | 2.053 | 29.42 | 288.5 | 191.9 | 3.31 |
| 10 | 2.588 | 46.77 | 458.7 | 305.1 | 5.261 |
| 8 | 3.264 | 74.37 | 729.3 | 485.2 | 8.367 |
| 6 | 4.115 | 118.2 | 1159.1 | 771.5 | 13.30 |
| 4 | 5.189 | 188.0 | 1843.7 | 1227 | 21.15 |
| 2 | 6.544 | 299.0 | 2932.2 | 1951 | 33.62 |
| 1 | 7.348 | 377.0 | 3697.1 | 2460 | 42.41 |
| 0 | 8.252 | 475.4 | 4662.1 | 3102 | 53.49 |
| 00 | 9.266 | 599.5 | 5879.1 | 3911 | 67.43 |
| 000 | 10.40 | 755.9 | 7412.8 | 4932 | 85.01 |
| 0000 | 11.68 | 935.2 | 9171.2 | 6219 | 107.2 |

circuit, such as small polyphase power, etc., service drops must not be smaller than no. 12 hard-drawn copper or equivalent, and service laterals must be not smaller than no. 12 copper or no. 10 aluminum.

The phrase **large enough to carry the load** requires elaboration. The various conductors of public-utility electric-supply systems are sized according to the calculations and decisions of the personnel of the specific public utility supplying the service drop or lateral. At the load end of the drop or lateral, the NEC rules apply, and from that point on into the consumer's premises, NEC rules are the governing authority. There is a discontinuity at this point in the calculation of combined load demand for electricity and allowable current (ampacity) of conductors, cables, etc. This discontinuity in calculations results from the fact that the utility company operates locally, whereas the NEC is a set of national standards and therefore cannot readily allow for regional differences in electrical coincident demand, ambient temperature, etc. The NEC's aim is the assurance of an electrically "safe" human environment. This will be fostered by following the NEC rules.

Service-entrance cables are conductor assemblies which bear the type codes **SE** (for overhead services) and **USE** (for underground services). Under specified conditions, these cables may also be used for interior feeder and branch-circuit wiring.

The service-entrance equipment must have the **capability of safely interrupting** the current resulting from a short circuit at its terminals. **Available short-circuit current** is the term given to the maximum current that the power system can deliver through a given circuit to any negligible-impedance short circuit applied at a given point. (This value can be in terms of symmetrical or asymmetrical, peak or rms current, as specified.)

In most instances, the available short-circuit current is limited by the impedance of the last transformer in the supply system. **Large power users**, however, must become aware of changes in the electricity supply system which, because of growth of system capacity or any other reason, would increase the short-circuit current available to their service-entrance equipment. If this current is too great, explosive failure can result.

Kilowatthour and sometimes demand-metering equipment are connected to the service-entrance conductors. Proceeding toward the utilization equipment, the power-supply system fans out into feeders and branch circuits (see Fig. 97). Each of the feeders, i.e., a run of untapped conductor or cable, is connected to the supply through a switch and fuses or a circuit breaker. At a point, usually near that portion of the electrical loads which are to be supplied, a **panel box** or perhaps a **load-center assembly** of switching and/or control equipment is installed. From this panel box or load-center assembly, circuits radiate; i.e., circuits are installed to extend into the area being served to connect electrical machinery or devices or make available electric receptacles connected to the source of electric power.

Each feeder and each branch circuit will have its own **overcurrent protection** and **disconnect** means in the form of a fuse and switch combination or a circuit breaker.

There is a provision in the NEC 1975 rules for the following types of feeder and branch circuit wiring:

**1. Open Wiring on Insulators** (NEC 1975, Art. 320). This wiring method uses approved cleats, knobs, tubes, and flexible tubing for the protection and support of insulated conductors run on or in buildings and not concealed by the building structure.

**2. Concealed Knob-and-Tube Work** (NEC 1975, Art. 324). Concealed knob-and-tube work may be used in the hollow spaces of walls and ceilings. It must not be used in commercial garages, theaters, and assembly halls having a capacity of more than 200 persons, motion-picture studios, or hazardous locations.

**3. Mineral-Insulated Metal-Sheathed Cable, Type MI** (NEC 1975, Art. 330). Type MI Cable contains one or more electrical conductors insulated with a highly compressed refractory mineral insulation and enclosed in a liquid- and gastight metallic tube sheath. Appropriate approved fittings must be used with it. It may be used for services, feeders, and branch circuits either exposed or concealed and dry or wet. It may be used in Class I, II, or III hazardous locations. It may be used for under-plaster extensions and embedded in plaster finish or brick or other masonry. It may be used where exposed to weather or continuous moisture, for underground runs and embedded in masonry, concrete or fill, in buildings in course of construction or where exposed to oil, gasoline, or other conditions. If the environment would cause destruction of the sheath, it must be protected by suitable materials.

**4. Aluminum-Sheathed Cable Type ALS** (NEC 1975, Art. 331). Aluminum-sheathed Type ALS cable is factory-assembled. It consists of one or more insulated conductors in an impervious, continuous, closely fitting tube of aluminum. It must be used with appropriate approved fittings. It may be used in exposed or concealed work and in dry or wet locations. Protection by suitable materials is required in environments containing strong chlorides, caustic alkalis, vapors of chlorine, or vapors of hydrochloric acid and when installed underground or buried in concrete or used in areas subject to severe corrosive influence.

**5. Copper-Sheathed Cable, Type CS** (NEC 1975, Art. 332). Type CS copper-sheathed cable is a factory assembly of one or more individually insulated conductors which are enclosed in a liquid- and gas-tight, closely fitting sheath of copper or brass. Use of this cable is permitted in wet or dry locations, either exposed or concealed. In corrosive locations the sheath must be properly protected. Care must be taken regarding bend radii when the cable is fished into place. Approved fittings must be used.

**6. Metal-Clad Cable, Type MC and AC Series** (NEC 1975, Art. 334). These are metal-clad cables, i.e., an assembly of insulated conductors in a flexible metal enclosure. Type MC are power cables and in the range up to 600 V are made in conductor sizes of no. 4 and larger for copper and no. 2 and larger for aluminum. Type AC are branch and feeder cables with armor of flexible metal tape. All AC types except ACL have an internal bonding strip of copper or aluminum in intimate contact with the armor for its entire length. Metal-clad branch circuit cable was formerly called BX. Metal-clad cables may generally be installed where not subject to physical damage, for feeders and branch circuits in exposed or concealed work, with qualifications for wet locations, direct burial in concrete, etc. The use of Type AC cable is prohibited (1) in motion-picture studios, (2) in theaters and assembly halls, (3) in hazardous locations, (4) where exposed to corrosive fumes or vapors, (5) on cranes or hoists except where flexible connections to motors, etc., are required, (6) in storage-battery rooms, (7) in hoistways or on elevators except (i) between risers and limit switches, interlocks, operating buttons, and similar devices in hoistways and in escalators and moving walkways and (*ii*) short runs on elevator cars, where free from oil, and if securely fastened in place, or (8) in commercial garages in hazardous locations. Type ACL (lead-covered) shall not be used for direct burial in the earth.

**7. Nonmetallic-Sheathed Cable, Types NM and NMC** (NEC 1975, Art. 336).    These are assemblies of two more insulated conductors (nos. 14 through 2 for copper, nos. 12 through 2 for aluminum) having an outer sheath of moisture-resistant, flame-retardant, nonmetallic material. In addition to the insulated conductors, the cable may have an approved size of uninsulated or bare conductor for grounding purposes only. The outer covering of NMC cable is also corrosion-resistant. This cable type has been commonly called **Nomex.** These cable types are used much the same as the AC cable but with different requirements for support, fastening, and mechanical protection.

**8. Shielded Nonmetallic-Sheathed Cable, Type SNM** (NEC 1975, Art. 337).    Type SNM is a factory-assembled cable consisting of two or more insulated conductors (nos. 14 through 2 copper and nos. 12 through 2 aluminum) in an extruded core of moisture-resistant material, covered with an overlapping spiral metal tape and wire shield and jacketed with an extruded moisture-, flame-, corrosion-, fungus-, and sunlight-resistant material. This cable is to be used (1) under appropriate ambient-temperature conditions and (2) in continuous rigid-cable support or in raceways. It can be used in some hazardous locations as defined by the NEC.

**9. Service Entrance Cable, Types SE and USE** (NEC 1975, Art. 338).    These cables, containing one or more individually insulated conductors, are primarily used for electric services. Type SE has a flame-retardant, moisture-resistant covering and is not required to have built-in protection against mechanical abuse. Type USE is recognized for use underground. It has a moisture-resistant covering, but not necessarily a flame-retardant one. Like the SE cable, USE cable is not required to have inherent protection against mechanical abuse. Under specified conditions, SE and USE cables can be used for feeders and branch circuits.

**10. Underground Feeder and Branch-Circuit Cable, Type UF** (NEC 1975, Art. 339).    This cable is made in sizes 14 through 4/0, and the insulated conductors are Types TW, RHW, and others approved for the purpose. As in the NM cable, the UF type may contain an approved size of uninsulated or bare conductor for grounding purposes only. The outer jacket of this cable shall be flame-retardant, moisture-resistant, fungus-resistant, corrosive-resistant, and suitable for direct burial in the ground.

**11. Other Installation Practices.**    The NEC details rules for nonmetallic circuit extensions and underplaster extensions. It also provides detailed rules for installation of electrical wiring in (*a*) rigid metal conduit (which may be used for all atmospheric conditions and locations with due regard to corrosion protection and choice of fittings), (*b*) rigid nonmetallic conduit (which is essentially corrosion-proof), in electrical metallic tubing (which is lighter-weight than rigid metal conduit), (*c*) flexible metal conduit, (*d*) liquidtight flexible metal conduit, (*e*) surface raceways, (*f*) underfloor raceways, (*g*) multioutlet assemblies, (*h*) cellular metal floor raceways, (*i*) structural raceways, (*j*) cellular concrete floor raceways, (*k*) wireways (sheet-metal troughs with hinged or removable covers), (*l*) flat,

Type FC, cable assemblies installed in a surface metal raceway (Type FC cable contains three or four no. 10 special stranded copper wires), (*m*) busways, and (*n*) cable-bus. Busways and cable-bus installations are permitted for exposed work only.

In all installation work, only approved outlets, switch and junction boxes, fittings, terminal strips, and dead-end caps shall be used, and they are to be used in an approved fashion (see NEC 1975, Art. 370).

Table 21 lists the allowable **current-carrying capacities** of copper conductors of various sizes. Table 22 lists the various **conductor insulations** approved by the NEC 1975 for conductors used in interior wiring. Table 20 relates AWG wire sizes to **metric units.** Dimensions and allowable fill of conduit and tubing are listed in Table 23. Table 24 lists the cross-sectional area of the various insulated conductors. **Derating factors** for cases where there are more than three conductors in a conduit are shown in a footnote to Table 21. Estimated full-load currents of motors can be taken from Table 25.

In all metallic protecting systems, such as conduit, armored cable, or metal raceways, joints and splices in conductors must be made only in junction boxes or other proper fittings; therefore, these fittings can be located only in accessible places and never concealed in partitions. Splices or joints in the wire must never be in the conduit piping, raceway, or metallic tubing itself, for the splices may become a source of trouble as a result of corrosion or grounding if water should enter the conduit.

**Switching Arrangements**    Small quick-break switches must be set in or on a metal box or fitting and may be of the push, tumbler, or rotary type. The following types of switches are used to control lighting circuits: (1) single-pole, (2) double-pole, (3) three-point or three-way, (4) four-way, in combination with three-way switches to control lights from three or more stations, (5) electrolier.

**AC Systems**    All conductors of an ac system must be placed in the same metallic casing so that their resultant magnetic field is nearly zero. If this is not done, eddy currents are set up causing heating and excessive loss. With single conductors in a casing, an excessive reactance drop may result.

**Service wires** are the conductors that bring the electric power into a building and should enter the building as near as possible to the service switch, so that when the switch is open all the electrical conductors and equipment inside the building will be dead. The service wires must be rubber- or thermoplastic-covered from the point of support on the outside of the building to the service switch or cutout and must be no. 6 wire or larger except for installations consisting of two-wire branch circuits where no. 8 wire may be permitted. A minimum of 100-A three-wire service is recommended for all single-family residences.

Generally, when the conductors from overhead lines enter a building, the wires are encased in rigid conduit equipped with a weather cap or a service entrance cable (type ASE armored or SE type, unarmored) may be attached directly to the building wall. The inner end of the service enters a metal service cabinet in which the service fuses and switch are located. Service conductors may also terminate at an air-break or oil-immersed switch in a metal case or on a panel board which is accessible to qualified persons only.

All underground service wires must be connected to the interior wiring through a blade of the service switch or circuit breaker and be fused or automatically interrupted at the service switch. A service switch controlling a three-wire dc or a single-phase system having a grounded neutral wire does not need to open that conductor.

The single-line diagram, Fig. 97, indicates a simplified interior arrangement of circuits and the necessary protection of the conductors and terminal load. Where a reduction is made in the wire size a protective device shall be installed to limit the conductor current to a safe value. Large motors and other terminal loads should also have overcurrent protection.

The fuse ratings and the setting of the protective devices for starting and for running protection of motors are given in Table 26. Tables 25 and 26 are taken from the 1975 NEC.

**Grounding** of direct- and alternating-current systems of 300 V and less is usually required. Inside a building the **grounded**

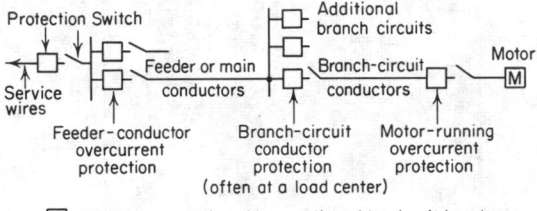

- -□- Fuses, non-adjustable or adjustable circuit breakers
- -\- The switches in the above diagrams are often a part of the circuit breaker or fuse box.

**Fig. 97** Motor and wiring protection. See Table 23 for selection of conductor sizes and protective devices.

**conductor** (one of the two conductors in two-wire system and the **neutral conductor** in a three-wire or a four-wire system) should have a **white or natural gray covering** throughout to distinguish it from the ungrounded conductors. This identified grounded conductor must not be fused or be opened unless the other conductors are opened simultaneously. **Green wires** only shall be used for grounding electrical equipment, such as motors, as well as conduits, armor, boxes, and such metallic enclosures. Four-wire circuits have **black, white, red, and blue** conductors.

DC systems need be grounded only at the generating stations because the grounded wire is electrically connected to one of the conductors in all the circuits throughout the system. In ac systems, since one section can be insulated from the other by a transformer, each section of 300 V or under is grounded at the individual services. The conductor grounding the ac system should not be less than no. 8 copper wire and must be without a joint or a splice and run from the supply side of the service switch.

The service conduit that protects the service wires on the outside of the building must be grounded by a wire at least as large as no. 8, run directly to ground.

The entire metallic system surrounding the conductors must be at ground potential. It is only necessary to ground the metallic system, including motors and other equipment, at one point, provided that each section makes a good electrical connection with the next.

Since January 1, 1973 **ground-fault circuit interrupters** have been required by the NEC for personnel protection from line to ground electric shock. Such circuit interrupters are required by the 1975 Code in branch circuits supplying the following: (1) residential receptacle outlets outdoors and in

bathrooms; (2) 120-V, 15- and 20-A receptacle outlets supplied by temporary wiring at construction sites; (3) electrical equipment, including power supply cords, used with storable pools; (4) receptacle outlets within 15 ft of a pool; and (5) fountain equipment.

Ground-fault circuit protection may be used at other locations and, if so used, will provide additional protection against line-to-ground shock hazard.

Ground fault-circuit interrupters monitor the current in the two conductors of a circuit. These two currents should be equal. If they are not equal, some current is leaking to ground, indicating a line to ground fault. If the difference between the two currents is 5 mA or more, the ground-fault circuit interrupter will automatically disconnect the faulted circuit in about 0.025 s.

**Overload Protection** A fuse or circuit breaker must be provided in all ungrounded conductors. Induction motors are usually protected by overload or thermal relays operating as automatic circuit breakers. To protect wiring properly, an automatic cutout must be installed at every point where a change is made in the size of the wire. Fuses or circuit breakers should not be placed either in a grounded line or in a ground wire. (See Fig. 97.)

**Fuses, cutout bases,** and **switches** are manufactured and change sizes, as follows: Edison plug (125 V only), 0 to 30 A; spring-clip cartridge (ferrule contact), 0 to 30, 31 to 60 A; knife-blade cartridge type, 61 to 100, 101 to 200, 201 to 400, 401 to 600 A; for 601 A and larger, knife-blade cartridge type with equal-size fuses in parallel may be used except for the protection of a branch motor circuit where a circuit breaker can be installed.

Since the rating of a fuse is only about 90 percent of the current that it will carry indefinitely, and since it may also take a few minutes before the heat due to slightly excessive current would be sufficient to melt the fuse wire and hence open the circuit, insulation may be permanently damaged if the fuses are larger than the current-carrying capacities given by Table 21.

**Demand Calculations for Building Feeder Sizes** The **demand factor** or **demand** is the ratio of maximum demand to the total connected load. This depends on the type of building, whether hotel, theater, factory, etc. The demand factor for any particular class of installation decreases as the floor area increases. Values of demand factors are found in the 1975 NEC Art. 220.

**Load centers** are panels or cabinets which act as distribution centers and which are supplied by feeder or main conductors, and from which the current to several branch circuits is taken. In each branch circuit there is usually a small combined switch and circuit breaker. Load centers for widely different types of circuits such as single-phase two-wire, single-phase three-wire, and three-phase are readily obtainable from several manufacturers.

## RESISTOR MATERIALS

For use in rheostats, electric furnaces, ovens, heaters, and many electrical appliances, a resistor material with high melting point and high resistivity which does not disintegrate or corrode at high temperatures is necessary. These requirements are met by the nickel-chromium and nickel-chromium-iron alloys. For electrical instruments and measuring apparatus, the resistor material should have high resistivity, low

**Table 21. Allowable Current-Carrying Capacities of Copper[a] Conductors, A**

(Not more than three conductors in raceway or cable. [b] Based on room temperature of 80°F or 30°C)

| Size, AWG-MCM | Max op. temp. 60°C (140°F)<br>Rubber, Type RUW[c] (nos. 14–2)<br>Thermoplastic, Types T,[d] TW[c]<br>Moisture-resistant, Type UF[c]<br>A | Max op. temp. 75°C (167°F)<br>Rubber, Types RH,[d] RHW[c], RUH[d] (nos. 14–2)<br>Thermoplastic, Types THW,[c] THWN[c]<br>Cross-linked synthetic polymer, Type XHHW[e]<br>Moisture- and heat-resistant, Type USE[c]<br>B | Max op. temp. 90°C (194°F)<br>Rubber, Type RHH[c]<br>Thermoplastic, Type THHN[d]<br>Fluorinated ethylene propylene, Types FEP,[d] FEPB[d]<br>Silicone asbestos, Type SA[d]<br>Asbestos and varnished cambric, Type AVB[d]<br>Cross-linked synthetic polymer, Type XHHW[d]<br>Switchboard wiring only, TA, TBS, SIS[h]<br>C |
|---|---|---|---|
| 14 | 15 | 15 | 25[f] |
| 12 | 20 | 20 | 30[f] |
| 10 | 30 | 30 | 40[f] |
| 8 | 40 | 45 | 50 |
| 6 | 55 | 65 | 70 |

| Size | | | |
|---|---|---|---|
| 4[a] | 70 | 85 | 90 |
| 3[a] | 80 | 100 | 105 |
| 2[a] | 95 | 115 | 120 |
| 1[a] | 110 | 130 | 140 |
| 0[a] | 125 | 150 | 155 |
| 00[a] | 145 | 175 | 185 |
| 000 | 165 | 200 | 210 |
| 0000 | 195 | 230 | 235 |
| 250 | 215 | 255 | 270 |
| 300 | 240 | 285 | 300 |
| 350 | 260 | 310 | 325 |
| 400 | 280 | 335 | 360 |
| 500 | 320 | 380 | 405 |
| 600 | 355 | 420 | 455 |
| 700 | 385 | 460 | 490 |
| 750 | 400 | 475 | 500 |
| 800 | 410 | 490 | 515 |
| 900 | 435 | 520 | 555 |
| 1,000 | 455 | 545 | 585 |
| 1,250 | 495 | 590 | 645 |
| 1,500 | 520 | 625 | 700 |
| 1,750 | 545 | 650 | 735 |
| 2,000 | 560 | 665 | 775 |

**Correction Factors for Room Temperatures over 86°F (30°C)**

| Temp, °F | 104 | 113 | 122 | 131 | 140 | 158 | 167 | 176 |
|---|---|---|---|---|---|---|---|---|
| Temp, °C | 40 | 45 | 50 | 55 | 60 | 70 | 75 | 80 |
| Column A | 0.82 | 0.71 | 0.58 | 0.41 | | | | |
| Column B | 0.88 | 0.82 | 0.75 | 0.67 | 0.58 | 0.35 | | |
| Column C | 0.91 | 0.87 | 0.82 | 0.76 | 0.71 | 0.58 | 0.50 | 0.41 |

[a] **For aluminum wire** the allowable carrying capacity of no. 12 is 15 A (75 percent of Cu), that of no. 10 is 25 A (83 percent of Cu). In general the allowable carrying capacity is 77 percent (rounded off to the nearest 5) from 12 to 4, 79 percent from 3 to 0000, 81 percent from 250 to 900 MCM, and 83 percent from 1,000 to 2,000 MCM.

[b] For four to six conductors the current is 80 percent, for seven to 24 conductors the current is 70 percent, for 25 to 42 conductors the current is 60 percent, and for 43 and above the current is 50 percent.

[c] For dry or wet locations.

[d] For dry locations.

[e] For wet locations.

The ampacities for Types FEP, FEPB, RHH, THHN, and XHHW conductors for sizes 14, 12, and 10 shall be the same as designated for 75°C conductors in this table. Because thermoplastic insulation, as distinct from thermosetting, may stiffen at temperatures below 14°F (−10°C), care should be used in its installation at such temperatures.

[f] For three-wire, single phase residential services, the allowable ampacity of RH, REH, RHW, THW, and XHHW copper conductors shall be for no. 4, 100 A; no. 3, 100 A; no. 2, 125 A; no. 1, 150 A; no. 1/0, 175 A; and no. 2/0, 200 A.

[g] TA—thermoplastic and asbestos; TB—thermoplastic; SIS—heat-resistant rubber.

**Table 22. Conductor Type and Application**

| Type letter | Insulation, trade name (See NEC 1975 or ANSI C1-1975, for complete information) | Environment | Outer covering[a] |
|---|---|---|---|
| Max operating temperature = 60°C (140°F) | | | |

| Type letter | Insulation, trade name (See NEC 1975 or ANSI C1-1975, for complete information) | Environment | Outer covering[a] |
|---|---|---|---|
| *Max operating temperature = 60°C (140°F)* | | | |
| RUW | Moisture-resistant latex rubber | Dry and wet | 1 |
| T | Thermoplastic | Dry | None |
| TW | Moisture-resistant thermoplastic | Dry and wet | None |
| TF | Thermoplastic-covered, solid or 7-strand | c,d | None |
| TFF | Thermoplastic-covered, flex. stranding | c,d | None |
| MTW | Moisture-, heat-, and oil-resistant thermoplastic machine-tool wiring (NFPA Stand. 79, NEC 1975, Art. 670) | Wet | None or nylon |
| UF | Moisture-resistant, underground feeder | Dry and wet | None |
| *Max operating temperature = 75°C (167°F)* | | | |
| RH | Heat-resistant rubber | Dry | 1,2 |
| RHW | Moisture- and heat-resistant rubber[e] | Dry and wet | 1,2 |
| RUH | Heat-resistant latex rubber | Dry | 1 |
| THW | Moisture- and heat-resistant thermoplastic | Dry and wet | None |
| THWN | Moisture- and heat-resistant thermoplastic | Dry and wet | Nylon |
| XHHW | Moisture- and heat-resistant cross-linked polymer | Wet | None |
| RFH-1&2 | Heat-resistant rubber-covered solid or 7-strand | b–d | None |
| FFH-1&2 | Heat-resistant rubber-covered flexible stranding | b–d | None |
| UF | Moisture-resistant and heat-resistant | Dry and wet | None |
| USE | Heat- and moisture-resistant | Dry and wet | 4 |
| *Max operating temperature = 85°C (185°F)* | | | |
| MI | Mineral-insulated (metal-sheathed) | Dry and wet | Copper |
| *Max operating temperature = 90°C (194°F)* | | | |
| RHH | Heat-resistant rubber | Dry | 1,2 |
| THHN | Heat-resistant thermoplastic | Dry | Nylon |
| THW | Moisture- and heat-resistant thermoplastic | f | None |
| XHHW | Moisture- and heat-resistant cross-linked synthetic polymer | Dry | None |
| FEP | Fluorinated ethylene propylene | Dry | None |
| FEPB | Fluorinated ethylene propylene | Dry | 3 |
| TFN | Heat-resistant thermoplastic covered, solid or 7-strand | c,d | Nylon |
| TFFN | Heat-resistant thermoplastic flexible stranding | c,d | Nylon |
| MTW | Moisture-, heat-, and oil-resistant thermoplastic, machine-tool wiring (NFPA Stand. 79, NEC 1975, Art. 670) | Dry | None or nylon |
| SA | Silicone asbestos | Dry | Asbestos or Glass |
| *Max operating temperature = 200°C (392°F)* | | | |
| FEP,FEPB | Fluorinated ethylene propylene Special applications | Dry | 5 3 |
| PF,PGF | Fluorinated ethylene propylene | c,d | None or glass braid |
| SF-2 | Silicone rubber, solid or 7-strand | c,d | Nonmetallic |
| *Max operating temperature = 250°C (482°F)* | | | |
| MI | Mineral-insulated (metal-sheathed), for special applications | Dry and wet | Copper |
| TFE | Extruded polytetrafluoroethylene, only for leads within apparatus, or within raceways connected to apparatus, or as open wiring (silver or nickel-coated copper only) | Dry | None |
| PTF | Extruded polytetrafluoroethylene, solid or 7-strand (silver or nickel-coated copper only) | c,d | None |

Source: NEC 1975, Tables 310–13 and 402–3; ANSI C1-1975.
[a]1: Moisture-resistant, flame-retardent nonmetallic; 2: outer covering not required when rubber insulation has been specifically approved for the purpose; 3; no. 14-8 glass braid, no. 6-2 asbestos braid; 4:moisture-resistant nonmetallic; 5:asbestos glass braid.
[b]Limited to 300 V.
[c]No. 18 and no. 16 conductor for remote controls, low-energy power, low-voltage power, and signal circuits; NEC 1975, Sec. 725-16.
[d]Fixture wire no. 18.
[e]For over 2,000 V, the insulation shall be ozone-resistant.
[f]Special applications within electric discharge lighting equipment. Limited to 1,000 V open-circuit volts or less.

temperature coefficient, and, for many uses, low thermoelectric power against copper. The properties of resistor materials are given in Table 27. Most of these materials are available in ribbon as well as in wire form. Cast-iron and steel wire are efficient and economical resistor materials for many uses, such as power-absorbing rheostats and motor starters and controllers. (See also Sec. 6.)

**Advance** has a low temperature coefficient and is useful in many types of measuring instrument and precision equipment. Because of its high thermoelectric power to copper, it is valuable for thermoelements and pyrometers. It is noncorrosive and is used to a large extent in industrial and radio rheostats. **Hytemco** is a nickel-iron alloy characterized by a high temperature coefficient and is used advantageously where self-regulation is required as in immersion heaters and heater pads. **Magno** is a manganese-nickel alloy used in the manufacture of incandescent lamps and radio tubes. **Manganin** is a copper-manganese-nickel alloy which, because of its very low temperature coefficient and its low thermal emf with respect to copper, is very valuable for high-precision electrical measuring apparatus. It is used for the resistance units in bridges, for shunts, multipliers, and similar measuring devices. **Nichrome V** is a nickel-chromium alloy free from iron, is noncorrosive, nonmagnetic, withstands high temperatures, and has high resistivity. It is recommended as material for heating elements in electric furnaces, hot-water heaters, ranges, radiant heaters, and high-grade electrical appliances. **Pure nickel** is used to satisfy the high requirements in the fabrication of radio tubes, such as the elimination of all gases and impurities in the metal parts. It has also other uses such as in incandescent lamps, for combustion boats, laboratory accessories, and resistance thermometers.

**Carbon** withstands high temperatures and has high resistance; its temperature coefficient is negative; it will safely carry about 125 A/in². Amorphous carbon has a resistivity of 3,800 and 4,100 $\mu\Omega\cdot$cm, retort carbon about 720 $\mu\Omega\cdot$cm, and graphite about 812 $\mu\Omega\cdot$cm. The properties of any particular kind of carbon depend on the temperature at which it was fired. Carbon for rheostats may best be used in the form of compression rheostats.

## MAGNETS

A **permanent magnet** is one that retains a considerable amount of magnetism indefinitely. Permanent magnets are used in electrical instruments, telephone receivers, loudspeakers, magnetos, tachometers, magnetic chucks, and for many purposes where a constant magnetic field or a constant source of magnetism is desired. The magnetic material should have high retentivity, a high remanence, and a high coercive force (see Fig. 14). These properties are usually found with hardened steel and its alloys and also in ceramic permanent magnet materials.

Since permanent magnets must operate on the molecular mmf imparted to them when magnetized, they must necessarily operate on the portion CDO of the hysteresis loop (see Fig. 14). The area CDO is proportional to the **stored energy** within the magnet and is a criterion of its usefulness as a permanent-magnetic material. In the left half of Fig. 98, are given the B-H characteristics of several permanent-magnetic materials; these include 5 to 6 percent tungsten steel (curve 1); 3½ percent chrome magnet steel (curve 2); cobalt magnet steel, containing 16 to 36 percent cobalt and 5 to 9 percent chromium and in some alloys tungsten (curve 3); and the carbon-free aluminum-nickel-cobalt-steel alloys called Alnico. There are many grades of Alnico; the characteristics of three of them are shown by curves 4, 5, 6. Their composition is as follows:

| Curve | Alnico no. | Composition, percent | | | | | |
|---|---|---|---|---|---|---|---|
| | | Al | Ni | Co | Cu | Ti | Fe |
| 4 | 5 | 8 | 14 | 24 | 3 | . . . | 51 |
| 5 | 6 | 8 | 14 | 24 | 3 | 1.25 | 49.75 |
| 6 | 12 | 6 | 18 | 35 | . . . | 8.0 | 33 |

## Table 23. Dimensions and Allowable Fill of Conduit and Tubing

| Trade size, in | ID, in | Area, in² | Allowable fill, in² of conductors (not lead-covered) | | |
|---|---|---|---|---|---|
| | | | One conductor, 53% fill | Two conductors, 31% fill | Over two conductors,* 40% fill |
| ½ | 0.622 | 0.30 | 0.16 | 0.09 | 0.12 |
| ¾ | 0.824 | 0.53 | 0.28 | 0.16 | 0.21 |
| 1 | 1.049 | 0.86 | 0.46 | 0.27 | 0.34 |
| 1¼ | 1.380 | 1.50 | 0.80 | 0.47 | 0.60 |
| 1½ | 1.610 | 2.04 | 1.08 | 0.63 | 0.82 |
| 2 | 2.067 | 3.36 | 1.78 | 1.04 | 1.34 |
| 2½ | 2.469 | 4.79 | 2.54 | 1.43 | 1.92 |
| 3 | 3.068 | 7.38 | 3.91 | 2.29 | 2.95 |
| 3½ | 3.548 | 9.90 | 5.25 | 3.07 | 3.96 |
| 4 | 4.026 | 12.72 | 6.74 | 3.94 | 5.09 |
| 4½ | 4.506 | 15.94 | 8.45 | 4.94 | 6.38 |
| 5 | 5.047 | 20.00 | 10.60 | 6.20 | 8.00 |
| 6 | 6.065 | 28.89 | 15.31 | 8.96 | 11.56 |

SOURCE: NEC 1975, p. 70-563.
*For conductor derating with more than three conductors see footnote b, Table 21.

**Table 24. Nominal Cross-Sectional Area of Rubber-Covered and Plastic-Covered Conductors**

|  | Size, AWG | | | | | |
|---|---|---|---|---|---|---|
|  | 18 | 16 | 14 | 12 | 10 | 8 |
| Type | Cross-sectional area, in² | | | | | |
| RFH-2 | 0.0167 | 0.0196 | (fixture wire) | | | |
| SF-2 | 0.0167 | 0.0196 | 0.0230 | (fixture wire) | | |
| RH | . . . | . . . | 0.0230 | 0.0278 | 0.0460 | 0.0854 |
| RHH, RHW | . . . | . . . | 0.0327 | 0.0384 | 0.0460 | 0.0854 |
| RHH, RHW (without outer covering) | . . . | . . . | 0.0206 | 0.0251 | 0.0311 | 0.0598 |
| THW | . . . | . . . | 0.0206 | 0.0251 | 0.0311 | 0.0598 |
| T, TW, RUH, RUW | . . . | . . . | 0.0135 | 0.0172 | 0.0224 | 0.0471 |
| TF | 0.0088 | 0.0109 | . . . | . . . | . . . | . . . |
| THWN, THHN | . . . | . . . | 0.0087 | 0.0117 | 0.0184 | 0.0373 |
| TFN | 0.0064 | 0.0079 | . . . | . . . | . . . | . . . |
| FEPB | . . . | . . . | 0.0087 | 0.0115 | 0.0159 | 0.0272 |
| FEP, TFE | . . . | . . . | 0.0087 | 0.0115 | 0.0159 | 0.0333 |
| PTF | 0.0052 | 0.0066 | 0.0087 | . . . | . . . | . . . |
| XHHW | . . . | . . . | 0.0131 | 0.0167 | 0.0216 | 0.0456 |

|  | Size, AWG | | | | | | | | |
|---|---|---|---|---|---|---|---|---|---|
|  | 6 | 4 | 3 | 2 | 1 | 1/0 | 2/0 | 3/0 | 4/0 |
|  | Cross-sectional area, in² | | | | | | | | |
| RH, RHH,* RHW* | 0.1238 | 0.1605 | 0.1817 | 0.2067 | 0.2715 | 0.3107 | 0.3578 | 0.4151 | 0.4840 |
| RUH, RUW | 0.0819 | 0.1087 | 0.1263 | 0.1473 | . . . | . . . | . . . | . . . | . . . |
| TW, T, THW | 0.0819 | 0.1087 | 0.1263 | 0.1473 | 0.2027 | 0.2367 | 0.2781 | 0.3288 | 0.3904 |
| TFE | 0.0467 | 0.0669 | 0.0803 | 0.0973 | 0.1385 | 0.1676 | 0.1974 | 0.2436 | 0.2999 |
| FEPB | 0.0716 | 0.0962 | 0.1122 | 0.1316 | . . . | . . . | . . . | . . . | . . . |
| FEP | 0.0467 | 0.0669 | 0.0803 | 0.0973 | . . . | . . . | . . . | . . . | . . . |
| THHN, THWN | 0.0519 | 0.0845 | 0.0995 | 0.1182 | 0.1590 | 0.1893 | 0.2265 | 0.2715 | 0.3278 |
| XHHW | 0.0625 | 0.0845 | 0.0995 | 0.1182 | 0.1590 | 0.1893 | 0.2265 | 0.2715 | 0.3278 |

|  | Size, MCM | | | | | |
|---|---|---|---|---|---|---|
|  | 250 | 500 | 750 | 1000 | 1500 | 2000 |
|  | Cross-sectional area, in² | | | | | |
| RH, RHH,* RHW* | 0.5917 | 0.9834 | 1.4082 | 1.7531 | 2.5475 | 3.2079 |
| TW, T, THW | 0.4877 | 0.8316 | 1.2252 | 1.5482 | 2.2748 | 2.9013 |
| THHN, THWN | 0.4026 | 0.7163 | 1.0623 | 1.3623 | . . . | . . . |
| XHHW | 0.4026 | 0.7163 | 1.0936 | 1.3893 | 2.0612 | 2.6590 |

SOURCE: NEC 1975, p. 70-563.
For general branch and feeder circuits the minimum conductor size is 14. Sizes 14 to 8 are solid wire. Sizes 6 and larger are stranded.
*RHH and RHW without covering have the same dimension as THW.

All the Alnicos can be made by the sand or the precision-casting (lost-wax) process, but the most satisfactory method is by the sintering process.

If the alloys are held in a magnetic field during heat-treatment, a **magnet grain** is established and the magnetic properties in the direction of the field are greatly increased. The alloys are hard, can be formed only by casting or sintering, and cannot be machined except by grinding.

The curves in the right half of Fig. 98 are "external energy" curves and give the product of $B$ and $H$. The optimum point of operation is at the point of maximum energy as is indicated at $A_1$ on curve 5.

Considering curve 5, if the magnetic circuit remained closed, the magnet would operate at point $B$. To utilize the flux, an air gap must be introduced. The air gap acts as a demagnetizing force, $H_1 (= B_1 A_1)$, and the magnet operates at point $A_1$ on the $HB$ curve. The line $OA_1$ is called the *air-gap line* and its slope is given by $\tan \theta_1 = B_1/H_1$ where $H_1 = B_g l_g/l_m$ and $B_1/B_g = A_g/A_m$, where $B_g$ = flux density in gap, G; $l_g$ = length of gap, cm; $l_m$ = length of magnet, cm; $A_g$, $A_m$ = areas of the gap and magnet, cm².

If the air gap is lengthened, the magnet will operate at $A_2$ corresponding to a lesser flux density $B_3$ and the new air-gap line is $OA_2$. If the gap is now closed to its original value, the magnet will not return to operation at point $A_1$ but will operate at some point $C$ on the line $OA_1$. If the air gap is varied between the two foregoing values, the magnet will operate along the **minor hysteresis loop** $A_2 C$. Return to point $A_1$ can be

**Table 25. Approximate Full-Load Currents of Motors, A**
(See NEC 1975 or ANSI C1-1975 for more complete information)

| hp | Three-phase ac motors, squirrel-cage and wound-rotor induction types | | | | Synchronous type, unity power factor | | | | Single-phase ac motors | | DC motors† | |
|---|---|---|---|---|---|---|---|---|---|---|---|---|
| | 230 V | 460 V | 575 V | 2,300 V | 220 V | 440 V | 550 V | 2,300 V | 115 V‡ | 230 V‡ | 120 V | 240 V |
| ½ | 2 | 1 | 0.8 | ... | ... | ... | ... | ... | 9.8 | 4.9 | 5.2 | 2.6 |
| ¾ | 2.8 | 1.4 | 1.1 | ... | ... | ... | ... | ... | 13.8 | 6.9 | 7.4 | 3.7 |
| 1 | 3.6 | 1.8 | 1.4 | ... | ... | ... | ... | ... | 16 | 8 | 9.4 | 4.7 |
| 1½ | 5.2 | 2.6 | 2.1 | ... | ... | ... | ... | ... | 20 | 10 | 13.2 | 6.6 |
| 2 | 6.8 | 3.4 | 2.7 | ... | ... | ... | ... | ... | 24 | 12 | 17 | 8.5 |
| 3 | 9.6 | 4.8 | 3.9 | ... | ... | ... | ... | ... | 34 | 17 | 25 | 12.2 |
| 5 | 15.2 | 7.6 | 6.1 | ... | ... | ... | ... | ... | 56 | 28 | 40 | 20 |
| 7½ | 22 | 11 | 9 | ... | ... | ... | ... | ... | 80 | 40 | 58 | 29 |
| 10 | 28 | 14 | 11 | ... | ... | ... | ... | ... | 100 | 50 | 76 | 38 |
| 15 | 42 | 21 | 17 | ... | ... | ... | ... | ... | ... | ... | ... | 55 |
| 20 | 54 | 27 | 22 | ... | ... | ... | ... | ... | ... | ... | ... | 72 |
| 25 | 68 | 34 | 27 | ... | 54 | 27 | 22 | ... | ... | ... | ... | 89 |
| 30 | 80 | 40 | 32 | ... | 65 | 33 | 26 | ... | ... | ... | ... | 106 |
| 40 | 104 | 52 | 41 | ... | 86 | 43 | 35 | ... | ... | ... | ... | 140 |
| 50 | 130 | 65 | 52 | ... | 108 | 54 | 44 | ... | ... | ... | ... | 173 |
| 60 | 154 | 77 | 62 | 16 | 128 | 64 | 51 | 12 | ... | ... | ... | 206 |
| 75 | 192 | 96 | 77 | 20 | 161 | 81 | 65 | 15 | ... | ... | ... | 255 |
| 100 | 248 | 124 | 99 | 26 | 211 | 106 | 85 | 20 | ... | ... | ... | 341 |
| 125 | 312 | 156 | 125 | 31 | 264 | 132 | 106 | 25 | ... | ... | ... | 425 |
| 150 | 360 | 180 | 144 | 37 | ... | 158 | 127 | 30 | ... | ... | ... | 506 |
| 200 | 480 | 240 | 192 | 49 | ... | 210 | 168 | 40 | ... | ... | ... | 675 |

*The values of current are for motors running at speeds customary for belted motors and motors having normal torque characteristics. Use name-plate data for low-speed, high-torque, or multispeed motors. For synchronous motors at 0.8 pf multiply the above amperes by 1.25, at 0.9 pf by 1.1. The motor voltages listed are rated voltages. Respective nominal system voltages would be 220 to 240, 440 to 480, and 550 to 600 V. For full-load currents of 208-V motors, multiply the above amperes by 1.10; for 200-V motors, multiply by 1.15.

†Ampere values are for motors running at base speed.

‡Rated voltage. Nominal system voltages are 120 and 240.

**Table 26. Maximum Rating or Setting of Motor Branch-Circuit Protective Devices and Starting-Inrush Code Letters**

| Type of motor | Percent of full-load current | | | | Code letter | kVA/hp with locked rotor |
|---|---|---|---|---|---|---|
| | Non-time delay fuse | Dual-element (time delay) fuse | Instantaneous breaker | Time-limit breaker | | |
| Single-phase, all types, no code letter | 300 | 175 | 700 | 250 | A | 0 − 3.14 |
| | | | | | B | 3.15 − 3.54 |
| AC motors: single-phase, polyphase squirrel-cage, or synchronous with full-voltage, resistor, or reactor starting: | | | | | C | 3.55 − 3.99 |
| | | | | | D | 4.0 − 4.49 |
| | | | | | E | 4.5 − 4.99 |
| | | | | | F | 5.0 − 5.59 |
| | | | | | G | 5.6 − 6.29 |
| | | | | | H | 6.3 − 7.09 |
| No code letter | 300 | 175 | 700 | 250 | J | 7.1 − 7.99 |
| F–V | 300 | 175 | 700 | 250 | K | 8.0 − 8.99 |
| B–E | 250 | 175 | 700 | 200 | L | 9.0 − 9.99 |
| A | 150 | 150 | 700 | 150 | M | 10.0 − 11.19 |
| | | | | | N | 11.2 − 12.49 |
| | | | | | P | 12.5 − 13.99 |
| AC squirrel-cage or synchronous motors with autotransformer starting: | | | | | R | 14.0 − 15.99 |
| | | | | | S | 16.0 − 17.99 |
| Not more than 30 A | | | | | T | 18.0 − 19.99 |
| No code letter | 250 | 175 | 700 | 200 | U | 20.0 − 22.39 |
| More than 30 A | | | | | V | 22.4 and up |
| No code letter | 200 | 175 | 700 | 200 | | |
| F–V | 250 | 175 | 700 | 200 | | |
| B–E | 200 | 175 | 700 | 200 | | |
| A | 150 | 150 | 700 | 150 | | |
| High reactance squirrel-cage: | | | | | | |
| Not more than 30 A | | | | | | |
| No code letter | 250 | 175 | 700 | 250 | | |
| More than 30 A | | | | | | |
| No code letter | 200 | 175 | 700 | 200 | | |
| Wound rotor, no code letter | 150 | 150 | 700 | 150 | | |
| DC motors: | | | | | | |
| No more than 50 hp | | | | | | |
| No code letter | 150 | 150 | 250 | 150 | | |
| More than 50 hp | | | | | | |
| No code letter | 150 | 150 | 175 | 150 | | |
| Low-torque, low-speed (450 r/min or lower) synchronous motors which start unloaded | 200 | 200 | 200 | 200 | | |

Source: NEC 1975, Tables 430-152 and 430-7(b).

accomplished only by remagnetizing and coming back down the curve from $B$ to $A_1$.

Alnico magnets corresponding to curves 4 and 5 are best adapted to operation with short air gaps, since the introduction of a long air gap will demagnetize the magnet materially. On the other hand, a magnet with a long air gap will operate most satisfactorily on curve 6 on account of the high coercive force $H_2$. With change in the length of the air gap, the operation will be essentially along that curve and the magnet will lose little of its original magnetization.

There are several other grades of Alnico with characteristics between curves 4, 5, and 6. Ceramic PM materials have a very large coercive force.

The steels for permanent magnets are cut in strips, heated to a red-hot temperature, and forged into shape, usually in a "bulldozer." If they are to be machined, they are cooled in mica dust to prevent air hardening. They are then ground, tumbled, and tempered. Alnico and ceramic types are cast and then finish-ground.

Permanent magnets are magnetized either by placing them over a bus bar carrying a large direct current, by placing them across the poles of a powerful electromagnet, or by an ampere-turn pulse.

the area of the core is determined by the pull. **Solenoid and plunger** is a solenoid provided with a movable iron rod or bar called a plunger. When the coil is energized, the iron rod becomes magnetized and the mutual action of the field in the solenoid on the poles created on the plunger causes the plunger to move within the solenoid. This force becomes zero only when the magnetic centers of the plunger and solenoid

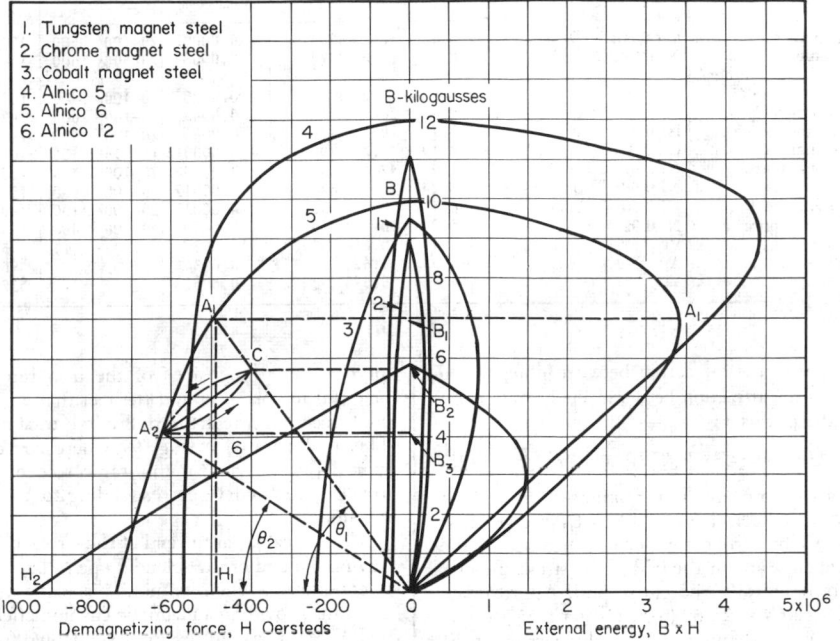

**Fig. 98** Characteristics of permanent magnet materials.

Unless permanent magnets are subjected to artificial aging, they gradually weaken until after a long period they become stabilized usually at from 85 to 90 percent of their initial strength. With magnets for electrical instruments, where a constant field strength is imperative, artificial aging is accomplished by mechanical vibration or by immersion in oil at 250°F for a period of a few hours.

In an **electromagnet** the magnetic field is produced by an electric current. The core is usually made of soft iron or mild steel because, the permeability being higher, a stronger magnetic field may be obtained. Also since the retentivity is low, there is little trouble due to the sticking of armatures when the circuit is opened. Electromagnets may have the form of simple solenoids, iron-clad solenoids, plunger electromagnets, electromagnets with external armatures, and lifting magnets, which are circular in form with a flat holding surface.

A **solenoid** is a winding of insulated conductor and is wound helically; the direction of winding may be either right or left. A **portative electromagnet** is one designed only for holding material brought in contact with it. A **tractive electromagnet** is one designed to exert a force on the load through some distance and thus do work. The **range** of an electromagnet is the distance through which the plunger will perform work when the winding is energized. For long range of operation, the plunger type of tractive magnet is best suited, for the length of core is governed practically by the range of action desired, and

coincide. If the load is attached to the plunger, work will be done until the force to be overcome is equal to the force that the solenoid exerts on the plunger. When the iron of the plunger is not saturated, the strength of magnetic field in the solenoid and the induced poles are both proportional to the exciting current, so that the pull varies as the **current squared.** When the plunger becomes highly saturated, the pull varies almost **directly** with the current.

The **maximum uniform pull** occurs when the end of the plunger is at the center of the solenoid and is equal to

$$F = CAnI/l \qquad \text{lb} \qquad (134)$$

where $A$ = cross-sectional area of plunger, in$^2$; $n$ = number of turns; $I$ = current, A; $l$ = length of the solenoid, in; and $C$ = pull, (lb/in$^2$)/(A-turn/in) $C$ depends on the proportions of the coil, the degree of saturation, the length, and the physical and chemical purity of the plunger. Table 28 gives values of $C$ for several different solenoids.

Curve 1, Fig. 99, shows the characteristic pull of an open-magnetic circuit solenoid, 12 in long, having 10,000 A-turns or 833 A-turns/in.

When a **strong pull** is desired at the end of the stroke, a stop may be used as shown in Fig. 100. Curve 2, Fig. 99, shows the pull obtained by adding a stop to the plunger. It will be noted that, except when the end of the plunger is near the stop, the stop adds little to the solenoid pull. The pull is made up of two

**Table 27. Properties of Metals, Alloys, and Resistor Materials**

| Material | Composition | Sp gr | Microohms-cm at 20 C | Ohms cir-mil-ft at 20 C | Temp coef of resistance per deg C | Temp range, deg C | Max safe working temp, deg C | Approx melting point, deg C |
|---|---|---|---|---|---|---|---|---|
| Advance......... | Cu 0.55; Ni 0.45 | 8.9 | 40 | 294 | ±0.00002 | 20–100 | 500 | 1210 |
| Comet.......... | Ni 0.30; Cr 0.05; Fe 0.65 | 8.15 | 95 | 570 | 0.00088 | 20–500 | 600 | 1480 |
| Bronze, commercial......... | Cu; Zn | 8.7 | 4.2 | 25 | 0.0020 | 0–100 | .... | 1015 |
| Hytemco........ | Ni 0.50; Fe 0.50 | 8.46 | 20 | 120 | 0.0045 | 20–100 | 600 | 1425 |
| Magno.......... | Ni 0.955; Mn 0.045 | 8.75 | 20 | 120 | 0.0036 | 20–100 | 400 | 1435 |
| Manganin....... | Cu 0.84; Mn 0.12; Ni 0.04 | 8.19 | 48.2 | 290 | ±0.000015 | 15–35 | 100 | 1020 |
| Monel metal..... | Ni 0.67; Cu 0.28 | 8.9 | 42.6 | 256 | 0.00198 | 20–100 | 425 | 1350 |
| Nichrome....... | Ni 0.60; Fe 0.25; Cr 0.15 | 8.247 | 112 | 675 | 0.00017 | 20–100 | 930 | 1350 |
| Nichrome V...... | Ni 0.80; Cr 0.20 | 8.412 | 108 | 650 | 0.00013 | 20–100 | 1100 | 1400 |
| Nickel, pure..... | Ni 0.99 | 8.9 | 10 | 60 | 0.0050 | 0–100 | 400 | 1450 |
| Platinum........ | Pt | 21.45 | 10.616 | 63.80 | 0.003 | ...... | .... | 1755 |
| Silver.......... | Ag | 10.5 | 1.622 | 9.755 | 0.00361 | ...... | .... | 960 |
| Tungsten........ | W | 19.3 | 5.523 | 33.22 | 0.00524 | ...... | .... | 3370 |

components: one due to the attraction between plunger and winding, the other to the attraction between plunger and stop. The equation for the pull is

$$P = AIn[(In/l_a^2 C_1^2) + (C/l)] \qquad \text{lb} \qquad (135)$$

where $A$ = area of the core, in²; $n$ = number of turns; $l_a$ = length of gap between core and stop; and $C, C_1$ = constants. At the beginning of the stroke the second member of the equation is predominant, and at the end of the stroke the first member represents practically the entire pull. Approximate values of $C$ and $C_1$ are $C_1 = 2,660$ (for $l$ greater than $10d$), $C = 0.0096$, where $d$ is the diameter of the plunger, in. In SI units

$$P = 1.7512 AnI\left(\frac{2.54nI}{l_a^2 C_1^2} + \frac{C}{l}\right) \qquad \text{N} \qquad (135a)$$

where A is in cm², $l$ and $l_a$ in cm, and the pull $P$ in N. All other quantities are unchanged.

The **range of uniform** pull can be extended by the use of conical ends of stop and plunger, as shown in Fig. 101. A stronger magnet mechanically can be obtained by using an iron-clad solenoid, Fig. 102, in which an iron return path is provided for the flux. Except for low flux densities and short

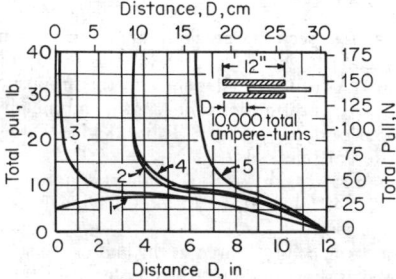

**Fig. 99**  Pull of solenoid on plunger. (1) Coil and plunger; (2) coil and plunger with stop; (3) iron-clad coil and plunger; (4) and (5) same as (3) with different lengths of stop.

air gaps the dimensions of the iron return path are of no practical importance, and the fact that an iron return path is used does not affect the pull curve except at short air gaps. This is illustrated in Fig. 99 where curves 3, 4, and 5 are typical pull curves for this same solenoid when it is made ironclad, each curve corresponding to a different position of the stop.

Mechanical jar at the end of the stroke may be prevented by leaving the end of the solenoid open. The plunger then comes to equilibrium when its middle is at the middle of the winding, thus providing a magnetic cushion effect. Electromagnets with external armatures are best adapted for **short-range** work, and the best type is the horseshoe magnet. The pull for short-range magnets is expressed by the equation

$$F = B^2 A/72,134,000 \qquad \text{lb} \qquad (136)$$

where $B$ = flux density, Mx/in²; and $A$ = area of the core, in².
In SI units

$$F = 397,840 B^2 A \qquad \text{N} \qquad (136a)$$

where the flux density $B$ is in Wb/m²; A = area, m²; and $F$ = force, N.

A greater holding power is obtained if the surfaces of the armature and core are not machined to an absolutely smooth contact surface. If the surface is slightly irregular, the area of contact $A$ is reduced but the flux density $B$ is increased approximately in proportion (if the iron is being operated below saturation) and the pull is increased since it varies as the square of the density $B$. Nonmagnetic stops should be used if it is desired that the armature may be released readily when the current is interrupted.

**Lifting magnets** are of the portative type in that their function is merely to hold the load. The actual lifting is performed by

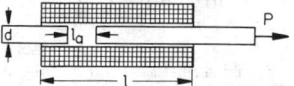

**Fig. 100**  Solenoid with stop.

the hoisting apparatus. The magnet is almost toroidal in shape. The coil shield is of manganese steel which is very hard and thus resists wear and is practically nonmagnetic. The holding power is given by Eq. (136), where $A$ = area of holding surface, in². It is difficult to calculate accurately the holding force of a lifting magnet for it depends on the magnetic characteristics of the load, the area of contact, and the manner in which the load is applied.

**Rapid action** in a magnet can be obtained by reducing the time constant of the winding and by subdividing the metal parts to reduce induced currents which have a demagnetizing effect when the circuit is closed. The movement of the plunger through the winding causes the winding and its bobbin to be cut by a magnetic field; if the bobbin is of metal and not slotted longitudinally, it is a short-circuited turn linked by a

**Fig. 101**  Conical plunger and stop.

changing magnetic field and hence currents are induced in it. These currents oppose the flux and hence reduce the pull during the transient period. They also cause some heating. Where it is found impossible to reduce the time constant sufficiently, an electromagnet designed for a voltage much lower than normal is often used. A resistor is connected in series which is short-circuited during the stroke of the plunger. At the completion of the stroke the plunger automatically opens the short circuit, reducing the current to a value which will not overheat the magnet under continuous operation. The extremely short time of overload produces very rapid action but does not injure the winding. The solenoids on many automatic motor-starting panels are designed in this manner.

When **slow action** is desired, it can be obtained by using solid cores and yoke and by using a heavy metallic spool or bobbin for the winding. A separate winding short-circuited on itself is also used to some extent.

**Sparking** at switch terminals may be reduced or eliminated by neutralizing the inductance of the winding. This is accomplished by winding a separate short-circuited coil with its wires parallel to those of the active winding. (This method can be used with dc magnets only.) This is not economical, since one-half the winding space is wasted. By connecting a capacitor across the switch terminals, the energy of the inductive discharge on opening the circuit may be absorbed. For the purpose of neutralizing the inductive discharge and causing a quick release, a small **reverse current** may be sent through the

coil winding automatically on opening the circuit. Sleeves of tin, aluminum, or copper foil placed over the various layers of the winding absorb energy when the circuit is broken and reduce the energy dissipated at the switch terminals. This scheme can be used for dc magnets only. **Sticking** of the parts of the magnetic circuit due to residual magnetism may be prevented by the use of nonmagnetic stops. In the case of lifting magnets subjected to rough usage and hard blows (as in a steel works), these stops usually consist of plates of manganese steel, which are extremely hard and non-magnetic.

**AC Tractive Magnets**  Because of the iron losses due to eddy currents, the magnetic circuits of ac electromagnets should be composed of laminated iron or steel. The magnetic circuit of large magnets is usually built up of thin sheets of

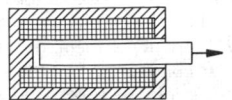

**Fig. 102**  Iron-clad solenoid.

sheet metal held together by means of suitable clamps. Small cores of circular cross section usually consist of a bundle of soft iron wires. Since the iron losses increase with the flux density, it is not advisable to operate at as high a density as with direct current. The current instead of being limited by the resistance of the winding is now determined almost entirely by the inductive reactance as the resistance is small. With the removal of the load the current rises to high values. The pull of ac magnets is nearly constant irrespective of the length of air gap.

In a **single-phase magnet** the pull varies from zero to a maximum and back to zero twice every cycle, which may cause considerable chattering of the armature against the stop. This may be prevented by the use of a spring or, in the case of a solenoid coil, by allowing the plunger to seek its position of equilibrium in the coil. **Chattering** may also be prevented by the use of a short-circuited winding or shading coil around one tip of the pole piece or by the use of polyphase. In a **two-phase magnet** the pull is constant and equal to the maximum instantaneous pull produced by one phase so long as the voltage is a sine function. In a **three-phase magnet** under the same conditions the pull is constant and equal to 1.5 times the maximum instantaneous pull of one phase. Should the load become greater than the minimum instantaneous pull, there will be chattering as in a single-phase magnet.

**Heating of Magnets**  The lifting capacity of an electromagnet is limited by the permissible current-carrying capacity of

**Table 28. Maximum Pull per Sq In. of Core for Solenoids with Open Magnetic Circuit**

| Length of coil $l$, in. | Length of plunger, in. | Core area $A$, sq in. | Total ampere-turns $I \times n$ | Max pull $P$, psi | $1,000 \times C$ | Length of coil $l$, in. | Length of plunger, in. | Core area $A$, sq in. | Total ampere-turns $I \times n$ | Max pull $P$, psi | $1,000 \times C$ |
|---|---|---|---|---|---|---|---|---|---|---|---|
| 6 | Long | 1.0 | 15,900 | 22.4 | 9.0 | 12 | Long | 1.0 | 11,200 | 8.75 | 9.4 |
| 9 | Long | 1.0 | 11,330 | 11.5 | 9.1 | 12 | Long | 1.0 | 20,500 | 16.75 | 9.8 |
| 9 | Long | 1.0 | 14,200 | 14.6 | 9.2 | 18 | 36 | 1.0 | 18,200 | 9.8 | 9.7 |
| 10 | 10 | 2.76 | 40,000 | 40.2 | 10.0 | 18 | 36 | 1.0 | 41,000 | 22.5 | 9.8 |
| 10 | 10 | 2.76 | 60,000 | 61.6 | 10.3 | 18 | 18 | 1.0 | 18,200 | 9.8 | 9.7 |
| 10 | 10 | 2.76 | 80,000 | 80.8 | 10.1 | 18 | 18 | 1.0 | 41,000 | 22.5 | 9.8 |

Source: From data by Underhill, *Elec. World*, 45, 1906, pp. 796, 881.

the winding, which, in turn, is dependent on the amount of heat energy that the winding can dissipate per unit time without exceeding a given temperature rise. Coils wound with wire having cotton insulation will, in general, be operating at a safe temperature if the average power expended does not exceed 0.5 W/in² of heat-dissipating surface.

**Design of Exciting Coil**   Let $n$ = number of turns, $l$ = mean length of turn, in ($l = 2\pi r$, where $r$ is the mean radius, in), $A$ = cross section of wire, cir mils. The resistance of 1 cir mil·ft of copper is practically 12 $\Omega$ at 60°C, or 1 $\Omega$/cir mil·in. Hence the resistance, $R = nl/A\ \Omega$; the current, $I = EA/nl$; the ampere-turns, $nI = EA/l$; the power to be dissipated, $P = E^2A/nl$ W. From the foregoing equations the cross section of wire and the number of turns can be calculated.

**Space Factor of Winding**   Space factor of a coil is the ratio of the space occupied by the conductor to the total volume of the coil or winding. Only in the theoretical case of uninsulated square or rectangular conductor may the space factor be 100 percent. For wire of circular section with insulation of negligible thickness, wound as shown in Fig. 103a, the space factor will be 78.5 percent. When the turns of wire are "bedded," as shown in Fig. 103b (the case in most windings, particularly

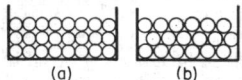

**Fig. 103**   Winding space factor.

with smaller wires), there is a theoretical gain of about 7 percent in space factor. Experiments have shown that in most cases this gain is about neutralized in practice by the flattening out of the insulation of the wire due to the tension used in winding. When wound in a haphazard manner, the space factors of magnet wires vary according to size, substantially as follows:

|  | *Double cotton covered* | | | | *Single cotton covered* | | |
|---|---|---|---|---|---|---|---|
| Size, AWG | 0 | 5 | 10 | 15 | 20 | 25 | 30 | 35 |
| Space factor, percent | 60 | 53.8 | 45.5 | 35.1 | 32.2 | 32 | 25.7 | 16 |

**Magnet wire** is a soft insulated copper wire of high conductivity. It can be obtained in square, rectangular, and circular section, but the round or cylindrical wire is used almost entirely in the smaller sizes. Ribbons are frequently used in the larger sizes. **Cotton covering** is used on large- and medium-sized wires where the space occupied by the cotton is small relative to the conductor cross section so that the space factor remains high. Cotton is also used where the insulation is to be impregnated with, or is to absorb, a liquid insulation. **Nylon** (replacing silk), used on the smaller sizes of wire, has a higher space factor than cotton. **Paper insulation,** consisting of one or more thin strips of dense high-quality paper applied helically over bare or enamel-covered conductors, is used sometimes with transformer windings where a high space factor is important. **Enameled** wire, a film-insulated copper conductor produced with oleoresinous enamels, is used particularly with the smaller wires. It is rated by the AIEE as Class A insulation. The wire is wound either layer upon layer or with paper between adjacent layers. **Enamel and cotton** and **enamel and nylon** are used in combination. The advantage lies in high dielectric strength, mechanical protection for the enamel, high space factor, and opportunity to impregnate if desired. Synthetic materials such as **Formvar** and **Formex** are being used as insula-

tion. Formvar is a vinyl acetal resin varnish and is a Class A material. It has film toughness, will withstand excessive elongation, has exceptional adherence to conductors, will not become brittle after prolonged exposure to maximum operating temperatures. **Glass-fiber** insulation such as **Fiberglas** is a continuous spun-glass fiber applied to bare or film-insulated (enamel or Formvar) magnet wire. It must be compound- or varnish-treated to give it abrasion resistance. The glass itself can withstand high temperature and resists acids. **Silotex** is a wire insulated with glass-fiber yarn over the conductor and affixed to it with silicone varnish. It is designated as Class H insulation. **Asbestos-covered** magnet wire is used where the temperature is high. It combines resistance to heat and abrasion, can resist mild acids, has good dielectric strength, and is fireproof.

Table 29 gives the diameters of magnet wire with the different types of insulation. For further data on electrical insulating materials, see Sec. 6.

## AUTOMOBILE IGNITION SYSTEMS

The ignition system in an automobile produces the spark which ignites the combustible mixture in the engine cylinders. This is accomplished by a high-voltage, or high-tension, spark between metal points in a spark plug. (A spark plug is an insulated bushing screwed into the cylinder head.) Spark plugs usually have porcelain insulation, but for some special uses, such as in airplane engines, mica may be used. There are two general sources for the energy necessary for ignition; one is the electrical system of the car which is maintained by the generator and the battery **(battery ignition),** and the other is a **magneto.** Battery ignition systems have traditionally operated electromechanically, using a spark coil, a high-voltage distributor, and low-voltage breaker points. Electronic ignition systems, working from the battery, became standard on U.S. cars in 1975. These vary in complexity from the use of a single transistor to reduce the current through the points to pointless systems triggered by magnetic pulses or interrupted light beams. Capacitor discharge into a pulse transformer is used in some systems to obtain the high voltage needed to fire the spark plugs.

**Battery ignition** is most widely used since it is simple, reliable, and low in cost, and the electrical system is a part of the car equipment. The high voltage for the spark is obtained from an ignition coil which consists of a primary coil of relatively few turns and a secondary coil of a large number of turns, both coils being wound on a common magnetic core consisting of either thin strips of iron or small iron wires. In a 6-V system the resistance of the primary coil is from 0.9 to 2 $\Omega$ and the inductance is from 5 to 10 mH. The number of secondary turns varies from 9,000 to 25,000, and the ratio of primary to secondary turns varies from 1:40 to 1:100.

The coil operates on the following principle. It stores energy in a magnetic field relatively slowly and then releases it suddenly. The power developed ($p = dw/dt$) is thus relatively large ($w$ = stored energy). The high emf $e_2$ which is required for the spark is induced by the sudden change in the flux $\phi$ in the core of the coil when the primary current is suddenly interrupted, $e_2 = -n_2(d\phi/dt)$, where $n_2$ is the number of secondary turns. For satisfactory ignition, peak voltages from 10 to 20 kV volts are desirable. Figure 104 shows the relation between the volts required to produce a spark and pressure with compressed air.

**Table 29. Diameters of Round Magnet Wire**

| Wire size | Nominal diam, bare, in | Nominal diam over insulation, in | | | | | | | |
|---|---|---|---|---|---|---|---|---|---|
| | | PE | HPE | SF, NE | HF, HNE | SCC | SCE | SVC, SX | SVSF |
| 8 | 0.1285 | ...... | ...... | 0.1306 | 0.1324 | 0.1359 | 0.1380 | 0.1345 | 0.1374 |
| 9 | 0.1144 | ...... | ...... | 0.1165 | 0.1182 | 0.1208 | 0.1229 | 0.1204 | 0.1233 |
| 10 | 0.1019 | 0.1039 | 0.1056 | 0.1039 | 0.1056 | 0.1074 | 0.1094 | 0.1069 | 0.1096 |
| 11 | 0.0907 | 0.0927 | 0.0943 | 0.0927 | 0.0943 | 0.0953 | 0.0973 | 0.0957 | 0.0984 |
| 12 | 0.0808 | 0.0827 | 0.0842 | 0.0827 | 0.0842 | 0.0854 | 0.0873 | 0.0858 | 0.0884 |
| 13 | 0.0720 | 0.0739 | 0.0753 | 0.0739 | 0.0753 | 0.0766 | 0.0785 | 0.0770 | 0.0796 |
| 14 | 0.0641 | 0.0660 | 0.0673 | 0.0660 | 0.0673 | 0.0687 | 0.0706 | 0.0691 | 0.0717 |
| 15 | 0.0571 | 0.0589 | 0.0602 | 0.0589 | 0.0602 | 0.0617 | 0.0635 | 0.0621 | 0.0646 |
| 16 | 0.0508 | 0.0525 | 0.0539 | 0.0525 | 0.0539 | 0.0554 | 0.0571 | 0.0558 | 0.0582 |
| 17 | 0.0453 | 0.0469 | 0.0483 | 0.0469 | 0.0483 | 0.0499 | 0.0515 | 0.0503 | 0.0526 |
| 18 | 0.0403 | 0.0418 | 0.0432 | 0.0418 | 0.0432 | 0.0449 | 0.0464 | 0.0453 | 0.0475 |
| 19 | 0.0359 | 0.0374 | 0.0387 | 0.0374 | 0.0387 | 0.0405 | 0.0420 | 0.0409 | 0.0431 |
| 20 | 0.0320 | 0.0334 | 0.0346 | 0.0334 | 0.0346 | 0.0366 | 0.0380 | 0.0370 | 0.0389 |
| 21 | 0.0285 | 0.0299 | 0.0310 | 0.0299 | 0.0310 | 0.0331 | 0.0345 | 0.0335 | 0.0354 |
| 22 | 0.0253 | 0.0266 | 0.0277 | 0.0266 | 0.0277 | 0.0299 | 0.0312 | 0.0303 | 0.0322 |
| 23 | 0.0226 | 0.0239 | 0.0249 | 0.0239 | 0.0249 | 0.0272 | 0.0285 | 0.0276 | 0.0294 |
| 24 | 0.0201 | 0.0213 | 0.0224 | 0.0213 | 0.0224 | 0.0247 | 0.0259 | 0.0251 | 0.0268 |
| 25 | 0.0179 | 0.0190 | 0.0201 | 0.0190 | 0.0201 | 0.0220 | 0.0231 | 0.0211 | 0.0227 |
| 26 | 0.0159 | 0.0170 | 0.0180 | 0.0170 | 0.0180 | 0.0200 | 0.0211 | 0.0191 | 0.0206 |
| 27 | 0.0142 | 0.0152 | 0.0161 | 0.0152 | 0.0161 | 0.0183 | 0.0193 | 0.0174 | 0.0189 |
| 28 | 0.0126 | 0.0136 | 0.0144 | 0.0136 | 0.0144 | 0.0167 | 0.0177 | 0.0158 | 0.0172 |
| 29 | 0.0113 | 0.0122 | 0.0130 | 0.0122 | 0.0130 | 0.0154 | 0.0163 | 0.0145 | 0.0159 |
| 30 | 0.0100 | 0.0109 | 0.0116 | 0.0109 | 0.0116 | 0.0141 | 0.0150 | 0.0132 | 0.0145 |
| 31 | 0.0089 | 0.0097 | 0.0105 | 0.0097 | 0.0105 | 0.0130 | 0.0138 | 0.0121 | 0.0133 |
| 32 | 0.0080 | 0.0088 | 0.0095 | 0.0088 | 0.0095 | 0.0121 | 0.0129 | 0.0112 | 0.0124 |
| 33 | 0.0071 | 0.0078 | 0.0085 | 0.0079 | 0.0085 | 0.0112 | 0.0119 | 0.0103 | 0.0115 |
| 34 | 0.0063 | 0.0069 | 0.0075 | 0.0070 | 0.0075 | 0.0104 | 0.0110 | 0.0095 | 0.0106 |
| 35 | 0.0056 | 0.0062 | 0.0067 | 0.0063 | 0.0067 | 0.0097 | 0.0103 | 0.0088 | 0.0099 |
| 36 | 0.0050 | 0.0055 | 0.0060 | 0.0056 | 0.0060 | 0.0087 | 0.0092 | 0.0082 | 0.0092 |
| 37 | 0.0045 | 0.0050 | 0.0055 | 0.0051 | 0.0055 | 0.0082 | 0.0087 | | |
| 38 | 0.0040 | 0.0044 | 0.0049 | 0.0045 | 0.0049 | 0.0077 | 0.0081 | | |
| 39 | 0.0035 | 0.0039 | 0.0043 | 0.0040 | 0.0043 | 0.0072 | 0.0076 | | |
| 40 | 0.0031 | 0.0034 | 0.0038 | 0.0036 | 0.0038 | 0.0068 | 0.0071 | | |

PE, plain enamel (single thickness)
HPE, heavy plain enamel (double thickness)
SF, single Formvar
NE, single nylon enamel
HF, heavy Formvar
HNE, heavy nylon enamel

SCC, single cotton covered
SCE, single cotton enamel
SVC, single Vitrotex cover
SX, single Silotex (silicone bond)
SVSF, single Vitrotex, single Formvar

Source: Anaconda Wire & Cable Co.

A battery ignition system for a four-cylinder engine is shown diagrammatically in Fig. 105. The primary circuit supplied by the battery consists of the primary coil $P$, a protective resistor, and a set of contacts, or "points" operated by a four-lobe cam, all in series. In order to reduce arcing and burning of the contacts and to produce a sharp break in the current, a capacitor $C$ of from 0.15 to 0.40 $\mu F$ is connected across the contacts. The energy, which would otherwise appear as an arc at the contacts, is stored in the capacitor ($ce^2/2$) and is dissipated when the contacts close again. The contacts, which are of pure tungsten, are operated by a four-lobe cam which is driven at one-half engine speed. A strong spring tends to keep the contacts open.

The secondary $S$ of the ignition coil is connected between the battery positive terminal and the distributor. The lobes of the cam force the contacts together long enough to build up a magnetic field in the coil, and then through the joint action of

the spring and the abrupt dropoff in the lobe of the cam, the primary current is suddenly interrupted, causing a high emf to be induced in the secondary $S$. The distributor arm connects the high-voltage terminal of the secondary to the spark plug of the cylinder which is firing at that instant. In a four-cylinder engine the firing order may be 1-3-4-2 or 1-2-4-3, the latter being shown in Fig. 105. The protective resistor in series with the primary of the ignition coil has a high temperature coefficient so that if the ignition switch is inadvertently left closed the resistor will heat up and limit the current. In this type of ignition system there is but a single spark for each explosion.

**High-Speed Distributors**  It is difficult to design the type of cam shown in Fig. 105 for eight- and even six-cylinder motors so that it operates well at high speeds. After each break takes place the cam follower must ride for considerable distance upon the next lobe before the contacts are closed. At high speed this does not give time for the flux in the iron core to build up to its full value. In the Delco-Remy distributor (Fig. 106) two breaker arms are connected in parallel; one coil and one capacitor are used. One set of contacts is open when the other is just breaking but closes a few degrees after the break occurs. This closes the primary of the ignition coil immedi-

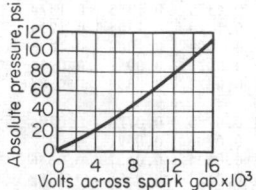

**Fig. 104** Pressure-voltage curve for spark plug.

ately after the break and increases the time that the primary of the ignition coil is closed and permits the flux in the iron to reach its full value. The interrupter shown in Fig. 106 is designed for an eight-cylinder motor.

Another method is to use one-half as many lobes as there are breaker arms and to use two sets of breaker arms so located that each pair interrupts the primary for only half the cylinders. This increases the time of contact. Both sets of breaker arms may operate through the same or different ignition coils. With a single ignition coil only a single distributor circuit is needed but the speed is limited since one pair of contacts cannot close until after the other opens. With two coils there is no such limitation.

The spark should advance with increase in engine speed so as to allow for the time lag in the explosion. To take care of this **automatically** most timers are now equipped with centrifugally operated weights which advance the breaker cam with respect to the engine drive as the speed increases.

## AUTOMOBILE LIGHTING AND STARTING SYSTEMS

Automobile lighting and starting systems initially operated at 6 V, but at present nearly all cars, except the smaller ones, operate at 12 V because larger engines, particularly V-8s, are now common and require more starting power. With 12 V, for the same power, the starting current is halved, and the effect of resistance in the leads, connections, and brushes is materially reduced. In some systems the positive side of the system is grounded, but more often the negative side is grounded. Six volts became standard about 1920.

A further development is the application of an **ac generator,** or alternator, combined with a rectifier as the generating unit rather than the usual dc generator. One advantage is the elimination of the commutator, made up of segments, which

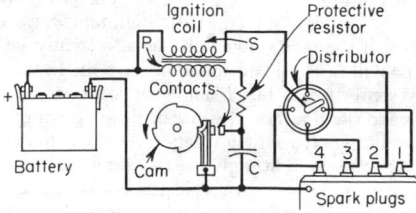

**Fig. 105** Battery-ignition system.

requires some maintenance due to the sparking and wear of the carbon brushes. With the alternator the dc field rotates, and brushes operating on smooth slip rings require almost no maintenance. Also, the system is greatly simplified by the fact that rectifiers are "one-way" devices, and the battery **cannot** deliver current back to the generator when its voltage drops below that of the battery. Thus, no cutout relay, such as is required with dc generators, is necessary. This ac development is the result of the development of reliable, low-cost germanium and silicon semiconductor rectifiers.

Figure 107 shows a schematic diagram of the Ford system (adapted initially to trucks). The generator stator is wound three-phase Y-connected, and the field is bipolar supplied with direct current through slip rings and brushes. The rectifier diodes are connected full-wave bridge circuit to supply the battery through the ammeter.

**Regulator** The function of the regulator is to control the generator current so that its value is adapted to the battery voltage which is related to the condition of charge of the battery (see Fig. 7). Thus, when the battery voltage drops (indicating a lowered condition of charge), the current should be increased, and, conversely, when the battery voltage increases (indicating a high condition of charge), the current should be decreased.

**Fig. 106** Delco-Remy eight-cylinder interrupter.

Neglecting for the moment the starting procedure, when the ignition switch is thrown to the normal "on" position at $c$, the coil actuating the field relay is connected to the battery + terminal and causes the relay contacts $a$ to close. This energizes the regulator circuits, and, if the two upper voltage regulator contacts $b$ are closed as shown, the rotating field of the alternator is connected directly to the battery + terminal, and the field current is then at its maximum value and produces a high generator voltage and large output current. At the same time the voltage regulator coil in series with the 0.3- and 14-$\Omega$ resistors is connected between the battery + terminal and ground. If the voltage of the battery rises owing to its higher condition of charge, the current to the voltage regulator coil increases, causing it to open the two upper contacts at $b$. Current from the battery now flows through the 0.3-$\Omega$ resistor and divides, some going through the 10-$\Omega$ resistor and dividing between the field and the 50-$\Omega$ resistor, and the remainder going to the voltage regulator coil. The current to the rotating field is thus reduced, causing the alternator output to be reduced. Because of the 0.3 $\Omega$ now in circuit, the current to the coil of the voltage regulator is reduced to such a value that it holds the center contact at $b$ in the mid, or open, position. If the battery voltage rises to an even higher value, the regulator coil becomes strong enough to close the lower contacts at $b$; this short-circuits the field, reducing its current almost to zero, and thus reducing the alternator output to zero. On the other hand when the battery voltage drops, the foregoing sequence is reversed, and the contacts at $b$ operate to increase the current to the alternator field.

As was mentioned earlier, the battery cannot supply current to the alternator because of the "one-way" characteristic of the rectifier. Thus, when the alternator voltage drops below that of the battery and even when the alternator stops running, its current automatically becomes zero. The alternator has a normal rectifier open-circuit voltage of about 14 V and a rating of 20 A.

**Starting** In most cases, for starting, the ignition key is turned far to the right and held there until the motor starts. Then, when the key is released, the ignition switch contacts assume a normal operating position. Thus, in Fig. 107, when the ignition-switch contact is in the starting position $S$, the starter relay coil becomes connected by a lead to the battery + terminal and thus becomes energized, closing the relay contacts. The starter motor is then connected to the battery to crank the engine. At the time that the contact closes it makes contact with a small metal brush $e$ which connects the battery + terminal to the primary terminal of the ignition coil through the protective resistor $R$. After the motor starts, the ignition switch contacts spring to the normal operating position $C$, and the starter relay switch opens, thereby breaking contact with the small brush $e$. However, when contact $C$ is closed, the ignition coil primary terminal is now connected through leads to the battery + terminal. The interrupter, the ignition coil, and the distributor now operate in the manner described earlier (see Fig. 105); the system shown in Fig. 107, is that for a six-cylinder engine.

The connection of **accessories** to the electric system is illustrated in Fig. 107 for the horn, head and other lights, and temperature and fuel gages.

### Magneto Ignition

**Principle of Magneto** A magneto is an electric generator in which the magnetic flux is provided by one or more permanent magnets. It is a self-contained unit and is used advantageously for ignition where a generator and a battery are not needed to supply power to other accessories. The design of magnetos was radically changed when Alnico, with its very high retentivity (Fig. 98), became available as a permanent-

magnet material. One method of utilizing Alnico magnets is to insert the bar magnets in the frame of the magneto (Fig. 108). The rotor is a soft-iron bobbin. A primary winding of relatively few turns and a secondary winding of a relatively large number of turns are wound over the laminated yoke $Y$. The position of the rotor shown in (a) provides a low reluctance path for the magnetic flux of the left-hand magnet and a high reluctance path for the magnetic flux of the right-hand magnet so that the flux goes through the yoke from left to right as shown. When the rotor turns one-eighth of a revolution, it becomes horizontal; obviously, each of the magnets acts in opposition relative to the yoke, and the flux therein becomes zero. In (b), which also shows the external electrical connections, the rotor is shown as having turned one-fourth of a revolution, or 90°, from its position in (a).

The rotor now provides a low-reluctance path for the right-hand magnet and a high-reluctance path for the left-hand one. Thus the magnetic flux now goes through the yoke from right to left. It follows that in each 90° interval the flux in the yoke undergoes a complete reversal. It will be recognized that this is an **inductor** type of ac generator.

In the diagram in (b), one end each of the primary and of the secondary are grounded together. The other end of the primary is connected to an insulated interrupter lever having a contact point $P$. This makes intermittent contact with the grounded contact point $P'$. The contact point $P$ is actuated by a cam which is driven by the same shaft as the magneto rotor and is in a definite relation to it. A switch $S'$ is provided to ground and thus short-circuit the secondary when it is desired to stop the engine. A capacitor $C$ is connected between the point $P$ and ground to absorb the energy of the spark which occurs when the contacts open.

With the contacts closed, the primary is short-circuited, and the varying flux in the core produced by the rotation of the rotor induces an alternating current in the winding which in

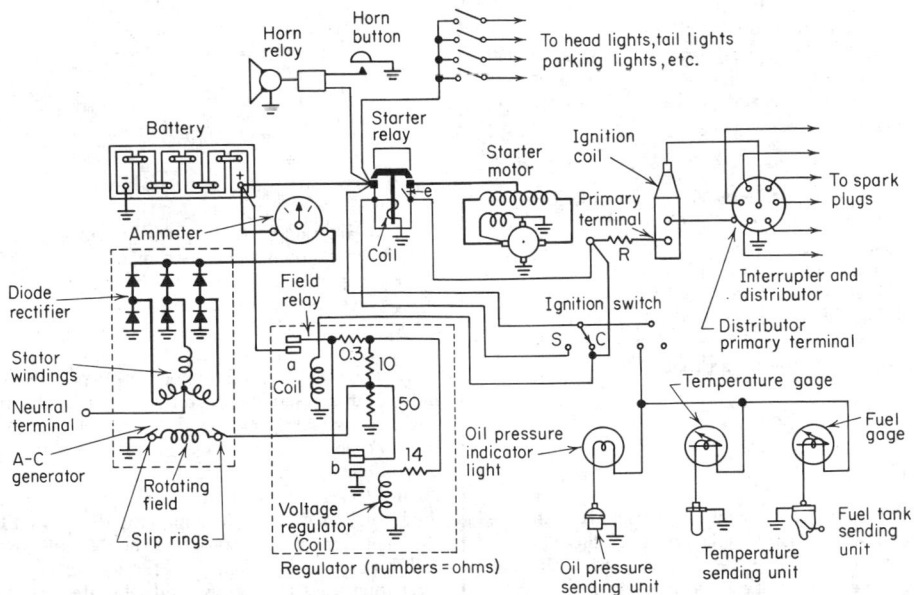

**Fig. 107** Schematic diagram of Ford lighting and starting system.

turn produces an alternating flux in the core. With the rotation of the rotor the current in the secondary rises cyclically to maximum values, and at these instants the cam causes the points $PP'$ to open suddenly, interrupting the current in the primary and thus causing a sudden collapse of the flux in the core. This induces a high-impulse emf in the secondary which

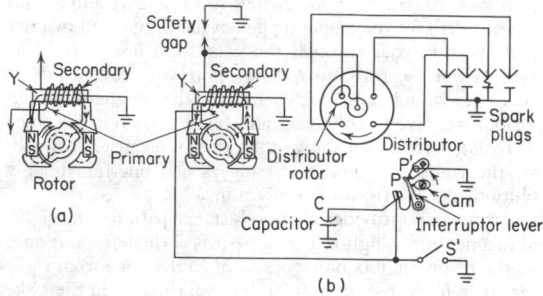

**Fig. 108** Magneto-ignition system.

is transmitted to the distributor and thence to the proper spark plug as shown.

On starting, the speed of the magneto may be so low that the emf is not sufficient to produce a hot spark. This difficulty can be met by **impulse starting,** in which the rotor is driven through a spring. During cranking the rotor is restrained from turning until the engine comes to the proper firing position, at which time the rotor is suddenly released. The energy stored in the spring produces a high, instantaneous, angular velocity to the rotor, resulting in a high emf and hot spark.

**Inductor-type magnetos,** having a large number of rotor poles and arranged differently from those shown in Fig. 108, are used for airplane-engine ignition. In another magneto design the rotor is a solid cylindrical Alnico magnet, permanently magnetized with an N and an S pole diametrically opposite. The frame is laminated, and there is a yoke with primary and secondary windings. When the rotor rotates, its N and S poles produce an alternating flux in the yoke which induces a short-circuit current in the primary winding and the method of producing the spark is then the same as with Fig. 108*b*.

# ELECTRONICS

## by Byron M. Jones

REFERENCES: "Reference Data for Radio Engineers," Howard Sams & Co. "Transistor Manual," General Electric Co. "SCR Manual," General Electric Co. "The Semiconductor Data Book," Motorola Inc. Terman, "Radio Engineering," McGraw-Hill. Fink, "Television Engineering," McGraw-Hill. Zworykin and Morton, "Television," Wiley. "Industrial Electronics Reference Book," Wiley.

The subject of electronics can be approached from the standpoint of the design of devices or the use of devices. For the practicing mechanical engineer, describing devices in terms of their external characteristics seems most likely to be profitable. The approach will be to describe devices as they appear to the outside world.

### Components

**Resistors, capacitors, reactors,** and **transformers** are described earlier in this section, along with basic circuit theory. These explanations are equally applicable to electronic circuits and hence are not repeated here. A description of additional components peculiar to electronic circuits follows.

A **rectifier,** or **diode,** is an electronic device which offers unequal resistance to forward and reverse current flow. Figure 1 shows the schematic symbol for a diode. The arrow beside

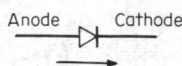

**Fig. 1** Diode schematic symbol.

the diode shows the direction of current flow. Current flow is taken to be the flow of positive charges, i.e., the arrow is counter to electron flow. Figure 2 shows typical forward and reverse volt-ampere characteristics. Notice that the scales for voltage and current are not the same for the first and third

quadrants. This has been done so that both the forward and reverse characteristics can be shown on a single plot even though they differ by several orders of magnitude.

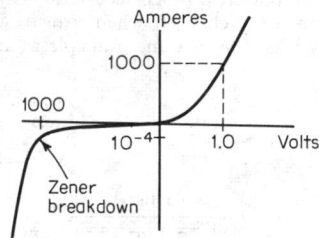

**Fig. 2** Diode forward-reverse characteristic.

**Diodes** are rated for forward current capacity and reverse voltage breakdown. They are manufactured with maximum current capabilities ranging from 0.05 A to more than 1,000 A. Reverse voltage breakdown varies from 50 V to more than 2,500 V. At rated forward current, the forward voltage drop varies between 0.7 and 1.5 V for silicon diodes. Although other materials are used for special-purpose devices, by far the most common semiconductor material is silicon. With a forward current of 1000 A and a forward voltage drop of 1 V, there would be a power loss in the diode of 1000 W (more than 1 hp). The basic diode package shown in Fig. 3 can dissipate about 20 W. To maintain an acceptable temperature in the diode, it is necessary to mount the diode on a **heat sink.** The manufacturer's recommendation should be followed very carefully to ensure good heat transfer and at the same time avoid fracturing the silicon chip inside the diode package.

The selection of **fuses** or **circuit breakers** for the protection of

**rectifiers** and rectifier circuitry requires more care than for other electronic devices. Diode failures as a result of circuit faults occur in a fraction of a millisecond. Special semiconductor fuses have been developed specifically for semiconductor

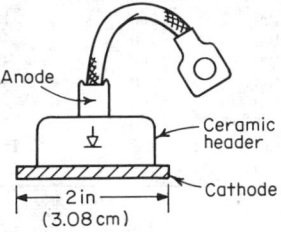

Anode

Ceramic header

Cathode

2 in
(3.08 cm)

**Fig. 3** Physical diode package.

circuits. Proper protective circuits must be provided for the protection of not only semiconductors but also the rest of the circuit and nearby personnel. Diodes and diode fuses have a short-circuit rating in $A^2 \cdot s$ ($I^2t$). As long as the $I^2t$ rating of the diode is greater than the $I^2t$ rating of its protective fuse, the diode and its associated circuitry is protected. Special circuit breakers are not available for the protection of diodes because of the nature of circuit breakers and the way they clear a fault. A circuit breaker does not clear a fault when its contacts open. The fault is cleared when the line voltage reverses at the end of the cycle of the applied voltage. This means that the **clearing time for a circuit breaker** is about ½ cycle of the ac input voltage. Diodes have a 1-cycle current rating corresponding to this. Additional line reactance must be provided to limit fault currents to this 1-cycle value in order to protect diodes and diode circuitry with a circuit breaker. The necessary reactance to limit fault currents is often provided by leakage reactance in the supply transformer for rectifier circuits.

A **thyristor**, often called a **silicon controlled rectifier (SCR)**, is a rectifier which blocks current in both the forward and reverse directions. Conduction of current in the forward direction will occur when the anode is positive with respect to the cathode and when the gate is pulsed positive with respect to the cathode. Once the thyristor has begun to conduct, the gate pulse can return to 0 V or even go negative and the thyristor will continue to pass current. To stop the cathode-to-anode current, it is necessary to reverse the cathode-to-anode voltage. The thyristor will again be able to block both forward and reverse voltages until current flow is initiated by a gate pulse. The schematic symbol for an SCR is shown in Fig. 4. The physical packaging of thyristors is similar to that of rectifiers with similar ratings except, of course, that the thyristor must have an additional gate connection.

The gate pulse required to fire an SCR is quite small compared with the anode voltage and current. Power gains in the range of $10^6$ to $10^9$ are easily obtained. In addition, the power loss in the thyristor is very low, compared with the

power it controls, so that it is a very efficient power-controlling device. Efficiency in a thyristor power supply is usually 97 to 99 percent. When the thyristor blocks either forward or reverse current, the high voltage drop across the thyristor accompanies low current. When the thyristor is conducting forward current after having been fired by its gate pulse, the high anode current occurs with a forward voltage drop of about 1.5 V. Since high voltage and high current never occur simultaneously, the power dissipation in both the on and off states is low.

The thyristor is rated primarily on the basis of its forward-current capacity and its voltage-blocking capability. Devices are manufactured to have equal forward and reverse voltage-blocking capability. Like diodes, thyristors have $I^2t$ ratings and 1-cycle surge current ratings to allow design of protective circuits. In addition to these ratings, which the SCR shares in common with diodes, the SCR has many additional specifications. Because the thyristor is limited in part by its average current and in part by its rms current, forward-current capacity is a function of the duty cycle to which the device is subjected. Since the thyristor cannot regain its blocking ability until its anode voltage is reversed and remains reversed for a short time, this time must be specified. The time to regain blocking ability after the anode voltage has been reversed is called the **turn-off time.** Specifications are also given for minimum and maximum gate drive. If forward blocking voltage is reapplied too quickly, the SCR may fire with no applied gate voltage pulse. The maximum safe value of rate of reapplied voltage is called the *dv/dt* rating of the SCR. When the gate pulse is applied, current begins to flow in the area immediately adjacent to the gate junction. Rather quickly, the current spreads across the entire cathode-junction area. In some circuits associated with the thyristor an extremely fast rate of rise of current may occur. In this event localized heating of the cathode may occur, with a resulting immediate failure or in less extreme cases a slow degradation of the thyristor. The maximum rate of change of current for a thyristor is given by its *di/dt* rating. Design for *di/dt* and *dv/dt* limits is not normally a problem at power-line frequencies of 50 and 60 Hz. These ratings become a design factor at frequencies of 500 Hz and greater. Table 1 lists typical thyristor characteristics.

A **triac** is a bilateral SCR. It blocks current in either direction until it receives a gate pulse. It can be used to control in ac circuits. Triacs are widely used for light dimmers and for the control of small universal ac motors. The triac must regain its blocking ability as the line voltage crosses through zero. This fact limits the use of triacs to 60 Hz and below.

A **transistor** is a semiconductor amplifier. The schematic symbol for a transistor is shown in Fig. 5. As shown Fig. 5 there are two types of transistors. *p-n-p* and *n-p-n*. Notice

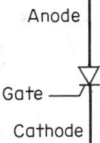

Anode

Gate

Cathode

**Fig. 4** Thyristor schematic symbol.

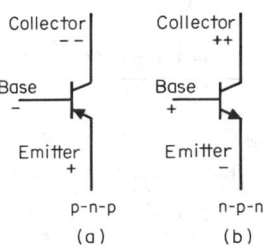

Collector
– –

Base
–

Emitter
+

p-n-p

(a)

Collector
+ +

Base
+

Emitter
–

n-p-n

(b)

**Fig. 5** Transistor schematic symbol.

**Table 1. Typical Thyristor Characteristics**

| Voltage | Current, A | | $I_{\frac{1}{2}}^2 t$, A·s | 1-cycle surge, A | $di/dt$, A/s | $dv/dt$, V/ε | Turn-off time, s |
|---|---|---|---|---|---|---|---|
| | rms | avg | | | | | |
| 400 | 35 | 20 | 165 | 180 | 100 | 200 | 10 |
| 1,200 | 35 | 20 | 75 | 150 | 100 | 200 | 10 |
| 400 | 110 | 70 | 4,000 | 1,000 | 100 | 200 | 40 |
| 1,200 | 110 | 70 | 4,000 | 1,000 | 100 | 200 | 40 |
| 400 | 235 | 160 | 32,000 | 3,500 | 100 | 200 | 80 |
| 1,200 | 235 | 160 | 32,000 | 3,500 | 75 | 200 | 80 |
| 400 | 470 | 300 | 120,000 | 5,500 | 50 | 100 | 150 |
| 1,200 | 470 | 300 | 120,000 | 5,500 | 50 | 100 | 150 |

that the polarities of voltage applied to these devices are reversed from one to the other. In many sizes matched *p-n-p* and *n-p-n* devices are available. The most common transistors have a collector dissipation rating of 150 to 600 mW. Collector to base breakdown voltage is 20 to 50 V. The amplification or gain of a transistor occurs because of two facts: (1) A small change in current in the base lead causes a large change in current in the collector and emitter leads. This current amplification is designated *hfe* on most transistor specification sheets. (2) A small change in base-to-emitter voltage can cause a large change in either the collector-to-base voltage or the collector-to-emitter voltage. Table 2 shows basic ratings for two matched audio amplifier transistors and one power transistor. There is a great profusion of transistor types so that the choice of type depends upon availability and cost as well as operating characteristics.

The gain of a transistor is independent of frequency over a wide range. At high frequency, the gain falls off. This cutoff frequency may be as low as 20 kHz for audio transistors or as high as 1 GHz for radio-frequency (rf) transistors.

The **unijunction** is a special-purpose semiconductor device. It is a pulse generator and widely used to fire thyristors and triacs as well as in timing circuits and waveshaping circuits. The schematic symbol for a unijunction is shown in Fig. 6. The device is essentially a silicon resistor. This resistor is connected to base 1 and base 2. The emitter is fastened to this resistor about halfway between bases 1 and 2. If a positive voltage is applied to base 2, and if the emitter and base 1 are at zero, the emitter junction is back-biased and no current flows in the emitter. If the emitter voltage is made increasingly positive, the emitter junction will become forward-biased. When this occurs, the resistance between base 1 and base 2 and between base 2 and the emitter suddenly switches to a very low value. This is a regenerative action, so that very fast and very energetic pulses can be generated with this device.

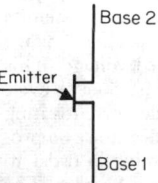

**Fig. 6**  Unijunction.

Before the advent of semiconductors, electronic rectifiers and amplifiers were **vacuum tubes** or **gas-filled tubes.** Some use of these devices still remains. If an electrode is heated in a vacuum, it gives up surface electrons. If an electric field is established between this heated electrode and another electrode so that the electrons are attracted to the other electrode, a current will flow through the vacuum. Electrons flow from the heated cathode to the cold anode. If the polarity is reversed, since there are no free electrons around the anode, no current will flow. This, then, is a vacuum-tube rectifier. If a third electrode, called a **control grid,** is placed between the cathode and the anode, the flow of electrons from the cathode to the anode can be controlled. This is a basic vacuum-tube amplifier. Additional grids have been placed between the cathode and anode further to enhance certain characteristics of the vacuum tube. In addition, multiple anodes and cathodes have been enclosed in a single tube for special applications such as radio signal converters.

If an inert gas, such as neon or argon, is introduced into the vacuum, conduction can be initiated from a cold electrode. The breakdown voltage is relatively stable for given gas and gas pressure and is in the range of 50 to 200 V. The **nixie** display tube is such a device. This tube contains 10 cathodes shaped in the form of the numerals from 0 to 9. If one of these cathodes is made negative with respect to the anode in the tube, the gas in the tube glows around that cathode. In this way each of the 10 numerals can be made to glow when the appropriate electrode is energized.

An **ignitron** is a vapor-filled tube. It has a pool of liquid mercury in the bottom of the tube. Air is exhausted from the enclosure, leaving only mercury vapor, which comes from the pool at the bottom. If no current is flowing, this tube will

**Table 2. Typical Transistor Characteristics**

| JEDEC number | Type | Collector-emitter volts at breakdown $BV_{CE}$ | Collector dissipation $P_c$ (25°C) | Collector current $I_c$ | Current gain $h_{fe}$ |
|---|---|---|---|---|---|
| 2N3904 | *n-p-n* | 40 | 310 mW | 200 mA | 200 |
| 2N3906 | *p-n-p* | 40 | 310 mW | 200 mA | 200 |
| 2N3055 | *n-p-n* | 100 | 115 W | 15 A | 20 |

block voltage whether the anode is plus or minus with respect to the mercury-pool cathode. A small rod called an **ignitor** can form a cathode spot on the pool of mercury when it is withdrawn from the pool. The ignitor is pulled out of the pool by an electromagnet. Once the cathode spot has been formed, electrons will continue to flow from the mercury-pool cathode to the anode until the anode-to-cathode voltage is reversed. The operation of an ignitron is very similar to that of a thyristor. The anode and cathode of each device perform similar functions. The ignitor and gate also perform similar functions. The thyristor is capable of operating at much higher frequencies than the ignitron and is much more efficient since the thyristor has 1.5 V forward drop and the ignitron has 15 V forward drop. The ignitron has an advantage over the thyristor in that it can carry extremely high overload currents without damage. For this reason ignitrons are often used as electronic "crowbars" which discharge electrical energy when a fault occurs in a circuit.

## Discrete-Component Circuits

Several common rectifier circuits are shown in Fig. 7. The waveforms shown in this figure assume no line reactance. The presence of line reactance will make a slight difference in the waveshapes and the conversion factors shown in Fig. 7. These waveshapes are equally applicable for loads which are pure resistive or resistive and inductive. In a resistive load the current flowing in the load has the same waveshape as the voltage applied to it. For inductive loads, the current waveshape will be smoother than the voltage applied. If the inductance is high enough, the ripple in the current may be indeterminantly small. An approximation of the ripple current can be calculated as follows:

$$I = \frac{E_{dc}\,PCT}{200\pi fNL} \tag{1}$$

where $I$ = rms ripple current, $E_{dc}$ = dc load voltage, $PCT$ =

| Type | Circuit | Output voltage waveform | $E_{dc}$ (avg) | Ripple fundamental frequency | % ripple | Peak inverse voltage |
|---|---|---|---|---|---|---|
| Half-wave $1\phi$ | 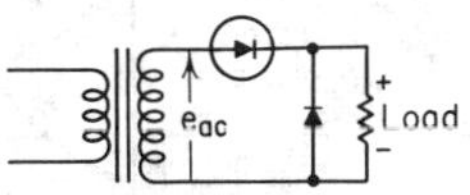 | 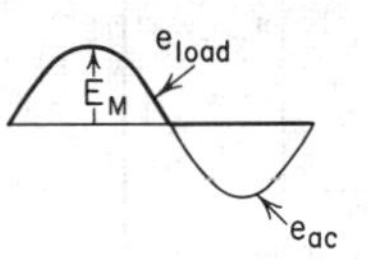 | 0.318 $E_M$ / 0.45 $E_{ac}$ | F | 121 | 3.14 $E_{dc}$ |
| Full-wave $1\phi$ | 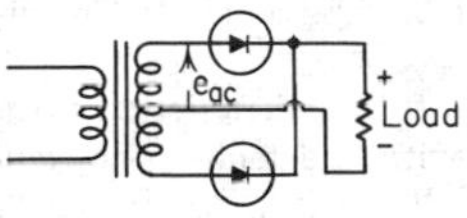 | 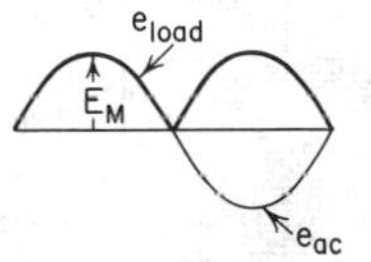 | 0.636 $E_M$ / 0.9 $E_{ac}$ | 2F | 48 | 3.14 $E_{dc}$ |
| Bridge $1\phi$ | 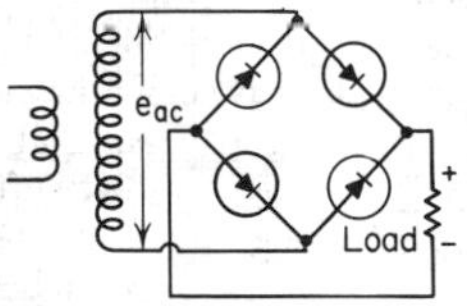 | 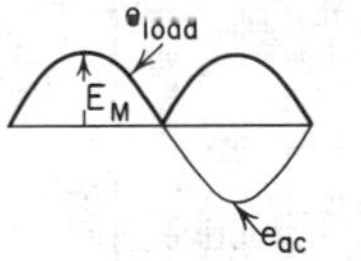 | 0.636 $E_M$ / 0.9 $E_{ac}$ | 2F | 48 | 1.57 $E_{dc}$ |
| Half-wave $3\phi$ | 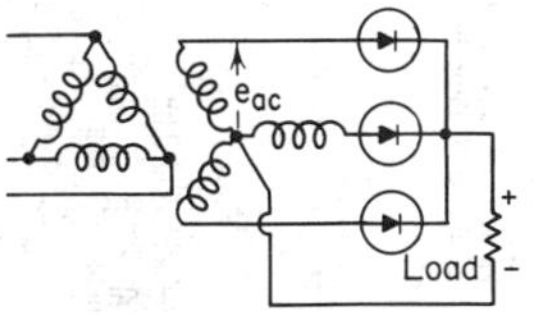 | 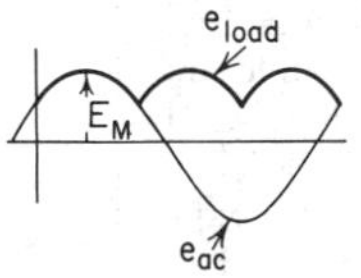 | 0.827 $E_M$ / 1.17 $E_{ac}$ | 3F | 18 | 2.09 $E_{dc}$ |

$E_M$ = maximun value of $e_{ac}$
$E_{ac}$ = effective value of $e_{ac}$
$E_{dc}$ = average value of d-c load voltage
F = line frequency
% ripple = 100 x rms ripple / $E_{dc}$

**Fig. 7** Comparison of rectifier circuits.

percent ripple from Fig. 7, $f$ = line frequency, $N$ = number of cycles of ripple frequency per cycle of line frequency, $L$ = equivalent series inductance in load. Equation (1) will always give a value of ripple higher than that calculated by more exact means, but this value is normally satisfactory for power-supply design.

Capacitance in the load leads to increased regulation. At light loads, the capacitor will tend to charge up to the peak value of the line voltage and remain there. This means that for either the single full-wave circuit or the single-phase bridge the dc output voltage would be 1.414 times the rms input voltage. As the size of the loading resistor is reduced, or as the size of the parallel load capacitor is reduced, the load voltage will more nearly follow the rectified line voltage and so the dc voltage will approach 0.9 times the rms input voltage for very heavy loads or for very small filter capacitors. One can see then that dc voltage may vary between 1.414 and 0.9 times line voltage due only to waveform changes when **capacitor filtering** is used.

Four different **thyristor rectifier circuits** are shown in Fig. 8.

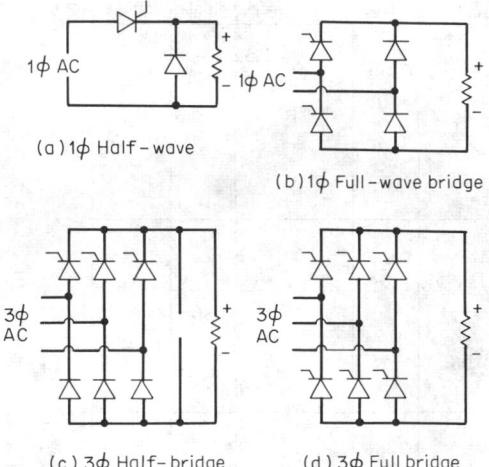

**Fig. 8** Basic thyristor circuits.

These circuits are equally suitable for resistive or inductive loads. It will be noted that the half-wave circuit for the thyristor has a rectifier across the load, as in Fig. 7. This diode is called a **freewheeling diode** because it freewheels and carries inductive load current when the thyristor is not conducting. Without this diode, it would not be possible to build up current in an inductive load. The gate-control circuitry is not shown in Fig. 8 in order to make the power circuit easier to see. Notice the location of the thyristors and rectifiers in the single-phase full-wave circuit. Constructed this way, the two diodes in series perform the function of a freewheeling diode. The circuit can be built with a thyristor and rectifier interchanged. This would work for resistive loads but not for inductive loads. For the full three-phase bridge, a freewheeling diode is not required since the carryover from the firing of one SCR to the next does not carry through a large portion of the negative half cycle and therefore current can be built up in an inductive load.

**Capacitance** must be used with care **in thyristor circuits**. A capacitor directly across any of the circuits in Fig. 8 will immediately destroy the thyristors. When an SCR is fired directly into a capacitor with no series resistance, the resulting $di/dt$ in the thyristor causes extreme local heating in the device and a resultant failure. A sufficiently high series resistor prevents failure. An inductance in series with a capacitor must also be used with caution. The series inductance may cause the capacitor to "ring up." Under this condition, the voltage across the capacitor can approach twice peak line voltage or 2.828 times rms line voltage.

The advantage of the thyristor circuits shown in Fig. 8 over the rectifier circuits is, of course, that the thyristor circuits provide variable output voltage. The output of the thyristor circuits depends upon the magnitude of the incoming line voltage and the phase angle at which the thyristors are fired. The control characteristic for the thyristor power supply is determined by the waveshape of the output voltage and also by the phase-shifting scheme used in the firing-control means for the thyristor. Practical and economic power supplies usually have control characteristics with some degree of non-linearity. A representative characteristic is shown in Fig. 9.

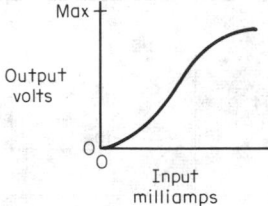

**Fig. 9** Thyristor control characteristic.

This control characteristic is usually given for nominal line voltage with the tacit understanding that variations in line voltage will cause approximately proportional changes in output voltage.

**Transistor amplifiers** can take many different forms. A complete discussion is beyond the scope of this handbook. The circuits described here illustrate basic principles. A basic **single-stage amplifier** is shown in Fig. 10. The transistor can be cut off by making the input terminal sufficiently negative. It can be saturated by making the input terminal sufficiently positive. In the linear range, the base of an $n$-$p$-$n$ transistor will be 0.5 to 0.7 V positive with respect to the emitter. The collector voltage will vary from about 0.2 V to $V_c$ (20 V, typically). Note that there is a sign inversion of voltage between the base and the collector; i.e., when the base is made more positive, the collector becomes less positive. The resistors in this circuit serve the following functions. Resistor $R1$ limits the input current to the base of the transistor so that it is not harmed when the input signal overdrives. Resistors $R2$ and $R3$ establish the transistor's operating point with no input signal. Resistors $R4$ and $R5$ determine the voltage gain of the amplifier. Resistor $R4$ also serves to stabilize the zero-signal operating point, as established by resistors $R2$ and $R3$. Usual prac-

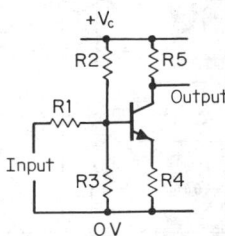

**Fig. 10** Single-stage amplifier.

tice is to design single-stage gains of 10 to 20. Much higher gains are possible to achieve, but low gain levels permit the use of less expensive transistors and increase circuit reliability.

Figure 11 illustrates a basic **two-stage transistor amplifier** using complementary *n-p-n* and *p-n-p* transistors. Note that the

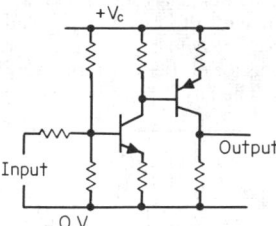

**Fig. 11** Two-stage amplifier.

first stage is identical to that shown in Fig. 10. This *n-p-n* stage drives the following *p-n-p* stage. Additional alternate *n-p-n* and *p-n-p* stages can be added until any desired overall amplifier gain is achieved.

Figure 12 shows the **Darlington** connection of transistors. This amplifier is used to obtain maximum current gain from two transistors. Assuming a base-to-collector current gain of

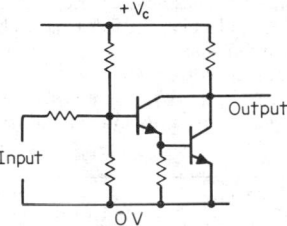

**Fig. 12** Darlington connection.

50 times for each transistor, this circuit will give an input-to-output current gain 2,500. This high level of gain is not very stable if the ambient temperature changes, but in many cases this drift is tolerable.

Figure 13 shows a circuit developed specifically to minimize temperature drift and drift due to power supply voltage changes. The **differential amplifier** minimizes drift because of the balanced nature of the circuit. Whatever changes in one transistor tend to increase the output are compensated by reverse trends in the second transistor. The input signal does not affect both transistors in compensatory ways of course, and so it is amplified. One way to look at a differential

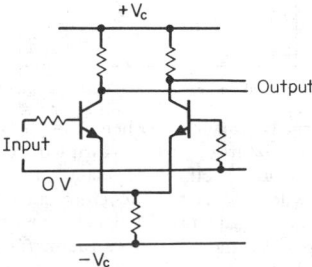

**Fig. 13** Differential amplifier.

amplifier is that twice as many transistors are used for each stage of amplification to achieve compensation. For very low drift requirements, matched transistors are available. For the ultimate in differential amplifier performance, two matched transistors are encapsulated in a single unit. **Operational amplifiers** made with discrete components frequently use differential amplifiers to minimize drift and offset. The operational amplifier is a low-drift, high-gain amplifier designed for a wide range of control and instrumentation uses.

**Oscillators** are circuits which provide a frequency output with no signal input. A portion of the collector signal is fed back to the base of the transistor. This feedback is amplified by the transistor and so maintains a sustained oscillation. The frequency of the oscillation is determined by parallel inductance and capacitance. The oscillatory circuit consisting of an inductance and a capacitance in parallel is called an *LC* **tank circuit.**

This frequency is approximately equal to

$$f = 1/2\pi\sqrt{CL} \qquad (2)$$

where $f$ = frequency, Hz; $C$ = capacitance, F; $L$ = inductance, H. A 1-MHz oscillator might typically be designed with a 20-$\mu$H inductance in parallel with a 0.05-$\mu$F capacitor. The exact frequency will vary from the calculated value because of loading effects and stray inductance and capacitance. The **Colpitts** oscillator shown in Fig. 14 differs from the

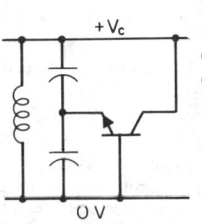

**Fig. 14** Colpitts oscillator.

**Hartley** oscillator shown in Fig. 15 only in the way energy is fed back to the emitter. The Colpitts oscillator has a capacitive voltage divider in the resonant tank. The Hartley oscillator has an inductive voltage divider in the tank. The **crystal oscillator** shown in Fig. 16 has much greater frequency stability than the circuits in Figs. 14 and 15. Frequency stability of 1 part in $10^7$ is easily achieved with a crystal-controlled oscillator. If the oscillator is temperature-controlled by mounting it in a small temperature-controlled oven, the frequency stability can be increased to 1 part in $10^9$. The resonant *LC* tank in the collector circuit is tuned to approximately the crystal frequency. The crystal offers a low impedance at its resonant frequency. This pulls the collector-tank operating frequency to the crystal resonant frequency.

As the desired operating frequency becomes 500 MHz and greater, **resonant cavities** are used as tank circuits instead of discrete capacitors and inductors. A rough guide to the relationship between frequency and resonant-cavity size is the wavelength of that frequency

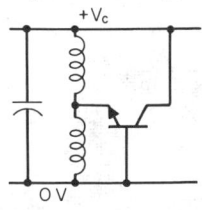

$$\lambda = 300 \times 10^6/f$$

where $\lambda$ = wavelength, m; $300 \times 10^6$ = speed of light, m/s; $f$ = frequency, Hz. The resonant cavities will be smaller than indicated by Eq. **(3)** because in general the cavity is either one-half or one-fourth wavelength and also, in general, the electromagnetic wave velocity is less in a cavity than in free space.

**Fig. 15** Hartley oscillator.

The operating principles of these devices are beyond the scope of this article. There are many different kinds of **microwave tubes** including **klystrons, magnetrons,** and **traveling-wave** tubes. All these tubes employ moving electrons to excite a resonant cavity. These devices serve as either oscillators or amplifiers at microwave frequencies.

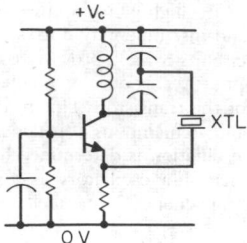

**Fig. 16**   Crystal-controlled oscillator.

**Lasers** operate at approximately visible-light frequency of 600 THz. This corresponds to a wavelength of 0.5 $\mu$m, or, the more usual measure of visible-light wavelength, $5 \times 10^3$ Å. Resonant cavities simply cannot be made small enough for these wavelengths. Electronic resonance in the atom serves as the tank circuit for these high frequencies. Quantum mechanics must be employed properly to explain these devices, but a practical understanding can be achieved without delving so deep. Most light is disorganized insofar as the axis of vibration and the frequency of vibration are concerned. When radiation along different axes is attenuated, as with a polarizing screen, the light wave is said to be polarized. When white light is filtered, or when the light source is not white, the light is frequency-limited, or colored. Colored light still has a relatively wide band of frequencies. The laser emits a very narrow band of frequencies, which are extremely stable, many times more stable than a crystal; therefore lasers are used as frequency standards. The narrow frequency band of lasers allows focusing the output into extremely small beams. This feature makes the laser attractive as a cutting tool and as an accurate surveying device. Extremely sharp focus and extremely high frequency make it attractive as a high-density communications carrier. Experimental work is being done with **phase-locked lasers,** which not only have a single frequency of output but have output oscillations in phase with each other. This degree of organization promises further commercial development of laser devices.

A radio wave consists of two parts, a **carrier,** and an information signal. The carrier is a steady high frequency. The information signal may be a voice signal, a video signal, or

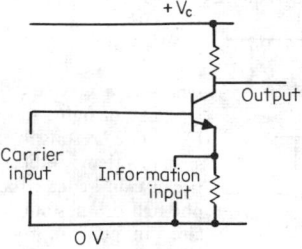

**Fig. 17**   AM modulator.

telemetry information. The carrier wave can be modulated by varying its amplitude or by varying its frequency. **Modulators** are circuits which impress the information signal onto the carrier. A **demodulator** is a circuit in the receiving apparatus which separates the information signal from the carrier. A simple amplitude modulator is shown in Fig. 17. The transistor is base-driven with the carrier input and emitter driven with the information signal. The modulated carrier wave appears at the collector of the transistor. An FM modulator is shown in Fig. 18. The carrier must be changed in frequency in

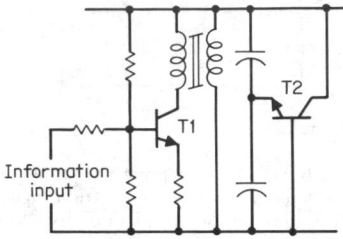

**Fig. 18**   FM modulator.

response to the information signal input. This is accomplished by using a saturable ferrite core in the inductance of a Colpitts oscillator which is tuned to the carrier frequency. As the collector current in transistor $T1$ varies with the information signal, the saturation level in the ferrite core changes, which in turn varies the inductance of the winding in the tank circuit and alters the operating frequency of the oscillator.

The **demodulator** for an AM signal is shown in Fig. 19. The diode rectifies the carrier plus information signal so that the

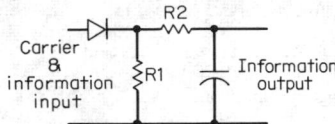

**Fig. 19**   AM demodulator.

filtered voltage appearing across the capacitor is the information signal. Resistor $R2$ blocks the carrier signal so that the output contains only the information signal. An **FM demodulator** is shown in Fig. 20. In this circuit, the carrier plus information signal has a constant amplitude. The information

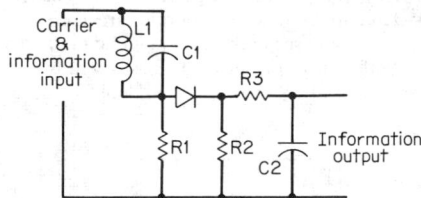

**Fig. 20**   FM discriminator.

is in the form of varying frequency in the carrier wave. If inductor $L1$ and capacitor $C1$ are tuned to near the carrier frequency but not exactly at resonance, the current through resistor $R1$ will vary as the carrier frequency shifts up and down. This will create an AM signal across resistor $R1$. The diode, resistors $R2$ and $R3$, and capacitor $C2$ demodulate this signal as in the circuit in Fig. 19.

The waveform of the basic **electronic timing circuit** is shown in Fig. 21 along with a basic timing circuit. Switch $S1$ is closed from time $t1$ until time $t3$. During this time, the transistor shorts the capacitor and holds the capacitor at 0.2 V. When switch $S1$ is opened at time $t3$, the transistor ceases to conduct

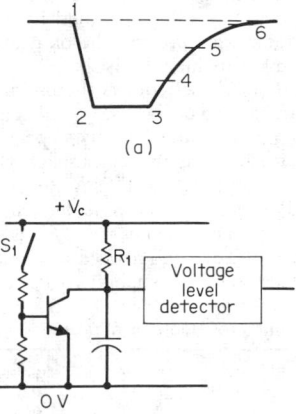

(a)

(b)

**Fig. 21**  Basic timing circuit.

and the capacitor charges exponentially due to the current flow through resistor $R1$. Delay time can be measured to any point along this exponential charge. If the time is measured until time $t6$, the timing may vary due to small shifts in supply voltage or slight changes in the voltage-level detecting circuit. If time is measured until time $t4$, the voltage level will be easy to detect, but the obtainable time delay from time $t3$ to time $t4$ may not be large enough compared with the reset time $t1$ to $t2$. Considerations like these usually dictate detecting at time $t4$. If this time is at a voltage level which is 63 percent of $V_c$, the time from $t3$ to $t4$ is one time constant of $R1$ and $C$. This time can be calculated by

$$t = RC$$

where $t$ = time, s; $R$ = resistance, $\Omega$; $C$ = capacitance, F. A timing circuit with a 0.1-s delay can be constructed using a 0.1-$\mu$F capacitor and a 1.0-M$\Omega$ resistor.

An improved timing circuit is shown in Fig. 22. In this circuit, the unijunction is used as a level detector, a pulse generator, and a reset means for the capacitor. The transistor is used as a constant current source for charging the timing capacitor. The current through the transistor is determined by resistors $R1, R2$, and $R3$. This current is adjustable by means of $R1$. When the charge on the capacitor reaches approxi-

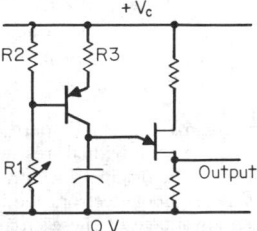

**Fig. 22**  Improved timing circuit.

mately 50 percent of $V_c$, the unijunction fires, discharging the capacitor and generating a pulse at the output. The discharged capacitor is then recharged by the transistor, and the cycle continues to repeat. The pulse rate of this circuit can be varied from one pulse per minute to many thousands of pulses per second.

A **multivibrator** circuit is shown in Fig. 23. This is a continuously toggling circuit. Transistors $T1$ and $T2$ conduct alter-

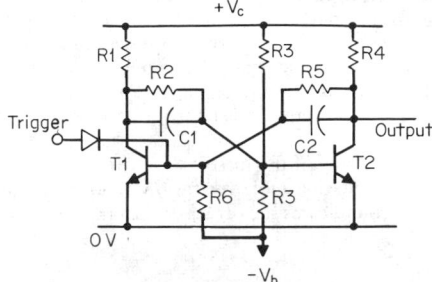

**Fig. 23**  Multivibrator circuit.

nately. With balanced resistors and capacitors, the circuit generates a square wave of voltage at the output. Assume that transistor $T2$ turns on. Its collector switches from $V_c$ to 0.2 V. Since capacitor $C2$ cannot discharge immediately, the base of transistor $T1$ is driven negative and $T1$ is turned off. Capacitor $C2$ begins to discharge through resistor $R6$. Capacitor $C1$ charges through resistor $R1$ and as long as the current through $C1$ is greater than the current through resistor $R3$, transistor $T2$ is kept in conduction. When the current in $C1$ decreases because of the increasing voltage across it, the current in $R3$ will turn transistor $T2$ off. This will cause capacitor $C2$ to charge, turning on transistor $T1$. Transistor $T1$ will stay on until capacitor $C2$ charges, and then $T2$ will again turn on. The circuit will continue to toggle in this way.

By changing the bias resistor $R3$ and connecting it to $+V_c$ transistor $T2$ will stay turned on until a positive external pulse is applied to the trigger input. The circuit will then cycle once, and when $T2$ turns on, it will remain in that state until

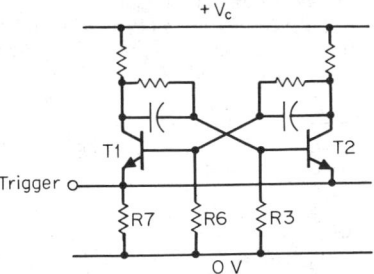

**Fig. 24**  Basic flip-flop.

another trigger pulse is applied. This form of the circuit is called a **monostable** multivibrator or simply a monostable.

A **flip-flop** is a circuit that assumes a state as directed by its inputs. This circuit is shown in Fig. 24. It is identical to the circuit in Fig. 23 except for the addition of resistor $R7$ and the reconnection of resistors $R3$ and $R6$ to 0 V rather than to $-V_b$. Transistors $T1$ and $T2$ will change state everytime a pulse appears at the trigger input. If $T1$ is on when a trigger pulse appears, it will turn off and $T2$ will turn on. If $T2$ is on, it will turn off and $T1$ will turn on when a trigger pulse appears. Figure 24 is a basic flip-flop. A more useful circuit for sequential machines, such as a computer, is the master-slave flip-flop. This circuit consists of two flip-flops in tandem. The master flip-flop can only be set up when the trigger pulse is on. During this time, the slave, or output flip-flop stays in its previous state. When the trigger pulse falls, the master flip-flop is inhibited from changing state and the slave is permitted to change to the state commanded by the master flip-flop. In this way, a command can be given to series of interconnected master-slave flip-flops at time $t$ without any change taking place at their outputs until time $t + 1$. This allows an orderly progression of logic commands to be paced by a series of clock pulses.

The most versatile of the master-slave circuits is called a **JK flip-flop.** The excitation table for the *JK* flip-flop is shown in Fig. 25. A zero indicates the off state; the one indicates the on

| Output | | | |
|---|---|---|---|
| $Q(t)$ | $Q(t+1)$ | $J$ | $K$ |
| 0 | 0 | 0 | X |
| 0 | 1 | 1 | X |
| 1 | 0 | X | 1 |
| 1 | 1 | X | 0 |

**Fig. 25** *JK* flip-flop excitation table.

state; the $X$ indicates the either on or off has no bearing on the result. It is an "I don't care" input. *JK* flip-flops are often incorporated in the arithmetic section of digital computers. Note that the $J$ input need have a specific value only when the flip-flop is in the off state and the $K$ input need have a specific value only when the flip-flop is in the on state at time $t$.

Binary logic requires additional circuits called **gates.** Six different types of gates are shown in Fig. 26. The AND gate has an output only when both its inputs are present. One might argue that the gate has an output of zero in all other cases, but it is customary when describing digital logic to call a zero "no output." The OR gate has an output whenever either or both of its inputs exist. The NOT gate is simply an inverter. Its output is what its input is not. The NAND gate and the NOR gate are

| $X$ | $Y$ | $Z$ |
|---|---|---|
| 0 | 0 | 0 |
| 0 | 1 | 0 |
| 1 | 0 | 0 |
| 1 | 1 | 1 |

*(a)* AND

| $X$ | $Y$ | $Z$ |
|---|---|---|
| 0 | 0 | 0 |
| 0 | 1 | 1 |
| 1 | 0 | 1 |
| 1 | 1 | 1 |

*(b)* OR

| $X$ | $Z$ |
|---|---|
| 0 | 1 |
| 1 | 0 |

*(c)* NOT

| $X$ | $Y$ | $Z$ |
|---|---|---|
| 0 | 0 | 1 |
| 0 | 1 | 1 |
| 1 | 0 | 1 |
| 1 | 1 | 0 |

*(d)* NAND

| $X$ | $Y$ | $Z$ |
|---|---|---|
| 0 | 0 | 1 |
| 0 | 1 | 0 |
| 1 | 0 | 0 |
| 1 | 1 | 0 |

*(e)* NOR

| $X$ | $Y$ | $Z$ |
|---|---|---|
| 0 | 0 | 0 |
| 0 | 1 | 1 |
| 1 | 0 | 1 |
| 1 | 1 | 0 |

*(f)* EXCLUSIVE OR

**Fig. 26** Combinational gate logic: *(a)* AND, *(b)* OR, *(c)* NOT, *(d)* NAND, *(e)* NOR, *(f)* EXCLUSIVE-OR.

equivalent to the combination of a NOT gate placed at the output of an AND gate and an OR gate, respectively. The EXCLUSIVE-OR gate is similar to the OR gate except that its output does not exist when both inputs are present.

The gates shown in Fig. 26 have two inputs (with the exception of the NOT gate). They can be extended to any number of inputs. The AND gate would have an output only when all its inputs are present; the OR gate would have an output when any of its inputs exist.

Analysis of digital logic requires the use of **Boolean algebra.** Boolean algebra has two operators, $\cdot$ , which indicates an AND operation, and +, which indicates an OR operation. The equals sign has the same meaning in Boolean algebra as in ordinary algebra. The symbol for "$X$ not" is $X'$ (or in some cases $\bar{X}$). The rules for Boolean algebra can be derived from set theory applied to a system in which only two numbers exist, i.e., zero and one. These rules are summarized in Huntington's postulates and De Morgan's theorem, and are listed in Table 3. To

**Table 3. Rules for Boolean Algebra**

| | |
|---|---|
| $X + 0 = X$ | $X \cdot 1 = X$ |
| $X + 1 = 1$ | $X \cdot X' = 0$ |
| $X + X' = 1$ | $X \cdot X = X$ |
| $(X')' = X$ | $X \cdot 0 = 0$ |
| $X + Y = Y + X$ | $X \cdot Y = Y \cdot X$ |
| $X + (Y + Z) = (X + Y) + Z$ | $X \cdot (Y Z) = (X \cdot Y)Z$ |
| $X + X \cdot Y = X$ | $X \cdot (X + Y) = X$ |

DeMorgan's theorem:
$(X + Y)' = X' \cdot Y'$                        $(X \cdot Y)' = X' + Y'$

facilitate the analysis of digital circuits and aid in the application of the rules given in Table 3, Karnaugh maps are used. A two-variable and a four-variable Karnaugh map are shown in Fig. 27 along with the algebraic expression represented by each map.

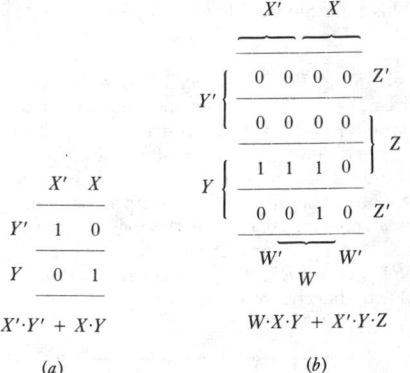

*(a)* $X' \cdot Y' + X \cdot Y$

*(b)* $W \cdot X \cdot Y + X' \cdot Y \cdot Z$

**Fig. 27** Karnaugh maps: *(a)* $X' \cdot Y + X \cdot Y$; *(b)* $W \cdot X \cdot Y + X' \cdot Y \cdot Z$.

## Integrated Circuits

The circuits described in the previous part were built with discrete components, i.e., individual resistors, capacitors, diodes, and transistors. The original incentive for integrated circuits was to reduce weight of electronic equipment in space vehicles. The added advantage of these circuits is that they cost much less to produce and are far more reliable. These two advantages allow the design of far more complex and far more sophisticated electronic equipment. There are three levels of circuit integration, **integrated circuits, MSI** (medium-scale integration), and **LSI** (large-scale integration). Integrated circuits include assemblages of parts into circuits such as gates, flip-flops, and amplifiers. MSI consists of single packages that contain complex circuits such as multidigit adders and multidigit code converters. LSI packages include very complex circuits such as accumulators and arithmetic processors. An electronic calculator that can add, subtract, multiply, divide, raise a number to a decimal power, evaluate sine, cosine, and tangent, and which has four memory registers and some additional functions is made with four LSI chips.

Table 4 shows a cost and complexity comparison between

**Table 4. Cost and Complexity Comparison of Discrete and Integrated Circuits**

|  | No. of gates per device | Cost per gate, $ | Tooling cost per gate, $ |
|---|---|---|---|
| Discrete circuit | 0.1 | 3.00 | 50.00 |
| Integrated circuit | 4 | 0.25 | 1200.00 |
| MSI | 200 | 0.05 | 50.00 |
| LSI | 10,000 | 0.01 | 0.50 |

discrete-component circuits and integrated circuits. The costs are, in general, proportional to the difference in reliability. This has brought the price of a four-function electronic calculator from $700 to under $40 in 3 years' time. Similar changes are taking place in other electronic equipment as well.

Integrated circuits are made with a refinement of the same methods used to make transistors. They are semiconductor devices. Dimensions within the integrated circuit are so small that electron microscopes are used for visual inspection. The circuits are produced by masking, etching, and diffusion techniques. Tooling for a new circuit costs $10,000 and more. For this reason, only high-volume applications can utilize integrated circuits. With the low cost of integrated circuits, many devices are being revised to use the circuits that are available. For instance, many devices that in the past have been made with analog circuits are now being made with digital circuits, since digital circuits were the first to be converted to integrated packages.

Analog circuits have also been integrated. High-performance operational amplifiers are available at low cost in integrated form. Entire sections of radio and television receivers have been designed in integrated packages. Electronic multipliers are also available. No attempt is made to list integrated circuits here because that list would rapidly become obsolete. Manufacturers of integrated circuits publish excellent data books which are free and kept current.

### Industrial Electronics

The power for dc motor armatures can be derived from thyristor circuits like those shown in Fig. 8. Single-phase bridge circuits are used for 5-hp drives and smaller. Three-phase bridge circuits are used for drives larger than 5 hp. A single set of six thyristors can supply power for about 300 hp. Above 300 hp, multiple sets of thyristors must be used in parallel. Mill drives have been built with more than 10,000 hp provided by thyristors.

The control of dc motors whether powered by thyristors or by dc generators is accomplished electronically. Control of individual drives can be accomplished by tachometer feedback or by armature voltage feedback. The speed-regulation accuracy for armature feedback is 5 percent; for tachometer feedback speed-regulation accuracy is from 0.1 to 1.0 percent. When two drives must be coordinated with each other, as in a continuous-web processing machine, they can be regulated to control torque, speed, position, draw, or a combination of these parameters. Torque controls can be achieved using dc motor armature current for a feedback signal. Speed-control signals are derived as for single motors. Position or draw control can be accomplished by using selsyn ties or dancer rolls. A **dancer roll** is a weight- or spring-loaded roll that rides on the web. It is free to move up and down, and as it does, a signal is taken from its position to serve as a feedback for the drive regulator before the dancer or after it.

Coordination of the motions of two or more drives requires tracking of the drives in both steady-state and transient conditions. Linearity of the control and feedback signals determine steady-state tracking. Provision must be made for both low-speed and high-speed matching signals. Transient matching requires that signals not only be the right magnitude but also arrive at the right time. An example will serve to illustrate this point. Suppose it is desired to have two drives with tachometer feedback which have a continuous web between them. One way to accomplish this would be to designate one drive as a master and the other as a slave. The tachometer on the master drive would serve as its own feedback signal and as the reference or command signal for the slave drive. The slave drive would have its own feedback from its own tachometer and so its regulator would try to minimize the difference between the two tachometer signals. On a transient basis the master drive will always start before the slave. An alternate and more common arrangement is to provide a common reference for both drives and let each drive receive its command signals at the same instant.

Digital computers are being used on line in mills and continuous processing industries. DC motors can be controlled by either analog or digital regulators. With the greatly reduced cost of integrated circuits, digital regulators are being increasingly used.

DC motors have been widely used for variable-speed applications because of their excellent characteristics. AC motors have been used primarily for constant-speed applications. The control schemes described above are equally applicable to ac motors (except of course for armature voltage and armature current feedback). If power circuitry is properly handled, the control of an ac motor is just as flexible and versatile as that of a dc motor.

AC motors can be supplied either from **phase-controlled circuits** or from **inverter circuits**. Phase control is a simple electronic circuit, but its use results in high losses in the ac motor. This limits the application of this type of drive to either a very limited speed range or to loads in which the torque required decreases rapidly as the speed decreases. Large pump drives and fan drives have been built using this form of ac motor control. Inverters can be designed so that excessive motor losses are not encountered. Inverters are quite complex and require auxiliary power components to commutate the thyristors. Cost and complexity have prevented the widespread use of inverter-powered ac motor drives.

Phase-control circuits are extensively used to control power flow to process heaters. Most industrial heating is done by gas because it is cheaper than electrical energy. In many applications, electric heat is needed or is sufficiently more convenient. Phase-controlled thyristors modulate the power to these heaters and provide smoother control than simple on-off control by contactors.

High frequencies can be generated by **thyristor inverter** circuits. This permits the use of thyristors for **induction heating** and supersonic cleaning. Thyristor supplies have been built with frequency output from 100 to 50,000 Hz. These power supplies can be controlled in frequency much more easily and rapidly than motor-alternator sets and so have added new capability to induction-heating apparatus.

**Dielectric heating** requires frequencies from 100 kHz to 1 MHz. Large vacuum-tube oscillators are used to generate these frequencies.

### Communications

The Federal Communications Commission (FCC) regulates the use of radio-frequency transmission in the United States. This regulation is necessary to prevent interfering transmissions of radio signals. Some of the frequency allocations are given in Table 5. The frequency bands are also classified as

**Table 5. Partial Table of Frequency Allocations**
(For a complete listing of frequency allocations, see "Reference Data for Radio Engineers," published by Howard Sams & Co.)

| Frequency, MHz | Utilization |
|---|---|
| 0.535–1.605 | Commercial broadcast band |
| 27.255 | Citizens' personal radio |
| 54–72 | Television channels 2–4 |
| 76–88 | Television channels 5–6 |
| 88–108 | Frequency-modulation broadcasting |
| 174–216 | Television channels 7–13 |
| 460–470 | Citizens' personal radio |
| 470–890 | Television channels 14–83 |

shown in Table 6. Very low frequencies are used for long-distance communications across the surface of the earth. Higher frequencies are limited to line-of-sight transmission. Because of bandwidth considerations, high frequencies are used for high-density communication links. Orbiting **satellites** allow the use of high-frequency transmission for long-distance high-density communications.

A **radio transmitter** is shown in Fig. 28. It consists of four basic parts: an rf oscillator tuned to the carrier frequency, an information-input device (microphone), a modulator to impress the input signal on the carrier, and an antenna to radiate the modulated carrier wave.

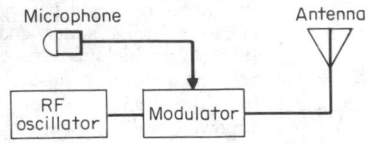

**Fig. 28**  Radio transmitter.

A **radio receiver** is shown in Fig. 29. This is called a **superheterodyne** receiver because it utilizes a frequency-mixing scheme. The tuned radio-frequency amplifier is tuned to receive the desired radio signal. The local oscillator is adjusted by the same tuning control to a lower frequency. The mixer produces an output frequency which is the difference between the incoming radio-signal frequency and the local-oscillator fre-

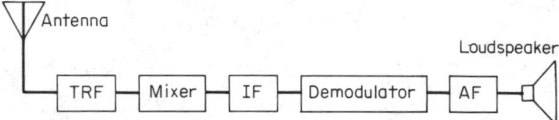

TRF  Tuned radio frequency amplifier

IF  Intermediate frequency amplifier

AF  Audio frequency amplifier

**Fig. 29**  Radio receiver.

quency. Since this difference frequency is constant for all tuning positions, the intermediate-frequency amplifier always operates with a constant frequency. This allows optimum design of the intermediate-frequency (IF) amplifiers since they are constant-frequency amplifiers. The IF frequency signal is modulated in just the same way as the radio signal. The demodulator separates this audio signal, which is then amplified so that the loudspeaker can be driven.

The term **radar** is **derived** from the first letters of the words "*ra*dio *d*etection *a*nd *r*anging." It is essentially an echo system in which the location of an object is determined by sending out short pulses of radio waves and observing and measuring the time required for their reflections or echoes to return to the sending point. The time interval is a measure of the distance of the object from the transmitter. The velocity of radio waves is the same as the velocity of light or 984 ft/$\mu$s, so that each microsecond interval corresponds to a distance of 492 ft. The

**Table 6. Frequency Bands**

| Designation | Frequency | Wavelength |
|---|---|---|
| VLF, very low frequency | 3–30 kHz | 100–10 km |
| LF, low frequency | 30–300 kHz | 10–1 km |
| MF, medium frequency | 300–3,000 kHz | 1,000–100 m |
| HF, high frequency | 3–30 MHz | 100–10 m |
| VHF, very high frequency | 30–300 MHz | 10–1 m |
| UHF, ultra-high frequency | 300–3,000 MHz | 100–10 cm |
| SHF, super-high frequency | 3,000–30,000 MHz | 10–1 cm |
| EHF, extremely high frequency | 30,000–300,000 MHz | 10–1 mm |

Wavelength in meters = 300/$f$ in megacycles.

direction of an object can be determined by the position of the directional transmitting and receiving antenna. Radio waves penetrate darkness, fog, and clouds, and hence are able to detect objects that otherwise would remain concealed. Radar can be used for the automatic "tracking" of objects such as airplanes.

A block diagram of a radar system is shown in Fig. 30. The transmitting system consists of an rf oscillator which is controlled by a modulator, or pulser, so that it sends to the antenna intermittent trains of rf waves of relatively high power but of very short duration, corresponding to the pulses

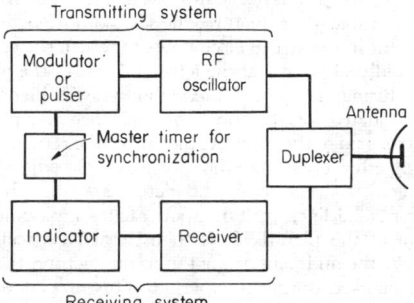

**Fig. 30**  Block diagram of radar system.

received by the modulator. The energy of the oscillator is transmitted through the duplexer and to the antenna through either coaxial cable or waveguides. The **receiver** is an ordinary heterodyne-type radio receiver which has high sensitivity in the band width corresponding to the frequency of the oscillator. For low frequencies the local oscillator is an ordinary oscillator for frequencies of 2,000 MHz; and higher a reflex **klystron** (hf cavity oscillator) is used. A common intermediate frequency is 30 MHz but 15 and 60 MHz are also frequently used.

In most radar systems the same antenna is used for receiving as for transmitting. This requires the use of a **duplexer** which cuts off the receiver during the intervals when the oscillator is sending out pulses and disconnects the transmitter during the periods between these pulses when the echo is being received.

The antenna is highly directional. By noting its angular position, the direction of the object may be determined. In the PPI (plan position indicator), the angle of the sweep of the cathode-ray beam on the screen of the oscilloscope is made to correspond to the azimuth angle of the antenna.

The receiver output is delivered to the indicator which consists of a cathode-ray tube or oscilloscope. The pulses which are received, corresponding to echoes from the target, must be synchronized with the sending pulses in order that the distance to the target may be determined. This is accomplished by synchronization of the sweep circuit of the oscilloscope with the pulses by the master timer.

**Displays**  Conversion of the received radar signals to usable display is accomplished by a cathode-ray oscilloscope. The simplest type, called the **A presentation,** is shown in Fig. 31a. When the pulser operates, a sawtoothed wave produces a linear sweep voltage (Fig. 31b), across the sweep plates of the cathode-ray tube; at the same time a transmitter pulse is impressed on the deflection plates and the return echoes appear as AM pulses, or "pips," on the screen, as shown in Fig. 31a. The distance on the screen between the transmitter

pulse and the pip caused by the echo is proportional to the distance to the target, and the screen can be calibrated in distance such as miles. [The return of the spot to its initial starting position, produced by the sweep interval *cd* (Fig. 31*b*)

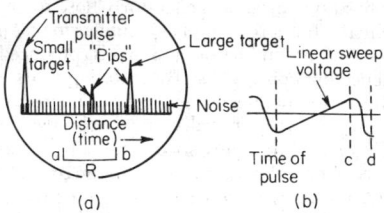

**Fig. 31**  Type A presentation.

is so rapid that it is not detectable by the eye.] The direction of the target may be determined by the angular position of the antenna, which can be transmitted to the operator by means of a selsyn. Different objects, such as airplanes, ships, islands, and land approaches, have characteristic pips, and operators become skilled in their interpretation. A bird in flight can be recognized on the screen. Also a portion of the scale such as *ab* can be segregated and amplified for close study of the characteristics of the pips.

**Plan Position Indicator (PPI)**  In the PPI (Fig. 32) the direction of a radial sweep of the electron beam is synchronized with the azimuth sweep of the antenna. The sweep of the beam is rotated continuously in synchronism with the antenna, and the received signals intensity-modulate the electron beam as it sweeps from the center of the oscilloscope screen radially outward. In this way the direction and range position of an object can be determined from the pattern on the screen of the oscilloscope, as shown in Fig. 32.

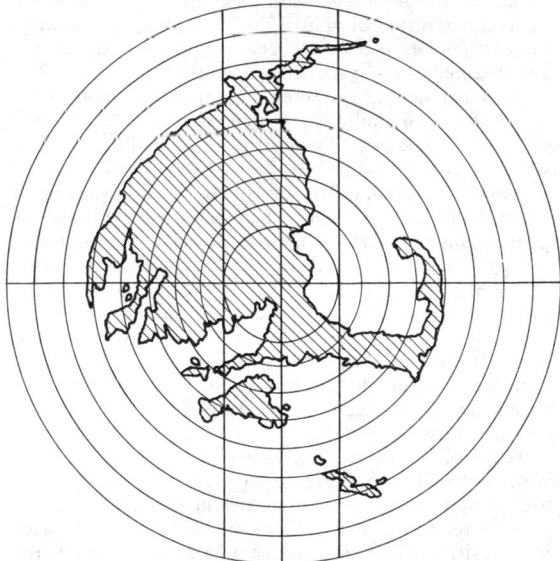

**Fig. 32**  Plan position indicator (PPI) of southeastern Massachusetts.

There are two methods by which the angular direction of the cathode spot is made to correspond with the angular position of the antenna. In one method, used on board ship, two magnetic deflecting coils are rotated around the neck of the tube in synchronism with the antenna, by means of a selsyn. In the other method, used on aircraft, two fixed magnetic deflecting coils at right angles to each other and placed at the neck of the tube are supplied with current from a small two-phase synchronous generator whose rotor is driven by the antenna. Thus a rotating field, similar to that produced by the stator of an induction motor, is produced by the magnetic deflecting coils. These two rotating fields, although produced by different means, are equivalent and cause the cathode beam to sweep radially in synchronism with the antenna. Circular coordinates spaced radially corresponding to distance are obtained by impressing on the control electrode short positive pulses synchronized with the transmitted pulse but delayed by time values corresponding to the desired distances. These coordinates appear as circles on the screen. Since the time of rotation of the antenna is relatively slow, it is necessary that a persistent screen be used in order that the operator may view the entire pattern. In Fig. 32 is shown a line drawing of a PPI presentation of Cape Cod, Mass., on a radar screen, taken from an airplane.

The applications of radar to war purposes are well known, such as detecting enemy ships and planes, aiming guns at them, and locating cities, rivers, mountains, and other landmarks in bombing operations. In peacetime, radar is used to navigate ships in darkness and poor visibility by locating navigational aids such as buoys and lighthouses, as well as protruding ledges, islands, and other landmarks. It can be similarly used in air navigation, as well as to operate altimeters for determining the height of the plane above ground. It is also used for aerial mapping.

There are also radio beacons, **shoran** (*short-ran*ge navigation) and **loran** ((*long-ran*ge navigation) by which ships or planes can locate their positions. In the **ground-controlled approach** (GCA) for airplanes, the ground operator picks up the plane on a PPI presentation at distances up to 30 mi, using a general surveillance radar, and gives instructions to the pilot by radio course and procedure. As the plane approaches the landing field, it is brought into vision on the screen of a high-resolution short-range radar, and the pilot is given continual detailed instructions as to the glide path which the plane is to follow until the landing is made.

**Television** is accomplished by systematically scanning a scene or the image of a scene to be reproduced and transmitting at each instant a current or a voltage which is proportional to the light intensity of the elementary area of the scene which at the instant is being scanned. The varying voltage or current is amplified, modulated on a carrier wave, and then transmitted as a radio wave. At the receiver the radio wave enters the antenna, is amplified, and demodulated to give a voltage or a current wave similar to the original wave. This voltage or current wave is then used to control the intensity of a cathode-ray beam which is focused on a fluorescent screen in a cathode-ray reproducing tube. The cathode-ray beam is caused to move over the screen in the same pattern as the scanning beam at the transmitter and in synchronism with it. Thus each small area of the receiver screen is illuminated instantaneously with light intensity corresponding to that of a similarly placed area in the original scene. This process is conducted so rapidly that owing to persistence of vision of the eye, the reproduction of each instantaneous scene appears to be a complete picture and the effect with successive scenes is similar to that produced by the projection of successive frames of a motion picture.

**Scanning and Blanking**  In the United States the ratio of width to height of a standard television picture is 4:3, and the picture is composed of 525 lines repeated 30 times a second, this last factor being one-half 60, the prevalent electric power frequency in the United States. The scanning sequence along the individual lines is from left to right and the sequence of the lines is from top to bottom. Also, interlacing is employed, the general method of which is shown in Fig. 33. The cathode-ray spot starts at 1 in the upper left-hand corner and is swept rapidly from left to right either by a sawtooth emf wave applied to the sweep plates or by the sawtooth current wave applied to the sweep coils of the tube. When the spot arrives at the right-hand side of the picture, the sawtooth wave of either emf or current in the sweep circuit acts to return the cathode-ray spot rapidly to point 3 at the left-hand side of the picture. However, during this period the cathode-ray is blanked, or entirely eliminated, by the application of a negative potential to the control grid of the tube. At the end of the return period, the blanking effect ceases and the spot appears at point 3, from which it again is swept across the picture and this process is repeated for 262.5 lines until the spot reaches a midpoint $C$ at the bottom of the picture. It is then carried vertically and rapidly to $B$, the midpoint of the top of the picture, the beam also being blanked during this period. This process of scanning is then repeated, a second set of lines corresponding to the even numbers 2, 4, 6 being established between the lines designated by the odd numbers. These lines are shown dashed in Fig. 33. This method or pattern of scanning is called **interlacing**. The two sets of lines taken together produce a frame of 525 lines, which are repeated 30 times each second. However, owing to interlacing, the flicker frequency is 60 Hz which is not noticeable, 50 Hz having been determined as the threshold of flicker noticeable to the average eye. In Fig. 33, for the sake of clarity, the distances between horizontal lines are greatly exaggerated and no attempt is made to maintain proportions.

**Frequency Band**  In order to obtain the necessary resolution of pictures, television frequencies must be high. In the United States, VHF frequencies from 54 to 88 MHz (omitting 72 to 76 MHz) and 174 to 216 MHz are assigned for television broadcasting. A UHF band of frequencies for commercial

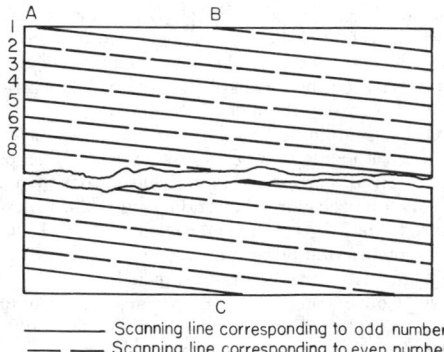

——— Scanning line corresponding to odd numbers
— — — Scanning line corresponding to even numbers

**Fig. 33**  Pattern of interlaced scanning.

television use is also allocated and consists of the frequencies of from 470 to 890 MHz (see also Tables 5 and 6).

In order to obtain the 525 lines repeated 30 times per second, a band width of 6 MHz is necessary. The video, or picture, signal with the superimposed scanning and blanking pulses is amplitude-modulated, amplified, and transmitted. The carrier frequency associated with the sound transmitter is 4.5 MHz higher than the video carrier frequency and is frequency-modulated with a maximum frequency deviation of 25 kHz.

In scanning motion-picture films a complication arises because standard film rate is 24 frames per second, while the television rate is 30 frames per second. This difficulty is overcome by scanning the first of two successive film frames twice and the second frame three times at the 60-Hz rate, making the total time for the two frames $\frac{1}{12}$ s ($\frac{2}{60}$ + $\frac{3}{60}$) or $\frac{1}{24}$ s average per frame.

**Kinescope**   The kinescope (Fig. 34) is the terminal tube in which the televised picture is reproduced. It is relatively simple, being not unlike the cathode-ray oscilloscope tube. It

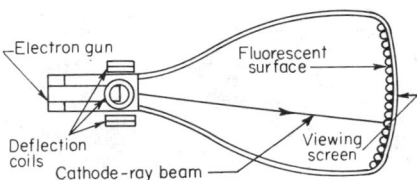

**Fig. 34**   Kinescope for television receiver.

has an electric gun operating at 8,000 to 20,000 V which produces an electron beam focused on a fluorescent surface within the front wall of the tube. The picture is viewed at the front wall. The horizontal and vertical deflections of the beam are normally controlled by deflection coils, as shown in Fig. 34.

**Television Receivers**   A block diagram for a television receiver is given in Fig. 35. It is in reality a superheterodyne receiver with tuned rf amplification, the separating of the sound and video or picture channels taking place at the inter-

mediate frequency in the mixer. The sound channel is then conventional, a discriminator being used to demodulate the FM wave (Fig. 20). The object of the dc restorer is to make the picture reproduction always positive, and it consists of applying a dc voltage at least equal in magnitude to the maximum values of the negative loops of the ac waves. The synchronizing pulses for both the vertical and the horizontal deflections are delivered by the dc restorer to an amplifier and the two pulses are then divided into the V and H components. The integrating and differentiating circuits are necessary to separate horizontal and vertical synchronizing signals.

As stated earlier, at any instant the magnitude of the current from the pickup tube varies in accordance with the light intensity of the part of the scene being scanned at that instant. This current is amplified and, together with the sound and synchronizing currents, is broadcast and received by the circuit shown in Fig. 35. The video current is detected by rectification, is amplified, and is then made to control the intensity of the kinescope electron beam. Tubes produce a scanning pattern, identical with that in the pickup tube, and these tubes are triggered by the synchronizing pulses which are transmitted in the broadcast wave. Hence, the original televised scene is reproduced on the fluorescent screen of the kinescope.

**Color-television** transmission is similar to black-and-white television, and the two signals must be compatible with each other. The kinescope for color TV has three electron guns, one for each primary color. The fluorescent screen has a matrix of three different colors of phosphor and a mask with many small holes in it. The intensity signals for each color are phase-shifted from each other so that the proper phosphors are excited by each electron stream at each mask point over the entire screen. A black-and-white signal does not have the same synchronizing signal as a color signal. The color receiver has circuits which recognize this state and switch it to black and white reception.

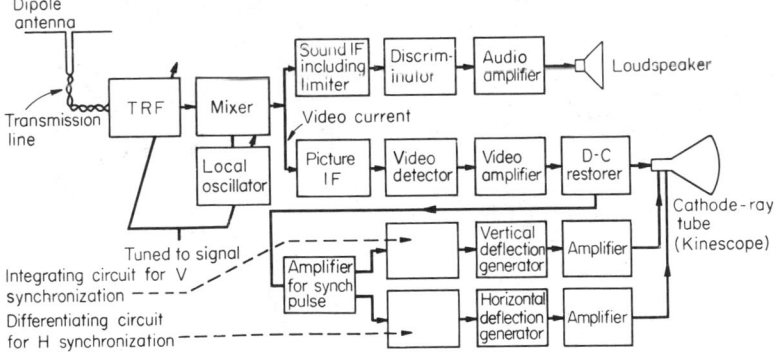

**Fig. 35**   Block diagram for television receiver; TRF—tuned radio frequency; IF—intermediate frequency.

# Instruments and Controls

**BY**

**ROBERT T. CORRY,** *Associate Professor of Mechanical Engineering, Polytechnic Institute of New York.*

**GUY LONGOBARDO,** *Corporate Staff, I.B.M. Corp.*

**ANNA KAZANJIAN LONGOBARDO,** *Manager of Plans & Programs, Systems Management Division, Sperry Rand Corp.*

**W. LUCAS GAILLARD,** *Practicing Surveyor, State of South Carolina.*

**W. LUCAS GAILLARD, JR.,** *Practicing Surveyor, State of South Carolina.*

### INSTRUMENTS
#### by Robert T. Corry

Introduction to Measurement ........................... 16-2
Counting Events ...................................... 16-2
Time and Frequency Measurement ....................... 16-3
Mass and Weight Measurement ......................... 16-3
Measurement of Linear and Angular Displacement ......... 16-4
Measurement of Area .................................. 16-7
Measurement of Fluid Volume .......................... 16-7
Force and Torque Measurement ......................... 16-8
Pressure and Vacuum Measurement ...................... 16-8
Liquid-Level Measurement ............................. 16-10
Temperature Measurement ............................. 16-10
Measurement of Fluid Flow Rate ....................... 16-14
Power Measurement ................................... 16-16
Electrical Measurements ............................... 16-16
Velocity and Acceleration Measurement ................. 16-20
Measurement of Physical and Chemical Properties ......... 16-21
Nuclear Radiation Instruments ......................... 16-22
Indicating, Recording, and Logging ..................... 16-23

### AUTOMATIC CONTROLS
#### by Guy Longobardo and Anna Kazanjian Longobardo

Introduction ......................................... 16-25

Basic Automatic-Control System ........................ 16-26
Transient Analysis of a Control System .................. 16-26
Block-Diagram Representation .......................... 16-28
Signal-Flow Representation ............................ 16-29
Steady-State Performance ............................. 16-30
Frequency Response .................................. 16-31
Graphical Display of the Frequency Response ............. 16-31
Stability and Performance of an Automatic Control ......... 16-34
Sampled-Date Control Systems ......................... 16-35
State-Space Concepts ................................. 16-37
General Design Procedure ............................. 16-37
Components .......................................... 16-37
Pneumatic Systems ................................... 16-38
Computer Control ..................................... 16-40

### SURVEYING
#### by W. Lucas Gaillard and W. Lucas Gaillard, Jr.

Linear Measurements ................................. 16-42
Leveling ............................................ 16-43
Transit Work ........................................ 16-45
Special Problems in Surveying and Mensuration ........... 16-48

# INSTRUMENTS

## Robert T. Corry

REFERENCES. ASME publications: "Instruments and Apparatus Supplement to Performance Test Codes (PTC 19.1–19.20)"; "Fluid Meters, pt. II, Application." ASTM, "Manual on the Use of Thermocouples in Temperature Measurement," STP 470A. ISA publications: "ISA Transducer Compendium," "Temperature: Its Measurement and Control in Science and Industry," "Flow: Its Measurement and Control in Science and Industry," vol. 1. Ambrosius et al., "Mechanical Measurement and Instrumentation," Ronald. Baker et al., "Temperature Measurement in Engineering, vols. I and II," Wiley. Beckwith et al., "Mechanical Measurements," Addison-Wesley. Benedict, "Fundamentals of Temperature, Pressure and Flow Measurements," Wiley. Carroll, "Industrial Process Measuring Instruments," McGraw-Hill. Carroll, "Industrial Instrument Servicing Handbook," McGraw-Hill. Considine, "Encyclopedia of Instrumentation and Control," McGraw-Hill. Considine, "Handbook of Applied Instrumentation," McGraw-Hill. Doebelin, "Measurement Systems: Application and Design," McGraw-Hill. Harris et al., "Shock and Vibration Handbook," McGraw-Hill. Holman, "Experimental Methods for Engineers," McGraw-Hill. Holzbock, "Instruments for Measurement and Control," Van Nostrand-Reinhold. Jones, "Instrument Technology," vol. I, "Measurement of Pressure, Level, Flow, and Temperature," Plenum Press. Lion, "Instrumentation in Scientific Research: Electrical Input Transducers," McGraw-Hill. Snell, "Nuclear Instruments and Their Uses," Wiley. Stout, "Basic Electrical Measurements," Prentice-Hall. "Instruments and Control Systems," monthly, Chilton Company.

## INTRODUCTION TO MEASUREMENT

An **instrument,** as referred to in the following discussion, is a device for determining the value or magnitude of a quantity or variable. The variables of interest are those which help describe or define an object, system, or process. Thus, in a manufacturing operation, product quality is related to measurements of its various dimensions and physical properties such as hardness and surface finish. In an industrial process, measurement and control of temperature, pressure, flow rates, etc., determine quality and efficiency of production.

Measurements may be direct, e.g., using a micrometer to measure a dimension, or indirect, e.g., determining moisture in steam by measuring the temperature in a throttling calorimeter.

Because of physical limitations of the measuring device and the system under study, practical measurements always have some error. The **accuracy** of an instrument is the closeness with which its reading approaches the true value of the variable being measured. **Precision** refers to the reproducibility of the measurements, i.e., with a fixed value of the variable, how much successive readings differ from one another. **Sensitivity** is the ratio of output signal or response of the instrument to a change in input or measured variable. **Resolution** relates to the smallest change in measured value to which the instrument will respond.

**Error** may be classified as systematic or random. Systematic errors are those due to assignable causes. These may be static or dynamic. Static errors are caused by limitations of the measuring device or the physical laws governing its behavior.

A static error is introduced in a micrometer reading, for example, when excessive pressure is applied to the spindle. Dynamic errors are caused by the instrument not responding fast enough to follow the changes in measured variable; e.g., a room thermometer does not show the correct temperature until several minutes after the temperature has stabilized to some steady value. Random errors are those due to causes which cannot be directly established because of random variations in the system.

**Standards** for measurement are established by the National Bureau of Standards. Secondary standards are prepared by very precise comparison with these primary standards and, in turn, form the basis for calibrating instruments in use. A well-known example is the use of precision gage blocks for the calibration of measuring instruments and machine tools.

There are three essential parts to an instrument: the **sensing element,** the **transmitting means,** and the **output or indicating element.** The sensing element responds directly to the measured quantity, producing a related motion, pressure, or electrical signal. This is transmitted by linkage, tubing, wiring, etc., to deflect a pointer or move a pen to indicate the value of the measurement on an appropriately calibrated scale or chart, or to store it in a memory device. The instrument may be actuated by mechanical, hydraulic, pneumatic, electrical, optical, or other energy medium. Often a combination of several energy modes is employed to obtain the accuracy, sensitivity, or form of output desired.

## COUNTING EVENTS

**Event counters** are used to measure the number of items passing on a conveyor line, the number of operations of a machine, etc. Coupled with time measurements, they yield measures of average rate or frequency. They find important application, therefore, in inventory control, production analysis, and in the sequencing control of automatic machines.

Choice of the proper counting device depends on the kind of events being counted, the necessary counting speed, and the disposition of the measurement; i.e., whether it is to be indicated remotely, used to actuate a machine, etc. Errors in the total count may be introduced by events being too close together or by too much nonuniformity in the items being counted.

The **mechanical counter** is shown in Fig. 1. Motion of the event being counted deflects the arm, which through an appropriate linkage advances the count register one unit. Alternatively, motion of the actuating arm may close an electrical switch which energizes a relay coil to advance the count register one step.

Where it is desired to avoid contact with the object being counted, the **photoelectric cell** is employed (Fig. 2). A count is registered each time the light beam focused on the photocell is interrupted. This generates an electric signal which is used to actuate a relay coil (as in the above example) or an electronic counter. This system can operate at considerably higher count speeds than the mechanical or electromechanical types.

Sensing methods based on electrical capacitance and magnetic effects are extremely sensitive and fast-acting. The **mag-**

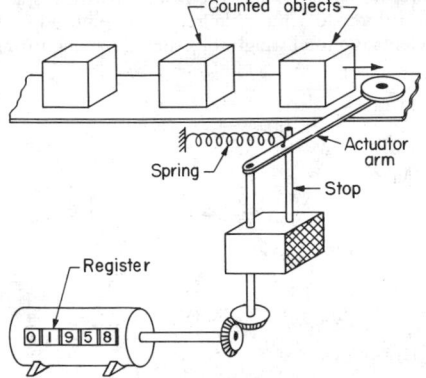

**Fig. 1**  Mechanical counter.

**netic pickup** responds to the motion of iron (or other magnetic material), and hence is particularly adaptable to counting machine operations.

The count is displayed by either a mechanical register as in Fig. 1, a dial-type register (as on the household watthour meter), or an electronic pulse counter with either number indicators or digital printing output. **Electronic counters** can operate accurately at rates exceeding 1 million counts per second.

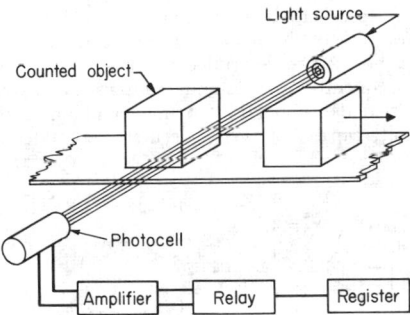

**Fig. 2**  Photoelectric counter.

## TIME AND FREQUENCY MEASUREMENT

Measurement of time is basic to time and motion studies, time program controls, and the measurements of velocity, frequency, and flow rate. (See also Sec. 1.)

**Mechanical devices** include the wide range of clocks, chronometers, and stop watches. In these, a spring tends to rotate the time indicator (clock hands) through a train of gears which is restrained by a hairspring and escapement assembly. The time measurement relates to the period of oscillation of the hairspring–balance-wheel combination. Since this period is affected somewhat by temperature, precise timepieces employ

a compensating element to maintain timing accuracies over long periods. Stop watches may be obtained to read to better than 0.1 s. The major limitation, however, is in the response time of the user.

**Electric timers** are simple, inexpensive, and readily adaptable to remote-control operations. The majority of these are ac synchronous motors geared in the proper ratio to the indicator. These depend for their accuracy on the frequency of the line voltage. Consequently, care must be exercised in using such devices for precise short-time measurements.

**Electronic timers** are started and stopped by electrical pulses and hence are not limited by the observer's reaction time. They may be made extremely accurate and capable of measuring to less than 1 $\mu$s. These measure time by counting the number of cycles in a high-frequency signal generated internally by means of a tuned oscillator, tuning fork, or quartz crystal.

There are a variety of timing devices designed to indicate or control to a fixed time. These include timers based on the charging time of a condenser, the flow of oil through a dashpot, or the release of air through a nozzle.

**Timing devices** can be **calibrated** by comparison with a standard instrument or by reference to the National Bureau of Standards timed radio signals and frequency standards.

The motion of fast-moving systems may be timed and studied by means of a **stroboscope**. This employs an electrically pulsed cold-cathode discharge tube. The frequency of the observed motion is measured by adjusting the stroboscopic frequency until the system appears to stand still. The frequency of the motion is then equal to the stroboscope frequency or an integer multiple of it.

Many other means exist for **measuring vibrational or rotational frequencies.** These include timing a fixed number of rotations or oscillations of the moving member. The sensing can be done by magnetic probes, phonograph-type pickups, and the like. The pulses can be counted by an electronic counter or displayed on an oscilloscope or oscillographic recorder and compared with a known frequency. Also used are reeds which vibrate when the measured oscillation excites their natural frequences, flyball devices which respond directly to angular velocity, and generator-type tachometers which generate a voltage proportional to the speed.

## MASS AND WEIGHT MEASUREMENT

Mass is the measure of the quantity of matter. The fundamental unit is the kilogram. The U.S. customary unit is the pound; 1 lb = 0.4536 kg (see Sec. 1, Measuring Units). Weight is a measure of the force of gravity acting on a mass (see the part of Sec. 4 titled Units of Force and Mass).

A general equation relating weight $W$ and mass $M$ is $W/g = M/g_c$, where $g$ is the local acceleration of gravity, and $g_c = 32.174$ lbm·ft/(lbf)(s²)[(1 kg·m/(N)(s²)] is a property of the unit system. Then $W = Mg/g_c$. The specific weight $w$ and the mass density $\rho$ are related by $w = \rho g/g_c$. Masses are conveniently compared by comparing their weights, and masses are often loosely referred to as weights. Indeed, almost all practical measures of mass are based on weight.

**Weighing devices** fall into two major categories: balances and force-deflection systems. The device may be batch or continuous weighing, automatic or manual. Accuracies are expected to be of the order of 0.1 to better than 0.0001 percent, depending on the type and application of the scale. Calibration is normally performed by use of standard weights (masses) with calibrations traceable to the National Bureau of Standards.

The **equal arm balance** compares the weight of an object with a set of standard weights. The laboratory balance shown in Fig. 3 is used for extreme precision and sensitivity. A chain poise provides fine adjustment of the final balance weight. The magnetic damper causes the balance to come to equilibrium quickly.

**Large weighing scales** operate on the same principle; however, the arms are unequal to allow multiplication between the tare and the measured weights. In this group are platform, track, hopper, and tank scales. Here balance is achieved by adjusting the position of one or more balance weights along a beam directly calibrated in weight units. In dial-indicating-type scales, balance is achieved automatically through the deflec-

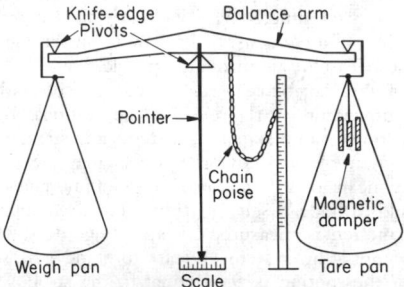

**Fig. 3** Laboratory balance.

tion of calibrated pendulum weights from the vertical. The deflection is greatly magnified by the pointer-actuating mechanism, providing a direct-reading weight indication on the dial.

Since the deflection of a spring (within its design range) is directly proportional to the applied force, a calibrated spring serves as a simple and inexpensive weighing device. Applications include the **spring scale** and **torsion balance**. These are subject to hysteresis and temperature errors and are not used for precise work.

**Other force-sensing elements** are adaptable to weight measurement. Of increasing importance are strain-gage load cells because they eliminate pivot maintenance and moving parts and provide an electrical output which can be used for direct recording and control purposes. Pneumatic pressure cells are also used with similar advantages.

In production processes, **continuous and automatic operating scales** are employed. In one type, the balancing weight is positioned by a reversible electric motor. Deflection of the beam makes an electrical contact which drives the motor in the proper direction to restore balance. The final balance position is translated by means of a potentiometer or digital encoding disk into a signal which is used for recording or control purposes.

The **batch-type scale** (Fig. 4) is adaptable to continuous flow streams of either liquids or solid particles. Material flows from the feed hopper through an adjustable gate into the scale hopper. When the weight in the scale hopper reaches that of the tare, the trip mechanism operates, closing the gate and opening the door. As soon as the scale hopper is empty, the

weight of the tare forces the door closed again, resets the trip, and opens the gate to repeat the cycle. The agitator rotates while the gate is open, to prevent the solids from packing. Also, a "dribble" (partial closing of the gate just before the mechanism trips) is employed to minimize the error from the falling column of material at the instant balance is achieved. Since each dump of the scale represents a fixed weight, a counter yields the total weight of material passing through the scale.

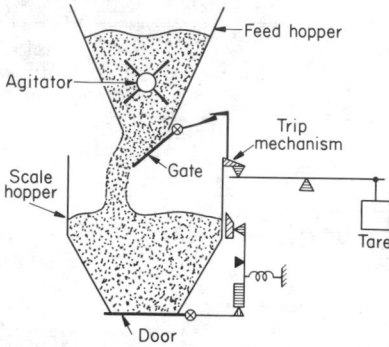

**Fig. 4** Automatic batch-weighing scale.

In **continuous weighers,** a section of conveyor belt is balanced on a weigh beam (Fig. 5). The belt is driven at a constant speed; hence, if the total weight is held constant, the weight rate of material fed through the scale is fixed. Unbalance of the weight beam causes the rate of material flow onto the belt to be changed in the direction of restoring balance. This is accomplished by a mechanical adjustment of the feed gate or by varying the speed of a belt or screw feeder drive.

If the density of the material is constant, **volume measurements** may be used to determine the mass. Thus, calibrated tanks are frequently used for liquids and vane and screw-type feeders for solids. Though often simpler to apply, these are not generally capable of as high accuracies as are common in weighing.

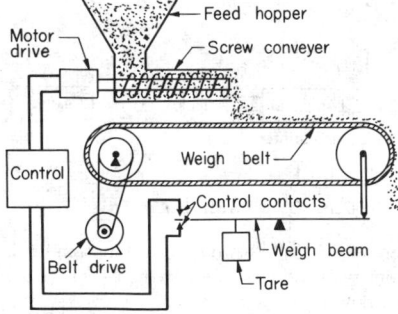

**Fig. 5** Continuous-weigh scale.

## MEASUREMENT OF LINEAR AND ANGULAR DISPLACEMENT

**Displacement-measuring devices** are employed to measure dimension, distances between points, and some derived quantities

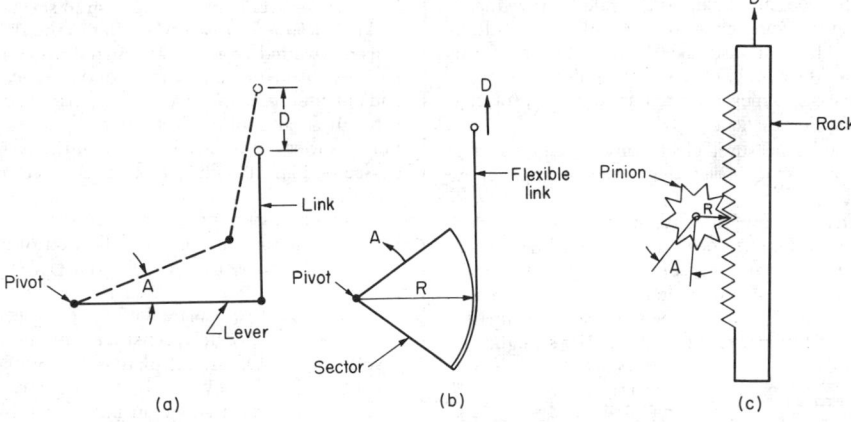

**Fig. 6**  Linear-rotary conversion mechanisms.

such as velocity, area, etc. These devices fall into two major categories: those based on comparison with a known or reference length and those based on some fixed physical relationship.

The **measurement of angles** is closely related to displacement measurements, and indeed, one is often converted into the other in the process of measurement. The common unit is the degree, which represents 1/360 of an entire rotation. The radian is used in mathematics and is related to the degree by $\pi$ rad = 180°; 1 rad = 57.3°. The grad is an angle unit = $\frac{1}{400}$ rotation.

Figure 6 illustrates some methods of **rotary to linear conversion.** Figure 6a is a simple link and lever, Fig. 6b is a flexible link and sector, and Fig. 6c is a rack-and-pinion mechanism. These can be used to convert in either direction according to the relationship $D = RA/57.3$, where $R$ = mean radius of the rotating element, in; $D$ = displacement, in; and $A$ = rotation, deg. (This equation holds for the link and lever of Fig. 6a only if the angle change from the perpendicular is small.)

**Comparative devices** are generally of the manual-indicating type and include ruled or graduated devices such as the machinist's scale, folding rule, tape measure, vernier caliper (Fig. 7), micrometer (Fig. 8), etc. These vary widely in their accuracy, resolution, and measuring span, according to their intended application. Since the readings are made manually, their accuracy is very dependent on the operator's care and skill.

The **vernier** and micrometer are two methods of increasing the sensitivity and precision of reading. The vernier (Fig. 7) is an auxiliary scale which slides along the main scale. It is

uniformly divided so that 10 subdivisions of the vernier correspond to exactly 9 subdivisions of the main scale. This means that each vernier subdivision is shorter than the main-scale subdivision by one-tenth the value of the main-scale subdivision. Accordingly, if the measurement (indicated by the position of the vernier index) falls between two scale subdivisions, the fraction of this interval is indicated by the number of the vernier division which coincides with a main-scale subdivision. Thus, in Fig. 7, the indicated reading is 1.33 in.

The **micrometer** magnifies small displacements by use of an accurately machined screw thread. In Fig. 8, one full rotation of the thimble moves the spindle 0.025 in. Since the thimble is divided into 25 divisions, each corresponds to $\frac{1}{1000}$ in. The displacement measurement is then the sum of the readings of the sleeve and thimble scales. Readings to 0.0001 in are made by estimating the last place or by addition of a vernier to the thimble scale. For a metric micrometer with 0.5-mm pitch thread and 50 divisions on the thimble, each division is 0.01 mm.

**Dial gages** are also used to magnify motion. A rack and pinion (Fig. 6c) converts linear into rotary motion, and a pointer moves over a calibrated scale.

Various modifications of the above-mentioned devices are available for making special kinds of measurements; e.g., **depth gages** for measuring the depth of a hole or cavity, **inside and outside calipers** (Fig. 7) for measuring the internal and external dimensions respectively of an object, **protractors** for angular measurement, etc.

For line production and inspection work, **go no-go gages** provide a rapid and accurate means of dimension measure-

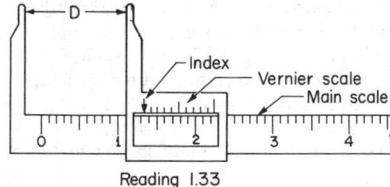

Reading 1.33

**Fig. 7**  Vernier caliper.

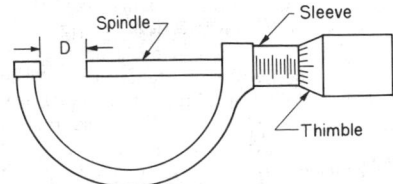

**Fig. 8**  Micrometer caliper.

ment and control. Since the measured values are fixed, the dependence on the operator's skill is considerably reduced. Such gages can be very complex in form to embrace a multidimensional object. They can also take the more general forms of the **feeler, wire, or thread-gage** sets. Of particular importance are **precision gage blocks,** which are used as standards for calibrating other measuring devices.

Displacement can be measured electrically through its effect on the resistance, inductance, or capacitance of an appropriate sensing element.

The **potentiometer** is comparatively inexpensive, accurate, and flexible in application. It consists of a fixed linear resistance over which slides a rotating contact keyed to the input shaft (Fig. 9). The resistance or voltage (assuming constant voltage across terminals 1 and 3) measured across terminals 1 and 2 is directly proportional to the angle $A$. For straight-line

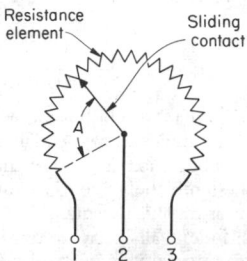

**Fig. 9**  Potentiometer.

motion, a mechanism of the type shown in Fig. 6 converts to rotary motion (or a rectilinear-type potentiometer can be used directly). (See also Sec. 15.) Versions with multiturns, straight-line motion, and special nonlinear resistance vs. motion are available.

The **synchro,** the **linear variable differential transformer (LVDT),** and the **E transformer** are devices in which the input motion changes the inductive coupling between primary and secondary coils. These avoid the limitations of wear, friction, and resolution of the potentiometer, but they require an ac supply and usually an electronic amplifier for the output. (See also Sec. 15.)

The **synchro** is a rotating device which is used to transmit rotary motions to a remote location for indication or control action. It is particularly useful where the rotation is continuous or covers a wide range. They are used in pairs, one transmitter and one receiver. For measurement of difference in angular position, the **control-transmitter** and **control-transformer** synchros generate an electrical error signal useful in control systems. A synchro differential added to the pair serves the same purpose as a gear differential.

The **differential transformer** consists of a primary and two secondary coils wound around a common core (Fig. 10). An armature (iron) is free to move vertically along the axis of the coils. An ac voltage is applied to the primary. A voltage is induced in each secondary coil proportional to the relative length of armature linking it with the primary. The secondaries are connected to oppose each other so that when the armature is centered, the output voltage is zero. When the armature is displaced off center by an amount $D$, the output will be proportional to $D$ (and phased to show whether $D$ is above or below the center). These devices are very linear,

require negligible actuating force, and have spans ranging from 0.1 to several inches (0.25 cm to several centimeters).

The **E transformer** is very similar to the above except that the coils are wound around a laminated iron core in the shape of an E (with the primary and secondaries occupying the center and outside legs respectively). The magnetic path is completed through an armature whose motion, either rotary or translational, varies the induced voltage in the secondaries, as in the device of Fig. 10. This, too, is sensitive to extremely small motions.

A method that is readily applied if a strain-gage analyzer is handy is the measure of the deflection of a cantilever spring with strain gages bonded to its surface (see Strain Gages, Secs. 5 and 16).

The **change of capacitance** with the displacement of the capacitor plates is extremely sensitive but limited to very small displacements. Often, one plate is fixed within the instrument; the other is formed by the object being measured. The capacitance can be measured by an impedance bridge or by determining the resonant frequency of a tuned circuit.

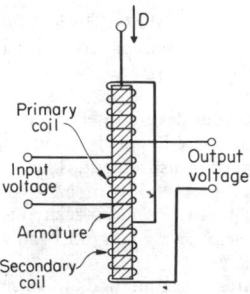

**Fig. 10**  Differential transformer.

Many optical instruments are available for obtaining precise measurements. The **transit and level** are used in surveying for measuring angles and vertical distances (see Sec. 16). A telescope with fine cross hairs permits accurate sighting. The angle scales are generally equipped with verniers. The **measuring microscope** permits measurement of very small displacements and dimensions. The microscope table is equipped with micrometer screws for sensitive adjustment. In addition, templates of scales, angles, etc., are available to permit measurement by comparison. The **optical comparator** projects a magnified shadow image of an object on a screen where it can readily be compared with a reference template.

Light can be used as a standard for the measurement of distance, straightness, and related properties. The wavelength of light in a medium is the velocity of light in vacuum divided by the index of refraction $n$ of the medium. For dry air $n - 1$ is closely proportional to air density and is about 0.000277 at 1 atm and 15°C for 550-nm green light. Since the wavelength changes about $+1$ ppm/°C, and about $-0.36$ ppm/mmHg, density gradients bend light slightly. A temperature gradient of 1°C/m (0.5°F/ft) will cause a deviation from a tangent line of about 0.05 mm (0.002 in) at 10 m (33 ft).

Optical equipment to establish and test alignment, plumb lines, squareness, and flatness includes **jig transits, alignment telescopes, collimators,** optical squares, mirrors, targets, and scales.

Interference principles can be used for distance measurements. An optical flat placed in close contact with a polished surface and illuminated perpendicular to the surface with a monochromatic light will show interference bands which are contours of constant separation distance between the surfaces. Adjacent bands correspond to separation differences of one-half wavelength. For 550-nm wavelength this is 275 nm (10.8 $\mu$in). This test is useful in examining surfaces for flatness and in length comparisons with gage blocks.

**Laser** beams can be used over great distances. Surveying instruments are available for measurements up to 40 mi (60 km). Accuracy is stated to be about 5 mm (0.02 ft) + 1 ppm. These instruments take several measurements which are processed automatically to display the distance directly. Momentary interruptions of the light beam can be tolerated.

A laser system for machine tools, measurement tables, and the like is available in modular form (Hewlett-Packard Co.). It can serve up to eight axes by using beam splitters with a combined range of 200 ft (60 m). Normal resolution of length is about one-fourth wavelength, with a digital display least count of 10 $\mu$in (0.1 $\mu$m). Angle-measurement display resolves 0.1 second of arc. Accuracy with proper environmental compensation is stated to be better than 1 ppm + 1 count in length measurement. Velocities up to 720 in/min (0.3 m/s) can be followed. Accessories are available for measuring straightness, parallelism, squareness and flatness, and for automatic temperature compensation. Various output options include displays and automatic computation and plots. The system can be used directly in measurement and control or to calibrate lead screws and other conventional measuring devices.

**Pneumatic gaging** finds an important place in line inspection and quality control. The device (Fig. 11) consists of a nozzle fixed in position relative to a stop or jig. Air at constant supply

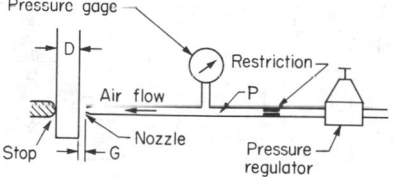

**Fig. 11**  Pneumatic gage.

pressure passes through a restriction and discharges through the nozzle. The nozzle back pressure $P$ depends on the gap $G$ between the measured surface and the nozzle opening. If the measured dimension $D$ increases, then $G$ decreases, restricting the discharge of air, increasing $P$. Conversely, when $D$ decreases, $P$ decreases. Thus, the pressure gage indicates deviation of the dimension from some normal value. With proper design, this pressure is directly proportional to the deviation, limited, however, to a few thousandths of an inch span. The device is extremely sensitive [better than 0.0001 in (0.003 mm)], rugged, and, with periodic calibration against a standard, quite accurate. The gage is adaptable to automatic line operation where the pressure signal is recorded or used to actuate "**reject**" or "**accept**" controls. Further, any number of nozzles can be used in a jig to check a multiplicity of dimensions. In another form of this device, the flow of air is measured with a rotameter in place of the back pressure.

The advent of automatically controlled machine tools has

brought about the need for very accurate displacement measurement over a wide range. Most commonly applied for this purpose is the calibrated **lead screw** which measures linear displacement in terms of its angular rotation. Several **digital** systems have recently been developed which greatly extend the resolution and accuracy limitations of the lead screw. In these, a uniformly spaced optical or inductive grid is displaced relative to a sensing element. The number of grid lines counted is a direct measure of the displacement (see above, Lasers).

Measurement of strip thickness or coating thickness is achieved by **X-ray** or **beta-radiation**-type gages (Fig. 12). A constant radiation source (X-ray tube or radioisotope) provides

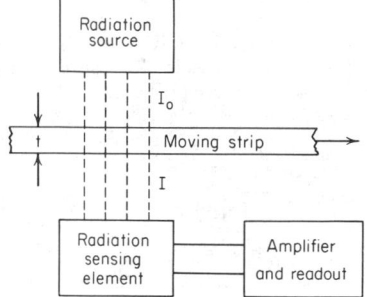

**Fig. 12**  Radiation-type thickness gage.

an incident intensity $I_0$; the radiation intensity $I$ after passing through the absorbing material is measured by an appropriate device (scintillation counter, Geiger-Müller tube, etc.). The thickness $t$ is determined by the equation $I = I_0 e^{-kt}$, where $k$ is a constant dependent on the material and the measuring device. The major advantage here is that measurements are continuous and nondestructive and require no contact. The method is extended to measure liquid level and density.

## MEASUREMENT OF AREA

Area measurements are made for the purpose of determining surface area of an object or area inside a closed curve relating to some desired physical quantity. Dimensions are expressed as a length squared; e.g., in² or m². The areas of simple forms are readily obtained by formula. The area of a complex form can be determined by subdividing into simple forms of known area. In addition, various numerical methods are available (see **Simpson's rule,** Sec. 2) for estimating the area under irregular curves.

Area measuring devices include various **flow integrators** (used with flowmeters) and the **polar planimeter.** The latter consists of two arms pivoted to each other. A tracer at the end of one arm is guided around the boundary curve of the area, causing rotation of a recorder wheel proportional to the area enclosed.

## MEASUREMENT OF FLUID VOLUME

For a liquid of known density, volume is a quick and simple means of measuring the amount (or mass) of liquid present. Conversely, measuring the weight and volume of a given quantity of material permits calculation of its density. Volume has the dimensions of length cubed; e.g., cubic metres, cubic feet. The volume of simple forms can be obtained by formula.

A volumetric device is any container which has a known and fixed calibration of volume contained vs. the level of liquid. The device may be calibrated at only one point (**pipette, volumetric flask**) or may be graduated over its entire volume (**burette, graduated cylinder, volumetric tank**). In the case of the tank, a sight glass may be calibrated directly in liquid volume.

Volumetric measure of continuous flow streams is obtained with the **displacement meter**. This is available in various forms: the nutating disk, reciprocating piston, rotating vane, etc. The **nutating-disk meter** (Fig. 13) is relatively inexpensive and hence is widely used (water meters, etc.). Liquid entering the meter causes the disk to nutate or "roll" as the liquid makes its way around the chamber to the outlet. A pin on the disk causes a counter to rotate, thereby counting the total number of rolls of the disk. Meter accuracy is limited by leakage past the disk and friction. The **piston meter** is like a piston pump operated backward. It is used for more precise measure (available to 0.1 percent accuracy).

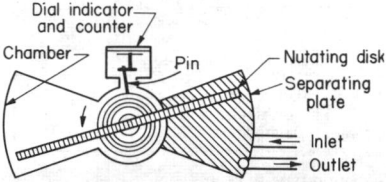

**Fig. 13**  Nutating-disk meter.

Volumetric gas measurement is commonly made with a **bellows meter.** Two bellows are alternately filled and exhausted with the gas. Motion of the bellows actuates a register to indicate the total flow. Various liquid-sealed displacement meters are also available for this purpose.

For precise volume measurements, corrections for temperature must be made (because of expansion of both the material being measured and the volumetric device). In the case of gases, the pressure also must be noted.

## FORCE AND TORQUE MEASUREMENT

**Force** may be measured by the deflection of an elastic element, by balancing against a known force, by the acceleration produced in an object of known mass, or by its effects on the electrical or other properties of a stress-sensitive material. The common unit of force is the pound (newton). **Torque** is the product of a force and the perpendicular distance to the axis of rotation. Thus, torque tends to produce rotational motion and is expressed in units of pound feet (newton metres). Torque can be measured by the angular deflection of an elastic element or, where the moment arm is known, by any of the force measuring methods.

Since weight is the force of gravity acting on a mass, any of the weight-measuring devices already discussed can be used to measure force. Common methods employ the deflection of springs or cantilever beams.

The strain gage is an element whose electrical resistance changes with applied strain (see Sec. 5). Combined with an element of known force-strain, motion-strain, or other input-strain relationship it is a transducer for the corresponding input. The relation of gage-resistance change to input variable can be found by analysis and calibration. Measure of the

resistance change can be translated into a measure of the force applied. The gage may be bonded or unbonded. In the bonded case, the gage is cemented to the surface of an elastic member and measures the strain of the member. Since the gage is very sensitive to temperature, the readings must be compensated. For this purpose, four gages are connected in a Wheatstone-bridge circuit such that the temperature effect cancels itself. A four-element unbonded gage is shown in Fig. 14. Note that as the applied force increases, the tension on two of the elements increases while that on the other two decreases. Gages subject to strain change of the same sign are put in opposite arms of the bridge. The zero adjustment permits balancing the bridge for zero output at any desired input. The $e_1$ and $e_2$ terminal pairs may be used interchangeably for the input excitation and the signal output.

The **piezoelectric** effect is useful in measuring rapidly varying forces because of its high-frequency response and negligible displacement characteristics. Quartz, rochelle salt, and barium titanate are common piezoelectric materials. They have the property of varying an output charge in direct proportion to the stress applied. This produces a voltage inversely proportional to the circuit capacitance. Charge leakage produces drifting at a rate depending on the circuit time constant. The voltage must be measured with a device having a very high input resistance. Accuracy is limited because of temperature dependence and some hysteresis effect.

Forces may also be measured with any of the pressure devices described in the next section by balancing against a fluid pressure acting on a fixed area.

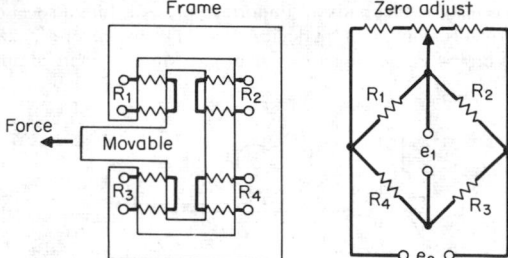

**Fig. 14**  Unbonded strain-gage board.

## PRESSURE AND VACUUM MEASUREMENT

Pressure is defined as the force per unit area exerted by a fluid. Pressure devices normally measure with respect to atmospheric pressure (mean value = 14.7 lb/in²), $p_a = p_g + 14.7$, where $p_a$ = total or **absolute** pressure and $p_g$ = **gage** pressure, both lb/in². Conventionally, gage pressure and vacuum refer to pressures above and below atmospheric, respectively. Common units are lb/in², inHg, ftH$_2$O, kg/cm², bars, and mmHg. The mean SI atmosphere is 1.013 bar.

Pressure devices are based on (1) measure of an equivalent height of liquid column; (2) measure of the force exerted on a fixed area; (3) measure of some change in electrical or physical characteristics of the fluid.

The manometer measures pressure according to the relationship $p = wh = \rho gh/g_c$, where $h$ = height of liquid of density $\rho$ and specific weight $w$ (assumed constants) supported by a pressure $p$. Thus, pressures are often expressed directly in terms of the equivalent height (head) of manometer liquid, e.g., inH₂O or inHg. Usual manometer fluids are water or mercury, although other fluids are available for special ranges.

The **U-tube manometer** (Fig. 15a) expresses the pressure difference $p_1 - p_2$ as the difference in levels $k$. If $p_2$ is exposed to the atmosphere, the manometer reads the gage pressure of $p_1$. If the $p_2$ tube is evacuated and sealed ($p_2 = 0$), the absolute value of $p_1$ is indicated. A common modification is the **well-type manometer** (Fig. 15b). The scale is specially calibrated to take into account changes of level inside the well so that only a

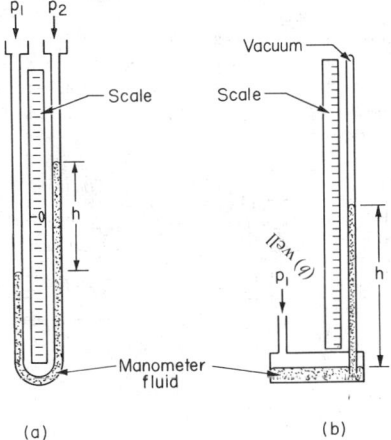

**Fig. 15**  Manometers: (a) U-tube type; (b) well type.

single tube reading is required. In particular, Fig. 15b illustrates the form usually applied to measurement of atmospheric pressure (**mercury barometer**).

The sensitivity of readings can be increased by inclining the manometer tubes to the vertical (**inclined manometer**), by use of low-specific-gravity manometer fluids, or by application of optical-magnification or level-sensing devices. Accuracy is influenced by surface-tension effects (reading of the meniscus) and changes in fluid density (due to temperature changes and impurities).

By definition, pressure times the area acted upon equals the force exerted. The pressure may act on a diaphragm, bellows, or other element of fixed area. The force is then measured with any force-measuring device, e.g., spring deflection, strain gage, or weight balance. Very commonly, the unknown pressure is balanced against an air or hydraulic pressure, which in turn is measured with a gage. By use of unequal-area diaphragms, the pressure can thus be amplified or attenuated as required. Further, it permits isolating the process fluid which may be corrosive, viscous, etc.

The **Bourdon-tube gage** (Fig. 16) is the most commonly used pressure device. It consists of a flattened tube of spring bronze or steel bent into a circle. Pressure inside the tube tends to straighten it. Since one end of the tube is fixed to the pressure inlet, the other end moves proportionally to the pressure difference existing between the inside and outside of the tube. The motion rotates the pointer through a pinion-and-sector mechanism. For amplification of the motion, the tube may be bent through several turns to form spiral or helical elements as are used in pressure recorders.

In the **diaphragm gage,** the pressure acts on a diaphragm in opposition to a spring or other elastic member. The deflection of the diaphragm is therefore proportional to the pressure. Since the force increases with the area of the diaphragm, very small pressures can be measured by the use of large diaphragms. The diaphragm may be metallic (brass, stainless steel) for strength and corrosion resistance, or nonmetallic (leather, neoprene, rubber) for high sensitivity and large deflection. With a stiff diaphragm, the total motion must be very small to maintain linearity.

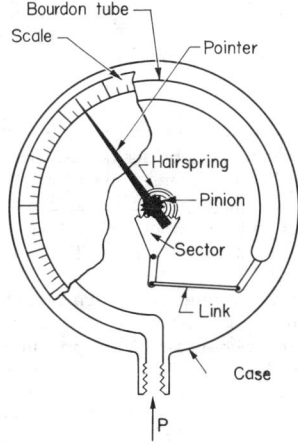

**Fig. 16**  Bourdon-tube gage.

The **bellows gage** (Fig. 17) is somewhat similar to the diaphragm gage, with the advantage, however, of providing a much wider range of motion. The force acting on the bottom of the bellows is balanced by the deflection of the spring. This motion is transmitted to the output arm, which then actuates a pointer or recorder pen.

The motion (or force) of the pressure element can be converted into an electrical signal by use of a differential transformer or strain-gage element or into an air-pressure signal through the action of a nozzle and pilot. The signal is then used for transmission, recording, or control.

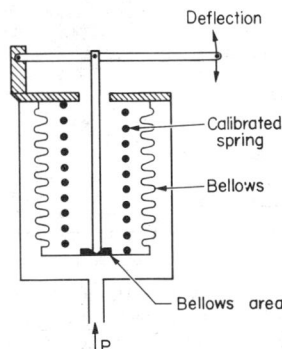

**Fig. 17**  Bellows gage.

The **dead-weight tester** is used as a standard for calibrating gages. Known hydraulic pressures are generated by means of weights loaded on a calibrated piston. The useful range is from 5 to 5,000 lb/in² (0.3 to 350 bar). For low pressures, the water or mercury manometer serves as a reference.

For many applications (fluid flow, liquid level), it is important to measure the **difference between two pressures.** This can be done directly with the manometer. Other pressure devices are available as differential devices where (1) the case is made pressuretight so that the second pressure can be applied external to the pressure element; (2) two identical pressure elements are mounted so that their outputs oppose each other.

Similar devices to those discussed are used to measure **vacuum,** the only difference being a shift in range or at most a relocation of the zeroing spring. When the vacuum is high (absolute pressure near zero) variations in atmospheric pressure become an important source of error. It is here that absolute-pressure devices are employed.

Any of the differential-pressure elements can be converted to an **absolute-pressure device** by sealing one pressure side to a perfect vacuum. A common instrument for the range 0 to 30 inHg employs two bellows of equal area set back to back. One bellows is completely evacuated and sealed; the other is connected to the measured pressure. The output is a bellows displacement, as in Fig. 17.

There are many instruments for high-vacuum work (0.001 to 10,000 $\mu$m range). These kinds of devices are based on the characteristic properties of gases at low pressures. The **McLeod gage** amplifies the pressure to be measured by compressing the gas a known amount and then measuring its pressure with a mercury manometer. The ratio of initial to final pressure is equal to the ratio of final to initial volume (for common gases). This gage serves as a standard for low pressures. It has also been mechanized for automatic operation.

The **Pirani gage** (Fig. 18) is based on the change of heat conductivity of a gas with pressure and the change of electrical resistance of a wire with temperature. The wire is electrically heated with a constant current. Its temperature changes with pressure, producing a voltage across the bridge network. The compensating cell corrects for room-temperature changes.

The **thermocouple gage** is similar to the Pirani gage, except that a thermocouple is used to measure the temperature difference between the resistance elements in the measuring and compensating cells, respectively.

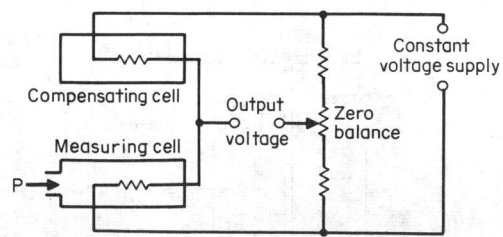

**Fig. 18** Pirani gage.

The **ionization gage** measures the ion current generated by bombardment of the molecules of the gas by the electron stream in a triode-type tube. This gage is limited to pressures below 1 $\mu$m. It is, however, extremely sensitive.

## LIQUID-LEVEL MEASUREMENT

Level instruments are used for determining (or controlling) the height of liquid in a vessel or the location of the interface between two liquids of different specific gravity. In large storage tanks the level is indicated by a **calibrated tape or chain** which is attached to a float riding the liquid surface. For measuring small changes in level, the **fixed displacer** is common (Fig. 19). The buoyant force is proportional to the volume of displacer submerged and hence changes directly with the level. The force is balanced by the air pressure acting in the bellows, which in turn is generated by the flapper and nozzle. A pressure gage (or recorder) indicates the level.

The level is often measured by means of a **differential-pressure meter** connected to taps in the top and bottom of the tank. As indicated in the discussion on manometers, the pressure difference is the height times the specific weight of the liquid. Where the liquid is corrosive or contains solids, then liquid seals, water purge, or air purge may be used to isolate the meter from the process.

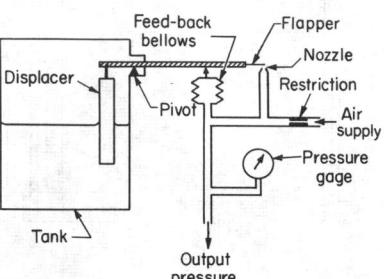

**Fig. 19** Displacer-type level meter.

For special applications, the dielectric, conducting, or absorption properties of the liquid can be used. Thus, in one model the liquid rises between two plates of a condenser, producing a **capacitance change** proportional to the change in level, and in another the **radiation** from a small radioactive source is measured. Since the liquid has a high absorption for the rays (compared with the vapor space), the intensity of the measured radiation decreases with the increase in level. An important advantage of this type is that it requires no external connections to the process.

## TEMPERATURE MEASUREMENT

The common temperature scales (Fahrenheit and Celsius) are based on the freezing and boiling points of water (see Sec. 4 for discussion of temperature standards, units, and conversion equations).

Temperature is measured in a number of different ways. Some of the more useful are as follows.

1. Thermal expansion of a gas (**gas thermometer**). At constant volume, the pressure $p$ of an (ideal) gas is directly proportional to its absolute temperature $T$. Thus, $p = (p_0/T_0)T$, where $p_0$ is the pressure at some known temperature $T_0$.

2. Thermal expansion of a liquid or solid (**mercury thermometer, bimetallic element**). Substances tend to expand with temperature. Thus, a change in temperature $t_2 - t_1$ causes a change in length $l_2 - l_1$ or a change in volume $V_2 - V_1$, according to the expressions

$$l_2 - l_1 = a'(t_2 - t_1)l_1 \qquad \text{or} \qquad V_2 - V_1 = a'''(t_2 - t_1)V_1$$

where $a'$ and $a'''$ = linear and volumetric coefficients of thermal expansion, respectively (see Sec. 4). For many substances, $a'$ and $a'''$ are reasonably constant over a limited temperature range. For solids, $a''' = 3a'$. For mercury at room temperature, $a'''$ is approximately $0.00018°C^{-1}$ ($0.00010°F^{-1}$).

3. Vapor pressure of a liquid (**vapor-bulb thermometer**). The vapor pressure of all liquids increases with temperature. The Clapeyron equation permits calculation of the rate of change of vapor pressure with temperature.

4. Thermoelectric potential (**thermocouple**). When two dissimilar metals are brought into intimate contact, a voltage is developed which depends on the temperature of the junction and the particular metals used. If two such junctions are connected in series with a voltage-measuring device, the measured voltage will be very nearly proportional to the temperature difference of the two junctions.

5. Variation of electrical resistance (**resistance thermometer, thermistor**). Electrical conductors experience a change in resistance with temperature which can be measured with a Wheatstone-bridge circuit. The relationship for platinum is very exact and hence serves as a primary standard over a wide temperature range. The Callendar equation is used in precise work:

$$t = 100 \frac{R_t - R_0}{R_{100} - R_0} + c(t - 100)t$$

Where $t$ = measured temperature, °C; $R_0$, $R_{100}$, $R_t$ = resistance of platinum element at 0, 100, and $t$°C, respectively; $c$ = const $\approx 1.5 \times 10^{-4}°C^{-1}$. **Thermistors** have relatively large and negative temperature coefficients of resistance = $-\beta/T^2$ from the equation $R = R_0 \exp\{\beta[(1/T) - (1/T_0)]\}$, where $R_0$ = resistance at absolute temperature $T_0$, commonly 298 K (=25°C, 77°F), $R$ = resistance at absolute temperature $T$ on the same scale, and $\beta$ = const, typically about 4000 K.

6. Change in radiation (**radiation and optical pyrometers**). A body radiates energy proportional to the fourth power of its absolute temperature. This principle is particularly adaptable to the measurement of very high temperatures where either the total quantity of radiation or its intensity within a narrow wavelength band may be measured. In the former type (radiation pyrometer), the radiation is focused on a heat-sensitive element, e.g., a thermocouple, and its rise in temperature is measured. In the latter type (optical pyrometer) the intensity of the radiation is compared optically with a heated filament. Either the filament brightness is varied by a control calibrated in temperature, or a fixed brightness filament is compared with the source viewed through a calibrated optical wedge.

Important relationships used in the design of these instruments are the Wien and Stefan-Boltzmann laws (in modified form):

$$\lambda_m = k_1/T \qquad q = k_2 \epsilon A(T_2^4 - T_1^4)$$

Where $\lambda_m$ = wavelength of maximum intensity, $\mu$m (nm); $q$ = radiant energy flux, Btu/h (Wa); $A$ = radiation surface, ft² (m²); $\epsilon$ = mean emissivity of the surfaces; $T_2$, $T_1$ = absolute temperatures of radiating and receiving surfaces, respectively, °R (K); $k_1 = 5215 \mu$m·°R (2898 $\mu$m·K); $k_2 = 0.173 \times 10^{-8}$ Btu/(h)(ft²)(°R⁴) [5.73 × 10⁻⁸ W/(m²)(K⁴)]. The emissivity depends on the material and form of the surfaces involved (see Sec. 4). Radiation sensors with scanning capability can produce maps, photographs, and television displays showing temperature-distribution patterns. They can operate with resolutions to under 1°C and at temperatures below room temperature.

7. Change in physical or chemical state (**Seger cones, Tempilsticks**). The temperatures at which substances melt or initiate chemical reaction are often known and reproducible characteristics. Commercial products are available which cover the temperature range from about 120 to 3600°F (50 to 2000°C) in intervals ranging from 3 to 70°F (2 to 40°C). The temperature-sensing element may be used as a solid which softens and changes shape at the critical temperature, or it may be applied as a paint, crayon, or stick-on label which changes color or surface appearance. For most the change is permanent; for some it is reversible. Liquid crystals are available in sheet and liquid form: these change reversibly through a range of colors over a relatively narrow temperature range. They are suitable for showing surface-temperature patterns in the range 20 to 50°C (68 to 122°F).

The most commonly used temperature device is the **mercury-in-glass thermometer**. As the temperature increases, the mercury in the bulb expands and rises through a fine capillary in the graduated thermometer stem. Useful range extends from −30 to 900°F (−35 to 500°C). In many applications of the mercury thermometer, the stem is not exposed to the measured temperature; hence a correction is required (except where the thermometer has been calibrated for **partial immersion**). Recommended formula for the correction $K$ to be added to the thermometer reading is $K_i = 0.00009D(t_1 - t_2)$, where $D$ = number of degrees of exposed mercury filament, °F; $t_1$ = thermometer reading, °F; $t_2$ = the temperature at about middle of the exposed portion of stem, °F. For Celsius thermometers the constant 0.00009 becomes 0.00016.

For **industrial applications** the thermometer or other sensor is encased in a metal protective well and case (Fig. 20). A threaded union fitting is provided so that the thermometer can be installed in a line or vessel under pressure. Ideally the

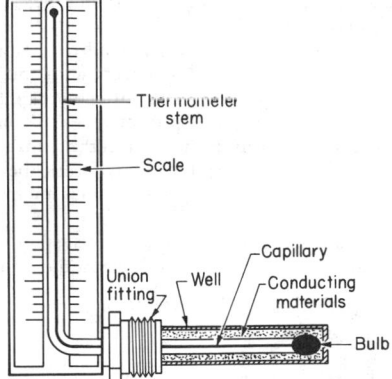

**Fig. 20** Industrial thermometer.

sensor should have the same temperature as the fluid into which the well is inserted. However, heat conduction to or from the pipe or vessel wall and radiation heat transfer may also influence the sensor temperature (see ASME PTC 19.3-1974 Temperature Measurement, on well design). An approximation of the conduction error effect is

$$T_{sensor} - T_{fluid} = (T_{wall} - T_{fluid})E$$

For a sensor inserted to a distance $L - x$ from the tip of a well of insertion length $L$, $E = \cosh [m(L - x)]/\cosh mL$, where $m = (h/kt)^{0.5}$; $x$ and $L$ are in ft (m); $h$ = fluid-to-well conductance, Btu/(h) (ft²)(°F) [J/(h) (m²)(°C)]; $k$ = thermal conductivity of the well-wall material, Btu/(h)(ft)(°F) [J/(h) (m)(°C)]; and $t$ = well-wall thickness, ft (m). Good thermal contact between the sensor and the well wall is assumed. For $(L - x)/L = 0.25$:

| $mL$ | 1 | 2 | 3 | 4 | 5 | 6 | 7 |
|------|-----|-----|------|-------|-------|-------|-------|
| $E$ | 0.67 | 0.30 | 0.13 | 0.057 | 0.025 | 0.012 | 0.005 |

Radiation effects can be reduced by a polished, low-emissivity surface on the well and by radiation shields around the well.

For **control** or **alarm indications** at fixed temperatures, the glass stem may be provided with electrical contacts such that when the mercury in the capillary rises to the contact point, an external relay circuit is energized.

Thermal expansion of a solid is employed in the simple **bimetal** (used in thermostats) and the bimetallic helix (Fig. 21). The bimetallic element is made by welding together two strips

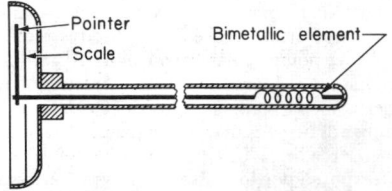

**Fig. 21**  Bimetallic temperature gage.

of metal having different coefficients of expansion. A change in temperature then causes the element to bend or twist an amount proportional to the temperature. A common bimetallic pair consists of invar (iron-nickel alloy) and brass.

A popular industrial-type instrument employs the deflection of a **pressure spring** to indicate (or record) the temperature (Fig. 22). The sensing element is a metal bulb containing some specific gas or liquid. The bulb connects with the pressure spring (in the form of a spiral or helix) through a capillary tube which is usually enclosed in a protective sheath or armor. Increasing temperature causes the fluid in the bulb to expand in volume or increase in pressure. This forces the pressure spring to unwind and move the pen or pointer an appropriate distance upscale.

The **bulb fluid** may be mercury (mercury system), nitrogen under pressure (gas system), or a volatile liquid (vapor-pres-

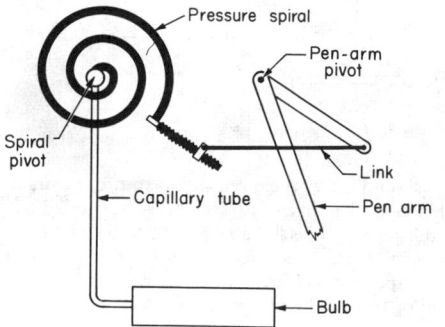

**Fig. 22**  Pressure-spring element.

sure system). Mercury and gas systems have linear scales; however, they must be compensated to avoid ambient temperature errors. The capillary may range up to 200 ft in length with, however, considerable reduction in speed of response.

For transmitting temperature readings over any distance (up to 1,000 ft), the **pneumatic transmitter** (Fig. 23) is better suited than the methods outlined thus far. This instrument has the additional advantages of greater compactness, higher response speeds, and generally better accuracy. The bulb is filled with gas under pressure which acts on the diaphragm. An increase in bulb temperature increases the upward force acting on the main beam, tending to rotate it clockwise. This causes the baffle or flapper to move closer to the nozzle, increasing the nozzle back pressure. This acts on the pilot, producing an increase in output pressure, which increases the force exerted by the feedback bellows. The system returns to equilibrium when the increase in bellows pressure exactly balances the effect of the increased diaphragm pressure. Since the lever ratios are fixed, this results in a direct proportionality between bulb temperature and output air pressure. For precision, compensating elements are built into the instrument to correct for the effects of changes in barometric pressure and ambient temperature.

**Electrical systems** based on the thermocouple or resistance thermometer are particularly applicable where many different temperatures are to be measured, where transmission distances are large, or where high sensitivity and rapid response are required. The thermocouple is used with high temperatures; the resistance thermometer for low temperatures and high accuracy requirements.

The **choice of thermocouple** depends on the temperature

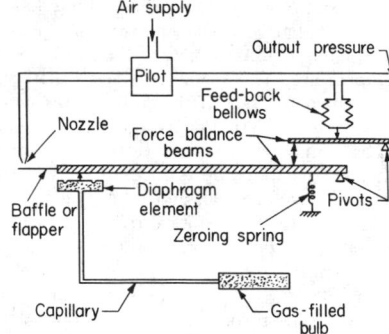

**Fig. 23**  Pneumatic temperature transmitter.

range, desired accuracy, and the nature of the atmosphere to which it is to be exposed. The temperature-voltage relationships for the more common of these are given by the curves of Fig. 24. Table 1 gives the recommended temperature limits for each kind of couple. The thermocouple voltage is measured by a millivoltmeter (deflection type) or potentiometer (null-balance type). Completion of the thermocouple circuit through the instrument immediately introduces one or more additional junctions. Common practice is to connect the thermocouple (hot junction) to the instrument with special lead wire (which may be of the same materials as the thermocouple itself). This assures that the **cold junction** will be inside the instrument case, where compensation can be effectively applied. A **deflection-type instrument** is shown in Fig. 25, where

**Table 1. Limits of Error on Standard Wires without Selection***

| ANSI symbol† | Materials and polarities | | °F: -150 | -75 | 32 | 200 | 530 | 600 | 700 | 1000 | 1400 | 2300 | 2700 |
|---|---|---|---|---|---|---|---|---|---|---|---|---|---|
| | Positive | Negative | °C: -101 | -59 | 0 | 93 | 277 | 316 | 371 | 538 | 760 | 1260 | 1482 |
| T | Cu | Constantan‡ | | 2% | 1.5°F (0.8°C) | | ¾% | | | | | | |
| E | Ni-Cr | Constantan | | | | | 3°F (1.7°C) | | | ¾% | | | |
| J | Fe | Constantan | | | | | 4°F (2.2°C) | | ¾% | | | | |
| K | Ni-Cr | Ni-Al | | | | | 4°F (2.2°C) | | ¾% | | | | |
| R | Pt-13%Rh | Pt | | | | | 5°F (2.8°C) | | | | ½% | | |
| S | Pt-10% Rh | Pt | | | | | 5°F (2.8°C) | | | | ½% | | |

*Closer tolerances are obtainable by selection and calibration. Consult makers' catalogs. Tungsten-rhenium alloys are in use up to 5000°F (2760°C). For cryogenic thermocouples see Sparks et al., Reference Tables for Low-Temperature Thermocouples, *Natl. Bur. Stand. Monogr.* 124.

†Individual wires are designated by the ANSI symbol followed by P or N; thus iron is JP.

‡Constantan is 55% Cu, 45% Ni. The nickel-chromium and nickel-aluminum alloys are available as Chromel and Alumel, trademarks of Hoskins Mfg. Co.

the millivoltmeter is similar in design to that in Fig. 39. A bimetal strip provides a zero shift of the galvanometer hairspring to compensate for changes in cold-junction temperature. Accordingly, the instrument scale is calibrated directly in thermocouple temperature. In the **potentiometer instrument,**

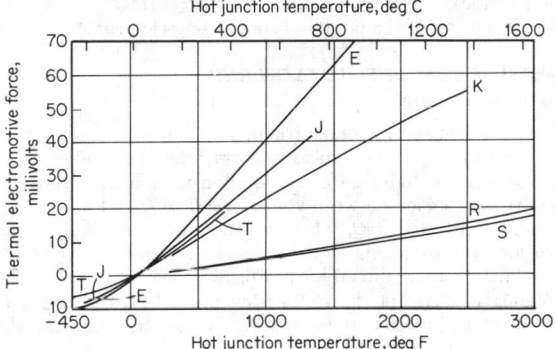

**Fig. 24** Thermocouple voltage-temperature characteristics (reference junction at 32°F, 0°C).

the thermocouple voltage is compared with the voltage of a precisely calibrated potentiometer (slide-wire). Any difference between the two is detected by a sensitive electronic amplifier or galvanometer. The error signal drives a servomotor which repositions the slide-wire contactor until balance is achieved.

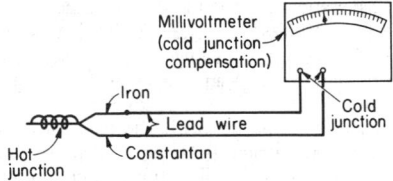

**Fig. 25** Temperature measurement with thermocouple and millivoltmeter.

The slide wire is mechanically linked to the indicator or recording pen. Cold-junction compensation is achieved by means of a temperature-sensitive resistor placed into the potentiometer circuit. The voltage applied across the potentiometer is checked periodically against a standard cell and

maintained constant by means of a self-standardizing circuit. (See also Sec. 15.) All-electronic voltmeters may also be used. A **zener diode** provides a reference voltage. The measured voltage may be displayed and/or recorded in analog or digital form. Automatic conversion from voltage to temperature can be built into the circuit, with choice among the ANSI standard types.

The **resistance thermometer** employs the same instrument described above, omitting the cold-junction compensation and the standardization circuit. The resistance bulb forms one leg of a Wheatstone bridge (Fig. 26), the slide wire forming the opposite leg. An upset of the balance of the circuit causes a current flow through the amplifier, which actuates the motor to drive the slide wire to a new position of balance. The relationship at balance is $R_T = (R_1/R_2)(R_s + R_3 + R_4)$, where $R_T$ = resistance of the designated arm of the bridge. (See also Sec. 15.)

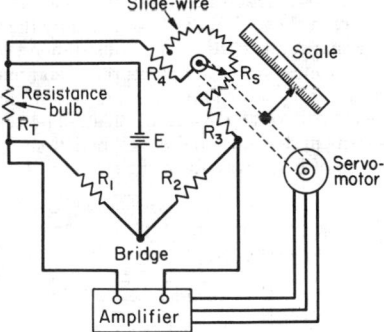

**Fig. 26** Resistance thermometer with self-balance potentiometer recorder.

The resistance bulb consists of a copper or platinum wire coil sealed in a protective metal tube. The **thermistor** has a very large temperature coefficient of resistance and may be substituted in low-accuracy, low-cost applications.

By use of a **selector switch,** any number of temperatures may be measured with the same instrument. The switch connects in order each thermocouple (or resistance bulb) to the potentiometer (or bridge circuit). When balance is achieved, the recorder prints the temperature value, then the switch advances on to the next position.

**Optical pyrometers** are applied to high-temperature measurement in the range 1000 to 5000°F (540 to 2760°C). One type is shown in Fig. 27. The surface whose temperature is to be measured (target) is focused by the lens onto the filament of a calibrated tungsten lamp. The light intensity of the filament is kept constant by maintaining a constant current flow. The intensity of the target image is adjusted by positioning the optical wedge until the image intensity appears exactly equal to that of the filament. A scale attached to the wedge is

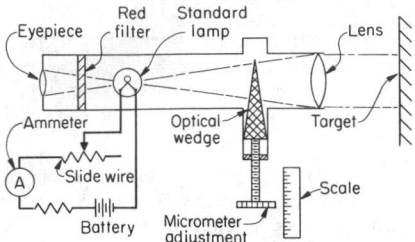

**Fig. 27**  Optical pyrometer.

calibrated directly in temperature. The red filter is employed so that the comparison is made at a specific wavelength (color) of light to make the calibration more reproducible. In another type of optical pyrometer, comparison is made by adjusting the current through the filament of the standard lamp. Here, an ammeter in series is calibrated to read temperature directly. **Automatic operation** may be had by comparing filament with image intensities with a pair of photoelectric cells arranged in a bridge network. A difference in intensity produces a voltage, which is amplified to drive the slide wire or optical wedge in the direction to restore zero difference.

The **radiation pyrometer** is normally applied to temperature measurements above 1000°F. Basically, there is no upper limit; however, the lower limit is determined by the sensitivity and cold-junction compensation of the instrument. It has been used down to almost room temperature. A common type of radiation receiver is shown in Fig. 28. A lens focuses the radiation onto a thermal sensing element. The temperature rise of this element depends on the total radiation received and the conduction of heat away from the element. The radiation

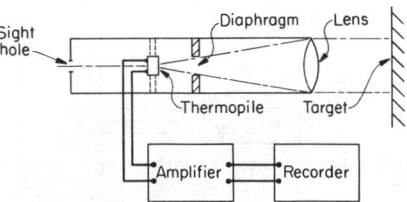

**Fig. 28**  Radiation pyrometer.

relates to the temperature of the target; the conduction depends on the temperature of the pyrometer housing. In normal applications the latter factor is not very great; however, for improved accuracy a compensating coil is added to the circuit. The sensing element may be a thermopile, vacuum thermocouple, or bolometer. The **thermopile** consists of a number of thermocouples connected in series, arranged so that all the hot junctions lie in the field of the incoming radiation; all the cold junctions are in thermal contact with the pyrometer housing so that they remain at ambient temperature. The

**vacuum thermocouple** is a single thermocouple whose hot junction is enclosed in an evacuated glass envelope. The **bolometer** consists of a very thin strip of nickel or platinum foil which responds to temperature in the same manner as the resistance thermometer. The **thermal sensing element** is connected to a potentiometer or bridge network of the same type as described for the self-balance thermocouple and resistance-thermometer instruments. Because of the nature of the radiation law, the scale is nonlinear.

**Accuracy of the optical- and radiation-type pyrometers** depends on:

1. Emissivity of the surface being sighted on. For closed furnace applications, blackbody conditions can be assumed (emissivity = 1). For other applications corrections for the actual emissivity of the surface must be made (correction tables are available for each pyrometer model). For measuring hot fluids, a target tube immersed in the fluid provides a target of known emissivity.

2. Radiation absorption between target and instrument. Smoke, gases, and glass lenses absorb some of the radiation and reduce the incoming signal. Use of an enclosed (or purged) target tube or direct calibration will correct this.

3. Focusing of the target on the sensing element.

## MEASUREMENT OF FLUID FLOW RATE

(See also Secs. 3 and 4)

Flow is expressed in volumetric or mass units per unit time. Thus gases are generally measured in ft³/min (m³/min) or ft³/h (m³/h), steam in lb/h (kg/h), and liquid in gal/min (l/min) or gal/h (l/h). Conversion between volumetric flow $Q$ and mass flow $m$ is given by $m = KpQ$, where $p$ = density of the fluid and $K$ is a constant depending on the units of $m$, $Q$, and $\rho$. Flow rate can be measured directly by attaching a rate device to a volumetric meter of the types previously described, e.g., a tachometer connected to the rotating shaft of the nutating-disk meter (Fig. 13).

Flow is most frequently measured by application of the principle of conservation of mechanical energy through conversion of fluid velocity to pressure (head). Thus, if the fluid is forced to change its velocity from $V_1$ to $V_2$, its pressure will change from $p_1$ to $p_2$ according to the equation (neglecting friction, expansion, and turbulence effects):

$$V_2^2 - V_1^2 = \frac{2g_c}{\rho}(p_1 - p_2) \tag{1}$$

where $g$ = acceleration due to gravity, $\rho$ = fluid density, and $g_c = 32.184$ lbm·ft/(lbf)(s²) [1.0 kg·m/(N)(s²)]. **Caution:** If the flow pulsates, the average value of $p_1 - p_2$ will be greater than that for steady flow of the same average flow (see Sec. 3, Fluids in Motion).

Local velocity is measured with the **pitot tube** (Fig. 29), consisting of two concentric tubes connected across a differential pressure-measuring device. The tubes extend into the flow stream such that the $L$ section is parallel to the direction of flow. The inner tube measures the impact pressure or velocity head. The outer tube has openings on the side (normal to the flow stream) so that it responds to the static pressure. Applying Eq. (1) and noting that, for a particle of fluid striking the impact opening, $V_1 = 0$, we have

$$V_2 = C_p \left[ \frac{2g_c(p_1 - p_2)}{\rho} \right]^{1/2} = C_p \left[ \frac{2g(\rho_m - \rho)h}{\rho} \right]^{1/2}$$

where $b$ = manometer reading, $\rho_m$ = density of manometer fluid, and $C_p$ = pitot-tube coefficient. The coefficient $C_p$ corrects for friction and turbulence effects and is determined by calibration. For the type illustrated, the value of $C_p$ ranges from 0.98 to 1.02. The **total volumetric flow** $Q$ is obtained by

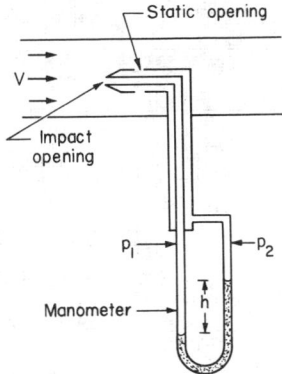

**Fig. 29**   Pitot tube.

integrating the local velocity over the entire area of the flow cross section. Thus, if $V$ is the average velocity and $A$ is the area, then $Q = AV$.

The **velocity distribution** depends on the flow rate and the shape of the flow channel. Hence, the average velocity is best determined by taking readings at selected points over the cross section, Table 2. When the pitot tube is preceded by a long section of straight pipe (over 50 pipe diameters), the average velocity runs normally about 0.83 times the velocity at the center of the pipe.

Average flow rate is more directly measured by inserting a calibrated restriction in the flow line, such as the **venturi tube** (Fig. 30). Since the volumetric flow is the same in both the pipe and the throat, the average velocity of the fluid must

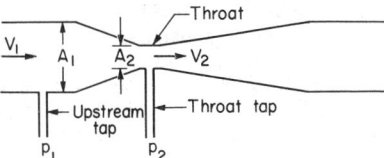

**Fig. 30**   Venturi tube.

increase as it passes through the throat (continuity principle). Thus, $Q = A_1 V_1 = A_2 V_2$, where subscripts 1 and 2 refer to the pipe and throat, respectively. The volumetric flow equa-

tion in terms of the pressure drop $p_1 - p_2$ then becomes,

$$Q = A_1 \beta^2 C_D Y \left[ \frac{2 g_c (p_1 - p_2)}{(1 - \beta^4) \rho} \right]^{1/2} \tag{2}$$

where $A_1 = \pi D_1^2 / 4$ (pipe cross-sectional area), $\beta = d_2/d_1$ (ratio of throat to pipe diameter), $C_D$ = discharge coefficient, $Y$ = compressibility factor, $\rho$ = density of the fluid. $C_D$ depends on the nature of the flow and design of the venturi (normal range 0.98 to 0.99). $Y$ is unity for liquid flow and for gas or steam flow where the pressure drop is not more than 1 to 2 percent of the static pressure.

**Flow nozzles** (Fig. 31) are modifications of the venturi used because of their lower cost or ease of installation. In general, the discharge coefficient approaches that of the venturi; but accurate values must be determined by test or obtained from the manufacturer.

The cheapest and most easily applied (hence, the most common) flow-sensing element is the **orifice** (Fig. 32). This is a thin metal plate with a sharp-edged hole accurately machined. The plate is installed between the flanges of the pipe so that (usually) the hole is concentric with the pipe. The orifice causes the flow lines to converge to much the same shape as the venturi of Fig. 30. The flow stream continues to converge a short distance downstream from the orifice plate, then diverges back to the full pipe diameter. The point of smallest

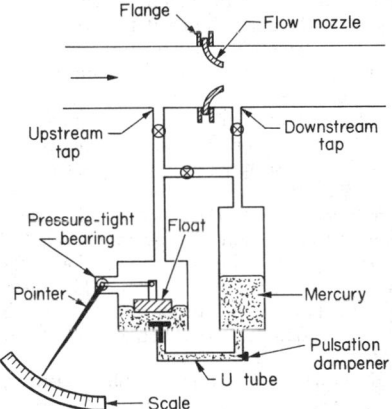

**Fig. 31**   Flow nozzle and mercury meter.

flow cross section is termed the **vena contracta**; this corresponds therefore to the point of lowest pressure.

A differential-pressure device is connected across pressure taps installed upstream and downstream from the orifice. Different **tap locations** are employed: flange taps are located 1 in (25 mm) from each side of the orifice plate and are installed in the flange; vena-contracta taps have the upstream tap 1 pipe

**Table 2. Layout Measurements for Pitot-Tube Traverses**

| Number of equal areas | Total number of readings | Distances from center of pipe to point of reading, percent of pipe diameter | | | | | |
|---|---|---|---|---|---|---|---|
| 3 | 12 | 20.4 | 35.3 | 45.5 | | | |
| 4 | 16 | 17.7 | 30.5 | 39.4 | 46.6 | | |
| 5 | 20 | 15.5 | 27.2 | 35.3 | 41.7 | 47.4 | |
| 6 | 24 | 14.5 | 25.0 | 32.3 | 38.2 | 43.3 | 47.9 |

diameter from the plate and the downstream tap at the vena contracta; pipe taps are located 2.5 and 8.0 pipe diameters upstream and downstream from the orifice, respectively.

Equation (2) applies to flow through orifices if $\beta$ is the ratio of orifice diameter to pipe diameter and $C_D$ is the flow coefficient of the orifice. The value of $C_D$ depends on $\beta$, the kind of taps used, and the Reynolds number. The Reynolds number, Re, is a dimensionless number which characterizes the flow and is given by the formula (in consistent units), $\text{Re} = \rho V d/\mu$, where $d$ = orifice diameter, $V$ = velocity through orifice, $\rho$ = fluid density, and $\mu$ = absolute fluid viscosity. The **expansion factor** for orifices with flange or vena-contracta taps is $Y = 1 - (0.41 + 0.35\beta^4)\,(p_1 - p_2)/p_1 k$, where $p_1$ and $p_2$ are upstream and downstream pressures, respectively, and $k$ = specific heat ratio of fluid [at 212°F(100°C) and 1 atm pressure, $k_{air} = 1.40$, $k_{steam} = 1.32$].

**Orifice coefficient values** vs. Reynolds number, for an orifice with vena-contracta taps, are given in Table 3. Note that $C_D$ ranges from 0.60 to 0.61 for most conditions. Values for flange taps are similar to those given in the table except for large orifice ratios. The values for pipe taps, however, deviate considerably (see ASME, "Fluid Meters").

The tabulated orifice coefficients apply only for **straight pipe** upstream and downstream from the orifice. In most cases, satisfactory results are obtained if there are no fittings closer than 25 pipe diameters upstream and 5 diameters downstream from the orifice. The upstream limitation can be reduced a bit by employing **straightening vanes.** Reciprocating pumps in the line may introduce serious errors and require special efforts for their correction.

A wide variety of differential pressure meters is available for measuring the orifice (or other primary element) pressure drop (see Sec. 16). The most frequently used types are illustrated in Figs. 31 and 32.

The **mercury meter** (Fig. 31) is basically a U-tube manometer (Fig. 15a) with the addition of a metal float which communicates to a pen or pointer assembly the position of the mercury level. The range of the meter is determined by the ratio of the diameters of the two manometer chambers. A needle valve in the U tube acts to dampen flow fluctuations to yield a smooth flow record.

Figure 32 shows the **diaphragm, or "dry," meter.** The orifice differential acts across a metal or rubber diaphragm, generating a force which tends to rotate the lever clockwise, moving the baffle toward the nozzle. This increases the nozzle back pressure, which acts on the pilot diaphragm to open the air supply port and increase the output pressure. This increases the force exerted by the feedback bellows, which generates a force opposing the motion of the main diaphragm. Equilibrium is reached when a change in orifice differential is exactly balanced by a proportionate change in output pressure. Often a damping device in the form of a simple oil dashpot is attached to the lever to reduce output fluctuations.

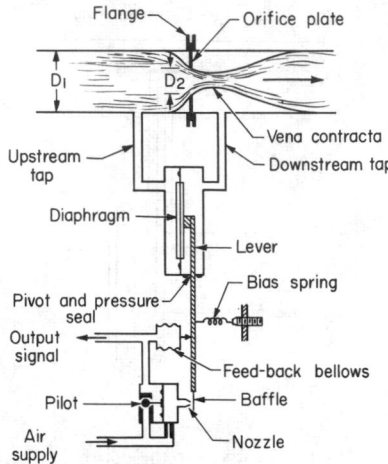

**Fig. 32** Orifice plate and diaphragm-type meter.

The flowmeter normally exhibits a square-root flow calibration [see Eq. (2)]. Some meters are designed to take out the square root by use of **cams, characterized floats or displacers (Ledoux bell)** or devices which describe a square-root behavior. These methods do not improve accuracy or performance but merely provide the convenience of a linear scale.

The meters described thus far are termed **variable-head** because the pressure drop varies with the flow, orifice ratio being fixed. In contrast, the **variable-area** meter maintains a constant pressure differential but varies the orifice area with flow.

The **rotameter** (Fig. 33) consists of a float positioned inside a tapered tube by action of the fluid flowing up through the tube. The flow restriction is now the annular area between the float and the tube (area increases as the float rises). The pressure differential is fixed, determined by the weight of the float and the buoyant forces. To satisfy Eq. (2) then, the annular area (hence the float level) must increase with flow rate. Thus the rotameter may be calibrated for direct flow

**Table 3. Orifice Flow Coefficients vs. Reynolds Number for Vena-Contracta Taps**

| Pipe size | | Orifice ratio $d_2/d_1$ | Reynolds number | | | | | | |
| cm | in | | 15,000 | 25,000 | 50,000 | 100,000 | 250,000 | 500,000 | 1,000,000 |
|---|---|---|---|---|---|---|---|---|---|
| 5 | 2 | 0.1 | 0.615 | 0.611 | 0.606 | 0.603 | | | |
| 10 | 4 | 0.1 | 0.606 | 0.603 | 0.599 | 0.597 | | | |
| 15–25 | 6–10 | 0.1 | 0.602 | 0.600 | 0.596 | 0.594 | | | |
| 5 | 2 | 0.2 | 0.607 | 0.604 | 0.600 | 0.598 | | | |
| 10–25 | 4–10 | 0.2 | 0.603 | 0.600 | 0.598 | 0.596 | | | |
| 5–25 | 2–10 | 0.3 | 0.606 | 0.603 | 0.601 | 0.599 | 0.597 | 0.596 | 0.596 |
| 5–25 | 2–10 | 0.4 | 0.611 | 0.607 | 0.605 | 0.603 | 0.601 | 0.600 | 0.600 |
| 5–25 | 2–10 | 0.5 | .... | .... | 0.610 | 0.607 | 0.605 | 0.604 | 0.603 |
| 5–25 | 2–10 | 0.6 | .... | .... | 0.614 | 0.611 | 0.609 | 0.608 | 0.607 |
| 5–25 | 2–10 | 0.7 | .... | .... | .... | 0.616 | 0.612 | 0.610 | 0.609 |

reading by etching an appropriate scale on the surface of the glass tube. The calibration depends on the float dimensions, tube taper, and fluid properties. The equation for volumetric flow is

$$Q = C_R(A_T - A_F)\left[\frac{2gV_F}{\rho A_F}(\rho_F - \rho)\right]^{1/2}$$

where $A_T$ = cross-sectional area of tube (at float position), $A_F$ = effective float area, $V_F$ = float volume, $\rho_F$ = float density, $\rho$ = fluid density, and $C_R$ = rotameter coefficient (usually between 0.6 and 0.8). The coefficient varies with the fluid viscosity; however, special float designs are available which are relatively insensitive to viscosity effects. Also, fluid density compensation can be obtained.

The **rotameter reading may be transmitted** for recording and control purposes by affixing to the float a stem which connects to an armature or permanent magnet. The armature forms part of an inductance bridge whose signal is amplified electronically to drive a pen-positioning motor. For pneumatic transmission, the magnet provides magnetic coupling to a pneumatic motion transmitter external to the rotameter tube. This generates an air pressure proportional to the height of the float.

The **area meter** is similar to the rotameter in operation. Flow area is varied by motion of a piston in a straight cylinder with openings cut into the wall. The piston position is transmitted as above by an armature and inductance bridge circuit.

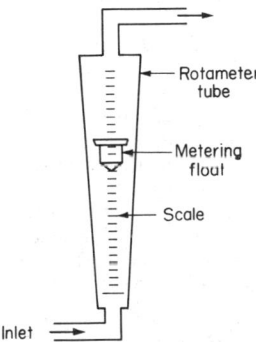

**Fig. 33**  Rotameter.

Primary elements for **flow in open channels** usually employ **weirs** or **open nozzles** to restrict the flow. Weir designs include the rectangular slot; the V notch; and for a linear-flow characteristic, the parabolically shaped weir (Sutro weir). The flow rate is determined from the height of the liquid surface relative to the base of the weir. This height is measured by a liquid-level device, usually float-actuated. A still well (float chamber or open standpipe) connected to the bottom of the weir or the nozzle tap is used to avoid errors in float displacement due to the motion of the flowing fluid or to the buildup of solids. (See also Sec. 3.)

There are many other kinds of flow instruments which serve special purposes of accuracy, response, or application. The **propeller type** (Fig. 34) responds linearly to the average velocity in the path of the propeller, assuming negligible friction. The propeller may be mechanically geared to a tachometer to indicate flow rate and to a counter to show total

quantity flow. The magnetic pickup (Fig. 34) generates a pulse each time a propeller tip passes. The frequency of pulses (measured by means of appropriate electronic circuitry) is then proportional to the local stream velocity. If the propeller occupies only part of the flow stream, an individual calibration is necessary and the velocity distribution must remain constant. The **turbine** type is similar, but is fabricated as a unit in a short length of pipe with vanes to guide the flow approaching the rotor. Its magnetic pickup permits hermetic sealing. A minimum flow is needed to overcome magnetic cogging and start the rotor turning.

The **metering pump** is an accurately calibrated positive-displacement pump which provides both measurement and control of fluid-flow rate. The pump may be either fixed volumetric displacement-variable speed or constant speed-variable displacement.

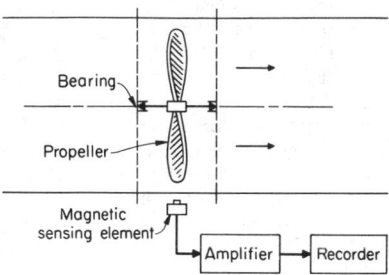

**Fig. 34**  Propeller-type flowmeter.

For air flow, a **vane-type meter (anemometer)** is often used. A mechanical counter counts the number of revolutions of the vane shaft over a timed interval. Instantaneous airflow readings are more readily obtained with the **hot-wire anemometer.** Here, a resistance wire heated by an electric current is placed in the flow stream. The temperature of the wire depends on the current and the rate at which heat is conducted away from it. This latter factor is related to the thermal properties of the air and its velocity past the wire. Airflow can be measured in terms of (1) the current through the wire to maintain a fixed temperature, (2) temperature of the wire for a fixed current, or (3) temperature rise of the air passing the wire for fixed current. The wire temperature is readily measured in terms of its resistance. The anemometer must be specially calibrated for the application. Lasers have also been applied to anemometer use.

The **electromagnetic flowmeter** is gaining importance because it has no moving parts and does not require any insertions in the flow stream. It is based on the voltage induced by the flow of charged particles of the fluid past a strong magnetic field. It is suitable for liquids having resistivities of 50 k$\Omega\cdot$cm or less. The **vortex-shedding meter** has a flow obstruction in the pipe; vortices form behind it at a rate nearly proportional to the volume flow rate. Vortex-formation-rate data give flow rate; a counter gives the integrated flow.

The effects of flow on sound propagation and jets transverse to the flow and flow-attachment instability related to the Coanda effect are also used in flowmeters. **Mass flowmeters** measure changes in momentum related to the mass flow rate.

Flowmeters measure rate of flow. To measure the total quantity of fluid flowing during a specified interval of time, the flow rate must be integrated over that interval. The integration may be done manually by estimating from the chart record the hourly flow averages or by measuring the area under the flow curve with a special square-root planimeter. **Mechanical integrators** use a constant-speed motor to rotate a counter. A cam converts the square-root meter reading into a linear displacement such that the fraction of time that the motor is engaged to the counter is proportional to the flow rate, resulting in a counter reading proportional to the integrated flow. **Electrical integrators** are similar in principle to the watthour meter in that the speed of the integrating motor is made proportional to the magnitude of the flow signal (see Sec. 15).

## POWER MEASUREMENT

Power is defined as the rate of doing work. Common units are the horsepower and the kilowatt: 1 hp = 33,000 ft·lb/min = 0.746 kW. The power input to a rotating machine in hp (W) = $2\pi nT/k$, where $n$ = r/min of the shaft where the torque $T$ is measured in lbf·ft (N·m), and $k$ = 33,000 ft·lbf/hp·min [60 N·m/(W)(min)]. The same equation applies to the power output of an engine or motor, where $n$ and $T$ refer to the output shaft. Mechanical power-measuring devices (**dynamometers**) are of two types: (1) those absorbing the power and dissipating it as heat and (2) those transmitting the measured power. As indicated by the above equation, two measurements are involved: shaft speed and torque. The speed is measured directly by means of a tachometer. Torque is usually measured by balancing against weights applied to a fixed lever arm; however, other force measuring methods are also used. In the **transmission dynamometer,** the torque is measured by means of strain-gage elements bonded to the transmission shaft.

There are several kinds of **absorption dynamometers.** The **Prony brake** applies a friction load to the output shaft by means of wood blocks, flexible band, or other friction surface. The **fan brake** absorbs power by "fan" action of rotating plates on surrounding air. The **water brake** acts as an inefficient centrifugal pump to convert mechanical energy into heat. The pump casing is mounted on antifriction bearings so that the developed turning moment can be measured. In the **magnetic-drag** or **eddy-current brake,** rotation of a metal disk in a magnetic field induces eddy currents in the disk which dissipate as heat. The field assembly is mounted in bearings in order to measure the torque.

One type of **Prony brake** is illustrated in Fig. 35. The torque developed is given by $L(W - W_0)$, where $L$ is the length of the brake arm, ft; $W$ and $W_0$ are the scale loads with the brake

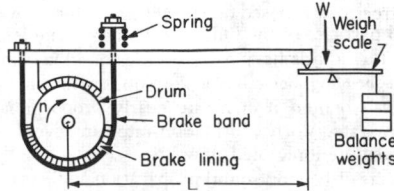

**Fig. 35** Prony brake.

operating and with the brake free, respectively. The brake horsepower then equals $2\pi nL(W - W_0)/33,000$, where $n$ is shaft speed, r/min.

In addition to eddy-current brakes, **electric dynamometers** include calibrated generators and motors and cradle-mounted generators and motors. In calibrated machines, the efficiency is determined over a range of operating conditions and plotted. Mechanical power measurements can then be made by measuring the electrical power input (or output) to the machine. In the electric-cradle dynamometer, the motor or generator stator is mounted in trunnion bearings so that the torque can be measured by suitable scales.

The **engine indicator** is a device for plotting cylinder pressure as a function of piston (or volume) displacement. The resulting $pv$ diagram (Fig. 36) provides both a measure of the work done in a reciprocating engine, pump, or compressor and a means for analyzing its performance (see Secs. 4, 9, and 14). If $A_d$ is

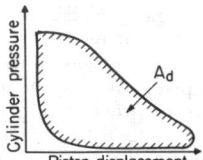

**Fig. 36** Indicator diagram.

the area inside the closed curve drawn by the indicator, then the indicated horsepower for the cylinder under test = $KnA_pA_d$ where $K$ is a proportionality factor determined by the scale factors of the indicator diagram, $n$ = engine speed, r/min, $A_p$ = piston area.

Completely mechanical indicators can be used only for low-speed machines. They have largely been superseded by electrical transducers using strain gages, variable capacitance, and piezoelectric principles which are suitable for **high-speed** as well as low-speed pressure changes (the piezoelectric principle has low-speed limitations). The usual diagram is produced on an oscilloscope display as pressure vs. time, with a marker to indicate some reference event such as spark timing or top dead center. Special transducers can be coupled to a crank or cam shaft to give an electrical signal representing piston motion so that a $p$-$v$ diagram can be shown on an oscilloscope.

## ELECTRICAL MEASUREMENTS
(See also Sec. 15)

Electrical measurements serve two purposes: (1) to measure the electrical quantities themselves, e.g., line voltage, power consumption, and (2) to measure other physical quantities which have been converted into electrical variables, e.g., temperature measurement in terms of thermocouple voltage.

In general, there is a sharp distinction between ac and dc devices used in measurements. Consequently, it is often desirable to transform an ac signal to an equivalent dc value, and vice versa. An ac signal is converted to dc (rectified) by use of **selenium rectifiers, silicon or germanium diodes, or electron-tube diodes.** Full-wave rectification is accomplished by the **diode bridge,** shown in Fig. 37. The rectified signal may be passed through one or more low-pass filter stages to smooth the wave form to its average value. Similarly, there are many ways of modulating a dc signal (converting it to alternating current). The most common method used in instrument applications is the so-called **chopper,** or **vibrator** (Fig. 38). An alternating voltage applied to the chopper coil causes the reed to vibrate at the same frequency (usually 60 or 400 Hz). The reed acts as the common terminal of a single-pole double-throw relay which alternates the direction of current flow through the primary coil of the transformer. This induces an alternating voltage (at line frequency) in the secondary coil of the transformer. The resulting output is proportional to the input signal and either in phase or 180° out of phase with the voltage on the chopper coil, depending on the input polarity. Solid-state dc to ac inverters operate on the same principle but without moving parts.

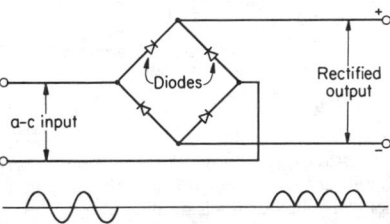

**Fig. 37**   Full-wave rectifier.

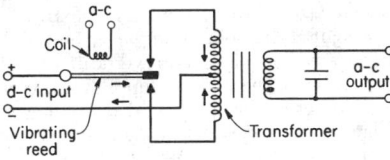

**Fig. 38**   Chopper modulator.

The **galvanometer** (Fig. 39) is the most common sensing device employed in dc measurement. The input signal is applied across a coil mounted in jeweled bearings or on a taut-band suspension so that it is free to rotate between the poles of a permanent magnet. Current in the coil produces a magnetic

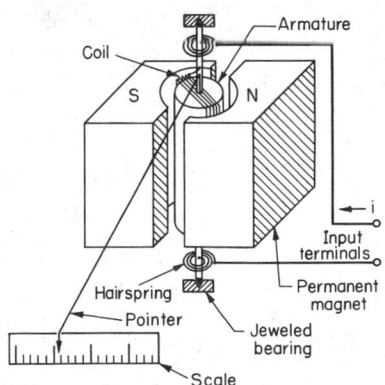

**Fig. 39**   D'Arsonval galvanometer.

moment which tends to rotate the coil. The rotation is limited, however, by the restraining torque of the hairsprings. The resulting deflection of the coil $\theta$ is proportional to the current $I$:

$$\theta = \frac{NBWL}{K}\,I$$

where $N$ = number of turns in coil; $W,L$ = coil width and length, respectively; $B$ = magnetic field intensity; $K$ = spring constant of the hairsprings. Galvanometer deflection is indicated by a balanced pointer attached to the coil. In very sensitive elements, the pointer is replaced by a mirror reflecting a spot of light onto a ground-glass scale; the bearings and hairspring are replaced by a torsion-wire suspension.

The galvanometer can be converted into a **dc voltmeter, ammeter, or ohmmeter** by application of Ohm's law, $IR = E$, where $I$ = current, A; $E$ = electrical potential, V; and $R$ = resistance, $\Omega$.

For a **voltmeter,** a fixed resistance $R$ is placed in series with the galvanometer (Fig. 40$a$). The current $i$ through the galvanometer is proportional to the applied voltage $E$: $i = E/(r + R)$, where $r$ = coil resistance. Different voltage ranges are obtained by changing the series resistance.

An **ammeter** is produced by placing the resistance in parallel with the galvanometer (Fig. 40$b$). The current then divides between the galvanometer coil and the resistor in inverse ratio to their resistance values ($r$ and $R$, respectively); thus, $i = IR/(r + R)$, where $i$ = current through coil and $I$ = total current to be measured. Different current ranges are obtained by using different shunt resistances.

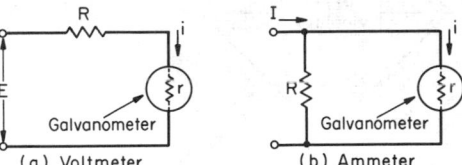

**Fig. 40**   ($a$) Voltmeter; ($b$) ammeter.

The common **ohmmeter** consists of a battery, adjustable rheostat, and ammeter in series with the resistance to be measured. The rheostat is adjusted to give zero reading with the resistance terminals short-circuited. The specially calibrated scale then gives the correct reading of the unknown resistor.

**Alternating current and voltage** must be measured by special means. A dc instrument with a rectifier input is commonly used in applications requiring high input impedance and wide frequency range. For precise measurement at power-line frequencies, the electrodynamic instrument is used. This is similar to the galvanometer except that the permanent magnet is replaced by an electromagnet. The movable coil and field coils are connected in series; hence they respond simultaneously to the same current and voltage alternations. The pointer deflection is proportional to the square of the input signal. The moving-iron-type instruments consist of a soft-iron vane or armature which moves in response to current flowing through a stationary coil. The pointer is attached to the iron to indicate the deflection on a calibrated non-linear scale. For measuring at very high frequencies, the **thermocouple voltmeter or ammeter** is used. This is based on the heating effect of the current passing through a fixed resistance $R$. Heat is liberated at the rate of $E^2/R$ or $I^2R$ W.

**DC electrical power** is the product of the current through the load and the voltage across the load. Thus it can be simply measured using a voltmeter and ammeter. AC power is directly indicated by the wattmeter, which is similar to the electrodynamic instrument described above. Here the field coils are connected in series with the load, and the movable coil is connected across the load (to measure its voltage). The deflection of the movable coil is then proportional to the effective load power.

**Precise voltage measurement** (direct current) can be made by balancing the unknown voltage against a measured fraction of a known reference voltage with a **potentiometer** (Fig. 9). Balance is indicated by means of a sensitive current detector placed in series with the unknown voltage. The potentiometer is calibrated for angular position vs. fractional voltage output. Accuracies to 0.05 percent are attainable, dependent on the linearity of the potentiometer and the accuracy of the reference source. The reference standard may be a Weston standard cell or a regulated voltage supply (based on diode characteristics). The balance detector may be a galvanometer or electronic amplifier.

Precision resistance and general impedance measurements are made with bridge circuits (Fig. 41) which are adjusted

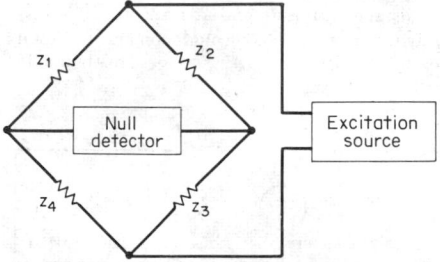

**Fig. 41**    Impedance bridge.

until no signal is detected by the null detector (bridge is balanced). Then $Z_1 Z_3 = Z_2 Z_4$. The basic Wheatstone bridge is used for resistance measurement where all the impedances ($Z$'s) are resistances ($R$'s). If $R_1$ is to be measured, $R_1 = R_2 R_4/R_3$ when balanced. A sensitive galvanometer for the null detector and dc voltage excitation are usual. All $R_2$, $R_3$, $R_4$ must be calibrated, and some adjustable. For general impedance measurement, ac voltage excitation of suitable frequency is used. The null detector may be a sensitive ac meter, oscilloscope, or, for audio frequencies, simple earphones. The basic balance equation is still valid, but it now requires also that the sum of the phase angles of $Z_1$ and $Z_3$ equal the sum of the phase angles of $Z_2$ and $Z_4$. As an example, if $Z_1$ is a capacitor, the bridge can be balanced if $Z_2$ is a known capacitor while $Z_3$ and $Z_4$ are resistances. The phase-angle condition is met, and $Z_1 Z_3 = Z_2 Z_4$ becomes $(1/2\pi f C_1) R_3 = (1/2\pi f C_2) R_4$ and $C_1 = C_2 R_3/R_4$. Variations on the basic principle include the Kelvin bridge for measurement of low resistance, and the Mueller bridge for platinum resistance thermometers.

Voltage measurement requires a meter of substantially higher impedance than the impedance of the source being measured. The vacuum-tube cathode follower and the field-effect transistor are suitable for high-impedance inputs. The following circuitry may be a simple amplifier to drive a pointer-type meter, or may use a digital technique to produce a digital output and display. Digital counting circuits are capable of great precision and are widely adapted to measurements of time, frequency, voltage, and resistance. Transducers are available to convert temperature, pressure, flow, length and other variables into signals suitable for these instru-

ments. The digital counter/timer and voltmeter are very versatile.

The charge amplifier is an example of an operational amplifier application (Fig. 42). It is used for outputs of piezoelectric transducers in which the output is a charge proportional to

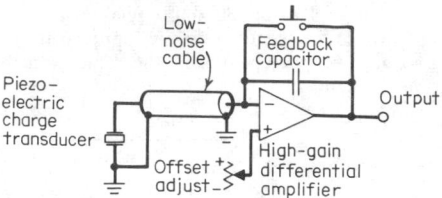

**Fig. 42**    Charge amplifier application.

input force or other input converted to a force. Several capacitors switchable across the feedback path provide a range of full-scale values. The output is a voltage.

The **cathode-ray oscilloscope** (Fig. 43) is an extremely useful and versatile device characterized by high input impedance and wide frequency range. An electron beam is focused on the phosphor-coated face of the cathode-ray tube, producing a visible spot of light at the point of impingement. The beam is deflected by applying voltages to vertical and horizontal deflector plates. Thus, the relationship between two varying voltages can be observed by applying them to the vertical and horizontal plates. The horizontal axis is commonly used for a linear time base generated by an internal sawtooth-wave generator. Virtually any desired sweep speed is obtainable as a calibrated sweep. Sweeps which change value part way across the screen are available to provide localized time magnifica-

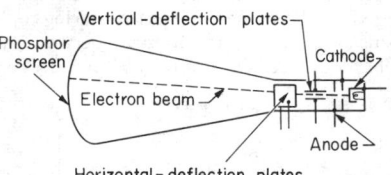

**Fig. 43**    Cathode-ray tube.

tion. As an alternative to the time base, any arbitrary voltage can be applied to drive the horizontal axis. The vertical axis is usually used to display a dependent variable voltage. **Dual-beam** and **dual-trace** instruments show two waveforms simultaneously. Special long-persistence and storage screens can hold transient waveforms for from seconds to hours. **Transient recorders** can store a waveform in digital or analog form in memories and play them back as desired. Other special features offered include digital display of voltage values and time-base calibration. Special accessories, some available as plug-ins, perform diverse functions such as providing complete circuitry for use with a strain-gage transducer and harmonic analysis.

## VELOCITY AND ACCELERATION MEASUREMENT

Velocity or speed is the time rate of change of displacement. Consequently, if the displacement measuring device provides an output signal which is a continuous (and smooth) function of time, the velocity can be measured by **differentiating** this signal either graphically or by use of a differentiating circuit. The accuracy may be very limited by noise (high-frequency fluctuations), however. More commonly, the output of an accelerometer is integrated to yield the velocity of the moving member. **Average speed** over a time interval can be determined by measuring the time required for the moving body to pass two fixed points a known distance apart. Here photoelectric or other rapid sensing devices may be used to trigger the start and stop of the timer. **Rotational speed** may be similarly measured by counting the number of rotations in a fixed time interval.

The **tachometer** provides a direct measure of angular velocity. One form is essentially a small permanent-magnet-type generator coupled to the rotating element; the voltage induced in the armature coil is directly proportional to the speed. The principle is also extended to rectilinear motions (restricted to small displacements) by using a straight coil moving in a fixed magnetic field.

Angular velocity can also be measured by magnetic drag-cup and **centrifugal-force** devices (flyball governor). The force may be balanced against a spring with the resulting deflection calibrated in terms of the shaft speed. Alternatively, the force may be balanced against the air pressure generated by a pneumatic nozzle-baffle assembly (similar to Fig. 23). The

Vibration velocity pickups may use a coil which moves relative to a magnet. The voltage generated in the coil has the same frequency as the vibration and, for sine motion, a magnitude proportional to the product of vibration frequency and amplitude. Vibration **acceleration pickups** commonly use strain-gage or piezoelectric elements to sense a force $F = Ma/g_c$. The maximum usable frequency of an accelerometer is about one-fifth of the pickup's natural frequency (see Sec. 5). The minimum usable frequency depends on the type of pickup and the associated circuitry. The output of an accelerometer can be integrated to obtain a velocity signal; a velocity signal can be integrated to obtain a displacement signal. The **operational amplifier** is a versatile element which can be connected as an integrator for this use.

**Holography** is being applied to the study of vibration patterns over the surface of turbine disks.

## MEASUREMENT OF PHYSICAL AND CHEMICAL PROPERTIES

Physical and chemical measurements are important in the control of product quality and composition. In the case of manufactured items, such properties as color, hardness, surface, roughness, etc., are of interest. Color is measured by means of a **colorimeter**, which provides comparison with color standards, or by means of a **spectrophotometer**, which analyzes the color spectrum. The **Brinell and Rockwell testers** measure surface hardness in terms of the depth of penetration of a hardened steel ball or special stylus. Testing machines with **strain-gage** elements provide measurement of the strength and elastic properties of materials. **Profilometers** are used to measure surface characteristics. In one type, the surface contour is magnified optically and the image projected onto a screen or viewer; in another, a stylus is employed to translate the surface irregularities into an electrical signal which may be recorded in the form of a highly magnified profile of the surface or presented as an averaged roughness-factor reading.

For liquids, attributes such as density, viscosity, melting point, boiling point, transparency, etc., are important. Density measurements have already been discussed. **Viscosity** is measured with a **viscosimeter**, of which there are three main types: flow through an orifice or capillary (Saybolt), viscous drag on a cylinder rotating in the fluid (MacMichael), damping of a vibrating reed (Ultrasonic) (see Secs. 3 and 4). **Plasticity** and consistency are related properties which are determined with special apparatus for heating or cooling the material and observing the temperature-time curve. The **photometer, reflectometer, and turbidimeter** are devices for measuring transparency or turbidity of nonopaque liquids and solids.

A variety of properties can be measured for determining chemical composition. **Electrical properties** include pH, conductivity, dielectric constant, oxidation potential, etc. **Physical properties** include density, refractive index, thermal conductivity, vapor pressure, melting and boiling points, etc. Of increasing industrial application are **spectroscopic measurements: infrared absorption spectra, ultraviolet and visible emission spectra, mass spectrometry,** and **gas chromatography.** These are specific to particular types of compounds and molecular configurations and hence are very powerful in the analysis of complex mixtures. As examples, infrared analyzers are in use to measure low-concentration contaminants in engine oils resulting from wear and in hydraulic oils to detect deterioration. **X-ray diffraction** has many applications in the analysis of crystalline solids, metals, and solid solutions.

Of special importance in the realm of composition measurements is the determination of **moisture content.** A common laboratory procedure measures the loss of weight of the oven-dried sample. More rapid methods employ electrical conductance or capacitance measurements, based on the relatively high conductivity and dielectric constant values for ordinary water.

Water vapor in air (**humidity**) is measured in terms of its physical properties or effects on materials (see also Secs. 4 and 12). (1) The **psychrometer** is based on the cooling effect of water evaporating into the airstream. It consists of two thermal elements exposed to a steady airflow; one is dry, the other is kept moist. (2) The **dewpoint recorder** measures the temperature at which water just starts to condense out of the air. (3) The **hygrometer** measures the change in length of such humidity-sensitive elements as hair and wood. (4) **Electric sensing elements** employ a wire-wound coil impregnated with a hygroscopic salt (one that maintains an equilibrium between its moisture content and the air humidity) such that the resistance of the coil is related to the humidity.

The **throttling calorimeter** (Fig. 44) is most commonly used for determining the moisture in steam. A sampling nozzle is located preferably in a vertical section of steampipe far removed from any fittings. Steam enters the calorimeter through a throttling orifice and into a well-insulated expansion chamber. The steam quality $x$ (fraction dry steam) is determined from the equation $x = (h_c - h_f)/h_{fg}$, where $h_c$ is the enthalpy of superheated steam at the temperature and pressure measured in the calorimeter, $h_f$ and $h_{fg}$ are, respectively, the liquid enthalpy and the heat of vaporization corresponding to line pressure. The chamber is conveniently exhausted to atmospheric pressure; then only line pressure and temperature of the throttled steam need be measured. The range of the throttling calorimeter is limited to small percentages of moisture; a **separating calorimeter** may be employed for larger moisture contents.

The **Orsat** apparatus is generally used for chemical analysis of flue gases. It consists of a graduated tube or burette designed to receive and measure volumes of gas (at constant temperature). The gas is analyzed for $CO_2$, $O_2$, $CO$, and $N_2$ by bubbling through appropriate absorbing reagents and measuring the resulting change in volume. The reagents normally employed are KOH solution for $CO_2$, pyrogallic acid and KOH mixture for $O_2$, and cuprous chloride ($Cu_2Cl_2$) for $CO$. The final remaining unabsorbed gas is assumed to be $N_2$. The most common errors in the Orsat analysis are due to **leakage**

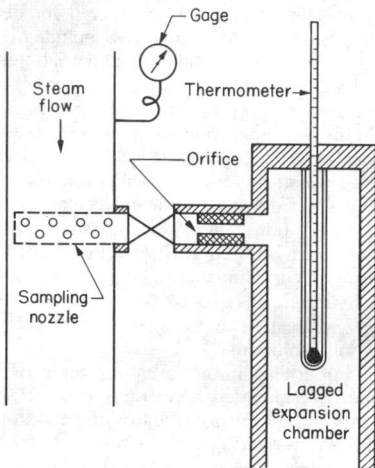

**Fig. 44**   Throttling calorimeter.

and poor **sampling.** The former can be checked by simple test; the latter factor can only be minimized by careful sampling procedure. Recommended procedure is the taking of several simultaneous samples from different points in the cross-sectional area of the flue-gas stream, analyzing these separately, and averaging the results.

There are many instruments for **measuring $CO_2$ (and other gases) automatically.** In one type, the $CO_2$ is absorbed in KOH, and the change in volume determined automatically. The more common type, however, is based on the difference in **thermal conductivity** of $CO_2$ compared with air. Two thermal conductivity cells are set into opposing arms of a Wheatstone-bridge circuit. Air is sealed into one cell (reference), and the $CO_2$-containing gas is passed through the other. The cell contains an electrically heated resistance element; the temperature of the element (and therefore its resistance) depends on the thermal conductivity of the gaseous atmosphere. As a result, the unbalance of the bridge provides a measure of the $CO_2$ content of the gas sample.

The same principle can be employed for analyzing other constituents of gas mixtures where there is a significant thermal-conductivity difference. A modification of this principle is also used for determining CO or other combustible gases by mixing the gas sample with air or oxygen. The combustible gas then burns on the heated wire of the test cell, producing a temperature rise which is measured as above.

Many other physical properties are employed in the determination of specific components of gaseous mixtures. An interesting example is the **oxygen analyzer,** based on the unique paramagnetic properties of oxygen.

## NUCLEAR RADIATION INSTRUMENTS

(See also Sec. 9.)

Nuclear radiation instrumentation is increasing in importance with two main areas of application: (1) measurement and control of radiation variables in nuclear reactor-based processes, such as nuclear power plants and (2) measurement of other physical variables based on radioactive excitation and tracer techniques. The instruments respond in general to electromagnetic radiation in the gamma and perhaps X-ray regions and to beta particles (electrons), neutrons, and alpha particles (helium nuclei).

**Gas Ionization Tubes**   The **ion chamber, proportional counter,** and **Geiger counter** are common instruments for radiation detection and measurement. These are different applications of the gas-ionization tube distinguished primarily by the amount of applied voltage.

A simple and very common form of the instrument consists of a gas-filled cylinder with a fine wire along the axis forming the anode and the cylinder wall itself (at ground potential) forming the cathode, as shown in Fig. 45. When a radiation particle enters the tube, its collision with gas molecules causes an ionization consisting of electrons (negatively charged) and positive ions. The electrons move very rapidly toward the positively charged wire; the heavier positive ions move relatively slowly toward the cathode. The above activity is detected by the resulting current flow in the external circuitry.

When the voltage applied across the tube is relatively low, the number of electrons collected at the anode is essentially equal to that produced by the incident radiation. In this voltage range, the device is called an **ion chamber.** The device

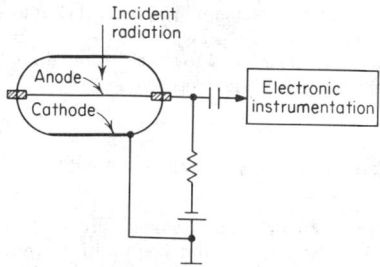

**Fig. 45**   Gas-ionization tube.

may be used to count the number of radiation particles when the frequency is low; when the frequency is high, an external integrating circuit yields an output current proportional to the radiation intensity. Since the amplification factor of the ion chamber is low, high-gain electronic amplification of the current signal is necessary.

If the applied voltage is increased, a point is reached where the radiation-produced ions have enough energy to collide with other gas molecules and produce more ions which also enter into collisions so that an "avalanche" of electrons is collected at the anode. Thus, there is a very considerable amplification of the output signal. In this region, the device is called a **proportional counter** and is characterized by the voltage or current pulse being proportional to the energy content of the incident radiation signal.

With still further increase in the applied voltage, a point of saturation is reached wherein the output pulses have a constant amplitude independent of the incident radiation level. The resulting **Geiger counter** is capable of producing output pulses up to 10 V in amplitude, thus greatly reducing the requirements on the external circuitry and instrumentation. This advantage is offset somewhat by a lower maximum counting rate and more limited ability to differentiate among the various types of radiation as compared with the proportional counter.

The **scintillation counter** is based on the excitation of a phosphor by incident radiation to produce light radiation which is in turn detected by a photomultiplier tube to yield an output voltage. The signal output is greatly amplified and nearly proportional to the energy of the initial radiation. The device may be applied to a wide range of radiations, it has a very fast response, and, by choice of phosphor material, it offers a large degree of flexibility in applications.

**Applications to the Measurement of Physical Variables** The ready availability of radioactive isotopes of long half-life, such as cobalt 60, make possible a variety of industrial and laboratory measuring techniques based on radiation instruments of the type described above. Most applications are based on (1) radiation absorption, (2) tracer identification, and (3) other properties. These techniques often have the advantages of isolation of the measuring device from the system, access to a variable not observable by conventional means, or measurement without destruction or modification of the system.

In the utilization of **absorption** properties, a radioactive source is separated from the radiation-measuring device by that part of the system to be measured. The measured radiation intensity will depend on the fraction of radiation absorbed, which in turn will depend on the distance traveled through the absorbing medium and the density and nature of the material. Thus, the instrument can be adapted to measuring thickness (see Fig. 12), coating weight, density, liquid or solids level, or concentration (of certain components).

**Tracer** techniques are effectively used in measuring flow rates or velocities, residence time distributions, and flow patterns. In flow measurement, a sharp pulse of radioactive material may be injected into the flow stream; with two detectors placed downstream from the injection point and a known distance apart, the velocity of the pulse is readily measured. Alternatively, if a known constant flow rate of tracer is injected into the flow stream, a measure of the radiation downstream is easily converted into a measure of the desired flow rate. Other applications of tracer techniques involve the use of tagged molecules embedded in the process to provide measures of wear, chemical reactions, etc.

Other applications of radiation phenomena include level measurements based on a floating radioactive source, level measurements based on the back-scattering effect of the medium, pressure measurements in the high-vacuum region based on the amount of ionization caused by alpha rays, location of interface in pipeline transmission applications, and certain chemical analysis applications.

## INDICATING, RECORDING, AND LOGGING

An important element of measurement is the display of the measured value in a form which the human operator can readily interpret. Two basic types of display are employed: analog and digital. **Analog** refers to a reading obtained from the motion of a pointer on a scale or the record of a pen moving over a chart. Digital refers to the reading displayed as a number, a series of holes on a punched card, or a sequence of pulses on magnetic tape. Further classification relates to indicating and recording functions. The **indicator** consists merely of a pointer moving over a calibrated scale. The scale may be concentric, as in the Bourdon gage (Fig. 16) or eccentric, as in the flowmeter (Fig. 31). There are also digital indicators which directly display or illuminate the specific digits corresponding to the reading. Obviously, use of the indicator is limited to cases where the variable of interest is constant during the measuring period, or at most, changes slowly.

The **recorder** is used where long-term trends or detailed variations with time are of interest, or where the response is too rapid for the human eye to follow. In the common **circular-chart recorder** (Fig. 46), the pointer is replaced by a pen which writes on a chart rotated by a constant-speed electrical or spring-wound clock. Various chart speeds are available from 1 r/min to 1 every 7 days. Up to four recording pens on a single chart are available (with a print-wheel mechanism, six color-identified records may be had). The **strip-chart recorder** shown in Fig. 47 is of the type used in electronic potentiometers, where the pen is positioned by a servomotor. A constant-speed motor drives the chart vertically past the pen, which deflects horizontally. **Multipoint recording** is achieved by replacing the pen with a print-wheel assembly. A selector switch switches the input signal from one variable to another at the same time that the print wheel switches from one number (or symbol) to another. The record of each variable appears then as a sequence of dots with an identifying numeral. Up to 16 different records may be recorded on a chart (with external switching, as many as 144 records have been applied). Miniature recorders with 3- and 4-in strip charts are gaining favor in process industries because of their compactness and readabil-

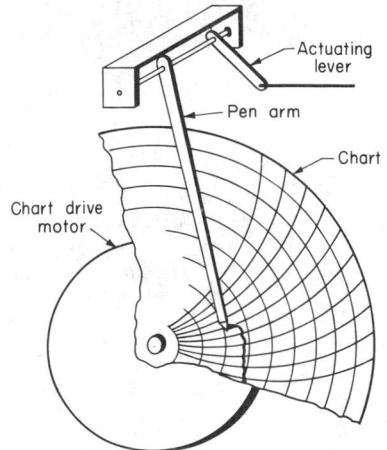

**Fig. 46** Circular-chart recorder.

ity. The pen may be pneumatically or electrically actuated. Maximum number of records per chart is two.

For **direct-writing recording** of high-speed phenomena up to about 100 Hz, a pen or stylus can be driven by a galvanometer. The chart is in strip form and is driven at a speed suitable

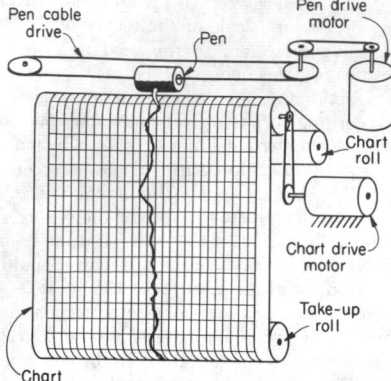

**Fig. 47**   Strip-chart recorder.

for the resolution needed. Recording may be done with ink and standard chart paper or heated stylus and special heat-sensitive paper. Mirror galvanometers projecting a spot of light onto a moving chart of light-sensitive paper can be used up to several kilohertz. A number of galvanometers can be used side by side to record several signals simultaneously on the same chart.

For higher frequencies a form of **magnetic recording** is common. Analog signals can be recorded by amplitude and frequency modulation. The latter is particularly convenient for playback at reduced speed.

**Digital signals** can be recorded in magnetic form. They can be recorded to any desired precision by using more bits to represent the data. Resolution is 1 part in $2^n$, where $n$ is the number

of bits used in straight binary form, less in binary-coded decimal, where 4 bits are used to encode each decimal digit. Digital recording and data transmission have the advantage that in principle error rates can be made as small as desired in the presence of noise by adding more bits which serve as checks in error correcting codes (Raisbeck, "Information Theory: An Introduction for Scientists and Engineers," M.I.T. Press).

Most physical variables are in analog form. Popular standards for the transmission of analog signals include 3- to 15-psig pneumatic signals, direct currents of 4 to 20 or 10 to 50 mA, 0 to 5 and 0 to 10 V. Suitable **signal conditioners** are needed to convert thermocouple outputs and the like to these levels. (Of course, if the instrument is specifically for the particular thermocouple, this conversion is not needed.) This standardization gives greater flexibility in interconnecting signal sources with indicators and recorders. Some transducers and signal conditioners are designed to receive their power over the same two wires used to transmit their output signals.

Often it is necessary to convert from analog to digital form (as for the input to a digital computer) and vice versa. The **analog-digital (A/D)** and **digital-analog (D/A)** converters provide these interfaces. They are available in various conversion speeds and resolutions. Resolution is specified in terms of the number of bits in the digital signal. The disk shown in Fig. 48 is angle-to-digital **encoder** of 7-bit resolution, or 1 part in $2^7 = 1$ part in 128. The form shown is straight binary and has the drawback that some angular positions will straddle one or more dividing lines and leave some of the bit values ambiguous, possibly even an uncertainty of 1 part in 2. The Gray code is a special form in which only one bit changes at a time, and the uncertainty is never more than 1 least significant bit. Digital circuitry can convert from one code to another, as needed, without ambiguity.

Data which have been stored in magnetic form can be recovered at any time by connecting the storage device to an electrically actuated typewriter, printer, or other read-out device. Modern **logging systems** have the measurements from

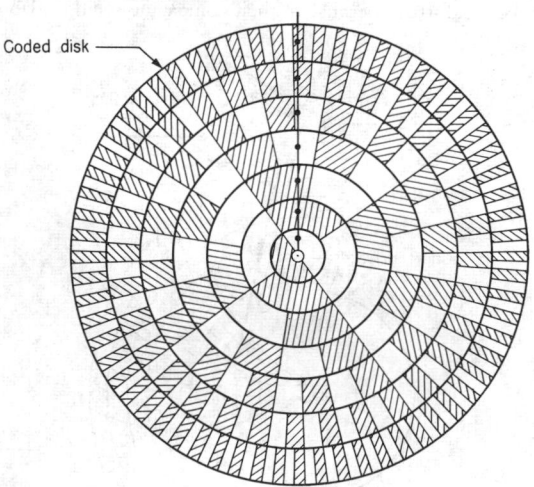

**Fig. 48**   Analog-to-digital converter.

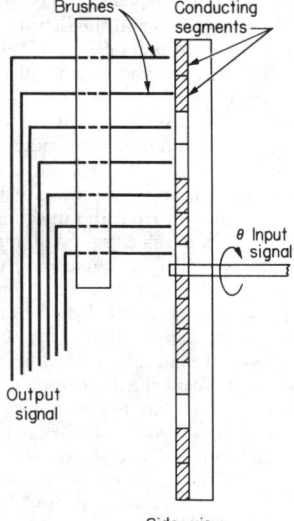

hundreds of different points in the process tabulated periodically. These systems may provide such additional features as the printing of deviations from the normal in red and the more frequent scanning of abnormal conditions. Computer elements are also used in conjunction with logging systems to compute derived variables (such as operating efficiency, system losses, etc.) and to apply corrections to measured variables, e.g., temperature and pressure compensation of gas-flow readings.

In quality control and time-motion studies, often a simple **on-off-type recorder** is sufficient for the purpose. Here, a pen is deflected when the machine or system is on and not deflected whenever the system is off. Pen actuation is usually by solenoid or other electromagnetic element.

# AUTOMATIC CONTROLS

### Guy Longobardo and Anna Kazanjian Longobardo

REFERENCES: Considine, "Process Instruments and Controls Handbook," McGraw-Hill. Thaler, "Elements of Servomechanism Theory," McGraw-Hill. Ahrendt, "Servomechanism Practice," McGraw-Hill. Porter, "An Introduction to Servomechanisms," Wiley. Chestnut and Mayer, "Servomechanisms and Regulating System Design," Wiley. Brown and Campbell, "Principles of Servomechanisms," Wiley. Truxal, "Automatic Feedback Control System Synthesis," McGraw-Hill. Gardner and Barnes, "Transients in Linear Systems," Wiley. Campbell, "Process Dynamics," Wiley. Hadley and Longobardo, "Automatic Process Control," Addison-Wesley. Murphy, "Control Engineering," Van Nostrand. Blackburn, Reethof, Shearer, "Fluid Power Control," Wiley. Truxal, "Control Engineers' Handbook," McGraw-Hill. Langill, "Automatic Control Systems Engineering," Prentice-Hall. Chang, "Synthesis of Optimum Control Systems," McGraw-Hill. Tsien, "Engineering Cybernetics," McGraw-Hill. Auslander, Takahashi, and Rabins, "Introducing Systems and Control," McGraw-Hill. Shinners, "Modern Control System Theory and Application," Addison-Wesley.

## INTRODUCTION

The purpose of an **automatic control** on a system is to produce a desired output when inputs to the system are changed. Inputs are in the form of commands, which the output is expected to follow, and disturbances, which the automatic control is expected to minimize. The usual form of an automatic control is a **closed-loop feedback control** which Ahrendt defines as "an operation which, in the presence of a disturbing influence, tends to reduce the difference between the actual state of a system and an arbitrarily varied desired state and which does so on the basis of this difference." The general theories and definitions of automatic control have been developed to aid the designer to meet primarily three basic specifications for the performance of the control system, namely, stability, accuracy, and speed of response.

The **nomenclature** of automatic control has been reviewed by both the ASME and the AIEE in recent years. Many terms adopted by the ASME naturally tend somewhat toward the vocabulary of the process-control engineer and in many cases are not sufficiently broad for general control application. The following terms and definitions have been selected to assist the reader and to serve as reference to a complex area of technology whose breadth crosses several professional disciplines.

**Automatic regulator,** an apparatus which measures the value of a quantity or condition which is subject to change with time, and operates to maintain within limits this measured value

**Controlled variable,** (types) that quantity or condition of the controlled system which is directly measured or controlled

**Throttling range,** that range of values through which the variable must change to cause the final control element to move from one extreme position to the other

**Set point,** the value of the controlled variable that it is desired to maintain

**Deviation,** the difference at any time between the controlled variable and the set point

**Corrective action,** a change in the flow of the control agent initiated by the measuring means of the automatic controller

**Control agent,** process energy whose flow is directly varied by the control element

**Self regulation,** that operating characteristic which inherently assists the establishment of equilibrium

**Two-position controller action,** that in which the final control element is moved immediately, from one extreme to the other of its stroke, at predetermined values of the variable

**Proportional-position controller action,** that in which there is continuous linear relation between the position of the final control element and the value of the controlled variable

**Floating controller action,** that in which there is a predetermined relation between the values of the controlled variable and the rate of motion of a final control element

**Proportional-speed floating controller action,** that in which there is a continuous linear relation between the rate of motion of the final control element and the deviation of the controlled variable

**Derivative controller action,** that in which there is a predetermined relation between a derivative function of the controlled variable and the position of a final control element

**Proportional plus floating controller action (proportional + reset),** that in which proportional-position and proportional-speed floating actions are additively combined

**Proportional band,** the range of scale values through which the controlled variable must pass in order that the final control element be moved through its entire range

**Floating speed,** the rate of movement of a final control element corresponding to a specified deviation

**Command signal,** the input which is established or varied by some means external to, and independent of, the feedback control system under consideration

**Disturbance,** a signal which tends to affect the value of the controlled variable

**Response time,** the time required for the controlled variable to reach a specified value after the application of a step input or disturbance

**Peak time,** the time required for the controlled variable to reach its first maximum following the application of a step input

**Rise time,** time required for the controlled variable to increase from one specified percentage of the final value to another, following the application of a step input

**Settling time,** the time required for the absolute value of the difference between the controlled variable and its final value to become and remain less than a specified amount following the application of a step disturbance

**Compensation,** a method of changing or maintaining the state of a system by employing means to offset the effects of disturbances without causal relationship between the error in the state of the system and the action of the compensating means

These definitions have been largely abstracted from Smith, "Automatic Control Engineering"; Ahrendt, "Servomechanism Practice"; and Ahrendt and Taplin, "Automatic Feedback Control."

## BASIC AUTOMATIC-CONTROL SYSTEM

The general components of a basic automatic-control system are shown in Fig. 1. Each block in the diagram represents a function which must be performed by the control. The opera-

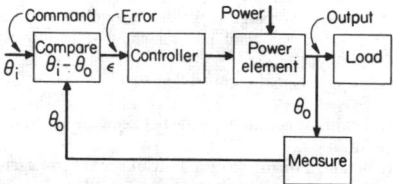

**Fig. 1**   Functional diagram of an automatic-control system.

tion may be explained as follows: (1) A command signal $\theta_i$ is applied to the input and compared with the instantaneous position of the output $\theta_o$. (2) The result of this comparison $\epsilon$, representing an error, is amplified by a controller and used to position a power element. (3) The power device in turn further amplifies the error signal to supply large amounts of power to the output or load to reduce the difference between $\theta_i$ and $\theta_o$.

## TRANSIENT ANALYSIS OF A CONTROL SYSTEM

The stability, accuracy, and speed of response of a control system are determined by analyzing the **steady-state** and the **transient** performance. It is desirable to achieve the steady state in the shortest possible time, while maintaining the output within specified limits. Steady-state performance is evaluated in terms of the accuracy with which the output is controlled for a specified input. The transient performance, i.e., the behavior of the output variable as the system changes from one steady-state condition to another, is evaluated in terms of such quantities as maximum overshoot, rise time, and response time (Fig. 2).

**Transient-Producing Disturbances**   An automatic control normally has only two places where disturbances can be expected: at the input or at the load. For a purely mechanical system the input disturbance may take the form of a periodic oscillation, a displacement, a velocity, or an acceleration. Disturbances at the output are usually load changes expressed as a torque or force quantity. Nonmechanical systems have disturbances expressed in different quantities; however, they are directly analogous to the mechanical system.

**The Basic Closed-Loop Control**   To illustrate some charac-

teristics of a basic closed-loop control, consider a mechanical, rotational system composed of a prime mover or motor, a total system inertia $J$, and a viscous friction $f$. To control the system's output variable $\theta_o$, a command signal $\theta_i$ must be supplied, the output variable measured and compared to the

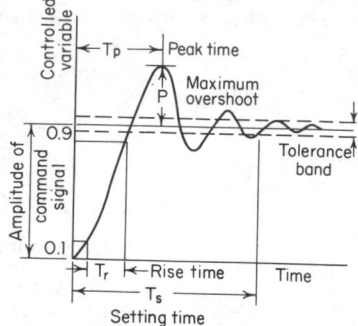

**Fig. 2**   System response to a unit step-function command.

input, and the resulting signal difference used to control the flow of energy to the load. The basic control system is represented schematically in Fig. 3.

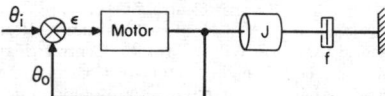

**Fig. 3**   A basic closed-loop control system.

The differential equation of this basic system is readily obtained from the idealized equations

$$\text{Load torque } T_L = J\frac{d^2\theta_o}{dt^2} + f\frac{d\theta_o}{dt} \tag{1}$$

$$\text{Developed torque } T_D = K\epsilon \tag{2}$$

$$\text{Error } \epsilon = \theta_i - \theta_o \tag{3}$$

The above equations combine to yield the system differential equation:

$$J\frac{d^2\theta_o}{dt^2} + f\frac{d\theta_o}{dt} + K\theta_o = K\theta_i \tag{4}$$

**Step-Input Response of a Viscous-Damped Control**   If the control system described in Fig. 3 by Eq. (4) is subjected to a step change in the input variable $\theta_i$, a solution $\theta_o = \theta_o(t)$ can be obtained as follows. (1) Let the ratio $\sqrt{K/J}$ be designated by the symbol $\omega_n$ and be called the **natural frequency.** (2) Let the quantity $2\sqrt{JK}$ be designated by the symbol $f_c$ and be called the **friction coefficient** required for critical damping. (3) Let $f/f_c$ be designated by the symbol $\zeta$ and be called the **damping ratio.** Equation (4) can then be written as

$$\frac{d^2\theta_o}{dt^2} + 2\zeta\omega_n\frac{d\theta_o}{dt} + \omega_n^2\theta_o = \omega_n^2\theta_i \tag{5}$$

For $\theta_i = 1$:

$$\theta_o = 0 \qquad \text{and} \qquad \frac{d\theta_o}{dt} = 0 \text{ at } t = 0$$

The complete solution of Eq. **(5)** is

$$\theta_o = 1 - \frac{e^{-\zeta\omega_n t}}{\sqrt{1-\zeta^2}}$$

$$\sin\left(\sqrt{1-\zeta^2}\,\omega_n t + \tan^{-1}\frac{\sqrt{1-\zeta^2}}{\zeta}\right) \qquad (6)$$

Equation **(6)** is plotted in dimensionless form for various values of damping ratio in Fig. 4. The curves for $\zeta = 0.1, 2,$ and $1$ illustrate the underdamped, overdamped, and critically damped case, where any further decrease in system damping

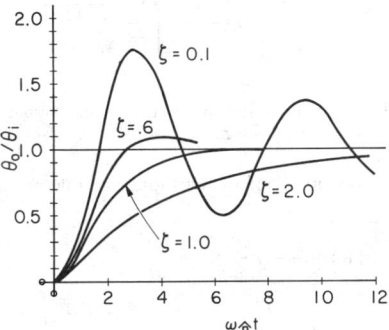

**Fig. 4** Transient response of a second-order viscous-damped control to unit-step input displacement.

would result in overshoot. Damping is a property of the system which opposes a change in the output variable.

The immediately apparent features of an observed transient performance are (1) the existence and magnitude of the maximum overshoot, (2) the frequency of the transient oscillation, and (3) the response time.

**Maximum Overshoot** When an automatic-control system is underdamped, the output variable overshoots its desired steady-state condition and a transient oscillation occurs. The first overshoot is the greatest, and it is the effect of its amplitude which must concern the control designer. The primary considerations for limiting this maximum overshoot are (1) to avoid damage to the process or machine due to excessive

excursions of the controlled variable beyond that specified by the command signal, and (2) to avoid the excessive settling time associated with highly underdamped systems. Obviously, exact quantitative limits cannot generally be specified for the magnitude of this overshoot. However, experience indicates that satisfactory performance can generally be obtained if the overshoot is limited to 30 percent or less.

**Transient Frequency** An undamped system oscillates about the final steady-state condition with a frequency of oscillation which should be as high as possible in order to minimize the response time. The designer must, however, avoid resonance conditions where the frequency of the transient oscillation is near the natural frequency of the system or its component parts.

**Rise Time $T_n$, Peak Overshoot $P$, Peak Time $T_P$**

These quantities are related to $\zeta$ and $\omega_n$ in Figs. 5 and 6. Some useful formulas are listed below (Murphy).

$$\omega_n T_n \approx 1.02 + 0.48\zeta + 1.15\zeta^2 + 0.76\zeta^3 \qquad 0 \le \zeta \le 1$$

$$\omega_n T_s = \left.\begin{matrix} 17.6 - 19.2\zeta & 0.2 \le \zeta \le 0.75 \\ -3.8 + 9.4\zeta & 0.75 \le \zeta \le 1 \end{matrix}\right\} \; 2\% \text{ tolerance band}$$

$$P = \frac{e^{-\pi\zeta}}{\sqrt{1-\zeta^2}} \qquad T_P = \frac{\pi}{\omega_n\sqrt{1-\zeta^2}}$$

Although these quantities are defined for a second-order system, they may be useful in the early design states of higher-order systems if the response of the higher-order system is dominated by roots of the characteristic equation near the imaginary axis.

**Derivative and Integral Compensation** (Thaler) Four common compensation methods for improving the steady-state performance of a proportional-error control without damaging its transient response are shown in Fig. 7. They are (1) error derivative compensation, (2) input derivative compensation, (3) output derivative compensation, (4) error integral compensation.

**Error Derivative Compensation** The torque equilibrium equation is

$$J\frac{d^2\theta_o}{dt^2} + f\frac{d\theta_o}{dt} = K_2\epsilon + K_1\frac{d\epsilon}{dt} \qquad (7)$$

Writing Eq. **(7)** in terms of the input and output variables yields

$$J\frac{d^2\theta_o}{dt^2} + (f + K_1)\frac{d\theta_o}{dt} + K_2\theta_o = K_2\theta_i + K_1\frac{d\theta_i}{dt} \qquad (8)$$

By adjusting $K_1$ and reducing $f$ so that the quantity $f + K_1$ is equal to $f$ in the uncompensated system, the system performance is affected as follows: (1) $\epsilon$ resulting from a constant-first-derivative input is reduced because of the reduction in viscous friction; (2) the transient performance of the uncompensated system is preserved unchanged.

**Derivative Input Compensation** The torque equilibrium equation is

$$J\frac{d^2\theta_o}{dt^2} + f\frac{d\theta_o}{dt} = K_2\epsilon + K_1\frac{d\theta_i}{dt} \qquad (9)$$

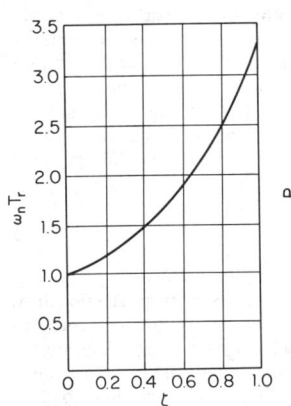

**Fig. 5** Rise time $T_r$ as a function of $\zeta$ and $\omega_n$.

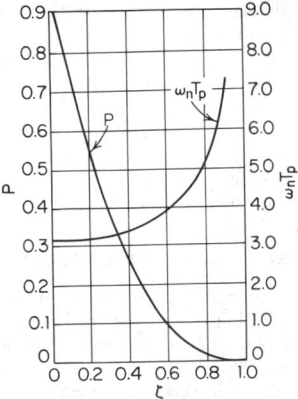

**Fig. 6** Peak overshoot $P$ and peak time $T_P$ as functions of $\zeta$ and $\omega_n$.

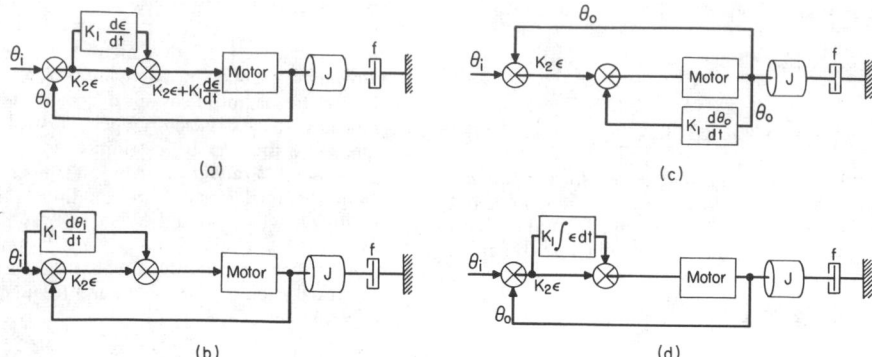

**Fig. 7** Derivative and integral compensation of a basic closed-loop system: (*a*) error derivative compensation; (*b*) input derivative compensation; (*c*) output derivative compensation; (*d*) error integral compensation. (*Thaler.*)

Writing Eq. (**9**) in terms of the input and output variables yields

$$J\frac{d^2\theta_o}{dt^2} + f\frac{d\theta_o}{dt} + K_2\theta_o = \theta_i + K_1\frac{d\theta_i}{dt} \qquad (10)$$

Examination of Eq. (**10**) yields the following information about the compensated system's performance: (1) since the characteristic equation is unchanged from that of the uncompensated system, the transient performance is unaltered; (2) the steady-state solution to Eq. (**10**) is

$$\theta_o = \theta_i - \frac{f}{K_2}\left(1 - \frac{K_1}{K_2}\right)\frac{d\theta_i}{dt} \qquad (11)$$

Therefore the input derivative signal can reduce the steady-state error by adjusting $K_1$ to equal $K_2$.

**Derivative Output Compensation**   The torque equilibrium equation is

$$J\frac{d^2\theta_o}{dt^2} + f\frac{d\theta_o}{dt} = K_2\epsilon \pm K_1\frac{d\theta_o}{dt} \qquad (12)$$

Writing Eq. (**12**) in terms of the input and output variables yields

$$J\frac{d^2\theta_o}{dt^2} + (f \pm K_1)\frac{d\theta_o}{dt} + K_2\theta_o = K_2\theta_i \qquad (13)$$

Examination of Eq. (**13**) yields the following information about the compensated system's performance. (1) Output derivative feedback produces the same system effect as the viscous friction does. This compensation therefore damps the transient performance. (2) Under conditions where $\theta_i = ct$, the steady-state error is increased.

**Error Integral Compensation**   Error integral compensation is used where it is necessary to eliminate steady-state errors resulting from input signals with constant first derivatives or under conditions of externally applied loads. The torque equilibrium equation is

$$J\frac{d^2\theta_o}{dt^2} + f\frac{d\theta_o}{dt} \pm \text{external load torque}$$
$$= K_2\epsilon + K_1\int_0^t \epsilon\, dt \qquad (14)$$

Writing Eq. (**14**) in terms of the input variable and the error yields

$$\text{External load torque} + J\frac{d^2\theta_i}{dt^2} + f\frac{d\theta_i}{dt}$$
$$= J\frac{d^2\epsilon}{dt^2} + f\frac{d\epsilon}{dt} + K_2\epsilon + K_1\int_0^t \epsilon\, dt \qquad (15)$$

At steady state for $t \gg 0$ and with a step change in the input derivative

$$\frac{d^2\theta_i}{dt^2} = \frac{d\epsilon}{dt} = \frac{d^2\epsilon}{dt^2} = 0 \qquad \frac{d\theta_i}{dt} = \text{const} \qquad (16)$$

Eq. (**15**) assumes the form

$$\pm \text{External load torque} + f\frac{d\theta_i}{dt} = K_2\epsilon + K_1\int_0^t \epsilon\, dt \qquad (17)$$

Since the sum of the load torque and the term $f(d\theta_i/dt)$ is finite, $\epsilon = 0$ for

$$\int_0^\infty \epsilon\, dt \to \infty \qquad (18)$$

If the integrating coefficient is a small number, additional torque produced by the integration action is developed very slowly, and, although the steady-state error is eventually eliminated, the transient performance is essentially unchanged. However, if $K_1$ is a large value, a large torque is produced in a short period of time, increasing the effect $T/J$ ratio, and thereby decreasing the damping. The general effects of error integral compensation within its useful range are (1) steady-state error is eliminated; and (2) transient response is adversely effected, resulting in increased overshoot and the attendant increase in response time.

**BLOCK-DIAGRAM REPRESENTATION**

A useful representation of the mathematical relationships defining the flow of information and energy through the control system is by means of a block diagram. In the diagram the components of the control system are considered as functional blocks in series and parallel arrangements according to their position in the actual control system. Each component is represented by its **transfer function,** the ratio of the Laplace transform of the output variable to the input variable with all initial conditions taken as zero.

For example, the overall transfer function of the control system described by Eq. (5) is obtained from Eq. (5) as

$$\frac{\theta_o(s)}{\theta_i(s)} = \frac{\omega_n^2}{s^2 + 2\zeta\omega_n s + \omega_n^2} \quad (19)$$

Two conditions are specified: (1) the components must be described by linear differential equations (or nonlinear equations linearized by suitable approximations), and (2) each block is unilateral. What occurs in one component may not affect the components preceding.

### Block-Diagram Algebra

The block diagram of a single-loop feedback-control system subjected to a command input $R(s)$ and a disturbance $U(s)$ is shown in Fig. 8.

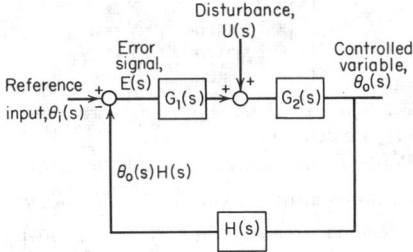

**Fig. 8**  Single-loop feedback control system.

When $U(s) = 0$ and the input is a reference change, the system may be reduced as follows:

$$E(s) = \theta_i(s) - \theta_o(s)H(s)$$
$$\theta_o(s) = E(s)[G_1(s)G_2(s)]$$

Therefore

$$\frac{\theta_o(s)}{\theta_i(s)} = \frac{G_1(s)G(s)}{1 + G_1(s)G_2(s)H(s)} \quad (20)$$

When $\theta_i(s) = 0$ and the input is a disturbance, the system may be reduced as follows:

$$E(s) = -\theta_o(s)H(s)$$
$$[E(s)G_1(s) + U(s)]\; G_2(s) = \theta_o(s)$$
$$\frac{\theta_o(s)}{U(s)} = \frac{G_2(s)}{1 + G_1(s)G_2(s)H(s)} \quad (21)$$

Equations (20) and (21) are in the form

$$\frac{\text{Response function}}{\text{Excitation function}} = \text{system function}$$

The **system function** is expressible as the ratio of two polynomials, $A(s)/B(s)$. The equation $B(s) = 0$ is the **characteristic equation.** When the excitation function is specified, inverse transformation of $\theta_o(s)$ yields $\theta_o(t)$, the transient response.

In many complex control systems, especially in the nonmechanical process-control field, auxiliary feedback paths are provided in order to adjust the system's performance. Figure 9a illustrates such a condition. In analyzing such a system it is usually best to combine secondary loops into the main control loop to form an equivalent series block and transfer function.

The system of Fig. 9a might be reduced in the following sequence.

1. Replace $K_3G_3(s)$ and $K_4H_1(s)$ with a single equivalent element

$$\frac{\theta_o}{\theta_2} = \frac{K_3G_3(s)}{1 + K_4H_1(s)K_3G_3(s)} = K_6G_6 \quad (22)$$

The result of this first reduction is shown in Fig. 9b.

2. Figure 9b can be treated in a similar fashion and a single block used to replace $K_2G_2$, $K_6G_6$, and $K_5H_2$

$$\frac{\theta_o}{\theta_1} = \frac{K_2G_2K_6G_6(s)}{1 + K_5H_2K_2G_2K_6G_6(s)} = K_7G_7 \quad (23)$$

The result of this second reduction is shown in Fig. 9c. The resulting open-loop transfer function is

$$\theta_o/\epsilon = K_1G_1K_7G_7(s) \quad (24)$$

The closed-loop or frequency response function is

$$\frac{\theta_o}{\theta_i} = \frac{K_1G_1K_7G_7(s)}{1 + K_1G_1K_7G_7(s)} \quad (25)$$

Equation (25) can, of course, be expanded to include the terms of the system's secondary loops.

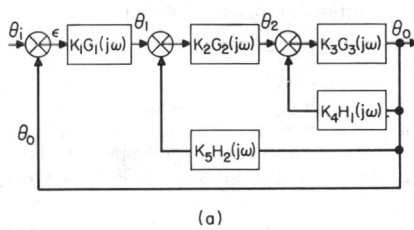

(a)

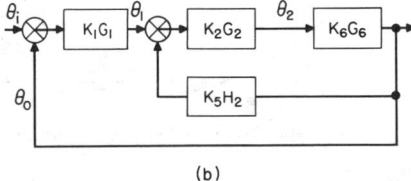

(b)

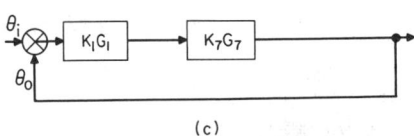

(c)

**Fig. 9**  Reduction of a closed-loop control system with multiple secondary loops.

### SIGNAL-FLOW REPRESENTATION

An alternate graphical representation of the mathematical relationships is the signal-flow graph. For complicated systems it allows a more compact representation and more rapid reduction techniques than the block diagram.

In Fig. 10, the nodes represent the variables $\theta_i$, $\epsilon$, $\theta_1$, . . . ,

$\theta_o$, and the branches the relationships between the nodes, of the system shown in Fig. 9. For example,

$$\theta_1(s) = \epsilon K_1 G_1(s) - K_5 H_2(s)\theta_o(s) \qquad \text{(26a)}$$

and

$$\epsilon = \theta_i - \theta_o \qquad \text{(26b)}$$

Signal-flow terminology follows:

**Source,** node having only outgoing branches, for example, $\theta_i$

**Sink,** node having only incoming branches, $\theta_o$

**Path,** series of branches with the same sense of direction, for example, *abcd*, *cdf*

**Forward path,** path originating at a source and ending at a sink, with no node encountered more than once, for example, *abcd*

**Path gain,** product of the coefficients along a path, for example, 1 $[K_1 G_1(s)]$ $[K_2 G_2(s)]$ $[K_3 G_3(s)]$

**Feedback loop,** path starting at a node and ending at the same node, for example, *bcdg*

**Loop gain,** product of coefficients along a feedback loop, for example, $[K_1 G_1(s)]$ $[K_2 G_2(s)]$ $[K_3 G_3(s)]$ $(-1)$

The overall gain of the system can be calculated from

$$G = \frac{\Sigma_i G_i \Delta_i}{\Delta} \qquad \text{(27)}$$

where $G_i$ = gain of $i$th forward path and

$$\Delta = 1 - \Sigma L_1 + \Sigma L_2 - \Sigma L_3 + \ldots + (-1)^k \Sigma L_k$$

where $\Sigma L_1$ = sum of the gains of each forward loop; $\Sigma L_2$ = sum of products of loop gains for nontouching loops (no node is common), taken two at a time; $\Sigma L_3$ = sum of products of loop gains for nontouching loops taken three at a time; $\Delta_i$ = value of $\Delta$ for signal flow graph resulting when $i$th path is removed.

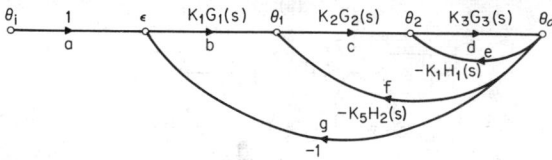

**Fig. 10**   Signal flow graph of the closed-loop control system shown in Fig. 9.

From Fig. 10 there is only one forward path, *abcd*.

$$\therefore G_1 = K_1 G_1(s)K_2 G_2(s)K_3 G_3(s)$$

Closed loops are *de*, *cdf*, and *bcdg*, with gains $-K_1 H_1(s)K_3 G_3(s)$,    $-K_2 G_2(s)K_3 G_3(s)K_5 H_2(s)$,    and $-K_1 G_1(s)K_2 G_2(s)K_3 G_3(s)$.

There are no nontouching closed loops;

$$\therefore \Delta = 1 + K_1 H_1(s)K_3 G_3(s)$$
$$+ K_2 G_2(s)K_3 G_3(s)K_5 H_2(s) + K_1 G_1(s)K_2 G_2(s)K_3 G_3(s)$$

There are no loops remaining if the forward path *abcd* is removed;

$$\therefore \Delta_1 = 1$$

Thus

$$\theta_o/\theta_i = K_1 G_1(s)K_2 G_2(s)K_3 G_3(s)/[1 + K_1 H_1(s)K_3 G_3(s)$$
$$+ K_2 G_2(s)K_3 G_3(s)K_5 H_2(s) + K_1 G_1(s)K_2 G_2(s)K_3 G_3(s)] \qquad \text{(28)}$$

which is identical with Eq. **(25)**.

## STEADY-STATE PERFORMANCE

The steady-state error of a control system can be determined by using the final value theorem (see Laplace Transforms, Sec. 2) in which

$$\theta_o(t) = s\theta_o(s) \qquad \text{provided } \theta_o(t) \text{ is stable}$$
$$t \to \infty \quad s \to 0$$

The classification of control systems according to the form of the open-loop transfer function facilitates the determination of the steady-state errors when the system is subjected to various inputs.

The open-loop transfer function $\theta_o(s)/\epsilon(s) = KG(s)$ may be written

$$KG(s) = \frac{\theta_o(s)}{\epsilon(s)} = \frac{K(1 + \tau_a s)(1 + \tau_b s)(1 + \tau_c s)\cdots}{s^N(1 + \tau_1 s)(1 + \tau_2 s)(1 + \tau_3 s)\cdots} \qquad \text{(29a)}$$

The **system type** is given according to the value of $N$ as

| | |
|---|---|
| Type 0 system | $N = 0$ |
| Type 1 system | $N = 1$ |
| Type 2 system | $N = 2$ |
| Type 3 system | $N = 3$ |

$$\text{(29b)}$$

**Error coefficients,** based on a system with unity feedback $[H(s) = 1]$, are defined as

$$\text{Positional error constant} = K_0 = \lim_{s \to 0} KG(s) \qquad \text{(30a)}$$
$$\text{Velocity error constant} = K_v = \lim_{s \to 0} sKG(s) \qquad \text{(30b)}$$
$$\text{Acceleration error constant} = K_a = \lim_{s \to 0} s^2 KG(s) \qquad \text{(30c)}$$

A summary of error coefficients for systems of different types is given in Table 1.

**Table 1. Summary of Error Coefficiencts**

| $N$ | $K_0$ | $K_v$ | $K_a$ |
|---|---|---|---|
| 0 | const | 0 | 0 |
| 1 | $\infty$ | const | 0 |
| 2 | $\infty$ | $\infty$ | const |

A summary of the errors for types 0, 1, and 2 systems, when subjected to various inputs, is given in Table 2.

The higher the system type, the better is the output able to follow the higher degrees of input. Higher-type systems,

**Table 2. Summary of Errors**

| Input error | $\theta_i(t) = A$ $\epsilon_0/A$ | $\theta_i(t) = vt$ $\epsilon_v/v$ | $\theta_i(t) = at^2$ $\epsilon_a/a$ |
|---|---|---|---|
| $N = 0$ | $1/(1 + K_0)$ | $\infty$ | $\infty$ |
| 1 | 0 | $1/K_v$ | $\infty$ |
| 2 | 0 | 0 | $1/K_a$ |

however, are more difficult to stabilize, and a compromise must be made between the steady-state error and the settling time of the response. (For a detailed treatment of steady-state errors, see Savant.) Another convenient means for classifying systems is according to controller action (see Pneumatic Systems).

## FREQUENCY RESPONSE

Although it is the time response of the control system that is of major importance, study of the effect on transient response of changes in system parameters, either in the process or controller, is more conveniently made from a **frequency-response** analysis of the system. The **frequency response** of a system is the steady-state output of the system to input sinusoids of varying frequency. The output for a linear system can be completely described in terms of the amplitude ratio of the output sinusoid to the input sinusoid and the phase of the output sinusoid to the input sinusoid. The amplitude ratio or gain, and phase, are functions of the frequency of the input sinusoid. For purposes of system analysis the frequency response function is more useful than that of the closed loop. Means for obtaining the closed-loop frequency response and evaluating transient performance from the open-loop frequency response are discussed below.

The frequency response can be obtained analytically from the transfer functions of the components, or system, by replacing $s$, the Laplace operator, with $j\omega$. Table 3 shows the freuqency responses for some common control-system elements.

The frequency response can also be obtained experimentally for systems not readily amenable to mathematical analysis by subjecting the system to input sinusoids of varying frequency.

## GRAPHICAL DISPLAY OF THE FREQUENCY RESPONSE

The transient performance characteristics of the control are conveniently obtained from curves of the open-loop frequency-response function. The most common methods of graphical presentation employed are (1) the **polar** or **Nyquist plot,** where the magnitude and phase angle of the direct-transfer function $G(j\omega)H(j\omega)$ are plotted as a vector with frequency $\omega$ as a parameter and (2) the **logarithmic** or **Bode plot,** where the phase angle and log of the magnitude of $G(j\omega)H(j\omega)$ are plotted against log $\omega$. The polar plot also enables the absolute stability of the system to be determined without the need for obtaining the roots of the characteristic equation. The logarithmic plot has the advantage of ease of plotting, especially in design, since the individual effects of cascaded elements can be gaged by superposition.

**Polar plots** for the frequency response of some common system components are shown in Table 4.

For a closed system the frequency response can be derived from the polar plot of the direct-transfer function $KG(j\omega)$. In Fig. 11, the amplitude ratio $\theta_o/\theta_i$ at any frequency $\omega$ is the ratio of the lengths of the vectors $O\beta$ and $\alpha\beta$. The angle formed by the vectors $\alpha\beta$ and $O\beta$ is the phase angle of the frequency response $\underline{/\alpha\beta O} = \underline{/\theta_o/\theta_i}(j\omega)$. The numerator of the direct-transfer function is a constant representing the gain of the control system. Changing the **gain** proportionally changes the length of each vector $O\beta$. This technique for deriving the frequency-response function from the direct-transfer function is not applicable if the control system has a transfer function in the feedback loop.

**Logarithmic Plots (Bode Diagrams)**   The direct-transfer-function equation can be plotted on logarithmic coordinates.

The first advantage of this method of data presentation is that the numerical computation of points on the curve is simplified by the fact that log magnitude vs. log $\omega$ can be approximated to engineering accuracy with straightline asymptotes. This

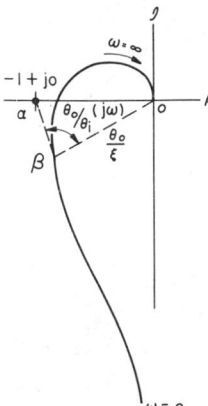

**Fig. 11**   Typical $(\theta_o/\epsilon)/j\omega)$ plot.

can be shown as follows. The generalized form of a direct transfer function is

$$KG(j\omega) = K \frac{(j\omega\tau_a + 1)(j\omega\tau_b + 1)\cdots}{(j\omega)^N(j\omega\tau_1 + 1)\cdots} \quad (31)$$

where there may be any number of terms in the numerator or denominator but where the denominator is of higher order. The exponent $N$ assumes positive integral values. Since the terms of Eq. (31) are complex, their magnitude and phase are computed separately. Taking logs of both sides of Eq. (31) yields

$$\log |KG| = \log |K| + \log |j\omega\tau_a + 1| + \cdots \\ - N \log |j\omega| - \log |j\omega\tau_1 + 1| \cdots \quad (32)$$

It is usually more convenient to express the logarithm in **decibels,** so each term of Eq. (32) is multiplied by 20. A decibel-to-amplitude-ratio conversion is given in Table 5.

$$20 \log |KG| = 20 \log |K| + 20 \log |j\omega\tau_a + 1| \cdots \\ - 20N \log |j\omega| - 20 \log (j\omega\tau_1 + 1) \cdots \quad (33)$$

The phase angle is computed as

$$\underline{/KG(j\omega)} = \underline{/K} + \tan^{-1}\omega\tau_a + \cdots \\ - (90N)° - \tan^{-1}\omega\tau_1 \cdots \quad (34)$$

The computational advantage of the logarithmic form of the amplitude function is quite apparent. For terms as $(j\omega)^N$, the decibel expression $-20N \log j\omega$ plots as a straight line. If $N = 1$, the slope is $-6$ dB/octave (one octave separates any two frequencies which are in a ratio of 2 or $\frac{1}{2}$). If $N = 2$, the slope is $-12$ dB/octave.

For such terms as $j\omega\tau_a + 1$ in the numerator of the transfer function, the decibel expression is $20 \log |j\omega\tau_a + 1|$.

For $\omega << 1$,               $|j\omega\tau_a + 1| = 1$   (35)
$\therefore 20 \log |j\omega\tau_a + 1| = 0$ dB   (36)
For $\omega >> 1$,               $|j\omega\tau_a + 1| = j\omega\tau_a$   (37)
$\therefore 20 \log |j\omega\tau_a|$ has a slope of $+6$ dB/octave   (38)

**Table 3. Frequency-Response Equations for Some Common Control-System Elements**

| Description | Transfer function $G(s)$ | Frequency response $G(j\omega)$ | Magnitude ratio | Phase angle |
|---|---|---|---|---|
| 1. Dead time | $\epsilon^{-T_L s}$ | $\epsilon^{-j\omega T_L}$ | 1 | $-\omega T_L$ radians |
| 2. First-order lag | $\dfrac{1}{Ts+1}$ | $\dfrac{1}{j\omega T+1}$ | $\dfrac{1}{\sqrt{\omega^2 T^2+1}}$ | $-\tan^{-1}(\omega T)$ |
| 3. Second-order lag | $\dfrac{1}{(Ts+1)(aTs+1)}$ | $\dfrac{1}{-a\omega^2 T^2 + j(1+a)\omega T + 1}$ | $\dfrac{1}{\sqrt{(1-a\omega^2 T^2)^2 + (1+a)^2\omega^2 T^2}}$ | $-\tan^{-1}\left[\dfrac{(1+a)\omega T}{1 - aT^2\omega^2}\right]$ |
| 4. Quadratic (underdamped) | $\left(\dfrac{s}{\omega_n}\right)^2 + \dfrac{2\zeta}{\omega_n}s + 1$ | $-\left(\dfrac{\omega}{\omega_n}\right)^2 + j2\zeta\dfrac{\omega}{\omega_n} + 1$ | $\sqrt{\left(1-\dfrac{\omega^2}{\omega_n^2}\right)^2 + 4\zeta^2\left(\dfrac{\omega}{\omega_n}\right)^2}$ | $-\tan^{-1}\left[\dfrac{2\zeta\dfrac{\omega}{\omega_n}}{1-\left(\dfrac{\omega}{\omega_n}\right)^2}\right]$ |
| 5. Ideal proportional controller | $K$ | $K$ | $K$ | 0 |
| 6. Ideal proportional-plus-reset controller<br><br>$T_i = \dfrac{1}{r}$<br><br>$r$ = reset rate | $K\left(1+\dfrac{1}{T_i s}\right)$<br>or<br>$K\dfrac{T_i s+1}{T_i s}$ | $K\left(1+\dfrac{1}{j\omega T_i}\right)$<br>or<br>$K\dfrac{j\omega T_i+1}{j\omega T_i}$ | $K\sqrt{1+\left(\dfrac{1}{\omega T_i}\right)^2}$ | $-\tan^{-1}\left(\dfrac{1}{\omega T_i}\right)$ |
| 7. Ideal proportional-plus-rate controller | $K(1+T_d s)$ | $K(1+j\omega T_d)$ | $K\sqrt{1+\omega^2 T_d^2}$ | $\tan^{-1}(\omega T_d)$ |
| 8. Ideal proportional-plus-reset-plus-rate controller | $K\left(1+T_d s+\dfrac{1}{T_i s}\right)$ | $K\left(1+j\omega T_d+\dfrac{1}{j\omega T_i}\right)$<br>or<br>$K\dfrac{j\omega T_i - \omega^2 T_d T_i + 1}{j\omega T_i}$ | $K\sqrt{(\omega T_i)^2 + (1-\omega^2 T_d T_i)^2}$ | $\tan^{-1}\left(\omega T_d - \dfrac{1}{\omega T_i}\right)$ |

Source: Considine, "Process Instruments and Controls Handbook," McGraw-Hill.

**Table 4. Polar Loci for Frequency Response of Common System Components**

| Type | Component | Polar loci plot in $G(j\omega)$ plane |
|---|---|---|
| A | First-order lag: $$G(j\omega) = \frac{1}{j\omega T + 1}$$ $$\|G(j\omega)\| = (\omega^2 T^2 + 1)^{-\frac{1}{2}}$$ $$\underline{/G(j\omega)} = -\tan^{-1}(+\omega T)$$ | |
| B | Dead time: $$G(j\omega) = \epsilon^{-j\omega T_L}$$ $$\|G(j\omega)\| = 1$$ $$\underline{/G(j\omega)} = -\omega T_L \text{ radians}$$ | |
| C | Second-order lag: $$G(j\omega) = \frac{1}{(j\omega T_1 + 1)(j\omega T_2 + 1)}$$ $$\|G(j\omega)\| = [(T_1^2\omega^2 + 1)(T_2^2\omega^2 + 1)]^{-\frac{1}{2}}$$ $$\underline{/G(j\omega)} = -\tan^{-1}(+\omega T_1) - \tan^{-1}(+\omega T_2)$$ | |
| 1 | Proportional controller: $$G(j\omega) = K$$ $$\|G(j\omega)\| = K$$ $$\underline{/G(j\omega)} = 0$$ | |
| 2 | Proportional-speed floating controller: $$G(j\omega) = \frac{1}{j\omega T_i}$$ $$\|G(j\omega)\| = \frac{1}{\omega T_i}$$ $$\underline{/G(j\omega)} = -90°$$ | |
| 3 | Proportional-plus-rate controller: $$G(j\omega) = K(1 + jT_d\omega)$$ $$\|G(j\omega)\| = K\sqrt{1 + \omega^2 T_d^2}$$ $$\underline{/G(j\omega)} = \tan^{-1}(+\omega T_d)$$ | |
| 4 | Proportional-plus-reset controller: $$G(j\omega) = K\left(1 + \frac{1}{j\omega T_i}\right)$$ $$\|G(j\omega)\| = K\sqrt{1 + \left(\frac{1}{\omega T_i}\right)^2}$$ $$\underline{/G(j\omega)} = -\tan^{-1}(+\omega T_i)$$ | |
| 5 | Proportional-plus-reset-plus-rate controller: $$G(j\omega) = K\left(1 + \frac{1}{j\omega T_i} + jT_d\omega\right)$$ $$\|G(j\omega)\| = K\sqrt{1 + \left(T_d\omega - \frac{1}{T_i\omega}\right)^2}$$ $$\underline{/G(j\omega)} = \tan^{-1}\left(T_d\omega - \frac{1}{T_i\omega}\right)$$ | |

SOURCE: Considine, "Process Instruments and Controls Handbook," McGraw-Hill.

**Table 5. Decibel-to-Amplitude-Ratio Conversion**

| Decibel | Amp-ratio | Decibel | Amp-ratio |
|---------|-----------|---------|-----------|
| 60 | 1,000.0 | 0.0 | 1.000 |
| 50 | 316.2 | −0.5 | 0.945 |
| 40 | 100.0 | −1.0 | 0.894 |
| 35 | 56.23 | −2.0 | 0.795 |
| 30 | 31.62 | −3.0 | 0.707 |
| 25 | 17.78 | −4.0 | 0.630 |
| 20 | 10.0 | −5.0 | 0.562 |
| 15 | 5.62 | −6.0 | 0.500 |
| 12 | 3.99 | −8.0 | 0.398 |
| 9 | 2.82 | −9.0 | 0.354 |
| 6 | 2.00 | −10.0 | 0.316 |
| 5 | 1.78 | −12.0 | 0.251 |
| 4 | 1.59 | −15.0 | 0.178 |
| 3 | 1.41 | −18.0 | 0.125 |
| 2 | 1.26 | −20.0 | 0.100 |
| 1 | 1.12 | | |
| 0.5 | 1.06 | | |

SOURCE: Thayer, "Elements of Servomechanism Theory," McGraw-Hill.

When $\omega = 1/\tau_a$, $20 \log |j\omega\tau_a| = 0$ dB. Therefore the 0-dB asymptote and the +6-dB/octave asymptote cross at $\omega = 1/\tau_a$. At this corner, it can be shown that the approximation yields a 3-dB error. This term $(j\omega\tau_a + 1)$ is plotted in Fig. 12. For denominator terms of the form $j\omega\tau_1 + 1$, the asymptotes are a horizontal line at 0 dB and a line sloping at −6 dB/octave. Quadratic terms of the form $(j\omega^2)\tau_b + j\omega A + 1$ do not lend themselves quite as well to asymptotic approximation. The low-frequency asymptote has a slope of 0 dB/octave and the high-frequency asymptote has a slope of ±12dB/octave. The magnitude and phase angle values at the corner frequency $\omega = 1/\tau_b$ depend on the value of the coefficient $A$. The magnitude error at this frequency cannot be conveniently approximated. The phase angle is −90°.

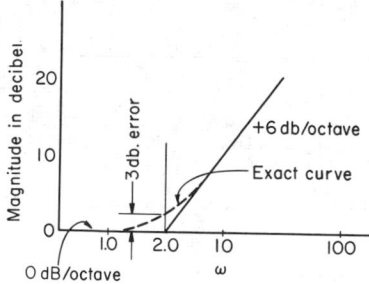

**Fig. 12**  Bode plot of term $(j\omega\tau_a + 1)$.

To obtain a logarithmic plot of a direct transfer function, the designer must first plot the asymptotic approximations to each term of the transfer function, add the ordinate values of each term at each value of $\omega$, and plot their sum.

## STABILITY AND PERFORMANCE OF AN AUTOMATIC CONTROL

An automatic-control system is stable if the amplitude of transient oscillations decreases with time and the system reaches a steady state. The stability of a system can be evaluated by examining the roots of the differential equation describing the system. The presence of positive real roots or complex roots with positive real parts dictates an unstable system. Any stability test utilizing the open-loop transfer function or its plot must utilize this fact as the basis of the test.

**The Nyquist Stability Criterion**  The $KG(j\omega)$ locus for a typical single-loop automatic-control system plotted for all positive and negative frequencies is shown in Fig. 13. The locus for negative values of $\omega$ is the mirror image of the positive $\omega$ locus in the real axis. To complete the diagram, a semicircle (or full circle if the locus approaches −∞ on the real axis) of infinite radius is assumed to connect in a positive sense, the + locus at $\omega \rightarrow 0$ with the negative locus at $\omega \rightarrow -0$. If this locus is traced in a positive sense from $\omega \rightarrow \infty$ to $\omega \rightarrow 0$, around the circle at $\infty$, and then along the negative-frequency locus the following may be concluded: (1) if the locus **does not** enclose the $-1 + j0$ point, the system is stable; (2) if the locus **does** enclose the $-1 + j0$ point, the system is unstable. The Nyquist criterion can also be applied to the log magnitude of $KG(j\omega)$ and phase−vs.−log $\omega$ diagrams. In this method of display, the criterion for stability reduces to the requirement that the log magnitude of $KG(j\omega)$ must cross the 0-dB axis at a frequency less than the frequency at which the phase curve crosses the −180° line. Two stability conditions are illustrated in Fig. 14. The Nyquist criterion not only provides a simple test for the stability of an automatic-control system but also indicates the degree of stability of the system by indicating the degree to which the $KG(j\omega)$ locus avoids the $-1 + j0$ point.

The concepts of **phase margin** and **gain margin** are employed to give this quantitative indication of the degree of stability of an automatic-control system. **Phase margin** is defined as the additional negative phase shift necessary to make the phase angle of the transfer function −180° at the frequency where the magnitude of the $KG(j\omega)$ vector is unity. Physically, phase margin can be interpreted as the amount by which the unity $KG$ vector has to be shifted to make a stable system unstable.

In a similar manner, **gain margin** is defined as the reciprocal of the magnitude of the $KG$ vector at −180°. Physically, gain margin is the number by which the gain must be multiplied to put the system to the limit of stability. Thaler suggests satisfactory results can be obtained in most control applications if

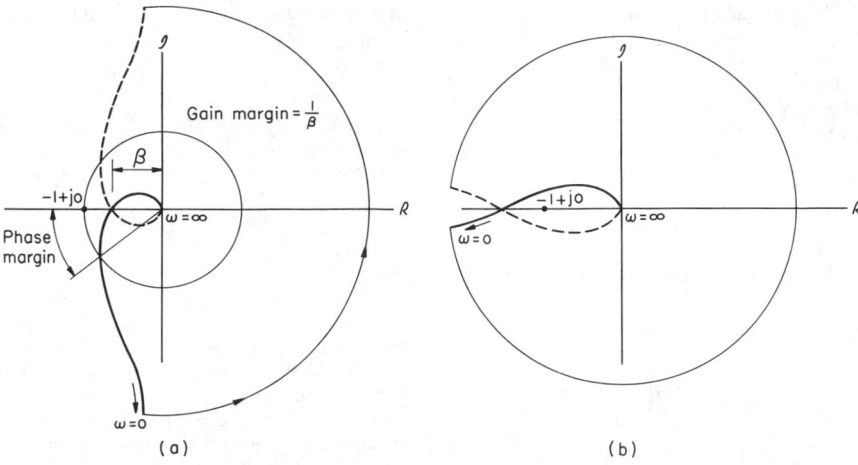

**Fig. 13** Typical $KG(j\omega)$ loci illustrating application of Nyquist's stability criterion: (a) stable; (b) unstable.

the phase margin is between 40 and 60° while the gain margin is between 3 and 10 (10 to 20 dB). These values will ensure a small transient overshoot with a single cycle in the transient. The margin concepts are qualitatively illustrated in Figs. 13 and 14.

**Routh's Stability Criterion**  The frequency-response equation of a closed-loop automatic control is

$$\frac{\theta_o}{\theta_i} = \frac{KG(j\omega)}{1 + KG(j\omega)} \qquad (39)$$

The characteristic equation obtained therefrom has the algebraic form

$$A(j\omega)^n + B(j\omega)^{n-1} + C(j\omega)^{n-2} + \cdots = 0 \qquad (40)$$

The purpose of Routh's method is to determine the existence of roots of this equation which are positive or which are complex with positive real parts and thus identify the resulting instability. To apply the criterion the coefficients are written alternately in two rows as

$$\begin{array}{cccc} A & C & E & G \\ B & D & F & H \end{array}$$

This array is then expanded to

$$\begin{array}{ccc} A & C & E & G \\ B & D & F & H \\ \alpha_1 & \alpha_2 & \alpha_3 \\ \beta_1 & \beta_2 & \beta_3 \\ \gamma_1 & \gamma_2 \end{array}$$

where $\alpha_1$, $\alpha_2$, $\alpha_3$, $\beta_1$, $\beta_2$, $\beta_3$, $\gamma_1$ and $\gamma_2$ are computed as

$$\alpha_1 = \frac{BC - AD}{B} \qquad \beta_1 = \frac{D\alpha_1 - B\alpha_2}{\alpha_1}$$

$$\alpha_2 = \frac{BE - AF}{B} \qquad \beta_2 = \frac{F\alpha_1 - B\alpha_3}{\alpha_1}$$

$$\alpha_3 = \frac{BG - AH}{B} \qquad \beta_3 = \frac{H\alpha_1 - Bo}{\alpha_1}$$

When the array has been computed, the left-hand column ($A$, $B$, $\alpha_1$, $\beta_1$, $\gamma_1$) is examined. If the signs of all the numbers in the left-hand column are the same, there are no positive real roots. If there are changes in sign, the number of positive real roots is equal to the number of changes in sign. It should be recognized that this is a test for instability; the absence of sign changes does not guarantee stability.

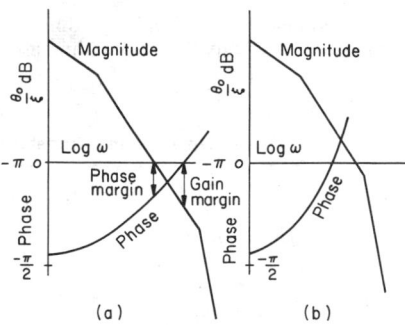

**Fig. 14**  Nyquist stability criterion in terms of log magnitude $KG(j\omega$ diagrams (a) stable; (b) unstable. (*Porter.*)

## SAMPLED-DATA CONTROL SYSTEMS

**Definition**  Sampled-data control systems are those in which continuous information is transformed at one or more points of the control system into a series of pulses. This transformation may be performed intentionally, e.g., the flow of information over long distances to preserve the accuracy of the data during the transmission, or it may be inherent in the generation of the information flow, e.g., radiating energy from a radar antenna which is in the form of a train of pulses, or the signals developed by a digital computer during a direct digital control of machine-tool operation.

Methods of analysis analogous to those for continuous-data systems have been developed for the sampled-data systems. Discussed herein are (1) sampling, (2) the $z$ transformation, (3) the $z$-transfer function, and (4) stability of sampled data systems.

**Sampling** The ideal sampler is a simple switch (Fig. 15)

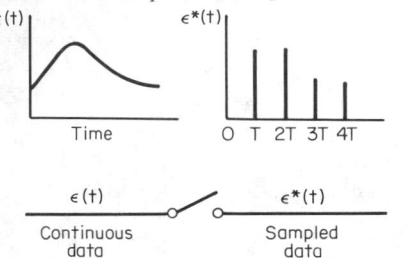

**Fig. 15** Ideal sampler, showing continuous input and sampled output.

which is closed only instantaneously and opens and closes at a constant frequency. The switch, which may or may not be a physical component in a sampled-data feedback system, indicates a sampled signal. Such a sampled-data feedback system is shown in Fig. 16. The error signal in continuous form is $\epsilon(t)$, and the sampled error signal is $\epsilon^*(t)$. Figure 15 shows the relationship between these signals in a graphical form.

**$z$ Transformation** In the analysis of continuous-data systems, it has been shown that the Laplace transformation can be used to reduce ordinary differential equations to algebraic equations. For sampled-data systems, an operational calculus, the $z$ transform, can be used to simplify the analysis of such systems.

Consider the sampler as an inpulse modulator; i.e., the sampling modulates an infinite train of unit impulses with the continuous-data variable. Then the Laplace transformation can be shown to be

$$F^*(s) = \sum_{n=0}^{\infty} f(nT)e^{-nTs} \tag{41}$$

(See A. W. Langill "Automatic Control Systems Engineering.")

In terms of the $z$ transform, this becomes

$$F^*(s) = F(z) = \sum_{n=0}^{\infty} f(nT)z^{-n} \tag{42}$$

where $z^{-n} = e^{-nTs}$.

Table 1 in Sec. 2, Laplace Transforms, lists the transforms for a number of continuous time functions. Table 6 in this subsection lists the $z$ transforms for some of these functions.

It should be noted that unlike the Laplace transforms, the $z$-transform method imposes the restriction that the sampled-data-system response can be determined only at the sampling instant. The same $z$ transform may apply to different time functions which may have the same value at the instant of sampling. The $z$-transform function is not defined, therefore, in a continuous sense, and the inverse $z$ transform is not unique.

**$z$ Transfer Function** The ratio of the sampled output function of a discrete network to the sampled input function is the

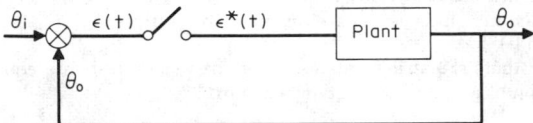

**Fig. 16** Sampled data feedback system.

$z$-transfer function. A discrete network is one which has both a sampled input and output. A table of block diagrams of a number of sampled-data control systems with their associated transfer functions is presented in Table 7.

**Stability of Sampled-Data Systems** The stability of sampled-data systems can be demonstrated utilizing frequency-response methods which have been discussed in this section.

**Table 6. Laplace and $z$ Transforms**

| Time function | $f(t)$ | $F(s) = \mathcal{L}[f(t)]$ | $F^*(z) = F(z)$ |
|---|---|---|---|
| Unit ramp function | $t$ | $\dfrac{1}{s^2}$ | $\dfrac{Tz}{(z-1)^2}$ |
| Unit acceleration function | $\dfrac{t^2}{2}$ | $\dfrac{1}{s^3}$ | $\tfrac{1}{2}T^2\dfrac{z(z+1)}{(z-1)^3}$ |
| Exponential function | $e^{-\alpha t}$ | $\dfrac{1}{s+\alpha}$ | $\dfrac{z}{z-e^{-\alpha T}}$ |
| Sinusoidal function | $\sin \beta t$ | $\dfrac{\beta}{s^2+\beta^2}$ | $\dfrac{z \sin \beta T}{z^2 - 2z \cos \beta T + 1}$ |
| Cosinusoidal function | $\cos \beta t$ | $\dfrac{s}{s^2+\beta^2}$ | $\dfrac{z(z - \cos \beta T)}{z^2 - 2z \cos \beta T + 1}$ |

See Stability and Performance of an Automatic Control. The Nyquist stability criterion again applies with the same conclusions relative to the $-1$ point, and the methods of generalizing the open- and closed-loop frequency response plots remain the same.

## STATE-SPACE CONCEPTS

State-space methods permit system analysis and design by study of a set of first-order differential equations rather than a single higher-order equation. This is convenient for solution by numerical methods, using the digital computer, and especially useful for systems with nonlinearities, time-varying characteristics, and multiple inputs and outputs.

For a system previously described [Eq. (4)] as

$$A_n \frac{d^n c}{dt^n} + A_{n-1} \frac{d^{n-1} c}{dt^{n-1}} + \cdots + A_0 c = u(t) \tag{43}$$

where $c(t)$ is the output and $u(t)$ the input, allowing $x_1 = c$, $x_2 = \dot{c}$, $x_3 = \ddot{c}$, ..., $x_n = c^{(n-1)}$ yields the state-variable representation

$$\dot{\mathbf{x}} = \mathbf{Px} + \mathbf{Bu} \tag{44}$$
$$c = \mathbf{Lx} \tag{45}$$

where $\mathbf{x}$ is the state vector and

$$\mathbf{x} = \begin{bmatrix} x_1 \\ x_2 \\ \cdot \\ \cdot \\ \cdot \\ x_{n-1} \\ x_n \end{bmatrix}$$

$$\mathbf{P} = \begin{bmatrix} 0 & 1 & 0 & \dots 0 \\ 0 & 0 & 1 & \dots 0 \\ \cdot & & & \dots \\ 0 & 0 & 0 & \dots 1 \\ -A_0/A_n & -A_1/A_n & -A_2/A_n & \dots A_{n-1}/A_n \end{bmatrix}$$

$$\mathbf{x} = \begin{bmatrix} \dot{x}_1 \\ \dot{x}_2 \\ \cdot \\ \cdot \\ \cdot \\ x_n \end{bmatrix} \quad \mathbf{B} = \begin{bmatrix} 0 \\ 0 \\ \cdot \\ \cdot \\ \cdot \\ 1/A_n \end{bmatrix} \quad \mathbf{L} = [1, \quad 0, \quad 0, \quad \dots, \quad 0] \tag{46}$$

The vector $\mathbf{P}$ is called the companion matrix.

**Transition Matrix** The transition matrix relates the transition of the system state at time $t_0 = 0$ to the state at some later time $t$.

From $\dot{\mathbf{x}} = \mathbf{Px} + \mathbf{Bu}$ can be derived

$$\mathbf{x}(t) = \mathbf{\Phi}(t)\mathbf{x}(0^+) + \int_0^t \mathbf{\Phi}(t - \tau)\mathbf{Bu}(\tau) \, d\tau \tag{47}$$

which is called the **state transition equation** of the system.

$\mathbf{\Phi}(t)$ is the transition matrix, calculable from

$$\mathbf{\Phi}(t) = \mathcal{L}^{-1}[(s\mathbf{I} - \mathbf{P})^{-1}] \tag{48}$$

$\mathbf{\Phi}(t)$ has the properties

$$\begin{aligned} \mathbf{\Phi}(0) &= \mathbf{I} \\ \mathbf{\Phi}(t_2 - t_0) &= \mathbf{\Phi}(t_2 - t_1)\mathbf{\Phi}(t_1 - t_0) \\ \mathbf{\Phi}(t + \tau) &= \mathbf{\Phi}(t)\mathbf{\Phi}(\tau) \\ \mathbf{\Phi}^{-1}(t) &= \mathbf{\Phi}(-t) \end{aligned} \tag{49}$$

Since $\mathbf{c}(t) = \mathbf{Lx}(t)$, the following is the system output in terms of the transition matrix:

$$\mathbf{c}(t) = \mathbf{L\Phi}(t)\mathbf{x}(0^+) + \int_0^t \mathbf{L\Phi}(t - \tau)\mathbf{Bu}(\tau) \, d\tau \tag{50}$$

See Shinners for digital-computer methods for determining $\mathbf{\Phi}(t)$, given $\mathbf{P}$.

## GENERAL DESIGN PROCEDURE

The initial performance specifications for an automatic control generally prescribe such quantities as the range of operation of the input variable and its derivatives, the maximum acceptable value of the steady-state error, and possibly other quantities, such as maximum settling time and peak overshoot. With preliminary knowledge of the nature of the input variable and the load, the designer integrates the components of the **basic** automatic-control system, develops the open-loop transfer function of this basic system, and examines its $G(s)$ locus. The system gain $K$ is then adjusted to satisfy the steady-state error requirements and the resulting locus $KG(s)$ is again examined for stability. If instability exists at the required gain, the $KG(s)$ locus is reshaped through the use of derivative or integral compensation by means of a phase-lead or phase-lag component to display acceptable phase and gain margins. A detailed discussion of gain adjustment and phase compensation may be found in the references.

## COMPONENTS

**Hydraulic systems** are used for rapid-response servomechanisms at high power levels. Operating system pressures are from 50 to 100 lb/in² for slower-acting systems and up to 5,000 lb/in² where lightweight and fast responses are required. Compared with **electrical systems** the major advantages are a rapid response in the large horsepower ranges and the capability of operating at high power-density levels since the fluid can transmit dissipated energy from the point of generation. Compared with **pneumatic systems,** hydraulic systems are faster because the fluid is essentially incompressible. Major disadvantages are vulnerability to dirt, since the components generally require close machining tolerances, and the danger of fire and explosion resulting from the flammability of the hydraulic fluids used (Blackburn).

The direction and volume of flow are controlled by **servo valves** in the system. They may be single-stage (pilot-operated) and mechanically or electrically actuated. A schematic of a spool-type four-way single-stage control piston and inertia load is shown in Fig. 17. Hydraulic fluid at constant pressure enters at the supply port. With displacement of the spool valve downward, for example, inflow to the top side of the piston moves the piston downward. Because of machining tolerances the spool dimensions are either larger (overlapped) or smaller (underlapped) than the port dimension. Underlapped valves permit leakage to the piston in the centered position; overlapped valves result in a dead zone, where motion $x$ results in no flow until a port is opened.

The transfer function of this circuit is given as (Truxal, "Control Engineers' Handbook")

$$\frac{y}{x} = \frac{C_1 \dfrac{1}{1 + \alpha(C_1/C_2)}}{s\left[\dfrac{VM}{2BA^2}\dfrac{1}{1 + \alpha(C_1/C_2)}s^2 + \dfrac{C_1m/C_2 + V\alpha/2BA^2}{1 + \alpha(C_1/C_2)}s + 1\right]}$$

where $C_1$ = servo velocity gradient, in/(s)(in), $C_2$ = servo force gradient, lb/in, $\alpha$ = viscous friction of load and piston, lb/(in)(s), $B$ = bulk modulus of fluid, lb/in² $M$ = mass of load and piston, lb/(in)(s²), $A$ = piston area, in², and $V$ = effective

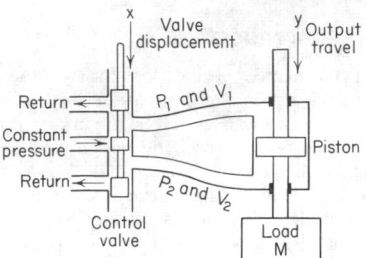

**Fig. 17**   Four-way valve-piston circuit. (*Truxal.*)

entrained fluid volume, in³ (one-half of total entrained volume between valve and piston).

The velocity of the output is proportional to the input resulting in a **velocity-control** servo. To convert this system to a **position-control** servo, mechanical, hydraulic, or electrical feedback may be employed. A valve-piston position servo with **mechanical feedback** is shown in Fig. 18. Any difference between the input $D$ and the piston position $y$ causes a motion $x$, which causes the piston to move in a direction opposite to $D$, that is, in a direction to reduce $x$. The lever ratio establishes the relationship between $y$ and $D$.

Most commercially available servo valves are two stages, permitting electrohydraulic action. The pilot stage can be operated by a low-power, short-travel electrical device, with a concomitant increase in flexibility. A typical pilot-operated servo valve is shown in Fig. 19. In this case the pilot is a double-flapper valve rather than a spool valve. (In general, small, accurate low-leakage spool valves are costly.) Upward movement of the flapper by the actuating motor results in increased pressure to the right end of the power spool. Hydraulic feedback occurs because of the increased flow across restrictor $a$. The power spool moves to the left until the unbalanced pressure is matched by the spring resistance. The disadvantage of this valve is the continual leakage flow through the flapper nozzle, but the torque motor has a low-power requirement and is inexpensive.

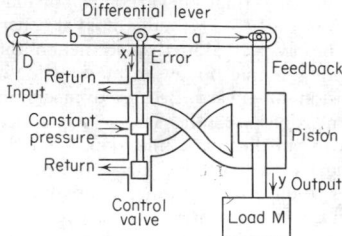

**Fig. 18**   Valve-piston position servo mechanical feedback. (*Truxal.*)

The major disadvantages of spool-type valves are (1) high cost, because of high-tolerance requirements between the valve lands, (2) high static friction and inertia, and (3) susceptibility to dirt. The **flapper** valve is less expensive to manufacture than a spool valve of equivalent characteristics and is not

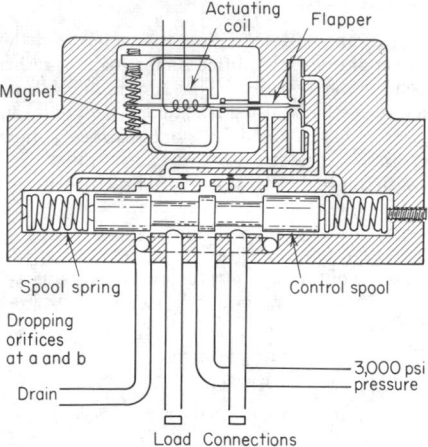

**Fig. 19**   Two-stage electrohydraulic servo valve. The first-stage is a four-way flapper valve with a calibrated pressure output driving a second-stage spring-loaded four-way spool valve. (*Moog Servocontrols, Inc.*)

so susceptible to damage by dirt particles. In Fig. 20, $P_1$, the supply pressure, is constant. Input motion of the flapper toward the nozzle increases $P_2$ and drives the piston toward the right. The steady-state characteristic $P_2$ vs. $x$ is shown in Fig. 21.

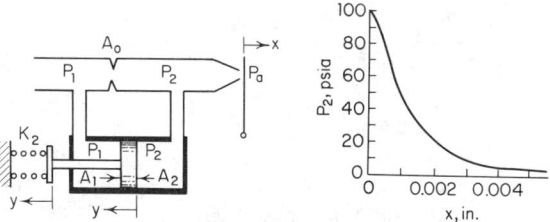

**Fig. 20**   Flapper valve. (*Raven.*)

**Fig. 21**   Equilibrium curve of $P_2$ vs. $x$ for a flapper valve. (*Raven.*)

## PNEUMATIC SYSTEMS

Pneumatic systems are widely used in industrial control. The advantage of pneumatic over hydraulic systems is the ready availability of air and the ability to discharge it indiscriminately into the atmosphere when it is used. The most common operating pressures are of the order of 20 psig; response times are considerably slower than those of hydraulic systems.

Important components in a pneumatic control system are the valve actuator, or motor, the valve, and the controller. The most commonly used **valve actuator** is the diaphragm motor in which the output pressure from the controller is counteracted

**Table 7. Typical Block Diagrams of Sampled-Data Control Systems and Their Transforms**

| System | Laplace transform of the output, $C(s)$ | z-transform of the output, $C^*(z)$ |
|---|---|---|
| 1. | $\dfrac{G_1(s)}{1 + G_1G_2^*(z)} R^*(z)$ | $\dfrac{G_1^*(z)}{1 + G_1G_2^*(z)} R^*(z)$ |
| 2. | $\dfrac{G_1(s)}{1 + G_1(s)G_2(s)} R^*(z)$ | $\left[\dfrac{G_1(s)}{1 + G_1(s)G_2(s)}\right]^* R^*(z)$ |
| 3. | | $\dfrac{G_1^*(z)}{1 + G_1^*(z)G_2^*(z)} R^*(z)$ |
| 4. | $G_1(s)\left[R(s) - G_2(s)\dfrac{RG_1^*(z)}{1 + G_1G_2^*(z)}\right]$ | $\dfrac{RG_1^*(z)}{1 + G_1G_2^*(z)}$ |
| 5. | $G_1(s)\left[R^*(z) - \dfrac{G_1^*(z)G_2^*(z)R^*(z)}{1 + G_1^*(z)G_2^*(z)}\right]$ | $\dfrac{G_1^*(z)}{1 + G_1^*(z)G_2^*(z)} R^*(z)$ |
| 6. | $G_1(s)\left[R^*(z) - G_2(s)\dfrac{R^*(z)G_1^*(z)}{1 + G_1G_2^*(z)}\right]$ | $\dfrac{G_1^*(z)}{1 + G_1G_2^*(z)} R^*(z)$ |
| 7. | $\dfrac{G_2(s)}{1 + G_1G_2G_1^*(z)} RG_1^*(z)$ | $\dfrac{G_2^*(z)}{1 + G_1G_2G_4^*(z)} RG_1^*(z)$ |
| 8. | $\dfrac{G_2(s)}{1 + G_2^*(z) + G_1G_2^*(z)} RG_1^*(z)$ | $\dfrac{G_2^*(z)}{1 + G_2^*(z) + G_1G_2^*(z)} RG_1^*(z)$ |

SOURCE: John G. Truxal (ed.), "Control Engineers' Handbook," McGraw-Hill.

not only by the spring but also by fluid forces at the valve body. The latter may cause serious deviation from linear static static behavior with deleterious control effects. High friction at the valve stem or large unbalanced fluid forces at the valve can be overcome with valve positioners, which are essentially proportional controllers.

The **control valve** is described by its lift-area relationship, where $L$ = lift and $A$ = area; e.g., linear valve, $L = kA$; equipercentage, $L = k \ln (A/A_0)$, where $A_0$ is area open to flow when lift is zero. This relationship is not necessarily the lift-flow characteristic of the valve when installed since the valve is but one component in a piping system in which pressure drops vary with the flow rate. The differential across the valve is usually taken as not less than one-third of the total losses of the system. Valves are generally selected according to how well they can compensate for nonlinearities in the system which result in a change in the character of the control response for a given controller setting when load or set-point

changes occur or when there is a variable overall pressure drop across the system. Equipercentage valves, for example, tend to control over widely varying operating conditions, since the change in flow is always proportional to the flow rate. The selection of the proper valve therefore depends on study of the particular system. Regardless of the type selected, however, the size of the valve should be such that pressure drop across the valve and not that across the connecting piping controls the flow.

The **controller** modifies the error signal in a desired manner to produce an output pressure which is used to actuate the valve motor. The several controller modes used singly or in combination are (1) the proportional mode in which $P_{out}(t) = K_c E(t)$, (2) the integral mode, in which $P_{out}(t) = 1/T_1 \int E(t)\,dt$, and (3) the rate mode, in which $P_{out}(t) = T_2\, dE(t)/dt$. In these expressions $P_{out}(t)$ = controller output pressure, $E(t)$ = input error signal, $K_c$ = proportional gain, $1/T_1$ = reset rate, and $T_2$ = rate time.

Since the proportional mode requires an error signal to change output pressure, set point and load changes in a proportionally controlled system are accompanied by a steady-state error inversely proportional to the gain. For systems which, because of stability considerations, cannot tolerate high gains, the integral mode added to the proportional will eliminate the steady-state error since the output from this mode is continually varying so long as an error exists. The addition of the integral mode to a proportional controller has an adverse affect on the relative stability of the control because of the 90° phase lag introduced.

The **rate mode,** called "anticipatory," can take large corrective action when errors are small but have a high rate of change. The mode resists not only departures from the set point but also returns and so provides a stabilizing action. Since the rate mode cannot control to a set point, it is not used alone. When used with the proportional mode, its stabilizing influence (90° phase lead; see Table 3) may allow an increase in gain $K_c$ and a consequent decrease in steady-state error.

**Controller Mechanisms**   A simple proportioning device is the pneumatic nozzle-flapper amplifier (Fig. 22 and 23). Since a typical nozzle area is 0.00002 in², the controller shown is not

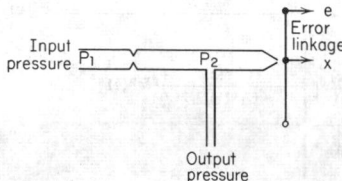

**Fig. 22**   Nozzle flapper amplifier.

capable of handling the large quantities of air that may be required for industrial use. For this purpose a second amplifier, or power relay (Fig. 24), is used as a second stage to the nozzle-flapper amplifier. As $P_i$ (from the nozzle flapper) increases, the bellows moves the valve against the supply seat, preventing flow from the supply and allowing air from the output to bleed to the atmosphere. When $P_i$ decreases, the reverse occurs.

The combination nozzle-flapper amplifier and power relay illustrated has high gain since small displacements of the flapper can result in the output traversing the full range of pressure available. Thus it can serve as an on-off controller.

In order to reduce the gain for processes in which a high gain would result in instability or too oscillatory a response, negative feedback is employed as in Fig. 24 to cancel part of the input signal. In this controller, as with the high-gain controller, the resistances and capacitances (although higher than those in the hydraulic counterpart) are still sufficiently small to be considered negligible.

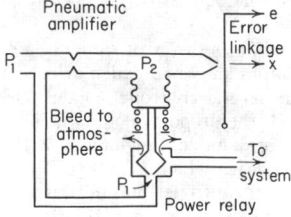

**Fig. 23**   Two-stage pneumatic amplifier.

The addition of an "integral" bellows (Fig. 25) to the low-gain proportional controller cancels the gain reduction brought about by the feedback bellows at a rate determined by the restriction $a$. That is, for a step-function input, the gain

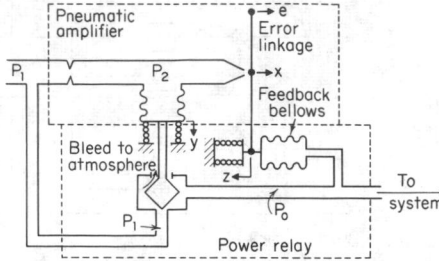

**Fig. 24**   Gain reduction of pneumatic amplifier by means of feedback bellows. *(Raven.)*

and hence the output pressure increases with time, thereby fulfilling the definition of a proportional plus integral controller.

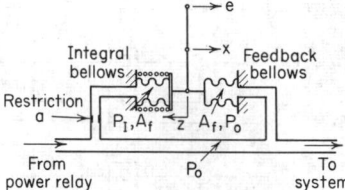

**Fig. 25**   Pneumatic amplifier with proportional plus integral action. *(Raven.)*

For a step input the rate mode requires an initial high gain which, as time proceeds, decays to the gain of the proportional controller. This is accomplished by delaying the feedback as in Fig. 26.

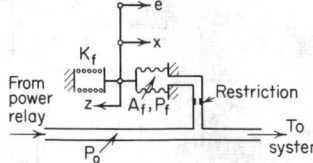

**Fig. 26**   Pneumatic amplifier with proportional plus rate action. *(Raven.)*

The selection of the proper controller can be made by methods previously outlined. Table 8 is offered as a general guide for preliminary design.

## COMPUTER CONTROL

Digital computers (Sec. 2) are being used with increasing frequency in the control of diverse processes. All but a few of the digital applications are supervisory or optimizing in nature; the computer, programmed to a model of the process, accepts measured data from conventional instruments, calculates optimum control settings for conventional controllers, and corrects them automatically. The computer need not be

**Table 8. Process Characteristics vs. Mode of Control**

| Number of process capacities | Process reaction rate | Process time lags | | Load changes | | Suitable mode of control |
| | | Resistance capacity (R-C) | Dead time (transportation) | Size | Speed | |
|---|---|---|---|---|---|---|
| Single....... | Slow | Moderate to large | Small | Any | Any | Two-position. Two-position with differential gap |
| | | | | Moderate | Slow | Multiposition. Proportional input |
| Single (self-regulating) | Fast | Small | Small | Any | Slow | Floating modes: Single speed Multispeed |
| | | | | | Moderate | Proportional-speed floating |
| Multiple..... | Slow to moderate | Moderate | Small | Small | Moderate | Proportional position |
| Multiple..... | Moderate | Any | Small | Small | Any | Proportional plus rate |
| Multiple..... | Any | Any | Small to moderate | Large | Slow to moderate | Proportional plus reset |
| Multiple..... | Any | Any | Small | Large | Fast | Proportional plus reset plus rate |
| Any......... | Faster than that of the control system | Small or nearly zero | Small to moderate | Any | Any | Wideband proportional plus fast reset |

SOURCE: Considine, "Process Instruments and Controls Handbook," McGraw-Hill.

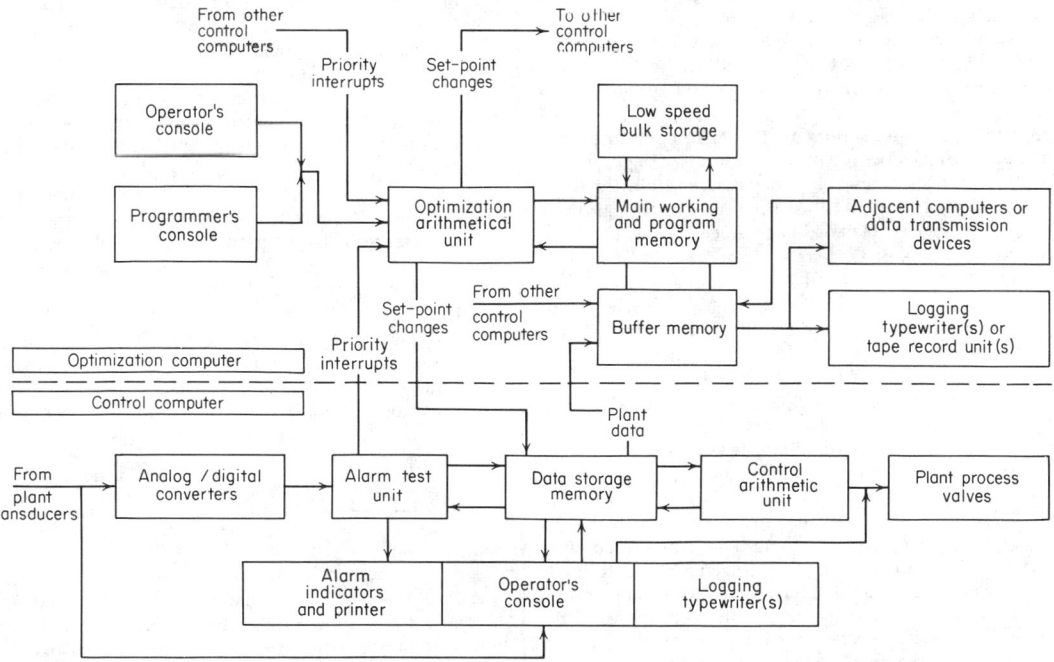

**Fig. 27**  Combined computer system for both direct digital and optimizing control. (*Chem. Eng., Mar.* 2, 1964.)

concerned only with optimizing the variables of a process for physical stability and quality but can also be used for economic optimization.

A second use of the digital computer is direct digital control (DDC) in which conventional automatic control instruments are directly displaced by a special-purpose digital computer time-shared among many control loops. The advantages of such a system are higher accuracy, more flexibility in incorpo-

rating advanced control techniques, and savings in control room costs because of compactness.

Applications of supervisory control computers are found in the electric utility industry, where they are applied to load-frequency control and automatic dispatch as well as closed-loop control; in the steel industry, where they are applied to rolling mills; and in the chemical industry, where they are used for closed-loop process control.

# SURVEYING

## by W. Lucas Gaillard and W. Lucas Gaillard, Jr.

REFERENCES: Johnson-Smith, "Theory and Practice of Surveying," Wiley. Breed and Hosmer, "Principles and Practice of Surveying," Wiley. Tracy, "Surveying," Wiley. Raymond, "Plane Surveying," American Book. Davis and Foote, "Surveying," McGraw-Hill. Rubey, Lommel, and Todd, "Engineering Survey," Macmillan. Kissam, "Surveying Practice," McGraw-Hill. Brinker and Taylor, "Elementary Surveying," International Textbook.

## LINEAR MEASUREMENTS

**Tapes** The linear measuring instrument most used is the **steel-ribbon tape.** For **surveying,** the **tape** is 100 or more ft long (30 m), graduated to feet, with 1 ft at the end graduated to tenths and sometimes hundredths. The **builder's tape** is graduated to feet, inches, and eights. **Linen tapes,** even when wire strands are woven in them, are useless for work of precision but serve for laying out road and railroad earthwork and the like. They are not sufficiently precise for laying out foundations or placing machinery.

**Variations in Tape Measurements Due to Temperature, Tension, and Sag** Steel tapes are usually of **standard length** at 68°F with a 10-lb pull for tapes 100 ft or less and 20-lb for tapes longer than 100 ft and in all cases supported throughout their length. They change 0.00000645 of their length per °F change in temperature. A 100-ft tape standard at 68°F (20°C) will be short 0.044 ft at 0°F (−12°C) and long 0.021 ft at 100°F (38°C).

A steel tape is usually standard for the pull necessary to straighten it when supported throughout its length. Additional pull stretches it at about the rate $1/28,000,000S$ of its length per pound of pull, $S$ being its cross-sectional area, in². For the common **sectional area** of 0.002 in², a pull of 10 lb will produce an elongation of 0.02 ft in 100 ft. The sectional area of a tape can be determined by dividing the weight of the tape in pounds by its length in inches and the quotient by 0.284.

When a tape is unsupported throughout its length, the measured distance is less than that indicated by about $C_s = l(wl/P)^2/24$, whereas $l$ = nominal length of unsupported tape; $w$ = weight of tape, lb per unit of length; and $P$ = pull, lb. A 100-ft tape weighing 0.624 lb pulled with 10 lb and unsupported throughout its length would in effect be shortened by 0.016 ft. Calling the correction for temperature $C_t$, for pull $C_p$, and for **sag** $C_s$, the distance between two points would be

given by the measured length $\pm\ C_t + C_p - C_s$. The **temperature correction** is **positive** or **negative** according as the temperature is **higher** or **lower** than standard. If a line of given length is to be measured from a given point, all the signs of the corrections are reversed (see also theory and tables for catenary curves, Sec. 2).

**To Measure a Horizontal Line on Sloping Ground** Most engineering linear measurements are horizontal or vertical. All land measurements mentioned in descriptions are horizontal. Vertical measurements may be made with tape or leveling instrument. When horizontal measurements are to be made on sloping ground, one of the two following methods is used: (1) The low end of the tape is raised to make the tape horizontal, the proper point on the tape or ground being transferred to the ground or tape by a plumb line held by the tapeman. When the slope is so steep that a full tape cannot be used, part of it is used—called **breaking the tape**—and it is better to pull the tape clear out, raising the several necessary sections consecutively, rather than to use one particular part of the tape several times. (2) The measurement is made along the slope, the angle of which or its rise in a tape length is determined, and the slope measurement reduced to the horizontal. In Fig. 1, the horizontal distance $B \approx S - R^2/2S$, with an error of about 0.0013 of 1 percent for a 10 percent slope and less for flatter slopes.

**To Measure or Lay Out an Angle with a Tape** In Fig. 2, from the apex of the angle $A$, measure equal distance $d$ along its sides and the distance $a$ between the ends of these measurements. Then $\sin \frac{1}{2}A = a/2d$. For small angles $A° \approx 57.3a/d$ or, if $d$ is 100 units, $A° \approx 0.573a \approx 4a/7$. Two lines separating at the rate of $n°$ separate in distance at the approximate rate of $7n/4$ units per hundred.

To lay out an angle, reverse the process. From the apex,

**Fig. 1**　　　　　　　**Fig. 2**

measure 100 units along one side; from the point obtained, describe an arc with a radius of 200 sin $\frac{1}{2}A$ or (approx) $7A/4$ units long; and from the apex with a radius of 100 units, describe an arc intersecting the first.

**To Lay Out a Right Angle**   In Fig. 3, from the point $A$ where the angle is to be, measure along one side a distance of $4n$ units; from the point thus obtained, swing an arc with radius of $5n$ units; and from $A$, swing an arc with a radius of $3n$ units to intersect the first arc at $B$; the line joining $A$ and $B$ is perpendicular to the $4n$ line.

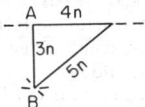

**Fig. 3**

## LEVELING

The **level**, shown in Fig. 4, consists of a telescope $EO$ resting in supports $YY$ attached to a bar $B$ and carrying a level bubble $L$. The bar $B$ is attached to a spindle $D$, which rests in a socket carrying the ball $J$ of a ball-and-socket joint. The bearing of the spindle carries the upper leveling plate $P$ and the socket of the ball joint is part of the lower leveling plate $P'$ which screws on to the tripod head $T$. By the leveling screws $S$ working through plate $P$ and resting on plate $P'$, the upper part of the instrument can be tipped with respect to $P'$. In the telescope, whose objective is at $O$ and eyepiece at $E$, there is a ring $R$ carrying two fine wires at right angles. This can be adjusted so that the intersection of the wires is in the optical axis of the telescope, and one wire can be made vertical when the other will be horizontal. Figure 4 shows a **Y level**, the supports $YY$ being in the form of Y's (wyes) in which the

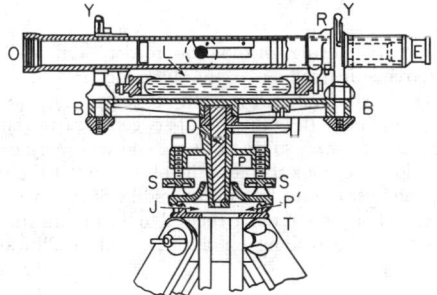

**Fig. 4**   Y level

telescope rests and from which it may be easily removed by opening clips on the Y's. In the Y level the telescope is usually an erecting one; i.e., it shows objects as they are. In another form of level, the supports are part of the telescope tube or are fastened rigidly to it and are held to the bar by screws. This is called a **dumpy level** and its (short) telescope (of large aperture) is usually inverting; i.e., it shows objects upside down and right side left. When the axis of the bubble tube and the line of sight of the telescope are parallel, the line of sight will be horizontal when the bubble is brought to the center of its tube. The level is **used with** some form of **graduated rod** read by a movable target or slide and vernier or self-reading, i.e., read directly by the level man.

**To set up the level,** the legs are planted firmly, with the lower leveling plate as nearly horizontal as practicable. The telescope and frame are swung over one diagonally opposite pair of leveling screws by which the bubble is brought to the middle of the tube. The frame is then brought over the other pair of screws and leveled; back to the first pair; to the second pair, etc., until the bubble remains in the middle of its tube for any position of the telescope. The screw which clamps the spindle motion should not be used in ordinary leveling operations. The eyepiece should be focused so that the wires are sharp against a blank ground as the sky or side of a light-colored building. It will not need changing so long as the same eye uses it. If the distance varies, the objective is focused for each object looked at.

To determine the difference in level between two points, set the level nearly midway between the points, hold a rod on one, look through the level and see where the line of sight, as defined by the eye and horizontal cross wire, cuts the rod, called **rod reading.** Move the rod to the second point and read. The difference of the readings is the difference in level of the two points. If it is impossible to see both points from a single setting of the level, one or more intermediate points, called **turning points,** are used. The readings taken on points of known or assumed elevation are called **plus sights,** those taken on points whose elevations are to be determined are called **minus sights.** The elevation of a point plus the rod reading on it gives the elevation of the line of sight; the elevation of the line of sight minus the rod reading on a point of unknown elevation gives that elevation. In Fig. 5, $I_1$ and $I_2$ are intermediate points between $A$ and $B$. The setups are numbered. Assuming $A$ to be of known elevation, the reading on $A$ is a + sight; the reading on $I_1$ from 1 is a − sight; the reading on $I_1$ from 2 is a + sight and on $I_2$ is a − sight. The algebraic sum of the plus and minus sights is the difference of level between $A$ and $B$. Target rods can usually be read by vernier to thousandths of a foot. In grading work the nearest tenth of a foot is good; in lining shafting the finest possible reading is none too good. It is desirable that the sum of the distances to the plus sights approximately equal the sum of the distances to the minus sights to ensure compensation of errors of adjustment. On a side hill this can be accomplished by zigzagging. When the direction of pointing is changed, if the bubble leaves the middle of its tube, the instrument should be releveled with the telescope in the direction of sight. If adjustments have been properly made little releveling will be necessary, but it must be remembered that **the bubble must be in the middle of its tube whenever a rod reading is taken.**

**To Make a Profile of a Line**   A bench mark is a point of reasonably permanent character whose elevation above some surface—as sea level—is known or assumed and used as a reference point for levels. The level is set up either on or a little off the line some distance—not more than about 300 ft (90 m)—from the starting point or a convenient bench mark

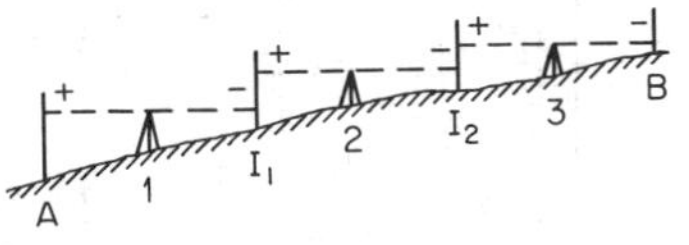

**Fig. 5**

(BM), as at K in Fig. 6. A reading is taken on the BM and added to the known or assumed elevation to get the height of the instrument, called HI. Readings are then taken at regular intervals (or **stations**) along the line and at such irregular points as may be necessary to show change of slope, as at B and C

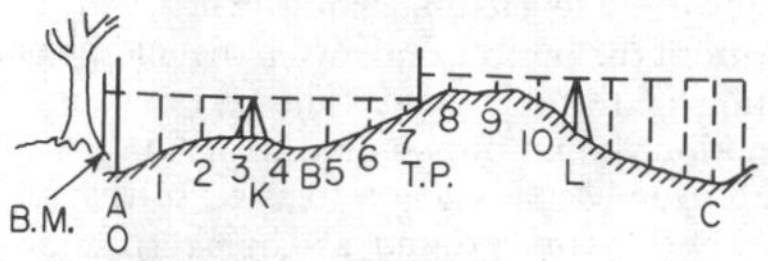

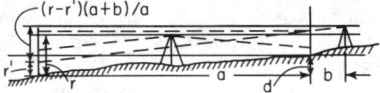

**Fig. 6**

between the regular points. The regular points are marked by stakes previously set "on line" at distances of 100 ft (30 m), 50 ft (15 m), or other distance suitable to the character of the ground and purpose of the work. When the work has proceeded as far as possible—not more than about 300 ft (90 m) from the instrument for good work—a **turning point** (TP) is taken at a regular point or other convenient place, the instrument moved ahead and the operation continued. The first reading on the BM and the first reading on a TP after a new setup are plus sights (+S); readings to points along the line and the first reading on a TP to be established are minus sights (−S).

The **notes** are taken in the following form. The elevation of a given point, both sights taken on it, and the HI determined from it all appear on a line with its station (Sta) designation. In plotting the **profile**, the vertical scale is usually exaggerated from 10 to 20 times.

**Adjustment of the Y Level** 1. Plumbing the Wire. Set up the level and bring the vertical wire to cover a suspended plumb line or the vertical corner of a building by rotating the wire ring if necessary after loosening its screws.

2. Line of Sight. Loosen the Y clips and by means of the leveling screws and the clamp and slow motion of the spindle bring the intersection of the wires to cover a minute distant point; carefully turn the telescope upside down in the Y's keeping the eye at the glass; if the intersection remains on the point, the line of sight is in adjustment; if not, bring the intersection halfway back to the point by the screws carrying the wire ring, and repeat until the distant point is covered by the intersection of the wires in either position of the telescope.

3. The Bubble Tube. Level the bubble over two sets of screws and carefully over one set; lift the telescope from the

Y's, turn it end for end and replace in the Y's; if the bubble returns to the center of the tube, its axis is parallel to the lower side of the telescope barrel and to its axis if the bearing rings are of equal diameter. For practically all work they are nearly enough so. If the bubble moves from the center of its tube, bring it halfway back by the adjusting nuts at the ends of the bubble tube, relevel, and repeat until the bubble remains in the center of its tube for both positions of the telescope.

4. The Y's. With the instrument leveled and more carefully over one set of screws, turn end for end on the vertical spindle and note whether the bubble remains in the middle of its tube; if not, adjust by the capstan nuts on the Y's at one end of the bar, bringing the bubble halfway back to the middle; relevel and repeat until the bubble remains in the middle through a complete revolution on the spindle. This last adjustment of the Y's is not essential to correct leveling, but is convenient in that when once set up the level requires no releveling for a change in the direction of pointing.

**Adjustment of the Dumpy Level** The dumpy level is adjusted by adjusting the bubble and afterward the line of sight to it.

1. The Bubble Tube. Set up and, having the instrument level over one set of screws, swing through 180° on the vertical axis; if the bubble moves from the center, bring it halfway back by the adjusting screws, relevel and repeat the test and adjustment until complete.

2. Line of Sight. Set the instrument midway between two stakes from 200 to 400 ft (60 to 120 m) apart, as in Fig. 7. With

$$(r-r')(a+b)/a$$

**Fig. 7**

the bubble in the center of the tube, read a rod on each stake. The difference in readings is the difference in level d of the two stakes. Remove the level and set it up near one of the stakes and in line with both. If set between the two stakes and close to the near stake so that the eye end will just clear a rod held on the stake, look through the object end at the rod and with a pencil get the reading in the middle of the small spot of light that will be seen; remove the rod to the distant stake, set a target to read the reading on the near rod + allowance for

|  | Left-hand page | | | Right-hand page |
| --- | --- | --- | --- | --- |

| Sta. | +S | H.I. | −S | Elev. |
| --- | --- | --- | --- | --- |
| B.M. | 6.42 | 506.42 | ......... | 500.0 |
| A = 0......... | | | 10.4 | 496.0 |
| 1............ | | | 8.2 | 98.2 |
| 2............ | | | 6.1 | 500.3 |
| +30......... | | | 5.5 | 0.9 |
| 3............ | | | 6.1 | 0.3 |
| 4............ | | | 7.9 | 498.5 |
| B = +40.... | | | 8.4 | 98.0 |
| 5............ | | | 7.5 | 98.9 |
| 6............ | | | 5.1 | 501.3 |
| 7............ | | | 3.2 | 3.2 |
| +10 T.P..... | 4.27 | 509.13 | 1.56 | 504.86 |
| 8............ | | | 2.2 | 506.9 |

This space for a heading, telling what the work is, who does it, and the date on which it is done.

This page is for remarks describing B.M.s and T.P.s or other important particulars.

earth's curvature ± *d* according as the far stake is the lower or higher; turn the telescope line of sight toward the distant rod and adjust the horizontal wire till it reads on the target when the bubble is centered. Earth's curvature is approximately 0.001 ft for 200 ft (0.3 mm for 60 m) distance and varies with the square of the distance. If the setup is a distance *b* outside the stakes which are *a* ft apart, read on the near stake, add earth's curvature *e*, and add or subtract *d* for the trial reading *r* on the distant rod; if the reading is not *r* but *r'*, move the target from *r'* by a distance $(r - r')(a \pm b)/a$, up if the result is plus, down if minus; by the wire adjusting screws bring the wire to read on the target, the bubble being kept in the middle of its tube. It is convenient to make $b = a/10$, so that $(a + b)/a = 11/10$. This is commonly called the **peg method**.

**Automatic (Self-Leveling) Level** With this instrument (Fig. 8) the circular level should be carefully centered. Thereafter

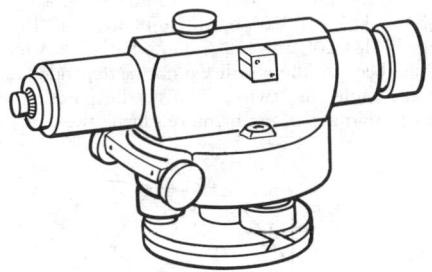

**Fig. 8** Automatic (self-leveling) level.

the pendulum device takes over so that no further leveling is required. Sometimes the pendulum sticks. To make sure that it is free after the circular level is approximately centered, turn one of the leveling screws quickly in one direction and back while looking through the telescope. If the line of sight heaves up and down once or twice, the pendulum is free. No special method is required for this instrument.

## TRANSIT WORK

**Transit** The essential parts of a surveyor's transit are shown in Fig. 9. The telescope *T* swings on axis *A* in standards *S* resting on plate *P* carrying verniers seen through openings in its upper side which permit readings on a graduated circle on plate *P'*; this plate can be turned on a spindle in a socket in the leveling head and clamped in position by the clamp *H*. The clamp *D* clamps the plates *P* and *P'* together, but it is still possible to move one on the other by the slow-motion, or tangent, screw *F*. A similar slow-motion screw attached to the clamp *H* serves to move the whole upper part of the instrument a little in a horizontal plane when the two plates are clamped together. The leveling head is like that of the level, but to the bottom of the spindle a hook is attached, from which a plumb line can be suspended for centering the transit over a point. For doing leveling and measuring vertical angles, there is a level under the telescope and a vertical circle *C* attached to the horizontal axis of the telescope and read by a vernier *V*. There is a clamp and slow-motion screw for the telescope axis of revolution. The telescope, though of shorter focal length, is like that of the level. **To set up the transit over a point,** plant the legs firmly in the ground with the plumb swinging as nearly as possible over the point. At the same time, the lower plate on the leveling head should be nearly level, as judged by eye. Loosen the leveling screws of the leveling head and shift the upper part on the lower plate till the bob swings over the required point. Bring the screws again to bearing—never tight,—swing the upper part of the instrument so that the plate bubbles are, respectively, parallel to the two diagonally opposite sets of leveling screws, and by the screws bring the bubbles to the center by leveling first one and then the other in turn until both are level. Focus the eyepiece of the telescope so that the cross wires are distinct against the sky or a light ground. Set the zeros of the verniers and graduated circle together by the clamp and slow-motion screws of the plates.

**To Produce a Straight Line** Set up the transit over one end of the line; with the lower motion clamp and tangent screw bring the telescopic line of sight to the other end of the line marked by a flag, a pencil, a pin, or other object; transit the telescope, i.e., plunge it by revolving on its horizontal axis, and set a point (drive a stake and "center" it with a tack or otherwise) a desired distance ahead in line with the telescopic line of sight; loosen the lower motion clamp and turn the instrument in azimuth until the line of sight can be again pointed to the other end of the line; again transit and set a point beside the first point set. If the instrument is in adjustment, the two points will coincide; if not, the point marking the projection of the line lies midway between the two established points.

**To Measure a Horizontal Angle** Set up the transit over the apex of the angle; with the lower motion bring the line of sight to a distant point in one side of the angle; unclamp the upper motion and bring the line of sight to a distant point in the second side of the angle, clamp and set exactly with the tangent screw; read the circle by the vernier for the angle turned.

**To Measure a Vertical Angle** Set up the transit over a point marking the apex of the angle *A* (see Fig. 10); by the lower motion and the motion of the telescope on its horizontal axis,

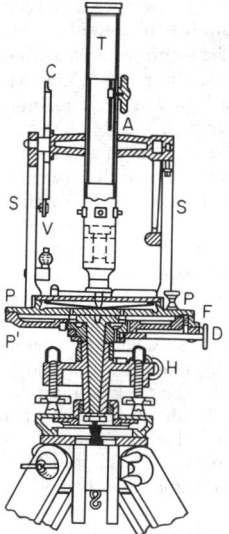

**Fig. 9** Surveyor's transit.

bring the intersection of the vertical and horizontal wires of the telescope in line with a point as much above the point defining the lower side of the angle as the telescope is above the apex; read the vertical circle; turn the telescope to a point which is the height of the instrument above the point marking the upper side of the angle and read the vertical circle. How to combine the readings to find the angle will be obvious.

**To Run a Traverse**   A traverse is a broken line marking the line of a road, bank of a stream, fence, ridge, or valley, or it may be the boundary of a piece of land. The bearing or azimuth and length of each portion of the line are determined, and this constitutes "running the traverse."

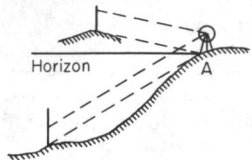

**Fig. 10**

The **bearing** of a line is the angle it makes east or west of a north and south line either true, magnetic, or assumed for the purpose of the survey. The bearing is read north or south so many degrees east or west, and never east or west so many degrees north or south. Thus, a line running only 1° north of east would have a bearing N89°E, and one running 1° south of east would have a bearing S89°E, etc.

To determine the bearing, set up over one end of the line, loosen the needle clamp, turn the telescope with its object end over the **fleur-de-lis** or north side of the compass box toward the farther end of the line, and read the needle, using the two letters between which its north end lies. It should be noticed that the compass-box letters E and W are reversed to make the reading agree with the telescope pointing.

The **azimuth** of a line is the angle the line makes with a north and south line, true, magnetic, or assumed, and differs from bearing in being measured always in one direction through 360°, while bearing is measured in each of four directions through 90°. Azimuth is measured to the right or clockwise. Astronomers use the south for zero azimuth. Surveyors, with some exceptions, use the north. Lines with bearings N88°E, S38°E, S70°W, and N60°W have azimuths of 88, 142, 250, and 300°, respectively.

To determine the azimuth of a line set up over one end, set the horizontal circle to read the azimuth of a known line through the point of setup (as the meridian or the preceding line of a traverse), and by the lower motion turn the line of sight in the direction of the known line; loosen the upper motion and set the line of sight in the direction of the required azimuth and read the circle. Always read the same vernier and the same row of figures—those inclined to the left—since the vernier reads with these when the telescope is turned clockwise. When the preceding line of the traverse is used for orienting the transit, the back azimuth should be set on the circle. The **back azimuth** is the azimuth read in a direction opposite to that in which the survey proceeds and is the forward azimuth plus 180°. If this gives more than 360°, subtract 360°. The distance may be measured with the tape or with the stadia.

In work with the transit, bearings are not usually read by the needle except for checking. Instead, the deflection angles from one course produced to the next are measured; one course—as the initial course—is taken as a meridian, and the bearings of the other courses are calculated with respect to the assumed meridian. The true or magnetic bearing of the first course can be determined, from which the bearings of all courses will be calculated from the true or magnetic meridian. To determine the **magnetic bearing** of the first course and to establish a meridian of reference, set up the transit over the initial point, let the needle swing free; with the zeros together turn the instrument on its vertical axis by the lower motion till the needle reads north, and set a point some distance away in the line of sight. The line ranged will be the magnetic meridian. By the upper motion set the telescope in the line of the initial course; the vernier will read the angle with the meridian, from which the bearing is calculated.

In Fig. 11, the bearing of $a$ is N40°E, of $b$ is N88°30′E, of $c$ is S49°20′E = 180° − (40° + 48°30′ + 42°10′), of $d$ is S36°40′W = 86° − 49°20′, or 40° + 48°30′ + 42°10′ + 86° − 180°, of $e$ is N81°20′W = 180° − (36°40′ + 62°). A meridian is established because the needle cannot be depended on to give exactly the same line twice. The needle pointing varies as much as 10 minutes of arc or more during the day.

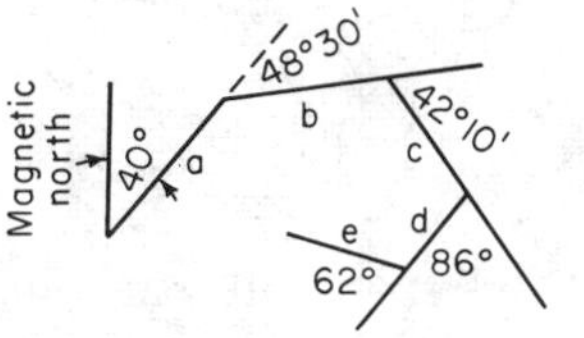

**Fig. 11**

**To Adjust a Transit**   The adjustment of the transit consists in: (1) making the plate bubbles parallel to the plates, i.e., perpendicular to the vertical axis; (2) making the line of sight perpendicular to the horizontal axis of revolution; (3) adjusting horizontal axis so that the line of sight may revolve in a vertical plane; (4) making the telescope bubble parallel to the line of sight; and (5) making the vernier of the vertical circle read zero when the line of sight is horizontal or determining the index error.

1. THE PLATE BUBBLES.   Set up the transit; when both plate bubbles are in the centers of their tubes turn the instrument on its vertical axis 180°, thus reversing the bubbles. If the bubbles remain in the centers of their tubes they are in adjustment; if not, raise or lower one end of one bubble tube with a small adjusting pin till the bubble seems to move halfway back to the center; do the same with the other tube, relevel with leveling screws, turn 180° to test correctness of work, and repeat until perfect.

2. LINE OF SIGHT.   Set up and fix the vertical wire on a suspended plumb line or corner of a vertical building. If the wire does not coincide with the vertical line, loosen all capstan screws carrying the wire ring, rotate the ring in the barrel by the screws until the wire is vertical, and then tighten the screws. By the lower motion and the vertical swing of the telescope fix the line of sight, as defined by the vertical wire, on a distant point about on a level with the instrument or the ground under it (do not clamp the horizontal axis); transit the telescope, i.e., plunge it on its horizontal axis, and, finding a minute point in the line of sight, note it carefully; turn in azimuth, i.e., on the vertical axis, until the line of sight covers

the first point sighted; transit and note whether line of sight covers the second point; if not, adjust the wires, moving the ring right or left as the case may be by the capstan-headed screws that carry it till the vertical wire seems to pass over one-fourth the distance between the two distant points; again set on the first point, transit and note a point in line—it will be neither of the points previously noted, but if the work has been completed at the first trial, the new point will lie midway between the two previously noted points—reverse in azimuth to the first point, transit, adjust if necessary, and repeat until the adjustment is complete. Stakes centered with pins 200 ft or more either side of the transit may be used for points.

3. THE HORIZONTAL AXIS. With the transit set up near a tall building, turn the line of sight on a plumb corner near the top; plunge the telescope and note if the line of sight follows down the edge of the building; if not, raise or lower one end of the horizontal axis—one end is adjustable—until the line of sight will revolve in a vertical plane. If no vertical line is available, set the line of sight on some high point, plunge and set a point on the ground; reverse in azimuth, transit, and set again on the high point and plunge; if the line of sight cuts the point set on the ground the horizontal axis is in adjustment; if not, adjust the axis until the line of sight cuts the same point below when plunged from the high point both direct and reversed.

4. LEVEL UNDER TELESCOPE. This is adjusted by the same peg method as described for the dumpy level, except that—the plate bubbles being centered—the wires are brought to the correct target reading by tipping the telescope, and the bubble is adjusted to bring it to the center, the wires being undisturbed.

5. VERTICAL CIRCLE. If the bubble tube is parallel to the line of sight and the latter is horizontal, the bubble will be in the center of its tube and the vertical circle should read zero. If it does not, the vernier may be moved slightly after loosening the screws that hold it. If not convenient, the reading may be noted and used as an index error. If the reading indicates a small angle of elevation, all angles of elevation will be read too large; i.e., the index error is to be subtracted; while depression angles will be read too small—the index error is to be added.

6. LINE OF SIGHT FOR LEVELING. For good leveling, the horizontal wire should be in the center of the telescope tube. If the eyepiece is nonadjustable, it will generally be sufficient to adjust the wire so that it appears to be in the field of view. This is not true of all instruments, and the only way to make the adjustment with certainty is to remove the telescope with its horizontal axis, place it in a pair of Y's made, for instance, by cutting notches in the ends of a wooden box of suitable size, and adjust the wires as in the Y level.

**To Measure Distances with the Stadia** In the transit telescope are two extra horizontal wires so spaced (when fixed by the maker) that they are $\frac{1}{100}$ of the focal length of the objective apart. When looking through the telescope at a rod held in a vertical position, 100 times the rod length $S$ intercepted between the two extra horizontal wires plus an instrumental constant $C$ is the distance $D$ from the center of the instrument to the rod if the line of sight is horizontal, or $D = 100 S + C$ (see Fig. 12). If the line of sight is inclined by a vertical angle $A$, as in Fig. 13, then if $S$ is the space intercepted on the rod and $C$ is the instrumental constant, the distance is given by the formula $D = 100 S \cos^2 A + C \cos A$. For angles less than 5 or 6°, the distance is given with sufficient exactness by $D = 100 S$. Although theory would indicate that distances can thus be

determined to within 0.2 ft, in practice it is not well to rely on a precision greater than the nearest foot for distances of 500 ft or less.

The instrumental constant is usually stated on a poster in the transit box. When not so given, it can be determined by measuring several distances on level ground, reading the dis-

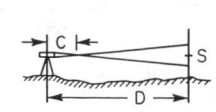

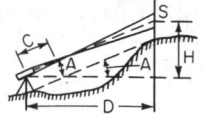

**Fig. 12**                    **Fig. 13**

tances with the stadia, as $D_1 = 100 S_1$, $D_2 = 100 S_2$, etc., subtracting the readings from the measurements and averaging the remainders. For most transits, the so-called instrumental constant is not constant but has an extreme variation of perhaps 0.1 ft. With different instruments its average value varies from about 0.75 to 1.25 ft. It is made up of the focal length of the objective and the distance from the objective to the horizontal axis.

For good work, the wire interval, i.e., the coefficient of the rod space, should be determined daily, as it may change slightly with atmospheric changes and may not always be 100. This can be done, if $C$ is known, by measuring $100 + C$ ft, $200 + C$ ft, etc., from the instrument, noting the rod intercepts. When $C$ is not known, measure two distances on level ground and read the intercepts $S$ and $S_1$; then, if $K$ is the coefficient of $S$, $C$ the instrumental constant, and $D$ and $D_1$ the two distances, $K = (D - D_1)/(S - S_1)$, and $C = D - KS$ or $D_1 - KS_1$. Several sets of readings should be taken and average results used.

The stadia wires are sometimes adjustable as to the space between them. When so, they are not in the same plane with the line and level cross wires, and hence are not seen with the same focusing of the eyepiece. Adjustable stadia wires should be tested daily and so set that 100 shall be the coefficient of the rod intercept. This can be done by laying off $100 + C$ ft from the center of the instrument and adjusting the wire to cover 1 ft on a rod held at the further end of the line.

**To Measure Differences of Level with Transit and Stadia** Measure the angle of elevation from the point of setup to the distant point required, according to the method already described, and read the rod intercept on a rod held **vertical** at the distant point. The rod intercept being $S$, its coefficient $K$, the instrumental constant $C$, angle of elevation or depression $A$, and difference of level $H$,

$$H = KS \cos A \sin A + C \sin A$$

If the rod is held normal to the line of sight,

$$H = (KS + C) \sin A$$

Tables of horizontal distances and differences of level for various vertical angles and a 100-unit rod are found in surveying textbooks.

**Contour Maps** A contour map is one on which the configuration of the surface is shown by lines of equal elevation called **contour lines.** In Fig. 14, contour lines varying by 10 ft in elevation are shown. $H$, $H$ are hill peaks, $R$, $R$ ravines, $S$, $S$ saddles or low places in the ridge $HSHSH$. The horizontal distance between adjacent contours shows the distance for a fall or rise of the contour interval—10 ft in the figure. A profile of any line as $AB$ can be made from the contour map as shown in the lower part of the figure. Conversely, a contour map may be made from a series of profiles, properly chosen. Thus, a profile line run along the ridge $HSHSH$ and radiating profile lines from the peaks down the hills and from the saddles down the ravines would give data for projecting points of equal elevation which could be connected for contour lines. This is the best method for making contour maps of very limited areas, such as city squares or very small parks. If the ground is not too much broken, the small tract is divided into squares and levels are taken at each square corner, and between two corners on some lines if necessary to get correct profiles.

**Contours with the Transit and Stadia**    When a large area of several hundred or more acres is to be contoured, or a long belt within which a railroad line is to lie, the best method is the transit and stadia method. Referring to Fig. 14, a traverse line

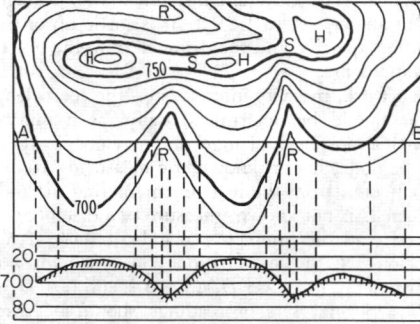

**Fig. 14**    Contour map.

would be run along the ridge by transit and stadia, establishing points in the saddles and on the peaks; from these, radial lines would be run, establishing points on the slopes of the hills; from each of these points a number of readings would be taken to slope-governing points, the azimuth, distance, and vertical angle being read, from which each point could be located in place and elevation.

## SPECIAL PROBLEMS IN SURVEYING AND MENSURATION

**Volume of Earth in Foundation and Area Grading**    The volume of earth removed from a foundation pit or in grading an area can be computed in several ways, of which two follow.

1. The area (Fig. 15) is divided into squares or rectangles, levels are taken at each corner before and after grading, and the volumes are computed as a series of prisms. If $A$ is the area (ft²) of one of the squares or rectangles—all being equal—and

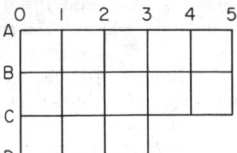

**Fig. 15**

$h_1$, $h_2$, $h_3$, $h_4$ are corner heights (ft) equal to the differences of level before and after grading, the subscripts referring to the number of prisms of which $h$ is a corner, then the volume in cubic yards is

$$Q = A(\Sigma h_1 + 2\Sigma h_2 + 3\Sigma h_3 + 4\Sigma h_4)/(4 \times 27)$$

In Fig. 15 the $h$'s at $A_0$, $D_0$, $D_3$, $C_5$, and $A_5$, would be $h_1$'s; those at $B_0$, $C_0$, $D_1$, $D_2$, $C_4$, $B_5$, $A_4$, $A_3$, $A_2$, and $A_1$ would be $h_2$'s; that at $C_3$ an $h_3$; and the rest $h_4$'s. The rectangles or squares should be of such size that their tops and bottoms are practically planes.

2. A large-scale profile of each line one way across the area is carefully made, as the $A$, $B$, $C$, and $D$ lines of Fig. 15, the final grade line is drawn on it, and the areas in excavation and embankment are separately measured with a planimeter or by estimation from the drawing. The excavation area of profile $A$ is averaged with that of profile $B$, and the result multiplied by the distance $AB$ and divided by 27 to reduce to cubic yards. Similarly, the material between $B$ and $C$ is found.

**To Pass an Obstacle**    Four cases are shown in Fig. 16. If the obstacle is large, as a building, (1) turn right angles at $B$, $C$, $D$, and $E$, making $BC = DE$ when $CD = BE$. All distances should be long enough to ensure sufficiently accurate sighting. (2) At $B$ turn the angle $K$ and measure $BC$ to a convenient point. At $C$ turn left = $360° - 2K$; measure $CD = BC$. At $D$ turn $K$ for line $DE$. $BD = 2BC \cos (180° - K)$. (3) At $B$ lay off a right angle and measure $BC$. At $C$ measure any angle to clear object and measure $CD = BC/\cos C$. At $D$ lay off $K = 90° + C$ for the line $DE$. $BD = BC \times \tan C$. If the obstacle is small, as a tree, (4) at $A$, some distance back, turn the small angle $a$ necessary to pass the obstacle and measure $AB$. At $B$ turn the angle $2a$ and measure $BC = AB$. At $C$ turn the small angle $a$ for the line $AC$, and transit, or turn the large angle $K = 180°$

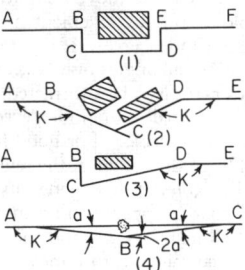

**Fig. 16**

$- a$. If $a$ is but a few minutes of arc, $AC = AB + BC$ with sufficient exactness. If only a tape is available, the right-angle method (1) above given may be used, or an equilateral triangle $ABC$ (Fig. 17) may be laid out, $AC$ produced a convenient distance to $F$, the similar triangle $DEF$ laid out, $FE$ produced to $H$ making $FH = AF$, and the similar triangle $GHI$ then laid out for the line $GH$. $AH = AF$.

**To Measure the Distance across a Stream**    To measure $AB$ (Fig. 18), $B$ being any established point, tree, stake, or building corner: (1) Set a transit over $A$; turn a right angle from $AB$ and measure any distance $AC$; set over $C$ and measure the angle $ACB$. $AB = AC \tan ACB$. (2) Set over $A$, turn any convenient angle $BAC'$ and measure $AC'$; set over $C'$ and measure $AC'B$. Angle $ABC' = 180° - AC'B - BAC'$. $BA = AC' \times \sin AC'B/\sin ABC'$. (3) Set up on $A$ and produce $BA$

any measured distance to *D*; establish a convenient point *C* about opposite *A* and measure *BAC* and *CAD*; set over *D* and measure *ADC*; set over *C*, and measure *DCA* and *ACB*; solve *ACD* for *AC*, and *ABC* for *AB*. For best results the acute angles of either method should lie between 30 and 60°.

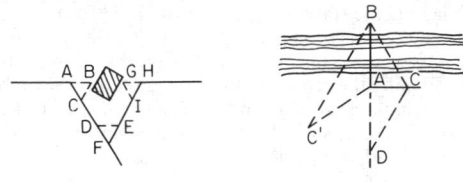

**Fig. 17**          **Fig. 18**

**To Measure a Visible but Inaccessible Distance** (as *AB* in Fig. 19) Measure *CD*. Set a transit at *C* and measure angles *ACB* and *BCD*; set at *D* and measure angles *CDA* and *ADB*.

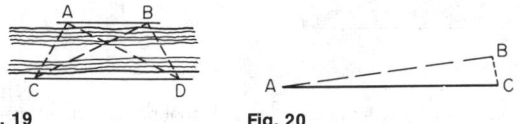

**Fig. 19**          **Fig. 20**

$CAD = 180° - (ACB + BCD + CDA)$. $AD = CD \times \sin ACD/\sin CAD$. $CBD = 180° - (BCD + CDA + ADB)$. $BD - CD \times \sin BCD/\sin CBD$. In the triangle *ABD*, $\frac{1}{2}(B + A) = 90° - \frac{1}{2}D$, where *A*, *B* and *D* are the angles of the triangle; $\tan \frac{1}{2}(B - A) = \cot \frac{1}{2}D(AD - BD)/(AD + BD)$; $AB = BD \sin D/\sin A = AD \sin D/\sin B$.

**Random Line** On many surveys it is necessary to run a random line from point *A* to a nonvisible point *B* which is a known distance away. On the basis of compass bearings, a line such as *AB* is run. The distances *AB* and *BC* are measured, and the angle *BAC* is found from its calculated tangent (see Fig. 20).

**To Stake out a Simple Horizontal Curve** A simple horizontal curve is composed of a single arc. Usually the curve must be laid out so that it joins two straight lines called tangents, which are marked on the ground by PT (points on tangent). These tangents are run to intersection, thus locating the PI (point of intersection). The plus of the PI and the angle *I* are measured (see Fig. 21). With these values and any given value of *R* (the radius desired for the curve), the data required for staking out the curve can be computed.

$$R = \frac{5,729.58}{D} \qquad L = 100\frac{\Delta}{D} \qquad T = R\tan\frac{\Delta}{2}$$

where *R* = radius, *T* = tan distance, *L* = curve length, *C* = long chord. In a sample computation, assume that $\Delta = 8°24'$ and a 2° curve is required. $R = 5,729.58/2 = 2,864.79$ ft (873 m); $L = (100) (8.40/2) = 420.0$ ft (128 m); $T = 2,864.79 \times 0.07344 = 210.39$ ft (64 m). The degree of the curve is always twice as great as the deflection angle for a chord of 100 ft (30 m).

**Setting Stakes for Trenching** A common way to give line and grade for trenching (see Fig. 22) is to set stakes *K* ft from the center line, driving them so that the near face is the measuring point and the top is some whole inch or tenth of a foot above the bottom grade or grade of the center or top of the pipe to be laid. The top of the pipe barrel is perhaps the better

line of reference. If preferred, two stakes can be driven on opposite sides and a board nailed across, on which the center-line is marked and the depth to pipeline given. When only one stake is used, a graduated pole sliding on one end of a level board at right angles is convenient for workmen and inspectors. On long grades, the grade stakes are set by "shooting in." Two grade stakes are set, one at each end of the grade, a transit is set over one, its height above grade determined, and a rod reading calculated for the distant stake such as to make the line of sight parallel to the grade line; the transit line of sight is then set at this rod reading; when the rod is taken to any intermediate stake, the height of instrument above grade less the rod reading will be the height of the top of the stake above grade. If the ground is uniform, the stakes may all be set at the same height above grade by driving them so as to give the same rod readings throughout.

**To Reference a Point** The point *P* (Fig. 23), which must be disturbed during construction operations and will be again required as a line point in a railway, pipeline, or other survey, is referenced as follows: (1) By setting the transit over it and setting four points, *A*, *B*, and *C*, *D* on two intersecting lines. When *P* is again required, the transit is set over *B* and, with foresight on *A*, two temporary points close together near *P* but on opposite sides of the line *DC* are set; the transit is then

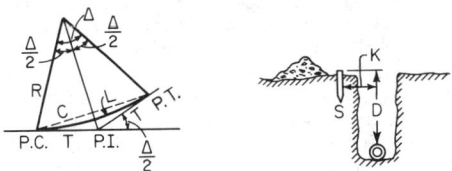

**Fig. 21**          **Fig. 22**

set on *D* and, with foresight on *C*, a point is set in the lines *DC* and *BA* by setting it in *DC* under a string stretched between the two temporary points on *BA*. (2) Points *A* and *E* and *C* and *F* may be established instead of *A*, *B*, *C*, *D*. (3) If the ground is fairly level and is not to be much disturbed, only points *A* and *C* need be located, and these by simple tape measurement from *P*. They should be less than a tape length from *P*. When *P* is wanted, arcs struck from *A* and *C* with the measured distance for radii will give *P* at their intersection.

**Foundations** The corners and lines of a foundation are preserved by setting stakes outside the area to be disturbed, as in Fig. 24. Cords stretched around nails in the stakes marking

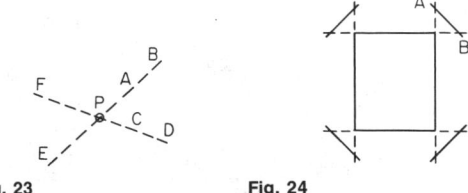

**Fig. 23**          **Fig. 24**

the reference points will give the referenced corners at their intersections and the main lines of the building. These corners can be plumbed down to the level desired if the height of the stakes above grade is given. It is well to nail boards across the stakes at *AB*, putting nails in the top edge of the board to mark the points A and B, and if the ground permits, to put all the boards at the same level.

**To Test the Alignment and Level of a Shaft**   Having placed the shaft hangers as closely in line as possible by the use of a chalk line, the shaft is finally adjusted for line by hanging plumb lines over one side of the shaft at each hanger and bringing these lines into a line found by stretching a cord or wire or by setting a transit instrument at one end and adjusting at each hanger till its plumb line is in the line of sight. The position of the line will be known either on the floor or on the ceiling rafters or beams to which the hangers are attached. If the latter, the transit may be centered over a point found by plumbing down, and sighted to a plumb line at the farther end.

To level the shaft, an ordinary carpenter's level may be used near each hanger, or, better, a pole with an improvised sliding target may be hung over the shaft at each hanger by a hook in one end. The target is brought to the line of sight of a leveling instrument set preferably about under the middle of the shaft, by adjusting the hanger.

When the hangers are attached to inclined roof rafters, the two extreme hangers can be put in a line at right angles to the vertical planes of the rafters by the use of a square and cord. The other hangers will then be put as nearly as possible without instrumental test in the same line. The shaft being hung, the two extreme hangers, which have been attached to the rafters about midway between their limits of adjustment, are brought to line and level by trial, using a transit instrument with a well-adjusted telescope bubble, a plumb line, and inverted level rod or target pole. Each intermediate hanger is then tested and may be adjusted by trial.

**To Determine the Verticality of a Stack**   If the stack is not in use and its top is accessible, a board can be fitted across the top, the center of the opening found, and a plumb line suspended to the bottom, where its deviation from the center will show any leaning. If the stack is in use or its top not accessible and its sides are battered, the following procedure may be followed. Referring to Fig. 25, set up a transit at any point T and measure the horizontal angles between vertical planes

tangent respectively to both sides of the top and the base and also the angle a to a second point $T_1$. On a line through T approximately at right angles to the chimney diameter, set the transit at $T_1$ and perform the same operations as at $T_1$, measuring also K and the angle b. On the drawing board, lay off K to as large a scale as convenient, and from the plotted T and $T_1$ lay off the several angles shown in the figure. By trial, draw circumferences tangent to the two quadrilaterals formed by the intersecting tangents of the base and top, respectively. The line joining the centers of these circumferences will be the deviation from the vertical in direction and amount. If the base is square, T and $T_1$ should be established opposite the middle points of two adjacent sides, as in Fig. 26.

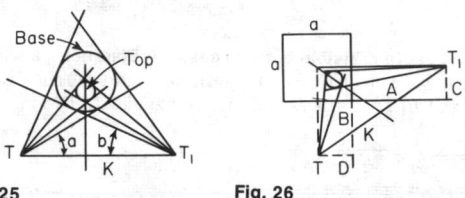

**Fig. 25**                                  **Fig. 26**

**Industrial Applications**   Optical tooling is providing the equipment necessary to meet the increase in the size of manufactured items and the decrease in tolerance limits. The precise measure required in both individual parts and the final assembly in mass production have made the toolmaker's old standbys obsolete.

**Basic Equipment**   The basic instruments used in industrial applications of surveying methods are alignment telescope or jig alignment telescope, jig transit, and precise level. The alignment telescope provides permanent reference lines on a jig or other structure. The jig transit establishes an absolutely vertical plane exactly where desired on line with two marks or precisely at right angles to other line of sight. The precise (tilting) level fixes a true horizontal plane at any desired height. All three instruments are self-checking and can be tested quickly and adjusted exactly.

Section **17**

# Industrial Engineering

BY

**B. W. NIEBEL** *Professor of Industrial Engineering, The Pennsylvania State University.*
**ADOLPH MATZ** *Professor Emeritus of Accounting, The Wharton School, University of Pennsylvania.*
**ROBERT W. KENNARD** *Manager, Systems Engineering, Engineering Department, E. I. du Pont de Nemours & Co.*
**WILLIAM ANTIS** *Technical Director, Maynard Research Council Incorporated, Pittsburgh, Pennsylvania.*
**ARTHUR J. FIEHN** *Vice President, Power Technology, Burns and Roe, Inc.*
**ASHOK AHUJA** *Manager, Power Generation Proposals, General Electric Co.*

**INDUSTRIAL ECONOMICS AND MANAGEMENT**
**by B. W. Niebel**

Plant Organization ........................................ 17-2
Process Planning ......................................... 17-3
Process Analysis ......................................... 17-4
Production Control ....................................... 17-5
Control of Materials ..................................... 17-7
Strategic Economic Evaluations ....................... 17-8
Wage Administration .................................... 17-9
Employee Relations ...................................... 17-9

**COST ACCOUNTING**
**by Adolph Matz**

Cost Determination and Control ........................ 17-10
Balance Sheet and Income Statement ................... 17-11
Types of Cost Systems .................................. 17-12
Cost Methods ............................................ 17-13
Elements of Costs ....................................... 17-13
Budgets and Standard Costs ............................ 17-14
Cost Analysis ........................................... 17-14
Capital-Expenditure Decisions ......................... 17-16
Linear Programming ..................................... 17-16

**ENGINEERING STATISTICS AND QUALITY CONTROL**
**by Robert W. Kennard**

Engineering Statistics and Quality Control ............ 17-19
Statistics and Variability .............................. 17-19
Characterizing Observational Data: The Average and
   Standard Deviation ................................... 17-19
Process Variability—How Much Data? .................. 17-20

Correlation and Association ............................ 17-21
Comparison of Methods or Processes ................... 17-22
Go/No-Go Data .......................................... 17-24
Control Charts .......................................... 17-25

**TIME AND MOTION STUDY**
**by William Antis**

Scope of Time and Motion Study ....................... 17-26
Aims of Time and Motion Study ........................ 17-26
Elements of Time and Motion Study ................... 17-26
Operation Analysis ...................................... 17-27
Principles of Motion Study ............................. 17-28
Standardizing the Job ................................... 17-28
Time-Study Observations ............................... 17-28
Performance Rating ..................................... 17-30
Allowances for Fatigue and Personal and Unavoidable
   Delays ................................................ 17-30
Developing the Time Standard .......................... 17-30
Time Formulas and Standard Data ..................... 17-32
Uses of Time Standards ................................. 17-32

**COST OF ELECTRIC POWER**
**by Arthur J. Fiehn**
**(with the assistance of Ashok Ahuja)**

Constructed Plant Costs ................................ 17-33
Fixed Charges .......................................... 17-37
Operating Expenses ..................................... 17-38
Overall Generation Costs ............................... 17-40
Transmission Costs ..................................... 17-41
Power Cost and Prices .................................. 17-42

# INDUSTRIAL ECONOMICS AND MANAGEMENT

## by B. W. Niebel

REFERENCES: Roscoe, "Project Economy," Irwin. Niebel, "Motion and Time Study," Irwin. Moore, "Manufacturing Management," Irwin. Folts, "Introduction to Industrial Management," McGraw-Hill. Bock and Holstein, "Production Planning and Control," Merrill. Mayer, "Production Management," McGraw-Hill. Maynard, "Handbook of Modern Manufacturing Management," McGraw-Hill. Eary and Johnson, "Process Engineering for Manufacturing," Prentice-Hall.

### Plant Organization

**Organization** generally is recognized as the foundation of management. The term, as it is used in industry and business, means the distribution of the functions of the business to the personnel logically qualified to handle them. It should be noted that the organization should be built around functions rather than individuals.

The majority of progressive concerns today are organized on a **line-and-staff** basis. The relationships usually are shown on an **organization chart,** which reveals the relationships of the major divisions and departments and the lines of direct authority from superior to subordinate. Lines of authority usually are shown as vertical lines. **Staff authority** frequently is indicated by a dotted line, which distinguishes it from direct authority. This same procedure is usually used to indicate committee relationships. Departments or activities are clearly identified within framed rectangles. The names of individuals responsible for a given department or activity often are included with their job organization titles. Although the organization chart shows the relationship of organization units, it does not clearly define the responsibilities of the individuals and the groups. Thus organization charts must be supplemented with carefully prepared job descriptions for all members of the organization. **Position descriptions** are written definitions of jobs enumerating the duties and responsibilities of each position.

A **line organization** comprises those individuals, groups, and supervising employees concerned directly with the productive operation of the business. The paths of authority are clearly defined, as each individual has but one superior from whom he or she obtains orders and instructions. This superior reports to but one individual, who has complete jurisdiction over his or her operation and supplies necessary technical information. In large- and middle-sized organizations, a pure line-type enterprise cannot exist because of the complexity of our business society.

A **staff organization** involves personnel, departments, or activities that assist the line supervisor in any advisory, service, coordinating, or control capacity. It should be noted that a staff position is a full-time job and is essentially the work of a specialist. Typical staff functions are performed by the company's legal department, controller, and production control. Figure 1 illustrates a typical line-staff activity.

Committees are used in some instances. A committee is a group of individuals which meets to discuss problems or projects within its area of assigned responsibility in order to arrive at recommendations or decisions. A **committee** operates on a staff basis. Although committees are time-consuming and frequently delay action, their use combines the experience and judgment of several persons, rather than a single individual, in reaching decisions.

The control of organization is the responsibility of two groups of management: (1) **administrative management,** which has the responsibility for determining policy and coordinating sales, finance, production, and distribution, and (2) **production management,** which has the responsibility for executing the policies established by administration.

In building an efficient organization, management should abide by certain principles, namely:

1. Clear separation of the various functions of the business should be established to avoid overlap or conflict in the accomplishment of tasks or in the issuance or reception of orders.

2. Each managerial position should have a definite location within the organization, with a written job specification.

3. There should be a clear distinction between line and staff operation and control.

4. A clear understanding of the authority under each position should prevail.

5. Selection of all personnel should be based on unbiased techniques.

6. A recognized line of authority should prevail from the top of the organization to the bottom, with an equally clear line of responsibility from the bottom to the top.

7. A system of communication should be well established and definitely known—it should be short, yet able to reach rapidly everyone in the organization.

Staff members usually have no authority over any portion of the organization that the staff unit assists. However, the department or division that is being assisted by the staff can make demands upon the staff to provide certain services. There are instances where a control type of staff may be delegated to direct the actions of certain individuals in the organization that they are servicing. When this takes place, the delegated authority may be termed **staff authority;** it is also frequently known as **functional authority** because its scope is determined by the functional specialty of the staff involved.

Many businesses today are finding it fruitful to establish "temporary organizations" in which a team of qualified individuals, reporting to management, is assembled to accomplish a mission, goal, or project, and then this organization is disbanded when the goal is reached.

**Good organization requires** that (1) responsibilities be clearly defined; (2) responsibility be coupled with corresponding authority; (3) a change in responsibility be made only after a definite understanding exists to that effect by all persons concerned; (4) no employee be subject to definite orders from more than one source; (5) orders not be given to subordinates over the head of another executive; (6) all criticism be made in a constructive manner and be made privately; (7) promotions, wage changes, and disciplinary action always be approved by the executive immediately superior to the one directly responsible; (8) employees whose work is subject to regular inspection or appraisal be given the facilities to maintain an independent check of the quality of their work.

### Process Planning

Process planning encompasses selecting the processes to be used in the most advantageous sequence, selecting the specific equipment to be employed, selecting the tooling to be used, and specifying the locating points of the special tools.

The **principal constraints** that must be considered in the selection of a given basic process (the first process used in the sequence that provides the evolvement of the finished design) are

1. Type and condition of raw material used
2. Size of raw material that the equipment can handle
3. Geometrical configurations that the equipment is capable of imparting to the raw material
4. Tolerance and surface finish capabilities of the equipment
5. Quantity of finished parts needed and their delivery requirements
6. Economy of the process

Once the basic process (casting, forging, pressing, weldment, etc.) that the design will be produced from has been determined, the process planner will need to determine the secondary operations required to transform the work to product specifications. Secondary operations can be classified into four categories. These are: critical operations, placement operations, tie-in operations, and protection operations.

**Critical manufacturing operations** are those that are applied to areas of the part where dimensional or surface specifications are sufficiently exacting to require quality control or are used for locating the workpiece in relation to other areas or mating parts. Critical surfaces almost always mate with other machined surfaces.

A critical manufacturing area can be recognized in one of three ways:

1. The surface or area represents a location to which other surfaces or areas are shown as having a relationship.
2. The surface or area has close dimensional tolerances. Close tolerances are usually thought of as those being within 0.005 in. Tolerances can be applied not only to size but also to roundness, straightness, and concentricity.
3. The surface or area has specified conditions such as surface finish, flatness requirements, or squareness requirements.

**Placement operations** are those whose method and sequence are determined principally by the nature and occurrence of the critical operations. Placement operations may be thought of as being of two types:

1. Those operations that take place to prepare for a critical operation: for example, it may be necessary to machine a surface to provide a suitable stable location for a subsequent critical operation.
2. Those operations that take place to correct the workpiece to return it to its required geometry or characteristic: for example, a press blanking operation may result in portions of the part curling so that it will be necessary to add a flattening placement operation.

**Tie-in operations** are those productive operations whose

sequence and method are determined by the geometry to be accomplished on the work as it comes out of a founding or critical operation in order to satisfy the specifications of the finished part. Tie-in operations may be thought of as those secondary productive operations which are necessary to produce the part, but which are not thought of as being critical. Tie-in operations usually are performed to standard machine tolerances and do not identify a mastering surface from which other surfaces are located to close tolerances.

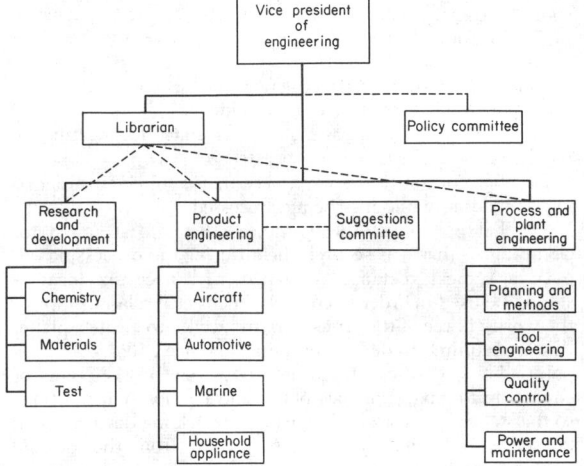

**Fig. 1** Organization chart illustrating the activities reporting to the vice president of engineering.

**Protection operations** are those operations, nonproductive in nature, that are performed to protect the product from the environment and handling during its progress through the plant and to the customer, and also those operations that control the product's level of quality. Broadly speaking, all protection operations may be classified as falling into one of three groups. These are:

1. Application of protective coatings
2. Inspection and test
3. Packing for shipment

**Protective coatings** may be applied at several stages in the manufacture of a product. Frequently a rust-preventive coating is applied to raw material as soon as it is received to protect it against corrosion in the processor's plant. Similarly, semifinished products may be given a protective coating before being sent to temporary storage. The finished product frequently is treated to protect it from the elements until it arrives safely at the customer's plant.

Protective and decorative coatings specified on the drawings are not operations falling under the classification of protective operations. The applications of these decorative and protective coatings are usually tie-in operations that are performed late in the process.

After all the critical, tie-in, placement, and protection operations have been identified, the best sequence should be determined. This can be done by considering the following:

1. Logical process order
2. Geometrical and dimensional control that the process is capable of maintaining

Regarding the **logical process order,** rough work involving heavy cuts and liberal tolerances should be performed early in the sequence while close-tolerance-type operations should be performed late in the sequence. In general, the final finishing of internal work is done in advance of external work. When internal work is performed, the logical sequence is: drilling, boring, recessing, reaming, and tapping. The logical sequence of external work is: turning, facing, grooving, forming, and threading.

In considering the **geometrical and dimensional control** that can be maintained, critical operations that establish locating points should be established early in the process. It is easier to maintain control from a large plane surface than from a curved, irregular, or small surface.

In providing the **special tooling** needed to perform the various operations, it is necessary to determine:

1. Those points or areas that are best suited for locating the workpiece while it is being processed
2. That portion of the workpiece that is suited to supporting or holding while it is being processed
3. That portion of the workpiece that is best suited for clamping so that it is securely held during the processing

In a typical rectangular-shaped workpiece six locators should be used in order to control dimensional relationships of the work. Three of the six **locators** are needed to locate a plane, two are required to determine a line, and one will determine a point. Thus work can be located positively by six points of contact in tooling. The locators should be chosen at positions so that when the work is clamped in the holding device, it will be stable and will not lift or rock away from the locators during the processing operation.

Locators should be placed on one of the two surfaces that identify a close tolerance. If close dimensions are shown to center lines, locators should be placed equally distant from either side of the center lines from which dimensioning takes place. If a surface requires a close tolerance relative to squareness, parallelism, or concentricity, three locators should be placed against the surface.

**Supports** should be provided at those areas where tool or clamping forces or the static weight of the part results in deflection of the work. To assure that all the work is held in a correct position throughout the process, it must be rigidly supported. The workpiece should not contact the support until the tooling or clamping forces are applied, since the only purpose of the support is to avoid or limit deflection and distortion.

It is necessary to provide adequate **holding force** so that all locators contact the workpiece during the processing cycle. It is usually most advantageous to place holding forces directly opposite locators and preferably on surfaces that do not require a fine finish. A nonrigid workpiece may require holding forces at several locations in order to hold the work against all locators.

### Process Analysis

**Process analysis** is a procedure for studying all productive and nonproductive operations for the purpose of optimizing cost, production output, or quality. The procedure is first to acquire all information related to the volume of the work that will be directed to the process under study, namely, the expected volume of business, the chance of repeat business, the life of the job, the chance for design changes, and the labor content of the job. This will determine the time and effort to be devoted toward improving the existing process or planning a new process.

Once an estimate is made of quantity, process life, and labor content, then all pertinent factual information should be collected on operations; facilities used for transportation and transportation distances; inspections, inspection facilities, and inspection times; storage, storage facilities, and time spent in storage; vendor operations, together with vendor prices; and all drawings and design specifications. When the information affecting cost and method is gathered, it should be presented in a form suitable for study, e.g., a **flow process chart.** This chart presents graphically and chronologically all manufacturing information. Studies should be made of each event with thought toward improvement. The recommended procedure is to take each step in the present method individually and analyze it with a specific approach toward improvement, considering the key points of analysis. After each element has been analyzed, the process should be reconsidered with thought toward overall improvement. The primary approaches that should be used when analyzing the flow chart include (1) purpose of operation, (2) design of parts, (3) tolerances and specifications, (4) materials, (5) process of manufacture, (6) setup and tools, (7) working conditions, (8) materials handling, (9) plant layout, and (10) principles of motion economy. (See also Sec. 12, Industrial Plants.)

**Purpose of Operation** Many operations can be eliminated if sufficient study is given the procedural process. Before accepting any operation as necessary, its purpose should be clearly determined and **checklist questions** should be asked to stimulate ideas that may result in eliminating the operation or some component of it. Typical checklist questions are: Can purpose be accomplished better in another way? Can operation be eliminated? Can operation be combined with another? Can operation be performed during idle period of another? Is sequence of operations the best possible?

**Design of Parts** Design should never be regarded as permanent. Experience has shown that practically every design can be improved. The analyst should consider the existing design to determine if it is possible to make improvements. In general, improvements can be made by (1) simplifying the design through reduction of the number of parts, (2) reducing the number of operations required to produce the design, (3) reducing the length of travel in the manufacture of the design, and (4) utilizing a better material in design.

**Tolerances and Specifications** These frequently can be liberalized to decrease unit costs without detrimental effects on quality; in other instances, they should be made more rigid to facilitate manufacturing and assembly operations. Tolerances and specifications must be investigated to ensure the use of an optimum process.

**Materials** Five considerations should be kept in mind relative to both the direct and the indirect material used in the process: (1) finding a less expensive material, (2) finding materials easier to process, (3) using materials more economically, (4) using salvage materials, and (5) using supplies and tools economically.

**Process of Manufacture** Improvement in the process of manufacture is perhaps the salient point, and possible improvements deserving special consideration include (1) mechanizing manual operations, (2) utilizing more efficient facilities on mechanical operations, (3) operating mechanical facilities more efficiently, and (4) when changing an operation, considering the possible effects on subsequent operations. There are almost always many ways to produce a given design, and better production methods are continually being developed. By systematically questioning and investigating the manufacturing process, more effective methods will be developed.

**Setup and tools** have such a dominant influence on economics that consideration must include quantity to be produced, chance for repeat business, amount of labor involved, delivery requirements of the customer, and capital needed to develop the setup and provide the tools. Specifically, consideration should be given to reducing the setup time by better planning in production control, designing tooling for the full-capacity utilization of the production facility, and introducing more efficient tooling.

**Good working conditions** are an integral part of an optimum process as they improve the safety record, reduce absenteeism and tardiness, raise employee morale, improve public relations, and increase production. Consideration should include (1) improved lighting; (2) controlled temperature; (3) adequate ventilation; (4) sound control; (5) promotion of orderliness, cleanliness, and good housekeeping; (6) arrangement for immediate disposal of irritating and harmful dusts, fumes, gases, and fogs; (7) provision of guards at nip points and points of power transmission; (8) installation of personnel-protection equipment; and (9) sponsorship and enforcement of a well-formulated first aid and safety program.

**Materials Handling**   The handling of materials is an essential part of each operation and frequently consumes the major share of the time. Materials handling adds nothing but cost to the product, and it should accordingly be reduced. When analyzing the flow process chart, keep in mind that the best-handled part is the least manually handled part. Whether distances of moves are large or small, points to be considered for reduction of time and energy spent in handling materials are (1) reduction of time spent in picking up material, (2) maximum use of mechanical handling equipment, (3) better use of existing handling facilities, (4) greater care in the handling of materials.

**Plant Layout**   Good process design requires good **plant layout.** This involves development of the workplace so that the location of the equipment introduces maximum economy during the manufacturing process. In general, plant layouts represent one or a combination of (1) product, or straight-line, layouts, and (2) process, or functional, arrangements. In the **straight-line layout,** machinery is located so the flow from one operation to the next is minimized for any product class. Thus it would not be unusual to see a surface grinder located between a milling machine and a turret lathe, with an assembly bench and plating tank in the immediate area. **Process, or functional, layout** is the grouping of similar facilities, e.g., all turret lathes in one section, department, or building.

The principal advantage of **product grouping** is lower materials-handling costs since distances moved are minimized. The major disadvantages are:

1. Since a broad variety of occupations are represented in a small area, employee discontent can readily be fostered.

2. Unlike facilities grouped together result in operator training becoming more difficult since no experienced operator on a given facility may be located in the immediate area to train new employees.

3. The problem of finding competent supervisors is increased due to the variety of facilities and jobs to be supervised.

4. Greater initial investment is required because of duplicate service lines such as air, water, gas, oil, and power lines.

5. The arrangement of facilities tends to give a casual observer the thought that disorder prevails. Thus it is more difficult to promote good housekeeping.

In general, the disadvantages of product grouping are more than offset by the advantage of low handling cost if production requirements are substantial.

**Process, or functional, layout** gives an appearance of neatness and orderliness and, consequently, tends to promote good housekeeping; new workers can be trained more readily, and it is easier to obtain experienced supervision since the requirements of supervising like facilities are not so arduous. The obvious disadvantages of process grouping are the possibilities of long moves and of backtracking on jobs that require a series of operations on diversified facilities. In planning the process, important points to be considered are: (1) For straight-line mass production, material laid aside should be in position for the next operation. (2) For diversified production, the layout should permit short moves and deliveries and the material should be convenient to the operator. (3) For multiple-machine operations, equipment should be grouped around the operator. (4) For efficient stacking, storage areas should be arranged to minimize searching and rehandling. (5) For better worker efficiency, service centers should be located close to production areas.

**Principles of Motion Economy**   The last of the primary approaches to process design is the analysis of the flow chart for the incorporation of basic principles of motion economy. When studying work performed at any work station, the engineer should ask: (1) Are both hands working at the same time and in opposite, symmetrical directions? (2) Is each hand going through as few motions as possible? (3) Is the workplace arranged so that long reaches are avoided? (4) Are both hands being used effectively, with neither being used as a holding device? In the event that "no" is the answer to any of these questions, then the work station should be altered to incorporate improvements related to motion economy. (See also Sec. 17, Time and Motion Study.)

## Production Control

Production control includes the scheduling of production; the dispatching of materials, tools, and supplies at the required time so that the predicted schedules can be realized; the follow-up of production orders to be sure that proposed schedules are realized; the maintenance of an adequate inventory to meet production requirements at optimum cost; and the maintenance of cost and manufacturing records to establish controls, estimating, and equipment replacement. Consideration must be given to the requirements of the customer, the available capacity, the nature of the work that precedes the production to be scheduled, and the nature of the work that succeeds the current work being scheduled.

**Scheduling** may be accomplished with various degrees of refinement. In low-production plants where the total number of hours required per unit of production is large, scheduling may adequately be done by departmental loading; e.g., if a department has a total of 10 direct-labor employees, it has 400 available work hours per week. Every new job is scheduled by departments giving consideration to the average number of available hours within the department. A refinement of this method is to schedule groups of facilities or sections, e.g., to schedule the milling machine section as a group. In high-production plants, detailed facility scheduling frequently is necessary in order to ensure optimum results from all facilities. Thus, with an 8-hour shift, each facility is recorded as having 8 available hours, and work is scheduled to each piece of equipment indicating the time that it should arrive at the work station and the time that the work should be completed.

Scheduling is frequently done on control boards utilizing commercially available devices, such as *Productrol, Sched-U-Graph,* and *Visi-trol.* These, in effect, are mechanized versions of **Gantt charts,** where schedules are represented by paper strips cut to lengths equivalent to standard times. The strips are placed in the appropriate horizontal position adjacent to the particular order being worked; delays are conspicuously marked by red signals at the delay point. Manual posting to a ledger maintains projected schedules and cumulative loads. The digital computer is successfully used as a scheduling facility.

A recent adaptation of the Gantt chart, **PERT** (Program Evaluation and Review Techniques), is finding considerable application to project-oriented scheduling (as opposed to repetitive-type applications). This prognostic management planning and control method graphically portrays the optimum way to attain some predetermined objective, usually in terms of time. The **critical path** (CPM = Critical Path Method) consists of that sequence of events in which delay in the start or completion of any event in the sequence will cause a delay in the project completion.

In using PERT for scheduling, three time estimates are used for each activity, based upon the following questions: (1) What is the earliest time (optimistic) in which you can expect to complete this activity if everything works out ideally? (2) Under average conditions, what would be the most likely time duration for this activity? (3) What is the longest possible time (pessimistic) required to complete this activity if almost everything goes wrong? With these estimates, a probability distribution of the time required to perform the activity can be made (Fig. 2). The activity is started, and depending on how successfully events take place, the finish will occur somewhere between $a$ and $b$ (most likely close to $m$). The distribution closely approximates that of the beta distribution and is used as the typical model in PERT. The weighted linear approximation for the expected mean time, using probability theory is given by

$$t_e = (a + 4m + b)/6$$

With the development of the project plan and the calculation of activity times (time for all jobs between successive nodes in the network, such as the time for "design of rocket ignition system"), a chain of activities through the project plan can be established which has identical early and late event times; i.e., the completion time of each activity comprising this chain cannot be delayed without delaying project completion. These are the critical events.

**Events** are represented by nodes and are positions in time representing the start and/or completion of a particular activity. A number is assigned to each event for reference purposes.

Each operation is referred to as an **activity** and is shown as an arc on the diagram. Each arc, or activity, has attached to it a

number representing the number of weeks required to complete the activity. Dummy activities, shown as a dotted line, utilize no time or cost and are used to maintain the correct sequence of activities.

The **time to complete** the entire project would correspond to the longest path from the initial node to the final node. In Fig. 3 the time to complete the project would be the longest path from node 1 to node 12. This longest path is termed **the critical path** since it establishes the minimum project time. There is at least one such chain through any given project. There can, of course, be more than one chain reflecting the minimum time. This is the concept behind the meaning of critical paths. The critical path method **(CPM) as compared to PERT** utilizes estimated times rather than the calculation of "most likely" times as previously referred to. Under CPM the analyst frequently will provide two time-cost estimates. One estimate would be for normal operation and the other could be for emergency operation. These two time estimates would reflect the impact of cost on quick-delivery techniques, i.e., the shorter the time the higher the cost, the longer the time the smaller the cost.

It should be evident that those activities that do not lie on the critical path have a certain flexibility. This amount of time flexibility or freedom is referred to as **float.** The amount of float is computed by subtracting the normal time from the time available. Thus the float is the amount of time that a noncritical activity can be lengthened without increasing the project's completion date.

Figure 3 illustrates an **elementary network portraying the critical path.** This path is identified by a heavy line and would include 27 weeks. There are several methods to shorten the project's duration. The cost of various time alternatives can be readily computed. For example, if the following cost table were developed, and assuming that a linear relation between the time and cost per week exists, the cost per week to improve delivery is shown.

| Activities | Normal | | Emergency | |
|---|---|---|---|---|
| | Weeks | Dollars | Weeks | Dollars |
| A | 4 | 4000 | 2 | 6000 |
| B | 2 | 1200 | 1 | 2500 |
| C | 3 | 3600 | 2 | 4800 |
| D | 1 | 1000 | 0.5 | 1800 |
| E | 5 | 6000 | 3 | 8000 |
| F | 4 | 3200 | 3 | 5000 |
| G | 3 | 3000 | 2 | 5000 |
| H | 0 | 0 | 0 | 0 |
| I | 6 | 7200 | 4 | 8400 |
| J | 2 | 1600 | 1 | 2000 |
| K | 5 | 3000 | 3 | 4000 |
| L | 3 | 3000 | 2 | 4000 |
| M | 4 | 1600 | 3 | 2000 |
| N | 1 | 700 | 1 | 700 |
| O | 4 | 4400 | 2 | 6000 |
| P | 2 | 1600 | 1 | 2400 |

The **cost of various time alternatives** can be readily computed.

27-week schedule—Normal duration for project; cost = $22,500

26-week schedule—The least expensive way to gain one week would be to reduce activity M or J for an additional cost of $400; cost = $22,900.

25-week schedule—The least expensive way to gain two weeks would be to reduce activities M and J (one week each) for an additional cost of $800; cost = $23,300.

24-week schedule—The least expensive way to gain three weeks would be to reduce activities M, J, and K (one week each) for an additional cost of $1,300; cost = $23,800.

23-week schedule—The least expensive way to gain four weeks would be to reduce activities M and J by one week each and activity K by two weeks for an additional cost of $1,800; cost = $24,300.

22-week schedule—The least expensive way to gain five weeks would be to reduce activities M and J by one week each and activity K by two weeks and activity I by one week for an additional cost of $2,400; cost = $24,900.

21-week schedule—The least expensive way to gain six weeks would be to reduce activities M and J by one week each and activities K and I by two weeks each for an additional cost of $3,000; cost = $25,500.

20-week schedule—The least expensive way to gain seven weeks would be to reduce activities M, J, and P by one week each and activities K and I by two weeks each for an additional cost of $3,800; cost = $26,300.

19-week schedule—The least expensive way to gain eight weeks would be to reduce activities M, J, P, and C by one week each and activity I by two weeks each for an additional cost of $5,000; cost = $27,500. (Note a second critical path is now developed through nodes 1, 3, 5, and 7.)

18-week schedule—The least expensive way to gain nine weeks would be to reduce activities M, J, P, C, E, and F by one week each and activities K and I by two weeks each for an additional cost of $7,800; cost = $30,000. (Note that by shortening time to 18 weeks, we develop a second critical path.)

### Control of Materials

Control of materials is critical to the smooth functioning of a plant. Raw materials and purchased parts must be on hand in the required quantities and at the time needed if production schedules are to be met. Unless management is speculating on raw materials, inventories should be at the lowest practicable level in order to minimize the capital invested and to reduce losses due to obsolescence, design changes, and deterioration. However, some minimum stock is essential if production is not to be delayed by lack of materials. The quantity for ordering replenishment stocks is determined by such factors as the lead time needed by the supplier, the reliability of the sources of supply in meeting promised delivery dates, the value of the materials, the cost of storage, and the risks of obsolescence or deterioration.

In many instances, plant management has the choice of

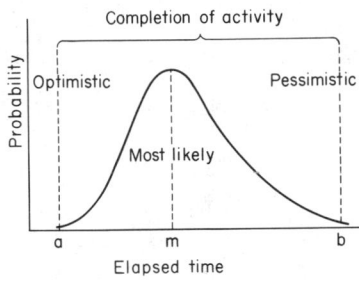

**Fig. 2** Probability distribution of time required to perform an activity.

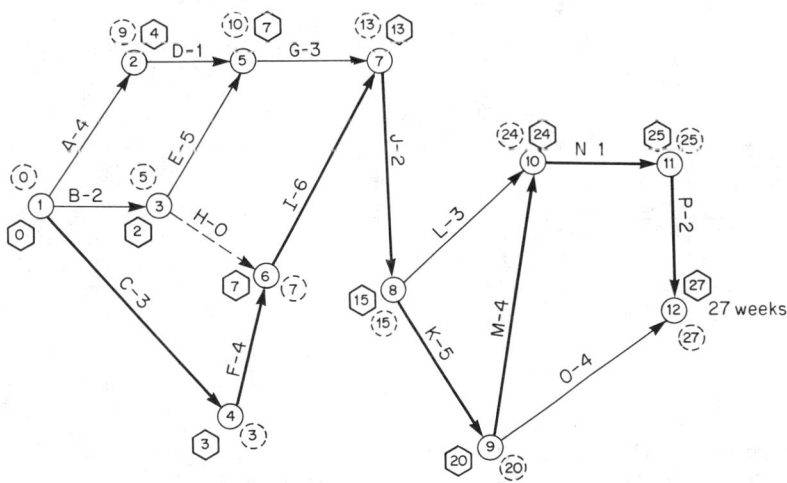

⬡ Earliest event time computed by logical procedure beginning with first event and working forward

⬠ Latest event time computed by logical procedure beginning with last event and working backwards

**Fig. 3** Network showing critical path (heavy line). Code numbers within nodes signify events. Connecting lines with directional arrows indicate operations that are dependent on prerequisite operations. Time values on connecting lines represent normal time in weeks. Hexagonals associated with events show the earliest time event. Dotted circles associated with events present the latest event time.

manufacturing the components used in its product or procuring them from outside suppliers. Where suppliers specialize in certain components, they may be able to reach high-volume operations and produce more economically than can the individual users. Procurement from outside suppliers simplifies the manufacturing problem within a plant and permits management to concentrate on the phases where it has critical know-how. Extreme quality specifications may preclude the use of outside suppliers. Likewise, if components are in short supply, the user may be forced to manufacture the units to ensure an adequate supply.

The control of raw materials and component parts may involve considerable clerical detail and many critical decisions. Systems and formulas will routinize this function, and larger companies use machines to process the vast amount of data.

Shrinkage throughout the manufacturing process may be a significant factor in materials control, scheduling, and dispatching. Spoilage rates at various stages in the process require that excess quantities of raw materials and component parts be started into the process in order to produce the quantity of finished product desired. If the original order has not allowed for spoilage, supplementary orders will be necessary; these are usually on a rush basis and may seriously disrupt the plant schedule.

**Production control** seeks optimum lot sizes with minimum total cost and adequate inventory. Figure 4 shows the time pattern, under an ideal situation, for active inventory. With assumed fixed cost per unit of output (except for starting and storage costs) and with zero minimum inventory, the optimum lot size $Q$ is given by $Q = \sqrt{ab/B}$, where $B$ = factor when storage space is reserved for maximum inventory = $0.5[1 - (d/r)](2s + ip)$; $b$ = starting cost per lot (planning and setup); $a$ = annual demand; $s$ = annual cost of storage per unit of product; $i$ = required yield on working capital; $p$ = unit cost of production; $d$ = daily demand; and $r$ = daily rate of production during production period.

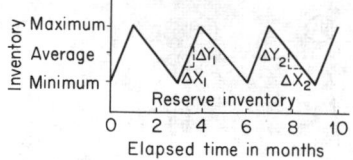

**Fig. 4**   Inventory time pattern. $\Delta Y_1/\Delta X_1$ = rate of increase of inventory; $\Delta Y_2/\Delta X_2$ = rate of decrease of inventory.

### Strategic Economic Evaluations

New equipment or facilities may be acquired for a variety of reasons: (1) Existing machines may be so badly worn that they are either beyond repair or excessively costly to maintain. (2) The equipment may be incapable of holding specified quality tolerances. (3) A technical development may introduce a process producing higher-quality products. (4) Changes in the product line may require new kinds of machines. (5) An improved model which reduces operating costs may come on the market.

A decision to invest in new capital equipment involves the risk that improved models of machines may become available and render the new equipment obsolete before its mechanical life has expired. Aggressive competitors who regularly modernize their plants may force other companies to adopt a

similar policy. The **paramount question** on strategic manufacturing expenditures is: Will it pay? Asking this question usually involves the consideration of alternatives. In comparing the economy of alternatives it is important that the engineer understand the concept of return on investment. Let:

$i$ = interest rate per period
$n$ = number of interest periods
$C$ = cash receipts
$D$ = cash payments
$P$ = present worth at the beginning of n periods
$S$ = lump sum of money at the end of n periods
$R$ = an end-of-period payment or receipt in a uniform series continuing for $n$ periods. The present value of the sum of the entire series at interest rate $i$ is equal to $P$.

Thus, \$1 $n$ years from now = $1/(1 + i)^n$

$$P = \frac{C_0 - D_0}{(1 + i)^0} + \frac{C_1 - D_1}{(1 + i)^1} + \frac{C_2 - D_2}{(1 + i)^2} \cdots \frac{C_n - D_n}{(1 + i)^n}$$

Then:

$$S = P(1 + i)^n \qquad \text{single payment}$$

$$P = S \frac{1}{(1 + i)^n} \qquad \text{single payment}$$

$$R = S \frac{i}{(1 + i)^n - 1} \qquad \text{uniform series, sinking fund}$$

$$R = P \frac{i(1 + i)^n}{(1 + i)^n - 1} \qquad \text{uniform series, capital recovery}$$

$$S = R \frac{(1 + i)^n - 1}{i} \qquad \text{uniform series, compound amount}$$

$$P = R \frac{(1 + i)^n - 1}{i(1 + i)^n} \qquad \text{uniform series, present worth}$$

For example, a new-type power-lift truck is being contemplated in the receiving department in order to reduce hand labor on a particular product line. The annual cost of this labor and labor extras such as social security taxes, industrial accident insurance, paid vacations, and various employees' fringe benefits are \$16,800 at present. An alternative proposal is to procure the new power equipment at a first cost of \$20,000. It is expected that this equipment will reduce the annual cost of the hand labor to \$6,900. Annual payments to utilize the new equipment include power (\$700); maintenance (\$2,400); and insurance (\$400).

It has been estimated that the need for this particular operation will continue for at least the next 8 years and that equipment will have no salvage value at the end of the life of this product line. Management requires a 15 percent rate of return before income taxes.

Since the burden of proof is on the proposed investment, its cost will include the 15 percent rate of return in the equivalent uniform annual cost of capital recovery. This computation approach assures that if the new method is adopted the savings will be at least as much before taxes as 15 percent per year.

The following is a **comparison of annual cost** of the present and proposed methods.

The **uniform equivalent cost** of the proposed method shows that nearly \$2,000 per year is saved in addition to the required minimum of 15 percent already included as a cost of investing in the new power-lift truck. Clearly the proposed plan is more economical.

*Present Method*

Labor + labor extras = $16,800
Total annual cost of the
present method = $16,800.00

*Proposed Method*

Equivalent uniform annual cost of capital recovery =
$(\$20,000)(0.15)(1 + 0.15)^8 = 20,000 \times 0.22285$

|  |  |
| --- | --- |
|  | = $ 4,457.00 |
| Labor and labor extras | $ 6,900.00 |
| Power | 700.00 |
| Maintenance | 2,400.00 |
| Insurance | 400.00 |
| Total uniform equivalent annual cost of proposed method | $14,857.00 |

## Wage Administration

Workers are compensated for their efforts in two principal ways: by hourly rates and by financial-type incentives. An hourly rate is paid to the worker for the number of hours worked and usually is not dependent upon the quantity or quality of the worker's output. Each worker is assigned a job title depending upon qualifications, experience, and skill. Under a structured system of wages, jobs are grouped into classifications, and a similar range of rates is applied to all jobs in a classification. The bottom of the range is paid to beginners, and periodic increases to the top of the range may be automatic or may depend upon the supervisor's appraisal of the individual's performance.

**Job evaluation** is a formal system for ranking jobs in classes. Each job is studied in relation to other jobs by analyzing such factors as responsibility, education, mental skill, manual skill, physical effort, and working conditions. A total numerical rating for job comparison is obtained by assigning points for each factor.

**Merit rating** is a point-scale evaluation of an individual's performance by a supervisor, considering such factors as quantity of output, quality of work, adaptability, dependability, ability to work with others, and attitude. These ratings serve as criteria for pay increases within a job classification.

The general level of hourly rates, or base rates, for a company should be determined with reference to the community level of rates. "Going rates" for the community are obtained from wage surveys conducted by the company, a local trade group, or a government agency. Generally, a company which offers wages noticeably lower than the community rates will attract less-competent and less-permanent employees.

Under **financial-type incentives**—or **piece rates,** as they are commonly called—the worker's compensation is dependent upon his or her rate of output. Ordinarily, there is a minimum hourly guarantee below which pay will not decline. Penalties for substandard work may reduce the worker's pay. In some instances, a maximum for incentive earnings is established. The incentive may be calculated so that only the individual's output affects his pay, or when the individual's output cannot be measured, a group incentive may be paid, where the pay of each member of the group is determined by his or her base rate plus the output of the group. Group incentives tend to promote cooperation among workers. The administration of incentive plans requires careful management attention if abuses are to be avoided. Restrictions on output and deteriorated standards may lead to higher unit labor costs rather than the anticipated lower costs.

A profit-sharing plan is a form of group incentive whereby each participating worker receives a periodic bonus in addition to regular pay, provided the company earns a profit. A minimum profit is usually set aside for a return on invested capital, and beyond this amount a percentage of profits goes into a pool to be shared by the employees. Many factors other than worker productivity affect profits, e.g., fluctuations in sales volume, selling prices, and costs of raw materials and purchased parts. To protect the workers against adverse developments outside their control, some plans give the workers a bonus whenever the actual payroll dollars are less than the normal amount expected for a given volume of production. Bonuses may be distributed quarterly or even annually and may consequently be less encouraging than incentives paid weekly and related directly to output.

## Employee Relations

Increasingly, management deals with collective bargaining units in setting conditions of work, hours of work, wages, seniority, vacations, and the like. The bargaining unit may be affiliated with a national union or may be independent and limited to the employees of a particular plant. A union contract is usually negotiated annually, although managements have sought longer-term agreements. Some managements, by making substantial concessions, have negotiated contracts covering a 3- to 5-year period. Under these contracts, the union frequently reserves the right to reopen the wage clauses annually.

**Grievance procedures** facilitate the processing of minor day-to-day disputes between workers and management. Grievances most commonly occur when the worker:

1. Thinks he or she is unfairly treated in matters of (a) pay rates, (b) promotion, (c) work assignment, (d) distribution of overtime, (e) seniority, or (f) disciplinary action.

2. Believes he or she is handicapped by (a) lack of clear policies or working rules; (b) inadequate supervision; (c) too many bosses; (d) supervisors who play favorites; (e) coworkers who are careless, inefficient, or uncooperative; or (f) lack of opportunity to show his or her ability.

3. Is dissatisfied with (a) general job conveniences (e.g., washrooms), (b) working conditions (e.g., light), (c) equipment and tools, (d) plant or office setup (e.g., working space), or (e) protection against job hazards and accidents.

A union steward acts as the employee's representative in discussing a complaint with the supervisor. If a settlement cannot be reached, the discussion may move to the general superintendent's level; the personnel manager is often involved at the various stages. Ways to reduce employee grievances are:

1. Make employees feel accepted; give them a sense of "belonging."

2. Make employees feel significant; give them recognition as people.

3. Make employees feel safe as to (a) job security, and (b) suitable working conditions.

4. Let employee experience the help of leadership.

5. Increase employees' knowledge about (a) the company, (b) its product(s), (c) their jobs, and (d) their next jobs.

6. Give employees fair and impartial treatment.

7. Give employees a chance to be heard: (*a*) Ask them for suggestions; acknowledge these suggestions; use where practicable and give credit. (*b*) Encourage them to discuss their problems and gripes; follow through if and as needed.

8. Aid employees to make their contribution to the solution of their problems.

9. Assist employees to develop pride in their work.

10. Recognize employees' status.

An outside arbitrator may be helpful when all internal grievance procedures have been exhausted.

**Selection of workers** must give attention to the suitability of applicants as well as their previous training. Interviewing and tests for qualities required on the job are necessary for good placement. Induction and instruction of new workers are equally important and should not be a secondary task of a busy supervisor; they may be a responsibility of the personnel manager, a member of the training staff, or of a subsupervisor. Induction includes an orientation to the total plant operations as well as to the duties of the specific job; introduction to the supervisor, subsupervisor, and coworkers; and an explanation of the employee's relations with the personnel department and with employee committees or groups that deal with management.

**Promotions** and recognition of accomplishment profoundly affect morale and require constant supervision by the personnel manager and the line managers. Training programs are used to prepare workers and apprentices for advancement; formal training programs for supervisors are used to develop the capabilities of members of management. The training sessions may be organized and conducted under the personnel manager, training staff member(s), or outside specialists.

Federal law and state laws in most states prohibit **discrimination** because of race, color, religion, age, sex, or national origin by most businesses and labor organizations. The **EEOC** (Equal Employment Opportunity Commission) polices these activities.

# COST ACCOUNTING

## By Adolph Matz

REFERENCES: Dickey, "Accountants' Cost Handbook," Ronald. Matz and Usry, "Cost Accounting—Planning and Control," South-Western Publishing. Neuner, "Cost Accounting," Irwin. Horngren, "Cost Accounting," Prentice-Hall. Shillinglaw, "Cost Accounting and Analysis," Irwin. Di Roccaferrara, Introduction to Linear Programming Processes," South-Western Publishing.

Cost accounting, an integral part of the management process, furnishes the costs of products, operations, or functions and compares actual costs and expenses with predetermined budgets and standards. It also provides data for special cost studies involving alternative choices regarding products, operations, and functions, thus enabling management to make decisions with respect to sales policies, production methods, purchasing procedures, financial plans, and capital structure. Such information is needed to assist management in (1) setting the company's profit goal, (2) establishing departmental targets which direct activities toward the achievement of the final goal, (3) measuring and controlling progress with the aid of budgets and standards, and (4) analyzing and deciding on adjustments and improvements to keep the entire organization moving forward in balance toward established company and profit objectives.

**Management and Its Functions**   To be successful, management must integrate its own knowledge, skills, and practices with the know-how and experience of those who are entrusted with the task of carrying out company objectives. Management, together with its employees and workers, can achieve its objectives through performance of the three managerial functions: (1) planning and setting objectives, (2) organizing, and (3) controlling.

**Planning** is a basic function of the management process. Without planning there is no need to organize or control. However, planning must precede doing, and the budget is the most important planning tool of an enterprise.

**Organizing** is essentially the establishment of the framework within which the required activities are to be performed, together with a list of who should perform them. Creation of an organization requires the establishment of organizational or functional units generally known as departments, divisions, sections, floors, branches, etc.

**Controlling** is the process or procedure by which management ensures operative performance which corresponds with plans. The control process is pictured diagrammatically in Fig. 1. Recognition of accounting as an important tool in the

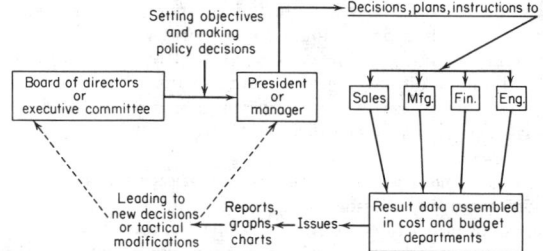

**Fig. 1**   Control circuit.

controlling phase is evidenced by the management position of the controller, the chief accounting officer. The controller, through the issuance of performance reports, points out the areas and jobs or tasks which require corrective action. These reports should make possible "management by exception."

The effectiveness of the control of costs depends upon proper communication through control and action reports from the accountant to the various levels of operating management. An organization chart is essential to the development of a cost system and cost reports which parallel the responsibilities of individuals for implementing management plans. The coordinated development of a company's organization with the cost and budgetary system will lead to "responsibility accounting."

**Annual Reports**  Accounting and cost systems are in the final analysis geared to the presentation of the financial and operating results at the end of the year. (See also Sec. 17, Industrial Economics and Management.) A balance sheet and income statement based on the published annual report of a company are presented below. Such statements are prepared in "conformance with generally accepted accounting principles," a phrase which is included in the certification statement thereto by the certified public accountant.

**Purposes of Cost Accounting**  Cost accounting is charged with the task of (1) determining costs and profit for an accounting period, (2) creating inventory values for costing and pricing purposes and, at times, controlling physical quantities, (3) aiding and participating in the creation and execution of budgets, (4) establishing costing methods and procedures that permit control and, if possible, reduction or improvement of costs, and (5) providing management with cost information in

### Balance Sheet (Illustrative)
### Black Carbon, Inc.
December 31, 19–
Assets

| | | | |
|---|---|---|---|
| Current Assets: | | | |
| Cash | | $ 5,050,000 | |
| Accounts Receivable (net) | | 6,990,000 | |
| Inventories: | | | |
| Raw Materials and Supplies | $ 1,000,000 | | |
| Work in Process | 1,800,000 | | |
| Finished Goods | 2,900,000 | 5,700,000 | |
| Investments | | 1,000,000 | |
| Deferred Charges | | 340,000 | |
| Total Current Assets | | | $19,080,000 |
| Property, Plant, and Equipment: | | | |
| Land | | $ 4,000,000 | |
| Buildings and Equipment | $75,500,000 | | |
| Less: Allowance for Depreciation | 47,300,000 | 28,200,000 | |
| Total Fixed Assets | | | 32,200,000 |
| Total Assets | | | $51,280,000 |

Liabilities

| | | | |
|---|---|---|---|
| Current Liabilities: | | | |
| Accounts Payable and Accruals | | $ 3,580,000 | |
| Provision for Income Taxes: | | | |
| Federal | $ 2,250,000 | | |
| State | 65,000 | 2,315,000 | |
| Total Current Liabilities | | | $ 5,895,000 |
| Long-term Debt | | | 5,300,000 |
| Total Liabilities | | | $11,195,000 |
| Stockholders Equity: | | | |
| Common Stock—no par value | | | |
| Authorized—2,000,000 shares | | | |
| Outstanding—1,190,000 shares | | $11,900,000 | |
| Earnings retained in the business | | 28,185,000 | |
| Total Stockholders' Equity | | | $40,085,000 |
| Total Liabilities and Stockholders' Equity | | | $51,280,000 |

### Income Statement (Illustrative)
### Black Carbon, Inc.
for the year 19—

| | | |
|---|---|---|
| Net Sales | | $50,087,000 |
| Cost of products: | | |
| Material, Labor, and Overhead (excluding depreciation) | $32,150,000 | |
| Depreciation | 5,160,000 | |
| Research and Development Costs | 260,000 | 37,570,000 |
| Gross Profit | | $12,517,000 |
| Less: Selling and Administrative Expenses | | 3,220,000 |
| Profit from Operations | | 9,297,000 |
| Other Deductions | | 305,000 |
| | | 8,992,000 |
| Other Income | | 219,000 |
| Income before Federal and State Income Taxes | | 9,211,000 |
| Less: Provision for Federal and State Income Taxes | | 4,055,000 |
| Total Net Income | | 5,156,000 |
| Dividends paid to shareholders | | 2,200,000 |
| Income retained in the business | | $ 2,956,000 |

connection with problems that involve choice from among two or more alternative courses.

**Classifications of Costs**  The purposes of cost accounting require classifications of costs so that they are recognized (1) by the nature of the item (a natural classification), (2) in their relation to the product, (3) with respect to the accounting period to which they apply, (4) in their tendency to vary with volume or activity, (5) in their relation to departments, (6) for control and analysis, and (7) for planning and decision making.

Direct material and direct labor may be listed among the items which have a **variable** nature. Factory overhead, however, must be carefully examined with regard to items of a variable and a fixed nature. It is impossible to budget and control factory-overhead items successfully without regard to their tendency to be fixed or variable; the division is a necessary prerequisite to successful budgeting and intelligent cost planning and analysis.

In general, **variable expenses** show the following characteristics: (1) variability of total amount in direct proportion to volume, (2) comparatively constant cost per unit or product in the face of changing volume, (3) easy and reasonably accurate assignments to operating departments, and (4) incurrence controllable by the responsible department head.

The characteristics of **fixed expenses** are (1) fixed amount within a relative output range, (2) decrease of fixed cost per unit with increased output, (3) assignment to departments often made by managerial decisions or cost-allocation methods, and (4) control for incurrence resting with top management rather than departmental supervisors. Whether an expense is classified as fixed or variable may well be the result of managerial decisions.

Some **factory overhead** items are semivariable in nature; i.e., they vary with production but not in direct proportion to the volume. For practical purposes, it is desirable to resolve each semivariable expense item into its variable and fixed components.

A factory is generally organized along departmental lines for production purposes. This factory departmentalization is the basis for the important classification and subsequent accumulation of costs by departments to achieve (1) cost control and (2) accurate costing. The departments of a company generally fall into two categories: (1) producing, or productive, departments, and (2) nonproducing, or service, departments. A producing department is one in which manual and machine operations are performed directly upon any part of the product manufactured. A service department is one that is not directly engaged in production but renders a particular type of service for the benefit of other departments. The expense incurred in the operation of service departments represents a part of the total factory overhead that must be absorbed in the cost of the product.

For **product costing,** the factory may be divided into departments, and departments may also be subdivided into cost centers. As a product passes through a cost center or department, it is charged with a share of the indirect expenses on the basis of a departmental factory-overhead rate. For cost-control purposes, budgets are established for departments and cost centers. Actual expenses are compared with budget allowances in order to determine the efficiency of a department and to measure the foreman's success in controlling his expenses.

Factory overhead, which is charged to a product or a job on the basis of a predetermined overhead rate, is considered indirect with regard to the product or the job to which the expense is charged. Service-department expenses are prorated to other service departments and/or to the producing departments. The prorated costs are termed **indirect departmental charges.** When all service-department expenses have been prorated to the producing departments, each producing department's total factory overhead will consist of its own direct departmental expense and the indirect (or prorated, or apportioned) charges. This total cost is charged to the product or the job on the basis of the predetermined factory-overhead rate.

A company's cost system provides the data required for establishing **standard costs** and for the preparation and operation of a **budget.**

The **budget** program enlists all members of management in the task of creating a workable and acceptable plan of action, welds the plan into a homogeneous unit, communicates to the managerial levels differences between planned activity and actual performance, and points out unfavorable conditions which need corrective action. The budget not only will help promote coordination of people, clarification of policy, and crystallization of plans, but with successful use will create greater internal harmony and unanimity of purpose among managers and workers.

The established **standard-cost** values for material, labor, and factory overhead form the foundation for the budget. Since standard costs are an invaluable aid in the process of setting prices, it is essential to set these standard costs at realistic levels. The measurement of deviations from established standards or norms is accomplished through the use of variance accounts.

Costs as a basis for planning are estimated costs which may be incurred if any one of several alternative courses of action is adopted. Different types of costs involve varying kinds of consideration in managerial planning and decision making.

**Types of Cost Systems**  The construction of a cost system requires a thorough understanding of (1) the organizational structure of the company, (2) the manufacturing procedure, and (3) the type of information which management requires of the cost system.

1. The organization chart gives a graphic picture of the ranking authority of superintendents, department heads, and foremen who are responsible for (a) providing the detailed information needed by the accounting division in order to install a successful system; (b) incurring expenditures in personnel, materials, and other cost elements, which the cost accountant must segregate and report to those in charge. The cost system with its operating accounts must correspond to organizational divisions of authority so that the individual supervisor, department head, or executive can be held "accountable" for the costs incurred in the department.

2. The manufacturing procedure and shop methods lead to a consideration of the type of pay (piece rate, incentive, day rate, etc.); the method of collecting hours worked; the control of inventories; the problem of costing tools, dies, jigs, and machinery; and many other problems connected with the factory.

3. The organizational setup on the one hand and the manufacturing procedure on the other form the background for the design of a cost system that is based on (a) recognition of the various cost elements, (b) departmentalization of factory and office, and (c) the chart of accounts.

Any cost system should be perfected so that it will (1) aid in the control and management of the company; (2) measure the efficiency of personnel, materials, and machines; (3) help in eliminating waste; (4) provide comparison within individual industries; (5) provide a means of valuing inventories; and (6) aid in establishing selling prices. The cost system's value is greatly enhanced when it is interlocked with a **budgetary control system.** When budget figures are based upon standard costs, the greatest benefit will be derived from such a combination.

Basically, two types of cost systems exist: (1) the **actual** (or **historical**) and (2) the **standard** (or **predetermined**). The actual cost system accumulates and summarizes costs as they occur and determines a final product cost after all manufacturing operations have been completed. The job is charged with actual quantities and costs of materials used and labor expended; the overhead or burden is allocated on the basis of some predetermined overhead rate. This predetermined overhead rate shows that even the so-called **actual system** does not entirely live up to its name. Under a standard cost system all costs are predetermined in advance of production. Both the actual (historical) and the standard cost system may be used in connection with either (1) the job-order cost method or (2) the process cost method.

**1. The Job-Order Cost Method**  When orders are placed in the factory for specific jobs or lots of product, which can be identified through all manufacturing processes, a job cost system is appropriate. This method has certain characteristics. A manufacturing order often corresponds to a customer's order, though sometimes a manufacturing order may be for stock. The customer's order may be obtained on the basis of a bid price computed from an estimated cost for the job. The goods in each order are kept physically separate from those of other jobs. The costs of a manufacturing order are entered on a job cost sheet which shows the total cost of the job upon completion of the order. This cost is compared with the estimated cost and with the price which the customer agreed to pay.

**2. The Process Cost Method**  When production proceeds in a continuous flow, when units of product are not separately identifiable, and when there are no specific jobs or lots of product, a process cost system is appropriate, for it has certain characteristics: work is ordered through the plant for a specific time period until the raw materials on hand have all been processed or until a specified quantity has been produced; goods are sold from the stock of finished goods on hand since a customer's order is not separately processed in the factory; the cost-of-production sheet is a record of the costs incurred in operating the process—or a series of processes—for a period of time. It shows the quantity produced in pounds, tons, gallons, or other units, and the cost per unit is obtained by dividing the total costs of the period by the total units produced. Performance is indicated by comparing the quantity produced and the cost per unit of the current period with similar figures of other periods or with standard cost figures.

**Elements of Costs**  The main items of costs are factory costs which include direct materials, direct labor and factory overhead, and selling and administrative expenses.

**Materials**  The cost of materials purchased is recorded from purchase invoices. When the materials are used in the factory, a decision must be made whether to charge them to operations at average prices, at costs based on the first-in, first-out method of costing, or at costs based on the last-in, first-out method of costing. Each method will lead to a slightly differ-

ent cost figure. Each situation must be studied individually to determine which practice will give a maximum of accuracy in cost figures with a minimum of accounting and clerical effort. Once the choice has been made, records must be set up to charge materials to operations based on requisitions. Indirect material is necessary to the completion of the product, but its consumption with regard to the final product is either so small or so complex that it would be futile to treat it as a direct-material item.

**Labor**  Labor also consists of two categories: direct and indirect. Direct labor, also called **productive labor,** is expended immediately on the materials comprising the finished product. Indirect labor, in contrast to direct labor, does not affect the construction or composition of the finished product. The term includes the labor of supervisors, shop clerks, general helpers, cleaners, and those employees engaged in maintenance work.

**Factory Overhead**  Indirect materials or factory supplies and indirect labor constitute an important segment of factory overhead. In addition, costs of fuel, power, small tools, depreciation, taxes on real estate, patent amortization, rent, inspection, supervision, social security taxes, health and accident insurance, workers' compensation insurance, and many others fall into this large category. These expenses must be collected and allocated to jobs, processes, and departments. Many expenses are definitely applicable to a specific department and are easily assigned thereto. Other expenses relate to the entire plant and must be prorated to departments on some suitable basis. For instance, heat might be prorated to departments on the basis of cubic feet of area occupied. The expenses of the service departments are prorated to the producing departments on some basis such as service rendered in the case of a maintenance department or so much dollar payroll processed in the case of a cost department.

The charging of factory overhead to jobs or products is accomplished by means of an overhead or burden rate. This rate is essentially a ratio computed to show the relationship of the total burden of a department to some other total figure for the department. For example, the total burden cost of a department may be divided by its direct-labor cost to give a percentage-of-direct-labor rate. This percentage applied to the direct-labor cost of a job or a product gives the amount of overhead chargeable thereto. Other common types of burden rates are the labor-hour rate (departmental expenses ÷ total direct-labor hours) and the machine-hour rate (departmental expenses ÷ total machine hours available). Labor rates are most commonly used. When, however, machines perform the greater amount of the work, machine-hour rates give better results. It must be clearly understood that these rates are computed in advance of production, generally at the beginning of the year. They are used throughout the fiscal period unless seasonal fluctuations or unusual changes in expense amounts necessitate the creation of a new rate. The determination of the overhead rate is closely tied up with overhead budgets.

**Departmental Classification**  As mentioned above, the establishment of departmental lines is important not only for costing purposes but also for budgetary control purposes. Departmental lines are set up in order to (1) segregate basically different processes of production, (2) secure the smoothest possible flow of production, and (3) establish lines of responsibility for control over production and costs. When the costing methods are designed to fit in with the departmentalization of factory and office, costs can be accumulated within a department with production being on either the job-order or process cost method.

**Budgets and Standard Costs**  A budget provides management with the information necessary to attain the following major objectives of budgetary control: (1) an organized procedure for planning; (2) a means for coordinating the activities of the various divisions of a business; (3) a basis for cost control. The planning phase provides the means for formalizing and coordinating the plans of the many individuals whose decisions influence the conduct of a business. Sales, production, and expense budgets must be established. Their establishment leads necessarily to the second phase of coordination. Production must be planned in relation to expected sales, materials and labor must be acquired or hired in line with expected production requirements, facilities must be expanded only as foreseeable future needs justify, and finances must be planned in relation to volume of sales and production. The third phase of cost control is predicated on the idea that actual costs will be compared with budgeted costs, thus relating what actually happened with what should have happened. To accomplish this purpose, a good measure of what costs should be under any given set of conditions must be provided. The most important condition affecting costs is volume or rate of activity. By predetermining, through the use of the flexible budget, the expenses allowed for any given rate of activity and comparing it with the actual expense, a better measurement of the performance of an individual department is achieved and the control of costs is more readily accomplished.

In the construction of overhead budgets the volume or activity of the entire organization as well as of the individual department is of considerable importance in their relationship to existing capacity. Capacity must be looked upon as that fixed amount of plant, machinery, and personnel to which management has committed itself and with which it expects to conduct the business. Volume or activity is the variable factor in business related to capacity by the fact that volume attempts to make the best use of the existing capacity. To find a profitable solution to this relationship is one of the most difficult problems faced by business management and the cost accountant who tries to help with appropriate cost data. Volume, particularly of a department, is often expressed in terms of direct-labor hours. With different rates of capacity, a different cost per hour of labor will be computed. This relationship can be demonstrated in the following manner:

per unit or hour. Variable overhead varies in total but remains fixed in relationship to the unit or hour.

**Standard Costs**  The budget, as a statement of expected costs, acts as a guidepost which keeps business on a charted course. Standards, however, do not tell what the costs are expected to be but rather what they will be if certain performances are attained. In a well-managed business, costs never exceed the budget. They should constantly approach predetermined standards. The uses of standard costs are of prime importance for (1) controlling and reducing costs, (2) promoting and measuring efficiencies, (3) simplifying the costing procedures, (4) evaluating inventories, (5) calculating and setting selling prices. The success of a standard cost system depends upon the reliability and accuracy of the standards. To be effective, standards should be established for a definite period of time so that control can be exercised and variances from standards computed. Standards are set for materials, labor, and factory overhead. When actual costs differ from standard costs with respect to material and labor, two causes can generally be detected. (1) The price may be higher or lower or the rate paid a worker may be different; the difference is called a material price or a labor rate variance. (2) The quantity of the material used may be more or less than the standard quantity or the hours used by the worker may be more or less. The difference is called material-quantity variance or labor-efficiency variance, respectively. For factory overhead, the computation is somewhat more elaborate. Actual expenses are compared not only with standard expenses but also with budget figures. Various methods are in vogue, resulting in different kinds of overhead variances. Most cost accountants compute a controllable and a volume variance. The controllable variance deals chiefly with variable expenses and measures the efficiency of the foreman's ability to hold costs within the budget allowance. The volume variance portrays fixed overhead with respect to the use or nonuse of existing capacity. It measures the success of management in its ability to fill capacity with sales or production volume. These two variances can be analyzed further into an expenditure and efficiency variance for the controllable variances and into an effectiveness and capacity variance for the volume variance. Such detailed analyses might bring forth additional information which would help management in making decisions. Of absolute importance for any cost system is the fact that the information must reach management promptly, with regularity, and in a report that is analytical, permitting quick

| Percentage of productive capacity........ | 60 % | 80 % | 100 % |
|---|---|---|---|
| Direct-labor hours...................... | 600 | 800 | 1,000 |
| Factory overhead | | | |
| Fixed overhead...................... | $1,200 | $1,200 | $1,200 |
| Variable overhead.................... | 600 | 800 | 1,000 |
| Total............................ | $1,800 | $2,000 | $2,200 |
| Overhead rate per direct-labor hour........ | $3.00 | $2.50 | $2.20 |

The existence of fixed overhead causes a higher rate at lower capacity. It is desirable to select that overhead rate which permits a full recovery of production costs by the end of the business cycle. The above tabulation reveals another important axiom with respect to fixed and variable overhead. Fixed overhead remains constant in total but varies in respect to cost

comparison with targets and goals. Only in this manner can management, which includes all echelons from the foreman to the president, exercise control over costs and therewith over profits.

**Cost Analysis**  The analytical phase of cost accounting has become more important and influential in the last few years.

Management must make many decisions, some of a short-range, others of a long-range nature. To base judgment upon good, reliable data and analyses is a major task for controllers and their staffs. Cost analysis comprises such matters as analysis of distribution costs, gross-profit analysis, break-even analysis, profit-volume analysis, differential-cost analysis, direct costing, capital-expenditure analysis, return on capital employed, and price analysis. A detailed discussion of each phase mentioned lies beyond the scope of this section, but a short description is appropriate.

**Distribution-cost** analysis deals with allocation of selling expenses to territories, customers, channels of distribution, products, and sales representatives. Once so allocated it might be possible to determine the most profitable and the least profitable commodity, product, territory, or customer. Standards have been introduced recently in these analyses. Some twenty years ago the Robinson-Patman Act, an amendment to Section 2 of the Clayton Act, gave additional impetus to the analytical phase of distribution costs.

**Gross-profit** analysis attempts to determine the causes for an increase or decrease in the gross profit. Any change in the gross profit is due to one or a combination of the following: (1) changes in the selling price of the products; (2) changes in the volume sold; (3) changes in the types of products sold, called the sales-mix; (4) changes in the cost elements. Cost elements are analyzed through budgetary control methods. Sales figures must be scrutinized to unearth the changes from the contemplated course and therewith from the final profit.

**Break-even** analysis, generally presented in the form of a break-even chart, constitutes one of the briefest and most easily understood devices for data presentation for policy-making decisions. The name "break-even" implies that point at which the company neither makes a profit nor suffers a loss from the operations of the business. A break-even chart can be defined as a portrayal in graphic form of the relation of production and sales to profit or, more briefly, a graphic variable income statement. The computation of the break-even point can be made by the following formula:

$$\text{Break-even sales volume} = \frac{\text{total fixed expenses}}{1 - \dfrac{\text{total variable expenses}}{\text{total sales volume}}}$$

EXAMPLE. Assume fixed expenses, $13,800,000; variable expenses, $27,000,000; total sales volume, $50,000,000. Computation: Break-even sales volume = $13,800,000/[1 − ($27,000,000/$50,000,000)] = $30,000,000.

Results can be obtained in chart form (Fig. 2).

**Cost-volume-profit** analysis deals with the effect that a change of volume, cost, price, and product-mix will have on profits. Managements of many enterprises attempt to stimulate the public to purchase their products by conducting intensive promotion campaigns in radio, press, mail, and television. The customer, however, makes the final decision. What management wants to know is which product or model will yield the most profitable margin; which is the least profitable; what effect a reduction in sales price will have on final profit; what effect a shift in volume or product-mix will have on product costs and profits; what the new break-even point will be under such changing conditions; what the effect of expected increases in wages or other operating costs on profit will be; what the effect will be on costs, profit, and sales volume

should there be an expansion of the plant. Cost-volume-profit analysis can also be presented graphically in a so-called *volume-profit-analysis graph*. Using the same data as in the break-even chart, a volume-profit analysis graph takes the form shown in Fig. 3.

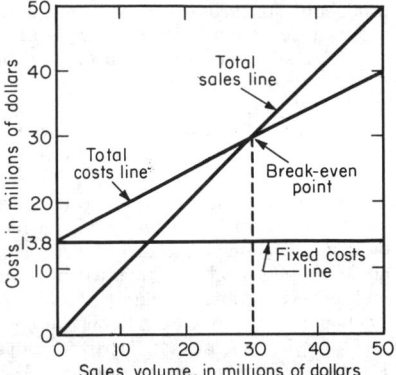

**Fig. 2** Illustrative break-even chart.

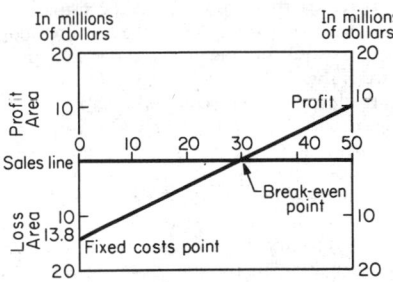

**Fig. 3** Illustrative cost-volume-profit analysis graph.

**Differential-cost** analysis treats differences, as the title suggests. These differences, also called alternative courses, arise when management wants to know whether or not to take business at a special price, to risk a decline in price of total sales, to sacrifice volume for price, to shut down part of the plant, or to enlarge plant capacity. While cost accountants generally use the term **"differential,"** economists speak of **"marginal"** and engineers of **"incremental"** costs in connection with such a study. As in any of the previously discussed analyses, the classification of costs into their fixed and variable components is absolutely essential. However, while in break-even analysis the emphasis rests upon the fixed expenses, differential-cost studies stress the variable costs. The differential-cost statement presents only the differences in the following manner:

|  | Present business | Additional business | Total |
|---|---|---|---|
| Sales | $100,000 | $10,000 | $110,000 |
| Variable costs | 60,000 | 6,000 | 66,000 |
| Marginal income | 40,000 | 4,000 | 44,000 |
| Fixed expenses | 30,000 | none | 30,000 |
| Profit | $ 10,000 | $ 4,000 | $ 14,000 |

This statement shows that additional business is charged with the variable expenses only because present business is absorbing all fixed expenses.

**Direct costing** is a costing method which charges the products with only those costs that vary directly with volume. Variable or direct costs such as direct materials, direct labor, and variable manufacturing expenses are examples of costs chargeable to the product. Costs that are a function of time rather than of production are excluded from the cost of the product. The only costs assignable to inventories are variable costs, and because they should vary in proportion to increases or decreases in production, the unit cost assigned to inventories should be uniform.

**Capital-Expenditure Decisions** The preparation of a capital-expenditure budget must be preceded by an analytical and decision-making process by management. This area of managerial decisions not only is important to the success of the company but also is crucial in case of errors. Financial requirements, present and anticipated costs, profits, tax considerations, and legal, personnel, and market problems must be studied and reviewed before making the final decision.

Four evaluation techniques are generally accepted as representative tools for decision making: (1) **payback-** or **payout-period method;** (2) **average-return-on-investment method;** (3) **present-value method;** and (4) **discounted-cash-flow method.** None of these methods serves every purpose or every firm. The methods should, however, aid management in exercising judgment and making decisions. Of significance in the evaluation of a capital expenditure is the time value of money which is employed in the present value and the discounted-cash-flow methods. The **present value** means that a dollar received a year hence is not the equivalent of a dollar received today, because the use of money has a value. For this reason, the estimated results of an investment proposal can be stated as a cash equivalent at the present time, i.e., its present value. Present-value tables have been devised to facilitate application of present-value theory.

In the present-value method the discount rate is known or at least predetermined. In the **discounted-cash-flow** method the rate is to be calculated and is defined as the rate of discount at which the sum of positive present values equals the sum of negative present values. The DCF method permits management to amortize corporate profits by selecting proposals with the highest rates of return as long as the rates are higher than the company's own **cost of capital** plus management's allowance for risk and uncertainty. Cost of capital represents the expected return for a given level of risk that investors demand for investing their money in a given firm or venture. However, when related to capital-expenditure planning, the cost of capital refers to a specific cost of capital from a particular financing effort to provide funds for a specific project or numerous projects. Such use of the concept connotes the marginal cost of capital point of view and implies linkage of the financing and investment decisions. It is, therefore, not surprising that the cost of capital differs, depending upon the sources. A company could obtain funds from (1) bonds, (2) preferred and common stock, (3) use of retained earnings, and (4) loans from banks. If a company obtains funds by some combination of these sources to achieve or maintain a particular capital structure, then the cost of capital (money) is the weighted average cost of each money source.

The **return-on-capital concept** aids management in making decisions with respect to proposed capital expenditures. This concept can also be used for (1) measuring operating performance, (2) profit planning and decision making, and (3) product pricing. The return on capital may be expressed as the product of two factors: the percentage of profit to sales and the rate of capital turnover. In the form of an equation, the method appears as

$$\frac{\text{Profit}}{\text{Sales}} = \text{percentage of profit to sales}$$

$$\frac{\text{Sales}}{\text{Capital employed}} = \frac{\text{turnover (times)}}{\text{of capital employed}} = \frac{\text{return on}}{\text{capital employed}}$$

Whether for top executive, plant or product manager, plant engineer, sales representative, or accountant, the concept of return on capital employed tends to mesh the interest of the entire organization. An understanding and appreciation of the return-on-capital concept by all employees help in building an organization interested in achieving fair profits and an adequate rate of return.

**Linear Programming**

At the heart of management's responsibility is the best or optimum use of limited resources including money, personnel, materials, facilities, and time. Linear programming, a mathematical technique, permits determination of the best use which can be made of available resources. It provides a systematic and efficient procedure which can be used as a guide in decision making.

As an example, imagine the simple problem of a small machine shop that manufactures two models, standard and deluxe. Each standard model requires 2 h of grinding and 4 h of polishing. Each deluxe model requires 5 h of grinding and 2 h of polishing. The manufacturer has three grinders and two polishers; therefore, in a 40-h week there are 120 h of grinding capacity and 80 h of polishing capacity. There is a profit of $3 on each standard model and $4 on each deluxe model and a ready market for both models. The management must decide on: (1) the allocation of the available production capacity to standard and deluxe models and (2) the number of units of each model in order to maximize profit.

To solve this linear programming problem, the symbol $X$ is assigned to the number of standard models and $Y$ to the number of deluxe models. The profit from making $X$ standard models and $Y$ deluxe models is $3X + 4Y$ dollars. The term *profit* refers to the **profit contribution,** also referred to as **contribution margin** or **marginal income.** The profit contribution per unit is the selling price per unit less the unit variable cost. Total contribution is the per-unit contribution multiplied by the number of units.

The restrictions on machine capacity are expressed in this manner: To manufacture one standard unit requires 2 h of grinding time, so that making $X$ standard models uses $2X$ h. Similarly, the production of $Y$ deluxe models uses $5Y$ h of grinding time. With 120 h of grinding time available, the grinding capacity is written as follows: $2X + 5Y \leq 120$ h of grinding capacity per week. The limitation on polishing capacity is expressed as follows: $4X + 2Y \leq 80$ h per week. In summary, the basic information is:

|  | Grinding time | Polishing time | Profit contribution |
|---|---|---|---|
| Standard model | 2 h | 4 h | $3 |
| Deluxe model | 5 h | 2 h | 4 |
| Plant capacity | 120 h | 80 h |  |

Two basic linear programming techniques, the graphic method and the simplex method, are described and illustrated using the above capacity-allocation–profit-contribution maximization data.

### Graphic Method

| Operations | Hours available | Hours required per model | | Maximum number of models | |
|---|---|---|---|---|---|
| | | Standard | Deluxe | Standard | Deluxe |
| Grinding | 120 | 2 | 5 | $\dfrac{120}{2} = 60$ | $\dfrac{120}{5} = 24$ |
| Polishing | 80 | 4 | 2 | $\dfrac{80}{4} = 20$ | $\dfrac{80}{2} = 40$ |

The lowest number in each of the two columns at the extreme right measures the impact of the hours limitations. The company can produce 20 standard models with a profit contribution of $60 (20 × $3) or 24 deluxe models at a profit contribution of $96 (24 × $4). Is there a better solution?

To determine production levels in order to maximize the profit contribution of $3X + $4Y when:

$$2X + 5Y \leqslant 120\,h \quad \text{grinding constraint}$$
$$4X + 2Y \leqslant 80\,h \quad \text{polishing constraint}$$

a graph (Fig. 4) is drawn with the constraints shown. The two-dimensional graphic technique is limited to problems having only two variables—in this example, standard and deluxe models. However, more than two constraints can be considered, although this case uses only two, grinding and polishing.

The **constraints** define the solution space when they are sketched on the graph. The solution space, representing the area of feasible solutions, is bounded by the corner points $a$, $b$, $c$, and $d$ on the graph. Any combination of standard and deluxe units that falls within the solution space is a feasible solution. However, the *best* feasible solution, according to mathematical laws, is in this case found at one of the corner points. Consequently, all corner-point variables must be tried to find the combination which maximizes the profit contribution: $3X + $4Y.

Trying values at each of the corner points:

$a = (X = \phantom{0}0, Y = \phantom{0}0);\ \$3\,(\phantom{0}0) + \$4\,(\phantom{0}0) = \$\phantom{00}0$ profit
$b = (X = \phantom{0}0, Y = 24);\ \$3\,(\phantom{0}0) + \$4\,(24) = \$\phantom{0}96$ profit
$c = (X = 10, Y = 20);\ \$3\,(10) + \$4\,(20) = \$110$ profit
$d = (X = 20, Y = \phantom{0}0);\ \$3\,(20) + \$4\,(\phantom{0}0) = \$\phantom{0}60$ profit

Therefore, in order to maximize profit the plant should schedule 10 standard models and 20 deluxe models.

**Simplex Method** The simplex method is considered one of the basic techniques from which many linear programming techniques are directly or indirectly derived. The method uses an iterative, stepwise process which approaches an optimum solution in order to reach an objective function of maximization (for profit) or minimization (for cost). The pertinent data are recorded in a tabular form known as the **simplex tableau.** The components of the tableau are as follows (see Fig. 5):

The **objective row** of the matrix consists of the coefficients of the objective function, which is the profit contribution per unit of each of the products.

The **variable row** has the names of the variables of the

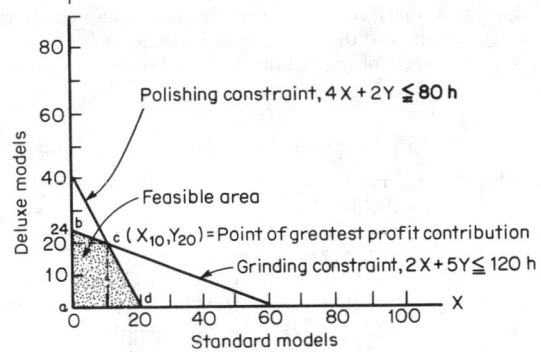

**Fig. 4** Graph depicting feasible solution.

problem including slack variables. **Slack variables** $S_1$ and $S_2$ are introduced in order to transform the set of inequalities into a set of equations. The use of slack variables involves simply the addition of an arbitrary variable to one side of the inequality, transforming it into an equality. This arbitrary variable is called **slack variable,** since it takes up the slack in the inequality. The simplex method requires the use of equations, in contrast to the inequalities used by the graphic method.

The **problem rows** contain the coefficients of the equations which represent constraints upon the satisfaction of the objective function. Each constraint equation adds an additional problem row.

The **objective column** receives different entries at each iteration, representing the profit per unit of the variables. In this first tableau (the only one illustrated due to space limitations) zeros are listed because they are the coefficients of the slack variables of the objective function. This column indicates that at the very beginning every $S_n$ has a net worth of zero profit.

The **variable column** receives different notations at each iteration by replacement. These notations are the variables used to find the profit contribution of the particular iteration. In this first matrix a situation of no (zero) production is considered. For this reason, zeros are marked in the objective column and the slacks are recorded in the variable column. As the iterations proceed, by replacements, appropriate values and notations will be entered in these two columns, objective and variable.

| | Mix | 0<br>Quantity | 3<br>$X$ | 4<br>$Y$ | 0<br>$S_1$ | 0<br>$S_2$ | Objective row<br>Variable row |
|---|---|---|---|---|---|---|---|
| 0 | $S_1$ | 120 | 2 | 5 | 1 | 0 | |
| 0 | $S_2$ | 80 | 4 | 2 | 0 | 1 | Problem rows |
| | | 0 | −3 | −4 | 0 | 0 | Index row |
| Objective<br>column | Variable<br>column | Quantity<br>column | | | | | |

**Fig. 5**  First simplex tableau and first solution.

The **quantity column** shows the constant values of the constraint equations.

Based on the data used in the graphic method and with a knowledge of the basic components of the simplex tableau, the first matrix can now be set up.

Letting $X$ and $Y$ be respectively the number of items of the standard model and the deluxe model that are to be manufactured, the system of inequalities or the set of constraint equations is

$$2X + 5Y \leqslant 120$$
$$4X + 2Y \leqslant 80$$

in which both $X$ and $Y$ must be positive values or zero ($X \leqslant 0$; $Y \leqslant 0$) for this problem.

The objective function is $3X + 4Y = P$; these two steps were the same for the graphic method.

The set of inequalities used by the graphic method must next be transformed into a set of equations by the use of slack variables. The inequalities rewritten as equalities are

$$2X + 5Y + S_1 = 120$$
$$4X + 2Y + S_2 = 80$$

and the objective function becomes

$$3X + 4Y + 0S_1 + 0S_2 = P \qquad \text{to be maximized}$$

The first tableau with the first solution would then appear as shown in Fig. 5.

The tableau carries also the first solution which is shown in the **index row**. The index row carries values computed by the following steps:

1. Multiply the values of the quantity column and those columns to the right of the quantity column by the corresponding value, by rows, of the objective column.
2. Add the results of the products by column of the matrix.
3. Subtract the values in the objective row from the results in step 2. For this operation the objective row is assumed to have a zero value in the quantity column. By convention the profit contribution entered in the cell lying in the quantity column and in the index row is zero, a condition valid only for the first tableau; in the subsequent matrices it will be a positive value.

In this first tableau the slack variables were introduced into the product mix, variable column, to find a *feasible* solution to the problem. It can be proven mathematically that beginning with slack variables assures a feasible solution. One possible solution might have $S_1$ take a value of 120 and $S_2$ a value of 80.

Index row:

| Steps 1 and 2: | Step 3: |
|---|---|
| $120(0) + 80(0) = 0$ | $0 - 0 = \phantom{-}0$ |
| $2(0) + \phantom{0}4(0) = 0$ | $0 - 3 = -3$ |
| $5(0) + \phantom{0}2(0) = 0$ | $0 - 4 = -4$ |
| $1(0) + \phantom{0}0(0) = 0$ | $0 - 0 = \phantom{-}0$ |
| $0(0) + \phantom{0}1(0) = 0$ | $0 - 0 = \phantom{-}0$ |

This approach satisfies the constraint equation but is undesirable since the resulting profit is zero.

It is a rule of the simplex method that the optimum solution has not been reached if the index row carries any negative values at the completion of an iteration in a maximization problem. Consequently, this first tableau does not carry the optimum solution since negative values appear in its index row. A second tableau or matrix must now be prepared, step by step, according to the rules of the simplex method.

**Duality of Linear Programming Problems and the Problem of Shadow Prices**  Every linear programming problem has associated with it another linear programming problem called its **dual.** This duality relationship states that for every maximization (or minimization) problem in linear programming, there is a unique similar problem of minimization (or maximization) involving the same data which describe the original problem. The possibility of solving any linear programming problem by starting from two different points of view offers considerable advantage. The two paired problems are defined as the dual problems because both are formed by the same set of data, although differently arranged in their mathematical presentation. Either can be considered to be the **primal;** consequently the other becomes its dual.

**Shadow prices** are the values assigned to one unit of capacity and represent economic values per unit of scarce resources involved in the restrictions of a linear programming problem. To maximize or minimize the total value of the total output it is necessary to assign a quantity of unit values to each input. These quantities, as cost coefficients in the dual, take the name of "shadow prices," "accounting prices," "fictitious prices," or "imputed prices" (values). They indicate the amount by which total profits would be increased if the producing division could increase its productive capacity by a unit. The shadow prices, expressed by monetary units (dollars) per unit of each element, represent the least cost of any extra unit of the element under consideration, in other words, a kind of marginal cost. The real use of shadow prices (or values) is for management's evaluation of the manufacturing process.

# ENGINEERING STATISTICS AND QUALITY CONTROL

## By Robert W. Kennard

REFERENCES: Brownlee, "Statistical Theory and Methodology in Science and Engineering," Wiley. Conover, Two $k$-sample slippage tests, *Journal of the American Statistical Association*, **63**: 614–626. Conover, "Practical Nonparametric Statistics," Wiley. Duncan, "Quality Control and Industrial Statistics," Richard D. Irwin. Gibbons, "Nonparametric Statistical Inference," McGraw-Hill. Olmstead, A corner test for association, *Annals of Mathematical Statistics*, **18**: 495–513. Owen, "Handbook of Statistical Tables," Addison-Wesley. Pearson, "Biometrika Tables for Statisticians," vol. 1, 2d ed., Cambridge University Press. Tukey, "A quick compact two sample test to Duckworth's specifications," *Technometrics*, **1**: 31–48. Wilks, "Mathematical Statistics," Wiley.

## ENGINEERING STATISTICS AND QUALITY CONTROL

Statistical models and statistical methods play an important role in modern engineering. Phenomena such as turbulence, vibration, and the strength of fiber bundles have statistical models for some of their underlying theories. Engineers now have available to them batteries of computer programs to assist in the analysis of masses of complex data. Many textbooks are needed to cover fully all these models and methods; many are areas of specialization in themselves. On the other hand, every engineer has a need for easy-to-use, self-contained statistical methods to assist in the analysis of data and the planning of tests and experiments. The sections to follow give methods that can be used in many everyday situations, yet they require a minimum of background to use and need little, if any, calculation to obtain good results.

## STATISTICS AND VARIABILITY

One of the primary problems in the interpretation of engineering and scientific data is coping with variability. Variability is inherent in every physical process; no two individuals in any population are the same. For example, it makes no real sense to speak of the tensile strength of a synthetic fiber manufactured under certain conditions; some of the fibers will be much stronger than others. Many factors, including variations of raw materials, operation of equipment, and test conditions themselves, may account for differences. Some factors may be carefully controlled, but variability will be observed in any data taken from the process. Even tightly designed and controlled laboratory experiments will exhibit variability.

Variability or variation is one of the basic concepts of statistics. Statistical methods are aimed at giving objective, quantitative, and reproducible ways of assessing the effects of variability. In particular, they aim to provide measures of the uncertainty in conclusions drawn from observational data that are inherently variable.

A second important concept is that of a **random sample**. To make valid inferences or conclusions from a set of observational data, the data should be able to be considered a random sample. What does this mean? In an operational sense it means that everything we are interested in seeing should have an equal chance of being represented in the observations we obtain. Some examples of what not to do may help. If machine setup is an important contributor to differences, then all observations should not be taken from one setup. If instrumental variation can be important, then measurements on the same item should not be taken successively—time to "forget" the last reading should pass. A random sample of $n$ items in a warehouse is not the first $n$ that you can find. It is the $n$ that is selected by a procedure guaranteed to give each item of interest an equal chance of selection. One should be guided by generalizations of the fact that the apples on top of a basket may not be representative of all apples in the basket.

## CHARACTERIZING OBSERVATIONAL DATA: THE AVERAGE AND STANDARD DEVIATION

The two statistics most commonly used to characterize observational data are the **average** and the **standard deviation.** Denote by $x_1, x_2, \ldots, x_n$ the $n$ individual observations in a random sample from some process. Then the average and standard deviation are defined as follows:

Average:

$$\bar{x} = \sum_{i=1}^{n} x_i / n$$

Standard deviation:

$$s = \left[ \sum_{i=1}^{n} (x_i - \bar{x})^2 / (n - 1) \right]^{1/2}$$

Clearly the average gives one number around which the n observations tend to cluster. And the standard deviation gives a measure of how the n observations vary or spread about this average. The square of the standard deviation is called the **variance.** If we consider a unit mass at each point $x_i$, then the variance is equivalent to a moment of inertia about an axis through $\bar{x}$. It is readily seen that for a fixed value of $\bar{x}$, greater spreads from the average will produce larger values of the standard deviation $s$. The average and the standard deviation can be used jointly to summarize where the observations are concentrated. Tchebysheff's theorem states: A fraction of at least $1 - (1/k^2)$ of the observations lie within $k$ standard deviations of the average. The theorem guarantees lower bounds on the percentage of observations within $k$ standard deviations of the average.

| Interval | Lower bound on % of measurements |
|---|---|
| $\bar{x} - 2s$ to $\bar{x} + 2s$ | 75% |
| $\bar{x} - 3s$ to $\bar{x} + 3s$ | 89% |
| $\bar{x} - 4s$ to $\bar{x} + 4s$ | 94% |

Since the average and the standard deviation are computed from a sample, they are themselves subject to fluctuation. However, if $\mu$ is the long-term average of the process and $\sigma$ is the long-term standard deviation, then:

Average $(\overline{x}) = \mu$ − process average
Average $(s) = \sigma$ − process standard deviation

Furthermore, the intervals $\mu \pm k\sigma$ contain the same percentage of all values, as do the intervals $\overline{x} \pm ks$ for the sample; that is, at least 89 percent of all the long-term values will be contained in the interval $\mu - 3\sigma$ to $\mu + 3\sigma$, etc.

### Range Estimate of the Standard Deviation

For $n \leq 20$ it is more convenient to compute the range $r$ to estimate the standard deviation $\sigma$. The range is $x_{(n)} - x_{(1)}$, where $x_{(n)}$ is the largest value in a random sample of size $n$ and $x_{(1)}$ is the smallest value. For example, if $n = 10$ and the observations are 310, 309, 312, 316, 314, 303, 306, 308, 302, 305, the range is $r = 316 - 302 = 14$. An estimate of the standard deviation $\sigma$ is obtained by multiplying $r$ by the factor $f_n$ in Table 1. The average value of $r \cdot f_n$ is $\sigma$. Thus, in

**Table 1. Average of Range** $\cdot f_n = \sigma$

| Sample size | $f_n$ | Sample size | $f_n$ |
|---|---|---|---|
| 2 | 0.8862 | 11 | 0.3152 |
| 3 | 0.5908 | 12 | 0.3069 |
| 4 | 0.4857 | 13 | 0.2998 |
| 5 | 0.4299 | 14 | 0.2935 |
| 6 | 0.3946 | 15 | 0.2880 |
| 7 | 0.3698 | 16 | 0.2831 |
| 8 | 0.3512 | 17 | 0.2787 |
| 9 | 0.3367 | 18 | 0.2747 |
| 10 | 0.3249 | 19 | 0.2711 |
| | | 20 | 0.2677 |

the example above, an estimate of $\sigma$ and a value that can be used for $s$ is $0.3249r = 0.3249(14) = 4.5486$.

### PROCESS VARIABILITY—HOW MUCH DATA?

Since the output of all processes is variable, one can make reasonable decisions about the output only if one can obtain a measure of how much variability or spread one can expect to see under normal conditions. Variability cannot be measured accurately with a small amount of data. Methods for assessing how much data are needed are given for two general situations.

### Specified Tolerances

A convenient statement about the variability or spread of a process can be based on the smallest and largest values in a random sample of the output. There are no practical limitations on its use. Suppose that we have a random sample of $n$ values from our process. Denote the values by $X_1, X_2, \ldots, X_n$. After obtaining the values we find the smallest, $X_{(1)}$, and the largest, $X_{(n)}$. Now we want to assess what percent of all future values that this process might generate will be covered by $X_{(1)}$ and $X_{(n)}$. In statistics, $X_{(1)}$ and $X_{(n)}$ are called **tolerance limits.** If the process generates bolts and $X$ is the diameter, then the engineering concept of tolerance and the statistical concept of tolerance are seen to be quite similar.

Let $p$ be the percentage of all the process values that on a long-term basis will be between $X_{(1)}$ and $X_{(n)}$. Let $P$ be a lower bound for this percentage $p$. Now consider the probability statement: Probability $(p \geq P) = C$. The quantity $C$ we call **confidence.** Since it is a probability its value is between 0 and 1. As $C$ approaches 1 our confidence in the percentage $P$ increases. The interpretation of $P$ and $C$ can be explained in terms of Table 2.

**Table 2**

| P, % | Confidence, $C$ | | | |
|---|---|---|---|---|
| | 0.995 | 0.99 | 0.95 | 0.75 |
| 99.9 | 7427 | 6636 | 4742 | 2692 |
| 99.5 | 1483 | 1325 | 947 | 538 |
| 99 | 740 | 661 | 473 | 269 |
| 98 | 368 | 330 | 235 | 134 |
| 97 | 245 | 219 | 156 | 89 |
| 96 | 183 | 163 | 117 | 67 |
| 95 | 146 | 130 | 93 | 53 |
| 90 | 71 | 64 | 46 | 26 |
| 80 | 34 | 31 | 22 | 13 |
| 75 | 27 | 24 | 18 | 10 |

*Note:* Sample size $r$ required to have a confidence $C$ that at least $P$ percent of all future values will be included between the smallest and largest values in a random sample.

Suppose that we take a random sample of size $n = 269$ values from our process output; the smallest value is 10 and the largest is 54. In Table 2 we see that 269 is the entry for $P = 99$ and $C = 0.75$. This tells us that at least $P = 99$ percent of all future values that this process will generate will be between 10 and 54, the smallest and largest values seen in the sample of 269. The confidence $C = 0.75 = \frac{3}{4}$ tells us that the chances are 3 out of 4 that our statement is correct. As we increase the sample size $n$, we increase the chances that our statement is correct. For example, if our sample size had been $n = 473$, then $C = 0.95$ and the chances are 95 in 100 that we are correct in making the statement that at least 99 percent of all process values will be between the 10 and 54 seen in the sample. Similarly, if the sample size had been 740, then the chances of being correct increase to 995 in 1000.

Further information on tolerance limits can be found in Wilks (1962) and Duncan (1965).

### Wear-Out and Life Tests

A special case of coverage occurs if our interest is in a wear-out phenomenon or a life test. For example, suppose we put a number of incandescent light bulbs on test; our interest is in the length of time to failure. Clearly we do not want to wait until all specimens fail to draw a conclusion; it might take an inordinate length of time for the last one to fail. From a practical point of view we would probably be interested in those that fail first anyway. If the sample size is properly chosen, there will be important information as soon as we obtain the first failure.

In a random sample of size $n$ let the failure times be $T_1 \geq T_2 \geq \cdots \geq T_n$. The value $T_1$ is thus the smallest value in a random sample of size $n$. Now let $q$ be the percentage of future units that can be expected to fail in a time less than $T_1$, the smallest value in the sample. As before we can make a probability statement about $q$. Let $Q$ be an upper bound to $q$. Then we can compute: Probability $(q \leq Q) = C$.

For example, suppose that we put a random sample of 299 items on test and the first one fails in time $T_1 = 151$ h. From Table 3 we see that 299 is an entry for $Q = 1$ and $C = 0.95$. Thus we can conclude that not more than 1 percent of future

**Table 3**

| Q, % | Confidence, C | | | |
|------|-------|------|------|------|
|      | 0.995 | 0.99 | 0.95 | 0.75 |
| 0.1  | 5296  | 4603 | 2995 | 1386 |
| 1    | 528   | 459  | 299  | 138  |
| 2    | 263   | 228  | 149  | 69   |
| 3    | 174   | 152  | 99   | 46   |
| 4    | 130   | 113  | 74   | 34   |
| 5    | 104   | 90   | 59   | 28   |
| 10   | 51    | 44   | 29   | 14   |
| 15   | 33    | 29   | 19   | 9    |
| 20   | 24    | 21   | 14   | 7    |
| 25   | 19    | 17   | 11   | 5    |

*Note:* Sample size $r$ required to have a confidence $C$ that fewer than $Q$ percent of future units will fail in a time shorter than the shortest life in the sample. For a more extensive table of values, see Owen (1962).

units should fail in a time less than 151 h. The confidence in the statement is 95 chances in 100 of being correct. Again referring to Table 3, we see that if $T_1 = 151$ were based on a sample of 528, then the confidence would be increased to 995 chances in 1000. Most importantly, Table 3 tells us that we need to test a very large sample if we want to have high confidence that only a small percentage of future units will fail in a time less than the smallest observed. The theory behind Table 3 can be found in Wilks (1962). For a more extensive table of values see Owen (1962). Or if $Q' = Q/100$, use

$$r = [\log(1 - C)]/[\log(1 - Q')]$$

## CORRELATION AND ASSOCIATION

One of the most common problems in the analysis of engineering data is to determine if a correlation or an association exists between two variables $X$ and $Y$, where the data occur in pairs $(X_i, Y_i)$. The "corner test of association" developed by Olmstead and Tukey (1947) is a quick and simple test to assist in making this determination.

### Corner Test

**Conditions for Use**  Each pair $(X_i, Y_i)$ should have been obtained independently; there are no other practical assumptions for its use. Of course, the user should consider the physical process generating the data when interpreting any correlation or association that is determined to exist.

**Procedure**

1. Make a scatter plot on graph paper of the data pairs $(X_i, Y_i)$, with the usual convention that $X$ is the horizontal scale and $Y$ the vertical.

2. Determine the median $X_m$ of the $X_i$ values. Determine the median $Y_m$ of the $Y_i$ values.

The median splits the data into two parts so that there is an equal number of values above and below the median. Let $N$ denote the total number of points. If $N$ is odd, then $N$ can be written as $2k + 1$ and the median is the $(k + 1)$st value as the values are ordered from the smallest to the largest. If $N$ is

even, then $N$ can be written as $2k$. Then the median is taken to be midway between the $k$th and $(k + 1)$st values.

3. Draw a vertical line through $X_m$.

4. Draw a horizontal line through $Y_m$.

5. The lines in (3) and (4) divide the graph into four quadrants. Label the upper right and lower left as plus. Label the upper left and lower right as minus.

6. Begin at the right side of the plot. Count the values, in order of decreasing $X$, until forced to cross the horizontal median $Y_m$. Give the count a plus sign if the values are in a plus quadrant, a minus sign if they are in a minus quadrant.

7. Repeat the procedure in (6), moving down from above until you have to cross the vertical median, moving from left to right until you have to cross the horizontal median, and moving up from below until you have to cross the vertical median.

8. Compute the algebraic sum of the four counts obtained in (6) and (7). Denote the sum by $T$.

**Test**  If $|T| \geq 11$, then there is evidence of correlation between $X$ and $Y$; ($|T|$ is the value of $T$ ignoring the sign.)

EXAMPLE.  Table 4 gives 33 pairs of values obtained from samples of a paper product. The $X$ coordinate is proportional to the reciprocal

**Table 4**

| i  | X     | Y    |
|----|-------|------|
| 1  | 10.45 | 4.1  |
| 2  | 13.81 | 2.7  |
| 3  | 12.22 | 1.6  |
| 4  | 9.05  | 4.3  |
| 5  | 17.86 | 2.6  |
| 6  | 14.54 | 0.1  |
| 7  | 19.99 | 3.7  |
| 8  | 8.73  | 3.5  |
| 9  | 4.66  | 5.3  |
| 10 | 13.88 | 3.9  |
| 11 | 5.10  | 4.4  |
| 12 | 3.98  | 4.1  |
| 13 | 8.12  | 6.3  |
| 14 | 12.26 | 6.6  |
| 15 | 10.30 | 6.5  |
| 16 | 5.40  | 11.9 |
| 17 | 10.39 | 5.8  |
| 18 | 9.65  | 3.8  |
| 19 | 7.44  | 5.4  |
| 20 | 10.70 | 7.6  |
| 21 | 13.38 | 6.0  |
| 22 | 13.00 | 10.4 |
| 23 | 13.90 | 10.7 |
| 24 | 11.94 | 9.4  |
| 25 | 14.11 | 10.7 |
| 26 | 0.93  | 12.9 |
| 27 | 3.18  | 12.5 |
| 28 | 13.13 | 6.5  |
| 29 | 13.45 | 11.7 |
| 30 | 12.70 | 9.6  |
| 31 | 15.95 | 8.5  |
| 32 | 7.30  | 16.6 |
| 33 | 7.78  | 8.8  |

*Notes:* Data are on paper samples. $X$ is proportional to reciprocal of light transmission. $Y$ is proportional to tensile strength.

of light transmitted by the sample. The $Y$ coordinate is proportional to tensile strength.

1. The 33 pairs of values are plotted in Fig. 1.

2. The median of $X$ values is $\bar{X}_m = 11.94$. The median of $Y$ values is $Y_m = 6.3$.

3 to 5. The medians are shown in Fig. 1, and the quadrants are labeled.

6 to 7. The counts are as follows:

| | | |
|---|---|---|
| Right to left: | −2 | (points at 19.99 and 17.86 on $X$) |
| Top to bottom: | −4 | (points at 16.6, 12.9, 12.5, 11.7 on $Y$) |
| Left to right: | −2 | (points at 0.93 and 3.18 on $X$) |
| Bottom to top: | −4 | (points at 0.1, 1.6, 2.6, 2.7 on $Y$) |

8. The algebraic sum of the counts is −12. Hence $T = -12$. And since $|T| \geq 11$, one can conclude that there is evidence of correlation or association between the variables $X$ and $Y$.

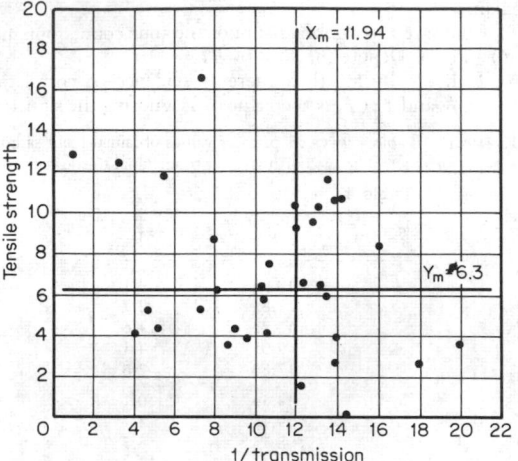

**Fig. 1** Plot of example data used in conjunction with the Corner Test.

## COMPARISON OF METHODS OR PROCESSES

A common problem in engineering investigations is that of using experimental or observational data to assess the performance of two processes, two treatment methods, or the like. Often one process or treatment is a standard or the one in current use. The other is an alternative that is a candidate to replace the standard. Sometimes it is cheaper, and one hopes to see no performance difference. Sometimes it is supposed to offer superior performance, and one hopes to see a measurable difference in the variable of interest. In either case we know that the variable of interest will have a distribution of values; and if the two processes are to be measurably different the distribution of values should not overlap too much. For an objective assessment we need to have some way to calibrate the overlap. A quick and easy-to-use test for this purpose is the **outside count** test developed by Tukey (1959).

### Two Methods—Outside Count Test

**Conditions for Use**   Given two groups of measurements taken under conditions 1 and 2 (treatments, methods, etc.), we identify the direction of difference by insisting that the two groups have minimum overlap. Use 1 to denote the group with the smaller number of measurements and let $N_1$ be the number of measurements for that group. Let $N_2$ be the number of measurements for the other group. The number of observations for each group should be about the same.

The conditions to be satisfied are:

$$4 \leq N_1 \leq N_2 \leq 30$$
$$N_2 \leq (4N_1/3) + 3$$

### Procedure

1. Count the number of values in the one group *exceeding* all values in the other.

2. Count the number of values in the other group *falling below* all those in the other.

3. Sum the two counts in (1) and (2). (It is required that neither count be zero. One group must have the largest value and the other the smallest.)

**Test**   If the sum of the two counts in (3) is 7 or larger, there is sufficient evidence to conclude that the two methods are measurably different.

EXAMPLE.   The following data represent the results of a trial of two methods for increasing the wear resistance of a grinding wheel. The data are proportional to wear:

Method 1: 13.06**, 9.52, 9.98, 8.83, 12.78, 9.00, 11.56, 8.10.*

Method 2: 8.44, 9.64, 9.94, 7.30, 8.74, 6.30*, 10.78**, 7.24, 9.30, 6.66.

The smallest value for each method is marked with an asterisk; the largest value is marked with two asterisks.

Count 1: The largest value is 13.06 for method 1. The values 13.06, 12.78, 12.21, and 11.56 for method 1 exceed the largest value for method 2, viz., 10.78. Hence the count is 4.

Count 2: The smallest value for method 1 is 8.10. For method 2 the values 7.30, 6.30, 7.24, and 6.66 are less than 8.10. Hence the count is 4.

Count 3: The total count is $4 + 4 = 8 > 7$.

The data support the conclusion that method 2 produces measurably less wear than method 1.

### Several Methods

The problem outlined in the preceding section can be generalized so that one can make a comparison of several processes, treatments, methods, or the like. Again, if there are differences among the methods, the values that we see should not overlap too much. We give you two easy-to-use tests. The first is for the situation where there is an equal amount of data for each method. For the second, the amount of data may differ. Each method will be demonstrated using the data in Table 5.

**Table 5**

| Sample No. | Supplier | | | |
|---|---|---|---|---|
| | 1 | 2 | 3 | 4 |
| 1 | 45.37 | 30.05 | 41.30 | 46.21 |
| 2 | 21.68 | 36.04 | 31.09 | 36.01 |
| 3 | 43.91 | 18.04 | 24.31 | 46.28 |
| 4 | 47.76* | 32.91 | 15.64 | 21.80 |
| 5 | 23.81 | 41.67 | 54.85* | 28.57 |
| 6 | 19.90 | 37.40 | 32.96 | 48.45 |
| 7 | 44.68 | 46.67* | 45.48 | 33.49 |
| 8 | 11.81 | 27.93 | 45.14 | 53.07* |
| 9 | 35.42 | 45.20 | 45.49 | 35.65 |
| 10 | 39.85 | 29.54 | 52.82 | 14.95 |

*Note:* The data are proportional to the time to failure of a standard cutting tool. Asterisks denote largest value in each group.

**Table 6. 95% Points for $k$-Sample Problems**

| | | | | $k$ | | | | |
|---|---|---|---|---|---|---|---|---|
| $n$ | 3 | 4 | 5 | 6 | 7 | 8 | 9 | 10 |
| 5 | 4 | 4 | 4 | 4 | 4 | 4 | 4 | 4 |
| 6 | 4 | 4 | 4 | 5 | 5 | 5 | 5 | 5 |
| 7 | 4 | 5 | 5 | 5 | 5 | 5 | 5 | 5 |
| 8 | 4 | 5 | 5 | 5 | 5 | 5 | 5 | 5 |
| 9 | 5 | 5 | 5 | 5 | 5 | 5 | 6 | 6 |
| 10 | 5 | 5 | 5 | 5 | 6 | 6 | 6 | 6 |
| 12 | 5 | 5 | 5 | 6 | 6 | 6 | 6 | 6 |
| 14 | 5 | 5 | 6 | 6 | 6 | 6 | 6 | 6 |
| 16 | 5 | 5 | 6 | 6 | 6 | 6 | 6 | 6 |
| 18 | 5 | 6 | 6 | 6 | 6 | 6 | 6 | 7 |
| 20 | 5 | 6 | 6 | 6 | 6 | 6 | 7 | 7 |
| 25 | 5 | 6 | 6 | 6 | 6 | 7 | 7 | 7 |
| 30 | 5 | 6 | 6 | 6 | 7 | 7 | 7 | 7 |
| 40 | 5 | 6 | 6 | 7 | 7 | 7 | 7 | 7 |
| >40 | 6 | 6 | 7 | 7 | 7 | 7 | 8 | 8 |

*Note:* $k$ is the number of groups; $n$ is the number of values per group. For other $n$, $k$, and % points see Conover (1968).

## Several Methods—Overlap Test

**Conditions for Use**   Independent data should be obtained for each of the $k$ methods. *The number of values n should be the same* for each method.

**Procedure**

1. For each of the $k$ methods, determine the *largest* value. Label it with an asterisk.

2. Scan the largest values. Label the group with the *largest largest* value as BIG. Label the group with the *smallest* largest value as SMALL, and its largest value as $S$.

3. In the group labeled BIG *count* the number of values that are larger than $S$, the largest value in SMALL. Denote this count by $C$.

4. Enter Table 6 for $n$ values of $k$ groups. If $C$ exceeds the tabled value, then the data support a conclusion that the methods are different. Otherwise, they do not.

EXAMPLE:   1. In Table 5 the largest value for each of the four groups is marked with an asterisk.

2. Group 3 is BIG. Group 2 is SMALL; the largest value in Group 2 is $S = 46.67$.

3. The number of values in Group 3 larger than 46.67 is 2, (52.82, 54.85).

4. Enter Table 6 with $n = 10$ and $k = 4$. The entry is 5 which is greater than 2. Hence, the data do not support a conclusion that the time to failure for the cutting tools of the four suppliers is measurably different.

## Several Methods—Rank Test

**Conditions for Use**   Independent data should be obtained for each of the methods. The amount of data for each method may be different.

**Procedure**

1. Let $n_i$ be the number of values in Group i.

2. Let $N = \sum_1^k n_i$ be the total number of values.

3. Rank each value from 1 to $N$ beginning with the smallest. (If there are ties among $t$ values, divide the successive ranks equally among them.)

4. Compute $r_i$, the sum of the ranks for the $i$th group.

[*Note:* For a check $\sum_1^k r_i = N(N + 1)/2$.]

**Test**

1. Compute

$$T = [12N/(N + 1)]\left[\sum_1^k (r_i^2/n_i) - 3(N + 1)\right]$$

2. Go to Table 7; find the entry under $k - 1$.

If $T$ exceeds the entry, then the data support the conclusion that the groups are different. Otherwise, they do not.

EXAMPLE:   We again use the data shown in Table 5. In Table 8 the numerical values representing times to failure have been replaced by their ranks. To facilitate such ranking it is convenient to order the values in each group from smallest to largest. Then all values are ranked from smallest to largest. In Table 8 the values have been reordered this way. The ranks are in parentheses.

**Table 7. Chi-Square Distribution**

| $k$ | $w$ | $k$ | $w$ |
|---|---|---|---|
| 1 | 3.841 | 16 | 26.30 |
| 2 | 5.991 | 17 | 27.59 |
| 3 | 7.815 | 18 | 28.87 |
| 4 | 9.488 | 19 | 30.14 |
| 5 | 11.07 | 20 | 31.41 |
| 6 | 12.59 | 22 | 33.92 |
| 7 | 14.07 | 24 | 36.42 |
| 8 | 15.51 | 26 | 38.89 |
| 9 | 16.92 | 28 | 41.34 |
| 10 | 18.31 | 30 | 43.77 |
| 11 | 19.68 | 40 | 55.76 |
| 12 | 21.03 | 50 | 67.50 |
| 13 | 22.36 | 60 | 79.08 |
| 14 | 23.68 | 70 | 90.53 |
| 15 | 25.00 | 80 | 101.9 |

*Note:* Entries are $P(W > w) = p = 0.05$. For other values of $k$ and $p$, see Pearson and Hartley (1962).

**Table 8.**

| | Supplier* | | | | | | | |
|---|---|---|---|---|---|---|---|---|
| 1 | | 2 | | 3 | | 4 | | |
| 11.81 | (1) | 18.04 | (4) | 15.64 | (3) | 14.95 | (2) | |
| 19.90 | (5) | 27.93 | (10) | 24.31 | (9) | 21.80 | (7) | |
| 21.68 | (6) | 29.54 | (12) | 21.09 | (14) | 28.57 | (11) | |
| 23.81 | (8) | 30.05 | (13) | 32.96 | (16) | 33.49 | (17) | |
| 35.42 | (18) | 32.91 | (15) | 41.30 | (24) | 35.65 | (19) | |
| 39.85 | (23) | 36.04 | (21) | 45.14 | (28) | 36.01 | (20) | |
| 43.91 | (26) | 37.40 | (22) | 45.48 | (31) | 46.21 | (33) | |
| 44.68 | (27) | 41.67 | (25) | 45.49 | (32) | 46.28 | (34) | |
| 45.37 | (30) | 45.20 | (29) | 52.82 | (38) | 48.45 | (37) | |
| 47.76 | (36) | 46.67 | (35) | 54.85 | (40) | 53.07 | (39) | |
| 180 | | 186 | | 235 | | 219 | | $r_i$ |
| 32400 | | 34596 | | 55225 | | 47961 | | $r_i^2$ |

*These are the data of Table 5 with the values for each supplier listed from smallest to largest. The values in parentheses are the ranks of the time to failure values from smallest to largest.

1. The number of values in each group is 10. Hence $n_i = 10$ for each value of $i$.

2. The total number of values $N = 40$.

3 and 4. The sum of the ranks $r_i$ is shown for each group. [Note that $\Sigma r_i = 820 = (40)(41)/2$.]

$$T = [12/N(N + 1)][\Sigma(r_i^2/10)] - 3(N + 1)$$
$$= [12/(40)(41)][170182/10] - 3(41)$$
$$= 124.523 - 123. = 1.523$$

Now go to Table 7 and obtain the entry under $k = 4 - 1 = 3$. The entry is 7.815, which is larger than 1.523. Hence, the data do not support a conclusion that the time to failure for the cutting tools for the four suppliers is measurably different.

## GO/NO-GO DATA

Quite often the data that we encounter will be **attribute** or **go/no-go data**; that is, we will not have quantitative measurements but only a characterization as to whether an item does or does not have some attribute. For example, if a manufactured part has a specification that it should not be shorter than two in, we might construct a template; and if a part is to meet the specification, it should not fit into the template. After inspecting a series of units with the template our data would consist of a tabulation of "gos" and "no-gos"—those that did not meet the specification and those that did.

If the items that are checked for an attribute are obtained by random sampling, the resulting data will follow what is known as the **binomial distribution.** Its standard form is as follows:

$p$ is the long term fraction of failures
$q = 1 - p$ is the long-term fraction of successes
$n$ is the size of the random sample.

Then the probability that our sample gives $x$ failures and $n - x$ successes is

$$f(x;n, p) = \binom{n}{x} p^x q^{n-x}; \quad x = 0,1,2, \ldots ,n$$

where $\binom{n}{x} = n!/x! \,(n - x)!$.

From $f(x; n, p)$, for a given $n$ and $p$, we can calculate the probability of $x$ failures in a sample size $n$. Similarly, by summing terms for different values of $x$, we can calculate the probability of having more than $w$ failures or fewer than $r$ failures, etc. Here we are not going to try to be so precise; rather we are going to try to show the general picture of the relationship between $x$, $p$, and $n$ by the use of examples and the graph in Figure 2.

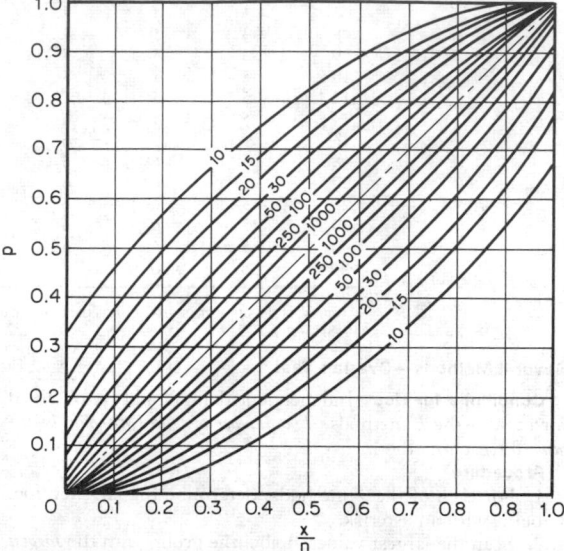

**Fig. 2** Binomial distribution, 95 percent confidence bands. (*Reproduced with the permission of the Biometrika Trust from C. J. Clopper and E. S. Pearson "The Use of Confidence or Fiducial Limits Illustrated in the Case of the Binomial,"* Biometrika, **26** (1934), p. 410.)

### Estimating the Failure Rate

In a manufacturing process a general index of quality is the fraction of items which fail to pass a certain test. Suppose that we take a random sample of size $n = 100$ from the process and observe $x_0 = 10$ failures. Clearly we have met the conditions for a binomial distribution and an estimate of $p$, the long-term failure rate is $\hat{p} = x_0/n = 10/100 = 0.1$. However, we would also like to know the accuracy of the estimate. In other words, if we operate the process for a long time under these conditions and obtain a large sample, what might be the value of $p$? One simple way to assess the estimate of $\hat{p}$ is to find values $p_1$ and $p_2$ $(p_1 < p_2)$ such that the following probabilities are satisfied for a fixed value of $\alpha$:

$$Pr[x \geq x_0 \,|\, p_1] = \alpha/2$$
$$Pr[x \leq x_0 \,|\, p_2] = \alpha/2$$

These values are the solutions for $p$ of the two equations.

$$\sum_{x=x_0}^{n} \binom{n}{x} p_1^x(1 - p_1)^{n-x} = \alpha/2$$
$$\sum_{x=0}^{x_0} \binom{n}{x} p_2^x(1 - p_2)^{n-x} = \alpha/2$$

General solutions for these equations for $\alpha = 0.05$ are shown in Fig. 2. If we go to Fig. 2 with $x/n = 0.1$ and read where the lines for $n = 100$ intersect the ordinate or $p$ scale, we see that $p_1 = 0.07$ and $p_2 = 0.18$. We can then state that we have

**Table 9. Factors for Control-Chart Limits**

| Sample size $n$ | For averages | | For ranges | | | | |
|---|---|---|---|---|---|---|---|
| | $A$ | $A_2$ | $d$ | $D_1$ | $D_2$ | $D_3$ | $D_4$ |
| 2 | 2.12 | 1.88 | 1.128 | 0 | 3.69 | 0 | 3.27 |
| 3 | 1.73 | 1.02 | 1.693 | 0 | 4.36 | 0 | 2.57 |
| 4 | 1.50 | 0.73 | 2.059 | 0 | 4.70 | 0 | 2.28 |
| 5 | 1.34 | 0.58 | 2.326 | 0 | 4.92 | 0 | 2.11 |
| 6 | 1.22 | 0.48 | 2.534 | 0 | 5.08 | 0 | 2.00 |
| 7 | 1.13 | 0.42 | 2.704 | 0.21 | 5.20 | 0.08 | 1.92 |
| 8 | 1.06 | 0.37 | 2.847 | 0.39 | 5.31 | 0.14 | 1.86 |
| 9 | 1.00 | 0.34 | 2.970 | 0.55 | 5.39 | 0.18 | 1.82 |
| 10 | 0.95 | 0.31 | 3.078 | 0.69 | 5.47 | 0.22 | 1.78 |

reasonable confidence (the probability is 0.95) that the long-term failure rate for the process is between 0.07 and 0.18.

**Estimating the Sample Size**

It should be evident that Fig. 2 can also be used to determine how large a sample is needed to estimate a proportion or a percentage with a specific accuracy or tolerance. Suppose that the proportion of interest is assumed to be around $p = 0.20$. Now enter Fig. 2 with $x/n = 0.20$. From the figure we see that if we take a sample of size 50, our estimate will have a range of about $\pm 0.10$ (actually $-0.10$, $+0.13$). On the other hand, if the sample size is 250, the estimate will have a range of about $\pm 0.06$.

Often one wants to compare the performance of two processes. As above, suppose that the rate $p$ for our process is 0.20. We have a modification that we want to test; however, to be economical the modification has to bring the rate $p$ down to 0.15 or less. If the modification is going to be assessed on the $p$'s for the standard and the modification, then we do not want the uncertainty in their estimates to overlap and the uncertainty should be less than half of 0.05 where $0.05 = (0.20 - 0.15)$. Figure 2 shows that we would have to use a sample size of at least 1000. This demonstrates that attribute sampling is effective only when the items and their characterization are not expensive. Otherwise, it is best to go to measurements where smaller sample sizes can be used to assess differences.

A more detailed exposition of the binomial distribution and its uses can be found in Brownlee (1970).

**CONTROL CHARTS**

When an industrial process is under control it is in a state of "statistical equilibrium." By equilibrium we mean that we can characterize its output by a fixed average $\mu$ and a fixed standard deviation $\sigma$. The variation in output is what one would expect to see from that $\mu$ and $\sigma$, as bounded by the values given in Tchebysheff's theorem, let us say. However, if control is lost, we tend to get a greater spread in values. In effect, the average $\mu$ or the standard deviation $\sigma$ is changing because of some cause. The causes of lack of control are manifold—it can be a change in raw materials, tool wear, instrumentation failure, operator error, etc. The important thing is that one wants to be able to detect when this lack of control occurs and take the appropriate steps to make corrections.

One of the most frequently used statistical tools for analyzing the state of an industrial process is control charts. The two most commonly used charts are those for the **average** and the **range**. The control chart procedure consists of these steps:

1. Choose a characteristic $X$ which will be used to describe the product coming from the process.

2. At time $t_i$, take a small number of observations $n$ on the process. The number $n$ should be small enough so that it is reasonable to assume that conditions will not change during the course of obtaining the observations.

3. For each set of $n$ observations, compute the average $\bar{x}_i$ and the range $r_i$ as defined in the section "Characterizing Observational Data."

There are two different control situations of interest.

4a. *Control standards given.* Suppose that from past operation of the process or from the need to meet certain specifications, a goal average $\mu^*$ and a goal standard deviation $\sigma^*$ are specified. Then we set up two charts as follows:

*Average chart:* Upper limit line: $\mu^* + A\sigma^*$
Central line:      $\mu^*$
Lower limit line: $\mu^* - A\sigma^*$

*Range chart:* Upper limit line: $D_2\sigma^*$
Central line:      $d\sigma^*$
Lower limit line: $D_1\sigma^*$

The values of $A$, $d$, $D_1$, and $D_2$ depend upon $n$ and can be found in Table 9.

Plot the values of $\bar{x}_i$ and $r_i$ obtained in (3) on the two charts. Whenever a value falls outside the limit lines, there is an indication of lack of control, and one is justified in seeking the causes for a change.

4b. *Control, no standards given.* Often one has no prior information about the process $\mu$ and $\sigma$, and one wants to determine if the process behaves as though it is in statistical equilibrium, and if not, take actions to get it there. In this case one has to determine the central lines for the charts from process data. To do this one first accumulates the data for 10 to 15 time periods as indicated in (2). Then two charts are set up as follows: Let $K$ be the 10 to 15 time periods observed. Compute an overall range $\overline{X} = \sum_{i=1}^{k} \bar{x}_i/K$ and the average range $\overline{R} = \sum_{i=1}^{k} r_i/K$. Set up charts with limits defined from:

*Average chart:* Upper limit line: $\overline{X} + A_2\overline{R}$
Central line:      $\overline{X}$
Lower limit line: $\overline{X} - A_2\overline{R}$

*Range chart:* Upper limit line: $D_4\overline{R}$
Central line:      $\overline{R}$
Lower limit line: $D_3\overline{R}$

The values of $A_1$, $D_3$, and $D_4$ depend upon $n$ and can be found in Table 9. The individual $\bar{x}_i$ and $r_i$ are plotted on the charts, and again a value outside the limits is an indication of lack of control and is justification for seeking the cause for a change.

### Charts for Go/No-Go Data

The control chart concept can also be used for attribute or go/no-go data. The procedures are, in general, the same as outlined for averages and ranges. Briefly, they are as follows:

1. Select a sample of size $n$ from the process; for best results $n$ should be in the range of 50 to 100.

2. Let $x_i$ denote the number of defective units in the sample of size $n$ at time $t_i$; then $\hat{p}_i = x_i/n$ is an estimate of the process fraction defective.

3. Set control limits for a standard fraction defective $p^*$ at

$$p^* \pm 3[p^*(1 - p^*)/n]^{1/2}$$

If no standard is given, then take $K = 10$ to 15 samples of size $n$ to get a good estimate of the fraction-defective $p$. Define $\bar{p} = \sum_{i=1}^{K} \hat{p}_i/K$. In this case set control limits at

$$\bar{p} \pm 3[\bar{p}(1 - \bar{p})/n]^{1/2}$$

4. Interpret a $\hat{p}_i$ outside the limits as an indication of a change worthy of investigation.

Further information on control charts can be found in Duncan (1965).

# TIME AND MOTION STUDY

## by William Antis

REFERENCES: ASME Standard Industrial Engineering Terminology. Lowry, Maynard and Stegemerten "Time and Motion Study," McGraw-Hill. Maynard and Stegemerten "Operation Analysis," McGraw-Hill. Barnes, "Motion and Time Study," Wiley. Mundel, "Motion and Time Study," Prentice-Hall. Morrow, "Time Study and Motion Economy," Ronald. Niebel, "Motion and Time Study," Irwin, Inc., Homewood, Ill. Nadler, "Motion and Time Study," McGraw-Hill. Chane, "Motion and Time Study," Harper. Maynard, Stegemerten, and Schwab, "Methods-Time Measurement," McGraw-Hill. Carroll, "How to Chart Timestudy Data," McGraw-Hill. Maynard, "Industrial Engineering Handbook," McGraw-Hill.

**Scope of Time and Motion Study**   Time study and motion study are two different procedures. When they are used by the industrial engineer, however, they are almost always used in conjunction with one another so that the combined name of the procedures as originally used by Frederick W. Taylor is still entirely appropriate.

According to ASME Standard Industrial Engineering Terminology, motion study is defined as

. . . the analysis of the manual and the eye movements occurring in an operation or work cycle for the purpose of eliminating wasted movements and establishing a better sequence and coordination of movements.

In the same publication, time study is defined as

. . . the procedure by which the actual elapsed time for performing an operation or subdivisions or elements thereof is determined by the use of a suitable timing device and recorded. The procedure usually but not always includes the adjustment of the actual time as the result of performance rating to derive the time which should be required to perform the task by a workman working at a standard pace and following a standard method under standard conditions.

Attempts have been made to separate the two functions and to assign each to a specialist. Although motion study deals with method and time study deals with time, the two are nearly inseparable in practical application work. The method determines the time required, and the time determines which of two or more methods is the best. It has, therefore, been found best to have both functions handled by the same individual. Although for convenience these individuals are usually called "time study specialists," they make various kinds of motion or methods studies as a part of their regular work.

**Aims of Time and Motion Study**   The aims of time and motion study are to subject each operation of a given piece of work to close analysis, so that every unnecessary operation may be eliminated and to determine the quickest and best method of performing each necessary operation; also to standardize equipment, methods, and working conditions; then and not until then to determine by scientific measurement the number of standard hours in which an average worker can do the job (Lowry, Maynard, and Stegemerten, "Time and Motion Study").

Time study specialists first devise the best practical method of doing a given job, a method which will result in necessary quality at the lowest cost. They then establish by time study the time which should be taken by a normally qualified operator to perform the task. Time study is a tool of precision measurement. It determines the time taken by the worker studied to perform a given piece of work when using the method employed when the study was taken. These data cannot be used to establish accurately the time for doing the same task by a method which differs from the one being studied, any more than measuring a shaft 2 in in diameter with micrometers will determine the diameter of another shaft of a different size.

Hence, the proper method for doing the job must be in use when the time study is made. Failure to recognize this or to act upon it is responsible for most of the inaccurate standards which are encountered.

**Elements of Time and Motion Study**   Figure 1 presents graphically the steps which must be taken to make a good time and motion study and shows their relation to each other and the order in which they must be performed.

The first step is the development of the method. Starting

with the drawing of the product, the operations which must be performed are determined and tools and equipment are specified. In large companies, this is usually done by a specialist called a process engineer. In smaller companies, processing is commonly done by the time study specialist.

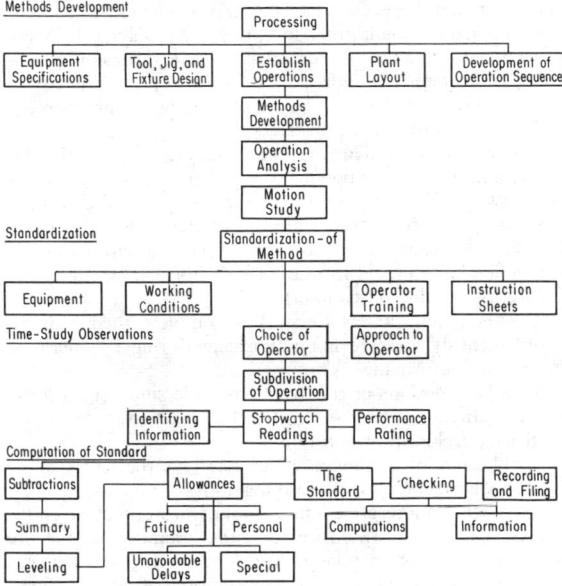

**Fig. 1** Graphic analysis of the elements of time and motion study.

Next, the detailed method by which each operation should be performed is developed. The procedures used for this are known as operation analysis and motion study. They will be described in more detail presently. When the method has been developed, conditions are standardized, and the operators are trained to follow the approved method.

At this time, not before, the job is ready for time study. Suitable operators are selected, the purposes of the study are carefully explained to them, and the time study observations are made. During the study, time study specialists rate the performance being given by operators either by judging the skill and effort they are exhibiting or by assessing the speed with which motions are made as compared with what they consider to be a normal working pace. Complete identifying information which will answer any questions which may be raised in the future is also recorded either before or after the stopwatch study is made.

The final step is to compute the standard. This involves "working up the study" in the manner which will be described presently, adding allowances for fatigue and personal, unavoidable, and special delays. When all data have been checked and all related papers have been carefully identified, the standard is ready for use for control, incentive payment, or other purposes.

**Operation analysis** is the procedure employed to study all major factors which affect a given operation. It is used for the purpose of uncovering possibilities of improving the method. The study is made by reviewing the operation with an open mind and asking either of oneself or others questions which

are likely to lead to methods-improving ideas. If this is done systematically, so that the possibility of overlooking factors which should be considered is minimized, worthwhile improvements are almost certain to result.

The 10 major factors explored during operation analysis, together with typical questions which should be asked about each factor, are as follows:

1. Purpose of operation
   a. Is the result accomplished by the operation necessary?
   b. Can the purpose of the operation be accomplished better in any other way?
2. Design of part
   a. Can motions be eliminated by design changes which will not affect the functioning of the product?
   b. Is the design satisfactory for automated assembly?
3. Complete survey of all operations performed on part
   a. Can the operation being analyzed be eliminated by changing the procedure or the sequence of operations?
   b. Can it be combined with another operation?
4. Inspection requirements
   a. Are tolerance, allowance, finish, and other requirements necessary?
   b. Will changing the requirements of a previous operation make this operation easier to perform?
5. Material
   a. Is the material furnished in a suitable condition for use?
   b. Is material utilized to best advantage during processing?
6. Material handling
   a. Where should incoming and outgoing material be located with respect to the work station?
   b. Can a progressive assembly line be set up?
7. Workplace layout, setup, and tool equipment
   a. Does the workplace layout conform to the principles of motion economy?
   b. Can the work be held in the machine by other means to better advantage?
8. Common possibilities for job improvement
   a. Can "drop delivery" be be used?
   b. Can foot-operated mechanisms be used to free the hand for other work?
9. Working conditions
   a. Has safety received due consideration?
   b. Are new workers properly introduced to their surroundings, and are sufficient instructions given them?
10. Method
   a. Is the repetitiveness of the job sufficient to justify more detailed motion study?
   b. Should full automation be considered?

General questions of the type given above or specific questions applying to a given department or industry may be added to the above list to provide a more complete set of questions (Maynard and Stegemerten, "Operation Analysis"). Experience has amply demonstrated that the persistent application of "the questioning attitude" will uncover numerous ideas for improving methods.

**Principles of Motion Study**  Operation analysis is a primary analysis which eliminates major inefficiencies. Motion study is a secondary analysis which refines the method still further. Motion study may and often does suggest further improvements in the factors considered during operation analysis, such as tools, material handling, design, and workplace layouts. In addition, it studies the human factors as well as the mechanical and sets up operations in conformance with the limitations, both physical and psychological, of those who must perform them.

The technique of motion study rests on the concept originally advanced by Frank B. and Lillian M. Gilbreth that all work is performed by using a relatively few basic operations in varying combinations and sequence. These Gilbreth Basic Elements have also been called "therbligs" and "basic divisions of accomplishment."

The basic elements together with their symbols (for definitions see ASME Industrial Engineering Terminology), grouped in accordance with their effect on accomplishment, are as follows:

*Group 1*
*Accomplishes*

| | |
|---|---|
| Reach, | R |
| Move, | M |
| Grasp, | G |
| Position, | P |
| Disengage, | D |
| Release, | RL |
| Examine, | E |
| Do, | DO |

*Group 2*
*Retards accomplishment*

| | |
|---|---|
| Change direction, | CD |
| Preposition, | PP |
| Search, | S |
| Select, | SE |
| Plan, | PL |
| Balancing delay, | BD |

*Group 3*
*Does not accomplish*

| | |
|---|---|
| Hold, | H |
| Avoidable delay, | AD |
| Unavoidable delay, | UD |
| Rest to overcome fatigue, | F |

Group 1 is the useful group of basic elements or the ones that accomplish work. They do not necessarily accomplish it in the most effective way, however, and a study of these elements will often uncover possibilities for improvement.

Group 2 contains the basic elements that tend to retard accomplishment when present. In most cases, they do this by slowing down the group 1 basic elements. They should be eliminated wherever possible.

Group 3 is the nonaccomplishment group. The greatest improvements in method usually come from the elimination of the group 3 basic elements from the cycle. This is done by rearranging the motion sequence, by providing mechanical holding fixtures, and by improving the workplace layout.

An operation may be analyzed into its basic elements either by observation or by making a micromotion study of a motion picture of the operation. The time and motion study books listed in the references describe various methods of doing this

(Lowry, Maynard, and Stegemerten, op. cit., Morrow, "Time Study and Motion Economy," Niebel, "Motion and Time Study," Nadler, "Motion and Time Study," Chane, "Motion and Time Study"). A study of the basic elements used for performing a job will almost always result in ideas for improvement. When the time required to make each motion is known, it is even easier to develop improvements. The methods time measurement procedure, or MTM (Maynard, Stegemerten, and Schwab, "Methods-time Measurement"), provides time standards for all types of industrially useful motions and is widely used for both methods-improvement and work-measurement purposes.

Methods improvement may be made on any operation by eliminating in so far as possible the group 2 and group 3 basic elements and by arranging the workplace so that the group 1 basic elements are performed in the shortest reasonable time. In doing this, certain laws of motion economy are followed. The following, derived from the laws originally stated by the Gilbreths, are the most important.

1. When both hands begin and complete their motions simultaneously and are not idle except during rest periods, maximum performance is approached.

2. When motions of the arms are made simultaneously in opposite directions over symmetrical paths, rhythm and automaticity develop most naturally.

3. The motion sequence which employs the fewest basic elements is the best for performing a given task.

4. When motions are confined to the lowest practical classification, maximum performance and minimum fatigue are approached. Motion classifications are: Class 1, finger motions; Class 2, finger and wrist motions; Class 3, finger, wrist, and forearm motions; Class 4, finger, wrist, forearm, and upper-arm motions; Class 5, finger, wrist, forearm, upper-arm, and body motions.

**Standardizing the Job**  When an acceptable method has been devised, equipment, materials, and conditions must be standardized so that the method can always be followed. Information and records describing the standard method must be carefully made and preserved, for experience has shown that, unless this is done, minor variations creep in which may in time cause a major problem. In the case of repetitive work, a job is not standardized until each piece is delivered to operators in the same condition, and it is possible for them to perform their work on each piece by completing a set cycle of motions, doing a definite amount of work with the same equipment under uniform working conditions.

The operator or operators must then be taught to follow the approved method. Operator training is always important if reasonable production is to be obtained, but it is an absolute necessity where methods have been devised by motion study. It is quite apparent that the operators cannot be expected to discover for themselves the method which the time-study specialist developed as the result of hours of concentrated study. They must, therefore, be carefully trained if they are to be expected to reach standard production. In addition, an accurate time study cannot be made until the operator is following the approved method with reasonable proficiency.

**Time Study Observations**  When the method has been made as efficient as is economically justified and when standardization has been accomplished, the job is ready for time study. The time study specialist can study any operator he or she wishes so long as that operator is using the accepted method. By applying what is known as the leveling procedure, the time study specialist will arrive at the same final time standard regardless of whether studying the fastest or the slowest worker. The time study specialist's work is made somewhat easier if he studies an intelligent, cooperative operator. If only one operator is doing the job, there is, of course, no choice.

The manner in which the operator is approached at the beginning of the study is important. This is particularly true if the operator is not accustomed to being studied. Time study specialists should be courteous and unassuming and should show a recognition of and a respect for the problems of the operator. They should be frank in their dealings with the operator and should be willing at any time to explain what they are doing and how they do it.

The first step in making time study observations is to subdivide the operation into a number of smaller operations which will be studied and timed separately. These subdivisions are known as elements, or elemental operations. Each element is exactly described in a few well-chosen words, which are recorded on the top of the time study form. Figure 2 shows how this is done. The beginning and ending points of these elements must be clearly recognizable so that the chances of overlapping watch readings will be minimized.

The timing is done with the aid of a stopwatch or, less frequently, with a special type of "time study machine." There are several types of stopwatches as well as several methods of recording watch readings in common use. The study illustrated by Fig. 2 was made using a decimal-hour stopwatch that reads directly in ten-thousandths of an hour. The readings were recorded using what is known as the continuous method of recording. In this method, the watch runs continuously from the beginning of the study to the end. Thus every moment of time is accounted for, something that may be important if the correctness of the study is ever questioned. The watch is read at the end of each elemental operation, and the reading is recorded in the "R" column under the proper element description. The elapsed time for each element is later secured by subtracting successive readings. This observation procedure gives results as accurate as any other and more accurate than some. (See also Sec. 16.)

Occasionally variations from the regular sequence of elemental operations occur. The time study specialist must be prepared to handle such situations when they happen. These

DATE 11-1-55
STUDY No. 1
SHEET No. 1
OF 1
SHEETS

| ELEMENTS | PICK UP PART FROM TABLE | | PLACE IN VISE | | TIGHTEN VISE | | START MACHINE | | RUN TABLE FORWARD 4" | | ENGAGE FEED | | MILL SLOT | | STOP MACHINE | | RETURN TABLE 5" | | RELEASE VISE | | LAY ASIDE PART IN TOTEPAN | | BRUSH VISE | |
|---|---|---|---|---|---|---|---|---|---|---|---|---|---|---|---|---|---|---|---|---|---|---|---|---|
| NUMBER | 1 | | 2 | | 3 | | 4 | | 5 | | 6 | | 7 | | 8 | | 9 | | 10 | | 11 | | 12 | |
| LINE | T | R | T | R | T | R | T | R | T | R | T | R | T | R | T | R | T | R | T | R | T | R | T | R |
| 1 | 06 | 06 | 06 | 12 | 23 | 35 | 03 | 38 | 08 | 46 | 02 | 48 | 72 | 120 | 16 | 36 | 13 | 49 | 12 | 61 | 09 | 70 | 11 | 81 |
| 2 | 05 | 86 | 08 | 94 | 21 | 215 | 03 | 18 | 09 | 27 | 03 | 30 | 76 | 306 | 13 | 19 | 12 | 31 | 14 | 45 | 07 | 52 | 06 | 58 |
| 3 | 07 | 65 | 07 | 72 | 21 | 93 | 02 | 95 | 10 | 405 | 03 | 08 | 70 | 78 | 12 | 90 | — | M | — | M | — | 513 | 08 | 21 |
| 4 | 05 | 26 | A 07 | 46 | 14 | 60 | 03 | 63 | 08 | 71 | 03 | 74 | 73 | 647 | 13 | 60 | 16 | 76 | 12 | 88 | 08 | 96 | 08 | 704 |
| 5 | 07 | 11 | 07 | 18 | 19 | 37 | 04 | 41 | 09 | 50 | 02 | 52 | 71 | 823 | 11 | 34 | 14 | 48 | 09 | 57 | 07 | 64 | 08 | 72 |
| 6 | 07 | 79 | 08 | 87 | 17 | 904 | 03 | 07 | 10 | 17 | 02 | 19 | 72 | 91 | 11 | 1002 | 13 | 15 | 11 | 26 | 09 | 35 | — | |
| 7 | 07 | 42 | 08 | 50 | 22 | 72 | 03 | 75 | 12 | 87 | 03 | 90 | 71 | 1161 | 16 | 77 | 14 | 91 | 09 | 1200 | 06 | 06 | 07 | 13 |
| 8 | 05 | 18 | 06 | 24 | 15 | 39 | 03 | 42 | B 10 | 64 | 02 | 66 | 76 | 1342 | 12 | 54 | 16 | 70 | 10 | 80 | 06 | 86 | 06 | 92 |
| 9 | 06 | 98 | 08 | 1406 | 12 | 18 | 02 | 20 | 13 | 33 | 03 | 36 | 74 | 1510 | 14 | 24 | 15 | 39 | 10 | 49 | 06 | 55 | 06 | 61 |
| 10 | 06 | 67 | 08 | 75 | 14 | 89 | 03 | 42 | 12 | 1604 | 04 | 08 | 71 | 79 | 12 | 91 | 16 | 1107 | 11 | 18 | 07 | 25 | 08 | 33 |
| 11 | 05 | 38 | 06 | 44 | 16 | 60 | 02 | 62 | 09 | 71 | 03 | 74 | 75 | 1849 | 13 | 62 | 12 | 74 | 09 | 83 | 08 | 91 | 07 | 98 |
| 12 | C 05 | 1918 | 08 | 26 | 19 | 45 | 03 | 48 | 12 | 60 | 04 | 64 | 71 | 2035 | 15 | 50 | 16 | 66 | 13 | 79 | 07 | 86 | 07 | 99/92 |
| 13 | 06 | 92/86 | 07 | 2106/99 | 14 | 20 | 04 | 24 | 13 | 37 | 02 | 39 | 75 | 2214 | 14 | 28 | 15 | 43 | 11 | 54 | 08 | 62 | 09 | 71 |
| 14 | (15) | 86 | 07 | 93 | 13 | 2306 | 03 | 09 | 12 | 21 | 04 | 25 | 71 | 96 | 13 | 2409 | 15 | 24 | 10 | 34 | 09 | 43 | 07 | 50 |
| 15 | 06 | 56 | 08 | 64 | 17 | 81 | 03 | 84 | 13 | 97 | 03 | 2500 | D 72 | 2694 | 16 | 2710 | 16 | 26 | 10 | 36 | 08 | 44 | 08 | 52 |

SUMMARY

| | 1 | 2 | 3 | 4 | 5 | 6 | 7 | 8 | 9 | 10 | 11 | 12 |
|---|---|---|---|---|---|---|---|---|---|---|---|---|
| TOTALS "T" | 0083 | 0109 | 0257 | 0044 | 0160 | 0043 | 1090 | 0201 | 0203 | 0151 | 0105 | 0106 |
| NO. OBSERVATIONS | 14 | 15 | 15 | 15 | 15 | 15 | 15 | 15 | 14 | 14 | 14 | 14 |
| AVERAGE "T" | 00059 | 00073 | 00171 | 00029 | 00107 | 00029 | 00727 | 00134 | 00145 | 00108 | 00075 | 00076 |
| MINIMUM "T" | 0005 | 0006 | 0012 | 0002 | 0008 | 0002 | 0070 | 0011 | 0012 | 0009 | 0006 | 0006 |
| MAXIMUM "T" | 0007 | 0008 | 0023 | 0004 | 0013 | 0004 | 0076 | 0016 | 0016 | 0014 | 0009 | 00011 |
| RATING (S.E.C.ACY.) | DCIDD → | | | | | | X → | | | | | |
| LEVELING FACTOR | 1.05 → | | | | | | X → | | | | | |
| L.F. x AVE. "T" | 00062 | 00076 | 00180 | 00030 | 00112 | 00029 | 00727 | 00141 | 00152 | 00114 | 00079 | 00079 |
| % ALLOWANCE | 15 | 15 | 15 | 15 | 15 | 15 | 10 | 15 | 15 | 15 | 15 | 15 |
| TIME ALLOWED | 00071 | 00087 | 00207 | 00035 | 00129 | 00033 | 00800 | 00162 | 00175 | 00131 | 00091 | 00091 |

| SYM | R | T | |
|---|---|---|---|
| A | 539 / 526 | 13 | Piece in vise wrong. |
| B | 1256 / 1244 | 12 | Machine stopped. |
| C | 1913 / 1898 | 15 | Clean hands |
| D | 2656 / 2534 | 122 | Piece pulled out of vise |
| E | | | |
| F | | | |
| G | | | |
| H | | | |
| I | | | |
| J | | | |
| K | | | |
| L | | | |
| M | | | |
| N | | | |
| O | | | |

| SKILL | | | EFFORT | | |
|---|---|---|---|---|---|
| | A1 A2 | SUPER | | A1 A2 | EXCESSIVE |
| | B1 B2 | EXCELLENT | | B1 B2 | EXCELLENT |
| | C1 C2 | GOOD | ✓ C1 | C1 C2 | GOOD |
| ✓ | D | AVERAGE | | D | AVERAGE |
| | E1 E2 | FAIR | | E1 E2 | FAIR |
| | F1 F2 | POOR | | F1 F2 | POOR |

| CONDITIONS | | | CONSISTENCY | | |
|---|---|---|---|---|---|
| | A | IDEAL | | A | PERFECT |
| | B | EXCELLENT | | B | EXCELLENT |
| | C | GOOD | | C | GOOD |
| ✓ | D | AVERAGE | | D | AVERAGE |
| | E | FAIR | | E | FAIR |
| | F | POOR | | F | POOR |

| GENERAL RATING FOR STUDY | SKILL | EFFORT | COND. | CONSY |
|---|---|---|---|---|
| | 0 | +.05 | 0 | — |

| STUDY STARTED | STUDY FINISHED | OVERALL TIME |
|---|---|---|
| 2:54 P.M. | 3:10½ P.M. | .275 HRS |

**Fig. 2** Face of time-study form.

variations may be divided into four general classes as follows: (1) elements performed out of order, (2) elements missed by the time study specialist, (3) elements omitted by the operator, (4) foreign elements.

The time study illustrated by Fig. 2 contains examples of each of these kinds of irregularity. Elements 12 and 1 on lines 12 and 13 were performed out of order. On line 3, the time study specialist missed obtaining the watch readings for elements 9 and 10. On line 6, element 12 was omitted by the operator. Foreign elements A, B, C, and D occurred during regular elements 2, 5, 1, and 7, respectively. A study of these examples will show how the time study specialist handles variations from the regular sequence of elements which occur during the making of a time study.

A time study to be of value for future use must tell the whole story of a job in such a way that it will be understood by anyone familiar with the time study procedure. This will not be possible unless all identifying and other pertinent information is recorded at the time the study is made. Records should be made to show complete identification of the operator; the part or assembly; the machines, tools, and equipment used; the operation; the department in which the operation was performed; and the conditions existing at the time the study was made. Sketches are generally a desirable part of this description. Figure 3 shows the information which would be recorded on the reverse side of the time study form illustrated by Fig. 2.

**Performance Rating**   The objective of a time study is to determine the time which a worker giving average performance will require to do the job under average or normal conditions. It is important to understand that when the time study specialist speaks of average performance, he or she is not referring to the mathematical average of all human beings, or even the average of all persons engaged in a given occupation. Average performance is established by definition and not statistically. It represents the time study specialist's conception of the normal, steady, but unhurried performance which may reasonably be expected from anyone qualified for the work. If sufficient inducement is offered by incentives or otherwise, this performance may be considerably surpassed.

If all operators available for study worked at the average performance level, the task of establishing a standard would be easy. It would be necessary merely to average the elapsed elemental times determined from time study and add an allowance for fatigue and personal and unavoidable delays. It is seldom, however, that a performance is observed which is rated throughout as average. Therefore, to establish a standard which represents the time which would be taken had an average performance been observed, it is necessary to use some method of adjusting the recorded elemental times when other than average performance is timed.

One of the well-known methods of doing this is the **leveling procedure.** When properly applied it gives excellent results. It must be correctly understood, of course, and the time study specialist who uses it must be thoroughly trained to apply it correctly.

The procedure recognizes that when the correct method is being followed, skill, effort, and working conditions will affect the level at which the operator works. These factors are judged during the making of the time study. Skill is defined as *proficiency at following a given method.* This is not subject to variation at will by the operator but develops with practice

over a period of time. Effort is defined as *the will to work.* It is controllable by the operator within the limits imposed by skill. Conditions are those conditions which affect the operator and not those which affect the method.

Definitions have been established for different degrees of skill, effort, and conditions. Numerical factors have been established by extensive research for each degree of skill, effort, and conditions. These are shown by Fig. 4. The algebraic sum of these numerical values added to 1.0 gives the leveling factor by which all actual elemental times are multiplied to bring them to the average or normal level. The leveling factor represents in effect the amount in percent which actual performance times are above or below the average performance level.

**Allowances for Fatigue and Personal and Unavoidable Delays**   The leveled elemental time values are net elapsed times adjusted to the average performance level. They do not provide for delays and other legitimate allowances. Something, therefore, must be added to take care of such things as fatigue, personal needs, delays outside the control of the workers, and special conditions of the work.

Fatigue allowances vary according to the nature of the work. Flat percentages are determined for each general class of work, such as bench work, machine-tool operation, hard physical labor, and so on. Personal allowances are the same for most classes of work. Unavoidable delay allowances vary with the nature of the work and the conditions under which it is performed. Peculiar conditions surrounding specific jobs sometimes require additional special allowances.

It is apparent, therefore, that the proper allowance factor to use can only be determined by a study of the class of work to which it is to be applied. Allowances are determined either by a series of all-day time studies or by a statistical method known as work sampling (Maynard, "Industrial Engineering Handbook"), or both. When an allowance factor has once been established, it is then applied to all time studies made on that class of work thereafter.

**Developing the Time Standard**   When time study observations have been completed, a series of calculations are made to develop the time standard. Elapsed times are determined by subtracting successive watch readings. Each subtraction is recorded between the two watch readings that determine its value. Elapsed time is noted in ink to ensure a permanent record and to distinguish it from the watch readings which are usually recorded in pencil. A study of Fig. 2 will show how subtractions are entered on the time study form and later summarized.

The several elapsed times for each element are next carefully compared and examined for abnormal values. If any are found, they are circled so that they can be distinguished and excluded from the summary.

The remaining elapsed times for each element are added and are averaged by dividing by the number of elapsed time readings. The results are average elapsed times which represent the time taken by the operator during that particular study. These times must be adjusted by multiplying them by a leveling factor to bring them to the average performance level. This factor is determined by the rating of skill, effort, and conditions made during the period of observation.

Each average elapsed time is multiplied by the leveling factor, except when the element is not controlled by the operator. An element that is outside the control of the opera-

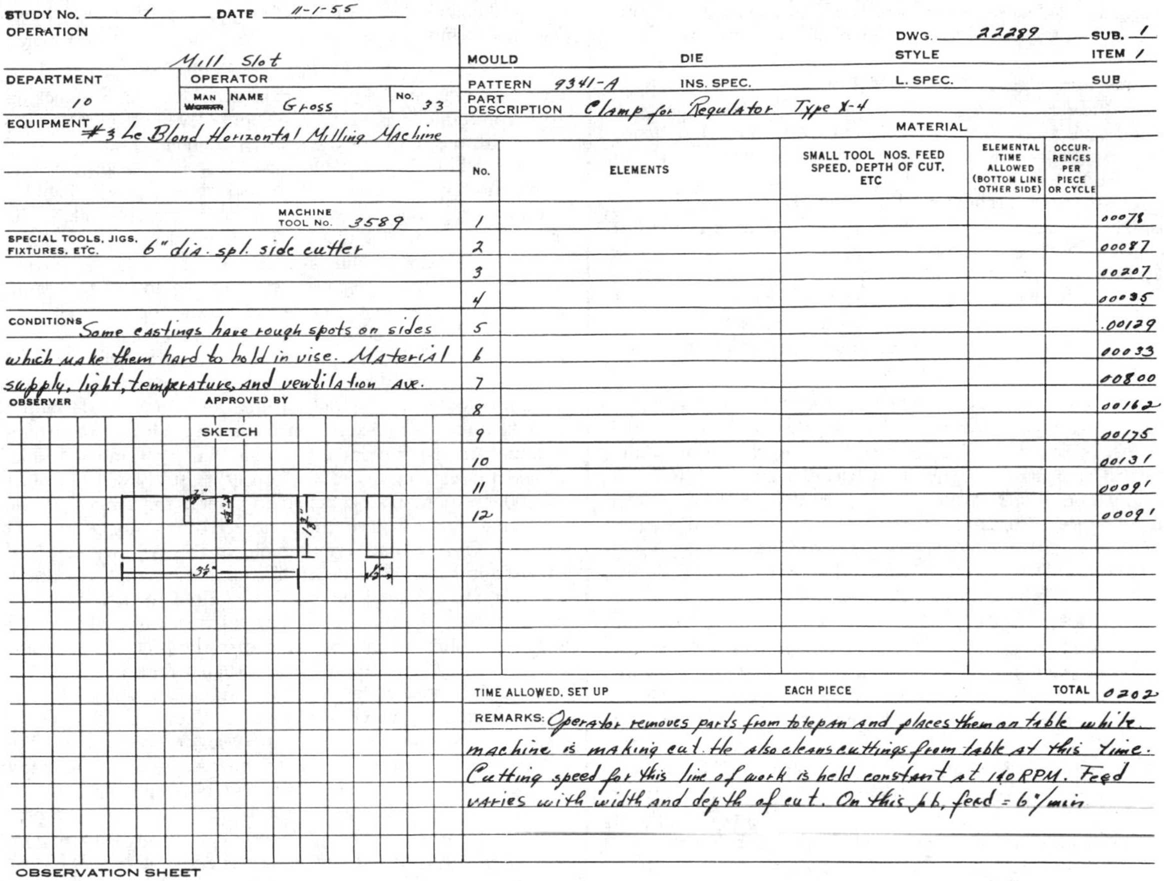

STUDY No. _1_    DATE _11-1-55_

OPERATION    _Mill Slot_

| DEPARTMENT | OPERATOR | | |
|---|---|---|---|
| 10 | MAN WOMAN | NAME _Gross_ | NO. _33_ |

MOULD          DIE

PART DESCRIPTION _Clamp for Regulator Type X-4_

PATTERN _9341-A_    INS. SPEC.

DWG. _22289_    SUB. _1_

STYLE    ITEM _1_

L. SPEC.    SUB

EQUIPMENT _#3 Le Blond Horizontal Milling Machine_

MATERIAL

| | | NO. | ELEMENTS | SMALL TOOL NOS. FEED SPEED, DEPTH OF CUT, ETC | ELEMENTAL TIME ALLOWED (BOTTOM LINE OTHER SIDE) | OCCUR-RENCES PER PIECE OR CYCLE | |
|---|---|---|---|---|---|---|---|
| MACHINE TOOL NO. _3589_ | | 1 | | | | | _.00078_ |
| SPECIAL TOOLS, JIGS, FIXTURES, ETC. _6" dia. spl. side cutter_ | | 2 | | | | | _.00087_ |
| | | 3 | | | | | _.00207_ |
| | | 4 | | | | | _.00035_ |
| CONDITIONS _Some castings have rough spots on sides_ | | 5 | | | | | _.00129_ |
| _which make them hard to hold in vise. Material_ | | 6 | | | | | _.00033_ |
| _supply, light, temperature, and ventilation Av._ | | 7 | | | | | _.00800_ |
| OBSERVER          APPROVED BY | | 8 | | | | | _.00162_ |
| | | 9 | | | | | _.00175_ |
| SKETCH | | 10 | | | | | _.00131_ |
| | | 11 | | | | | _.00091_ |
| | | 12 | | | | | _.00091_ |

TIME ALLOWED, SET UP          EACH PIECE          TOTAL _.0202_

REMARKS: _Operator removes parts from totepan and places them on table while machine is making cut. He also cleans cuttings from table at this time. Cutting speed for this line of work is held constant at 140 RPM. Feed varies with width and depth of cut. On this job, feed = 6"/min_

OBSERVATION SHEET

**Fig. 3**  Back of time-study form.

tor, such as element 7 in Fig. 2 which is a cut with power feed, should not be leveled, because it is unaffected by the ability of the operator. As long as the proper feed and speed are used, the time for performing this element will be the same whether the worker is an expert or a learner.

If workers were able to work continuously, the leveled time would be the correct value to allow for doing the operation studied, but constant application to the job is neither possible nor desirable. In the course of a day, there are certain to be occasional interruptions and delays, for which due allowance must be made in establishing the final standard. Therefore,

each elemental time is increased by an allowance which covers time that will be consumed by personal and unavoidable delays, fatigue, and any special factors that may affect the job.

The numbers and descriptions of the elemental operations together with their allowed time are transcribed on the back of the time study form as shown by Fig. 3. The number of times an elemental operation occurs on each piece or cycle of the operation is taken into account, and the total time allowed for each element is determined and recorded. The final standard for the operation is the sum of the amounts recorded in the "time allowed" column. When all computations have been

| Skill | | | Conditions | | | Effort | | |
|---|---|---|---|---|---|---|---|---|
| +0.15 | A1 | Superskill | +0.06 | A | Ideal | +0.13 | A1 | Excessive |
| +0.13 | A2 | | +0.04 | B | Excellent | +0.12 | A2 | |
| +0.11 | B1 | Excellent | +0.02 | C | Good | +0.10 | B1 | Excellent |
| +0.08 | B2 | | 0.00 | D | Average | +0.08 | B2 | |
| +0.06 | C1 | Good | −0.03 | E | Fair | +0.05 | C1 | Good |
| +0.03 | C2 | | −0.07 | F | Poor | +0.02 | C2 | |
| 0.00 | D | Average | | | | 0.00 | D | Average |
| −0.05 | E1 | Fair | | | | −0.04 | E1 | Fair |
| −0.10 | E2 | | | | | −0.08 | E2 | |
| −0.16 | F1 | Poor | | | | −0.12 | F1 | Poor |
| −0.22 | F2 | | | | | −0.17 | F2 | |

**Fig. 4**  Leveling factors for performance rating.

checked and all supporting records have been properly identified and filed, the task of developing the time standard is complete.

**Time Formulas and Standard Data**  On repetitive work, time study is a satisfactory tool of work measurement. A single time study may be sufficient to establish a standard which will cover the work of one or more operators for a long period of time.

As quantities become smaller, however, the cost of establishing standards by individual time study increases until at length it becomes prohibitive. In the extreme case, where products are manufactured in quantities of one, it would require at least one time study specialist for each operator if standards were established by detailed time study, and the standards would not be available until after the jobs had been completed.

In order to simplify the task of setting standards on a given class of work and in order to improve the consistency of the standards, standard data are frequently used by time study specialists (Carroll, "How to Chart Timestudy Data"). A compilation of standard data in its simplest form is merely a list of all the different elements that have occurred during all the time studies made on a given class of work, with representative time values for each element. Every element that differs even slightly from any other element has its own time value.

When a job comes into the shop on which no standard has previously been established, time study specialists analyze the job either mentally or by direct observation and determine the elements required to perform it. They then select time values from the standard data for each element. Their sum gives the standard for the job.

This method, although a decided improvement from a time, cost, and consistency standpoint over individual time study, is capable of further refinement and improvement. On a given class of work, certain elements will be performed—for example, "pick up part"—on every piece produced, while others—such as "secure in steady rest"—will be performed only when a piece has certain characteristics. In some cases, the performance of a certain element will always require the performance of another element, e.g., "start machine" will always require the subsequent performance of the element "stop machine." Then again, the time for performing certain elements—for example, "engage feed"—will be the same regardless of the characteristics of the part being worked upon, while the time for performing certain other elements—like "lay part aside"—will be affected by the size and shape of the part.

Thus it is possible to make certain combinations and groupings which will simplify the task of applying standard data. Time study specialists construct various charts and tables which they still call standard data, or, in the ultimate refinement, develop time formulas (Lowry, Maynard, and Stegemerten, op. cit.). A time formula is a convenient arrangement of standard data which simplifies their accurate application. Much of the analysis which is necessary when applying standard data is done once and for all at the time the formula is derived. The job characteristics which make the performance of certain elements or groups of elements necessary are determined, and the formula is expressed in terms of these characteristics.

Figure 2 illustrates a detailed time study made to establish a standard on a simple milling machine operation. The same standard can be derived much more quickly from the following time formula:

$$\text{Curve 1} + \text{Table 1} = \text{each piece time}$$

Curve 1 combines the times for the variable elements "pick up part from table," "place in vise," and "lay aside part in totepan" with the times for the constant elements "tighten vise," "start machine," "run table forward 4 in," "engage feed," "stop machine," "release vise," and "brush vise." Table 1 combines the times for "mill slot" and "return table." The standard time for milling a slot in a brass clamp of any size is computed by determining the variable characteristics of the job from the drawing—in this case, the volume of the clamp and the perimeter of the cut—and adding together the time read from curve 1 and the time read from Table 1.

The amount of time which the use of time formulas will save the time study specialist is readily apparent. It takes a certain amount of time and no little know-how to develop a time formula, but once it is available, the job of establishing accurate standards becomes a simple, fairly routine task. The time required to make and work up a time study will be from 1 to 100 or more hours, depending upon the length of the operation cycle studied. The time required to establish a standard from a time formula will, in the majority of cases, range from 1 to 15 min, depending upon the complexity of the formula and the amount of time required to determine the characteristics of the job. Where all necessary information may be obtained from the drawing of the part, the standard may generally be computed in less than 5 min.

**Uses of Time Standards**  Time study data can be of value to nearly every function of a business. They are used directly by those who plan, schedule, supervise, or direct the work to which the data apply. In slightly different form, time study data can be useful to the product and tool design departments, the accounting departments, the sales department, the processing department, the methods department, and others. Even the president will find time standards of value as he or she develops policies, plans for the future of his or her company, and makes decisions on the day-to-day conduct of his or her company's business.

Some of the more common uses of time standards are in connection with

1. Wage incentive plans
2. Plant layout
3. Plant capacity studies
4. Production planning and control
5. Standard costs
6. Budgetary control
7. Cost reduction activities
8. Product design
9. Tool design
10. Top-management controls
11. Equipment selection
12. Bidding for new business
13. Machine loading
14. Effective labor utilization
15. Material-handling studies

In short, wherever it is necessary to know the time required to do a given piece of work, time standards will be of value. Since it costs money to establish a time standard, any company will be well advised to endeavor to get the most for the money it spends by making full use of its time standards.

# COST OF ELECTRIC POWER

**by Arthur J. Fiehn**
**(with the assistance of Ashok Ahuja)**

REFERENCES: Federal Power Commission, National Power Surveys; Steam-Electric Plant Construction Cost and Annual Production Expenses, 1973; Hydroelectric Plant Construction Cost and Annual Production Expenses, 1973. *Electrical World*, Annual Steam Station Cost Surveys. *Power Engineering*, "Capital Cost Calculations for Future Power Plants," compiled from AEC Source Material, January 1973. ASME, Annual Reports on Diesel and Gas Engines Power Costs. Edison Electric Institute, Annual Statistical Bulletins. "Handy-Whitman Index of Public Utility Construction Costs," Whitman, Reguardt & Assoc. Grant, "Principles of Engineering Economy," Ronald Press.

This chapter is primarily concerned with the economics of electric power production by central station generating plants. The cost of generation by stationary thermal and hydroelectric plants makes significant contribution to the price of electric service provided by interconnected and pooled electric utility systems. For the single isolated plant, often assigned specific industrial or commercial loads, generation costs can comprise a major portion or can establish the total price of power. Generation costs consist of operating expenses—including the cost of fuel, operating labor, and plant maintenance—and fixed charges on investment, i.e., the cost of investment capital, depreciation or amortization expenses, and taxes and insurance.

Interconnected power supply systems must price power to include transmission costs, distribution costs, and commercial expenses. They effect price savings by scheduling the installation and operation of a range of plant types to optimize overall generation costs. Generally, efficient plants burning lowest-cost fuels, or operating on available year-round natural water flows, are assigned to continuous base-load operation at or near full capacity. Plants bearing low unit kilowatt investment costs are installed to provide **peaking capacity.** Operation of these units during short daily periods of peak load limits their energy output, and thus minimizes characteristic cost penalties associated with poorer station efficiency or a requirement for higher-priced gas or distillate fuels. Large interconnected utilities may also achieve price benefits by accommodating the staggered installation of large-size units which carry lower unit kilowatt investments, by supplementing capacity and reserve needs of neighboring systems, and by the daily and seasonal exchange of off-peak, low-incremental-cost energy. Historically, there have been significant regional differences in the price of power. These differences have reflected such factors as availability of hydroelectric energy, the cost of fossil fuels, labor costs, ownership type and investment composition, local tax structure, and the opportunities pooling and coordination provide to exploit the economies of scale.

Cost or price is a basic characteristic of power supply. Along with relative abundancy, reliability, and high quality of service, low prices have encouraged the widespread use of electric energy that has come to be associated with our national way of life. Industrial use of electric energy is particularly heavy in the electroprocess and metallurgical industries, and the price of electricity has a significant effect on end-product cost. In many manufacturing and process industries, quality of service in terms of voltage regulation, frequency control, and reliability is a major concern. Uninterrupted supply is of crucial importance in many process industries where a power failure may entail material waste or damage to equipment, in addition to loss of production revenue. Because of the flexibility and convenience of electric service, increases in industrial, residential, and commercial consumption are anticipated despite inflationary trends, price rises brought on by fuel shortages, environmental considerations, and energy conservation efforts.

## Constructed Plant Costs

A **central station** serving a utility system is, in present practice, designed to meet not only the existing and prospective loads of the system in which it is to function, but also the pooling and integration obligations to adjacent systems. Service requirements which establish station size, type, location, and design characteristics ultimately affect cost of delivered power. Selection of plant type and the overall philosophy followed in design must accommodate a combination of specific demands for high operating efficiency, minimum investment, high reliability and availability, maximum reserve capability margins, rapid load change and quick start capabilities, or service adaptability as spinning reserve. For any given plant, design will account for factors such as subsoil and meteorological conditions, temperature and quantity of available water supply, transmission access and capability, fuel delivery and storage needs, outages for maintenance and nuclear fuel loading, and equipment redundancy. In addition, plant siting and design will be significantly affected by legal restrictions on effluents which may have adverse impact on the natural environment.

In the case of **nuclear plants,** siting must take into account the proximity of population centers and the size of exclusion areas. Reactor plant design must bear the investments required to limit radioactive releases to acceptable levels and to provide safeguard systems which protect against accidents. Conventional thermal plants may be sited adjacent to fuel supplies or in proximity to load centers, thereby increasing transmission costs on the one hand or fuel delivery charges on the other. Depending primarily on climate, plants may be enclosed or constructed outdoors. Spare auxiliaries can be installed to improve reliability, and increased investments in sophisticated heat cycles and controls, or for improved equipment performance, can achieve higher efficiencies. Internal-combustion-engine plants are sensitive to loss of capacity as elevation increases above sea level. Gas-turbine outputs are further restricted as ambient air temperatures rise. Hydroelectric sites are frequently far removed from load centers and thus involve added costs for extensive transmission facilities.

**Table 1. Typical Installed Plant Costs, 1975 Price Level**

| Plant description | | | Total investment cost, $ per net electrical kW | | |
|---|---|---|---|---|---|
| Type | Nominal size, MW | Fuel | Enclosed | Semienclosed | Outdoors |
| Conventional fossil | 600 | Coal | 300–360 | 290–350 | 270–330 |
| | | Oil | 280–340 | 270–330 | 250–310 |
| | | Gas | 260–320 | 250–310 | 230–290 |
| Light-water reactor | 1200 | Nuclear | 480–620 | | |
| Packaged combined cycle | 300 | Natural gas/ Distillate | | 210–260 | 190–240 |
| Combustion gas turbine | 60 | Natural gas/ Distillate | | | 120–160 |
| Diesel engine | 8 | Diesel | 210–240 | | 180–210 |

Notes: (*a*) Once through cooling system with conventional surface discharge. (*b*) Land and land-right costs not included. (*c*) Air pollution control equipment not included. (*d*) First unit of a multiunit station. (*e*) Transmission and switchyard costs not included. (*f*) Outdoor combustion gas turbines and diesel engines are assumed to be weather-protected. Semienclosed combined cycle assumes steam plant enclosure.

Also, hydro facilities may provide for flood control, navigation, or recreation as by-products of power production. In such instances, total cost should be properly allocated to the various product elements of the multipurpose project.

An **industrial power plant** provided to meet the requirements of an isolated load entails design considerations and exhibits cost characteristics which differ from those of a central station assigned an integrated role within a connected generating system. Often industrial facilities must economically accommodate both base and peak load requirements. They may be designed to provide for on-site reserve capacity or spinning reserve. Frequency control and voltage regulation must be viewed as special problems because of the limited capability of a single plant to meet load changes.

Table 1 provides typical installed cost data for alternate-type thermal generating plants. The figures represent cost of facilities in place, including interest during construction. In all cases, however, the cost of land, fuel in storage, or loaded nuclear fuel is not included. Cost ranges apply for 1975 plant completion. Escalation effects for plants to be completed beyond this date may be extrapolated in accordance with anticipated cost trends for labor, material, and equipment. Historical cost trends, as experienced in the power industry, which can be helpful in forecasting future costs, may be determined by use of the figures in Table 2. Escalation can be expected to significantly affect plant costs on future projects, especially in view of the 8- to 10-year engineering and construction periods presently envisioned for nuclear facilities, and the corresponding 4 to 5 years required for fossil steam-electric plants. In addition to rising equipment, construction labor, and material prices, major factors influencing the upward trend in plant costs include increased investment in environmental control systems, an emphasis on improved quality assurance and plant reliability, and a concern for safety, particularly in the nuclear field.

**Steam-Electric Plants** Conventional fossil plant investment costs given in Table 1 are for 600 MW nominal unit size. A range of costs accommodates the alternate plot plan, foun-

dation, and condensing water designs dictated by plant site conditions. It also accounts for the variety of cycle arrangements and steam conditions selected to economically balance investment and operating costs for individual plants. Present-day parameters, popular in the face of current economics, call for drum-boiler regenerative reheat cycles at initial steam pressures of 1800 and 2400 psig with superheat and reheat temperatures of 1000°F. Similar temperature levels are

**Table 2. U.S. Cost Trends for Electric Plant Construction, North Central Region**

| January 1 | Total: steam generating plant | Total: hydro generating plant |
|---|---|---|
| 1949 | 100 | 100 |
| 1950 | 105 | 104 |
| 1955 | 136 | 135 |
| 1960 | 163 | 160 |
| 1961 | 161 | 159 |
| 1962 | 160 | 160 |
| 1963 | 160 | 163 |
| 1964 | 163 | 168 |
| 1965 | 167 | 170 |
| 1966 | 171 | 173 |
| 1967 | 177 | 179 |
| 1968 | 186 | 189 |
| 1969 | 194 | 199 |
| 1970 | 212 | 218 |
| 1971 | 226 | 232 |
| 1972 | 247 | 253 |
| 1973 | 253 | 260 |
| 1974 | 275 | 284 |
| 1975 | 347 | 353 |

Note: Cost trends for other regions follow the North Central region index with minor deviation of up to about 4 percent.
SOURCE: "Handy-Whitman Index of Public Utility Construction Costs," compiled and published by Whitman, Reguard & Associates, 1304 St. Paul St., Baltimore, Md.

employed for 3500 psig initial steam pressure supercritical reheat and double reheat cycles which require higher investment outlays in exchange for efficiency or heat rate improvements of 5 to 10 percent. Increased investment costs also apply for more generous boiler-furnace sizing and the fuel and ash-handling facilities required for liquid and solid fuels. Investment increments are also required to provide partial enclosure of facilities in the form of turbine building and furnace structure enclosures and to fully enclose the boiler by providing extended housing to weather-protect ductwork and breeching. Investments are for an initial unit plant which bears costs for site development and office and shop facilities. Costs for plant additions contemplate savings of 5 and 10 percent for successive units.

**Nuclear Plants** Primarily because of site-related design variations in expensive containment structures, a wide range in light-water reactor-plant investment cost is reported. In general, reactor plants require considerably higher investments than do fossil-fueled plants, reflecting the need for leakproof reactor-pressure containment structures, radiation shielding, and a host of reactor-plant safety-related devices and back-up equipment. Also, light-water reactor plants operate at lower initial steam conditions than do fossil-fueled plants, and because of their poorer conversion efficiency, require larger steam flows and increased equipment sizes at added investment. Over a 12-year period of initial commercial development, nuclear-plant costs have escalated dramatically. This is a consequence of rising price levels in general, but importantly reflects stricter licensing criteria applied by the United States Government in the areas of system safeguards, containment structure integrity, and control of radioactive releases.

Potential competition to both pressurized and boiling light-water reactor plants is provided by the helium-gas-cooled reactor, which produces steam in a reheat cycle configuration at initial steam conditions and thermal cycle efficiencies matching those of drum-type fossil units. Sufficient data is not yet available to establish cost levels for the commercially sized units. However, 1973 projected cost for Southern California Edison Company's 770-MW gas-cooled reactor plants, scheduled for operation in 1983 but subsequently cancelled, were reported at $615 per kW.

Both light-water reactor and conventional fossil plants evidence declining unit cost with size. The economy of scale has been demonstrated most strikingly in the nuclear field, where orders for units at the 1200-MW size level predominate. Large fossil units of tandem compound design have reached the 600- to 900-MW power level, and cross compound units rated at 1300 MW are being commissioned. The utilization of large-size units has been made possible by the interconnection of utility transmission and distribution systems, many of which have grown to levels capable of absorbing large individual units without assuming economic penalties for the reserve equipment required to assure service continuity in the event of unscheduled outages. Representative investment costs for nuclear- and fossil-fueled steam stations are plotted in Fig. 1 as a function of size.

**Combined Cycle Plants** Packaged combined cycle plants are offered by a number of vendors, generally in the 100- to 600-MW size range. These plants consist of multiple installations of open-cycle gas turbines arranged to exhaust to steam generators which also may be equipped for supplementary

firing of light oil or gas. Steam produced is supplied to a conventional nonreheat steam cycle. Advantages of combined cycles are lower unit investment costs, attractive thermal performance, increased flexibility which allows independent operation of the gas-turbine portion of the plant, shorter installation schedules, reduced thermal discharge to the aquatic environment, and the reduction in sulfur oxide and particulate emissions characteristic of distillate and gaseous fuels.

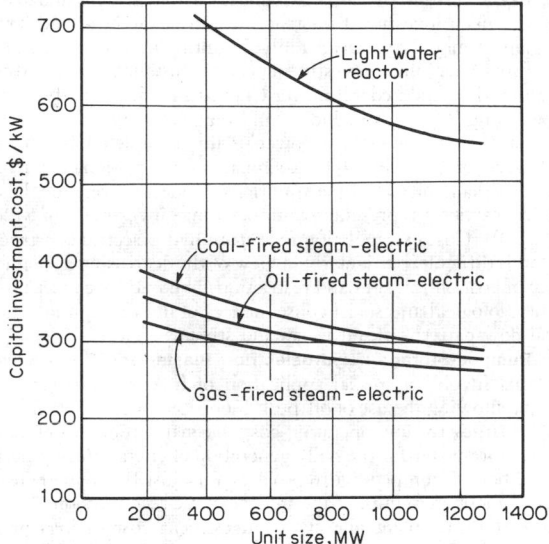

**Fig. 1**  Representative investment versus unit size. *Note:* Costs are for 1975 price levels, and environmental protection systems are not included.

Crude- and heavy-oil firing of gas turbines is now proposed by various manufacturers of this equipment. These proposals contemplate the addition of treatment plants to inhibit corrosive effects of the fuels on turbine internals. Reported investment increments required for crude or residual fuel-oil treatment range between $5 and $10 per kW of combined cycle capacity.

**Combustion Gas Turbine and Diesel Engine Plants** Combustion gas turbines and diesel engines show significant investment savings as compared to base-load steam electric plants. They are characterized by quick start abilities and, in the case of diesel engines, relatively high efficiency. Fluctuations in investment costs for internal-combustion thermal plants are primarily caused by variations in the fuel-storage and noise-attenuation designs provided individual plants. Total investment savings in this equipment is largely attributable to factory manufacture and assembly of major components coupled with limited on-site construction. As in the case of combined cycle plants, heavy oil may be considered for gas turbine use. Here, fuel treatment costs must be assigned against combustion turbine output only.

**Hydroelectric Plants** Hydroelectric generation offers unique advantages. Fuel, a heavy contributor to thermal-plant operating costs, is eliminated. Also, hydro facilities last longer than do other plant types, and thus they carry lower depreciation rates. They result in lower maintenance and operating expenses, eliminate air and thermal discharges, and, because of relatively simple design, exhibit attractive availability and forced outage rates. Quick start capability and rapid response to load change ideally suit hydro turbines to spinning reserve and frequency control assignments.

The constructed cost of a hydroelectric station is essentially site-dependent. Overall costs fluctuate significantly with variations in dam costs, with intake and discharge system requirements, with the pondage provided to firm up capacity, and with the cost of relocation of facilities within areas inundated. For a given investment in structures, available head and flow quantity may vary considerably, resulting in a wide range of outputs and unit investment costs. Installed plant costs reported by the Federal Power Commission for recently completed hydro plants include a mid-range $386 per kW investment for the 140-MW Keowee Plant, completed by Duke Power Company in 1971. A similarly sized station, the Don Pedro Plant of California's Turlock Modesto Irrigation District, carries a representative upper-range investment of $726 per kW. Cost extrapolation for future hydroelectric construction is difficult, particularly in view of the decreasing availability of economical sites and restrictions imposed by concern for the ecological and social consequences of disrupting the natural flow patterns of our rivers and streams.

**Pumped-Storage Hydroelectric Plants**  Pumped-storage plants involve a special application of hydroelectric generation, allowing the use of off-peak energy supplied at incremental charges by low-operating-cost thermal stations to elevate and store water for the daily generation of energy during peak load hours. Pumped hydro projects must justify the inefficiencies of storage pumping and hydroelectric reconversion of off-peak thermal plant energy by investment cost savings over competing peaking plants. Installation of a pumped hydro station calls for a suitable high head site which minimizes required water storage and upper and lower reservoir areas, and an available makeup source to supply the evaporative losses of the closed hydraulic loop. Despite the added complications of installing both pumping and generating units, or of utilizing reversible motor-generator pump turbines, costs for pumped hydro stations generally fall below those for conventional hydroelectric stations. Lower installed costs are the result of elimination of dams, extensive pondage, and the siting need for appreciable natural water flow. The Federal Power Commission reports a 1970 installed cost of $151 per kW for Pennsylvania Electric's 422-MW Seneca Plant. The 1965 Yards Creek Project in western New Jersey carries an investment chage of only $89 per kW, reflecting lower price levels of a decade ago.

**Geothermal Plants**  Geothermal generation applies the earth's heat by abstracting it from steam or hot water found below the earth's outer crust. Prevalent in geological formations underlying the western United States, geothermal energy is predominantly unexploited, but is receiving increased attention in view of escalating demands on limited worldwide fossil fuel supplies. Because natural geothermal heat supplants fuel, the atmospheric release of combustion products is eliminated. Nevertheless, obnoxious gases and chemical residues, usually contained in geothermal steam and hot water, must be dissipated. There is a current lack of significant cost data covering geothermal plants. Only one major commercial United States facility exists, the 500-MW

Geysers Plant in northern California, now undergoing staged expansion to a projected multiunit capacity of 900-MW in 1978. Because boiler and associated fuel-handling facilities are eliminated, investment in generating plant proper is perhaps half to three-fourths the cost of comparable fossil-fueled units. Overall investment chargeable to geothermal facilities, however, includes significant exploration and drilling costs, which are site-dependent and elude accurate prediction (see also Sec. 9).

**Environmental Considerations**  Environmental protection has become a dominant factor in the siting and design of new power generating stations. 1970 amendments to the Federal Clean Air Act, and Federal Water Pollution Control Act amendments passed in 1972 have greatly strengthened government regulatory powers over air and water pollution. It is abundantly clear that both stack emissions to the atmosphere and thermal discharges to natural water courses must be significantly reduced in order to meet increasingly stringent environmental criteria. In general, the regulatory emphasis is on new sources. However, in many cases older plants will be required to reduce emission levels to achieve legislated ambient air quality standards and to control thermal discharges by the use of closed cooling systems allowing heat dissipation to the atmosphere (see also Sec. 18).

Control of **air pollution** by fossil-fueled generating sources involves the reduction of particulates, sulfur oxides, and nitrogen oxides in flue gas emissions. Particulate collection is currently achieved principally by the use of high-efficiency electrostatic precipitators. Control of sulfur dioxide can be effected through the use of low-sulfur fuels. The use of generally higher cost low-sulfur fuels, however, can add significantly to plant operating costs. For this reason, a major effort has been devoted to the development of stack gas cleanup systems. Although considerable controversy exists as to the practicality of present design in terms of availability for continuous service, it is anticipated that environmental pressures coupled with decreasing economic availability of low-sulfur fuels will result in the near-term commercial application of stack gas scrubbers. Table 3 shows the investment increment required to limit stack gas emission. The estimate is based on relatively limited data, but is indicative of the magnitude of the investment cost penalty involved in stack gas sulfur removal. Tabulated figures reflect system installed cost, plus the penalty in restricted plant capability due primarily to increased auxiliary power to overcome system draft losses.

Control of nitrogen oxides is attained primarily through modifications to flame propagation and combustion processes and by flue gas recirculation. These features can be readily incorporated in modern steam-generator designs without appreciable cost impact.

In order to avoid plant discharges of waste heat to the aquatic environment, evaporative-type closed cooling cycles may be employed in place of conventional, natural-water-course-cooled, once-through condensing designs (see Sec. 9). Closed-loop cooling systems in current use employ evaporative **cooling towers,** cooling ponds, and spray canals. Use of these systems entails investment penalties consisting of net increases in equipment and facilities cost, and unit kilowatt investment penalties incurred by losses in peak capability due to the lower conversion efficiency normally experienced with evaporative cooling. Table 3 shows representative facility cost

**Table 3. Representative Incremental Investment Costs and Capability Losses for Environmental Control Systems**

| Plant type / Control system | Incremental investment cost $ per kW, 1975 price level | | | Percent loss in plant kW capability | | |
|---|---|---|---|---|---|---|
| | Conventional, 600 MW | LWR, 1200 MW | Comb. cycle, 300 MW | Conventional, 600 MW | LWR, 1200 MW | Comb. cycle, 300 MW |
| Precipitators | 15–22 | | | 0.2–0.4 | | |
| Sulfur-oxide removal | 60–80 | | | 3–5 | | |
| Evaporative cooling towers: | | | | | | |
| Natural draft | 8–12 | 11–16 | | 2–6 | 2–6 | |
| Mechanical draft | 4–7 | 6–11 | 3–6 | 3–8 | 3–8 | 1–4 |
| Canal spray modules | 10–17 | 14–23 | 6–10 | 3–8 | 3–8 | 3–8 |
| Cooling ponds | 8–12 | 12–20 | 4–8 | 2–6 | 2–6 | 2–6 |

*Notes:* (a) LWR = Light-water reactor. (b) Costs show incremental increases over and above normal costs for conventional once-through cooling. (c) Cooling-system investment costs do not include pond lining or makeup system costs. No penalty is assessed for water consumptive losses. (d) Land and land-right costs are not included. (e) Electrostatic precipitator installed costs are based on a removal efficiency of 99 percent. (f) Sulfur-oxide removal-system costs assume throwaway-type wet limestone/lime scrubbing systems. Costs do not include waste-disposal facilities. (g) Mechanical towers assume wooden construction. Cost of concrete towers is currently approximately 50 percent higher. (h) Combined cycles assume the use of low-sulfur, low-ash distillate fuels which meet air pollution criteria. (i) Natural draft towers are not economically feasible for 300-MW combined cycle units.

penalties and peak capability restrictions for various closed-cycle cooling systems. Penalties for fitting light-water reactor plants with evaporative systems are higher than for conventional fossil plants. The lower efficiency of light-water nuclear plants requires the dissipation of larger quantities of heat and also results in more severe capability restrictions.

More advanced closed cooling loop designs anticipate the use of dry and wet/dry cooling towers. These represent possible alternates to conventional evaporative systems where makeup water is in short supply or where visible vapor plumes or ice formed by vapor discharge present hazards. The penalties for dry-tower cooling are significantly higher than those for conventional evaporative designs. Large, more costly water-to-air heat-transfer surfaces are required, and characteristically high condensing pressures result in severe restrictions in plant capability.

Although no large-size dry towers are currently in operation in the United States, estimates indicate incremental investment cost penalties for conventional fossil-fueled plants in the range of $25 to $40 per kW for mechanical draft towers, and of $55 to $80 per kW for dry natural-draft designs. Similarly equipped light-water reactor plants bear dry-tower investment cost penalties approximately 40 percent higher than the above. Wet/dry towers show promise for practical application, combining the advantages of both wet- and dry-tower designs. Wet/dry towers incur added investment penalties for closed-surface heat transfer only to the extent necessary to eliminate visible plume or reduce makeup requirements. Investment costs fall between those for wet and dry towers.

**Fixed Charges**

Costs that are established by the amount of capital investment in plant and which are fixed regardless of production level are termed **fixed charges**. Annual fixed charges are ordinarily expressed as a percentage of investment and include interest or the cost of money, funds applied to amortize investment or to allow for replacement of depreciated plant, and charges cover-ing property taxes and insurance. Additionally, fixed charges may include an interim replacement allowance to cover the cost of replacement of items of plant equipment not expected to last the full life of the plant.

The **cost of money** employed for utility-plant expansion depends upon money-market conditions in general and upon the attitude of investors with regard to a particular utility enterprise or specific project. Funds for investor-owned utility expansion are derived from both the risk capital (equity) and debt capital (bond) markets. Utility bonds command a return of 9 to 10 percent in the current market, while equity capital returns between 10 and 12 percent are typical. Prevailing return rates for investments in utility plant facilities are influenced by the rate-setting practices of public-utility regulatory agencies on the one hand, and by supply and demand factors on the other.

Public-utility facilities owned by state and municipal government organizations are generally financed by long-term revenue bonds, most of which qualify for tax-free-income status. Interest rates currently fall between 7 and 8 percent and reflect the tax relief on interest income enjoyed by the bond holders. Generating facilities are often financed by industrial concerns whose primary business is the production of a manufactured product. In these instances, the annual cost of money invested in power facilities will be established by considering alternate investment of the required funds in manufacturing plant. Inasmuch as returns on equity capital invested in manufacturing industries are usually on the order of 12 to 20 percent, rates of return for industrial power plants will tend to be set at these higher levels.

Transmission and distribution cooperatives qualifying under provisions of the Rural Electrification Act of 1936 present a special case. Direct treasury financing at a preferential interest rate, traditional in the past, is now limited to the partial funding of projects. A major portion of future cooperative investment requirements will be met by long-term loans of private capital at anticipated interest rates of 8 to 9 percent.

Corporate federal taxes are levied on equity capital income. Thus, corporate earnings on equity investment must exceed the return paid the investor by an amount sufficient to cover the tax increment. Current federal tax rates of 48 percent apply. Therefore, revenue requirements for a given return to the investor will amount to slightly less than twice the given rate (see Table 4).

The amortization of debt capital or the provision for **depreciation** over the physical life of utility-plant facilities may be effected by several methods. Straight-line depreciation requiring uniform charges in each year over a predetermined period of useful service is commonly applied because of its simplicity. The percentage method of depreciation assumes a constant percentage decrease in the value of capital investment from its value the previous year, thereby resulting in annual depreciation charges which progressively diminish. The sinking-fund method, popularly used by engineers for economic analysis, assumes equal annual payments which, when invested at a given interest rate, will accumulate the capital value of facilities less their salvage value over a predetermined useful service life. Representative life of alternate utility facilities is illustrated in the following table.

| Facility | Representative useful service life, years |
|---|---|
| Steam-electric generating plant | 30 |
| Hydroelectric plant | 50 |
| Gas turbine generator | 25 |
| Diesel electric generator | 20 |
| Nuclear plant | 30 |
| Transmission and distribution plant | 40 |

Use may be made of interest tables which show, for any rate of interest and any number of years, the equal annual payment (sinking-fund) rate which will amortize an investment and additionally will yield an annual return on investment equal to the interest rate.

$$\text{Equal annual payment} = \frac{i(1 + i)^n}{(1 + i)^n - 1}$$

where $n$ = number of years of life and $i$ = interest rate or rate of return. (See Table 5 and Grant, "Principles of Engineering Economy," 4th ed., Ronald Press, New York).

Property taxes and property insurance premiums are normally established as a function of plant investment and thus are properly included as fixed charges. Property taxes vary with the location of installed facilities and with the rates levied by the various governmental authorities having jurisdiction. In general, public power authorities will be free of taxes, although public enterprises often render payments to government in lieu of taxes. Annual property-tax rates for private enterprise will amount to perhaps 2 to 4 percent of investment, while property insurance may account for annual costs of 0.2 to 0.4 percent.

A representative makeup of fixed charges on investment in a conventional steam-electric station having a useful service life of 30 years is shown for various classes of ownership in Table 6.

## Operating Expenses

**Fossil Fuels** Currently, fossil fuels contribute approximately three-quarters of the primary energy consumed by the United States in the production of electric energy. Predictions are that in about a dozen years nuclear fuel will emerge as the single largest source of energy used for power production. Nevertheless, generating-station demands for fossil fuels, particularly coal, are continuing to increase as the electric-utility and industrial power markets grow. Price comparisons of fossil fuel are generally made on the basis of delivered cost per million Btu. This cost includes mine-mouth or well-head price, plus the cost of delivery by pipeline or carrier. Price comparisons must recognize that solid and, to a lesser extent, liquid fuels require plant investments and operating expenditures for fuel receipt, storage, handling and processing facilities, and ash collection and removal. High-transportation-cost contributions on a Btu basis will be incurred by high moisture- and ash-content coals with low heating values. This explains the popularity of firing lignite and subbituminous coals at mine-mouth generating plants.

**Coal** represents our most abundant indigenous energy resource, with enough economically recoverable supplies at current use rates to last well into the next century. About half of the recoverable coal reserves, however, have a sulfur content which, without costly stack gas cleanup, will exceed federal emission standards. Low-sulfur coals are found chiefly in the low-load areas of the mountainous West, and delivered cost at the major markets east of the Mississippi must include high transportation charges. Increasingly, coal production will bear the cost of more rigid enforcement of stringent deep-mine safety regulations and the charges associated with strip-mine land restoration. Delivered price will depend upon transportation economies as may be effected by unit train haulage or by pumping in slurry pipelines. A long history of market losses to oil, gas, and most recently nuclear fuels in the domestic heating, rail transport, and power generation fields has left the coal industry in an unfavorable financial condition. Substantially higher mine-mouth prices are likely in order to ensure the funding of needed expansion in production. Delivered price levels for coal fuel vary with plant location. Plants conveniently located with respect to eastern coal reserves report delivered coal prices in the general range of 80 to 140¢ per million Btu, depending upon sulfur and ash content. Mine-mouth plants in North Dakota utilize strip-mine lignite at current prices of 25 to 40¢ per million Btu. Low-sulfur western coals supplied to Commonwealth Edison stations in the Chicago area command delivered prices in the neighborhood of 100¢ per million Btu.

The domestic supply of **petroleum** is now outstripped by nationwide demand. If current trends continue, the United States will be even more dependent upon overseas sources to meet growing energy demands. Free-market needs for foreign imports of oil are predicted to reach over 50 percent of the total consumed in the United States by the mid-1980s. The consequences of this trend in terms of balance-of-payment deficits are enormous, and dependence on foreign supplies from politically unstable areas in the Middle East and North Africa has already resulted in supply shortages. Residual and distillate oil prices have risen principally because of short supply. Pressure by the major oil producers on worldwide-

**Table 4. Comparison of Interest or Return and Federal Income Taxes on Private and Public Projects**

| | Investor-owned utility, % | Government-owned utility, % | Industrial/commercial ownership, % |
|---|---|---|---|
| Distribution of investment | | | |
| Equity capital (stocks) | 40 | 0 | 100 |
| Debt capital (bonds) | 60 | 100 | 0 |
| Total | 100 | 100 | 100 |
| Rate of return or interest | | | |
| On stocks | 11.0 | 0.0 | 12.0 |
| On bonds | 9.0 | 7.0 | 0.0 |
| Income subject to federal income tax (FIT) | | | |
| Average rate of return | 9.8* | 7.0 | 12.0 |
| Deduction for interest | 5.4† | 7.0 | 0.0 |
| Net taxable income | 4.4 | 0.0 | 12.0 |
| Return on equity before 48% FIT | | | |
| 4.4%‡/(100% − 48%) | 8.46 | | |
| 12%/(100% − 48%) | | | 23.08 |
| FIT as percentage of capital | | | |
| 8.46% − 4.4% | 4.06 | | |
| 23.08% − 12% | | | 11.08 |
| Summary revenue requirements | | | |
| Equity return before FIT | 8.46 | 0.0 | 23.08 |
| Bond interest | 5.40§ | 7.0 | 0.00 |
| Total | 13.86 | 7.0 | 23.08 |

*Average return is 11 percent equity return on 40 percent of investment plus 9 percent bond interest on 60 percent of investment.
†Deduction is 9 percent bond interest on 60 percent of investment.
‡Return on equity equals 11 percent equity return to stockholders on 40 percent of investment.
§Return on debt equals 9 percent bond interest on 60 percent of investment.

market price levels can be expected to continue this trend. Blends of low-sulfur oils delivered during the last months of 1975 to generating plants along the eastern seaboard from overseas sources were priced upwards of 160¢ per million Btu, reaching as high as 240¢ per million Btu for the 0.3 percent sulfur fuel required for firing in some metropolitan areas.

During this same period distillate oil commanded a nationwide price ranging approximately from 220 to 270¢ per million Btu.

Consumption of **natural gas** as a power fuel is not expected to keep pace with future growth in energy demands, primarily due to restricted availability of domestic supplies and the high

**Table 5. Equal Annual Rate which Will Amortize Original Investment over Estimated Life* and Yield on Annual Return on Investment Equal to Interest Rate**

| Life, yr | Annual rate of interest or return | | | | | |
|---|---|---|---|---|---|---|
| | 6% | 7% | 8% | 10% | 12% | 14% |
| 10 | 0.13587 | 0.14238 | 0.14903 | 0.16275 | 0.17698 | 0.19171 |
| 15 | 0.10296 | 0.10979 | 0.11683 | 0.13147 | 0.14682 | 0.16281 |
| 20 | 0.08718 | 0.09439 | 0.10185 | 0.11746 | 0.13388 | 0.15099 |
| 25 | 0.07823 | 0.08581 | 0.09368 | 0.11017 | 0.12750 | 0.14550 |
| 30 | 0.07265 | 0.08059 | 0.08883 | 0.10608 | 0.12414 | 0.14280 |
| 35 | 0.06897 | 0.07723 | 0.08580 | 0.10369 | 0.12232 | 0.14144 |
| 40 | 0.06646 | 0.07501 | 0.08386 | 0.10226 | 0.12130 | 0.14075 |
| 50 | 0.06344 | 0.07246 | 0.08174 | 0.10086 | 0.12042 | 0.14020 |
| 75 | 0.06077 | 0.07044 | 0.08025 | 0.10008 | 0.12002 | 0.14001 |

*Amortization is by sinking fund earning interest at indicated rate.

**Table 6. Derivation of Fixed-Charge Rate for Conventional Fossil-fueled Steam-Electric Plant with 30-Year Economic Life**

|  | Investor-owned utility, % | Government-owned utility, % | Industrial/commercial ownership, % |
|---|---|---|---|
| Rate of return or interest* | 9.80 | 7.00 | 12.00 |
| Amortization or depreciation* | 0.63 | 1.10 | 0.41 |
| Federal income tax | 4.06 | 0 | 11.08 |
| Local taxes (or payment in lieu of taxes) | 2.00 | 2.00 | 2.00 |
| Insurance | 0.30 | 0.30 | 0.30 |
| Total | 16.79† | 10.40 | 25.79 |

*Sum of sinking-fund amortization or depreciation rate (at given rate of return or interest) and interest or return rate is listed in Table 5.

†Assuming initial plant investment of $300/kW and 7500 h per year operation, the annual fixed charges in mils per kilowatthour will be $300 \times 0.1679 \times 1000 \times 1/7500$ of full load = 6.72 mils/kWh.

cost of overseas transportation in liquefied form. Because it is clean-burning, convenient to handle, and generally requires smaller and cheaper furnaces, natural gas is premium rated; its limited supply is best suited competitively for consumption by residential and commercial users and to meet industrial process needs. Present-day power-plant use of natural gas is generally restricted to existing boiler plants located in the gas-producing areas of the south-central and western states and to gas turbines operating on off-peak gas-fuel supplies. FPC data for 1975 reports gas-fuel prices for firm supplies in the range of 70 to 150¢ per million Btu. Price levels of natural gas, severely regulated in the past by government controls imposed at the well, have been rising sharply in response to current supply-and-demand factors. A continuation of this trend should further discourage the use of gas as a generating-station fuel.

**Nuclear Fuel** Reactor-plant fuel costs present a special case. Actually, the initial core loading which will support operation of a nuclear plant over its early years of life requires a single purchase prior to commissioning of the plant. For comparison with fossil-fuel prices, therefore, nuclear-fuel-cycle costs, including first-core investment, periodic charges for reload fuel, and spent-fuel shipment and processing costs are ordinarily extrapolated at assigned load factors over the life of the plant and are converted to an economic equivalent expressed in cents per million Btu of released fission heat. Nuclear-fuel costs are not only influenced by ore prices and by fuel fabrication and processing costs, they are also sensitive to investment costs and to uranium-enrichment charges levied for this government service. Future costs are subject to inflationary pressures and could reflect changes in government policy or the assumption of enrichment operations by private industry. Charges of 70 to 90¢ per million Btu are representative of the equivalent or levelized fuel prices extrapolated for nuclear plants for which designs are now being initiated.

**Operating Costs for Fuel** Fuel-price contributions to energy-generation costs will reflect start-up and furnace-banking losses and will depend upon plant efficiencies which, at low loads, show considerable departure from the best-point performance achieved at or near full unit loadings. These factors significantly affect the operating expenses of load following utility system units as well as plants assigned fluctuating demands in manufacturing or industrial service. As an example, calculated performance for a nominal 600-MW 2400-

psig oil-fired regenerative reheat steam unit shows a best-point heat rate of 9,400 Btu per net kWh at rated net output. Load reduction yields heat rates of 10,100 and 12,800 Btu per kWh at net MW loadings of 300 and 150, respectively. Typical operating-expense ranges for given fuel prices and estimated full-load heat rates may be determined directly where continuous operation at or near unit rating is assumed (Table 7).

**Operating Labor and Maintenance Costs** In addition to fuel costs, operating expenses include labor costs for plant operation and maintenance plus charges for operating supplies and maintenance materials, general administrative expenses, and other costs incidental to normal plant operation. Operating labor and maintenance costs vary considerably with unit size, operating regimen, plant-design conditions, type of facility, and the local labor market. Representative figures appear in Table 8. Nuclear liability insurance costs are included under reactor-plant operating expenses. For units in the 1000- to 1200-MW size range operating at high-capacity factor, these charges may be in the order of 0.04 mils per kWh. They finance a program of government indemnity, coupled with private insurance, for public damage that could arise from a nuclear incident.

**Environmental Controls** Systems and equipment applied for air- and water-pollution abatement generally carry increased fuel and maintenance labor and materials costs. Reductions in plant output resulting from the higher condensing pressures associated with cooling-tower operation, or the added auxiliary power for gas clean-up systems, lower plant efficiency and increase fuel consumption. Operating-cost increments for major environmental protection systems are listed in Table 9. The reader is cautioned that these estimates are based on the rather limited data now available and are subject to significant variation for individual installations, particularly in the case of sulfur-oxide removal systems.

### Overall Generation Costs

The total cost of power generation may now be estimated by reference to the preceding material assuming type of ownership, capital structure, plant type, fuel, and loading regimen. Table 10 comprises an illustrative tabulation of the factors determining the overall generation cost of an investor owned nuclear generating facility.

It should be noted that capacity factor, or the ratio of

**Table 7. Operating Expense for Fuel for Representative Heat Rates and Fuel Prices.**

| Type of plant | Nominal size, MW | Typical heat rate Btu/kWh | Fuel price, ¢/MBtu | Fuel cost, mils/kWh |
|---|---|---|---|---|
| Conventional fossil | 600 | 9,400 | 200 | 18.8 |
| Light-water reactor | 1200 | 10,200 | 80 | 8.16 |
| Combined cycle | 300 | 8,800 | 250 | 22.0 |
| Combustion gas turbine | 60 | 13,000 | 250 | 32.5 |
| Diesel engine | 8 | 9,000 | 250 | 22.5 |

*Note:* Operating fuel cost, mils/kWh $= \dfrac{\text{Btu}}{\text{kWh}} \times \dfrac{¢}{\text{Btu}} \times 10$.

**Table 8. Representative Operating and Maintenance Costs (1975 Price Level)**

| Type of plant | Nominal size, MW | Fuel | Operating and maintenance costs, mils/kWh |
|---|---|---|---|
| Conventional fossil | 600 | Coal | 0.7–1.2 |
| | | Oil | 0.6–0.8 |
| | | Gas | 0.4–0.7 |
| Light-water reactor | 1200 | Nuclear | 0.7–1.2 |
| Combined cycle | 300 | Natural gas | 1.2–1.8 |
| | | distillate | 1.5–2.2 |
| Conventional hydro | 300 | | 0.2–0.6 |
| Combustion gas turbine | 60 | Natural gas | 2.0–3.0 |
| | | distillate | 2.5–3.5 |
| Diesel engine | 8 | Distillate | 4.0–6.0 |

*Notes: (a)* Unit kWh costs include labor, maintenance materials, operating supplies, and incidental expenses. *(b)* Costs shown for conventional light-water reactor and combined cycle plants assume base-load operation. Combustion turbines and diesel engine costs assume operating in the range of 500 to 1500 h per year.

average to peak load carried by a given generating facility, will have significant effect on generation expenses. In addition to the effects of part-load operation on fuel costs as previously discussed, capacity factor will determine the plant output quantity which will support fixed charges. High-capacity factor operation will spread fixed charges over a larger number of kilowatthours of output, thereby reducing unit generation costs. During its initial life, a thermal plant is usually operated at high-capacity factor. As inevitable obsolescence brings newer and more efficient equipment into service, a unit's base-load position on the load duration curve is relinquished, and capacity factor tends to drop. This decline in capacity factor must be acknowledged in estimating output and generation costs over the life of a given facility.

Most modern electric utilities incorporate computerized systems designed to economically dispatch power generated at each production plant feeding the load. Individual generating unit loads are assigned in a manner that can be demonstrated to result in minimum overall cost, i.e., at each system power level, load is shared between units so that all operate at the same incremental production cost. Telemetered data reflecting system load and generation is transmitted to central dispatch computers by multiplexing via power-line carrier, microwave, or telephone lines, and the communication schemes include channels for transmitting load-adjustment commands developed by the computer to on-line generating units. Loading instructions account for the unit-production efficiencies and transmission losses associated with each assignment.

**Transmission Costs**

Because of a growing scarcity of urban sites and an increasing emphasis on environmental protection and nuclear safety, it has become more and more difficult to site major power stations at the centers of load. As a consequence, added cost of transmission, along with attendant resistive power losses, add significantly to the overall cost of service. Because of the distances involved, these costs are generally greatest for nuclear facilities, mine-mouth stations, and hydroelectric plants where dam locations govern. Transmission-plant investment also reflects a trend toward interconnection of neighboring utilities. Designed to improve service reliability by the pooling of reserves and to effect savings by capacity and energy interchanges, such interties must carry substantial ratings so that emergency power transfers can be accommodated without exceeding system stability limits. Based primarily on data reported in the National Power Survey, 1975 costs for major overhead transmission ties (1000 MW and up) are estimated to range between $150,000 and $250,000 per circuit mile. Investments are sensitive to factors of climate and topography, and routings near urban areas can increase right-of-way costs significantly. Where underground transmission is elected, installed investment could escalate to as much as 10 times the cost of overhead lines. According to the 1970 National Power Survey, transmission costs comprised 13 percent of the energy prices paid by consumers in 1968. As the effects of remote plant siting and system integration continue to be felt, pressure will be exerted to further increase this percentage. The cost of transmission as a portion of total energy price, however, will be significantly affected by the relative escalation of transmission, as opposed to production costs.

The choice of **transmission voltage** level and whether alternat-

**Table 9. Estimated Operating and Maintenance Cost Increments in Mils/ kWh for Environmental Control Systems**

| Control system | Plant type Conventional, 600 MW | Light-water reactor, 1200 MW | Combined cycle, 300 MW |
|---|---|---|---|
| Evaporative cooling towers | | | |
| Mechanical draft | 0.007–0.015 | 0.01–0.02 | 0.004–0.01 |
| Hyperbolic natural draft | 0.004–0.015 | 0.005–0.02 | |
| Canal spray modules | 0.015–0.03 | 0.02–0.04 | 0.01–0.02 |
| Cooling ponds | 0.004–0.015 | 0.005–0.02 | 0.002–0.01 |
| Electrostatic precipitator | 0.05–0.30 | | |
| Sulfur oxide removal | 1.0–3.0 | | |

*Notes:* (*a*) Cost increments shown for all plants are for operation at high capacity factors. (*b*) Sulfur-oxide removal-system costs assume throwaway-type wet limestone/lime scrubbing systems. Costs associated with waste disposal are not included.

**Table 10. Total Cost of Generating Power**

| Plant type | 1200 MW Light-water reactor |
|---|---|
| Plant net heat rate | 10,200 Btu/kWh |
| Generating costs, mils/kWh | |
| Nuclear fuel @ 80¢/MBtu | 8.16 |
| Operating and maintenance | 1.00 |
| Fixed charges @ 16% per annum | 11.00 |
| Total Operating costs | 20.16 |

*Notes:* (*a*) Operating and maintenance includes nuclear liability insurance. (*b*) Fixed charges based on assumed initial plant investment of $550/kW and 8000 h per year of full load operation. (*c*) Values shown are typical and could vary significantly for individual plants.

ing current or direct current is used for bulk power transfer depends on the amount of power transmitted, the transmission distance involved, and at each voltage level, the cost of line and substation equipment. Voltages generally employed are 230, 345, 500, and 765 kV ac and ± 400 kV dc. As the trend is to ever-larger blocks of power, still higher ac transmission voltages are being examined in an effort to improve economy and to hold rights-of-way to manageable limits. Where long distances in the order of 400 mi or more are encountered, dc transmission becomes economically attractive. For shorter lines, however, savings in fewer conductors and lighter transmission towers are nullified by the high cost of dc/ac conversion equipment at the terminals (see Sec. 15).

**Power Cost and Prices**

The price of power delivered by a utility system must account for the production costs at each of its generating stations. As previously noted, these costs depend upon labor rates, fuel prices, and material charges. They reflect investment levels in generating plant and the fixed-charge rates established by funding patterns, type of ownership, and expected equipment service life. Overall system production costs are affected by the investment requirements of specific mixes of generating equipment types, and by the manner in which load is shared by units, i.e., how production is allocated between highly

efficient base-load stations and the lower-cost peaking equipment which normally runs for only a few hours each day. Also, important cost reductions are achieved by controlling natural and stored water flows to allow optimized sizing and scheduling of hydroelectric output, thereby reducing needs for thermal peaking capacity and decreasing the generation requirements of high-fuel-cost fossil plants.

Power prices cover the investment charges, maintenance costs, and capacity and energy losses chargeable to transmission and distribution plant. They further include the administrative costs incurred to maintain corporate enterprise, and the commercial expense of submitting and collecting bills.

Generally, power prices must provide a return to cover the average cost of power production throughout a given system. Rate schedules and supply contracts, however, are drawn to reflect the reduced cost of off-peak energy produced by available generating units during periods of low system load. Additionally, prices for high-load-factor service often recognize the cost reductions effected by spreading fixed charges over increased units of energy output. Large blocks of capacity and energy supplied for industrial use are often priced by establishing an annual charge for capacity which equals the fixed charges on investment in committed generation and transmission plant, plus charges for energy representing the sum of the variable kilowatthour production costs for fuel, maintenance, and operation.

Historically, utilities have applied rate schedules which promote consumption by applying progressively lower rates to blocks of increased energy usage. Rationale for such pricing is the savings that load growth can realize through economies of scale, as well as the improved utilization of existing utility plant. Most recently, however, regulatory pressure, reflecting a policy of minimizing the industry's impact on the environment and the critical need to conserve high-cost imported fuel, has favored a marginal cost pricing system more nearly reflecting the actual cost of production and transmission of a particular user's supply of power. Rate setting under this conservationist approach calls for flat rather than reduced rates as usage increases and for high unit energy charges during peak load periods.

Facilities designed for the **dual-purpose production of power**

**and process steam** permit investment savings, principally in steam generation plant, which result in combined production charges falling below the total cost of separate single-purpose production of power and of steam. Such savings can permit proportionate decreases in the prices ordinarily charged for separate single-purpose production of each of the products, or they may be assigned in total to reduce the price of one or the other by-product. This latter option is often exercised in the case of dual-purpose water-production plants arranged for seawater flash evaporation, using power-turbine extraction as a process steam source. Where severe shortages of fresh water exist, social considerations favor the total assignment of dual-purpose savings to the water product. Thus, minimal prices for desalted product water are achieved, while dual-purpose power is marketed at prices competing with single-purpose power generation costs. Similarly, assignment of the total savings of dual-purpose power and process steam production to the power product may justify on-site industrial-plant power generation in preference to outside purchases of higher-priced utility-system power supplies.

Where hydro facilities supply power in combination with irrigation, flood control, navigation, or recreational benefits, power costs are largely sensitive to the allocation of investment charges against each of the multipurpose project functions. Should generation be treated as a by-product, power prices can be reduced drastically so as to reflect equipment operating expenses and the limited fixed charges covering investment in only the generating plant itself.

Cheap fuel, advances in design, the economy of scale, and the economic application of alternate generating unit types have been responsible for past downward trends in the price of electric service. More recently, however, these trends have been reversed by inflationary effects on plant costs, by rising interest rates, and by the increased prices commanded by fossil and nuclear fuels (Table 11). Additionally, a dwindling number of favorable plant sites, licensing delays, and the

**Table 11. Average Revenue per kWh Sold, Total U.S. Electric Utility Industry**

| | Cents per kWh | | |
|---|---|---|---|
| | | Light and Power | |
| Year | Residential | Small | Large |
| 1974 | 2.83 | 2.85 | 1.55 |
| 1973 | 2.38 | 2.30 | 1.17 |
| 1972 | 2.29 | 2.22 | 1.09 |
| 1971 | 2.19 | 2.12 | 1.03 |
| 1970 | 2.10 | 2.01 | 0.95 |
| 1969 | 2.09 | 1.99 | 0.91 |
| 1968 | 2.12 | 2.00 | 0.90 |
| 1967 | 2.17 | 2.04 | 0.90 |
| 1966 | 2.20 | 2.06 | 0.89 |
| 1965 | 2.25 | 2.13 | 0.90 |
| 1964 | 2.31 | 2.19 | 0.91 |
| 1963 | 2.37 | 2.28 | 0.93 |
| 1962 | 2.41 | 2.37 | 0.96 |
| 1961 | 2.45 | 2.35 | 0.97 |
| 1960 | 2.47 | 2.46 | 0.97 |
| 1959 | 2.51 | 2.38 | 0.96 |
| 1958 | 2.54 | 2.43 | 0.97 |
| 1957 | 2.56 | 2.44 | 0.94 |
| 1956 | 2.61 | 2.47 | 0.92 |
| 1955 | 2.65 | 2.50 | 0.94 |
| 1954 | 2.70 | 2.51 | 0.99 |
| 1953 | 2.74 | 2.51 | 1.00 |
| 1952 | 2.77 | 2.54 | 1.01 |
| 1951 | 2.81 | 2.54 | 1.00 |

SOURCE: Edison Electric Institute.

added cost of environmental impact controls have combined to cause further upward pressure on power prices, and increases are expected to extend into the foreseeable future.

Section **18**

# Environmental Control

BY

**PASCAL M. RAPIER** *Consulting Engineer, Richmond, CA.*
**KENNETH A. ROE** *Chairman and President, Burns and Roe, Inc.*

The Principle of Environmental Control .................... 18-2
National Policy .......................................... 18-2
An Overview of Ecology ............................... 18-2
Cost/Benefit Balancing ................................ 18-3
Control of Thermal Discharges ........................ 18-4
Wastewater Control ................................... 18-5

Atmospheric Pollution and Gas Cleaning .................. 18-8
Sources of Air Pollution................................ 18-8
Radioactive Waste Management ......................... 18-18
Solid Waste Handling.................................. 18-21
Occupational Safety and Health ........................ 18-22
Fire Protection ....................................... 18-27

# ENVIRONMENTAL CONTROL

## by Pascal M. Rapier and Kenneth A. Roe

REFERENCES: National Environmental Policy Act of 1969, Public Law 91-190. Schurr: "Energy, Economic Growth and the Environment," John Hopkins Univ. Press. "Handbook of Environmental Control," vols. I–IV, Chemical Rubber Co.

### THE PRINCIPLE OF ENVIRONMENTAL CONTROL

Nature has provided two almost inexhaustible sumps for maintaining a steady-state environment on earth. The first of these is the 4 K background temperature of absolute space, which nature uses for heat rejection to close its heat balances. The second is the oceans, which serve to close the material balances of its cyclic processes by accepting the combined runoffs of the continents. The greatest engineering progress comes when people control their environmental activities so as to take maximum advantage at minimum cost of these sumps and of nature's cyclic processes. This is the **basic principle** upon which the science of **environmental control** is founded.

### NATIONAL POLICY

The National Environmental Policy Act of 1969, Public Law 91-190 (NEPA), concerns air, water, and land quality and conservation of natural resources. Section 2 of NEPA declares a national policy which will encourage productive and enjoyable harmony between people and their environment and which will enrich their understanding of ecological systems and natural resources.

Section 101(b) (2) provides for the "widest range of beneficial uses of the environment without degradation, risk to health or safety, or other undesirable consequences."

Section 101(b) (5) provides for achieving "a balance between population and resource use which will permit high standards of living" (e.g., cost/benefit balancing).

Section 101(b) (6) provides for enhancing the quality of renewable resources and the maximum attainable recycling of depletable resources (e.g., developing breeder reactors to extend nuclear fuel reserves).

Section 102(c) requires that all agencies of the Federal Government shall file environmental impact statements. [In principle, this would allow cost/benefit balancing of the effects of their significant actions to be made available to the public by detailing in a published document for each case: (*a*) its adverse environmental effects, (*b*) its alternatives, (*c*) its enhancement of long-term environmental productivity, and (*d*) its irreversible and irretrievable commitments of natural resources.]

### AN OVERVIEW OF ECOLOGY

(See also Solar Energy, Sec. 9)

Ecology is the study of the **biosphere,** a hypothetical system comprising the surface of the earth and all the subsystems necessary to maintain a steady-state life-support system. The **thermal cycle** is open, isenthalpic, and irreversible. Solar energy is utilized, becomes degraded, and is finally rejected into outer space by means of long-wave radiation through the optical window of the earth's atmosphere. Major quasicyclic processes in the biosphere, necessary for life support, are the **hydrological cycle,** the **carbon cycle,** the **nitrogen cycle,** the **potassium cycle,** the **phosphorous cycle,** and the **sulfur cycle.** These are almost closed cycles, which are occasionally interrupted and renewed by geological upheavals.

In these cycles, vegetation and animal wastes decay, thereby furnishing nutrients to land, air, and water. New green plants grow, animals eat them, and the recycling continues. Ecologically, the bacteria which decompose all these wastes are called *destroyers*, the green plants which employ photosynthesis are called *producers*, and animals which eat the plants are called *consumers*. A hypothetically bounded small portion of the biosphere is called an **ecosystem.** All ecosystems are **open,** as their essential cycles extend into the biosphere. Ecological problems arise in ecosystems when people disturb the essential cycles.

An ecologically idealized community would be one in which the ecosystem, including the surrounding agricultural area would operate steady-state, almost closed cycle. Tucson, Arizona, approaches this state, with its crops being irrigated with secondary sewage effluent, so that water, as well as all the plant nutrients, are recycled. Minerals entering the ecosystem through the Santa Cruz River, and through the burning of fossil fuels in the area, are almost counterbalanced by exports of crops from the area and by irrigation runoff into the normally dry Gila River bed. These activities conform to nature's naturally occurring cycles, and maintain the fertility of the soil.

In larger, heavily overpopulated communities, however, people's activities are counter to nature's, and so humanity suffers. The earth is denuded of local vegetation, and crops are imported. All the nitrogen, potassium, sulfur, and phosphorous contained in the agricultural commodities and fossil fuels, instead of being recycled to the farmlands, are ultimately handled by the waste disposal systems of the communities and dumped into rivers and harbors, thereby reaching the oceans and being lost to the continents, despite the recycling and soil fixation methods intended by nature to prevent such losses. Severe pollution problems result also, mainly from nonpoint sources and storm runoffs which are not amenable to control, no matter how much money is spent. But the major damage done by people's failure to recycle wastes is depletion of farmlands. It is irreversible and irretrievable, and it will continue, regardless of whether society is industrialized, as long as people persist in living in large cities.

**Environmental control,** the engineering approach to ecology, seeks to subdue and to utilize nature's ecological cycles to the maximum extent possible in order to serve people's needs, thereby conserving natural energy and mineral resources, and to replenish desirable local flora and fauna populations by agriculture and cultivation to provide adequate food, clothing, and shelter. Environmental control seeks to extend depletable fuel supplies by utilizing them most wisely and efficiently, substituting, where possible, solar and nuclear energy and other replenishable or nearly inexhaustible supplies, as necessary to conserve them. Also, replenishable substitutes for other depletable supplies and natural resources are sought, as well as recycling means for scarce and irretrievable substances. Environmental control seeks to conserve land and water quality by diversion into adequately controlled drainage canals of concentrated runoffs, e.g., from irrigation, municipalities, industries and mining, so that these nonrecyclable, unsalvageable mixtures are ultimately discharged into the oceans, where they blend inconsequentially with the vastly larger amounts of runoff from the continents, occurring naturally as part of the hydrological cycle. The public standard of living is highest by diversion into adequately controlled drainage minimum cost and effort for the consumer. This requires that, on a national scale, productivity of consumer items be maximized and losses minimized by avoiding wastes and by utilizing replenishable cyclic processes.

## COST/BENEFIT BALANCING

REFERENCES: Summary of UWAG-EEI Comments on EPA's Proposed Regulations under 304, 306, and 316(a) of Public Law 92-500, the Federal Water Pollution Control Act Amendments of 1972, June 1974. Effluent Limitations Guidelines for Steam Electric Power Generating Point Source Category 40 CFR, Chapter I, Subchapter N, Part 423.

### Cost/Benefit Calculations

In order to assess the feasibility [under Section 304b (1B) of the FWPCA Amendments of 1972] of any environmental control technology or augment, both the public cost and benefit of its operation must be considered. For demonstrating feasibility, this dimensionless cost/benefit ratio has the limits

$$0 \leqslant C/B \leqslant 1$$

The upper limit of unity means that $1 spent in the incremental public operational costs of the augment will yield $1 of return in public benefits. If an augment is required for which the cost/benefit ratio is greater than unity, it is unfeasible. (It is more feasible to accept the pollution.) If an augment is required for which the cost/benefit ratio is negative, the public benefit is negative and constitutes a public detriment. The more that is spent on such an augment, the worse the environmental degradation becomes.

EXAMPLE. Once-through vs recirculating cooling systems: A 1000-MW(e) steam/electric generating plant is operating with once-through cooling on a river. The heat rate is 10,300 Btu/kWh (0.33 thermal efficiency), with about one-half of the heat, or 5150 Btu/kWh, being discharged into the river. The public benefit of this thermal enrichment is a stable and abundant fish population, which is estimated to offset all public detriments except occasional fish kills caused by emergency shutdowns. These average each year not more than 8 h duration and $45,000 public detriment.

Proper environmental management to provide emergency heating of the water during unexpected outages will avoid the fish kills with an estimated 90 percent confidence factor. A cooling tower retrofit will reduce the thermal discharge by 90 percent but will substitute some undesirable atmospheric, noise, and stream pollution conditions. Neglect these problems, but otherwise compare the feasibility of an open-recirculation cooling tower retrofit with that of proper environmental management.

*First alternative:* Environmental management of existing plant. The annual public detriment for the fish kills is $0.045/kW(e) of installed capacity. For avoiding these by emergency heating of the river water, let the estimated operating cost to the consumer be $1 per million Btu. The annual environmental management cost for an 8-h outage per year becomes:

$$8 \text{ h} \times \$.00515/\text{kWh} = \$0.0412/\text{kW(e)}$$

The cost/benefit ratio becomes $0.412/$.045 or 0.916, so that 91.6¢ spent will yield $1 return in public benefits. This is a wise investment. For the existing plant the possibility of a fish kill is reduced to one-tenth its original value, and the remaining annual public detriment becomes $0.0045 per kW(e) of installed capacity.

*Second alternative:* Retrofit of an open recirculating cooling tower. The viability of a cooling tower retrofit must be assessed on the basis of this reduced public detriment for the existing plant. The estimated annual operating cost to the public for a cooling tower augment, based on public operational overhead, a 1970 overall plant investment, and a 3 percent incremental capacity loss, due to the retrofit, is $20/kW(e) of installed capacity. Dividing this by the $0.0045 expected public benefit yield a cost/benefit ratio of 4444. This means that $4444 must be spent (and the cost passed on to consumer) for every dollar of public benefit. As this is clearly not in the public interest, it is unfeasible. It is cheaper to replenish the fish supply. Moreover, the 3 percent capacity loss of power generation due to the augment results in a 3 percent wastage of fuel and a 3 percent increase in the total thermal discharge to the environment. Energy, in contrast to the fish population, is a nonreplenishable natural resource. It must be conserved, and it must be considered when assessing the public detriment. Valued at 3 percent of the heat rate, $1 per million Btu and 0.9 load factor, this waste of natural resources will cost $2.435 per kW(e) annually.

Loss of cooling water by evaporation in the cooling tower is also a public detriment that can be conservatively valued at 25¢ per 1000 gal. At 5150 Btu/kWh thermal discharge rate and 0.9 evaporative load factor, this depletion of a natural resource will cost:

$$0.9 \times 5150 \times \$0.25 \times 24 \times 365 \div 8.33 \times 10^6 = \$1.219/\text{kW(e)}$$

Subtracting these two detriments, amounting to $3.654, from the annual public benefit of $0.0045 for conserving the fish population yields a negative public benefit due to the augment of −$3.65. The overall cost/benefit of this installation therefore becomes

$$\$20/(-\$3.65) = -5.48$$

The negative sign indicates that money spent to improve the environment by this means simply makes matters worse. The expected public benefit is not possible, and therefore the technology is not economically achievable.

### Direct Water Cooling

For large-scale power generation, natural surface cooling in open water bodies is most cost-effective. The intake and outfall should be so oriented as to minimize possibility of heat recirculation and short-circuiting, which can usually be done by locating the outfall sufficiently down-current of the intake.

Surface discharge of the condenser water as a hot, buoyant plume requires the least area of water surface for heat dissipation, yields the largest value for the overall surface heat-transfer coefficient $K$, and hence minimizes both the cost and environmental impact of the facility.

The surface heat-transfer coefficient $K$ combines the heat-loss contributions due to long-wave radiation, evaporation, and convection into a function which can be reliably estimated by two independent equations (J. E. Edinger, 1974):

$$K = H_s(T_s - T_{dp})^{-1} \tag{1}$$

$$K = \frac{T_s - 183.16}{20} + (0.47 + B)f(W) \tag{2}$$

where $K$ = overall surface heat-transfer    $Wm^{-2}\cdot K^{-1}$
   coefficient,

   $H_s$ = gross 24-h average sunshine    $Wm^{-2}$
   factor,

   $T_s$ = water surface temperature,    K
   ambient,

   $T_{dp}$ = dew-point temperature,    K
   $W$ = wind speed,    m/s
   $f(W)$ = wind-speed function, 9.2 +    $Wm^{-2}\cdot mmHg^{-1}$
   $.46W^2$ (Brady 1969),

   $B$ = slope of vapor-pressure    $mmHg/K$
   curve at $T_s$,

The derivative of the 1967 ASME steam table saturation line, B, is given accurately over the temperature range 0 to 50°C by the Rapier log-hyperbolic curve fit

$$B = \frac{5301.734}{T_s^2} \exp\left(20.94673 - \frac{5301.734}{T_s}\right)$$

The overall heat transfer coefficient $K$ can be most accurately obtained from (1), since all the variables can be directly measured; then the wind-speed function $f(W)$ can be determined for the particular site by substitution of $K$ into (2). On the other hand, if $T_s$ is not known, both $T_s$ and $K$ can be approximated by iteration between the two equations, after adopting a reasonable wind-speed function, such as the Brady parabola, given above.

Once the value of $K$ has been determined and the temperature rise through the condensers has been decided upon, the initial temperature difference or excess above ambient lake surface temperature at the discharge point, and the final temperature difference or excess above ambient at the plant intake, can be readily calculated, since the effective area $A_e$ in square meters from the discharge point up to the intake determines the log-mean temperature difference, as given by the equation

$$\text{LMTD, °C} = \frac{\text{thermal discharge rate, watts}}{KA_e}$$

At locations where limited quantities of water are available, some heat recirculation occurs, and particularly during summer months can cause temperatures in excess of 32°C at the inlet of the plant. When considering such a location, trade-offs should be made to determine whether it is more economical to accept a higher temperature at the inlet or to install auxiliary cooling means such as towers or spray ponds.

EXAMPLE. During July, a proposed inland lake site has a gross sunshine factor of 387.3, a dew point of 13.9°C, and a windspeed of 4.07 m/s. Iteration yields 24.2°C for the water surface temperature and 36.4 $Wm^{-2}\cdot K^{-1}$ for the heat-transfer coefficient. If thermally loading the lake will cause a temperature excess of more than 8° at the intake, trade-offs should be considered here.

## CONTROL OF THERMAL DISCHARGES

REFERENCES: "Water Resources and Waste Heat," *Power Engineering*, pp. 26–33, June, 1973.

The total solar flux reaching the earth is 182 trillion kW. Of this, about one-sixth, or 32.4 trillion kW, is utilized in the hydrological cycle, whereby fresh water distilled from the oceans is utilized on the continents by the plant and animal kingdoms and finally drains back into the ocean to complete the cycle. The entire thermal demand estimated for the United States by 1980 is about 5 billion kW, very small compared with the solar flux in the earth's biosphere. The temperature of the earth's biosphere is stabilized by reradiating the solar flux as long-wave radiation back into absolute space through the "optical window" (7 to 13 $\mu$m wavelength) of the earth's atmosphere. (See also Solar Energy, Sec. 9.) It remains almost unaffected by the total environmental activities of people and constitutes the one natural and unchangeable thermal background for making comparisons when evaluating the environmental impact of thermal discharges. Thus, for example, if once-through cooling for 25,000-MW(e) power generation were added to Lake Michigan, the average surface temperature rise would be limited to ¼°F by long-wave radiation from the surface of the lake.

The Joint Committee on Atomic Energy, 93rd Congress (1973), anticipated that by 1980 the total heat dissipated into the environment by the generation and use of energy in the United States will be equivalent to 43.2 million bbl of oil per day. Of this, the combined heat loss caused by electric and industrial power generation will be 11.5 million bbl per day, or 26.6 percent, which includes all the point sources. Distributed and diffuse sources, such as automotive, residential, commercial, and industrial heating and cooling systems, will constitute the balance, account for 73.4 percent of the heat dissipation, and create "heat islands" around every major metropolis. About half the heat loss from power generation is presently discharged into bodies of water via nonconsumptive once-through cooling systems and does not contribute to the "heat island." The use of consumptive evaporative cooling for this would add to the atmospheric discharges in the "heat island" and raise the ambient temperatures in the island. By 1978, increased urbanization in the Los Angeles Basin will raise the ambient temperature an estimated 5°F. The surface temperatures of water bodies in the Basin would be raised over 5°F, which exceeds guideline limitations for thermal discharges. By the year 2000, if evaporative cooling towers are required for all the incremental power generating capacity until then, the New York-Philadelphia "heat island" will suffer a mean ambient temperature increase of about 15°F, and the cooling towers will consume twice the flow of the entire Colorado River Basin, 31,000 ft³/s, or 3000 times their 1970 consumption of water.

For the year 2000, the percentages of total heat discharged to the environment from various human activities have been estimated as shown in Table 1.

The largest thermal discharge, 31.75 percent, is primarily attributable to residential and commercial heating and venti-

**Table 1. Heat into Environment**

| Thermal discharge source | Coolant | Total heat, % |
|---|---|---|
| Residential and commercial | Atmospheric and water | 31.75 |
| Industrial | Atmospheric and water | 27.00 |
| Transportation (all forms) | Atmospheric | 22.20 |
| Electric power generation | Atmospheric and water | 14.30 |
| Electric power consumption | Atmosphere | 4.75 |
| Total heat into environment | | 100.00 |

lating systems, with air-conditioning units making the most undesirable contributions to the environmental "heat island" effect. Better thermal insulation and reduction of electrical loads (such as by replacing electrical heating units with gas- or oil-fired units and atmospheric air conditioners with once-through, water-cooled units) could substantially reduce the undesirable side effects.

The second largest thermal discharge is from industry. Because of keen marketplace competition, thermal processes have been selected to provide the most favorable cost/benefit ratio for the consumer. Regulatory interference here, except to prevent malpractices, can do more public harm than good. Once-through nonevaporative cooling, where practicable, will produce the maximum public benefit.

Transportation is the third largest contributor to the "heat island" effect and the one with the worst side effects. Thermal discharge to city streets could be reduced to one-third its present value and petroleum stocks conserved if all-electric transportation could be substituted economically for the internal-combustion engine. This would be particularly advantageous in the summertime.

The smallest contributor to the "heat island" effect is electric power generation, which accounts for 7.15 percent of the total atmospheric discharge, with another 7.15 percent thermal discharge to water. Irrigation canals, navigation canals, and artificial waterway complexes designed to include treated municipal effluents could, if implemented in the future, provide sufficient water for a significant number of thermal power generating units, save energy, and reduce the atmospheric discharge near cities by as much as 7.15 percent.

In short, the environmental impact of heat discharged by human activities can be minimized and fuel conserved by selecting processes having the highest practicable thermodynamic efficiencies and the lowest practicable heat-sink temperatures. To prevent raising ambient temperatures, means must be provided to increase the emissions of nocturnal long-wave radiation by geodetic surfaces in the locale.

**Use of Cooling Towers**   (See Sec. 9, Cooling Towers)

REFERENCES: Nomographs for Thermal Pollution Control Systems, EPA 660/2-73-004, September, 1973.

In locales where water supplies are inadequate to provide nonconsumptive once-through cooling, evaporative open-recirculating cooling towers become necessary. These consume water at about 2 percent of the recirculation rate, but reduce circulation into rivers or streams by over 95 percent even when employing low concentrating cycles. Nomographs are available from EPA which provide cost comparisons of the cooling systems sampled by Table 2. Selection of the most feasible cooling system depends on specific site conditions. Economics dictate once-through cooling where possible, wet cooling towers or ponds where water supplies are limited, and dry cooling towers where water is unavailable. All of these will contribute significantly to the "heat island" effect unless specifically designed to provide buoyant plumes that will puncture inversion layers. In this regard, see SDEL procedures under "Air Pollution."

Wet cooling towers substitute water pollution for thermal pollution by concentrating blowdown, require chemical treatment, and release heavy-metal corrosion products into the blowdown stream. Their operation requires an optimized water chemistry which balances the acid treatment and cycles of concentration to provide nonaggressive water with a slightly positive Langlier index, and to retard the buildup of scale. The Langlier (saturation) index is the difference between the actual measured pH and the calculated $pH_s$ at saturation with $CaCO_3$. A zero Langlier index balances pH, alkalinity, and calcium hardness, and establishes the threshold for alkaline scale formation. (For further information on this and other water treatments, see Betz, "Handbook of Industrial Water Conditioning," Betz Laboratories.) For environmental reasons, pretreatments based upon zinc, chromate, and phosphate inhibitors should generally be avoided, while posttreatments to remove pollutants and corrosion products from the blowdown may have to be employed. With all constraints considered, the most economical design results from a series of trade-offs. Dry cooling towers are justified only where water shortages preclude the use of evaporative coolers, as their cost, inefficiency, and environmental impact from abnormal heating of the air are otherwise excessive.

Cooling towers may be required by environmental regulations, even though completely unjustifiable at the site. "Zero discharge" of blowdown may also be required, with high concentrating cycles and "perpetual storage" of the resultant salty wastes. The salty plume will cause additional difficulties. Blowdown volume can be reduced by concentrating to 10 percent solids using vapor compression. The power requirement for this is 90 kWh/kgal evaporated. Further concentration requires open kettles. (For further information, see Kolflat: "Cooling Tower Practices," *Power Engineering*, pp. 32–39, January 1974, and Boles and Levin: "Treat Cooling Tower Blowdown," *Power*, pp. 68–76, August 1974.)

## WASTEWATER CONTROL

REFERENCES: Twenty-seven Industries' Comments on EPA's Proposed Effluent Limitations Guidelines Amending 40 CFR Chapter I, Subchapter N, 1974. The Federal Water Pollution Control Act Amendments of 1972, PL 92-500, S. 2770, October 1972. Lund: "Industrial Pollution Control Handbook," McGraw-Hill. Pollution Control Technology, Research and Education Association, 342 Madison Avenue, New York, N. Y.

### Industrial Categorization by the 1972 FWPCA Amendments

The Federal Water Pollution Control Act (FWPCA) Amendments of 1972, Public Law 92-500, preempt all existing federal, state, and local regulations by authorizing guidelines for "effluent discharges," rather than by establishing "receiving water quality."

Title III, Section 306(b) (1) (A) categorizes sources. "Source" means any building, structure, facility, or installa-

**Table 2. Cooling System Augments**

| Cooling systems | Relative capital cost | Power consumption |
|---|---|---|
| Natural draft wet towers | 3.2 | Low |
| Mechanical draft wet towers | 1 | High |
| Spray ponds | 2.3 | High |
| Cooling ponds | 1.4 | Low |
| Natural/mechanical draft dry towers | 7.3/6.7 | Medium/high |

tion from which there is or may be the discharge of pollutants for which effluent limitations guidelines and revisions will be developed. These will be published periodically in the Federal Register, under the 40 CFR part numbers shown by Table 3. The limitations are to be achieved by July 1, 1977, for "best practicable control technology currently available" (BPCTCA) and by July 1, 1983, for "best available technology economically achievable" (BATEA).

Both BPCTCA and BATEA guidelines must meet feasibility and achievability criteria under cost/benefit testing and will be modified periodically to reflect current needs and technology.

For example, it has been estimated that to achieve "zero discharge" of pollutants from all point sources by 1985, as implied by BATEA, will require expenditures equal to seven times the $1 trillion U.S. annual gross national product. Yet this would be meaningless, since the major pollution is from nonpoint sources, not readily amenable to control.

Except where pretreatment guidelines have been established by EPA under Section 307(b), industries with effluent discharge into municipal systems are not regulated. Such waste becomes part of the municipal wastewater flow and thus a municipal responsibility.

### Permit and Application for Industrial Discharges

The Permit Program for Industrial Discharges under the provisions of the "Refuse Act of 1899" was previously administered by the Army Corps of Engineers. The Act of 1972 in effect transferred control of the Discharge Permit Program from the Army Corps of Engineers to EPA. The Permit Program is enforced by the state, but each state program must either be approved or is subject to takeover by EPA.

A Permit Program Application plus an environmental report [see NEPA-102(c) for criteria] is filed by the discharger with the state, which then refers it to the Regional EPA permit section for review. The state draws a permit that

**Table 3. 27 Industries Categorized under 40 CFR, Chapter I, Subchapter N**

| | Industrial groups | Part nos. | Stds. of performance |
|---|---|---|---|
| 1 | Food and grain processing | 405–409 | In general, BPCTCA 1977 guidelines limit pH* between 6 and 9, BOD,* TSS,* and COD,* to the ppm range, toxic substances, e.g., Ba, Pb, Hg, Ag, As, Cr⁺⁶, F, to nontoxic concentration. BATEA 1983 guidelines limit these further on nonprocess streams, and seek "zero discharge" of process streams. |
| 2 | Textiles | 410 | |
| 3 | Feed lots; meat products | 412; 432 | |
| 4 | Chemical; petroleum metallurgy | 410–424 | |
| 5 | Power generation | 423 | |
| 6 | Leather, glass, and rubber products | 425–427 | |
| 7 | Wood products, paper, and pulp | 428–431 | |

*BOD (biochemical oxygen demand); TSS (total suspended solids); COD (chemical oxygen demand); pH (log of the reciprocal hydrogen ion concentration).

includes limitations, compliance schedules, and state monitoring requirements, making the permit forms available to the public 30 days prior to issuing the permit. Hearings on the permit may cause much delay. After this procedure is completed, the permit may be issued for a maximum of 5 years.

### Environmental Considerations for Construction Projects

**Introduction**  The environmental considerations which follow should be taken into account by industry when preparing an environmental report for construction projects, and before an effluent discharge permit can be granted. Cost/benefit factors for applicable environmental control technologies and regulations limiting effluent discharges should be obtained and included in the record of public hearings and should lead to rational decision-making based on a total short- and long-term project cost to the public consumer in relation to the worth of anticipated public benefits. (*Note:* Cost/benefit ratios for meeting effluent discharge regulations can be obtained from the comments submitted to EPA by the 27 categories of industries listed by Table 3. Public costs generally range from 100 to 1000 times greater than social or environmental benefits. The engineer will also find these reports a valuable source of design information for environmental projects.)

It should be recognized that the considerations identified should be individually tailored to respond to the unique guidelines for effluent discharge encountered on a particular project and in a particular state.

**General Construction Projects**

1. What site selection criteria will be used? Will they include cost/benefit balancing of all environmental worth factors such as the conservation of land and fuel reserves, air and water quality, noise, impacts on residents of the area, fish, wildlife, and vegetation? Will alternative sites and alternative orientations of the plant on the selected site be considered?

2. What disposition will be made of solid and liquid residues (ashes and chemical and industrial wastes)? Does the disposal method include adequate cassetting or neutralization to minimize the danger of soil or water pollution? What steps are planned to contain and reclaim ash dumps and chemical and oil spills to avoid pollution of surface and ground water by runoff? What provision has been made for the disposal of waste-concentrated brines? If waste disposal into water bodies is planned, what will be the dollar value of its detriment to aquatic life? To what degree will tidal action and currents dilute plant effluents? What disposition will be made of potentially harmful and toxic corrosion products occurring in the plant effluent? Can they be injected into underground aquifers? What provision will be made for controlling the release of waste material into water bodies? If additional units are constructed, what will be the total load of waste materials? What downstream environmental effects can be anticipated with respect to humans, crops, forests, and wildlife? With what cost effectiveness can any harmful effects be minimized by dilution and toxic materials be discharged below toxic concentrations to merge into the normal background of compositions?

3. What public benefits and detriments will thermal effluents have on the receiving waters? What temperature increase can be anticipated? Is there sufficient water motion in the receiving bodies to dissipate heat effectively? Has the use of cooling towers been explored? What is the probable cost of

their (cooling towers) public detriment, incurred by the added consumptive use of water and energy, by producing undesirable fog and weather conditions through the dissipation of waste heat, and by producing undesirable blowdown concentrations?

4. What impact will the impoundment for a water supply have in terms of the destruction of agricultural, forest lands, and habitats for fish and wildlife? What measures are planned to mitigate the loss of natural habitats for fish and wildlife? To what degree will archaeological and scenic values be affected? How will the reservoir and downstream flow affect water quality parameters, i.e., temperature, dissolved oxygen, nutrients, nitrogen concentration, hydrogen sulfide, and color?

5. How vulnerable is the facility to surface subsidence, earthquakes, hurricanes, war, insurrection, and other catastrophes? What is the probable annual cost (e.g., liability insurance cost) of the environmental degradation which could be expected from such catastrophes? What preventive and remedial safeguards for downstream inhabitants will be incorporated in plant design and construction? What provisions are made for training plant operators in environmental protection?

6. Have the environmental consequences of transportation, pipeline, and power transmission been considered in site selection? What steps are planned to avoid soil erosion and the silting of streams as pipelines, transmission facilities, and access roads are constructed?

7. What provision has been made for recycling and for industrial by-product development associated with the facility? (See also "Water Resources," Sec. 6.) What impact will that activity have on the environment?

### Process Selection, Recycling, and Ultimate Disposal of Residual Wastes

(See also Water Resources, Sec. 6)

In the design of an industrial installation, segregation of process and nonprocess streams, together with good housekeeping practices and diversion of storm waters will generally result in minimum volumes of water requiring special treatments before being discharged. Wastewater treatment processes involved are: chemical and physical treatments for removal of undesirable waste products; suspended and dissolved solids from process waste streams; removal of corrosion products and carbonaceous wastes (biological treatment) from nonprocess streams; oil-water separations; suspended-solids removal from storm runoffs; good housekeeping and protected stockpiling to prevent contamination of runoffs; and blending of treated effluents and impoundment to meet local end-of-pipe regulations (see Table 4).

Recycling of consumer and commercial wastes back to their sources (e.g., crankcase oil, glycol, fat, photographic and plating wastes, paper, rags, rubber, metals, and irrigational and fertilizer values of municipal wastes), following the good housekeeping practices of industry, is a very cost-effective method for reducing municipal stream pollution. However, within most industries recycling of salvageable material is almost always practiced to the point of optimum return, so that proposed benefits from further recycling in an effort to obtain "zero discharge" are minimal at best. Recycling, or "closed cycle" water reuse schemes consume water and produce blowdown streams of concentrated pollutants in contrast to "open cycle" schemes, whereby solids have traditionally been discharged at normal and below toxic levels, within the assimilative capacity of the receiving bodies of water. The latter, and not the former, conserves water and enhances the naturally occurring ecological cycles—provided, of course, that abnormal toxic substances are not discharged, but are removed by a posttreatment for ultimate disposal.

Descriptions, flowsheets, and costs for a variety of industrial water-use and recovery schemes are found in "Water Reuse, Industry's Opportunity" (1973), published by the AIChE.

Water reuse by consumptive processes always results in a stream of concentrated residual wastes for which an ultimate disposal site must be found. Thus, for brine from saline-water conversion processes, there are four major avenues: (1) ocean

### Table 4. Removal Process Guide for Selected Contaminants

| Treatment processes | pH, acidic, alkaline | Suspended and precipitated solids | Bacteria, odor, organic, color | Nutrients (N, P) | Herbicides, pesticides, CN⁻ | TDS | Oils, solvents | Heavy metals $Cr^{+6}$ |
|---|---|---|---|---|---|---|---|---|
| Chemical* | C | B | A | A | A | A, B | B | A, B |
| Clarification† and settling | | G, D, I | D, E, H | H, D | | D | F, H, D | D, I |
| Selected‡ biological | | J, K | J, K | K | | J | | |
| Desalting§ | | O, M | L, N | | L | O, M | L, N | |
| Sludge¶ dewatering | P | P | | | | | P | P |

*A, selected chemical treatment; B, coagulation; C, neutralization.
†D, selected clarification; E, aeration, oxidation; F, flotation, decantation; G, filtration; H, lagooning; I, sedimentation.
‡J, secondary (e.g., aerobic, anaerobic); K, tertiary (e.g., carbon adsorption; extended aeration); anaerobic denitrification.
§L, selected desalting and brine blowdown; M, membranes; N, ion exchange; O, distillation.
¶P, dewatering (e.g., gravity, centrifugal, vacuum filtration; drying).

discharge; (2) underground discharge into saline aquifers; (3) dry lake beds, liner protected; and (4) solidification and use for landfill. Reference to current regulatory and trade publications should be made to determine which of these is permissible for a given situation.

## ATMOSPHERIC POLLUTION AND GAS CLEANING

REFERENCES: "Laws: Current Legislation," *Chemical Engineering*, June 21, 1971. Victor Sussman, "New Priorities in Air Pollution Control," *Journal Air Pollution Control Association*, **21**(4):201, April 1971.

### Abstract of Federal Air Pollution Control Legislation

**Air Pollution Control Act, 1955**   Provided for research, technical assistance, and training activities by HEW, as well as leadership and support to the states and local governments, whose authority it redefined primarily as responsibility to Congress for air pollution control.

**Clean Air Amendments, 1965**   Authorized federal control of new motor vehicle emissions. Extended federal control to include abatement of international pollution originating in the United States.

**Air Quality Act of November 21, 1967**   Under Public Law 90-148 the states are held responsible for establishing local air quality standards consistent with federal criteria.

**Air Quality Act Amendments of 1970**   Under Public Law 91-604, EPA can require offending sources to be shut down immediately. Unsatisfactory state enforcement programs can be taken over and run by the EPA. New plants and new additions to old ones must have the best available pollution control technology, and meet emission standards, including zero levels for hazardous substances. A 90 percent reduction from baseline in automotive emissions by 1975 is required.

**Some Existing Federal Standards**   Following passage of the 1970 Amendments, EPA published in the Federal Register national ambient air quality standards covering six pollutants (see Table 5).

**Effect of Energy Crisis on Legislation**   Because of the developing energy crisis, by late 1974 Congress began considering new legislation to postpone the National Environmental Policy Act (NEPA) and to modify the Clean Air Act so that the energy shortage problem could be alleviated. The energy shortage is expected to continue for the next decade; thus, the prudent environmental trend is toward conservation of natural resources by the most productive use of energy, materials, and naturally occurring ecological cycles, and away from any pollution control alternatives that tend to squander these resources.

## SOURCES OF AIR POLLUTION

REFERENCES: Magil, Holden, and Ackley, "Air Pollution Handbook," McGraw-Hill. Gibbs, "Clouds and Smoke," Churchill. White, "Industrial Electrostatic Precipitation," Addison-Wesley. Morgensen, "Fan Engineering," Buffalo Forge. Dallavalle, "Micrometrics," Pitman. Stern, "Air Pollution," Academic Press. "Air Pollution Engineering Manual," J. A. Danielson, EPA-Research Triangle Park, N.C., May 1973.

### Domestic and Municipal Sources of Air Pollution (See Tables 7 and 8)

**Air pollutants** are usually classified as gases, vapors, and **particu-**

**Table 5. Ambient Air Quality Standards**

| Pollutant | Primary, enforcement by summer 1975 | Secondary, no time limit on enforcement |
|---|---|---|
| Particulates, $\mu g/m^3$ | | |
| annual geometric mean | 75 | 60 |
| max. 24-h conc.* | 260 | 150 |
| Sulfur oxides, $\mu g/m^3$ | | |
| annual arith. aver. | 80(0.03 ppm) | 60(0.02 ppm) |
| max. 24-h conc.* | 365(0.14 ppm) | 260(0.1  ppm) |
| max. 3-h conc.* | | 1,300(0.5  ppm) |
| Carbon monoxide, $mg/m^3$ | | |
| max. 8-h conc.* | 10(9 ppm) | 10 |
| max. 1-h conc.* | 40(35 ppm) | 40 |
| Photochemical oxidants, $\mu g/m^3$ | | |
| 1-h max.* | 160(0.08 ppm) | 160 |
| Hydrocarbons, $\mu g/m^3$ | | |
| max. 3-h conc.* 6–9 am | 160(0.24 ppm) | 160 |
| Nitrogen oxides, $\mu g/m^3$ | | |
| annual arith. aver. | 100(0.05 ppm) | 100 |
| 24-h max. aver. | | |

*Note:* All values subject to periodic revision.
*Not to be exceeded more than once a year.

**late matter** (see Table 6 and Fig. 1). Particulate matter includes: **Dusts**—a loose term applied to solid particles predominantly larger than colloidal and capable of temporary gas suspension. Dusts do not tend to flocculate except under electrostatic forces; they do not diffuse but settle under the influence of gravity. Dusts result from such operations as crushing, grinding, drilling, screening, and blasting. **Fumes**—the solid particles generated by condensation from the gaseous state, generally after volatilization from melted substances, and often accompanied by a chemical reaction such as oxidation. For example, zinc vapor will evaporate from the surface of heated liquid metal and will then condense upon contact with ambient air to form small, fluffy particles of zinc oxide. Fumes are usually less than 1 $\mu$m in size, although they may coalesce or flocculate to form larger particles. **Smokes**—gas-borne particles (usually less than 0.5 $\mu$m) resulting from incomplete combustion of materials such as wood, coal, and oil. **Mists** and **plumes**—loose terms applied to dispersions of liquid particles (0.1 to 2.5 $\mu$m), the dispersion being of low concentration and the particles of **large size.**

## Low-Emission Automotive Propulsion Systems
(See Sec. 9, Automobile Engines)

The automobile is a major source of air pollution and a significant contributor to photochemical smog (caused by the products of photoreactions in the atmosphere—organic peroxides, peracids, hydroxy peracids, perozyacyl nitrate, and other compounds). For every 1,000 gal of gasoline consumed by an automobile engine, there are discharged the following air pollutants, in pounds: carbon monoxide, 3,000; hydrocarbons, 200 to 400; nitrogen oxides, 50 to 150; aldehydes, 5;

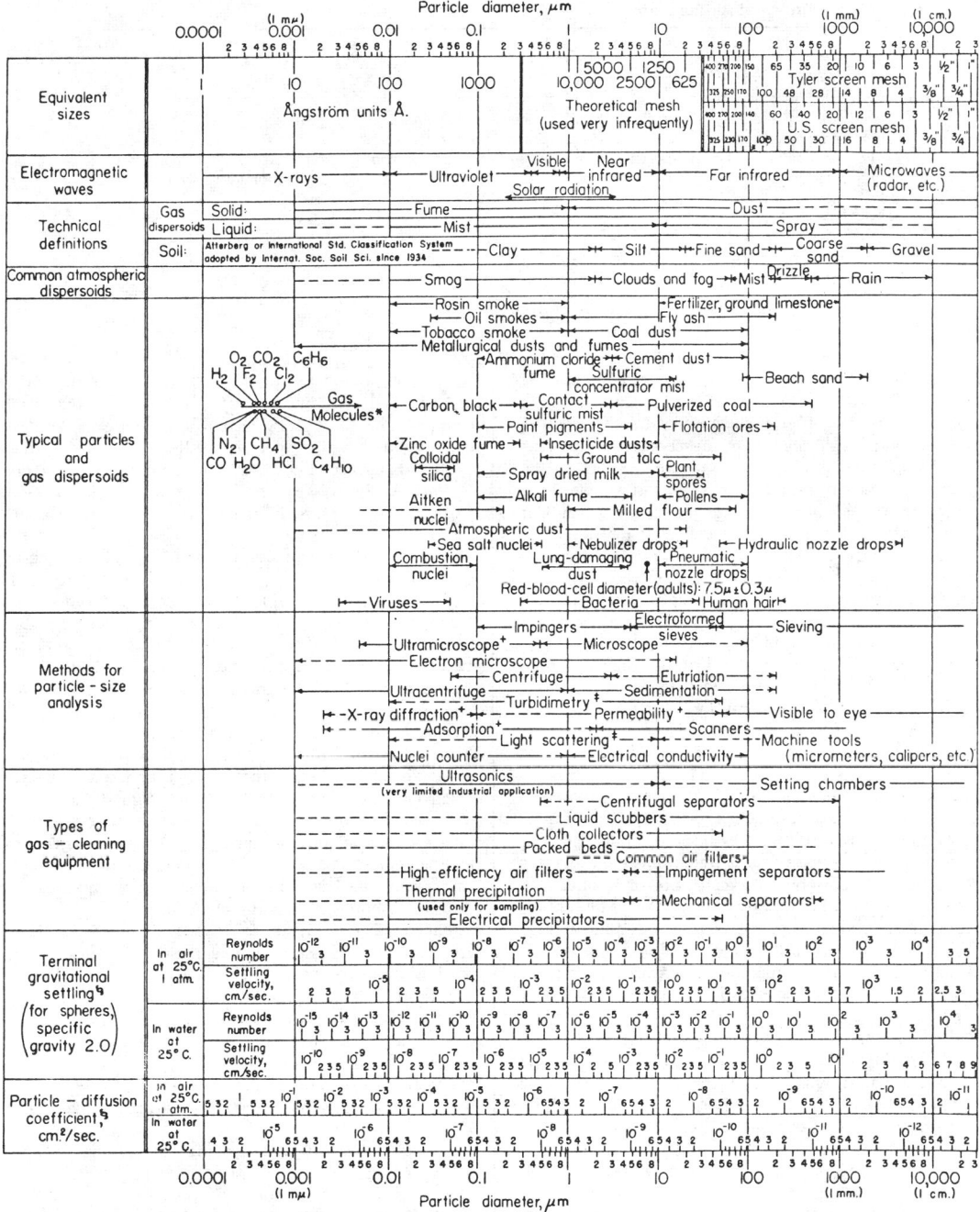

**Fig. 1**  Sizes and characteristics of airborne particles. [*From C. E. Lapple, Stanford Res. Inst. Jour., Vol. 5, p. 94 (Third Quarter 1961). Reprinted by permission.*]

\* Molecular diameters calculated from viscosity data at 0° C.

⁺ Furnishes average particle diameter but no size distribution.

‡ Size distribution may be obtained by special calibration.

⁹ Stokes–Cunningham factor included in values given for air but not included for water.

**Table 6. Principal Air Pollutants**

| Gases | Vapors | Particulate matter | | | |
| | | Dusts | Fumes | Smokes | Mists |
|---|---|---|---|---|---|
| Acid gases | Alcohols | Alumina | Metallic | Ash | Acid |
| Carbon monoxide | Esters | Calcium | halogens | Cinders | Chromic |
| Hydrogen chloride | Hydrocarbons | fluoride | Metallic oxides | Organic | Phosphoric |
| Hydrogen fluoride | Ketones | Cement | Silicon | compounds | Sulphuric |
| Hydrogen sulphide | Mercaptans | Coal | tetrafluoride | Soot | Organic |
| Nitrogen oxides | | Grain | | | chemicals |
| Sulphur dioxide | | Limestone | | | Oil |
| Alkaline gases | | Metal | | | Salt spray |
| Ammonia | | Ore | | | |
| | | Rock | | | |
| | | Wood | | | |

sulphur compounds, 5 to 10; organic acids, 2; ammonia, 2; solids, 0.3.

During the continuing energy shortage in the next decade, automobiles must provide maximum fuel economy consistent with cleanliness. Two obvious methods are to increase the engine compression ratio and reduce the size and weight of the car. Another is to suspend and revise the Clean Air Act of 1970, to provide both the time and incentives to develop more efficient, cleaner engines. All these can be done, so that the alternatives listed below can be considered for the future.

**Conventional Gasoline Engines**  Equipped with a **thermal converter,** the high-compression engine using leaded gasoline will operate at 10 percent less efficiency than the same engine unequipped. This is an excellent combination, since by refinements in ignition and in carburetion, leading to higher compression ratios, the 10 percent loss can be recovered. Other advantages of the thermal converter are low capital cost and maintenance, plus reduced emissions over base values as follows: hydrocarbons, 98.2%; CO, 94.4%; and NO$_x$, 82%.

Other emission systems, including the **catalytic converter,** possess fewer advantages, increase fuel consumption by about 30 percent, and are more difficult to maintain. The catalytic converter is less reliable because it is fragile and the converter material can be easily poisoned. It creates more environmental problems than it solves, because of its dangerously high operating temperature, emissions of colloidal Pt, Pd, and sulfates, and its irretrievable commitments of natural resources. It requires nonleaded gasoline which becomes scarce during fuel shortages, and lower compression ratios than can be used with the thermal converter.

**Fuel-Injection Gasoline Engines**  A typical fuel-injection engine of about 1700-cm$^3$ displacement and burning 91 octane fuel, when compared with a conventional two-carburetor engine of the same displacement and using the same gasoline, will demonstrate an improved gasoline economy of about 17 percent and substantially less emissions because of improved combustion. When equipped with a thermal converter, it will still be about 7 percent more efficient than the conventional engine. The cost of the fuel-injection system, however, is substantially greater than for conventional carburetion.

**Stratified Charge Gasoline Engines**  In the stratified-charge engine, a rich mixture is ignited in a precombustion chamber by a sparkplug and is then fed into the main combustion chamber, where it ignites a much leaner mixture of gas and air. By this means, relatively low peak combustion temperatures, a slow rate of combustion, and an overabundance of

free air are obtained. The respective results are low emissions of NO$_x$, hydrocarbons, and CO, without misfiring. Stratified-charge engines in general have lower compression ratios and, consequently, higher specific gasoline consumption than conventional engines. They cost more than conventional engines.

**Wankel Rotary Engines**  Licensed by Curtiss-Wright Corp., these engines, built somewhat like a rotary compressor, have less than half as many moving parts and half the weight and are about one-third the size of the equivalent conventional gasoline engine. They require add-on pollution control hardware similar to the conventional gasoline engine, and with conventional carburetion they have about the same fuel economy. With high pressure, or diesel, fuel injection their efficiency approaches that of the diesel engine.

**Two-Stroke Diesel Engines**  (See Sec. 9.) With its high compression ratio (up to 21 on Mercedes-Benz 240-D) and complete combustion, the two-stroke diesel engine is an excellent performer. Emissions are quite low. It is highly efficient, even when equipped with a thermal converter. Its apparent disadvantages, such as high weight, high initial cost, noise, slow acceleration, and difficult cold-weather starting, have been worked on and overcome, so that the passenger-car diesel engine is now available from several manufacturers. Use of more abundant diesel fuel relieves some of the pressure produced by shortages of high-octane gasoline.

**Phillips (Stirling Cycle) Engines**  (See Secs. 4 and 9)  These are external-combustion engines, using helium or freon as a working fluid. Consequently, they require no pollution control devices, since combustion occurs in a highly efficient burner system. In the Stirling Cycle, the engine, too, is highly efficient and has a low specific fuel consumption. With its heat exchangers, the system is somewhat bulky and requires expensive construction techniques. Nevertheless, the engine holds considerable promise because it can utilize almost any kind of fuel.

**Steam (Rankine Cycle) Engines**  Steam engines are bulky and extremely hard to cool and protect from freezing. Their efficiency, at best, is about 75 percent that of the average gasoline engine. When tested in automobiles, they delivered less than 10 miles per gallon. They show little promise for the future.

**Gas Turbines**  Much effort has been invested to develop gas-turbine-driven vehicles. However, numerous problems and high cost have all but ruled these highly efficient devices out of consideration.

**Hydrogen Fuel Cells**  (See Sec. 9)  In 1974 a compact fuel

**Table 7. Pounds of Contaminants Discharged Daily from Domestic Sources in a Metropolitan Area of 100,000 Persons**

| Principal contaminants | Fuel—domestic heating | | | Domestic incineration | | |
|---|---|---|---|---|---|---|
| | Coal | Oil | Gas | Backyard burning | Household incinerator | Apartment incinerator |
| $SO_x$ | 42,000 | 17,000 | 0.4 | 180 | 1 | 12 |
| $NO_x$ | 8,000 | 6,000 | 6 | 90 | 1,150 | 30 |
| $H_2S$ | 1,000 | 500 | 0.1 | | | 24 |
| $NH_3$ | 2,000 | 800 | 0.3 | 345 | | 24 |
| HCl | 2,000 | 500 | 0.3 | | | |
| Aldehydes | 2,000 | 800 | 1 | 600 | 8,400 | 72 |
| Organics | 20,000 | 4,000 | 1 | 42,000 | 12,000 | 1,800 |
| Organic acids | 30,000 | 12,000 | 1 | 225 | 1,900 | 4,800 |
| Solids | 200,000 | 800 | 0.1 | 3,400 | 16,500 | 4,000 |
| Total of above categories, lb | 307,000 | 42,400 | 10 | 46,800 | 40,000 | 10,700 |
| Total of above categories, kg | 139,100 | 19,210 | 4.5 | 21,200 | 18,120 | 4,850 |

Pounds per day per 100,000 persons using each category of heating and refuse disposal

SOURCE: Eliassen, Domestic and Municipal Sources of Air Pollution, *Proc. National Conference on Air Pollution*, 1958.
*Notes:*
Each column shows estimates of the pollutants released to the atmosphere if the entire population used the noted fuel or method of incineration.
$SO_x$—oxides of sulphur—$SO_2$ and $SO_3$.
$NO_x$—oxides of nitrogen.
Aldehydes measured as formaldehyde.
Organic acids measured as acetic acid.
Total includes only above categories and does not imply total contamination in the ambient atmosphere.

**Table 8. Sources of Air Pollution and Typical Loss Rates**

| Class | Aerosols | Gases and vapors | Typical loss rates |
|---|---|---|---|
| Combustion processes | Dust, fume | $NO_2$, $SO_2$, CO, organics, acids | 0.05–1.5% by weight of fuel |
| Stationary engines | Fume | $NO_2$, CO, acids, organics | 4–7% by weight of fuel (hydrocarbons) |
| Petroleum operations | Dust, mist | $SO_2$, $H_2S$, $NH_3$, CO, hydrocarbons, mercaptans | 0.25–1.5% by weight of material processed |
| Chemical processes | Dust, mist, fume, spray | Process-dependent ($SO_2$, CO, $NH_3$, acids, organics, solvents, odours, sulphides) | 0.5–2% by weight of material processed |
| Pyro- and electro-metallurgical processes | Dust, fume | $SO_2$, CO, fluorides, organics | 0.5–2% by weight of material processed |
| Mineral processing | Dust, fume | Process-dependent ($SO_2$, CO, fluorides, organics) | 1–3% by weight of material processed |
| Food and feed operations | Dust, mist | Odorous materials | 0.25–1% by weight of material processed |

SOURCE: Rose et al., Prevention and Control of Air Pollution by Process Changes and Equipment, W.H.O. Rept. Air Pollution, *Monograph Series* 46-307–343.

system was developed that reacts gasoline with other liquids to produce hydrogen. In addition, a packaged industrial power plant was developed which reforms fossil fuel into hydrogen and, using fuel cells, delivers 26 MW of net power at a 9,000 Btu/kWh heat rate. Thus, it is now within the range of practicality to power electrically driven automobiles with compact, light-weight fuel-cell generating systems that operate at a 39 percent overall thermal efficiency, and are pollution-free. Considerable developmental effort will have to be made, however, to design the automobile. This system, if successfully developed, should easily outperform the huge, heavy, and underpowered electric storage battery systems in the current generation of electric vehicles.

CLASSIFICATION OF SOLID POLLUTANTS

(See also Sec. 3)

The **micrometer** ($10^{-6}$ m), abbreviated $\mu$m, is customarily used to measure fine particles. Particles over 10 $\mu$m ($10^{-5}$ m) are classified by Gibbs as **dusts**; between 0.1 and 10 $\mu$m, **clouds**; between 0.001 and 0.1 $\mu$m, **smokes**; below 0.001 $\mu$m, molecular dimensions. The law governing the **terminal velocity of the settling of particles** under the influence of gravity varies with the size of the particle. In the turbulent region (above 2,000 $\mu$m)

$$V_t = Ks^{1/2}p^{1/2}D^{1/2} = k_1\sqrt{SD}$$

In the intermediate region (between 1,000 and 100 $\mu$m), the law is $V_n = K's^{2/3}\rho^{-1/3}\eta^{-1/3}D = k_2s^{2/3}D$. In the streamline region (2 to 50 microns), Stokes' law holds: $V_s = K''sD^2\eta^{-1} = k_3sD^2$. From 0.1 to 1.0 $\mu$m, Cunningham's correction must be applied to Stokes' law:

$$V_c = V_s(1 + 1.72\lambda/D) = V_s(1 + 0.172/D)$$

Below 0.1 $\mu$m, the movement due to molecular shock (Brownian motion) exceeds that due to velocity. In these equations, $V$ is the terminal velocity, fpm; $D$, particle diameter, $\mu$m; $p$, gas pressure, atm; $s$, specific gravity; $\rho$, gas density, g/cm$^3$; $\eta$, gas viscosity, P; $\lambda$, mean free path of the gas molecules, $\mu$m. The equations for falling in standard air are given below, where values of the constants give velocities in m/s. For irregular shapes, $k_1 = 0.142$, $k_2 = 0.00259$, and $k_3 = 1.98 (10)^{-5}$. For spheres, $k_1 = 0.243$, $k_2 = 4.11 (10)^{-3}$, and $k_3 = 3.0 (10)^{-5}$.

**The Tyler standard screen scale** is related to particle size as below:

| Meshes per in | 10 | 20 | 35 | 48 | 64 | 100 | 150 | 200 | 325 |
|---|---|---|---|---|---|---|---|---|---|
| Micrometer scale | 1650 | 830 | 420 | 300 | 220 | 150 | 110 | 74 | 44 |

Diameters of the more common gas molecules range from about 0.0003 to 0.00045 $\mu$m; their mean free path is from 0.06 to 0.2 $\mu$m at atmospheric pressure.

The size ranges of **particles** in typical **aerosols** and **industrial dusts** are, in micrometers: raindrops, 500 to 5,000; mist, 40 to 500; fog, 1 to 40; tobacco smoke, 0.01 to 0.15; oil smoke, 0.03 to 1.0; pigments, 1 to 7; fly ash, 3 to 70; carbon black, 0.04 to 0.2; pulverized coal, 10 to 400; foundry dusts, 1 to 200; cement, 10 to 150; metallurgical fumes, 0.1 to 100; sprayed zinc dust (condensed), 2 to 15; normal impurities in quiet outdoor air, less than 1; dust particles causing silicosis, below 10; pollens, 20 to 60; plant spores, 10 to 30; and bacteria, 1 to 15.

Normal **city air** carries around 0.0006 grains of suspended matter per cubic foot (1.37 mg/m$^3$), which is a practical limit for most industrial gas cleaning; the amount of dust in normal air in manufacturing plants frequently is as high as 0.002 gr/ft$^3$ (4.58 mg/m$^3$). The amount of dust in **blast-furnace gas** after passing the first dust catcher is of the order of 10 gr/ft$^3$ (22.9 g/m$^3$), the same as raw hot producer gas. All dust-content figures are based on air volumes at 60°F and 1 atm. (15.6°C and 101,000 N/m$^2$).

### Cleaning Apparatus

The removal of suspended matter is accomplished by the following methods. Each particle case must be studied to find the most desirable method. Economic considerations require that cleaning should not be any more thorough than necessary.

**Gravity Separation** This method is applicable only to the larger-sized suspended particles (100 $\mu$m and over). A long horizontal chamber is built, in which the gases are slowed down to a velocity which will allow the particles to settle to the bottom of the chamber. Even gas flow should be maintained throughout the chamber. Hoppers or a drag scraper are used to collect the settled-out material. The settling rate can be calculated by Stokes' equation above, and the size of the settling chamber for a given particle size determined.

**Inertia Separation** An effort is made in inertia separators to magnify the settling tendency of solid or liquid particles in gases by increasing the velocity of the gas and by providing for rapid changes in direction which, by inertia, cause the particles to leave the gas stream. Baffle chambers are used for this purpose with a zigzag movement of the gas.

Of all inertia separators, the **cyclone** type constitutes the most widely used industrial dust collector. The gas is passed tangentially into a vertical cylinder with a conical bottom. The gas follows a spiral path, with most of the separation taking place in the smaller sections of the cyclone. Cyclones can be used when particles of over 5 $\mu$m diam are involved. **Multiclones** have a large number of parallel, small cyclone units. The gas pressure drop is about four times the entrance velocity head (1 to 4 in w. g.). The dust content of the cleaned gas is seldom less than 1 gr/ft$^3$ (2.29 g/m$^3$).

In the **Rotoclone** apparatus, the impeller is a concave disk with curved blades along which the dust slips and is discharged over the edge of the disk into a ring-shaped dust chamber. The main air stream leaves in a wide scroll on the side of this chamber. The apparatus is usually preceded by a cyclone for removal of coarse dust. By spraying water in the entering air, the precipitation efficiency of dust removal can be materially raised. Particles of minimum size 8 $\mu$m are claimed to be removed by the dry apparatus, 1 $\mu$m by the wet apparatus. The power for dust separation cannot readily be separated from that for compressing the gases. The power requirement is 3 bhp per 1,000 ft$^3$/min for an outlet pressure of 10 in w.g.; in the water-sprayed machine, the required power is at least 25 percent more.

**Static spray scrubbers** are usually of the tower type, the gas passing upward countercurrently to the descending liquid. Sets of sprays are placed in the top zone, with various materials used in layers to channel and mix the gas and water. Hurdles, cylindrical tiles, and random-packed ceramic tiles or metal spirals are common packing materials. With a gas flow-rate of about 350 ft$^3$/min per ft$^2$ of cross-sectional area and a water rate of 25 gal per Mcf of gas, a cleanliness of 0.1 to 0.3 gr/ft$^3$ can be obtained with blast furnace, coke oven, and producer gas. The pressure drop through a tower scrubber is in the range of 4 to 10 in w. g. The scrubber process is used as the primary cleaning and cooling stages before the cleaning of gases.

**Dynamic Spray Scrubbers** Contact between water droplets and suspended matter in gas is improved by mechanical agitation of gas and water. The *Theisen* disintegrator (Fig. 2) is an example of a modern type of dynamic spray scrubber. It consists of a substantial cast-iron casing enclosing two stationary and one motor-driven rotor baskets. The water is injected by gravity to the center of the perforated cones. Centrifugal force distributes the water over the bar system. The rotation atomizes the water which agglomerates with the dust, wetting it to effect its removal. The disintegrator is always followed by a moisture eliminator which removes the wetted particles from the gas. The range of cleanliness is between 0.02 and 0.005 grains of dust per cubic foot. In addition, any desired pressure increase up to 16 in w. g. can be obtained. The disintegrator is not sensitive to the quality of the incoming gas. It will clean just as well with 0.30 gr/ft$^3$ in the entering gas as with 0.10 gr. It has been installed mostly on blast-furnace gas cleaning systems. In recent years, it has been successful on waste-gas cleaning for the oxygen steelmaking process and electric-furnace waste-gas cleaning systems. A similar dynamic scrubbing effect can be obtained by spraying water into the "eye" of the final exhaust fan and dewatering with a wet cyclone. Water requirements range from 4 to 8 gal per 1,000 ft³/min of gas, and a power demand of from 4 to 10 kW per 1,000 ft³/min of gas.

The other example of dynamic spray scrubbers is the **Pease-Anthony Venturi.** In this system, the gas is forced through a Venturi throat in which the gas is mixed with high-pressure water sprays. A tank after the Venturi is needed to cool and eliminate the moisture. Cleanliness in the range of 0.1 to 0.3 gr/ft³ has been reported.

Filtration is comparatively little used for cleaning fuel gases; it is in extensive use in cleaning air and waste gases. Materials used for gas filtration are ordinarily thickly woven cotton or wool cloth for temperatures up to 250°F; for higher temperatures, metal cloth or woven glass cloth is advisable. The gases filtered should be well above their dewpoint, as condensation on the filter cloth plugs the pores. If necessary, saturated gas should be reheated. The cloth is frequently made into "bags," tubes of 6 to 30 in in diameter and up to 30 ft long, suspended from a steel framework (baghouse). The gas inlet is at the lower end through a header to which the bags are connected in parallel; the outlet is through a housing surrounding all the bags. At frequent intervals, the operation of all or part of each unit is interrupted while the bags are rapped or shaken by a mechanical device to dislodge the accumulated dust which drops back into the gas-inlet header, and is removed by a conveyor screw. If necessary, the gas flow is temporarily reversed through the bags to assist loosening of the dust. The dust content can be reduced to 0.01 gr/ft³ or less at reasonable cost. The apparatus also is used for the recovery of valuable solids from gases.

**Electrical Precipitation**   Gas cleaning by this method is the most efficient system known. The principle of electrical precipitation, although involving highly technical physioelectrical phenomena, is generally familiar to all. If suspended particles are placed in a high-voltage electric field, they receive an electric charge and move toward one or the other of the electrodes between which the electric field is established. If one electrode is a pipe or flat plate and the other electrode is a wire axially suspended in the center of the pipe or between two plates, the field is strongest around the wire and weakest near the surface of the pipe or plate. The charged dust particles move from the strong part of the field toward the weak part of the field.

Two types of precipitators are manufactured. **The vertical-round type** is usually built in a cylindrical tank having a header sheet near the inside top. In this header sheet are nested pipes which act as the collecting electrodes. The high-voltage ionizing electrodes are twisted square steel rods, about ¼ in in size, supported from above the header sheet and hung axially in the collecting electrode pipes. Water is introduced above the header sheet and flows over weir rings at the tops of the pipes to form a water film on the inside of the pipes. **The horizontal-plate type** has vertical parallel plates as collecting electrodes. The discharge electrodes are a series of wires suspended equidistantly between the plates, and can be operated as either a "wet" or "dry" unit. When used as a wet precipitator, weir boxes are built on the top of each plate to produce a wetting of both sides of the plate, washing the collected particles down the plate. As a dry precipitator, a rapping device operated on a time cycle shakes the plates, causing the collected particles to fall into a hopper under the plates. The dry dust is gathered from the hoppers by means of gates or drag scrapers.

The electrical apparatus provided with a precipitator to produce a dc potential of 30,000 to 90,000 V can be mechanical, vacuum tube, or a selenium rectifier. Power packs are self-contained cabinet-type units, built in sizes from 2½ to 50 kVA. Rectifiers are mounted in a substation building close to the precipitator or can be mounted in weatherproof enclosures on the roof of the precipitator. Controls can be at a distance because the wiring is of low potential.

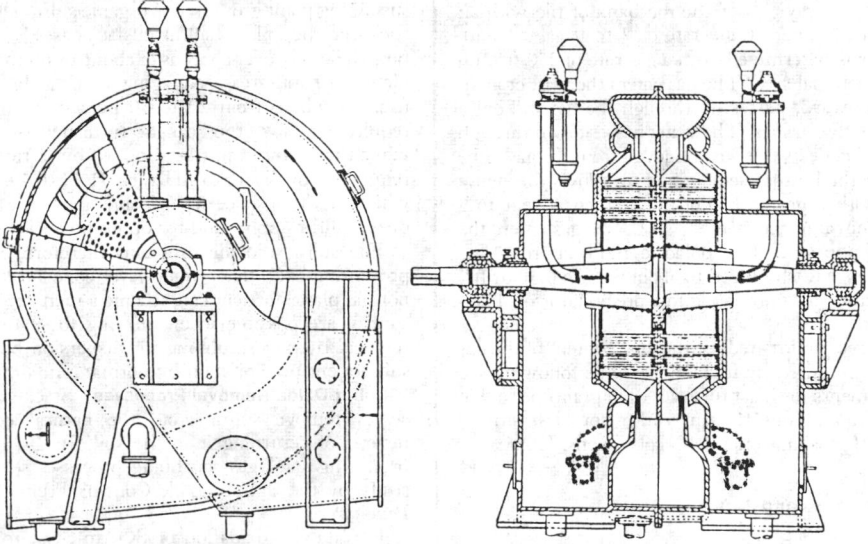

**Fig. 2**   Theisen disintegrator air washer.

The efficiency of dust removal by precipitators is expressed as log $(1-E) = t \log K$, where $E$ = the efficiency as fraction removal; $t$ = the time the gas is in the electrical field, s; and $K$ = an apparatus constant varying from 0.05 to 0.80 ("Cottrell Electrical Precipitators," Western Precipitator Corp.; see also Schmidt and Anderson, *Elec. Eng.*, 1938). The above function implies that small variations of $t$ cause considerable variations of $(1-E)$, the fraction of dust left in the gas. This indicates a sensitivity of the precipitators to overload. (See Table 12 for common applications).

Precipitators working on coke oven gas for tar removal, under favorable temperature and moisture conditions, show outlet figures of 0.005 gr of dry tar per ft$^3$ of gas with inlet of 0.25 gr/ft$^3$. The outlet tar content increases rapidly upon overloading the apparatus.

Precipitators for blast furnace gas reduce the dust to about 0.01 gr/ft$^3$. Electrical precipitators are used for removal of suspended matter from gases and air from boiler gases, metallurgical fumes, cement and lime dusts, etc. Precipitators can be built for dust in practically all particle sizes, as well as mist. They cannot collect material in the gaseous or vapor state. Precipitators are not applicable where explosive gas or dust is involved in the presence of air or oxygen in explosive proportions because of the hazard of ignition by an electrical spark.

**Blast-Furnace Gas Cleaning** Modern equipment for this service represents a field in which gas cleaning has been developed to a high degree of efficiency. The method described here is used on most modern blast-furnace gas-cleaning plants.

The **dust catcher** is a large cylindrical steel tank, brick-lined, about 30 to 40 ft in diameter. The dust catcher is fundamentally a settling chamber. Dust is dropped out of the gas by reducing the velocity of the flow. Gas enters at 400°F approx and with 10 to 20 grains of dust per cubic foot. A good dust catcher will clean the gas to 3 grains of dust per cubic foot.

The **primary washer** is a cylindrical steel shell 16 to 21 ft in diameter, with a conical bottom section. Four to five banks of ceramic, round, or cross-partition tile are arranged in the tower. At the top, sprays cover the top bank of tile with an even distribution of water at the rate of 7 to 9 gal/ft$^2 \cdot$ min). Size of the tower is determined with a gas rate of 350 ft$^3$/min per ft$^2$ of cross-sectional area. The gas enters the washer at an angle of 45° downward, passes up through the tile and out a main at the top of the washer. The water passes down through the tile banks and collects the dust removed from the gas. The water collects in the bottom section and overflows through a pipe arranged with a gas seal. The water then is sent to a thickener for removal of the iron-ore dust. The gas enters the washer at 400°F approx and is cooled to 100°F approx. The dust content of the gas is reduced to about 0.15 to 0.30 gr/ft$^3$, which is satisfactory for entrance into a precipitator or Theisen disintegrator.

Electrostatic precipitators reduce the dust content of the gas to the cleanliness required by its final use. The following are common requirements for blast furnace gas in grains per cubic foot at standard conditions (1 atm, 70°F): hot blast stoves, 0.01; boilers, 0.01; soaking pits, 0.10; coke ovens, 0.005.

### Effect of Sulphur Gases and Particulates on Vegetation

REFERENCES: Kamprath, E. J., "Possible Benefits from Sulfur in the Atmosphere." *Combustion*, **44**(4):16, 17, 1972; Grennard, Alf, and Ross, Frazer, "Progress Report on Sulfur Dioxide," *Combustion*, **45** (7):4–9, 1974; Wagner, Hon. A. J., "TVA's Proposed Clean Air Strategy," *Public Utilities Fortnightly*, 34–37, June 6, 1974.

**Damage from Sulphur Fumigation** Vegetation is damaged by abnormal concentrations (above 0.3 ppm) of sulphur gases at ground level. Users of sulphur-bearing fuels may become involved with damage claims where ground-level concentrations are of consequent increasing importance. The contaminants causing damage to vegetation are sulphur trioxide (SO$_3$), sulphur dioxide (SO$_2$), and particulate materials.

**Particulate materials** on leaf surfaces have no harmful effect other than to exclude sunlight, and to that extent, retard growth.

**Sulphur trioxide** is hygroscopic and occurs at ground level as sulphuric acid mist. Plant damage by sulphuric acid aerosol has been achieved in the laboratory but has not been observed in the field. Sulphur trioxide, therefore, is not a consideration under field conditions.

**Sulphur dioxide** has the greatest effect on plants. Concentrations greater than 0.3 ppm will damage the leaves of sensitive vegetation if maintained for over 5 h. Animals and humans are more resistant than plants. Concentrations above 0.5 to 2.0 ppm are recognizable by smell and taste. The concentration and the duration of the fumigation determine the degree of injury. (See Table 9). Environmental conditions of temperature, humidity, soil moisture, soil fertility, nutrient supply, light intensity, age of plants, and moist-leaf conditions have a marked influence on the effect of sulfur dioxide on vegetation.

**Benefits from Sulphur in the Atmosphere** In normal atmospheric concentrations below 0.3 ppm, as occurs in metropolitan areas 95 percent of the time, sulfur dioxide is an essential plant nutrient. Vegetation utilizes it for growth by absorption from the atmosphere through leaves, and from rain-saturated earth through the roots. Fossil fuels, by virtue of their origin as former vegetation contain from 0.4 to over 4 percent sulfur (Table 11). To maintain the natural ecological sulfur cycle for sustaining plant growth, this organic sulfur should be recycled back into the soil. As sulfur in stack gases is brought down by rainfall largely over a radius of about 15 mi from steam-electric plants in rural areas, averaging about 5 lb/acre-year (0.55 g/ m$^2$), anywhere from 20 to 50 percent of the annual sulfur requirement for crops can be furnished by stack emissions, while the balance must be supplied by fertilizers. Agriculture requires annually 10 to 40 lb/acre (1.11 to 4.44 g/m$^2$) of sulfur, which, in the absence of stack emissions, must be furnished by direct sulfur application for optimum yields of crops.

The sulfur dioxide problem, therefore, is simply one of preventing excessive ground-level concentrations, whereby a normal plant nutrient can become a menace. Two methods of control are possible when burning fossil fuels: sulfur oxide removal by wet scrubbing, or dispersion by tall stacks with subsequent precipitation by normal rainfall.

**Sulfur Oxide Removal Processes** Stack-gas scrubbing processes remove sulfur from the natural ecological cycles, instead of returning it to the soil. In 1973 the seven most promising stack-gas scrubbing processes selected for further study by the Sulfur Oxide Control Technology Assessment Panel were

1. Catalytic oxidation of SO$_2$ to SO$_3$, to recover sulfuric acid as a saleable by-product

**Table 9. Susceptibility of Vegetation to Injury by Sulphur Dioxide (from Field Observations)**

| Maximum susceptibility | Moderate susceptibility | Resistant |
|---|---|---|
| WEEDS | | |
| Plantain | Bracken fern | Goldenrod |
| Ragweed | Wild carrot | Small-leaved milkweed |
| Smartweed | Wild grape | Knotweed |
| Dewberry | Sweet clover | Milkweed |
| Greenbriar | | |
| Dandelion | | |
| Galinsoga | | |
| Pigweed (redroot) | | |
| WILDFLOWERS | | |
| Greenbriar | Blackberry | Goldenrod |
| Galinsoga | Witch hazel | Small-leaved milkweed |
| Dandelion | Huckleberry | Elderberry |
| | Blueberry | Mountain laurel |
| | Bracken fern | Yarrow |
| | Viburnum | Knotweed |
| | Wild carrot | Joe-Pye weed |
| | Sweet clover | Milkweed |
| FARM CROPS | | |
| Alfalfa | Alsike clover | Potato |
| Oats | Corn | |
| Buckwheat | Cabbage | |
| Barley | | |
| GARDEN VEGETABLES | | |
| Beet | Huckleberry | Onion |
| Endive | Blueberry | Cucumber |
| Bean | Tomato | Corn |
| Peas | Parsley | Potato |
| Brussels sprouts | Cauliflower | |
| Sunflower | Cabbage | |
| LANDSCAPE MATERIALS | | |
| Hawthorn | Huckleberry | Hazelnut |
| Sunflower | Blueberry | Mountain laurel |
| Cosmos | Nasturtium | |
| Sweet William | Dahlia | |
| | Gladiolus | |
| | Willow | |
| TREES | | |
| White pine | Larch | Red pine |
| Hemlock | Scotch pine | Blue spruce |
| Hawthorn | Norway spruce | Black locust |
| Black birch | Sumac | Black oak |
| Yellow birch | Ash | Red oak |
| | Wild cherry | White oak |
| | Domestic apple | Sugar maple |
| | Willow | Swamp red maple |
| | | Norway maple |

2. Wet lime/limestone scrubbing, which produces calcium sulfite sludge for disposal

3. Alkali scrubbing without regeneration, which produces 18 percent sodium sulfite brine for disposal

4. Alkali scrubbing with lime regeneration, which produces calcium sulfite sludge for disposal

5. Alkali scrubbing with thermal regeneration, which recovers $SO_2$ gas by-product, and a sodium sulfite/sulfate brine for disposal

6. Magnesium oxide scrubbing, with off-site sulfuric acid manufacture and regeneration of magnesium oxide by others

7. Alkali scrubbing, plus electrodialytic regeneration to recover $SO_2$ gas and a purge stream of impure brine

A comparison of these systems is given by Table 10. The sulfur dioxide emissions limitations (SDEL) case, with its low-cost, negligible environmental impact and its enhancement of the normally occurring sulfur ecological cycle, is technically justifiable with a cost benefit ratio of 0.19. The other designs are bulky, costly, noisy, and highly experimental. They consume energy, water, fuel, and chemicals, and require direct sulfur application to surrounding agricultural areas, plus waste sludge disposal. These adverse environmental impacts usually result in negative total public benefits, unjustifiable except under unusually favorable circumstances.

By far the most difficult and costly environmental hazard produced by sulfur removal is the volume of high chemical-oxygen-demand (COD) liquid and solid wastes produced. By 1980, 48 million tons of $SO_2$ are expected to be removed annually by 80 percent recovery of 100,000 MW installed flue-gas desulfurization capacity, and will yield 300 million tons annually of calcium sulfite sludge. Over a 20-year period, this would cover 160 mi² (400 km²) of land 10 ft (3 m) deep. This contaminated thixotropic sludge is environmentally unacceptable for landfill or for any other conceivable use. Reduction to elemental sulfur for permanent underground storage by an inverse Frasch process would reduce the volume of wastes, but the cost is exorbitant.

### SULFUR DIOXIDE EMISSION LIMITATION (SDEL) PROGRAMS

To avoid the water pollution difficulties and the wet sludge disposal difficulties introduced by wet scrubbing, Tennessee Valley Authority and others have developed various SDEL programs. TVA's Chairman, the Honorable Aubrey J. Wagner, states:

Not only is TVA's plan the only way in which health and welfare standards can be achieved in time to meet the Clean Air Act deadlines, but as an air pollution control technique, it is also the most reliable and least costly and has the fewest adverse effects on energy conservation and other aspects of the environment.*

SDEL methods employ combinations of five or more sensing, dispersing, and limitation approaches simultaneously.

1. Very tall stacks (1,000 ft) for dispersing $SO_2$ gases high in the atmosphere, thereby minimizing unacceptable ground-level concentrations and occurrences. [See the Holland equation for stack height, below.]

2. Stadium-type cooling tower arrays, surrounding the stacks. These arrays, 600 ft in diameter or more, create large, buoyant, nondispersive, rising hot-air masses, which puncture existing inversion layers and distribute their contents high in the stratosphere, over a wide area. [See Brown and Sneck: "Cooling Tower Plume Rise," *Proc. Amer. Power Conference*, 33:546–553, 1971.]

3. Computers, which quickly correlate meteorological and plant operational data, so as to anticipate and prevent unacceptable ground-level concentrations.

4. Sensors and monitors, which sense and record automatically the local ambient $SO_2$ concentrations, and pH's, thereby ensuring the effectiveness of the control operations. (Limestone application is made directly to lakes in the area to raise pH.)

5. Emission limitations, accomplished by fuel-switching between high- and low-sulfur fuels during normal power plant operations, and by local power-generation curtailment during emergency situations, when even switching to low-sulfur fuels might not prevent unacceptable ground-level concentrations.

6. Daily local plant weather observations and meteorological measurements by means of light aircraft and balloons.

*Quoted with permission. *Public Utilities Fortnightly*, 1974.

Table 10. Comparison of SO$_2$ Control Systems

| Process | Requirements | Throwaway or recovery | Additional cost of power generation, % | SO$_2$ control efficiency, % | Additional plant capacity investment, % |
|---|---|---|---|---|---|
| SDEL* | Fuel-switching and tall stacks | | 1–2.35 | 99+ | 1–2.8 |
| Low-sulfur fuel (coal or oil) | | | 28–85 | Varies | — |
| Dry limestone injection | Limestone (200% of stoichiometric) | Throwaway CaSO$_3$/CaSO$_4$ | 10–14 | 22–45 | 10–12 |
| Wet lime/ limestone scrubbing | Lime (100–120% stoich.); limestone (120–150% stoich.) | Throwaway CaSO$_3$/CaSO$_4$ | 19–37 | 60–85 | 17–30 |
| MgO scrubbing | MgO Carbon and fuel (for regeneration and drying) | Recover conc. H$_2$SO$_4$ | 25–50 | 90 | 20–35 |
| Catalytic oxidation (add-on) | V$_2$O$_5$ catalyst heat | Recover dilute H$_2$SO$_4$ | 25–45 | 85–90 | 25–40 |
| Na$_2$SO$_3$ scrubbing | Sodium makeup heat | Recover conc. H$_2$SO$_4$ or sulfur | 25–50 | 90 | 25–40 |
| Double alkali | Sodium makeup plus lime (100–130% stoich.) | Throwaway CaSO$_3$/CaSO$_4$ | 17–35 | 90 | 15–30 |

BASIS: 80 percent plant load factor. Yearly fixed charges are 18 percent of capital investment. Cost ranges are for 200-MW to 1000-MW plant sizes. Costs of sludge disposal ponds not included.
SOURCES: Burns and Roe Data, 1974, "TVA's Clean air Strategy," *Public Utilities Fortnightly*, **34**: June 6, 1974.

These complete the picture, and by allowing high-sulfur fuels to be burned most of the time, thereby conserve scarce low-sulfur fuels for domestic and other ground-level uses not employing SDEL methods.

As opposed to wet scrubbers for SO$_2$ control, SDEL methods are estimated to require less than one-tenth the total investment cost and one-thirteenth the annual operating cost, exclusive of wet-sludge disposal costs.

## ATMOSPHERIC DISPERSION OF POLLUTANTS FROM STACKS

**Stacks** can provide effective atmospheric dispersion of gaseous and particulate pollutants with acceptable ground-level concentrations. Theoretical and empirical formulas are available to estimate the dispersion of air borne pollutants continuously emitted from stacks. Effective stack height of a plume is given by $H = h_s + h_r$. **The Holland formula**, recommended for calculating $h_r$, is $h_r = (1.5vd + 4.09 \times 10^{-5}Q_h)/u$. A ratio of stack height $h_s$ to building height of 2½ to 1 or more is commonly used to avoid entrapment of the plume in the vortex of adjacent buildings and the associated high values of ground-level concentration $X_g$. Stack-exit velocity $v$ should be 14 to 28 m/s (45 to 90 fps) to minimize plume entrapment in stack vortices. The lower the ratio of stack inside to outside diameter at its top, the higher the necessary $v$.

**The Cramer equation,** recommended for calculating $X_g$, is $X_g = 10^6 Q/(\pi u x^b \sigma_A \sigma_E) \exp [H^2/2\sigma_E^2 x^{2p}]$. The equation is for sampling intervals of approximately 20 min, with $X_g$ varying approximately inversely as the one-fifth power of the interval.

Simplified approximate formulas for $X_{mg}$ and $x_m$ are

$$X_{mg} = 2Q 10^6 \sigma_E/\pi euH^{1.9}\sigma_A \qquad \text{and} \qquad x_m = H/1.35\sigma_E$$

Atmospheric dispersal characteristics and hence the values of $X_g$, $X_{mg}$, and $x_m$ are influenced by thermal stratification and atmospheric turbulence. A superadiabatic (adiabatic) (subadiabatic) lapse rate; i.e., a drop in temperature with altitude greater than (equal to) (less than) 5.4°F per 1,000 ft (9.84 K/km) results in an unstable (neutral) (stable) atmosphere with much (moderate) (very little) vertical mixing.

**Notation**

| Symbol | Definition | System of consistent units |
|---|---|---|
| $b$ | Atmospheric-dispersal parameter | Dimensionless |
| $d$ | Stack-exit diameter | m |
| $H$ | Effective stack height | m |
| $h_r$ | Rise of plume | m |
| $h_s$ | Stack height | m |
| $p$ | Atmospheric-dispersal parameter | Dimensionless |
| $Q$ | Emission rate of pollutant | cu m per sec |
| $Q_h$ | Heat emission rate | cal per sec |
| $q$ | Atmospheric-dispersal parameter | Dimensionless |
| $u$ | Mean wind speed | m per sec |
| $v$ | Stack-exit velocity | m per sec |
| $X_g$ | Ground-level concentration of pollutant at a distance, $x$, from base of stack | ppm by vol |
| $X_{mg}$ | Maximum ground-level concentration of pollutant at a distance, $x_m$, from base of stack | ppm by vol |
| $x$ | Downwind distance from emission source in direction of mean wind | m |
| $x_m$ | Distance from base of stack to maximum ground-level concentration | m |
| $\sigma_A$ | Basic diffusion parameter related to azimuth plume angle in radians | $m^{1-q}$ |
| $\sigma_E$ | Basic diffusion parameter related to elevation plume angle in radians | $m^{1-p}$ |

The recommended parameters for use in the Cramer equations under the above conditions of thermal stratification are:

| Thermal stratification | $\sigma_A$, radians | $\sigma_E$, radians | $b$ | $p$ | $q$ |
|---|---|---|---|---|---|
| Unstable....... | 0.39 | 0.13 | 2 | 1.1 | 0.9 |
| Neutral........ | 0.21 | 0.07 | 1.85 | 1.0 | 0.85 |
| Stable.......... | 0.12 | 0.04 | 1.6 | 0.8 | 0.8 |

The above formulas deal only with steady-state conditions. There are transient meteorological conditions which can result in significant ground-level concentrations of pollutants and which cannot readily be analyzed by computations. These conditions include the breakup of a temperature inversion and the slow, steady buildup of pollutants during stagnations accompanied by deep ground fogs in a valley. For large power plants, high stacks (400 ft and more) (122 m), high stack-exit velocities [45 to 90 ft/s (13.7 to 27.4 m/s)], and adequate exit temperatures [250 to 300°F (394 to 422 K)] are usually effective in avoiding such fumigations.

STACK EMISSIONS

Emission of sulphur-bearing gases and solids from a 1,000-MW fossil-fuel-fired power station is shown in Table 11. (See also Tables 12 and 13 for supplementary data.) Reducing the remaining solids in stack emissions from bituminous-coal-fired (excluding stoker firing) power boilers to 0.02 gr/standard ft³ (45.8 mg/m³) and less results in a clear stack discharge. This requires collection efficiencies of 99+ percent, attainable with commercial equipment. These high collection-efficiency requirements demonstrate the need for optical instruments which will readily determine stack-emission quality. The **Ringelman chart** is frequently used to evaluate stack emission, but it is a crude and inaccurate method. A good indicator is a stack emission invisible to the naked human eye. With oil firing, present practice is to control stack emissions by furnace design, combustion control, use of additives, and multiple cyclone-type collectors located in the low- or high-temperature zones of the flue-gas system. For future plants of 1,000 MW and larger, use of high-temperature electrostatic precipitators located ahead of the air preheater may be considered.

**Air Pollution Control in Various Industries**

(See Tables 12 and 13)

**Industrial sources** of air pollution and typical loss rates are given in Table 8.

**Non-ferrous-metal** smelting (e.g., copper, zinc, lead) and thermal operations (e.g., blast furnace, sintering, converters) are sources of air pollution and of potentially recoverable metallic materials. A large variety of collection equipment has been tried. Experience has proved that precipitators, cloth filters, and gas washers are best. By such means, Chattanooga, TN, was able to achieve a 67 percent cleanup of air pollution particulates between 1969 and 1974.

**Paper industry** sources of air pollution are (1) recovery-furnace gases and (2) lime-kiln exhaust gases. **Recovery-furnace gases** in the sulphate and soda processes of manufacture emit solids [2 to 6 gr/standard ft³ (4.6 to 13.7 g/m³)] which are mainly sodium salts. The small particle size is particularly bothersome because of the large, obscuring plume in the stack tail. Chemically, the solids are destructive to painted and finished surfaces (e.g., automobiles, structures). Valuable constituents are recovered for return to the cycle. Electrostatic precipitators with 95 to 99 percent efficiency are favored for collection and return of solids. Corrosion-resistant materials (glazed tile, special cements) must be used to reduce maintenance. **Lime kilns** are used to calcine calcium carbonate and recover it as CaO for use in the process. The consequent air pollutant is recovered by gas washers and mechanical collectors.

**Cement-industry** sources of air pollutants are (1) clinker kilns and (2) bagging and other mechanical operations. The **kilns** are the major source of air pollution, with large, unattractive exhaust plumes depositing cement-making solids on the landscape and on painted surfaces. Damage is reduced by the use of electrostatic precipitators with the wet process and cyclones plus cloth filters with the dry process. **Bagging and other mechanical losses** are reduced by use of cloth filters following cyclone separators.

**Table 11. Emission of Sulphur-Bearing Gases and Solids from a 1,000,000-kW (1,000-MW) Fossil-Fuel-Fired Power Station**

| Fuel | Mid-Western bit coal | Eastern bit coal | Central Ill. bit coal | High-sulphur fuel oil | Low-sulphur fuel oil |
|---|---|---|---|---|---|
| Firing method | Pulv.-coal round burners | Pulv.-coal round burners | Cyclone burners | Round burners | Round burners |
| Sulphur content of fuel, % | 4.14 | 1.15 | 4.65 | 2.41 | 0.40 |
| Ash content of fuel, % | 18.16 | 7.59 | 15.00 | 0.10 | 0.10 |
| Sulphur-bearing gases leaving boiler: | | | | | |
|   Sulphur dioxide, lb per h | 69,000 | 15,100 | 97,500 | 23,200 | 3,620 |
|   Sulphur trioxide, lb per h | 3,450 | 755 | 4,875 | 1,160 | 180 |
|   Total sulphur-bearing gases, lb per h | 72,450 | 15,855 | 102,375 | 24,360 | 3,800 |
| Solids in flue gas leaving boiler: | | | | | |
|   Sulphur compounds, lb per h | 4,600 | 920 | 920 | 2,520 | 420 |
|   Carbon, lb per h | 8,280 | 2,760 | 1,190 | 4,200 | 4,200 |
|   Fly ash, lb per h | 118,680 | 38,640 | 20,250 | 420 | 420 |
|   Total solids, lb per h | 131,560 | 42,320 | 22,360 | 7,140 | 5,040 |
| Collection efficiency,* % | 99+‡ | 99+‡ | 99+‡ | † | † |
| Solids in flue gas leaving stack, lb per h | 360 | 360 | 15 | † | † |

*Based on clear stack.
†See text.
‡Total efficiency whether combination mechanical and electrostatic units or electrostatic alone.
Note: 1 lb per hr = 7.55(10)$^{-3}$ kg/s.

**Carbon-black** production utilizes electrostatic precipitators, cloth filters, and scrubbers to collect the black from the gas stream and to control air pollution.

**Iron-and-steel-industry** sources of air pollution are open-hearth, oxygen-furnace, electric-furnace, and sintering operations. The dispersoids in these operations are mostly oxides of iron in particle sizes below 10 $\mu$m and with loadings ranging to 15 gr/standard ft$^3$ (34.3 g/m$^3$). Dust emittance is acceptably controlled generally by use of precipitators and scrubbers for open-hearth and basic oxygen processes, of cloth filters and scrubbers for electric furnaces, and of precipitators, scrubbers, and filters for sintering.

**Petroleum-refining** sources of air pollution are mainly gaseous in nature. Solid dispersoids from the fluid-cracking processes are removed by electrostatic precipitators and/or cyclone-type collectors.

**Miscellaneous Dusts** Four industrial dusts have been classified as causing severe respiratory ailments: asbestos, causing asbestosis; crystalline silica, causing silicosis; cotton and cottonseed dust, a powerful allergen causing bronchial asthma and byssonosis; and lead, causing plumbism.

In mining operations and foundries, cyclonic collectors, followed by baghouses, have been found most effective. Likewise, in textiles and in cottonseed-oil pressing, baghouses following the lint-collecting cyclones can remove almost 100 percent of the hazardous pollutant.

**Large central incineration plants** can produce odors, noxious gases, and particulates (see Sec. 7). Odors and noxious gases can be minimized by carefully controlled combustion, using modern travelling-grate continuous systems. Furnace temperatures between 1600 and 2200°F (1150 to 1480 K) are recommended. Particulates can be minimized by well-managed covered conveying of dust-bearing refuse, incinerator residues and fly ash, and the use of electrostatic precipitators and wet scrubbers to comply with increasingly stringent air quality standards. (See Table 13.) Gaseous emissions of SO$_x$, CO, and NO$_x$ do not cause problems with continuous incineration, as their concentrations are small, but HCl is of some concern and is expected to come under closer scrutiny. Stack height should be sufficient to minimize ground effects of any residual pollution (see Sec. 7). Municipal refuse can also be fired into waste-heat utility boilers, a fuel conservation measure which will become increasingly important in electric power generation.

## RADIOACTIVE WASTE MANAGEMENT

References: "Additional Testimony of Dr. Walton A. Rodger on Behalf of Consolidated Utility Group," Rulemaking Hearing RM50-2, "Effluents from Light Water-Cooled Nuclear Power Reactors," Nov. 9, 1973. Ayers, "Decontamination of Nuclear Reactors and Equipment," The Ronald Press Co., New York. "The Problem Beyond Removal," *Proceedings from Electrical World Conference Waste Disposal in Utility Environmental Systems*, McGraw-Hill, Oct. 1973. Etherington, "Nuclear Engineering Handbook," McGraw-Hill. Code of Federal Regulations 10CFR50, Appendix I, and 10 CFR-20.

### General Considerations

Under existing legislation and authorization the Nuclear Regulatory Commission (NRC) has the primary responsibility for developing and regulating waste management methods and practices, sites of operation, and enforcement of all applicable standards, including those developed adjunctively with EPA. Radioactive wastes are generally divided into two broad categories: (1) those from nuclear reprocessing plants, containing highly radioactive fission products together with small amounts of plutonium, called **high-level radioactive wastes,** and (2) those from commercial nuclear power-generating facilities, called **low-level radioactive wastes.** The high-level radwastes,

**Table 12. Common Applications of Industrial Precipitators**

| Industry | Application | Gas-flow range, ft³/min* | Temp. range, °F | Percent weight of dust (below 10) | Usual efficiencies, % | Dust conc. range in ppm† by wt in air | Temp. range, K |
|---|---|---|---|---|---|---|---|
| Electric power | Fly ash from pulverized-coal-fired boilers | 50,000–750,000 | 270–600 | 25–75 | 95–99+ | 760–9,500 | 405–589 |
| Portland cement | Dust from kilns | 50,000–1,000,000 | 300–750 | 35–75 | 85–99+ | 950–29,000 | 422–672 |
| | Dust from dryers | 30,000–100,000 | 125–350 | 10–60 | 95–99 | 1,905–29,000 | 325–450 |
| | Mill ventilation | 2,000–10,000 | 50–125 | 35–75 | 95–99 | 9,500–98,000 | 283–325 |
| Steel | Cleaning blast-furnace gas for fuel | 20,000–100,000 | 90–110 | 100 | 95–99 | 38–950 | 306–317 |
| | Collecting tars from coke-oven gases | 50,000–200,000 | 80–120 | 100 | 95–99 | 190–1,905 | 300–322 |
| | Collecting fume from open-hearth and electric furnaces | 30,000–400,000 | 300–700 | 95 | 90–99 | 95–5,700 | 422–644 |
| Non-ferrous metals | Fume from kilns, roasters, sintering machines, aluminum potlines, etc. | 5,000–1,000,000 | 150–1100 | 10–100 | 90–98 | 5,700–95,000 | 339–867 |
| Pulp and paper | Soda-fume recovery in kraft pulp mills | 50,000–200,000 | 275–350 | 99 | 90–95 | 950–7,600 | 408–450 |
| Chemical | Acid mist | 2,500–20,000 | 100–200 | 100 | 95–99 | 38–1,905 | 311–367 |
| | Cleaning hydrogen, CO₂, SO₂, etc. | 5,000–20,000 | 70–200 | 100 | 90–99 | 19–1,905 | 295–367 |
| | Separate dust from vaporized phosphorus | 2,500–7,500 | 500–600 | 30–85 | 99+ | 19–1,905 | 533–589 |
| Petroleum | Powdered catalyst recovery | 50,000–150,000 | 350–550 | 50–75 | 90–99.9 | 190–4,800 | 450–561 |
| Rock products | Roofing, magnesite, dolomite, etc. | 5,000–200,000 | 100–700 | 30–45 | 90–98 | 950–4,800 | 311–644 |
| Gas | Tar from gas | 2,000–50,000 | 50–150 | 100 | 90–98 | 19–3,800 | 283–339 |
| Carbon black | Collecting and agglomerating carbon black | 20,000–150,000 | 300–700 | 100 | 10–35 | 57–9,500 | 422–644 |
| Gypsum | Dust from kettles, conveyors, etc. | 5,000–20,000 | 250–350 | 95 | 90–98 | 2,850–9,500 | 394–450 |

SOURCE: Holden and Ackley, "Air Pollution Handbook," McGraw-Hill.
*ft³/min = 4.72(10)⁻⁴ m³/s.
†1 ppm = 0.525 gr/1000 ft³ = 2 mg/m³, in air

together with all the legal aspects of nuclear-power-plant licensing and operation, are briefly discussed in Sec. 9. The remaining category, "low-level radioactive wastes," consists of three types of effluents: gases, liquids, and solids.

**Gases** From the buildings which contain the fuel-handling system, auxiliary equipment, and waste systems, ventilating air, filtered if necessary, is released through a vent which is monitored for radioactive gases, iodine, and airborne particulates. Automatic controls will shut down the ventilating system, thereby stopping any release of airborne activity if the monitor set points, which regulate emissions to stay within established MPC limits at the site boundary, are exceeded. The steam-jet ejector of PWRs is included in this vent.

Ventilating air for the containment vessel, which holds the reactor, primary pumps, and steam generators, is drawn when it is necessary for inspections or for other reasons to purge the vessel. **Purge air** is filtered through high-efficiency-particulate-air (HEPA) and through charcoal filters before release. In case monitor set points are exceeded, purging is automatically curtailed.

The third source of gases is the nitrogen cover gas of the volume control system PWR, and the steam-jet ejector system in the BWR. Krypton, xenon, and iodine are produced within the reactor and mix with the cover gas. To release them, the principle employed is to "delay and decay" by compressing them into decay tanks of about 45 days' storage capacity in the PWR case, and into decay pipes of 20- to 60-min storage capacity in the BWR case. All releases are carefully monitored to be within established MPC limits.

To provide further time for decay in the BWR case, an augment on the BWR air ejector can be shown to be justifiable by a cost/benefit analysis based upon the worth factor of $1,000 per person-rem of population dose. On this basis, off-gas reduction systems employing catalytic recombiners alone or with cryogenics and carbon are now required.

**Liquids** Liquid radwaste streams from boiling water reactors may be classified into two major systems. In one of these, relatively high radioactivity, low conductivity water is reclaimed by filtration and deionization, and is stored for recycling into plant makeup systems. In the other, relatively low radioactivity, high conductivity water from floor and laboratory drains, laundry drains, makeup demineralizer

**Table 13. Characteristics of Air- and Gas-Cleaning Devices**

| General class | Specific type | Device most suitable for | Removable contaminants | Optimum size particle μm | Limits of gas temperature °F | Opt. conc. ppm by wt* | Lim. gas temp. K |
|---|---|---|---|---|---|---|---|
| Odor adsorbers | Shallow bed | | Malodors, gases | (Molecular) | 0–100 | <1.9 | 256–311 |
| Air washers | Spray chamber | | | >20 | 40–700 | | |
| | Wet cell | | | >5 | 40–700 | <9.5 | 278–644 |
| Electro. precip., low-voltage | Two-stage, plate | Atmospheric air cleaning | Lints, dusts, pollens, tobacco smoke | <1 | 0–250 | | |
| | Two-stage, filter | | | <1 | 0–180 | <1.9 | 256–394 |
| Air filters, viscous-coated | Throwaway | | | >5 | 0–180 | <3.81 | 256–356 |
| | Washable | | | >5 | 0–250 | <3.81 | 256–394 |
| Air filters, dry-fiber | 5–10 μm | | | >3 | 0–180 | <1.9 | 256–356 |
| | 2–5 μm | | | >0.5 | 0–180 | <1.9 | 256–356 |
| Absolute filters | Paper | | Special† | <1 | 0–1800 | <1.9 | 256–1256 |
| Industrial filters | Cloth bag | | | >0.3 | 0–180‡ | >190 | 256–356 |
| | Cloth envelope | | | >0.3 | 0–180‡ | >190 | 256–356 |
| Electro. precip., high-voltage | Single-stage, plate | | | <2 | 0–700 | >190 | 256–644 |
| | Single-stage, pipe | | | <2 | 0–700 | >190 | 256–644 |
| Dry inertial collectors | Settling chamber | | Dusts, fumes, smokes, mists | >50 | 0–700 | >9,520 | 256–644 |
| | Baffled chamber | | | >50 | 0–700 | >9,520 | 256–644 |
| | Skimming chamber | | | >20 | 0–700 | >1,905 | 256–644 |
| | Cyclone | | | >10 | 0–700 | >1,905 | 256–644 |
| | Multiple-cyclone | | | >5 | 0–700 | >1,905 | 256–644 |
| | Impingement | | | >10 | 0–700 | >1,905 | 256–644 |
| | Dynamic | Stack gas cleaning | | >10 | 0–700 | >1,905 | 256–644 |
| Scrubbers§ | Cyclone | | | >10 | 40–700 | >1,905 | 256–644 |
| | Impingement | | | >5 | 40–700 | >1,905 | 256–644 |
| | Dynamic | | | >10 | 40–700 | >1,905 | 256–644 |
| | Fog | | | <2 | 40–700 | >190 | 256–644 |
| | Pebble bed | | | >5 | 40–700 | >190 | 256–644 |
| | Multidynamic | | | <1 | 40–700 | >190 | 256–644 |
| | Venturi | | | <2 | 40–700 | >190 | 256–644 |
| | Submerged nozzle | | | >2 | 40–700 | >190 | 256–644 |
| | Jet | | | <5 | 40–700 | >190 | 256–644 |
| Incinerators Afterburners | Direct | | Gases, vapors, malodors | Any | 2000 | Combustible | 1367 |
| | Catalytic | | | (Molecular) | 1000 | any | 811 |
| Gas absorbers | Spray tower | | | (Molecular) | 40–100 | >1.9 | 278–311 |
| | Packed column | | | (Molecular) | 40–100 | >1.9 | 278–311 |
| | Fiber cell | | | (Molecular) | 40–100 | >1.9 | 278–311 |
| Gas adsorbers | Deep bed | | | (Molecular) | 0–100 | >1.9 | 256–311 |

SOURCE: Jorgensen, "Fan Engineering," Buffalo Forge Co. By permission.
*Based on std. air @ 0.075 lb/ft³ (1.2 kg/m³).
†Bacteria, radioactive, or highly toxic fumes.
‡500°F for glass (553 K).
§Reheating of scrubbed stack is necessary to avoid plumes.

regeneration, and equipment decontamination is treated by various means, such as filtration, waste concentration, and deionization, is monitored for acceptable MPC limits and is then discharged or recycled for in-plant uses, depending upon its quality. Pressurized water reactors can be characterized by three major liquid streams. The first of these consists of borated, relatively high radioactivity demineralized water from the primary cooling loop. During the feed-and-bleed operating mode, borated water is removed and replaced with fresh water, which slowly reduces the amount of chemical shim (boric acid) present. A boric acid evaporator makes it possible to recover boric acid and to recycle a large portion of the distillate. The second major stream consists of floor drains and equipment drains from the containment vessel and the auxiliary building and from the radiochemistry laboratory, which go to a large **waste holdup tank.** These are subsequently transferred to an evaporator, and the distillate therefrom is deionized, passed through a radiation monitor, and then either recycled or released to the discharge canal. With both types of reactor, the heavy ends, crud, and dirt concentrate in the evaporator feed tank. This waste is pumped to the radwaste **drumming station,** where it is solidified with Portland cement or chemical grout. The third of the major streams is soapy water from laundry and showers. Because of foaming difficulties, this stream is not always processed through an evaporator, but instead is filtered or put through a reverse osmosis unit, monitored for radioactivity, and if satisfactory, released into the discharge canal.

**Solids** Solid wastes for BWRs and PWRs are handled similarly. Evaporator and filter sludges are pumped to the radwaste drumming station and solidified with Portland cement or chemical grout in 55-gal (or larger) drums. Spent ion-exchange resins that are not regenerated are sent to an accumulation tank for decay and are then flushed into a shipping cask and dewatered for removal to a burial ground. They may also be solidified. Balable wastes are drummed but, if highly radioactive, are set in concrete. Large, heavy, junk equipment items can sometimes be stored on-site more economically than cutting them up and packaging them for off-site shipment.

**Environmental Releases** For routine releases of radioactivity into the environment to meet the "as low as practicable" limitation, the maximum off-site dose to an individual from liquid effluents is 5 mrem per year under Appendix I of 10CFR-50, with a 10-mrem gamma whole body and a 20-mrem beta dose from gaseous effluents also being allowed.

New plants are capable of meeting these criteria. With an older plant, there may be cases where the radwaste systems will have to be augmented. In these cases, a cost benefit analysis, based upon the worth of a person-rem in terms of abated population degradation, should be made to determine which augments, if any, are economically justifiable. Exemplary data from the final environmental statement for Appendix I (Wash 1258) list worth factors per person-rem which vary from a low of $12 to a high of $990.

Reference to current federal and trade publications must be made to keep abreast of the latest developments.

## SOLID WASTE HANDLING

REFERENCES: "Pollution Control Technology, Solid Waste Disposal," Research & Educational Assn., New York. Schwing and Puntenney, "Denver Plan: Recycle Sludge to Feed Farms," *Water and Wastes Engineering*, p. 24, Sept. 1974.

**Solid waste** includes solids discarded permanently, or temporarily, and materials suspended in air or water. Also forming part of this group are wet solids with insufficient liquid content to be free-flowing. The solid waste produced in the United States, excluding sewage, reached 200 million tons per year in 1975, and could reach 250 and 475 million tons respectively in the 1980s and 2000s. The increasing accumula-

tion of dry waste which results from the high standard of living of modern society will deplete existing economically exploited natural resources and further aggravate the already costly waste disposal problem. Product and energy recovery, therefore, constitute the preferable direction to follow rather than elaborate means of disposal. There is, however, an uncontrollable growth factor for solid residuals which is sheltered by the present environmental control legislation: unpredictable amounts of dry waste generated by air and water pollution control. Environmental regulations have so far been oriented toward zero discharge levels for contaminants in all pollution control areas, regardless of their interfaces. As a consequence, proposed solutions of air and water problems simply transfer the pollution from those media to the solid state.

Federal environmental legislation in this area consists of "The Solid Disposal Act," Title II of Public Law 89-272 (1965), the Amendments of 1970, and the Federal Environmental Pesticide Control Act of 1972. Further regulations will stem from a proposed "Hazardous Waste Management Act" (1974). Additional federal regulations are published in the *Federal Register* (for example, "Thermal Processing and Land Disposal of Solid Waste," **39**:158, parts 240 and 241).

### Solid Waste Disposal

**Solid waste handling** involves two main steps: collection and disposal. **Solid waste conditioning** involves recovering those materials of value already present while conditioning the rest for conversion, recycling, or ultimate disposal. **Solid waste segregation** prior to collection into garbage, combustible wastes, and noncombustible wastes, is practiced in some communities as a beneficial preconditioning step.

The principal solid disposal methods used are **incineration, landfill,** and **ocean disposal.** Many mining and processing operations leave a process residue on the premises since disposal cost elsewhere is prohibitive. Examples are red mud from refining bauxite and sulfite sludge from $SO_2$ stack-gas scrubbing processes.

**Sanitary landfilling** requires spreading and compacting the waste and covering it with soil. This system is applicable both to biodegradable and nondegradable types of wastes. Careful attention must be given to possible subsidence, leaching-contamination of the ground water and gas formation in landfilling. Topsoil disposal is limited to biodegradable organic materials.

**Ocean dumping** is regulated and takes advantage of the ocean-bottom trench configuration. Industrial wastes are disposed at sea in large amounts in containers, in bales, or in bulk form. The cost benefit of ultimate ocean disposal is quite favorable compared with other disposal alternatives, but must be compatible with ecological considerations.

**Incineration** is a volume reduction process rather than solid waste disposal, resulting in a residue which still must be ultimately disposed of. Heating values of solid wastes range from 5,000 Btu/lb for ordinary trash and for industrial wastes to about 15,000 Btu/lb. The recent growth of incineration is affected by higher fuel costs, construction costs, and added expense for air pollution control.

Another means of solid waste reduction is **composting,** not widely used because of the limited market for its products, but ultimately necessary to maintain a rural-urban ecological balance.

## Product Recovery

**The average solid waste** produced in the United States on a 100-lb daily basis is roughly 6 lb industrial (from process), 5 lb municipal, and 89 lb mine tailings, smelter slags, dredging spoil, and agricultural wastes. Materials salvaged from solid waste recycled back to their sources not only reduce disposal costs but also conserve farmland values and natural resources. Recovery of metal cans and paper are outstanding examples of materials recycling.

**Materials separation** from solid waste may be performed by physical or chemical means. The most common types of equipment used are material separators (varies with application: magnetic, thermal, flotation, size reduction—shredders, crushers, grinders, and pulverizers), materials flow (conveyors, lifters), mixers, and blenders. (See Sec. 10.)

**Basic chemical conversion processes** for industrial waste are hydrogenation, oxidation, acetylation, cross-linking cellulosic materials. For more detail, see Engdahl, "Solid Waste Processing—A State of the Art Report on Unit Operations and Processes," Public Health Service Publication No. 1856, Supt. of Documents, Washington, D.C.

**Biochemical processing** of wastes from the food and chemical industries is common practice. An example is recovery of B-complex vitamins and animal feed from solid brewery wastes.

**Pyrolysis** of organic wastes can economically utilize them as supplementary fuels depending upon waste types and waste energy conversion processes. Because of fuel shortages, new efforts are being made in that direction to process municipal refuse as well as industrial waste.

A breakthrough in **dry sludge production** from municipal sewage has been achieved with the development of activated sludge processes using high-purity oxygen. These processes reduce the volume of activated sludge by two-thirds and improve its dewatering characteristics. In Denver, for example, it is now economically feasible to recycle dried sludge to the neighboring farmlands more cheaply than to incinerate it or to haul it away. In this manner, 100 percent of the nutrient values of fertilizer plus all the trace elements can be recycled, and $11.34/dry ton was saved (in 1974) in hauling costs to the disposal site.

Table 14 itemizes the 1974 cost factors at Denver from which cost benefit ratios have been developed. (See Cost/Benefit.) The cost benefit for air-dried sludge is particularly favorable, inasmuch as for every 20¢ invested, the public consumer receives $1.00 in benefits.

## OCCUPATIONAL SAFETY AND HEALTH

REFERENCES: "General Industry Guide for Applying Safety and Health Standards," 29CFR 1910 OSHA, U.S. Dept. of Labor, 1973. 29CFR 1910.1–1910.309, Fed. Register, **37**:202, October 18, 1972. "Accident Prevention Manual for Industrial Operations," National Safety Council. "Manual of Accident Prevention in Construction," Associated General Contractors of America, Inc. American Conference of Governmental Industrial Hygienists, "Industrial Ventilation—A Manual of Recommended Practice." Blake, "Industrial Safety," Prentice-Hall. DeReamer, "Modern Safety Practices," Wiley. Heinrich, "Industrial Accident Prevention," McGraw-Hill. Patty (Ed.), "Industrial Hygiene and Toxicology," Interscience. Simonds and Grimaldi, "Safety Management," Irwin. Stubbs (Ed.), "Handbook of Heavy Construction," McGraw-Hill.

Worker's compensation laws in this country started in 1911 and have now been enacted in all the states. The principle involved is that the worker injured or disabled in industry should be enabled, through proper medical treatment, to return to wage-earning capacity as promptly as possible and, while incapacitated, should receive compensation in lieu of wages, regardless of fault. The expense of medical treatment and compensation should properly be borne by industry and become a part of the cost of its products. The laws generally provide that workers injured in industry shall be furnished the necessary medical treatment, and, in addition, compensation based on a percentage of their average weekly wages, payable periodically. Dependents of employees killed in industry are likewise compensated (see Table 15).

Many states have supplemented their worker's compensation laws by providing comparable benefits in cases of incapacity or death due to occupational disease. Some states make these provisions applicable only in case of specified occupational diseases; others make them of general application to cases of any disease directly attributable to the employment. At the present time, all states and the District of Columbia provide for compensation benefits in occupational-disease cases either by enlarging the scope of the worker's compensation law, by separate legislative enactment, or by judicial construction.

The enactment of worker's compensation laws has been followed in many jurisdictions by more stringent provisions relating to factory inspections for the prevention of accidents in industry and of occupational disease.

The enactment of worker's compensation and occupational-disease laws has increased materially the cost of insurance to

**Table 14. Cost/Benefit Evaluation of Recycled Sewage Sludge as Fertilizer (Denver)***

| | Total cost delivery and application, $/dry ton | Current worth of nutrients, $/dry ton | Disposal cost savings, $/dry ton | Total benefit (worth and savings), $/dry ton | Cost/benefit |
|---|---|---|---|---|---|
| Air-dried (trucked) | $ 4.00 | $ 8.37 | $11.34 | $19.71 | 0.20 |
| Liquid (5% solids) (trucked) | 14.00 | 12.20 | 11.34 | 23.54 | 0.595 |
| Liquid (5% solids) (pipelined) | 9.00 | 12.20 | 11.34 | 23.54 | 0.382 |

*1975 data.

Table 15. Types Of Occupational Accidents, Compensable Injuries

| Source of injury | All disabilities, % | Fatal and permanent total disabilities, % | Permanent partial disabilities, % | Temporary total disabilities, % |
|---|---|---|---|---|
| Handling objects, manual.............. | 22.6 | 13.9 | 9.6 | 28.5 |
| Falls............................... | 20.4 | 17.4 | 18.5 | 21.2 |
| Same level...................... | 10.4 | 4.8 | 9.2 | 11.0 |
| Different level.................... | 10.0 | 12.6 | 9.3 | 10.2 |
| Struck by falling, moving objects....... | 13.6 | 9.3 | 19.3 | 11.1 |
| Machinery........................ | 10.2 | 3.1 | 19.2 | 6.3 |
| Vehicles........................... | 7.1 | 20.7 | 7.1 | 6.9 |
| Motor........................... | 5.0 | 18.0 | 4.3 | 5.2 |
| Other............................ | 2.1 | 2.7 | 2.8 | 1.7 |
| Stepping on, striking against objects... | 6.9 | 2.3 | 5.6 | 7.6 |
| Hand tools........................ | 6.1 | 1.5 | 8.1 | 5.3 |
| Electricity, heat, explosives........... | 2.5 | 7.7 | 2.2 | 2.6 |
| Harmful substances................. | 2.5 | 8.2 | 1.1 | 3.0 |
| Elevators, hoists, conveyors........... | 2.2 | 3.6 | 3.8 | 1.5 |
| Engines, motors.................... | 0.4 | 0.7 | 0.7 | 0.2 |
| Other............................. | 5.5 | 11.6 | 4.8 | 5.8 |
| Total.......................... | 100.0 | 100.0 | 100.0 | 100.0 |

SOURCE: *"Accident Facts,"* 1965 ed., National Safety Council, Chicago, Ill.

industry. The increased cost and the certainty with which it is applied have put a premium on accident-prevention work. This cost can be materially reduced by the installation of safety devices. Experience has shown that approximately 80 percent of all industrial accidents are preventable.

The logical time to install safety devices is when new machines are being built, while general construction work is being done, or when alterations or repairs are being made; results can be accomplished with a minimum of expense and delay at the time plans and specifications are being prepared.

**Checking for Safety** In order to make sure that the question of safety will not be overlooked, it is well to have all plans, specifications, and drawings checked for safety, making special provision for this in each set of specifications and in the title plate of each drawing.

### Buildings[1]

**Plant Arrangement** A fundamental factor in effective accident prevention is the provision of ample ground space in the plant site to reduce crowding of buildings, congestion of plant traffic, unusual fire hazards, unsafe yard conditions, etc. One-story buildings have definite advantages in regard to fire hazards, building collapse, and natural lighting.

Where a plant is located near a main line of a railroad, consideration for safe access by employees and vehicular traffic should be given serious attention. Safe passageways from one building to another are also important. Blind corners, doorways opening onto yard railway tracks, etc., should be avoided as much as possible, and, where such conditions must necessarily exist, safety railings, gates, signs, or gongs should be installed to warn pedestrians of danger.

Because of their inherent hazards of fire and explosion, the storage, handling, and utilization of *flammable liquids* should be carefully controlled. The National Fire Codes, promulgated by the NFPA, are considered authoritative in establishing minimum safeguards necessary to control these hazards. It is necessary in any event to conform to the requirements of any applicable state and local code.

[1]See 29CFR 1910, Subparts D, E, F.

Buildings in which **dusty operations** are carried on should be so designed as to present a minimum area of projections, ledges, and resting places for dust accumulations.

**Floors, Stairways, Aisles, etc.** Probably the most important factor to be considered in connection with floors, stairways, etc., concerns **slipperiness**. Floors and stairs should be free from projecting nails, boltheads, etc., as noiseless as possible, wear well, and be strong enough to carry safely any static or moving load. The weight of modern industrial machinery and material handling equipment should be carefully considered in checking floor-load calculations. Floors and stairways should be kept clear of unnecessary obstructions over which workers may trip.

**Spilling** of oil, water, acid, etc., should be prevented to eliminate slipping hazards. Splash guards, drip collectors, etc., can be designed in many instances to reduce slippage. Excessive spillage of dusty materials onto the floor is sometimes taken care of by installing floor gratings beneath which pits or conveyor systems are located to collect the falling material.

**Ample aisle space** is very important, especially in foundries where workers carry ladles of molten metal and where there is considerable shop traffic involving power-driven trucks, etc. Aisles should be clearly marked off to assist in keeping them clear. One-way traffic is often advantageous.

Stairways should be provided with **handrails** on both sides, and an intermediate middle handrail should be installed on stairways over 88 in in width. Nonslip treads on stairs are desirable. Stairways should be adequately lighted (see Sec. 12).

**Exits and Fire Escapes** (see Life Safety Code NFPA) As far as practicable, all doors should open outward or with the natural direction of egress; they must not block passageways from other floors or parts of the building.

For factories, not less than two means of exit should be provided on every floor, including basements, of all buildings or sections; these exits should be separated in such a manner that they are not liable to be cut off by a single local fire. The location of stairways around or adjacent to passenger elevators is undesirable unless there is separation by fire walls.

**Outside fire escapes** are inferior to stairways as means of egress. Where they are used, they should be located on blank walls or arranged so that persons on them will be protected from flames issuing from windows or openings underneath by use of wired glass in standard metal frames, fire doors, etc. Outside fire escapes, to provide maximum protection for persons using them during fire, should be enclosed in noncombustible towers which will protect against weather, smoke, or fire, with access through or over some intermediate balcony or structure to the building proper.

**Lighting** (see Sec. 12)    Adequate light has a definite bearing on the prevention of accidents. Workrooms should be well lighted to reduce eye strain and the possibility of permanent eye impairment and also to remove any danger of employees falling over obstructions or being caught in machinery in darkened areas.

**Ventilation** (see Sec. 12)    A lack of adequate ventilation in a workroom tends to bring on fatigue and reduces the alertness of workmen thus making them more susceptible to accidents. Where injurious dusts or noxious vapors are encountered, it is necessary to provide for their removal by the installation of adequate local exhaust systems (see Sec. 12) and ACGIH "Industrial Ventilation" and ASA Standard, "Fundamentals Governing the Design and Operation of Local Exhaust Systems").

**Identification of Piping** (see Sec. 8)    It is desirable that a plan of identifying the contents of various pipelines be adopted so that in case of an emergency it will be possible to determine quickly the service of all pipelines involved.

### Mechanical Guarding in Machine Design[2]

The most logical time at which to consider the safeguarding of a machine is during its design. In this stage, features of safe operation can be incorporated so that there will be a minimum of specific guarding required on the finished machine. The following points are fundamental considerations concerning accident prevention which should be taken into account in machine design.

Care should be taken in arranging clearances of moving parts to avoid shearing or crushing points in which hands or other parts of operator's body might be caught or injured.

Arrangements should be made so that adjustments, inspections, and hand lubrications can be done safely.

Machines should be so designed that operators are not required to stand in an uncomfortable position, reach over moving parts, or exert themselves in awkward positions.

Machines should be designed so that there will be little danger of the operator tripping over parts of the frame or striking against projecting parts during normal operation movements.

Careful attention should be given to strength of all parts whose failure might result in injury to operator.

All guards, covers, or enclosures should be designed strong enough to prevent the possibility of their giving way and permitting an accident in case the operator should fall or be thrown against them.

**Point-of-Operation Guarding**    The point of operation on a machine is taken to be that zone where the work of the machine is actually performed and where the operator, by manipulating the material being processed, is exposed to a hazard from moving parts of the machine. Guards for point-of-operation protection are placed on the machines as additional equipment. The first requirement for a successful guard of this type is that it shall be convenient and not interfere with the operator's movement or affect the output of the machine. The following statements describe several basic principles which may be utilized in point-of-operation guarding.

Where possible, the **danger point** should be completely **covered** by a barrier or enclosure before the hazardous operation of the machine begins. This may be accomplished for example on a treadle-controlled machine by having the treadle operate the guard which when in proper location will in turn operate the machine.

Operator's hands may be kept out of dangerous positions by installing **starting devices** so located that, to operate the machines, both hands of the operator must be out of the danger zone.

**Feeding devices** may be used so that material to be processed is placed in a feeding mechanism at a point where there is no exposure to moving parts. Special holders or feeding tongs can also be used to place work in hazardous positions.

**Electronic controls,** which operate by the interruption of a light beam or other energy source to protect the danger zone, may be used to start and stop machines.

**Electrical interlocks** on guarding devices may be utilized in operating circuits so that unless a guard is in proper position the circuit is open and no current will flow until the machine is safely protected by the guard being brought into proper position.

**Automation** has minimized the hazards associated with the manual handling of stock and has eliminated the need for repetitive exposure at the point of operation, but the urgency of making repairs or adjustments introduces the need for special precautions covering maintenance work, such as locking out the power source.

### Electrical Equipment[3]

In considering electrical equipment for industrial establishments, it should be borne in mind that the **hazard** to human life **increases with increase in voltage,** and economy in transmission wiring and copper parts, which is achieved through the use of high-voltage motors, may be offset by increased danger of accident. For small motors, lights, and general service inside industrial plants, installations of 110 or 220 V are recommended.

All **switches, fuse boxes, terminals, starting rheostats, motors,** etc., located within 8 ft of a floor or working platform, should be enclosed or guarded in such a manner as will prevent accidental contact with live parts, irrespective of voltage. **Switches** should be arranged so that they can be locked in the open position, to guard against a switch being thrown in accidentally while workers are at work on the lines or equipment that the switch controls. **Equipment for operation at 550 V and higher** should be isolated from other operating equipment, in separate rooms or enclosures, with provision for locking up these enclosures. **All metallic cases, frames,** and **supports** of such equipment should be permanently grounded; foundation bolts should not be depended upon for this purpose, but substantial ground conductors should be used. It is preferable to have these ground wires accessible for inspection. **All low-voltage**

---

[2]29CFR 1910, Subpart 0.

[3]See 29CFR 1910, Subpart S.

secondary circuits of 300 V or less (such as light, motor, and meter circuits) should be permanently grounded whenever a neutral point is available for the purpose. Secondary circuits of 250 V or less should be permanently grounded even though a neutral point is not available. Whenever grounding has to be omitted, the frame or casing of the apparatus should be permanently grounded, and all live parts of the secondary circuit should be shielded to prevent accidental contact therewith. It is also desirable to have floors or platforms adjacent to switchboards built of nonconducting material, or suitably insulated. In the absence of such construction, rubber mats may be used.

Portable electric lamps, tools, and machines are subjected to severe operating conditions. Many electric-shock fatalities have occurred on electric portables, even with 110-V lighting circuits. Cable used to service portable equipment should be high-grade rubber-covered cord designed for the service, not ordinary twisted lamp cord. Its mechanical attachment to both the portable and the attachment plug or source of power should be designed to prevent sharp bending or chafing that would break down the insulation. All portable lamps should have nonmetallic sockets (such as rubber, composition, or porcelain). The heavier portable tools and machines cannot practically have their metallic parts insulated from contact by the operator, and protection from shock due to accidental charge thereon must be provided by means of a special grounding wire. Such protective grounding should be in the form of an additional wire in the cable that feeds the portable, one end permanently connected to the machine frame and the other connecting to ground through an additional pole in both the attachment plug and the receptacle to which the device is to be attached.

### Occupational-Disease Prevention[4]

In any industrial operation where a toxic material is being processed in such a manner that those persons engaged in or working near the operation are exposed to appreciable quantities of dusts, fumes, vapor, or gas, it is important that adequate control measures be adopted. The following statements cover the major considerations involved in the application of effective control to industrial occupational disease.

Contaminants (See 29CFR 1910, 93.) The physical and chemical characteristics of a contaminant should be known. In the case of dusts or fumes, the chemical nature, particle size, solubility, etc., should be determined. For gases or vapors, the composition, vapor pressure, flash point, etc., are important factors. In all atmospheric contamination, the quantity of material in the worker's breathing zone must be known before the degree of hazard can be evaluated. The chemical characteristics are important in the selection of materials to be used in the construction of any control equipment where corrosion, etc., might be factors.

A careful investigation should be made to determine accurately the sources from which the contaminant is being produced or from which it is being dispersed. The most common types of dust-producing operations are crushing, screening, grinding, polishing, etc. Dispersion of dust is encountered in practically all dry-handling operations of fine materials. Vapors and fumes are produced by chemical processes and reactions and are most commonly found in connection with the use of solvents.

[4]See 29CFR 1910, Subpart G.

A number of testing instruments and procedures have been developed for determining quantitatively the concentrations or amounts of various toxic material in the atmosphere of a work room (see ACGIH, "Air Sampling Instruments" and reports of U.S. Public Health Service and U.S. Bureau of Mines; see also AIHA Hygienic Guide Series and ACGIH, "Annual Threshold Limits"). In general, the seriousness of any exposure involving a health hazard is directly proportional to the dosage (concentration and length of time of the exposure). Engineering control should be directed to the reduction of these two factors.

Local Ventilation at the Source of Contaminant Removal by means of exhaust hoods, enclosures, etc., so located as to prevent the escape into occupied areas of any appreciable amount of contaminant at its source is the most effective method of control. For details of hood design, piping, collectors, fan characteristics, and for general principles of exhaust design system, see Sec. 12 and ANSI Standard, "Fundamentals Governing the Design and Operation of Local Exhaust Systems"; AFS, "Engineering Manual for Control of In-plant Environment in Foundries."

Natural ventilation has a limited application in industrial occupational-disease control. It is important with a natural-ventilation system to maintain close supervision of adjustments as required by changes in temperature, wind directions, etc. Regular tests of air content should be made to check on the degree of dilution being obtained. This method of control is not recommended for exposures where severe hazards exist owing to extreme toxicity or high concentrations.

Isolation Under some circumstances, the isolation of a hazardous operation, physically, or in point of time, is indicated. For example, a hazardous operation may be carried on in a separated room in which all contamination can be confined or it may be carried on outside of regular working hours when no one except the persons engaged in the operation will be present. By isolating an operation, the number of persons exposed to any accompanying hazard may be reduced to a minimum.

Process Revision The possibility of substituting a less toxic material should be borne in mind; e.g., high-boiling petroleum naphtha for benzol in rubber cements, thinners, etc.; dolomite lime for quartz in foundry-parting compounds. It may be possible to reduce the concentrations of objectionable contaminants and the time of exposure by changing handling operations, by substituting mechanical feeds or conveyors for manual operations, or by installing automatic machinery or controls so as to dispense with the presence of an attendant. An automatic temperature- or pressure-control device, in connection with a chemical process, may reduce materially the time of an operator's exposure as well as prevent the production of excess or undesirable fumes or gases. An enclosed feed-conveyor system handling dry material and automatically weighing it may be utilized to eliminate a severe exposure that would result from hand shoveling.

Physical Hazards In addition to exposures to toxic gases, vapors, dusts, fumes, etc., there are several physical conditions such as abnormal pressures, temperatures, and humidities as well as radiation (including ultraviolet; infrared; x-ray; $\alpha$, $\beta$, and $\gamma$ rays from radioactive substances; and radiation from handling radioactive isotopes), all of which may, in case of excessive exposure, prove detrimental to the health of exposed persons. Evaluation depends on physical methods of measurement. Protective measures include shielding from radiation and control of exposure time rather than ventilation, as in the case of atmospheric contaminations.

**Ambient Noise Control** (See also Sec. 12) Outer property noise propagation is regulated by EPA, under provisions of the Noise Control Act of 1972: Section 6 applies to new-product noise emission standards; Section 17 is concerned with noise emission regulations for railroads and Section 18 for motor carriers. EPA is directed to publish noise emission guidelines in the Federal Register. The regulations for air-craft-airports are promulgated by the Federal Aviation Administration (FAA) consistent with Section 611 of the Federal Aviation Act of 1958 and guidelines proposed by EPA.

**Occupational Noise Exposure** Mandatory protection of employees against noise is required by 29CFR1910.95 as follows:

(*a*) Protection against the effects of noise exposure shall be provided when the sound levels exceed those shown in Table 16 when measured on the A scale of a standard sound-level meter at slow response. When noise levels are determined by octave band analysis, the equivalent A-weighted sound level may be determined from Fig. 3. Octave band sound pressure levels may be converted to the equivalent A-weighted sound level by plotting them on this graph and noting the A-weighted sound level corresponding to the point of highest penetration into the sound-level contours. This equivalent A-weighted sound level, which may differ from the actual A-weighted sound level of the noise, is used to determine exposure limits from Table 16.

(*b*) (1) When employees are subjected to sound levels exceeding those listed in Table 16, feasible administrative or engineering controls shall be utilized. If such controls fail to reduce sound levels within the levels of Table 16, personal protective equipment shall be provided and used to reduce sound levels within the levels of the table. (2) If the variations in noise level involve maxima at intervals of 1 s or less, it is to be considered continuous. (3) In all cases where the sound levels exceed the values shown herein, a continuing, effective hearing conservation program shall be administered.

When the daily noise exposure is composed of two or more

**Table 16. Permissible Noise Exposure***

| Duration per day, h | Sound level, slow response, dBA |
|---|---|
| 8 | 90 |
| 6 | 92 |
| 4 | 95 |
| 3 | 97 |
| 2 | 100 |
| 1½ | 102 |
| 1 | 105 |
| ½ | 110 |
| ¼ or less | 115 |

*29CFR 1910.95.

periods of noise exposure of different levels, their combined effect should be considered, rather than the individual effect of each. If the sum of the fractions $C_1/T_1 + C_2/T_2 + \cdots\cdots + C_n/T_n$ exceeds unity, then the mixed exposure should be considered to exceed the limit value. $C_n$ indicates the total time of exposure at a specified noise level, and $T_n$ indicates the total time of exposure permitted at that level. Exposure to impulsive or impact noise should not exceed 140 dB peak sound-pressure level.

**Personal Respiratory Protection** (See 29CFR1910, 134) Where it is not practicable to control air contamination in the breathing zone of workers by adequate exhaust ventilation, etc., and it is necessary for personnel to be exposed to harmful amounts of dust, smoke, fumes, or gases or to work in an atmosphere with a deficiency of oxygen, personal respiratory equipment should be provided. Such equipment is not, in general, suited for prolonged daily use because of inherent discomfort and inconvenience to the wearer. For emergency or temporary situations or until effective control of contamination can be developed and applied, personal respiratory protection should be given very careful consideration where harmful exposures are encountered (see "The Respirator Manual," prepared jointly by the American Industrial Hygiene Association and the American Conference of Governmental Industrial Hygienists).

The U.S. Bureau of Mines (Schedules 19 and 21) has developed certain standards for approval of respiratory protection devices. The equipment may be (1) **supplied-air respirators** (hose type), devices that supply clean, respirable air to the wearer through a hose line extending from a source outside the contaminated zone; (2) **supplied-oxygen respirators** (self-contained type of oxygen-breathing apparatus), devices that supply oxygen to the wearer from a source of supply that he or she carries as part of the respirator; or (3) **filter-type respirators,** which may be chemical filter respirators that remove the harmful constituents from air passing through them by chemical reactions, absorption, and adsorption, including ordinary gas masks and chemical respirators; or which may be mechanical-filter respirators that remove the harmful constituents from air by mechanical filtration, including dust, mist, or fume respirators.

**Safety Codes and Published Material. OSHA** The American National Standards Institute (ANSI) publishes a series of

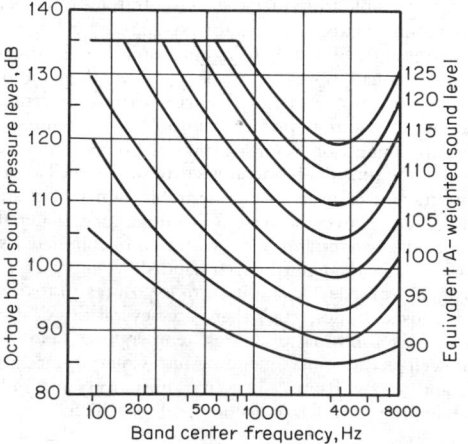

**Fig. 3** Equivalent sound-level contours.

ANSI Standards. State and local ordinances must be complied with as required.

The National Safety Council publishes a series of Industrial Data Sheets and Safe Practice Pamphlets covering the safeguarding of industrial operations, in addition to literature, posters, films, and other safety educational material. The annual publication "Accident Facts" reports current trends in accident experience on an overall basis.

United States government agencies, such as the Bureau of Mines, the Bureau of Labor Statistics, the National Bureau of Standards, and the Division of Labor Standards publish statistical information concerning industrial-accident occurrence and accident prevention. The National Fire Protection Association (NFPA) promulgates fire codes and publishes information of all types relating to fire protection.

**The Williams Steiger Act of 1970** created the **Federal Occupational Safety and Health Administration** (OSHA), which has now codified and published blanket standards for national enforcement in all public safety and health areas. Updated revisions will continue to appear in the Federal Register under Title 29 CFR1910 and 1518, and these standards make all preexisting standards obsolete.

**Radiation Health Physics** (OSHA: Code of Federal Regulations, Title 29, 29CFR1910.96, Occupational Exposure and Health Standards)    **Radiation,** which accompanies all nuclear reactions, can be advantageous or disadvantageous, depending generally on the manner in which contact is made and on the amount of energy involved. In its application to humans, the common terminology is **radiation dose,** which is measured in units of **roentgen (r), rad,** and **rem.** Each of these units is applicable to a different type of dose: (1) roentgen, to exposure dose; (2) rad, to absorbed dose; and (3) rem, to dose equivalent (DE), formerly called **RBE dose.** The roentgen is a physical unit applicable only to x- and gamma-ray radiation. A measurement in roentgens is a determination of the ionization produced in air under specified conditions. The rad, also a physical unit, is a measure of the energy absorbed from radiation per unit mass of irradiated material. Unlike the roentgen, the rad is applicable to all types of ionizing radiation. A roentgen (of x- or gamma rays), however, is not biologically equivalent to a rad of other types of radiation; neither is a rad of one type of radiation biologically equivalent to a rad of a different type of radiation. Therefore, a dose in rads must have certain modifying factors applied to it in order to relate it to biological effect. For radiation-protection purposes, use is made of **dose equivalent,** which is defined as the product of absorbed dose (in rads), quality factor (QF), dose distribution factor (DF), and other necessary modifying factors. The unit of dose equivalent is the rem. For purposes of radiation protection, any of the following external doses may be considered to be equivalent to a dose of 1 rem: (1) 1 R of x- or gamma-ray radiation; (2) 1 rad of x- , gamma-, or beta-ray radiation; (3) 0.1 rad of neutrons or high-energy protons; and (4) 0.05 rad of particles heavier than protons and with sufficient energy to reach the lens of the eye.

**Exposure** to radiation may be divided into two general types: (1) external exposure, which is that resulting from radiation sources external to the body; and (2) internal exposure, which is that resulting from radionuclides within the body. Radiation-protection standards are established for both types of exposure. Limits for external exposure are given in units of rem (or rad) per unit time. Basic limits are usually for a control period of ¼ year and for 1 year. Local operational rules may specify control limits for shorter exposure times (day, week, month), such limits being based on the basic limits. Radiation-protection standards for controlling internal exposures are based on the same basic limits as those for external exposures. Controls, however, are effected by use of maximum-permissible-concentration (MPC) values established for each of the various radionuclides, for both air and water. Many factors (physical, chemical, biological, and physiological) are involved in determining the MPC of a radionuclide. As such, the MPCs of the radionuclides have a very wide range of values. MPCs are normally expressed in units of microcuries, $\mu$ci, per $cm^3$ of air (or water). Generally, MPC values are based on a 40-h work week for radiation workers and on a 168-h week for the general population.

**Monitoring** is the periodic or continuous determination of the presence and extent of ionizing radiation and radioactive contamination. It provides information for radiation protection and normally is assigned to a health physics group, which establishes schedules, executes or supervises the operation, and evaluates their results. It is divided into these special types: personnel, area, source, surface, air, and water.

**Regulations** governing exposure to radiation are promulgated in the form of AEC Regulations: Title 10, Chapter 1, Part 20. Reference to the current full statement of Part 20 is mandatory for a full understanding of the regulations. As a general guide, material and equipment must be shielded and the exposure time limited so that no worker will receive a radiation dose during a calendar quarter in excess of 1¼ rem on whole body, head and trunk, active blood-forming organs, lenses of eyes, or gonads; 18¾ rems on hands and forearms, feet and ankles; and 7½ rem on skin of whole body. **Additional regulations** governing all forms of occupational radiation exposure are found in the U.S. Department of Labor Ionizing Radiation and Health Standards (29CFR1910.96(r)), issued under authority of Sections 6 and 8(g) of the Williams Steiger (OSHA) Act of 1970. Nuclear power plant safety and accident prevention are basically mechanical design criteria. (See also Sec. 9.)

## FIRE PROTECTION

REFERENCES: "Fire Protection Handbook," "National Fire Codes," "Fire Inspection Manual," National Fire Protection Assn. "Handbook of Industrial Loss Prevention," McGraw-Hill. "Uniform Building Code 1976," International Conference of Building Officials, Building Offic. and Code Admin. International (BOCA), 1975, Fire Insurance Assn. (FIA), Factory Mutual (FM), Code of Federal Regulations: 29CFR1518, July 1976, and 29CFR1910, July 1976, Jan. 28, 1977.

### Importance of Fire Protection (29CFR1910, Subpart L)

The profitable use or availability for use of facilities is the aim of all business, whether industrial, mercantile, professional, scientific, or educational. Destruction of or damage to the facilities cripples the attainment of that purpose.

Monetary compensation for destroyed property and for resulting lost profits is obtained by the purchase of insurance. Reputation, goodwill, and other intangible factors may be irreparably damaged. Loss of life is not recoverable. Loss by fire is largely an avoidable, nonproductive tax resulting from carelessness, ignorance, apathy, or incompetence. The annual fire loss in the United States, (refer to NFPA) has increased steadily since 1964 at about 8 to 12 percent per year, reflecting the population-economic growth and inflation. It amounted to $3.1 billion in 1973. The number of losses has exceeded 2 million. Building fires, 13 percent of which are attributable to smoking, have made up about 80 percent of the **amount** of loss but only about 40 percent of the **number.**

**Fire-loss Prevention** Neither insurance organizations nor legally established regulatory bodies assume the responsibility of industry for conserving its own resources and operating its facilities safely, although they may detect deficiencies or a lack of compliance with regulations and may serve in a consulting capacity. Fire-loss prevention is an indispensable element in industry and business. It exists only with top management direction and the support of labor.

The designation **fire protection** usually encompasses the entire field of **prevention of loss by fire,** including both the causes for the occurrence of fires and methods for minimizing their consequences. Included along with fire are other destructive agencies such as explosion, lightning, electric current, wind, earthquake, nuclear excursions, and radioactive contamination. OSHA and building codes mandate minimum standards of protection to prevent injury and loss of life. Higher standards to protect investment are a worthwhile management option.

**Fire brigades** (29CFR1910.164) are essential to the development and maintenance of an effective fire protection program at every job site. By prompt, immediate response and notification of the local fire department, every effort should be made to bring the fire quickly under control during the early minutes of an outbreak. The immediate availability of the correct fire protection and suppression equipment is essential.

**Fire protection engineering** involves the application of sound engineering principles to the reduction of loss by fire and related hazards. The Society of Fire Protection Engineers has established a well-defined scope for fire-protection engineering practice, both in building design and in safe operating practices.

**Limitation of Loss** With large businesses it is difficult to limit single destructive occurrences to an amount that will not be catastrophic to the owner or to others having a financial interest. Principles and methods for preventing large losses by fire are well known to fire protection engineers.

**Sources of Fire-Protection Information** Much specific, detailed, and technical information is given in the standards, codes, and rules of the NFPA, Factory Mutual Engineering Division, Factory Insurance Assoc., American Insurance Association, ANSI, Underwriters Laboratories Inc., and may also be found in the procedures of many insurance companies and inspection bureaus.

The NFPA "Handbook of Fire Protection" and the Factory Mutual "Handbook of Industrial Loss Prevention" are especially comprehensive references. The most complete and readily available source of standards is the "National Fire Codes," prepared and published by the NFPA, currently in ten volumes as follows: 1—Flammable Liquids; 2—Gases: 3—Combustible Solids, Dusts, and Explosives; 4—Building Construction and Facilities; 5—Electrical; 6—Sprinklers, Fire Pumps and Water Tanks; 7—Alarms and Special Extinguishing Systems; 8—Portable and Manual Fire-control Equipment; 9—Occupancy Standards and Process Hazards; 10—Transportation. Many of the standards have been adopted by insurance inspection and advisory groups, such as the American Insurance Association, and by Federal, state and munici-

pal regulatory bodies and have been incorporated into building codes and other regulations. The Occupational Safety and Health Administration, (OSHA) U.S. Dept. of Labor, is promulgating blanket national standards for fire protection, which appear in the Federal Register under Title 29, Code of Federal Regulations for the working place, and Part 1518 during construction. Where applicable, these supersede and make obsolete all existing standards.

### Construction

Building construction and the protection of buildings and their contents against fire are frequently governed by local building codes and ordinances and by insurance standards. When new construction or changes are planned, the property owner should have the advice of a competent fire-protection engineer and should consult local authorities and insurance carriers to avoid delay and the possibility of expensive changes later.

Some construction features which have increasing importance in fire-loss prevention are (1) the trend toward large fire areas which present high values to possible loss and make manual fire-fighting difficult; (2) blank wall, air-conditioned, and artifically lighted buildings which interfere with ready access; and (3) mechanization of materials handling, resulting in conveyor systems that interconnect areas and in larger storage areas and high-piled stock which make fires difficult to extinguish. These cause increased cost of fire protection and emphasize the need for automatic fire-control methods, sources of dependable water supplies, and special extinguishing systems. A measure that should be adopted where practical is the use of outdoor, totally enclosed process equipment, with shelters provided only for control rooms, laboratories, and some maintenance functions.

**Types of Construction** **Fire-resistive** refers to types that withstand considerable fire without serious damage, such as reinforced concrete or protected steel. **Noncombustible** refers to any construction that contains no elements of burnable material but which may be structurally damaged by fire, such as unprotected metal. **Combustible** means structures entirely of combustible materials or having combustible elements of such character and distribution that a fire can spread and contribute fuel so that severe damage results. Combustible types of construction are frequently subdivided into (1) "heavy timber," also called "plank on timber," "mill" or "slow burning" this has masonry walls with floors and roof of plank on heavy timbers; (2) "ordinary construction" this has masonry walls with floors and roof made of boards on joists and is sometimes called "quick burning"; (3) "wood frame" this has all of its elements of wood, except that the exterior may be surfaced with a noncombustible sheathing.

Many types of construction have composite elements that may include combustible materials. Insulation, acoustic materials, and surface treatments may aid the start and contribute to the rapid spread of fire. A roof of interlocking metal sheets, with asphalt on its upper surface as part of a vapor barrier or as an adhesive for insulation or weather surfacing, can, if initially heated by a local interior exposing fire, furnish gaseous fuel through the joints to produce a spreading damaging fire. Owing to the difficulty and cost of protection, enclosed spaces at roofs, ceilings, in walls, and below floors should be avoided. Important steel structural members and steel supports for heavy equipment that may be exposed to severe fire should have heat-insulating protection. Economic tradeoffs, based on the probability of exposure, should be made.

**Fire Performance of Building Materials** The kind and amount of fire protection needed are governed by the fire performance of materials used in construction and by the combustibility of building contents. Completely non-combustible construction avoids the need for sprinkler protection, provided that contents are also certain to be non-combustible. Knowledge of the fire performance of materials, based upon tests made by qualified laboratories, is essential before they are adopted (see Sec. 6). Fire performance is in two categories, fire resistance and fire hazard.

**Fire resistance** measures the susceptibility of materials to damage by exposure to fire and is usually measured as the time period of exposure, without significant damage, to a standard fire exposure as specified by standard fire tests, such as "ASTM E-119 Fire Tests of Building Construction and Materials," and expressed as a **Fire-Resistance Classification** in hours as, ½ hr, 2 hr, 6 hr, etc.

**Fire-hazard** indicators are also in two generally recognized categories used by the Underwriters Laboratories. The **listing** of specific appliances and materials indicates that the products comply with the Laboratories' test requirements, with particular reference to fire-preventive and fire-protective capabilities when the manufacturer's instructions and limitations of use have been followed. The **Fire-Hazard Classification** of materials and products is related mainly to burning characteristics. It has a numerical basis determined by (1) flame spread, (2) fuel contributed, (3) smoke developed. A Fire-hazard Classification of 0 indicates a material with a fire performance equivalent to that of asbestos-cement board. A value of 100 corresponds to the behavior of red oak lumber. Building codes and other regulations generally prescribe limiting values for Fire-hazard Classifications.

**Protection against Exposure Fires** Protection of buildings or other structures against fires in nearby property must sometimes be provided. A practical barrier against a conflagration is the presence of fire-resistive buildings along the exposed side. Usually the most severe exposure is localized, and protection against it may be provided by a blank brick or concrete wall, by wired-glass metal-frame windows, or by open sprinklers alone or in combination. The relative value of various safeguards against exposure fires has been estimated to be roughly blank brick or concrete walls, 100; tin-clad shutters, 60; wired-glass windows, metal frames, glass block (4 in. minimum thickness), 40; wired-glass windows, wooden frames, 20; plain glass windows, 5; additional value of open sprinklers with any of the foregoing, except blank fire walls, 30; open windows, 0.

**Horizontal Cutoffs** Provide effective fire walls between important buildings, and subdivide large building areas in order to limit the probable maximum damage from a single fire.

**Vertical Cutoffs** Enclose stairs, elevator wells, vertical conveyors, chutes, and other floor openings with adequate fire-resistive walls, having fire doors or their equivalent at openings to prevent the rapid spread of fire and heat upward from floor to floor.

**Isolation of Hazards** Cut off hazardous occupancies by fire-resistive partitions or fire walls or, if the degree of hazard warrants, isolate them in separate buildings. Make provision for adequate ventilation and for explosion vents where needed.

**Floor Leakage** Provide drained, watertight floors over large values susceptible to water damage.

**Storehouses and Vaults** Provide suitable storehouses for large quantities of combustible raw stock, tools, and patterns, and for valuable finished goods. Provide reliable vaults for the storage of business records and valuable drawings.

### Safeguards during Building Construction
(29CFR 1518, Subpart F)

A building under construction is more vulnerable to fire than the same building completed and in use with normal fire protection. As construction progresses, concentrations of readily combustible materials appear at new locations from day to day. So do many potential ignition sources—temporary heaters, welders' torches, riveters' forges, roofers' tar kettles, and workers' discarded matches or cigarettes. If fire starts, wall and floor openings create draft and help flames to spread rapidly. Many serious construction fires have taken place. Most can be prevented. Guard against cutting and welding hazards by proper supervision of operations. Plan in advance for effective notification of, and coordinated firefighting operations with, any available public fire department. (See 29CFR 1518.24.)

Start installation of the fire-protective equipment—including water supplies, yard mains, and hydrants—as soon as construction of the building starts. Make layouts for sprinklers as soon as building drawings are completed, and let contracts without delay. It is sometimes possible to arrange for limited temporary protection by providing connections for hose from the piping system which furnishes water for construction uses. Have hose and nozzles available as soon as the yard piping, hydrants, and water supplies are ready. Distribute ample hand extinguishing equipment throughout the premises, including contractor's temporary buildings.

**Temporary Heating** Arrange any needed heat safely. Use coke as fuel for salamanders, rather than wood trash or rubbish. Oil-fired heaters, if well-designed and properly located, are safer. Unit heaters are still better and are recommended if an adequate steam supply is available. Locate salamanders or oil heaters away from woodwork or tarpaulins. Keep the floor clean and free from combustible material. Flame-proofed, water-repellent tarpaulins are available, and their use is advised. The tarpaulin-salamander-wooden-form combination has caused many serious fires, particularly in reinforced concrete construction work during freezing weather. Scaffolding constructed of metal, or of fire-retardant treated wood, largely avoids a serious fire hazard. Locate construction sheds a safe distance from a building under erection, never inside.

**Temporary Storage** Keep combustible storage out of buildings under construction, as far as possible and at least 10 ft (3.05 m) from structures when stored outside. Under no condition erect canvas tents or wooden shelters inside. A safe location and good fire protection are important for valuable equipment or machinery, delivered before the building is ready for its installation.

### Automatic Sprinklers (29CFR1910.159)

**Advantages of Automatic Sprinklers** Automatic sprinklers are the most dependable and effective means of fire protection. The advantages are (1) automatic sprinklers go into operation soon after a fire starts, and before it has gained dangerous proportions; (2) the automatic sprinklers that open are over, and in the vicinity of, the fire; (3) fires at all locations, including out-of-the-way places, are controlled as effectively as fires easily seen and reached; (4) sprinklers operate where fire fighters could not enter or would be driven out by heat or smoke; (5) sprinklers are always ready; (6) only sprinklers that are needed to control a fire operate, so that water is used to better advantage than from hose streams.

**Where Sprinklers Are Needed** Automatic sprinklers are needed for complete protection where there is an appreciable amount of combustible material in either building construction or in building contents, e.g., (1) throughout buildings having floors or roofs of sufficient combustibility to contribute to a spreading fire, whether or not the contents are combustible; (2) in wholly noncombustible buildings where contents are combustible, including the storage or use of flammable liquids; (3) in concealed spaces, such as attics and under low roofs, if they contain any combustibles, including heat insulation and exposed electrical wiring; (4) in vacant spaces beneath a combustible first floor of a building, unless the floor is completely tight, the space sealed against entry, and no possible sources of ignition are present; (5) in dryers, large ducts, closets, and small offices, unless there is no fuel for a fire in either the construction or the contents; (6) under wide storage shelves or work tables over 4 ft in width having any combustible stock, and under canopies over platforms where combustible materials may be present.

**Sprinklers over Electrical Equipment** If the construction is fire-resistive or adequately fireproofed, sprinklers can generally be omitted in electric-generator rooms and over switchboards. If voltages exceed 600, it is generally better to replace combustible roofs or ceilings with noncombustible construction or to protect all exposed woodwork with metal lath and cement plaster, rather than to install sprinklers. Metal shields or hoods may be used to protect generators, switchboards, or other important electrical equipment from possible water damage.

**Automatic-Sprinkler Equipment** The responsibility for the design of automatic-sprinkler systems should be given only to experienced and responsible parties. Approval of preliminary layouts and working plans is usually required by municipal or insurance inspection bureaus before installation is started. Sprinkler installation is a well-established trade by itself. Standards for the Installation of Sprinkler Systems are given in complete detail in a publication under that name by the National Fire Protection Assoc.

**Dry-Pipe Systems** The ordinary wet-pipe sprinkler system cannot be used where subject to temperatures below freezing. A dry-pipe system must be substituted. In the dry-pipe system the sprinkler piping contains air under pressure instead of water. When a sprinkler is opened by fire, the air pressure falls and water is admitted automatically by the operation of a dry-pipe valve.

**Special Sprinkler Systems** The deluge system is a special type of sprinkler equipment frequently used in hazardous occupancies, such as airplane hangars and storage areas for flammable liquids or materials, where a flash fire could spread before the regular automatic sprinklers became operative, and where the prompt discharge of a large amount of water over a considerable area is needed. The deluge system has sprinkler heads with the fusible element removed. Water is controlled by a quick-opening valve (deluge valve) operated by heat-sensitive elements distributed over the area protected. Piping and heads are arranged as in a standard sprinkler system, with larger pipe sizes.

**Preaction systems** are actuated by heat-sensitive devices, smoke detectors, or ionization detectors, which trip a controlling valve and admit water to the otherwise dry sprinkler piping before the regular closed automatic-sprinkler heads operate. The system has the advantage of giving an alarm before sprinklers operate. Accidental opening of automatic sprinklers or mechanical breakage of piping does not result in the discharge of water.

**Deluge** and other systems actuated by heat-sensitive or other special devices introduce additional mechanical equipment and lose the simplicity of the standard automatic-sprinkler system. Some additional skilled maintenance is required. Various arrangements for mechanically supervising the condition of such equipment can be provided.

**Nonfreeze sprinkler** systems are small and are sometimes used in relatively unimportant locations where it is impracticable to provide heat. They consist of special piping connections to wet-pipe systems, with a nonfreezing solution in the exposed piping. Glycerin is acceptable for this purpose. They should be limited to systems with less than 20 heads. It is better to provide heat, to connect to existing dry-pipe systems, or to provide small dry-pipe valves. Nonfreeze systems are preferable to shutting off the sprinklers and draining the piping during the winter. No portion of an automatic-sprinkler system connected to a public water supply should be filled with a nonfreeze liquid without determining that the arrangement will comply with water department health regulations.

**Outside sprinklers** are used for protection against fire from outside sources. Windows in outside walls present the most frequent problem. Outside sprinklers may be operated either manually or automatically. The effectiveness of these systems depends on the water-supply valve opening in time. They can sometimes be adapted to automatic control by a thermostatically operated deluge valve.

**Types of Automatic-Sprinkler Heads** Approved automatic-sprinkler heads are well standardized as to the rate of water discharge, pattern of discharge, and temperature operating characteristics. The essential elements of an automatic sprinkler are a nozzle, a closing device that is released at a definite temperature, and a deflector that produces the desired distribution pattern. Temperature ratings appropriate for the location must be selected. The ratings for normal atmospheric temperatures, not substantially above 100°F, are 135 to 165°F. Ratings for other ceiling temperature ranges are as follows:

| Max ceiling temp | | Sprinkler rating | |
|---|---|---|---|
| °F | K | °F | K |
| 101–150 | 311–338 | 175–212 | 352–373 |
| 151–225 | 339–380 | 250–286 | 394–414 |
| 226–300 | 381–422 | 325–360 | 436–455 |

Heads of higher rating are obtainable for ambient temperatures up to 500° F. Corrosion-resistant heads, usually wax-coated, are obtainable for use in corrosive atmospheres.

**Sprinkler-Flow Alarms** The flow of water from a sprinkler system may be made to sound an alarm which gives warning of both a fire and accidental leakage from the system. This alarm may be either a hydraulic gong or an electric bell, or both. If the property has no constant attendance, or if added security is desired, central-station supervisory service is available in many industrial centers. This service transmits water-flow alarms to public fire departments and can perform numerous other supervisory functions. Preaction alarms give advance notice of hazard.

**Location and Spacing of Automatic Sprinklers**  As the purpose of an automatic-sprinkler system is to protect both building and contents, the water-distribution pattern and the quantity of water applied must be held within close limits. Sprinklers customarily are installed so that the discharge from one head is allowed for each 60 to 130 ft² of floor area. The distance between adjacent heads should not exceed 15 ft. Spacing averages 80 to 120 ft² per head with 8 to 12 ft between heads.

Recent developments in sprinkler-system design for the more severe hazards produced by high piled combustible goods or flammable liquids are leading to standards including a **discharge density** in terms of gallons per minute per square foot of floor area. In this consideration building contents, rather than building construction, are the determining factor. Densities called for have commonly been in a range of 0.15 to 0.3 gal/min per ft² of floor area. High-piled storage of combustible materials, such as rolled paper, imposes a severe hazard which may call for a water density of 0.6 gal/min per ft², or more. NFPA standards (1973) call for intermediate levels of sprinkler heads in high rack storage areas. Open and deluge sprinklers require hydraulically designed systems based on friction loss instead of standard pipe schedules (see NFPA No. 13, No. 13a, and No. 231c).

The number of heads, to be supplied through pipes of standard sizes, with ordinary hazards and reasonably uniform water discharge, is as follows:

| Size of pipe, in | 1 | 1¼ | 1½ | 2 | 2½ | 3 | 3½ | 4 | 5 | 6 | 8 |
|---|---|---|---|---|---|---|---|---|---|---|---|
| Max no. of sprinklers | 2 | 3 | 5 | 10 | 20 | 40 | 65 | 100 | 160 | 275 | 400 |

Not more than eight sprinklers should be on one branch line on either side of a cross main. For deluge systems, with which all heads are open and must be supplied simultaneously, and for especially hazardous occupancies where a large proportion of the automatic heads are expected to operate, larger pipe sizes are required as follows:

| Size of pipe, in | 1 | 1¼ | 1½ | 2 | 2½ | 3 | 3½ | 4 | 5 | 6 |
|---|---|---|---|---|---|---|---|---|---|---|
| Max no. of sprinklers | 1 | 2 | 5 | 8 | 15 | 27 | 40 | 55 | 90 | 150 |

## Water Supplies for Fire Protection

(See Sec. 3 for hydraulics, Sec. 8 for piping, Sec. 9 for engines, Sec. 14 for pumps)

Insurance companies or their inspection and engineering bureaus should be consulted concerning water-supply requirements for fire service. In addition to providing water for automatic sprinklers, the supply should be adequate for hydrants and hose streams. Common water supplies are public water systems, gravity tanks or private reservoirs, and fire pumps taking suction from above-ground suction tanks, rivers, or ponds. Two or more independent supplies are usually needed for larger properties. Strong, dependable public water systems alone are adequate only for small plants, of good construction, having safe occupancies, and not dangerously exposed.

The "primary" water supply automatically maintains pressure on the property fire system at all times. The "secondary" supply supplements it as needed. The **primary supply** should not be less than 500 gpm available at a pressure that will be effective for automatic sprinklers at the highest plant elevation. A **secondary water supply** serves several purposes. The most important usually is to provide more water, frequently at higher pressure than is available from the constant primary supply. A secondary supply is also of value in maintaining protection at any time the primary supply may be interrupted due to public supply deficiencies or to the necessity of taking tanks or reservoirs out of service for repairs and maintenance.

High water demands for special hazard protection, especially where large areas must have deluge-system protection, can seldom be met by connections to public water systems. Fire pumps and suction reservoirs can give the added volume of water at desirably high pressure for the expected duration of the total demand, which is made up of the anticipated sprinkler demand and an allowance for the number of hose streams likely to be used. Large airplane hangars, large building areas containing flammable liquids, or very extensive properties that could have a general fire, approaching conflagration proportions, can have an estimated total demand of 5,000 gal/min or even more.

**Gravity Water Tanks**  The smallest gravity tank on a tower considered advisable for fire protection is 25,000 gal which may be suitable for the protection of small properties. Water from tanks of such limited size should be reserved for automatic sprinklers only. Tanks of 50,000 to 100,000 gal capacity may provide either the primary or secondary supplies for large properties of moderate hazard. Tanks combining fire service and domestic or industrial supplies in a single structure are allowable, provided that the amount needed for the fire supply cannot be withdrawn into the industrial-use system.

Gravity tanks should be at such elevation that the bottom of the tank will be at least 35 ft above the highest sprinkler to be supplied. The bottom of large tanks, intended to provide water for hose streams as well as sprinklers, should be at least 75 ft, and preferably 100 ft, above ground level.

**Pressure Tanks**  Hydropneumatic steel tanks can be used for primary water supplies for demands of short duration, such as needed to bring automatic fire pumps to full operation. Under some conditions they can be used to facilitate the automatic control of the automatic fire pumps by providing a rapid drop in the control pressure. High cost and limited capacity deter the general use of pressure tanks.

**Private Fire Pumps**  Well-located fire pumps with ample suction supplies and capable of maintaining high pressure over a long period provide a very satisfactory secondary supply. Centrifugal fire pumps of approved design are in common use. Practically all new installations use centrifugal pumps driven by electric motors, gasoline or diesel engines, or steam turbines. The electric-motor-driven centrifugal pump is most desirable because of simplicity of control and operation. Gasoline-engine drives are used to a limited extent where other reliable sources of power are not available, or as auxiliary units in connection with other pumps.

Adequate suction and reliable power supply are essential for fire pumps. The suction supply should be sufficient to operate the pump at rated capacity for at least 1½ h at the smaller plants, and an inexhaustible suction supply is desirable for good protection of the largest industrial properties.

An automatically controlled, electrically driven centrifugal fire pump is sometimes used as a booster pump when the public water is of good volume but too low in pressure for direct use in sprinkler systems.

Fire pumps are used in capacities of 500, 750, 1,000, 1,500, and occasionally 2,000 or 2,500 gal/min. The 750 and 1,000 gal/min sizes are most common.

**Fire-Department Connections** For properties where the public water supply is at comparatively low pressure but adequate in quantity, fire-department connections to the plant fire-service piping can serve as a valuable auxiliary for pumpers to deliver water into the fire system at high pressures. In city properties, particularly, fire-department connections are usually a part of the sprinkler system.

**Underground Water Piping for Fire Service**

Complete specifications or standards covering all of the main features of underground piping for fire service are given in NFPA National Fire Codes, vol. 6, "Sprinklers, Fire Pumps, and Water Tanks," and the Factory Mutual "Handbook of Industrial Loss Prevention." The "Handbook of Cast-Iron Pipe," published by the Cast-Iron Pipe Research Association, contains much useful information.

**Underground Mains** Buried underground piping should be so located that hydrants and control valves are at a safe distance from possible falling building walls. A complete loop system around buildings or groups of buildings, with multiple supplies preferably connected at opposite sides and with prudently located division valves, affords the best hydraulic characteristics for heavy flows and for freedom from impairments. Pipe trenches should allow careful laying and uniform support for the pipe. Foreign materials should be kept out of the pipe. Pipe should be anchored at bends, dead ends, and branch connections. The system should be hydraulically tested for 2 h at a minimum pressure of 200 lb/in² or at least 50 lb/in² in excess of the maximum static pressure if it is to be more than 150 lb/in². Leakage at joints must be less than specified amounts. Underground piping should be thoroughly flushed before it is connected to indoor piping. Pipe should be buried well below the deepest frost penetration as shown by weather charts. In the coldest areas a cover of at least 5 ft is necessary. In the southern states 2½ ft may be adequate, except that under roadways a greater cover may be needed for protection against traffic loads. Clearance to prevent breakage by settlement is needed at foundation walls.

**Soil Corrosion** Rapid external corrosion of cast-iron pipe may be expected if the mains pass under coal piles, through cinder fill, or where pickling liquors, acids, alkalies, or salts penetrate the soil. Cast-iron pipe covered with heavy coating of asphalt, or Transite pipe should be used to offset soil corrosion in such locations. Backfill should be of clean sand or gravel. (See Sec. 6.)

**Type of Pipe** Pipe for underground use is usually cast iron or asbestos-cement. Ordinarily the working water-pressure rating does not need to exceed 150 lb/in². Representative specifications for pipe suitable for fire service are: "Specifications for Cast-Iron Pit-cast Pipe" (ANSI A21.2 or AWWA C102), "Specifications for Cast-Iron Pipe Centrifugally Cast in Sand-Lined Molds" (ANSI 21.8 or AWWA C108); asbestos-cement pipe should conform to "Tentative Standard Specifications for Asbestos-Cement Water Pipe" (AWWA C400) or "Commodity Federal Specification SS-P-55/A." Specific approval from insurance companies or inspection bureaus is usually required for the use of asbestos-cement pipe (see Sec. 8).

**Outside Hydrants and Hose**

**Location of Hydrants** Where space permits, outdoor hydrants and hose provide a supplement to automatic sprinklers and afford means for fighting fires in combustible yard storage, in railroad cars and vehicles, and small combustible unsprinkled sheds. Recommended space between hydrants varies from 150 to 300 ft, depending upon the type of buildings and the character of outdoor combustibles. Hose for outdoor use, 2½ in in size, is made of woven cotton or modern synthetic fiber and rubber-lined.

**Standpipes and Inside Hose** (29CFR 1910.158)

Standpipe systems furnish the best means of obtaining effective fire streams in the upper stories of buildings. They are designed for small hose streams used by the occupants and for large hose streams to be used by public or plant fire departments. Outlets may be designed to supply both. In buildings of unusual height, standpipes are sometimes supplied by a series of fire pumps and tanks at different elevations.

Small hose is of particular value in areas with hazardous occupancies, such as opener, picker, spinning, and woodworking rooms. Hose should be 1½-in cotton, rubber-lined, or unlined linen, with ⅜- or ½-in nozzles. Spray-type nozzles are usually best for the hazards mentioned. The water supply should preferably be taken from a connection independent of the sprinkler piping so that hose streams will be available when the sprinklers are shut off after a fire or during sprinkler repairs or changes. Small hose, for fire service only, in locations of ordinary hazards, may be connected to wet-pipe sprinkler systems, but in no case should hose connections be made to sprinkler lines smaller than 2½ in diam. (See Sec. 8.)

**Special Forms of Fire Protection**

Special types of protection are adapted to the control of unusual hazards, such as flammable liquids. Equipment should be secured from makers specializing in the form of protection required, but such use does not supplant general building protection by automatic sprinklers.

**Water Spray** A dense strong spray of water from suitably designed nozzles is effective for controlling fires in flammable liquids of moderate hazard, for unusually flammable solid materials, and for surface fires of ordinary combustible materials. Such a system may be most appropriate for the protection of transformers and other oil-filled electric equipment and for systems handling fuel or lubricating oil under pressure. The entire zone to be protected must be within reach of the strong spray. Water pressures of 50 lb/in² prevail with properly designed equipment. (See Sec. 3.)

**Foam** Foams for fire control provide a blanketing action to exclude air and in some cases to give useful insulating effect. Foams are commonly designated as "chemical foam" and "air foam," depending upon their service or method of production. For flammable-liquid fires, foam is available for manual or automatic application with a selection of characteristics appropriate for a wide variety of conditions. The amount of foam varies from ½ ft³ to several cubic feet per square foot of surface protected. Rate of application and means of distribution largely determine its effectiveness. A sprinkler system, designated as "Foam-water," provides an initial discharge of very fluid foam backed up by a sprinkler discharge of water. Spe-

**Table 17. Portable Extinguishers by Hazard Classification**

| Extinguisher type | Class A general purpose | Class B flammable liquids | Class C electrical fires | Class D combustible metal |
|---|:---:|:---:|:---:|:---:|
| Soda-acid | x | | | |
| Foam | x | | | |
| Loaded stream | x | x | | |
| Dry chemical | | x | x | |
| CO₂ (plastic horn) | | x | x | |
| Vaporizing liquid | | x | x | |
| Dry powder | | | | x |

SOURCE: NFPA Handbook, Vol. 8.

cial **alcohol-type** foam is needed for application to alcohols, alcohol-type liquids, and organic solvents, all of which seriously break down the commonly used foams. **High-expansion** foams are receiving increasing attention and use. They are easily produced in a **high-expansion foam generator** by blowing air through a screen wet with a continuous spray of water having a bubble-producing additive. The foam is very light and fluid and can be applied to fill completely and quickly a room or other enclosed area of considerable size. One gallon of foam-producing liquid can form as much as 1,000 gal of foam. Ordinary foam-producing liquids yield only about 10 gal of foam per gal of liquid.

**Carbon Dioxide** (29CFR 1910.161)  For (1) flammable liquids; (2) electric equipment, such as large enclosed electric generators; and (3) hazards where a space-filling effect by an inert atmosphere is needed, or where a nonwetting extinguishing agent is desired. Carbon dioxide for fire extinguishing is available in small cylinders for manual use; or in banks of large cylinders, or in refrigerated storage tanks for piped-extinguishing systems.

**Dry chemical** (29CFR 1910.160) extinguishers and extinguishing systems are mainly used for flammable liquids and electrical fires. They are effective also on surface fires in combustible fibers. Multi-purpose dry-chemical extinguishers have special ingredients which make them suitable for fires in ordinary combustibles.

**Portable Hand Extinguishers** (29 CFR 1910.157)

**Hand-operated extinguishers** and small hose are effective when employees are on hand to attack fires immediately after discovery. They are frequently used to put out the final vestiges of fires brought under control by automatic sprinklers. Common types of hand extinguishers are the fire pail and water; the hand pump tank; and extinguishers called soda-acid, nonfreeze, foam, carbon dioxide, dry chemical, and vaporizing liquid. Dry compounds, applied by shovel or scoop from bulk containers, are available for fires in combustible metals, such as magnesium, powdered aluminum, zirconium, sodium, and potassium. Grouped by hazards, the hand extinguishers are selected from four classifications, as shown by Table 17.

# Refrigeration, Cryogenics, Optics, Patents, Trademarks, and Copyrights, and Miscellaneous

BY

**FRED P. ROSLYN**  *Consulting Engineer.*

**K. D. WILLIAMSON, JR.**  *Associate Group Leader—Controlled Thermonuclear Research (Cryogenics), Los Alamos Scientific Laboratory.*

**F. J. EDESKUTY**  *Associate Group Leader—Cryogenics, Los Alamos Scientific Laboratory.*

**F. G. BRICKWEDDE**  *Evan Pugh Research Professor Emeritus of Physics, The Pennsylvania State University.*

**HEARD K. BAUMEISTER**  *Senior Engineer, International Business Machines Corporation.*

**EDWARD TAYLOR NEWTON**  *Counsellor at Law; Senior Partner, Newton, Hopkins and Ormsby, Atlanta, Georgia.*

### MECHANICAL REFRIGERATION
#### by Fred P. Roslyn

Refrigeration Machines and Processes .................... 19-3
Properties of Refrigerants ............................... 19-3
Overall Cycles ......................................... 19-6
Components of Compression Systems .................... 19-7
Absorption Systems .................................... 19-11
Air Machines .......................................... 19-13
Thermoelectric Cooling ................................ 19-13
Methods of Applying Refrigeration .................... 19-13
Refrigerant Piping ..................................... 19-15
Cold Storage .......................................... 19-16
Ice Making ............................................ 19-21
Skating Rinks.......................................... 19-21
Refrigeration in the Chemical Industries ............... 19-21
Deep Refrigeration .................................... 19-22

### CRYOGENICS
#### by K. D. Williamson, Jr., F. J. Edeskuty, and F. G. Brickwedde

Notation ............................................... 19-23
Refrigeration Methods ................................. 19-23
Gas Liquefaction ...................................... 19-25
Applications of Cryogenics ............................ 19-26
Properties of Solids at Low Temperatures ............. 19-26
Properties of Cryogenic Fluids (Cryogens) ........... 19-31
Instrumentation ....................................... 19-33
Insulation ............................................. 19-35
Safety ................................................ 19-39

### OPTICS
#### by Heard K. Baumeister

Index of Refraction .................................... 19-40

Dispersion .......................................... 19-40
Refraction .......................................... 19-40
Aperture ........................................... 19-40

## PATENTS, TRADEMARKS, AND COPYRIGHTS
### by Edward Taylor Newton

Patents for Inventions ................................ 19-41
Trademarks ......................................... 19-43

Copyrights .......................................... 19-44

## MISCELLANEOUS

Sizes of Type ....................................... 19-45
Copy Preparation and Copy Fitting .................... 19-45
Proofreading ........................................ 19-45
Proofreaders' Marks ................................. 29-45
Special Alphabets ................................... 19-46

# MECHANICAL REFRIGERATION
## by Fred P. Roslyn

REFERENCES: Publications of the ASRE and ASHRAE; "Air-conditioning Refrigerating Data Books"; *Refrigerating Engineering;* ARI Equipment Standards; ASHRAE Guide and Data Book. Merkel and Bosnjakovic, "Diagrammen und Tabellen zur Berechnung der Absorptionskältemaschinen," Springer, Berlin. Jordan and Priester, "Refrigeration and Air Conditioning," Prentice-Hall.

## REFRIGERATION MACHINES AND PROCESSES

(For general theory of refrigeration, see Sec. 4; for air-conditioning uses, see Sec. 12; for cryogenics, see Sec. 19.)

Refrigeration is a special aspect of heat transfer and involves the production and utilization of below-atmospheric temperatures by a number of practical processes. Substances are cooled when their heat is transferred, via a temperature drop, to solid, liquid, or gaseous media which are naturally or artificially colder, their lower temperature stemming from radiation, sensible- or latent-heat physical effects, or endothermic chemical, thermoelectric, or even magnetic effects. Effects, such as cold streams, melting ice, and sublimating solid carbon dioxide, are included.

Of particular practical use is the achievement of lowered temperature for a circulating fluid (**refrigerant**) on a continuous basis, with heat-absorption capability, by (1) expansive flow of a fluid through a pressure drop in a restriction ("expansion valve") *without* production of external work, i.e., throttling (Joule-Thomson) effect, and (2) expansive flow of a fluid (usually a gas) through a pressure drop in a machine (expander) *with* production of external work.

The integrated systems for continuous cold effect are best recognized through the similarity of format in their closed flow circuits, particularly in (1) vapor-compression and (2) gas-compression systems. Both these systems contain a high-level and a low-level heat exchanger and two pressure difference devices, a pressure generator and a pressure-reducing element, typical of energy-transport systems. In the vapor-compression system the heat exchangers are characterized by latent-heat effects; in the gas compression, by sensible-heat effects primarily. Although some cases may entail discharge of some circulant at low pressure and low temperature with subsequent fresh make-up, the closed systems are more usual. Generally, the low-level-exchanger refrigeration effect is achieved through energy input, such as work input to a compressor.

With continuous recirculation of the cycle fluid, the low-temperature heat pickup must be pumped, via energy input, to a higher level so as to discharge it to atmospheric temperature. In this sense the system acts as an **energy-transport** device or **heat pump** or, from another viewpoint, as a temperature transformer. This philosophy of energy transport may equally well be applied to more remote systems, such as thermoelectric. It is, of course, a manifestation of the principles of the Second Law, and the effectiveness of the system's operation may be evaluated numerically.

**Coefficient of performance** (see Sec. 4) is the ratio of useful refrigeration to the work supplied, either directly or indirectly, as energy capable of doing work. The ratio of the coefficient of performance of any actual cycle to that of the Carnot cycle operating between the same two temperature levels is called the **relative efficiency** of the cycle. Relative efficiencies must always be less than unity, but coefficients of performance can be severalfold greater than one.

**Refrigeration capacity** is also of interest; the capability of a system is defined in terms of an arbitrary measure of capacity, the "ton."

**Units of Refrigeration** In the United States, a **standard ton of refrigeration** corresponds to a heat absorption at a rate of 288,000 Btu/day or 200 Btu/min. The heat absorption per day is approximately the heat of fusion of 1 ton of ice at 32°F. The **standard rating** of a refrigerating machine, using a condensable vapor, is the number of standard tons of refrigeration it can produce under the following conditions: (1) liquid only enters the expansion valve and vapor only enters the compressor of a compression-type unit or the absorber of an absorption system; (2) the liquid entering the expansion valve is subcooled 9°F and the vapor entering the compressor or absorber is superheated 9°F, these temperatures to be measured within 10 ft of the compressor cylinder or absorber; (3) the pressure at the compressor or absorber inlet corresponds to a saturation temperature of 5°F; and (4) the pressure at the compressor or absorber outlet corresponds to a saturation temperature of 86°F.

The **British unit of refrigeration** corresponds to a heat-absorption rate of 237.6 Btu/min with inlet and outlet pressures corresponding to saturation temperatures of 23 and 59°F, respectively. On the European continent a unit of refrigeration capacity called the **frigorie** is used. A frigorie is approximately equivalent to 50 Btu/min, or ¼ of a standard ton of refrigeration.

## PROPERTIES OF REFRIGERANTS

(For detailed properties, see Sec. 4)

**Refrigerants** are the transport fluids which convey the heat energy from the low-temperature level to the high-temperature level, where it can, in terms of heat transfer, give up its heat. In the broad sense, gases involved in liquefaction processes or in gas-compression cycles go through low-temperature phases and hence may be termed "refrigerants," in a way similar to the more conventional vapor-compression fluids.

According to ANSI Code B79.1-1968, refrigerants are **designated by number**, to wit, "5.1.3 The identifying number of the refrigerant, or the word 'Refrigerant' or both, may be preceded by the manufacturer's trademark or trade name"; also, "5.1.4 Examples are as follows: 'Isotron' 12 or 'Isotron' Refrigerant 12; 'Genetron' 12 or 'Genetron' Refrigerant 12; 'Freon' 12 or 'Freon' Refrigerant 12 or 'Freon' 12 Refrigerant or 'F-12' Refrigerant."

One of the attributes that must be considered for vapor-compression systems is the **normal boiling point** since this is of concern in the selection of a fluid which will be above atmospheric pressure on the low side and hence free from the threat of inward air leaks. Table 1 shows some common fluids in order of ascending boiling point; included are halocarbon compounds, such as hitherto named "freons," hydrocarbons and sulphur compounds, and inorganic compounds.

The desirable thermal properties of a refrigerant are (1) convenient evaporation and condensation pressures, (2) high critical and low freezing temperatures, (3) high latent heat of evaporation and high vapor specific heat, (4) low viscosity and high film heat conductivity. Desirable practical properties include (1) low cost, (2) chemical and physical inertness under operating conditions, (3) noncorrosiveness toward ordinary construction materials, and (4) low explosive hazard both alone and mixed with air. The refrigerant should be nonpoisonous and nonirritating and should not cause deterioration in the lubricant used. Leakage should be detectable by simple tests, easily performed.

The *specific volume of refrigerant* to be handled is important with reciprocating compressors, as it determines the size of the compressor; but with centrifugal compression a large volume is not objectionable and may be a positive advantage for small units. A large compression ratio is undesirable in reciprocating compressors from the standpoint of clearance losses and may make the use of compound compression necessary.

A comparison of various refrigerants based on *ideal performance* is given in Table 2. The results are for the **standard temperature range** of 5 to 86°F and also for certain other ranges. For these calculations, zero piston clearance and expansion through a throttle valve were assumed. The use of an expansion cylinder instead of a valve would have yielded work (available for supplying some of the work of compression), but this is always a negligible quantity, except when the compression pressure approaches the critical pressure, as for carbon dioxide. Theoretical *coefficients of performance* may be obtained by dividing 4.72 by the theoretical horsepower given in the table.

**Ammonia** is used extensively in large installations, industrial and commercial. It is toxic and because of corrosive action must be kept out of contact with copper or copper-bearing alloys. It has a high latent heat of evaporation and convenient pressure-specific volume relations. It is not miscible to any large extent with lubricating oil. **Carbon dioxide** was used for a long time as a safety refrigerant; exposure to it in a confined space is not dangerous unless concentrations are high. With cooling water at 70°F, the condensing pressure of $CO_2$ is high, and since its critical temperature is 87.8°F condensation will not occur at water temperatures above this. Power consumption is high on $CO_2$ machines. **Sulphur dioxide** has rapidly gone out of use even in the small household units where it was previously widely employed. It is extremely corrosive unless absolutely anhydrous and although not dangerous in the quantities used in a household unit (2 to 3 lb), it may be dangerous in larger installations. It is extremely irritating even in small amounts. **Methyl chloride** ($CH_3Cl$), an anesthetic in amounts of 5 to 10 percent by volume, has been used in air-cooled units of either moderate or small sizes. It is miscible with minerals oils; small amounts of moisture in a methyl chloride system will cause trouble by freezing in expansion valves. The series of fluoro-chloro hydrocarbons known as **Freons** or **Genetrons** are increasingly used both in small and medium-sized units. Many compounds of this type are commercially produced; those which are nontoxic, nonirritating, and nonflammable are commonly employed. They include **F-11**, trichloromonofluo-

### Table 1. Selected Refrigerants and Gases in Low-Temperature Applications

| Refrigerant | Refrigerant No. | Symbol | Molecular weight | B.P. deg F | Critical point Temp, deg F | Critical point Press, psia |
|---|---|---|---|---|---|---|
| Helium | 704 | He | 4.0 | −452 | −450 | 33 |
| Hydrogen | 702 | $H_2$ | 2.0 | −423 | −400 | 188 |
| Nitrogen | 728 | $N_2$ | 28.0 | −320 | −232 | 492 |
| Air | 729 | ......... | 29.0 | −312 | −220.3 | 547 |
| Oxygen | 732 | $O_2$ | 32.0 | −297 | −182 | 730 |
| Methane | 50 | $CH_4$ | 16.0 | −258.9 | −115.8 | 673 |
| Carbon tetrafluoride | 14 | $CF_4$ | 88.0 | −198.4 | −49.9 | 542 |
| Ethylene | 1,150 | $C_2H_4$ | 28.0 | −155.0 | 48.8 | 732 |
| Ethane | 170 | $C_2H_6$ | 30.0 | −127.5 | 90.1 | 708 |
| Nitrous oxide | 744A | $N_2O$ | 44.0 | −127.0 | 96.5 | 1,050 |
| Monochlorotrifluoromethane | 13 | $CClF_3$ | 104.5 | −114.6 | 83.9 | 561 |
| Carbon dioxide | 744 | $CO_2$ | 44.0 | −109.3 | 87.8 | 1,071 |
| Propane | 290 | $C_3H_8$ | 44.1 | −44.2 | 202 | 661 |
| Monochlorodifluoromethane | 22 | $CHClF_2$ | 86.5 | −41.4 | 204.8 | 716 |
| Ammonia | 717 | $NH_3$ | 17.0 | −28.0 | 271.4 | 1,657 |
| Dichlorodifluoromethane | 12 | $CCl_2F_2$ | 120.9 | −21.6 | 233.6 | 597 |
| Methylchloride | 40 | $CH_3Cl$ | 50.5 | −10.8 | 289.4 | 969 |
| Isobutane | 601 | $C_4H_{10}$ | 58.1 | 10.3 | 272.7 | 537 |
| Sulphur dioxide | 764 | $SO_2$ | 64.1 | 14.0 | 314.8 | 1,142 |
| Butane | 600 | $C_4H_{10}$ | 58.1 | 31.3 | 306 | 550 |
| Dichlorotetrafluoroethane | 114 | $CClF_2\text{-}CClF_2$ | 170.9 | 38.4 | 294.3 | 474 |
| Dichloromonofluoroethane | 21 | $CHCl_2F$ | 102.9 | 48.1 | 353.3 | 750 |
| Trichloromonofluoromethane | 11 | $CCl_3F$ | 137.4 | 74.8 | 388.4 | 635 |
| Ethyl-ether | 610 | $C_4H_{10}O$ | 74.1 | 94.3 | 522.1 | 381 |
| Methylene chloride | 30 | $CH_2Cl_2$ | 84.9 | 103.6 | 480 | 670 |
| Trichlorotrifluoroethane | 113 | $CCl_2F\text{-}CClF_2$ | 187.4 | 117.6 | 417.4 | 495 |
| Water | 718 | $H_2O$ | 18.0 | 212 | 706.1 | 3,226 |

**Table 2. Ideal Performance of Refrigerants for Various Temperature Ranges**

| Refrigerant and (number) | Operating temperature range, deg F | Suction pressure, psia | Head pressure, psia | Ratio of head to suction pressure | With dry and saturated suction vapor, per ton | | | Temperature at end of compression, deg F | Type of compressor |
|---|---|---|---|---|---|---|---|---|---|
| | | | | | Weight of vapor, lb per min | Piston displacement, cfm | Theoretical hp | | |
| Air (729)............... | 5–86 | 14.7 | 73.5 | 5.0 | 7.02 | 82.3 | 2.82 | 277.0 | Recip. and exp. cyl |
| Water (718)............ | 32–86 | 0.0885 | 0.6152 | 6.95 | 0.1957 | 647.0 | 0.618 | 332 | Centrif. |
| | 32–100 | 0.0885 | 0.9492 | 10.73 | 0.1985 | 656.2 | 0.819 | 420 | or |
| | 40–100 | 0.1217 | 0.9492 | 7.80 | 0.1978 | 483.4 | 0.687 | 366 | ejector |
| Carbon dioxide (CO₂) (744)................ | 5–86 | 332.0 | 1,043.0 | 3.14 | 3.61 | 0.960 | 1.827 | 160.3 | Recip. |
| Ammonia (NH₃) (717).... | 5–86 | 34.27 | 169.2 | 4.94 | 0.421 | 3.44 | 0.99 | 209.8 | Recip. |
| | 20–100 | 48.21 | 211.9 | 4.40 | 0.421 | 2.49 | 0.94 | 212.8 | |
| | 40–100 | 73.32 | 211.9 | 2.89 | 0.427 | 1.70 | 0.65 | 176.0 | |
| Freon 11 (CCl₃F) (11).... | 5–86 | 2.931 | 18.28 | 6.20 | 3.058 | 37.0 | 0.94 | 112.7 | Centrif. |
| | 20–100 | 4.342 | 23.60 | 5.44 | 3.086 | 26.2 | 0.89 | 122.1 | |
| | 40–100 | 7.032 | 23.60 | 3.36 | 2.945 | 16.08 | 0.63 | 114.2 | |
| Freon 12 (CCl₂F₂) (12)... | 5–86 | 26.51 | 107.9 | 4.07 | 3.916 | 5.82 | 1.00 | 100.2 | Recip. |
| | 20–100 | 35.75 | 131.6 | 3.68 | 4.054 | 4.54 | 0.97 | 112.5 | or |
| | 40–100 | 51.68 | 131.6 | 2.55 | 3.880 | 3.07 | 0.67 | 108.0 | centrif. |
| Freon 21 (CHCl₂F) (21).. | 5–86 | 5.243 | 31.23 | 5.96 | 2.364 | 20.87 | 0.94* | 99.0* | Rotary |
| | 20–100 | 7.699 | 40.04 | 5.20 | 2.404 | 15.61 | 0.94* | 110.4* | |
| | 40–100 | 12.32 | 40.04 | 3.25 | 2.315 | 10.19 | 0.68* | 105.9* | |
| Freon 22 (CHClF₂) (22).. | 5–86 | 43.02 | 174.5 | 4.06 | 2.926 | 3.65 | 1.03 | 131.7 | Recip. |
| | 20–100 | 57.98 | 212.6 | 3.67 | 3.023 | 2.83 | 0.99 | 152.7 | |
| | 40–100 | 83.72 | 212.6 | 2.54 | 2.936 | 1.93 | 0.68 | 131.0 | |
| Methylene chloride (CH₂Cl₂) (30) | 5–86 | 1.28 | 10.07 | 8.56 | 1.485 | 74.0 | 0.96* | 205.1* | Centrif. |
| | 20–100 | 1.92 | 13.25 | 6.90 | 1.520 | 47.72 | 0.91* | 157.1* | |
| | 40–100 | 3.38 | 13.25 | 3.92 | 1.493 | 27.76 | 0.63* | 167.7* | |
| Methyl chloride (CH₃Cl) (40) | 5–86 | 21.15 | 94.70 | 4.48 | 1.331 | 5.95 | 0.96 | 178.1 | Recip. |
| | 20–100 | 29.16 | 116.7 | 4.00 | 1.363 | 4.51 | 0.90 | 184.4 | |
| | 40–100 | 43.25 | 116.7 | 2.69 | 1.342 | 3.07 | 0.62 | 157.0 | |
| Sulphur dioxide (SO₂) (764) | 5–86 | 11.81 | 66.45 | 5.63 | 1.415 | 9.08 | 0.97 | 191.4 | Recip. |
| | 20–100 | 17.18 | 84.52 | 4.92 | 1.453 | 6.52 | 0.92 | 193.4 | |
| | 40–100 | 27.10 | 84.52 | 3.12 | 1.444 | 4.17 | 0.63 | 162.6 | |
| Propane (C₃H₈) (290)..... | 5–86 | 42.1 | 155.3 | 3.69 | 1.653 | 4.10 | 1.35* | 92.9* | Recip. |
| | 20–100 | 55.5 | 187.0 | 3.37 | 1.730 | 3.29 | 1.32* | 103.9* | |
| | 40–100 | 78.0 | 187.0 | 2.40 | 1.646 | 2.26 | 0.90* | 101.5* | |
| Ethane (C₂H₆) (170)..... | 5–86 | 236.0 | 675.9 | 2.87 | 3.41 | 1.82 | 2.180 | 105 | Recip. |
| Ethyl chloride (C₂H₅Cl) (160) | 5–86 | 4.65 | 27.10 | 5.83 | 1.405 | 24.0 | 0.95* | 106.3* | Rotary |
| | 20–100 | 6.80 | 34.79 | 5.12 | 1.425 | 17.19 | 0.92* | 116.1* | |
| | 40–100 | 10.79 | 34.79 | 3.22 | 1.375 | 10.73 | 0.63* | 109.9* | |

*Values may be slightly in error.

romethane. (CCl₃F); **F-12**, dichlorodifluoromethane (CCl₂F₂); **F-13**, monochlorotrifluoromethane (CClF₃); **F-21**, dichloromonofluoroethane (CHCl₂F); **F-22**, monochlorodifluoromethane (CHClF₂); **F-113**, trichlorotrifluoroethane (CCl₂FCClF₂); and **F-114**, dichlorotetrafluoroethane (C₂Cl₂F₄). F-12 is widely used for air conditioning and general commercial cooling. Methyl chloride, having similar physical properties, has been largely supplanted by F-12 in this field. Refrigerants containing chlorine as part of the molecule are readily detected in minute amount by passing them through a copper gauze kept hot by the essentially colorless flame of burning methyl alcohol. Even traces of such refrigerants in air give a readily detected test due to the intense green color imparted to the flame by their presence.

For industrial work, **butane** (C₄H₁₀), **propane** (C₃H₈), and, at low temperatures, **ethane** (C₂H₆) are used. **Dieline** (dichloroethylene, C₂H₂Cl₂) and **Carrene** (dichloromethane, CH₂Cl₂)

have been used to some extent, usually in centrifugal compressors.

The pressure-temperature relations for the saturated vapors of many of the commercially important refrigerants are given in Fig. 1. If the temperature at which the refrigeration is desired and the temperature at which heat can be discarded (condenser temperature) are known, the chart is convenient for determining for any chosen refrigerant the pressures which must be maintained. For instance, if the use of methyl chloride is contemplated in an air-cooled cabinet food freezer located in a room where the air temperature may rise to 90°F, the compressor discharge must be at least 100 psia and if the cabinet cooling coil is to be held at −30°F, compressor suction will be 9 psia.

The large volumes required for F-11, dieline, and water vapor can be handled satisfactorily by centrifugal compressors. When the evaporating pressure is below atmospheric, as

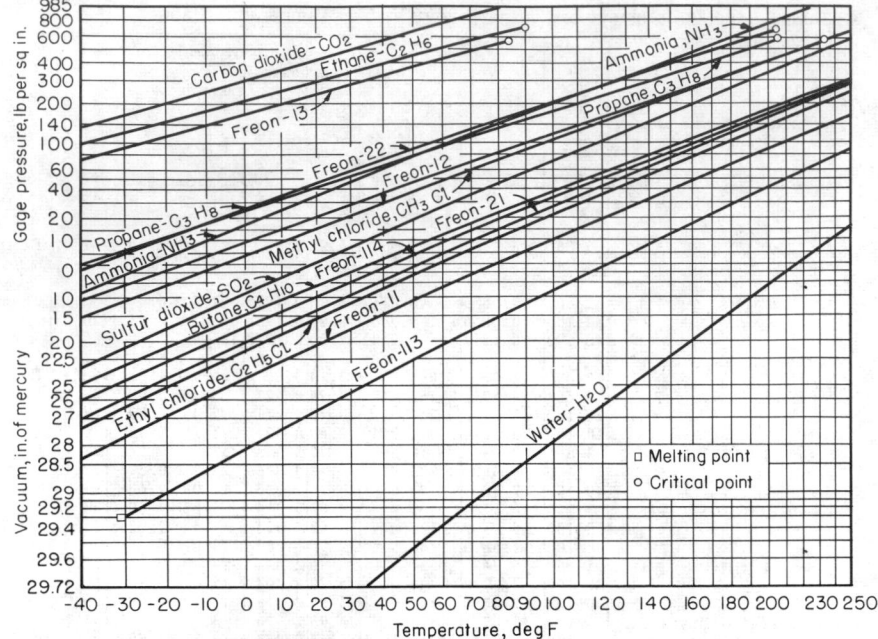

**Fig. 1** Pressure-temperature relations for saturated vapors of refrigerants.

in the case of dieline, ethyl chloride, sulfur dioxide, water vapor, and butane, air leaks are likely to be excessive; where the refrigerant is nearly odorless (Freon 12 and methyl chloride), leaks into the atmosphere are difficult to detect, and loss of the refrigerant may be excessive.

## OVERALL CYCLES

Figure 2 represents the simple closed-circuit **vapor-compression**

system. The upper heat exchanger is a vapor condenser (with some superheat), and after the throttling expansion valve, the lower one is the evaporator in which the refrigerant liquid at reduced pressure and temperature evaporates with the inward refrigeration heat flow. The refrigerant vapor is elevated by the compressor to a higher pressure and condensing temperature so that it will liquefy in its transfer of heat to the atmospheric level. Figures 3 to 5 illustrate the cycle.

Figure 6 is the counterpart of Fig. 2 and represents the

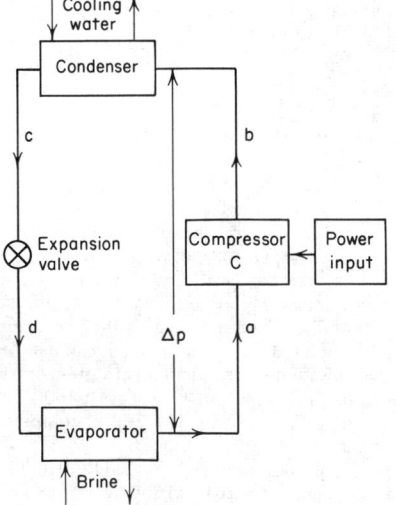

**Fig. 2** Simple closed-circuit vapor-compression system.

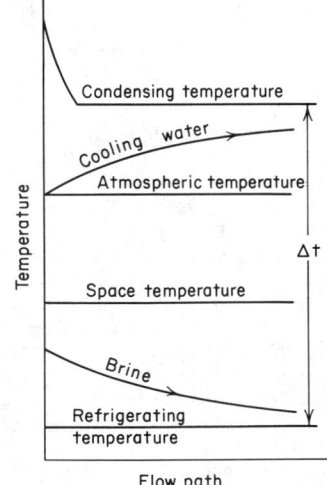

**Fig. 3** Comparative temperature relations for fluids of a simple system.

closed sensible-heat **gas-compression** cycle, with an expander (either displacement or turbine-type) parallel to a gas compressor which may provide a compressor-work "assist" in addition to its exhaust stream of cold gas. This is the refrigeration version of the Brayton cycle (see Sec. 4), the upper exchanger serving to cool the hot compressed gas, and the lower exchanger handling the sensible-heat refrigeration effect. In place of the expander, an expansion valve could be used; this is particularly effective on very low-temperature open-cycle operations, such as with gas liquefaction processes incorporating secondary "cold-regenerating" heat exchangers. An extended *Ts* diagram (i.e., Fig. 7 for air) shows (1) expander stages 3–4 (ideal, isentropic) and 3–4' (real, polytropic), and (2) throttling (isenthalpic process), 5–6 (high level), and 7–8 (low level). (See Deep Refrigeration, below, for modifications.)

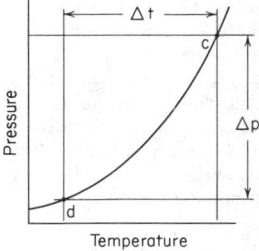

**Fig. 4** Pressure-temperature curve illustrating overall differences.

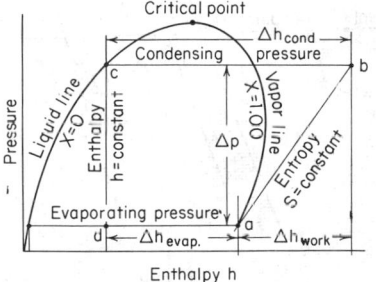

**Fig. 5** Pressure-enthalpy diagram for simple ideal refrigeration cycle.

## COMPONENTS OF COMPRESSION SYSTEMS

A refrigeration system is an energy-transport complex of assorted components. It may be a conventional system in which the same fluid, the refrigerant, recirculates continuously in a closed circuit, or it may entail a partially open system with discharge of some of the processed fluid either as a liquefied gas or as a solidified gas, with replacement makeup, e.g., by air liquefaction and solid carbon dioxide production. Components include vapor and gas compressors; liquid pumps; heat-transfer equipment (gas coolers, intercoolers, aftercoolers, exchangers, economizers); vapor condensers and the counterpart evaporators; liquid coolers and receivers; expanders; control valves and pressure-drop throttling devices (capillaries, refrigerant-mixture separating chambers, stream-mixing chambers); and connecting piping and insulation.

Compression may be single stage (Fig. 2) or multiple stage. The **compressors** may be displacement or kinetic. The former

include piston **reciprocating** machines in which the refrigerant, as vapor, is raised in pressure, usually with some simultaneous cooling. These machines may be horizontal or vertical, single acting or double acting; the vertical arrangement, with multiple cylinders in automotive arrangement and running at high rotative speeds, is common. Of utmost importance in the realized performance of a displacement compressor is the real volumetric effectiveness, or "efficiency," which measures the extent of utilization of the machine displacement volume. It is influenced by cylinder clearance volume, pressure ratio, nature of the gas, pressure drops through the valves and suction heating, and the equivalent compression exponent $n$. [The numerical value of $n$ expresses the effect of heat transfer to or from the gas during compression or expansion in a cylinder. Normally it lies between $n = 1.0$ and $n = k$ (adiabatic), such as for the ideal (isentropic).]

**Air-cooled** compressors are used where discharge temperatures are low, such as with Refrigerant 12; **water cooling,** where discharge temperatures are high, as with ammonia. **Oil separators** return to the compressor the lubricating oil carried over by the refrigerant vapors. **Rotary compressors** of the vane, eccentric, gear, and screw types are in use.

**Kinetic** compressors include high-speed centrifugal and axial flow machines, usually multistaged, and jet-entrainment devices. Centrifugal machines are especially adapted to high-volume flow (>500 ft³/min). (See also Sec. 14.)

The entire machine-compression operation may be replaced by a secondary absorber-pump-generator system, in which the complex is known as an absorption refrigeration system.

**Dual** (or **multiple-effect**) **compression** may be used when refrigeration at two temperatures is desired. The compressor takes vapor from a lower temperature expansion coil during the first part of its intake stroke, and from a higher temperature expansion coil at or near the end of the stroke. The mixture is then compressed and condensed.

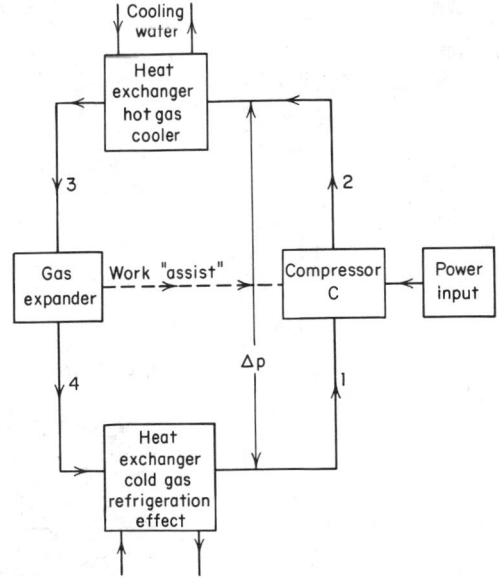

**Fig. 6** Closed sensible-heat gas-compression cycle.

In **wet compression,** cooling is obtained during compression by spraying liquid refrigerant into the compressor cylinder. The desuperheating of the compressed vapors results in better heat transfer in condensers and more nearly isothermal compression; the disadvantages are reduced compressor capacity and the problem of control of the amount of injection. Wet compression is not widely used.

Table 3 shows some ideal performance values for a single-stage ammonia system under different conditions.

**Condensers** are usually shell-and-tube type, with the refrigerant passing outside the tubes. Older industrial installments still use **double-pipe condensers.** In double-pipe condensers, gas flows between the two pipes while water passes through the inner. The outside pipe diameter may be 2 in; the inner, 1¼ in. In some instances, the pipes are exposed to the atmosphere either with or without water drip. In an **evaporative condenser** the refrigerant vapor is condensed as it passes through tubes over which water is sprayed; the water is then evaporated by air flowing over the wet tubes. In this way, cooling-water requirements are reduced to from 5 to 15 percent of the water requirements of a nonevaporative condenser with no water reuse. The evaporative condenser combines in a single unit a refrigerant condenser and the atmospheric cooling tower or spray pond which is required if the cooling water is to be reused. Air-cooled finned-tube condensers with forced ventilation are widely used on small units, and shell-and-coil or double-tube water-cooled condensers on medium units.

**Evaporators** may have finned or plain surfaces. Defrosting may be automatic or manual. **Flooded** evaporators are operated practically full of liquid refrigerant, the level being controlled by a float valve. **Wet-expansion** evaporators are operated with a level approaching that for flooded operation; **dry-expansion** (once-through) units operate with an indefinite amount of liquid in the evaporator. Flooded or wet operation gives high heat-transfer rates, but requires larger amounts of refrigerant than dry operation. Pumped recirculation of refrigerant is used on some flooded evaporator systems to promote heat transfer, and also on shell and tube evaporators (liquid chillers), e.g., in water-vapor systems. For a low-cost unit operating in limited space under conditions of motion (as on boats, trains, etc.) dry operation is preferred.

**Controls** are required on the liquid level of the refrigerant and on the temperature of the refrigerated space. The liquid control regulates the flow of refrigerant into the evaporator and also serves as the pressure barrier between the high operating pressure of the condenser and the lower operating pressure of the evaporator. It may take the form of a **capillary tube** between the condenser (high side) and the evaporator (low

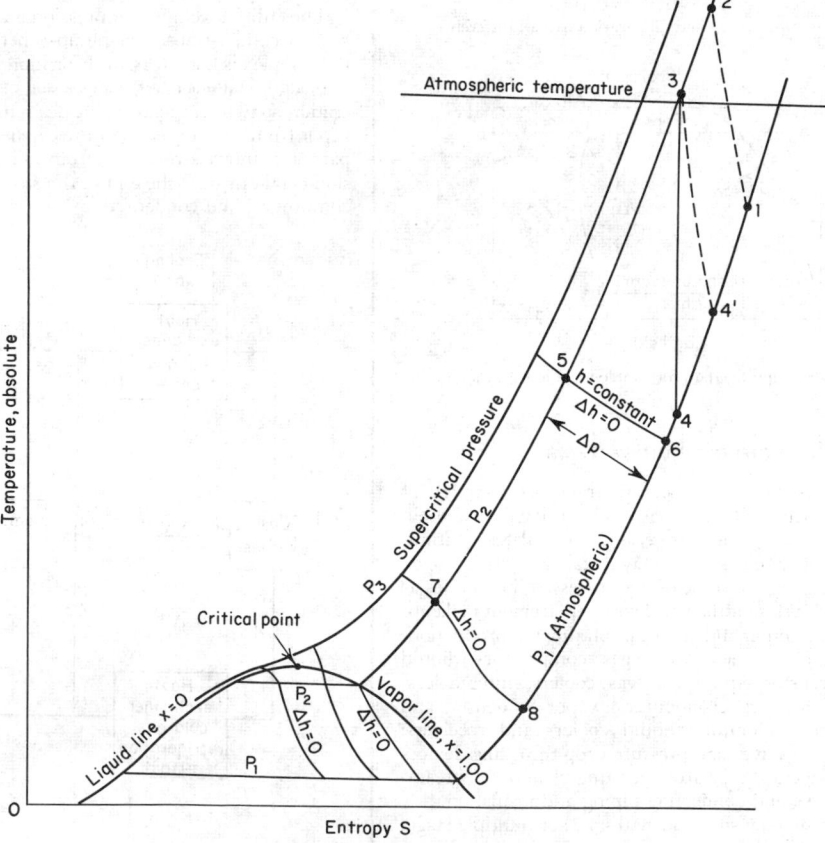

**Fig. 7**  Extended temperature-entropy diagram illustrating processes.

**Table 3. Theoretical Horsepower and Theoretical Volume of Dry Ammonia Gas Pumped per Minute to Produce 1 Ton of Refrigeration**

| Suction pressure and temperature | | Condenser pressure, psig (temperature, deg F) | | | | | | | | | |
|---|---|---|---|---|---|---|---|---|---|---|---|
| | | 103 (65) | | 127 (75) | | 153 (85) | | 182 (95) | | 215 (105) | |
| Psig | Deg F | Hp | Cfm | Hp | Cfm | Hp | Cfm | Hp | Cfm | Hp | Cfm |
| 4 | −20 | 1.058 | 5.84 | 1.205 | 5.96 | 1.361 | 6.09 | 1.525 | 6.23 | 1.691 | 6.43 |
| 6 | −15 | 0.997 | 5.35 | 1.145 | 5.46 | 1.300 | 5.58 | 1.461 | 5.70 | 1.546 | 5.83 |
| 9 | −10 | 0.903 | 4.66 | 1.045 | 4.76 | 1.193 | 4.86 | 1.347 | 4.97 | 1.435 | 5.08 |
| 13 | −5 | 0.818 | 4.09 | 0.954 | 4.17 | 1.094 | 4.25 | 1.244 | 4.35 | 1.321 | 4.44 |
| 16 | 0 | 0.735 | 3.59 | 0.865 | 3.66 | 1.002 | 3.74 | 1.147 | 3.83 | 1.219 | 3.91 |
| 20 | 5 | 0.666 | 3.20 | 0.795 | 3.27 | 0.928 | 3.34 | 1.066 | 3.41 | 1.138 | 3.49 |
| 24 | 10 | 0.592 | 2.87 | 0.726 | 2.93 | 0.854 | 2.99 | 0.991 | 3.06 | 1.060 | 3.12 |
| 28 | 15 | 0.541 | 2.59 | 0.664 | 2.65 | 0.792 | 2.71 | 0.922 | 2.76 | 0.994 | 2.82 |
| 33 | 20 | 0.474 | 2.31 | 0.592 | 2.36 | 0.715 | 2.41 | 0.842 | 2.46 | 0.903 | 2.51 |
| 39 | 25 | 0.410 | 2.06 | 0.523 | 2.10 | 0.599 | 2.15 | 0.767 | 2.20 | 0.829 | 2.24 |
| 45 | 30 | 0.351 | 1.85 | 0.461 | 1.89 | 0.576 | 1.93 | 0.694 | 1.97 | 0.759 | 2.01 |
| 51 | 35 | 0.300 | 1.70 | 0.410 | 1.74 | 0.521 | 1.77 | 0.640 | 1.81 | 0.701 | 1.85 |

side), in which case it is nonadjustable. Plugging due to dirt is a common difficulty. Capillary-tube liquid control is largely confined to relatively small units assembled and charged at the factory and particularly for hermetically sealed systems. The **constant-pressure expansion valve,** maintaining a constant evaporator pressure, and the **thermal expansion valve,** maintaining a constant superheat leaving the evaporator, are standard liquid controls for most commercial applications. A **low-side float** liquid control, used with a flooded evaporator operating at evaporator (low) pressure, consists of a float-operated valve to admit liquid refrigerant to the evaporator in accordance with demand so that a constant liquid level is held in it. A **high-side float** liquid control is often used with a single flooded evaporator; the float operating the valve between the evaporator and the condenser is in a float chamber containing liquid refrigerant at the condenser (high-side) pressure. As the liquid level in the float chamber falls, the valve closes, thus preventing hot gaseous refrigerant from passing from the high to the low side as is possible when using a capillary or a low-side float.

### Vapor-Compression Circuits

**Compound Low-Temperature Systems** Multistaging is advantageous as it reduces operating temperatures by intercooling and power requirements if the fluid superheats considerably on adiabatic compression, as with ammonia; with F-12 refrigerant, propane, and butane, the superheat is negligible (see Figs. 8 to 11).

**Gas Intercooler Plus Liquid Injection** Since gas temperature in the intercooler is limited by cooling water temperature, further favorable reduction of gas temperature, prior to entrance to the second stage, may be accomplished by direct injection, via an expansion valve, of refrigerant liquid (Fig. 8).

**Gas Intercooler Plus Flash Intercooler** With a flash intercooler separator located midway between successive expansion valves, the gas leaving the water-cooled intercooler passes through the flash intercooler, gets cooled by direct contact with the liquid, and, augmented by the flash vapor and at reduced temperature, enters the second stage (Fig. 9).

**Low-Temperature Booster Plus Single Stage** For the addition of a second lower-temperature evaporator to an existing single-stage system, a booster for the low-temperature vapor may be employed, its output feeding into the single-stage

suction. The booster is a centrifugal or rotary compressor, suitable for handling the large vapor volumes encountered at low temperatures (and pressures); the reciprocating compressor is more advantageous at high pressures (Fig. 10).

**Cascade Systems with Segregated Circuits** Here two separate circuits are employed with the low-stage vapor condensed in the second-stage evaporator. Two different refrigerants may be appropriately chosen, a lower-boiling-point refrigerant serving for the first stage. Thus, for low temperatures ($< -50°F$), $CO_2$ may be employed in the lower stage with $NH_3$ in the second stage condensing the $CO_2$. Or the

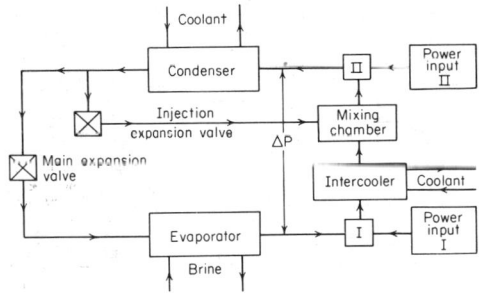

**Fig. 8** Two-stage system with conventional intercooler plus refrigerant injection.

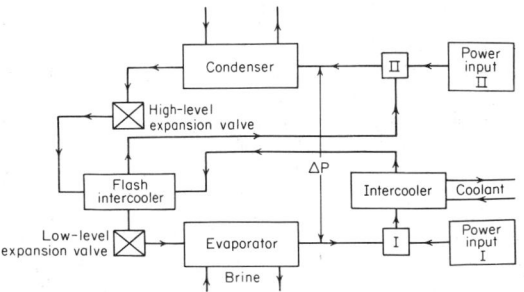

**Fig. 9** Two-stage system with conventional intercooler plus flash intercooler.

same refrigerant may be used in both stages, limiting the migration of the lubricant to the low-temperature realm (Fig. 11).

**Water-vapor refrigeration** is necessarily limited to near-atmospheric temperature levels; it involves large suction volumes at high vacuum and utilizes a nontoxic refrigerant; this type of

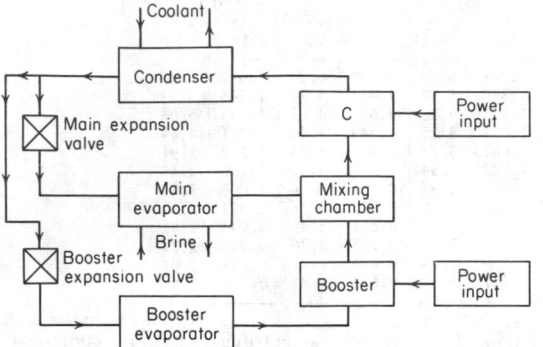

**Fig. 10** Two-stage system incorporating low-temperature booster circuit.

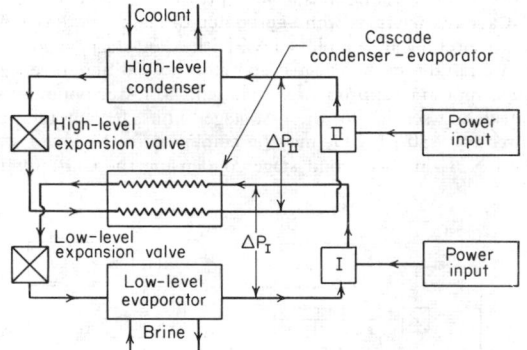

**Fig. 11** Two-stage system in cascade form.

refrigeration was frequently used in the early days of space air-conditioning. It utilizes (1) centrifugal compressors or (2) steam-jet compressors, entraining low-density water vapor from the evaporator. The centrifugal water-vapor compressor, usually multistage, must operate at high speeds (7,000 to 10,000 rpm) with a geared electric motor or steam-turbine drive. Operation at high vacuum (Fig. 2) requires an air pump at the condenser to eliminate leakage effects.

**Steam-jet refrigeration** systems are used where cooling to temperatures above 32°F is desired. Applications include industrial air conditioning; and cooling of city gas to condense out tar and other objectionable impurities, of gas absorbers to increase efficiency of absorption, of reaction units where heat removal and temperature control are important during chemical transformations, and of wort and mash in the brewing and other fermentation industries particularly in the summer months. Jet refrigeration has been used for cooling passenger trains and is coming into popularity for marine installations, e.g., cooling of banana boats and large passenger vessels. The system is a compression-type refrigerator: it uses water as a refrigerant, a part of which is evaporated to produce cooling of

the remainder; steam-jet ejectors to compress the water vapor resulting from evaporation, and a condenser, either the surface type, where refrigerant water is to be reused, or the barometric type, where refrigerant water can be discarded. The advantages of the system are: low installation cost; economical operating cost, particularly if low-cost, low-pressure steam and adequate condensing water are available; the absence of moving parts except for small liquid pumps; safety, since no noxious or toxic refrigerants are used; ability to carry considerable overload with only a small rise in refrigerating temperature level; and elimination of heat-transfer surfaces. (See Havemeyer, *Chem. Eng.*, Sept. 1948, pp. 103–106.) For considerable differences in temperatures between condensing-water temperature and refrigerating temperatures (20°F or more), cooling of the refrigerant water by evaporation in two stages operating at different pressures (and temperatures) will usually give better operating economy, but at somewhat higher installation cost.

Figure 12 shows the variation of refrigerating capacity with variation of the chilled-water temperature, capacity increasing as chilled-water temperature rises.

**Household compression machines** are designed for continuous automatic operation and for conservation of the charges of refrigerant and oil. These units are almost universally motor-driven air-cooled compressors, the principal exception being the Electrolux-Servel absorption unit (see below). The compressor may be hermetically sealed, with the compressor and motor enclosed in the same casing, or may use a shaft seal, embodying some form of sylphon bellows, with an outside coil spring at the place where the crankshaft passes out of the casing. The compressor may be reciprocating, but rotary types are being increasingly used either with a floating piston ring driven by an eccentric or with sliding blades. Lubrication is by internal forced-feed circulation in most cases. The condenser may be of the radiator, coil, or plate type, cooled by natural draft, by a fan on the main motor, or by a separate motor. The refrigerant is usually Freon 12. Refrigerant feed control to the evaporator may be (1) float feed (flooded system), (2) pressure-actuated diaphragm valve, or (3) by fixed orifice. The fixed orifice may be either a plate orifice or a capillary tube. The evaporator is usually made of stainless

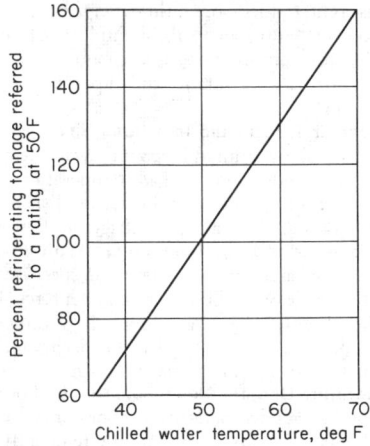

**Fig. 12** Variation of refrigerating capacity with the jet-chilled water temperature in steam refrigeration.

steel. Controls are generally pressure-operated thermostatic switches. The power consumption of an average box (say 6 ft³) is about 20 kWh per month. The heat gain of a refrigerator cabinet of the same size is about 3.3 Btu/h·°F.

## ABSORPTION SYSTEMS

Absorption refrigeration systems are essentially vapor-compression plants (Fig. 2) with the powered compressor replaced by a thermally activated arrangement (Fig. 13) where the basic elements are absorber, pump, heat exchanger, throt-

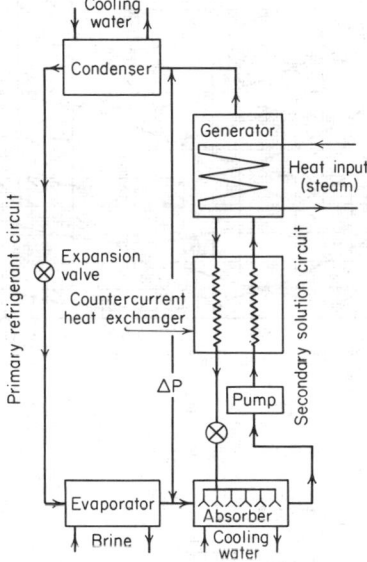

**Fig. 13**  Elemental absorption system circuit.

tle valve, and generator. Applications of this type of system are (1) ammonia-water system, for general industrial application, (2) household refrigerators (Platen, Munters, Electrolux), and (3) water and lithium bromide, for air conditioning.

Many combinations of soluble gas and solvent liquid have been proposed for absorption refrigeration but ammonia-water systems alone have been used industrially. Few of these are now in operation, except for the small domestic Electrolux machines (see below), probably because they were bulky and difficult to operate efficiently.

In the absorption system, liquid ammonia vaporizes in the evaporator, producing refrigeration. The ammonia gas goes to an **absorber,** where it is absorbed by a weak solution of ammonia in water (weak liquor) at a low temperature, the heat of solution being abstracted by water (or air) and a strong liquor being formed. The strong liquor is pumped to the **generator,** where its temperature is raised by steam coils and a mixture of water vapor and ammonia vapor is evolved. This gaseous mixture passes up through the **analyzer,** where it meets the relatively cool strong liquor entering the generator and loses some of its vapor content. From the top of the analyzer ammonia and water vapor pass to the **rectifier,** often merely a cooling coil or tubular heat interchanger, where cooling water condenses more of the water vapor, the water flowing back to the analyzer and the generator and the residual ammonia gas passing to the condenser. The ammonia gas after liquefaction in the condenser goes through an expansion valve to the

evaporator. Weak liquor flowing from the generator to the absorber passes through a heat exchanger, where it heats the strong liquor coming from the absorber to the generator. Efficient heat exchange is necessary in absorption units; the steam requirement may be reduced to one-third that necessary with no heat exchange. (See Stickney, *Ice and Refrig.*, May to Aug., 1936, for an analysis of the effect on the absorption cycle of the use of heat exchangers.) **Two-stage** and even **three-stage** absorption systems have been proposed. Staged operation makes possible a greater temperature difference between cooling water and refrigeration temperature, but increases heat requirements almost in proportion to the number of stages. With cooling water at 80°F refrigeration may be produced at −94°F in a two-stage system; with three stages, the refrigeration may be at −200°F or even lower.

Some of the properties of ammonia-water solutions are given in Fig. 14 (from Stickney, New Tables and Chart for Ammonia Solutions, *Refrig. Eng.*, Oct., 1935). The enthalpy of liquid water is assumed 0 at 32°F, the enthalpy of liquid ammonia is assumed 0 at −40°F. Six values can be read from any point, namely, gage pressure; temperature, °F; composition (weight of ammonia per lb of mixture) and enthalpy (Btu per lb) in both the liquid and vapor phases.

> EXAMPLE.  A solution containing 30 percent by weight $NH_3$ at 120°F has a pressure of 17 psig; the enthalpy of the liquid solution is 22 and of the vapor 693 Btu/lb; the vapor in equilibrium with the solution will be 96.3 percent $NH_3$ by weight.

For a second-law analysis of the parts of the absorption cycle, the availability tables of Scatchard (*Refrig. Eng.*, **58**, 1947, pp. 413–419) will be found convenient.

**Electrolux-Servel Absorption Process** (Platen-Munters Patent)  The Electrolux-Servel ammonia absorption process eliminates the use of pumps or other moving parts. The gas pressure is uniform throughout the hermetically sealed system, the difference between the vapor pressure of ammonia in the condenser and that in the evaporator being compensated by the presence of hydrogen; the sum of the partial pressures of the hydrogen and of the ammonia vapor in the evaporator is equal to the sum of the partial pressures in the condenser. The general arrangement is shown in Fig. 15.

The most prominent heat-activated absorption system for air conditioning is that involving **water and lithium bromide,** a "thermal compression" water-vapor cycle for temperatures above 32°F. This parallels the water-vapor systems (Fig. 13) in which the elevation of pressure to the condenser is accomplished either by a high-volume centrifugal or by a steam-jet entrainment device. Whereas in the ammonia and water system, ammonia vapor alternately dissolves in or is disengaged from an aqueous ammonia solution, with ammonia subsequently the cycle refrigerant, here water vapor parallels the ammonia; it dissolves in (absorber) or disengages from (generator) the lithium bromide solution, and water subsequently is the cycle refrigerant.

**Adsorption systems** using solids to adsorb the refrigerant consist of the same elements as absorption systems except that no rectifier or analyzer is needed since the solids used are nonvolatile under operating conditions. Since solids are troublesome to transfer, they are usually held in stationary beds; continuous refrigeration is accomplished by installing duplicate beds of solids, serving intermittently as adsorber and generator through a system of switching valves. (See also Sec. 12.)

**Silica gel,** a hard glassy granular solid made by precipitating

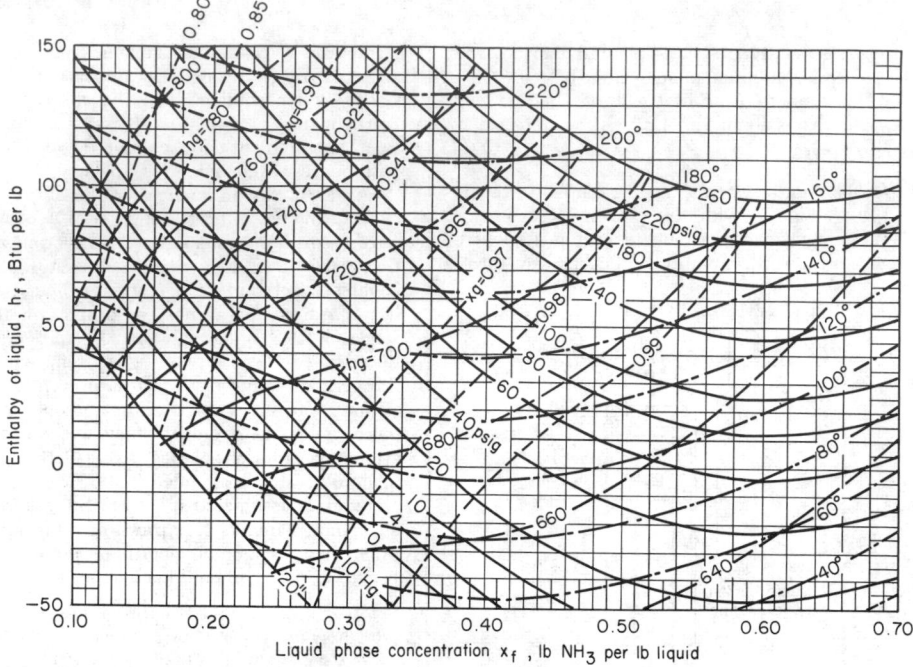

**Fig. 14** Physical properties of ammonia-water solutions.

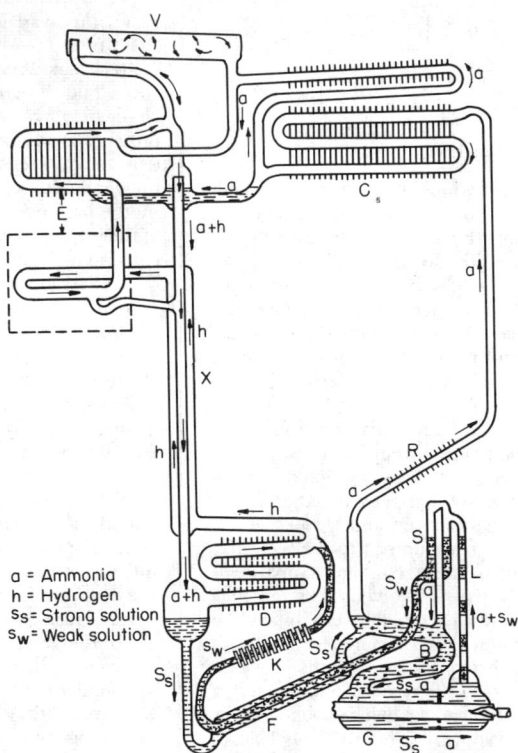

a = Ammonia
h = Hydrogen
$s_S$ = Strong solution
$s_W$ = Weak solution

**Fig. 15** Diagrammatic layout of Electrolux-Servel absorption machine: B, analyzer; C, condenser; D, absorber; E, evaporator; F, exchanger; G, generator; K, forecooler; R, rectifier; S, liquid-vapor separator; V, hydrogen reserve vessel; X, heat exchanger.

$SiO_2$ from sodium silicate solution with acid, has been used in adsorption systems. It is extremely porous; adsorbed vapors may be readily driven off by heating. It will absorb 25 to 35 percent of its own weight of sulphur dioxide, the usual refrigerant. The gel (8–20 mesh) is packed with about 50 percent voids. Ammonia is unsatisfactory because of interaction with the gel. A typical adsorber consists of ¾-in steel tubes welded into headers and filled with granular gel, the total weight of gel being 1,000 lb. It has an ice-melting effect of 1 to 1½ lb per 24 h per lb of silica gel; this may be raised to 4 lb/lb by forced circulation. Fuel consumption is about 135 lb of propane per ton of refrigeration. Silica gel is used in the chemical industry as a dehydrating agent, and also in some air-conditioning and high-temperature cooling units because of its ability to absorb water. The water is eliminated and the gel restored to its original activity by raising its temperature.

## AIR MACHINES

**Compressed-air machines** in which air is compressed, cooled, and then adiabatically expanded in an engine while doing work are obsolete for usual industrial refrigeration because of bulk and inefficiency. Their coefficient of performance is less than unity as compared with 4 or 5 for vapor machines. Operation may be closed cycle, in which the air is reused, or open cycle, in which the air is discarded after expansion. Compression ratios of about 3 seem best. In the **dense-air** closed system, the pressure at the compressor intake averages 75 and at the outlet 225 to 250 psig. The use of super-atmospheric pressure throughout the cycle decreases the size, and the reuse of dried air eliminates the ice formation on engine valves which occurs with open-system operation. Air cycles are used for liquefaction of the so-called permanent gases and to an increasing extent for comfort cooling in high-speed planes where the small lightweight equipment and the safety of the cooling medium are important. High-speed (90,-000 to 100,000 r/min) gas turbines are used to extract work from moderately compressed (5 to 100 psig) air.

## THERMOELECTRIC COOLING

Recently thermoelectric cooling, utilizing the Peltier effect with dissimilar metallic conductors (and variously called thermoelectric refrigeration, Peltier-effect cooling, or electronic cooling), has attained practicality in units of small sizes (less than 1 ton).

The elements of a thermoelectric cooler are quite simple and are equivalent to those of a thermoelectric generator (see Sec. 9). For development of various output capacities, the number of basic elements is proportionately increased; they all act in parallel, and their outputs add together. **Semiconductor** materials, such as bismuth-telluride-selenide and bismuth-antimony-telluride alloys, are employed.

With an input to the circuit of low-voltage direct current and with heat continuously abstracted at one junction at room temperature, the other junction will become cold. Essentially each of the two junctions becomes an "activity cell," in which at the lower refrigeration temperature level, heat is converted into electrical effect, and at the higher atmospheric temperature, the electrical effect is converted into heat. This establishes the basis of "heat-pump" operation: the lower-temperature heat requirement must be supplied, essentially as in the evaporator of a vapor-compression system, by heat with-drawal via heat transfer, from all connected surroundings. The energy-transport circuit thus parallels the machine-activated circuit of Fig. 2. The amount of heat for a single-component circuit depends on the circulating electric current, on the properties and dimensions of the two conductors ($p$ and $n$ types), and on the resultant temperature difference between the high side room temperature and the low side input heat that is divided into two groups, useless and useful. The **useless heat** is the heat conduction between the high and low temperature through the thermoelements plus the current-generated resistance-loss heat within the elements that flows to the colder junction. The **useful heat** is the refrigeration effect in a regular evaporator. All the heat must in turn be discharged via the Peltier effect at the higher room temperature, as in the condenser of a vapor-compression system. Ultimate performance of the circuit depends on the voltage-generating nature of the two dissimilar conductors and on their electrical resistance, thermal conductivity, and physical dimensions. The different properties are mathematically combined into a **figure of merit,** $Z$.

## METHODS OF APPLYING REFRIGERATION

In **direct expansion systems** the evaporator is placed in the space which is to be cooled; in **brine systems** a brine is cooled by contact with the evaporator surface, and the cooled brine goes to the space which is to be refrigerated. Brine systems require 40 to 60 percent more surface than do direct expansion; they have an equalizing effect due to the large heat capacity of cold brine, they are safer (particularly if the refrigerating effect must be carried considerable distance or widely distributed), and they permit closer temperature regulation than is possible with direct expansion. Brine systems are recommended for larger cold-storage plants. If two temperatures are to be held, the lower may be by direct expansion, the higher by brine cooling. Development of better controls and newer piping methods has made direct expansion more attractive than previously.

**Brines** used for industrial refrigeration are usually aqueous solutions of either calcium chloride or sodium chloride. Calcium chloride should not contain over 0.2 percent magnesia, calculated as magnesium chloride. The purest economically available salt should be used in making sodium chloride brines; magnesia and sulphates are especially undesirable as they cause sludge formation. Calcium chloride brines are recommended down to $-45°F$; sodium chloride brine should not be used below $0°F$. Brines should be chemically neutral: acidic brines attack ferrous materials, alkaline brines attack zinc, ammonia in brine (resulting from leaks in the ammonia system) is especially harmful to most nonferrous metals. Corrosion by brine is increased by the presence of oxygen, air, or carbon dioxide and by galvanic action between dissimilar metals. The contact of brine with air should be minimized. Corrosion inhibitors are widely used, a satisfactory one being about 100 lb of sodium chromate or dichromate per 1,000 $ft^3$ of calcium chloride brine or 200 lb per 1,000 $ft^3$ of sodium chloride brine.

The properties of sodium and calcium chloride brines are given in Tables 4 to 7.

The density of brine is measured by a **salinometer** (or **salometer**), which is a simple hydrometer the indications on which are 4 times greater than on the corresponding Baumé scale (see Sec. 1).

## Table 4. Properties of Sodium Chloride Solutions
(For variation of sp gr with temperature see Table 6)

| Parts of NaCl by weight in 100 parts of the solution | Specific gravity at 60 F | Deg Baumé | Weight per gal, lb | Weight per cu ft, lb | Freezing point, deg F | Specific heat at | | | | |
|---|---|---|---|---|---|---|---|---|---|---|
| | | | | | | 14 F | 32 F | 50 F | 68 F | 86 F |
| 6 | 1.044 | 6.06 | 8.71 | 65.1 | 25.5 | ....... | 0.924 | 0.927 | 0.929 | 0.932 |
| 8 | 1.058 | 8.00 | 8.82 | 66.0 | 22.9 | ....... | 0.902 | 0.906 | 0.909 | 0.912 |
| 10 | 1.073 | 9.91 | 8.95 | 66.9 | 20.2 | ....... | 0.882 | 0.887 | 0.890 | 0.893 |
| 12 | 1.088 | 11.78 | 9.08 | 67.8 | 17.3 | ....... | 0.865 | 0.869 | 0.873 | 0.876 |
| 14 | 1.104 | 13.63 | 9.22 | 68.8 | 14.1 | ....... | 0.848 | 0.853 | 0.857 | 0.859 |
| 16 | 1.119 | 15.45 | 9.33 | 69.8 | 10.6 | 0.827 | 0.834 | 0.839 | 0.842 | 0.844 |
| 18 | 1.135 | 17.25 | 9.47 | 70.8 | 6.7 | 0.815 | 0.821 | 0.825 | 0.828 | 0.830 |
| 20 | 1.151 | 19.02 | 9.60 | 71.8 | 2.4 | 0.804 | 0.809 | 0.813 | 0.815 | 0.817 |
| 22 | 1.167 | 20.78 | 9.74 | 72.8 | −2.5 | 0.794 | 0.798 | 0.801 | 0.803 | 0.804 |
| 24 | 1.184 | 22.51 | 9.88 | 73.8 | +1.4 | 0.784 | 0.788 | 0.791 | 0.792 | 0.793 |

It is undesirable to use a **strength of solution** of salt greater than is necessitated by its freezing temperature, as the specific heat (Tables 4 and 5) decreases as the concentration of the brine increases, and consequently the stronger the brine, the less heat a given amount of it is able to convey between certain definite temperatures and the more power is required to pump the brine. Moreover, brine which is too strong may cause clogging of pipes, etc., by depositing salt. On the other hand, if the solution is too weak it may not be able to withstand the temperature existing in the expansion coil, so that a layer of thin ice will form around the latter and interfere with the absorption of heat from the brine. The surface of the expansion coils in the brine tank should be inspected from time to time to see if any ice has formed on them. In larger plants, it is customary to use a solution with a freezing point not less than 10°F below the lowest temperature which will be obtained in the operation of the plant. In smaller isolated plants and where careful supervision is not ensured, it is customary to make the solution as strong as possible without being unstable, usually 1.240 to 1.250 sp gr.

**Brine coolers** may be of three types, shell-and-tube, shell-and-coil, and double pipe. The shell-and-tube type is the most widely used, the brine flowing through the tubes which are surrounded by the evaporating refrigerant. Tubes may be arranged for multipass operation. The effective heat-transfer surface varies from 8 to 15 ft² per ton, varying with temperature and brine velocity. A submerged coil in an open brine tank is used for ice making by the can process.

The **double-pipe cooler** is usually of 2-in inner or brine-flow pipe and 3-in outer pipe. The commercial rating is 15 to 20 ft length of coil per ton of refrigeration.

The **shell-and-tube cooler** is used with closed heads and is erected both vertically and horizontally; brine flows through the tubes and ammonia is in the shell. It is made in sizes from 1 to 350 tons with ratings of 8 to 15 ft² effective surface per ton, varying with the temperature and brine velocities; tubes 1

## Table 5. Properties of Calcium Chloride Solutions
(For variation of sp gr with temperature see Table 6)

| Parts of CaCl₂ by weight in 100 parts of the solution | Specific gravity at 60 F | Deg Baumé | Weight per gal, lb | Weight per cu ft, lb | Freezing point, deg F | Specific heat at | | | | | |
|---|---|---|---|---|---|---|---|---|---|---|---|
| | | | | | | −4 F | 14 F | 32 F | 50 F | 68 F | 86 F |
| 6 | 1.050 | 7.0 | 8.76 | 65.52 | 28.0 | ....... | ....... | ....... | ....... | ....... | ....... |
| 8 | 1.069 | 9.33 | 8.926 | 66.70 | 24.2 | ....... | ....... | 0.882 | 0.887 | 0.892 | 0.897 |
| 10 | 1.087 | 11.57 | 9.076 | 67.83 | 21.4 | ....... | ....... | 0.853 | 0.858 | 0.863 | 0.868 |
| 12 | 1.105 | 13.78 | 9.227 | 68.95 | 18.2 | ....... | ....... | 0.825 | 0.831 | 0.836 | 0.842 |
| 14 | 1.124 | 15.96 | 9.377 | 70.08 | 14.4 | ....... | ....... | 0.799 | 0.805 | 0.811 | 0.817 |
| 16 | 1.143 | 18.12 | 9.536 | 71.26 | 9.9 | ....... | 0.768 | 0.775 | 0.781 | 0.787 | 0.792 |
| 18 | 1.162 | 20.24 | 9.703 | 72.51 | 4.7 | ....... | 0.745 | 0.752 | 0.759 | 0.764 | 0.769 |
| 20 | 1.182 | 22.32 | 9.853 | 73.63 | −1.0 | ....... | 0.723 | 0.731 | 0.738 | 0.744 | 0.749 |
| 22 | 1.202 | 24.38 | 10.04 | 75.0 | −7.3 | 0.695 | 0.704 | 0.711 | 0.718 | 0.724 | 0.729 |
| 24 | 1.223 | 26.41 | 10.21 | 76.32 | −14.1 | 0.678 | 0.686 | 0.693 | 0.900 | 0.706 | 0.712 |
| 26 | 1.244 | 28.41 | 10.38 | 77.56 | −22.0 | 0.663 | 0.670 | 0.677 | 0.683 | 0.690 | 0.696 |
| 28 | 1.265 | 30.39 | 10.56 | 78.94 | −32.0 | 0.649 | 0.656 | 0.662 | 0.669 | 0.675 | 0.682 |
| 30 | 1.287 | 32.34 | 10.75 | 80.35 | −46.0 | 0.638 | 0.643 | 0.648 | 0.655 | 0.661 | 0.668 |

### Table 6. Specific Gravities of Brines
(To change to lb/ft³ multiply by 62.43; to change to lb/gal multiply by 8.35)

| Parts by weight of salt in 100 parts of brine | Sodium chloride | | | | Calcium chloride | | | | Magnesium chloride | | | |
|---|---|---|---|---|---|---|---|---|---|---|---|---|
| | Temperature, deg F | | | | | | | | | | | |
| | 14 | 32 | 50 | 68 | 14 | 32 | 50 | 68 | 14 | 32 | 50 | 68 |
| 6 | ...... | 1.046 | 1.044 | 1.041 | ...... | 1.053 | 1.051 | 1.049 | ...... | 1.053 | 1.051 | 1.049 |
| 8 | ...... | 1.061 | 1.059 | 1.056 | ...... | 1.071 | 1.069 | 1.066 | ...... | 1.070 | 1.069 | 1.067 |
| 10 | ...... | 1.077 | 1.074 | 1.071 | ...... | 1.090 | 1.087 | 1.084 | ...... | 1.089 | 1.087 | 1.084 |
| 12 | ...... | 1.093 | 1.090 | 1.086 | ...... | 1.108 | 1.106 | 1.103 | 1.108 | 1.107 | 1.105 | 1.102 |
| 14 | ...... | 1.108 | 1.105 | 1.101 | ...... | 1.127 | 1.124 | 1.121 | 1.127 | 1.126 | 1.123 | 1.121 |
| 16 | 1.128 | 1.124 | 1.121 | 1.116 | 1.150 | 1.147 | 1.144 | 1.140 | 1.147 | 1.145 | 1.142 | 1.139 |
| 18 | 1.144 | 1.140 | 1.136 | 1.132 | 1.170 | 1.167 | 1.163 | 1.159 | 1.166 | 1.164 | 1.161 | 1.158 |
| 20 | 1.161 | 1.157 | 1.152 | 1.148 | 1.190 | 1.187 | 1.183 | 1.179 | 1.186 | 1.183 | 1.181 | 1.178 |
| 22 | 1.178 | 1.173 | 1.169 | 1.164 | 1.211 | 1.208 | 1.203 | 1.199 | 1.206 | 1.203 | 1.201 | 1.197 |
| 24 | 1.195 | 1.190 | 1.185 | 1.180 | 1.233 | 1.229 | 1.224 | 1.219 | 1.226 | 1.224 | 1.221 | 1.218 |

### Table 7. Weight of Commercial Calcium Chloride in Brine

| Sp gr | 1.10 | 1.12 | 1.14 | 1.16 | 1.18 | 1.20 | 1.22 | 1.24 | 1.26 | 1.28 | 1.30 | 1.32 |
|---|---|---|---|---|---|---|---|---|---|---|---|---|
| Wt per gal, lb | 1.41 | 1.70 | 2.00 | 2.30 | 2.59 | 2.90 | 3.20 | 3.50 | 3.83 | 4.13 | 4.46 | 4.78 |
| Wt per cu ft, lb | 10.55 | 12.72 | 14.96 | 17.20 | 19.37 | 21.69 | 23.94 | 26.18 | 28.62 | 30.89 | 33.36 | 35.75 |

Specific gravity is at 60°F for both brine and water. The weights are of 73 to 75 percent solid calcium chloride per gal of brine at 60°F. For flake 77 to 80 percent calcium chloride multiply the weights given by 0.94.

to 2½ in arranged multipass. This type of cooler has largely displaced all other types in recent installations (Table 8).

## REFRIGERANT PIPING
(See also Sec. 8)

It is important to the proper operation of a refrigeration system that the piping or mains interconnecting the compressors, condensers, evaporators, and receivers be properly sized. This piping must be considered in three categories, viz., liquid lines, suction lines, and discharge lines. Essentially, pipeline sizing is governed by the permissible frictional pressure drop in each piping system. Excessive pressure drops penalize compressor efficiencies and may affect control-valve operation adversely. Liquid-line velocities for most refrigerants are in the order of 60 to 400 ft/min, suction lines from 700 to 4,600 ft/min, and discharge lines from 1,000 to 5,000 ft/min. Pressure drops vary approximately as the square of the velocity (or tonnage) and directly as the length of the piping. For liquid lines, when evaporator is located above condenser,

the following pressure drops in pounds per square inch per foot of static lift should be allowed: ammonia, 0.26; Freon 12, 0.57; Freon 22, 0.51; Freon 11, 0.64. (See also Secs. 3 and 4.)

**Ammonia Mains** For the average installation, Table 9 shows the maximum tons of refrigeration normally allowed for various sizes of standard pipe, based on 100 ft equivalent length of piping (measured length plus allowance for valves and fittings). The gas pressure drops per 100 ft equivalent length upon which Table 9 is based are as follows: suction lines at 5 psig = 0.25 lb/in², suction lines at 20 psig = 0.50 lb/in², suction lines at 45 psig = 1 lb/in², and discharge lines = 1 lb/in².

Standard-weight (schedule 40) steel pipe is used for ammonia mains, except for liquid lines 1½ in and smaller where extra-strong (schedule 80) pipe is used. Joints may be either screwed, flanged, or welded, but welding is preferred.

**Freon Mains** Maximum tons refrigerant for various pressure drops per 100 ft equivalent length of Freon 12 mains are given in Table 10. For average conditions, allowable pressure drops in Freon 12 suction lines are 0.5 to 1.0 lb/in² below 0°F

### Table 8. Capacity of Multipass Shell-and-Tube Brine Coolers, Flooded
(Tons refrigeration)

| Diameter of shell, in. | Length of shell, ft | Velocity of brine through cooler, fpm | | | | | | | | | | |
|---|---|---|---|---|---|---|---|---|---|---|---|---|
| | | 75 | | | | 200 | | | | 400 | | | |
| | | Total brine, gpm | Mean temp diff, brine and ammonia, deg F | | | Total brine, gpm | Mean temp diff, brine and ammonia, deg F | | | Total brine, gpm | Mean temp diff, brine and ammonia, deg F | | |
| | | | 10 | 15 | 20 | | 7½ | 12½ | 17½ | | 5 | 10 | 15 |
| 26 | 6 | 55 | 5.7 | 8.6 | 11.5 | 140 | 7.74 | 12.9 | 18.1 | 290 | 7.15 | 14.3 | 21.2 |
| 26 | 9 | 55 | 8.5 | 12.8 | 17.2 | 140 | 11.5 | 19.3 | 26.0 | 290 | 10.7 | 21.4 | 32.0 |
| 26 | 12 | 55 | 11.4 | 17.2 | 22.9 | 140 | 15.5 | 25.8 | 36.2 | 290 | 14.3 | 28.6 | 42.9 |
| 34 | 9 | 90 | 13.6 | 20.5 | 27.4 | 230 | 18.5 | 30.7 | 43.1 | 440 | 17.1 | 34.1 | 51.1 |
| 34 | 12 | 90 | 18.1 | 27.3 | 36.4 | 230 | 24.6 | 41.0 | 57.4 | 440 | 22.6 | 45.3 | 68 |
| 34 | 18 | 90 | 27.2 | 41.0 | 54.8 | 230 | 37.0 | 62.0 | 86.2 | 440 | 34.1 | 68.2 | 101 |
| 42 | 12 | 190 | 33.1 | 50 | 66.5 | 510 | 45.0 | 74.8 | 105 | 970 | 41.5 | 83 | 123 |
| 42 | 18 | 190 | 50 | 75 | 100 | 510 | 67.4 | 112 | 157 | 970 | 62.2 | 124 | 187 |

**Table 9. Maximum Tons Refrigeration for Ammonia Mains**
(100 ft equivalent pipe length)

| Pipe size, in. | Suction line | | | Discharge line | Liquid line | |
|---|---|---|---|---|---|---|
| | Suction pressure, psig | | | | Condenser to receiver | Receiver to system |
| | 5(−17.2 F) | 20(5.5 F) | 45(30 F) | | | |
| ⅜ | . . | . . | . . | . . | 2.5 | 12.0 |
| ½ | 0.6 | 1.1 | 2.0 | 3.1 | 6.0 | 20.0 |
| ¾ | 1.2 | 2.2 | 4.1 | 6.0 | 14.0 | 75.0 |
| 1 | 2.2 | 4.0 | 7.5 | 11.4 | 24.0 | 137 |
| 1¼ | 4.4 | 8.0 | 15.0 | 22.4 | 50.0 | 245 |
| 1½ | 6.4 | 11.8 | 21.6 | 30.9 | 77.0 | 400 |
| 2 | 12.1 | 22.2 | 42.0 | 62.0 | 140 | 850 |
| 2½ | 19.1 | 35.5 | 65.0 | 97.5 | 220 | 1,475 |
| 3 | 31.5 | 59.0 | 108 | 160 | 375 | 2,400 |
| 3½ | 46.6 | 87.5 | 156 | 238 | 540 | 3,500 |
| 4 | 64.0 | 118 | 240 | 330 | 740 | |
| 5 | 117 | 208 | 385 | 560 | 1,320 | |
| 6 | 175 | 306 | 600 | 905 | 2,030 | |
| 8 | 362 | 650 | 1,200 | 1,810 | 4,200 | |
| 10 | 640 | 1,180 | 2,160 | 3,200 | | |
| 12 | 940 | 1,850 | | | | |

SOURCE: From ASRE, Air-Conditioning Refrigerating Data Book, 1955–1956 ed., based on ARI Equipment Standards; see current ASHRAE, Guide and Data Book, for more complete tables.

evaporator temperature, 1.0 to 1.5 lb/in² from 0 to 25°F, and 2.0 to 2.5 lb/in² from 25 to 50°F. Discharge-line pressure drops upon which Table 10 is based are approximately 1 lb/in² per 100 ft equivalent length. Pressure drops of 1 to 4 lb/in² are permissible. Liquid-line pressure drops of 3 to 5 lb/in² per 100 ft equivalent length are normal, with 10 lb/in² usually considered maximum. To facilitate oil return in vertical suction lines with evaporator located below compressor, the velocity in the vertical section should be 1,000 ft/min or more. Suction-line capacities in Table 10 may be increased 9 percent for 50°F evaporator temperature, and will decrease approximately 7 percent for each 10°F below 40°F.

Freon 22 piping will, in general, be smaller for the same tonnage than Freon 12 piping. Suction lines will handle about one-third more tonnage for the same pressure drop. Liquid and discharge lines can be the same as for the equivalent tonnage of Freon 12, or slightly smaller.

Freon piping may be standard-weight steel pipe, but copper tubing is used in most cases. Medium-weight type L copper tubing is normally used for land installations, and the heavier type K copper tubing is used for marine work. Joints in the copper tubing may be flared compression fittings for tubing ¾ in OD and smaller. However, hard solder (silver-base alloy melting above 1000°F) is preferred for most tubing connections. Copper tubing is almost always used for liquid lines. Steel pipe may be used for the larger suction and discharge lines but should be sandblasted on the inside and carefully cleaned before installation.

**Mains for Other Refrigerants** For sizing piping for methyl chloride, sulphur dioxide, carbon dioxide, and other refrigerants, reference may be made to the latest edition of the ASHRAE Guide & Data Book. (For additional pipe sizing tables and charts, see Hendrickson, *Refrig. Eng.*, Oct., 1946, Determination of Refrigerant Pipe Size.)

## COLD STORAGE

**Insulation** For tables of conductivities, see Sec. 4. Insula-

tion used in refrigeration is often compound. The insulating value of filling material varies nearly inversely as its specific gravity. It increases to a point where the material is so loose as to permit air circulation. With most material, a specific gravity of 0.160 seems to be the limit. When any fibrous material, so far in use, is arranged so that it weighs much under this, the effect of an open air space is obtained. This does not apply to **material arranged in layers** of different densities transversely to the direction of heat flow. **Dryness of insulation** is of great importance. Insulation until the heat loss falls to 2 Btu per ft² per °F temp diff per 24 h has been found to be economic in most cases in temperate climates and for plants of average efficiency.

A satisfactory insulation for the **walls, ceilings, floors, partitions, etc.,** of cold-storage buildings consists of corkboard of good quality and medium density with sides and edges dipped in hot asphalt or coated with asphaltic emulsion or other cold-setting adhesive, to seal the surfaces against moisture penetration. Interior finish is usually two ⅛-in-thick coats of mastic plaster, or two ¼-in-thick coats of portland-cement plaster. In self-supporting cork partition walls, ½-in-thick portland-cement mortar is often used as a strengthening core between layers. Vapor seal should be on the warm side.

The standard thickness for temperatures down to 32°F is two layers of 2-in corkboard; to this is added 1 in for each 15°F below (see Table 11). **Piping, fittings, shell vessels, etc.,** should be covered with molded covering and fitting covers in which the cork as molded weighs 1.1 lb per board ft approx. This molded covering should have the inner and outer surfaces sealed with a rubberized or asphaltic mastic and made up with the joints thoroughly cemented and filled. Standard **brine pipe covering** for temperatures down to 0°F varies from 2 to 3 in in thickness with the diameter of the pipe. "Special thick" brine covering for temperatures below 0°F is 3 to 4 in thick. Final treatment may be plaster, canvas, paint, or sheet metal.

Either aluminum or stainless steel, about 0.022 in thick, applied in multiple layers, with ⅝-in air spaces between layers, is now used as insulation in low-temperature test rooms.

**Table 10. Maximum Tons Refrigeration for Freon 12 Lines**

| Line sizes, in. | Suction-line (based on 105 F condensing temp) pressure drop, psi per 100 ft equivalent length at 40 F saturation | | | | | | Discharge-line condensing temp | | Liquid-line (based on 105 F condensing temp; 25 F evaporating temp) pressure drop, psi per 100 ft equivalent length (type L tubing) | | | |
|---|---|---|---|---|---|---|---|---|---|---|---|---|
| | ½ | 1 | 2 | 3 | 4 | 5 | 115 F | 90 F | 3 | 5 | 10 | 20 |
| ⅜ O.D. | .... | .... | .... | .... | .... | .... | .... | .... | 0.88 | 1.14 | 1.80 | 2.58 |
| ½ O.D. | 0.14 | 0.20 | 0.28 | 0.35 | 0.41 | 0.45 | .... | .... | 2.89 | 3.64 | 5.56 | 8.50 |
| ⅜ IPS | 0.17 | 0.24 | 0.34 | 0.42 | 0.49 | 0.54 | | | | | | |
| ⅝ O.D. | 0.25 | 0.35 | 0.51 | 0.62 | 0.73 | 0.81 | 1.43 | 1.15 | 4.86 | 6.81 | 10.2 | 15.8 |
| ½ IPS | 0.35 | 0.45 | 0.65 | 0.79 | 0.93 | 1.03 | 1.87 | 1.50 | 4.86 | 6.81 | 10.2 | 15.8 |
| ⅞ O.D. | 0.55 | 0.76 | 1.10 | 1.34 | 1.58 | 1.75 | 2.97 | 2.38 | 10.5 | 14.1 | 21.8 | 33.0 |
| ¾ IPS | 0.68 | 0.94 | 1.35 | 1.65 | 1.92 | 2.12 | 3.26 | 2.62 | 9.73 | 12.6 | 18.5 | 27.0 |
| 1⅛ O.D. | 1.26 | 1.80 | 2.57 | 3.17 | 3.76 | 4.15 | 5.05 | 4.05 | 21.4 | 28.2 | 41.3 | 60.8 |
| 1 IPS | 1.43 | 2.01 | 2.89 | 3.54 | 4.17 | 4.60 | 5.29 | 4.25 | 21.4 | 28.2 | 41.3 | 60.8 |
| 1⅜ O.D. | 2.21 | 3.12 | 4.45 | 5.50 | 6.38 | 7.05 | 7.72 | 6.19 | 36.9 | 48.1 | 70.5 | 101 |
| 1¼ IPS | 2.70 | 3.82 | 5.37 | 6.72 | 7.68 | 8.48 | 9.16 | 7.35 | 36.9 | 48.1 | 70.5 | 101 |
| 1⅝ O.D. | 3.40 | 4.78 | 6.79 | 8.42 | 9.77 | 10.8 | 10.92 | 8.75 | 62.0 | 80.2 | 114 | 160 |
| 1½ IPS | 4.05 | 5.75 | 8.10 | 10.12 | 11.6 | 12.8 | 12.5 | 10.0 | 62.0 | 80.2 | 114 | 160 |
| 2⅛ O.D. | 6.12 | 8.60 | 12.1 | 15.1 | 17.4 | 19.2 | 19.2 | 15.3 | | | | |
| 2 IPS | 7.66 | 10.9 | 15.3 | 19.2 | 22.2 | 24.5 | 20.6 | 16.5 | 124 | 161 | 231 | 328 |
| 2⅝ O.D. | 12.0 | 17.1 | 24.0 | 30.1 | 34.6 | 38.2 | 32.2 | 25.9 | | | | |
| 2½ IPS | 12.0 | 17.1 | 24.0 | 30.1 | 34.6 | 38.2 | 32.2 | 25.9 | 230 | 297 | 426 | 607 |
| 3⅛ O.D. | 19.1 | 27.2 | 38.2 | 47.8 | 55.0 | 60.7 | 51.5 | 39.8 | | | | |
| 3 IPS | 20.9 | 29.4 | 42.3 | 51.8 | 60.0 | 66.2 | 54.5 | 43.8 | 364 | 469 | 676 | 972 |
| 3⅝ O.D. | 27.8 | 39.7 | 55.7 | 69.8 | 80.3 | 88.7 | 72.0 | 57.6 | | | | |
| 3½ IPS | 30.2 | 43.2 | 61.0 | 76.1 | 87.0 | 96.0 | 78.8 | 62.3 | 539 | 704 | 1,005 | 1,430 |
| 4⅛ O.D. | 38.6 | 55.2 | 78.0 | 97.3 | 111 | 123 | 95.8 | 77.1 | | | | |
| 4 IPS | 40.7 | 58.6 | 83.0 | 103 | 118 | 130 | 101.6 | 81.6 | 753 | 972 | 1,385 | 1,945 |
| 5 IPS | 71.3 | 100 | 141 | 176 | 203 | 224 | 171.5 | 137.8 | | | | |
| 6 IPS | 126 | 183 | 257 | 322 | 366 | 403 | 266 | 214 | | | | |
| 8 IPS | 211 | 297 | 422 | 523 | 602 | 664 | 461 | 370 | | | | |
| 10 IPS | 352 | 503 | 712 | 887 | 1 024 | 1,130 | 725 | 582 | | | | |
| 12 IPS | 550 | 780 | 1,106 | 1,373 | 1 582 | 1,748 | 1,041 | 836 | | | | |

SOURCE: From ASRE, Air-Conditioning Refrigerating Data Book, 1955–1956 ed., based on ARI Equipment Standards; see current ASHRAE, Guide and Data Book, for more complete tables.

**Table 11. Cold-Storage Usage and Transmission Factors**

| | Room volume, cu ft | | | | | | | | | |
|---|---|---|---|---|---|---|---|---|---|---|
| | 20 | 50 | 100 | 500 | 1,000 | 2,000 | 5,000 | 10,000 | 50,000 | 100,000 |
| Usage load* | | | | | | | | | | |
| Average............ | 4.68 | 2.28 | 1.61 | 1.21 | 1.10 | 0.835 | 0.403 | 0.240 | 0.178 | 0.173 |
| Heavy.............. | 5.51 | 3.55 | 2.52 | 1.87 | 1.67 | 1.29 | 0.625 | 0.408 | 0.305 | 0.295 |
| Storage temp, deg F..... | 40 and above | | 25–40 | | 15–25 | | 0–15 | | −25–0 | |
| Insulation thickness, in.. | 3 | | 4 | | 5 | | 6 | | 8 | |
| Theoretical $U$ factor†.. | 0.086 | | 0.066 | | 0.055 | | 0.046 | | 0.035 | |
| Practical $U$ factor‡... | 0.111 | | 0.083 | | 0.066 | | 0.055 | | 0.046 | |
| Piping $U$ factor§..... | 2.5 | | 2.2 | | 2.0 | | 1.8 | | 1.6 | |

*Btu per cu ft per 24 h per °F temp diff between outside and inside. Usage load includes 10°F or less of product cooling, and infiltration losses, etc., but does not include any freezing of product or fan motor loads when unit coolers are used. Values are for normal conditions and should not be used for unusual product loads.

†Theoretical $U$ factor is based on corkboard ($k = 0.30$) with plaster on both sides, Btu per ft² per h per °F temp diff.

‡Practical $U$ factor allows for inefficiency of joints and structural supports in insulation.

§Piping $U$ factor is Btu/ft² outside pipe surface per h per °F temp diff between cooling medium and air (gravity circulation). Allowance has been made for frosting at the lower temperatures.

The outer layer may be $\frac{1}{16}$-in-thick or heavier for mechanical protection.

The scope of insulation has enlarged greatly, with cellular plastics (prefoamed or foamed-in-place) such as polystyrene and polyurethane, with powders (vacuum and gas-filled cavities), and with preformed insulated panels for field assembly into complete refrigerated warehouse constructions.

**Cold-Storage Temperatures**  A great deal of research and experience are required to obtain authoritative information on optimum temperatures and humidities for various products in cold storage. Table 12 is representative of good practice. The safe storage period depends upon the product and the storage temperature, and operational techniques vary greatly. Modern cold-storage warehouses of the larger concerns are cooled by brine which is furnished at two different temperatures only. The higher temperature for the mild-temperature warehouses is 10 to 12°F, and the low-temperature brine for the freezers is −10 to −12°F. All temperatures above that of the brine are obtained by regulating the amount of brine circulated in any particular set of coils. In the low-temperature warehouses, the piping is arranged for two classes of service: (1) **sharp freezers**, where the goods which are to be frozen are kept while their temperature is brought down quickly to the holding temperature (say in from 6 to 10 h) after which they are stored in (2) **holding rooms** where the desired temperature is maintained.

The system of cooling which is now being installed in the highest type of warehouses consists of a coil room containing the necessary brine coils, through which the air from the different rooms is circulated by a pressure blower. The inlet and outlet of each room are so arranged that the cooled circulating air will cover the entire room in its transit; this is usually accomplished by having the cold-air inlet in the center of the room and two return outlets—one at each end of the room. The piping ratio for the coil rooms in this system, assuming a high-grade insulation with 2 to 4 Btu transmission per ft² per 24 h per °F temp diff, should be 1 ft² of external pipe surface to 15 ft³ of space to be cooled (with brine at −10 to −12°F) for warehouses carrying temperatures of zero and below, and 1 ft² of pipe surface to 24 ft³ of space to be cooled (with brine at 10 to 12°F) in warehouses carrying mild temperatures of from 30 to 40°F. Another system is used in which the

blower and coils are supplemented by coils in the rooms. With this arrangement, it is possible to reduce temperatures quickly and hold them with the coils.

The average coil transmission in cold-storage rooms without forced circulation of the air is about 2 Btu per ft² of outside metal surface per h per °F temp diff with horizontal piping and 2.5 Btu with vertical piping. When forced air circulation is used, the transmission rate will increase to 20 Btu or more. In **brine circulation,** the brine, at same compressor back pressure, has a higher temperature than the ammonia, and consequently 1½ times as much pipe or more is used in brine circulation as in direct expansion for a given back pressure.

**Piping of Rooms**  The **size of pipe** usually employed for piping rooms varies from 1 to 2 in with either brine circulation or direct expansion.

The extra cost of liberal piping allowance will often be offset by the consequent improvement in the efficiency of operation of the compressor. An expansion valve should be provided for every 500-ft length of 1-in pipe, every 650 ft of 1¼-in pipe, and every 1,000 ft of 2-in pipe when direct expansion is used.

Values of **overall coefficients of heat transfer** in Btu/h·ft²·°F for refrigerating practice are given in Tables 11 and 13. (See Sec. 4).

**Brine circulation** is generally preferred to direct expansion so as to avoid danger from escaping ammonia or other refrigerant in case the pipes should leak. An advantage of the brine system is that there is always a considerable mass of refrigerated brine which can be drawn on in case the machinery should have to be stopped for any reason. In small plants, the general machinery may be stopped at night and only the brine pump be kept going to distribute the surplus refrigeration which has been accumulated in the brine during the day. Brine piping must consist of two lines, a flow and a return, usually of the same size. Brine storage is seldom used in large plants because of its bulk, its first cost, and the practical inability to store much refrigeration.

The **brine coils** in each cooled space are in parallel across the supply and return pipes. It has been common practice to allow 100 to 120 running feet of **pipe** per circuit for low temperatures, and 400 to 440 for high temperatures. The **tons refrigera-**

**Table 12. Product Storage Data**
(Bush Manufacturing Co., West Hartford, Conn.)

| Product | Quick-freeze temp, deg F | Storage temp | | Humidity, % R.H. | Specific heat | | Latent heat | Freezing point | Respiration, Btu per lb per day |
|---|---|---|---|---|---|---|---|---|---|
| | | Long | Short | | Above freezing | Below freezing | | | |
| Apples............. | —15 | 30–32 | 38–42 | 85–88 | 0.86 | 0.45 | 121.0 | 28.4 | 0.75 |
| Asparagus......... | —30 | 32 | 40 | 85–90 | 0.95 | 0.44 | 134.0 | 29.8 | |
| Bacon, fresh....... | .... | 0–5 | 36–40 | 80 | 0.55 | 0.31 | 30.0 | 25.0 | |
| Bananas........... | .... | 56–72 | 56–72 | 85–95 | 0.80 | 0.42 | 108.0 | 28.0 | 4.18 |
| Beans, green....... | .... | 32–34 | 40–45 | 85–90 | 0.92 | 0.47 | 128 | 29.7 | 3.3 |
| Beans, dried...... | .... | 36–40 | 50–60 | 70 | 0.30 | 0.237 | 18 | | |
| Beef, fresh, fat..... | —15 | 30–32 | 38–42 | 84 | 0.60 | 0.35 | 79 | | |
| Beef, fresh, lean.... | —15 | 30–32 | 38–42 | 85 | 0.77 | 0.40 | 100 | | |
| Beets, topped...... | .... | 32–35 | 45–50 | 95–98 | 0.86 | 0.47 | 129 | 31.1 | 2.0 |
| Blackberries....... | —15 | 31–32 | 42–45 | 80–85 | 0.89 | 0.46 | 125 | 28.9 | |
| Broccoli.......... | .... | 32–35 | 40–45 | 90–95 | 0.92 | 0.47 | 130 | 29.2 | |
| Butter............ | +15 | ..... | 40–45 | ..... | 0.64 | 0.34 | 15 | 15.0 | |
| Cabbage.......... | —30 | 32 | 45 | 90–95 | 0.93 | 0.47 | 130 | 31.2 | |
| Carrots, topped.... | —30 | 32 | 40–45 | 95–98 | 0.87 | 0.45 | 120 | 29.6 | 1.73 |
| Cauliflower........ | .... | 32 | 40–45 | 85–90 | 0.93 | 0.47 | 132 | 30.1 | |
| Celery............ | —30 | 31–32 | 45–50 | 90–95 | 0.95 | 0.48 | 135 | 29.7 | 2.27 |
| Cheese............ | +15 | 32–38 | 39–45 | ..... | 0.70 | | | | |
| Cherries.......... | .... | 31–32 | 40 | 80–85 | 0.87 | 0.45 | 120 | 26.0 | 6.6 |
| Chocolate coatings.. | .... | 45–50 | ..... | ..... | 0.3 | | | | |
| Corn, green........ | .... | 31–32 | 45 | 85–90 | 0.80 | 0.43 | 108 | 29.0 | 4.1 |
| Cranberries........ | .... | 36–40 | 40–45 | 85–90 | 0.90 | 0.46 | 124 | 27.3 | |
| Cream............ | .... | 34 | 40–45 | ..... | 0.88 | 0.37 | 84 | | |
| Cucumbers........ | .... | 45–50 | 45–50 | 80–85 | 0.97 | 0.47 | 137 | 30.5 | |
| Dates, cured....... | .... | 28 | 55–60 | 50–60 | 0.83 | 0.44 | 104 | | |
| Eggs, fresh........ | —10 | 30–31 | 38–45 | ..... | 0.76 | 0.40 | 98 | 31.0 | |
| Eggplants......... | .... | 45–50 | 46–50 | 85–90 | 0.94 | 0.47 | 132 | 30.4 | |
| Flowers........... | .... | 35–40 | ..... | 85–90 | | | | | |
| Fish, fresh, iced.... | —15 | 25 | 25–30 | ..... | 0.82 | 0.41 | 105 | 30.0 | |
| Fish, dried........ | .... | 30–40 | ..... | 60–70 | 0.56 | 0.34 | 65 | | |
| Furs............. | .... | 32–34 | 40–42 | 40–60 | | | | | |
| Furs, to shock..... | .... | 15 | 15 | | | | | | |
| Grapefruit......... | .... | 32 | 32 | 85–90 | 0.91 | 0.46 | 126 | 28.4 | 0.5 |
| Grapes............ | .... | 30–32 | 35–40 | 80–85 | 0.86 | 0.44 | 116 | 27.0 | 0.5 |
| Ham, fresh........ | .... | 28 | 36–40 | 80 | 0.68 | 0.38 | 87 | | |
| Honey........... | .... | 31–32 | 45–50 | ..... | 0.35 | 0.26 | 26 | | |
| Ice cream........ | —20 | ..... | 0–10 | ..... | 0.5–0.8 | 0.45 | 96 | | |
| Lard............. | .... | 32–34 | 40–45 | 80 | 0.52 | 0.31 | 90 | | |
| Lemons........... | .... | 55–58 | ..... | 80–85 | 0.92 | 0.46 | 127 | 28.1 | 0.4 |
| Lettuce........... | .... | 32 | 45 | 90–95 | 0.96 | 0.48 | 136 | 31.2 | 8.0 |
| Liver, fresh........ | .... | 32–34 | 36–38 | 83 | 0.72 | 0.42 | 94 | | |
| Lobster, boiled..... | .... | 25 | 36–40 | ..... | 0.81 | 0.42 | 105 | | |
| Maple syrup....... | .... | 31–32 | 45 | ..... | 0.24 | 0.215 | 7.0 | | |
| Meat, brined...... | .... | 31–32 | 40–45 | ..... | 0.75 | 0.36 | 75.0 | | |
| Melons............ | .... | 34–40 | 40–45 | 75–85 | 0.92 | 0.35 | 115 | 28.5 | 1.0 |
| Milk.............. | .... | 34–36 | 40–45 | ..... | 0.92 | 0.46 | 124 | 31.0 | |
| Mushrooms........ | .... | 32–35 | 55–60 | 80–85 | 0.93 | 0.47 | 130 | 30.2 | |
| Mutton........... | .... | 32–34 | 34–42 | 82 | 0.81 | 0.39 | 96 | 29.0 | |
| Nut meats......... | .... | 32–50 | 35–40 | 65–75 | 0.30 | 0.24 | 14 | 20.0 | |
| Oleomargarine..... | .... | 34–36 | ..... | ..... | 0.65 | 0.34 | 35 | 15.0 | |
| Onions............ | .... | 32 | 50–60 | 70–75 | 0.91 | 0.46 | 120 | 30.1 | 1.0 |
| Oranges........... | .... | 32–34 | 50 | 85–90 | 0.90 | 0.46 | 124 | 27.9 | 0.7 |
| Oysters........... | .... | ..... | 32–35 | ..... | 0.85 | 0.45 | 120.0 | | |
| Parsnips.......... | —30 | 32–34 | 34–40 | 90–95 | 0.82 | 0.45 | 120.0 | 28.9 | |
| Peaches, fresh...... | .... | 31–32 | 50 | 85–90 | 0.90 | 0.46 | 126 | 29.4 | 1.0 |
| Pears, fresh........ | .... | 29–31 | 40 | 85–90 | 0.86 | 0.45 | 118 | 28.0 | 6.6 |

**Table 12. Product Storage Data—(Continued)**

| Product | Quick-freeze temp, deg F | Storage temp | | Hu-midity, % R.H. | Specific heat | | Latent heat | Freezing point | Respira-tion, Btu per lb per day |
|---|---|---|---|---|---|---|---|---|---|
| | | Long | Short | | Above freezing | Below freezing | | | |
| Peas, green........ | .... | 32 | 40–45 | 85–90 | 0.80 | 0.42 | 108 | 30.0 | |
| Peas, dried........ | .... | 35–40 | 50–60 | ..... | 0.28 | 0.22 | 14 | | |
| Peppers.......... | .... | 32 | 40–45 | 85–90 | 0.94 | 0.47 | ..... | 30.1 | 2.35 |
| Pineapples, ripe.... | .... | 40–45 | 50 | 85–90 | 0.88 | 0.45 | 122 | 29.9 | |
| Plums............. | .... | 31–32 | 40–45 | 80–85 | 0.88 | 0.45 | 123 | 28.0 | |
| Pork, fresh........ | .... | 30 | 36–40 | 85 | 0.60 | 0.38 | 66 | 28.0 | |
| Potatoes, white.... | −30 | 36–50 | 45–60 | 85–90 | 0.77 | 0.44 | 105 | 28.9 | 0.85 |
| Poultry, dressed.... | −10 | 28–30 | 29–32 | ..... | 0.80 | 0.41 | 99 | 27 | |
| Pumpkins......... | .... | 50–55 | 55–60 | 70–75 | 0.92 | 0.47 | 130 | 30.2 | |
| Quinces........... | .... | 31–32 | 40–45 | 80–85 | 0.90 | .... | ..... | 28.1 | |
| Raspberries........ | .... | 31–32 | 40–45 | 80–85 | 0.89 | 0.46 | 125 | 30.0 | 3.3 |
| Sardines, canned... | .... | ..... | 35–40 | | 0.76 | 0.410 | 101 | | |
| Sausage, fresh..... | .... | 31–36 | 36–40 | 80 | 0.89 | | | | |
| Sauerkraut........ | .... | 33–36 | 36–38 | 85 | 0.91 | 0.47 | 128 | 26 | |
| Squash........... | .... | 50–55 | 55–60 | 70–75 | 0.92 | 0.47 | 130 | 29.3 | |
| Spinach.......... | .... | 32 | 45–50 | 85 | 0.94 | 0.48 | 132 | 30.8 | |
| Strawberries....... | −15 | 31–32 | 42–45 | 80–85 | 0.92 | 0.48 | 129 | 30.0 | 3.3 |
| Tomatoes, ripe..... | .... | 40–50 | 55–70 | 85–90 | 0.95 | 0.48 | 135 | 30.4 | 0.5 |
| Turnips........... | .... | 32 | 40–45 | 95–98 | 0.93 | 0.40 | 137 | 30.5 | 1.0 |
| Veal............. | −15 | 28–30 | 36–40 | ..... | 0.71 | 0.39 | 91 | 29 | |

tion produced by brine at various temperature differences and rates of pumping may be calculated approximately as follows: tons refrigeration = gal/min × °F range/28.

**Air Conditioning of Low-Temperature Storage Rooms** Bacterial growth and chemical decomposition are retarded by lowering the **temperature** of perishable goods. Too low a temperature will freeze the goods and may result in spoilage. A holding temperature above the freezing point of the article is best for storage conditions. The maintenance of a proper relative **humidity** in a storage space holding unwrapped goods is as important as the maintenance of the proper temperature. High humidity favors the growth of mold and bacteria. Low humidities rob the product of moisture resulting in losses in value through impaired appearance and lost weight. **Air motion** is of importance in maintaining uniform conditions throughout the storage space. Too high air velocities cause excessive drying, and stagnant air through high humidities will cause mold. An optimum air motion falling between stagnation on one side and excessive drying on the other must be selected.

Low-temperature conditioning for storage purposes is preferably accomplished by the use of **cold-diffuser methods.** The cold diffuser connected to or located in the storage space consists of a fan and cooling means. The cooling means may be either a brine or cold-water spray or a surface-type cooler using brine, or a volatile refrigerant such as ammonia, Freon, etc. Unitary equipment, performing the functions of the cold diffuser, is commercially available. For the requirements for various perishable products, see ASHRAE Guide and Data Books and Table 12.

**Table 13. Overall Coefficients of Heat Transfer**
(Btu/f².h.°F)

| | | | |
|---|---|---|---|
| Can ice-making piping | | Brine coolers | |
|   Old-style feed, non-flooded........... | 12–15 |   Shell and tube...................... | 45–100 |
|   Flooded.......................... | 20–40 |   Double pipe........................ | 150–300 |
|   High-velocity raceway trunk coils..... | 80–110 | Cooling coils | |
| Ammonia condensers | |   Boiling refrigerant to air in unit coolers. | 4–8 |
|   Submerged (obsolete)................ | 30–40 |   Water to air in unit coolers........... | 5–9 |
|   Atmospheric, gas entering at top...... | 60–65 |   Brine to unagitated air.............. | 2.2–2.8 |
|   Atmospheric, drip or bleeder........ | 125–200 |   Direct expansion to unagitated air*... | 1.6–2.5 |
|   Flooded.......................... | 125–150 | Water cooler, shell and coil............ | 15–25 |
|   Shell and tube..................... | 150–300 | Water cooler, shell and tube........... | 50–150 |
|   Double pipe....................... | 150–250 | Water cooler, shell and finned tube | |
| Baudelot coolers, counterflow, atmos-pheric type | |   (Freon)........................... | 30–150 |
|   Milk coolers...................... | 75 | Liquid-ammonia cooler, shell and coil | |
|   Cream coolers..................... | 60 |   accumulator...................... | 45 |
|   Oil coolers....................... | 10 | Air dehydrator | |
| Water coolers | |   Shell and coil (brine in coil) 1st coil... | 5.0 |
|   Direct expansion.................. | 60–150 |     2d coil.... | 3.0 |
|   Flooded.......................... | 100–200 |   Double pipe........................ | 6–7 |
| | |   Superheat remover, shell and tube...... | 15–25 |

Forced circulation of the air increases the coefficient to 1½ to 2½ times the values for still air. One inch of frost decreases the value 25 percent.

*U factor increases to 3.3 at 15°F temp diff with ammonia recirculation systems.

**Quick freezing** is freezing in 2 h or less; as employed for foods the temperatures may be as low as −20 to −50°F. The methods of freezing include direct immersion in cold brine or a brine spray with the commodity held in a metal container. Another procedure is the use of a freezing tunnel approximately 50 ft long, equipped with a stainless-steel conveyor belt. The commodities to be frozen pass through this tunnel in 15 to 30 min. The speed of the belt is variable, and the heat transfer is accomplished by high-velocity air circulation at a number of points in the tunnel, the circulation being transverse to the belt. The circulating air is cooled by brine sprays which maintain the required temperature and the high humidity necessary to prevent shrinkage. In another process, packages of the material to be quick-frozen are clamped between steel platens containing evaporating refrigerant. Successful results are obtained with temperatures as high as 0°F. Storage may be at −5 to −10°F and transportation at 10°F or lower.

**Cold-storage lockers** for individual family use are used in large numbers, particularly in suburban and rural areas. Lockers are rented usually on a yearly basis. A locker plant may have several hundred rental steel lockers, 72 × 20 × 30 in, 6 ft³ capacity each, placed in rooms at about 0°F; a room for cutting and wrapping the meat; and a sharp freezer room held at about −20°F. Two compressors are frequently specified or a single compressor using automatic control valves on the separate rooms. Ammonia and Freon 12, operating under direct expansion, are commonly used.

## ICE MAKING

Ice refrigeration is economical for some applications. The use of natural ice is fast disappearing. Manufactured ice is made by several methods. In the can system, galvanized cans containing 300 to 400 lb of water are immersed in brine for 25 to 40 h for freezing. Air dissolved in the water separates as bubbles causing opacity unless the water is agitated during the early stages of cooling, and any dissolved impurities come out of solution to form a central off-color core in the ice block unless the core is removed and replaced with fresh water just before freezing. The time and investment required for making ice in cans are often uneconomical. In the **Flakice** method (Crosby Field, *Trans. ASME*, May, 1951, pp. 347–357) refrigerant is sprayed against the inner wall of a flexible, slowly rotating cylinder which is partially immersed in a tank of water. As the cylinder rotates, a thin layer of ice forms on the outside and then is broken off as further rotation causes the cylinder surface to flex under the action of an internal cam. For a patent history of small ice machines of this type, refer to Crosby Field, *Refrig. Eng.*, **58**, Dec., 1950, p. 1163. In the **Pak-Ice** process (Taylor, *Refrig. Eng.*, **22**, 1931, p. 307) liquid ammonia is evaporated in an annular space formed by two concentric cylinders, the inner being corrugated. Water sprayed on the inner corrugated surface quickly freezes in a thin layer about 0.01 in thick and is continually removed by mechanical scrapers. In an **extrusion method** (Watt, *Trans. ASME*, 1949) a slightly tapered circular or rectangular vertical cylinder with large end up is open at the top, but closed by a ram at the bottom. This cylinder has a jacket in which refrigerant is evaporated directly against the cylinder walls. Water enters the cylinder and freezes to a tapered block at the start of the operation. As soon as this occurs, the ram lifts (about ¼ in), shearing the ice block from the cold cylinder walls. Water

admitted to this space quickly freezes, the ram again operates and the block of ice is again lifted, and shearing occurs again between the freshly formed ice and the cylinder walls, but not at the fresh ice and old ice interface. In this way, a continuous cylinder of ice is formed by the series of nested ice shells. The installation and upkeep costs are small and so is the space required per ton of ice produced; a pilot-plant 1 ton unit requires 3 ft² of floor space.

The refrigeration tonnage required for making 1 ton of ice varies from 1.4 with 50°F inlet water temperature to about 1.7 if the water to be frozen enters as high as 80°F.

## SKATING RINKS

Rinks for ice skating, hockey, or curling vary in size from about 400 ft² area to 16,000 ft² or more. Construction of the ice floor is important. For a low-cost, efficient unit, brine pipes are laid directly in sand, which is kept wet during freezing to increase heat transfer. As freezing progresses, a layer of ice is built up by spraying water on the surface. The expense of maintaining such a floor in a satisfactory level condition may be considerable, and corrosion of the pipes, from the outside in, in contact with the wet sand may be excessive, particularly during shutdown periods. Moreover, it is not possible to use such a floor for other purposes. When a floor for diversified activities is needed, it is usual to provide a lower layer of insulation, primarily to prevent sweating on ceilings under the floor, surmounted by a concrete slab in which the steel pipes carrying the brine are imbedded and on which the ice layer is formed. This construction protects the pipe from external corrosion, affords fast freezing, and makes possible use of the floor in a few hours for purposes where an ice surface is not desired. Floor pipes are usually 1 or 1¼ in diam spaced 3 to 6 in on centers; care must be exercised to ensure a uniform distribution of brine. Brine flow may be 10 gal/min per ton of refrigeration, corresponding to about 1 gal/min per 10 to 20 ft² of piping surface. In a few installations, cold brine is sprayed directly against the under side of a steel floor on which the ice layer is frozen. Close control of ice surface temperature is necessary for good skating, usually 27°F + 1.0. Refrigeration capacity should be from 0.4 to 0.85 ton per 100 ft² of floor surface, although the load may be much higher for open-air rinks or other unusual conditions.

## REFRIGERATION IN THE CHEMICAL INDUSTRIES

The use of refrigeration in petroleum and other industries is widespread. In petroleum processing, refrigeration is used (1) to control vapor pressure of highly volatile constituents (methane, ethane, propane, and butane), such control being necessary during distillation, during processing, and for recovery of gasoline fractions from natural gas; (2) to shift the solubility relationship so that undesired constituents such as asphalt and wax in lubricating oils may be removed by precipitation; (3) to produce selective chemical reaction such as occurs when sulphuric acid is used to remove gumforming constituents from light fuels or when an alkylate fuel fraction is formed by combining a low molecular weight unsaturated with a similar saturated hydrocarbon. Aklylate is a valuable and important constituent of aviation fuels.

For (1), temperatures ranging from −35°F or lower to as

high as 50 or 60°F are required, the refrigerant being ammonia, propane, or butane. For the higher temperatures spray-cooled water or jet refrigerating units (see above) are sometimes satisfactory. For (2), it is possible to use propane as a combined solvent and refrigerant, evaporation of the propane causing the required cooling, temperature level (approximately −40°F) being controlled by the pressure held on the equipment. The propane vapor is recompressed, condensed, and reused as in any compression cycle. As much as ⅓ ton of refrigeration per barrel of lubricating oil dewaxed is needed. For (3), temperatures of 20 to 60°F are general, and ammonia is a satisfactory and widely used refrigerant.

The trend in the petroleum industry, as recovery of the lightest petroleum fractions is more widely practiced, and as better and more carefully prepared products are demanded, is toward more widespread use of refrigeration and toward lower temperatures. It is to be noted that, even at low temperatures, use of refrigeration is economical if efficient heat-transfer equipment is used to cool the incoming warm streams to the temperature of the leaving cold streams; in the ideal case, refrigeration is required only for the removal of heat liberated at the low temperature.

### DEEP REFRIGERATION

**Dry ice** (solid $CO_2$) is useful as a refrigerating medium in special cases. The main objection to the general use of dry ice is its cost, but the ease in its handling, its low temperature (−110°F at atmospheric pressure), its noncorrosive and nontoxic properties, its high latent heat, and the absence of liquid drip make it desirable. The heat absorbed per pound of solid $CO_2$ during sublimation at atmospheric pressure and −110°F is 245 Btu approx. The specific heat of the gas at constant pressure is about 0.2.

Carbon dioxide is obtained commercially either by fermentation or by burning. The gas is compressed, usually in three stages, and cooled to atmospheric temperature, thereby forming a liquid at 900 to 1,000 psia; the liquid is then throttled to below 5.3 atm pressure. During throttling, part of the $CO_2$ solidifies and is compressed to form a dense ice. The fraction that returns to the vapor phase is recompressed.

**Production of Solid Carbon Dioxide**  Figure 16 shows a simple cycle and Fig. 17 the elemental Ph diagram for the

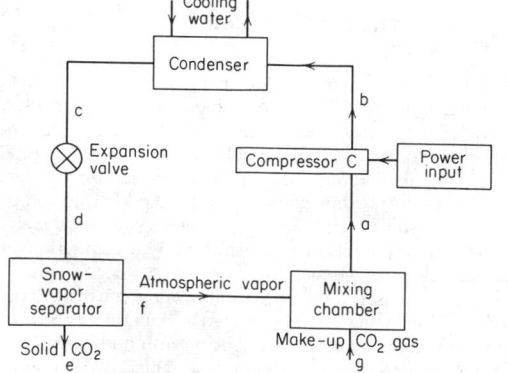

**Fig. 16**  Elemental dry-ice production circuit.

production of solid carbon dioxide. From the high pressure at atmospheric temperature the liquid is throttled (isenthalpic) to atmospheric pressure, the resultant point being within the sublimation solid-vapor zone below the triple-point realm. In this resultant throttling expansion, part of the $CO_2$ solidifies; it is then removed in the separator from the residual vapor and is

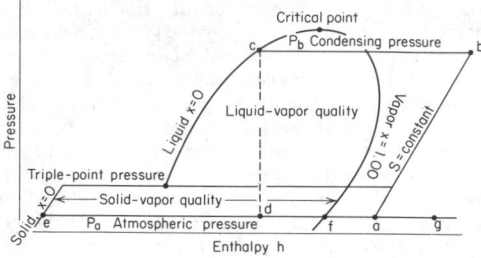

**Fig. 17**  Pressure-enthalpy diagram illustrating dry-ice production circuit.

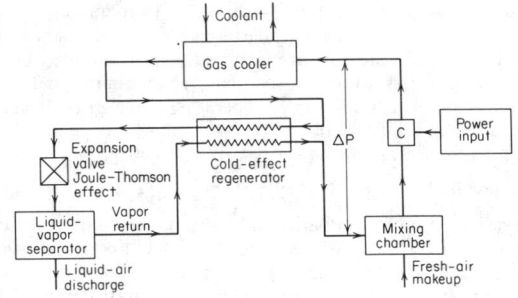

**Fig. 18**  Elemental Linde-system circuit.

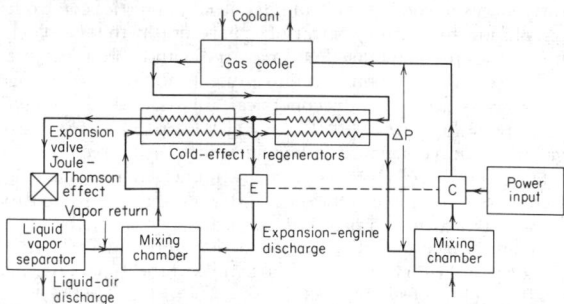

**Fig. 19**  Elemental Claude-system circuit.

compacted to form a dense $CO_2$ ice. The fraction representing the vapor phase is augmented by fresh makeup gas to continue the production cycle.

Beyond the manufacture of dry ice, operations at deep refrigerations, such as air liquefaction (at −312°F approx), lead into **cryogenics** (see Sec. 19). Illustrative systems are (1) Linde (throttling) and (2) Claude (expander). In the elemental **Linde system** (Fig. 18), by virtue of the "cold-effect" regenerator, the process air is reduced in temperature by the return fraction of the air from the separator. The throttling process thus ends in the "wet zone" at atmospheric pressure with generation of low-temperature liquid air that flows from the separator. In the **Claude system** (Fig. 19), use is made of an expander (whose work output reduces the net work requirement of the compressor) and of an expansion valve with the Joule-Thomson effect.

# CRYOGENICS

### by K. D. Williamson, Jr., F. J. Edeskuty, and F. G. Brickwedde

(Work supported by the U.S. Atomic Energy Commission.)

REFERENCES: Timmerhaus (ed.), "Advances in Cryogenic Engineering" (annual), Plenum Press. Vance and Duke, "Applied Cryogenic Engineering," Wiley. Scott, "Cryogenic Engineering," Van Nostrand. Vance (ed.), "Cryogenic Technology," Wiley. Mendelssohn, "Cryophysics," Interscience. McClintock, "Cryogenics," Reinhold. Scott, Denton, and Nicholls (eds.), "Technology and Uses of Liquid Hydrogen," Pergamon. Hoare, Jackson, and Kurti, "Experimental Cryophysics," Butterworth. White, "Experimental Techniques in Low-temperature Physics," Oxford. Johnson (ed.), "Compendium of Properties of Materials at Low Temperature," WADD Technical Report 60-56, U.S. Dept. of Commerce. Durham, McClintock, and Reed, "Cryogenic Materials Data Handbook," PB 171-809, U.S. Dept. of Commerce. Cook (ed.), "Argon, Helium, and the Rare Gases," Interscience. The National Bureau of Standards, Cryogenic Engineering Laboratory at Boulder, Colorado, maintains the Cryogenic Data Center which is an excellent source of information about cryogenics. Perry, Chilton, and Kirkpatrick, "Chemical Engineers' Handbook," McGraw-Hill. Barron, "Cryogenic Systems," McGraw-Hill. Bailey, "Advanced Cryogenics," Plenum. Harelden, "Cryogenic Fundamentals," Academic.

## Notation

Some symbols in this section differ from the notation of engineering practice but are retained to facilitate use of the prevalent scientific sources of information.

$A$ = area, ft$^2$ (m$^2$)
b.c.c. = body centered cubic lattice structure
$C$ = specific heat, Btu/lbm·°R (J/kg·K)
$C_p, C_v$ = specific heats at constant pressure and volume respectively, Btu/lbm·°R (J/kg·K)
f.c.c. = face centered cubic lattic structure
$H$ = enthalpy, Btu/lbm (J/kg)
$H$ = magnetic field, G($T$)
h.c.p. = hexagonal close-packed lattice structure
$\kappa$ = thermal conductivity, Btu/h·ft·°R (W/m·K)
$\sigma$ = electrical conductivity (reciprocal of resistivity) or pair separation for zero potential in Lenard-Jones equation
$s$ = Boltzmann's constant
$L$ = length, ft (m)
LH$_2$ = liquid hydrogen (H$_2$)
LHe = liquid helium (He)
LN$_2$ = liquid nitrogen (N$_2$)
LNG = liquefied natural gas
LO$_2$ = liquid oxygen
$n$ = number of moles
$P$ = pressure, psia or torr (vacuum) (Pa)
$R_e$ = electrical resistance, $\Omega$
$T$ = temperature, °R (K)

$V$ = volume, ft$^3$ (m$^3$)
$\alpha$ = thermal linear-expansion coefficient, °R$^{-1}$, or accommodation coefficient
$\eta$ = viscosity, lbm/ft·h (N·s/m$^2$)
$\rho$ = density, lbm/ft$^3$ (kg/m$^3$)
$\Delta T$ = temp difference, °R (K)

*Denotes property divided by quantum parameter (Table 4) which is not the critical point value.

**Cryogenics** is the study, production, and utilization of low temperatures. Cryogenic temperatures have been defined so ambiguously that "upper limits" to the cryogenic range from 216 to 396°R (120 K to 220 K) may be found in the literature. In this section the cryogenic range for a given property is considered to embrace the scale between absolute zero and the temperature above which the property has the expected or normal behavior. Cryogenics thus embraces the unusual and unexpected variations which appear at low temperatures and make extrapolations of properties from ambient to low temperatures unreliable.

Progressively lower temperatures become increasingly difficult to attain in practice. As the working temperature of a refrigerator is lowered, the work required to transfer a given amount of heat increases as demonstrated by the Carnot limitation, to wit, $W = Q[(T_1 - T_2)/T_2]$, where $W$ is the work required to extract the heat $Q$ at a low temperature $T_2$ and reject it at a higher temperature $T_1$ (see Sec. 4). The actual work is always greater than this because of inefficiencies of mechanical equipment, thermal losses associated with finite temperature differences in heat exchangers, and heat leaks from the surroundings to the cold equipment.

### Refrigeration Methods

(See also Sec. 4 and Mechanical Refrigeration, this section.)

**Cryogenic refrigerators** may be classified by (1) the functions they perform (e.g., the delivery of liquid cryogens, the separation of mixtures of gases, and the maintenance of spaces at cryogenic temperatures, (2) their refrigerating capacities and (3) the temperatures they reach. Large industrial-sized plants (*a*) deliver LNG (~120 K), LO$_2$ (~83 K), LN$_2$ (~77 K), LH$_2$ (~20 K), and LHe (4 K), (*b*) separate gaseous mixtures, e.g., the constituents of the atmosphere, H$_2$ from petroleum refinery gases, H$_2$ and CO from coke oven and coal-water gas reactors, and He from natural gas, and (*c*) provide refrigeration to maintain spaces at low temperatures. For the latter, i.e., (*c*), units have been built to give 7 kW of refrigeration at 20 K, 2.5 kW at 4 K, and 350 W at 1.85 K.

An important area of refrigerator development that is being commercially exploited now is laboratory-sized cryogenic refrigerators for laboratory research and development at LHe temperatures (<5 K). These refrigerators are used for many different purposes, refrigerating for example: high-field superconducting electromagnets, LHe bubble chambers for high-energy (nuclear) particle research, experimental superconducting power transmission lines, experimental superconducting electric generators and motors and superconducting magnets for levitation of railroad trains. Another area of commercial exploitation that is important for the progress of cryogenic physics research is the development of refrigerators for the continuous production of refrigeration at temperatures below 1 K. These include L³He evaporation refrigerators and L³He-L⁴He dilution refrigerators that reach temperatures of 0.4 and 0.003 K, respectively.

Of various methods of refrigeration, the most commonly used to produce temperatures as low as 1 K are (1) the evaporation of a volatile liquid (referred to as the **cascade** method when applied in several successive stages using progressively lower-boiling liquids), (2) Joule-Thomson (isenthalpic) expansion of a compressed gas, and (3) an adiabatic (isentropic) expansion of a compressed gas in an engine (reciprocating or turbine) or from a bomb or cylinder through a throttling valve (Simon expansion). Using L³He, method (1) is capable of reaching temperatures as low as ~0.4 K. Numerous refrigeration cycles have been devised utilizing various combinations of the above three methods. The final stage of refrigeration in plants that deliver liquid cryogens is generally a Joule-Thomson (isenthalpic) expansion. **Expansion engines** (reciprocating and turbine) are commonly used for producing the refrigeration needed to reach the final stage. Turbine expanders are preferred for the large refrigerators because their efficiencies (70 to 85 percent) are higher than for reciprocating expanders. The efficiencies of the heat exchangers for counterflowing "cold" and "warm" gases are an important factor in the overall refrigerator efficiency. In industrial plants heat exchanger efficiencies reach 98 percent. Their design represents a compromise between the attainment of a high heat-transfer coefficient, on the one hand, and a low resistance to the flow on the other.

Figure 1 typifies a commonly employed modified Brayton cycle which uses an expansion engine for generating the refrigeration. The theoretical figure of merit (the ratio of the work $W_c$ done by the compressor to the heat $Q$, transferred to the refrigerant at the low temperature) is for this cycle

$$\frac{W_c}{Q} = \frac{RT_1 \ln (P_1/P_5)}{M(H_4 - H_3)}$$

where $M$ is the molecular weight of the gas whose enthalpy per unit mass is $H$. The other symbols are defined by reference in Fig. 1.

The **Stirling refrigerating engine** and modifications of it are used in a number of commercial makes of laboratory and miniature-sized refrigerators. The Stirling-cycle refrigerator is well suited to (1) the liquefaction of air on a laboratory scale (~7 l/h), (2) the recondensation of evaporated liquid cryogens, and (3) the refrigeration of closed spaces where the refrigeration load is not very large. These laboratory scale refrigerators supply ~1 kW of refrigeration at ~80 K and ~2 kW at 160 K. They are used with liquid air fractionating columns for the production of LN₂ (~7 l/h) and LO₂ (~5 l/h).

The Stirling-cycle refrigerator consists of a piston for compressing isothermally the working fluid (usually He), and a displacer operating in the same cylinder with the piston. The displacer and piston are connected to the same electrically driven shaft but displaced in phase by 90°. The displacer pushes compressed gas isochorically from the warm region where it was compressed, through the regenerator into the cold region where the compressed gas is expanded isothermally doing work on the piston and producing refrigeration. The regenerator consists of a porous mass (packed metal wool of high heat capacity) in which a steep temperature gradient is established between the warm region of compression and the cold region of expansion. The displacer returns the gas after expansion to the region for compression through the regenerator in which it is warmed to the temperature of the isothermal compression. The refrigerant (He) is recycled. The regenerators in the larger refrigerators are usually placed around the outside of the engine cylinders, but in small capacity, lower temperature refrigerators they are put inside the displacers.

In Stirling refrigerators that reach temperatures of 17 K and produce 1 W of refrigeration at 25 K, the compressed He is expanded in two or three stages at temperatures intermediate between room temperature and the lowest temperature reached. There is only a single piston for compression. The expansion in stages allows part of the heat that leaks into the coldest region of a single stage engine to be absorbed and removed at a higher (intermediate) temperature where the efficiency for transferring heat is greater.

The **Vuilleumier refrigerating engine** is a modification of the Stirling refrigerator. It *resembles* two single Stirling engines placed back to back. One of the engines operates as a heat engine and the other as a refrigerator. The lower temperature at which the heat engine discharges its waste heat, is also the top temperature at which the refrigerator engine discharges its waste heat. Hence, the Vuilleumier refrigerator operates at elevated, intermediate and low temperature levels which may be 800, 300 (ambient), and 90 K, respectively. Each engine has a cylinder, a displacer, and a regenerator, and the two engines are connected to a common crankshaft but displaced by 90°. Very little external power is needed to operate the two displacers which are operated in sinusoidal motion, displaced 90° in phase. The working fluid (He) is recycled.

The **Gifford-McMahon refrigerator** is another modification of

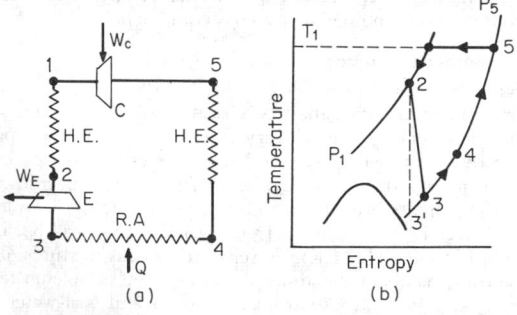

**Fig. 1**   (*a*) Schematic of a modified (isothermal compression) Brayton cycle, and (*b*) the process path. $W_c$ is the energy (work) to drive the compressor C, and Q, the heat absorbed by the refrigerator from the refrigerated area R.A., where the working fluid passes from state 3 to state 4. The heat-exchanger (H.E.) processes are isobaric. The dashed line 2–3' is the ideal isentropic expansion; 2–3 represents the actual expansion in the expander E.

the Stirling-cycle refrigerator. It may be single stage for refrigeration at higher temperatures (~80 K), or multistage for lower temperatures. Each stage has a cylinder, a displacer and a regenerator. The displacer pushes gas (usually He) under a very small head of pressure from the top of the cylinder, where the temperature is ambient, through the regenerator, into the bottom of the cylinder which is the cold region of the refrigerator. Compressed He is supplied from an external source—the refrigerator has no piston for compressing gas. Only a small source of external energy is needed for driving the displacer which has very little work to do in overcoming the forces of friction. The valves that control the admission of compressed He to the top of the cylinder, and the expansion of the cold compressed He from the bottom of the cylinder are externally operated by a drive mechanism. The cycling rate is <2 Hz. The expansion of the cold He is of the adiabatic-Simon-bomb type. Expanded gas is discharged from the refrigerator.

A method for reaching temperatures lower than 0.3 K utilizes a paramagnetic salt which is isothermally magnetized at a higher temperature (e.g., ~1.5 K) followed by an adiabatic (isentropic) reduction or removal of the magnetic field. Although materials can be cooled by a single adiabatic "demagnetization" to very low temperatures (0.001$_5$ K), this method has not as yet been successfully exploited for maintaining constant low temperatures. The continuously operating paramagnetic salt refrigerator marketed for a time has been superseded by the boiling L$^3$He and the L$^3$He-L$^4$He dilution refrigerators whose practical low limits are 0.4 K and 0.003 K, respectively.

**$^3$He-$^4$He dilution refrigerators** for reaching temperatures between 0.01 and 0.3 K and generating continuously as much as 750 ergs/s of refrigeration are commercially available. Refrigeration results from the solution of L$^3$He in L$^4$He, the heat of solution being negative. Figure 2 is a schematic diagram of a dilution refrigerator. $^3$He vapor from the "still" at ~1 K (the upper operating temperature of the refrigerator) is collected and returned by a pump to the "mixing" chamber

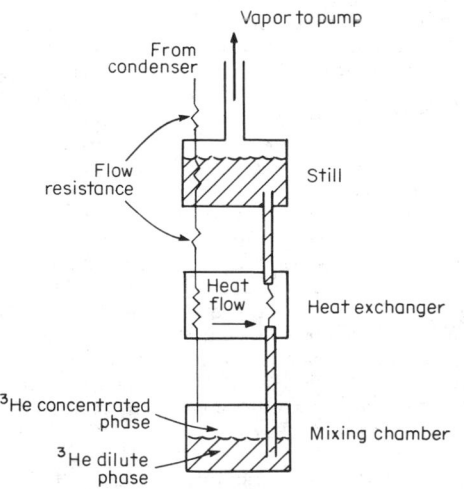

**Fig. 2** Schematic diagram of the mixing chamber and still of a $^3$He-$^4$He dilution refrigerator.

where solution takes place. The $^3$He arrives at the mixing chamber as L$^3$He near the mixing chamber temperature (the lower operating temperature of the refrigerator) after flowing through a heat exchanger counter to the flow of cold $^3$He-$^4$He solution on its way to the "still." In the electrically heated still, $^3$He is evaporated from the solution. The evaporated $^3$He is withdrawn by the collecting pump (diffusion) that returns the $^3$He to the mixing chamber. At 1 K the vapor pressure of L$^4$He is negligible. L$^4$He in the still, freed of $^3$He, returns to the mixing chamber by superfluid flow, building up in the mixing chamber an osmotic pressure that drives $^3$He atoms towards the still against the forces of viscous flow.

In the mixing chamber there is a two-phase separation of $^3$He and $^4$He. A layer of nearly pure L$^3$He of lower density rides on a heavier L$^4$He-rich layer containing ~6 percent $^3$He. At temperatures lower than 0.87 K liquid solutions of $^3$He and $^4$He separate into two liquid phases, one $^3$He-rich and the other $^4$He-rich, in thermodynamic equilibrium. At 0 K, the $^3$He-rich phase is 100 percent $^3$He, whereas the $^4$He-rich phase (in equilibrium with L$^3$He phase) contains ~6 percent $^3$He.

In the mixing chamber, L$^3$He enters the upper $^3$He-rich phase; solution of $^3$He in $^4$He takes place at the interphase boundary. The lower (denser) $^4$He-rich phase connects with the still.

Starting at 50 mK (a temperature reached in a $^3$He-$^4$He dilution refrigerator), temperatures in the 2- to 3-mK range are attainable with a **Pomaranchuk refrigerator.** Refrigeration is generated by compression of a mixture of liquid and solid phases of $^3$He at pressures in excess of 28.9 atm, the minimum $P$ at which these phases can coexist in equilibrium. The entropy of solid $^3$He exceeds the entropy of L$^3$He, which is contrary to the normal behavior for other substances. This occurs in $^3$He because of its nuclear magnetic properties. The $^3$He nucleus is magnetic and at temperatures in the mK range and above, in solid $^3$He, the nuclear moments are randomly oriented whereas in L$^3$He, at temperatures lower than ~0.3 K, the nuclear moments are paired or partially paired, antiferromagnetically. An adiabatic increase of $P$ converts liquid to solid with a reduction in $T$, whereas an isothermal increase of $P$ results in an absorption of heat. $^3$He, condensed, is confined in a container with flexible metal walls in the "cold" region of a $^3$He-$^4$He dilution refrigerator. The $^3$He container is surrounded with L$^4$He which serves as the pressure transmitting fluid.

### Gas Liquefaction

A conventional method of gas liquefaction utilizes a gas compressor, a countercurrent heat exchanger, and a throttling valve through which the gas expands isenthalpically (Joule-Thomson). After expansion, the cold gas returns through the heat exchanger, in which it exchanges heat with the countercurrent compressed gas flowing to the throttling valve. It leaves the heat exchanger near the temperature of the entering compressed gas. The temperature decreases at the throttling valve until the condensing temperature is reached, and then liquefaction occurs at a rate determined by the rate of refrigeration.

The rate of liquefaction $x$ in pounds per hour is $x = \{[H$ (expanded *gas* out) $- H$ (compressed *gas* in)]$w - q\}/[H$ (compressed *gas* in) $- H$ (*liquid* following valve)], where the $H$'s are enthalpies in Btu per pound for the final heat exchanger, $w$ is the flow rate in the same units as $x$, and $q$ is the heat leak in Btu per hour from outside into the heat exchanger. For an ideal heat exchanger the expanded gas leaves at the temperature of the entering compressed gas. In practice, there is a small difference in temperature which represents a small loss of refrigeration. For ideal gases, $H$ is independent of $P$ for a given $T$, and hence no liquefaction of an ideal gas results. The $H$'s of air, $O_2$, $N_2$, $A$, $CO_2$, the hydrocarbons, and the normal refrigerants decrease with increasing pressure at ambient $T$, except at very high $P$'s, and these gases are liquefiable by this kind of isenthalpic (Joule-Thomson) expansion. The $H$'s of $H_2$ and He at ambient $T$, however, increase with increasing $P$, even at low $P$'s, and hence an auxiliary mechanism is required for the liquefaction of these gases. In one method, the stream of compressed gas is split in the liquefier, and a part goes to an engine in which it is cooled by an isentropic expansion with the performance of work. This part of the flow, thus cooled, is sent to a heat exchanger, where it flows counter to the other fraction of compressed gas, cooling it to a temperature below the inversion temperature (see Sec. 4). This flow of compressed gas, thus precooled, is run through another and final heat exchanger with a throttling valve at its lower end, with the result that the quantity $x$ is liquefied.

Another method of precooling $H_2$ and He below their inversion temperatures makes use of a boiling liquid cryogen through which the compressed gas flows in a heat exchanger. $LN_2$ is used for precooling $H_2$, and $LH_2$ for He.

### Applications of Cryogenics

Gases, such as $O_2$, $N_2$, natural gas, $H_2$, and He, are liquefied for transportation. Shipments have been made of $LH_2$ by trailer trucks in quantities of $13 \times 10^3$ gal (50 m³) for thousands of miles and by railway tank cars holding as much as $28 \times 10^3$ gal (100 m³). LHe is regularly shipped in quantities from 25 gal (0.1 m³) to 10,000 gal (40 m³). Liquid cryogens are used as liquid refrigerants. The aerospace, steel, and bottled-gas industries are the principal users of liquefied cryogens. Important applications as coolants can be found in vacuum technology, electronics, biology, medicine, and metal forming.

In recent years superconductivity has found industrial applications in the form of superconducting magnets. Readily available are magnets of Nb-Ti with fields up to 80 kG (8 T) with large magnetic field volumes (1 m³). Higher field magnets [up to 150 kG (15 T)] of $Nb_3SN$ are available with smaller magnetic field volumes ($10^{-4}$ m³). Development of ac and dc superconducting electric power transmission lines as well as magnetic electrical energy storage and train levitation are being developed.

Cryogenics is also becoming increasingly important in the storage and transport of energy. Large quantities of natural gas are routinely transported as a liquid (LNG). Peak shaving in municipal gas systems is accomplished by storage of LNG. If, due to depletion of fossil fuels, hydrogen becomes an important, synthetic portable fuel, cryogenic hydrogen will play an important role. An already obvious potential use for $LH_2$ is aircraft fuel. This may be extended to surface transportation applications.

### Properties of Solids at Low Temperatures

(See also Sec. 4)

**Specific heats** of solids in general decrease with decreasing $T$, becoming zero at 0°R (Fig. 3). The downward approach to $C_p$

= 0 at 0°R is interrupted for some substances (principally compounds) by "bumps" on the curve (excess $C_p$ that rises to a maximum and then decreases). Paramagnetic salts undergoing transitions to either a ferromagnetic or an antiferromagnetic state are examples. This excess specific heat of a paramagnetic

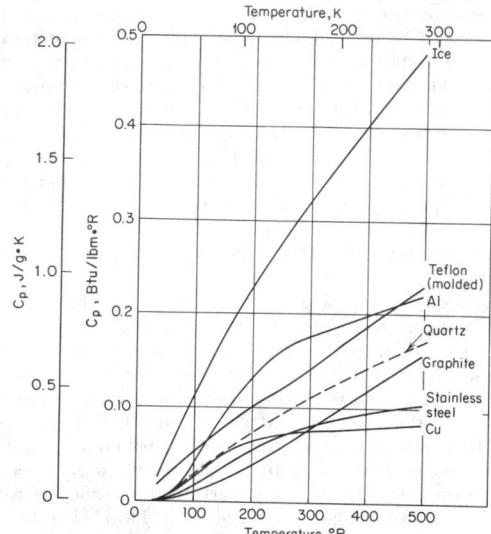

**Fig. 3**  Specific heats of solids at low temperatures.

salt is connected with the effectiveness of the salt for reaching low temperatures by the method of adiabatic demagnetization. There are also transitions in solids from a more orderly to a less orderly arrangement of atoms and molecules in the lattice that give rise to excess $C_p$. The transition in solid ortho- and normal-$H_2$ below 20°R (11 K) is an example. In Fig. 3, $C_p$'s are plotted for various materials.

**Heat is transferred** in dielectrics by lattice vibrations, or waves. In good electrical conductors, heat is transferred principally by the conduction electrons, and thermal and electrical conductivities are related by the Wiedemann-Franz law: $\kappa/\sigma T$ = const. This means that the ratio of the thermal and electrical conductivities at a given $T$ is approximately the same for the good conductors. In the poor electrical conductors, alloys for example, both lattice waves and conduction electrons play important parts in the transfer of heat. The $\kappa$'s of "pure" dielectrics and "pure" metals rise with increasing $T$ from $\kappa$ = 0 at 0°R (proportional to $T$ for metals and to $T^3$ for dielectrics), reach a maximum, normally between 10 and 100°R (5 and 55 K), and then decrease to a value approximately independent of $T$ (Fig. 5 and ice in Fig. 4). The $\kappa$'s of alloys (Fig. 4) are an order of magnitude smaller than for "pure" metals and *do not* exhibit the maxima characteristic of the "pure" metals at low $T$. Lattice disorder introduced by alloying, even in small amounts, and working a metal, even a pure metal, reduces $\kappa$. Annealing, in general, raises $\kappa$.

At 0°R, an absolutely pure and perfect single crystal of metal would have zero electrical resistance, $R_e = 0$. **Electrical resistance** arises from the scattering of the conduction electrons as they move through the lattice of metal ions under the influence of an externally applied electric field. Scattering arises for two reasons: (1) the amplitudes of the thermal vibrations of the lattice which increase with $T$ at a rate proportional to $\sqrt{C}$ of the metal, and (2) the imperfections in the regularity of the lattice as caused by impurity atoms (solid-solution alloys included), lattice vacancies, dislocations, and

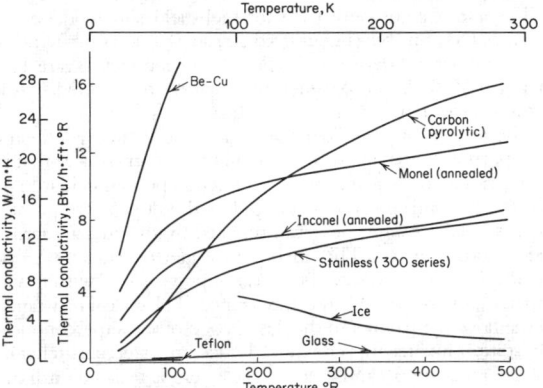

**Fig. 4**  Thermal conductivity of solids at low temperatures, part 1.

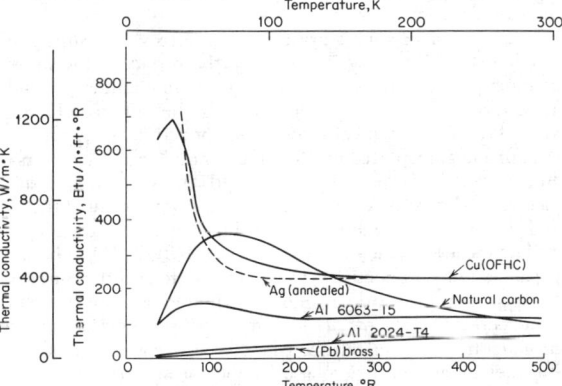

**Fig. 5**  Thermal conductivity of solids at low temperatures, part 2.

grain boundaries. Resistances for "pure" metals increase with $T$ and are roughly proportional to $C : T$, except at very small $T$'s where they are proportional to $T^5$ for absolutely pure crystals. The resistance due to impurities and imperfections is very roughly independent of $T$ (Matthiessen's rule). For very pure metals, $R_e$ is approximately constant below 20°R (11 K). Magnetic impurities can give rise to a minimum in resistivity [usually below 36°R (20 K)] called the Kondo effect. Resistivity in $\Omega$·cm (Fig. 6) is the $R_e$ of a cm cube.

Some metals, including elements (none from column 1 of the Mendeleeff table or from the ferromagnetic elements), intermetallic compounds, and alloys exhibit the phenomenon of **superconductivity.** Their electrical resistance is zero from 0°R to a transition $T$ at which normal resistance (value extrapolated from higher $T$) appears. In pure, single crystals the transition range $\Delta T$, from $R_e = 0$ to the normal value, may be as small as a few millidegrees. The highest known transition $T$ is about 23 K for $Nb_3Ge$. Closed circuits consisting entirely of superconducting metals can support persistent, resistanceless

currents without an external source of voltage. A superconducting circuit maintains constant the value of the total flux enclosed by the circuit at the time it entered the superconducting state. Hence, superconducting metals influence magnetic fields in their environment. Sometimes their use in the construction of apparatus and equipment for $T$'s lower than 15°R (8 K) has been avoided for this reason. Lead brasses and some solders, in particular Pb-Sn alloys, become superconductors.

**Superconductivity** is characterized by (1) perfect electrical conductivity ($\sigma \rightarrow \infty$), and (2) the Meissner effect, which is $B$ (the internal magnetic induction) is zero when $H$ (external) $\neq$ 0. Persistent currents at the surface of a specimen shield the interior from $H$ (external). $H$ parallel to the surface of a specimen is continuous at the surface and falls off exponentially below the surface. The penetration depth of $H$ for perfect (chemically and mechanically) specimens is of the order of $5 \times 10^{-8}$ m for $T \ll T_0$ (the superconducting transition temperature). The penetration depth is larger for alloys, specimens with lattice imperfections, and all superconductors near their superconducting transition temperatures $T_0$.

The normal-to-superconducting transition temperature $T_c$ is a function of $H$ (external): $T_c = T_0[1 - (H_c/H_0)]^{1/2}$ where $H_0$ is $H_c$ at 0 K. The superconducting state exists when $T < T_c$ and $H < H_c$, and the normally resisting state when $T > T_c$ or $H > H_c$.

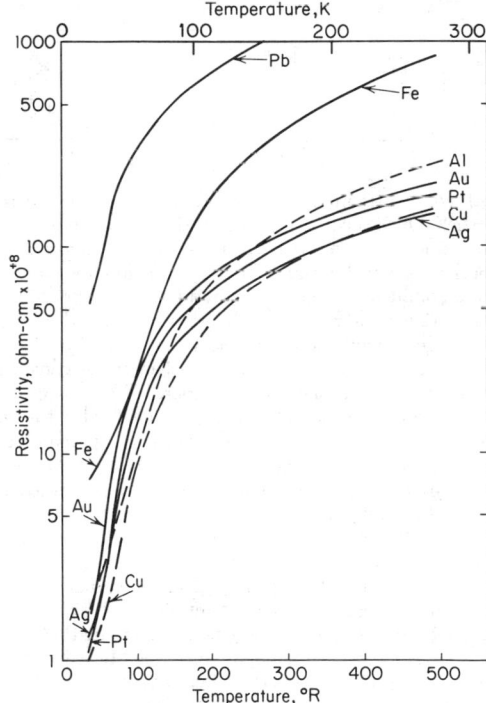

**Fig. 6**  Electrical resistivity of solids at low temperatures.

Tables 1 and 2 are only representative. The number of superconductors including compounds and alloys runs into the hundreds. The superconductors in their normal states are

**Table 1. Transition Temperatures $T_0$ and Critical Magnetic Fields $H_0$ of Some Type I Superconductors**

| Superconductor | $T_0$ (K) | $H_0$ (A/m)* |
|---|---|---|
| Nb | 9.25 | $1.57 \times 10^5$ |
| Pb | 7.23 | $6.4 \times 10^4$ |
| V | 5.31 | $8.8 \times 10^4$ |
| Hg | 4.154 | $3.3 \times 10^4$ |
| Sn | 3.722 | $2.4 \times 10^4$ |
| In | 3.405 | $2.2 \times 10^4$ |
| Al | 1.175 | $8.4 \times 10^3$ |
| Mo | 0.916 | $7.2 \times 10^3$ |
| Zr | 0.53 | $3.7 \times 10^3$ |
| Ti | 0.39 | $8.0 \times 10^3$ |
| W | 0.0154 | $9.2 \times 10$ |
| | | |
| Na Bi | 2.2 | The elements, in pure |
| $Au_2Bi$ | 1.7 | state at normal |
| Cu S | 1.6 | pressure, are not |
| | | superconductors. |

*To convert to oersteds, divide by 79.57.

**Table 2. Transition Temperatures $T_0$ and Upper Critical Fields $H_{c2}$ of Some High Field, Type II Superconductors**

| Superconductor | $T_0$ (K) | $H_{c2}$ (A/m)* | $T$ (K) for $H_{c2}$ |
|---|---|---|---|
| $Al_{0.75}Ge_{0.25}Nb_3$ | 18.5 | $3.4 \times 10^7$ | 4.2 |
| $Nb_3Sn$ | 18.0 | $1.9 \times 10^7$ | 4.2 |
| $N_{0.93}Nb$ | 15.9 | $1.3 \times 10^7$ | 0 |
| $Ga\ V_3$ | 14.8 | $1.9 \times 10^7$ | 0 |
| NbZr | 10.8 | $7.4 \times 10^6$ | 0 |
| $Nb_{0.2}Ti_{0.8}$ | 7.5 | $6.4 \times 10^6$ | 4.2 |
| $Ti_{0.6}V_{0.4}$ | 7.0 | $8.8 \times 10^6$ | 2 |

*To convert to oersteds, divide by 79.57.

classed with substances that are metallic in their electrical properties. Of the chemical elements, 38 are known to become superconductors. Some that are not superconductors at normal pressures have high-pressure allotropic modifications that are superconductors, e.g., Bi, Si, and Ge at pressures of ~25, 130, and $120 \times 10^8$ Pa, respectively. Many of the superconducting alloys have nonsuperconducting constituents, and there are superconducting compounds all of whose constituent elements in their pure state are nonsuperconductors (see Table 2). It is interesting but as yet unexplained why the good conductors Cu, Ag, and Au do not become superconductors to the lowest temperatures practically realizable (~0.01 K).

There are two **types of superconductors.** Type I superconductors exhibit a complete Meissner effect for $0 \leqslant H \leqslant H_c$; i.e., $B = 0$ until $H_c$ is reached when penetration of the specimen by $B$ becomes complete. The elemental metals in a pure and mechanically perfect state are Type I superconductors. Alloys (intermetallic compounds, solid solutions, and mixed phases) and work-hardened (unannealed) metals are Type II superconductors. For Type II superconductors, the Meissner effect is complete only for $0 \leqslant H \leqslant H_{c1}$, where $H_{c1}$ is smaller than $H_c$. For $H_{c1} \leqslant H \leqslant H_{c2}$, the magnetic field penetrates the body of the specimen in the form of dispersed bundles of a fixed quantized amount of magnetic flux, called fluxons, that pass through filamentary regions of the specimen where the specimen is normally resisting. Hence, the state $H_{c1} \leqslant H \leqslant H_{c2}$ is called the *mixed state* of a Type II superconductor. The num-

ber of fluxon bundles and the volume fraction of the specimen that is normally resisting increase with $H$ from zero at $H_{c1}$ to complete penetration of the field $B$ and the restoration of the normal state at $H_{c2}$. $H_{c2}$ is a function of $T$ (see Table 2); it increases as $T$ decreases, more rapidly at higher $T$'s ($< T_0$), and at 0 K, $dH_{c2}/dT = 0$.

The alloys commonly used for high-field superconducting magnets ($Nb_3Sn$, Nb-Ti, and Nb-Zr, see Table 2) are Type II superconductors. $H_c$'s for Type I superconductors are less than $1.6 \times 10^5$ A/m. Magnet alloy wires are strained (work hardened) to increase their $H_{c2}$ values.

The fluxons in the mixed state of a Type II superconductor transporting a current are acted upon by a Lorentz force that is proportional to $\mathbf{i} \times \mathbf{B}$ and acts in a direction perpendicular to $\mathbf{B}$, and to $\mathbf{i}$ (the current intensity). Unless the fluxons are pinned to crystal lattice sites they are propelled by the current across the superconductor. This involves the performance of work by the current and results in (1) a power loss within the superconductor, and (2) the appearance of electrical resistance to the flow of current and the destruction of the superconducting state. Fluxons are pinned by lattice imperfections (chemical and mechanical) and their displacement is resisted until the Lorentz force exceeds a break-away value, freeing the fluxons to move. Lattice imperfections are introduced in the wires for high-field superconducting magnets to pin fluxons as well as to increase $H_{c2}$.

A current exceeding a critical value destroys the superconducting state. The magnetic field, at the surface of the superconductor, generated by the current acts in conjunction with an external magnetic field in limiting the superconducting state. Hence, the critical value of an applied field depends on the current transported by the superconductor and vice versa. In general, the larger the applied field, the smaller $I_c$ is, and vice versa. For superconducting magnets, it is essential that $I_c$ and the applied magnetic field be simultaneously large.

The **coefficient of linear expansion** $\alpha$ is $(1/L)(dL/dT)$. Thermal expansion of asymmetric crystals differs along different crystal axes. For an isotropic solid, the volume or cubical-expansion coefficient $(1/V)(dV/dT) = 3\alpha$. Expansion coefficients are in general approximately constant at ambient temperatures, but they all approach zero at 0°R, and the approach to zero is tangential to the $T$ axis ($d\alpha/dT = 0$ at 0°R). Some expansion coefficients are negative at low temperatures, e.g., stainless steel, some Invars, and fused quartz. The expansivities of some crystals are negative in some directions even though their volume coefficients are positive. Cold-working may produce differences in expansivity in different directions. Annealing restores isotropy. Figure 7 shows the relative change in length, the integral of $\alpha$ from $T$ to 528°R (293 K).

Because the cold interiors of cryogenic equipment must at times be warmed to ambient temperature, provisions have to be made for those changes in dimensions of the interior that result from large changes in $T$. Even for equipment of ordinary size, these changes can be too large for accommodation within the elastic limits of the materials of construction or of the structure. In such situations, flexibility has to be designed into the structure. Bellows and U bends are commonly employed to obtain this flexibility in insulated transfer lines for liquid cryogens.

The **mechanical properties** important for the design and construction of cryogenic equipment are the same as for other temperatures (see also Sec. 5). Frequently, the choice of materials for the construction of cryogenic equipment will depend upon other considerations besides mechanical strength, e.g., lightness (density or weight), thermal conductivity (heat transfer along structural support members), and thermal expansivity (change of dimensions when cycling between ambient and

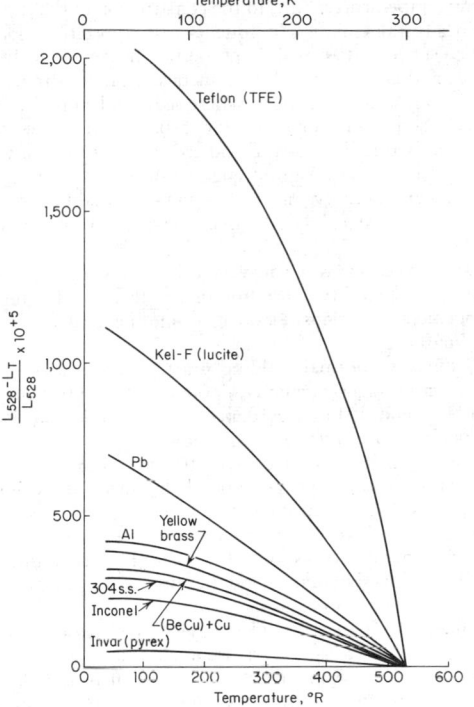

**Fig. 7** The relative change in length from $T$ to 528 °R (293 K), equal to $\int_t^{528°R} \alpha dT$. Bracketed materials are within ~ 5 percent of the curve shown.

low temperatures). Frequently, mechanical properties at low temperature are significantly different from properties at ambient temperature. This makes room-temperature data unreliable for engineering use at low temperatures.

It is not possible to make generalizations that would not have numerous exceptions about the temperature variations of the mechanical properties. In this discussion, Figs. 8 to 10 are only illustrative guides. There is no substitute for test data on a truly representative specimen when designing for the limit of effectiveness of a cryogenic material or structure. Just as the mechanical properties at ambient temperatures are dependent upon the impurities (metallic and non-metallic), their chemical nature and concentration, the thermal history of the specimen, the amount and kind of working (microstructure and dislocations of the lattice), and the rate of loading and type of stress (uni-, bi-, and triaxial), so also are the changes, quantitative and qualitative, in the mechanical properties when changing the temperature from ambient to low temperatures.

Metals, non-metallic solids (glass), plastics, and elastomers are discussed in order. The metals are classed by their lattice (crystal) symmetry.

The **f.c.c. metals** and their alloys are most commonly used for the construction of cryogenic equipment. Al, Cu, Ni, their

alloys, and the austenitic stainless steels of the 18-8 type (300 series) are f.c.c. They do not exhibit an impact (or a notched-tensile) ductile-to-brittle transition at low temperatures. As a general rule, which has some exceptions, the mechanical properties of these metals improve as $T$ is reduced: (1) Young's modulus at 40°R (22 K) is 5 to 20 percent larger than at 530°R (294 K), (2) the yield strength at 40°R (22 K) is considerably greater than the strength at 530°R (294 K). (Cu is an exception; see Fig. 9), and (3) the fatigue properties at low $T$ are improved. There is a large difference between the yield strength and the ultimate tensile strength of these metals and alloys, especially when they have been annealed. Pb (f.c.c.) and In (face centered tetragonal) are used for low-$T$ deformable gaskets because of their creep properties. Low-$T$ creep data are meager, but the rate of creep decreases with $T$.

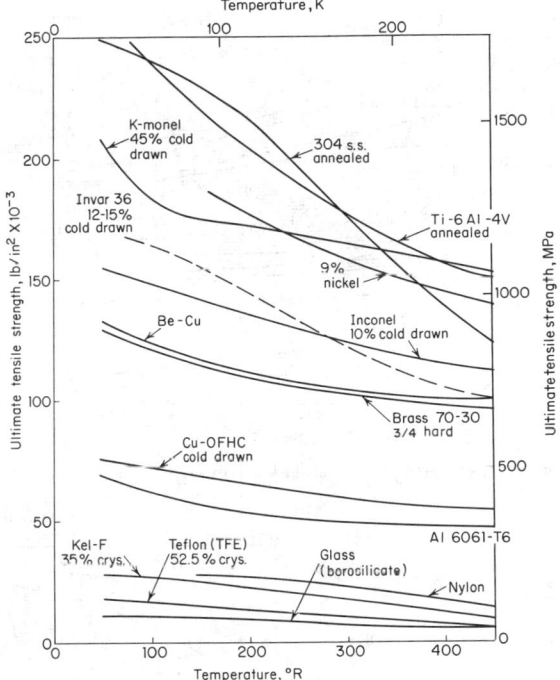

**Fig. 8** Ultimate tensile strength of solids at low temperatures.

Beta brass (f.c.c.) is ductile down to 7°R, though it is like all alloys of Cu in being less ductile than Cu itself. Thin brass is sometimes porous, and this limits its usefulness for high-vacuum enclosures at low temperatures. Free-machining brass normally contains Pb, which even in low concentrations can render the brass superconducting at LHe temperatures. In the superconducting state, it may then affect the magnetic field in its vicinity (see electrical properties, above).

The **b.c.c. metals and alloys** are normally classed as undesirable. These include Fe, the martensitic steels (low carbon and the 400 series of stainless steels), Mo, and Nb. If not brittle at room temperature, they have a ductile-to-brittle transition at low $T$. Working can induce the austenite-to-martensite transition in some steels. AISI 301 austenitic stainless is an example. It remains moderately "tough" at very low $T$ and is a valuable material for construction, but cold-drawing, in excess of 70 percent strain, induces a partial transformation of structure and reduces the elongation and notched-strength ratio to nearly zero. This same type of reduction in toughness is observed in the "heat-treatable" stainless steels. Improved

tensile properties can be obtained by cold-working and by heat-treating. These alloys usually have decreased toughness at low temperatures. Alloys of V, Nb and Ta, although not f.c.c., behave well as regards brittleness at low temperatures. These alloys have the advantage of being suitable for use at high $T$. Carbon steels, though brittle, have found special uses at low temperatures, e.g., in the construction of the expansion engines of the Collins He liquefier and cryostat.

The **h.c.p. metals** exhibit properties intermediate between those of the f.c.c. and b.c.c. metals; e.g., Zn undergoes a brittle-to-ductile transition, whereas Zr and pure Ti do not. Ti and some Ti alloys, having a h.c.p. structure, remain moderately ductile at low $T$ and are excellent for many applications. They have high strength-to-weight and strength-to-thermal-conductivity ratios. The low-$T$ properties of Ti and its alloys are extremely sensitive to even small amounts of O, N, H, and C.

**Brittle materials,** when in very thin sheets, can have a high degree of flexibility, and this makes them useful for some cryogenic applications. **Flexibility** cannot be used as a criterion for ductility.

Ordinarily, normal welding practices are observed in the construction of cryogenic equipment. These practices, however, are modified in accordance with available knowledge of the performance of the metals at low $T$.

**Nonmetal** materials for construction are in many cases brittle, or they are susceptible to brittle fracture. The strength of **glass,** measured at a constant rate of loading, increases on going to low $T$. Failure occurs at a lower stress when the glass surface contains cracks and abrasions. The strength of glass can be improved by tempering the surface, i.e., by putting the surface under compression.

The **plastics** increase in strength as $T$ is decreased, but this is accompanied by a rapid decrease in elongation in a tensile test and a decrease in impact resistance. Teflon and the glass-reinforced plastics (e.g., glass-reinforced epoxy resin) retain

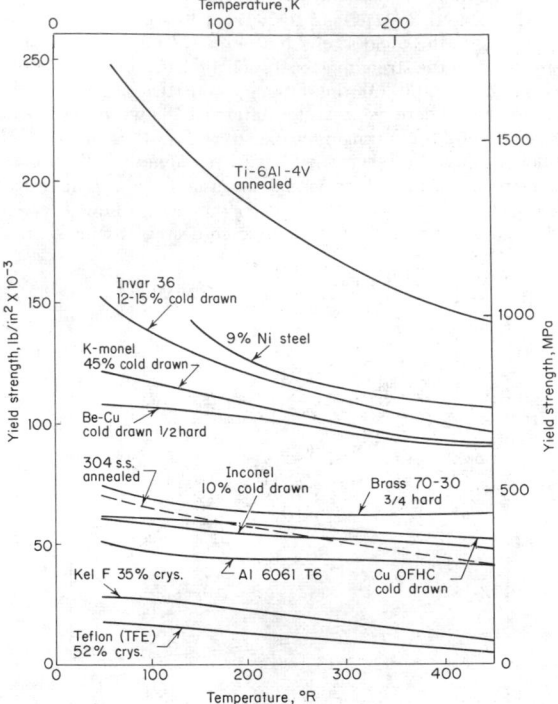

**Fig. 9**   Yield strength of solids at low temperatures.

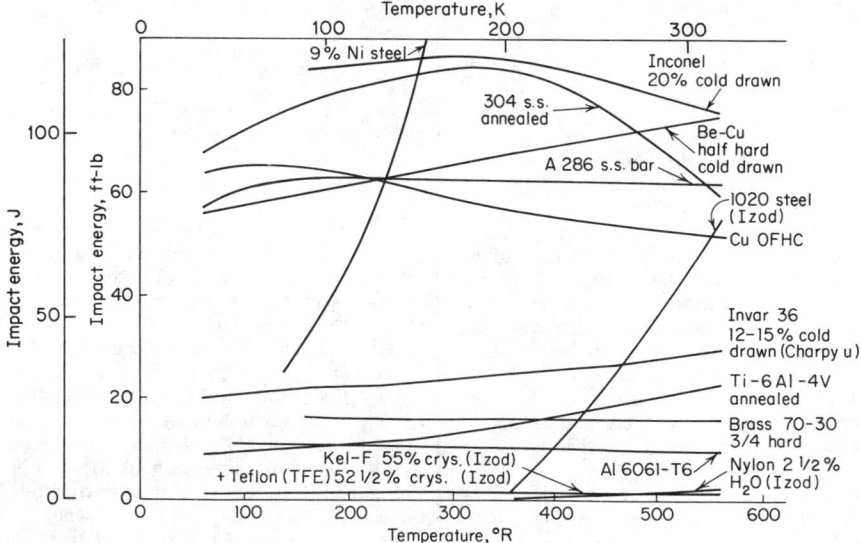

**Fig. 10**   Impact energy for solids at low temperatures (Charpy V unless noted). For Kel-F, nylon, and Teflon, the impact energy units are foot-pound per inch of notch width.

appreciable impact resistance as $T$ is lowered. Teflon, which is polytetrafluroethylene, can be deformed plastically at $T$'s as low as 7°R (3.9 K). The amount is considerably less than at room temperature, but it is enough to make Teflon very useful for some cryogenic applications. The glass-reinforced epoxies, besides having appreciable impact resistance at low $T$'s, also have high strength-to-weight and strength-to-thermal-conductivity ratios.

All the **elastomers** become brittle at low $T$. However, elastomers like natural rubber, nitrile rubber, Viton A, and plastics such as Mylar and nylon that become brittle at low $T$ can be used for static seal gaskets when *highly compressed* at room temperature, prior to cooling.

### Properties of Cryogenic Fluids (Cryogens)

Table 3 indicates the **temperature ranges** accessible with liquid cryogen baths. There are two inaccessible ranges: (1) between helium and hydrogen [9.36 to 24.9°R (5.2 to 13.8 K)] and (2) between neon and oxygen [80 to 97.9°R (44.4 to 54.4 K)]. The limiting temperatures are set by the critical and triple points. Pumping on a cryogen bath to lower the pressure results in lower temperatures ultimately reaching the triple point; except for helium, further pumping leads to solidification and in practice poor heat-transfer. Raising the bath pressure results in higher boiling temperatures with the limit set by the critical temperature, above which liquid and vapor phases cannot coexist. Cryogen baths are most frequently operated near atmospheric pressure.

The algebraic sign of the Joule-Thomson coefficient determines whether a gas cools or warms upon free expansion. Figure 11, based on the law of corresponding states, allows rough calculations of inversion temperatures and pressures. Free (Joule-Thomson) expansion inside the "cooling" region results in cooling, i.e., refrigeration; outside this region, heating results. Upper inversion temperatures at 14.7 psia (0.101 MPa) are given in Table 3.

**Volumetric latent heats** (Table 3) decrease with the lower-boiling cryogens and emphasize the importance of the insulation in storing and handling LH₂ and LHe.

Two forms of **hydrogen** and **deuterium** exist: **ortho** and **para**. The ortho and para molecules differ in the relative orientation of the nuclear spins of the two atoms composing the diatomic molecule. In the ortho form of $H_2$ the nuclear spins of the two atoms in the molecule are parallel (in the same direction), whereas in the para form the spins are antiparallel. The

relative orientations for ortho and para $D_2$ differ from those for $H_2$. The thermodynamic equilibrium composition of ortho and para varieties is temperature dependent as shown in Fig. 12.

In liquid hydrogen, the uncatalyzed ortho-para reaction is second order (proportional to the square of the o-$H_2$ concentration) with a rate constant of 0.0114/h. Thus, in the course of uncatalyzed liquefaction, normal LH₂ (75 percent ortho) is produced. Since the heat of conversion of o-$H_2$ to p-$H_2$ at the normal boiling point is 302 Btu/lbm (702 J/g) (greater than the heat of vaporization; see Table 3), long-term storage of normal LH₂ is impractical. Therefore, in the commercial production of LH₂ a catalyst is used to produce LH₂ with more than 95 percent para which for practical purposes is equivalent to the equilibrium composition (99.8 percent para).

For pressures of no more than 150 lb/in² (1.03 MPa) and temperatures at least twice the critical temperature, the ideal gas law ($PV = nRT$) enables one to calculate $PVT$ data with sufficient accuracy for many engineering purposes. For cases in which the ideal gas law is not adequate or in which experimental data are not available, the procedures outlined in Sec. 4 may be used. The theoretical prediction of $PVT$ data for

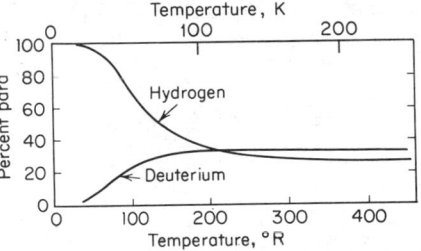

**Fig. 12** Thermodynamic equilibrium composition of ortho and para varieties of $H_2$ and $D_2$: percent ortho = 100 − percent para.

liquids is considerably more difficult than for gases. Consequently no universal equation analogous to the perfect gas law exists for extensive ranges of $P$ and $T$.

**Reduced quantum mechanical correlations** of saturated liquid densities, viscosities, and thermal conductivities for several cryogens are shown in Figs. 13 to 15. Dashed lines represent areas void of experimental data.

Utilization of Figs. 13 to 15 requires knowledge of the constants $\epsilon$ and $\sigma$ for the Lennard-Jones function for the intermolecular potential energy $\varphi(r)$ of a pair of molecules separated by a distance $r$, so that $\varphi(r) = 4\epsilon[(\sigma/r)^{12} - (\sigma/r)^6]$. Here $\epsilon$ is the depth of the minimum in the potential energy of a pair of molecules, and $\sigma$ is the separation of the pair at which $\varphi = 0$. To facilitate use of Figs. 13 to 15, all required constants and conversion factors have been combined and are listed in Table 4.

**Thermal-expansion coefficient** $(1/V)(\partial V/\partial T)_P$ and **compressibility** $(1/V)(\partial V/\partial P)_T$ data for the cryogenic liquids are obtained as derivatives of $PVT$ data. Liquids hydrogen and helium have large thermal-expansion and compressibility coefficients (see Table 5).

For gases, kinetic theory predicts an increase of **viscosity** with rising temperature as shown in Fig. 16 ($\eta \sim \sqrt{T}$). Normally, the **thermal conductivities** of liquids decrease with increasing temperature, whereas the thermal conductivities of gases increase. Figure 17 presents thermal conductivity for

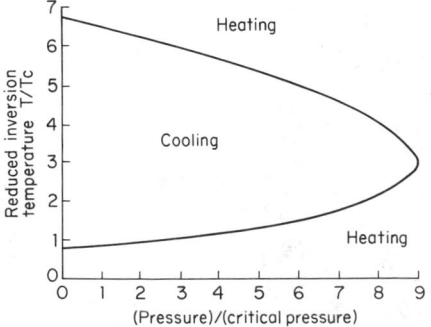

**Fig. 11** Reduced inversion temperature versus reduced pressure.

**Table 3. Common Cryogen Properties**

| Cryogen | Boiling point, °R | Triple-point Temp, °R | Triple-point Pressure, psia | Critical temp, °R | Critical pressure, psia | Upper Inversion temp,* °R | Heat of vaporization* Btu/lbm | Heat of vaporization* Btu/ft³ | Liquid density,† lbm/ft³ | Vapor density,† lbm/ft³ | Gas density,‡ lbm/ft³ |
|---|---|---|---|---|---|---|---|---|---|---|---|
| He³ | 5.7 | | | 6 | 17 | 72 | 3.6 | 13.4 | 3.72 | 1.50 | 0.0084 |
| He⁴ | 7.6 | | | 9.36 | 33.2 | 92 | 8.8 | 68.6 | 7.80 | 1.04 | 0.0111 |
| H₂ (equilib.) | 36.5 | 24.9 | 1.02 | 59.4 | 187.7 | 368 | 192 | 853 | 4.42 | 0.0837 | 0.00561 |
| D₂ (normal) | 42.6 | 33.7 | 2.49 | 69.0 | 240 | | 131 | 1,343 | 10.25 | | 0.0112 |
| Ne | 48.7 | 44.2 | 6.26 | 80.0 | 395.3 | 486 | 37.1 | 2,790 | 75.35 | 0.58 | 0.056 |
| N₂ | 139.2 | 113.7 | 1.86 | 227.27 | 492.3 | 1,115 | 85.9 | 4,294 | 50.4 | 0.28 | 0.078 |
| Air | (141.8) | | | | | | 88.22 | 4,813 | 54.56 | 0.28 | 0.0807 |
| A | 157.1 | 150.8 | 9.97 | 271.3 | 709.8 | 1,300 | 70.2 | 6,135 | 87.4 | 0.36 | 0.111 |
| F₂ | 151 | (96.4)§ | | 259.3 | 808 | | 74.1 | 6,965 | 94 | | 0.106 |
| O₂ | 162.3 | 97.9 | 0.022 | 278.59 | 736.3 | 1,605 | 91.5 | 6,538 | 71.24 | | 0.089 |
| CH₄ | 201.1 | 163.2 | 1.69 | 343.27 | 673 | | 219.2 | 5,804 | 26.5 | 0.115 | 0.0448 |

Note: To convert °R to K, multiply °R by 0.556; to convert psia to MPa, multiply by 0.00689; to convert Btu/lbm to J/g, multiply by 2.326; to convert Btu/ft³ to J/m³, multiply by 3.73 × 10⁴; to convert lbm/ft³ to Kg/m³, multiply by 16.01.
*14.7 psia.
†At normal boiling point.
‡At 14.7 psia and 492°R.
§Melting point, 14.7 psia.

several gases as a function of temperature at 14.7 psia (0.1013 MPa).

Figure 18 is a graph of **vapor pressures** of the common cryogens.

Figure 19 presents **specific heats** at constant pressure for several of the cryogenic liquids, while Fig. 20 presents the specific heats for liquid hydrogen and helium along the saturation curve. Specific heats of gases are given in Fig. 21.

## Instrumentation

(See also Secs. 15 and 16)

The usual instrumentation problems of engineering practice are further complicated by low-temperature problems which require special calibration procedures.

**Pressures** are measured with normally used apparatus, e.g., bourdon gages, transducers, and manometers. If the measuring device is located outside the cryogenic environment and insulated from it by a thermal barrier, no special problems are introduced. Occasionally response time requirements necessitate the placing of the transducer in the cryogen space. Then the problems of temperature compensation and calibration must be considered.

**Level measurements** currently utilize several methods. The differential-pressure method can in principle be used with all cryogens. With cryogens of low densities and low heats of vaporization (e.g., $H_2$ and He) pressure oscillation occurs in the liquid leg. In the case of $LH_2$ these can be eliminated by the proper design of the liquid leg and the introduction of He gas which does not condense.

**Direct weighing** is also used to determine "levels," although difficulties are encountered owing to the large tare-to-cryogen weight ratios (as large as 10 to 1 for a full container in the case of $H_2$) and extraneous loadings introduced by permanently connected piping, frost, and wind. Weighing systems have nevertheless been successfully used on 50,000-gal (190 m³) $LH_2$ dewars with an accuracy in level changes of 250 gal (1 m³).

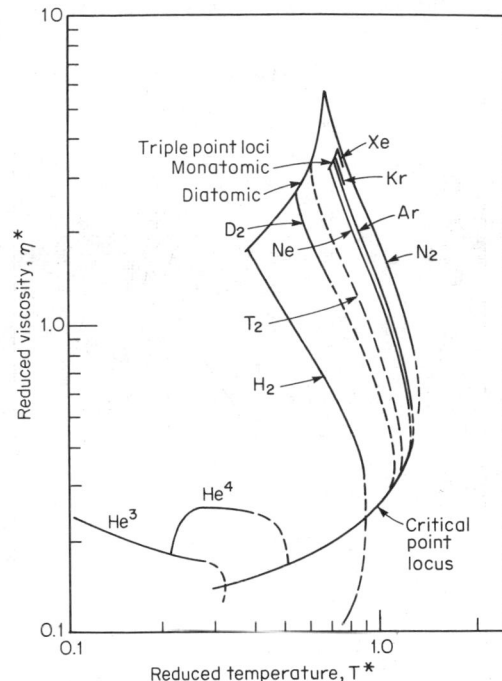

**Fig. 14**  Reduced viscosity vs. reduced temperature along the saturation curve for cryogenic liquids. Values of $T^*$ are obtained by dividing the desired temperature (°R) by the value of $T/T^*$ given in Table 4. Reduced viscosity $\eta^*$ must be multiplied by $\eta/\eta^*$ given in Table 4 to obtain viscosity in lbm/ft·h. Multiply lbm/ft·h by 0.000413 to obtain Pa·s.

With the **capacitance gage,** the sensing element consists of a concentric tube capacitor which can be calibrated to give liquid-level measurements accurate to a few tenths of a percent. An inexpensive, accurate **point-level sensor** is easily fabricated from a carbon resistor, typically $1/10$W, 1,500 Ω. Its use depends upon the fact that its resistance is a strong function of temperature (Fig. 22). A small current (50 to 80 mA) is passed through the resistor to heat it. Since heat from the resistor is dissipated more readily in the liquid than in the gas, the temperature of the resistor undergoes a step increase when the resistor environment changes from liquid to gas, whereas the resistance decreases stepwise.

**Flow measurements** utilize orifices, venturis, and turbine-type meters. Various devices for direct mass-flow measurement exist, but none are widely used. "Quality" meters have been built to determine the percentage liquid in two-phase flow. The determination of quality is complicated by nonequilibrium of temperature in the gas phase.

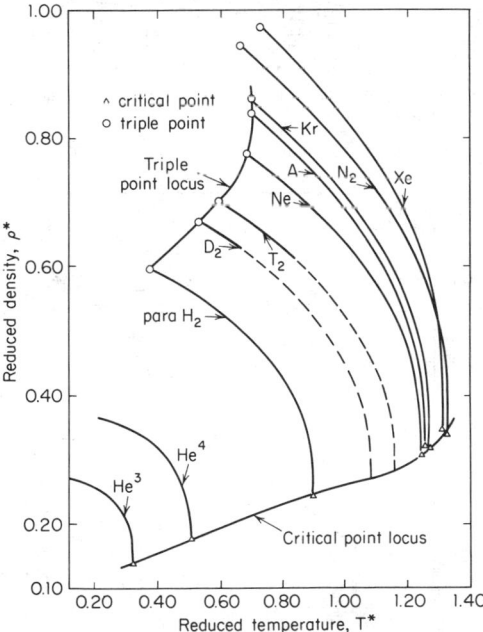

**Fig. 13**  Reduced density vs. reduced temperature along the saturation curve for cryogenic liquids. Values of $T^*$ are obtained by dividing the desired temperature (°R) by the value $T/T^*$ given in Table 4. Reduced density $\rho^*$ must be multiplied by $\rho/\rho^*$ given in Table 4 to obtain density in lb/ft³. Multiply lbm/ft³ by 16.01 to obtain kg/m³ and °R by 0.556 to obtain K.

**Temperature measurements** are usually made with thermocouples, resistance thermometers, and vapor-pressure thermometers.

The **vapor-pressure thermometer** depends upon a vapor-pressure-temperature relationship (see Fig. 18). In general, a small cavity is filled with a gas which has a condensation temperature in the vicinity of the temperature to be measured. If sufficient gas is present, the pressure will be that of the liquid vapor pressure at the coldest part of the measuring system.

Maximum speed of response is obtained if the total quantity of gas is minimized to permit only a thin film of liquid to form. In the case of $H_2$, a catalyst (e.g., iron hydroxide) to promote ortho-para conversion should be included in the bulb for an accurate measurement since the vapor pressure of $H_2$ is dependent upon its ortho-para composition (see above).

The pressure at the surface of a liquid cryogen in a dewar may, under conditions of thermal equilibrium, be used to measure the temperature of the cryogen and of apparatus immersed in it. Corrections for the pressure of the hydrostatic head of liquid may be required. If using this method, it should be realized that large vertical temperature gradients can exist in an unstirred liquid cryogen in a vessel with good insulation.

**Thermocouples** are favored for the measurement of temperatures because of their low cost, ease of application, and rapid response. The thermoelectric powers (temperature sensitivities) of the thermocouples commonly used at higher $T$'s decrease with decreasing temperature, and spurious emfs generated in wires of non-uniform composition are troublesome. The proper use of a reference junction at a known fixed temperature close to the temperature being measured is often advantageous for accuracy. Variability in composition of thermocouple alloys makes individual calibrations necessary for

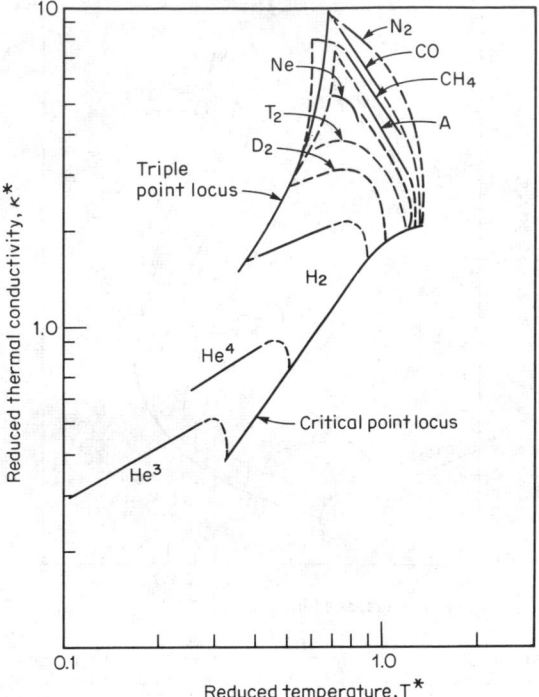

**Fig. 15** Reduced thermal conductivity vs. reduced temperature along the saturation curve for cryogenic liquids. Values of $T^*$ are obtained by dividing the desired temperature (°R) by the value of $T/T^*$ given in Table 4. Reduced thermal conductivity $\kappa^*$ must be multiplied by $\kappa/\kappa^*$ given in Table 4 to obtain thermal conductivity in Btu/h·ft·°R. (Multiply by 1.73 to obtain W/m·K.)

**Table 4. Multiplication Factors for Use with Figs. 13 to 15**

| Substance | $T/T$, °R | $\rho/\rho$, lbm/ft³ | $\eta/\eta$, lbm/ft-hr | $\kappa/\kappa$, Btu/(h·ft·°R) |
|---|---|---|---|---|
| Helium 3 | 18.4 | 18.71 | 0.0311 | 0.0204 |
| Helium 4 | 18.4 | 24.84 | 0.0359 | 0.0178 |
| Para-hydrogen | 66.1 | 8.06 | 0.0360 | 0.0363 |
| Deuterium | 63.4 | 16.22 | 0.0500 | 0.0257 |
| Tritium | 62.1 | 24.37 | 0.0607 | 0.0210 |
| Neon | 64.1 | 100.66 | 0.1300 | 0.0128 |
| Nitrogen | 171.1 | 57.39 | 0.1382 | 0.0098 |
| Argon | 215.6 | 104.84 | 0.2183 | 0.0109 |
| Krypton | 297.4 | 177.1 | | 0.0080 |
| Xenon | 397.8 | 197.4 | | |

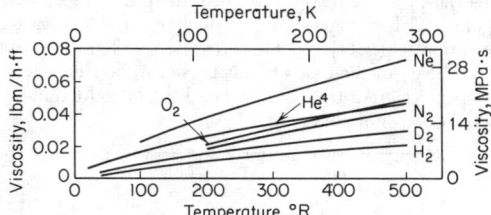

**Fig. 16** Viscosity of selected gases at low temperatures and 14.7 psia (0.1013 MPa).

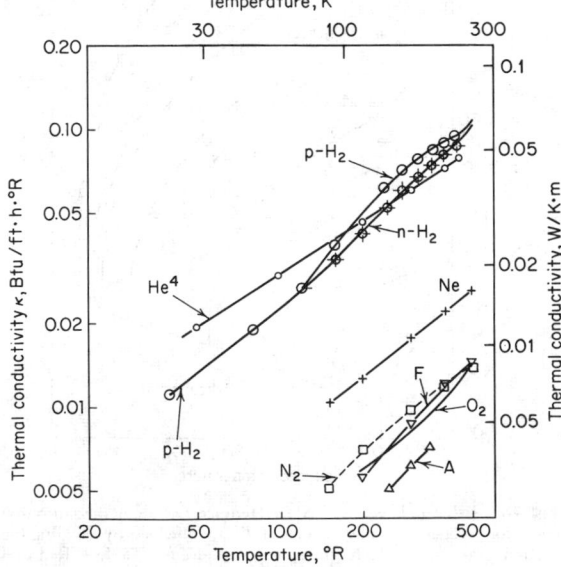

**Fig. 17** Thermal conductivity of gases at low temperatures and 14.7 psia (0.1013 MPa).

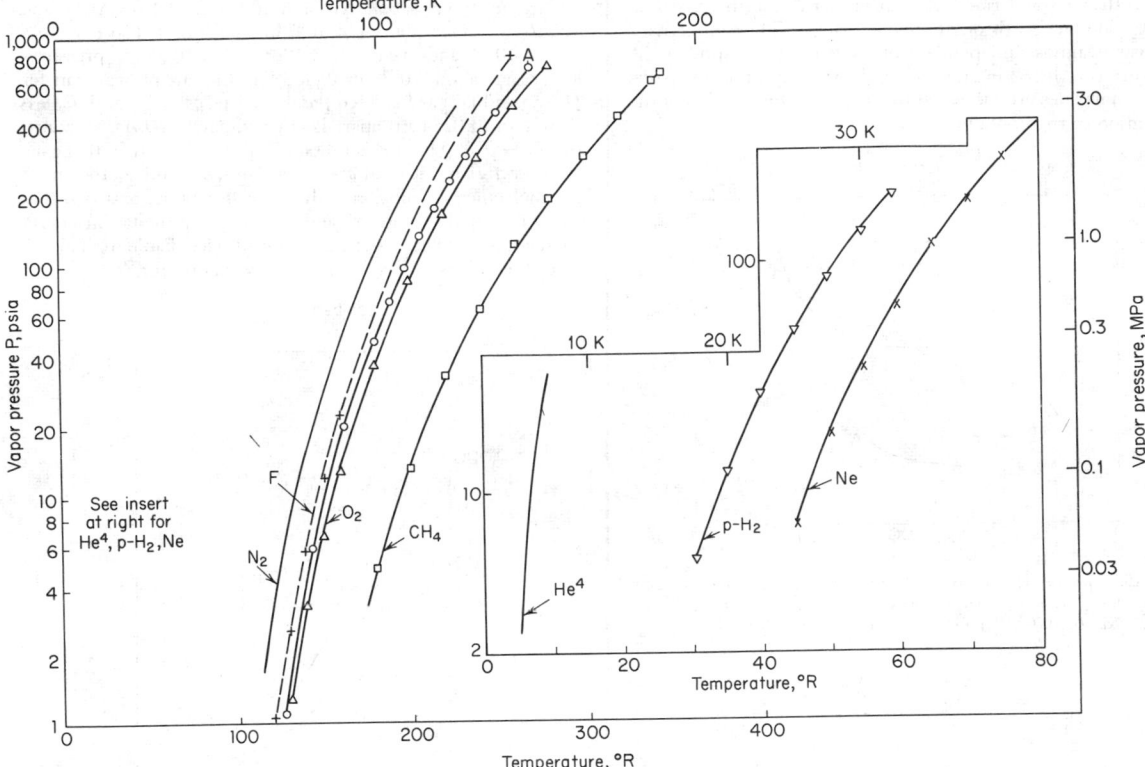

**Fig. 18**  Vapor pressures of the common cryogens.

accurate results. Figure 23 gives typical thermocouple emf-versus-temperature curves for some common thermocouples.

### Insulation

(See also Sec. 6)

The degree of thermal isolation required at low $T$'s is normally greater than at elevated $T$'s because it is more costly to remove heat leaking into a low-temperature system than to replace heat lost at elevated $T$'s, as demonstrated by the Carnot cycle. Other differences between insulating for low and elevated temperatures are:

(1) Condensation of moisture in low-temperature insulation is possible. When this occurs, the conductance of the insulation is significantly increased. This problem is avoided by a vapor barrier on the outside of the insulation.

(2) Condensation of the atmosphere in the insulation and surface washing by the condensed liquid are possible when the insulating surfaces are below 150°R (83 K). This can result in a large added transfer of heat. It is avoided by using an outer cover or surface impermeable to air, and evacuating the insulating space or replacing the atmosphere in it with a noncondensable gas.

At low temperatures, as at other temperatures, the fundamental modes of heat transfer are conduction and radiation, and it is against these that insulation is used. Convection of heat is practically eliminated by the insulation.

*Low-T insulation* categories are (1) high vacuum, with or without multiple radiation shields, (2) powders, (3) rigid foams, and (4) low-conductivity solids, such as balsa wood and corkboard.

**Table 5. Compressibility and Thermal-Expansion Coefficients. Coefficients for Liquid Nitrogen, Hydrogen, and Helium at 14.7 psia (0.101 MPa)**

| Cryogen | $T$, °R (K) | Compressibility, $1/V(\partial V/\partial P)_T$, 1/psia 1/(MPa) | | Thermal expansion, $1/V(\partial V/\partial T)_P$, 1/°R (1/K) | |
|---|---|---|---|---|---|
| Helium 4 | 6 (3.3) | 0.0013 | $(1.89 \times 10^{-5})$ | 0.071 | (0.127) |
| Nitrogen | 139 (77) | 0.000022 | $(3.1 \times 10^{-6})$ | 0.0032 | (0.0058) |
| Para-hydrogen | 37 (20) | 0.0001 | $(1.45 \times 10^{-5})$ | 0.01 | (0.018) |
| Water | 540 (300) | 0.0000036 | $(5 \times 10^{-7})$ | 0.000115 | (0.00019) |

**(1.i) Heat is transferred across an evacuated space** by radiation and by conduction through the residual gas. The conductivity of a gas is almost independent of its density (pressure) as the pressure is reduced in an evacuated space until the free paths of the molecules are increased to an appreciable fraction of the separation of the containing walls.

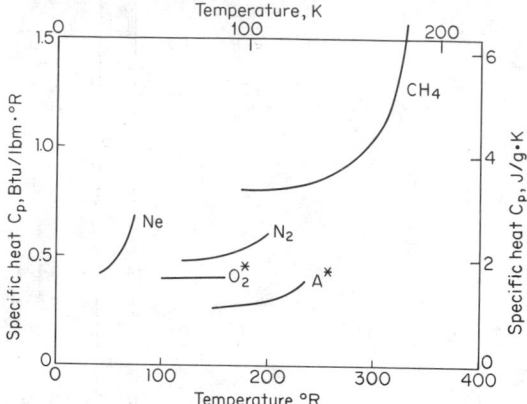

**Fig. 19** Specific heat at constant pressure ($C_p$) for liquid cryogens at saturation pressure. Asterisk means $C_p$ evaluated along an isobar at 14.7 psia (0.1013 MPa) instead of saturation pressure.

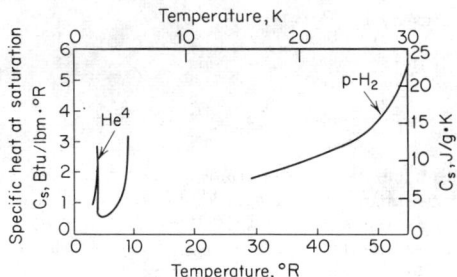

**Fig. 20** Specific heats at saturation $C_s$ along the saturation curve vs. temperature for $^4$He and para hydrogen.

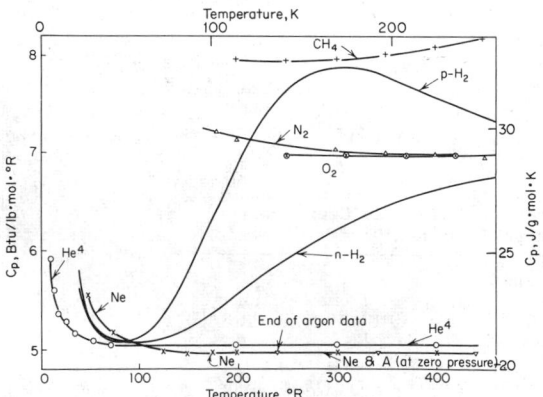

**Fig. 21** Specific heat, $C_p$, of cryogen gases at 14.7 psia (0.1013 MPa) vs. temperature.

The molecular *mean* free path at 1 atm (0.1013 MPa) and 492°R (273 K) in air is $4 \times 10^{-6}$ in ($10^{-7}$ m); in $H_2$, $6 \times 10^{-6}$ ($1.5 \times 10^{-7}$); and in He, $10 \times 10^{-6}$ ($2.5 \times 10^{-7}$). At pressures lower than about $10^{-3}$ torr (0.133 Pa), the rate of heat transfer $Q$ by residual gas between parallel surfaces at $T_1$ and $T_2$ less than 1 in (0.0254 m) apart is approximately $(Q/A) \approx$ (const) $\alpha P(T_2 - T_1)$, where $P$ is pressure *measured* in torr with a gage at *ambient* $T$ and $\alpha$ is an *over-all* accommodation coefficient of gas molecules in collision with the walls. Thus, $\alpha = \alpha_1\alpha_2/[\alpha_2 + \alpha_1(1 - \alpha_2)]$, where $\alpha_1$ and $\alpha_2$ are accommodation coefficients of gas molecules at the two walls (see Table 6). Table 7 gives the values of the (const) in the equation for $Q/A$.

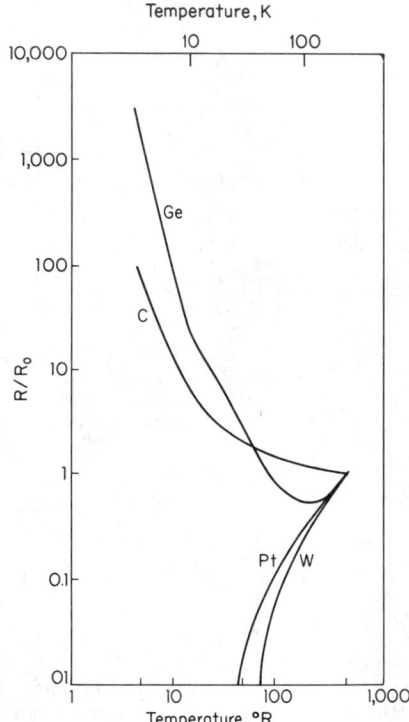

**Fig. 22** Ratio of resistance $R$ to the resistance $R_0$ at 492°R (273 K) vs. temperature for several resistance thermometers. (Values for germanium and carbon are representative only.)

The **rate of radiative heat transfer** between parallel surfaces at $T_1$ and $T_2 > T_1$, whose emissivities are $e_1$ and $e_2$, respectively, is $Q = e_1e_2s/[e_2 + (1 - e_2)e_1](T_2^4 - T_1^4)$, where $s$ is the Stefan-Boltzmann constant whose value is $1.712 \times 10^{-9}$ Btu/ft²·h·°R⁴ ($5.67 \times 10^{-8}$ W/m²K⁴) (see also Sec. 4). Table 8 contains emissivities for 540°R (300 K) thermal radiation; these for most engineering purposes may be considered equivalent to minimal values of the $e$'s for very carefully prepared and cleaned surfaces. Organic coatings have high emissivities approaching unity (>0.9). Mechanically polished surfaces have higher $e$'s than the minimal values. Handling a surface transfers an absorbing (high $e$) coating to it.

**(1.ii) Refrigerated Radiation Shield** A $LN_2$ (140°R) (78 K) cooled surface interposed in the insulating vacuum space of an

LH$_2$ or an LHe container reduces the heat transferred by radiation to the LH$_2$ or LHe by a factor of about 250 when compared with that transferred by a surface of the same emissivity at room temperature without the interposition of the refrigerated shield. Nearly all the heat radiated by the

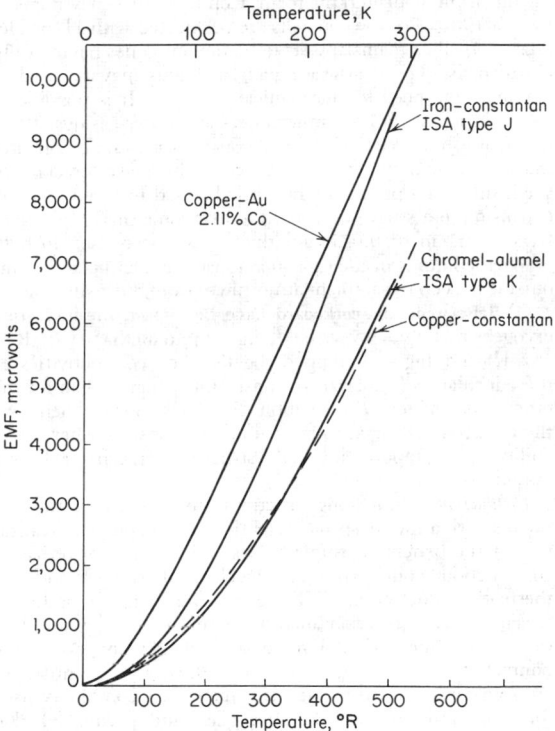

**Fig. 23**  EMF vs. temperature for several thermocouples.

surface at room temperature is absorbed by the LN$_2$, and this results in evaporation of the relatively inexpensive LN$_2$.

(1.iii)  **A floating radiation shield**  is an opaque layer (e.g., a metal sheet), having surfaces of low emissivity suspended in the vacuum space with a minimum of thermal contacts with surfaces that are warmer or cooler. If the emissivity of the surfaces of the floating shield is the same as the emissivity of the vacuum-space walls, the floating shield decreases the *radiative* heat transfer by a factor of 2. For *m* floating shields arranged in series between the inner cold and outer warm surfaces, the rate of *radiative* heat transfer becomes $\dot{Q}_m$ (*m* floating shields) = $[1/(m + 1)]\dot{Q}_0$ where $\dot{Q}_0$ = heat transfer for $m = 0$.

(1.iv)  **Multilayer (Super) Insulation**  The principle of multiple floating radiation shields has been extended to the use of many thin metal layers (up to 75 layers per inch or 3000 layers per meter) separated by thin thermal insulation. Best results have been obtained with Al foil separated by glass-fiber paper. Other materials have been used successfully, e.g., nylon nets as separators and aluminized Mylar with no spacer material. The *apparent* thermal *conductivity* of Al foil multilayer insulation spaced with glass fiber paper is variable depending on the number of layers of foil, the thickness of the paper, and the

thickness and compacting of the insulating layer. For $\Delta T = (540 - 36)°$R or $(300 - 20)$ K, apparent mean conductivities range from 2.0 to $4.0 \times 10^{-5}$ Btu/h·ft·°R (3.5 to $7 \times 10^{-5}$ J/s·m·K). For $\Delta T = [540 - 140°$R $(300 - 78)$ K], the values are about one-third larger. Using nylon nets, glass fabric or glass fibers for separating the Al-foil increases the conductance from three to seven times. The above conductivities are for the direction normal to the foil surfaces. Lateral conductivities are many thousands of times larger. An advantage of aluminized Mylar is its reduced lateral conductivity. Another advantage is its relatively low density (one-third to one-half of other multilayer insulations). If the residual gas pressure in multilayer insulation is less than $10^{-4}$ torr, the conductance is practically independent of the residual gas pressure. At $10^{-3}$ torr (0.133 Pa), the conductance is increased by the order of 50 percent.

(2.i)  **Powder insulation** consists of finely divided materials. Conductances of evacuated powders may vary by as much as 100 percent with variations of particle size and *apparent* density (packing) of the powder. The most commonly used powders are perlite, expanded SiO$_2$ (aerogel), calcium-silicate, diatomaceous earth, and carbon black. The conductance of

**Table 6. Approximate Values of Accommodation Coefficients, $\alpha$**

| Temp, °R (K) | He | H$_2$ | Air |
|---|---|---|---|
| 540 (300) | 0.3 | 0.3 | 0.8–0.9 |
| 140 (78) | 0.6 | 0.5 | 1 |
| 35 (19) | 0.6 | 1 | 1 |

**Table 7. Constant in the Gas-Conduction Equation**
$Q/A \approx (\text{const})\,\alpha P(T_2 - T_1)$

| Gas | $T_2$ and $T_1$, °R (K) | Const, Btu/ h·ft²·torr·°R [J/ (N·s·K)] |
|---|---|---|
| N$_2$ | <700 (389) | 28 (21,100) |
| O$_2$ | <540 (300) | 26 (19,600) |
| H$_2$ | 540 and 140 (300 and 78) | 93 (70,200) |
| H$_2$ | 140 and 36 (78 and 20) | 70 (52,800) |
| He | Any | 49 (37,000) |

$T_1$—inner wall (cold).
$T_2$—outer wall (hot).

**Table 8. Minimal Values of Emissivity of Metal Surfaces at Various Temperatures for 540°R (300 K) Thermal Radiation**

| Surfaces | Surface temp, °R (K) | | |
|---|---|---|---|
| | 7 (3.9) | 140 (78) | 540 (300) |
| Copper | 0.005 | 0.008 | 0.018 |
| Silver | 0.0044 | 0.008 | 0.02 |
| Aluminum | 0.011 | 0.018 | 0.03 |
| Chromium | | 0.08 | 0.08 |
| Nickel | | 0.022 | 0.04 |
| Brass | 0.018 | | 0.035 |
| Stainless steel 18-8 | | 0.048 | 0.08 |
| 50 Pb-50 Sn solder | | 0.032 | |
| Glass, paints, carbon | | | >0.9 |
| Nickel plate on copper | | 0.033 | |

powder insulation is practically independent of gas pressure down to 10 torr (1.333 kPa), at which pressure it begins to decrease rapidly with decreasing pressure as the free path of the gas molecules becomes comparable with the space between powder particles. The most rapid decrease of conductance occurs between 1 and $10^{-1}$ torr (133 and 13.3 Pa). At $10^{-3}$ torr (0.133 Pa) the conductance reaches practically the lower limit. Even at $10^{-2}$ torr (1.33 Pa) the conductance is quite low. For this reason, good mechanical pumps are adequate for the evacuation of powder insulation (fine mesh filters are needed to protect the pump from being damaged by the powder). When evacuated to $10^{-4}$ torr (1.33 $\times$ $10^{-2}$ Pa) or less, the apparent mean thermal conductivity for perlite [540°R (300 K) to 140°R (78 K)] ranges from 0.6 to 1.2 mW/m·K. For nonevacutated perlite filled with He or $H_2$ an apparent mean thermal conductivity of 11.5 mW/m·K can be used. With $N_2$ a typical value is 3.0 mW/m·K.

The principal function of the powder is to impede the transfer of heat by radiation. Powders are used when radiation would constitute an important heat leak. If an insulating vacuum is bounded by highly reflecting walls, such as clean Cu, Ag, or Al, adding powder may result in an increase of heat transfer because of the paths of solid conduction through the powder.

(2.ii) **Al powder or flakes added to a powder insulation** reduces the radiative transfer through the insulation. Powdered Al is more effective than flaked Al. The conductance of highly evacuated perlite insulation [$\Delta T = 540 - 140$°R (300 − 78 K)] may be reduced 40 percent by the addition of 25 percent of Al powder. A similar addition of Al powder to Santocel and Cab-O-Sil (diatomaceous earth) reduces the conductance by 70 percent. Metal powders are most effective with those powders that are partially transparent to infrared radiation of wavelengths longer than 3 $\mu$m.

**(3) Rigid-Foam Insulation**  The foams most used in cryogenic applications are the more or less closed-cell foams. Table 9 gives thermal conductivities of selected samples of insulating foams. The conductivity of a foam is dependent on the conduction through the intracellular gas, on the transfer

**Table 9. Apparent Mean Thermal Conductivities of Selected Foams, Balsa Wood, and Corkboard**

| Material | Density lbm/ft³ (kg/m³) | Thermal conductivity,* Btu/h· ft·°R (W/K· m) |
|---|---|---|
| Polystyrene foam | 2.4 (38) | 0.019 (0.033) |
| | 2.9 (46) | 0.015 (0.026) |
| Epoxy resin foam | 5.0 (80) | 0.019 (0.033) |
| Polyurethane foam | 5.0–8.7 (80–139) | 0.019 (0.033) |
| Rubber foam | 5.0 (80) | 0.021 (0.036) |
| Silica foam | 10.0 (160) | 0.032 (0.055) |
| Glass foam | 8.7 (139) | 0.020 (0.035) |
| Balsa wood, across grain | 7.3 (117) | 0.027 (0.047) |
| | 8.8 (141) | 0.032 (0.055) |
| | 20.0 (320) | 0.048 (0.083) |
| Corkboard, no added binder | 5.4 (86) | 0.021 (0.036) |
| | 7.0 (112) | 0.0225 (0.039) |
| | 10.6 (170) | 0.025 (0.043) |
| | 14.0 (224) | 0.028 (0.048) |
| Corkboard with asphalt binder | 14.5 (232) | 0.027 (0.047) |

*Test space pressure = 14.7 psia (0.101 MPa).

by thermal radiation, and on solid conduction. Temperatures below the condensing $T$ of the encased gas condense it and improve the insulation. Gases encased when polymer foams are blown gradually diffuse out of the cells and are replaced by ambient gases. The conductivities of many freon and $CO_2$ blown foams increase in time as much as 30 percent because of the diffusion of air into the foam. Conductivities may increase by a factor of 3 or 4 when cells are permeated with $H_2$ or He. Glass and silica foams appear to be the only ones having fully closed cells. The **structural rigidity** of foams may be used to eliminate the need for mechanical supports. It is possible to use rigid foams without inner liners and outer casings. Polymer foams have relatively high **thermal expansions,** even greater than the same polymer without voids. This makes cracking of the insulation a problem when it is bonded to metal. Sliding expansion joints have been used to overcome this. Mylar film bags (0.001 in or 0.025 mm thick) have been used to hold cryogenic liquids inside rigid-foam insulation and plastic films outside to keep the ambient atmosphere from the insulation.

(4) **Balsa wood and corkboard** have been used for insulating cryogens that are not very cold, e.g., liquid methane (201°R or 112 K) and higher-boiling cryogens. The conductivities of these insulators (Table 9) are considerably higher than for the other types of low-$T$ insulation discussed above. Their use, therefore, is restricted to special applications, as where their ability to support internal structures mechanically is important.

*Conduction* of heat along structural supports passing through the insulation merits special attention for an adequate realization of the benefits of superior insulation. Ideal materials of construction would have high mechanical strength and low thermal conductivity. In Table 10, the yield strengths, in tension, of construction materials are compared with their thermal conductivities. This kind of comparison favors the nonmetallic materials in Table 10. When large quantities of cryogens are handled in the field, metals are commonly used for structural members because glass and plastics (Teflon excepted) are brittle at low $T$'s.

Minimizing the transfer of heat through the supporting structure offers opportunities for ingenuity and inventiveness in design. The transfer is minimized by lengthening the supporting members. A stack of thin metal disks or washers, utilizing the contact resistance between adjacent disks, will increase thermal resistance without increase in length of support. A stack of 0.0008-in (0.02-mm) stainless steel disks under a compressive stress of 1,000 lb/in² (6.895 MPa), in vacuum, has the same thermal resistance as a solid rod 50 times the length of the stack.

**Insulation Selection**  Because of their smaller heats of vaporization and greater cost of production, the lowest-$T$ cryogens (LHe and LH$_2$) ordinarily merit more effective insulation than the higher-boiling cryogens like LO$_2$ and LN$_2$, and these in turn are ordinarily better insulated than other cryogens boiling at still more elevated $T$'s. The choice of insulation ordinarily involves the conditions of use of the equipment, as well as considerations of the tolerable loss of refrigeration and the cost of the insulation and its installation. If equipment is to have intermittent, rather than long, uninterrupted operating periods, the insulation heat capacity and time lag in reaching steady state are important. An insulation with large heat capacity (e.g., powders) will evaporate large quantities of refrigerant and take a long time to reach steady state. Powders

**Table 10. Comparison of Materials for Support Members**

| Material | $E_u$,**<br>kpsi (MPa) | $\kappa$,**<br>Btu/h·ft·°R (W/K·m) | $E_y/\kappa$<br>kpsi·h·ft·°R/Btu<br>(MN·K/W·m) |
|---|---|---|---|
| Aluminum 2024 | 55 (379) | 47 (81) | 1.17 (4.68) |
| Aluminum 7075 | 70 (483) | 50 (87) | 1.4 (5.55) |
| Copper, ann | 12 (83) | 274 (474) | 0.044 (0.18) |
| Hastelloy† "B" | 65 (448) | 5.4 (9.3) | 12 (48) |
| Hastelloy† "C" | 48 (331) | 5.9 (10.2) | 8.1 (32) |
| "K" Monel‡ | 100 (690) | 9.9 (17.1) | 10.1 (40) |
| Stainless steel 304, ann | 35 (241) | 5.9 (10.2) | 5.9 (24) |
| Stainless steel (drawn 210,000 psi) | 150 (1034) | 5.2 (9.0) | 29 (115) |
| Titanium, pure | 85 (586) | 21 (36.3) | 4.0 (16) |
| Titanium alloy (4A1-4Mn) | 145 (1000) | 3.5 (6.1) | 41.4 (164) |
| Dacron§ | 20 (138) | 0.088 (0.15)¶ | 227 (920) |
| Mylar§ | 10 (69) | 0.088 (0.15)¶ | 113 (460) |
| Nylon§ | 20 (138) | 0.18 (0.31) | 111 (445) |
| Teflon§ | 2 (14) | 0.14 (0.24) | 14.3 (58) |

**$E_y$ is the yield stress, and $\kappa$ is the average thermal conductivity between 36 and 540°R (20 and 300 K).
†Haynes Stellite Co.
‡International Nickel Co.
§E.I. du Pont de Nemours & Company
¶Room-temperature value.

are compacted by mechanical shocks and vibrations. Continued mechanical loading will compact both multilayer insulation and powders. Weight and bulkiness of the insulation can be important. The heat transported by the mechanical supports and vents connecting the cold interior with the warm exterior becomes the minimum refrigeration loss, attainable only with perfect insulation of the rest of the system. A large loss of refrigeration through supports and vents ordinarily reduces the attractiveness of costly insulation having the lowest conductance.

## Safety

Precautions are needed for the safe handling and storage of cryogens because (1) air can contaminate the cryogen, creating a potential explosive, either directly as when air mixes with $H_2$ or $CH_4$ or indirectly by transforming an inert cryogen like $LN_2$ into an oxidant (a potential hazard for combustible insulation), and (2) moisture and air can freeze on cold surfaces, clogging vents and preventing normal operation of valves.

**Vent systems** should be sized for sufficient exhaust velocities to prevent back-diffusion of air into the cryogen space. Storage at slightly elevated pressure is preferable to prevent inleakage of contaminants. Condensation can also be a problem on poorly insulated lines carrying low-boiling cryogens; frost is not a hazard but the oxygen-enriched condensed air is hazardous in the presence of combustibles. The insulation of liquid vent lines is a necessary precaution. Where this is not possible, appropriately placed guttering can prevent the contact of any condensed air with combustible objects.

Any space, either containing a cryogen or being refrigerated by a cryogen, should be protected with proper **safety-relief mechanisms**, i.e., relief valves or rupture disks. Such space includes less obvious ones such as the insulating vacuum space surrounding a cryogenic fluid or a section of line which can trap liquid between two valves. A heat leak to trapped liquid confined without a vent for escape of vapor may develop pressures sufficient to burst the containing vessel. Gas pressures necessary to maintain liquid density at ambient temperature are for helium, 18,000 lb/in² (124 MPa); for hydrogen, 28,000 lb/in² (193 MPa); and for nitrogen, 43,000 lb/in² (296 MPa).

The **selection of structural** materials calls for consideration of the effects of low temperatures on the properties of those materials. A number of otherwise suitable materials become brittle at low temperature (Fig. 10). Materials must perform satisfactorily over the complete range from cryogenic to room temperature, with hydrogen-embrittlement particularly significant. While hydrogen-embrittlement is not believed to be a problem at liquid-hydrogen temperatures, it has been shown to occur at room temperature. The 300-series stainless steels, aluminum alloys and BeCu, among other alloys, are more resistant to hydrogen embrittlement.

In calculating allowable stresses at room temperature, yield and tensile strengths should be used because (1) pressure testing of containers at ambient temperature is frequently necessary, and (2) in the case of large vessels, large unknown temperature gradients can exist above the liquid level both in the ullage space and in the walls of the vessel. The 300-series stainless steels are normally austenitic (f.c.c.). However, in some 300-series stainless steels austenite is partially transformed by cold-working and possibly by temperature cycling to martensite. As martensite is brittle at low $T$'s, the ductility of these steels is reduced by this kind of treatment. For this reason, it is recommended as an added safety factor that room temperature strengths of the 300-series stainless steels be used for the design of structural members.

Careful **stress analysis** is essential and must include stresses due to (1) thermal contraction of equipment, and (2) radial, axial, and circumferential temperature gradients caused by non-uniform cooling rates. The latter are frequently encountered in the cool-down of cryogenic transfer lines (see Sec. 5).

Cryogenic equipment should be **purged** before being placed in service to remove unwanted condensables and gases that could form explosive mixtures. Commonly used methods are (1) evacuating and back-filling with purge gas and (2) flowing of purge gas through the system.

The cryogen can be a direct **hazard to personnel.** Cryogen "burns" can result from direct contact with either the cryogen

or uninsulated equipment containing the cryogen. The large evolution of gases associated with cryogenic spills can result in asphyxiation. Asphyxiation is a hazard in entering a warmed cryogenic vessel. Further safety details are found in the references.

# OPTICS
## by Heard K. Baumeister

REFERENCES: AIP Handbook, McGraw-Hill. Conrady, "Applied Optics and Optical Design," Dover. Jenkins and White, "Fundamentals of Optics," McGraw-Hill.

**Index of Refraction**    **Visible light** is a small band in the electromagnetic spectrum between 4,000 and 7,000 Å wavelengths with about 5,600 Å (green-yellow) seeming the brightest to the average person. (Å, an angstrom, $= 10^{-10}$ metres.) The **velocity** of light in a vacuum $c$ is taken as $2.998 \times 10^8$ m/s ($9.836 \times 10^8$ ft/s; 186,300 mi/s). The velocity is lower in all substances, and the ratio of the velocity of light in a vacuum $c$ to that in a substance $c'$ is defined as the **index of refraction,** $n = c/c'$. The index of refraction in air at standard conditions is 1.0003; it is often assumed to be unity, and the reference (air or vacuum) must be determined before using published values if the slight differences are significant in the application. Most substances, gases, liquids, and solids, have indexes between 1 and 2 (see AIP Handbook).

**Dispersion**    The index of refraction of substances (including optical glass) is not constant but varies with the color of the light. The dispersion $v$ of glass is normally taken as $v = (n_D - 1)/(n_F - n_C)$, where $n_C$, $n_D$, and $n_F$ are the indexes of refraction for the three spectral lines 6,563 Å, 5893 Å, and 4,861 Å, respectively, obtained from hydrogen ($n_C$ and $n_F$) and sodium ($n_D$) discharge tube light sources. Most optical glasses have dispersions between 35 and 65. For special purposes, the dispersion may be determined for other colors (see AIP Handbook for further data).

**Refraction**    A ray of light undergoes refraction at a surface separating two regions of different indexes of refraction (see Conrady). For rays intersecting a plane surface interface, **Snell's law** holds (Fig. 1) that $n \sin I = n' \sin I'$.

For rays that are directed toward a point on the axis of a spherical interface (Fig. 2) and that are close to the axis (paraxial rays), the location of the intersection of the rays with the axis after refraction may be found from $n'/l' = (n' - n)/r + n/l$.

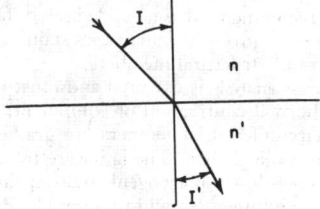

**Fig. 1**    Refraction at a plane surface.

For two spherical surfaces close together (thin lens in air) the relationship is $1/l'_2 - 1/l_1 = (n - 1)(1/r_1 - 1/r_2) = 1/f$. If the focal length and magnification are assumed, the object and image distances are $l_1 = f(1 - m')$ and $l'_2 = f(1 - m')/m'$.

The **magnifications** $m'$ for such systems are respectively $m' = h'/h = nl'/n'l$ and $m' = h'_2/h_1 = l'_2/l_1$. A word of caution is appropriate: If the image is inverted with respect to its object, the magnification is a negative number. Fractional magnifications indicate an actual reduction in size.

The **focal length** $f$ of a thin lens is the distance from the image to the lens when the object is infinitely far away. The focal length for a single surface, a spherical mirror, and a thick lens is more difficult to define and interpret (see Jenkins and White), but it is analogous to that for a thin lens and is indicative of the lens power and image size. The focal lengths of commercial lenses for use in air vary between 3 mm for microscope objectives to 100 ft (30.5 m) for large observatory telescope objectives.

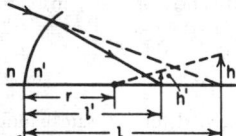

**Fig. 2**    Refraction at a spherical surface; all parameters shown in their positive sense.

**Aperture**    The larger the **aperture** of an optical element, the brighter will be the image and the better the resolving power, if the rest of the system (including the eye) is capable of accommodating all the light. The aperture of telescope objectives and camera-projection lenses is usually described by the $f$ **number,** i.e., the ratio of the focal length to the clear diameter. Commercially available lenses usually fall between $f/1.2$ and $f/15$. The aperture of microscope objectives is usually described by the **numerical aperture NA;** that is, $n \sin U$, where $n$ is the index of refraction of the object space and $U$ is the half angle of the vertex of the cone of light from the object that passes through the lens. Commercially available lenses usually fall between NA/1.40 and NA/0.08.

**Photographic film speeds** are usually given as an ASA exposure index. Black-and-white films are available commercially in the ASA range from 25 to 3,000. Color films fall in the slower end of this range. A rule of thumb for exposing films on a bright, sunny day is to set the aperture of the camera at $f/16$ and the shutter speed at 1/ASA exposure index sec.

The **energy flow** through an optical system is difficult to calculate accurately but is amenable to rapid rough estimation. The details must be arranged to suit the particular application but a representative estimation can be given here. The fundamental unit is the energy flow on, off, or through (watts) per unit area (metre$^2$) per unit solid angle (steradian) per unit wave length (metre); the spectral brightness, spectral luminance, spectral irradiance per steradian, spectral emittance per steradian etc. [W/(m$^2\cdot$sr$\cdot$m)]. The simplified fundamental concept is that the spectral brightness is transmitted throughout an ideal optical system unchanged in magnitude not only from object to image planes but from one aperture plane to the next. Estimating the limiting solid angle for the system and its associated area yield a convenient system constant. If the limiting field area is $A_1$ and the adjacent limiting aperture area is $A_2$ and their axial separation is $l$ then the system constant is $A_1 A_2/l^2$. (The solid angle is either $A_1/l^2$ or $A_2/l^2$, as convenient.) Multiplication of the input (source) spectral brightness by this system constant yields the ideal maximum energy transmitted per unit band width. Suitable reduction for reflection, absorption, and scattering must be made to obtain the output. Since these, as well as the input, are functions of the wavelength (color), the total passband must be divided into several narrow bands, individually treated, and summed to yield the total output.

EXAMPLE. Assume the spectral brightness for a narrow band at $6 \times 10^{-7}$ m wavelength to be $2.9 \times 10^{11}$ W/(m$^2\cdot$sr$\cdot$m). Let the source area $A_1$ be $1 \times 10^{-4}$ m$^2$ and the entrance area $A_2$ be $2.5 \times 10^{-3}$ m$^2$, separated a distance $l$ of 0.1 m. The system constant is then $2.5 \times 10^{-5}$ m$^2\cdot$ sr. The ideal maximum energy per unit band width is then $(2.9 \times 10^{11})(2.5 \times 10^{-5}) \approx 7.25 \times 10^6$ W/m. If the optical system has 10 air-glass surfaces, four that get dirty with a transmission of 0.8 each and six sealed, coated, and clean at 0.98 each, then the efficiency due to reflection would be about $(0.8)^4(0.98)^6 \approx 0.36$. If the glass were some 0.05 m thick then the efficiency due to absorption would be about 0.98. For a narrow color band $0.1 \times 10^{-7}$ m wide centered at the same $6 \times 10^{-7}$ m color for which the input was taken and for which the efficiencies are applicable, the total power transmitted by the system would be $(7.25 \times 10^6)(0.36)(0.98)(0.1 \times 10^{-7}) \approx 2.6 \times 10^{-2}$ W.

# PATENTS, TRADEMARKS, AND COPYRIGHTS
## by Edward Taylor Newton

### PATENTS FOR INVENTIONS

REFERENCES: Deller's Edition, "Walker on Patents," Baker, Voorhis & Co., Inc. Robinson, "The Law of Patents," Little, Brown. Ellis, "Patent Assignments and Licenses," Baker, Voorhis & Co., Inc. Rules of Practice of U.S. Patent and Trademark Office, 37 C.F.R., Title 1.

### United States of America

[Section references (§) refer to Patent Act of 1952, 35 U.S.C.]

**What Subject Matter Is Patentable**  Any new and useful process, machine, manufacture, or composition of matter, or any new and useful improvement thereof, or any asexually reproduced distinct and new variety of plant, including cultivated sports, mutants, hybrids, and newly found seedlings, other than a tuber-propagated plant found in an uncultivated state. "Process" means art or method and includes a new use of a known process, machine, manufacture, composition of matter, or material. "Machine" includes every mechanical device, usually having relatively moving parts. "Manufacture" means articles other than machines, such as pencils and chairs for example. A "composition of matter" includes anything resulting from the chemical union or mechanical mixture of two or more substances.

**Conditions for Patentability; Novelty and Loss of Right to Patent**  A person shall be entitled to a patent unless (1) the invention was known or used by others in this country or patented or described in a printed publication in this or a foreign country before the invention thereof by the applicant for patent, or (2) the invention was patented or described in a printed publication in this or a foreign country or in public use or on sale in this country more than 1 year prior to the date of the application for patent in the United States, or (3) the applicant has abandoned the invention, or (4) the invention was first patented or caused to be patented by the applicant or his or her legal representatives or assigns in a foreign country prior to the date of the application for patent in this country on an application filed more than 12 months before the filing of the application in the United States, or (5) the invention was described in a patent granted on an application for patent by another filed in the United States before the invention thereof by the applicant for patent, or (6) the applicant did not invent the subject matter sought to be patented, or (7) before the applicant's invention thereof the invention was made in this country by another who had not abandoned, suppressed, or concealed it (§102); and a patent may not be obtained if the differences between the subject matter sought to be patented and the prior art are such that the subject matter as a whole would have been obvious at the time the invention was made to a person having ordinary skill in the art to which said subject matter pertains (§103). In determining priority of invention there shall be considered not only the respective dates of conception and reduction to practice of the invention, but also the reasonable diligence of the one who was first to conceive and last to reduce to practice, from a time prior to conception by the other (§102). It is necessary that an applicant not withhold from the Patent Office evidence of pertinent prior art known to him or her, and especially such prior art as may anticipate or render obvious the invention claimed.

**Who May Apply for Patent**  The original inventor or his or her personal representative, or inventors jointly if more than one. The existence of joint invention can be determined only by the facts of each case. The only general rule is that two or more persons are properly joined as applicants for patent if the conception grew spontaneously out of conversation between them or if in working together they both contributed to the claimed invention. The *owner* of an invention, by assignment and sale from the inventor, may not *apply* for patent, but may receive the patent as assignee.

**Term of Patent**  Seventeen years from the date of issue. No extensions are granted without congressional action. The **patent grant** gives to the patentee the right to exclude others from making, using, or selling the patented subject matter during the life of the patent.

The **patent application** should be prepared by a competent solicitor. For information concerning the forms and rules, obtain from the Commissioner of Patents and Trademarks, Washington, D.C., a copy of the Rules of Practice of the U.S. Patent and Trademark Office.

**Proceedings in the U.S. Patent and Trademark Office**  Each application for patent found to be correct in form is examined in its turn. Rejections of claim or requirements of amendment must be answered by the applicant (or the applicant's attorney) within 6 months after the date of the official communication, or a shorter period specified by the Patent Office, or the application will be held to be abandoned.

**Interferences**  When an application for patent is found to interfere with, i.e., to present or claim substantially the same invention as (1) another pending application, (2) a patent issued on a date less than 1 year prior to the date of the application in question, (3) a reissued patent, of which the original was granted on a date less than 1 year prior to the date of the application in question, the two interfering cases are impleaded in an action in the Patent Office called an interference, the purpose of which is to determine which of two or more rival claimants is the original and/or first inventor. The practice before the Patent Office in interference is of such a character that an interference should be conducted only by legal counsel equipped with special experience in that practice.

The first of one or more applicants to **file** applications on the same invention is prima facie the first inventor. Proving that the invention was made **before** the effective filing dates of the respective applications requires evidence generally corroborated by one or more witnesses who had **direct** knowledge of, and understood, what was being done and when, during the making of the invention. A daily **log book** for the inventor and witnesses is desirable, and any pertinence of running entries to the invention should be **clear,** as references to sketches, descriptions, etc., already in the book, and to such items as drawings, photographs, completed samples, and test reports elsewhere which should be preserved.

**Cost of Obtaining a Patent**  Patent Office fees and time for payment are fixed by law but attorneys' fees vary. In general, the better the protection afforded by the patent, the more it will cost.

**Design Patents**  The inventor of a new, original, and ornamental design for an article of manufacture may obtain a patent for such design. Design patents are granted for $3\frac{1}{2}$ years, 7 years, or 14 years. Fees vary with the duration. In other respects, the requirements for design patents are substantially similar to those for mechanical patents. Design patents protect only the ornamental design of an article, i.e., its shape or ornamentation, as distinguished from its mechanical construction or its functional or utilitarian attributes.

**Sale or assignment of patent,** to be binding, must be by an instrument in writing. To provide for constructive notice to all, an assignment must be recorded in the Patent Office within 3 months of its date.

**Sales of Fractional Interests in Patents**  Undivided fractions or territorial division of patent rights are advisable only in special cases, because with the best intentions the interests of joint or territorial owners are liable to interfere.

**Licenses** are permissions by the patentee to make, use, or sell the thing patented, and broadly speaking a license may carry any imaginable provisions the parties agree on, subject to certain restrictions upon agreements in restraint of trade, upon fixing resale prices, upon limiting the use of patented articles after their sale, upon limiting the manufacture, use, or sale of articles not covered by the patent, and other restrictions of like character, as to which a lawyer should be consulted. License contracts should be reduced to writing. License by oral arrangement, or by implication from circumstances, is as binding on the parties as if reduced to writing, the difference lying in difficulty of proof.

**Relations of employer and employee,** when the latter makes an invention. Unless by express contract, or by implication from circumstances, the employee agrees that his or her patentable inventions should be the property of the employer, the employer has no title or claim to the invention or to the patent for it. However, if the employee uses the employer's time and/or materials in making the invention, the employer has a shop right (a nonexclusive royalty-free license). The scope of such a shop right, or implied license, must be determined from the nature of the employer's business, the character of the invention involved, the circumstances that created it, and the relation, conduct, and intention of the parties. The extent of such shop right depends upon the circumstances of each case.

Many employers require or attempt to secure from their employees an express agreement that inventions shall belong to the employer, and a prospective employee should carefully read and understand any such agreement before signing.

**Infringements of Patent**  As the grant gives the right to prevent others from making, selling, or using, so an unlicensed manufacture, sale, or use of the thing patented is an infringement, and the maker, seller, or user may be sued in a District Court of the United States. **Suit** may be **brought** in the judicial district where the defendant resides, or where the defendant has committed acts of infringement and has a regular and established place of business. In all but very exceptional cases, the courts will not grant **preliminary injunctions** in a suit on a patent that has not previously been held valid in another contested litigation.

**Marking Patented Articles**  Unless the patentee has given particular notice of his or her patent, or general notice by marking the patented articles "Patent" followed by the number of his patent, he or she may not recover damages for infringement. If the patented articles themselves cannot be marked, the mark may be affixed to the packages in which they are contained. Marking requirements do not apply to process patents or to patents under which the patentee has not manufactured.

**False Marking**  Whoever, without the consent of the patentee, marks upon, or affixes to, or uses in advertising in connection with anything made, used, or sold by him, the name or any imitation of the name of the patentee, the patent number, or the words "patent," "patentee," or the like, with the intent of counterfeiting or imitating the mark of the patentee or of deceiving the public and inducing them to believe that the thing was made or sold by or with the consent of the patentee; or

Whoever marks upon, or affixes to, or uses in advertising in connection with any unpatented article, the word "patent" or any word or number importing that the same is patented, for the purpose of deceiving the public; or

Whoever marks upon, or affixes to, or uses in advertising in connection with any article, the words "patent applied for," "patent pending," or any word importing that an application for patent has been made, when no application for patent has been made, or if made is not pending, for the purpose of deceiving the public—

Shall be fined not more than $500 for every such offense.

Any person may sue for the penalty, in which event one-half shall go to the person suing and the other to the use of the United States (§292).

**Damages for Infringement**  §284 provides that the successful claimant in a patent-infringement suit may be awarded "damages adequate to compensate for the infringement, but in no event less than a reasonable royalty for the use made of the invention by the infringer, together with interest and costs as fixed by the Court," and "the Court may increase the damages up to three times the amount found or assessed." §285 provides that "the Court in exceptional cases may award reasonable attorney fees to the prevailing party." The measure of damages may be the established royalty if licenses have been granted under the patent, or it may be the profit which the claimant lost due to the infringing competition. In any event the claimant is entitled to a reasonable royalty. In the case of a design patent §289 provides that the infringer shall also be "liable to the extent of his total profit on the infringing sales and not less than $250."

**Reissues of Patent**  If through inadvertence, accident, or mistake the original patent was defective, or claimed too much or too little, the error may be repaired by surrender of the patent and reissue, provided the reissue be for the same invention as was disclosed as such by the original and be applied for without unreasonable delay. No reissued patent shall be granted enlarging the scope of the claims of the original patent unless applied for within 2 years from the grant of the original patent (§251). The term of a reissued patent expires on the day that the original patent would have expired.

**Cooperation with Attorney**  The attorney who prepares a patent application should thoroughly **understand** both how the invention is made and how it works and try to anticipate what the inventor's competitor may wish to do; so discussion with the attorney is advisable at the outset. The action of the Patent Office will be based exclusively on the written record (Rule 2). An application contains a description and usually a drawing which essentially must enable one skilled in the art, to which the invention most closely pertains, to make and use the invention without appreciable experimentation. Neither the drawing nor the description may be substantially added to after being filed without adequate support somewhere in the application as **originally** filed, but the claims in a pending application may be timely amended. Each valid claim in a patent is a legal definition of what the patent owner may exclude others from making, using, or selling.

**Patents as Technical Literature**  The disclosures in the more recently issued patents are generally well detailed and reliable sources of information. Copies may be obtained readily at 50¢ each when identified by patent number, and they may be ordered in groups according to subclass, both according to patents already issued and according to future patents that may be issued. For details write Commissioner of Patents and Trademarks, Washington, D.C. 20231.

**Foreign Countries**

**International Convention for the Protection of Industrial Property**  The important provision of the Convention is that any person who has duly applied for a patent in one of the contracting states shall have a right of priority for 12 months in making application in the other states. Such subsequent application is unaffected by any acts accomplished in the interval, as, for example, by publication of the invention or by the working of it.

The laws and regulations in foreign countries differ so much with respect to their requirements as to novelty and patentability, the effect of a public disclosure of the invention either in the country in question or elsewhere, who may obtain a patent, the term of the patent, the cost of a patent, working of the patent, taxes, importation, compulsory licenses, revocation, and other particulars that it is not practicable to make a useful summary in the space allocated to this note. An inventor interested in foreign protection should consult a patent attorney.

## TRADEMARKS

REFERENCES: Callman, "Unfair Competition, Trademarks and Monopolies," Callaghan and Co., Chicago. McCarthy, "Trademarks and Unfair Competition," Lawyers Cooperative, Bancroft-Whitney. Vandenburgh, "Trademark Law and Procedure," Bobbs-Merrill. Gilson, "Trademark Protection and Practice," Matthew Bender.

**United States of America**

[Section references (§) refer to Lanham Act of 1946, and 15 U.S.C., both section numbers being given]

**What Is a Trademark?**  The term *trademark* includes any word, name, symbol, or device or any combination thereof adopted and used by manufacturers or merchants to identify their goods and distinguish them from those manufactured or sold by others (§45, §1127). A *service mark* is defined as a mark used in the sale or advertising of services to identify the services of one person and distinguish them from the services of others (§45, §1127).

**How Rights in a Mark Are Obtained**  Unlike a patent, but somewhat like a copyright, ownership of a mark is obtained by use. Subsequent to use within a state, a mark may be registered in that state. Forms for registering marks under state law are obtainable from the Secretary of State of each state. Subsequent to use in interstate or foreign commerce, a mark may be registered in the United States Patent and Trademark Office by filing an application in the form prescribed (37 C.F.R. Pt. IV). After federal registration the symbol ® may be used with the mark. Prior to federal registration, the symbols **TM** for a trademark or **SM** for a service mark may be used to give notice of a claim of common law or state law rights.

**Tests for Registrability and Infringement**  A mark may be registrable on the Principal Register only if when applied to the goods or used in connection with the services, it is not likely to cause confusion or mistake or to deceive. The same test is used to determine whether a mark infringes on an earlier mark. Marks that are descriptive, misdescriptive, geographical, or primarily merely a surname may not be registrable on the Principal Register, if at all, until after 5 years of substantially exclusive and continuous use in interstate commerce (§2[f], §1052[f]). Marks not registrable on the Principal Register may be registrable on the Supplemental Register if they are capable of becoming distinctive as to the applicant's goods or services (§23, §1091).

**Term of Registration**  Federal registrations remain in force for twenty (20) years unless cancelled. However, unless an affidavit is filed during the sixth year after registration showing that the mark is still in use, the Commissioner of Patents and Trademarks will cancel the registration of the mark (§8, §1058). Within the sixth year after registration on the Principal Register a registered mark may become more secure from legal challenge, i.e., "incontestable" upon the filing of the required affidavit (§15[3], §1065[b]). During the last 6 months of the registration term an application for renewal may be filed (§9, §1059). The terms of state registrations vary.

**Preliminary Search**  Before adopting a mark, it is advisable to have a search made in the United States Patent and Trademark Office to determine whether the mark under consideration would conflict with any pending application or mark registered for use on or in connection with any similar goods or services.

**Cost of Registering a Mark**  Government fees and time for payment are fixed by law (§31, §1113) but attorneys' fees vary.

**Effect of Federal Registration**  Federal registration of a mark affords nationwide protection, and once the certificate has issued, no person can acquire any additional rights superior to those obtained by the federal registrant. Federal registration of a mark establishes federal jurisdiction in an infringement action, can be the basis for treble damages, and is admissible as evidence of trademark rights. Registration on the Principal Register constitutes constructive notice, constitutes prima facie or conclusive evidence of the exclusive right to use the mark in interstate commerce, may become incontestable, and may be recorded with the United States Treasury Department to bar importation of goods bearing an infringing trademark.

**Assignment of a Mark**  It is only in conjunction with the goodwill of the business or that portion thereof with which the mark is associated that the mark can be assigned. Assignments of registered marks or of applications for registration should be recorded in the United States Patent and Trademark Office within 3 months after the date of assignment (§10, §1060).

### Foreign Countries

**Registration**  In general, most foreign countries require the registration of a mark in compliance with local requirements for trademark protection. In most countries registration is compulsory and provides the sole basis for protection of a trademark. A few countries afford common-law protection to unregistered marks. Because the laws and regulations in foreign countries regarding trademark registration vary so much, it is not practicable to summarize the requirements for registration in the space allocated to this note. Anyone interested in foreign protection of a mark should consult an attorney.

## COPYRIGHTS

REFERENCES: Nimmer, "Nimmer on Copyright," Matthew Bender. Copyright Office Regulations, 37 C.F.R., Chapter 2.

### United States of America

[Section references (§) refer to Copyright Act of 1976, 17 U.S.C.]

**Subject Matter of Copyright**  Copyright protection subsists in works of authorship in literary works, musical works including any accompanying words, dramatic works, including any accompanying music, pantomimes and choreographic works, pictorial, graphic, and sculptural works, motion pictures and other audio-visual works, and sound recordings, but not in any idea, procedure, process, system, method of operation, concept, principle, or discovery [§102].

**Method of Registering Copyright**  After publication (sale, placing on sale, public distribution), registration may be effected by an application to the Register of Copyrights along with the fee ($10.00) and two (2) copies of the writing (§401; 407; 409). The Copyright Office (Register of Copyrights, Library of Congress, Washington, D. C. 20540) supplies without charge the necessary forms for use when applying for registration of a claim to copyright.

**Form of Copyright Notice**  The notice of copyright required (§401) shall consist either of the word "copyright," the abbreviation "Copr.," or the symbol ©, accompanied by the name of the copyright owner and the year date of first publication.

**Who May Obtain Copyright**  The name of the author (creator) of the work is normally used in the copyright notice, even though an assignee may file the application for registration. Every assignment of copyright should be recorded in the Copyright Office within 3 months after its execution within the United States (§30).

**Duration of Copyright**  Copyright in a work created on or after January 1, 1978 subsists from its creation and, with certain exceptions, endures for a term consisting of the life of the author and fifty years after the author's death (§302). Copyright in the work created before January 1, 1978, but not theretofore in the public domain or copyrighted, subsists from January 1, 1978 and endures for the term provided by §302. In no case, however, shall the term copyright in such a work expire before December 31, 2002; and, if the work is published on or before December 31, 2002, the term of copyright shall not expire before December 31, 2027 (§303). Any copyright, the first term of which is subsisting on January 1, 1978, shall endure for twenty-eight years from the date it was originally secured, with certain exceptions (§304). All terms of copyright provided by §302–304 run to the end of the calendar year in which they would otherwise expire (§305).

**Infringement of Copyright**  A violation of any of the exclusive rights (§1) by a person not licensed could lead to liability (§101) for infringement. Actions for infringement of copyright are brought in the United States District Courts.

**Consulting Attorneys**  An attorney should be consulted before publication unless the author or proprietor has previously registered copyrights. A mistake could cause the copyright to be lost if the work is published without proper notice.

### Foreign Countries

**Universal Copyright Convention**  Under this Convention a U.S. citizen may obtain a copyright in most countries of the world simply by publishing within the United States using the prescribed notice, namely, © accompanied by the name of the copyright proprietor and the year of first publication placed in such manner and location as to give reasonable notice of claim of copyright. While the term "Copyright" on the notice is adequate for copyright under U.S. law, only the © is recognized under the Convention.

## MISCELLANEOUS
### (Staff Contribution)

#### Sizes of Type

The unit of height of a line of printer's type is the "point." One point = $\frac{1}{72}$ in. Six-point type is consequently $\frac{6}{72}$ = $\frac{1}{12}$ in high. The sizes generally employed in books and periodicals are 6-, 7-, 8-, 9-, 10-, 11-, and 12-point. These are shown below as set in Janson, the type face used in this book.

| | |
|---|---|
| 6-point | 10-point |
| 7-point | 11-point |
| 8-point | |
| 9-point | 12-point |

This handbook is set in 8-point type (8-point Janson) with a 1-point lead (space) between each line, with 124 text lines per page, and with a type-page width of 36½ picas. One printed page of this book is equivalent to 5.23 pages of elite typewritten manuscript and 7.42 pages of pica typewritten manuscript (see Copy Preparation and Copy Fitting below).

#### Copy Preparation and Copy Fitting

It is recommended that manuscripts be typed on 8½- by 11-in paper with 1½-in margins at the top and left side, with 1-in margins at the right and bottom, and with each page containing 26 lines, double-spaced. Some typewriters have pica type and write 10 characters per inch; others have elite type and write 12 characters per inch. One manuscript page typed as specified is equivalent to 260 words if typed in pica type and 312 words if typed in elite (assuming that each word averages six characters).

### This is pica type
### 1234567890

### This is elite type
### 123456789012

One way to find approximately how many lines of type will be equivalent to lines of manuscript copy is to type to the specifications given above at least enough lines from the particular printed article or book to fill one typewritten page. This will show the relationship between typewritten lines and lines set in type.

Another method is to take a number of lines from the printed article or book and type them line for line on the typewriter. This will also give the approximate number of lines of type that will be equivalent to the manuscript lines.

For a useful introduction to copy-fitting methods, including the more precise "character-count" method, see the article on copy fitting in Melcher and Larrick, "Printing and Promotion Handbook," 3d ed., McGraw-Hill.

#### Proofreading

When correcting proof, use ink or pencil of a different color from the marks already on the proof. Put your marks in the right or left margin, whichever is nearer to the word corrected. If there are several corrections in a single line, place them in order from left to right, separated by a slant line (e.g., tr/cap/,). If the same correction is made several times in the same line with no intervening correction, make your correction once in the margin, followed by an appropriate number of slant lines. For instance, if you add an *s* to three words in a line, write in the margin s///. When you wish to insert words, put a caret at the point of insertion and write the additional words in the margin. To delete material without substituting anything, cross it out and put a delete sign in the margin. When you delete words and substitute other words, cross out the unwanted material, use a caret within the line, and write the new material in the margin; the delete sign is then unnecessary.

Occasionally you may decide that material you have deleted should be restored. Place a row of dots under the crossed-out material, cross off the delete sign in the margin, and write "stet" (let it stand).

| | |
|---|---|
| ✄ | Delete |
| | Delete and close up |
| □ | Quad (one em) space |
| ⊔ | Move down |
| ⊓ | Move up |
| ⊏ | Move to left |
| ⊐ | Move to right |
| eq # | Equalize spacing between words |
| × | Broken letter |
| ¶ | Begin a new paragraph |
| no ¶ | No new paragraph |

| | |
|---|---|
| lc | lowercase Word |
| cap | Capital letter |
| sc | SMALL CAPITAL LETTER |
| bf | **Boldface** type |
| ital | *Italic* type |
| rom | Roman (type) |
| wf | Wrong font |
| # | Insert space |
| ⌢ | Close up |
| ⊙ | Turn letter |
| ⊙ | Period |

| | |
|---|---|
| *stet* | Let type stand ~~as set~~ |
| ? | Verify or (supply) information |
| tr | Transpose letters or marked words |
| sp | Spell out (abbrev) or (7) |
| ⌣ | Push space down |
| = | Straighten type |
| ‖ | ‖ Align type |
| *run in* | Run in material on same line |
| bu | Change ((x/y)) to built-up fraction |
| sh | Change (x/y) to shilling fraction |
| ⌄ | Set (s) as subscript |
| ⌃ | Set (s) as exponent |

| | |
|---|---|
| ⌃ | , Comma |
| ⌄ | ' Apostrophe |
| ⌄/⌄ | " " Quotation marks |
| ;/ | ; Semicolon |
| ⊙ | : Colon |
| ?/ | ? Question mark |
| !/ | ! Exclamation mark |
| =/ | - Hyphen |
| ⊥ | – En dash |
| ⊥ | — Em dash |
| ⊥ | Two-em dash |
| ⊂/⊃/ | ( ) Parentheses |
| [/⊐/ | [ ] Brackets |
| {/}/{} | Braces |

## Special Alphabets

### GREEK ALPHABET

The printer identifies Greek letters by means of the numbers in parentheses.

Note that the capital Greek letters A, B, E, Z, H, I, K, M, N, O, P, T, X used by the printer are like the English roman letters and may not be suitable as symbols. If used, however, they should be identified by marginal note so that they will not be improperly set in italic.

Note that the form of many boldface Greek symbols is markedly different from the style of the lightface characters.

| | | |
|---|---|---|
| Alpha | A | $\alpha$ (49), $\alpha$ (25) |
| Beta | B | $\beta$ (26), $\beta$ (53) |
| Gamma | $\Gamma$ (3) | $\gamma$ (27) |
| Delta | $\Delta$ (4) | $\delta$ (28), $\partial$ (56) |
| Epsilon | E | $\epsilon$ (29) |
| Zeta | Z | $\zeta$ (30) |
| Eta | H | $\eta$ (31) |
| Theta | $\Theta$ (8) | $\theta$ (32), $\vartheta$ (50) |
| Iota | I | $\iota$ (33) |
| Kappa | K | $\kappa$ (34), $\varkappa$ (54) |
| Lambda | $\Lambda$ (11) | $\lambda$ (35) |
| Mu | M | $\mu$ (36) |
| Nu | N | $\nu$ (37) |
| Xi | $\Xi$ (14) | $\xi$ (38) |
| Omicron | O | o (39) |
| Pi | $\Pi$ (16) | $\pi$ (40) |
| Rho | P | $\rho$ (41) |
| Sigma | $\Sigma$ (18) | $\sigma$ (42), $\varsigma$ (52) |
| Tau | T | $\tau$ (43) |
| Upsilon | $\Upsilon$ (20) | $\upsilon$ (44) |
| Phi | $\Phi$ (21) | $\phi$ (51), $\varphi$ (45) |
| Chi | X | $\chi$ (46) |
| Psi | $\Psi$ (23) | $\psi$ (47) |
| Omega | $\Omega$ (24) | $\omega$ (48) |

SPECIAL GREEK CHARACTERS

| | | | |
|---|---|---|---|
| No. 83M 8 | $\Theta$ | No. 55 | ϝ |
| No. 83M 29 | ε | No. 59 | ϛ |

BOLDFACE GREEK
ALPHABETS

SCRIPT

Available in capital letters only.

GERMAN

Script and German symbols may be indicated by circling the corresponding English letters each in a distinctive color.

Capitals: $\mathbf{\Gamma\;\Delta\;\Theta\;\Lambda\;\Xi\;\Pi\;\Sigma\;\Phi\;\Psi\;\Omega}$

Lower case: $\mathbf{\alpha\;\beta\;\gamma\;\delta\;\epsilon\;\zeta\;\eta\;\theta\;\iota\;\kappa\;\lambda\;\mu\;\nu\;\xi}$ $\mathbf{o\;\pi\;\rho\;\sigma\;\tau\;\upsilon\;\phi\;\chi\;\psi\;\omega\;\alpha\;\vartheta\;\varphi\;\varsigma\;\varkappa\;\partial}$

Lightface: $\mathcal{A\;B\;C\;D\;E\;F\;G\;H\;I\;J\;K\;L}$ $\mathcal{M\;N\;O\;P\;Q\;R\;S\;T\;U\;V\;W\;X\;Y\;Z}$

Boldface: $\boldsymbol{\mathcal{B\;E\;E\;F\;H\;I\;M}}$

Capitals: $\mathfrak{A\;B\;C\;D\;E\;F\;G\;H\;I\;K\;L\;M}$ $\mathfrak{N\;O\;P\;Q\;R\;S\;T\;U\;V\;W\;X\;Y\;Z}$

Lower case: $\mathfrak{a\;b\;c\;d\;e\;f\;g\;h\;i\;j\;k\;l\;m\;n\;o\;p}$ $\mathfrak{q\;r\;s\;t\;u\;v\;w\;x\;y\;z}$

# Index

Abacá (manila fiber), **6**-151
Abampere (def), **15**-3
Ablators (heat-resisting materials), **6**-170
Abrams method (proportioning cement), **6**-184
Abrasive cutoff saws, **13**-66
Abrasive paper, **6**-140
Abrasive tools (wood sanding), **13**-83
Abrasive wheels (*see* Grinding wheels)
Abrasives, **6**-138, **6**-139
ABS plastic, **6**-162
Absolute-pressure gages, **16**-10
Absolute temperature scale, **4**-3
Absorber in refrigeration system, **19**-11
Absorptance (heat, def), **4**-71
Absorption:
  coefficient of (sound), **12**-140
  of gases by water, **6**-7
Absorption dynamometer, **16**-18
Absorption machine, **19**-11 to **19**-13
Absorption refrigerators, **4**-46
Absorptivity (heat, def), **4**-71
Acceleration (def), **3**-11
  angular (def), **3**-14
  composition of, **3**-12, **3**-15
  conversion (table), **1**-5
  equivalents (table), **1**-51
  of gravity, **1**-40, **3**-2
  resolution of, **3**-13
  units of, **1**-33
Acceleration resistance (train), **11**-30
Acceleration-time curve (kinematics), **3**-12
Accelerometers, **5**-75, **16**-21
Accidents, occupational (table), **18**-23
  prevention of, **18**-22 to **18**-25
Accounting (*see* Cost accounting)
Accumulators, hydraulic, **8**-41
Acetone (solvent), **6**-169
Acetylene welding, **13**-33 to **13**-36
Acid number of lubricating oil, **6**-199
Acid-resisting cast iron, **6**-49
Ackerman steering gear (automobiles), **11**-11
Acme screw threads, **8**-11, **8**-12
Acoustic theory (aerodynamics), **11**-79 to **11**-84
Acoustics in industrial plants, **12**-9
  (*See also* Sound)
Acrylic fibers, **6**-54
Acrylic resins (plastics), **6**-161
Active power (a-c circuits, def), **15**-4
Actuators, hydraulic, **8**-41
ACV (air cushions vehicles), **11**-56
Addendum (gear teeth), **8**-99
Addition:
  algebraic, **2**-16
  arithmetical, **2**-2
  of complex quantities, **2**-23
  of vectors, **2**-63
Adhesion, coefficient of (locomotives), **11**-24
Adhesives, **6**-140, **6**-141
  properties and uses of (table), **6**-143 to **6**-148
  resins, performance of (table), **6**-142

Adiabatic expansion of gases, **4**-19
Admiralty brass (table), **6**-66
Admittance (electric circuits, def), **15**-5, **15**-23
Adsorption, refrigeration, **19**-11
Advance (resistor alloy), **15**-73
  table, **15**-78
Aerodynamics (def), **11**-59
  axes in, **11**-59
  coefficients in, **11**-60
  hypersonic (def), **11**-58
  speed in, **11**-59
    subsonic, **11**-60
    (*See also* Airfoils)
    supersonic (def), **11**-58
    (*See also* Supersonic and hypersonic
      aerodynamics)
  transonic (def), **11**-58
Aeronautics (*see* Airplanes)
Afterburner in jet engine, **11**-86
Aftercondensers, **9**-67
Agde-Damm coal test, **7**-38
Agglomeration test (coal), **7**-2, **7**-38
Agglutination test (coal), **7**-38
Aggregates for concrete, **6**-179
Aging of steel (hysteresis), **15**-16, **15**-17
  after temper rolling, **6**-28
Air:
  cleaners, characteristics of (table), **18**-20
  composition of, **6**-10
  dehumidification of, **4**-43
    by surface cooler, **4**-44
  density of (def), **4**-31
    at altitudes (table), **11**-59
    formula, **14**-50
  in ducts, friction in (tables), **12**-99, **12**-100
  enthalpy-entropy diagram, **14**-37
  heat-transfer coefficients to or from, **4**-63 to
    **4**-70
  humidification of (spray chamber), **4**-44
  internal energy of (table), **4**-56
  liquefaction of, **19**-22
  moist, properties of, **4**-28 to **4**-45
    charts, **4**-12, **4**-43
  moisture vs. dewpoint (table), **12**-75
  pressure of, at altitudes (table), **11**-59
  saturated, vapor pressure of (table), **12**-75
  solubility of, in water, **6**-7
  surface conductance (table), **12**-86
  surface resistance (table), **12**-86
  temperature at altitudes (table), **11**-59
  temperature-entropy chart for, **4**-25
  thermal conductivity of (table), **4**-62, **4**-65
  velocity pressure, **14**-51
  water vapor mixtures, **4**-28 to **4**-49
Air chambers of pumps, **14**-7
Air compression, **14**-32
  adiabatic (chart), **14**-37
  effect of altitude on, **14**-31
  multistage: power required in, **14**-32
    theory of, **4**-22

Air compression (*Cont.*):
  power required for, **14**-32
    chart, **14**-33
  theory of, **4**-21
  work of, **4**-22
Air compressors, **14**-30 to **14**-44
  centrifugal (*see* Centrifugal compressors)
  compression efficiency of, **14**-33
  efficiency of, **14**-41
  fuel-injection for Diesel engines, **9**-98
  lubrication of, **14**-37
  oils for, **6**-205
  ratio of compression, **14**-32
  regulators for, **14**-38
  rotary, **14**-39 to **14**-44
  thrust loading, **14**-32
  types of, **14**-31, **14**-32
  unloaders for, **14**-38
  uses for, **14**-31
  valves for, **14**-36
Air conditioning, **12**-76 to **12**-116
  absorption systems, water and lithium
      bromide, **19**-11
  air distribution, **12**-76
  air infiltration (tables), **12**-65, **12**-66
  atmospheric cooling for, **11**-98
  of automobiles, **11**-17, **11**-18
  chilled-water systems, water distribution in,
      **12**-98
  of cold storage rooms, **19**-20
  comfort indexes for: ASHRAE comfort
      chart, **12**-105, **12**-116
    effective temperature (chart), **12**-115
    temperature-humidity index, **12**-105, **12**-
      106
  cooling load formulas, **12**-76
  duct air velocities (table), **12**-98
  ducts, friction of air in (tables), **12**-99, **12**-100
  ductwork for, **12**-83, **12**-98
  fans for, **12**-98
  heat gain: from appliances (table), **12**-96, **12**-97
    from electric motors (table), **12**-97
    from lighting fixtures (charts), **12**-77
    from occupants (table), **12**-96
    from people (table), **12**-98
    through roofs (tables), **12**-82, **12**-83
    solar, factors for (table), **12**-90 to **12**-92
    through sunlit walls (table), **12**-78 to **12**-
      80
    from various processes (table), **12**-103
  inside design conditions (table), **12**-97
  moisture infiltration, **12**-72
  moisture load in, **12**-72
  outdoor air requirements (table), **12**-67
  outside design temperatures (charts), **12**-80,
      **12**-81
  overall heat-transfer coefficients (table), **12**-
      64
  power input to (table), **12**-111
  processes, **4**-44
  psychrometric chart for, **12**-81
  refrigeration, **4**-45
  return-intake air velocities (table), **12**-97
  supply-air rate, **12**-76
  supply-air temperature, **12**-76
  water-cooling equipment for, effectiveness
      of (table), **12**-104
Air ejectors, **9**-66 to **9**-71
  capacities of (table), **9**-68 to **9**-70
  hogging, **9**-68
  intercondensers, **9**-67
  materials for, **9**-68

Air ejectors (*Cont.*):
  performance of, **9**-67
  priming, **9**-68
  steam jet pressure, **9**-67
  table, **9**-68, **9**-71
  types of, **9**-66, **9**-67
Air-lift pumps, **14**-14, **14**-15
Air meters, **16**-14 to **16**-18
Air motor hoist, **10**-15
Air pollution, **18**-8
  characteristics of (chart), **18**-9
  classification of, **18**-11, **18**-12
  control of, in industries, **18**-17, **18**-18
  from diesel engines, **9**-111
  dispersion from stacks, **18**-16, **18**-17
  effects on vegetation, **18**-14, **18**-15
  emission standards for passenger cars and
      light trucks (table), **9**-113
  from gasoline engines, **9**-110 to **9**-112
  from heavy-duty engines, **9**-113
  from internal-combustion engines, **9**-109 to
      **9**-113
  from light-duty engines, **9**-112
  particle sizing in, **18**-11, **18**-12
  by power plants, **17**-36
  principal pollutants (table), **18**-10
  sources of, **18**-8 to **18**-11
    domestic, **18**-8
    industrial (table), **18**-11
    municipal (table), **18**-11
  sulfur effects, **18**-14 to **18**-17
  from transportation engines (table), **9**-110
  (*See also* Environmental control)
Air preheaters (boilers), **9**-22
Air resistance:
  of automobiles, **11**-3, **11**-57, **11**-58
  of ships, **11**-57, **11**-58
  of trains, **11**-57, **11**-58
Aircraft (def), **11**-58
Aircraft jet propulsion, **11**-84 to **11**-98
  efficiency and power of, **11**-92, **11**-93
  equations, notation, **11**-90 to **11**-92
  fuels, **7**-17, **7**-18
  ramjet, **11**-85
    performance of, **11**-93 to **11**-96
  rocket engines, **11**-87 to **11**-90
    electric, **11**-89
    electromagnetic, **11**-89
    electrostatic, **11**-89
    electrothermal, **11**-89
    liquid propellants, **11**-87
    nuclear heat-transfer, **11**-89
    performance of, **11**-96 to **11**-98
    propellant combinations (table), **11**-88
    scramjet, **11**-85
    solid propellants, **11**-88
  thermal air-jet systems, **11**-85 to **11**-87
  thrust equations, **11**-92
  turbofan, **11**-86
    performance of, **11**-96
  turbojet, **11**-86
    performance of, **11**-93 to **11**-96
Airflow:
  through orifices, **4**-46
  in pipes, **4**-66
Airfoils, **11**-61 to **11**-71
  aerodynamic forces on, **11**-60
  angle of attack (def), **11**-60
  aspect ratio (def), **11**-60
  boundary layer control, **11**-64
  camber line of, **11**-62
  characteristics of, **11**-61

Airfoils (*Cont.*):
 drag (def), **11**-60, **11**-61
 dynamic pressure (def), **11**-61
 flaps and slots in, **11**-63, **11**-64
 induced drag (def), **11**-61
 lift (def), **11**-60, **11**-61
 moment (def), **11**-61
 pressure distribution on, **11**-64
 profile drag, **11**-61
 sections, properties of (charts), **11**-63
 stagnation point (def), **11**-61
 stalling angle, **11**-61
 transonic, **11**-63
 wing section, selection of, **11**-62, **11**-63
Airplanes (def), **11**-58
 aerodynamic forces on, **11**-60 to **11**-71
  supersonic and hypersonic, **11**-73 to **11**-79
 ceiling, **11**-65
 climbing rate, **11**-65
 climbing time, **11**-66
 compass (gyroscopic), **3**-23
 control, **11**-71, **11**-72
 dimensions of (table), **11**-67
 drag, **11**-66 to **11**-71
  (*See also* Airfoils; Drag)
 dynamic stability, **11**-71
 engines, **9**-80
  carburetors, **9**-98
  cooling, **9**-105
  drag, **11**-70
  tables, **9**-81
 flaps and slots, **11**-63, **11**-64
  table, **11**-65
 floats (seaplane) drag, **11**-71
 fuselage, drag, **11**-70
 gas turbines, **9**-117
 gyrocompass, **3**-23
 jet propulsion (*see* Aircraft jet propulsion)
 landing speed, **11**-64
 parasite drag (def), **11**-66
 performance of, **11**-64 to **11**-66
  with jet thrust, **11**-66
  typical (table), **11**-67
 power, available, **11**-65, **11**-66
  table, **11**-66
 propellers, **11**-98 to **11**-104
  aerodynamic theories, **11**-98 to **11**-100
   axial-momentum, **11**-98, **11**-99
   blade element, **11**-99
   vortex, **11**-99
  blades, **11**-102, **11**-103
   construction of, **11**-102, **11**-103
  centrificial loads, **11**-102
  coefficients for, **11**-100
   power, **11**-100
   thrust, **11**-100
   torque, **11**-100
  control, **11**-103
   beta, **11**-104
  counterweights, **11**-102
  design, **11**-102, **11**-103
  ENT (emergency negative thrust), **11**-104
  excitation factor, **11**-102
  fatigue strength, **11**-102
  feathering, **11**-103
  gyroscopic forces on, **3**-23, **11**-104
  hubs, **11**-102
  mechanical design, **11**-102 to **11**-104
  performance of, **11**-100, **11**-101
   characteristics of, **11**-100, **11**-101
   compressibility effects, **11**-101
   noise, **11**-101

Airplanes, propellers, performance of (*Cont.*):
   reverse thrust, **11**-101
   static thrust, **11**-100
  performance characteristics of, **11**-100,
    **11**-101
   advance ratio, **11**-100
   velocity coefficient, **11**-100
  pitch, change mechanism for, **11**-103
  resonance, **11**-102
  reversing, **11**-103
  slipstream contraction, **11**-98
  store damage, **11**-102
  vibratory stresses in, **11**-102
 range of (table), **11**-67
 speed of (table), **11**-67
 stability, **11**-71
 stalling angle, **11**-64
 stalling speed, **11**-64
 static stability, **11**-71
 struts, drag, **11**-19
 tail surfaces, drag, **11**-70
 turn-indicator, **3**-23
 wings: delta, **11**-80
  drag, **11**-63, **11**-69
  flaps and slots, **11**-63, **11**-64
  scale effect, **11**-62
   chart, **11**-63
  sections, **11**-62, **11**-79
  (*See also* Airfoils)
 wire (streamline), drag (chart, table), **11**-70
Airslide conveyor, **10**-63
Albumin glue, **6**-141
Alclad alloys (aluminum), **6**-78, **6**-117
Alcohol, **6**-168
 air for combustion (table), **4**-55
 compressibility, **6**-9
 denatured, **6**-156, **6**-168
 freezing preventive (table), **4**-5
 as fuel, **7**-20
 heat of combustion (table), **4**-54
 products of combustion (table), **4**-55
 solvents, **6**-168
 specific gravity and density (table), **6**-8
 wood, **6**-156, **6**-168
Algebra:
 of complex quantities, **2**-22 to **2**-25
 elementary, **2**-15 to **2**-25
 solution of equations, **2**-18 to **2**-25
Alignment charts, **2**-58 to **2**-62
Alkalinity of boiler water, **9**-31
Allowance (def), **8**-45
 for various types of fit, **8**-45
  table, **8**-45, **8**-46
Alloys:
 aluminum, **6**-75 to **6**-81
 antifriction, **6**-92, **6**-93
 cast-iron, composition and properties of
   (tables), **6**-49
 for castings, **13**-7, **13**-8
 copper, **6**-63, **6**-68
  for bearings (table), **6**-93, **6**-94
 creep rates (table), **5**-11
 fusible (table), **6**-90
 hardenable, **6**-38
 for high temperature, **6**-92 to **6**-100
  superalloys (tables), **6**-97 to **6**-99
 iron-nickel, **6**-41
  magnetic properties of (table), **6**-41
 lead, **6**-89 to **6**-91
 of low thermal expansion, **6**-41
 magnesium, **6**-81 to **6**-85
 nickel, **6**-70 to **6**-75

Alloys (*Cont.*):
  nonferrous, **6**-59 to **6**-106
    hardening, **6**-62
    heat-treatment, **6**-62
  resistivity (table), **15**-6
  resistor materials, **15**-69 to **15**-73
  specific gravity and density (table), **6**-7
  steel (*see* Steel, alloys)
  tin, **6**-87
  titanium, **6**-87
  zinc, **6**-85 to **6**-87
Alnico (magnet material), **15**-73
Alpha iron (def), **6**-17
Alpha particle, atomic, **9**-121
Alphabet:
  German, **19**-47
  Greek, **19**-46, **19**-47
  script, **19**-47
Alternating-current instruments, **15**-25, **15**-26
Alternating currents (*see* Circuits, alternating-current; Currents, alternating; Electric motors; Generators)
Alternation, electric current (def), **15**-21
Alternators, **15**-36 to **15**-40
Altitude:
  correction factors for, due to air volume (table), **12**-103
  variation of atmospheric pressure and temperature with (table), **11**-59
Alumina brick, **6**-171
Aluminum:
  alloys, **6**-75 to **6**-81
    aging, **6**-78
    anodizing treatment of castings, **6**-81
    castings, composition and properties of (tables), **6**-76, **6**-78
    designations, **6**-75
    die-casting, **6**-75
      table, **6**-78
    heat-treatment, **6**-78
      table, **6**-81
    machining, **6**-78
    magnesium, **6**-75
    mold-casting (table), **6**-76, **6**-77
    protection, **6**-81, **6**-116
    riveting, **6**-81
    silicon, **6**-75
    strength at high temperatures (table), **6**-80
    welding, **13**-45 to **13**-47
      and soldering, **6**-81
    wrought, **6**-75, **6**-76
      composition and properties of (table), **6**-78
  brass: composition and properties of (table), **6**-65
    condenser tubes, **6**-112
  bronze, **6**-68
    for bearings, **6**-93
    composition and properties of (table), **6**-65
  conductors, **6**-81, **15**-6
    table, **6**-82, **15**-59
  corrosion resistance of, **6**-81
  paint, **6**-119
  painting and lacquering, **6**-81
  pipe (table), **8**-173
  plastic range chart, **13**-15
  soldering, **6**-81
  tubing, **8**-169
  welding, **6**-81, **13**-45 to **13**-47
Alundum (abrasive), **6**-140
Amatol (explosive, table), **7**-26
American farm windmill, **9**-165

American Standard pipe threads, **8**-12, **8**-19, **8**-20
American Standard screw threads, **8**-9 to **8**-15
American Standard wire gage, **15**-6
  table, **6**-45
American wire gage, copper wire sizes (table), **15**-7
Ammeters, **15**-24
Ammonia, **4**-26
  aqua-, properties of (chart), **4**-36
  pipes for, valves and fittings, **8**-188
  properties of: chart, **4**-27
    saturated (table), **4**-33, **4**-34
    solutions (chart), **19**-12
    superheated (table), **4**-35
  as refrigerant, **4**-51, **19**-4
Ammonia absorption machines, **19**-11 to **19**-13
Ammonia compressors, **19**-7, **19**-10
  condenser pressure (table), **19**-9
  dual, **19**-7
  horsepower per ton of refrigeration (table), **19**-9
  multiple effect of, **19**-7
  performance of (table), **19**-5
  volume of gas per ton of refrigeration (table), **19**-9
  wet compression, **19**-8
Ammonia condensers, **19**-8
  double-pipe, **19**-8
  evaporative, **19**-8
Ammonia dynamites, **7**-25
Ammonia gelatins in explosives, **7**-25
Amortisseur windings (a-c armatures), **15**-39
Ampere (def), **1**-32, **15**-3
Ampere-turn (def), **15**-2
Amplidyne (d-c generator), **15**-31
Amplifier:
  analog computer, **2**-73
  differential (electronic circuit), **15**-89
  radio, **15**-94
  transistor, **15**-85, **15**-86, **15**-88, **15**-89
Analog computer, **2**-73 to **2**-75
Analog-to-digital converter, **16**-24
Analysis, dimensional, **3**-49 to **3**-52
Analytical geometry, **2**-30 to **2**-41
Analyzer, absorption refrigeration, **19**-11
Anchor ring, volume and area, **2**-14
Ancit process (coke making), **7**-41
Anemometers, **16**-17
Angle:
  of attack (aerodynamics, def), **11**-60
    induced, **11**-61
  of repose, friction, **10**-2
  of slide (piled material, def), **10**-2
  of stall (airfoils, def), **11**-61
Angle valves, **8**-188
Angles:
  analytical geometry formulas, **2**-30
  bisection of, **2**-9
  complementary (def), **2**-25
  congruent (def), **2**-25
  conversion tables for, **1**-11, **1**-12, **1**-30
  dihedral, **2**-8, **2**-14
  half, functions of, **2**-27
  laying out with tape, **16**-42
  measurement of (surveying), **16**-45
    units for, **2**-25
  multiple, functions of, **2**-27
  negative, functions of, **2**-27
  radii of gyration for two (tables), **12**-41, **12**-42
  solid, **2**-14

Angles (*Cont.*):
  steel, used as beams, **12**-33
    tables, **12**-39 to **12**-42
  structural steel (tables), **12**-39 to **12**-42
  sum and difference of, **2**-27
  supplementary (def), **2**-25
  of a triangle, relations between, **2**-27
  trigonometric formulas, **2**-25 to **2**-27
Angular acceleration, **3**-14
Angular displacement, **3**-14
Angular momentum (def), **3**-21
Angular velocity, **3**-14
  conversion factors (table), **1**-50
Animals, power output of, **9**-162
Annealing:
  of nonferrous metals, **6**-62
  of steel, **6**-21 to **6**-23
Annual reports, **17**-11
Annuity tables, **1**-26 to **1**-29
Annulus:
  area of, **2**-11, **2**-12
  number of contiguous circles in, **2**-11
Anodes, sacrificial, **6**-108
Anodic reaction in corrosion, **6**-106
Anomalistic year, **1**-41
Anthracite (*see* Coal, anthracite)
Antifriction alloys, **6**-91 to **6**-94
Antifriction curve, Schiele's, **2**-41
Antigudermannian, **2**-30
Antilogarithms, **2**-4
Antimonal lead, **6**-89
Antimony, alloys, **6**-89
Antioch process (molding), **13**-3
Antirolling gyroscope (ships), **3**-23
Anvils, steam-hammer, **13**-25
Aperture (optics), **19**-40
API (American Petroleum Institute) scale
  (specific gravity) conversion tables,
  1-42, **1**-43
Apothecaries' liquid measure, **1**-31
Apothecaries' weight, **1**-32
Apparent power (a-c circuit, def), **15**-4, **15**-22
Apron conveyors, **10**-50
  table, **10**-51
Aqua ammonia (chart), **19**-12
Arc:
  circular: center of gravity of, **3**-7
    length of (construction), **2**-10
  of contact (belts), **8**-53
  sin, arc tan, etc., **2**-27
Arc furnaces, **7**-60, **7**-64 to **7**-66
Arc welding, electric, **13**-28 to **13**-33
Arch beams, **5**-46
Arches:
  furnace, **6**-175
  masonry, laying of, **12**-22
Archimedean spiral, **2**-40
Are, **1**-34
Area meters, fluid flow, **16**-17
AREA railway line clearances, **11**-31
Areas:
  centers of gravity for, **3**-7, **3**-8
  of circles (tables), **1**-4, **1**-5
  conversion tables, **1**-46
  equivalent (table), **1**-46
  measurement of, **1**-31, **16**-7
  methods of calculating, **2**-11, **2**-12, **2**-48, **2**-49
  moments of inertia, **3**-9
  plane: centers of gravity, **3**-7, **3**-8
    by graphics, **3**-11
    moment of inertia, **3**-9
    by graphics, **3**-11

Areas (*Cont.*):
  of similar figures, **2**-8
  of solids, **2**-12 to **2**-15
  units of, **1**-31
Arithmetic, **2**-2 to **2**-7
  fixed point, **2**-65
  floating point, **2**-65
Arithmetic mean, **2**-17
Arithmetical progression, **2**-17
Armature reactance and resistance, a-c
  generators, **15**-37
Armature reaction, **15**-34, **15**-37
Armco ingot iron, **6**-13
Armored cable (interior wiring), **15**-67
Arms, flywheel, **8**-50
Asbestos, **6**-153
  friction, **3**-25
  insulation products, **6**-170
  magnet wire insulation, **15**-80
  specific gravity and density of (table),
    **6**-9
  thermal conductivity of (table), **4**-63
Asbestos-cement pipes, **6**-114, **8**-172
Asbestos shingles, **6**-164
Ash:
  coal, **9**-7 to **9**-10
    tables, **7**-4
  combustibles, **9**-13
    heat loss due to (chart), **9**-14
  content in fuel oil, **9**-9
    analysis (table), **9**-9
  deposits, **9**-8, **9**-9
  effect of fuel additives on, **9**-8
  fusibility, **7**-7
    table, **7**-8
  removal by shot cleaning, **9**-9
  uses of, **9**-10
  viscosity of, **9**-8
Ash collectors, **9**-10
Ash conveyors, pneumatic, **10**-62
Ashlar masonry (table), **12**-23
ASHRAE comfort chart, **12**-105, **12**-116
Aspect ratio (aerodynamics, def), **11**-60
Asphalt, **6**-163
  roofing, **6**-163, **6**-164
Asphaltum, specific gravity and density
  (table), **6**-9
Assay ton (def), **1**-32
ASTM specifications:
  alloy-steel castings (table), **6**-56
  brick (table), **6**-149
  cast iron, **6**-50
  cement, **6**-178, **6**-179
  copper wire, **6**-63
    table, **6**-63
  gypsum plaster, **6**-182
  malleable iron, **6**-51
  nodular iron, **6**-51
  standard sieves (proportioning concrete,
    table), **6**-180
  steel castings, **6**-54, **6**-55
Astroid, **2**-40
Astronautics, **11**-104 to **11**-129
  (*See also* Space)
Asymptote:
  of hyperbola, **2**-35, **2**-36
  of tractrix, **2**-41
Asynchronous generator, **15**-45
Ata (pressure unit, def), **14**-31
Atgas process (gas making), **7**-46
Atmosphere:
  control in industrial furnaces, **7**-51 to **7**-53

Atmosphere (*Cont.*):
  impurities in (chart), **18**-9
  protective gas (heat treating, table), **7**-52
  standard (NACA, def), **11**-58
    table, **11**-59
  upper, data on (NACA), **11**-58, **11**-59
  variation with altitude (table), **11**-59
Atom (def), **6**-3, **9**-120
Atomic energy, **9**-119 to **9**-121
  alloys for, **6**-100 to **6**-103
  beryllium as absorber, **6**-103
  control rods, **6**-102, **6**-103
    boron, **6**-102
    cadmium, **6**-103
    gadolinium, **6**-103
    hafnium, **6**-103
  coolants, metallic, **6**-101
  fuels, **6**-102
  fusion systems, **9**-135, **9**-136
  metals for, **6**-100 to **6**-103
  moderators (table), **6**-101
  nuclear properties, **6**-100
  reactor component requirements (table), **6**-100
  slow neutron absorption (table), **6**-101
  sodium, resistance to (table), **6**-102
  stainless steel, uses of, **6**-103
  zircoloys, uses of, **6**-103
  zirconium, uses of, **6**-103
Atomic number (def), **6**-7
  of elements (table), **6**-3, **6**-4
  of metals (table), **6**-60, **6**-61
Atomic power, **9**-119 to **9**-136
  boilers for (*see* Nuclear boilers)
  cycle for, **9**-122
  fuel enrichment costs, **9**-132
  operating and maintenance costs, **9**-132
  plant cost, **9**-130 to **9**-133
  plant licensing, **9**-133, **9**-134
  plant safety, **9**-133
Atomic reactors (*see* Reactors)
Atomic weights:
  of elements (table), **6**-3, **6**-4
  of metals (table), **6**-60, **6**-61
Attemperation of steam, **9**-23
  submerged-type, **9**-23
Attraction, laws of, **3**-22
Atu (pressure unit, def), **14**-31
Austenite (def), **6**-17
  grain size, **6**-20
    influence of alloys (table), **6**-21
Austenitic hardenable alloys, **6**-36
Autogiro (def), **11**-58
Autoignition temperatures (gaseous fuels), **9**-93
Automatic control:
  basic system, **16**-26
  block-diagrams, **16**-28, **16**-29
    algebra, **16**-29
    representation, **16**-28
  closed-loop, **16**-26
  command signal, **16**-25
  compensation (def), **16**-26
    derivative, **16**-27
      error, **16**-27
      input, **16**-27
      output, **16**-28
    integral, **16**-27
      error, **16**-28
  components, **16**-37, **16**-38
  by computer, **16**-40, **16**-42
  definition of terms, **16**-25, **16**-26
  design procedure, **16**-37

Automatic control (*Cont.*):
  electrical systems, **16**-37
  error coefficients, **16**-30
  frequency response, **16**-31
    equations for common elements (table), **16**-32
    graphical display, **16**-31 to **16**-34
      Bode diagrams, **16**-31
    logarithmic plots, **16**-31
    Nyquist plots, **16**-31
    polar plots, **16**-31
    polar loci (table), **16**-33
  hydraulic systems, **16**-37
  modes vs. process characteristics (table), **16**-41
  nomenclature, **16**-25
  peak overshoot, **16**-27
  peak time, **16**-27
  pneumatic systems, **16**-30 to **16**-40
    mechanisms, **16**-40
  process characteristics for (table), **16**-41
  rise time, **16**-27
  sampled-data systems, **16**-35 to **16**-37
    stability of, **16**-36
  servo valves in, **16**-37
  signal flow representation, **16**-29, **16**-30
  stability and performance of, **16**-34, **16**-35
    gain margin (def), **16**-34
    Nyquist stability criterion, **16**-34
    phase margin (def), **16**-34
    Routh's stability criterion, **16**-35
    steady-state performance, **16**-30
  state-space concepts, **16**-37
  transient analysis of system, **16**-26 to **16**-28
    overshoot, **16**-27
    transient frequency, **16**-27
    viscous-damped, **16**-26
      response to step-input, **16**-26
  transition matrix for, **16**-37
  z transformation, **16**-36
Automatic pilot (gyroscopic), **3**-23
Automatic regulator (def), **16**-25
Automobile engines, **9**-82, **9**-83
  in buses, **9**-83
  compression, **9**-107
  compression ratio (table), **9**-84, **9**-85
  cooling, **9**-83, **9**-104, **9**-105
  cylinders, **9**-82, **9**-88 to **9**-92
  firing orders, **9**-100
  foreign, **9**-83
    table, **9**-85
  fuel consumption (charts), **11**-5
  ignition, **9**-100, **15**-80 to **15**-82
  lubrication, **9**-105, **9**-106
  passenger-car data, **9**-82
  pollution from, **18**-10
  power (table), **9**-84, **9**-85
  valve timing, **9**-2
Automobiles, **11**-3 to **11**-18
  acceleration, **11**-4
  air conditioning, **11**-17, **11**-18
    compressors, **11**-18
  brakes, **11**-13 to **11**-17
    adjustment, **11**-15
    caliper disk, **11**-16
    drums, **11**-16
    fade, **11**-15
    grab, **11**-15
    hydraulic, **11**-14
    internal expanding, **11**-14, **11**-15
    parking, **11**-14
    pedal pressure, **11**-17

Automobiles, brakes (*Cont.*):
    power, **11**-16, **11**-17
        booster, **11**-17
        Hydra-Vac, **11**-17
    self-energizing, **11**-15
    service, **11**-14
    shoes, **11**-15
    stopping distance, **11**-13
  characteristics purchased, **11**-3
  cooling, **11**-17, **11**-18
  dimensions of body, **11**-3
  engines (*see* Automobile engines)
  foreign, **9**-82
  front suspensions, **11**-9
  fuel consumption, **11**-5
    charts, **11**-5
  heating, **11**-17
  rear axles, **11**-9
    differentials, **11**-8
    semifloating, **11**-9
  rear suspensions, **11**-9
  registration in the United States, **11**-3
  resistance, **11**-3
    air, **11**-3, **11**-57, **11**-58
    tires, **11**-3
  starting and lighting systems, **15**-82 to **15**-84
  steels for, influence of heat-treatment, **6**-27
  steering, **11**-11 to **11**-13
    Ackermann, **11**-11
    power, **11**-11 to **11**-13
      failure, effects of, **11**-12
      oil pumps, **11**-12
      road feel, **11**-11, **11**-12
    recirculating ball, **11**-11
    worm and roller, **11**-11
  tires (table), **11**-4
    inflation pressure (table), **11**-4
  traction required, **11**-3
  transmissions, **11**-5 to **11**-9
    automatic, **11**-7 to **11**-9
      Hydramatic, **11**-8
    fluid couplings, **11**-5
    friction clutches, **11**-5
    manual, **11**-7
      gear ratios, **11**-4
      synchromesh, **11**-7
    overdrives, **11**-7
    torque converter, **11**-8
  trucks (*see* Trucks)
  turning radius, **11**-3
  ventilation, **11**-17
  wheel: alignment, **11**-10
    camber (def), **11**-10
    caster (def), **11**-10
    toe-in (def), **11**-10
Autotransformer, **15**-42, **15**-43
  for squirrel-cage motor, **15**-43
Available heat (def), **4**-15
  in steam engine cycle, **4**-28
Aviation (def), **11**-58
Aviation fuels, **7**-17, **7**-18
Avogadro's number (def), **4**-11
Avoirdupois weight, **1**-32
AWG (American Wire Gauge), **15**-6
Axial fans, **14**-49 to **14**-53
  characteristic curves, **14**-53
  efficiency of (chart), **14**-55
  formulas, **14**-50, **14**-51
  performance curves for, **12**-102
  specific speed (chart), **14**-55
Axial-flow pumps, **14**-16, **14**-24

Axis:
  of inertia, principal (def), **3**-9
  of oscillation, **3**-19
Axles:
  automobile, **11**-9
  railroad car, standard (table), **11**-29
Azimuth (def), **16**-46
  back (def), **16**-46

Babbitt linings in large bearings, **8**-126 to **8**-128
Babbitt metal (table), **6**-92
Babcock coefficient of friction (steam flow), **4**-50
Back plaster, **6**-182
Bagasse (fuel, table), **7**-13
Bainite (def), **6**-19
Baking varnishes, **6**-121
Balance:
  dynamic, **3**-18
  standing, **3**-17
Balancer, motor-generator, **15**-62, **15**-63
  for three-wire systems, **15**-63
Balances:
  equal arm, **16**-4
  spring, **16**-4
  torsion, **16**-4
Balancing, **3**-17
  of machines, **5**-70 to **5**-72
    reciprocating, **5**-72
    rotating, **5**-70
    steam turbines, **9**-45
Ball bearings (*see* Bearings, ball)
Ball-mill grindability test for coal, **7**-9
Ball valves (pump), **14**-9
Balsa, **6**-170
Banana oil, **6**-168
Band brakes, **8**-42
  friction of, **3**-31
Band saws, metal, **13**-66
Bar iron and steel (table), **6**-46
Barn (unit of nuclear cross-section, def), **9**-127
Barometers, mercury, **16**-9
Barometric condensers, **9**-65
Barrel:
  standard (def), **1**-70
  volume of, **2**-14
Batteries, **15**-10 to **15**-14
  dry, **15**-11, **15**-12
    block assembly, **15**-11
    button-type, **15**-11
    effect of temperature, **15**-11
    efficiency of, **15**-11
    flashlight, **15**-11
    large, **15**-11
    radio B, **15**-11
    Ruben-cell, **15**-11
  electrodes, **15**-10
  electromotive force, **15**-10
  Exide Ironclad, **15**-13
  ignition system, **15**-80
  polarization, **15**-10
  poles, **15**-10
  storage, **15**-12 to **15**-14
    capacity of (table), **15**-13
    care of, **15**-14
    charging, **15**-12 to **15**-13
    copper-oxide, zinc, caustic-soda, **15**-10, **15**-11
    Edison, **15**-12, **15**-14
    efficiency of, **15**-13

Batteries, storage (*Cont.*):
　Nicad, **15**-12, **15**-14
　pasted-plate type, **15**-12
　Planté, **15**-12, **15**-13
　portable, **15**-13
　rechargeable, **15**-14
　specific gravity of electrolyte, **15**-12
　stationary, **15**-13
　voltage, **15**-12
Baumé scale (specific gravity), conversion
　tables, **1**-42, **1**-43
Beam lengths, mean (radiation), **4**-77
　table, **4**-78
Beams, **5**-22 to **5**-37
　angles used as, **12**-33
　bending moment, **5**-22
　cantilever (def), **5**-22
　connections in steel-framed structures, **12**-
　　34
　　table, **12**-44
　constrained, **5**-35
　continuous, **5**-35, **5**-36
　　uniformly loaded (table), **5**-37
　and crank mechanisms, **8**-3
　curved, strength of, **5**-45, **5**-46
　deflection of, **5**-27 to **5**-30, **5**-33, **12**-32
　　Castigliano's theorem, **5**-36, **5**-46
　　curve for, **5**-28, **5**-34
　　diagram, **5**-24
　　formulas, **5**-28
　　as function of stress, **5**-33
　　graphical method, **5**-33
　　Maxwells' theorem, **5**-36
　design of, **5**-22 to **5**-37
　　factors governing, **5**-31
　elastic curve, **5**-28
　I (*see* I beams)
　internal moment beyond elastic limit, **5**-26
　internal resilience, formula, **5**-34
　loads and reactions, **5**-22
　　diagram, **5**-33
　　oblique, **5**-23
　　rolling or moving, **5**-35
　moment and shear diagrams, **5**-24 to **5**-27, **5**-
　　33, **5**-35, **5**-36
　moments of inertia of various sections
　　(table), **5**-30 to **5**-33
　neutral axis in (def), **5**-23
　neutral plane and line of (def), **5**-23
　radii of gyration of various sections, **5**-30 to
　　**5**-33
　rectangular, uniformly loaded, safe loads
　　(table), **5**-28
　reinforced-concrete, **12**-50 to **12**-53
　resilience, **5**-34
　　per unit volume (table), **5**-18
　section modulus (def), **5**-23
　　tables, **5**-30 to **5**-33
　sections, properties of (tables), **5**-30 to **5**-33
　shear, **5**-22
　　transverse, **5**-18
　shear diagram, **5**-33
　simple (def), **5**-22
　slope diagram, **5**-33
　steel: allowable stresses (table), **12**-31
　　deflections (tables), **5**-29, **12**-32
　　maximum safe load on, **5**-23
　　　table, **5**-29
　　properties of (tables), **12**-35 to **12**-38
　　proportion of, **12**-30
　　safe loads, **5**-23, **12**-32
　　　table, **5**-29

Beams, steel (*Cont.*):
　　short, calculation of, **12**-32
　　supports for, **12**-34
　　web connections, **12**-34
　　　table, **12**-44
　stiffness of, **5**-31, **5**-70
　strength of (formula), **5**-23
　uniform cross-section: bending moments
　　(table), **5**-24 to **5**-27
　　vertical shear (table), **5**-24 to **5**-27
　uniform strength, **5**-36
　　table, **5**-28 to **5**-40
　vibration, **5**-73, **5**-74
　wooden, **12**-23, **12**-24
　　properties of (table), **12**-25
　　safe loads (table), **5**-28
Bearing, determination of (in surveying), **16**-
　46
Bearing metals, **6**-91 to **6**-94
　aluminum alloys, **6**-92
　babbitt (table), **6**-92
　cadmium-base, **6**-92
　copper-base (table), **6**-93
　metal-powder-sintered (oil-impregnated,
　　table), **6**-95
　miscellaneous, **6**-93, **6**-94
　porous, **6**-93
　(*See also* Brass; Bronze)
Bearing pressures of soils and rock (table), **12**-
　20
Bearings, **8**-120 to **8**-143
　ball, **8**-136
　　annulus calculation, **2**-11
　　friction, **3**-28
　　(*See also* rolling contact, *below*)
　closures, **8**-141
　conical, friction, **3**-30
　efficiencies, **3**-29
　friction, **3**-30, **3**-31, **8**-123
　gas-lubricated, **8**-133 to **8**-135
　　compressibility parameter, **8**-133
　　thrust, **8**-133
　　whirl stability, **8**-133
　guide, **8**-120
　hardwood, **8**-130
　journal, **8**-120 to **8**-130
　　allowable pressures (table), **8**-122
　　babbitt linings, **8**-126
　　　thickness, **8**-126
　　bushings, **8**-126
　　clearance, **8**-122
　　elements, **8**-126 to **8**-130
　　film thickness, **8**-121
　　friction, **3**-30
　　　variation with clearance, **8**-123
　　graphite-lubricated, **8**-129
　　heat dissipation from, **8**-124
　　length-diameter ratios, **8**-122
　　load distribution, **8**-127
　　lubrication, **8**-120
　　mean pressures (table), **8**-122
　　mountings, **8**-129
　　oil grooves, **8**-127
　　　for various load directions, **8**-127, **8**-128
　　oilless, **8**-129
　　porous-metal, **8**-130
　　propeller shaft, **11**-54
　　seals, **8**-128
　　turbulence of fluid film, **8**-123
　　types of, **8**-129
　Kingsbury thrust, **8**-130 to **8**-133
　　friction, **3**-31

Bearings (*Cont.*):
  lignum vitae, **8**-130
  line-shaft, **8**-129
  lubrication, **6**-203, **6**-205
    complete, **8**-120
    failure, **8**-120
    film thickness, **8**-121
    hydrostatic oil lift, **8**-126
    minimum oil feed, **8**-125
    mixed, **8**-120
    pressure feed, **8**-124
    semifluid, **8**-120
    starting and stopping, **8**-126
    wicks, **8**-125
  materials (*see* Bearing metals)
  mine car, **10**-23
  Mitchell, **8**-131
  mounting, **8**-129, **8**-141
  needle, **8**-137
  oilers for, **8**-125
  for oscillatory motion, **8**-128
  plain, **8**-120 to **8**-135
  porous, **6**-93, **8**-130
  roller, **8**-137
    coefficient of friction, **3**-28
    efficiency of, **3**-28
    friction, **3**-28
    railway, resistance of, **11**-30
    (*See also* rolling contact, *below*)
  rolling contact: AFBMA standards, **8**-136
    ball and roller, **8**-136 to **8**-143
    closures, **8**-141
    components, **8**-136
      rings, **8**-136
      rolling elements, **8**-136
      separators, **8**-136
    fits (table), **8**-141
    friction, **8**-140
    load-capacity life, **8**-138
    loads and ratings, **8**-138 to **8**-140
    lubrication, **8**-142
      table, **8**-142
    mountings, **8**-141
    thrust bearings, **8**-137
    types of, **8**-136 to **8**-138
  shielded (rolling contact), **8**-141
  sliding, **8**-133
  step, friction, **3**-31
  thrust, **8**-120, **8**-130 to **8**-133
    ball, **8**-137
    capacities of (table), **8**-135
    film thickness, **8**-131
    friction, **3**-31
    friction coefficient, **8**-131
    gas-lubricated, **8**-133
    grooves, **8**-130
    hydrostatic step, **8**-131
    Kingsbury, **8**-131
    roller, **8**-137
    rolling contact, **8**-136 to **8**-143
    segmental, pivoted, **8**-131
Beaufort scale of wind force (table), **9**-167
Bêché pneumatic forge hammer, **13**-26
Bell-and-spigot pipe (table), **8**-168
Bellows gages, **16**-9
Bellows gas meter, **16**-8
Belts, **8**-51 to **8**-58
  arc of contact, **8**-53
  conveyor, **10**-55, **10**-56
  drives, **8**-51 to **8**-58
    arrangements, **8**-52
    efficiency, **3**-29

Belts, drives (*Cont.*):
    friction, **3**-32
    idler pulleys, **10**-56
    power, **8**-53
  high-strength, **10**-56
  joints, **8**-52
  leather, **8**-51, **8**-52
    capacity factors (table), **8**-51
    design of, **8**-53
    power transmitted by, **8**-53
    strength of, **8**-51
    weight of, **8**-51
  lengths, open and crossed, **8**-52
    formulas, **8**-52
    graphical methods of determining, **8**-53
  power transmission by, **8**-53
  rubber, **8**-52
    minimum pulley diameters (table), **8**-55
    power ratings (tables), **8**-54
  slack removal, **8**-52, **8**-58
  tension, **8**-52
  tightener, **8**-52, **8**-58, **10**-57
  V (*see* V-belts)
  weights, **8**-52
Bench mark (surveying, def), **16**-43
Bending:
  elastic limit in, **5**-27
  of metals, **13**-19
    allowance for, **13**-19
    loads for, **13**-19
    machines for, **13**-19, **13**-26, **13**-27
  theory of (beams), **5**-23, **5**-27 to **5**-33, **5**-36, **5**-45 to **5**-49
  and torsion of shafts, **5**-20
Bending moment of beams, **5**-22
Bendix automobile brakes, **11**-15
Bendix-Weiss universal joint, **8**-37
Bends:
  pipe, **5**-61 to **5**-63
  rope, **8**-89, **8**-97
Benzene, **7**-20
  as solvent, **6**-168
  (*See also* Hydrocarbons)
Benzol (solvent), **6**-168
Bergbau-Forschung process (coke making), **7**-41
Bernoulli differential equation, **2**-50
Bernoulli's theorem (hydraulics), **3**-43, **3**-45
Beryllium, **6**-102
Beryllium copper, **6**-63
Bessel's formula for computing probable error, **2**-21
Bessemer steel, **6**-12, **6**-15
Beta particle (def), **9**-121
Beta rays, **9**-119
Bethell process (wood preservation), **6**-137
Betz momentum theory (windmills), **9**-164
Bevel gears (*see* Gears, bevel)
Beveloid gears, **8**-111
Bi-Gas process (gas making), **7**-46
Billet (steel, def), **6**-16
Billet reheating furnaces, heat transmission in, **4**-82
Binary numbers, **2**-65, **2**-66
Binders, core, **13**-6
Binding energy (nuclear), **9**-120
Binomial coefficients (tables), **1**-8, **2**-18
Binomial theorem, **2**-17
Bipropellants (rocket), **7**-35
Birefringent coatings, **5**-60
Birmingham wire and sheet-metal gages, **6**-45

Bisection:
    of angle, **2**-9
    of line, **2**-9
Bit (computers), **2**-65
Bituminous coal (see Coal, bituminous)
Black powder (explosive), **7**-24
Black-surface enclosures (radiant-heat), **4**-73
Blackbody (def), **4**-71
Blades, fan, centrifugal, **14**-50
Blasius equation (friction-drag coefficient), **11**-66
Blast furnace (def), **6**-12
Blast-furnace gas:
    cleaning, **18**-14
    flame temperatures (table), **4**-57
Blasting gelatin, **7**-25
Bleeding cycle (turbines), **4**-28
Blister copper, **6**-63
Block brakes, **8**-41
Blocks, pulley, **8**-8
Bloom (steel, def), **6**-16
Blowers, **14**-49 to **14**-57
    for draft, **14**-57
    Roots, **14**-43
    rotary, **14**-42 to **14**-44
    spiral-axial, **14**-43
    for superchargers, **9**-102
    for ventilation, **14**-56
    (See also Centrifugal compressors; Fans)
Blue gas, **7**-42
    results from manufacture of (table), **7**-42
BM-AGA coal test, **7**-38
Board drop hammer, **13**-25
Board measure (def), **1**-32
Boat spikes (table), **8**-94
Bode diagrams, **16**-31
Boghead coal (def), **7**-5
Bohr theory of atoms, **6**-7
Boiled oil, **6**-119
Boiler furnaces:
    cinder and fly-ash recovery, **9**-7
    combustion of, **9**-12 to **9**-14
    controls, **9**-23, **9**-24
    design and construction of, **9**-16 to **9**-20
    draft loss, **9**-26
    heat transfer, **9**-20
    heat transmission, **4**-80
    mechanical draft, **9**-26
    oil and gas firing, **9**-22
    pulverized-coal: firing, **9**-11 to **9**-13
        slag-tap, **9**-7
    stokers, **9**-10
    walls, **9**-16 to **9**-20
    waste fuels, **9**-7
    water-cooled, **9**-14 to **9**-16
Boilers, **9**-7 to **9**-35
    acid cleaning, **9**-34
    air preheaters, **9**-22 to **9**-23
    boil-out, **9**-33
    carbon monoxide, **9**-16
    caustic embrittlement, **6**-111
    chemical cleaning, **9**-34
    circulation in, **9**-24 to **9**-26
        natural, **9**-14, **9**-24
    codes for, **9**-34
    combined circulation, **9**-14, **9**-26
    controls, **9**-24
    corner-fired, **9**-12
    corrosion in, **6**-109 to **6**-111, **9**-29
    draft loss through, **9**-26
    dry storage, **9**-34
    economizers, **9**-21, **9**-22

Boilers (Cont.):
    efficiency of, **9**-28
        calculation by heat balance, **9**-27, **9**-28
    emergency operation of, **9**-33
    explosions, **9**-33
    external cleaning, **9**-34
    feed pumps, **14**-7
    feedwater (see Feedwater)
    foaming, **9**-29
    forced-circulation, **9**-14, **9**-25
    fuels for (table), **9**-7
    furnace (see Boiler furnaces)
    heat balance, **9**-27, **9**-28
    heat losses, **9**-27
    heating, **12**-70, **12**-71
    high-temperature-water, **9**-15
    hydrogen damage to, **6**-111
    injectors, **14**-14
    inspection and maintenance, **9**-34
    integral furnace, **9**-14
    normal operation, **9**-33
    nuclear (see Nuclear boilers)
    once-through type of, **9**-14, **9**-25, **9**-33
    package, **9**-14
    performance of, **9**-26 to **9**-28
        acceptance tests, **9**-26
        component tests, **9**-28
        draft loss, **9**-26
        guarantee, **9**-26
        heat-adsorption calculation, **9**-27
        thermal losses, **9**-27
    radiant, **9**-14
    rating (def), **9**-26
    recovery, **9**-16
    safety interlocks, **9**-33
    start-up, **9**-33
    supercritical, **9**-14, **9**-25
    superheaters, **9**-20, **9**-21
    surface combustion applied to, **4**-58
    tubes: film coefficients for, **4**-63 to **4**-70
        radiation factors for rows of (chart), **4**-74
        standard dimensions (table), **8**-167
    types of, **9**-14 to **9**-16
    universal pressure, **9**-14
    waste-heat, **9**-15
    water (see Feedwater)
    water-tube, **9**-14 to **9**-16
Boiling point:
    of common substances, **4**-6
    of hydrocarbons, **4**-6
    of liquids (table), **4**-12
    of metals (table), **6**-60, **6**-61
Bolometer, **16**-14
Bolted joints, design of, **8**-24, **8**-25, **8**-27
Bolts:
    carriage, **8**-24
    drift, pulling resistance, **12**-28
    head, standard dimensions (table), **8**-20, **8**-21
    materials for, **8**-24
        strength of (tables), **8**-30
    and nuts, thread standards, **8**-9 to **8**-30
    for steel-framed structures, **12**-31
    stove (table), **8**-27
    stress due to tightening, **8**-25
    tension due to tightening nuts on, **3**-30
    for timber trusses, **12**-27
        table, **12**-27
    U.S. Standard: safe loads (table), **8**-28
        strength of (table), **8**-30
        wrench openings (table), **8**-20
Bonderizing (protection of iron), **6**-116

Bonus wage systems, **17**-9
Boolean algebra (circuit analysis), **15**-92
Booster, electric converters, **15**-49
Borers, wood, **13**-82
Boring machines, **13**-59, **13**-60
Boring mills, **13**-59, **13**-60
Boron carbide (abrasive), **6**-139
Bosch fuel-injection pump, **9**-99
Bottom dump wagons, **10**-28
Boundary layer (def), **3**-6, **11**-66
    aerodynamics, **11**-66
    fluid flow, **3**-52
Boundary-layer control (BLC), **11**-64
Bourdon pressure gages, **16**-9
Bower-Barff process, **6**-116
Bow's notation (trusses), **12**-18
Box coking test for coal, **7**-38
Brakes, **8**-41 to **8**-44
    automobile and truck, **11**-13 to **11**-17
    band, **8**-42
    block, **8**-41
    cone, **8**-43
    disk, **8**-43
    dynamometers, **16**-18
    eddy-current, **8**-44
        dynamometer, **16**-18
    electric, **8**-44
    fan, **16**-18
    friction, **3**-21, **3**-25, **3**-31
        horsepower absorbed by, **3**-21
    hoisting machinery, **10**-18
    hydraulic (automobile), **11**-14
    internal, **8**-44
    multidisk, **8**-44
    post-type (power hoists), **10**-16
    prony, **16**-18
    railway, **11**-30
    vacuum, **11**-17
Braking:
    of railway trains, **11**-30
    regenerative (def), **10**-11
Brass, **6**-68
    composition and properties of (tables), **6**-64
        to **6**-69
    dezincification of, **6** 112
    leaded, **6**-68
        table, **6**-64 to **6**-67
    machinability of, **6**-70
    naval, **6**-68
    pipe (table), **8**-171
    plastic range (chart), **13**-15
    press forgings, mixture for, **13**-21
    rolled, mechanical properties of (table), **6**-69
    season cracking, **6**-68, **6**-70, **6**-112
    strength at high temperatures, **5**-12
    (*See also* Copper alloys)
Brass-aluminum alloy for condenser tubes, **6**-68
Brayton cycle, **4**-20, **4**-22, **9**-114
Braze welding, **13**-36
Brazing, **13**-42
    aluminum, **13**-47
    filler metals (table), **6**-91
Break-even chart, cost accounting, **17**-15
Brick, **6**-141, **6**-149
    ASTM specifications (table), **6**-149
    cement, **6**-149
    chrome, **6**-171
    common, **6**-149
    facing, **6**-149
    fire (*see* Firebrick)
    grades (table), **6**-149

Brick (*Cont.*):
    high-alumina, **6**-171
    magnesite, **6**-171
    manufacture, **6**-149
    masonry: specific gravity and density of
        (table), **6**-8
        strength of, **12**-22
        table, **12**-23
    mortar for (table), **6**-183
    paving, **6**-149
    properties of (table), **6**-98, **6**-149
    sand-lime, **6**-149
    silica, **6**-171
    silicon carbide, **6**-172
Brickwork:
    lateral support, **12**-22
    laying and bonding, **12**-22
    mortar for (table), **6**-183
    strength of, **12**-22
Bridges:
    corrosion of, **6**-114
    steel cables for (table), **8**-92
    wind pressure on girders of, **12**-14
Briggsian logarithms (def), **2**-16
Brightness (illumination, def), **12**-117
Brine coolers, **19**-14
    double-pipe, **19**-14
    shell-and-tube, **19**-14
Brinell hardness test, **5**-13, **16**-21
Brinell number (hardness, def), **5**-13
    for heat-treated steels (table), **6**-23
    for metals (table), **5**-3
Brines:
    circulation, **19**-18
    corrosion of metals in, **6**-109
    properties of (tables), **19**-14, **19**-15
    for refrigerating plants, **19**-13 to **19**-15
British thermal units (Btu), **1**-31, **4**-8
    mechanical equivalent, **1**-34
British unit of refrigeration (def), **19**-3
Brittleness, impact tests for, **5**-7
Broaches, **13**-65, **13**-66
Broaching, **13**-65, **13**-66
Broaching machines, **13**-66
Bronze, **6**-68, **6**-69
    aluminum, **6**-68
        for bearings, **6**-93
        tables, **6**-65, **6**-67
    cupronickels, **6**-69
        tables, **6**-65, **6**-67
    manganese, **6**-68
    phosphor, **6**-68
        table, **6**-65, **6**-67
    silicon, **6**-68
        table, **6**-65, **6**-67
    strength at high temperatures (table), **5**-12
    table, **15**-6
    tin, **6**-68
        tables, **6**-65, **6**-67
    various, uses of, **6**-68
    welding, **13**-36
Brown and Sharpe wire gage:
    scheme of, **15**-6 to **15**-7
    table, **6**-45
    (*See also* American wire gage)
Brown coat (plastering), **6**-182
Btu (*see* British thermal units)
Bucket carriers, **10**-50 to **10**-53
    (*See also* Conveyors)
Bucket elevators, **10**-53 to **10**-55
    casings, **10**-54
    power requirements (chart), **10**-53

Buckets:
  clamshell, **10**-13 to **10**-14
    capacities of (table), **10**-13
  dragline: scraper, **10**-13
    self-filling, **10**-13
  grab, **10**-13 to **10**-14
    electric, **10**-13
  in impulse hydraulic turbines, **9**-144
  orange-peel, **10**-13
  in steam turbines, **9**-45
Buckingham's II theorem (dimensional analysis), **3**-50
Buckling of columns, **5**-41
Budget, **17**-12
Buffing, **13**-69
Buhrstones, **6**-139
Building blocks, **12**-22
Building construction, **12**-2 to **12**-62
  fire-resisting, **18**-28
  foundations, **12**-20 to **12**-21
  industrial, **12**-2 to **12**-11
  materials (see Building materials)
  partitions: sound-adsorption coefficients (table), **12**-141
    sound-transmission loss (table), **12**-140
  reinforced concrete (see Reinforced concrete)
  steel (see Steel-framed structures)
  steel for, allowable stresses in (tables), **12**-30
  structural design, **12**-11 to **12**-62
Building materials:
  fire resistance of, **12**-42, **18**-29
  specific gravity and density of (table), **6**-8
  thermal conductivity of (tables), **4**-63, **4**-64, **12**-84 to **12**-85
  weight of (tables), **18**-29
Building stone, **6**-158
  specific gravity and density of (table), **6**-8
Buildings:
  closed, wind forces on, **12**-13
  earthquake forces in, **12**-11
  exits and fire escapes, **18**-23
  fire protection, **18**-27 to **18**-33
  heating (see Heating)
  industrial (see Industrial plants)
  safety provisions, **18**-23 to **18**-25
  soundproofing, **12**-139
  ventilation (see Ventilation)
  wind pressure on, **12**-12, **12**-13
Bulk flo conveyor, **10**-46
Bulk modulus (def), **3**-35
  of various liquids (table), **3**-35
Bulk modulus of elasticity (def), **5**-18
  of metals (table), **5**-5
Bulldozers, wheeled, **10**-27
Buoyancy, **3**-40
  center of, **3**-40
  for ships, **11**-39
Burners, **9**-12, **9**-13
  air-atomized, **9**-12
  for combustion furnaces, **7**-48
  corner-fired, **9**-12
  gas, **9**-12
    blast-furnace, **9**-13
    coke-oven, **9**-13
  oil, **9**-12
  pulverized coal, **9**-12
    primary air for, **9**-12
    secondary air for, **9**-12
Bus bars for switchboards, **15**-53
Buses, engines in, **9**-3
  (See also Automobiles)
Bushel, U.S. and imperial (def), **1**-31

Bushings, dimensions of (table), **8**-128
Butane:
  properties of (chart), **4**-28
    table, **4**-39
  for refrigerant (table), **19**-4
Butterfly valves for hydraulic turbines, **9**-149
Butyl rubber, **6**-166
BWG (Birmingham wire gage), **6**-45
Byte (computers), **2**-65

Cable length (nautical unit), **1**-31
Cable tramways (see Cableways)
Cables:
  for airplanes (table), **11**-70
  aluminum, electrical properties of (table), **15**-59
  armored (interior wiring), **15**-67
  copper (table), **15**-7, **15**-8
    resistance and reactance of (table), **15**-8, **15**-63
  electric: insulation, **15**-60, **15**-61, **15**-67, **15**-68
    paper for, **6**-153
    underground power, **15**-60, **15**-61
      carrying capacity (table), **15**-60
  galvanized steel wire (tables), **8**-87, **8**-90, **8**-91
  nonmetallic sheathed (interior wiring), **15**-68
  service entrance (interior wiring), **15**-68
Cableways, **10**-34 to **10**-40
  deflection, **10**-35
  factor of safety, **10**-35
  hoisting and conveying, **10**-34
  slackline, **10**-39, **10**-40
    table, **10**-40
  speeds and loads, **10**-35
  supporting towers, **10**-36
  tramways, **10**-35 to **10**-37
    cables and traction ropes, **10**-36
    loading and discharge terminals, **10**-36
    power required, **10**-37
      stresses, **10**-36
  transporting, **10**-34
  types of, **10**-35
Cadmium, **6**-103, **6**-115
Cadmium plating, **6**-115
Caisson method for foundations, **12**-20
Calcium carbide furnace (electric), **7**-69
Calcium carbonate, saturation index, **6**-117
Calcium chloride:
  brine, properties of (tables), **19**-14, **19**-15
  as freezing preventive, **6**-156
Calculators, **2**-75
Calculus, **2**-42 to **2**-53
  rules for differentiation, **2**-43
  table of integrals, **2**-46 to **2**-49
Calendars, **1**-41
Calipers, **16**-5
  micrometer, **16**-5
  vernier, **16**-5
Calite (heat-resistant noncorroding alloy), **6**-118
Calorie, IT (def), **4**-8
Calorific value (see Heat value)
Calorimeters:
  separating, **16**-22
  steam, **16**-21
  throttling, **16**-21
Calorizing (metal protection), **6**-116
Camber, line, of airplane wing section, **11**-62

Cams, **8**-4 to **8**-7
  acceleration of (chart), **8**-5
  design of, **8**-6
  diagrams, **8**-4
  jerk (chart), **8**-5
  pitch line, **8**-5
  velocity of (chart), **8**-5
Candela (unit of luminous intensity, def), **1**-33, **12**-116
Cannel coal (def), **7**-5
Cantilever beam (def), **5**-22
  vibration of, **5**-74
  (*See also* Beams)
Capacitance:
  condenser, **15**-18, **15**-19
  electrical, unit of (def), **15**-4
Capacitances:
  in parallel, **15**-18
  in series, **15**-18
Capacitive circuit, **15**-20
Capacitive reactance (def), **15**-4, **15**-21
Capacitor, **15**-18, **15**-19
  (*See also* Condensers, electric)
Capacity and volume equivalents (table), **1**-47
  conversion table, **1**-47
Capillary attraction, **3**-36
  in tubes (chart), **3**-37
Car:
  box-body dump, **10**-23
  gable-bottom, **10**-23
  hopper-bottom, **10**-23
  industrial, **10**-22, **10**-23
  mine, **10**-23
    frictional resistance (table), **10**-23
  railway, **11**-28, **11**-29
    air resistance, **11**-57, **11**-58
    roller bearing resistance, **11**-30
    wheels and axles for, **11**-28
      table, **11**-29
  rocker side-dump, **10**-23
  scoop dump, **10**-23
  shakers, **10**-25
  unloading machinery, **10**-24, **10**-25
Car dumps:
  cross-over, **10**-24
  rotary: gravity, **10**-24
    power, **10**-24
  unit-train, **10**-24
Car positioners (railroad), **10**-25
Car retarders (railroad), **11**-33
Carat, metric (def), **1**-32
Carbide:
  cemented, for tools, **6**-104 to **6**-106
  composition, properties, uses (table), **6**-105
  tungsten, **6**-105
Carbolon (abrasive), **6**-139
Carbon:
  air required for combustion, **4**-56
  heat of combustion (table), **4**-54
Carbon-arc cutting of metals, **13**-37
Carbon dioxide:
  dissociation, **4**-54
  emissivity (chart), **4**-79
  measurement of, **16**-22
  properties of (table), **4**-38
  as refrigerant (chart, table), **19**-5, **19**-6
  solid, **19**-22
Carbon monoxide:
  flame temperature and dissociation (table), **4**-57
  measurement of, **16**-22
Carbon-residue test for lubricating oils, **6**-199
Carbon resistor, **15**-73

Carbon steel (*see* Steel, carbon)
Carbon tetrachloride as solvent, **6**-168
Carbonic acid (*see* Carbon dioxide)
Carbonization of coal, **7**-37 to **7**-41
  apparatus for, **7**-39 to **7**-41
  coal-chemical recovery in, **7**-40
    heat balances (table), **7**-40
  coals for, **7**-37
  coking process, **7**-38
  gas yield (table), **7**-39
    variation during (table), **7**-39
  heat required, **7**-39
  high-temperature, yields from, **7**-37
    table, **7**-39
  plastic zone, **7**-38
  processes, **7**-40, **7**-41
  temperature effects during, **7**-39
    gradients in (diagram), **7**-38
  thermal efficiency of, **7**-39
Carborundum (abrasive), **6**-139
Carburetors, **9**-97, **9**-98
Carburetted water gas, **7**-42
Carburizing of steel, **6**-23
Cardioid, **2**-40
Carnot cycle, **4**-14, **4**-20
Carrene (refrigerant), **19**-5
Carriage bolts, **8**-24
Carrier wave, television, **15**-96
Carriers:
  open-top, **10**-50
  pivoted-bucket, **10**-51
  V-bucket, **10**-50
    table, **10**-52
Cartridge brass (table), **6**-64, **6**-66
Casehardening of steel, **6**-23 to **6**-25
Casein in glues, **6**-141
Casings, centrifugal fan, **14**-50
Casks, volumes of, **2**-14
Cast iron, **6**-47 to **6**-51
  alloys, composition and mechanical properties of (tables), **6**-49
  carbon in, **6**-47
  classification of, **6**-47
  coefficient of expansion (table), **4**-7
  columns (tables), **5**-42, **5**-43
  composition of, **6**-47
    tables, **6**-49
  corrosion of, **6**-109
  corrosion resistance of, **6**-49
  cutting speeds (lathe, tables), **13**-58
  damping capacity (vibrations), **6**-48
  definition of, **6**-12
  ductile, **6**-50
  graphite in, **6**-47
  gray, **6**-47 to **6**-49
    mechanical properties of, **6**-48
      table, **6**-48
  machining, **6**-49
  malleable, **6**-49
  mechanical properties of, **6**-48
    tables, **6**-49
  nodular, **6**-47, **6**-50
  pipe (*see* Pipe, cast-iron)
  shrinkage, **6**-50
  silicon in, effect of, **6**-47
  specifications of, **6**-49
  strength of (tables), **6**-48
    at high temperature (table), **5**-11
  tapping depth, **8**-27
  welding, **13**-44, **13**-45
  white, **6**-47
  (*See also* Castings)

Cast steel:
  properties of (charts), **6**-53
  strength at high temperatures (table), **5**-11
  welding, **13**-44
Castable mixes for refractories, **6**-174
Castigliano's theorem (beam deflections), **5**-36
Castings:
  alloys, **13**-7, **13**-8
  aluminum (table), **6**-76, **6**-77
  centrifugal, **6**-58
  ceramic mold, **6**-58
  chilled-iron, **6**-47
  cold-chamber machines, **13**-4
  copper, **6**-70
  design of, **6**-58, **13**-9
  die, **13**-4
  gooseneck machine, **13**-4
  graphite mold, **6**-58
  gray-iron, **6**-47, **6**-49
    composition and mechanical properties
      of, **6**-49, **13**-7
    tolerances of, **6**-50
  inspection of, **13**-8, **13**-9
  investment molding, **6**-57
  iron, **6**-47 to **6**-52
    alloy, composition and mechanical
      properties of (table), **6**-49
    ASTM specifications for, **6**-49
  magnesium alloy, **6**-82
  malleable iron, **6**-47, **6**-51, **6**-52
    ASTM specifications for, **6**-51
    composition and mechanical properties
      of, **6**-51, **13**-7
    processing methods, **6**-52
    specifications for, **6**-51
    tolerances for, **6**-52
    uses of, **6**-47
  methods, design and cost features of (table),
    **13**-5
  nickel and nickel-alloy, composition and
    properties of (tables), **6**-71 to **6**-73
  permanent-mold, **13**-3
  sand, **13**-2, **13**-6
    basic steps, **13**-2
    patterns, **13**-2
    processes, **13**-3
  sand-blast cleaning of, **13**-8
  shell-mold, **6**-58
  steel, **6**-47, **6**-52 to **6**-58
    allowances for machine finish (table), **6**-57
    alloy, **6**-52
      composition and properties of (table),
        **6**-56
      specifications for (ASTM, table), **6**-55,
        **6**-56
    classification of, **6**-47
    composition of, **6**-52
    corrosion resistance of, **6**-54
    design of, **6**-58
    ductility, **6**-53
    endurance limit, **6**-54
    heat resistance of, **6**-54
    impact resistance of, **6**-54
    machinability of (table), **6**-55
    melting practice, **6**-55
    properties of, **6**-52, **13**-7
      charts, **6**-53
    specifications for, **6**-54
    tolerances of, **6**-55
      table, **6**-57
    uses of, **6**-58
  vacuum, **6**-15, **6**-96

Castings, steel (*Cont.*):
  wear resistance of, **6**-54
  weight range, **6**-52
  welding, **13**-44
  (*See also* Foundry practice and equipment;
    Molding)
Castor oil, properties of (table), **6**-202
Catenary, **2**-36 to **2**-38
  tables, **2**-37
Cathode-ray tubes, **2**-68, **2**-74
Cathodic reaction in corrosion, **6**-106
Cauchy-Euler equation, **2**-51
Caustic embrittlement in boilers, **6**-111,
  **9**-29
Cavalieri's theorem, **2**-15
Cavitation:
  limits of hydraulic turbines, **9**-147
  in power steering pumps, **11**-13
  in screw propellers, **11**-51
Ceiling (airplanes), **11**-65
Cells, electric (*see* Batteries)
Cellulose acetate, **6**-161
Cellulose acetate butyrate, **6**-161, **6**-162
Celsius:
  to Fahrenheit (conversion table), **4**-3, **4**-4
  to Kelvin (eq), **4**-3
Cement, **6**-177, **6**-178
  brick, **6**-149
  high-early-strength, **6**-177
  masonry, **6**-178
  mortar, **6**-181
  natural, **6**-178
  plaster, **6**-182
  portland, **6**-177, **6**-178
    ASTM specifications for (tables), **6**-178
    specific gravity and density of (table),
      **6**-8
    strength of (tables), **6**-178
    sulphate-resisting, **6**-178
  pozzolana, **6**-178
  setting time, **6**-178
  surfaces, painting of, **6**-120
  tests for, **6**-178
  waterproofed, **6**-178
  white, **6**-178
Cement-lined pipe, **8**-170
Cemented-carbide tools, **6**-104 to **6**-106
Cementite (def), **6**-17
Center:
  of buoyancy, **3**-40
  of curvature, **2**-45
  of gravity, **3**-7, **3**-8
    by experiment, **3**-8
    of lines, **3**-7
    of plane areas, **3**-7
      by graphics, **3**-11
    of solids, **3**-8
  of mass, **3**-7
  of percussion, **3**-18, **3**-19
Centesimal measure of angles, **2**-25
Centigrade (*see* Celsius), **4**-2
Centipoise (unit of viscosity, def), **3**-35
Centistoke (unit of kinematic viscosity), **3**-36
Central heating, **12**-62 to **12**-76
  (*See also* Heating)
Centrifugal casting, **13**-4
  of iron pipe, **8**-164
Centrifugal compressors, **14**-39 to **14**-42
  axial thrust, **14**-42
  efficiency of, **14**-40
  fluid flow in, **3**-49
  (*See also* Turbocompressors)

Centrifugal fans, **14**-49 to **14**-57
  for air conditioning: equations for, **12**-98
    fan laws for (table), **12**-102
  applications of, **14**-56, **14**-57
  blade form, **14**-50
  blowback in, **14**-56
  casings, **14**-50
  characteristics of, **14**-52
    charts, **14**-53
  compressibility factor (def), **14**-51
  efficiency of, **14**-51
  fluid flow in, **3**-49
  formulas, **14**-50, **14**-51
  head (def), **14**-51
  impeller types of, **14**-49, **14**-50
  laws of, **14**-54, **14**-55
  parallel operation, **14**-54
  performance curves for, **12**-101
  power output (def), **14**-51
  pressures in (def), **14**-51
  puffing in, **14**-56
  pulsation, **14**-56
  relative characteristics of (table), **12**-103
  series operation, **14**-54
  sound power level, **14**-51
  stability of, **14**-55, **14**-56
  system characteristics of, **14**-52
  system matching, **14**-53, **14**-54
Centrifugal force (def), **3**-17
Centrifugal pumps, **14**-15 to **14**-30
  application of, **14**-25 to **14**-28
  axial-flow, **14**-16, **14**-24
  axial thrust in, **14**-18
  balance drums and disks, **14**-18, **14**-20
  bearings, **14**-21
  casings, **14**-17, **14**-18
  cavitation limits, **14**-26
  characteristic curves, **14**-26, **14**-28
    influence of fluid properties on, **14**-25
    influence of viscosity on, **14**-25
  classification of, **14**-15 to **14**-17
  condensate-injection sealing, **14**-21
  couplings, **14**-21
  diffusers, **14**-16
  effect of speed change, **14**-25
  efficiency of (chart), **14**-28
  fluid flow in, **3**-49
  head-capacity curves (charts), **14**-28, **14**-29
  hydraulic balancing devices, **14**-18
  hydraulic theory, **14**-23 to **14**-27
  impellers (def), **14**-16, **14**-18
    unbalanced forces on, **14**-18
    velocity diagram, **14**-27
  lattice-effect coefficient (chart), **14**-27
  maintenance of, **14**-28 to **14**-30
  materials for, **14**-22
    table, **14**-23
  mechanical construction, **14**-15 to **14**-22
  mechanical seals, **14**-21
  mixed-flow, **14**-23
  motors, **15**-52
  mountings, **14**-21
  multistage, **14**-17, **14**-18
  nomenclature of, **14**-15 to **14**-17
  NPSH (def), **14**-25
  packing, **14**-20
  parallel operation of, **14**-26
  parts of, recommended names of (table), **14**-18
  performance of, **14**-22 to **14**-28
  priming, **14**-27
  series operation of, **14**-26

Centrifugal pumps (*Cont.*):
  shaft sleeves, **14**-20
  shafts, **14**-20
  similarity laws, **14**-25
  specific speed (def), **14**-22
    upper limits, **14**-26
  stuffing boxes, **14**-20
  suction specific speed, **14**-26
  system friction losses, **14**-26
  system-head curve, **14**-29
  theory of, **14**-23 to **14**-27
  vertical-shaft, construction, **14**-21, **14**-22
  volute, **14**-16
  water-seal packing, **8**-144
  wearing rings, **14**-18
Centripetal force (def), **3**-17
Centrode, **3**-15
Centroids (def), **3**-7
  of lines, **3**-7
  of plane areas, **3**-7
  of solids, **3**-8
Ceramic products, **6**-150
Ceramics, **6**-149, **6**-150, **13**-11
  manufacture of, **6**-150
  properties of, **6**-150
Cetane number (fuels, def), **7**-19, **9**-97
Cgs system of units, **1**-40
Chain (unit of length, def), **1**-31
  binding, high-strength, **10**-6
    data on (table), **10**-7
  Flite, **10**-8
    pocket-wheel specifications for (table), **10**-8
  high-test, ASTM specifications for (table), **10**-6
  proof coil, ASTM specifications for (table), **10**-6
Chain block:
  differential, **8**-8
  triplex, **8**-8
Chain conveyors, **10**-50 to **10**-55
  (See also Conveyors, bucket)
Chain drives, **8**-59 to **8**-69
  efficiency, **3**-32
  inverted-tooth (silent), **8**-68
  length of chain, **8**-61
  power ratings, **8**-61
    tables, **8**-65 to **8**-67
  roller, **8**-59 to **8**-69
Chain-grate stokers, **9**-10
Chain hoists, **10**-14
  data for (table), **10**-14
Chaining uphill, sag and stretch, **2**-38
Chains:
  alloy, **10**-4
    data on (table), **10**-5
  ASTM specifications for (table), **10**-5
  BBB, **10**-6
  conveyor, **10**-8
  end fittings for, **10**-7
  friction in, **3**-32
  high-test, **10**-5
  master links for, **10**-7
  master rings for, **10**-7
  pocketwheels (table), **10**-8
  power transmission, **10**-8
  proof coil, **10**-6
  roller, **8**-59 to **8**-69
    dimensions and speeds (table), **8**-64
    length calculations, **8**-61
    power ratings (tables), **8**-65 to **8**-67
    service factors (table), **8**-67

Chains, roller (*Cont.*):
  sprocket wheels, **8**-67, **8**-68
    diameter formulas, **8**-67
    teeth, **8**-67, **8**-68
  silent, **8**-68
    power ratings (tables), **8**-68, **8**-69
    service factors (table), **8**-69
  sling, **10**-4, **10**-5
  special, **10**-8
  strength of, **10**-6, **10**-7
    tables, **10**-5 to **10**-7
  welded-link, **10**-7
  wheel, **10**-7
  working load (tables), **10**-5 to **10**-7
  wrought-iron, **10**-4
Channels:
  open (hydraulic), **3**-66 to **3**-68
    roughness coefficients (table), **3**-68
  steel, properties of (table), **12**-33
  structural steel, rivet gages, **12**-45
Characteristic of a logarithm, **2**-4
Characteristic curves:
  of fans, **14**-53, **14**-55
  of pumps, **14**-26, **14**-28
Charcoal as fuel, **7**-13
Charpy impact test, **5**-15
Charts:
  alignment of, **2**-58 to **2**-62
  contour-line, **2**-58
  parallel or proportional, **2**-61, **2**-62
Chemag process (oxide coating), **6**-116
Chemical compounds (def), **6**-3
  calculations of, **6**-7
Chemical elements (def), **6**-3
  table, **6**-3, **6**-4
Chemical equipment, corrosion of, **6**-113
Chemical fire extinguishers, **18**-33
Chemical symbols (table), **6**-3, **6**-4
Chemistry, **6**-3 to **6**-11
Chézy formula (hydraulics), **3**-67
Chi-square distribution (statistics, table), **17**-23
Chicago pneumatic simplate valve
    (compressor), **14**-36
Chimneys:
  compression at base, **5**-44
  draft, **9**-26
  gases (*see* Flue gas)
  (*See also* Stacks)
Chrome brick, **6**-171
Chromium, **6**-96
Chromium plating, **6**-115
Chromium steel alloys (*see* Steel, chromium)
Chromizing, **6**-25, **6**-116
Chutes, **10**-48 to **10**-50
  for bulk materials, **10**-48
  for lumpy materials, **10**-48
  spiral, **10**-49, **10**-50
  for unit loads, **10**-49
Cinder and fly ash, **9**-7 to **9**-10, **9**-13
Cinder concrete, corrosion of pipes in, **6**-114
Cippoletti trapezoidal weir, **3**-65
Circle:
  angles in, **2**-8
  arc of, **2**-10
  area of, **2**-12
    tables, **1**-4, **1**-6, **1**-7
  circumference of, **2**-12
    table, **1**-2, **1**-7
  constructions of, **2**-9 to **2**-11
  equations of (analytic geometry), **2**-31
  involute of, **2**-40
  radical axis of, **2**-8, **2**-31

Circle (*Cont.*):
  segments of (tables), **1**-6, **1**-7
  theorems on, **2**-8, **2**-12
Circuit breakers, **15**-54, **15**-55, **15**-69
Circuits (electric), **15**-8 to **15**-10
  alternating-current: four-phase, **15**-23
    interior wiring, **15**-66 to **15**-69
    parallel, **15**-22 to **15**-24
    polyphase power: advantages of, **15**-24
      measurement of, **15**-26
    quarter-phase, **15**-23
    solution of problems, **15**-22
    three-phase, **15**-23, **15**-26
    two-phase, **15**-23
    voltage drop, **15**-64
      Mershon diagram, **15**-56, **15**-58
    wiring calculations, **15**-64 to **15**-66
  branch, voltage drop, **15**-64
  capacitivity, **15**-21
    reactance, **15**-21
  carrying capacity of conductors, **15**-60, **15**-66
  critical damping resistance of, **15**-20
  dielectric, **15**-18, **15**-19
  direct-current, wiring calculations, **15**-64, **15**-65
  effective resistance, **15**-22
  impedance, **15**-21
  inductive, **15**-19
    reactances, **15**-21
    time constant, **15**-20
  integrated, **15**-92, **15**-93
  loop, **15**-62
  magnetic, **15**-14
  natural frequency, **15**-22
  network, **15**-64
  parallel, **15**-8, **15**-62
  phase difference, **15**-22
  resonance, **15**-22
  series, **15**-8, **15**-61
  series-parallel, **15**-62
  street-lighting, **15**-62
  three-wire: a-c, **15**-63, **15**-64
    balancer set for, **15**-62
    d-c, **15**-62
    resistance and reactance (table), **15**-63
Circular arc (*see* Circle)
Circular cutters, **13**-19
Circular inch (def), **1**-31
Circular measure, **1**-32
  of angles, **2**-25
    tables, **1**-11, **1**-12, **1**-30
Circular mil (def), **1**-31, **15**-5
Circular pitch (gears, def), **8**-98
Circular saws, **13**-66
Circulation (aerodynamics, def), **11**-61
Clad steels, **6**-24
Clairaut, differential equation of, **2**-50
Clapeyron equation, **4**-14
Claude's system of liquefying air, **19**-23
Clay for firebricks, **6**-171
Cleaning, sonic, **12**-142
Cleansers, **6**-150, **6**-151
Clearance in steam engines, **9**-38
Climb of airplanes, **11**-65, **11**-66
Clock motors, **15**-48
Cloud point (petroleum oils, def), **6**-198
Cloudburst hardness test, **5**-14
Clouds (particles in suspension, def), **18**-11
  of large black particles, **4**-78
Clutches, **8**-38 to **8**-41
  allowable pressures in (table), **8**-39

Clutches (*Cont.*):
  band, **8**-38
  cone, **8**-38
  disk, **8**-38
  dry-disk (automobile), **11**-5
  friction, **8**-38
    for automobiles, **11**-5
    coefficients (table), **8**-39
  jaw, **8**-38
  multiple-disk (automobile), **11**-5
  overrunning, **8**-38
  positive, **8**-38
  rim, **8**-38
  torque transmissible by, **11**-6
CO Acceptor process (gas making), **7**-45
Coach screws (table), **8**-23
Coal, **7**-2 to **7**-11
  agglomerating index, **7**-2
  analyses (tables), **7**-3, **7**-4
    proximate, **7**-2, **7**-5
      formulas, **7**-2
    ultimate, **7**-7
  anthracite (def), **7**-5
    in chain-grate stokers, **9**-10
    pulverizing, **9**-11
  ash in, **7**-7, **9**-7 to **9**-10
    composition of (table), **7**-7
    fusibility, **7**-7
    table, **7**-8
  banded, **7**-5
  bituminous, **7**-5, **7**-8
    coking quality, **7**-37
  boghead (def), **7**-5
  bulk density of, **7**-9
  caking (def), **7**-8
  calorific value, **7**-7
    Dulong's formula for, **7**-7
    Parr formulas for, **7**-2
    tables, **7**-2 to **7**-4
  cannel (def), **7**-5
  carbonization (*see* Carbonization of coal)
  classification of, **7**-2 to **7**-5
    tables, **7**-3, **7**-4, **7**-6
  cleaning, **7**-9
  coking characteristics of, **7**-37
  coking process, **7**-38
  combustion of, **9**-12 to **9**-14
    spontaneous, **7**-11
  common banded (def), **7**-5
  density and specific gravity of, **7**-8
  firing in boiler furnaces, **9**-12 to **9**-14
  fixed carbon in, **7**-6
  free-swelling, **7**-8
  free-swelling test for, **7**-38
  for gas manufacture (table), **7**-39
  gasification of, **7**-41 to **7**-47
    underground, **7**-24
  grindability, **7**-9, **9**-11
    tests, **7**-9
  gross calorific value (def), **7**-7
  heat value (*see* Coal, calorific value)
  for industrial heating, **7**-47
  lignitic (def), **7**-5
    analyses of (tables), **7**-2 to **7**-4
  meta-anthracite, **7**-5
  mineral-matter-free basis of classifying, **7**-2
  mining, **7**-9
  moisture in, **7**-5, **7**-6
  net calorific value (def), **7**-7
  piled, specific gravity and density of (table), **6**-9
  preparation of, **7**-9, **7**-10

Coal (*Cont.*):
  processing, **7**-11
  production of, **7**-11
  proximate analyses of (def), **7**-5
    tables, **7**-2, **7**-4
  pulverized, **9**-11, **9**-12
    air supply for burning, **9**-12
    burners for, **9**-12
    direct system of firing, **9**-11
    fineness of, **9**-11
    fly-ash recovery in burning, **9**-10
    mills for, **9**-11
  purchase of, under specifications, **7**-10
  rank (def), **7**-2
    mined in various states (table), **7**-6
    tables, **7**-2 to **7**-4, **7**-6
  reserves, **9**-4
  sampling, **7**-10
  semianthracite (def), **7**-5
  slurry pipeline, **11**-134 to **11**-135
  specific gravity of, **7**-8
    table, **6**-9
  specifications for, **7**-10, **7**-11
  splint (def), **7**-5
  spontaneous combustion, **7**-10
  storage, **7**-10
  subbituminous, **7**-5
    tables, **7**-2 to **7**-4
  sulfur in, **7**-7
    tables, **7**-4
  transportation, **7**-10
  ultimate analysis of, **7**-7
    methods, **7**-7
    tables, **7**-4
  volatile matter in, **7**-4
    tables, **7**-4
Coal-chemical ovens, **7**-39, **7**-40
Coal gas, flame temperatures (table), **4**-57
Coal-in-oil suspensions as fuel, **7**-20
Coal tar as fuel, **7**-20
Coaltek (coal charging to coke ovens), **7**-41
Cocks, **8**-189
Coefficients:
  of area expansion, **4**-3
  binomial (table), **2**-18
  of contraction, **3**-68
  of discharge for liquids through orifices, **3**-63, **3**-68
  of excess air in combustion (def), **4**-51
  of expansion (def), **4**-3
    at low temperatures, **19**-28
    tables, **4**-7, **4**-8
  film, **4**-63 to **4**-70
  of form (ships), **11**-37
  of friction, **3**-24 to **3**-28, **4**-50
  Joule-Thomson, **4**-14, **4**-51
  of performance (def), **4**-20, **4**-45
  of restitution, **3**-21
  of rigidity (def), **5**-18
  of thermal expansion of metals (table), **6**-11
  of velocity, **3**-68
  for steam, **4**-49
Cohesion of liquids, **3**-36
Coils, pipe, **8**-155
Coining (metal squeezing), **13**-22
Coke, **7**-11, **7**-12, **7**-37 to **7**-41
  analyses of (table), **7**-11
  blast-furnace, **7**-11
  domestic, **7**-11
  foundry, **7**-11
  high-temperature, **7**-11
  low-temperature, **7**-11

Coke (*Cont.*):
  medium-temperature, 7-11
  petroleum, 7-12
  pitch, 7-12
  specific gravity and density of (table), 6-9
  water-gas, 7-12
Coke-oven gas (table), 7-39
Coke ovens, 7-39, 7-40
  by-products of, 7-40
  heat balances of (table), 7-40
Coking properties of coal, 7-37, 7-38
  tests, 7-38
Cold-air refrigerating machines, 19-13
Cold-diffuser conditioning of storage rooms, 19-20
Cold forging of metals, 13-21
Cold storage, 19-16 to 19-21
  insulation requirements, 19-16
  lockers, 19-21
  product data (table), 19-19, 19-20
  rooms: air conditioning, 19-20
    piping, 19-18
  temperatures, 19-18
Cold-working of metals, 13-15
  influence on corrosion, 6-108
  nonferrous metals, 6-59
Colebrook equation for friction, 3-54
Collapsing pressure of tubes, formula, 5-50
Collimators, 16-6
Collision, laws of, 3-21
Colloidal fuels, 7-20
Cologarithm (def), 2-4
Color scale for temperatures of iron and steel, 4-6
Color television, 15-97
Color test for lubricating oils, 6-199
Color vision, 12-117
Colorimeters, 16-21
Colors:
  for electrical wiring, 15-69
  for identifying piping, 8-199
Colpitts oscillator, 15-89
Columbium, 6-96
Columns:
  cast-iron, 12-38
    strength of (tables), 5-42, 5-43
  classification (long and short), 5-41
  eccentrically loaded, stresses in, 5-42
  ends, forms of, 5-41
  long: critical load, 5-41
    Euler's formula, 5-41
    slenderness ratio, 5-41
    strength of (table), 5-42
  reinforced-concrete, 12-53 to 12-55
  round-ended, strength of, Euler's formula
    (table), 5-42
  short (table), 5-43
    formulas, 5-42
  steel: allowable unit stresses (table), 5-42
    pipe, 12-42
      concrete-filled, 12-55
      table, 12-45
    properties of (tables), 12-35 to 12-38
    proportions of, 12-30
    strength of (tables), 5-42, 5-43
  timber, 12-24
  wooden, 12-24
    working stresses (table), 12-26
  wrought-iron pipe, strength of (tables), 5-42
Combinations and permutations, 2-18
Combined flexure and longitudinal force, 5-42
  and torsion, 5-20
Combined stresses, 5-19 to 5-21

Combustion, 4-51 to 4-58
  air required by gaseous fuels, 4-51
  in boiler furnaces, 9-10 to 9-13
  of coal, 9-10 to 9-13
    spontaneous, 7-10
  control of, automatic, 9-23
  dissociation of gases, 4-54
  equations, 4-51
  excess air for (charts), 4-58
  flame temperatures (table), 4-57
  of gaseous fuels, 4-51 to 4-55
    products (table), 4-55
    temperature (table), 4-57
  of gases: air required, 4-51
    products, 4-51
  heat of formation (def), 4-53
  heat value of fuels, 4-53
  heats of (def), 4-52
    table, 4-52
  incomplete, loss due to, 4-57, 9-13
  of liquid fuels, 4-55
  products of, 4-51
    table, 4-55
  of solid fuels, 4-56 to 4-58
    air required, 4-56
    excess air for (charts), 4-58
  surface, 4-58
  temperatures attained in, 4-53
  volume contraction due to, 4-52
Combustion chambers of engines, 9-88 to 9-92
Combustion furnaces (*see* Furnaces)
Combustion products, radiation from, 4-77 to 4-80
Combustion turbines (*see* Gas turbines)
Come-alongs, 10-14
Comfort indexes, 12-104 to 12-113
Common logarithms, 2-17
Commutating-pole motors, d-c, 15-34
Commutation in d-c generators and motors, 15-33, 15-34
Commutator motors, a-c, 15-47
Compensation (automatic control, def), 16-26
Compensator (transformer), 15-42
  starting, a-c motor, 15-44
  for three-wire generator, 15-63
Compiler, 2-70
Complex quantities:
  algebra of, 2-22
  standard forms of, 2-24
Composite steels, 6-24
Compound, chemical, 6-5
  calculation of composition, 6-7
Compound interest (tables), 1-26, 1-27
Compound motors, electric, 15-34
Compound-wound generators, 15-30
Compressed-air machinery, lubrication of, 6-205
  (*See also* Air compressors; Centrifugal compressors)
Compressibility of liquids, 6-9
  table, 4-7, 6-10
Compressible fluids, flow of, 4-46 to 4-51
Compression:
  of air (*see* Air compression)
  of saturated and superheated vapors, 4-24
  in steam engines, effect on efficiency, 9-38
Compression couplings, 8-35
Compression-ignition (engines), 9-92
Compression machines, vapor, theory of, 4-45
Compression ratios of internal-combustion engines (def), 9-78
  permissible, 9-95, 9-96

Compression testing, **5**-5
Compressors, **14**-30 to **14**-44
  accessories for, **14**-38
  air (*see* Air compressors)
  ammonia, **19**-7 to **19**-11
  centrifugal (*see* Centrifugal compressors)
  compression efficiency, **14**-33
  cylinders, **14**-34, **14**-35
    cooling, **14**-39
    displacement, **14**-34
    sizing, **14**-34
    wear, **14**-36
  efficiencies, **14**-32, **14**-39, **14**-40
  for gas turbines, **9**-116
  helical screw, **14**-43
  line sizing equations, **14**-38
  liquid-ring, **14**-43
  lubrication, **14**-37, **14**-38
  mechanical efficiency, **14**-32
  motors for, **15**-47, **15**-53
  multistage sizing, **14**-35
  piston rings, **14**-36
  piston-rod racking, **14**-36
  pulse bottles, **14**-38
  relief valves, **14**-38
  spiral-axial, **14**-43
  temperature rise, **14**-33
  types of, **14**-31
  valve resistance, **14**-34
    table, **14**-34
  valves, **14**-36
  volumetric efficiency, **14**-34
  warm-up factor, **14**-33
Computation:
  graphical methods, **2**-58
  numerical, **2**-2
Computers, **2**-64 to **2**-75
  analog, **2**-64, **2**-73 to **2**-75
    components, **2**-73
  applications of, **2**-73
  arithmetic, **2**-65, **2**-66
  automatic control by, **2**-68
  characteristics of, **2**-67
  comparison of analog and digital, **2**-74
  compiler, **2**-70
  digital, **2**-65 to **2**-74
    components, **2**-66 to **2**-68
  engineering calculations with, **2**-71, **2**-72
  hybrid, **2**-65
  input, **2**-68
  letters in, **2**-66
  memory, **2**-67
  mini-, **2**-67
  multiprocessor, **2**-65
  numbers in, **2**-65, **2**-66
  operating systems for, **2**-69
  operation, **2**-67
  output, **2**-68
  program, **2**-67
  program-preparation systems for, **2**-69, **2**-70
  programming languages for, **2**-70 to **2**-73
  simulation with, **2**-71
  size of, **2**-61
  software, **2**-68
  solutions with, **2**-71, **2**-72
  speed of, **2**-67
  subroutine, **2**-67
  systems routines, **2**-69
  tape, **2**-68
  timesharing, **2**-69
  types of, **2**-64
  word (def), **2**-65

Concatenation of induction motors, **15**-45
Concrete, **6**-183 to **6**-189
  admixtures, **6**-180, **6**-181
  aggregate for, **6**-179, **6**-180
  air-entrained, **6**-187
  clay in sand, **6**-188
  compaction, **6**-187
  consistency, **6**-186
  curing, **6**-187
  dry rodded weight, **6**-185
  effect of freezing, **6**-188
  effect of oils and acids, **6**-188
  effect of sea water, **6**-188
  fineness modulus of (def), **6**-180
  forms for, **6**-186
  masonry, specific gravity and density
    (table), **6**-8
  for masonry units, **6**-187
  materials, per cubic yard (table), **6**-185
  mica in sand, **6**-188
  mixing, **6**-186
  painting, **6**-120
  piles, **12**-21, **12**-59
  pipe, **8**-172
  placement, **6**-186
  proportioning, **6**-183, **6**-184
  quality control, **6**-186
  ready-mixed, **6**-186
  reinforced (*see* Reinforced concrete)
  for reinforced work, proportions of, **12**-48
  sand for, **6**-179
  slump test (table), **6**-185
  strength of, **6**-187, **6**-188
    variation with age (table), **6**-188
  water for, **6**-180
  water content (table), **6**-184
  watertightness, **6**-187
  weight, **6**-187
Concrete mixers, **6**-186
Condensate pumps, **9**-64
Condensation, **9**-59 to **9**-66
Condensers:
  ammonia, **19**-8
  circulating water, United States
    temperatures (map), **9**-62
  cooling water, air in (table), **9**-67
  direct-contact, low-level, **9**-65
  electric, **15**-18, **15**-19
    capacitance of, **15**-18
    synchronous, **15**-47
  evaporative, **4**-45
  refrigeration, **19**-8
  steam, **9**-59 to **9**-66
    air-cooled, **9**-64, **9**-65
    air ejectors, **9**-66, **9**-71
      two-stage, **9**-67
    barometric, **9**-67
    direct-contact, **9**-65, **9**-66
    ejectors for, **9**-66 to **9**-71
    noncondensibles, removal of, **9**-66
    surface-type, **9**-59 to **9**-64
      backwash systems for, **9**-60
      calculations, **9**-60 to **9**-64
        of size, **9**-60 to **9**-62
      circulating-water pumps, **9**-63
      condensate pumps, **9**-64
      configurations, **9**-60
      expansion in, **9**-60
      materials for (table), **9**-64
      performance curves, **9**-63
      proportions of (table), **9**-63
      sizing, **9**-61, **9**-62

Condensers, steam, surface-type (*Cont.*):
  tube-bundles, **9**-60
  tubes: cleanliness, **9**-60
    corrosion, **6**-112
    heat-transfer rates, **4**-68
    special alloys for, **6**-112
  water-boxes, **9**-60
  steam tables for (table), **4**-31
  (*See also* Cooling towers)
Condensing water, cooling of, **9**-71 to **9**-74
Conductances:
  a-c (def), **15**-22
  electrical (def), **15**-4
  thermal, **4**-61
    conversion table, **1**-54
Conduction:
  and convection, **4**-61 to **4**-63
  heat transmission by, **4**-59 to **4**-70
Conductivity:
  electric, **15**-4
  thermal (tables), **4**-60 to **4**-65
    conversion table, **1**-54
    of insulation (table), **4**-64, **4**-65
    of liquids and gases (table), **4**-62
    in materials for low temperatures (table), **4**-64
    of metals (table), **4**-60, **4**-61
    of miscellaneous solids (table), **4**-63
    (*See also* Heat transmission)
Conductors, electric, **15**-5 to **15**-10
  aluminum (table), **6**-82
  copper alloys for (table), **6**-63
  cross-sectional area (table), **15**-74
  current-carrying capacities (table), **15**-70
  economical cross section, **15**-56
  estimating resistances and weights, **15**-6
  materials for, **15**-5, **15**-6
  properties of metals and alloys (table), **15**-6
  specific resistance, **15**-5
  temperature coefficients (def), **15**-5
    table, **15**-5
  types and applications of (table), **15**-72
  (*See also* Cables)
Conduits for interior wiring (table), **15**-73
Cone pulleys, **8**-52
Cones, surface and volume of, **2**-13
Conic sections, **2**-31 to **2**-36
Conical pendulum, **3**-17
Conjugate diameters:
  of ellipse, **2**-33
  of hyperbola, **2**-35
Conjugate hyperbolas (def), **2**-35
Conjugate imaginaries, **2**-24
Conradson carbon-residue test for lubricating oils, **6**-199
Conservation:
  of energy, **3**-3, **3**-20, **4**-13, **9**-120
  of mass, **3**-3
  of matter, **9**-120
  of momentum, **3**-3, **3**-21
Consolidation Coal process (coke making), **7**-41
Constant-entropy expansion:
  of gases, **4**-19
  of vapors, **4**-24
Constant-pressure expansion:
  of gases, **4**-19
  of vapors, **4**-24
Constant-temperature expansion:
  of gases, **4**-19
  of vapors, **4**-24

Constant velocity joints, **8**-37
Constant-volume expansion:
  of gases, **4**-19
  of vapors, **4**-24
Constrained beams, **5**-35
Construction (*see* Building construction)
Containerization, **10**-25
Continuity equation (in fluid flow), **3**-41, **4**-46
Continuous beams, **5**-35, **5**-36
Continuous conveying, **10**-2
Contour-line charts, **2**-58
Contour lines (maps, def), **16**-47
Contour mapping (surveying), **16**-48
Contracts, **12**-10
Control, automatic (*see* Automatic control)
Control charts (quality control), **17**-25
Controllable-pitch propellers:
  airplane, **11**-102, **11**-103
  marine, **11**-51
Controlled variable (automatic control), **16**-25
Convection:
  coefficients of, **4**-63 to **4**-68
  and conduction, **4**-61 to **4**-63
  heat transmission by, **4**-67
  natural, **4**-67
Conversion of mass to energy, **9**-120
Conversion tables:
  acceleration, **1**-51
  angular measure: decimals of degrees, minutes, and seconds, **1**-30
    degrees-radians, **1**-11, **1**-12, **1**-32
    velocity, **1**-50
  area, **1**-31, **1**-46
  conductance, thermal, **1**-54
  conductivity, thermal, **1**-54
  density, **1**-42, **1**-43, **1**-54
  energy, **1**-52
  flow of heat, **1**-54
  heat, **1**-52
  kinematic viscosity, **3**-36
  length, **1**-31, **1**-44, **1**-45
  logarithms (base 10 and base *e*), **1**-24
  mass, **1**-48
  power, **1**-53
  pressure, **1**-49
  specific gravity and density, degrees API and Baumé, **1**-42, **1**-43
  temperature, **4**-3, **4**-4
  velocity, **1**-50
  volume, **1**-31, **1**-47
  weight, **1**-48
  work, **1**-52
Converters:
  phase, **15**-47
  synchronous, **15**-49, **15**-50
    switchboard equipment for, **15**-53
Conveying, **10**-40 to **10**-63
  bucket, **10**-50 to **10**-55
    (*See also* Conveyors, bucket)
  cableways, **10**-34, **10**-35
    (*See also* Cableways; Conveyors)
  continuous (def), **10**-2
Conveyors, **10**-40 to **10**-63
  apron, **10**-50
    table, **10**-51
  belt, **10**-55 to **10**-60
    arrangements, **10**-55, **10**-60
    automatic scales, **10**-63, **10**-64
    belts for, **10**-55, **10**-56
    capacity (table), **10**-58
    drive calculations, **10**-58
    drives, **10**-56

Conveyors, belt (*Cont.*):
 electromagnetic separators, **10**-59
 feeders, **10**-60, **10**-61
 horsepower required, **10**-57
 idler pulleys, **10**-56
 life, **10**-56
 magnetic pulleys, **10**-59
  table, **10**-60
 maximum slope (table), **10**-57
 portable, **10**-60
 pulleys, **10**-56, **10**-59
 sectional, **10**-60
 shuttle, **10**-59
 sliding, **10**-60
 slope for various materials (table), **10**-57
 speed (table), **10**-59
 take-ups, **10**-57
 trippers, **10**-57, **10**-59
 width required for lumps (table), **10**-58
bucket, **10**-50 to **10**-55
 open-top, **10**-50
 Peck carrier, **10**-53
 pivoted, **10**-51, **10**-53
  capacities (table), **10**-54
 power requirements (chart), **10**-53
 V-bucket, **10**-50
  table, **10**-52
Bulk flo, **10**-46
carrying, **10**-50 to **10**-63
chutes, **10**-48 to **10**-50
continuous-flow, **10**-46
drag-chain, **10**-44
drives for, **10**-56 to **10**-59
feeders for, **10**-60, **10**-61, **10**-64
flight, **10**-44 to **10**-47
 arrangements for, **10**-47
 capacity and lump size of (table), **10**-45
 chain pull, equation for, **10**-45
 design of, **10**-46
 inclined, power required, **10**-46
 scraper type of, **10**-44
 suspended-chain type of, **10**-44
 suspended-flight type of, **10** 44
friction of materials on (table), **10**-46
gravity roller, **10**-61
hydraulic, **10**-63
noncarrying, **10**-44 to **10**-50
oscillating, **10**-60
overhead, **10**-41 to **10**-44
 components of, **10**-43, **10**-44
 control, **10**-43
 drop-lift, **10**-44
 power-and-free, **10**-42 to **10**-44
 powered, **10**-40 to **10**-42
 safety guards, **10**-41, **10**-42
 switching, **10**-42, **10**-43
  controls, **10**-43
 track for, **10**-41
 transfer devices, **10**-42
 trolleys, **10**-41
 turns, **10**-41
package, **10**-49
platform, **10**-61
pneumatic, **10**-61 to **10**-63
reciprocating plate feeder for, **10**-60
Redler, **10**-46
ribbon, **10**-48
roller, **10**-61
runaround, **10**-46
scraper, **10**-44

Conveyors (*Cont.*):
 screw, **10**-47, **10**-48
  cut-flight, **10**-48
  paddle, **10**-48
  power required, **10**-48
  short-pitch, **10**-48
  speeds and capacities of (tables), **10**-49
  variable-pitch, **10**-48
 selection of (table), **10**-48
 spiral, **10**-48
 suction for, **10**-62
 trolley, **10**-41, **10**-42
 type preferred for various material (table), **10**-3
 vertical, **10**-44
 vertical interfloor, **10**-44
 vibrating feeders, **10**-61
 weighers, **10**-63, **10**-64
 wheel, **10**-61
Coolants (cutting fluids), **13**-54
Coolers:
 brine, **19**-14
 capacity of (table), **19**-15
 heat-transfer coefficients (table), **19**-20
Cooling:
 effect of throttling, **4**-51
 of internal-combustion engines, **9**-104, **9**-105
 thermoelectric, **19**-13
Cooling ponds, **9**-74
Cooling sprays, **9**-74
Cooling towers, **4**-45, **9**-71 to **9**-74
 approach (def), **9**-72
 cost evaluation, **9**-73
 costs of (table), **17**-37
 drift in, **9**-72
 dry, **9**-74
 height of, **9**-73
 hyperbolic, **9**-72
 makeup, **9**-72
 materials for, **9**-74
 mechanical-draft, **9**-72
 performance calculations, **9**-72, **9**-73
 performance curves, **9**-74
 range (def), **9**-72
 wet-bulb temperatures (map), **9**-73
Coordinates:
 polar, **2**-31, **2**-57
 rectangular, **2**-30, **2**-54
 transformation of, **2**-31
Copper, **6**-63 to **6**-70
 alloys (*see* Copper alloys)
 bus bars, **15**-53
 cable, resistance and reactance of (table), **15**-57
 commercial grades, **6**-61
  production and properties of, **6**-63
 conductors, current-carrying capacities (table), **15**-70
 deoxidized, **6**-63
 for electrical purposes, **6**-63
 extruded, **6**-63
 lake, **6**-63
 oxygen-free, **6**-63
 resistivity of, **15**-5
 tough-pitch, **6**-63
 tubing, **8**-169
 welding of, **6**-69, **13**-47
Copper alloys, **6**-63 to **6**-68
 for bearings, **6**-92
  table, **6**-93, **6**-94
 brazing, **6**-69

Copper alloys (*Cont.*):
  conductor, **15**-6
    table, **6**-63
  copper-base castings, **6**-70
    composition, properties of (table), **6**-71
  copper-nickel (tables), **6**-65, **6**-67
  copper-silicon (table), **6**-65, **6**-67
  copper-tin, **6**-68
  copper-zinc, **6**-63
  corrosion resistance, **6**-70
  dezincification, **6**-112
  effects of temperature, **6**-70
  fabrication, **6**-69
  machining of, **6**-70
  for sea-water service, **6**-68
  strength of (tables), **6**-64 to **6**-67
  welding, **6**-69, **13**-47
  wrought, composition and properties of
    (table), **6**-64 to **6**-67
  (*See also* Brass; Bronze)
Copper pipe, **8**-169
Copper plating, **6**-115
Copper tubing, **8**-169
Copper wire (table), **6**-63
  alloys for (table), **6**-64 to **6**-67
  ASTM specifications (table), **6**-63
  insulation, **15**-67
  reactance (table), **15**-63
  resistance (tables), **15**-5, **15**-7, **15**-8, **15**-63
    temperature coefficient of, **15**-5
  weight (table), **15**-7
Copperweld (electrical conductor), **15**-6
Copy (printing):
  fitting, **19**-45
  preparation of, **19**-45
Copyrights, **19**-44
Cord (lumber measure, def), **1**-31, **7**-13
Cordage, **6**-151
Core binders, **13**-6
Core blowers, **13**-7
Core boxes, **13**-2
Core drier, **13**-7
Core molding, **13**-3
Core ovens, **13**-7
Core sands, **13**-6
Coremaking methods, **13**-7
Cores (def), **13**-2
Coriolis acceleration, **3**-15
Cork, specific gravity and density (table), **6**-8
Corner test (statistics), **17**-21
Corona (electrical discharge), **15**-59
Corporation, annual reports, **17**-11
Corrective action (automatic control), **16**-25
Corrosion, **6**-106 to **6**-117
  in boilers, **6**-109 to **6**-111, **9**-29
  in bridges, **6**-114
  in brine, **6**-109
  in chemical plants, **6**-112, **6**-113
  in concrete structures, **6**-114
  of condenser tubes, **6**-112
  electrochemical, **6**-106
    electrode potentials, **6**-107
    electrolyte in, **6**-103
    mechanism, **6**-106
    metal in, **6**-107
    overvoltage (def), **6**-107
    oxygen in, **6**-107
  electrolytic, **6**-108
  factors: inhibiting, **6**-108
    stimulating, **6**-107
  of iron and steel, **6**-108
  of metal stacks, **6**-114

Corrosion (*Cont.*):
  of metals under stress, **6**-108
  minimizing, **6**-114 to **6**-117
    platings, **6**-114 to **6**-117
    water treatments, **6**-117
    zinc coatings, **6**-114, **6**-115
  of pipes, **6**-113
    ferrous, **6**-113
    nonferrous, **6**-113
    protective coatings, **6**-113
    protective linings, **6**-114
    underground, **6**-113
  protection against, **6**-114 to **6**-117
  of pumps, **6**-112
  resistance to (*see* Corrosion resistance)
  in steam-generating plants, **6**-109, **6**-112
    copper, **6**-110
    creep cracks, **6**-111
    in economizer, **6**-111
    gases in steam, **6**-110
    hydrazine, **6**-110
    hydrogen embrittlement, **6**-111
    pH, effect of, **6**-110
    scale, **6**-109
    in superheater, **6**-111
  stress, **6**-108
Corrosion fatigue of metals, **5**-10
Corrosion number (lubricating oils), **6**-199
Corrosion resistance:
  of alloys, **6**-108, **6**-109
  in chemical plants, **6**-113
  of polished surfaces, **6**-108
Corrosion-resisting alloys, steel, **6**-36
  table, **6**-56
Corrosiron (corrosion-resisting iron alloy), **6**-109
Corrugated sheets, **12**-20, **12**-42
Corundum (abrasive), **6**-139
Cosecant (trigonometry), **2**-25
  graph, **2**-55
  tables, **1**-17, **1**-18
Cosine (trigonometry), **2**-25
  graph, **2**-55
  law of, **2**-28
  table, **1**-13, **1**-14
Cosslett process (protection of iron), **6**-116
Cost accounting, **17**-10 to **17**-18
  break-even, **17**-15
  control by, **17**-11
  factory overhead, **17**-13
  labor, **17**-13
  linear programming in, **17**-16 to **17**-18
  materials, **17**-13
  methods of, **17**-13
  for product costing, **17**-12
  purposes of, **17**-11
  standard costs, **17**-14
  types of systems, **17**-12
Cost analysis, **17**-14, **17**-15
Cotangent (trigonometry), **2**-25
  graph, **2**-55
  tables, **1**-15, **1**-16
Cotter pins, **8**-31
Cottered joints, **8**-21
Cotton:
  fiber (tables), **6**-155
  mercerized, **6**-154
Cottonseed oil (table), **6**-202
Coulomb (def), **15**-3
Counters (event measurement), **16**-2
  pickups for, **16**-3
Counterweights for elevators, **10**-20

Couples, **3**-4
  composition of, **3**-4
  displacement of, **3**-4
  moment of, **3**-4
  rotation moments of, **3**-4
Coupling coefficient (electric circuits), **15**-20
Couplings, **8**-35 to **8**-38
  clutch (*see* Clutches)
  compression, **8**-35
  constant velocity, **8**-37
  fire-hose: American National Standard (table), **8**-200
    screw threads, **8**-199
  flanged, **8**-35
  flexible, **8**-35 to **8**-37
    chain, **8**-36
    Falk, **8**-36
    Fast, **8**-35
    Oldham, **8**-35
    Steelflex, **8**-36
    Waldron, **8**-36
  fluid, **8**-37, **8**-38
    for automobiles, **11**-5, **11**-6
    torque capacity (eq) **8**-37
    torque converter, **8**-37
  ribbed clamp, **8**-35
  rigid, **8**-35
  rubber, **8**-36
  sleeve (pipe), **8**-155
  slider, double, **8**-35
  universal, Hooke's joint, **8**-36
Coversed sine (trigonometry), **2**-26
Cracked gasoline, **7**-15
Cramer equation for pollution calculations, **18**-16
Cranes:
  capacities of (table), **10**-31, **10**-33, **10**-34
  column jib, **10**-30
  derricks, **10**-30
  efficiency of, **3**-29
  gantry, **10**-30
  hand-power, **10**-28
  hooks, **10**-7
  industrial (table), **10**-31
  jib, **10**-30
  lifting speeds (table), **10**-31
  locomotive, **10**-30
    data on (table), **10**-33
  motors for, **15**-51
  overhead, **10**-28 to **10**-30
    special purpose, **10**-30
  power shovels as, **10**-37
  rotary, **10**-30
  telescoping boom, **10**-37
  traveling, **10**-28 to **10**-30
    electric, **10**-29
  truck, **10**-32, **10**-34
  yard jib, **10**-30
Crank angles:
  and piston positions (tables), **8**-70, **8**-71
  and piston velocities (table), **8**-72
Crank mechanism, **8**-3
Crankshafts, **8**-49
  for diesel-marine engines, **11**-47, **11**-48
  (*See also* Shafts)
Creep:
  of low-chromium steel (table), **5**-11
  of metals (def), **5**-10, **6**-95
  testing, **5**-11, **5**-12
  of zinc and zinc alloys, **6**-85

Creep rates:
  of iron (table), **5**-11
  of metals and alloys (table), **5**-11
  of steel (chart, table), **5**-11, **5**-12
Crescent beams, **5**-46, **5**-47
Critical cooling rate (steel), **6**-19
Critical damping resistance, **15**-20
Critical data for various gases (table), **4**-23
Critical path method of scheduling, **12**-2, **17**-6
Critical pressure (def), **4**-22
  in gas or vapor flow, **4**-47
Critical speeds of shafts, **5**-72, **5**-73
Critical state for gases (def, table), **4**-23
Critical temperature (def), **4**-22
  of iron (def, chart), **6**-17
Crocus (abrasive), **6**-139
Cross-section paper:
  logarithmic, **2**-55
  semilogarithmic, **2**-56
Cross-section symbols for drafting, **8**-97
Crude oils:
  distillates (chart), **7**-15
  as fuels (table), **7**-14
  refining, **7**-14
Crushed steel (abrasive), **6**-139
Cryogenic refrigerators, **19**-23
Cryogenics, **19**-23 to **19**-40
  applications of, **19**-26
  cryogen hazards, **19**-40
  definition of, **19**-23
  equipment materials, **19**-26 to **19**-31, **19**-39
    stress analysis, **19**-30
  gas liquefaction for, **19**-25
  instruments for, **19**-33 to **19**-35
  insulation for, **19**-35 to **19**-39
    selection of, **19**-38
  level measurements, **19**-33
    by capacitance gage, **19**-33
    by weighing, **19**-33
  temperature measurements, **19**-34
  vent systems for, **19**-39
Cryogens, **19**-31
  compressibility, **19**-31
  liquid density, quantum mechanical correlation, **19**-31
  properties of (table), **19**-32
  specific heat (chart), **19**-36
  thermal conductivity, **19**-31
  thermal expansion coefficient, **19**-31
  vapor pressure (chart), **19**-35
  viscosity, **19**-31
  volumetric latent heats (table), **19**-32
Crystalline alumina (abrasive), **6**-139
Crystolon (abrasive), **6**-139
Crystoplastic (def), **13**-10
Cube roots, **2**-3
Cubes, summation of series of, **2**-17
Cubic equation, **2**-19
Cubic measure, conversion table, **1**-47
Cubical expansion:
  coefficient of (def), **4**-3
  table, **4**-8
Cupronickel, **6**-69
  tables, **6**-65, **6**-67
Current regulator, **15**-82
Current transformers, **15**-27
Currents:
  alternating, **15**-21 to **15**-24
    active or energy (def), **15**-22
    circuits (*see* Circuits)
    reactive or wattless (def), **15**-22

Currents, alternating (*Cont.*):
   waves: average value, **15**-21
     effective value, **15**-21
     form factor, **15**-21
     phasor representation, **15**-21
     vector representation, **15**-21
   electric, unit of (def), **15**-3
   heat developed by, **15**-10
   wattless, **15**-22
   (*See also* Electric motors)
Curvature, radius and center of, **2**-45
Curve resistance (train), **11**-30
Curved beams, **5**-45, **5**-46
   eccentrically curved (table), **5**-48
Curves:
   empirical, equations for, **2**-55
   equations and construction, various, **2**-39 to **2**-41
   railway, radii of (table), **11**-32
   representation of various functions by, **2**-54
Curvilinear motion, **3**-13, **3**-14
Cutters:
   circular, **13**-19
   milling, **13**-62
Cutting:
   oxyacetylene (table), **13**-37
   oxygen, **13**-37
     of metals under water, **13**-37
   stainless steel, **13**-37, **13**-38
   thermal, of metals, **13**-36 to **13**-38
   (*See also* Metal cutting)
Cutting fluids (machining), **13**-54
Cutting-off machines (metal), **13**-66
Cutting oils, **6**-206
Cutting processes, gears, **13**-62 to **13**-64
Cutting tools, **13**-49 to **13**-54
   action of, **13**-49
   boring mill, **13**-62
   broaches, **13**-65
   cemented carbide, **13**-52
   ceramic, **13**-53
   chip formation by, **13**-49, **13**-51
   cutting angle of, **13**-49
   diamond, **13**-54
   failure of, **13**-50
   fluids for, **13**-54
   grinding of, **13**-54
   heat-treatment of, **13**-54
   for lathe: life of, **13**-50
     power requirements of, **13**-50
     shapes for, **13**-57
   materials for, **13**-51 to **13**-54
   milling, **13**-61, **13**-62
   nomenclature of, **13**-49, **13**-57
   nonferrous, **13**-52
   nose radius of, **13**-57
   planers, **13**-64, **13**-65
   rake of, **13**-49, **13**-57
   reamers, **13**-60, **13**-61
   relief angle of, **13**-49, **13**-57
   saws, metal-cutting, **13**-66
   screw machine, **13**-59
   shaper, **13**-65
   shapes of, **13**-57
   side-cutting edge angle, **13**-57
   slotter, **13**-65
   steel, **13**-51, **13**-52
   Stellite, **13**-53
   surface finish by, **13**-51
   threading, **13**-61
   wood, **13**-80 to **13**-84
   (*See also* Machine tools)

Cyaniding (heat-treatment), **6**-24
Cycle:
   alternating-current (def), **15**-21
   ideal gas, **4**-20, **4**-28
   for perfect gases, **4**-20 to **4**-22
   steam, **4**-27, **4**-28
     bleeding, **4**-28
     Rankine, **4**-27
     reheating, **4**-28
Cyclogiro wind rotor, **9**-166
Cycloid, **2**-39
Cyclone furnaces, **9**-13
Cyclone separators, **18**-12
   for dust exhaust systems, **18**-12
Cylinder-boring machines, **13**-59
Cylinder grinders, **13**-67
Cylinders:
   air resistance (table), **11**-69
   of automobile engines, **9**-88 to **9**-92
   collapsing pressure, **5**-50
   hollow, volume of, **2**-13
   of internal-combustion engines, **9**-88 to **9**-92
   offset, **8**-69
   oval hollow, strength of, **5**-51
   rolling down a plane, **3**-18
   shrink-fit effects, **5**-51
   of steam engines, jackets, effect on economy, **9**-37
   stiffening rings, **5**-50
   surface and volume of, **2**-12
   thick, **5**-51
   thin-walled, **5**-49

Dacron rope, **6**-151
Dalton's law for gases, **4**-18
Damping:
   of vibrations, **5**-67 to **5**-72
   viscous, **16**-26
Damping ratio (automatic control), **16**-26
Darcy formula in pipeline flow, **11**-133
Darlington connection (electronic circuit), **15**-89
Darrieus wind rotor, **9**-166
D'Arsonval galvanometer, **15**-24
Data logging, **16**-23 to **16**-25
Davis Plastometer, **7**-38
Day, definitions of, **1**-41
Dead-weight gage tester, **16**-9
Deaeration:
   of feedwater, **9**-30, **9**-77, **9**-78
   of water, **6**-110
Deaerators, **9**-77, **9**-78
Decibel (sound, def), **12**-137
Decimal equivalents (table), **1**-30
Decimal point, position of, **2**-2
Dedendum (gear teeth), **8**-99
Definite integrals, **2**-49
Deflection:
   angular, under torsion (table), **5**-41
   of beams (*see* Beams, deflection of)
Deformation (def), **5**-16
   plastic flow theory, **5**-54
   of spheres and cylinders under compression, **5**-49 to **5**-51
Degrees:
   API (table), **1**-42
   Baumé (table), **1**-43
   from radians (conversion table), **1**-11
   spherical (def), **2**-14
Dehumidification of air, **4**-43
   (*See also* Air conditioning)

de Lavaud cast-iron pipe, metal mold process, **13**-4
Delco-Remy distributors, **15**-82
Delta connections, three-phase circuit, **15**-23
Delta iron (def), **6**-17
Demand factor of buildings (electric power), **15**-69
Demodulation of radio waves, **15**-90
De Moivre's theorem (complex quantities), 2-24
De Morgan's theorem (Boolean algebra), **15**-92
Denier (unit of fineness of fibers), **6**-156
Density (def), **1**-42, **3**-24
  of air (formula), **14**-50
  conversion table, **1**-54
  equivalents (table), **1**-54
  of metals (table), **6**-7, **6**-60, **6**-61
  of varieties of iron and steel (table), **6**-44
  of various substances (table), **6**-7 to **6**-9
  of water (table), **6**-9, **6**-10
Depreciation of power equipment, **17**-38
Derivative:
  calculus, **2**-42, **2**-43
  compensation (automatic control), **16**-27, **16**-28
  partial, **2**-43
Derricks, **10**-30, **10**-32
Design stress (mechanics), **5**-22
Detergents, **6**-151
Determinants, **2**-22
Detonation, **7** 16
  (See also Knock)
Deuterium as cryogen, **19**-31
Deviation in automatic control (def), **15**-25
Dew point, **4**-31
Dewpoint recorders, **16**-21
Dextrin baked cores, **13**-6
Diametral pitch (gears, def), **8**-98
Diamond (abrasive), **6**-139, **13**-67
Diamond cutting tools, **13**-54
Diaphragm gages, **16**-9
Diaphragms to replace packing, **8**-147
Diatomaceous earth (abrasive), **6**-131
Dichlorodifluoromethane (Freon 12), **19**-5
  chart, **19**-6
  tables, **19**-17
Die blocks:
  materials for, **13**-21, **13**-22
  proportions of, **13**-22
Die casting, **13**-4
  alloys, zinc-base (table), **6**-85
Dielectric (def), **15**-18
Dielectric circuit, **15**-18, **15**-19
Dielectric constant (def), **15**-4
  of insulating materials (table), **15**-19
Dielectric heating, **7**-64, **15**-94
Dieline (refrigerant), **19**-5
Dies:
  resistance to shearing in (table), **13**-18
  threading, **13**-61
Diesel cycle, **4**-20, **9**-78
Diesel-electric drive, locomotives, **11**-21 to **11**-26
  table, **11**-22
Diesel engines, **9**-78 to **9**-114
  analysis of engine process, **9**-106 to **9**-109
  combustion chambers, **9**-88 to **9**-92
  cooling systems, **9**-104, **9**-105
  cycle for, **4**-20, **9**-78
  efficiency of: indicated, **9**-106
    theoretical, **9**-106
    volumetric, **9**-107

Diesel engines (Cont.):
  exhaust-gas analysis of (table), **9**-109
  exhaust temperatures, **9**-108
  fuel injection, **9**-98 to **9**-100
    methods, **9**-99
    nozzles, **9**-99
  fuel lines, **9**-99
  fuel sprays, **9**-99
  fuels (see Diesel fuels)
  governing, **9**-104
  intake manifolds, **9**-101
  locomotive (see Locomotives, diesel-electric)
  lubrication, **6**-204
  marine, **9**-86
    data on (table), **9**-88, **9**-90
  permissible compression ratios, **9**-95
  pollution from, **18**-10
  pump injection, **9**-99
  stationary, **9**-86
    data on (table), **9**-88, **9**-89
  supercharging, **9**-102, **9**-103
  tractor, **9**-85
  trucks and buses, **9**-83, **9**-85
Diesel fuels, **7**-18, **7**-19, **9**-97
  additives, **7**-19
  distillation, curves, **7**-15
  grades, **7**-18
  specifications for (table), **7**-19
Differences, first, second, third, etc., **2**-17, **2**-43
Differential calculus, **2**-42 to **2**-44
Differential chain block, **8**-8
Differential equations:
  Laplace transforms, **2**-53
  solutions of, **2**-49 to **2**-53
Differentials, **2**-42, **2**-43
  in automobiles, **11**-8
  in calculus, **2**-42, **2**-43
Differentiation:
  of functions of complex variable, **2**-24
  rules for, **2**-42
  of vectors, **2**-63
Dihedral (def), **11**-72
Dimensional analysis, **3**-49 to **3**-52
  theorems, **3**-50
Dimensions:
  of common quantities (table), **3**-51
  of a quantity (def), **3**-50
Dings magnetic separators, **10**-59
Diode (electronics), **15**-84
  free-wheeling, **15**-88
  gas, **15**-86
  vacuum, **15**-86
Diphenyl, thermal properties of (table), **4**-26
Direct-current circuits, wiring calculations for, **15**-64, **15**-65
  (See also Electric motors, direct-current)
Direct-current generators (see Generators, direct-current)
Direct-current instruments, **15**-24
Direct energy conversion, **9**-149 to **9**-152
  fuel cells, **9**-151
    notable surface of action in, **9**-151
  magnetohydrodynamic generation, **9**-152
  photovoltaic generation, **9**-152
  thermionic generation, **9**-150
    Fermi level, **9**-150
    work-function barrier, **9**-150
  thermoelectric generation, **9**-150
    figure of merit for, **9**-150
Directrix:
  of catenary, **2**-36
  of ellipse, **2**-32

Directrix (*Cont.*):
  of hyperbola, **2**-34
  of parabola, **2**-31
Discharge coefficients for flow of liquids, **3**-61
    to **3**-64
Disk brakes, **8**-43, **11**-16
Disk clutches, **8**-38, **11**-5
Disk wheels, steam turbine, **5**-55, **5**-56
Disks:
  air resistance (table), **11**-68
  nutating, liquid meter, **16**-8
  rotating, **5**-55, **5**-56
    with central hole, **5**-55
    with noncentral hole, **5**-56
  vibration, **5**-75
Dispersion (optical), **19**-40
Displacement (motions), **3**-11
Displacement of ships (def), **11**-36
Displays (radar), **15**-95
Dissociation:
  in combustion of gases, **4**-54
  of gaseous fuels and explosion temperatures
    (table), **4**-57
Distance, measurement of, **16**-4 to **16**-7
    analytical geometry formulas, **2**-30
    surveying, **16**-42, **16**-47, **16**-49
Distillation of crude oils (chart), **7**-15
Distribution systems, electric (*see* Electric
    power distribution)
Distributors (electric ignition), **15**-81, **15**-82
Disturbance (automatic control, def), **16**-25
Divergence of nozzles for steam flow, **4**-49
Division:
  algebraic, **2**-16
  arithmetical, **2**-2
  of complex quantities, **2**-23
  by logarithms, **2**-4
  by slide rule, **2**-5
Dodecahedron, **2**-8, **2**-14
Dolomite, composition of, **6**-171
Domestic refrigerating machines, **19**-10
Dowtherm, **4**-26
  thermal properties of (table), **4**-33
Dozers, **10**-27
Draft, **9**-26
  forced, **9**-26
  induced, **9**-26
  natural, **9**-26
  of ships (def), **11**-37
  stack effect (table), **9**-27
Draft loss, **9**-26
Draft tubes (hydraulic turbines), **9**-143
Drafting:
  geometrical constructions in, **2**-9
  standard cross sections in, **8**-97
Drag (aerodynamics):
  of floats, **11**-71
  of fuselage, **11**-70
  hydraulics, **3**-52
  induced, **11**-61
  interference, **11**-71
  of nacelles, **11**-70
  parasite, **11**-66
  profile, **11**-61
  supersonic (*see* Supersonic and hypersonic
    aerodynamics)
  of tail surfaces, **11**-70
  of various bodies (table), **11**-68
  of wings, **11**-61, **11**-69
Drag coefficients (table), **11**-68
Drag forces, **11**-66 to **11**-71
Drag-link mechanism, **8**-3

Dragging of materials, **10**-2
Dragline buckets:
  scraper, **10**-13
  self-filling, **10**-13
Dragline excavators, **10**-23, **10**-37, **10**-39
Drain pipe (table), **8**-178
Drainage fittings, screwed (table), **8**-189
Dram or drachm:
  apothecaries' liquid measure (def), **1**-32
  apothecaries' weight (def), **1**-32
  avoirdupois weight (def), **1**-32
Drawbar horsepower, **11**-24
Drawbar pull (locomotives, tables), **11**-20, **11**-
    22
Drawing of metals, **13**-19 to **13**-21
  pressure in, **13**-19
  various shapes, **13**-20
  work done in, **13**-20
Dredges, **10**-38, **10**-39
  diesel, **10**-39
  digging ladders for, **10**-38
  elevator, **10**-38
  hydraulic, **10**-38, **10**-39
  placer, **10**-38
    capacities of (table), **10**-38
Driers for paints, **6**-118
  (*See also* Evaporators)
Drift bolts, pulling resistance of, **12**-28
Drilling:
  sonic, **12**-142
  various metals (table), **13**-61
Drilling machines, **13**-60
  sizes of, **13**-60
Drills:
  for pipe taps (table), **8**-20
  tap, sizes of (table), **8**-32
  twist, **13**-60, **13**-61
Drop-forge dies, **13**-21, **13**-22
Drop forging, **13**-21
Drop hammers, **13**-21, **13**-24, **13**-25
Drop presses, **13**-24
Drop-weight test, **5**-7
Dropping point (greases), **6**-202
Drums:
  computer, **2**-67
  hoisting: rope, **10**-9, **10**-10
    wire rope, **8**-85
Dry batteries (*see* Batteries, dry)
Dry-bulb temperature, **4**-28
Dry cells, **15**-11
Dry ice (solid $CO_2$), **19**-22
Dry measure, **1**-31
Dual (linear programming), **17**-18
Ducts:
  air friction in (tables), **12**-99, **12**-100
  air velocities in (table), **12**-98
  design methods for, **12**-83, **12**-98
  pressure loss in, **12**-83
  return-inlet air velocities (table), **12**-97
Dulong and Petit's rule (specific heat), **4**-10
Dulong's formula (calorific value of coal), **7**-7
Dumbwaiters, **10**-20
Dumet wire, **6**-41
Dump cars, **10**-23
Dumpy level, **16**-43, **16**-44
Duralumin, **6**-78
Duriron (corrosion-resisting iron alloy), **6**-49,
    **6**-109
Dust (def), **7**-26, **18**-8, **18**-11
  explosive characteristics of (table), **7**-28 to **7**-31
  removal of: by electrostatic collectors, **9**-10
    by mechanical separators, **9**-10

Dust (*Cont.*):
  screen scale (table), **18**-12
  size of particles, **18**-12
  velocity of settling (chart), **18**-9
Dust collectors, **9**-10
Dust explosions, **7**-25 to **7**-33
  building design to prevent, **7**-32
  characteristics of (table), **7**-28 to **7**-31
  composition of atmosphere and dust, **7**-27, **7**-31
  concentration of dust, **7**-27
  definition of, **7**-26
  effects of inerts, **7**-27
  explosibility index of (def), **7**-31
  factors affecting, **7**-27, **7**-31
  fineness of dust, **7**-27
  ignition source, **7**-27, **7**-32
  ignition temperatures for, **7**-31
  inerted atmosphere to prevent, **7**-32
  maximum pressure in, **7**-31
  minimum concentrations for, **7**-31
  minimum energy to ignite, **7**-31
  oxygen concentrations for, **7**-31
  pressure attained by (chart), **7**-33
  pressure-rise rates in, **7**-31
  prevention of, **7**-32, **7**-33
  relative hazards, **7**-31
  relief venting for, **7**-32
  turbulence in, **7**-27
  of various materials (table), **7**-28 to **7**-31
Duty of pumps, **14**-8
Dynamic braking, **11**-23
Dynamic electricity (def), **15**-18
Dynamic unbalance (def), **5**-71
Dynamics, **3**-16 to **3**-23
  of fluids, **3**-41 to **3**-46
Dynamite, **7**-24
Dynamometers, **16**-18
Dyne (def), **1**-40

*E* and *e*, value of, **1**-19, **2**-16
E transformer, **16**-6
Earth, packed and loose, specific gravity and density of (table), **6**-8
Earthmoving equipment, **10**-26 to **10**-28
Earthquake forces in buildings, **12**-11
Eccentric angle in ellipse, **2**-38
Eccentric loads:
  on circular rings (table), **5**-44, **5**-45
  on cylinders, **5**-43
  on short blocks, **5**-43
  on various cross sections, **5**-42 to **5**-44
Eccentricity:
  of ellipse, **2**-32
  of hyperbola, **2**-35
Ecology (*see* Environmental control)
Economic lot size, **17**-8
Economics, industrial, **17**-2 to **17**-10
Economizers, **9**-21, **9**-22
  corrosion of, **9**-22
  size of, **9**-22
  types of, **9**-22
Ecosystems, **18**-2
Eddy-current brakes, **8**-44, **16**-18
Eddy-current losses, **15**-17
Eddy-current testing of materials, **5**-83
Edison storage cell, **15**-12, **15**-14
Effective temperatures (wind chill, table), **12**-113
Efficiency:
  Carnot cycle, **4**-14
  of machine elements (table), **3**-29

Efflorescence on brickwork, **12**-23
Ejectors:
  air (*see* Air ejectors)
  high-vacuum, **14**-44
    table, **14**-46
Elastic constants of metals (table), **5**-5
Elastic limit (def), **5**-2, **5**-18
  in flexure, **5**-27
  proportional (def), **5**-2
Elastic solid (def), **3**-33
Elasticity (def), **5**-18
  modulus of (def), **5**-18
    for metals (table), **5**-5, **6**-60, **6**-61
  theory of, **5**-47 to **5**-49
Elastomers (rubberlike substances), **6**-165 to **6**-167, **13**-11
  comparative properties of (table), **6**-167
  properties of, at low temperatures, **19**-31
Electric apparatus:
  efficiency calculations, **15**-51
  rating, **15**-50
  symbols for, **15**-9
  temperature limits, **15**-51
  (*See also* Electric instruments)
Electric-arc welding (*see* Welding)
Electric brakes, **8**-44
Electric circuits (*see* Circuits)
Electric conductivity, **15**-4
Electric current (*see* Current)
Electric-discharge machining of metals, **13**-71
Electric energy:
  measurement of, **15**-26, **15**-27
  units of (def), **15**-4
Electric equipment, safety devices for, **18**-24
Electric-furnace steel (def), **6**-12, **6**-15
Electric furnaces, **7**-60 to **7**-69
  arc, **7**-64 to **7**-66
    arcs in, **7**-65
    charges, **7**-65
    electrode consumption, **7**-68
    rating and sizes of (table), **7**-66
    reactance, **7**-66
    refractories for, **7**-65
    regulation characteristics of (chart), **7**-66
    single-phase, **7**-66
    submerged-arc, **7**-69
    temperature in, **7**-65
    three-phase, **7**-65
  atmosphere: artificial, **7**-61, **7**-62
    natural, **7**-61
  for cast iron, **6**-49
  dielectric heating, **7**-64
  energy consumption of (table), **7**-68
  high-temperature, resistors for, **7**-62
  induction, **7**-60, **7**-66 to **7**-68
    core-type, **7**-67
    coreless, **7**-67
    frequency for, **7**-67
    operation, **7**-66
    sizes of (table), **7**-67
  induction heating for, **7**-64
  for refining steel, **6**-15
  resistance, **7**-60, **7**-69
    with fixed or movable electrodes, **7**-69
    uses of, **7**-69
  resistor, **7**-60 to **7**-64
    bath heating, **7**-62
    heating chamber, **7**-60
    high-temperature, **7**-62
    losses from, **7**-63
    operating efficiency, **7**-63
    ovens, **7**-63

Electric furnaces, resistor (*Cont.*):
  resistors for, 7-62, **15**-69, **15**-73
  sizes of, 7-63
  temperature regulation, 7-63
  for tempering, 7-63
  ventilation of, 7-64
  types of, 7-60
Electric generators (*see* Generators)
Electric ignition systems (*see* Ignition systems)
Electric instruments, **15**-24 to **15**-30, **16**-18 to **16**-20
  alternating-current, **15**-25
  ammeter, **16**-19
  direct-current, **15**-24
  galvanometers, **16**-19
  high-voltage, **15**-28
  impedance bridge, **16**-20
  ohmmeter, **16**-19
  oscilloscope, **16**-20
  potentiometer, **15**-29, **15**-30, **16**-13, **16**-20
  transformers, **15**-27
  voltmeter, **16**-19
  Wheatstone bridge, **15**-29, **16**-20
Electric insulating materials, **6**-151 to **6**-153
Electric lamps, **12**-118 to **12**-120
Electric locomotives, **10**-21, **10**-22
  (*See also* Locomotives)
Electric measurements, **15**-24 to **15**-30
Electric meters, **15**-24 to **15**-30
Electric motors:
  adjustable bases for, **8**-58
  alternating-current, **15**-43 to **15**-49
    branch-circuit maximum ratings (table), **15**-76
    capacitor, **15**-46
    commutator, **15**-47
    heat gain from (table), **12**-97
    induction, **15**-43 to **15**-46
      autotransformer, **15**-44
      breakdown torque, **15**-43
       concatenation, **15**-45
      efficiency of (table), **15**-46
      polyphase, **15**-43
      power and weight (table), **15**-46
      power factor (table), **15**-46
      resistor starters, **15**-45
      rotors, **15**-43
      single-phase, **15**-46
       starting, **15**-46
      slip-ring, **15**-45
      speed control, **15**-45
      squirrel-cage, **15**-43 to **15**-45
       classes of, **15**-43
      starting compensator, **15**-44
      switchboard equipment, **15**-54
      wiring calculations, **15**-65
    industrial applications, **15**-51, **15**-53
    synchronous, **15**-48, **15**-49
      performance of (table), **15**-48
      power and weight of (table), **15**-48
      power factor of, **15**-47
    synchronous-induction, **15**-48
  balancer set for three-wire system, **15**-62
  direct-current, **15**-32 to **15**-36
    armature reaction, **15**-34
    commutating pole, **15**-33, **15**-34
    commutation, **15**-34
    compound, performance of (table), **15**-34
    cumulative compound, **15**-34
    full-load currents (table), **15**-75
    fundamental equations, **15**-32
    Ilgner speed control, **15**-35

Electric motors, direct-current (*Cont.*):
    industrial applications, **15**-51 to **15**-53
    series, **15**-33
      speed control, **15**-36, **15**-93
    shunt, **15**-33
      armature resistance control, **15**-35
      control by changing impressed voltage, **15**-35
      field current control, **15**-35
      speed control, **15**-35, **15**-93
      stabilizing windings for, **15**-33
      starters, **15**-33
    speed and torque characteristics (chart), **15**-33
    speed control, **15**-35, **15**-36
      regulation of (def), **15**-33
      traction, **15**-36
      Ward-Leonard, **15**-35
  industrial applications, **15**-51, **15**-52
  oils, **6**-205
  selection of, **15**-52
  universal, **15**-47
Electric power (def), **15**-4
  active (def), **15**-4
  apparent (def), **15**-4
  cost of, **17**-33 to **17**-43
  formulas, **15**-10
  measurement of, **15**-26, **15**-27
  reactive (def), **15**-4
Electric power distribution, **15**-61 to **15**-64
  circuits, **15**-61 to **15**-64
  demand factor (buildings), **15**-69
  feeders and mains, **15**-63
  load centers, **15**-69
  networks, a-c, **15**-64
  service wires, **15**-68
  systems, **15**-61
  three-wire: a-c, **15**-62
    d-c, **15**-61
  wire resistance and reactance of (table), **15**-63
  wiring calculations, **15**-64 to **15**-66
Electric power plants (*see* Power plants)
Electric power transmissions, **15**-55 to **15**-61
  corona in, **15**-59
  d-c, high-voltage, **15**-58
  lines: Mershon diagram, **15**-56, **15**-58
    overhead, reactance of (table), **15**-57
    voltage drop, **15**-56
  symmetrical system, **15**-56
  underground cables, **15**-60, **15**-61
    ampere rating (table), **15**-60
Electric resistivity of metals (table), **6**-60, **6**-61
Electric resistor materials, **15**-69, **15**-73
  table, **15**-78
Electric sheet steel, **6**-38
Electric steel, **6**-15
Electric switchboards, **15**-53 to **15**-55
Electric switches, **15**-54
Electric traction, **11**-19 to **11**-21
Electric vehicles, storage batteries for, **15**-13
Electric waves:
  average value, **15**-21
  form factor, **15**-21
Electric welding (*see* Welding)
Electric wiring (*see* Wiring)
Electrical engineering, **15**-1 to **15**-97
Electrical symbols (table), **15**-3, **15**-9
Electrical units, **15**-2 to **15**-5
  table, **15**-3
Electricity:
  dynamic (def), **15**-18
  static (def), **15**-18

Electrochemical machining, **13**-71
Electrodes, coated, for arc welding, **13**-29 to **13**-33
Electrodynamometer, **15**-25
Electrohydraulic forming, **13**-20, **13**-21
Electrolux Servel process of refrigeration, **19**-11, **19**-13
Electrolysis, corrosion due to, **6**-108
Electrolytic iron:
    composition of (table), **6**-13
    properties and uses of, **6**-12
Electromagnetic system of units, **15**-2
Electromagnets, **15**-73 to **15**-80
    alternating-current, **15**-79
    exciting coil, **15**-80
    heating of, **15**-79
    lifting, **15**-78
    polyphase, **15**-79
    sparking of, **15**-79
    time constant in, **15**-20
    tractive, **15**-79, **15**-80
    wire for, **15**-79, **15**-81
Electromotive force:
    induced, direction of, **15**-17
    series of metals, **6**-107
    unit of, **15**-3
Electron (def), **6**-7
Electron capture (def), **9**-121
Electron tubes, **15**-84 to **15**-88
Electronic components, **15**-84 to **15**-87
Electronics, industrial, **15**-93, **15**-94
Electrostatic collectors (fly ash), **9**-10
Electrostatic system of units, **15**-2
Electrotype metal (table), **6**-89
Elements:
    chemical (def, table), **6**-3, **6**-4
    periodic table of, **6**-6
    physical properties of (table), **6**-60, **6**-61
Elevator dredges, **10**-38
Elevators, **10**-19 to **10**-21
    automatic control of, **10**-20
    belt-and-bucket, **10**-54
    bucket, **10**-53 to **10**-55
        capacity of (table), **10**-54
        casings, **10**-54
        continuous, **10**-54
        gravity discharge, **10**-54
        table, **10**-55
        supercapacity of, **10**-54
    car mileage of, **10**-20
    controls for, **10**-20
    counterweights for, **10**-20
    drives for, **10**-19
    gearless, **10**-19
    hydraulic, **10**-19
    loads for, **10**-20
    motors for, **10**-19, **15**-51
    power consumption and efficiency, **10**-20
    push-button, **10**-20
    traction-type, **10**-19
Elinvar (steel alloy of low expansion), **6**-41
Ellipse:
    area and perimeter of, **2**-12
    constructions of, **2**-33, **2**-34
    properties of, **2**-32
    quadrant, center of gravity, **3**-8
Ellipsoid, volume of, **2**-14
Elliptic integrals, **2**-49
Elongation due to tension (table), **5**-3
Embossing of metals, **13**-22
Emery (abrasive), **6**-139
Emery wheels (see Grinding wheels)

Emissivity (radiant-heat, def), **4**-71
    of air spaces (table), **12**-87
    of carbon dioxide (chart), **4**-79
    gas, **4**-77
    of various surfaces (table), **4**-73, **12**-87
    of water vapor (chart), **4**-79
Emittance (def), **4**-71
Empirical curves:
    equations for, **2**-54, **2**-55
    plotting, **2**-54
Employee relations, **17**-9
    grievance procedures, **17**-9
    promotions, **17**-10
Emulsions, cutting fluid (machining), **13**-54
Enamel coatings, **6**-118
Energy (def), **3**-2, **3**-20
    atomic (see Atomic energy)
    conservation of, **3**-3, **3**-20, **3**-44
        thermodynamics, **4**-13
    conversion: direct (see Direct energy conversion)
        to mass (formula), **9**-120
    conversion table for, **1**-52
    electric (see Electric energy)
    equivalents (table), **1**-52
    free (def), **4**-15
        Helmholtz (def), **4**-15
    internal: of perfect gas, **4**-18
        table, **4**-56
    kinetic (def), **3**-20
    law of conservation of, **3**-3, **3**-20
    mass equivalence of (formula), **9**-120
    of moving fluid, **3**-43
    nuclear (see Atomic energy)
    potential (def), **3**-20
    solar (see Solar energy)
Energy requirements (table), **9**-6
    from coal, **9**-6
    from natural gas, **9**-6
    from nuclear sources, **9**-6
    from petroleum, **9**-6
Energy sources, **9**-3 to **9**-6
    geothermal, **9**-5
        (See also Geothermal power)
    hydroelectric, **9**-5
    hydrogen, **9**-6
    irreplaceable: heat of the seas, **9**-169
        reserves, **9**-3, **9**-4
            coal, **9**-4
            natural gas, **9**-3
            nuclear, **9**-4, **9**-5
            petroleum, **9**-3
            shale oil, **9**-4
            tar sand, **9**-4
    replaceable: solar, **9**-5
        (See also Solar energy)
        tidal, **9**-5
        (See also Tidal energy)
        vegetation, **9**-163
        waves, **9**-168, **9**-169
        wind, **9**-5, **9**-166 to **9**-168
    thorium, **9**-5
    uranium, **9**-4
    wood, **9**-5
Engineering statistics, **17**-19 to **17**-26
Engines:
    airplane (see Airplanes, engines)
    automobile (see Automobile engines)
    balancing of, **5**-72
    cranks, **8**-69 to **8**-71
    cycles: internal-combustion, **4**-20
        steam, **4**-27, **4**-28

Engines (*Cont.*):
  diesel (*see* Diesel engines)
  hot-air, **9**-163
  inertia factors (table), **8**-73
  internal-combustion (*see* Internal-combustion engines)
  jet (*see* Aircraft jet propulsion)
  marine (*see* Marine engines)
  piston accelerations, **8**-73
  piston positions for various crank angles (table), **8**-70, **8**-71
  reciprocating, balancing of, **5**-72
  reciprocating parts: inertia force of, **8**-70
    motion of, analysis of, **8**-69
    velocity and acceleration of, **8**-69, **8**-70
  steam (*see* Steam engines)
  tangential pressure factors (table), **8**-72
Enthalpy (def), **4**-12
  of gases, **4**-53
  of perfect gas, **4**-18
Enthalpy-entropy chart:
  for air, **14**-37
  for steam, **4**-25, **4**-26
Enthalpy-pressure chart (*see* Pressure-enthalpy chart)
Entropy (def), **4**-14
  of ideal gas, **4**-18
Environmental control:
  air- and gas-cleaning devices for (table), **18**-20
  air-cleaning apparatus for, **18**-12 to **18**-14
  air pollution, sources of, **18**-8 to **18**-11
    table, **18**-11
  condenser water cooling, **18**-3
  construction project considerations for, **18**-6
  cost/benefit calculations, **18**-3
  cyclone separators for, **18**-12
  federal legislation on: air pollution, **18**-8
    water pollution, **11**-43, **18**-5
  industrial discharge permits, **18**-6
  national policy, **18**-2
  process selection for, **18**-7
  of radioactive waste, **18**-18 to **18**-21
  recycling, **18**-7
  scrubbers for, **18**-13, **18**-14
  size of airborne particles (chart), **18**-9
  standards for industrial discharges (table), **18**-6
  of sulfur gases and particulates, **18**-14 to **18**-77
  sulfur oxide removal for, **18**-14, **18**-15
  thermal discharges, **18**-4
  in various industries, **18**-17, **18**-18
  waste disposal, **18**-7
  wastewater, **18**-5 to **18**-8
Epicyclic gear trains, **8**-7, **8**-8
Epicycloid, **2**-39
Epitrochoid, **2**-40
Epoxy plastic (table), **6**-164
Equations:
  algebraic, **2**-18 to **2**-22, **2**-54 to **2**-58
  Cauchy-Euler equidimensional, **2**-51
  circle, **2**-31
  cubic, **2**-19
  differential, **2**-49, **2**-50
    linear: constant coefficients, **2**-51
      undetermined coefficients, **2**-51
    methods of solving, **2**-50
  ellipse, **2**-33
  for empirical curves, **2**-54
  exponential, **2**-19, **2**-54
  legitimate operations on, **2**-18

Equations (*Cont.*):
  linear, **2**-18, **2**-54
  parabola, **2**-31
  quadratic, **2**-18
  radical axis, **2**-31
  simultaneous, **2**-20 to **2**-22
  of state (thermodynamics), **4**-17
  straight-line, **2**-30
  tangent to circle, **2**-31
  tangent to parabola, **2**-31
  trial-and-error solution, **2**-19
  trigonometric, **2**-19, **2**-54
  types of, **2**-18
Equilibrium (def), **3**-4
  forces in (mechanics), **3**-5, **3**-6
Equilibrium diagram, iron–iron carbide (chart), **6**-17
Erg (def), **1**-36, **15**-4
Error:
  absolute and relative, **2**-2
  compensation (automatic control), **16**-27, **16**-28
  mean square, **2**-21
  in measurement, **16**-2
  probable, **2**-21, **2**-22
Escalators, **10**-21
Esters (solvents), **6**-168
Ethane:
  as refrigerant, **19**-5
  vapor pressure (chart), **4**-28
Ethyl alcohol:
  as fuel, **7**-20
  as solvent, **6**-168
Ethyl chloride, thermal properties of, **4**-38
Ethylene, vapor pressure (chart), **4**-28
Ethylene glycol (antifreeze), properties of, **6**-156
Euler number, **3**-48
Euler's equations of motion, **3**-22
Euler's formula for long columns, **5**-41
Eutectoid (steel), **6**-18
Evaporation temperature of common liquids (table), **4**-12
Evaporative condensers, **4**-45, **19**-8
Evaporative cooling of condensing water, **9**-71 to **9**-74
Evaporators, **19**-8
  heat transmission in, **4**-68
    maximum (table), **4**-70
  vapor binding in, **4**-69
Everdur, **15**-7
Evolute of a curve (def), **2**-39, **2**-46
Excavating machines, **10**-37 to **10**-40
Excavation, volume computation in, **16**-48
Excavators, dragline, **10**-39, **10**-40
Excess air for combustion (charts), **4**-58
  coefficient of, **4**-51
Exchangers, heat (*see* Heat exchangers)
Exciter, switchboard equipment for, **15**-54
Exhaust gas:
  analysis of (table), **9**-108
  apparatus for, **16**-22
Exide ironclad battery, **15**-13
Expansion:
  of bodies by heat, **4**-3
  coefficients of (table), **4**-7, **4**-8, **6**-60, **6**-61
    of petroleum products, **7**-15
  in compound engines, **9**-37
  of functions in series, **2**-44, **2**-45
  of gases (formulas and tables), **4**-19 to **4**-22
  of pipelines, stresses due to, **5**-61 to **5**-67
  ratio of, effect on engine economy, **9**-36

Expansion (*Cont.*):
   of saturated and superheated vapors
     (formulas), **4**-22 to **4**-25
   tests of, for coal, **7**-38
   thermal: coefficients of (def), **4**-3
     of liquids (tables), **4**-8
     of metals (table), **4**-7
     of various materials (table), **4**-8
Expansion joints for steam pipes (table), **8**-194
Explosions:
   detonation rates of (tables), **7**-26
   of dusts (*see* Dust explosions)
Explosive D, **7**-25
   table, **7**-26
Explosive forming, **13**-20
Explosives, **7**-24, **7**-25
   ballistic mortar strength of (table), **7**-26
   dynamite, **7**-24
   military (table), **7**-25
   permissible in mines, **7**-25
Exponential equations, solution of, **2**-19
Exponential functions, **2**-24
   of *e* (table), **1**-19
   graph of, **2**-55
   series for, **2**-44
Exponential values of *e* and −*e* (table), **1**-19
Exponents (algebra), **2**-16
Extraction turbines, **9**-47 to **9**-56
Extrusion of metals, **13**-22
Eyebolts, proportions and strength of (table), **8**-22, **8**-23

Fabrics, **6**-154, **6**-156
   impregnated (insulation), **6**-153
Face brick, **6**-149
Facing by welding, **13**-42
Factories (*see* Industrial plants)
Factoring (algebra), **2**-16
Factory accounts (*see* Cost accounting)
Fahrenheit:
   conversion to Celsius (eq), **4**-2
     table, **4**-3, **4**-4
   conversion to Rankine (eq), **4**-3
Failure, theories of, **5**-53, **5**-54
Falk flexible couplings, **8**-36
Fans:
   axial-flow, **14**-49, **14**-50
   centrifugal (*see* Centrifugal fans)
   classification of, **14**-49, **14**-50
   forced-draft, **9**-26
   induced-draft, **9**-26
   laws of, **14**-51
   propeller, **14**-49, **14**-50
   (*See also* Blowers)
Farad (capacitance, def), **15**-4
Fast flexible couplings, **8**-35
Fathom (def), **1**-31
Fatigue:
   endurance limit in (table), **5**-9
   of metals, **5**-9, **5**-10
     effects of corrosion, **5**-10, **6**-111
     effects of notches, **5**-10
     effects of overstressing and
      understressing, **5**-10
Fatigue failure, **5**-9, **5**-10
   of shafts, **8**-48
     notch sensitivity, **8**-48
Faure storage battery, **15**-12
Feedback loop (automatic control), **16**-30
Feeder panels, equipment for, **15**-54

Feeders:
   belt conveyor, **10**-60, **10**-61
   of electric distribution systems, **15**-63
Feedwater, **9**-30, **9**-31
   alkalinity, **9**-31, **9**-32
   amines for, **9**-31
   blowdown, **9**-29, **9**-33
   corrosion from impurities, **6**-111, **6**-112
   deaeration, **6**-110, **9**-77
   evaporation, **9**-30
   filtering, **9**-30
   heaters, **9**-74 to **9**-78
     deaerators, **9**-77
      spray-type, **9**-78
      tray-type, **9**-78
     desuperheating section, **9**-77
     drain-cooling section, **9**-77
     friction loss (table), **9**-77
     heat-transfer rates, **9**-75
     high-pressure, **9**-75
     length of, **9**-74
     low-pressure, **9**-75
     materials for, **9**-77, **9**-78
     open, **9**-78
     performance calculations, **9**-76, **9**-77
     pressure drop in, **9**-74
     regenerative cycle for, **4**-28
     tube-wall thickness (table), **9**-77
     types of, **9**-74
     vent condensers for, **9**-78
   heating, regenerative (turbines), **9**-56
   impurities in, **9**-29
     effect in boiler, **9**-31, **9**-32
     limits, ABAI (table), **9**-33
   make-up, **9**-29
   oil and grease in, **9**-29
   oxygen removal, **9**-30
   pH, **9**-31, **9**-32
     hydrogen-ion measurement, **9**-31
     indicators for, **9**-32
     table, **9**-32
   sampling and analysis, **9**-32
   scale from, **9**-29
   settling of, **9**-30
   silica in, **9**-32
   sludge, **9**-29
   softening, **9**-30
   treatment of, **9**-29 to **9**-33
     amines, use of, **9**-31
     cation exchange, **9**-30
     chemical, **9**-30
     demineralization, **9**-30
     hardness elimination, **9**-32
     lime-soda process, **9**-30
     of raw water, **9**-29, **9**-30
     solids removal, **9**-30, **9**-32
     Zeolite process, **9**-30
Felts, **6**-154
   roofing, **6**-164
Ferrite (def), **6**-17
Ferritic hardenable alloys, **6**-38
Ferroalloy furnace (electric), **7**-69
Fiberglas, electric insulation, **15**-80
Fibers, **6**-153 to **6**-156
   animal, **6**-154
   for cordage, **6**-151
   creep of, **6**-154
   denier (unit of fineness), **6**-156
   elastic properties of (table), **6**-155
   glass, **6**-153
     table, **6**-155
   heat endurance, **6**-154

Fibers (*Cont.*):
  identification of, **6**-154
  inorganic, **6**-153
  nylon, **6**-154
    table, **6**-155
  properties of (table), **6**-155
  synthetic, **6**-154
  uses of, **6**-154, **6**-155
  vegetable, **6**-154
Field of force, **3**-22
Field intensity (magnetic, def), **15**-2, **15**-15
Film coefficients, **4**-63 to **4**-70
  for boiling liquids, **4**-68
  combined convection and radiation, **4**-69
  condensing vapors, **4**-68
  extended surfaces, **4**-67
  factors influencing, **4**-63
  forced-circulation evaporators, **4**-69
  for gases, **4**-64
  heat transmission (def), **4**-61
  for high-velocity gases, **4**-65
  laminar flow, **4**-67
  natural convection, **4**-67, **4**-68
  over tubes, **4**-65
  turbulent flow, **4**-64
Film speed, photographic, **19**-40
Filmogen (paint, def), **6**-118
Filters, capacitor, **15**-88
Filtration of sound, **12**-139
Financial arithmetic, **2**-7
Fineness modulus (concrete, def), **6**-180
Fire extinguishing, hand apparatus for, **18**-33
Fire hose, **18**-32
  couplings for: American National Standard
      (table), **8**-200
    screw threads, **8**-199
Fire point of oil (def), **6**-198
Fire protection, **18**-23 to **18**-33
  building construction for, **18**-28, **18**-29
  hose (*see* Fire hose)
  hydrants for, **18**-32
  information sources on, **18**-28
  by isolation, **18**-29
  loss limitation in, **18**-28
  pumps for, **18**-31
  resistant materials for, **18**-29
  sprinkler equipment for, **18**-29 to **18**-31
  standpipes for, **18**-32
  with temporary heating, **18**-29
  with temporary storage, **18**-29
  underground piping for, **18**-32
  water supply for, **18**-31
Fire pumps, **18**-31
Fire resistance of building materials, **12**-42,
    **18**-29
Firebrick, **6**-171
  for boiler furnaces, **9**-16
  coatings for, **6**-174
  heat losses and heat-storage capacity of
      (table), **6**-175
  manufacture of, **6**-171
  mortars for, **6**-173, **6**-174
  plastics and ramming mixtures for, **6**-174
  selection of, **6**-175, **6**-177
  shapes and sizes of, **6**-172, **6**-173
  (*See also* Refractories)
Fireclay refractories, **6**-171, **6**-172
  table, **6**-172
Fission of nuclei (*see* Nuclear fission)
Fits, **8**-45 to **8**-47
  allowances and tolerances, **8**-45
  classification of, **8**-45, **8**-46

Fits (*Cont.*):
  force, **8**-46
  locational clearance, **8**-45
  locational interference, **8**-46
  press, **8**-47
    pressures required in making, **8**-47
    stresses due to, **8**-47
    torsional holding of, **8**-47
  running, allowances for (table), **8**-45
  shrink, allowances for (table), **8**-46
    stresses due to, **8**-47
  transition, **8**-46
Fittings, pipe (*see* Pipe fittings)
Flame gouging, **13**-37
Flame hardening (heat-treatment), **6**-24
Flame propagation in gaseous fuels (chart), **7**-23
Flame radiation, **4**-77
Flame speed in internal-combustion engines,
    **9**-93
Flame temperatures of gaseous fuels (table), **4**-57
Flame travel in internal-combustion engines,
    **9**-108
Flammability of fuels:
  gaseous, **7**-23
    tables, **7**-23
  limits of (table), **7**-23
    LeChâtelier's equation for, **7**-23
Flanged fittings:
  cast-iron, **8**-174 to **8**-176
  steel, **8**-177 to **8**-186
Flanges:
  cast-iron, **8**-174 to **8**-177
  couplings, **8**-35
  steel, **8**-179
  unions, **8**-186
Flash point of lubricating oils (def), **6**-198
Flash welding, electric resistance, **13**-40
Flats of bolts and nuts (table), **8**-21
Flax, properties of (tables), **6**-155, **6**-156
Flexible metal hose and tubing, **8**-199, **8**-200
Flexure, theory of (beams), **5**-23 to **5**-27
Flight (*see* Aerodynamics)
Flight conveyors, **10**-44 to **10**-48
Flip-flop (electronic circuit), **15**-91
  *JK* circuit, **15**-92
Floors:
  industrial plant, **12**-8
  loads on, **12**-11
  plank, safe load on, and deflection (table),
      **12**-24
  reinforced-concrete, **12**-55 to **12**-58
  safety provisions for, **18**-23
  trusses, **12**-14, **12**-15
  wood, **12**-23
Flow:
  of air: measurement of, **16**-14 to **16**-18
    in pipes, **4**-49, **4**-50
    in pneumatic conveyors, **10**-62
  coefficients of friction, **4**-50
  of compressible fluids, **4**-48 to **4**-51
    through nozzles, **4**-48, **4**-49
    through orifices, **4**-46 to **4**-48, **16**-16
    in pipes, **4**-49, **4**-50
      friction loss, **4**-49
  of cryogens, measurement of, **19**-33
  of gases: fundamental equations, **4**-46
    measurement of, **16**-14 to **16**-18
  of heat, **4**-59 to **4**-82
    conversion table for, **1**-54
    (*See also* Heat transmission)

Flow (*Cont.*):
  of liquids: dimensional analysis of, **3**-49 to **3**-52
    laminar, **3**-52
    measurement of, by metering pump, **16**-17
      (*See also* Flowmeters)
    through orifices, **3**-63, **3**-64
    in pipes, **3**-53 to **3**-57
    turbulent, **3**-52
      (*See also* of water, *below*)
  of metal at high temperatures (creep), **5**-10 to **5**-12
  of steam: through labyrinth packings, **9**-46
    through nozzles, **4**-48, **4**-49
    through orifices, **4**-46, **4**-47
    in pipes, **4**-49, **4**-50
      friction loss due to, **4**-49
      resistance of fittings to, **4**-50
    saturated, **4**-47, **4**-49
  of water: measurement of (*see* Flowmeters)
    through notches, **3**-66
    through nozzles, **3**-61
    in open channels, **3**-66 to **3**-68
    in pipes, **3**-53 to **3**-60
      Chézy equation for, **3**-67
      Manning formula for, **3**-67
    venturi meter for, **3**-61
    over weirs, **3**-65, **3**-66
Flow diagrams in industrial plants, **12**-3
Flowmeters, **3**-60, **3**-64, **16**-8, **16**-14 to **16**-18
  electromagnetic, **16**-17
  integrators, **16**-18
  nozzle, **16**-15
  propeller, **16**-17
Flue gas:
  analysis of, **4**-51, **4**-57, **4**-58
    apparatus for, **16**-22
    chart, **12**-73
  cleaning of, **9**-10
  heat loss in, **9**-27
Fluid coupling (*see* Couplings)
Fluids:
  coefficients of friction (formula), **4**-50
  flow (*see* Flow)
  ideal (def), **3**-34
  for machining, **13**-54
  measurement of, **1**-31
    (*See also* Flowmeters)
  newtonian (def), **3**-34
  nonnewtonian (def), **3**-34
  throttling of, **4**-50, **4**-51
  viscosity of (chart), **3**-36, **4**-50
  volume, measurement of, **16**-7, **16**-8
Fluorescent lamps, **12**-119
Flux density (magnetic, def), **15**-2, **15**-15
Fly ash:
  sintering of, **9**-8
  utilization of, **9**-10
Flywheel effect, **3**-11
Flywheels, **8**-49 to **8**-51
  arms, design of, **8**-49, **8**-50
  for energy storage, **9**-170, **9**-171
  of flat-rolled steel plates, **8**-50
  high-speed, stresses in rims, **8**-50
  rims, **8**-50
  split rim, **8**-50
  stresses in, **8**-50
  test data (table), **8**-51
  vehicles powered by, **9**-171
  weight determination, **8**-72
  Wittenbauer's analysis, **8**-73
FMC coke process, **7**-41

Foaming of boilers, **9**-29
Focal length of lenses, **19**-40
Focus:
  of ellipse, **2**-32
  of hyperbola, **2**-34
  of parabola, **2**-31
Foot (def), **1**-31
Foot-lambert (def), **12**-117
Footings:
  for foundations, **12**-20
  reinforced-concrete, **12**-58
Force (def), **3**-2
  centrifugal, **3**-17
  centripetal, **3**-17
  components of, **3**-3, **3**-5
  composition of, **3**-5
  definition of units, **1**-40
  equilibrium of, **3**-3 to **3**-6
  equivalents, **1**-40
  external and internal (def), **3**-3
  field of, **3**-22
    intensity of (def), **3**-22
  fundamental equation of, **3**-2
  line of (def), **3**-22
  polygon, **3**-6
  resultant of, **3**-3
  supporting, **3**-6, **3**-7
  system of, in equilibrium, **3**-6
  unbalanced, **3**-16 to **3**-20
  units of, **1**-40, **4**-8
Force fits, **8**-46
Forced-circulation boilers, **9**-14, **9**-25
Forced draft, **9**-26
Forging of metals, **13**-15 to **13**-23
Forging dies, **13**-21, **13**-22
Forging hammers, **13**-21, **13**-24 to **13**-26
Forging steel (table), **6**-26
  uses of, **6**-33
Fork trucks, **10**-25, **10**-26
Form factor, a-c waves, **15**-21
Formation, heats of (def), **4**-53
Formcoke (coke making), **7**-41
Formex, magnet wire insulation, **15**-80
Forms for reinforced concrete, **12**-61, **12**-62
Formvar, magnet wire insulation, **15**-80
Forsterite (refractory), **6**-172
Fortran, **2**-70
Fossil fuels, reserves of, **9**-4
Foucault-current losses, **15**-17
Foundations, **12**-20 to **12**-23
  bearing pressure on soils, **12**-20
  caisson methods for, **12**-20
  excavation, computations in, **16**-48
  footings for, **12**-20
  pile, **12**-21
  setting stakes for, **16**-49
  spread (reinforced-concrete columns), **12**-59
Foundry practice and equipment, **13**-2 to **13**-9
  cast iron, **6**-49
  casting inspection in, **13**-8
  cleaning equipment for castings in, **13**-8
  melting processes used in, **6**-49, **6**-55, **13**-8
    (*See also* Castings; Molding)
Four-cycle engines, **9**-79
Fourier series, **2**-45
Fourier's law (heat conduction), **4**-60
Fractionating high-vacuum pumps, **14**-46
Fractions (algebra), **2**-16
  decimal values (table), **1**-30
Fracture:
  at low stress, **5**-6 to **5**-9
    under tension and compression, **5**-7 to **5**-9

Fracture mechanics, 5-7 to 5-9
Frame of reference (def), 3-2
Framed structures, steel (*see* Steel-framed structures)
Francis turbine, 9-136 to 9-143
Free energy (thermodynamics), 4-15
Free-swelling index test for coal, 7-8, 7-38
Freezing points of liquids and mixtures, 4-5, 4-6
Freezing preventives, 6-156
Freight cars, 11-28
  axles for (table), 11-29
  dimensions of (table), 11-28
  train resistance, 11-29, 11-30
  wheels for (table), 11-29
Freon, properties of (tables and chart), 19-5, 19-6
Freon 11, properties of (table), 4-39
Freon 12, properties of (table), 4-39
Freon 21, properties of (table), 4-40
Freon 22, properties of (table), 4-40
Freon lines, maximum refrigeration for (table), 19-17
Frequency:
  a-c (def), 15-3, 15-21
  natural, 15-21, 15-22
Frequency bands (table), 15-94
Frequency modulation (radio), 15-90
Frequency response of automatic controls, 16-31
  graphical display, 16-31 to 16-34
Fresh air requirements for various locations, 12-67
Friction, 3-24 to 3-32
  coefficients of (def, tables), 3-24 to 3-26, 3-28
    clutches (table), 8-39
    fluid flow in pipes, 4-50
    journal bearings, 3-30, 8-123
    rolling (table), 3-28
    static and sliding (def), 3-24
      tables, 3-25 to 3-27
    tables, 3-25, 3-26
    thrust bearings, 3-31, 8-131
    for various materials, 3-25 to 3-28
    of various materials on steel (table), 10-46
  dry (def), 3-24
  of liquids in pipes, 3-53 to 3-57
  of machine elements, 3-28 to 3-32
  rolling (def), 3-28
  skin (aerodynamics), 11-66
  of steam engines, 9-36
Friction brake, horsepower absorbed by, 3-21
Friction clutches (automobile), 11-5
Friction factors for pipes (liquid flow), 3-53 to 3-57
Friction head (liquids), 3-43
Friction loss in pipe fittings (table), 3-58
Friction resistance of ships, 11-40, 11-41
Friction sawing (metal), 13-66
Frigorie, British unit of refrigeration (def), 19-3
Froude's law (ship resistance), 11-41
Froude's number, 3-48, 3-49, 3-68, 11-37, 11-40
Frustum of cone or pyramid, volume of, 2-13
Fuel cells, 9-151, 18-10
Fuel oils, 7-14 to 7-21
  analyses of (table), 7-14
  API and specific gravity of (tables), 7-14 to 7-16, 7-19
  ash content (table), 7-19, 9-9

Fuel oils (*Cont.*):
  combustion in boiler furnaces, 9-12
  composition and properties of, 7-20
  crude, 7-14, 7-15
  heat of vaporization, 7-15
    table, 7-16
  heat values (table), 7-14, 7-16
  specific heat, 7-15
  specifications for (table), 7-19
  (*See also* Fuels, liquid)
Fuels, 4-51 to 4-58, 7-2 to 7-24, 7-47, 7-48
  aviation turbine, 7-17
    table, 7-19
  for boilers, 9-7
  briquets, 7-13
  by-product (table), 7-13
  colloidal, 7-20
  combustion of (*see* Combustion)
  consumption of (automobiles), 11-5
  cost in power plants, 17-38
  diesel (*see* Diesel fuels)
  flammability of, 7-23
    table, 7-23
  fossil, reserves of, 9-3, 9-4
  gaseous, 7-21 to 7-24
    analysis of, 7-22
    autoignition temperatures, 9-93
    combustion of, 4-51 to 4-55
      air required, 4-51
        table, 4-55, 4-57
      in boiler furnaces, 9-12
      dissociation in, 4-55
        table, 4-57
      products of (table), 4-52, 4-55
      temperature, 4-53
      volume contraction, 4-52
    composition of (table), 7-22
    flame propagation, 7-23
    flame speed, 9-93
    flammability, 7-23
      table, 7-23
    heat value (table), 4-54
    for industrial heating, 7-48
    manufacturing processes, 7-41 to 7-47
    odorization of, 7-21
    physical constants of, 7-22
    specifications for, 7-21
  heat of formation, 4-53
  heat value (def), 4-52, 4-53, 7-7
    high and low (def), 4-53
  high-octane, 7-17, 7-18
  for industrial heating, 7-47, 7-48
  for internal-combustion engines, 7-15 to 7-20, 9-92
  jet, 7-17
    table, 7-19
  knock rating, 9-96, 9-97
  knock suppressors (table), 9-95
  liquid, 7-14 to 7-21
    ash content, 9-9
    combustion of, 4-51 to 4-55
      air required (table), 4-55
      products of (table), 4-55
    heat values (tables), 4-54, 7-14, 7-16
    from hydrogenation of coal, 7-20
    for industrial heating, 7-47
    molecular weight (table), 4-51
    properties of, 7-14, 7-15
    synthetic, 7-20, 7-21
    volatility of (def), 9-92
  nuclear, 9-132
    costs of, 9-131

Fuels (*Cont.*):
  oils (*see* Fuel oils)
  peat, 7-12
  products of combustion of (table), 4-55
  rocket, 7-33 to 7-37
  solid: burning equipment for (table), 9-7
    combustion of, 4-56 to 4-58
      air required, 4-57
      products of, 4-57
    specific gravity and density (table), 6-9
  wood, 7-12 to 7-14
Full-cell process (wood preservation), 6-137
Fuller-Kinyon system, 10-62
Function generator, 2-73
Functions:
  algebraic, series for, 2-44, 2-45
  of complex variable, 2-22 to 2-25
  graphical representation of, 2-54
  hyperbolic (tables), 1-22 to 1-24, 2-29
    series for, 2-44
  implicit, 2-43
  trigonometric, 2-25 to 2-27
    series for, 2-44
    tables, 1-13 to 1-18
  of two or more variables, 2-44
Furlong (unit of length, def), 1-31
Furnaces:
  arches, 6-175
  boiler (*see* Boiler furnaces)
  combustion, 7-47 to 7-53
  cyclone, 9-13
  electric (*see* Electric furnaces)
  for electric steel, 6-15
  forced convection, 7-48
  heat transmission in, 4-80 to 4-82
  incineration (*see* Incinerators)
  induction, 6-15
  industrial heating, 7-48 to 7-53
    automatic control, 7-51
    classification of, 7-48
    construction of, 7-50, 7-51
    efficiency of (table), 7-52
    firing methods, 7-48
    flue areas and velocities (tables), 7-50
    fuel requirements (table), 7-52
    fuels, 7-47, 7-48
    heat balance (table), 7-51
    heat conservation, 7-50
    heat losses, 7-49, 7-50
    heat transfer, 7-48
    heating and soaking time, 7-49
    material handling, 7-48
    metal parts for, 7-50
    protective gas atmospheres, 7-51 to 7-53
      table, 7-52
    recuperators and regenerators, 7-50, 7-51
    size of, factors influencing, 7-48 to 7-50
    temperature control, 7-51, 7-63
    temperatures (table), 7-52
    useful heat in, 7-49
    ventilation, 7-64
  liquid-bath, 7-48
  melting (casting), 13-8
  muffle, 7-48
  open-hearth, 6-15
    for steel castings, 6-55
  oven, 7-48
  pressure, 9-26
  radiant-tube fired, 7-48
  radiation, 4-80 to 4-82
  recirculating, 7-48
  refractories for (*see* Refractories)

Furnaces (*Cont.*):
  resistors, materials for (table), 15-78
  suction, 9-26
  for surface combustion, 4-58
  walls of, 6-174, 9-16 to 9-20
    air-cooled, 6-174
    expansion joints, 6-174
    heat losses and heat storage (table), 6-175
  water-cooled, 9-16 to 9-20
    construction of, 9-20
    heat transfer, 9-20
Fuses, 15-69
  sizes for motor branch circuits (table), 15-76
Fusible alloys, 6-89 to 6-91
  table, 6-90
Fusion:
  heat of (table), 4-12
  nuclear (*see* Nuclear fusion)
Fusion point of refractories (table), 6-176
FWPCA (Federal Water Pollution Control
    Act), 11-43, 18-5

*G* and *g*, value of, 1-40, 3-2
Gage blocks, 16-6
Gages:
  absolute pressure, 16-10
  bellows, 16-9
  Bourdon-tube, 16-9
    dead-weight tester for, 16-9
  depth, 16-5
  dial, 16-5
  diaphragm, 16-9
  go no-go, 16-5
  ionization, 16-10
  pneumatic, 16-7
  pressure, 16-8 to 16-10
  railway track, 11-31
  sheet metal (table), 6-45
  for steel angles (table), 12-44
  strain, 5-58, 5-59
    in vibration, 5-76
  thermocouple, for vacuum measurement,
    16-10
  thread, 16-6
  U-tube, 16-9
  using beta radiation, 16-7
  using X-rays, 16-7
  wire (table), 6-45
Gallon (def), 1-31
Galvanized conductors (electrical), 15-6
Galvanized surfaces, paints for, 6-120
Galvanizing, 6-85, 6-114, 6-115
  electrolytic, 6-114, 6-115
Galvanometer, 15-24
Gamma function, 2-49
Gamma iron (def), 6-17
Gamma rays, 9-121, 9-129
  irradiation by, 9-129
Gang saws, 13-80
Gangue of iron ore, 6-12
Gantry cranes, 10-30
Gantt charts (scheduling), 17-6
Garnet (abrasive), 6-139
Gas:
  adiabatic expansion, 4-19
  blue, 7-42
  change of state, 4-18, 4-19
  chromatography, 7-22
  cleaning (*see* Gas cleaning)
  combustion (*see* Combustion)
  compressibility factors (chart), 4-18

Gas (*Cont.*):
  compression, power required for (chart), 14-33
  compressors (*see* Air compressors; Centrifugal compressors)
  constant-pressure expansion, 4-19
  constant-volume expansion, 4-19
  constants (table), 4-17
  critical state (def, table), 4-17
  density of (table), 4-17, 4-23
  enthalpy, 4-18, 4-53
  entropy, 4-18
  expansion, 4-18 to 4-20
    with variable specific heat, 4-19
  flow of (*see* Flow, of gases)
  flue (*see* Flue gases)
  heat-transfer coefficients (table), 4-62
  heat transmission, 4-61 to 4-82
  ideal: changes of state, 4-18 to 4-20
    equation of state, 4-17
    ideal cycles with, 4-20 to 4-22
    laws of, 4-16
  internal energy, 4-18, 4-53
    table, 4-56
  isentropic expansion, 4-19
  isothermal expansion, 4-19
  liquefaction, 19-23
  mixtures of, 4-18
    partial pressures, 4-18
    specific heat, 4-18
    total pressure (Dalton's law), 4-18
  natural (*see* Natural gas)
  perfect (*see* Gas, ideal)
  polytropic expansion of, 4-19
  producer (*see* Producer gas)
  properties of, 4-16
  radiation from, 4-77 to 4-82
  solubility in water (table), 6-7
  specific heat (table), 4-10, 4-17
  specific volume (table), 4-17
  thermal conductivity (table), 4-62
  thermal expansion of, 4-3
  viscosity (chart), 4-50
  waste, heat in, 7-49
  water (*see* Water gas)
Gas cleaning, 18-8
  apparatus (table), 18-20
  of flue gas, 9-10
  stack gases, 9-10
Gas cutting of metals, 13-36 to 13-38
Gas cycles, 4-20 to 4-22
  for nuclear reactors, 9-122
Gas engines (*see* Internal-combustion engines)
Gas making, 7-41 to 7-47
  processes of, 7-42 to 7-47
Gas meters, 16-8, 16-14 to 16-18
Gas oil (petroleum distillate), 7-18
Gas pipe (*see* Pipe)
Gas producers (*see* Gas making; Producer gas)
Gas turbines, 9-114 to 9-119
  aircraft, 9-118
  aircraft jet engine, 9-117
  applications of, 9-118, 9-119
  characteristics of, 9-117, 9-118
  closed, 9-115
  combustors, 9-116
    heat release rate, 9-116
  components, 9-116
  compressors, 9-116
  cycles, 4-22, 9-114, 9-115
  effect of inlet temperatures (chart), 9-115
  effect of pressure ratio (chart), 9-115, 9-116

Gas turbines (*Cont.*):
  fan engine, 9-117
  fuels, 9-116
  gas pipeline, 9-118
  inlet temperatures, 9-114
  intercooling, 9-114
  locomotive, 9-119, 11-27
  marine, 9-119
  multiple-shaft type of, 9-117
  open, 9-115
  peaking units, 9-118
  power stations, cost of, 17-33
  pressure ratio, 9-114
    effect of (charts), 9-115, 9-116
  regenerators, 9-114, 9-116
  reheating, 9-114
  ship propulsion, 9-119
  single-shaft type of, 9-117
  steam combination plants, 9-118
  superalloys for, 6-96
  types of, 9-115, 9-116
  working fluids, 9-116
Gas welding (*see* Welding)
Gaseous fuels (*see* Fuels, gaseous)
Gases, natural, pipeline transmission, 11-130 to 11-132
Gasification of coal, underground, 7-24
Gaskets, 8-143, 8-144
  ammonia fittings, 8-188
  compressibility (chart), 8-146
  O-ring, 8-144
Gasoline, 7-15 to 7-17, 9-92
  additives, 7-16
  antiknock requirements for (table), 9-95
  aviation, 7-17
    grades of (table), 7-18
    specifications for (table), 7-18
  condensation temperature of (table), 9-102
  cracked, 7-15
  distillation curves, 7-15
  heat value of (table), 7-14, 7-16
  knock characteristics of, 7-16
  knock rating, 9-96, 9-97
  knock suppressors (table), 9-95
  motor, ASTM specifications (table), 7-17
  natural, 7-16
  octane number (table), 7-17
  properties and specifications (table), 7-17
  reformed, 7-15
  road ratings, 9-97
  straight-run, 7-15
  vapor pressure (table), 7-17
Gasoline engines, 9-78 to 9-114
  (*See also* Internal-combustion engines)
Gate valves, 8-188
Gauss (magnetic flux density, def), 15-2, 15-15
Gear-cutting processes, 13-62 to 13-64
Gear pumps, 14-12
Gear ratio (def), 8-99
Gear-shaving machines, 13-64
Gearing, 8-98 to 8-120
Gears:
  AGMA standards for, 8-111
  AGMA strength and durability ratings, 8-112
  automobile, 11-4, 11-7 to 11-9
  backlash, 8-102
  bevel, 8-106 to 8-110
    cutting processes for, 13-63, 13-64
    definition of terms, 8-106, 8-107
    efficiency of, 3-30
    Gleason, 8-108

Gears, bevel (*Cont.*):
    hypoid, **8**-106
    mounting surface, **8**-107
    recommended backlash (table), **8**-109
    registering surface, **8**-107
    skew, **8**-106
    spiral, **8**-109
        dimensions of (table), **8**-109
    straight, **8**-108, **8**-109
        dimensions of (table), **8**-109
    Zerol, **8**-110
        dedendum formulas (table), **8**-111
        dimensions of (table), **8**-110
Beveloid, **8**-111
Buckingham equation: for durability, **8**-112
    dynamic, **8**-112
        factor (table), **8**-114
center distance, **8**-100
contact ratio (charts), **8**-103
cutters for, **13**-63
design standards, **8**-111
durability, **8**-112
efficiencies, **3**-30
endurance limits for (table), **8**-117, **8**-118
greases, **6**-205
grinding, **13**-64
helical, **8**-104 to **8**-106
    axial pitch (def), **8**-105
    calculations for, **8**-106
    contact ratio (def), **8**-100
    crossed axis, **8**-105, **8**-106
    formulas (table), **8**-106
    friction, **3**-30
    parallel shaft, **8**-105
    skew shafts, **8**-105
Helicon, **8**-111
herringbone, **8**-105
hypoid, **8**-106
    figure, **11**-9
Lewis formula for strength, **8**-111
    refinement, **8**-112
    table, **8**-113
long and short addendum, **8**-102
lubricants for (table), **8**-116, **8**-119
lubrication, **6**-204
materials for (table), **8**-116, **8**-117
mesh ratio (def), **8**-99
metric: American equivalents (table), **8**-101, **8**-102
    design equations for (table), **8**-105
    ISO specifications, **8**-100
    tooth proportions and standards, **8**-100
miter, **8**-108
nomenclature for, **8**-98, **8**-99
noncircular, **8**-111
nonspur standards (table), **8**-99
oil, **6**-200, **6**-204
pitch: circular (def), **8**-98
    diametral (def), **8**-98
pitch circle of (def), **8**-99
pressure angle of (def), **8**-99
profile shifted, **8**-102
rolling of, **13**-64
service factors (table), **8**-115
shaving, **13**-64
speed ratio, **8**-99
spiral, bevel, **8**-109, **8**-110, **13**-64
Spiroid, **8**-111
spur, **8**-100 to **8**-104
    contact ratio (def), **8**-100
    friction, **3**-30
    metric, design equations for (table), **8**-105

Gears, spur (*Cont.*):
    over-pins measurements, **8**-100
    pitch (def), **8**-98
    speed ratios, **8**-99
    teeth: proportions of, **8**-98
        table, **8**-99
    teeth: nomenclature for, **8**-98
    proportions of (table), **8**-99
    thickness of, **8**-100
    testing radius, **8**-102
    trains: bevel, **8**-8
        epicyclic, **8**-7
    worm, **8**-110, **8**-111
        double enveloping, **8**-110
        efficiency of, **3**-30
        friction, **3**-30
        nonreversibility of, **8**-111
        single enveloping, **8**-110
        teeth, proportions of (table), **8**-99
        velocity ratio (def), **8**-111
Geiger counters, **16**-22, **16**-23
Gelatin in explosives, **7**-25
GEM (ground effect machines), **11**-56
General Motors two-cycle engines, **9**-103
Generators:
    absorption refrigeration, **19**-11
    alternating-current, **15**-36 to **15**-40
        armature reactance, **15**-34
        armature resistance, effective, **15**-37
        booster, **15**-49
        classes of, **15**-36
        construction of, **15**-36
        design of, **15**-36
        efficiencies of (tables), **15**-40
        excitation for, **15**-38
        frequency, **15**-37
        ground resistors for, **15**-40
        hunting, prevention of, **15**-39
        induced emf of, **15**-37
        parallel operation, **15**-39
        performance of (tables), **15**-40
        regulation of, **15**-38
        synchronous, **15**-36 to **15**-40
        voltage regulation, **15**-37 to **15**-39
    asynchronous, **15**-45
    automobile, **15**-82
    direct-current, **15**-30 to **15**-32
        armature reaction in, **15**-34
        balancer set for three-wire system, **15**-63
        commutating poles in, **15**-34
        commutation in, **15**-34
        compound-wound, **15**-30
            performance of (table), **15**-31
        induced emf of, **15**-30
        parallel operation, **15**-31, **15**-32
        series, **15**-30
        shunt, **15**-30
        as synchronous converters, **15**-49, **15**-50
        three-wire, **15**-64
    double-current, **15**-49
    efficiency calculations, **15**-51
    induction, **15**-45
    switchboard equipment, **15**-53
    synchronous (*see* Generators, alternating-current)
Genetrons (refrigerants), **19**-4
Geometric constructions:
    ellipse, **2**-33, **2**-34
    hyperbola, **2**-36
    parabola, **2**-32
    various, **2**-9 to **2**-11, **2**-32 to **2**-36

Geometric mean, **2**-17
  construction for, **2**-9
Geometric progression, **2**-17, **8**-201
Geometry:
  analytical, **2**-30 to **2**-41
  elementary, **2**-7 to **2**-15
Geothermal power, **9**-5, **9**-159 to **9**-162
  electrical energy conversion, **9**-160 to **9**-162
    processes, **9**-160 to **9**-162
  environmental considerations, **9**-160
  exploration technology, **9**-160
  sources of, **6**-159
German alphabet, **19**-47
Gibbs function (thermodynamics), **4**-15
Giesler coal test, **7**-38
Gifford-McMahon refrigerator (cryogenics),
  **19**-24
Gilbert (magnetomotive force, def), **15**-2, **15**-15
Gilbreth's micromotion study, **17**-28
Girders, steel, proportions of, **12**-30
  (*See also* Beams)
Glass, **6**-157, **6**-158
  composition of, **6**-157
  fiber, **6**-157
    properties of (table), **6**-155
  for industrial plants, **12**-7
  as insulation, **6**-157
  properties of, **8**-158
    at low temperatures, **19**-30
  safety, **6**-157
  shading coefficients for (tables), **12**-93 to **12**-95
  specific gravity and density of (table), **6**-8
  types of, **6**-157, **6**-158
  wire, **6**-157
  wool, **6**-170
Glass blocks, **6**-157
Gleason bevel gears, **8**-108
Globe valves, **8**-188
Glues, **6**-140, **6**-141
  for plywood and laminated wood, **6**-135
  (*See also* Adhesives)
Glulam (glued-laminated wood), **6**-132
Glycerine as freezing preventive (table), **4**-6
Glycerol as freezing preventive, **6**-156
GMR Stirling thermal engine, **9**-163
Gold, uses of, **6**-103
Golden section, **2**-9
Governing of internal-combustion engines, **9**-104
Governors:
  hydraulic turbine, **9**-147 to **9**-149
  steam turbine, **9**-47
Grab buckets, **10**-13, **10**-14
Grad (angular measure, def.), **16**-5
Grade resistance (train), **11**-30
Grading, volume of earth, computation of, **16**-48
Grain alcohol as fuel, **7**-20
Grain size, austenitic (metallography), **6**-20
Granite, composition of, **6**-158
Granodizing (protection of iron), **6**-116
Graphical methods of computation, **2**-58 to **2**-62
Graphical relations in beams, **5**-33, **5**-34
Graphical representation of functions, **2**-54 to **2**-62
Graphical solution of trusses, **12**-18
Graphical statics, problems, **3**-6, **3**-7
Graphite:
  bearings lubricated by, **8**-129
  as lubricant, **3**-25
  specific gravity and density of (table), **6**-9

Graphitization of pipe, **8**-197
Graphitizing furnace (electric), **7**-69
Graphs, equations for, **2**-54
Grashof's formula for steam flow, **4**-47
Gravel:
  for concrete, **6**-179
  specific gravity and density of (table), **6**-9
Gravitation, law of, **3**-3
Gravity:
  acceleration of, **1**-40
  center of, **3**-7, **3**-8
  equation for, **3**-2
  specific (*see* Specific gravity)
  standard (def), **1**-40
  standard acceleration of, **1**-40
Gravity conveyors, **10**-61
Gray iron (def), **6**-12
  castings, **6**-49
    composition and mechanical properties of (tables), **6**-48
    specifications for (ASTM), **6**-49
    tolerances, **6**-50
Gray-King assay, **7**-38
Greases, **6**-202, **6**-203
  characteristics of (table), **6**-203
  consistency tests, **6**-202
  dropping point, **6**-202
  numbers, **6**-202
  removers, **6**-150
  synthetic fluid, **6**-200
Greek alphabet, **19**-46, **19**-47
Gregorian calendar, **1**-41
Grid (electron tube), **15**-86
Grinding:
  allowances for, **13**-68
  cross feeds, **13**-68
  of cutting tools, **13**-54
  depth of cut in, **13**-68
  surface speeds, **13**-68
  work speeds, **13**-68
Grinding machines, **13**-66 to **13**-69
  centerless, **13**-66
  cylinder, **13**-67
  drill, **13**-67
  internal, **13**-67
  surface, **13**-67
  types of, **13**-66
  universal, **13**-67
Grinding wheels, **6**-139, **6**-140, **13**-67 to **13**-69
  abrasives, selection of, **13**-67
  bonding processes, **6**-139, **6**-140
  dressing, **13**-68
  grades, **6**-140
  grain size, **6**-140
  speeds, **6**-140, **13**-68
  truing, **13**-68
Grindstones, friction, **3**-27
Gripsprings, **8**-31
Grog (calcined clay), **6**-171
Grooving of steam boilers, **6**-110
Ground-controlled approach (radar), **15**-96
Grounding transformer secondary, **15**-64
Grounds for interior wiring, **15**-69
GR-S rubber, **6**-166
  table, **6**-167
Gudermannian function, **2**-30
Gunter's measure (surveying), **1**-31
Gust load, **12**-12
Gutta percha, **6**-167
Guy ropes, galvanized (table), **8**-91
Gypsum plaster, **6**-182
Gyration, radius of (def), **3**-9

Gyrocompass, 3-23
Gyroscope:
    applications of, 3-23
    theory of, 3-22, 3-23

H-S diagram, description of, 4-23
    (See also Mollier diagram)
Hack saw, 13-66
Hadfield's manganese steel, 6-42
Half-life (def), 5-81
Half-life of nuclei (def), 9-121
Hammers:
    drop-forge, 13-25
    energy of blows (tables), 13-26
    pneumatic, 13-26
    steam, 13-25
Hand (unit of length, def), 1-31
Hand-operated squeezer and molding
        machines, 13-4
Hangers, shaft, 8-129
Hankinson's formula for strength of wood, 6-
    127
Hard coal (def), 7-5
Hard-facing materials, 13-42
Hardenability of steel (def), 6-19
    Jominy test, 6-19
Hardening:
    cold work, 6-13, 6-16, 6-17
    nonferrous alloys, 6-59, 6-62
    by precipitation, 6-62
    steel, 6-18
Hardgrove machine grindability test for coal,
    7-9
Hardinge feeder-weigher, 10-64
Hardness:
    of materials (def), 5-12
    Mohs scale of, 1-40, 5-12
    of steel (table), 6-23
    tests for, 5-12 to 5-14
    of wood, 5-14
Hardwood bearings, 8-130
Harmonic mean, 2-17
Harmonic motion, 3-14
Harmonic vibration, 5-67
Hartley oscillator, 15-89
Hastelloy (tables), 6-71, 6-72, 6-97, 6-99
    uses of, 6-74
Haulage:
    by locomotives, 10-21, 10-22
    off-highway, 10-26 to 10-28
    wire rope for, 10-9
        table, 8-86
    (See also Conveying)
Hawsers, galvanized steel (tables), 8-91
Hayward grab bucket, 10-13
    electric, 10-14
    table, 10-13
Head:
    loss due to fittings (table), 3-58
    potential (hydraulics), 3-43
    pressure (hydraulics), 3-43
    velocity (hydraulics), 3-43
Head gates for hydraulic turbines, 9-149
Heading operations, steel for, 13-21
Heat:
    available (def), 4-15
    of combustion (table), 4-59
    conduction (see Heat transfer; Heat
        transmission; Thermal conductivity)
    conversion table for, 1-52
    developed by electric current, 15-10

Heat (Cont.):
    of earth, power from, 9-159 to 9-163
    expansion by, 4-3
    of formation (tables), 4-53
    of fusion: of materials (table), 4-12
        of metals (table), 6-60, 6-61
    latent, 4-11
        of vaporization (table), 4-12
    measurement of, 4-8
    mechanical equivalent of, 4-8
    of seas, power from, 9-168, 9-169
    specific (see Specific heat)
    units of, 1-52, 4-8
    of vaporization of liquids (table, chart), 4-
        11, 4-12
Heat capacity (def), 4-8
Heat-engine cycles, 4-20 to 4-22
Heat exchangers:
    power plant, 9-59 to 9-78
    tubes (table), 8-166
Heat flow, conversion table, 1-52
Heat insulators (see Insulators)
Heat pumps, 12-102 to 12-105, 19-3
    cycles for (charts), 12-114
    heat sources and sinks (table), 12-112
Heat-resisting alloy castings (table), 6-56
Heat-resisting stainless steels (table), 6-37
Heat transfer:
    coefficients of (def), 4-60
        to or from air, 4-65 to 4-68
        to or from boiling liquids, 4-58
        combined convection and radiation, 4-69
        to or from condensing vapors, 4-68
        in refrigerating plant (table), 19-20
        for scale deposits (table), 4-66
        to or from steam (chart), 4-68
        to or from water, 4-64, 4-65
        (See also Film coefficients)
    in dielectrics, 19-26
    across evacuated space, 19-36
    overall coefficients for windows (table), 12-
        95
    through pipe insulation, 4-69
        chart, 4-70
    radiation shield for, refrigerated, 19-36
    radiative, 19-36
    in steam boilers, 9-18
    U values resulting from additional
        insulation (tables), 12-88
Heat transmission:
    coefficients of, 4-60
        (See also Film coefficients; Heat transfer)
    in condensers, 4-68
    by conduction, 4-60 to 4-69
        and convection, 4-61 to 4-69
        laws of, 4-60
    by convection, 4-60 to 4-69
        and conduction, 4-61 to 4-69
        natural, 4-67
        and radiation, 4-69
    conversion table, 1-52
    in evaporators, 4-69
    with extended surfaces, 4-67
    in furnaces (calculations), 4-80 to 4-82
    to high-velocity gases, 4-65
    with laminar flow, 4-67
    mean temperature difference, 4-62, 4-63
    nomenclature for, 4-59, 4-60
    by radiation, 4-60, 4-70 to 4-82
    between steam and boiling liquids (table), 4-
        70
    with streamline flow, 4-67

Heat transmission (*Cont.*):
    in surface condensers, **9**-62, **9**-63
    temperature gradient, **4**-62
    units of, **4**-59, **4**-60
Heat-treatment:
    of nonferrous alloys, **6**-62, **6**-78
    of steel (*see* Steel)
Heat value (def), **4**-52, **4**-53
    of by-product fuels (table), **7**-13
    of coal (tables), **7**-3, **7**-4
    of coke (table), **7**-11
    of fuels (def), **4**-52, **7**-7
        Dulong's formula for, **7**-7
        high and low (def), **4**-52
        per unit of gaseous combustible, **4**-53
        table, **4**-54
        (*See also* Calorific value)
    of gaseous fuels (table), **7**-22
    of hydrocarbons (table), **4**-54
    of petroleum products, **7**-14, **7**-15
        table, **7**-14
    of various liquid fuels, **7**-18, **7**-19
    of various solid fuels, **7**-13
    of wood, **7**-13
        tables, **7**-13
Heaters:
    chimneys for, **12**-71
    fuel consumption of, **12**-71
Heating, **12**-62 to **12**-76
    air infiltration (tables), **12**-65, **12**-66
    boiler loads for, **12**-70
    coils, film coefficients for, **4**-64
    dielectric, **7**-64, **15**-94
    ductwork for, **12**-83, **12**-98
    flue gas calculation (chart), **12**-73
    fuel consumption for, **12**-71
    fuel-consumption constants for (table), **12**-73
    hot water (*see* Hot-water heating)
    industrial plants, **12**-19
    inside design conditions (table), **12**-63
    moisture infiltration, **12**-72
    outdoor air requirements (table), **12**-67
    outdoor design temperatures, correction
        factors (table), **12**-72
    overall heat-transfer coefficients (tables), **12**-
        64
    radiators (*see* Radiators)
    stack sizing calculations for (tables), **12**-73
    steam (*see* Steam heating)
    system efficiency (table), **12**-72
Heave (ships, def), **11**-40
Heavy water, **9**-127
Hectare (def), **1**-37
Helical gears, **8**-104 to **8**-106
Helicon gears, **8**-111
Helicopters, **11**-72, **11**-73
    definition of, **11**-58
Helix, **2**-41
Helmholtz free energy (thermodynamics), **4**-
    15
Helve hammers, **13**-25
Hemp, **6**-151, **6**-154
    properties of (table), **6**-155
Henequen fiber, **6**-151
Henry (inductance, def), **15**-3, **15**-4, **15**-19
Herbert hardness tests, **5**-13, **5**-14
Herringbone gears, **8**-105
Hexagon (construction), **2**-10
High heat value of fuels, **4**-72
High-speed steel, **6**-33
    composition and uses of (tables), **6**-36
    tools, **13**-51

High-vacuum pumps, **14**-44 to **14**-48
    applications of, **14**-48
    baffles and traps, **14**-47
    compression ratio (def), **14**-45
    conductance, **14**-47
    cryopumps, **14**-47
    cryosorption, **14**-47
    diffusion, **14**-45, **14**-46
    forepressure, **14**-45
    fractionating, **14**-46
    gas-ballast, **14**-47
    gas flow at low pressure, **14**-47
    getter-ion, **14**-46
    installation of, **14**-47
    Knudsen number, **14**-48
    mercury-vapor, **14**-46
    net pumping speed, **14**-47
    outgassing rate, **14**-48
    pressure range (table), **14**-46
    Roots-type, **14**-48
    rotary mechanical, **14**-44
    roughing time, **14**-48
    speed (def), **14**-44
    sputter-ion, **14**-46
    throughput (def), **14**-45
    turbomolecular, **14**-47
    types and sizes of (table), **14**-46
    ultimate pressure (def), **14**-44
    ultrahigh vacuum, **14**-45
    vapor contamination, **14**-47
High-voltage measurement, **15**-28
Hildebrand function, heat of vaporization
    (chart), **4**-11
Hill climbing (optimizing), **2**-71
Hipernik (magnetic alloy), **6**-41
Hitches, rope, **8**-89
Hitenso bronze (electrical conductor), **15**-7
Hobbing process, **13**-63
Hodograph (def), **3**-13
Hogging (ships, def), **11**-37
Hohmann transfer (space maneuver), **11**-113
Hoisting:
    cableways, **10**-34 to **10**-37
    drums, **10**-9, **10**-10
    load-suspension devices, **10**-4, **10**-11, **10**-12
    mechanisms, **8**-8
    speeds, **10**-19
    wire rope for, **8**-85, **8**-86, **10**-9
Hoisting machinery:
    drums, **10**-9, **10**-10
    sheaves, **10**-10
    tackle blocks, **10**-10
Hoists, **10**-14 to **10**-16
    air motor, **10**-15
    brakes for, **10**-15, **10**-18
    chain, **10**-14
    electric, **10**-14
        table, **10**-17
    lever-operated, **10**-14
        table, **10**-15
    mine, **10**-16 to **10**-19
        balanced (def), **10**-17
        brakes for, **10**-18
        electrical equipment, **10**-19
        motors for, **10**-19
        speeds, **10**-19
    monorail, **10**-28
        table, **10**-29
    pneumatic, **10**-15
    power: motors, **15**-51
        wire rope, **10**-9
    skip, **10**-18

Holding mechanisms, **10**-11, **10**-12
Holland formula (for stack plume), **18**-16
Hollow building blocks (table), **12**-23
Honing, **13**-70
Hooke's joint (coupling), **8**-36
Hooke's law, **5**-18
Horsepower (def), **1**-37
Horses, work of (table), **9**-162
Hose:
    American Standard coupling threads, **8**-199
    couplings for, American Standard (table), **8**-200
    fire, **18**-32
        American National: couplings, **8**-199
            dimensions of (table), **8**-200
            screw thread, **8**-199
    flexible metal, **8**-199
    pressure, **8**-199, **8**-200
        for hydraulic fluids, **8**-41
    rubber-lined, **8**-199
    SAE standard (table), **8**-41
Hot-air engines, **9**-163
    cycles for, **4**-20 to **4**-22
    internally focusing regenerative, **9**-163
    solar, **9**-163
    for space power, **9**-163
Hot drawing, **13**-20
Hot forging of metals, **13**-11
Hot-water heating:
    boiler load, **12**-70
    design water temperatures, **12**-62
    expansion of water in (chart), **12**-70
    expansion tank design, **12**-70
Hot-working of metals, **13**-11
Hotchkiss drive (automobile), **11**-9
Household refrigerating machines, **19**-10
Humidification of air, **4**-44
    (*See also* Air conditioning)
Humidity, **4**-28 to **4**-32
    chart, **4**-41 to **4**-43
        low temperature, **4**-42
        medium temperature, **4**-43
    measurement, **4**-21
        by chemical analysis, **4**-31
    molal (def), **4**-31
    relative (def), **4**-30
    specific (def), **4**-31
Huntington's postulates (Boolean algebra), **15**-92
Huygens' approximation to length of circular arc, **2**-12
Hydramatic transmission (automobiles), **11**-8
Hydrated lime, **6**-179
Hydraulic brakes, **11**-14 to **11**-17
Hydraulic conveyors, **10**-63
Hydraulic couplings, **8**-37, **8**-38, **11**-5, **11**-6
Hydraulic governors, **9**-147
Hydraulic grade line (def), **3**-43
Hydraulic hoists, friction of, **3**-27
Hydraulic jacks, **10**-17
    efficiency of, **3**-29
Hydraulic power transmission, **8**-39 to **8**-41
    fluids for, **6**-207
Hydraulic presses, **13**-23
Hydraulic radius (table), **3**-42
Hydraulic torque converter (automobiles), **11**-6
Hydraulic turbines, **9**-136 to **9**-149
    air valves, **9**-143
    auxiliaries for, **9**-149
    cavitation, **9**-147
        plant sigma for preventing, **9**-147
    classification of, **9**-137

Hydraulic turbines (*Cont.*):
    Francis (*see* reaction, *below*)
    fundamental formulas, **9**-137
    general arrangements (table), **9**-137
    governors, **9**-147 to **9**-149
        electric, **9**-148
    head limits (table), **9**-137
    homologous turbines, **9**-147
    impulse, **9**-137, **9**-143 to **9**-145
        basic dimensions, **9**-143
        blade forces in, **3**-45
        buckets, **9**-144
        efficiency of, **9**-143
        general arrangement (table), **9**-137
        housing, **9**-143
        jet deflectors, **9**-145
        nozzle, **9**-145
        regulation, **9**-145
        runaway speed, **9**-145
        runners, **9**-144
        setting, **9**-144
        table, **9**-137
        specific speeds, **9**-143
        speed selection, **9**-143
        tailwater level, **9**-144
        usual head limits (table), **9**-137
    model tests of (chart), **9**-147
    number of units in plant, **9**-138
    Pelton-type (*see* impulse, *above*)
    pressure regulators, **9**-149
    propeller-reaction: adjustable blade runners, **9**-138
        Kaplan-type, **9**-138, **9**-139
        specific speed, **9**-138
        thrust, **9**-141
        usual head limits (table), **9**-137
    proportionality laws for, **9**-137, **9**-147
    reaction, **9**-138 to **9**-143
        axial-flow, **9**-139
            types of, **9**-139
        bearings for main shaft, **9**-142
        case velocities (chart), **9**-142
        characteristics of, **9**-141
        design of, **9**-139
        draft tubes, **9**-143
        general arrangement (table), **9**-137
        number of runners (table), **9**-137
        runners, **9**-141
        setting (table), **9**-137
        specific speeds, **9**-138
        spiral case, **9**-142
        stay ring, **9**-142
        thrust bearings, **9**-141
            pressure in, **9**-141
        usual head limits (table), **9**-137
        wearing rings, **9**-141
    reversible pump, **9**-145, **9**-146
    runaway speed, **9**-141, **9**-145, **9**-146
    selection of, **9**-137
    specific speed (def), **9**-137
    speed regulation, **9**-147 to **9**-149
        requirements, **9**-148
    speed selection, **9**-138
    surge tank, **9**-149
    tests of, **9**-149
    types of: characteristic, **9**-137
        general arrangement (table), **9**-137
        selection of, **9**-137
        usual head limits (table), **9**-137
    vacuum breakers, **9**-143
    water hammer in penstocks, **9**-148
    wicket gates and mechanisms, **9**-142

Hydraulic weighing systems, **5**-14
Hydra-Vac power brake, **11**-17
Hydrazine (rocket fuel), **7**-36
Hydrocarbons:
  air for combustion of (table), **4**-55
  boiling points of (table), **4**-6
  heats of combustion of (table), **4**-54
  products of combustion of (table), **4**-55
  vapor pressures of (chart), **4**-28
Hydroelectric power stations, cost of, **17**-36
Hydrofoil craft, **11**-35, **11**-55
  lift-drag ratios (chart), **11**-55
Hydrogasification processes (gas making), **7**-47
Hydrogen:
  air for combustion of (table), **4**-55
  as cryogen, **19**-31
  flame temperature and dissociation (table), **4**-57
  as fuel, **9**-169
  heat value (table), **4**-54
  liquid (rocket fuel), **7**-36
  products of combustion of (table), **4**-55
Hydrogen gas, **9**-6
Hydrogen peroxide as rocket fuel, **7**-36
Hydrogenation of coal, **7**-41 to **7**-47
Hydrometers, **3**-40
  scale for, **1**-42
Hydrophones, **12**-137
Hydrostatics, **3**-37 to **3**-40
Hygas process (gas making), **7**-45
Hygrometer, **4**-31, **16**-21
Hyperbola:
  area of, **2**-12
  conjugate, **2**-34
  constructions for, **2**-36
  equilateral, **2**-36
  properties of, **2**-34 to **2**-36
Hyperbolic functions (sinh, cosh, tanh), **2**-24, **2**-29
  of complex variable, **2**-24
  graphs of, **2**-30
  series for, **2**-44
  tables, **1**-22 to **1**-24
Hyperbolic types of power functions, graph, **2**-54
Hypereutectoid (steel), **6**-18
Hypersonic aerodynamics (*see* Supersonic and hypersonic aerodynamics)
Hypocycloid, **2**-39
Hypoeutectoid (steel), **6**-18
Hypoid gearing, **8**-106
Hypotrochoid, **2**-40
Hysteresis, magnetic, **15**-16, **15**-17
Hysteresis loop, **15**-17
Hysteresis loss:
  in magnetic metals (table), **15**-17
  Steinmetz's law of, **15**-17
Hytemco (resistor alloy), **15**-73
  table, **15**-78

*I* and *i* (imaginary number), properties of, **2**-22
I beams:
  deflection of (table), **12**-32
  safe loads for (tables), **12**-33 to **12**-43
  standard, properties of (tables), **12**-34
  structural steel rivet gages (tables), **12**-44, **12**-45
  wide-flange, properties of (tables), **12**-35 to **12**-38

Ice:
  heat of fusion (table), **4**-12
  specific gravity and density (table), **6**-8
Ice making:
  can system, **19**-21
  extrusion method, **19**-21
  Flakice method, **19**-21
  Pak-ice method, **19**-21
Ice-skating rinks, piping for, **19**-21
Icosahedron, **2**-8, **2**-14
Ideal gas law, **3**-44
Ideal radiator (def), **4**-71
Identity, algebraic (def), **2**-18
Ignition coil, **15**-80 to **15**-82
Ignition lag (internal-combustion engines), **9**-93
Ignition quality of fuels, **9**-97
Ignition systems:
  in automobile engines, **9**-100
  battery, **15**-80 to **15**-83
  electrical, **15**-80 to **15**-84
  generators for, **15**-80
  magneto, **15**-83, **15**-84
  spark gaps for, **9**-100
  voltage of, **15**-80
Ignition temperatures (gaseous fuels), **9**-93
Ignitron (voltage regulation for rectifiers), **11**-19
Ilgner method of motor speed control, **15**-35
  for mine hoists, **10**-19
Illuminance (def), **12**-117
Illumination, **12**-116 to **12**-135
  calculation of, **12**-126 to **12**-135
  calculation sheet for, **12**-134
  coefficients of utilization of, **12**-126, **12**-127
  color of, **12**-118
  dimming systems for, **12**-135
  fixtures (*see* Lighting fixtures)
  glare in, **12**-125
  industrial, **12**-9
  prescribing of, **12**-120 to **12**-126
  recommended: for various activities (table), **12**-126
    for various locations (table), **12**-126
  visual comfort criteria, **12**-125
  (*See also* Lighting)
Imaginary quantities, **2**-22 to **2**-25
Imaginary roots of equations, **2**-19
Impact, **3**-21, **3**-22, **5**-46, **5**-47
  collinear, **3**-21
  definition of, **3**-21, **5**-64
  of elastic and inelastic bodies, **3**-21
  oblique, **3**-21
  stress due to, **5**-47
  of water jet against plate, **3**-22
Impact failure, **5**-6 to **5**-9
Impact tests, **5**-7
Impedance (electric circuits), **15**-5, **15**-20
  synchronous, **15**-37
Implicit functions, **2**-43
Impregnating compounds (motor coils), **6**-152
Impulse turbines (*see* Hydraulic turbines; Steam turbines)
Incentive wage systems, **17**-9
Inch:
  circular (def), **1**-31
  miner's (def), **1**-31
Incinerators, **7**-53 to **7**-60
  air pollution control of, **7**-54, **18**-21
  air supply to, **7**-55
  facilities for, **7**-53, **7**-54
  fuel used in, **7**-53

Incinerators (*Cont.*):
  furnace design, **7**-54 to **7**-56
  furnace types of, **7**-53
  heat balance, **7**-56
  heat calculations for, **7**-55
  heat recovery from, **7**-57
  materials balance in (table), **7**-56
  plant cross-section (figure), **7**-59
  plant design, **7**-53, **7**-54
  salvage from, **7**-57
  waste recovery plants (table), **7**-57, **7**-58
Inclined planes, laws of, **3**-16
Income statement, **17**-11
Inconel, **6**-74
  composition and properties of (tables), **6**-72,
    **6**-96, **6**-97
  strength of, at high temperatures (tables), **6**-
    73, **6**-74, **6**-97
Indeterminate forms (calculus), **2**-45
Index of refraction, **19**-40
Indicator diagrams of steam engines, **9**-36
Indicators, engine, electric, **16**-18
    high-speed, **16**-18
Induced angle of attack (aerodynamics, def),
  **11**-61
Induced drag (aeronautics), **11**-61
Induced emf, **15**-19
  direction of, **15**-17
Inductance (electrical circuits), **15**-19, **15**-20
  units of (def), **15**-4, **15**-19
Induction (electrical circuits), **15**-19
Induction furnaces, **7**-66 to **7**-68
Induction generator, **15**-45
Induction hardening of steel, **6**-24
Induction heaters, **7**-60, **7**-64, **15**-94
Induction motors, **15**-46 to **15**-49
Induction watt-hour meter, **15**-27
Inductive circuits, **15**-4, **15**-19
Inductive reactance, **15**-4, **15**-21, **15**-22
Industrial accidents, prevention of, **18**-22 to
  **18**-27
Industrial accounting, **17**-10 to **17**-18
  balance sheets, **17**-11
Industrial cars, **10**-22, **10**-23
Industrial electronics, **15**-93, **15**-94
Industrial engineering, **17**-2 to **17**-43
Industrial heating furnaces, **7**-48 to **7**-53
Industrial management, **17**-2 to **17**-10
Industrial organization, **17**-2 to **17**-4
Industrial plants, **12**-2 to **12**-11
  acoustics, **12**-9
  aisle space (accident prevention), **18**-23
  automatic control (*see* Automatic control)
  building materials, **12**-6 to **12**-9
  buildings, **12**-5, **12**-6
  contracts, **12**-10, **12**-11
  conveyors, **12**-4
  cost estimates, **12**-10
  equipment replacement, **17**-8
  floor-space requirements, **12**-3
  floors, **12**-8
  flow diagrams for, **12**-3, **12**-4
  heating and ventilating, **12**-9
  lighting, **12**-9, **12**-118
  location, **12**-2
    of equipment, **12**-4
  material handling, **12**-3 to **12**-5
  partitions, **12**-8
  planning, **12**-2 to **12**-4
  plans and specifications, **12**-10
  plumbing, **12**-9
  process charts, **12**-4

Industrial plants (*Cont.*):
  reinforced-concrete construction, **12**-7
  roof profiles, **12**-5, **12**-6
  roofs, **12**-7
  routing diagrams, **12**-3
  safety provisions, **18**-23
  sequence of operations, **12**-3
  site selection, **12**-2, **12**-5
  skylights and monitors, **12**-9
  stairs, **12**-8
  steel-frame construction, **12**-7
  truss construction, **12**-5
  ventilation, **12**-9
  wall construction, **12**-7
  windowless, **12**-9
  windows, **12**-7
  wood construction, **12**-6
Inertia (def), **3**-2
  force of reciprocating engine parts, **8**-69 to
    **8**-74
  moment of (*see* Moment, of inertia)
  principal axis of (def), **3**-9
  product of (def), **3**-9
  radius of (def), **3**-9
Inflammability (*see* Flammability)
Inflection, point of, **2**-43, **2**-44
Infrared testing of materials, **5**-83
Infusorial earth (abrasive), **6**-139
Ingot iron, **6**-12, **6**-13
  composition of (table), **6**-13
  effect of cold rolling on (chart), **6**-13
  mechanical properties (table), **6**-14
  physical constants, **6**-13
  uses of, **6**-13
Ingots, steel, **6**-15
  defects in, **6**-15, **6**-16
Injectors, **14**-14
Inspection of castings, **13**-8
Instant center, **8**-4
Instantaneous axis, **3**-15, **8**-4
Instruments, **16**-2 to **16**-25
  accuracy of (def), **16**-2
  for cryogenics, **19**-33 to **19**-35
  electrical (*see* Electric instruments)
  error in, **16**-2
  gages (*see* Gages)
  nuclear radiation, **16**-22, **16**-23
  precision of (def), **16**-2
  pressure measuring, **16**-8 to **16**-10
  recording, **16**-23 to **16**-25
  resolution of (def), **16**-2
  sensitivity of (def), **16**-2
Insulating materials, electrical, **6**-151 to **6**-153
  properties of (table), **15**-19
  thermal conductivity of (tables), **4**-64, **4**-
    65, **12**-84, **12**-85
Insulating paper, **6**-153, **15**-61
Insulating varnishes, **6**-152
Insulation:
  additional, U values resulting from (tables),
    **12**-88
  electric, **15**-18, **15**-19
    classes of, **15**-51
    copper wire, **15**-67
    impregnated fabrics, **6**-153, **15**-60
    impregnated paper, **15**-61
    for magnet wire, **15**-80, **15**-81
    resistance: measurement of, **15**-28, **15**-29
      underground cable, **15**-60, **15**-61
  thermal, **6**-169 to **6**-171
    of brine piping, **19**-16
    of cold-storage rooms, **19**-16

Insulation, thermal (*Cont.*):
  for cryogenic temperatures, **6**-169
  for cryogenics, **19**-35 to **19**-39
  for high temperatures, **6**-170
  for moderate temperatures, **6**-170
  multilayer, **19**-37
  of pipes, heat transmission through, **4**-69
  powder, **19**-37
  reflective, **6**-170
  for refrigeration, heating, and air
    conditioning, **6**-169
  rigid-foam, **19**-38
Insulators:
  conductivity of (tables), **4**-64, **4**-65
  electric properties of (tables), **15**-19
  glass, **6**-157
  heat transmission through, **4**-69
Integral calculus, **2**-46 to **2**-49
Integral compensation (automatic control), **16**-
    27, **16**-28
Integrals:
  approximate computation of, **2**-49
  definite, **2**-48, **2**-49
  double, **2**-49
  elliptic, **2**-49
  indefinite (table), **2**-46 to **2**-48
  probability, **2**-49
Integraph, mechanical, **2**-49
Integrated circuits, **15**-92, **15**-93
  cost of (table), **15**-93
Integrators, electronic, **2**-73
Intensity:
  field, unit of, **3**-22
  of radiation (heat transfer), **4**-71
  of sound (def), **12**-137
Intercondensers, **9**-67
Interest, compound, tables for, **1**-26 to **1**-29
Interference drag, **11**-71
Interference fits, **8**-46
Interior wiring (*see* Wiring)
Internal brakes, **8**-44
Internal-combustion engines, **9**-78 to **9**-114
  air cooling, **9**-105
  air lines, **9**-101
  air pollution from, **9**-109 to **9**-113
  airplane (*see* Airplanes, engines)
  analysis of engine process of, **9**-106 to **9**-109
  automobile (*see* Automobile engines)
  back pressure, **9**-102
  carburetion, **9**-97, **9**-98
    for maximum economy, **9**-98
    for maximum power, **9**-97
    metering characteristics, **9**-98
  combustion chambers, **9**-88 to **9**-92
    divided, **9**-92
    open, **9**-92
  combustion knock, **9**-92 to **9**-97
  combustion process, **9**-108
  combustion roughness, **9**-91
  compression-ignition type of, **9**-79, **9**-92
  compression process, **9**-107
  compression ratios, **9**-78
  cooling systems, **9**-104, **9**-105
  cycles, **4**-20, **9**-78 to **9**-80
  deviations from ideal processes, **9**-107 to **9**-
    109
  diesel (*see* Diesel engines)
  dissociation, **4**-54
  dual-fuel, **9**-79
  exhaust-gas: analysis of, **9**-108, **9**-109
    temperatures, **9**-108
  exhaust manifolds, **9**-102

Internal-combustion engines (*Cont.*):
  exhaust process, **9**-108
  expansion process, **9**-108
  flame travel, **9**-108
  four-cycle, **9**-79
  fuel injection, **9**-98 to **9**-100
    pumps, **9**-99
  fuel lines, **9**-101
  fuels, **7**-14 to **7**-21, **9**-92 to **9**-97
    antiknock compounds (table), **9**-95
    autoignition temperatures, **9**-93
    cetane number (def), **9**-97
    flame speed in, **9**-93
    gasoline (*see* Gasoline)
    ignition quality, **9**-97
    ignition temperatures of (tables), **9**-94
    injection for, **9**-98 to **9**-100
    kerosine (*see* Kerosine)
    knock rating, **9**-96, **9**-97
    knock suppression, **9**-94
    knock suppressors (table), **9**-95
    lead susceptibility, **9**-97
    octane number (def), **9**-96
    permissible compression ratios, **9**-95
    road ratings, **9**-97
    vapor lock, **9**-101
    volatility of, **9**-92
  gas lines, **9**-101
  horsepower, **9**-107
  idling jet, **9**-98
  ignition systems (*see* Ignition systems)
  indicators, high-speed, **16**-18
  intake manifolds, **9**-101
  knock, combustion of, **9**-92 to **9**-97
    (*See also* fuels, *above*)
  L-head, **9**-91
  locomotive, **9**-86
  lubrication, **6**-203, **6**-204, **9**-105, **9**-106
  marine, **9**-86
  mean effective pressure, **9**-106
  mixture distribution, **9**-101
  muffle pits, **9**-102
  mufflers, **9**-102
  oil cooling, **9**-105
  outboard (table), **9**-90
  pistons (*see* Pistons)
  precombustion chambers, **9**-92
  pumping work, **9**-107
  regulation (speed and load), **9**-104
  scavenging, **9**-103, **9**-104
  spark advance, **9**-94, **9**-100
  spark-ignition type of, **9**-78, **9**-88 to **9**-92
  stationary, **9**-86
  stratified-charge spark-ignition, **9**-113
  supercharging, **9**-102, **9**-103
  tractor, **9**-85, **9**-86
  truck and bus, **9**-83
  two-cycle, **9**-79
  volumetric efficiency, **9**-107
    table, **9**-108
  water cooling of, **9**-104
Internal-combustion power stations, cost of, **17**-35
Internal energy of gases, **4**-18, **4**-53
  table, **4**-56
International Aluminum Standard, **15**-5
International Copper Standard, **15**-5
Interpolation, **2**-17, **2**-18
Interpoles in d-c generators and motors, **15**-34
Invar (low-expansion alloy), **6**-41
Inventions, patents for, **19**-41 to **19**-43
Inverse hyperbolic functions, **2**-30
  graphs, **2**-56

Inverse trigonometric functions, **2**-27
  graphs, **2**-55
Inversion temperature (Joule-Thomson effect), **4**-51
Investment castings, **6**-57
Investment process (molding), **13**-3
Involute:
  of circle, **2**-40
  of curve (def), **2**-46
Involute gear teeth, **8**-78
Involute splines, **8**-31
Ionization, effects of, on materials, **9**-125
Iridium, uses of, **6**-103
Iron:
  alloys: corrosion-resisting, **6**-108
    machinability, **13**-50
  alpha (def), **6**-17
  Armco ingot, **6**-13
  cast (*see* Cast iron)
  castings (*see* Castings)
  chromium, welding, **13**-43
  classification of, **6**-12
  corrosion of (*see* Corrosion)
  creep rates (table), **5**-11
  delta (def), **6**-17
  densities of (table), **6**-44
  electrolytic, composition of (table), **6**-13
  foundry practice (*see* Foundry practice and equipment)
  gamma (def), **6**-17
  ingot, **6**-12
    composition of (table), **6**-13
    properties of, **6**-13
  malleable, **6**-47, **6**-51, **6**-52
  passivating of, **6**-117
  pig, **6**-12
    table, **6**-13
  pure, **6**-12
  Russia, **6**-116
  sintered, strength of (table), **6**-104
  specific heat, mean (table), **4**-10
  and steel, **6**-12 to **6**-58
  strength at high temperature (table), **5**-11
  wrought (*see* Wrought iron)
Iron–iron carbide equilibrium diagram, **6**-17
Iron ore, sources of, **6**-12
Ironing (thinning of metal in bending), **13**-19
Irradiation, effects of, **9**-129
  (*See also* Radiation)
Isentropic expansion:
  of gases, **4**-19
  of vapors, **4**-24
Isobutane temperature-entropy chart, **4**-25
Isoclinics (stress analysis), **5**-57
Isostatics (stress analysis), **5**-57
Isothermal expansion:
  of gases, **4**-19
  of vapors, **4**-24
Isotope, **6**-7, **9**-120
IT calorie (International Steam Table) (def), **1**-35, **4**-8
  mechanical equivalent of, **1**-35

Jacks, **10**-15, **10**-16
Jaw clutches, **8**-38
Jet, water, impact of, **3**-22, **3**-45
Jet condensers, **9**-66
Jet pressure against blade or vane, **3**-45
Jet propulsion (*see* Aircraft jet propulsion)
Jib cranes, **10**-30
Jig boring machines, **13**-59

Joints:
  in belts, **8**-52
  cottered, **8**-31
  knuckle, **8**-31
  method of (stress determinations), **3**-6
  pin, friction in, **3**-31
  pipe (*see* Pipe, joints)
  riveted (*see* Riveted joints)
  toggle, **8**-8
  universal, **8**-36
Joists, steel, **12**-42
Jolt molding machines, **13**-5
Jominy hardenability test, **6**-19
Joule (unit of work, def), **1**-33, **15**-4
Joule cycle, **4**-20, **4**-22, **9**-114
Joule-Thomson coefficient (def), **4**-14, **4**-51
Joule-Thomson effect (throttling), **4**-51
Joule's law (electric currents), **15**-10
Journal bearings (*see* Bearings)
Journals (*see* Shafts)
Jute fiber, properties of (table), **6**-155

Kanthal (heat-resisting alloy, table), **6**-96
Kaolin in firebrick, **6**-172
Kaplan-type turbines (hydraulic), **9**-138
Kármán vortices, **12**-12
Karnaugh maps (circuit analysis), **15**-92
Keene's cement (plaster), **6**-182
Kelly ball test for concrete, **6**-186
Kelvin (unit of temperature, def), **1**-32
Kelvin absolute temperature scale, **4**-3
  conversion to degrees Celsius (eq), **4**-3
Kelvin double bridge, **15**-29
Kelvin's law for economical conductor size, **15**-56
Kepler's law (satellites), **11**-112
Kern (def), **5**-44
Kerosine, **7**-18
Keys, **8**-30, **8**-31
  feather, **8**-30
  flat, **8**-30
  gib-head (table), **8**-34
  sunk (table), **8**-34
  taper, **3**-28, **8**-30
    table, **8**-34
  Woodruff (table), **8**-32
Kilocalorie (def), **1**-37, **1**-40
  mechanical equivalent of, **1**-52
Kilogram (unit of mass, def), **1**-32
Kilovars (reactive current, def), **15**-4, **15**-22
Kilowatt (def), **15**-4, **15**-10
Kilowatt-hour (def), **15**-4, **15**-10
Kinematic viscosity, **3**-35
Kinematics, **3**-11 to **3**-20
  definition of, **3**-11
  of fluids, **3**-40, **3**-41
  (*See also* Mechanism)
Kinescope (television), **15**-97
Kinetic friction (def), **3**-24
Kingsbury thrust bearings, **8**-131
  coefficient of friction, **3**-31
Kirchhoff's laws:
  electrical circuits, **15**-8
  radiation (def), **4**-71
Kirsten-Boeing ship propellers, **11**-52
Klystron, **15**-90, **15**-95
Knock (detonation), **7**-16
  suppression of, **9**-94
Knock rating of gasolines, **7**-16, **9**-96, **9**-97
  table, **7**-17
Knot (nautical unit, def), **1**-31

Knots in rope, **8**-89
Knuckle joints, **8**-31
Knudsen number, **14**-48
Koppers-Totzek process, **7**-43, **7**-47
Kopp's approximation (specific heat), **4**-10
  constants for (table), **4**-10
Kort nozzle system (ship propulsion), **11**-52
Kraft paper, **6**-160

Labyrinth seal, **8**-147
  for steam turbines, **9**-46
Lac (resin), **6**-121
Lacquers, **6**-118, **6**-121
  composition and uses of, **6**-121
Lag screws:
  holding power of, **12**-28
  lead holes for, **12**-28
  proportions of (table), **8**-23, **12**-30
Lake copper, **6**-63
Lambert surfaces (radiant heat, def), **4**-73
Lamé's formula for collapse of thick tubes, **5**-51
Laminar flow, **3**-52, **3**-54, **3**-56
Laminated wood, **6**-132
Laminating of plastics, **13**-23
Lamps:
  ballasts for, **12**-120
    circuits (chart), **12**-121
  economics of installations, **12**-135
  electric, **12**-118 to **12**-120
  energy distribution from (fig), **12**-135
  filament, **12**-118
    bases for, **12**-118
    bulbs, **12**-118
    characteristics of (table), **12**-122
  fluorescent: characteristics of (table), **12**-123
    life of (table), **12**-123
    preheat circuits for, **12**-119
  high-intensity-discharge, **12**-120
    characteristics of (table), **12**-124
    figure, **12**-122
  high-pressure sodium vapor, **12**-120
    figure, **12**-122
  metal halide, **12**-120
    figure, **12**-122
  starting circuits for, **12**-119, **12**-120
  vapor, **12**-120
Lanchester-Prandtl theory (aerodynamics), **11**-61
Land measure, units of, **1**-31
Landfill, **18**-21
Lang-lay wire rope, **8**-85, **10**-9
Laplace inverse transforms, **2**-53
Laplace transformation calculus, **2**-53
Laplace transforms, **2**-53
  table, **2**-52
Lapping, **13**-69
Lapping machines, **13**-66
Lard oil, properties of (table), **6**-202
Larson-Miller parameter (creep), **6**-95
Lasers, **15**-90
  for machine alignment, **16**-7
Latent heats (def), **4**-11
  of fusion (table), **4**-12
  of vaporization (table), **4**-12
Latex, **6**-167
Lathes, **13**-56 to **13**-59
  cutting tools for (see Cutting tools)
  gear calculations for thread cutting, **13**-56
  multispeeds for, **13**-57
  sizes of, **13**-56

Lathes (*Cont.*):
  speed ratios for, **13**-56
  turning recommendations (table), **13**-58
  turret, **13**-59
Latus rectum:
  of parabola, **2**-31
  semi-, of ellipse, **2**-33
Law:
  of cosines, **2**-28
  of sines, **2**-28
Lay cables, **15**-7
  copper (table), **15**-8
Layouts, industrial, **17**-5
L.B.P. (length between perpendiculars, def), **11**-36
Lead:
  alloys, **6**-89 to **6**-91
    for bearings (table), **6**-92
  chemical, **6**-89
    specifications (table), **6**-89
  corroding, **6**-89
    specifications (table), **6**-89
  specifications (table), **6**-89
  tetraethyl (for gasoline), **7**-16
  in type metals (table), **6**-89
Lead-antimony alloys, **6**-89
Lead pipe (table), **8**-175
Lead solders, melting points of (table), **6**-91
Lead tubing (table), **8**-174
League (unit of length, def), **1**-31
Leakage:
  loss through steam turbines, **9**-46
  magnetic, **15**-16
Least squares, method of, **2**-21, **2**-22
Least work, principle of, **5**-36
Leather:
  in belts (see Belts)
  friction of, **3**-26
  specific gravity and density of, **6**-8
Leclanché cell, **15**-10
Ledoux bell flowmeter, **16**-16
Leidenfrost point (def), **4**-68
Lemniscate, **2**-41
Length:
  conversion table for, **1**-44, **1**-45
  equivalents (table), **1**-44
  of plane figures, **2**-11, **2**-12
  units of (def), **1**-31, **1**-32
Lenses, **19**-40
Leveling, **16**-43 to **16**-45
  with transit and stadia, **16**-47
Levels (surveying), **16**-6
  adjustment of, **16**-44
  setup and use of, **16**-43
Lever (mechanism), **8**-3
Lewis formulas for strength of gear teeth, **8**-111
Lift (aerodynamics, def), **11**-60, **11**-61
  in hydraulics, **3**-52
Lift coefficient (def), **11**-61
Lift-slab construction, **12**-7
Lift trucks, **10**-25, **10**-26
Lifting magnets, **10**-12, **15**-78
Lifting tongs, **10**-11
Light:
  units of, **12**-116, **12**-117
  velocity of, **19**-40
  (See also Illumination)
Light meters, **12**-118
Lighting, **12**-117 to **12**-135
  artificial, **12**-9
  economics of, **12**-135

Lighting (*Cont.*):
heat from, **12**-135
of industrial plants, **12**-9, **12**-126
natural, **12**-9
of offices, **12**-126
(*See also* Illumination)
Lighting design, cavity reflectances (table), **12**-128, **12**-129, **12**-132
Lighting fixtures, **12**-118 to **12**-120
heat gain from (charts), **12**-77
Lighting systems for automobiles, **15**-82
Lignite, **7**-5
analyses of (table), **7**-3, **7**-4
combustion of, **9**-7, **9**-10
Lignum vitae bearings, **8**-130
Lime:
common (quicklime), **6**-179
hydrated, **6**-179
magnesium, **6**-179
mortar, **6**-179
specific gravity and density of (table), **6**-8
Lime-soda process (feedwater treatment), **9**-30
Limestone, composition of, **6**-158
Linde liquid-air system, **19**-22
Line-shaft bearings, **8**-129
Linear differential equations, **2**-50, **2**-51
Linear equations, solution of, **2**-18
Linear expansion, coefficient of (def), **4**-3
table, **4**-7
Linear function, graph of, **2**-54
Linear measurements, **16**-4 to **16**-7
surveying, **16**-42
Linear programming, **2**-71
to minimize costs, **17**-16 to **17**-18
Lines:
bisection of, **2**-9
center of gravity of, **3**-7
equations of, **2**-18
of force (def), **3**-22
directions of (magnetic), **15**-17
geometric constructions, **2**-9 to **2**-11
straight, **2**-54
Link (unit of length, def), **1**-31
Linkages, **8**-3, **8**-4
Linotype metal (table), **6**-89
Linseed oil, **6**-118
Liquid measure, **1**-31
Liquid meters, **16**-7, **16**-8, **16**-14 to **16**-18
Liquids (def), **3**-37
boiling, heat-transfer coefficients to or from, **4**-68
coefficient of thermal expansion of, **4**-3
tables, **4**-7, **4**-8
compressibility of, **6**-9
densities of (table), **3**-34
evaporation temperatures of (table), **4**-12
flow of (*see* Flow)
freezing points of (table), **4**-5
general properties of, **3**-34 to **3**-37
latent heat of (table), **4**-12
mechanics of, **3**-33 to **3**-70
in motion, properties of, **3**-40 to **3**-46
pressure, law of, **3**-37
at rest, properties of, **3**-37 to **3**-40
specific gravity and density of (table), **6**-8, **6**-9
specific heat of (table), **4**-9
surface tension of (table), **3**-37
thermal conductivity of (table), **4**-62
thermal expansion of (table), **4**-8
vapor pressure of (table), **4**-12
Liter (def), **1**-32

Ljungström air preheater, **9**-24
Loaders:
crawler, **10**-27
wheeled, **10**-27
Loading:
by power shovel, **10**-37
safety factors for, **5**-22
Lobar-type pumps (rotary), **14**-12
Lock nuts, **8**-21
Locomotive cranes, **10**-30, **10**-72
data on (table), **10**-33
Locomotives, **11**-19 to **11**-28
air resistance, **11**-58
diesel-electric, **11**-21 to **11**-26
batteries, **11**-23
braking, **11**-23
air, **11**-23
dynamic, **11**-23
characteristics of (table), **11**-22
controls, **11**-23
propulsion, **11**-23
wheel slip detection, **11**-23
engines, **11**-21
governors, **11**-21
generators, **11**-23
horsepower, **11**-24
brake, **11**-24
drawbar, **11**-24
indicated, **11**-24
rail, **11**-24
performance of, **11**-24 to **11**-26
adhesion, **11**-24
efficiency (thermal), **11**-24
speed-tractive effort, **11**-24
traction motors, **11**-24
characteristics of, **11**-24
field shunting, **11**-25
maximum speed, **11**-25
saturation curves, **11**-25
transition, **11**-25
diesel-hydraulic, **11**-26, **11**-27
hydrodynamic brakes, **11**-27
efficiency of, **3**-29
electric, **11**-19 to **11**-21
characteristics of (table), **11**-20
drives, **11**-19
ignitron rectifier, **11**-19
speed-tractive effort (curves), **11**-21
storage battery, **10**-22
traction motors, **11**-19
electric haulage, **10**-21, **10**-22
mine, **10**-21, **10**-22
engines, **9**-86
table, **9**-89
gas-turbine, **9**-119, **11**-27
comparison with diesel-electric, **11**-27
mine, **10**-21, **10**-22
haulage capacity of (table), **10**-22
steel for, ASTM specifications (table), **6**-27
storage-battery, **10**-22
Logarithmic cross-section paper, **2**-55
Logarithmic function, **2**-54
complex variable, **2**-24
graph, **2**-55
series for, **2**-44
Logarithmic mean temperature difference, **4**-62, **4**-63
Logarithmic scale (slide rule), **2**-5
Logarithmic spiral, **2**-41
Logarithms, **2**-16
common (table), **1**-9, **1**-10
conversion (base 10 and base $e$), **1**-24

Logarithms (*Cont.*):
   fundamental properties, **2**-4
   Napierian (table), **1**-20, **1**-21
      multiples of (table), **1**-24
   natural (table), **1**-20, **1**-21
   theory of, **2**-16
   use in computation, **2**-3 to **2**-5
Logging of data, **16**-23 to **16**-25
Loop (automatic control), **16**-26
Loop circuit, **15**-62
Loose materials, weight of (table), **10**-2
Loran (navigation by radar), **15**-96
Lost-wax process (molding), **13**-3
Loudness level, **12**-138
Low heat value of fuels (table), **4**-54
Lowry process (wood preservation), **6**-137
LOX (liquid-oxygen explosive), **7**-25
LPG (liquefied petroleum gas):
   in manufactured gas, **7**-21
   specifications for (table), **7**-22
Lubricants, **6**-196 to **6**-207
   coefficients of friction, **3**-25
   disposal of used, **6**-207
   extreme-pressure, **6**-200
   fats, **6**-202
   for gears (table), **8**-119
   greases, **6**-202, **6**-203
   for industrial applications, **6**-206
      additive requirements (table), **8**-206
   for internal-combustion engines, **6**-204
   laboratory-engine tests for, **6**-204
   for machining operations, **6**-206
   marine, **6**-204
   oiliness (def), **6**-200
   oils (*see* Oil, lubricating)
   for press work, **13**-20
   properties of, **6**-196
   solid, **3**-25, **6**-200
   sulphated-residue test, **6**-199
   synthetic (table), **6**-201
   tests, **6**-199
   viscosity, **3**-36, **6**-196, **6**-197
   (*See also* Lubrication)
Lubrication, **6**-203 to **6**-207
   of air compressors, **6**-205, **14**-37, **14**-38
   of aircraft engines, **6**-204
   of automobile engine, **9**-105, **9**-106
   of bearings, **6**-205, **8**-120 to **8**-126
   boundary (def), **3**-24
   complete (def), **3**-24, **8**-120
   fluid, **8**-120
   of gears, **8**-204
   greasy (def), **3**-24
   incomplete, **3**-24
   of internal-combustion engines, **6**-203, **6**-204, **9**-105, **9**-106
   and journal friction, **3**-30
   of refrigerating machinery, **6**-205
   semifluid, **8**-120
   of steam turbines, **6**-204
   systems, **6**-203
   of textile machinery, **6**-205
   of thrust bearings, **8**-130, **8**-131
   viscous (def), **3**-24
   (*See also* Lubricants)
Lucite (plastic), **6**-161, **6**-162
Lüders lines (rolled steel), **5**-4, **6**-26
Lumber, **6**-122 to **6**-138
   allowable unit stresses (table), **6**-128 to **6**-135
   commercial standards, **6**-138
   preservative-treated, allowable stresses, **6**-138

Lumber (*Cont.*):
   stress-graded, **6**-127
   structural glued-laminated, **6**-132
   (*See also* Plywood; Wood)
Lumen (unit of luminous flux, def), **12**-116
Luminaires, **12**-120
   dirt depreciation factors (table), **12**-133
      chart, **12**-133
   lenses for, **12**-120
   reflectors for, **12**-120
   utilization coefficients (table), **11**-130, **11**-131
Luminance (def), **12**-117
Luminous efficiency (chart), **12**-117
Lune, **2**-14
Lurgi process (gas making), **7**-42
L.W.L. (length on load waterline, def), **11**-36


McAdams and Sherwood coefficient of friction (fluid flow), **4**-50
Mach angle (aerodynamics), **11**-75, **11**-76
Mach cone (aerodynamics), **11**-76, **11**-80
Mach number (aerodynamics), **11**-60, **11**-73
Mach number (fluid flow), **3**-48
Mach-Zehnder interferometer, **11**-84
Machinability (def), **13**-5
Machine elements, **8**-9 to **8**-147
   efficiencies of (table), **3**-29
   rotary and reciprocating, **8**-69 to **8**-74
Machine screws:
   driving recesses for (fig), **8**-27
   head types for, **8**-13, **8**-16
   heads, standards for (tables), **8**-21, **8**-22
   threads, **8**-9 to **8**-11
      standards for (tables), **8**-10 to **8**-12
Machine-shop practice, **13**-2 to **13**-79
Machine tools, **13**-48 to **13**-73
   automation, **13**-55
   boring machines, **13**-59
   broaching, **13**-65
   chatter, **13**-51
   control by computer, **2**-71
   cutting fluids for, **13**-54, **13**-55
   cutting-off, **13**-66
   direct numerical control of, **17**-55
   drilling, **13**-60
   gear cutting, **13**-62 to **13**-64
   grinders, **13**-66 to **13**-69
   lathes (*see* Cutting tools; Lathes)
   milling, **13**-61, **13**-62
   mist cooling, **13**-54
   numerical control of, **13**-55, **17**-55
   planers, **13**-64, **13**-65
   reamers, **13**-60
   saws, metal-cutting, **13**-65, **13**-66
   screw machines, **13**-59
   shapers, **13**-64, **13**-65
   slotters, **13**-65
   tool shapes for, **13**-57 to **13**-59
   tools for (*see* Cutting tools)
   turret lathes, **13**-57, **13**-59
   variable-speed transmission, **13**-57
   woodworking, **13**-80 to **13**-84
Machinery steel, **6**-30
Machines:
   guarding of, **18**-24
   point of operation (def), **18**-24
   safety devices for, **18**-24
Machining:
   by chemical milling, **13**-72
   electrical discharge, **13**-71

Machining (*Cont.*):
  electrochemical, **13**-71
  of plastics, **13**-70
  recommendations, **13**-56, **13**-62
  ultrasonic, **13**-72
  (*See also* Machine tools; Metal cutting)
Maclaurin's series, **2**-44
McLeod gage (high-vacuum), **16**-10
McQuaid-Ehn test (austenitic grain size), **6**-20
Magnaflux inspection of castings, **13**-8
Magnesite brick, **6**-171
Magnesium, **6**-81 to **6**-85
  alloys, **6**-81 to **6**-85
    casting mechanical properties (table), **6**-83
    forgings, **6**-82
    joining, **6**-82
    machining, **6**-82
    properties of (tables), **6**-83, **6**-84
    resistance to corrosion, **6**-85
    surface protection, **6**-85
    welding, **6**-82
Magnesium chloride, brine properties of
    (table), **19**-15
Magnet wire, **15**-80
  diameters of (table), **15**-81
Magnetic bearing (surveying), **16**-46
Magnetic circuit, **15**-14
  Ohm's law of, **15**-15
Magnetic coercive force (def), **15**-17
Magnetic drag dynamometer, **16**-18
Magnetic field intensity (def), **15**-2, **15**-15
Magnetic flux (def), **15**-2, **15**-14, **15**-15
Magnetic flux density (def), **15**-2, **15**-15, **15**-16
Magnetic flux direction, **15**-17
Magnetic forming, **13**-21
Magnetic hysteresis, **15**-16
Magnetic leakage, **15**-16
  coefficient of, **15**-16
Magnetic permeability (def), **15**-2, **15**-15
Magnetic permeance (def), **15**-2, **15** 15
Magnetic pole (def), **15**-2
Magnetic potential (def), **15**-2
Magnetic pulleys (conveyors), **10**-59
Magnetic reluctance (def), **15**-3, **15**-15
Magnetic reluctivity (def), **15**-2, **15**-15
Magnetic remanence (def), **15**-17
Magnetic symbols (table), **15**-15
Magnetic tape (computers), **2**-67
Magnetic units, **15**-2, **15**-3, **15**-15
  table, **15**-15
Magnetism, **15**-14 to **15**-18
Magnetization curves, **15**-15, **15**-16
Magnetohydrodynamics, **9**-152
Magnetomotive force (def), **15**-2, **15**-15
Magnetos for ignition systems, **15**-83, **15**-84
Magnets, **15**-73 to **15**-80
  lifting, **10**-12, **15**-78
  permanent, **15**-73
  specialty, **10**-12
Magnification of optical system, **19**-40
Magnifying power of lenses, **19**-40
Magno (resistor alloy), **15**-73
  table, **15**-78
Mains for electric distribution systems, **15**-63
Malleable cast iron, **6**-12, **6**-47
Malleable castings (*see* Castings)
Man:
  conversion efficiency of, **9**-162
  energy output of, **9**-162
    in bursts, **9**-162
  physiological limit of, **9**-162
Management, industrial, **17**-2 to **17**-10

Manganese steel (table), **6**-31, **6**-32
Manganin (resistor alloy), **15**-73
  table, **15**-78
Manila fiber, properties of (table), **6**-155
Manila rope, U.S. specification for (table), **6**-152
Manning formula for flow in open channels, **3**-67
Manometers, **3**-38, **16**-9
  amplification of pressures in, **16**-9
  inclined, **3**-38, **16**-9
  U-tube, **3**-38, **16**-9
  well-type, **3**-38, **16**-9
Mantissa of logarithm, **2**-4
Manufactured gas, **7**-21
Manufacturing plants (*see* Industrial plants)
Mapping, contour, **16**-47, **16**-48
Maraging steel, **6**-29
Marble, composition of, **6**-158
Marine compass, **3**-23
Marine engineering, **11**-35 to **11**-58
  (*See also* Ships)
Marine engines:
  diesel, **9**-86
  gas turbines, **9**-119
  internal-combustion, **9**-86
Marine steam turbines, **9**-48
Martensite, **6**-18
Martensitic hardenable alloys, **6**-38
Masonry:
  ashlar and rubble, **6**-158, **12**-23
  cement, **6**-181
  construction, **12**-22, **12**-23
    in industrial plants, **12**-7
  mortar for (tables), **6**-183
  specific gravity and density (table), **6**-8
  stone, properties of (table), **6**-158
  terms for, **6**-158
  working compression (table), **12**-23
Mass (def), **3**-2
  center of, **3**-7
  conversion to energy, **9**-130
  conversion table for, **1**-48
  energy equivalence of, **9**-120
  equivalents (table), **1**-48
  law of conservation of, **3**-3, **3**-44
  relativistic, **9**-120
  units of, **4**-8
    defined, **1**-32
  variable, **3**-22
Mass defect, nuclear, **9**-120
Materials:
  angle of repose of (def), **10**-2
  general properties of, **6**-3 to **6**-11
  handling, codes and specifications for, **10**-2
  handling systems, **10**-2
  loose, weight of (table), **10**-2
  mechanical properties of, **5**-2 to **5**-4, **5**-22
    at low temperatures, **19**-28
  methods of moving, selection of, **10**-2
  preferred conveyors for (table), **10**-3, **10**-4
Materials testing, **5**-14, **5**-15
Mathematical models, **2**-71
Mathematical tables, **1**-2 to **1**-30
Mathematics, **2**-2 to **2**-64
Matter:
  conservation of, **9**-120
  particles of, **9**-119
Maxima and minima, **2**-43 to **2**-45
Maxwell (magnetic flux, def), **15**-2, **15**-14
Maxwell's relations (thermodynamics), **4**-15
  table, **4**-16

Maxwell's theorem (beam deflections), **5**-36
Mean:
    arithmetic, **2**-17, **17**-19
    geometrical, **2**-16, **2**-17
        construction, **2**-9
    harmonic, **2**-17
    proportional, **2**-16, **2**-17
        construction, **2**-9
Mean camber line (airfoils, def), **11**-62
Mean effective pressure, **4**-20
    in steam engines, **9**-36
Mean specific heats (*see* Specific heat)
Mean square error (def), **2**-21
Mean temperature difference in heat
        transmission, **4**-62
Mean value theorem, **2**-49
Measurements, **16**-2 to **16**-25
    absolute systems (mechanics), **3**-2
    acceleration, **16**-20, **16**-21
    angle, **16**-4, **16**-5
    area, **16**-7
    counting, **16**-2, **16**-3
    electrical, **15**-24 to **15**-30, **16**-18 to **16**-20
    fluid flow, **16**-14 to **16**-18
        integrators for, **16**-18
    fluid volume, **16**-7, **16**-8
    force, **16**-8
    frequency, **16**-3
    gravitational systems (mechanics), **3**-2
    heat, **4**-8 to **4**-11
    humidity, **16**-21
    linear, **16**-4 to **16**-7
    liquid level, **16**-10
    mass, **16**-3, **16**-4
    nuclear radiation, **16**-22, **16**-23
    physical and chemical properties of, **16**-21,
        **16**-22
    plasticity, **16**-21
    power, **16**-18
    pressure, **16**-8 to **16**-10
        differences, **16**-9
    standards for, **16**-2
    temperature, **4**-2, **4**-3, **16**-10 to **16**-14
    of thickness, **16**-4 to **16**-7
    time, **16**-3
    torque, **16**-8
    vacuum, **16**-8 to **16**-10
    velocity, **16**-20, **16**-21
    viscosity, **16**-21
    weight, **16**-3, **16**-4
Measures and weights:
    metric (*see* Metric measures and weights)
    U.S., **1**-31, **1**-32
Measuring instruments (*see* Instruments;
        Meters)
Mechanical draft, **9**-26
Mechanical efficiency of steam engines, **9**-36
Mechanical equivalent of heat (def), **1**-35, **4**-8
Mechanical movements, **8**-3 to **8**-8
Mechanical properties of materials, **5**-2 to **5**-15
Mechanical refrigeration (*see* Refrigeration)
Mechanical seals, **8**-146, **8**-147
Mechanics:
    of fluids, **3**-33 to **3**-70
    of materials, **5**-16 to **5**-61
    Newtonian, **2**-63
    Newton's laws of, **3**-2
    of solids, **3**-2 to **3**-23
Mechanism, **8**-3 to **8**-8
Meehanite (cast iron), **6**-47, **6**-48
Megger (measurement of insulation
        resistance), **15**-28

Megohmit (electrical insulation), **6**-152
Megotalc (electrical insulation), **6**-152
Melamine-formaldehyde (plastic), **6**-163
Melting points:
    of brazing alloys (table), **6**-92
    of fusible alloys (table), **6**-90
    of metals (table), **6**-11, **6**-60, **6**-61
    of nonmetallic elements (table), **4**-5
    of reactor materials (table), **4**-5
    of refractories (table), **6**-173, **6**-176
    of resistor alloys (table), **15**-78
    of solders (table), **6**-91
    of various solids (table), **4**-5
Melting pots, resistor heating of, **7**-63
Membranes, vibration of, **5**-74
Mensuration, **2**-7 to **2**-15
Mercury:
    properties of (table), **4**-32
    uses of, **6**-87
Mercury thermometers, **16**-10
Mercury vapor lamps, **12**-120
Merrick weightometer, **10**-63
Mershon diagram for a-c voltage drop, **15**-56
    chart, **15**-58
Mesh connections, three-phase circuit, **15**-23
Metacenter, **3**-40
    of ship (def), **11**-39
Metal cutting, **13**-48 to **13**-73
    basic mechanics of, **13**-49 to **13**-51
    carbon-arc, **13**-37
    chips, **13**-49
    forces in, **13**-49
    gas, **13**-36, **13**-37
    power requirements, **13**-50
    principles of, **13**-49 to **13**-51
    tools for (*see* Cutting tools; Machine tools)
    vibration effects, **13**-51
    under water, **13**-37
Metal working (*see* Metalworking)
Metallurgy, powder, **6**-104
    (*See also* Alloys; *specific metals*)
Metals:
    antifriction, **6**-91 to **6**-94
    for atomic-energy applications, **6**-100 to **6**-
        103
    atomic numbers (table), **6**-60, **6**-61
    atomic weights (table), **6**-60, **6**-61
    b.c.c., for cryogenic equipment, **19**-29
    bearing (*see* Bearing metals)
    bending of, **13**-19
    boiling points (table), **6**-60, **6**-61
    calorizing of, **6**-116
    chemical symbols (table), **6**-60, **6**-61
    coefficients of thermal expansion, **4**-3
        tables, **4**-7, **6**-60, **6**-61
    coining of, **13**-22
    cold-working of, **13**-15
    corrosion of (*see* Corrosion)
    creep of (table, chart), **5**-11, **5**-12
    for cryogenic equipment, **19**-29, **19**-30
    crystal lattice of (table), **6**-60, **6**-61
    density (table), **6**-7, **6**-11, **6**-60, **6**-61
    drawing of, **13**-19, **13**-20
    elastic constants (table), **5**-5
    electrical resistivity (table), **6**-60, **6**-61, **15**-6
    embossing of, **13**-22
    emissivity (table), **4**-73
    extrusion of, **13**-22
    fatigue of, **5**-9, **5**-10
    f.c.c., for cryogenic equipment, **19**-29
    flame machining of, **13**-37
    forging of, **13**-10 to **13**-23

Metals (*Cont.*):
  h.c.p., for cryogenic equipment, **19**-30
  heat of fusion (tables), **4**-12, **6**-60, **6**-61
  heat content at various temperatures (chart),
    **7**-49
  for high-temperature use, **6**-94 to **6**-100
  hot-working of, **13**-11
  jewelry, **6**-103
  machinability, **13**-50
  melting point (table), **4**-61, **6**-60, **6**-61
  modulus of elasticity (table), **5**-5, **6**-60, **6**-61
  molten, properties of (table), **4**-61
  nonferrous, **6**-59 to **6**-106
  physical constants (table), **6**-60, **6**-61
  plastic range chart, **13**-15
  powdered, fabrication by, **6**-104
  properties of (table), **6**-7, **6**-11
  relaxation test, **5**-12
  resistivity (table), **15**-6, **15**-7
  resistor materials, **15**-69 to **15**-73
  shearing of, **13**-17, **13**-18
  sizing of, **13**-20
  solution pressures of (def), **6**-107
    order of, **6**-107
  specific gravity and density (table), **6**-7
  specific heat (table), **6**-60, **6**-61
  spinning of, **13**-17
  spraying processes, **6**-115
  squeezing of, **13**-21, **13**-22
  stamping of, **13**-22
  strength of (table), **5**-5
    at high temperatures, **5**-11
  structure, working, **13**-9, **13**-10
  surface finish, **13**-73 to **13**-79
  swaging of, **13**-21
  thermal conductivity (tables), **4**-60, **4**-61, **6**-60, **6**-61
  true-stress-strain curves, **13**-16
  types of, composition and properties of
    (table), **6**-89
  welding of (*see* Welding)
Metalworking, **13**-9 to **13**-23
  equipment for, **13**-23 to **13**-27
Meters, **16**-2 to **16**-25
  air, **16**-14 to **16**-18
  for conveyors, **10**-63, **10**-64
  displacement, **16**-8
  electric, **15**-24 to **15**-30
  gas, **16**-14 to **16**-18
  liquid, **16**-14 to **16**-18
  nutating disk, **16**-8
  piston, **16**-8
  venturi, **3**-61, **3**-62, **16**-15
  (*See also* Measurements)
Methane (*see* Hydrocarbons)
Methanol (wood alcohol), **6**-168
Method of joints (stresses in structures), **3**-6
Method of sections (stresses in structures), **3**-7
Methyl alcohol, **6**-168
  as freezing preventive (table), **6**-156
  as fuel, **7**-20
  as solvent, **6**-168
Methyl chloride:
  as refrigerant, **19**-4
  thermal properties of, **4**-38
Metre (unit of length, def), **1**-22
Metric measures and weights, **1**-40
  conversion tables and equivalents, **1**-34 to **1**-39
  prefixes, **1**-34
Metric units (*see* SI units)
Mho (conductance, def), **15**-4

Mica, **6**-152
Micanite (electrical insulation), **6**-152
Microfarad (capacitance, def), **15**-4
Micrometers, **16**-5
Micron (def), **1**-37
Microwave, testing of materials, **5**-83
Mil, circular (def), **1**-31
Mil-foot (def), **15**-5
Mill-construction floors, **12**-23
Milling cutters, **13**-62
Milling machines, **13**-61, **13**-62
  types of, **13**-61, **13**-62
Millivoltmeters, **15**-24
Mills, pulverized coal, **9**-11
Millstones, **6**-139
Mine cars, **10**-23
Mine hoists, **10**-16 to **10**-19
Mine locomotives, **10**-21, **10**-22
Mineral spirits (solvent), **6**-168
Minerals, specific gravity and density (table), **6**-9
Miner's inch (def), **1**-31
Minicomputers, **2**-67
Minim (apothecaries' liquid measure, def), **1**-31
Minimum (point on a curve), **2**-44
Minute (circular measure, def), **1**-32
Minutes to decimals of a degree (table), **1**-30
Mitchell bearing, **8**-131
Miter gears, **8**-108
Mixing of two atmospheres, **4**-44
Mixtures:
  of air and water vapor, **4**-28 to **4**-45
  of liquid and vapor (thermodynamics), **4**-23
Mobilgas Economy Run, **11**-5
Models:
  mathematical, **2**-71
  for plant layout, **12**-4
  theory of, **3**-46, **3**-47
  for wind-pressure tests, **12**-12, **12**-13
Moderators (for slowing neutrons), **9**-125, **9**-128
  table, **9**-128
Modulators:
  amplitude modulation (AM), **15**-90
  frequency-modulation (FM), **15**-90
Modulus:
  bulk, of elasticity (def), **5**-18
    table, **5**-5
  of elasticity (def), **5**-18
    of metals (tables), **5**-5, **6**-60, **6**-61
    table, **5**-5
  of resilience (def), **5**-18
  of rigidity (shearing) of metals (def), **5**-18
    tables, **5**-5
  of rupture (def), **5**-26
  of strain hardening (def), **13**-15
  Young's (def), **5**-2
    table, **5**-5
Mohr's circle:
  in moments of inertia, **3**-9, **3**-10
  in stress analysis, **5**-19
Mohr's strain circle, **5**-20
Mohs scale (hardness), **1**-40, **5**-12
Moisture:
  in air, measurement of, **4**-31
  in steam, measurement of, **16**-21
Mol (def), **1**-32, **4**-11
Molal specific heat, **4**-11
Molding, **13**-3 to **13**-7
  centrifugal process, **13**-4
  De Lavaud process, **13**-4

Molding (*Cont.*):
  dry and green sand, **13**-3
  investment process, **13**-3
  lost-wax process, **13**-3
  process, **13**-3
  sand, **13**-3, **13**-5, **13**-6
  sand preparation, **13**-6
  Shaw process, **13**-3
  for steel castings, **6**-57, **6**-58
  (*See also* Castings; Cores)
Molding machines, **13**-4, **13**-5
  flasks, **13**-5
  jolt and squeeze, **13**-5
  sand slinger, **13**-5
  vibrators, **13**-5
Molds:
  graphite, **13**-4
  permanent, **13**-3
  plaster, **13**-3
Mole (unit of amount of substance, def),
  **1**-32
Molecular weight:
  of common gases (tables), **4**-51, **4**-55
  of liquid fuel (tables), **4**-51, **4**-55
Molecule (def), **6**-3
Mollier diagram:
  description of, **4**-23
  for sodium, **4**-27
  for steam, **4**-25, **4**-26
  use of, **4**-23
Molybdenum alloys, **6**-98
Moment:
  in aerodynamics, **11**-61
  of a couple, **3**-34
  of a force, **3**-4
  of inertia, **3**-8 to **3**-11
    of areas (def), **3**-10
    of beam sections (tables), **5**-30 to **5**-33
    of built-up sections, **3**-10
    determined experimentally, **3**-19
    about parallel axes, **3**-9
    of plane area, **3**-10
      by graphics, **3**-11
    polar (def), **3**-9
    principal, **3**-9
    referred to any axis, **3**-10
    of solid body (def), **3**-8
    of solids (table), **3**-10
  method of, in trusses, **12**-17
  of momentum, **3**-21
  of resistance (beams, def), **5**-23
  and shear in beams, **5**-34
  unit of rotation (def), **3**-4
Moment diagram for beams, **5**-24 to **5**-27
Momentum (def), **3**-21
  angular (def), **3**-2
  law of conservation of, **3**-3, **3**-21
  linear (def), **3**-2
  moment of, **3**-21
Monel, **6**-70 to **6**-75
  castings, **6**-73
  composition and properties of (table), **6**-72
    to **6**-74, **15**-78
  heat treatment, **6**-73
  machining, **6**-73
  magnetic properties of, **6**-74
  mechanical properties of (table), **6**-72
    at high temperature (table), **6**-73
    at low temperature, **6**-74
  welding, **13**-45
Monopropellants (rocket), **7**-35
Monorail hoists, **10**-14

Monorails, **10**-28, **11**-33
  trolley dimensions (table), **10**-29
Monotron hardness test, **5**-13
Monotype metal (table), **6**-89
Monte-Carlo, **2**-72
Moody formula for hydraulic turbine
  efficiency, **9**-147
Morse standard taper pins (table), **8**-34
Mortar, **6**-180 to **6**-183
  for brickwork, **6**-181
    amount required (table), **6**-183
  cement, **6**-181
  for firebrick, **6**-173, **12**-22
  lime, **6**-180
  lime and cement, **6**-180
  for masonry construction, **12**-22
  for plastering, **6**-181
  sand for, **6**-179
  for stonework and masonry, **6**-181
    amount required (table), **6**-183
  strength of (table), **6**-182
  weight of materials per cubic yard (table), **6**-181
Motion:
  angular, **3**-14
  under constant force, **3**-16
  curvilinear, **3**-13 to **3**-20
  harmonic (simple), **3**-14
  plane (def), **3**-14
  polygon of, **3**-12
  rectilinear, **3**-20
  relative, **3**-15
  of rigid bodies, **3**-12 to **3**-20
  uniform, **3**-12
  uniformly accelerated or retarded, **3**-12
Motion study, **17**-26 to **17**-32
  aims of, **17**-26
  elements of, **17**-26
  motion classification, **17**-28
  operation analysis of, **17**-27
  principles of, **17**-28
  scope of, **17**-26
  time standards, **17**-28, **17**-30 to **17**-32
Motor (*see* Airplanes, engines; Automobile
  engines; Electric motors)
Motor-generator balancer set, **15**-62
Motor vehicles (*see* Automobiles; Trucks)
Muffle furnaces, **7**-48
  (*See also* Furnaces)
Mufflers (exhaust), **9**-102, **12**-139
Multiple-effect compression (refrigeration),
  **19**-7
Multiplication:
  algebraic, **2**-16
  arithmetical, **2**-2
  of complex quantities, **2**-23
  by logarithms, **2**-4
  by slide rule, **2**-6
  tables, references for, **2**-2
  of vectors, **2**-63
Multivibrator (electronic circuit), **15**-91
Muntz metal (copper-zinc alloy, table), **6**-64, **6**-66
Muscular energy, **9**-162
Mutual inductance, **15**-4, **15**-20

Nails:
  cut steel (table), **8**-95
  holding power of, **12**-29
  lateral resistance in wood, **12**-27
    table, **12**-29

Nails (*Cont.*):
  points for, **8**-97
  wire (tables), **8**-94, **8**-95
Naphtha, solvent, **6**-168
Napierian logarithms, **2**-16
  multiples of (table), **1**-24
  table, **1**-20, **1**-21
Napier's formula for steam flow, **4**-47
Natural cement, **6**-178
Natural gas (table), **7**-22
  demand, **9**-6
  flame temperatures (table), **4**-57
  reserves (table), **9**-3, **9**-4
Natural logarithms, **2**-16
  table, **1**-20, **1**-21
Nautical units of length and speed, **1**-31
Naval brass (table), **6**-65, **6**-67
Neoprene (synthetic rubber), **6**-166
Neutral axis (beams, def), **5**-23
Neutron balance, **9**-127
Neutron diffractometry, **9**-121
Neutron-level instruments, **9**-129
Neutron radiography, **9**-121
Neutrons, **6**-7, **9**-119
  mass of, **9**-119
  thermal, absorption and scattering cross
    sections (table), **9**-128
Newton (magnetic repulsion, def), **15**-2
Newtonian fluid (def), **3**-34
Newtonian mechanics, **2**-63, **3**-2
Newton's laws (mechanics), **3**-2
Nicad battery, **15**-14
Nichrome (resistor alloy), **15**-73
  table, **15**-78
Nickel, **6**-70 to **6**-75
  alloys, **6**-70 to **6**-75
    castings, composition and properties of
      (tables), **6**-70 to **6**-73
    copper (*see* Monel)
    heat-resisting, **6**-95, **6**-96
    heat treatment of, **6**-73
    magnetic properties, **6**-74
    mechanical properties of (table), **6**-72
    strength at high temperatures (table), **5**-
      12, **6**-74
    welding of, **13**-37, **13**-43, **13**-45
  composition and properties of (table), **6**-71,
    **6**-72, **15**-73
  electroplating of, **6**-74, **6**-115
  high-temperature properties of (table), **6**-74
  production of, **6**-70
  properties of (tables), **6**-72 to **6**-74
  resistor, **15**-73
Nickel plating, **6**-115
Nickel silver, **6**-69
  table, **6**-65, **6**-67
Nickel steel (*see* Steel, nickel)
Nimonic (heat-resisting alloy, table), **6**-97
Niobium, **6**-96
  uses of, **6**-97
Nitriding (heat-treatment), **6**-24
  steels for (table), **6**-24
Nitrile rubber, **6**-166
Noise, **12**-136 to **12**-143
  control of, **12**-139 to **12**-142, **18**-26
    by damping, **12**-139
    by design changes, **12**-139
    by filtration, **12**-139
    by isolation, **12**-139
    by quieting, **12**-140
    by shielding, **12**-139
  in ducts, versus friction drop (table), **12**-100

Noise (*Cont.*):
  occupational exposure to, **18**-26
  permissible exposures (table), **12**-143
  reduction, coefficient (def), **12**-140
  safe levels of, **12**-143
  (*See also* Sound)
Nomograms, **2**-58 to **2**-62
Nonferrous metals, **6**-59 to **6**-106
  annealing of, **6**-62
  casting, **6**-59
  cold-working of, **6**-59
  effect of temperature on, **5**-12
  effects of environment, **6**-62
  precipitation hardening of, **6**-62
  (*See also* Alloys; *specific metals*)
Notation, algebraic, **2**-15
Notch sensitivity in fatigue of metals (def), **5**-
  10, **8**-48
Notches (*see* Weirs)
Nozzles, **3**-44, **3**-61, **3**-62
  divergence of, **4**-49
  flow: of compressible fluids through, **4**-46 to
    **4**-49
    formula for, **3**-62
    of steam through, **4**-46 to **4**-49, **9**-40 to **9**-
      42
  for flow measurement, **16**-15
  fuel-injection (diesel engines), **9**-99
  hydraulic turbines, **9**-145
  steam, design of, **9**-40
Nuclear boilers, **9**-34, **9**-35
  BWR system, **9**-34
  chemical poisons from, **9**-35
  chemical reactions in, **9**-35
  codes and specifications for, **9**-35
  design considerations, **9**-34
  economics of, **9**-35
  hazards, **9**-35
  heat transfer in, **9**-34
  PWR system, **9**-34
  radioactivity, **9**-35
Nuclear energy (*see* Atomic energy)
Nuclear fission, **9**-121
  energy available, **9**-121
Nuclear fuels, **9**-129
  costs of, **9**-131 to **9**-133
    chemical reprocessing, **9**-132
    enrichment, **9**-132
    fabrication, **9**-132
    fuel-burnup, **9**-132
    shipping charges, **9**-132
  properties of (table), **9**-126
Nuclear fusion, **9**-120
  energy released in, **9**-120
Nuclear particles:
  binding energy of, **9**-120
  elementary, **9**-119
Nuclear physics, **9**-119 to **9**-121
Nuclear power (*see* Atomic power)
Nuclear radiation instruments, **16**-22,
  **16**-23
Nuclear reactors (*see* Reactors)
Nucleus:
  cross section of, **9**-127
  fission of, **9**-121
  half-life of, **9**-121
  radioactivity of, **9**-121
Numbers, preferred, **8**-201
Nutating disk, flowmeter, **16**-8
Nuts:
  force required to tighten or loosen, **3**-30
  lock, **8**-21

Nuts (*Cont.*):
  materials for, **8**-24
    strength of (table), **8**-28
    tests for, **8**-24
  standard threads for, **8**-10, **8**-11
  wrench openings for (table), **8**-20
Nuveyor pneumatic ash handling, **10**-62
Nylon, **6**-151, **6**-154
  fiber properties of (table), **6**-155
  plastic (table), **6**-161, **6**-162
Nyquist stability criterion (automatic control), **16**-34

Obelisk, volume of, **2**-13
Occupational accidents (table), **18**-23
Occupational disease, prevention of, **18**-25 to **18**-27
Octagon (construction), **2**-10
Octahedron, **2**-8, **2**-14
Octane number of gasolines (def), **9**-96
  table, **7**-17
Oersted (magnetic field intensity, def), **15**-2, **15**-15
Offset cylinders, **8**-60
Ohm (def), **15**-4
Ohm's law, **15**-8
  of magnetic circuits, **15**-15
Oil:
  acidity, tests for, **6**-199
  additives for, **6**-199, **6**-204
  animal, **6**-202
    properties of (table), **6**-202
  blown, **6**-118
  burners for, **9**-12
  castor, **6**-118
  cloud point, **6**-198
  coal suspensions in, as fuel, **7**-20
  compressibility, **6**-9, **6**-198
  core binder, **13**-6
  crankcase, SAE viscosity number (table), **6**-197
  crude, as fuel, **7**-14
  cutting fluid (machining), **6**-206, **13**-54
  density of, **6**-198
  extreme pressure, **6**-200
  fish, **6**-118, **6**-202
  flash and fire points (def), **6**-198
  fuel (*see* Fuel oils)
  in gas manufacture, **7**-46
  heat-transfer coefficients to or from (chart), **4**-67
  hydrocarbon (paints), **6**-118
  insulating, specifications for, **6**-152
  linseed, **6**-118
  lubricating, **6**-196 to **6**-200
    additives for, **9**-106
    gravity of, **6**-198
    SAE classifications of, **9**-106
    tests of, **6**-198 to **6**-200
  oiliness of (def), **6**-200
  oiticica, **6**-118
  for paint, **6**-118
  petroleum (*see* Petroleum)
  pour point, **6**-198
  slushing, **6**-117
  soybean, **6**-118
  specific gravity of, **6**-198
  specific gravity and density of (table), **6**-8, **6**-9
  synthetic, **6**-200
  tung, **6**-118

Oil (*Cont.*):
  vegetable, **6**-202
    properties of (table), **6**-202
  viscosity of, **3**-35, **6**-197
    index of (def), **6**-198
    recommended (table), **6**-197
    temperature variation of, **6**-198
  world reserves of crude (table), **9**-3
Oil burners (*see* Burners)
Oil cooling, internal-combustion engines, **9**-105
Oil engines (*see* Diesel engines; Internal-combustion engines)
Oil grooves, arrangement of, **8**-127
Oil paint, **6**-118
Oil Pollution Act of 1961, **11**-43
Oil quenching, **6**-22
Oil-sand cores, **13**-6
Oil seals for turbocompressors, **14**-41
Oilless bearings, **8**-129
Oilstones, **6**-139
Oiticica oil, **6**-118
Once-through boilers, **9**-14, **9**-25
Open channels:
  critical values for, **3**-68
  flow of water in, **3**-66 to **3**-68
  roughness coefficient of (table), **3**-68
  specific energy in flow in, **3**-68
Open-hearth furnace, operation of, **6**-15, **6**-55
Open-hearth steel (def), **6**-12, **6**-15
Optical comparators, **16**-6
Optical pyrometers, **16**-11, **16**-13, **16**-14
Optics, **19**-40, **19**-41
Optimization:
  using computers, **2**-71
  using linear programming, **17**-16 to **17**-18
Ores, specific gravity and density of (table), **6**-7
Organization, industrial, **17**-2
Orifice meters, **3**-62 to **3**-64, **16**-15
Orifices, **16**-15, **16**-16
  diaphragm meters for, **16**-16
  discharge coefficients for, **3**-63
    table, **3**-64, **16**-16
  expansion factors for, **16**-16
  flow: of air through, **4**-46, **4**-47
    of compressible fluids through, **3**-63, **3**-64, **4**-46, **4**-47
    of gas through, **4**-47
    of steam through, **4**-47
  formulas for, **4**-47
  pipe, **3**-63
  standard, **3**-63
  tap locations for, **3**-63, **16**-15
Orsat apparatus (flue-gas analysis), **16**-22
Oscillating, rolling (metalworking), **13**-17
Oscillation, axis of, **3**-19
Oscillators, **15**-89, **15**-90
Oscilloscope, **16**-20
OSHA (Occupational Safety and Health Act), **18**-27
Oswatitsch shock diffuser, **11**-78
Ottawa sand for concrete, **6**-179
Otto cycle, **4**-20
Otto-cycle engines (*see* Internal-combustion engines)
Outboard engines, **9**-86, **11**-52
  table, **9**-90
Outside count test (statistics), **17**-22
Oven furnaces, **7**-48
Ovens:
  for carbonizing coal, **7**-39, **7**-40
  core, **13**-7

Ovens (*Cont.*):
    electric (*see* Electric furnaces)
    resistor, **7**-95
Overdrive (automobiles), **11**-7
Overhead in cost accounting, **17**-12
Overhead cranes (*see* Cranes, overhead)
Overhead trackage, **10**-40 to **10**-44
Overlap test (statistics), **17**-23
Oxidation tests for lubricating oils, **6**-199
Oxide coating of iron (corrosion protection), **6**-116
Oxyacetylene flame cutting, **13**-37
Oxyacetylene welding, **13**-36
Oxygen:
    corrosion in boilers, **6**-107, **9**-30
    liquid (rocket fuel), **7**-36
    in water, removal of, **6**-110, **9**-30
Oxygen analyzer, **16**-22
Oxyhydrogen welding, **13**-36

Pack rolling, **13**-16
Package conveyors, **10**-49
Packing, **8**-144 to **8**-146
    carbon, for steam turbines, **9**-46
    conical ring, **8**-144
    diaphragms to replace, **8**-147
    dynamic, **8**-144
    friction of, **3**-27
    jamb, **8**-145
    labyrinth, for steam turbines, **9**-46
    liquid seal, **8**-144
    O-ring, **8**-144
    oil seals, **8**-146
    plastic, **8**-146
    ring, for steam turbines, **9**-46
    shaft, **8**-145
    soft, **8**-145
    steam turbine, **9**-46
Painting, **6**-120, **6**-121
    aluminum, **6**-120
    application, **6**-121
    concrete, **6**-120
    copper, **6**-120
    exterior, **6**-120
    galvanized iron, **6**-120
    interior, **6**-120
    magnesium, **6**-120
    masonry, **6**-120
    plastics, **6**-121
    spray method, **6**-121
    spreading rates (table), **6**-121
    steel, **6**-120
    tin plate, **6**-120
    water tanks, **6**-120
    wood products, **6**-120
Paints and protective coatings, **6**-118 to **6**-121
    aluminum, **6**-119
    antifouling (ships), **6**-119
    bituminous, **6**-119
    calcimine, **6**-119
    casein, **6**-119
    copper powder, **6**-119
    definition of, **6**-118
    destroying agencies of, **6**-121
    driers for, **6**-118
    drying oils in, **6**-118
    electrical, **6**-119
    emulsion, **6**-119
    enamel, **6**-118
    fire-retardant, **6**-119
    fungicides added to, **6**-119

Paints and protective coatings (*Cont.*):
    heat radiation of, **6**-119
    heat-reflecting, **6**-119
    heat-resistant, **6**-119
    ingredients of, **6**-118, **6**-119
    lacquer, **6**-118, **6**-121
    latex, **6**-119
    luminous, **6**-119
    paste, **6**-119
    pigments in, **6**-119
    plastic, hot, **6**-119
    plasticizers for, **6**-118
    resins in, **6**-119
        natural, **6**-119
        synthetic, **6**-119
    ship-bottom, **6**-119
    silicone, **6**-119
    solvents, **6**-119
    thinners for, **6**-119
    traffic, **6**-119
    varnish, **6**-118, **6**-121
    water-repellent, **6**-120
    water-thinned, **6**-118
    wood preservative, **6**-120
Palladium, uses of, **6**-103
Pallets:
    in industrial plants, **10**-26, **12**-4
    lift trucks for, **10**-25, **10**-26
    loads on, **10**-26
Panhandle formula (natural gas pipelines), **11**-130
Paper, **6**-159, **6**-160
    bleaching of, **6**-160
    coatings, **6**-160
    converting, **6**-160
    grades of, **6**-159
    insulating, **6**-153
    plastic-fiber, **6**-160
    pulping, **6**-159
    refining, **6**-160
Pappus, theorems of, **2**-14, **2**-15
Parabola:
    area of, **2**-12
    constructions of, **2**-32
    properties of, **2**-31, **2**-32
Parabolic type of power function, graph of, **2**-54
Paraboloid of revolution, volume of, **2**-14
Parallel charts, **2**-61
Parallel circuits, **15**-8, **15**-62
    a-c, **15**-23
Parallel operation of alternators, **15**-39
Parallelogram:
    area of, **2**-11
    of motion, **3**-3, **3**-12
Parallelopiped of motion, **3**-3
Parametric equations, **2**-31
Parasite drag (aerodynamics, def), **11**-66 to **11**-71
Parasitic currents, **15**-17
Parkerizing (protection of iron), **6**-116
Parr formulas (mineral-matter-free coal), **7**-2
Partial derivatives, **2**-43
Partial pressures, **4**-18
Particle clouds, **4**-77 to **4**-80
Particles of matter, **9**-119
Partitions:
    industrial plant, **12**-8
    reinforced-concrete, **12**-59
Pascal's law, **3**-37
Passenger cars (railroad), **11**-28
    train resistance, **11**-30, **11**-57, **11**-58

Passivating of iron surfaces, **6**-117
Patents, **19**-41 to **19**-43
  cost of, **19**-42
  infringement of, **19**-42
    damages for, **19**-43
  interferences, **19**-42
  International Convention for the Protection
    of Industrial Property, **19**-43
  licenses, **19**-42
  procedures, **19**-42
  reissues of, **19**-43
  requirements, **19**-41
  sales or licensing, **19**-42
  term of, **19**-41
  who may apply for, **19**-41
Patterns, **13**-2, **13**-3
  distortion allowance, **13**-2
  draft, **13**-2
  machine-finish allowance, **13**-2
  making, **13**-3
  for molding, **13**-2, **13**-3
  shrinkage allowance, **13**-2
  types of, **13**-2, **13**-3
Paving, brick, **6**-149
Pearlite (steel), **6**-18
Peat, **7**-12
  moisture content of, **7**-12
  properties of, **7**-12
  specific gravity and density of, **6**-9
Peck, dry measure (def), **1**-31
Peck carrier buckets, **10**-53
Peltier effect, **9**-150
Pendulum:
  compound, time of oscillation, **3**-19
  conical, **3**-17
  simple, **3**-17
Penetrameters (radiographic), **5**-82
Perch (unit of length, def), **1**-31
Percussion, center of, **3**-18
Perfect gases, **4**-16 to **4**-22
Perfectly diffuse surface, **4**-73
Periodic table of elements, **6**-6
Permalloy, **6**-41
  magnetic properties (table), **6**-41
Permanent set (def), **5**-2
Permeability curves, **15**-15, **15**-16
Permeance, magnetic (def), **15**-2, **15**-15
Perminvar, **6**-41
Permutations and combinations, **2**-18
Personnel management, **17**-9, **17**-10
PERT (Program Evaluation and Review
  Techniques) for scheduling, **17**-6
Peter's formula for computing probable error,
  **2**-21
PETN (explosive), **7**-25
  table, **7**-26
Petroleum:
  analyses of (table), **7**-14
  API and specific gravity (table), **7**-14
  as fuel, **7**-14, **7**-15
    (See also Fuel)
  heat values (table), **7**-14
  oils, **6**-196 to **6**-200
    (See also Oil)
  refining, **7**-14
  reserves, **9**-3, **9**-4
  specific gravity and density (table), **6**-8,
    **6**-9
  spirits (paint thinner), **6**-119
Pewter, **6**-89
pH value:
  definition of, **6**-107, **9**-31

pH value (*Cont.*):
  indicators for, **9**-31
  relationship to hydrogen in concentration
    (table), **9**-32
Phase converter, **15**-47
Phase difference in alternating currents, **15**-22
Phenol formaldehyde adhesive, **6**-141
Philips hot-air engine, **9**-163
Phon (loudness level of sound, def), **12**-138
Phono-electric bronze (electrical conductor), **15**-6
Phosphate coatings (protection of iron), **6**-116
Phosphor bronze, **6**-68
  tables, **6**-65, **6**-67
Photoelasticity, **5**-57, **5**-58
  material-fringe value, **5**-57
  oblique incidence, **5**-58
  stress separation, **5**-57
  three-dimensional, **5**-58
Photometers, uses of, **16**-21
Photostress, **5**-60
Pi ($\pi$), multiples of (tables), **1**-2, **1**-12
Picric acid (military explosive, tablc), **7**-26
Piece-rate wage systems, **17**-9
Piezoelectric force measurement, **16**-8
Piezoelectric transducers, **5**-76, **5**-82
Pig iron (def), **6**-12
  production of, **6**-12
  for steel making (table), **6**-13
Pigments (paints), **6**-119
Pile, atomic (see Reactors)
Piles:
  capping, **12**-21
  concrete, **12**-21, **12**-59
  driving, **12**-21
  foundations, **12**-21
  safe loads for, **12**-21
  spacing of, **12**-21
  wood, **12**-21
Pinions (see Gears)
Pins, taper (table), **8**-31
Pipe:
  aging of, **3**-54
  allowable stresses for (tables), **8**-150 to **8**-155
  alloy, **8**-149
    butt welding, **8**-195
  aluminum (table), **8**-173, **8**-174
  anchoring of, **8**-197
  asbestos-cement, **6**-114, **8**-172
  bends, expansion, formulas and table, **5**-61
    to **5**-67
  block tin (table), **8**-175
  branch flow, compound, **3**-59
  brass (table), **8**-171
  butt-welded, **8**-195
  carbon-molybdenum steel, codes for, **8**-149
  carbon-steel, codes for, **8**-149
  cast-iron, **8**-155 to **8**-169
    allowances for water hammer, **8**-149
    bell-and-spigot (tables), **8**-168
      ends, **8**-163
    centrifugally cast, **8**-164
    for fire protection, **18**-32
    fittings for (see Pipe fittings)
    flanged (table), **8**-180, **8**-181
    flexible-joint (table), **8**-169
    for gas supply, **8**-166
    joints for, **8**-164
    pit-cast, **8**-166
    soil (table), **8**-171
    thickness of (tables), **8**-168 to **8**-171
    Universal, **8**-167
    for water supply, **8**-167

Pipe (*Cont.*):
cement-lined, **8**-170
chromium-molybdenum steel, codes for, **8**-149
code designations, **8**-149
coils, **8**-155
center-to-center dimensions (table), **8**-164
collapsing pressure of, **5**-50
concrete, **8**-172
copper (table), **8**-171
corrosion of, **6**-113, **6**-114
in cinder concrete, **6**-114
coverings, **8**-198, **8**-199
heat transmission through, **4**-69
curves, resistance of (hydraulics), **3**-58
distance between supports, **8**-198
drain (table), **8**-178
fittings for (table), **8**-189
expansion, thermal (table), **8**-196
expansion joints for (table), **8**-194
fabrication methods, **8**-148
cupping and drawing, **8**-148
electric-fusion welding, **8**-148
electric-resistance welding, **8**-148
electric submerged-arc welding, **8**-148
forged, turned, and bored, **8**-148
hollow-forged, **8**-148, **8**-149
fittings (*see* Pipe fittings)
flanges (*see* Pipe flanges)
flow: of compressible fluids in, **4**-49, **4**-50
general formulas for, **3**-57 to **3**-60
of steam in, **4**-50
friction factor for (table), **3**-58
friction loss of water in (tables), **12**-104, **12**-105
hard rubber, **8**-171
insulation, **8**-198, **8**-199
heat transmission through, **4**-69
chart, **4**-70
joints: expansion, **8**-193
flanged, **8**-182
gaskets for, **8**-143
unions, **8**-186
lead (table), **8**-175
lead-lined, **8**-170
materials: corrosion of, **6**-113
strength of, ASTM specifications (table), **8**-162
molding machines, **13**-4, **13**-5
orifices (*see* Orifices)
plastic (tables), **8**-176, **8**-177
plugs (table), **8**-193
plywood, **8**-172
in pneumatic conveyors, **10**-62
polyethylene (tables), **8**-176, **8**-177
reinforced-concrete, **8**-172
resistance parameters (fluid flow), **3**-57
return main, flow capacities of (table), **12**-68
riser, flow capacities of (table), **12**-68
roughness in (table), **3**-55
rubber-lined, **8**-170
schedule designations, **8**-149
screw threads, **8**-12, **8**-13
American Standard straight (table), **8**-19, **8**-20
American Standard taper (table), **8**-19, **8**-20
table, **8**-19, **8**-20
seamless-copper-lined, **8**-170
sewer (table), **8**-178
sleeve couplings for, **8**-155
spiral, **8**-155

Pipe (*Cont.*):
steam: ANSI code, **8**-154, **8**-155
ANSI specifications (table), **8**-150 to **8**-155
flow rates in (table), **12**-68
steel, **8**-148 to **8**-155
ANSI specifications, **8**-154, **8**-155
API specifications (table), **8**-164
table, **8**-150 to **8**-155
butt welding, **8**-195
as column, **12**-42
concrete-filled, as column, **12**-55
high-pressure steam, **8**-154
high-temperature, **8**-149, **8**-154
properties of (table), **8**-156 to **8**-161
sizes of (tables), **8**-150 to **8**-163
spiral welded, **8**-155
thickness of (tables), **8**-156 to **8**-161, **8**-163
stress relieving after welding, **8**-195
supports, **8**-197, **8**-198
cast-iron rollers for, **8**-197
constant-support hangers, **8**-197
location of, **8**-198
stresses in line caused by, **5**-61 to **5**-67
sway braces, **8**-198
variable-spring hangers, **8**-198
taps, drills for (table), **8**-20
threads for, **8**-12, **8**-13
American Standard: straight (table), **8**-19
taper (table), **8**-19
Dryseal threads (table), **8**-20
engagement for tight joint (table), **8**-187
tap drills for (table), **8**-20
tin-lined, **8**-170
Transite (asbestos-cement), **8**-172
unions, **8**-186
vitrified (table), **8**-176
water hammer in, **3**-69, **3**-70
welding of, **8**-194 to **8**-197
wood-lined, **8**-170
wood stave, **8**-172
wrench torques for, **8**-177
wrought-iron, **8**-148 to **8**-155
ANSI specifications, **8**-154 to **8**-155
butt welding of, **8**-195
thickness of (tables), **8**-163
Pipe fittings, **8**-174 to **8**-199
ammonia, **8**-188
cast-iron, **8**-167
elbows, **8**-176
flanged (tables), **8**-180
ANSI, **8**-230
drilling templates (table), **8**-180
laterals, **8**-176
long turn, **8**-188
pressure ratings, **8**-174
reducing, **8**-176
sizes of, **8**-174
table, **8**-188, **8**-189
water pipe, **8**-167
drainage, screwed (table), **8**-189
head losses in (table), **12**-108 to **12**-110
for high-pressure and high-temperature steam, **8**-154
tables, **8**-182, **8**-183
loss factors for (table), **3**-58
railing, **8**-187
resistance to fluid flow, **4**-50
screwed, **8**-186 to **8**-188
brass, **8**-187
malleable-iron (table), **8**-187
soldered-joint (table), **8**-190

Pipe fittings (*Cont.*):
  steel: American standard, **8**-177 to **8**-186
    bolting, **8**-179
      material for, **8**-179
    butt-welded (tables), **8**-191, **8**-192
    materials, **8**-179
    pressure-temperature ratings (table), **8**-182
    reducing fittings, **8**-182
    ring joints for, **8**-179
    side outlet fittings, **8**-182
    valve dimensions, **8**-182
    welding neck, **8**-182
  for welding (table), **8**-191, **8**-192
Pipe flanges:
  cast-iron: American Standard, **8**-174 to **8**-176
    bolting of, **8**-174
    bolts, **8**-175
    dimensions of, **8**-174
    drilling templates for (tables), **8**-180, **8**-181
    facing of, **8**-174
    inspection limit, **8**-174
    nuts for, **8**-175
    screwed companion, **8**-176
    spot facing of, **8**-176
  faces of, **8**-186
    male and female, **8**-186
    plain straight, **8**-186
    raised, **8**-186
    serrated, **8**-186
    tongue-and-groove, **8**-186
    V grooves, **8**-186
  steel: American standard, **8**-177 to **8**-186
    bolting, **8**-179
    companion (table), **8**-185
    dimensions of (table), **8**-183, **8**-185
    drilling templates for (table), **8**-184
    facing of (table), **8**-183, **8**-185
    fitting dimensions of, **8**-179
    metal thickness, **8**-179
    properties of (table), **8**-182
    ring joints for, **8**-179
    sizes of, **8**-179
    temperature ratings (table), **8**-182
    thickness of metal for, **8**-179
    welded, **8**-182
Pipelines:
  API specifications (table), **8**-164
  coal, **11**-134, **11**-135
    cost of transportation, **11**-135
    economics and design of, **11**-134
    velocity in, **11**-135
    wall thickness of, **11**-135
  compression, **11**-131
  deflection (formulas), **5**-63
  design of, **11**-131, **11**-133
  design pressures, **11**-131
  economic analysis, **11**-132 to **11**-134
  flexure, **5**-61 to **5**-67
  flow equations, **11**-130, **11**-133, **11**-135
  horsepower required, **11**-131, **11**-133, **11**-135
  natural gas, **11**-130 to **11**-132
  oil and oil products, **11**-133, **11**-134
  prime movers, **11**-130
  pumps, **11**-133, **11**-135
  for solids, **11**-134, **11**-135
  station spacing, **11**-133
  stresses: due to flexure, **5**-61 to **5**-67
    formulas for, **5**-63
Pipetran (pipeline computer program), **11**-30

Piping:
  ammonia (table), **19**-16
  brine, **19**-18
    for rooms, **19**-18
  colors for identification, **8**-199
  expansion and flexibility, **8**-193
  Freon, **19**-15
    maximum refrigeration (tables), **19**-17
  for other refrigerants, **19**-16
  power systems, allowable stresses (table), **8**-150 to **8**-155
  standards for, **8**-147, **8**-148
Pirani gage (high-vacuum), **16**-10
Pistons:
  acceleration of, **8**-69, **8**-70
  for compressors, **14**-36
  inertia force of, **8**-70
  packing, **8**-144
  positions for various crank angles (table), **8**-70, **8**-71
  tangential factors (table), **8**-72
  velocity and acceleration, **8**-69, **8**-70
Pitch, specific gravity and density of, **6**-9
Pitch of gears:
  circular (def), **8**-98
  diametral (def), **8**-98
Pitch of sound, **12**-138
Pitch (ships, def), **11**-40
Pitot tube, **3**-64, **16**-14
  traverse for (table), **16**-15
  velocity distributions, **16**-15
Placer dredges, **10**-38
Plan position indicator (radar), **15**-95
Planck's constant, **9**-120
Planck's law, **4**-71
Plane, inclined, **3**-16
Plane area:
  center of gravity, **3**-7, **3**-8
  moments of inertia, **3**-9
Planers:
  metal, **13**-64, **13**-65
  wood, **13**-81
  motors for, **15**-52
Planets, data on (table), **11**-108
Planimeters, **2**-49, **16**-7
Planning:
  in industrial management, **17**-3 to **17**-9
  of industrial plants, **12**-2
Planté storage battery, **15**-12
Plants, industrial (*see* Industrial plants)
Plasma-arc cutting, **13**-38
Plastering, mortars for, **6**-182
Plastic, ideal (def), **3**-34
Plastic design, **5**-21, **5**-22
Plastic modulus, **5**-54
Plastic packings, **8**-146
Plastic range chart (metals), **13**-15
Plasticity (def), **13**-10 to **13**-15
Plasticizers (in paints), **6**-118
Plastics, **6**-160 to **6**-162
  laminating of, **13**-23
  machining of, **13**-70
  molding, **13**-22, **13**-23
  properties of (tables), **13**-12 to **13**-14
    at low temperatures, **19**-30
  thermoplastic, **6**-160 to **6**-162
    definition of, **13**-11
    properties and uses of (table), **6**-161, **6**-162
  thermosetting, **6**-162, **13**-11
    fillers for, **6**-162
    properties and uses of (table), **6**-163
  working, **13**-9 to **13**-27

Plate-bending machines, **13**-26, **13**-27
Plate-straightening machines, **13**-26, **13**-27
Plates:
    circular (chart), **5**-52
    elliptical (chart), **5**-52
    flat, **5**-52
    rectangular, **5**-52
    strength of, **5**-52, **5**-53
    vibration, **5**-74
    wind pressure, **12**-13
Platinite (low-expansion alloy), **6**-41
Platinum, uses of, **6**-103
Plexiglas, **6**-161, **6**-162
Pliofilm, **6**-167
Plug-cocks, **8**-189
Plumbing, industrial-plant, **12**-9, **12**-10
Plutonium, use of, in atomic energy work, **6**-102
Plywood, **6**-135, **6**-136
    allowable stresses (table), **6**-135
    back of (def), **6**-135
    center of (def), **6**-135
    crossbands of (def), **6**-135
    Douglas fir, **6**-135, **6**-136
        working stresses (table), **6**-135
    exterior, **6**-135
    face of (def), **6**-135
    glues for, **6**-136
    grade designation, **6**-136
    interior, **6**-135
    marine, **6**-136
    moments of inertia (table), **6**-137
    section moduli (table), **6**-137
    shrinkage, **6**-136
    strength of, **6**-136
Pneumatic conveyors, **10**-61 to **10**-63
Pneumatic hammers, **13**-26
Pneumatic hoists, **10**-15
Pneumatic riveters, **13**-27
Pohlé air-lift pump, **14**-15
Point (printers' type measure, def), **19**-45
Poise (unit of viscosity, def), **3**-35
Poisson's ratio (def), **5**-16
    values of, **5**-16
        for metals (table), **5**-5, **6**-11
Polar coordinates, **2**-31, **2**-57
Polar moment of inertia (def), **3**-9
Polarization in batteries, **15**-10
Pole (unit of length, def), **1**-31
Polishing, **13**-69, **13**-70
Polishing wheels, **13**-69
Pollution, air (see Air pollution)
Polyester fibers, **6**-154
Polyethylene as electrical insulation, **15**-60
Polyethylene plastic, **6**-161
    pipe (tables), **8**-176, **8**-177
Polygons:
    area of, **2**-12
    construction of, **2**-10
    of forces, **3**-6
    regular, **1**-8
    table of, **1**-8
Polyhedra, **2**-8, **2**-14
Polymethyl methacrylate plastic, **6**-162
Polynomial, **2**-19
Polyphase induction motors, **15**-43
Polyphase power:
    advantages of, **15**-24
    transformation, **15**-42
Polyphase squirrel-cage motors, **15**-43 to **15**-46
Polyphase wattmeter, **15**-26
Polypropylene plastic, **6**-161

Polystyrene plastic, **6**-162
Polytropic expansion:
    determination of exponent, **4**-19
    values of *n* for water vapor (table), **4**-20
Polyurethane foams, **6**-162
Polyvinyl resins (plastics, table), **6**-161
Pomaranchuk refrigerator (cryogenics), **19**-25
Porous-metal bearings, **6**-93, **8**-130
Port and Waterways Safety Act of 1972, **11**-43
Portland cement, **6**-177, **6**-178
    (See also Cement)
Positron (radiation), **9**-121
Potential difference, unit of (magnetic, def),
    **15**-2
Potential energy (def), **3**-20
Potential transformers (instruments), **15**-27,
    **15**-28
Potentiometers, **15**-29, **16**-6, **16**-13, **16**-20
    in analog computer, **2**-73
Pound (def), **1**-32
Pound mol (def), **4**-11
Poundal (def), **1**-38
Pour point of petroleum oils (def), **6**-198
Powder, compacting, **13**-22
Powder metallurgy, **6**-104
Powdered coal, **9**-11, **9**-12
Power (def), **3**-20
    atomic (see Atomic power; Reactors)
    conversion tables for, **1**-53
    cost of (see Power costs)
    distribution (see Electric power distribution)
    electric (see Electric power)
    measurement of, **16**-18
        one- and multiphase, **15**-26, **15**-27
    muscular, **9**-162
    nuclear (see Atomic power)
    solar (see Solar energy)
    sources (see Energy sources)
    tidal, **9**-168
    transmission: by belts, **8**-51 to **8**-59
        by chain drive, **8**-59 to **8**-86
        electric (see Electric power transmission)
        hydraulic, **8**-39 to **8**-41
        wire rope for (table), **8**-86
    units of, **1**-33
    wave, **9**-168, **9**-169
Power costs, **9**-130 to **9**-133, **17**-33 to **17**-43
    environmental (table), **17**-42
    of environmental controls, **17**-40
    fixed charges, **9**-130, **9**-131, **17**-37, **17**-38
        cost of money in, **17**-37
        depreciation, **17**-38
    fuel: coal, **17**-38
        natural gas, **17**-39, **17**-40
        nuclear, **17**-40
        petroleum, **17**-38
        table, **17**-41
    index for North Central region of the
        United States (table), **17**-34
    for maintenance, **17**-40
    operating and maintenance (table), **17**-41
    operating expenses, **17**-38 to **17**-41
        nuclear, **9**-132, **17**-40
    for operating labor, **17**-40
    overall, **9**-132, **17**-40, **17**-41
        index (table), **17**-34
    plant: combined cycle, **17**-35
        gas-turbine, **17**-35
        geothermal, **17**-36
        hydroelectric, **17**-35
        industrial, **17**-34
        internal-combustion, **17**-35

Power costs, plant (*Cont.*):
    nuclear, **9**-130 to **9**-133, **17**-33, **17**-35
    pumped storage, **17**-36
    steam-electric, **17**-34
    trends (table), **17**-34
    prices, **17**-42
        table, **17**-43
    total (table), **17**-42
    for transmission, **17**-41
Power factor:
    a-c (def), **15**-4, **15**-22
    of induction motors (table), **15**-46
    of loads, approximate values, **15**-65
    measurement of, **15**-27
Power function, graphs, **2**-54
Power plants:
    atomic (*see* Atomic power, plant cost)
    costs of, **17**-33 to **17**-43
        (*See also* Power costs)
    environmental considerations, **17**-36
    heat rates, **9**-54
    installed costs (table), **17**-34
    multipurpose, **17**-42, **17**-43
    profits, comparison of private and public
        (table), **17**-39
    sulfur emission from (table), **18**-18
    useful service life (table), **17**-38
Power presses (*see* Presses)
Power pumps, **14**-2 to **14**-7
Power shovels, **10**-37
Power spectral density (def), **11**-110
Power steering (automobiles), **11**-11 to **11**-13
Powers:
    algebraic, and roots, **2**-16
    arithmetical, **2**-3
    of complex numbers, **2**-23
    of *e* (table), **1**-19
    by logarithms, **2**-5
Pozzolana cement, **6**-178
Prandtl-Glauert rule (supersonic flow), **11**-79
Prandtl number in acoustic theory, **11**-82
Precipitation number (lubricating-oil tests), **6**-199
Precipitators:
    applications of (table), **18**-19
    electrical (gas cleaning), **18**-13, **18**-20
Precise length equivalents (table), **1**-44
Preferred numbers, **8**-201
Preheaters (boilers), **9**-22, **9**-23
    corrosion of, **9**-23
    types of, **9**-22
Press fits, **8**-46
    torsional holding ability, **8**-47
Presses:
    double-action, **13**-24
    frames for, **13**-23, **13**-24
    hydraulic, **13**-21, **13**-23, **13**-24
    power, **13**-23, **13**-24
        bearing pressures, **13**-23
        energy required, **13**-23
        flywheel capacity (formula), **13**-23
        shaft capacities (table), **13**-24
    screw, **13**-24
    types of, **13**-24
Pressure:
    absolute (def), **3**-37
    between bodies with curved surfaces, **5**-51, **5**-52
    conversion table for, **1**-49
    critical: flow of compressible fluids, **4**-47
        for gases (table), **4**-23
    effect on volume of water (table), **6**-10

Pressure (*Cont.*):
    gage (def), **3**-37
    mean effective, **4**-20
    measurements, **16**-8 to **16**-10
        at low temperatures, **19**-33
    partial, **4**-18
    total, in air (def), **4**-28
    units of, **1**-33
Pressure drop in pipes, **4**-49, **4**-50
    (*See also* Flow)
Pressure-enthalpy chart:
    for ammonia, **4**-27
    use of, **4**-24
Pressure equivalents (table), **1**-49
Pressure hose, **8**-199, **8**-200
Pressure-volume diagram, description of, **4**-19
Prestone (antifreeze), properties of, **6**-156
Prevention:
    of accidents, **18**-23 to **18**-25
        codes for, **18**-26
    of occupational disease, **18**-25
Primary cells, **15**-10 to **15**-12
    (*See also* Batteries)
Printers' type, sizes of, **19**-45
Prismoidal formula, **2**-15
Prisms, surface and volume of, **2**-12
Probability integral, **2**-49
Probable error, **2**-21, **2**-22
    table, **1**-25
Process analysis, **17**-4, **17**-5
Processes, automatic control of (*see* Automatic control)
Producer gas, flame temperatures (table), **4**-57
Producers, gas, **7**-41 to **7**-47
Product of inertia, **3**-9
Production control and planning, **17**-5 to **17**-9
Products of combustion of fuels, **4**-51
    table, **4**-55
Profile drag (aerodynamics, def), **11**-61, **11**-69
Profile surveying, **16**-43
Profilometers, **16**-21
Profit-sharing plan, **17**-9
Progression:
    arithmetical, **2**-17
    geometrical, **2**-17
Projectile, motion of, **3**-17
Projection electric-resistance welding, **13**-39
Prony brake, **16**-18
Proofreaders' symbols, **19**-45, **19**-46
Proofreading, **19**-45
Propane:
    commercial (table), **7**-22
    properties of (chart, table), **4**-28, **4**-39, **19**-5, **19**-6
    as refrigerant, **19**-5
Propellants, rocket (*see* Rocket fuels)
Propeller fans, **14**-49, **14**-50
Propeller shafts (ships), **11**-53, **11**-54
Propeller-type turbines (hydraulic), **9**-137 to **9**-143
Propellers:
    airplane (*see* Airplanes, propellers)
    ship (*see* Screw propellers)
Proportion (algebra), **2**-16
Proportional charts, **2**-61, **2**-62
Proportional elastic limit (def), **5**-2
Propulsion:
    jet (*see* Aircraft jet propulsion)
    of ships, **11**-44 to **11**-55
Protective coatings (*see* Paints and protective coatings)
Protons, **6**-7, **9**-119

Proximate analysis of coals, **7**-5
Psychrometer, wet-bulb, **4**-31, **16**-21
Psychrometric chart, **4**-40, **4**-41
  low-temperature, **4**-42
  medium-temperature, **4**-43
  normal temperatures, **12**-81
  in SI units, **12**-87
Pullers, **10**-14
Pulleys, **8**-49 to **8**-51
  arms, design of, **8**-49, **8**-50
  for belting, minimum diameter (table), **8**-51, **8**-55
  block, **8**-8
    efficiency of, **3**-32
  cone, **8**-52
  friction of, **3**-29, **3**-32
  magnetic (conveyors), **10**-59
  rims, design of, **8**-50
  step, **8**-52
  for V belts (table), **8**-55
Pulp for papermaking, **6**-159
Pulsation welding (electric resistance), **13**-40
Pulverized coal, **9**-11, **9**-12
Pulverized lime, **6**-179
Pulverizers (coal), **9**-11
  capacities of, **9**-11
  direct-firing system, **9**-11
  high-speed, **9**-11
  medium-speed, **9**-11
  slow-speed, **9**-11
  storage system, **9**-11
Pumice (abrasive), **6**-139
Pumped storage hydro power, **9**-145
Pumps, **14**-2 to **14**-30
  air-lift, **14**-14, **14**-15
  allowable stresses in, **14**-10
  boiler-feed, turbines for, **9**-47
  centrifugal (*see* Centrifugal pumps)
  classification of, **14**-2, **14**-15
  cushion chambers for, **14**-7
  direct-acting, **14**-7, **14**-8
  double-screw, **14**-14
  efficiencies of, **14**-8
  fire protection, **18**-31
  fuel injection for engines, **9**-99
  gear, **14**-12
  heat (*see* Heat pumps)
  high-vacuum (*see* High-vacuum pumps)
  internal gear, **14**-13
  materials for various liquids (table), **14**-12
  metering, **16**-17
  motors for, **15**-52
  nomenclature, **14**-2, **14**-15
  piston, **14**-2 to **14**-10
    packing for, **8**-144
  plunger, **14**-6
  power (def), **14**-2
  radial plunger, **14**-12
  reciprocating (*see* Reciprocating pumps)
  rotary, **14**-10
    clearances, **14**-10
    eccentric piston, **14**-11
    guided-vane, **14**-11
    lobar type of, **14**-12
    swinging-vane, **14**-11
  screw, **14**-13
  standard fitted, **14**-10
  swash-plate, **14**-12
  vacuum (*see* High-vacuum pumps)
  valves, **14**-8 to **14**-10
  vane, **14**-11

Pumps (*Cont.*):
  vertical turbine pump, **14**-22
  very-high-pressure, **14**-4
Punching of metals, **13**-17
Purifiers, steam, **9**-28
PVC (polyvinyl chloride) electrical insulation, **15**-60
P-V-T relations, **4**-17
Pyramids, surface and volume of, **2**-12
Pyranometer (solar radiation meter), **9**-154
Pyrex glass, **6**-157
Pyrheliometer (solar radiation meter), **9**-154
Pyrolysis (coke), **7**-14
Pyrometers:
  accuracy of, **16**-14
  optical, **16**-11, **16**-13
  radiation, **16**-11, **16**-14

Quadrant, trigonometric, **2**-25
Quadratic equations, **2**-18
Quadrature, **2**-49
Quadrature current (def), **15**-22
Quadrilateral, area of, **2**-11, **2**-12
Quality control, statistical, **17**-19
Quantum field theory, **9**-119 to **9**-121
Quarry machines, **10**-37
Quarter-phase a-c circuit, **15**-23
Quartz, fused, **6**-158
Quenching of steels, **6**-18, **6**-22
Quenching mediums, selection of, **6**-22
Quick freezing, **19**-21
Quick-return motion, Whitworth, **8**-3
Quicklime, **6**-179
Quiet rooms, **12**-139
Quieting (noise control), **12**-140

Raceways for interior wiring, **15**-68
Racks, cutting processes for, **13**-63
Radar, **15**-94, **15**-95
  loran, **15**-96
  plan position indicator, **15**-95
  shoran, **15**-96
Radians (def), **1**-33
  measure of angles, **2**-25
  table, **1**-11
Radians-degrees (conversion tables), **1**-11, **1**-12
Radiant-heat transmission, **4**-60, **4**-70 to **4**-82
Radiation, **4**-70 to **4**-82
  blackbody (def), **4**-71
  and convection, combined coefficients, **4**-69
  definition of, **4**-60
  emissive power (def), **4**-71
  emissivity, of surfaces (table), **4**-73
  from flames and gases, **4**-77 to **4**-80
    beam length for, **4**-77
    table, **4**-78
  from furnaces, **4**-80, **7**-49
  gray-surface (def), **4**-75
  heat transmission by, **4**-70 to **4**-82
  intensity of, **4**-71
  laws of, **4**-71
  of nonblack surfaces, **4**-75 to **4**-77
  nuclear: damage caused by, **9**-129
    measurement of, **9**-129, **9**-130
      instruments for, **16**-22, **16**-23
      proportional counter for, **16**-23
      scintillation counter for, **16**-23
  OSHA exposure standards, **18**-27
  to or from polished metals, **4**-71
  pyrometers, **16**-11, **16**-14

Radiation (*Cont.*):
  refractory, **4**-72
  solar (*see* Solar energy)
  Stefan-Boltzmann law of, **4**-71
  between surfaces, view factors for (charts), **4**-74
  view factors for, **4**-73
Radical axis of circles, **2**-8, **2**-31
Radicals and exponents (algebra), **2**-16
Radio batteries, **15**-11
Radio-frequency allocations (table), **15**-94
Radio-frequency bands (table), **15**-94
Radio receivers, **15**-94
  superheterodyne, **15**-94
Radio transmitters, **15**-94
Radioactive tracers, **16**-23
Radioactive transformation (chemistry), **6**-5
Radioactivity, **9**-35, **9**-121, **9**-129
Radioisotopes, **16**-23
  table, **9**-135
Radius:
  of curvature, **2**-45
  of gyration (def), **3**-9
    for beam sections (tables), **5**-30 to **5**-33
    of two angles (tables), **12**-41, **12**-42
  of inertia (def), **3**-9
  of railway curves (table), **11**-32
Railing fittings, **8**-187
Rails:
  friction on, **3**-25
  railway (table), **11**-32
Railway engineering, **11**-19 to **11**-34
  braking, **11**-30, **11**-31
  car retarders, **11**-33
  car wheels and axles, **11**-28
    tables, **11**-29
  cars, **11**-28, **11**-29
    air resistance, **11**-30, **11**-57, **11**-58
  clearance, **11**-31
  curves (table), **11**-32
  engines (*see* Locomotives)
  rails, **11**-32
    table, **11**-32
  resistance, starting, **11**-30
  track, **11**-31, **11**-32
    gage, **11**-31
    spacing, **11**-31
  train resistance, **11**-29, **11**-30, **11**-57, **11**-58
    curve and grade, **11**-30
    inherent, **11**-29
Rake of cutting tools, **13**-49, **13**-57
Ram-jet system of aircraft propulsion, **11**-85
Ramie fiber (table), **6**-155
Ramsbottom coking test for lubricating oils, **6**-199
Rank test (statistics), **17**-23
Rankine absolute temperature scale, **4**-3
  conversion to Fahrenheit degrees (eq), **4**-3
Rankine cycle:
  efficiencies, **4**-27
  in nuclear reactor, **9**-122
Rankine formula for retaining walls, **12**-22
Rateau formula for steam flow, **4**-47
Rating of electrical apparatus, **15**-50 to **15**-52
Ratio (algebra), **2**-16
  to divide a line in extreme and mean, **2**-9
Rayleigh's method (dimensional analysis), **3**-50
Raymond concrete piles, **12**-21
Rayon fibers (table), **6**-155, **6**-156
RDX (military explosive), **7**-25
  table, **7**-26

Reactance:
  capacitive (def), **15**-3, **15**-21, **15**-23
  inductive (def), **15**-3, **15**-21, **15**-22
  table, **15**-57
Reaction turbines, hydraulic, **9**-138 to **9**-143
Reactive current (def), **15**-22
Reactive power (a-c circuit, def), **15**-4
Reactors, **9**-121 to **9**-136
  controls, **9**-129
  coolants, **9**-125 to **9**-127
    gas, **9**-127
    liquid metals, **9**-127
    properties of (table), **9**-127
  design of, **9**-127 to **9**-130
    breeding ratio, **9**-128
    control, **9**-128
      of criticality, **9**-129
    coolant temperature, **9**-129
    coolant velocity, **9**-129
    cross sections, **9**-127
      for thermal neutrons (table), **9**-128
    diffusion theory, **9**-128
    gamma-ray attenuation, **9**-129
    high power density, **9**-129
    hot-spot and hot-channel factors, **9**-128
    instrumentation, **9**-129, **9**-130
    irradiation effects, **9**-129
    mechanical, **9**-129
    multiplication factor, **9**-128
    neutron attenuation, **9**-129
    neutron balance, **9**-127
    shielding, **9**-129
    sources of heat, **9**-128
    thermal, **9**-128, **9**-129
    thermal design criteria, **9**-128, **9**-129
  materials for, **9**-125 to **9**-127
    for control, **9**-125
    damage to, **9**-125
    fertile, **9**-125
    fissionable, **9**-125
    fuel, **9**-125
    fuel diluents, **9**-125
    melting points of (table), **4**-5
    metals, **9**-125
    moderating, **9**-125
    properties of (table), **6**-102, **9**-126
  neutron-level instruments for, **9**-129
  nuclear, heat from, **4**-28
  power, **9**-122 to **9**-124
    boiling water, **9**-122
    fast breeder, **9**-123
    gas-cooled, **9**-123
    liquid-metal, **9**-123
    pressurized-water, **9**-123
  power cycle for, **9**-122
  power generators, **9**-135
  research, **9**-121
  safety, **9**-133
  teaching, **9**-121
  test, **9**-122
Reamers, **13**-60, **13**-61
Rear axles, automobiles, **11**-9
Reciprocals, **2**-3
Reciprocating machines, balancing of, **5**-72
Reciprocating pumps, **14**-2 to **14**-10
  air chambers, **14**-7
  direct acting, steam, **14**-7
    power requirements, **14**-7, **14**-8
  ends for, **14**-7
  flow and acceleration, **14**-6
  heads of, **14**-5
  high-speed, **14**-2, **14**-6

Reciprocating pumps (*Cont.*):
  horizontal, **14**-3
  low-speed, **14**-4
  materials for (table), **14**-10
  motors for, **15**-52
  packing for, **8**-144 to **8**-146
  plunger load, **14**-8
  pressure pulsation in, **14**-7
  quintuplex, **14**-3
  slip of, **14**-8
  steam ends of, **14**-7
  suction lift of, **14**-5
  triplex, **14**-3
  valves, **14**-8 to **14**-10
  for very high pressure, **14**-4
Rectangles, area of, **2**-11
Rectifiers:
  absorption refrigeration, **19**-11
  electric, **15**-84 to **15**-88
    circuits, **15**-87, **15**-88
      semiconductor, **15**-87, **15**-88
  full-wave, **15**-86, **16**-19
  gas diode, **15**-86
  half-wave, **15**-86
  ignitron, **15**-86
Recuperator for industrial heating furnaces, **7**-51
Red lead (paint), **6**-120
Redler continuous-flow conveyor, **10**-46
Reduced values (thermodynamics), **4**-17
Reduction of area in tensile test, **5**-3
Referencing a point, **16**-49
Reflectance of heat, **4**-73
Reflective heat insulation, **6**-170
Reflectivity:
  of air spaces (table), **12**-87
  of various surfaces (table), **12**-87
Refraction (optics), **19**-40
Refractories, **6**-171 to **6**-177
  for arc furnaces, **7**-65
  for boiler furnaces, **9**-16
  brick shapes and sizes, **6**-172, **6**-173
  castable, **6**-174
  coatings for, **6**-174
  composition of (table), **6**-172
  fiber, **6**-177
  furnace: heat radiation from (chart), **7**-50
    selection of, **6**-175
  heat losses and heat storage capacities (table), **6**-175
  high-purity, properties of (table), **6**-176
  insulating, **6**-171
  metal, properties of (table), **6**-99
  mortars for, **6**-173
  physical properties of (table), **6**-173, **6**-176
  plastics and ramming mixtures for, **6**-174
  pure-oxide, **6**-177
  for space vehicles, **6**-177
  specific heats (tables), **6**-173, **6**-176
  types of, **6**-171, **6**-172
Refrigerants, **19**-3 to **19**-7
  ideal performance of (table), **19**-5
    standard temperature range, **19**-4
  identifying number, **19**-3
  pressure-temperature relations (chart), **19**-6
  properties of (tables), **4**-27, **4**-33 to **4**-40, **19**-4, **19**-5
  (*See also* Ammonia)
Refrigerating machines, **19**-6 to **19**-16
  absorption, **19**-11 to **19**-13
    processes in, **19**-11 to **19**-13
    Servel, **18**-19

Refrigerating machines (*Cont.*):
  adsorption, **19**-11
  coefficient of performance, **4**-45, **19**-3
  cold-air, **19**-13
  compression, **19**-7 to **19**-13
    ammonia volume per ton of refrigeration (table), **19**-9
    controls for, **19**-8
    horsepower per ton of refrigeration (ammonia, table), **19**-9
    household, **19**-10
  condensers for, **19**-8
  lubrication of, **6**-205
  oil for, **6**-205
  rating of (def), **19**-3
  unit of capacity of (def), **19**-3
Refrigerating plants:
  brine, **19**-13 to **19**-16
  piping of cooling and storage rooms, **19**-15 to **19**-21
  skating rinks, **19**-21
Refrigeration, **4**-45, **4**-46, **19**-3 to **19**-23
  absorption system of, **4**-46
  brine system of, **19**-13
  cascade system, **19**-9
  in chemical industry, **19**-21
  cold-storage, **19**-16 to **19**-21
  compound low-temperature, **19**-9
  in cryogenics, **19**-23 to **19**-25
    engines for, **19**-24, **19**-25
  cycles, **19**-6, **19**-7
    gas-compression, **19**-7
  deep, **19**-22
  direct-expansion system of, **19**-13
  gas intercoolers, **19**-9
  ice making, **19**-21
  low-temperature booster, **19**-9
  multiple-effect, compression in, **19**-7
  power required for, **4**-45
  quantity of fluid circulated, **4**-45
  quick freezing (cold storage), **19**-21
  unit of (def), **19**-3
  vapor-compression circuits, **19**-9 to **19**-11
  vapor-compression machines (theory), **4**-45
  water-vapor, **19**-10
Refuse Act of 1899, **11**-43
Regenerative braking (def), **10**-11
Regenerative cycle:
  feedwater heating, **4**-28
  for turbines, **4**-28, **9**-56
Regenerators:
  for industrial heating furnaces, **7**-51
  tile sizes for, **6**-172
Register ton (def), **1**-31
Reheaters, **9**-20, **9**-21
Reheating cycle for steam turbines, **4**-28, **9**-56
Reinforced concrete, **12**-46 to **12**-62
  beams, **12**-50 to **12**-53
    shear forces in, **12**-51
  bearing plates, **12**-59
  bearing walls, **12**-59
  columns, **12**-53 to **12**-55
    footings, **12**-58
    spiral reinforcement, **12**-53
    tied, **12**-53
  concrete for, **12**-48
  construction, of industrial plants, **12**-7
  contraction and expansion joints, **12**-61
  dead load, **12**-49
  definition of, **12**-48
  dowels, **12**-59
  floor systems, **12**-55 to **12**-58

Reinforced concrete (*Cont.*):
footings, **12**-58, **12**-59
for columns, **12**-58
combined, **12**-58
forms, **12**-61, **12**-62
design, **12**-61
liners, **12**-61
removal, **12**-62
joints, **12**-61
construction, **12**-61
joist floors, **12**-55
live load, **12**-49
load factors for, **12**-49
loads, **12**-49
moduli of elasticity, **12**-48
notation, **12**-47
partitions, **12**-59
pipe, **8**-172
precast units, **12**-60
prestressed, **12**-48, **12**-59, **12**-60
allowable stresses in, **12**-60
protection of reinforcement, **12**-49
reinforcing steel, **12**-42, **12**-48
dimensions (table), **12**-42
retaining walls, **12**-59
slabs, **12**-55
steel for, **12**-48
sizes of, **12**-48
T beams, **12**-55
two-way slabs, **12**-55
walls, **12**-59
Relative capacitivity (def), **15**-4
Relative humidity, **4**-30
Relative permeability (def), **15**-2, **15**-15
Relative roughness (fluid flow), **3**-53
Relativistic mass, **9**-120
formula for, **9**-120
Relaxation test of metals (def), **5**-12
Relief angle of cutting tools, **13**-49, **13**-57
Reluctance, magnetic (def), **15**-3, **15**-15
Reluctivity (def), **15**-2, **15**-15
Remanence (magnetic hysteresis, def), **15**-17
Replacement of equipment, **17**-8
Reports, annual, **17**-11
Repose, angle of, **10**-2
Residuals (errors of observation), **2**-22
tables of, **1**-25
Resilience (def), **5**-18
of beams, **5**-34
of beams and springs (table), **5**-19
modulus of (def), **5**-18
ultimate, **5**-18
Resin-type adhesives (table), **6**-142
Resinoid grinding wheels, **6**-140
Resins in varnishes, **6**-118
Resistance:
effective (electric circuits, def), **15**-22
electric, **15**-4
of conductors, specific, **15**-5, **15**-6
measurement of, **15**-28 to **15**-30
of copper lay cables (table), **15**-8
of copper wire (table), **15**-7
of insulating materials (table), **15**-19
of materials (table), **15**-70 to **15**-72
of parallel circuits, equivalent, **15**-8
temperature coefficient of (table), **15**-5,**15**-6
unit of (def), **15**-4
welding (*see* Welding, electric resistance)
of resistor alloys (table), **15**-78
of ships, **11**-40, **11**-41, **11**-58
traction, of automobiles, **11**-3, **11**-57, **11**-58
of trains, **11**-29, **11**-30, **11**-57, **11**-58

Resistance brazing, **13**-42
Resistance furnaces (*see* Electric furnaces, resistance)
Resistance strain gages, **5**-58, **5**-59
Resistance welding (*see* Welding, electric resistance)
Resistivity, electric, of metals (def), **15**-5, **15**-6
tables, **6**-60, **6**-61, **15**-6
Resistor furnaces, **7**-60 to **7**-63
Resistor materials, **15**-69, **15**-76
table, **15**-78
Resistor ovens, **7**-63
Resistors for electric furnaces, **7**-62, **15**-69
Resonance, electric circuits, **15**-22
Resonance vibration, **5**-68
Respirators, use of, in hazardous occupations, **18**-26
Respiratory protection, **18**-26
Response time (automatic control, def), **16**-25
Restitution, coefficient of (def), **3**-21
Retaining walls, design of, **12**-21, **12**-59
Reynolds number, **3**-47, **3**-49, **3**-56, **11**-41, **11**-60, **11**-62
chart for, **3**-55
Rheostats, materials for, **15**-69
table, **15**-78
Rhodium, uses of, **6**-103
Rhombus, area of, **2**-11
Ribbon, area of, **2**-12
Richardson conveyor weighing scale, **10**-64
Rigging, wire rope (tables), **8**-91
Right-hand rule (for conductor current), **15**-17
Right-hand rule (generators), **15**-17
Rigid bodies:
dynamics of, **3**-16 to **3**-23
forces supporting, **3**-5
instantaneous axis of, **3**-15
motion of, **3**-11 to **3**-16
statics of, **3**-3 to **3**-11
Rigidity:
coefficient of (def), **5**-18
modulus of, for metals (table), **5**-5
Ringelman chart, **18**-17
Rings, strength of, **5**-45
Rise time (automatic control, def), **16**-33
Riveted joints:
design of, **12**-31, **12**-34, **12**-38
punched vs. drilled plates, **8**-30
Riveters, **13**-27
Rivets:
conventional signs for, **8**-33
forms and proportions of, **8**-30
gages, standard, **12**-32
tables, **12**-45
for steel-framed structures, **12**-31, **12**-34, **12**-38
Roads:
friction on, **3**-27, **3**-28
rolling resistance on, **11**-3
Rocker (mechanism), **8**-3
Rocket engines, **11**-87 to **11**-90
chemical, **11**-87, **11**-90
electric, **11**-89
electromagnetic, **11**-89
electrostatic, **11**-89
electrothermal, **11**-89
fuels for (*see* Rocket fuels)
liquid-bipropellant, **11**-87, **11**-88
nuclear, **11**-128
nuclear heat-transfer, **11**-89
photon, **11**-129
solid propellant, **11**-88, **11**-89

Rocket fuels, **7**-33 to **7**-37
  alcohol, **7**-36
  ammonia, **7**-36
  bipropellant, **7**-35
  boron hydrides, **7**-36
  characteristics of, **7**-35 to **7**-37
    table, **7**-36
  classification by physical state, **7**-34
  combinations of, **11**-88
  composite, **7**-34
  diergolic (def), **11**-87
  fluorine (table), **7**-36
  future developments, **7**-36, **7**-37
  hydrazine, **7**-36
  hydrocarbons with liquid oxygen, **7**-36
  hydrogen peroxide, **7**-36
  hypergolic (def), **7**-35, **11**-87
  ignition of, **7**-35
  liquid, **7**-36, **11**-87
  liquid hydrogen, **7**-36
  liquid oxygen, **7**-36
  monopropellants, **7**-35
  multicomponent, **7**-36
  propyl nitrate, **7**-36
  slush hydrogen, **7**-36
  solid, **7**-34, **11**-88
    case-bonded, **11**-88
  specific impulse of (table), **7**-36
  UDMH, **7**-36
  various (table), **7**-36
Rocket nozzles, **11**-77
Rockets, nuclear, **11**-128, **11**-129
    nozzle thermodynamics, **11**-128
    reactor analysis, **11**-128
    space vehicle applications of, **11**-129
    specific impulse of, **11**-128
Rockwell hardness test, **5**-13, **16**-21
Rockwool (heat insulator), **6**-170
Rod (unit of length, def), **1**-31
Roll (ships, def), **11**-40
Roll-pins, **8**-33
Roller bearings (*see* Bearings, roller)
Roller chains (*see* Chains, roller)
Roller conveyors, **10**-61
Rolling:
  cold, **13**-16
  contour, **13**-17
  operations, **13**-16, **13**-17
  shape, **13**-17
Rolling friction, **3**-28
Rolling loads on beams, **5**-39
Rolling motion, **3**-18
Rolling surface (mechanism), **8**-7
Roof(s):
  flashings for, **12**-8
  heat gain through (table), **12**-82, **12**-83
  for industrial plants, **12**-5, **12**-7
  insulation for, **12**-8
  joist and mill construction, **12**-23
  live loads on, **12**-11
  masonry, coefficients of heat transmission
    (table), **12**-89
  metal, corrosion of, **6**-114
  trusses (*see* Trusses)
  wind pressure on, **12**-13
Roofing:
  asphalt, **6**-163
  elastomers for, **6**-165
  materials for, **6**-163 to **6**-165
  metallic, **6**-164
Roofing cement, **6**-164
Root-mean-square value of a-c wave, **15**-21

Roots:
  algebraic, **2**-16
  arithmetical, **2**-3
  of complex numbers, **2**-23
  of equations, **2**-18
    imaginary, **2**-19
  by logarithms, **2**-5
  by slide rule, **2**-6
Roots blower, **9**-102, **14**-44
Rope, **6**-151
  braided, **6**-151
  drums for hoists, **3**-32, **10**-9
  friction, **3**-32
  haulage, **10**-9
  knots, hitches, and bends, **8**-89
  lay of (def), **6**-151
  manila, U.S. specification (table), **6**-152
  materials for, breaking strength of (table), **8**-93
  nylon, **6**-151
  polyester, **6**-151
  sheaves, **10**-10
  wire, **10**-9, **10**-10
Rotameter (for fluids), **16**-16
Rotary blowers, **14**-42 to **14**-44
Rotary car dumps:
  gravity, **10**-24
  power, **10**-24
Rotary cranes, **10**-30
Rotary pumps, **14**-10, **14**-11
Rotary shafts (*see* Shafts)
Rotating machines, balancing of, **5**-70, **5**-71
Rotating shaft, critical speed of, **5**-72
Rotation of solid bodies about axes, **3**-18, **3**-22
Rotation moments of a couple, **3**-4
Rouge (abrasive), **6**-139
Roughness (metal surfaces, def), **13**-76
Roughness factors for open channels, **3**-67
  table, **3**-68
Routh's stability criterion (automatic control),
  **16**-35
Rubber and rubberlike materials, **6**-165 to **6**-167
  balata, **6**-167
  buna S, **6**-166
  butyl, **6**-166, **6**-167
  chlorinated, **6**-167
  cyclized, **6**-167
  derivatives, **6**-167
  elastomers (def), **6**-165
    comparative properties of (table), **6**-167
    uses of, **6**-166
  GR-S, **6**-166, **6**-167
  gutta-percha, **6**-167
  latex, **6**-167
  natural, **6**-165, **6**-167
  neoprene, **6**-166, **6**-167
  nitrile, **6**-166, **6**-167
  pigments in, **6**-166
  polysulfide, **6**-166
  polyurethane, **6**-166
  properties of, **6**-166
  on roads, friction of, **3**-27
  silicone, **6**-153, **6**-166, **6**-171
  specific gravity and density (table), **6**-8
  thiokol, **6**-167
  vulcanization, effect of, **6**-165
Rubber belts (*see* V-belts)
Rubber bonding of grinding wheels, **6**-140
Rubber-die forming, **13**-20
Rubber hydrochloride, **6**-167
Rubble masonry (table), **12**-23

Ruben cell, **15**-11
Rueping process (wood preservation), **6**-137
Rupture:
  modulus of (def), **5**-26
  unit work, **5**-18
Russell movable-wall oven, **7**-38
Russia iron, **6**-116

Σ (sigma) function for moist air (table), **4**-32
*S-N* diagram (fatigue analysis), **5**-9
SAE viscosity number:
  for crankcase oils, **6**-197
  for transmission and axle lubricants (table),
    **6**-198
Safety, **18**-22 to **18**-25
  in cryogenics, **19**-39
  of grinding wheels, **13**-68
Safety codes, **18**-27
Safety devices for machines, **18**-24
Safety factor (def), **5**-22
Safety glass, **6**-157
Sag in chaining uphill, **2**-38
Sagging (ships, def), **11**-37
Saginaw power-steering system, **11**-12
Salinometer (def), **19**-13
Salometer (def), **19**-13
Salt, common, as freezing preventive, **6**-156
Salts, inorganic, solubility of (table), **6**-5
Sampling (statistics), **17**-19, **17**-20, **17**-25
Sand:
  for concrete, **6**-179
  core, **13**-6
  molding, **13**-3, **13**-7
  specific gravity and density, **6**-8
Sand blast, **13**-70
Sand castings, making, **13**-3
Sand-lime brick, **6**-149
Sand slinger (molding machine), **13**-5
Sanders, wood, **13**-83, **13**-84
Sandstone, composition of, **6**-158
Saponification number (organic oils), **6**-202
Saran (fiber), **6**-154
Sash cord, wire (table), **8**-92
Saturated ammonia, properties of (table), **4**-33,
  **4**-34
Saturated Dowtherm A, properties of (table),
  **4**-33
Saturated steam:
  properties of (tables), **4**-29 to **4**-31
  vapor (def), **4**-22
Saturated sulfur dioxide, properties of (table),
  **4**-37
Saturation of air in spray chamber, **4**-44
Saunders air-lift pump, **14**-15
Saws, **13**-66
  metal-cutting, **13**-66
  wood, **13**-80, **13**-81
    classification of, **13**-80
    gullet-feed index, **13**-81
    materials, **13**-81
    motors for, **15**-52
    power requirements (table), **13**-82
    teeth, **13**-80
Sawtooth roof trusses, **12**-6
Saybolt-Furol viscometer, **3**-36
Saybolt Universal viscometer, **3**-36
Scalar (def), **2**-63
Scalar product, **2**-63
Scale deposits, heat-transfer coefficients
  (table), **4**-66
Scale effect (aerodynamics), **11**-63

Scales, **16**-3, **16**-4
  automatic, **16**-4
    for conveyors, **10**-63, **10**-64
  batch-type, **16**-4
  continuous, **16**-4
  spring, **16**-4
Scanning (television), **15**-96
Scavenging two-cycle engines, **9**-103, **9**-104
Scheduling:
  by computer, **12**-2
  flow of materials, **17**-7
  production, **17**-5
  using CPM (critical path method), **12**-2, **17**-
    6
  using PERT, **17**-6
Schiele's antifriction curve (tractrix), **2**-41
Schlieren optical system (compressible flow
  visualization), **11**-84
Scleroscope hardness test, **5**-13
Scott tap (transformers), **15**-42
Scrap iron and steel (foundry), **6**-49
Scraper conveyors, **10**-44 to **10**-47
Scrapers, **10**-27
Scratch coat (plastering), **6**-182
Scratch hardness (minerals), **5**-12
Screen scale, Tyler (for particle sizes, table), **7**-
  12
Screw conveyors, **10**-47, **10**-48
Screw jacks, **10**-15
Screw machines, **13**-59
  free-cutting steel stock for (AISI and SAE),
    **6**-31
Screw presses, **13**-24
Screw propellers (ship), **11**-49 to **11**-52
  blade shapes, **11**-49
  cavitation, **11**-51
  controllable and reversible pitch, **11**-51
  design of, **11**-50
  developed area, **11**-49
  fitting to shaft, **11**-54
  partially submerged, **11**-52
  pitch, **11**-49
  projected area, **11**-49
  shafts for, **11**-54
  slip ratio, **11**-50
  slip velocity, **11**-50
  super cavitating, **11**-52
  tandem and contrarotating, **11**-52
  ventilated, **11**-52
  wake velocity, **11**-50
Screw pumps, **14**-13, **14**-14
Screw threads:
  Acme (table), **8**-18
  American Standard (tables), **8**-10 to **8**-15
    coarse (tables), **8**-10
    drill sizes for (table), **8**-32
    8-thread series (table), **8**-13
    extra-fine (table), **8**-12
    fine (tables), **8**-11
    16-thread series (table), **8**-14
    special-pitch (UN), **8**-9
    12-thread series (table), **8**-14
  British Association (table), **8**-17
  Dryseal pipe, **8**-13
  efficiency of, **3**-29, **3**-30
  fits for, standard, **8**-9
  French (metric, table), **8**-17
  high-strength (table), **8**-18
  inserts, **8**-27
  international (metric) standard (table), **8**-17
  limiting dimensions of, **8**-9
  machine, **8**-9 to **8**-16

Screw threads (*Cont.*):
 machines for cutting, **13**-61
 major diameter (def), **8**-9
 metric (table), **8**-17
 minor diameter (def), **8**-9
 pipe, American, **8**-12, **8**-13
 pitch diameter (def), **8**-10
 power transmission, **8**-11
 Sellers, **8**-11
 square, friction of, **3**-29
 tap drill sizes for (table), **8**-32
 tolerances for, **8**-9
 Unified thread standards, **8**-9
 U.S. standard, **8**-9
 (*See also* American Standard, *above*)
 Whitworth standard (table), **8**-16
Screws, **8**-9 to **8**-22
 cap (table), **8**-22
 coach (table), **8**-23
 cutting, gear calculations, **13**-56
 lag: holding power, **12**-28
  table, **8**-23
 machine (*see* Machine screws)
 self-tapping (tables), **8**-26, **8**-27
 V-thread, friction, **3**-29, **3**-30
 wood (table), **8**-24
  holding power, **12**-28
Script alphabet, **19**-47
Scrubbers, gas, **18**-12
Scruple (apothecaries' weight, def), **1**-32
Sea as source of tidal energy, **9**-5, **9**-168, **9**-169
Seals for shafts, **14**-41
 controlled gap, **8**-147
 mechanical, **8**-146
Seam welding, electric resistance (table), **13**-39
Seamless steel tubing (tables), **8**-165 to **8**-167
Season cracking of brass, **6**-68, **6**-70, **6**-112
Seawater, specific gravity and density of (table), **6**-8
 (*See also* Water, desalinization)
Secant:
 graph, **2**-55
 tables, **1**-17, **1**-18
Second (circular measure, def), **1**-32
Second (unit of time, def), **1**-32
Section modulus, of beams (def), **5**-23
 table, **5**-30 to **5**-32
Sector:
 of circle: area of, **2**-12
  center of gravity of, **3**-8
 of sphere: center of gravity of, **3**-8
  volume and area of, **2**-14
Seebeck effect, **9**-150, **15**-25
Seger cones, **16**-11
Segments:
 of circle, **2**-12
  center of gravity of, **3**-8
  table, **1**-6, **1**-7
 of sphere, **2**-13
  center of gravity of, **3**-8
Self-inductance, unit of (def), **15**-4, **15**-19
 coefficient of (def), **15**-4
Self-tapping screws (table), **8**-27
Selsyns, **15**-50
Semiconductor diodes, **15**-84
Semigels (explosives), **7**-25
Semilogarithmic cross-section paper, **2**-56
Separating calorimeter, **16**-22
Separators:
 for dust exhaust systems, **18**-12
 steam, **9**-28

Series:
 expansion of functions in, **2**-12, **2**-44, **2**-45
 Fourier's, **2**-45
 reversing of, **2**-45
 semigeometric, **8**-201
 summation of, **2**-17
 Taylor's and Maclaurin's, **2**-44
Series circuits, **15**-18
 a-c solution of problems, **15**-22
Series generators, **15**-30
Series motors, **15**-33
Servel process of refrigeration, **19**-11
Set point (automatic control, def), **16**-25
Setscrews, **8**-16
 holding power of (table), **8**-23
 points and heads for, **8**-29
Settling time (automatic control, def), **16**-26
SEV (surface-effect vehicles), **11**-56
Sewer pipe, **8**-171
Sexagesimal measure of angles, **2**-25
Shadow prices (linear programming), **17**-18
Shaft seals for compressors, **14**-41
Shafting, marine, **11**-53, **11**-54
Shafts, **8**-48, **8**-49
 alignment of, **16**-50
 bending moments in, **8**-49
 combined torque and bending in, **8**-49
 couplings for (*see* Couplings)
 critical speed of, **5**-72 to **5**-74
 failure due to fatigue, **8**-48
 fits, **8**-48 to **8**-47
 hangers for, **8**-129
 keys for, **8**-30, **8**-31
 leveling of, **16**-50
 packing for, **8**-144 to **8**-146
 resilience of (table), **5**-19
 ship propeller, **11**-53, **11**-54
 splined, **8**-31, **8**-35
 stiffness of, **8**-49
 strength of, **8**-48, **8**-49
 torsion of, **5**-37, **5**-40, **8**-48
  combined with other stresses, **5**-19, **8**-48
  table, **5**-41
 torsional vibration of, **5**-67
 vibration of, **5**-67 to **5**-74
Shale oil:
 as fuel, **7**-20
 reserves of, **9**-4
Shape factor in plastic design (def), **5**-21
Shapers, **13**-64, **13**-65
Shaw process (molding), **13**-3
Shear:
 and bending, **5**-18
 and bending moment in beams, **5**-34
 diagram for beams, **5**-24 to **5**-27
 maximum pressure in, **13**-17
 modulus of elasticity in, **5**-18
Shearing in dies, resistance to (table), **13**-18
Shearing of metals, **13**-17, **13**-18
 allowance for shaving, **13**-18
 power for, **13**-18
Shearing modulus of metals (table), **5**-5
Shearing stress (def), **5**-17
 at end of beams, **5**-36
 for various cross sections, **5**-18
Shears, metal-cutting, **13**-18
Sheaves:
 efficiency of rope and chain, **3**-32
 grooves for wire-rope transmission, **8**-85
 hoisting, rope, **8**-85
 stresses in, **8**-50

Sheaves (*Cont.*):
for V belts (tables), **8**-55 to **8**-59
wire-rope, **8**-85
Sheet-metal gages (table), **6**-45
Shellac grinding wheels, **6**-140
Shellac varnish, **6**-121
Shells, drawing of (metalworking), **13**-20
Sherardizing process, **6**-114
Shielding:
of atomic power plants, **9**-129
sound, **12**-139
Shingles:
asbestos, **6**-164
asphalt, **6**-163
slate, **6**-164
wood, **6**-164
Shipping measure, **1**-31, **1**-32
tons, United States and British, **1**-31, **1**-32
Ships, **11**-35 to **11**-55
air resistance of, **11**-41, **11**-58
antirolling gyroscope for, **3**-23
boilers, **11**-44
bottom paints for, **6**-119
catamaran, **11**-35
center of buoyancy of (def), **11**-39
center of gravity of, **11**-39
combined propulsion plants for, **11**-49
dead weight of (def), **11**-36
definitions of power and speed, **11**-44
designed load waterline (DWL), **11**-36
diesel engines for, **11**-47, **11**-48
displacement of (def), **11**-36
draft of (def), **11**-37
drag, designed, **11**-37
electric drives for, **11**-53
engineering constraints, **11**-42 to **11**-44
engines (*see* Marine engines)
environment, **11**-35
environmental constraints, **11**-43
form coefficients of (table), **11**-37
gas-turbine drive for, **9**-119
gas turbines for, **11**-48
high-performance, **11**-35, **11**-55
hull forms, **11**-35
Kort nozzles for, **11**-52
lengths of (def), **11**-36
line shafts for, **11**-54
metacenter of, **11**-39
molded beam (def), **11**-36
motions, **11**-40
nuclear, **9**-134
powering, **11**-40 to **11**-42
propeller shafts for, **11**-53, **11**-54
propulsion plants, **11**-44 to **11**-49
reduction gears for, **11**-53
resistance of, **11**-40, **11**-41
frictional, **11**-40
wavemaking, **11**-40
screw propellers for, **11**-49 to **11**-52
stability of, **11**-39, **11**-40
steam plants for, **11**-44 to **11**-47
steam turbines for, **11**-45
structural steel for, ASTM specifications
(table), **6**-27
structure of, **11**-37 to **11**-40
submarines, **11**-35
tonnage of (def), **11**-36
trim of, **11**-37
water jet propulsion for, **11**-52
Shoran (navigation by radar), **15**-96
Shore scleroscope (hardness testing), **5**-14
Shovels, power, **10**-37

Shrink fits, **8**-46, **8**-47
Shrinkage:
of castings (table), **4**-8
of refractories (table), **6**-173
of rings on thick cylinders, **5**-51
Shunt generators, **15**-30
parallel operation, **15**-31
Shunt motors, **15**-33
starters for, **15**-33
SI units (The International System of Units):
base and supplementary (defs), **1**-32, **1**-33
conversion tables for, **1**-34 to **1**-39
prefixes for (table), **1**-34
Side-cutting-edge angle (tools), **13**-57
Sidereal time, **1**-41
Sidewalks, moving, **10**-24
Siemans, **1**-33, **15**-4
Sigma function for moist air (table), **4**-32
Signal bronze, **15**-6
Significant figures, number of, **2**-2
Silastic (silicone rubber), **6**-145
Silent chains, **8**-68
Silica in steam, **9**-32
Silica brick, **6**-171
Silica gel, refrigerant system, **19**-11
Silicate grinding wheels, **6**-140
Silicon, softening effect of, in cast iron, **6**-47
Silicon carbide (abrasive), **6**-139, **13**-67
Silicon carbide furnace (electric), **7**-69
Silicon carbide refractory, **6**-172
Silicon controlled rectifier (SCR), **15**-85
Silicones, **6**-170
greases, **6**-171
oils, **6**-170
paint, **6**-120
resins, **6**-171
rubber, **6**-153, **6**-166, **6**-171
varnish, **6**-152, **6**-171
water repellents, **6**-170
Silk fiber (table), **6**-155, **6**-156
Silver, uses of, **6**-103
Silver solders, **6**-91, **13**-42
Similar figures (geometry), **2**-7
Simplex concrete piles, **12**-21
Simplex method (optimization), **17**-17
Simpson's rule for areas and volumes, **2**-12, **2**-15, **2**-49
Simulation on computers, **2**-71
Simultaneous equations:
contour-line charts for, **2**-58
solution of, **2**-20 to **2**-22
by determinants, **2**-22
Sine (trigonometry), **2**-25
graph, **2**-55
tables, **1**-13, **1**-14
Sine waves (a-c circuits), **15**-21
Sines, law of, **2**-28
Single-phase induction motors, **15**-46
Single-phase power measurement, **15**-25
Sinking fund, table, **1**-28
Sintered copper and iron (table), **6**-104
Sintered steel, strength of (table), **6**-105
Sintering, **13**-22
of powdered metals, **6**-104
Siphons, **3**-60
Sisal, properties of (tables), **6**-155
Skating rinks, ice, piping for, **19**-21
Skelp (steel, def), **6**-16, **8**-148
Skim coat (plastering), **6**-182
Skin friction (aerodynamics), **11**-66
Skip hoists, **10**-18
Skylights for industrial plants, **12**-9
Slab (steel, def), **6**-16

Slabs (*see* Reinforced concrete)
Slackline cableways, **10**-39, **10**-40
Slag:
  in boilers, **9**-7 to **9**-10
  sintering strength test, **9**-8
  specific gravity and density (table), **6**-8
  on superheaters, **9**-23
Slate, **6**-158, **6**-165
Sleds, friction of, **3**-27
Slenderness ratio of columns, **5**-41
Slide rule, **2**-5 to **2**-7
Slider couplings, double, **8**-35
Sliding bearings, **8**-133
Sliding-block linkage, **8**-3
Sliding friction (def), **3**-24
  tables, **3**-25, **3**-26
Slings (wire rope), **8**-89
Slip (of pumps), **14**-8
Slip ratio of screw propellers, **11**-50
Slip rings for a-c motors, **15**-45
Slope of a curve (def), **2**-42
Slotters, **13**-65
Slug (def), **1**-39
Slump test for concrete, **6**-186
Slushing oils, **6**-117
Smoke (def), **18**-8, **18**-11
SNAP (nuclear power system), **9**-135
Snatch block (def), **10**-10
Snell's law (optics), **19**-40
Soap, **6**-151
Sodium chloride brine, properties of
  (tables), **19**-13, **19**-14
Sodium vapor, Mollier diagram for, **4**-
  27
Soil pipe, cast-iron (table), **8**-171
Soils:
  corrosive effect on pipes, **6**-113
  safe bearing power of, **12**-20
Solar energy, **9**-5, **9**-153 to **9**-158
  applications of, **9**-157, **9**-158
  collectors, **9**-155 to **9**-157
    flat-plate, **9**-156
    heat-transfer coefficients (table), **9**-
      157
  for cooking, **9**-158
  direct conversion, **9**-158
  equilibrium temperatures, **9**-156
  furnaces, **9**-158
  for house heating, **9**-157
  incident-angle determination, **9**-154
  instruments: pyranometer, **9**-154
    pyrheliometer, **9**-154
  opaque surface, absorptance and admittance
    of, **9**-155
  power from, **9**-158
  radiation intensity, **9**-154
  for refrigeration, **9**-158
  for stills, **9**-157
  transparent materials for, solar-optical
    properties, **9**-154
  utilization of, **9**-153
    heliochemical (def), **9**-154
    helioelectrical (def), **9**-154
    heliothermal (def), **9**-154
    processes, **9**-153
  for water heaters, **9**-157
Solar time, **1**-41
Solder, alloys (table), **6**-91
Soldered-joint, fittings (table), **8**-190
Soldering:
  sonic, **12**-142
  with ultrasonics, **12**-142

Solders, **6**-90
  brazing (table), **6**-91, **6**-92
  silver, **13**-42
Solenoids, **15**-77 to **15**-80
  magnetic pull of (table), **15**-79
Solids:
  buoyancy of, **3**-40
  center of gravity of, **3**-8
  melting points of (table), **4**-5
  moments of inertia of, **3**-8 to **3**-10
    approximate, **3**-10
  rotation about axes, **3**-17, **3**-18, **3**-22
  specific gravity and density of (table), **6**-7 to
    **6**-9
    by immersion, **3**-40
  specific heats of (table), **4**-9
  surface and volume of, **2**-12 to **2**-15
  thermal conductivity of (tables), **4**-60, **4**-61
  thermal expansion of (table), **4**-8
  volume of, by immersion, **3**-40
Solubility:
  of gases in water (table), **6**-7
  of inorganic substances in water (table), **6**-5
  of salts in water (table), **6**-5
Solution pressures of metals (def), **6**-107
Solutions, specific heat of, **4**-11
Solvents, **6**-168, **6**-169
  alcohols as, **6**-168
  chlorinated, **6**-168
  esters as, **6**-168
  hydrocarbons as, **6**-168
  ketones as, **6**-169
  organic (cleaners), **6**-150
  petroleum, **6**-168
  for plastic materials, **13**-10
  for polymeric materials, **6**-168
Sonar (underwater detection), **12**-142
Sone (loudness level of sound, def), **12**-138
Sonotone battery, **15**-14
Soot blowers, **9**-9, **9**-10
Soot luminosity, **4**-77
Sound, **12**-136 to **12**-143
  absorption, coefficient of (def), **12**-140
    table, **12**-141
  attenuation, **12**-136
  audible range of, **12**-136
  beat-frequency oscillator, **12**-138
  contact transducers, **12**-138
  industrial applications, **12**-141 to **12**-143
  intensity of, **12**-137
  levels (table), **12**-139
  loudness, **12**-138
  masking of, **12**-138
  perception of, **12**-137
  pitch of, **12**-138
  production and reception of, **12**-137, **12**-138
  quality of, **12**-138
  specific acoustic impedance, **12**-137, **12**-140
  spectrum (table), **12**-136
  speed of, **3**-35
  testing of materials by, **12**-142
  timbre of, **12**-138
  total absorption, **12**-140
  transducers, **12**-137
  transmission loss in building partitions
    (table), **12**-140
  velocity of, in various media (table), **12**-136
  vibration pickup, **12**-138
  wavelength of, **12**-136
  of whistles and sirens, **12**-137
  (*See also* Noise)
Sound analyzer, **12**-138

Sound-level meter, **12**-138
Sources and sinks (radiation), **4**-75
Soybean oil (paints), **6**-118
Space, **11**-104 to **11**-129
    effects due to: meteoroids, **11**-122
        radiation, **11**-122
            table, **11**-123
        vacuum, **11**-122
            table, **11**-123
    effects on lubrication, **11**-122
    entry from, **11**-114 to **11**-117
    nuclear-powered systems for (table), **9**-136
    orbital mechanics, **11**-112, **11**-113
    planets in, atmospheres of (table), **11**-115
    satellites, **11**-104
        energy of, **11**-114
        lifetime of, **11**-113
        orbits, perturbations of, **11**-113
    solar system: astronomical constants, **11**-104
            to **11**-107
        data on bodies in (tables), **11**-108
Space factor in winding magnets, **15**-80
Space flight:
    hyperbolic excess velocity, requirements
            (tables), **11**-114
    mechanics of, **11**-110 to **11**-114
        interplanetary, **11**-114
        lunar, **11**-113
    trajectories, performance optimized, **11**-115
    (See also Space probes; Space vehicles;
        Spacecraft)
Space flights (table), **11**-105
Space lattice (metals), **13**-9
Space probes, **11**-104
    table, **11**-105
Space-time curve (kinematics), **3**-11
Space vehicles:
    ablative systems, **11**-115
    damage, by meteoroids, **11**-122
    environments, **11**-107 to **11**-110, **11**-122
    lubricants for, **11**-122
    noise, effects of, **11**-109, **11**-110
    nuclear motors for, **9**-135
    nuclear rockets for, **11**-128
    propulsion, **11**-127 to **11**-129
    radiative-heat-transfer systems, **11**-115
    stresses and stability in, **11**-124, **11**-125
    structures of, **11**-119
        analysis of, **11**-122 to **11**-124
    vibrations, **11**-110, **11**-125 to **11**-127
    (See also Space; Space flight; Space probes;
        Spacecraft)
Spacecraft:
    control, **11**-117 to **11**-119
    engines for, **9**-163
    materials, **11**-119 to **11**-122
        high-temperature, **11**-120
        intermediate-temperature, **11**-120
        low-temperature, **11**-119 to **11**-121
            properties of (table), **11**-121
    operation of, **11**-105
    welding, **11**-120
    (See also Space; Space flights; Space probes;
        Space vehicles)
Span (unit of length, def), **1**-31
Spark coil for ignition system, **15**-81
Spark gaps for spark plugs, **9**-100
Spark-ignition (engines), **9**-79, **15**-81
Spark plugs for ignition systems, **15**-81
Spark voltage for ignition systems, **15**-82
Specific gravity (def), **1**-42, **3**-35
    of brines (tables), **19**-14, **19**-15

Specific gravity (*Cont.*):
    of calcium chloride solutions (tables), **19**-14,
        **19**-15
    of liquids (conversion tables): API, **1**-42
        Baumé, **1**-43
    of magnesium chloride brines (tables), **19**-15
    of sodium chloride solutions (table), **19**-14,
        **19**-15
    of solids by immersion, **3**-40
    of various substances (table), **6**-7 to **6**-10
    of water at various temperatures (table), **6**-9
Specific heat (def), **4**-8 to **4**-11
    of alloys (table), **4**-9
    of gas mixtures, **4**-11
    of gases, **4**-11
        table, **4**-10
        variable, **4**-19
    humid (def), **4**-32
    of iron, mean (table), **4**-10
    of liquids (table), **4**-9
    mean molal (chart), **4**-11
    of metals (table), **6**-11, **6**-60, **6**-61
    of mixtures, **4**-11
    of refractories (tables), **6**-173, **6**-176
    of solids (table), **4**-9
        at low temperatures (chart), **19**-26
    of solutions, **4**-11
    of various materials (table), **4**-9
    of water (table), **4**-9
Specific resistance (def), **15**-5
Specific speed:
    of centrifugal pumps (def), **14**-22
    of hydraulic turbines (def), **9**-137, **9**-138
Specific volume of ideal gases (eq), **3**-34
Specific weight of fluids (def), **3**-35
Spectrophotometers, **16**-21
Speed control:
    of d-c motors, **15**-35, **15**-36
    of induction motors, **15**-45
Spelter, **6**-85
Sphere gaps (high-voltage measurement), **15**-
    28
Spheres:
    drag coefficient of (chart), **11**-68
    and flat plates, deformation of, under
        compression, **5**-52
    hollow: strength of, **5**-51
        volume of, **2**-13
    solid, deformation under compression, **5**-52
    surface and volume of, **2**-13
    theorems on, **2**-8
Spherical degree (def), **2**-14
Spherical excess, **2**-29
Spherical sector, surface and volume of, **2**-14
Spherical segments, areas of (formula), **2**-13
Spherical triangles:
    area of, **2**-14, **2**-29
    solutions of, **2**-29
Spherical wedge, surface and volume of, **2**-14
Spherical zone, surface and volume of, **2**-13
Spheroid, volume of, **2**-14
Spikes (tables), **8**-92, **8**-95
    for timber construction, **12**-28
Spiral:
    Archimedean, **2**-40
    logarithmic, **2**-41
Spiral bevel gears, dimensions of (table), **8**-109
Spiral chutes, **10**-49, **10**-50
Spiral conveyors, **10**-48
Spiral gears, **8**-109, **8**-110
Spiral pipe, **8**-155
Spiroid gears, **8**-111

Splines, **8**-31, **8**-35
  dimensions of fittings for (table), **8**-36
  involute, **8**-31
  parallel-side, **8**-35
  proportions of (table), **8**-36
Splint coal (def), **7**-5
Split-ring piston packing, **8**-144
Spontaneous combustion of coal, **7**-10
Spot welding, **13**-38
  by arc, **13**-33
  machines for, **13**-38
  steel (table), **13**-39
Spray chamber (air conditioning), **4**-44
Spray painting, **6**-121
Spray ponds, **9**-74
Spreader stokers, **9**-10
Springs, **8**-74 to **8**-85
  allowable working stresses, **8**-77
  brass for (table), **6**-69
  coiled, **8**-76
  compound (leaf or laminated), **8**-75
  conical, **8**-77
  cylindrical-helical, **8**-76, **8**-77
    strength and deflection of (table), **8**-81 to
      **8**-83
  deflection, **8**-74 to **8**-85
    work done in, **8**-74
  elliptic, **8**-75
  flat-leaf, **8**-75
  helical, **8**-76
  leaf, **11**-9
  materials for, properties of (tables), **8**-78 to
      **8**-80
  rectangular plate, **8**-74
  resilience (table), **5**-19
  semielliptic, **8**-75
  single-leaf flat, strength and deflection of
      (table), **8**-75
  spiral-coiled, **8**-76
  steel, **6**-35
    type and heat-treatment of (table), **6**-36
    wire for, **6**-35
  stiffness, **5**-70
  straight-bar torsion, **8**-76
  time of vibration, **8**-74
  torsion, **8**-76, **8**-77
  triangular plate, **8** 74
Sprinklers, **18**-29 to **18**-31
  alarms for, **18**-30
  deluge, **18**-30
  dry pipe systems for, **18**-30
  over electrical equipment, **18**-30
  heads for, **18**-30
    alloys for, **6**-90
  location and spacing, **18**-30, **18**-31
  nonfreeze systems for, **18**-30
  outside, **18**-30
  preaction, **18**-30
  tanks for, **18**-31
  temperature rating (table), **18**-30
Sprocket wheels:
  for chain drives, **8**-67, **8**-68
  chains for, **8**-59 to **8**-67
  diameters (formulas), **8**-67
  teeth, design of, **8**-67
Spur gears, **8**-89 to **8**-106
Square roots, **2**-3, **2**-26
Squares, summation of series of, **2**-17
Squaring shears, **13**-18
Squirrel-cage motors, **15**-43 to **15**-47
Stability:
  of airplanes, **11**-72

Stability (*Cont.*):
  of automatic control systems, **16**-34, **16**-35
  of floating bodies, **3**-40
  of ships, **11**-39, **11**-40
  of submerged bodies, **3**-40
Stack effect:
  air flow due to, **9**-26
  table, **9**-27
Stack emissions (*see* Air pollution)
Stack gases (*see* Flue gas)
Stacks:
  metal, corrosion of, **6**-114
  verticality determination of, **16**-50
Stadia:
  distance measurements with, **16**-47
  leveling with, **16**-47
Stagnation (fluids, def), **3**-52
Stagnation point (aerodynamics, def), **11**-61
Stagnation point (fluid flow), **3**-52
Stainless alloys, **6**36 to **6**-42
Stainless iron, **6**-38
Stainless steel (*see* Steel, stainless)
Stairs, industrial plant, **12**-8
Stairways, safety provisions, **18**-23
Stalling angle (aerodynamics, def), **11**-61
Standard cubic foot (def), **14**-31
Standard deviation (statistics, def), **17**-19
Standard time, **1**-41
  U.S. zones, **1**-41
Standard ton (refrigeration, def), **19**-3
Standard wire rope, **8**-85
Standpipes for fire protection, **18**-33
Star connections, three-phase circuits, **15**-23
Starches in adhesives, **6**-141
Starting boxes for shunt motors, **15**-33
Starting compensator for a-c motor, **15**-44
Starting devices for a-c motors, **15**-44, **15**-45
Starting resistance (trains), **11**-30
Starting systems for automobiles, **15**-83
Static electricity (def), **15**-18
Static friction (tables), **3**-25 to **3**-27
Static load, safety factors for, **5**-22
Static unbalance (def), **5**-70
Statically determinate and indeterminate
      supports (def), **3**-6
Statics:
  of framed structures, **12**-11 to **12**-19
  graphical, **3**-6, **3**-7
    of trusses, **12**-14 to **12**-19
  of rigid bodies, **3**-3 to **3**-7
Stationary engines (internal-combustion), **9**-86
Statistical quality control, **17**-19
Statistics, **17**-19 to **17**-26
  average (def), **17**-19
  confidence, **17**-20
  correlation, **17**-21, **17**-22
  for life tests, **17**-20
  standard deviation (def), **17**-19
  variance (def), **17**-19
Steady-flow process (def), **4**-13
Steam:
  attemperation, **9**-23
  constant-pressure expansion, **4**-24
  constant-volume expansion, **4**-24
  critical pressure, **9**-25
  enthalpy-entropy charts, **4**-25, **4**-26
  flow rates in pipes (tables), **12**-68
  isentropic expansion, **4**-24
  isothermal expansion, **4**-24
  moisture in, measurement of, **16**-21
  Mollier chart for, **4**-25, **4**-26
  properties of (tables), **4**-29 to **4**-31

Steam (*Cont.*):
  purification, **9**-28
  quality of, **4**-22
    determination of, **16**-21
  reheating, **9**-20, **9**-21, **9**-56
  saturated (charts), **4**-24, **4**-25
    pressure-temperature conversion (table), **9**-63
    tables, **4**-29 to **4**-31
  silica removal, **9**-28
  specific heat (table), **4**-10
    chart, **4**-11
  superheated (tables), **4**-30
    isentropic expansion, **4**-24
  tables, **4**-29 to **4**-31
  temperature control, **9**-23, **9**-24
  temperature-entropy chart, **4**-24
  wet: flow of, through nozzles, **4**-47 to **4**-49
    quality of, **4**-22, **16**-21
Steam boilers (*see* Boilers)
Steam calorimeters, **16**-21, **16**-22
Steam condensers (*see* Condensers)
Steam cycles, **4**-28
Steam emulsion test for lubricating oils, **6**-199
Steam ends of pumps, **14**-7
Steam engines, **9**-35 to **9**-38
  clearance, **9**-38
  compound, **9**-37
  compounding, **9**-37
  economy factors, **9**-37
  efficiency, **4**-27
  friction horsepower, **9**-30
  indicator diagrams for (theoretical), **9**-36
  indicators for, **16**-18
  marine (*see* Marine engines)
  mean effective pressure, **9**-36
  mechanical efficiency of, **9**-36
  operation of, condensing, **9**-37
  Rankine cycle for, **9**-38
  Rankine efficiency ratio, **9**-38
  ratio of expansion, **9**-36
  separation of inlet and outlet ports, **9**-37
  steam jackets for, **9**-37
  steam rates, **9**-37
  superheating, effects of, **9**-37, **9**-38
  theory of, **4**-27
  uniflow, **9**-37
  valve and port sizes, **9**-38
Steam flow:
  effect of fittings on, **4**-50
  in nozzles, **4**-47 to **4**-49, **9**-39
  in orifices, **4**-47, **4**-48
    velocity coefficients for, **4**-47, **4**-49
  in pipes, **4**-49, **4**-50
  throttling by, **4**-50, **4**-51
  in turbine buckets, **9**-40 to **9**-45
  wire drawing from, **4**-50, **4**-51
Steam hammers, **13**-25
Steam jackets, **9**-37
Steam-jet, refrigeration, **19**-10
Steam lines, expansion joints for, **8**-193
Steam nozzles:
  design of, **4**-49
  flow-through, **4**-48, **4**-49, **9**-40
Steam pipes (*see* Pipe; Piping)
Steam power plants, cost of, **17**-34
  (*See also* Power costs)
Steam pumps, **14**-7, **14**-8
Steam purifiers, **9**-28
Steam reheaters, **9**-20, **9**-21
Steam separators, **9**-28

Steam turbines, **9**-38 to **9**-59
  advantages of, **9**-40
  back-pressure, **9**-47
  balancing of, **9**-45
  blades: erosion of, **9**-43
    materials for, **9**-46
    maximum speed of, **9**-45
    vibrations of, **9**-45
  bleeding cycle for, **4**-28, **9**-56
  bolting materials, **9**-46
  buckets, flow of steam through, velocity diagrams, **9**-39, **9**-45
  casing materials, **9**-46
  classification of, **9**-38, **9**-39
  clearances in, **9**-46
  cross-compound, **9**-43
  damage to, due to water induction, **9**-57, **9**-58
  diagram efficiency of, **9**-40
  disk wheels, **5**-55
  divided flow, **9**-43
  efficiency of: correction factors, **9**-53
    table, **9**-54
  engine efficiency of (def), **9**-52
  extraction, **9**-48
    in regenerative heating, **9**-56
  feedwater treatment, **9**-59
  governors for, **9**-47
  heat rates (charts), **9**-58
    effect of regenerative heating on, **9**-56
    representative (table), **9**-55, **9**-56
  helical flow, **9**-47
  high-temperature bolting for, **9**-46
  impulse, **9**-39
    bucket-velocity diagram, **9**-39
  impurities in steam, **9**-58, **9**-59
  internal efficiency (def), **9**-42
  labyrinth packings for, **9**-46
  leakage of steam through, **9**-46
  leakage loss, **9**-41
  leaving loss, **9**-43
  loss due to moisture in steam, **9**-43
  low-capacity, **9**-47
  low-pressure, elements of, **9**-42 to **9**-45
  lubrication of, **6**-204
  marine, **9**-48
  mechanical drive, **9**-48, **9**-53
  modern, large central station, **9**-49 to **9**-51
  multirow stages, **9**-41
  nozzles: design of, **9**-40
    efficiency of, **9**-40
    mouth area, **9**-40
    theoretical work, **9**-40
    throat area, **9**-40
    velocity coefficient, **9**-40
  oils, **6**-204, **9**-58
  operating pressures and temperatures for, **9**-56
  packings for, **9**-46
  performance of, **9**-52 to **9**-57
  performance calculations, **9**-53
  radius ratio of, **9**-39
  reaction, **9**-39
    bucket-velocity diagram for, **9**-39
    (*See also* Hydraulic turbines, reaction)
  regenerative cycle for, **4**-28, **9**-56
  reheating cycle for, **4**-28
  rotative speeds of, **9**-45
  rotor materials, **9**-46
  silica deposits on, **9**-32
  stage design, **9**-39
  stage efficiency, **9**-40

Steam turbines (*Cont.*):
　　starting and loading, **9**-57
　　steam bled (table), **9**-57
　　steam-path design, **9**-42
　　steam rate of (def), **9**-52
　　　theoretical (table), **9**-52
　　superposition, **9**-47
　　tandem, **9**-43
　　thrust bearings for, **9**-46
　　turning gears for, **9**-46
　　velocity ratio, effect on efficiency (chart), **9**-42
Steam washers, **9**-28
Steel (def), **6**-12
　　aging: of magnet, **15**-77
　　　methods of decreasing, **6**-28
　　AISI designations, **6**-30
　　AISI properties (tables), **6**-34, **6**-35
　　AISI specifications (tables), **6**-23, **6**-30 to **6**-40
　　alloys, **6**-12
　　　AISI specifications (tables), **6**-23, **6**-30 to **6**-40
　　　castings, **6**-47, **6**-52 to **6**-58
　　　corrosion- and heat-resisting, **6**-36, **6**-54
　　　　AISI specifications for (table), **6**-37, **6**-38
　　　　properties of (table), **6**-39, **6**-40
　　　furnace for (electric), **6**-15
　　　gas cutting of, **13**-36 to **13**-38
　　　hardenability of, **6**-19
　　　hardness, influence on: of elements (table), **6**-21
　　　　of heat-treatment (tables), **6**-23, **6**-34, **6**-35
　　　　of specimen size (tables), **6**-35
　　　magnetic properties of (table), **6**-41
　　　precipitation-hardening, **6**-38
　　　properties of (tables), **6**-39, **6**-40
　　　strength at high temperatures (table), **5**-11
　　　welding, **13**-38 to **13**-42
　　angles (*see* Angles, steel)
　　　gages for (table), **12**-44
　　annealing of, **6**-18
　　ASTM specifications for, **6**-25
　　　tables, **6**-27, **6**-28
　　austenitic manganese, **6**-42
　　bars, round and square, weight (table), **6**-46
　　basic oxygen, **6**-14
　　beams (*see* Beams, steel)
　　　deflection of (table), **12**-32
　　Bessemer (def), **6**-12
　　　manufacture of, **6**-15
　　boiler, ASTM specifications for (table), **6**-27
　　boron, **6**-30
　　capped (def), **6**-15
　　carbon (def), **6**-12
　　　AISI specifications (table), **6**-30, **6**-31
　　　applications of (table), **6**-26
　　　castings, **6**-52
　　　hardness, influence of heat-treatment (table), **6**-23
　　　uses of, **6**-31, **6**-32
　　　welding, **13**-38 to **13**-42
　　carbon tool, **13**-52
　　carburized: annealing, **6**-23
　　　hardening, **6**-23
　　carburizing, **6**-23
　　casehardening, **6**-23 to **6**-25
　　castings (*see* Castings, steel)
　　chromium, stainless (high-chromium), **6**-36, **6**-38
　　　AISI specifications (tables), **6**-37, **6**-38

Steel (*Cont.*):
　　clad, **6**-24
　　classification of, **6**-12
　　cobalt, for tools, **13**-51
　　cold-drawn: physical properties of (table), **6**-35
　　　uses of, **6**-33
　　cold-rolling of, **6**-16, **6**-28
　　cold-working of, effect of, **6**-16
　　columns, **5**-40 to **5**-44
　　　table, **12**-35 to **12**-38
　　commercial, applications of (table), **6**-26
　　composite, **6**-24
　　concrete reinforcement bars, **12**-48
　　conductors, electrical, **15**-6
　　constitution of, **6**-17, **6**-18
　　construction (*see* Steel-framed structures)
　　copper-covered (high-voltage transmission), **15**-6
　　corrosion of (*see* Corrosion)
　　corrosion-resisting, **6**-36
　　creep rates for (chart, table), **5**-11, **5**-12
　　cutting speeds, **13**-53
　　cyaniding, **6**-24
　　deep-hardening, **6**-22
　　　liquids for quenching, **6**-22
　　defects in, **6**-16
　　　removal of, **6**-16
　　densities of (table), **6**-44
　　deoxidizing, **6**-15
　　drawing, **6**-19
　　electric, **6**-15
　　　furnace (def), **6**-12
　　electric sheet, **6**-38
　　extrusion, **6**-17
　　flat-rolled, weight of (table), **6**-44
　　forging, **6**-33
　　　effect of, **6**-16
　　foundry practice (*see* Foundry practice and equipment)
　　free-cutting, **6**-31
　　　AISI specifications (table), **6**-31
　　　uses of, **6**-31
　　galvanized sheets (table), **6**-42
　　gas cutting of, **13**-36 to **13**-38
　　for gears, yield strengths (table), **8**-120
　　H-type, **6**-30
　　hammering, effect of, **6**-16
　　hardenability (def), **6**-19
　　　effect of alloys on (table), **6**-21
　　hardened, tempering of, **6**-22
　　hardening of, **6**-24
　　　effect of surface condition on, **6**-22
　　heat-resistant alloys, properties of (table), **6**-96
　　heat-resisting, **6**-36
　　heat-treatment, **6**-18 to **6**-22, **13**-54
　　　carbon, temperatures of (table), **6**-21
　　　quenching mediums, **6**-22
　　　relation of design to, **6**-22
　　　time of, and cooling rate, **6**-21
　　high-speed, **6**-33, **13**-52
　　　(*See also* Tool steel)
　　hot-working, effect of, **6**-16
　　hysteresis loss in (chart), **15**-17
　　I beams (table), **12**-34
　　inclusions in, **6**-16
　　ingots, defects in, **6**-16
　　joists, **12**-42
　　Jominy test, **6**-19
　　killed (def), **6**-15

Steel (*Cont.*):
  low-carbon, **6**-25, **6**-26
    AISI specifications (table), **6**-30
    composition of (tables), **6**-13, **6**-30
    low-alloy, uses of, **6**-29
    mechanical properties of (tables), **6**-27
    uses of, **6**-25 to **6**-33
  Lüders lines in, **6**-26
  machinability, **13**-50
  machinery, **6**-30
    composition of (tables), **6**-30 to **6**-33
    influence of heat-treatment (tables), **6**-35
    influence of specimen size (tables), **6**-35
  McQuaid-Ehn test, **6**-30
  magnet (permanent), **15**-73
    chart, **15**-77
  magnetic properties of alloys (table), **6**-41
  manganese, **6**-42
    AISI specifications (table), **6**-32, **6**-33
  manufacture of, **6**-13 to **6**-16
  maraging, **6**-29
  mechanical treatment, **6**-16, **6**-17
  members, structural, properties of, **12**-23 to **12**-46
  microscopic structure, **6**-17
  molybdenum: AISI specifications (table), **6**-32, **6**-33
    for tools, **13**-51
  NE-type, **6**-30
  nickel, AISI specifications (table), **6**-32, **6**-33
  nickel-chromium, AISI specifications (table), **6**-32, **6**-33
  nickel-molybdenum, AISI specifications (tables), **6**-32, **6**-33
  nitriding, **6**-24
  nonmagnetic, **6**-41
  normalizing of, **6**-19
  open-hearth (def), **6**-12, **6**-15
  patenting, **6**-18
  pipes (*see* Pipe)
  piping in (cavities), **6**-15
  plastic range chart for, **13**-15
  pressing, effect of, **6**-16
  quenching (heat-treatment), **6**-22
  railroad, ASTM specifications (table), **6**-27
  reinforcing, for concrete, **12**-48
  rimmed (def), **6**-15
  rolled, weight (table), **6**-43, **6**-44
  rolling, effect of, **6**-16
  scabs in, **6**-16
  scrap, **6**-49
  seams in, **6**-16
  segregation in, **6**-16
  semikilled (def), **6**-15
  shallow-hardening (def), **6**-22
    liquids for quenching, **6**-22
  shapes of, properties of (tables), **12**-30 to **12**-45
  sheet and strip (table), **6**-43, **6**-44
  silicon (table), **6**-41
  silicon-manganese, AISI specifications (table), **6**-32, **6**-33
  sintered, strength of (table), **6**-105
  spheroidizing, **6**-18
  spring, **6**-35
  springs (vibration absorption), **5**-72
  stainless, **6**-36 to **6**-38
    compositions of (table), **6**-37, **6**-38
    corrosion-resistance of, **6**-36, **6**-109
    cutting by flame, **13**-37
    passivity of, **6**-109
    properties of (tables), **6**-37 to **6**-40

Steel, stainless (*Cont.*):
    strength at high temperatures (table), **5**-11
    uses of, **6**-39 to **6**-42
  strength of: AISI (tables), **6**-34, **6**-35
    AISI alloys (tables), **6**-39, **6**-40
    ASTM (table), **6**-27
    low-alloy (table), **6**-28
  stretcher strains in, **6**-26
  strip: mechanical properties of (table), **6**-26
    temper rolling, **6**-26
  structural: ASTM specifications, **6**-27, **12**-28, **12**-30
    fire-resistance of, **12**-46
    for locomotives, ASTM specifications (table), **6**-27
    materials for, **12**-28, **12**-30
    paints for, **6**-120
    properties of (tables), **12**-30 to **12**-45
    for ships, ASTM specifications (table), **6**-27
    specifications for (table), **6**-27
    welding, **12**-32
  structure of, **6**-17, **6**-18
  temper rolling of, **6**-28
  tempering of, **6**-19, **6**-22
  thermomechanical treatment of, **6**-25
  tool (*see* Tool steel)
  tubes, electric resistance welded, **13**-38
  tubing (tables), **8**-166, **8**-167
  for turbine blading, **9**-46
  Ugine-Sejournet process, **6**-17
  welding (*see* Welding)
  wire, properties of (table), **6**-43
Steel construction (*see* Steel-framed structures)
Steel-framed structures, **12**-28 to **12**-46
  allowable stresses (table), **12**-30
  beam connections (table), **12**-44
  beams and girders, proportions of, **12**-30
  bolts, **12**-31
  bracing, **12**-32
  columns, **12**-30
  compression members, **12**-30
  corrosion, **6**-114
  fire resistance (table), **12**-46
  gages for rivets, **12**-32
    tables, **12**-45
  lacing bars, **12**-32
  pins, **12**-31
  riveted connections, **12**-31
  rivets, **12**-31
  specifications, **12**-28, **12**-29
  tension members, **12**-30
  tie plates, **12**-31
  welding, **12**-32
Steel wire:
  strength of (table), **6**-43
  weight of (table), **6**-43
Steel-wire gage (table), **6**-45
Steering gear, automobile, **11**-11 to **11**-13
Stefan-Boltzmann law (radiation), **4**-71
Steinmetz coefficients (table), **15**-17
Steinmetz law of hysteresis, **15**-17
Stellite (cobalt alloy tool material), **13**-51
Step bearing, **8**-31
  friction of, **3**-31
Step function (automatic control), **16**-26
Step pulleys, **8**-52
Steradian (def), **1**-33, **2**-14
Steregon (def), **2**-14
Stereotype metal (table), **6**-89
Stiffness of beams, **5**-31
Stirling cycle, **4**-21

Stirling refrigerating engine for cryogenics, **19**-24
Stoke (unit of kinematic viscosity), **3**-36
Stokers, **9**-10
   chain-grate, **9**-10
   spreader, **9**-10
   traveling-grate, **9**-10
   underfeed, **9**-10
Stokes' law for settling of dust, **18**-12
Stokes' theorem (vector analysis), **2**-64
Stone:
   building, **6**-158
   properties of (table), **6**-159
   specific gravity and density of (table), **6**-8, **6**-9
Stonework, **12**-22
   mortar for, **6**-181
   (*See also* Masonry)
Storage:
   in computers: disk, **2**-67
      drum, **2**-67
      tape, **2**-67
   of inflammable liquids, **18**-23
Storage batteries (*see* Batteries, storage)
Stove bolts, **8**-24
Stovewood, **7**-13
Straight line, equation of, **2**-30
   (*See also* Lines)
Straightening operations (metals), **13**-19, **13**-26, **13**-27
Strain:
   elastic, **5**-54
   plastic, **5**-54
Strain gages, **5**-58, **5**-59, **16**-8
   foil, **5**-59
   in rosettes, **5**-59
   transverse sensitivity, **5**-59
Strain hardening:
   coefficient of, **5**-54
   modulus of, **13**-15
Strands, wire, **8**-85
Strap hammers, **13**-25
Streamline form (aerodynamics, def), **11**-66
   drag of, **11**-69
Streams, flow measurement of (*see* Weirs)
Strength:
   of bolts (tables), **8**-30
   of metals (table), **5**-5, **6**-11
   tensile (def), **5**-2
   yield (def), **5**-2
Stress (def), **5**-4, **5**-16
   analysis of, **5**-57 to **5**-61
      by birefringent coating, **5**-60
      by brittle-coating, **5**-59
      by holography, **5**-60
   combinations of, **5**-5, **5**-19 to **5**-21
   corrosion, **5**-9, **6**-109
   creep rates due to (table), **5**-11
   in framed structures, diagrams, **12**-16 to **12**-19
   longitudinal, **5**-16
   method of joints, **3**-6
   moment (def), **5**-23
   in pipelines due to flexure, **5**-61 to **5**-67
   polygon, **3**-6
   shearing, **5**-17, **5**-18
   simple, **5**-16 to **5**-19
   in static plane structure, **3**-6, **3**-7
   tangential, **5**-17
   true, **5**-4
   in turbine disk wheels, **5**-55, **5**-56
   volume change due to, **5**-18

Stress concentration, **5**-6
   factors of (charts), **5**-6
Stress-cycle diagram (endurance), **5**-9
Stress-optic law, **5**-57
Stress-rupture test, **5**-12
Stress-strain diagram, **5**-2 to **5**-4
Stresscoat (strain indicator), **5**-59
Stretcher strains, **5**-4, **6**-26
Stripping by power shovel, **10**-37
Stroboscopes, **16**-3
Strouhal number (vibrations in fluids), **3**-47, **3**-53
Structural steel (*see* Steel, structural)
Structural timber (*see* Timber)
Structures:
   statically determinate and indeterminate (def), **3**-6
      stress in, **3**-6, **3**-7
   steel-framed (*see* Steel-framed structures)
   stresses in, by method of sections, **3**-7
   wind pressure on, **12**-12 to **12**-14
Struts, drag of (aerodynamics), **11**-69
Stubs gage (table), **6**-45
Stucco, **6**-183
Stud partitions, safe loads on (table), **12**-26
Studs, drilling and tapping cast iron for (table), **8**-30
Stuffing boxes, friction of, **3**-27
Sturm's equation for collapse of cylinders, **5**-49
Submerged bodies:
   forces on, **3**-38, **3**-39
   loss of weight in, **3**-40
Submerged openings, flow through, **3**-68, **3**-69
Submerged surfaces, pressure on, **3**-38
Submergence of air-lift pumps, **14**-14
Subsonic velocity (def), **11**-58
Subtraction:
   algebraic, **2**-16
   arithmetical, **2**-2
   of complex quantities, **2**-23
Subway trains, **11**-34
Suction lift of pumps, **14**-5
Sulfur:
   in coals (table), **7**-4
   effect on vegetation, **18**-14
   heat of combustion (table), **4**-54
Sulfur dioxide:
   effect on vegetation, **18**-14
      table, **18**-15
   properties of (table), **4**-37
   as refrigerant, **19**-4
Sulfuric acid, freezing temperatures of (table), **15**-12
Sulfurized oils as cutting fluid (machining), **13**-54
Sulfurous acid (*see* Sulfur dioxide)
Sulzer two-cycle engines, **9**-104
Summation:
   of series by differences, **2**-17
   of squares and cubes, **2**-17
Summer (analog computer), **2**-73
Superalloys for high-temperature use (tables), **6**-97
Supercharging (engines), **9**-102, **9**-103
Superconductivity, **19**-27
   of metals, **19**-27
Superconductors (cryogenics), **19**-28
   columbium alloys, **6**-97
Superfinish (honing process), **13**-70
Superheated ammonia, properties of (table), **4**-35
Superheated steam:
   properties of (chart), **4**-24
   tables, **4**-30

Superheated sulfur dioxide (table), **4**-37
Superheated vapors, expansion of, **4**-24, **4**-25
Superheaters, **9**-20, **9**-21
 allowable stress (table), **9**-21
 characteristic steam temperature, **9**-20
 continuous-tube, **9**-20
 convection, **9**-20
 draft-loss in, **9**-21
 fouling of, **9**-21
 hairpin, **9**-20
 integral, **9**-20
 mass flow in (table), **9**-22
 nondraining, **9**-20
 pendant, **9**-20
 pressure drop in, **9**-21
 radiant, **9**-20
 self-draining, **9**-20
 separately fired, **9**-20
 steam flow rates, **9**-21
 tube spacing, **9**-21
 tubes for (table), **9**-21
Superheterodyne reception (radio), **15**-94
Supermalloy, properties of (table), **6**-41
Supersaturation, theory of, **4**-49
Supersonic and hypersonic aerodynamics, **11**-73 to **11**-79
 acoustic theory of, **11**-79 to **11**-84
 aerodynamic heating, **11**-81
 airfoil section, **11**-79
 area rule, **11**-81
 base drag, **11**-81
 density measurement, **11**-84
 diffusers, **11**-77
  Oswatitsch, **11**-78
  subsonic, **11**-77
  supersonic, **11**-77
 force measurements, **11**-83
 gas dynamics relations, **11**-73
  table, **11**-74
 hypersonic flow (def), **11**-73
 lift: of axially symmetric body, **11**-80
  coefficient, **11**-79
  curves for wings: delta plan-form, **11**-80
   rectangular, **11**-80
   swept-back, **11**-80
  interference, **11**-81
 Mach angle, **11**-75
 Mach cone, **11**-76, **11**-80
 Mach number (def), **11**-73
 Mach wave, **11**-75
 nozzles, **11**-76
  converging, **11**-77
  ordinates for (table), **11**-77
  rockets, **11**-77
 Prandtl-Glauert rule in, **11**-79
 pressure measurements, **11**-82
 recovery factor (temperature), **11**-82
 shock polar, **11**-75
 shock relations (table), **11**-76
 shock tubes, **11**-78
 shock waves, **11**-74, **11**-75
 skin friction drag, **11**-81
 stagnation temperature, **11**-82 to **11**-84
 subsonic flow (def), **11**-73
 supersonic flow (def), **11**-73
  past a cone, **11**-75
  chart, **11**-76
 transonic flow, **11**-73
 wind tunnels, **11**-78, **11**-79
 (*See also* Aircraft jet propulsion)
Surface of various solids, **2**-12 to **2**-15
Surface combustion, **4**-58

Surface condensers, **9**-59 to **9**-64
Surface finish (*see* Surface texture)
Surface temperature, measurement of, **16**-13, **16**-14
Surface tension (def) **3**-36
 of liquids (table), **3**-37
Surface texture:
 design criteria, **13**-74
 design requirements, **13**-75
 flaws in (def), **13**-77
 lay of (def), **13**-77
 lay symbols (table), **13**-78
 measurement of, **13**-77
 of production processes (chart), **13**-79
 roughness of, **13**-76
 symbols of, **13**-75, **13**-76
 waviness of (def), **13**-77
Surge tanks, **9**-149
Surveying, **16**-42 to **16**-50
 leveling, **16**-43 to **16**-45
 mapping, **16**-47
 measurement of inaccessible distances, **16**-49
 profile, **16**-43
 referencing a point in, **16**-49
 special problems in, **16**-48 to **16**-50
 stadia, **16**-47
 transit work, **16**-45 to **16**-48
 traverse, **16**-46
Surveyor's measure, **1**-31
Susceptance (electric circuit, def), **15**-5, **15**-22
Sutro weir, **16**-17
Swaging of metals, **13**-21
Swaging machines, rotary, **13**-27
Swash-plate pumps, **14**-12
SWG (wire gage, table), **6**-45
Swinging-block linkage, **8**-3
Switch gear (electric), **15**-54
Switchboards, **15**-53 to **15**-55
 bus bars for, **15**-53
 panels, **15**-53, **15**-54
  equipment for standard, **15**-53, **15**-54
 wiring diagrams for, **15**-54
Switches, electric, **15**-54, **15**-68, **15**-69
Symbols:
 algebraic, **2**-15
 chemical (table), **6**-3, **6**-4
 cross-section, **8**-97
 for electrical apparatus, **15**-9
 for electrical units (table), **15**-3
 proofreaders', **19**-45, **19**-46
 welding, **13**-38
Synchromesh (automobile), **11**-7
Synchronous condenser, **15**-47
Synchronous converter, **15**-49, **15**-50
 switchboard equipment for, **15**-53
Synchronous generators (*see* Generators, alternating-current)
Synchronous impedance, **15**-37, **15**-38
Synchronous motors, **15**-47 to **15**-49
Synchros, **16**-6
Synthane process (gas making), **7**-46

Tachometers, **16**-21
Tacks, wire (table), **8**-97
Tail shafts (ships), **11**-54
Tallow oil (table), **6**-202
Tangent (trigonometry), **2**-25
 to circle, **2**-8, **2**-31
 to ellipse, **2**-33
 graph, **2**-55

Tangent (trigonometry) (*Cont.*):
   to hyperbola, **2**-35
   to parabola, **2**-31
   tables, **1**-15, **1**-16
Tangential stress, **5**-17
Tanks, water, for fire protection, **18**-31
Tantalum, uses of, **6**-98
Tantiron (corrosion-resisting iron alloy), **6**-49,
   **6**-109
Tape:
   corrections for temperature, tension, and
      sag, **16**-42
   magnetic (computers), **2**-67
   measurements with, **16**-42
   surveyor's and builder's, **16**-42
Taper keys, friction of, **3**-28
Taper pins (table), **8**-34
Taps:
   pipe, drills for (table), **8**-20
   sizes of (table), **8**-32
Tar:
   coal, as fuel, **7**-20
   for roofing, **6**-163
   specific gravity and density of (table), **6**-9
Tar sands, reserves of, **9**-4
Taylor's diagram for three-blade propellers,
   **11**-51
Taylor's propeller coefficients (ship
   propellers), **11**-50
Taylor's series, **2**-44
Tees, structural steel, properties of (tables),
   **12**-43, **12**-44
Teflon:
   electrical insulation, **6**-153
   plastic, **6**-161, **6**-162
Television, **15**-96, **15**-97
   antenna, **15**-96
   color, **15**-97
   frequency band, **15**-96
   interlacing, **15**-96
   kinescope, **15**-97
   receivers, **15**-97
   scanning and blanking, **15**-96
Temperature:
   absolute scale, **4**-3
   of adiabatic saturation, **4**-32
   air volume correction factors due to (table),
      **12**-103
   coefficient of resistance (def, table), **15**-5
   of combustion, **4**-53
      table, **4**-57
   conversion, tables, **4**-3, **4**-4
   creep rate, effect on (table), **5**-11
   critical: of gases (def), **4**-22
      tables, **4**-17, **4**-23
   dry-bulb, **4**-31
   effect on volume of water (table), **6**-10
   flame, with gaseous fuels (table), **4**-57
   gradient (heat flow), **4**-62
   inferred absolute zero, **15**-5
   inversion (Joule-Thomson effect), **4**-51
   of iron or steel by color, **4**-6
   limits of electrical apparatus, **15**-51
   measurements of, **4**-2, **4**-3, **16**-10 to **16**-14
      pneumatic transmitter, **16**-12
      of surfaces, **16**-11, **16**-13, **16**-14
   stresses, deformation by, **5**-17
   wet- and dry-bulb (def), **4**-31
   wet-bulb, thermodynamic, **4**-32
Temperature-entropy chart:
   for air, **4**-25
   description of, **4**-23

Temperature-entropy chart (*Cont.*):
   for isobutane, **4**-25
   for steam, **4**-24
Temperature-humidity index, **12**-105, **12**-106
Tempering (def), **6**-19
   furnaces, **7**-63
   of hardened steel, **6**-22
   steel (def), **6**-19
   (*See also* Steel, heat-treatment)
Tempilsticks, **16**-11
Templates for plant layouts, **12**-4
Tensile strength of metals at high
   temperatures (table), **5**-11
Tension, elongation and contraction due to, **5**-2
Termites, destruction of wood by, **6**-126
Terne plate, **6**-89, **6**-115
Test specimens, **5**-15
Testing:
   by acoustic signature analysis, **5**-84
   compression, **5**-5
   by eddy-current methods, **5**-83
   by electrified particles, **5**-81
   by gamma rays, **5**-81
   by infrared methods, **5**-83
   by magnetic particles, **5**-81
   of materials, **5**-14, **5**-15
   methods of, **5**-76
      table, **5**-77, **5**-80
   by microwave methods, **5**-83
   nondestructive, **5**-76 to **5**-84
   by penetrants, **5**-81
   of porous objects, **5**-81
   by radiation, **5**-81
   radiographic, **5**-81, **5**-82
      dangers in, **5**-82
      intensifying screens for, **5**-82
      penetrameters for, **5**-82
      standards for, **5**-82
   ultrasonic, **5**-82, **5**-83
      frequencies, **5**-82
      materials used on, **5**-82
      standards for, **5**-83
   by X-rays, **5**-81
Testing machines, **5**-14, **5**-15
   accuracy and calibration, **5**-15
   grips for, **5**-14
   specimens, **5**-15
   (*See also* Testing)
Tetraethyl lead (knock suppressor, table), **7**-18, **9**-94
Tetrahedron, **2**-8, **2**-14
Tetryl explosive, **7**-25
   table, **7**-26
Tex (fiber fineness unit), **6**-156
Textile machinery, lubrication of, **6**-205
Therbligs (motion study, table), **17**-28
Thermal conductance, **4**-61
   conversion table, **1**-54
Thermal conductivity (def), **4**-60
   of air (table), **4**-62
   of building materials (tables), **4**-63, **4**-64
   conversion table, **1**-54
   of gases (table), **4**-62
   of insulating materials (tables), **4**-64, **4**-65
   of liquids (table), **4**-62
   of metallic elements (table), **6**-60, **6**-61
   of metals (table), **4**-60, **4**-61, **6**-11
   of miscellaneous substances (table), **4**-63
   of nickel-chromium with iron (table), **4**-61
   of refractories (tables), **4**-65
   of steam (table), **4**-62

Thermal insulation (*see* Refractories)
Thermal properties of substances, **4**-2 to **4**-12
Thermal unit:
 British (def), **1**-34, **4**-8
 IT calorie (def), **1**-35, **4**-8
Thermistors (resistance thermometer), **16**-11, **16**-13
Thermocouples, **15**-25, **16**-11, **16**-12
 at low temperatures, **19**-34
 temperature-millivolt relations (chart), **16**-13
 vacuum, **16**-14
Thermodynamics:
 cycle (def), **4**-12
 efficiency of steam engines, **4**-28
 equations of state, **4**-16
 first law of, **4**-13
 flow process (def), **4**-13
 of gases, variable specific heat, **4**-19
 mixtures of gases and vapors, **4**-28 to **4**-45
 notation and symbols of, **4**-12
 process (def), **4**-12
  reversible and irreversible (def), **4**-13
  steady-flow, **4**-13
 second law of, **4**-14
 of vapors, **4**-22 to **4**-28
Thermoelectric cooling, **19**-13
 figure of merit for, **19**-13
Thermometers, **4**-2, **16**-10 to **16**-13
 bimetallic elements, **16**-10, **16**-12
 calibration of, **4**-3
  standards for (table), **4**-5
 gas, **16**-10
 hydrogen, **4**-2
 for industrial applications, **16**-11
 mercury, **4**-2, **16**-11
 partial immersion, **16**-11
 readings, conversion of (tables), **4**-3, **4**-4
 resistance, **16**-11, **16**-13
 scales for, **4**-2, **4**-3
 stem exposure correction for, **16**-11
 vapor bulb, **16**-11
 vapor pressure, **19**-34
Thermopile, **16**-14
Thermoplastic substances, **6**-153
Thermosetting plastics, **6**-153, **6**-162
Thermosiphon (water-cooled engines), **9**-105
Thin-plate orifice meters, **16**-15, **16**-16
 flow coefficients for (table), **16**-16
Thinners for paint, **6**-119
Thiokol (synthetic rubber), **6**-167
Thoma coefficient (for hydraulic turbines), **9**-147
Thorium, **9**-6
Threading dies, **13**-61
Threading machines, **13**-61
Threads, screw (*see* Screw threads)
Three-phase a-c circuits, **15**-23
Three-phase alternators, **15**-36
Three-phase power, measurement of, **15**-26
Three-phase transformers, **15**-42
Three-wire d-c generator, **15**-63
Three-wire distribution circuits, a-c, **15**-63, **15**-65
Throttling, **4**-50, **4**-51
 calorimeter, **16**-21
 efficiency loss due to, **4**-51
 range (automatic control, def), **16**-25
Thrust, jet engines, equations for, **11**-92, **11**-93
Thrust bearings (*see* Bearings, thrust)
Thrustor for wire rope brake, **10**-11

Thyristors, **15**-85
 capacitance in circuits, **15**-88
 characteristics of (table), **15**-86
 inverter circuit, **15**-94
 rectifier circuits, **15**-88
Tidal power, **9**-5, **9**-168
 multiple-basin, **9**-168
 risks and uncertainties, **9**-168
 single-basin, **9**-168
Tile, roofing, **6**-164
Timber, **6**-122 to **6**-138
 allowable stresses in (tables), **6**-128 to **6**-131, **12**-26
 beams, **12**-23, **12**-24
 columns, **12**-24 to **12**-27
 connections, **12**-27, **12**-28
 connectors for, **12**-27
  allowable loads (table), **12**-29
 construction, **12**-23 to **12**-28
 fire resistance of, **12**-46
 floors, properties of (table), **12**-23
 specific gravity and density of (table), **6**-8
 trusses, **12**-15
 (*See also* Wood)
Timbre (of sound), **12**-138
Time (def), **3**-2
 measure of, **1**-41, **11**-112
 sidereal (def), **1**-41
 solar and apparent solar day, **1**-41
 standard, **1**-41
Time standards (work measurements):
 development of, **17**-30, **17**-31
 uses of, **17**-32
Time study, **17**-26 to **17**-32
 fatigue allowances in, **17**-30
 form for, **17**-29, **17**-31
 observations in, **17**-28 to **17**-30
 performance rating by, **17**-30
 standards development from, **17**-30 to **17**-32
Time zones, **1**-41
Timers, **16**-3
Timesharing (computers), **2**-69
Timing circuit (electronic), **15**-91
Tin, **6**-87
 alloys, **6**-87
 for bearings (table), **6**-92
 pipe, block (table), **8**-175
Tin plate, **6**-87, **6**-120
Tin plating, **6**-115
Tin solders, melting points (table), **6**-91
Tinned surfaces, paints for, **6**-120
Tires:
 details of (table), **11**-4
 effect of temperature on rolling resistance of, **11**-3
 friction of, **3**-27, **3**-28
Tirrill regulator (a-c generators), **15**-38
Titanium:
 alloys, **6**-87
  composition of (table), **6**-87
  properties of (table), **6**-88
 production of, **6**-87
TNT explosive, **7**-25
 table, **7**-26
Tobin bronze, plastic range chart for, **13**-15
Toggle joint, **8**-8
Tolerance (def), **8**-45
Toluene, solvent, **6**-168
Ton:
 assay (def), **1**-32
 metric (def), **1**-39
 register (def), **1**-31

Ton (*Cont.*):
  shipping (def), **1**-31
  standard (refrigeration, def), **19**-3
Tongs, lifting, **10**-11
Tonnage of ships (def), **11**-36
Tool steel, **6**-33, **6**-35, **13**-51 to **13**-53
  cast alloys, **6**-96
    compositions of (table), **6**-99
  classification of (table), **6**-33
  cold-work, **6**-33
  composition and uses of (tables), **6**-36
  heat-treatment, **13**-54
  high-speed (table), **6**-36
  hot-work, **6**-33
  properties of (table), **13**-52
  shock-resisting, **6**-33
  special-purpose, **6**-35
  tungsten alloy, **6**-35
  water-hardening, **6**-33
Tools:
  carbide, **6**-104 to **6**-106
  machine (*see* Machine tools)
Toothed and worm gearing efficiency, **3**-29
Torque of electric motor, **15**-32
Torque converter, hydraulic, **11**-6, **11**-8
Torr (def), **14**-44
Torsion, **5**-37, **5**-40
  of circular sections, **5**-37
  of composite sections, **5**-40
  external twisting moment, effect of, **5**-37
  of noncircular sections, **5**-40
  of shafts, **5**-37, **5**-40, **8**-48, **8**-49
  of various cross sections (table), **5**-41
Torsion balance, **16**-4
Torsional vibration of shafts, **5**-67
Torus, surface and volume of, **2**-14
Towers, wind pressure on, **12**-14
Towing of materials, **10**-2
Track, railway, **11**-31, **11**-32
  curves, **11**-31
Track jacks, **10**-16
Trackage, overhead, **10**-40
  monorail, **10**-28, **11**-33, **11**-34
    table, **10**-29
Traction, motors (railway), **11**-19, **11**-23, **11**-24
Tractive electromagnet, **15**-79
Tractive force, locomotive (table), **11**-20, **11**-22
Tractors, **10**-26
  crawler, **10**-26
  engines, **9**-85, **9**-86
  wheeled, **10**-26
Tractrix, **2**-41
Trademarks, **19**-43, **19**-44
Trains (*see* Railway engineering)
Tramways, cable, **10**-35 to **10**-37
  (*See also* Cableways, tramways)
Transducers (sound), **12**-137
Transfer functions in automatic control, **16**-31
Transformers, **15**-40 to **15**-42
  compensator, **15**-42
  data on, **15**-42
  differential, **16**-6
  E, **16**-6
  efficiency of, **15**-41
  grounding secondary of, **15**-64
  instrument, **15**-27, **15**-28
  insulation, chlorinated hydrocarbons, **6**-153
  leakage flux in, **15**-41
  polyphase, connections, **15**-42
  regulation of, **15**-41

Transformers (*Cont.*):
  Scott tap for, **15**-42
  T connection for, **15**-42
  testing of, **15**-41
  theory of, **15**-40
  three-phase, **15**-42
  V connection for, **15**-42
Transient analysis (*see* Automatic control, transient analysis of system)
Transients (electrical), **15**-19 to **15**-21
Transistors, **15**-85, **15**-86
  characteristics of (table), **15**-86
Transit, **16**-6, **16**-45 to **16**-48
  adjustment of, **16**-46, **16**-47
  angular measurements with, **16**-45
  and stadia, leveling with, **16**-47
Transmission:
  of electric power (*see* Electric power transmission)
  of heat (*see* Heat transmission)
Transmission dynamometers, **16**-18
Transmission mechanism, automobile, **11**-5 to **11**-9
Transmission shafting, design of, **8**-48, **8**-49
Transportation by cableways, **10**-34 to **10**-37
Trapezoid:
  area of, **2**-11
  center of gravity of, **3**-7
Trapezoidal weir, Cippoletti, **3**-65
Traveling cranes, **10**-28 to **10**-30
Trenching, setting stakes for, **16**-49
Triac, **15**-85
Trial-and-error solution of an equation, **2**-19
Triangles:
  plane: lengths and areas, **2**-11, **2**-28
    solution of, **2**-28
    theorems on, **2**-7
  polar, **2**-8
  spherical: solution of, **2**-29
    theorems on, **2**-14, **2**-29
Triangular notch weirs, **3**-65
Trigonometric equations, solution of, **2**-19
Trigonometric functions, **2**-25
  of a complex variable, **2**-24
  graphs of, **2**-55
  inverse, **2**-27
  series for, **2**-45
Trigonometric tables, **1**-13 to **1**-18
Trigonometry, **2**-25 to **2**-30
  solution: of plane triangles, **2**-28
    of spherical triangles, **2**-29
Triplex chain block, **8**-8
Tripoli (abrasive), **6**-139
Trippers (conveyors), **10**-57, **10**-59
Trochoid, **2**-39
Trolley trackage, overhead, **10**-28
Tropical year, **1**-41
Troy weight, **1**-32
Truck cranes, **10**-32 to **10**-34
  table, **10**-33
Trucks:
  dump, **10**-28
  engines, **9**-83, **9**-85
    (*See also* Automobile engines)
  fork, **10**-25
  lift, **10**-25, **10**-26
  (*See also* Automobiles)
Trusses:
  analytical determination of stresses, **12**-16 to **12**-18
  Bow's notation for, **12**-18
  floor, **12**-15

Trusses (*Cont.*):
  graphical determination of stresses, **12**-18, **12**-19
  roof, **12**-15
  stresses in, determination of, **12**-15 to **12**-19
  timber, joints for, **12**-27, **12**-28
  weight of (formula), **12**-15
Tubes, **8**-148, **8**-155
  boiler (*see* Boilers, tubes)
  cathode-ray, **2**-68, **2**-74
  collapsing pressure of, **5**-49
  condenser (*see* Condensers, steam, surface-type)
  electron, **15**-86
  film coefficients for (heat transmission), **4**-63 to **4**-69
  radiation factors for rows of (chart), **4**-74
  strength of, **5**-49, **5**-51
  (*See also* Pipe)
Tubing:
  aluminum, **8**-169
  boiler (table), **8**-167
  brass, **8**-169
  condenser (table), **8**-166
  copper (table), **8**-172
  heat exchanger (table), **8**-166
  lead (table), **8**-174
  metallic, for interior wiring (table), **15**-66 to **15**-68
  plastic, **8**-169
  seamless steel, weight (table), **8**-165
  steel (tables), **8**-166, **8**-167
  (*See also* Piping)
Tubular heat exchangers, formulas for, **4**-64
Tung oil, **6**-118
Tungsten, uses of, **6**-100
Tungsten carbide alloys, **6**-104 to **6**-106
  menstruum process for, **6**-105
  properties of (table), **6**-105
Tungsten steel, **6**-35
Tungsten tools, **13**-52
Turbidimeters, **16**-21
Turbine disks, stresses in, **5**-55
Turbines:
  gas (*see* Gas turbines)
  hydraulic (*see* Hydraulic turbines)
  steam (*see* Steam turbines)
Turbo-blowers, **14**-39 to **14**-42
Turbocompressors, **14**-39 to **14**-42
  adiabatic efficiency, **14**-40
  adiabatic heat, **14**-40
  axial, **14**-39
  characteristics of (chart), **14**-40
  efficiency of, **14**-40
  head per stage, **14**-41
  performance chart, **14**-41
  pressure coefficient, **14**-40
  seals for, **14**-41
  sonic velocity (eq), **14**-40
  specific diameter (eq), **14**-40
  specific speed (eq), **14**-40
  thrust pressure, **14**-42
Turbojet system of aircraft propulsion, **11**-86, **11**-87
Turbomachines (*see* Turbocompressors)
Turbulent flow, **3**-52, **3**-54
Turn indicator for airplanes, **3**-23
Turning-block linkage, **8**-3
Turpentine (paint thinner), **6**-119
Turret lathes, **13**-57, **13**-58
Twist drills, **13**-60
  angles, **13**-60
  sizes of (table), **13**-61

Twisting moments on shafts, **5**-37, **8**-48, **8**-49
Two-cycle engines, **9**-79, **9**-83, **9**-86 to **9**-89, **9**-92, **9**-103, **9**-104, **9**-107
Two-phase a-c circuits, **15**-23
Tyler standard screen scale, **18**-12
Type metals (table), **6**-89
Type sizes:
  printers, **19**-45
  typewriter, **19**-45

U-tube manometers, **16**-9
UGI-CCR catalytic reforming, **7**-47
Ugine-Sejournet steel extrusion process, **6**-17
Ultimate analysis of coal (tables), **7**-4
Ultrasonic testing (*see* Testing, ultrasonic)
Ultrasonic transducers, **5**-76, **5**-82
Ultrasonics, **12**-141 to **12**-143
  generation, by whistle, **12**-142
  industrial applications, **12**-141
  metal machining by, **12**-142, **13**-72
  for testing materials, **12**-142
  (*See also* Sound)
Underfeed stokers, **9**-10
Underground gasification of coal, **7**-24
Underground pipes, corrosion of, **6**-113
Underwater cutting of metals, **13**-37
Ungula, surface and volume of, **2**-13
Unified screw-thread standards, **8**-9
  (*See also* Screw threads)
Uniflow engines, **9**-37
Unijunction (semiconductor), **15**-86
Unions, pipe, **8**-186
Unit pole (magnetism), **15**-15
Unit-train unloader, **10**-24
U.S. standard gages for sheet metal (table), **6**-45
Units:
  absolute and gravitational systems, **3**-2
    table, **1**-40
  cgs system, **1**-40
  of common variables (table), **3**-51
  electrical and magnetic, **15**-2 to **15**-5
  heat (def), **4**-8
  magnetic (table), **15**-15
  systems, **1**-40
Unloader:
  airslides for, **10**-63
  rotary cardumper, **10**-24
Upper atmosphere, **11**-58, **11**-59
Uranium:
  enrichment, **9**-124
  fabrication, **9**-124
  natural, **9**-124
  properties of (table), **6**-102
  recycle, **9**-124
  reprocessing and reconversion, **9**-124
Urea-formaldehyde, adhesive, **6**-141, **6**-146

V-belts, **8**-54 to **8**-59
  arc of contact factors (tables), **8**-55, **8**-60
  cogged, **8**-58
  cross sections (table), **8**-60
    for required horsepower (chart), **8**-57
  on flat pulleys, **8**-57
  groove dimensions, **8**-56, **8**-60
  horsepower ratings (tables), **8**-54, **8**-57 to **8**-64
  length correction factors (table), **8**-61
  light-duty, **8**-55
  minimum pulley diameters (table), **8**-55, **8**-56

V-belts (*Cont.*):
  multiple drives, **8**-56
  quarter-turn drives, **8**-57
  ribbed, **8**-58
  service: factors (tables), **8**-50, **8**-60
    heavy, **8**-60
  sheave dimensions (table), **8**-56
  small diameter factors (table), **8**-61
  tensioning of drive, **8**-58
V-bucket carriers, **10**-50, **10**-51
  table, **10**-52
V-guides, friction in, **3**-28
V-notches, flow of water through, **3**-66, **16**-17
Vacuum, measurement of, **14**-44, **16**-8 to **16**-10
Vacuum gage, **16**-10
Vacuum pumps:
  for condensers, **9**-71
  (*See also* High-vacuum pumps)
Vacuum tubes (radio), **15**-86
Valence of chemical elements (table), **6**-3, **6**-4
Valves, **8**-188, **8**-189
  ammonia, **8**-188
  angle, **8**-188
  ball (pump), **14**-9
  cocks, **8**-189
  compressor, **14**-36
  conical wing (pump), **14**-8
  control, hydraulic power, **8**-39
  disk (pump), **14**-8
  disks, **8**-144
  gate, **8**-188
  globe, **8**-188
  head losses in (table), **12**-106
  high-pressure and high-temperature, steam, **8**-154, **8**-155
  hydraulic turbines, **9**-149
  materials (pump, table), **14**-10
  motion, duplex steam pump, **14**-7
  reciprocating pump, **14**-8 to **14**-10
    bearing pressures for (table), **14**-11
    velocity through seats of, **14**-9
  seats for, **8**-144
  wedge gate, **8**-188
van der Waals equation, **4**-17
Vanes, curved, force of liquid stream against, **3**-45
Vapor, **4**-22 to **4**-27
  compressibility factors (chart), **4**-18
  condensing, heat-transfer coefficients (chart), **4**-68
  properties of (tables and charts), **4**-23 to **4**-41
  saturated (def), **4**-22
  superheated (def), **4**-22
  temperature-entropy charts, **4**-24, **4**-25
  thermodynamics equations, **4**-22, **4**-23
Vapor-compression machines (refrigeration), **4**-45
Vapor pressures (def), **4**-11
  of hydrocarbons (chart), **4**-28
  of liquids (table), **4**-12
  of water (table), **4**-41
Vapor refrigerating machines, **19**-6
Vaporization:
  Hildebrand function for enthalpy of, **4**-11
  latent heats of (table), **4**-12
Vara (def), **1**-31
Variance (statistics, def), **17**-19
Variation (algebra), **2**-16
Varnish, **6**-118, **6**-121
  airplane dopes, **6**-121
  catalytic, **6**-121

Varnish (*Cont.*):
  chemical resistant, **6**-121
  flat, **6**-121
  floor, **6**-121
  insulating, **6**-152
  lac, **6**-121
  lacquer, **6**-121
  oleoresinous, **6**-118, **6**-121
  pyroxylin, **6**-121
  shellac, **6**-118, **6**-121
  silicone, **6**-152, **6**-171
  spar, **6**-121
  spirit, **6**-118
  water, **6**-121
Vars (reactive current, def), **15**-22
Vector (def), **2**-63, **3**-2
  analysis of, **2**-63, **2**-64
  differential operators, **2**-64
  differentiation, **2**-63
  products, **2**-63
  representation, **2**-64
Vehicles, road resistance of, **3**-27, **3**-28
Velocity (def), **3**-11
  acoustic (def), **3**-35
  angular (def), **3**-14
  coefficient, in turbine nozzles, **4**-49
  composition of, **3**-12, **3**-13
  conversion table for linear and angular, **1**-50
  diagrams, pumps, **14**-27
  distribution in pipes, **3**-54
  equivalents (table), **1**-47
  head (hydraulics), **3**-43
  hypersonic (dcf), **11**-58
  of light, **19**-40
  ratios (mechanism), **8**-4
  resolution of, **3**-12
  of sound: in ideal gases, **3**-35
    in various media (table), **3**-35, **12**-136
  subsonic (def), **11**-58
  supersonic (def), **11**-58
  transonic (def), **11**-58
  units of, **1**-33
Velocity-time curve (kinematics), **3**-11
Vena contracta (def), **16**-15
  taps, orifice coefficients for (table), **16**-16
Ventilation (*see* Air conditioning)
Venturi meters, **3**-61, **3**-62
  ASME coefficients for (table), **3**-62
  factors for (table), **3**-63
Vernier calipers, **16**-5
Versed sine (trigonometry), **2**-26
  graph, **2**-55
Vertex of parabola, **2**-31
Vessels (*see* Ships)
Vibration, **5**-67 to **5**-76
  absorption of, **5**-72
  balancing, **5**-70 to **5**-72
  damped, **5**-67
  energy method, **5**-68
  forced, **5**-68
  free, **5**-67
  isolation, **5**-69
  Kármán vortices, **12**-12
  lateral, of bars, **5**-73, **5**-74
  logarithmic decrement, **5**-67
  measurement of, **5**-75, **5**-76
  of membranes, **5**-74
  pickup (sound), **12**-139
  of plates, **5**-74, **5**-75
  of shafts, **5**-69, **5**-73
  in ships, **11**-42, **11**-43
  of simple systems, **5**-69

Vibration (*Cont.*):
  in structures, **5**-68
  torsional, of shafts, **5**-69
  of turbine blades, **9**-45
Vibrator (molding), **13**-5
Vickers hardness test, **5**-13
View factor (def), **4**-73
Vinyl plastic (table), **6**-162
Viscometers, **3**-36
Viscosimeters, **16**-21
Viscosity (def), **3**-35, **6**-196
  of cryogens, **19**-31
  dynamic (def), **3**-35
  of gases (chart), **4**-50
    table, **3**-36
  kinematic, **3**-35
    conversion, formulas, **3**-36
  of liquids (table), **3**-36
  of lubricating oils, test for, **6**-196
  of oils, temperature variation of (chart), **6**-197
  units of (def), **3**-35, **3**-36
Viscosity index (chart), **6**-198
Viscous damping in automatic control systems, **16**-26
Vision, **12**-117
Vitallium (heat-resisting alloy, table), **6**-99
Vitrified grinding wheels, **6**-139
Vitrified pipe (table), **8**-178
Voith-Schneider propulsion system (ships), **11**-52
Volatility of liquid fuel (def), **9**-92
Volt (def), **15**-3
Volt-ampere (def), **15**-22
Voltage drop:
  in a-c circuits, **15**-64
  Mershon diagram for, **15**-58
Voltage regulation:
  of a-c generators, **15**-37 to **15**-40
  with synchronous converters, **15**-49
Voltage regulator, **15**-82
  Tirrill, **15**-38
Voltmeters, **15**-24, **15**-25, **15**-28
Volume:
  and capacity equivalents (table), **1**-47
  change under stress, **5**-18
  conversion tables for, **1**-47
  measures of, **1**-31
  of similar figures, **2**-8
  of solids by immersion, **3**-40
  units of (def), **1**-31 to **1**-33
  of various solids, **2**-12 to **2**-15
Volumetric efficiency:
  of compressor, **14**-34
  of internal-combustion engines, **9**-107
Volute pumps, **14**-16, **14**-17
Vortex theory for propellers, **11**-99
Vuilleumier refrigerating engine (cryogenics), **19**-24
Vulcanized rubber, **6**-165


Wage systems (industrial management), **17**-9
Wakes in fluid streams, **3**-53
Walls:
  building: heat gain through (tables), **12**-78 to **12**-80
    thickness of (table), **12**-20
  corrugated sheeting for, **12**-20, **12**-42
  masonry, coefficients of heat transmission (table), **12**-89

Walls (*Cont.*):
  reinforced-concrete, **12**-20, **12**-59, **12**-60
  retaining, design of, **12**-21
  sound-absorption coefficients (table), **12**-141
  sound-transmission loss (table), **12**-140
Wankel engines, **9**-80
Ward-Leonard method of speed control, **15**-35
  for mine hoists, **10**-19
Washburn & Moen wire gage (table), **6**-45
Washers:
  for bolts, dimensions of (table), **8**-25
  steam, **9**-28
Waste disposal, solid, **18**-21, **18**-22
Waste recovery, **18**-22
  of sewage as fertilizer (table), **18**-22
Water, **6**-189 to **6**-195
  agricultural (def), **6**-190
  boiler (*see* Feedwater)
  compressibility of (tables), **4**-7, **6**-10
  critical pressure, **9**-25
  deaeration of, **6**-110
  desalinization of, **6**-192 to **6**-195
    costs of, **6**-195
    by electrodialysis, **6**-194
    by flash distillation, **6**-193
  expansion of (chart), **12**-70
  for fire protection: sprays, **12**-32
    supply, **12**-31
  flow of: in open channels, **3**-67
    through orifices, **3**-62 to **3**-64
    in pipes, **3**-53 to **3**-60
    from tank openings, **3**-68
    over weirs, **3**-65, **3**-66
  hardness (def), **6**-190
  heat-transfer coefficients to or from, **4**-64 to **4**-66
  household use of, average, **6**-189
  index (def), **6**-189
  industrial, **6**-191
    for cooling, **6**-191
    usage, **6**-191
      variance in (table), **6**-191
  ions in sea water (table), **6**-192
  measurement(s), **1**-31, **6**-190
  pollution, **6**-192
  potable (def), **6**-190
    specifications (table), **6**-190
  quality (def), **6**-190
  as reactor coolant, **9**-127
  resources, **6**-189
  runoff, **6**-189
  solubility: of gases in (table), **6**-7
    of inorganic substances in (table), **6**-5
  specific gravity and density (tables), **6**-10
  specific heat (table), **4**-9
  temperatures, for circulating water, U.S. (map), **9**-62
  thermal conductivity (table), **4**-62
  thermal expansion of, **12**-62
  treatment (table), **6**-192
    (*See also* Feedwater)
  use of (table), **6**-189
  volume, in pipes and tubes (table), **12**-70
Water brake, **16**-18
Water gas (table), **7**-42
  carbureted, **7**-42
    composition of (tables), **7**-42
  flame temperature (table), **4**-57
Water hammer (def), **3**-69
  in penstocks, **9**-148
  in pipelines, **3**-69
  relief devices for, **3**-69

Water jet, impact, **3**-45, **3**-52
Water meters, **16**-7, **16**-8, **16**-14 to **16**-18
Water power, **9**-168, **9**-169
    (*See also* Hydraulic turbines)
Water seal (packings), **8**-144
Water-tube boilers, **9**-14 to **9**-21
Water turbines (*see* Hydraulic turbines)
Water vapor:
    in air (charts), **4**-41 to **4**-43
    emissivity of (chart), **4**-79
    permeability, of various building materials
        (table), **12**-74
    specific heat, **4**-10
Waterproofed cement, **6**-114, **6**-178
Waterproofed concrete, **6**-187
Watt (def), **1**-39, **15**-4, **15**-10
Watt-hour (def), **15**-4
Watt-hour meters, **15**-26, **15**-27
Watt-second (def), **15**-4
Wattless current (def), **15**-22
Wattmeters, **15**-25 to **15**-27
    polyphase, **15**-26
Wave-making resistance (ships), **11**-40
Wave motors, **9**-169
Wave power, **9**-168
Wavelength:
    of radio waves (table), **15**-94
    of sound, **12**-136
Waves:
    alternating-current, **15**-21
    radio, **15**-94
Weber (magnetic flux, def), **15**-2
Weber's (dimensionless) number, **3**-48, **3**-49
Wedge:
    rectangular, volume of, **2**-13
    spherical, volume of, **2**-14
Wedges, friction of, **3**-28
Weighers for conveyors, **10**-63, **10**-64
Weighing scales, **16**-3, **16**-4
Weightometer, Merrick, **10**-63
Weights:
    atomic (table), **6**-3, **6**-4
    conversion table for, **1**-48
    corresponding to degrees: API (table),
        **1**-42
        Baumé (table), **1**-43
    per cubic foot of various materials (table), **6**-
        7 to **6**-9
    equivalents (table), **1**-48
    fundamental equation of, **3**-2
    of materials (table), **10**-2
    and measures: metric (*see* Metric measures
        and weights)
        U.S., **1**-31, **1**-32
Weirs, **3**-65, **3**-66, **16**-17
    Cippoletti, **3**-65, **3**-66
    contracted, **3**-65
    hyperbolic, **3**-65
    parabolic, **3**-65, **16**-17
    rectangular-notch, **3**-65, **3**-66, **16**-17
    Sutro, **3**-65, **16**-17
    trapezoidal, **3**-65
    triangular-notch, **3**-65, **3**-66, **16**-17
    V-notch, **3**-65, **3**-66, **16**-17
Welding, **13**-27 to **13**-48
    aluminum and alloys, **13**-45 to **13**-47
    arcs, **13**-28
    braze, **13**-36, **13**-45
    cast iron, **13**-44
    cast steel, **13**-44
    chrome iron and steels, **13**-43
    codes for, **13**-27, **13**-28

Welding (*Cont.*):
    copper and copper alloys, **13**-47
    drafting symbols for, **13**-29, **13**-38
    electric arc, **13**-28 to **13**-33
        currents and speeds (table), **13**-32, **13**-34,
            **13**-35
        electrodes for, **13**-29 to **13**-32
        metal grooves, proportions of, **13**-27
        semiautomatic machines for, **13**-33
        shielded arc, **13**-29, **13**-44
        spot, **13**-33
        voltage for, **13**-29
    electric flash, **13**-40
    electric resistance, **13**-38 to **13**-42
        current and energy for (table), **13**-39, **13**-
            41
        electrodes, **13**-38, **13**-39
        flash, **13**-40
        percussion, **13**-40
        projection, **13**-39, **13**-40
        pulsation, **13**-40
        seam, **13**-38, **13**-39
        spot, **13**-38, **13**-39
        upset, **13**-40
    electrodes: classifications of, **13**-29 to
        **13**-32
        coverings for, **13**-29
        current ranges for (table), **13**-31
        properties of (table), **13**-29
        tungsten, **13**-33
    electron-beam, **13**-47
    Electroslag, **13**-48
    exotic metals, **13**-47
    explosive, **13**-48
    facing by, **13**-42
    filler metals for, **13**-33
    forehand and backhand, **13**-36
    fundamentals of, **13**-27
    gas, **13**-33 to **13**-36
        fluxes for, **13**-36
        by machine, **13**-36
    gas metal-arc, production data (table), **13**-
        33, **13**-34
    joints for, **13**-28
    laser-beam, **13**-48
    magnesium and alloys, **13**-47
    mash, **13**-39
    monel metal, **13**-45
    nickel alloys, **13**-45
    oxyacetylene, **13**-36
    oxyhydrogen, **13**-33
    pipes, **8**-194 to **8**-197
    plasma-arc, **13**-47
    preheating, **8**-195
    procedures for, **13**-43 to **13**-48
    shielding, **13**-29
    solid-state, **13**-48
    sonic, **12**-142
    specifications for, **13**-27
    steel and steel alloys, **13**-43 to **13**-45
    structural steel, **12**-32
    submerged arc, **13**-32
        table, **13**-32
    symbols for, **13**-29, **13**-38
    tungsten: electrodes for, **13**-33
        production data (table), **13**-34
    wrought iron, **13**-45
Welds:
    metal, strength of, **13**-27, **13**-28
        at low temperatures, **19**-30
    types of, **13**-27
Weston cells, **15**-12

Wet-bulb psychrometer, 4-31
Wet-bulb temperature, 4-28
  isolines (map), 9-73
Wetted perimeter (hydraulics), 3-42
Wheatstone bridge, 15-29
  for strain gages, 5-58
Wheels:
  and axles (railroad), 11-28, 11-29
  friction coefficients for, 3-27
    tables, 3-27, 3-28
White cast iron (def), 6-12
White coat (plastering), 6-182
White iron (def), 6-47
White lead (paints), 6-119
Whitworth quick-return motion, 8-3
Whitworth screw threads (table), 8-16
Wide-flange beams, properties of (tables), 12-36 to 12-38
Willans line vs. fuel consumption, 11-5
Williams and Hazen formula in pipeline flow, 9-37, 11-133
Wind:
  Beaufort scale (table), 9-167
  gusts, 12-12
  power in, 9-166
  velocities of, 12-12
    in the United States (table), 9-167
Wind-chill index, 12-107
  table, 12-113
Wind pressure:
  on bridge girders, 12-14
  on buildings, 12-13
  distribution of, 12-12
  on roofs, 12-13
  on structures, 12-12 to 12-14
  on towers, 12-14
Wind tunnels, 11-78
  intermittent, 11-78
  shock tubes for, 11-78
  supersonic, 11-78
  transonic, 11-79
Wind turbines, 9-164
  blade element theory, 9-165
  cross-wind-axis, 9-165
  cyclogiro rotor, 9-166
  Darrieus rotor, 9-166
  general momentum theory, 9-164
Windmills (see Wind turbines)
Window glass, 6-157
Windows:
  glass, shading coefficients for (tables), 12-93 to 12-95
  for industrial plants, 12-7
Wings:
  aircraft (see Airfoils)
  airplane (see Airfoils; Airplanes, wings)
Winkler process (gas making), 7-43
Wire:
  airplane, 11-70
  copper (see Copper wire)
  insulated, types of, 15-67, 15-68
    table, 15-70, 15-71
  magnet (table), 15-81
  steel: gage of (table), 6-45
    strength and weight of (table), 6-43
Wire drawing (fluid flow), 4-50, 4-51
Wire gage:
  American (table), 15-7
  table, 6-45
Wire glass, 6-157
Wire nails (tables), 8-94, 8-95

Wire rope, 8-85 to 8-89, 10-9 to 10-11
  brakes for, 10-11
  classification numbers, 8-85
  clips (tables), 8-89
  coarse-laid, standard, 8-86
  core, 8-85
  drums for, 10-9
  fittings, 8-89, 10-9
    efficiencies of (table), 8-90
  flat (table), 8-86
    flattened strand (table), 8-87
  galvanized: bridge rope (table), 8-92
    bridge strand (table), 8-92
    guy (tables), 8-91
    hawser (table), 8-91
    mast-arm (table), 8-90
    mooring line (table), 8-91
    rigging (tables), 8-91
    running (table), 8-91
    special grades (table), 8-90
    strands (table), 8-87, 8-90, 8-92
  handling of, 8-85
  haulage (table), 10-10
  hoisting, 10-9
    drums, 10-9
    extra-pliable (table), 8-84
    nonrotating (table), 8-87
    nonspinning, 8-88
    standard (table), 8-84
  idlers for, 10-10
  IWRC, 8-85
  lang-lay, 8-85, 10-9
  left-lay, 8-85
  materials for, 8-85
  for power hoists, 10-9
  power-transmission (table), 8-86
  sash cords (table), 8-92
  seizings, 8-89
    tables, 8-88
  sheaves for, 10-10
  sizes of, 8-85
  slings, 8-89
  smooth-coil track strand (table), 8-92
  steel-clad, 8-88
  strength of, 8-85
  tackle, efficiencies (table), 10-11
  tiller (table), 8-92
  track cables, 10-9
    table, 10-10
  working loads for, 8-85
Wire strand, galvanized (tables), 8-87, 8-90, 8-92
Wire tacks (table), 8-97
Wiring:
  calculations: for a-c circuits, 15-65, 15-66
    for d-c circuits, 15-64
  diagrams, generator switchboard, 15-54
  interior, 15-66 to 15-69
    approved methods of, 15-67, 15-68
    cable for, 15-67, 15-68
    conduit and tubing for, 15-68
    grounds for, 15-69
    insulation for, 15-68
      table, 15-70 to 15-72
    open, 15-67
    protective devices for, 15-69
    switches, 15-68, 15-69
  three-phase a-c circuits, 15-65
Wittenbauer's analysis for fly-wheel performance, 8-73

Wolfram (*see* Tungsten)
Wood, **6**-122 to **6**-138
  allowable stresses for (tables), **6**-128 to **6**-135, **12**-26
  beams, **12**-23, **12**-24
  chemical composition of, **6**-122
  classification of, **6**-122
  columns, **12**-24 to **12**-27
  commercial standards for, **6**-138
  construction, **12**-23 to **12**-28
    of industrial plants, **12**-6
  damping in, **6**-127
  decay of, **6**-126
  density and specific gravity, **6**-123
  electrical properties of, **6**-125
  fatigue properties of, **6**-127
  fire-retardant treatment for, **6**-125, **6**-138
  flooring, safe load and deflection (table), **12**-24
  as fuel, **7**-12, **7**-13
    analyses of (table), **7**-13
    heat value of (table), **7**-13
  fungi in, **6**-126
  glue-laminated, **6**-132
    allowable stresses in, **6**-132
    curvature factor, **6**-132
    glues for, **6**-132
    preservative treatment for, **6**-132
    (*See also* Plywood)
  glues for, **6**-132
  hardness test, **5**-14
  hardwood (def), **6**-122
  heartwood (def), **6**-122
  heat of combustion, **6**-125
  heat value, **6**-125
  ignition temperatures, **6**-125
  insects, attack by, **6**-126
  marine organisms, attack by, **6**-126
  moisture relations of, **6**-122
  old, strength of, **6**-126
  preservative treatments, **6**-136 to **6**-138
    allowable stresses after, **6**-138
    against biological action, **6**-136
    methods of, **6**-137
    preparation for, **6**-137
  preservatives, **6**-136
  properties of (table), **6**-124
  rheological properties of, **6**-126
  sanding recommendations (table), **13**-84
  sapwood (def), **6**-122
  shrinking or swelling of, **6**-123
    treatments to prevent, **6**-123
  soapy, friction of, **3**-25
  softwood (def), **6**-122
  sound transmission in, **6**-126
  specific gravity and density (tables), **6**-8, **6**-123, **6**-124
  strength of, **6**-125 to **6**-132
    bolted (table), **12**-27
    effect of age, **6**-126
    effect of heat, **6**-125
    effect of moisture (table), **6**-123, **6**-124
    table, **6**-124, **6**-127 to **6**-131, **6**-133 to **6**-135
    for various grain slopes (table), **6**-127
  thermal conductivity of, **6**-125
  trusses, joints for, **12**-27
  walls: safe loads on (table), **12**-26
    stud, **12**-19
  weight of (tables), **6**-8

Wood alcohol, **6**-168, **7**-20
Wood pulp in paper manufacture, **6**-159
Wood screws, **8**-22
Wood-stave pipe, **8**-172
Wood waste as fuel, **7**-13
Woodruff keys (table), **8**-32
Wood's metal, **6**-90
Woodworking machines, **13**-80 to **13**-84
  motors for, **15**-52
Wool fibers (tables), **6**-155, **6**-156
Word (computers, def), **2**-65
Work (def), **3**-2, **3**-20
  computation of: diagram, **3**-20
    rule for, **3**-20
  conversion tables for, **1**-52
  of friction, **3**-28
  muscular, **9**-162
  units of (def), **1**-33, **3**-20
Work hardenability, determination of, **5**-14
Worm gears (*see* Gears, worm)
Worm hoist, **8**-8
Worthington feather valve (compressor), **14**-36
Wrench openings, standard (table), **8**-20
Wronskian determinant, **2**-51
Wrought-aluminum alloys, **6**-75
Wrought iron (def), **6**-12
  bars, round and square, weight of (table), **6**-46
  composition of (table), **6**-13
  mechanical properties of (table), **6**-14
  pipe (*see* Pipe)
  shapes of, finished, **6**-12
  uses of, **6**-12
  welding, **13**-44
Wrought-magnesium alloys (table), **6**-84


X-ray diffraction, uses of, **16**-21
Xylene, solvent, **6**-168


Y connections, three-phase circuits, **15**-23
Y level (surveying), **16**-44
Yard (def), **1**-31
Yarns (tables), **6**-155
Yaw (airplanes), **11**-72
Yaw (ships, def), **11**-40
Year, definitions of, **1**-41
Yield point (def), **5**-3
Yield strength (def), **5**-2
  of metals (table), **5**-3
Yield-tensile ratio, **13**-15
Young's modulus (def), **5**-2, **5**-18


z-transfer function (automatic control), **16**-36
z transformation (automatic control), **16**-36
Zeolite process (feedwater treatment), **9**-30
Zinc:
  alloys, **6**-85 to **6**-87
  coatings for metals, **6**-114, **6**-115
  commercial: composition of (ASTM specifications, table), **6**-85
    production of, **6**-85
    uses of, **6**-85
  corrosion resistance of, **6**-86
  die castings, **6**-86
    aging, **6**-86

Zinc, die castings (*Cont.*):
 composition and properties of (table), **6**-86
 dust, in paint, **6**-120
 effect of temperature on, **6**-86
 fabrication of, **6**-85
 galvanized sheet steel (table), **6**-42
 galvanizing, **6**-85
 painting of, **6**-120
 rolled, **6**-85
 spelter, **6**-85
 uses of, **6**-85

Zinc (*Cont.*):
 white, in paint, **6**-119
 wrought, **6**-85
Zinc plating, **6**-114, **6**-115
Zircoloys, **6**-113
Zirconium, **6**-87, **6**-103
 mechanical properties of (table), **6**-102
  absorption of slow neutrons (tables), **6**-101
 in nuclear technology, **6**-103
 uses of, **6**-87
Zones of spheres, area of (formulas), **2**-13